D0930163

DICTIONARY OF SCIENCE AND TECHNOLOGY

ENGLISH - FRENCH

DICTIONNAIRE DE SCIENCE ET TECHNOLOGIE

ANGLAIS - FRANÇAIS

DICTIONNAIRE DE SCIENCE ET TECHNOLOGIE

ANGLAIS - FRANÇAIS

compilé par

A.F. DORIAN

ELSEVIER SCIENTIFIC PUBLISHING COMPANY
AMSTERDAM - OXFORD - NEW YORK
1979

DICTIONARY OF SCIENCE AND TECHNOLOGY

ENGLISH-FRENCH

compiled and arranged by

A.F. DORIAN

ELSEVIER SCIENTIFIC PUBLISHING COMPANY
AMSTERDAM - OXFORD - NEW YORK
1979

Distribution of this book is being handled by the following publishers

ELSEVIER SCIENTIFIC PUBLISHING COMPANY
335 JAN VAN GALENSTRAAT
P.O. BOX 211, 1000 AE AMSTERDAM, THE NETHERLANDS

ELSEVIER/NORTH-HOLLAND INC.
52 VANDERBILT AVENUE
NEW YORK, NEW YORK 10017

Library of Congress Cataloging in Publication Data

Dorian, Angelo Francis.
Dictionary of science and technology.

Added t.p.: Dictionnaire de science et technologie.
1. Science--Dictionaries. 2. Technology--Dictionaries. 3. English language--Dictionaries--French.
I. Title. II. Title: Dictionnaire de science et technologie.
Q123.D672 503 79-18507
ISBN 0-444-41829-6

ISBN 0-444-41829-6

Printed in The Netherlands

PREFACE

There are many, not a few first-class English/French dictionaries available; yet, they often seem sadly inadequate to fulfil the expectations of scholars, scientists and students who move in the ever growing vastness of science and technology. In this sense, this work is a humble attempt at filling in some of the gaps which are so frequently a source of irritation and of frustration. The 150.000 terms in it cover a large number of fields, as far as possible screened through a balanced selection, so as to make it a valid and easy-to-use instrument of work.

This is, in fact, the main scope of the present dictionary, especially in view of the fact that all scientific and technological subjects, owing to their very complexity, are inevitably intertwined: mechanics invade the territory of nucleonics, electronics, telecommunications, astronautics and so on; nor is it possible to divorce -and this is but an example- telecommunications from electricity, electronics and building construction; metallurgy from mineralogy and petrology; mining and oilfield technology from geology; medicine from the guiding principles of radiations and isotopes and so on, practically ad infinitum. Of course, it would be sheer presumption, not to say absurd, to suggest that every single term of every single subject will be found in this work. It can only be assured that the very best has been done, within the author's ability, to avoid teasing the user in his search.

All entries are placed in alphabetical order and each term is followed by the indication of the subject to which it is related. In many instances a short definition is added, so as to avoid misinterpretations or errors. When such indication is "general", this simply means that the term is thus translated in all its applications.

Finally, the author would like to seek the user's indulgence for all imperfections which may be found in his work and to express the hope that the dictionary, as it is, may be of real, easy and satisfactory use.

A.F. Dorian

FREFACE

Il y a beaucoup et de très bons dictionnaires anglais-français disponibles; cependant ils sont souvent insuffisants pour répondre aux demandes des savants, des scientifiques, et des étudiants qui travaillent dans le champ toujours plus grand de la science et de la technologie. En ce sens, ce travail est une humble tentative pour combler quelques vides qui sont si fréquemment une source d'irritation et de frustration. Les 150.000 termes couvrent un grand nombre de domaines, et ayant fait l'objet d'une sélection harmonieuse, ils en font un instrument de travail valable et facile à utiliser.

C'est en fait le principal but de ce dictionnaire, surtout si l'on considère que tous les sujets scientifiques et technologiques, en raison de leur grande complexité, sont inévitablement mêlés: la mécanique envahit le territoire du nucléaire, de l'électronique, des télécommunications, de l'astronautique, etc., et il n'est pas possible non plus de séparer — et ce ne sont que des exemples — les télécommunications de l'électricité, l'électronique et le bâtiment; la métallurgie de la minéralogie et de la pétrologie; l'extraction minière et la technologie pétrolière de la géologie; la médecine des principes fondamentaux des radiations et des isotopes, etc., pratiquement à l'infini. Naturellement ce serait pure présomption, même absurde de suggérer que l'on trouvera dans ce travail chaque terme concernant chaque domaine. On peut assurer que l'auteur a fait de son mieux pour éviter à l'utilisateur des tracas dans sa recherche.

Tous les termes sont en ordre alphabétique et chacun est suivi de l'indication du sujet auquel il se rapporte. Dans beaucoup de cas une courte définition est ajoutée, pour éviter de mauvaises interprétations ou erreurs. Quand une telle indication est "générale", cela veut dire simplement que le terme est traduit dans toutes ses acceptions.

Enfin, l'auteur demande l'indulgence de l'utilisateur pour toutes les imperfections qui peuvent se trouver dans son travail, et exprime l'espoir que ce dictionnaire, tel qu'il est, sera d'un usage réel, facile et satisfaisant.

A.F. Dorian

ABBREVIATIONS

acoust	acoustics	acoustique
adj	adjective	adjective
aero	aeronautics	aéronautique
agric	agriculture	agriculture
anat	anatomy	anatomie
arch	architecture	architecture
astr	astronomy	astronomie
astronaut	astronautics	astronautique
auto	automobiles	automobiles
automat	automation	automation
biochem	biochemistry	biochimie
biol	biology	biologie
bookbind	bookbinding	reliure
bot	botany	botanique
brew ind	brewing industry	brasserie
carp	carpentry	charpenterie
ceram	ceramics	céramique
chem	chemistry	chimie
cin	cinema	cinématographie
comm	commerce	commerce
comput	computers	calculateurs
constr	construction work	bâtiment
contr	control	commande automatique
cryst	crystallography	cristallographie
cytol	cytology	cytologie
dent	dentistry	dentisterie
draw	drawing	dessin
el	electricity	électricité
el acoust	electroacoustics	électro-acoustique
el chem	electrochemistry	électro-chimie
el metall	electrometallurgy	électrométallurgie
electron	electronics	électronique
fin	finance	finances
firearms	firearms	armes à feu
fishing ind	fishing industry	pêche
food	food industry	industrie alimentaire
gas ind	gas industry	industries du gaz
gen	general	général
genet	genetics	génétique
geogr	geography	géographie
geol	geology	géologie
glass man	glassmaking	verrerie
horol	horology	horlogerie

hydr	hydraulics	hydraulique
impl	implement	outil
ind chem	industrial chemistry	chimie industrielle
instr	instruments	instruments
insur	insurance	assurance
leather ind	leather industry	travail de cuir
leg	legal	légal
mach tool	machine tools	machines-outils
meas	measures	mesures
mech	mechanics	mécanique
med	medicine	médecine
met	meteorology	météorologie
metall	metallurgy	métallurgie
min	mineralogy	minéralogie
mining	mining	industrie minière
mus	music	musique
naut	nautical	marine
nucl	nucleonics	technique nucléaire
oil ind	oil industry	industrie pétrolifère
opt	optics	optique
packag	packaging	emballage
paint	painting	peinture
paper man	papermaking	fabrication du papier
photo	photography	photographie
phys	physics	physique
physiol	physiology	physiologie
plast ind	plastics industry	matières plastiques
plumb	plumbing	plomberie
print	printing	imprimerie
radiat	radiation	rayonnement et radio-activité
radio	radio	radio
radiol	radiology	radiologie
railw	railways	chemins de fer
rubber ind	rubber industry	industrie du caoutchouc
shipbuild	shipbuilding	architecture navale
shoe man	shoe manufacture	cordonnerie
soil	soil and soil mechanics	sol et mécanique des sols
statist	statistics	statistique
sugar ind	sugar industry	industrie sucrière
surv	surveying	levé de plans
telecomm	telecommunications	télécommunications
telev	television	télévision
text	textiles	textiles
tool	tools	outils
transp	transport	transports
vet	veterinary science	art vétérinaire
zool	zoology	zoologie

A

A - [el] ampère *m*
[chem] argon *m*
[phys] (abbrev for absolute temperature) température *f* absolue
[geom] aire *f*, superficie *f*, surface *f*
A-AMPLIFIER - [el acoust] (amplifier used with a high-fidelity microphone) amplificateur *m* classe A
A-BATTERY - [el] (used for cathode heating) batterie *f* de chauffage
A-BOMB - [nucl] bombe *f* atomique
A-DIGIT SELECTOR - [telecomm] sélecteur *m* primaire
A-DISPLAY - [radar] indicateur *m* linéaire (type A)
A-FRAME - [constr] chevalet *m*, bâti *m* en pyramide
A-INDICATOR - s.A-display
A-LEVEL [impl] niveau *m* triangulaire
A-POLE - [telecomm] appuis *m* en A
A-POSITION - [telecomm] groupe *m* de départ
A-POWER SUPPLY - [electron] source *f* d'alimentation de chauffage
A-SERVICE AREA - [radio] (the area surrounding a broadcasting station where the field strength exceeds ten millivolts per metre) zone *f* autour d'une station où l'intensité du champ est supérieure à dix millivolts par mètre
A-STANDARD - [constr] chaise *f* de sol
A-SWITCHBOARD - [telecomm] positions *f*pl de départ, tableau *m* manuel, positions *f*pl A
a-a - [geol] (a type of lava) aa *m*
A.A. - [aero] (abbrev for anti-aircraft) contre-avion, antiaérien
AAMCOLOSE - [chem] dextrine
A.B.POWER PACK - [electron] (the power source for electronic tubes operated by battery) bloc *m* d'alimentation
ABAC - [radar] (a graphic system of lines on a chart corresponding to an equation which can be solved without further calculation) abaque *m*, monogramme *m*
ABACA - [text] abaca *m*, chanvre *m* de Manille
~FLOWER - [bot] fleur *f* d'Abaca
ABACK - [naut] coiffé, masqué, bordé au vent
ABACUS - [instr] (primitive instrument used to add or subtract mechanically) abaque *m*, boulier-compteur *m*
[mining] augette *f*, sébile *f*, batée *f*
[arch] abaque *m*, tailloir *m*
ABAMPERE - [el] (c.g.s. electromagnetic unit) ampère *m* absolu (u.é.m. de courant électrique)
ABAMURUS - [constr] (a supporting wall or buttress) contrefort *m*, mur *m* de soutènement
ABANDON, to - [gen] abandonner, évacuer, renoncer à
ABANDONMENT - [gen] abandon *m*, cession *f*, évacuation *f*
ABATE, to - [gen] diminuer, réduire, s'apaiser
[med] faire cesser
[met] (of the wind) tomber
[comm] rabattre
ABATIA - [chem] abatia *f* (utilisée pour la teinture en noir)
ABATING - [metall] réduction *f* de trempe
ABATTOIR - [constr] (public slaughter-house) abattoir *m*
ABAXIAL - [phys etc] abaxial
ABB - [text] trame *f*
ABBCITE - [min] (explosive used in coal mines) abbcite *f* (dynamite au nitrate d'ammoniaque)
ABBE REFRACTOMETER - [instr] (instrument for the direct measurement of the refractive index of liquids) refractomètre *m* Abbe
ABCOULOMB - [el] (c.g.s. unit of quantity of electricity) coulomb *m* absolu (u.é.m. C.G.S. de quantité d'électricité)
ABDOMEN - [anat] abdomen *m*
ABDOMINAL BELT - [med] ceinture *f* abdominale
ABDUCENT - [anat] abducteur *m*
ABDUCT, to - [gen] enlever
ABDUCTION - [gen] abduction *f*, enlèvement *m*
ABEL CLOSED TESTER - [ind chem] (an apparatus used for the determination of the flash point of petroleum) appareil *m* à coupe fermée (détermination du point d'inflammabilité; pétroles etc)
~FLASH POINT APPARATUS - s.Abel closed tester
ABELITE - [chem] (an explosive) explosif *m* à base de trinitrotoluène et de nitrate d'ammoniaque
ABELMOSCHUS - [bot] (used for perfumes) abelmosch *m*, ambrette *f*
ABEL'S REAGENT - [chem] used for the microanalysis of carbon steel) réactif *m* d'Abel
ABERRANT - [gen] (showing characteristics different from type) aberrant, anormal
ABERRATION - [gen phys opt] aberration *f*, déviation *f*
[telev] (distortion of image) distortion *f*
~CROWN - [opt] cercle *m* d'aberration (chromatique)
~OF LIGHT - [opt] aberration *f* de la lumière
ABERROMETER - [instr] aberromètre *m*
ABERROSCOPE - [instr] aberroscope *m*
ABFARAD - [el] (electrostatic capacitance of c.g.s. unit) farad *m* absolu (u.é.m. C.G.S. de capacité)
ABHENRY - [el] (inductance of c.g.s.unit) henry *m* absolu (u.é.m. d'inductance)

ABHESION - [ind chem] (loss of adhesion) perte *f* d'adhérence
ABIE - [chem] huile *f* de pin
ABIES - [bot] abies *m*, sapin *m* blanc
~BARK - [bot] (used in tanning) écorce *f* de sapin
ABIETATE - [chem] abiétate *m*
ABIETIC ACID - [chem] acide *m* abiétique
ABIETIN - [chem] abiétine *f*
ABIETINIC ACID - s. abietic acid
ABILITY - [gen] capacité *f*, pouvoir *m*, habileté *f*
~TO FLOW - [phys] fluidité *f*
ABIOCHEMISTRY - [chem] chemie *f* minérale
ABIOGENESIS - [genet] abiogénèse *f*
ABLACTATION - [med] ablactation *f*
ABLATE, to - [gen] enlever, retrencher
ABLATING MATERIAL - [astronaut] (material having low thermal conductivity, high melting point, specific heat and latent heats of fusion and evaporation, used for the outer surface of a space capsule) matériau *m* d'ablation
ABLATION - [gen] ablation *f*
~[med] (removal of tissue by surgery) ablation *f*, élimination *f*
ABLUTION - [ind chem] ablution *f*
ABMHO - [el] (conductance of c.g.s. unit) u.é.m.-C.G.S. de conductance
ABNORMAL - [gen] anormal
~GLOW DISCHARGE - [electron] (glow discharge in which the working voltage increases with the increase of the current) décharge *f* luminescente anormale
~STEEL - [metall] acier *m* à texture anormale
ABNORMALITY - [gen] anomalie *f*, caractère *m* anormal
ABOHM - [el] (resistance of C.G.S. unit) ohm *m* absolu
ABOPON - [chem] abopon *m* (borophosphate complexe)
ABORTION - [med] avortement *m*
ABOUT-SLEDGE - [impl] marteau *m* à frapper devant, marteau *m* à devant
ABOVE - [gen] au dessus de, ci-dessus
~AVERAGE - [gen] supérieur à la moyenne
~GROUND - [gen] au dessus du sol, superficiel
[mining] (in open cut) à ciel ouvert, au jour
~LOCK - [hydr] en amont de l'écluse
~PLATEN DEVICE - [comput] (in a tabulating machine) guidage *m* du tambour supérieur
ABRADANT - [gen & metall] abrasif *m* (en poudre)
ABRADE, to - [gen] écorcher, user par frottement
ABRADER - [mech & metall] abrasif *m*, dispositif *m* abrasif
ABRASION - [gen] abrasion *f*, usure *f* par frottement
[med] (skin or mucous membrane rubbed away) écorchure *f*
~HARDNESS - [metall] dureté *f* de meulage
~MACHINE - [mech] appareil *m* d'essai de résistance à l'abrasion
~MARKS - [metall] marques *fpl* de meulage
~PATTERN - [photo] état *m* de surface (après abrasion)
~RESISTANCE - [mech] résistance *f* à l'abrasion
~SERVICE TEST - [ind] essai *m* de mesure de résistance à l'abrasion
~TEST - [ind] essai *m* de résistance à l'abrasion, essai *m* d'usure par frottement, essai *m* à la meule
ABRASION TESTER - [mech] s. abrasion machine, usomètre *m*
~TESTING MACHINE - s. abrasion machine
~WEAR - [mech] usure *f* par abrasion, usure *f* par frottement
ABRASIVE - [gen] abrasif
~BAND MACHINE - [mech] machine *f* à meuler à bandes
~BELT - [metall] bande *f* abrasive
~BELT GRINDING MACHINE - s. abrasive band machine
~FABRIC - [gen] toile *f* abrasive
~HARDNESS - [metall] dureté *f* d'abrasion
~MATERIAL - [gen] abrasif *m*
ABREAST - [gen & aero] (en ligne) de front, par le travers
[el] en dérivation
~CONNECTION - [el] couplage *m* parallèle
~OF THE BEAM - [constr] à angle droit
~OPPOSITE - [naut] à contrebord
ABRIDGE, to - [gen] abréger, diminuer, réduire
ABRIN - [chem] abrine *f*, toxalbumine *f*
ABROCOL - [chem] monostéarate *m* de glycérine
ABROGATE, to - [gen & leg] abroger
ABROGATION - [gen & leg] abrogation *f*
ABROTANINE - [chem] (toxic alkaloid) abrotanine *f*
ABS - (abbrev for absolute) absolu
ABSCESS - [med] abscès
~WALL - [med] paroi *f* d'un abscès
ABSCISSA - [math] (the distance of a point from the axis of ordinates) abscisse *f*
~AXIS - [radar] (the line on the screen of a cathode-ray tube along which the electron ray moves) axe *m* horizontal
ABSCISSION - [gen] abscission *f*, excission *f*
ABSENCE - [gen] absence *f*
~OF CURRENT - [el] absence *f* de courant
~OF PATTERN - [autom] absence *f* de plan, absence *f* de modèle
ABSENTEE - [ind etc] absentéiste *m* & *f*
ABSENTEISM - [ind etc] absentéisme *m*
ABSINTHIN - [chem] absinthine *f*
ABSOLUTE - [gen] absolu
[chem] pur, absolu
~ADDRESS - [comput] (label assigned to a specific register in the storage) adresse *f* absolue
~AGE - [nucl] (the age of some minerals determined by the natural radioactivity) âge *m* absolu
~ALCOHOL - [chem] (water-free alcohol) alcool *m* absolu, alcool *m* anhydre
~ALTIMETER - [instr] altimètre *m* radio-électrique
~ALTITUDE - [aero] (height above the earth's surface) altitude *f* absolue
~APPARATUS - [instr] appareil *m* gradué en valeurs absolues
~ASSAY - [nucl] dosage *m* absolu
~CALIBRATION - [instr] étalonnage *m* absolu
~CEILING - [aero] (maximum height which can be reached in normal conditions) plafond *m* théorique
~CODING - [comput] (coding which uses absolute addresses) codage *m* en adresses absolues
~DRIFT - [electron] dérive *f* absolue
~ELECTRICAL UNIT - [el] (ampere: unit of electric current; volt: unit of potential difference between two points of a conductor; ohm: unit of electrical resistance; coulomb: unit of quantity of electricity) unité *f* électrique absolue
~ELECTRICAL UNIT SYSTEM - [el] système *m* d'uni-

tés électriques absolues
ABSOLUTE ELECTRODE POTENTIAL - [el] (electric potential in the electrode metal less that of the solution) potentiel *m* absolu d'électrode
~ELECTROMETER - [instr] (used for absolute measurement of potential) électromètre *m* absolu (Kelvin)
~ERROR - [instr] (of an instrument) erreur *f* absolue
~FILM - [cin] cinéma *m* pur
~FORCE - [phys] force *f* absolue
~FREQUENCY METER - [instr] fréquencemètre *m* absolu
~GALVANOMETER - [instr] (used for absolute measurement of current) galvanomètre *m* absolu
~GAUSS - [el] (unit of magnetic potential) gauss *m* absolu
~GILBERT - [el] (unit of magnetomotive force) gilbert absolu
~HUMIDITY - [met] (quantity of water vapour in a cubic metre of atmosphere) humidité *f* absolue
~INSENSITIVITY - [instr] insensibilité *f* absolue
~INSTRUMENT - [instr] (instrument measuring a quantity in absolute units) instrument *m* de mesure gradué en valeurs absolues
~LUMINOSITY - [phys] luminosité *f* absolue (totale)
~MAGNITUDE - [astr] (the apparent magnitude of a star at the distance of ten units used for the measurement of stellar distances) grandeur *f* absolue
~MEASUREMENT INSTRUMENT - [instr] appareil de mesure absolue
~METHOD OF MEASUREMENT - [instr] méthode de mesure absolue
~MOTION - [phys] mouvement *m* absolu
~OERSTED - [el] (unit of magnetic intensity) oersted *m* absolu
~PARALLAX - [opt] parallaxe *f* absolue
~PERMEABILITY - [el] (magnetic flux density divided by the magnetic field intensity in relation to an isotropic medium) perméabilité *f* absolue
~PERMITTIVITY - [el] permittivité *f* absolue, constante *f* diélectrique absolue
~PITCH - [acoust] diapason *m* absolu
~POTENTIAL - [el] (true potential difference between metal and solution) potentiel *m* absolu
~PRESSURE - [phys] pression *f* absolue
~PROGRAMMING - [comput] (programming with the use of absolute addresses) programmation *f* en adresses absolues
~SENSITIVITY - [instr] (ratio of the deflection change to the change in the quantity) sensibilité *f* absolue
~SPEED - [phys] vitesse *f* absolue
~SPEED DROP - [mech] (numerical value of the absolute speed variation between two working conditions) chute *f* absolue de vitesse
~SPEED RISE- [mech] élévation *f* absolue de vitesse
~SPEED VARIATION - [mech] (algebraic variation of the speed between two working conditions) variation *f* absolue de vitesse
~STOP LIGHT - [railw] feu *m* non franchissable
~STOP SIGNAL - [railw] signal *m* d'arrêt absolu
~SYSTEM OF UNITS - [phys] système *m* d'unités absolues
~TEMPERATURE - [phys] (the temperature measured in relation to the absolute zero) température absolue
~UNIT - [phys] (unit in terms of basic unit) unité *f* absolue
ABSOLUTE VACUUM - [phys] (perfect vacuum) vide *m* absolu
~VALUE - [math] valeur *f* absolue
~VALUE REPRESENTATION - [comput] (of complex numbers) représentation *f* de la valeur absolue
~WEIGHT - [phys] (weight of a body in a vacuum) poids *m* absolu
~ZERO - [phys] (temperature equal to -273.1°C) zéro *m* absolu
ABSOLUTELY REFRACTORY STATE - [el] état *m* absolument réfractaire
ABSORB, to - [gen] absorber
[mech] amortir
ABSORBABILITY - [chem] absorptivité *f*
ABSORBED - [gen & chem] absorbé
~DOSE - [nucl] dose *f* absorbée
~ENERGY - [phys] énergie *f* absorbée
~NEUTRON - [nucl] neutron *m* absorbé
ABSORBENT - [chem] absorbant *m*
~COTTON - [ind chem] coton *m* hydrophile
ABSORBER - [el] (in an electrical circuit) absorbeur *m*
[oil ind] colonne *f* d'absorption, tour *f* d'absorption
~CIRCUIT - [el] circuit *m* absorbant
~VALVE - [radio] (valve used in an absorption modulator) tube *m* absorbant
ABSORBING - [gen] absorbant
~AGENT - [chem] absorbant *m*
~CAPACITY - [phys & chem] pouvoir *m* absorbant
~CIRCUIT - [el] circuit *m* absorbant
~COMPLEX - [soil] (the ion-absorbing material of the soil) complexe *m* absorbant
~FILTER - [radio] filtre *m* absorbant
~MATERIAL - [ind chem] substance *f* absorbante
~MEDIUM - [nucl] milieu *m* absorbant
[chem] s. absorbing agent
~POWER - [phys] pouvoir *m* absorbant
~ROD - [nucl] barre *f* absorbante, tige *f* absorbante
~SHUTTER - [constr] (or vacuum shutter, vacuum form shutter; timber shuttering in a dam) coffrage *m* absorbant, coffrage *m* sous vide
ABSORBIT - [chem] charbon *m* activé, charbon *m* actif
ABSORBITIVITY - [phys] (of radiant heating) absorptivité *f*
ABSORPTIOMETER - [instr] (used to measure the solubility of gas in a liquid) absorptiomètre *m*
ABSORPTION - [gen & chem] absorption *f*
~APPARATUS - [ind chem] appareil *m* d'absorption
~APPLIANCE - [ind chem] dispositif *m* absorbeur
~BAND - [opt] (dark interruption of the spectrum of white light) bande *f* d'absorption
~BOTTLE - [oil ind] flacon *m* d'absorption
~CAPACITY - s. absorbing capacity
~CIRCUIT - s. absorbing circuit
~COEFFICIENT - [chem] coefficient *m* d'absorption
~COIL - [oil ind] serpentin *m* pour l'absorption
~COLUMN - [ind chem] colonne *f* absorbante, colonne *f* d'absorption
~COMPLEX - s. absorbing complex
~CONTROL - [nucl] réglage *m* par absorption
~CROSS-SECTION - [nucl] section *f* efficace d'absorption
~CURVE - [chem] courbe *f* d'absorption
~DISCONTINUITY - [chem] discontinuité *f* d'absorption
~DYNAMOMETER - [instr] amortisseur *m* à moulinet
~EDGE - [phys] (as of X-rays) limite *f* d'absorption
~EFFECT - [phys] (gettering effect) effet *m* getter

ABSORPTION EXTRACTION - [chem] extraction *f* par absorption
~FACTOR - [chem] s. absorption coefficient
~FLASK - s. absorption bottle
~FREQUENCY METER - [instr] ondemètre *m* à absorption
~LIMITING FREQUENCY - [radio] fréquence *f* limite d'absorption
~LINE - [phys] (of a spectrum) raie *f* d'absorption
~LOSS - [chem] perte *f* par absorption
~METER - [instr] absorptiomètre *m*
~METHOD - [radio] méthode *f* par absorption
~MODULATION - [radio] modulation *f* par absorption
~OF A NEUTRON - [nucl] absorption *f* d'un neutron
~OF CHARGED PARTICLES - [el] absorption *f* de porteurs électrisés
~OF HEAT BY EVAPORATION - [phys] absorption *f* de chaleur par vaporisation
~OF HYDROGEN BY METALS - [metall] absorption *f* de l'hydrogène par les métaux
~OF NOISE - [acoust] absorption *f* du bruit, absorption des sons
~OF OXYGEN BY COAL - [chem] absorption *f* d'oxygène par la houille
~OF THERMAL NEUTRON - [nucl] absorption des neutrons thermiques
~OF WATER - [chem] absorption *f* d'eau
~OIL - [chem] huile *f* d'épuration, huile *f* de lavage
~PIPETTE - [ind chem] pipette *f* d'absorption
~PLANE - [chem] plan *m* d'absorption
~PLANT - [oil ind] installation *f* de récupération des gaz
~POINT - [chem] point *m* d'absorption
~PROCESS - [phys & chem] (physical or chemical change of the absorbent during process) modification *f* de l'absorbant au cours du processus d'absorption
~REFRIGERATOR - [mech] réfrigérateur *m* à absorption
~RESISTANCE - [chem] résistance *f* à l'absorption
~SPECTROMETRY - [phys] spectrométrie *f* d'absorption
~SPECTRUM - [opt] spectre *m* d'absorption
~TOWER - s. absorption column
~TRAP - [ind chem] piège *f* d'absorption
~WAVEMETER - s. absorption frequency meter
ABSORPTIVE - [gen] absorbant
~ATTENUATOR - [radio] atténuateur *m* d'absorption
~POWER - [phys] (that part of the incident radiation absorbed by the surface) pouvoir *m* absorbant
ABSTAT UNIT - [el] (absolute electrostatic unit) unité *f* électrostatique absolue
ABSTRACT, to - [chem] distiller, extraire
ABSTRACT - [gen] résumé *m*
~CODE - [comput] pseudo-code *m*
~HEAT, to - [mech] soustraire de la chaleur, éliminer de la chaleur
~MATHEMATICS - [math] mathématique *f* pure
~MECHANICS - [mech] mécanique *f* rationnelle
~NUMBER - [math] nombre *m* abstrait
~SET - [cin & telev] décoration *f* de studio
ABSTRACTING - [gen] extrait *m*, extraction *f*, relevé *m*, résumé *m*
ABSTRACTION OF HEAT - [mech] élimination *f* de chaleur, soutirage *m* de chaleur
ABUNDANCE RATIO - [nucl] rapport *m* isotopique
ABUT, to - [gen & constr] buter, s'appuyer contre
~AGAINST - [mining] disparition *f* brusque d'une couche
ABUTMENT - [constr] arc-boutant *m*, butée *f*, coulée *f* (pont), pied-droit *m*, pilier *m* (d'une voûte) [dent] dent *f* de soutien
~CHEEK - [constr] chambrante *f*
~HINGE - [constr] gond *m* de chambrante
~LINE - [constr] ligne *f* de poussée
~PIER - [constr] (support at end of bridge) arc-boutant *m*, pile *f*
~PRESSURE - [constr] poussée *f*
~WALL - [constr] pied-droit *m*, culée *f*, arc *m* boutant
ABUTTAL - s. abutment [dent]
ABUTTING - [gen] adjacent, aboutissant
~END - [carp] (of a surface) about *m*
~JOINT - [carp] assemblage *m* en about
ABVOLT - [el] (unit of electromotive force) volt *m* absolu (u.é.m. C.G.S. de force électromotrice)
ABYSSAL - [gen] abyssal
~AREA - [geol] région *f* abyssale
~DEPOSITS - [geol] dépôts *m* pl abyssaux
~ROCKS - [geol] (deep-seated rocks) roches *f* pl abyssales, roches *f* pl plutoniques (ou plutoniennes)
ABYSSINIAN DRIVEN WELL - [hydr] puits *m* abyssin, puits *m* instantané
~GOLD - [metall] alliage *m* de cuivre et de zinc (légèrement plaqué d'or)
A.C. - [el] (alternating current) courant *m* alternatif
[comput] (automatic computer) calculatrice *f* électronique automatique
[aero] (approach control) contrôle *m* d'approche
A.C. AMMETER - [el] ampèremètre *m* pour courant alternatif
A.C. BALANCER - [radio] bobine *f* égalisatrice
A.C. BRIDGE - [el] pont *m* à courant alternatif
A.C. CURRENT - [el] courant *m* alternatif
A.C. DUMP - [el] chûte *f* de tension alternative
A.C. ERASING HEAD - [el acoust] tête *f* d'effacement à courant alternatif
A.C. GENERATOR - [el] alternateur *m*
A.C. HYSTERESIS LOOP - [el] boucle *f* d'hystérésis dynamique
A.C. INTERRUPTION - [el] interruption *f* du réseau d'alimentation (alternative)
A.C. LINE FREQUENCY - [el] fréquence *f* de ligne
A.C. MAGNETIC BIASING - [el] polarisation *f* magnétique alternative
A.C. MAINS - [el] réseau *m* de distribution de courant alternatif
A.C. METER - [instr] appareil *m* de mesure de courant alternatif
A.C. MOTOR - [el] moteur *m* à courant alternatif
A.C. SYSTEM - [el] installation *f* à courant alternatif
A.C. VOLTAGE - [el] tension *f* du courant alternatif
A.C.-D.C. - [el] CA-CC, tous courants, universel
A.C.-D.C. MOTOR - [el] moteur *m* à tous courants
A.C.-D.C. RECEIVER - [radio] récepteur *m* tous courants
A.C.-D.C. TUBE - [radio] tube *m* universel
A.C.-D.C. VALVE - s. A.C.-D.C. tube
ACANTHITE - [min] acanthite *f* (sel d'argent)
ACANTHUS - [bot] acanthe *f*
ACARDITE - [chem] diphénylurée *f*
ACARICIDE - [chem] acaricide *m*

ACCELERATE, to - [gen] accélérer
ACCELERATED AGEING TEST - [phys à metall] essai *m* de vieillissement accéléré
~LIGHT AGEING - [phys] essai *m* de vieillissement accéléré à la lumière
~MOTION - [phys] mouvement *m* accéléré
~WEATHERING TEST - [phys] essai *m* de vieillissement accéléré aux agents atmosphériques
ACCELERATING - [gen] accélérateur, d'accélération
~ABILITY - [mech] accélération *f*, reprise *f*
~ANODE - [electron] anode *f* accélératrice
~CHAMBER - [nucl] chambre *f* accélératrice
~ELECTRODE - [el] (electrode kept at high positive potential) électrode *f* accélératrice
~FIELD - [electron] champ *m* accélérateur
~GRID - [electron] grille *f* d'accélération
~JET - [mech] (carburettor) compensateur *m* (du carburateur)
~PERIOD - [phys] période *f* d'accélération
~POTENTIAL - [electron] (potential of the accelerating electrode) potentiel *m* d'accélération, potentiel de reprise
~PUMP - [mech] pompe *f* d'accélération
~RELAY - [el] relais *m* d'accélération
~TUBE - [electron] tube *m* accélérateur
~VOLTAGE - [el] tension *f* accélératrice
ACCELERATION - [gen] accélération
~CONSTANT - [mech] constante *f* d'accélération
~CONTROL UNIT - [mech] régulateur *m* d'accélération
~CURVE - [mech] courbe *f* d'accélération
~DUE TO GRAVITY - [mech] accélération *f* de la pesanteur
~ERROR - [mech] erreur *f* d'accélération
~FACTOR - [mech] facteur *f* d'accélération
~GRID - [el] grille *f* accélératrice
~INDICATOR - [instr] indicateur *m* d'accélération
~INSTRUMENT - [instr] appareil *m* indicateur d'accélération
~JET - [mech] compensateur *m*
~LAG - [mech] espace *m* d'accélération
~METER - [instr] accéléromètre *m*
~MISALIGNMENT - [mech] désalignement *m*
~OF GRAVITY - [mech] accélération *f* de la pesanteur
~OF OSCILLATORY MOTION - [mech] accélération d'un mouvement oscillatoire
~OF THE CURRENT - [el] accélération *f* du courant
~PUMP - [mech] pompe *f* d'accélération
~SPACE - [mech] espace *m* d'accélération
~TRANSDUCER - [el] transducteur *m* d'accélération, capteur *m* d'accélération
~VOLTAGE - [el] tension *f* anodique; (electron)tension *f* entre cathode et anode
ACCELERATOR - [gen & mech] accélérateur *m*
~PEDAL - [auto] pédale *f* d'accélérateur
~PEDAL SPINDLE - [auto] axe *f* de pédale d'accélérateur
~PUMP - [auto] pompe *f* d'accélérateur
~PUMP JET - [auto] gicleur *m* de pompe d'accélération
~PUMP PISTON - [auto] piston *m* de pompe d'accélération
~TUBE - [electron] tube *m* accélérateur
ACCELERATORS - [chem] (substances which hasten the speed of a reaction) accélérateurs *m* pl
ACCELEROMETER - [instr] accéléromètre *m*
ACCENT - [gen] accent *m*
ACCENTUATION - [photo] accentuation *f*, renforcement *m*
[radio & telev] (in audio systems) accentuation *f*, amplification *f* sélective
ACCEPT - [gen] accepter, admettre
ACCEPTABLE QUALITY LEVEL - [ind] (of manufactured items) niveau *m* de qualité acceptable
ACCEPTANCE - [gen & comm] réception *f*, recette *f*
~FLIGHT - [aero] vol *m* de recette
~FLIGHT TEST - [aero] essai *m* de recette en vol
~GAUGE - [instr] calibre *m* d'essai de recette
~TEST - [ind] essai *m* de recette, essai *m* de reception
ACCEPTATION - [gen & comm] acceptation *f*, accord *m*, agrément *m*
ACCEPTOR - [phys] accepteur *m*, élément *m* accepteur
~ATOM - [nucl] atome *m* accepteur
~CIRCUIT - [electron] circuit *m* accepteur
~IMPURITY - [electron] (in a semiconductor) impureté *f* acceptrice
~LEVEL - [electron] (in the energy diagram of anextrinsic semiconductor) niveau *m* accepteur
ACCESS - [gen] accès *m*, entrée *f*
~CYCLE - [comput] (of a core memory) cycle *m* d'accès
~DOOR - [constr] porte *f* de visite, porte *f* d'accès
~EYE - [constr] orifice *m* d'accès, regard *m*
~PANEL - [constr] panneau *m* de visite
~ROAD - [constr] voie *f* d'accès
~TIME - [comput] (memory) temps *m* d'accès (mémoire), temps *m* de montée, durée *f* d'établissement (impulsion, signal)
ACCESSIBILITY - [gen] accessibilité *f*
ACCESSIBLE - [gen] accessible
~HEARTH - [metall] sole *f* accessible
~SHAFT - [metall] cuve *f* accessible
~TERMINAL - [el & telecomm] borne *f*
ACCESSORY - [gen] accessoire
~APPARATUS - [gen & mech] appareil *m* auxiliaire
~DRIVE - [auto] commande *f* auxiliaire
~EQUIPMENT - [gen] appareillage *m* auxiliaire
~GEAR TRAIN - [mech] train *m* d'engrenage auxiliaire
~MINERALS - [min] minerais *m* pl accessoires
~PLATE - [genet] plaque *f* accessoire
ACCIDENT - [gen] accident *m*
~AT WORK - [ind] accident *m* du travail
~FREE BONUS - [auto insur] prime *f* de non-accident
~INSURANCE - [insur] assurance *f* contre les accidents
~PREVENTION - [ind] prévention *f* des accidents
ACCIDENTAL - [gen] accidentel, fortuit
~COINCIDENCE - [phys] coïncidence *f* accidentelle
~COINCIDENCE CORRECTION - [autom] correction *f* de coïncidences accidentelles
~EARTH - [el] mise *f* à la terre accidentelle
~ERROR - [gen] erreur *f* accidentelle
~PRINTING - [el acous] effet *m* d'echo, effet *m* d'empreinte magnétique
ACCLIMATION - s. acclimatization
ACCLIMATIZATION - [gen] acclimatisation *f*
ACCLIVITY - [geol] acclivité *f*, montée *f*
ACCOMMODATION [gen] ajustement *m*, adaptation *f*
[physiol, opt etc] accommodation *f*
[el] aménagement *m* électrique

ACCOMMODATION COEFFICIENT FOR CONDENSATION - [phys] (condensation coefficient; the ratio between the number of molecules condensing on a surface per second and the total number incident per second) coefficient *m* de condensation, probabilité *f* de condensation
~LADDER - [naut] échelle *f* de coupée
~OF THE CREW - [naut] logement *m* de l'équipage
~ROAD - [constr] chemin *m* de terre
~TRAIN - [railw] (US) train *m* omnibus
ACCOMPANIMENT - [gen] accompagnement *m*, accessoires *mpl*
ACCOMPANYING MINERAL - [min] minérai *m* associé
ACCORD - [gen] accord *m*, consentement *m*
[mus] accord *m*
ACCORDANCE - [gen] conformité *f*
ACCORDING SCALE - [gen] à l'échelle de
ACCORDION - [mus] accordéon *m*
~DOORS - [constr] portes *fpl* repliantes
~FOLDED - [gen] plié accordéon
ACCOUNT - [gen] compte *m*, calcul *m*
~BOOK - [comm] livre *m* (de comptes)
~CARD - [comput] carte *f* compte
~FOR, to - [gen & comm] justifier, rendre compte
~STATED - [comm] arrêté *m* de compte
ACCOUNTABILITY - [gen] responsabilité *f*
ACCOUNTANCY - [comm] comptabilité *f*, tenue *f* des livres
ACCOUNTANT - [comm] comptable *m*, expert comptable *m*
ACCOUNTING - s.accountancy
~DEPARTMENT - [gen] service *m* de la comptabilité
~MACHINE - [mech] machine *f* comptable
ACCOUNTS PAYABLE - [comm] dettes *fpl* passives
~PAYABLE REMITTANCE - [comput] extrait *m* de compte créditeurs
~RECEIVABLE - [comm] dettes *fpl* actives
~RECEIVABLE STATEMENT - [comput] extrait *m* de compte débiteurs
ACCOUTERMENT - s.accoutrement
ACCOUTREMENT - [gen] équipement *m* (militaire)
ACCRA COPAL - [min] copal *m* dur (d'Afrique)
ACCRETION - [gen] accroissement *m*
[med] synechie *f*, adhérence *f*, bride *f*
[hydrol] (ground water) enrichissement *m* (de la nappe), raveinement *m*, renforcement *m*
ACCROIDES GUM - [chem] (a red gum soluble in hot alcohol) gomme *f* acroïde (ou accroïde)
ACCUMULATE - [gen] accumuler, s'amasser
~HEAT, to - [phys] accumuler de la chaleur
ACCUMULATING COUNTER - [comput] compteur *m* additif, compteur *m* totalisateur
~SPEED - [comput] vitesse *f* (ou rapidité) d'addition, vitesse *f* de totalisation (ou de sommation)
~STIMULUS - [biol] stimulus *m* d'accumulation
ACCUMULATION - [gen] accumulation *f*
~OF CHARGE - [el] accumulation *f* de charge
~TIME - [nucl] temps *m* d'accumulation (requis pour accumuler un quantum d'énergie rayonnante avant de la libérer)
ACCUMULATOR - [el] (storage battery) accumulateur *m*
~ACID - [chem] acide *m* pour accumulateur
~BATTERY - [el] batterie *f* d'accumulateurs
~BOX - [el] accumulator plates and electrolyte container) bac *m* d'accumulateur
ACCUMULATOR CAN - s.accumulator box
~CELL - [el] élément *m* d'accumulateur
~CHARGE - [el] charge *f* d'un accumulateur
~CONTENTS - [comput] contenu *m* d'un accumulateur
~GRID - [el] (a grid forming a plate) grille *f* (d'une plaque d'accumulateur)
~JAR - s.accumulator box
~PLATE - [el] plaque *f* d'accumulateur
~RECTIFIER - [el] redresseur *m* pour accumulateurs
~STILL - [ind chem] réservoir *m* de décantation
~SUBSTATION - [el] sous-station *f* d'accumulateurs
~SUITABLE FOR RAPID DISCHARGE - [el] accumulateur *m* à décharge rapide
~SWITCH - [el] commutateur *m* d'accumulateurs
~TANK - [hydr] réservoir *m* d'eau
[nucl] réservoir *m* d'emmagasinage
~TRACTION [el] (system of electric traction fed by batteries) traction *f* électrique (par batteries)
~VEHICLE - [el] (electrically propelled vehicle) véhicule *m* électrique (à accumulateurs)
~VESSEL - s.accumulator box
ACCURACY - [gen] exactitude *f*, justesse *f*, précision *f*
~GRADE - [instr] (degree of accuracy) degré *m* de précision
~IN MEASUREMENT - [meas] précision *f* de la mesure
~OF FIRE - [gen] (of firearms) précision *f* du tir
~TEST - [gen] (e.g. of firearms) essai *m* de précision, essai *m* d'exactitude
ACCURATE - [gen] exact, juste, précis
~CURRENT RANGE OF A METER - [el] (effective current range of a meter) plaque *f* utile d'un appareil de mesure
~GRINDING - [mech] rectification *f* de précision
~RANGE MARKER - [surv] marqueur *m* de distance précise
~SCANNING - [radar] analyse *f* de précision, exploration *f* précise
A.C.E. - [comput] (automatic computer engine) calculatrice *f* électronique automatique
ACE - [phys etc] (particle) particule *f*
ACENAPHTHENE - [chem] acénaphtène *m*
ACENAPHTHENEDIONE - [chem] acénaphténédione *m*
ACENAPHTHYLENE - [chem] acénaphtylène *m*
ACENTRIC - [gen] sous centre, acentrique
~INVERSION - [genet] (of a chromosome section) inversion *f* acentrique
ACERATION - [metall] (steeling) aciérage *m*, aciération
ACERDESE - [min] (manganite) acerdèse *f*, manganite *f*
ACERDOL - [chem] (calcium permanganate) permanganate *m* de calcium
ACESCENT - [gen] acescent, acide
ACETAL - [chem] acétal *m*
ACETALDEHYDE - [chem] acétaldéhyde *m*, aldéhyde *m* acétique, éthanal *m*
~OXIME - [chem] oxime *m* d'acétaldéhyde
ACETALDOXIME - [chem] acétaldoxime *m*
ACETAMIDE - [chem] (a primary amide) acétamide *m*
ACETAMIDINE - [chem] acétamidine *f*
ACETANILIDE - [chem] (aniline and glacial acetic acid) acétanilide *f*
ACETANISOL - [chem] acétanisol *m*
ACETANNIN [chem] acide *m* diacétyltannique
ACETATE - [chem] (salt of acetic acid) acétate *m*

ACETATE DISK - [el magn] (a recording disk made of various acetate compounds) disque *m* à l'acétate
~DOPE - [ind chem] vernis *m* cellulosique
~DYE - [text] colorant *m* pour la rayonne acétate
~FILM - [photo] (non-inflammable film) pellicule *f* cellulosique (à base d'acétate de cellulose)
~OF ALUMINIA - [chem] acétate *m* d'alumine
~OF ALUMINIUM - [chem] acétate *m* d'aluminium
~OF BARIUM - [chem] acétate *m* de baryte
~OF IRON - [chem] acétate *m* de fer
~OF LEAD - [chem] acétate *m* de plomb
~OF LIME - [chem] acétate *m* de calcium
~OF POTASSIUM - [chem] acétate *m* de potasse
~PROCESS - [text] procédé *m* à l'acétate
~RAYON - [text] (a cellulosic fibre) rayonne *f* acétate
ACETIC - [chem] acétique
~ACID - [chem] (important raw material, obtained from wood, acetylene or alcohol; the acid of vinegar) acide *m* acétique
~ALDEHYDE - [chem] aldéhyde *m* acétique, éthanal *m*
~ANHYDRIDE - [chem] anhydride *m* acétique
~ETHER - [chem] (ethyl acetate) éther *m* acétique, acétate *m* d'éthyle
~FERMENTATION - [chem] (the fermentation of alcohol solutions by oxidation) fermentation *f* acétique
~OXIDE - s.acetic anhydride
~NITRILE - s.acetonitrile
ACETIMETER - [ind chem] acétimètre *m*, acétomètre
ACETIN - [chem] (any of three liquid acetates formed when glycerol and acetic acid are heated together) acétine *f*, monoacétate *m* de glycérine
~ACETIC BLUE - [chem] bleu *m* d'acétine
ACETOACETIC ACID - [chem] acide *m* acéto-acétique
~ETHER - [chem] (used in perfumery etc) acétylacétate *m* d'éthyle
ACETOACETOANILIDE - [chem] acétoacétanilide *f*
ACETOBROMIDE - [chem] bromoacétamide *f*
ACETOETHYLAMIDE - [chem] éthylacétamide *f*
ACETOL - [chem] (obtained indirectly from acetone) acétol *m*
ACETOLYSIS - [chem] (reaction analogous to hydrolysis in which acetic acid has the role of water) acétolyse *f*
ACETOMETER - s.acetimeter
ACETOMETHYLANILIDE - [chem] acétométhylanilide
ACETONAPHTHALIDE - [chem] acétonaphtalide *f*
ACETONATE, to - [chem] convertir en combinaison acétonique
ACETONATE - [chem] acétonate *m*
ACETONE - [chem] (a very important solvent; also a basis for organic synthesis) acétone *m*, propanone *f*
~AZINE - [chem] azine *f* acétonique
~CARBOXYLIC ACID - [chem] acide *m* acétoacétique
~CHLORIDE - [chem] chlorure *m* d'acétone
~CHLOROFORM - [chem] acétone-chloroforme *m*
~EXTRACT - [chem] extrait *m* acétonique
~FLUORIDE - [chem] fluorure *m* d'acétone
~OIL - [chem] (a solvent for cellulose esters) huile *f* d'acétone
~OXIME - [chem] acétoxime *m*
~SOLUBLE MATTER - [chem] substance *f* soluble dans l'acétone
ACETONITRILE - [chem] (a solvent, also used in organic synthesis) acétonitrile *m*
ACETONYLACETONE - [chem] acétonylacétone *f*
ACETOPHENONE - [chem] acétophénone *f*
ACETOXIME - [chem] acétoxime *f*
ACETYL - [chem] (the acetyl group is the radical of acetic acid) acétyle *m*
~ANILINE - [chem] acétylaniline *f*
~BROMO ANILINE - [chem] acétyl-bromo-aniline *f*
~BURNER - [ind chem] brûleur *m* à l'acétylène
~CYLINDER - [ind chem] bouteille *f* d'acétylène
~GENERATOR - [el] générateur *m* d'acétylène
~ISATIN - [chem] acétylisatine *f*
~METHYLUREA - [chem] acétylméthylurée *f*
~NUMBER - [chem] indice *m* d'acétyle
~PEROXIDE - [chem] peroxyde *m* d'acétyle
~SALICYLIC ACID - [chem] (aspirin) acide *m* acétylsalicylique
~VALUE - s.acetyl number
ACETYLACETANILIDE - [chem] (a plastifier) acétylacétanilide *m*
ACETYLACETONE - [chem] acétylacétone *f*
ACETYLAMINOPHENOL - [chem] acétylaminophénol *m*
ACETYLANILIDE - [chem] éthylacétanilide *f*
ACETYLATIVE DESULFATION - [chem] passage *m* d'un ester sulfurique à un ester acétique
ACETYLCHOLINE - [chem] acétylcholine *f*
ACETYLENE - [chem] (colourless poisonous gas, prepared by the action of water on calcium carbide. Used for welding, lighting, acetic acid synthesis etc) acétylène *f*
~BLACK - [ind chem] noir *m* d'acétylène
~BUNSEN BURNER - [metall] brûleur *m* de Bunsen à l'acétylène
~FLAME - [metall] flamme *f* acétylénique
~GAS - [metall] gaz *m* d'acétylène
~GENERATOR - [ind chem] générateur *m* d'acétylène
~LAMP - [ind chem] lampe *f* à acétylène
~OXYGEN BLOW-PIPE FLAME - [metall] flamme *f* de chalumeau oxy-acétylénique
~WELDING - [metall] soudure *f* autogène
ACETYLENIC ALCOHOL - [chem] alcool *m* acétylènique
~LINKAGE - [chem] liaison *f* acétylénique
ACETYLITH - [chem] carbure *m* de calcium
ACETYLTHIOUREA - [chem] acétylthiourée *f*
ACETYLUREA - [chem] acétylurée *f*
ACETYLVANILLIN - [chem] acétylvanilline *f*
ACHESON FURNACE - [metall] four *m* électrique pour la fabrication du carborundum
ACHILLEA OIL - [chem] essence *f* d'achillée
ACHROMATIC - [opt] achromatique
~FIGURE - [genet] (the mitotic spindle and cell structures which do not stain with usual microtechnical dyes) fuseau *m* achromatique
~LENS - [opt] lentille *f* achromatique, objectif achromatique
~POINT - [telev] (in colour television) point *m* blanc
~PRISM - [opt] prisme *m* achromatique
~THRESHOLD - [telev] seuil *m* achromatique
ACHROMATICITY - [opt] achromatisme *m*
ACHROMATISATION - [opt] achromatisation *f*
ACHROMATISM - s. achromaticity
ACHROMATIZE, to - [opt] (to deprive of colour) achromatiser
ACICULAR - [gen] aciculaire, en forme d'aiguille
~CAST IRON - [metall] fonte *f* aciculaire, fonte *f* baitinique
~HABITUS - [crystal] faciès *m* aciculaire

ACID - [chem] acide *m*
~AMIDES - [chem] amides *f*pl acides
~BATH - [chem] bain *m* acide
~BESSEMER STEEL - [metall] acier *m* Bessemer
~BRIGHT DIP - [chem] solution *f* de polissage
~BRITTLENESS - s.acid embrittlement
~BLOW CASE - [ind chem] monte-jus *m* (pour acides)
~BRONZE - [metall] (used in chemical industries) alliage *m* contenant 78 p.c. de cuivre
~BROWN - [text] colorant *m* pour laines
~CASEIN - [chem] caséine *f* obtenue sous l'influence d'un acide
~CHAMBER - [ind chem] chambre *f* de plomb
~COLOUR - [dyes] couleur *f* acide
~CONCENTRATION - [chem] concentration *f* d'acide
~DIPPING - [metall] décapage *m* au bain acidulé
~DYE - [dyes] colorant *m* acide
~EGG - [ind chem] (closed vessel for pumping liquids, especially corrosives by means of compressed air) monte-jus *m*
~ELECTRIC PROCESS - [metall] procédé *m* de fabrication de l'acier à l'arc électrique avec laitier acide
~EMBRITTLEMENT - [metall] fragilité *f* par décapage
~ETCHING - [metall] décapage *m* à l'acide
~ESTER - [chem] (acid compound with an alkyl radical replacing the hydrogen) ester *m* acide
~FAST - [chem] résistant aux acides, inattaquable aux acides, antiacide
~FASTNESS - [chem] résistance *f* aux acides
~FERMENTATION - [chem] fermentation *f* acide
~-FIXING BATH - [photo] (to obtain rapid cessation of development) acide *m* fixateur
~FUCHSINE - [chem] fuchsine *f* acide
~FUME - [chem] vapeur *f* acide
~FUNNEL - [ind chem] entonnoir *m* à acide
~GREEN - [chem] vert *m* acide
~HEARTH - [metall] four *m* Martin
~LATEX - [chem] latex *m* acide
~LEAD ARSENATE - [chem] biarséniate *m* de plomb
~LEAD SULPHATE - [chem] sulfate *m* acide de plomb
~LINING- [metall] revêtement *m* intérieur acide (siliceux), garnissage *m* acide
~LIQUOR - [chem] bain *m* acide, liqueur *f* acide
~LITHIUM CARBONATE - [chem] bicarbonate *m* de lithium
~LITHIUM OXALATE - [chem] oxalate *m* acide de lithium
~MAGENTA - [chem] fuchsine *f* acide
~MAGNESIUM SULPHATE - [chem] bisulfate de magnésium
~MEDIUM - [chem] milieu *m* acide
~NUMBER - [chem] indice *m* d'acide
~OF ANTS - [chem] acide *m* formique
~OF APPLES - [chem] acide *m* maléique
~OF LEMON - [chem] acide *m* citrique
~OF MILK - [chem] acide *m* lactique
~OF SUGAR - [chem] acide *m* oxalique
~OPEN HEARTH FURNACE - [metall] four *m* Martin
~PHOSPHATE - [chem] superphosphate *m*
~PICKLING SOLUTION - [chem] solution *f* de décapage acide
~PIG IRON - [metall] fonte *f* acide
~PROCESS - [metall] procédé *m* acide
~PROCESS REGENERATION - [ind chem] régénération *f* à l'acide
~-PROOF - [chem & metall] inattaquable aux acides
~-PROOF BOX - [ind chem] boîte *f* à l'épreuve des acides
ACID-PROOF FLOORTILE - [constr] carreau *m* inattaquable aux acides (revêtement de sol)
~-PROOF SULPHURIC PUTTY - [ind chem] mastic *m* sulfo-asphaltique résistant aux acides
~PUMP - [ind chem] pompe *f* pour acides
~RADICAL - [chem] (a molecule of the acid with the hydrogen) radical *m* acide
~RECLAIM - [ind chem] régénération *f* à l'acide
~RESISTANCE - [chem] résistance *f* aux acides
~RESISTANT COATING - [ind chem] revêtement *m* inattaquable par les acides
~-RESISTING - s.acid fast
~-RESISTING ALLOY - [metall] alliage *m* résistant aux acides
~-RESISTING RUBBER - [rubber] caoutchouc *m* antiacide, caoutchouc *m* résistant aux acides
~ROOM - [ind chem & metall] dépôt *m* des acides
~ROSEINE - [chem] fuchsine *f* acide
~RUBIN - [chem] fuchsine *f* acide
~SALT - [chem] (a salt derived from the replacement of part of the hydrogen) sel *m* acide
~SEAL PAINT - [ind chem] vernis *m* inattaquable aux acides
~SHELLING DRUM - [ind chem] bac *m* de décantation d'acide
~SIZE - [ind chem] colle *f* à la résine
~SLAG - [metall] scorie *f* acide, laitier *m* acide
~SLUDGE - [chem] boue *f* acide
~SOIL - [soil] sol acide
~SOLUTION - [chem] solution *f* acide, électrolyte *m*
~SOLVENT - [chem] solvant *m* acide
~STAGE - [chem] fermentation *f* acide, phase *f* acide
~STEEL - [metall] (a steel obtained through an acid process) acier *m* acide
~TEST - [ind chem] essai *m* à l'acide, macrographie *f*
~THALLIUM SULPHATE - [chem] sulfate *m* acide de thallium
~TO LITMUS - [chem] réaction *f* acide au tournesol
~TOWER - [ind chem] colonne *f* à acide
~TREATMENT - [chem] traitement *m* à l'acide (sulfurique
~VALUE - s.acid number
~VALVE - [ind chem] vanne *f* pour acides
~VAPOUR - [chem] vapeur *f* acide
~YELLOW - [dyes] jaune *m* acide
ACIDIC CONTENT - [chem] (the degree of acidity) acidité *f*, teneur *f* en acide
~OXIDE [chem] oxyde *m* acide
ACIDIFEROUS - [chem] acidifère
ACIDIFIABLE - [chem] acidifiable
ACIDIFICATION - [chem] acidification *f*
ACIDIFIED WATER [chem] eau *f* acidulée
ACIDIFIER - [chem] acidifiant *m*
ACIDIFY, to - [chem] acidifier
ACIDIMETER - [instr] acidimètre *m*
ACIDIMETRY - [chem] acidimétrie *f*
ACIDITY - [chem] (the measure of acid content of a solution) acidité *f*
~MEASUREMENT - [chem] mesure *f* de la concentration (d'acide)
ACIDIZE, to - [chem] acidifier
ACIDIZING OF WELLS - [oil ind] traitement *m* à l'acide chlorhydrique
ACIDNESS - [chem] acidité *f*
ACIDOMETER - [instr] acidomètre *m*

ACIDULATE, to - [chem] aciduler
ACIDULATED WATER - [chem] eau *f* acidulée
ACIERABLE - [metall] aciérable
ACIERAGE - [metall] aciération *f*, aciérage *m*
ACIERATE, to - [metall] (to steel) aciérer
ACINOSE - [min] granuleux
ACKNOWLEDGEMENT SIGNAL - [sign] signal *m* d'aperçu, accusé *m* de reception
ACLINAL - s.aclinic
ACLINIC - [phys] aclinique
~LINE - [phys] (also called magnetic equator) ligne *f* isoclinique
ACME THREAD - [mech] filet *m* trapézoïdal
ACNODE - [math] acnode *m*
ACONINE - [chem] aconéine *f*
ACONITE - [chem] (the dried tuberous root of the plant) aconit *m*
ACONITIC ACID - [chem] acide *m* aconitique
ACONITINE - [chem] (an alkaloid of unknown constitution) aconitine *f*
ACORN - [bot] gland *m*, pomme *f* (de girouette)
~CUP - [mech] cupule *f*
~NUT - [mech] écrou *m* borgne
~OIL - [chem] huile *f* de glands
~TUBE - [electron] (a colloquial term for thermionic valve) tube-gland *m*
~VALVE - s.acorn tube
ACOUMETER - [instr] acoumètre *m*
ACOUPHONY - [acoust] acouphonie *f*
ACOUSTIC - [acoust] acoustique
~ABSORPTION - [acoust] absorption *f* acoustique
~ADMITTANCE - [el acoust] admittance *f* acoustique
~BACKING - [acoust] (e.g. of a studio) revêtement *m* acoustique
~BAFFLE - [acoust] écran *m* acoustique
~BASS - [acoust] basse *f* harmonique
~BRIDGE - [acoust] pont *m* acoustique
~BURGLAR ALARM - [el acoust] avertisseur *m* sonore antivol
~COEFFICIENT REPRODUCTION - [acoust] résonance *f* acoustique
~COMPLIANCE - [acoust] élasticité *f* acoustique
~CONSTRUCTION - [constr] construction *f* insonorisée
~DELAY LINE - [telecomm] ligne *f* à retard acoustique
~DISPERSION - [acoust] dispersion *f* acoustique
~DUCT - [anat] conduit *m* auditif
~FEEDBACK - [radio etc] (positive feedback loop, generally involving a loudspeaker and a microphone in which sound is fed back and amplified) réaction *f* acoustique
~GENERATOR - [el acoust] radiateur *m* acoustique
~HORN - [radio] pavillon *m* acoustique
~IMPEDANCE - [el acoust] (the ratio between sound pressure amplitude and volume-velocity amplitude) impédance *f* acoustique
~INSULATION OF ROCK WOOL - [acoust] isolation *f* phonique en laine de roche
~MEMORY - [comput] mémoire *f* acoustique
~OSCILLATION - [acoust] vibration *f* acoustique, oscillation *f* acoustique
~PICK-UP - [el acoust] lecteur *m* acoustique, pick-up *m* acoustique, phonocapteur *m*
~PLASTER - [acoust] (a plaster with good acoustic qualities) enduit *m* acoustique
~POWER - [acoust] puissance *f* acoustique
ACOUSTIC PRESSURE - [phys] pression *f* acoustique
~RADIATING ELEMENT - [radio] (a vibration element supplying acoustic energy, e.g. sound waves) radiateur *m* acoustique
~RADIATOR - s.acoustic radiating element
~RADIOMETER - [instr] radiomètre *m* acoustique, sonomètre *m* acoustique
~REACTANCE - [el acoust] (of sound waves) réactance *f* acoustique
~REFLECTION COEFFICIENT - [acoust] coefficient *m* de réflection acoustique
~REFLECTION FACTOR - s.acoustic reflection coefficient
~REFRACTION - [acoust] réfraction *f* acoustique
~RESISTANCE - [acoust] (of sound waves) résistance *f* acoustique
~SHOCK ABSORBER - [telecomm] amortisseur *m* acoustique, amortisseur *m* de bruit
~STIFFNESS - [acoust] raideur *f* acoustique
~STORE - s.acoustic memory
~STRAIN GAUGE - [meas] jauge *f* de contrainte acoustique
~TILE - [constr] carreau *m* insonorisant
~TREATMENT - [constr] insonorisation
~VIBRATION - [acoust] vibration *f* acoustique, oscillation *f* acoustique
~WAVE - [acoust] onde *f* acoustique
ACOUSTICAL - [acoust] acoustique
~CORRECTION - [acoust] correction *f* acoustique
~SYSTEM - [el acoust] système *m* acoustique
ACOUSTICIAN - [el acoust] ingénieur *m* du son
ACOUSTICS - [acoust] acoustique *f*
ACQUIRE, to - [gen] acquérir
ACQUIRED CHARACTER - [genet] caractère *m* acquis
ACQUISITION - [radar] acquisition *f*, localisation *f*
~AND TRACKING RADAR - [radar] radar *m* de localisation et de poursuite
A.C.R. - [radar] (Approach Control Radar) radar *m* de contrôle d'approche
ACRANIA - [med] acranie *f*
ACRE - [meas] (agricultural measure) acre *f*, arpent *m*
ACREAGE - [meas] superficie *f* en acres
ACRIBOMETER - [instr] (a pair of callipers used for the measurement of very small objects) acribomètre *m*
ACRID - [gen] âcre
ACRIDINE - [chem] (analogous to anthracene) acridine *f*
ACRIFLAVINE - [chem] acriflavine *f*
ACROBATIC FLYING - [aero] acrobatie *f* aérienne
ACROCARPOUS - [bot] (with fruit at the end of the branch) acrocarpe
ACROCENTRIC - [genet] (relating to a subterminal centromere) acrocentrique
ACROCEPHALIA - [anat] (congenital deformity of the skull) acrocéphalie *f*
ACROCYANOSIS - [med] acrocyanose *f*
ACRODERMATITIS - [med] acrodermatite *f*
ACROGAMY - [bot] acrogamie *f*
ACROGENOUS - [bot] acrogène
ACROPATHY - [med] acropatie *f*
ACROPHOBIA - [med] acrophobie *f*
ACROSPIRE - [bot] (the spiral plumule in a germinating grain) acrospire
ACROSPORE - [bot] acrospore *f*
ACROSS-LINE - [el] en court-circuit

ACROSS-LINE START - [el] démarrage *m* par directe connexion au réseau (moteurs à induction)
~THE GRAIN - [mech] à contre sens, à contre fil, à rebours
~THE LINE - [el] à connexion directe au réseau
ACROTER - [arch] acrotère *m*
ACRYLATE - [chem] (salt or ester of acrylic acid) acrylate *m*
ACRYLIC - [chem] acrylique
~ACID - [chem] (highly reactive acid belonging to the oleic acid series) acide *m* acrylique
~FIBRE - [ind chem] (synthetic fibre) fibre *f* acrylique
~RESINS - [plastics] (obtained through the polymerization of esters or amides of acrylic acid) résines *f* pl acryliques
~RUBBER - [rubber] (synthetic rubber partially made from acrylonitrile) caoutchouc *m* acrylique
ACT, to - [gen] agir, fonctionner, réagir sur
ACT - [gen & leg] acte *m*, action *f*, décret *m*, loi *f*
~OF GOD - [gen] cas *m* de force majeure
ACTH - [biol] (abbrev for adrenocorticotrophic hormone) ACTH (hormone adrénocorticotrophique)
ACTINIA - [zool] (genus of sea anemones) actinie *f*
ACTINIC - [chem] (having photochemical properties or effect) actinique
~LIGHT - [photo] lumière *f* actinique
~PHOTOMETER - [opt] photomètre *m* actinique
~RADIATION - [el magn] (electro-magnetic radiation capable of producing chemical effects on photo-sensitive materials) radiation *f* actinique
ACTINIDES - [nucl] (series of heavy radioactive metallic elements) actinides *m* pl
ACTINISM - [phys] (property of radiant energy) actinisme *m*
ACTINIUM - [chem] (symbol Ac; radioactive element found as a constituent of uranium ores) actinium *m*
~EMANATION - [chem] actinon *m*
~SERIES - [chem] famille *f* de l'actinium
ACTINOBACILLOSIS - [vet] actinobacillose *f*
ACTINOGRAPH - [photo] actinographe *m*
ACTINOLITE - [min] (a source of asbestos) actinolithe *f*, actinote *f*
ACTINOMETER - [instr] (instrument for the measurement of the actinic power of a luminous source) actinomètre *m*
ACTINOMETRY - [phys] (the measure of the chemical effect of light) actinométrie *f*
ACTINOMORPHIC - [phys] (radially symmetrical) actinomorphe
ACTINON - [chem] (radon isotope produced by the disintegration of actinium) actinon *m*
ACTINOURANIUM - [nucl] actinouranium *m*
ACTION - [gen] action *f*, acte *m*, operation *f*
[mech] fonctionnement *m*, opération *f*
[chem] action *f*
[leg] procès *m*, plainte *f*
~FINDING - [telecomm] recherche *f* automatique d'une ligne (par un sélecteur)
~IMPULSE - [telecomm] recherche *f* d'une ligne (par un sélecteur à impulsions)
~HOMING - [telecomm] recherche *f* automatique de la position de repos (sélecteur)
~LINE - [electron] ligne *f* explorée, ligne *f* analysée
~OF THE BLAST - [metall] allure *f* de la soufflerie
ACTION PERIOD - [electron] temps *m* d'exploration, temps *m* d'analyse
~PHASE - s. action period
~POTENTIAL - [el biol] courant *m* d'action nerveuse
~SIGNAL - [autom] signal *m* d'action
~SPOT - [electron] spot *m* explorateur
~TURBINE - [mech] turbine *f* à action, turbine *f* à impulsion
[nucl] intégrale *f* de phase
ACTIVATE, to - [chem] activer
ACTIVATED - [chem] activé
~CARBON - [chem] (form of carbon produced by heating vegetable matter in the absence of air) charbon *m* activé, charbon *m* actif
~CATHODE - [electron] (an activated thermionic cathode) cathode *f* activée
~CHARCOAL - [chem] charbon *m* activé
~FILAMENT - [electron] (used as electron emitter) filament *m* activé
~MOLECULE - [chem] (a molecule of which one or more of the atoms is or are excited) molécule *f* activée
~PETROLEUM COKE - [ind chem] coke *m* de pétrole activé
~SLUDGE - [chem] (a sewer disposal method used to increase the effectiveness of ordinary aeration) boue *f* activée
~SLUDGE PLANT - [ind chem] installation *f* à boues activées
~WATER - [nucl] eau *f* radioactivée
ACTIVATING ISOTOPE - [nucl] isotope *m* d'activation
ACTIVATION - [nucl] (the process of inducing radioactivity) activation *f*
~ANALYSIS - [chem] analyse *f* par activation
~CROSS-SECTION - [nucl] section *f* d'activation
~ENERGY - [nucl] énergie *f* d'activation
~METHOD - [nucl] méthode *f* d'activation
~OVERVOLTAGE - [el chem] surtension *f* d'activation
ACTIVATORS - [chem] (compounds capable of converting zymogens, proenzymes or proferments to active enzymes) activants *m* pl
ACTIVE - [gen] actif, efficace
~AERIAL - [radio] (exciter in the USA) antenne *f* active
~AREA - [electron] surface *f* active
~BALANCE RETURN LOSS - [el] attenuation *f* (circuit à amplificateur)
~CARBON - s. activated carbon
~CELL - [el] élément *m* chargé
~CIRCUIT - [comput] (containing transistors) circuit *m* actif
~COIL - [el] enroulement *m* actif
~COMPONENT - [el] composante *f* active, composante *f* en phase
~CORE - [nucl] coeur *m* de la pile
~CURRENT - [el] (the component of the current vector which is in phase with the voltage) courant *m* actif
~DEPOSIT - [nucl] dépôt *m* radioactif
~ELECTRODE - [el] électrode *f* active
~ENERGY - [phys] énergie *f* active
~ENERGY METER - [instr] (instrument for the measurement of electrical energy in watthours) wattheuremètre *m*, compteur *m* d'énergie active
~FAILURE - [geol] rupture *f* active (par poussée)

ACTIVE FIELD PERIOD - [telev] temps *m* utile de trame
~HOMING - [radar] auto-guidage *m*
~HYDROGEN - [phys] (the atomic form of hydrogen produced when molecular hydrogen is heated to a temperature of 2500°C under a pressure of 1mm of mercury) hydrogène *m* actif
~INGREDIENT - [chem] principe *m* actif
~IRON - [metall] fer *m* actif
~LATTICE - [nucl] réseau *m* actif
~MASS - [phys] (molecular concentration) masse *f* active
~MATERIAL - [el chem] (material participating in the chemical changes in the cells of an accumulator) matière *f* active
[nucl] matière *f* fissile
~MUD - [chem] boue *f* active
~NETWORK - [el] réseau *m* actif, réseau *m* sous tension
~NITROGEN - [chem] (unstable form of nitrogen produced by a silent electric discharge) azote *m* actif (naissant)
~OXYGEN - [chem] oxygène *m* actif (naissant)
~POWER - [el] (the true power in a circuit) puissance *f* active
~PRESSURE - [mech] pression *f* effective
~PRINCIPLE - [chem] principe *m* actif
~PRODUCT - [nucl] (the result of the radioactive decay of a radio-nuclide) produit *m* radioactif
~PROFILE - [mech] profil *m* actif
~RETURN LOSS - [el] atténuation *f*
~SATELLITE REPEATER - [telev] répéteur *m* de satellite actif
~SECTION - [nucl] section *f* active, milieu *m* actif
~SOIL PRESSURE - [constr] poussée *f* active des terres
~STORAGE (area) - [nucl] dépôt *m* de déchêts radioactifs
~SULPHUR - [chem] soufre *m* actif
~SURFACE OF SLIDING - [soil] surface *f* active de glissement
~TRANSDUCER - [el] transducteur *m* actif
~VALENCE - [nucl] (the valence exhibited by an element in any particular compound) valence *f* normale
~VOLTAGE - [el] (in a.c. circuits) tension *f* en phase
ACTIVITY - [phys] (the thermodynamic or ideal concentration of a substance) activité *f*
~COEFFICIENT - [chem] (ratio of activity) coefficient *m* d'activité
~CURVE - [phys] (graph showing the activity of a radioactive source in function of time) courbe *f* d'activité
~FILE - [comput] fichier *m* activités
ACTUAL - [gen] actuel, effectif, réel
~ADDRESS - [comput] adresse *f* effective
~BREAKING LOAD - [mech] charge *f* ultime, charge *f* de rupture
~DRAFT - [text] étirage *m* effectif
~EFFICIENCY - [mech] rendement *m* effectif
~ENERGY - [phys] puissance *f* réelle, énergie *f* cinétique
~LEVEL - [meas] niveau *m* réel
~LOAD FACTOR - [aero] coefficient *m* de charge effective
~MONITOR - [telev] moniteur *m* de sortie
ACTUAL POWER - [el] puissance *f* effective, puissance active
[mech] puissance *f* au frein
~SIZE - [gen] grandeur *f* naturelle
~VALUE - [gen] valeur *f* effective
[instr] valeur *f* instantanée
ACTUATING ARM - [mech] bras *m* de commande, bras *m* de manoeuvre
~CAM - [mech] came *f* de commande, excentrique *m* de commande
~LINKAGE - [mech] tringlerie *f* (de direction)
~PINION - [mech] pignon *m* de commande
~PRESSURE - [mech] (pressure at which an actuator operates) pression *f* de fonctionnement
~ROD - [mech] tige *f* de commande
~TRANSFER FUNCTION - [mech] fonction *f* de transfert de commande
~VOLTAGE - [el] tension *f* de fonctionnement
ACTUATION - [gen] commande *f*, entraînement *m*, mise *f* en action
ACTUATOR - [mech] (a servomotor producing a limited output motion) moteur *m* électrique d'asservissement, vérin *m*
ACUITY - [opt] acuité *f* visuelle
~MATCHING - [telev] (in colour television) adaptation *f* de l'acuité visuelle
~METER - [opt] appareil *m* de mesure de l'acuité visuelle
[acoust] sonomètre *m*
~OF HEARING - [acoust] acuité *f* auditive
ACUMEN - [bot] pointe *f*
ACUMINATE - [gen] acuminé, pointu
ACUTE - [gen & geom] aigu
~ANGLE - [geom] angle *m* aigu
~EXPOSURE - [nucl] exposition *f* brève
ACUTENESS - [gen] acuité *f*
A.C.W. - [radio] (alternating continuous waves) ondes *f* pl entretenues alternatives
ACYCLIC - [gen] acyclique
[bot] (with parts of flower arranged in spirals) acyclique
~PROCESS - [comput] processus *m* apériodique, processus *m* acyclique
ACYL - [chem] (organic acid radical) acyle *m*
ACYLATION - [chem] acylation *f*
AD - [gen] (abbrev for advertisement) annonce *f*
ADALIN - [chem] adaline *f*
ADAM'S APPLE - [anat] pomme *f* d'Adam
ADAMANT - [gen] adamantin, dur
ADAMANTINE DRILL - [mining] couronne *f* à grenaille d'acier, sondeuse *f* à grenaille
ADAMINE - [min] adamine *f*
ADAPT, to - [gen] adapter, ajuster
ADAPTABILITY - [gen] convertibilité *f*, faculté *f* d'adaptation
ADAPTATION - [gen] adaptation *f*
ADAPTER - [el] adapteur *m*, raccord *m* adaptateur
[mech] rallonge *f*, connexion *f* tubulure *f* de raccordement
~BOX - [el] boîte *f* de jonction
~FLANGE - [mech] bride *f* de réduction
~FOR MILLING MACHINES - [mach tools] manchon réducteur *m* à doigts d'entraînement
~PART - [metall] (of a mould) partie *f* mobile
~PLATE - [mech] adapteur *m*, plaque *f* adaptrice
~PLUG - [el] fiche *f*
~SKIRT - [astronaut] jupe *f* de raccordement

ADAPTER TRANSFORMER - [el] transformateur *m* intermédiaire
~UNIT - [mech] pièce *f* de liaison
ADAPTIVE COLOUR SHIFT - [telev] change *m* adaptatif de couleur
~MODIFICATION - [genet] modification *f* d'adaptation
~PEAK - [genet] pic *m* d'adaptation
ADCOCK AERIAL - [radio] (a directional receiving aerial) antenne *f* Adcock
ADD, to - [math] additionner, ajouter
ADD CARRY - [comput] (when the count changes from 9 to 0) retenue *f* d'addition
~INSTRUCTION - [comput] instruction *f* d'addition, ordre *m* d'addition
~-SUBTRACT COUNTER - [comput] compteur *m* réversible, additionneur-soustracteur *m*
ADDED - [gen] ajouté, supplémentaire
~IRON - [metall] fer *m* à ajouter, addition *f* de fer
ADDEND - [math] additeur *m*
~REGISTER - [comput] registre *m* totalisateur
ADDENDUM - [mech] saillie *f*
~ANGLE - angle *m* de saillie
~CIRCLE - [mech] cercle *m* de tête
~LINE - s.addendum cercle
ADDER - [comput] additionneuse *f*, machine *f* à calculer
[telev] circuit *m* combinateur, mélangeur *m*
[zool] vipère *f*
ADDER'S WORT - [bot] renouée *f* bistorte
ADDING COUNTER - [comput] compteur *m* additif, compteur *m* totalisateur
~MACHINE - [mech] additionneuse *f*, machine *f* arithmétique
~OF ORE - [metall] addition *f* de minerai
ADDITION - [math] addition *f*
~POLYMERIZATION - [plastics] polymérisation *f* d'addition
~PRODUCT - [chem] produit *m* d'addition
ADDITIONAL - [gen] additionnel, supplémentaire
~EQUIPMENT [gen] appareils *mpl* périphériques, appareils *mpl* auxiliaires
~NOISE - [acoust] bruit *m* de fond
~SET - s.additional equipment
~TANK - [oil ind] réservoir *m* supplémentaire
ADDITIVE - [gen] additif
~COLOUR SYSTEM - [telev]système *m* additif de couleurs
~COMPOUND - [chem](compound obtained through additive reactions) produit *m* d'addition
~GENOTYPE VALUE - [genet] valeur *f* héréditaire
~MIXING - [telev] mélange *m* additif
~PRIMARIES - [photo etc] couleurs *fpl* primaires de synthèse additive
~PROCESS - [photo] (of colour photography) procédé *m* d'addition
~REACTION - [chem] réaction *f* d'addition
~REVERSAL MATERIAL - [photo] surfaces *fpl* additives réversibles
ADDITIVITY EFFECT - [nucl] (of irradiation) effet *m* cumulatif de l'irradiation
ADDITRON - [comput] (special type of computer tube) additron *m*
ADDRESS - [gen] adresse *f*
[comput] (a symbol which denotes a particular position for the storage of information in the storage device of an electronic computer) adresse *f*

ADDRESS BLANK - [comput] (open space in an address) blanc *m* d'adresse
~CARD - [comput] carte-adresse *f*
~CODE - [comput] code-adresse *m*, code *m* d'adresse
~DECODER - [comput] décodeur *m* d'adresses
~FILE - [comput] registre *m* d'adresses
~PART - [comput] section *f* d'adresse
~PATTERN - [comput] modèle *m* d'adresse
~-READ-WIRE - [comput] fil *m* de lecture d'adresse
~REGISTER - s.address file
~RESEARCH- [comput] recherche *f* d'adresse
~SELECTION SWITCH - [comput] sélecteur *m* d'adresses
~WIRE - [comput] fil *m* d'adresse
ADDRESSABLE - [comput] accessible, adressable
ADDRESSED MEMORY [comput] mémoire *f* adressée
~STORAGE - s.addressed memory
ADDUCT - [chem] produit *m* d'addition
ADDUCTOR MUSCLE - [anat] muscle *m* adducteur
ADENASE - [chem] adénase *f*
ADENOID - [anat] adénoïde
ADENOMA - [med] adénome *m*
ADENOTOME - [instr] adénotome *m*
ADF - (abbrev for automatic direction finder) radiogoniomètre *m* automatique
~LET-DOWN - [aero] atterrissage *m* au radiogoniomètre automatique
ADHERE, to - [gen] adhérir
ADHERENCE - [gen] adhérence *f*, adhésion *f*
~TIME - [chem] durée *f* d'adhérence, temps *m* de retention
ADHERENT - [gen] adhérent
ADHERING MOULDING SAND - [metall] sable m de moulage adhérent
ADHESION - [gen] adhésion *f*, adhérence *f*
[el] (the mutual force between two magnetic forces) adhésion *f*
~COEFFICIENT - [el] coefficient *m* d'adhérence
ADHESIVE - [gen] adhésif *m*, colle *f*
[adj] adhésif, gommé
~CAPACITY - [ind chem] adhérence, adhésivité *f*
~CEMENT - [ind chem] colle *f*
~COEFFICIENT - s.adhesion coefficient
~FABRIC - [text] tissu *m* adhésif
~FILM - [ind chem] pellicule *f* adhésive
~FORCE - [chem] adhérence
~PAPER - [paper] papier *m* gommé
~POWER - s.adhesive force
~STRENGTH - [mech] force *f* d'adhésion
~TAPE - [gen] ruban *m* isolant
[el] chatterton *m*
~WEIGHT - [mech] poids *m* adhérent
ADHESIVENESS - [gen & paint] adhérence *f*, adhésivité
ADHESIVITY - s.adhesiveness
ADIABATIC - [phys] (incapable of gaining or losing heat) adiabatique
~CHANGE - [phys] changement *m* adiabatique
~COMPRESSION - [phys] compression *f* adiabatique
~COOLING - [phys] refroidissement *m* adiabatique
~CURVE - [th dyn] courbe *f* adiabatique
~EFFICIENCY - [th dyn] (ratio of the work performed by a pound of steam to the energy indicated by the adiabatic drop of heat) rendement *m* adiabatique
~EQUATION - [th dyn] équation *f* adiabatique
~EXPANSION - [th dyn] détente *f* adiabatique
~FLOW - [th dyn] écoulement *m* adiabatique

ADIABATIC HEAT DROP - [th dyn] diminution *f* adiabatique de chaleur
~LINE - [th dyn] ligne *f* adiabatique
~MOTION - [phys] mouvement *m* adiabatique
~PRESSURE DROP - [th dyn] diminution *f* adiabatique de pression
~PROCESS - [th dyn] processus *m* adiabatique
ADIAPHORESIS - [med] adiaphorèse *f*
ADIE BAROMETER - [instr] baromètre *m* Adie
ADION - [nucl] (an ion absorbed on a surface)adion *m*
ADIPAMIDE - [chem] adipamide *f*
ADIPATE - [chem] (salt or ester of adipic acid) adipate *m*
ADIPIC ACID - [chem] acide *m* adipique
ADIPOCELLULOSE - [biochem] (cellulose associated with suberin in the cell walls of cork tissue) adipocellulose *f*
ADIPOCERE - [biochem] (waxy substance of fatty acids and calcium soaps due to chemical changes in dead animal fat and muscle when deprived of contact with air) adipocère *f*
ADIPOLYSIS - [chem] adipolyse *f*
ADIPOSE - [gen] adipeux
~TISSUE - [biol] (connective tissue) tissu *m* adipeux
ADIPOSITY - [med] adiposité *f*
ADIPOSURIA - [med] adiposurée *f*
ADIT - [gen] accès *m*
[mining] galerie *f* d'accès (à flanc de coteau), galerie *f* d'écoulement, fenêtre *f* d'accès à une galerie
~CUT MINING - [mining] exploitation *f* à flanc de coteau
~END - [mining] fond *m* de fendue
~ENTRANCE - [mining] entrée *f* de fendue
~LEVEL - [mining] niveau *m* de la fendue
~ROCK - [mining] roche *f* encaissante
ADJACENT - [gen] adjacent, attenant, contigu
~ANGLE - [geom] angle *m* adjacent
~CHANNEL - [telev] canal *m* adjacent
~CHANNEL SELECTIVITY - [telev] sélectivité *f* adjacente
~CHROMINANCE TRAP - [telev] filtre *m* de suppression du signal de chrominance du canal adjacent
~PICTURE CARRIER - [telev] porteuse *f* image adjacente
~PICTURE CARRIER SPACING - [telev] écart *m* entre porteuses
~SOUND CARRIER - [telev] onde *f* porteuse de son adjacent
~VISION CARRIER - [telev] onde *f* porteuse image adjacente
ADJOINING - [gen] contigu
~ROCK - [geol] roche *f* encaissante
ADJOINT FUNCTION - [nucl] fonction *f* adjointe
ADJUGATE DETERMINANT - [math] déterminant *m* adjoint
ADJUNCT - [gen] adjoint, auxiliaire
ADJUST, to - [gen] régler
[mech] ajuster, caler
[instr] régler, tarer, compenser
ADJUSTABLE - [gen] ajustable, réglable, variable
~ARM - [mech] bras *m* réglable
~BAFFLE PLATE - [mech] déflecteur *m* réglable
~BEARING - [mech] palier *m* réglable, coussinet *m* réglable
~BEVEL - [mech] cone *m* réglable
ADJUSTABLE BLADE - [mech] pâle *f* orientable
~BRUSH - [el] balai *m* réglable
~BUSHING - [mech] bague *f* réglable
~CLAMP - [mech] collier *m* réglable
~CONDENSER [el] condensateur *m* variable plat
~DISCHARGE PUMP - [mech] pompe *f* à débit variable
~DISCHARGER - [mech] déchargeur *m* mobile
~DOG - [mach tools] butée *f* enregistrable
~FRICTION DAMPER - [mech] amortisseur *m* à friction réglable
~GAUGE - [instr] calibre *m* réglable
~GIB - [mach tool] lardon *m* de guidage
~GUIDE - [mech] guide *f* réglable
~KNOB - [mech] bouton *m* de réglage
~LEVEL - [impl] niveau *m* réglable
~MIRROR - [photo] miroir *m* orientable
~PIPE WRENCH - [mech] clef *f* à tubes réglable
~PITCH PROPELLER - [aero] hélice *f* à pas variable
~POINT - [comput] virgule *f* réglable
~PROP - [mining] étançon *m* métallique flexible
~RATCHET GEAR FOR AUTOMATIC FEED - [mach tool] groupe *m* d'encliquetage enregistrable pour commande alimentation automatique
~REAMER - [mech] alésoir *m* réglable
~RESISTANCE - [el] résistance *f* variable
~RING - [mech] anneau *m* réglable, bague *f* réglable
~RIVETTING MACHINE - [mech] riveteuse *f* réglable
~SCREW - [mech] vis *f* de réglage
~SET SQUARE - [impl] équerre *f* à bras mobile
~SPEED MOTOR - [el] moteur *m* à vitesse réglable
~SUPPORT CLAMPING LEVER [mach tool] levier *m* de blocage support registrable
~TABLE - [mach tool] table *f* mobile
~TRANSFORMER - [el] transformateur *m* réglable
~TRIPOD - [photo] trépied *m* à branches coulissables, trépied *m* télescopique
~VENT FLAP - [metall] fermeture *f* de ventilation
~WRENCH - [mech] clef *f* à molette
ADJUSTED - [gen & mech] reglé
[meas] compensé, taré
ADJUSTER - [mech] dispositif *m* de réglage
ADJUSTING - [mech etc] mise *f* au point, réglage, tarage *m*
~APPARATUS - [mech] dispositif *m* de réglage
~CAM - [mech] excentrique *m* de réglage
~DEVICE - s.adjusting apparatus
~[photo] (of an exposure meter) dispositif *m* de réglage
~GIB - s.adjustable gib
~HAND WHEEL - [mech] volant *m* de réglage
~KNOB - [mech] bouton *m* de réglage
~LEVER - [mech] levier *m* de réglage
~MARK - [photo] point *m* de repère
~NUT - [mech] écrou *m* de serrage, écrou *m* de réglage
~SCREW - [mech] vis *f* de réglage, vis *f* de rappel
~SLEEVE - [mech] manchon *m* de réglage
~VOLTAGE - [el] tension *f* de réglage
~WINDING - [mech] enroulement *m* de réglage
ADJUSTMENT - [gen & mech] ajustement *m*, compensation *f*
[opt] mise *f* au point
[meas] compensation *f*, tarage *m*
[contr] commande *f* de régulation
~PLUG - [mech] bouchon *m* de réglage

ADJUSTMENT SCREW - [mech] vis *f* de réglage, vis *f* de rappel
~SETTING OF THE ZERO - [instr] mise *f* à zéro
~SHAFT - [mech] axe *m* de rotation, tige *f* de réglage
ADJUTAGE - [hydr] (tube for the discharging of water) ajutage *m*
ADMEASUREMENT - [meas] mesure *f*
ADMINISTRATION - [gen] administration *f*
ADMIRALTY METAL - [metall] laiton *m* 70-29 à 1 p.c. d'étain, bronze *m* de canons
ADMISSIBLE CHARACTER - [comput] signe *m* admissible
~LOAD - [mech] charge *f* (maximum) autorisée
ADMISSION - [gen] admission *f*, accès *m*
[mech] admission *f*, adduction *f*, entrée *f*, introduction *f*
~PORT - [mech] ouverture *f* d'admission
~VALVE - [mech] (or slide valve) soupape *f* d'admission
ADMITTANCE - [el] (flowing of current under the action of a difference in potential) admittance *f*
~CHART - [el] diagramme *m* d'admittance
ADMITTING PORT - [mech] orifice *m* d'admission
ADMIXTURE - [chem] addition *f*, dosage *m* mélange *m*
ADOBE - [constr] (sun-dried brick consisting of mud and straw) adobe *m*
[soil] (a soil with a high content of colloidal clay) adobe *m*
ADRENAL - [anat] surrénal
~CORTEX - [anat] substance *f* corticale surrénale, écorce *f* de la surrénale
~GLANDS - [anat] capsules *f*pl surrénales
ADRENALINE - [biochem] (the principal blood-pressure raising hormone of the medulla of the adrenal glands; epinephrine) adrénaline *f*
ADRIFT - [naut] à la dérive
ADSORB, to [chem] (to take up gases, vapours or dissolved substances at the surface of a solid) adsorber
ADSORBABILITY - [chem] adsorbabilité *f*
ADSORBABLE - [chem] adsorbable
ADSORBATE - [chem] (adsorbed substance) substance *f* adsorbée
ADSORBED LAYER - [chem] couche *f* adsorbée
ADSORBENT - [chem] (substance having the ability to take up other substances on its surface) adsorbant *m*
ADSORBING AGENT - [chem] matériau *m* adsorbant
ADSORPTION - [chem] (the process by which a solid can take up a liquid or a gas at its surface) adsorption *f*
~CHARCOAL - [ind chem] charbon *m* adsorbant, charbon actif
~CYCLE - [nucl] cycle *m* d'adsorption
~PUMP - [mech] pompe *f* à sorption (par un getter massif thermiquement régénérable)
~RECOVERY - [ind chem] récupération *f* par adsorption
~TRAP - [ind chem] piège à adsorption
ADTEVAC PROCESS - [med] déshydratation *f* sous vide
ADULARESCENCE - [min] adularescence *f*
ADULTERATE - [gen] adultérer, falsifier
ADULTERATING AGENT - [ind chem] agent *m* adultérant
ADULTERATION - [ind chem] adultération *f*, altération, dénaturation *f*, falsification *f*
ADVANCE, to - [gen] avancer, progresser
[mech] mettre de l'avance
ADVANCE - [gen] avance *f*, développement *m*, progrès *m*, crue *f* (d'un glacier)
~BOREHOLE - [mining] trou *m* d'avancement
~DIAMETER RATIO - [aero] argument *m* de similitude (d'une hélice)
~FIRE - [auto] avance *f* à l'allumage
~HEADING - [mining] front *m* de taille
~NOTICE - [comm & leg] préavis *m*
~REGULATOR - [mech] régulateur *m* d'avance (à l'allumage)
~THE BRUSHES, to - [el] caler les balais
~THE SPARK , to - [auto] mettre de l'avance à l'allumage
~WINDING - [contr] enroulement *m* de commande
ADVANCED IGNITION - [auto] avance *f* à l'allumage
~OPENING - [auto] avance *f* à l'ouverture (des soupapes)
~SPARK - [auto] s. advanced ignition
ADVANCING AND RETREATING MINING - [mining] exploitation *f* en avant et en retour
ADVANTAGE FACTOR - [nucl] facteur *m* de flux, facteur *m* d'irradiation optimale
ADVECTION - [met] (horizontal heat transfer caused by air currents in the atmosphere) advection *f*
~FOG - [met] brouillard *m* d'advection
ADVENTITIOUS - [gen] accidentel, fortuit
[bot] adventice
ADVERTISEMENT - [gen] annonce *f* publicitaire, publicité *f*
ADVERTISING - [gen] publicité *f*, réclame *f*
~AGENCY - [comm] agence *f* de publicité
ADVISORY - [gen] consultatif
~BOARD - [comm] conseil *m* consultatif
ADYNAMIA - [med] adynamie *f*
ADYNAMIC - [phys] adynamique
ADZE - [impl] (cutting tool with a curved blade fixed at 90° to the handle) herminette f, erminette *f*
AEOLATION - [geol] érosion *f* éolienne
AEOLIAN DEPOSIT - [soil] (sediment or deposit laid by the wind) dépôt *m* éolien
~VIBRATION - [acoust] (musical sound emanating from a string exposed to an air stream) vibration *f* due au vent
AEOLIPILE - [phys] (used to explain the reaction on the air of jets of steam from a closed container) éolipyle *f*
AEOLOTROPIC - [phys] (with physical properties varying in relation to the direction in which measurements are taken) anisotropique
AEOLOTROPY - [phys] anisotropie *f*
AERATE, to - [gen] aérer, ventiler
[ind chem] (mixing air with gas in a gas burner) gazéifier
[hydr] (water) aérer (de l'eau)
AERATED - [gen & chem] aéré, gaséifié, ventilé
~MUD - [ind chem] boue *f* aérée
~PLASTICS - [plastics] mousse *f* plastique
~WATER - [chem] eau *f* aérée, eau *f* gazeuse
AERATION - [gen & constr] aération *f*, ventilation *f*
[chem] barbottage *m* (d'air), gazéification *f*
~TANK - [ind chem] bassin *m* d'activation, bassin *m* d'aération
AERATING APPARATUS - [min] diviseur-aérateur *m*
AERATOR - [impl] aérateur *m*, ventilateur *m*

[ind chem] dispositif *m* de barbottage, gazéificateur
[min] appareil *m* diviseur du sable, diviseur-aérateur *m*
AERIAL - [gen] aérien
[radio & telev] antenne *f*
~ARCH - [geol] voûte *f* anticlinale dénudée, pli *m* dénudé, selle *f* aérienne
~ARRAY - [radio] réseau *m* d'antennes
~ASSEMBLY - [radio] ensemble *m* d'antennes
~ATTENUATOR - [radio] atténuateur *m* d'antenne
~BEACON - [aero] radio balise *f*
~BOOSTER - [radio] (aerial amplifier) amplificateur *m* d'antenne
~BUOY - [aero] bouée *f* aéronautique
~CABLE - [telecomm] (cable on a wire between telegraph poles) câble *m* aérien
[radio] conducteur *m* d'antenne
~CABLEWAY - [transp] blondin *m*
~CAMERA - [photo] appareil *m* de photographie aérienne
~CAPACITOR - [radio] condensateur *m* d'antenne
~CASING - [radio] boîtier *m* antenne
~CHANGE-OVER SWITCH - [radio] commutateur *m* d'antenne
~CIRCUIT - [radio] circuit *m* d'antenne
~CIRCUIT BREAKER - [radio] disjoncteur *m* d'antenne
~COIL - [radio] (the first coil in a receiver through which the aerial current flows) bobine *f* d'antenne
~CONTROL TABLE - [radio] tableau *m* de commande d'antenne
~COUPLING - [radio] couplage *m* d'antenne
~CROSS-TALK - [radio] diaphonie *f*
~CROSSING - [el] conduite *f* aérienne, traversée *f* aérienne
~CURRENT - [radio] courant *m* d'antenne
~DAMPING - [radio] amortissement *m* de l'antenne
~DIRECTIVITY - [radio] (the property by which it radiates more strongly in some direction than in others) directivité *f* d'antenne
~DISCHARGER - [el] limiteur *m* de tension
~DUPLEXING - [radio] utilisation *f* d'une seule antenne pour l'émission et la réception
~EARTHING SWITCH - [el] commutateur *m* de mise à la terre (ou à la masse)
~EFFICIENCY - [radio] (ratio between power supplied to an aerial and power radiated) rendement *m* de l'antenne
~ELEVATION PAWL - [radio] crochet *m* de hissage d'antenne
~ENERGY - [radio] puissance *f* d'antenne
~EXCITATION - [radio] excitation *f* de l'antenne
~FEED IMPEDANCE - [radio] impédance *f* d'entrée d'antenne
~FEEDBACK - [radio] réaction *f* d'antenne
~FEEDER - [radio] alimentation *f* de l'antenne
~FERRY - [transp] pont *m* transbordeur
~FIELD GAIN - [telev] gain *m* d'antenne
~FRAME - [radio] cadre *m*
~FROG - [el] conduite *f* aérienne, traversée *f* aérienne
~FUSE - [el] limiteur *m* de tension
~GAIN - s.aerial field gain
~IMAGE - [opt] (image formed in space and not received on a surface) image *f* aérienne, image *f* virtuelle
~IMPEDANCE - [radio] impédance *f* d'antenne
~INPUT - [radio] entrée *f* d'antenne

AERIAL INSULATOR - [radio] isolateur *m* d'antenne
~LADDER - [gen] échelle *f* de sauvetage
~LEAD - [radio] descente
~LENS - [radio] lentille *f* d'antenne
~LINE - [mining] canalisation *f* aérienne, ligne *f* aérienne
~LOADING INDUCTANCE - [radio] bobine *f* d'accord d'antenne
~LOSS - [radio] perte *f* d'antenne
~MAST - [radio] mât *m* d'antenne, poteau *m* d'antenne, pylone *m* d'antenne
~METAL - [metall] [alloy of aluminium and lithium) alliage *f* léger
~MOSAIC - [photo] mosaïque *f* de photographies aériennes
~NETWORK - s.aerial array
~NOISE - [radio] bruit *m* d'antenne, souffle *m* d'antenne
~PHOTOGRAPH - [photo] (photograph taken from an aircraft) photographie *f* aérienne
~PHOTOGRAPH INTERPRETATION - [surv] interprétation *f* des photographies aériennes
~PLANT - [bot] épiphyte *f*, plante *f* aérienne
~POWER GAIN-[radio] (directive gain) facteur *m* de directivité
~RAILWAY - [transp] chemin de fer *m* aérien ou surélevé
~REACTANCE - [radio] réactance *f* d'antenne
~REPEAT DIAL - [radio] cadran *m* indicateur (d'antenne)
~RESISTANCE - [radio] résistance *f* d'antenne
~ROPEWAY - [transp] câble *m* aérien, voie *f* à câbles
~SERIES CAPACITOR - [radio] condensateur *m* en série dans l'antenne
~SIGNAL - [radio] signal *m* d'antenne
~SKETCHMASTER - [photo] appareil *m* de dessin par projection de clichés aériens
~SOCKET - [radio] douille *f* d'antenne
~SOUNDING LINE - [acoust] altimètre *m* acoustique
~SURVEY - [surv] levé *m* aérophotogrammétrique
~SWITCH - [radio] commutateur *m* d'antenne
~SWITCHING - [radio] commutation *f* de lobe
~TERMINAL - [radio] (the terminal introduced into the socket) borne *f* d'antenne
~TOWER - [radio] pylone *m*
~TRIANGULATION - [surv] triangulation *f* aérienne
~TRIMMER - [radio] trimmer *m* d'antenne
~TRUNK - [radio] pipe *f* d'entrée d'antenne
~TUNING CAPACITOR - [radio] (variable capacitor used to adjust the natural frequency of an aerial circuit to the desired value) bobine *f* de syntonisation d'antenne
~TUNING COIL - s.aerial tuning inductor
~TUNING INDUCTOR - [radio] (variable inductor used like an aerial tuning capacitor) bobine *f* de syntonisation d'antenne
~VELOCITY - [nucl] vitesse *f* dans l'air
~WINCH - [radio] rouet *m* d'antenne
AERIFEROUS - [gen] aérifère
AERIFICATION - [phys & chem] aérification *f*
AERIFORM - [phys] aériforme, gazeux
AEROASTRO MEDICINE - [astronaut] médecine *f* aérospatiale
AEROBATICS - [aero](trick or "stunt" flying) acrobatie *f* aérienne
AEROBE - [biol] (a micro-organism which can live and develop only in the presence of oxygen) aérobie *f*

AEROBIC CELL - [biol] aérobie *f*
AEROBOAT - [aero] hydravion *m*
AEROCAMERA - [photo] appareil *m* de photographie aérienne
AEROCARTOGRAPHY - [surv] cartographie *f* aérienne
AERODONETICS - [aero] technique *f* du vol à voile
AERODROME - [aero] aérodrome *m*, aéroport *m*
AERODUCT - [astronaut] statoréacteur *m* à énergie électrique atmosphérique
AERODYNAMIC - [gen] aérodynamique
~BALANCE - [aero] (a condition in which the resultants of aerodynamic forces on the blades of propellers consist only of thrust about the axis of rotation and torque about such axis) équilibrage *m* aérodynamique, balance *f* aérodynamique
~BALANCED SURFACE - [aero] gouverne *f* compensée, surface *f* aérodynamique
~CENTRE - [aero] (the point about which the rate of change of the pitching moment in relation to incidence is of zero value) foyer *m* (du profil d'aile), centre *m* de poussée (du profil d'aile)
~INDUCTION - [phys] induction *f* aérodynamique
~LIFT- [aero] (the component of the resultant force due to airflow, taken along the lift axis) portance *f* aérodynamique
~MEAN CHORD - [aero] (of an imaginary aerofoil) corde *f* aérodynamique moyenne (de référence)
~STIFFNESS - [aero] (stiffness due to aerodynamic forces) rigidité *f* aérodynamique
~TWIST - [aero] (the variation of the zero-lift line along the span of an aerofoil) torsion *f* aérodynamique, vrillage *f* aérodynamique
~VOLUME - [aero] (total volume of lighter-than-air craft) volume *m* aérodynamique
AERODYNAMICS -[phys] (the study of those forces which act on bodies moving in the air) aérodynamique *f*
AERODYNE - [aero] (heavier-than-air flying vehicle) aérodyne *m*
AEROELASTIC - [aero] aéroélastique
~DIVERGENCE - [aero] (instability due to the rate of change of aerodynamic forces) divergence *f* aéroélastique
AEROELASTICITY - [mech] (the mechanics dealing with the results of interaction between aerodynamic forces and elastic reactions) aéroélasticité *f*
AEROEMBOLISM - [med] (formation of nitrogen bubbles in blood and tissue) aéroembolie *f*
AERO-ENGINE - [mech] moteur *m* d'avion
AEROFOIL - [aero] (bodies or surfaces producing aerodynamic reactions) surface *f* portante, plan *m*
~SECTION - [aero] profil *m* d'aile, profil *m* de la surface portante
AEROGEL - [chem] aérogel *m*
AEROGRAPH - [met] météorographe *m*
AEROGRAPHY - [met] (the recording of various magnitudes simultaneously, e.g. temperature, pressure, humidity) météorographie *f*
AEROLITE - [geol] (a meteoric stone) aérolithe *f*
AEROLOGY - [met] (the study of those strata of the atmosphere which are not influenced by conditions on the surface) aérologie *f*
AEROMAGNETIC PROFILE - [aero] profil *m* aéromagnétique
AEROMECHANICS [aero] mécanique *f* aéronautique
AEROMETEOROGRAPH - [instr] (instrument fitted in an aircraft to record atmospheric conditions) météorographe *m*
AEROMETER - [instr] (instrument for measuring the weight and density of air and other gases) aéromètre *m*
AEROMETRY - [phys] aérométrie *f*
AEROMODELLING - [aero] aéromodélisme *m*
AEROMOTOR - s.aero-engine
AEROMYCIN - [chem] aéromycine *f*
AERONAUTICAL - [aero] (relating to the science of flight) aéronautique
~ENGINEERING- [aero] industrie *f* aéronautique
~FIXED SERVICE - [telecomm] service *m* fixe aéronautique
~LIGHT - [sign] feu *m* aéronautique
~MILE - [meas] mille *m* aéronautique
~MOBILE SERVICE - [radio] service *m* mobile aéronautique
~RADIONAVIGATION - [aero] radionavigation *f* aérienne
~TELECOMMUNICATION LOG - [aero] service *m* de télécommunication aéronautique
~TELECOMMUNICATION SERVICE - [aero] cahier *m* de veille
AERONAUTICS - [aero] (the science relating to aerial flights) aéronautique *f*
AEROPHOBIA - [med] aérophobie *f*
AEROPHONE - [acoust] (a primitive device used as an amplifier) aérophone *m*
AEROPHOTOCARTOGRAPHY - [surv] photocartographie *f* aérienne
AEROPHOTOGRAM - [photo] photogramme *m* aérien
AEROPHOTOGRAMMETRY - [surv] photogrammétrie *f* aérienne
AEROPHOTOGRAPHIC MOSAIC - [photo] mosaïque *f* de photographies aériennes
AEROPHOTOGRAPHY - [photo] photographie *f* aérienne
AEROPHYTE - [bot] (a plant growing on another without being a parasite as it takes no material from the plant to which it is attached) aérophyte *f*
AEROPLANE - [aero](power-driven heavy-than-air flying machine) aéroplane *m*, avion *m*
~AERIAL - [radio] antenne *f* d'avion
~EFFECT - [radio] (interference with radio reception due to the passage of an aircraft) effet *m* d'avion; (error of bearing caused by the aerial) effet *m* dû à la composante polarisée horizontale de l'onde directe provenant d'un émetteur en altitude
~FLUTTER - [telev] (interference caused by reflection from an aircraft) interférence *f* d'avion
~TRIM - [aero] assiette *f*, stabilité *f*
AEROSCOPE - [instr] (instrument for determining the composition of air) aéroscope *m*
AEROSOL - [chem] (a colloïdal system of fluid and solid particles -fog or smoke- in a gaseous dispersion medium; used in bomb forms for dispersion of insecticides) aérosol *m*
~BOMB - [ind chem] bombe *f* à aérosols
AEROSPACE - [astronaut] domaine *m* aérospatial
AEROSPACECRAFT - [astronaut] engin *m* aérospatial
AEROSPHERE - [met] (the whole envelope of gases surrounding the earth) aérosphère
AEROSTAT - [aero] (lighter-than-air craft) aérostat
AEROSTATIC BALANCE - [aero] balance *f* aérostatique
~LIFT - [aero] portance *f* aérostatique

AEROSTATICS - [phys] (the study of gases not in motion) aérostatique *f*
AEROSTATION - [aero] (the science of operating lighter-than-air craft) aérostation *f*
AEROTECHNICAL - [aero] aérotechnique
AEROTHERAPY - [med] aérothérapie *f*
AEROTHERMODYNAMIC - [th dyn] aérothermodynamique
~DUCT - [aero] (ramjet) tuyère *f* thermopropulsive, tuyère *f* de réacteur
AEROTHERMOMETER - [instr] aérothermomètre *m*
AEROTRIANGULATION - [surv] triangulation *f* aérienne
AEROTYRE - [aero] pneumatique *m* pour avions
AESTHESIOMETER - [instr] (to measure a person's sensibility to the touch) esthésiomètre *m*
AETHRIOSCOPE - [instr] éthrioscope *m*
AETIOLOGY - [med & biol] (the study of the causes of diseases or abnormal conditions) étiologie *f*
AF - [acoust] (abbrev for audio frequency) basse fréquence, fréquence *f* acoustique
AF AMPLIFICATION STAGE - [radio] étage *m* basse fréquence (amplification)
AFC - [radio] (Automatic Frequency Control) régulation automatique de fréquence
AFC CHARACTERISTICS - [radio] caractéristiques *fpl* de régulation automatique de fréquence
AFC CORRECTION RATIO [radio] taux *m* de correction de régulation automatique de fréquence
AFC HOLDING RANGE - [radio] gamme *f* effective de régulation automatique de fréquence
AFC THRESHOLD - [radio] seuil *m* de régulation automatique de fréquence
AFFECT, to - [gen] affecter, influencer, toucher
AFFERENT - [anat] (of nerves etc) afférent
~BLOOD VESSELS [anat] vaisseaux *mpl* afférents
AFFIDAVIT - [leg] déclaration *f* écrite enregistrée sous serment, déposition *f* sous serment
AFFINATION - [chem] affinage *m*
~CENTRIFUGE - [sugar ind] turbine *f* à affiner
AFFINE, to - [chem] affiner, purifier
AFFINITY - [chem] (the force which causes substances to combine chemically) affinité *f*
AFFIX, to - [gen] fixer, attacher, joindre
AFFLUENT - [geogr] confluent *m*
AFFORESTATION - [agric] reboisement *m*, soumission *f* au régime forestier
AFFUSION - [med] affusion *f*
AFLAME - [gen] en feu, en flamme
AFLOAT - [naut] à flot, à la surface
AFT - [naut] à l'arrière
~-WIND - [naut] vent *m* en poupe, vent *m* arrière
AFTER - [gen] après, d'après
[naut] de l'arrière
~ANNEALING - [metall] (reheating for the elimination of internal stresses) recuit *m*
~-BAKE - [plastics] après-cuisson *f*, étuvage *m*
AFTERBIRTH - [zool] (part of the placenta) arrière-faix *m*, délivre *m*
AFTERBLOW - [metall] sursoufflage *m*
AFTERBURNING - [aero] (of jet engines) post-combustion *f*
AFTER CABIN - [naut] chambre *f* de l'arrière
AFTER-CONTRACTION - [metall] retrait permanent
AFTERCOOLER - [metall] refroidisseur *m* après combustion, refroidisseur *m* final
AFTERCOOLING - [metall] refroidissement après combustion
AFTERDAMP - [mining] (gas produced by the combustion of coal-gases) gaz *mpl* délétères
AFTER-DEFLECTION FOCUSING - [electron] focalisation *f* après déviation
AFTER DRAFT - [naut] pêche *f* par l'arrière
AFTER-EFFECT - [gen] effet *m* résiduel, suites *fpl*
AFTER-EXPANSION - [metall] dilatation *f* permanente
AFTERFLAMING - [astronaut] post-combustion *f*
AFTERFLOW - [metall] fluage *m* postérieur
AFTERFRACTIONATING TOWER - [oil ind] tour *f* de fractionnement secondaire
AFTERGASES - s. afterdamp
AFTERGLOW - [phys] persistance *f* de luminescence
[metall] incandescence *f* résiduelle
[electron] persistance *f* d'écran, phosphorescence *f*
~[radar] durée *f* de persistance
~TIME - s. afterglow [radar]
AFTERGROWTH - [biol] recrudescence *f*
AFTERHEAT - [nucl] chaleur *f* résiduelle
AFTERHOLD - [naut] cale *f* arrière
AFTER-IMAGE - [opt] image *f* rémanente
[telev] traînage *m*
AFTERPEAK - [naut] arrière-bec *m*
AFTERRIPENING - [bot] post-maturation *f*
[photo] (of emulsions) refonte *f*, maturation *f* chimique
AFTER-RUNNING - [ind chem] queue *f* de distillation
AFTERSHOCK - [metall] secousse *f* consécutive
AFTER-TASTE - [gen] arrière-goût *m*
AFTERTREATMENT - [ind chem] traitement *m* complémentaire
Ag - [symbol for silver] argent *m*
AGAMIC - [genet] (reproduced without male) agame
~REPRODUCTION - [bot] reproduction *f* des plantes agames
AGAR-AGAR - [chem] agar-agar *m*
AGARICINE - [chem] agaricine *f*
AGARIC MINERAL - [min] agarice *m*
AGATE - [min] agate *f*
~MORTAR - [ind chem] (small bowl-shaped vessel made from agate used for crushing very hard material by hand) mortier *m* d'agate
~FOR BURNISHING - [min] agate *f* à brunir
AGATED - [min] agaté
AGAVE - [bot] agave *m*
AGC - [radio] (Automatic Gain Control) régulation *f* automatique de sensibilité (anti-fading)
AGE, to - [gen] vieillir
[brew] mûrir
AGE - [gen] âge *m*
AGE-DIFFUSION EQUATION - [phys] équation *f* de diffusion en fonction de l'âge
~EQUATION - [nucl] équation *f* de l'âge
~HARDEN - [metall] vieillir, augmenter la dureté par traitement thermique
~HARDENING - [metall] vieillissement *m* d'un alliage (traitement thermique), durcissement *m* par vieillissement structural, trempe *f* par vieillissement, (precipitation hardening) durcissement *m* par précipitation
~OF A CHARGE - [metall] degré *m* de cuisson
~-RESISTING - [metall] résistant au vieillissement
AGED - [gen] vieilli
AGEING - [metall] vieillissement *m* (trait. thermi-

que)
AGEING CRACK - [metall] fissure *f* de vieillissement
~TEST - [ncul] essai *m* de vieillissement
AGENCY - [gen] agence *f*, action *f*
AGENDA - [gen] ordre *m* du jour, programme *m*
AGENT - [chem] agent *m*
AGGLOMERATE, to - [phys] agglomérer
AGGLOMERATE - [gen bot geol] agglomérat *m*
~BASE - [ind chem] agglomérant *m*
AGGLOMERATED CAKE - [metall] gâteau *m* aggloméré
~COAL - [min] charbon *m* aggloméré
~PLANT - [min] installation *f* d'agglomération
AGGLOMERATION - [gen] agglomération *f*
AGGLOMERATIVE - [gen] agglomératif
AGGLUTINANT - [ind chem] (a cementing substance) agglutinant *m*, liant *m*
AGGLUTINATE, to - [ind chem] agglutiner
AGGLUTINATING POWER - [ind chem] pouvoir *m* agglutinant
AGGLUTINATOR - [ind chem] agglutinant *m*
AGGLUTININ - [chem] agglutinine *f*
AGGRADATION - [geol] alluvionnement *m*
[soil] ensablement *m*, engravement *m*
AGGRADATIONAL DEPOSIT - [geol] dépôt *m* alluvionnaire
AGGREGATE - [gen] ensemble, global, total
[constr] aggloméré *m*
~BREAKING - [mech] charge *f* de rupture théorique
~RECOIL - [nucl] recul *m* moléculaire, groupe *m* de recul
AGGREGATES - [constr] agrégats *mpl*; (rounded) agrégats *mpl* roulés
AGGREGATION - [gen] agrégation *f*
[nucl] amas *m*
AGGREGATIVE FLUIDIZATION - [chem] bullage *m* (formation de bulles)
AGGRESSIVENESS OF GROUND WATER - [geol] agressivité *f* de l'eau phréatique
AGING - s. ageing
AGIO - [comm] agio *m*
AGITATE, to - agiter, remuer
AGITATING PAN - [sugar ind] bac *m* à agitation, bac *m* mélangeur
~TANK - [metall] cuve *f* d'agitation
AGITATOR - [impl](container in which crushed material is agitated) agitateur *m*
AGOMETER - [instr] agomètre *m*
AGONIC LINE - [phys] (line of zero magnetic declination) ligne *f* agonique
AGONISIS - [genet] (or certation; the competition of male elements of different genotype for the fertilization of available female elements) certation *f*
AGRAVIC - [phys] sans gravité
AGREEMENT - [gen] accord *m*, arrangement *m*, convention *f*
[comm] contrat *m*
AGRICULTURAL - [agric] agricole
~BELT - [agric] zone *f* agricole
~CHEMISTRY - [agric] chimie *f* agricole
~ENGINEERING - [agric] technique *f* agricole, génie *m* rural
~HOLDING - [agric] domaine *m* agricole, ferme *f*
~IMPLEMENT - [agric] matériel *m* agricole
~LIME - [soil] chaux *f* agricole
~PESTICIDE - [chem] anticryptogamique *m*
~TRACTOR - [mech] tracteur *m* agricole
AGRIMOTOR - s. agricultural tractor
AGROLOGY - [agric] agrologie *f*
AGRONOMY - [agric] agronomie *f*
AGROUND - [naut] au sec, échoué
A.H. - [metall] (abbrev for air-hardening) trempe *f* à l'air
a.h. - [el] (abbrev for ampere-hour) ampèreheure *m*
a.h.m. - [el] (abbrev for ampere-hour-meter) ampère-heure-mètre *m*
AID - [gen] aide *f*, secours *m*
~STATION - [gen] poste *m* de secours
AIGRETTE - [el] (brush discharge) (décharge en) aigrette *f*
AILANTUS - [bot] ailante *m*
AILERON - [aero] (wing flap controlled by the pilot) aileron *m*
~ANGLE - [aero] (the angle which the chord of an aileron makes with the chord of the fixed surface corresponding to it) angle *m* de déplacement de l'aileron
~CONTROL CABLE - [aero] câble *m* de commande
~DEFLECTION - [aero] braquage *m*
~DRAG - [aero] traînée *f*
~LEVER - [aero] guignol *m* d'aileron
~ROLL - [aero] tonneau *m*
~TAB - [aero] tab *m* d'aileron
AIM - [gen] but *m*, dessein *m*, objet *m*
AIMING - [firearms] pointage *m*, visée *f*
~APPARATUS - [photo] (in aerial photography) appareil *m* de visée
~DEVICE - [firearms] dispositif *m* de pointage
~MECHANISM - [firearms] mécanisme *m* de pointage
~POINT - [gen] point *m* de mire
AIR - [gen & phys] air *m*
[mus] air *m*, mélodie *f*
~ACCUMULATOR - s. air-hydraulic accumulator
~ADJUSTING SCREW - [mech] vis *f* de réglage de l'air
~AGITATION LEACH - [ind chem] lessivage *m* par brassage, lixiviation *f* par barbotage
~ALARM - [acoust] alerte *f* aérienne
~BACK REST - [gen] dossier *m* pneumatique
~BAFFLE - [mech] (surface designed to modify the flow of a stream of air) chicane *f*, déflecteur *m*
~BALLAST - [mech] (or gas ballast; the quantity of gas admitted to the pump chamber before and during the compression stroke) lest *m* d'air, injection *f* d'air
~BEACON - [aero] radio-balise *f*
~BELT - [metall] boîte *f* à vent (du cubilot)
~BLAST - [gen] jet *m* d'air, circulation *f* forcée d'air
[mech] (injection of oil fuel into the cylinder of a diesel engine by a blast of high-pressure air) injection *f* à air comprimé
~BLAST CIRCUIT BREAKER - [el] (a switch whose arc is extinguished by a blast of air) disjoncteur *m* à soufflage d'arc (par air comprimé)
~BLAST COOLING - [th eng] refroidissement *m* par circulation forcée d'air
~BLAST TRANSFORMER - [el] (cooled by forces circulation of air) transformateur *m* à refroidissement par air
~BLASTING - [metall] nettoyage *m* à jets d'air, refroidissement *m* par air
~BLEED - [mech] prise *f* d'air
~BLEED ELBOW - [mech] coude *m* de piquage
~BLEEDING VALVE - [mech] soupape *f* à désaération

AIR BLOWING - [ind chem] (process to change the properties of some vegetable oils by forcing large volumes of air through them) soufflage *m*
~BLOWN ASPHALT - [ind chem] asphalte *m* soufflé
~BOTTLE - [mech] bouteille *f* d'air comprimé
~BOX s. air belt
~BRAKE - [mech] (mechanical brake actuated by air pressure) frein *m* à air comprimé, frein *m* pneumatique
[aero] (system of reducing air speed by control surfaces) aérofrein *m*
~BRATTICE - [mining] cloison *f* d'aérage
~BREAK SWITCH - [el] disjoncteur *m* à l'air libre
~BREAK-UP - [phys] désintégration *f* dans l'air
~BREATHER - [aero] véhicule *m* à moteur aérobie
~BREATHING ENGINE - [mech] (engine which obtains its oxygen from the air) moteur *m* oxygéné à air
~BRICK - [build] (perforated brick built into a wall) brique *f* perforée (ventilation)
~BRIDGE - [mining] crossing *m* d'aérage
~BRUSH - [impl] (for spraying a colour or coating on to a surface by compressed air) aérographe *m*, pistolet *m* à peinture
~BUBBLE - [phys] (small body of air in a mass of liquid or solid material) bulle *f* d'air
~BUFFER - [gen] tampon *m* atmosphérique
[mech] amortisseur *m* à air
~CABLEWAY - [transp] blondin *m*
~CAPACITOR - [el] (a capacitor in which the dielectric is air) condensateur *m* à air
~CASING - [th eng] chemise *f*, enveloppe *f*
~CELL - [motor] (a small auxiliary combustion chamber) chambre *f* de réserve d'air
[el] (US for air-depolarized cell) pile *f* à dépolarisation par l'air
~CHAMBER - [hydr] chambre *f* de détente
[mech] chambre *f* de la pompe à schlamms
~CHANGE - [gen] changement *m* d'air, ventilation *f*
~CHANNEL - [gen] conduit *m* d'air
~CHILL - [metall] trempe *f* à l'air, durcissement à l'air
~CHISEL - [mech] burin *m* pneumatique
~CHOKE - [mech] volet *m* d'air
~CHUCK - [mech] mandrin *m* pneumatique
~CIRCULATION - [phys] (natural or induced movement of air) circulation *f* d'air
~CIRCULATION AUTOCLAVE - [ind chem] autoclave *m* à circulation d'air
~CLAMP - [motor] bride *f* de serrage à air comprimé
~CLASSIFICATION - [ind chem] (separation and classification of solid particles by means of an air stream) classification *f* par air
~CLEANER - [motor] épurateur *m*, filtre *m* à air
~CLEANING FILTER UNIT - [mach tool] groupe *m* filtre dépuration de l'air
~CLUTCH - [mech] friction *f* pneumatique
~CLUTCH FORGING MACHINE - [mech] machine *f* à forger à embrayage pneumatique
~COCK - [motor] (robinet) purgeur *m* d'air
~COMPRESSOR - [mech] compresseur *m*
~COMPRESSOR INTAKE THROTTLE - [mech] (automatic valve on the intake of a compressor to control delivery conditions) volet *m* d'admission d'air au compresseur
~COMPUTER - [aero] calculateur *m* de bord

AIR CONDENSER - [mech] condenseur *m*
~CONDITION, to - [th eng] climatiser
~CONDITIONING - [th eng] climatisation *f*
~CONDITIONING CENTRE - [constr] centre *m* de climatisation, conditionneur *m* d'air
~CONDITIONING PLANT - [th eng] installation *f* de climatisation, conditionneur *m* d'air
~CONDUCTION - [acoust] conduction *f* de l'air
~CONDUIT - [gen] conduit *m* d'air
~CONSIGNMENT - [aero comm] expédition *f* (de marchandises) par avion
~CONTENT - [soil] indice *m* d'aération
~CONTROL DISK - [burners] disque *m* fileté
~COOLED - [gen] à refroidissement par air
~COOLED TUBE - [electron] tube *m* à refroidissement par air
~COOLING - [th eng] (abstraction of heat by means of a current of air) refroidissement *m* par air
~COOLING PLANT - [ind] installation *f* de refroidissement par air
~CORE - [el] (as a single layer solenoid) noyau *m* à air
~CORE COIL - [el] bobinage *m* à air
~CORE MAGNETIC CIRCUIT - [el] (without ferromagnetic core) circuit *m* à noyau à air
~CORED MAGNET - [el] aimant *m* sans fer
~CORRIDOR - [aero] couloir *m* aérien
~CRASH - [aero] accident *m* d'aviation, catastrophe *f* aérienne
~CROSSING - [mining] crossing *m* d'aérage
~CURRENT - [met] courant *m* d'air
~CUSHION - [phys] coussin *m* d'air, tampon *m* d'air
~DAMPER - [th eng] registre *m* d'air
~DAMPING - [mech] amortissement *m* par air comprimé
~DENSITY - [phys] densité *f* de l'air
~DISPLAY - [aero] fête *f* aérienne
~DISTANCE - [aero] distance *f* à vol d'oiseau
~DISTANCE RECORDER - [instr] (an instrument making a continuous record of air distances flown) enregistreur *m* de distance aérienne
~DISTRIBUTION - [metall] répartition *f* du vent
~DOME - [diesel eng] dôme *m* d'air
~DOSE - [radiat] dose *f* dans l'air
~DRAUGHT - [heat] tirage *m*
~DRIED - [gen] séché à l'air
~-DRIED BRICK - [constr] brique *f* crue
~-DRIED SHEETS - [rubber] (raw plantation rubber dried in the air) feuilles *f* pl séchées à l'air
~DRIFTER DRILL - [ming] perforateur *m* pneumatique à colonne
~DRILL - [tool] perforatrice *f* à air comprimé
~DRILLING - [mining] forage *m* à air comprimé
[oil ind] forage *m* à l'air
~DROP HAMMER - [tool] mouton *m* pneumatique
~DRUM - [ind] réservoir *m* d'air comprimé
~DRYING - [ind] siccatif *m* à l'air
~DRYING LACQUER - [paint] (lacquer producing a dry film on exposure to ambient temperatures) vernis *m* séchant à l'air
~DUCT - [ind] canal *m* aérien, conduit *m* d'air, manche *f* d'aspiration d'air
~DUCTING - [ind] adduction *f* d'air
~EJECTOR - [ind] éjecteur *m* d'air
~ENGINE - [mech] moteur *m* à air
~-ENTRAINED CONCRETE - [constr] béton *m* à air occlus

AIR ENTRAINING AGENT - [constr] (for concrete) entraîneur *m* d'air
~ESCAPE - [ind] soupape *f* d'échappement, détendeur *m*
~EXHAUSTER - [mech] ventilateur *m* aspirant
~FACTOR - [gas ind] facteur *m* d'air
~FEED - [mining] avancement *m* à air comprimé
~FEED STOPER - [mech] perforatrice *f* télescopique
~FILM - [telev] (in video recording) couche *f* intermédiaire d'air
~FILTER - [mech] (deviced fitted in the air intake to clean the air) filtre *m* à air
[photo] tamis *m* antipoussière
~FIN COOLER - [mech] refroidisseur *m* d'air à ailettes
~-FLOW - [mech] debit *m* d'air
~-FLOW METER - [instr] débitmètre *m* d'air
~FLUE - [th eng] conduit *m* d'air
~FOG - [photo] voile *m*
~FUEL RATIO - [motor] rapport *m* air-carburant
~GALLERY - [mining] galerie *f* de ventilation
~GAP - [el] (a short break in the ferro-magnetic portion of a magnetic circuit) entrefer *m*
~GAP CHOKE - [mech] bobine *f* d'induction avec entrefer
~GAP CLEARANCE - [el] (length of the break) longueur *f* de l'entrefer
~GAS - [th eng] gaz *m* d'air
~GAUGE - [instr] calibre *m* pneumatique
~-GLASS SURFACES - [photo] surfaces *f* pl air-verre
~GRATING - [build] grille *f* de ventilation
~-GROUND COMMUNICATION - [aero] (two-way communication between aircraft and ground) communication *f* au sol
~GUN - [armam] fusil *m* à air comprimé
[mech] marteau-piqueur *m*
~HAMMER - [impl] marteau *m* pneumatique
~HARDENING - [metall etc] trempe *f* à l'air
~HARDENING STEEL - [metall] acier *m* trempé à l'air
~HEADING - [mining] galerie *f* de ventilation
~HEATER - [th eng] réchauffeur *m* d'air, aérotherme *m*
~HOIST - [mech] élévateur *m* à air comprimé
~HOLE - [gen] évent *m*, prise *f* d'air, trou *m* d'air
[mining] ouverture *f* d'entrée d'air, arrivée *f* d'air
~HORN - [acoust] avertisseur *m* à air comprimé, trompe *f*
~HOSE - [naut] manche *f* à air
~HUMIDIFIER - humidificateur *m* d'air
~-HYDRAULIC ACCUMULATOR - [plastics] accumulateur *m* aérohydraulique
~INJECTION - s. air blast
~INJECTION MACHINE - [diecasting] machine *f* à pression directe de l'air sur le métal
~INJECTION STARTER - [motor] démarreur *m* à air comprimé
~INLET - [mech] (admission of air) entrée *f* d'air, rentrée *f* d'air, remise *f* à l'air
~INLET CONTROL - [mech] commande *f* d'admission d'air
~INLET VALVE - [mech] soupape *f* d'admission d'air, robinet *m* d'entrée d'air, vanne *f* de rentrée d'air
~INSULATION - [el] isolement *m* par air

AIR INTAKE - [mech] prise *f* d'air, entrée *f* d'air
~INTAKE CASING - [aero] (casing through which the air passes to a compressor) carter *m* d'admission d'air
~INTAKE COOLER - [motor] refroidisseur *m* de prise d'air
~INTAKE GUIDE-VANE - [aero] aubage *m* directeur d'admission d'air
~INTAKE PIPE - [motor] prise *f* d'air (carburateur)
~INTERCHANGER - [aero] (in a wind-tunnel) échangeur *m* d'air
~JET - [oil ind] injecteur *m* à air
~JIG - [min] crible-classeur *m* à air comprimé
~LAG - [aero] angle *m* de chute théorique
~LAW - [leg] droit *m* aéronautique
~LAYER - [met] couche *f* d'air
~LEAK - [el] disposition *f* par ionisation, fuite *f* d'air
~LEG - [mech] support *m* à air comprimé
~LENS - [opt] lentille *f* aérienne (interstice entre deux lentilles)
~LEVEL - [instr] niveau *m* à bulle
[mining] galerie *f* d'aérage
~LEVER - [mech] levier *m* d'admission d'air
~LICENCE - [aero] licence *f* de navigation aérienne
~LIFT - [mech] ascenseur *m* à air comprimé, élévateur *m* à air comprimé
[mining] extraction *f* par air comprimé
[aero] pont *m* volant
~LIFT PUMP - [mech] (a compressed-air ejector) éjecteur *m* à air comprimé
~LINE - [aero] ligne *f* aérienne
~LINER - [aero] avion *m* de ligne, avion *m* de transport
~LOCK - [naut] écluse *f*, sas *m* pneumatique
[hydr] (condition in a liquid flow system caused by air, e.g. an air bubble in a gravity fuel feed line) gaz *m* d'aérage
~LOCK - [aero] poche *f* d'air
~LOG - [aero] journal *m* de bord
~MAIL - [aero] poste *f* aérienne
~MAIN - [ind] circuit *m* d'air, tuyauterie *f* de soufflage
~MECHANIC - [aero] mécanicien *m* d'aviation
~MONITOR - [nucl] (for measuring radioactivity in the air) détecteur *m* de la radioactivité de l'air
~NAVIGATION - [aero] navigation *f* aérienne
~NET - [aero] réseau *m* aérien
~NIPPLE - [mech] tubulure *f* d'air
~NOZZLE - [mech] injecteur *m* d'air
~OPERATED - [mech] à commande pneumatique
~-OPERATED ELECTRIC VALVES - [mach tool] électro-soupapes *f* pl pneumatiques
~OUTLET - sortie *f* d'air
~PASSAGE - [mech] conduit *m* d'air
~PHOTOGRAPHY - [aero] photographie *f* aérienne
~PIPING - [mech] circuit *m* d'air
~PIT - [aero] trou *m* d'air
~PLOT - [aero] (position of aircraft as determined on a chart) trace air *m*
~POCKET - [aero] trou *m* d'air, courant *m* d'air descendant
[metall] retassure *f*, soufflure *f*, poche *f* d'air
[brew] bulle *f* d'air.
~POSITION - [aero] position *f* en vol
~PREHEATING - [th eng] préchauffage *m* de l'air
~PRESSURE - [mech] pression *f* d'air (comprimé)

AIR PRESSURE BRAKE - [mech] frein *m* à air comprimé
~PUMP - [mech] pompe *f* à air, compresseur *m*
~PUMP BELL - [mech] cloche *f* de la pompe à air
~PUMP DISC - [mech] plateau *m* de la pompe à air
~PUMP PISTON - [mech] piston *m* pneumatique
~PUMP PLATE - s. air pump disc
~PUMP RECEIVER - s. air pump bell
~PUMPING - [mech] pompage *m* à air
~RAID - [aero] raid *m* aérien
~RAMMER - [tool] pilon *m* pneumatique
~RECEIVER - [th eng] réservoir *m* à air (comprimé)
~-REFINED STEEL - [metall] acier *m* affiné par le vent
~RELEASE - [mech] échappement *m* d'air
~RELEASE VALVE - [mech] soupape *f* d'échappement d'air
~RELEASE VENT - [mech] orifice *m* d'échappement, évent *m*
~REMOVAL - [gen] désaération *f*, ventilation *f*
~RENEWAL - [aero] (replacement of vitiated air in the cabin) renouvellement *m* d'air
~RESERVOIR - [mach tool] (for air-operated clutches) réservoir *m* de l'air (pour embrayages pneumatiques)
~RESISTANCE - [aero dyn] résistance *f* de l'air
~SADDLE - [geol] voûte *f* anticlinale dénudée
~SCOOP - [mech] prise *f* d'air
[aero] manche *f* à air
~SEASONING - [bot] maturation *f* à l'air, séchage *m* à l'air
~SEPARATION COLUMN - [chem] (separation of nitrogen from oxygen) colonne *f* de séparation d'air
~SHAFT - [mining] puits *m* d'aération, conduit *m* d'aération
~SHIP - [aero] aéronef *m*, dirigeable *m*
~SHOWER - [astr] gerbe *f* atmosphérique
~SHUTTER - [th eng] régulateur *m* d'admission d'air
~SICKNESS - [med] mal *m* de l'air
~SIFTING - [min] épuration *f* pneumatique, épuration *f* à air comprimé
~SLACKED LIME - [chem] chaux *f* fusée, chaux *f* éteinte *m* à l'air, chaux *f* délitée
~SLACKING - [chem] extinction *f* de la chaux
~SLIT - [mining] recoupe *m* de ventilation
~SLUICE - [mining] sas *m* à air, sas *m* d'aérage
~SNIFFLER - [mech] (air intake valve) reniflard *m*
~SPACE - [el] entrefer *m*, isolement *m* par air
[aero] espace *m* aérien
~SPACED COIL - [el] bobinage *m* à air
~SPEED - [aero] (the speed of the aircraft in respect to the ambient air)vitesse *f* vraie
~SPEED METER - [aero] anémomètre *m*
~SPINNER - [mech] centrifuge *m* à air
~SPRING - [mech] amortisseur *m* à air
~STARTING - [mech] démarrage *m* à l'air comprimé
~STATION - [aero] aéroport *m*
~STRANGLER - [mech] régulateur *m* d'admission d'air
~SURVEY - [surv] relevé *m* photogrammétrique
~SWEETENING - [metall] affinage *m*
~SWEPT - [gen] refroidi par air
~TANK - [naut] caisson *m* de décompression
[mech] chambre *f* de compression
~TIME - [telecomm] temps *m* d'émission
~-TO-GROUND COMMUNICATION - [aero] communication *f* air-sol
AIR-TO-SURFACE VESSEL - [radar] radar *m* de bord pour la recherche d'objets à la surface de la mer
~TRAFFIC - [aero] circulation *f* aérienne
~TRAIN - [aero] train *m* aérien
~TUBE - chambre *f* à air
~VENT - [th eng] évent *m*, prise *f* d'air, ouverture *f* de ventilation
~VENTURE CHOKE TUBE - [mech] diffuseur *m* Venturi
~WALL IONIZATION CHAMBER - [nucl] chambre *f* d'ionisation à paroi d'air
~WASHER - [th eng] épurateur *m* d'air
~WIRE - [radio] fil *m* aérien d'une antenne
AIRBAG - [rubber] airbag *m*, sac *m* à air, sac *m* à cuisson
~BUFFER - [rubber] machine *f* à raper les airbags
~MOULD - [rubber] moule *m* pour sac à cuisson (ou airbag)
AIRBAGGING MACHINE - [rubber] (to shape tires) presse *f* à galber, conformateur *m* de pneumatiques
AIRBORNE - [aero] (carried by aerial transport) aéroporté
~DUST - [air pollution] poussière *f* en suspension dans l'air
~EARLY WARNING - [radar] radar *m* (aéroporté) de veille avancée
~ELECTRONICS - [electron] (electronic equipment fitted to an aircraft) appareillage *m* électronique de bord
~GUN SIGHT RADAR - [radar] radar *m* de direction de tir à pointage automatique
~INTERCEPTION - [radar] interception *f* par radar aéroporté
~PARTICULATE - [nucl] macromolécule *m* en suspension dans l'air
~RADAR - [radar] radar *m* de bord
~RADIOACTIVITY - [nucl] radioactivité *f* atmosphérique, radioactivité *f* de l'air (des matières en suspension dans l'air)
AIRCRAFT - [aero] (any aerial vehicle) aéronef *m*, aéroplane *m*, avion *m*
~BONDING - [el] connexion *f* électrique entre les différents éléments métalliques destinées à augmenter l'effet de masse
~FITTER - [mech] monteur *m*, ajusteur *m*
~FLUTTER - [telev] (due to passing aircraft) vibration *f* parasite
~FUEL - [aero] carburant *m* pour avions
~GEAR - [aero] engrenage *m* pour avions
~GROUNDING - [aero] mise *f* au sol, immobilisation *f* des appareils
~HUNTING - [aero] (uncontrolled oscillation of approximately constant amplitude) flottement *m*, oscillation *f* (lente)
~INTAKE DUCT - [aero] canalisation *f* d'entrée d'air
~INTERCEPTION - [radar] interception *f* par radar de veille au sol
~LANDING FLARE - [aero] fusée *f* d'atterrissage
~MARKINGS - [aero] marques *f* pl d'immatriculation
~PLOTTER - [aero] calculateur *m* de route
~PREPARATED FOR SERVICE WEIGHT - [aero] poids *m* en ordre de marche
~PROXIMITY INDICATOR - [aero] indicateur *m* de

risque de collision
AIRCRAFT RADIO STATION - [aero] (airborne radio station) local *m* radio de bord
~TAIL WARNING - [radar] radar *m* de queue (avion)
~TENDER - [navy] ravitailleur *m* d'aviation
~WIRELESS OPERATOR - [aero] radio-télégraphiste *m* de bord
AIRDROP, to - [aero] (the operation of dropping troops etc) parachuter
AIRDROP - [aero] lancement *m* en parachutes
AIRFIELD - [aero] terrain *m* d'aviation
~BEACON - [aero] balise *f* lumineuse, radiobalise *f*
AIRFLOW - [aero] écoulement *m* de l'air, débit *m* d'air
~METER - [instr] débitmètre *m*
AIRFOAM - [rubber] caoutchouc *m* mousse
~CUSHIONING - [rubber] rembourrage *m* de caoutchouc mousse
AIRFOIL - [aero] profil *m* d'aile, plan *m* aérodynamique
~PROFILE - [aero] profil *m* d'aile
~SECTION - [aero] profil *m* d'aile, section *f* d'aile
AIRFRAME - [aero] (the complete structure of an aircraft without its means of propulsion) cellule *f*
~BONDING LEAD - [aero] câble *m* de mise à la masse
~DE-ICING - [aero] dégivrage *m* de la cellule
~RESONANCE - [aero] résonance *f* de la cellule
AIRFREIGHT - [aero] transport *m* aérien, cargaison *f*, fret *m*
~CARGO - [aero] fret *m*, cargaison *f* (transportée par air)
AIRING - [gen] aération *f*, aérage *m*, ventilation *f*
~CUPBOARD - [ind] séchoir *m*
AIRLEG - [mech] servomoteur *m* pneumatique
AIRLESS - privé d'air, exempt d'air
~INJECTION - [mech] injection *f* mécanique
AIRLIFT - [aero] pont *m* aérien, chargement *m*, cargaison *f*
[mech ind] pompe *f* à air comprimé
AIRLINE - [aero] ligne *f* aérienne
~DISTANCE - [aero] (the length of a great circle passing through two given points on the surface of the earth) distance *f* de grand cercle
AIRLINER - [aero] avion *m* de ligne, avion *m* de transport
AIRMAN - [aero] aviateur *m*
AIRMARK, to - [aero] repérer, signaler (pour la navigation aérienne)
AIRPLANE - [aero] aéronef *m*, aéroplane *m*, avion *m*
~BUDDY TANK - [aero] (US) réservoir *m* auxiliaire
AIRPORT - [aero] aéroport *m*
~DANGER BEACON - [radar] radiophare *m* de danger
AIRPROOF - [gen] étanche à l'air
~CLOTH - [text] tissu *m* imperméable à l'air
~PAPER - [paper] papier *m* imperméable à l'air
AIRSCREW - [aero] hélice *f*
~AXIS - [aero] axe *m* d'hélice
~BLADE - [aero] pale *f* d'hélice
~BOSS - [aero] moyeu *m* d'hélice
~DISK - [aero] disque *m* d'hélice
~PITCH - [aero] pas *m* d'hélice
~SHAFT - [aero] arbre *m* d'hélice
~SLIP - [aero] recul *m* de l'hélice
~THRUST - [aero] poussée *f* de l'hélice
~TURBINE - [aero] turbopropulseur *m*
AIRSHAFT - [mining] puits *m* de ventilation
AIRSHIP - [aero] aéronef *m*, dirigeable *m*
~GONDOLA - gondola *f*, nacelle *f*
~HULL - [aero] coque *f*
~KEEL - [aero] grille *f*
AIRSICKNESS - [med] mal *m* de l'air
AIRSTART - [astronaut] démarrage *m* en vol
AIRSTOP TUBE - [rubber] chambre *f* à air increvable
AIRSTRIP - [aero] piste *f* d'atterrissage
AIRTAXI - [aero] avion-taxi *m*
AIRTIGHT - étanche à l'air, hermétique
~COVER - [gen] couvercle *m* étanche
~JOINT - [mech] fermeture *f* hermétique (imperméable) au vent
~SEAL - [mech] scellement *m* étanche à l'air
AIRVANE - [astronaut] palette *f* directrice
AIRWAY - [aero] ligne *f* aérienne, route *f* aérienne
AIRWORTHINESS - [aero] navigabilité *f*
~CERTIFICATE - [aero] certificat *m* de navigabilité
AIRWORTHY - [aero] en bon état de navigabilité
AISLE - [constr] bas-côté *m*, nef *f* latérale, coursive *f*, nef *f*
AJUTAGE s. adjutage
Al - [chem] (aluminium) Al (aluminium)
ALABANDITE - [min] alabandine *f*
ALABASTER - [min] albâtre *m*
ALAR - [zool] alaire
ALARM - [gen] alarme *f*, alerte *f*, signal *m*, avertisseur *m*
~BELL - [sign] sonnerie *f* d'alarme, timbre *m* avertisseur
~CLAPPER - [horol] battant *m* de sonnerie
~CLICK - [horol] cliquet *m*
~CLOCK - [horol] réveil-matin *m*
~DEVICE - [acoust] dispositif *m* d'alarme, dispositif *m* avertisseur
~FUSE - [el] fusible *m* avertisseur
~HAND - [horol] aiguille *f* de réglage de sonnerie
~LAMP - [light] voyant *m* avertisseur
~POWER WHEEL - [horol] roue *f* matrice de sonnerie
~SETTING KNOB - [horol] bouton *m* de réglage
~SIGNAL - [sign] signal *m* d'alarme, signal *m* avertisseur
~WHISTLE - [sign] sifflet *m* d'alarme
ALASKA - [text] (mohair or wool and cotton yarn) tissu *m* de laine et coton
ALATE - [zool] (especially of shells) ailé
[bot] (stems with decurrent leaves) ailé
ALBEDO - [astr] (light-reflecting power of non luminous heavenly bodies) albédo *m*
[phys] pouvoir *m* de reflexion
[nucl] (measure of the reflecting force of surfaces for neutrons) albédo *m*
ALBERTITE - [min] (black solid bitumen) albertite *f*
ALBIAN - [geol] (a stage of the cretaceous system) albien
ALBINO - [biol] albinos *m*
ALBION METAL - [metall] feuille *f* de plomb étamée
ALBITE - [min] (silicate of sodium and aluminium with potash and lime) albite *f*
ALBUGINEOUS TISSUE - [anat] tissu *m* albuginé
ALBUMEN - [bot & zool] albumen *m*
~PAPER - [ind chem] papier *m* à l'albumine, papier *m* albuminé
ALBUMIN - [chem] (simple protein soluble in pure water) albumine *f*
~DENATURATION - [chem] dénaturation *f* des corps

albuminoïdes
ALBUMIN PROCESS - [photo] (an obsolete method of printing in photography) procédé *m* à l'albumine
ALBUMINATE - [chem] (alkali albumin compounds) albuminate *m*
ALBUMINOID - [chem] (insoluble protein) albuminoïde *m*
~DEGENERATION - [ind] dégenérescence *f* granuleuse
ALBUMINOMETER - [instr] albuminomètre *m*
ALBUMINOUS - [chem] albumineux
~MATTER - [chem] matières *f* pl albumineuses
ALBUMOSE - [chem] albumose *m*
ALBURNUM - [bot] aubier *m*
ALCALI - [chem] s. alkali
ALCLAD - [metall] (an alloy) alclad *m*, védal *m* (alliage d'aluminium)
ALCOHOL - [chem] alcool *m*
~ACID - [chem] acide-alcool *m*
~CONTENT - [chem] teneur *f* en alcool
~DE-ICING - [mech] dégivrage *m* à l'alcool
~MOTOR FUEL - [mech] mélange *m* alcoolisé
~THERMOMETER - [instr] thermomètre *m* à l'alcool
ALCOHOLATE - [chem] alcoolate *m*
ALCOHOLIC - [chem] alcoolique
~FERMENTATION - [ind chem] (production of alcohol from sugar and yeast) fermentation *f* alcoolique
ALCOHOLICITY - [chem] alcoolicité *f*
ALCOHOLIZATION - [chem] alcoolisation *f*
ALCOHOLOMETER - [instr] (to measure the alcohol content in acqueous solutions) alcoolomètre *m*
ALCOVE - [constr] alcove *f*, niche *f*, rotonde *f*
ALDEHYDASE - [chem] aldéhydase *f*
ALDEHYDE - [chem] (hydrocarbon derivatives containing the carbonyl group) aldéhyde *f*
~ACID - [chem] acide *m* aldéhydique
~AMMONIA - [chem] aldéhydate *m* d'ammoniaque
~CONDENSATION - [chem] condensation *f* aldéhydique
~GROUP - [chem] groupe *m* aldéhydique
~RESIN - [chem] résine *f* aldéhydique
ALDEHYDIC ACID - [chem] acide *m* aldéhydique
ALDER - [bot] aulne *m*
ALDERCOPSE - [bot] aulnaie *f*
ALDIMINE - [chem] (condensation product of phenol) aldimine *f*
ALDOL - [chem] (condensation product of acetaldehyde) aldol *m*
ALDOSE - [chem] (mono-saccharose of an aldehydic constitution) aldose *m*
ALE - [brew] (a beer flavoured with hops) bière *f* blonde
ALECOST OIL - [chem] essence *f* de balsamite
ALEMBIC - [chem impl] alambic *m*
ALERT - [acoust] alerte *f*, alarme *f*
ALEUROMETER - [instr] aleuromètre *m*
ALEURONE - [bot] aleurone *m*
ALEWIFE - [zool] alose *f*
ALEXANDRITE - [min] (a variey of chrysoberyl) alexandrite *f*
ALEXIA - [med] (inability to understand written language) alexie *f*
ALFA GRASS - [bot] alfa *m*
ALFALFA - [bot] (purple medick) luzerne *f*, éragrostide *f*
ALGA - [bot] algue *f*
ALGAE BLOOM - [hydrol] fleur *f* d'eau, prolifération *f* d'algues
~CONTROL - [hydrol] destruction *f* des algues, lutte *f* contre la prolifération des algues
ALGAL LAYER - [bot] couche *f*
ALGEBRA - [math] algèbre *m*
ALGEBRAIC - [math] algébrique
~ADDER [comput] addeur *m* algébrique
~ANGLE - [math] angle *m* algébrique
~AREA - [math] aire *f* algébrique
~CALCULUS - [math] calcul *m* algébrique
~COMPLEX - [math] quantité *f* complexe algébrique
~CURVE - [math] courbe *f* algébrique
~EQUATION - [math] équation *f* algébrique
~EXPRESSION - [math] expression *f* algébrique
~GEOMETRY - [math] géométrie *f* algébrique
~INTEGER - [math] intégrale *f* algébrique
~NUMBER - [math] nombre *m* algébrique
~SUM - [math] somme *f* algébrique
ALGESIA - [med] (sensitivity to pains) algésie *f*
ALGID - [gen] algide
ALGIDITY - [phys] algidité *f*
ALGINATE - [chem] alginate *m*
ALGODONITE - [min] algodonite *f*
ALGORISM - [math] algorithme *m*
ALGRAPHY - [print] algraphie *f*
ALICYCLIC - [chem] alicyclique, cyclanique
ALIDADE - [instr] alidade *f*, pinnule *f*
ALIGHT, to [gen] descendre, atterrir, se poser
ALIGHT - [gen] en feu, en flammes
~ON WATER, to - [aero] amerrir, amerir
ALIGHTING - [gen] descente *f*
[aero] amerissage *m*, atterrissage *m*
~GEAR - [aero] train *m* d'atterrissage
~RUN - [aero] longueur *f* de roulement à l'atterrissage
ALIGN, to - [gen] aligner, redresser
~BY SPIRIT LEVEL, to-[gen] aligner au niveau d'eau
ALIGNED - [gen] aligné, mise en ligne
~GRID VALVE - [electron] (a thermionic valve) tube *m* à grilles alignées
ALIGNMENT - [gen] alignement *m*
[meas] alignement *m*, mise *f* dans l'axe, centrage *m*
[radar] alignement *m*
~ANTENNA ARRAY - [radio] antenne *f* à éléments alignés
~CHART - [math] abaque *m*, nomogramme *m*
~ERROR - [acoust] erreur *f* d'alignement
~GENERATOR - [telev] oscillateur *m* de service
~OF LIGHTS - [naut] (maritime signals) alignement *m* de feux
~OF NUCLEI - [phys] orientation *f* sans polarisation des noyaux
ALIMENTARY - alimentaire
ALIPHATIC - [chem] aliphatique, de la série grasse
~ACID - [chem] acide *m* aliphatique, acide *m* gras
~ALCOHOL - [chem] alcool *m* aliphatique
~HYDROCARBON - [chem] carbure *m* aliphatique
ALIQUOT PART - [math] partie *f* aliquote
ALITE - [chem] alite *f*
ALIVE - [gen] vivant, en vie
[el] sous tension, actif
ALIZARIN - [chem] (important natural and synthetic dye) alizarine *f*
~BLACK - [chem] noir *m* d'alizarine
~BROWN - [chem] brun *m* d'alizarine

ALIZARIN GREEN - [chem] vert *m* d'alizarine
~VIOLET - [dyes] violet *m* d'alizarine
~YELLOW - [dyes] jaune *m* d'alizarine
ALKALESCENCY - [chem] alcalescence *f*
ALKALESCENT - [chem] alcalescent
ALKALI - [chem] (a substance which is alkaline in acqueous solution) alcali *m*
~CELLULOSE - [chem] alcali-cellulose *f*
~FAST - [dyes] résistant aux alcalis
~GRANITE - [geol] granit *m* basique
~LIQUOR - [ind chem] eau *f* résiduaire ammoniacale
~METAL - [ind chem] (the monovalent metals in the first group of the periodic table) métal *m* alcalin
~PROCESS - [chem] régénération *f* à l'alcali, régénération *f* à la soude
~RECLAIM - [rubber] régénération *f* à l'alcali, régénération *f* à la soude
~RECLAIMING PROCESS - [rubber] régénération *f* à l'alcali, régénération *f* à la soude
~WASH - [ind chem] lavage *m* à la soude
ALKALIFIABLE - [chem] alcalifiable
ALKALIFY, to - [chem] alcaliniser, alcaliser
ALKALIMETER - [instr] (to measure the proportion of alkali in a solution) alcalimètre *m*
ALKALIMETRICAL - [chem] alcalimétrique
ALKALIMETRY - [chem] alcalimétrie *f*
ALKALINE - [chem] alcalin
~ACCUMULATOR - [el] accumulateur *m* alcalin
~AIR - [chem] ammoniac *m*
~CLEANING - [el chem] (removal of grease by an alkaline solution) dégraissage *m*
~EARTH - [chem] terre *f* alcaline
~EARTH METAL - [ind chem] (all divalent metals in the second group of the periodic table) métal *m* alcalinoterreux
~PHOTOCELL - [el] cellule *f* photoélectrique alcaline
~SOLUTION - [chem] solution *f* alcaline
~STORAGE BATTERY - [el] accumulateur *m* alcalin
~TIN PLATING - [el chem] étamage *m* électrolytique
ALKALINITY - [chem] alcalinité *f*
ALKALIPROOF - [chem] résistant aux alcalis
ALKALIZATION - [chem] alcalinisation *f*, alcalisation *f*
ALKALIZE, to - [chem] alcaliniser, alcaliser
ALKALOID - [chem] alcaloïde *m*
ALKALOIDAL - [chem] alcaloïde
ALKANE - [chem] alcane *f*
ALKANNIN - [chem] alcannine *f*
ALKARGEN - [chem] acide *m* cacodylique
ALKARSINE - [chem] oxyde *m* de cacodyle
ALKYD - [chem] alkyde
~PLASTICS - [plastics] alkyde *m*
~RESIN - [chem] résine *f* alkyde, glyptal *m*
ALKYL - [chem] (generic term for monovalent paraffin hydrocarbon radicals) alcoyle *m*
~HALIDE - [chem] halogénure *m* d'alcoyle, alcoylhalogène *m*
ALKYLAMINE - [chem] alcoylamine *f*.
ALKYLATION - [ind chem] (process in petroleum refining to obtain combination of olefins and isoparaffin hydrocarbons) alkylation *f*, alcoylation *f* [chem] alcoylation *f*, alkylation *f*
~ACID - [chem] acide *m* pour alkylation
~CONTACTOR - [chem] réactif *m* d'alcoylation
ALKYLENE - [chem] alcoylidène *m*

ALL - tout, complètement
~ABACK - [naut] complètement masqué
~-BAND AMPLIFIER - [telev] amplificateur *m* à large bande
~-BAND TUNER - [telev] sélecteur *m* pour tous les canaux
~BOTTOM SOUND - [acoust] son *m* sourd
~BRASS UNION - [metall] raccord *m* en laiton
~CHANNEL DECODER - [comput] décodeur *m* à plusieurs canaux
~CLEAR - [acoust] fin *f* d'alerte
~-CROP TRACTOR - s. all-purpose tractor
~-FLYING TAIL - [aero] ensemble *m* gouverne de profondeur - stabilisation mobile
~GAS BURNER - [burners] brûleur *m* universel
~-GEAR HEADSTOCK - [mach tools] poupée *f* à monopoulie et boîte de vitesse
~-GRAIN TYPE THRESHER - [agric] batteuse *f* mixte à céréales et à graines fourragères
~HANDS - [gen & naut] tout l'équipage *m*, tous ensemble
~-IN TARIFF - [el] tarif *m* simple à compteur unique
~-INERTIA GUIDANCE - [astronaut] guidage *m* totalement inertiel
~MAINS CURRENT - [el] courant *m* universel
~MALLEABLE IRON UNION - [metall] raccord *m* en fonte malléable
~-MINE IRON - [min] fer *m* virginal
~-ON ALL-OFF CONTROL - [burners] régulation *f* par tout ou rien
~-ON ALL-OFF CONTROLLER - [telecontr] régulateur *m* par tout ou rien
~-OUT SEARCH - [aero] recherche *f* mettant tout en oeuvre
~-FILTER - s. all-pass network
~-PASS NETWORK - [el] filtre *m* passe-tout
~PURPOSE - d'usage général, universel
~-PURPOSE CAMERA - [telev] caméra *f* universelle
~-PURPOSE TRACTOR - [agric] tracteur *m* universel, polyculteur *m*
~-RELAY SELECTOR - [telecomm] sélecteur *m* à relais
~-RELAY SYSTEM - [telecomm] système *m* tout à relais
~ROUND SLEWING - [mech] (cranes) à orientation totale
~SLIMING - [min] broyage *m* fin
~STEPS CONTROL - [comput] commande *f* de tous les cycles, contrôle *m* de toutes les opérations
~TOPSOUND - [acoust] son *m* aigu
~TRUNKS BUSY REGISTER - [telecomm] compteur *m* d'occupation totale
~UP WEIGHT - [aero] poids *m* total
~WAVE RADIO SET - [radio] récepteur *m* radio toutes ondes
~-WEATHER - [gen] tout temps
ALLANIC ACID - [chem] acide *m* allanique
ALLANTOIN - [chem] allantoïne *f*
ALLAY, to - [gen] alléger, appaiser, calmer
ALLELE - [genet] (having alternate or constrasting Mendelian characters) allèle
~SHIFT - [genet] modification *f* par la présence d'un allèle
ALLELOMORPH - s. allele
ALLENE - [chem] allène *m*
ALLERGIC - [med] allergique
ALLERGY - [med] allergie *f*

ALLETTE - [constr] membrette *f* (arche)
ALLEVIATOR - [mech] équilibreur *m* de pression
ALLEY - [gen] allée *f*, passage *m*, ruelle *f*
ALLEYWAY - [gen] corridor *m*
[naut] coursive *f*
ALLIACEOUS - [bot] alliacé
ALLIGATOR, to - [paint] (to crack along many lines forming a pattern like the skin of an alligator) lézarder
ALLIGATOR - [zool] alligator *m*
[metall] presse *f* à cingler, presse *f* à former
[mech] clef *f* à tubes
~CLIP - [impl] (hinged spring clip with serrated jaws) pince *f* crocodile
~CRACKING - [paint] fendillement *m*, lézarde *f*
~SHEARS - [impl] cisaille *f* à balance
~SKIN - [metall] effet *m* de pelure d'orange
~SQUEEZER - [metall] macque *f*, presse *f* à cingler
~WRENCH - [impl] pince *f* universelle crocodile
ALLIGATORING - [paint] fendillement *m*, lézarde *f*
ALLOBAR - [phys] (form of an element with an atomic weight different from that of the form occurring naturally) allobare *f*
ALLOCATE, to - [gen] affecter, assigner, distribuer, répartir
ALLOCATION - [gen] allocation *f*, attribution *f*, distribution *f*, répartition *f*
~OF FREQUENCIES - [radio] allotissement *m* des fréquences
ALLOCHROMATIC - [phys] allochromatique
~MINERAL - [min] minéral *m* allochromatique
ALLOCHTONOUS - [geol] allochtone
ALLOCLASITE - [min] alloclase *f*
ALLOGAMY - [bot] fécondation *f* croisée
ALLOGENE - [genet] allogène
ALLOIOBIOGENESIS s. alloiogenesis
ALLOIOGENESIS - [genet] (alternation of sexual and parthogenetic generations) alloiogénèse *f*
ALLOISOMERY - [chem] (condition of substances having the same structure but with different spatial arrangement of their atoms) allomérie *f*
ALLOMERIC - [chem] allomérique
ALLOMERISM - [chem] s. alloisomery
ALLOMORPHISM - [chem] (condition of substances having the same chemical composition but different crystalline forms) allomorphisme *m*
ALLOPATRIC - [genet] (relating to different areas) allopatrique
ALLOPHANE - [min] allophane
ALLOSE - [chem] allose *m*
ALLOSOME - [genet] (atypical chromosome) allosome *m*
ALLOT, to - [gen] assigner, distribuer, répartir
ALLOTHERMIC - [zool] allothermique
ALLOTMENT - [gen] attribution *f*, distribution *f*
[agric] lotissement *m*
ALLOTROPE - [chem] allotrope *m*
ALLOTROPIC - [chem] allotropique
~MODIFICATION - [chem] transformation *f* allotropique
ALLOTROPY - [chem] (the existence of an element in two or more forms, like diamond and graphite) allotropie *f*
ALLOW, to - [mech] prévoir une tolérance
ALLOWABLE - admis, admissible, légitime
~BEARING VALUE OF SOIL - [soil] pression *f* admissible sur le sol porteur
ALLOWABLE LOAD - [gen] charge *f* admissible
ALLOWANCE - [gen] indemnité *f*, attribution *f*
[comm] remise *f*, réduction *f*, bonification *f*, rabais *m*
[mech] tolérance *f*, jeu *m*, surépaisseur *f* d'usinage
ALLOWED TRANSITION - [nucl] transition *f* permise
ALLOY - [metall] alliage *m*
~CAST IRON - [metall] fonte *f* alliée
~CLADDING - [metall] revêtement *m* d'alliage
~CONSTITUENT - [metall] constituant *m* d'un alliage
~FOR ELECTRIC RESISTORS - [metall] alliage *m* pour rhéostats
~JUNCTION TRANSISTOR - [transm] transistor *m* à raccord en alliage
~OF HIGH PERCENTAGE - [metall] alliage *m* à haute teneur
~OF LOW PERCENTAGE - [metall] alliage *m* à faible teneur
~PLATE - [metall] revêtement *m* d'alliage
~STEEL - [metall] acier *m* allié, acier *m* spécial
~STRUCTURE - [metall] structure *f* des alliages
ALLOYING ELEMENT - [metall] élément *m* d'addition
ALLSPICE - [bot] piment *m*
ALLUVIAL - [geol] alluvial, alluvionnaire
~CLAIM - [mining] concession *f* de terrains d'alluvion
~CLAY - [geol] argile *f* alluviale, glaise *f* alluviale
~CONE - [geol] cône *m* de déjection
~DEPOSIT - [geol] dépôt *m* alluvionnaire
~DIGGING - [mining] exploitation *f* des alluvions, orpaillage *m*
~FAN - [geol] cône *m* de déjection
~MEADOW SOIL - [soil] (gleyed forest soil) sol *m* alluvial fluviatile de formation récente
~PLAIN - [geol] plaine *f* alluvionnaire
~SLIME - [geol] limon *m* alluvial
~SOIL - [geol] terrain *m* alluvionnaire
~TERRACE - [geol] (drift terrace) terrasse *f* alluviale, terrasse *f* d'accumulation
ALLUVIATION - [geol] alluvionnement *m*
[soil] atterrissement *m*
ALLUVION - [geol] alluvion *m*
ALLUVIUM - [geol] alluvion *m*, terre *f* d'alluvions
[soil] atterrissement *m*
~DRY VANNING - [min] vannage *m* à sec des alluvions
ALLYL - [chem] allyle *m*
~ALCOHOL - [chem] alcool *m* allylique
~AMINE - [chem] allylamine *f*
~MERCAPTAN - [chem] allylmercaptan *m*
~PLASTICS - [plastics] plastiques *m* pl allyliques
ALLYLENE - [chem] allylène *m*
ALMANDITE - [min] almandite *f*
ALMOND - [bot] amande *f*
~GREEN - [ind chem] vert *m* malachite
~OIL - [ind chem] huile *f* d'amandes
~SECTION IRON - [metall] fer *m* à feuille de saule
~TREE - [bot] amandier *m*
ALNICO - [metall] (aluminium, nickel, cobalt and copper alloy) alnico *m* (alliage aluminium-nickel-cobalt-cuivre)
ALOE - [bot] aloes *m*
ALOIN - [chem] aloïne *f*
ALONG THE STRIKE - [mining] en direction
ALP - [geogr] alpe *f*

ALPACA - [text] alpaga *m*
ALPHA - (first letter of Greek alphabet) alpha
~BOMBARDMENT - [nucl] bombardement *m* par particules apha
~BRASS - [metall] (copper-zinc alloy) laiton *m* alpha
~BRONZE - [metall] (alpha solid solution of tin in copper) bronze *m* alpha
~BURST DETECTOR - [nucl] détecteur *m* d'explosion alpha
~CHAMBER - [nucl] (counting chamber for the detection of alpha particles) chambre *f* alpha
~COUNTER - [phys] compteur *m* alpha
~DECAY - [phys] désintégration *f* alpha
~DISINTEGRATION ENERGY - [nucl] énergie *f* de désintégration alpha
~EMITTER - [phys] émetteur *m* alpha
~IRON - [metall] (polyform of iron, stable below 90 centigrades) fer *m* alpha
~PARTICLE - [phys] particule *f* alpha
~PARTICLE BINDING ENERGY - [nucl] énergie *f* de liaison des particules alpha
~RAY SPECTRUM - [phys] spectre *m* de rayonnement alpha
~RAYS - [nucl] rayons *m* pl alpha
~URANIUM - [phys] uranium *m* alpha
ALPHABETIC INTERPRETER - [comput] traductrice *f* alphabétique, traductrice *f* alphanumérique
~PUNCH - [comput] perforatrice *f* alphabétique
~TELEGRAPHY - [telecomm] télégraphe *f* alphabétique
~VERIFIER - [comput] vérificatrice *f* alphabétique, vérificatrice *f* alphanumérique
ALPHABETICAL ACCOUNTING MACHINE - [comput] tabulatrice *f* alphabétique
~NOTATION - [mus] notation *f* alphabétique
~OFFICE CODE - [telecomm] indicatif *m* litéral
ALPHACELLULOSE - [ind chem] alpha-cellulose *f*
ALPHAMERIC - [comput] alphanumérique
ALPHATOPIC - [nucl] alphatopique
ALPHATRON GAUGE - [nucl] alphatron *m* (appellation commerciale), jauge *f* à ionisation
ALTAZIMUTH - [instr] (navigational instrument) altazimut *m*
[adj] altazimutal
ALTER, to - changer, modifier, transformer
ALTERATION - changement *m*, modification *f*, révision *f*
~OF ANGLE - [metall] (angular slip) déformation *f* angulaire
ALTERING - changement *m* , modification *f*
~OF COURSE - [naut] changement *m* de route, changement *m* de cap
ALTERNANCE - [radio] alternance *f*
ALTERNATE - alternatif, alternant, alterné
ALTERNATE ANGLES - [geom] angles *m* pl alternes
~CHANNEL INTERFERENCE - [radio] interférence *f* du canal alternant
~ROUTE - [telecomm] voie *f* de déroutement
ALTERNATING BURST - [telev] (in colour television) salve *f* alternante
~COMPONENT OF CURRENT - [el] composante *f* alternative de courant
~CURRENT - [el] courant *m* alternatif
~-CURRENT ARC - [el] arc *m* à courant alternatif
~-CURRENT BRIDGE - [el] pont *m* à courant alternatif
~-CURRENT DIALLING - [telecomm] sélection *f* à distance par impulsions de courant alternatif
ALTERNATING-CURRENT PULSE - [el] impulsion *f* de courant alternatif
~-CURRENT PULSING - [transm] sélection *f* par courants alternatifs
~-CURRENT VOLTAGE - [el] tension *f* alternative
~FIELD - [el] champ *m* alternatif
~LIGHT - [signal] feu *m* alterné
~QUANTITY - [el] grandeur *f* alternative
~STRESS - [mech] (stress repeated at intervals, ranging from tension to compression) effort *m* pulsatoire, contrainte *f* alternée
ALTERNATION - [el] alternance *f*
ALTERNATIVE FUEL ENGINE - [mech] moteur *m* à alimentation mixte
ALTERNATOR - [el] alternateur *m*, génératrice *f* (de courant alternatif)
~DISC SET - [el] générateur *m* à étincelle tournante
~TRANSMITTER - [radio] (utilizing power generated by a radiofrequency alternator) émetteur *m* alternateur
ALTERNOMOTOR - [el] moteur *m* à courant alternatif
ALTIGRAPH - [instr] altimètre *m* enregistreur
ALTIMETER - [instr] (an aneroid barometer measuring altitude by the decrease of atmospheric pressure) altimètre *m*
ALTIMETRY - [gen] altimétrie *f*
ALTITUDE - altitude *f*, élévation *f* (au dessus du niveau de la mer)
[astr] hauteur *f* (d'un astre)
~ANGLE - [astr] angle *m* d'élévation
~CHAMBER - [aero] chambre *f* à vide
~CONTROL - [aero] commande *f* altimétrique
~PRESSURE CHAMBER - [astronaut] (high-altitude test chamber) chambre *f* de haute altitude, chambre *f* d'altitude
~RECORDER - [instr] altimètre *m* enregistreur
~SWITCH - [instr] (an instrument making or breaking an electrical circuit at a pre-set height) interrupteur *m* altimétrique
~VALVE - [aero] (valve fitted to the carburettor of an aero-engine and correcting the strength of the mixture as air density falls) soupape *f* altimétrique
ALTO-CUMULUS - [met] (rounded mass of cloud at heights between 10,000 and 25,000 feet) alto-cumulus *m*
ALTOMETER - [instr] théodolite *m*
ALTO-STRATUS - [met] alto-stratus *m*
ALTROSE - [chem] (stereo-isomeric form of glucose) altrose *m*
ALULA - [zool] (the bastard wing) alule *f*, aile *f* bâtarde
ALUM, to - [text] (to treat with alum) aluner
ALUM - [chem] (alums: a large number of isomorphous compounds) alun *m*
~EARTH - [min] alumine *f* calcinée, oxyde *m* d'aluminium, terre *f* alumineuse
~LEATHER - [leather] cuir *m* aluné, cuir *m* blanc, cuir *m* mégis
~MINE - s. alum pit
~PIT - [min] aluminière *f*, alumière *f*
~SLATE - [geol] schiste *m* alumineux
ALUMEL - [min] alumel *m*
ALUMEN - [metall] alliage *m* d'aluminium
ALUMINA - [chem] (the trioxide of aluminium obtained as mineral corundum) alumine *f*

ALUMINA BRICK - [ovens] brique *f* réfractaire
~WHITE - [chem] blanc *m* d'alumine (charge)
ALUMINATE - [chem] aluminate *m*
ALUMING - [text] aluminage *m*
[leather] alunage *m*
ALUMINITE - [chem] aluminite *f*, aluminaire *f*
~PROCESS - [ind chem] oxydation *f* anodique de l'aluminium
ALUMINIUM - [metall] (light, ductile metal with high electrical conductivity and resistance to corrosion) aluminium *m*
~ACETATE - [chem] acétate *m* d'aluminium
~ALLOY - [metall] (general term for those alloys whose basis is aluminium) alliage *m* d'aluminium
~ALLOY CASTING - [metall] pièce *f* coulée en alliage d'aluminium
~ANODE CELL - [el] pile *f* à anode d'aluminium
~ARRESTER - [el] parafoudre *m* à pointe d'aluminium
~BLACK - [chem] noir *m* d'aluminium
~BONDING - [metall] métallisation *f*, revêtement *m* d'aluminium
~BRASS - [metall] laiton *m* d'aluminium
~BROMIDE - [chem] bromure *m* d'aluminium
~BRONZE - [metall] bronze *m* d'aluminium, cupro-aluminium *m*
~BRONZE POWDER - [ind chem] poudre *f* d'aluminium
~CELL RECTIFIER - [el] redresseur *m* à anode d'aluminium
~CHLORIDE - [chem] chlorure *m* d'aluminium au chrome
~FLAKE - [ind chem] blanc *m* d'alumine
~FOAM - [metall] mousse *f* d'aluminium, écume *f* d'aluminium
~HYDROXIDE - [chem] hydroxyde *m* d'aluminium, alumine *f*
~INGOT - [metall] aluminium *m* en lingot
~MAGNESIUM ALLOY - [metall] alliage *m* d'aluminium au magnésium, alumag *m*
~MANGANESE ALLOY - [metall] alliage *m* d'aluminium au manganèse
~NICKEL ALLOY - [metall] alliage *m* d'aluminium au nickel
~ORE - [min] minerai *m* d'aluminium
~OXIDE - [chem] alumine *f*, corindon *m*, oxyde *m* d'aluminium
~PAINT - [ind chem] peinture *f* d'aluminium
~PAPER - [paper ind] papier *m* métallisé, papier *m* d'aluminium
~PLATE - [metall] tôle *f* d'aluminium
~POWDER - [ind chem] poudre *f* d'aluminium, aluminium *m* en poudre
~SHEATED CABLE - [metall] câble *m* à gaîne d'aluminium
~WELDING - [metall] soudure *f* à l'aluminium
~WIRE - [metall] fil *m* d'aluminium
~WOOL - [metall] laine *f* d'aluminium
ALUMINIZING - [ind chem] (coating with aluminium) aluminage *m*, métallisation *f* à l'aluminium
ALUMINOGRAPHY - [print] algraphie *f*
ALUMINOSIS - [med] aluminose *f*
ALUMINOTHERMIC PROCESS - [chem] (reduction of metallic oxides by fine aluminium powder) méthode *f* aluminothermique
~WELDING - [metall] soudure *f* aluminothermique
ALUMINOTHERMY - [metall] (welding process) aluminothermie *f*
ALUMINOUS - [chem] alumineux
~CEMENT - [mas] ciment *m* alumineux
ALUMSHALE - [geol] schiste *m* alumineux
ALUMSTONE - [min] alunite *f*
ALUNITE - [min] alunite *f*
ALUNOGEN - [min] alunogène
ALURGITE - [min] alurgite *f*
ALVEOLAR - [bot] alvéolé, alvéolaire
A.M. - (abbrev for anti-meridian) du matin, avant midi, après minuit
A.M. - (abbrev for amplitude modulation) modulation *f* d'amplitude
A.M. TUNER - [radio] dispositif *m* d'accord de modulation d'amplitude
AMALGAM - [metall] amalgame *m*
~DISTILLING FURNACE - [metall] four *m* pour distiller l'amalgame
~POT RETORT - [metall] cornue *f* pour distiller l'amalgame
AMALGAMATE, to - [gen & chem] amalgamer
AMALGAMATING BARREL - [metall] amalgamateur *m*
~BATH - [metall] bain *m* d'amalgamation
~CASK - [metall] amalgamateur *m*
AMALGAMATION - [gen & chem] amalgamation *f*
~IN BARRELS - [metall] amalgamation *f* aux tambours tournants
~IN PANS - [metall] amalgamation *f* aux cuves
AMALGAMATOR - [ind chem] amalgamateur *m*
AMATEUR BAND - [radio] bande *f* réservée aux radioamateurs
AMAZONITE - [min] amazonite *f*
AMBA - [geol] amba *m*
AMBER - [min] ambre *m*
~ACID - [chem] acide *m* succinique
~BROWN - [ind chem] brun *m* ambré
~MICA - [min] mica *m* ambré, phlogopite *f*
~OIL - [min] huile *f* de succin
~OPAL - [min] opale *f* ambrée
~TREE - [bot] liquidambar *m*
~VARNISH - [ind chem] vernis *m* au succin
~WHITE - [ind chem] blanc *m* ambré
~YELLOW - [ind chem] jaune *m* ambré
AMBERGRIS - [chem] ambre *m* gris
AMBIENT - [gen] ambiant
~AIR - [gen] air *m* ambiant, atmosphère *f* ambiante
~LIGHT - [telev] (normal room lighting illuminating the screen from outside) éclairage *m* ambiant, lumière *f* ambiante
~TEMPERATURE - [therm] température *f* ambiante
AMBIGUITY - [radar] (a condition in which navigational co-ordinates define more than one point) ambiguité *f*
AMBIPOLAR - [el] ambipolaire
~DIFFUSION - [el] (diffusion in a plasma with ions of both signs) diffusion *f* ambipolaire
AMBIT - [gen] circonférence *f*, bornes *f* pl, étendue *f*
[telev] (outline of an object) contour *m*
AMBIVALENCE - [med] ambivalence *f*
AMBLYACOUSIA - [acoust] dureté *f* d'oreille
AMBO - [build] ambon *m*
AMBOCEPTOR - [biochem] ambocepteur *m*
AMBRA SHEETING - [photo] filtre *m* ambré
AMBRETTE - [chem] ambrette *f*
~SEED OIL - [ind chem] essence *f* de graines d'ambrette
AMBRITE - [min] ambrite *f*

AMBROSIA - [bot] (a type of fungi used as food by some beetles) ambroisie *f*
AMBULACRUM - [zool] ambulacre *m*
AMBULANCE - [gen] ambulance *f*, hôpital *m* de campagne
~COACH - [railw] voiture *f* sanitaire
~STATION - [med] poste *m* d'ambulance
~TRAIN - [railw] train *m* sanitaire
AMBULANT CLINIC - [med] clinique *f* ambulante
AMEIOSIS - [genet] (the elimination of one of the meiotic divisions without reduction of chromosomes) améiose *f*
AMELIORATION - [gen] amélioration *f*
AMEND, to - [gen] amender, corriger, modifier
AMEND - [gen] amendement *m*, modification *f*, rectification *f*
AMENDE - [leg] amende *f*
AMENDED FORMULA - [chem] formule *f* modifiée
AMENT - [bot] chaton *m*, iule *m*
AMERICAN ASHES - [chem] carbonate *m* de potasse (brut)
~BRIGGS STANDARD - [mech] filet *m* américain (pour tubes)
~CLOTH - [text] toile *f* cirée
~MELTING POINT - [oil ind] point *m* de fusion américain
~TERN - [zool] sterne *m* d'Amérique, hirondelle *f* des mers arctiques
~VERMILLION - [chem] chromate *m* basique de plomb
AMERICIUM - [chem] (element No 95; radioactive) américium *m*
AMETABOLISM - [biol] (development with no apparent metamorphosis) amétabolisme *m*
AMETHYST - [min] (form of quarz; a semi-precious stone) améthyste *f*
AMETROPIA - [opt] amétropie *f*
AMIANTHUS - [min] (a fine quality of asbestos) amiante *m*
AMICABLE NUMBER - [math] (two numbers, either of which is the sum of the aliquots of the other) nombre *m* amiable
AMICRON - [phys] (smallest particle visible) ultramicron *m*
AMICTIC - [genet] (incapable of being fertilized) amictique
AMIDASE - [biochem] amidase *f*
AMIDATED COTTON - [text] coton *m* amidonné
AMIDE - [chem] (a compound derived from the replacement of the hydroxyl by an amido group) amide *f*, amidure *m*
AMIDO - [chem] (containing the NH_2 group) amido
AMIDOACETIC - [chem] amidoacétique
AMIDOGEN - [chem] amidogène
AMIDONAPHTHOL RED - [ind chem] rouge *m* amidonaphtol
AMIDOXIME - [chem] amidoxime *m*
AMIDPULVER - [chem] amidon *m* en poudre
AMIDSHIP - [naut] au milieu du navire, par le travers
~SHELL - [naut] coque *f* au maître couple
AMINE - [chem] (one of the organic derivatives of ammonia by the replacement of one or more atoms of hydrogen by an organic radical or radicals) amine *f*
AMINO - [chem] amino
AMINOACETIC ACID - [chem] acide *m* aminoacétique
AMINOACID - [chem] (group of fatty acids in which an amino group has replaced an atom of hydrogen in the hydrocarbon radical) amino-acide *m*
AMINOANTHRAQUINONE - [chem] aminoanthraquinone *m*
AMINOAZOBENZENE - [chem] amino-azobenzène *m*
AMINOAZOTOLUENE - [chem] amino-azotoluène *m*
AMINOBENZALDEHYDE - [chem] aminobenzaldéhyde *m*
AMINO-COMPOUND - [chem] composé *m* aminé, amino-composé *m*
AMINODIMETHYLANILINE - [chem] aminodiméthylaniline *f*
AMINODIMETHYLBENZENE - [chem] aminodiméthylbenzène *m*
AMINODIPHENYLAMINE - [chem] aminodiphénylamine *f*
AMINO GROUP - [chem] amines *f* pl
AMINO-KETONE - [chem] aminocétone *f*
AMINOPHEN - [chem] aniline *f*
AMINOPHENACETIN - [chem] aminophénacétine *f*
AMINOPHENOL - [chem] aminophénol *m*
AMINOPLAST - [ind chem] aminoplaste *m*
AMINOPURINE - [chem] aminopurine *f*
AMINOPYRIDINE - [chem] aminopyridine *f*
AMINOTHIAZOLE - [chem] aminothiazole *m*
AMINOTHIOPHENE - [chem] aminothiophène *m*
AMITOSIS - [genet] (division of a cell without regular distribution of chromosomes) amitose *f*
AMIXIA - [genet] (absence of interbreeding) amixie *f*
AMMETER - [instr] ampèremètre *m*
~SHUNT - [el] (conductor of low resistance) shunt *m* d'ampéremètre
AMMINE - [chem] (complex inorganic compounds formed by the addition of ammonia molecules to a salt or like compounds) ammine *f*
AMMINO - [chem] (including ammonia molecules) ammino
~COMPOUND - [chem] ammino-composé *m*, composé *m* amminé
AMMONIA - [chem] (a colourless gas) ammoniac *m*, ammoniaque *f*
~COMPRESSOR - [th eng] compresseur *m* à ammoniac
~CONTENT - [chem] teneur *f* en ammoniaque
~FERTILIZER - [agric] engrais *m* pl ammoniacaux
~HARDENING - [metall] nitruration *f*
~INJECTION - [aero eng] (injection of ammonia into the compressor inlet of a gas turbine to obtain increased power) injection *f* d'ammoniaque
~LEAK TESTING - [chem] détection *f* de fuites à l'ammoniaque
~LIQUOR - [chem] (the water distillate obtained as a by-product of gas or coke manufacture; a source of ammonia) eau *f* ammoniacale, liqueur *f* ammoniacale
~PLANT - [ind chem] atelier *m* de fabrication d'ammoniaque
~PROCESS - [chem] procédé *m* Solvay
~RECOVERY PLANT - [ind chem] fabrique *f* d'ammoniaque
~SOLUTION - [chem] solution *f* d'ammoniaque
~TREATMENT - [ind chem] procédé *m* par l'ammoniaque
~WASTE - [ind chem] eaux *f* pl résiduaires ammoniacales

AMMONIA WATER - [chem] (s. ammonia liquor) eau *f* ammoniacale
AMMONIAC - [chem] ammoniacal
~PLANT - [bot] dorème *m* ammoniac
AMMONIACAL - [chem] ammoniacal
~LATEX - [rubber ind] latex *m* ammonié, latex *m* ammoniacal
AMMONIAK - [chem] chlorure *m* d'ammonium
AMMONIATE, to - [ind chem] ammonier, traiter à l'ammoniaque
AMMONIATION - [rubber ind] ammoniation *f*
AMMONIFICATION - [chem] ammonisation *f*
AMMONIUM - [chem] ammonium *m*
~ACETATE - [chem] acétate *m* d'ammonium, esprit *m* de Mindérérus
~ALGINATE - [chem] alginate *m* d'ammonium
~ALUM - [chem] alun *m* d'ammoniaque
~ARSENATE - [chem] arséniate *m* d'ammonium
~BICARBONATE - [chem] bicarbonate *m* d'ammonium
~BROMIDE - [chem] bromure *m* d'ammonium
~CASEINATE - [chem] caséinate *m* d'ammonium
~CHLORIDE - [chem] chlorure *m* d'ammonium, sel *m* ammoniac
~FLUORIDE - [chem] fluorure *m* d'ammonium
~HYDROSULPHIDE - [chem] sulfhydrate *m* d'ammonium
~HYDROXIDE - [chem] hydroxyde *m* d'ammonium, alcali *m* volatil
~MOLYBDATE - [chem] molybdate *m* d'ammonium
~NITRATE - [chem] nitrate *m* d'ammonium
~OXALATE - [chem] oxalate *m* d'ammonium
~PHOSPHATE - [chem] phosphate *m* d'ammonium
~SALT - [chem] sel *m* ammoniac
~SULPHATE - [chem] sulfate *m* d'ammoniaque
~SULPHIDE - [chem] sulfure *m* d'ammonium
~THIOCYANATE - [chem] thiocyanate *m* d'ammonium
~THIOSULPHATE - [chem] thiosulfate *m* d'ammonium
AMMONIURIA - [med] ammoniurie *f*
AMMONIZATION - [chem] ammonisation *f*
AMMONIZE, to - [chem] ammoniser
AMMONOLYSIS - [chem] (decomposition by means of ammonia) décomposition *f* par l'ammoniaque
AMMUNITION, to - approvisionner (en munitions)
AMMUNITION - munitions *f* pl
~BOX - [armam] caisson *m* à munitions, chargeur *m*
~CHEST - [mil] caisse *f* de munitions
AMNESIA - [med] (loss of memory) amnésie *f*
AMNESTY - [leg] amnistie *f*
AMNIOGRAPHY - [med] amniographie *f*
AMOEBA - [biol] amibe *f*
AMOEBIC DYSENTERY - [med] dysenterie *f* amibienne
AMOEBOCYTE - [biol] (colourless blood corpuscle) leucocyte *m*
AMORPHOUS - [adj] (devoid of orderly structure) amorphe
~MATERIAL - [nucl radiat] (material without orderly arrangement of the atoms) matière *f* amorphe
AMOUNT - [gen] quantité *f*
[comm] montant *m*, somme *f*
[tech] teneur *f*, contenu *m*
~BROUGHT FORWARD - [comm] report *m* à nouveau
~OF DEFLECTION - [phys] grandeur *f* de la déviation
~OF MODULATION - [radio] taux *m* de modulation
~OF SLUDGE PRODUCED - [ind chem] rendement *m* en boues

AMP - [el] (abbrev for ampere) ampère *m*
AMPANGABEITE - [chem] (iron niobate and titanate; uranium and rare earths) ampangabéite *f*
AMPERAGE - [el] (the current flowing in a circuit) intensité *f*
AMPERE - [el] (unit of current) ampère *m*
~ARC - [el] ampère-arc *m*
~CONDUCTORS - [el] (product of the peripherical conductors of the winding by the current which circulates in these conductors) ampère-conducteurs *m* pl
~GAUGE - [el] ampèremètre *m*
~HOUR - [el] (a practical unit of quantity of electricity) ampère-heure *m*
~HOUR CAPACITY - [el] (capacity of a battery measured in ampere-hours) capacité *f* en ampère-heures
~HOUR EFFICIENCY - [el] rendement *m* d'un accumulateur
~HOUR METER - [el] ampèreheure-mètre *m*
~TURN - [el] (unit of magnetomotive force) ampère-tour *m*
~TURN GAIN - [ampl] (in magnetic amplifiers) gain *m* en ampère-tours
AMPEROMETER - [instr] ampèreheure-mètre *m*
AMPEROMETRIC COIL - [el] enroulement *m* ampèreheuremétrique
AMPERSAND - [print] (the sign '&') abréviation typographique de ET ('&')
AMPHIBIAN - [tool] amphibie
~PLANE - [aero] (aircraft capable of alighting on land or water) avion *m* amphibie
AMPHIBIOUS VEHICLE - [auto] véhicule *m* amphibie
AMPHIBOLE - [min] (large group of rock forming silicates) amphibole *f*
AMPHIBOLITE - [min] (a crystalline rock) amphibolite *f*
AMPHIGONY - [genet] (sexual reproduction with the participation of two gametes) amphigonie *f*
AMPHIMIXIS - [genet] (the union of germ cells in sexual reproduction) amphimixie *f*
AMPHIPROTIC - [chem] (with protophillic and protogenic properties) amphotère
AMPHITHEATRE - [build] amphithéatre *m*
AMPHOLYTE - [chem] (amphoteric electrolyte) ampholite *m*
AMPHOLYTICS - [ind chem] composés *m* pl ampholytes
AMPHOLYTOID - [chem] (a colloid with acid and basic properties) colloïde *m* amphotère
AMPHOTERIC - [chem] (exhibiting both acid and basic properties) amphotère
~ION - [nucl] (an ion which has a positive and a negative charge) ion *m* amphotère
~SURFACTANT - [deterg] agent *m* tensio-actif amphotère
AMPLICORDER - [instr] ondemètre *m*
AMPLIDYNE - [el] (rotary magnetic or dynamic-electric amplifier giving high power gain, used in servomechanisms) amplidyne *f*
~DRIVE - [radar] (special form of D.C. generator) système *m* amplidyne
AMPLIFICATION - [ampl] (the process of amplifying a small signal by suitable circuitry) amplification *f*
~CONSTANT - [radio] facteur *m* d'amplification
~FACTOR - [electron] (a figure of merit for a ther-

mionic valve: the maximum gain in voltage theoretically obtainable from it) coefficient *m* d'amplification
AMPLIFICATION STAGE - [radio] étage *m* amplificateur
AMPLIFIER - [el] (an arrangement of thermionic valves to increase the power of electric currents) amplificateur *m*
[photo] (an additional lens) amplificateur *m*
~BAY - [telev] (in a studio) baie *f* d'amplificateurs
~CHAIN - [comput] chaîne *f* d'amplificateurs, chaîne *f* amplificatrice
~FEEDBACK - [el] (return of energy from the output to the input of an amplifier) réaction *f* d'amplificateur
~METADYNE - [el] métadyne *f* à amplificateur
~RACK - [telev] bâti *m* d'amplificateurs
~ROOM - [acoust] centre *m* d'amplification
~STAGE - [comput] étage *m* amplificateur
AMPLIFY, to - amplifier
AMPLIFYING EQUIPMENT - [acoust] équipement *m* d'amplification
~KLYSTRON - [electron] klystron *m* amplificateur
~SIGNAL CONVERTER - [instr] convertisseur-amplificateur *m*
~VIBROGRAPH - [instr] (vibrograph recording curves of greater amplitude than actual vibrations) vibrographe *m* amplificateur
~WINDING - [el] (of a metadyne) enroulement *m* amplificateur
AMPLITUDE - [phys] (the greatest value of a quantity periodically varying during a cycle) amplitude *f*
~CHANGE SIGNALLING - [radio] formation *f* de signaux par modulation d'amplitude
~CHARACTERISTIC - [telev] caractéristique *f* d'amplitude, courbe *f* de réponse
~COMPASS - [astr] boussole *f* d'amplitude
~CONTROL - [electron] réglage *m* de l'amplitude
~CORRECTOR - [electron] circuit *m* correcteur d'amplitude
~DISCRIMINATOR - [instr] circuit *m* de discrimination d'amplitude
~DISTORTION - [electron] (distortion of wave form) distortion *f* harmonique, distorsion *f* d'amplitude
~EXCURSION - [telev] excursion *f* d'amplitude
~FACTOR - [el] facteur *m* de crête
~FADING - [radio] (fading in which all portions of the modulated wave are attenuated equally) fading *m* d'amplitude
~FILTER - [telev] (valve circuit in a television receiver) filtre *m* d'amplitude
~GATE - [radio] (a transducer) découpeur *m*
~INCREASE - [telev] gain *m* d'amplitude
~LIMITER - [radar] (a circuit which clips the signal) écréteur *m*
~LINEARITY - [telev] linéarité *f* d'amplitude
~LOPPER - [radar] (a circuit clipping a signal) écréteur *m*
~MODULATED OSCILLATION - [radio] oscillation *f* modulée en amplitude
~MODULATION - [radio] (the process of combining a signal with a carrier wave so that the carrier amplitude varies in accordance with the amplitude of the signal) modulation *f* d'amplitude
~MODULATION NOISE - [radio] bruit *m* de modulation d'amplitude
~OF A WAVE - [radio] amplitude *f* d'une onde
AMPLITUDE OF ACCOMMODATION - [opt] amplitude *f* d'accommodation
~PEAK - [telev] crête *f* d'amplitude
~RATIO - [radio] rapport *m* d'amplitude
~RESONANCE - [radio] résonance *f* d'amplitude
~RESPONSE - [radio] réponse *f* en amplitude
~SCALE FACTOR - [electron] échelle *f* des amplitudes
~SLICER - [radio] s. amplitude gate
AMPOULE - [ind chem] ampoule *f*
AMPUTATION - [gen] amputation *f*
[el] (of a circuit) exclusion *f*
A.M.S.L. - [top] (above mean sea level) au dessus du niveau de la mer
AMU - [nucl] (abbrev for atomic mass unit) nombre *m* de masse
AMYGDALA - [anat] (one of the palatal tonsils) amygdale *f*
AMYGDALIN - [chem] (glucoside obtained from bitter almonds, peach kernels, etc) amygdaline *f*
AMYGDALOID - [bot] (almond-like) amygdaloïde
AMYGDALOSIDE - [chem] amygdaloside *f*, amygdaline *f*
AMYGDOPHENINE - [chem] amygdophénine *f*
AMYL - [chem] amyle *m*
~ACETATE - [chem] (used as a solvent for nitrocellulose) acétate *m* d'amyle
~ALCOHOL - [chem] (a mixture of isomers used as solvents) alcool *m* amylique
~ANILINE - [chem] amylaniline *f*
~BORATE - [chem] borate *m* d'amyle
~GROUP - [chem] composés *m* pl amyliques
~NITRITE - [chem] nitrite *m* d'amyle
~OXIDE - [chem] oxyde *m* d'amyle
~PHENOL - [chem] amylphénol *m*
~SALICYLATE - [chem] salicylate *m* d'amyle
AMYLACEOUS - [bot] (starchy) amylacé
AMYLAMINE - [chem] amylamine *f*, manoamylamine *f*
AMYLASE - [chem] (enzyme capable of hydrolizing starch) amylase *f*
AMYLENE - [chem] (a term for pentene) amylène *f*
~DICHLORIDE - [chem] bichlorure *m* d'amylène
~HYDRATE - [chem] hydrate *m* d'amylène
AMYLIC - [chem] amylique
~FERMENTATION - [chem] fermentation *f* amylique
AMYLIFEROUS - [chem] amylifère
AMYLIN - [chem] amyline *f*
AMYLO FERMENTATION PROCESS - [chem] (use of certain moulds yielding diastase and fermentation enzymes to obtain alcohol from starchy substances) processus *m* de fermentation amylique
AMYLODEXTRIN - [biochem] amylodextrine *f*
AMYLOGEN - [chem] amylogène
AMYLOGENESIS - [bot] (formation of starch in the cells of a plant) amylogénèse *f*
AMYLOID - [chem] (starch-like compounds produced by treatment of cellulose with concentrated sulphuric acid) amyloïde
~BODY - [zool] (concretion in the prostate gland) amyloïde *m*
~DEGENERATION - [med] (formation of amyloid in the small arteries) dégénérescence *f* amyloïde
AMYLOLEUCITE - [bot] amyloleucite *m*
AMYLOLYSIS - [chem] amylolyse *f*
AMYLON - [chem] amylone *m*, méthylisobutylcétone *f*

AMYLOPECTIN - [biochem] amylopectine *f*
AMYLOPSIN - [chem] (a starch hydrolizing enzyme) amylopsine
AMYLOSE - [chem] (a glucose polymer from which plants produce starch) amylose *f*
AMYLUM - [chem] (synonym for starch) amidon *m*
AMYOTROPHY - [med] (atrophy of muscles) amyotrophie *f*
A-N RADIO RANGE - [radar] radioguidage *m* à signaux (morse) équilibrés
ANA - [chem] (prefix in chemical names, meaning a condensed double aromatic nucleus substituted in the 1.5 positions) ana-
ANABATIC - [met] (of a wind which flows up a hillside and is caused by diurnal insolation) anabatique
~WIND - [met] vent *m* anabatique
ANABIOSIS - [zool] (the return to life after apparent death) anabiose *f*
ANABOLISM - [biol] (chemical changes in living organisms) anabolisme *m*
ANABOLITE - [biol] anabolite *m*
ANACIDITY - [med] anacidité *f*
ANAEMIA - [med] (decrease of the amount of haemoglobin in the blood) anémie *f*
ANAEROBE - [biol] (organism living in complete absence of oxygen) anaérobie *f*
ANAEROBIC BACTERIA - [bact] anaérobie *f*
~DECOMPOSITION - [biol] (limited decomposition of organic materials owing to the absence of oxygen) décomposition *f* anaérobie
~RESPIRATION - [biol] respiration *f* anaérobie
ANAEROBIOSIS - [biol] (the existence without oxygen) anaérobiose *f*
ANAESTHESIA- [med] (loss of sensibility; also the science of administering anaesthetics) anesthésie *f*
ANAESTHETIC - [med] anesthésique
ANAESTHETIZATION - [med] administration *f* d'un anesthésique
ANAFLOW - [met] (air stream on a mountain side giving rise to anabatic winds) flux *m* anabatique
ANAGLYPH - [photo] anaglyphe *m*
ANAKINESIS - [biol] (the restoration of energy) anakinésie *f*
ANAKINETIC - [biol] anacinétique
ANAL - [anat] anal
ANALCIME - [min] (natural hydrated silicate of aluminium and sodium) analcime *f*
ANALCITE - [min] (alternative term for analcime) analcite *f*
ANALGESIA - [med] (loss of sensibility to pain) analgésie *f*
ANALGESIC - [med] analgésique
ANALGESINE - [chem] analgésine *f*, antipyrine *f*
ANALLATIC - [surv] anallatique
~LENS - [opt] (special lens in a tacheometric telescope reducing the additive constant for the tacheometer to zero) objectif *m* anallatique
~TELESCOPE - [surv] (a telescope used for tacheometric purposes) lunette *f* anallatique
ANALLATISM - [surv] anallatisme *m*
ANALOG - [comput] analogique
ANALOGOUS TRANSISTOR - [electron] (a transistor with practically the same characteristics as a three electrode tube) transistor *m* équivalent
ANALOGUE - [comput] analogique
~COMPUTER - [comput] (a computer which calculates by using physical analogs of the variables) calculatrice *f* analogique
ANALOGUE SIGNAL - [autom] (a signal with progressively varying values) signal *m* analogique
ANALOG-TO-DIGITAL CONVERSION UNIT - [comput] convertisseur *m* analogique-numérique, traducteur *m* analogique-numérique
ANALYSIS - analyse *f*
~CERTIFICATE - [comm ind chem] certificat *m* d'analyse
~FILTER - [telev] (in colour television filter fitted in front of a camera to intercept, field by field, the scene) filtre *m* analyseur de couleurs
~IN A DISSOLVED STATE - [chem] analyse *f* par voie humide
~IN DRY STATE - [chem] analyse *f* par voie sèche
~IN DRY WAY - [chem] analyse *f* par voie sèche
~IN WET WAY - [chem] analyse *f* par voie humide
~OF VARIANCE - [ind chem] analyse *f* de variance
ANALYST - [ind chem] (chimiste) analyste *m*, expert *m*
ANALYTIC - [chem] analytique
~CHEMISTRY - [chem] chimie *f* analytique
~CONTROL EQUIPMENT - [comput] appareillage *m* de contrôle de précision
ANALYTICAL BALANCE - [ind chem] balance *f* de précision, balance *f* de laboratoire
~FINDINGS - [gen] expertises *f* pl analytiques
~GEOMETRY - [math] géométrie *f* analytique
~REAGENT - [chem] réactif *m* d'analyse
~WEIGHT - [ind chem] poids *m* de précision
ANALYZE, to - [gen & chem] analyser
ANALYZER - analyseur *m*, contrôleur *m* universel (radiodépannage)
ANAMORPHIC - [gen & opt] anamorphotique
~LENS - [opt] objectif *m* anamorphotique
ANAMORPHOSIS - [opt] (distorted picture) anamorphose *f*
ANAPHASE - [genet] (stage in nuclear division) anaphase *f*
ANAPHORESIS - [chem] (the migration of suspended particles towards an anode under the influence of an electric field) anaphorèse *f*
ANASTIGMATIC - [opt] anastigmatique
~DEFLECTION YOKE - [electron] bobine *f* de balayage anastigmatique
~LENS - [opt] (compound lens combination so corrected that astigmatism and curvature of that field are almost eliminated over a large area in the image plane) objectif *m* anastigmatique
ANASTOMOSIS - [physical] (cross connection between arteries etc) anastomose *f*
ANATASE - [min] (titanium dioxide) anatase *m*, bioxyde *m* de titane
ANATOMY - anatomie *f*
ANCHOR, to - [naut] mouiller, jeter l'ancre
[constr] ancrer, fixer
ANCHOR - [naut] ancre *f*
[el] induit *m*
[el constr] ancrage *m*
[plastic] culot *m* d'injection
~APEAK - [naut] ancre *f* à pic
~ARM - [naut] patte *f* d'ancre
~AT LONG STAY - [naut] ancre *f* à long pic
~AT SHORT STAY - [naut] ancre *f* à pic
~AWEIGH - [naut] ancre *f* dérapée
~BERTH - [naut] poste *m* de mouillage
~BOLT - [mech] boulon *m* d'ancrage

ANCHOR BUOY - [naut] bouée *f* de mouillage, bouée *f* de corps mort
~CHOCK - [naut] support *m* d'ancre
~CLAMP - [mech] crampon *m*
~CROWN - [naut] diamant *m* de l'ancre
~CURTAIN - [constr] rideau *m* d'ancrage
~DAVIT - [naut] bossoir *m*
~DECK - [naut] pont-teugue *m*
~DUES - [naut] droits *m* pl de mouillage
~EAR - [el] oreille *f* d'ancre
~ESCAPEMENT - [mech] échappement *m*
~FLUKE - [naut] patte *f* de l'ancre
~ICE - [soil] glace *f* de fond
~LIGHT - [naut] feu *m* de mouillage.
~LINK - [mech] tige *f* d'ancrage
~LOOP - [constr] (a u-bolt in timber forms for dams) boucle *f* d'ancrage
~PLATE - [mech] plaque *f* d'ancrage, plaque *f* de fixation
~POLE - [el] poteau d'ancrage
~PULL - [constr] traction *f* du tirant d'ancrage
~RING - [naut] organeau *m*
~ROD - [constr] (a tie rod) tirant *m* de retenue
~ROPE - câble *m* de retenue
~SCREW s. anchor bolt
~SHACKLE - [naut] cigale *f*
~SHANK - [naut] verge *f* de l'ancre
~SHOE - [naut] semelle *f*, savatte *f* de l'ancre
~SLIPPER - [naut] semelle *f* d'ancre
~STOCK - [naut] jas *m* de l'ancre
~TOWER - [el] pylone *m* d'ancrage
~WELL - [naut] puits *m*
ANCHORAGE - [naut] mouillage *m*, ancrage *m*
[constr] ancrage *m*, bloc *m* d'ancrage, point *m* d'attache
[med] fixation *f*
~CHAIN - [naut] chaîne *f* de mouillage
~CHAMBER - [build] chambre *f* d'ancrage
~DUES - [naut] s. anchor dues
~GROUND - [naut] mouillage *m*
~OF A CHAIN - [mech] ancrage *m* de chaîne
~PILLAR - [constr] pylone *m* d'ancrage
~PIN - [mech] goupille *f* de fixation
ANCHORED - [gen & naut] ancré, mouillé
~BEARING - [mech] support *m* ancré
ANCHORING - [naut] mouillage *m*
~SLIT - [photo] (for films) fente *f* d'accrochage
~WIRE - [el] câble *m* d'ancrage
ANCHOVY - [zool] anchois *m*
ANCHYLOSIS - [med] (fixation of joints by fibrous bands) ankylose *f*
ANCILLARY - [gen] accessoire, auxiliaire, secondaire
~CIRCUIT - [telecomm] circuit *m* auxiliaire
~EQUIPMENT - appareillage *m* auxiliaire
~GRID - [telecomm] grille *f* auxiliaire
~JACK - [el] jack *m* d'entr'aide
~WORK - [hydr constr] travail *m* de soutien
ANCLE - cheville *f*
ANCON - [build] ancon *m*
AND - [comput] (circuit) ET
~-CIRCUIT - [comput] circuit *m* ET, circuit *m* à coïncidences, circuit *m* intersection
~-NOT GATE - [comput] circuit *m* intersection-négation
~-OR CIRCUIT - [comput] circuit *m* ET-OU
ANDALUSITE - [min] (naturally occurring aluminium silicate) andalousite *f*
ANDERSON BRIDGE - [el] (comparison of capacitance and inductance) pont *m* d'Anderson
ANDERSONITE - [min] (rare mineral containing over one third uranium) andersonite *f*
ANDESITE - [geol] (an igneous rock) andésite *f*
ANDIRON - [impl] chenet *m*, landier *m*
ANDRABITE - [min] andrabite *f*
ANDROGENESIS - [genet] (development in which the embryo contains only paternal chromosomes owing to the failure of the egg nucleus to participate in fertilization) androgénèse *f*
ANECHOIC ROOM - [acoust] chambre *f* sourde
ANELASTICITY - [metall] anélasticité *f*
ANELECTRIC - [el] (old fashioned term of a body which is not electrified through friction) anélectrique
ANELECTROTONOUS - [adj] anélectrotonus
ANEMIA - s. anaemia
ANEMOBIOGRAPH - [instr] anémobiographe *m*
ANEMOCLINOGRAPH - [instr] anémoclinographe *m*
ANEMOGAMY - [genet] (pollination by wind) anémogamie *f*
ANEMOGRAM - [instr] (curve recorded by an anemograph) anémogramme *m*
ANEMOGRAPH - [instr] anémographe *m*, anémomètre *m* enregistreur
ANEMOGRAPHY - [instr] (the recording of the speed and direction of winds) anémographie *f*
ANEMOLOGY - [meter] (the study of winds) anémologie *f*
ANEMOMETER - [instr] anémomètre *m*
ANEMOMETRY - [instr] anémométrie
ANEMOPHILY - [bot] (pollination by the wind) anémophilie *f*
ANEMOSCOPE - [instr] anémoscope *m*
ANEMOSTAT - [instr aero] anémostat *m*
ANEMOTROPISM - [biol] anémotropisme *m*
ANERGIA - [med] anergie *f*
ANEROID - [instr] (not containing a fluid) anéroïde
~BAROMETER - [instr] (measuring atmospheric pressure) baromètre *m* anéroïde
~MANOMETER - [instr] (pressure gauge) manomètre *m* anéroïde
ANESON - [chem] chloroforme *m* d'acétone
ANESTHESIA - [med] anesthésie *f*
ANESTHETIC - [chem] anesthésique
ANEURYSM - [med] (dilatation of an artery) anévrisme
ANGEL - [radar] écho *m* parasite
ANGIITIS - [med] angésite
ANGINA PECTORIS - [med] angine *f* de poitrine
ANGIOGRAPHY - [med] (vasography) angiographie
ANGIOSPERMS - [genet] (plants producing seeds enclosed in an ovary) angiospermes *m* pl
ANGLE - [geom] (the inclination of one line to another)
[geom] angle *m*
[metall] cornière *f*, profilé *m*
~AT THE CENTRE - [geom] angle *m* au centre
~AT THE CIRCUMFERENCE - [geom] angle *m* inscrit
~BAR - [metall] cornière *f*, profilé *m*
~BEAD - [build] (small round moulding) bourrelet *m* de moulure
~BEND - [plumb] (pipe or tube) coude *f*, tube *m* coudé, raccord *m* d'angle
~BETWEEN THE SIDES - [mech] angle *m* entre parois

ANGLE BEVEL GEARS - [mech] engrenages *m* pl conico-obliques
~BLADE - [mech] (of earth moving machines) lame *f* orientable
~BOARD - [carp] (used as a gauge to give required angle to planed boards) aisselier *m*
~BRACE - [tool] foret *m* à angle
[carp] renfort *m* d'angle, lien *m*
[constr] contrefiche *f* de ferme de toit
~BRACKET - [mech] (mechanical part uniting two or more other parts meeting at an angle) équerre *f* de fixation, cornière *f* d'angle
~BRANCH - [network equip] té *m* oblique
~BRICK - [build] brique *f* d'angle
~BUILD UP - [mining] accroissement *m* d'angle
~BUTT STRAP - [mech] couvre-joint *m* oblique, éclisse *f* oblique
~COCK - [mech] robinet *m* coudé
~COMPRESSOR - [network equip] compresseur *m* à cylindres perpendiculaires
~CORRELATION - [radiat] corrélation *f* angulaire
~CRANE - [mach tool] grue *f* à support triangulaire
~DRAFT - [text] armure *f* à retour
~DROP OFF - [mining] réduction *f* d'angle
~FISHPLATE - [railw] éclisse *f* oblique
~HEAD - [plastic] tête *f* oblique
~HINGE - [mech] gond *m* d'angle
~IRON - [metall] cornière *f*
~IRON STIFFENER - [mech] raidisseur *m*, cornière *f* de renforcement
~-JAW TONGS - [impl] tenaille *f* à cornières
~JOINT - [metall] joint *m* d'angle, assemblage *m* d'angle
~LEVER - [impl] levier *m* coudé
~LEVER SHEARS - [impl] cisailles *f* pl coudées
~MODULATION - [mod] modulation *f* de phase
~MOULDING PRESS - [mach] presse *f* de moulage à tête d'équerre
~OF A SINE WAVE - [mod] angle *m* de phase
~OF ACTION - [tools] angle *m* d'attaque
~OF AIRSCREW SETTING - [aero] angle *m* de calage de l'hélice
~OF APPROACH INDICATOR - [instr] (light arranged to show a required approach path in a vertical plane) indicateur *m* d'angle d'approche
~OF ARRIVAL - [el magn waves] angle *m* d'arrivée
~OF ATTACK - [aero] (the angle between the chord of an aerofoil and the direction of the undisturbed flow) angle *m* d'attaque
~OF BACK OF TOOTH - [mech] angle *m* du pied de la dent
~OF BALANCE - [aero] angle *m* d'équilibrage
~OF BANK - [aero] angle *m* d'inclinaison latérale
~OF BASE FRICTION - [constr] angle *m* de frottement de la base
~OF BEDDING - [mining] angle *m* de stratification
~OF BENDING - [mech] angle *m* de flexion
~OF BOSH - [th eng] angle *m* d'étalage (d'un haut-fourneau)
~OF BRUSH LAG - [mech] angle *m* de décalage (des balais)
~OF CENTRE - [mach tool] angle *m* de pointe
~OF CHORDWISE INCIDENCE - [aero] (the angle of the direction of undisturbed flow with the chord of an aerofoil) angle *m* d'incidence
~OF CLIMB - [aero] angle *m* de montée
~OF CONTACT - angle *m* de contact

ANGLE OF CONTINGENCE - [geom] angle *m* de contingence
~OF CONTROL LEVER - [aero] angle *m* du levier de commande
~OF COUNTERSINKING - [mech] angle *m* de fraisage (d'une entrée)
~OF CRAG - [aero] angle *m* de route "en crabe"
~OF CURRENT FLOW - [th eng] angle *m* d'écoulement
~OF CURVATURE - [geom] angle *m* de courbure
~OF DECLINATION - angle *m* de déclinaison
~OF DEFLECTION - [phys] angle *m* de déviation
~OF DEPARTURE - [artill] angle *m* de tir
[el magn waves] angle *m* de projection
~OF DEPRESSION - [top] angle *m* de dépression
~OF DEPTH - [shooting] angle *m* de profondeur
~OF DESCENT - [aero] angle *m* de descente
~OF DEVIATION - [phys & opt] (the angular change in direction of a ray of light and other electromagnetic radiation when entering a different medium) angle *m* de déviation
~OF DIVERGENCE - [electron] angle *m* de divergence
~OF DOWNWASH - [aero] angle *m* vertical de déviation
~OF DRIFT - [aero] angle *m* de dérive
~OF ECCENTRICITY - [math] angle *m* d'excentricité
~OF ELEVATION - [top] hauteur *f* angulaire
[geol] angle *m* d'inclinaison
~OF ELONGATION - [astr] angle *m* d'élongation
~OF EMERGENCE - [opt] angle *m* d'émergence
~OF FLOW - [gen] angle *m* d'écoulement
~OF FRICTION - [mech] angle *m* de frottement
~OF GLIDE - [aero] angle *m* de vol plané
~OF GRADIENT - [top] angle *m* d'inclinaison
~OF HEEL - [aero] angle *m* de bande, angle *m* d'inclinaison latérale
~OF INCIDENCE - [phys] angle *m* d'incidence
~OF INCLINATION - [surv] angle *m* d'inclinaison
~OF LAG - [el] angle *m* de retard (de phase)
~OF LAP - [mech] angle *m* de recouvrement
~OF LEAD - [el] angle *m* d'avance (de phase), angle *m* de calage
[mech] (of a slide valve) angle *m* d'avance à l'admission
~OF LOCK - [mech] angle *m* de braquage
~OF LOSS - [el] angle *m* de pertes
~OF OBLIQUITY - [mech] angle *m* de pression
~OF PARALLAX - [opt] angle *m* de parallaxe
~OF PHASE DIFFERENCE - [el] écart *m* angulaire
~OF PHASE DISPLACEMENT - [el] déphasage *m*
~OF PITCH - [aero] angle *m* de tangage
~OF POLARIZATION - [opt] angle *m* de polarisation
~OF POSITIVE STAGGER - [aero] angle *m* de décalage en avant
~OF PRESSURE - [mech] angle *m* de pression
~OF PROPELLER PITCH - [aero] angle *m* de pas d'une hélice
~OF RAKE - [cin] (angle of projector axis with horizontal) angle *m* d'attaque
~OF REFLECTION - [opt] angle *m* de réflexion
~OF REFRACTION - [opt] angle *m* de réfraction
~OF RELIEF - [mech] angle *m* de tranchant, angle *m* d'incidence
~OF REPOSE - [build] angle *m* d'éboulement, angle *m* naturel de repos
~OF REPOSE OF THE NATURAL SLOPE - [constr] (in works for berthing) angle *m* du talus naturel
~OF ROLL - [aero] angle *m* de roulis

ANGLE OF ROTATION - [mech] angle *m* de rotation
~OF RUPTURE - [phys] angle *m* de cisaillement
~OF SAFETY - [ballist] angle *m* de sécurité
~OF SCATTERING - angle *m* de dispersion
~OF SEPARATION - [geom] angle *m* de séparation
~OF SHAFTS - [mech] angle *m* des arbres
~OF SHEARING RESISTANCE - [mech] angle *m* de frottement
~OF SIDESLIP -[aero] angle *m* de dérapage
~OF SIGHTING - [artill] angle *m* de site, angle *m* de visée
~OF SITUATION - [astr] angle *m* de position
~OF SLIDE - [build] angle *m* minimal de glissement
~OF SLOPE - [build] angle *m* de talus
[constr] (angle of inclination) angle *m* de déclivité *f*, angle *m* d'inclinaison *f*
~OF STABILITY - [mech] angle *m* de friction
~OF STALL - [aero] angle *m* d'incidence critique
~OF STREAMWISE INCIDENCE - [aero] angle *m* d'incidence (dans la direction du flux)
~OF SWEEP-BACK - [aero] angle *m* de flêche
~OF THE V-BELT - [mech] angle *m* de courroie trapézoïdale
~OF THE V-GROOVE - [mech] angle *m* de gorge
~OF THREAD - [mech] angle *m* de filetage
~OF TORSION - [mech] angle *m* de torsion
~OF TRIM - [aero] angle *m* d'assiette
~OF TWIST - [aero] angle *m* de torsion
~OF VIEW - [opt] champ *m* angulaire
~OF WING SETTING - [aero] angle *m* de calage de l'aile
~OF YAW - [aero] angle *m* de lacet
~OF ZOOM - [aero] angle *m* de cabré
~PIECE - [plumb] piece *f* coudée, raccord *m* angulaire
~PIN - [mech] goupille *f* oblique, broche *f* oblique
[metall] doigt *m* de démoulage
[diecasting] doigt *m* de démoulage, broche *f* oblique
~PLATE - [mech] (machine tools) équerre *f*, plan *m* incliné
~PLATE - (a plate bent to an angle) gousset *m* en équerre
~RAFTER - [build] chevron *m* d'arête
~SHOT - [cin] bascule *f*
~SLEEKER - [metall] spatule *f* (outil à biscauter les arêtes de moule) lissoir *m*
~SPLICE BAR - [metall] éclisse à cornière
~STEEL - [metall] cornière *f* d'acier
~STRAGGLING - [nucl] (the variation in the direction of particles whose initial paths are parallel) dispersion *f* angulaire
~SUPPORT - [telecomm] support *m* d'angle
~TEMPLATE - [mech] gabarit *m* circulaire
~TIE - [mech] renfort *m* d'angle
~TRACKING - [radar] poursuite *f* continue
~TRANSMISSION - [mech] renvoi *m* d'angle
~VALVE - [network equip] vanne *f* à disque, vanne *f* à tampon d'équerre
~VARIABLE - [phys] variable *f* angulaire
~WITH UNEQUAL FLANGES - [mech] cornière *f* à ailes inégales
ANGLEDOZER - [mach] bulldozer *m* à lame orientable
ANGLER - [zool] poisson-grenouille *m*, baudroie *f*
ANGLESITE - [min] (a common lead ore) anglesite *f*
ANGORA GOAT - [text] chèvre *f* angora

ANGOSTURA - [bot] angusture *f*, angosture *f*
~BARK OIL - [chem] essence *f* d'angusture
ANGSTROM - [el] (spectroscope unit) Angström *m*
ANGUILLIFORM - [gen] anguilliforme
ANGULAR - [gen] angulaire , d'angle
~ACCELERATION - [mech] (acceleration of a rotating body, expressed in angular measurements) accélération *f* angulaire
~ADVANCE - [mech mach] angle *m* d'avance
~APERTURE - [camera] angle *m* d'ouverture
~ARTERY - [anat] artère *f* angulaire
~BEARING - [mach] roulement *m* à billes angulaire
~BEVEL GEAR - [mech] engrenages *m* pl conico-obliques
~CABLE SOCKET - [el] douille *f* coudée, prise *f* de courant coudée
~CUTTER - [mach] fraise *f* angulaire
~DEFLECTION - [phys] déviation *f* angulaire
~DEPENDENCE OF SCATTERING - [nucl] dépendance *f* angulaire de la diffusion
~DEVIATION - [mech] déviation *f* angulaire
~DEVIATION LOSS - [acoust] perte *f* de déviation angulaire
~DIAMETER - [opt] angle *m* visuel
~DISPLACEMENT - [gen mech & phys] écart *m* angulaire
~DISTANCE - [math] distance *f* angulaire
~DRILLING - [mech] forage *m* oblique
~EXTRUDER HEAD - [mech mach] tête *f* oblique d'extrudeuse
~FIELD - [el] champ *m* angulaire
~FLIGHT - [build] rampe *f*
~FREQUENCY - [el] (frequency expressed in terms of the angular velocity of the corresponding vector) fréquence *f* angulaire, pulsation *f*
~GAUGE - [mech] calibre *m* angulaire
~GEAR - [mech] (gearing transmitting motion between shafts at an angle to each other) engrenage *m* conique, renvoi *m* d'angle
~HEIGHT - [radio] hauteur *f* angulaire
~IMPULSE - [phys] impulsion *f* angulaire
~INDEXING - [mech] division *f* angulaire
~KINETIC ENERGY - [nucl] énergie *f* critique angulaire
~MAGNIFICATION - [opt] grossissement *m* angulaire
~MEMBER - [mech] (drive mechanism of the speedometer) renvoi *m* d'angle
~MILL - [mech mach] fraise *f* conique
~MOMENTUM - [phys] moment *m* cinétique, moment *m* angulaire, spin *m*
~MOTION - [mech] déplacement *m* angulaire
~PHASE DIFFERENCE - [el] déphasage *m*
~-POSITION ENCODER - [comput] codeur *m* de rotation, traducteur *m* de position angulaire à sortie numérique
~RADIUS - [math] rayon *m* angulaire
~SAND - [metall] sable *m* à grains anguleux
~SECTION - [mech] domaine *m* angulaire, secteur *m* d'angle
~SPEED - [astronaut] vitesse *f* angulaire
~SPREAD - s. angular section
~VEIN - [anat] veine *f* angulaire
~VELOCITY - [mech] pulsation *f*, vitesse *f* angulaire
~WIDTH OF SPOT - [telev] (aperture of the beam) largeur *f* du faisceau
ANGULARITY OF CONNECTING ROD - [mech] obliqui-

té *f* d'une bielle
ANHARMONIC - [math] anharmonique
ANHIDROSIS - [med] anhidrose *f*, anidrose *f*
ANHYDRATION - [ind chem] (to dehydrate quickly in food processing) anhydrisation *f*
ANHYDRIDE - [chem] anhydride *m*
ANHYDRITE - [chem] anhydrite *f*
ANHYDROUS - [chem] anhydre
~ALCOHOL - [ind chem] alcool *m* anhydre, alcool *m* absolu
~LANOLIN - [chem] lanoline *f* anhydre
~LIME - [chem] chaux *f* vive
~MAGNESIUM CHLORIDE - [chem] chlorure *m* de magnésium anhydre
~SODIUM CARBONATE - [chem] carbonate *m* de sodium anhydre
ANIL - [bot] anil *m*, indigotier *m*
ANILIDE - [chem] anilide *m*
ANILINE - [chem] aniline *f*
~BLACK - [ind chem] noir *m* d'aniline
~BLUE - [ind chem] bleu *m* d'aniline
~DYE - [ind chem] colorant *m* azoïque
~PINK - [ind chem] rosaniline *f*
~POINT - [chem] point *m* d'aniline
~PRINTING - [print] impression *f* à l'aniline
~RED - [ind chem] fuchsine *f*, rouge *m* d'aniline
~RESIN - [ind chem] résine *f* à l'aniline
~RUBBER PLATE PRINTING - [print] impression *f* à l'aniline, flexographie *f*
~SALT - [chem] sel *m* d'aniline
~SULPHATE - [chem] sulfate *m* d'aniline
~VIOLET - [ind chem] violet *m* d'aniline
~YELLOW - [ind chem] jaune *m* d'aniline
ANIMAL - [gen] animal *m*
~BLACK - [ind chem] noir *m* animal
~CHARCOAL - [ind chem] charbon *m* animal, noir *m* animal
~DRAWN - [gen] à traction animale
~ELECTRICITY - [physiol] électricité *f* animale
~GLUE - [ind chem] colle *f* animale
~HUSBANDRY - [gen] élevage *m*
~OIL - [ind chem] huile *f* animale
~SIZING - [paper ind] encollage *m* à la colle animale
ANIMATED - [gen] animé
~CARTOONS - [cin] dessins *m pl* animés, animation *f* dessinée
ANIMATION - [cin] animation *f*
~TIMING - [cin & telev] synchronisation *f* de l'animation
ANIMATIONS - [telev] (mechanical devices supplying with apparent movement inanimate subjects) dispositifs *m pl* mécaniques d'animation
ANIMATOR - [cin] animateur *m*
ANION - [phys] (a negatively charged atom molecule or group of molecules) anion *m*
ANIONIC - [phys & chem] anionique
ANIONICS - [chem] composés *m pl* anioniques
ANISE - [bot] anis *m*
ANISEED - [bot] graine *f* d'anis
~OIL - [ind chem] essence *f* d'anis
ANISIDINE - [chem] anisidine *f*
ANISOCORIA - [med] anisocorie *f*
ANISOL - [chem] anisol *m*, éther *m* méthylique
ANISOMERY - [bot] (condition of a flower) anisomérie *f*
ANISOPHYLLY - [bot] anisophyllie *f*
ANISOPLEURAL - [zool] anisopleural
ANISOSPORES - [zool] (spores of two kinds found at the same time) anisospores *f*
ANISOTROPIC - [chem] anisotropique
~CONDUCTIVITY - [el] conductivité *f* anisotropique
~METAL - [met] métal *m* anisotrope
~STRESS CHANGE - [phys] variation *f* de contrainte anisotropique
ANISOTROPY - [phys] anisotropie *f*
~OF THE SOIL - [soil] anisotropie *f* des sols, éolotropie *f*
ANKERITE - [min] ankerite *f*
ANKLE - [gen] cheville *f*
~BONE - [anat] astragale *f*
~JOINT - [med] (of an artificial leg) attache *f* du pied, articulation *f* de la cheville
~PATCH - [shoe man] protège-cheville *m*
ANKYLOSIS - [med] ankylose *f*
ANNABERGITE - [min] arséniate *m* de nickel
ANNATTO - [ind chem] (vegetable dye used in food processing) annatto *m*
ANNEAL, to - [metall] (heating followed by slow cooling) recuire
ANNEALED - [metall] recuit
~CAST IRON - [metall] fonte *f* malléable
~GLASS - [glass] verre *m* recuit
~TUBING - [metall] tube *m* de fonte malléable
ANNEALING - [metall] recuit *m*
~BOX - [metall] boîte *f* de recuit
~CHAMBER - [metall] chambre *f* à recuire, chambre *f* de recuit
~COLOURS - [metall] couleurs *f pl* de recuit
~FURNACE - [metall] four *m* à recuire
~OVEN - [glass ind] four *m* à recuire le verre
~POT - [metall] pot *m* de recuit
ANNERODITE - [min] (mineral containing niobium and erbium) anérodite *f*
ANNEX - [gen] annexe *f*
ANNIHILATION - [nucl] (meeting of antiparticles and their conversion into photons) annihilation *f*, fission *f* de particules avec dématérialisation
~FORCE - [nucl] force *f* d'annihilation
~GAMMA QUANTUM - [nucl] photon *m* gamma, photon *m* d'annihilation
~PHOTON - [nucl] photon *m* d'annihilation (ou de dématérialisation)
~RADIANT - [nucl] radiation *f* d'annihilation, radiant *f* de dématérialisation
ANNOUNCER - [rad] announceur *m*, présentateur *m*, meneur *m* de jeu
ANNUAL EQUATION - [astr] équation *f* annuelle
~LOAD DIAGRAM - [el] (load of a generating station over the period of one year) diagramme *m* de charge annuelle
~PLANT - [bot] plante *f* annuelle
~RAINFALL - [met] hauteur *f* annuelle des précipitations
~RING - [bot] couche *f* annuelle
~VARIATION OF COMPASS - [phys] (the yearly magnetic variation of a compass) déclinaison *f* annuelle du compas
ANNUITY - [gen & leg] annuité *f*, rente *f* annuelle
ANNULAR - [gen] annulaire
[aero eng] entrée *f* d'air
~ARCH - [build] arc *m* outrepassé
~AUGER - [tool] tarière *f* à couronne, sondeuse *f* à couronne
~BALL BEARING - [mech] roulement *m* à billes annu-

laire
ANNULAR BOILER - [th eng] chaudière *f* annulaire
~BORER - [mech impl] (tool employed in rock-boring and cutting an annular channel in the rock) mèche *f* annulaire
~COMBUSTION CHAMBER - [aero] chambre *f* de combustion à prise d'air annulaire
~ECLIPSE - [astr] (central eclipse of the sun) éclipse *f* annulaire
~ELECTRODE - [el] électrode *f* annulaire
~GAP - [mech] (between nozzle and casing wall of a diffusion pump) aire *f* de diffusion, espace *m* annulaire
~GEAR - [mech] couronne *f* dentée
~GRILLED TUBE - [mech] tube *m* ondulé pour raccords flexibles
~GROOVE - [mech] rainure *f* annulaire
~MAGNET - [phys] aimant *m* annulaire, aimant *m* torique
~MANIFOLD - [aero eng] prise *f* d'air annulaire
~MICROMETER - [instr] (the flat ring in the focal plane of telescope) micromètre *m* à anneau
~NOZZLE - [mech] (of a diffusion pump) diffuseur *m* à jet déflecté
~OPENING - [mech] ouverture *f* annulaire
~SEAT - [mech] siège *m* annulaire, portée *f* annulaire
~TANK - [gas ind] (gasholders) cuve *f* annulaire, réservoir *m* annulaire
~VAULT - [constr] voûte *f* outrepassée
~WATER JACKET - [therm eng] enveloppe *f* double (à circulation d'eau)
ANNULATE - [bot] (ring shaped) annelé
ANNULET - [constr] annelet *m*, armille *f*, filet *m*
ANNULUS - [geom] anneau *m* sphérique, tore *m*
[aero eng] partie *f* annulaire
ANNUNCIATOR - [el] (a system of indicators operated by relays) indicateur *m*, voyant *m* lumineux
~DROP - [telecomm] volet *m* d'annonciateur
ANODE - [electron] (an electrode whose task it is to collect electrons and use the resultant current in an external circuit) anode *f*
~ANGLE - [electron] angle *m* de l'anode, angle *m* de foyer
~BATTERY - [el] batterie *f* anodique
~BEND - [electron] courbure *f* anodique
~BEND DETECTION - [electron] (thermion. valve) détection *f* plaque
~BEND DETECTOR - [electron] détecteur *m* plaque
~BEND VOLTMETER - [instr] (type of vacuum tube voltmeter) voltmètre *m* à redressement par l'anode
~BREAKDOWN VOLTAGE - [electron] (the value of the anode voltage which causes conduction across the main gap) tension *f* d'amorçage
~BRIDGE - [el] cavalier *m* d'anode
~BRIGHTENING - [metall] (a process of electrodeposition) polissage *m* anodique
~BUTT - [electron] (an anode partially consumed) anode *f* partiellement détruite
~CAP - [electron] (part of ignitron) corne *f* capuchon *m* d'anode
~CATHODE CAPACITANCE - [electron] capacité *f* anode-cathode
~CHOKE COIL - [electron] contrôle *f* d'anode
~CIRCUIT - [el] circuit *m* anodique
~CLAMP - [el] (part of electronic tube) pince *f* d'anode
ANODE CONDUCTANCE - [electron] conductance *f* (d'un tube)
~CONVERTER - [el] convertisseur *m* anodique
~CURRENT - [electron] (the current flowing from the cathode to the anode in an electronic tube) courant *m* anodique
~CURRENT RECTIFICATION - [radio] redressement *m* d'un courant anodique
~DARK SPACE - [electron] (dark area near the surface of the anode) espace *m* sombre anodique
~DECOUPLING RESISTOR - [radio] résistance *f* de découplage dans un circuit anodique
~DIFFERENTIAL RESISTANCE - [electron] résistance *f* d'anode (dans un tube)
~DISSIPATION - [electron] dissipation *f* anodique
~DROP - [electron] chute *f* anodique
~EFFECT - [electron] (increase in potential difference between the anode and the electrolyte) effet *m* d'anode
[el chem] effet *m* d'anode
~EFFICIENCY - [radio] rendement *m* d'un amplificateur
~FEED RESISTANCE - [transm] résistance *f* de charge d'anode
~FIN - [el] (in an electronic tube) ailette *f* d'anode
~FOLLOWER - [telev] circuit *m* à charge anodique, inverseur *m* de polarité à contre-réaction
~GLOW - [electron] (luminous area at the near end of the positive column facing the anode) lumière *f* anodique, lumière *f* positive
~GRID - [electron] grille *f* anodique
~GRID CAPACITY - [electron] capacité *f* grille-plaque
~HUM - [electron] (the hum voltage between anode and cathode) ronflement *m* d'anode
~IMPEDANCE - [electron] résistance *f* de charge d'anode
~INDUCTANCE - [electron] inductance *f* (bobine) d'anode, inductance *f* anodique
~INPUT POWER - [electron] puissance *f* d'entrée
~KETING - [radio] manipulation *f* dans la plaque
~LAYER - [electron] (molten metal forming a cathode) couche *f* métallique servant d'anode
~LOAD - [electron] charge *f* d'anode
~MODULATION - [electron] modulation *f* dans l'anode
~MUD - [chem] (the residue left on the anode in electrolytic refining) dépôt *m* sur une anode
~NECK - [electron] (the neck of the bulb) col *m* anodique
~NEUTRALIZATION - [electron] (method of neutralizing an amplifier) neutralisation *f* du circuit anodique
~PEAK CURRENT - [electron] intensité *f* anodique de crête
~PICKLING - [electron] traitement *m* désoxydant (des anodes), décapage *m*
~PLATE - [electron] (the material of the anode) anode *f* (lame métallique)
~PLUG - [electron] fiche *f* de tension anodique
~POWER INPUT - [electron] alimentation *f* anodique
~PROCESS - [ind chem] procédé *m* anodique
~PULSE MODULATION - [modul] modulation *f* dans l'anode par impulsions
~REGION - [electron] (the group of regions including positive column, anode glow and anode dark

space) zone *f* anodique
ANODE RESISTANCE - [electron] résistance *f* d'anode
~REST CURRENT - [radio] courant *m* anodique permanent
~SCRAP - [electron] résidu *m* d'anode
~SCREENING GRID - [electron] grille- écran *f* d'anode
~SIDE CAP - [electron] (part of X-ray tubes) capuchon *m* d'anode
~SLIME - [el chem] s. anode mud
~SLOPE CONDUCTANCE - [electron] pente *f*
~SPLUTTERING - [electron] pulvérisation *f* anodique
~STEM - [electron] (metallic rod on X - ray tube) tige *f* anodique
~STOPPER - [radio] éliminateur *m* d'oscillations parasites
~STRAP - [electron] (part in electronic tube) bande *f* de court-circuit d'anode
~STRIP - [electron] (part of the anode in an electronic tube) collier *m* d'anode
~TERMINAL - [radio] borne *f* d'anode
~-TO-GRID CAPACITANCE - [thermion] capacité *f* plaque-grille
~VOLTAGE - [el] tension *f* anodique
ANODIC - [electron] anodique
~AREA - [electron] (of stray current) zone *f* de sortie de courants parasites
~BEHAVIOUR - [radio] comportement *m* anodique
~CLEANING - [el chem] dégraissage *m* (par électrolyse)
~COATING - [el chem] revêtement *m* électrolytique
~INHIBITORS - [el chem] (inhibitors retarding the anodic reaction) inhibiteurs *m* pl anodiques
~NEUTRAL - [electron] neutre *m* anodique
~OXIDATION - [el chem] (for aluminium or light alloys) oxydation *f* anodique
~PASSIVATION - [el chem] passivation *f* anodique
~POLARISATION - [electron] polarisation *f* anodique
~REACTION - [el chem] réaction *f* anodique
~TRANSFORMER - [electron] transformateur *m* anodique
~TREATMENT - [el chem] traitement *m* anodique
ANODIZATION - [el chem] anodisation *f*
ANODIZE, to - [el chem] soumettre à un traitement anodique
ANODIZING - [el chem] traitement *m* anodique
ANODOLUMINESCENCE - [electron] anodoluminescence *f*
ANODYNE - [med] (an analgesic) anodin *m*
ANOLYTE - [el chem] anolyte *m*
ANOMALOUS - [gen] anormal, anomal, irrégulier
~CATHODE FALL - [electron] chute *f* cathodique anormale
~DISPLACEMENT CURRENT - [el] courant *m* résiduel
~MAGNETIC MOMENT - [nucl] moment *m* magnétique anomal
~VALENCE - [nucl] (exceptional valence of an element in certain compounds) valence *f* anomale
ANOMALY - [gen] (a difference from the characteristics of the type) anomalie *f*, irrégularité *f*
ANOPSIA - [med] anopsie *f*
ANOPTIC SYSTEM - [telev] (television system without the use of optical means) système *m* anoptique
ANORECTAL - [anat] anorectal
ANORTHITE - [min] (a silicate of calcium and aluminium) anorthyte *f*
ANOTRON - [electron] (cold cathode rectifier valve) anotron *m* (soupape à cathode froide)
ANOXIC CELL - [biol] (living cell to which oxygen is not accessible) cellule *f* anaérobie
ANSWER - [gen] réponse *f*, solution *f*
~-BACK CODE - [radio] indicatif *m*
~ON A CIRCUIT, to-[telecomm] s'annoncer sur une ligne
~PRINT - [cin] (the first positive shown) première copie *f*, copie *f* destinée au producteur
~SIGNAL - [telecomm] accusé *m* de réception, signal *m* d'aperçu
ANSWERING INTERVAL - [telecomm] délai *m* de réponse
~JACKFIELD - [telecomm] tableau *m* de jacks locaux
~PLUG - [telecomm] fiche *f* d'écoute
~WAVE - [radio] onde *f* de réponse
ANT - [zool] fourmi *f*
~EATER - [zool] fourmilier *m*
~-HILL - [zool] fourmilière *f*
~OIL - [chem] furfurol *m*
ANTA - [constr] (square pilaster placed at the side of a door) ante *f*
ANTACID - [chem] antacide
ANTAGONIST SPRING - [mech] ressort *m* de rappel
ANTECHAMBER - [constr] antichambre *f*
[mech mach] chambre *f* de précombustion
ANTEFIX - [constr] (ornament applied to the end of roof tiles) antéfixe *f*
ANTENNA - [radio telev] (a means to transmit or receive radiowaves) antenne *f*, aérien *m*
~ARRAY - [radio] (a group of antennas having radiating or receiving properties) aérien *m*, réseau *m* d'antennes
~ATTENUATOR - [radio] (device decreasing the strength of the signal) atténuateur *m* d'antenne
~BOOSTER - [radio] (device increasing the strength of the signal) amplificateur *m* d'antenne
~CABLE - [telecomm] conducteur *m* d'antenne
~CAPACITOR - [radio] condensateur *m* d'antenne
~CHANGE OVER SWITCH - [radio] (a switch in the antenna circuit used for switching from transmission to reception and vice-versa) commutateur *m* d'antenne (émission-réception)
~CHOKE - [radio] bobine *f* d'arrêt (d'antenne)
~CIRCUIT - [radio] circuit *m* d'antenne
~CIRCUIT BREAKER - [radio] interrupteur *m* d'antenne, limiteur *m* de tension
~CONDENSER - [radio] condensateur *m* d'antenne
~CONDUCTOR - [radio] conducteur *m* d'antenne
~CONNECTION - [radio] (the terminal at the receiving set end) borne *f* d'antenne
~CONTROL BOARD - [radio] (U.S. - a panel with the switches relating to the antenna) tableau *m* de commande d'antenne
~COUPLING - [radio] couplage *m* d'antenne
~CROSS-TALK - [radio] diaphonie *f*
~CURRENT - [radio] courant *m* d'antenne
~DECREMENT - [radio] amortissement *m* de l'antenne
~DOWNLEAD - [radio] descente *f* d'antenne
~EARTHING - [radio] mise *f* à la terre d'une antenne
~EFFECT - [radio] (defective definition due to electrical asymmetry of the circuit) effet *m* d'antenne

ANTENNA EFFICIENCY - [radio] rendement *m* d'une antenne
~ELEVATION PAWL - [radio] crochet *m* de hissage d'antenne
~ENERGY - [radio] puissance *f* d'antenne
~EXCITATION - [radio] excitation *f* de l'antenne
~FEED - [radio] alimentation *f* de l'antenne, attaque *f* de l'antenne
~FEEDER - [radio] feeder *m*, ligne *f* d'alimentation d'une antenne
~FIELD GAIN - [telev] gain *m* d'antenne (télévision)
~FRAME - [radio] (the structure on which the conductor is wound) cadre *m*
~GAIN - [radio] gain *m* d'antenne
~GROUNDING SWITCH - [radio] (connecting the antenna to the ground) commutateur *m* antenne-terre
~IMPEDANCE - [radio] impédance *f* d'antenne
~INDUCTANCE - [radio] bobine *f* d'antenne, self *f* d'antenne
~INSULATOR - [radio] isolateur *m* d'antenne
~JACK - [radio] (socket connecting the antenna to the receiving set) douille *f* d'antenne
~LEAD - [radio] descente *f* d'antenne
~LEAD-IN - [radio] conducteur *m* d'antenne, descente *f* d'antenne
~LENS - [radar] (dielectric material concentrating the beam of radar antennas) lentille *f* d'antenne
~LOADING - [radio] bobine *f* (d'accord) d'antenne
~LOSS - [radio] (decrease of power in transmission) perte *f* de l'antenne
~LOSS DAMPING - [radio] amortissement *m* des pertes d'antenne
~MOUNT - [radio] (a rotating antenna with reflector for microwave radar) aérien *m* tournant
~NOISE - [radio] (noise due to interference picked up by the antenna) bruit *m* d'antenne, souffle *m* d'antenne
~PICK-UP - [radio] s. antenna noise
~POWER - [radio] (the power fed to a transmitting antenna) puissance *f* d'antenne
~RADIATION RESISTANCE - [radio] résistance *f* de rayonnement de l'antenne
~REACTANCE - [radio] réactance *f* d'antenne
~REEL - [radio] (used in aircraft) rouet *m* d'antenne
~REPEAT DIAL - [radio] (dial connected to a rotary antenna to show directive data) cadran *m* indicateur d'antenne
~RESISTANCE - [radio] résistance *f* d'antenne
~RESONANCE CURVE - [radio] courbe *f* de résonance d'une antenne
~SERIES CAPACITOR - [radio] condensateur *m* en série dans l'antenne
~SERIES CONDENSER - [radio] condensateur *m* en série dans l'antenne
~SHORTENING CONDENSER - [radio] condensateur *m* en série dans l'antenne
~SOCKET - [radio] s. antenna jack
~STABILIZATION - [radio] stabilisation *f* d'antenne
~STRAND - [radio] fil *m* d'antenne forsadé
~SYSTEM - [radio] système *m* d'antenne, aérien *m*
~TERMINAL - [radio] borne *f* d'antenne
~TOWER - [radio] (the structure supporting the transmitting antenna) pylone *m*, mât *m*
~TRUNK - [radio] entrée *f*, tubulaire (d'antenne)
~TUNING CAPACITOR - [radio] condensateur *m* d'accord d'antenne
~TUNING CONDENSER - [radio] s. antenne tuning capacitor
ANTENNA WINCH - [radio] (used in aircraft) rouet *m* d'antenne
~WIRE CHANGE-OVER SWITCH - [radio] commutateur *m* d'antenne
~YARD - [radio] (on a ship's mast) vergue *f* d'antenne
ANTERIOR - [gen] antérieur
~AQUEOUS CHAMBER - [anat] chambre *f* antérieure
~POSTERIOR VIEW - [radar] projection *f* antéro-postérieure
ANTEROOM - [constr] antichambre *f*, vestibule *m*
ANTEVERSION - [med] antéversion *f* (utérine)
ANTHELIX - [biol] anthélix *m*
ANTHESIS - [biol] anthèse *f*
ANTHION - [chem] persulfate *m* de potassium
ANTHOBIOLOGY - [bot] anthobiologie *f*
ANTHOPHOROUS - [bot] anthophore
ANTHOPHYLLITE - [min] (metasilicate of magnesium and iron used as a filler for paints) anthophyllite *f*
ANTHRACENE - [chem] (an intermediate in the production of some dyestuffs) anthracène
~ACID BLACK - [ind chem] noir *m* acide d'anthracène
~BLUE - [ind chem] bleu *m* d'alizarine
~BROWN - [ind chem] anthragallol *m*
~CHROME BROWN - [ind chem] brun *m* d'anthracène au chrome
~OIL - [chem] huile *f* anthracénique
ANTHRACIPHEROUS - [geol] anthracifère
ANTHRACIN GREEN OIL - [chem] huile *f* verte anthracénique
ANTHRACITE - [min] (a variety of hard coal) anthracite *m*
~BLACK - [ind chem] noir *m* anthracite
~COAL - [min] anthracite *m*, houille *f* sèche
~PIG IRON - [metall] fonte *f* à l'anthracite
ANTHRACITIZATION - [gas ind] anthracitisation *f*
ANTHRACNOSE - [bot] anthracnose
ANTHRACOSIS - [med] (produced by inhalation of coal dust) anthracose *f*
ANTHRAFLAVIC ACID - [chem] (intermediate for the production of dyes) acide *m* anthraflavique
ANTHRAFLAVINE - [chem] (an anthraquinone dyestuff) anthraflavine *f*
ANTHRAFLAVONE - [chem] anthraflavone *m*
ANTHRAGALLOL - [chem] anthragallol *m*
ANTHRAPURPURIN - [chem] anthrapurpurine *f*
ANTHRAQUINONE - [chem] (an intermediate for several dyestuffs) anthraquinone *m*
ANTHRARUFIN - [chem] anthrarufine *f*
ANTHRAX - [med] (infection due to the anthrax bacillus which can be communicated by animal to man) anthrax *m*
ANTHRONE - [biol] anthrone *m*
ANTHROPOID - [zool] anthropoïde
~APE - [zool] anthropoïde *m*
ANTHROPOLOGIST - [gen] anthropologiste *m*
ANTHROPOLOGY - [gen] anthropologie *f*
ANTHROPOMETER - [instr] anthropomètre *m*
ANTHROPOMETRY - [gen] anthropométrie
ANTHROPOMORPHIC - [gen] anthropomorphe
ANTHROPOPHAGOUS - [gen] anthropofage
ANTHROPOPHOBIA - [gen] anthropophobie *f*
ANTHROPOZOIC - [adj] anthropozoïque
ANTI-ACID COAT - [gen] couche *f* anti-acide
ANTIAGER - [ind chem] agent *m* anti-vieillissement

ANTIAIRCRAFT - [aero] antiaérien, contre-avion
ANTIARCING SCREEN - [el] écran *m* anti-arc
ANTIATOM - [nucl] antiatome *m*
ANTIBACTERIAL - [med] antibactérien
ANTIBIOSIS - [biol] (condition of mutual antagonism in living organisms) antibiose *f*
ANTIBIOTIC - [chem] (complex organic compounds with remarkable antimicrobial properties) antibiotique *m*
ANTIBLOW - [build] résistant au souffle (d'une explosion)
ANTIBODY - [med] (a substance introduced into the plasma, which is antagonistic to bodies or substances injurious to the organism) anticorps *m*
ANTICAKING - [ind chem] anti-agglutinat
ANTICAPACITANCE SWITCH - [radio] commutateur *m* à faible capacité
ANTICATALYST - [chem] (substance preventing or reducing the action of a catalyst) catalyseur *m* négatif
ANTICATHODE - [phys] (the electrode in a vacuum tube reflecting the rays emitted from the cathode) anticathode *f*
ANTICER - [auto, aero etc] (device or arrangement for preventing formation of ice or detaching it after formation) dispositif *m* anti-givre
ANTICHLOR - [chem] (a substance neutralizing free chlorine after bleaching) antichlore *m*
ANTICLINAL - [geol] anticlinal
~CORE - [geol] noyau *m* anticlinal
~FAULT - [geol] faille *f* anticlinale
~FLEXURE - [geol] flexure *f* anticlinale
~FOLD - [geol] anticlinale *m*
~LIMB - [geol] flanc *m* anticlinal
~RIDGE - [geol] crête *f* anticlinale
ANTICLINE - [geol] (a fold with the convex side upward) anticlinal *m*
~CREST - [geol] crête *f* anticlinale
ANTICLOCKWISE - [gen] dans le sens inverse des aiguilles d'une montre
~MOTION - [mech] mouvement *m* d'avancement à gauche
ANTICLOTTER - [chem] anticoagulant
ANTICLUTTER - [radar] (a device to eliminate or reduce the gain by decreasing the amount briefly after the transmission of the pulse) éliminateur *m* de retour de mer (signaux parasites), suppresseur *m* des parasites, régleur du couplage à faible constante de temps
~DEVICE - [radar] (s.anticlutter) dispositif *m* anti-retour de mer
ANTICOAGULANT - [ind chem] anticoagulant *m*
ANTICOINCIDENCE - [nucl] (a count which occurs in a detector but is not accompanied by a count in other detectors) anticoïncidence *f*
~COUNTER - [nucl] (a system of counters and circuits recording a count only if a ionizing particle passes through some of the counters but not through others) compteur *m* à anticoïncidence
ANTICOLLISION - [radar] (the prevention of collision in the course of movements by ships or aircraft) anticollision
~RADAR - [radar] radar *m* anticollision
ANTICONTAMINATION CLOTHING - [nucl] (clothing supplied to people working in the immediate vicinity of a reactor) vêtements *m* protecteurs
ANTI-CORROSION ENAMEL - [metall] laque *f* anticorrosive, laque *f* antirouille
ANTI-CORROSION PROTECTIVE COATING- s.anti-corrosion enamel
ANTICORROSIVE - [gen] anticorrosif
~AGENT - [metall] anticorrosif *m*
~PAINT - [ind chem] peinture *f* anticorrosive
ANTICREASE - [text] infroissable
~FINISH - [text] apprêt *m* infroissable
ANTI-CREEP BAFFLE - [mech] baffle *m* contre les migrations superficielles
~-CREEP SHIELD - [metall] barrière *f* écran contre les migrations superficielles
ANTICREEPER - [railw] dispositif *m* d'encrage des rails
ANTICRYPTOGAMIC - [ind chem] anticryptogamique
ANTICYCLONE - [met] (area of high barometric pressure in the atmosphere) anticyclone *m*
ANTICYCLONIC GLOOM - [met] (dimness of light caused by the persistence of dense stratocumulus during anticyclonic conditions) obscurité *f* anticyclonique
ANTI-DAZZLE - [gen] (any arrangement preventing or reducing dazzling conditions) anti-éblouissant
~SCREEN - [auto] écran *m* anti-éblouissant
~VISOR - [auto] para-soleil, visière *f* anti-éblouissante
ANTI-DAZZLING - [opt] anti-éblouissant
ANTIDIAZO COMPOUNDS - [chem] (stereoisomers of the diazo compounds in which the nitrogen-attached groups are remote from each other) antidiazoïque
ANTIDIFFUSION SCREEN - [radar] grille *f* anti-diffusante
ANTIDIMMING - [opt instr] anti-buée
ANTI-DIP - [instr] plongeur *m* mobile
ANTIDOTE - [gen] antidote *m*
ANTI-DRAG - [aerodyn] (possessing the quality of reducing drag) anti-trainée
~WIRE - [aero] hauban *m* de recul
ANTI-DRIBBLE DEVICE - [mech] (device preventing the slow escape of small quantities of liquid) dispositif *m* contre dégouttement
ANTI-DRUMMING - [auto] anti-bruit
~COMPOUND - [ind chem] pâte *f* anti-bruit, induit *m* anti-drumming
ANTIEMETIC - [med] antiémétique
ANTIENZYME - [chem] antienzyme *m*
ANTIFADING - [radio] antifading *m*
~DEVICE - [radio] dispositif *m* antifading
~FREQUENCY MODULATION - [modul] modulation *f* de fréquence antifading
ANTIFERMENT - [chem] antiferment *m*
ANTIFERMENTATIVE - [chem] antifermentatif
ANTIFIRE - [gen] anti-incendie, ignifuge
~PAINT - [ind chem] peinture *f* ignifuge
ANTI-FLUCTUATOR - [gas ind] antifluctuateur *m*, réservoir *m* amortisseur
ANTIFLUTTER - [aero] antiflutter
ANTIFOAMING - [chem] agent *m* antimousses
ANTIFOGGING - [gen] anti-buée
~AGENT - [photo] agent *m* anti-voile
ANTIFOULING - [ind chem] antisalissures
~COMPOSITION - [ind chem] (special toxic paints for submerged metalwork to prevent accumulation of organisms) peinture *f* antisalissures (couche de fond)
~PAINT - [ind chem] peinture *f* antisalissures, pein-

ture *f* de carène
ANTIFREEZE - [gen] antigel *m*
~PUMP - [mech] pompe *f* antigel
ANTI-FREEZER - [gasholders] dispositif *m* de chauffage des gorges
ANTIFREEZING - [gen] antigel
ANTIFRICTION - [mech] antifriction
~BEARING - [mech] coussinet *m* régulé
~METAL - [metall] métal *m* antifriction, régule *m*
ANTIFROST - [chem] antigel *m*
ANTIFROTHING - [chem] antimousses
ANTIGELLING AGENT - [ind chem] agent *m* antigélifiant
ANTIGLARE - [opt] anti-éblouissant
~GLASS - [auto] verre *m* anti-éblouissant
ANTI-g SUIT - [astronaut] vêtement *m* anti-g
ANTIGRAVITY - [phys] antigravité *f*
ANTIHALATION - [photot] anti-halo *m*
ANTIHISTAMINE - [med] (syntetic substance inhibiting the physiological activity of histamine) antihistamine *f*
ANTIHUM DEVICE - [radio] dispositif *m* anti-ronfle
ANTI-INCRUSTATOR - [gen boilers] désincrustant
ANTI-INDUCTION NETWORK - [radio] réseau *m* anti-inductif
ANTIKNOCK - [ind chem] anti-détonant *m*
~ADDITIVE - [ind chem] (a compound added to fuel to prevent pre-ignition) additif *m* anti-détonant
~DOPE - [ind chem] mélange *m* anti-détonant
~FUEL - [ind chem] carburant *m* anti-détonant
~VALUE - [ind chem] pouvoir *m* anti-détonant
ANTILIFT WIRE - [aero] (a wire designed to oppose forces acting in an opposite direction to a lift) câble *m* de retention
ANTILOGARITHM - [math] (the number corresponding to a given logarithm) cologarithme *m*
ANTILOGOUS POLE - [el] pôle *m* opposé
ANTIMAGNETIC - [phys] antimagnétique
ANTIMATTER - [phys] antimatière *f*
ANTI-MIGRATION SHIELD - s. anti-creep shield
ANTIMONIAL ALLOY - [metall] alliage *m* d'antimoine
~LEAD - [metall] plomb *m* aigre, plomb *m* antimonial
ANTIMONIC - [chem] antimonique, stibique
ANTIMONIDE - [chem] antimoniure *m*
ANTIMONITE - s. antimony glance
ANTIMONY - [chem] (an element, symbol Sb. Brittle, silvery, poisonous metal used for hardening lead, and in various alloys) antimoine *m*
~BLENDE - [min] kermésite *f*
~GLANCE - [min] (or antimonite, stibnite; natural trisulphide of antimony, the main source of the metal) stibine *f*, stibnite *f*
~OCHRE - [min] cervantite *f*, antimonocre *m*
~OXIDE - [chem] (used in paints) oxyde *m* d'antimoine
~OXYCHLORIDE - [chem] (a starting material) oxychlorure *m* d'antimoine
~POTASSIUM TARTRATE - [chem] (a dyeing mordant, insecticide and emetic) tartrate *m* antimonico-potassique
~SULPHIDE - [chem] antimoine *m* cru, antimoine *m* sulfuré
~WHITE - s. antimony oxide
ANTINEURALGIC - [med] antinévralgique
ANTINEUTRON - [nucl] anti-neutron *m*
ANTINODE - [phys & acoust] ventre *m*, antinoeud *m*
ANTINOISE - [acoust] anti- bruit
~MICROPHONE - [acoust] microphone *m* anti-bruit
~PAINT - [ind chem] peinture *f* insonorisante
ANTINOUS RELEASE - [camera] déclencheur *m* souple
ANTIOXIDIZER - [chem] anti-oxydant
ANTIPARALLAX - [gen] sans parallaxe
~MIRROR - [el] miroir *m* sans parallaxe
ANTIPARALLEL - [geom & nucl] antiparallèle *f*
ANTIPARTICLE - [nucl] (elementary particle of negative energy) antiparticule *f*
ANTIPHASE - [el] opposition *f* de phase
ANTIPHLOGISTIC - [chem] antiphlogistique
ANTIPODE - [geogr] antipode *m*
ANTIPOLARIZING WINDING - [el] enroulement *m* antipolarisant
ANTIPROTON - [nucl] (a proton with negative charge) anti-proton *m*
ANTI-PULSATOR - [gas ind] anti-pulsateur
ANTIPYRETIC - [chem] fébrifuge, antipyrétique
ANTIPYRINE - [chem] (used in medicine as a febrifuge) antipyrine *f*
ANTIQUARIAN - [paper] format *m* de papier à dessin (134x79 cms)
[gen] antiquaire *m*
ANTIQUE - [paper man] (good, rough-surfaced paper) antique
ANTIRABIC - [gen] antirabique
ANTIRATTLE SPRING - [mech] ressort *m* antivibreur, ressort *m* amortisseur
ANTIREDEPOSITION AGENT - [deterg] agent *m* d'antiredéposition
ANTI-REFLECTION COATING - [opt] (on optical lenses) couche *f* antiréfléchissante
ANTIRESONANCE - [phys & el] (condition of maximum impedance) anti-résonance
ANTIRIVELLING AGENT - [ind chem] agent *m* contre la formation des plis
ANTIROLLING - [naut & aer] anti-roulis
~TANK - [naut] caisson *m* anti-roulis
ANTIRUST - [gen] anti-rouille
ANTI-RUST COAT - [metall] couche *f* anti-rouille
ANTISAG BAR - [build] barre *f* de soutien
ANTISCALE - [ovens] antitartre *m*
ANTISCORCH - [ind chem] antigrilleur
ANTISCORCHER - [rubber ind] agent *m* antigrilleur, retardateur *m* de grillage
ANTISEISMIC - [build] résistant aux tremblements de terre
ANTISEIZE - [gen] anti-grippant
ANTISEPSIS - [med] (the destruction of bacteria by chemical agents) antisepsie *f*
ANTISEPTIC - [med] antiseptique
ANTISHATTER COMPOSITION - [glass ind] intercouche *f* pour verre de sécurité
ANTISHOCK MOUNTING - [auto] montage *m* amortisseur, dispositif *f* antichoc
ANTISHIMMY DAMPER - [aero] (device reducing rapid mechanical oscillations of a castoring wheel in a landing gear) anti-shimmy *m*
ANTI-SIDETONE - [radio] dispositif *m* anti-local
ANTISKID - [rubber ind] antidérapant
ANTI-SKID - [metall] (of surface iron plate) tôle *f* striée
~CHAIN - [auto] chaîne *f* antidérapante
ANTI-SKID UNIT - [aero] (mechanism designed to prevent locking of landing gear wheels when brakes are applied after touch-down) dispositif *m* limiteur

de patinage
ANTISKINNING AGENT - [ind chem] (agent preventing the formation of skin on surface coatings) agent *m* anti-peaux
ANTISLIP METAL - [metall] métal *m* antidérapant
~TYRE - [auto] pneu(matique) *m* antidérapant
ANTISPIN PARACHUTE - [aero] (parachute attached to wing tips or tail of an aircraft to combat spin) parachute *m* anti-vrille
ANTISPLASH - [gen] anti-éclaboussures
ANTISQUEAK - [gen & auto] anti-bruit
ANTISTALLING DEVICE - [aero] correcteur *m* de perte de vitesse
ANTISTATIC AERIAL - [radio] (designed to avoid interference) antenne *f* antistatique
~AGENT - [ind chem] agent *m* antistatique
~ANTENNA - [radio] (anti-interference antenna) antenne *f* antistatique
~RUBBER - [rubber ind] caoutchouc *m* antistatique
ANTISUBMARINE - [gen & nav] anti-sous-marine
~DETECTOR - [instr] détecteur *m* anti-sous-marin, asdic *m*
ANTI-SUCTION VALVE - [mech] clapet *m* de sécurité à minimum (de pression)
ANTISURGE BAFFLE - [aero] (a baffle plate in a tank to prevent undesirable movements of the contents) déflecteur *m* régulateur de débit
ANTISWEEP - [naut] antidragage
~MINE - [nav] mine *f* antidragage
ANTISYMMETRICAL FLUTTER - [aero] flutter *m* asymétrique
ANTISYNCLINAL AXIS - [geol] axe *m* antisynclinal
ANTITELESCOPING - [railw] anti-télescopage *m*
ANTITHEFT - [gen] (mechanism to prevent theft) antivol
ANTITHERMIC - [phys] antithermique
ANTITHETIC - [gen] antithétique
ANTITOXIN - [med] (substance produced by the organism which prevents the poisonous action of toxins) antitoxine *f*
ANTITRACKING VARNISH - [ind chem] (special varnish applied to insulating materials to prevent the formation of carbon track under high potential differences) vernis *m* antitrace
ANTITRADES - [meter] (westerly winds occurring at high level above the trade winds) contre-alizés *mpl*
ANTI-TRANSMIT RECEIVE SWITCH - [radar] (a device automatically switching the aerial from transmitter to receiver) tube *m* à décharge permettant l'émission et la réception sur une seule antenne
ANTITRIPTIC WIND - [met] (weak winds of short duration, blowing along the pressure gradient) vent *m* antitriptique
ANTIVACUUM - [electron] (counter-pressure exerted in the evacuation of vessels by a diffusion pump) anti-dépression
ANTI-VIBRATION MICA - [electron] mica *m* amortisseur, mica *m* anti-vibration
~MOUNTING - [aero] montage *m* amortisseur, montage *m* antivibreur
ANTI-VORTEX BAFFLE - [mech] déflecteur *m* anti-tourbillon
ANTIMONIAL - [chem] antimonié, stibié
~GLASS - [chem] verre *m* d'antimoine
~LEAD - [metall] (lead containing between 4 and 6 p.c. of antimony) plomb *m* antimonial, plomb *m* dur
~LEAD GRID - [el] grille *f* en plomb antimonial
ANTIMONIATE - [chem] (a compound produced by the reaction of antimonic acids with aqueous solutions of potassium hydroxide) antimoniate *m*
ANTIMONIC - [chem] antimonique, stibique
~ACID - [chem] acide *m* antimonique
~ANHYDRIDE - [chem] anhydride *m* antimonique
~CHLORIDE - [chem] chlorure *m* d'antimoine
~OXIDE - [chem] oxyde *m* d'antimoine
ANTIMONIDE - [chem] antimoniure *m*
ANTIMONINE - [chem] antimonine *f*
ANTIMONIOUS - [chem] antimonieux, stibieux
ANTIMONIPHEROUS - [chem] antimonière
ANTIMONITE - [min] (stibnite. Natural trisulphide of antimony) antimonite *m*, sulfure *m* d'antimoine
ANTIMONIUM BLENDE - [min] kermésite *f*
ANTIMONY - [chem] (white, metallic, brittle and poisonous element whose symbol is Sb, used for hardening lead and in several alloys) antimoine *m*
~ALLOY - [metall] alliage *m* d'antimoine
~ASH - [chem] cendre *f* d'antimoine
~BLACK - [ind chem] noir *m* d'antimoine
~BLOOM - [chem] oxyde *m* d'antimoine
~CROCUS - [chem] vermillon *m* d'antimoine
~CRUDE - [min] sulfure *m* d'antimoine
~ELECTRODE - [metall] électrode *f* à l'antimoine
~GLANCE - [min] stibnite *f*, stibine *f*
~GLASS - [chem] verre *f* d'antimoine
~GREY - [chem] gris *m* d'antimoine
~HALIDE - [chem] halogénure *m* d'antimoine
~HYDRIDE - [chem] hydrure *m* d'antimoine
~NEEDLES - [chem] antimoine *m* aciculaire
~ORE - [min] minerai *m* d'antimoine
~PENTACHLORIDE - [chem] pentachlorure *m* d'antimoine
~PENTOXIDE - [chem] anhydride *m* antimonique
~REGULUS - [min] régule *f*
~RUBBER - [rubber ind] caoutchouc *m* au sulfure d'antimoine
~SAFFRON - [chem] vermillon *m* d'antimoine
~SALT - [chem] sel *m* d'antimoine
~SULPHATE - [chem] sulfate *m* d'antimoine
~TRICHLORIDE - [chem] trichlorure *m* d'antimoine
~TRIOXIDE - [chem] anhydride *m* antimonieux, trioxyde *m* d'antimoine
~TRISULPHIDE - [chem] sulfure *m* d'antimoine
~YELLOW - [chem] jaune *m* de Naples, antimoniate *m* de plomb
~WHITE - [chem] blanc *m* d'antimoine
ANTIMONYL - [chem] (the monovalent radical SbO) antimonyle *m*
~POTASSIUM TARTRATE - [chem] tartrate *m* émétique d'antimoine
ANTLER - [zool] (annual outgrowth of bony material from frontal bone of deer) andouiller *m*
ANURESIS - [med] (diminution or suppression of urinary secretion) anurèse *f*, anurie *f*
ANVIL - [gen & mech & firearms] enclume *f*
[anat] (an ossicle in the ear) enclume *f*
~BLOCK - [impl] billot *m* d'enclume, souche *f* d'enclume
[mech] (of a power hammer) chabotte *f*
~CHISEL - [impl] tranche *f*
~CONTACT - [el] contact *m* fixe d'un interrupteur
~CUSHION - [impl] souche *f* d'enclume, semelle *f* d'enclume
~CUTTER - [mech] tranchet *m* d'enclume, casse-fer *m*

ANVIL DROSS - [metall] scories *f* pl de forge, battitures *f* pl
~FACE - [impl] table *f* de l'enclume
~PALLET - [mech] (of a power hammer) tas *m* inférieur
~STAKE - [impl] tas *m*
~-TONGS - [mech] pinces *f* pl de forgeron
~VICE - [mech] étau *m* à enclumette
AORTA - [anat] aorte *f*
AORTIC INSUFFICIENCY - [med] insuffisance *f* aortique
~MURMUR - [med] souffle *m* aortique
~VALVE - [anat] valvule *f* aortique
APATITE - [min] (natural calcium phosphate; a source of phosphorus and phosphoric acid) apatite *f*
APEPSIA - [med] apepsie *f*
APERIODIC - [phys] (having no natural frequency) apériodique
~AERIAL - [radio] (designed to have no natural frequency) antenne *f* apériodique
~CIRCUIT - [radio] circuit *m* apériodique
~COMPASS - [instr] compas *m* apériodique
~CURRENT - [el] courant *m* apériodique
~ELEMENT - [instr] équipage *m* apériodique
~FILTER - [mech] filtre *m* apériodique
~OSCILLATION - [phys] oscillation *f* apériodique
~REGENERATION - [radio] régénération *f* apériodique
APERIODICITY - [phys] (the quality of being free from natural frequency of oscillation, as in a compass) apériodicité *f*
APERTOMETER - [instr] (to measure numerical aperture of an objective) apertomètre *m*
APERTURAL EFFECT - [cinema] cercle *m* de confusion
APERTURE - [gen & opt] (any opening through which radiation may pass) ouverture *f*
[radar] (the effective size of the scanning spot in a television cathode-ray tube) dimension *f* du spot ouverture *f*
[cin] (in film projector) fenêtre - image *f*
~ANGLE - [opt] angle *m* d'ouverture
[radar] largeur *f* de faisceau, dimension *f* du spot
~COMPENSATION - [telev] (reduction of aperture distortion) compensation *f* d'ouverture
~CONTROL - [photo] commande *f* du diaphragme
~DISK - [telev] disque *f* perforé
~DISTORTION - [telev] définition *f* insuffisante, distortion *f* d'ouverture
~ILLUMINATION - [radio] répartition *f* du champ dans l'ouverture
~IMPEDANCE - [phys] impédance *f* d'ouverture, effet *m* d'extrémité
~LENS - [electron] lentille *f* électronique
~MASK - [telev] masque *m* d'ombre
~OF THE BEAM - [telev] largeur *f* du faisceau, dimension *f* du spot
~PLATE - [camera] diaphragme *m*, plaque *f* du couloir
~RATIO - [opt] ouverture *f* relative
[radar] ouverture *f* relative
~STOP - [telev] diaphragme *m*
APETALOUS - [bot] (devoid of petals) apétale
APEX - [gen] (the highest point, as in a triangle) apex *m*, point *m* culminant, sommet *m*
~DRIVE - [radio] alimentation *f* médiane (antennes)
~INDUCTOR - [telecomm] inducteur *m* d'extrémité de câble sous-marin
APEX OF CONE ANGLE - [mech] (of gears) sommet *m* du cône primitif
APHAGIA - [med] aphagie *f*
APHANITE - [petr] aphanite *f*
APHASIA - [med] (lack of faculty of expressing thoughts in words) aphasie *f*
APHELION - [astr] (that point in the orbit of a planet, espec, the earth, when it is furthest from the sun) aphélie *m*
APHIDAN - [zool] aphidien
APHIS - [zool] aphidé *m*, aphis *m*
APHONIA - [med] (loss of voice) aphonie *f*
APHOTIC - [bot] (able to grow with very little or no light) aphotique
APHOTOMETRIC - [bot] (does not react to light) aphotométrique
APHROMETER - [instr] (to measure carbonic acid contents in wine) aphromètre *m*
APHTHA - [med] (a small ulcer in the mouth) aphte *f*
APHYLLOUS - [bot] (devoid of leaves) aphylle
A.P.I. - (Air Position Indicator) indicateur *m* de position en vol
A.P.I. - (American Petroleum Institute) Institut Américain du Pétrole
APIARY - [gen] rucher *m*
APICAL - [gen] (relating to an apex) apical, disposé au sommet
~BODY - [zool] corps *m* apical
~CELL - [bot] cellule *f* apicale
~GROWTH - [bot] croissance *f* apicale
~ORGAN - [eool] organe *m* apical
APICITIS - [med] apicite *f*
APICULATE - [bot] apiciforme
APICULTURE - [gen] apiculture *f*
APIOLE - [chem] apiol *m*
APLACENTAL - [zool] aplacentaire
APLANATIC - [opt] (optical system which produces images free from spherical aberration) aplanétique
~LENS - [opt] objectif *m* aplanétique, aplanat *m*
~POINT - [photo] point *m* aplanétique
~PROJECTION - [opt] projection *f* aplanétique
~REFRACTION - [opt] réfraction *f* aplanétique
~SURFACE - [geom] réfraction *f* aplanétique
APLANATISM - [opt] aplanétisme *m*
APLASIA - [med] (faulty structural development) aplasie *f*
APLIC ROCKS - [geol] aplites *f* pl
APLITE - [geol] (an igneous rock) aplite *m*
APNEA - [gen] (transient cassation of respiration) apnée *f*
APNEUSIS - [physiol] (need for oxygen) apneusie *f*
APNEUSTIC CENTRE - [zool] (the part of the brain controlling the inflation of the lungs) centre *m* apneutique
APNOEA - s. apnea
APOATROPINE - [chem] apoatropine *f*
APOCARPOUS - [bot] apocarpé
APOCHROMATIC - [opt] apochromatique
~LENS - [opt] (with spherical and chromatic aberrations corrected as completely as possible) objectif *m* apochromatique
APODEME - [zool] apodème *m*
APODOUS - [zool] (without feet) apode
APOGEE - [astr] (that point in the orbit of the moon when it is most remote from the earth) apogée *m*
APOGENY - [bot] (sterility) apogénie *f*

APOGINY - [bot] (sterility in female organs) apogynie *f*
A-POLE - [telecomm] poteau *m* double de ligne
APOLLONIAN CIRCLES - [math] cercles *m* pl d'Apollonios
~SPHERES - [math] sphères *f* pl d'Apollonios
APOMECOMETER - [instr] (to measure elevation or distance) apomécomètre *m*
APONEUROSIS - [med] aponeurose *f*
APOPETALOUS - [bot] (lacking petals) apopétale
APOPHYLAXIS - [med] apophylaxie *f*
APOPHYLLITE - [min] apophyllite *f*
APOPHYSARY - [anat] apophysaire
APOPHYSIS - [anat] apophyse *f*
APOPLECTIC - [med] apoplectique
APOPLEXY - [med] (sudden loss of consciousness and paralysis) apoplexie *f*, congestion *f* cérébrale
APOROGAMY - [bot] aporogamie *f*
APOSAFRANINE - [chem] aposafranine *f*
APOSEMATIC COLORATION - [zoöl] (warning coloration) coloration *f* aposématique
APOTHECARY'S LIQUID MEASURES - [ind chem] mesures *f* pl utilisées en pharmacie
APOTHEM - [geom] apothème *m*
APOTROPOUS - [bot] apotrope
APPARATUS - [gen, mech, el etc] appareil *m*, dispositif *m*, équipement *m*, mécanisme *m*
~GLASS - [glass man] verre *m* de laboratoire
~ROOM - [telecomm] salle *f* des relais
APPAREL - [gen] habillement *m*, vêtements *m* pl
[naut] équipement
APPARENT - [gen] apparent, évident
~ALTITUDE - [surv] altitude *f* apparente
~ASTRONOMIC HORIZON - [astr] (the Great Circle upon which a plane perpendicular to the direction of gravity and tangential to the earth' surface cuts the celestial sphere) horizon *m* mathématique
~CANDLE POWER - [phys] intensité *f* lumineuse apparente
~DEPRESSION OF HORIZON - [astr] (angular difference between the true mathematical and the visible horizon) dépression *f* de l'horizon
~DIP - [phys] inclinaison *f* apparente
~EFFICIENCY - [el] rendement *m* apparent
~FREEZING POINT - [phys] point *m* de congélation apparent
~HEAVE - [geol] rejet *m* horizontal transversal apparent
~HORIZON - [astr] (visible horizon) horizon *m* apparent, horizon *m* visible
~LOAD - [el] charge *f* apparente
~MOTION - [phys] mouvement *m* apparent
~PITCH - [mech] pas *m* apparent
~POWER - [el] puissance *f* apparente
~RESISTANCE - [el] impédance *f*
~SLIP - [mech] recul *m* apparent
[geol] rejet *m* incliné, glissement *m*
~SOLAR DAY - [astr] jour *m* solaire
~SPECIFIC GRAVITY - [hydrol] densité *f* relative apparente
~STRATIGRAPHICAL GAP - [geol] rejet *m* apparent parallèle aux couches
~STRATIGRAPHICAL OVERLAP - [geol] recouvrement *m* stratigraphique apparent
~THROW - [geol] rejet *m* vertical
APPEARANCE - [gen] apparence *f*, aspect *m*
[leg] comparition *f*
APPEARANCE CHECK - [gen] contrôle *m* extérieur
~POTENTIAL - [electrom] potentiel *m* d'apparition
APPENDAGE - [gen & build] accessoire *m*, dépendance *f*, annexe *f*
APPENDECTOMY - [med] appendicectomie *f*
APPENDICITIS - [med] appendicite *f*
APPENDIX - [gen & mech] appendice *m*
APPLAUSEOGRAPH - [instr] appareil *m* de mesure des applaudissements
APPLE - [bot] pomme *m*
~ACID - [chem] acide *m* malique
~CANKER - [bot] chancre *m* du pommier
~FRUIT MINER - [bot] teigne *f* du pommier
~MILDEW - [bot] oidium *m* du pommier
~TUBE - [telev] (a colour display tube) tube *m* dit pomme, tube *m* de télévision à couleurs à écran rayé
APPLETON LAYER - [met] (the second stratum of the Heaviside layer, altitude approximately 180 Km) couche *f* d'Appleton
~REGION - [Appleton Layer] couche *f* d'Appleton
APPLIANCE - [gen] accessoire *m*, appareil *m*, dispositif *m*, équipement *m*, instrument *m*, matériel *m*
APPLICATION - [gen] application *f*, demande *f*, requête *f*
APPLICATOR - [print] encreur *m*
[nucl] (device containing radio-active material) applicateur *m*, dispositif *m* pour traitement local
APPLIED - [gen & mech] appliqué
~CHEMISTRY - [chem] chimie *f* appliquée
~ELECTRICITY - [el] électricité *f* appliquée
~IMPULSE - [electron] impulsion *f* appliquée
~LOAD - [mech] charge *f* appliquée
~POWER - [el] puissance *f* appliquée
~PRESSURE - [el] (potential difference applied between the terminals of an electric circuit) tension *f* appliquée
~POTENTIAL - [el] potentiel *m* appliqué
~SHOCK - [acoust] choc *m* appliqué
~STRESS - [mech] effort *m* appliqué
APPLY, to - [gen & techn] (e.g. the brakes) appliquer, faire une demande, solliciter
APPORTIONMENT - [gen] distribution *f*, répartition *f*, lotissement *m*
APPOSITION - [bot] (growth in thickness of the cell wall) apposition *f*
APPRAISAL - [gen & comm] appréciation *f*, estimation *f*, évaluation *f*
~WELL - [ind chem] puits *m* d'essai
APPRAISE, to - [gen & comm] estimer, évaluer, priser
APPRAISER - [leg & comm] estimateur *m*, priseur *m*
APPRENTICE - [gen] apprenti *m*
APPRENTISHIP - [gen] apprentissage *m*
APPROACH, to - [gen] approcher
APPROACH - [gen etc] approche *f*, abord *m*, accès *m*, voie *f* d'accès
[aero] (the movement of an aircraft nearing a landing point) approche *f*
~AND LANDING AIDS - [aero] (system designed to aid pilots in approach and landing along a predetermined path) aides *f* pl d'approche et d'atterrissage
~AREA - [aero] aire *f* d'approche
~BEACON - [aero] radiophare *m* d'atterrissage
~CHART - [aero] carte *f* d'approche
~CONFIGURATION - [aero] (configuration of an air-

craft during the approach phase of a flight, e.g. with flaps, landing gear etc down) position *f* en vol d'approche
APPROACH CONTROL - [radar-aero] contrôle *m* d'approche
~CONTROL RADAR - [radar-aero] (A.C.R.) radar *m* de contrôle d'approche
~CUTTING - [mining] taille *f* d'accès
~LIGHT - [aero] feu *m* d'atterrissage
~LIGHTING - [railw] signal *m* (lumineux) d'approche
~LOCKING - [railw] (route with approach locking) blocage *m* d'approche
~LOCKING DEVICE - [railw] blocage *m* d'approche
~MARKER-BEACON - [radar] radiophare *m* d'atterrissage
~NAVIGATION - [aero] (the navigation of an aircraft during approach) vol *m* d'approche
~PATH - [aero] circuit *m* d'approche
~RECEIVER - [radio-aero] (a radio receiver designed to interpret approach beacon signals) récepteur *m* d'approche
~SPEED - [aero] vitesse *f* d'approche
~SURFACE - [aero] aire *f* d'approche
~TRACK - [railw] voie *f* d'accès
APPROBATION REPORT - [comm] procès-verbal *m* d'agrément
APPROJECTION - [radio] projection *f* antéro-postérieure
APPROPRIATE, to - [gen] s'approprier, (finance) approprier
APPROPRIATE - [gen] approprié, convenable, propre
APPROPRIATION - [gen] affectation *f* de fonds, appropriation *f*
APPROVED - [gen] approuvé, ratifié, homologué
~MAXIMUM RATE OF DESCENT - [aero] vitesse *f* de descente maximale autorisée
APPROXIMATE - [gen & comm] approximatif, approché
APPROXIMATION - [gen] approximation *f*, valeur *f* approchée
APPURTENANCE - [gen] appartenance *f*, droit *m* accessoire (immeubles)
APPURTENANCES - [gen] accessoires *m* pl, appareils *m* pl, appareillage *m*
APRICOT - [bot] abricot *m*
APRON - [hydr] radier *m* (d'un bassin) chape *f*
[mech] tablier *m*
[mech mach] trainard *m* (de tour), tablier *m* (d'alimentation)
[aero] (that part of the surface of an airfield immediately in front of a hangar or other buildings) aire *f* de manoeuvre
[astronaut] tablier *m* de protection
[naut] contre-étrave *f*
[text] tablier *m*
~CLOTH - [text] tissu *m* pour tabliers
~CONVEYOR - [mech] convoyeur *m* à éléments articulés
~FEEDER - [mech] alimentateur *m* à courroie
~TANK - [cin] (tank for the processing of films) cuve *f* (pour le développement)
A.P.S. WEIGHT - [abbrev for Aircraft prepared for service weight] poids *m* en ordre de marche
APSE - [build] abside *m*
APSIDAL LINE - s. apsis
APSIS - [astr] (the line joining the points in the orbit of a planet at which it is nearest to and furthest from the sun) apside *f*
APYRETIC - [gen] apyrétique
AQUA AMMONIA - [chem] solution *f* aqueuse d'ammoniaque
~FORTIS - [chem] (dilute nitric acid) eau-forte *f*
~REGIA - [chem] eau *f* régale
~VITAE - (a spirit) eau *f* de vie
AQUADAG - [ind chem] (an aqueous preparation of graphite used for forming a conducting internal coating in vacuum tubes) aquadag *m* (revêtement interne pour tubes à vides)
AQUALUNG - [impl] scaphandre *m* autonome
AQUAMARINE - [min] aigue marine *f*
AQUARELLE - [aer] aquarelle *f*
AQUARIUM - [gen] aquarium *m*
[acoust] salle *f* de mixage
~REACTOR - [nucl] (also called swimming pool reactor) pile *f* piscine, réacteur *m* piscine
AQUARIUS - [astr] le Verseau *m*
AQUATIC FAUNA - [zool] faune *f* aquatique
~FLORA - [bot] flore *f* aquatique
AQUATINT - [phot] aquatinte *f*
AQUEDUCT - [hydr] aqueduc *m*, réseau *m* de distribution
AQUEOUS - [gen & chem] aqueux
~DESIZING - [glass fibre] désensimage *m* aqueux
~HOMOGENEOUS REACTOR - [nucl] réacteur *m* homogène à suspension aqueuse
~SHEATH - [soil] film *m* d'eau, coquille *f* d'eau
~SOLUTION - [chem] solution *f* aqueuse
AQUICLUDE - [hydr] couche *f* imperméable
AQUIFER - [hydr] formation *f* aquifère, couche *f* aquifère, nappe *f* aquifère
~LAYER - (in a gasholder) couche *f* aquifère
AQUIFUGE - [geol] formation *f* aquifuge, formation *f* hydrophobe
AQUINITE - [chem] chloropicrine
AQUO ION - [phys] (a hydrated ion) ion *m* hydrate
AQUOLYSIS - [chem] aquolyse *f*
ARABIC GUM - [ind chem] gomme *f* arabique
ARABINOSE - [chem] (a pentose used as a culture medium) arabinose *f*
ARABITOL - [chem] arabitol *m*
ARABLE - [gen] arable, labourable
~SOIL - [agric] terre *f* arable
ARACHIS - [bot] arachide *f*
~OIL - [ind chem] (peanut oil) huile *f* d'arachide
ARAGONITE - [min] (a form of crystalline calcium carbonate) aragonite *f*
ARBITRARY PARAMETER - [comput] paramètre *m* arbitraire
~-PRECISION MULTIPLICATION - [comput] multiplication *f* à precision arbitraire
ARBITRATION - [gen] arbitrage *m*
ARBITRATOR - [leg] arbitre *m*
ARBOR - [mech] (a shaft or spindle) arbre *m*, axe *m*, broche *f*
[mech mach] mandrin *m* de tour
~BRACE - [mech] porte-mandrin *m*
~FOR SHELL-END MILL - [mach tools] arbre *m* porte-fraise, arbre *m* porte-taille
~SUPPORT - [mach tool] (of a milling machine) support *m* d'arbre porte-fraise
ARBOREAL - [bot] d'arbre, arboricole
ARBORESCENT - [bot] arborescent
~CRYSTALS - [crystal] cristal *m* en forme de sapin
ARBORICULTURE - [agric] arboriculture *f*

ARBUTIN - [chem] (glucoside obtained from the bear berry) arbutine *f*

ARC - [math] (any part of algebraic curve) arc *m*
[astr] arc *m*
[el] (a luminous discharge of electricity across a gas) arc *m*

~AND RESISTANCE FURNACE - [metall] four *m* à arc et à résistance

~BACK - [electron] (backfire; failure of the rectifying action resulting in a strong current flow in the inverse direction) allumage *m* en retour, retour *m* d'arc

~BAFFLE - [electron] (in mercury-pool tubes) écran *m* déflecteur d'arc

~BAR - [build] barre *f* d'un arc

~BLOW - [el] (the energy contained in an arc) souffle *m* de l'arc

~BOUTANT - [constr] arc-boutant *m*

~BRAZING - [metall] soudobrasage *m* à l'arc

~CATHODE - [electron] (a cathode with a self-supporting electron emission with a weak voltage drop) cathode *f* à arc

~CONTROL DEVICE - [el] dispositif *m* de commande d'arc

~CONVERTER - [radio] oscillateur *m* à arc

~CUTTING - [metall] coupage *m* à l'arc

~CUTTING MACHINE - [mech mach] machine *f* de découpage à l'arc

~DISCHARGE - [electron] décharge *f* en arc

~DISSOCIATION - [metall] dissociation *f* à l'arc

~DROP - [electron] chute *f* dans l'arc

~DROP LOSS - [electron] (the power which is dissipated in the arc) pertes *f pl* dans l'arc

~DROP VOLTAGE - [electron] chute *f* de tension dans l'arc

~DURATION - [el] durée *f* d'arc

~ELIMINATOR - [el] dispositif *m* d'extinction d'arc

~EXTINGUISHER - [el] dispositif *m* d'extinction d'arc

~FURNACE - [metall] (furnace heated by an electric arc) four *m* à arc

~GAP - [el] entrefer *m*

~HEATING - [electro heat] (heating generated by electric arc) chauffage *m* par arc

~HORN - [el] corne *f* de protection, électrode *f* de garde

~IMAGE FURNACE - [metall] four *m* à arc indirect

~LAMP - [gen & el] (electric lamp using an electric arc between two carbon electrodes as source of light) lampe *f* à arc

~LENGTH - [welding] longueur *f* de l'arc

~MELTING - [metall] fusion *f* à l'arc

~OF ACCESS - [mech] arc *m* d'avance

~OF ACTION - [mech] arc *m* d'engrènement

~OF APPROACH - [mech] (arc on the pitch circle of a gear wheel when two teeth are in contact and approach the pitch point) arc *m* d'accès

~OF FOLDING - [geol] arc *m* de plissement

~OF RECESS - [mech] (as above, but when the two teeth in contact recede from the pitch point) arc *m* de recul

~OF VIBRATION - [mech] arc *m* d'oscillation

~-OVER - [el] (arc discharge across an insulator or an air gap) décharge *f* extérieure (isolateurs), fuite *f*

~PIECE - [naut] voûte *f* d'arcasse

~PITCH - [mech] (of gear) pas *m* circulaire

ARC QUENCHING - [el] extinction *f* d'arc

~RECTIFIER - [el] soupape *f* à arc, redresseur *m* d'arc

~RESISTANCE - [el] résistance *f* à l'arc

~RESISTANCE FURNACE - [metall] four *m* à résistance de l'arc

~SHIELD - [el] anneau *m* de garde (isolateurs)

~SPECTRUM - [el] (spectrum produced at the temperature of the electric arc) spectre *m* de l'arc

~SPLITTER - [el] sectionneur *m* d'arc

~STREAM VOLTAGE - [el] chute *f* de tension d'arc

~SUPPRESSION COIL - [el] (a choke coil inserted in a circuit to suppress arcing) bobine *f* d'équilibrage

~TRACKING - [el] cheminement *m* d'arc

~TRANSMITTER - [radio] s. arc converter

~TUBE - [el] s. arc converter

~VOLTAGE - [el] (the total voltage across an electric arc) tension *f* d'arc

~WELDER - [metall] soudeuse *f* à l'arc

~WELDING - [metall] (welding in which the heat is produced by one or more arcs either between one or more electrodes and the workpiece or between electrodes) soudure *f* à l'arc

~WELDING FLUX - [metall] flux *m* de soudure à l'arc

~WELDING SET - [mech] poste *m* de soudage à l'arc

~WELDING WITH STEAM - [welding] soudage *m* à l'arc avec vapeur

ARCADE - [build] arcade *f*, galerie *f*

ARCATOMIC WELDING - [metall] soudage *m* arcatomique

ARCH, to - [gen] arquer, voûter, se voûter
[min] (of ore) s'arc-bouter en voûte

ARCH - [gen & constr] arc *m*, cintre *m*, voûte *f*
[of a bridge] arche *f*
[metall] (of a furnace] voûte
[railw] (of a locomotive) boîte à fumée
[geol] selle *f*, voûte *f*
[mining] stot *m*, estau *m*

~ABUTMENT - [constr] culée *f*

~ACTION - [constr] effet *m* de voûte, arc-boutement *m*

~BARREL - [constr] (an inclined barrel arch for a multiple-arch dam) contrefort *m*

~BEARING - [mech] appui *m* à arc

~BEND - [geol] charnière *f* anticlinale

~BRICK - [constr] coin *m*

~BUTTRESS - [constr] arc-boutant *m*

~CENTERING - [constr] cintre *m* de l'arc

~CORE - [geol] noyau *m* du pli

~CROWN - (of an oven) voûte *f* (d'un four)

~DAM - [hydr] barrage-voûte *m*

~FALSEWORK - [build] armement *m* de l'arc

~GRAVITY DAM - [hydr] barrage *m* poids-voûte

~LID - [constr] couvercle *m* de la voûte

~LIMB - [geol] flanc *m* supérieur (d'un pli couché)

~OF FOOT - [anat] cambrure *f* du pied

~PIPE - [gas install] chevalement *m*

~SPAN - [constr] (in a dam) écartement *m* intérieur des contreforts

~SPRINGING LINE - [constr] naissance *f*

~STONE - [constr] vousseau *m*, voussoir *m*

~SUPPORT - [shoe man] cambrure *f*

~THRUST - [constr] poussée *f* de l'arc

~TRUSS - [constr] arc *m* en treillis

ARCHED - [gen] en arc, cintré, voûté, incurvé
[shoe man] cambré

ARCHED CALVERT - [hydr] ponceau *m* voûté, rigole *f* voûtée
~CATENARY SUPPORT - [railw] caténaire *m* en arc
ARCHETYPE - [arch] (original type from which others derive) archétype *m*
ARCHIMEDEAN DRILL - [mech] (a drill in which the movement of a nut on a helix gives an alternating rotary motion to the bit) forêt *m* à vis d'Archimède
~SCREW - [impl] (an ancient device to lift water) vis *f* d'Archimède
~SPIRALS - [radiation] spirales *f* pl d'Archimède
ARCHIPELAGO - [geogr] archipel *m*
ARCHITECTURE - [gen] architecture *f*
ARCHITRAVE - [build] architrave *f*
ARCHIVOLT - [build] (ornament around the face of an arch) archivolte *m*
ARCHLUTE - [mus] (a large lute with extra bass strings and a long neck) archiluth *m*
ARCHWAY - [gen] passage *m* voûte, voûte *f*
ARCING - [el] (the occurrence of electric arcs) amorçage *m* d'arc
~BACK - [electron] allumage *m* en retour, retour *m* d'arc
~CONTACT - [el] contact *m* de coupure (pare-étincelles)
~GROUND - [el] arc *m* à la terre (isolement défectueux)
~VOLTAGE - [el] tension *f* d'arc
ARCOM - [telecomm] (Arctic Communication Satellite Earth Station) station *f* terrestre arctique de communication
ARCONOGRAPH - [instr] oscillographe *m*
ARCTIC - [geogr] arctique
~CIRCLE - [geogr] cercle *m* polaire arctique
~FRONT - [met] front *m* arctique
~SEA SMOKE - [met] (sea smoke observed in the North Polar region) fumée *f*, brouillard *m* d'évaporation
ARDOMETER - [instr] (type of total radiation pyrometer) ardomètre *m*
ARE - [meas] (100 sq metres, i.e. 119.6 sq yds) are *m* (surface)
AREA - [gen] aire *f*, superficie *f*, surface *f*
[top] zone *f*, région *f*
[build] aire *f*
~CONTROL - [aero] (air traffic control service operating over a given area) contrôle *m* de région
~DENSITY - [phys] densité *f* superficielle
~EXPANSION RATIO - [phys] rapport *m* géométrique de détente
~MONITOR - [radiation] (device for detecting or measuring the level of radiation in a given area) détecteur *m* local
~MONITORING - [nucl] contrôle *m* local
~OF COUNTERPRESSURE - [phys] aire *f* de contre-pression
~OF CRANE - [mech] portée *f* d'une grue
~OF HIGH PRESSURE - [meter] zone *f* de haute pression, anticyclone *m*
~OF LOW PRESSURE - [meter] zone *f* de basse pression
AREAL GEOLOGY - [geol] géologie *f* régionale
ARENA - [gen] arène *f*
~STAGE - [build] scène *f* circulaire
~THEATRE - [build] théâtre *m* en rond, théâtre *m* en arène, arène *f*
ARENACEOUS - [gen] arénacé, arénifère
[bot] (growing well in sandy soil) arénacé
ARENACEOUS LIMESTONE - [min] calcaire *m* arénacé
~ROCKS - [geol] roches *f* pl arénacées
~TEXTURE - [geol] structure *f* psammitique
AREOMETER - [instr] (instrument for the determination of specific gravities) aréomètre *m*
ARGENTAN - [nickel-silver] argentan *f*
ARGENTIC OXIDE - [chem] (an oxyde of silver) oxyde *m* d'argent
ARGENTIFEROUS - [min] argentifère
ARGENTITE - [min] (important ore of silver) argentite *f*, argyrose *m*, argyrite *f*
ARGENTOMETER - [instr] (used to establish the strength of silver nitrate solutions) argentimètre *m*
ARGENTOUS OXIDE - [chem] (lower oxide of silver) oxyde *m* argenteux
ARGIL - [min] argile *m*
ARGILLACEOUS - [min] argileux
~CEMENT - [build] ciment *m* argileux
~ROCK - [min] roche *f* argileuse
~SANDSTONE - [min] grès *m* argileux
~SLATE - [geol] schiste *m* argileux
ARGININE - [chem] (a widely distributed amino-acid) arginine *f*
ARGOL - [chem] (crude potassium acid tartrate recovered from wine vats) tartre *m* brut
ARGON - [chem] (an element, symbol Ar. One of the inert gases) argon *m*
~-ARC WELDING - [metall] soudage *m* à l'arc en atmosphère d'argon
ARGUMENT - [math] argument *m*
[el] argument *m* (d'une impédance)
ARID - [gen met] aride
ARIES - [astr] (a constellation) le Bélier *m*
ARIL - [bot] (outgrowth on a seed) arille *f*
ARITHMETIC - [gen] arithmétique
~PROGRESSION - [math] progression *f* arithmétique, série *f* arithmétique
~SERIES - [math] série *f* arithmétique, progression *f* arithmétique
ARITHMETICAL CHECK - [comp] vérification *f* arithmétique
~ELEMENT - [comp] organe *m* de calcul
~MEAN - [math] moyenne *f* arithmétique
~ORGAN - [comp] organe *m* de calcul
~PROGRESSION - [math] progression *f* arithmétique
~SHIFT - [comput] décalage *m* arithmétique
~VALUE - [math] valeur *f* absolue
ARITHMOMETER - [mech] (machine performing multiplications by successive additions) arithmomètre *m*
ARM - [anat] bras *m*
[mach] bras *m*, levier *m*, fléau *m* de balance
[gearing] levier *m* d'entraînement
[naut] patte *f* d'ancre
[mining] pied-droit
~BRACES - [mech mach] bretelles *f* pl
~COAL CUTTER - [mining] haveuse *f* à barre
~DOWN SWITCH - [mach tool] interrupteur *m* pour commande descente bras
~EXTENTION BRACKET - [telecomm] prolongement *m* de traverse
~FILE - [tool] lime *f* plate-à-main
~HOIST - [mech] palan *m* à bras, élévateur *m* (hydraulique) à bras
~LIFTING SWITCH - [mach tool] interrupteur *m* pour commande montée bras

ARM LOCKING LEVER - [mach tool] levier *m* pour le blocage du bras
~MOORING - [naut] amarrage *m* en patte d'oie
~OF A WHEEL - [spoke] rayon *m*
~PIT - [anat] aisselle *f*
~REPEATER - [railw] sémaphore *m*
~-REST TABLE - [aero] tablette *m* pour accoudoir
~SAW - [metall] scie *f* à main
~SEAT - [telecomm] semelle *f* (placée entre la traverse et le poteau)
~SUPPORT - [mach tool] support *m* du bras, appui *m* de bras (bretelles)
ARMAMENT - [gen] armement *m*
ARMATURE, to - [el] monter un induit
ARMATURE - [el & gen] induit *m* (machine tournante), armature *f* (relais etc), rotor *m*
[constr] semelle *f* d'encrage pour haubans
~AXIS - [el] axe *m* de l'induit
~BANDS - [el] forettes *f* pl d'induit
~BEARING - [el] palier *m* d'induit
~CASING - [el] blindage *m* d'induit
~CLAMP - [el] borne *f* d'armature
~CLEARANCE - [el] libre trajet *m* de l'armature
~COIL - [el] enroulement *m* d'induit
~CORE - [el] noyau *m* d'induit
~CURRENT - [el] courant *m* d'induit
~DUCT - [el] canal *m* de ventilation (d'un induit), rainure *f* de refroidissement de l'induit
~FILED - [el] champ *m* d'induit
~OF A RELAY - [el] armature *f* d'un relais
~REACTION - [el] (the reaction of the field produced by the current flowing in the armature winding of a generator or motor on the main field of the machine) réaction *f* d'induit
~REED - [el] ressort *m* d'armature
~SHAFT - [mech] arbre *m* d'induit
~SPIDER - [el] croisillon *m* (machines tournantes)
~TESTER - [impl] ronfleur *m*, vibreur *m*
~WINDER - [el] machine *f* à bobiner les induits
ARMOUR, to - [metall] cuirasser
ARMOUR - [gen] armure *f*, cuirasse *f*
[armam] engins *m* pl blindés, unités *f* pl blindées
~CLAD - [gen] blindé, cuirassé
~CLAMP - [mech] collier *m* de câble
~GLAND - [mech & el] collier *m* de câble
~GRIP - [mech] collier *m* de câble, serre-câble *m*
~-PIERCING - [armam] (of a bullet or shell) perforant
~-PIERCING BULLET - [armam] balle *f* perforante
~-PIERCING SHELL - [armam] obus *m* perforant
~PLATE, to - [gen & armam] blinder, cuirasser
~PLATE - [mech & mach] plaque *f* de blindage, plaque *f* de cuirasse
~PLATE ROLLING MILL - [metall] laminoir *m* à blindages
~PLATING - [armam] blindage *m*, cuirasse *f*
ARMOURED - [gen] blindé, cuirassé
[el] armé
~CABLE - [el] câble *m* armé
~CAR - [mech] voiture *f* blindée
~CONDUIT - [mech] tube *m* armé (pour canalisations électriques)
~HOSE - [gen] (hose protected against mechanical damage by a spiral wire winding) tube *m* souple armé, tuyau *m* protégé
ARMOURER - [gen] armourier *m*
ARMOURING - [gen & cables] (steel wires woven on submarine cables) blindage *m*, cuirasse *f*, armure *f* (câbles etc)
ARMOURY - [gen] magasin *m* d'armes, armurerie *f*
ARMRACK - [gen] râtelier *m* d'armes
ARMURE - [text] armure *f*
ARMY - [gen] armée *f*
~CLOTH - [text] tissu *m* militaire
ARNATTO - [bot] rocou *m*
ARNICA - [bot] arnica *f*
~ROOT OIL - [ind chem] essence *f* d'arnica
AROMA - [gen] arome *m*
AROMATIC - [gen] aromatique
~ACID - [chem] (acids in which the carboxyl group is attached to an aryl radical) acide *m* aromatique
~ALCOHOL - [chem] alcool *m* aromatique
~BASE - [petr] pétrole *m* brut
~CHEMISTRY - [chem] chimie *f* aromatique
~COMPOUND - [chem] (a term formerly in general use for closed-chain organic compounds as opposed to the so-called aliphatic or open-chain compounds) composé *m* aromatique
~HYDROCARBON - [chem] hydrocarbure *m* aromatique
~HYDROGENATION - [chem] (hydrogenation in the naphtalene series) hydrogénation *f* aromatique
~PROPERTIES - [chem] (the characteristic properties of aromatic compounds) propriété *f* aromatique
~TAR - [oil ind] résidu *m* aromatique
~VINEGAR - [chem] (obtained by the distillation of copper diacetate) acide *m* aromatique
AROMATICS - [chem] carbures *m* pl aromatiques
AROUND - [gen] autour de, à l'entour (for prices and quantities) environ
AROUSAL RESPONSE - [physiol] réaction *f* d'excitation
ARRAIGNEMENT - [leg] mise *f* en accusation, acte *m* d'accusation, interpellation *f* de l'accusé
ARRANGE, to - [gen] aménager, arranger, disposer, mettre en ordre, ordonner, ranger
ARRANGEMENT - [gen] arrangement *m*, aménagement *m*, dispositif *m*, disposition *f*, montage *m*
[comm] accord *m*, compromis *m*, entente *f*, transaction *f*
[math] arrangement *m*
[mus] adaptation *f*, arrangement *m*
[telecomm] montage *m*
~OF THE GRAINS - [soil] arrangement *m* des grains
ARRAS - [text] tapisserie *f*, tenture *f*
ARRAY - [gen] disposition *f*, ordre *m*
[radio] (a complex aerial system comprising various elements giving directional properties) réseau *m* d'antennes
[comput] disposition *f* en rangées, alignement *m*
ARREARS - [gen & comm] arriérés *m* pl, arrérages *m* pl
ARREST, to - [gen] arrêter
[mech] arrêter, empêcher, freiner
ARREST - [gen & mech] arrestation *f*, arrêt *m*
ARRESTER - [el] (lightning arrester) éclateur *m*, parafoudre *m*
~CABLE - [aero] (wire rope forming the principal part of an arrester system, for checking aircraft quickly after landing in a restricted area such as the flight deck of an aircraft carrier) brin *m* d'appontage
~HOOK - [aero] (hook fixed to an aircraft intended

for alighting on the flight deck of aircraft carrier) crosse *f* d'appontage
ARRESTER WIRE - [aero] s.arrester cable
~GEAR - [aero] (device used to facilitate the alighting of an aircraft in a restricted area) dispositif *m* d'arrêt
ARRIS - [mech] arête *f*
~FILLET - [build] (small strip of wood used beneath tiles) congé *m* en V
~GUTTER - [build] (V-shaped gutter, generally of wood) gouttière *f* en V
~RAIL - [carp] (rails with triangular cross-sections) guide *m* à section triangulaire
~TILE - [build] tuile *f* faîtière
ARRIVAL - [gen] arrivée *f*
~NOTE - [comm & railw] avis *m* d'arrivée
~PLATFORM - [railw] quai *m* d'arrivée, quai *m* de débarquement
~TIME - qgen] heure *f* d'arrivée
ARROW - [gen] flêche *f*
[surv] fiche *f*
~ENGINE - [mech] moteur *m* en W
~HEADED DRILL - [mech] mèche *f* plate, foret à langue d'aspic
~POINT BRACING - [mech] entretoisement *m* en triangle
ARROWHEAD AUGER - [gas ind] crochet *m* à déboucher
~TWILLED FABRIC - [text] tissu *m* croisé
ARSENAL - [gen] arsenal *m*
ARSENATE - [chem] arséniate *m*
ARSENIC - [chem] (an element, symbol As. Silvery grey, brittle, poisonous, used to harden lead and in the production of various alloys) arsenic *m*
~ACID ANHYDRIDE - [chem] anhydride *m* arsénique
~ANHYDRIDE - [chem] anhydride *m* arsénique
~ANTIDOTE - [chem] antidote *m* de l'arsenic
~BLOOM - [min] arsénolite *f*
~DISULPHITE - [chem] bisulfite *m* d'arsenic
~GLASS - [chem] oxyde *m* d'arsenic, verre *m* d'arsenic
~HALIDE - [chem] halogénure *m* d'arsenic
~HYDRIDE - [chem] hydrure *m* d'arsenic
~ORANGE - [chem] bisulfite *m* d'arsenic
~ORE - [min] minerai *m* d'arsenic
~OXIDE - [chem] oxyde *m* d'arsenic
~PENTOXIDE - [chem] pentoxyde *m* d'arsenic
~TRIOXIDE - [chem] anhydride *m* arsénieux
~TRISULPHIDE - [chem] trisulfure *m* d'arsenic
ARSENICAL - [chem] arsenical
~ACID - [chem] acide *m* arsénique
~COBALT - [min] cobalt *m* arsenical
~COPPER - [metall] cuivre *m* arsenical
~IRON ORE - [min] minerai *m* de fer arsénieux
~NICKEL - [metall] nickel *m* arsenical
~PAPER - [pap man] papier *m* arsénié
~PYRITES - [min] pyrite *f* arsenicale
ARSENIDE - [chem] arséniure *m*
ARSENIFEROUS - [min] arsénifère
ARSENIOUS - [chem] arsénieux
~ACID - [chem] acide *m* arsénieux
ARSENITE - [chem] (salt of arsenious acid) arsénite *f*, arsénolithe *f*
ARSENOBENZENE - [chem] arsénobenzène *m*
ARSENOLITE - [chem] arsénolite *m*, arsénite *m*
ARSENOPYRITE - [min] (natural sulpharsenide of iron; an important ore of arsenic) arsénopyrite *f*, pyrite *f* arsenicale
ARSINE - [chem] (a poisonous colourless gas used as a synthetic intermediate) arsine *f*, hydrogène *m* arsénié
ARSINIC - [chem] arsinique
ARSON - [gen] incendie *m* volontaire
ART - [gen] art *m*
~DIRECTOR - [cinema] chef-décorateur *m*
~PAPER - [pap man] papier *m* couché
~STILL - [cinema] photo *f* d'artiste
ARTERIAL DRAINAGE - [construct] (one main channel collecting the flow from several branch drains) collecteur *m*
~HIGHWAY - [gen] autoroute *m*, voie *f* à grande circulation
~PIPE - [hydr] conduite *f* secondaire
~ROAD - s.arterial highway
ARTERIOGRAM - [radio] artériogramme *m*
ARTERIOGRAPHY - [radio] artériographie *f*
ARTERIOLOGY - [med] artériologie *f*
ARTERIOPATHY - [med] artériopathie *f*
ARTERIOSCLEROSIS - [med] (hardening or stiffening of the arteries) artériosclérose *f*
ARTERY - [anat] artère *f*
ARTESIAN AQUIFER - [hydr] puits *m* artésien, nappe *f* artésienne
~BORED WELL - s.artesian well
~FLOW - [hydr] débit *m* artésien
~HEAD - [hydr] charge *f* artésienne, pression *f* artésienne
~WELL - [geol] puits *m* artésien
ARTHRITIQUE - [med] arthritique
ARTHRITISM - [med] arthrite *f*
ARTHROGRAPHY - [radio] arthrographie *f*
ARTIC ESSENCE - [chem] chlorure *m* de méthyle
ARTICHOKE - [bot] artichaut *m*
ARTICLE - [gen & comm] article *m*, clause *f*
~-CARD - [comput] carte-article *f*
ARTICLES - [leg] stipulations *f*pl d'un contrat
ARTICULATED - [mech] articulé
~ARM - [mech] bras *m* articulé
~BLADE - [aer & mech] pale *f* articulée
~CAR - [railw] wagon *m* articulé, voiture *f* articulée
~CHUTE - [constr] (or elephant trunk) trompe *f* d'éléphant
~CONNECTING ROD - [mech] bielle *f* articulée, bielle *f* secondaire
~COUPLING - [mech] joint *m* articulé
~CROSS SHAFT - [mech] arbre *m* transversal articulé
~JACK - [mech] cric *m* articulé
~JOINT - [mech] joint *m* articulé
~LANDING GEAR - [aer] (type of landing gear in which the wheel is carried by a lever pivoted to the main leg) train *m* d'atterrissage articulé
~LEVER - [mech] levier *m* articulé
~LOCOMOTIVE - [railw] locomotive *f* articulée
~MAT - [auto] tapis *m* en segments de caoutchouc
~NIPPLE - [mech] raccord *m* articulé
~SIX WHEELER - [mech] véhicule *m* articulé à six roues
~TRAIN - [railw] train *m* articulé
~VEHICLE - véhicule *m* articulé
ARTICULATING LINK - (U.S. universal joint) joint *m* articulé, joint *m* universel
ARTICULATION - [gen & mech] articulation *f*
~REDUCTION - [acoust] perte *f* de netteté

ARTIFICIER - [milit] artificier *m*
ARTIFICIAL - [gen] artificiel
~AERIAL - [rad] (a system of components so arranged as to have the electrical characteristics of those of an aerial but incapable of radiating energy) antenne *f* fictive
~AGEING - [phys] (production of effects similar to those caused by the passage of time, by artificial means) vieillissement *m* artificiel
~ANTENNA - s. artificial aerial
~ATMOSPHERE FURNACE - [metall] (thermal treatment is effected in an atmosphere other than air) four *m* à atmosphère artificielle
~BLACK SIGNAL - [rad] (corresponding in amplitude or frequency to the black signal) signal *m* de noir normal
~DIELECTRIC - (electro magn, wavea) diélectrique *m* artificielle
~DISINTEGRATION - [phys] désintégration *f* artificielle
~EAR - [acoust] oreille *f* artificielle
~EARTH - [el] prise *f* de terre artificielle, mise *f* à la masse artificielle
~ECHO - [radar] (echo obtained from an artificial target) écho *m* artificiel
~FEEL SYSTEM - [aero] (arrangement for imparting to control a degree of resistance to movement by the pilot) dispositif *m* de sensation masculaire
~FERTILIZERS - [ind chem] engrais *m* pl chimiques
~FLAGS - [construct] dalles *f* pl en simili-pierre
~HARBOUR - [construct] (by means of breakwaters) port *m* artificiel
~HORIZON - [radar] (an instrument designed to show the altitude of an aircraft in respect to the horizon) horizon *m* artificiel
~HORSEHAIR - [text] crin *m* artificiel
~INSEMINATION - [zool] insémination *f* artificielle
~KNOP - [text] bouton *m* artificiel
~KNOP YARN - [text] fil *m* boutonneux imprimé
~LEATHER - [leather ind] simili-cuir *m*
~LIGHT - [photo] lumière *f* artificielle
~LIGHT EXPOSURE - [photo] prise *f* de vue à la lumière artificielle
~LIMB - [mech] appareil *m* de prothèse
~MAGNET - [el] aimant *m* artificiel
~MARBLE - [build] simili-marbre *m*
~MOUTH - [electro acoust] (loudspeaker mounted in a baffle) bouche *f* artificielle
~PARCHMENT - [paper man] parchemin *m* végétal, papier *m* suffurisé
~POLLINATION - [bot & zool] pollination *f* artificielle, fécondation *f* artificielle
~RADIOACTIVITY - [phys] (of artificially produced radioelements) radioactivité *f* artificielle
~RESIN MOULDING PRESS - [plast] moule *m* à presser la matière synthétique
~RESPIRATION - [med] respiration *f* artificielle
~SILK - [text] soie *f* artificielle
~SILK YARNS - [text] filés *m* pl de soie artificielle
~SKIN - [leather ind] simili-cuir *m*
~STONE - [build] simili-pierre *f*, pierre *f* artificielle, pierre *f* reconstituée, aggloméré *m*, parpaing *m*
~TARGET - [radar] (device for reflecting radar signals) cible *f* artificielle
~TRANSMISSION LINE - [radar] (a circuit designed to simulate the characteristics of a transmission line) ligne *f* artificielle, ligne *f* fictive
ARTIFICIAL VOICE - [electro- acoust] (complex sound with a spectrum corresponding to that of the human voice) voix *f* artificielle
~WHITE SIGNAL - [radio] signal *m* de blanc normal
~WOOL - [text] laine *f* artificielle
ARTILLERY - [gen] artillerie *f*
~TRACTOR - [mech] tracteur *m* pour l'artillerie
~WHEEL - [mech] roue *f* type artillerie
ARTISAN - [gen] artisan *m*
ARYL - [chem] (indicating aromatic monovalent hydrocarbon radicals) aryle *m*
ARYLAMINE - [chem] (amino derivatives of the aromatic series) arylamine *f*
AS-CAST - [metall] brut de fonderie
~-ROLLED - [metall] brut *m* de laminage
ASA FETIDA OIL - [ind chem] essence *f* d'ase fétide
ASARUM OIL - [ind chem] essence *f* d'asaret
ASBESTOS - [min] (fibrous mineral composed of complex metallic silicates) amiante *m*, asbeste *m*
~BOARD - [gen] carton *m* d'amiante
~BRAKE BAND - [mech] collier *m* de frein en amiante
~BRAKE LINING - [mech] garniture *f* de frein en amiante
~CEMENT - [construct] (weather proof and fire resisting building material) fibrociment *m*, amiante-ciment *m*
~CEMENT PIPE - [construct] tube *m* en fibrociment
~CEMENT ROOFING - [construct] matériau *m* de couverture en fibrociment
~CLOTH - [text] (fire-resisting fabric) toile *f* d'amiante, tissu d'amiante
~CORD - [mech] cordon *m* d'amiante
~CUT - [text] titre *m* de l'amiante
~FABRIC - [constr] tissu *m* d'asbeste
~FELT - [text] feutre d'amiante
~JOINTING - [mech] joint *m* à l'amiante
~LUMBER - [build] fibrociment *m*
~MILLBOARD - [paper man] carton *m* d'amiante
~PACKING - [mech] garniture *f* d'amiante, joint *m* en amiante
~PAPER - [paper man] papier *m* d'amiante
~SHEET - [gen] feuille *f* d'amiante
~WASHER - [mech] joint *m* en amiante, rondelle *f* en amiante
~WEAVING - [text] tissage *m* de l'amiante
~WOOL - [metall] laine *f* d'amiante
~YARNS - [text] filés *m* pl d'amiante
~YARN WITH COTTON CORE - [text] fil *m* d'amiante à mèche (ou âme) de coton
ASBESTOSIS - [med] (lung disease caused by inhalation of asbestos particles) silicose *f* (provoquée par l'amiante)
ASCARID - [zool] ascaride *m*
ASCEND, to - [gen] monter, s'élever
ASCENDING COLON - [anat] colon *m* ascendant
~FLUE - [metall] rampant *m* ascendant
~HORIZONTAL SLICING - [mining] exploitation *f* par tranches horizontales ascendantes
~WARP - [text] chaîne *f* montante
~WATER - [hydrol] eau *f* ascendante, eau *f* hypogène
ASCENSION - [gen] ascension *f*
~HANDLE - [auto] rampe *f* de montée
ASCENSIONAL FORCE - [phys] force *f* ascensionnelle
ASCENT - [gen] ascension *f*, montée *f*
[mining] montée *f*, remonte *f*

ASCENT PATH - [astronaut] trajectoire *f* de montée
A-SCOPE - [instr] (an oscilloscope tube used in an A-display) tube *m* pour indicateur de type A
ASCORBIC ACID - [chem] (vitamin C, a very important food factor) acide *m* ascorbique
ASDIC - [instr] (for the detection of submarines) (GB) Asdic *m*, (US) Sonar *m*
ASEISMATIC - [construct] résistant aux tremblements de terre
~BUILDING - [build] bâtiment *m* résistant aux tremblements de terre
ASEISMIC - [geol] asismique
ASEPSIS - [med] (the exclusion of putrefactive organisms) asepsie *f*
ASEXUAL - [biol] asexué, asexuel
~REPRODUCTION - [biol] (any reproduction which does not depend on normal or modified sexual processes) reproduction *f* asexuée
ASH - [gen] cendre *f*
[bot] frêne *m*
~BOAT - [naut] bette *f* à escarbilles
~-CAN - [gen] poubelle *f*
[cinema-colloq] projecteur *m* de 1000 watts
~CELLAR - [ovens] cave *f* aux cendres, soute *f* aux cendres
~COAL - [min] charbon *m* riche en cendres
~CONTENT - [chem] (the proportion of incombustible residue in organic substances, especially fuels) teneur *f* en cendres
~DAMPING VALVE - [mech] vanne *f* pour le mouillage des cendres
~DUMP - [construct] décharge *f* pour cendres
~EJECTOR - (of boilers) éjecteur *m* d'escarbilles
~HOPPER - [ovens] trémie *f* à cendres
~LOCK CHAMBER - [gas ind] sas *m* à mâchefer
~PAN - [gen] cendrier *m*
~PAN WATER CHANNEL - (in an oven) bac *m* à eau de cendres, cuve *f* à mâchefer
~PATH PLOUGH - (in an oven) soc *m* déscendeur
~PERCENTAGE - [chem] teneur *f* en cendres
~PICKING STICK - [text] fouet *m* de chasse
~PIT - [gen] cendrier *m*
~POCKET - [gas ind] poche *f* à mâchefer
~PRECIPITATOR - [ind chem] précipitateur *m* de cendres
~SHOOT - [naut] manche *f* à escarbilles
~TRAY - [gen impl] cendrier *m*
~TREE - [bot] frêne *m*
~WASHINGS - [ind chem] eau *f* de lavage des cendres
A-SHELTER - [gen] abri *m* anti-atomique
ASHING - [plastics man etc] (the process of cleaning up moulding with a buffing wheel and pumice) cendrage *m* (d'une moule)
ASHLAR - [construct] (accurate squaring of the stones) pierre *f* de taille, moellon *m* d'appareil
~ARCH - [construct] arc *m* en pierre de taille
~CONSTRUCTION - [construct] construction *f* en pierre de taille
~FACING - [constr] parement *m* en pierre de taille
ASHORE - [gen] (towards or on the shore) à terre, sur la côte
[naut] (aground or stranded) échoué
A.S.I. - [instr] (Air Speed Indicator) anémomètre *m*
ASIDERITE - [geol] asidérite *f*
ASKEW - [gen] (placed obliquely) de biais, de côte
ASMI - [radar] s. Aerodrome Surface Movement Indicator
ASP - [zool] vipère *f*, aspic *m*
ASPARAGINE - [chem] (monoamide of aminosuccinic acid, found in asparagus etc) asparagine *f*
ASPARAGUS - [bot] asperge *f*
ASPARTIC ACID - [chem] acide *m* aspartique
ASPECT - [gen] aspect *m*, apparence *f*
~RATIO - [aero] (the ratio of the square of the span of an aerofoil to the gross area) allongement *m* (d'une aile)
[telev] rapport *m* hauteur-largeur (image)
ASPEN - [bot] tremble *m*, peuplier *m* tremble
ASPERGILLIN - [chem] aspergilline *f*
ASPHALT, to - [gen] asphalter
ASPHALT - [oil] (bituminous substance) asphalte *m*
~BASE - [min] pétrole *m* bitumineux
~BLOCKS - [gen] pans *mpl* d'asphalte
~CAKES - [construct] pans *mpl* d'asphalte
~CEMENT - [construct] ciment *m* asphaltique
~CONCRETE - [construct] béton *m* asphaltique, béton *m* à base de bitume
~COVERING - [construct] revêtement *m* d'asphalte, revêtement *m* bitumineux
~DEPOSITS - [min] gisements *mpl* asphaltiques
~EMULSION - [ind chem] émulsion *f* bitumineuse
~GROUT - [construct] injection *f* de bitume
~GROUTED MACADAM - [road construct] macadam *m* bitumé, macadam *m* à pénétration de bitume
~MASTIC - [constr] mastic *m* d'asphalte
~OIL - [min] pétrole *m* asphaltique
~PAINT - [ind chem] peinture *f* à base de bitume
~PLANT - [gen] fabrique *f* d'asphalte
~POWDER - [construct] asphalte *m* en poudre
~ROCK - [geol] roche *f* asphaltique
~SEEPAGE - [geol] suintement *m* bitumineux
~STOCK - [min] produit *m* asphaltique de base, matière *f* première asphaltique
~TILE - [constr] carreau *m* d'asphalte, plaque *f* d'asphalte *m*, dalle *f* d'asphalte *m*
ASPHALTED - [gen] asphalté
~FELT - [text] feutre *f* bituminé
ASPHALTENES - [chem] asphaltènes *fpl*
ASPHALTIC - [gen] asphaltique
~BITUMEN -[construct] bitume *m* asphaltique
~CARDBOARD - [build] carton *m* bituminé
~CEMENT - [ind chem] ciment *m* asphaltique
~LIMESTONE -[geol] calcaire *m* asphaltique
~MASTIC - [ind chem] (a stopping material) mastic *m* d'asphalte
~MORTAR - [construct] mortier *m* asphaltique
~SAND - [geol] sable *m* asphaltique
ASPHALTITE - [min] (mineral bitumen) asphaltite *f*
ASPHERIC SURFACE - [opt] (surface of a lens or mirror which has been slightly modified from the true spherical form to correct aberrations) surface *f* asphérique
ASPHERICAL - [gen] asphérique
ASPHERICITY - [phys] asphéricité *f*
ASPHIXIA - [med] asphyxie *f*
ASPIRATION - [med] (drawing in a breath) aspiration *f*
ASPIRATOR - [ind chem] (device for sucking air from an apparatus) aspirateur *m*
ASPIRIN - [chem] (acetylsalicylic acid) aspirine *f*
ASS - [zool] âne *m*

ASSARTING - [agric] (land clearing) défrichement *m*, essouchage *m*
ASSAY, to - [gen & chem] analyser, essayer, titrer
ASSAY - [gen] essai *m*, (of precious metals) essai *m*
[chem] analyse *m*, vérification *f*
~BALANCE - [impl] balance *f* d'essai
~BUTTON - [metall] bouton *m* d'essai
~CRUCIBLE - [metall] creuse *m* d'essai
~FURNACE - [metall] four *m* à coupellation
ASSEMBLAGE - [math] famille *f* , ensemble *m*
~OF CURVES - [math] famille *f* de courbes
ASSEMBLE, to - [mech] monter
[comput] assembler
ASSEMBLER - [mech] monteur *m*, ajusteur *m*
ASSEMBLING - [mech] assemblage *m*, montage *m*
~BAY - [mech] atelier *m* de montage
~JIG - [gen] gabarit *m* d'assemblage
~LINE - [mech] chaîne *f* de montage
~NUT - [mech] écrou *m* d'assemblage
~PROCEDURE - [gen] montage *m*, procédé *m* de montage
~SHOP - [mech] atelier *m* de montage
ASSEMBLY - [mech] montage *m*, groupe *m*, rassemblement *m*, ensemble *m*
[nucl] groupement *m*
~BAY - s. assembling bay
~BOLT - [mech] boulon *m* d'assemblage
~CORE - [nucl] milieu *m* multiplicateur
~DRAWING - [gen] plan *m* de montage
~LINE - s. assembling line
~STAND - [impl] banc *m* de montage
ASSESS, to - [gen & comm] estimer, établir, évaluer
(of expenses) estimer
ASSESSMENT - [gen & comm] estimation *f*, évaluation *f*, répartition *f*
ASSET - [comm & fin] valeur *f* active, capital *m*
ASSETS - [fin & comm] capital *m*, disponibilités *f*pl, actif *m*
ASSIGN, to - assigner, donner, céder
ASSIGNMENT - [gen] affectation *f*, attribution *f*, allocation *f*
[leg] cession *f*, concession *f*
ASSIMILATE, to - [gen] assimiler
ASSIMILATION - [gen] assimilation *f*
ASSISTANT - [gen] adjoint, assistant, auxiliaire
~-DIRECTOR - [cin & telev] assistant *m* de réalisation
ASSISTED RUNNING - [railw] marche *f* en double traction
ASSISTER - [aero] s. automatic pilot
ASSOCIATED AIR MASS - [met] (an air mass moving at the same velocity as the canopy in relation to the airstream) masse *f* d'air associée
~CORPUSCULAR EMISSION - [radiation] émission *f* corpusculaire associée
~WAVE - [nucl] onde *f* associée
ASSOCIATION - [gen] association *f*
[comm] association *f*, société *f*
[soil] (soils associated geographically on one relatively uniform parent material) association *f* (de sols)
ASSOCIATIVE LAW - [math] propriété *f* d'association
ASSONANCE - [acoust] assonance *f*
ASSORTMENT - [gen & comm] assortiment *m*, classement *m* (par sortes)

ASTABLE MULTIVIBRATOR - [radio] multivibrateur *m* astable
ASTASIA - [med] (inability to stand or walk) astasie *f*
ASTATIC - [phys] astatique
~APPARATUS - [instr] (a moving-magnet apparatus in which the magnets form an astatic system) appareil *m* astatique
~ARM - [phys] bras *m* astatique
~COIL - [el] enroulement *m* astatique
~GALVANOMETER - [instr] galvanomètre *m* astatique
~METER - [el] instrument *m* de mesure astatique
~MICROPHONE - [acoust] microphone *m* omni-directionnel, microphone *m* anti-directionnel
~PAIR - [el] équipage *m* astatique
~POINT - [el] point *m* astatique
~SYSTEM - [instr] (arrangement of two or more magnetic needles on only one suspension, so that there is no resultant torque on the suspension in a uniform magnetic field) système *m* astatique
ASTATICISM - [phys] astatisme *m*
ASTATINE - [chem] (a radioactive element, symbol At) astate *m*
ASTATIZING - [instr] rendre astatique
ASTERISK - [print] astérisque
ASTERISM - [astr] astérisme *m*, constellation *f*
[phys] (one of the characteristic effects which can be observed in X-ray spectrograms) astérisme *m*
ASTERN - [naut] vers l'arrière, sur l'arrière
~GUIDES - [naut] filoirs *m*pl de l'arrière
~MOVEMENT - [naut] erre *f* en arrière
ASTEROID - [astr] astéroïde *m*
ASTHENIA -.[med] (loss of muscular strenght) asthénie *f*
ASTHENOPIA - [med] (weakness of eye muscles) asthénopie *f*
ASTHMA - [med] (difficulty of breathing due to spasm of the bronchial muscles) asthme *f*
ASTIGMATIC - [opt] astigmate
ASTIGMATISM - [opt] (defect in an optical system) astigmatisme *m*
[radar] (failure of focussing electrons in different axial planes at the same point) astigmatisme *m*
ASTON DARK SPACE - [electron] espace *m* sombre d'Aston
ASTRAGAL - [carp] astragale *m*
ASTRAKAN - [text] (dressed skins of very young lambs) astrakhan *m*
[text] (pile fabric imitating natural) astrakhan *m* (imitation)
ASTRAL - [astr] (relating to the stars) astral
~OIL - [chem] kérosène *m*
ASTRIDE - [gen] à cheval, à califourchon
ASTRINGENT - [gen & mech] astringent *m*
ASTRIONICS - [astr] (science of measurement and control in astronautics) mesure *f* et régulation *f* en astronautique
ASTRO-COMPASS - [instr] astro-compas *m*
ASTRODOME - [aero] coupole *f* vitrée (navigateur)
ASTROGRAPHICS - [astr] (the science of mapping the heavens) astrographie *f*
ASTRO-HATCH - [aero] (door in an astrodome) écoutille *f* pour observation astrale
ASTROLABE - [instr ancient] astrolabe *m*
ASTRONAUT - [astronaut] (pilot or member of the crew of a space-vehicle) astronaute *m*
ASTRONAUTICS - [astronaut] (the science of travel

in interplanetary or interstellar space) astronautique *f*
ASTRO-NAVIGATION - s. astronomical navigation
ASTRONOMER - [gen] astronome *m*
ASTRONOMIC LATITUDE - [astr] (latitude reckoned N. or S. of the celestial equator i.e. the ecliptic) latitude *f* géographique
ASTRONOMICAL - [gen] astronomique
~ALTITUDE - [aero] hauteur *f* astronomique
~DATE - [astr] date *f* astronomique
~DIAL - [astr] quadrant *m* astronomique
~HORIZON - [astr] horizon *m* astronomique, horizon *m* géocentrique
~NAVIGATION - [nav] (the science of navigating by observation of celestial bodies) navigation *f* astronomique*f*
~OBSERVATION - [astr] observation *f* astronomique
~REFRACTION - [astr] réfraction *f* astronomique
~STATION - [astr] station *f* astronomique, observatoire *m*
~TELESCOPE - [astr] télescope *m* astronomique
~TRIANGLE - [astr] triangle *m* boréal, triangle *m* austral
~TWILIGHT - [astr] crépuscule *m* astronomique
~UNIT - [nav] (the mean distance of the earth from the sun, used as a unit in measurements within the solar system) unité *f* astronomique
~VERTICAL - [astr] verticale *f* astronomique
~ZENITH - [astr] (the point at which a plumb line extended upwards from the earth's surface would intersect the celestial sphere) zénith *m* astronomique
ASTRONOMY - [astr] astronomie *f*
ASTROPHOTOGRAPHY - [astr] photographie *f* des astres
ASTROPHYSICS - [astr] astrophysique *f*
ASTYLAR - [constr] astyle
ASYLUM - [gen] asile *m*, hospice *m*
ASYMMETER - [instr] asymètre *m*
ASYMMETRIC - [gen] asymétrique
~ANTICLINE - [geol] anticlinal *m* asymétrique
~DEFLECTION - [electron] déviation *f* asymétrique
~LENS - [photo] objectif *m* asymétrique
~MODULATION - [radio] modulation *f* asymétrique
~SIDEBAND - [radio] bande *f* latérale résiduelle
~SYNTHESIS - [chem] synthèse *f* asymétrique
ASYMMETRICAL CELL - [el] élément *m* à conductibilité unilatérale
~CONDUCTIVITY - [electron] conductivité *f* unidirectionelle
~DISTORTION - [radio] distortion *f* asymétrique
~FLUTTER - [aero] battement *m* asymétrique
~FOLD - [geol] plis *m* déjeté
~HETEROSTATIC CIRCUIT - [electron] montage *m* hétérostatique dissymétrique
~INDUCTION MOTOR - [el] moteur *m* à induction dissymétrique
~NETWORK - [telecomm] réseau *m* asymétrique
~TWO-WIRE SYSTEM - [telecomm] ligne *f* asymétrique à deux conducteurs
ASYMMETRY - [gen] (the condition of being irregular in form) asymétrie *f*
~POTENTIAL - [electron] (the potential difference between the outer and inner surfaces of a hollow electrode) potentiel *m* d'asymétrie
ASYMPTOTE - [math] (the limiting case of a tangent in which the point of contact is infinitely distant from the origin) asymptote
ASYMPTOTIC - [math] asymptotique
~CURVE - [math] courbe *f* asymptotique
ASYNCHRONOUS - [gen] asynchrone
[el] (term used of an electrical machine in which rotational speed is not a constant function of the supply frequency) asynchrone
~ALTERNATOR - [el] alternateur *m* asynchrone
~COMPUTER - [comput] (automatic computer in which the performance of an operation begins as a result of a signal confirming the completion of the previous operation) calculateur *m* asynchrone
~CONDENSER - [electron] condensateur *m*
~IMPEDANCE - [electron] impédance *f* asynchrone
~MACHINE - [el] machine *f* asynchrone
~MOTOR - [el] moteur *m* asynchrone
~SPARK-GAP - [el] éclateur *m* asynchrone
~VIBRATOR - [el] vibreur *m* asynchrone
ASYSTOLISM - [physiol] asystolie *f*
A/T (ampere-turn) ampère-tour *m*
AT ANCHOR - [naut] au mouillage
~GRADE - [gen] à niveau
~RANDOM - [gen] au hasard
A.T.C. - (Aerial Tuning Condenser) condensateur *m* d'accord d'antenne
[aero] (Air Traffic Control)service *m*de contrôle de traffic aérien
ATHERMANOUS - [therm] (does not transmit radiant heat) athermane, athermique
ATHERMIC - [therm] athermique
ATHODYD - (abbrev for Aero Thermo Dynamic Duct) (s. Ramjet) tuyère *f* de réacteur
ATHWART - [gen] en travers, par le travers, transversal
[naut] par le travers (du navire)
A.T.I. - (abbrev for Aerial Tuning Inductor) inducteur *m* d'accord d'antenne
ATLAS - [gen] atlas *m*
[paper man] format *m* atlas, format *m* journal
ATMOMETER - [instr] (instrument for measuring evaporation) atmomètre *m*
ATMOSPHERE - [gen & astr] (the whole of the body of gases surrounding a celestial body) atmosphère *f*
[meas] atmosphère *f*
ATMOSPHERIC - [gen] atmosphérique
~ABSORPTION - [gen] absorption atmosphérique
~ACOUSTICS - [acoust] (the study of sound propagation in the atmosphere) acoustique *f* atmosphérique
~AGENTS - [met] agents *m* pl atmosphériques
~BRAKING - [astronaut] freinage *m* atmosphérique
~BURNER - [gas ind] brûleur *m*
~CRACKING - [gen] gerçures *f*pl atmosphériques, craquelage *m* atmosphérique
~DISCHARGE - [el] (electric discharge in the atmosphere) décharge *f* atmosphérique
~DISPERSION - [light] dispersion *f* atmosphérique
~DISTURBANCES - [radio] perturbations *f*pl atmosphériques
~DUCT - [radio] (a layer in the atmosphere capable of conducting radio-frequency waves) couche *f* ionisée (dans la troposphère), conduit *m* troposphérique
~ELECTRICITY - [el] (electricity developed in the atmosphere from meteorological causes) électricité *f* atmosphérique

ATMOSPHERIC ENTRY - [astronaut] entrée *f* dans l'atmosphère
~EQUILIBRIUM - [met] équilibre *m* atmosphérique
~EXCESS PRESSURE - [phys] surpression *f* atmosphérique
~INSTABILITY - [met] instabilité *f* atmosphérique
~INTERFERENCE - [radio etc] parasites *m* pl atmosphériques
~INVERSION - [met] (condition in which the temperature of the atmosphere increases with height) inversion *f* de température
~MOISTURE - [met] état *m* hygrométrique de l'air
~NOISE - [radio] parasite *m* atmosphérique
~POLLUTION - [air pollution] pollution *f* de l'air
~PRECIPITATION - [met] précipitation *f* atmosphérique
~PRESSURE - [met] pression *f* atmosphérique
~-PRESSURE CAPSULE - [instr] (an airtight capsule containing gas at atmospheric pressure) capsule *f* à pression atmosphérique
~RADIATION - [radiat] rayonnement *m* atmosphérique
~RADIATION ATTENUATION - [nucl] atténuation *f* atmosphérique du rayonnement
~RADIOWAVE - [electro magn waves] onde *f* réfléchie
~REFRACTION - [radar] (bending of radar waves in the atmosphere) réfraction *f* atmosphérique
~ROCK - [geol] roche *f* éolienne
~SHELL - [astronaut] couche *f* atmosphérique
~SOUND REFRACTION - [radar] (bending of sound waves in the atmosphere) réfraction *f* atmosphérique du son
~STABILITY - [met] stabilité *f* atmosphérique
~STEAM ENGINE - [mech mach] machine *f* (à vapeur) atmosphérique
~TOPPING - [oil ind] distillation *f* atmosphérique
~TRANSMITTANCE - [nucl] transmittance *f* atmosphérique
~TUNNEL - [aero] (type of wind-tunnel in which the return circuit is at atmospheric pressure) soufflerie *f* à retour atmosphérique
ATMOSPHERICS - [radio] (effects in radio reception caused by electrical disturbances in the atmosphere) perturbations *f* pl atmosphériques, parasites *m* pl
ATOLL - [geogr] atoll *m*
ATOM - [phys] (the smallest part of an element, in the neutral electric state, which can enter into chemical combination) atome *m*
~CHAIN - [phys] chaîne *m* atomique
~PHYSICS - [nucl] physique *f* atomique
~SMASHER - [nucl] accélérateur *m* de particules
ATOMIC ABSORPTION - [nucl] absorption *f* atomique
~ABSORPTION COEFFICIENT - [nucl] coefficient *m* d'absorption atomique
~ACCELERATOR - [nucl] accélérateur *m* de particules
~ARC WELDING - [metall] soudage *m* à l'hydrogène atomique
~ARRANGEMENT - [nucl] (the arrangement of the atoms in a molecule) disposition *f* atomique
~BATTERY - [nucl] (a battery which converts radioactive radiation into electric energy for the alimentation of transistors) batterie *f* atomique (émission B)
~BEAM - [nucl] rayonnement *m* atomique
~BLAST - [nucl] explosion *f* atomique
ATOMIC BOMB - [nucl] bombe *f* atomique
~BOMBARDEMENT - [nucl] bombardement *m* atomique
~BOND - [nucl] (valence linkage between atoms which consist of a pair of electrons) liaison *f* atomique
~CENTRE - [nucl] (alternative term for atomic core) noyau *m* atomique
~CHARGE - [nucl] (the electrical charge of an ion equal to the number of electrons which it has gained or lost on ionization, multiplied by the charge on one electron) charge *f* atomique
~CLOCK - [instr] horloge *f* atomique
~CONTROL - [nucl] réglage *m* nucléaire
~CORE - [nucl] (an atom which has lost its valence electrons and has only closed electron shells round the nucleus) noyau *m* d'atome
~DISINTEGRATION - [phys] désintégration *f* atomique
~ENERGY - [nucl] (colloq. nuclear energy released in a quantity which can be of interest for practical purposes) énergie *f* atomique
~FIELD - [nucl] (the region surrounding the atom) champ *m* nucléaire
~FISSION - [phys] fission *f* nucléaire
~FORM FACTOR - [nucl] facteur *m* de forme
~FREQUENCY - [nucl] fréquence *f* atomique
~HEAT - [nucl] (the amount of heat required to increase the temperature of one gram-atom of an element by one centigrade) chaleur *f* atomique
~HYDROGEN - [phys] hydrogène *m* atomique
~MASS - [nucl] (the mass of a neutral atom of a nuclide) masse *f* atomique
~MASS UNIT - [nucl] unité *f* de masse atomique (UMA)
~MIGRATION - [nucl] (the transfer of the valence bond of an atom from one to another atom in a molecule) migration *f* atomique
~NUCLEUS - [nucl] (the positively charged core) noyau *m* atomique
~NUMBER - [nucl] (the number of an element arranged with others in order of progressive atomic weight) numéro *m* atomique
~ORBIT - [nucl] (the wave function of the electron in the atom) arbite *m* atomique
~PARTICLE - [nucl] particule *f* atomique
~PHOTOELECTRIC EFFECT - [nucl] (occurring when an electron is ejected photoelectrically from one of the inner orbits of an atom of a gas) photo-ionisation *f*
~PILE - [nucl] pile *f* atomique, réacteur *m* nucléaire
~PLANE - [nucl react] (a plane which passes through the atoms of a crystal space lattice) plan *m* réticulaire
~POLARIZATION - [phys] polarisation *f* atomique
~POWER PLANT - [nucl] centrale *f* d'énergie atomique
~RADIUS - [nucl] rayon *m* de l'atome
~RATIO - [nucl] rapport *m* atomique
[phys] rapport *m* des nombres d'atomes
~REACTOR - [nucl] (apparatus in which nuclear fission may be sustained in a chain reaction) réacteur *m* nucléaire
~REFRACTION - [phys] (contribution of a gram-atom of an element to the molecular refraction of a compound) réfraction *f* atomique
~SCATTERING - [nucl] (the scattering of X-rays of a given frequency into a given direction) diffusion *f* atomique

ATOMIC SPECTRUM - [nucl] (the spectrum of radiation emitted by an excited atom) spectre *m* atomique
~STOPPING POWER - [nucl] (the loss of energy per atom) section *f* efficace d'arrêt
~STRUCTURE - [nucl] structure *f* atomique
~SUBMARINE - [armam] sous-marin *m* atomique (ou nucléaire)
~THEORY - [phys] théorie *f* atomique
~TRANSMUTATION - [nucl] (the changing of an atom into an atom of a different atomic number, that is to say not an atom of a different element) transmutation *f* atomique
~VOLUME - [nucl] (volume occupied by one gram-atom of an element) volume *m* atomique
~WARHEAD - [armam] ogive *f* atomique, ogive *f* nucléaire
~WEAPON - [armam] arme *f* atomique
~WEIGHT - [nucl] (relative weight of an atom of an element, expressed in atomic weight units) poids *m* atomique, masse *f* atomique
~WEIGHT UNIT - [nucl] (1 atomic weight unit: 1.660×10^{-24} g) unité *f* de masse atomique
ATOMICITY - [chem] (the number of atoms in the molecule of a given element) atomicité
ATOMIZATION - [ind chem] pulvérisation *f*, vaporisation *f*
ATOMIZE, to - [gen] atomiser, pulvériser
ATOMIZER - [mech] vaporisateur *m*, pulvérisateur *m*
[med] pulvérisateur *m*
[for disinfection] pulvérisateur *m*, vaporisateur *m*
[of a burner] brûleur *m*, atomiseur *m*
~NEEDLE - [mech] aiguille *f* de pulvérisateur
~PASSAGE - [mech] diffuseur *m*
ATOMIZING - [phys] pulvérisation *f*, vaporisation *f*
ATONAL - [acoust] atonal, atone
~INTERVAL - [mus] intervalle *m* atonal
ATONALITY - [acoust] (lack of tonality) atonalité *f*
ATONY - [med] (loss of muscular tone) atonie *f*
ATOXIC - [chem] atoxique
ATREPSIA - [med] atrepsie *f*
ATRESIA - [med] (narrowing of part of channels in the body) atrésie *f*, occlusion *f*
ATRIUM - [arch] atrium *m*
ATROPHY - [med] (wasting of cells or organs in a body) atrophie *f*
ATROPIC ACID - [chem] acide *m* atropique
ATROPINE - [chem] (medically important, poisonous alkaloid) atropine *f*
AT SIGHT - [comm] à vue
ATTACH, to - [gen] attacher, fixer, lier, relier, unir
ATTACHABLE - [gen] à montage rapide, qui peut être fixé
~CORE OF EXTRUDER HEAD - [mech] mandrin *m* amovible d'extrudeuse
ATTACHING PARTS - [mech] pièces *f* pl de fixation
~THE WARP THREADS - [text] nouage *m* des fils de chaîne
ATTACHMENT - [gen] attache *f*, fixation *f*, action *f* d'attacher
[mech] mode *m* de fixation, accessoire *m*
[leg] saisie *f*, contrainte *f* par corps
~COEFFICIENT - [electron] (the number of negative ions being formed by one electron moving along a path of 1 cm in field direction) coefficient *m* d'adhésion, coefficient *m* d'attachement
~FLANGE - [mech] flasque *m* de fixation, bride *f* de fixation
ATTACHMENT SCREW - [mech] vis *f* de fixation
~UNITS - [mech] accessoires *m* pl, organes *m* pl de fixation
ATTACK, to - [gen & chem] attaquer
ATTACK - [gen] attaque *f*, assaut *m*
[acoust] attaque *f*, intonation *f*
~TIME - [electron] temps *m* de mise au point
ATTAR - [chem] essence *f* de rose
ATTEMPERATION - [brewing] contrôle *m* de la température
ATTEMPERATOR - [gen] désurchauffeur *m*
[brew] réfrigérant *m*
ATTEND, to - [gen] assister à
[med] soigner
[ind] faire attention à, surveiller
ATTENDANCE - [gen] service *m*, assistance *f*, présence *f*
ATTENTION - [gen] attention *f*, soin *m*
ATTENUATE, to - [gen & radio] atténuer
ATTENUATED - [bot] acuminé, ténu
ATTENUATION - [gen & radio] atténuation *f*, affaiblissement
~BOX - [el] atténuateur *m*, affaiblisseur *m*
~COMPENSATION - [el] compensation *f* de l'atténuation
~CONSTANT - [el] constante *f* d'affaiblissement
~CORRECTION - [el] correction *f* de l'affaiblissement
~EQUALIZER - [rad] (network to compensate for attenuation-frequency distortion) réseau *m* équilibreur
~EQUALIZER-SHUNT TYPE - [el] réseau *m* équilibreur en dérivation
~FACTOR - [el] (the real part of the propagation coefficient) facteur *m* d'affaiblissement
~FREQUENCY DISTORTION - [rad] distortion d'amplitude/frequence, distortion *f* d'affaiblissement (appareil électro-acoustique)
~MEASUREMENT - [radio] mesure *f* d'équivalent (de transmission)
~MEASURING SET - [instr] appareil *m* de mesure d'atténuation
~RANGE - [radio] gamme *f* de valeurs d'atténuation
~RATIO - [radio] (the extent of the propagation ratio) affaiblissement *m*, taux *m* d'affaiblissement
ATTENUATOR - [radio] (arrangement of resistances introducing a specified quantity of attenuation into the circuit in which it is inserted) atténuateur *m*, affaiblisseur *m*
~PAD - [el] unité *f* d'atténuation
ATTIC - [build] mansarde *f* grenier *m*
[arch] attique *m*
ATTITUDE - [aero] (the position of an aircraft in relation to horizontal and vertical fixed axes) position *f* en vol
[gen] attitude *f*
~CONTROL - [astronaut] commande *f* d'attitude
~GYRO - [astronaut] gyroscope *m* d'attitude
~INDICATOR - [aero] indicateur *m* de position en vol
~INSTRUMENTS - [instr] (instruments for the indication of attitude) indicateurs *m* pl de position en vol
~JET - [astronaut] jet *m* de gouverne d'attitude
~OF FLIGHT - [aero] position *f* en vol
~RELATIVE TO GROUND - [aero] position *f* par rap-

port au sol
ATTORNEY - [leg] mandataire *m*, fondé *m* de pouvoir [at law] avoué *m*
ATTRACTED DISC ELECTROMETER - [instr] (form of absolute electrometer with circular metal discs held parallel to a larger metal disc by a spring) électromètre *m* absolu de Kelvin
ATTRACTION - [gen & phys] attraction *f*
~PARTICLE (of cells) (central granule within the centrosome) centriole *f*
~SPINDLE (of cells) (the terminal section of the achromatic spindle) fuseau *m* achromatique
ATTRACTIVE FORCE - [phys] (force acting on the particle, so that its acceleration is in the direction of the agency causing the force) force *f* d'attraction
ATTRITION - [gen & phys] attrition *f*, usure *f* par frottement
~TEST - [gen] (a test for the determination of wear resisting properties) essai *m* d'attrition
A-TYPE POLE - [el] poteau *m* double
A.U. - (abbrev. for Astronomical Unit) (the mean distance between the earth and the sun) unité *f* astronomique
Au - [chem] (the symbol for gold) or (Au) *m*
A/U - [el] (Angstrom Unit) Angstrom
AUBERGINE - [bot] aubergine *f*
AUCTION - [gen & comm] enchère *f* adjudication *f*
~ROOM - [gen] salle *f* des ventes
AUCTIONEER - [gen] commissaire-priseur *m*
AUDIBILITY - [acoust] perceptibilité *f* d'un son, audibilité *f*
~RANGE - [acoust] gamme *f* des fréquences audibles
AUDIBLE - [acoust] audible, perceptible à l'oreille
~DOPPLER ENHANCER - [radar] (observation by aural reception) traducteur *m* audible de l'effet Doppler
~INDICATOR - [comput] avertisseur *m* acoustique
~MACHMETER - [instr] machmètre *m* sonore
~RINGING TONE - [acoust] signal *m* de retour d'appel
~SIGNAL - [acoust] signal *m* sonore, signal *m* audible
~SPECTRUM - [acoust] (diagrammatic representation of frequency ranges compared with the limits of the human ears) spectre *m* des fréquences audibles
~TEST - [telecomm] test *m* audible
~WARNING SIGNAL - [railw] avertisseur *m* sonore
AUDIENCE - [gen] audience *f*, public *m*, spectateurs *mpl*
AUDIO - [radio] acoustique, basse fréquence
~-AMPLIFICATION - [radio] (amplification of a signal at frequencies of audible sound) amplification *f* des basses fréquences
~CARRIER - [telev] porteuse *f* son
~CENTRE FREQUENCY - [telev] fréquence *f* nominale de la porteuse son
~-CONTROL PANEL - [radio] (control panel for audio-frequency circuits) panneau *m* de contrôle des circuits basse fréquence
~FEEDBACK CIRCUIT - [electron] retour *m* d'écoute
~-FOLLOW-VIDEO OPERATION - [telev] opération *f* à présélection consécutive d'image et son
~-FREQUENCY - [acoust] (AF) (any frequency within that of audible sound waves (30 c/s to 15 kc/s approx.)) fréquence *f* acoustique, basse fréquence *f*
AUDIO-FREQUENCY AMPLIFIER - [el] (circuit designed to amplify a signal at audio-frequency) amplificateur *m* basse fréquence
~-FREQUENCY BAND - [acoust] gamme *f* des fréquences acoustiques (ou audibles)
~-FREQUENCY BEAT NOTE - [acoust] son *m* fondamental à fréquence acoustique
~-FREQUENCY PEAK LIMITER - [radio] écrèteur *m* basse fréquence
~-FREQUENCY SIGNAL GENERATOR - [radio] (test equipment designed to produce signals of controlled frequencies within the audio range) générateur *m* basse fréquence
~-FREQUENCY SPECTRUM ANALYSIS - [radio] analyse *f* spectrale des basses fréquences
~-FREQUENCY TELEPHONY - [telecomm] téléphonie *f* à fréquences vocales
~-FREQUENCY TRANSFORMER - [el] (ferro-magnetic core transformer, designed for coupling audio-frequency circuits) transformateur *m* basse fréquence
~-GAIN - [radio] (gain obtained in audio-frequency amplification) gain *m* d'amplification basse fréquence
~RECORDING - [el acoust] enregistrement *m* d'impulsions sonores
~SIGNAL - [telev] signal *m* audio (basse fréquence)
~SPECTRUM - [acoust] spectre *m* de fréquences acoustiques
~TAPE - [el acoust] bande *f* magnétique (pour enregistrement sonore)
~TAPE RECORDING - [el acoust] enregistrement *m* sur bande magnétique
~TRANSMITTER - [telev] émetteur *m* son pour télévision
AUDIOGRAM - [acoust] audiogramme *m*
AUDIOLOGY - [acoust] (the science of hearing and sound) audiologie *f*
AUDIOMETER - [instr] audiomètre *m*, sonomètre
AUDIOMETRIC - [acoust] audiométrique
AUDIO MIXER - [telev] (a means in which the signals of two or more microphones are mixed) table *f* de mixage
AUDION - [radio] (the original name for triode amplifying valves) audion *m*
AUDIO-OSCILLATION - [radio] oscillation *f* à fréquence acoustique
~-OSCILLATOR - [radio] générateur *m* basse fréquence
~RANGE - [acoust] gamme *f* des fréquences audibles
~-TRANSFORMER - [el] (audio-frequency transformer) transformateur *m* basse fréquence
AUDIT - [comm] apurement *m* (de comptes), vérification *f*
AUDIT, to - [comm] apurer (des comptes), vérifier
AUDITION - [gen] audition *f*, faculté *f* d'entendre
AUDITION, to - [gen & entert] donner une audition, auditionner
AUDITOR - [comm] commissaire *m* aux comptes, expert-comptable *m*, auditeur *m* (conférence etc)
AUDITORIUM - [arch] salle *f* (de concerts etc)
~NOISE - [cinema] bruit *m* de salle
AUDITORY - [acoust] auditif
~NERVE - [acoust] nerf *m* auditif

AUDITORY PERSPECTIVE - [acoust] perspective *f* sonore
~SENSATION AREA - [acoust] zone *f* d'audibilité
AUGER - [impl] foret *m*, mêche *f*, tarère *f*
~BIT - [mech] mèche *f* hélicoïdale, mèche *f* à bois
AUGER'S BORE - [mech] trou *m* de sondage
AUGITE - [min] (complex aluminous silicate of calcium, iron and magnesium) pyroxène *m*
AUGMENTATION DISTANCE - [nucl] (the distance of the extrapolated boundary from the true boundary of a nuclear reactor) écart *m* d'extrapolation
AUGMENTER CONDENSER - [el] condensateur *m* auxiliaire
~TUBE - [astronaut] tuyau *m* d'augmentation de poussée
AURAL - [physiol] auditif, de l'oreille, sonore
AURAMINE - [chem] (a dyestuff) auramine *f*
AURANTIA - [photo] (reddish dye for sensitizing emulsions) aurantia *f*
AURANTINE - [chem] aurantine *f*
AUREOLE - [light & met] auréole *f*
AUREOMYCIN - [chem] auréomycine *f*
AURIC ACID - [chem] (the acidic oxide of gold) acide *m* aurique
AURICLE - [anat] auricule *m*, pavillon *m* de l'oreille
AURICULAR - [anat] auriculaire
~CANAL - [anat] conduit *m* auriculaire
~FINGER - [anat] auriculaire *m*
AURICULATE - [bot] auriculé
AURICYANIC ACID - [chem] acide *m* cyanoaurique
AURIFEROUS - [geol] aurifère
~DEPOSIT - [geol] dépot *m* aurifère
~PYRITE - [min] pyrite *f* aurifère
~ROCK - [geol] roche *f* aurifère
AURIN - [chem] aurine *f*
AURISCOPE - [instr] otoscope *m*
AURORA - [met] (luminous effect in the sky) aurore *f*
~ASTRALIS - [met] (aurora occurring round the South Polar region) aurore *f* australe
~BOREALIS - [met] (aurora occurring round the North Polar region) aurore *f* boréale
AUSTEMPERING - [metall] trempe *f* étagée, trempe *f* baïnitique interrompue
AUSTENITE - [ind chem] (a solid solution of carbon in iron, occurring in steel manufacture) austénite *f*
AUSTENITIC - [metall] austénitique
~STEEL - [metall] acier *m* austénitique
~STRUCTURE - [gen] (characteristic structure of austenitic steel) structure *f* austénitique
AUSTENITIZE, to - [metall] rendre austénitique
AUSTRAL - [geog] austral
AUSTRALITE - [min] (spheroidal forms of green and black glass probably of cosmic origin) australite *f*
AUSTRIAN CINNABAR - [chem] chromate *m* de plomb basique
AUTHOR'S PROOFS - [print] épreuves *f* pl en seconde, épreuves *f* pl en placards
AUTOAERIAL - [auto] antenne *f* d'automobile
AUTOAGGLUTINATION - [chem] autoagglutination *f*
AUTOALIGNMENT - [radio] autoalignement *m*
AUTOBOAT - [naut] embarcation *f* à moteur
AUTOCAR - [auto] autocar *m*, automobile *f*
AUTO-CATALYSIS - [chem] (catalysis caused by a product of the reaction itself) auto-catalyse *f*
AUTOCATALYTIC - [chem] auto-catalyseur
AUTOCATALYZE, to - [chem] auto-catalyser
AUTOCHROME - [photo] plaque *f* autochrome
AUTOCHTONOUS - [zool] autochtone
AUTOCLAVE - [gen & chem] autoclave
AUTOCLAVING - [chem] passage *m* à l'autoclave
AUTOCOLLIMATING SPECTROGRAPH - [instr] spectrographe *m* à autocollimation
AUTOCOLLIMATION - [opt] autocollimation *f*
AUTOCOLLIMATOR - [instr] autocollimateur *m*
AUTOCOMBUSTION - [phys] autocombustion *f*
AUTOCOMPOUNDED TRANSFORMER - [el] transformateur *m* autocompoundé
AUTOCONVECTIVE LAPSE RATE - [meter] gradient *m* de température par autoconvection
AUTOCOUND - [nucl] comptage *m* automatique
AUTOCOUNTER - [instr] compteur *m* automatique
AUTOCOUPLING - [mech] couplage *m* automatique
AUTOCYCLE - [mech] bicyclette *f* à moteur, vélomoteur *m*
AUTODROME - [gen] autodrome *m*
AUTODYNE - [radio] autodyne *f*
~CIRCUIT - [el] (circuit in which the function of an oscillator and a detector are combined) circuit *m* autodyne
~OSCILLATOR - [el] oscillateur *m* à réaction
AUTO-EMISSION - [radio] (field emission; emission due solely to high-voltage gradients at the emitting surface) émission *f* par champ électrique
~-EXPOSURE - [photo] temps *m* de pose automatique
AUTOFECONDATION - [zool] autofécondation *f*
AUTOFEED - [mech] à avance automatique, à alimentation automatique
AUTOGAMY - [zool] autogamie *f*
AUTOGENEOUS - [gen] autogène
~IGNITION TEMPERATURE - [metall] température *f* d'autoallumage
~SOLDERING - [metall] soudure *f* autogène
~WELDING - [metall] (welding in which only the material of the parts to be welded is used to effect the junction) soudure *f* autogène
AUTOGENESIS - [bot] autogenèse *f*
AUTOGIRO - [aero] (a rotorcraft with freely revolving rotors) autogyre *m*
AUTOHETERODYNE RECEIVER - [radio] (a receiver operating on the beat reception principle and embodying an autodyne oscillator) émetteur *m* à autohétérodyne
AUTO-HOIST - [mech] élévateur *m* à colonne
AUTOIGNITION - [mech] auto-allumage *m*
~-INDUCTION - induction *f* propre
AUTOINDUCTIVE COUPLING - [radio] couplage *m* autoinductif
AUTOIONIZATION - [phys] autoionisation *f*
AUTO-LIFT - [mech] pont *m* élévateur
AUTOLUMINESCENCE - [phys] autoluminescence *f*
AUTOLYSIS - [zool] autolyse *f*
AUTOMANUAL SWITCHBOARD - [telecomm] central *m* mixte
AUTOMATE, to - [gen] automatiser, rendre automatique
AUTOMATIC - [gen] automatique
~ADVANCE - [mech] avance *f* automatique
~ADVANCE REGULATOR - [mech] régulateur *m* d'avance automatique
~AIMING - [radar] pointage *m* automatique
~ARC LAMP - [el] lampe à arc automatique
~ARC WELDING - [metall] soudure *f* à l'arc continue

AUTOMATIC AREA - [telecomm] réseau *m* téléphonique automatique
~ASSEMBLY - [autom] montage *m* automatique
~BATTERY CHARGER - [el] chargeur *m* de batterie automatique
~BIAS - [electron] (automatic grid bias) polarisation *f* automatique de grille
~BLANKET - [el] couverture *f* (électrique) chauffante
~BLOCK - [railw] bloc *m* automatique
~BOOST CONTROL - [aero] contrôleur *m* automatique de pression d'admission
~BRAKE - [mech] frein *m* automatique
~BRAKING DEVICE - [mech] dispositif *m* automatique de freinage
~BREAKER - [el] disjoncteur *m* automatique
~BRIGHTNESS CONTROL - [telev] (circuit for the automatic control of the illumination) régulation *f* automatique de brillance
~BUCKET - [mech impl] benne *f* automatique
~CARRIAGE - [comput] (typewriter carriage automatically controlled by information and programme) chariot *m* à commande automatique
~CASE SHIFT - [radio] inversion *f* automatique
~CAT HEAD - [mining] poulie *f* automatique du câble de cabestan
~CENTRE-PUNCH - [mech] pointeau *m* à centrer automatique
~CHECK - [comput] vérification *f* automatique
~CHEQUE SORTER - [autom] (automatic sorting of up to 1,000 cheques per minute) trieuse *f* de chèques automatique
~CHOKE - [mech] diffuseur *m* automatique
~CIRCUIT BREAKER - [el] disjoncteur *m* automatique
~CLEARING - [telecomm] signal *m* automatique de fin de communication
~CLUTCH - [mech] embrayage *m* automatique
~CODING - [comput] codage *m* automatique
~COIL WINDER - [el] bobineuse *m* automatique
~COMPUTER - [comput] calculateur *m* automatique, ordinateur *m* automatique
~CONTRAST CONTROL - [telev] commande *f* automatique de contraste
~CONTROL - [mech, el etc] commande *f* automatique, régulation *f* automatique
~CONTROL-CABLE TENSIONER - [mech] tendeur *m* automatique de câble de commande
~CONTROL VALVE - [comput] vanne *f* de régulation automatique
~COUPLER - [railw] attelage *m* automatique
~COUPLING - [railw] attelage *m* automatique [mach] accouplement *m* automatique
~CUT-OUT - [el] interrupteur *m* automatique
~CUT-OUT VALVE - [mech] clapet *m* de court-circuit automatique
~CYCLE - [mech etc] cycle *m* automatique
~DECIMAL POINT COMPUTER - [comput] (a computer programming itself with fully automatic decimal point alignment) calculatrice *f* à alignement automatique de la virgule
~DIRECTION FINDER - [radar] (direction-finding system in which the incoming signals control the goniometer motors) radiogoniomètre *m* automatique
~DISCONNECTION - [mech] débrayage *m* automatique
~DISSOLVE - [telev] caméra *f* fade, fondu *m* automatique
~DOOR OPERATOR - [mech] dispositif *m* d'ouverture automatique des portes
AUTOMATIC DRILL - [mech] perceuse *f* automatique
~EJECTION - [plastics] éjection *f* automatique
~ELEVATION MEASUREMENT - [radar] (the automatic measurement of the height of an object by means of a device operated by the echo) mesure *f* automatique de l'altitude
~EXCHANGE - [telecomm] central *m* téléphonique automatique
~FEED - [mech] avance *f* automatique, alimentation *f* automatique
~FEED ADJUSTMENT - [mach tool] poignée *f* pour réglage alimentation automatique
~FEED CONTROL - [mach tool] poignée *f* pour embrayage alimentation automatique
~FEEDER - [print] margeur *m* automatique
~FEEDING DEVICE - [mech comp] dispositif *m* d'alimentation automatique
~FIELD PHASING - [telev] mise *f* en phase automatique des trames
~FIRE ALARM - [instr]avertisseur *m* automatique d'incendie
~FLOOD GATE - [hydr] vanne *f* automatique (de décharge)
~FLUSHING - [hydr] chasse *f* automatique
~FOAM INSTALLATION - [oil mining] système *m* automatique à mousse contre l'incendie
~FOCUSING - [telev] focalisation *f* automatique
~FREQUENCY CONTROL - [el] (A.F.C. an arrangement designed to maintain the frequency of an oscillator at or near the required value) régulation *f* automatique de fréquence
~FULL-COP MOTION - [text] mécanisme *m* d'arrêt automatique de cannette
~GAIN CONTROL - [electron] (A.G.C.) (circuits in a radio receiver which cause the gain to vary substantially inversely as the magnitude of the radio-frequency input, so as to maintain a generally constant output for a given modulation value) antifading *m*, contrôle *m* automatique de gain
~GAIN STABILISATION - (AGS) s.automatic gain control
~GAUGE EQUIPMENT - [meas] appareils *mpl* de mesure automatique
~GRAB - [mining] griffe *f* automatique, cuiller *f*, benne-drague *f*
~GRID BIAS - [electron] polarisation *f* automatique de grille
~HANDLING EQUIPMENT - [comp] installation *f* de manutention automatique
~HUNTING - [telecomm] recherche *f* automatique, sélection *f* automatique
~INSPECTION DEVICES - [autom] appareils *mpl* de contrôle automatiques
~INTERLOCK - [comput] (operation automatically preventing the reading of an item which has already been the subject of an operation) verrouillage *m* automatique
~LADLING - [diecasting] alimentation *f* automatique
~LATHE - [mech] (machine-tool which performs a repeated cycle of turning operations automatically) tour *m* automatique
~LIGHTER - [gas ind] allumeur *m* automatique
~LOADER - [autom] chargeur *m* automatique
~LOCKING - [mech] blocage *m* automatique, verrouillage *m* automatique
~LUBRICATION - [mech] lubrification *f* automa-

tique, graissage *m* automatique
AUTOMATIC MACHINE CONTROLS - [auto] commandes *f*pl automatiques
~MILLING MACHINE - [mech] fraiseuse *f* automatique
~MIXTURE CONTROL - [mech] (a device for automatically adjusting the fuel-air ratio of the mixture) correcteur *m* automatique de mélange
~NUMBERING TRANSMITTER - [radio] (transmitter automatically transmitting serial numbers before each telegram) émetteur *m* à numérotation automatique
~OBSERVER - [aero] appareil *m* enregistreur automatique
~OVERLOAD RESET SWITCH - [el] disjoncteur *m* à maxima à réenclenchement automatique
~PACKAGING - [auto] emballage *m* automatique
~PARACHUTE - [aero] parachute *m* à ouverture automatique
~PEAK LIMITER - [telev] limiteur *m* de crête (blanc) automatique
~PHASE CONTROL - [telev] (a synchronization method) mise *f* en phase automatique
~PILOT - [aero] (automatic control device to keep the aircraft on set course and in level flight) pilote *m* automatique
~PRINTER - [telecomm] téléimprimeur *m*
~PROCESS CONTROL - [contr] commande *f* à programme, cycle *m* automatique programmé
~PROGRAMMING - [comput] programmation *f* automatique
~PROPELLER - [aero] hélice *f* à vitesse constante
~PUNCH - [mech mach] perforatrice *f* automatique, poinçonneuse *f* automatique
~PUNCHING MACHINE - [mech] perforatrice *f* automatique, poinçonneuse *f* automatique
~RADIOMONITOR - [radio] comparateur *m* automatique de signaux
~RECORD PLAYER - [mus] phonographe *m* automatique, tourne-disque *m* automatique
~REELING APPARATUS - [text] bobineuse *f* automatique
~RELAY OPERATED EXCHANGE - [telecomm] central *m* automatique à relais
~RELEASE - [mech] déclenchement *m* automatique [photo] obturateur *m* à action différée
~RELIEF VALVE - [el] soupape *f* de sûreté automatique, clapet *m* de surpression automatique
~RESET - [comput] remise *f* à zéro automatique
~RESHUTTLING LOOM - [text] métier *m* à tisser automatique
~ROUTING TRANSMISSION - [radio] transmission *f* télégraphique par bandes perforées
~SCALES - [impl] balance *f* automatique.
~SCREW CUTTING LATHE - [mech] tour *m* à fileter automatique
~SCREW MACHINE - [mech] tour *m* automatique
~SELECTIVITY CONTROL - [radio] régulation *m* automatique de sélectivité
~SEQUENCE-CONTROLLED COMPUTER - [comput] calculatrice *f* équipée d'un programme de commande, calculatrice *f* à programme enregistré
~SEQUENCE MANUFACTURE - [autom] fabrication *f* en séquence automatisée
~SHARPENING MACHINE FOR TOOTH CUTTERS - [mach tool] affûteuse *f* automatique pour couteaux à denter

AUTOMATIC SHIFT - [mech] changement *m* automatique
~SHUTTER - [camera] obturateur *m* automatique
~SHUTTLE CHANGING - [text] changement *m* automatique de navette
~SKIP - [mech] benne *f* automatique
~SPARK ADVANCE - [auto] avance *f* automatique
~SPOOLER - [text] bobineuse *f* automatique
~STABILIZER - [aero] (an automatic pilot set to provide increased aerodynamic stability) stabilisateur *m* automatique
~STARTER - [auto] démarreur *m* automatique
~STEEL - [metall] acier *m* à décolletage rapide
~STOP - [mech] arrêt *m* de fin de course automatique
~STOP DEVICE - [railw] dispositif *m* d'arrêt automatique
~SUBSTATION - [el] (substation with rotating machinery, started and stopped automatically) sous-station *f* automatique
~SWEEP-FREQUENCY IMPEDANCE METER - [instr] appareil *m* automatique à mesurer l'impédance et la fréquence de balayage
~TAP CHANGING EQUIPMENT - [el] (voltage regulating device) régulateur *m* automatique de tension
~TELEPHONE - [telecomm] téléphone *m* automatique
~TESTING - [autom] (the testing of the quality of goods by automatic means) contrôle *m* automatique
~THREAD MILLING MACHINE - [mech] machine *f* à fileter automatique
~THRUST - [mech] avance *f* automatique
~TIME STAMP - [instr] horodateur *m* automatique
~TIMER - [aero] déclencheur *m* automatique d'ouverture (parachute)
~TIMING - [auto] avance *f* automatique
~TIPPER - [mech] culbuteur *m* automatique
~TITRATING APPARATUS - [instr] appareil *m* de titration automatique
~TOLL EXCHANGE - [telecomm] central *m* interurbain automatique
~TONE COMPENSATION - [radio] correction *f* automatique de tonalité
~TRACKING - [radar] poursuite *f* automatique
~TRAIN STOP - [railw] dispositif *m* d'arrêt automatique
~TRANSMISSION - [auto] transmission *f* automatique
~TRANSMITTER - [radio] émetteur *m* automatique
~TRANSVERSE FEED - [mech] avance *f* transversale automatique
~TRIP - [mech] rupteur *m* automatique
~TRUNK EXCHANGE - [telecomm] central *m* interurbain automatique
~TUNING - [radio] accord *m* automatique
~UNCOUPLING - [mech] décrochage *m* automatique
~VOLTAGE REGULATOR - [el] régulateur *m* automatique de tension
~VOLUME COMPRESSOR - [radio] compresseur *m* de volume automatique
~VOLUME CONTRACTOR - [radio] compresseur *m* de volume automatique
~VOLUME CONTROL - [radio] (A.V.C.) régulation *f* automatique de puissance
~VOLUME EXPANDER - [radio] expanseur *m* automatique (des contrastes)
~VOLUME RANGE REGULATOR - [radio] régulateur *m* automatique de puissance
~WEFT SUPPLY - [text] distribution *f* automatique des cannettes

AUTOMATIC WELDING - [metall] soudage *m* automaque

AUTOMATION - [gen] (the technique of rendering processes self-regulating and self-cheking, especially by electronic control) automatisation *f*

~CONTROL GEAR - [autom] dispositif *m* régulateur pour l'automatisation

~PLAN - [autom] (plan of the details which lead to automation) plan *m* d'automatisation, macroprogramme *m*

AUTOMATISM - [med] (psyco-patology. Automatic act performed without the full cooperation of the person) automatisme *m*

AUTOMOBILE - [auto] automobile *f*, véhicule *m* automobile

~BULB - [el] lampe *f* pour automobile

~ENGINE - [mech] moteur *m* d'automobile

~PLANT - [mech ind] usine *f* automobile

~RADIO - [auto] récepteur *m* radio pour automobile

~VALVE GEAR - [mech] commande *f* de soupapes

AUTOMOTIVE - [auto] automoteur, automobile

AUTONAVIGATOR - [aero] (an automatic navigation computer system) calculateur *m* automatique de navigation

AUTONOMOUS - [gen] (self-controlling) autonome

AUTONOMY - [gen] autonomie *f*

~OF OPERATION - [mech] autonomie *f* de fonctionnement

AUTOOXIDATION - [chem] autooxydation *f*

AUTOPATROL - [mech mach] (an earth grading machine) niveleuse *f*

AUTOPHONY - [med] autophonie *f*

AUTOPILOT - [aero] pilote *m* automatique

~ENGAGEMENT - [aero] enclenchement *m* du pilote automatique

~RELEASE - [aero] désenclenchement *m* du pilote automatique

~RESPONSE TIME - [aero] (the period required for an automatic pilot to respond to information received by it) temps *m* de réponse du pilote automatique

AUTOPLASMA - [zool] (a medium obtained from the same plasma of the animal from which the tissue was originally taken) autoplasma *m*

AUTOPLASTIC TRANSPLANTATION - [zool] autoplastie *f*

AUTOPLASTY - [med] autoplastie *f*

AUTOPOLYMERIZATION - [chem] autopolymérisation *f*

AUTOPSY - [med] autopsie

AUTOPULSE MAGNETIC FUEL PUMP - [mech] pompe *f* d'alimentation à commande électromagnétique

AUTO RADAR PILOT - [radar] (a mechanism to position a radar map on a navigational chart) report *m* d'une image radar (P.P.I.) sur une carte géographique

AUTORADIOGRAPH - [phys & chem] autoradiographe *m*

AUTORADIOGRAPHY - [phys & chem] autoradiographie *f*

AUTO-RECLOSE BREAKER - [el] disjoncteur *m* à réenclenchement automatique

AUTO-REPEATER - [electron] (device designed to obtain direct current which is proportional to the power in an a.c. circuit) générateur *m* de tension de signal

AUTOROTATION - [aero] autorotation *f*

AUTOSCALER - [instr] échelle *f* de comptage automatique

AUTOSCOPY - [med] autoscopie *f*

AUTOSET LEVEL - [surv] (level for rapid operations, with a quick levelling head) niveau *m* à auto-alignement

AUTOSOME - [cells] (typical chromosome) autosome *m*

AUTO-STABILISER - [aero] (device for improving the stability of an aircraft while it is still under the control of the pilot) autostabilisateur *m*

AUTOSTARTER - [mech] démarreur *m* automatique

AUTOSTEREOGRAM - [radiation] auto-stéréogramme *m*

AUTOSYN - [mech] transmetteur *m* autosyn, synchrotransmetteur *m* autosyn

~TAKE-OFF - [radio] synchrotransmetteur *m* autosyn, autosyn *m*

AUTO-TAB - [aero] (automatically-operated tab) tab *m* (de compensation) automatique

AUTOTITRATOR - [instr] instrument *m* de titration automatique

AUTOTOMY - [zool] (voluntary separation of parts of the body) autotomie *f*

AUTO-TRANSFORMER - [el] autotransformateur *m*

AUTOTROPHIC - [bot] autotrophe

~BACTERIA - [bot] (bacteria able to utilise carbondioxide in assimilation) bactérie *f* autotrophe

AUTOTROPISM - [bot] (tendency to grow in a straight line) autotropisme *m*

AUTOTRUCK - [mech] camion *m*

AUTOTYPE - [photo] phototypographie *f*

AUTOVALVE - [el] (lightning diverter) parafoudre *m*

AUTUMN - [astr] automne *m*

~OVERTURN - [hydr] (of lakes etc) circulation *f* automnale

AUTUMNAL EQUINOX - [met] équinoxe *m* d'automne

AUTUNITE - [min] (natural hydrated phosphate of calcium and uranium; a source of uranium) autunite *f*, uranite *f*

A.U.W. - [aero] (All Up Weight) poids *m* total

AUXANOMETER - [instr] (instrument for the measure of the rate of growth of plants) auxanomètre *m*

AUXILIARY - [gen] auxiliaire

~AIR VALVE - [mech] soupape *f* d'air auxiliaire, clapet *m* d'air auxiliaire

~ANODE - [electron] anode *f* auxiliaire

~ATTACHMENT - [mech] accessoire *m*

~BEAM - [text] rouleau *m* auxiliaire, ensouple *m* auxiliaire

~DIESEL GENERATOR - [el] générateur *m* diesel auxiliaire, groupe *m* électrogène de secours

~ELECTRODE - [el] électrode *f* de comparaison

~GENERATING PLANT - [el] groupe *m* électrogène de secours

~GRID - [radio] grille *f* auxiliaire

~HOPPER - (of ovens) trémie *f* intermédiaire

~JACK - [el] jack *m* d'entr'aide

~JACQUARD - [text] jacquard *m* auxiliaire

~JET - [mech] gicleur *m* auxiliaire

~LANDING GEAR - [aero] train *m* d'atterrissage auxiliaire

~LIFTER - [text] lice *f* auxiliaire

~LINES - [telecomm] lignes *f* pl d'extension

~MOTOR - [mech] moteur *m* auxiliaire, servomoteur *m*

~PARACHUTE - [aero] parachute *m* auxiliaire

~POWER PLANT - [el] groupe *m* électrogène auxiliai-

re
AUXILIARY POWER STATION - [el] sous-station *f* de secours
~RELAY - [el] relais *m* auxiliaire
~RIGGING LINES - [aero] (in a parachute, to distribute the load evenly) suspentes *fpl* auxiliaires
~ROTOR - [aero] (of helicopter) rotor *m* auxiliaire
~SCREW BOAT - [naut] voilier *m* à moteur auxiliaire
~SHAFT - [mech] arbre *m* auxiliaire
~SHUTTLE - [text] navette *f* auxiliaire
~SPRING - [mech] ressort *m* compensateur
~SPRING BRACKET - [mech] support *m* de ressort compensateur
~TANK - [gen & auto] réservoir *m* auxiliaire
~WINDING - [el] enroulement *m* auxiliaire
~WINDSHIELD WIPER - [auto] essuie-glace *m* auxiliaire
AUXIOMETER - [instr] (for the measurement of the magnifying power of a lens) auxomètre *m*
AUXOCHROMES - [ind chem] (used for the improvement of colouring properties) auxochromes *mpl*
AVAILABILITY - [gen] disponibilité
~STORE - [comput] mémoire *f* de disponibilité
AVAILABLE - [gen] disponible
~CONVERSION GAIN - [radio] amplification *f* de conversion disponible
~ENERGY - [el] énergie *f* disponible
~LOAD - [el] charge *f* disponible
~MACHINE TIME - [comput] temps *m* machine disponible, temps *m* d'exploitation disponible
~MAGNIFICATION - grossissement *m* utile
~NOISE POWER - [radio] puissance *f* de bruit disponible
~POWER - [acoust] puissance *f* disponible (maximum)
~POWER EFFICIENCY - [acoust] rendement *m* disponible
~POWER RESPONSE - [acoust] réponse *f* à la puissance
~SIGNAL-TO-NOISE RATIO - [radio] rapport *m* signal-bruit disponible
AVALANCHE - [gen] avalanche *f*
[nucl] avalanche *f*, ionisation *f* en chaîne
~BREAKDOWN - [el] charge *f* d'avalanche
~VOLTAGE - [el] tension *f* d'avalanche
AVAST, to - [naut] tenir bon
AVAST - [naut] tiens bon
A.V.C. - [radio] (Automatic Volume Control) régulation *f* automatique de puissance
AVENS - [bot] (a medicinal plant) benoîte *f*
AVENTURINE FELDSPATH - [min] (variety of sodic plagioclase, also called sunstone) pierre *f* de soleil
~GLASS - [glass man] (glass containing fragments of glittering material separated from the main body) aventurine *f* artificielle
~QUARTZ - [min] (quartz charged with minute inclusions of mica, or iron oxide) aventurine *f* naturelle
AVERAGE, to - [gen] calculer une moyenne, établir une moyenne
AVERAGE - [gen] moyenne *f*
[naut] avarie
[adj] moyen
~ADJUSTER - [leg] répartiteur *m* d'avaries, dispacheur *m*
~AGREEMENT - [leg] compromis *m* d'avarie
AVERAGE BEHAVIOUR - [nucl] (the behaviour or a nuclear reactor over a specified period) comportement *m* moyen
~BIAS - [el] tension *f* moyenne d'une électrode
~BINDING ENERGY - [nucl] (per nucleon) énergie *f* de liaison globale
~BOND - [leg] compromis *m* d'avaries, acte *m* de compromis
~BRIGHTNESS - [telev] brillance *f* moyenne
~CALCULATING SPEED - [comput] vitesse *f* de calcul moyenne
~CIRCULATION - [publish] tirage *m* moyen
~CURRENT - [el] (average value of a current taken over half a cycle) courant *m* moyen
~EXCITATION ENERGY - [nucl] énergie *f* moyenne d'excitation
~HEAD - [hydr] hauteur *f* de chute moyenne
~LIFE - [leg & mech] durée *f* moyenne de vie
~NOISE FACTOR - [radio] facteur *m* moyen de bruit
~OUTPUT - [el] rendement *m* moyen
~PER HOUR - [gen] moyenne *f* horaire
~PRESSURE - [phys] pression *f* moyenne
~PULSE AMPLITUDE - [radio] amplitude *f* moyenne d'impulsion
~SPEECH POWER - [acoust] puissance *f* acoustique moyenne
~SPEED - [mech etc] vitesse *f* moyenne
~STATEMENT - [leg] règlement *f* d'avaries, dispache *f* d'avarie
~TEST - [mech mach] essai *m* moyen
~VALUE - [math] valeur *f* moyenne
AVIATION - [aero] aviation *f*
~BEACON - [radio] radiophare *m* pour la navigation aérienne
~COMPASS - [instr] compas *m* pour avions
~FUEL - [combust] carburant *m* pour l'aviation
~SPIRIT - [aero] essence *f* d'avion
AVIATOR - [gen] (pilot of an aircraft) aviateur *m*
~SUIT - [aero] combinaison *f* pressurisée
AVIGATION - [aero] (term for aerial navigation(obsolete)) navigation *f* aérienne
AVITAMINOSIS - [med] (lack of vitamins) avitaminose *f*
AVOCADO OIL - [ind chem] essence *f* d'avocat
AVOGADRO CONSTANT - [chem] (the number of molecules in a grammolecule) nombre *m* d'Avogadro
~HYPOTHESIS - [chem] (equal volumes of different gases at the same temperature and pressure contain the same number of molecules) hypothèse *f* d'Avogadro
AVOGADRO'S LAW - [phys] loi *f* d'Avogadro
~NUMBER - [chem] s. Avogadro's constant
AVOIRDUPOIS - [meas] mesures *fpl* de poids du commerce (G.B.)
AWARD - [gen & leg] arbitrage *m*, sentence *f* arbitrale, adjudication *f*, récompense *f*
AWASH - [naut] à fleur d'eau
AWEIGH - [naut] dérapée (ancre)
AWL - [impl] alène *f*, poinçon *m*
AWN - [agric] barbe *f* (céréales etc)
~CUTTER - [agric impl] ébarbeuse *f*
AWNING - [text] tente *f*, bâche *f*, vélum *m*
~DECK - [naut] pont-abri *m*
~STANCHION - [naut] chandelier
AWNLESS - [agric] sans barbe
AWU - [nucl] (Atomic Weight Unit) unité *f* de masse atomique

AXE - [impl] hache *f*
~HOLE - [impl] trou *m* d'axe
AXES - [cryst] (lines of reference in crystal structures, which intersect at the centre of the crystal, normally three in number) axes *m* pl
~CROSSING DEVICE - [mech] dispositif *m* de croisement des cylindres
AXIAL - [gen] axial
~ARMATURE RELAY - [el] relai *m* à armature mobile
~CLEARANCE - [mech] jeu *m* axial
~COMPRESSOR - compresseur *m* axial
~CORD - [aero] (a central rigging line in a parachute) suspense *f* centrale
~DUCTS - [el] canaux *m* pl de ventilation axiaux
~ENGINE - [mech] moteur *m* axial
~FLOW - [mech] écoulement *m* axial
[aero] (helicopter; the component of the air flow which is at right angles to the tip-path plane) flux *m* axial
~FLOW COMPRESSOR - [mech] compresseur *m* axial
~FLOW ENGINE - [mech] (an engine in which the flow of gas is parallel with the centre-line of the rotor) moteur *m* à flux axial
~FLOW SUPERCHARGER - [mech] (a supercharger employing an axial flow compressor) compresseur *m* à flux axial
~FLOW TURBINE - [mech] turbine *f* à écoulement axial
~GIRDER - [aero] poutre *f* axiale
~HEAD - [mech mach] (of extruder) tête *f* axiale
~INCIDENCE - [acoust] (of a microphone) incidence *f* axiale
~MAGNIFICATION - [opt] grossissement *m* axial
~PITCH - [mech] pas *m* axial
~RESPONSE - [acoust] réponse *f* axiale, efficacité *f* axiale
~SWINGING ELEMENT - [mech] élément *m* d'essieu oscillant
~THRUST - [mech] poussée *f* axiale, butée *f* axiale
~VECTOR - [phys] (in three-dimensional space) vecteur *m* axial (espace à 3 dimensions)
AXIALLY MOVABLE COMB - [text] peigne *m* à déplacement axial
AXILLARY - [anat] axillaire
AXIOM - [math] axiome *m*
AXIOMETER - [instr] (instrument indicating the position of the rudder) axiomètre *m*
AXIS - [gen] axe
~CYLINDER - [zool] (the excitable core of a medullated nerve-fibre) cylindraxe *m*
~OF A WELD - [welding] axe *m* du cordon de soudure
~OF AFFINITY - [chem] axe *m* d'affinité
~OF BUOYANCY - [hydr] axe *m* de poussée
~OF CENTRES - [mech mach] axe *m* des pointes
~OF DEPRESSION - [geol] axe *m* synclinal
~OF ELEVATION - [geol] axe *m* anticlinal
~OF GROOVES - [metall] ligne *f* médiane des cannelures
~OF LENS - [opt] axe *m* optique
~OF ORDINATES - [math] axe *m* des ordonnées
~OF OSCILLATION - [phys] axe *m* d'oscillation
~OF POLARIZATION - [opt] axe *m* de polarisation
~OF REVOLUTION - [astr] axe *m* de révolution
~OF ROTATION - [mech] axe *m* de rotation
~OF SCAN - [radar] axe *m* de balayage
~OF SPIN - [mech] axe *m* de rotation
~OF SYMMETRY -[phys] (an immaginary line about which the elements of a body are symmetrically disposed) axe *m* de symétrie
AXIS OF X - [math] axe *m* des abscisses, axe des X
~OF Y - [math] axe *m* des ordonnées, axe des Y
AXLE - [mech] axe
[auto] pont *m* arrière
[railw] essieu
~ARM - [mech] fusée *f* d'essieu
~BAR - [auto] corps *m* d'essieu
~BASE - [mech] empattement *m*
~BEAM - [mech] s. axle bar
~BEARING - [mech] portée *m* d'axe, palier *m* d'axe
~BOOMS - [aero] longerons *m*
~BOX - [mech] boîte *f* à graisse, boîte *f* d'essieu
~BREAKAGE - [mech] rupture *f* d'essieu
~CAP - [mech] chapeau *m* de moyen, chapeau *m* d'essieu
~COLLAR - [mech railw] collier *m* d'essieu
~DRIVEN ALTERNATOR - [railw] alternateur *m* couplé aux essieux
~DRIVEN GENERATOR - [railw] génératrice *f* couplée aux essieux
~END - [mech] bout *m* d'arbre
~HOUSING - [auto] carter *m* de pont (arrière)
~JOURNAL - [mech] arbre *m*, axe *m*
~LATHE - [railw] tour *m* pour essieux
~LOAD - [mech] charge *f* sur un essieu
~NUT - [mech] écrou *m* d'arbre (ou d'axe)
~OFFSET - [mech] (the length of a line which is normal to both the castor axis and the wheel axis) déport *m*, désaxage *m*
~SEAT - [railw] portée *f* d'essieu
~SHAFT - [mech] arbre *m*, axe *m*, tourillon *m* d'essieu, pont *m* demi-arbre, demi-essieu *m*, arbre *m* porteur
~SHAFT BEARING - [mech] palier *m* de demi-arbre
~SWINGING ELEMENT - [mech] élément *m* d'essieu oscillant
AXLETREE - [mech] essieu *m*, axe *m*, pont *m*
~SPINDLE - [mech] fusée *f* d'essieu
AXONOMETRIC - [geom] axonométrique
~PROJECTION - [geom] projection *f* axonométrique
AXONOMETRY - [geom] axonométrie *f*
AYRTON SHUNT - [el] shunt *m* universel
AZALEA - bot] azalée *f*
AZEL DISPLAY - [radar] indicateur *m* azimut-élévation
AZELAIC ACID - [chem] acide *m* azélaïque
AZEOTROPIC - [chem] azéotrope, azéotropique
~DISTILLATION - [nucl] distillation *f*, azéotropique, distillation *f* fractionnée
~ENTRAINER - [chem] entraîneur *m* azéotropique, agent *m* de déplacement azéotropique
~MIXTURE - [chem] (e.g. chlorhydrin and water) mélange *m* azéotropique
AZEOTROPISM - [chem] azéotropisme *m*
AZIDE - [chem] (the salts of hydrazoic acid) azoture *m*, azothydrure *m*
AZIMUTH - [nav & aero] (the arc of the horizon which a vertical plane passing through a celestial body makes with the meridian of the point of observation) azimut *m*
~ANGLE - [aero] angle *m* azimutal
~CIRCLE - [astr] cercle *m* de relèvement
~COMPASS - [instr] compas *m* de relèvement
~DIAL - [instr] (the dial of an instrument on which the azimuth angles are read) cadran *m* azimutal

AZIMUTH DIRECTION - [radar] (direction in the plane of the horizon) direction *f* azimutale, relèvement *m*
~FINDER - [instr] (electronic navigational aid) indicateur *m* de relèvement
~INDICATING GONIOMETER - [instr] radiogoniomètre *m* azimutal
~RING - [naut] cercle *m* de relèvement
~STABILIZED P.P.I. - [radar] indicateur *m* panoramique (P.P.I.) à alignement automatique
AZIMUTHAL - [astr & instr] azimutal
~CONTROL - [hel] (the lever by which the pilot controls the cyclic pitch) commande *f* azimutale
~PROJECTION - [math] projection *f* azimutale
AZINDONE BLUE - [chem] indamine *f*
AZINE - [chem] azine *f*
AZO COMPOUNDS - [chem] composés *m* pl azoïques
~DYES - [ind chem] (azobenzine derivatives) colorante *m* pl azoïques
AZOBENZENE - [chem] (an intermediate for dyestuffs and rubber chemicals) azobenzène *m*
AZOCYANIDE - [chem] azocyanure *m*
AZODERMIN - [chem] azodermine *f*
AZOFLAVIN - [chem] azoflavine *f*
AZOFORMIC - [chem] azoformique
AZOIC - [chem] azoïque
AZOIC ERA - [geol] ére *f* azoïque
AZOLE - [chem] pyrrol *m*
AZOMETHANE - [chem] azométhane *m*
AZON - [radar] (azimuth only) missile *m* radioguidé en azimut, "azon" *m*
AZONAL SOIL - [soil] sol *m* peu évolué
AZOOSPERMIA - [zool] azoospermie *f*
AZOPROTEINS - [chem] azoprotéines *f* pl
AZORES HIGH - [met] hautes pressions *f* pl des Açores
AZORUBINE - [ind chem] azorubine *f*
AZOSULPHONIC - [chem] azosulfonique
AZOTATE - [chem] nitrate *m*, azotate *m*
AZOTE - [chem] (term occasionally used for nitrogen) azote *m*
AZOTIC - [chem] nitrique, azotique
~ACID - [chem] acide *m* nitrique
AZOTITE - [chem] nitrite *m*
AZOTIZE, to -[chem] azoter
AZOTOMETER - [instr] azotomètre *m*
AZULENE - [chem] azulène *m*
AZURE - [col] bleu *m* d'azur
~STONE - [min] lapis-lazuli *m*
AZURITE - [min] (basic copper carbonate) azurite *f*
AZYGOTE - [genet] (individual produced by haploid parthenogenesis) azygote *m*

B

b - [el] (the symbol for susceptance in an a.c. circuit) symbole de la susceptance dans un circuit de courant alternatif
B - [el] (the symbol for flux density) symbole de l'induction magnétique (densité de flux)
B - [chem] (the symbol for boron) symbole du Bore
B - [light] (symbol for brightness) symbole de la brillance
B - [telev] (blue primary) bleu *m* primaire
B BATTERY - [el] (battery required for the anodic current of a thermionic valve) batterie *f* anodique
B CLASS MODULATOR - [radio] modulateur *m* équilibré
B-DISPLAY - [radar] indicateur *m* radar type B
B ELIMINATOR - [electron] (power source which eliminates the anode battery) appareil *m* de tension anodique
B LINE - [comput] (machine register in which a number can change an address) registre *m* de base, registre *m* index
B-POSITION - [telecomm] groupe *m* d'arrivée, groupe *m* B
B SCOPE - [radar] (radar display) indicateur *m* radar type B
B-Y AXIS - [telev] (blue colour difference axis) axe *m* B-Y
B-Y SIGNAL - [telev] (blue colour difference signal) signal *m* B-Y
Ba - [chem] (symbol for barium) symbole du Baryum
BABBITT, to- [metall] garnir d'antifriction, garnir de métal antifriction, réguler
BABBITT - [metall] (s. also babbitt metal) métal *m* antifriction, régule *m*
~ALLOYS - [metall] métal *m* blanc, métal *m* antifriction
~LINING - [metall] garniture *f* de métal antifriction
~MELTING FURNACE - [metall] four *m* pour la fusion des alliages antifriction
~METAL - [metall] (antifriction alloy of tin, antimony, copper and lead used as an antifriction metal for bearings) métal *m* antifriction, régule *m*
BABBITTED GUIDE - [mech] guide *m* garni de métal antifriction
BABBLE - [acoust] (unwanted disturbing sounds in telecomm. system) diaphonie *f* multiple
BABS - [radio] (Beam-Approach Beacon System) dispositif de radioguidage (approche sans visibilité)
BABULUM OIL - [ind chem] huile *f* de pied de boeuf
BABY ANVIL - [impl] enclumette *f*
~BESSEMER STEEL - [metall] acier *m* Bessemer de petit convertisseur
BABY CAN - [light] petit réflecteur *m* diffuseur
~DRILL - [impl] (used in quarrying) barre *f* de carrière
~KEG-LIGHT - [cinema] petit projecteur *m* (500 W)
~SPOT - [cinema] s. baby keg-light
BACILLUS - [biol] (micro-organism characterized by rod-shaped form) bacille *m*
BACITRACIN - [chem] (an antibiotic) bacitracine *f*
BACK, to - [gen] appuyer, soutenir, reculer, faire reculer
[constr] renforcer, épauler
[auto] faire marche arrière
[met] (of a wind; the change of direction anticlockwise) devier dans le sens direct
[mining] boiser
[bookbind] endosser, passer en presse
[comm] (cheques etc) avaliser, endosser
[naut] empenneler (une ancre), scier, nager à culer
BACK - [gen] dos *m*, arrière *m*, dossier *m*, verso *m*, inverse *m*
[adv] en arrière, dans le sens contraire
[text] envers *m*
[books] dos *m*
[large vat] bac *m*
[mining] plafond *m* de mine, amont pendage *m*
[horol] fond *m*
[constr] extrados *m*, tête *f* de pilon
[carp impl] (of a saw) dosseret *m*
~AND FORTH MOTION - [mech] mouvement *m* de va-et-vient, mouvement *m* d'avance et de recul
~AND FRONT PITCH OF WINDING - [el] pas *m* partiel d'un enroulement en tambour
~ARM - [mech] bras *m* de contrepoids
~AXLE - [mech] pont *m* arrière, essieu *m* arrière
~AXLE CASING - [mech] carter *m* de pont arrière
~AXLE CENTRE - [mech] contre-pointe *f* (tour)
~BALANCE - [mech] (a counterweight) contrepoids *m* de queue
~BEAM - [text mach] ensouple *m* d'ourdissage
~BENDING TEST - [metall etc] essai *m* de pliage en sens inverse
~BIAS - [electron] (the illumination of the rear surface in a TV camera tube) éclairage *m* de l'arrière côté
~-BLAST FORGE - [metall] forge *f* à tuyère en bout
~BRICK - [space heat equip] dalle *f* réfractaire, dalle *f* calorifuge
~BUMPER - [railw] traverse *f* d'arrière
~BUSHING - [mech] coussinet *m* arrière
~CAGE - [build] (of stairs) crinoline *f*

BACK CATCH BAR MOTION - [text] mouvement *m* horizontal de la barre mouvante
~CENTRE - [mach tool] contrepointe *f* (tour), pointe *f* de la poupée mobile (d'un tour)
~CENTRE SOCKET - [mech] douille *f* de contrepointe
~CLOTH - [text] toile *f* de renforcement
[plast ind] feuille *f* ensouple, bande *f* ensouple
~COMB - [text] peigne *m* arrière
~COMING - [mining] exploitation *f* en rabattant
~CONDUCTANCE - [el](the reciprocal of the forward resistance) conductance *f* inverse
~CONE ANGLE - [mech] angle *m* de cone complémentaire
~CONNECTED - [el] à contact arrière
~CONTACT - [telecomm] contact *m* arrière, contact *m* de repos
~COUPLING - [mech] renvoi *m*, retour *m* de force
~COVER - [print] couverture *f*, jaquette *f*, couvre-livre *m*
~COVERING - [text] housse *f* pour dossier
~CRANK PUMPING - [petr ind] pompage *m* combiné
~CROSS BREEDING - [genet] reproduction *f* par rétrocroisement
~CROSSING - [biol] (cross between a hybrid and one of its parents) croisement *m*
~CURRENT - [el] (reverse current through the depletion layer of a transistor) courant *m* de retour, contre-courant *m*
~DIFFUSION - [mech] (of gas) diffusion *f* en retour reflux *m* de gaz par diffusion
~DIFFUSION OF ELECTRONS - [nucl] rétrodiffusion *f* (des électrons), diffusion *f* en arrière des électrons
~DRIVE - [mech] marche *f* arrière
~ECHO - [radar] (echo due to a minor back lobe of a radar beam) écho *m* arrière
~ELECTROMOTIVE FORCE - [electron] (electromotive force generated by a motor in opposition to the supply voltage) force *f* contre-électromotrice
~-EMISSION - [electron] (an unwanted effect in vacuum rectifier tubes, caused by the evaporation of barium particles from the cathode towards the anode) contre-émission *f*
~-FILL STOPING - [mining] exploitation *f* par remblayage
~FILLING - [build] matériau *m* de remblayage, remblaiement *m* (tranchée de pipe-line)
~FIRE - [mech el] retour *m* de flamme, arc *m* en retour
~FIRING - [el] retour *m* de flamme, arc *m* en retour
~-FLASH - [metall] cannelure *f* à ébarbures
~-FLOW - [th eng] (of the boiler water) refoulement *m*
~FOCAL PLANE - [photo] plan *m* focal arrière
~FOCUS - [photo] distance *f* du foyer à la surface arrière d'une lentille
~GEAR - [mech] engrenage *m* intermédiaire, harnais *m*
~GEAR SHAFT - [mech mach] arbre *m* d'engrenages réducteurs
~-GEARED FOOT-LATHE - [tool] tour *f* à engrenages marchant au pied
~GEARS - [mech mach] engrenages *m* réducteurs
~GREY FABRIC - [text] étoffe *f* pour doublier
~-HAND WELDING - [metall] soudure *f* à droite (en arrière)

BACK HOLE - [metall] trou *m* de décharge
~IRON - [metall] contrefer *m*, fer *m* de dessous
~JOINT - [mining] limet *m*
~KICK - [mech] retour *m* de flamme, retour *m* à l'allumage
~-LASH - [mech] jeu *m* de dentures, jeu *m* d'engrènement
~LATH - [mining] palplanche *f*
~-LIGHT - [cinema] (projection lamp illuminating the back of the scene) projecteur *m* de fond
~-LIGHTED - [photo] en contrejour
~LINING - [print] endossage *m*
~LOBE - [radar] (the lobe which is radiated towards the rear of the intended direction) lobe *m* postérieur
~LOT - [cinema] (large surface surrounding studios for exterior shots) terrain *m* pour extérieurs
~MIGRATION - [mech] (from a vapour pump into a vacuum system) migration *f* en retour (dans une pompe à jet de vapeur)
~MUTATION - [genet] mutation *f* reverse, mutation *f* renversée
~NUT - [mech] contre-écrou *m*, écrou *m* de blocage
~OF A VAULT - [constr] (extrados) extrados *m* de voûte, crête *f* d'une voûte
~OF GALLERY - [mining] toit *m* de galerie
~OF SEAT - [gen] dossier *m* de siège
~OFF, to - [mech] dégager, dépouiller, détalonner, dévisser
[oil ind] décrocher dans le puits
~OUT, to - [gen] sortir à reculons
[auto] sortir en marche arrière
~OUTLET RETURN - [gas ind] culotte *f*
~PACK - [telev] (to house the electronic components) boîtier *m* dorsal
~-PACK PARACHUTE - [aero] parachute *m* dorsal
~PAY - [comm] rappel *m*, arrérages *m*, arrière *m*
~PEDAL BRAKE - [mech] frein *m* à rétropédalage
~PICK - [text] trame *f* d'envers, trame *f* de dessous
~-PITCH - [el] (the pitch of the winding of a motor or generator at the end of the armature opposite to the commutator) pas *m* de l'enroulement de la section arrière (d'un induit)
~PLATE - [mech] contre-plaque *f*
~POINTER - [nucl] index *m* de Back
~POLLINATING - [genet] rétrocroisement *m*, croisement *m* en retour
~PORCH - [telev] (part of the video signal at black level after each of the synchronization pulses) palier *m* arrière (sur un graphique)
~PRESSURE - [mech] contre-pression *f*
[phys] contre-pression *f*
~PRESSURE BELL - [rubber ind] couvercle *m* de fermeture de presse à cloche
~PRESSURE TURBINE - [el] turbine *f* à contre-pression
~PRESSURE VALVE - [mech] clapet *m* de retenue
~PUTTY - [constr] mastic *m* de vitrier
~-RACKING - [text] rélargissement *m*, augmentation *f*, recul *m*
~RESISTANCE - [el] (the contact resistance opposing the forward current) résistance *f* inverse
~REST - [gen] lunette *f* de tour
[auto] dossier *m* rembourré
~ROLLER - [text mach] cylindre *m* arrière
~-RUN - [gas ind] (water gas) fabrication *f* à contre

courant (gaz à l'eau)
BACK RUN - [mech] (of a film) marche *f* arrière
~SADDLE - [mining] cloche *f* du toit
~SAW - [impl] scie *f* à dossière, scie *f* à tenon
~-SCATTERING - [nucl] s. backscattering
~-SCRATCHER - [impl] gratte-dos *m*
~SEAT - [auto] siège *m* arrière
~SHAFT - [text] arbre *m* du chariot
~SHIELD - [auto] lunette *f* arrière
~SHUNT - [railw] refoulement *m*, manoeuvre *f* en arrière
~-SHUNT KEYING - [radio] (a way of keying a transmitter whose radio-frequency energy is fed to the antenna when the telegraph key is closed) manipulation *f* par suppression de l'onde porteuse
~SIDE PUMPING - [oil ind] pompage *m* latéral
~SLOPE - [mech] (aiming device) hausse *f*
~SPACER - [mech] (of typewriter) rappel *m* de chariot
~SPEED - [mech] redoublement *m* d'engrenages
~SQUARE - [impl] équerre *f* à chapeau
~-STEAM - [th eng] contrevapeur *f*
~-STEP SEQUENCE - [metall] (a welding technique) soudage *m* en arrière
~STOP - [telecomm] butée *f*, arrêt *m*
[mech] cliquet *m*
~STOPING - [mining] abattage *m* montant, exploitation *f* en gradins renversés
~STREAMING - [mech] (of a pump) flux *m* inverse, reflux *m*
~SWEEP - [aero] (or sweep-back; the angle between a line at right angle to the plane of symmetry and the plan projection of a specified spanwise line in the wing) dièdre *m* horizontal, flèche *f* arrière
~SWEPT WING - [aero] aile *f* en flêche
~TIMBER - [mining] boisage *m* du toit
~-TO-BACK CHANNEL SECTIONS - [mech] profilés *mpl* en U placés dos à dos
~-TO-BACK CONNEXION - [el] montage *m* en opposition
~-TO-BACK METHOD - [mech & el] méthode *f* d'opposition
~-TO-BACK TEST - [mech] essai *m* par la méthode d'opposition
~-UP LIGHT - [auto] phare *m* de marche arrière
~UP ROLL - [metall] cylindre *m* de soutien
~WALL - [auto] panneau *m* arrière
~WAVE - [telecomm] (electromagnetic waves radiated during the spacing interval of the telegraphic code) onde *f* de contre-manipulation
~WELD - [metall] soudure *f* de renforcement
~WHEEL - [auto] roue *f* arrière
~WINDING - [gen & of films] ré-enroulement *m*
~WINDOW - [auto] lunette *f* arrière
BACKACTER - [mech] rétrocaveuse *f*
~SHOVEL - [mech] pelle *f* rétrocaveuse
BACKBLOWING - [hydr] (the surging of a well) décolmatage *m* d'un puits
BACKBOARD - [gen] dossier *m*
BACKBONE - [anat] colonne *f* vertébrale
~TYPE OF FRAME - [auto] châssis-poutre *m*, châssis *m* à poutre centrale
BACKCLOTH - [photo] toile *f* de fond
BACKDRAUGHT - [heat] refoulement *m*, contre-tirage *m*
BACKDROP - [text] arrière-plan *m*
BACKED FABRIC - [plast ind] tissue *m* renforcé, tissu *m* doublé, tissu *m* enduit
BACKED FILM - [photo] film *m* antihalo
~METAL CUTTING SAW - [impl] scie *f* à métaux à dossière
~OFF CUTTER - [mech] fraise *f* à profil invariable
~-OFF MILLING-CUTTER - [mach tool] fraise *f* à denture à dépouille
~STAMPER - [acoust] matrice *f* renforcée
~-UP TRAIN - [railw] train *m* refoulé
BACKFILL,to-[oil ind] remblayer (une tranchée de pipe-line)
BACKFLOW - [mech] (reversed flow in the tail-pipe of a pulsejet) reflux *m*, retour *m*, refoulement *m*, contre-courant *m*
~PIPE - [hydr] conduit *m* de refoulement, conduit *m* de retour
~PREVENTER - [hydr] clapet *m* de retenue
BACKGROUND - [gen] fond *m*, arrière-plan *m*
[nucl] mouvement *m* propre
[cinema] fond *m* sonore
[radio] bruit *m* de fond
~ADAPTATION - [genet] adaptation *f* au milieu
~BRIGHTNESS - [telev] (the brightness on the screen independently of the pictures on it) luminosité *f* du fond
~CONTROL - [telev] (in colour television) contrôle *m* de la luminosité de fond
~COUNTING RATE - [nucl] (the counting rate caused by background radiation) taux *m* de comptage des coups parasites, taux *m* de comptage du mouvement propre
~COUNTS - [nucl] (counts originating from agencies other than those which must be detected) coups *m* parasites
~FADE-IN - [telev] apparition *f* graduelle d'un fond
~FADE-OUT - [telev] disparition *f* graduelle d'un fond
~LOUDSPEAKER - [radio] haut-parleur *m* de fond
~NOISE - [radio] (due to a large number of disturbances) bruit *m* de fond
~PROJECTION - [telev] (projection of a scenery used as background during the taking of a picture) projection par transparence
~RADIATION - [nucl] (ionizing radiation other than that which is measured) rayonnement *m* dû au mouvement propre
~RETURNS - [radar] échos *m* de sol
BACKHEATING - [electron] (failure of the rectifying action resulting in a substantial current flow in the inverse direction) surchauffage *m*
BACKHOE - [mech] pelle *f* rétrocaveuse
BACKING - [gen] appui *m*, renforcement *m*, soutien *m*, support *m*
[auto] marche *f* arrière
[naut] empennelage *m* (d'une ancre), nage *f* à culer
[met] déviation *f* du vent, recul *m* du vent (dans le sens direct)
[photo] (of emulsion) couche *f* dorsale
[cinema] décor *m* de fond, enduit *m* anti-halo
[text] toile *f* du fond
~BAR - [metall] (in welding) bande-support *f*
~BRICK - [build] brique *f* de remplissage
~CLOTH - [rubber ind] tissu *m* support
~CONDENSER - [vacuum techn] refroidisseur *m* pour le vide primaire
~DEAL - [constr] bois *m* de garnissage

BACKING LINE - [vacuum techn] conduite *f* de vide primaire
~LINE VALVE - [mech] robinet *m* de vide primaire, vanne *f* à vide primaire
~MEMORY - [comp] (part of interior storage) mémoire *f* additionnelle
~MOVEMENT - [railw] refoulement *m*
~OFF - [mech] dégagement *m*, dépouillement *m*, détalonnage *m*
~-OFF BRAKE - [mech] frein *m* de dépointage
~-OFF CAM -[text] came *f* de dépointage
~-OFF CHAIN - [text] plan *m* incliné, dispositif *m* pour régler le dépointage
~-OFF LATHE - [mach tool] tour *m* à dépouiller
~PICK - [text] trame *f* supplémentaire
~PLATE - [mech mach] plaque *f* d'appui, pièce *f* de renfort, plateau *m* de frein
[plast ind] plateau *m* de fixation
~PRESSURE - [vacuum techn] prévide *m*, vide *m* primaire
~PRESSURE GAUGE - [mech] jauge *f* à vide primaire
~PUMP - [vacuum techn] pompe *f* à prévide, pompe *f* primaire
~SAND - [constr] sable *m* de remblayage
~SIDE TRAP - s. backing condenser
~SPACE - [vacuum techn] volume *m* à prévide, espace *m* primaire
~STORAGE - [comput] mémoire *f* d'appoint, mémoire *f* d'intermédiaire
~THREAD - [text] fil *m* de fourrure
~-UP PLATE - [mech] plaque *f* d'appui, contre-plateau *m*
~UP RING - [metall] support *m* à anneau
~WEFT - [text] trame *f* inférieure
~WHEEL - [text] roue *f* pour doubler
BACKLASH - [el] (s. also reverse grid current) courant *m* inverse de grille
[mech] jeu *m*
BACKPLATE - [telev] (of a TV camera tube) contre-électrode *f*
BACKSCATTER - [radar] s. backscattering
~FACTOR - [nucl] facteur *m* de rétrodiffusion
BACKSCATTERING - [nucl] (the occurrence of radiation from the surface of a material through which it entered, due to its collision with atoms in the material and reflection from these atoms) rétrodiffusion *f*, diffusion *f* rétrograde
BACKSPACE, to - [comput] rappeler en arrière (pas à pas), ramener en arrière (pas à pas)
BACKSTAIR - [constr] escalier *m* de service
BACKSTAY - [gen] dossier *m*
[naut] galhauban *m*
[mining] chambrière *f*
[radio] hauban *m* (d'antenne)
BACKSTEP WELDING - [metall] soudure *f* de gauche à droite, soudage *m* en arrière
BACKSTITCH - [text] point *m* arrière
BACKSTREAMING - [mech] (of pump fluid vapour) reflux *m* (de la vapeur motrice)
BACKSTROKE - [mech] course *f* de retour, contre-coup *m*, choc *m* en retour
BACKTITRATION - [ind chem] titrage *m* en retour
BACKTWIST - [mech] (cables) contre-torsion *f*
BACKUP SAFETY DEVICE - [nucl] (operating in case of failure of the normal safety rods) protection *f* de réserve (ou de secours)
BACKWARD - [gen] rétrograde, en arrière, arrière

BACKWARD ACTING REGULATOR - [radio] régulation *m* à réaction
~LEAD - [el] (or backward shift) changement *m* de la position des balais
~MOTION - [mech] mouvement *m* rétrograde, mouvement *m* de recul
~RECALL SIGNAL - [radio] signal *m* de rappel
~ROTATION - [ind chem] marche *f* en arrière (ou rétrograde)
~SCATTER - [ballist] rétrodiffusion *f*
~SHIFT - [el] (adjustment of the brushes of a commutator motor or generator in the opposite direction from that of the rotation, to reduce sparking) changement *m* de la position des balais
~SIGNALLING - [radio] rappel *m* de la station origine par le destinataire
~TILT - [aero] (blade tilt in an aftward direction) déport *m* de la pale dans le plan d'avancement de l'avion *f*
~TRANSFER ADMITTANCE - [radio] transadmittance *f* anode-grille
~WAVE - [electron] (a wave in a travelling-wave tube opposite to the direction of the stream motion of the electrons) onde *f* rétrograde, onde *f* inverse
~-WAVE OSCILLATOR - [instr] oscillateur *m* à onde rétrograde, oscillateur *m* à onde inverse
~-WAVE TUBE - [electron] (travelling wave tube used as a backward wave oscillator) tube *m* à ondes réfléchies (ou de retour)
~WELDING - [metall] soudure *f* à gaz à droite
BACKWASH - [ind chem] (water passed through a filter in a direction opposite to that of the normal flow, for cleaning purposes) remous *m*, lavage *m* à contre-courant
[naut] remous *m*, renvoi *m* de mer
BACKWASHER - [text] dégraisseur
BACKWASHING - [ind chem] lavage *m* par retour de courant
[text] dégraissage *m*, lavage *m* à dos, lavage sur pied
BACKWATER, to - [naut] nager à culer, scier
BACKWATER - [geogr] bras *m* de décharge (rivière)
[hydr] eau *f* arretée (par un bief etc), remous *m* (en amont d'un barrage)
~CURVE - [hydr] courbe *f* de remplissage
~GATE - s. backflow preventer
~PUMP - [paper man] pompe *f* à eaux résiduaires (eaux noires)
~SLOPE - [hydr] courbe *f* de retenue
~TANK - [paper man] cuve *f* de récuperation pour eaux noires
BACKWAVE - [hydr] (hydraulic pump) ressaut *m* hydraulique
BACON - [food ind] bacon *m*, lard *m* fumé
BACONER - [agric] porc *m* à bacon
BACTERIA - [biol] bactéries *f*
BACTERIAL ACTION - [biol] action *f* bactérienne
~COAGULATION - [ind chem] coagulation *f* bactérienne
~CULTURE - [biol] culture *f* bactérienne
BACTERICIDE - [biol] bactéricide
BACTERIOLYSIS - [chem] (decomposition of bacteria) bactériolyse *f*
BACTERIOSCOPY - (microscopic study of bacteria) bactérioscopie *f*
BACTERIUM - [biol] bactérie *f*
BACULIFORM - [genet] bacilliforme, en forme de

bâtonnet
BAD CONTACT - [el] mauvais contact *m*, faux contact *m*
~GEOMETRY - [nucl] (an arrangement of equipment which causes errors in nuclear physics measurements) géométrie *f* mauvaise
~RUNNER - [railw] mauvais rouleur *m*
~WEATHER CIRCUIT - [aero] (a special course to avoid adverse weather conditions) route *f* de sécurité
BADGE - [auto] emblème *m*, insigne *m*
BADGER - [zool] blaireau *m*
[impl] pinceau *m* en poil de blaireau
BAFFLE - [acoust radio] déflecteur *m*, écran *m* acoustique, parois *f*
[mech] déflecteur *m*, chicane *f*
[hydr] diaphragme
[electron] (auxiliary member in the arc path without external connection) chicane *f*
~BLANKET - [cinema] (felt and muslin sheet around a set to absorb sound) tenture *f* insonorisante, écran *m* insonorisant
~BOARD - [min] plaque *f* protectrice, plaque *f* de déviation
~PLATE - [mech] (a plate designed to prevent undesirable movement in a fluid, e.g. wave-motion, in a vessel subject to oscillation) déflecteur *m*, plaque *f* de déviation, chicane *f*, cloison *f* de réservoir, surface *f* de rebondissement
BAFFLED PISTON - [auto] piston *m* à déflecteur
~SPEED - [mech] (of a pump) débit *m* massique (d'une pompe) avec baffle
~THROUGHPUT - s. baffled speed
BAFFLER - [th eng] coupe-tirage *m*
BAFFLES - [plast ind] déflecteurs *m*
BAG, to - [gen] ensacher, mettre en sac
BAG - [gen] sac *m*, enveloppe *f*, bourse *f*
[mining] poche *f*
~AND SPOON - [mech] drague *f* à sac
~FILLER - [mech] ensacheuse *f*
~FILTER - [ind chem] manche *f* en toile à voile
[sugar ind] (a mechanical filter) filtre *m* à poches
~LEATHER - [leather ind] cuir *m* pour valises
~MOULDING - [plast ind] moulage *m* au sac
~PAPER - [paper man] papier *m* pour sacs
~SEALER - [mech] machine *f* à souder les sachets
~STRINGS - [gen] lacets *m*, cordons *m*
BAGASSE - [ind chem] (crushed sugar cane from which the sugar has already been extracted) bagasse *f*
BAGGAGE - [gen] bagage *m*
~CAR - [railw] fourgon *m* à bagages
~TROLLEY - [railw] chariot *m* à bagages
BAGGING - [text] toile *f* à sac
~MACHINE - [mech mach] ensacheuse *f*
~OFF - [gas ind] (network equip) mise *f* en place d'un ballon, ballonnement *m*
~-UP MACHINE - [rubber ind] presse *f* à galber, conformateur *m*
BAGPIPE - [mus] cornemuse *f*, binion *m*, musette *f*
BAIL, to - [naut] écoper
[mining] puiser par cuiller, curer
[leg] cautionner, se porter caution pour, déposer (des biens) sous contrat
BAIL - [leg] (security for prisoner's appearance) cautionnement *m*, caution *f*, garant *m*
[mech] arceau *m*, étrier *m* de suspension
[metall] étrier *m* (support de poche)
[oil ind] anse *f*
[naut] écope *f*
BAIL CLAMP - [mech] étrier *m*
BAIL OUT, to - [aero] sauter (en parachute)
BAILER - [naut impl] écope *f*
[mech] cuiller *f*, curette *f*, soupape *f*, sonde *f* à percussion
[leg] garant *m*
[oil ind] cuiller *f*
BAILIFF - [gen] régisseur *m*, intendant *m*
[leg] huissier *m*
BAILING - [mining] extraction *f* par skip
[oil ind] puisage *m*, épuisement *m*, dénoyage *m*
~CRUSHER - [min] broyeur *m* à boulets
~IRON - [metall] fer *m* en loupe
~LADLE - [impl] cuiller *f*
~LINE - [oil ind] câble *m* de curage
~TANK - [impl] cuffat *m* d'épuisement
~TEST - [mining] essai *m* par curage
~WELL - [oil ind] puits *m* en puisage
BAIN-MARIE - [gen] bain marie *m*
BAINITE - [metall] bainite *f*
BAIT-POKE - [mining] musette *f*, sac *m* de mineur
BAIZE - [text] reps *m*, serge *f*
BAKE, to - [gen] cuire (au four), étuver (une moule le), cuire (a brick)
BAKE - [ind] cuisson *f* au four, cuite *f* au four
~OUT, to - [metall] étuver, récuire (dans un four par chemise chauffante)
BAKED - [ind] cuit au four
~CLAY - [ind] terre *f* cuite
~COAL - [min] charbon *m* cuit
~CORE - [mech] noyau *m* étuvé
BAKELITE - [ind chem] (synthetic resin widely used for electrical components etc) bakélite *f*
~BOND - [ind chem] liant *m* résine
~BONDED - [plast ind] lié à la bakélite
BAKELIZE, to - [ind chem] bakéliser
BAKEOUT - [metall] dégazage *m* au four
~CLAMP - [metall] étrier *m* d'étuvage
~FURNACE - s. baking oven
BAKERY - [food ind] boulangerie *f*
BAKING - [ind] (the process of heating in special ovens or stoves for finishing varnished articles etc) cuisson *f*, cuite *f* au four, étuvage *m* (moule)
~CHERRY COAL - [min] houille *f* maigre collante
~CLAMP - s. bakeout clamp
~COAL - [min] charbon *m* collant
~OVEN - [metall] four *m* d'étuvage
~POWDER - [food] levure *f*
~SODA - [chem] bicarbonate *m* de soude
~TRAY - [impl] plaque *f* à pâtisserie
~TROUGH - [carp] pétrin *m*
~VARNISH - [ind chem] (type of varnish which is finished by baking after normal air-drying) vernis-émail *m*
BAKUIN - [oil ind] (Russian origin) huile *f* lubrifiante Bakuin
BALANCE, to - [gen & mech] équilibrer, compenser, stabiliser, balancer
[comm] solder un compte, balancer
BALANCE - [gen] équilibre *m*, équilibrage *m*, stabilité *f*, compensation *f*
[impl] balance *f*
[mech] contrepoids *m*
[horol] balancier *m* de montre

[aero] équilibre
[aero mech] équilibrage *m*, compensation *f*
[comm] solde *m* d'un compte, bilan *m*
[aerodyn] équilibre *m* aérodynamique
BALANCE ARC - [horol] arc *m* balancier
~ARM - [horol] bras *m* de balancier
~BAR - [hydr] balancier *m* (porte d'écluse), flèche *f*
~BEAM - [mech] fléau *m* de balance, verge *f* de balance
~BEAM METER - [instr] compteur *m* à balancier
~BOB - [mech] contre-balancier *m*
~BOX - [mech mach] caisson-contrepoids *m* (grue)
~BRIDGE - [horol] pont *m* de balancier
~CAR - [impl] chariot *m* contrepoids
~CARD - [comput] carte *f* stock
~ELECTROMETER - [instr] électromètre *m* balance
~GATE - [hydr] porte *f* d'écluse à compensation (balancier)
~GEAR - [mech] différentiel *m*
~HORN - [aero] (a balance area localized at the extremity of a control surface) volet *m* correcteur à corne
~METHOD - [gen & el] méthode *f* de compensation, méthode *f* de zéro
~OF HEAT - [th eng] bilan *m* thermique
~OF PAYMENT - [comm & finance] balance *f* de paiements
~OF TORSION - [phys] balance *f* de torsion
~OF TRADE - [comm & finance] balance *f* commerciale
~OUT, to [gen] (to neutralize forces by counterbalancing them with others) équilibrer
~PLOUGH - [agric impl] charrue-balance *f*
~POINT - [meas] prise *f* de dérivation
~POINT INDICATOR - [instr] indicateur *m* d'équilibre
~RETURN LOSS - [radio] atténuation *f* (de retour), affaiblissement *m* d'équilibrage
~ROPE - [mining] câble *m* d'équilibre
~SASH - [constr] châssis *m* ouvrant
~SHEET - [comm] bilan *m*
~SPRING - [horol] ressort *m* spirale
[mech] ressort *m* compensateur
~SPRING COLLET - [horol] bague *f* de ressort compensateur
~STAFF - [horol] axe *m* de balancier
~TAB - [aero] (a tab attached to a control surface and designed to reduce the force needed to operate it) volet *m* de compensation
~THEORY - [phys] théorie *f* de l'équilibre
~TRUCK - s. balance car
~WEIGHT - [mech] contrepoids *m*, masse *f* d'équilibrage
~WHEEL - [horol] balancier *m* (de montre), roue *f* de rencontre
[mech] volant *m* compensateur
~WITH UNEQUAL ARMS - [text] romaine *f* à bras inégaux
BALANCED - [gen mech el] équilibré, compensé, stabilisé
[comm] soldé, balancé
~AERIAL - [radio] (an aerial in which the voltage at all transversal planes are equal in magnitude and opposite in polarity in respect to the ground) antenne *f* compensée, antenne *f* équilibrée
~AERIAL CUPOLA - [metall] cubilot *m* équilibré
~AILERON - [aero] (an aileron in which the axis of deflection is so placed that the aerodynamic forces on it are balanced) aileron *m* compensé
BALANCED AMPLIFIER - [radio] (an amplifier circuit with input and output connections balanced to ground) amplificateur *m* compensé
~ANTENNA - [radio] s. balanced aerial
~ARMATURE - [acoust] induit *m* tétrapolaire
~BACKFLOW VALVE - [hydr] clapet *m* de retenue
~BEAM RELAY - [el] relais-balance *m*
~BLAST CUPOLA - [metall] cubilot *m* à air soufflé équilibré
~CIRCUIT - [el] circuit *m* équilibré, circuit *m* symétrique
~CONTROL SURFACE - [mech] (a control surface so designed that the aerodynamic forces about the axis of deflection are balanced) gouverne *f* compensée
~CURRENTS - [waveguides] (a current flowing in the two conductors of a balanced line, which are equal in magnitude and opposite in direction) courants *m* équilibrés
~DETECTOR - [radio] détecteur *m* équilibré
~DOOR - [oven] porte *f* équilibrée
~FLUE GAS HEATER - [gas ind] appareil *m* de chauffage à tirage équilibré
~FLUE SYSTEM - [gas ind] circuit *m* étanche
~LEVER ARM - [text] bras *m* de levier équilibré
~LINE - [waveguides] (transmission line consisting of two conductors whose currents are equal in magnitude and opposite in direction) ligne *f* équilibrée
~LINE SYSTEM - [waveguides] système *m* de ligne équilibrée
~LOOP - [telecomm] cadre *m* équilibré
~MIXER - [waveguides] (hybrid junction with crystal detectors in a pair of uncoupled arms) mélangeur *m* équilibré
~MODULATOR - [electron] (a circuit used to generate the side bands of an amplitude modulation wave and to suppress the carrier) modulateur *m* équilibré
~NEEDLE VALVE - [hydr] soupape *f* à aiguille compensée
~NETWORK - [el] réseau *m* équilibré
~OSCILLATOR - [radio] (oscillator with the impedance centres of the tank circuits at earth potential and the voltages between either end and the centres of these equal in magnitude and in phase opposition) oscillateur *m* compensé
~OUTPUT - [comput] sortie *f* symétrique, signal *m* de sortie symétrique
~POLYPHASE SYSTEM - [el] (of circuits) système *m* polyphasé symétrique
~QUADRIPOL - [el] quadripôle *m* équilibré
~RUDDER - [naut] gouvernail *m* compensé
~SURFACE - [aero] (a surface in which the hinge is aerodynamically balanced) gouverne *f* compensée, surface *f* compensée
~TERMINATION - [waveguides] charge *f* équilibrée
~THREE-PHASE SYSTEM - [el] système *m* triphasé équilibré
~THRUST - [phys] poussée *f* équilibrée
~TRANSMISSION LINE - [radio & el] ligne *f* de transmission équilibrée
~TWO TERMINAL PAIR NETWORK - [el] quadripole *m* équilibré
~WIRE CIRCUIT - [radio] circuit *m* symétrique

BALANCER - [gen & mech] balancier *m*, compensateur *m*, équilibreur *m*, vernier *m* d'accord (gonio)
~BOOST - [el] groupe *m* compensateur à élévateur de tension
~SET - [el] groupe *m* compensateur
BALANCING - [gen & mech] équilibrage *m*, compensation *f*, stabilisation *f*, répartition *f* des masses
~BLADE - [cinema] (blades of shutters eliminating flickering on the screen) volet *m* de compensation
~CAPACITANCE - [radio] capacité *f* de neutralisation
~CONDENSER - [el] condensateur *m* de neutralisation
~DYNAMO - [el] dynamo *f* compensatrice
~FLAP - [aero] aileron *m*
~GATE PIT - [hydr] (of a sector gate) fosse *f* d'équilibrage
~MACHINE - [mech] (a mechanical device for effecting the balancing of mechanical parts) machine *f* à équilibrer
~MAGNETIC STRIPE - [el acoust] piste *f* magnétique de compensation
~MAIN - [gas ind] barillet *m* compensateur, barillet *m* d'équilibrage
~MASS - [meas] masse *f* d'équilibrage
~NETWORK - [el] réseau *m* d'équilibrage
~OF OPERATOR'S LOAD - [telecomm] équilibre *m* des groupes
~REPEATING COIL - [telecomm] translateur *m* de l'équilibreur
~RESERVOIR - [hydr] réservoir *m* compensateur, réservoir *m* régulateur du débit
~SPEED - [mech] vitesse *f* normale de marche
~SURFACE - [aero] s. balanced surface
~TABULATOR - [comput] tabulatrice *f* de vérification
~TAIL LOAD - [aero] (air or inertial load on the tail of an aircraft to obtain equilibrium under specified flying conditions) charge *f* d'équilibrage de l'empennage horizontal, charge *f* compensatrice de queue
~TANK - [hydr] bassin *m* de compensation
~WEIGHT - [mech etc] contrepoids *m*, masselotte *f* (weight added to a moving part to give mechanical balance) masse *f* d'équilibrage
BALATA - [bot] (the coagulated latex of the bullet tree growing in Latin America) balata *m*
~GUM - [ind chem] balata *f*, gomme *f* balata
BALAUSTINE - [bot] balaustier *m*
BALCONET - [constr] petit balcon *m*
BALCONY - [constr] balcon *m*, (theatre) galerie *f*
BALD EAGLE - [zool] pygargue *m*
~SPOT - [dyes] zone *f* passive
~TYRE - [rubber ind] pneumatique *m* à bande de roulement usée (ou lisse)
BALE, to [gen & comm] emballer
BALE - [gen & comm] balle *f*, ballot *m*
~BREAKER - [text mach] machine *f* à ouvrir les balles
~COATING - [packag] revêtement *m* des balles
~COATING SOLUTION - [ind chem] dissolution *f* protectrice pour balles
~ELEVATOR - [mech] élévateur *m* à balles
~HOOP CUTTER - [impl] pince *f* à ouvrir les balles
~OPENER - [impl] ouvreuse *f* de balles
~-OUT FURNACE - [metall] four *m* fixe
~PICKER - s. bale opener
BALE PLUCKER - [text] éplucheur *m* de balles
~PRESS - [text] presse *f* à mettre en balles
~PRESSING MACHINE - [mech mach] presse *f* à paqueter, presse *f* à balles
~SCUTCHER - s. bale opener
~SPLITTING MACHINE - [mech mach] trancheuse *f* pour balles
~TIES - [mech] fils *m* métalliques pour mise en balles
BALEEN - [zool] fanon *m* de baleine
BALER - [gen] presse *f* à balles
[mech mach] presse *f* à paqueter
[agric] presse *f* à fourrage
BALING - [text] mise *f* en balles
~BAND - [metall] feuillard *m* d'emballage
~WIRE - [metall] fil de fer *m* pour emballage
BALK - [constr] grosse poutre *f*, solive *f*
[agric] (a boundary strip) lisière *f* d'un champ
BALKING - [el mach] défaillance *f*
BALL, to - [metall] agglomérer
BALL - [gen] balle *f*, bille *f*, boule *f*, sphère *f*
[text] pelote *f*
[metall] loupe *f* (de fer fondu)
[min] tas *m*
~AND DISC INTEGRATOR - [el] totalisateur *m* à sphère et à plateau
~-AND-SOCKET DEVICE - [mech] joint *m* à rotule
~-AND-SOCKET GEAR SHIFTING - [mech] changement *m* de vitesse à levier à rotule
~-AND-SOCKET JOINT - [mech] joint *m* à rotule
~BEARING - [mech] roulement *m* à billes, palier *m* à billes
~BEARING GUIDE BUSH - [metall] (in sheet metal forming) guidage *m* par billes
~BEARING HOUSING - [mech] boîtier *m* de roulement
~BEARING HUB - [mech] boîtier *m* de roulement
~BEARING RACE - [mech] moyen *m* à roulement à billes
~BEARING RETAINER RING - [mech] jonc *m* d'arrêt de roulement, bague *f* de retenue de roulement
~BEARING SEAT - [mech] boîtier *m* de roulement
~BLUE - [chem] bleu *m* d'empois, bleu *m* des blanchisseuses
~BOBBIN - [text] bobine *f* pour le ruban de coeur
~BURNISHING - [metall] brunissage *m* à la bille
~CAGE - [mech] cage *f* de roulement à billes
~CASTOR - [mech] roulette *f* orientable
~-CATCH - [mech] loquet *m* à ressort, loquet *m* à bille
~CHECK - [mech] clapet *m* à bille, clapet *m* de retenue à bille
~CHECK HOUSING - [mech] boîtier *m* de clapet à bille, logement *m* de clapet à bille
~CHECK NOZZLE - [mech] buse *f* anti-retour
~CHECK VALVE - [mech] clapet *m* à bille, clapet *m* de retenue à bille
~CIRCUIT SCREW - [mech] vis *f* à recirculation de billes
~CLAY - [geol] (used for pottery) argile *f* figuline, rognons *m* d'argile
~COCK VALVE - [mech] robinet *m* à flotteur, soupape *f* à flotteur
~CUP - [mech] cuvette-rotule *f*, coussinet *m* sphérique
~-END ROD - [mech] tige *f* à tête sphérique
~-ENDED MAGNET - [instr] aimant *m* à pôles sphériques

BALL FLOAT LEVER - [mech & hydr] bras *m* de flotteur sphérique
~FLOAT LEVER CONTROL - [mech] régulateur *m* à flotteur sphérique
~GATE - [metall] jet *m* de coulée rond
~GEAR CHANGE LEVER - [mech] levier *m* de changement de vitesses à rotule
~GRINDER - [mech] broyeur *m* à boulets
~GRIP - [mech] poignée *f* à pommeau
~GUDGEON - [mech] tourillon *m* à bille
~HAMMER - [impl] marteau *m* à panne ronde
~HARDNESS TESTER - [mech] appareil *m* d'essai à la bille (dureté)
~HEADED AIMING SCREW - [mech] vis *f* de réglage à tête ronde
~HEADED BOLT - [mech] boulon *m* à tête ronde
~JOINT - [mech] joint *m* sphérique
~JOINT AXLE - [mech] pont *m* à joint sphérique
~JOURNAL - [mech] tourillon *m* à bille
~LIGHTNING - [met] éclair *m* en boule, boule *f* de feu
~LUMP - [min] loupe *f*
~-MILL - [mech mach] broyeur *m* à boulets
~MOULDING PRESS - [mech] presse *f* à mouler les billes
~OF FIRE - [nucl] (bright sphere of luminous gases following a nuclear explosion) boule *f* de feu
~OF WARP - [text] pelote *m* d'ourdissage
~PANE - [impl] panne *f* ronde
~-PEEN HAMMER - [mech] marteau *m* à panne ronde
~PEN - [impl] crayon *m* à bille
~PIN - [mech] tourillon *m* à bille, pivot *m* sphérique
~PIVOT - [mech] pivot *m* sphérique, tourillon *m* à bille
~POINT - [mech] à bille
~RACE - [mech] chemin *m* de roulement des billes
~REAMER - [mech] alésoir *m* sphérique
~RECEPTION - [telev] (programme relayed to another transmitter to be re-broadcast) réception *f* par relais de retransmission
~RETAINER - [mech] cage *m* de roulement à billes
~ROLLER SCREW - [mech] vis *f* de galet à billes
~-SCREW - [mech] (a system in which a screw and a nut have half-round grooves which are filled with hardened steel balls, forming a continuous ball-bearing taking the place of the more solid threads) vis *f* à rotule
~SEAT - [mech] cuvette *f* de rotule, alvéole *f* de bille
~SHAKER CONVEYOR - [min] goulotte *f* oscillante à billes
~SHAPED STOPPER - [mech] obturateur *m* sphérique
~SIZING MACHINE - [text] pareuse *f*
~SOCKET - [mech] cuvette *f* de rotule, alvéole *f* de bille
~SODA - [chem] carbonate *m* de soude
~SPUNT - [mech] gorge *f* de guidage pour billes, cuvette *f* de roulement
~SQUEEZER - [metall] presse *f* à cingler
~STEAMING MACHINE - [text] machine *f* à vaporiser les pelotes
~TESTER - [instr] duromètre *m* à bille, appareil *m* de mesure de dureté à la bille
~THRUST - [mech] butée *f* à billes
~THRUST BEARING - [mech] butée *f* à billes
~THRUST TEST - [metall] billage *m*
~TRACK - [mech] chemin *m* de roulement des billes

BALL TUBE MILL - [min] broyeur *m* à boulets tubulaires
~-TYPE UNIVERSAL JOINT - [mech] joint *m* sphérique universel
~VALVE - [mech] clapet *m* à bille, clapet *m* de retenue à bille
~VISCOSIMETER - [rubber ind] viscosimètre *m* à chute de bille
~WARP BEAMING MACHINE - [text] dresseuse *f* mécanique pour chaîne en boyau
~WARP SIZING - [text] encollage *m* de la chaîne en boyau
~WINDING MACHINE - [text] pelotoneuse *f*, peloteuse *f*

BALLAST, to - [naut & aero] lester
[road& railw] empierrer, ballaster

BALLAST - [gen & aero] (any substance intentionally used to increase the weight of a vehicle or device) lest *m*, ballast *m*, charge *f*
[road & railw] empierrement *m*, ballast *m*
[concrete] cailloutage *m*
~BOXING - [road works] caisson *m* à ballast
~CAR - [railw] wagon *m* de terrassement
~CHIPS - [railw works] gravillons *m* (concassés) empierrement *m*
~CLEANER - [mech mach] (for railway works) machine *f* à laver le ballast
~CRIBBER - [mech mach] (for railway and road works) scarificatrice *f*
~HAMMER - [impl] marteau *m* à concasser
~HOPPER TRUCK - [mech] (road and railway works) benne *f* trémie
~LAMP - [el] lampe *f* ballast, ballast *m*
~PIT - [geol] ballastière *f*
~PUMP - [naut] pompe *f* des ballasts
~RAKE - [road works] rateau *m*
~RESISTANCE - [el] résistance *f* de charge
~RESISTOR - [el] (a self-controlling resistor used to compensate for variations in line voltage) résistance *f* de charge
~SCARIFIER - [mech mach] (road & railway works) scarificatrice *f*
~SCREENING MACHINE - [mech mach] machine *f* à cribler le ballast
~ SIDE-TIPPING TRUCK - [railw works] benne *f* de terrassement basculante
~TANK - [mech] (between diffusion pump and backing pump) réserve *f* de vide, réservoir *m* tampon
~TANK - [naut] ballast *m* (sous-marin etc)
~TUBE - [electron] (a ballast resistor in a gas filled tube, used to maintain constant current) tube *m* à résistance de charge
~WATER - [gen & naut] lest *m* liquide

BALLASTING - [naut & aero] lestage *m*
[constr] remblai *m*, remblayage *m*
[road & railw] empierrement *m*, ballastage *m*
~MATERIAL - [road & railw works] empierrement *m*, ballast *m*, ballastage *m*
~UP - [naut] lestage *m*

BALLING - [text] empelotage *m*, mise *f* en pelote
[metall] ballage *m* (du fer)
~AND REWINDING THE WARP - [text] empelotage *m* et rebobinage de la chaîne
~MACHINE - s. ball winding machine
~ROLLER - [text] rouleau *m* livreur, cylindre *m* dérouleur, rouloir *m*

BALLISTIC - [ballist] balistique

BALLISTIC CIRCUIT BREAKER - [el] disjoncteur *m* ultra-rapide
~CONE - [ballist] cône *m* balistique
~DENSITY - [rocketry] densité *f* balistique
~FACTOR - [of measuring instrument] (ratio between first and permanent deflection) facteur *m* balistique
~GALVANOMETER - [el] galvanomètre *m* balistique
~MISSILE - [ballist] engin *m* balistique, engin *m* à longue portée
~MISSILE EARLY WARNING SYSTEM - [ballist] (BMWS - system for giving warning of the approach of ballistic missiles with minimum delay) radar *m* d'alerte avancée anti-missile
~PATH - [ballist] trajectoire *f* balistique
~PENDULUM - [phys] pendule *m* balistique
~REENTRY - [astronaut] rentrée *f* balistique
~VEHICLE - [astronaut] véhicule *m* balistique
~WIND - [astronaut] vent *m* balistique
BALLISTICS - [ballist] balistique *f*
BALLISTITE - [min] balistite *f*
BALLISTOCARDIOGRAPHY - [med] balistocardiographie *f*
BALLONET - [aero] ballonet *m* compensateur
BALLOON - [aero] (lighter-than-aircraft, consisting of a flexible envelope filled wholly or partially with a gas of less specific gravity than of the atmosphere) ballon *m*, aérostat *m*
[print] ballon *m*
~ANCHOR - [aero] ancre *f* de ballon (ou d'aérostat)
~BARRAGE - [aero] barrage *m* de ballon captif
~BASKET - [aero] nacelle *f*
~BED - [aero] ancrage *m* de ballon captif
~CHECKING RING - [text] anneau *m* anti-ballon
~FABRIC - [text] tissu *m* pour ballons
~FLYING CABLE - [aero] câble *m* de retenue
~FORESAIL - [naut] foc *m* de ballon
~JIB - [naut] foc *m* ballon
~SEPARATING PLATE - [text] séparateur *m* anti-ballon
~SEPARATOR - s. ballon separating plate
~SHED - [aero] hangar *m* à ballon
~SILK - [text] soie *f* pour ballons
~-SONDE - [met] (a free unmanned balloon, carrying instruments, used to investigate atmospheric conditions) ballon-sonde *m*
~-TYRE - [rubber ind] pneu-ballon *m*
BALLOONING - [aero] s. bouncing
BALLOONIST - [aero] aérostier *m*, aéronaute *m*
BALLOT - [gen] scrutin *m*, vote *m*
BALLS OF MASSECUITE - [sugar ind] (fragments of a dense ball of sugar crystals mixed with mother liquor obtained by evaporation) grumeaux *mpl* de masse cuite
BALLSTUD - [mech] boulon *m* à tête ronde
BALLUTE - [astronaut] (parachute brake) ballute *m* (frein de parachute)
BALM - [gen] baume *m*
~OIL - [chem] essence *f* de mélisse
BALSA MEAL - [explos] farine *f* de balsa
~WOOD - [bot] balsa *m*
BALSAM - [chem] baume *m*
~FIR - [bot] sapin *m* baumier
~TANSY OIL - [chem] essence *f* de balsamite (ou de baume-coq)
BALSAMIFEROUS - [bot] balsamifère
BALSAMINE - [bot] balsamine *f*
BALUN - (abbrev. for BALanced-to-UNbalanced Transformer) (device which can be switched from a symmetrical to an asymmetrical system, obtaining at the time transformation of impedance) transformateur *m* symétrique-asymétrique
BALUSTER - [constr] balustre *m*, parapet *m* à balustres, fuseau *m* de rampe
BALUSTRADE - [build] balustrade *f*
BAMBOO - [bot] bambou *m*
~-REED - [bot] roseau-canne *m*, bamboo *m*
BAN, to - [gen] interdire, mettre au ban
BAN - [gen] ban *m*, interdiction *f*, interdict *m*
BANANA - [bot] banane *f*
~OIL - [ind chem] acétate *m* d'amyle
~PLUG - [el] fiche *f* banane
~TUBE - [telev] (colour display device) tube *m* banane (tube cathodique analyseur à tambour à lentilles)
BAND - [gen] bande *f*, courroie *f*, ruban *m*
[radio] bande *f*, plage *f*
[metall] frette *f*, bande *f*, bride *f*, collier *m*
[of a wheel] bandage *m*
[mining] filon *m* mince
[of shell] ceinture *f*
[print] nerf *m*, nervure *f*
[build] ruban *m*
[comput] groupe *m* de pistes
~BLADE - [impl] lame *f* de scie (à ruban)
~BOX - [paper man] carton *m* à chapeaux
~BRAKE - [mech] frein *m* à collier, frein *m* à ruban
~CHAIN - [meas] chaîne *f* à ruban
~CLUTCH - [mech] embrayage *m* à collier (ou à ruban)
~CONVEYER - [mech] transporteur *m* à courroie, convoyeur *m* à courroie
~COUPLING - [mech] accouplement *m* à collier (ou à ruban)
~DRIVE - [text] transmission *f* par corde
~EDGE - [phys] bord *m* d'une bande d'absorption
~ELIMINATION FILTER - [radio] (a wave filter with a single attenuation band) filtre *m* éliminateur de bande
~FILTER - [el] filtre *m* de bande
~HEATER - [plast ind] élément *m* de chauffage annulaire
~IRON - [mech] feuillard *m*
~LOSSES - [el] pertes *f* dues à des courants parasites
~NIPPERS - [impl] pince *f* de relieur
~-PASS/LOW-PASS ANALOGY - [radio] analogie *f* passe-bande/passe-bas
~PASS COIL - [el] filtre *m* passe-bande
~PASS FILTER - [radio] (a filter with negligible attenuation at all frequencies) filtre *m* passe-bande
~-PASS TUNING - [radio] syntonisation *f* à passe-bande
~PRESSURE LEVEL - [radio] niveau *m* de pression acoustique dans une bande déterminée
~PULLEY - [mech] poulie *f* à courroie
~RESPONSE - [radio] réponse *f* de bande
~ROPE - [text] câble *m* plat
~SAW - [impl] scie *f* à ruban
~SAW SHARPENER - [mech] affeteuse *f* pour scies à ruban
~SELECTOR - [radio] sélecteur *m* de gammes
~SHOCK ABSORBER - [auto] amortisseur *m* à ressort

à bandes
BAND SPECTRUM - [radio] spectre *m* de bande
~-SPREADING - [radio] étalement *m* de bande
~SWITCHING - [radio] commutation *f* de gammes d'ondes
~WHARVE - [text] noix *f* à corde
~WHEEL - [mech] poulie *f* à courroie
BANDAGE, to - [gen] bander
BANDAGE - [gen] bandage *m*, bande *f*, bandeau *m*
~WINDING MACHINE - [text] enrouleuse *f* pour pansements
BANDED CLAY - [geol] argile *f* rubanée
~COAL - [min] charbon *m* lité, charbon *m* barré
~VEIN - [mining] filon *m* zoné
BANDELET - [build] bandelette *f*
BANDING - [text] cordelette *f*
[geol] structure *f* de bande
~MACHINE - [mech mach] machine *f* à coudre les cahiers (de livres)
[text] machine *f* à commettre
BANDWIDTH - [radio] largeur *f* de bande
~CHARACTERISTIC - [radio] caractéristique *f* de l'onde passante
~COMPRESSION - [telev] compression *f* de largeur de bande
BANG-BANG CONTROL - [astronaut] commande *f* par tout ou rien
BANISTER - [constr] balustre *m*
BANISTERS - [constr] rampe *f* d'escalier, main *f* courante
BANJO - [mus] banjo *m*
~AXLE - [mech] pont *m* banjo
~CONNECTION - [mech] raccord *m* banjo
~UNION - [mech] raccord *m* banjo
BANK, to - [fin] mettre *m* en banque, déposer (dans une banque)
[road constr] relever, remblayer
[aero] viser sur l'aile, s'incliner sur l'aile
BANK - [gen] banc *m*, rampe *f*
[geol] rive *f*, berge *f*, banc *m* (de sable etc)
[comm] banque *f*
[mech] banc *m*, batterie *f*, assemblage *m*, groupe *m*, rang *m*, rangée *f*, faisceau *m*, rampe *f*
[mining] (the working face) front *m* de taille, front *m* d'abatage
(the surface around the opening of a shaft)carreau *m* de la mine
[el electron & etc] (row of push-button switches) portant *m*
[aero] inclinaison *f* de l'avion, virage *m* incliné
[cinema] rampe *m* de projecteurs
[roads] talus *m*, banquette *f*, remblais *m*, berge *f*
[min] (auriferous gravel) barre *f*
[print] banc *m*, rang *m*
[geogr] banc *m* de sable (etc), bord *m*
~AND BOND PAPER - [paper man] papier *m* coquille
~ANGLE - [aero] (the angle at which an aircraft is inclined during banking) angle *m* de virage
~CABLE - [telecomm] câble *m* à plusieurs âmes, câble *m* à plusieurs conducteurs
~DISCOUNT - [fin] escompte *m* en dehors
~HEAD - [mining] recette *f* supérieure du plan incliné
~INDICATOR - [aero instr] indicateur *m* de virage
~MULTIPLE - [telecomm] (the multiple connecting the bank of contacts in automatic telephone services) champ *m* de multiplage du banc de contacts
BANK OF BUTTONS - [electron etc] réglette *f* de touches, rangée *f* de touches
~OF CAPACITORS - [el] banc *m* de condensateurs
~OF CONTACTS - [telecomm] banc *m* de contacts
~OF CYLINDERS - [mech] cylindres *m* en ligne
~OF LIGHTS - [cinema] rampe *f* de projecteurs, herse *f*
~OF RUBBER - [on a mill] bourrelet *m* de caoutchouc
~OF TRANSFORMERS - [el] transformateur *m* montés en parallèle
~OVERDRAFT - [fin] découvert *m*
~PASS - [mech] écartement *m* de cylindres
~PROOF MACHINE - [comput] sélectionneuse *f* de chèques, machine *f* de comptabilisation des chèques
~RATE - [fin] taux *m* d'escompte
~REJECTOR FILTER - [el] filtre *m* éliminateur de bande
~RIDER - [mining] freineur *m*
~STEEL - [metall] feuillard *m* (d'acier)
~THE FIRE, to - [metall] coucher les feux
~TURN - [aero] virage *m*
~UP, to [constr] remblayer, surhausser, relever
~-UP WATER LEVEL - [hydr] surélévation *f* d'un plan d'eau
BANKED - [road & railw] surélevé
~BOILER - [ovens] chaudière *f* à feu couvert
~CURVE - [road works] courbe *f* relevée, virage *m* relevé
~EARTH - s. ballasting
~FIRE - (of ovens) feu *m* couvert, feu *m* réduit
~WINDING - [el] bobinage *m* à plusieurs couches entrelacées
BANKER - [gen] banquier *m*
BANKET - [min] conglomérat *m* aurifère
~STRUCTURE - [geol] structure *f* en plaquettes
BANKING - [aero] (inclination of an aircraft in respect to the lateral axis) virage *m* incliné, inclinaison *f* latérale
[road] relèvement *m*, remblayage *m*
[mining] atterrissage *m* de la cage
[railw] marche *m* en double traction, double traction *f*
~AT REAR - [railw] renfort *m* en queue de train
~PINS - [horol] ergots *m* d'arrêt
~UP - of ovens) couvrage *f* de feu, mise *f* en veilleuse
BANKSMAN - [oil ind] receveur *m*, porion *m* de surface
BANNER - [gen] bannière *f*, étendard *m*, pavillon *m*
~CLOTH - [text] étamine *f*
~CLOUD - [met] (cloud trailing from a mountain top in a flag-like form) nuage *f* en forme de drapeau
BANNOCK - [fodder] fourrage *m*
BANNS - [leg] bans *m*
BANQUET - [constr] (a continuous footing) banquette *f*
BAPTISTRY - [arch] baptistère *m*
BAR, to - [gen] barrer
(a steam engine) amorcer
BAR - [gen] barre *f*
[metall] barre *f*, barreau *m*, mise *f* (d'un paquet)
[constr] fer *m* à béton
[min] barre *f* à mine, fleuret *m*,
(a cross vein) filon *m* croiseur
[phys] (unit of pressure) bar *m*

[telev] (black line used for testing, or appearing as interference) barre *f*
[leg] barreau *m*, barre *f*
[mus] mesure *f*
[geol] barre *f*
[mech] barre *m* de commande; (a mounting) affût *m*; (a tamping rod) bourroir *m*
[gold] lingot *m*
BAR AGITATOR - [mech] agitateur *m* à barres
~AND TUBE TURNING MACHINE - [mech] tour *m* à barres et à tubes
~BENDER - [mech] machine *f* à cintrer les ronds (les barres)
~BENDING - [mech] cintrage *m* des ronds (des barres)
~BURNER - [th eng] brûleur *m* rectiligne
~CHANNELER - [mining] (in quarrying) trancheuse *f*
~-COAL-CUTTING MACHINE - [mining] haveuse *f* à barre coupante
~CODE - [comput] (machine readable code, in which the information is represented by standardized bars) code *f* de barres
~CONSTRUCTION - [constr] construction *f* triangulée, construction *f* à noeuds rigides
~COPPER - [el] cuivre *m* en barres
~CROPPING MACHINE - [mech] (machine for shearing or cropping lengths of metal bar) tronçonneuse *f*
~DOWN, to - [mining] abattre par des pinces
~GOLD - [metall] or *m* en barres
~GUIDE - [text] porte-guide *m* de doublage
~HANGER - [mech] appareil *m* de suspension à barre
~HOLE LEAK DETECTION - [gas ind] sondage *m* de recherche de fuites
~-HOLES - [mach tool] mortaises *f*pl
~IRON - [metall] fer *m* en barres
~KEEL - [shipbuild] quille *f* massive
~LATHE - [mach tool] tour *m* à barres
~LIGHT - [light] (a lamp park) parc *m* de projecteurs
~MAGNET - [el] barreau *m* aimanté, aimant *m* droit
~MARK - [rubber ind] bande *m* de treillis
~MILL - [metall] laminoir *m* marchand, train *m* de gros fers marchands
~MOULD - [plast ind] moule *f* à empreintes mobiles
~SCRAPER - [impl] grattoir *m*
~SCREEN - [impl] crible *m* à barreaux
[hydr] grille *f* à barreaux
~SHEARS - [metall] cisailles *f*pl à barres
~STRAIGHTENING MACHINE - [mech] machine *f* à redresser les barres
~TEST SURVEY - [gas ind] s. bar-hole leak detection
~TIMBERING - [tunnel constr] boissage *m*
~TIN - [metall] étain *m* en saumons
~TOOL HOLDER - [impl] porte-outil *m* à barre
~-TYPE TRANSFORMER - [el] transformateur *m* à barres
~WINDOW - [constr] fenêtre *f* grillée
~WOUND ARMATURE - [el] induit *m* en barres
BARB - [gen] ardillon *m* de crochet, barbillon *m* d'hameçon, barbe *f*, bavure *f*
BARBATE - [bot] barbé, aristé
BARBED - [metall] barbelé (fil de fer), aristé; (in foundry work) avec bavures
~AUGER - [impl] crochet *m* à deboucher
BARBED WIRE - [metall] fil *m* de fer barbelé, ronce *f* artificielle
~WIRE CUTTER - [impl] cisailles *f* à fil de fer barbelé
~WIRE FENCE - [gen] haie *f* de fil de fer barbelé, barrière *f* en fil de fer barbelé
BARBER'S RASH - [med] sycosis *m*, mentagre *f*
BARBETTE - [build] barbette *f*
~TOWER - [build] tour *f* à barbette
BARBIERITE - [min] barbiérite *f*
BARBITAL - [chem] (a derivative of urea with uses in medicine) acide *m* diéthylbarbiturique
~SODIUM - [chem] acide *m* sodico-diéthylbarbiturique
BARBITURATES - [chem] barbituriques *m*
BARBITURIC ACID - [chem] (a condensation agent in plastics manufacture; also an intermediate in the synthesis of pharmaceuticals) acide *m* barbiturique
BARBWIRE - [gen & agric] fil *m* de fer barbelé
BARE - [electr] (not insulated) nu
[constr] (of a shingle etc) pureau *m*
~-BACK PACKING - [packag] emballage *m* "bare back"
~CARBON - [el] carbone *m* nu
~DUCK - [text] toile *f* de revêtement extérieur
~DUCK BELT - [mech] courroie *f* de transmission à toile apparente
~ELECTRODE - [el] (a type of welding electrode which is not provided with a slag-forming preparation) électrode *f* non enrobée
~MEASURES - [mining] couches *f*pl stériles
~REACTOR - [nucl] (reactor without a surrounding reflector) réacteur *m* sans réflecteur, pile *f* sans réflecteur
~WIRE - [el] fil *m* nu, conducteur *m* nu
BAREBACK - [mech] tracteur *m* sans semi-remorque
BAREFACED TENON - [mech and carp] tenon *m* bâtard
BAREFOOT COMPLETION - [mining] nu-pied *m*
BARGAIN, to - [comm] négocier, traiter, faire marché
BARGAIN - [gen & comm] marché *m*, affaire *f*, bonne affaire *m*, occasion *f*
BARGE - [naut] péniche *f*, chaland *m*, mahonne *f*, bugalet *m* à munitions, bette *f* à saletés, gabare *f*, barque *f* de cérémonie
~COURSE - [constr] cordon *m*
BARGEMAN - [naut] batelier, marinier *m*
BARGEMASTER - [naut] patron *m* de chaland, patron *m* de gabare
BARILLA - [chem] (impure sodium carbonate) barille *f*
BARING - [mining] déblaiement *m* de terrains de recouvrement, dépouillement *m*, décapelage *m*
BARINGS - [geol] havrit *m*
BARITE - [min] s. barytes
BARITONE - [mus] baryton *m*
BARIUM - [chem] (an element, symbol Ba. A silvery, malleable, poisonous metal with uses in the production of alloys and in medicine) baryum *m*
~ACETATE - [chem] (a mordant in textile dyeing) acétate *m* de baryum
~ARSENATE - [chem] arséniate *m* de baryum
~ARSENIDE - [chem] arséniure *m* de baryum
~BICHROMATE - [chem] bichromate *m* de baryum
~BORIDE - [chem] borure *m* de baryum

BARIUM BROMATE - [chem] bromate *m* de baryum
~BROMIDE - [chem] bromure *m* de baryum
~CARBIDE - [chem] carbure *m* de baryum
~CARBONATE - [chem] (occurring naturally as witherite) carbonate *m* de baryum
~CHLORATE - [chem] (a constituent of explosives and textile chemicals) chlorate *m* de baryum
~CHLORIDE - [chem] chlorure *m* de baryum
~CHROMATE - [chem] chromate *m* de baryum
~CHROME - [chem] s. barium chromate
~CITRATE - [chem] citrate *m* de baryum
~CONCRETE - [radiat] (radiation protective material) béton *m* au baryum
~CYANIDE - [chem] cyanure *m* de baryum
~DICHROMATE - [chem] bichromate *m* de baryum, dichromate *m* de baryum
~DIOXIDE - [chem] bioxyde *m* de baryum
~ENEMA TEST - [radiol] examen *m* après lavement (opaque) baryté
~FERROCYANIDE - [chem] ferrocyanure *m* de baryum
~FLINT GLASS - [glass ind] verre *m* flint au baryum
~FLUORIDE - [chem] fluorure *m* de baryum
~GETTER - [electron] getter *m* au baryum
~HYDROXIDE - [chem] hydroxyde *m* de baryum
~HYPOPHOSPHATE - [chem] hypophosphate *m* de baryum
~IODATE - [chem] iodate *m* de baryum
~IODIDE - [chem] iodure *m* de baryum
~LACTATE - [chem] lactate *m* de baryum
~MANGANATE - [chem] manganate *m* de baryum
~MEAL - [med] (for radiographic purposes) bouillie *f* barytée
~MEAL EXAMINATION - [med] examen *m* de l'appareil digestif après ingestion de bouillie barytée
~MIXER - [radiol] malaxeur *m* pour bouillie barytée
~MONOXIDE - [chem] monoxyde *m* de baryum
[min] baryte *f*
~NITRATE - [chem] nitrate *m* de baryum
~NITRITE - [chem] nitrite *m* de baryum
~OLEATE - [chem] oléate *m* de baryum
~OXIDE - [chem] oxyde *m* de baryum
~PEROXIDE - [chem] peroxyde *m* de baryum
~PHOSPHITE - [chem] phosphite *m* de baryum
~PLASTER - [radiol] (protective building material) plâtre *m* baryté
~PROPIONATE - [chem] propionate *m* de baryum
~SALICYLATE - [chem] salicylate *m* de baryum
~SELENATE - [chem] séléniate *m* de baryum
~SILICATE - [chem] silicate *m* de baryum
~STEARATE - [chem] stéarate *m* de baryum
~SULPHATE - [chem] sulfate *m* de baryum
~SULPHIDE - [chem] sulfure *m* de baryum
~SULPHITE - [chem] sulfite *m* de baryum
~SWALLOW EXAMINATION - [chem] examen *m* pendant l'ingestion d'un repas baryté
~TARTRATE - [chem] tartrate *m* de baryum
~THIOCYANATE - [chem] thiocyanate *m* de baryum
~THIOSULPHATE - [chem] thiosulfate *m* de baryum
~TITANATE - [chem] titanate *m* de baryum
~TUNGSTATE - [chem] tungstate *m* de baryum
~YELLOW - [chem] jaune *m* de baryum
BARK, to - [zool] aboyer
BARK - [bot] écorce *f*
[metall] couche *f* intermédiaire décarburée
[naut]trois-mâts barque *m*, barque *f*
[animal] aboiement *m*
~FLAP - [bot] languette *f* d'écorce, lambeau *m*
BARK ROT - [bot] maladie *f* de l'écorce
~SHAVING - [rubber ind] pellicule *f* d'écorce, copeau *m* d'écorce
~STRIPPING MACHINE - [mech] écorceuse *f*
~TANNERY - [leather ind] tannerie *f* à l'écorce
BARKED WOOD - [paper ind] bois *m* écorcé, bois *m* pelard
BARKEN, to - [ind chem] taniser
BARKER - [mech] écorceuse *f*
BARKHAUSEN EFFECT - [radio] effet *m* Barkhausen
~OSCILLATOR - [radio] oscillateur *m* de Barkhausen, oscillateur *m* à grille positive
BARKING - [acoust] aboiement *m*
[agric] écorçage *m*
~LATHE - [paper ind] machine *f* à écorcer les grumes
BARKPEELING - [forestry] écorçage *m*
~MACHINE - [mech mach] écorceuse *f*
BARLEY - [bot] (a cereal used in making beer and spirits) orge *m*
~CORN - [bot] grain *m* d'orge
~FLOOR - [brew ind] grenier *m* à orge
~GRADER - [brew ind] trieuse *f* à orge
~MALT - [agric] malt *m* d'orge
~MEAL - [agric] farine *f* d'orge, farine *f* d'orge fourragère
BARM - [baking] (froth on malt liquor during fermentation) levure *f* de bière
~BEER - [brew ind] bière *f* de levure
~YEAST - [brew ind] levure *f* de bière
BARN - [gen] grange *f*
[nucl] barn *m*
~-FEEDING - [agric] alimentation *f* à l'auge, alimentation *f* à l'étable
BARNACLE - [zool] bernacle *f*, oie *f* marine
BARNDOOR GATE - [auto] porte *f* arrière rabattable latéralement
BARNDOORS - [cinema-colloq] (small movable screens) écrans *mpl* opaques
BARNEY - [mining] chariot *m* contrepoids d'un plan incliné, chariot *m* butoir
BARNSTORM, to [aero] piloter de manière acrobatique
BARNYARD ANIMALS - [zool] animaux *m* de basse-cour
~MANURE - [agric] fumier *m* d'étable
BAROCYCLONOMETER - [instr] barocyclonomètre *m*
BARODYNAMICS - [phys] (in the construction of heavy works) dynamique *f* des structures pesantes
BAROGRAM - [instr] barogramme *m*
BAROGRAPH - [met] (a recording barometer) barographe *m*, baromètre *m* enregistreur
BAROLUMINESCENCE - [met] baroluminescence *f*
BAROMETER - [met] (instrument for measuring the pressure of the atmosphere) baromètre *m*
~READING - [instr] hauteur *f* barométrique
BAROMETRIC - [instr] barométrique
~ALTIMETER - [instr] altimètre *m* barométrique
~GRADIENT - [met] gradient *m* barométrique
~PRESSURE CONTROL - [mech] commande *f* barométrique, régulateur *m* barométrique
~SWITCH - [el] (switch controlled by the pressure of the atmosphere) interrupteur *m* barométrique
~TENDENCY - [met] (the changes in barometric pressure occurring during a given time, usually 3 hours, before the observation) tendance *f* barométrique

BAROQUE - [arch] baroque *m*
BAROSCOPE - [met] (an instrument giving approximate indications of changes in atmospheric pressure) baroscope *m*
BAROSTAT - [instr] commande *f* barostatique, barostat *m*
~FUEL CONTROL - [mech] (a system of gas turbine fuel control in which a barostat capsule operates the relief valve) commande *f* barostatique d'admission (de carburant)
BAROSTATIC RELIEF VALVE - [mech] commande *f* barostatique de pression d'admission
BAROTHERMOGRAPH - [instr] (a combined instrument recording several magnitudes simultaneously e.g. temperature, pressure, humidity) barothermographe *m*
BAROTRAUMA - [astronaut] barotraumatisme *m*
BAROTROPIC - [phys] (of a fluid, when its density varies with the pressure) barotropique
~VORTICITY EQUATION - [phys] équation *f* de vorticité barotropique
BARQUE - [naut] trois-mâts barque *m*
BARQUENTINE - [naut] trois-mâts goélette *m*
BARRACK - [gen] baraque *f*
BARRACKS - [constr] caserne *f*
BARRAGE -[hydr] barrage *m*
~BALLOON [aero] (a small captive-kite balloon used as a defence against low-flying aircraft) ballon *m* de protection
~FIRE - [milit oper] tir *m* en barrage
~FIXE - [hydr] barrage *m* fixe
~JAMMING - [radio] (various frequencies jammed simultaneously) brouillage *m* simultané sur plusieurs gammes
~MOBILE - [hydr] barrage *m* mobile
BARRED BOX - [metall] châssis *m* à barres
BARREL - [gen] baril *f*, tonneau *m*, fût *m*
[firearms] canon *m*
[mech] cylindre *m*, corps *m* de pompe
[ovens] (of a boiler) corps *m* cylindrique
[mech] (of a capstan) cloche *f*
[mech](of a crab-winch) tambour *m*
[mech] (of a syringe) canon *m*
[mech] (of a screw] tige *f*
[opt] (of a lens) barillet *m*
[naut] cloche *f* de cabestan, tambour *m* de treuil
[unit of meas] baril *m*
~ARBOR - [horol] arbre *m* à barillet
~ARCH - [build] voûte *f* en berceau
~ATTEMPERATOR - [brew ind] réfrigérant *m*
~-BOLT - [mech] verrou *m* à coquille
~CHURN - [agric] baratte *f* rotative
~CONVERTER - [metall] convertisseur *m* horizontal
~DISTORTION - [telev] (distortion of the picture which bulges at the sides) distortion *f* en tonneau
~DRAIN - [build] buse *f*
~DRUM - [horol] barillet *m* (de mouvement d'horlogerie)
~ELECTRO-PLATING - [electro-metall] (mechanical plating in which the cathodes are kept loosely in a rotating container) revêtement *m* électrolytique au tonneau
~FINISHING - [electro-metall] polissage *m* au tonneau
~NIPPLE - [mech] mamelon *m* double
~OF A PUMP - [mech] corps *m* de pompe
~ORGAN - [mus] orgue *m* de Barbarie
BARREL PLATING - [metall] métallisation *f* électrolytique au tonneau, rotogalvanostégie *f*
~POLISHING - [metall] polissage *m* au tonneau, tonnelage *m*
~PROCESS - [min] (an ore-dressing process) procédé *m* du baril
~ROLL - [aero] (an aerobatic manoeuvre in which the aircraft follows a spiral path while at the same time revolving about the longitudinal axis) tonneau *m*
~SHAPED DOUBLE FLANGED BOBBIN - [text] bobine *f* bombée
~SHUTTER - [cinema] obturateur *m* cylindrique
~-SLING - [mech] élingue *f* de barrique, patte *f* à futaille
~SPANNER - [mech] clé *f* à tube
~TYPE CRANKCASE - [mech] (used in internal combustion engines) carter *m* d'une seule pièce, carter *m* tubulaire
~TYPE ENGINE - [mech] moteur *m* pour transporteur à courroie
~-VAULTED - [arch] à voûte en berceau
~WINDING - [el] enroulement *m* en manteau
BARREN - [gen biol & bot] stérile
~SAND - [geol] sable *m* stérile
~WELL - [oil ind] sondage *m* stérile
BARRETTER - [el] (ballast resistor enclosed in a gas-filled vessel) tube *m* régulateur fer-hydrogène, bolomètre *m*
BARRICADE - [gen & milit] barricade *f*
BARRICO - [naut] baril *m* d'eau douce
BARRIER - [gen] obstacle *m*, barrière *f*
[el] isolant *m*
[in circuit breakers] écran *m* anti-arc
[cinema] (the black track between sound track and pictures) ligne *f* de séparation
[genet] isolation *f*, barrière *f*
~DIFFUSION METHOD - [nucl] (separation of isotopes in which a gas which is an isotopic mixture is allowed to escape) méthode *f* par barrière de diffusion
~FREQUENCY - [electromagn waves] fréquence *f* de coupure
~HEIGHT - [nucl] (maximum energy of a Coulomb barrier) hauteur *f* de barrière
~LAYER - [el] (double electrical layer formed at the surface contacts between two substances) couche *f* d'arrêt, couche *f* de barrage
~LAYER CAPACITANCE - [el] capacité *f* d'une couche d'arrêt
~PILLAR - [mining] pilier *m* de limite
~REEF - [geol] récif *m* en barrière
~SPRING - [hydr] source *f* de barrage
~TO FISSION - [nucl] barrière *f* de fission
BARRING - [mining] boisage *m* du toit
~DOWN - [mining] abattage *m* au moyen de pinces
~ENGINE - [mech] (equipment for turning the fly-wheels of large engines) moteur *m* démarreur, vireur *m*
~GEAR - [mech] (equipment for moving heavy machinery) engin *m* de levage pour gros matériel
~MOTOR - [mech] moteur *m* démarreur, vireur *m*
~WHEEL - [mach] (rope pulley) volant *m* à gorge, poulie *f* à corde
BARRISTER - [leg] avocat *m*
BARROW - [impl] brouette *f*
[mining] (a heap of dead ore) halde *f*, charrée *f*,

terril *m*
BARROW PUMP - [hydr] pompe *f* mobile
BARTER - [gen & comm] échange *m*, troc *m*
BARTIZAN - [arch] bretèche *f*, échauguette *f*
BARYCENTRE - [geom] barycentre *m*
BARYE - [phys] (the absolute cgs unit of pressure) barye *f*, microbar *m*
BARYSPHERE - [geol] barysphère *f*
BARYTA - [chem] (barium oxide) baryte *f*
~GREEN - [ind chem] vert *m* de baryte
~YELLOW - [ind chem] jaune *f* de baryte
BARYTES - [min] (heavy spars. Natural calcium phosphate and the most important source of calcium compounds) barytine *f*
BARYTOCALCITE - [min] barytocalcite *f*
BASAL - [gen] fondamental, basique, de base
~ANAESTHESIA - [med] (anaesthesia acting as basis for further deeper anaesthesia) anesthésie *f* de base
~CELL - [bot] (uninucleate cell) cellule *f* basale
~CORPUSCLE - [bot] corpuscule *m* basal
~METABOLIC RATE - [zool] (rate of oxygen consumption in a resting organism) taux *m* métabolique basal
~PLANES - [cryst] plans *m* basaux
BASALT - [min] (a fine-grained igneous rock of dark colour) basalte *m*
BASALTIC TUFF - [geol] tuf *m* basaltique
BASCULATION - [med] redressement *m* de l'utérus rétrodévié
BASCULE - [mech] bascule *f*
~BARRIER - [railw] barrière *f* à bascule
~BRIDGE - [build] pont *m* basculant, pont *m* à bascule
~GATE - [hydr] vanne *f* à bascule, porte *f* d'écluse
~-TYPE SETTLING CONE - [metall] cône *m* décanteur basculant
BASE - [gen] base *f*, fondation *f*, assise *f*
[el] culot *m*, base *f*, embase *f*, patère *f*
[chem] base *f*
[motor & engine] bâti *m*, socle *m*
[mach tool] support *m*, pied *m*, bâti *m*
[comput] (for magnetic film) support *m*
[constr](basement) soubassement *m*
[electron] (of an electronic tube) base *f*, culot *m*
[impl] (of a plummer-block) patin *m*, semelle *f*
[top] base *f*
[adj] (of metals) commun, vil
~ANGLE - [geom] angle *m* à la base
~-BAND - [radio] bande *f* de base
~BEARING - [mech] palier *m* de vilebrequin
~BOARD - [constr] plinthe *f*
~BULLION - [metall] plomb *m* brut
~CIRCLE - [mech] cercle *m* de base
~COLLECTOR JUNCTION - [el] jonction *f* base-collecteur
~COMPOUND - [rubber ind] mélange *m* de base
~CONE - [mech] (of gears) cône *m* de base
~CONGLOMERATE - [geol] conglomérat *m* de base
~COURSE - [constr] couche *f* de fondation, couche *f* de base
~CURRENT - [el] courant *m* base
[electron] (of a transistor) courant *m* de base, courant *m* base
~DIAMETER - [mech] diamètre *m* du cercle de base
~DRAG - [astronaut] traînée de culot
~DYE - [dyes] couleur *f* de fond
~ELBOW - [hydr] coude *f* à pied (ou à patin)
BASE ELECTRODE - [metall] électrode *f* base
~EMITTER JUNCTION - [radio] jonction *f* émetteur-base
~EXCHANGE CAPACITY - [soil] capacité *f* d'échange des bases
~EXPLOSION - [mech] explosion *f* dans le carter moteur
~IRON - [metall] fonte *f* de base
~LEAD - [el] connexion *f* de base
~LINE - [top] base *f*, ligne *f* de terre
~LOAD - [el] charge *f* minimale
~MATERIAL - [gen] matière *f* première, corps *m* simple
~METAL - [metall] (generally copper, lead, tin or zinc: a metal having a relatively low negative electrode potential) métal *m* basique
~MIX - [rubber ind] mélange *m* de base
~MOULDING - [build] moulure *f* de base
~MOUNTING - [el] montage *m* sur pied
~OF A A DAM - [hydr] assise *f* d'un barrage
~OF BOBBIN - [text] base *f* de bobine
~OF NATURAL LOGARITHM - [math] (the base of natural or Napierian logarithms: 2.7182818) base *f* logarithmique (log. népérien)
~OF RIM - [rubber] base *f* de jante
~OF THE RAIL - [railw] patin *m* de rail
~OF TRIANGULATION - [top] base *f* de triangulation
~ORE - [min] minerai *m* commun, minerai *m* pauvre
~PLATE - [mech] embase *f*, plaque *f* d'appui, plaque *f* base
[mech] plaque *f* de fondation
~REGION - [electron] (the region of a transistor into which minority carriers are injected) base *f*, galette *f* de base
~REGISTER - [comput] (indexing register) registre *m* de base, registre *m* index
~SATURATION - [soil] (exchange capacity saturated with metallic cations) saturation *f* en bases
~SIGNAL - [electron] signal *m* de base
~SLAB - [constr] radier *m* de fondation, semelle *f*, dalle *f* de fondation
~SURGE - [nucl] (outward rolling cloud in a subsurface explosion) nuage *m* de base
~TANK - [aero] réservoir *m* principal
~TRIM - [build] moulure *f* d'embase
~UNIT - [mech] plaque *f* de fondation, plaque *f* d'appui
~VEIL - [photo] voile *m* de fond
~WALL - [constr] mur *m* de soubassement
~WELL RIM - [rubber ind] jante *f* à base creuse
~WIDTH - [tyres] largeur *f* de la base (de jante)
~YEAR - [statist] année *f* de base
BASEBALL FINGER - [med] doigt *m* en marteau
BASEBOARD CAMERA - [photo] chambre *f* à base
BASEMENT - [build] soubassement *m*, sous-sol *m*, cave *f*
[mech mach] embase *f*, base *f*
~MEMBRANE -[anat] couche *f* sousépithéliale
BASH, to-[mining] remblayer
BASHER - [light] (in a studio) bol *m*, réflecteur *m*
BASIC - [gen] fondamental, de base
[chem] basique
~ALUM - [chem] alum *m* basique
~BLUE - [ind chem] bleu *m* basique
~CARBONATE - [chem] carbonate *m* basique
~CIRCUIT - [electron] montage *m* de principe, circuit *m* fondamental

BASIC DUTY - [comm] (in the EEC) droit *m* de base
~DYESTUFF - [ind chem] colorant *m* basique
~ELEMENT - [automat] (element performing a necessary and specified function in a sequence of operations) équipement *m* de base
~GROUP - [radio] (group of basic circuits) groupe *m* primaire de base
~HOLE - [mech] trou *m* de base
~INDUSTRY - [comm] industrie *f* de base
~INSTRUCTION - [comput] instruction *f* de base
~LATTICE - [nucl] réseau *m* fondamental
~LEAD CHROMATE - [chem] chromate *m* de plomb basique
~LEAD SULPHATE - [chem] sulfate *m* de plomb basique
~MACHINE TIME - [comput] cycle *m* de base
~MANURE - [agric] fumier *m* de base
~MATERIAL - [ind chem & plast] matière *f* première, matière *f* de base
~METALLIC RECTIFIER - [electron] (rectifier, each rectifying element of which consists of a single metallic-rectifier cell) redresseur *m* métallique cellulaire
~NETWORK - [radio] réseau *m* fondamental
~NOISE - [radio] bruit *m* propre
~OPEN HEARTH STEEL - [metall] acier *m* Martin (obtenu au four basique)
~OXIDE - [chem] (an oxide having the properties of a base or one which forms a hydroxide with water) oxyde *m* basique
~PIG - [metall] fonte *m* basique
~PRICE - [comm] prix *m* de base
~PROCESS - [metall] (steel making process) procédé *m* au four basique
~RATE - [comm] salaire *m* de base
~REFRACTORY - [metall] (heat resisting material used to line furnaces) garnissage *m* basique, revêtement *m* basique
~SALT - [chem] (a normal salt combined with the specific amount of the base) sous-sel *m*
~SIZE - [mech] cote *f* nominale
~SLAG - [ind chem] (by-product of steel manufacture with uses as fertilizer) scorie *f* basique
~STEEL - [metall] acier *m* basique
~SUPERGROUP - [radio] groupe *m* secondaire de base
~TOLERANCE - [mech] tolérance *f* de base
~WEIGHT - [aero] (aircraft weight including all fixed operating equipment and trapped fuel and oil, but not expendable load) poids *m* brut, poids *m* en ordre de marche
~WIRING - [comput] connexion *f* de principe
BASICITY - [chem] (the number of hydrogen atoms in an acid which can be replaced by metal atoms) basicité *f*
BASIFY, to [chem] rendre basique
BASIL - [leather ind] (tanned, not yet dyed sheepskin) basane *f*
BASILAR - [gen zool & bot] basilaire
BASILISK - [zool] basilic *m*
BASIN - [gen] bassin *m*, réservoir *m*, cuvette *f*
[hydr] bassin *m*
[naut] bassin *m*,darse *f*
[geol] bassin *m* d'un fleuve
[metall] bassin *m* de coulée
~AND GATE - [metall] entonnoir *m* de coulée, bassin *m* de coulée
~FLOOR - [hydr] radier *m* de réservoir
BASIN TRIAL - [naut] essai *m* au bassin
~WITH HANDLE - [metall] casserole *f*
BASIS - [gen] base *f*, principe *m*
BASKET - [aero] (car suspended beneath a balloon for passengers etc) nacelle *f*
[gen] corbeille *f*, panier *m*
[sugar refining] (wire mesh lining in the centrifuge to retain the crystals) panier *m*
~CELLS - [med] cellules *f* en panier de Boll
~COIL - [radio] bobinage *m* en fond de panier
~HANDLED ARCH - [build] arc *m* en anse de panier
~TROLLEY - [text] chariot *m* à panier
~-TYPE CORE BARREL - [mining] tube-carottier *m* avec arrache-carotte
~WEAVE - [text] armure *f* panama
BASOID - [soil] (soil constituents displaying basic properties) basoide
BASQUE - [metall] revêtement *m* d'un four
BASS - [mus] basse *f*
[bot] liber *m* , filasse *f*, raphia *m*
[geol] argile *f* compacte
~-BAR - [mus] barre *f* d'harmonie
~BOOST - [acoust] (amplification to the lower frequencies) amplification *f* des basses fréquences
~CLEF - [mus] clef *f* de fa
~CORRECTOR - [rad & acoust] (a circuit correcting the lower frequencies) circuit *m* correcteur des graves
~FIBRE - [text] (piassava) fibre *f* de piassava
~FREQUENCIES - [acoust] graves *m*
~HORN - [mus] cor *m* de basset
~TUNING - [acoust] accord *m* des graves
BASSETITE - [min] bassetite *f*
BASSOON - [mus] basson *m*
BASSWOOD - [bot] tilleul *m*
BAST - [bot] s. bass
~BAND - [text] ruban *m* filasse
BASTARD - [gen zool etc] bâtard
[geol] roche *f* massive
~COAL - [min] charbon *m* dur
~CUT - [mech] taille *f* bâtarde
~FALLOW - [agric] jachère *f* intermittente
~FILE - [mech] (a file intermediate in "cut" or depth of tooth between "second-cut" and "smooth") lime *f* bâtarde
~FOUNT - [print] bâtarde *f*
~GRANITE - [geol] gneiss *m*
~PITCH - [mech] pas *m* bâtard
~STATUARY - [arch] marbre *m* statuaire veiné
~THREAD - [mech] filet *m* bâtard
~VAT - [text] cuve *f* bâtarde
~WING - [zool] (quill feathers borne on the first digit of the wing) alule *f*
BASTE, to - [text] bâtir, faufiler
[in the kitchen] arroser de jus (ou de graisse)
BASTIAN METER - [instr] ampèreheure-mètre *m* à électrolyse
BASTING - [text] bâti, faufilage *m*
~COTTON - [text] coton *m* à bâtir, fil *m* à bâtir
BASTION - [arch] bastion *m*
BASTITE - [min] bastite *f*
BASTOSE - [chem] cellulose *f* de jute
BAT - [gen] batte *f*, battoir *m*, palette *f*
[zool] chauve-souris *f*
[build] briqueton *m*
[geol] schiste *m* bitumineux compact
[min] (clay in a coal deposit) gore *f*, gord *m*

BAT-BOLT - [mech] boulon *m* de scellement à crans
~EAR - [anat] creille *f* en anse
BATCH, to - [gen ind chem etc] doser, mélanger
[rubber ind & plast ind] traiter en lots
BATCH - [gen] groupe *m*, lot *m*, fournée *f*, série *f*
[comm] lot *m* de marchandise, quantité *f* produite
[metall] charge *f*, lot *m* d'un produit
[ind chem] dose *f*, opération *f* discontinue
[glass man] charge *f*
~-BOARD - [auto] cloison *f* de fractionnement
~BOOKING - [telecomm] demande *f* de communication en série
~BOOKINGS - [telecomm] liste *f* de demandes de communications en série
~-BULK PROCESSING - [comput] traitement *m* échelonné des données, traitement *m* par étapes des données
~CARBONIZATION - [sugar ind] carbonatation *f* discontinue
~CHARGER - [glass man] chargeuse *f*
~COKE STEEL - [metall] four *m* de coke intermittent
~DISTILLATION - [ind chem] (a special technique in which the still is charged with all the substances to be distilled before distillation is begun) distillation *f* intermittente
~FEEDER - [mech] distributeur-doseur *m*
~OFF, to - [ind process] prélever en quantité limitée, répartir en lots
(to take small quantities from a process machine for treatment in a separate batch) enlever du malaxeur (en petites quantités)
~-OFF MILL - [mech] mélangeur *m* à cylindres
~PROCESS - [ind chem] procédé *m* discontinu
~PROCESSING - [comput] (operations preceding the updating of the master records) pré-groupement *m*
[photo] façonnage *m* en grandes quantités
~PRODUCTION - [comm & ind] production *f* discontinue, production *f* en séries limitées
~SAMPLING - [comm etc] prélèvement *m* par lots
~STILL - [ind chem] (for the distillation of benzol) installation *f* de distillation discontinue, distillateur *m* discontinu
~TEST - [mech] (test made on specimens from a given batch) essai *m* par prélèvement
~-TYPE FURNACE - [metall] four *m* discontinu, four *m* dormant
BATCHER - [ind chem] doseur *m*
~PLANT - [chem & ind applications] installation *f* de dosage
BATCHING - [gen & naut] groupage *m*
[ind chem] dosage *m*
~PLANT - [ind chem] dispositif *m* doseur
~ROLLER - s. balling roller
BATCHWISE - [gen & comm] en discontinu, par lots de manière intermittente
BATEA - [min] (flat cone-shaped pan of wood or iron) battée *f*, sébile *f*, auge *f*
BATH - [gen] bain *m*
[chem] bain *m*
[metall] bain *m*
~-HEATER - [th eng] chauffe-bain *m*
~ITCH - [med] dermatite *f* des piscines
~LUBRIFICATION - [mech] graissage *m* par barbotage
~-TUB - [build] baignoire *f*
BATHESTHESIA - [med] sensibilité *f* profonde

BATHOCHROMES - [chem] bathochromes *m*
BATHOLITH - [geol] (igneous rock) batholithe *f*
BATHOMETER - [instr] (instrument designed for deepsea soundings) bathymètre *m*
BATHROOM - [build] salle *f* de bain
BATHYHYPERESTHESIA - [med] hyperesthésie *f* profonde
BATHYHYPOESTHESIA - [med] perte *f* de sensibilité profonde
BATHYMETER - [instr] s. bathometer
BATHYMETRY - [meas] bathymétrie *f*
BATHYPNEA - [med] respiration *f* profonde
BATHYSCAPH - (diving apparatus capable of resisting very great pressure) bathyscaphe *m*
BATHYSMAL - [zool] abyssal
BATHYSPHERE - (bathyscaph built in spherical shape) bathysphère *f*
BATHYTHERMOGRAPH - [instr] (temperature recorder in the sea) bathythermographe *m*
BATIK, to - [text] imprimer au batik, batiker
BATIK PRINTING - [text] impression *f* batik
BATING - [leather ind] (steeping of light skins in fermenting solution) chipage *m* des peaux
[mining] abaissement *m* du plan d'excavation
BATISTE - [text] batiste *f*
~RIBBON - [text] ruban *m* de batiste
BATON - [mus] baton *m* (de chef d'orchestre)
BATSWING BURNER - [mech] bec *m* papillon
BATT - [geol] schiste *m* bitumineux compact
BATTEN, to - [naut] fermer les panneaux, condamner les panneaux
BATTEN - [gen & carp] couvre-joint *m*, baguette *f*, latte *f*
[text] battant *m*
[constr] volige *f*, grosse latte *f*, tringle *f*
[theatre] herses *f* d'éclairage
[naut] barre *f*, latte *f*, claire-voie *f*, vaigrage *m* à claire-voie
~DOOR - [carp] porte *f* à claire-voie, porte *f* à traverses
~FLOOR - [build] parquet *m* planchéié
~PIN - [text] pivot *m* de battant
~PLATE - [build] traverse *f* de liaison, étrésillon
~SPRING - [text] ressort *m* de battant
BATTENED PARTITION - [build] cloison *f* en lattes
BATTER, to - [gen & mech] battre, canonner
[constr] terrasser, taluter
BATTER - [build] inclinaison *f* (sur la verticale), fruit *m* (d'un mus)
[constr] talus *m*, escarpe *f*
[hydr] (in dam works) fruit *m*
[print] caractère *m* écrasé
[cooking] pâte *f* lisse
~LEVEL - [surv] (form of clinometer) clinomètre *m*
~PILE - [build] (pile driven in at an angle to the vertical) pieu *m* porteur incliné
~POST - [build] pieu *m* incliné
BATTERED - [mech] (bruised or deformed by blows) délabré, bossué
BATTERING RAM - [impl] bélier *m*
BATTERY - [gen] (a number of pieces of equipment grouped together) batterie *f*, série *f*
[el] (a group of two or more primary cells or accumulators) batterie *f*, pile *f*, accumulateur *m*
~ACID - [chem] (sulphuric acid of a strength which is usable in accumulators) acide *m* pour accumulateurs

ATTERY BOOSTER - [el] (a motor generator set used to supply the necessary extra voltage to enable a battery to be charged from a source of voltage equal to the normal output voltage of the battery) survolteur *m*
~BOX - [el] bac *m* d'accumulateur
~CAPACITY - [el] puissance *f* utile d'un accumulateur
~CARRIER - [mech] chariot *m* porte-batteries
~CARRYING STRAP - [impl] sangle *f* pour le transport des batteries
~CART - [el] (a small vehicle containing an accumulator battery, used for starting aircraft engines on the ground) chariot *m* porte-batteries
~CELL - [el] (an individual cell of a battery) élément *m* d'accumulateur
~CHARGER - [el] chargeur *f* de batterie (ou d'accumulateur)
~COMPARTMENT - (in a factory or in a submarine) compartiment *m* des accumulateurs, salle *f* des accumulateurs
~COVER - [mech] couvercle *f* de batterie
~CUT-OUT - [el] (an automatic switch used to disconnect a battery being charged when the charging voltage falls too low) interrupteur *m* de batterie
~-DRIVEN - [el mech] alimenté par batterie
~ELIMINATOR - [radio] (arrangement for supplying power from mains to radio sets, usually operated by batteries) appareil *m* remplaçant une batterie
~FILLER - [auto etc] dispositif *m* de remplissage pour batterie
~FILLING - [el] bouchon *m* de remplissage de batterie
~HUM - [radio] bruit *m* d'alimentation
~IGNITION - [el] allumage *m* par batterie
~LIFTER CARRIER - [mech auto] sangle *m* de levage et de transport pour batteries
~LOOP - [el] bouclage *m* par batterie
~MANGANESE - [chem] bioxyde *m* de manganèse
~MASTER SWITCH - [el] interrupteur *m* principal de batterie
~OF TESTS - [ind phys] série *f* de tests
~PLATE - [el] plaque *f* d'accumulateur
~PLATE TERMINAL - [el] borne *f* de connexion
~PULSING - [radio] signalisation *f* par impulsions de tension
~SUPPLY BRIDGE - [el] pont *m* d'alimentation
~SUPPLY CIRCUIT - [el] pont *m* d'alimentation
~SUPPLY RELAY - [el] relais *m* d'alimentation
~SUPPORT - [mech] support *m* de batterie
~SYRINGE - [auto] seringue *f* pour batterie
~TESTER - [instr] vérificateur *m* de batterie
~TRAY - [impl] berceau *m* de batterie
~TRIODE - [electron] (triode used in battery sets) triode *f* pour récepteur (fonctionnant sur batterie)
~VENT PLUG - [mech] bouchon *m* à orifice de ventilation
~VOLTAGE - [el] tension *f* de la batterie
BATTLE - [gen] bataille *f*, combat *m*
~CRUISER - [naut milit] croiseur *m* de bataille
~FATIGUE - [med] traumatisme *m* des combâts
BATTLEAXE - [impl] hache *f* d'armes
BATTLEGROUND - [gen] champ *m* de bataille
BATTLEMENT - [constr] (indented parapet at the top of a wall) crénelage *m*, créneau *m*, merlon *m*
BATTLESHIP - [naut milit] cuirassé *m*
BATWING AERIAL - [telev] antenne *f* croisée multiple
BAUD - [telecomm] (in telegraphy, the unit of speed of transmission equal to twice the number of dots per second) baud *m*
BAULK - [mining] étranglement d'une couche
~END - [text] extrémité *f* de peigne
~OF REED - [text] coronelle *f*, jumelle *f* du peigne
BAUXITE - [min] (an important ore of aluminium. Also used as a filler in plastics and rubber) bauxite *f*
~KILN - [min] four *m* pour bauxite
BAVENITE - [min] bavénite *f*
BAWLING - [acoust] (sustained shouting) hurler, brailler, hurlement *m* prolongé
BAY - [geog] baie *f*, anse *f*
[constr](any division of an arcade or space between two columns in a building) travée *f*, baie *f*, travée *f* de façade
[in a factory] atelier *m*, section *f*, hall *m*
[in a shop] section *f*
[aero] travée *f* d'aile
[in a hangar] travée *f*
[of a window] baie *f*, balcon *m*
[hospital & naut] infirmerie
[bot] laurier *m*
~FRONT - [mech] panneau *m* avant (de semi-remorque)
~OF A WALL - [build] masque *m* de mur, panneau *m*
~QUOIN - [build] pan *m* rectangulaire coupé
~TREE - [bot] laurier *m*
~WINDOW - [build] (internal recess formed when a wall projects outside the general line) fenêtre *f* en saillie, fenêtre *f* à balcon, bretèche *f*
BAYING - [acoust] aboiement *m*
BAYONET - [gen & el] baïonnette *f*
~CAP - [el] culot *m* à baïonnette
~CATCH - [mech] fermeture *f* à baïonnette, joint *m* à baïonnette
~HOLDER - [el] douille *f* à baïonnette
~JOINT - s. bayonet catch
~LOCK - s. bayonet catch
~MOUNT - [mech] support *m* à baïonnette
~PLUG - [el] bouchon *m* à baïonnette
~PLUNGER - [mech] piston *m* plongeur à baïonnette
[diecasting] plongeur *m* à baïonnette
~SOCKET - [el] (a lampholder in which the contacts consist of spring-loaded pins) douille *f* à baïonnette
B.D.C. - [mech] (abbrev. for Bottom Dead Centre) point *m* mort inférieur
B.D.V. - [el] (abbrev. for Break Down Voltage) tension *f* de percement, tension *f* de rupture
BEACH, to - [naut] (to haul out of the water on to a sloping shore) échouer, mettre au sec
BEACH - [top] plage *f*, grève *f*, rivage *m*
~-GRASS - [agric] roseau *m* des sables
~HEAD - [milit] tête *f* de pont
BEACHED - [naut] échoué, mis au sec
BEACHING - [aero] mise *f* à terre, mise *f* au sec
~GEAR - [aero] (a wheeled carriage used for beaching a sea-plane) dispositif *m* de mise à terre
BEACON, to - [gen] baliser, échelonner les feux (ou les radiophares)
BEACON - [gen] phare *m*, balise *f*, projecteur *m* fanal *m*
[navig] balise *f*, repère *m*, marque *f* feu *m*
[radio] radiophare *m*

BEACON FIRE - [aero] feu *m* de balisage
~MODULATOR - [radar] (modulator used to supply the signal to the radar beacon) modulateur *m* de balise radar
~SKIPPING - [nav] silence *m* de balise répondeuse
~STEALING - [nav] subtilisation *f* de balise répondeuse
~TRACKING - [nav] poursuite *f* par balise
BEACONED - [navig] (provided with beacon) balisé
BEAD, to [mech] (to form a bead or beading) appliquer une baguette (un joint, un cordon), dudgeonner (un tube)
BEAD - [mech] bourrelet *m*, talon *m*, cordon *m*
[chem] perle *f*
[build] bourrelet *m*, baguette *f* (moulure convexe), congé *m*
[metall] passe *f* de soudure, cordon *m* de soudure
[rubber tyres] talon *m*
[plastics] cordon *m*
~BASE - [of tyre] (the lower surface of a tyre bead) base *f* de talon
~CORE - [rubber ind] armure *f* de talon
~EFFECT - [text] effet *m* perlé
~FLIPPER - [mech mach] machine *f* à fabriquer les armures de talon, machine *f* à enrober les armures de talon
~FRACTURE - [rubber ind] cassure *f* du talon
~HEEL - [rubber ind] pointe *f* extérieur du talon
~IN THE YARN - [text] grosseur *f*
~OF A TYRE - [rubber ind] (the thickened portion of a pneumatic tyre, which engages the rim of the wheel) talon *m* de pneumatique
~OF RIM - [rubber ind] rebord *m* de jante
~SEPARATING DISC - [text] disque *m* séparateur de perles
~SEPARATOR - [text] séparateur *m* de perle
~TEST - [metall] essai *m* à la perle
~THERMISTOR - [instr] thermistor *m* à perle
~THREAD - [text] coton *m* perlé
~TOE - [rubber ind] pointe *f* intérieure du talon
~TRANSISTOR - [electron] transistor *m* à perle
~TRIMMING MACHINE - [mech mach] machine *f* à ébarber les talons
~TWIST - [text] fil *m* retors perlé
~WARP - [text] chaine *f* perlée
~WEAVING - [text] tissage *m* perlé
~WEFT - [text] trame *f* perlée
~WELD, to - [metall] (weld laid down in the form of a single pass) faire un depôt en une passe
~WELD - [metall] soudure *f* en cordon, soudure *f* à cordon
~WIRE - [of tyre] (round steel wire incorporated in the bead to retain shape) armure *f* de talon
~WORK - [carp] bourrelet *m* de moulure, baguette *f*
~YARN - [text] fil *m* perlé simple
BEADED BRAID - [text] galon *m* perlé
~EDGE RIM - [auto] jante *f* à rebord
~EDGE TYRE - [rubber] pneumatique *m* à talon
~IRON - [metall] fer *m* pour clôtures
~MATERIAL - [text] tissu *m* perlé
~RIBBON - [text] ruban *m* perlé
~SCREEN - [photo] écran *m* perlé
~TEXTURE - [metall] structure *f* en chapelet
~TUBE - [plumb] tube *m* à bord rabattu
~VEIN - [mining] filon *m* en chapelot
BEADING - [rubber ind] talon *m*, bourrelet *m*
[build] baguette *f*, congé *m*, bourrelet *m* de moulure
[mech] bordage *m*, dudgeonnage *m*
[metall] (deposition of weld metal without oscillating the electrode) chenille *f*, cordon *m* de soudure
[metall] (or necking, the production of a channel in the wall of a hollow component) rainurage *m* (à la molette), nervurage *m*
BEADING MACHINE - [mech mach] machine *f* à border
~PLANE - [impl] rabot *m* à moulures
~TOOL - [metall] (or curling tool; an edge rolling tool) outillage *m* à rouler
BEADLESS TYRE - [rubber ind] pneumatique *m* à bord droit
BEAK - [metall] (of an anvil) bigorne *f*
~IRON - [mech] bec *m* d'enclume, bigorne *f*
~OF ANVIL - [mech] (the tapering pointed projection at one end of an anvil) bec *m* d'enclume
BEAKER - [impl] bécher *m*, vase *m* à filtration chaude
BEAM, to - [gen] émettre des rayons, rayonner, émettre (un signal))
[text] enrouler sur ensouple, emplier
BEAM - [gen] poutre *f*, poutrelle *f*, madrier *m*, fléau *m* de balance
[build] poutre *f*, profilé *m*
[of light] faisceau *m*, rayon *m*
[radio] (transmission restricted to a narrow path) émission *f* dirigée
[radar] faisceau *m*
[navig] faisceau *m* lumineux
[shipbuild] barrot *m*
[text] rouleau *m*, ensouple *f* d'un métier
[naut] bau *m*, largeur *f* d'un bâtiment
~AERIAL - [radar] (an aerial system of which the directional qualities are high) antenne *f* directionnelle
~ALIGNMENT - [telev] (when the scanning beam is moving down the centre of the tube) alignement *m*
~ANGLE - [radio] angle *m* du faisceau d'émission
~APPROACH - [radar] (the operation of bringing in an aircraft along a radio beam) approche *f* radioguidée
~APPROACH BEACON SYSTEM - BABS - [radar] (radar navigational aid giving lateral guidance and distance from alighting point during approach) système "BABS" (aide à la navigation pour le radioguidage de l'approche)
~ARRAY - [radar] (an arrangement of radiators to form a special aerial having high directional properties) antenne *f* directionnelle
~BAR - [text] baguette *f* de l'ensouple, verdillon *m*
~BEARING - [text] support *m* de rouleau
~BENDER - [telev] aimant *m* de piège d'ions
~BENDING - [telev] déplacement *m* du faisceau
~CARRIER - [text] porteur *m* d'ensouples
~CARRIER SLIDE - [text] glissière *f* du support d'ensouple
~CEILING - [constr] plafond *m* à poutres apparentes
~COMPASS - [impl] compas *m* à verges, compas *m* à trusquin
~COUPLING - [electron] couplage *m* électronique
~CREEL - [text] porte-rouleau *m*
~CURRENT - [electron] (the part of a C.R.T. electron beam which reaches the flourescent screen) courant *m* de faisceau
~CUT-OFF - [telev] tension *f* de coupure du faisceau
~CUTTER - [impl] machine *f* à détalonner

BEAM DEFLECTION - [radar] (the deflection of the electron beam in a C.R.T.) déviation *f* de faisceau
~DEFLECTION VALVE - [radiat] tube *m* à faisceau dirigé
~DIRECTION INDICATOR - [radiat] indicateur *m* de direction du rayonnement
~DISC - s. beam flange
~EFFECT - [acoust] effet *m* directionnel
~END PLATE - [text] plaque *f* du bout de l'ensouple
~ENGINE - [metall] machine *f* à balancier
~FLANGE - [text] disque *m* d'ensouple
~FLANGE WITH CORRUGATIONS - [text] plateau *m* d'ensouple avec côtes
~FLANGE WITH COUPLING SLEEVE - [text] plateau *m* d'ensouple avec manchon de serrage
~FOCUS - [telev] finesse *f* du faisceau
~FOCUSING - [electron] (the focusing of the beam in a C.R.T.) concentration *f* du faisceau
~-FORMING ELECTRODES - [radar] (electrodes used to concentrate an electron beam into the desired form) électrodes *fpl* de focalisation
~GATE - [telev] blocage *m* du faisceau
~HOLDER - [text] levier *m* à auge
~HOLE - [nucl] (a hole cut through the shield to permit the escape of a beam of radiation for experimental purposes) trou *m* de faisceau, canal *m* d'irradiation, canal *m* d'expérimentation
~HOUSE - [leather ind] (of a tannery) atelier *m* de chaux
~JITTER - [electron] fluctuation *f* du faisceau
~KNEE - [naut] courbe *m* de barrot
~MACHINE - [mech mach] machine *f* à balancier
~MAGNET - [electron] (used in three-gun picture tubes) aimant *m* de convergence
~MODULATION - [radar] (modulation of the beam current in a C.R.T. by applying the signal voltage between cathode and control grid) modulation *f* du courant de faisceau
~OF LIGHT - [gen] faisceau *m* lumineux
~ORE - [min] limonite *f* pisolitique
~PATTERN - [acoust] diagramme *m* directionnel de rayonnement
~PENTODE - [electron] pentode *f* à faisceau dirigé
~POWER TUBE - [electron] (designed for use in a power-output stage) tube *m* à faisceaux électroniques dirigés
~POWER VALVE - [electron] s. beam power tube
~RADIOSTATION - [radio] émetteur *m* dirigé
~RECEPTION - [radio] réception *f* directionnelle
~RIDER - [astronaut & aero] engin *m* spatial, avion *m* guidé par faisceau
~RIDING GUIDANCE - [ballistic & radio] (the control of a missile or a/c by means of beamed radio transmissions along the beam it travels) système *m* de guidage à faisceaux dirigés
~SHAFT - [text] arbre *m* d'ensouple
~SPLITTER - [opt] diviseur *m* optique du faisceau
~-SPLITTING PLATES - [radar] (electrodes in a C.R.T. used to divide the beam into two separate streams to give independent traces) électrodes *f* divergentes
~STABILIZATION - [radar] stabilisation *f* du faisceau
~SUPPORTED AT BOTH ENDS - [builds] poutre *f* soutenue aux deux extrémités
~SUPPRESSION - [radar] (interruption of the electron beam in a C.R.T. by imposing a large negative potential on the grid) suppression *f* du faisceau
BEAM SWITCHING - [radar] variation *f* d'orientation d'un faisceau
~SYSTEM - [radar] émission *f* dirigée
~TEST - [metall] essai *m* de flexion
~TETRODE - [radar] (four-electrode valve using beam forming electrodes to produce a space charge near the anode, thus reducing secondary emission) tétrode *f* à faisceau dirigé
~THREAD - [text] fil *m* de chaîne
~TILT - [ant] inclinaison *f* du diagramme
~TRANSMISSION - [radar & radio] (transmission of radio signals by means of highly directional aerial systems, for communications between given points) émission *f* dirigée
~TRAP - [électron] électrode *f* collectrice du faisceau
~VALVE - [radio] (a thermionic valve in which secondary emission is reduced by means of beaming electrodes, s. beam tetrode) tube *m* à faisceaux électroniques dirigés
~VOLTAGE - [electron] (voltage between cathode and anode in a C.R.T.) tension *f* entre anode et cathode
~WARPING - [text] ourdissage *m* direct
~WARPING MACHINE - [text] ourdissoir *m*
~WELL - [petr ind] puits *m* exploité au balancier
~WIDTH - [radio] (the angular width of a radio beam between lines marking the limits of an agreed field intensity) largeur *f* de faisceau
BEAMER - [text] monteur *m* de chaîne, ourdisseur *m*
BEAMHOUSE - [leather ind] atelier *m* de chevalage
BEAMING - [text] montage *m* de chaînes, ourdissage *m*
~DEVICE - [telecomm] dispositif *m* de concentration
~FRAME - [text] ourdissoir *m*
~MACHINE - [text] ensouple *m*, ourdissoir *m*
~ROLLER - [text] rouleau *m* livreur, rouloir *m* nappe, cylindre *m* enrouleur
BEAMS AND RAFTERS - [metall] poutrages *mpl*
BEAN - [bot] haricot *m*, fève *f*
~SEED BEETLE - [agric] (a pest) coccinelle *f* des haricots
~SEPARATOR - [agric mach] trieuse *f* de haricots
~WEEVIL - s. bean seed beetle
BEANSTALK - [bot] tige *f* de haricot (ou de fève)
BEAR, to - [gen] soutenir, supporter, contenir
[navig] se diriger vers
BEAR - [zool] ours *m*
[fin] baissier *m*, jouer à la baisse *f*
[metall] loup *m*, bloc *m*, renard *m*
~AWAY, to - [naut] laisser porter, arriver, abattre
~CAT - [oil ind] puits *m* à grand débit, puits *m* à haute production
~FRAME - [mech] bâti *m* en col de cygne
[mining] cadre *m* à col d'oie
BEARD - [gen] barbe *f*
[print] talus *m* d'un caractère, blanc *m*
[bot] arête *f* d'épi, glume *m*
[carp] arête *f* vive
[mining] travers-banc *m*
BEARDED - [gen & bot] barbu, aristé
~NEEDLE - [text] aiguille *f* à barbe (ou à bec)
BEARER - [gen] porteur *m*
[carp] solive *f* transversale, appui *m*, support *m*, tasseau *m* de support
~CABLE - [telecomm] câble *m* porteur
BEARERS - [mech mach] (in a printing machine)

colonne *f*, porte-page *m*

BEARING - [gen] rapport *m*, aspect *m* (d'une question), portée *f* (d'un argument)
[build constr] appui *m*, surface *f* d'appui, portée *f*
[radio] (the angle of direction of an arriving radio wave) relèvement *m* radiogoniométrique
[surv] relèvement *m*
[mech] coussinet *m*, palier *m*, roulement *m*
[naut] relèvement *m*
[min] gisement *m* d'une veine, direction *f* d'une veine

~AREA - [mech] surface *f* de portage, surface *f* portante, aire *f* portante

~AXLE - [mech] essieu *m* porteur, arbre *m* porteur

~BASE - [mech] base *f* d'appui

~BED - s. bearing course

~BLOCK - [mech] imposte *f*, sommier *m*

~BRACKET - [mech] chaise *f* de palier, chevalet *m* de palier, porte-coussinet *m*

~BRASS - [mech] coussinet *m* antifriction

~BRONZE - [metall] bronze *m* à coussinets

~BUSH - [mech] coussinet *m* cylindrique, coussinet *m*

~CAP - [mech] couvercle *m*, chapeau *m* de palier

~CAPACITY - [mech] capacité *f* de charge, portance *f*

~CHAIR - [mech] sabot *m* d'appui

~COMPASS - [surv] compas *m* de relèvement

~CONE - [mech] roulement *m* conique

~COURSE - [constr] couche *f* portante

~CURRENT - [el] courant *m* porteur

~DISTANCE - [build] portée *f*

~DOOR - [mining] porte *f* d'aérage

~EXTRACTOR - [mech] extracteur *m* de coussinets

~FACE - s. bearing area

~FOOT - [mech] pied *m* de montant

~FRICTION - [mech] frottement *m* d'appui (ou de palier)

~HOUSING - [mech] siège *m* de coussinet

~-IN - [metall] profondeur *f* de la saignée

~LINE - [mech] (in grinding surface roughness) ligne *f* portante

~LINE FRACTION - [mech] fraction *f* portante du profil

~LINING - [mech] revêtement *m* de coussinet, métal *m* antifriction, blanc *m* à coussinets, régule *m*

~LOAD - [mech] charge *f*

~LOCKING RING - [mech] bague *f* d'arrêt de coussinet

~METAL - [metall] (alloys, such as bronze or white metal, used in bearings to reduce friction) métal *m* antifriction

~OF BRIDGE - [build] culée *f*

~OF SPINDLE - [mech mach] palier *m* d'axe, appui *m* de pivot

~PACKING - [mech] joint *m* de palier, garniture *f*

~PARTITION - [build] cloison *m* portante, cloison *f* d'appui

~PILE - [constr] pieu *m* porteur

~PLATE - [constr] plate-forme *f* de support

~PLATE BED - [railw] assise *f* des plaques d'appui

~POTENTIOMETER - [instr] (in radar, arranged in the support of the aerial to transmit angular information) potentiomètre *m* sinus-cosinus, potentiomètre *m* de gisement pour écran type B

~PRESSURE - [mech] (the specific load on a bearing surface) pression *f* d'appui

BEARING PROJECTOR - [aero] projecteur *m* d'atterrissage

~PULLEY - [mech] poulie *f* portante

~RESOLUTION - [radar] discrimination *f* de relèvement

~RETAINER - [mech] arrêtoir *m* de roulement, jonc *m* d'arrêt de roulement

~ROD - [mech] support *m* du balancier, bielle *f* de suspension

~ROLLER - [mech] rouleau *m* de roulement (ou de palier), galet *m* de roulement

~SHELL - [mech] coquille *f* de coussinet

~SPRING - [mech] ressort *m* de suspension

~STAND - [mech] chaise *f* de palier, chevalet *m* de palier

~STANDARD - [mech] support *m* à colonne, montant *m*

~STOP - [mining] cloison *f* d'aérage

~STRATUM - [constr] strate *f* portante

~STRENGTH - [mech] résistance *f* de portée

~STRESS - [constr] pression *f* de contact, contrainte *f* au contact

~SUPPORT - [mech] support *m* de coussinet

~SURFACE - [mech] (the part of a bearing which is in direct contact with the moving part it supports, or viceversa) surface *f* de roulement, surface *f* de palier, surface *f* portante, surface *f* d'appui

~TEST - [constr] essai *m* de chargement, essai *m* de charge

~TIMBER - [constr] poutre *f* portante, cadre-porteur *m*

~TRANSMISSION UNIT - [radar] (mechanism transmitting data to a reading scale) indicateur *m* de relèvement

~WALL - [build] mur *m* porteur, mur *m* d'appui

~WITH OIL RINGS - [mech] coussinet *m* avec anneaux de lubrification

BEARINGS - [aero & nav] position *f*

BEAT, to - [gen] battre, frapper, avoir des battements
[text] frapper le battant
[of cotton] battre (le coton)

BEAT - [acoust phys el etc] battement *m*
[hor] tic-tac *m*
[med] pulsation *f*, battement *m*
[radio] (a cyclic change in amplitude caused by the combination of two different frequencies) battement *m*, interférence *f*
[geol] affleurement *m*

~FREQUENCY - [radio] (the frequency of the amplitude variations by the combination of two different frequencies) fréquence *f* de battement

~FREQUENCY OSCILLATOR - [radio] (arrangement for producing approximately sinusoidal oscillations by combining signals of different frequencies) oscillateur *m* à battements

~INDICATOR - [instr] indicateur *m* de battements

~NOTE - [radio] note *f* de battement

~OSCILLATOR - [radio] (valve oscillator used to furnish local oscillations in beat reception) oscillateur *m* à battement

~RECEIVER - [radio] récepteur *m* hétérodyne, superhétérodyne *m*

~RECEPTION - [radio] (reception by combining incoming signals with a local oscillation of a different frequency to produce an audio-frequency signal) réception *f* de battements, réception *f* en

hétérodyne
BEATEN GOLD - [metall] or *m* battu, or *m* martelé, or *m* en feuilles
BEATER - [text] fouloir *m* (de foulon), gaulette *f*, agitateur *m*
[impl] bourroir *m*
[paper man] pilon *m*
[rubber ind] (used in latex-foam industry) fouetteuse
~ARM - [text] fouet *m*, bras *m* de chasse, battant *m*
~-BAR THRESHER - [agric] batteuse *f* à battes
~BLADE - [text] règle *f* du volant batteur
~FEED ROLLER - [text] cylindre *m* alimentateur du batteur
~MIXER - [rubber ind] mélangeur-agitateur *m*
~ROLLER - [text] cylindre *m* du volant
~SHAFT - [mech] arbre *m* de l'agitateur
~TUB - [paper man] cuve *f* de pilon
BEATING BOARD FRAME - [text] cloison *f* à claire-voie
~BRUSH - [text] escoubette *f*, balai *m* de battage
~CURRENT - [radio] courant *m* de battement, courant *m* oscillatoire
~ENGINE - [paper man] pilon *m*
~IN OF CABLE - [telecomm] préparation *f* de l'enveloppe d'un câble pour la faire adhérer aux conducteurs
~IRON - [metall] sabot *m* de bocard
~MACHINE - [rubber ind] (in latex foam industry) fouetteuse *f*, batteuse *f*, machine *f* de fabrication de mousse
~OSCILLATOR - [el] oscillateur *m* hétérodyne
~UP - [text] serrage *m*, battage *m*
~UP BY PRESSURE - [text] battage *m* par pression
~UP NEEDLE - [text] aiguille *m* de pressage
~UP TOOL - [text] battoir *m* de trame
BEAUFORT NOTATION - [met] (a code of letters corresponding to meteorological conditions) notation *f* Beaufort
~SCALE - [met] (scale of wind velocity in steps from 0 to 12; 0 to 71 knots) échelle *f* Beaufort
BEAUMONTAGE - [ind chem] (composition for stopping holes in castings or the like) mastic *m* à reboucher, badigeon *m*
BEAVER - [zool] castor *m*
[comm] fourrure *f* de castor
~CLOTH - [text] castorine *f*
~COTTON - [text] flanelle *f* de coton
~HYDE - [leather ind] peau *f* de castor
~PLUSH - [text] feutre *f* imitation castor
BEAVERTAIL - [radar] (ground based air-transportable radar for height finding) radar *m* à faisceau étalé, radar *m* d'altitude
~AERIAL - [radar] (special type of aerial used chiefly for determining height) antenne *f* à faisceau étalé
BEAVERTEEN - [text] molleton *m* de coton, velours *m* de coton
BECALM, to - [naut] encalminer, déventer
BECALMED - [naut] encalminé
BECK - [impl] (large vat used in dyeing, soapmaking, brewing etc) baquet *m*
BECKET - [naut] ganse *f*, ringot *m*, erse *f*, garcette *f*
BECKING - [metall] (forging operation) élargissement *m*, dégorgement *m* à cloc
~BAR - [metall] dégorgeoir *m*
BECKIRON - [tool] bigorne *f*, bec *m* (d'enclume)
BECKITE - [min] (a chemically precipitated form of silica) beckite *f*
BECKMANN APPARATUS - appareil *m* de Beckmann
~MOLECULAR REARRANGEMENT - s. Beckmann rearrangement
~REARRANGEMENT - [chem] (the transformation of ketoximes to amides by the intermolecular rearrangement) transposition *f* moléculaire de Beckmann
~THERMOMETER - [instr] (a special type of mercurial thermometer provided with a large bulb and thus possessing high sensitivity over a limited range) thermomètre *m* différentiel de Beckmann
BECQUEREL CELL - [photo el] pile *f* photochimique
~EFFECT - [photo el] effet *m* Becquerel
~RAYS - [phys] (obsolete term including the three types of radiation emitted by radio-active substances) radioactivité *f*
BED, to - [mech] sceller (une poutre dans un mur), asseoir (fondations)
BED - [gen] lit *m*
[build] fondation *f*, bâti *m*
[geol] assise *f*, couche *f*, banc *m*, lit *m*
[geol] (of a layer) couche *f*, strate *f*
[mech] support *m*, bâti *m* de moteur, banc *m* de tour, sommier *m* (de machine), table *f* (de raboteuse), banc *m*, base *f* d'appui
[agric] planche *f*, carré *m*
[metall] couche *f* de coulée
[text] (of a full-fashioned knitting machine) cornière *f* du métier, barrette *f* de la barre à aiguilles
[min] gisement *m*, couche *f*
[ind chem] couche *f* filtrante, lit *m*
[print] marbre *m* de presse
[shipbuild] souille *f*
~BEARERS - [mech] glissières *f*
~-BUG - [zool] punaise *f* des lits
~CHARGE - [metall] fausse charge *f*, charge *f* de base
~COKE - [metall] coke *m* pour lit de fusion
[min] coke *m* d'allumage
~DIE - [mech] matrice *f*, perçoir *m*
~DOWEL - [build] pieu *m*, cheville *f*, tirefond *m*
~FUEL - (ovens) couche *f* de base de combustible
~IN, to - [metall] mettre sur couche
~IRRIGATION - [agric] irrigation *f* par rigoles
~JOINT - [build] joint *m* horizontal, joint *m* de lit, joint *m* d'assise
~LENGTH - [mech] longueur *f* de banc (de tour etc)
~LINEN - [text] linge *m* de lit
~MAT - [text] descente *f* de lit
~MOULDING - [build] corniche *f*, entablement *m*
~OF CLAY - [soil] couche *f* d'argile
~PAN - [impl med] bassin *m*
~PLATE - [mech] plaque *f* de base, socle *m*, plaque *f* d'assise
[in ovens] plaque *f* de fond, sole *f*
[mech mach] plaque *f* de fondation, base *f* commune
[auto] plaque *f* d'appui
~PLATE BEARING - [auto] coussinet *m* de tête de bielle
~SHEET - [text] drap *m* de lit
~SIZE - [mech & gen] dimension *f* de la base (ou du banc, du bâti etc)

BED SLIDE - [mech mach] glissière *f*, coulisse *f* de lit
~SPREAD - [text] couvre-lit *m*, dessus *m* de lit
~-STONE - [build] pierre de fondation, pierre *f* de soubassement
[in mills] gisante *f*, meule *f* dormante
~TICKING - [text] toile *f* coutil à matelas
~-TYPE MILLING MACHINE - [mech mach] fraiseuse *f* horizontale
~VEIN - [geol] filon-couche *m*
BEDDED - [geol] stratifié
~-IN MOULDING - [metall] troussage *m*
~ORE DEPOSITS - [min] formations *f*pl métallifères
~ROCK - [geol] roche *f* sédimentaire
~SET - [brew ind] bouture *f* racinée
~STRUCTURE - [geol] structure *f* stratifiée
BEDDING - [gen] literie *f*, fournitures *f*pl d'un lit
[agric] labour *m* en billons
[geol] stratification *f*, gisement *m*
[build] enrochement *m*, scellemtent *m*, assiette *f*, matériau *m* d'enrochement
[cables] gaine *f* fibreuse imprégnée
[metall] lit *m* de fusion, (in foundry) paillasse *f*
~CULTIVATION - [agric] billonnage *m*
~FAULT - [geol] fissure *f*, rejet *m*
[in records] défaut *m* d'impression
~-IN - [mech] (the process of fitting a bearing to its shaft by scraping and testing) ajustage *m* de précision
~PLANE - [geol] plan *m* de stratification
~OF ENGINE IN RUBBER CUSHIONS - [auto] suspension *f* d'un moteur sur tampons de caoutchouc
BEDFAST - [med] alité
BEDRIDDEN - [med] alité, cloué au lit
BEDROCK - [geol] roche *f* en place, roche *f* de fond, assise *f* rocheuse
BEDROOM - [gen & build] chambre *f* à coucher
BEDSIDE RUG - [text] descente *f* de lit, saut *m* de lit
BEE - [zool] abeille *f*
~-VEIL - [agric] masque *m* d'apiculteur
BEEBREAD - [zool] pâtée *f* de pollen et de miel (donné au couvain), gateau *m* de miel
BEECH - [bot] hêtre *m*
~NUT OIL - [ind chem] huile *f* de faînes
BEEF - [food ind] viande *f* de boeuf
BEEFING-UP - [plast ind] (operation for reinforcing the core of a laminated sheet) renforcement *m*
BEEHIVE - [zool] rûche *f*
BEEKEEPER - [agric] apiculteur *m*
BEEP - [acoust] (sound signal which is used to determine the speed of the tape) signal *m*, top *m*
BEER - [brew] bière *f*
[text] (unit for warp threads) écheveau *m* de fil
~BARREL - [brew] baril *m* à bière, tonneau *m* à bière
~FROM THE WOOD - [brew ind] bière *f* en fût (en tonneau)
~HEART - [med] coeur *m* de buveur de bière
~LOSS - [brew ind] freinte *f* (de bière)
~SCALE - [brew ind] tartre *m* (de bière)
~STILL - [ind chem] (still used for the production of intermediate crude alcohol from fermented liquor) appareil *m* à distiller la bière
~WELL - [ind chem] (tank used to receive fermented liquor from the fermenters) cuve *f* à liquide fermenté
BEESWAX - [ind chem] (wax obtained from the honeycomb of the bee) cire *f* d'abeille
BEET - [bot] betterave *f*
~AND WATER WHEEL - [sugar ind] roue *f* élévatrice mixte
~DIGGER - [agric] arracheuse *f* de betteraves
~FEEDER - [sugar ind] régulateur *m* d'alimentation en betteraves
~FLY - [agric] (a pest) mouche *f* de la betterave, pégomye *f*
~HARVESTER - [agric] récolteuse *f* de betteraves
~HOPPER - [sugar ind] trémie *f* à betteraves
~LIFTER - [agric] souleveuse *f* de betteraves
~PICK-UP LOADER - [agric] ramasseur-chargeur *m* de betteraves
~PULP DRIER - [sugar ind] séchoir *m* à pulpes
~RASP - [sugar ind] râpe *f* à betteraves
~SCREW - [sugar ind] hélice *f* à betteraves
~SLICING - [sugar ind] découpage *m* des betteraves
~SUGAR - [ind chem] (sucrose obtained from sugarbeets) sucre *m* de betteraves
~TAIL CATCHER - [sugar ind] séparateur *m* de queues, ramasse-radicelles *m*
~WASHER - [agric] laveur *m* de racines
~WHEEL - [sugar ind] roue *f* à betteraves
BEETLE, to - [ind] (to make into pulp) beetler, pilonner, écraser
BEETLE - [text mach] (row of wooden hammers falling on rolls of cloth as they revolve) mailloche *f*
[zool] coléoptère *m*, cafard *m*, blatte *f*
[impl] (heavy hammer or mallet) maillet *m*, mailloche *f*, masse *f* en bois
[plast mat] (proprietary) matière *f* plastique thermodurcissable (appellation commerciale)
~-HEAD - [tools] mouton *m*
BEETROOT - [bot] betterave *f*
BEFORE - [gen] avant, devant
~AND AFTER TEST - [nucl] essai *m* comparatif (après traitement)
~THE BEAM - [naut] sur l'avant du travers
BEFOUL, to - [gen & ind] salir, souiller
BEGET, to - [gen] engendrer, procréer
BEGIN, to - [gen] commencer, débuter, amorcer
BEGOHM - [el] mille megohms, gigaohm *m*
BEHAVE, to - [gen & mech] (se) comporter, fonctionner
BEHAVIOUR - [gen] comportement *m*, tenue *f*
[mech] fonctionnement *m*, allure *f* d'une machine
~FLEXIBILITY - [genet] flexibilité *f* de comportement
BEL - [acoust] (unit ten times the size of a decible) bel *m*
BELAY, to - [naut] amarrer, fixer, tourner (un cordage)
BELAYING - [naut] amarrage
~PIN - [naut] cabillot *m*, taquet *m*
~RACK - [naut] ratelier *m* de cabillots
BELCHERING - [metall] (or bulging; production of an outward channel in the wall of a hollow component) nervurage *m* (à la presse)
BELEMNOID - [med] apophyse *f* styloïde
BELFRY - [build] beffroi *m*, clocher *m*, clocheton *m*
BELGIAN SLAG - [metall] scorie *f* Thomas
B ELIMINATOR - [electron] (power source which eliminates the anode battery) appareil *m* de tension anodique
BELITE - [chem] belite *f*
BELL, to - [gen] munir d'une cloche, évaser
[metall] claquer
BELL - [gen] (sound emitting metal device) cloche *f*

sonnerie *f*
[metall] cloche *f*, cône *m* de fermeture
[plumb] évasement *m*
[naut] cloche *f*, sonnerie *f*
[mus] (opening of a wind instrument) pavillon *m*
BELL AND SPIGOT JOINT - [plumb] joint *m* à cordon et emboîtement
~AND SPIGOT PIPE - [plumb] tuyau *m* à emboîtement et cordon
~ARMATURE - [el] induit *m* en cloche, rotor *m* en cloche
~-BUOY - [nav] (a buoy fitted with a bell actuated by the wave motion used to mark a shoal or the like) bouée *f* à cloche
~BUTT JOINT - [plumb] joint *m* à érasement
~BUTTON - [el] bouton *m* de sonnette (ou de sonnerie)
~CAGE - [build] campanile *m*
~CENTRE PUNCH - [impl] pointe *f* auto-centreuse
~CHUCK - [mech mach] mandrin *m* à vis
~CONTROL - [mech] commande *f* à sonnette
~CONTROL SYSTEM - [mech] commande *f* à sonnette
~-CRANK - [mech] (a device for converting linear movement of a rod or cable to related linear movement in a different direction) levier *m* coudé, levier *m* de changement de direction, genouillère *f*
~CRANK CAM LEVER - [text] levier *m* coudé à galet
~CRANK LEVER - [mech] levier *m* coudé à renvoi, pièce *f* coudée
~CRUSHER - [min] broyeur *m* à cloche
~FOUNDING - [metall] fonte *f* des cloches
~FOUNDRY - [metall] fonderie *f* de cloche
~FRAME - [build] campanile *m*
~GABLE - [build] campanile *m*
~GLASS - [glass man] cloche *f* en verre
~HAMMER - [text] battant *m* de fuseau de dentellière
~HARP - [mus] (acient string instrument) psaltérion *m*
~HOLE - [plumb] emboîtement *m* (de tuyaux)
~JAR - [glass man] s.bell glass
~JOINT - [plumb] joint *m* à emboîtement
~MANOMETER - [instr] manomètre *m* à cloche
~METAL - [metall] métal *m* de cloches, bronze *m* de cloche
~METAL ORE - [min] stannine *f*
~-MOUTH - [plumb] évasement *m*, égueulement *m*
~-MOUTHED - [plumb] évasé, à trompe, en entonnoir
~NOZZLE - [mech] tuyère *f* en cloche
~OF PIPE - [plumb] extrémité *f* femelle d'un tuyau
~PUSH - [el] (button shaped switch) bouton *m* de sonnette (ou de sonnerie)
~SHAPED - [gen] en forme de cloche, campanulé
~SHAPED FUNNEL - [build] entonnoir *m* en forme de cloche
~TOWER - [build] campanile *m*
~TRANSFORMER - [el] transformateur *m* de sonnerie
~TYPE FURNACE - [ovens] four-cloche *m*
~WIRE - [el] fil *m* de sonnerie
BELLADONNA - [chem] (a plant from which atropine and hyoscyamine are extracted) belladone *f*
BELLING - [mech plumb] évasement *m*
BELLITE - [ind chem] (explosive mixture) bellite *f*
BELLMOUTH - [mech] s.belling
~INTAKE - [hydr] trompe *f* d'entrée
BELLOW, to - [zool] mugir
[fig] hurler

BELLOWS - [impl] (flexible device resembling domestic bellows, to enclose and protect moving parts while allowing axial movement) soufflet *m*, accordéon *m* (protecteur) en caoutchouc
[ind chem] compensateur *m* de dilatation à soufflet
[aero & gen engines) (bellows which are sensitive to pressure) capsule *f* anéroïde
~ADAPTER - [photo] soufflet *m* auxiliaire
~CAMERA - [photo] appareil *m* à soufflet, chambre *f* à soufflet
~MURMUR - [med] bruit *m* de souffle
~SEAL - [mech] soufflet *m* en caoutchouc, accordéon *m* (protecteur en caoutchouc)
~SHEETING - [text] tissu *m* caoutchouté pour soufflets
~SOUND - s.bellows murmur
~TYPE PRESSURE GAUGE - [instr] (type of pressure gauge in which the moving element consists of a bellows connected to a pointer) - manomètre *m* à capsule, capsule *f* manométrique
BELLY, to [gen] gonfler, (se) bomber, s'enfler
BELLY - [gen anat] ventre *m*, abdomen *m*
[mech ovens] ventre *m*
[metall] calotte *f*
[naut] (said of a sail) - creux *m* d'une voile, ventre *m* d'un bateau
~-BAND - [impl] (of a harness) sous-ventrière *f*
~-BOUND - [med] constipé
~-BRACE - [mech] (device for rotating a drill or carpenter's bit, feed being obtained by the pressure of the user's body) - vilebrequin *m*, chignole *f*
~LANDING - (emergency landing, the a/c grounding on the under-side of the fuselage, when undercarriage cannot be lowered because of a fault) - atterrissage *m* sur le ventre, atterrissage *m* forcé
~PIPE [metall] conduit *m* du vent
BELONG, to - [gen] appartenir
BELOW - [gen] au dessous, sous
[naut] en dessous, en bas
~DECK - [naut] en dessous, en bas
BELT - [gen] ceinture *f*, courroie f, bande *f*
(a flexifle belt used to transmit power) - convoyeur *m* à courroie
(a flexible belt used to transport material) - transporteur *m* à courroie
[constr] (a projecting course of stones or bricks) bandeau *m* saillant
[shipbuild] ceinture *f* cuirassée
[geogr] région *f*, ceinture *f*, zone *f*
~CARRIER - [impl] transporteur *m* à courroie, chaîne *f* transporteuse
~CHARGING MACHINE - [in ovens] chargeuse *f* à courroie
~CLAMP - [mech] agrafe *f*
~CLAW - [mech] s. belt clamp
~CONE - [text] cône *m* de poulie, poulie *f* étagée, poulie *f* à gradins
~CONVEYOR - [impl] transporteur *m* à courroie, convoyeur *m* à courroie
~COURSE - [arch] s. belt
~DRESSING -[mech] enduit *m* pour courroie
[ind chem] (resinous preparation applied to driving belts to prevent slip) apprêt *m* pour courroie
~DRIVE - [mech] transmission *f* par courroie
~DRIVEN - [mech] entraîné par courroie
~DRIVEN MIXER - [mech mach] mélangeur *m* à courroie

BELT DRIVEN PUMP - [mech] pompe *f* commandée par courroie
~DRIVEN RIVETER - [mech mach] riveuse *f* à courroie
~DRIVING - [text] commande *f* à courroie
~FASTENER - [mech] agrafe *f* (pour courroie), attache-courroie *m*
~FEEDING - [mech] alimentation *f* par courroie
~FINISHER - [mech mach] (abrasion machine consisting of a band of abrasive-coated textile material running over rollers) meuleuse *f* à courroie, polisseuse *f* à bande, ponceuse *f* à bande
~FORK - [mech impl] fourchette *f* de débrayage (courroies)
~GEAR - [mech] renvoi *m* à courroie, transmission *f* à courroie
~GRINDING - [mech] polissage *m* par bandes (abrasives)
~GUARD - [mech] protection *f* des courroies
~GUIDE - [mech] guide *m* pour courroie, fourche *f* de guidage
~HOOL - [mech] fourchette *f*
~H.P. - [Belt Horse Power] (as in tractors) puissance *f* à la poulie
~JOINT - [mech] jonction *f* de courroies
~LINE - [town planning] ligne *f* de ceinture (chemin de fer etc)
~LINK - [mech] agrafe *f* (pour les extrémités d'une courroie)
~LOOM - [text] métier *m* à tisser les sangles
~OF ARMOUR - [naut shipbuild] ceinture *f* cuirassée, cuirasse *f*
~OF CALMS - [met] (belt of light variable winds extending between 10 and 15 degrees north and south of the equator) zone *f* des calmes équatoriaux
~OF CONE PULLEY - [mech] courroie *f* pour poulie étagée (ou à gradins)
~PRODUCTION - [mech] production *f* à la chaîne
~PULLEY - [mech] (in earth moving equip., i.e. tractors)poulie *f* pour courroie
~PUNCH - [mech impl] perce-courroie *m*
~SANDER - [mech mach] s. belt finisher
~-SANDING - [mech] polissage *m* à bande, ponçage *m* à bande
~SAW - [impl] scie *f* à ruban
~SCALES - [impl] (automatic continuous weighing machine) bande *f* convoyeuse-peseuse
~-SCANNER - [telev] (endless belt of opaque material with scanning holes) analyseur *m* à bande
~SHIFTER - [mech] fourchette *f* de courroie, embrayeur *m*, monte-courroie *m*
~SHIPPER- [mech] fourche *f* de débrayage
~SLING - [gas-pipe install] élingue *f* plate
~SLIP - [mech mach] glissement *m* de la courroie
~SLIPPER - [impl text] monte-courroie *m* embrayeur
~SPEEDER-[mech] poulies-cônes *f*pl, cônes *m*pl alternes
~STRETCHER - [impl] tendeur *m* de courroie
~SURFACER - s. belt finisher
~TENSION - [mech] tension *f* de courroie
~TIGHTENER - [mech] tendeur *m* de courroie, galet *m* tendeur
~TONGS - [mech] pince *f* à courroie
~TRACTION-[mech] traction *f* de la courroie
~WEAVING - [text] tissage *m* des courroies
BELTED - [rubber ind] ceinturé
~TYPE CABLE - [telecomm] conducteur *m* électrique plat
BELTING - [gen] (materials from which driving belts are made) matériaux *m* pour courroies, courroies *m* de transmission *f*
~DUCK - [text] tissu *m* pour courroie
~HEALDS - [text] lisses *f* pour courroies
~LEATHER - [leath ind] cuir *m* pour courroies
BELVEDERE - [arch] belvédère *m*, mirador *m*
BENCH - [gen] banc *m*, gradin *m*, banquette *f*
[impl] établi *m*, marbre *m* d'ajusteur
[mining] gradin *m*, strate *f*
[build] (horizontal ledge on the side of an enbankment) accotement *m*, berme *f*
[glass man] banc *m* (de verrier)
~AND BENCH [mining] exploitation *f* par tranches
~ASSEMBLY - [mech] montage *m* sur table
~DRILLING MACHINE - [mech] perceuse *f* d'établi
~ENGINE LATHE-[mach tool] tour *m* d'établi
~FILING MACHINE - [mech mach] limeuse *f* d'établi
~GRAFTING - [agric] greffe *f* sur table
~HAMMER - [mech] marteau *m* d'établi
~HOOK - [carp] valet *m* d'établi
~LATHE - [mech] (a small lathe designed to be mounted on a workbench) tour *m* d'établi
~MARK - [top] repère *m*, borne *f*, repère *m* de nivellement
~MICROMETER - [instr] micromètre *m* d'établi
~PHOTOMETER - [instr] photomètre *m* de banc
~PLANE - [carp] colombe *f* à joindre
~SCALE CRAKING UNIT - [ind chem] unité *f* de pyrolyse de laboratoire
~SHAPING MACHINE - [mach tool] étau-limeur *m*, limeuse *f*
~SHEARS - [mech] cisailles *f* d'établi
~STOP - [carp] griffe *f* d'établi
~STOPING - [mining] exploitation *f* par gradins, abattage *m* en gradins
~TERRACE - [soil] (a terrace with a steep drop on the down-hill side) gradin *m* (à surface se determinant par un talus abrupt)
~-VICE - [mech] (vice designed for mounting on a workbench) - étau *m* d'établi
BENCHING - [min] (concrete sloped up from the foundation) exploitation *f* en gradins
[mining] exploitation *f* par gradins, abattage *m* en gradins
BEND,to - [gen & mech] cintrer, courber, couder, fléchir, plier, arquer, flamber, (se) gauchir
[of ropes] étalinguer, abouter deux cordages
[naut] frapper (un cordage), enverguer (une voile)
[of anchor] saisir, amarrer
BEND - [gen] courbe *f*, courbure *f*, coude *m*, pliage *m*, flexion *m*
[plumb] raccord *m* coudé, coude *m*
[of roads] tournant *m*, courbe *f*, virage *m*
[mech] courbure *f*, cintrage *m*, cambrure *f*
[of rope] noeud *m*, oeil *m*, ganse *f*
~AERIAL - [radio] antenne *f* coudée
~OF GROUND - [geol] pli *m* de terrain
~OVER,to - [gen & mech] replier, recourber, rabattre
~TEST (COATINGS) - (a test for coating, in which the coated part is bent to determine the adhesion of the coating to it) - essai *m* de pliage (ou de flexion)
BENDABLE - [gen & metall] flexible, souple, pliable,

faussable
BENDER - [mech] machine *f* à cintrer, cintreuse *f*, plieur *m*, cintreur *m*
BENDING - [mech] courbure *f*, ployage *m*, cintrage *m*, flexion *f*
[of roads] courbe *f*, tournant *m*, virage *m*
~AND FOLDING - [plast] ployage *m*, cintrage *m*
~ANGLE - [mech] angle *m* de courbure
~BLOCK - [carp] bloc *m* pour le cambrage du bois
~COEFFICIENT - [metall] coefficient *m* de pliage
~FATIGUE TEST - [text] essai *m* de rupture au pliage
~JAWS - [metall] mâchoires *f*pl de pliage
~LINE - [metall] ligne *f* de flexion
~MACHINE - [mech mach] plieuse *f*, cintreuse *f*, machine *f* à plier
~MOMENT - [phys] (moment tending to distort a structural element from the straight or from its proper curvature) moment *m* de flexion, moment *m* fléchissant
~OF THE SPRING - [mech] flexion *f* du ressort
~-OFF PRESS - [metall] presse *f* à chanfreiner
~PRESS - [metall] presse *f* à cintrer, cintreuse *f*
~RADIUS - [gen] rayon *m* de courbure
~ROLLS - [mech] cylindres *m* de cintrage
~SHOE - [gas install] sabot *m* de cintrage
~SLAB - [impl] plateforme *f* à dresser, marbre *m*
~SPEED - [mech] vitesse *f* de pliage (ou de cintrage)
~STIFFNESS - [phys] rigidité *f* de flexion
~STRAIN - [metall] effort de flexion, effort *m* transversal, travail *m* à la flexion
~STRENGTH - [phys] résistance *f* à la flexion
~STRESS - [metall] effort *m* de flexion, effort *m* fléchissant
~STRESS TESTER - [mech] machine *f* pour mesurer les efforts de flexion
~TEMPLATE - [mech] mandrin *m* pour cintrer
~TEST - [metall] essai *m* de flexion
~TORQUE - s.bending moment
~VIBRATION - [metall] vibration *f* de flexion
BENDS - [med] aérémie *f*, mal *m* des caissons, aéroembolisme *m*
BENEATH - [gen] sous, au dessous
BENEFICIATE, to[min] (treatment for the improvement of the metallurgical use of minerals) enrichir, réduire
BENEFICIATION - [min] enrichissement *m*, réduction *f*
BENEWORTH - [chem] belladone *f*
BENGAL FIRE - [signal] feu *m* de Bengale
~LIGHT - [signal] feu *m* de Bengale
BENGALINE - [text] bengaline *f*
BENIGN - [gen & med] bénin
~NEOPLASMA - [nucl] (new growth of cells) néoplasme *m* bénin
~TUMOR - [med] (tumor which does not metastasize) tumeur bénigne
BENITOITE - [min] bénitoïte *f*
BENNET - [bot] (a medicinal plant) benoîte *f*
BENT - [gen mech etc] coudé, courbé, faussé, fléchi, gauchi
[build] chevalet *m*, portique *m*
~ENGRAVING PLATE - [print] plaque *f* courbée pour taille douce
~GOUGE - [carp] gouge *f* à bec de corbin
~LEVER - [mech impl] levier *m* coudé, levier *m* brisé, levier *m* à renvoi
BENT NEEDLE - [text ind] aiguille *f* tordue
~POINTER - [instr] aiguille *f* courbée, index *m* courbé
~RAY - [opt] rayon *m* réfracté
~RIM - [mech] cercle *m* à rebord coudé
~SHAFT - [mech] arbre *m* fléchi
~SOCKET WRENCH - [impl] clé *f* à tube coudé
~TOOL - [tool] outil *m* coudé
BENTHOS - [geol] faune *f* et flore *f* du fond de la mer
BENTHOSCOPE - [instr] benthoscope *m*
BENTONITE - [min] (naturally occurring colloidal clay used as a rubber filler) bentonite *f*
~SLURRY - [soil] coulis *m* de bentonite
BENZAL CHLORIDE - [chem] chlorure *m* de benzilidène
~GREEN - [ind chem] vert *m* malachite
BENZALDEHYDE - [chem] benzaldéhyde *m*, aldéhyde *m* benzoïque
BENZALKONIUM - [chem] benzalkonium *m*
BENZANTHRONE - [chem] benzanthrone *m*
BENZAZOL - [chem] indole *m*
BENZEDRINE - [chem] benzédrine *f*
BENZENE - [chem] benzène *m*
~CARBOXYLIC ACID - [chem] acide *m* carboxylique
~HYDROCARBONS - [chem] hydrocarbures *m* de benzène
~RING - [chem] noyau *m* du benzène
~SERIES - [chem] série *f* benzénique
~SULPHONIC ACID - [chem] acide *m* benzène sulfonique
BENZENYL - [chem] benzényle
BENZIDINE - [chem] benzidine *f*
~BASE - [chem] benzidine *f* (base)
~BLUE - [ind chem] bleu *m* de benzidine
~TRANSFORMATION - [chem] transposition *f* benzidinique
BENZINE - [oil ind] (a liquid hydrocarbon derived from mineral petroleum) benzine *f*
BENZOATE - [chem] benzoate *m*
BENZOAZURINE - [chem] benzoazurine *f*
BENZOFLAVINE - [chem] benzoflavine *f*
BENZOGUANIMINE - [chem] benzoguanimine *f*
BENZOIC ACID - [chem] acide *m* benzoïque
BENZOIN - [chem] benzoïne *f*
~GUM - [ind chem] benjoin *m*
BENZOL - [chem] benzol *m*, benzène *m*
~BLACK - [ind chem] (amorphous carbon from benzol flames) noir *m* de benzol
~EXTRACTION - [ind chem] débenzolage *m*
~EXTRACTION BY REFRIGERATION - [ind chem] débenzolage *m* par réfrigération
~SCRUBBER - [ind chem] épurateur
BENZOLISM - [med] benzolisme *m*
BENZOLIZE - [ind chem] benzoliser
BENZONITRILE - [chem] benzonitrile *m*
BENZOPHENONE - [chem] benzophénone *f*
BENZOPURPURIN - [chem] violet *m* hexaméthylé
BENZOQUINONE - [chem] benzoquinone *f*, benzotriazole *f* azimide
BENZOTRICHLORIDE - [chem] chlorure *m* de benzényle
BENZOTRIFLUORIDE - [chem] benzotrifluorure *m*
BENZOYL - [chem] benzoyle *m*
~ACETYL PEROXIDE - [chem]peroxyde *m* de benzoyle
BENZOYLATION - [ind chem] benzoylation *f*
BENZYL - [chem] benzyle *m*, dibenzyle *m*

BENZYL ALCOHOL - [chem] alcool *m* benzylique
~BENZOATE - [chem] benzyl-benzoate *m*
~BLUE - [chem] bleu *m* de benzyle
~BROMIDE - [chem] bromure *m* de benzyle
~CELLULOSE - [ind chem] cellulose *f* benzylique
~CHLORIDE - [chem] chlorure *m* de benzyle
~CINNAMATE - [chem] cinnamate *m* de benzyle
BENZYLANILINE - [chem] benzaniline *f*
BENZYLIDENE - [chem] benzylidène *m*
BEQUEATH, to - [leg] léguer
BERBERAMINE - [chem] berbéramine *f*
BERG CRYSTAL - [min] cristal *m* de roche
BERGAMIOL - [chem] bergamiol *m*
BERGAMOT - [bot] bergamote *f*
~OIL - [ind chem] essence *f* de bergamote
BERGERON THEORY - [met] (an explanation of the formation of rain, put forward by the meteorologists Bergeron and Finseisen) théorie *f* de Bergeron
BERGWIND - [met] (steady katabatic wind blowing down a valley before sunrise) bergwind *m*
BERI-BERI - [med] béri-béri *m*
BERILLIOSIS - [med] bhérylliose *f*
BERKELIUM - [chem] berkélium *m*
BERLIN BLUE - [ind chem] bleu *m* de Berlin
BERM - [build](horizontal ledge on the side of an embankment) berme *f*, banquette *f*, accotement *m*
[of a road] berme *f*, berge *f*
~DITCH - [build] (channel cut along a berm for the drainage of excessive water) fossé *m* latéral, caniveau *m*
BERME - [mining] berme *f*
BERNOULLI'S LAW - [phys] loi *f* de Bernoulli
BERRY - [bot] baie *f*
BERTH - [gen] place *f*, emplacement *m*, couchette *f*, cabine *f*
[naut] poste *m* de mouillage, poste *m* d'amarrage
BERTRANDITE - [min] bertrandite *f*
BERYL - [min] (a silicate of beryllium and aluminium) béryl *m*
~GREEN - [ind chem] aigue-marine *f*
BERYLLATE - [chem] béryllate *m*
BERYLLIA - [min] oxyde *m* de béryllium
BERYLLIUM - [chem] (light steely metallic element) béryllium *m*
~ACETATE - [chem] acétate *m* de béryllium
~BRONZE - [metall] bronze *m* au béryllium
~COPPER - [min] cupro-béryllium *m*
~COPPER ALLOY - [metall] cuivre *m* au béryllium
~METAPHOSPHATE - [ind chem] métaphosphate *m* de béryllium
~MODERATED REACTOR - [nucl] réacteur *m* modéré au béryllium
~MODERATOR - [nucl] modérateur *m* au béryllium
~OXIDE - [chem] oxyde *m* de béryllium
~REACTOR - [nucl] réacteur *m* au béryllium
~SILICATE - [chem] silicate *m* de béryllium
BERYL OXIDE - [ind chem] oxyde *m* de béryl
BESIDE - [gen] à coté, auprès de
BESOM - [impl] balais *m* de jonc (ou de bruyère)
BESPOKE TAILORING - [text] atelier *m* de tailleur sur mesure
BESSEL FUNCTION - [el] (the basic function in the Bessel Zero Method) fonction *f* de Bessel
~ZERO METHOD - (a method for determining frequency variation in F.M. transmission) méthode *f* du pont de Bessel
BESSEMER PROCESS - [metall] (process of steel-making in which air is blown through iron to eliminate certain elements, after which required constituents are added in a special alloy) procédé *m* Bessemer
BESSEMER STEEL - [metall] acier *m* Bessemer
BEST - [gen] meilleur, optimal
~BAR - [metall] fer *m* ébauché
~CLIMBING ANGLE - [aero] (optimum value of climbing angle) angle *m* de montée optimal
~COAL - [min] charbon *m* en gros morceaux
~FOUNDRY PIG IRON - [metall] fonte *f* spéciale pour pièces de machines
~IRON - [metall] fer *m* doux
~TAP - [min] oxyde *m* magnétique pur de fer
BESTOW, to - [gen] accorder, donner, conférer, octroyer
BESTRIDE - [gen] à cheval, à califourchon
BETA - [phys] bêta, particule *f* bêta
~ABSORPTION GAUGE - [nucl] jauge *f* d'épaisseur (à rayonnement bêta)
~ACTIVITY - [radiat] radioactivité *f* bêta
~BRASS - [metall] laiton *m* B (bêta)
~CELLULOSE - [paper man] bêta B-cellulose *f*
~CONTROL - [aero] (system of propeller pitch control in which the pilot selects the pitch directly in the range between normal fin pitch and maximum reverse) commande *f* bêta
~COUNTER - [nucl] compteur *m* de particules bêta
~DECAY - [nucl] désintégration *f* bêta
~DECAY ELECTRON - [nucl] électron *m* de désintégration bêta
~DISINTEGRATION ENERGY - [nucl] énergie *f* de désintégration bêta
~EMITTER - [nucl] source *f* de particules bêta
~FUNCTION - [math] fonction *f* bêta
~IRON - [metall] fer *m* bêta
~PARTICLE DISINTEGRATION - [nucl] désintégration *f* bêta
~RAY - [phys] rayon *m* bêta (électrons négatifs)
~-RAY ELECTROSCOPE - [instr] électroscope *m* pour rayons bêta
~RAY EMISSION - [nucl] émission *f* de rayons bêta
~RAY PLAQUE - [nucl] (radium container) plaque *f* radifère
~RAY SPECTRUM - [nucl] spectre *m* bêta
~SCREEN - [nucl] écran *m* pour rayonnement bêta
~SENSITIVE - [nucl] sensible aux rayons bêta
~TRANSFORMATION - [phys] transformation *f* bêta
~URANIUM - [nucl] uranium *m* bêta
~URANOPHANE - [min] uranophane *m* bêta
~URANOTIL - s.uranophane
BETAFITE - [min] bétafite *f*
BETAINE - [chem] bétaine *f*
~FORMULA - [chem] formule *f* de la bétaine
BETANAPHTHOL - [chem] bêtanaphtol *m*
BETATOPIC - [nucl] bêtatopique
BETATRON - [electron] (device for the acceleration of electrons) bêtatron *m*
BETEL BIT - [agric] chique *f* de bétel
~CHIP - s.betel bit
~NUT - [bot] (a stimulant plant) noix *f* d'arec
BETTERMENT - [gen] amélioration *f*
[fin] plus-value *f*, amélioration *f*
BETWEEN - [gen] entre
~CENTRES - [mech] (in lathes) distance *f* entre pointes
~DECKS - [naut] entrepont *m*, faux-pont *m*

BETWEEN SEASON HEATING - [th eng] chauffage *m* de demi-saison
BEV - [electron] gigaélectron-volt *m*
BEVATRON - [nucl] (powerful accelerator of atomic particles) bévatron *m*
BEVEL, to - [mech] (to form a surface at an angle to another, more or less than 90°) chanfreiner, biseauter
BEVEL - [carp etc] facette *f*, biseau *m*, chanfrein *m*, fausse équerre *f*
[mech] biseau *m*, conicité *f*, cone *m*
[print] biseau *m*
~COUPLING - [mech] accouplement *m* à cône
~CUT - [mech] chanfrein *m*
~DIFFERENTIAL - [mech] différentiel *m* à engrenages coniques
~DRIVE GEAR - [text] pignon *m* conique de commande
~DRIVE PINION - [mech] pignon *m* conique
~-EDGED - [mech etc] taillé en biseau
~GEAR - [mech] engrenage *m* conique (ou d'angle), engrenage *m* à biseau
~GEAR CUTTING MACHINE - [mech mach] machine *f* à tailler (ou rectifier) les engrenages coniques
~GEAR DIFFERENTIAL - [mech] différentiel *m* à engrenages coniques
~GEAR DRIVE - [mech] transmission *f* à engrenages coniques
~GEAR GENERATION MACHINE - [mech mach] machine *f* à tailler les engrenages coniques
~GEARING - [mech] transmission *f* à engrenages coniques
~HEAD - [metall] cordon *m* conique
~JOINT - [mech] assemblage *m* en fausse coupe
~PINION - [mech] pignon *m* conique
~PROTRACTER - [instr] goniomètre *m*
~RULE - [impl] fausse *f* équerre
~SEATED VALVE - [mech] soupape *f* à siège conique
~STICK - [print] stylet *m*
~WELD - [metall] (a weld between two bevelled structural members) soudure *f* chanfreinée
~WHEEL - [mech] pignon *m* conique, roue *f* conique
~WHEEL REVERSING GEAR - [mech] inverseur *m* de marche à engrenages coniques
BEVELLED - [metall] biseauté, chanfreiné
~COGGING - [mech] assemblage *m* en biseau
~HALVING - [carp] traverse *f* trapéziforme
BEVELLER - [mech mach] biseauteuse *f*
BEVELLING - [gen] biseautage *m*, chanfreinage *m*, angle *m* d'équerrage
BEVERAGE - [gen] boisson *f*
BEZEL - [mech] (annular part surrounding the transparent cover of an instrument) biseau *m* de boîtier chaton *m*, portée *f* de bague
~OF HEAD LAMP - [aero] visière *f* de phare
B.F.O. - (Beat Frequency Oscillator) oscillateur *m* à battements
B - H CURVE - (magnetization curve) courbe *f* de saturation magnétique
B.H.P. - (Brake Horse Power) puissance *f* au frein
BIAS, to - [gen] influencer, rendre partial
BIAS - [gen] prévention *f*, tendance *f*
[adv] obliquement
[text] biais *m*
[mech] arme *m* d'un ressort
[electron] (steady direct voltage between grid and cathode in a thermionic valve) tension *f* de grille
[electron] (in video recording) prémagnetisation *f*
[telecomm] tension *f* de polarisation
[instr] erreur *f* systématique (due au décentrement)
~CONSTRUCTION - [text] confection *f* en tissu coupé en biais
~CUTTING MACHINE - [text] machine *f* à coupe transversale
~FABRIC - [rubber ind] tissu *m* en fil biais
~LIGHTING - [electron] éclairage *m* arrière
~METER - [instr] appareil *m* de mesure de polarisation
~POINT - [electron] point *m* de charge
~RESISTOR - [electron] (resistor used to fix grid bias) résistance *f* de polarisation de grille
~TAPE - [text] bande *f* en biais
~VOLTAGE - [electron] (voltage applied to produce grid bias) tension *f* de polarisation
BIASED AUTOMATIC GAIN CONTROL - [radio] antifading *m* retardé
~PROTECTIVE SYSTEM - [el] dispositif *m* de protection polarisé
~RELAY - [electron] (relay so adjusted as to operate on one polarity only) relais *m* polarisé
~TELEGRAPH DISTORSION - [radio] distortion *f* asymétrique, distortion *f* biaise
BIASING - [electron] (in video recording, polarization of a recording head in magnetic-tape recording) polarisation *f*, prémagnétisation *f*
~COIL - [comput] (of a magnetic core) enroulement *m* de polarisation, enroulement *m* de prémagnétisation
~PULSE - [telecomm] impulsion *f* de commande
BIAX MAGNETIC ELEMENT - [electron comp] élément *m* magnétique biaxe
BIAXIAL - [gen & phys] biaxial, biaxe
~CRYSTAL - [cryst] (crystal having two optic axes) cristal *m* biaxe
~FATIGUE TEST - [metall] essai *f* de fatigue biaxiale
BIAZZI PROCESS - [ind chem] (a method of continuous nitration) procédé *m* de nitration continue (de Biazzi)
BIB-VALVE - [mech] (a type of screw-down valve with a downward curved discharge) vanne *f* à robinet de purge
BIBASIC - [gen] bibasique
BIBB - [shipbuild] collier *m* des élonges de chouque
BIBCOCK - [mech] (a small plug-cock having the discharge downward) robinet *m* coudé, robinet *m* à bec courbé
BIBENDUM RIM - [rubber ind] jante *f* Michelin
BIBULOUS - [med] spongieux
BICAPSULAR - [bot] bicapsulaire
BICARBONATE - [chem] bicarbonate *m*
~OF SODA - [chem] bicarbonate *m* de soude
BICEPS - [anat] biceps *m*
BICHLORIDE - [chem] bichlorure *m*
BICHROMATE - [chem] bichromate *m*
~CELL - [el] pile *f* au bichromate, élément *m* au bichromate de potasse
BICHROMATED GELATINE - [chem] gélatine *f* au bichromate, gélatine *f* chromatée
BICHROME - [chem] bicolore, bichromate *m* de potassium
BICIPITAL - [anat] bicipital, bicéphale, bicipité
BICKERIN - s.beak-iron
BICONCAVE - [gen] biconcave
~SPHERE - [opt] sphère *f* biconcave

BICONICAL AERIAL - [radio] (aerial consisting of two conical conductors with common axis and vortex, excited at the latter) antenne *f* biconique
~ANTENNA - [radio] s.biconical aerial
~HORN - [radio] (aerial system) cornet *m* double
BICONVEX - [gen] biconvexe
BICYCLE - [mech] bicyclette *f*
~LAYOUT - [aero] (an arrangement of a landing gear in which there are two main wheels, one following the other, the wings being supported by outrigger wheels or skids) train *m* d'atterrissage, atterrisseur *m* axial
~TUBE - [rubber ind] chambre *f* d'air (de bicyclette)
BICYCLIC - [chem] bicyclique
~TERPENE - [ind chem] terpène *m* bicyclique
BID, to - [gen] offrir, inviter, faire une enchère, ordonner
BID - [gen & comm] offre *f*, enchère *f*, mise *f*
BIDDER - [gen & comm] enchérisseur *m*
BIDDING - [gen] commandement *m*, invitation *f*
[comm] enchère *f*, mise *f* aux enchères
BIDIRECTIONAL - [gen] bidirectionnel, alternatif
~MICROPHONE - [acoust] microphone *m* bidirectionnel
~PULSES - [modul] (pulses of which a few rise in one direction and the remainder in the other direction) impulsions *f* bidirectionnelles
~TRANSDUCER - [radio] transducteur *m* bidirectionnel
BIDWELL CELL - [el] pile *f* au sélénium
BIENNIAL - [gen] biennal, bisannuel
~PLANT - [bot] plante *f* bisannuelle
BIFID - [zool] (forked: divided into two lobes) bifide
BIFILAR - [gen el etc] bifilaire
~BRIDGE - [meas instr] (a bridge for the measurement of the coefficient of self-induction) pont *m* bifilaire
~OSCILLOGRAPH - [instr] oscillographe *m* bifilaire
~PENDULUM - [phys] pendule *m* bifilaire
~SUSPENSION - [el] suspension *f* bifilaire
~WINDING - [el] (method of winding so as to eliminate self-induction) enroulement *m* bifilaire (anti-inductif)
BIFOCAL - [opt] bifocal, à double foyer
~BULB - [el] lampe *f* à deux filaments
BIFUNCTIONAL STRUCTURAL UNIT - [chem] (of a polymer) motif *m* structural bifonctionnel (d'un polymère)
BIFURCATED RIVET - [mech] (a type of rivet with a divided shank used for light work) rivet *m* fendu
BIFURCATING BOX - [telecomm] (connecting box joints between one two-core and two single-core cables) boîte *f* de dérivation
BIFURCATION - [gen & road or railw] bifurcation *f*, embranchement *m*
BIG-END - [mech] (the end of the connecting-rod in a reciprocating engine) tête *f* de bielle
~BEARING - [mech] coussinet *m* de tête de bielle
~BEARING SHELL - [mech] coquille *f* de coussinet de tête de bielle
~BOLT - [mech] boulon *m* de tête de bielle
~BUSH - [mech] coussinet *m* de tête de bielle
~DOWN MOULD - [metall] lingotière *f* droite
~HALF BEARING - [mech] demi-coussinet *m* de tête de bielle
~HEAD - [mech] tête *f* de bielle
BIG KEEP - [mech] (the separate piece at the end of the connecting rod which forms the retaining part for half the bearing) chape *f* de tête de bielle
~KNOCK - [mech] (characteristic sound produced by a big-end bearing having excessive clearance) cliquetis *m* de tête de bielle
~OF CONNECTING ROD - [mech] tête *f* de bielle
~UP MOULD - [metall] lingotière *f* renversée
BIG SPOT - [cinema lighting] (large rotary lamp) projecteur *m* intensif
BIGENERIC CROSS - [genet] croisement *m* bigénérique
BIGG - [agric] (four-row barley) orge *f* commune, orge *f* à quatre rangs
BIGHT - [geogr] baie *f*, anse *f*, crique *f*, sinuosité *f*
[of ropes] mou *m* (d'un cordage), double *m* d'un cordage
BIGNESS - [gen] grandeur *f*, grosseur *f*
BIGRID - [electron] lampe *f* bigrille
~VALVE - [electron] (four electrode thermionic valves with two control grids) valve *f* bigrille
BIKE - [mech] (colloquial for bicycle) vélo *m*, bicyclette *f*
BILATERAL - [gen] bilatéral
~AREA TRACK - [electro acoust] (variable area track with the two edges of the central area modulated in accordance with the signals) piste *f* double
[cinema] piste *f* sonore double
~FIT - [mech] jeu *m* bilatéral
~LIMITS - [mech] (upper and lower limits of bilateral fit system) tolérances *fpl* bilatérales
~RECORDING - [cinema] (system of recording with high sensitivity obtained by increasing or decreasing the width of the illuminated area from both sides at the same time) enregistrement *m* bilatéral
~TRANSDUCER - [radio] transducteur *m* bidirectionnel
BILE - [physiol] bile *f*
~ACID - [med] acide *m* biliaire
~DUCTS - [anat] canaux *m* biliaires
~PIGMENT - [med] pigment *m* biliaire
~SALTS - [physiol] sels *m* biliaires
~STONES - [med] calcul *m* biliaire
BILGE, to - [anat] faire eau, crever (une navire)
BILGE - [naut] sentine *f*, cale *f*
[shipbuild] bouchain *m*
[of barrels and casks] ventre *m* d'un baril, bouge *m* (de barrique)
~BLOCK - [naut] anguiller *m*, tin *m* latéral
~KEEL - [shipbuild] quille *f* de roulis
~PIECE - [shipbuild] s.bilge keel
~PUMP - [shipbuild] pompe *f* de cale
~STRAKE - [shipbuild] virure *f* de bouchain
~STRINGER - [shipbuild] serre *f* de bouchain
~WATER - [naut] eau *f* de cale
~WAYS - [shipbuild] (blocks of timber under a vessel) couettes *f* courantes
~WELL - [shipbuild] archipompe *f*, puisard *m*
BILGED - [naut] qui a une voie d'eau, défoncé
BILIARY - [anat] biliaire
BILINEAR - [gen] bilinéaire
BILL, to - [comm] facturer
[advert] afficher
BILL - [gen] compte *m*, facture *f*, addition *f*
[comm] compte *m*, calcul *m*, traite *f*, lettre *f* de change, effet *m*

[advert] affiche *f*, écriteau *m*, placard *m*, programme *m*, prospectus *m*
[leg] projet *m* de loi
[zool] bec *m*
[naut] bec *m* de l'oreille de l'ancre
[impl] serpe *f*, serpette *f*
BILL BOOK - [comm] livre *m* d'effets
~OF ENTRY - [comm] déclaration *f* en douane
~OF EXCHANGE - [fin] lettre *f* de change
~OF HEALTH - [med] patente *f* de santé
~OF INDICTMENT - [leg] acte *m* d'accusation
~OF LADING - [comm] connaissement *m*
~OF MATERIALS - [comm & ind] spécification *f*, devis *m* quantitatif, bordereau *m* matières
~OF QUANTITIES - [build] (quantities of materials needed for a specified work and description) devis *m* quantitatif
BILLBOARD - [naut] renfort *m* d'écubier
BILLET, to - [gen] loger, cantonner
BILLET - [gen] emploi *m*, place *f*, situation *f*, billet *m* de logement
[metall] (iron or steel bloom drawn into a short bar before further treatment) billette *f*, lingot *m*, larget *m*
[plast ind] billette *f*
[carp] bille *f*, bûche *f*, rondin *m*
~ROLL - [mech] cylindre *m* ébauché
~ROLLING MILL - [metall] (mill reducing ingots to billets) train *m* à billettes, laminoir *m* à billettes, train *m* ébaucheur
~STEEL - [metall] acier *m* en billette
BILLHOOK - [agric impl] serpe *f*, serpette *f*
BILLI-CAPACITOR - [radio] (a variable capacitor of a few pico-farads, used for very fine adjustments) condensateur *m* ajustable de petite capacité
~-CONDENSER - [radio] (variable condenser with a maximum capacity of a few micro-microfarads, used for fine tuning) condensateur *m* ajustable
BILLIARD - [gen] billard *m*
~CLOTH - [text] (a very fine quality, perfectly smooth woollen cloth) drap *m* de billard
BILLIETITE - [min] (secondary mineral containing approximately 70 p.c. of uranium) billiétite *f*
BILLING - [comm] facturation *f*
~OFFICE - [comm] service *m* du portefeuille (-effets)
BILLION - [math] trillion *m* (GB), milliard *m* (US)
BILLOW - [naut] grande vague *f*, lame *f* de fond
BILLY - [gen] gamelle *f*, baton *m*
[text] métier *m* en gros, banc *m* à broches
[slubbing bill] boudineuse *f*
~GOAT - [zool] bouc *m*
BILOBATE - [med] bilobé
BILOBED - [anat] bilobé
BILOCULAR - [bot] (consisting of two chambers) biloculaire
BIMETAL - [metall] (composed of two different metals) bimétal, bimétallique
~FUSE - [el] (a fuse consisting of two different metals, e.g. lead coated copper wire) fusible *m* bimétallique
~STRIP - [el] (strip composed of thinner strips of two different metals so as to deform when its temperature changes) bilame *m*
~THERMAL SWITCH - [el] (switch operated by a change of temperature) interrupteur *m* à bilame
BIMETALLIC INSTRUMENT - [instr] (the deformation of a bimetallic element is transmitted to the moving element) instrument *m* bimétallique
BIMETALLIC STARTER - [mech] starter *m* à bilame
~STRIP - [mech] lame *f* bimétallique
BIMOLECULAR REACTION - [chem] (a reaction in which there is interacting of two molecules) réaction *f* bimoléculaire
BIMOTORED - [mech mach] à deux moteurs, bimoteur
BIN - [gen] coffre *m*, huche *f*, bac *m*, coffre *m* à grain, compartiment *m*, casier *m*
[in factories] casier *m*, caisse *f*, caisson *m*, réservoir *m*, trémie *f*
~CARD - [comput] carte *f* stock
BINANT - [instr] (component part of Hoffman electrometer) segment *m* chargé
BINARY - [math] (involving the integer two) binaire
[astr] (double star in which the two components revolve about their common centre of mass) étoile *f* double
[chem] binaire
~ALLOY - [metall] alliage *m* binaire
~CELL - [comput] (an element capable of storing a unit of information) élément *m* binaire, élément *m* de mémoire binaire
~CODE - [comput] code *m* binaire
~-CODED DECIMAL QUOTATION - [electron comp] notation *f* décimale en code binaire, notation *f* décimale codée sous une forme binaire
~-CODED DECIMAL SYSTEM - [comput] système *m* décimal codé binaire
~COIL - [comput] cellule *f* binaire, élément *m* binaire
~COMMAND - [comput] ordre *m* par tout ou rien
~COMPOUND - [chem] composé *m* binaire, mélange *m* binaire
~CONVERTER - [el] (a.c. to d.c. converter with a three-phase winding and a d.c. exciting winding) convertisseur *m* binaire
~COUNTER - [radio] (used to give one output pulse for each pair of input pulses) compteur *m* binaire
~DECIMAL CODE - [comput] code *m* binaire décimal
~DIGIT - [comput] (digit in the binary scale of notation) bit *m*, chiffre *m* binaire
~DIGITAL COMPUTER - [comput] (computer which operates with numbers represented by ones and zeros) calculatrice *f* digitale binaire
~DIVIDER - s, binary counter
~FISSION - [chem] (the division of the nucleus into two daughter nuclei) fission *f* binaire
~FUEL - [ind chem] carburant *m* binaire
~GRANITE - [geol] granit *m* à deux micas
~NOTATION - [electron comput] (writing of number in the scale of two) notation *f* binaire
~NUMBER - [math] (written in binary notation) nombre *m* binaire
~OPERATION - [comput] (computer operation with two variables, or with variables in binary notation) opération *f* binaire
~POINT - [comput] (marking the place between integral and fractional powers of two) virgule *f* binaire
~PUNCH - [comput] perforation *f* en code binaire
~PUNCHED CARD - [comput] carte *f* perforée en code binaire
~REPRESENTATION - [comput] représentation *f* binaire

BINARY SCALE - [math] échelle *f* binaire
~SCALER - [math] échelle *f* binaire
~SEARCH METHOD - [comput] (to find definite data in a sequence) méthode *f* de recherche en code binaire
~SHIFT REGISTER - [comput] registre *m* binaire à décalages
~STAR - [astr] étoile *f* double
~-TO-DECIMAL CONVERSION - [math] conversion *f* binaire-décimale
~-TO-DECIMAL CONVERTER - [comput] convertisseur *m* binaire-décimal, décodeur *m* binaire-décimal,
BINAURAL - [acoust] (pertaining to both ears) binaural, binauriculaire
~EFFECT - [acoust] (the effect which is achieved by hearing a sound with both ears) effet *m* stéréophonique
~RECORDER - [acoust] (tape recorder with two separate recording channels) enregistreur *m* stéréophonique, enregistreur *m* binauriculaire
~STETHOSCOPE -[med] stéthoscope *m* binauriculaire
BIND, to - [gen] attacher, lier
[comm] lier (par contrat), engager, obliger, tenir
[chem] lier, relier, fixer, s'agglomérer, s'agglutiner
[mech mach] se coincer, gripper, gommer
[med]constiper, bander, ligaturer
[of books] relier
BIND - [min] schiste *m* bitumineux
[mining] amincissement *m* en coin
BIND WITH IRON, to - [mech] ferrer
BINDER - [ind chem] liant *m*, agglomérant *m*
[agric mach] lieuse
[constr] poutre *f* maîtresse, poutre *f* de traverse
~AND REAPER - [agric] moissonneuse-lieuse *f*
~CORE - [metall] portée *f*
~DENT - [text] garde *f* du peigne
~HEALD FRAME - [text] lame *f* à coulisses
~LEVER - [mach tool] levier *m* de blocage
~SHEETING - [agric mach] (for a harvesting machine) feuille *f* pour moissonneuse-lieuse
~TWINE - [text] fil *m* pour lieuse
BINDERS - [ind chem] (natural or synthetic resins, cellulose or protein compounds) liants *m*, agglomérants *m*
BINDING - [gen] lien *m*, fixation *f*, serrage *m*, frettage *m*, reliure *f* (de livre)
[chem] (of gas in a solid or liquid) captage *m*, absorption *f*
[text] passementerie *f*, garniture *f*, entrelacement, liage *m*
[mech] grippage *m*, gommage *m*, coincement *m*
[of agreements or contracts] obligatoire, qui lie, qui engage
~AGENT - [ind chem] (generic term denoting natural and synthetic agents which retain their adhesiveness after thickening) liant *m*, agglomérant *m*, agglutinant *m*
~BAND - [el] ruban *m* à border
~CLAMP - [mech] serre-fils *m*
~CORE - [metall] portée *f* du noyau
~ENERGY - [phys] (the energy per unit charge required to remove an electron from the atom to an infinite distance) énergie *f* de liaison
~ENERGY FORMULA - [nucl] formule *f* de l'énergie de liaison
BINDING GRAVEL - [build] couche *f* de liaison, assise *f* de liaison
~-IN WIRE - [el] fil *m* de ligature
~MATERIAL - [mech mach] liant *m*, agglomérant *m*, matière *f* agglutinante
~PICK - [text] trame *f* de fond
~PLAN - [text] liage *m*
~POINT - [text] point *m* de liage, pointé *m*
~-POST - [el] (a type of terminal in which the wire passes through a hole and is clamped by a set screw) borne *f* à vis, serre-fil *m*
~POWER - [chem] pouvoir *m* adhésif
~RIG - [metall] anneau *m* de serrage
~SCREW - [mech] vis *f* de pression
~STRIP - [gen] bande *f* gommée
~TWINE - [text] fil *m* de ligature
~WARP - [text] chaîne *f* de piqûre, chaîne *f* de liage
~WEFT - [text] trame *f* de liage
~WIRE - [metall] fil *m* de ligature
~YARN - [text] fil *m* frisé
BINDWEB - [med] névroglie *f*
BINDWEED - [bot] liseron *m*
BINE - [bot] sarment *m*, tige *f*, liane *f*
BINET'S AGE - [med] âge *m* intellectuel, âge *m* de Binet
BINEUTRON - [nucl] (in theory, a neutral particle with a mass near two) bineutron *m*
BINIT - [electron comput] bit *m*, chiffre *m* binaire
~-TEN - [comput] binaire-décimal
BINNACLE - [naut] habitacle *m* (du compas)
BINOCULAR - [opt] binoculaire
~CAMERA - [photo] (camera with two lenses for stereoscopic images) appareil *m* photographique binoculaire, caméra *f* binoculaire
~VISION - [opt] vision *f* binoculaire
BINODE - [electron] (a three electrode thermionic tube with one cathode and two anodes) diode-triode *f*, duodiode-triode *f*, binode *f*
BINOMIAL - [math] (an algebraic expression having two terms) binôme *m*
~COEFFICIENT - [math] coefficient *m* binomial
~DISTRIBUTION - [math] distribution *f* binomiale
~EXPANSION - [math] développement *m* binomial
~LAW - [math] loi *f* binomiale
~SERIES - [math] série *f* binomiale
~THEOREM - [math] théorème *m* du binôme
~TWIST - [waveguide] torsion *f* binomiale, torsade *f* binomiale
BINORMAL - [math] binormal
BINOVULAR - [med] biovulaire
~TWINS - [med] (twins from two separate ova) jumeaux *m* biovulaires, faux jumeaux *mpl*
BINUCLEAR - [med] binucléaire
BINUCLEATE - [med] binucléée
BIOAERATION - [hydr] (system of sewage purification by oxidation) aération *f* biochimique (épuration)
BIOASSAY - [biol] (test to determine the power of a drug) essai *m* biologique
BIOBLAST - [genet] (minute mass of amorphous protoplasm with formative powers) bioblaste *m*
BIOCATALYST - [chem] biocatalyseur *m*
BIOCHEMICAL - [biochem] biochimique
~OXYGEN DEMAND - [chem] (sewage purification) demande *f* biochimique d'oxygène (D.B.O.)
BIOCHEMISTRY - [chem] (the chemistry of living

things) biochimie *f*

BIOCHORE - [genet] (group of similar biotopes) biochore *m*

BIOCHROMOTHERAPY - [med] biochromothérapie *f*

BIOELECTRIC NULL - [electrobiology] (potential of an area of a tissue with an electric symmetry the potential of which does not appreciably alter during stimulation) zéro *m* bioélectrique

BIOELECTRICITY - [el] bioélectricité *f*

BIOELEMENT - [phys] bioélément *m*

BIOFILTER - [biochem] lit *m* percolateur, lit *m* bactérien

BIOFILTRATION - [biochem] filtration *f* biologique, épuration *f* par lit bactérien

BIOGENERIC CROSS - [genet] croisement *m* biogénérique

BIOGENESIS - [biol] (the generation of life from living beings) biogenèse *f*

BIOGENOUS - [bot & med] (parasitic) parasitique

BIOGENY - s.biogenesis

BIOGRAPHY - [gen] biographie *f*

BIOLOGICAL - [gen] biologique

~CHEMISTRY - [chem] (the chemistry of living organisms) biochimie *f*

~GEOGRAPHY - [biol] biogéographie *f*

~HALF-LIFE - [nucl] (the time during which a living tissue, organ etc eliminates one half of a specified substance in it) période *f* biologique

~HOLE - [nucl] (cavity in a nuclear reactor to place animals or plants near the active section) chambre *f* d'essais biologiques

~OXIDATION - [chem] oxydation *f* biologique

~PROCESS - [biol] procédé *m* biologique, action *f* biologique

~PURIFICATION OF SEWAGE - [hydr] épuration *f* biologique des eaux d'égout

~RESPONSE - [phys] réaction *f* biologique

~SEWAGE DISPOSAL WORKS - [hydr] installation *f* de clarification biologique

~SHIELD - [nucl] (shield designed to reduce the intensity of radiation to an amount which is physiologically acceptable) écran *m* de protection

~SHIELDING - [nucl] protection *f* biologique

~SLIME - [hydr] pellicule *f* biologique, film *m* biologique

~SLUDGE - [ind chem] boue *f* activée

~TREATMENT - [ind chem] procédé *m* de traitement biologique

BIOLOGY - [biol] (the science of all living organisms) biologie *f*

BIOLUMINESCENCE - [phys] (production of light by living organisms) bioluminescence *f*

BIOMECHANICS - [biochem] (the study of the principles and relations of biological activity) biomécanique

BIOMOTOR - [med] appareil *m* pour la respiration artificielle

BIOMUTANT - [genet] biomutant *m*

BIOPHYSICAL - [biol] biophysique

BIOPHYSICS - [phys] (the physics of biological processes) biophysique *f*

BIOPOESIS - [genet] biopoèse *f*

BIOPSY - [med] (examination of tissues removed from a body for diagnostic purposes) biopsie *f*

BIOSATELLITE - [astronaut] satellite *m* biologique

BIOSPECTROMETRY - [instr] biospectrométrie *f*

BIOSPHERE - [geogr] biosphère *f*

BIOTHERAPY - [med] biothérapie *f*

BIOTIN - [chem] (Vitamin H) biotine *f* (Vitamine H)

BIOTITE - [min] (rock-forming mineral belonging to the mica group) biotite *f*

BIPACK FILM - [cinema] (film for the bipack system) film *f* à deux couches

~SYSTEM - [cinema] (a system for colour photography with a film having two sensitive layers) assemblage *m* de deux bandes

BIPHASE - [el] (synonym for two-phase) biphasé

BIPHENYL - [chem] biphényle *m*

BIPLANE - [aero] biplan

BIPOLAR - [el] (having two poles) bipolaire

~CAM - [mech] came *f* bipolaire

~DYNAMO - [el] dynamo *f* bipolaire

~ELECTRODE - [electron] (in an electro-plating bath, an electrode without connexion to anode or cathode) électrode *f* bimétallique

~GERMINATION - [bot] (germination of a spore by the formation of two germ tubes) germination *f* bipolaire

BIPRISM - [photo] (a prism with a very obtuse angle) biprisme *f*

BIPROPELLANT - [mech] (of rockets) bi-combustible, à deux constituants

~ROCKET - [astronaut] fusée *f* à diergol

BIQUADRATE - [math] bicarré, puissance quatre

BIQUADRATIC EQUATION - [math] équation *f* bicarrée

BIQUINARY CODE - [electron comput] code *m* biquinaire

~NOTATION - [electron comput] (scale of notation where the base is alternately three and five) notation *f* biquinaire

BIRADIAL SYMMETRY - [zool] (condition in which radial and part bilateral symmetry appears in a body) symétrie *f* biradiale

BIRAMOUS - [zool] (with two branches) à deux branches

BIRCH - [bot] bouleau *m*

BIRD - [zool] oiseau *m*

~ARM - [med] bras *m* atrophié

~FEVER - [vet] (synonym for fowl cholera) choléra *f* aviaire, pasteurellose *f*

~GUARDS - [gen] épouvantail *m*

BIRD'S BEAK ORNAMENT - [build] bec *m* d'oiseau

~-EYE - [text] (a pattern) oeil *m* de perdrix

~EYE MAPLE - [bot] érable *m* à broussin, érable *m* madré

~EYE VIEW - [gen & surv] vue *f* à vol d'oiseau [cinema] plongée

BIRDCAGE BOBBIN - [text] roue *f* de dévidoir

~REEL - [text] dévidoir *m* à bobine

BIRDSMOUTH ATTACHMENT - [mech] joint *m* à chevron, emboîtement *m* à griffe

BIREFRACTIVE - [opt] biréfringent

BIREFRINGENCE - [opt] (the double bending of light by crystalline minerals) biréfringence *f*

BIROTATION - [chem] (the change of the optical activity of a fresh solution of an active substance) birotation *f*

BIRTH - [gen] naissance *f*, parturition *f* [med] accouchement *m*, naissance *f*

~CANAL - [med] voies *f* génitales

~-CONTROL - [med] régulation *f* des naissances

~MARK - [med] naevus *m*, envie *f*, tache *f* de naissance

BIRTH PALSY - [med] paralysie *f* de parturition
~RATE - [gen] natalité *f*
~TRAUMA - [med] trauma *m* obstétrique
BISCUIT - [food] biscuit *m*
[pottery] (pottery ware after preliminary firing but before glazing) biscuit *m*
[diecasting] galette *f*, pastille *f*
[plast ind] (injection moulding) grappe *f* (de moulage en injection)
~BAGGING - [text] toile *f* à filet
~OF RUBBER - [text] galette *f* de caoutchouc
~OVEN - [pottery] four *m* à porcelaine
BISE - [met] (dry northerly wind occurring in southern France) bise *f*
BISECT, to - [gen] (to devide into two parts, espec, equal parts) diviser en deux, bisegmenter
[math] partager en deux
BISECTOR - [opt] (the line which divides an angle into two equal parts) bissectrice *f*
BISECTRIX - [opt] bissectrice *f*
BISEXUAL - [zool] (with male and female sexual organs) bissexué, bissexuel, hermaphrodite
BISILICATE - [chem] silicate *m* double, bisilicate *m*
BISMANOL - [metall] (magnetic alloy of bismuth and manganese) alliage *m* magnétique, bismuth-manganèse
BISMITE - [chem] bismuthite *f*, oxyde *m* de bismuth
BISMUTH - [chem] (an element, symbol Bi. A hard, brittle grey metal used in low melting point alloys, medicine etc) bismuth *m*
~AMMONIUM CITRATE - [chem] (an antacid employed in medicine) citrate *m* de bismuth-ammoniaque
~CHROMATE - [chem] (a pigment) chromate *m* de bismuth
~GLANCE - [min] (an ore of bismuth) bismuthine *f*
~MEAL - [radio] bouillie *f* opaque en bismuth
~NITRATE - [chem] (intermediate for other bismuth salts) nitrate *m* de bismuth
~OCHRE - [chem] (natural form of bismuth trioxide; also called bismite) bismuthocre *m*
~OXYCHLORIDE - [chem] (a pigment; also used in medicine) oxychlorure *m* de bismuth, blanc *m* de perles
~SODIUM TARTRATE - [chem] (used in medicine) tartrate *m* double de sodium et de bismuth
~SPAR - [min] bismuthite *f*
~SPIRAL - [instr] (instrument using the variation of the resistivity of bismuth to measure field strength) sonde *f* au bismuth
[el] sonde *f* bismuthique
~SUBCARBONATE - [chem] (an x-ray contrast medium; also used in medicine, cosmetics etc) sous-carbonate *m* de bismuth
~SUBGALLATE - [chem] (astringent used in medicine) sous-gallate *m* de bismuth
~SUBNITRATE - [chem] (an astringent white powder used in medicine, cosmetics, ceramics etc) nitrate *m* basique de bismuth, blanc *m* de fard, sous-nitrate *m* de bismuth
~TELLURIDE - [chem] (a component of semi-conducteurs) tellurure *m* de bismuth
~TRIOXIDE - [chem] sesquioxyde *m* de bismuth
~WHITE - [chem] blanc *m* de bismuth
~YELLOW - [chem] jaune *m* de bismuth
BISMUTHIC ACID - [chem] acide *m* bismuthique
BISMUTHYL - [chem] bismuthyle *m*
BISON - [zool] bison *m*
BISQUE - [pottery] biscuit *m*, porcelaine *f* blanche sans couverte
~OVEN - s,biscuit oven
BISTABLE CIRCUIT - [electron] circuit *m* bistable, circuit *m* à deux états stables
~MULTIVIBRATOR - [telev] (a relaxation generator changed from a stable to a quasistable state) générateur *m* flip et flop
~TRIGGER CIRCUIT - [radio] (trigger circuit with two stable or quasistable states) circuit *m* de déclenchement bistable, circuit *m* basculeur
~UNIT - [electron comput] élément *m* binaire
BISTOURY - [med instr] bistouri *m*
BISTRE - [ind chem] (a brown pigment obtained from beechwood soot) bistre *m*
BISULPHATE - [chem] bisulfate *m*
BISULPHATES - [chem] (the acid salts of sulphuric acid) bisulfates *m*
BISULPHIDE - [chem] bisulfure *m*
BISULPHITE - [chem] bisulfite *m*
BISULPHITES - [chem] (the acid salts of sulphurous acid) bisulfites *m*
BISYNCHRONOUS MOTOR - [mech] (a synchronous motor designed to run also at twice synchronous speed) moteur *m* synchrone à deux vitesses
BIT - [gen] fragment *m*, morceau *m*, pièce *f*, tranche *f*
[tools cutting edge] mêche *f*, foret *m*, tranchant *m*, taillant *m*
[min] (the cutting edge of steel for rock drilling) fleuret *m*, outil *m* à mèche, bit *m*
[of keys] panneton *m*
[for horses] mors *m*
[paper man] filigrane *m*
[electron comput] bit *m*, chiffre *m* binaire, unité *f* d'information
~BRACE - [mech] vilebrequin *m*
~BREAKER - [mining] débloqueur *m* de trépan
~-DENSITY - [comput] densité *f* de bits, densité *f* de digits binaires
~FILE - [mech] lime *f* pour mèche (ou foret)
~HOOK - [mining] caracole *f* à trépan
~SAMPLE - [mining] échantillon *m* de trépan
~TRAFFIC - [comput] (traffic with reference to the number of bits sent over a transmission line) trafic *m* de bits
BITARTRATE - [chem] bitartrate *m*, tartrate *m* acide
BITCH - [zool] chienne *f*, femelle *f*
BITE, to - [gen & mech] mordre, attaquer, corroder, inciser, prendre, crocher
BITE - [gen] morsure *f*, coup *m* de dent
[metall] (corrosion of metal) attaque *f*, trace *f* de corrosion
[mech] jeu *m*, intervalle *m* (entre cylindres), écartement *m*
[mech] (friction surface) adhérence *f*, surface *f* de pression
[med] morsure *f*, piqûre *f*
BITER SHAFT - [text] arbre *m* à ailettes
BITING ANGLE - [ballist] angle *m* d'attaque minimum
~TEST - [brew ind] (to test the hardness of the grain) essai *m* de dureté du grain (en le mordant)
BITS FOR HANDBRACE - [mech] mèche *f* pour vilebrequin
BITT, to - [naut] bitter un câble (une aussière)
BITT - [naut] bitte *f* d'amarrage, bollard *m*

BITTER - [gen] amer, âpre, piquant
[naut] tour *m* de bitte
~ALMOND OIL - [chem] essence *f* d'amandes amères, benzaldéhyde *m*, aldéhyde *m* benzoïque
~CASSAVA - [bot] tapioca *m*
~EARTH - [min] magnésie *f*
~ORANGE - [bot] orange *f* amère
~SALT - [chem] sulfate *m* de magnésie
~SPAR - [min] (alternative term for dolomite, a natural carbonate of calcium and magnesium) chaux *f* carbonatée magnésifère, dolomie *f*
~SUBSTANCE OF HOPS - [brew ind] principe *m* amer (du lupulin), lupuline *f*
BITTERING POWER - [brew ind] pouvoir *m* d'amertume
BITTERN - [chem] (the residual liquor remaining from the evaporation of sea water after the removal of salt crystals) eaux-mères *mpl*
BITTY - [dyes] (of dyes and varnishes) grumelé
BITUMEN - [chem] (non mineralized substances of coal, lignite etc) bitume *m*, asphalte *m* minéral
~-COATED STEEL MAIN - [plumb] tuyau *m* en tôle plombée et bitumée
~COATING - [constr] revêtement *m* de bitume, enduit *m* bitumineux
~EMULSION - [constr] émulsion *f* bitumineuse
~PLASTICS - [plast] plastiques *m* à base de bitume, matière *f* plastique tirée du bitume
~SHEETING - [road constr] revêtement *m* de bitume
~SPRINKLER - [road constr] bitumeuse *f*
~VARNISH - [ind chem] vernis *m* à base de bitume
BITUMINIZE, to - [road constr] bitumer, bituminer
BITUMINOSIS - [med] bituminose *f*
BITUMINOUS - [gen] bitumineux
~COAL - [min] (term denoting a large class of coals used as fuels and in the production of gas) charbon *m* bitumineux, charbon *m* gras
~PAVEMENT - [constr] (in road construction) revêtement *m* bitumineux (d'une route)
~PEAT - [geol] tourbe *f* grasse
~PITCH - [min] poix *m* d'asphalte, brai *m* d'asphalte
~SCHIST - [min] schiste *m* bitumineux
~SHALE - [min] (shaly sandstone) schiste *m* bitumineux
BITUMIZATION - [road constr] bitumage *m*
BITUMIZED PAPER - [paper man] papier *m* bitumé
BIURET - [chem] (analytical reagent) biuret *m*, amide *f* allophanique
BIVALENCE - [chem] (the property of having a valence of two) bivalence *f*
BIVALENT - [chem] (of a valence of two) bivalent
~INFORMATION - [comput] information *f* bivalente
BIVALVE - [zool] (with the shell in the form of two plates) bivalve *m*
BIVARIANT - [chem] (having two degrees of freedom) bivariant
BIVINYL - [chem] bivinyle *m*
BIVOUAC - [gen] bivouac *m*
BLACK, to - [metall] noircir (un moule)
[radio] perturber
BLACK - [gen] noir, obscur, sombre
[print] noir
~AFTER WHITE - [telev] (defect in which an unnatural black line marks the right-hand edge of objects on the screen) noir après blanc
~ALUM - [ind chem] (water purification mixture) alun *m* noir
~-AND WHITE - [photo] noir et blanc
BLACK-AND-WHITE RECEPTION - [telev] réception *f* en noir et blanc
~ANILINE INK - [ind chem] encre *m* au noir d'aniline
~ANNEALING - [metall] recuit *m* intermédiaire
~ASH - [chem] carbonate *m* de soude
~BEETLE - [zool] blatte *f*, cafard *m*, cancrelat *m*
~BODY - [phys] (a body which when heated to incandescence emits a light spectrum) corps *m* noir, radiateur *m* intégral
~BODY CONSTANT - [phys] constante *f* du corps noir
~BODY RADIATION - [phys] (the quantity and quality of the radiation emitted by a black body with no reflecting power) rayonnement *m* du corps noir
~BODY TEMPERATURE - [phys] (the temperature at which a black body emits the same radiation as that of the given radiating body at a given temperature) température *f* du corps noir
~BOX - [mech] (colloquial term for any all-enclosed self-controlling device) boîte *f* noire, appareil *m* automatique scellé
~CHALK - [ind chem] (a mixture of black ivory and clay) sauce *f*, craie *f* noire
~CLIPPER - [telev] écrêteur *m* du noir, limiteur *m* du noir
~COMPRESSION - [telev] (amplitude compression of the signals corresponding to the black areas of the picture on the screen) compression *f* de l'échelle des luminances dans le noir
~COPPER - [min] cuivre *m* noir
~COTTON SOIL - [soil] régur *m*, tirs *m*
~CRUSHING - [telev] noircissement *m* des gris
~CYANIDE - [chem] (mixture containing approximately 50 p.c. of calcium cyanide) cyanure *m* noir
~-DAMP - [min] (residual gas after a fire-damp explosion, chiefly consisting of carbon dioxide) grisou *m*
~DIAMOND - [min] (a crystalline form of carbon, similar to diamond but without crystal form) diamant *m* noir, carbonado *m*
~DROP - [cinema] (black velvet used to lend night effect) tenture *f* noire, rideau *m* noir
~EARTH WAX - [min] ozokérite *f*
~EGYPTIAN WARE - [pottery] (black, fine-grained stoneware) poterie *f* de grès
~HEAD - [vet] (a contagious typhlo-hepatitis) tête *f* noire, typhlo-hépatite *f*
~-HEART MALLEABLE CAST-IRON - [metall] (s.also malleable cast-iron) fonte *f* malléable, malléable *f* à coeur noir
~IRON PIPE - [plumb] tube *m* de fer noir, tuyau *m* de fer noir
~JACK - [min] (colloq. term for Sphalerite, zinc blende) blende *f*, minerai *m* de zinc, fausse galène *f*
~JAPAN - [paint] (fast-drying black varnish, almost transparent) vernis *m* à l'asphalte, laque *f* à l'asphalte
~-LEAD - [min] (commercial graphite, one of the two crystalline forms of carbon) graphite *m*, mine *f* de plomb, plombagine *f*
~-LEAD CRUCIBLE - [metall] creuset *m* en graphite
~-LEG OF POTATOES - [agric] (a plant disease) jambe *f* noire
~-LEG OF SUGAR BEET - [agric] (a plant disease) pythium *m* de la betterave

BLACK LETTER - [print] caractère *m* gothique
~LEVEL - [telev] niveau *m* du noir
~LIFT - [telev] décollement *m* du noir
~LIGHT - [phys] lumière *f* noire
~LINE CANKER - [rubber ind] (a disease of rubber) black-stripe *f*, black-thread *f*, maladie *f* des raies noires
~LIQUOR - [ind chem] (waste liquid) eau *f* noire
~MALT - [brew ind] (malt specially kilned to obtain a dark colour, as in the "stout" beer) malt *m* grillé
~MANGANESE - [min] pyrolusite *f*
~MICA - [min] biotite *f*
~NEGATIVE - [telev] tension *f* négative pour le noir
~OIL - [min] huile *f* minérale
~OPAL - [min] (Australian type of opal) opale *f* noire
~ORE - [min] pyrite *f* cuivreuse oxydée
~OXIDE OF MANGANESE - [chem] bioxyde *m* de manganèse
~PEAK - [telev] (peak black in G.B.) (the maximum excursion of the signal in the black direction) crête *f* du noir
~PIG-IRON - [metall] fonte *f* grise
~PINE - [bot] pin *m* noir, pin *m* d'Autriche
~PLATE - [metall] tôle *f* noire
~POSITIVE - [telev] tension *f* positive pour le noir
~POWDER - [ind chem] (explosive) poudre *f* noire
~PRECIPITATE - [chem] chlorure *m* mercureux
~RUST - [bot] (in barley) rouille *f* (noire)
~SATURATION - [telev] (a distortion of the picture signal caused by the compression of the blackest range) saturation *f* du noir
~SCREEN - [telev] filtre *m* optique plastique
~SCREEN TELEVISION SET - [telev] (system designed to increase picture contrasts by fitting a black or grey glass filter in front of the cathode-ray tube) téléviseur *m* avec filtre optique
~SHADING - [telev] voile *m*
~SPOTTER - [telev] (special diode in the circuit to act against brief interferences) diode *f* antibrouillage
~STONE - [min] schiste *m* charbonneux
~STRAP - [ind chem] huile *f* émulsionnée noire
~TELLURIUM - [min] (a natural sulpho-telluride of gold and lead) tellure *m* natif-auro-plombifère
~TONGUE - [med] (nutritional disease affecting dogs) langue *f* noire pileuse, glossophytie *f*
~TRANSMITTER - [radio] émetteur *m* clandestin
BLACKBAND - [min] fer *m* carbonaté lithoïde
BLACKBERRY - [bot] mûre *f*
BLACKEN, to - [gen] noircir
BLACKENED TIPS - [agric] (in barley) bases *f* brunes des grains
BLACKENING - [metall] passage *m* au noir
~BATH - [photo] bain *m* de noircissement
BLACKER THAN BLACK REGION - [telev] (the part of the signal near the black level used for synchronization signals) ultranoir *m*
BLACKING - [metall] (carbonaceous material) noir *m* de fonderie
~HOLE - [metall] défaut *m* (en forme de trou) dû au noir
BLACKOUT - [gen] obscurcissement *m*
[air-raid protection] extinction *f* des lumières, blackout *m*
[telecomm] évanouissement *m* total des communications radio
[cin] (in projection) obscurcissement *m*, éclipse *f*
[telev] (replacement of the signal by another of constant amplitude) suppression *f* du faisceau
[med] syncope *f*, éblouissement *m*
[telev] circuit *m* de suppression
BLACKOUT EFFECT - [electron] (temporary loss of sensitivity of a vacuum tube) étourdissement *m*
~LEVEL - [telev] (the level of the signal during the blackout interval) (blanking level in G.B.) niveau *m* de suppression
~PULSE - [telev] (a pulse which blacks out the traces of the scanning beam) impulsion *f* de suppression
~SIGNALS - [telev] (wave used to effect blacking-out) signal *m* de suppression de faisceau
~TIME - [comput] temps *m* des blancs
BLACKROT - [agric] (a plant disease) nervation *f* noire
BLACKSMITH - [gen] forgeron *m*, maréchal-ferrant *m*
BLACKWATER FEVER - [med] hématurie *f*
BLACKWORK - [metall] taillanderie *f*, grosse serrurerie *f* non meulée
BLADDER - [gen] vessie *f*, bulle *f*
[anat] vessie *f*
[of ball] chambre *f* à air
~FOR TYRE SHAPING - [rubber ind] bag *m* de galbage (des pneumatiques)
~STONE - [med] calcul *m* vésical
~WORM - [med] ver *m* cystique
BLADE, to - [mech] (to remove, e.g. with a bulldozer) enlever (avec une lame)
BLADE - [gen] lame *f*, couteau *m*, feuille *f*
[aero] (one of the aerofoils forming the propulsive elements of a propeller) pale *f* d'hélice
[of a rotary compressor] ailette *f*, aube *f*
[of tools] lame *f*, couteau *m*
[of a turbine] aube *f*
[of a mixer] palette *f*
[of earth moving equip] lame *f*
[moving part of knife switch] couteau *m*
[of a shutter in a camera] lame *f*
[of an oar] pelle *f*
~ACTIVITY FACTOR - [aero] (a factor expressing the power-absorbing capacity of a propeller blade) capacité *f* d'absorption d'une pale, profondeur *f* relative d'une pale
~ANGLE - [aero] (the angle between the chord of a propeller blade and the plane of rotation) angle *m* de calage, angle *m* de pale
~AZIMUTH ANGLE - [hel] (the angle, measured in the direction of rotation, and in plan, between the downwind position and a line through the lag hinge and the rotor hub centres) angle *m* d'incidence azimutale, angle *m* azimutale de poussée de l'hélice
~BACK - [aero] extrados *m*, dos *m* de la pale
~COOLING - [gas turbine] (means of extracting heat from blades of a gas turbine) refroidissement *m* des aubes (ou des ailettes)
~DAMPER - [hel] (damping device acting against unwanted motion of a rotor blade about the lag hinge) amortisseur *m* de pale
~FACE - [aero] face *f* de la pale, intrados *m*
~FINGER - [text] palette *f* du couteau
~GUIDE - [mech] guide *m* de couteau
~LOADING - [aero] (quantity obtained by dividing the rotor thrust by the total blade area) charge *f*

sur la pale, charge *f* de pale
BLADE-LOCKING - [gas turbine] (method of securing blades to rotor in gas turbines) emmanchement *m* des ailettes
~ROOT - [gas turbines] (the base of a blade which is engaged in the rotor) queue *f* d'ailette, emmanchement *m*
[mech] queue *f* d'ailette, embase *f* d'ailette
~SECTION - [aero] profil *m* de section de pale
~SPACING SYSTEM - [gas turbine] (the system determining the distance between the blades of turbine or compressor rotors) dispositif *m* d'écartement des ailettes
~SURFACE - [aero] aire *f* de la pale
~SWEEP - [aero] (if one line be drawn through the centroids of the blade sections, and another tangentially to it from the propeller axis, the angle between these, projected upon the plane of rotation, is termed the blade sweep) déport *m* dans le plan de rotation, compensation *f* dans le plan de rotation
~TANG - [aero] emplanture *f* de pale, embase *f* d'ailette
~TILT - [hel] (the angle between the line drawn through the centroids of the blade sections, and the plane of rotation) angle *m* de calage de pale
~TIP - [aero] extrémité *f* de pale
~TOOL - [text] outil *m* en forme de couteau
~TYPE CIRCUIT BREAKER - [el] interrupteur *m* à couteaux
~TYPE ROTARY BOOSTER - [gas ind] compresseur *m* à palettes
~WHEEL - [mech] roue *f* à aubes
~WIDTH RATIO - [aero] rapport *m* de largeur de l'hélice
BLADED EXHAUSTER - [gas install] extracteur *m* à palettes
~SHUTTER - [photo] obturateur *m* à pales
BLADING - [mech] aubage *m*, ailetage *m*
BLANC DE CHINE - [cer] (brilliant white glaze) blanc *m* de Chine
~FIXE - [paint] (a filler consisting of barium sulphate) blanc *m* fixe
BLANCHING LIQUID - [ind chem] solution *f* d'hypochlorite de chaux
~WATER - [chem] eau *f* de blanchiment
BLANK, to - [hydr etc] (to cut off flow through an orifice by inserting an umperforated disk over it) obturer, masquer, boucher, couvrir
[metall] (of a metal sheet) découper
BLANK - [metall] (a part which has only received a general but not a final form) pièce *f* brute, ébauche *f* de forme, flan *m*, fromage *m*
[print] espace *m* en blanc, blanc *m*, tiret *m*, blocage *m*
[ind] (of any industrial product) pièce *f* ébauchée (non finie, non usinée, brute)
[plumb] (without an opening) borgne, brut
[plast ind] flan *m*, feuille *f* préparée, ébauche *f*
[glass man] (of optical glass) ébauche *f*
[comput] (blank column of punched card) blanc *m*, perforation *f* manquante
~AND CENTRE PUNCH DIE - [tool] outil *m* à centrer et à découper
~AND PIERCE DIE - [tool] outil *m* à découper et à perforer
~ANNEALING - s.close annealing
BLANK CARD - [text] carton *m* blanc, carton *m* plein
[comput] carte *f* vierge, carte *f* non perforée
~CARTRIDGE - [explos] cartouche *f* à blanc
~CHEQUE - [comm] chèque *m* en blanc
~COLUMN DETECTOR - [comput] détecteur *m* de colonnes vierges
~DOOR - [build] (a bricked up imitation door) fausse porte *f*
~FIXE - s.blanc fixe
~FLANGE - [mech] bride *f* aveugle, bride *f* d'obturation, bride *f* pleine, fausse bride *f*
[an unpierced disk of metal] flan *m* brut, bride *f* brute
~GROOVE - [acoust] (in mechanical recording) sillon *m* sans modulation
~HOLDER - [mech] presse-flan *m*, serve-flan *m*
~INSTRUCTION - [comput] (instruction which does not operate the computer) instruction *f* nulle
~NUT - [mech] écrou *m* borgne
~OF THE CARD - [text] blanc *m* du carton, plein *m* du carton
~-OFF FLANGE - s.blank flange
~-OFF PRESSURE - [mech] (ultimate, or limiting pressure) pression *f* finale, vide *m* limite
~PICTURE SIGNAL - [telev] signal *m* vidéo avec signal de suppression
~SPACE - [print] blanc *m*, espace *m* en blanc
~STOCK - [ind chem] mélange *m* de base, composé *m* de base
~TEST - [metall] essai *m* à blanc
~TYPE - [print] espacement *m*
~WALL - [build] fausse paroi *f*, mur *m* nu
~WINDOW - [build] (bricked up imitation window) fausse-fenêtre *f*
BLANCKED OFF - [of mines] bouché, fermé, obturé
BLANKET, to - [radio] (the action of some powerful signal which makes it impossible for a receiving set to receive the desired signals) supprimer un signal
[naut] déventer, masquer
BLANKET - [text] (thick woollen or cotton fabric) couverture *f*
[paper man] feutre *m*
[print] (the material covering the impression surface on a rotary press) blanchet *m*
[road constr] revêtement *m* hydrocarboné, revêtement *m* au bitume
[mining] couche *f*, filon-couche *m* horizontal
[nucl] (a layer of material external to the core) manteau *m*, couverture *f*
~CYLINDER - [of rotary press] cylindre *m* recouvert d'un blanchet
~FEED - [ovens] alimentation *f* par bande transporteuse
~FOR CALICO-PRINTING - [text] blanchet *m* pour indiennage
~FOR LETTER-PRESS PRINTING - [print] blanchet *m* pour la typographie
~MATERIAL - [nucl] (material used at the ends of the fuel rods, consisting of natural uranium) matériau *m* pour manteau (ou couverture)
~PILE - [text] (cloth for bed-covers) tissu *m* pour couvertures
~VEIN - [mining] - s.blanket (mining)
BLANKETING - [radio] s.blank, to
~FREQUENCY - [radio] (electromagn waves) fréquence *f* d'occultation

BLANKING - [mech] découpe *f* d'ébauche, découpe *f* des flans
[telev] s.blackout
[radio] (making a channel non effective for a predetermined interval) suppression *f* d'un canal
~COVER - [mech] (a special cover for closing an aperture) tampon *m*, tape *f*, couvercle *m* de protection
~DIE - [plast ind] outil *m* à découper
~LEVEL - [telev] niveau *m* de suppression
~MACHINE - [mech mach] machine *f* à découper
~OFF PLATE - [mech] plaque *f* d'obturation, couvercle *m* de protection
~PEDESTAL - s.blanking pulse
~PLATE - [mech] plaque *f* d'obturation, couvercle *m* de protection
~PULSE - [telev] (pulse blanking out the bright traces of the scanning beam) impulsion *f* de suppression de faisceau
~SIGNAL - [telev] - s.blackout signal
BLAST, to - [gen] faire exploser, faire sauter à la mine, pétarder (des roches), faire sauter
[metall] dessabler
[missile] lancer
BLAST - [explos] (the ignition or detonation of an explosive charge) explosion *f*
[metall] (air blown into a furnace) air *m* insufflé, vent *m*, vent *m* de la soufflerie
[met] bouffée *f* de vent
[mech] (of steam) jet *m* de vapeur
~AIR - [gen] jet *m* d'air, courant *m* d'air
~BOX - [metall] boîte *f* à vent
~BURNER - [ind chem] brûleur *m* à mélange surpressé
~CHAMBER - [a] chambre *f* de combustion
~CHEST - [med] lésion *f* thoracique due à une explosion
~CONNECTION - [metall] porte-vent *m*
~COOLING - [metall] refroidissement *m* par air forcé
~EJECTION - [mech] éjection *f* pneumatique
~FIRING - [mining] sautage *m*
~FURNACE - [metall] (smelting furnace in which air blast is used) haut-fourneau *m* (à soufflage d'air)
~-FURNACE BLOWING ENGINE - [metall] soufflante *f* pour haut-fourneau
~-FURNACE CEMENT - [metall] ciment *m* de haut-fourneau, ciment *m* de laitier
~-FURNACE DUST - [ind chem] (dust derived from blast furnace gases and used as a source of potash) poussière *f* de haut-fourneau
~FURNACE GAS - [fuels] (by-product of iron smelting) gaz *m* de haut-fourneau
~-FURNACE GAS ENGINE - [metall] moteur *m* à gaz pour haut-fourneau
~-FURNACE SINTER - [metall] aggloméré *m* pour haut-fourneau
~FURNACE SLAG - [metall] (material containing the gangue from the ore combined with the flux, drawn off during smelting) laitier *m* de haut-fourneau
~-GATE - [jet engine] (device for controlling the pressure of the air supplied by a turbo-supercharger) vanne *f* réglage du vent
~GATE - [metall] registre *m* du vent, régulateur *m* du vent
~-HOLE - [min] coup *m* de mine, trou *m* de mine
~INJURY - [med] blessure *f* due à une explosion, lésion *f* due à une explosion
BLAST INLET - [th eng] (in furnaces) tuyère *f* arrivée *f* d'air
~LOADING - [nucl] (force of an object borne of the air blast of an explosion) force *f* de l'onde de choc
~MAIN - [mech] (the main blast air pipe supplying air to a furnace) conduite *f* de vent, tuyauterie *f* de soufflage
~PIPE - [in steam locomotives] tuyau *m* d'échappement
[in ovens] tuyère *f*
~PURGE - [in water gas] purge *f* à l'air
~ROASTING - [metall] (roasting in which this is accompanied by sintering) grillage *m* à l'air, calcination *f*, frittage *m*
~SCALING LAW - [nucl] loi *f* du calcul de l'onde de choc (ou de pression)
~TUBE - [metall] tubulure (du portevent)
~VALVE - [ind chem] (heating of retorts) vanne *f* de soufflage, vanne *f* d'air
~WAVE - [phys] (pressure pulse of air) onde *m* de pression
BLASTED ORE - [min] minerai *m* abattu
BLASTER - [metall] sableuse *f*
~CAP - [mining] détonateur *m*
BLASTINE - [chem] mélange *m* de perchlorate d'ammonium et de T.N.T.
BLASTING - [explos] explosion *f*
[min] tir *m* de mine, abattage *m*, travail *m* aux explosifs
[metall] dessablage *m*
[acoust] onde *f* de choc
[of a microphone] poussée *f* d'intensité
~AGENT - [explos] (substance or mixture for blasting) explosif *m*, agent *m* explosif
~CAP - [explos] capsule *f* fulminante, détonateur *m*
~CHARGE - [explos] charge *f* explosive
~FUSE - [explos] (a slow burning composition delaying the firing for a specified time) mèche *f*
~GELATINE - [explos] dynamite gomme *f*, gélatine *f* détonante
~MACHINE - [min] exploseur *m*
~OIL - [explos] nitroglycérine *f*
~PAPER - [paper man] papier *m* pour tir de mine
~PLANT - [min] installation *f* d'abattage
~POWDER - [explos] poudre *f* de mine
BLASTOCYTE - [genet] blastocyte *m*
BLASTODERM - [zool] blastoderme *m*
BLASTOGENESIS - [biol] (transmission of inherited features through germ-plasm) blastogenèse *f*
BLASTOGENIC - [biol] (of germ-plasm) blastogénique
BLASTOID - [geol] blastoïde
BLASTOPHERE - [zool] blastophère *f*
BLASTOPHORE - [med] blastophore
BLASTOPORE - [zool] blastopore
BLAU GAS - [chem] propane *m*
BLAZE, to - [gen] flamber, s'enflammer
[agric] (marking of trees) griffer, marquer
BLAZE - [gen] flamme *f*, feu *m*, flambée *f*
~OFF, to - [metall] (the operation of tempering hardened steel by dipping it in oil) tremper à l'huile, recuire par le flambage
BLAZING OFF - [metall] trempe *f* à l'huile
BLEACH, to - [ind chem] décolorer, blanchir
[paper man] blanchir
BLEACH - [ind chem] décoloration *f*

[paper man] blanchiment *m*
BLEACH LIQUOR - [ind chem] solution *f* de chlorure de calcium
BLEACHED COTTON - [text] tissu *m* de coton blanchi
BLEACHERS - [arch] (section of a stadium) places *f*pl découvertes (d'un stade)
BLEACHING - [ind chem] (the elimination of natural colours by chemical action or by the action of light) décoloration *f*, blanchiment *m*
[text] blanchiment *m*, blanchissement *m*
[paper man] blanchiment
[photo] blanchissement
~AGENT - [ind chem] décolorant *m*, produit *m* à blanchir, agent *m* de blanchiment
~ASSISTANTS - [ind chem] (substances aiding the action of bleaching agents) auxiliaires *m*pl de blanchiment
~BATH - [ind chem] bain *m* de blanchiment (ou de blanchissement)
~CHEST - [paper man] caisse *f* de blanchiment
~CLAY - [ind chem] terre *f* à blanchir
~EARTH - [ind chem] terre *f* à blanchir
~EFFICIENCY - [ind chem] pouvoir *m* blanchissant, pouvoir *m* de blanchissement
~LIQUOR - s.bleach liquor
~MACHINE - [paper man] machine *f* pour le blanchiment
~POWDER - [ind chem] (a synonym for chlorinated lime) poudre *f* à blanchir, chlorure *m* de chaux
~PROCESS - [ind chem] procédé *m* de blanchiment
~RANGE - [text ind] blanchisserie *f*, installation *f* de blanchiment
~TOWERS - [paper man] tours *f* de blanchiment
~VAT - [text] cuve *f* de blanchiment
BLEATING - [acoust] (the sound made by sheep) bêlement *m*
BLED - [med] saigné
[print] rogné à l'excès
~EDGES - [photo] (of a print) pleins bords *m*pl
~STEAM - [th eng] vapeur *f* soutirée
BLEED, to - [gen] saigner, perdre du sang
[hydr] (the draining of air from a hydr. system) purger
[mech] (draining of oil from brake systems) purger
[text] déteindre, s'étendre au lavage, dégorger
[th eng] soutirer, prélever
[book-binding] tailler
[construct] (of bitumen) décharger
~AIR - [mech] air *m* de rinçage, air *m* de balayage
~-BURN POWER AUGMENTATION - [mech] (method of obtaining extra power for take-off in gas turbines by bleeding off air from a high pressure stage of the compressor and passing it to an auxiliary combustion chamber) surcharge *f* par post-combustion, augmentation *f* de puissance par purge d'air
~GAS - [mech] gaz *m* de balayage
~OFF, to - [mech] (to allow undesired air to escape from a fluid) purger
~SCREW - [mech] (a screw which can be used for bleeding off by loosening it) vis *f* de purge
~THROUGH - [gen & chem] infiltration *f*
~VALVE - [mech] (a valve to allow bleeding) robinet *m* de purge
~WHITE, to - [gen] saigner à blanc
BLEEDER - [mech] robinet *m* de vidange, robinet *m* purgeur, trop-plein *m*
BLEEDER CONDENSING TURBINE - [mech] turbine *f* à condensation et prise de vapeur
~PIPE - [in natural gas] torche *f*
~RESISTANCE - [el] résistance *f* de fuite
~RESISTOR - [telev] (leak resistance forming a steady load on the supply) résistance *f* de fuite, diviseur *m* de tension
~SCREW - [mech] s.bleed screw
~VENT - [mech] (for the extraction of unwanted air) mise *f* à l'air libre, évent *m*
BLEEDING - [oil ind] séparation *f*
~CORE - [mining] carotte *f* exsudante
BLEMISH, to - [gen] tacher, souiller, endommager
[electron comp] défaut *m* de mémoire
BLEMISH - [gen] défaut *m*, défectuosité, imperfection *f*, tache *f*
[el magn] défaut *m* dans une couche magnétique
BLEND, to - [gen & ind chem] mélanger, mêler
[mech] raccorder, joindre
[text] mélanger
BLEND - [gen] mélange *m*, alliage *m*
[ind proc] mélange *m* dosé
[text] (in spinning) mélange *m*
~TANK - [rubber ind] bac *m* de mise en mélange, cuve *f* de mise en mélange
BLENDE - [min] (zinc blende, zinc sulphide found in metalliferous veins) blende *f*
BLENDED FABRIC - [text] tissu *m* mélangé
~FUEL - [oil ind] combustible composé
~WOOL - [text] laine *f* de mélange
~YARN - [text] fil *m* mixte
BLENDER - [mech mach] mélangeur *m*
BLENDING - [gen] mélange *m*, alliage *m*, union *m*, fusion *f*, mise en mélange
[ind] mélange *m*, mise *f* en mélange
[of paints] mélange *m* des couleurs, mariage *m* des couleurs
[oil ind] mélange *m*
[brew ind] coupage *m* des vins, mise *f* en cuvée
~AGENT - [chem] (distillation) adjuvant *m*
~CHARACTERS - [genet] caractères *m*pl mêlés
~INHERITANCE - [genet] hérédité *f* à facteurs multiples
~OF SOUNDS - [radio & telev]fusion *f* de sons, mixage *m* de sons
~PLATE - [ind chem] plaque *f* doseuse
~TABLE - s.blending plate
~WORTS - [brew ind] apport *m* de moût (pendant la fermentation)
BLENNORRHAGIA - [med] blénnorragie *f*, gonorrhée
BLEPHARAL - [med] blépharique
BLEPHARITIS - [med] blépharite *f*
BLEPHAROPLASTY - [med] blépharoplastie *f*
BLEPHAROSTAT - [med instr] blépharostat *m*
BLEPHAROTOMY - [med] blépharotomie *f*
BLIMP - [photo] (camera equip) cabine *f*, blindage *m*, insonorisant *m* de caméra
[aero] (small non-rigid airship) saucisse *f*, vedette *f* aérienne
BLIMPED CAMERA - [cinema] caméra *f* silencieuse
BLIND, to - [gen] aveugler, éblouir
[paint] rendre opaque, opacifier
BLIND - [gen] rideau *m*, store *m*, persienne *f*
[impl] (for horses) oeillères *f*pl
[photo] (of a shutter) rideau *m*
[med] aveugle

BLIND ABUTMENT - [build] faux arc-boutant *m*
~ANGLE - [gen] angle *m* mort
~APPROACH BEAM SYSTEM - [radar] (ground based navigation beacon) dispositif *m* d'approche sans visibilité directionnel
~ARCADE - [constr] fausse arcade *f*
~ARCH - [constr] fausse voûte *f*
~AREA - [constr] (sunken space round the basement of a building) soubassement *m*
[radar] zone *f* de silence
~BLOCKING - [print] dorure *f* à froid
~COAL - [min] charbon *m* sec
~COPPER - [min] cuivre *m* ampoulé
~DIVE - [aero] piqué *m*
~DRIFT - [mining] galerie *f* en cul-de-sac
~END CYLINDER - [mech] cylindre *m* borgne
~FEEDER - [metall] masselotte *f* borgne
~FLANGE - s. blank flange
~FLYING - [aero] (flight without the aid of visual observation, as in darkness or fog) vol *m* sans visibilité, vol *m* en P.S.V.
~FLYING EQUIPMENT - [aero] (the set of instruments for blind flying) instruments *m* de vol sans visibilité
~FLYING INSTRUMENTS - [instr] (instruments specified by regulations as being indispensable for instrument (blind) flying) instruments *m* de vol sans visibilité
~FLYING SHIELD - [aero] (device to preclude external vision when flying by instruments only) écran *m* pour vol sans visibilité
~GUT - [anat] caecum *m*
~HOLE - [metall] (a hole which does not pass right through the part in which it is made) trou *m* borgne
~JOINT - [geol] schistosité *f* confuse
~LANDING - [aero] (landing by the use of instruments only) atterrissage *m* sans visibilité
~LEVEL - [mining] galerie *f* d'exhaure
~LODE - [mining] (a lode without outcrop on the surface) filon *m* sans affleurement
~MORTISE - [carp] fausse mortaise *f*
~NAVIGATION - [aero] vol *m* sans visibilité, vol *m* aux instruments
~NUT - [mech] écrou *m* plein, écrou *m* borgne
~SECTOR - [radio] (region in the normal service area of a transmitter where signal strength is greatly reduced by selection effects) secteur *m* mort
~SHAFT - [mining] puits *m* intérieur
~SHELL - [explos] (an exploded shell) obus *m* qui n'éclate pas
~SHOT - [radio] s. blind sector
~SPOT - [opt] papille *f* optique
~SUPERVISION - [telecomm] surveillance *f* par rentrée en écoute
~TOOLING - [impl] (for bookbinding) outils *m* pour la dorure à froid
~TRACK - [railw] voie *f* en impasse
BLINDAGE - [radio] blindage *m*
BLINDING - [constr] ensablement *m*, couche *m* de sable
BLINDNESS - [med] cécité *f*
B-LINE - [comput] (machine register in which a variable is put to change another variable) registre *m* de base, registre *m* index
BLINK - [gen & med] battement *m* de paupière, clignotement *m*
BLINK MICROSCOPE - [instr] (instrument in which two photographic plates of the same area are viewed simultaneously) microscope *m* à clignotement
BLINKER - [signal] feu *m* clignotant
BLINKING - [gen & med] battement *m* de paupière, clignotement *m*, clignotement *m* des yeux
[signal] feu *m* clignotant, signal *m* intermittent
BLIP - [radar] (sudden deflection of the scanning spot by a received signal) top *m* d'écho
BLISTER, to - [metall] s'écailler
BLISTER - [gen & med] ampoule *f*, cloque *f*, boursouflure *f*
[casting] soufflure *f*, cloque *f*, bulle *f*, vésicule *f*
[paper man] bulle *f*
[rubber ind] bulle *f* d'air, poche *f* d'air
[radio] carter *m* profilé, coupole *f*, dôme *m*
[metall] cuivre *m* ampoulé
[paint] cloque *f*, boursouflure *f*
~BAR - [metall] (wrought iron bar impregnated with carbon) barre *f* cémentée
~RUST - [bot] (a plant disease) rouille *f* du groseillier
~SERUM - [med] sérosité *f* d'ampoule, sérum *m* d'ampoule, liquide *m* vésiculaire
~STEEL - [metall] (wrought iron bars impregnated with carbon, formerly the only available steel) acier *m* boursouflé, acier *m* de cémentation, acier *m* de cémentation à soufflures
BLISTERING - [paint] (defect arising when the painted surface is exposed to direct heating) cloquage *m*, gondolage *m*, boursouflage *m*
~GAS - [chem] (chemical warfare) gaz *m* vésicant
BLISTERS - [aero] (fitted on the sides of bombers for aiming purposes) couples *f*, dômes *m*, renflements *mpl*
BLIZZARD - [met] (strong, very cold, snow-laden wind) blizzard *m*, tempête *f* de neige
BLOAT - [vet] (tympanite) tympanite *f*
BLOATING TENDENCY - [rubber ind] tendance *f* à gonfler
BLOB - [glass man] larme *f* de verre
[met] hétérogénéité *f* atmosphérique locale
BLOBS - [paint] (a paintwork defect) grumeaux *mpl*
BLOCK, to - [gen] arrêter, bloquer, obstruer, barrer
[chem] arrêter, (of a reaction) empêcher (une réaction)
BLOCK - [gen] bloc *m*, encombrement *m*, obstruction *f*
[mech] (of a pulley) poulie *f*
[mech] patin *m*, sabot *m* de frein
[int comb eng] bloc cylindres *m*
[naut] poulie *f*, palan *m*
[print] cliché *m*
[paint] (the mounting of a printing plate) bloc *m*
[mining] (stop used to prevent the rolling back of a wagon) cale *f*, coin *m*
[aero] (device placed before the wheels of an aircraft to prevent movement) cale *f*
[carp] (in wood engraving) planche *f*, bois *m*
~ACCESS - [comput] (the transfer of numbers in groups or blocks from one location to another) transfert *m* en bloc
~ADDRESS - [comput] (address of selected numbers adresse *f* en bloc
~AERIAL - [radio] (system of aerials which may be

used by a number of receivers in the same building) antenne *f* collective
BLOCK AND TACKLE - [naut] palan *m*
~ANTENNA - s.block aerial
~APPARATUS - [railw] appareillage *m* de cantonnement, dispositif *m* de fermeture
~BOOK - [print] livre *m* de gravures sur bois
~BRAKE - [mech] (brake consisting of a block of cast-iron forced against the rim of the wheel) frein *m* à sabot
~CAST CYLINDER - [mech] groupe-cylindres *m* monobloc
~CAVING - [mining] abattage *m* par le tir après sous-cavage
~CHAIN - [mech] chaîne *f* à maillons pleins, chaîne *f* à blocs
~CIRCUIT - s.block diagram
~-CLIP - [mech] louve *f* à pince, pinces *f*pl à genouillère
~CLUTCH - [mech] (friction clutch with blocks forced inwards) embrayage *m* à sabot
~COEFFICIENT - [shipbuild] volume *m* de carène sur parallélépipède circonscrit
~CONSTRUCTION - [aero] (of parachutes) assemblage *m* de bandes parallèles
~COURSE - [build] assise *f*
~DIAGRAM - [el] (a diagram in which groups of components or circuits are indicated by rectangles and not shown on detail) diagramme *m* synoptique
[comput] schéma *m* de fonctionnement
~EQUIPMENT - [railw] dispositif *m* de fermeture (cantonnement)
~-FAULT - [geol] faille *f* tabulaire
~FITTING - [naut] accessoires *m* de pouliage
~GAP - [comput] (space between two blocks in magnetic tape storage systems) distance *f* entre deux blocs
~ICE - [gen] glace *f* en pains
~INTERLOCKING SYSTEM - [railw] dispositif *m* de fermeture avec verrouillage de sécurité
~IRON - [metall] fonte *f* en bloc, fonte *f* en lingot
~LETTERS - [print] majuscules *f* d'imprimerie
~MACHINE - [mech mach] poulierie *f*
~-MAKER - [print] photograveur *m*, clicheur *m*
~OF HOUSES - [constr] pâté *m* de maisons, immeuble *m*, ensemble *m* d'habitations
~OF INFORMATION - [comput] bloc *m*
~OF PLATES - [el] bloc *m* de plaques
~OUT, to - [mining] découper en massifs d'abattage
[photo] (the background etc) silhouetter, masquer
~PATTERNS EFFECT - [genet] effet *m* de combinaison en bloc
~PAVEMENT - [constr] pavage *m* (routes)
~PLANER - [print] machine *f* à planer les blocs
~POLYMER - [chem] (a polymer with a molecule made up of alternate sections of one chemical composition separated by sections of another chemical composition) copolymère *m* bloc
~POST - [railw] poste *m* de blocage
~POST IN ADVANCE - [railw] poste *m* de blocage aval, poste *m* de cantonnement aval
~POST IN REAR - [railw] poste *m* de blocage amont
~PRESS - [mech] presse *f*
~PRINT - [print] gravure *f* sur bois
~PRINTING - [text] impression *f* à la planche (ou à la main)
BLOCK PROCESS - [print] phototypographie *f*
~PROFILE MAT - [rubber ind] tapis *m* formé de blocs articulés
~PULL - [print] première épreuve *f* d'un cliché
~PULLEY - [mining] moufles *f*pl
~RATE - [comm] tarif *m* à tranches
~REPEATER - [railw] répétiteur *m* de fermeture
~SCREW - [mech] vis *f* de verrouillage
~SHEARS - [mech] cisaille *f* à bras
~SHEAVE - [mech] (of a pulley) rouet *m*
~SIGNAL BOX WITH AXLE COUNTER - [railw] poste *m* de blocage (ou de block) à comptage d'essieux
~SIGNALS - [railw] signaux *m* de fermeture, disques *m* de fermeture
~SLICING MACHINE - [mech] (in plastic industry) trancheuse *f*
~SORT - [comput] (the sorting out of a large number of data) classification *f* en bloc
~SPEED - [aero] (value of speed of an aircraft derived from block-to-block time) vitesse *f* cale à cale
~SPRING - [horol] ressort *m* de blocage
~SYSTEM - [railw] block-système *m*, cantonnement *m*
~SYSTEM WITH SIGNALS NORMALLY AT CLEAR - [railw] cantonnement *m* à signaux normalement ouverts
~SYSTEM WITH SIGNALS NORMALLY AT DANGER - [railw] cantonnement *m* à signaux normalement sur la position arrêt
~TERMINAL - [telecomm] boîte *f* de jonction, prise *f* multiple
~TIME - [aero] (the period between first movement of an a/c under pilot's control and its coming to rest at the end of the flight) temps *m* de vol, temps *m* de "cale à cale"
~TIN - [metall] (pure tin) étain *m* en saumons
~-TO-BLOCK TIME - s.block time
~WORK - [metall] gros ouvrage *m* de fer
BLOCKADE, to - [gen] bloquer, faire le blocus
BLOCKADE - [gen] blocus *m*
BLOCKAGE - [med] anesthésie *f* locale
BLOCKED FLUTE - [mus] flûte *f* douce, flûte *f* à bec
~IMPEDANCE - [el] (of an electrochemical transducer) impédance *m* en circuit ouvert
~SHADOWS - [photo] ombres *f*pl bouchées
BLOCKER - [telecomm] dispositif *m* de blocage
BLOCKETTE - [comput] (group of words belonging together) sous-bloc *m*
BLOCKING - [gen] blocage *m*, engorgement *m*, calage *m*
[railw] cantonnement *m*
[metall] (forging) enrayage *m*, ébauchage *m*
[bookbinding] gauffrage *m*
[carp] (securing together two pieces of board) calage *m*
[ovens] (of furnaces) arrêt *m*
[ind chem] (of a crucible) engorgement *m* de creuset
[glass man] blocage *m*
[radar] (valve overdriving strong signals which paralyse the receiver circuits) blocage *m*
[mining] calage *m*
[plast ind] adhérence *f* de contact, blocking *m*, adhérence *f* mutuelle
~AND EMBOSSING PRESS - [print & bookbinding]

presse *f* à gauffrer (et à dorer)
BLOCKING BATTERY – [telecomm] batterie *f* d'arrêt
~BIAS – [electron] tension *f* de blocage
~CAPACITOR – [el] (asymmetrical cell used to prevent a flow of current in a specified direction) condensateur *m* d'arrêt
~CHARACTERISTIC – [radio] caractéristique *f* de blocage
~CIRCUIT – [el] circuit *m* de blocage
~CONDENSER – [radio] condensateur *m* d'arrêt
~DEVICE – [metall] (metal plating) dispositif *m* de placage
[mech] dispositif *m* de verrouillage, dispositif *m* de blocage
~DIRECTION – [telecomm] sens *m* de blocage, sens *m* d'arrêt
~GATE CIRCUIT – [comput] circuit *m* bloqueur
~LAYER – [telecomm] couche *f* d'arrêt
~LAYER PHOTOCELL – [telecomm] cellule *f* photoélectrique à couche d'arrêt
~LAYER RECTIFIER – [telecomm] redresseur *m* à couche d'arrêt
~LEVER – [mech] levier *m* de blocage
~MACHINE – s.block machine
~MAGNET – [telecomm] aimant *m* d'arrêt
~MEDIUM – [radiat] (a material with adequate radiation opacity) matériau *m* absorbant, milieu *m* absorbant
~OF SHEETS – [rubber ind] adhérence *m* de contact entre feuilles
~OF THOUGHT – [med] inhibition *f* de la pensée
~OSCILLATOR – [radio] (an electronic tube oscillator in which blocking occurs for a time determined by the circuit constants after a cycle of oscillation) oscillateur *m* de blocage, oscillateur *m* surcouplé
~-OUT – [photo] (the use of indian ink or other pigments to cover parts of negatives) silhouettage *m*
[mining] découpage *m* en massif d'abattage
~PAWL – [mech] cliquet *m* de blocage
~PERIOD – [electron] (the interval between the initiation and the establishment of conduction) temps *m* de blocage
~PLATE – [metall] plaque *f* de réduction
~RATIO – [radio] (the ratio of the response to an excitation) taux *m* de blocage
~RELAY – [el] relais *m* de verrouillage
~SIGNAL – [radio] signal *m* de blocage
~VOLTAGE RATING – [electron] (permissible peak inverse voltage) tension *m* inverse de crête admissible
BLOCKS – [telecomm] (telegr & teleph installation) ancrages *m*
BLOCKY COKE – [fuel] coke *m* massif
BLONDIN – [mech] (cable-way) blondin *m*
BLOOD – [physiol] sang *m*
~ALBUMEN GLUE – [ind chem] colle *f* à l'albumine
~ALBUMIN – [med] (brown amorphous albuminous product obtained from blood serum) albumine *f* de sang
~AQUEOUS BARRIER – [med] barrière *f* sanguino-aqueuse
~BANK – [med] banque *f* du sang
~BLACK – [ind chem] (pigment made by charring blood) charbon *m* de sang
~-BRAIN BARRIER – [med] barrière *f* cérébro-spinale hématique
BLOOD CELL – [physiol] hématoblaste *f*
~CHANNEL – [med] voie *f* sanguine
~CLOT – [med] caillot *m* sanguin
~CORPUSCLES – [med] globules *mpl* sanguins
~COUNT – [med] numération *f* globulaire
~CRYSTALS – [med] sang *m* cristallisé
~GROUP – [physiol] groupe *m* sanguin
~LETTING – [med] saignée *f*
~PLASMA – [physiol] plasma *m* sanguin
~PRESSURE – [med] pression *f* sanguine, tension *f* artérielle
~PRESSURE CUSHION – [rubber ind] coussinet *m*
~PRESSURE FOLLOWER – [instr] (electronic equipment recording and controlling human blood pressure) appareil *m* électronique de mesure de la pression sanguine
~RELATION – [genet] rapport *m* consanguin
~-ROT – [bot] (used for dyes) tormentille *f*
~SERUM – [physiol] sérum *m* sanguin
~STONE – s.bloodstone
~STREAM – [med] circulation *f* du sang
~-SUCKER – [zool] sangsue *f*
~TRANSFUSION – [med] transfusion *f* sanguine
~VESSEL – [physiol] vaisseau *m* sanguin
~VOLUME – [med] volume *m* du sang
~-WORT – s.blood rot
BLOODLESS – [gen & med] exsangue
BLOODSTONE – [min] sanguine *f*, jaspe *m* sanguin
BLOOM, to – [opt] (lenses) métalliser sous vide
[metall] battre, marteler
BLOOM – [gen] fleur *f*, floraison *f*
[metall] (intermediate product in the rolling of steel) bloom *m*, masse f de fer cinglé, brame *f*
[constr] efflorescence *f*
[rubber & leath ind] efflorescence *f*
[oil ind] fluorescence *f*
[glass man] loupe *f*, masse *f* de verre fondu
[paint] voile *m*
[met] (on glass panes, caused by atmosphere) buée *f*
[telev] (defect in the receiver causing the picture to lose brightness and expand like a flower) hyperluminosité *f*, flou *m* d'image
~-FREE – [rubber ind] non efflorescent
~IRON – [metall] fer *m* en blooms
~MILL – [metall] train *m* à blooms, ébaucheur *m*
~OIL – [ind chem] (type of rosin oil obtained in the destructive distillation of rosin) huile *f* de résine
~SIDE – [leather ind] coté *m* du poil de la peau
BLOOMERY – [metall] bloomerie *f*, affinerie *f*
~FURNACE – [metall] four *m* à loupes
BLOOMING – [gen] floraison *f*
[paint] voile *m*, louchissement *m*
[rubber ind] efflorescence *f*
[metall] dégrossissage *m*, ébauchage *m*
[opt] (optics coating) revêtement *m* de surfaces optiques
~MILL – [metall] train *m* à blooms, ébaucheur *m*, train *m* dégrossisseur
BLOOMS – [metall] blooms *mpl* carrés
BLOOP – [cinema] (low frequency sounds caused by defective blooping patches, q.v.) bruit *m* causé par un collage
BLOOPING NOTCH – [cinema] (defect in a sound splice) défaut *m* dans un raccord sonore
~PATCH – [cinema] (sections introduced over spli-

ces in the positive sound track to eliminate noises) éliminateur *m* de bruit de collage
BLOSSOM - [gen] fleur *f*, floraison *f*
[telev] s. bloom
[geol] affleurement *m* oxydé
BLOTCH - [gen] tache *f*, éclaboussure *f*
[text] à motif flou
[med] rougeur *f*, tache *f* rouge, pustule *f*
~PRINTING - [text] impression *f* de surface
BLOTCHING - [photo] formation *f* de cratères
BLOTTER - [gen] buvard *m*, papier *m* buvard
~PRESS - [mech] filtre-presse *m*
BLOTTING PAPER - [paper man] papier *m* buvard
BLOW, to - [gen] souffler
[el] (of a fuse) fondre, sauter
[glassman] souffler (le verre)
[mining] faire sauter
[oil ind] nettoyer un puits (par injection d'air etc)
BLOW - [gen] coup *m*
[metall] soufflage *m*, charge *f* de soufflage
[metall] (casting defect) creux *m*
[casting cavity] soufflure *f*, ampoule *f*
[soil] soufflure *f*, soulèvement *m* local
[mining] (sudden inrush of gas) dégagement *m* instantané de gaz
~AND BLOW MACHINE - [glass man] soufflante *f* continue, soufflerie *f* continue
~BACK - [mech] retour *m* au carburateur
~-BY - [mech] fuite *f* des gaz
~CASE - [ind chem] monte-jus *m*
~CASING - [metall] corps *m* de soufflante
~-COCK - [mech] robinet *m* de vidange, robinet *m* d'extraction
~DOWN, to - [in boilers] vider, faire l'extraction, purger, vidanger
~-DOWN - [mech] purge *f*
~-DOWN LINE - [metall] (in boilers etc) tuyauterie *f* de vidange (ou de purge)
~-DOWN PLANT - [mech] installation *f* à clapet à sûreté placé en bas
~-DOWN VALVE - [mech] (in the lower part of a boiler) robinet *m* de purge
~-FLY - [zool] mouche *f* à viande , lucilie *f*
~FORMING - [plast ind] soufflage *m*, moulage *m* par soufflage
~GAS - [gas ind] (water gas) gaz *m* de soufflage
~GUN - [impl] pistolet *m* à air
~HOLE - [metall] soufflure *f*
~IN, to - [metall] (of furnaces)mettre en feu, insuffler (de l'air)
~-LAMP - [mech] (device for producing a hot Bunsen-type flame from petrol or paraffin) lampe *f* à souder
~MOULDING - [plast ind] moulage *m* par soufflage
~NOZZLE - [plast ind] buse *f* de soufflage
~OF PIPE - [mech] tuyauterie *f* de purge
~OFF, to - [plumb] (in boilers) purger, vidanger
~OFF - [auto] (of tyres) éclatement
~-OFF COCK - [mech] robinet *m* de purge
~-OFF PLUG - [plumb] bouchon *m* de purge, bouchon *m* de vidange
~-OFF VALVE - [mech] clapet *m* de sûreté
~OUT, to - [el] (of fuses etc) sauter, fondre
~OUT - [auto] (of tyres) éclatement *m* (d'un pneumatique)
[oil ind] (of gas) dégagement *m* instantané de gaz
[el] (of a fuse) fusion *f*
BLOW-OUT ACTION - [el] soufflage *m*
~-OUT COIL - [el] (magnetic blow-out) bobine *f* de soufflage
~-OUT CORE - [el] noyau *m* de soufflage
~-OUT MAGNET - [el] aimant *m* de soufflage
~-OUT PATCH - [rubber ind] pièce *f*, rustine *f*
~-OUT PREVENTER - [oil ind] (natural gas installations) vanne *f* d'éruption
~-OUT PROOF TYRE - [rubber ind] pneu *m* increvable
~PIPE - [welding] chalumeau *m*
[metall] (oxyhydrogen blowing) chalumeau *m* oxhydrique
[glass man] (metal tube for glass blowing) canne *f* de souffleur
~PIPE TEST - [ind chem] essai *m* au chalumeau
~-PIPE WELDING - [metall] soudure *f* au chalumeau
~PURGE - [gas ind] (in water gas) purge *f* à l'air
~ROOM - [text] salle *f* de battage
~-RUN - [gas ind] (water gas) fabrication *f* avec incorporation des gaz de soufflage
~THROUGH GAP - [ind chem] (gas passing through a metal bath) barbotage *m*
~TORCH - [metall] lampe *f* à souder, chalumeau *m*
~UP, to - [gen] faire exploser, exploser, sauter, éclater
[photo] agrandir
BLOWER - [gen] soufflerie *f*, ventilateur *m*, compresseur *m*
[metall] (in furnaces) soufflante *f*
[min] échappement *m* de gaz, puits *m* de gaz naturel
~AIR-FLOW - [aero] débit *m* de compresseur
~CASING - [aero] carter *m* du rotor, logement *m*
~DEVICE - [text] dispositif *m* de soufflage
~-GENERATOR - [railw] groupe *m* générateur-ventilateur
~PIPE - [aero] prise *f* d'air
~TURBINE SHAFT - [aero] arbre *m* de compresseur à turbine
~VALVE - [text] soupape *f* de soufflement
BLOWING - [gen & ind] soufflage *m*, soufflement *m*
[rubber ind] gonflement *m*
~AGENT - [ind chem] (compound decomposing with the evolution of gas, used to produce cellular rubber or plastics) agent *m* gonflant
~AN EXTRUDED TUBE - [plast ind] extrusion-soufflage *f* de gaines, procédé *m* de soufflage de gaines
~AN EXTRUDED TUBULAR FILM - s. blowing an extruded tube
~APPARATUS - [gen] machine *f* soufflante, soufflerie *f*
~CONE - [geol] hornito *m*
~CURRENT - [el] (in fuse links, to indicate the current which will cause the links to melt) courant *m* de court circuit
~DOWN ENGINE - [mech] (machine) soufflante *f*
~FAN - [metall] ventilateur *m* soufflant
~FURNACE - [glass man] four *m* de réchauffage
~IRON - [glass man] (metal tube used for blowing glass) canne *f* de souffleur
~MACHINE - [glass man] soufflante *f*, soufflerie *f*
~MOULD - [glass man] moule *m* pour soufflage
~OUT - [gen & metall] extinction *f*, mise *f* hors feu (four)
~PIPE - s. blowing iron
~ROOM MACHINE - [text] machine *f* de nettoyage, machine *f* d'ouverture et de battage
~TUBE - s. blowing iron

BLOWINGS - [text] déchets *mpl* de coton
BLOWN - [mech] (of a supercharged petrol engine) gonflé
~BITUMEN - [oil ind] (product obtained by high-temperature air-blowing of residual mineral bitumens) bitume *m* soufflé
~CASTING - [metall] (casting with inclusion of blow-holes) pièce *f* moulée qui présente des soufflures
~EXTRUSION METHOD - [plast ind] procedé *m* par extrusion et soufflage
~EXTRUSION MOULDING - [plast ind] moulage *m* par extrusion et soufflage
~FILM - [plast ind] film *m* soufflé
~FLAP - [aero] (a flap over the upper surface of which air or gas is blown to increase its effect) volet *m* soufflé
~FUSE - [el] (a fuse which has melted, thus interrupting the circuit) fusible *m* fondu, fusible *m* sauté
~GLASS - [glass man] verre *f* soufflé
~INGOT - [metall] s. blown casting
~LATEX FOAM - [rubber ind] mousse *f* de latex soufflé
~LINSEED OIL - [ind chem] (linseed oil partially polymerized by agitation with air at high temperatures) huile *f* de lin soufflé
~OIL - [ind chem] (oil subjected to partial oxidation by blowing air through it) huile *f* soufflée
~PERIPHERY - [aero] (a part of the periphery of the canopy blown between two rigging lines as the parachute opens and thus inflates inside out) double coupole *f*
~POINT - [metall] (welding) joint *m* soudé au chalumeau
~PULP - [paper man] pâte *f* soufflée
BLUBBER - [zool] graisse *f* de baleine
BLUE, to - [gen & metall] (the operation of producing oxide-colouring of a steel moulding by overheating) bleuir
BLUE - [a primary colour] bleu, azure
~ADDER - [telev] (in colour television) circuit *m* mélangeur pour le bleu
~ANNEALED - [metall] recuit (au) bleu
~ANNEALING - [metall] bleuissage *m*, recuit *m* bleu
~ASBESTOS - [min] (silicate of sodium and iron which is found in the Asbestos mountains in South Africa) amiante *f* bleue
~ASHES - [ind chem] cendres *fpl* bleues
~-BEAM MAGNET - [telev] aimant *m* du faisceau pour le bleu
~-BELL - [bot] campanule *f*, jacinthe *f* des prés
~BLACK - [paint] (ivory black) bleu *m* noir
~BLINDNESS - [med] acyanopsie *f*, acyanoblepsie *f*
~-BOTTLE - [zool] mouche *f* à viande
~BRITTLE RANGE - [metall] intervalle *m* de fragilité à la chaleur bleue
~BRITTLENESS - [metall] (the lack of malleability in iron and steel between 200 and 400 centigrades) fragilité *f* à la chaleur bleue
~COPPER - [ind chem] bleu *m* cuivre
~COPPERAS - [chem] vitriol *m* bleu, sulfate *m* de cuivre
~DIP - [electro metall] (solution with mercury compound used to deposit mercury on a metal by immersion before silver-plating) solution *f* mercurielle
~DISEASE - [med] (also called the Rocky Mountain fever) maladie *f* bleue, cyanose *f*
BLUE FILTER - [photo] filtre *m* bleu
~FLAME - [th eng] (a non luminous flame) flamme *f* bleue
~FOX - [zool] renard *m* bleu
~FRIT - [paint] (mixture of copper and calcium silicates, a pigment) fritte *f* bleue
~GAS - [water gas] (mixture of carbon monoxide and hydrogen) gaz *m* d'eau
~GLASS - [glass man] (glass of a special blue tint used to determine colour values) verre *m* fumé
~GLOW - [el] fluorescence *f*
~GROUND - [min] (decomposed agglomerate containing plutonic rock-fragments and diamonds) brèche *f* diamantifère, gîte *m* de diamants
~HEAT - [metall] chaleur *f* bleue
~HIGHS - [telev] hautes fréquences *fpl* pour le bleu
~IRON EARTH - [min] vivianite *f*
~JOHN - [min] (a massive variety of the mineral fluorite, to be found in Derbyshire, an English county) fluorite *f* bleue (ou violette) du Derbyshire
~LEAD - [chem] (industrial term for metallic lead, to distinguish it from white lead, etc) plomb *m* bleu
~LIGHT - [cinema] lumière *f* bleue
~LOWS - [telev] basses fréquences *fpl* pour le bleu
~MALACHITE - [min] azurite *f*
~METAL - [metall] (in zinc refining by distillation, a bluish dust consisting of small grains of zinc coated with oxide) poudre *m* de zinc, zinc *m* en poudre
~OIL - [ind chem] huile *f* bleue
~OXIDE - [ind chem] mélange *m* de zinc et d'oxyde de zinc
~PAPER - [ind chem] (blue-process paper) papier *m* prussiate
~PEACH - [min] tourmaline *f* à grain fin
~PETER - [naut] pavillon *m* de partance
~POT - [metall] creuset *m* en mine de plomb, creuset *m* en graphite
~POWDER - s blue metal
~-PRINT PAPER - [paper man] papier *m* héliographique (au ferroprussiate)
~-PRINTER - [print] machine *f* à tirer les calques
~PRINTING - [print] (a reproductive process for drawings) photocopie *f* bleue, bleu *m*, calque *m*
~PULP - [ind chem] pâte *f* bleue
~SALT - [chem] sulfate *m* de nickel cristallisé
~SAND - [pottery] émail *m* au cobalt
~SHIFT - [telev] décalage *m* du canevas bleu
~SHORTNESS - s. blue-brittleness
~SPOT - [med] tache *f* bleue
~STEEL - [metall] (blued steel) acier *m* bleui
~STUFF - [geol] kimberlite *f*
~SYSTEM - [mech & el] (one of a set of systems designated by colours, as blue green, red etc for identification) circuit *m* bleu, réseau *m* bleu
~TOURMALIN - [min] indicolite *f*, tourmaline *f* bleue
~VERDITER - [paint] (a blue pigment consisting chiefly of very basic copper carbonate) bleu *m* de Brême
~VITRIOL - [chem] (commercial term for copper sulphate) vitriol *m* bleu, sulfate *m* de cuivre
~WATER GAS - [ind chem] (mixture of carbon monoxide and hydrogen in almost equal proportions. It is produced by passing steam over heated coke)

gaz *m* d'eau
BLUE WATER GAS SET - [ind chem] (water gas) installation *f* à gaz d'eau
~WHALE - [zool] baleine *f* bleue
BLUED - [metall] bronzé, bleui
~STEEL - [metall] acier *m* bleui
BLUEING - [metall] (to form a protective coating) bleuissage *m*, bronzage *m*
[text] bleutage *m*
[pottery] (the whitening of yellow lead glaze) bleuissement *m*
BLUELINE - [print] tirage *m* bleu
BLUEPRINT - [print] photocopie *f* bleue, bleu *m*, calque *m*
BLUESTONE - [chem] sulfate *m* de cuivre
BLUFF - [geogr] à pie, promontoire *m* à pie
~BODY - [aero] (a body which often offers a broad front to gas flow) fuselage *m* renflé
~BOW - [naut] proue *f* renflée
~-RACKING - [text] pousser à vide, commander à vide
~SINKER - [text] platine *f* d'abattage à vide
~SLIDER - [text] clavette *f* avec talon
~WHEEL - [text] roue *f* réductrice
BLUISH - [paint] bleuâtre
BLUNDERBUSS - [mus] tromblon *m*
BLUNGE, to - [pottery] (to mix clay with water for pottery making) malaxer
BLUNGER - [pottery ind] malaxeur *m* à couteaux
BLUNT - [gen] épointé, émoussé, contondant
~FILE - [impl] lime *f* à section uniforme
~NOSE - [mech] ogive *f* arrondie
~NOSE RASP - [impl] râpe *f* ronde
~TOOL - [tool] outil *m* émoussé, outil *m* épointé
BLUNTNESS - [gen & of tool] état *m* émoussé, manque *m* de tranchant
BLUR - [gen] tache *f*
[print] macule *f*, tache *f*, frison *m*
[acoust] son *m* diffus
[photo] flou *m*
[radiation] flou *m* cinématique
~CIRCLE - [opt] halo *m*
BLURRED - [gen] flou, indécis, confus
~PAN - [cinema] (distorted picture) panoramique *m* déformé
~PICTURE - [telev & cinema] image *f* floue, flou *m* d'image
~PRINT - [print] épreuve *f* maculée
BLURRING - [telev] (the reduction of the apparent sharpness of the reproduced scene) flou *m* d'image
~BY MOTION - [photo] flou *m* de bougé
~REGION - [telev] zone *f* floue
BLUSHING - [paint] (cloudy film appearing on a recently varnished surface) voile *m*, louchissement *m*
~TENDENCY - [paint] tendance *f* au louchissement
B.M.E.P. - [mech] (Brake Mean Effective Pressure) pression *f* moyenne efficace
B.M.E.W.S. - [ballist] (Ballistic Missile Early Warning System) système *m* d'alerte avancée pour les engins balistiques
BOAR - [zool] sanglier *m*
~SKIN - [leather ind] peau *m* de sanglier, cuir *m* de sanglier
BOARD, to - [gen] monter (dans un train, dans un véhicule) monter à bord
[metall] planchéier
[naut] aborder, accoster, embarquer, monter à bord
BOARD - [gen] planche *f*, madrier *m*, panneau *m*, ais *m*
[carp] planche *f* de bois
[impl] batte *f*
[naut] (side of a ship) bordé *m*, bord *m*
[comm] conseil *m* d'administration
[paper man] carton *m*
[auto] marche-pied *m*, plancher *m*
~-AND-PILLAR METHOD - [mining] exploitation *f* par chambres et piliers
~COAL - [min] charbon *m* fibreux
~CUTTING AND ROUND CORNERING MACHINE - [paper man] machine *f* à couper le carton et à rogner les angles
~CUTTING MACHINE - [mech] machine *f* à couper le carton
~DECATIZING - [text] décatissage *m* à plaques
~-FLANGED BEAMS - [metall] poutrelles *f*pl à larges ailes
~FOOT - [meas] (unit measure for timber) pied *m* cube (mesure des bois de construction)
~FRAME - [text] cloison *f* à claire- voie
~MACHINE - [paper man] machine *f* à carton
~OF DIRECTORS - [comm] conseil *m* d'administration
~OF MANAGEMENT - [comm] conseil *m* de gérance
~PRESS - [text] presse *f* à cartons
~SIGNAL - [telecomm] signal *m* à disque
BOARDED - [gen] en planches, planchéié
~FENCE - [build] palissade *f* en planches
~PARTITION - [build] cloison *f* en planches
~SHUTTERING - [build] volet *m* en planches
BOARDING - [build] planchéiage *m*
[naut] accostage *m*, abordage *m*
BOARDS - [mining] voies *f*pl d'exploitation
[light] (in a studio) herse *f* lumineuse
BOARDY - [text] (of a fabric) carteux, rigide, dur
~FEEL - [text] raide au toucher
BOAST, to - [mining] (quarry work) dégrossir
BOASTER - [tool] ébauchoir *m*, burin *m* de carrier
BOASTING - [carp] ébauchage *m*
[mining] taille *f* de pierres
BOAT, to - [gen] transporter par mer
BOAT - [naut] barque *f*, canot *m*, embarcation *f*, bateau *m*, bâtiment *m*, navire *m*
~ACROSS, to - [naut] traverser en bateau
~AMPHIBIAN - [aero] aéronef *m* amphibie
~COMPASS - [instr] compas *m* d'embarcation
~DAVITS - [naut] bossoir *m* d'embarcation
~DECK - [naut] pont *m* des embarcations
~FENDER - [naut] défense *f*, débordoir *m*
~-HOOK - [naut] gaffe *f*
~SEAPLANE - [aero] (flying boat) hydravion *m* à coque
~SLINGS - [naut] saisines *f*
~STRINGER - [naut] serre *f*, gouttière *f*, hiloire *m*
~TRAIN - [railw] train-paquebot *m*
~VARNISH - [paint] (highly water-resistant varnish) vernis *m* pour embarcations
BOAT'S PAINTER - [naut] amarre *f*
BOATMAN - [naut] batelier *m*, canotier *m*
BOATSWAIN - [naut] maître *m* d'équipage, maître *m* de manoeuvre
BOATSWAIN'S CHAIR - [naut] chaise *f* de calfat
BOB - [gen] bouchon *m* (de pêcheur)
[mech] entille *f* (d'un pendule), (of a beam engine)

balancier *m*
[impl] contrepoids *m*, masselotte *f*, plomb *m* (de fil à plomb), disque *m* de feutre
[metall] masselotte *f* latérale, poids *m* curseur
BOB-PUNCH - [mech] pointeau *m*
~WEIGHT - [mech] (weight counterbalancing the moving part of a machine) contrepoids *m*, balancier *m*
BOBBIN - [gen text & el] bobine *f*, revêtement *m* dépolarisant
[electron] carcasse *f* de bobine
~CARRIAGE - [text] dispositif *m* d'enroulage
~CENTRIFUGE - [text] essoreuse *f* pour bobines
~CLEANING MACHINE - [text] machine *f* à nettoyer les tubes
~CORE - [comput] (tape-wound core in which the ferro-magnetic tape has been wrapped on a form, or bobbin, which furnishes the mechanical support to the tape) noyau *m* bobiné
~CREEL - [text] râtelier *m* à bobines
~FEELER - [text] tâteur *m* de bobine, tâteur de canettes
~GAUGE - [text] gabarit *m* de tube
~HANGER - [text] dispositif *m* pour bobines suspendues
~HEIGHT - [text] (axial diameter of the bobbin) longueur *f* de bobine
~INTERNAL DIAMETER - [comput] (static magnetic storage) diamètre *m* intérieur de la bobine
~LACE LEVER - [text] battant *m* de fuseau de dentellière
~LACE WORK - [text] travail *m* aux fuseaux de dentellière, travail de dentelles a.f.
~LOADER - [text] alimentation *f* de bobines, alimentation de canettes
~NET - [text] tulle *m*
~OVERALL DIAMETER - [comput] (static magnetic storage) diamètre *m* hors-tout de la bobine
~SKIP - [text] panier *m* à bobines, panier à canettes
~WASTE - [text] déchêts *m pl*
~WHEEL - [text] bobinoir *m* à main
~WINDER - [text] bobineuse *f*
BOBBINET - [text] dentelle *f* au fuseau
BOBBING - [radar] fluctuation *f* d'écho
BOBINE - [el & tech] bobine *f*
BOBSLED - s. bob-sleigh
BOBSLEIGH - [transp] bob-sleigh *m*
BOBSTAY - [naut] sous-barbe *f* de beaupré
BOBTAIL - [mech] tracteur *m* sans sa semi-remorque
B.O.D. - [Biochemical oxygen Demand] D.B.O. (demande *f* biochimique d'oxygène)
BOD - [metall] (ball of clay used to close the tap-hole of a furnace or cupola) bouchon *m* d'argile destiné à obturer le trou de coulée
BODICE - [med] corset *m*
BODKIN - [tool] passe-lacets *m*, poinçon *m*, pointe *f*
BODY, to - [gen] donner du corps, épaissir
BODY - [gen] corps *m*
[auto] carrosserie *f*, caisse *f* coque *f*
[aero] fuselage *m*
[constr] corps *m* de bâtiment; (of a church) nef *f* centrale
[mech] tige *f*, queue *f*; (of a connecting-rod) corps *m*
[of a spindle] socle *m* de tour
[plumb] (of a tap) cannelle *f*
[pump] corps *m* de pompe
[naut] corps *m* d'un vaisseau, coque *f*
[truck] benne *f*
[min] gisement *m*, massif *m*
[print] corps *m*, force *f* du corps
[phys] solide *m*
[oil ind] consistance *f*, viscosité *f*
[text] (shedding motion) plaque *m* de fixation
[railw] caisse *f*
[of a saw] châssis *m*
[of a screw] noyau *m* de vis
BODY APRON - [radiat] (worn by X-ray operator) tablier *m* opaque
~AXES - [aero] (a system of co-ordinates originating in the G.C. of the a/c body and extending in three dimensions) axes *m pl* d'un aéronef (origine au centre de gravité)
~BACK PANEL - [auto] panneau *m* arrière de carrosserie
~BELT - [text] ceinture *f*
~BOTTOM - [auto] dessous *m* de caisse, plancher *m*
~-BOUND COMMUTATOR - [el] commutateur *m* blindé
~BRACKET - [auto] attache *f* de carrosserie
~BUILDER - [auto] carrossier *m*
~BUILDING - [auto] carrosserie *f*
~CAPACITANCE - [el] (capacitance introduced into an electric circuit by the proximity of the human body) effet *m* de main
~CAVITY - [anat] (the perivisceral cavity in which the viscera lie) cavité *f* périviscérale
~CEILING - [auto] pavillon *m*
~CELL - [biol] cellule *f* somatique
~CEMENT - [ind chem] mastic *m* à l'acétate de cellulose
~CENTRED CUBIC STRUCTURE - [crystall] structure *f* cubique à corps centrés
~CENTRED LATTICE - [phys] (lattice points situated at the centre of unit cells) réseau *m* à corps centrés
~CROSS SHAFT - [auto] arbre *m* transversal de benne
~FITTING - [aero] garniture *m* de fuselage, aménagement *m* de fuselage
~FORCES - [phys] (those forces which act on, and are proportional to, the mass of a fluid element, e.g. gravity) forces *f pl* de volume, forces *f pl* massiques
~FRAME - [gas ind] carcasse *f*
~FRONT PANEL - [auto] panneau *m* avant de carrosserie
~GUIDE - [auto] guide *m* de benne
~HEAD - [auto] avant *m* de benne
~HINGE - [auto] articulation *f* de basculement
~IMAGE - [med] schéma *m* du corps humain
~INNER PANEL - [auto] panneau *m* intérieur de carrosserie
~INSULATION - [auto] isolement *m* de la carrosserie
~INTEGRAL WITH FRAME - [auto] carrosserie *f* intégrée au châssis
~LAG - [text] carton *m* de fond
~LIFTING JACK - [auto] élévateur *m* pour benne, cric *m* pour carrosserie
~LINEN - [text] linge *m* de corps
~-MAKER - [auto] carrossier *m*
~OF A BLAST FURNACE - [metall] cuve *f*, ventre *m* de haut four
~OF A BOILER - [th eng] corps *m* de chaudière

ODY OF A CARBURETTOR - [auto] corps *m* de carburateur, corps *m* principal de carburateur
OF A PISTON - [mech] corps *m* de piston
OF HEADLAMP - [auto] boîtier *m* de phare
~OF OIL - [oil ind] consistance *f* de l'huile
OF REVOLUTION - [phys] solide *m* de révolution, corps *m* de révolution
~OF ROLL - [metall] table *f* du cylindre
OF SCREW - [mech] corps *m* de vis, noyau *m* de vis
OF SOIL - [soil] masse *f* de sol
OF TOOL - [mech] tige *f* d'un outil, queue *f* d'un outil
OUTER PANEL - [auto] panneau *m* extérieur de carrosserie
OVERHANG BEYOND REAR AXLE - [auto] porte-à-faux de la carrosserie
PAINT SHOP - [auto] atelier *m* de peinture de carrosseries
~PLAN - [shipbuild] plan *m* latitudinal, projection *f* transversale
[metall] section *f* transversale
POLISH - [ind chem] produit *m* de lustrage pour carrosserie, cire *f* pour carrosserie
~REAR PANEL - [auto] panneau *m* arrière de carrosserie
~ROLL - [auto] rebord *m* roulé de benne
~SECTION RADIOGRAPHY - [radiat] (radiograph showing structures at one particular level in a body or object) radiotomographie *f*, tomographie *f*, stratigraphie *f*
~SHELL - [auto] caisse *f*, coque *f*
~SHOP - [auto] atelier *m* de carrosserie
~SIDE - [auto] panneau *m* latéral de carrosserie
~SIDEWALL - [auto] panneau *m* latéral de carrosserie
SIZE - [print] force *f* de corps
~SPRING - [auto] ressort *m* de carrosserie
~STAMPINGS - [auto] pièces *f*pl embouties pour carrosserie
STIFFENING PANEL - [auto] panneau *m* de renforcement pour carrosserie
STRESS - [mech] tension *f* interne
~SUBRAME - [auto] soubassement *m* de carrosserie
~SUPPORT - [auto] support *m* de carrosserie
TEMPERATURE - [med] température *f* du corps
TRIMMING - [auto] finition *f* de la carrosserie
TRUSS-ROD - [railw] tirant *m* de caisse
TUBE - [opt] (of a lens) barillet *m*
TYPE - [print] caractères *m*pl de texte
~VENTILATION - [auto] ventilation *f* de la carrosserie
~WASHER - [mech] rondelle *f* d'épaulement d'essieu
WEAR PLATE - [auto] plateau *m* de protection (benne)
WITH CONCEALED HOOD - [auto] carrosserie *f* à capote escamotable
~WORK - [auto] carrosserie *f*
ODYING - [gen] épaississement *m*
~FABRIC RUBBER - [rubber ind] adhérence *f* du caoutchouc au tissu
~SPEED - [chem] (rate of increase of viscosity) taux *m* d'accroissement de la viscosité
TEST - [rubber ind] essai *m* d'adhérence
UP - [paint] donner du corps
OG - [gen] fondrière *f*, marécage *f*, marais *m*
BODY - [min] gîte *m* de minerai des prés
~COAL - [min] houille *f* des marais

BOG DRAINAGE - [soil] dessèchement *m* des marais
~FORMATION - [soil] formation *f* des terrains tourbeux
~IRON ORE - [min] limonite *f*, fer *m* limoneux, ferro-hydride *f*, hématite *f* brune
~LIME - [geol] (earthy impure calcium carbonate deposited in ponds and lakes) craie *f* lacustre, travertin *m*
~MANGANESE - [chem] hydroxyde *m* impur de manganèse
BOGIE - [mech] (a group of two or more wheels forming a pivoting supporting unit of a long rigid vehicle) bogie *m*
[railw] bogie *m*
~BEAM - [aero] (the beam member which connects the wheels or groups of wheels in a landing-gear bogie) entretoise *f* transversale de train
~COACH - [railw] voiture *f* à bogie
~ENGINE - [railw] locomotive *f* de terrassement
~FRAME - [railw] châssis *m* de bogie
~HEARTH FURNACE - [metall] four *m* à sole mobile, four *m* à chariot
~LANDING GEAR - [aero] (a type of landing gear in which wheels are grouped in bogies) train *m* d'atterrisage à bogies
~PIVOT - [aero] (pivot or spindle on which a bogie assembly turns) axe *m* de train
~SIDE FRICTION BLOCK - [railw] patin *m* pour bogie
~TRUCK - [mech] plateforme *f* de bogie
~UNDERCARRIAGE - [aero] (an assembly of alighting gear comprising a pair of wheels one behind the other, pivoted to a central strut) train *m* d'atterrissage à bogies
~UNIT - [mech] bogie *m*
~WAGON - [railw] wagon *m* à bogies, voiture *f* à bogies
BOHR ATOM - [el] (model of the atom according to the conception of Bohr and Sommerfield, in which the electrons move around the nucleus in circular or elliptical orbits) atome *m* de Bohr
BOIL, to - [gen] bouillir, faire bouillir, entrer en ébullition, bouillonner
BOIL - [gen] ébullition *f*, bouillonnement *m*
[med] furoncle *m*
[acoust] bruit *m* parasite
[glass man] bulle *f*
[metall] travail *m* du bain
[hydr] (a spring) source *f*
~DOWN, to - [ind chem] réduire, concentrer, réduire par ebullition
~ERUPTION PERIOD - [metall] période *f* de bouillonnement et de projections
~TO GRAIN, to - [sugar ind] grainer
BOILED - [gen] bouilli
[ind chem] cuit
~LINSEED OIL - [ind chem] (linseed oil brought to a temperature of 210/315 c. and mixed with driers) huile *f* de lin cuite
~OIL - [ind chem] huile *f* cuite
~SILK - [text] soie *f* dégommée
BOILER - [th eng] (steam generator consisting of water-drums and tubes exposed to the heat of a furnace and producing circulation) chaudière *f*, bouilleur *m*
[ind chem] (pressure vessel designed to produce vapour from liquid by the application of heat) chaudière *f*

BOILER BACK- [th eng] fond *m* de chaudière
~BARREL - [th eng] corps *m* cylindrique (chaudière)
~BEAR - [tool] poinçonneuse *f* à main (pour tôles de chaudière)
~BEARER - [th eng] chevalet *m* de chaudière
~BURNER - [th eng] brûleur *m*
~CAPACITY - [th eng] (weight of steam a boiler can evaporate when at full output) débit *m* de vapeur d'une chaudière
~CASING - [mech] casing *m* (tôles recouvrant la maçonnerie de la chaudière)
~CLAMP - [th eng] presse *f* de chaudronnier
~COCK - [th eng] robinet *m* pour chaudière (à vapeur)
~COMPOSITIONS - [ind chem] (chemicals introduced into the boiler feed-water to prevent scale formation etc) anti-incrustants *mpl*
~COVERING - [th eng] (lagging of boiler) calorifugeage *m*
~CRADLE - [naut] (longitudinal strength members of a ship forming the shell-plating stiffeners) berceau *m* de chaudière
~CRAMP - s boiler clamp
~DRUM - [th eng] collecteur *m* de chaudière
~DUTY - [th eng] débit *m* d'une chaudière
~EFFICIENCY - [th eng] rendement *m* d'une chaudière
~END - [th eng] fond *m* de chaudière, bout *m* de chaudière
~END PLATE - [th eng] s boiler end
~FEED PUMP - [th eng] pompe *f* d'alimentation, pompe *f* alimentaire
~FEED REGULATOR - [th eng] régulateur *m* d'eau d'alimentation
~FEED-WATER - [th eng] eau *f* d'alimentation de chaudière
~FEED WATER CONTROL - [th eng] régulateur *m* d'eau d'alimentation (chaudières)
~FEEDING - [th eng] alimentation *f*, alimentation *f* d'une chaudière
~FITTINGS AND MOUNTINGS - [th eng] (boiler valves and gauges) accessoires *m*; garnitures *f* pour chaudières
~FLOAT - [th eng] flotteur *m* d'alarme
~FLUE - [th eng] carneau *m* de chaudière
~FLUID - [ind chem] anti-incrustant *m*, désincrustant *m*
~FRONT - [th eng] façade *f* de chaudière
~FURNACE - [th eng] foyer *m*, chambre *f* de combustion
~GRATE - [th eng] grille *f* de foyer
~HEAD - [th eng] fond *m* de chaudière
~HEADER - [th eng] collecteur *m*
[metall] collecteur *m* de chaudière
~HEATING SURFACE - [th eng] surface *f* de chauffe
~HORSEPOWER - [th eng] équivaut à la chaleur de vaporisation de 15,63 Kg d'eau à l'heure à 100 centigrades, soit à la pression atmospherique: 8405 Kcal à l'heure
~HOUSE - [th eng] salle *f* des chaudières, bâtiment *m* des chaudières
~JACKET - [th eng] tôle *f* d'enveloppe extérieure
~LAGGING - [metall] enveloppe *f* d'une chaudière
~MAKER - [th eng] chaudronnier *m*
~-MAKER'S HAMMER - [impl] (used for caulking and scaling) marteau *m* de chaudronnier
~MAKING - [th eng] chaudronnerie *f*

BOILER OUTPUT - [th eng] débit *m* de vapeur d'une chaudière
~PLATE - [th eng] tôle *f* de chaudière
~PRESSURE - [th eng] (the pressure at which steam is generated in a boiler) pression *f* de la chaudière
~PROVER - [impl] pompe *f* pour épreuves de chaudière
~-RIVET - [th eng] clou *m* à river
~ROOM - [th eng] chaufferie *f*, chambre *f* de chauffe
~SCALE - [th eng] (a hard deposit of magnesium, iron and calcium carbonates formed in boilers in which hard water is used) incrustation *f*, tartre *m*, calcin *m*
~-SCALING HAMMER - [impl] marteau *m* à piquer les chaudières
~SHELL - [th eng] enveloppe *f* de chaudière, corps *m* cylindrique (de chaudière)
~SHELL RINGS - [th eng] joints *mpl* de chaudière
~SHOP - [mech] atelier *m* de chaudières
~SHOP TOOLS - [tool] outils *mpl* pour chaudières
~-SMITH - [th eng] chaudronnier *m*
~STAY - [th eng] (tubes or rods supporting the flat surfaces of a boiler to prevent bursting caused by internal pressure) tirant *m*
~STEEL - [metall] acier *m* pour chaudières
~TEST - [th eng] (hydraulic pressure test; also an efficiency test) épreuve *m* d'une chaudière
~-TEST PLATE - [th eng] timbre *m* de chaudière
~TRIAL - [th eng] (a test relating to the efficiency of the boiler) timbrage *m* d'une chaudière
~TUBE - [th eng] (tubes of steel which are part of the heating surface) (tube) bouilleur *m*, tube *m* de chaudière
~WASTE LIQUOR - [ind chem] lessive *f* épuisée du bouilleur
~WATER CONDITIONING SYSTEM - [th eng] installation *f* de traitement de l'eau d'alimentation
~WATER DRUM - [th eng] ballon *m* de chaudière
~WITH CORRUGATED FLUE TUBE - [th eng] chaudière *f* à carneau ondulé
~WITH CROSSED TUBES - [th eng] chaudière *f* à tubes transversaux (ou à bouilleurs transversaux)
~WITH ECCENTRIC FLUE - [metall] chaudière *f* à tube-foyer excentrique
~WITH HORIZONTAL TUBES - [th eng] chaudière *f* à tubes horizontaux (ou à bouilleurs horizontaux)
~WITH REMOVABLE NEST OF TUBES - [th eng] chaudière *f* à tubes démontables
~WITH STEPPED FLUE - [metall] chaudière *f* à tube-foyer échelonné
~-WORK - [th eng] chaudronnerie *f*
BOILING - [phys] (the evolution of bubbles of vapour induced by the increase in temperature, or a reduction in pressure) ébullition *f*
[text] (the removal of sericin-natural gum - from silk yarn) dégommage *m*, cuisson *f*
[text] (yarn hanks) cuisson *f*
[paper man] décreusage *m* (des fibres), lessivage *m*
~AGENT - [paper man] lessive *f*
~BURNER - [gas ind] (gas cookers) brûleur *m* de dessus
~FERMENTATION - [ind chem] (very rapid fermentation with abundant evolution of gas) fermentation *f* avec bouillonnement
~FLASK - [ind chem] (a spherical glass flask with

wide neck, used to boil liquids) flacon *m* à essai
OILING KIER - [text] cuve *f* à bouillir
~OFF - s boiling [text]
~-OFF BATH - [text] bain *m* de dégommage
~OUT - [text] (cleaning out of cotton) nettoyage *m* du coton
~PAN - [text] (in size mixing apparatus) marmite *f*, chaudière *f* de cuisson, bac *m* à cuisson
~PLATE - [ind chem] (a metal plate which is heated for heating flasks etc) plaque *f* chauffante
~POINT - [phys] (the temperature at which the vapour pressure of a liquid exceeds that of the ambient pressure and bubbles of vapour are formed) point *m* d'ébullition
~POINT CONSTANT - [phys] constante *f* du point d'ébullition
~POINT ELEVATION - [phys] (the raising of the boiling point of liquids by substances in solution) élévation *f* du point d'ébullition
~-POINT THERMOMETER - [instr] thermomètre *m* à ébullition
~RANGE - [ind chem] courbe *f* d'ébullition
~RESISTANT - [rubber ind] résistant *m* l'ébullition
~RING - [el] fourneau *m* électrique, réchaud *m* électrique
~TEMPERATURE - [phys] température *f* d'ébullition
~WATER REACTOR - [nucl] (reactor with solid fuel elements and water as coolant and moderator) réacteur *m* à eau bouillante
OLD - [gen] assuré, audacieux, hardi
[print] gras
[ind chem] (in natural resins) (a term indicating that the substance in question is in fairly large pieces) en piéces
~FACE - [print] (caractère) gras *m*
OLE - [bookbinding] (a reddish variety of clay used for guilt edges) bol *m*, terre *f* bolaire
[bot] tronc *m*, fût *m*
OLIDE - [met] (a meteorite which explodes in the atmosphere) bolide *m*
OLL - [bot] (the fruit of the cotton plant) capsule *f* (du cotonnier)
~ROT - [bot] (in cotton plants) moisissure *f* du cotonnier
~WEEVIL - [zool] (in cotton plants) anthonome *m* des cotonniers
~WEEVIL HANGERS - [oil ind] bridage *m* à "boll weevil"
~WORM - [zool] (the larvae of a moth which feeds on the bolls of the cotton plant) ver *m* du coton (Heliothis armigera)
OLLARD - [naut] bitte *f*, bollard *m*
~PULL TEST - [naut] essai *m* de traction sur les amarres
OLOGNA PHOSPHATE - [chem] sulfure *m* de barium
~PHOSPHORUS - [chem] sulfate *m* de baryte (phosphorescent)
OLOGRAPH - [instr] bolographe *m*, tracé *m* du bolomètre
OLOMETER - [instr] (high sensitivity temperature measuring instrument, designed to deal with feeble radiations, as from celestial bodies) bolomètre *m*
~MOUNT - [instr] monture *m* de bolomètre
OLOMETRIC INSTRUMENT - [instr] appareil *m* (thermique) à résistance
OLSTER - [gen] coussin *m*, traversin *m*, rembourrage *m*
[constr] racinal *m*, soús-poûtre *f*, semelle *f*, chapeau *m*
[text mach] collet *m*
[mech] embase *f*, patin *m*, sabot *m*, plateau *m* mobile
[metall] (of die-plate) matrice *f*
[plast ind] châssis *m*, manteau *m* de moule, frette *f*
[tool] (a punching tool) perçoir *m*
[railw] (rocking steel frame) sellette *f* de wagon, sommier *m* de caisse, traverse *f* dansante
[naut] coussin *m* (de capelage)
BOLSTER EDDY - [met] (an eddy of approximately cylindrical form with a generally horizontal axis) tourbillon *m* transversal, turbulence *f* transversale
~PLATE - [mech] plateau *m* porte-matrice
[diecasting] (or die plate) plateau *m* porte-moule
~RAIL - [text mach] chariot *m*
~SPRING - [railw] ressort *m* de suspension
BOLT, to - [mech] (to secure by means of bolts) boulonner, fixer au moyen de boulons, cheviller
[ind] (to sift) bluter
BOLT - [mech] boulon *m*, goupille *f*, cheville *f*
[lock] verrou *m*, pêne *m* (de serrure)
[firearm] culasse *f* mobile, fermeture *f* de culasse
[text] pièce *f* de toile
~AND NUT - [mech] boulon *m* avec écrou
~CLIPPER - s.bolt croppers
~CROPPERS - [mech] (powerful short bladed shears, hand or power operated designed for cutting, or cropping, metal rods or bolts) coupe-boulons *m*
~CUTTER - s.bolt croppers
~FLANGE - [mech] bride *f* boulonnée
~HANDLE - [firearm] levier *m*
~HANGER SOCKET - [constr] (in concrete) alvéole *f* pour boulon à charpente
~HEAD - [mech] tête *f* de boulon
~-HEADER - [mech] boulonnière *f*
~HOLE - [mech] trou *m* de boulon
~LOAD - [build] charge *f* d'un boulon
~MAKING MACHINE - [mech] machine *f* à faire les boulons
~NUT - [text] écrou *m*
~OF SPRING - [mech] boulon *m* étoquiau
~ROD - [mech] corps *m* de boulon
~-SCREWING AND NUT-TAPPING MACHINE - [mech] machine *f* à tarauder (tiges et écrous)
~SCREWING MACHINE - [mech] machine *f* à fileter les boulons
~WASHER - [mech] rondelle *f*
~WITH RECESSED HEAD - [mech] boulon *m* à tête fraisée, boulon *m* à tête noyée
BOLTED CONNECTION - [carp] assemblage *m* par boulons
~FLANGE - s.bolt flange
BOLTER - [impl] (piece of cloth used for sifting) tamis *m*, blutoir *m*
BOLTING - [mech] (screw coupling) vissage *m*, raccord *m* vissé
~CLOTH - [text] (a fine cloth used for sizing ground materials) toile *f* à tamis, étamine *f*
~SILK - [text] (a fine silk cloth) gaze *f* de soie
BOLTROPE - [naut] ralingue *f*
BOLTZMANN CONSTANT - [phys] (ratio between the mean total energy of a molecule and its temperature) constante *f* de Boltzmann
BOLTZMANN'S EQUATION - [nucl] (the particle conservation equation) équation *f* de Boltzmann

BOLUS - s. bole
~MATERIAL - [radiat] (material used in radiotherapy to fill up void space) matière *f* de remplissage Bolus
BOMB, to - [gen] bombarder
BOMB - [gen & impl] bombe *f*
[mining] mesureur *m* de pression de couche
[instr] (in a calorimeter) bombe *f* calorimétrique
~BAY - [aero] panneau *m* de soute à bombe
~CALORIMETER - [instr] (used to determine the calorific values of fuels) bombe *f* calorimétrique
~CORE - [nucl] (the active central part of an atom bomb) partie *f* centrale (d'une bombe atomique)
~DEBRIS - [nucl] (the debris of a nuclear bomb after explosion) débris *mpl* radioactifs, retombées *fpl* radioactives
~PROOF - [gen] à l'épreuve des bombes
~RACK - [aero] lance-bombes *m*
~RELEASE - [aero] dispositif lance-bombes *m*
~RELEASE CONTROL - [aero] commande *f* de lance-bombes *m*
~RELEASE GEAR - [aero] dispositif lance-bombe *m*
~SIGHT - [aero] viseur *m*
~STICK - [aero] chapelet *m* de bombes
BOMBARD, to - [nucl] bombarder
BOMBARDED PARTICLE - [nucl] (a particle which is hit by another particle) particule *f* bombardée, particule *f* cible
BOMBARDING PARTICLE - [nucl] (particle coming into collision with another particle) particule *f* projectile, particule *f* incidente
BOMBARDMENT - [nucl] (impact of particles or radiations for the production of other particles and radiations) bombardement *m*
~BY IONS - [nucl] bombardement *m* ionique
~DAMAGE - [nucl] (the damage which is caused by bombardment by particles) dommages *mpl* dûs au bombardement
~OF ELECTRONS - [electron] bombardement *m* électronique
BOMBARDON - [mus] contrebasse *f* (à vent)
BOMBER - [aero] bombardier *m*
BOND, to - [build] liaisonner, appareiller
[electron] mettre à la masse
BOND - [gen] lien *m*, attache *f*, liaison *f*, joint *m*
[comm] obligation *f*, titre *m*, bon *m*, contrat *m*, cautionnement *m*, engagement *m*
[ind chem] liaison *f*, collure *f*, liant *m*, aggloméran *m*
[el] connexion *f*, soudure *f*
[build] appareil *m* (disposition des briques etc)
[concrete constr] adhérence *f* (de l'armature dans le béton)
[mech] (laying of parts in overlapping courses) superposition *f*
[metall] joint *m*, agglomérat *m*
[mining] révolution *f* de machine d'extraction
[railw] (in electric railways) rail-bond *m*
[paint] (adhesion between a coating and a surface) liaison *f*
~DIRECTION - [nucl] (direction of covalent bonds with respect to the bonded atoms) direction *f* de la valence
~ENERGY - [ind chem] énergie *f* de liaison
~MOMENT - [nucl] (the electromagnetic dipole moment of a chemical bond between atoms) moment *m* de liaison
BOND RING DOUBLE - [nucl] liaison *f* double nucléaire
~-STONE - [build] (a long stone used as a header through a wall) pierre *f* de liaison
~STRENGTH - [phys] (the degree of stability of the linkage between atoms within a molecule) cohésion *f*, force *f* de liaison, résistance *f* d'adhésion
~STRESS - [constr] tension *f* dans la couche de collage
~TEST - [ind chem] essai *m* d'adhérence
~TESTING MACHINE - [mech] machine *f* pour essai de cohésion, machine *f* pour essai d'adhérence
BONDED DIODE - [electron] diode *f* à microjonctio[n]
~MASONRY - [constr] maçonnerie *f* en liaison
~SEALS - [mech] joints *m* renforcés
~STRAIN GAUGE - [instr] (a strain gauge made of wires) jauge *f* extensométrique
~WAREHOUSE - [comm] entrepôt *m* en douane
BONDERIZING - [metall] bondérisation *f*
BONDING - [el] (electrical connection between parts to ensure that they are at the same potential) mise *f* à la masse
[constr] appareillage *m*
[mining] boisage *m*
~AGENT - [ind chem] (substance used to bond materials together) liant *m*, adhésif *m*
~CLIP - [el] attache *f* de mise à la masse
~JUMPER - [el] (a conductor, usually of copper braid, used to make an electrical connexion between adjacent parts of a structure) fil *m* de mise à la masse
~LAYER - [rubber ind] couche *f* de liaison
~OF RAILS - [railw] connexion *f* de rail de contact
~PROPERTY - [ind chem] propriété *f* agglomérante
~RIBBON - [el] (for prevention of corrosion) tress[e] *f* de mise à la masse
~STRIP - [el] s. bonding ribbon
~SYSTEM - [el] circuit *m* de mise à la masse
~TAB - [mech] (tongue of metal for bonding purposes) languette *f* de raccordement
~WIRE - [el] fil *m* de mise à la masse
BONE, to - [gen] désosser
BONE - [anat] os *m*
[fish] arête *f*
[min] charbon *m* schisteux
~ASH - [chem] (a fertilizer and a source of superphosphate) cendre *f* d'os, poudre *f* d'os, terre *f* d'os
~-BED - [geol] bone-bed *m*
~BLACK - [ind chem] (a black pigment) noir *m* anim[al]
~BRECCIA - [geol] brèche *f* osseuse
~CARTILAGE - [med] osséine *f*
~-CAVE - [geol] caverne *f* à ossements
~CELL - [med] cellule *f* osseuse
~CHARCOAL - [ind chem] (charcoal obtained from bones) charbon *m* d'os
~CHINA - [pottery] (type of porcelain fluxed with bone ash) porcelaine *f* anglaise, porcelaine *f* d'o[s]
~COAL - [min] charbon *m* barré, schiste *m* houiller
~CONDUCTION - [acoust] conduction *f* osseuse
~CONDUCTION HEADPHONE - [el acoust] ostéophone *m*, récepteur *m* à conduction osseuse
~CONDUCTION RECEIVER - s. bone conductor headphone
~CRUSHER - [mech] (machine for reducing bones to powder) broyeur *m* d'os
~EARTH - s. bone ash

ONE GLASS - [glass man] opaline *f*
GLUE - [ind chem] colle *f* d'os
MARROW - [physiol] moelle *f* osseuse
MEAL - [ind chem] farine *f* d'os
OIL - [ind chem] (oil containing a mixture of pyridine compounds) huile *f* d'os
PHOSPHATE - [chem] phosphate *m* de chaux
PORCELAIN - s. bone china
SEEKER - [nucl] (compounds or ions migrating into bones) ion *m* ostéotrope
TURQUOISE - [min] (fossil bones or teeth used as a gemstone) fausse turquoise *f*, odontolite *f*
ONING - [survey] (locating and driving in pegs in line) nivellement *m*
ROD - [surv] (T-shaped rods used for boning) nivelette *f*, voyant *m*
ONMARTINI CATERPILLAR GEAR - [aero] (type of caterpillar L/g in which the track consists of a pneumatic tyre having the form of an endless belt) train *m* d'atterrissage à pneumatique Bonmartini
ONNET - [auto] capot *m*
[mech] couvercle *m*, chapeau *m* de valve, capot *m*
[mining] (of a cage) toit *m* protecteur (d'une cage d'extraction)
[mining] (timbering work) écoin *m*
[metall] calotte *f*, coupole *f*
[naut] (added to sails) bonnette *f* maillée
~DOOR - [auto] panneau *m* ouvrant de capot
~FASTENER - [auto] fermeture *f* de capot, verrouillage *m* de capot
~FASTENING CLIP - [auto] fermeture *f* de capot
~LATCH - [auto] crochet *m* de fermeture de capot
~LOUVRE - [auto] fentes *f*pl d'aération de capot, persienne *f* d'aération
~RELEASE - [auto] ouverture *f* de capot
~SIDE PIECE - [auto] panneau *m* latéral de capot
ONUS - [gen & comm] gratification *f*, boni *m*, prime *f*
~RATE - [comm] taux *m* d'indemnité
OOBY HATCH - [naut] écoutille *f*, capot *m*
~TRAP - [milit] mine *f* piégée
OOK, to - [gen] réserver, retenir, louer, enregistrer
OOK - [gen] livre *m*, registre *m*
~BACK ROUNDING MACHINE - [mech] presse *f* à condosser
~BINDER - [bookbinding] relieur *m*
~BINDERY - [print] atelier *m* de reliure
~BINDING - [gen] reliure *f*
~BINDING WIRE - [bookbinding] fil *m* pour reliure
~DESIGNER - [bookbinding] metteur *m* en page
~FACE - [print] caractères *m* pour textes courants, caractères *m* de texte
~-FORM MASK - [photo] cache *m* double (s'ouvrant comme un livre)
~INK - [print] encre *f* typo, encre *f* d'imprimerie
~JACKET - [bookbinding] couverture *f*, jaquette *f*
~-KEEPING - [comm] comptabilité *f*
~MUSLIN - [text] organdi *m*
~REVIEW - [gen] critique *f* d'un livre
~-SEWING - [bookbinding] couture *f* des cahiers
~-SEWING MACHINE - [mech] (in book binding) machine *f* à coudre les cahiers
~SIZE - [print] format *m*
~TRIMMING MACHINE - [mech] (in bookbinding) machine *f* à rogner
~TYPE - [print] caractères *m*pl de texte, caractéres *m*pl pour textes courants
BOOK TYPE PRESS - [mech] presse *f* à charnière horyzontale
~VALUE - [comm] valeur *f* d'inventaire
BOOKING - [gen] enregistrement *m*, inscription *f*, location *f*
[surv] (recording field observations) enregistrement *m*
~DATA - [comput] données *f* de réservation
~OF CALLS - [telecomm] enregistrement *m* des demandes de communication
~OFFICE - [railw] guichet *m* (de vente de billets)
~TIME - [telecomm] (of a call) heure *f* de la demande de communication
BOOKLET - [gen] livret *m*, opuscule *m*, brochure *f*
BOOLEAN CALCULATION - [electron] calcul *m* de Boole
BOOM, to - [gen & acoust] gronder, ronfler
BOOM - [acoust] grondement *m*, ronflement *m*
[gen build] (a long beam) longeron *m*
[aero] (flange of a built up girder) longeron *m*, poutrelle *f*
[naut] bôme *m*, gui *m*, tangon *m*
[mech] (main spar of hoist) flêche *f*, volée *f*
[constr] (of a girder) chapeau *m* (d'une poutre), plate-bande *f*
[naut] (barrier of logs to prevent the passage of a ship) barrage *m* mobile d'estuaire, estacade *f* flottante
[impl] (spar attached to a pole to lengthen it) rallonge *f*
[radio etc] (a telescopic support for a microphone) girafe *f*
[comm] "boom" *m*, emballement *m*, montée *f* rapide
~CAT - [oil ind] tracteur *m* portetubes
~HOIST - [oil ind] treuil *m* à antenne
~IRON - [naut] blin *m* (de bout-dehors)
~MEMBER - [build] barre *f* de membrure
~PALTE - [build] tôle *f* pour bordages
~STACKER - [mech] chargeuse *f* à flèche, élévateur *m* à flèche
BOOMER - [radio] (colloq) "boomer" *m*, haut-parleur *m*
BOOMINESS - [radio & telev] son *m* de tonneau
BOOMING - [acoust] ronflement *m*, grondement *m*
[mining] poussage *m* (par palplanches)
BOOMY - [gen & acous] sombre (voix, son etc)
BOON - [bot] chènevotte *f*, tige *f* ligneuse (du chanure)
BOORT - s bort
BOOST, to - [el] (to increase the total voltage in a circuit by connecting an additional source of voltage) survolter, surélever (le potentiel)
[gen] suralimenter, augmenter, aider
[mech] pousser
BOOST - [gen] augmentation *f*, aide *f*
[el] (the increase of the total voltage in a circuit by connecting an additional source of voltage) survoltage *m*, suralimentation *f*
[mech] (in intern. comb. eng; the amount by which the induction pressure of a supercharged engine exceeds atmospheric pressure) surpression *f*, poussée *f* additionnelle, surpuissance *f*, pression *f* additionnelle, pression *f* de suralimentation
~CONTROL - [mech] (a device for limiting the manifold pressure to a pre-set level) commande *f* de

la pression d'admission
BOOST CONTROL OVER-RIDE - [mech] (boost control cut-out) surpression *f* d'admission
~FEED - [mech] (of engines) suralimentation *f*, alimentation *f* forcée
~GAUGE - [instr] (sensitive pressure gauge indicating the measure of boost used in a supercharged aeroengine) indicateur *m* de pression d'admission
~PRESSURE - [mech] (the manifold pressure produced by the boost system) pression *f* de suralimentation, pression *f* d'admission
BOOSTER - [el] (additional source of voltage) survolteur *m*, générateur *m* autorégulateur
[of pressure and generally of power] élévateur *m*, dispositif *m* de renfort
[mech] servo-commande *f*
[mech] (pump or blower increasing pressure in a fluid) pompe *f* d'alimentation
[radio] (additional amplifier between aerial and receiver) préamplificateur *m* d'antenne
[railw] locomotive *f* de tête, (ou locomotive *f* de queue)
[aero] dispositif *m* élévateur de pression
~AMPLIFIER - [radio] (an amplifier used in audio circuits to improve signal noise ratio) amplificateur *m* de puissance
~BATTERY METERING - [el] mesure *f* par batterie tampon
~BRAKE - [mech] servo-frein *m*
~CIRCUIT - [electron] récupération *f* d'énergie
~COIL - [el] (induction coil supplied from a battery) survolteur *m*
~COMPRESSOR - [mech] surpresseur *m*
~DYNAMO - [el] dynamo *f* survoltrice
~DIODE - [electron] diode *f* de récupération
~DOSE - [med] dose *f* d'entretien
~ENGINE - [astronaut] moteur *m* de lancement
~EXPANDER - [mech] (expansion engine coupled to a compressor used in liquid oxygen manufacture) machine *f* de détente avec surpresseur
~FAN - [el] ventilateur *m* de surpression
~FUEL PUMP - [mech] (booster used in a fuel system) pompe *f* d'alimentation
~INJECTION - [med] injection *f* de rappel, injection *f* de soutien
~LIGHT - [cinema] projecteur *m* d'appoint
~LOCOMOTIVE - s. booster [railw]
~MAGNETO - [mech] magnéto *f* auxiliaire
~MECHANISM - [mech] mécanisme *m* d'asservissement
~PUMP - [mech] pompe *f* d'alimentation, pompe *f* de suralimentation*f*
~RAM - [mech] piston-plongeur *m* additionel
~ROCKET - [aero] (in a ramjet, for initial acceleration) fusée *f* auxiliaire
~STATION - [gas ind] poste *m* de surpression, poste *m* de recompression
[oil ind] station *f* de relais
[telev] station-relais *m* de diffusion
~SWITCH - [el] commutateur *m* auxiliaire
~TRANSFORMER - [el] (a transformer with the secondary winding in series with a circuit to add, or oppose voltage provided by another source) transformateur *m* survolteur (ou dévolteur)
~VENTURI TUBE - [mech] (a Venturi tube placed within the throat of another Venturi) petit venturi *m*, petit tube *m* de venturi

BOOSTING - [gen, phys & el] augmentation *f*, suralimentation *f*, surcompression *f*
~BATTERY - [el] batterie *f* tampon
~CHARGE - [el] charge *f* additionnelle
~MAIN - [el] ligne *f* auxiliaire, ligne *f* supplémentaire
~VOLTAGE - [el] surtension *f*
BOOT - [gen] chaussure *f*, bottine *f*, brodequin *m*
[auto] coffre *m* à bagages, malle *f* arrière
[mech] (a protective shield for hydraulic brake cylindres) gaine *f* de protection
[mech] (a protective piece of rubber) manchon *m* protecteur
[metall] (in a blast furnace) trémie *f*, hotte *f*
[constr] (of a rain-water down-pipe) dauphin *m*
[aero] (colloq for inner tube of an A/c tyre) chambre *f* à air (de pneumatique)
~JACK - [mining] fourche *f* à cliquet
~LACE - [text] lacet *m* à chaussure, lacet *m* de soulier
~LID - [auto] couvercle *m* de malle (ou de coffre)
~MACHINE - [mech] machine *f* pour chaussures
~-TOPPING - [shipbuild] bande *f* de flottaison
~-TREE - [impl] embauchoir *m*
BOOTH - [gen] (market stall etc; also temporary enclosure) baraque *f* foraine, cabine *f*
BOOTLEG - [mining] culot *m* de mine
~PACKER - [mech] presse-étoupe *m*
BOOTLEGGING - [rubber ind] décollement *m* des plis
BOOTSTRAP - [comput] (coded instructions at the beginning of an input tape employed to insert routine into the computer) instructions *f*pl initiales
~PILE LOAD TEST - [constr] essai *m* de chargement au vérin
~AMPLIFIER - [radio] (an amplifier circuit designed to change the potential of the input source in relation to earth by an amount equal to the output signal) amplificateur *m* à contre-réaction
~CIRCUIT - [radar & telev] circuit *m* auto-élévateur
~CYCLE - [el] (turbines) cycle *m* "bootstrap"
BOOTY - [gen] butin *m*
BORA - [met] (cold N.E. wind occurring in the Adriatic region) borée *m* (vent de N.E. froid en Adriatique)
BORACIC ACID - [chem] acide *m* borique
BORACITE - [min] boracite *f*
BORAGE - [bot] bourrache *f*
BORAL - [nucl] (material which absorbs neutrons in a nuclear reactor) boral *m*
~SHIELD - [nucl] (special shield against thermal neutrons) écran *m* en boral
BORANES - [chem] (fuels for rocket propulsion) boranes *m*pl
BORATE - [chem] (boric oxyde) borate *m*
BORAX - [min] (mineral obtained through evaporation of the waters of alkaline lakes) borax *m*
~HEAD - [ind chem] perle *f* de borax
~PENTAHYDRATE - [chem] (a weedkiller) pentahydrate *m* de borax
BORAZONE - [chem] (crystallized boron nitride) nitrure *m* de bore, borure *m* d'azote
BORD - [mining] chambre *f*, recoupe *f*
~-AND-PILLAR METHOD - [mining] exploitation *f* par chambres et piliers
BORDEAUX MIXTURE - [ind chem] (mixture of lime, copper sulphate and water used as an insecticide) bouillie *f* bordelaise, bouillie *f* cuprique

BORDER, to - [metall] (to form a divergent lip on the end of a tube) border, rabattre, marger
BORDER - [gen] bord *m*, lisière *f*, bordure *f*, marge *f*
[print] vignette *f*, baguette *f*, marge *f*
[geogr] frontière *f*, confins *m*
[text] galon *m*, bordure *f*
~CURVE - [draw] (in diagrams) courbe *f* limite
~HEM - [text] ourlet *m* de bordure, ourlet *m* de bordé
~LAG - [text] carton *m* des bordures
~-LINE - [geogr & gen] frontière *f*, ligne *f* de séparation
~-LINE CASE - [med] cas *m* limite, cas *m* indéterminé
~PLANE - [carp] rabot *m* pour moulures
~THREAD - [text] fil *m* à lisière
BORDEREAU - [comm] (list of documents) bordereau *m*
BORDERED CARPET - [text] tapis *m* à bordure
~PIPE - [mech] (a pipe having the end formed with a divergent lip) tube *m* à collerette rabattue, tube *m* bordé
BORDERING - [gen] limitrophe, contigu, bordure *f*
BORDERLAND - [geogr] pays *m* limitrophe, région *f* limitrophe
BORE, to - [gen] forer, percer, perforer
[mech] (to enlarge or finish a hole by internal turning) aléser
[mining] (of a mine gallery) forer, sonder, creuser
[hydr] (of a well) forer
[constr] (of a tunnel) forer, percer
BORE - [gen] forage *m*
[mech] chambre *f* cylindrique, forage *m*
[firearm] calibre *m*, âme *f* (d'une arme)
[mech] (diameter of a cylinder in a motor) alésage *m*, diamètre *m* interieur
[met] (tide wave running rapidly up a river) mascaret *m*, raz de marée *m*
[oil ind] sondage *m*, trou *m* de sonde
~-BIT - [mech] tranchant *m*, fleuret *m*
~-CORE - [mining] carotte *f*, carotte -témoin *f*
~FRAME - [mech] chevalet *m* de forage
~GAUGE - [instr] calibre *m* pour diamètre intérieur
~-HOLE - [min] trou *m* de sonde, sondage *m*, trou *m* de mine
~HOLE PUMP - [mining] pompe *f* de forage, pompe *f* de fonçage
~MEAL - [mining] farine *f* de sondage
~OF A PIPE - [mech] diamètre *m* intérieur
~OF NOZZLE - [mech] calibre *m* du gicleur (ou de l'ajutage)
~ON THE LATHE, to - [mech] percer au tour
~OUT, to - [mech] creuser, forer, foncer (un puits), percer, aléser (un cylindre)
~PIT - [mining] puits *m* de recherche
~PROFILOMETER - [instr] profilomètre *m* d'alésage
~-ROD - [mining] tige *f* de sonde
BOREAL - [met] boréal
BORED PLATEN PRESS - [mech] presse *f* à plateaux forés
~ROLL - [mech] cylindre *m* foré
BORER - [tool] alesoir *m*, foret *m*, tarière *f*, vrille *f*, perçoir *m*
[in mining] fleuret *m*, perforatrice *f*, sonde *f*, burin *m*

BORIC - [chem] borique
~ACID - [chem] acide *m* borique
~OXIDE - [chem] anhydride *m* borique
BORIDE - [chem] borure *m*
BORING - [mech] alésage *m*, perçage *m*, percement *m*
[mining] (the process of drilling holes into ground or rock) sondage *m*, forage *m*
~AND MORTISING MACHINE - [mech] mortaiseuse *f* à mèche
~AND SURFACING LATHE - [mech tool] tour *m* vertical
~BAR - [mining] barre *f* de sondage, sondeuse *f*
[mach tool] (bar or mandrel mounted between centres in a lathe) barre *f* d'alésage, arbre *m* porte-foret
[mech] (regrinder for cylinder) alésoir *m*
~BEETLES - [bot] insectes *m* térébrants
~BIT - [tool] tranchant *m* de mèche, mèche *f*, alésoir *m*, ébauchoir *m*, dégorgeoir *m*
~BY ROTATION - [mining] sondage *m* rotatif, forage *m* rotatif
~CHISEL - [mech] trépan *m*, burin *m*
~CHUCK - [mech] manchon *m* porte-foret
~CUTTER - [tool] lame *f* d'aléseuse, lame *f* d'alésage
~HEAD - [mech] tête *f* de sondage
~HOSE - [mech] tuyau *m* pour forage
~MACHINE - [mech] foreuse *f*, perforatrice *f*, perceuse *f*, aléseuse *f*
~MILL - [mach tool] (a heavy type of boring machine) aléseuse-fraiseuse *m*, tour vertical
~RIG - [mining] tige *f* de forage, tige *f* du carottier
~ROD - [impl] barre *f* d'alésage, arbre *m* porte-mèche, barre *f* de sondage
~SPINDLE - [mech] broche *f* porte-mèche, mandrin *m* porte-outil
~STAY - [mech] barre *f* d'alésage
~STAY CAP BEARING - [mech] coussinet *m* de la barre d'alésage
~STAY UPRIGHT - [mech] montant *m* de la barre d'alésage
~TEST - [mining] sondage *m* de recherche
~TOOL - [tool] outil *m* à aléser, barre *f* de sondage, outil *m* de forage
~TOWER - [mining] tour *f* de sondage
~WINCH - [mech] treuil *m* de foreuse
BORINGS - [mech] alésures *f*pl copeaux *m*pl
BORNEO CAMPHOR - s borneol
BORNEOL - [chem] (translucent substance with camphor-like smell used in perfumery and medicine) bornéol *m*
BORNITE - [min] (a natural sulphide of iron and copper; a valuable ore of the latter) bornite *f*
~DETECTOR - [radio] (crystal detector: steel point in contact with a bornite crystal) détecteur *m* à cristal de bornite
BORNYL ACETATE - [chem] (a derivative of borneol used as a plasticizer for nitrocellulose and in perfumery) acétate *m* de bornyle
~ALCOHOL - [chem] bornéol *m*
~CHLORIDE - [chem] chlorure *m* de bornyle
BORNYLAMINE - [chem] bornylamine *f*
BOROETHANE - [chem] boroéthane *m*, biborane *m*
BOROFLUORIDES - [chem] borofluorures *m*pl
BOROLANITE - [geol] borolanite *f*

BORON - [chem] (an element, symbol B, used in glass and ceramics production etc) bore *m*
~AZIDE - [chem] azoture *m* de bore
~BROMIDE - [chem] bromure *m* de bore
~CARBIDE - [ind chem] (compound of carbon and boron, highly resistant to heat and used for gas turbine blades) carbure *m* de bore
~CHAMBER - [ind chem] chambre *f* au bore
~COUNTER - [ind chem] compteur *m* à bore
~COUNTER TUBE - [nucl] tube-compteur *m* à bore
~DETECTOR - [nucl] détecteur *m* à bore
~HYDRIDE - [chem] (an alternative term for Diborane) hydrure *m* de bore
~IODIDE - [chem] iodure *m* de bore
~IRON - [metall] ferrobore *m*
~NITRIDE - [chem] (a high temperature refractory) nitrure *m* de bore
~STEEL - [metall] acier *m* au bore
BORONIC COPPER - [metall] cuivre *m* au bore
BOROSILICATE - [chem] borosilicate *m*
BORROW, to - [gen] emprunter
~PIT - [build] (site outside the works whence extra material for filling is taken) ballastière *f*, emprunt *m* de terre
BORT - [min] (diamond of confused crystalline construction, used for industrial purposes) bort *m* égrisé *m*, égrisée *f*
BOSH - [metall] (of blast furnace) étalage *m* (d'un haut-fourneau)
BOSON - [nucl] boson *m*
BOSS - [mech] bossage *m*, renflement *m*, bosse *f*, portée *f*
[mech] (of a shaft) moyeu *m*
[mach] (of a press) matrice *f*, estampe *f*
[metall] saillie *f*
[geol] (of intrusive rock) massif *m* d'injection
[min] (ore stamping) tête *f*, surcharge *f*
[build] auge *f* à mortier
[aero] (of a propeller) moyeu *m*
[med] bosse *f*, protubérance *f*
~JOINT BOLT - [mech] boulon *m* de moyeu
~OF COGWHEEL - [mech] moyeu *m* de roue dentée
~OF SCREW - [naut] moyeu *m* de l'hélice
~WITH STRENGTHENING RIBS - [mech] moyeu *m* à nervures de renforcement
BOSSAGE - [constr] (roughly dressed stones) bossage *m*
BOSSED ROLLER - [mech] cylindre *m* à boutons
BOSSING UP - [mech] (the operation of forming a boss) estampage *m* en relief
BOSUN - [naut] maître *m* d'equipage, maître *m* de manoeuvre
BOT - [zool] larve *f* d'oestre
BOTANY - [bot] botanique *f*
~WOOL - [text] (merino) laine *f* Botany
BOTFLY - [zool] oestre *m*, mouche *f* des chevaux
BOTH - [gen] deux, tous les deux
~-WAY JUNCTION - [telecomm] jonction *f* à deux directions, à double sens
BOTT - [metall] (foundry) bouchon *m* d'argile pour obturer le trou de coulée, tampon *m*
~UP, to - [metall] boucher
BOTTING - [metall] (of furnace cupola) obturation *f* du trou de coulée (avec un bouchon d'argile)
BOTTLE, to - [brew ind] (to draw off into bottles) mettre en bouteille, embouteiller
BOTTLE - [gen] bouteille *f*, flacon *m*, fiole *f*
[metall] (metal cylinder containing fluid under pressure) bouteille *f*
BOTTLE BATTERY - [el] pile *f* bouteille
~BOBBIN - [text] bobine *f* à bouteille
~BRUSHEAD - [impl] tête *f* de brosse à bouteille
~-CAP - [gen] capsule *f* (de bouteille)
~CAPPER - [mech] capsuleuse *f*
~CLEANER - [mech] rince-bouteille *m*, machine *f* à rincer les bouteilles
~CLEANSER - s. bottle cleaner
~CRATE - [impl] caisse *f* à bouteilles
~FILLING MACHINE - [mech] machine *f* à remplir les bouteilles
~GLASS - [glass man] verre *m* à bouteille, verre *m* vert
~GREEN - [dye] vert-bouteille *m*
~JACK - [tool] vérin *m* à bouteille
~SCREW - [impl] tendeur *m*
~SOAKER - s, bottle cleaner
~STOPPER - [gen] bouchon *m* (pour bouteilles)
~WASHER - [mech] lave-bouteille *m*, rince-bouteille *m*
~WRAPPER - [gen] paillon *m* à bouteille
BOTTLED BEER - [brew] bière *f* en bouteille
BOTTLEMAKER CATARACT - [med] cataracte *f* des verriers
BOTTLENECK - [gen] goulot *m* (de bouteille)
[ind] (in industrial production) col *m* de bouteille
[town planning] (of traffic) embouteillage *m*, goulot *m* d'étranglement
[constr] (road build) éntranglement *m*
BOTTLING - [ind] embouteillage *m*, mise *f* en bouteilles, soutirage *m*
~DEPARTMENT - [brew ind] canetterie *f*
~MACHINE FILLING PLUG - [rubber ind] raccord *m* de soutireuse
~STORE - [brew ind] atelier *m* de mise en bouteille
~TANK - [brew ind] (tank under pressure from which bottles are filled in breweries) réservoir *m* sous pression
BOTTOM, to - [radio] (a thermionic valve is bottomed when the potential applied to its grid reduces the anode current to zero) bloquer
[carp] mettre un fond
[mech] (of the piston touching the end of the cylinder) toucher le fond, buter sur le fond
[naut] toucher le fond
BOTTOM - [gen] fond *m*, base *f*, socle *m*
[ind chem] bas *m*
[metall] (of a furnace) sole *f*
[mining] mur *m*, fond *m* de galerie
[mining] (of a pit) fond *m* de la fouille
[shoe man] cuir *m* pour semelles
~ANCHORED CORE - [metall] noyau *m* fixé par le bas
~APRON - [text] lamière *f* inférieure
~ARM - [mech] bras *m* inférieur
~BACKING PLATE - [metall] plaque-support *f*
~BALANCE END PIECE - [horol] contre-pivot *m* dessous de balancier
~BAR JACQUARD - [text] jacquard latéral
~BEARING - [instr] support *m*, crapaudine *f*
~BEND - s. anode bend
~BOARD - [shipbuild] vaigrage *m* du fond
~BOARD TAPPET - [text] excentrique *m* de la planche à collets
~BOOM - [build] membrure *f* inférieure, élément *m* inférieur

BOTTOM BOOM MEMBER - [constr] barre *f* de membrure inférieure
~BOX - [text] boîte *f* à cannettes
~BUCKLE - [mech] (of a saw) bride *f* inférieure à oeil
~CARRIAGE - [gas ind] guidage *m* de la partie inférieure de la dernière levée
~CASK - [brew ind] foudre *f* à la rangée inférieure
~CASTING - [metall] coulée *f* en source
~CLOTH - [text] tissu *m* dessous, tissu *m* d'envers, toile *f* d'envers
~CLOTH WEAVE - [text] armure *f* de tissu de dessous
~CURB - [gas ind] cornière *f* de renforcement (bas de jupe)
~DEAD CENTRE - [mech] point *m* mort inférieur
~DEPOSIT - [gen] dépôt *m*, sédiment *m*, dépôt *m* de fond, vase *f* de fond
~DIE - [metall] matrice *f* inférieure
~DISCHARGE - [ind chem] échappement *m* par le fond, vidange *f* par le fond
~DOOR - [metall] (of a furnace) porte *f* de fond
~DRAUGHT - [brew ind] insufflation *f*
~EJECTION - [plast ind] ejection *f* par le fond
~ELECTRODE - [metall] (arc furnaces) (electrode located in the lower part of the furnace) électrode *f* de sole
~FERMENTATION - [brew ind] fermentation *f*, basse fermentation *f* avec dépôt
~FLASH - [metall] ébarbure *f* inférieure (d'un lingot)
~FORCE - [plast ind] matrice *f* inférieure
~FORCE PRESS - [rubber ind] presse *f* à piston inférieur
~FRACTION - [ind chem] fraction *f* de queue
~FRICTION WASHER - [mech] disque *m*
~GATE - [of furnaces] ouverture *f* de fond
~GUIDE PLATE - [text] contre-plaque *f*
~HALF MOULD - [rubber ind] demi-moule *m* inférieur
~HEADING - [mining] galerie *f* d'avancement de fond
~HOLE - [mining] coup *m* de fond, mine *f* de relevage
~-HOLE PRESSURE - [oil ind] pression *f* de fond
~ICE - [soil] glace *f* de fond
~JOURNAL - [mech] pivot *m* inférieur, tourillon *m* inférieur
~LAYER - [constr] (in road building) couche *f* inférieure
~LAYER OF YEAST - [brew] levure *f* de couche inférieure
~LEASE - [text] envergeure *f* inférieure, envergeure *f* du bas
~LIFT - [mining] jeu *m* de fond
~LINING BASE - [metall] revêtement *m* de fond, tôle *f* de fond
~MAN - [gas ind] déluteur *m*
~MOUNTING PLATE - [mech] plateau *m* inférieur
~OF SHUTTLE BOX - [text] fond *m* de la boîte à navette
~OF TOOTH - [text] pied *m* de dent
~OUTLET - [mech] vidange *f* de fond, tuyau *m* de vidange
~PAINT - [naut] peinture *f* sous-marine, peinture *f* de carène
~PART - [metall] partie *f* de dessous, demi-moule *m* inférieur
~PILLAR - [mining] pilier *m* de protection de puits

BOTTOM PLANE - [aero] aile *f* inférieure
~PLATE - [plast ind] plateau *m* inférieur, plaque *f* de dessous
[metall] plaque *f* de fond, armature *f* du dessous
~PLATING - [naut] bordé *m* de carène
~PLUG - [plast ind] (base of matrix) fond *m* rapporté
[mining] tampon *m* d'avance
~POURING LADLE - [metall] poche *f* à quenouille
~PRINT - [text] impression *f* préalable, impression *f* première
~PULLER - [mech] tirefond *m*
~RAKE - [mach tool] angle *m* de dépouille inférieur, angle *m* d'incidence
~RAM PRESS - [rubber ind] presse *f* à piston inférieur
~-ROAD BRIDGE - [constr] pont *m* à tablier inférieur
~ROLL - [metall] (of a rolling mill) cylindre *m* femelle
~RUNNER - [mech] longeron *m* inférieur de carrosserie
~SEDIMENTS - [oil ind] fonds *mpl* de réservoir
~SHED - [text] foule *f* d'en bas
~SHEDDING MACHINE - [mach] machine *f* à pas inférieur
~SHOT - [text] liage *f* de la trame de fond
[mining] coup *m* de relevage
~SIDE - [mech] (of bevel gears) flanc *m* inférieur
~SPROCKET - [mech] (sprocket transporting a projected film towards the take-up spool) tambour *m* denté inférieur
~STEP - [build] premier palier *m*
~-STOPING - [mining] abattage *m* descendant
~SWAGE - [tool] (a bottom rounding tool) étampe *f* de dessous
~WATER - [geol] nappe *f* superficielle
~YEAST - [ind chem] (yeast vegetating at the bottom of a fermenting vat) levure *f* basse
BOTTOMING - [constr] empierrement *m* (d'une route)
[electron] (the clamping of the plate potential of a pentode) blocage *m*
[electron] (the operation of an electronic device which determines the lowest instantaneous potential of the output electrode) coinçage *m*
~BATH - [text] bain *m* de piétage, bain de mordançage
~COLOUR - [text] couleur *f* de fond
~HOLE - [glass man] four *m* de réchauffage
~TAP - [mech] taraud *m* de finition, taraud *m* finisseur
BOTTOMLANDS - [geol] terres *fpl* d'alluvion
BOTTOMLESS - [gen] sans fond, insondable
BOTTOMS - [oil ind] fonds *m*
[ind chem] produits *mpl* de queue, résidu *m*
BOTULISM - [med] botulisme *m*
BOUCHERIZING - [telecomm] imprégnation *f* des poteaux par le procédé Bouchery
BOUGE - [brew ind] (of a cask) bombement *m*
BOUGH - [bot] branche *f*, rameau *m*
BOUGIE - [med] sonde *f*, cathéter *m*
~DECIMAL - [el & light] bougie *f* internationale
BOUGIRAGE - [med] sondage *m*, dilatation *f* à la sonde
BOULDER - [geol] galet *m*, gros cailloux *m*, bloc *m* erratique
~CLAY - [geol] argile *f* à blocaux, argile *f* erratique

BOULDER FLINT - [min] silex *m*
~PAVING - [constr] pavage *m* en galets
BOULE - [min] (small mass of synthetic sapphire, ruby etc) pierre *f* artificielle, pierre *f* synthétique
BOULTER - [fishing] (multiple fishing line) ligne *f* munie de plusieurs hameçons
BOUNCE, to - [phys] rebondir, sauter
[aero] rebondir
BOUNCE - [acoust] qualité *f* acoustique
~FLASH - [light] éclair *m* réfléchi
~LIGHTING - [light] éclairage *m* indirect
~PLATE - s. baffled plate
BOUNCING - [gen] rebond *m*
[telev] instabilité *f* verticale, (of the picture) instabilité *f* vertical de l'image, sautillement *m* vertical
~PIN - [mech] aiguille *f* sauteuse
~PUTTY - [ind chem] (silicone with high impact resilience) mastic *m* élastique
BOUND - [gen] saut *m*, rebond *m*
[surv] limite *f*, borne *f*
[ind chem] lié, combiné
~ATOM - [nucl] atome *m* lié
~CHARGE - [el] (inducted electrostatic charge) charge *f* latente
~ELECTRON - [phys] (an electron bound to an atomic nucleus) électron *m* lié
~IN PAPER BOARDS - [bookbind] cartoné
~RUBBER - [ind chem] (rubber in chemical combination) caoutchouc *m* lié
~SULPHUR - [ind chem] (the proportion of sulphur contained in a vulcanized rubber) soufre *m* combiné
~VORTEX - [phys] (virtual vortex formed by the vortex sheet surrounding the surface of a body) tourbillon *m* théorique
~WATER - [bot] eau *f* liée, eau *f* absorbée
BOUNDARY - [gen] limite *f*, borne *f*, frontière *f*
[geol] ligne *f* de séparation
[crystall] (the contact between adjacent crystals in a metal) face *f* limite
~CONDITION - [phys] condition *f* aux limites
~CONTRAST - [electron] valeur *m* de seuil de luminance
~-FAULT - [geol] faille-limite *f*
~LAYER - [phys & aero] (the layer of fluid adjacent to the surface of an aerofoil) couche *f* limite, couche *f* laminaire
~LAYER BLADE COOLING - [aero] (method of cooling the blades of a gas turbine) refroidissement *m* des ailettes par la couche limite
~LAYER BLOWING - [aero] (the control of a boundary layer by injection of air or gas) contrôle *m* d'une couche limite par injection
~LAYER CONTROL - [aero] (artificial control of a boundary layer) commande *f* de la couche limite
~LAYER SEPARATION - [aero] (the separation of a flow having a boundary layer) décollement *m* de la couche limite
~LAYER SUCTION - [aero] aspiration *f* de la couche limite
~LAYER THICKNESS - [aero] épaisseur *f* de la couche limite
~LAYER TRANSITION - [aero] (the transition from laminar to turbulent flow in a boundary layer) transition *f* de la couche limite
BOUNDARY LIGHTS - [aero] (for landing areas) feux *mpl* de balisage
~LINE - [surv] ligne *f* de démarcation
~LUBRICATION - [mech] (state of partial lubrication that may exist between two surfaces) lubrification *f* superficielle
~MARKERS - [aero] balises *fpl* de délimitation
~MONUMENT - [top] pierre *f* de bornage, borne *f*
~POTENTIAL - [biol] (the potential difference through any chemical or physical discontinuity) potentiel *f* de contact
~STRESS - [mech] contraintes *fpl* au contour
~VALUE - [math] valeur *f* limite
~WAVELENGTH - [radio] longueur *f* d'onde limite
~ZONE - s. boundary layer
BOUNDLESS - [gen] sans borne, illimité, infini
BOURDON GAUGE - [mech] (type of pressure gauge) manomètre *m* de Bourdon
~TUBE - [mech] (curved blind tube) tube *m* de Bourdon, manomètre *m* de Bourdon
BOURETTE - [text] bourette *f*
~YARN - [text] fil *m* de bourette de soie
BOURGEOIS - [print] petit romain *m* (corps 8)
BOURNONITE - [min] (natural sulphide of copper, antimony and lead) bournonite *f*
BOUSE, to - [naut] palanquer
BOUT - [gen] tour *m*, reprise *f*
[med] accès *m*, attaque *f*, crise *f*
BOUTON TERMINAL - [med] terminaison *f* nerveuse en plaque
BOVINE- [gen] bovin
BOW - [gen] arc *m*, arceau *m*, archet *m* (de violin), réverence *f*, salut *m*
[naut] (of boats) avant *m*, proue *f*
[naut] (of the rudder in a boat) étrave *f*
[el] archet *m*
[el] (current collector in electrical vehicles) archet *m* de prise de courant
[auto] s. bow of head
[mech] (of a saw) cadre *m* de scie
~CAP - [aero] (of airship) capuchon *m* d'avant, chapeau *m* de proue
~CHOCK - [naut] chaumard *m*, galoche *f*
~COMPASS - [impl] compas *m* à balustre, compas *m* à pompe, petit balustre *m*
~DRILL - [impl] foret *m* à arçon, foret *m* à archet
~FAST - [naut] amarre *f* de l'avant
~HAWSE HOLES - [naut] écubiers *mpl* de proue
~HEAVY - [naut] lourd du nez, chargé sur l'avant
~LIGHT - [naut] fanal *m* de proue
~LINE - [shipbuild] arc *m*, courbure *f*
~OAR - [naut] rameur *m* de sête, brigadier *m*
~OF HEAD - [auto] partie *f* frontale de la carrosserie
~PEN - [impl] compas *m* à balustre à encre, tireligne *m* de compas
~SAW - [impl] scie *f* à chantourner, scie *f* à étrier (ou à archet)
~SEPARATOR - [auto] entretoise *f* d'arceau
~SHUTTLE - [text] navette *f* à arceau
~SIDE - [naut] tribord *m*
~SPRING TENSION - [mech] tendeur *m* à ressort à arc
~STIFFENER - [naut & aero] renfort *m* d'étrave, entretoise *f* de l'avant
~SUPPORTER - [auto] support *m* d'arceau
~TROLLEY - [el railw] archet *m* de prise de courant

OW WAVE - [naut] lame *f* d'étrave, vague *f* de proue
~WINDOW - [build] fenêtre *f* en rotonde, fenêtre *f* en saillie
OWDEN CABLE - [mech] câble *m* flexible Bowden
~CABLE STOP PIECE - [auto] arrêtoir *m*, serre-câble *m*
~WIRE - [mech] fil *m* flexible
BOWDLERIZE, to - [gen] expurger, émasculer
OWEL - [anat] intestins *mpl*, viscères *fpl*, boyaux *mpl*, entrailles *fpl*
BOWENITE - [min] (massive, finely granular form of serpentine) bowenite *f*
BOWER - [gen] charmille *f*, tonnelle *f*
[naut] ancre *f* de bossoir
BOWHAIR - [mus] (hair used for the bow of a violin) crin *m*
BOWING UNDER A LOAD - [nucl] (of fuel elements) flexion *m*
BOWK - [mining] (a hoisting bucket) cuffat *f*
BOWL - [gen] bol *m*, coupe *f*, cuvette *f*, écuelle *f*, jatte *f*
[mech] bol *m*
[meas] (of scales) plat *m*, bassin *m*
[instr] (of a compass) cuvette *f*
[mech] (motor fuel filter) faux-carter *m*
[text] galet *m* de carte, roulette *f* de carte
[light] (of a lamp) culot *m*
[geol] entonnoir *m*
[paper mach] rouleau *m*, cylindre *m*
[plumb] (of a pipe) fourneau *m*
[plastics] cylindre *m* de calandre
~METAL - [metall] antimoine *m* impur
~MILL - [ind chem] moulin *m*
~OF CAM LEVER - [text] galet *m* du levier
~PATTERN ROLLER - [text] rouleau *m*
BOWLEG - [med] jambe *f* arquée (ou bancale)
BOWLINE - [naut] bouline *f*
~BIRD - [naut] patte *f* de bouline, andaillot *m*
~BRIDLE - [naut] attache *f* de bouline
~CRINGLE - s. bowline bird
~HITCH - [naut] noeud *m* de chaise
~KNOT - [naut] noeud *m* de chaise (simple)
~TOGGLE - [naut] cabillot *m* de bouline
BOWR - s. bort
BOWS - [horol] anneaux *mpl*
BOWSPRIT - [naut] beaupré
BOWSTRING SUSPENSION - [el railw] prise *f* de courant à archet
BOX, to - [gen] boxer, souffleter
[ind] mettre en caisse, mettre en boîte
[naut] masquer (une voile)
BOX - [gen] boîte *f*, coffret *m*, caisse *f*, coffre *m* boîtier *m*
[el] (of an aneroid barometer) tambour *m*
[mech] (a housing) niche *f*
[railw] (for the axle) boîte *f* d'essieu
[mining] berline *f*
[auto] cadre *m*, box *m* (de garage)
[metall] (in foundries) chassis *m* (à mouler)
[naut] (of a paddle wheel) tambour *m* (de roue à aubes
[railw] cadre *m*, cabine *f* de signaleur, poste *m* d'aiguillage
[build] cabine *f*, loge *f*
[hydr] (irrigation) vanne *f*
[of a pump] corps *m*
[theatre] loge *f*
[agric] (in a stable) stalle *f*, box *m*
[comput] unité *f*
BOX A WHEEL, to - [gen] cercler une roue
~AERIAL - [radio] antenne *f* à boîte
~ANNEALING - [metall] recuit *m* en vase clos
~AUGER - [mech] tarière *f* à caisson
~BAFFLE - [acoust] enceinte *f* acoustique
~BAR - [metall] barre *f* d'ancrage
~BEAM - [build] poutre *f* à caissons
~BEAM SECTION - [build] section *f* à caissons
~BILL - [mining] instrument *m* de repêchage des tiges de sonde
~BODY - [auto] carrosserie *f* en caisson
~BRACE - [auto] ceinture *f* de benne
~CALF - [leather ind] box-calf *m*, veau *m* chromé
~CAMERA - [photo] appareil *m* à chambre rectangulaire, "boîte" *f*
~CAR - [railw] wagon *m* couvert
~CART - [impl] diable *m* à caisses
~CLOTH - [text] "box cloth" *m*
~COLUMN - [build] colonne *f* à section à caisson
~CONNECTING-ROD END - [mech] tête *f* de bielle à cage fermée
~CONNECTOR - [impl] boîte *f* de jonction
~COUPLING - [mech] accouplement *m* à manchon
~CULVERT - [hydr] conduit *m* à section rectangulaire, aqueduc *m* à section rectangulaire
~DAM - [hydr] batardeau *m* à double coffrage
~DOCK - [naut] dock *m* flottant
~DRAIN - [build] égout *m* à section rectangulaire
~ELEVATOR - [mech] monte-charge *m*
~FOR STEERING GEAR - [auto] boîtier *m* de direction
~GIRDER - [aero] poutre *f* à caisson, poutre *f* tubulaire
~GUTTER - [constr] gouttière *f* en U, rigole *f* en U, chéneau *m* encaissé
~HARDENING - [metall] cementation *f* en vase clos, cémentation *f* en châssis
~LONGERON - [mech] longeron *m* caisson
~MALTING - [brew ind] maltage *m* en caisse
~METAL - [metall] métal *m* à coussinets
~MOULDING - [metall] moulage *m* en chassis
~NAILING MACHINE - [mech mach] machine *f* à clouer les caisses
~NEGATIVE PLATE - [el] (storage battery) plaque *f* negative
~NUT - [mech] écrou *m* borgne, écrou *m* à chapeau
~OFFICE - [gen] bureau *m* de location, caisse *f*, guichet *m*
~PLATE - [telecomm] plaque *f* à caissons
~RECEIVER - [telecomm] récepteur *m* montre, recepteur *m* en forme de montre
~RELAY - [el] relais *m* à boîte
~RULE - [print] filet *m*
~SEAT - [auto] baquet *m*, siège *m* enveloppant
~SECTION FRAME - [mach tool] bâti *m* en forme de caisson
~SECTION MEMBER - [auto] traverse *f* à section caisson
~SPANNER - [impl] clé *f* à tube, clé *f* à douille
~SPAR - [aero] longeron *m* caisson
~STAPLE - [mech] gâche *f*
~SWITCH - [el] interrupteur *m* rotatif, commutateur *m* à boîte
~TOE - [shoe man] bout *m* rapporté
~TOLL BLADE - [tool] lame *f* pour porte-outil multi-

ple
BOX TYPE CROSSHEAD - [mech] crosse *f* à caisson
~TYPE ELECTRIC FURNACE - [metall] four *m* électrique à mouffle
~TYPE FRAME - [auto] chassis-caisson *m*
~WAGON - [railw] wagon *m* couvert, fourgon *m* (à bestiaux)
~WRENCH - [mech] s.box spanner
BOXBOARD - [paper man] carton *m* pour boîtes
BOXER'S EAR - [med] oreille *f* en choux-fleur
BOXING - [ind] emballage *m*
[build] (part of window frame for the folded shutter) chambranle *m*
[constr] ensablement *m*, lit *m* de charge, ballastage *m*
~MACHINE - [mech] machine *f* à mettre en boîte
~TENON - [mech] tenon *m* de raccordement
BOXWOOD - [bot] buis *m*, bois *m* de buis
~SHUTTLE - [text] navette *f* en buis
BRACE, to - [gen] amarrer, ancrer, consolider
[naut] brasser
[build] entretoiser, haubanner, renforcer
[constr etc] armer, moiser, contreventer
[oil ind] étayer
BRACE - [naut] bras *m* de vergue
[naut] (rudder gudgeon) pentures *fpl*, ferrures *fpl*
[aero] (to oppose deforming stresses) croisillon *m*, suspente *f*
[build] entretoise *f*, attache *f*, tirant *m*, croisillon *m*; (diagonal member for wooden roof-truss) contre-fiche *f*; (diagonal member in iron roof-truss) bielle *f*; (for a lattice truss) croisillon *m*; (in a trussed frame) écharpe *f*
[carp] vilebrequin *m*, moise *f*
[mech] support *m*, armature *f*, entretoise *f*, tirant *m*
[mining] bouche *f* de puits de mine, (a kind of platform) plancher *m* de manoeuvre
[mech] (ratchet brace) cliquet *m*
[print] accolade *f*
[meas] (a pair) couple *m*, paire *f*
~BACK, to - [naut] brasser à culer
~BIT - [tool] mèche *f* de vilebrequin
~ENDS - [print] grisés *mpl*
~HEAD - [mining] manivelle *f*, manche *f* de manoeuvre
~JAWS - [carp] mors *m*
~ROD - [mech] tringle *f* de renforcement
BRACED - [gen & mech] renforcé, entretoisé
~ARCH - [build] arceau *m* à croisillons
~GIRDER - [build] poutre *f* à croisillons
~LOOM FRAMING - [text] fixation *f* du cadre du métier
~PURLIN - [build] panne *f*, filière *f* à croisillon
~TIMBERING - [mining] boisage *m* contrefiche
BRACELET - [gen & mech] bracelet *m*
BRACES - [gen & mech] bretelles *fpl*
[build] contreventement *m*, entretoises *fpl*
[oil ind] croisillons *mpl* de la tour
BRACHIAL - [zool] brachial
~ARTERY - [anat] artère *f* brachiale
~PLEXUS - [anat] plexus *m* brachial
BRACHIALGIA - [med] névralgie *f* du plexus brachial, brachialgie *f*
BRACHYCEPHALY - [met] brachycéphalie *f*
BRACHYPNEA - [physiol] brachypnée *f*
BRACING - [el] ancrage *m*
[build] entretoisage *m*, renforcement *m*, étrésillonnement *m*
[constr] (metal frame-work) contreventement *m*
[mech] (to prevent deforming stresses) renforcer, haubanner
~CROSS - [build] croisillon *m* de renforcement
~MEMBER - [build] poutre *f* de contreventement
~WIRE - [mech] fil *m* tenseur de rappel, hauban *m*
BRACINGS - [oil ind] s.braces
BRACKEN - [bot] fougère *f* arborescente, fougère *f* à l'aigle
BRACKET, to - [gen] mettre entre parenthèses
[firearm] (of guns) encadrer, prendre en fourchette
BRACKET - [gen] support *m*, console *f*, potence *f*
[mech] patte *f* de fixation, support *m*
[carp] (of shelves) tasseau *m*
[firearm] (of guns) fourchette *f*, flasque *m* d'affut
[print] crochet *m*, parenthèse *f*, accolade *f*
[el] collier *m* de fixation, étrier *m*
[instr] applique *f*
[mining] plancher *m* de manoeuvre
~BALUSTER - [build] balustre *m* sur console
~CRAB - [mech] treuil *m* d'applique
~CRANE - [mech] grue *f* à console
~-HANGER - [mech] palier *m* à potence
~POLE - [telecomm] poteau *m* à potence
~RIM - [metall] couronne *f* de support, marâtre *f*
~SCAFFOLD - [build] échafaudage *m* sur supports en équerre
~SEAT - [auto etc] strapontin *m*
BRACKETING - [mech] système *m* de support, fixation *f*, palier *m*
[aero] (flying along a beam by skirting the twilight zones on each side alternatively) vol *m* en zig-zag le long d'un faisceau d'ondes radioélectriques
[photo] mise *f* au point (sur distance moyenne)
BRACKISH - [gen] saumâtre
BRACKISHNESS - [gen] caractère *m* saumâtre
BRACT - [bot] bractée *f*
BRACTEAL LEAF - [bot] feuille *f* bractéale
BRACTEATE - [bot] bractifère, à bractées
BRAD - [carp] pointe *f*, clou *m* à tête perdue, clou *m* étêté
BRADAWL - [impl] alène *f* plate, poinçon *m*, pointe *f* carrée
BRADS - [metall] pointes *fpl* étêtées
BRADYCARDIA - [med] bradycardie *f*
BRADYPEPSIA - [med] bradypepsie *f*
BRADYSEISM - [geol] bradyséisme *m*
BRADYTROPHY - [med] bradytrophie *f*
BRAID, to - [gen] tresser, natter, galonner, soutacher, passementer, guiper
BRAID - [text] tresse *f*, galon *m*, ganse *f*, passement *m*, natte *f*
[mech] tresse *f*, enroulement *m* hélicoidal
~PHOTOGRAPHING - [rubber ind] empreinte *f* du tissu d'enroulement (sur la surface d'un tuyau)
~ROPE - [gen] corde *f* tressée, tresse *f*
BRAIDED - [el] sous tresse
~HOSE - [rubber ind] tuyau *m* à armature en tissu guipé
~TOW ROPE - [rope man] remorque *f* tressée
~WIRE - [mech] fil *m* tressé, fil *m* guipé
BRAIDER - [text] machine *f* à tresser
BRAIDING - [gen] tressage *m*

[el] (covering of braided metal) gaine *f* tressée, guipage *m*
BRAIDING BOBBIN - [text] bobine *f* de tressage
~HEAD - [text] tête *f* de tressage
~LOOM - [text] métier *m* de passementerie
~MACHINE - [text] machine *f* à tresser, machine *f* à lacets, tresseuse *f* mécanique
BRAIL, to - [naut] carguer
BRAIL - [naut] cargue *m*, cargue-point *m*
BRAIN - [anat] cerveau *m*
~BLADDER - [anat] vésicule *m* cérébrale
~DISEASE - [med] maladie *f* du cerveau, affection *f* cérébrale
~FEVER - [med] fièvre *f* cérébrale
~STEM - [anat] axe *m* cérébro-spinal
BRAIZE - [mining] poussière *f* de charbon
BRAKE, to - [mech] freiner
[text] briser, broyer, teiller, macquer, piler
BRAKE - [mech] frein *m*
[text] (to beat linen) brisoir *m*, broie *f*, macque *f*, tillotte *f*
[mech] (for the extraction of juices) presse *f*
[naut] (pump handle) brimbale *f*, levier *m*
~ACTION - [mech] action *f* du frein
~ACTUATING CAM - [mech] came *f* de commande de frein
~ADJUSTER - [mech] registre *m* du frein
~ADJUSTING - [mech] réglage *m* du frein
~ADJUSTMENT SCREW - [mech] vis *f* de réglage du frein
~ASSEMBLY - [mech] ensemble *m* de freinage
~BACKING PLATE - [mech] disque *m* porte-frein
~BAND - [mech] ruban *m* de frein, bande *f* de frein, collier *m* de frein
~BAND LINING - [mech] garniture *f* de la bande de frein
~BEAM - [mech] sommier *m* de frein
[auto] levier *m* de commande de frein
~BLOCK - [mech] patin *m* de frein, sabot *m* de frein, coussinet *m* de freinage
~BOLT - [mech] boulon *m* de frein
~BOOSTER - [mech] servo-frein *m*
~CABIN - [railw] cabine *f* de frein
~CABLE - [mech] câble *m* de frein
~CAM - [mech] came *f* de frein
~CHEEK - [mech] mâchoire *f* de frein
~COMPENSATING DEVICE - [mech] dispositif *m* compensateur du frein
~CYLINDER - [mech] cylindre *m* de freinage
~CYLINDER TANK - [railw] réservoir *m* du cylindre de freinage
~DISC - [mech] disque *m* de frein
~DRUM - [mech] tambour *m* de frein, poulie *f* de frein
~DRUM LINER - [mech] revêtement *m* du tambour de frein
~EQUALISER - [auto] compensateur *m* de frein
~EQUALISER SHAFT - [auto] arbre *m* d'équilibrage des freins
~FACING - [mech] garniture *f* de frein
~FIELD - [electron] champ *m* retardateur
~FIELD VALVE - [electron] (electronic valve for the prevention of the passage of secondary electrons between the screen grid and the cathode) tube *m* à champ de freinage
~FLANGE - [auto] disque *m* porte-frein
~FLUID - [auto] fluide *m* pour frein, fluide *m* hydraulique pour freins
BRAKE FORCE - [phys] force *f* de freinage
~FRICTION PAD - [mech] patin *m* du disque de frein
~GOVERNOR - [mech] régulateur *m* à frein
~HAND LEVER - [auto] levier *m* de frein à main
~HANDLE - [auto] poignée *f* de levier de frein
~HANGER - [railw] bielle *f* de suspension de la tête de frein
~HEAD - [railw] tête *f* de frein
~HOP - [mech] broutement *m* au freinage
~HORSEPOWER - [mech] (B.H.P.) (effective or useful horsepower) puissance *f* au frein
~HOSEPIPE - [railw] tube *m* flexible pour l'accouplement des freins
~HOUSING - [auto] carter *m* de frein
~HUB - [mech] moyeu *m* de frein
~INDUCTION COIL - [mech] bobine *f* d'induction de frein
~INTERMEDIATE CONTROL - [mech] commande *f* intermédiaire de frein
~INTERMEDIATE SHAFT - [mech] arbre *m* intermédiaire de frein
~LATCH - [mech] butée *f* d'arrêt de frein
~LEVER - [mech] levier *m* du frein
~LEVER SEGMENT - [auto] cremaillère *f* de levier de frein
~LINE - [auto] tubulure *f* de frein, canalisation *f* de frein
~LINING - [auto] garniture *f* de frein
~LINKAGE - [auto] tringlerie *f* de frein
~LOAD - [mech] charge *f* de frein
~LOCOMOTIVE - [railw] locomotive-frein *f*
~MAGNET - [el] électro (aimant) *m* de démarrage de frein, électro *m* de frein
~MEAN EFFECTIVE PRESSURE - [mech] pression *f* moyenne effective au frein
~METER - [auto] appareil *m* d'essai des freins
~NUT - [mech] écrou *m* de frein
~OIL - [ind chem] liquide *m* (hydraulique) pour freins
~ON DIFFERENTIAL SHAFT - [mech] frein *m* sur l'arbre du différentiel
~ON TRANSMISSION SHAFT - [mech] frein *m* sur l'arbre de transmission
~OPERATING - [mech] commande *f* du frein
~OPERATING LEVER - [mech] levier *m* de commande du frein
~PARACHUTE - [aero] (parachute used to reduce the ground speed of an aircraft after landing) parachute *m* de freinage
~PAWL - [mech] cliquet *m* d'arrêt de frein
~PEDAL - [auto] pédale *f* de frein
~PIPE - [railw] conduite *f* principal de frein
~PISTON - [mech] piston *m* de frein
~PISTON CUP - [mech] coupelle *f* de piston de frein
~PLATE - [mech] disque *m* porte-sabot
~POWER - [mech] puissance *f* du frein, puissance *f* au frein, puissance *f* de freinage
~PULL CABLE - [auto] câble *m* de commande du frein
~PULL ROD - [mech] tringle *f* de commande de frein
~PULLEY - [mech] poulie *f* de frein
~QUADRANT - [auto] secteur *m* denté d'immobilisation du frein à main
~RELEASE SPRING - [mech] ressort *m* de rappel des freins
~RELINER - [impl] outil *m* pour le changement de garniture de frein

BRAKE RELINING - [auto] changement *m* de la garniture de frein
~RIGGING - [mech] tringlerie *f* de frein
~ROD - [auto] tige *f* de frein
~SERVO - [auto] servo-frein *m*
~SHAFT - [mech] barre *f* d'accouplement des freins, tige *f* de raccordement des freins
~SHOE - [auto] patin *m*, sabot *m* de frein, machoires *f*pl, segments *m*
~SHOE LINING - [auto] garniture *f* de mâchoires de frein
~SPIDER - [mech] disque *m* porte-patin
~SPRING - [mech] ressort *m* de frein
~STOP - [mech] butée *f* de frein
~SUPPORT OF RETARDER - s. brake beam
~TEST - [mech] essai *m* des freins
~TESTER - [mech] appareil *m* d'essai des freins
~THERMAL EFFICIENCY - [mech] rendement *m* (d'un moteur) en chevaux-vapeur
~TOGGLE - [auto] commande *f* des mâchoires de frein
~TOGGLE LEVER - [auto] commande *f* de l'écartement des mâchoires (freins)
~TOGGLE SHAFT - [mech] axe *m* du levier de commande des mâchoires de frein
~TORQUE - [mech] couple *m* de freinage, réaction *f* de torsion du frein
~VALVE - [mech] soupape *f* de frein
[railw] soupape *m* de mécanicien
~VAN - [railw] wagon-frein *m*, fourgon *m*
~WATER - [mech] (in turbines) eau *f* de contrejet, contrejet *m* de freinage
~WHEEL - [railw] volant *m* de frein à main
~WHEEL ARC LAMP - [el] lampe *f* à arc à réglage automatique
BRAKESMAN - [railw] serre-frein *m*, garde-frein *m*
~CABIN - [railw] guérite *f* du serre-frein
BRAKING - [mech] freinage *m*, serrage *m* des freins
~ACTION - [mech] effet *m* de freinage
~COUPLE - s. braking torque
~CYLINDER - [mech] cylindre *m* de frein
~DISTANCE-[gen] distance *f* nécessaire pour s'arrêter
~EFFECT - [mech] effet *m* des freins
[aero] puissance *f* de freinage
~EFFICIENCY - [mech] rendement *m* d'un frein
~LINKS - [mech] bielles *f*pl de sécurité
~MAGNET - s. brake magnet
~NOTCHES - [el] (positions of the handle of a drum-type controller applying a measure of electric braking) crans *m*pl de freinage
~PITCH - [aero] (setting of propeller pitch giving a reversal thrust) pas *m* de freinage
~POWER - [mech] puissance *f* de freinage
~PROPELLER - [aero] hélice *f* freinante
~RATIO - [mech] coefficient *m* de freinage
~SURFACE - [mech] surface *f* de freinage
~TORQUE - [mech] (of a meter) couple *m* de freinage
~VOLTAGE - [el] tension *f* de freinage
~WEIGHT - [mech] poids *m* du frein, poids de freinage
~WITH THE MOTOR - [mech] freinage *m* par le moteur, frein-moteur *m*
BRALE - [metall] presse *f* Brinell
BRAMBLE - [bot] ronce *f* sauvage, murier *m* des haies
BRAMBLING FINCH - [zool] pinçon *m* des Ardennes

BRAN, to - [text] ébrouer (la laine)
BRAN - [bot] son *m*, remoulage *m*
BRANCH, to - [gen] ramifier, bifurquer, suddiviser
[el] brancher sur
BRANCH - [gen] branche *f*, rameau *m*
[plumb] (of piping) tubulure *f* de connexion, dérivation *f*
[metall] (of oven) dérivation *f*
[build] branche *f* d'une voûte
[instr] (of compass) branche *f* de compas
[comput] aiguillage *m* (d'un programme)
[nucl] (product or series of products from one mode of decay of a radioactive nuclide showing two or more such modes) embranchement *m*
~ABSCISSION - [bot] (shedding of branches by plants) chûte *f* des branches
~BILL - [impl] serpe *f*
~BOX - [el] boîte *f* de branchement
~CIRCUIT - [el] dérivation *f*
~EXCHANGE - [telecomm] bureau *m* secondaire (téléphone)
~GATE - [metall] attaque *f* de coulée multiple
~HOUSE - [comm] succursale *f*, filiale *f*
~LINE - [railw] ligne *f* d'intérêt local, ligne *f* secondaire, embranchement *m*
~MAIN - [hydr etc] conduit *m* de dérivation, canalisation *f* de dérivation
~OF TEE - [gas ind] (network equip) tubulure *f*
~OFF, to - [gen] bifurquer, s'embrancher sur, se diviser (en plusieurs branches), se ramifier
[el] déconnecter
~PIPE - [mech] (special pipe having one or more branches) dérivation *f*, canalisation *f* de dérivation, tubulure *f* de branchement
~POINT - [el] branchement *m*, connexion *f*
~PROGRAMM - [comput] programme *m* d'aiguillage
~RAILWAY - s. branch line
~SPUR - [telecomm] ligne *f* de dérivation
~STRIP - [telecomm] baratte *f* de dérivation
~SWITCH - [el] interrupteur *m* de dérivation
~TENDRIL - [bot] vrille *f*
~WIRE - [el] fil *m* dérivé
BRANCHED - [gen] branchu, ramifié
[mech] (of pipes) à embranchements, piqué sur
~FLUE - [gas ind] conduit *m* à gaine commune (à départs individuels)
~SPARK - [el] étincelle *f* multiple
~VEIN - [mining] filon *m* ramifié
BRANCHIA - [zool] (respiratory organ in aquatic animals) branchies *f*pl, oules *f*pl
BRANCHIAL - [zool] branchial
~ARCH - [zool] arc *m* branchial
~DUCT - [zool] (ventral respiratory bronchus) conduit *m* branchial
BRANCHING - [gen] bifurcation *f*, branchement *m*, ramification *f*
[el] dérivation *f*
[nucl] (the occurrence of two or more modes by which an active nuclide is subject do decay) branchement *m*
~APPARATUS - [el] appareil *m* de dérivation
~CABLE - [el & telecomm] câble *m* de dérivation
~CURRENT - [el] courant *m* de dérivation
~DRAIN - [plumb] canalisation *f* d'évacuation secondaire
~FRACTION - [nucl] (ratio of the number of atoms disintegrating by a particular mode to the total

number of disintegrating atoms) fraction *f* d'embranchement
BRANCHING JACK - [telecomm] (jack without breaking contacts) jack *m* sans contacts de rupture
~OFF - [railw] embranchement *m*, bifurcation *f*
~OFF POINT - [railw] branchement *m* de voie, raccordement *m*, changement *m* de voie
~OUT AT A RIGHT ANGLE - [gen & mech] embranchement *m* à angle droit
~RATIO - [nucl] (the ratio of two specified branching fractions) rapport *m* d'embranchement
~TRACK - s. branching line
BRANCHIOMA - [med] tumeur *f* bronchiale
BRAND - [gen] marque *f*
[comm] marque *f*, qualité *f*, sorte *f*, étiquette *f*,
[agr] (torch of twisted straw) brandon *m*
[agr] (of animals) marque *f* faite au fer rouge
~IRON - [impl] fer *m* (chaud) à marquer
BRANDER - [text] griffe *f*, boîte *f* à couteaux
BRANDERING - [build] fixation *f* des lattes (sous le plâtre)
BRANDING - [gen] impression *f* au fer chaud, marque *f*
~IRON - s. brand iron
BRANDISH, to - [gen] brandir
BRANDY - [gen] cognac *m*, eau-de-vie *f*
~NOSE - [med] couperose *f* du nez
BRANNERITE - [min] brannerite *f*
BRASQUE - [metall] brasque *f*
BRASS, to - [metall] laitonner
BRASS - [metall] laiton *m*, cuivre *m* jaune
[mech] bronze, coussinet *m*, coquille *f*, demi-coussinet *m*
[min] pyrite *f*
~BED - [horol] bâti *m* en laiton
~BOBBIN - [text] bobine *f* plate
~CHILL - [med] fièvre *f* des fondeurs
~CURVES - [metall] réglettes *fpl* courbées en laiton
~FOIL - [metall] laiton *m* en feuilles, clinquant *m*
~FOUNDRY - [metall] fonderie *f* de laiton, robinetterie *f*
~FRACTIONAL WEIGHTS - [mech] poids *mpl* en lame de cuivre
~FURNACE - [metall] four *m* de fusion pour le laiton
~GAUZE - [metall] toile *f* de cuivre
~HAMMER - [impl] marteau *m* à panne en laiton
~INSTRUMENTS - [mus] cuivres *mpl*
~PLATING - [metall] cuivrage *m*, laitonnage *m*
~SHEET - [metall] feuille *f* de laiton
~SOLDER - [metall] soudure *f* au cuivre
~TUBE - [metall] tube *m* de cuivre, tuyau *m* de laiton
~WIRE - [metall] fil *m* de laiton
~-WIRE GAUZE - [metall] toile *f* en fil de laiton
~WOOL - [metall] laine *f* de cuivre
BRASSIL - [min] charbon *m* pyriteux
BRASSING - [metall] laitonnage *m*
BRASSWORK - [metall] cuivrerie *f*
BRASSY - [min] houille *f* pyriteuse
BRATTICE -[mining](partition for diverting air) cloison *f* d'aération
~CLOTH - [mining] toile *f* d'aérage
BRATTICING - [constr] cloisonnage *m*
BRAUNITE - [min] (ore of manganese found in India) braunite *f*
BRAWNY - [gen] musculeux, charnu
~ARM - [med] callosité *f* du bras
BRAXY - [vet] charbon *m*, fièvre *f* charbonneuse
BRAY, to - [gen] braire
[ind proc] broyer, concasser, piler, réduire en poudre
BRAYER - [print] (a hand ink-roller) encreur *m* à main
~ROLL - [print mach] rouleau *m* encreur
BRAYING - [acoust] braiment *m*
BRAYTON CYCLE - [phys] (cycle of constant pressure used in early gas turbines) cycle *m* de Brayton
BRAZE, to - [metall] braser, souder au laiton, se braser
BRAZE - [metall] joint *m* brasé
~WELDING - [metall] (special brazing method) soudobrasage *m*
BRAZIER - [impl] braséro *m*, chaudronnier *m*, dinandier *m*
BRAZIL WAX - [ind chem] s. carnauba wax
BRAZING - [metall] (the process of joining two sections of metal by fusing a layer of brass between the adjoining surfaces) brasage *m* fort, brasage *m*, soudo-brasure *f*
~CRUCIBLE - [metall] creuset *m* pour soudure
~FLUX - [metall] (agent used in brazing to promote fusion) fondant *m*, flux *m*
~LAMP - [metall] lampe *f* à braser
~PINCERS - [impl] pinces *fpl* pour souder
~SEAM - [metall] cordon *m* de soudure, ligne *f* de brasage
~SOLDER - [metall] alliage *m* pour soudure
~SPELTER - s. brazing solder
~TORCH - [metall] (gas torch used in welding) chalumeau-braseur *m*
BREACH - [gen] bris *m*, rupture *f*
[build] brèche *f*
BREAD - [gen] pain *m*
~-CRUST BOMB - [geol] bombe *f* craquelée
~TASTE - [brew ind] goût *m* de pain
BREADBOARD - [electron] montage *m* sur table
~CIRCUIT - [electron] montage *m* expérimentel d'un circuit
BREADFRUIT - [bot] fruit *m* de l'arbre à pain
BREADTH - [gen] largeur *f*, ampleur *f*
~EXTREME - [shipbuild] largeur *f* au fort
~FACTOR - [el] (distribution factor) facteur *m* de distribution
~OF BEAM - [naut] largeur *m* au maître bau
BREAK, to - [gen] briser, casser, rompre
[el] interrompre, couper
[physiol] (of voices) muer, casser
[paper man] désagréger, défibrer
[mus] changer de ton
[telecomm] interrompre, couper
[comput] (temporary break of a series of operations) interrompre
BREAK - [gen & mech] bris *m*, fracure *f*, cassure *f*, rupture *f*, brèche *f*, percée *f*, brisure *f*
[el] interruption *f*, rupture *f*, rupteur *m*
[min] (to crush or pound) concasser
[telecomm] déconnexion *f*, défaut *m* de continuité
[shipbuild] coupée *f*, décrochage *m*
[build] rupture *f*, renfoncement *m*, brisis *m*
[paint] casser (une émulsion) , mucilages *mpl*
[geol] dislocation *f*, fracture *f*, faille *f*, rupture *f*, cassure *f*
[print] alinéa *m*
~A MOULD, to - [plast ind] ouvrir un moule
~ADRIFT, to - [naut] partir à la dérive

BREAK ARC - [el] arc *m* de rupture
~BACK, to - [constr] faire un renfoncement
~BEFORE MAKING CONTACT - [telecomm] contacts *mpl* échelonnés dans l'ordre Repos-Travail
~BULK, to - [naut] désarrimer, commencer le déchargement
~BULKHEAD - [naut] cloison *m* de séparation
~CONSTRICTION - [genet] constriction *f* de rupture
~CONTACT - [telecomm] contact *m* de rupture
[el] contact *m* de repos
~CONTACT UNIT - [el] contact *m* de rupture
~DOWN, to - [gen] démolir, abattre, diviser, rompre
[chem] dissocier, décomposer, fractionner
[el] rompre, percer
[comm] (of accounts) analyser dans le détail
[mech] (of machines) tomber en panne
~DOWN BY MILLING, to - [rubber ind] plastifier, rendre plastique
~DOWN THE ORE, to - [mining] abattre le minerai
~FORMATION - [brew ind] flocculation *f*
~GROUND, to - [gen & build] commencer à labourer, donner le premier coup de pioche, défricher, entamer le travail
[naut] déraper
~IMPULSE - [telecomm] (impulse formed by the interruption of a current in a circuit) impulsion *f* provoquée par l'interruption d'un circuit, impulsion *f* d'ouverture
~IN, to - [mech] (of engines) roder
[print] (to insert illustrations etc) intercaler, insérer
~IN THE SUCCESSION - [geol] lacune *f* stratigraphique
~INTO, to - [gen] envahir
~-IRON - [impl] (of a plane) contre-fer *m*
~JACK - [telecomm] jack *m* de rupture
~JOINT - [build] joint *m* décalé, joint *m* brisé
~KEY - [el] interrupteur *m* à touches
~LENGTH - [telecomm] longueur *f* d'interruption
~LINE - [print] dernière ligne *f* (d'un alinéa)
~LOAD - [phys] charge *f* de rupture
~LOCATOR - [instr] localisateur *m* d'interruption
~MOORINGS, to - [naut] briser ses amarres
~OFF, to - [mining] abattre
~OUT, to - [gen] s'échapper, s'évader, éclater
~-OUT CATHEAD - [oil ind] cabestan *m* automatique de premier vissage
~OUT INTO, to - [gen] pénétrer (dans), résulter
[math] décomposer (en)
~PIN - [mech] goupille *f* de sécurité
~POINT - [electron] (in a gamma correction circuit) point *m* de flexion
~POINT INSTRUCTION - [comput] instruction *f* d'arrêt (pour contrôle)
~PULSE - [telecomm] impulsion *f* d'ouverture
~ROLL - [mech] (in a mill) cylindre *m* broyeur
~SHEER, to - [naut] faire une embardée, déboîter
~SURFACE, to - [naut] faire surface, émerger
~THE CIRCUIT, to - [el] interrompre, couper (un circuit)
~THROUGH - [nucl] seuil *m* d'absorption
[mining] boyau *m* de mine
~-THRUST - [geol] chevauchement *m* anticlinal
~UP, to - [gen] démolir, mettre en morceaux, disperser
[ind] diviser, broyer, fragmenter, morceler
[constr] (the ground) ameublir (le sol), défoncer

BREAK UP THE SAND, to - [metall] diviser le sable
BREAKABLE - [gen] fragile, cassable, brisable
BREAKAGE - [gen & comm] avarie *f*, bris *m*, cassure *f*, casse *f*, fracture *f*, rupture *f*, colis *m* démoli, fragmentation *f*
[naut] espace *m* perdu (dans les cales)
BREAKAWAY - [gen] sécession *f*, désertion *f*, évasion *f*, dérive *f*
[mech] (start of engine) démarrage *m*
[cinema](breakaway scenery) décor *m* qui doit s'effondrer
~CURRENT - [el] (the current which is used by an electric starter) intensité *f* au démarrage
~OF FLOW - [phys] (the detachment of flow from a surface with which it has been in contact) décollement *m* des filets d'air
BREAKDOWN - [gen] rupture *f*, interruption *f*, décomposition *f*
[mech] panne *f*, dérangement *m*, arrêt *m*
[el] (sudden passage of current through an insulator when voltage exceeds a prescribed value) claquage *m* percement *m*, décharge *f* disruptive
[naut] avarie *f*, panne *f*
[ind chem] décomposition *f*, fractionnement *m*
[comm] analyse *f* des comptes
[brew ind] désagrégation *f*
[med] dépression (nerveuse), effondrement *m*
~BENCH - [electron] (for the measurement of the breakdown voltage of rectifier tubes) banc *m* de mesure pour tubes redresseurs (ou soupapes électroniques)
~CRANE - [mech] (portable jib crane carried on a railway truck or lorry) grue *f* de dépannage; (if carried in a lorry) camion-grue *m*
~DRAWING - [eng] plan *m* de détail, dessin *m* détaillé
~LORRY - [mech] dépanneuse *f*
~SET - [el] groupe *m* de secours
~SIGNAL - [radio & telecomm] signal *m* d'interruption
~TORQUE - [el] couple *m* maximum (d'un moteur)
~VAN - [railw] voiture *f* de secours
~VISCOSITY - [oil ind] viscosité *f* de rupture
~VOLTAGE - [el] (voltage required to break down a cable or an insulator) tension *f* de rupture, tension *f* de percement, tension *f* disruptive
[el] (the lowest voltage at which an electrochemical valve fails) tension *f* de claquage
BREAKER - [naut] (of the sea) brisant *m*, vague *f* déferlante
[el] (circuit breaker) interrupteur *m*
[paper man] (washing roll fitted with knives) broyeur *m*, défibreuse *f*
[min] concasseur *m*, broyeur *m*
[naut] (water cask) baril *m* de galère
~ARM - [el] linguet *m* mobile de rupteur
~CAM - [el] came *f* de rupteur
~CARD - [text] carde *f* de dégrossissage
~CONTACT - [el] contact *m* de rupteur
~DRUM - [mech] cylindre *m* de défibreuse, tambour *m* de défibreuse
~LINING - [text] garniture *f* du briseur
~MOUTH - [mech] mâchoire *f* du concasseur
~PICKER - [text] batteur *m* dégrossisseur
~PLATE - [el] platine *f* de rupteur
[plast ind] disque *m* perforé
~-STATION - [mining] banc *m* de cassage

BREAKER STRIP - [rubber ind] (of a tyre, layers of rubberised fabric between thread and casing for the spreading of the load) entreplis *m* de gomme, "breakerstrip" *m*
BREAKING - [gen] rupture *f*, fracture *f*, bris *m*, cassage *m*
[el] interruption *f*, coupure *f*
[chem] (of an emulsion) recombination of the dispersed phase of an emulsion) floculation *f* (d'une émulsion)
[min] désagrégation *f*
~ANGLE - [text] angle *m* de pliage
~BOLT - [mech] boulon *m* de sécurité
~BUCKTHORN - [bot] aune *m* noir
~CAPACITY - [el] (the capacity of breaking an electric circuit under certain conditions) pouvoir *m* de coupure, puissance *f* de rupture
~CHAMFER - [metall] chanfrein *m* de rupture
~CURRENT - [el] (the maximum current which can be interrupted by a switch without damage) courant *m* coupé par un pôle (d'appareil d'interruption)
~DOWN - [gen] abattage *m*, démolition *m*, débitage *m*
[min] broyage *m*
[mining] abattage *m*, éboulement *m*
~-DOWN PASS - [metall] cylindre *m* dégrossisseur
~DOWN TIMBER - [wood ind] débitage *m* des bois
~ENERGY - [chem] énergie *f* de dissociation
~ENGINE - [paper man] pilon *m*
~GROUND - [mining] abattage *m*
~IN - [auto etc] rodage *m*, dressage *m*
~-IN HOLE - [mining] mine *f* d'empiétage
~IN PERIOD - [auto] période *f* de rodage
~JOINT - [build etc] (the arrangement of joints between parts so that they are not in line with each other) alternance *f* des joints
~LINK - [mech] liaison *f* mécanique de rupture
~LOAD - [mech] charge *f* de rupture
~OF COUPLINGS - [railw] rupture *f* d'attelages
~OF MERES - [bot] (masses of algae appearing rapidly in small fresh water meres) apparition *f* d'algues
~OUT POINT - [telecomm] point *m* de dérivation, embranchement *m*
~POINT - [gen & phys] limite *f* critique de resistance, point *m* de rupture
[el] contact *m* fixe
[constr] (in tar) point *m* de fragilité.
[telecomm] (a sudden distortion of sound and noise making transmission faulty) point *m* critique
~SCUTCHER - [text] broyeuse-teilleuse *f*
~STRENGTH - [mech] (stress required to break materials) résistance *f* à la rupture
~STRESS - s. breaking strength
~TEST - [mech] essai *m* de rupture
~THE FLAX - [text] broyage *m*, teillage *m*
~THROUGH - [mech] (through drilling) pérforation *f*, percée *f*
~TOOL - [tool] outil *m* pour le démontage
~UP - [gen] fragmentation *f*, morcellement *m*
[of ice, e.g. in rivers) débâcle
~UP OF THE FIRE - [th eng] tombée *f* du feu
BREAKOFF - [min] séparation *f*, éloignement *m* de la terre
[mining] recoupe *f*, galerie *f* de ventilation
BREAKOUT - [metall] (from a mould) coulage *m*; (from a blast furnace) chat *m*, pissée *m*

BREAKSTAFF - [impl] branloire *f*
BREAKTHROUGH - [mining] boyau *m* de mine
BREAKWATER - [constr] (barrier resisting the force of the waves) brise-lames *m*, môle *m*, jetée *f*
BREAM, to - [naut] flamber, brusquer
BREAM - [zool] brème *f*
BREAST - [anat] sein *m*, poitrine *f*, mamelle *f*, poitrail *m*
[build](the section of the wall between the window and the floor) allège *f* de fenêtre, mur *m* à hauteur d'appui
[metall] (of a furnace) ventre *m* de haut-fourneau
[carp] rambarde *f*
[mining] (the coal face being worked on) front *m* de taille, front *m* d'abattage, taille *f*, chambre *f*
[text] (front roller of a carder) premier cylindre *m*
[agric] (of a plough) versoir *m*
~-AND-PILLAR - [mining] exploitation *f* par poche et pilier
~-BAND - [naut] traversier *m*
[gen] (in a harness) tablier *m*
~BEAM - [text] (a guide at the front of a loom) poitrinière *f*
~BEAM BAR - [text] axe *m* de poitrinière
~BEAM LEVER - [text] levier *m* de poitrinière
~BEAM PLATE - [text] plaque *f* de poitrinière
~BEAM REGULATOR - [text] régulateur *m* de poitrinière
~BEAM WHEEL - [text] roue *f* de poitrinière
~BOARD - [text] planchette *f* de protection
~BONE - [anat] sternum *m*, bréchet *m* (d'oiseau)
~BORER - [tool] vilebrequin *m*
~COLLAR - [gen] (in a harness) tablier *m*
~DOOR - [metall] porte *f* de préchauffage
~DRILL - [tool] vilebrequin *m*, chignole *f*, porte-foret *m* à conscience
~FAST - [naut] traversier *m*
~HARNESS - [impl] bricole *m*
~HOLE - [metall] (in a furnace for the removal of slags) porte *f* de décrassage, tour *m* à laitier, trou *m* à crasse
[mining] (in a mine gallery) coup *m* de mine vers le toit
~-HOOK - [naut] guirlande *f*
~LINING - [build] (the panelling of the "breast") parement *m* de parapet, revêtement *m* du parapet
~MOULDING - [build] moulure *f* de parapet
~-PLATE - [gen] plastron *m*, poitrail *m*
[mech] (of a breast drill) plaque *f* conscience
~PROTHESIS - [rubber ind] (artificial breast) faux sein *m*, sein *m* artificiel
~-PUMP - [rubber ind] tire-lait *m*
~ROLL - [paper man] cylindre *m* antérieur
~ROLLER - [text](in a calender, a roller placed in front of another at the same level) cylindre *m* frontal
~-ROPE - s. breast-band
~STOPE - [mining] chantier *m* d'abattage en gradins gradin *m* droit, chantier *m* d'avancement frontal
~STOPING - [mining] (frontal excavation of ore from a vein, reef or lode) exploitation *f* en gradins abattage *m* de front
~SUMMER - [build] poitrail *m*, linteau *m* de baie, sommier *m*
~TRANSMITTER - [radio] microphone *m* thoracique
~WALL - [build] (breast high wall) allège *f* de fenê-

tre, mur *m* à hauteur d'appui, mur *m* de soutènement
BREAST WHEEL - [hydr] (obsolete type of water wheel) roue *f* de côté
~-WORK - [build] parapet *m*, garde-corps *m*
[naut] rambarde *f*
BREASTING - [hydr] coursier *m*
BREATH - [gen] haleine *f*, souffle *m*, respiration *f*
[of wind] souffle *m*
BREATHE, to - [physiol] respirer
BREATHER - [build etc] cheminée *f* d'aération
[mech] (of a motor) reniflard *m*, évent *m*
[oil ind] soupape *f* de respiration
~CAP - [auto] capouchon *m*, chapeau *m* de reniflard
[oil ind] goulotte *f* de décharge
~PIPE - [mech] évent *m*, tuyau *m* d'aspiration
BREATHING - [physiol & gen] respiration *f*, souffle *m*, ventilation *f*
[el] (in a transformer core) ventilation *f*, canal *m* de ventilation
[plast ind] (the period of a moulding cycle in which gases and water vapour escape from a moulding) dégazage *m*
~APPARATUS - [med] appareil *m* respiratoire
~CYCLE - [ind chem] cycle *m* de dégazage
~FABRIC - [text] (cotton backed vinyl material with a degree of porosity) tissu *m* aéré
~HOSE - [med] (in oxygen apparatuses) tube *m* respiratoire
~ROOT - [bot] (in mangroves and other plants growing in mud) racine *f* aérienne
~STROKE - s. breathing cycle
BRECCIA - [geol] (coarse clastic rock, mostly consisting of fragments of preexisting rocks) brèche *f*
BRECCIATED - [gen & min] brecciolaire
BRECCIATION - [geol] structure *f* anguleuse, structure *f* brecciolaire
BREECH - [anat] derrière *m*, fondement *m*, culasse *f*
~ACTION - [firearms] (of a gun) mécanisme *m* de culasse
~BAND - [impl] courroie *f* de reculement
~DELIVERY - [med] accouchement *m* avec présentation par le siège
~EXTRACTION - [med] s. breech delivery
~-LOADER - [firearms] arme *f* se chargeant par la culasse
~-LOADING - [firearms] chargement *m* par la culasse
~PLUG - [firearms] (of a gun) obturateur *m*
~WELL - [firearms] (of a gun) logement *m* de l'obturateur
BREECHBLOCK - [firearms] (of a gun) obturateur *m*
BREECHES - [text] culotte *f*, pantalons *mpl*
~PIECE - [gas ind] culotte *f*
~PIPE - s. breeches piece
BREECHING - [th eng] (of a boiler) culottes *fpl* de la cheminée
~PIECE - [plumb] raccord *m* à culotte
~STRAP - [leather ind] (of a harness) courroie *f* de reculement, avaloire *f*
BREED, to - [gen] engendrer, procréer, élever, faire saillir, faire couvrir, couver
BREED - [gen] race *f*
BREEDER - [gen] éleveur *m*
[nucl] pile *f* couveuse, pile *f* surgénératrice
~BLANKET - [nucl] enveloppe *f* régénératrice, milieu *m* fertile
BREEDER END - [nucl] (tube of pure uranium at both ends of tubular enriched uranium elements) élément *m* d'uranium pur
~REACTOR - [nucl] réacteur *m* autorégénérateur, réacteur *m* surgénérateur
BREEDING - [gen] reproduction *f*, multiplication *f*
[agric] (of an animal used for breeding purposes) reproducteur
[agric] (agricultural selection) sélection *f*
[nucl] (the process of generating nuclear fuel, as uranium from thorium etc. by the absorption of neutrons) régénération *f*, surgénération *f*
~COCOON - [text] cocon *m* pour l'élevage
~GAIN - [nucl] (the breeding ratio minus one) gain *m* de surgénération
~RATIO - [nucl] rapport *m* de surgénération
BREEZE - [met] (a light wind, between 2 and 6 in the Beaufort scale) brise *f*
[min] (fuel & build material) (small broken coke, or other material, e.g. furnace clinker) petit coke *m*, grésillon *m*
~BLOCKS - [meas] parpaings *mpl*
~CONCRETE - [build] (concrete made of 3 parts coke breeze, 1 part sand and 1 part cement) béton *m* de scories
BREEZING - [glass man] lit *m*, couche *f*
[cinema] (a projected image which is not clear) image *f* floue
B-REGISTER - [comput] (machine register in which a number can change an address) registre *m* de base, registre *m* index
BREGMA - [med] bregma *m*
BREMEN BLUE - [paint] (greenish-blue pigment consisting chiefly of copper hydroxide) bleu *m* de Brème
~GREEN - [paint] (pigment consisting of green copper carbonate) vert *m* minéral
BREMSSTRAHLUNG - [radiat] (German term generally accepted in English. The production of electromagnetic radiation by the acceleration of a charged particle) bremsstralhung *m*, rayonnement *m* de freinage
BREUNNERITE - [min] (used for the manufacture of bricks) breunerite *f* (ou breunnerite)
BREVE - [mus] brêve *f*
BREW, to - [brew] brasser (la bière)
BREW - [brew] brassage *m* de la bière
~-HOUSE - [brew ind] salle *f* de brassage
~KETTLE - [brew ind] bouilleur *m*
BREWER - [brew ind] brasseur *m*
BREWER'S BARLEY - [brew ind] orge *m* pour la bière
~DRAFT - [brew ind] drêche *f*
~GRAINS - [brew ind] résidus *mpl* de fabrication
~YEAST - [brew ind] levure *f* de bière
BREWERY - [brew ind] (factory for the preparation of beer) brasserie *f*
BREWING - [brew ind] (the series of processes for the making of beer) brassage *m*
~BARLEY - s. brewer's barley
~INDUSTRY - [brew ind] brasserie *f*
~LIQUOR - [ind chem] eau *f* pour la brasserie
~ROOM - [brew ind] salle *f* de brassage
BREWSTER PROCESS - [ind chem] (a process for extracting acetic acid from the distillate obtained by the destructive distillation of wood) procédé *m* de Brewster

BRIAR - [bot] (used for the manufacture of smoking pipes) bruyère *f* (arborescente)
BRICK, to - [build] briqueter, murer, maçonner
BRICK - [build] brique *f*
[comput] (a programme with a number of sequences of operations which occur frequently) sous-programme *m*
~ARCH - [build] voûte *f* en briques
~-BUILT - [build] construit en briques
~CLAY - [geol] terre *f* à briques, argile *f* à briques
~COURSE ON EDGE - [build] assise *f* de champ, rang *m* de briques posées verticalement
~DRIER - [constr] four *m* à briques
~DUST - [build] poussière *f* de brique
~EARTH - [geol] terre *f* à briques, argile *f* à briques
~FACING - [build] revêtement *m* de briques, parement *m* en briques
~FLOOR - [build] pavage *m* en briques
~FLOORING - s. brick floor
~FORK - [impl] (fitted to a lift truck) fourche *f* à briques pour élévateur
~FURNACE - s. brick kiln
~HAMMER - [impl] marteau *m* de maçon
~-KILN - [constr] four *m* à briques
~-LAYER - [build] maçon *m* en briques, briqueteur *m*
~-LAYER'S ITCH - [med] gale *f* du ciment
~-LAYING - [build] maçonnerie *f* en briques, briquetage *m*, pose *f* des briques
~MOULDING MACHINE - [mech] machine *f* à mouler les briques
~NOGGING - [build] maçonnerie *f* de briques avec entretoises
~-ON-EDGE COPING - [build] couronnement *m* en briques posées verticalement
~-ON-END COURSE - [build] assise *f* de champ, rang *m* de briques posées verticalement
~PAVEMENT - [build] pavage *m* en briques
~PRESS - [mech] presse *f* à briques
~RED - [paint] rouge *m* brique
~ROAD - [constr] pavage *m* en briques
~TROWEL - [impl] truelle *f* à briques
~VENEER - [biild] révêtement *m* en briques
~WALL - [build] mur *m* en briques
~-WORK - [build] construction *f* en briques, maçonnerie *f* de briques, briquetage *m*
BRICKYARD - [build] briqueterie *f*
BRIDGE, to - [constr] jeter un pont, construire un pont, ponter
[el] monter en pont
[mining] coffrer le toit *m*
BRIDGE - [arch] pont *m*
[el] (arrangement of electrical elements in an equivalent quadrilateral form, one of the diagonals being connected to a source of current and the other to a measuring instrument) pont *m* (de mesure)
[th eng] (of boilers) autel *m* de chaudière, pont *m*
[mech] (of cranes) pont *m*, portique *m*
[mining] croisement *m* de courants d'air
[metall] autel *m*
[naut] passerelle *f*
[dent] bridge *m*
[mus] chevalet *m* (de violon)
~ARM - [build] bras *m* (de pont)
~ARMS - [el] branches *fpl* d'un pont
BRIDGE BALANCE - [build] équilibre *m* d'un pont
~BEARING - [build] appui *m* d'un pont
~-BUILDING - [build] construction *f* de ponts
~CLEARANCE - [bridge build] tirant *m* d'air sous un pont
~CONTACT - [el] contact *m* en pont
~CRANE - [mech] grue *f* à portique, grue *f* à pont, pont-grue *m*
~DECK - [naut] passerelle *f* découverte
~DUPLEX SYSTEM - [radio] réseau *m* duplex équilibré en pont
~FEEDBACK - [radio] (feedback in which the returned energy is independent of the impedance of the load) réaction *f* équilibrée
~FUSE - [el] fusible *m* en pont
~GIRDER - [constr] poutre *f* de pont
~GUIDE - [metall] pont *m*
~-HEAD - [gen] tête *f* de pont
~HOUSE - [naut] roufle *m*, rouf *m*
~KEY - [constr] clef *m* de voûte
~INTERSECTION - [roads] croisement *m* à saut de mouton
~MILL - [mech] fraiseuse *f* à portique
~NETWORK - [el] réseau *m* en pont
~OF A VALVE - [mech] support *m* de soupape
~OSCILLATOR - [el] oscillateur *m* à pont
~PIECE - [el] pont *m* polaire (d'accumulateur)
~PIER - [constr] pile *f* de pont
~PILE - [constr] palée *f*
~PIPE - [gas ind] chevalement *m*
~PLATE - [railw] plaque *f* de jonction
~RECEIVER - [radio] récepteur *m* à pont
~RECTIFIER - [el] (full-wave rectifier connected as in a bridge circuit) redresseur *m* en pont
~SEAM WELD - [mech] soudure *f* continue à couvre-joint
~SIGNAL BOX - [railw] cabine *f* de signalisation sur portique
~SLEEPER - [constr] traverse *f* de pont
~SWITCH - [text] interrupteur *m* du déclencheur
~TEST - [constr] essai *m* d'un pont
~TRANSFORMER - [el] (a single transformer with three windings) transformateur *m* différentiel
~TRANSITION - [radio] transition *f* série-parallèle par pont
~TRUSS - [build] (structural framework supporting a bridge) ferme *m*
~-TYPE CAPACITANCE ANALYZER - [instr] (instrument for measuring capacity by means of a bridge circuit) capacimètre *m* à pont
~WALL - [glass man] pont *m*
~WITH WIND TIES - [constr] pont *m* à contreventement
BRIDGED T FILTER - [el] (filter section consisting of a T-network, used for phase compensation) filtre *m* en T
~T NETWORK - [radio] (a T network with the series impedances bridged by a fourth impedance) réseau *m* en T
BRIDGING - [el] shuntage *m*
[metall] (of furnaces) couronnement *m*, accrochage *m* au cubilot
[photo] (in photogrammetry) triangulation *f* aérienne
~AMPLIFIER - [el] amplificateur *m* de pont
~COIL - [el] bobine *f* de dérivation
~CONNECTION - [radio] (a parallel connection) pont

m

BRIDGING FIBRILS - [bot] fibrilles *f*pl
~GAIN - [radio] gain *m* résultant d'un montage en pont
~JACK - [telecomm] jack *m* sans contact de rupture
~LOSS - [radio] perte *f* provenant d'un montage en pont
~PIECE - [carp] (short piece of timber inserted to reduce lateral distortion) traversière *f*
BRIDGMAN SEAL - [hydr] (a type of seal for the piston of hydraulic actuators) joint *m* étanche Bridgman
BRIDLE - [impl] bride *f*
[mech] bride *f*
[anat] frein *m*, filet *m* (de la langue)
[horol] bride *f*
~CABLE - [telecomm] câble *m* de raccordement entre lignes aériennes et souterraines
~PATH - [gen] piste *f* cavalière, allée *f* cavalière
~ROAD - [constr] route *f* cavalière, chemin *m* muletier
~WIRE - [telecomm] s.bridle cable
BRIEF, to - [leg] (to retain the services of a barrister) confier une cause à un avocat
[to prepare a report for counsel] instruire
BRIEF - [leg] (a summary of the facts prepared for the counsel) dossier *m* (d'une procédure), instructions *f*pl
[adj] bref, concis, succint
BRIER - [bot] bruyère *f* arborescente
BRIG - [naut] brick *m*
~-RIGGED - [naut] grée en brick
~SCHOONER - [naut] brick-goélette *f*
BRIGANTINE - [naut] brigantin *m*
BRIGHT - [adj] clair, lumineux, brillant, vif, poli
~ACID DIPPING - [metall] brillantage *m* acide
~ANNEALING - [metall] recuit *m* brillant
~CHERRY-RED HEAT - [metall] chaleur *f* rouge clair
~COAL - [min] charbon *m* brillant
~COLD-ROLLED - [metall] laminé au blanc
~COPPER PLATING - [metall] cuivrage *m* brillant
~DIP - [metall] (solution designed to produce a bright surface on a metal object) solution *f* de polissage, solution *f* de brillantage
~DRAWN - [metall] étiré brillant
~EMITTER - [electron] s.bright-emitting cathode
~-EMITTING CATHODE - [electron] (a cathode generally in the form of a tungsten filament) cathode *f* à filament chauffé visible
~ENAMEL PAPER - [paper man] papier *m* couché
~FINISH - [metall] brillantage *m*
~HARDENING - [metall] trempe *f* brillante
~INTERVAL - [met] éclaircie *f*
~LIGHTS - [cinema] (said of a picture with vast areas white) blancs *m*, playes *f*pl lumineuses
~NICKEL - [metall] solution *f* pour nickelage brillant
~NICKEL PLATING - [metall] nickelage *m* brillant
~PLATING - [metall] revêtement *m* (électrolytique) brillant
~RADIATION - [heat] rayonnement *m* lumineux
~SILVER PLATING - [metall] argenture *f* brillante
~STEEL - [metall] acier *m* bruni
~STOCK - [oil ind] (high viscosity lubricating oil obtained from the residues of petroleum refining) huile *f* lubrifiante à grande viscosité
~THREAD - [text] fil *m* brillanté
BRIGHTENER - [electro plating] (additive to an electrolyte to give more brilliant deposits) produit *m* à polir, brillanteur *m*
BRIGHTENING - [gen & metall] brillantage *m*
~AGENT - [text] agent *m* d'azurage
~PULSE - [radar] (square pulse of a duration which is equal to the swept time of the time base) impulsion *f* de grande intensité
~SCREEN - [photo] réflecteur *m*
BRIGHTNESS - [gen & telev] éclat *m*, luminosité *f*, brillant *m*, brillance *f*
[text] aspect *m* brillant
~CONTROL - [telev radar] (device designed to control the brightness of a television or radar picture) réglage *m* de luminosité
~DIFFERENCE - [telev] contrastes *m*pl des brillance
~METER - [instr] luxmètre *m*
~MgO - [paints] (the apparent brilliance of a white pigment compared to pure Magnesium oxide) blancheur *f*
~OF IMAGE - [telev] luminosité *f* de l'image
~OF THE SPOT - [opt] luminosité *f* du spot
BRILLIANCE - [opt] brillance *f*
[telev] luminosité *f*
[radar] (the luminosity of a cathode ray tube screen) luminosité *f*
~CONTROL - s.brightness control
BRILLIANCY - [gen] brillant *m*, éclat *m*, brillance *f*
BRILLIANT - [gen] brillant, éclatant
[print] corps trois et demi
~FINDER - [cinema] (view finder of glass of high clarity) viseur *m* clair
BRILLIANTINE YARN - [text] fil *m* brillanté
BRIM - [gen] bord *m*
BRIMFUL - [gen] plein jusqu'au bord
BRIMSTONE - [min] (old name for sulphur) soufre *m* natif
~LIVER - [med] foie *m* de soufre
~YELLOW - [text] jaune soufre
BRINDLED - [gen & zool] tacheté, tavelé, bringé
~BRICKS - [build] briques *f*pl striées
BRINE - [chem] (solution of common salt in water) eau *f* salée, saumure *f*
~PIT - [geol] saline *f*
~PUMP - [hydr] pompe *f* à saumure
~SPRING - [geol] source *f* salée, source *f* saline
~TANK - [brew ind] générateur *m* de glace
~WELL - [hydr] puits *m* d'extraction de saumure
BRINELL HARDNESS - [phys] (a value for the hardness of a material obtained by measuring the area of the indentation made by a hard steel ball loaded to a standard degree) dureté *f* Brinell
~MACHINE - [mech] machine *f* pour l'essai de dureté Brinell
~TEST - [mech] (system of hardness testing in which a ball of hardened steel or tungsten carbide is used to indent the test piece) essai *m* de dureté Brinell
BRINELLING - [mech] empreinte *f* produite par une bille
BRING, to - [gen] porter, apporter, conduire, amener
~ABOUT, to - [gen] causer, déterminer, occasioner provoquer
~-BACK, to - [mining] exploiter en rabattant
~BACK TO ZERO - [instr] remettre à zéro
~DOWN, to - [gen & aero] abattre, faire descendre
~FORWARD, to - [fin] reporter

BRING INTO PHASE, to - [el] mettre en phase, synchroniser
~INTO SERVICE, to - [telecomm] mettre en service
BRINY - [adj] saumâtre, salé
BRIQUETTE - [min] briquette *f*, aggloméré *m*
~CEMENT - [build] agglomérant *m* pour briquettes
BRIQUETTED COAL - s. briquette
BRIQUETTING - [min] (of coke etc) agglomération *f*, fabrication *f* de briquettes
~MACHINE - [build] (a machine for forming powdered, granular or fibrous materials into blocks by mechanical pressure) installation *f* d'agglomération
BRISTLE - [bot] (very stiff hair) soie *f*
[el] (a relatively thick monofilament) filament *m*
[plast ind] crin *m*
~BRUSH - [impl] brosse *f* en soies naturelles
~HAIR - [text] jarre *m*
~STENCIL BRUSH - [impl] pinceau *m* à soies courtes
BRISTOL BOARD - [paper man] (cardboard of fine quality) carton *m* Bristol, bristol *m*
~DIAMONDS - [min] (small crystals of quartz occurring in the district of Bristol); diamants *mpl* de Bristol
BRITANNIA JOINT - [metall] joint *m* soudé sans manchon
~METAL - [metall] (an alloy used especially for tableware) métal *m* anglais; alliage *m* Britannia
BRITISH ASSOCIATION SCREW THREAD - [B.A.S.T.] filet *m* anglais
~CANDLE - [meas] (in photometry) bougie *f* anglaise
~GUM - [ind chem] "british gum" *f*, (fécule de manioc,ou farine de maïs, utilisée comme apprêt ou comme épaississant)
~THERMAL UNIT - [B.T.U.] (the quality of heat absorbed by one pound of water in rising one degree Fahrenheit) unité *f* anglaise de quantité de chaleur (1 B.T.U.= 0,252 kilo calorie)
BRITTLE - [gen & mech] (easily broken or shattered) fragile, cassant
~DIABETES - [med] diabète *m* instable
~FAILURE - [metall] résilience *f*
~FRACTURE - [metall] rupture *f* par fragilité
~LACQUER - [metall] vernis *m* craquelant
~POINT - [meas] point *m* de fragilité
~SILVER ORE - [min] (stephanite: a natural silver antimony sulphide and an ore of silver) stéphanite *f*, argent *m* fragile, psaturose *m*
~TEMPERATURE - [plast ind] température *f* de fragilité
BRITTLENESS - [gen] (the quality of being easily fractured) fragilité *f*
BROACH, to - [mech] (to enlarge or internally finish a hole by a special tool) aléser, mandriner, équarrir
[naut] venir en travers
BROACH - [mech] (tool for enlarging a hole to exact dimensions) alésoir *m*, broche *f*
[arch] (a pyramid at the projecting corner of square tower) flêche *f*, aiguille *f*
~GRINDER - [mech] machine *f* à affûter les broches
~SHARPENING MACHINE - [mech] machine *f* à affûter les broches
BROACHING - [mech] alésage *m*, mandrinage *m*
[mining] battage *m* au large
~BIT - [tool] alésoir *m*
BROACHING MACHINE - [mech] machine *f* à brocher
BROAD - [adj] large, ample, vaste
[cinema] (box of a rectangular shape with a row of incandescent lamps) banc *m* d'éclairage, herse *f*
[cinema-US] (incandescent lamps arranged as a unit in a studio) banc *m* de lampes dans un studio
~-AXE - [impl] hache *f* à large fer
~BEAM ABSORPTION - [radiat] (absorption measured under conditions in which scattered radiation is not excluded from the measuring mechanism) absorption *f* à large faisceau
~-BEAM HEADLAMP - [el] projecteur *m* grand angle
~DIMENSION - [electron] (the dimension of a cross section of a waveguide determining the critical frequency) dimension *f* critique, côte *f* large
~-GAUGE LINE - [railw] voie *f* à grand écartement
~IRRIGATION - irrigation *f* superficielle
~LIGHT - [light] (a diffuse light in a studio) lumière *f* diffuse
~PULSE - [telev] (one of the pulses, usually repetitive at twice the line frequency, forming the frame-synchronizing pulse of a television wave-form) impulsion *f* de grande durée
~SPECTRUM ANTIBIOTIC - [med] antibiotique *m* à large spectre
~TUNING - [radio] accord *m* étalé
~WAVES - [radio] ondes *fpl* amorties
BROADBAND - [telev] large bande *f*
~AMPLIFIER - [radio] (radio-frequency amplifier characterized by a very large pass-bandwidth) amplificateur *m* à large bande
BROADCAST, to - [radio]transmettre radiodiffuser
[agric] semer à la volée
BROADCAST - [radio] radiodiffusion *f*, émission *f* en l'air
[agric] semaille *f* à la volée
~CHANNEL - [radio] bande *f* d'émission
~FREQUENCY MONITOR - [radio] stabilisateur *m* à haute fréquence
~NETWORK - [radio] réseau *m* de radiodiffusion
~RECEIVER - [radio] récepteur *m* radio
~SEEDER - [agric] semeuse *f* à la volée
~SOWING - [agric] semaille *f* à la volée
~SYSTEM - [telecomm] (a system in telegraphy whereby a central station can communicate with a number of stations at the same time) trasmission *f* télégraphique à destinataires multiples
~TELEGRAPHY - [telecomm] transmission *f* télégraphique à destinataires multiples
~TRANSMITTER - [radio] émetteur *m* radio
BROADCASTER - [radio] appareil *m* émetteur
BROADCASTING - [radio] radio-diffusion *f*, émission *f* en l'air
~OF FERTILIZERS - [agric] distribution *f* des engrais (par dispositif placé sur le semoir)
~REPEATER - [radio] émetteur-relais *m* de radiodiffusion
BROADCLOTH - [text] drap *m* fin
BROADEN, to - [gen] élargir, étendre
BROADLEAF TREE - [bot] arbre *m* à larges feuilles, arbre *m* latifolié
BROADLOOM - [radar] (airborne magnetron jammer) brouilleur *m* anti-radar aéroporté
BROADSIDE - [naut] flanc *m*, travers *m*
[light] réflecteur *m* diffuseur, projecteur *m* grand angle, projecteur *m* diffuseur

[advert] (large folder) dépliant *m*
[print] (broadsheet) in-plano *m*, feuille *f* imprimée
BROADSIDE ARRAY - [radio] (a system of two elements fed with equal currents) réseau *m* d'antenne à grande ouverture
BROCADE - [text] brocart *m*
~WARP - [text] chaine *f* de brocart
BROCADED ALTAR CLOTH - [text] nappe *f* d'autel brochée
~GAUZE - [text] gaze *f* brochée
BROCADES FOR CHURCH HANGINGS - [text] brocarts *mpl* pour tentures d'église
BROCADING - [text] fabrication *f* du brocart
BROCATELLE - [text] brocatelle *f*
[constr] brocatelle *f*
BROCHANTITE - [min] (natural basic copper sulphate: a minor ore of copper) brochantite *f*, waringtonite *f*
BROCHE FABRIC - [text] tissu *m* broché
~FABRIC FOR UPHOLSTERY - [text] tissu *m* broché pour tapisserie d'ameublement
BROCHURE - [print] brochure *f*, plaquette *f*
BROGGERITE - [min] (uranium mineral with a large proportion of thorium) broggérite *f*, thoruraninite *f*
BROILER - [zool] poulet *m* à rôtir
BROKE - [paper man] chûtes *fpl* de papier
BROKEN - [gen] brisé, cassé, rompu
[mech] en panne, en avarie
~-BACKED - [naut] arqué, cassé
~BARLEY - [brew ind] brisures *fpl* d'orge
~BREAST - [med] abscès *m* de la glande mammaire
~CLOUD - [met] (clouds occurring in separate masses) nuages *m* pl en masses isolées
~COAL - [min] anthracite *f* de 60 à à 100 mm
~COKE - [min] coke *m* cassé
~DOWN - [gen] avarié, défectueux, en panne, raté
~END - [text] (fabric defect) fil *m* rompu, fente *f*
~FOLD - [geol] pli *m* faillé
~GRAINS - s.broken barley
~HEMP - [text] chanvre *m* teillé, chanvre *m* broyé
~JOINTS - [build] joints *mpl* alternés
~LINE - [gen] ligne *f* brisée
~LINE PROFILE - [rubber ind] sculpture *f* en ligne brisée
~LINE TREAD - s. broken line profile
~ORE - [min] minerai *m* abattu
~PASS - [text] rentrage *m* à pointe et à retour
~RANGE - [build] opus *m* incertum
~STONE - [constr] pierre *f* cassée
~STONE BALLAST - [railw] ballast *m* de pierre cassée
~TWILL - [text] sergé *m* brisé; (of a cloth) étoffe *f* satinée
~WEATHER - [met] temps *m* incertain
~WEAVE - [text] armure *f* satin
~WEFT FINDER - [text] dispositif *m* de recherche de la duite
~WHITE - [paint] (white coating in which the pure whiteness has been slightly modified by the addition of a small quantity of another pigment) blanc *m* cassé
~WIND - [vet] pousse *f*
~WIRE LOCKING DEVICE - [railw] appareil *m* contrôleur de rupture de fil
BROKENWINDED - [vet] poussif
BROKER - [comm] courtier *m*, agent *m* de change, commissaire-priseur *m*
BROKERAGE - [comm] courtage *m*
BROKES - [text] laine *f* de qualité inférieure
BROMAL - [chem] bromal *m*
BROMATE - [chem] bromate *m*
BROMATOLOGY - [med] bromatologie *f*
BROMATOXISM - [med] intoxication *f* alimentaire
BROMCRESOL PURPLE - [chem] pourpre *m* de bromocrésol
BROMHIDROSIS - [med] bromhydrose *f*
BROMIC ACID - [chem] acide *m* bromique
BROMIDE - [chem] (bromides are salts of hydrobromic acid, important in medicine and photography) bromure acid *m*
~OF SILVER - [chem] bromure *m* d'argent
~PAPER - [photo] papier *m* au bromure
~PENCIL - [impl] crayon *f* pour retouche (des épreuves sur papier au bromure)
~PRINT - [photo] épreuve *f* sur papier au bromure
~PROCESS - [photo] (a silver bromide emulsion on paper for printing or enlarging) procédé *m* au (gélatino) bromure d'argent
BROMIDROSIS - [med] bromidrose *f*
BROMINATE, to - [chem] bromer
BROMINATION - [chem] (the replacement of bromine in organic compounds) bromuration *f*, bromation *f*
BROMINE - [chem] (symbol Br; atomic number 35; a non metallic liquid element) brome *m*
~ACNE - [med] acné *f* bromique
~NUMBER - [chem] indice *m* de brome
~PENTAFLUORIDE - [chem] (a synthesis intermediate and component of liquid rocket fuels) pentafluorure *m* de brome
~SOLIDIFICATION - [chem] solidification *f* du brome, agglomération *f* du brome
~THERMAL VALUE - [chem] (the amount of temperature rise observed when one c.c. of bromide is quickly stirred into a solution of one gramme of vegetable oil in 10 c.c. of chloroform) équivalent *m* thermique
~WATER - [chem] (a solution of bromine in water) eau *f* de brome
BROMISM - [med] bromisme *m*
BROMOACETONE - [chem] (a tear gas and synthetic intermediate) bromoacétone *f*
BROMOAURIC ACID - [chem] acide *m* aurobromique
BROMOBENZENE - [chem] (a solvent, synthesis intermediate and component for motor fuels) bromobenzène *m*, bromure *m* de phényle
BROMOBENZYL CYANIDE - [chem] cyanure *m* de bromobenzyle
BROMOCHLOROETHANE - [chem] (a synthesis intermediate and filter for fire extinguishers) bromochloroéthane
BROMOCHLOPHENOL - [chem] bromochlorophénol *m*
BROMOCRESOL PURPLE - [chem] pourpre *m* de bromocrésol
BROMOCYANIDE - [chem] bromocyanure *m*
BROMODERMA - [med] bromodermie *f*
BROMODIETHYLACETYLUREA - [chem] (hypnotic used in medicine) bromodiéthylacétylurée *f*, adaline *f*
BROMOETHYLENE - [chem] bromoéthylène *m*
BROMOFORM - [chem] (a sedative used in medicine and a synthesis intermediate) bromoforme *m*,

tribromométhane *m*
BROMOHYDRATE - [chem] bromohydrate *m*
BROMOPHENOL BLUE - [chem] (an acid-base indicator) bleu *m* de bromophénol
BROMOSTYROL- [chem] (a perfumery intermediate) bromostyrol *m*
BROMOTHYMOL BLUE - [ind chem] (indicator for acid bases) bleu *m* de bromothymol
BROMYRITE - [min] (an ore of silver: natural silver bromide) bromyrite *f*, bromargyrite *f*
BRONCHADENITIS - [med] adéno-bronchite *f*
BRONCHIAL - [med] bronchial, bronchique
~ASTHMA - [med] asthme *m* bronchique
~TREE - [anat] arbre *m* bronchique
BRONCHILOQUY - [med] bronchophonie *f*
BRONCHIOLE - [med] bronchiole *f*
BRONCHIOLITIS - [med] bronchiolite *f*
BRONCHITIS - [med] bronchite *f*
BRONCHOGENIC - [med] bronchogène
BRONCHIOGRAM - [radiat] bronchogramme *m*
BRONCHOGRAPHY - [med] (the radiological examination of the bronchial system) bronchographie *f*
BRONCHOLITH - [med] broncholithe *m*, calcul *m* des bronches
BRONCHOPLEGIA - [med] bronchoplégie *f*
BRONCHOPNEUMONIA - [med] broncho-pneumonie *f*
BRONCHOSCOPE - [instr] bronchoscope *m*
BRONCHOSCOPY - [med] bronchoscopie *f*
BRONZE - [metall] (a general term for alloys of copper and tin in various proportions) bronze *m*
~BUSHING - [mech] bague *f* de bronze
~HAMMER - [impl] marteau *m* à panne de bronze
~-LINED CYLINDER - [mech] (a working cylinder e.g.of a hydraulic press, fitted with a bronze lining) cylindre *m* chemisé en bronze
~PAINT - [paint] peinture *f* dorée, peinture *f* bronzée
~POWDER - [metall] bronzine *f*, poudre *f* de bronze
~DIABETES - [med] diabète *m* bronzé, cirrhose *f* pigmentaire diabétique
BRONZING - [paint] bronzage *m*
~MACHINE - [mech] machine *f* à bronzer
~MEDIUM - [paint] vernis *m* bronzant
~SHEET - [metall] toile *f* à bronzer
BRONZITE - [min] bronzite *f*
BROOD, to - [zool] couver
BROOD - [zool] (set of offsprings produced from the same batch of eggs, or from the same birth) couvée *f* volée *f*
~-HEN - [zool] couveuse *f*
~SOW - [zool] truie *f* mère
BROODER - [zool] couveuse *f* artificielle, éleveuse *f*
BROODING - [zool] incubation *f*, couvée *f*
~TIME - [tool] durée *f* d'incubation
BROODY - [zool] qui demande à couver
BROOKLET - [geogr] ruisselet *m*
BROOM - [impl] balai *m*
[bot] (a medicinal plant) genêt *m* (à balai)
~-CORN - [bot] millet *m*
~FIBRE - [text] fibre *f* de genêt
~-MILLET - [bot] sorgho *m*
BROTH - [gen] bouillon *m*
[chem] bouillon *m* de culture
BROUGHT FORWARD - [fin] reporté
BROW - [naut] jouttière *f* de hublot
[geol]front *m* de charriage
[mining] front *f* d'une nappe de charriage, galerie *f* inclinée, descenderie *f*
BROWN - [colour] brun, marron
~ACETATE - [chem] acétate *m* gris
~COAL - [min] lignite *f*
~CREPE - [rubber ind] (a variety of crepe rubber) crêpe *m* brun
~DRAWING - [text] laine *f* des hanches
~-EYE MOTH - [zool] noctuelle *f* potagère
~FAT TISSUE - [anat] glande *f* interscapulaire
~HAEMATITE - [min] hématite *f* brune, limonite *f*
~HOLLAND - [text] toile *f* écrue
~IRON - [chem] oxyde *m* ferrique
~ORE - [min] limonite *f*
~-OUT - (or BROWNOUT; US for dimming of lights) mise *f* en veilleuse, éclairage *m* réduit
~POWDER - [metall] zinc *m* en poudre
~-PRINT - [photo] tirage *m* sépia, épreuve *f* sépia
~ROOT - [bot] sclerotinia *f*
~ROOT DISEASE - [bot] (a plant disease) maladie *f* brune
~SALT - [chem] sulfate *m* de nickel cristallisé
~SALTS - [chem] résidus *mpl* d'évaporation de lessive de potasse
~SIENNA - [paint] terre *f* de Sienne
~SPAR - [min] ankérite *f*
~STONE - [plast ind] bioxyde *m* de manganèse
~TIPPED - [bot] à pointes brunes
BROWNIAN MOVEMENT - [phys] (rapid random movement imparted to particles in a solution by bombardment of the molecules of the liquid) mouvement *m* Brownien
BROWNING OF CONES - [brew ind] brunissement *m* des cônes
BROWNISH - [adj] brunâtre
BRUCE AERIAL - [ant] (an aerial shaped like an inverted V) antenne *f* de Bruce
~ANTENNA - s.bruce aerial
BRUCELLOSIS - [med] brucellose *f*
BRUCINE - [min] (naturally occurring alkaloid with intensely bitter taste, used for denaturing alcohol) brucine *f*
BRUCITE - [min] (natural magnesium hydroxide used as refractory material) brucite *f*, hydrophyllite *f*, shepardite *f*
BRUISE, to - [gen] meurtrir, contusionner, écraser
BRUISE - [gen] meurtrissure *f*, contusion *f*
BRUISING - [agric] dégâts *mpl* causés par le vent
BRUNSWICK BLACK - [paint] (term used for inferior black japan) vernis *m* à base de bitume
~BLUE - [paint] bleu *m* de Brunswick
BRUSH, to - [gen] brosser, frôler, affleurer
BRUSH - [gen] brosse *f*
[impl] pinceau *m*
[metall] brosse *f* à nettoyer
[opt] faisceau *m*
[el] (a part designed to conduct current from a fixed point to a moving part) balais *m*
~-AERATOR - [el] aérateur *m* à brosse, rouleau *m* de balai-brosse
~BINDER - [paint] (product used to cement the bristles of paint brushes unto the mounting) colle *f* pour pinceaux
~BURN - [med] écorchure *f*, abrasion *f*
~-COAT, to - [paint] appliquer au pinceau (à la brosse)
~COLLAR - [el] collier *m* porte-balai

BRUSH COMPARE CHECK - [comput] contrôle *m* de balai
~DAMPENER - [impl] mouilleur *m* pour pinceaux
~DINAMO - [el] dynamo *f* à balais
~DISCHARGE - [electron] (a silent luminous electric discharge) aigrette *f*
~DISPLACEMENT - [el] (the electrical angle defining the relative position of the brushes and the neutral plane) angle *m* de calage des balais
~-END - [el] extrémité *f* de balais
~GEAR - [el] porte-balais
~-HOLDER - [el] (part designed to hold and support a brush in a motor or generator) porte-balais *m*
~-HOLDER COLUMN - [el] pivot *m* de porte-balais
~-HOLDER STUD - [el] tige *f* de porte-balais
~LIFTING DEVICE - [el] dispositif *m* de levage des balais
~ON, to - [paint] (to apply a coating by means of a brush) appliquer au pinceau, enduire au pinceau
~ORE - [min] minerai *m* de fer en forme de stalactites
~PAINTING - [paint] peinture *f* à la brosse
~-PLATE - [el] frotteur *m*, lame *f* de contact
~PLATING - [metall] (plating by means of a brush loaded with electrolyte and arranged to surround the anode) revêtement *m* électrolytique au tampon
~-POLISHING - [paper man] lissage *m* à la brosse
~PRINTING - [text] impression *f* à la brosse
~-PROOF - [print] épreuve *f* à la brosse, morasse *f*
~RING - s. brush collar
~-ROCKER - [el] (an assemblage of parts on which the brush-holders are fixed relatively to each other so as to provide a circumferential displacement of the assembly) couronne *f* porte-balais, collier *m* porte-balais
~ROLL - [impl] (a cylindrical brush arranged to rotate on its longitudinal axis) balais *m* cylindrique
~SCRAPER - [impl] (a scraping device in the form of a stiff brush) racleur *m* à brosses
~SHIFT - [el] (in a commutator; the modification of the angular position of the brushes) décalage *m* des balais
~SPARKLING - [el] étincelle *f*
~SPREADING MACHINE - [paint] machine *f* à étendre la peinture à la brosse
~STILL - [ind chem] (a fractionating still) installation *f* de distillation à brosse fractionnante
~TRIMMER - [mech] machine *f* à tailler les balais
~-TYPE SPREADER ROLL - [impl] (a cylindrical brush used for spreading fluid material) rouleau *m* à peinture
~-WARE - [gen] brosserie *f*
~-WEAR - [el] (wear occurring in a brush) usure *f* du balais
~YOKE - [el] armature *f* de porte-balais
BRUSHABILITY - [paint] (quality in a liquid coating of being easily applied with a brush) brossabilité *f*, aptitude *f* à l'application au pinceau
BRUSHING - [gen etc] (designed to be applied with a brush) au pinceau, à la brosse
[mining] recoupage *m*
[agric] badigeonnage *m*
~LACQUER-[paint] (type of varnish compounded for application by brush as distinct from spraying) lacque *f* au pinceau
~MACHINE - [mech] machine *f* à brosser
~PAINT - [paint] (paints especially compounded for application with a brush) peinture *f* appicable à la brosse
BRUSHITE - [min] brushite *f*
BRUSHWOOD - [bot] brousailles *f*pl, brindilles *f*pl
BRUSSELS SPROUT - [bot] choux *m* de Bruxelles
BRUTE - [cinema] (colloq. - an arc light of 225 A) lampe *f* à arc (225 A)
BRUXISM - [med] grincement *m* des dents (dans le sommeil)
B.S. - [British Standard] norme *f* britannique
B SCOPE - [radar] (radar display) indicateur *m* radar type B
B.S.I. - [British Standard Institute] Institut *m* Britannique de Normalisation
B -STAGE - [plast ind] (transition stage through which a phenol-formaldehyde thermosetting resin passes) état *m* B
B-STAGE RESIN - [chem] (intermediate stage in the formation of a cured phenol formaldehyde resin, at which the resin softens but does not melt on heating and does not dissolve in alcohol and acetone) résine *f* à l'état B
B STATION - [nav] (the slave station of a hyperbolic navigation system) station B *f*
B SWITCHBOARD - [telecomm] tableau *m* pour la position B
B-TUBE - [comput] (cathode-ray storage tube) tube *m* de mémoire de base
BUBBLE, to - [el] (of a battery while being charged) bouillonner
[chem] bouillonner, barboter
BUBBLE - [phys] bulle *f*
[ind chem] bulle *f*, boursoufflement *m*, soufflure *f*
~BUCKET - [oil ind] (a bucket-shaped container to collect stones during drilling operations) seau *m*, godet *m*, seau *m* à pierres
~CAP - [nucl] coupelle *f* à bulles, calotte *f* de barbotage
~CAP WASHER - [oil ind] laveur *m* à barbotage
~CHAMBER - [phys] (vessel filled with transparent liquid so overheated that a moving ionizing particle causes violent boiling) chambre *f* à bulles
~DECK - [oil ind] plateau *m* à barbotage
~FERMENTATION - [brew ind] fermentation *f* à bouillonnement
~GAUGE - [phys] (small liquid-containing trap in a gas line for the measuring of the rate of gas flow by counting the number of gas bubbles through the liquid) débimètre *m* à bulles (de gaz)
~OVERVOLTAGE - [el] (overvoltage associated with visible evolution of gas) surtension *f* due aux bulles de gaz
~PLATE - [ind chem] (a tray or plate with bubble caps and overflow pipe in a distillation column) plateau *m* de colonne à distiller
[oil ind] plateau *m* à calottes de barbotage
~POINT - [oil ind] pression *f* de bulle
~PROTRACTOR - [instr] rapporteur *m* à niveau
~RAFT - [metall] (a device designed to give a visual demostration of the dislocation properties in metals) modèle *m* à bulles de savon
~SEXTANT - [instr] (a sextant in which a spirit bubble takes the part of an artificial horizon) sextant *m* à bulle
~TOWER - [oil ind] laveur *m* à barbotage
~TRAY - [ind chem] (perforated tray fitted with bubble caps and fixed in a bubble tray column and

containing a liquid through which the gas is bubbled) plateau *m* de colonne (à distiller)
BUBBLE TRAY COLUMN - [ind chem] (cylindrical structure containing bubble trays in which a gas is brought into intimate contact with a liquid) colonne *f* à plateaux de barbotage, laveur *m* à barbotage
~TUBE - [instr] niveau *m* à bulle
~-TYPE SEPARATION - [ind chem] séparation *f* par barbotage
~-TYPE WASHER - s. bubble tower
BUBBLER - [chem] barboteur *m*
BUBBLES - [rubber ind] (bubbles occurring in the rubber when the inner surface is dry) bulles *f*pl, "bubbles" *f*pl
BUBBLING - [ind chem] bouillonnement *m*, barbotage *m*, dégazage *m* d'un bain (métallurgie)
[chem] (in a fluidized bed) bullage *m*
[paint] formation *f* de bulles
[aero] formation *f* de tourbillons
~HOOD - [ind chem] cloche *f* de barbotage
~RALE - [med] râle *m* bulleux
~-THROUGH - [metall] (when a gas passes through a metal bath) barbotage *m*
BUBBLY - [paint] bulleux, plein de bulles
BUBONIC PLAGUE - [med] peste *f* bubonique
BUBONOCELE - [med] bubonocèle *f*, hernie *f* inguinopubienne
BUCCAL - [med] buccal
~CAVITY - [anat] cavité *f* buccale
BUCHMANN AND MEYER PATTERN - [el acoust] [also called Christmas-tree pattern; method of measuring the maximum lateral velocity of a recorded signal) méthode *f* du faisceau lumineux réfléchi
BUCK, to - [el] agir en opposition, réduire la tension
BUCK - [zool] daim *m*, chevreuil *m*, mâle *m*
[impl] chevalet *m*, chêvre *f*
[glass man] plaque *f* fixe
[auto] (a wooden model) maquette *f* en bois
~-RAKE - [impl] rateau *m*
~-SAW - [impl] scie *f* de long, scie *f* de carrier
BUCKBEAN - [bot] trèfle *m* d'eau, ményanthe *m*
BUCKET - [gen] (an open, slightly tapering, cylindrical vessel, fitted with a swinging handle, used to contain and convey small quantities of material, especially liquids) seau *m*, baquet *m*
[impl] (in earth moving machinery) benne *f*, godet *m*
[naut] baille *m*, moque *f*
[mech] (of a belt conveyor) godet *m*, auget *m*
[mech] (the valved piston of a pump) piston *m* à clapet
[mech] (of a steam turbine) aube *f*, ailette *f*
[electron] (specially formed electrode in a velocity modulated tube) collecteur *m*, électrode *f* collectrice
[comput] (expression designating a location in a storage device) adresse *f*
~CHAIN - [tool] chaîne *f* à godets, noria *f*
~CHAIN EXCAVATOR - [mech] (earth moving mach) excavateur *m* à godets
~CONVEYOR - [mech] (handling equip) transporteur *m* à godets
~DREDGER - [mech] (earth moving mach) drague *f* à godets
BUCKET ELEVATOR - [mech] (handling equip) élévateur *m* à godets
~FRONT LOADER - [mech] chargeur *m* à benne frontale
~LIP - [mech] trousse *f* coupante (de benne)
~LOADER - [mech] (earth moving machinery) chargeur *m* à godets
~PISTON - [electron] (contact piston in waveguides, by which the actual contact occurs in a region of low current and high voltage, thus minimizing the effects of contact resistance) piston *m* à lames de contact
~PLUNGER - s. bucket piston
~PUMP - [mech] pompe *f* élévatoire
~SEAT - [auto] baquet *m*, siège *m* en baquet
~WHEEL - [hydr] roue *f* à augets
BUCKING - [el] en opposition
~BAR - [mech] contre-bouterolle *f*, tas *m* à river
~CIRCUIT - [electron] (compensation circuit for photocells etc) circuit *m* de compensation
~COIL - [electron] (coil used in a bucking circuit) bobine *f* de compensation
~CURRENT - [el] courant *m* en opposition
BUCKLE, to - [gen] (to deform usually under compression) gauchir, flamber, tordre
BUCKLE - [mech] (device for attaching flat parts together or for securing the end of metal ribbons) agrafe *f*, fermoir *m*, boucle *f*
[metall] (a bend in bar or sheet) gondolement *m*, courbure *f*
[mech] (of a pump) piston *m* de pompe
BUCKLED PIPE - [plumb] tube *m* ondulé au cintrage
~PLATE - [el chem] (battery plate distorted by lateral curvature caused by heavy discharge or other incorrect treatment) plaque *f* gondolée
~SURFACE - [metall] surface *f* déformée
~TYRE - [rubber ind] (buckling due to the mould being too small) carcasse *f* plissée, pneumatique *m* refoulé
BUCKLING - [mech & metall] (deformation due to compressive stress) déformation *f*, flambage *m*, flambement *m*, gauchissement *m*
[electron] (mechanical distortion of the grid wires of an electronic tube) déformation *f*
[telecomm] (of cables) courbe *f*, courbure *f*
[naut] devers *m*
[aero] (deformation in the structure of an aircraft) flexion *f*, flambage *m*
~FAILURE - [constr] rupture *f* par flambage
~LOAD - [mech] (load imposed on an element in such a way as to produce an axially compressive stress) charge *f* de flambage
~STRENGTH - [phys] (capacity to resist axial compressive load withouth failure or deformation) résistance *f* au flambage
~STRESS - [phys] (axial compressive stress) contrainte *f* de flexion, contrainte *f* de flambage
~TEST - [metall] (bending of sheet metal) essai *m* de flexion
BUCKRAM - [text] bougran *m*
BUCKSKIN - [leather ind] peau *m* de daim
BUCKSTAY - [gen] (ovens, retorts, etc) armature *f*, montant *m*
BUCKTHORN - [bot] nerprun *m*
BUCKWHEAT - [agric] sarrasin *m*, blé *m* noir
BUCKY SCREEN - [radiat] (anti-diffusion screen, i.e. a barrier allowing only primary radiation to

pass) écran *m* anti-diffusant
BUD, to - [bot] bourgeonner, greffer (en écusson)
BUD - [bot] bourgeon *m*, bouton *m* (de fleur)
~SCALE - [bot] écaille *f* de bourgeon
~SPORT - [agric] variété *f* anormale, variation *f* sportive, mutation *f*
~VARIATION - s. bud sport
BUDDING - [agric] (insertion of buds to obtain artificial propagation) bourgeonnement *m*, greffe *f* par oeil détaché
[rubber ind] greffe *f*, greffage *m*
~KNIFE - [impl] écussonnoir *m*
BUDDLE - [min] (shallow annular pit for concentrating finely crushed ores) lavoir *m*
BUDDLED ORE - [min] minerai *m* lavé
BUDDLING - [min] lavage *m* de minerai
BUDDY TANK - [aero] réservoir *m* supplémentaire
BUDGET, to - [fin] budgéter, préparer un budget
BUDGET - [fin] budget *m*
~LOCK - [auto] serrure *f* de boîte à gants
~YEAR - [fin] exercice *m* financier
BUDGRAFT, to - s. to bud
BUDGRAFTED PLANTATION - [bot] plantation *f* d'arbres greffés
BUDWOOD - [bot] bois *m* de greffe
BUFF, to - [metall] (to finish metal surfaces by contact with rapidly rotating disks of cloth) polir (au buffle)
BUFF - [impl] (revolving disc consisting of layers of cloth used for polishing metals) disque *m* en buffle, polissoir *m*
~BOB - [impl] disque *m* de polissage
~LEATHER - [leather ind] cuir *m* pour disque de polissage
BUFFALO - [zool] buffle *m*
BUFFER - [impl] polisseuse *f*
[mech] (a device for absorbing impact shocks) amortisseur *m*, tampon *m*
[electron] (isolating circuit) circuit *m* intérmédiaire, tampon *m*
[comput] (mechanism linked to an output device for the assembling of information from a storage) mémoire *f* tampon, mémoire *f* intermédiaire, mémoire *f* de transit
[ind chem] tampon *m*
~ACTION - [gen] effet *m* tampon, action *f* tamponnante, effet *m* tamponnant
~AMPLIFIER - [radio] (a circuit designed to prevent load fluctuations of subsequent stages affecting the characteristics of preceding circuits) amplificateur *m* intermédiaire
~BAR - [auto] lame *f* de pare-choc
[railw] traverse *f* avant
~BATTERY - [el] (set of batteries connected in parallel with a direct current generator to smooth out variations in voltage and current) batterie *f* tampon
~BOLT - [railw] tige *m* de tampon, axe *m* de tampon
[text] tige *m* d'arrêt
~CAPACITOR - [electron] condensateur *m* shunt
~CAPACITY OF THE SKIN - [physiol] pouvoir *m* tampon de la peau
~CASE - [railw] corps *m* de tampon
~CASTING - [railw] enveloppe *f* fixe de tampon
~CIRCUIT - [electron] (an isolated circuit designed to prevent reaction from a driven circuit upon the driving system) circuit *m* tampon
BUFFER COATING - [paint] (coating applied to prevent defects such as bleeding occurring when the second coat is put on) couche *f* tampon
~CONICAL SPRING - [mech] ressort *m* amortisseur
~FUNCTION - [comput] (function synchronizing two or more systems)fonction *f* intermédiaire
~GEAR - [railw] butoir *m*, heurtoir *m*
~GENE - [genet] gène *m* tampon
~HEAD - [railw] tête *f* de tampon
~-HOLDER - [horol] fixateur *m* de tampon amortisseur
~IMPACT - [railw] (head-on collision) tamponnement *m*
~LAYER - [ind chem] couche *f* tampon
~LEATHER - [text] cuir *m* de butée
~MIXTURE - [chem] mélange *m* tampon
~PLUNGER - [railw] tige *f* de tampon, axe *m* de tampon
~REAGENT - [chem] réactif *m* tampon
~RESISTANCE - [el] résistance *f* tampon
~REST - [rubber ind] (a rubber cap at the base of a billiard cue) tampon *m* (de bout de queue de billard)
~ROD - s. buffer plunger
~SALTS - [chem] (compounds which reduce changes in the pH value of the solution concerned when an acid or alkali is added to it) produits *m* stabilisateurs, agents *mpl* stabilisants
~SCREW - [text] vis *f* de tampon
~SHANK - s. buffer plunger
~SOLUTION - [chem] solution *f* tampon
~SPACE - [el] espace *m* intermédiaire
~SPRING - [mech] (a spring used to absorb mechanical impact or shock) ressort *m* amortisseur
~STAGE - [electron] (an electronic tube amplifying stage) étage *m* intermédiaire
[electron] (valve circuit between two parts of a system to protect the early stages against variations in the later stages) étage *m* tampon
~STEM - s. buffer bolt
~STOP - [auto] (mainly in the US) butoir *m* de pare-choc
[railw] butoir *m*, heurtoir *m*, tampon *m* d'arrêt
~STORAGE - [comput] mémoire *f* intermédiaire mémoire *f* tampon
~TUBE - [electron] (tube preventing reaction between two adjacent amplifier tubes in a circuit) tube *m* intermédiaire
~VALVE - s. buffer tube
~VOLUME SPRING - [railw] ressort *m* spiral conique pour tampon
~WAGON - [railw] (shock absorbing truck) wagon *m* tampon
BUFFERED SOLUTION - [chem] solution *f* tamponnée
BUFFERING - [gen] action-tampon *f*
~ACTION - [gen] effet *m* tampon
~EFFECT - s. buffering action
BUFFETING - [aero] (the effect of eddying flow of an aircraft) vibrations *fpl*, chocs *mpl*, secousses *fpl*
BUFFING - [metall] (finishing process for metals etc using rapidly revolving discs of similar material) polissage *m*, avivage *m*, bufflage *m*, lustrage *m*
[railw] (the action of the buffer) choc *m*
~COLOURING - [metall] polissage *m*
~GEAR - s. buffer gear
~LATHE - [metall] touret *m* à aviver
~LOAD TEST - [railw] essai *m* de résistance au choc

BUFFING MACHINE - [mech] polisseuse *f*, brosseuse-polisseuse *f*
~RASP - [impl] meule *f* à brosser, meule *f* à polir
~WHEEL - [impl] meule *f* à polir, disque *m* en buffle
BUG - [zool] insecte *m*, punaise *f*
[comput] (interference during a computer operation) perturbation *f*
[mech] (colloq: a defect) défaut *m*, pépin *m*
~KEY - [telecomm] manipulateur *m* semi-automatique
~LIGHT - [naut] feu *m* intermittent
BUGGY - [impl] (a truck for heavy materials) ; chariot *m*
[colloq: of a hoist] benne *f*
[mining] wagonnet *m*
BUGLE - [mus] clairon *m*
~WEED OIL - [chem] essence *f* de lycope
BUGLER - [mus] clairon *m*
BUHRSTONE - [min] meule *f*
BUILD, to - [gen] bâtir, construire, édifier
BUILD - [gen] construction *f*, carrure *f*, conformation *f*, taille *f*
~IN, to - [gen] murer, boucher, incorporer
~UP, to - [constr] construire, édifier, bâtir
[mech] (of pressure) faire monter (la pression)
~-UP - [gen & ind] accumulation *f*
~-UP OF A SELF EXCITED GENERATOR - [el] (the whole of the transient phenomena preceding the establishment of permanent values of electromotive force and exciting current in a self-excited generator, when it is started) amorçage *m* d'une génératrice auto-excitatrice
~-UP TIME - [telev] (duration of the output signal of a network) temps *m* de montée, durée *f* d'établissement
BUILDER'S HOIST - [build] appareil *m* de levage, monte-charge *m*
~LEVEL - [instr] (a spirit level tube set in a long wooden straight edge) riveau *m* d'eau
~STAGING - [constr] (a strong type of scaffold) échafaudage *m*
~TAPE - [impl] (steel or linen measuring tape contained in a circular case) mètre *m* à ruban
BUILDING - [gen] bâtiment *m*, construction *f*, édifice *m*
[the operation of building] construction *f*
~BAG - [rubber ind] mandrin *m* souple
~BERTH - [shipbuild] cale *f* de construction
~BOARD - [build] (used for lining walls and ceilings) bois *m* en planche pour la construction
~CERTIFICATE - [comm] (document made out by the architect to enable contractors to be paid for the work completed) certificat *m* de construction
~CODE - [gen & leg] règlements *mpl* de la construction, code *m* de la construction
~CONTRACTOR - [gen] entrepreneur *m*
~CRANE - [mech] grue *f* de chantier
~DRUM - [rubber ind] tambour *m* de confection (des pneumatiques)
~FORMS - [build] coffrage *m*
~FRAME - [constr] charpente *f*, ossature *f*
~GROUND - [constr] chantier *m* de construction, terrain *m* à bâtir
~HARDWARE - [build] serrurerie *f* de bâtiment
~LAND - [constr] terrain *m* à bâtir
~LIFT - [constr] monte-charge *m*
~LINE - [constr] (the line-beyond which a building may not be erected) alignment *m*
BUILDING MATERIALS - [constr] matériaux *mpl* de construction
~-OUT CAPACITOR - [el] (used to increase the capacitance of an electric circuit to a specified value) condensateur *m* supplémentaire
~-OUT NETWORK - [radio] (type of matching network) réseau *m* supplémentaire
~-OUT SECTION - [telecomm] complément *m* d'une section de pupinisation, longueur *f* d'amenée
~PAPER - [paper man] papier *m* fort (utilisé comme isolant thermique)
~PIT - [constr] fouille *f*, excavation *f*, souille *f*
~PLOT - [town planning] lot *m* à bâtir
~SITE - [constr] site *m*, chantier *m*
~SKELETON - s. building frame
~SLIP - [shipbuild] cale *f* de construction, cale *f* de lancement
~TIMBER - [constr] bois *m* de charpente
~-UP - [gen] developpement *m*, augmentation *f*
[metall] épaississement *m*, tassement *m*, chevauchement *m*
[telev] (of the picture) synthèse *f* de l'image
[metall] (electro-plating for the purpose of increasing the dimensions of a piece) dépôt *m* électrolytique épais
~-UP SEQUENCE - [metall] (to form a welded joint) rechargement *m*
~-UP TIME - s. build-up time
~-UP TIME - [radio & telev] période *f* transitoire initiale
BUILT-IN - [adj] (incorporated or forming a permanent part of something) incorporé
[constr] encastré
[mech] (of valve seat) rapporté
[carp] (of furniture) encastré
~-IN AERIAL - [ant] (an aerial which is fitted inside a set) antenne *f* incorporée
~-IN ANTENNA - s. built-in aerial
~-IN ARCH PROTECTOR - [shoe man] semelles *fpl* orthopédiques incorporées
~IN CHECK - [comput] (a mechanism for the automatic checking of the information) vérification *f* automatique
~-IN FURNITURE - [build] mobilier *m* encastré
~-IN JACK - [auto] cric *m* incorporé
~-IN PRODUCER - [gas ind] gazogène *m* incorporé
~-IN RADIATION HEATING - [el] (method of space heating, whereby all surfaces in a room are heated by means of heated elements built into the structure) chauffage *m* rayonnant par éléments incorporés
~-UP - [gen constr & mech] composé
~-UP AREA - [town planning] zone *f* urbaine, agglomeration *f*
~-UP BEAM - [constr] poutre *f* composée, poutre *f* rapportée
~-UP CONSTRUCTION - [gen] construction *f* composée
~-UP GIRDER - s. buil-up beam
~-UP MEMBER - [constr] élément *m* prefabriqué
BULB - [el](electric lamp; the envelope of a vacuum tube) ampoule *f*
[electron] (envelope of electronic tube) ampoule *f*
[mech] (any structure resembling a plant bulb in its shape) bulbe *m*
[bot] bulbe *m*
[instr] (of a measuring instrument) bulbe *m*,

boule *f*, cuvette *f*, reservoir *m*
[electron] (a thermionic valve) tube *m* thermoïonique
BULB ANGLE - [mech] cornière *f* à boudin
~BAR - [metall] (rolled bar with the section thickened along one edge) fer *m* à boudin
~BEAM - s.bulb bar
~BLACKENING - [electron] (black deposit in a bulb of a vacuum tube) noircissement *m* d'un tube
~BLOWING MACHINE - [mach] (automatic machine designed to blow glass bulbs) machine *f* à souffler les ampoules (ou les tubes)
~CONDENSER - [chem] réfrigérant *m* à boules
~CRADLE - [el] support *m*
~CUTTING MACHINE - [mech] (used to manufacture bulbs of electronic tubes) machine *f* à couper les tubes
~EXPOSURE - [photo] pose *f* à un temps
~GROWING - [agric] culture *f* de bulbes
~IRON-s.bulb bar
~MOULD - [electron] (a part of the bulb blowing machine) moule *m* à ampoules, moule *m* à tubes
~-NIPPLE TAP - [plumb] robinet *m* porte-caoutchouc
~RECTIFIER - [el] soupape *f* électrique
~RING - [electron] (the lower end of the bulb) collet *m* de tube
~SOCKET - [el] support *m* (de lampe ou de tube)
~-T - [metall] fer *m* en T à boudin
~-T BAR - [metall] fer *m* en T à boudin
~TUBE - [ind chem] (glass tube with one or more bulbs blown on it) tube *m* à boules
BULBAR - [med] bulbaire
BULBIFORM - [med] bulbiforme
BULBIPHEROUS - [bot] bulbifère
BULBOCAPNINE - [chem] (a naturally occurring alkaloid used as a sedative in medicine) bulbocapnine *f*
BULBOUS - [gen] (swollen like a bulb) bulbeux
~BOW - [shipbuild] étrave *f* à bulbe
BULGE - [gen] renflement *m*, bombement *m*, soulèvement *m*
[shipbuild] bulge *m*, caisson *m* pare-torpilles
~DIE - [metall etc] matrice *f* à repousser
~SYSTEM - [shipbuild] construction *f* à bulge
BULGED - [gen] (forming a protuberance) bombé, renflé, incurvé
[shipbuild] (outward bent) bombé, renflé
BULGING - [constr] gonflement *m* latéral
BULK, to - [gen] entasser, amasser, grouper
BULK - [gen] grandeur *f*, grosseur *f*, masse *f*
[naut] (in ships) chargement *m* arrimé
[phys] (space occupied by structures) encombrement *m*, volume *m*
~CONCRETE - [build] béton *m* en masse
~DENSITY - [phys] (relation between the weight and the volume of a material) densité *f* apparente
~ERASER - [el] (device designed to erase a complete reel of magnetic tape) tête *f* d'effacement instantané (de bobine)
~FACTOR - [mech] (the ratio of the density of a moulded article to the powder density of the moulding material) facteur *m* de compression, facteur *m* de contraction
[ind chem] facteur *m* de contraction
~GETTER - [metall] (getter metals in solid form) getter *m* massif
~GOODS - [gen] marchandises *f*pl en vrac
BULK INFORMATION - [comput] grande quantité *f* de données (d'informations)
~-INJECTION CARBURETTOR - [mech] (type of carburettor designed to inject fuel direct into the air supply before it reaches the cylinders) carburateur *m* à injection
~-INJECTION PUMP - [mech] (a direct injection pump) pompe *f* à injection directe
~-LOAD CHASSIS - [mech] chassis *m* pour charges volumineuses
~MATERIAL - [gen] (material in a loose condition, not packed in any way) marchandises *f*pl en vrac
~MEMORY - [comput] grande mémoire *f*, mémoire *f* de grande capacité
~MODULUS - [phys] (the change in the volume of materials caused by the application of pressure) module *m* d'élasticité cubique
~POLYMERIZATION - [chem] copolymérisation *f* en masse
~SHIPMENT - [gen] (a consignment of material in loose form, not packed in any way) transport *m* en vrac
~STORAGE - [gen] stockage *m* en vrac
~TEST - [nucl] (test shield sample with high attenuation) essai *m* de masse, essai *m* de volume
~VELOCITY - [phys] vitesse *f* de propagation dans le milieu
BULKAGE - [med] aliments *m*pl de lest
BULKED YARN - [text] fil *m* gonflant, fil mousse, filé *m* high-bulk, filé texturisé
BULKHEAD - [shipbuild & aero] (transverse partition in a ship or aircraft) cloison *f* étanche; (the ribs in the fuselage structure) fausse *f* nervure, couple *m* de fuselage
[constr] (masonry or timber partition to support earth etc as in a waterfront, a tunnel etc) cloison *f* étanche, bâtardeau *m*, rideau *m* de palplanches
~DECK - [shipbuild] pont *m* de cloisonnement
~PACKING - [packag] garniture *f* d'étanchéité
~RIB - [shipbuild aero] (structural member forming a stiffening element in a bulkhead) couple *m* de fuselage
BULKING - [gen] augmentation *f* de volume, gonflement *m*, expansion *f*
~AGENT - [rubber & plast ind] charge *f*
~FACTOR - [soil] coifficient *m* de foisonnement
~TANK - [ind] (a vessel used to contain material in bulk for storage and mixing) réservoir *m* de stockage (et de mélange)
BULKY - [gen] (of a large size in relation to weight) volumineux, gros, encombrant
BULL - [zool] taureau *m*
[fin colloq] bévue *f*
~ENGINE - [mining] machine *f* à maîtresse-tige
~FOR BREEDING - [zool] taureau *m* reproducteur
~FOR SERVICE - s.bull for breeding
~GEAR - [mech] engrenage *m* de grande taille
~HEAD - [constr] brique *f* pour voûtes
[railw] champignon *m*
~-HEAD RIVET - [mech] rivet *m* à tête fraisée et goutte-de-suif
~-HEADED FISH PLATE - [railw] éclisse *f* à champignon
~-HEADED RAIL - [railw] rail *m* à double champignon
~NOSE - [oil ind] boulon *m* à haute pression
~NOSE PLIERS - [impl] pinces *f*pl universelles

BULL-NOSED BRICK - [build] brique *f* arrondie sur un côté
~-NOSED STEP - [build] (half or a quarter round at the end) marche *f* à coins arrondis
~PUMP - [mining] pompe *f* à maîtresse-tige
~RING - [el] (metal ring designed to form the junction of three or more straining wires) cercle *m* de garde
[mech] (supporting ring in a piston) anneau *m* de soutien
~ROD - [mining] tige *f* de sondage
~STRETCHER - [build] brique *f* à angles arrondis
~SWITCH - [cinema] (a switch controlling the lights in a studio) commutateur *m* principal
~WHEEL - [mining] treuil *m* de forage, tambour *m* de forage
BULL'S EYE - [firearms] centre *f*, mouche *f* de cible
[met] (a small patch of clear sky at the centre of a cyclonic storm) oeil *m*
[shipbuild] hublot *m*; (a wooden block) moque *f*
~EYE CLOUD - [met] (small isolated cloud seen at the beginning of a bull's eye squall) nuage *f* "d'oeil de beuf"
~EYE PACKING - [shipbuild] garniture *f* d'étanchéité de hublot
BULLACE - [agric] béloce *f*, prune *f* sauvage
BULLDOG - [metall] scories *fpl* de revêtement (pour soles de fours)
~CASING SPEAR - [mining] navette *f*
BULLDOZE, to - [constr] niveller, préparer au bulldozer
BULLDOZER - [mech] (earth moving mach) bulldozer *m*, tracteur *m* à refouler
[metall] presse *f* à forger à plusieurs poinçons
BULLDOZING - [mining] fragmentation *f* de blocs de minerai
BULLET - [firearms] balle *f*
~AMPLIFIER - [el] amplificateur *m* à cartouche
~(TYPE) CHECK VALVE - [mech] clapet *m* en forme de balle de fusil
BULLETIN - [gen] bulletin *m*, communiqué *m*
~BOARD - [gen] tableau *m* d'affichage
BULLFINCH - [zool] bouvreuil *m*
BULLION - [comm] or *m* en lingot(s), argent *m* en lingot(s)
BULLOCK - [zool] boeuf *m*
BULLOUS - [med] bulleux
BULRUSH - [bot] scirpe *m*, joinc *m* des marais
BULWARK - [constr] boulevard *m*, rempart *m*, bastingage *m*
[naut] pavois *m*
~STAY - [naut] jambe *f* de force de pavois
BUMBLE-BEE - [zool] bourdon *m*
BUMBOAT - [naut] bateau *m* à provisions
BUMMER - [transp] chariot *m* à deux roues
[min] (a worker dealing with the conveyors) opérateur *m* de transporteur
BUMP, to - [gen] cogner, heurter, frapper, buter
[auto] (impact from the rear) tamponner
BUMP - [gen] choc *m*, coup *m*, heurt *m*, secousse *f*, collision *f*, cahot *m*
[aero] (sudden upward movement of the aircraft) trou *m* d'air
[acoust] (a low frequency sound, heard in the reproduction of a sound track) boum *m*
~AGAINST, to - [auto etc] buter contre, heurter

BUMP ARTIST - [cinema] (an actor who is standing-in for risky parts) cascadeur *m*
~STOP - [auto] butée *f* de suspension
~YARN - [text] (for carpets) fil *m* de fourrure
BUMPER - [auto] pare-choc *m*
[railw] tampon *m*, butée *f* élastique
[naut] défense *f*
[mech] machine *f* à mouler à secousses
~BAR - [auto] lame *f* de pare-choc
~CLAMP - [auto] attache *f* de pare-choc
~CROP - [agric] récolte *f* très abondante, récolte *f* exceptionnelle
~GROMMET - [mech] anneau *m* isolant pour pare-chocs
~GUARD - [auto] butoir *m* de pare-choc
~PAD - [mech] butée *f* de suspension, butée *f* élastique
~PLATE - [diecasting] plaque *f* d'éjection
~ROD - [auto] lame *f* de pare-choc
BUMPING - [railw] (bumpers at the end of a track) buttoir *m*, tampon *m*, heurtoir *m*
[mech] (pressure bursts due to overheating of pump fluid) instabilité *f* de surchauffage, ébullition *f* par secousses
~CONVEYOR - [min] transporteur *m* à secousses
~TABLE - [min] trieuse *f* à secousses
BUMPKIN - [naut] (a boom) bout-dehors *m* de tapecul
BUMPY AIR - [aero] (air in a condition of eddy or turbulence causing irregular disturbance in the motion of an aircraft) air *m* instable
BUNA - [chem] buna *m*
~N - [ind chem] (a copolymer of butadiene and acrylonitrile) buna N *m*
~S - [chem] (copolymer of butadiene and styrene) buna S *m*
BUNCH, to - [gen] grouper, lier, mettre en tas, mettre en bouquet
BUNCH - [gen] botte *f*, bouquet *m*, touffe *f*
[agric] (of grapes) grappe *f*
[text] (of ribbons) flot *m* de rubans; (of feathers) touffe *f* de plumes, houppe *f*
[electron] (bunch of electrons) groupe *m*, paquet *m*, faisceau *m*
~LIGHT - [light] (set of lamps illuminating a stage) banc *m* d'éclairage
~OF ORE - [min] poche *f* de minerai
~PLATING - [agric] plantation *f* en touffes
BUNCHED CIRCUIT - [telecomm] circuit *m* groupé
BUNCHER - [electron] (system of electrodes in a klystron designed to break up the electron stream into discrete groups) résonateur *m* d'entrée
~SPACE - [electron] (the section of the tube which follows the acceleration space. This is comprised between the input-resonator grids) espace *m* de modulation
BUNCHING - [electron] (in a velocity-modulated electron stream, the action producing an alternating convection current) groupement *m* des electrons
~ANGLE - [electron] (the average transit angle in a specified drift space) angle *m* de groupement
~MACHINE - [mech] machine *f* à retordre les fils métalliques
~OF CABLE CONDUCTORS - [telecomm] (for testing purposes) groupement *m* des conducteurs en faisceau
~OF IONS - [nucl] (the joining of ions in a number

of groups) groupage *m* d'ions
BUNCHING PARAMETER - [electron] (waveguides) paramètre *m* de groupement
BUNCHY - [min] irregulier
BUNDLE - [gen] paquet *m*, ballot *m*, groupe *m*, botte *f*, tas *m*; (of papers or documents) liasse *f*
[anat] faisceau *m*
~OF LINES OF FORCE - [phys] faisceau *m* de lignes de force
~OF RAYS - [geom] faisceau *m*
~PRESS - [packag] presse *f* à paqueter
BUNDLING - [gen] groupement *m* en faisceaux
BUNG, to - [ind chem] mettre sous pression
BUNG - [brew ind etc] (wooden plug to close the opening in a cask) bondon *m*, tampon *m*, tape *f*, obturateur *m*
~BUSH - [brew ind etc] obturateur *m*, bouchon *m*
~HOLE - [brew ind etc] bonde *f*
~PLUG - [brew ind etc] bondon *m*, tampon *m*
~STAVE - [brew ind etc] douve *f* du bondon
BUNGALOW - [constr] (one-storey house, frequently with a veranda) "bungalow" *m*, ville *m* à véranda
BUNGEE - [aero] (a spring device for aircraft controls) vérin *m* à ressort
~LOOP - [aero] (a system of rubber cords to provide spring effect in the landing gear) dispositif *m* amortisseur
BUNGHOLE - [brew ind etc] (circular opening in a cask through which it is filled and emptied) bonde *f*
BUNGING APPARATUS - [ind chem] régulateur *m* de pression, soupape *f*
~PRESSURE - [brew ind] pression *f* dans la bouteille
BUNION - [med] oignon *m*
BUNKER, to - [naut] charbonner
BUNKER - [gen] (large receptacle for storing solid or liquid fuel in bulk) réservoir *m*, caisson *m*; (storage room for coal) soute *f*
[railw] (at railway stations or depots) fosse *f* à combustible
[naut] soute *f* à charbon, soute *f* à combustible
~BULKHEAD - [naut] cloison *f* de soute
~CAPACITY - [naut] capacité *f* des soutes (à combustible), volume *m* des soutes
~FUEL - [fuel] mazout *m*
~HATCH - [naut] panneau *m* de soute
BUNKERING - [naut] charbonnage *m*
BUNS - [gen] (large loaf-like masses of material for further processing) pains *mpl*
BUNSEN BURNER - [ind chem] (device for burning gas mixed with air, for heating chemical apparatus) bec *m* Bunsen
~CELL - [ind chem] (a double-fluid primary cell. Positive electrode: carbon; negative: zinc; electrolyte: dilute suphuric acid; depolarizer: concentrated nitric acid. Voltage 1,9v. Gives off nitric fumes) pile *f* Bunsen
~COEFFICIENT - [phys chem] (the volume of gas under standard conditions which is absorbed by a unit volume of gas solution) coefficient *m* de Bunsen
~FLAME - [chem] flamme *f* de Bunsen
~SCREEN - [opt] (a photometer screen) écran *m* photométrique de Bunsen
BUNT - [aero] (a manoeuvre consisting of a partial inverted loop) looping *m* à l'envers
[naut] (of a sail) fond *m* d'une voile
[bot] (a parasitic fungus) carie *f* du froment
BUNTING - [text] étamine *f*
BUNTLINE - [naut] (a rope used for hauling the sail) cargue-fond *m*
BUNTON - [impl] (a piece of square timber; divider poussard *m*
BUOY, to - [naut] baliser
BUOY - [naut] (a buoyant structure used as anchorage) bouée *f*, balise *f*
~ROPE - [naut] orin *m*
BUOYAGE - [naut] balisage *m*
BUOYANCY - [phys] (the vertical force which a fluid exerts on a body, wholly or partially submerged in it, when the overall specific gravity of the body is less than that of the fluid) poussée *f* (d'un liquide), force *f* ascensionnelle
[naut] (the capacity to flow) flottabilité *f*
[aero] force *f* ascensionnelle
[hydr] poussée *f* hydrostatique
~BAG - [rubber ind] sac *m* flottant
~BALANCE - [instr] balance *f* de précision pour la mesure de la densité des gaz
~CHAMBER - [naut] (in submarines) compartiment *m* des régulateurs d'immersion
~CUSHION - [rubber ind] coussin *m* flottant
~TANK - [naut] caisse *f* d'assiette
~WEIGHT - [aero] portance *f*, flottabilité *f*
BUOYANT - [gen] flottable, léger
BUR - [bot] (a weed) gratteron *m*, glouteron *m*, bardane *f*
[mech] capsule *f* épineuse
~-CRUSHER ROLLER - [text] cylindre *m* échardonneur, cylindre briseur des chardons
BURAN - [met] (a cold north-east wind peculiar to Russia and Siberia) buran *m*
BURBLE - [phys] (turbulent flow resulting from the breakdown of streamline flow round a body, especially on the upper surface of an aircraft wing) remous *m*
BURBLING POINT - [aero] point *m* de decollement, point *m* de discontinuité
BURDEN, to - [gen] alourdir, charger, aggraver; (of machines) charger
BURDEN - [gen] contenance *f*, charge *f*, fardeau *m*, poids *m*
[comm] port *m*, maximum *m* de charge
[mining] (waste material) épaisseur *f* de morts terrains
[naut] (the weight of the cargo) charge *f*, contenance *f*
[el] (the load) charge *f*
[metall] (of blast furnaces; ratio of ore coke in the charge) charge *f*, laitier *m*
[metall] (the metal charge in a furnace) charge *f* métallique; (the charge of ore and flux of a furnace without fuel) charge *f*
~OF AN INSTRUMENT - [meas] charge *f* totale d'un instrument
BUREAU - [gen] bureau *m*, office *m*, service *m*
~OF STANDARDS - [comm] bureau *m* des poids et measures
BURETTE - [ind chem] (graduated glass column with a stopcock at the lower end, for use in titration) burette *f*, éprouvette *f* graduée
BURGLAR ALARM - [el] dispositif *m* antivol
BURGUNDY PITCH - [paint] (an oleoresin obtained from the Norway spruce) poix *f* de Bourgogne
BURIAL GROUND - [gen] cimetière *m*
[nucl] (an area used for the burial of radio-active

objects) dépôt *m* de rebuts radio-actifs
BURIED AERIAL - [radio] (an aerial which is buried into the earth) antenne *f* enterrée
~OUTCROP - [geol] affleurement *m* masqué
BURIN - [tool] burin *m*
BURLAP, to - [gen] envelopper dans de canevas
BURLAP - [text] (coarse woven jute fabric used for packing etc) toile *f* d'emballage, gros canevas *m*
BURLING - [text] époutiage *m*, épinçage *m*
~IRON - [text] épincette *f*
~MACHINE - [text] machine *f* à époutier, ratineuse *f*
BURMITE - [min] (an amber-like mineral whitout succinic acid) burmite *f*
BURN, to - [gen] brûler, flamber, faire une combustion
[ind chem] cuire
[metall] (plates in boilers etc) brûler
[rubber ind] cuire, griller
BURN - [gen & med] brûlure *f*
[telev] (the retention of the image by the screen in a cathode-ray tube) rémanence *f*
~IN, to - [gen] graver au feu
~MARKS - [metall] (marks caused by overheated air trapped in a mould) traces *f*pl de brûlure
~-OFF - [glass man] (cutting off unwanted section by fusing the glass) fonte *f* du verre superflu
[metall] usure *f*
~-OFF MACHINE - [glass man] machine *f* à fondre (le verre superflu)
~OUT - [electron] (the failure of a cathode ray tube, a vacuum tube etc) claquage *m*
[telev] perte *f* de gradation
[nucl] épuisement
~THROUGH, to - [gen & metall] brûler, griller, fondre
~-UP - [nucl] (the destruction of atoms by neutrons) taux *m* de combustion
~UP FACTOR - [nucl] coefficient *m* de consommation
BURNER - [th eng] brûleur *m*
[impl] bec *m*
~CAP - [th eng] (in burners) calotte *m* de brûleur, chapeau *m* de brûleur
~FLOW DISTRIBUTOR - [mech] (unit providing each burner of a multiburner gas turbine with the correct amount of fuel) distributeur *m* de combustible
~HEAD - [th eng] (burners) tête *f* de brûleur, corps *m* de brûleur
~HOLDER - [mech] porte-brûleur *m*
~PORT - [th eng] (burners) orifice *f* de formation de la flamme
~RAIL - [gas ind] (in a domestic gas cooker) rampe *f*
~THROAT - [th eng] (burners) col *m* du mélangeur
BURNING - [gen] brûlure *f*
[mining] élargissement *m* du fond de trou de mine
[mech] cuite *f*, cuisson *f*, combustion *f*, fusion *f*
[metall] (overheating of steel during working, thus affecting its original qualities) surchauffage *m* (de l'acier)
[US-radio] (noise due to irregular current) bruit *m* de microphone
[med] brûlure *f*
~-BUSH - [bot] (with uses in perfumery) grande pimprenelle *f*
~FEET - [med] paresthésie *f* ariboflavinique des pieds
BURNING-IN - [photo] postlumination *f*
~-IN KILN - [glass man] four *m* de recuit
~-OFF - [photo] (of prints) flambage *m* d'épreuves
~-OFF SHAFT - [mech] (in a meter) axe *m* de blocage
[glass man] s. burn-off
~POINT - [chem] point *m* d'inflammation
~RATE - [ind chem] (the speed at which combustion proceeds in a substance) vitesse *f* de combustion
~UP - [photo] (colloq. - over exposure) surexposition *f*
~VELOCITY - [phys] (the rate at which combustion is propagated in relation to the unburnt mixture) vitesse *f* de combustion
~VOLTAGE - [el] (the minimum voltage between cathode and anode at which the discharge in a cold-cathode tube can be sustained) tension *f* minimale de décharge
BURNISH, to - [metall] (to finish or harden a metal surface by the application of a metal roller or rod) brunir, polir
BURNISHED - [metall] bruni
BURNISHER - [mech] (tool for burnishing) brunissoir *m*, polissoir *m*
BURNISHING - [metall] (the smoothing or hardening of metal surfaces) brunissage *m*, polissage *m*
~BRUSH - [text] brosse *f* pour lissage, polissoire *f*
~FILLET - [text] ruban *m* à polir
~MACHINE - [mech mach] polisseuse *f*, brunisseuse
BURNOUT - [el] court-circuit *m*
[mech] (in a motor) fin *f* de la combustion
~OF A CRYSTAL RECTIFIER - [electron] (change in the properties of a crystal rectifier owing to voltage overload) brûlure *f* d'un redresseur à cristal
~-PROOF - [electron] (of a cathode) résistant au brûlage, à l'épreuve de fusion
BURNT - [gen] brûlé
[ind chem] cuit
[metall] surchauffé
~BALLAST - [build] argile *f* cuite
~DEPOSIT - [el plating] (rough electro-deposit caused by an excess of current density) dépôt *m* brulé
~LIME - [chem] oxyde *m* de calcium, chaux *f* (vive)
~METAL - [metall] métal *m* sarchauffé
~OUT - [el] brûlé, claqué, détruit
[mech] (of a bearing) fondu
~-OUT BEARING - [mech] palier *m* coulé, coussinet *m* fondu
~-OUT LAMP - [el] (an electric lamp the filament of which has been destroyed by excessive current) lampe *f* grillée, lampe *f* claquée
~SIENNA - [paint] (pigment obtained by calcining raw Sienna) terre *f* de Sienne brûlée
~UMBER - [paint] (deep brown pigment obtained by calcining umber, consisting of manganese) terre *f* d'ombre brûlée
BURR, to - [mech] (to produce a burr) ébarber
BURR - [metall] (a rough irregular edge caused by working or wear) bavure *f*, barbe *f*, barbure *f*
[impl] (in dentistry) fraise *f*
[mech] rosette *f*
[tool] burin *m* triangulaire, alésoir *m* cannelé
~ROLLER - [text] cylindre *m* échardonneur
~STONE - [geol] pierre *f* meulière
BURRING - [text] chardonnage *m*

BURRING MACHINE - [mech] ébarbeuse *f*
[text] machine *f* à chardonner, échardonneuse *f*
BURROW, to - [mining] prospecter
BURROW - [geol] roche *f* stérile
BURRS - [text] (seed vessels found in wool and removed before scouring) graterons *mpl*
BURSITIS - [med] bursite *f*
BURST, to - [gen] éclater, exploser
BURST - [gen & expl] éclatement *m*, explosion *f*
[firearms] rafale *f*
[nucl] (shower of cosmic ray particles over a large area) gerbe *f* extensive
[telecomm] bruit *m* impulsif, bruit *m* impulsionnel
[telev] (a colour burst) signal *m* de synchronisation de couleur, salve *f* de référence
[min] rejettement *m*
[mech] (of a grinding wheel) rupture *f*
~AMPLIFIER - [telev] amplificateur *m* de salve
~GATE - [telev] porte *f* déclenchant le signal de synchronisation de la sousporteuse de chrominance
~KEYING PULSE - [telev] impulsion *f* en phase avec le signal de synchronisation de couleur
BURSTING - [gen] explosion *f*, éclatement *m*, crevaison *f*
~CHARGE - [gen] charge *f* d'explosion
~PRESSURE - [phys] (pressure at which a closed pressure vessel will explode) pression *f* d'explosion
~STRENGTH - [phys] (the capacity of a vessel to resist internal pressure) résistance *f* à l'éclatement
~VELOCITY - [gen] vitesse *f* d'explosion
BURTON - [naut] petit *m* palan
BURY, to - [gen] ensevelir, enterrer, inhumer, enfouir
BURYING METHOD - [ind chem] méthode *f* d'enfouissement des boues
BUS - [auto] autobus *m*
[comput] (a path over which the information is moved) barre *f* omnibus
~-BAR - [el] (a metal bar or rod to which a number of leads are connected) barre *f* omnibus
~-BAR MOUNTED TRANSFORMER - [el] (a current transformer fixed on a bus-bar acting both as primary and support) transformateur *m* monté sur une barre omnibus
~-BAR SECTIONALIZING SWITCH - [el] sectionneur *m* placé sur une barre omnibus
~-BAR VOLTAGE - [el] tension *f* à la barre omnibus
BUSH, to - [mech] (to insert a bush) baguer, manchonner, mettre un coussinet
BUSH - [mech] (a ring or short tube inserted in a hole in another part to receive a shaft, a bolt etc) bague *f*, coussinet *m*, manchon *m*, douille *f*
[el] manchon *m*, revêtement *m* isolant
[plast ind] buse *f*
~BEAN - [bot] haricot *m* blanc
~-HAMMER, to - [build] (to dress a surface of stone by means of a special hammer having rows of points on its striking face) boucharder
~-HAMMER - [tool] boucharde *f*, marteau *m* rustique
~METAL - [metall] alliage *m* pour coussinets et paliers
~OF JACK - [el] douille *f* de jack
~PILLER - [agric] machine *f* à égruger
~ROLLER - [tool] boucharde *f* à rouleau
~TREE - [bot] arbrisseau *m*, arbuste *m*
BUSH VETCH - [bot] vesce *f*
BUSHEL - [meas] (35.210 l) boisseau *m*
BUSHING - [mech] manchon *m*, douille *f*, bague *f*, coussinet *m*, fourrure *f*
[el](conductor with an insulating unit) borne *f* traversée, isolateur *m* de traversée
~INSULATOR - [el] isolateur *m* de traversée
~TRANSFORMER - [el] (a magnetic circuit carrying a winding fitted over an insulated bushing to act as a current transformer) transformateur *m* de traversée
BUSINESS - [gen & comm] métier *m*, operation *f* commerciale
~DAY - [comm] jour *m* ouvrable
~HOURS - [gen & comm] heures *fpl* d'ouverture, heures *fpl* ouvrables
BUSS - [naut] embarcation *f* pour la pêche; (herring fishing-boat) embarcation *f* pour la pêche aux harengs
BUST - [anat] buste *m*, gorge *f*, poitrine *f*
[tool] (a riveting hammer) marteau *m* à river
BUSTER - [tool] (pneumatic hammer to break up foundations etc) marteau *m* pneumatique
BUSULPHAN - [chem] (a cytotoxic agent used in medicine for the treatment of leukaemia) busulphan *m*
BUSY - [gen] actif, occupé
[telecomm] occupé
~-FLASH SYSTEM - [telecomm] (a signal operating a lamp to indicate that the outlet or the subscriber is busy) signal *m* d'occupation
~LINE - [telecomm] ligne *f* occupée
~RELAY - [telecomm] relais *f* d'occupation
~TEST - [telecomm] contrôle *m* d'occupation d'un circuit
BUTABARBITAL SODIUM - [chem] (a hypnotic used in medicine) butabarbital *m* sodique
BUTACAINE SULPHATE - [chem] (a local anaesthetic) butacaïne *f*
BUTADIENE - [chem] (butadiene - 1,3; vinylethylene erythene, bivinyl, divinyl B; a major raw material for the production of butadiene/styrene rubbers and other rubbers, water-based latex paints, and a starting material for adiponitrile) butadiène *m*
BUTANE - [chem] (n-butane, butyl hydride; an important starting material in organic synthesis, a refrigerant aerosol propellant and fuel for industrial and domestic use) butane *m*
~-AIR MIXTURE - [gas] air *m* butané
BUTONATE - [chem] (an insecticide) butonate *m*
BUTOPYRONOXYL - [chem] (an insect repellant) butonate *m*, butopyronoxyl *m*
BUTOXYETHYL LAURATE - [chem] (a solvent and plasticizer) laurate *m* de butoxyéthyle
BUTT, to - [gen] abouter, abutter, rabouter
[metall] mettre bout à bout
BUTT - [gen] bout *m*, gros bout *m*
[metall] (of a joint) bout *m*
[firearms] crosse *f*
[carp] about *m*
[mech] aboutement *m*
~-END OF A RAIL - [railw] extremité *f* d'un rail
~JOINT - [mech] raccord *m* à bride lisse
~PULLER - [text] extracteur *m* de talons
~-SEAM - [shoe man] soudure *f* bout à bout
~-SEAM WELD - [metall] soudage *m* en bout en ligne continue

JTT SPLICE - [cin & telev] collage m de bout

STRAP - [mech] couvre-joint m

-WELDING - [metall] (welding of two abutting ends without overlapping) soudage m bout à bout, soudage m en bout

-WELDING MACHINE - [mech] (machine for welding parts end-to-end by the passage of a current from one to the other) machine f à souder en bout

WELDING WITH PRESSURE - [metall] (butt welding in which the simultaneous effects of static pressure and heat flow cause an important widening of the welded surface) soudage m en bout avec refoulement

JTTER - [food] beurre m

BEAN - [bot] haricot m beurre

BUR - [bot] (a weed) pétasite m commun

CHURNER - [impl] baratte f

-FAT - [zool] matières fpl grasses du lait

-FAT CONTENT - [chem] teneur m en matières grasses

MOULD - [impl] moule m à beurre

MOULDING MACHINE - [agric] machine f à mouler le beurre

-PAT MACHINE - [mech] machine f à mouler les médaillons de beurre

JTTERCUP - [bot] bouton m d'or, renoncule f des champs

JTTERFLY - [zool] papillon m
[mech] (of a gate valve) papillon m
[cinema] écran m diffuseur

NUT - [mech] (a nut provided with two projecting ears or lugs so as to be easily turned with the fingers) écrou m à oreilles, écrou m papillon

-VALVE - [mech] (a valve consisting of an elliptical metal plate pivoted on its minor axis and mounted in a cylindrical passage so as to control the flow therein) vanne f papillon

JTTERING - [build] (the operation of spreading mortar on the edges of a brick) étalement m
[metall] (the process of precoating the faces of prepared plates with weld metal) beurrage m

TROVEL - [impl] truelle f

JTTERMILK - [zool] babeurre m

JTTERY - s. butter-like

JTTOCK - [anat] fesse f
[naut] arcasse f, arrière m arrondi, fesses mpl

LINE - [shipbuild] (longitudinal sectional line through the stern's form) section f verticale longitudinale

PLANES - [shipbuild] (longitudinal sectional planes drawn through a ship's form) plans mpl de la section verticale longitudinale

JTTON - [gen] bouton m
[el] (of a bell) bouton-poussoir m
[metall] (metal remnants attached to a weld during tests) culot m
[mus] (the pin bearing the pull of the strings of a violin) bouton m de queue de violòn

BASE - [el] (a standard seven-pin valve base, all in glass, designated B7G) culot m de tube à 7 broches

HEAD BOLT - [mech] boulon m à tête en goutte de suif

-HEADED SCREW - [mech] vis f à tête bombée

KNOB - [mech] bouton m

SWITCH - [el] interrupteur m à bouton-poussoir

JTTONHOLE - [text] boutonnière f

JTTONWOOD - [bot] platane m d'Amerique

BUTTRESS, to - [constr] arc-bouter

BUTTRESS - [constr] (support or structure to resist a thrust) contrefort m
[rubber ind] (the shoulder of a tyre) épaulement m (d'un pneumatique)

~THREAD - [mech] (special thread used for unidirectional thrusts, one face slipping at 45°, the other being at right angles to the axis) filet m trapézoïdal, filetage m à appui

~WALL - [build] (a thrusting block in the construction of ovens) massif m de butée

BUTTSTOCK - [firearms] crosse f

BUTYL - [chem] butyle m

~ACETOACETATE - [chem] (synthesis intermediate in the production of pharmaceuticals and dyestuffs and a solvent for synthetic resins) acétylacétate m de butyle

~ACETOXYSTEARATE - [chem] (a component of textile processing and a plastics plasticizer) acétoxystéarate m de butyle

~ACETYL RICINOLEATE - [chem] (a detergent, plasticizer, lubricant and emulsification agent) acétyl ricinoléate m de butyle

~ACRYLATE - [chem] (an intermediate for copolymers for use in protective coatings) acrylate m de butyle

~ALCOHOL - [chem] alcool m butylique

~BENZENESULPHONAMIDE - [chem] (a plasticizer for a number of synthetic resins and an intermediate in the production of pharmaceutical and dyestuffs) butylbenzènesulfonamide f

~BENZOATE - [chem] (solvent for cellulose derivatives, plasticizer and ingredient of perfumes) benzoate m de butyle

~BENZYL PHTHALATE - [chem] (a plasticizer for cellulose derivatives and polyvinyl resins) phtalate m de butyle et de benzyle

~BENZYL SEBACATE - [chem] (a plastics plasticizer) sébacate m de butyle et de benzyle

~CHLORAL - [chem] (a sedative used in medicine) butylchloral m

~DECYL PHTHALATE - [chem] (a plastics plasticizer) phtalate m de butyle et de décyle

~ETHER - [chem] (solvent for the extraction of fats) éther m butylique

~FORMATE - [chem] (a synthesis intermediate and solvent for cellulose esters and natural and synthetic resins. Also has uses in perfumery) formate m de butyle

~LACTATE - [chem] (a synthesis intermediate and solvent for a number of synthetic and natural resins; also has applications in perfumery and dry cleaning) lactate m de butyle

~LAURATE - [chem] (a plasticizer) laurate m de butyle

~MYRISTATE - [chem] (a plasticizer and textile lubricant) myristate m de butyle

~OLEATE - [chem] (a component of waterproofing compounds, solvent and plasticizer) oléate m de butyle

~PERBENZOATE - [chem] (a polymerization catalyst for styrene and acrylic plastics) perbenzoate m de butyle

~PERMALEIC ACID - [chem] (a polymerization catalyst) acide m butylpermaléique

~-PHOSPHORIC ACID - [chem] (a polymerisation agent and component of textile chemicals) acide m

butylphosphorique
BUTYL PHTHALYL BUTYL GLYCOLATE - [chem] (a plasticizer for cellusose derivatives and a number of other plastics) glycolate *m* de butyle et phtalylbutyle
~RUBBER - [rubber ind] (a synthetic rubber produced from isobutene) butylcaoutchouc *m*
~STEARAMIDE - [chem] (a synthesis intermediate) stéaramide *f* de butyle
~STEARATE - [chem] (a component of lubrificants, cosmetics, pharmaceuticals and polishes, and a plasticizer for rubber) stéarate *m* de butyle
BUTYLENE - [chem] butylène *m*
~OXIDE - [chem] (intermediate for the production of polymers) oxyde *m* de butylène
~OXIDE MIXTURES - [chem] (mixtures of 1,2- and 2,3- butylene oxides and an intermediate for a number of polymers) mélange *m* d'oxydes de butylène
BUTYNEDIOL - [chem] (a crop defoliant, corrosion inhibitor and polymerization agent) butynédiol *m*
BUTYRACEOUS - [chem] butyracé
BUTYRALDEHYDE - [chem] (a synthesis intermediate) butyraldéhyde *m*, butanal *m*
BUTYRIC ACID - [chem] (an intermediate in the production of emulsifying agents, disinfectants, flavours and perfumes, pharmaceuticals and varnishes) acide *m* butyrique
BUTYRIN - [chem] butyrine *f*
BUTYROLACTONE - [chem] (a solvent for resins and a synthesis intermediate) butyrolactone *m*
BUTYRONITRILE - [chem] (a synthesis intermediate for pharmaceutical and veterinary compounds) butyronitrile *m*
BUTYROMETER - [instr] (for the measurement of fat contents in milk and butter) butyromètre *m*
BUY, to - [gen] acheter, acquérir
BUYER - [gen] acheteur *m*, acquéreur *m*
BUYING OFFICE - [comm] service *m* des achats
BUYS BALLOT'S LAW - [met] (law stating that the atmospheric pressure decreases towards the right and increases towards the left of an observer facing the wind in the Northern hemisphere, while in the Southern hemisphere the reverse is the case) loi *f* de Buy Ballot
BUZZ, to - [gen] bourdonner, vrombir
[aero] survoler à basse altitude
~TRACK - [cinema] piste *f* d'essais de bruits
BUZZER - [telecomm] (an electrical device which emits a buzzing sound when a current passes through, used for giving warnings or the like) vibreur *m*, ronfleur *m*, trembleur *m*; (a telegraphy receiver instrument which consists of armature and electromagnet) récepteur *m* acoustique
~COIL - [el] bobine *f* de trembleur
~PRACTICE SET - [telecomm] récepteur *m* acoustique
~SIGNAL - [telecomm] signal *m* acoustique par ronfleur
~WAVEMETER - [instr] ondemètre *m* à vibreur
BUZZING - [acoust] (a sibilant hum) bourdonnement *m*, ronflement
B.W.G. - [Blue Water Gas] gaz *m* à l'eau, gaz *m* bleu

B-WIRE SYSTEM - [telecomm] utilisation *f* de la ter pour la transmission des impulsions
BY-AIR - [commun & transp] par voie aérienne
BY-LAW - [leg] arrêté *m*
BY-PASS, to - [electron] ponter, shunter
BY-PASS - [phys mech el etc] (arrangement to allo fluid or electric current to pass something by a different path) dérivation *f*
[plumb] conduit *m* de dérivation
[constr] route *f* d'évitement
~A SYSTEM, to - [phys el mech] (to conduct a flow by a path alien to the system in question) dériver, by-passer, contourner
~CAPACITOR - [el] (a capacitor connected in paral lel with a component or circuit to allow currents c certain frequencies to avoid it) condensateur *m* de découplage, condensateur *m* de fuite
~CONTROL VALVE - [mech] vanne *f* régulatrice de dérivation
~DILUTION RATIO - [el] (ratio of primary to secondary air mass flow in a by-pass turbine unit) rapport *m* de dérivation
~ENGINE - s. by-pass gas turbine
~FILTER - [el] (filter circuit arranged as a by-pass filtre *m* de dérivation
~FILTERING - [el] (arrangements of two or more filters in parallel liquid circuits, so that they can be used alternatively, for maintenance and cleaning) filtrage *m* en dérivation
~GOVERNOR - [oil ind] régulateur *m* de retour
~LINE - [mech] (a by-pass pipe) dérivation *f*, conduite *f* de dérivation, by-pass *m*
~PIPE - s. by-pass line
~RATIO - s. by-pass dilution ratio
~REGULATOR - s. by-pass governor
~TURBINE - [aero] (type of turbine engine in which part of the air taken in is used in the turbine and part delivered into the jet pipe: such air may be used as an afterburner) turbine *f* à dérivation
~VALVE - [mech] (a valve so arranged as to cause the fluid which controls it to flow past some part of its normal path, e.g. to allow a liquid to avoid a filter through which it usually passes) vanne *f* c dérivation, robinet *m* by-pass
BY-PASSED - [gen] dérivé
BY-PATH - [telecomm] découplage *m*, dérivation *f*
~CONDENSER - [el] condensateur *m* de découplage
~SYSTEM - [telecomm] système *m* de libération de organes sélecteurs
BY-PRODUCT - [gen] sous-produit *m*, produit *m* dér
~CIRCUIT - [el] (a superposed circuit) circuit *m* supplémentaire
~OVEN - [ind chem] four *m* à récupération de sous produits
BY-ROAD - [constr] route *f* secondaire, chemin *m* vicinal
BY SEA - [transp] par voie maritime, par mer
BY-WAY - [constr] route *f* secondaire, voie *f* indirecte
BYE-CHANNEL - [constr] (channel at the end of a dam) canal *m* de dérivation
BYRE - [agric] étable *f* à vaches

C

- [chem] (symbol for Carbon) symbole du Carbone *m*
[met] (symbol for Cloudy) nuageux
-AMPLIFIER - [el] (amplifier following the B-amplifiers in broadcasting studios) amplificateur *m* de classe C
-BATTERY - [el] (power supply required for the polarization of thermionic valves) batterie *f* de polarisation de grille
-BIAS - [el] polarisation *f* de grille
-NETWORK - [el] réseau *m* en C
-NUMBER THEORY - [phys] (field theory in which wave functions are functions of position and time) théorie *f* des nombres C
a - [chem] (symbol for Calcium) symbole du Calcium *m*
AB - [railw] cabine *f*, abri *m* (du mécanicien)
[colloq for taxi] taxi *m*, voiture *f* de place
CONTROL LEVER - [mech] levier *m* de commande placé dans la cabine
GUARD - [auto] bouclier *m* de protection de la cabine
LIGHT - [auto] éclairage *m* intérieur de cabine
-TYRE SHEATHING - [el etc] gaine *f* en caoutchouc dur
ABANE - [aero] (system of struts in an aircraft to which wings of an early type were braced) cabane *f*
ABBAGE, to - [metall] (the pressing of scraps) paqueter la ferraille, empaqueter les chutes
ABBAGE - [bot] chou *m*
OIL - [chem] essence *f* du sureau
SEED OIL - [ind chem] (used in the manufacture of soap) huile *f* de chou
ABBAGING - [metall] paquetage *m* de la ferraille
PRESS - [mech mach] presse *f* à paqueter la ferraille, machine *f* à empaqueter les chutes
ABBLE, to - [metall] (the operation of cutting iron bars in lengths) tronçonner
ABIN - [aero] (enclosed crew and passenger space in an aircraft) carlingue *f*, cabine *f*
[mech] (of a crane etc) cabine *f*, guérite *f*
[railw] (control room) poste *f* de manoeuvre
ALTIMETER - [instr] (altimeter fitted in the cabin) altimètre *m* de cabine
PRESSURE - [aero] (prevailing air pressure within the cabin) pression *f* de la cabine
SUPERCHARGER - [aero] (rotary air-compressor used to produce pressure in an aircraft cabin) compresseur *m* de cabine
ABINET - [gen] meuble *m* à tiroir, cabinet *m*, classeur *m*, armoire *f*, coffret *m*

CABINET AIR HEATER - [th eng] aérotherme *m*, réchauffeur *m* d'air
~BURNISHER - [carp] brunissoir *m*
~DRYER - [text] séchoir *m* à cellules, armoire-sécheuse *f*
~FILE - [carp] lime *f* d'ébéniste, écouane *f*, écoine *f*
~LOUDSPEAKER - [radio] haut-parleur *m* à coffret
~-MAKER - [carp] ébéniste *m*, menuisier *m* en meubles
~ORGAN - [mus] harmonium *m* américain
~PATTERN MULTIPLE SWITCH-BOARD - [el] commutateur *m* multiple
~RASP - [carp] râpe *f* d'ébéniste
~SCRAPER - [carp] grattoir *m*, raclette *f*
~SHELL - [carp ètc] enveloppe *f*, coffrage *m*
~SIZE - [photo] format *m* album
~WORK - [carp] ébénisterie *f*
CABLE, to - [telecomm] télégraphier, envoyer un câble, câbler
[metall] couper les barres de fer
[arch] rudenter
[naut] amarrer (avec un câble)
CABLE - [el] (insulated electrical conductor, usually stranded, which may contain more than one separate conductor) câble *m*
[text] (strong, thick rope of hemp or other fabrics) câble *m*, aussière *f*, corde *f*, amarre *f*, cordage *m*
[naut] chaîne *f* d'ancre, câble *m*
~ANGLE INDICATOR - [instr] (instrument designed to show the vertical angle between the longitudinal axis of a towed glider and the towing cable) indicateur *m* de l'angle du câble de remorque
~ARMOURING MACHINE - [mech] machine *f* à armer les câbles, armeuse *f*
~ASSIGNMENT RECORD - [telecomm] registre *m* des câbles
~ATTENUATION - [telecomm] affaiblissement *m* dû au câble
~BELT CONVEYOR - [mech] transporteur *m* à courroie et câble
~BOND - [el] connexion *f* de gaine métallique
~BOX - [el] boîte *f* de jonction
~BRACKET - [telecomm] équerre *f* pour support à câbles
~BREAK - [el etc] (a complete disconnection in a cable) rupture *f* d'un câble
~BRIDGE - [constr] pont *m* suspendu
~-BRUSH - [impl] hérisson *m* à cordes
~BUOY - [naut] bouée *f* pour câble sous-marin
~CARRIER - [el] support *m* de câbles
~CHAMBER - [telecomm] chambre *f* de répartition,

trémie *f* d'entrée de câbles
CABLE CHUTE - [telecomm] cheminée *f* d'ascension
~CIRCUIT - [el] circuit *m* câblé
~CIRCUIT DIAGRAM - [el] schéma *m* des circuits
~CLAMP - [mech] collier *m* de câble, serre-câble *m*
~CLAMPING BELT - s. cable clamp
~CLEATS - [mech] crampons *mpl* de fixation
~CLIP - [el] (device for securing cables, in the form of a loop of thin strip metal) serre-câble *m*, collier *m* de câble
~CODE - [telecomm] (variation of the Morse, used on submarine cable) code *m* pour câble
~CODE DIRECT PRINTER - [telecomm] (type of telegraph printer) téléimprimeur *m*
~COMPONENT - [telecomm] (pairs of conductors in a cable with a common characteristic) paire *f*
~COMPOUND - [el] brai *m* de câble, compound *m*
~CONDUCTOR - [el] (the metal conductor in an electric cable) conducteur *m*
~CONDUCTOR SPLICE - [el] épissure *f*
~CONNECTION - [el] connexion *f* de câble, raccord *m* de câble, attache *f* de câble
~CONNECTOR - [el] coupleur *m* électrique
~CORE - [el etc] âme *f* d'un câble
~CORRECTION - [el acoust] correction *f* d'atténuation de câble
~COUNT - [telecomm] nombre *m* de mots
~COUPLER - [el] joint *m* rapide pour câbles
~COUPLING - [el] couplage *m* par câble
~COVERING MACHINE - [mech] (a machine for applying the insulation over the conductors) machine *f* à guiper les câbles
~CURRENT TRANSFORMER - [el] (a magnetic circuit forming a current transformer) transformateur *m* de câble
~DETECTOR - [instr] (instrument designed to determine the path of an underground cable) appareil *m* localisateur de câble
~DISTRIBUTION BOX - [telecomm] boîte *f* de jonction
~DISTRIBUTION HEAD - [el] tête *f* de câble
~DISTRIBUTION POINT - [el & telecomm] point *m* de concentration
~DRILLING - [oil ind] forage *m* au câble, sondage *m* à la corde
~DRUM - [el] bobine *f* de câble
[of ropes] touret *m*
~DUCT - [el] (a channel for electric cables in the frame of a machine or in a wall etc) gouttière *f*, caniveau *f*, conduit *m*, canalisation *f*
~DUCT RING - [el] anneau *m* de guidage
~END BOX - [el] manchon *m*
~END INSULATOR BOX - [el] manchon *m* d'extrémité de câble
~EXTRUSION - [plast ind] (the operation of sheathing an electric cable with plastic material extruded round it) gainage *m* par extrusion
~FAULT - [el] défaut *m* de câble
~FAULT LOCALIZATION - [el] localisation *f* des défauts de câble
~FILL - [el] (ratio of the number of pairs in use to the total number of pairs in a cable) remplissage *m* de câble
~FILLER - [el] brai *m* de câble, matière *f* de remplissage de câble
~FOR FOUR-WIRE WORKING - [el] câble *m* pour circuit à quatre fils
~FOR TWIN BAND TELEPHONY - [telecomm] câble *m* pour téléphonie à deux bandes
CABLE FORM - [telecomm] forme *f* de câblage
~GLAND - [el] manchon *m* d'entrée
~GREASE - [ind chem] lubrifiant *m* pour câbles, graisse *f* pour câbles
~GRIP - [impl] cosse *f* de câble, manchon *m* à mailles
~GROMMET - [el] anneau *m* isolant
~GUIDE - [impl] glissière *f*
~HARNESS - [el] (pre-assembled set of electric leads) conducteurs *mpl* montés à l'avance
~HAUL - [transp] traîneur *m* à câble
~HOIST - [transp] élévateur *m* à câble, monte-charge *m*
~HOOK - [mech] équerre *f* pour support de câbles
~HOSE - [mech] gaine *f* pour câble
~HUT - [telecomm] chambre *f* de répartition
~JOINT - [el] épissure *f* de câble, jonction *f*, joint *m*
~JUNCTION INSULATOR SLEEVE - [el] manchon *m* de jonction
~-LAID - [naut] (a set of three three-strand ropes commis en grelin
~LAY UP - [telecomm] câblage *m*, constitution *f* de l'âme du câble, pas *m* de câblage, rotation *f*, torsion *f*
~LAYING - [el] pose *f* de câble
~-LAYING SHIP - s. cable ship
~LENGHT COMPENSATION - [telev] compensation *f* de longueur de câble
~LIFT - [mech] élévateur *m* à câble
~LINK - [el] câble *m* de raccord, liaison *f* par câble
~LOCKER - [naut] puits *m* aux chaînes
~LOOP - [el] noeud *m* dans un câble, boucle *f* dans câble
~LUG - [el] cosse *f* de câble, oeillet *m* de câble
~MAKE-UP - [telecomm] composition *f* d'un câble spécification *f* de câblage
~MANHOLE - s. cable hut
~MARKER - [el] borne *f* de repérage
~OF THE BRAKE - [auto] câble *m* de commande de frein
~OUTLET - [el] raccord *m* de sortie
~PLUG - [mech] fiche *f* pour câble
~PROTECTION PIPE - [el & telecomm] tuyau *m* de protection pour câble
~PROTECTIVE SLEEVE - [mech] douille *f* de protection pour câble
~RACK - [el] herse *f*, panneau *m* de fixation de câble, bâti *m* pour câbles
~RAILWAY - [transp] funiculaire *m*, téléphérique *m*
~RECORD - [el] registre *m* des câbles
~REEL - [el & telecomm] bobine *f* à câble, touret *m* tambour *m* d'enroulement
~REEL TRAILER - [mech] remorque *f* à bobines
~RIG - [oil ind] appareil *m* de forage au câble
[mining] appareil *m* de forage à la corde
~RING FOR AERIAL CABLES - [mech] anneau *m* porte câble
~RUN - [telecomm] chemin *m* de câble
~SADDLE - [el] support *m* pour câble
~SEAL - [el] joint *m* d'étanchéité de câble
~SEALING BOX - [el] boîte *f* d'extrémité unipolaire
~SHAFT - [telecomm] cheminée *f* d'ascension
~SHEATH - [el] (external envelope protecting a group of insulated conductors) gaine *f* de câble,

armature *f* de câble
ABLE SHIELD - [el] manchon *m* d'entrée de canalisation
SHIP - [naut] câblier *m*
SHOE - [el etc] oeillet *m* de câble, tête *f* de câble
SILK - [text] fil *m* câblé
SLEEVE - [el] manchon *m* de jonction
SOCKET - [el] cosse *f* de câble
SPLICE - [el] épissure *f* de câble
-STITCH DESIGN - [text] dessin *m* de nattes, torsade *f*
STRANDING MACHINE - [mech] toronneuse *f*
SYSTEM - [mining] forage *m* au câble, sondage *m* à la corde
TANK - [el] bac *m* d'essai pour câbles
TAPE - [el] ruban *m* pour câbles
TENSION - [phys] tension *f* d'un câble
TERMINAL - [el] tête *f* de câble, embout *m* protecteur
TERMINATION - [telecomm] boîte *f* de coupure, tête *f* de câble
TESTING CAR - [el] (vehicle carrying the necessary instruments to carry out tests on cables) véhicule *m* laboratoire pour câbles
TESTING TRUCK - s. cable testing car
TIER - [naut] soute *f* aux câbles
TOOL - [impl] (used in mining) outil *m* de forage au câble
VAULT - [el] trémie *f* d'entrée de câbles
VULCANIZING PAN - [impl] (a process vessel for vulcanizing the covering of electric cables) bac *m* de vulcanisation pour câbles
WAX - [el] cire *f* isolante pour câbles
-WAY - [transp] transporteur *m* aérien
WINCH - [mech] entraîneur *m* à câble
WITH FLEXIBLE COATING - [el] câble *m* à gaine souple
WORKS - [ind] fabrique *f* de câbles, câblerie *f*
YARN - [text] s. cable silk
ABLE'S LENGTH - [meas] (one tenth of a nautical mile) encâblure *f* (185,2 m)
ABLEGRAM - [telecomm] câblegramme *m*, télégramme *m*
ABLEWAY EXCAVATOR - [mech] grue *f* à câble, blondin *m*
ABLING - [text] (the production of cabled yarn, obtained through a double-twisting operation) retordage *m*
[arch] cordelière *f*, torsade *f*
[el] câblage *m*
ABOOSE - [naut] cuisine *f*
[railw] (separate truck at the end of a goods train carrying the staff) fourgon *m* de queue
ABOT PROCESS - [ind chem] (carbon black manufacturing process in which deposition is effected from natural gas flames on large plates with moving scrapers) procédé *m* cabot
QUILT - [acoust] (acoustic absorbing material between sheets of paper or canvas) matériau *m* isolant thermo-acoustique
ABOTAGE - [shipping] (coastal shipping for trade purposes) cabotage *m*
ABRERITE - [min] cabrérite *f*
ACAO - [bot] cacao *m*
BEANS - [bot] (the seeds of the cacao tree, used in the manufacture of chocolate) graines *m* de cacao
BUTTER - [chem] (the fat obtained from cacao beans, used in soap, sweets and pharmaceuticals) beurre *m* de cacao
CACAO OIL - s. cacao butter
~SHELL - [bot] gousse *f* de cacao
CACHINNATION - [med] fou rire *m*, rire *m* hystérique
CACKLING - [acoust] (the sound made by hens) caquetage *m*
CACODYL - [chem] (arsenical radical) cacodyle *m*
~OXIDE - [chem] oxyde *m* cacodylique
CACODYLATE - [chem] cacodylate *m*
CACODYLIC ACID - [chem] (herbicide and intermediate for pharmaceuticals) acide *m* cacodylique
CACOGENESIS - [genet] (inability to produce viable hybrids; also racial deterioration) cacogénèse *f*
CACOSMIA - [med] (a bad smell) cacosmie *f*, odeur *f* fétide
CACTUS - [bot] cactus *m*
CADASTRAL - [top] cadastral
~CONTROL - [top] triangulation *f* cadastrale
~MAP - [top] carte *f* cadastrale
CADASTRE - [surv] cadastre *m*
CADAVERINE - [chem] (colourless syrupy liquid) cadaverine *f*
CADDICE - [text] cadis *m*
CADDIS - s. caddice
CADE OIL - [ind chem] (thick brown liquid obtained by dry distillation of juniper wood) huile *f* de cade
CADENCE - [telecomm] (the audible signal given to the operator) appel *m*
CADION - [chem] (a reagent) cadion *m*
CADMIUM, to - [metall] cadmier
CADMIUM - [chem] (an element, symbol Cd. Soft bluish metal used in bearing metals, pigment manufacture etc) cadmium *m*
~ACETATE - [chem] (used for dyeing and for the preparation of ceramic glazes) acétate *m* de cadmium
~BROMIDE - [chem] (used in photography) bromure *m* de cadmium
~CARBONATE - [chem] carbonate *m* de cadmium
~CARMINE - [paint] rouge *m* de cadmium
~CELL - [chem] (a standard cell consisting of an amalgamate cadmium anode covered with crystals of cadmium sulphate, and a mercury cathode covered with solid mercurious sulphate) pile *f* Weston
[electron] (a vacuum photo-electric cell with a cadmium-coated cathode) cellule *f* photoélectrique au cadmium
~CHLORIDE - [chem] (photographing chemical; a starting material in the production of cadmium sulphide and an ingredient of cadmium plating baths) chlorure *m* de cadmium
~COPPER - [metall] (alloy of copper with 0.7 to 1.0 p.c. of cadmium) cuivre *m* au cadmium
~CUT-OFF - [nucl] (the sharp drop of the neutron absorption cross section of cadmium at higher energies) limite *f* de capture du cadmium
~ELECTRODE - [electron] électrode *f* en cadmium
~GREEN - [paints] (pigment; a mixture of cadmium yellow and viridian) vert *m* de cadmium
~HYDROXIDE - [chem] (used in cadmium plating) hydroxyde *m* de cadmium
~IODIDE - [chem] (an analitical reagent and photographic chemical) iodure *m* de cadmium
~LAMP - [light] lampe *f* à vapeur de cadmium
~-NICKEL STORAGE BATTERY - [el] (alkaline storage

battery in which the positive electrode is principally of nickel hydroxide and the negative chiefly of cadmium alloy) accumulateur *m* nickel-cadmium
CADMIUM NITRATE - [chem] nitrate *m* de cadmium
~OXIDE - [chem] (a catalyst and a pigment) oxyde *m* de cadmium
~PLATED - [ind chem] cadmié
~PLATING - [metall] cadmiage *m*
~RED - [paint] (red pigment consisting of a mixture of cadmium sulphide and selenide) rouge *m* de cadmium
~RED LINE - [light] (the source selected for comparing optical wavelength with standard lengths) ligne *f* rouge du cadmium
~REGULATOR - [nucl] (regulating rod containing cadmium) régulateur *m* au cadmium
~RICINOLATE - [chem] (a heat and light stabilizer for PVC copolymers) ricinoléate *m* de cadmium
~SELENIDE - [chem] (a red pigment used in rubber and in paints) séléniure *m* de cadmium
~SILVER - [metall] alliage *m* argent-cadmium
~STANDARD CELL - s. cadmium cell
~STRIP - [nucl] (a strip-like control element made of cadmium in a reactor) bande *f* de cadmium
~SULPHATE - [chem] (analytical reagent) sulfate *m* de cadmium
~SULPHIDE - [chem] (used in paints; a pigment) sulfure *m* de cadmium
~TEST - [el] (method of testing lead-acid storage cells) essai *m* au cadmium
~TUNGSTATE - [chem] (a catalyst and a component of fluorescent X-ray screens) tungstate *m* de cadmium
~YELLOW - [paint] (an alternative name for cadmium sulphide) cadmium *m* sulfuré, jaune *m* de cadmium
CADUCOUS - [bot] (falling from the plant in a short time) caduc, caduque
CAENOGENESIS - [genet] (introduction of adaptive characters or structures during development) cénogénèse *f*
CAESIUM - [chem] (also Cesium) (an element, symbol Cs. Soft silvery metal decomposing in water. Used in photoelectric cells and in electronics) césium *m*
~CELL - [el] (a photo-electric cell in which the cathode is a thin layer of caesium deposited on exceedingly small globules of metallic silver. This cell is particularly sensitive to infra-red radiation) cellule *f* photo-électrique au césium
~OXYGEN CELL - [el] (a modification of the caesium cell in which sensitivity to red light is increased) cellule *f* photoélectrique au césium à atmosphère d'oxygène
C.A.F. - [abbrev. for Cost & Freight] coût-fret
CAFFEINE - [chem] (an alkaloid derived from tea or coffee and produced synthetically. Used in medicine and as an ingredient of soft drinks) caféine *f*
CAGE, to - [gen] verrouiller
CAGE - [gen] cage *f*
[build] palissade *f*
[constr] ossature *f* d'un bâtiment
[mech] (device for spacing out the balls in a ball-bearing) cage *f* de roulement
[electron] (the electrode system of a thermionic valve) cage *f*
[of a lift] cage *f* d'ascenseur
[mining] cage *m*
CAGE AERIAL - [radio] (type of aerial consisting of a cage-like arrangement of wires connected in parallel) antenne *f* en cage
~ANTENNA - s. cage aerial
~BAIL - [mining] attache *f* de câble à la cage d'extraction
~BOX - [mining] cage *f* d'extraction
~CENTRIFUGE - [mech] essoreuse *f* verticale, centrifuge *f* verticale
~COIL - [el] bobine *f* à cage
~DIPOLE - [ant] (a dipole with elements consisting of rods spaced equally in a circle) dipôle *m* en cag
~GRID - [electron] (grid with the shape of a cage) grille *f* en forme de cage
~OF THE DIFFERENTIAL - [mech] boîtier *m* de différentiel
~RELAY - [el] relais *m* inductif
~SAFETY APPARATUS - [mining] parachute *m*
~SEATS - [mining] clichage *m* pour cages
~-TYPE NEGATIVE PLATE - [el] (a negative plate fo a battery made up by riveting two lead grids together and placing the active material in the space between them) plaque *f* négative à grille
~-TYPE VALVE - [mech] soupape *f* à corbeille
~WHEEL EXTENSION - [mech] (used to increase traction) barillets *mpl* de jumelage
~WINDING - [el] (also called squirrel cage winding cage *f* d'écureuil
[mining] extraction *f* par cages
~WORK - [mech] grillage *m*
CAGER - [mining] clicheur *m* (de wagons)
CAGING - [mech] blocage *m*
[mining] clichage *m* (des wagons)
~DEVICE - [mech] dispositif *m* de blocage (gyroscope)
CAIMAN - [zool] caïman *m*
CAIQUE - [naut] caïque *m*
CAISSON - [hydr] caisson *m*, bâtardeau *m*
[naut] (a boat) bateau-porte *m*
[build] (section of ceiling) caisson *m*
~CEILING - [constr] plafond *m* à compartiments
~FOUNDATION - [constr] fondation *f* sur caissons (cylindriques)
~SICKNESS - [med] mal *m* des caissons
~SYSTEM - [constr] fondation *f* à caissons non-perdus
CAKE, to - [ind chem] (to collect in discrete masses) se concrétionner, floculer, se prendre en gâteau, s'agglutiner
CAKE - [food] gâteau *m*
[of soap] pain *m* de savon
[ind chem] (a compressed substance) pain *m*, gâteau *m*
[an oil cake] tourteau *m*
[metall] lingot *m* de départ
~ALUM - [chem] sulfate *m* d'aluminium
~MOULDING MACHINE - [mech] machine *f* à moule les pains
~OF CINDER - [metall] gâteau *m* (ou plaque) de scories
~OF RUBBER - [rubber ind] bloc *m* de caoutchouc
~TOGETHER, to - [oil ind] se coller, (se) cailler, se lier
~WAX - [el acoust] (thick disc of wax on which a recording can be registered) disque *m* (vierge) en cire

AKING - [gen] agglomération *f*, mise *f* en blocs
[paint] (setting of paint in a caked mass) agglomération *f*, concrétion *f*, floculation *f*
CAPACITY - [metall] pouvoir *m* collant
˜COAL - [min] charbon *m* cokéfiant
ALABAR BEAN - [bot] fève *f* de Calabar
ALABASH - [bot] calebasse *f*
ALAMIFEROUS - [bot] calamifère
ALAMINE - [min] (carbonate of zinc occurring in calcareous rocks, veins and beds; also called smithsonite in England and hemophormite in the USA) calamine *f*
ALAMUS - [zool] (the hollow part of a feather) calamus *m*, rotin *m*
˜OIL - [ind chem] essence *f* de calamus
ALANDRIA - [gas ind] (in relation to evaporators) élément *m* chauffant
[nucl] cuve *f*
[sugar ind] chambre *f* de vapeur
ALATHIDE - [bot] capitule *m*
ALAVERITE - [min] calavérite *f*
ALCANEITIS - [med] calcanéite *f*
ALCANEUS - [med] calcanéum *m*, pied *m* bot talus
ALCAR - [glass man] arche *f*
[metall] four *m* à recuire, fourneau *m* à calciner
˜AVIS - [zool] petit hippocampe *m*
ALCAREOUS - [chem] (containing compounds of calcium) calcaire
˜BINDING - [metall] ciment *m* calcaire
˜CEMENTING MATERIAL - [constr] ciment *m* calcaire
˜CLAY - [geol] (very fine-grained rock, clay or loam, with an addition of calcium carbonate) argile *f* calcaire
˜MARL - [geol] calcaire *m* marneux, marne *f* calcaire
˜ROCK - [geol] (sedimentary rocks containing calcium carbonate) roche *f* calcaire
˜SANDSTONE - [geol] grès *m* calcaire
˜SPAR - [min] calcite *f*, spath *m* calcaire
˜TUFF - [geol] tuf *m* calcaire
˜WATER - [chem] eau *f* calcaire
ALCEMIA - [med] calcémie *f*
ALCEOLARIA - [bot] calcéolaire *f*
ALCIC LIVER OF SULPHUR - [chem] foie *m* de soufre calcique
ALCICOLE - [bot] (flourishing on grounds rich in calcium carbonate) calcicole
˜PLANTS - [bot] plante *f* calcicole
ALÇICOSIS - [med] (a lung disease caused by the inhalation of marble dust) chalicose *f*, phtisie *f* des tailleurs de pierre
ALCIFEROL - [chem] vitamine *f* D_2, calciférol *m*
ALCIFEROUS - [chem] (producing or containing calcium salts) calcifère
˜SANDSTONE - [geol] grès *m* calcifère
ALCIFICATION - [zool & bot] calcification *f*
ALCIFUGE - [bot] (intolerant of limy soil) calcifuge
˜PLANTS - [bot] plantes *mpl* calcifuges
ALCIFY, to - [gen] calcifier; convertir en carbonate *m* de chaux
ALCIGENOUS GLANDS - [zool] (glands producing a limy secretion) glandes *fpl* calcigènes
ALCIMETER - [instr] calcimètre *m*
ALCIMINE - [paint] (made up of whiting and glue with water) badigeon *m*

CALCINATION - [chem] (the process of subjecting materials to the effects of prolongued heating) calcination *f*, cuisson *f*, frittage *m*, grillage *m*
[metall] (the elimination of water and carbon dioxides from ores) calcination *f*
CALCINE, to - [chem] calciner
CALCINED ALUMINIA - [chem] alumine *f* calcinée, alumine *f* anhydre
˜BARYTA - [chem] baryte *f* anhydre
˜COPPER METAL - [metall] matte *f* de cuivre grillée
˜FLINT - [geol] silex *m* calciné
˜MAGNESIA - [chem] magnésie *f* calcinée
˜PYRITES - [metall] résidu *m* de pyrites grillées, cendres *fpl* de pyrites
˜SODA - [chem] carbonate *m* neutre de calcium anhydre
CALCINING - [chem] calcination *f*
˜FURNACE - [metall] four *m* à calciner
CALCINOSIS - [med] calcinose *f*
CALCIOSAMARSKITE - [min] (a rare ore, containing uranium and thorium) calciosamarskite *f*
CALCIOTHORITE - [min] (a radioactive product of thorite containing calcium) calciothorite *f*
CALCIPEXIS - [med] fixation *f* de calcium
CALCIPHOBE - s. calcifuge
CALCIPRIVIA - [med] perte *f* de calcium
CALCIPYELITIS - [med] pyélite *f* calculeuse
CALCITE - [min] (calcium carbonate in crystalline form) calcite *f*, spath *m* calcaire
˜GRATING SPACE - [cryst] (the distance between the diffracting spaces in calcite crystals) réseau *m* de diffraction de la calcite
CALCITHERAPY - [med] calcithérapie *f*
CALCIUM - [chem] (an element, symbol Ca. Soft white metal used in the production of intermediates and in metallurgy) calcium *m*
˜ACETATE - [chem] (a dyeing mordant and an intermediate in the production of acetic acid and derivatives) acétate *m* de calcium
˜ACETYLSALICYLATE - [chem] (an antipyretic, used in medicine) acétylsalicylate *m* de calcium
˜ACRYLATE - [chem] (binding agent for foundry sands; also an intermediate for ion-exchange resins) acrylate *m* de calcium
˜ALGINATE - [chem] (used in the production of surgical dressings) alginate *m* de calcium
˜ALUMINATE - [chem] (an ingredient of cement; also a refractory material) aluminate *m* de calcium
˜AMMONIUM NITRATE - [chem] (a fertilizer) nitrate *m* de chaux et d'ammoniaque, ammonitre *m*
˜ARSENATE - [chem] (a disinfectant and insecticide) arseniate *m* de calcium
˜ARSENITE - [chem] (a component of bactericides and insecticides) arsénite *m* de calcium
˜BICARBONATE - [chem] bicarbonate *m* de calcium
˜BISULPHITE - [chem] (a bleaching agent for textiles, a disinfectant, preservative and germicide) bisulfite *m* de calcium
˜BROMIDE - [chem] (a food and wood preservative; also a sedative used in medicine and a component of photographic chemicals) bromure *m* de calcium
˜CARBIDE - [chem] (a source of acetylene gas for industrial use and a starting material for a number of organic syntheses) carbure *m* de calcium
˜CARBONATE - [min] (a common mineral widely distributed as chalk and limestone) carbonate *m* de calcium, carbonate *m* de chaux

CALCIUM CHLORATE - [chem] (a herbicide, photographic chemical and a component of pyrotechnic compounds) chlorate *m* de calcium, chlorate *m* de chaux

~CHLORIDE - [chem] (a drying agent and component of refrigerant and dustproofing mixtures) chlorure *m* de calcium, chlorure *m* calcique

~CHLORIDE CYLINDER - [ind chem] (a small glass column packed with calcium chloride along which gases are passed to extract water from them) tube *m* à chlorure de calcium

~CHROMATE - [chem] (a yellow pigment) chromate *m* de calcium

~-CHROMIUM GARNET - [min] grenat *m* chromo-calcareux

~CITRATE - [chem] (an intermediate in the manufacture of citric acid) citrate *m* de calcium

~CYANAMIDE - [chem] (a fertilizer and herbicide; also used in iron and steel production) cyanamide *f* calcique, cyanamide *f* de chaux

~CYANIDE - [chem] (a rodenticide and pesticide. Principally used in the extraction of gold and other metals from their ores) cyanure *m* de calcium

~DEHYDROACETATE - [chem] (a fungicide) déhydroacétate *m* de calcium

~FLUORIDE - [chem] (raw material in the manufacture of hydrofluoric acid and a flux in metallurgy) fluorure *m* de calcium

~GLUCONATE - [chem] (used in medicine as a source of calcium) gluconate *m* de calcium

~GLYCEROPHOSPHATE - [chem] (a plastics stabilizer; also used in medicine) glycérophosphate *m* de calcium

~HYDRIDE - [chem] (an analytical reagent and portable source of hydrogen) hydrure *m* de calcium, hydrolithe *m*

~HYPEROXIDE - [chem] (slaked lime, used in sugar and coal-gas purification, in glass production and in the manufacture of plasters and mortars) hydroxyde *m* de calcium

~HYPOCHLORITE - [chem] (a bleaching agent, disinfectant and fungicide) hypochlorite *m* de calcium

~HYPOPHOSPHITE - [chem] (used in medicine as a tonic) hypophosphite *m* de calcium

~IODIDE - [chem] (used in medicine and photography) iodure *m* de calcium

~LACTATE - [chem] (used in medicine as a source of calcium and in food processing) lactate *m* de calcium

~LAEVULINATE - [chem] (used in medicine) lévulinate *m* de calcium

~LINOLEATE - [chem] (an emulsifying, stabilizing and waterproofing agent) linoléate *m* de calcium

~MAGNESIUM CHLORIDE - [chem] (a reagent and intermediate for textile chemicals and dyes) chlorure *m* de magnésium et de calcium

~NAPHTENATE - [chem] (a component of lake pigments, waterproofing agents, varnishes and waxes) naphténate *m* de calcium

~NITRATE - [chem] (a rubber processing chemical and a component of explosives and fertilizers) nitrate *m* de calcium

~OXALATE - [chem] (a starting material for oxalic acid and organic oxalates) oxalate *m* de calcium

~OXIDE - [chem] (quicklime. Widely used in industrial processes) chaux *f* vive

~PALMITATE - [chem] (a lubricant oil additive and waterproofing agent) palmitate *m* de calcium

CALCIUM PANTOTHENATE - [chem] (used in medicine as a nutritional supplement) pantothénate *m* de calcium

~PERMANGANATE - [chem] (an oxidant and disinfectant) permanganate *m* de calcium

~PEROXIDE - [chem] (an oxidizing agent, seed disinfectant and bleach for oils) bi-oxyde *m* de calcium, peroxyde *m* de calcium

~PHOSPHATE - [chem] (white crystalline powder, insoluble in alcohol) phosphate *m* de calcium, phosphate *m* de chaux

~PHOSPHIDE - [chem] (an ingredient of signal flares) phosphure *m* de calcium

~PLUMBATE - [chem] (an oxidizing agent) plombate *m* de calcium

~PYROPHOSPHATE - [chem] (a dietary supplement and ingredient of polishes) pyrophosphate *m* de calcium

~RESINATE - [chem] (limed resin. A component of paint driers, tanning agents and waterproofing compounds) résinate *m* de chaux, résinate *m* de calcium

~RICINOLEATE - [chem] (a stabilizer for polyvynil chloride and a component of lubricant additives) ricinoléate *m* de calcium

~SALYCILATE - [chem] (used in medicine) salicylate *m* de calcium

~SILICATE - [chem] (an absorbent and a filter in rubber compounding) silicate *m* de calcium

~SILICON - [chem] silico-calcium *m*

~STANNATE - [chem] (an intermediate for ceramic glazing) stannate *m* de calcium

~STEARATE - [chem] (a lubricant, stabilizer for plastics, waterproofing agent and a component of surface coatings) stéarate *m* de calcium

~SULPHAMATE - [chem] (a flameproofing agent for textiles) sulfamate *m* de calcium

~SULPHATE - [chem] (a pigment, filler and drier for paints etc. Plaster of Paris is the anhydrous commercial form) sulfate *m* de calcium

~SULPHIDE - [chem] (a depilatory and an ingredient of luminous paints) sulfure *m* de calcium

~SULPHITE - [chem] (a disinfectant, food preservative and antichlor in textile processing) sulfite *m* de calcium

~THIOCYANATE - [chem] (a solvent for polyacrylates and cellulose, and a textile finishing agent) thiocyanate *m* de calcium

~TUNGSTATE - [chem] (an ingredient of fluorescent paints) tungstate *m* de calcium

~UREA - [agric] urée *f* calcique

CALCIURIA - [med] calciurie *f*

CALCOGRAPHY - [draw] calcographie *f*

CALCOID - [med] néoplasme *m* de la pulpe dentaire

CALCSPAR - s. calcite

CALCULATE, to - [gen] calculer

CALCULATED LENGTH - [text] longueur *f* théorique

~LENGTH OF HANKS - [text] longueur *f* théorique des écheveaux

~WEIGHT - [text] poids *m* théorique

~WEIGHT OF YARN - [text] poids *m* théorique du fil

~ZENITH DISTANCE (C.Z.D.) - [astr] (the Great Circle distance between the geographical position of the body under observation and the dead reckoning position of the observer) distance *f* zénithale calculée

CALCULATING DISK - [impl] disque *m* à calculer
~MACHINE - [mech] (a machine which can perform one or more of the basic arithmetical operations) machine *f* à calcul, calculatrice *f*
~PUNCH - [comput] (machine which reads a punched card, performs a series of operations and punches the result on the cards) unité *f* de lecture et de perforation, lectrice-perforatrice *f*
CALCULATION - [math] calcul *m*, compte *m*
~CHART - [math] abaque
~OF BLAST - [metall] calcul *m* du vent (ou de l'air)
~OF EARTH THRUST - [constr] calcul *m* de la poussée du sol
~OF HEALDS - [text] calcul *m* des lisses
~OF THE WEIGHT OF MATERIAL - [text] calcul *m* de poids de la pièce
CALCULATOR - [mech] (a calculating machine) calculatrice *f*
CALCULUS - [math] calcul *m*
[med] calcul *m*
~OF PROBABILITIES - [math] calcul *m* des probabilités
~OF VARIATIONS - [math] calcul *m* des variations
CALEFACTION - [phys] caléfaction *f*
CALENDAR - [gen] calendrier
~DAYS - [astr] jours *m* successifs
~MONTH - [astr] mois *m* civil
~WORK - [horol] mouvement *m* de calendrier
~YEAR - [astr] année *f* civile
CALENDER, to - [text ind. paper man etc] calandrer
CALENDER - [mech] (a mill with heated rollers used to impart close dimensional tolerance and high finish to sheet material) calandre *f*, laminoir *m*
[paper man] calandre *m*
~BOWL - [mech] cylindre *m* de calandre
~COATER - [mech] (machine with rollers for applying a coating) calandre *f*
~COATING - [mech] (a coating applied to sheet material by means of a calender) revêtement *m* par calandrage
~COLOURED - [paper man] coloré en surface
~CUTS - [paper man] (blemishes or defects caused by the calender) défauts *mpl* dûs au calandrage
~FELT - [text] feutre *m* pour calandre
~FINISH - [text] finissage *m* à la calandre, apprêt *m* à la calandre
~FRICTION - [rubber ind] rapport *m* de vitesse des cylindres de calandre à friction
~GRAIN - [rubber] effet *m* de calandrage
~PROFILE - [rubber] profile *m* de calandre
~ROLLER - [mech] (rollers designed to feed timber into a machine) rouleaux *mpl* d'alimentation
~TRAIN - [mech] (a series of calenders used to perform sequential operations in a single process) train *m* de cylindres à calandrer
CALENDERED - [gen] calandré
~LINEN - [text] toile *f* calandrée
~PAPER - [paper man] papier *m* couché
~SHEET - [rubber ind] (sheet rubber treated by passing it through pressure rollers) feuille *f* calandrée
[plast ind] feuille *f* laminée
CALENDERING - [gen] (the operation of reducing thickness) laminage *m*
[text paper man etc] calandrage *m*
CALEOMETER - [instr] (measuring the heat loss from a calibrated wire) caléomètre *m*

CALF - [anat] mollet *m*
[zool] veau *m*
[leather ind] veau *m*, vachette *f*
~LINE - [oil ind] (drilling) câble *m* de manoeuvre
~PAPER - [paper man] papier *m* simili-cuir
~PEN - [agric] enclos *m* pour veaux
~SCOURS - [vet] diarrhée *f* des veaux
~-SKIN - [leather ind] veau *m*, vachette *f*, cuir *m* de veau
~WHEEL - [mining] treuil *m* pour tubage
CALIBER - [mech] (the bore, or internal diameter of a pipe) calibre *m*, diamètre *m* intérieur
CALIBRATE, to - [mech] (the operation of determining the absolute values corresponding to a graduation on an instrument) calibrer, étalonner, graduer, régler
[the operation of adjusting the graduation] tarer
[the operation of measuring a caliber] vérifier un calibre
~RESISTANCES, to - [el] étalonner les résistances, vérifier les résistances
CALIBRATED - [gen] calibré, étalonné, gradué, taré
~AIR-SPEED - [aero] vitesse *f* corrigée
~ATTENUATOR - [electron] (power measuring device using a thermistor) atténuateur *m* calibré
~CHAIN - [metall] chaîne *f* calibrée
~DIAL - [instr] cadran *m* gradué
~FOR AIR - [instr] (of a gauge) étalonné pour l'air, calibré pour l'air (jauge)
~ORIFICE - [mech] orifice *m* calibré, trou *m* calibré
~SLIDE-WIRE - [instr] (a component of potentiometers) fil *m* à contact glissant étalonné
CALIBRATING GAS - [gas ind] gaz *m* de référence
~TABLE - [impl] banc *m* d'étalonnage, banc *m* de tarage
CALIBRATION - [instr] calibrage *m*, étalonnage *m*, graduation *f*, réglage *m*, tarage *m*
~ACCURACY - [instr] précision *f* d'étalonnage
~BATTERY - [el] (battery of a constant voltage feeding circuits in calibrating instruments) pile *f* d'étalonnage
~BOARD - [el] panneau *m* d'étalonnage
~CAPACITOR - [el] (a capacitor whose value is accurately known) condensateur *m* étalon
~CIRCLE - [radar] (circles on the screen of a cathode-ray tube to facilitate the reading of the bearing) cercle *m* gradué, cercle *m* de distance, cercle *m* d'étalonnage
~CIRCUIT - [el] (reference circuit specifying certain values) circuit *m* d'étalonnage
~CURVE - [instr] courbe *f* d'étalonnage
~ERROR - [instr] erreur *f* d'étalonnage
~FACTOR - [instr] facteur *m* d'étalonnage, facteur *m* de calibrage
~FREQUENCY - [el] (the frequency which is set for calibrating a signal generator) frèquence *f* étalon
~INSTRUMENT - [instr] (reference measuring instrument) appareil *m* de mesure étalon
~OSCILLATOR - [el] (oscillator used to calibrate instruments) oscillateur *m* d'étalonnage
~PRESSURE - [instr] pression *f* d'étalonnage
~PULSE - [electron] impulsion *f* de calibrage
~READING - [instr] (of a measuring instrument) signal *m* d'étalonnage, lecture *f* d'étalonnage
~RESISTANCE - [el] (a resistance whose value is accurately known) résistance *f* étalon

CALIBRATION RING - [radar] s. calibration circle
[mech] cercle *m* gradué
~TEST - [instr] contrôle *m* de l'étalonnage
~TEST BENCH - [el] (test unit comprising instruments for calibration) banc *m* d'étalonnage
~UNIT - [radar] générateur *m* de cercles de distance
~VALUE - [instr] (of a gauge head) valeur *f* d'étalonnage
CALICHE - [min] (mineral of variable composition containing sodium nitrate, sulphate and chloride, potassium nitrate etc) caliche *m*
CALICO - [text] calicot *m*, blanc *m* de coton
~FOR LINING - [text] calicot *m* pour doublure
~WEAVE - [text] armure *f* toile
CALIFORNIAN JADE - [min] (used as ornamental stone) californite *f*
~ONYX - [min] (a brown tinted aragonite) aragonite *f* brune
CALIFORNITE - s. californian jade
CALIFORNIUM - [chem] (a name suggested for element 98; symbol Cf) californium *m*
CALIGATION - [med] obscurcissement *m* de la vue
CALIPER, to - [mech] (to measure with calipers or with a caliper gauge) calibrer
CALIPER - [horol] (the size of a watch movement) calibre *m*
~GAUGE - [instr] (a measuring instrument with a graduated bar carrying a fixed and a sliding jaw) jauge *f* à coulisse, calibreur *m*
~SQUARE - [instr] pied *m* à coulisse, jauge *f* à coulisse
CALIPERS - [instr] (measuring device consisting of two sliding or pivoting arms) compas *m* à calibrer
CALK, to - s. caulk, to
CALKED ENDS - [constr] (the ends of built-in iron ties) extrémités *f*pl matées de tirants
~SEAMS - [shipbuild] coutures *f*pl calfatées
CALKING - s. caulking
~CHISEL - s. caulking chisel
~TOOL - s. caulking tool
CALL, to - [telecomm] appeler
[naut] faire escale
CALL - [telecomm] appel *m*, communication *f*
~ADDRESS - [comput] (instruction address of the input instruction of a subroutine) adresse *f* d'appel
~ANNOUNCER SYSTEM - [telecomm]indicateur *m* des numéros demandés
~-BACK TELEPHONE APPARATUS - [telecomm] appareil *m* téléphonique mixte
~BOX - [telecomm] cabine *f* téléphonique
~CIRCUIT - [telecomm] ligne *f* d'ordres, circuit *m* de liaison entre opérateurs
~CIRCUIT KEY - [telecomm] touche *f* de circuit de liaison
~CIRCUIT METHOD - [telecomm] méthode *f* de la ligne d'ordres
~CIRCUIT OPERATION - [telecomm] exploitation *f* avec ligne d'ordres
~COUNT - [telecomm] nombre *m* d'appels
~-COUNT RECORD - [telecomm] compte *m* du nombre d'appels
~-COUNTING METER - s. call meter
~DISTRIBUTOR - [telecomm] distributeur *m* d'appels
~FILL - [telecomm] coefficient *m* d'occupation, coefficient *m* d'utilisation
CALL FILLED FOR LATER COMPLETION - [telecomm] appel *m* différé
~FINDER - [telecomm] sélecteur *m* d'appel, chercheur *m* d'appel
~FOR TENDERS - [comm] (an invitation to submit detailed offers for a contract) appel *m* d'offres
~-IN - [comput] (the transfer of the control of a digital computer from a main routine to a subroutine) appel *m*
~IN A STATION, to - [telecomm] faire entrer une station dans un réseau
~INDICATOR - [telecomm] volet *m* d'appel
~INDICATOR POSITION - [telecomm] position *f* à indicateur d'appel
~KEY - [telegraph] touche *f* d'appel
~LETTER - [comm] avis *m* d'appel de fonds
~LETTERS - [radio] indicatif *m*
~LOCATOR - [telecomm] localisateur *m* d'appels
~METER - [telecomm] compteur *m* d'abonné, compteur *m* d'appels, compteur *m* de statistique
~MINUTE - [telecomm] communication-minute *f*
~NOTIFICATION - [telecomm] avis *m* d'appel
~NUMBER- [comput] (set of characters identifying a subroutine) numéro *m* d'appel
~OF HAND - [telecomm] appel *m* en instance
~ORDER TICKET - [telecomm] fiche *f* de communication, fiche *f* de conversation
~SIGN - [telecomm] indicatif *m* d'appel
~SIGNAL - [telecomm] indicatif *m* d'appel
~SIGNAL DEVICE - [railw] dispositif *m* d'appel
~THE ROLL, to - [telecomm] faire l'appel
~TICKET - [telecomm] fiche *f* de conversation, fiche *f* de communication
~UP, to - [telecomm] téléphoner, appeler
~WITH REQUEST FOR CHARGES - [telecomm] demande *f* de communication avec indication de taxe
~WORD - [comput] (call number corresponding to a machine word) mot *m* d'appel
CALLAINITE - [min] (rare green phosphate of aluminium) collaïnite *f*
CALLAITE - [min] (mineral turquoise) callaïte *f*
CALLAN CELL - [el] (a primary cell, similar to the Bunsen cell, but whith a cast-iron positive electrode) pile *f* de Callan
CALLAUD CELL - [el] (a primary cell in which the zinc sulphate solution floats on the upper and lower parts of the container) pile *f* impolarisable de Callaud
CALLED AND CALLING SUBSCRIBER'S RELEASE - [telecomm] libération *f* au raccrochage des deux correspondants
~EXCHANGE - [telecomm] central *m* d'arrivée
~LINE - [telecomm] ligne *f* demandée
~SUBSCRIBER - [telecomm] abonné *m* demandé
~SUBSCRIBER RELEASE - [telecomm] remise *f* en circuit de la ligne de l'abonné appelé
CALLING CORD - [telecomm] cordon *m* d'appel
~DROP - [telecomm] volet *m* d'indicateur d'appel
~EQUIPMENT - [telecomm] équipement *m* d'abonné
~FREQUENCIES - [radio] (frequencies reserved for special cases in the radioservice) fréquence *f* d'appel
~JACK - [telecomm] jack *m* local
~LAMP - [telecomm] voyant *m* d'appel
~LINE - [telecomm] ligne *f* appelante
~MAGNETO - [telecomm] magnéto *f* d'appel
~PLUG - [telecomm] fiche *f* d'appel

CALLING POSITION - [telecomm] position *f* d'appel
~RELAY - [telecomm] relais *m* d'appel
~SEQUENCE - [comput] séquence *f* d'appel
~SIGNAL - [telecomm] indicatif *m* d'appel, indicateur *m* d'appel
~SUBSCRIBER - [telecomm] demandeur *m*
~SUBSCRIBER RELEASE - [telecomm] remise *f* en circuit de la ligne du demandeur
~WAVE - [telecomm] impulsion *f* d'appel
CALLIPER - s. caliper
CALLIPERS - s. calipers
CALLOSE - [chem] (a carboydrate) callose *f*
CALLOSITY - [med] callosité *f*, cal *m*, durillon *m*
CALLOUS - [med] calleux
CALLUS - [med & agric] cal *m*, calus *m*
~PLATE - [bot] (pad of callose) callosité *f*
CALM, to - [gen] calmer, apaiser, tranquilliser, modérer
CALM - [gen] calme, tranquille, posé
[met] (free from wind disturbance) calme, sans vent; (atmospheric condition when no wind is present) bonace *f*
CALOMEL - [chem] (mercurous chloride, used in medicine) calomel *m*
[min] (also called Horn Quicksilver; a naturally occurring mercurious chloride) calomel *m*
~ELECTRODE - [el] (a mercury electrode contained in a half-cell and immersed in a solution of potassium chloride saturated with mercurous chloride) électrode *f* au calomel
~HALF-CELL - [el chem] (half-cell with mercury electrode and potassium chloride electrolyte of specified concentration, saturated with mercurous chloride) électrode *f* au calomel
CALORESCENCE - [phys] (the production of visible light by energy derived from invisible radiation) calorescence *f*
CALORIE - [phys] (or calory; the quantity of heat required to raise the temperature of 1 kg of water 1 centigrade) calorie *f*
CALORIFIC - [phys] calorifique
~CAPACITY - [phys] chaleur *f* spécifique, capacité *f* calorifique
~EFFECT - [phys] effet *m* calorifique
~INTENSITY - [phys] (the highest temperature obtainable with the combustion of a specified fuel) châleur *f* maximale de combustion
~POWER - [phys] (the heat produced by the complete combustion of a specified weight of fuel) pouvoir *m* calorifique
~REQUIREMENT - [phys] besoin *m* calorifique
~VALUE - s. calorific power
CALORIFIER - [the eng] calorifère *m*
CALORIGENIC - [med] calorigène
CALORIMETRIC - [phys] calorimétrique
~BOMB - [instr] bombe *f* calorimétrique
~COEFFICIENT - [th dyn] coefficient *m* calorifique
~TEST - [phys] essai *m* calorimétrique
[metall] détermination *f* calorimétrique
CALORIMETRY - [phys] calorimétrie *f*
CALORIZE, to - [metall] (the operation of rendering steel or iron resistant to oxidation by a treatment with aluminium) caloriser
CALORIZING - [metall] calorisation *f*
CALORY - s. calorie
CALOTTE - [build] (small dome in the ceiling of a room) calotte *f*
CALUTRON - [nucl] (an isotope separator of the electromagnetic type) calutron *m*
CALVARIA - [med] calotte *f* du crâne
CALVE, to - [agric] mettre bas (un veau), vêler
CALVING - [agric] vêlage *m*, vêlement *m*
~PEN - [agric] enclos *m* réservé au vélage
CALX - [anat] talon *m*
[metall] résidu *m* de calcination, cendre *f* métallique
CALYX - [bot] calice *m*, vase *f* de tulipe
CAM - [mech] (a curved surface or profile designed to produce a specific motion to a part forced to follow it during its movement) came *f*, excentrique *m*, élément *m* de course, serrure *f* d'aiguilles, partie *f* de la serrure, came d'aiguille
~ACTUATED - [mech] commandé par came
~AND BOWL PICKING MOTION - [text] mécanisme *m* de chasse à excentrique et rouleau
~AND ROLLER HOIST - [mech] (hydraulic hoist in which a horizontal cylinder thrust is converted into a vertical thrust) vérin *m* hydraulique à rampe et galet
~ANGLE DEGREE - [mech] degré *m* de l'angle de came
~BAR - [mech] barre *f* courbe, barre incurvée
~BIT - [text] piège *f* réglable de la came
~BOWL - [mech] roulette *f*
~BOX - [mech] boîte *f* à cames
~BOX RING - [text] plaque *f* à cames du plateau
~CHANNEL - [mech] chemin *m* de la serrure
~CIRCLE - [mech] cercle *m* primitif de la came
~DISC - [mech] excentrique *m*
~DRIVE - [mech] commande *f* à came
~FINGER - [mech] doigt *m* de démoulage, broche *f* oblique
~FOLLOWER - [mech] (the part which follows the curve of the cam) basculeur *m*, poussoir *m*, commande *f* de came
~FOR CROSSING LAYER - [text] excentrique *m* pour couche croisée
~FOR VARYING HEIGHT OF LAYERS - [text] excentrique *m* faisant varier la hauteur des couches
~GROOVE - [text] rainure *f* excentrique
~HUMP - [mech] bossage *m* de came
~INCLINE - [mech] rampe *f* de came
~INTERMITTENT MOVEMENT - [cinema] (mechanism moving the film intermittently) avance *f* à excentrique
~LEVER WITH TOOTHED SECTOR FOR CUTTING MOTION - [text] levier *m* d'excentrique à secteur denté pour mécanisme de coupe
~LIFT - [mech] élévation *f* de la came
[text] course *f* de l'excentrique
~LOBE - [mech] bossage *m* de came
~LOCK - [mech] arrêt *m* de came, blocage *m* de came
~MOTION - [text] mouvement *m* d'excentrique, mouvement par excentrique
~NOSE - [mech] bossage *m* de came
~-PARACHUTE - [mining] parachute *m* à excentriques
~PIECE FOR OPEN SHEDDING - [text] section *f* de l'excentrique commandant le pas ouvert
~PLATE - [mech] disque-guide *m*, disque *m* de guidage, came *f* de commande
~PRESS - [mech] (press actuated by a cam) presse *f* excentrique
~PROFILE - [mech] (the geometrical outline of a cam) profile *m* de la came

CAM RAMP - s. cam incline
~RING - [mech] anneau *m* d'excentrique
~ROD - [mech] tige *f* d'excentrique
~ROLLER - [text] rouleau *m* tâteur
~SHAFT - [text] arbre *m* à cames
~SLEEVE - [mech] (in radial engines) tambour *m* à cames
~SPINDLE - [mech] axe *m*, pivot *m* de came
~STROKE AS THE SLAY BEATS UP - [text] course *f* de l'excentrique pendant le coup de battant
~STUD - [mech] boulon *m* d'excentrique
~THROW - [text] course *f* de l'excentrique
~VICE - [impl] étau *m* à came
CAMBER, to - [gen] cambrer, bomber, cintrer, donner de la flêche
CAMBER - [gen] cambrure *f*, courbure *f*
[aero] (the amount by which the curve of an aerofoil rises along the span) courbure *f*, flêche *f*
[of a road] bombement *m*
[metall] (of furnaces) ventre *m*
[of a beam etc] flêche *f*
[shipbuild] (the convexity of a deck line in a transverse section) bouge *m*
[auto etc] (of the wheels) carrossage *m*
~ANGLE - [auto etc] angle *m* de carrossage
~LINE - [aero] (also called median line: a line drawn through an aerofoil, every point of which is equidistant from the upper and lower surfaces when the aerofoil is measured at right angles to the line) ligne *f* médiane
~OF THE SPRING - [mech] cambrure *f* du ressort
~SLIP - [constr] courbe *f* en bois
CAMBERED BOWL - [mech] (a roller whose surface is curved in the sense of its longitudinal axis) cylindre *m* à cambré
~BRIDGE - [naut] pont *m* en pente
CAMBERING OF A TYRE - [rubber ind] galbe *m* d'un pneumatique, galbage *m* d'un pneumatique
CAMBIFORM CELL - [bot] cellule *f* cambiforme
CAMBIUM - [bot] (a layer of cells giving rise to daughter cells) cambium *m*
~LAYER - [bot] couche *f* de cambium
CAMBRIC - [text] (fine linen cloth) batiste *f*
~MUSLIN - [text] percale *f*
CAMBRIDGE ROLLER - [agric] rouleau *m* à disques
CAMEL - [zool] chameau *m*
~HAIR - [text] poil *m* de chameau
~HAIR BELT - [impl] courroie *f* en poil de chameau
CAMEO - [min] (a shell carved in relief) camée *m*
CAMERA - [photo] appareil *m* photographique, chambre *f* photographique
[telev] (a unit comprising a lens, a tube and a vision aplifier to pass on the pulses from the tube to the transmitter)caméra *f*
~ALIGNMENT - [telev] (when the scanning beam moves down the centre of the tube) alignement *m* du faisceau
~AMPLIFIER - [telev] (one of two amplifiers in the television camera) amplificateur *m* de caméra
~APERTURE - [photo] (diameter of the part of the lens which is used to take a photograph) ouverture *f* du diaphragme
~BOOTH - [cinema] (a soundproof booth for the cinematographer and his equipment) cabine *f* de projection insonorisée
~CHAIN - [telev] (the camera, the control unit and the power supply) chaîne *f* de caméra
CAMERA CHANNEL - [telev] appareillage *m* de prise de vues, voie *f* de caméra
~CONTROL - [telev] contrôle *m* des voies de caméra
~CONTROL ROOM - [telev] salle *f* de contrôle d'image, salle *f* de contrôle de video
~COVERAGE - [telev] (the area covered by the camera) angle *m* de prise de vue
~CRANE - [telev] (a platform on which the camera is raised) chariot *m* élévateur, grue *f* de caméra
~DOLLY - [telev] (a support on wheel) chariot *m* pour caméra
~EASEL - [photo] (copy holder) porte-copie *m*, porte-modèle *m*
~GRIP - [photo] poignée *f*
~HOOD - [telev] paresoleil *m*
~LIFTING PLATTFORM - [telev] chariot *m* élévateur, plateau *m* élévateur
~LINE-UP - [telev] ajustage *m* de la caméra
~LINES - [cinema] (the area covered by the lens of the camera) limites *fpl* de netteté
~MONITOR - [telev] (the person controlling the picture of a camera in a studio) moniteur *m* de caméra
~PRINTER - [photo] chambre *f* de copie
~SCANNING AMPLITUDE - [telev] amplitude *f* d'analyse de la caméra
~SHIFTING - [telev] panoramique *m* rapide
~SHOOTING - [telev] prise *f* de vues
~SIGNAL - [telev] (a signal from a television camera) signal *m* vidéo, signal *m* de caméra
~SPECTRAL RESPONSE - [telev] sensibilité *f* chromatique de la caméra
~STATION - [photo] point *m* de prise de vue
~TUBE - [telev] (an electron-beam tube for converting an optical image into a television signal without any mechanical means) tube *m* de prise de vues, tube *m* analyseur
CAMERAMAN - [photo] photographe *m*
[cinema] opérateur *m*
CAMISOLE - [med] camisole *f* de force
CAMITE - [chem] camite *f*
CAMOMILE - [bot] camomille *f*
CAMOUFLAGE, to - [gen] camoufler
CAMOUFLAGE - [gen] camouflage *m*
~NET - [text] filet *m* de camouflage
CAMOUFLAGED - [gen] camouflé
CAMP, to - [gen] camper
CAMP - [gen] camp *m*, campement *m*
CAMPAIGN - [gen] campagne *f*
[glass man] (life of a melting unit) campagne *f*
CAMPANIFORM - [gen] campaniforme
CAMPANILE - [arch] (a bell tower) campanile *m*
CAMPANULA - [bot] campanule *f*
CAMPBELL METHOD - [phys] (a method of measurring low pressures) méthode *f* de Campbell
CAMPHANE - [chem] (a saturated terpene hydrocarbon) camphane *m*
CAMPHENE - [chem] (an intermediate for synthetic and camphor substitutes) camphène *m*
CAMPHOR - [chem] camphre *m*
~OIL - [chem] (substitute for oil of turpentine in the production of surface coating) essence *f* de camphre
~TREE - [bot] camphrier
~WATER - [chem] eau *f* camphrée
~WEED - [bot] ambroisie *f*
CAMPHORATED OIL - [food] huile *f* camphrée

CAMPHORIC ACID - [chem] (a plasticizer for celluloid) acide *m* camphorique
CAMPING - [gen] campement *m*
[holiday camping] camping *m*
CAMPYLITE - [min] (natural chloroarsenate of lead) campylite *f*
CAMSHAFT - [mech] (a shaft carrying one or more cams) arbre *m* à cames
~BEARING - [mech] palier *m* d'arbre à cames
~CASING - [mech] carter *m* d'arbre à cames
~CONTROLLER - [el] commutateur *m* à cames
~GEAR - [mech] (the gear fixed on the camshaft to take the drive from the crankshaft) pignon *m* d'arbre à cames
~GRINDER - [mech] (machine tool designed for the grinding of camshafts) rectifieuse *f* d'arbre à cames
~GRINDING MACHINE - s. camshaft grinder
~LEVER - [mech] levier *m* de levée
~SPROCKET - [mech] pignon *m* d'arbre à cames
~TIMING GEAR - [mech] pignon *m* de distribution
CAN, to - [nucl] (the operation of forming an envelope for a fuel bar) gainer
CAN - [gen] bidon *m*, boîte *f*, bac *m*, boîtier *m*
[cin] (metal container for films) boîte *f*
[text] pot *m*
[nucl] (a container for fuel bars) gaine *f*, enveloppe *f*
[radio etc] (ear-phone used for monitoring) écouteur *m* de contrôle
~ANODE - [electron] anode *f* cylindrique
~COILER - [text] (a revolving can in which slivers from the comb are laid in a double spiral form) pot *m* tournant, empoteur *m*, appareil *m* empoteur
~EMPTYING CREEL - [text] râtelier *m* vide-pots
~FEED - [text] alimentation *f* par pots
~FEED CREEL - [text] râtelier *m* d'alimentation à pots
~GILL BOX - [text] étirage *m* vide-pots, intersecting *m* vide-pots
~HOIST - [mech] élévateur *m* à bidons
~HOOK - [naut] patte *f* à futailles
~INTERSECTING GILL BOX - [text] étirage *m* préparatoire
~MOVER - [text] rouleuse *f* de pots
~OPENER - [impl] ouvre-boîte *m*
~SEALING MACHINE - [mech] machine *f* à fermer les boîtes (métalliques)
~SPINNING SYSTEM - [text] procédé *m* de filature en pots
~TRUCKER - [text] transporteur *m* de pots
~-TYPE COMBUSTION CHAMBER - [mech] chambre *f* de combustion cylindrique
~WINDER - [text] bobinoir *m* à pots
CANADA BALSAM - [chem] (balsam of fir) baume *m* du Canada
~THISTLE - [bot] (a weed) chardon *m* des champs
~TURPENTINE - s. Canada balsam
CANADIAN ASBESTOS - [min] (orthosilicate of iron and magnesium) chrysolite *f*
~DRILLING - [mining] forage *m* canadien (à percussion)
~POPLAR - [bot] peuplier *m* du Canada
~SHELF - [geol] bouclier *m* canadien
CANADOL - [chem] éther *m* de pétrole
CANAL - [gen & hydr] (an artificial water channel) canal *m*
[naut] canal *m*
[bot] (elongated intercellular space] canal *m*
CANAL BRIDGE - [constr] aqueduc *m*
~NEEDLE TRICK - [text] canal *m* de transport, rainure *f*
~RAYS - [phys] (positive rays) rayons *mpl* canaux
~REGULATOR - [hydr] canal *m* de prise d'eau
~SYSTEM - [zool] (water vascular system) système *m* vasculaire
~TOLL - [comm] droits *mpl* de péage
CANALICULAR - [anat] canaliculaire
CANALICULUS - [anat] canalicule *m*
CANALIZATION - [hydr etc] (the construction of canals; also a system of canals conveying water, gas etc) canalisation *f*
CANALIZE, to - [gen] canaliser
CANANGA OIL - [chem] essence *f* de cananga
CANARD - [aero] (an aircraft in which the longitudinal stability and control surfaces are in front of the main plane) appareil *m* avec le gouvernail à l'avant, canard *m*
CANARIES - [el acoust] (disturbances in the sound track) bruits *mpl* parasites
CANARY BIRD - [zool] canari
~GRASS - [bot] phalaris *m*, alpiste *m*
~LITHARGE - [paints] (a bright yellow form of litharge, i.e. lead monoxide) litharge *m* jaune
~WHITEWOOD - [bot] tulipier *m* d'Amérique
CANCEL, to - [gen] annuler, résilier, rescinder, révoquer, supprimer, décommander
CANCEL HEY - [telecomm] touche *f* d'annulation
~OUT, to - [math] s'annuler, s'éliminer, se contrebalancer
CANCELLATE - [anat] réticulé
CANCELLATION - [gen] annulation *f*, résiliation *f*, révocation *f*, retrait *m*
[leg] abrogation *f*
~FEE - [leg] frais *mpl* d'annulation
~OF A BOOKING - [telecomm] annulation *f* d'une demande de communication
~OF A CALL - [telecomm] s. cancellation of a booking
~OF LEASE - [leg] résiliation *f* d'un bail
~OF THE ROUTE - [railw] libération *f* d'un itinéraire
CANCELLED CALL - [telecomm] demande *f* de communication annulée
CANCELLOUS - [med] réticulé, d'aspect réticulé
CANCER - [med] (malignant neoplasm) cancer *m*, carcinome *m*
CANCEROUS - [med] cancéreux
CANCH - [mining] caniveau *m*, entaillement *m*
~HOLE - [mining] trou *m* de mine horizontal
CANCRIFORM - [med] cancériforme
CANCROID - cancroïde
CANCRUM - [med] ulcère *f* à propagation rapide
CANDELA - s. candle
CANDELABRUM - [gen] candélabre *m*
CANDELILLA WAX - [chem] (a natural wax used in surface coatings, in polishes and as a substitute for carnauba wax) cire *f* de candelille
CANDID CAMERA - [photo] chambre *f* pour prises de vues à la dérobée
~SNAPSHOT - [photo] instantané *m* à la dérobée, instantané *m* à l'insu du sujet
CANDLE - [gen] chandelle *f*, bougie *f*
[el] (the unit of luminous intensity) bougie *f*

CANDLE COAL - [mining] houille *f* grasse
~FILTER - [impl] filtre *m* à bougie
~GREASE - [ind chem] suif *m*
~-HOUR - [el] bougie-heure *f*
~MOULDING - [ind] moulage *m* de bougies
~POWER - [el] (the luminous flux emitted by a source of light, expressed in terms of international candle and new candle) puissance *f* lumineuse, intensité *f* en bougies
~-TREE - [bot] cirier *m*
CANDLING - [agric] mirage *m* des oeufs
~DEVICE - [agric] mire-oeufs *m*
CANDY - [food] sucre *m* candi
CANE - [bot] canne *f*, jonc *m*, rotin *m*
[impl] (of a switch) badine *f*
[glass man] canne *f*
~-APPLE TREE - [bot] arbousier *m*
~JUICE - [brew ind] sirop *m* de canne
~MAT - [text] natte *f* en roseau
~MESH CEILING - [constr] claie *f* en roseau, canisse *f*
~SUGAR - [sugar ind] (sucrose derived from sugar cane) sucre *m* de canne
~SUGAR-MILL - [sugar ind] sucrerie *f*
~TRIPOD - [photo] pied-canne *m*
~WEAVE - [text] treillis *m* de roseau
CANELLA - [bot] (cinnamon) cannelle *f*
~OIL - [chem] essence *f* de cannelle
CANESCENT - [bot] (with a hoary appearance) grisâtre
CANICULA - [astr] canicule *f*
CANINE - [zool] canin
[anat] (the canine tooth) canine *f*
~DISTEMPER - [vet] maladie *f* des chiens
~PLAQUE - s. canine distemper
CANISTER - [gen] boîte *f* en fer blanc, boîte *f* métallique
CANKER - s. cancrum
~OF THE HOOF - [vet] crapaud *m* du sabot
CANNED - [gen] en boîte
[nucl] gainé, blindé
~BEER - [brew ind] bière *f* en boîte
~MEAT - [food] viande *f* en conserve
~MUSIC - [el acoust] (recorded music) musique *f* enregistrée
~PUMP - [mech] pompe *f* à carter, pompe *f* blindée, pompe *f* cuirassée
CANNEL COAL - [min] a variety of coal which burns with a bright flame) cannel *m*, houille *f* grasse
CANNERY - [food] fabrique *f* de conserves, conserverie *f*
CANNIBALISM - [vet] (the action of eating parts of their own body or other animals' by animals lacking minerals or protein) cannibalisme *m*
CANNIBALIZE, to - [ind] (colloq. To recover component parts of a machine for use in another) récupérer des pièces (sur du matériel endommagé)
CANNING - [food] (a method of food preservation) mise *f* en conserve, mise *f* en boîte
[nucl] the operation of sheathing the fuel slugs to avoid contamination) gainage *m*
~INDUSTRY - [food] industrie *f* des conserves alimentaires
~MACHINE - [brew ind] machine *f* pour mise en boîte (de la bière)
~PLANT - [oil ind] département *m* de remplissage et confection des produits pétrolifères

CANNON - [gen] canon *m*
~BONE - [anat] canon *m* (de la jambe du cheval)
~PINION - [horol] chaussée *f*
~PLUG - [el] prise *f* de courant encastrée
CANOE - [naut] canoë *m*, pirogue *f*
CANON - [gen] règle *f*, critère *m*, canon *m*
CAÑON - s. canyon
CANOPY - [arch] (part projecting from a wall or roof) auvent *m*, marquise *f* gâble *m*, dais *m*
[aero] (the main supporting surface of a parachute) calotte *f*
[aero] (the transparent covering of the cockpit) dôme *m* (d'habitacle), verrière *f*
[el] couvercle *m* de boîte de dérivation
[auto] protège-cabine *m*
CANT, to - [gen] (to incline an object to the horizontal plane) incliner
[naut] (of ships) chavirer (une embarcation pour la réparer)
CANT - [gen] (inclination to the horizontal) inclinaison *f*
~BAY - [build] fenêtre *f* en pan coupé
~BOARD - [constr] planche *f* de pied, volige *f* de pied
~DOG - s. cant hook
~HOOK- [impl] croc *m* à levier, grappin *m*
~OF THE RAIL - [railw] surhaussement *m* (du rail extérieur)
~OF THE TRACK - [railw] surélévation *f*, devers *m*
CANTBOARD - [constr] coyau *m*
CANTED OVER - [gen] incliné
~SHOT - [cin](photography obtained by rolling the camera) bascule
CANTEEN - [gen] cantine *f*, restaurant *m*
CANTHARIDIN - [chem] (a pharmaceutical product) cantharidine *f*
CANTILEVER - [constr] (a beam or girder fixed at one end only) encorbellement *m*
~ARCHED BRIDGE - [constr] pont *m* à poutres en console
~ARM - [constr] (a projecting arm of a cantilever bridge) poutre *f* en console
~BRAKE - [mech] frein *m* cantilever
~BRIDGE - [constr] pont *m* cantilever, pont *m* en encorbellement
~CRANE - [mech] (straight steel truss resting on a central support, used for the removal of excavated material) grue *f* cantilever
~FORM - [constr] coffrage *m* suspendu (en porte-à-faux)
~GIRDER - [constr] poutre *f* en console
~LOW WING - [aero] aile *f* (basse) cantilever
~MAST - [oil ind] mât *m* de forage cantilever
~RACK - [gen] ratelier *m* à consoles
~ROOF - [constr] toit *m* en porte-à-faux
~SHEET PILING - [constr] rideau *m* de palplanches en porte-à-faux
~SPRING - [mech] (a laminated spring anchored to the frame of an automobile) ressort *m* cantilever, ressort *m* en porte-à-faux
~SUSPENSION - [auto] suspension *f* cantilever
~WING - [aero] aile *f* cantilever
CANTING - [gen] inclinaison f, dévoiement *m*
~FENCE - [mech] guide *f* inclinable
~OF THE RAIL - [railw] surhaussement *m* du rail, surélévation *f* du rail
~STRIP - [constr] (projecting sloping element

around a building to deflect water from the wall) solin *m*
CANVAS - [text] toile *f*
~BELT - [mech] courroie *f* en toile
~CONVEYOR - [mech] (e.g. of a drying machine) toile *f* de transport sans fin
~COVERING - [text] couverture *f* en toile
~FOR EMBROIDERY - [text] canevas *m*
~MATTRESS - [text] toile *f* à matelas
~FOR SAILS - [text] toile *f* à voile
~OF A SURVEY - [surv] canevas *m*
~PAPER - [paper man] papier *m* toilé
CANYON - [geol] canyon *m*
CAP, to - [rubber ind] (of tyres) rechaper, refaire la bande de roulement
[glass man] couronner
CAP - [gen] bonnet *m*, casquette, béret *m*
[mech] chapeau *m*, couvercle *m*, collerette *f*
[el] chapeau *m* de protection; (of a lamp) culot *m*
[mech] (cap holding a spring in a valve) chapeau *m*
[firearms] (part of the cartridge) capsule *f*, amorce *f*
[auto] bouchon *m* de radiateur
[shoe man] bout *m* rapporté
[mining] chapeau *m*, chapeau *m* de cadre
[glass man] couronne *f*
[shipbuild] chouque *m* de mâture
[ind chem] capsule *f*
[arch] chapiteau *m* de colonne
[rubber ind] (the tread of a tyre) bande *f* de roulement
[print] (abbreviation for capital letter) majuscule *f*
~CELL - [bot] (cell surmounting the antheridium of a fern) cellule *f* apicale
~-COLLAR GASKET - [mech] joint *m* d'arbre à section angulaire, joint *m* à chapeau
~END - [el] capot *m* de couverture, capot *m* de protection; (for cable) bouchon *m* de tête de câble
~FILLER - [electron] (designed to fill the caps of electronic tubes with cement) machine *f* a remplir les culots
~FLANGE - [mech] bride *f* à chapeau, bride *f* à ergot
~HEAD SCREW - [mech] vis *f* à tête
~KEY - [impl] clef *f* fermée
~LAMP - [mining] lampe *f* frontale, lampe *f* de casque
~NUT - [mech] écrou *m* borgne
[plumb] (of pipes) chapeau *m* fileté
~PAPER - [paper man] papier *m* bulle, papier *m* (gris épais) d'emballage
~PIECE - [mining] chapeau *m* de cadre
~SCREW - [mech] vis *f* à tête cubique
~SHORE - [naut] épontille *f* de chouque
~SLEEVE - [electron] (part of the oscillator tube) gaine *f* de culot
~SPINDLE - [text] broche *f* à cloche
~SPINNING FRAME - [text] métier *m* à filer à cloches
~STONE - [constr] chaperon *m* (de toit), pierre *f* de faîte
CAPABILITY - [gen] capacité *f*, qualification *f* aptitude *f*
~OF SUBLIMATION - [chem] capacité *f* de sublimation
CAPACITANCE - [el] (the property of an electric nonconductor permitting the storage of energy) capacitance *f*
CAPACITANCE ALTIMETER - [instr] (instrument designed to measure the height of an aircraft in terms of its capacitance to earth) altimètre *m* à capacitance
~BEAM SWITCHING - [radar] commutation *f* capacitive du faisceau
~BOX - [el] (assembly of electrical resistors of a specified nominal value mounted in a box) boîte *f* de capacités
~BOX WITH PLUGS - [el] boîte *f* de capacités à fiches
~BRIDGE - [el] pont *m* de capacités
[instr] pont *m* de mesure de la capacité
~COEFFICIENT - [el] coefficient *m* de capacitance
~DEVIATIONS - [el] différences *fpl* de capacité
~DIVIDER - [el] diviseur *m* (de tension) capacitif
~METER - [el] capacimètre *m*, faradmètre *m*
~OF A CAPACITOR - [el] (the charge on one of the capacitor plates divided by the potential difference between them) capacité *f* d'un condensateur
~OF A CONDUCTING BODY - [el] (the charge of the conducting body divided by its potential) capacité *f* électrique d'un conducteur
~OF A CONDUCTOR - [el] (the charge of a conductor divided by its potential) capacité *f* d'un conducteur
~OF THE DIAPHRAGM - [el] capacité *f* du diaphragme
~PER UNIT LENGTH - [el]capacité *f* unitaire
~RELAY - [electron - (an electronic relay whose operation depends on a small change of capacitance) relais *m* capacitif
~STRAIN GAUGE - [instr] jauge *f* de contrainte à capacité
CAPACITIVE - [el] capacitif
~COUPLING - [el] (circuits associated with one another by a capacitance which is mutual) couplage *m* capacitif
~FEED-BACK - [el] réaction *f* capacitive
~HEATING - [el] (method of heating a body by making it part of a capacitor) chauffage *m* capacitif
~LOAD - [el] charge *f* capacitive
~POST - [electron] (extending across a waveguide) saillie *f* capacitive
~REACTANCE - [electron] (of a waveguide component) réactance *f* capacitive
~SAWTOOTH GENERATOR - [el] (generator of sawtooth voltage across a capacitor) générateur *m* capacitif de dents de scie
~TUNING - [radio] (form of tuning in which the capacitance is varied) réglage *m* capacitif
~WINDOW - [electron] (conducting diaphragm extending into a waveguide) fenêtre *f* capacitive
CAPACITIVITY - [el] (term sometimes used for permittivity) permittivité *f*, constante *f* diélectrique
CAPACITOR - [el] (an electrical circuit component designed to provide capacity) condensateur *m*
~INPUT FILTER - [el] (a power supply wave filter) filtre *m* à entrée capacitive
~LOUDSPEAKER - [el acoust] haut-parleur *m* électrostatique
~MOTOR - [el] moteur *m* à condensateur
~OF THE GRID CIRCUIT - [el] condensateur *m* de grille
~PICK-UP - [el acoust] (pick-up depending for its operations on the variations of its capacitance) tête *f* de lecture électrostatique
CAPACITY - [gen] capacité *f*, aptitude *f*
[phys] (quantity) contenance *f*

[el] (the output of an electrical apparatus) puissance *f* capacité *f*
[hydr] (the output of a pump) débit *m*
[mech] (of a motor) rendement *m*, puissance *f*
[th eng] (of boilers) production *f* de vapeur
[railw] (of railway cars and generally vehicles) charge *f* admise, charge *f* utile
CAPACITY BALANCE - [el] couplage *m* capacitif
~COUPLING - s. capacitive coupling
~DEVIATIONS - [el] différences *f*pl de capacité
~EXCEEDING NUMBER - [comput] (any number which is larger then the maximum number stored by the computer in any register) nombre *m* dépassant la capacité de mémoire
~LOAD - [constr] charge *f* maximale
~MICROPHONE - [el] (microphone depending for its operation on the variations of the capacitance of a capacitor) microphone *m* électrostatique
~OF A DAM - [hydr] capacité *f* de retenue
~OF A DISTRIBUTION SYSTEM - [gen] capacité *f* d'un réseau
~OF A POWER STATION - [el] puissance *f* d'une centrale
~OF A STORAGE BATTERY - [el] capacité *f* d'un accumulateur
~PER UNIT LENGTH - [el] capacité *f* unitaire
~REACTANCE - s. capacitive reactance
~UNBALANCE - [el] déséquilibre *m* de capacité
CAPE - [geogr] cap *m*
[text] cape *f*, capot, mantelet, pélerine *f*
~BOLT - [mech] boulon *m* à chapeau, boulon *m* à tête
CAPEL - [min] silex *m* corné
CAPER - [bot] câprier *m*, câpre *f*
CAPILLACEOUS - [bot] capillacé
CAPILLARECTASIA - [med] dilatation *f* des capillaires
CAPILLARITY - [phys] (a phenomenon caused by surface tension occurring in fine bore tubes or channels) capillarité *f*
CAPILLARY - [gen] (very-thin, hair-like; of a very small diameter) capillaire
[gen] tube *m* capillaire, capillaire *m*
~ACTION - s. capillarity
~ACTIVITY - [phys] activité *f* capillaire
~ATTRACTION - [metall] (occurring in the brazing process) attraction *f* capillaire
~BED - [anat] réseau *m* capillaire
~CONSTANT - [phys] constante *f* capillaire
~CORRECTION - [phys] (correction applied to mercury barometers etc for the effect of capillarity on the height of the column) correction *f* de capillarité
~DRAG - [phys] vide *m* d'adhésion du mercure, vide *m* collant
~ELECTROMETER - [instr] électromètre *m* capillaire
~ELEVATION - [hydr] ascension *f* capillaire
~FLOW - [phys] (the phenomenon of the ascent or descent of fluids in tubes of microscopic calibre, due to the relative attraction of the molecules of the fluid for each other and for those of the walls of the tube) capillarité *f*, écoulement *m* capillaire
~FORCE - [phys] capillarité *f*
~FRINGE - [hydr] frange *f* capillaire
~INTERSTICE - [geol] interstice *m* capillaire
~LEAK - [phys] fuite *f* capillaire
~PRESSURE - [phys] (the pressure which is caused by capillary action) pression *f* capillaire
CAPILLARY RISE - [phys] ascension *f* capillaire, montée *f* par capillarité
~SILVER - [photo] argent *m* capillaire (filamenteux)
~SOIL WATER - [bot] eau *f* de capillarité
~TENSION - [phys] tension *f* capillaire
~THEORY OF SEPARATION - [phys] théorie *f* de la séparation par diffusion gazeuse
~TUBE - [instr] tube *m* capillaire
~VESSELS - [med] vaisseaux *m*pl capillaires, capillaires *m*pl
~VISCOSIMETER - [instr] viscosimètre *m* capillaire
~WATER - [hydr] eau *f* capillaire
CAPITAL - [gen] capital *m*
[geogr] capitale *f*
[fin] capital *m*
[arch] (of a column) chapiteau *m*
[print] (of a letter) majuscule *f*
~ACCOUNT - [fin] compte *m* de capital
~INVESTMENT - [fin] valeur *f* d'investissement
~LOCK KEY - [mech] (typewriter part) touche *f* fixe des majuscules
~SHIP - [naut] cuirassé *m*
~STOCK - [fin] capital *m* versé
CAPITALIST - [gen] capitaliste
CAPITALIZATION - [fin] placement *m* de capital, financement *m*, capitalisation *f*
CAPITALIZE, to - [gen & fin] capitaliser
CAPITALIZED CUSHION GAS - [gas ind] gaz-coussin *m* injecté
CAPITATE - [bot] en capitule, capité
CAPON - [zool] chapon *m*
CAPONIZE, to - [vet] chaponner, châtrer (un poulet)
CAPPED STEEL - [metall] acier *m* effervescent
CAPPELENITE - [min] cappelenite *f*
CAPPING - [carp] corniche *f*, chape *f*
[mech] capsulage *m* des bouteilles
[geol] couche *f* supérieure
[constr] (head beam) chapeau *m*
~KNIFE - [impl] longrine *f* à chapeau
CAPRIC ACID - [chem] (a plasticizer and synthetic resins intermediate) acide *m* caprique
CAPRICORN - [astr] capricorne *m*
CAPRIFICATION - [bot] (the fertilization of the flowers of fig-trees by fig insects) caprification *f*
CAPREOLATE - [bot] capréolate *m*
CAPROCK - [gas ind] couverture *f*, terrain *m* de recouvrement
CAPROIC ACID - [chem] (starting material for rubber and resin chemicals) acide *m* caproïque
CAPRYLATE - [chem] caprylate *m*
CAPRYLIC ACID - [chem] (organic intermediate and plastics plasticizer) acide *m* caprylique
CAPSICUM - [bot] (Cayenne pepper) piment *m*
CAPSIZE, to - [gen] capoter
[naut] chavirer, cabaner
CAPSTAN - [naut] cabestan *m*, guindeau *m*
[el acoust] (the spindle, or shaft, rotating against the tape on recording and playback) arbre *m* d'entraînement
~BAR - [naut] barre *f* de cabestan, levier *m* de cabestan
~BARREL - [naut] cloche *f* de cabestan
~IDLER - [el acoust] (in recording machines) rouleau *m* presseur
~LATHE - [tools] (a lathe in which the required tools are mounted radially) tour *m* revolver

CAPSTAN PARTNER - [naut] étambrai *m* de cabestan
~PULLER - [plast ind] (mechanism for pulling off coated wire from the extruder) éjecteur *m*
~SCREW - [mech] vis *f* à tête ronde et forée
~SPINDLE - [naut] mèche *f* de cabestan
CAPSTRIP - [aero] chapeau *m* de nervure
CAPSULE - [gen] capsule *f*
[el acoust] pastille *f* de microphone
[astronautics] capsule *f*
~STACK - [ind chem] pile *f* de capsules
CAPTAIN DRESSER - [mining] maître-bocardeur *m*
CAPTAN - [chem] (a fungicide, used in plastics and surface coating) captan *m*
CAPTANCE - [el] réactance *f* de capacité
CAPTION - [print] légende *f*, en-tête *m*
[cinema] sous-titre *m*, légende *f*
CAPTIVE - [mech] (of any part which is attached loosely to an assembly, so that it cannot be lost when removed from its position) prisonnier *m*
~BALLOON - [aero] ballon *m* captif
~FOUNDRY - [metall] fonderie *f* integrée
~SCREW - [mech] (a screw so made that when unscrewed completely, it is still attached to the assembly) vis *f* prisonnière
~TEST - [astronaut] essai *m* au point fixe
~THUMBSCREW - [mech] (captive screw provided with a knurled head) vis *f* prisonnière à molette
CAPTURE, to - [gen] capturer, prendre
CAPTURE - [gen] capture *f*, prise *f*
~CROSS-SECTION - [nucl] (the cross-section which is effective for the capture) section *f* efficace de capture
~EFFICIENCY - [nucl] (ratio of time available during a frequency modulation cycle for starting particles into stable orbits to the total time for repetition of the frequency modulation cycle) rendement *m* de capture
~GAMMA RADIATION - [nucl] rayonnement *m* gamma de capture
~RANGE - [telev] plage *f* de rattrapage
~REACTION - [nucl] réaction *f* de capture
~SPOT - [electron] (also called trapping spot) zone *f* de capture
~-TO-FISSION RATIO - [nucl] rapport *m* capture-fission
CAR - [gen] véhicule *m*, voiture *f*, chariot *m*
[auto] automobile *f*
[aero] (the structure hung below the envelope of a balloon to contain passengers, instruments and other load) carlingue *f*, nacelle *f*
[mining] (for transport in mines) wagonet *m*, chariot *m*
[transp] (of a cableway) cabine *f*
[railw] voiture *f*, wagon *m*
~AERIAL - [auto] antenne *f* pour automobile
~ANTENNA - s. car aerial
~BATTERY - [auto] accumulateur *m* pour véhicules
~BODY - [auto] carrosserie *f*
~BOTTOM RESISTOR FURNACE - [el] four *m* électrique à sole mobile
~CASTINGS - [auto] pièces *fpl* coulées (pour automobiles)
~CHASSIS - [mech] châssis *m* (de véhicule)
~FRAME - [auto] châssis *m* (de véhicule)
~HEATER - [auto] appareil *m* chauffrage pour automobile
~ICER - [mech] (a hoist designed to rise vertically to supply ice to refrigerator cars) élévateur *m* à déplacement parallèle
~LIFT - [auto] pont *m* élévateur pour automobiles
~LOAD - [railw] charge *f* admise
~PLATE - [auto] plaque *f* d'immatriculation
~POLISH - [auto] cire *f* pour automobiles
~RADIO - [auto] récepteur *m* radio pour voiture
~SHAKER - [railw] déchargeur *m* à secousses
~SHED - [auto] garage *m*
~SHIELD - [radio] (the prevention of interference generation in cars) dispositif *m* antiparasites
~WAX - s. car polish
~WITH OBSERVATIONS PLATFORM - [railw] wagon *m* à plateforme
CARACOLE - [build] escalier *m* en spirale
CARACOLITE - [min] caracolite *f*
CARAMEL - [sugar ind] caramel *m*
[brew ind] caramel *m* (colorant pour bière)
CARAT - [meas] (standard of weight for gold and precious stones) carat *m*
CARAVEL - [naut] caravelle *f*
~PLANKING - [shipbuild] bordage *m* franc
CARAWAY - [bot] cumin *m*, carvi *m*
~OIL - [ind chem] essence *f* de carvi
CARBAMIC ACID - [chem] acide *m* carbamique
CARBAMIDE - [chem] (a diuretic, fertilizer, starting material for synthetic resins) carbamide *f*
~PEROXIDE - [chem] (oxidizing, bleaching and polymerizing agent) percarbamide *f*, peroxyde *m* de carbamide
~PHOSPHORIC ACID - [chem] (a flame proofing agent) acide *m* uréo-phosphorique
CARBAMYL CHLORIDE - [chem] (used in the synthesis or organic acids) chlorure *m* de carbamyle
CARBANILIDE - [chem] diphénylcarbamide *f*
CARBAZOLE - [chem] (an intermediate for rubber chemicals) carbazol *m*
CARBENE - [chem] (a free radical with paired electrons and divalent carbon) carbène *m*
CARBIDE - [chem] carbure *m*
~CARBON - [metall] carbone *m* de cémentation
~LAMP - [light] lampe *f* à acétylène
~STEEL - [metall] acier *m* au carbone
~STRINGERS - [metall] (defects appearing on metal surfaces) défauts *m* de cémentation
[oil ind] alignement *m* des carbures
~-TIPPED SAW - [impl] (a type of saw with tungsten carbide tips on its teeth) scie *f* à pastilles de carbure rapportées
~TOOL - [mech] (tool made of tungsten carbide) outil *m* au carbure
~TOOL GRINDING MACHINE - [mech] (special machine for grinding tungsten carbide tools) machine *f* à affûter les outils au carbure
CARBIDING - [chem] carburation *f*
CARBINE - [firearms] carabine *f*
CARBINOL - [chem] (a synonym for methyl alcohol) carbinol *m*, alcool *m* methylique, méthanol *m*
CARBO - [chem] carbo-
CARBOCYCLIC COMPOUNDS - [chem] (also called isocyclic compounds, i.e. those in which the closed chain or ring contains only carbon atoms) composés *mpl* homocycliques
CARBOHYDRASE - [chem] (an enzyme which breaks down carbohydrates) carbohydrase *f*
CARBOHYDRATES - [chem] (compounds which contain carbon, hydrogen and oxygen, the latter in

such proportion as to form water) hydrates *mpl* de carbone, glucides *m*
CARBOLATED - [chem] phéniqué
CARBOLIC ACID - [chem] acide *m* phénique
CARBOLISM - [med] empoisonnement *m* par le phénol
CARBOMYCIN - [chem] (an antibiotic) carbomycine *f*
CARBON - [chem] (a non-metallic element, symbol C) carbone *m*
~ANODE - [electron] (anode made of graphite, used in thermionic valves designed to work at high anode temperatures) anode *f* au carbone
~ARC - [el] arc *m* au carbone
~ARC CUTTING - [metall] découpage *m* à l'arc
~ARC LAMP - [el] lampe *f* à arc au carbone, lampe *f* à arc
~ARC WELDING - [metall] (arc welding process where the arc is maintained between carbon electrodes or between one or more carbon electrodes and the piece) soudage *m* à l'arc au carbone
~ASSIMILATION - [bot] assimilation *f* du carbone
~BLACK - [ind chem] noir *f* de fumée
~BLOCK - [ind chem] (a rectangular block of gas carbon on which specimens are placed for blowpipe analysis) bloc *m* de carbone
~BLOCK PROTECTOR - [el] paratonnerre *m* à charbon strié
~BLOCK PROTECTOR - [el] (voltage discharge gap) paratonnerre *m* à charbon
~BORER - [mining] foret *m* à charbon
~BRICK - [metall] brique *f* de carbone
~BRUSH - [el] balais *m* en carbone
~CHAIN - [chem] chaîne *f* carbonée
~-COMBUSTION CELL - s. carbon-consuming cell
~-CONSUMING CELL - [el] (primary cell acting by voltaic oxidation of carbon) pile *f* au carbone
~CONTENT - [chem] teneur *f* en carbone
~COPY - [gen] double, copie *f*
~CYCLE - [bot] (the circulation of carbon in nature) cycle *m* du carbone
[nucl] (a series of thermonuclear reactions with the release of energy that is probably of the sun) cycle *m* du carbone
~DEPOSIT - [mech] (motors) calamine *f*
~DIAMOND - [min] carbonado *m*
~DIOXIDE - [chem] (gas used as an intermediate, a fire extinguisher, a refrigerant and an inert atmosphere for chemical reactions) anhydride *m* carbonique, gaz *m* carbonique
~DIOXIDE FLASK - [ind chem] bouteille *f* de gaz carbonique
~DIOXIDE LEAKAGE - [nucl] (sometimes occurring in nuclear power plants at various points) fuite *f* de gaz carbonique
~DIOXIDE SNOW - [chem] neige *f* carbonique, carboglace *f*
~DISULPHIDE - [chem] (a solvent for rubber, fats, sulphur, resins and waxes) bisulfure *m* de carbone, anhydride *m* sulfocarbonique
~ELECTRODE - [el, heat] (used for arc furnaces; electrode of amorphous carbon, ready for use) électrode de carbone
~FILAMENT LAMP - [el] lampe *f* à filament de carbone
~FILTER - [ind chem] filtre *m* à charbon
~FIN - [astronaut] ailette *f* en graphite, directrice *f* en graphite
CARBON-FUNCTIONAL SILICONE - [chem] compound *m* silicone au carbone
~GRANULE - [el acoust] grenaille *m* de charbon
~GRANULE MICROPHONE - [el acoust] microphone *m* à charbon
~GRANULE TRANSMITTER - [el acoust] microphone *m* à charbon
~HEXACHLORIDE - [chem] hexachlorure *m* de carbone
~HYDRIDES - [chem] (synonym for hydrocarbons) hydrocarbures *m*
~ISOTOPE RATIO - [nucl] (the relative value of an isotope in a sample of carbon) proportion *f* des isotopes du carbone
~LIGHTNING PROTECTOR - [el] paratonnerre *m* au carbone
~LINING - [ind chem] chemise *f* de graphite, revêtement *m* de graphite
~LINKAGE - [chem] liaison *f* carbonée
~MICROPHONE - [el acoust] (microphone depending for its operation on variations of electrical resistance of carbon contacts) microphone *m* à charbon
~MONOXIDE - [chem] (poisonous gas) oxyde *m* de carbone
~NITROGEN CYCLE - [phys] (Bethe's theory on the interchange of carbon and nitrogen in nature) cycle *m* solaire de Bethe
~OIL - [chem] kérosène *m*
~OXYCHLORIDE - [chem] (phosgene) oxychlorure *m* de carbone, phosgène *m*, chlorure *m* de carbonyle
~PAPER - [paper man] papier *m* carbone
~PILE REGULATOR - [el] régulateur *m* de tension à pile au carbone
~PREVENTIVE FUEL - [mech] (motors) carburant *m* décalaminant
~PROCESS - [photo] procédé *m* au charbon
~PROTECTOR - [el] paratonnerre *m* à condensateur
~RESIDUE - [mech] (in motors) calamine *f*
~RESTORATION - [metall] régénération *f* (des pièces décarburées)
~SEPARATION - [chem] séparation *f* du carbone
~STEEL BILLET - [metall] billette *f* d'acier au carbone
~STEEL BLOOM - [metall] bloom *m* d'acier au carbone
~STEEL CASTING - [metall] pièce *f* coulée en acier au carbone
~-STEEL VESSEL - [nucl] (the close container of the reactor core, made of carbon steel) récipient *m* en acier au carbone
~TETRACHLORIDE - [chem] (a degreasing agent, a solvent and a refrigerant) tétrachlorure *m* de carbone
~TISSUE - [photo] papier *m* au carbone
~TRACK POTENTIOMETER - [instr] (type of potentiometer in which the adjustable resistor consists of a circular path of carbon traversed by a radial contact arm) potentiomètre *m* à résistance de carbone
~TRANSMITTER - [el acoust] microphone *m* à charbon
CARBONADO - [min] (impure aggregate of small diamond crystals, used for industrial purposes) carbonado *m*
CARBONATE, to - [chem] (to saturate a liquid with carbon dioxide under pressure) saturer de gaz

carbonique
[to burn] carboniser
[ind chem] (the operation of transforming into a carbonate) transformer en carbonate, carbonater
CARBONATE - [chem] carbonate *m*
~BALANCE - [chem] équilibre *m* des (bi)carbonates
~HARDNESS - [chem] dureté *f* des carbonates
~OF CALCIUM - [chem] carbonate *m* de calcium, carbonate *m* de chaux
~OF LIME - [chem] carbonate *m* de chaux
~OF POTASH - [chem] carbonate *m* de potasse
~WATER - [chem] eau *f* calcaire
CARBONATED JUICE - [sugar ind] jus *m* carbonaté
CARBONATING SODA - [chem] soude *f* à la chaux
~TANK - [sugar ind] chaudière *f* à carbonater
~UNIT - [ind chem] installation *f* de carbonatation
CARBONATION JUICE PUMP - [sugar ind] pompe *f* à jus trouble
~JUICE RECYCLING - [sugar ind] recyclage *m* du jus carbonaté
CARBONATOR - [ind chem] (a vessel for the treatment of basic lead acetate with carbon dioxide) appareil *m* de carbonisation
CARBONIC ACID - [chem] (the weak acid formed by the solution of carbon dioxide in water) acide *m* carbonique
~ACID GAS - [chem] gaz *m* carbonique, anhydride *m* carbonique
~ACID PUMP - [mech] pompe *f* à gaz
~ANHYDRIDE - [chem] s. carbonic acid gas
CARBONIFEROUS - [geol] carbonifère
~IGNEOUS ROCK - [geol] roche *f* ignée carbonifère
~LIMESTONE - [geol] calcaire *m* carbonifère
CARBONITE - [expl] carbonite *f*
CARBONITRIDING - [metall] carbonitruration *f*
CARBONIUM ION - [chem] (an electrophilic, positively charged ion, the carbon atom of which has only six valence electrons) ion *m* carbonium, carbo-cation *m*
CARBONIZATION - [chem] (the destructive distillation of organic substances) carbonisation *f*, houillification *f*
[text] carbonisation *f*
[mech] (of i.c. engines) calaminage *m*
~BY DRY PROCESS - [text] carbonisation *f* par voie sêche
~BY GAS CIRCULATION - [gas ind] distillation *f* par des gaz de balayage
~BY WET PROCESS - [text] carbonisation *f* par voie humide
~INDEX - [ind chem] degré *m* de cuisson
~OF COAL - [ind chem] carbonisation *f* du charbon, distillation *f* du charbon
~PROCESS - [text] (the destruction of vegetable matter in wool by steeping it in a weak solution of sulphuric acid) carbonisation *f*
~WITH STEAMING - [ind chem] distillation *f* humide
~WITHOUT STEAMING - [ind chem] distillation *f* sèche
CARBONIZE, to - [gen ind chem & text] carboniser
CARBONIZED FUEL - [metall] combustible *m* carbonisé
~STEEL - [metall] acier *m* cémenté
CARBONIZING - [chem] carbonisation *f*
~BATH - [ind chem] bain *m* de carbonisation
~CONDITIONS - [ind chem] régime *m* de distillation, allure *f* de distillation
CARBONIZING FURNACE - [metall] four *m* de cémentation
~LIQUOR - [ind chem] bain *m* de carbonisage
~METHOD - [text] procédé *m* de carbonisage
~POT - [metall] creuset *m* de cémentation
~PROGRAMME - [ind chem] programme *m* des défournements
~RANGE - [text] installation *f* de carbonisation
~STOVE - [text] (used for carbonisation by dry process) four *m* de carbonisation, carboniseuse *f*
~TIME - [ind chem] durée *f* de distillation
CARBONYL - [chem] carbonyle *m*
~CYANIDE - [chem] cyanure *m* de carbonyle
~GROUP - [chem] (the group formed by carbon monoxide acting as a radical) groupement *m* carbonyle
~IRON POWDER - [metall] (iron powder of high purity, used in the electronic industry) poudre *f* de fer carbonyle
CARBORUNDUM - [metall] (proprietary name, silicon carbide used as an abrasive) carborundum *m*
~BRICK - [ind chem] brique *f* de carborundum
~DETECTOR - [radio] détecteur *m* à carborundum
~PASTE - [ind chem] pâte *f* au carborundum
~POWDER - [ind chem] (carborundum reduced to fine powder, used as an abrasive) poudre *f* de carborundum
~SLIP - [mech] (small flat strip of carborundum used for sharpening tools) lamelle *f* de carborundum
~WHEEL - [impl] meule *f* en carborundum
CARBOXYL - [chem] carboxyle *m*
CARBOXYLASE - [chem] carboxylase *f*
CARBOXYLATION - [chem] carboxylation *f*
CARBOXYLIC - [chem] (term denoting compounds with acid properties) carboxylique, carbonique
CARBOY - [ind chem] bombonne *f*, tourie *f*, ballon *m* d'acide
~FILTER - [ind chem] filtre *f* pour bombonne
~TIPPER - [mech] (a device for tilting a carboy) vide-tourie *m*
~WAGON - [railw] wagon *m* citerne
CARBROMAL - [chem] adaline *f*, carbromal *m*
CARBURAN - [chem] (uranium-lead compound) carburane *m*
CARBURATE, to - [chem] carburer
CARBURATION - [metall] carburation *f*
CARBURETTED IRON - [metall] fer *m* carburé
~WATER-GAS - [ind chem] (water gas enriched with hydrocarbon gases) gaz *m* à l'eau carburé
~WATER-GAS SET - [gas ind] ligne *f* de gaz à l'eau carburé
CARBURETTOR - [mech] (also called carburetter; a mechanism for mixing air and volatile fuel) carburateur *m*
~AIR SCOOP - [mech] prise *f* d'air réglable
~BODY - [mech] corps *m* du carburateur
~COVER - [mech] couvercle *m* de carburateur
~COVER GASKET - [mech] joint *m* de couvercle de carburateur
~FLOAT - [mech] flotteur *m*
CARBURIZATION - [metall] carburation *f*
[metall] (heat-treatment) carburation *f*, cémentation *f*
CARBURIZE, to - [metall] carburer, cémenter
CARBURIZER - [metall] (carbon containing material used in adding carbon to steel or iron) cément *m*

CARBURIZING - [metall] (introduction of carbon into surfaces by heating) carburation *f*, cémentation *f*
~FLAME - [metall] flamme *f* de cémentation
~FURNACE - [metall] four *m* de cémentation
~MATERIAL - [metall & ind chem] cément *m*
CARBYLAMINE - [chem] carbylamine *f*
CARCASE - s. carcass
CARCASS - [build] carcasse *f*, charpente *f*
[gas ind] (obsolete; body of an installation) installation *f* intérieure
[rubber ind] carcasse *f* (de pneumatique)
~BREAK - [rubber ind] rupture *f* de la carcasse
~PLY - [rubber ind] pli *m* de carcasse
~RECLAIM - [rubber ind] régénéré *m* de carcasses
~SAW - [impl] scie *f* à dos
CARCEL LAMP - [illum] lampe *f* Carcel
CARCINEMIA - [med] cachexie *f* cancéreuse
CARCINOGENESIS - [med] (production of cancer) carcinogénèse *f*, cancérogénèse *f*
CARCINOGENIC - [med] (capable of producing cancer) cancérigène, carcinogène
CARCINOGENS - [chem] (substances inducing malignant growths in living tissues) substances *fpl* cancérigènes
CARCINOID - [med] carcinoîde
CARCINOMA - [med] (a malign cancer tumor) carcinome *m*, cancer *m*
CARCINOTRON - [electron] (magnetron type of electron tube in which a beam of electrons is moved perpendiculary to electric and magnetic fields) carcinotron *m*
CARD, to - [text] carder
CARD - [gen] carte *f*
[text] (strip of cardboard in a weaving mill, lifting or depressing the threads to form the desired pattern) carte *f*, carton *m*
~ASSEMBLY - [compass] rose *f* d'un compas
~BIN - [electron] (the receptacle into which selected cards are deposited) récepteur *m* de cartes, casier *m* à cartes
~CABINET - [comput] fichier *m*
~CAN - [text] boîte *f* (pot de carde)
~CAPACITY - [comput] (number of characters which can be punched into a card) capacité *f* de la carte
~CHAIN - [text] chaîne *f* des cartons
~CLOTH - [text] garniture *f* de carde, drap *m* de carde
~CLOTHING - s. card cloth
~COLUMN - [comput] (one of a number of columns in a punch card) colonne *f*
~COUNTER - [comput] compte-cartes *m*
~COVER - [text] couvercle *m* protecteur
~CRADLE - [text] support *m* des cartes
~CUTTER - [text] perforatrice *f* de cartes
~CUTTING - [text] perforation *f* des cartes
~CYLINDER - [text] tambour *m* de la carde
~DELIVERY - [text] (for wool) sortie *f* de carde
~ENGINE - [text] (for wool) carde *f*
~FEED - [comput] alimentation *f*, dispositif *m* d'alimentation
~FEEDING - [comput] (the action of inserting cards one by one into the computer) alimentation *f*
~FIELD - [comput] (a set of columns, one or more in a given number of cards) zone *f* de carte perforée
CARD FILE - [comput] fichier *m*, cartothèque *f*
~FILLET - [text] ruban *m* pour carde
~FILM - [text] voile *m* de carde
~FITTER - [text] régleur *m* de cardes
~FIXER - [text] régleur *m* de carde
~FLAT - [text] chapeau *m* de carde
~FLEECE - [text] voile *m* de carde
~FLY - [text] duvet *m* de carde
~FOUNDATION - [text] soubassement *m* de garniture
~GRINDER - [text] aiguiseur *m* de cardes
~GUIDE - [text] guide-cartons *m*
~HOLDER - [comput] fichier *m*
[text] support *m* de carte
~HOOK - [text] aiguille *f* de la garniture de carde
~HOPPER - [comput] (a pocket stacking cards after they have gone through the machine) récepteur *m* de cartes
~INDEX - [comput] fichier *m*
~LACER - [text] attache *f* des cartes
~LACING - [text] attache *m* des cartes
~LEVELLING - [text] nivelage *m* des cardes
~OF WARPING - [text] feuille *f* d'ourdissage, carte *f* d'ourdissage
~PERFORATING MACHINE - [text] machine *f* à piquer les cartons
~PITCH - [text] division *f* des cartons, numérotage *m* des cartons
~POCKET - s. card bin
~PROGRAMMED COMPUTER - [comput] calculatrice *f* électronique à programme sur cartes perforées
~PULLEY - [text] poulie *f* à gorge
~PUNCH - [comput] perforatrice *f*
~MINDER - [text] cardeur *m*
~READER - [comput] (a mechanism causing the information to be read) lectrice *f* de cartes perforées
~REPEATER - [text] machine *f* à copier les cartes
~ROLLER - [text] cylindre *m* de carde
~ROOM - [text] atelier *m* de cardage
~RUN - [comput] passage *m* (des cartes)
~SETTER - [text] régleur *m* des cardes
~SHEET - [text] garniture *f* de carde
~SHEETS - [text] plaques *fpl* pour carde
~SLIVER - [text] ruban *m* de carde
~STACKER - [comput] récepteur *m* de cartes
~STAMPING MACHINE - [text] matrice *f* à cartons, appareil *m* d'étampage de cartons
~STRIPPING PLANT - [text] dispositif *m* à nettoyer la carde, dispositif de débourrage de la carde
~TENTER - [text] cardeur *m*
~THISTLE - [text] chardon *m* à foulon
~TO TAPE CONVERTER - [comput] (a mechanism for the transfer of punched card data to a tape) convertisseur *m* de transfert cartes-bandes perforées
~WASTE - [text] déchet *m* de carde
~WEB - [text] voile *m* de carde
~WIRE - [text] aiguillage *m* de la carde
CARDAMON - [bot] cardamone *m*
~OIL - [chem] essence *f* de cardamone
CARDAN - [mech] cardan *m*
~DRIVEN - [mech] à cardan
~JOINT - [mech] joint *m* à la cardan
~SHAFT - [auto] arbre *m* de transmission
~SHAFT HOUSING - [auto] logement *m* d'arbre de transmission
CARDANIC - [mech] à la cardan
~SUSPENSION - [mech] suspension *f* à la cardan

CARDBOARD - [paper man] carton *m*
~BENDING MACHINE - [mech] plieuse *f* à carton
CARDED - [text] cardé
~COTTON - [text] coton *m* cardé
~SLIVER - [text] mèche *f* cardée
~WOOL - [text] laine *f* cardée
~YARN - [text] fil *m* de laine cardée
CARDER - [text] cardeuse *f*
CARDIAC - [med] cardiaque
~ADYNAMIA - [med] adynamie *f* cardiaque
~ASTHMA - [med] asthme *f* cardiaque
~CYCLE - [med] cycle *m* cardiaque
~FAILURE - [med] insuffisance *f* cardiaque
~IMPULSE - [med] battement *m* de coeur perceptible au cinquième espace intercostal à la gauche du sternum
~INSUFFICIENCY - [med] insuffisance *f* cardiaque
~MURMUR - [med] souffle *m*
~MUSCLE - [anat] muscle *m* cardiaque
~NERVES - [anat] nerfs *mpl* cardiaques
~PLEXUS - [anat] plexus *m* cardiaque
~SPHINCTER - [anat] sphincter *m* cardiaque
CARDIALGIA - [med] cardialgie *f*
CARDIGAN CAM - [text] came *f* façon-métier
~LOCK - [text] s. cardigan cam
CARDINAL - [gen] cardinal
~NUMBER - [math] nombre *m* cardinal
~PLANES - [light] plans *mpl* cardinaux
~POINTS - [astr] points *mpl* cardinaux
~STIMULUS - [opt] excitation *f* repère
~VEINS - [anat] veines *fpl* cardinales
CARDING - [text] cardage *m*
~BEATER - [text] volant *m* cardeur
~COMB - [text] peigne *m* à carder
~CYLINDER - [text] cylindre *m* à tambour
~ENGINE - [text] carde *f* (laine)
~FILLET - [text] ruban *m* pour carde
~MACHINE - [text] cardeuse *f*, carde *f*
~SURFACE - [text] surface *f* de carde
~WILLOW - [text] loup *m* cardeur
CARDIOACCELERATOR - [med] cardio-accélérateur (stimulant l'action cardiaque)
CARDIOCENTESIS - [med] ponction *f* cardiaque
CARDIOCIRRHOSIS - [med] maladie *f* de coeur associée à une cirrhose du foie, pseudo-cirrhose *f* péricardique
CARDIOGRAM - [med] cardiogramme *m*
CARDIOGRAPH - [med] cardiographe *m*
CARDIOID - [math] cardioïde *f*
~CAM - [mech] came *f* à cardioïde
~CONDENSER - [el] condensateur *m* à cardioïde
~CURVE - [el] courbe *f* cardioïde
~DIAGRAM - [radio] (heart shaped diagram used in direction finding systems) diagramme *m* cardioïde
CARDIOINHIBITORY - [med] cardio-inhibiteur
CARDIOKINETIC - [med] stimulant l'action du coeur
CARDIOLOGY - [med] cardiologie *f*
CARDIOLYSIS - [med] cardiolyse *f*
CARDIOMEGALY - [med] cardiomégalie *f*
CARDIOMETER - [med] cardiomètre *m*
CARDIOMYOLIPOSIS - [med] dégénérescence *f* graisseuse du muscle cardiaque
CARDIOMYOTOMY - [med] cardiomyotomie *f*
CARDIONECROSIS - [med] gangrène *f* du coeur
CARDIONEUROSIS - [med] cardionévrose *f*
CARDIOPALMUS - [med] palpitation *f* du coeur
CARDIOPATHY - [med] cardiopathie *f*
CARDIOSCLEROSIS - [med] cardio-sclérose *f*
CARDIOSPASM - [med] cardiospasme *m*, phrénospasme *m*
CARDIOSURGERY - [med] chirurgie *f* du coeur
CARDIOTOMY - [med] cardiotomie *f*
CARDITIONER - [comput] (card reconditioner) reconditionneur *m* de cartes
CARDITIS - [med] cardite *f*
CARDOON - [bot] cardon *m*, chardonnette *f*
CAREEN, to - [naut] carêner
CAREEN - [naut] carène *f*
CAREENING - [naut] carénage *m*, abattage *m* en carène
CARET - [typ] signe *m* d'omission
CARGO - [gen] chargement *m*, cargaison *f*
~BATTENS - [naut] vaigrage *m* à claire-voie
~BOAT - [shipbuild] cargo *m*
~BOOM - [naut] mât *m* de charge
~CAPACITY - [naut] volume *m* des cales
~COMPARTMENT - [aero] (space in an aircraft designed to receive cargo) soute *f* (à marchandises)
~GEAR - [naut] appareillage *m* de manutention
~GLIDER - [aero] planeur *m* de transport
~HOIST - [mech] (mechanical device for raising cargo into an aircraft or ship) élévateur *m*
~PLANE - [aero] avion *m* de transport, avion *m* cargo
~ROCKET - [astronaut] fusée-cargo *f*
~SHIP - [naut] cargo *m*
~TIE-DOWN - [aero] système *m* de fixation (du fret)
CARIES - [med] (decay of teeth) carie *f*
~OF BONES - [med] ostéite *f* chronique
CARINA - [zool] carène *f*, bréchet *m*
~OF VAGINA - [anat] colonne *f* antérieure du vagin
CARINATE - [zool] (shaped like the keel of a boat) caréné, en forme de carène
~ANTICLINE - [geol] anticlinal *m* caréné
~FOLD - [geol] anticlinal *m* caréné, isoclinal *m*
CARIOUS - [med] carié
~TUBE - [chem] tube *m* scellé
CARLINA OIL - [chem] essence *f* de carline
CARLING - [shipbuild] entremise *f* des baux, élongis *m*
CARMINATIVE - [med] (relieving gastric flatulence) carminatif
CARMINE - [dyes] (a red pigment obtained from alizarin ground in oil) carmin *m*
~PAPER - [ind chem] papier *m* carmin
~RED - [dyes] rouge *m* carmin
CARMINIC ACID - [chem] acide *m* carminique
CARMINITE - [min] carminite *f*
CARNALLITE - [min] (natural potassium magnesium chloride, an important source of potash salts) carnallite *f*
CARNATION - [bot] oeillet *m*
[photo] (flesh tint) carnation *f*
~CLOVER - [bot] trèfle *m* incarnat
~OIL - [chem] essence *f* d'oeillet
CARNAUBA PALM - [bot] palmier *m* à cire (du Brésil)
~WAX - [chem] (hard, naturally-occurring wax used in some surface coatings) cire *f* de carnauba
CARNIC ACID - [chem] acide *m* carnique
CARNIFICATION - [med] carnification *f*, carnisation *f*
CARNIVOROUS - [gen] (flesh eating) carnivore
~PLANT - [bot] (plant which catches and digests

insects) plante *f* carnivore
CARNOSITY - [med] excroissance *f* de chair
CARNOT'S CYCLE - [phys] (ideal cycle of four reversible changes in the physical conditions of a substance) cycle *m* de Carnot
~REAGENT - [chem] (used for the determination of potassium) réactif *m* de Carnot
~THEOREM - [th dyn] principe *m* de Carnot
CARNOTITE - [min] (natural uranium potassium vanadate, bright yellow in colour; an important source of uranium and radium) carnotite *f*, vanadate *m* hydreux
CAROB - [bot] caroube *f*
~TREE - [bot] caroubier
CAROTENE - [chem] (naturally-occurring vitamin A precursor) carotène *m*
CAROTENEMIA - [med] caroténémie *f*, carotinémie *f*
CAROTID - [anat] carotide
~ARTERY - [anat] artère *f* carotide
~BODY - [anat] ganglion *m* carotidien
~GLAND - [anat] ganglion *m* carotidien
~SINUS - [anat] sinus *m* carotidien
CARP - [zool] carpe *f*
CARPAL - [anat] carpien, os *m* carpien
CARPEL - [bot] carpelle *m*
CARPENTER - [gen] charpentier *m*
~RATCHET BRACE - [carp] vilebrequin *m* à cliquets
CARPENTER'S BENCH - [carp] établi *m* de menuisier
~GLUE - [ind chem] colle *f* pour bois
~RIG - [oil ind] chevalement *m* de sondage
~YARD - [carp] atelier *m* de charpentage, chantier *m* de charpentage
CARPENTRY - [carp] charpentage *m*, menuiserie *f*
CARPET, to - [gen] recouvrir d'un tapis
CARPET - [gen] tapis *m*
[constr] (bituminous carpenting in road works) tapis *m*, couche *f* de surface
~BACKING - [rubber ind] dos *m* de tapis caoutchouté
~FELT - [text] feutre *m* pour tapis
~KNOTTED MECHANICALLY - [text] tapis *m* noué mécaniquement
~LATEXING - [text] apprêt *m* sur envers de tapis
~LOOM - [text ind] métier *m* à tisser les tapis
~MOQUETTE - [text] moquette *f*
~MOUNTING - [rubber] pièce *m* moulée de caoutchouc antivibration
~SHEARING MACHINE - [text] tondeuse *f* pour tapis
~SHUTTLE - [text] navette *f* pour tapis
~STEAMING PLANT - [text] installation *f* de vaporisage de tapis
~STRIP - [text] bande *m* de moquette
~UNDERLAY - [gen] assise *f* de feutre
~WARP PILE - [text] poils *mpl* de chaîne de tapis
~WEAVING - [text] tissage *m* des tapis
~WOOL - [text] laine *f* pour tapis
~YARN - [text] fil *m* pour tapis
CARPHOLOGY - [med] (the movements of a delirious patient) carphologie *f*, crocidisme *m*
CARPOPEDAL CONTRACTION - [med] tétanie *f* chez les enfants due à la dentition et aux vers
CARPOPHORE - [bot] carpophore *m*
CARPOPHYL - [bot] carpophylle *m*
CARPOXENIA - [genet] (the effect of pollen on the carpel tissue) carpoxénie *f*
CARPUS - [anat] (the region of the fore-limb) carpe *m*
CARRAGHEEN - [bot] mousse *f* d'Irlande, mousse *f* perlée
~MOSS - s. carragheen
CARREAU - [med] carreau *m* (tuberculose des ganglions mésentériques)
CARRIAGE - [vehicle] véhicule *m*, voiture *f*
[railw] voiture *f*, wagon *m*
[mech] chariot *m*
[text] (carriage of a spinning machine) chariot *m*
[text] (mule) chariot *m* à broches (selfacting)
[artill] (of a gun) affût *m*
[of a typewriter] chariot *m*
[a support for the steps of a wooden stair] support *m*
~BODY - [auto] carrosserie *f*
~BOGIE - [railw] bogie *m*, boggie *m*
~BOLT - [mech] boulon *m* de carrosserie
~CAM - [text] partie *f* de la serrure, came *f* d'aiguille
~DRAFT - [text] étirage *m* par le chariot
~DRAWING-UP WORM - [text] scroll *m* de rentrée
~DRIVE - [text] commande *f* du chariot
~FREE - [comm] franc de port
~GANGWAY - [mining] galerie *f* principale de roulage
~GREASE - [ind chem] graisse *f* pour voitures
~GUIDE - [mech] guidage *m* du chariot
~LOCK KEY - [mech] (typewriter) touche *f* de blocage du chariot
~RELEASE - [mech] (typewriter) déblocage *m* du chariot
~RETURN - [mech] (typewriter) retour *m* du chariot
~STOCK - [railw] dotation *f* en voitures (ou wagons)
~TIPPER - [text] mécanisme *m* à basculer le chariot, culbuteur *m* de chariot
~TWIST - [text] torsion *f* par la sortie du chariot
~TYRE - [rubber ind] bandage *m* plein
~WITH SPENT BOBBIN - [text] chariot *m* à bobine vide
CARRIAGEWAY - [constr] (the road surface) chaussée *f*
~JOINTING CHAMBER - [telecomm] chambre *f* sous chaussée
~JOINTING MANHOLE - [telecomm] chambre *f* sous chaussée
~MARKINGS - [constr] (roads) lignes *fpl* de signalisation, bandes *fpl* de signalisation
CARRICK BEND - [naut] noeud *m* de vache
CARRIED FORWARD CALL - [telecomm] demande *f* de communication déposée la veille
CARRIER - [gen] porteur *m*
[comm] transporteur *m*, voiturier *m*, camionneur *m*
[med] (the carrier of bacteria who has not the disease caused by such bacteria) porteur *m* de germes
[el] (the conductor bringing the charge to a main conductor) porteur *m*
[chem] entraîneur *m*
[ind chem] (dyeing accelerator) véhiculeur *m*, accélérateur *m*, carrieur *m*
[mech] (in a machine) plaque *f* support
[mech] (of a lathe) bride *f*
[plast ind] (the spider supporting the torpedo in an extrusion head) bride *f*
[mech] (a conveyor) transporteur *m*
[radio] (a wave which is suitable for modulation) porteuse *f*, onde *f* porteuse
[nucl] (an element associated with traces of isoto-

pes of the same element which may be radioactive) entraîneur *m*
[text] soutien, appui, support *m*
[photo] (a frame for holding the negative) cadre *m*
[electron] (a mobile conduction electron in a semiconductor) porteur *m* électrisé
CARRIER AMPLITUDE - [radio] (amplitude of a carrier) amplitude *f* de l'onde porteuse
~AMPLITUDE REGULATION - [radio] changement *m* de fréquence de l'onde porteuse
~ARM FOR FEELER ROLLER - [text] bras *m* de rouleau palpeur
~BALANCE - [telev] équilibrage *m* de la sousporteuse de couleur
~BASED AIRCRAFT - [aero] (based on aircraft carrier) avion *m* embarqué
~BED - [geol] roche *f* magasin primaire, roche-relais *f*
~-BORNE AIRCRAFT - [aero] avion *m* embarqué
~BRACKET - [mech] bras *m* de support
~CATALYST - [chem] catalyseur *m*, support *m*
~CHANNEL - [telecomm] canal *m* à courant porteur
~CHROMINANCE SIGNAL - [telev] signal *m* de chrominance
~COMPOUND - [nucl] (bulk material containing the radioactive atoms) composé *m* entraîneur
~COMPRESSION - [radio] (undesired change in the amplitude of a carrier) compression *f* de la porteuse
~CONTROLLED APPROACH RADAR - [radar] C.C.A. *m*, radar *m* d'appontage
~CURRENT CHANNEL - s. carrier channel
~CURRENT PROTECTION - [el] protection *f* par courants porteurs
~CURRENT TELEPHONY - [telecomm] téléphonie *f* par courants porteurs
~DENSITY - [electron] (the number of mobile charge carriers per cubic centimetre) densité *f* des porteurs
~DISC - [mech] disque *m* transporteur
~FILTER - [radio] filtre *m* pour onde porteuse
~FOR THE PISTON RING - [auto] outil *m* de mise en place de segments de piston
~FOR THE SWIVEL SHUTTLE - [text] support *m* de navette à brocher
~-FREE - [nucl] (of radioisotopes of an element in very small quantities) sans entraîneur
~-FREE ISOTOPE - [nucl] isotope sans entraîneur
~-FREE TRACER - [nucl] traceur *m* sans entraîneur
~FREQUENCY - [radio] (the frequency of an unmodulated carrier wave) fréquence *f* porteuse
~FREQUENCY OSCILLATOR - [telecomm] oscillateur *m* à fréquence porteuse
~-FREQUENCY PEAK PULSE POWER - [radio] (average of power over the carrier-frequency carrier occurring at the peak of the pulse of power) puissance *f* de pointe de la fréquence porteuse
~-FREQUENCY PULSE - [modul] impulsion *f* de modulation de la fréquence porteuse
~-FREQUENCY RANGE - [radio] gamme *f* de fréquence porteuses
~-FREQUENCY STABILITY - [radio] (the ability of a transmitter to maintain the specified average frequency) stabilité *f* de la fréquence porteuse
~GAS - [gas ind] gaz *m* porteur, gaz *m* vecteur
~GENERATOR - [telecomm] génératrice *f* de courants porteurs
CARRIER INTERVAL - [telecomm] intervalle *m* entre courants porteurs
~ISOLATING CHOKE COIL - [radio] bobine *f* d'arrêt de la porteuse
~LINE LINK - [radio] liaison *f* par lignes à paires symétriques
~LINE SECTION - [radio] section *f* de ligne à paires symétriques
~METAL - [electron] métal *m* porteur, support *m*
~MOBILITY - [electron] mobilité *f* des porteurs
~MODULATION PERCENTAGE - [telev] taux *m* de modulation de l'onde porteuse
~NOISE - [radio] (unwanted variations of a radio-frequency signal) bruit *m* de la porteuse
~NOISE LEVEL - [radio] niveau *m* de bruit de la porteuse
~OFFSET - [telev] décalage *m* de la porteuse
~PIGEON - [zool] pigeon *m* voyageur
~REPEATER - [telecomm] amplificateur *m* pour courants porteurs
~RING - [text] bague *f* intérieure sans épaulement, anneau *m* porte-guide-fil
~ROCK - s. carrier bed
~ROD - [text] butée *f* d'arrêt des guide-fils
~ROLLER - [text] rouleau *m* entraîneur
~ROLLER COVERED WITH PLUSH - [text] rouleau *m* entraîneur recouvert de feutre
~SENTENCE - [telecomm] phrase *f* de liaison
~SHIFT - [electron] décalage *m* de la fréquence de l'onde porteuse (par rapport à la fréquence nominale de travail)
~SLIDE BAR - [text] barre *f* des guide-fils, barre à jeteurs
~SPRING - [text] ressort *m* de support
~STOP - [text] butée *f* de guide-fil, arrêt *m* de guide-fil
~STORAGE - [electron] (semi-conductors) accumulation *f* des porteurs
~SUPPRESSION - [radio] (method of radio communication in which the carrier wave is not transmitted, or is reduced to a pilot) suppression *f* de l'onde porteuse
~SYNCHRONIZATION - [radio] synchronisation *f* du courant porteur
~TAP CHOKE COIL - [radio] (a choke coil which isolates the carrier) bobine *f* d'arrêt à prises multiples
~TELEGRAPHY - [telecomm] (carrier transmission of telegraph signals) télégraphie *f* par courants porteurs
~TELEPHONY - [telecomm] (transmission of signals by modulating a carrier wave)- téléphonie *f* par courants porteurs
~TELEPHONY EQUIPMENT - [telecomm] appareillage *m* pour téléphonie par courants porteurs
~TERMINAL EQUIPMENT - [telecomm] appareillage *m* d'extrémité de ligne à courants porteurs
~-TO-NOISE RATIO - [radio] rapport *m* porteuse-bruit
~TRANSMISSION - [radio] (transmission by means of a modulated carrier wave) transmission *f* par modulation d'onde porteuse
~TRYCYCLE - [transp] triporteur *m*
~WAVE - [radio] (a wave easily modulated) onde *f* porteuse, porteuse *f*
~WAVE SUPPLY - [telecomm] alimentation *f* de

l'onde porteuse
CARRIER WAVE SUPPLY EQUIPMENT - [telecomm] alimentation *f* de l'onde porteuse
~WAVE TELEGRAPHY - s. carrier telegraphy
CARRIERS - [el] (electrically charged particles; also electrons in semi-conductors) porteurs *mpl* électrisés
CARRION BEETLE - [zool] néocrophore *m*
~CROW - [zool] corneille *f* noire
CARROLLITE - [min] (a sulphide of cobalt which crystallizes in the cubic system) carrollite *f*
CARROT - [bot] carotte *f*
CARROUSEL - [el] (a switch which makes it possible to choose the right voltage) commutateur-adapteur *m* de tension
CARRY, to - [gen] porter, conduire, transporter
CARRY - [comput] (the process of transferring the carry digit to the next higher column) report *m*
~ALONG, to - [gen] emporter, entraîner
~ALONG WITH, to - [auto] traîner
~AWAY, to - [gen] emporter, enlever
~FORWARD, to - [math] reporter
~IN, to - [comput] (the introduction of information into a computer) introduire
~INTO EFEECT, to - [gen] mettre en pratique
~OFF HEAT, to - [phys] (to remove heat by conduction, convection or radiation) éliminer de la chaleur
~ON, to - [gen] continuer, poursuivre
~ONE'S WAY, to - [naut] conserver son erre
~OUT, to - [gen] exécuter
~OVER, to - [gen & math] reporter
~-OVER - [math] report *m*
~THE TRAFFIC, to - [telecomm] écouler le traffic
~THROUGH, to - [gen] conduire à bonne fin
CARRYING - [gen] portant
~AXLE - [railw] essieu *m* porteur
~CABLE - [constr] câble *m* porteur
~CAPACITY - [gen] charge *f* utile, contenance *f*
[el] (current carrying capacity) capacité *f* limite
~CAPACITY OF A LINE - [railw] capacité *f* d'une ligne
~CAPACITY OF THE TRACK - [railw] capacité *f* limite d'un rail conducteur
~COMPANY - [gen] entreprise *f* de transport
~CURRENT - [el] courant *m* porteur
~EFFECT - [telecomm] traînage *m*
~HANDLE - [gen] poignée *f* de transport
~SLEEVE - [auto] manchon *m* support
~STRENGTH - [mech] (of vehicles) résistance *f* des véhicules
~VOLTAGE - [el] tension *f* porteuse
~WIRE - [el] câble *m* porteur
CART - [gen] charrette *f*, chariot *m*, fourgon *m*
~HOUSE - [constr] remise *f*
~MANURE, to - [agric] transporter le fumier
~-WHEEL, to - [aero] faire tonneau
CARTAGE - [transp] transport *m* par voiture, charroi *m*, camionnage *m*
CARTER - [gen] charretier *m*
~PROCESS - [ind chem] (method of manufacturing white lead in which the lead is reduced to fine particles by melting it and subjecting it to air-blast, before corrosion in revolving barrels) procédé *m* de Carter
CARTESIAN - [math] cartésien
~COORDINATE AXES - [math] (the axes of reference of a rectangular Cartesian system) axes *mpl* de coordonnées cartésiennes
CARTESIAN CURVES - [geom] courbes *fpl* cartésiennes
~DEVIL - [phys] ludion *m*
~OVAL - [phys] ovale *m* de Descartes
CARTILAGE - [anat] cartilage *m*
CARTILAGINOUS - [gen] (hard & tough) cartilagineux
CARTING - [transp] charriage *m*
CARTOGRAPHER - [surv] cartographe *m*
CARTOGRAPHY - [surv] (the science of preparing maps and charts) cartographie *f*
CARTON - [packag] boîte *f* en carton
CARTOON - [draw] (drawing or painting of patterns for decorations) carton *m*
[cinema] dessin *m* animé
~CAMERA - [cinema] caméra *f* pour dessins animés
CARTOUCHE - [arch] (ornamental block supporting the eaves) cartouche *f*
CARTRIDGE - [firearms] cartouche *f*, gargousse *f*
[mech] (int comb engines) cartouche *f* de lancement
[photo] (holder for roll films for daylight loading) bobine *f*
[impl] (of a filter) cartouche *f* de filtre
[el acoust] tête *f* de lecture
[nucl] (a fuel element which consists of a rod of atomic fuel hermetically sealed into a metal container) cartouche *f* de combustible
~BELT - [milit] cartouchière *f*, bande-chargeur *f*
~BRASS - [metall] (copper-zinc alloy of high ductility) laiton *m* pour douilles
~CASE - [firearms] douille *f*
~CHAMBER - [firearms] chambre *f*
~FUSE - [el] (a type of electrical fuse in which the fusible element is enclosed in an insulating casing, the whole being changed after operation) fusible *m* à cartouche
~HEATER - [rubber] réchauffeur *m* à cartouche
~POUCH- s. cartridge box
~STARTER - [el] (aircraft engine starter consisting of a small turbine driven by the gases of combustion from a cordite cartridge, the associated reduction gearing and cartridge holder) démarreur *m* à cartouche
~TYPE FILTER - [el] (type of filter in which the separating element forms a removable and replaceable cartridge) filtre *m* à cartouche
~UNIT - [el] cartouche *f*
CARUNCLE - [med] caroncule *f*
CARVACROL - [chem] (an ingredient of fungicides and perfumes) carvacrol *m*
CARVE, to - [gen] sculpter, découper
[text] égratigner, érafler, rayer
CARVEL - s. caravel
~-BUILT - [shipbuild] bordé à franc-bord
~JOINT - [shipbuild] joint *m* à franc-bord
~-PLANKED - s. carvel-built
~WORK - [shipbuild] bordage *m* franc
CARVING - [gen] sculpture *f*
CARVONE - [chem] (a flavouring agent) carvone *f*
CARYATID - [arch] caryatide
CASCADE, to - [el] monter en cascade
CASCADE - [met] (a mass of spray or dense vapour thrown out at the base of a waterspout) cascade *f*
[el] (method of connecting two circuits or units so

that the output of one is used as the input of the second) montage *m* en cascade, montage *m* en série
[ind chem] (a repetitive system of separation or purification) cascade *f*
CASCADE ACCELERATOR - [ind chem] accélérateur *m* en cascade
~AMPLIFIER - [radio] (valve amplifier containing two or more stages connected in cascade) amplificateur *m* en cascade
~CAPACITOR - [telecomm] condensateur *m* en cascade
~CARRY - [comput] report *m* accéléré
~CIRCUIT - [el] montage *m* en cascade
~COMBINATION - [control] (multiple system in which a number of controllers regulate the input of other controllers) réglages *mpl* en cascade
~CONNECTION - [el] montage *m* en cascade
~CONTROL - [el] réglages *mpl* en cascade
~CONVERTER - [el] convertisseur *m* en cascade
~COOLER - [el] réfrigérant *m* à cascade
~GENERATOR - [nucl] (high-voltage apparatus to obtain high-speed particles) générateur *m* en cascade
~LIMITER - [electron] limiteur *m* à deux étages
~OF SEPARATING UNITS - [nucl] (a series of isotope separating units) cascade *f* d'appareils de séparation
~PARTICLE - [nucl] corpuscule *m* d'une gerbe en cascade
~RECTIFIER - [telecomm] redresseur *m* en cascade
~RECTIFIER CIRCUIT - [electron] montage *m* en cascade
~SEQUENCE - [electron] (series of events in an electrical circuit caused by electronic devices) succession *f* de cascade
~SHOWER - [nucl] (a shower of cosmic rays which is started when a high-energy electron produces one or more protons of comparable energies when it passes through matter) gerbe *f* en cascade, avalanche *f*
~-STREAMER - [impl] tamis *m* en cascade
~SYSTEM - [constr] système *m* en chaîne
~THEORY - [phys] (the theory of multiplication of electrons, positrons and y-rays when a cosmic ray particle passes through matter) théorie *f* des cascades
~TUBE - [electron] (a high voltage vacuum tube producting hard X-rays or high-speed ion beams) tube *m* à rayons X à éléments séparés
~VANES - [aero] (guides used to direct the airstream round curves in a wind-tunnel) aubages *mpl* redresseurs de l'écoulement
CASCADES - [aero] (a series of fixed vanes designed to guide the airstream round bends in a wind-tunnel) aubages *mpl* redresseurs
CASCARILLA OIL - [ind chem] essence *f* de cascarille
CASE - [gen] cas *m*
[leg] cause *f*, affaire *f*, procès *m*
[container] caisse *f*, boîte *f*, boîtier *m*
[print] casse *f*
[of dry cells] chemise *f*
[of a storage battery] bac *m*
[print] (a metal plate to which is applied a layer of wax to serve as matrix) forme *f*
[geol] fissure *f* aquifère
~BAY - [build] entrevous *m*
CASE CARBURIZING - [metall] cémentation *f*
~CHILLED - [metall] coulée *f* en coquille
~FINDING - [med] dépistage *m*
~FOR STEERING GEAR - [auto] boîtier *m* de direction
~-HARDEN, to - [metall] (the operation of producing a hard surface layer on a non-hardening type of steel by heating the latter in contact with certain carbon compounds) cémenter
~-HARDENED - [metall] cémenté
~-HARDENED SURFACE - [metall] couche *f* cémentée, couche *f* carburée
~HARDENING FURNACE - [metall] four *m* de cémentation
~-HARDENING POT - [metall] caisse *f* de cémentation
~-HARDNESS - [metall] durcissement *m* superficiel, dureté *f* de cémentation
~LEATHER - [leather] cuir *m* pour valises
~RACK - [print] cassier *m*
~RATE - [med] taux *m* de la mortalité
~REPORT - [med] observation *f*, protocolle *m*
~SCREW - [horol] vis *f* de fermeture de boîte
~SHIFT - [telecomm] (change over of the translating mechanism of a telegraph machine from figures to letter case) inversion *f*
~SPRING - [horol] ressort *m* de retenue de boîte
~VELVET - [text] velours *m* pour étuis
CASEATE - [chem] caséate *m*
[adj] caséeux
CASEIC - [chem] caséique
CASEIFICATION - [med] caséification *f*
CASEIN - [chem] (a phosphoprotein, the main constituent of milk, and the raw material of casein plastics) caséine *f*
~FIBRE - [ind chem] fibre *f* de caséine
~GLUE - [ind chem] (a cheap adhesive made by mixing acid casein with calcium and sodium compounds) colle *f* de caséine, colle *f* à froid
~PAINT - [paints] peinture *f* à la caséine
~PLASTICS - [plast ind] (strong non flammable thermoplastics with poor water resistance) matières *fpl* plastiques à base de caséine
CASEINOGEN - [chem] (a phosphoprotein occurring in milk, from which casein can be obtained by acidification) caséinogène *m*
CASEMAKING DRUM - [rubber ind] tambour *m* de confection (de pneumatiques)
~MACHINE - [rubber ind] machine *f* à confectionner les pneumatiques
CASEMENT - [constr] (of a window; hinged to open about one of its vertical edges) châssis *m* de fenêtre (à deux battants)
~CLOTH - [text] toile *f* pour rideaux
~STAY - [constr] crochet *m* à pitons
CASEOUS - [chem] caséeux
CASH, to - [gen] encaisser
CASH - [gen] argent *m* liquide
~BALANCE - [comm] solde *m* de caisse
~BOOK - [comm] livre *m* de caisse
~CREDIT - [fin] crédit *m*
~-DESK - [comm] caisse *f*
~-DESK MAT - [rubber ind] plaque *f* pour caisse
~IN HAND - [comm] fonds *mpl* en caisse
~ON DELIVERY - [comm] (C.O.D.) payement *m* à la livraison
~ON HAND - s. cash in band
~PAYMENT - [comm] payement *m* au comptant

CASH REGISTER - [comm] caisse *f* enregistreuse
~-SALE - [comm] vente *f* au comptant
~WITH ORDER - [comm] payement *m* à la commande
CASHEW - [bot] anacardier *m*
~RESIN - [plast ind] (thermosetting resin produced from the phenolic fraction of cashew nut shell oil) résine *f* d'acajou
CASHMERE - [text] (fabric made from the winter coat of a Cashmere goat) cachemire *m*
~SHAWL - [text] cachemire *m* de l'Inde
~TWILL - croisé *m* de cachemire, cachemirette *f*
~WOOL - [text] laine *f* de cachemire
~YARN - [text] fil *m* de cachemire
CASING - [metall] (structure, usually of sheet metal, enclosing a part or unit) enveloppe *f*, logement *m*, carter *m*, boîtier *m*, corps *m*
[constr] (a framework] coffrage *m*, cadre *m*
[mech] (of motors) carcasse *f*
[text] (the sorting of qualities) classement *m*
[auto] enveloppe *f* de pneumatique
[mining] (the lining of a drill hole) tubage *m*, colonne *f* de tubage
[constr] (lining of tunnels) coffrage *m* de galerie
[metall] (the metal supporting the refractory walls of a furnace) armature *f* métallique doublant la maçonnerie, casing *m*
~BLOCK - [mining] moufle *m* mobile (pour forage au câble)
~BOWL - [mech] cloche *f* de repêchage à coins
~BREAK-UP - [rubber ind] rupture *f* de carcasse
~CAN - [nucl] (a thin cartridge container) gaine *f*
~CLAMP - [horol] bride *m* de fixation
[mining] collier *m* de serrage, collier *m* de retenue pour tubes de sondage
~COLLAR - [mining] manchon *m* de tubage
~-COUPLING BOLT - [mech] (used in aircraft construction) boulon *m* d'assemblage
~CUTTER - [mining] coupe-tube *m*
~DOG - [mining] arrache-tubes *m*, arrache-cuvelage *m*
~ELEVATOR - [mining] élévateur *m* à tubes, élévateur *m* de tubage
~FLOAT - [mining] flotteur *m* à tube
~+GRAB - [mech] accroche-tube *m*
~-GUN - [mining] (used on oil-well drilling) perforateur *m* de tubage
~HANGER - [mining] dispositif *m* de suspension de tubage
~-HEAD - [oil ind] (in oil wells) tête *f* de tubage
~HEAD GAS - [oil ind] (term used for gasoline obtained from natural gases escaping from an oil-well, used for blending with aircraft fuel) essence *f* obtenue par dégazolinage des gaz sortant du puits
~HOOK - [mining] crochet *m* de levage, crochet *m* à tubes
[oil ind] crochet *m* à émerillon
~IN - [bookbind] cartonnage *m*, finissure *f*
~JACKET - [text] (of a roller) chemise *f* du rouleau, manchon *m*
~KNIFE - [mining] coupe-tubes *m*
~LINE - [oil ind] (in oil well drilling) câble *m* de levage, câble *m* de treuil de manoeuvre, câble *m* de tubage
~PLY - [rubber ind] pli *m* de carcasse
~PROTECTOR - [oil ind] (in oil wells) manchon *m* protecteur de tubage, manchon *m* de protection
~RING - [horol] cercle *m* d'emboîtage

CASING RIPPER - [mining] incise-tubes *m*
~SHOE - [mining] sabot *m* de cuvelage, sabot *m* de tubage
~SHUTTERING - [rubber ind] pneumatique *m*, enveloppe *f*
[mech] (of motors etc) carter *m*
[constr] caisson *m* de coulée, coffrage *m*
~SPIDER - [mining] support *m* de tubage à coins
~STRING - [oil ind] colonne *f* de tubage
~SWAGE - [mining] redresse-tubes *m*
~TONGS - [mining] clefs *mpl* de tubage
CASK - [gen] tonneau *m*, baril *m*, fût *m*
~BEER - [brew ind] bière *f* en tonneau
~BODY - [brew ind] carcasse *f* de tonneau
~BUNG - [brew ind] bonde *f*
~DRIVING MACHINE - [brew ind] machine *f* à cercler
~RINSING DEPARTMENT - [brew ind] atelier *m* de lavage des fûts
~-WAGON (WINE) - [railw] wagon-citerne *m*
~WASHER - [brew ind] lave-fûts *m*
CASKET - [gen] écrin *m*, coffret *m*
[nucl] récipient *m* de transport
~LENSES - [opt] (lens combination) trousse *f* d'objectifs
CASQUE - [telecomm] casque *m*
CASSAVA - [bot] cassave *f*, manioc *m*
CASSE-PAPER - [paper man] papier *m* cassé
CASSEL'S YELLOW - [chem] (commercial lead oxychloride) jaune *m* de Cassel
CASSETTE - [photo] chassis *m* à plaques (ou films)
CASSIA OIL - [ind chem] (an essential oil used as flavouring agent and in perfumery) essence *f* de cassie, essence *f* de cannelle de Chine
CASSIOPEIUM - [chem] (earth metallic element, symbol Lu, atomic number 71) lutécium *m*
CASSITERITE - [min] (important ore of tin) cassitérite *f*
CAST, to - [gen] jeter, lancer
[metall] (to form articles by pouring melted material into a mould) fondre; couler
[naut] laisser porter, abattre
CAST - [gen] jet *m*
[metall etc] pièce *f* moulée, moulage *m* au plâtre
[constr] (with plaster) plâtre *m*, moulage *m* au plâtre
~ABOUT, to - [naut] virer
~ANCHOR, to - [naut] jeter l'ancre, mouiller
~-BRASS - [metall] laiton *m* coulé
~COLD, to - [metall] couler à froid
~CONCRETE - [constr] (portland cement concrete poured into special formwork to determine its final shape) ciment *m* fondu
~FILM - [plast ind] (films of cellulose acetate, nitrate, triacetate, ethyl cellulose or polystyrene produced by casting a solution on to an endless band, followed by drying and stripping) film *m* coulé, fouille *f* mince coulée
~GATE - [metall] jet *m*, attaque *f*
~FURNACE - [metall] fourneau *m* de fonderie
~GATE - [metall] jet *m*, attaque *f*
~GLASS - [opt] verre *m* coulé
~GREEN, to - [metall] couler à vert, mouler à vert
~HEATER - [plast ind] (annular extruder barrel heater in which resistance wires are cast) collier *m* chauffant
~HOLLOW, to - [metall] couler à noyau

CAST HORIZONTALLY, to - [metall] couler à plat
~-IN - [metall] venu de fonte (avec)
~-IN INSERT - [metall] pièce *f* noyée
~IN MOULD, to - [metall] couler en moule
~IN ONE PIECE - [metall] coulé en bloc, coulé monobloc
~IN OPEN SAND, to - [metall] couler à découvert
~INTO INGOTS, to - [metall] couler en lingots, lingoter
~-IRON - [metall] (iron-carbon alloy containing substantial amounts of graphite and cementite, and thus unsuitable for forging, rolling or the like) fonte *f* (de moulage)
~IRON CASTINGS - [metall] pièce *f* de fonte, fonte *f* moulée
~IRON CHIP - [metall] éclat *m* de fonte
~-IRON FRAME - [rubber ind] cadre *m* en fonte, châssis *m* en fonte
~IRON FURNACE - [metall] cubilot *m*
~IRON PIPE - [metall] tuyau *m* de fonte
~-IRON SLEEVE - [metall] manchon *m* en fonte
~-IRON TEMPERED SURFACE ROLL - [metall] cylindre *m* coulé en coquille
~LINE - [soil] courbe *f* cumulative
~METAL FRAME - [mus] (parts of the pianoforte mechanism) sommier *m* métallique
~MOULDING - [metall] (a moulding produced without the use of pressure and at relatively low temperatures) pièce *f* moulée directement (sans utilisation de la pression), moulage *m* par la coulée
~OFF, to - [naut] larguer, laisser porter
[print] évaluer le nombre de pages imprimées etc, évaluer un manuscrit
~-OFF CAM - [mech] came *m* d'abattage
~-OFF POSITION - [mech] position *f* de dépointage
~OFF TYRE - [rubber ind] pneumatique *m* mis au rebut
~OFF VORTEX - [phys] (a vortex which is left behind in the ambient fluid when a body is put into motion) tourbillon *m* de départ
~ON BAR - [metall] barreau *m* attenant
~ON END, to - [metall] couler en source
~-ON FLANGE - [metall] manchon *m* venu de fonte, tubulure *f* venue de fonte
~-ON PIN - [metall] tenon *m* venu de fonte
~OUT, to - [gen] rejeter
~RESIN - [ind chem] résine *f* coulée
~SCRAPS - [metall] rebuts *mpl* de fonderie
~SHADOW - [photo] ombre *f* portée
~SOFT, to - [metall] couler en fonte à faible teneur de carbone
~STEEL - [metall] acier *m* moulé
~STEEL PLAT - [metall] plaque *f* en acier moulé
~STEEL PUNCH - [metall] perforatrice *f* en acier coulé
~STEEL SPOKED WHEEL - [auto] roue *f* à rayons en acier coulé
~STEEL WIRE HEALD - [text] lisse *f* à fil d'acier
~THROUGH THE PLUG HOLE, to - [metall] couler en source
~TOOTH - [metall] dent *f* brute (de fonte)
~WHITE, to - [metall] couler en fonte blanche
CASTABILITY - [metall etc] coulabilité *f*
~TEST - [metall] essai *m* de coulabilité
CASTABLE - [metall] coulable, moulable
CASTANETS - [mus] castagnettes *f*
CASTELLANUS - [met] (a type of altocumulus showing marked turret-like upward projections which appear to be in lines and connected to a common base) castellanus *m*
CASTELLATED NUT - [mech] (a nut slotted across the top, so that a cotter-pin can be inserted in a hole in the bolt, thus securing the nut against undesired rotation) écrou *m* crénelé
CASTER, to - [mech] (in a landing gear; to swivel so as to follow the true direction on the ground of an aircraft's movement) pivoter
CASTER - [ind] couleur *m*, fondeur *m*, mouleur *m*
[mech] (a small roller attached to the feet of furniture items and which can be swivelled) roulette *f*, roue *f* pivotante
[print] (a type casting machine) machine *f* à fondre les caractères
[aero & auto] (the tendency of a wheel which is free to swivel about a vertical or near vertical axis, to set itself in line with the direction of motion of the structure which it supports) chasse *f* de direction
~ACTION - [auto etc] chasse *f* de direction
~ANGLE - [mech] angle *m* de chasse
~EFFECT ON STEERING - [auto] effet *m* de rappel, effet *m* de l'angle de chasse
~LENGTH - [aero] (of landing gear) convergence *f* de la roue
~LOCK - [mech] (device for locking a castering wheel or landing gear when desired) blocage *m* de roue orientable
~OFFSET - [auto] projection *f* de la chasse
~TRAILER - [mech] remorque *f* à quatre roues
~-UP - [mining] (in coal mining) déschisteur *m*
CASTERING LANDING GEAR - [aero] (landing gear so mounted that the wheels can swivel to follow the true direction of the aircraft when landing yawed in a crosswind) train *m* d'atterrissage orientable
~WHEEL - [aero] (in a landing gear) roue *f* orientable, roue *f* pivotante
CASTING - [metall etc] (the operation of casting) coulée
[any article obtained by the process of casting] pièce *f* coulée
[plast ind & glass man] coulée *f*
[theatre & cinema] distribution *f*
[print] (the pouring of electro-type metal on the thinned shells) coulage *m*
~AREA - [metall] aire *f* de coulée
~BAY - [metall] chantiers *mpl* de coulée
~BED - [metall] (of a blast furnace; the area on which sand moulds are made to receive molten iron when the furnace is tapped) lit *m* de coulée
~BOX - [metall] châssis *m*
~BRUSH - [metall] brosse *f* pour fontes brutes
~CLEANING MACHINE - [mech] ébarbeuse *f* pour pièces coulées
~CRANE - [metall] grue *f* de fonderie
~CRITICAL AREA - [metall] zone *f* critique de coulée
~DIE - [metall] matrice *f* pour moulage sous pression
~DIRECTOR - [cinema] directeur *m* de la distribution
~FORM - [metall] moule *m*
~GATE - [metall] jet *m*, attaque *f*
~HEAD - [metall] masselotte *f*, tête *f* de lingotière
~HOUSE - [gen] fonderie *f*

CASTING IN CHILL - [metall] pièce *f* coulée en coquille
~IN FLASKS - [metall] moulage *m* en châssis
~IN OPEN - [metall] pièce *f* coulée à découvert
~IN SAND - [metall] pièce *f* coulée en sable
~LADLE - [metall] poche *f* de coulée
~LATEX - [rubber ind] pièce *f* moulée en latex
~LINE - [rubber ind] ligne *f* de coulée
~MACHINE - [print] machine *f* à fondre les caractères
[metall] (of blast furnaces; machine which receives molten iron from the furnace and forms pigs automatically) machine *f* à couler
~MATRIX - [metall] matrice *f* de coulée
~NET - [fishing] épervier *m*
~ON - [text] montage *m* des mailles (tricot)
~ON FLAT - [metall] coulée *f* à plat
~PATTERN - [metall] modèle *m*
~PIT - [metall] fosse *f* de coulée
~PIT CRANE - [metall] grue *f* de fosse de coulée
~RESINS - [plast ind] (liquid polymerizable resins, containing hardening agents, which can be poured into moulds and hardened with or without the use of heat) résines *f* de coulée
~SEAM - [metall] bavure *f* (de fonte)
~SHOP - [metall] fonderie *f*
~SKIN - [metall] peau *f* (d'une pièce moulée)
~STRAINS - [metall] déformations *fpl* sous tensions internes
~STRAND - [metall] convoyeur *m* de coulée
~TEMPERATURE - [metall] température *f* de coulée
~WAX - [metall] cire *f* de fonderie
~WHEEL - [metall] (large wheel on a vertical spindle, fitted with moulds which are brought successively under a spout delivering iron or slag) roue *f* de coulée
~WITH BLOWHOLES - [metall] pièce *f* coulée comportant des soufflures
CASTINGS BREAKER - [metall] casse-fonte *m*
CASTLE - [arch] château *m*
[nucl] (a housing for a counter and the radioactive material for assay, so designed as to reduce extraneous radiation) château *m* (de plomb)
~NUT - [mech] écrou *m* crénelé
~WHEEL - [horol] pignon *m* crénelé
CASTNER CELL - [el] (electrolytic cell with mercury cathode used for the production of chlorine and caustic soda from brine) pile *f* de Castner
CASTOR - s. caster
CASTOR - [chem] castoréum *m*
~OIL - [ricinus oil] (starting material for plasticizers, certain nylons, and alkyd resins for surface coatings) huile *f* de ricin
~OIL PLANT - [bot] ricin *m* commun
CASTOREUM - [chem] castoréum *m*
CASTORIN - [chem] castorine *f*
CASTRATE, to - [vet] castrer, émasculer
CASTRATION - [vet] castration *f*
CASUAL - [gen] fortuit, accidental
CASUALTY - [gen] victime *f*
~CLEARING STATION - [med] centre *m* d'évacuation
CAT, to - [naut] caponner (l'ancre)
CAT - [zool] chat *m*
[naut] capon *m*
[shipbuild] petit *m* bâtiment gréé à taille-vent
~-CRACKING - [colloq] (term for catalytic cracking) cracking *m* catalytique
CAT DAVIT - [naut] bossoir *m* de capon
~DIRT - [mining] charbon *m* pyriteux
~FACES - [timber] noeuds *m*
~FALL - [naut] garant *m* de capon
~GOLD - [min] mica *m* aurifère
~HOLES - [naut] écubiers *m*
~-LINE - [oil ind] câble *m* de cabestan
~-MINT OIL - [chem] essence *f* de cataire
~SILVER - [min] argent *m* de chat, mica *m*
~TACKLE - [naut] palan *m* de capon
~-WALK - [cinema] (colloq. for an elevated platform for lights etc) passerelle *f*
[aero] (of an airship) coursive *f*, couloir *m* de communication
[constr] passerelle *f*
CAT'S EYE - [mining] (a lamp) oeil *m* de chat
[glass man] (elongated bubble caused by extraneou[s] matter) occlusion *f*, bouillon *m*
~EYES - [light] (in road works, glass reflectors to show the middle line of the road, or curbstones etc) cataphotes *mpl*, cabochons *mpl* lumineux
~PAW - [naut] noeud *m* de gueule de raie
~WHISKER - [radio] (the elastic wire making the contact in crystal detectors) chercheur *m*, spirale *f* métallique
CATA - [chem] (prefix denoting that the substance contains a condensed double isocyclic nucleus substituted in the 1.7 positions) cata-
CATABATIC - [phys] (relating to a downward motion) catabatique
CATABOLISM - [biol] (the total of the disruptive metabolic processes) catabolisme *m*
CATACLASTIC - [geol] cataclastique
CATACOMB - [arch] catacombe *f*
CATADIOPTRIC - [opt] catadioptrique
CATAGENESIS - [genet] (regressive evolution) catagénèse *f*
CATALAN FURNACE - [metall] four *m* catalan; bas-foyer *m*
CATALASE - [chem] (enzyme present in plant and animal tissues and which will decompose hydrogen peroxide) catalase *f*
CATALEPSY - [med] catalepsie *f*
CATALOGUE - [gen] catalogue *m*
CATALYSIS - [chem] (change in the rate of a reaction induced by the presence of small quantities of a substance which itself remains unchanged) catalyse *f*
~CRACKING - [chem] (the breaking of a carbon-carbon bond with the aid of a catalyst; an importan[t] process in petroleum refining) cracking *m* catalytique
~OF NUCLEUS - [biochem] catalyse *f* du germe
CATALYST - [chem] (a substance which accelerates a reaction without taking part in it) catalyseur *m*
~CHAMBER - [chem] chambre *f* de catalyse
~DISTILLATION - [ind chem] distillation *f* catalytique
~LIFT - [ind chem] (in oil refining) élévateur *m* de catalyseur
~VAPOUR DISENGAGER - [ind chem] (in oil refining) séparateur *m* catalyseur-vapeur
CATALYTIC AGENT - [ind chem] catalyseur *m*
~BOMB - [ind chem] (high-pressure vessel used in catalytic synthesis of ammonia) bombe *f* catalytiq[ue]
~COMBUSTION RADIANT UNIT - [th eng] radiateur

m à combustion catalytique
CATALYTIC CONVERSION - [chem] (of carbon monoxide) conversion f catalytique
~CRACKING - s. catalysis cracking
~CYCLE STOCKS - [oil ind] charges fpl de catalyseur
~EXCHANGE REACTION - [nucl] (chemical exchange reaction by means of a catalyst) réaction f d'échange catalytique
~POISON - [chem] (term used for an inhibitor of catalytic processes) poison m
~REACTOR - [ind chem] (a vessel containing catalyst material in which a reaction involving it is carried on) réacteur m catalytique
~REFORMING - [chem] (the use of heat, pressure and a catalyst to isomerize hydrocarbon molecules) régénération f catalytique
CATALYZE, to - [chem] catalyser
CATAMARAN - [naut] catamaran m
CATAPHORESIS - [chem] cataphorèse f
CATAPHORIA - [med] tendance f au déplacement vers le bas de l'axe visuel
CATAPULT, to - [aero] catapulter
CATAPULT - [aero] (a device for launching an aircraft at flying speed) catapulte f
[gen] fronde f
CATAPULTING - [aero] catapultage f
CATARACT - [hydr] cataracte f
[opt] (opacity of the lens of the eye) cataracte f
CATARRH - [med] catarrhe f
CATASTASIS - [med] remise f en place d'un organe déplacé
CATCH, to - [gen] prendre, attraper, accrocher, enclencher
[naut] prendre (le vent)
CATCH - [mech] (spring bolt for securing doors) loquet m, dispositif m d'arrêt
[a clamp] collier m, fixation f
[text] (of a knitting machine) platine f auxiliaire, poussoir m
~-ALL - [paper ind] ramasse-pâte m
~AN IMAGE, to - [photo] saisir une image, recueillir une image
~BAR - [text] barre f à poignée, barre f mouvante
~BAR MOTION - [text] mécanisme m de la barre mouvante
~BASIN - [build] collecteur m, chambre f de décantation (égout)
[metall] bassin m épurateur
~BOX - [ind chem] (used for chemical purification) cuve f de sûreté
~CROP FODDER - [agric] fourrage m en culture dérobée
~CROP GROWING - [agric] culture f dérobée
~DRAIN - [metall] caniveau m de drainage
~FEEDER - [hydr] canal m d'irrigation
~FINGER - [text] griffe f, doigt m de retenue
~FOR PICKING ARM - [text] cliquet m de bras de chasse
~IN, to - [mech] enclencher
~LEVER - [text] levier m à cliquet
~LIGHTS - [photo] accents mpl de lumière
~OF THE HOOK - [text] bec m de crochet
~PIN - [text] goupille f réceptrice
~PIT - [constr] tranchée f drainante
~PIT GULLY - [constr] fosse f d'écoulement m, regard m d'égout m
CATCH-POT - [ind chem] séparateur m (piège m à condensat)
~PROPS - [mining] étais mpl provisoires
~THREAD - [text] fil m de retenue, fil de raccroc
~WATER DRAIN - [hydr] canal m de drainage
CATCHER - [mech] arrêt m
[electron] (electrode in a klystron designed to draw energy from electron bunches) résonateur m de sortie
[a mechanism used for separating dust particles in pipes] capteur m poussière
[mining] parachute m pour tubes
CATCHMENT AREA - [geogr] bassin m hydrographique
~BASIN - s. catchment area
CATCHPIT - [agric] puits m perdu
CATCHPLATE - [mech] plateau m toc (de tour)
CATECHOL - [chem] (important for its derivatives guaiacol and adrenaline) catéchol m, pyrocatéchol m
CATECHU - [photo] (dark brown dyeing extract used for toning platinum prints) cachou m
~BROWN - [dyes] cachon
CATEGORY - [gen] catégorie f, classe f
CATENARIAN ARCH - [arch] (arch with the shape of an inverted catenary) arc m caténaire
CATENARY - [phys] (the curve assumed by a chain or heavy flexible cord attached at its ends only to two points) caténaire f
~CONTACT-LINE CONSTRUCTION - [telecomm] ligne f caténaire
~LINKAGE - [chem] (the linking of molecules occurring when amine acids unite to form polypeptides) liaison f à chaîne
~SUPPORT - [railw] support m de ligne caténaire
~SUSPENSION LINE - [transp] ligne f caténaire
CATENATION - [genet] (formation of rings or chains by chromosomes) caténation f
CATENOID - [math] alysséide f
CATERING - [gen] alimentation f, approvisionnement m
CATERPILLAR - [mech] (a device designed to increase the tractive of road vehicles or tractors) chenille f (CATERPILLAR: Reg. trade mark)
[zool] chenille f
~BELT - s. caterpillar
~DRIVE - [mech] traction f à chenilles
~GATE - [eng] vanne-wagon f
~GRADER - [mech] niveleuse f à chenilles
~TRACK - s. caterpillar
~TRACTOR - [mech mach] tracteur m à chenilles
CATERWAULING - [acoust] (noise made by cats) miaulement m de chat
CATFORMING - s. catalytic reforming
CATGUT - [mus] (used for strings) boyau m
[med] catgut m
CATHEAD - [naut] cabestan m, petit treuil m
CATHEDRAL - [arch] cathédrale f
~GLASS - [glass man] verre m cathédrale, vitraux mpl sertis de plomb
CATHETER - [med] cathéter m
CATHETERIZE - [med] cathétériser
CATHETOMETER - [instr] (designed to measure vertical distances not exceeding a few centimeters) cathétomètre m
CATHETRON - [el] (grid controlled mercury-arc rectifier) redresseur m à vapeur de mercure à

grilles commandées
CATHEXIS - [med] concentration *f* mentale
CATHODE - [el] (the negative pole of an electrical system) cathode *f*
[electron] (an electrode which is the primary source of electron emission) cathode *f*
~ANCHOR - [electron] (in electronic tubes) patte *f* de cathode
~BEAM - [electron] faisceau *m* cathodique
~BIAS - [electron] (biasing potential applied to the cathode of an electron tube) polarisation *f* de cathode
~BORDER - [electron] (the distinct surface of separation between the cathode dark space and the negative glow) lisière *f* cathodique
~CLEANING - [metall] (electrolysis) dégraissage *m*
~COMPENSATION - [telev] (a video frequency stage) compensation *f* cathodique
~CONTAMINATION - [electron] (chemical effect of residual gases) contamination *f* de la cathode
~COPPER - [metall] cuivre *m* électrolytique
~-COUPLED AMPLIFIER - [ampl] amplificateur *m* à couplage cathodique
~-COUPLED CIRCUIT - s. cathode follower
~CUP - [electron] (a metal device for focussing the electron beam) cupole *f* de focalisation
~CURRENT - [electron] courant *m* cathodique, intensité *f* cathodique
~DARK SPACE - [electron] (region of faint luminescence following the cathode glow in which the electrons do not excite the gas but only ionize it) espace *f* sombre-cathodique
~DEPOSIT - [ind chem] dépôt *m* cathodique
~DROP - s. cathode fall
~EFFICIENCY - [el] (current efficiency in a specified cathodic process) rendement *m* cathodique
~EVALUATION - [electron] évaluation *f* de la cathode
~FALL - [electron] (also: cathode potential fall. The difference of potential due to the space charge near the cathode) chute *f* cathodique
~FOLLOWER - [electron] (a radio circuit in which the load is in the cathode circuit of an electronic tube) circuit *m* à charge cathodique
~GLOW - [electron] (glow discharge occurring in a vacuum discharge tube when the applied voltage reaches a sufficiently high level) lueur *f* cathodique
~-GRID CAPACITY - [electron] capacité *f* grille-cathodique
~GROUNDED CIRCUIT - [el] circuit *m* à cathode mise à la masse
~HEATING TIME - [electron] (the time required for the cathode to attain a specified condition) temps *m* de chauffage
~HUM - [electron] (the hum voltage between cathode and earth) ronflement *m* de la cathode
~KEYING - [modul] manipulation *f* dans le retour commun
~LAYER - [metall] (layer of molten metal acting as an anode in a fused-electrolyte cell) couche *f* métallique servant de cathode
~LUMINESCENCE - [electron] (caused by cathode or beta-rays) cathodoluminescence *f*
~MERCURY - [ind chem] mercure *m* cathodique
~MODULATION - [radio] (modulation effected in a transmitting valve by injecting a signal voltage into the cathode circuit) modulation *f* cathodique
CATHODE NECK - [electron] (of the tube which surrounds the cathode) col *m* cathodique
~POISONING - s. cathode contamination
~PULSE MODULATION - [radio] modulation *f* cathodique par impulsions
~RAY - [electron] (the stream of negatively charged particles, or electrons, emitted from the surface of the cathode in a rarefied gas) rayon *m* cathodique
~-RAY CAMERA - [telev] caméra *f* à tube cathodique
~RAY DIRECTION FINDER - [radar] (direction finder apparatus in which the bearing required is shown visually on the screen of a cathode ray tube without the need for a goniometer) radiogoniomètre *m* à oscilloscope
~RAY OSCILLOGRAPH TUBE - [instr] (cathode ray tube used to give visual representation of instantaneous values of the currents and voltages fed to the tube) oscillographe *m* cathodique
~-RAY SCREEN - [electron] écran *m* fluorescent
~RAY TRACE - [electron] oscillogramme *m*
~RAY TUBE - [electron] (an electron tube of generally conical form in which an electron beam generated at one end is deflected by coils or plates supplied with electric quantities. The beam impinges on a fluorescent screen on the large end of the tube and thus produces a trace which gives a visual indication of the quantities supplied to the deflector elements) tube *m* à rayons cathodiques
~RAY TUBE DISPLAY - [telev] (the presentation of a received signal on the screen) information *f* cathodique
~-RAY TUBE HARNESS - [electron] dispositif *m* de fixation pour tube à rayons cathodiques
~-RAY TUBE SCALE - [electron] graduation *f* pour tube à rayons cathodiques
~-RAY VOLTMETER - [instr] voltmètre *m* à rayons cathodiques
~RAYS - [electron] (beam of electrons emitted from the cathode of a vacuum tube) rayons *mpl* cathodiques
~REGION - [electron] (the group of regions extending from the cathode to the Faraday dark space) domaine *m* cathodique
~RESISTANCE - [electron] (a resistance in the cathode circuit of an electronic tube) résistance *f* de la cathode
~SCREEN - [electron] écran *m* cathodique
~SPARKING - [electron] (flashing of the cathode of a vacuum tube) scintillement *m* cathodique
~SPOT - [electron] (the luminous part of an arc cathode surface on which the electron stream is concentrated) tache *f* cathodique
~SPUTTERING - [electron] (erosion of the cathode of a vacuum tube under ionic bombardment) pulvérisation *f* cathodique
CATHODIC BOMBARDMENT - [electron] bombardement *m* cathodique
~CLEANING - [metall] (method of electrolytic cleaning in which the object being treated is the cathode) dégraissage *m* cathodique
~ETCHING - [mteall] nettoyage *m* par bombardement *m* ionique, effluvation *f*, nettoyage *m* ionique
~INHIBITORS - [el] (substances which modify the cathodic reaction) inhibiteurs *mpl* cathodiques
~PICKLING - [metall] (method of electrolytic pickling in which current is passed through the

solution to the metal being treated) décapage *m* cathodique
CATHODIC POLARIZATION - [el] (polarization occurring at the cathode) polarisation *f* cathodique
~PROTECTION - [metall] (suitable cathodic polarization to give protection to a metal) protection *f* cathodique
~REACTION - [el] (a reaction involving transfer of possitive charges from the electrolyte to the electrode) réaction *f* cathodique
CATHODOLUMINESCENCE - s. cathode luminescence
CATHODOPHOSPHORESCENCE - [electron] (the result of cathode-ray bombardment as in cathodoluminescence) cathodophosphorescence *f*
CATHOLYTE - [metall] (in electroplating, the electrolyte surrounding the cathode) catholyte *f*
CATION - [nucl] (an atom, molecule or group of molecules carrying a positive charge) cation *m*
CATIONIC - [nucl] cationique
~REAGENTS - [chem] (flotation in which the active component is the positive ion) réactifs *mpl* cationiques
~SOAP - [chem] (a surface-active agent) savon *m* inversé, savon *m* acide
CATLIN - [med] couteau *m* à imputations
CATLINE - [mining] câble *m* de cabestan
CATMINT - [bot] cataire *f*
CATOPTRICS - [opt] catoptrique *f* (étude de la lumière réfléchie)
CATTLE - [zool] bétail *m*, gros bétail *m*
~BOWL - [agric] mangeoire *f*, abreuvoir *m*
~CAR - [railw] wagon *m* à bestiaux
~CLIPPER - [agric] tondeuse *f*
~GRUB INFESTATION - [vet] infestation *f* par les vers
~GUARD - [railw] (fitted in front of a locomotive) chasse-pierres *m*
[telecomm] fil *m* de protection contre les animaux
~HIDE - [leather ind] peau *f*
~ON THE HOOF - [agric] bétail *m* sur pied
~PLAGUE - [vet] perte *f* bovine
~REARING - [agric] élevage
~TRAIN - [railw] train *m* à bestiaux
~TRUCK - [transp] bétaillère
~WAGON - s. cattle car
~WEIGHING PLATFORM - [agric] bascule *f* pour bestiaux
CAUDATE - [zool] caudé
CAUDICLE - [bot] caudicule *f*
CAUL - [carp] (a sheet of metal, plywood or other material used as a cover plate to equalize pressure in making plywood) lamelle *f*, couche *f* mince
[obstetr] coiffe *f*
CAULDRON - [gen & ind] chaudron *m*
[mining] effondrement *m* circulaire
CAULESCENT - [bot] (having a stalk or a steam) caulescent
CAULIFEROUS - [bot] caulifère, caulescent
CAULIFLOWER - [bot] chou-fleur *m*
~HEAD - [metall] (of feedhead) boursouflement *m* (de masselotte)
CAULINE - [bot] (growing from the stem) caulinaire
CAULK, to - [mech] (to make tight by hammering the exposed edges of riveted joints) mater
[naut] calfater
~A RIVET, to - [mech] mater un river
CAULKED SEAM - [shipbuild] couture *f* matée
CAULKER - [shipbuild] (a caulking specialist) calfat *m*
[tool] matoir *m*
CAULKING - [shipbuild] calfatage *m*
[mech] matage *m*
~CHISEL - [tool] matoir *m*, ciseau *m* de calfat
~FELT - [naut] ploc *m*
~HAMMER - [impl] marteau *m* à mater, matoir *m*
~IRON - [mech] (blunt-ended chisel-like tool used for caulking joints) matoir *m*
~RING - [mech] anneau *m* de matage
CAULOCARPOUS - [bot] caulocarpe
CAUSEWAY - [constr] (road carried by an embankment or a wall) chaussée *f* surélevée, route *f* sur digue
CAUSTIC - [chem] caustique
~ALKOHOL - [chem] soude *f* à l'alcool
~BARYTA - [chem] hydrate *m* de barium, baryte *m* caustique
~-CALCINED MAGNESITE - [chem] magnésite *f* calcinée
~CRACKING - [metall] fragilité *f* caustique
~DIP - [metall] bain *m* caustique
~EMBRITTLEMENT - s. caustic cracking
~LIME - [chem] (the residue of calcium oxide) chaux *f* caustique, chaux *f* vive
~LIQUOR - [chem] soude *f* caustique diluée
~LYE OF SODA - [chem] lessive *f* de soude
~POTASH - [chem] (a term for potassium hydroxide) potasse *f* caustique
~SILVER - [chem] nitrate *m* d'argent
~SODA - [chem] (sodium hydroxide) soude *f* caustique
~SODA CELL - [el] (of a primary battery, a cell in which the electrolyte consists mainly of a solution of sodium hydroxide) pile *f* à hydrate de sodium
~SURFACE - [light] (a surface to which rays of light are tangential after reflection or refraction at another surface) surface *f* caustique
~WASH - [ind chem] lavage *m* caustique
CAUSTICITY - [chem] causticité *f*
CAUSTICIZER - [ind chem] caustifiant *m*
CAUSTICIZING - [ind chem] caustification *f*
CAUTERIZATION - [med] cautérisation *f*
CAUTERIZER - [med] cautère *m*
CAUTERY - [med] cautère *m*
~ELECTRODE - [med] électrode *f* de cautérisation
CAUTIOUS RUNNING - [railw] marche *f* à vue
CAVE, to - [oil ind] (of a well) s'effondrer, s'ébouler
CAVE - [gen] caverne *f*, grotte *f*
[nucl] (a heavily shielded compartment in which highly radioactive material can be handled (by remote control) fosse *f*
[oil ind] éboulement *m* d'un puits
~IN, to - [gen] s'ébouler, s'éffondrer, foudroyer
~-IN - [gen] éboulement *m*, affaissement *m*
CAVERN - [gen] caverne *f*, grotte *f*
CAVERNILOQUY - [med] bronchophonie *f* caverneuse
CAVERNITIS - [med] inflammation *f* des corps caverneux
CAVERNOUS - [gen] caverneux
~BREATHING - [med] souffle *m* caverneux
CAVETTO - [arch] (a hollow moulding) cavet *m*, gorge *f*

CAVING - [mining] (the caving-in of a well) affaissement *m*, éboulement *m*
[mining] (method of mining) foudroyage *m*
~FORMATION - [mining] terrain *m* boulant
CAVITATION - [phys] (the formation of a cavity on the downstream side of a body moving in a fluid (e.g. a propeller blade or a pump vane) in which the pressure is below that of the vapour pressure of the liquid) cavitation *f*
CAVITY - [gen] cavité *f*
[metall] (of a die) empreinte *f*
[metall] soufflure; (in a mould) empreinte *f* d'un moule
~COUPLED FILTER - [radio] (wave filter consisting of two identical chambers coupled with a third chamber) filtre *m* à cavités résonantes couplées
~DEPTH - [plast ind] profondeur *m* de l'empreinte
~FILLING - [mining] remplissage *m* de fissures
~FREQUENCY METER - [electron] (variable resonant cavity used for measuring the frequency of electromagnetic waves) fréquencemètre *m* à cavité résonante
~MAGNETRON - [electron] (magnetron in which resonant cavities in the body of the anode act as tuned circuits) magnétron *m* à cavité
~PLUG - [plast ind] (another term for a mould insert) prisonnier *m*
~RADIATION - [phys] rayonnement *m* de corps noir
~RESONATOR - [electron] (resonant cavity in the anode-body of a magnetron) cavité *f* résonante
~RETAINER PLATE - [plast ind] (the fixing plate on which the mould is mounted) porte-empreinte *m*
~SHEET LATEX FOAM - [rubber ind] mousse *f* de latex en plaques
~SIDE PART - [plast ind] (term used in the U.S.A. for the stationary part of a mould) demi-moule *m* fixe
~WALL - [build] mur *m* à éléments creux
~WALLS - [build] (two walls held together by wall ties) mur *m* creux
CAWK - [min] barytine *f*
CEASE, to - [gen] cesser, discontinuer
CEASING - [gen] cessation *f*
CECAL - [anat] caecal
CECUM - [anat] caecum *m*
CEDAR - [bot] cèdre *m*
~NUT OIL - [chem] huile *f* de bois de cèdre
~-WOOD OIL - [chem] (poisonous, volatile oil with uses in medicine and perfumery) essence *f* de bois de cèdre
CEIL, to - [gen] plafonner
[build] plafonner, lambrisser
[shipbuild] vaigrer
CEILING - [gen] plafond *m*
[build] (the upper surface of a room) plafond *m* plafonnage *m*
[aero] plafond *m*
[met] plafond *m*, (height of the level of a bank of clouds) plafond *m*
[artill] flêche *f*
[shipbuild] vaigrage *m*, plafond *m*
[railw] (the upper surface of a coach) plafond *m*
[astronaut] (of rockets) altitude *f* maximale
~BOARDING - [build] bois *m* de plafonnage
~DRILLING MACHINE - [mech] perceuse *f* à plafonds
~FAN - [el] ventilateur *m* de plafond
CEILING HANGER - [constr] (ceiling joist) gite *f* pour plafond suspendu
~HEATING - [th eng] chauffage *m* par le plafond
~HEIGHT INDICATOR - [instr] indicateur *m* de hauteur de plafond
~LAMP - [el] plafonnier *m*
~LIGHT PROJECTOR - [instr] (used to throw a light beam on clouds) projecteur *m* pour déterminer la hauteur des nuages
~REED - [constr] roseau *m* pour plafonds
~ROSE - [build] rosace *f* de plafond
~ROSETTE - s. ceiling rose
~SWITCH - [el] interrupteur *m* de plafond
~VOLTAGE - [el] (the maximum voltage produced by a generator) tension *f* de sortie maximale
CEILOMETER - [met] projecteur *m* pour déterminer la hauteur des nuages
CEL - [cinema] bande *f* en celluloïd, cel *m*
CELADON - [ceram] (porcelain of a pale grey-green colour) vert *m* céladon
CELERY - [bot] céleri *m*
CELESTE - [dyes] bleu céleste
CELESTIAL - [astr] céleste
~BODY - [astr] corps *m* céleste
~EQUATOR - [astr] (the Great Circle of the celestial sphere) équateur *m* céleste
~FIX - [aero] point *m* astronomique
~HORIZON - [astr] horizon *m*
~LATITUDE - [astr] latitude *f* céleste
~LONGITUDE - [astr] longitude *f* céleste
~MERIDIANS - [astr] (great Circles on the celestial sphere) méridiens *mpl* célestes
~NAVIGATION - [astr] navigation *f* astronomique
~POLES - [astr] (the two poles in which the earth's axis cuts the celestial sphere) pôles *mpl* célestes
~SPHERE - [astr] (imaginary sphere of infinite radius having its centre within the solar system) sphère *f* céleste
CELESTITE - [min] (mineral consisting essentially of strontium sulphate, occurring in association with rocksalt and gypsum) célestine *f*
CELIAC - [anat] coeliaque
~PLEXUS - [anat] plexus *m* solaire
CELIALGIA - [med] coelialgie *f*
CELIOMA - [med] tumeur *f* abdominale
CELIOSCOPE - [med] appareil *m* pour examen visuel direct de la cavité abdominale
CELITE - [chem] célite *f*
CELITIS - [med] inflammation *f* abdominale
CELL - [gen] cellule *f*, alvéole *m*
[el chem] (a single galvanic cell in practical form) cellule *f*
[el] (of a battery) élément *m*, pile *f*, accumulateur *m*
[auto] (of a radiator) élément *m*
[comput] élément *m* de mémoire
~AMPLIFICATION - [electron] amplification *f* par ionisation
~BOX - [el] (of a battery) bac *m*
~CAVITY - [el] (the container formed by the cell lining for holding the fused electrolyte) creuset *m*
~CONNECTOR - [el] (conductor to connect adjacent cells in a storage battery) connecteur *m* (entre deux accumulateurs)
~CONSTANT - [el] (relation between the resistance of a cell and that of the electrolyte; it is depended only on the geometrical form of the cell) constante

f de cellule
CELL CONTAINER - s. cell box
~COUNT - [biol] (numerical value for cell population in a culture) numération *f* des cellules
~FILTER - [brew ind] filtre *m* à chambre
~GROWTH - [biol] croissance *f* des cellules
~PRESSURE - [el] tension *f* d'un élément [soil] pression *f* de soutien, contrainte *f* de soutien, pression *f* d'étrainte
~SAP - [biol] suc *m* cellulaire
~STRUCTURE - [phys] structure *f* cellulaire
~TESTER - [el] (a portable voltmeter) voltmètre *m* pour accumulateurs
~TISSUE - [gen] tissu *m* cellulaire
~WALL - [gen] parois *f* cellulaire
CELLAR, to - [agric & brew] encaver, mettre en chai
CELLAR - [build] cave *f*, cellier *m*, chai *m* [mining] avant-puits *m*
~FOUNDATION - [constr] fondation *f* des caves
CELLARAGE - [brew ind] mise *f* en cave, mise *f* en chai
CELLARING - [text] (the seasoning of wool tops) séchage *m*, conditionnement *m*
CELLARIUS VESSEL - [ind chem] (stoneware vessel of a saddle-shape, used for cooling or absorbing gases or acids etc) vase *m* en grès en forme de selle
CELLATED - [biol] cellulé
CELLIFORME - [biol] celluliforme
CELLO - [mus] violoncelle *m*
CELLOPHANE - [plast ind] (transparent film produced from wood pulp) "cellophane" *m*, cellulose *f* régénérée de viscose
CELLOSOLVE - [chem] (a colourless liquid used as solvent in the plastics industry) éther *m* éthylglycolique, "cellosolve" *m* (nom commercial)
CELLUCOTTON - [paper man] ouate *f* de cellulose
CELLULAR - [gen] (composed of cells) cellulaire
~APPEARANCE - [photo] (of the negative) fausse reticulation *f*
~BODY - [biol] protoplasme *m*
~CLOTH - [text] tissu *m* cellulaire
~COFFERDAM - [constr] gabions *mpl* de palplanches, batardeau *m* cellulaire
~CONCRETE - [build] (concrete with introduction of air bubbles to obtain a low unit weight) béton *m* cellulaire
~DOUBLE BOTTOM - [shipbuild] double-fond *m* cellulaire
~GRID- [photo] diaphragme *m* à rayons
~-LIKE - [gen] alvéolaire
~RUBBER - [rubber ind] caoutchouc *m* cellulaire
~SPORE - [bot] (multicellular body set free like a spore in which each cell can germinate separately) spore *f* cellulaire
~SWITCHBOARD - [telecomm] table *f* d'opérateur à cellule
~TISSUE - [zool] tissu *m* cellulaire
CELLULASE - [biochem] callulase *f*
CELLULATED - [zool] cellulé, celluleux
CELLULATION - [zool] cellulation *f*
CELLULIFUGAL -[physiol] cellulifuge
CELLULIPETAL - [physiol] cellulipète
CELLULITH - [paper man] (replaces ebonite; dried and ground wood paste) cellulithe *f*
CELLULOID - [plast ind] (thermoplastics based on cellulose nitrate and camphor) celluloïd *m*
CELLULOID BASED ADHESIVE - [ind chem] (used in shoe manufacturing) adhésif *m* à base de celluloïd
~TECHNIQUE - [ind chem] (a process of treatment comprising mixing, block-pressing, slicing and polishing) technique *f* de fabrication de celluloïd
CELLULOSE - [bot] (the chief structural material of plants and consisting of a long, regular chain of glucose units) cellulose *f*
~ACETATE (CA) - [ind chem] (basic material in the production of a number of thermoplastics, acetate fibre and transparent films, acétate *m* de cellulose
~ACETATE BUTYRATE (CAB CELLULOSE ACETOBUTYRATE) - [chem] (for the production of transparent film and thermoplastic moulding compounds) acéto-butyrate *m* de cellulose
~ACETATE PROPIONATE - [chem] (a cellulose derivative with properties similar to those of cellulose acetate butyrate) acéto-propionate *m* de cellulose
~ACETATE RAYON - [text] rayonne *f* à l'acétate
~BASE - [text] base *f* de cellulose
~ESTERS - [chem] (cellulose derivatives based on esterification products of cellulose with an inorganic or organic acid. They include cellulose acetate nitrate, propionate, acetatebutyrate and triacetate and have wide applications) esters *mpl* cellulosiques
~ETHER - [chem] (the partial or complete etherification of the hydroxyl groups of a cellulose molecule) éther *m* cellulosique
~ETHERS - [chem] (cellulose derivatives based on the etherification products of cellulose and organic derivatives. They include ethyl cellulose, methyl cellulose and sodium carboxymethyl cellulose) éthers *mpl* cellulosiques
~FIBRE - [text] fibre *m* de cellulose, fibre cellulosique
~FINISH - [paint] vernis *m* cellulosique
~LACQUER - [chem] (the product of dissolved nitrocellulose or acetyl-cellulose in a suitable solvent) laque *f* cellulosique
~NITRATE - [chem] (thermoplastic used in the production of several moulded articles. It is highly flammable) nitrate *m* de cellulose
~NITRATE DISK - [el acoust] (recording disk, made of metal, glass or paper, coated with a lacquer compound) disque *m* laqué
~PLASTICS - [plast ind] (plastics derived from cellulose) matières *fpl* plastiques cellulosiques
~PROPIONATE - [chem] (thermoplastic material) propionate *m* de cellulose
~SPONGE - [ind chem] (highly absorbent regenerated cellulose sponge) éponge *f* en cellulose
~TRIACETATE - [chem] (completely esterifield cellulose resin used in the production of tapes, surface coatings and fibres) triacétate *m* de cellulose
~XANTATE - [chem] (intermediate in the production of viscose rayon) xanthate *m* de cellulose
~YARN - [text] fil *m* de cellulose
CELLULOSICS - [ind chem] (resins based on cellulose) résines *fpl* cellulosiques
CELOSIA - [bot] célosie *f*
CELSIUS SCALE - [meas] (a synonym for centigrade scale) échelle *f* centigrade, échelle *f* Celsius
~THERMOMETER - [instr] thermomètre *m* centigrade
CELTIUM - [min] (metallic element, symbol Hf, atomic number 72) celtium *m*, Hafnium *m*
CEMBRA PINE - [bot] cembro *m*, cembre *m*

CEMENT, to - [ind chem] (to attach by means of an adhesive) cimenter, coller
[build] (to use cement in masonry) cimenter
[metall] cémenter
CEMENT - [ind chem build etc] (a substance used for uniting or binding materials) ciment *m*
[metall] cément *m*
[rubber ind] adhésif *m*, dissolution *f*
~BLOCK - [constr] parpaing *m*, bloc *m* en béton *m*
~BOND - [gen] adhérence *f* du ciment
~BRICK - [build] parpaing *m*
~CAKE - [constr] galette *f* de ciment *m* pur
~BURNING - [ind] cuisson *f* du ciment
~CLAY - [geol] (clay rock used for the manufacture of cement) argile *m* à ciment
~CHURN - [ind chem] agitateur *m* pour adhésifs
~CONCRETE - [build] béton *m* de ciment
~COPPER - [metall] (impure copper) cuivre *m* impur
~DUCT - [telecomm] canalisation *f* en ciment
~FACING - [build] enduit *m* de ciment
~FACTORY - [gen] cimenterie *f*
~FILLET FLAUNCHING- [constr] filet *m* de mortier *m*
~FLOOR - [build] sol *m* en ciment
~GOLD - [metall] or *m* cémentatoire
~GROUT - [build] lait *m* de ciment, coulis *m*
~GROUTING - [constr] (cement injection) injection *f* de ciment
~GUN - [build] canon *m* à ciment, guniteuse *f*
~HEAD - [oil ind] tête *f* de cimentage
~HOPPER - [build] malaxeur de ciment
~KILN - [build mach] four *m* à ciment
~LAYER - [gen] cimentier *m*
~MIXER - [mech mach] bétonneuse *f*, bétonnière *f*
~MIXING - [build] malaxage *m* du ciment
~PLANT - s. cement factory
~PLASTERING - [build] enduit *m* de ciment
~RETARDER- [ind chem]agent *m* ralentisseur de prise
~SAND RATIO - [build]rapport *m* sable/ciment
~SLURRY - [metall] cément*m* liquide
[oil ind] pâte *f* de ciment, lait *m* de ciment
~STONE - [build] pierre *f* à ciment, calcaire *m* à ciment
~TILE - [constr] carreau *m* de ciment *m*
~WORKER -[build] cimentier *m*, cimenteur *m*
~WORKS - s. cement factory
CEMENTATION - [metall] (the increasing of the carbon content of steel by heating in contact with carbon compounds, as in case-hardening) cémentation *f*
[mining] (of a well) cimentation *f*
~PROCESS - [constr] méthode *f* de cimentation des terrains
~STEEL - [metall] acier de cémentation
CEMENTATORY - [metall] cémentatoire
CEMENTED - [gen] cimenté, collé
~BELT JOINT - [mech] joint *m* de courroie collé
~CARBIDES - [metall] (powdered carbides of tungsten, tantalum or titanium mixed with powdered cobalt or nickel, compressed and sintered) carbures *mpl* frittés
~COAT - [metall] couche *f* cémentée
~LENSES - [opt] lentilles *mpl* accollées
~SHOE - [shoe man] chaussure *f* à semelle collée
~SOCKET JOINT - [mech] joint *m* à rotule collé
~STEEL - [metall] acier *m* de cémentation
~SURFACE - [opt] surface *f* accollée
CEMENTED TUBE - [plast ind] tube *m* collé
CEMENTING - [build] cimentaire
[metall] cémentation *f*
~FLOAT COLLAR - [oil ind] soupape *f* flottante pour tubes
~MACHINE - [mech] machine *f* à étendre les adhésifs
~MATERIAL - [ind chem] adhésif *m*, colle *f*, ciment *m*
~OF LENSES - [opt] accollage *m* de lentilles
CEMENTITE - [metall] (a carbide of iron, constituent of steel and cast iron, brittle and very hard) cémentite *f*
CEMENTUM - [med] cément *m* (d'une dent)
CENOTIC - [med] drastique, purgatif
CENSOR KEY - [telev] interrupteur *m* de canal
CENSORSHIP - [gen & leg] censure *f*
CENSURE - s. censorship
CENSUS - [gen] recensement *m*
CENT - [nucl] (a unit of reactivity) cent *m*
[acoust] (in music) centième *m*
CENTAL - [meas] cent livres anglaises
CENTERING - [mech] (the adjusting of work in a lathe) centrage *m*
[surv] (setting in line) alignement *m*
[build] cintre *m* (de voûte)
~AND CLAMPING RING - [mech] anneau *m* de centrage et collier *m* de serrage
~AND FACING MACHINE - [mech] machine *f* à centrer et à surfacer
~AND MILLING MACHINE - [mech] machine *f* à centrer et à fraiser
~CONTROL - [telev] réglage *m* du cadrage
~MACHINE - [mech] machine *f* à centrer
~OF VAULTS - [build] cintre *m* de voûte
~RIBS - [rubber ind] (of tyres) nervures *fpl* de centrage
~RIM - [acoust] (in a loudspeaker) bord *m* de centrage
~RING - [mech] anneau *m* de centrage
CENTERLESS GRINDER - [mech] rectifieuse *f* sans centres
~GRINDING MACHINE - s. centerless grinder
~INTERNAL GRINDER - [mech] machine *f* à rectifier les surfaces intérieures sans centres
CENTIGRADE - [meas] centigrade
~SCALE - [meas] (thermometric scale of which 0 is the temperature of melting ice and 100 that of boiling water at standard atmospheric pressure) échelle *f* centigrade
~THERMOMETER - [instr] thermomètre *m* centigrade
CENTIGRAM - [meas] centigramme *m*
CENTILITER - [meas] centilitre *m*
CENTIMETRE - [meas] centimètre *m*
~-GRAMME-SECOND UNIT (C.G.S. Unit) - [el] unité *f* C.G.S. (centimètre-gramme-seconde)
CENTIMETRIC LOCAL OSCILLATOR - [instr] (an ultra-high frequency local oscillator) oscillateur *m* local à haute fréquence
~RADAR - [radar] (radar operated on centimetric waves) radar *m*centimétrique
~WAVES - [radio] (waves in the 300 to 30000 Mc/s band) ondes *fpl* centimétriques
CENTIPOISE - [ind chem] (a measure of viscosity, one hundredth of a poise) centipoise *f*
CENTISTOKE - [phys] (one hundredth of a stoke) centistoke *m*

CENTO - [mus] centon *m*
CENTRAL - [gen] central
[telecomm] central *m* téléphonique
~BARRIER - [nucl] (high central rise in a nuclear potential) barrière *f* centrale
~BATTERY - [el] batterie *f* centrale
~BODY - [genet] centrosome *m*
~COMPUTER - [comput] (computer as a unit, without any additional equipment) calculatrice *f* centrale
~CONTROL ROOM - [telev] centre *m* de commutation, centre *m* distributeur de modulation
~CONTROL SYSTEM - [gen] système *m* de commande centralisée
~FEED - [radio] alimentation *f* médiane
~FIELD - [opt] champ *m* central
~FORCE - [phys] (a force acting along the line joining two centres) force *f* centrale
~GEAR CHANGE LEVER - [auto] levier *m* de changement de vitesse central
~GIRDER - [shipbuild] carlingue *f*
~HEATING - [th eng] chauffage *m* central
~MONITORING POSITION - [radio] position *f* centrale de contrôle
~PILOT FOR AUTOMATIC IGNITION - [gas ind] veilleuse *f* centrale d'allumage
~PLANE OF THE LENS - [opt] plan *m* moyen de l'objectif
~POTENTIAL - [nucl] (nuclear potential which is spherically symmetric) potentiel *m* de forces centrales, potentiel *m* central
~RAY - [radiat] (radiation propagated along the axis of the cone of radiation used) rayon *m* normal
~REGISTER - [comput] (register making possible a quick transfer between other registers) registre *m* central
~RUNNER - [aero] (single skid beneath the fuselage of a glider to act as landing gear) patin *m* central
~SHUTTER - [photo] obturateur *m* central
~SPINDLE - [genet] fuseau *m* central
~STOP - [photo] diaphragme *m* central
~TAPPING - [el] prise *f* médiane
~THROTTLE - [aero] manette *f* centrale des gaz
~TUBE FRAME - [auto] châssis *m* à tube central
~TUBULAR BACKBONE FRAME - [auto] châssis *m* à poutre centrale tubulaire
CENTRALIZED - [gen] centralisé
~LUBRICATION - [auto] graissage *m* centralisé
CENTRALISER - [tool] outil *m* à centrer
CENTRALIZER ARM - [auto] levier *m* de compensation
CENTRE, to - [gen & opt] centrer
[surv] (to set up an instrument vertically above a station point) collimater
CENTRE - [gen] centre *m*
[build] (a timber frame as temporary support for the construction of an arch or dome) cintre *m*
[mech] pointe *f*
[med] (nerve centre) ganglion *m*, plexus *m*, centre *m*
~AISLE COACH - [railw] voiture *f* à couloir central
~AND COUNTERSINK - [mech] machine *f* à centrer et à fraiser
~ANGLE - [geom] angle *m* au centre
~ARBOR - [horol] axe *m* de centre
~BAY - [build] travée *f* centrale
~BIT - [mech] (wood boring tool with a projecting central point and two side wings) mèche *f* à trois pointes, foret *m* centré
CENTRE BODY PILLAR - [auto] montant *m* central
~BOLT - [mech] boulon *m* de centrage
~CAP - [mech] chapeau *m* de pivot
~COMB - [text] peigne *m* du milieu, peigne central
~-COUPLED LOOP - [electron] (coupling loop in a multicavity magnetron) boucle *f* de couplage central
~DISTANCE - [mech] entr'axes *m*
~-DOT, to - [metall] amorcer un trou au pointeau
~DRILL - [mech] foret *m* à centrer
~EXPANSION - [radar] information *f* à repère annulaire de zéro
~FEED - [radio] (an aerial in which the feeder is applied to the centre of the aerial wire) alimentation *f* médiane, attaque *f* médiane
~-FINDER - [gen] centreur *m*
~FRAME - [text] plaque *f* intermédiaire, bâti *m* intermédiaire
~HOLES - [telecomm] (holes in the telegraph tape enabling it to be moved forward or backward) perforations *f* centrales (pour l'avancement)
~KEELSON - [shipbuild] carlingue *f*
~LATHE - [mech] tour *m* à pointes
~LINE - [draw] axe *m*
[draw] (the line on a drawing indicating the centre of the object) ligne *f* médiane
[aero] (a line in an aerofoil such that any point in it is equidistant from the upper and lower surfaces, the measurements being made at right angle to the line itself) axe *m*
~-LINE CAMBER - [aero] (the relation of the greatest height of the centre line of an aerofoil above the chord line to the length of the chord) cambrure *f* de l'axe
~-MARK - [mech] coup *m* de pointeau
~OF A LENS - [opt] centre *m* optique
~OF ACCELERATION - [phys] centre *m* d'accelération
~OF ACTION - [met] (any of the several areas of high and low barometric pressure which changes little in location and exists throughout a season or an entire year) centre *m* barométrique
~OF AREA - [phys] (the centre of mass of a two-dimensional region) centre *m* de superficie
~OF ATTRACTION - [astr] centre *m* d'attraction, centre *m* de gravité
~OF BUOYANCY - [phys] (the point at which the buoyancy of a body may be assumed to act) centre *m* de carène,
[aero] centre *m* de volume, centre *m* de portance
~OF CURVATURE - [opt] centre *m* de courbure
~OF DISPLACEMENT - [phys] centre *m* de poussée
~OF EFFORT (OF THE SAILS) - [naut] point *m* vélique
~OF FLOTATION - [phys] centre *m* de flottaison
~OF FLOW - [phys] milieu *m* du courant, centre *m* du courant
~OF FRICTION - [mech] centre *m* de frottement
~OF GRAVITY - [phys] (point at which all the gravitational forces on a body are assumed to act) centre *m* de gravité
~OF GYRATION - [phys] centre *m* de rotation
~OF IMMERSION - s. centre of buoyancy
~OF INERTIA - [phys] centre *m* de masse
~OF INVERSION - [cryst] (symmetry element possessed by certain crystals, whereby the crystal

can be brought into self-coincidence) centre *m* d'inversion
CENTRE OF LATERAL RESISTANCE - [naut] centre *m* de dérive
~OF LIFT - [aero] centre *m* de portance
~OF LINE - [naut] ligne *f* de quille
~OF MASS - s. centre of gravity
~OF MASS SYSTEM - [phys] système *m* de centre de masse
~OF MOTION - [mech] point *m* d'appuit
~OF OSCILLATION - [mech] centre *m* d'oscillation
~OF OSSIFICATION - [anat] centre *m* d'ossification
~OF PERCUSSION - [mech] centre *m* de percussion
~OF PERSPECTIVE - [math] centre *m* de perspective
~OF PRESSURE - [aero] (a point on a reference line of an aerofoil about which the pitching moment is zero) centre *m* de pression
~OF ROTATION - [mech] centre *m* de rotation
~OF SENSITIVITY - [photo] centre *m* de sensibilité
~OF SYMMETRY - [cryst] centre *m* de symétrie
~OF THRUST - [phys] centre *m* de poussée
~OF VOLUME - [mech] (the centre of gravity of a homogeneous region) centre *m* de volume
~PIER - [constr] (in a gas-holder) pylone *m* central, pylone *m* de cuve
~PILLAR - [mech] montant *m* central, pylone *m* central
~PIN - [mech] cheville *f* ouvrière, goupille *f*
~PINION - [horol] pignon *m* de centre
~PIVOT - [horol] pivot *m*
[railw] (of a bogie) tourillon *m*
~POINT - [mech] pointe *f*
~PUNCH - [mech] (a tool with a hardened conical point) pointeau *m*
~RAIL - [railw] crémaillère
~REAMER - [mech] alésoir *m* central
~RING - [mech] anneau *m* de centrage
~SECTION - [aero] (the middle part of an aircraft's fuselage) section *f* centrale
~SELVEDGE - [text] lisière *f* centrale
~SPEAR - [oil ind] harpon *m*
~SQUARE - [impl] équerre *f* à centrer
~-STABLE RELAY - [el] relais *m* polarisé avec position centrale
~STAFF - [horol] axe *m* de centre
~STOP MOTION - [text] casse-trame *m* à fourchette
~STRIKING - [build] décintrement *m*
~STUB - [radio] (a stub at the middle point of a conducting element) téton *m* central
~STUD - [horol] tenon *m* de centre
~TAP - [el] prise *f* médiane
~-TO-CENTRE - [mech] entre-axe *m*
~TRAVERSE BEAM - [text] ensouple *f* centrale
~WEFT STOP MOTION - [text] casse-trame *m* central
~WHEEL - [horol] roue *f* de centre
~WHEEL COCK - [horol] barrette *f* de roue de centre
~WHEEL SPINDLE - [horol] axe *m* de roue de centre
~ZERO INSTRUMENT - [instr] (an instrument in which the zero point is at the middle of the scale) instrument *m* à zéro au centre de l'échelle
CENTREBOARD - [naut] (a sliding keel) dérive *f*
CENTRELESS - s. centerless
CENTRIFUGAL - [mech] (moving or directed outward) centrifuge
~BLENDER - [mech] (mixing machine operating by centrifugal action) mélangeur *m* centrifuge
CENTRIFUGAL BLOWER - [mech] (a radial flow compressor, in which centrifugal forces provide the compression effect) compresseur *m* centrifuge
~BRAKE - [mech] (automatic brake used on heavy machinery and in which the high speed of the rope drum is checked by outward forced revolving shoes) frein *m* centrifuge
~CAST IRON PIPE - [gas ind] tuyau *m* en fonte centrifugée
~CASTER - [auto] chasse *f* de direction centrifuge
~CASTING - [metall] (a casting made by rotating the mould rapidly while the metal is solidifying, to increase its density) pièce *f* coulée centrifugée, coulée *f* centrifuge
~CASTING DIE - [metall] coquille *f* de centrifugation
~CLARIFYING - [ind chem] (the process of clearing a suspension by centrifugal action) clarification *f* par centrifugation
~CLUTCH - [mech] (a clutch in which the frictional elements are engaged by the action of bob-weights, so as to be free when stationary and to engage when a certain speed is reached) embrayeur *m* centrifuge
~CONCENTRATION - [ind chem] concentration *f* par centrifugation
~DREDGING PUMP - [hydr] pompe *f* centrifuge de dragage
~DRIER - [el mech] essoreuse *f* centrifuge
~EFFECT - [phys] effet *m* centrifuge
~EXHAUSTER - [gas ind] extracteur *m* centrifuge
~FAN - [mech] surpresseur *m* centrifuge
~FILTER - [mech] filtre *m* centrifuge
~FLOUR SIFTER - [mech] tamiseur *m* centrifuge
~FORCE - [phys] (a radially outward force) force *f* centrifuge
~FORCE TACHOMETRE - [instr] tachymètre *m* centrifuge
~GOVERNOR - [mech] régulateur *m* centrifuge
~IMPELLER - [mech] compresseur *m* centrifuge
~LATEX CLARIFIER - [rubber ind] (machine for clarifying rubber latex using a centrifugal action) centrifugeuse *f* à clarifier le latex
~PRESSURE CASTING - [metall] coulée *f* sous pression centrifuge
~PUMP - [mech] (pump in which increase of pressure is obtained by centrifugal forces) pompe *f* centrifuge
~SIFTER - [agric] tamiseur *m* centrifuge
~STARTER - [mech] (used with small induction motors) démarreur *m* centrifuge
~STRAINER - [paper man] épurateur *m* centrifuge
~SUGAR - [ind chem] (sugar as produced from the centrifuging plant, before decolorizing) sucre *m* lavé, sucre *m* sortant des turbines
~SUPERCHARGER - [aero] compresseur *m* centrifuge
~SWITCH - [el] commutateur *m* centrifuge
~THICKENING - [bot] (deposition of layers of wall material on the outside of a cell wall) épaississement *m* centrifuge
CENTRIFUGATE - [gen] centrifugé
CENTRIFUGATION - [gen] centrifugation *f*
~TUNNEL - [mech] galerie *f* de centrifugation
CENTRIFUGE, to - [mech & ind chem] centrifuger, séparer par centrifugation
CENTRIFUGE - [mech] (machine in which separation is effected by centrifugal action in a rapidly rotating vessel) centrifugeuse *f*

[sugar ind] turbine *f* centrifuge, essoreuse *f*
CENTRIFUGE BASKET - [sugar ind] panier *m* de turbine
~FEEDING TROUGH - [sugar ind] malaxeur *m* distributeur
~METHOD - [nucl] (the method of separating isotopes by centrifuges) séparation *f* par centrifugation
~PLOUGH - [sugar ind] charrue *f*, déchargeur *m*
~STEAM WASHING - [sugar ind] clairçage *m* à la vapeur dans la turbine
CENTRIFUGED - [gen] centrifugé
~LATEX - [rubber ind] latex *m* centrifugé
CENTRIFUGING - [mech & ind chem] centrifugation *f*
CENTRING - [telev & radar] (the adjustment of the beam of the cathode-ray tube, so that when the deflectors are not excited, it impinges on the central point of the screen) centrage *m*, cadrage *m* [mech] centrage *m*
~CONTROL - [telev & radar] réglage *m* de cadrage, réglage *m* de centrage
~DEVICE - [photo] dispositif *m* de centrage
~DIAPHRAGM - [photo] diaphragme *m* de centrage
~SCREWS - [mech] (in an extruder die; screws placed radially in a die body for locating the mandrel) vis *f*pl de centrage
CENTRIPETAL - [phys] (moving or directed inward, i.e. toward the centre) centripète
~ACCELERATION - [mech] (directed towards the centre of curvature of the path) accélération *f* centripète
~FAULT - [geol] faille *f* normale
~FORCE - [phys] force *f* centripète
~PRESS - [mech] presse *f* centripète
~THICKENING - [bot] (deposition of layers of wall material on the inside of a cell wall) épaississement *m* centripète
CENTROCLINAL DIP - [geol] (structure in which rocks converge to a central point) pendage *m* périclinal
CENTRODE - [phys] (the path of the instantaneous centre of a plane figure in plane motion) lieu *m* du centre, parcours *m* d'un centre
CENTROID - [phys] (the centre of mass of a two-dimensional homogeneous region) s'appliquant au centre de gravité
CENTROMERE DISTANCE - [genet] distance *f* au centromère
~MISDIVISION - [genet] fausse division *f* du centromère
~SHIFT - [genet] déplacement *m* du centromère
CENTROOSTEOSCLEROSIS - [med] ostéosclérose *f* de la cavité centrale de l'os
CENTROPLASM - [bot] protoplasme *m* de la centrosphere
CENTROSOME - [cytol] centrosome *m*
CENTURY - [gen] siècle *m*
~PLANT - [bot] agave *m* d'Amérique
CEPHALGIA - [med] céphalée *f*, céphalalgie *f*
CEPHALIC - [zool] céphalique
~INDEX - [zool] indice *m* céphalique
CEPHALOCENE - [med] céphalocène *f*
CEPHALOGRAM - [med] céphalogramme *m*
CEPHALOID - [biol] céphaloïde
CEPHALOMETRY - [med] céphalométrie
CEPHALOPOD - [zool] céphalopode *m*
CERA ALBA - [paint] (bleached beeswax) cire *f* blanchie, cire *f* d'abeille raffinée
CERA FLAVA - [paint] (yellow or unbleached beeswax) cire *f* jaune, cire *f* d'abeille brute
CERACEOUS - [bot] (with a superficial resemblance to beeswax) céracé
CERAMEL - s. cermet
CERAMETALLIC - [metall] métal *m* céramique
CERAMIC - [gen] (made of a ceramic material) céramique
~CAPACITOR - [el] (component of radio circuits) condensateur *m* céramique
~CAPILLARY - [electron] (component part of measuring devices in electronics) tube *m* capillaire en céramique
~COATED - [metall & gen] à revêtement céramique, recouvert de céramique
~COATING - [metall] revêtement *m* céramique
~DISK - [electron] (disk made of ceramic material used in electronic tubes) disque *m* en céramique
~FERROMAGNETIC MATERIAL - [metall] (compounds with ferro-magnetic characteristics combined with a ceramic structure) matériau *m* ferromagnétique céramique
~FUEL - [nucl] combustible *m* céramique
~INSULATOR - [el] isolateur *m* en céramique
~KILN - [metall] four *m* céramique, four *m* de céramique
~-METAL SEAL - [metall] (ceramic-to-metal seal) scellement *m* métal-céramique, joint *m* céramique-métal
~REACTOR - [nucl] (reactor with fuel and moderator assemblies of ceramic materials) réacteur *m* céramique
CERAMICS - [gen] (the art and science of making articles of baked clay or comparable materials) céramique *f*
CERAMOPLASTICS - [plast ind] (combinations of synthetic mica with high melting point glass or inorganic crystalline binder) plastocéramiques *f*pl
CERARGYRITE - [chem] (natural silver chloride; an ore of silver) cérargyrite *f*
CERASIN - [chem] cérasine *f*
CERATE - [med] cérate *m*
CERAUNOMETRE - [instr] (measuring the intensity of a lightning flash) céraunomètre *m*, céraunographe *m*
CERE - [zool] (the soft skin covering the base of the upper beak in birds) cire *f* (du bec)
CEREAL - [bot] (any edible grain) céréale *f*
CEREBELLAR - [anat] cérébelleux
CEREBELLUM - [anat] cervelet *m*
CEREBRAL - [anat] cérébral
~ANEMIA - [med] anémie *f* cérébrale
~PALSY - [med] paralysie *f* par encéphalopathie
CEREBRASTHENIA - [med] cérebrasthénie *f*, neurasthénie *f* cérébrale
CEREBROMALACIA - [med] cérébromalacie *f*, ramollissement *m* cérébral
CEREBROPATHIA - [med] encéphalopathie *f*, cérébropathie *f*
CEREBROSCLEROSIS - [med] cérébro-sclérose *f*
CEREBROSPINAL - [med] cérébrospinal
CERESINE WAX - [chem] (purified ozokerite used in candle manufacture) cérésine *f*
CERIC - [chem] cérique
~AMMONIUM NITRATE - [chem] (analytical reagent and oxidizing agent) nitrate *m* d'ammonium et de

cérium
CERIC HYDROXIDE - [chem] (starting material for cerium salts) hydroxyde *m* cérique
~OXIDE - [chem] (a sensitizer and stabilizer in glass manufacture) oxyde *m* cérique
CERITE - [min] (important ore of cerium, essentially a hydrate silicate of cerium) cérite *f*
CERIUM - [chem] (one of the rare-earth elements, symbol Ce) cérium *m*
~DIOXIDE - [chem] bioxyde *m* de cérium
~FLUORIDE - [chem] fluorure *m* de cérium
~NAPHTENATE - [chem] (waterproofing agent and drier for paints and inks) naphténate *m* de cérium
CERMET - [el] (a class of materials consisting of ceramic substances in a fine state of division, bonded with a metallic medium) cermet *m*
CERNUOUS - [bot] courbé
CEROMA - s. cere
CEROPLASTY - [gen] (wax-modelling) céroplastique *f*
CEROUS - [chem] céreux
~HYDROXIDE - [chem] (a constituent of glass pigments and ceramic glazes) hydroxyde *m* céreux
CERTIFICATE - [gen] certificat *m*, attestation *f*
~OF AIRWORTHINESS - [aero] (certificate issued by the competent authorities, accepting an aircraft as fit for service in all respects) certificat *m* de navigabilité
~OF CARRIAGE - [railw] certificat *m* de transport
~OF CLASSIFICATION - [naut] acte *m* d'immatriculation
~OF COMPLIANCE - [gen] certificat *m* de conformité
~OF CONVEYANCE - s. certificate of carriage
~OF NATIONALITY - [naut] acte *m* de nationalité
~OF ORIGIN - [comm] certificat *m* d'origine
~OF RECEIPT - [naut] certificat *m* de chargement
~OF REGISTRY - s. certificate of classification
~OF SHIPMENT - [naut] certificat *m* d'ambarquement
~OF SURVEY - [naut] certificat *m* de visite
~OF TONNAGE - [naut] certificat *m* de jauge
CERTIFICATION - [comm etc] certification *f*
~MARK - [comm] estampille *f* de conformité
CERTIFIED - [gen & leg] authentifié, certifié conforme, légalisé
CERTIFY, to - [comm etc] authentifier, légaliser, homologuer
CERUMEN - [zool] (a waxy substance secreted by the ceruminous glands) cérumen *m*
CERUMINOUS GLAND - [zool] glandes *fpl* cérumineuses
CERUSE - [paint] céruse *f*
CERUSITE - [min] (natural lead carbonate) cérusite *f*
CERVICAL - [anat] cervical
~FLEXURE - [anat] flexion *f* cervicale
~TRIANGLE - [anat] triangle *m* cervical
CERVICECTOMY - [med] (removal of the cervix uteri) excision *f* du col de l'utérus
CERVICIS - [med] cervicite *f*
CESAREAN SECTION - [med] opération *f* césarienne
CESIUM - s. caesium
CESSATION - [gen] interruption *f*, cessation *f*
~OF CURRENT - [el] interruption *f* du courant, coupure *f* du courant
CESSPIPE - [plumb] canalisation *f* d'évacuation
CESSPIT - [sewage] fosse *f* d'aisance, puits *m* perdu
~EMPTIER - [transp] véhicule *m* de vidange
CESSPOOL - s. cesspit
~DEPOSIT - [build] fosse *f* d'aisance, puits *m* perdu
CETACEUM - [chem] huile *f* de blanc de baleine
CETANE - [chem] cétane *m*
~NUMBER - [chem] (a performance criterion; the percentage of cetane in a mixture of cetane and alpha-methylnaphtalene which will give the same performance as the fuel to which the number refers) indice *m* de cétane
CETIN - [chem] (a major constituent of spermaceti, used in soap and candle manufacture and as a base for ointments) cétine *f*
CETYL ALCOHOL - [chem] (a derivative of palmitic acid, used in perfumery, cosmetics and medicine) alcool *m* cétilique
~MERCAPTAN - [chem] (a rubber processing chemical) mercaptan *m* cétilique
CEYLON ISINGLAS - [chem] mousse *f* de Ceylan
C.G.S. SYSTEM - [meas] systéme *m* C.G.S. (centimètre-gramme-seconde)
CHABAZITE - [min] chabasite *f*
CHADLESS PERFORATION -[comput](tape perforation in which the holes are only partially through, so that the chads, i.e. cuttings hinge on the tape) perforation *f* incomplète
CHADS - [gen] (of punched cards or tape) déchets *mpl* de perforation
CHAFE, to - [gen] frotter, écorcher, érailler
[to cause heat by friction] échauffer par frottement
CHAFER - [rubber ind] (in tyres) bande *f* de renfort (du talon)
~FABRIC - [rubber ind] bande *f* de renfort (du talon)
CHAFERY - [metall] chaufferie *f*
CHAFF - [agric] balle *f*, paille *f* hachée
[radar] (US term, metallized strip of paper) rubans *mpl* métalliques anti-radar
~CLOUD - [radar] (quantity of interference strips of metallized paper) masse *f* de rubans métallisés
CHAFING - [metall] chauffage *m* par frottement
~SLEEVE - [auto] manchon *m* de guidage
CHAIN, to - [gen] enchaîner
[surv] chaîner
CHAIN - [gen] chaîne *f*
[text] chaîne *f*
[chem] (a series of linked atoms, usually characterizing an organic molecule) chaîne *f*
[draw] ligne *f* de points et de traits
[mech] (a series of interconnected metal links, forming a flexible cable) chaîne *f* de transmission
[telev] chaîne *f* de stations de télévision
[naut] (holding the deadeyes of the shrouds) porte-haubans *m*
[meas] mesure *f* de longueur (d'environ 20 mètres), chaînée *f*
~ADJUSTER - [mech] tendeur *m* de chaîne
~ANCHORING - [gen] ancrage *m* de chaîne, fixation *f* de chaîne
~-AND-BUCKET TRENCHER - [mech] excavatrice *f* à godets
~BARREL - [mech] tambour *m* pour chaîne
~BLOCK - [naut] palan *m* à chaîne
[mech] moufle *f* à chaîne
~BOLT - [mech] boulon *m* de chaîne
~-BOND - [build] chaînement *m*
~BOWL - [text] poulie *f* à chaîne
~BRAKE - [mech] (in text mach) frein *m* à chaîne
~BRANCHING - [chem] ramification *f* en chaîne

CHAIN BRIDGE - [constr] pont *m* suspendu à chaînes
~BUCKET ELEVATOR - [mech] élévateur *m* à godets
~CABLE - [naut] chaîne *f* d'ancre
~CABLE STOPPER - [naut] étrangloir *m*
~CARRIER - [text] entraîneur
~COAL CUTTER - [mining] haveuse *f* à chaîne
~COGWHEEL - [mech] pignon *m* de chaîne
~CONVEYOR - [transp] transporteur *m* à chaîne
~COUPLING - [railw] attelage *m* à chaînes
~COVER - [mech] (in bicycles) carter *m* couvre-chaîne
~DECAY - [nucl] (the process of radioactive transformation in radioactive series) désintégration *f* en chaîne, décomposition *f* en chaîne
~DISINTEGRATION - s. chain decay
~DREDGER - [mech] drague *f* à godets
~DRIVE - [mech] commande *f* par chaîne, transmission *f* par chaîne
~DRIVEN - [mech] commandé par chaîne
~DRUM - s. chain barrel
~ELEVATOR - [transp] élévateur *m* à chaîne
~EXTRACTOR - [mech] extracteur *m* à chaîne
~FEED MOTION - [text] mécanisme *m* d'avance de la chaîne
~FISSION YIELD - [nucl] rendement *m* de fissions en chaîne
~GEAR - [mech] engrenage *m* à chaîne
~GEARING - [mech] transmission *f* a chaîne
~GRATE STOKER - [boilers] grille *f* mécanique, grille *f* à chaîne
~GUARD - [mech] couvre-chaîne
~GUIDE - [mech] guide-chaîne *m*
~HARROW - [agric] herse *f* articulée
~HOIST - [mech] palan *m* à chaîne, treuil *m* à chaîne
~HOME BEAMED - [radar] (a chain of radar stations pooling information on in-flying aircraft) radar *m* de sol à grande portée pour la détection aérienne
~HOME LOW - [radar] radar *m* de sol *m* grande portée pour la détection des avions volant à basse altitude
~HOOK - [horol] croc *m* de chaîne, croc *m* de câble
~INSULATOR - [el] isolateur *m* à chaîne
~INTERMITTENT WELDS - [metall] soudures *fpl* intermittentes
~IRON - [constr] (chain link) anneau *m* (ou maillon *m*) d'une chaîne
~JACK - [mech] tendeur *m* pour chaîne
~LEASE - [text] envergeure *f* basse
~LENGTH - [nucl] (the measure characterizing the chains of atoms which theoretically constitute the molecules of certain polymers) longueur *f* de chaîne
~LIGHTNING - [met] (several flashes of lightning occurring in very rapid succession) éclair *m* sinueux
~LINK - [mech] maillon *m*
~LOCKER - [naut] puits *m* aux chaînes
~LOOM - [text] métier *m* à ourdir
~MAGAZINE - [impl] râtelier *m* à chaîne
~NEEDLE - [text] aiguille *f* pour métier chaîne
~OF CARDS - [text] chaîne *f* sans fin, jeu *m* de cartons
~OF CONTACTS - [telecomm] chaîne *f* de contacts
~OF LAGS - [text] lattis *m*
~OF STATIONS - [telecomm] chaîne *f* de stations
~PILLAR - [mining] pilier *m* de protection de galerie
~PIN - [mech] goupille *f* de chaîne

CHAIN PIPE - [naut] manchon *m* de puits à chaînes, écubier *m* de pont
~PIPE WRENCH - [tool] clé *f* à chaîne, serre-tubes *m* à chaîne
~PITCH - [mech] pas *m* de chaîne
~PLATE - [shipbuild] cadène *f* de haubans
~PROCESS - [nucl] processus *m* en chaîne
~PROFILE - s. chain tread
~PULL - [mech] tension *f* de la chaîne
~PULLEY - [mech] roue *f* à chaîne
~PULLEY BLOCK - [mech] palan *m* à chaîne
~PUMP - [mech] pompe *f* à chapelet, pompe *f* à godets
~RAILWAY - [mech] transporteur *m* à chaîne
~REACTING PILE - [nucl] pile *f* à réaction en chaîne
~REACTION - [chem] (a reaction which is self-propagating) réaction *f* en chaîne
~REAXTOR - s. chain reacting pile
~RIVETING - [mech] rivure *f* parallèle, rivure *f* en vis-à-vis
~RULE - [math] règle *f* conjointe
~SAW - [impl] scie *f* articulée
~SHACKLE - [naut] maillon *m* de chaîne
~SHEAVE - [mech] poulie *f* à chaîne, roue *f* à chaîne
~SLING - [mech] chaîne *f* de suspension
~SPROCKET - [mech] pignon *m* pour chaîne
~STAYS - [mech] (in motorbicycles) bases *fpl*
~STITCH - [text] point *m* de chaînette
~STOPPER - [naut] stoppeur *m*
~STRAIGHTENER - [ind chem] (in rubber ind) inhibiteur *m* de ramification
~STRETCHING DRIVE - [mech] s. chain tightener
~STUD - [mech] tête *f* de la chaîne
~SURVEY - [surv] chaînage *m*
~TACKLE - [mech] palan *m* à chaîne
~TIGHTENER - [mech] tendeur *m* de chaîne
~TONGS - [impl] pince *f*, clé *f* à chaîne, serre-tubes *m* à chaîne
~TRANSMISSION - s. chaîn drive
~TREAD - [rubber ind] (of tyres) sculpture *f* à chaîne
~TRIPOD - [photo] pied-chaîne *m*, chaînette *f*
~-TYPE CONTINUOUS DIFFUSER - [sugar ind] diffusion *f* continue à chaînes
~VICE - s. chain tongs
~WALE - [shipbuild] porte-haubans *m*
~WELL - s. chain locker
~WHEEL - [mech] roue *f* à chaîne, pignon *m* pour chaîne
~WHEEL GEAR - [mech] renvoi *m* à chaîne
~WINDING - [el] enroulement *m* à chaîne
~WRENCH - s. chain tongs
CHAINE- [text] s. chain
CHAINED BLOCKS - [build] blocs *mpl* enchaînés, chaînage *m*
CHAINING - [surv] chaînage *m*, chaînée
~-UP - [text] enchaînement *m*, liaison *f*
CHAINLESS - [text] sans chaîne
CHAIR - [gen] chaise *f*, siège *m*
[railw] (the cast-iron support spiked to the sleeper) coussinet *m*, chaise *f* de rail
[glass man] (team of glassworkers) équipe *f*
~LIFT - [transp] télesiège *m*, élévateur *m* à siège
~PASS - [metall] cannelures *fpl* pour plaques (ou selles)
~RAIL - [carp] barreau *m* (de chaise) antebois *m* (d'une salle), planche *f* du lattage (à la hauteur

des dossiers des chaises)
CHAIR TIP - [carp] embout *m* de pied de chaise
~UPHOLSTERY - [gen] capitonnage *m* pour chaises
~WITH ONE JAW - [railw] coussinet *m*
CHALCEDONY - [min] (a variety of silica, occurring in cavities in lava) calcédoine *f*
CHALCOCITE - [min] (natural copper sulphide; not an important ore of copper, but frequently occurring in association with the major sources of this metal) chalcosine *f*
CHALCOGRAPHY - [print] chalcographie *f*, gravure *f* sur cuivre
CHALCOPYRITE - [min] (natural sulphide of copper and iron, a good ore occurring in bright yellow masses, often iridescent as the result of superficial tarnish) chalcopyrite *f*
CHALCOSINE - s. chalcocite
CHALCOSIS - [med] dépôt *m* de particules de cuivre dans les tissus
CHALCOSTIBITE - [min] (natural copper antimony sulphide) chalcostibine *f*, chalcostibite *f*
CHALCOTRICHITE - [min] (a variety of cuprite) chalcotrichite *f*
CHALDRON - [meas] mesure *f* à charbon
CHALICOSIS - [med] chalicose *f*, silicose *f*, phtisie *f* des tailleurs de pierre
CHALINOPLASTY - [med] opération *f* pour réconstituer un frein à la langue
CHALK - [min] (naturally occurring calcium carbonate; it is used as a filler and extender for plastics, rubber, paints, in medicine etc) craie *f*
~FOG - [photo] voile *m* calcaire
~FORMATION - [geol] terrain *m* crétacé
~LINE - [surv] tringle *f*
~MARK - [metall] (on a laminate; loose powdery mark consisting of pigment and filler particles and caused by breakdown of the binding medium) marque *f* de chaulage
~OVERLAY - [paper man] papier *m* baryté, papier *m* porcelaine
~PUTTY - [ind chem] mastic *m*
~-STONE - [med] concretion *f* calcaire, tophus *m* des arthritiques
~TEST - [metall] essai *m* à la craie
~WATER - [chem] eau *f* calcaire
CHALKINESS - [gen] nature *f* crayeuse
CHALKING - [paint] (disintegration of pigmented coatings by the action of the weather etc) farinage *m* (en surface)
[metall] pulvérisation *f*
~MACHINE - [rubber ind] machine *f* à talquer
CHALKY - [gen] crayeux, crétacé
~SANDSTONE - [geol] grès *m* calcaire
~SPOTS - [photo] taches *fpl* calcaires
CHALYBEATE - [chem] (containing salts of iron in solution, especially of mineral springs) ferrugineux
CHALYBITE - [min] (natural carbonate of iron. Also called siderite and spathic Iron Ore) sidérite *f*
CHAMBER - [gen] (a room) chambre *f*
[phys] espace *m*, chambre *f*
[firearms] (the section of the bore housing the charge) chambre *f*
[ind chem] (large, box-shaped compartment lined with lead for chemical reactions in the manufacture of sulphuric acid) chambre *f* de plomb
[in retorts etc] compartiment *m*, chambre *f*
[mech] corps *m*, cage *f*, chapelle *f*, boîte *f*, carter *m*
[mech] (of an injector) cheminée *f*, mélangeur (d'injecteur)
[mech] (a narrow flat surface formed in place of the angle between two adjacent surfaces) pan *m* coupé
[mining] taille *f* (in blasting), fourneau *m* de mine
[hydr] (of a canal-lock) chambre *f* d'écluse, coffre *m*
~ACID - [chem] (acid as it leaves the lead chambers in sulphuric acid manufacture) acide *m* sulfurique des chambres de plomb
~-AND-PILLAR - [mining] exploitation *f* par chambres et piliers
~BLAST - [mining] sautage *m* par chambres
~-BORED ROLLER - [plast ind] (a calender roller with an enlarged cylindrical chamber on the centreline) cylindre *m* de calandre à chambre centrale
~CRYSTALS - [chem] (crystals of nitrosulphuric acid which may appear on the walls of sulphuric acid chambers) cristaux *m* d'acide sulfonitrique
~FILTER PRESS - [ind chem] filtre-presse *m*
~LEVEL TUBE - [instr] (in surveying) niveau *m* à tube
~OF A RESERVOIR - [hydr] compartiment *m* d'un réservoir
~OF A SOURCE - [hydr] chambre *f* de captage d'une source
~PRESS - [mech] filtre-presse *m* à plateaux
CHAMBERED CORE - [metall] noyau *m* chambré
~VEIN - [mining] filon *m* en forme de chambres
CHAMBERING - [mining] agrandissement *m* du fond d'un trou de mine
CHAMBERS OF THE EYE - [anat] chambres *fpl* oculaires
~OF THE HEART - [anat] cavités *fpl* du coeur
CHAMFER, to - [mech] (to form a small bevel, especially at the end of a rod or bolt) biseauter, chanfreiner, abattre (les angles)
CHAMFER - [carp] (the surface produced by bevelling the edges) chanfrein *m*, biseau *m*, onglet *m*
[mech] biseau *m*
[the operation of bevelling] biseautage *m*, chanfreinage *m*
~CUT - [carp] onglet *m*
~PLANE - [tool] rabot *m* à chanfreiner
~TEETH - [mech] dents *fpl* biseautées
CHAMFERED - [carp & mech] chanfreiné, biseauté
CHAMFERING - [carp] biseautage *m*, chanfreinage *m*
[mech] (the cutting of grooves) cannelure *f*
~MACHINE - [mech] (machine for producing bevelled edges) chanfreineuse *f*
~TOOL - [tool] outil *m* à chanfreiner
CHAMOIS - [zool] chamois *m*
~LEATHER - [leather ind] peau *m* de chamois
CHAMOISE, to - [gen] chamoiser
CHAMOISING - [leather ind] chamoisage *m*, tannage *m* à l'huile
CHAMOISITE - [min] (a silicate of iron) chamoisite *f*
CHAMOMILLE - [bot] camomille *f*
CHAMOTTE - [metall] chamotte *f*, argile *f* réfractaire
~BRICK - [constr] brique *f* de chamotte *f*

CHAMOTTE MOULDING - [metall] moulage *m* en chamotte
CHAMPION TOOTH - [carp impl] dent *m* de scie à taille croisée
CHANCEL - [arch] (in a church) choeur *m*
CHANCRE - [med] chancre *m*
CHANCRIFORM - [med] chancriforme
CHANCROID - [med] chancre *m* mou, chancroïde *m*
CHANDELIER - [el] lustre *m*
CHANDELLE - [aero] (abrupt climbing turn to near stalling, so as to use the momentum of the aircraft to obtain higher rate of climb) chandelle *f*
CHANDLER - [gen] (candle holder) chandelier *m*
[comm] (supplier of goods to a ship) ship chandler *m*, fournisseur *m*
CHANGE, to - [gen] changer, modifier, altérer
[mech etc] changer, remplacer
CHANGE - [gen] changement *m*, altération *f*, variation *f*, modification *f*, change *m*
~BOBBIN - [text] canette *f* de changement, bobine *f* de changement
~BOX - [text] boîte *f* à plusieurs trames, boîte à plusieurs navettes
~-CAN MIXER - [ind chem] (type of paddle mixer in which the container can be easily removed and exchanged) mélangeur *m* à récipient interchangeable
~CARD - [text] carton *m* de changement
~GEAR - [mech] changement *m* de vitesse
~GEARS - [mech] engrenages *mpl* de changement de vitesse
~MOTION - [mech] dispositif *m* de changement
~OF CONNECTION - [el] commutation *f*
~OF DEVIATION - [instr] (in a compass) variation *f* de la déviation
~OF GRADIENT - [surv] variation *f* de pente, variation *f* de dénivellation
~OF ROUTE - [railw etc] déviation *f* de parcours, changement *m* d'itinéraire
~OF STATE - [phys] (the transformation from one to another of the three states of matter: gaseous, liquid or solid) changement *m* d'état
~OF STEEP WATER - [brew ind] changement *m* du bain de macération
~-OVER - [gen] changement *m*, transposition *f*, changement *m* de position
[el] commutation *f*
[cin] (the operation of changing from one projector to another) passage *m*, changement *m* d'appareil
~-OVER CONTACT UNIT - [el] contact *m* triple de commutation
~-OVER CUE - [cin] (a mark on a film, indicating that the end of the reel is near) marque *f* de fin de bobine
~-OVER SWITCH - [el] (a switch designed to change a connection or a set of connections from one circuit to another) commutateur *m*
~-OVER TIME - [el] (the period of time between the break of one connection and the switching of another) temps *m* de commutation
~-OVER WINCH - [mech] (in the heating of retorts) treuil *m* inverseur
~POINT - [surv] point *m* de canevas
~-POLE MOTOR - [el] moteur *m* à plusieurs polarités
~RINGING - [mus] carillon *m* à permutations, sonnerie *f* à permutations
CHANGE SHAFT - [auto] arbre *m* de changement de vitesse
~WHEEL - [mech] roue *f* de rechange
[mach tool] roue *f* de filetage (ou à fileter)
~WHEELS - [mech] (the gear wheels through which the lead screw of a screw cutting lathe is driven from the mandril) engrenages *mpl* de changement de vitesse
CHANGEABLE BORING BAR - [mech] barre *f* d'alésage interchangeable
~EFFECT - [text] effet *m* changeant
CHANGED - [gen] changé
~OVER - [el] commuté
CHANGING - [gen] changeant, variable
~BOX - [photo] châssis-magasin *m*
~CRANK - [mech] manivelle *f* interfixable
~DEVICE - [mech] dispositif *m* de changement
~LOAD - [el] charge *f* variable
~NOTE - [mus] note *f* d'appoggiature
~ZERO POSITION - [instr] position *f* variable du zéro
CHANNEL, to - [gen] tailler en caniveau, évider, échancrer
[carp] canneler, rainurer
[mining] haver
CHANNEL - [gen] canal *m*
[el] (telecomm) circuit *m*
[metall] (standard form of rolled steel section) profilé *m* en U, fer *m* en U
[hydr] (a duct or conduit) canal *m*, conduit *m*
[geogr] canal *m*, détroit *m*
[geol] filon *m* de roche
[geol] (of a river) lit *m*
[naut] porte-haubans *m*
[mech] (a groove) cannelure *f*, rainure *f*, conduit *m*
[radio] (the range of frequencies occupied by a modulated transmission) canal *m*
[el acoust] (in recording) chaîne *f* d'enregistrement
[constr] (of a road) fossé *m*, rigole *f*
[control] (part of a control system appropriated to a specific control surface, e.g. rudder channel) canal *m*, conduit *m*
[comput] (a path along which information may flow or be stored) canal *m*
[comput] (a circular path through a delay line store) trajectoire *f* circulaire
[el acoust] (a complete set of recording equipment) chaîne *f* d'enregistrement
~BANDWIDTH - [telev] largeur *f* de bande du canal
~BED - [geol] couche *f* de gravier
~BLACK - [ind chem] (a variety of carbon black used in rubber compounding and as a pigment) noir *m* au tunnel
~EQUIPMENT - [telecomm] équipment *m* multivoies
~FLAT-TILE - [build] tuile *f*
~FORM - [metall] (a flat surface having two others at right angles to it on the same side, used in rolled metal members) chenal *m* en U
~GROUP - [radio] groupe *m* de canaux
~IRON - [metall] fer *m* en U
~LIGHTS - [signal] (luminous beacons on an alighting channel) feux *mpl* de chenaux
~LODE - [mining] canal *m* filonien
~OF ASCENT - [geol] cheminée *f* volcanique, évent

m
CHANNEL PATCH - [gen] pièce *f* en U
~PIPE - [build] canalisation *f* à section semi-circulaire
~RAIL - [railw] rail *m* à gorge
~RIB - [aero] nervure *f* en U
~SECTION - [metall] (rolled or extruder metal bar having the form of a through, used in structural work) profilé *m* en U
~SELECTION - [radio] (adjustment of reveiver to respond to the desired channel) sélection *f* de canal
~SELECTOR - [radio] sélecteur *m* de canal
~STEEL - [metall] profilé *m* en U
~SUPERGROUP - [radio] | groupe *m* secondaire de canaux
~TRANSLATING EQUIPMENT - [radio] modulateur *m* démodulateur
~WITH SLOPING SIDES - [metall] fer *m* à section trapézoïdale
CHANNELED - [gen] cannelé, rainuré
~INSOLE - [shoe man] semelle *f* striée
~PLATE - [metall] plaque *f* rainurée
~SPECTRUM - [phys] (a spectrum in which interference bands are visible) spectre *m* de réseau
CHANNELLING - [gen] cannelure *f*, rainure *f*
[nucl] (the additional transmission of particles through a medium containing voids) effet *m* de canalisation
~[oil ind] canalisation *f*
~EFFECT - s. channelling
~EFFECT FACTOR - [nucl] facteur *m* d'inhomogénéité
~SWITCH - [el] commutateur *m* de direction
CHANNELLIZING ISLAND - [constr] (in roadwork) terre-plein *m*, refuge *m*
CHANTER - [mus] (a component part of the bagpipe) chalumeau *m*
CHANTRY - [arch] (in a church) chantrerie *f*, chapelle *f* de fondation
CHAPEL - [arch] chapelle *f*
CHAPLET - [arch] moulure *f* en perles, chapelet *m*
[metall] support *m* d'âme (de moule)
CHAPPED - [gen & med] crevassé
~SKIN - [med] peau *f* crevassée
CHAPTALIZATION - [agric] (of wine) chaptalisation *f*
CHAPTALIZE - [agric] chaptaliser (le moût)
CHAPTER - [gen] chapitre *m*
[build] salle *f* capitulaire
CHAPTERS - [horol] (in watch or clock) chiffres *m* romains
CHAPTREL - [arch] petit *m* chapiteau
CHAR, to - [gen] carboniser
[constr] charruer, ciseler la pierre
CHAR - [ind chem] (a general term for product of carbonization of an organic material) noir *m* animal charbon *m* animal
[min] (a synonym for charcoal) charbon *m* de bois
CHARACTER - [gen] caractère *m*, qualité *f*
CHARACTERISTIC - [gen] (any special feature) caractéristique *f*
[radar] (group of signals by radio beacons, consisting of a code sign for identification) signal *m* codé
~ANODE VOLTAGE - [electron] (the theoretically smallest value of the anode voltage of a magnetron, at which an oscillation can start) tension *f* d'amorçage caractéristique
CHARACTERISTIC CONDUCTIVITY - [phys] (of a photoelectric cell) amplification *f* (d'une cellule)
~CURVES - [el] (curves showing the relations between quantities used in the study of machines and electrical apparatus) courbes *fpl* caractéristiques
~DATA - [electron] (set of values of current, voltages etc) données *fpl* caractéristiques
~DISTORTION - [electron] distortion *f* caractéristique
~EQUATION - [electron] (the relation between the photoelectric current, the anode voltage and the luminous flux) équation *f* caractéristique
~IMPEDANCE - [electron] impédance *f* caractéristique
~INDUCTION - [electron] (the theoretically smallest value of the induction in a magnetron, at which oscillation can start) induction *f* caractéristique
~INSTANTS - [electron] (instants limiting intervals of modulations) instants *mpl* significatifs
~RADIATION - [phys] (characteristic X-radiations emitted as a result of absorption of X-rays of higher frequency) raie *f* caractéristique
~SURFACE - [aero] (also called Mach surface; in supersonic flow, a surface which corresponds to the wavefront of an infinitesimal disturbance) surface *f* de Mach
~TELEGRAPH DISTORTION - [radio] distortion *f* caractéristique
~WAVE IMPEDANCE - [electron] (waveguides) impédance *f* caractéristique d'onde
~X-RAYS - [phys] (electromagnetic radiation resulting from a rearrangement of the electrons in the inner shell of atoms) rayons *mpl* X caractéristiques
CHARACTERISTICS - [gen] caractéristiques *fpl*
[math] caractéristique *f* (d'un logarithme)
[electron] caractéristique *f*
~OF A GAS - [phys] constante *f* d'un gaz
CHARACTRON - [electron] caractron *m*
CHARCOAL - [min] (a form of coal produced by the destructive distillation of wood) charbon *m* de bois
[draw] fusain *m*
~BLACKING - [metall] noir *m* végetal
~-BURNER - [gen] charbonnier *m*
~CRAYON - [draw] fusain *m*
~FILTER - [ind chem] filtre *m* à charbon
~FURNACE - [ind chem] carbonisateur *m*
~HEARTH - [metall] feu *m* brasqué
~PIG IRON - [metall] fonte *f* au charbon de bois
CHARGE, to - [gen] charger
[a battery] charger
[metall] charger
[comm] charger, imputer (un compte), mettre une dépense à un compte
CHARGE - [gen] charge *f*, commission *f*
[el] (the quantity of electricity in a body) charge *f*
[comm] frais *mpl*, prix *m*
[fin] privilège *m* d'hypothèque
[leg] accusation *f*, acte *m* d'accusation
[mech] (intern comb motors; of a mixture) charge *f*
[metall&plast ind] (the measured amount of material placed in a mould, machine etc, for each non continuous operation) charge *f*
~ACCOUNT - [accounts] compte *m* débité, compte-courant *m*

CHARGE BALL CHECK PLUG - [auto] clapet *m* de retenue à bille
~BRIDGE - [metall] pont *m* de chargement
~CARRIER - [electron] (a particle carrying an electric charge) porteur *m* électrisé
~CAVITY - [metall] (in foundry) cavité *f* d'alimentation
~COKE - [metall] coke *m* de fusion
~CONJUGATION - [el] (the operation of changing the signs of all electric charges and fields in a system) conjugaison *f* des charges
~DISSIPATION - [el] dissipation *f* de charge
~-EXCHANGE PHENOMENON - [electron] (the phenomenon in which a positive ion having sufficient kinetic energy is neutralized by colliding with a molecule and capturing an electron from it) phénomène *m* d'échange
~-GAUGE - [metall] (in furnaces) indicateur *m* de charge
~IMAGE - [telev] (the total of the charged particles of the luminescent substance on an insulating surface in a cathode-ray tube) image *f* de potentiel
~INDEPENDENCE - [phys] indépendance *f* par rapport à la charge
~INVARIANCE - [phys] (nucleon-nucleon interactions are supposed to be invariant under rotations in isotopic spin space) invariance *f* de la charge
~LEVEL - [metall] niveau *m* de chargement
~-LIMITING DEVICE - [gen] limitateur *m* de charge
~-MASS RATIO - [el] (the quotient of the electric charge by the mass) charge *f* spécifique d'un porteur électrisé)
~METERING DEVICE - [metall] (in foundry) dispositif *m* de dosage, dispositif *m* de réglage de l'alimentation
~MULTIPLICATION - [electron] (due to the collision of electrons with gas molecules under the action of an electric field) multiplication *f* de charge
~NEUTRALIZATION - [electron] neutralisation *f* de la charge
~OF A CAPACITOR - [el] (the quantity of electricity carried by one of the plates) charge *m* d'un condensateur
~OF A SECONDARY CELL - [el] charge *f* d'un accumulateur
~OF ORE - [metall] (the measured quantity of ore introduced in a blast furnace) charge *f* de minerai
~OF THE ELECTRON - [electron] (the elementary quantity of negative electricity associated with the electron) charge *f* de l'électron
~ON THE LATEX PARTICLE - [rubber ind] charge *f* de la particule de latex
~PER CHAMBER - [ind chem metall] charge *f* par chambre
~RESISTANCE - [el] résistance *f* de charge
~RESISTANCE FURNACE - [metall] four *m* à résistance
~STOCK - [oil ind] pétrole *m* à traiter
~-STORAGE TUBE - [electron] (a storage tube, in which information is in the form of electric charges) tube *m* à memoire électrostatique
~SWITCH - [el] commutateur *m* de charge
~THE KILN, to - [brew ind] charger un four, charger un séchoir
~TRANSFER - [electron] (passage of an electron between a bonding and an antibonding orbital) transfert *m* de la charge

CHARGE TRANSFER SPECTRUM - [phys] (a charge transfer) spectre *m* de transfert de charge
~TRESTLE - [metall] (in ovens) estacade *f* de déchargement
CHARGEABLE CALL - [telecomm] conversation *f* taxée
~DISTANCE - [railw] distance *f* taxable
~MINUTES - [telecomm] minutes *fpl* taxées
~ROUTE - [railw] parcours *m* taxé, acheminement *m* taxé
~TIME INDICATOR - [telecomm] (measuring the length of a telephone conversation) chronotaximètre *m*, indicateur *m* de durée
~TIME LAMP - [telecomm] lampe *f* indicatrice de durée
~WEIGHT - [railw] poids *m* taxé
CHARGED - [gen] chargé
[ind chem] (of a liquid) gazéifié
~MAIN - [gas ind] conduite *f* en charge
~MINUTES - [telecomm] minutes *fpl* taxées
~PARTICLE - [electron] (the general name given to electrically charged particles) particule *f* chargée
~PARTICLE CARRIER - s. charge particle
CHARGER CONE VALVE - [mech] cône *m* de chargement
~-READER - [nucl] (a mechanism for charging and reading pocket chambers) lecteur *m* de charge
CHARGING - [el] (battery etc) charge *f*; (the operation of loading) chargement *m*
~AND DELIVERY HOSE - [mech] tube *m* d'alimentation et de distribution
~APPARATUS - [metall] appareil *m* de chargement
~BAR - [metall] flèche *f* mobile horizontale
~BELL - [metall] (in ovens) cône *m* de chargement
~BOARD - [el] panneau *m* de charge
~BOX - [metall] récipient *m* de chargement, cuiller *f* de chargement
~BY DISTANCE - [telecomm] taxation *f* selon la distance
~BY TIME - [telecomm] taxation *f* selon la durée
~CAR WEIGHBRIDGE - [meas] bascule *f* de chargement
~CLIPS - [el] pinces *fpl* de charge
~CONE - [metall] cône *m* de glissement
~CONTROL LAMP - [el] lampe *f* témoin de charge
~CURRENT - [el] courant *m* de charge
~CURRENT SOURCE - [el] alimentation *f* de courant de charge
~DENSITY - [ind chem] densité *f* de chargement
~DOOR - [metall] (in ovens) porte *f* de chargement
~END - [metall] bande *f* collectrice, brin *m* collecteur
~FACE - [nucl] (the side of the reactor in which fuel is introduced) front *m* de chargement, face *f* de chargement
~GALLERY - [metall] (the platform round the throat of a blast furnace, from which the charge is tipped in) pont *m* chargeur, plateforme *f* de chargement
~HOLE - [of boilers] orifice *m* de chargement, bouche *f* de chargement
~HOPPER - [metall] (in ovens) trémie *f* de chargement
~LADLE - [metall] poche *f* à couler
~MACHINE - [mech] chargeuse *f*, chargeur *m* mécanique, enfourneuse *f* mécanique
~ON THERMAL BASE - [gas ind] facturation *f* à

l'unité calorifique
CHARGING PAN - [metall] benne *f* de chargement
~PANEL - [el] tableau *m* de charge
~PLATFORM - [metall] plateforme *f* de chargement
~PLUG - [auto] prise *f* de charge
~POTENTIAL - [el] (of a battery) potentiel *m* de charge
~PUMP - [mech] pompe *f* de remplissage
~RATE - [el] (the current at which a battery should be charged, usually in amperes) régime *m* de charge
~RECTIFIER - [electron] redresseur *m* de charge
~RESISTOR - [el] (a resistor used in series with a capacitor to regulate the time taken to charge it) résistance *f* de charge
~SCOOP - [mech] cuiller *f*, cuillère *f*
~SET - [mech] (gas turbines) groupe *m* alimentaire d'air
~SKIP - [mech] benne *f* de charge, skip *m*
~TRAY - [ind chem] chargeur *m*
~TURBINE - [mech] (gas turbines) turbine *f* d'alimentation
~VOLTAGE - [el] (of an accumulator) tension *f* de charge
~VOLTAGE FLUCTUATIONS - [el] variations *fpl* de la tension de charge
CHARLE'S LAW - [phys] (also called Gay-Lussac's Law; the volume of a given mass of any gas under constant pressure increases by 1/273 rd of its volume at 0 centigrades for each degree of temperature increase) loi *f* des gaz parfaits
CHARLOCK - [bot] sanve *f*, moutarde *f* des champs
CHARPY IMPACT TEST - [metall] essai *m* de choc de Charpy
CHARRED - [gen] carbonisé, brulé
CHARRING - [gen] carbonisation *f*
~HAMMER - [impl] boucharde *f*
CHART - [top] (a map for the use of navigators) carte *f* marine
[draw] (a graphic; of curves etc) diagramme *m*, graphique *m*
[statist] (indications in tabular form) table *f*, schéma *m*
~COMPARISON UNIT - [radar] ensemble *m* de report d'image radar P.P.I. sur carte géographique
~OF A RECORDING INSTRUMENT - [instr] graphique *m*, diagramme *m* (d'enregistrement)
~PAPER - [instr] papier *m* millimétré (pour enregistreur)
CHARTACEOUS - [bot] cartacé
CHARTER, to - [naut & aero] affréter, noliser
CHARTER - [naut & aero] affrètement *m*
[leg] charte *f*, contrat *m*
~PARTY - [naut & aero] contrat *m* de nolisement, contrat *m* d'affrètement
CHARTERED - [naut & aero] affrété
CHARTHOUSE - [naut] chambre *f* des cartes
CHARTOGRAPH - s. cartography
CHARTROOM - s. charthouse
CHASE, to - [gen] chasser, poursuivre
[mech] fileter, ciseler
[mech] (to cut a groove) rainurer
[mech] (to cut a thread) fileter, tarauder
[constr] (to cut a groove in concrete or masonry to accomodate a pipe, conduit etc) encastrer
[arch] (to cut ornamentally) ciseler
CHASE - [gen] chasse *f*, poursuite *f*
[constr] (a groove in concrete or masonry to accomodate pipes etc) passage *m* pour canalisations
[artill] volée *f* (d'un canon)
[print] châssis *m* de mise en page
[plast ind] châssis *m* (manteau de moule), frette *f*
CHASER - [tools] peigne *m* à fileter
[min] moulin *m* chilien
~MILL - [ind chem] (mill used for mixing white lead with linseed oil in paint manufacture) mélangeur *m* pour couleurs
CHASING - [mining] exploration *f* du filon en direction
[text] ciselure *f*
[arch etc] ciselage *m*
~CALENDER - [text] calandre *f* pour chasing
~MACHINE - [text] machine *f* à fileter au peigne
CHASM - [gen] gouffre *m*, abîme *m*, chasme *m*
CHASMUS - [med] bâillement *m*
CHASSIS - [auto] châssis *m*
[aero] châssis *m*, cadre *m*
[the carriage of a gun] châssis *m* d'affût
[electron] (a metal frame on which the components of a circuit or circuits are mounted) châssis *m*
[metall] (die casting machine frame) bâti *m*
[radio] châssis *m*
~-BODY CONSTRUCTION - [auto] construction *f* à châssis et carrosseries séparés
~CAB - [auto] châssis-cabine *m*
~FITTINGS - [auto] accessoires *mpl* de châssis
~FRAME - [auto] châssis *m*
~LUBRIFICATION - [auto] graissage *m* de châssis
~SIDE MEMBERS - [auto] longerons *mpl* de châssis
CHASTE TREE OIL - [chem] essence *f* de gattilier
CHATTER, to - [mech] (to vibrate continuously) vibrer, cogner, brouter
CHATTER - [mech] (any continuous rattling sound caused by vibration of a component part not rigidly supported or secured) vibration *f*, broutage *m*
~MARK - [mech] (marks left on a workpiece by the cutter of a machine tool, when such a cutter is not supported rigidly enough and is therefore liable to chatter) trait *m* de broutage
CHATTERING - [el] (term used especially for rattling sounds in an electric motor) cliquetis *m*
CHATTERTON'S COMPOUND - [ind chem] (an insulating compound based on gutta percha) chatterton *m*
CHAULMOOGRA OIL - [chem] huile *f* de chaulmoogra
CHECK, to - [gen] arrêter, freiner, modérer, vérifier, contrôler
[naut] (of ropes) choquer
CHECK - [gen] arrêt *m*, butée *f*, échec *m*, contrôle *m*, vérification *f*
[metall] (a crack in a casting) craquelure *f*
[paint] (appearance of cracks in a coating) fendillement *m*
[glass man] craquelure *f*
[text] carreau *m*
~BOLT - [mech] boulon *m* d'arrêt, boulon *m* de butée
~CHAINS - [auto] chaînes *fpl* d'arrêt, chaînes *fpl* de sécurité
~CIRCUIT - [comput] circuit *m* de contrôle
~LAMP - [el] lampe *f* de contrôle
~PATTERN - [text] dessin *m* quadrillé, dessin à carreaux, dessin gaufré
~ROD - [mech] déclic, verrou *m*, cliquet *m* d'arrêt
~SAMPLE - [gen] échantillon *m* étalon

CHECKED CUT - [auto] coupe f (de segment) à recouvrement
~DIGIT - [comput] chiffre m de vérification
~LOCK - [carp] loquet m à ressort de retenue
~LOOM - [text] métier m à navettes multiples
~METER - [instr] compteur m divisionnaire
~NUT - [mech] contre-écrou
~OF CHARGEABLE MINUTES - [telecomm] comptage m des minutes de conversation
~PATTERN - [text] motif m à carreaux, dessin m quadrillé
~PATTERN FABRIC - [text] étoffe f à carreaux, tissu m écossais
~PIN - [mech] goupille f d'arrêt
~PLOT - [agric] parcelle f témoin
~POINT - [gen] point m de référence, point m de repère
~PROBLEM - [comput] (especially to indicate whether there is a fault in the computer) programme m de vérification, programme m d'essai
~RAIL - [railw] contre-rail m
~ROD - [mech] barre f de retenue, tige f de retenue
~ROUTINE - s. check problem
~SPRING - [mech] ressort m de retenue
~-STRAP - [auto] sangle f de portiére
[text] lanière f d'arrêt, bride f de choc
~SUM - [comput] (the sum of the amounts punched on cards) somme f de vérification
~TEST - [gen] essai m de vérification, essai m de contrôle
~VALVE - [mech] (a valve which allows flow in one direction only) vanne f d'arrêt, clapet m de retenue
CHECKED - [gen & text] à carreaux, quadrillé
~CUT - [auto] coupe f (de segment) à recouvrement
~OPERATION - [comput] opération f vérifiée
CHECKER, to - [gen] quadriller, diviser en carreaux
CHECKER - [gen] employé m du contrôle, vérificateur m, contrôleur m, pointeur m, marqueur m
[opt] chercheur m
[metall] chambre f
~BOARD PATTERN - [telev] mire f à damier
~BRICK - [constr] (in water gas installations) brique f d'empliage
~COAL - [mining] anthracite f en morceaux
~-WORK - [metall] (in ovens) chambre f (à air, à gaz)
CHECKERED - [gen & metall] guilloché, gaufré
[text] à carreaux, à damier, quadrillé
~LINING - [text] tissu m écossais pour doublure
~PLATE - [metall] tôle f striée, tôle f goufrée
CHECKING - [gen] contrôle m, vérification f, essai m, mesure f
[metall] (in a die) craquelure f du moule (par la chaleur)
[paint] (defect in paint, taking the form of fissuring in all directions) fendillement m
~-CODE TIME - [comput] (the time which is required to check the effect of a problem on the machine) temps m de vérification de la programmation
~FOR GAS LEAKS - [gas ind] recherche f des fuites
~STATION - [control] point m de référence
CHEEK - [anat] joue f
[mech] mâchoire f d'étau
[build] (one of the sides of an opening) montant m
[firearms] joue f
[in impl & tools] flasque m
[metall] (in foundry, the middle section of a flask) chape f (châssis intermédiaire)
CHEEK BONE - [anat] pommette f
~OF REED - [text] garde f de peigne, jumelles fpl
CHEEKING - [leather ind] amincissage m
CHEEPING - [acoust] piaulement m
CHEESE - [foood] fromage m
[text] (yarn wound on a cheese) fromage m, bobine f soleil
[metall] (cylindrical forging with convex sides) loupe f
~AERIAL - [radio] (a mirror aerial shaped like a semi-circular section of a cylindrical cheese) antenne f à cornet
~AND CONE DYEING - [text] appareil m de teinture pour bobines
~ANTENNA - s. cheese aerial
~BOLT - [mech] boulon m à tête (ronde)
~BOX - [oil ind] (US term; cylindrical retort used in the distillation of kerosene) cornue f à kérosène
~-CLOTH - [text] gaze f
[photo] écran m diffuseur
~CUTTER - [cinema] écran m opaque
~DAIRY - [agric] fromagerie f
~-HEAD SCREW - [mech] vis f à tête ronde
~-HEADED - [mech] à tête ronde
~HOPPER - [zool] ver m du fromage
~MOULD - [food ind] moule m à fromage
~PRESS - [food ind] presse f à fromage
~SKIPPER - s. cheese hopper
CHEESINESS - [paint] (of a coating which has not dried properly and is soft and sticky) séchage m imparfait
CHEESY - [paint] caséeux, mou
CHEETAH - [zool] guépard m
CHEILECTOMY - [med] chéilectomie f
CHEILITIS - [med] chéilite f
CHEILOPLASTY - [med] chéiloplastie f, chiloplastie f
CHELATE - [chem] (a compound containing a metallic ion attached by coordinate links to two or more atoms in the molecule) noyau m chélaté, composé m chélaté
~COMPOUND - [chem] composé m de chélation
CHELATING AGENT - [chem] agent m de chélation
CHELATION - [chem] chélation f
CHEMICAL - [chem] chimique
~ACTION - [chem] action f chimique
~ACTIVITY - [chem] chimisme m
~ADDITIVES - [chem] adjuvants mpl chimiques
~AFFINITY - [chem] affinité f chimique
~AGE - [nucl] (the age calculated from chemical analysis in a uranium containing mineral) âge m chimique
~AGENT - [chem] agent m chimique
~ANALYSIS - [chem] analyse f chimique
~ASSAY - [chem] essai m chimique
~ATTACK - [chem] attaque f chimique
~BALANCE - [impl] (a delicate and very sensitive weighing apparatus) balance f de laboratoire
~BENCH - [chem] banc m d'expériences, paillasse f
~BINDING EFFECT - [phys] (the dependance of the neutron cross-section of a material on the chemical binding of the atoms of the material itself) effet m de liaison chimique

CHEMICAL BOND - [chem] liaison *f* chimique
~CHANGE - [chem] réaction *f* chimique
~COLOURING - [metall] (production of colours on metal surfaces by chemical or electro-chemical means) brunissage *m* chimique
~COMBINATION - [chem] combinaison *f* chimique
~COMPOUND - [chem] composé *m* chimique
~CONDITIONING PROCESS - [ind chem] procédé *m* de traitement chimique
~DECOMPOSITION - [chem] décomposition *f* chimique
~DEVELOPMENT - [photo] développement *m* chimique
~DIFFUSION - [chem] (movement of a neutral element of a phase from one part of the phase to another) diffusion *f* chimique
~DIP METHOD - [metall] procédé *m* de revêtement par immersion
~DIP TIN PROCESS - [metall] procédé *m* d'étamage par immersion
~DOSIMETER - [instr] (a self indicating mechanism for measuring the radiation exposure dose) dosimètre *m* chimique
~EFFECT - [chem] effet *m* chimique
~ENERGY - [chem] énergie *f* chimique
~ENGINEERING - [gen] génie *m* chimique
~EQUATION - [chem] (symbolic representation of a chemical reaction) équation *f* chimique
~EQUIVALENT - [phys] (the atomic mass divided by the valency) équivalent *m* chimique
~EXCHANGE - [nucl] (process by which isotopes of the same element in two different molecules exchange place) échange *m* chimique
~FADE - [cin] fondu *m* chimique
~FOAMING - [plast ind] (production of cellular plastics by the incorporation of an agent which yields gas under heat) fabrication *f* chimique des mousses
~FOCUS - [opt] foyer *m* chimique
~FOG - [photo] voile *m* chimique
~FORMULA - [chem] (symbolic expression of a molecule of a substance) formule *f* chimique
~FUNCTION - [chem] fonction *f* chimique
~IMPLEMENTS - [chem & ind chem] accessoires *mpl* de laboratoire, matériel *m* de laboratoire
~IMPURITY - [phys] (when an atom is found in a crystal which is alien to the crystal itself) impureté *f*
~INERTNESS - [chem] (absence of combining power or other form of chemical activity) inactivité *f*, inertie *f* chimique
~INTERMEDIATE - [chem] intermédiaire *m*
~LEAD - [chem] (almost pure lead used for resistant linings in chemical vessels) plomb *m* chimique
~MACHINING - [metall] (the use of chemical action to shape or finish metal components) usinage *m* chimique
~MILLING - [metall] (also called etch-machining; the removal of material from a workpiece by chemical means) fraisage *m* par attaque à l'acide
~PASSIVATION - [chem] (approximately perfect passivity produced in a metal by chemical action) passivation *f* chimique
~PASSIVITY - [phys] (a marked decrease in the reactivity of certain metals when treated by certain reagents) passivité *f* chimique
~PICKLING - [metall] (in electro-plating; removal of oxides etc from metal by chemical process) décapage *m* chimique
CHEMICAL PLANT - [chem] usine *f* de produits chimiques
~PLATING - [metall] (deposition of metal without the use of external electric supply) revêtement *m* chimique
~PRECIPITATION - [chem] précipitation *f* chimique
~PROCESSING - [chem] (treatment of a substance by chemical processes) traitement *m* chimique
~PROPERTIES - [chem] propriétés *fpl* chimiques
~PULP - s. chemical wood pulp
~PURIFICATION - [ind chem] purification *f* chimique
~RECOVERY - [chem] récupération *f* par voie chimique
~REPROCESSING - [chem] traitement *m* chimique de récupération
~RESISTANCE - [chem] (the quality of withstanding chemical action) résistance *f* aux actions chimiques
~RIPENING - [photo] maturation *f* chimique, refonte *f*
~ROCK WEATHERING - [geol] altération *f* chimique des roches, désintégration *f* chimique des roches
~SEPARATION - [chem] (the separation of chemical substances by means of chemical processes) séparation *f* chimique
~STABILITY - [chem] (not easily decomposed by chemical action) stabilité *f* chimique
~STONEWARE - [ind chem] (stoneware specially producted to resist acid and mildly alkaline solutions) grès *m* résistant aux acides
~STRIPPING - [metall] (removal of a metal from a surface by chemical means) décapage *m* chimique
~TRACER - [chem] indicateur *m* chimique
~TREATMENT - [chem] purification *f* chimique, épuration *f* chimique
~VALENCE - [chem] valence *f* chimique
~VALUE - [chem] s. chemical valence
~WASTE - [chem] déchets *mpl* chimiques
~WOOD-PULP - [paper man] (pulp made from wood by one of the sulphite, sulphate or soda processes) pâte *f* chimique
~WORKS - [chem] usine *f* de produits chimiques
CHEMICALIZE - [chem] traiter selon des procédés chimiques
CHEMICALLY FORMED ROCK - [geol] roche *f* d'origine chimique
~INERT - [chem] chimiquement inert
~PURE - [chem] chimiquement pur
CHEMICALS - [chem] produits *mpl* chimiques [oil ind] correctifs *mpl*
CHEMILUMINESCENCE - [phys] (light obtained without heat in certain chemical reactions) chimiluminescence *f*
CHEMISM - [chem] mécanisme *m* d'une réaction, chimisme *m*
CHEMISORPTION - [chem] (adsorption in which a chemical reaction takes place only at the surface of the adsorbent) adsorption *f* chimique
CHEMISTRY - [chem] chimie *f*
CHEMOCEPTOR - [biochem] récepteur *m* de cellule vivante pouvant fixer des corps chimiques, chimiorécepteur *m*
CHEMOLYSIS - [chem] lyse *f* chimique
CHEMORECEPTOR - [zool] chimiorécepteur *m*
CHEMOSIS - [med] chémosis *f*

CHEMOSMOSIS - [chem] (chemical reactions occurring through semipermeable membranes) réactions *fpl* osmotiques
CHEMOSURGERY - [med] cautérisation *f* chimique
CHEMOSYNTHESIS - [chem] chimiosynthèse *f*
CHEMOTAXIS - [bot] chimioactisme *m*, chimiotropisme *m*
CHEMOTHERAPY - [med] chimiothérapie *f*
CHEMOTROPISM - [bot & zool] chimiotropisme *m*
CHEMPURE TIN - [chem] étain *m* chimiquement pur
CHEMURGY - [chem] agrotechnie *f*
CHENILLE - [text] (silk or worsted tufted cord, used for embroidery) chenille *f*
~CUTTING MACHINE - [mech] machine *f* à couper la chenille
CHENOPODIUM OIL - [mech] (an antithelmintic used in medicine) essence *f* de chénopode
CHEQUE - [fin] chèque *m*
~PAPER - [paper man] (a specially treated paper so that it may reveal any attempt of tampering with the printing on it) papier *m* pour chèques
~PLATE - [metall] (hard steel plate with a raised pattern of diagonal lines, used as floor plating to prevent slipping) tôle *f* gaufrée
CHEQUERBOARD FREQUENCY - [telev] fréquence *f* de damier
CHEQUERED - s. checkered
[text] en damier
CHERALITE - [min] (a rare ore, containing uranium and thorium) chéralite *f*
CHEROFOBIA - [med] peur *m* morbide de la gaité
CHERRY - [bot] cerise *f*
[wood ind] cerisier *f*
[mech] (a small milling cutter) fraise *f* ronde
~COAL - [metall] houille *f* grasse à longue flamme
~LAUREL OIL - [ind chem] essence *f* de laurier-cerise
~RED - [dyes] rouge cerise
CHERVIL - [bot] cerfeuil *m*
CHESNY PROCESS - [ind chem] (a method of obtaining magnesium and its salts from sea water) procédé *m* Chesney
CHESS - [naut] (a plank on a pontoon) madrier *m* de ponton
CHESSBOARD-LIKE - [text] s. chequered
CHESSYLITE - [min] (a synonym for Azurite; natural hydrated basic carbonate of copper; deep azure blue in colour) azurite *f*
CHEST - [gen] caisse *f*, coffre *m*, boîte *f*
[anat] poitrine *f*, thorax *m*
~MICROPHONE - [el acoust] microphone *m* cravate, microphone *m* plastron
~OF DRAWERS - [gen] commode *f*
~TONE - s. chest voice
~VOICE - [mus] (the lowest register of the voice) voix *f* de poitrine
CHESTNUT - [bot] châtaigne *f*
~OAK - [timber] rouvre *m*
~TREE - [bot] châtaignier *m*
CHEVAL-GLASS - [glass man] (a mirror) psyché *f*
CHEVAUX-DE-FRISE - [gen] (iron or timber bars with projecting spikes) chevaux *mpl* de frise
CHEVET - [arch] chevet *m* (d'une église)
CHEVIOT - [text] (a long-wool, medium-lustre bread, adapted to hilly country; it produces the wool used for Scottish tweeds) cheviot *m*
~CLOTH - [text] cheviotte *f*
CHEVIOT WOOL - [text] laine *f* "cheviot"
CHEVRON - [arch] chevron *m*
~BAFFLE - [ind chem] baffle *m* à chevron, baffle *m* à écrans angulaires multiples
~PACKING - [mech] garniture *f* en V
CHIASM - [genet] entrecroisement *m*
CHIASTOLITE - [min] (variety of andalusite) chiastolite *f*, macle *f*
CHICK - [zool] poussin *m*
CHICKEN - [zool] poulet *m*
[telev] signal *m* errant vers le noir
~BREEDING - [agric] élevage *m* de volaille
~-POX - [med] varicelle *f*
CHICKLING VETCH - [bot] gesse *f*
CHICKWEED - [bot] mouron *m* des oixaux
CHICLE - [ind chem] (gum obtained from the sapodilla, the main ingredient in chewing-gum) chiclé *m*
CHICORY - [bot] chicorée *f*, endive *f*
CHIEF - [gen] chef *m*, premier, principal
~CAMERAMAN - [cinema] chef opérateur *m*
~RAY - [opt] (the ray passing through an optical system so as to intersect the axis at the plane of the aperture stop) rayon *m* principal
~STATION - [railw] gare *f* principale, station *f* principale
CHIFFON - [very fine, soft silk material] chiffon *m*
~TWIST - [text] (single yarn of raw silk) fil *m* de soie grège
~VELVET - [text] velours *m* chiffon
CHIGGER - [zool] puce-chique *f*
CHIGOE - [zool] puce *f* pénétrante
CHILBLAIN - [med] engelure *f*
CHILD-BED FEVER - [med] fièvre *f* puerpérale
~-BIRTH - [med] accouchement *m*
CHILDHOOD - [gen] enfance *f*
CHILE BAR - [metall] barre *f* de cuivre impur
~SALTPETER - [chem] nitre *m* du Chili, nitrate *m* de soude
CHILEAN MILL - [min] (a large type of edge-runner mill) broyeur *m* à minerai
CHILL, to - [gen] refroidir, glacer
[metall] (casting in iron moulds) couler en coquille
[metall] (to obtain greater hardness by cooling) tremper (à l'air)
CHILL - [gen] froid *m*
[metall] moulage *m* en coquille, solidification *f* blanche, trempe *f*
[metall] (the component part of a sand mould) refroidisseur *m*; (of a sand mould) coquille *f*
[metall] (an iron mould, used to accelerate cooling) coquille *f*
~-BACK - [ind chem] (in heat-processing varnish or resin composition any additive which reduces the temperature of the mix) réfrigérant *m*
~BLOCK - [metall] (chill test-piece) éprouvette *f* de trempe
~BOX - [ind chem] armoire *f* frigorifique
~-CAST - [metall] moulé *m* en coquille
~CAST INGOT - [metall] gueuset *m* (moulé en coquille)
~COIL - [metall] (foundry) ressort *m* refroidisseur
~CRYSTALS - [metall] cristaux *mpl* de trempe
~DEPTH - [metall] profondeur m de trempe
~-HARDEN, to -[metall] couler en coquille, mouler en coquille

CHILL HAZE - [brew ind] turbidité *f* due au froid
~MARK - [glass man] ride *f*
~MOULD - [metall] coquille *f*
~NAIL - [metall] (foundry) clou *m* de cheval refroidisseur
~-PROOF - [brew ind] résistant au froid
~-PROOFING COMPOUND - [ind chem] enduit *m* protecteur
~ROLL - [metall & plast ind] (a cooler roller round which extruded sheets are passed to reduce the temperature) cylindre *m* de refroidissement
~VALUE - [metall] valeur *f* de refroidissement
CHILLED - [metall] trempé, coulé en coquille
[gen] réfrigéré
~-CASTING - [metall] fonte *f* en coquille
~IRON - [metall] (cast-iron in moulds chiefly of metal, so that the surface of the casting is white and hard, while the inner part is grey) fonte *f* moulée en coquille
~IRON WHEEL - [metall] roue *f* en fonte moulée en coquille
~MEAT - [food ind] viande *f* congelée
~ROLL - [metall] cylindre *m* moulé en coquille
~SPOTS - [metall] zones *fpl* trempées
~STEEL - [metall] acier *m* coulé en coquille, acier *m* à trempe glacée
CHILLER - [metall] (part of a cast-iron mould) refroidisseur *m*
CHILLING - [gen] réfrigération *m*, refroidissement *m*
[metall] (the operation of hardening by sudden cooling) trempe *f* en coquille, coquillage *m*
[glass man] voile *m*
[paint] (loss of lustre in a newly-finished surface, caused by a draught of cold air) voile *m*
[photo] (of the gelatine) coagulation *f* (de la gélatine), prise en gelée *f*
~EFFECT - [metall] effet *m* de trempe
~INJURY - [agric] dommage *m* dû au froid
CHIME - [acoust] carillon *m*
[horol] sonnerie *f*, carillon *f*
CHIMING OF THE BELLS - [mus] carillonner, sonner en carillon
CHIMNEY - [gen] cheminée *f*
[mining] colonne *f* de minerai
~BACK - [build] dos *m* de cheminée
~BOARD - [build] manteau *m* de cheminée
~BREAST - [build] revêtement *m* du conduit de fumée, [constr] sailli *m* des conduites *fpl* de fumée
~CAN - [build] faîte *m*
~CAP - [build] capuchon *m* de cheminée, mitron *m*, recouvrement *m* de cheminée *f*
~COPING - [build] capuchon *m* de cheminée
~CORNER - [arch] coin *m* de cheminée, coin *m* du feu
~DRAUGHT - [gen] tirage *m* de la cheminée
~FLUE - [build] conduit *m* de fumée
~JACK - [build] mitre *f* de cheminée
~JAMBS - [build] (the upright sides of a fireplace opening) montants *mpl* de cheminée
~LOSS - [th eng] perte *f* de chaleur (par la cheminée)
~PIECE - [build] (the dressing around a fireplace) manteau *m* de cheminée
~POT - [build] (the pipe, generally of metal, fitted to the smoke outlet at the top of the flue) pot *m* de cheminée, mitre *f* de cheminée
~ROCK - [geol] pyramide *f* coiffée, demoiselle *f*
CHIMNEY SHAFT - [build] âme *f* de la cheminée
~STACK - [build] (a unit comprising a number of flues) tuyaux *mpl* de cheminée
~-SWEEP - [gen] ramoneur *m*
~TOP - [build] couronnement *m* de cheminée, chapiteau *m* de cheminée
CHIMONOPHILOUS - [bot] (growing principally in winter months) hivernal
CHIMPANZEE - [zool] chimpanzé *m*
CHIN - [anat] menton *m*
~COUGH - [med] coqueluche *f*
~REST - [med] mentonnière *f*
CHINA - [pottery] porcelaine *f*
~BARK - [bot] (a medicinal plant) écorce *f* de cinchonia
~CEMENT - [ind chem] mastic *m*
~CLAY - [chem] (a decomposition product of felspar, essentially hydrated aluminium silicate. Also called Kaolin; very important in ceramics) kaolin *m*
~CRAPE - [text] crêpe *m* de Chine
~CRUCIBLE - [impl] creuset *m* en porcelaine
~GRASS - [text] china grass *f*, ramie *f*
~INK - [draw] encre *m* de Chine
~JUTE - [text] jute *m* de Chine
~PAPER - [paper man] papier *m* de Chine
~ROOT - s. China bark
~SILVER - [metall] (a copper, tin, nickel and silver alloy used by jewellers) alliage *m* de cuivre, d'étain, de nickel et d'argent
~STONE - [geol] pétunsé *m*
~WOOD OIL - [chem] huile *f* de bois de Chine, huile *f* de tung
CHINAWARE - [pott] porcelaine *f*
CHINCHILLA - [zool] chinchilla *m*
CHINE - [shipbuild aero] (the intersection of the bottom of a float or amphibial hull with the side or deck) quille *f* d'angle, bordé *m* de coque
[anat] échine *f*
[geogr] (of a mountain) crête *f*
CHINESE OIL - [ind chem] essence *f* de cannelle de Chine
~RED - [paint] (a basic lead chromate pigment) rouge *m* de chrome
~WAX - [chem] cire *f* de Chine
~WHITE - [paint] (white pigment consisting of zinc oxide) blanc *m* de zinc
~WOOD BLOCK - [mus] (used for jazz music) caisse *f* chinoise
CHINK - [gen] fente *f*, lézarde *f*, crevasse *f*
CHINKING - [acoust] tintement *m*
CHINSE, to - [naut] étouper, calfater
CHINSING - [naut] calfatage *m*
~IRON - s. caulking iron
~TOOL - s. caulking tool
CHIP, to - [carp etc] tailler par éclats, hacher
[mech] (by means of a chisel) buriner, enlever au ciseau
[metall] (the elimination of oxidation) piquer (la rouille)
[gen] (of china, glass etc) ébrécher
[metall] (the elimination of burrs) ébarber
CHIP - [gen] éclat *m*, copeau *m*, écaille *f*, rognure *f*, bavure *f*
[a defect in pottery and glass] brêche *f*, brisure *f*
[of glasses] éclat *m*
[fragments of resins] granule *m*, petit cube *m*, chip

HIP AXE - [impl] doloire *f*
BASKET - [impl] panier *m* à copeaux
BOARD - [paper man] carton *m* de mauvaise qualité
BREAKER - [tool] brise-copeaux *m*
PAN - [impl] panier *m* à copeaux
SCREEN - [paper man] crible *m*, tamis *m*
HIPPER - [paper man] (a machine designed to slice pulp) défibreuse *f*
[metall] machine *f* à décriquer
[impl] marteau *m* à air comprimé
HIPPING - [gen] taille *f* par éclats, burinage *m*
[metall] (the removal of defects from metal surfaces) burinage *m*
[mech] (the removal of burrs) bavurage *m* ébarbage *m*
[glass man] dégrossissage *m*
[paints] écaillage *m*, effritement *m*
HAMMER - [metall] marteau *m* à ébarber, marteau *m* à buriner
KNIFE - [impl] (for pipes) couteau *m* à couper le plomb
HIPS - [brew] (as used in the brewing industry) copeaux *mpl* de clarification
HIROGRAPHY - [gen] chirographie *f*
HIROPLASTY - [med] opération *f* plastique sur la main
HIROPODIST - [med] pédicure *m*
HIROPODY - [med] (the treatment of foot ailments) chirurgie *f* pédicure
HIRPING - [acoust] pépiement *m*, gazouillement *m*
HISEL, to - [gen] buriner, tailler au ciseau
HISEL - [tool] ciseau *m*, burin *m*, tranche *f*
~AUGER - [oil ind] tarière *f* à tranchant
~CHIPPING - [metall etc] ébarbage *m* au ciseau, burinage *m*
~FOR COLD METAL - [tool] ciseau *m* à froid
~FOR HOT METAL - [tool] ciseau *m* à chaud
~OFF, to - [metall] découper au ciseau
~STEEL - [metall] acier *m* à burins, acier *m* à ciseau
HIT, to - [brew] (of hops) germer
~MALT - [brew] malt *m* incomplètement germé
HITIN - [zool] chitine *f*
HITINOUS - [med] chitineux
HITTING COUCH - [bot] (relative to brewing industry) couche *f* de germination
HIVES - [bot] ciboulette *f*
HLAMIDOSPORE - [bot] (thick walled fungal spore) chlamydospore *f*
HLOANTHITE - [min] (arsenic of nickel) chloantite *f*
HLOASMA - [med] chloasma *f*, masque *m* des femmes enceintes
HLORACETALDEHYDE - [chem] (a fungicide and intermediate) chloracétaldéhyde *m*
HLORAL - [chem] (trichloracetaldehyde. A synthesis intermediate for DDT and pharmaceuticals) chloral *m*
~FORMAMIDE - [chem] (a hypnotic used in medicine) formamide *f* de chloral
~HYDRATE - [chem] (a hypnotic used in medicine) hydrate *m* de chloral
HLORAMINATION - [chem] procédé *m* aux chloramines, procédé *m* chlore-ammoniaque, traitement *m* au chlore et à l'ammoniaque
HLORAMINE - [chem] (an intermediate in the production of hydrazine) chloramine *f*
CHLORAMINE-T - [chem] (a disinfectant and antiseptic used in medicine) chloramine *f* T
CHLORAMPHENICOL - [chem] (an antibiotic) chloromycétine *f*
CHLORANIL - [chem] (an intermediate for dyestuffs and a fungicide) chloranile *f*
CHLORARGYRITE - [min] (natural silver chloride, an ore of silver) chlorargyrite *f*
CHLORATE - [chem] (chlorates are the salt of chloric acids. Very strong oxidizing agents) chlorate *m*
CHLORAURIC - [chem] chloraurique
CHLORAZIDE - [chem] (a compound formed by the combination of three nitrogen atoms with one chlorine atom; gaseous at N.T.P. and highly unstable) chlorazide *f*
CHLORBENSIDE - [chem] (an insecticide) chlorbenside *f*
CHLORBUTANOL - [chem] (a sedative used in medicine and plasticizer for cellulose derivatives) chlorobutanol *m*
CHLORDANE - [chem] (an insecticide) chlordane *m*
CHLORETHANE - [chem] éthane *m* monochloré
CHLORHYDRIA - [med] chlorhydrie *f*
CHLORIC ACID - [chem] (a monobasic acid, which combines with bases to form chlorates) acide *m* chlorique
CHLORIDE OF LIME - [chem] (properly called chlorinated lime; a powerful disinfectant, bleacher and deodorizer. Also called Bleaching Powder) chlorure *m* de chaux, chaux *f* chlorée
~OF POTASH - [chem] chlorure *m* de potassium
CHLORIDES - [chem] (compounds formed by the combination of metals directly with chlorine or derived by the action of hydrochloric acid to form salts) chlorures *mpl*
CHLORINATE, to - [gen] traiter par le chlore, chlorurer
CHLORINATED LIME - [chem] (a white powder obtained by absorbing chlorate in slaked lime. Chlorinated lime is used as a bleaching agent, disinfectant and deodorizer) chaux *f* chlorée
~POLYPROPYLENE - [chem] (a polymer used in paper and fabric coating) polypropylène *m* chloré
~RUBBER - [chem] (natural or synthetic rubbers to which 60 p.c. or more of chlorine has been added) caoutchouc *m* chloré
CHLORINATOR - [ind chem] appareil *m* à chlore, chloreur *m*, chloromètre *m*
CHLORINE - [chem] (an element, symbol Cl. Heavy, greenish, poisonous gas. Chlorine is an important industrial chemical and is used in the manufacture of a number of organic and inorganic compounds; in metallurgy, as a bleaching agent and in water purification) chlore *m*
~ABSORPTIVE PROPERTIES - [chem] capacité *f* d'absorption de chlore
~ACNE - [med] éruption *f* acnéiforme causée par le chlore
~AZIDE - [chem] azoture *m* de chlore
~COMBINING CAPACITY - [chem] s. chlorine absorptive properties
~DEMAND - [chem] besoin *m* en chlore, dose *f* nécessaire de chlore, demande *f* en chlore
~DIOXIDE - [chem] (reddish-yellow, explosive gas used as a bleaching agent and deodorant) bioxyde *m* de chlore

CHLORINE DOSAGE - [chem] dose f de chlore, dosage m de chlore
~MONOXIDE - [chem] (dissolves in water to form hydrochlorous acid) monoxyde m de chlore
~PEROXIDE - [chem peroxyde m de chlore
~TRIFLUORIDE - [chem] trifluorure m de chlore
~WATER - [chem] (a solution of chlorine in water used as a disinfectant, and in laboratory) eau f chlorée

CHLORISONDAMINE CHLORIDE - [chem] (a hypotensive agent used in medicine) chlorure m de chlorisondamine

CHLORITE - [chem] chlorite f
~SCHIST - [geol] (a schist largely consisting of mineral chlorite) chloritoschiste m

CHLORITIC MARL - [geol] (thin bed of fossiliferous green marl) marne f chloritique
~QUARTZITE - [geol] quartzite f chloritique

CHLORNAPHTALENES - [chem] naphtalènes mpl chlorès

CHLOROACETIC ACID - [chem] (an insecticide and an intermediate for pharmaceuticals) acide m chloroacétique

CHLOROACETONE - [chem] (intermediate for insecticides, antioxidants, photographic chemicals, and pharmaceuticals) chloroacétone f

CHLOROACETYL CHLORIDE - [chem] (a synthesis intermediate) chlorure m de chloracétyle

CHLOROAZODIN - [chem] (a bactericide used in medicine) chloro-azodine f

CHLOROBENZENE - [chem] (intermediate for a number of organic compounds and a solvent for resins, rubbers and paints) chlorobenzène m

CHLOROBENZOIC ACID - [chem] acide m chlorobenzoïque

CHLOROBROMIDE - [chem] chlorobromure m

CHLOROBUTANOL - [chem] (a plasticizer for cellulose derivatives) chlorobutanol m

CHLORODIFLUOROACETIC ACID - [chem] (a catalyst for condensation reactions) acide m chlorodifluoracétique

CHLORODIFLUOROMETHANE - [chem] (a refrigerant and propellant for aerosols) chloro-difluorométhane m

CHLOROETHYLENE - [chem] éthylène m monochloré

CHLOROFORM - [chem] (an anaesthetic, solvent and intermediate) chloroforme m

CHLOROGUANIDINE HYDROCHLORIDE - [chem] (a drug used in the treatment of malaria) chlorhydrate m de chloroguanidine

CHLOROHYDRIN - [chem] (a synthesis intermediate and a solvent for a number of resins and gums) chlorhydrine f

CHLOROHYDROQUINONE - [chem] (synthesis intermediate, bactericide and photographic chemical) chlorohydroquinone f

CHLOROMALEIC ANHYDRIDE - [chem] (a catalyst for certain synthetic resins) anhydride m chloromaléique

CHLOROMETER - [instr] (apparatus for the determination of chlorine) chloromètre m

CHLOROMETHANE - [chem] chlorométane m, méthane m monochloré

CHLOROMETHYLPHOSPHONIC DICHLORIDE - [chem] (an intermediate for plasticizers, synthetic resins, and flameproofing agents) bichlorure m chloro-méthylphosphonique

CHLOROMYCETIN - [chem] (an antibiotic) chloromycétine f

CHLORONAPHTHALENE OILS - [chem] (generic name for chlorinated naphthalene compounds. The have many uses as solvents, plasticizers and in th rubber industry) chloronaphtalènes mpl
~WAXES - [chem] (rubber and resin plasticizers and solvents) cires fpl de chloronaphtalène

CHLORONITROACETOPHENONE - [chem] (a synthes intermediate, bacteriostat and fungistat) chloronitroacétophénone f

CHLOROPHAELITE - [min] (mineral closely related to chlorite) chlorophaleite f

CHLOROPHENOL - [chem] (a solvent and an interm diate for dyes and pharmaceuticals) chlorophénol
~RED - [chem] (acid base showing colour change from yellow to red) rouge m de chlorophénol

CHLOROPHYLL - [chem] (the green pigment in plants, necessary for photosynthesis. It is used as a colouring agent, in medicine and as a vulcan zation accelerator) chlorophylle f

CHLOROPHYLLASE - [bot] (enzime occurring in association with chlorophyll in plants) chlorophyllase f

CHLOROPICRIN - [chem] (a poison gas and an inter mediate for dyestuffs and fungicides) chloropicrin f

CHLOROPLAST - [bot] (a chlorophyll containing plastid) chloroplaste m

CHLOROPLATINATE - [chem] chloroplatinate m

CHLOROPLATINIC ACID - [chem] (known commercially as platinum chloride) acide m chloroplatinique

CHLOROPRENE - [chem] (the monomer for neopren synthetic rubber) chloroprène m

CHLOROPROCAINE HYDROCHLORIDE - [chem] (a local anaesthetic used in medicine) chloroydrate de chloroprocaîne

CHLOROQUINE - [chem] (an anti-malarial drug) chloroquine f

CHLOROSIS - [bot] (caused by a deficiency of chlorophyl) chlorose f

CHLOROSULPHONIC ACID - [chem] (an intermedia in the manufacture of saccharin) chlorhydrine f sulfurique, acide m chloro-sulfonique

CHLOROTHEN CITRATE - [chem] (an anti-histamine drug) citrate m de chlorothène

CHLOROTHIOPHENOL - [chem] (a rubber chemical and plasticizer) chlorothiophénol m

CHLOROTHYMOL - [chem] (a disinfectant) chlorothymol m

CHLOROTIC - [chem] chlorotique

CHLOROTOLUENE - [chem] (a solvent and intermediate for dyes and other organic compounds) chlor toluène m

CHLOROTRIFLUOROETHYLENE - [chem] (monomer for a number of plastics) chlorotrifluoroéthylène

CHLOROTRIFLUOROMETHANE - [chem] (a refrigerant) chlorotrifluorométhane m

CHLOROUS ACID - [chem] acide m chloreux

CHLORPHENOL - s. chlorophenol

CHLORPROMAZINE - [chem] (a tranquillizer used in medicine) chlorpromazine f

CHLORTETRACYCLINE - [chem] (an antibiotic ; also used for stimulating growth by adding it to animal feed) chlorotétracycline f

CHLORUREMIA - [med] rétention f des chlorures de

urine dans le sang
OCK - [gen] cale f, coin m
nech] chaise f, empoise f, coussinet m de
urillon
uto & aero] (a block placed against the wheel
prevent movement) cale f
aut] taquet m, cabrion m
OCOLATE - [food] chocolat m
OIR - [arch & mus] choeur m
RGAN - [mus] orgue m de choeur, orgue m d'ac-
ompagnement
OKE, to - [gen] étouffer, étrangler, suffoquer
nech] (int comb eng) fermer l'arrivée d'air
hys] (in combustion) étouffer, éteindre
OKE - [mech] (a butterfly valve or an equivalent
vice provided in the carburation system to redu-
the air intake and give a specially rich mixture
r starting) diffuseur m
l] (a coil designed to act as a choke) bobine f
arrêt
en] (imperfection in a container) étranglement

lectron] (a discontinuity in a waveguide surface)
ège m
MPLIFIER - [radio] amplificateur m à impédance
ORE - [firearms] choke-bore m, étranglement m
OIL - [radio] bobine f d'arrêt
ONTROL MODULATION - [radio] modulation f à
urant constant
OUPLING - [radio] (method of coupling valves,
ing a choke coil as the anode load of the first
lve) couplage m par bobine d'arrêt
AMP - [chem] (a general term for any suffoca-
ng mixture of gases occurring in coal mines)
yde m de carbone, grisou m
NPUT FILTER - [el] (a power-supply wave filter
which the first element is a series choke) fil-
e m à impédances
OINT - [electron] (waveguides) jonction f à piège
ODULATION - [radio] (modulation of transmitter
ode current by means of a choke in the high tenon feed common to transmitter and modulator
lves) modulation f à courant constant, modula-
on f Heising
STON - [electron] (waveguides) piston m à piège
UNGER - s. choke piston
BE - [mech] (the venturi in the air-passage
a carburettor) diffuseur m
ED - [bot] orobanche f
OKED FLANGE - [radio] (a flange in whose
rface a ditch is cut to prevent the passage of
ided waves within a limited frequency range)
ide f à piège
OKER BAR - [plast ind] (metal bar running across
extruder die for thickness control and for mini-
izing stagnation) barre f de contrôle
OKES - [med] suffocation f, pneumatose f de
compression
OKING - [gen] étouffement m, étranglement m
nech] (in a gas turbine) étranglement m, obtura-
on f
last ind] (of an extruder) engorgement m
OIL - [el] bobine f d'arrêt
OLANGIOMA - [med] tumeur f des voies biliaires
OLANGITIS - [med] inflammation f d'un conduit
liaire
OLECYSTATONIA - [med] atonie f de la vésicule
biliaire
CHOLECYSTITIS - [med] cholécystite f
CHOLEMIA - [med] cholémie f
CHOLEPATHIA - [med] cholépathie f
CHOLERA - [med] choléra m
~VIBRIO - [bact] bacille m du choléra
CHOLERESIS - [med] cholérèse f
CHOLERETIC - [med] choléretique
CHOLESTERASE - [chem] cholestérase f
CHOLESTEROL - [chem] (the principal animal sterol: used in medicine as an emulsifying agent) cholestérol m
CHOLIC ACID - [chem] (a bile acid used in medicine and in the manufacture of pharmaceuticals) acide m cholique
CHOLINE - [chem] (a component of the lecithins; a dietary factor essential for some mammals, but not for man) choline f
CHOLINERGIC - [med] cholinergique
CHOLOCHROME - [med] pigment m biliaire
CHOLOLITH - [med] calcul m biliaire
CHOLURIA - [med] cholurie f
CHOMOPHYTE - [bot] (of plant growing on rock ledges) chomophyte
CHONDRAL - [anat] (pertaining to cartilage) chondral, cartilagineux
CHONDRIFICATION - [anat] cartilaginification f
CHONDRIN - [anat] (a firm substance forming the ground substance of the cartilage) chondrine f
CHONDRITE - [geol] (type of stony meteorite which contains aggregates of minerals) chondrite f
CHONDROADENOMA - [med] chondroadénome m
CHONDROMATOSIS - [med] chondromatose f
CHONDROSIS - [med] formation f de cartilage, tumeur f cartilagineuse
CHONDROTOMY - [med] chondrotomie f
CHOOSE - [gen] choisir, adopter, opter
CHOP, to - [gen] couper, fendre, hacher, tailler
CHOP - [impl] mors m
[metall] impureté f enfoncée
[paper man] longueur f de coupe
[comm] marque f de fabrique
~-OUT DIE - [plast ind] (a die for cutting out blanks) outil m de découpage
CHOPPED - [gen] (sheared into short lengths) haché, coupé en petits morceaux
~COTTON CLOTH - [text] (cotton fabric cut finely for use as a filler) tissu m de coton
~STRANDS - [text] (rovings cut into short pieces for filling) flocons mpl de coton
CHOPPER - [impl] couperet m, hacheur m, hachoir m
[el] (a device making and breaking contact periodically) vibreur m
~AMPLIFIER - [radio] amplificateur m à interrupteur
~BAR - [radio] (a conducting bar in continuous recording, forming the second electrode in conjunction with a helix) couteau m, barre f d'impression
~BAR RECORDING - [meas] (method of registering measurements) enregistrement m par pression périodique
~-BLOWER - [agric] machine f hacheuse à souffleur
~DISC - [electron] (a toothed disc rotating and causing the light signal on the photocell to be interrupted at a required frequency) disque m interrupteur perforé
CHOPPING BIT - [impl] trépan m

CHOPPING MACHINE - [mech] (machine for dividing material into small pieces by shearing action) hacheur *m*, hacheuse *f*
CHOPPY SEA - [met] mer *f* hachée, lames *fpl* courtes
CHORD - [gen] corde *f*
[mus] accord *m*
[geom] corde *f*
[aero] (straight line drawn parallel to the air stream through leading and trailing edge centres of curvature in an aerofoil) corde *f*, profondeur *f* de l'aile
[constr] (the flange of a large girder) semelle *f* de poutre
~INCIDENCE - [aero] angle *m* d'attaque
~LENGTH - [aero] (the length of the projection of the profile of an aerofoil on the chord) profondeur *f* d'un profile d'aile
~LINE - [aero] ligne *f* moyenne
~MEMBERS OF TRUSS - [constr] membrures *fpl*
~POSITION - [aero] (the position of the chord line as defined by coordinates the origin of which is a fixed point in the plane of symmetry) position *f* de la corde
CHORDAL - [gen & mech] à la corde, de la corde
~ADDENDUM - [mech] (of gears) saillie *f* de la dent à la corde
~PITCH - [mech] (of gears) pas *m* à la corde
~THICKNESS - [mech] (of gear teeth) épaisseur *f* à la corde
CHORDITIS - [med] chordite *f*, inflammation *f* des cordes vocales
CHORDWISE PATTERN - [aero] disposition *f* parallèle à la corde
CHORIOMA - [med] néoplasme *m* du chorion
CHORION - [med] chorion *m*
CHORIOPETALOUS - [bot] (polypetalous) choripétale
CHORISIS - [bot] (fission into two or more lobes) chorise *f*
CHOROGRAPHY - [geogr] chorographie *f*
CHOROID - [anat] choroïde *f*
CHOROIDITIS - [med] choroïdite *f*
CHORUS - s. choir
CHRISTMAS TREE - [oil ind] (a complex system of tubes and accessories at the head of an oil well casing) arbre *m* de Noël
~TREE AERIAL - [radio] (a fishbone aerial) antenne *f* directionnelle en arête de poisson
~TREE ANTENNA - s. Christmas tree aerial
~TREE PATTERN - [acoust] (on the illuminated surface of a record) méthode *f* du faisceau lumineux réflechi
CHROMA - [opt] (alternative name for chromaticity, describing the extent to which a colour departs from grey) chromaticité *f*
~CODER - [telev] convertisseur *m* de système
~KEY - [telev] mixage *m* de signaux d'image à activation électronique par différence de couleur
CHROMALLUMINIUM - [metall] chrome-aluminium *m*
CHROMATE - [chem] (the salts corresponding to chromium trioxide) chromate *m*
~GELATIN - [photo] gélatine *f* bichromatée
~RECOVERY - [ind chem] récupération *f* des solutions de chrome
~TREATMENT - [metall] (the operation of applying a protective skin to magnesium alloy parts) chromage *m*
CHROMATE WASTES - [ind chem] déchets *mpl* de chrome
CHROMATIC - [gen etc] chromatique *m*
~ABERRATION - [opt] (lack of coincidence of the component colour images formed by a lens) aberration *f* chromatique
~ADAPTATION - [bot] (a variation in coloration du to the amount of light to which a plant is exposed) adaptation *f* chromatique
~ADDITION - [ind chem] addition *f* chromatique
~CHORD - [mus] accord *m* chromatique
~CIRCLES - [opt] cercles *mpl* chromatiques
~COMPONENT - [telev] composante *f* chromatique
~DEFINITION - [telev] netteté *f* de la transition colorée
~EFFECT - [photo] effet *m* chromatique
~HARP - [mus] harpe *m* chromatique
~INTERCHANGE - [genet] échange *m* de chromatide
~POLARIZATION - [light] polarisation *f* chromatic
~SCALE - [mus] gamme *f* chromatique
~SENSITIVITY - [photo] sensibilité *f* chromatique
~SPLITTING - [telev] division *f* chromatique
~TEST - [opt] essai *m* des couleurs
CHROMATICITY - s. chroma
~ABERRATION - [opt] erreur *f* de teinte
~COORDINATES - [opt] coordonnées *f* trichromatiques
~DIAGRAM - [opt] (plane diagram formed by plotting one of the three chromaticity coordinates against another) diagramme *m* colorimétrique
~FLICKER - [telev] papillotement *m* chromatique
CHROMATICNESS - [opt] sensation *f* de chroma
CHROMATID BREAK - [genet] (the transverse brea of the chromatic threads after the longitudinal division of the chromosome) division *f* de la chromatide
CHROMATIN - [biochem] (the cell-nucleus materi from which the chromosomes are formed) chromtine *f*
CHROMATING - s. chromate treatment
CHROMATOGRAM - [chem] chromatogramme
CHROMATOGRAPHIC ADSORPTION -[chem](the application of the adsorption of solutes from solution) adsorption *f* chromatographique
~ANALYSIS - [chem] analyse *f* chromatographique
CHROMATOGRAPHY - [opt] chromatographie *f*
[chem] (a separation technique based on selectiv absorption) cromatographie *f*
CHROMATOPLASM - [genet] (the peripheral regic of the protoplast, containing the pigments of the cell) substance *f* constituant les cellules conjonctives qui produisent du pigment
CHROMATOPTOMETER - [instr] (instrument designed to measure the colour sensitivity of the eye chromatomètre *m* optique
CHROMATOSCOPE - [opt] chromatoscope *m*
CHROMATRON - [telev] (post-deflection focus tube chromatron *m*, tube *m* tricolore à canon unique
CHROME, to - [metall] chromer
[photo] (a negative) passer au bichromate *m* de potassium
CHROME - [mech] chrome *m*
~ALUM - [chem] (potassium chromium sulphate used in tanning etc)alun *m* de chrome
~BROWN - [text] (used for the dyeing of wools) brun *m* de chrome

ROME CHECK - [metall] craquelure *f* (du chrome)
COMPLEX - [chem] complexe *m* chromifère
YESTUFF - [ind chem] colorant *m* au chrome
REEN - [dyes] (a pigment made by mixing Prusian blue and yellow lead chromate) vert *m* de chrome
RON ORE - s. chromite
EATHER - [leather ind] (leather tanned with salts of chromium) cuir *m* chromé
IORDANT - [ind chem] mordant *m* au chrome
NICKEL - [metall] nickel-chrome *m*
NICKEL STEEL - [metall] acier *m* au nickel-chrome
RANGES - [chem] (pigments consisting of basic lead chromate) orangés *mpl* de chrome, chromates *pl* basiques de plomb
PLATING PLANT - [gen] installation *f* de chromage
RINTING - [text] impression *f* au chrome
ED - [chem] (a lead chromate pigment) rouge de chrome
CARLETS - [dyes] (complex pigments consisting of co-precipitated chromate, molybdate and sulphate of lead) orangés *mpl* de chrome
PINEL - [min] (a name for picotite, a dark coloured spinel containing iron, magnesium and chromium) picotite *f*, spinelle *f* de chrome
TEEL - [metall] acier *m* chromé, acier *m* au chrome
ANNING - [leather ind] tannage *m* aux sels de chrome
ANADIUM STEEL - [metall] acier *m* au chrome-vanadium
ELLOW - [dyes] (alternative term for yellow chrome. A pigment obtained by precipitating lead chromate with potassium chromate) jaune *m* de chrome
IROMEL - [metall] alliage *m* nickel-chrome
IROMESTHESIA - [med] association *f* des couleurs avec les sons, les mots etc
IROMHIDROSIS - [med] chromhydrose *f*, chromidrose *f*
IROMIC - [chem] chromique
ACETATE - [chem] (a tanning agent and a mordant for textiles) acétate *m* chromique
ACID - [chem] (orange, caustic crystals; used in rubber and textile industries, ceramic glazes etc as an oxidizing agent and catalyst) acide *m* chromique
ACID ANODIZING - [chem] (anodizing using an electrolyte of chromic acid solution) traitement *m* de surface par oxydation anodique à l'acide chromique
ACID MIST - [chem] (haze of chromic acid solution droplets which may occur over a chromic acid anodizing bath) vapeur *f* d'acide chromique
ANHYDRIDE - [chem] anhydride *m* chromique
CHLORIDE - [chem] (a polymerization catalyst for olefins, component of plating baths etc) chlorure *m* chromique
LUORIDE - [chem] (mothproofing agent and mordant for textiles) fluorure *m* chromique
HYDROXIDE - [chem] (a green pigment for surface coating) hydroxyde *m* chromique
RON - [chem] fer *m* chromé, sidérochrome *m*
OXIDE - [chem] (a green pigment for surface coating and ceramics; also a catalyst in organic reactions) sesquioxyde *m* de chrome
CHROMIC PHOSPHATE - [chem] (a green pigment for dyes) phosphate *m* chromique
~SALTS - [chem] (the salts in which chromium is trivalent; usually stable and blue or violet in colour) sels *mpl* de chrome
~SULPHATE - [chem] (green pigment for paints and ceramic glazes) sulfate *m* chromique
CHROMIDIUM - [bot] (algal cell in the thallus of a lichen) chromidium *m*
CHROMINANCE - [opt] (the colorimetric difference between a colour and a standard colour of equal luminance) chrominance *f*
~CANCELLATION - [telev] suppression *f* de chrominance
~CARRIER - [telev] porteuse *f* couleur, sousporteuse *f* couleur
~CARRIER MODIFIER - [telev] modificateur *m* de la porteuse couleur
~CHANNEL - [telev] canal *m* de chrominance
~DEFINITION - [telev] résolution *f* chromatique
~LUMINANCE FACTOR - [telev] facteur *m* chrominance-luminance
~MODULATOR - [telev] modulateur *m* couleur
~PRIMARY SIGNAL - [telev] signal *m* du primaire de chrominance
~VECTOR - [telev] vecteur *m* de chrominance
~VIDEO SIGNAL - [telev] signal *m* d'une image en couleur
CHROMING TIME - [metall] durée *f* de chromatage
CHROMIOLE - [cryst] (a deeply staining granule) chromiole *m*
CHROMITE - [min] (chrome iron ore. An ore of chromium) chromite *f*, fer *m* chromé
CHROMIUM - [chem] (an element, symbol Cr. Atomic Number 24. Hard white metal used in the production of stainless and other alloys) chrome *m*
~ACETATE - [chem] acétate *m* de chrome
~AMMONIUM SULPHATE - [chem](a tanning agent) sulfate *m* de chrome et d'ammonium
~CARBIDE - [chem] carbure *m* de chrome
~CHLORATE - [chem] chlorate *m* de chrome
~DIOXIDE - [chem] bioxyde *m* de chrome
~NAPHTENATE - [chem] (a chromium soap used to reduce the tendency to chalking in titanium dioxide paints) naphténate *m* de chrome
~PAPER - [paper man] papier *m* pour chromolithographie
~-PLATE, to - [metall] (electro-plating) chromer
~-PLATED - [metall] chromé
~PLATING - [metall] (the electro deposition of a thin layer of chrome for protection and decoration) chromage *m*
~POTASSIUM SULPHATE - [chem] (a tanning agent, textile mordant, and photographic chemical) basichrome *m*, alun *m* de chrome
CHROMIZING - [metall] (a treatment of steel obtained through the absorption of chromium) chromisation *f*, cémentation *f* par le chrome
CHROMOBLAST - [zool] (embryonic cell developing into a chromatophore) chromoblaste *m*
CHROMOCENTRE - [biochem] (accumulation of chromatin) chromocentre *m*
CHROMOGEN - [chem] (coloured compound containing chromophores) chromogène *m*
CHROMOGENIC - [opt] (producing or giving rise to

colour) chromogène
CHROMOISOMERISM - [chem] (the possession of identical chemical composition by substances of different colours) chromo-isomérisme *m*
CHROMOLITOGRAPHY - [print] chromolithographie *f*
CHROMOMETER - [instr] (an instrument for the comparison of the colour of a substance with a standard one) chromomètre *m*
CHROMOPHILIC - [biochem] (readily stained, as applied to specimens in microscopy) chromophile
CHROMOPHOBIC - [biochem] (difficult to stain) chromophobe
CHROMOPHORES - [chem] (certain groups which produce the colours in dyestuffs when attached in sufficient quantities to hydrocarbon radicals) chromophores *mpl*, chromotophores *mpl*
CHROMOPHORIC ELECTRONS - [phys] (electrons in the double bonds of the chromophoric groups) électrons *mpl* chromophores
CHROMOPHOTOGRAPHY - [photo] chromophotographie *f*
CHROMOPLASMA - [genet] chromoplasma *m*
CHROMOPLAST - [genet] chromoplaste *m*
CHROMOSCOPE - [telev] (a cathode-ray tube for colour television; its screen consists of four layers of phosphor) chromoscope *m*
CHROMOSOME - [biochem] (a deeply staining body into which the chromatin becomes condensed) chromosome *m*
~ABERRATION - [genet] (rearrangement of chromosome parts as the result of breakage and reunion of broken ends) aberration *f* chromosomique
~ARM - [genet] (one of the two sections of a chromosome to which the spindle fibre is attached) bras *m* d'un chromosome
~BREAK - [genet] rupture *f* chromosomique
~CYCLE - [genet] (the whole of the changes in the chromosomes) cycle *m* chromosomique
~DELETION - [genet] (the loss of a section of a chromosome) déficience *f* chromosomique
~EXCHANGE - [genet] échange *m* chromosomique
~INVERSION - [genet] inversion *f* chromosomique
~MAP - [genet] (diagram showing the position of the genes in the chromosome) carte *f* des chromosomes
~MOTTLING - [genet] coloration *f* segmentaire alternée, mottling *m*
~RING - [genet] chromosome *m* en anneau
~SET - [genet] garniture *f* chromosomique, génome *m* chromosomique
CHROMOSPHERE - [astr] (the layer of the sun's atmosphere) chromosphère *f*
CHROMOTHERAPY - [med] chromothérapie *f*
CHROMOTROPY - s. chromoisomerism
CHROMOTROPISM - [chem] chromotropisme *m*
CHROMOTYPE - [print] chromotypie *f*, chromotypographie *f*
CHROMOUS CHLORIDE - [chem] (a reducing agent) chlorure *m* chromeux
~SALTS - [chem] (salts of chromium in the divalent form; their aqueous solutions are blue and they are strong reducing agents) sels *mpl* chromeux
CHROMULE - [bot] (general term indicating plant pigments) chromule *m*
CHROMYL CHLORIDE - [chem] (a dyestuff intermediate and oxidizing and chlorinating agent) chlorure *m* de chromyle
CHRONAXIA - [med] chronaxie *f*
CHRONAXIMETER - [instr] (a device measuring th minimum time required for the excitation of an excitable structure) chronaximètre *m*
CHRONIC - [gen] chronique
CHRONOGRAPH - [instr] (an electrically controlle instrument writing the times) chronographe *m*
CHRONOMETER - [instr] chronomètre *m*
CHRONOPHOTOGRAPHIC CAMERA - [photo] chamb *f* chronophotographique, pistolet *m* photographiqu
CHRONOSCOPE - [instr] (electronic measuring instrument measuring the velocity of bullets) chronoscope *m*
CHRYSAROBIN - [chem] (a parasiticide in medicir chrysarobine *f*, acide *m* chrysophanique officinal
CHRYSALIS - [zool] (the pupa of some insects) chrysalide *f*
~STAGE - [zool] phase *f* de chrysalide
CHRYSENE - [chem] (a high fraction in coal-tar distillation; it shows red-violet fluorescence) chrysène *m*
CHRYSOBERYL - [chem] (aluminate of beryllium) chrysobéryl *m*
CHRYSOCOLLA - [min] (a natural hydrate silicate of copper usually occurring as an incrustation and sometimes in thin seams) chrysocolle *f*, chrysocole *f*
CHRYSOIDINE - [chem] chrysoïdine *f*, jaune *m* acide RS
CHRYSOPHENINE - [chem] chrysophénine *f*, jaune *m* diamine CP
~DYE - [ind chem] colorant *m* de la chrysophénine
CHRYSOTHERAPY - [med] chrysothérapie *f*, aurothérapie *f*
CHRYSOTILE - [min] (an important variety of asbestos) chrysotile *f*
CHUBASCO - [met] (violent squall occurring on tl West coast of South America in tropical and subtropical latidudes) chubasco *m*
CHUCK, to - [gen] jeter, lancer
[mech] (the operation of fixing a tool or a workp ce to the spindle) monter (en le mandrin), serre en mandrin, fixer sur le tour, mandriner
CHUCK - [mech] mandrin *m*
[mach tool] (of a planing machine) mors *m*
[mech] (a kind of wedge) cale *f*
[tool] (a drilling bit-holding device) porte-foret
~JAWS - [mech] (the part of the chuck which bear on and grips the workpiece) mors *m* de mandrin
CHUCKER - [mech] tour *m* multibroche automatiqu
CHUCKING LATHE - [mach tool] (machine tool in which the workpiece is held and driven by a chuck tour *m* à mandriner
~REAMER - [mach tool] alésoir *m* à queue
CHUGGING - [mech] combustion *f* pulsée instable grondement *m* de la combustion
CHUNK - [gen](thick solid lump) gros morceau *m*, tronçon *m*
CHUNKY - [gen] massif
~SCRAP - [gen] gros débris *m*
CHURN, to - [ind chem etc] (the operation of agit ting violently for some manufacturing process) battre, brasser
[metall] (in founding) pomper
[agric] baratter
CHURN - [impl] agitateur *m*
[agric] baratte *f*

URN DRILL - [impl] (used in mining) sonde *f* ercutante, trépan *m* pour sondage au câble
UTE - [hydr] (an inclined channel for water) hute *f* d'eau, conduite *f*, couloir *m* d'évacuation, oulotte *f*
min] (inclined trough for gravity-conveyance of re etc) plan *m* incliné
aero] parachute *m*
OOR - [mining] porte *f* de cheminée
RYER - [text] séchoir *m* à tunnel
EED - [gen] alimentation *f* par plan incliné
ATE - [hydr etc] vanne *f* réglable
build] trappe *f* de cheminée
INING -[metall] (special material used to pro- ect the inside of a trough against wear caused by ny substance sliding through it) revêtement *m* rotecteur (de plan incliné)
LOPE - [constr] pente *f* de coulage, pente *f* de eversement *m*
YLE - [zool] (lymph containing the result of the igestive processes) chyle *m*
YLIFICATION - [physiol] chylification *f*
YMIFICATION - [physiol] chymification *f*
CADA - [zool] cigale *f*
CATRIX - [med & bot] (the scar left after the ealing of a wound; or left on a plant after shed- ing one of its members) cicatrice *f*
LIA - [zool] (eyelashes in mammals and the arbicels of the feather in birds) cils *mpl*, cils *mpl* ibratiles
LIARY - [anat] ciliaire
MUSCLE - [anat] muscle *m* ciliaire
LIUM - [bot] (hair-like appendage to a spore) il *m*
MENT FONDU - [build] (trade name for a type f quick setting cement) ciment *m* fondu, ciment alumineux
MOLITE - [geol] cimolite *f*
NCHING - [cin] (the operation of tightening a oll film) enroulement *m* serré
NCHONA - s. China bark
ARK - [ind chem] (dried bark of several varie- ies of cinchona; a source of quinine and related ompounds) écorce *f* de cinchona, quinquina *m* jaune
NCHONIDINE - [chem] (a pharmaceutical inter- ediate) cinchonidine *f*
NCHONINE - [chem] cinchonine *f*
NCHOPHEN - [chem] (an antipyretic and analgesic sed in medicine) cinchophène *m*
NCTURE- [arch] (plain ring or fillet round a olumn, to separate the shaft from the capital and he base) ceinture *f*, filet *m*, moulure *f*
NDER - [gen] scorie *f*, mâchefer *m*, cendre *f*, scarbille *f*
metall] (the slag in furnaces) scorie *f*, laitier *m*, rasse *f*
ANK - [impl] crassier *m*
HARGING - [metall] addition *f* de scories
ONCRETE - [build] béton *m* de mâchefer
RAME - [railw] pare-étincelles
metall] capteur *m* de cendres
NOTCH - [metall] (in blast furnace) trou *m* à aitier, tuyère *f* à laitier
AP - s. cinder notch
NECAMERA - [photo] caméra *f*, caméra *f* de prise e vue
NEMA - [build] cinéma *m*

CINEMA LAMP - [cin] lampe *f* de projection
CINEMATOGRAPH - [cin] projecteur *m* cinématographique
~CAMERA - s. cinecamera
CINEMATOGRAPHY - [cinema] cinématographie *f*
CINERIN - [chem] (an insecticide derived from pyrethrum flowers) cinérine *f*
CINNABAR - [min] (natural sulphide of mercury, an important ore of that metal, bright red in colour, occurring as acicular crystals, or massive) cinabre *m*
CINNAMATE - [chem] cinnamate *m*
CINNAMIC ALCOHOL - [chem] (an ingredient of perfumes) alcool *m* cinnamique
~ALDEHYDE - [chem] (perfumery and flavouring agent) aldéhyde *m* cinnamique
~ETHER - [chem] éther *m* éthylcinnamique
CINNAMON OIL - [chem] (flavouring and perfumery agent) essence *f* de cannelle
~STONE - [min] grenat *m* jaune
CINNAMYL CINNAMATE - [chem] cinnamate *m* de cinnamyle, styracine *f*
CINQUEFOIL - [bot] potentille *f* rampante, quintefeuille *f*
CIPHER, to - [gen] chiffrer, calculer
CIPHER - [gen] chiffre *m*
[math] chiffre *m*, zéro *m*
~BOOK - [gen] code *m*
~DISC - [gen] disque *m* de chiffrement
~SIGNAL - [telecomm] signal *m* chiffré, message *m* chiffré
~TUNNEL - [build] (false chimney built for symmetrical effect) fausse cheminée *f*
CIPOLIN - [build] (white marble with green streaks) cipolin *m*
CIRCLE - [gen & geom] cercle *m*
[geogr] méridien *m*, parallèle *m*
[astr] orbite *m*, révolution *f*
~BRICK - [build] brique *f* radiale
~COEFFICIENT - [el] (denoting the leakage factor of an induction motor) coefficient *m* de dispersion
~CUTTING MACHINE - [mech] (a machine for cutting circular disks from sheet material) machine *f* à tailler circulaire
~DIAGRAM - [el] (a vector diagram to indicate particular conditions of electric circuits) diagramme *m* circulaire
~-DOT MODE - [comput] (a mode of storage of binary digits) mode *m* cercle-point
~IN, to - [photo] (to decrease the aperture of the lens gradually) fermer l'objectif graduellement
~OF ALTITUDE - [astr] cercle *m* de hauteur
~OF COMPASS - [instr] rose *f* de compas, cadran *m* de boussole
~OF CONFUSION - [photo] (also called apertural effect; the area of a focused bright point of light which determines the best possible definition) cercle *m* de confusion
~OF CONVERGENCE - [math] cercle *m* de convergence
~OF CURVATURE - [geom] (circle touching a curve at one point and having its centre at the centre of curvature of the curve at the point) cercle *m* de courbure, cercle *m* osculateur d'une courbe
~OF DIFFUSION - [opt] diffuseur *m*
~OF LEAST CONFUSION - [photo] cercle *m* de diffusion minimale, cercle *m* de la moindre aberration

CIRCLE OF POSITION - [astr] (a circle surrounding a point on the earth directly below a celestial body, at all points of which the altitude of the body is the same) cercle *m* de position
~OF RIGHT ASCENSION - [astr] cercle *m* d'ascension droite
~OF STRESS - [soil] (Mohr's diagram) cercle *m* des contraintes de Mohr, cercle *m* de Mohr
~OUT, to - [photo] (to increase gradually the lens aperture) ouvrir l'objectif graduellement
~RAILWAY - [railw] ligne *f* circulaire, chemin de fer *m* de ceinture
~SHEAR - [mech] cisaille *f* circulaire
~SHIPPING - [naut] navigation *f* orthodromique
~STAMPING - [mech] estampage *m* de pièces circulaires
~SYSTEM - [aero] système *m* de cercle
CIRCLIP - [mech] segment *m* d'arrêt, jonc *m* d'arrêt
~PLIERS - [mech] tenaille *f* à circlip, tenaille à Seeger
CIRCUIT - [gen] circuit *m*
[el] (a path designed for an electric current) circuit *m*
~ANALYZER - [el] (to measure two or more electrical quantities in a circuit) analyseur *m* de circuit
~BREAKER - [el] (term usually employed for an automatic switch to open a circuit in emergency such as overload) interrupteur *m*, coupe-circuit *m*
~CLOSER - [el] conjoncteur *m*
~EFFICIENCY - [electron] (the efficiency of the output circuit of an electron tube) rendement *m* du circuit
~ELEMENT - [el] (component part of a circuit) élément *m* d'un circuit
~GAP ADMITTANCE - [electron] (the admittance of the circuit at a gap when the electron stream is absent) admittance *f* de l'espace d'interaction
~LAYOUT RECORD - [el] spécification *f* sommaire d'un circuit
~MODEL - [telecomm] circuit *m* d'essai
~NODE - [el] noeud *m* de circuit
~NOISE - [el] (noise currents occurring in a circuit due to thermal agitation of electrons) bruit *m* de ligne
~NOISE LEVEL - [radio] (the ratio of the circuit noise at any point in a transmission system to a standard amount of noise used as reference) niveau *m* de bruit de ligne
~NOISE METER - [instr] psophomètre *m*
~SELF-INDUCTION - [el] induction *f* propre d'un circuit
CIRCUITRY - [el] circuits *mpl*, montage *m*
CIRCULAR - [gen & geom] circulaire *f*
~AERIAL - [radio] antenne *f* circulaire
~ANTENNA - s. circular aerial
~APERTURE DIFFRACTION - [opt] diffraction *f* d'une ouverture circulaire
~ARC - [geom] arc *m* de cercle
~BLAST FURNACE - [metall] haut-fourneau *m* circulaire
~BLAST-MAIN - [metall] conduite *f* circulaire de vent
~BRAIDER - [text] machine *f* à tresser circulaire
~BRUSH - [impl] brosse *f* circulaire
~CHART - [meas] (the paper disk used for the registration of a recording meter results) diagramme *m* circulaire d'enregistrement
CIRCULAR COMB - [text] peigne *m* circulaire
~COVERAGE - [electron] omnidirectionnel
~CROSS SECTION - [geom] coupe *f* circulaire
~-CUT FILE - [tool] lime-fraiseuse *f* à main
~-DIE STOCK - [metall] filière *f* à lunettes rondes
~DOME - [arch] coupole *f* de plein cintre
~ELECTRIC WAVE - [electron] (transverse electric wave for which the lines of electric force form concentric circles) onde *f* électrique circulaire
~FLANGING PRESS - [mech] machine *f* à border
~FLASH - [photo] (ring flash) éclair *m* circulaire
~GEAR - [mech] engrenage *m* cylindrique
~GIRDER - [constr] poutre *f* circulaire
~GRATING - [electron] filtre *m* circulaire
~GRINDING - [mech] rectification *f* cylindrique
~GROOVE - [mech] rainure *f* ou gorge *f* annulaire
~INTERLOCK MACHINE - [text] métier *m* circulaire interlock
~JOIST - [constr] poutre *f* circulaire, poutre *f* de répartition *f*
~KILN - [brew ind] four *m* de séchage circulaire
~KNIFE - [impl] couteau *m* circulaire, couteau-disque *m*
~KNITTING - [text] tricotage *m* circulaire
~LETTER - [gen & comm] lettre *f* circulaire, circulaire *f*
~LINKING MACHINE - [text] remailleuse *f* circulaire
~LOOM - [text] métier *m* circulaire à mailleuses
~LOOPING MACHINE - [text] remailleuse *f* circulaire
~MAGNETIC WAVE - [electron] (waveguides) onde *f* magnétique circulaire
~MEASURE - [geom] mesure *f* des angles
~MICROMETER - [meas] micromètre *m* à anneau
~MIL - [meas] (a unit of area for the cross-section of a wire) 0,0005067 mm^2
~MILLING - [mech] fraisage *m* circulaire
~NOTE - [comm] lettre *f* de crédit circulaire
~PIER - [build] pilier *m* circulaire
~PITCH - [mech] (the distance between similar points on adjacent teeth of a gear-wheel measured along the pitch circle) pas *m* circonférentiel
~PLANE - [impl] rabot *m* courbe
~POLARIZATION - [opt] polarisation *f* circulaire
~SAW - [impl] (a saw consisting of a steel disk on the periphery of which teeth are cut) scie *f* circulaire
~SAW SHARPENER - [mech] machine *f* à affûter les scies circulaires
~SCANNING - [radar] (also called spiral scanning, a type of cathode-ray tube scan in which the spot moves spirally over the screen) balayage *m* circulaire
~SEAM - [metall] soudure *f* circulaire
~SLIDE-VALVE - [meas] (used in meters) tiroir *m* cylindrique, tiroir *m* à pistons
~SLOT - [mech] rainure *f* circulaire
~SLOT BURNER - [metall] brûleur *m* à orifice circulaire
~SLOTTING - [mech] rainure *f* circulaire
~SWEEP - [radar] (the sweep form on the screen of a cathode-ray tube which is produced by two sets of sinusoidal voltages 90 degrees apart in phase, applied to two complete sets of deflectors)

balayage *m* circulaire
:IRCULAR TIMEBASE - [radar] (timebase producing circular scanning) base *f* de temps circulaire
~TIPPING SKIP - [mech] benne *f* basculante
~TOOTH THICKNESS - [mech] épaisseur *f* circulaire de dent
~TRACE - [radar] (the trace produced by circular scanning) trace *f* circulaire
~TRACK - [railw] (of a railway turntable) cercle *m* de roulement
~TRIMMER - [impl] cisaille *f* circulaire
~WAVEGUIDE - [electron] guide *m* d'ondes cylindrique
:IRCULARITY - [geom etc] circularité *f*
:IRCULARLY POLARIZED WAVE - [phys] (electromagnetic waves) onde *f* à polarisation circulaire
:IRCULATING - [gen] circulation *f*
~BOILER - [th eng] chaudière *f* à circulation
~CAPITAL - [fin] capitaux *mpl* circulants, capital *m* disponible
~FRACTION - [math] fraction *f* périodique
~FUEL SYSTEM - [nucl] (in a nuclear reactor) système *m* à combustible circulant
~GAS - [metall] (used in low temperature carbonisation) gaz *m* de recirculation, gaz *m* de recyclage
~MEMORY - [comput] (a mechanism using a delay line which stores information in a train of pulses or waves) mémoire *f* à circulation
~PUMP - [mech] (a pump used to maintain movement of a fluid in a closed circuit) pompe *f* de circulation
~REACTOR - [nucl] (a nuclear reactor in which the fissionable material circulates through the core) pile *f* à combustible circulaire
~REFLUX - [oil ind] reflux *m* de circulation
~SIGNAL - [radio] (signal travelling once or more round the earth) signal *m* circulant
~WATER - [auto] eau *f* de refroidissmeent
~WAVE - [phys] (electromagnetic waves) onde *f* circulante
IRCULATION - [gen] circulation *f*
[print] (of newspapers) tirage *m*
[aero] (the integral of the component of the fluid velocity along any closed path with respect to the distance round that path) circulation *f*
[mech] (the circulation of oil in a motor) circulation *f*
[el] (the line integral of a vector round a closed curve) circulation *f* (d'un vecteur)
~BOILER - s. circulating boiler
~DEGASSING - [metall] dégazage *m* par circulation (coulée de l'acier sous vide)
~LOOP - [nucl] (a duct for fluid) boucle *f* de circulation, circuit *m* fermé
~OF ELECTROLYTE - [el] (the constant flow of electrolyte through the cell) circulation *f* de l'électrolyte
~OVEN - [th eng] four *m* à courant circulaire
IRCULATOR - [electron] (a passive waveguide junction of generally four arms) circulateur *m*
IRCUMCIRCLE - [geom] cercle *m* circonscrit
IRCUMCISION - [med] circoncision *f*
IRCUMFERENCE - [geom] circonférence *f*
IRCUMFERENTIAL FORCE - [phys] force *f* tangentielle
~GROOVE) [mech] rainure *f* périphérique
~PITCH - [mech] pas *m* circonférentiel, pas *m* circulaire
CIRCUMFERENTIAL RIB - [rubber ind] (of tyres) nervure *f* longitudinale
~RIB PROFILE - [rubber ind] (in tyres) profil *m* à nervures à la circomférence
~SPEED - [phys] vitesse *f* périphérique, vitesse *f* circonférentielle
~WELD - [metall] soudure *f* circulaire
CIRCUMFERENTOR - [instr] graphomètre *m*
CIRCUMNAVIGATION - [naut] circumnavigation *f*
CIRCUMNUTATION - [bot] (the rotation of the tip of a stem, so that it traces a helical curve) circumnutation *f*
CIRCUMPOLAR - [astr] circumpolaire
CIRCUMSCRIBE, to - [geom] circonscrire
CIRCUMVALLATION - [constr] rempart *m*, retranchement *m*
CIRCUMVOLUTION - [gen] circonvolution *f*
CIRQUE - [geol] (a large curve hollow in mountain regions, often excavated by ice) cirque *m*
CIRRHOSIS - [med] cirrhose *f*
CIRRIFORM - [met] (having the general form of a cirrus cloud) cirriforme
CIRROCUMULUS - [met] (small billowed cirrus-type cloud, composed of ice crystals) cirrocumulus *m*
CIRROSE - [bot] (curved) cirreux
CIRROSTRATUS - [met] (more or less transparent cloud-veil, composed of ice crystals) cirrostratus *m*
CIRSECTOMY - [med] excision *f* d'une partie de veine variqueuse
CIS - [chem] (prefix denoting an isomer having certain groups of atoms on the same side of the plane) cis-
CISLUNAR - [astr] cislunaire
CISSING - [paint] (a defect of paint or varnish taking the form of blistering due to imperfect adhesion) retrait *m* avec formation de raies (ou de gouttes)
CISSOID - [math] cissoïde *f*
CISTERN - [gen] (a vessel used to store liquids) citerne *f*, bâche *f*, caisse *f*, cuve *f*
[instr] (of measuring instrument) cuvette *f*
[railw] (railway truck used for the transport of liquids) wagon-citerne *m*
[hydr] (of a water closet) réservoir *m* de chasse
~BAROMETER - [instr] baromètre *m* à cuvette
~CLINKER - [constr] (engineering brick) brique *f* très cuite (pour des citernes)
CISTUS - [bot] (aromatic plant) ciste *m*
CITRAL - [chem] (ingredient for flavours and perfumes) citral *m*
CITRATES - [chem] (the salts of citric acid) citrates *mpl*
CITRIC ACID - [chem] (mordant and sequestering agent and an ingredient of soft drinks, food etc) acide *m* citrique
CITRIFORM - [bot] (shaped like a lemon) en forme de citron
CITRIN - [med] citrine *f*, vitamine *f* P
CITRINE - [bot] (lemon coloured) citrin, jaune verdâtre
[min] (a yellow, cheap variety of quartz) citrine *f*, topaze *f* occidentale, prime *f* de topaze
CITRON - [bot] (afruit) cedrat *m*
CITRONELLA OIL - [ind chem] (insect repellent and

intermediate in the production of some perfumes) essence *f* de citronnelle
CITRONELLAL - [chem] (a perfumery intermediate) citronnellal *m*
~HYDRATE - [chem] laurine *f*
CITRONELLOL - [chem]citronnellol *m*, alcool *m* citronnellique
CITRONELLYL ACETATE - [chem] acétate *m* de citronnellyle
CITRUS GROVE - [agric] plantation *f* d'agrumes
CIVET - [chem] (a fixative in perfumery) civette *f*
CIVETONE - [chem] civettone *f*
CIVIL - [gen] civil
~AVIATION - [aero] aviation *f* civile
~DAY - [astr] (mean solar day) jour *m* solaire moyen
~ENGINEERING - [constr] génie *m* civile
~TIME - [gen] temps *m* civil
~YEAR - [astr] (tropical year) année *f* civile
CLABBER, to - [oil ind] coaguler
CLACK - [mech] (a check valve admitting water from a feed-pump to the boiler of a locomotive) soupape *f* à charnière, clapet *m*, soupape *f* à clapet
[acoust] claquement *m*
~BOX - [mech] boîte *f* à clapet, chapelle *f*
~VALVE - s. clack
CLAD - [gen & mech] vêtu, revêtu, recouvert
CLADDING - [nucl] revêtement métallique
~WALL - [constr] mur *m* de revêtement *m*
CLAIM - [gen] demande *f*, revendication *f*, réclamation *f*
[insur] demande *f* d'indennité
[mining] concession *f*
~LICENCE - [mining] titre *m* de concession
CLAIRCE - [sugar ind] clairce *f*, claircée
CLAISEN FLASK - [ind chem] (a special flask for vacuum distillation provided with a separate neck for a thermometer in addition to the outlet) ballon *m* de Claisen
CLAM - [metall] coquille *f*
~SHELL BUCKET - s. clamshell bucket
CLAMMINESS - [gen & med] moiteur *f* froide, état *m* collant
CLAMP, to - [mech] (to secure in position by means of mechanical pressure) fixer, bloquer, serrer, brider, caler
CLAMP - [gen] crampon *m*, bride *f* de serrage
[mech] pince *f*, serre-joint *m*, presse *f*, clampe *f*, pince *f* d'arrêt
[auto] collier *m*, attache *f*
[min] tas *m* de minerai à griller
[mining] (in boring) agrafe *f* de manoeuvre
[metall] presse *f* de coulée
[constr] étreignoir *m*
[mach tool] bride *f* de fixation, mâchoire *f*, mordache *f*
[carp] (wooden frame used to hold parts together) presse *f*, serre-joint *m*, valet *m* d'établi
[shipbuild] bauquière *f*, plat-bord *m*
[oil ind] (oil wall casing) clamp *m*
[el] serre-fil *m*, serre-câble *m*
[build] (stack of dried raw bricks for burning) pile *f* de briques crues
[med] (in surgery) agrafe *f*, pince *f*
~FLANGE - [mech] bride *f* à griffe
~FOR LOCKING HEADSTOCK - [mach tool] fermeture *f* pour blocage poupée
CLAMP JAWS LEVER - [mach tool] levier *m* commande à main pour mâchoires de blocage
~JOINT - [mech] joint *m* à griffes
~DIELECTRIC CONSTANT - [el] (dielectric constant of a material when placed under mechanical stress) permittivité *f* d'un matériau soumis à une contrainte mécanique
~NUT - [mech] écrou *m* de serrage
~PULSES - [el] impulsions *fpl* de clamp, impulsions *fpl* de verrouillage
~RING - [mech] (in instruments) collier *m* de serrage
~SCISSORS - [text] pince *f* coupe-fil, dispositif *m* de pinçage
~SCREW - [mech] vis *f* tendeur, vis *f* de serrage
~SLIDE BRACKET - [mach tools]console *f* chariot avec mâchoires de blocage
~TERMINAL - [mech] borne *f* de fixation, collier *m* raccord
CLAMPED SHEET - [gen] (sheet of material fixed in a frame) feuille *f* fixée sur un cadre
CLAMPER - [electron] coupleur *m* électronique
CLAMPING - [mech] blocage *m*, bridage *m*, calage *m*
[metall] élevatage *m* (du moule)
[telev] clamp *m*, verrouillage *m* du niveau
~ARM - [mech] bras *m* de serrage
~BAR - [metall] traverse *f* de serrage
~BATTEN - [constr] latte de liaison *f* provisoire pour huisseries
~BOLT - [mech] boulon *m* de fixation
~CIRCUIT - [electron] (an electronic circuit holding the amplitude extremes of an electronic waveform) circuit *m* de blocage
~DEVICE - [mech] verrouillage *m*, dispositif *m* de fixation (ou de serrage)
~DIODE - [electron] (a diode used to clamp a voltage at a point in the circuit) diode *f* de blocage
~DOG - [of a timber-sawing machine] griffe *f*
~EFFECT - [text] effet *m* de pinçage
~FINGERS - [mech] griffes *fpl* de serrage
~FRAME - [impl] cadre *m* de fixation
~HANDLE - [mach tools]griffe *f* de blocage
~LEVER - [mech] (a lever used to exert pressure on a part to retain it in place) manette *f* de serrage, levier *m* de serrage
~LUG - [mech] mâchoire *f*
~MECHANISM - [mech] blocage *m*
~NUT - [mech] écrou *m* de serrage, écrou *m* de blocage
~PLATE - [mech] plaque *f* de serrage, (ou de fixation)
~PLATEN - [plast ind] (a plate used to hold a mould closed) plateau *m* de fixation
~PRESSURE - [mech] pression *f* de pinçage
~RING - [mech] collier *m* de serrage, anneau *m* de serrage
~SCREW - [el] (a screw for holding a conductor) vis *f* de blocage, vis *f* de fixation
~SHOE - [auto] sabot *m* de serrage, patin *m* de fixation
~SPIKE - [impl] crampon *m*
~SURFACE - [mech] surface *f* d'appui
~TIME - [ind chem] (the time during which it is necessary to apply pressure to a glued joint while the glue hardness) temps *m* de fermeture (de la presse)

CLAMPING WASHER - [mech] rondelle *f* de fixation
CLAMPS - [text] pinces *fpl*
CLAMSHELL BUCKET - [mech] benne *f* preneuse (ou à deux mâchoires), benne *f* piocheuse
~EXCAVATOR - [mech] excavateur *m* à benne preneuse (ou à deux mâchoires)
CLAP, to - [gen] applaudir, battre des mains
CLAP BOARD -[carp] bardeau *m*
~-SILL - [hydr] (also called mitre-sill; the raised part of the bed of a canal lock) seuil *m* d'écluse
~-STICK - s. clappers
CLAPP MARK - [cin] repère *m*
~PICTURE - s. clapp mark
~STICK - s. clapper
CLAPPER - [mech] (a slotted tool head) tête *f* porte-outil
~BOX - [mach tool] (carried on the saddle of a planing machine) logement *m* du porte-outil, support *m* oscillant de l'outil
CLAPPERS - [cinema] (component part of a sound picture recorder) claquoir *m*
CLARAIN - [chem] (separate constituent of bright coal) clarain *m*
CLARENDON - [print] normande *f*, caractère *m* gros
CLARET-COLOURED CANKER - [rubber ind] patch canker *m*
CLARIFICATION- [gen] clarification *f*
[chem] (the operation of freeing from impurities and make clear a liquid) clarification *f*, purification *f*, défécation *f*
~FILTER - [ind chem] filtre *m* de clarification
~PLANT - [hydr] installation *f* de clarification, station *f* d'épuration des eaux d'égout
CLARIFIER - [gen] clarificateur *m*
[chem] (clarifying agent) claire *f* (sucre), colle *f* (vins)
[metall] chaudière *f* à clarification
~PLATE - [brew ind] plaque *f* de clarification
CLARIFY, to - [gen] clarifier, éclaircir
[chem] clarifier, purifier, défequer, coller (le vin)
CLARIFYING REDUCER - [photo] faiblisseur *m* clarificateur
CLARINET - [mus] clarinette *f*
CLARION - [mus] clairon *m*
CLARK CELL - [chem] (a standard cell having zinc and mercury electrodes, the latter pasted with mercurous sulphate; the electrolyte is a saturated zinc sulphate solution) pile *f* de Clark
~PROCESS - [chem] (a process for obtaining the partial softening of water) procédé *m* de Clark
CLARKEITE - [min] (a mineral containing potassium, sodium, lead and uranium) clarkéite *f*
CLASH, to -[gen] se choquer, se heurter
CLASH - [gen] choc *m* violent, collision *f*, heurt *m*
~GEAR - [mech] engrenages *mpl* baladeurs
CLASHING OF GEARS - [mech] grincement *m* des dentures
CLASHPANS - [mus] cymbales *f*
CLASMATOCYTE - [zool] clasmatocyte *m*
CLASP, to - [gen] agrafer, serrer, fermer, joindre
[mech] serrer, s'agrafer
CLASP - [gen] agrafe *f*, fermoir *m*
[mech] fermoir *m*
~BRAKE - [railw] (the application of two brake shoes to each wheel) frein *m* à deux sabots
~KNIFE - [impl] couteau *m* de poche, couteau *m* pliant
CLASP LOCK - [mech] serrure *f* à fermeture automatique
~NAIL - [mech] (a square cut nail with two pointed projections sinking into the wood) clou *m* à crampons
~NUT - [mech] (a nut split diametrically into halves) écrou *m* à mâchoires
~ROD - [text] baguette *f* de serrage
CLASS, to - [gen] classer, classifier, ranger
[naut] coter
CLASS - [gen] classe *f*, catégorie *f*, sorte *f*, genre *m*
~A AMPLIFIER - [radio] (amplifier in which anode current flows during the whole signal cycle) amplificateur *m* de classe A
~A MODULATION - [radio] modulation *f* de classe A
~A OPERATION - [radio] (operation of a valve as class A amplifier) fonctionnement *m* en amplificateur de classe A
~AB AMPLIFIER - [radio] (amplifier so biased that anode current flows during more than half the signal cycle) amplificateur *m* de classe AB
~B AMPLIFIER - [radio] (amplifier so biased that anode current flows only during positive-going half cycles of the signal) amplificateur *m* de classe B
~B MODULATION - [radio] modulation *f* de classe B
~B OPERATION - [radio] (operation of a valve as class B amplifier) fonctionnement *m* en amplificateur de classe B
~C AMPLIFIER - [radio] (amplifier biased so far below cut-off that anode current only flows during much less than half the signal cycle) amplificateur *m* de classe C
~C OPERATION - [radio] (operation of a valve as a class C amplifier) fonctionnement *m* en amplificateur de classeC
~OF ACCURACY - [instr] (a classification of instruments, whose precision is characterized by the same number indicating the upper limit of error) classe *f* de précision
CLASSICAL ANHARMONIC MOTION - [phys] (the motion of an anharmonic oscillator as treated by classical mechanism) mouvement *m* anharmonique classique
~ELECTRON RADIUS - [electron] rayon *m* classique de l'électron
~FLUTTER - [aero] (flutter due only to coupling between two or more degrees freedon) battement *m* classique
~SCATTERING - [nucl] diffusion *f* classique
~SYSTEM - [electron] système *m* non quantifiè, système *m* classique
CLASSIFICATION - [gen etc] classification *f*
~LAMP - [railw] feu *m* de position
~OF ORE - [min] désignation *f* du minerai
~YARD - [railw] gare *f* de triage
CLASSIFIER - [min] (a machine which separates the product of the ore-crushing plate into two portions, consisting of particles of different size) classeur *m*, trieur *m*
~BOX - s. classifier
CLASSIFY, to - [gen] classifier
CLASSIFYING AND DEWATERING MACHINE - [min] classificateur *m* débourbeur

CLASSIFYING SCREEN - [min] crible-classeur *m*
CLASSING - [gen] classification *f*
CLASTIC - [geol] (of rocks formed of fragments of preexisting rocks) clastique, deutogène
~ROCKS - [geol] roches *fpl* clastiques
CLATHRATE - [chem] (a type of compound in which one kind of molecule is enclosed within the structure of another) clathrate *m*, composé *m* en cage, composé *m* d'insertion
CLATTER, to - [acoust] faire du bruit
[mech] (the noise caused by a broken part) cliqueter
CLATTERING - [acoust] bruit *m*, fracas *m*, vacarme *m*
CLAUDE AMMONIA PROCESS - [chem] (a process in which the uncombined pases from the first pass are treated in a series of other catalyzers) procédé *m* de Claude
CLAUDICATION - [med] claudication *f*
CLAUSTROPHOBIA - [med] claustrophobie *f*
CLAVA - [bot] clavaire *f*
CLAVATE - [zool] (shaped like a club) clavé
CLAVICLE - [zool] clavicule *f*
CLAVICULATE - [zool] claviculé
CLAW - [gen] griffe *f*, ongle *m*, serre *f* d'oiseau, crochet *m*, mâchoire *f*
[mech] (of belts) griffe *f* de fixation
[mech] (in hoists) griffe *f*
[zool] (of a lobster) pince *f*
[cin] (component part of the mechanism for intermittent movement) griffe *f*
~BAR - [impl] (used for extracting nails) pied *m* de biche
~BOLT - [mech] boulon *m* à griffe
~BUCKET - [mech] benne *f* à dents
~CHISEL - [impl] ciseau *m* à ébaucher
~CHUCK - [mech] mandrin *m* à griffes
~CLUTCH - s. claw coupling
~COUPLING - [mech] (a shaft coupling in which the flanges on each engage through teeth cut in the opposing surfaces) embrayage *m* à griffes, accouplement *m* à griffes
~END - [mech] crochet *m*, griffe *f*, crabot *m*
~HAMMER - [impl] (a hammer with a bent and split pen) marteau *m* à panne fendue
~HAND - [med] main *f* en griffe
~MOVEMENT - [cin] (intermittent movement in which the film is moved on by two claws engaging the perforation) entraînement *m* à griffe
~SLIPPING - [cinema] glissement *m* des griffes
CLAY - [min] (a general term for sedimentary or residual material consisting primarily of hydrated aluminium silicates with varying amounts of impurities. It is very important as the basis of ceramic products) argile *f*
~AUGER - [mining] tarière *f* à glaise, sonde *f*
~-BAND - [mining] (claylike substance in coal, easily detached from the roof of the gallery, thus requiring a strengthening of the roof) claya *m*
~BASE MUD - [oil ind] boue *f* à base d'argile
~BEARING - [geol] argilifère
~BLANKET - [soil] masque *m* en argile
~CEMENT - [build] ciment *m* argileux
~CORE - [hydr] (of a dam) massif *m* en argile corroyée
~COURSE - [mining] salbande *f*, lisière *f*
~CRUCIBLE - [impl] creuset *m* en argile
~CUTTER - [metall] malaxeur *m* à argile
CLAY DAM - [hydr] barrage *m* en terre
~EMBANKEMENT - [constr] remplissage *m* d'argile, remblai *m* d'argile
~FIELD - [geol] argilière *f*, glaisière *f*
~FLOOR - [build] sol *m* en terre battue
~FLUX - [metall] erbue *f*, herbue *f*
~FURNACE - [metall] petit four *f* en terre réfractaire
~GALL - [metall] tache *f* d'argile
~GRAINS - [soil] grains *mpl* d'argile
~GROUTING - [soil] injection *f* d'argile
~IRON ORE - [min] minerai *m* argileux
~IRONSTONE - [geol] (nodular beds of clay and iron compounds) minerai *m* de fer argileux
~LEVEE - [hydr] digue *f* en argile
~MARL - [geol] marne *f* argileuse
~MORTAR - [constr] mortier *m* d'argile *f*, placage
~-PIT - [geol] argilière *f*, glaisière *f*, carrière *f* d'argile
~PUDDLE - [constr] corroi *m* d'argile
~PUDDLE WALL - [hydr] (of a dam) s. clay core
~-REVIVIFYING SYSTEM - [metall] régénérateur *m* de sable
~ROCK - [geol] roche *f* argileuse
~SEAL - [soil] scellement *m* d'argile, couche *f* argileuse d'étanchéité
~SEAM - [mining] veine *f* d'argile
~SHALE - [min] schiste *m* argileux, argile *f* schisteuse, argile *f* feuilletée, ardoise *f*
~SOIL [geol] sol *m* argileux
~TAMPING - [mining] bourre *m* d'argile
~WORK - [constr] construction *f* en pisé réfractaire
CLAYED ORE - s. clay iron ore
CLAYEY - [geol] argileux
~MARL - [geol] marne *f* argileuse
~SOIL - [soil] terres *fpl* argileuses
CLAYING - [metall] glaisage *m*
~BAR - [metall] tarière *f* à argile
CLAYISH - [geol] argileux
CLEADING - [th eng] (the operation of lining a vessel or a pipe with non conducting material; also the material itself) enveloppe *f* calorifuge, revêtement *m* isolant, chemise *f*
CLEAN, to - [gen] nettoyer, épurer, purifier
[metall] (to finish a surface smoothly) ébarber ébavurer
CLEAN - [gen] clair, pur, net
~ANCHORAGE - [naut] mouillage *m* sain
~BOMB - [nucl] (a hydrogen bomb with a smallest possible amount of radioactive fall-out) bombe *f* propre
~BOX - [ind chem] cuve *f* d'épuration complète
~CASTING - [metall] pièce *f* moulée ébarbée
~CONFIGURATION - [aero] (outline of an aircraft free from projections and forms of low aerodynamic efficiency) contour *m* sans saillie
~CUT- [gen & mech] coupure *f* nette, taille *f* nette
~EFFECTS - [telev] musique *f* de fond sans commentaire
~HARDENING - [metall] trempe *f* brillante
~HEMP - [text] chanvre *m* blanc
~OFF, to - [carp] replanir
~OUT, to - [gen] écurer (ou curer) débourber
~REACTOR - [nucl] (reactor without induced radio activity) réacteur *m* froid
~SHEDDING - [text] (in a jacquard loom) pas *m* oblique

CLEAN SPIRIT - [chem] alcool *m* rectifié
~TIMBER - [timber] bois *m* uni, bois *m* net (blanchi)
~TONE - [acoust] son *m* pur
~-UP - [el] (the improvement in the vacuum occurring in an electric discharge tube) pompage *m*, dégazage *m* final, absorption *f* de gaz, effet *m* getter
[mining] (of gold) récolte *f* de l'or
~UP TIME - [instr] rémanence *f* d'un détecteur de fuites
~WEEDING SYSTEM - [agric] sarclage *m*, désherbage *m*
CLEANED - [gen] nettoyé, épuré, blanchi
CLEANER - [metall & plast ind] (a small brass tool used to clean the surface of a mould) racloir *m*, crochet *m* à ramasser
[mech] curette *f*
[ind chem] (a compound or mixture used in degreasing) dégraissant *m*, solvant *m*, détersif *m*
CLEANING - [gen] nettoyage *m*, assainissement *m*
[metall]ébarbage *m*
[metall] (of wires) décapage *m*
[mech] ébavurage *m*, finissage *m*
[ind chem] dégraissage *m*
[text] époutiage *m*, épinçage *m*
[min] (of coals) classification *f*
~AND PRIMING MACHINE - [mech] (used in pipe laying) brosseuse-vernisseuse *f*
~BARREL - [metall] tonneau *m* dessableur, tambour *m* dessableur
~BRUSH - [text] brosse *f* nettoyeuse
~CLOTH - [text] chiffon *m* de nettoyage
~CROP - [agric] culture *f* nettoyante
~EYE -[constr] (a screwed plug in drain pipes etc to make access possible in order to clear a stoppage) regard *m*
~HINGE - [mech] charnière *f* en coude
~HOLE [mech] (in engines) regard *m* de nettoyage
~OF ORE - [min] débourbage *m*
~OFF - [carp] replanissage *m*
~RAKE (BAR) GRATE -[impl] (used in ovens etc) grille *f* à peigne
~SHOP - [metall] atelier *m* de dessablage
~SPIRIT - [chem] essence *f* pure, essence *f* de nettoyage
~STATION - [naut] poste *m* de propreté, poste *m* de lavage
~STRAINER - [mech] filtre *m* épurateur
CLEANOUT RIG - [oil ind] installation *f* de nettoyage d'un puits
CLEANSE, to - [min & metall] assainir, curer, débourber
CLEANSER - [gen & metall] nettoyeur *m*, ébarbeur *m*
CLEANSING - [min etc] assainissement *m*, curage *m*, débourbage *m*
~AGENT - [chem] agent *m* de purification, agent *m* dégraissant
~DRUM - [metall] tambour *m* de nettoyage
~SOLUTION - [chem] solution *f* de lavage
~SYSTEM - [brew ind] méthode *f* de fermentation en fût
CLEAR, to - [gen] éclaircir, déconnecter, purifier, dégager
[naut] dégager, parer
[met] (of the sky) se dégager
[photo] (of a negative) affaiblir
[leg] dédouaner
[plumb] (theoperation of eliminating a stoppage in a pipe) déboucher
~[comput] (to reset at zero) remettre à zero, liberer
[comm] (of a sale in shops) solder, liquider
[aero] autoriser à décoller
CLEAR - [gen] clair, limpide, net
[traffic & telecomm] libre
~ABREAST - [naut] en route libre sur le côté
~-BACK SIGNAL - [radio] signal *m* de fin de communication
~FELLING - [timber] coupe *f* blanche
~-FORWARD SIGNAL - [radio] signal *m* de fin de communication
~GAS - [gas ind] (tar-free gas induced between carbonization and gasification zones in a coal gasification plant) "clear gas" *m*
~HEIGHT - [railway and road constr] hauteur *f* libre
~INTERLACING EFFECT - [text] effet *m* clair d'entrecroisement
~MESSAGE - [telecomm] message *m* clair
~OCTANE - [chem] octane *m* sans additifs
~POSITION - [railw] signal *m* effacé, voie *f* libre
~SKY - [met] ciel *m* clair, ciel *m* dégagé
~SPAN - [constr] (the horizontal distance between the inner ends of the two bearings at the ends of a beam) ouverture *f*
~SPOT FOCUSING - [photo] mise *f* au point sur disque central lumineux
~STUFF - [timber] (timber free from knots and other blemishes) bois *m* sans noeuds
~THE POINTS, to- [railw] franchir les aiguilles
~TIMBER - s. clear stuff
~TOP - [text] peigné *m* nettoyé
~UP, to - [gen] éclaircir
[met] (of the sky) se dégager, s'éclaircir
[brew ind] éclaircir
~VARNISH - [paint] vernis *m* incolore
~WEATHER - [met] temps *m* clair
CLEARANCE - [gen] dégagement *m*, éclaircissement *m*
[mech] (the distance between two elements) jeu *m*, espace *m* libre, dépouille *f*, dégagement *m*
[constr]espace *m* libre
[comm] liquidation *f*, réalisation *f* du stock, dédouanement *m*
[tools] (the angular backing-off of a tool) dégagement *m*, dépouille *f*
[naut] permis *m* d'entrée
[aero] (the authorization to land) autorisation *f* (de décoller ou d'atterrir)
[mech] (of a slide valve) découvert *m*
[mining] (in the headframe) hauteur *f* de sécurité
[metall] jeu *m* de coiffage
~ANGLE - [mech] (the angular backing-off of a tool) angle *m* de dépouille, angle *m* d'incidence
~CAR - [railw] wagon *m* gabarit
~CIRCLE - [mech] (of a gear) cercle *m* limite
~DIAGRAM - [metall]gabarit *m*
~FIT - [mech] accouplement *m* mobile
~GAUGE - [mech] calibre *m* de tolérance
~LAMP - [auto] feu *m* de position
~OF FAULTS - [gen] réparation *f*, remise *f* en état
~OF POLE LINES - [telecomm] hauteur *f* libre
~PRINT - [metall] fausse portée *f* (du noyau)
~RING- [mech] anneau *m* de tolérance
~SPACE - [mech] (internal combustion engines) espace *m* de montée, chambre *f*; (of a pump) corps *m*

CLEARANCE SPACE - [mech] (in a steam cylinder) espace *m* mort, espace *m* nuisible
[el] (in a dynamo) intervalle *m*
~TEST [med] épreuve *f* d'élimination de l'urée sanguine
~VOLUME - [mech] (in reciprocating engines or compressors, the volume between the piston and the adjacent end of the cylinder when the crank is on the dead centre) espace *m* nuisible
~WASHER - [mech] rondelle *f* d'égalisation
CLEARER - [tool] curette *f*
[text] cylindre *m* débourreur
~BEARING - [aero] relèvement *m* de sécurité
~BELT - [text] courroie *f* de renversement de la marche
~BOARD [text] plaque *f* guide-fils
~GUIDE PLATE - [text] platine *f* nettoyeuse
~PLATE - [text] planchette *f* guide-fils, guide-fil *m* à fente
~ROLLER - [text] rouleau *m* nettoyeur
~ROLLER PLUSH - [text] peluche *f* pour rouleau-nettoyeur
CLEARING - [gen] clarification *f*
[text] nettoyage *m*
[mech] (the space between gear teeth) creux *m* de dents
[constr] déblaiment *m*, débroussaillage *m*
[geogr] (in a wood) clairière *f*, éclaircie *f*
[railw] libération *f* de la voie
[fin] acquittement *m* de dettes, compensation *f* de chèques
~BATH - [photo] bain *m* de clarification
~CAM - [text] came façon-métier
~CIRCUIT - [railw] circuit *m* d'ouverture de signaux
~DEVICE - [comput] (device resetting to zero) dispositif *m* de remise à zéro
~FIELD - [nucl] electrostatic field across the gas space in a cloud chamber of ions formed at undesired times) champ *m* de déionisation, champ *m* de balayage
~IRON - [impl] perçoir *m*, débouchoir *m*
~PULSE - [telecomm] impulsion *f* d'interruption
~SIGNAL BELL - [telecomm] sonnerie *f* de fin de conversation
~SIGNAL BUZZER - [telecomm] ronfleur *m* de signal de fin de conversation
~THE BALLAST - [railw] (from the track) nettoyage *m* de la voie
CLEARNESS - [gen] clarté *f*, netteté *f*, transparence *f*
CLEAT - [mech] (in the track of a caterpillar) crampon *m*
[carp] (a strip of wood fixed to another for strengthening purposes) tasseau *m*, languette *f*
[build] échantignole *f*
[el] serre-fil *m*, isolateur *m* à gorge
[naut] taquet *m*
~INSULATOR - [el] serre-fil *m*
~OF THE COAL - [mining] feuille *f* du charbon, limets *mpl* de la houille
CLEATED SOLE [shoe man] semelle *f* à dessin
CLEAVABILITY - [geol] clivage *m*
CLEAVABLE - [gen] clivable
~ROCK - [geol] roche *f* clivable
CLEAVAGE - [gen] fissure *f*, scission *f*, division *f*
[geol] clivage *m*
[chem] (the division of a complex protein molecule into simpler molecules) scission *f*
[chem] (in hydrolitic processes) dissociation *f*
[phys] (the property of crystals to break along definite planes) clivage *m*
~FRACTURE - [metall] fragilité *f* à la rupture
~LOADING - [phys] charge *f* de clivage
~NUCLEUS - [biol] (the nucleus of the fertilized ovum) noyau *m* de segmentation
~PLANE - [geol] plan *m* de clivage
~PRODUCT - [brew ind] produit *m* de scission
CLEAVE, to - [gen] fendre, refendre, cliver
[geol] se cliver, cliver
CLEAVERS - [bot] (a weed) gaillet *m* accrochant, grateron *m*
CLEAVING - [gen & geol] fendage *m*, clivage *m*, refente *f*
~CHISEL - [impl] rabattoir *m*
~SAW - [impl] (also called pit-saw; large, two-handed rip-saw used forcuttinglogs) scie *f* à refendre
CLEFT - [gen] fissure *f*, crevasse *f*, fente *f*
~GRAFTING - [agric] griffe *f* à fente
~PALATE - [med] palais *m* fendu
~WELD - [metall] soudure *f* à gueule de loup
CLEIDOTOMY - [med] (the cutting of the clavicles to ease delivery in a difficult labour) cléidotomie *f*
CLEISTOCARP - [bot] cléistocarpe *m*
CLEISTOGAMY - [bot] (the production of small flowers, in which selfpollination occurs) cléistogamie *f*
CLEMATIS - [bot] clématite *f*
CLENCH, to - [gen] serrer, contracter
~A NAIL, to - [mech] mater un clou
CLEPSYDRA - [horol] (the most ancient mechanism for indicating intervals of time) clepsydre *f*
CLERESTORY - [arch] clair-étage *m*, lanterneau *m*
~RING - [constr] (eye of a cupola) anneau *m* de lanterneau
CLEVEITE - [min] (uraniumcompound with rare earth and helium contents) clévéite *f*
CLEVIS - [mech] manille *f*, maillon *m*, attache *f*, crochet *m* de sûreté, crochet *m* à ressort, mousqueton *m*
[naut] manille *f*
~PIN - [mech] axe *m* à chape, broche *f* d'étrier
CLEW, to - [naut] (of sails) carguer (les voiles)
CLEW - [naut] araignée *f* de hamac
[naut] (the ring at a corner of a sail) point *m* d'écoute (de voile)
~GARNET - [naut] cargue-point (de basse voile) *m*
~LINE - [naut] cargue-point (de haute voile) *m*
~ROPE - [naut] faux point *m*
CLICHE - [print] (a stereotype or electrolyte plate) cliché *m*
CLICK - [gen] clic *m*, bruit *m* sec, cliquetis *m*
[horol] (a pawl permitting rotation in one direction only) cliquet *m*
[mech] cliquet *m*
[the noise of the pawl] cliquetis *m*
[radio] (circuit noise caused by interference) bruit *m*
~AND RATCHET WHEEL - [mech] roue *f* à rochet
~-BEETLE - [agric] (a pest) élatère *m*
~GAUGE - [meas] (used for the measure of moderate pressures) manomètre *m* à détente
~POST - [horol] support *m* de cliquet

LICK SOLE - [horol] semelle *f* de cliquet
SPRING - [mech] cliquet *m*, ressort *m* à cliquet
[horol] ressort *m* de cliquet
STOP - [mech] diaphragme *m* à encliquetage
WHEEL - [mech] roue *f* à cliquet, roue *f* à rochet, roue *f* à chien
LICKER - [leather ind] machine *f* à couper les peaux
PRESS - [mech] (a press designed to cut out blanks from plastic sheet material) machine *f* à découper
LICKING MACHINE - s. clicker press
SPRING - [horol] cliquet *m* ressort
LIFDEN BLUE - [zool] papillon *m* adonis
LIFF - [geogr] falaise *f*, escarpement *m*
LIMATIC - [gen & met] climatique
CHAMBER - [ind chem] chambre *f* climatique
FACTOR - [bot] (the climatic conditions determining the life of a plant community) facteur *m* climatique
TEST - [ind chem] essai *m* climatique
ZONES - [met] (the earth's zones according to climates)zones *f*pl climatiques
LIMATOLOGY - [met](the study of climates and its causes) climatologie *f*
LIMATOTHERAPY - [med] climatothérapie *f*
LIMAX - [gen] point *m* culminant, apogée *m*, comble *m*
LIMB, to - [gen] monter, gravir, grimper, escalader
[aero] prendre de l'altitude
LIMB - [gen] ascension *f*, montée *f*, rampe *f*
[aero] vol *m* ascendant, montée *f* en altitude
~CUTTING - [mech] fraisage *m* en avalant
~INDICATOR - [instr] (an instrument showing the rate of change of altitude) variomètre *m*
~MILLING - s. climb cutting
LIMBING - [gen] escalade *f*, montée *f*, ascension *f*
~ABILITY - [auto] pouvoir *m* de traction en cote
~BOOT - [shoe man] chaussure *f* de montagne
~-FILM EVAPORATOR - [ind chem] appareil *m* d'évaporation à grimpage
~IRON - [impl] crampon *m*
~POWER - [aero] puissance *f* ascensionnelle
~SHAFT - [aero] puits *m* de montée
~SPEED INDICATOR - s. climb indicator
~TAKE-OFF - [aero] décollage *m* cabré
~TURN - [aero] montée *f* en virage
~-WAY - [mining] compartiment *m* des échelles
LINCH, to - [gen] rabattre, aplatir
[mech] rabattre, river
LINCH - [mech] rivet *m*
[naut] étalingure *f*, agrafe *f*, crampon *m*
LINCHER -[mech] (a machine to join belts etc) agrafeuse *f*
~TYRE - [rubber ind] pneumatique *m* à talon
LINCHING - [mech] rivetage *m*; (of belts) agrafage *m*
LING, to - [gen] s'agripper, se cramponner, s'attacher
[naut] (the wind) serrer (le vent)
LINGSTONE - [bot] alberge *f*
LINGY - [gen] collant
LINICAL PATTERN - [med] description *f* clinique
~THERMOMETER - [instr] (mercury thermometer used for an accurate measure of the temperature of the human body) thermomètre *m* médical

CLINK - [metall] crique *f* interne, fracture *f* interne
~-STONE - [geol] (also called phonolite; a fine-grained igneous rock) phonolite *f*
CLINKER - [gen]mâchefer *m*, clinker*m* scories *f*pl
[metall] (iron slag) crasse *f* de fonte
[constr] brique *f* de mâchefer, brique *f* vitrifiée
~BAR - [impl] ringard *m*
~-BUILT - [shipbuild] bordé à clins
~DOOR - [metall] (of furnaces) porte *f* de décrassage
~PLATING - [shipbuild] bordage *m* à clins
~RAKE - s. clinker bar
~SLICE - s. clinker bar
~WORK - s. clinker plating
CLINKERING - [metall] (of furnaces) décrassage *m*
~COAL - [min] charbon *m* produisant beaucoup de scories
~DOOR - s. clinker door
~TOOL - s. clinker bar
~UP OF A GENERATOR - [gas ind] engorgement *m* du gazogène
CLINKERS - [gen](fused ash in furnaces etc) mâchefer *m*, scories *f*pl
[constr] (portland cement; the calcined material delivered from the rotary kiln before grinding and sizing) clinker *m*
CLINOCHLORE - [min] (a variety of chlorite) clinochlore *m*
CLINOGRAPH - [instr] (to measure the inclination of drilling) clinographe *m*, clitographe *m*
CLINOID - [med] clinoïde
CLINOMANIA - [med] clinomanie *f*
CLINOMETER - [instr] (instrument measuring angles of inclination) clinomètre *m*, indicateur *m* de pente
CLINORHOMBOIDAL SYSTEM - [crystal]système *m* triclinique, système *m* doublement oblique
CLINOSCOPE -[instr] clinoscope *m*
CLINOTHERAPY - [med] clinothérapie *f*
CLIP, to - [gen] tailler, tondre, couper, cisailler
[metall] ébarber
[mech] (to hold firmly in a grip) serrer, agrafer
CLIP - [gen] attache *f*
[mech] agrafe *f*, attache *f*, pince *f*, serre *f*, grip *m*, tenaille *f* d'attelage
[agric] (of sheep) toison *f*, tonte *f*
[railw] serre-rail *m*, crapaud *m*
[firearms] chargeur *m*
[el] borne *f*
[cin] chute f
[photo] griffe *f*
~ANGLE - [mech] cornière *f* d'assemblage
~BUFFER - [mech] garniture *f* de collier de fixation
~-ON - [gen] à mâchoires
~PULLEY - [mech] poulie *f* d'adhérence
~SCREW - [mech] (screws used to adjust the two verniers of the vertical circle of a theodolite) vis *f* antagoniste
~TOGETHER, to - [gen]agrafer
~WOOL - [text] laine *f* de tonte
CLIPPER - [naut]clipper *m*
[el] écrêteur *m*
[telev] séparateur *m*
[nucl] circuit *m* d'écrêtage
~AMPLIFIER - [el] amplificateur *m* limiteur
~BOW - [shipbuild] avant *m* fin
~CIRCUIT - [radio] (a circuit designed to prevent

the maximum signal level to exceed a predetermined value) circuit *m* écrêteur
~DIODE - [el] (diode used in a clipper circuit diode *m* pour circuit écrêteur
~LIMITER - [radio] découpeur *m*
~TUBE - [electron] tube *m* à seuil
CLIPPING - [text] tonte *f*
[acoust] distortion *f*
[metall] ébarbage *m*
~BASE - [mech] plaque *f* d'ébarbage, base *m* d'appui de machine à ébarbage
~CIRCUIT - [el] circuit *m* limiteur
~KNIFE - [metall] couteau *m* à tailler les bordures
~LEVEL - [el] niveau *m* d'écrêtage
~MACHINE - [mech] machine *f* à découper
~TIME - [radio] (the time constant of the clipper circuit) constante *f* de temps de circuit écrêteur
~TOOL - [tool] outil *m* d'ébarbage
CLIPRING - [mech] collier *m* de fixation
CLOACA - [constr] (a term for sewer) cloaque *m*, égout *m*
[zool] cloaque *m*
CLOAKROOM - [build] vestiaire *m*, toilettes *f*pl, consigne *f*
CLOCHE CIRCUIT - [telev] circuit *m* cloche
CLOCK - [instr] horloge *f*, pendule *f*
~CASE - [horol] boîtier *m*, caisse *f*
~CONTROLLED GOVERNOR - [mech] régulateur *m* à programme
~FREQUENCY- [comput] fréquence *f* pilote
~HAMMER - [impl] marteau *m* de cloche, marteau *m* d'horloge
~IGNITION - [mech] allumeur *m* horaire
~METER - [el] compteur *m* à mouvement d'horlogerie, compteur *m* à balancier
~SPRING - [mech] ressort *m* spiral
~STAR [astr] (stars whose position is known and help in the determination of time) étoile *f* de position continue
~-WATCH - [horol] montre *m* à sonnerie
CLOCKWISE - [gen] (in the same direction as the rotation of the hands of the clock) dans le sens des aiguilles d'une montre
~POLARIZED WAVE - [phys] (electromagnetic waves) onde *f* à polarisation elliptique à droite
~ROTATION - [gen] rotation *f* dans le sens des aiguilles d'une montre
CLOCKWORK - [horol] mouvement *m* d'horlogerie
[mech] (train of wheels) rouage *m*, engrenage *m*
CLOD - [soil] (earthy clay found near a coal seam) motte *f*, corps *m* d'argile, motte *f* de terre
~BREAKER - [agric] brise-mottes *m*, rouleau *m* brise-mottes
CLODDY - [soil] rempli de mottes, se cassant en mottes
CLOG, to - [gen] entraver, empêcher, obstruer
[plumb] (of a pipe) encrasser, colmater, s'encrasser
[mech] (of a grinding wheel) encrasser
CLOG - [gen] entrave *f*, empêchement *m*
[mining] écoin *m*
[plumb] (of a pipe) engorgement *m*, obstruction *f*
CLOGGED - [gen] bouché, obstrué, engorgé
CLOGGING - [gen] obstruction *f*, colmatage *m*
~UP MATERIALS - [geol] colmatants *m*pl
CLOISTER - [arch] cloître *m*
~VAULT - [arch] voûte *f* à pavillon

CLONAL PROGENY - [bot] (the entire stock of plant obtained by means of vegetative multiplication fro one original seedling, e.g. budding, grafting etc) clône *m*
~SEEDLING RUBBER - [rubber ind] caoutchouc *m* provenant de "seedlings" clonaux
CLONE - s. clonal progeny
CLONIC- [med] clonique
CLONISM - [med] clonisme *m*, convulsions *f*pl cloniques
CLONUS - [med] clonus *m*
CLOSE, to - [gen] fermer, clore
CLOSE - [gen] fermé, étroit, proche, rapproché
[text] (of texture) serré
~A CIRCUIT, to - [el] fermer un circuit
~A RIVET, to - [carp] refouler un rivet
~AN ELECTRIC CIRCUIT - [el] mettre dans le circuit fermer un circuit
~ANNEALING - [metall] recuit *m* en vase clos
~BOILING POINT - [metall] coupe *f* courte
~BURNING COAL - [min] houille *f* à courte flamme
~BUTTRESS - [constr] contrefort *m*, éperon *m*
~CEILING - [shipbuild] vaigrage *m* continu
~CONTACT GLUE - [ind chem] (an adhesive giving a satisfactory bond only with surfaces not more than a fraction of a millimiter apart) colle *f* de contact
~-CONTROL BOMBING - [radar] (a bombardment from a short distance which is controlled by radar bombardement *m* radioguidé
~-COUPLE TRUSS - [constr] ferme *f* à tirant
~COUPLING - [telecomm] couplage *m* serré
~CRIBBING - [mining] boisage *m* jointif, boisage *m* serré
~CROPPED - [gen & zool] coupé ras, tondu de près
~FITTING - [gen] ajusté, collant
~FOLD - [geol] pli *m* resserré
~GRAIN - [metall] grain *m* serré
~-GRAIN CAST IRON -[metall] fonte *f* compacte
~GRAINED - [metall] à grain serré
[brew ind] à grain fin
~-GRAINED WOOD - [timber] bois *m* compact, bois plein
~-HAULED - [naut] au plus serré
~-IN - [aero] pincement *m* des roues
~IN FALL-OUT - [nucl] retombées *f*pl radioactives primaires
~JOINT - [mech] joint *m* étanche
~LATHING - [mech] lattage *m* jointif
~LAY OF THREADS - [text] torsion *f* serrée des fils
~-MOUTH TONGS - [impl] tenaille *f* plate fermée
~NIPPLE - [plumb] (a short length of pipe threaded from end to end) mamelon *m* simple
~-PACKED METAL - [metall] métal *m* à structure compacte
~-PACKED STRUCTURE - [crystal] structure *f* compacte
~PILING - [constr] palplanches *f*pl jointives
~PLANTING - [agric] plantation *f* dense
~POLLINATION - [bot] pollination *f* entre plantes voisines
~PRESSING UP - [text] tramage serré
~RIVET JOINT - [mech] rivure *f* étanche
~RUNNING FIT - [mech] ajustage *m* serré
~SCANNING - [telev] (scanning system with a very small light beam and a large number of lines) balayage *m* serré

OSE SET - [text] chaîne *f* serrée
ETS - [mining] (of timbering) cadres *mpl* jointifs
HOT - [photo] gros plan *m*, plan *m* très rapproché
PACED TRIODE - [electron] (triode with a very mall grill-cathode spacing) triode *f* à faible disance entre grille et cathode
PIRAL SPRING - [mech] ressort *m* de rappel
TRING - [build] limon *m* droit
TALKING MICROPHONE - [el acoust] microphone *m* le bouche
TALKING RESPONSE - s. close-talking sensitivity
TALKING SENSITIVITY - [el acoust] (of a microhone for specified frequencies) efficacité *f* paraphonique
HE WIND - [naut] serrer le vent
IMBERING - s. close cribbing
OLERANCE - [mech] tolérance *f* serrée
UP - [photo & cin] (shot of a detail in image size) ros plan *m*, vue *f* rapprochée
UP FOCUSING - [photo] mise *f* au point sur plan approché
JP ILLUSTRATION - [print] illustration *f* d'un détail
UP LENS - [photo] bonnette *f* d'approche
UP VIEW - s. close-up
UP WINDING - [el] enroulement *m* à spires jointives
WOVEN - [text] serré (tissu à contexture serrée)
OSED - [gen] fermé, bouché, obturé
RMOURING - [el etc] armure *f* à recouvrement
ASSEMBLY - [mech] (method of assembly in which parts are brought together as soon as ready and pressure is applied subsequently) assemblage *m* avant pression
ASSEMBLY TIME - [mech] (the time between the assembly of glued surfaces and the application of pressure) temps *m* d'assemblage avant pression
-CELL PLASTICS - [plast ind] (cellular plastics in which the cavities are not connected with one another or with the atmosphere) matières *fpl* plastiques à cellules fermées
CELLS - [plast ind] (cavities in a foam material without communication with the atmosphere) cellules *fpl* fermées
-CHAIN COMPOUND - [chem] composé *m* à chaîne fermée
CIRCUIT - [el] circuit *m* fermé
-CIRCUIT ALARM SYSTEM - [telecomm] dispositif *m* d'alarme à circuit fermé
-CIRCUIT ARMATURE - [el] induit *m* à circuit fermé
CIRCUIT ARRANGEMENT - [telecomm] montage *m* en circuit fermé
CIRCUIT GRINDING - [constr] (of cement) broyage *m* en circuit fermé
-CIRCUIT OPERATION - [telecomm] (a single-circuit signalling method) fonctionnement *m* en courant constant
-CIRCUIT RADIATOR - [th eng] radiateur *m* à circuit étanche
-CIRCUIT SIGNALLING - [radio] transmission *f* par courant constant
-CIRCUIT TELEVISION- [telev] télévision *f* en circuit fermé, télévision *f* industrielle
CIRCUIT VOLTAGE - [el] (potential difference between the terminals of a cell or battery when current is flowing) tension *f* de service
WORKING - [telecomm] système *m* à courant constant
~COIL WINDING - [el] enroulement *m* d'induit fermé
~COMMUNITY - [bot] (a plant community occupying the ground without leaving any space withoutvegetation) végétation *f* dense
~CORE - [el] noyau *m* fermé
~CORE TRANSFORMER - [el] transformateur *m* à noyau fermé
~CUP - [ind chem] recipient *m* à pétrole fermé
~CYCLE - [control] circuit *m* fermé
~EAVES - [constr] gouttières *fpl* en saillie
~FOLD -FAULT - [geol] pli-faille *m* fermé
~FRAME PRESS - [mech] presse *f* à bâti fermé
~-GRAINED - [gen & metall] à grain fin, dense
~-IN PRESSURE - [oil ind] pression *f* statique d'un puits
~IRON CIRCUIT TRANSFORMER - [el] transformateur *m* à noyau fermé
~-JET WIND TUNNEL - [aero] (type of wind tunnel in which the experimental section is provided with rigid walls) soufflerie *f* à veine guidée
~LOOP - [el] circuit *m* fermé
~LOOP CONTROL - [control](control method using a feed-back loop) réglage *m* à boucle fermée, réglage *m* à reaction
~-LOOP CIRCUIT [mech] (a control method for gas turbines in which a special device provides the correct amount of fuel for any given setting of the pilot's throttle lever) alimentation *f* en circuit fermé
~MAGNETIC CIRCUIT - [el] circuit *m* magnétique fermé
~MOULD - [metall] moule *m* fermé
~OSCILLATING CIRCUIT - [el] circuit *m* oscillant fermé
~PASS - [hydr] canal *m* fermé
[metall] cannelure *f* emboîtée
~PATH - [control] contour *m* fermé
~PILE - [text] poil *m* serré
~PRESSURE - [oil ind] pression *f* de gisement
~REED - [text] peigne *m* fermé
~SCOOP - [mining] cuillière *f* fermée
~SHAFT - [mining] puits *m* aveugle
~SHED - [text] pas *m* fermé, pas *m* clos
~SHED DOBBY - [text] mécanique *f* d'armure à pas fermé
~SHELL - [nucl] (an electron shell containing the full number of electrons according to the Bohr theory) couche *f* électronique saturée
~STRING - s. close string
~STUB - [radio] (of aerials) têton *m* de court-circuit
~SUBROUTINE - [comput] sous-programme *m* fermé
~SWITCH - [railw] signal *m* fermé
~TO TRAFFIC - [gen] fermé à la circulation
~TRAVERSE - [surv] contour *m* fermé
~-TUBE TEST - [chem] essai *m* au tube fermé
~TURBINE-CHAMBER - [mech] bâche *f* fermée, enveloppe *f*
~-TYPE HOTPLATE - [th eng] plaque *f* coupe-feu
~VASCULAR BUNDLE - [bot] (vascular bundle which cannot increase in diameter) faisceau *m* vasculaire fermé
~WORK - [mining] ouvrage *m* souterrain, exploitation *f* souterraine
CLOSELY - [gen] étroitement, hermétiquement
~PACKED LATTICE - [nucl] réseau *m* très compact
~WOVEN BACK - [text] envers *m* serré

CLOSED WOVEN FABRIC - [text] tissu *m* serré
CLOSENESS OF WINDING - [el] densité *f* d'enroulement
CLOSER - [constr] (a brick cut to make up a course) brique *f* de fin de rangée
[gen] appareil *m* de fermeture
CLOSING - [gen] fermeture *f*, clôture *f*
[adj] dernier, final, qui ferme
~CAPSULE - [mech] capsule *f* de fermeture
~CYLINDER - [metall] (in die casting) cylindre *m* de fermeture
~ERROR - [surv] erreur *f* de fermeture du contour
~HEAD - [mech] (of a nail) tête *f* matée
~JOINT - [plumb] joint *m* par rapprochement
~RELAY - [electron] conjoncteur *m*
~TIME - [mech] (the time taken by a press to close completely) temps *m* de fermeture
~TRAVEL - [mech] (the distance through which the press moves in closing completely) course *f* de la fermeture
~VALVE - [mech] vanne *f* obturatrice, robinet *m* obturateur
CLOSURE - [gen] clôture *f*, fermeture *f*
~OFMOULD - [plast ind] the operation of bringing the mould into the closed condition) fermeture *f* d'un moule
CLOT, to - [gen] former des grumeaux, se grumeler
[rubber ind] former des grumeaux
[agric] (milk products) cailler, coaguler
CLOT - [gen] grumeau *m*, caillot *m*
CLOTH - [text] tissu *m*, étoffe *f*, drap *m*, toile *f*
[naut] (the breadth of the canvas) laize *f*
~ANALYSIS - [text] décomposition *f* du dessin
~BEAM - [text] ensouple *f* d'avant
~BEAM LOCKING - [text] arrêt *m* de l'ensouple d'avant
~BOUND - [book bind] relié en toile
~DESIGN- [text] dessin *m* du tissu
~DOFFER - [text] coupeur *m* de pièces
~EXPANDER - [text] élargisseur *m* de tissu
~FACE- [text] endroit *m* du tissu
~FILTER - [impl] filtre *m* en toile
~FINISHING - [text] apprêt *m* façon drap
~GOVERNING - [text] guide *m* de nappe
~GUIDE- [text] conducteur *m* de nappe de tissu
~HALL - [comm] halle *f* aux draps
~INSPECTING MACHINE - [text] machine *f* à contrôler le poids
~LOOKER - [text] vérificateur *m*
~MAKER - [text] drapier *m*
~MANGLE - [text] calandre *f* pour linge
~MANUFACTURE - [text] fabrication *f* de tissus, tissage *m*
~MILL - [text] fabrique *f* de tissus
~OIL - [ind chem] huile *f* d'ensimage
~PASTE - [ind chem] colle *f*
~POLISHING BAND - [mech] endless strip of cloth running over pulleys and used for polishing) toile *f* pour polissage
~ROLL - [text] rouleau *m* de tissu
~ROLLER - [text] cylindre *m* d'enroulement
~SCRAY - [text] accumulateur *m* de tissu
~TENSION - [text] tension *f* du tissu
~TESTING - [text] contrôle *m* des tissus
~WEAVER - [text] tisserand *m*
~WHEEL - [impl] disque *m* souple, disque *m* en tissu

CLOTH WIDTH - [text] largeur *f* du tissu
~WITH PLAIN SURFACE - [text] tissu *m* uni, tissu *m* lisse
~WITH SIMILAR FACES - [text] tissu *m* reversible
~WITH WARP EFFECTS - [text] tissu *m* à effets de chaîne
~WITH WEFT EFFECTS - [text] tissu *m* à effets de trame
CLOTHE, to - [gen] vêtir, habiller, revêtir
CLOTHES - [gen] vêtements *mpl*, habits *mpl*
~DRYER - text] séchoir *m* à linge
~HANGER - [impl] cintre *m*, bois *m*
~PIN - [impl] pince *f* à linge
~WRINGER - [impl] essoreuse *f*
CLOTHING - [gen] vêtements *mpl*, habillement *m*
[text] (card) garniture
[plumb etc] chemise *f*, enveloppe *f*
~INDUSTRY - [text] industrie *f* du vêtement, industrie *f* d'habillement
~NEEDLE - [text] aiguille *f* de garniture
~PINCERS - [text] pince *f* à monter les garnitures
~WIRE - s. clothing needle
~WOOL - [text] laine *f* cardée
CLOTTED - [gen] coagulé, caillé
CLOTTING - [med] coagulation *f*, consolidation *f*
[agric] (milk products) caillement *m*
~ENZYMES - [chem] enzymes *fpl* causant le caillement
~TIME - [med] temps *m* de coagulation
CLOUD, to - [gen] couvrir, voiler
[text] chiner
CLOUD - [met] nuage *m*
[chem] turbidité *f*
~AMOUNT - [met] (the proportion of the sky covered with clouds, measured in eighths or oktas) nébulosité *f*
[telev] contrastes *mpl* trop marqués
~AND COLLISION WARNING SYSTEM - [radar] (airborne primary-radar installation designed to give warning of objects ahead) radar *m* météorologique
~AND POUR POINT - [oil ind] point *m* de trouble et de fluage
~BANNER - [met] (a cloud streaming from a mountain peak in a flag-like form) nuage *m* en bannière
~BASE - [met] (the under-side of a cloud layer) plafond *m*
~-BURST - [met] (sudden and very heavy rainfall) averse *f*, trombe *f* d'eau
~-BURST HARDENING - [metall] trempe *f* au jet
~CANNON - [met] canon *m* paragrêle
~CAP - [met] (cloud forming a cap over another cloud or over the summit of a mountain) capuchon *m* de nuages
~CEILOMETER - [instr] projecteur *m* pour la détermination de la hauteur des nuages
~CHAMBER - [nucl] (device designed to show the path of a charged particle by the fog droplets formed by its passage) chambre *f* de Wilson
~COLUMN - [nucl] (the visible column of smoke from the point of explosion of a nuclear weapon) colonne *f* de fumée
~DETECTION RADAR - [radar] (centimetric-wave radar combined with a special scanning system, to give warning of dangerous cloud formations) radar *m* météorologique
~HEIGHT - [met] (the altitude of the cloud base

bove the ground surface) hauteur *f* des nuages
OUD LAYER - [met] couche *f* de nuages
OF ELECTRONS - [phys] nuage *m* d'électrons
POINT - [ind chem] (the temperature at which solids begin to separate from an oil in the process of cooling) point *m* de trouble
PULSE - [electron] impulsion *f* de la charge spatiale
SEARCHLIGHT - [met] (light ray projector used to measure the height of a cloud by trigonometry) projecteur *m* permettant de déterminer la hauteur des nuages
STREETS - [met] nuages *mpl* rectilignes
-TRACK INTERPRETATION - [nucl] (the study of the cloud-tracks in a cloud chamber) interprétation *f* de la trajectoire
WARNING - s. cloud detection radar
OUDINESS - [met] nébulosité *f*
[oil ind] aspect *m* trouble
OUDING [paint] voile *m*
[med] obscurcissement *m*
[brew ind] turbidité *f*
OUDY - [met] nuageux, couvert
[text] (of uneven shade) strié
LIQUID - [ind chem] liquide *m* trouble
WEB - [text] (web of fibres from the doffer of a carding engine uneven in density) voile *m* de densité irrégulière
LOUT, to - [gen] battre
LOUT - [gen] chiffon *m*
[impl] (a nail with a large thin flat head) clou *m* à tête platte
LOVE - [bot] caïeu *m*, clou *m* de girofle
HITCH - [naut] demi-clef *f* à capeler
OF GARLIC - [agric] gousse *f* d'ail
OIL - [ind chem] (yellow, aromatic liquid used in perfumery, medicine and confectionery) essence *f* de girofle
LOVER - [bot] trèfle *m*
HAY - [agric] foin *m* de trèfle, luzerne *f*
-LEAF - [constr] (in road constr) croisement *m* en as de trèfle
-LEAF AERIAL - [radio] (aerial of the magnetic dipole type consisting of four small loops connected in parallel across a coaxial line) antenne *f* à éléments multiples
-LEAF ANTENNA - s. clover-leaf aerial
MIXTURE HAY - [agric] foin *m* mixte graminées-légumineuses
LUB, to - [gen] frapper, battre (avec un gourdin)
LUB - [gen] massue *f*, gourdin *m*
[build] club *m*, cercle *m*
FOOT - [med] pied *m* bot
-HAUL - [naut] virer vent devant en mouillant l'ancre sous le vent
MOSS - [bot] lycopode *m* en massue, soufre *m* végétal
-ROOT - [bot] (plant disease) hernie *f* (des choux)
RUSH - [bot] scirpe *m*
-SHAPED - [gen] claviforme, clavé
-TOOTH ESCAPEMENT - [horol] échappement *m* à ancre
WHEAT - [agric] blé *m* nain
LUCKING - [acoust] gloussement *m*
LUE - [gen] indication *f*, indice *m*
LUMP - [gen] groupe *m*
[min] masse *f* d'argile
[print] lingot *m*
CLUMP BOOT - [shoe man] botte *f*
CLUMPS - [print] (large metal spacing material) lingots *mpl*
CLUSIUS COLUMN - [nucl], (apparatus employed in the separation of isotopes by their thermal diffusion) colonne *f* Closius
CLUSTER - [gen] groupe *m*, bouquet *m*, grappe *f*
[el] (a group of lamps) faisceau *m*
[aero] (a group of parachutes carrying a single load) groupe *m*, grappe *f*
[ind chem] agglomérat *m*, amas *m*, houppe *f*
[metall] grappe *f*
[auto] (instrument panel board) tableau *m* de bord
~BURNER - [gas ind] multibec *m*, bec *m* multiple
~GEAR - [mech] pignon *m* multiple
~OF ATOMS - [nucl] faisceau *m* d'atomes
~PINE - [bot] pin *m* maritime
~SWITCH - [el] interrupteur *m* circulaire
CLUSTERED - [arch] en faisceau
~COLUMN - [arch] colonne *f* en faisceau
CLUTCH - [gen] prise *f*, griffe *f*, serre *f*
[mech] (a device by which two shafts may be connected or disconnected, both in rest and motion) embrayage *m*, accouplement *m*
~ARC LAMP - [light] lampe *f* à arc à commande automatique
~BRAKE - [auto] frein *m* sur l'embrayage
~CASING - [auto] carter *m* d'embrayage
~CONE - [auto] cône *m* d'embrayage
~CONTROL LEVER - [mech] levier *m* de commande de la friction
~COUPLING - [mech] accouplement *m* par friction
~COVER - [auto] cloche *f* d'embrayage
~DISC - [auto] disque *m* d'embrayage
~DISC HUB - [auto] moyeu *m* de disque
~DISENGAGEMENT - [mech] ressort *m* de débrayage
~FLUID FLY-WHEEL - [auto] accouplement *m* hydraulique
~FORK - [mech] embrayeur *m*, fourchette *f* d'embrayage
~HOUSING - [mech] carter *m* d'embrayage
~HUB - [mech] moyeu *m* d'embrayage
~INSERTS - [mech] (in a motorbicycle) pièces *mpl* de l'embrayage
~LEVER - [mech] levier *m* d'embrayage
~MAGNET- [el] électro-aimant *m* d'accouplement
~OF STARTING HANDLE - [mech] dent *f* de loup de la manivelle
~OUT - [auto] débrayer
~PEDAL - [mech] pédale *f* d'embrayage
~PLATE - [mech] disque *m* d'embrayage
~PRESSURE- [mech] pression *f* d'embrayage
~PRESSURE PLATE - [mech] plateau *m* de pression d'embrayage
~PRESSURE SPRING - [auto] ressort *m* de pression
~RELEASE BEARING - [mech] butée *f* d'embrayage
~RELEASE FORK - [mech] fourchette *f* d'embrayage
~SHAFT - [mech] arbre *m* d'embrayage
~SLIP - [auto] patinage *m* de l'embrayage
~SPINDLE - [mech] arbre *m* d'embrayage
~SPRING - [mech] ressort *m* d'embrayage
~THROW-OUT SLEEVE - [mech] manchon *m* de débrayage
~THRUST BEARING - [mech] butée *f* d'embrayage
~WHEEL FOR ALARM - [horol] pignon *m* coulant de sonnerie

CLUTCH WITH MULTIPLE LAMINATED DISC - [auto] embrayage *m* à disques multiples
CLUTCHLESS - [mech] sans embrayage
~GEARSHIFT - [mech] changement *m* de vitesse automatique
CLUTTER - [acoust] tapage *m*, bruit *m*
[radar] (confused unwanted echoes) retour *m* de mer, échos *mpl* parasites
CLYPEATE - [bot] (shaped like a shield) clypéacé
CLYPEUS - [bot] (a parasitic formation in a leaf) clypéus *m*
C NETWORK - [radio] (network composed of three impedances connected in series) réseau *m* en C
C NEUTRONS - [nucl] (neutrons of which the energy level is such as to make cadmium absorption possible) neutrons *mpl* épicadmiques
CNIDOBLAST - [zool] (a stinging cell) cnidoblaste *m*
COACERVATE - [bot] (massed in a small heap) coarcevé
COACERVATION - [chem] (the production, by coagulation of hydrophilic sol, of a liquid phase, often appearing as viscous drops instead of forming a continuous liquid phase) coacérvation *f*
COACH - [auto] voiture *f*, autocar *m*
[railw] voiture *f*, wagon *m*
[horse-drawn carriage] carrosse *m*
~-BUILDER - [auto] carrossier *m*
~-BUILDING - [auto] carrosserie *f*
~ROLL - [paper man] presse *f* humide
~-SCREW - [mech] (large wood screw with square head) tire-fond *m*
~SCREW SPANNER - [impl] clé *f* pour tire-fonds, clé *f* carrée
~WITH AUTOMATIC COUPLINGS - [railw] voiture *f* à attelage automatique
~WITH CENTRE GANGWAY - [railw] voiture *f* à couloir central
~WORK - [auto] carrosserie *f*
~WRENCH - [impl] clé *f* anglaise
COACTION - [ecology] (correlation between plants, or plants and animals) coaction *f*
COADAPTATION - [biol] (correlated adaptation in organs or organism) coadaptation *f*
COAGEL - [chem] (gel obtained by coagulation) coagel *m*
COAGENT - [gen] aide *m*, adjoint *m*
COAGULANT - [chem] (substance which initiates formation of relatively large particles in a finely divided suspension, thus hastening settling out. Also a substance which assists in the formation of a gel) agent *m* coagulant, coagulant *m*
~CASTING PROCESS - [rubber ind] procédé *m* de moulage utilisant un coagulant
~DIPPING RUBBER - [rubber ind] immersion *f* dans un bain de latex contenant un coagulant
COAGULATE, to - [gen] coaguler
COAGULATING BATH - [ind chem] bain *m* coagulant
~TANK - [ind chem] cuve *f* de coagulation
COAGULATION - [chem] (chemical change inducing transition from a liquid to a semisolid or gel-like state) coagulation *f*
~BASIN - s. coagulating tank
~VALUE - [chem] (the concentration of a coagulant) indice *m* de coagulation
COAGULATIVE - [chem] coagulateur
COAGULATOR - [chem] coagulant *m*
COAGULOMETER - [instr] (apparatus measuring the time required for the coagulation of a fluid, particularly blood) coagulomètre *m*
COAGULUM - [chem] coagulum *m*
COAL, to - [gen] charbonner
COAL - [min] (combustible carbonaceous materia formed from the remains of prehistoric plants) charbon *m*
~BARGE - [naut] chaland *m* à charbon, péniche *f* à charbon, charbonnière *f*
~BASIN - s. coal bed
~BED - [geol] bassin *m* houiller
~-BLASTING - [mining] tir *m* au charbon
~BOX - [mech] (of a forge) garde-frasil *m*
~BRASS - [min] inclusion *f* de pyrite dans la houil
~BREAKER - [mech] broyeur *m* à charbon
~BRICK - [min] briquette *f*
~BRIQUETTE - s. coal brick
~BUCKET - [railw] benne *f* à charbon
~BUNKER - [railw] soute *f* à charbon
[ind] silo *m*, trémie *f* à charbon
~BURNING - [th eng] chauffe *f* au charbon
~CAR - [transp] chariot *m* à charbon
~CARBONIZATION - [ind chem] (conversion of coa to coke, tars, gases and other products, by heating processes) carbonisation *f*, distillation *f*
~CHARGE - [th eng] charge *f* de charbon
~CHARGING CAR - [th eng] chariot *m* à charbon
~CHUTE - [mining] (in a mine) cheminée *f* à charbon
[th eng] couloir *m* à charbon
~CLASSIFICATION - [min] classification *f* de charb
~CLASSIFICATION BY SIZE - [min] granulométrie des charbons
~CONVEYOR - [mech] transporteur *m* à charbon
~CRACKER - s. coal breaker
~CRUSHER - s. coal breaker
~CUTTER - [mech] haveuse *f*
[transp] (coal barge) chaland *m* à charbon
~CUTTING MACHINE - s. coal cutter
~DEPOSIT - [min] gisement *m* houiller
~DISCHARGE TRESTLE - [th eng] estacade *f* de déchargement
~DRAWING - [mining] extraction *f* du charbon
~DRESSING - [th eng] conditionnement *m* du charbc préparation *f* du charbon (mécanique)
~DRIFT - [mining] fendue *f*
~DUMP - [gen] silo *m*, tas *m* de charbon
~DUST - [gen] poussière *f* de charbon, poussière *f* de houille
~DUST FURNACE - s. coal furnace
~ELECTRODE - [el] électrode *f* de carbone
~EXTRACTION - [mining] extraction *f* du charbon
~FACE - [mining] front *m* de taille (du charbon)
~FEED - [th eng] dispositif *m* de chargement du sas à charbon
~FEED CHAMBER - [th eng] sas *m* à charbon
~FEED HOPPER - [th eng] trémie *f* intermédiaire
~FIELD - s. coal bed
~FIRING - [mech] chauffe *f* au charbon
~FURNACE - [metall] four *m* à charbon pulvérisé
~GAS - [ind chem] (a mixture of combustible gase obtained by the destructive distillation of coal) gaz *m* de houille
~GASIFICATION - [gas ind] (treatment of coal to obtain industrial gases) gazéification *f* du charbo
~GETTING - [mining] abattage *m* du charbon

AL GRAB - [railw] benne *f* à charbon
RINDING - [th eng] pulvérisation *f* du charbon, royage *m* du charbon
GRIT - [geol] grès *m* houiller
AMMER - [impl] marteau *m* à charbon
HANDLING PLANT - [railw] installation *f* de nanutention du charbon
EAVER - [mech] élévateur *m* à charbon
OLD - [naut] soute *f* à charbon
OPPER - [fuel] trémie *f* à charbon
YDROGENATION - [ind chem] hydrogénation *f* du harbon
EVELER - [mining] machine *f* à tasser le charbon
IGHTER - s. coal barge
OCK CHAMBER - [th eng] sas *m* à charbon
IEASURES - [min] couches *fpl* de charbon
IILL - [min] broyeur *m* à charbon
IINE - [min] mine *f* de charbon, houillère *f*
IINING - [mining] exploitation *f* du charbon ou de la houille), charbonnage *m*
IAPHTHA - [oil ind] huile *f* de houille
IL - [min] huile *f* lourde de roche, naphte *m* ninéral
ICK - [mining] pic *m* à la veine (ou au charbon)
IPE - [mining] veine *f* irrégulière de charbon
PIT - [mining] puits *m* houiller, fosse *f*
REPARATION - [th eng] conditionnement *m* du harbon, préparation *f* du charbon
ULVERIZER - [mech] broyeur *m* à charbon
OAD - [mining] galerie *f* au charbon
CREEN - [impl] crible *m* à charbon, tamis *m* à harbon
CUTTLE - [impl] seau *m* à charbon
EAM - [min] veine *f*, couche *f*, filon *m*, isement *m* houiller
HED - [gen] hangar *m* à charbon
IZING - [min] granulométrie *f* dû charbon
SLACK - [mining] charbon *m* menu
TORAGE - [gen] dépôt *m* de charbon
TORAGE GROUND - [gen] parc *m* à charbon *f*
TORAGE YARD - s. coal storage ground
TORE - s. coal storage ground
AR - [ind chem] (black viscous liquid obtained by the destructive distillation of coal; an important raw material for pharmaceuticals, dyes, solvents, etc) goudron *m* de houille, coaltar *m*
AR OIL - [ind chem] huile *f* de goudron
AR PITCH - [ind chem] brai *m* de huille, brai *m* de goudron, poix *f* de houille
AR SOAP - [ind chem] savon *m* coaltar
ALVE - [fuel] robinet *m* à charbon
VEIN - s. coal bed
WALL - [mining] front *m* de taille d'une bouillère
WASHERY - [min] laverie *f* à charbon
WATER GAS - [gas ind] (a mixture of water-gas and carbonization gas obtained in a single operation from coal or lignite) gaz *m* double
WEIGHING MACHINE - [impl] bascule *f* à charbon
OALER - [naut] navire *m* charbonnier
OALESCE, to - [phys etc] (to merge together into a single whole) s'associer, se combiner, s'unir, se fondre
OALESCENCE - [phys] coalescence *f*, combinaison *f*, fusion *f*
[bot & zool] (growing together) coalescence *f*
OALESCENT - [bot & zool] coalescent
OALIFICATION - [min] carbonification *f*

COALING - [mining] charbonnage *m*
~DOOR - [th eng] gueulard *m*, porte *f* du foyer
~STAGE - [railw] rampe *f* à charbon
~STATION - [naut] port *m* à charbon
COALITE - [min] (trade name for a type of smokeless fuel; also called semi-coke) coalite *f*, semi-coke *m*
COAMING - [naut] hiloire *f*
[build] bord *m* exhaussé
COANDA EFFECT - [phys] (the tendency of a jet of gas to follow the contour of a surface near which it is discharged) effet *m* Coanda
COARCTATE - [zool] coarcté
COARCTATION - [med] constriction *f*, coarctation *f*
COARSE - brut
[min] brut
[mech] grossier
[text] rude, grossier
~ADJUSTMENT - [instr] (device which gives only an approximate setting) réglage *m* approximatif
~AGGREGATE - [constr] (gravel or crushed stone used with concrete) gros agrégat *m*
~CONTROL - [control] réglage *m* approximatif
~COPPER - [metall] cuivre *m* brut, cuivre *m* noir
~COUNT SPINNING - [text] filature *f* de gros numéros
~FEED - [mech] grande avance *f*
~-FINE ACTION - [control] action *f* approche-précision
~-FILE RELAY - [control] (an electric relay used to operate the amplifier of a remote-position-control servo-mechanism from a coarse data transmission) relais *m* approche-précision
~FLOCCULENT - [brew ind] à gros flocons
~GRAINDED -[metall] à gros grain, à grain grossier
~GRAINED WHEEL - [impl] meule *f* à gros grain
~GRINDING - [mech] dégrossissage *m* à la meule, broyage *m* grossier
~HAIR SHEEP - [zool] mouton *m* à laine commune
~HECKLE - [text] peigne *m*
~METAL - [metall] matte *f*
~PITCH - [aero] grand pas *m*
~PLASTER OF PARIS - [build] plâtre *m* de hourdis
~QUALITY - [gen] qualité *f* inférieure
~RADIOLOCATION - [radar] radio-repérage *m* approximatif
~REGULATING ROD - [nucl] (control rod in a nuclear reactor, suitable for coarse control) barre *f* de réglage grossier
~SALT - [min] gros sel *m*
~SAND - [geol] sable *m* grossier
~SCANNING - [telev] (the scanning of an image by using a light spot of a comparatively large diameter) balayage *m* approximatif
~SCREEN - [min] tamis *m* à larges mailles
~SILT - [geol] (sediments with individual grains of a diameter of not less than 0,05mm) dépôt *m* grossier, boue *f*
~SKINNED - [agric] à peau épaisse
~SLUDGE - [brew ind] coagulation *f* à chaud
~STUFF - [build] (a mixture of mortar and hair forming a first coat for walls) mortier *m* de chaux et de fibre
~TEETH - [mech] dents *fpl* non dégrossies
~TEXTURE - [metall] texture *f* à gros grain
~TUNING - [radio] réglage *m* approximatif
~VACUUM - [phys] vide *m* grossier, vide prélimi-

naire, vide primaire
COARSE WOOLHAIR - [text] poil *m* mal relevé
~YARN - [text] fil *m* grossier
COARSELY GROUND MALT - [brew ind] malt *m* broyé gros
COARSENESS - [gen] grossièreté *f*, grosseur *f*, gros grain *m*
[gen] manque *f* de précision
[ind chem etc] granulation *f*
COAST, to - [aero & astronaut] (to travel by inertia, without the use of propelling power) se déplacer par inertie
[aero] descendre en vol plané
[naut] suivre la côte, caboter
[auto] marcher *m* moteur débrayé, couper l'allumage
COAST - [geogr] côte *f*
[naut] (a point of the compass) aire *f*
~-LINE - [geogr] côte *f*, littoral *m*, ligne *f* de côte
~STATION - [radio] (radiostation on or near the coast) station *f* cotière
COASTAL - [geogr] côtier
~REFRACTION - [radio] (refraction to which a radio wave is subject in passing a coastline) réfraction *f* d'une onde sur la côte
COASTER - [naut] caboteur *m*
~BRAKE - [mech] (in bicycles) frein *m* à rétropédalage
COASTING - [naut] cabotage *m*, navigation *f* en vue des côtes
[aero] vol *m* plané
[auto] descente *f* en roue libre
[geogr] littoral *m*
~FLIGHT - [astronaut] vol *m* balistique
~NAVIGATION - [naut] cabotage *m*, navigation *f* en vue des côtes
~TRADE - [naut] cabotage *m*
COAT, to - [gen] enduire, couvrir, revêtir, appliquer (une couche)
COAT - [text] habit *m*, pardessus *m*
[ind] couche *f*, enduit *m*, revêtement *m*
[paints] couche *f*, application *f*
[constr] (in road construction) enduit *m*, revétement *m*
[bot] enveloppe *f*, tunique *f*, peau *m*
[naut] (the canvas round the mast in the section near the deck) braie *f*
[ind chem] dépôt *m*
[zool] (of animals) robe *f*, pelage *m*
[med] paroi *f*
~HANGER HOOK - [auto] crochet *m* porte-vêtement
~LINING - [text] doublure *f* de manteau
~OF LACQUER - [auto] couche *f* de peinture, couche *f* de vernis
~WITH NICKEL, to - [metall] nickeler
COATED - [gen & ind] recouvert, enduit, enrobé
~ABRASIVE - [tools] abrasif *m* rapporté
~BOARD - [paper man] carton *m* glacé
~ELECTRODE - [el] électrode *f* enrobée
~FABRIC - [plast ind] (a fabric to which a coating of plastic or natural or synthetic rubber has been applied) tissu *m* caoutchouté, tissu *m* enduit
~LENS - [opt] (lenses which are coated with a low reflectance film) objectif *m* traité, objectif *m* bleuté
~MAGNETIC TAPE - [el acoust] bande *f* à couche magnétique

COATED PAPER - [paper man] papier *m* glacé, papier *m* enduit (encollé)
~SIDE - [paper man] face *f* émulsionnée
~TAPE - [acoust] (coating of a powdered ferromagnetic material on a non magnetic base for a tape) bande *f* à couche d'oxyde
COATING - [gen] revêtement *m*
[text] étoffe *f* pour paletot, tissu *m* pour paletot
[text] (fabric) enduction *f* (tissu)
[ind chem] (the application of a coating compound to a material) couche *f*, enrobage *m*
[chem] (composition of natural or synthetic resins or rubbers for the coating or impregnation of cloth, paper, wood etc) revêtement *m*, enduit *m*, enduction *f*
[paints] couche *f*
[paper man] glacis *m*
[el] (the metallic sheets or films used as plates of a condenser) armature *f*
[electron](a layer of special material imposed on the cathode of an electron tube to improve its emission qualities) enrobage *m*
~ACTIVATION - [electron] (the operation of giving an emitter the best possible emittivity) activation *f* de l'enrobement (ou de revêtement)
~CONDUCTIVITY - [electron] (the conductivity of a cathode coating) conductivité *f* de l'enrobement
~FILM - [plast ind] (plastic film used in the production of a variety of laminates) pellicule *f* protectrice
~FINISH - [text] apprêt *m* de couverture
~HEAD - [photo] (component part of the film coating machine) tête *f* d'application de l'émulsion
~KNIFE - [plast ind] (a doctor knife; a blade used to spread coating material) couteau *m* de peintre
~MACHINE - [mech] machine *f* à revêtement
~MATERIALS - [ind chem] (compositions based on natural or synthetic resins used to impart special properties to a base material, such as fabric, paper, wood etc) matériaux *mpl* de revêtement
~MIXTURE - [ind chem] masse *f* de revêtement
~PAN - [impl] bac *m* de revêtement
~PAPER - [paper man] papier *m* à glacer
~PISTOL - [impl] pistolet *m* pneumatique
~ROLLER - [impl] rouleau *m* d'enduction
~SIDE - [photo] côte *m* émulsion
~TESTER - [el] appareil *m* de contrôle de revêtement
COAXIAL - [geom] coaxial
~AERIAL - [radio] antenne *f* à feeder coaxial
~ANTENNA - s. coaxial aerial
~-CABLE - [radio & telev] (type of cable for the transmission of radio frequency power consisting of concentric inner and outer conductors insulated from one another) câble *m* coaxial
~CIRCUIT - [telecomm] circuit *m* coaxial
~ELECTRODE SYSTEM - [el] (an electrode system which is geometrically coaxial but electrical unsymmetrical) système *m* d'électrodes concentrique
~ELECTROMAGNETIC BRAKE - [mech] frein *m* électromagnétique coaxial
~FEEDER - s. coaxial cable
~GEAR - [mech] transmission *f* coaxiale
~LINE - [el] (a transmission line in which one conductor surrounds the other completely) câble *m* coaxial
~LINE OSCILLATOR - [el] oscillateur *m* à lignes

:oaxiales
OAXIAL PAIR - [electron] paire *f* coaxiale
PROPELLERS - [aero] (propellers rotating about the same centre-line, having separate drive running in opposite senses) hélices *fpl* coaxiales
ROTORS - [mech] rotors *mpl* coaxiaux
STUB - [electron] (in waveguides) a length of non-dissipative coaxial line) téton *m* coaxial
SWITCH - [el] (a type of switch having constant impedance characteristics) interrupteur *m* coaxial
TRANSMISSION LINE - [electron] (in waveguides; line consisting of two coaxial cylindrical conductors) ligne *f* coaxiale
OB - [build] brique *f* crue
COAL - [min] gaillette *f*
WALL - [build] (a wall built of clay and straw with lime and earth) mur *m* en torchis (on en pisé)
-WEB - [zool] toile *f* d'araignée
OBALT - [chem] (an element, symbol Co. A greyish, hard, ductile, magnetic metal used in the production of alloys and steel, catalysts, paint additives and pigments) cobalt *m*
60 - [nucl] (a radio-active isotope of cobalt used as a trace element, X-ray source, and in radiation therapy) cobalt *m* 60
BLOOM - [min] (natural cobalt arsenate, crystallizing in the monoclinic system) érythrite *f*
BLUE - [chem] (a blue pigment consisting of mixtures of alumina and cobalt oxide) bleu *m* de cobalt
CARBIDE - [chem] (a catalyst used in the Fischer-Tropsch synthesis of hydrocarbons) carbure *m* de cobalt
DRIERS - [paint] (used for their powerful drying action) siccatifs *mpl* au cobalt
GLANCE- [min] (also called cobaltite; a natural sulphide and arsenide of cobalt; an important ore of cobalt) cobaltine *f*, cobalt *m* gris
GLASS - [chem] verre *m* de cobalt, bleu *m* de Saxe
GREEN - [paint] (pigment obtained by high-temperature calcination, chiefly consisting of oxides of zinc and cobalt) vert *m* de cobalt
MORDANT - [chem] mordant *m* de cobalt
OXIDE - [chem] (used to produce a deep-blue colour in glass) sesquioxyde *m* de cobalt
-PLATING - [chem] cobaltage *m*
POTASSIUM NITRITE - [chem] (a yellow pigment for glass and ceramics) nitrite *m* cobalti-potassique
STEEL - [metall] acier *m* au cobalt
TETRACARBONYL - [chem] (a catalyst and an antiknock additive for petrol) cobalt-carbonyle *m*
TRIFLUORIDE - [chem] (fluorinating agent) tétrafluorure *m* de cobalt
ULTRAMARINE - [paint] (bright blue pigment consisting primarily of oxides of cobalt and aluminium) bleu *m* d'azur de cobalt
VIOLET - [paint] (a term used generically for cobalt phosphates, cobalt arsenite, cobalt ammonium arsenate etc and for mixtures of these compounds) violet *m* de cobalt
YELLOW - [paint] jaune *m* de cobalt
:OBALTIC HYDROXIDE - [chem] (a starting material in the production of cobalt salts) hydroxyde *m* de cobalt
~OXIDE - [chem] (used in ceramic glazes and as a pigment) sesquioxyde *m* de cobalt
COBALTITE - s. cobalt glace
COBALTOUS ACETATE - [chem] (a catalyst and an ingredient of varnish driers) acétate *m* cobalteux, acétate *m* de cobalt
~AMMONIUM SULPHATE - [chem] (a catalyst and a pigment for ceramic glazes) sulfate *m* d'ammonium cobalteux
~ARSENATE - [chem] (a pigment for glass) arséniate *m* cobalteux
~CARBONATE - [chem] carbonate *m* de cobalt
~CHLORIDE - [chem] (dyeing mordant, trace element in animal feeds, catalyst and absorbent for ammonia) chlorure *m* cobalteux, protochlorure *m* de cobalt
~CHROMATE - [chem] (a pigment for ceramics) chromate *m* cobalteux
~HYDROXIDE - [chem] (intermediate in the production of driers for surface coatings) hydroxyde *m* cobalteux
~LINOLEATE - [chem] (a drier for surface coatings) linoléate *m* de cobalt
~NAPHTENATE - [chem] (drier for paints and varnishes) naphténate *m* cobalteux
~NITRATE - [chem] (ceramics pigment, feed additive, starting material for pigments and catalyst) nitrate *m* cobalteux, protonitrate *m* de cobalt
~OLEATE - [chem] (drier for surface coatings) oléate *m* cobalteux
~OXALATE - [chem] (starting material in the production of catalysts)oxalate *m* cobalteux
~OXIDE - [chem] (a catalyst, pigment, feed additive, and constituent of colours for glass and ceramics) oxyde *m* cobalteux
~PHOSPHATE - [chem] (pigment for glass and ceramics) phosphate *m* cobalteux, rose *m* de cobalt
~RESINATE - [chem] (drier for paints and varnishes) résinate *m* cobalteux
~SULPHATE - [chem] (a catalyst, soil and feed additive, component of plating baths and pigment for ceramics) sulfate *m* cobalteux
~TUNGSTATE - [chem] (an anti-knock additive and pigment and drier for surface coatings) tungstate *m* cobalteux
COBBED ORE - [min] mineral *m* scheidé
COBBLE - [gen] galet *m*, cailloux *m*
[naut] barque *f* de pêche
COBBLER - [gen] cordonnier *m*
COBBLER'S WAX - [ind chem] poix *f* de cordonnier
COBBLES - [geol] moellons *mpl*, galets *mpl*, cailloux *mpl* roulés
COBBLESTONE - [gen] pavé *m*
~PAVEMENT - [road constr] chaussée *f* en pavés arrondis
~ROAD - [constr] route *f* pavée (en pavés arrondis)
COBBLING - [gen] cordonnerie *f*
[text] (redyeing fabric to perfect their shade) teinture *f* de finissage
COBS - [radar] (distortion of image due to frequency modulation interference) distortion *f* d'image en forme de cloche (causée par un brouillage de modulation de fréquence)
COBWEBBING - [paint] (defect of a spraying composition which causes it to form individual web-like threads when ejected from the spraying gun) formation *f* de filaments
[packag] (a protective cover) emballage *m* "cocon"
COBWORK - [build] construction *f* en torchis (on en

pisé)
COCA - [bot] coca *f*
[ind chem] (the dried leaves of the coca, a source of cocaine) coca *f*
~LEAVES - [bot] feuilles *f*pl de coca
COCAINE - [chem] (an alkaloid obtained from species of erythroxylon and used as a local anaesthetic in medicine) cocaïne *f*
COCCID - [bot] (a weed) cochenillier *m*
COCCUS - [bact] (a minute spherical bacterium) coccus *m*
COCHINCHINA WAX - [chem] beurre *m* d'Irvingia
COCHINEAL - [chem] (the desiccated bodies of the female insects of the coccus cacti) cochenille *f*
~CACTUS - [bot] cochenillier *m*, nopal *m*
~CARMINE - [chem] (a pigment)carmin *m* de cochenille
COCK, to - [firearms] armer (le chien)
[photo] armer
[mech] être non-aligné
[agric] mettre en meule (le foin)
COCK - [zool] coq *m*
[agric] meule *f* de foin
[firearms] chien *m*
[hydr] (a water valve) robinet *m*
[horol] (pivot carrier) pont *m* de balancier
[mech] aiguille *f* de balance
~BODY - [hydr] corps *m* de robinet
~-DISK - [mech](disk or pad of moderately elastic material to obturate element of small hand-operated valve) garniture *f* de robinet
~METAL - [metall] bronze *m* pour robinetterie
~-PLUG HEAD - [gas ind] (in gas meters) carré *m* de manoeuvre
~-UP - [print] (a large initial extending above the first line) lettre *f* supérieure, lettre *f* initiale
COCKCHAFER - [surv] (triangle of error enclosed by three position lines laid down on a chart) chapeau *m*
COCKEREL - [zool] jeune coq *m*, cochelet *m*
COCKING - [mech] non-alignement *m*
[photo] armement *m*
~KNOB - [photo] armement *m*, bouton *m* d'armement
~RING CATCH - [mech] levier *m* de verrouillage
COCKLE, to - [text] onduler, friser, crêper
[photo] incurver, s'incurver, cintrer
COCKLE - [ceramics] four *m*
[metall] ondulation *f*
~STAIR - [build] (a winding stair) escalier *m* en colimaçon
COCKPIT - [aero] (the compartment for the pilot) cockpit *m*, carlingue *f*
[shipbuild] cockpit *m*, habitacle *m*
[agric mach] cabine *f* de commande
~COWLING - [aero] (protective structure over a cockpit) capot *m* de carlingue
~ENCLOSURE - [aero] dôme *m* d'habitacle
COCKROACH - [zool] blatte *f*, cancrelas *m*, cafard *m*
COCKS HEAD - [bot] sainfoin *m* des prés, éparcet *m*
COCO-MATTING - [text] tapis *m* de coco
COCOA - s. cacao
~MOTH - [zool] teigne *f* du cacao
~PALM - [bot] cocotier *m*
~PALM BUTTER - [ind chem] beurre *m* de coco
~-PLUM - [bot] icaque *f*
~POWDER - [expl] poudre *f* chocolat

COCONUT ACID - [chem] (fatty acids mixture derived from coconut oil and used in soap manufacture and detergents) acide *m* retiré de la noix de coco
~FIBRE - [bot] fibre *f* de coco
~MILK - [food] lait *m* de coco
~OIL - [ind chem] huile *f* de coprah, huile *f* de co
~PALM - [bot] cocotier *m*
COCOON, to - [gen] coconner, filer son cocon
COCOON - [zool] cocon *m*
~BREEDING - [zool] élevage *m* de cocons
~CROP - [text] récolte *f* des cocons
~FILAMENT - [text] filament *m* de cocon
~REELING - [text] dévidage *m* des cocons
COCOONED - [packag] sous emballage "cocon", sous cocon
COCOONING - [packag] (the process of enclosing an article in a closed cell of sheet material by a spraying process for protection) emballage *m* "cocon", mise *f* sous cocon
[text] (of silk worm) coconnage *m* (de la soie)
COCOONIZATION - s. cocooning
CO-CHANNEL INTERFERENCE - [radio] (interferenc in a channel caused by a transmitter operating in the same channel) brouillage *m* sur le même canal
CO-CONDENSATION - [chem] co-condensation *f*
CO-CURRENT FLOW - [plast] (the parallel flow in the same direction of two streams within a systen écoulement *m* parallèle
COCTION - [med] coction *f*
COD - [zool] cabillot *m*, morue *f*
[metall] (a sand projection in a mould) projectior *f* de sable
~LIVER OIL - [chem] (pale yellow oil obtained from the livers of various species of codfish. Used in medicine as a source of vitamine A and D and as a leather dressing) huile *f* de foie de morue
CODAN - [telecomm] éliminateur *m* de bruit à onde porteuse pilote
CODE, to - [comput] (or to encode; the expression of information in symbols for the computer) coder
CODE - [gen] code *m*
[telecomm] (system of signals corresponding to units of information) code *m*
[comput] (the system of symbols in a computer and the rules for associating them) code *m*
~BEACON - [radio] (beacon transmitting known co signals by which it can be recognized) balise *f* à occultations codées
~CHARACTER - [radio] signal *m* télégraphique
~CHECK - [comput] vérification *f* de code
~CHECKING TIME - [comput] temps *m* de vérification de codage
~CONVERTER - [comput] (a device for the conversi of information from one code into another) convertisseur *m*
~ELEMENT - [telecomm] (a single element of a code used in signalling) élément *m* de code
~LETTER - [telecomm] indicatif *m* littéral
~LIGHT - [radio] (a luminous beacon giving signals in Morse) balise *f* à occultations codées
~OF PRACTICE - [gen comm etc] code *m* profession règle *m* de conduite
~PULSE TRAIN - [radio] (modulated transmission pulses) impulsions *f*pl codées
~RINGING - [telecomm] appel *m* codé

CODE SELECTOR - [telecomm] sélecteur *m* de préfixe
~SIGN - [radar] signal *m* codé
~-TELEGRAMM - [telecomm] télégramme *m* chiffré
~TRANSLATION - [telecomm] décodage *m*
~TRANSMITTER - [telecomm] émetteur *m* télégraphique
CODECLINATION - [nav] complément *m* de la déclinaison
CODED DECIMAL - [comput] (notation by which every decimal digit is converted into a pattern of binary one and zero) code *m* décimal binaire
~-DECIMAL DIGIT - [comput] chiffre *m* en code décimal binaire
~PROGRAMME - [comput] programme *m* codé
CODEINE - [chem] (an alkaloid of the morphine group) codéine *f*
CODER - [telecomm] (a device for encoding information for transmission) modulateur *m*, changeur *m* de code
CODEX - [pharm] codex *m*
CODICIL - [leg] codicille *m*
CODIFY - [gen & leg] codifier
CODING - [telecomm] (the application of a code) codage *m*
~LINE - [comput] (a single instruction) instruction *f* codée
~PULSE MULTIPLE - [radar] codage *m* des impulsions
~SECTION - [comput] (series of instructions of a programme section) série *f* d'instructions codées
COEFFICIENT - [phys etc] (an expression denoting the degree to which a certain property is possessed by a body or substance) coefficient *f*, facteur *m* [math] (a numerical constant used as a multiplier to a variable quantity) coefficient *m*, facteur *m*
~OF ABSORPTION - [phys] coefficient *m* d'absorption
~OF ADHESION - [phys] coefficient *m* d'adhérence
~OF AMPLIFICATION - [electron] (the ratio of input to output in a given system, expressed in the same unit) coefficient *m* d'amplification
~OF ATTENUATION - [phys] facteur *m* d'affaiblissement
~OF COMPRESSIBILITY - [phys] coefficient *m* de compressibilité
~OF CONDENSATION - [phys] coefficient *m* de condensation
~OF CONSOLIDATION - [soil] coefficient *m* de consolidation
~OF CONTRACTION - [phys] (the ratio of the smallest sectional area of the jet of liquid issuing from an orifice to the area of the orifice) coefficient *m* de contraction
~OF CORRECTION - [phys] coefficient *m* de correction
~OF COUPLING - [radio] (the relation between the mutual impedance elements of two coupled circuits to the square root of the product of their total impedance) coefficient *m* de couplage
~OF DETECTION - [radio] coefficient *m* de détection
~OF DIFFERENTIAL TONES - [telecomm] coefficient *m* de sons différentiels
~OF DIFFUSION - [phys] coefficient *m* de diffusion
~OF DISCHARGE - [mech] (the ratio of the discharge through an orifice to the discharge computed by assuming uniform flow) coefficient *m* d'écoulement
COEFFICIENT OF DISPLACEMENT - [shipbuild] coefficient *m* de déplacement
~OF EARTH PRESSURE AT REST - [soil] coefficient *m* de pression des terres au repos
~OF ELASTIC RECOVERY - [phys] coefficient *m* de détente élastique
~OF ELECTROLYTIC DISSOCIATION - [chem] (the relation between the number of molecules dissociated electrochemically to the total number of molecules in the solution in question) coefficient *m* de dissociation électrolytique
~OF ELONGATION - s. coefficient of linear expansion
~OF EQUIVALENCE - [metall] (used in the conversion of quantities of aluminium, iron and manganese into equivalent quantities of zinc, in relation to their effect on the constitution of brass) coefficient *m* d'équivalence
~OF EXPANSION - [phys] (the fractional expansion per degree rise of temperature) coefficient *m* de dilatation
~OF EXCELLENTE - [gen] coefficient *m* de qualité
~OF FINENESS OF WATER-PLANE - [shipbuild] (the area of a ship's load water plane divided by the product of breadth and length) coefficient *m* de finesse du plan de flottaison
~OF FRICTION - [mech] coefficient *m* de frottement
~OF HARMONIC DISTORTION - [electron] coefficient *m* de distortion non-linéaire
~OF HEAT TRANSFER - [phys] coefficient *m* de transmission de chaleur
~OF HEAT TRANSMISSION - s. coefficient of heat transfer
~OF HYSTERESIS - [phys] coefficient *m* d'hystérésis
~OF IMPACT - [phys] coefficient *m* d'impact
~OF INDUCED MAGNETIZATION - [phys] coefficient *m* d'aimantation (susceptibilité magnétique)
~OF INDUCTIVE COUPLING BETWEEN TWO CIRCUITS [el] (the ratio of mutual inductance to the geometric mean of the two self-inductances) facteur *m* de couplage inductif de deux circuits
~OF INTERNAL FRICTION - [mech] coefficient *m* de frottement interne
~OF KINEMATIC VISCOSITY - [phys] (the value of the tangential force per unit area) coefficient *m* de viscosité cinématique
~OF KINETIC FRICTION - [mech] (the ratio of the tangential force required to sustain motion to the normal force pressing the two surfaces together) coefficient *m* de frottement en mouvement
~OF LINEAR ABSORPTION - [phys] coefficient *m* d'absorption linéaire
~OF LINEAR EXPANSION - [phys] (the change in length of a material per unit length in temperature) coefficient *m* de dilatation linéaire
~OF MUTUAL INDUCTION - [el] coefficient *m* d'induction mutuelle
~OF NON-LINEAR DISTORTION - [radio] coefficient *m* de distortion non-linéaire
~OF OCCUPATION - [telecomm] coefficient *m* d'occupation
~OF PERMEABILITY - [phys] coefficient *m* de pérméabilité
~OF RECOMBINATION - [phys] (expressing the rate of recombination of ions in a gas) coefficient *m* de recombinaison
~OF REDUCTION - [metall] coefficient *m* de réduction

COEFFICIENT OF REFLECTION - [phys] (the ratio of the light reflected from a surface to that falling on it) coefficient *m* de réflexion
~OF RESISTANCE) [mech] coefficient *m* de résistance
~OF RESTITUTION - [nucl] (in a collision of two bodies involving particles) coefficient *m* de restitution
~OF RIGIDITY - s. coefficient transverse elasticity
~OF RUN-OFF - [hydr] coefficient *m* d'écoulement
~OF SAFETY - [mech] facteur *m* de sécurité
~OF SELF-INDUCTION - [el] (self inductance) inductance *f* propre, coefficient *m* d'induction propre
~OF SLIDING FRICTION - [mech] coefficient *m* de glissement
~OF STATIC FRICTION - [mech] (the ratio of the maximum tangential force required to cause motion to the normal force pressing the two surfaces together) coefficient *m* de frottement au repos
~OF SURFACE TENSION - [phys] (the force per unit length which seems to act across lines drawn in the surface of the fluid) coefficient *m* de tension superficielle
~OF THERMAL CONDUCTIVITY - [phys] (the time rate of heat conduction per unit area per degree rise of temperature) coefficient *m* de conductivité calorifique
~OF TRANSVERSE ELASTICITY - [mech] (the elastic modules which corresponds to a shear stress on a pair of orthogonal planes) module *m* d'élasticité au cisaillement
~OF TWIST - [text] coefficient *m* de torsion
~OF UTILIZATION - [light] (ratio of the useful light to the total output) coefficient *m* d'utilisation
~OF VELOCITY - [mech] coefficient *m* de vitesse d'écoulement
~OF VISCOSITY - [phys] (the value of the tangential force per unit area) coefficient *m* de viscosité
~OF VOLUME ELASTICITY - [phys] module *m* de volume
~OF VOLUME EXPANSION - [soil] coefficient *m* de gonflement spécifique à la détente
~OF VOLUMETRIC EXPANSION - [phys] coefficient *m* de dilatation cubique
~OF VULCANIZATION - [rubber ind] coefficient *m* de vulcanisation
~POTENTIOMETER - [instr] (used in circuitry of computers to adjust coefficient) potentiomètre *m* de réglage des coefficients
COELIAC - [med] coeliaque
COELOSTAT - [instr] (an instrument designed to reflect continuously the same region of the sky into the field of a telescope) coelostat *m*
COENURE - [zool] cénure *m*
COERCIMETER - [instr] (instrument measuring the coercive force) coercimètre *m*
COERCIVE - [gen] coercitif
~FORCE - [el] (the magneitizing force required to eliminate the residual magnetism of a substance) force *f* coercitive
COERCITIVITY - [el] (static magnetic storage) coercivité *f*
COENZYME - [chem] (an organic compound which in combination with a protein, can form an enzyme system) co-enzyme *m*
COFFEE - [gen] café *m*
~BEAN - [bot] grain *m* de café
COFFEE BERRIES OIL - [ind chem] huile *f* de café
~DRYER - [mech] machine *f* à sécher le café
~GRADING - [ind proc] (the classification of coffee classification *f* du café
~GRINDER - s. coffee mill
~MILL - [impl] moulin *m* à café
~PLANTATION - [agric] plantation *f* de café
~WASHER - [mech] machine *f* à laver le café
COFFER - [gen] coffre *m*
[arch] caisson *m*
[mining] coffrage *m*, revêtement *m* d'un puits
[hydr] (a lock chamber) bassin *m*, sas *m* d'écluse
COFFERDAM - [build] (a temporary wall to keep water out of a site) cofferdam *m*, bâtardeau *m*
COFFERING - [min] coffrage *m*
COFFIN - [gen] cercueil *m*
[nucl] (colloq for a thick-walled container, usuall of lead, used for the transport of radioactive material) château *m* de plomb
COFFINITE - [min] (an important ore of uranium, essentially hydrated uranium silicate) coffinite *f*
COG, to - [carp] tenonner, caler
[metall] ébaucher le fer
[mech] denter, endenter
COG - [carp] adent *m*, tenon *m*
[mech] dent *f*, denture *f*
[a separate wooden tooth in gear wheel] dent *f* rapportée
[in bicycles] multiplication *f*
~-WHEEL - [mech] (a wheel with cogs or teeth) pignon *m*, roue *f* dentée, roue *f* d'engrenage, rouage *m*
COGGED BLOOM - [metall] lingot *m* dégrossi (ou ébauché)
~V-BELT - [mech] (a V-belt formed with teeth on th inner surface) courroie *f* trapézoïdale à dents
COGGING - [metall] (the preparation of forging and rolling an ingot) ébauchage *m*
~MILL - [metall] laminoir *m* ébaucheur (ou à blooms
~ROLL - [metall] cylindre *m* ébaucheur
COHERE - [gen] adhérer, cohérer, s'agglomérer
COHERENT - [chem] (possessing the property of sticking together in masses without special treatment) cohérent
~INTERRUPTED WAVES - [radio] ondes *fpl* cohérentes interrompues
~OSCILLATOR - [radio] (having a fixed phase relationship to another reference oscillator) oscillateur *m* cohérent
~PULSE RADAR - [radar] radar *m* à impulsions synchronisées
~RADIATION - [radiat] radiation *f* cohérente
~SCATTERING - [nucl] (scattering of particles with definite phase relationship between incoming and scattered waves) diffusion *f* cohérente
COHERER - [radio] (earlier type of detector-magnetic waves) cohéreur *m*
~TYPE ACOUSTIC SHOCK REDUCER - [telecomm] cohéreur *m* protecteur
COHESION - [phys] (the attraction between the molecules of a fluid, which causes the formation of drops and films of liquids) cohésion *f*
~ENERGY - [phys] (the energy required to break up a solid or fluid into its constituent atoms or molecules) énergie *f* de cohésion
~PRESSURE - [phys] pression *f* de cohésion
COHESIONAL RESISTANCE-[phys] résistance *f* co-

hésive
:OHESIONLESS SOIL - [soil] sol *m* non cohérent, sol *m* pulvérulent
OHESIVE - [gen] susceptible de cohésion
:OHESIVENESS - [gen] cohésion *f*
.OHOBATE, to - [chem] cohober, enrichir par distillations successives
:OHOBATION - [chem] cohobation *f*
OIL, to - [gen & mech] enrouler, bobiner
[of a rope] lover, enrouler
:OIL - [gen & ind] rouleau *m*, bobine *f*, roue *f*
[el] bobine *f*, bobinage *m*, enroulement *m*
[of a pipe] serpentin *m*
[one ring of a spiral] spire *f*
[of rope] rouleau *m* de corde, glène *f* de filin
[el] (a group of turns connected in series in a distributed winding with no commutator) bobine *f* élémentaire
~AERIAL - [radio] cadre *m*
~CAPACITY - [el] capacité *f* propre de bobine
~CORE MATERIAL - [el] matériau *m* pour noyaux
~CRADLE - [mech] berceau *m* de rouleau
~DIAMETER - [mech etc] diamètre *m* d'une spire
~DOWN, to - [naut] lover
~DRIVE - [acoust] (transducer represented by a coil carrying an alternating current) transducteur *m* à enroulement
~FORMER - [el] carcasse *f*, mandrin *m* de bobine
~HEATING - [th eng] (a form of panel-heating) chauffage *m* par radiation à serpentin
~IGNITION - [auto] (high-tension supply for sparking plugs in automobiles) allumage *m* haute tension
~-LOADED CABLE - [telecomm] câble *m* pupinisé
~-LOADED CIRCUIT - [telecomm] circuit *m* pupinisé
~LOADING - [el] pupinisation *f*
~OF CABLE - [el] bobine *f* de câble, touret *m* de câble
~OF WIRE - [el] couronne *f* de câble
~PACK - [el] bloc *m* de bobinages
~PAN - [ind chem] cuite *f* à serpentines
~PITCH - [el] (the number of teeth which separate the slots in which the two sides of a cell are placed) pas *m* d'une bobine
~POT - [el] pot *m* pour bobinage
~SET - [el] circuit *m* de démarrage
~SPACING - [el] pas *m* de pupinisation
~SPRING - [mech] ressort *m* spiral, ressort *m* hélicoïdal
~SYSTEM - s. coil set
~TAP - [el] prise *f* d'enroulement
~TESTING - [electron] (high-frequency method for testing the quality of the vacuum in electronic tubes) passage *m* à la bobine, essai *m* à haute frequence
~TOUCHING - [mech] (condition of a spring under compression, when the adjacent coils are in mechanical contact) ressort *m* spiral comprimé
~WINDER - [el] bobineuse *f*
~WINDING - [mech] solénoïde *m*, bobinage *m* enroulement *m*
~WINDING MACHINE - [mech] bobineuse *f* automatique
~WITH INSULATED WIRE - [el] bobine *f* à fil isolé
COILED - [gen] bobiné, spiralé
~-COIL FILAMENT - [el] (spiral filament coiled into a further coil to reduce radiation losses) filament *m* bispiralé
COILED-COIL LAMP - [el] (electric filament lamp with a coiled-coil filament) lampe *f* à filament bispiralé
~LOOP - [text] maille *f* d'arrêt
~PIPE - [mech] serpentin *m*, tuyau *m* spiralé
~TUBE - [mech] serpentin *m*
~-UP MOLECULE - [chem] molécule *f* enroulée
COILER - [mech] bobineuse *f*
[text] (in cotton spinning; mechanism for the carding engine and designed to deliver the sliver in coil into the coiler cans) bobineuse *f*
~BOX - [text] (in cotton spinning) assembleur *m*
~CAN - [text] (a slowly rotating cylinder in which sliver is delivered in coils) pot *m* tournant
COILING - [gen el] enroulement *m*
~CAN - [text] pot *m*, pot-enrouleur *m*
COIN, to - [metall] battre monnaie, frapper de la monnaie
COIN - [gen] pièce *f* de monnaie
~BOX - [telecomm] dispositif *m* d'encaissement
~BOX RELAY - [telecomm] relais *m* de commande d'encaissement
~RADIO - [radio] récepteur *m* à prépaiment
~RELAY - s. coin box relay
~RETURN - s. coin relay
COINAGE - [ind] frappe *m* de la monnaie
~BRONZE - [metall] bronze *m* pour pièce de monnaie
COINCIDENCE - [gen] coïncidence *f*
[nucl] (counts occurring in two or more detectors simultaneously or almost simultaneously) coïncidence *f*
~ARRAY - [nucl] dispositif *m* à coïncidence
~CIRCUIT - [electron] (producing suitable output pulse only when each circuit reveives pulses simultaneously) circuit *m* à coïncidence
~CORRECTION - [nucl] correction *f* de temps mort
~COUNTER - [instr] compteur *m* à coïncidence
~COUNTING - [instr] comptage *m* par coïncidence
~-CURRENT SELECTION - [comput] sélection *f* par coïncidence de courants, sélection *f* de cellules
~GATE - [comput] (gate whose output is energized only when every input is in its specified state) circuit *m* ET
~RANGE FINDER - [instr] télemètre *m* à coïncidence
COINING - [metall] frappe *f* de la monnaie
COIR - [text ind] fibre *f* de coco, coir *m*, kair *m*
~MAT - [text] paillasson *m*, tapis-brosse *m*
COKE, to - cokéfier
COKE - [gas ind] (the carbonaceous residue remaining in the retort after the destructive distillation of coal) coke *m*
~BAGGER - [gen] ensacheur *m*
~BED - [metall] paillasse *f* de coke, première couche *f* du lit de fusion
~BENCH - [th eng] aire *f* de refroidissement
~BIN - [th eng] caisse *f* à coke
~BLAST FURNACE - [metall] haut-fourneau *m* au coke
~BOX - [th eng] s. coke bin
~BREAKER - [mech] (a special type of crushing machine designed to treat coke) concasseur *m* à coke
~BREEZE - [build] (small broken coke used in the manufacture of breeze concrete) grésillon *m*, poussier *m*, fraisil *m* de coke
~BUTTON - [th eng] bouton *m* de coke
~CAKE - [th eng] saumon *m* de coke, gateau *m* de

coke
COKE CHARGE - [th eng] charge *f* de coke
~DRUSS - [gas ind] coke *m* en petits morceaux
~DUST - [gas ind] poussier *m*
~EMPTYING PIT - [gas ind] fosse *f* à coke
~END - [gas ind] côte *m* de coke
~EXTRACTOR - [th eng] extracteur *m* de coke
~FORK - [th eng] fourche *f* à coke
~GRADING PLANT - [gas ind] installation *f* de triage de coke
~GRINDING PLANT - [mech] (a plant for sorting coke into specified sizes) installation *f* de classification du coke
~GROUND - [gas ind] parc *m* à coke
~GUIDE - [th eng] guide-coke *m*
~IRON - [metall] fonte *f* au coke
~MILL - [mech] (a machine for reducing coke to powder) broyeur *m* à coke
~OVEN - [gas ind] (a large type of oven in which high-temperature carbonization of hard-caking bituminous coal is carried out) four *m* à coke
~OVEN GAS - [gas ind] gaz *m* de cokerie
~OVEN PLANT - [gas ind] cokerie *f*
~PUSHING AND LEVELING MACHINE - [mech] déformeuse-repaleuse *f*
~QUENCHING - [gas ind] extinction *f* du coke
~QUENCHING CAGE - [gas ind] benne *f* d'extraction du coke
~QUENCHING SKIP - s. coke quenching cage
~QUENCHING TOWER - [gas ind] chevalet *m* d'extinction
~-QUENCHING WATER - [gas ind] eau *f* d'extinction du coke
~SCRUBBER - [gas ind] ruisseleur *m* à coke
~SIDE - [gas ind] côte *m* de coke
~SPLIT - [metall] charge *f* de coke
~STORAGE GROUND - [ind] parc *m* à coke
~YELD - [th eng] rendement *m* en coke
COKED-UP - [mech] (of an engine or other device in which carbon deposit has reached a level high enough to impair its functioning) encrassé, calaminé
COKES - [metall] (thin sheets, formerly made by coating wrought iron in a coke furnace) fer *m* blanc
COKING - [gas ind] (the process of converting coal into coke) cokéfaction *f*
[mech] (in motors and engines) calaminage *m*, encrassement *m*
~BLEND - [gas ind] pâte *f* à coke
~CHAMBER - [gas ind] cellule *f* de cokéfaction
~COAL - [gas ind] charbon *m* cokéfiant, charbon *m* à coke
~PLANT - [gas ind] cokerie *f*
~POWER - [gas ind] pouvoir cokéfiant
~PRESSURE - [gas ind] poussé *f* d'une charge
~TIME - [gas ind] durée *f* de distillation
COL - [met] (narrow corridor of comparatively low pressure between two anticyclones) col *m*
COLA NUT - [bot] noix *f* de cola
~TREE - [bot] cola *m*, kola *m*
COLANDER - [impl] passoire *f*
COLATITUDE - [astr] colatitude *f*
COLCHICINE - [chem] (an alkaloid used in medicine for the treatment of gout) colchicine *f*
COLCHOTAR - [chem] (iron oxide) colcotar *m*
COLD - [gen] froid *m*
[med] rhume *m*

COLD AIR DRYING - [agric] (of cereals etc) séchage *m* par air froid
~AIR MASS - [met] (relatively cold in respect to near-by masses) masse *f* d'air froid
~AREA - [nucl] (where there is negligible contact with radioactive chemicals or radiations) région *f* de faible radioactivité
~ASPHALT - [ind chem] émulsion *f* de bitume
~BEND TEST - [metall & plast ind] (method of measuring the flexibility at low temperatures) essai *m* de flexion à froid, essai *m* de pliage à froid
~-BLAST VALVE - [metall] (in a blast furnace) valve *f* à vent froid
~-BLEACHING - [paper man] blanchiment *m* à froid
~-BLOODED - [zool] (with a body temperature which depends on the ambient temperature) à sang froid
~BREAK - [brew ind] coagulation *f* à froid
~BRITTLE - [metall] cassant à froid
~CAP - [ind chem] baffle *m* à chapeau, baffle-chapeau *m*
~CASTING - [metall] coulée *f* à froid
~CATHODE - [electron] (electron-tube cathode emitting electrons without application of heat) cathode *f* froide
~-CATHODE DISCHARGE LAMP - [el] (an electric discharge lamp in which the cathode is not heated) lampe *f* à décharge à cathode froide
~CATHODE OSCILLOGRAPH - [instr] oscillographe *m* à cathode froide
~CATHODE RECTIFIER - [electron] (rectifier for low-frequency alternating currents) redresseur *m* à cathode froide
~CHAMBER MACHINE - [metall] (in die casting) machine *f* à chambre froide
~CHARGE - [metall] (of furnaces) charge *f* solide
~CHECK - [paint] fendillement *m* à froid
~CHISEL - [impl] (a chisel used for cutting metals) ciseau *m* à froid, burin *m* en pointe de diamant
~COINAGE - [metall] frappe *f* à froid
~COINING PRESS - [metall] presse *f* à frapper à froid
~CRACKING - [metall] (of a casting) tapure *f*
~CREEP - [rubber ind] fluage *m* à froid
~CURE - [plast ind] (the curing or hardening of synthetic resins at ambient temperatures) vulcanisation *f* à froid
~-CUT VARNISH - [paint] vernis *m* fait à froid
~CUTTER - [impl] tranche *f* à froid
~-CUTTING - [paint] (production of a solution by cold mixing) mélange *m* à froid
~-DRAW, to - [metall] étirer à froid
~-DRAWING - [metall] (the process of producing bars or wire without the application of heat to the material) étirage *m* à froid
~-DRAWN - [metall] étiré à froid
~-DYEING METHOD - [text] méthode *f* de teinture à froid
~ELECTRON - [nucl] electron *m* lent
~EMISSION - [electron] autoémission *f*
~EXTRUSION - [plast ind] extrusion *f* à froid
~EXTRUSION PLUNGER - [plast ind] piston *m* de forçage pour extrusion à froid
~FLEXIBILITY - [rubber & plast] flexibilité *f* à basse température, souplesse *f* à basse température
~FLOW - [plast ind] (the "unmoulding" of plastics material, due to static stress at ambient temperatures) fluage *m* à froid

COLD FLOW TESTER - [instr] (used to measure the tendency of a material to cold flow) appareil *m* de mesure du fluage à froid
~FORMING - [metall] forçage *m* à froid
~FRONT - [met] (discontinuity at the forward edge of an advancing cold air mass before which warmer air is displaced) front *m* froid
~GALVANIZING - [metall] galvanisation *f* à froid
~GRINDING PROCESS - [text] défibrage *m* à froid
~HAMMER, to - [metall] écrouir
~-HEAD, to - [metall] refouler à froid
~-HEADED - [metall] refoulé à froid
~-HEADING - [metall] (the formation of rivet-heads without heating) refoulement *m* à froid
~JUNCTION - [el] (of a thermocouple) coudure *f* froide
~JUNCTION COMPENSATION - [control] compensation *f* de soudure froide
~-LAP - [metall] (a fault in metal casting due to metal from different sprues failing to unite in meeting) repli *m*, tourbillon *m*
~-LIGHT MIRROR - [telev] miroir *m* à lumière froide
~LIQUID CHILLER - [ind chem] réfrigérant *m* à liquide
~MASHING - [brew ind] (to obtain a diastatic extract) extraction *f* à froid
~-MIRROR REFLECTOR - [telev] réflecteur *m* à revêtement dichroïque
~MOULDING - [plast ind] (a process by which an article is shaped in a mould under high pressure at ambient temperature) moulage *m* à froid
~NEUTRON - [nucl] (also called slow neutron; having low kinetic energy) neutron *m* lent
~OCCLUSION - [met] (having more conspicuous cold frontal than warm frontal characteristics) occlusion *f* froide
~PIT - [mining] puits *m* d'aérage
~POLYMERIZATION - [chem] (polimerization occurring at ambient temperatures) polymérisation *f* à froid
~-PRESS, to - [text] catir, satiner à froid
~PRESSING - [el acoust] (a defective section in a record, due to bad definition of track, the material having failed to reach an adequate temperature) pressage *m* à froid
~PRESSURE WELDING - [metall] soudage *m* par pression à froid
~PRODUCTION - [th eng] production *f* du froid
~REACTOR - [nucl] (without induced radioactivity) réacteur *m* non-empoisonné
~RECTIFYING - s. cold rolling
~RIVETING - [mech] rivetage *m* à froid
~TOLL, to - [metall] écrouir
~ROLLED - [metall] laminé à froid
~ROLLED SHEET - [metall] tôle *f* laminée à froid
~ROLLED STEEL - [metall] acier *m* laminé à froid
~-ROLLING - [metall] (the rolling of metal at near-ambient temperature) laminage *m* à froid, écrouissage *m*
~ROOM - [gen] chambre *f* froide
~RUBBER - [rubber ind] (synthetic rubber manufactured at a temperature not exceeding 10 centigrades) caoutchouc *m* froid
~SAW - [impl] scie *f* à froid
~SEPARATION UNIT - [oil ind] (natural gas installation) unité *f* de separation à froid
COLD SET - [impl] ciseau *m* à froid, tranche *f* à froid
~-SETTING ADHESIVE - [ind chem] (adhesives whose maximum strength is obtained without heat) colle *f* à froid, colle *f* durcissable à froid
~-SETTING GLUE - [ind chem] (synthetic resin hardening at ambient temperatures) colle *f* à froid
~-SETTING LACQUER - [paint] (synthetic resin lacquer drying at ambient temperatures) lacque *f* séchant à froid
~-SHORT - [metall] (brittle at normal temperatures) cassant à basse température
~-SHORT IRON - [metall] fer *m* cassant à basse température
~-SHORTNESS - [metall] fragilité *f* à basse température
~SHRINK FIT - [metall] (tolerance dimensions for shrinking parts together by cooling the internal parts rather than heating the external one) ajustage *m* par retrait
~SHUT - s. cold-lap
~SIZING - [text] encollage *m* à froid
~SLUDGE - [ind chem] turbidité *f* due au froid [brew ind] dépôt *m* dû au froid
~-SLUG WELL - [metall] puits *m* de coulée
~SPRAYING - [paint] pulvérisation *f* à froid
~STARTING - [mech] (of motors) démarrage *m* à froid
~STORAGE - [th eng] conservation *f* par le froid
~STORE - [th eng] chambre *f* frigorifique
~STRETCH - [plast ind] (artificial lengthening of extruded filaments to improve their tensile strength-étirage *m* à froid
~STRIP MILL - [metall] laminoir *m* à bandes pour froid
~TAR - [ind chem] émulsion *f* de bitume
~TEST - [mech etc] essai *m* à froid
~TIME - [metall] (the interval between heating times) temps *m* d'interruption
~TRAP - [nucl] piège *m* à condensation [ind chem] piège *m* refroidi, piège *m* réfrigérant
~-TREATING - [gen] traitement *m* à froid
~-VATTING METHOD - [ind chem] procédé *m* de dissolution à froid
~VULCANIZATION - [rubber ind] vulcanisation *f* à froid
~WAVE - [met] (rapid marked fall of temperature during the cold season) vague *f* de froid
~WELL - [plast ind] (the space which is left in an injection mould to trap the cold slug) canal *m* de retenue de la carotte
~-WORKING - [metall] travail *m* à froid, allure *f* froide
COLEMANITE - [min] (naturally occurring hydrated calcium borate, a source of boric acid and borates) colemanite *f*
COLEOPTERA - [zool] coléoptères *m*
COLEOPTEROID - [bot] (a seed or a fruit looking like a beetle) coléoptéroïde
COLEOPTILE - [bot] (in a seedling of grass, the first leaf) coléoptile *m*
COLEOPTOSIS - [med] coléoptose *f*, prolapsus *m* du vagin
COLEOSPASTIA - [med] vaginisme *m*
COLESEED - [bot] colza *m*, graine *f* de colzan
COLIBACILLOSIS - [vet] (an infection) colibacillose

f
COLIFORME - [med] cibriforme
COLINERGIC - [med] colinergique
COLLAGE - [agric] (in the wine industry) collage m
COLLAGEN - [biol] (a sclero-protein and a major constituent of bones and connective tissue) collagène m
COLLAGENOUS - [med] collagène m
COLLAPSE, to - [gen] s'affaisser, s'écrouler, s'effondrer
[build] s'écrouler, s'effondrer
[mech] (of collapsible objects) se replier
[chem] s'arrêter
[cin] (of a film) se casser
COLLAPSE - [gen] écroulement m
[med] (extreme prostration) collapsus m, prostration f
[build] écroulement m, effondrement m, affaissement m
[chem] arrêt m
[cin] (of a film) rupture f, cassure f
[oil ind] écrasement m d'un tubage
~OF LUNG - [med] affaissement m (de la plèvre du poumon)
~OF YEAST HEAD - [brew ind] chute f
~SINK - [geol] formé par effondrement
COLLAPSED - [gen] écroulé, effondré
[mech] (of flexible objects) replié, démonté
COLLAPSERS - [plast ind] (sheet metal guides closing the walls of blown tubing together before the nip rolls) barrettes fpl de pression
COLLAPSIBILITY - [metall] (friability of the cores, after casting) friabilité f
COLLAPSIBLE - [mech] pliant, démontable, repliable, télescopique, extensible
[gen] (of flexible objects) rentrable, rabattable
~AERIAL - s. collapsible mast
~BIT - [oil ind] trépan m à effacement
~CAMERA - [photo] chambre f pliante
~CONTAINER - [railw] caisse f basculante
~LENS - [photo] objectif m télescopique
~LOAD - [mech] effort m de compression axiale
~MANDREL - [mech] (designed to be collapsed for removal from the completed duct) mandrin m amovible
~MAST - [radio] (the mast of the aerial which can be easily dismounted) mât m démontable
~PROP - [mining] étançon m télescopique
~RACER - [text] dévidoir m à bras mobile
~REEL - [text] dévidoir m à bras tombant
~ROOF - [auto] toit m ouvrant
~RYCE - s. collapsible racer
~STAND WITH PIPE VICE - [mech] établi m démontable avec étau à tube
~TAP - [tool] taraud m à expansion
~VIEWFINDER - [photo] (folding viewfinder) viseur m pliant
COLLAPSOTHERAPY - [med] (the treatment of lung disease by injected air) collapsothérapie f
COLLAR - [gen] collier m, col m, collet m, frette f
[mech] anneau m d'arrêt, collier m, bague f, bride f, collerette f
[metall] (in rolling mills) cordon m
[mech] (of a valve spring) coupelle f de ressort de soupape; (of an axle) champignon m, talon m, (d'essieu)
[mining] cadre m de la surface, collet m, embase f
[metall] (a section of larger diameter which separates the grooves in rolls) frette f, collier m
[railw] (of axle) collet m
[arch] (around a column) collier m, collerette f
COLLAR BEAM - [carp] entrait m retroussé, faux entrait m
~BEARING - [mech] palier m à cannelures
~-BONE - [anat] clavicule f
~BUSTER - [mech] brise-manchon m
~COLLET - [mech] (in valves) manchon m d'arrêt
~FINDER - [oil ind] localisateur m de joint
~FLANGE - [mech] bride f à cornière, bride f à chapeau, bride f à ergot
~-HEADED SCREW - [mech] (a screw with a collar under the head) vis f à collier
~NUT - [mech] écrou m à collet
~OF A LENS - [photo] embase f d'un objectif
~RING FLANGE - [mech] s. collar flange
~TIE - [carp] s. collar beam
~VELVET - [text] velours m pour cols
~VORTEX - [mech] (the motion of a closed circular vortex line of concentrated vorticity) tourbillon m annulaire
COLLARED CASING - [oil ind] tube m manchonné
~PIN - [metall] goujon m à embase
COLLARGOL - [chem] collargol m
COLLATE, to - [gen & bookbind] collationner, assembler (les cahiers)
[comput] (the operation of combining two sequences of items of information) interclasser
COLLATE - [comput] interclassement m
COLLATERAL - [gen] collatéral, additionnel, accessoire, subsidiaire
~SERIES - [nucl] (radioactive decay series) branches fpl collatérales
COLLATING - [gen] mise f en ordre, assemblage m
[bookbing] collationnement m, collationnure f
~TABLE - [bookbind] table f d'assemblage
COLLATOR - [comput] (a machine with two card feeds, four card pockets and three stations where a card may be placed in order with regard to other cards) interclasseuse f
COLLECT, to - [gen] rassembler, réunir, collecter, recueillir
[mech] (dismantling instruction implying that removed parts are to be set aside in a group for re-use on reassembling) démonter selon les instruction d'assemblage
COLLECTING - [min] échantillonnage m, rassemblement m (de minéraux)
~BAND - [el] lame f collectrice
~DITCH - [agric] canal m collecteur
~DRAIN - [agric] collecteur m
~FLUE - [th eng] carneau m collecteur
~GRID - [electron] grille f d'arrêt
~GROOVE - [mech] canal m collecteur, gouttière f
~MAIN - [th eng] (coke ovens) barillet m
~[oil ind] conduite f principale
~PIPE - [mech] tube m collecteur
~RING - [el] bague f collectrice
~ROLLER - [text] rouleau m collecteur
~SYSTEM - [th eng] réseau m de collecte
~TANK - [hydr] puits m collecteur
~TROUGH - [hydr] rigole f collectrice, goulette f collectrice
~TUN - [brew ind] cuve f collectrice

COLLECTING VAT - [min] (used for ore-dressing) bac *m* collecteur
~VESSEL - s. collecting tun
~ZONE - [telev] (a frequency range in which an oscillator can be synchronized by a synchronization signal) plage *f* d'accrochage
COLLECTION - [gen] collecte *f*, rassemblement *m*, récupération *f*, collection *f*
COLLECTIVE - [gen] collectif
~PITCH - [aero] (in helicopters; blade pitch angle applied to all the blades of a rotor independently of their relative position) pas *m* commun, pas *m* collectif
~PITCH CONTROL - [aero] commande *f* de pas collectif
~PITCH LEVER - [aero] (in helicopters; control lever for the regulation of the collective pitch) levier *m* de commande de pas collectif
COLLECTIVIZATION - [agric] collectivisation *f*
COLLECTOR - [gen & comm] receveur *m*, encaisseur *m*
[mech] (set of slip rings on an electrical machine) bague *f*
[auto] collecteur *m*
[electron] (the electrode in a transistor or crystal diode in which the current is modified by the control signal) collecteur *m*
~CURRENT - [el] (in semi-conductors) courant *m* collecteur
~DRAIN - [hydr] drain *m* collecteur
~FUNNEL - [ind chem] entonnoir *m* collecteur
~JUNCTION- [electron] jonction *f* de collecteur
~PLATES - [el] (metal plates inserted on the cell lining to reduce the electric resistance between cell lining and current leads) joints *mpl* conducteurs, plaques *fpl* collectrices
~RING - [el] (one of the rings mounted on the rotor shaft from which they are insulated) bague *f* collectrice
~RING HUB - [el] manchon *m* de bague collectrice
~SHOE - [el] (a metal shoe used to maintain contact with the conductor rail) patin *m*, sabot *m*
~WELL - [hydr] puits *m* horizontal, puits *m* à drains horizontaux
COLLET - [mech] douille *f* de serrage, pince *f* américaine, bague *f*, collier *m*
~CHUCK - [mech] (a type of chuck in precision lathes etc in which the workpiece is held in a small split collar) mandrin *m*, pince *f* de serrage, fermeture *f* pince
~HAND-LEVER CHUCK - [mach tool] levier *m* pour commande à main, fermeture pince *f*
COLLETED HAIRSPRING - [horol] spiral *m* virolé
COLLICULUS - [zool] (a small prominence) petite éminence *f*, crête *f* urétrale
COLLIDE, to - [gen] se choquer, se rencontrer, se heurter
[auto] entrer en collision, heurter
[nucl] (of atoms) entrer en collision
[naut] aborder, entrer en collision
COLLIDING PARTICLE - [phys] (a particle colliding with another particle) particule *f* incidente
COLLIER - [naut] charbonnier *m*, bateau *m* charbonnier
COLLIER'S LUNG - [med] anthracose *f* pulmonaire
COLLIERY - [min] mine *f* de charbon, houillère *f*
COLLIGATIVE PROPERTY - [phys] (depending on the number rather than on the quality of the molecules) propriété *f* colligative
COLLIMATE, to - [gen & phys] aligner
COLLIMATED - [gen] aligné
COLLIMATION - [gen] collimation *f*
[phys] (the process by which a divergent beam of energy is converted into a parallel beam) collimation *f*
~ERROR - [surv] (error produced in levelling work) erreur *f* de collimation
~LINE - [surv] (the imaginary line through the centre of the object glass and the intersection of the cross-hairs in the diaphragm) ligne *f* de collimation
COLLIMATOR - [opt] collimateur *m*
COLLINEAR - [gen] en ligne droite
~ARRAY - [radio] (a series of dipoles so arranged as to form a single line) aérien *m* linéaire à plusieurs éléments
~DISTORTION - [photo] déformation *f* collinéaire
~EQUATION - [opt] équation *f* collinéaire
COLLINEATION - s. collimation
COLLIQUATION - [med] colliquation *f*
COLLIQUATIVE - [med] colliquatif
COLLISION - [gen & auto] collision *f*, choc *m*, heurt *m*
[of trains] tamponnement *m*, collision *f*
[naut] collision *f*, abordage *m*
[nucl] (close approach between two particles, or photons etc giving rise to an interchange of energy) choc *m*, collision *f*
[phys] (the removal of electrons from an atom through collision with another particle) choc *m*
~BULKHEAD - [shipbuild] cloison *f* d'abordage
~COEFFICIENT - [nucl] (the coefficient of restitution in a two-body collision involving particles) coefficient *m* de choc
~COURSE - [naut] route *f* de collision
~COURSE INDICATION - [radar] signalisation *f* de route de collision
~CROSS SECTION - [nucl] section *f* efficace de collision
~DENSITY - [nucl] (the number of collisions per unit volume per unit time) efficacité *f* neutronique
~EXCHANGE - [nucl] échange *m* par choc
~EXCITATION - [electron] (the excitation of a gas by collision with moving charged particles) excitation *f* par chocs
~FREQUENCY - [phys] (electromagnetic waves; the frequency of collision of free electron with surrounding molecules in the ionosphere) fréquence *f* de collision, fréquence *f* des chocs
~INTEGRAL - [phys] intégrale *f* de collision
~IONIZATION - [phys] (the ionization of atoms or molecules of a gas or a vapour by collision with other particles) ionisation *f* par chocs
~NUMBER - [phys] (the average number of collisions of an ion when it diffuses out of the cavity) nombre *m* de collisions
~OF THE FIRST KIND - [nucl] (the collision of an excited atom with an atom resulting in a transfer of energy) choc *m* de première espèce
~OF THE SECOND KIND - [nucl] (the collision of an excited atom with a slow particle) choc *m* de deuxième espèce
~PROBABILITY - [phys] probabilité *f* de collision
~WARNING - [radar] système *m* avertisseur anti-

collision
COLLOCHEMISTRY - [med] chimie *f* des colloïdes
COLLODION - [chem] (a solution of cellulose nitrate in alcohol and ether) collodion *m*
~COTTON - [ind chem] coton-collodion *m*
~FILAMENT - [text] fil *m* de collodion
~SILK - [text] rayonne *f* de nitrocellulose
COLLOID - [chem] (a substance easily entering the colloidal stage) colloïde *m*
~BALL-MILL - [ind chem] moulin *m* à colloïdes
~COAL - [chem] charbon *m* colloïdal
~EQUIVALENT - [phys] (the number of atoms sharing unit charge) équivalent *m* colloïdal
~GOITRE - [med] (enlargement of the thyroid gland due to accumulation of iodine-containing colloid) poitre *m* colloïdal
~MILL - [ind chem] (high-speed, low-clearance mill capable of reduction to particle sizes of less than 1,0 micron) moulin *m* à colloïdes
~RECTIFIER - [el] redresseur *m* à cathode
COLLOIDAL - [chem] colloïdal
~CHEMISTRY - [chem] chimie *f* des colloïdes
~CLAY - [geol] argile *f* colloïdale
~ELECTROLYTE - [el] (an electrolyte whose ions are of colloidal dimensions) électrolyte *m* colloïdal
~EQUIVALENT - [phys] (term denoting the number of molecules per unit electric charge) équivalent *m* colloïdal
~FILAMENT - [light] (metal filament with the use of colloidal substances) filament *m* colloïdal
~FUEL - [min] (mixture of fuel oil and finely ground coal) combustible *m* colloïdal
~GRAPHITE - [ind chem] (graphite used in the dispersion of lubricating oil) graphite *m* colloïdal
~METALS - [chem] métaux *mpl* colloïdaux
~PARTICLES - [phys] (electrically charged particles in the dispersing phase of a colloidal solution) particules *fpl* colloïdales
~SOLUTION - s. colloid
~STATE - [phys] (the level of subdivision of matter at which the size of the particles lies between the type of solution termed molecular and that prevailing in coarse suspension) état *m* colloïdal
~STRUCTURE - [chem] solution *f* colloïdale
~SYSTEM - [phys] (in a multiphase system at least one dispersed phase uniformly distributed through the dispersion medium) système *m* colloïdal
COLLOIDITY - [chem] colloïdité *f*
COLLOTYPE - [print] collotypie *f*, phototypie *f*
COLLUTORY - [med] collutoire *m*
COLLUVIARIUM - [build] (access opening in an aqueduct) colluviarum *m*
COLOCYNTH - [bot] coloquinte *f*
COLOENTERITIS - [med] entéro-colite *f*
COLOGARITHM - [math] (the logarithm of a reciprocal of a number) cologarithme *m*
COLOGNE EARTH - [paint] terre *f* (brune) de Cologne
~SPIRIT - [ind chem] (a pure grade of ethyl alcohol) eau *f* de Cologne
COLON - [print] deux points
[anat] colon *m*
COLONIAL - [gen] colonial
~SPIRIT - [paint] (commercial methyl alcohol) alcool *m* méthylique
COLONNADE - [arch] colonnade *f*
COLONOPEXY - [med] colopexie *f*
COLONY - [gen] colonie *f*
[phys] (of crystals) amas *m* (de cristaux)
[bact] colonie *f* de bactéries
[agric] (taking over a ground) installation *f*, plantation *f*
~SHED - [agric] poulailler *m* transportable
COLOPHONIC ACID - [chem] acide *m* colophonique
COLOPHONY - [chem] (obsolete term for rosin) colophane *f*
COLOPTOSIS - [med] coloptose *f*
COLORADO BEETLE - [zool] doryphore *m*
~TICK FEVER - [med] fièvre *m* de la tique du Colorado
COLORADOITE - [min] (mercuric telluride) coloradoïte *f*
COLORANTS -[dyes] (dyes and pigments) colorants *mpl*
COLORATION - [print] (tone) ton *m*, teinte *f*
[text] coloration, teinture, salissure *f*, tonalité *f*
COLOROFIC - [paint] colorant
COLORIMETRIC - [meas] colorimétrique
~ANALYSIS - [chem] analyse *f* colorimétrique
~METHOD - [ind chem] méthode *f* colorimétrique
~PURITY - [opt] facteur *m* de pureté d'une couleur
COLORIMETRY - [meas] (analytical technique for the comparison of colour intensities) colorimétrie *f*
COLORIZATION - [gen] coloration *f*, coloris *m*
COLOTOMY - [med] (a hole cut into the colon) colotomie *f*
COLOUR, to - [gen] colorer
COLOUR - [gen] couleur *f*
[opt] couleur *f*
[ind chem] colorant *m*, teinture *f*
~ADJUSTMENT - [photo] mise *f* au point de la couleur
~BALANCE - [telev] équilibrage *m* de couleurs
~BAR - [telev] barre *f* colorée
~BLEEDING - [paint] exsudation *f* d'un colorant, migration *f* d'un colorant
~BLEND - [dyes] mélange *m* de couleurs
~BLINDNESS - [med] (lack of spectral colour sensations of the eye) daltonisme *m*, achromatopsie *f*
~BREAK-UP - [telev] dissociation *f* des couleurs, décomposition *f* des couleurs
~BRIGHTNESS - [telev] brillance *f* de couleur
~CARRIER - [opt] (chromophore) chromophore *m*
~CAST - [telev] prédominance *f* d'une couleur primaire
[photo] coloration *f*
~CELL - [biol] cellule *f* pigmentaire
~CENTRE - [phys] (centre to which electrons are attached; the deep colour taken by an alkali halide crystal heated in an atmosphere of the alkali vapour) centre *m* de couleur
~CHANGE - [chem] virage *m*
~CODE - [electron] (code for indicating the characteristics of components by coloured markings) code *m* des couleurs
~CODER - [telev] générateur *m* de signaux
~COMPARATOR - [instr] (a device for the comparison of a colour with a specified specimen) comparateur *m* de couleurs
~CONTRAST - [photo] contraste *m* des couleurs
~-CONVERSION FILTER - [photo] (colour-correction filter) filtre *m* compensateur d'éclairage
~CONVERTER - [telev] convertisseur *m* du signal de couleur

COLOUR CORRECTION - [telev] correction *f* de couleur
~COUPLER - [photo] (colour-forming agent) copulant *m* chromogène, formateur *m* de couleur
~DECODER - [telev] décodeur *m* couleur
~DEPRECIATION - [chem] altération *f* des couleurs
~DENSITY - [opt] facteur *m* de pureté colorimétrique
~DIFFERENCE - [opt] différence *f* de couleur
~DIFFERENCE SIGNAL - [telev] signal *m* de différence de couleur
~DISCRIMINATION - [opt] (the perception of colour differences) séparation *f* des couleurs
~DISK - [telev] disque *m* à secteurs colorés, disque *m* chromatique
~DISPLAY TUBE - [telev] tube *m* à image couleur
~DISTORTION - [opt] distortion *f* des couleurs
~DOCTOR - [impl] racle *f*
~EDGING - [telev] frange *f* colorée, couleur *f* fausse dans les bords
~EMBOSSING - [text] impression *f* en relief en couleur
~FADING - [dyes] (fading of colour pigment through light, temperature or chemical action) décoloration *f*
~-FAST - [text] bon teint, grand teint
~FASTNESS - [dyes] (resisting bleaching) solidité *f* d'une couleur
[text] solidité *f*
~FATIGUE - [opt] (changes in the sensation when the eye is tired) fatigue *f* rétinienne
~FIDELITY - [photo & cin] (the degree of faithful reproduction of the colours in the original scene) fidélité *f* des couleurs
~FIELD - s. colour frame
~FIELD CORRECTOR - [telev] aimant *m* d'uniformisation de la trame à couleurs
~FILM - [photo] film *m* en couleurs
~FILTER - [ind chem] (a sheet of translucent material effecting colour separation) filtre *m* coloré
~FILTER DISK - [telev] disque *m* à filtre chromatique
~FLASH - [telev] éclat *m* de couleur
~FLICKER - [telev] scintillement *m* des couleurs
~FLOTATION - [paint] flottaison *f*
~FRAME - [telev] (subdivision of the colour picture by scanning once in each of the three primary colours) trame *f* primaire
~FRINGERS - [photo] franges *fpl* d'interférence
~FRINGING - [telev] effet *m* de franges colorées
~GATE - [telev] porte *f* de signal couleur
~GRADING - [opt] étalonnage *m* des couleurs
~GRID - [telev] grille *f* de couleur
~GRINDING MILL - [ind chem] (for pigments) broyeur *m* à pigments
~IMAGE SEPARATION - [telev] séparation *f* des images en couleurs primaires
~INDEX - [opt] indice *m* de couleur
~KILLER - [telev] dispositif *m* de suppression de la couleur
~LACQUER - [dyes] vernis *m* coloré
~LEVEL - [telev] (modulated level corresponding to a fixed tone of the image) niveau *m* de couleur
~-LIGHT SIGNAL - [el] feu *m* coloré
~LOCK - [telev] enchaînement *m* du signal couleur
~MATCH - [opt] équilibrage *m* colorimétrique
~-MATCHING - [dyes] (careful adjustment of colours) appariement *m* des couleurs
COLOUR-METER - [meas] kelvinomètre *m*, thermocolorimètre *m*
~MIGRATION - [phys] (the movement of material of one colour into the structure of another of different colour) migration *f* de couleur
~MILL - [rubber ind] moulin *m* à couleurs
~MIXTURE - [opt] mélange *m* de couleurs
~MIXTURE CURVE - [opt] courbe *f* de mélange des couleurs
~MIXTURE FUNCTION - [opt] fonction *f* de mélange des couleurs
~MONITOR - [telev] récepteur *m* de contrôle couleur, moniteur *m* d'image couleur
~NOISE - [telev] bruit *m* du signal couleur
~NUMBER - [ind chem] s. colour index
~OF HIGH ORDER - [paint] couleur *f* d'ordre supérieur
~OVERLOAD - [telev] sursaturation *f* de la couleur
~PERCEPTION - [opt] perception *f* des couleurs
~PHASE - [telev] angle *m* de phase du signal de chrominance
~PICTURE - [telev] image *f* couleur
[photo] photo *f* en couleurs, photo *f* couleur
~PICTURE SIGNAL - [telev] signal *m* video couleur
~PICTURE TUBE - [electron] (a cathode-ray tube for colour television) chromoscope *m*, tube *m* image couleur
~PIGMENT - [chem] pigment *m* colorant
~PLATING - [metall] coloration *f* électrolytique
~PRIMARIES - [opt] (primary colours) couleurs *m pl* primaires, primaires *mpl*
~PRODUCING - [biol] chromogène
~PURITY COIL - [telev] bobine *f* de pureté de couleur
~RANGE - [dyes] gamme *m* des couleurs, gamme *f* de couleurs
~RECEIVER - [telev] récepteur *m* de télévision couleur, téléviseur *m* de couleur
~REGISTRATION - [telev] superposition *f* des couleurs
~RESPONSE - [phys] sensibilité *f* chromatique
[telev] réponse *f* de couleur
~RETENTION - [dyes] (colour fastness) solidité *f* de la couleur
~SAMPLING - [electron] commutation *f* électronique de couleurs
~SAMPLING FREQUENCY - s. colour sampling rate
~SAMPLING RATE - [telev] (the number of times per second each primary colour is sampled) fréquence *f* de commutation des couleurs
~SAMPLING SEQUENCE - [telev] séquence *f* de commutation des couleurs
~SATURATION - [telev] saturation *f* de couleur
~SCALE - [paint] échelle *f* de teintes, gamme *f* des couleurs, échelle *f* des couleurs
~SCHEME - [paint] combinaison *f* des couleurs, agencement *m* des couleurs
~SCREEN - [opt] (a colour filter designed to exclude certain frequencies of light from a reaction system) écran *m* coloré, écran *m* orthochromatique
[cin] (a coloured gelatine screen to change the colour of the irradiated scene) écran *m* diffuseur coloré
~SELECTIVE MIRRORS - [telev] miroirs *mpl* sélectifs de couleurs
~-SENSITIVE EMULSION - [photo] émulsion *f* sensible

aux couleurs
COLOUR SENSITIVITY - [meas] sensibilité *f* spectrale
~SENSITIZATION - [photo] sensibilisation *f* chromatique
~SEPARATION - [opt] division *f* chromatique, séparation *f* chromatique
~SEQUENCE - [telev] séquence *f* chromatique
[dyes] suite *f* des couleurs
~SHADE - [paint] teinte *f*
~SHADING - [telev] virage *m* de teinte
~SIGNAL - [telev] signal *m* de chromaticité
~SLIDE - [photo] diapositive *f* couleur
~SPACE - [opt] espace *m* chromatique
~STABILITY - [dyes] (stability of colour properties) stabilité *f* des couleurs
~STIMULUS - [opt] stimulus *m* chromatique, stimulus *m* coloré
~SUPERIMPOSITION - s. colour registration
~TAPE - [instr] ruban *m* encreur (pour enregistreur)
~TELEVISION - [telev] télévision *f* en couleurs
~TEMPERATURE - [phys] (the temperature of an incandescent "black body" emitting the same radiation as the colour in question) température *f* de couleur
~TEMPERATURE METER - [instr] appareil *m* de mesure de la temperature de couleur
~THRESHOLD - [opt] (any measure of the degree of colour discrimination) seuil *m* chromatique, seuil *m* de sensation chromatique
~TRANSMISSION - [telev] transmission *f* en couleur
~TRIANGLE - [photo] (the arrangement of colours with the primaires at the corners) diagramme *m* colorimétrique, triangle *m* des couleurs
~TROUGH - [text] cuve *f* à couleurs
~VALENCE - [chem] valence *f* des couleurs
~VALUE - [paint] valeur *f* chromatique
~VAT - [text] récipient *m* pour colorants, cuve *f* pour colorants
~VISION - [opt] vision *f* chromatique
~WASH - [paint] badigeon *m*
~WASH PAINT - s. colour wash
~YELD - [chem] rendement *m* tinctorial
COLOURED TRACER SERVING - [el] fil *m* de couleur de repérage (des conducteurs)
~TRACER THREAD - s. coloured tracer
COLOURING -[metall](the production of a specified colour on metal surfaces by chemical or electrochemical action) coloration *f*
~CAPACITY - [paint] pouvoir *m* colorant
~MATTER - [paint] matière *f* colorante
~OF FLAME - [chem] (used in blow-pipe tests) coloration *f* de la flamme
~POWER - [paint] pouvoir *m* colorant, pouvoir *m* tinctorial
~STRENGHT - [dyes] s. colouring power
COLOURLESS - [gen] incolore
~GLASS - [glass man] verre *m* incolore
COLOURS IN OIL - [paint] (highly concentrated mixtures of pigments in oil) pigments *mpl* à l'huile
COLPATRESIA - [med] atrésie *f* du vagin, occlusion *f* du vagin
COLPECTASIA - [med] dilatation *f* du vagin
COLPECTOMY - [med] colpectomie *f*
COLPITIS - [med] (inflammation of the vagina) colpite *f*, inflammation *f* vaginale
COLPITTS OSCILLATOR - [radar] (oscillator circuit consisting of a triode in which the grid-cathode and anode-cathode paths are capacitive and the anode-grid path is inductive) oscillateur *m* de Colpitts
COLPOCELE - [med] colpocèle *f*, hernie *f* vaginale
COLPOPOLYPUS - [med] polype *m* vaginal
COLT - [zool] poulain *m*
COLTER - s. coulter
~KNIFE - [agric mach] coutre *m*
COLTSFOOT - [bot] (a weed) tussilage *m*
COLUMBITE - [min] (niobate and tantalate of manganese and iron; the principal source of tantalum) columbite *f*
COLUMBIUM - [metall] (metallic element, symbol Cb or Nb, occurring in various rare minerals) niobium *m*
COLUMN - [arch] colonne *f*, pilier *m*, montant *m*
[anat] (the spinal cord and any other columnal structures) colonne *f*
[print] colonne *f*
[ind chem] colonne *f*, tour *f*
[mech](of a balance; the vertical element supporting the beam) montant *m*
[mech] (of a press, the vertical pillar connecting the head to the base) montant *m*, bâti *m*
[mach tools] montant *m*, colonne *f*
[metall] pied *m*
~BASE - [metall] (in welding) base *fpl* de colonnes
~CONSTRUCTION - [constr] construction *f* des poteaux
~DRILL - [min] perforatrice *f* à colonne
~DRILLING MACHINE - [mech] perceuse *f* à colonne
~FEED - [mech] avance *f* du montant
~FLOODING - [chem] engorgement *m* de la colonne
~HEAD - [chem] tête *f* de colonne
~OF AIR - [chem] colonne *f* d'air
~OF FLUID - [oil ind] colonne *f* de liquide
~OF WATER - [hydr] colonne *f* d'eau
~PACKING - [chem] remplissage *m* de la colonne
~PRESS - [mech] (a press whose head is carried on two or more vertical columns) presse *f* à colonnes
~RADIATOR - [th eng] radiateur *m* tubulaire (à éléments)
~STILL - [ind chem] colonne *f* de distillation
~STRENGTH - [mech] résistance *f* au flambage
~STRIPPER - [ind chem] colonne *f* de distillation
~WAY - [mech] chemin *m* de déplacement
COLUMNAR CRYSTALS - [metall] (elongated crystals at right angle to the surface of the mould) cristaux *f* à structure basaltique
~EPITHELIUM - [zool] (consisting of prismatic columnar cells) épithélium *m* colomnaire
~IONIZATION - [nucl] (ionization so dense, that no external field can prevent recombination) ionisation *f* en colonne
~RECOMBINATION - [nucl] (recombination occurring before the ions have left the track where ionization takes place along a column) recombinaison *f* colomnaire
COLUMNING - [med] columnisation *f*, tamponnement *m* du vagin
COLURES - [astr] (the great circles passing through the poles of the celestial equator) colures *m*
COLZA - [bot] colza *m*
~CAKE - [agric] (animal feeding) tourteau *m* de colza
~OIL - [paint] (used as an illuminant) huile *f* de

colza
COMA - [med] coma *m*
[opt] (aberration of a lens with spherical surfaces) coma *m*
[telev] (image defect) coma *m*
COMATOSE - [med] comateux
COMB, to - [gen & text ind] peigner
COMB - [mech] peigne *m*, peigne *m* à fileter
[text ind] peigne *m*
[build] ligne *f* de partage des eaux, faîte *m*, faîtage *m*
[zool] crête *f*
~BAR - [text] tige *f* de peigne
~BIT - [min] barroir *m*
~COLLECTOR - [el] collecteur *m* à peigne
~CYLINDER - [text] peigne *m* circulaire
~FILTER - [radio] (a wave filter) filtre *m* de bande à peigne
~-GATE - [metall] attaque *f* de coulée à peigne
~IN TWO PARTS - [text] peigne *m* en deux parties
~-LIKE - [gen] crénelé, denté
~NEEDLE - [text] aiguille *f* du peigne, crochet *m* du peigne, dent *f* du peigne
~NOILS - [text] duvet *m* de peigne
~PIN - [text] dent *f* de peigne, boulon *m* du peigne
~POLES - [el] (in electric machines) pôles *mpl* à peigne
~STRIP - [text] barrette *f* à aiguilles
~-STRUCTURE - [geol] structure *f* crêtée
~TOOTH - [text] s. comb needle
~WASTE - [text] blousse *f*, freinte *f*
COMBED COTTON YARN - [text] fil *m* de coton peigné
COMBER - [text ind] peigneuse *f*
COMBINATION - [gen] combinaison *f*, association *f*
[chem] (the formation of a compound) combinaison *f*, composé *m*, combiné *m*
~ARRAY - [electron] réseau *m* complexe
~AUTOMATIC CONTROLLER - [el] boucles *fpl* de réglage couplées
~BEAM - [build] poutre *f* composée
~BIT - [oil ind] trépan *m* pour roches à deux taillants à queue de carpe
~BOILER - [metall] chaudière *f* combinée (ou mixte)
~BURNER - [th eng] brûleur *m* mixte
~CHECKER BOARD - [genet] échiquier *m* de croisement ment
~CHUCK - [mech] mandrin *m* universel (ou à combinaison)
~COLOURS - [zool] (effects produced by pigments combining with structural colours) combinaison *f* de couleurs
~DIE - [metall] moule *m* à empreintes multiples
~DIES - [metall] matrices *fpl* multiples d'estampage et de découpe
~DRILL AND COUNTERSINK - [mach tools] mêche *f* à percer et à fraiser
~FLUE - [th eng] conduit *m* unitaire
~LATHE - [mach tools] tour *m* à combinaisons
~LEVER - [mech] (of a valve-gear) levier *m* d'avance
~LOCK - [mech] serrure *f* à combinaisons
~MICROPHONE - [acoust] (consisting of two or more microphones) microphones *mpl* combinés
~OF COLOUR AND WEAVE - [text] effet *m* de carreaux par combinaison des couleurs et de l'armure
~OF RADIAL AND TANGENTIAL GUIDING - [mech] guidage *m* mixte
COMBINATION OVEN - [th eng] four *m* combiné, four *m* compound
~PIPE AND NUT WRENCH - [impl] clé à tubes et à écrous
~PLANER - [tool] rabot *m* à filet
~PLANT - [mech] installation *f* de dégazolinage
~PLIERS - [mech] (used for gripping various types of objects) pinces *fpl* universelles
~PRINCIPLE - [phys] (Ritz's principle; the addition or subtraction of the wave numbers of two spectral lines often give the value of the wave number of another line in the same spectrum) principe *m* de combinaison
~PRINTING - [photo] assemblage *m* de photographies
~REAMER - [mining] trépan *m* aléseur avec foret au fond
~RIG - [oil ind] (in well sinking) installation *f* de forage combiné
~SCALE - [meas] (composed of two or more concentric scales) échelle *f* composée
~SOCKET - [mining] douille *f* combinée
~SQUARE - [impl] équerre *f* universelle
~STRING - [mining] tubage *m* combiné
~SWITCH - [el] interrupteur *m* à combinaison
~TONE - [acoust] (supplementary tone produced when two tones are sounded simultaneously) son *m* de combinaison
~TONE DISTORTION - [radio] distortion *f* d'intermodulation
~TURRET LATHE - [mach tools]tour *m* semi-automatique
~UNIT - s. combination rig
~YARN - [text] fil *m* fantaisie
COMBINATIONAL - [gen] composé, mixte
COMBINATIONS - [math] (the different groupswhich can be formed with a number of items) combinaisons *fpl*
COMBINATORIAL ANALYSIS - [math] analyse *f* combinatoire
COMBINE, to - [gen] combiner, associer, se combiner
COMBINE - [agric] moissonneuse-batteuse *f*
~BALER - [agric] rateau *m* ramasseur et presse combinés
~HARVESTER - [agric] moissonneuse-batteuse *f*
~SEED AND FERTILIZER DRILL - [agric] semoir *m* et distributeur d'engrains
~WITH BINDER ATTACHMENT - [agric] moissonneuse lieuse *f*
COMBINED - [gen & chem] combiné, composé, mixte
~AERIAL - [radio] antenne *f* commune
~BEAM FLANGE AND BRAKE RING - [text] flange *f* à frein, plateau *m* à frein
~CARBON - [metall] (the carbon present as iron-carbide) carbone *m* combiné, carbone *m* de cémentation
~COMPRESSIVE AND BENDING STRESS - [mech] effort *m* combiné de compression et de flexion
~COUPLING - [el] couplage *m* mixte
~CROSS TWILL - [text] croisé *m* combiné
~CURVE AND SIDE PLANER - [impl] raboteuse *f* pour surfaces planes et courbes
~DEGENERATION - [med] sclérose *f* combinée (de la moelle)
~DIAGRAM - [mech] (of a multicylinder engine) diagramme *m* totalisé

COMBINED DISTRIBUTION FRAME - [telecomm] (used when the number of circuits does not justify individual frames) répartiteur *m* mixte
˜DRAW AND BUFFING-GEAR - [railw] appareil *m* de traction et de répulsion
˜DRILLING AND TAPPING MACHINE - [mach tools] perceuse-taraudeuse *f*
˜FUEL BURNER - [th eng] brûleur *m* mixte
˜FUSE AND CUT-OUT - [telecomm] fusible *m* et disjoncteur combiné
˜GAS AND ELECTRICITY COOKER - [th eng] cuisinière *f* combinée
˜GRINDER AND SIEVE - [mech] broyeur *m* tamiseur
˜LINE AND HALFTONE BLOCK - [print] trait-simili *m*, cliché *m* combiné
˜LINE AND RECORDING (C.L.R.) - [telecomm] frafic *m* interurbain rapide
˜LISTENING AND SPEAKING KEY - [telecomm] clé *f* d'écoute et de conversation
˜ROLLER AND LEVER TACKLE WEAVES - [text] armure *f* composé
˜STOPING - [mining] abattage *m* combiné
˜SULPHUR - [chem] soufre *m* combiné
˜SYSTEM - [hydr] (sewage serving drains and rain-water at once) réseau *m* unitaire
˜TIMBERING - [constr] charpente *f* composée
˜WATER - [th eng] (boilers) eau *f* de constitution
˜WEAVE - [text] (when regular weaves are combined to form new ones) armure *f* composée
COMBING - [text] peignage *m*, blousse *f*
[paint] (graining with special tool) glacis *m* veiné (au peigne)
˜CARD - [text] carde *f*
˜MACHINE - [text] peigneuse *f*
˜WASTE - [text] déchets *mpl* de peigne, blousse *f*
˜WAVE - [naut] vague *f* déferlante
˜WORKS - [text] peignerie *f*
COMBINING EQUIVALENT - [chem] (an element or radical weight in any way equivalent to a unit weight of hydrogen) nombre *m* proportionnel
˜VOLUME - [metall] volume *m* de combinaison
˜WEIGHT - [metall] poids *m* relatif de combinaison
COMBS FOR FLAX - [text] peignes *mpl* pour lin
COMBURENT - [chem] comburant *m*
COMBUSTIBILITY - [gen] combustibilité *f*
COMBUSTIBLE - [gen] combustible *m*, matière *f* inflammable
˜SHALE - [min] schiste *m* combustible, transmanite *f*
COMBUSTION - [phys] combustion *f*
˜AGENT - [chem] comburant *m*
˜AT CONSTANT PRESSURE - [metall] combustion *f* à pression constante
˜BOAT - [ind chem] (small container of great heat resistance used to hold substances for ignition) coupelle *f*
˜BOMB - [chem] bombe *f* à combustion
˜CHAMBER - [mech] (in boiler furnaces and in internal combustion engines) chambre *f* de combustion, chambre *f* d'explosion
˜CHAMBER HEAD - [mech] (in a jet engine) tête *f* de chambre de combustion
˜CHAMBER OUTER CASING - [mech] enveloppe *f* de chambre
˜CHARACTERISTICS - [th eng] caractéristiques *fpl* de combustion
˜CONTROL - [mech] (the control of the rate of combustion in a furnace) régulation *f* de la combustion
COMBUSTION DIAGRAM - [th eng] diagramme *m* de combustion
˜EFFICIENCY - [mech] rendement *m* de la combustion
˜ENGINE - [mech] moteur *m* à combustion interne
˜EQUATION - [phys] équation *f* de combustion
˜FLUE - [th eng] carneau *m* de chauffage
˜GLASS - [metall] verre *m* à fusion
˜HEAT - [nucl] chaleur *f* de combustion
˜INDEX - [ind chem] (the ratio of carbon monoxide to carbon dioxide in flue gas) indice *m* de combustion
˜LAG - [mech] délai *m* d'inflammation
˜POTENTIAL - [phys] potentiel *m* de combustion
˜PRODUCTS - [th eng] produits *mpl* de combustion
˜RESIDUAL PRODUCTS - [th eng] résidus *mpl* de combustion
˜SHAFT - [metall] cheminée *f* de combustion, puits *m* de combustion
˜SPACE - [mech] chambre *f* de combustion
˜STARTER - [mech] démarreur *m* à cartouche
˜STROKE - [mech] course *f* de détente
˜TEMPERATURE - [th eng] (the maximum temperature of a specified fuel with atmospheric oxygen under atmospheric pressure) température *f* de combustion
˜TUBE FURNACE - [metall] (furnace used for the determination of carbon contents in steel) four *m* à analyse
˜TUBING - [ind chem] (tube of special glass to resist high temperatures) tubes *mpl* réfractaires
˜UNIT - [mech] (in a jet engine) ensemble *m* de combustion
˜VELOCITY - [phys] vitesse *f* de propagation, vitesse *f* de déflagration, vitesse *f* de combustion
˜ZONE - [mech] (the part of the combustion chamber in which burning takes place) zone *f* de combustion
COMBUSTIVE - [mech] comburant
COMBUSTOR - [mech] (in a jet engine) chambre *f* de combustion
COME, to - [gen] arriver, venir
˜ALONG - [mining] treuil *m* manuel
˜ABOUT, to - [naut] virer de bord
˜DOWN, to - [gen] descendre
[comm] baisser
[chem] précipiter
˜IN, to - [gen] entrer
[el] entrer *m* en fonctionnement
˜INTO FORCE, to - [gen] entrer en vigueur
˜OFF, to - [gen] se détacher, tomber
[chem] se dégager
˜OUT, to - [gen] sortir, se dégager
˜OVER, to - [gen] traverser
[chem] distiller, déborder
˜UP, to - [gen] monter, gravir
[naut] lofer, serrer le vent
COMET - [astr] (a small mass member of the solar system) comète *f*
COMING - s. coaming
˜INTO STEP - [el] (of a synchronous machine; the phenomenon by which a synchronous machine acquires synchronism with another synchronous machine without being mechanically coupled) accrochage *m*

COMMA - [print] virgule *f*
[acoust] (pitch error) comma *mf*
COMMAND - [gen] commande *f*, commandement *m*, ordre *m*
~GUIDANCE - [constr] (missile control system based on the information to the missile from an external source) système *m* de guidage à télécommande
~RESOLUTION - [contr] (the maximum permissible change in command without altering the ultimately controlled variable) tolérance *f* de réglage
COMMENSURABLE - [gen & math] mesurable
~QUANTITIES - [math] quantités *fpl* mesurables
COMMENTARY - [gen] commentaire *m*
[radio] (a verbal report for immediate use) reportage *m*, commentaire *m*
[cin & tel] commentaire *m*
COMMENTATOR - [gen & radio] commentateur *m*
COMMERCE - [gen] commerce *m*
COMMERCIAL - [gen] commercial
[telev & radio] (paid broadcast and or televised publicity) annonce *f* publicitaire
~AVIATION - [aero] aviation *f* civile
~BRASS - [metall] (copper-zinc alloy with 10 percent zinc) bronze *m* marchand
~COPPER - [metall] cuivre *m* marchand
~COMPUTER - [comput] ordinateur *m* de gestion
~FILM - [cin & telev] film *m* publicitaire
~IRON - [metall] fer *m* marchand
~LOAD - [comm] charge *f* marchande
~PRODUCTION - [comm] (on commercial lines) production *f* industrielle
~TANK - [el chem] (electrolytic cell producing refined metal) cuve *f* électrolytique finale
~TELEVISION - [telev] télévision *f* commerciale
~VEHICLE - [auto] véhicule *m* utilitaire, véhicule *m* commercial
~WELL - [oil ind] puits *m* rentable
COMMINGLE, to - mélanger, méler, se mélanger
COMMINUTE, to - [min] (to crush coarse material into finer particles) broyer finement, concasser, pulvériser, porphyriser
COMMINUTED - [med] (crushed into small fragments) pulvérisé, porphyrisé, grenaillé
~FRACTURE - [med] fracture *f* comminutive
COMMINUTION - [min] (the operation of breaking, crushing or grinding ores to reduce size) grenaillage *m*, concassage *m*, pulvérisation *f*, porphyrisation *f*, broyage *m*
COMMINUTOR - [mech] (a machine for crushing coarse material to a finer particle size) broyeur *m*
COMMISSION - [gen] commission *f*
[leg] (the authority to act) délégation *f*
~COMBING - [text] peignage *m* à façon
~SPINNING - [text] filature *f* à façon
~WEAVING - [text] tissage *m* à façon
COMMISSIONING - [comm] (of a machine etc) prise *f* en charge
COMMISSURE - [zool] (a joint) commissure *f*
[build] (a joint courses of bricks) joint *m* entre les rangées de briques
COMMITMENT - [gen] engagement *m*
[leg] emprisonnement *m*, incarcération *f*
COMMIXTURE - [gen] mélange *m*
COMMODITY - [gen] whatever is of use to human beings) produit *m*, denrée *f*, marchandise *f*, matière *f* première
COMMON, to - [el] (to connect several electric leads to the same circuit point) brancher ensemble
COMMON - [gen] commun, ordinaire, mutuel, courant
~AERIAL - [radio] (an aerial system which may be used by a number of persons in the same building) antenne *f* commune
~BASE - [electron] (transistor operation in which the base is common to input and output circuits) base *f* commune
~BATTERY - [el] (also called central battery; a large battery for telephone and telegraph circuits) batterie *f* commune, batterie *f* centrale
~BRICK - [build] (bricks of common quality used for a variety of purposes) brique *f* ordinaire
~CATHODE - [electron] (cathode in a multiple electronic tube which is common to all electrode systems) cathode *f* commune *f*
~CHORD - [mus] accord *m* parfait
~COLLECTOR - [electron] (transistor operation in which the emitter is common to input and output circuits) collecteur *m* commun
~DENOMINATOR - [math] commun denominateur *m*
~DIVISOR - [math] commun diviseur *m*
~EMITTER - [electron] (transistor operation in which the emitter is common to input and output circuits) émetteur *m* commun
~FACTOR - [math] facteur *m* commun
~FREQUENCY BROADCASTING - [radio] (the use of the same carrier frequency by more than one transmitter) émission *f* sur fréquence commune
~GABLE ROOF - [build] toit *m* ordinaire à deux pans
~-IMPEDANCE COUPLING - [el] couplage *m* direct
~INFORMATION CARRIER - [comput] porteur *m* d'information commun
~ION EFFECT - [phys] effet *m* d'ion commun
~LEAD - [metall] (commercial lead below the purity standard of corroding lead) plomb *m* marchand
~LIMB - [geol] flanc *m* médian
~LIME - [chem] (impure calcium oxide, made by calcining limestone or chalk) chaux *f* vive
~LOGATITHM - [math] logarithme *m* décimal, logarithme *m* vulgaire
~MEASURE - s. common divisor
~METAL - [metall] métal *m* commun
~-MODE REJECTION - [radio] (the ability of certain amplifiers to eliminate a common-mode signal and to respond to an out-of-phase signal) mode *m* commun de rejection
~-MODE REJECTION QUOTIENT - [amplif] facteur *m* de mode commun de réjection
~-MODE SIGNAL - [electron] (signal applied equally to the inputs of a balance amplifier stage or other differential devices) signal *m* commun, signal *m* en phase
~MULTIPLE - [math] commun multiple *m*
~RAFTER - [build] (subsidiary rafter supporting the roof covering) chevron *m* de ferme
~REACTANCE - [electron] induction *f* mutuelle
~RETURN - [el] (the single lead forming the return circuit for more than one separate circuit) conducteur *m* de retour commun
~ROSIN - [chem] colophane *f*
~SALT - [min] sel *m* de cuisine, sel *m* commun

COMMON SPRUCE - [bot] épicéa *m*
~WALL - [constr] mur *m* mitoyen
~WIRE - [el] fil *m* neutre
~WIRE NAIL - [carp] clou *m* ordinaire
COMMONING-STRIP - [el] (metal strip to which a number of leads are connected in common) barrette *f* de connexion
COMMUN TRUNK - [telecomm] réseau *m* commun
COMMUNICATE, to - [gen etc] communiquer, transmettre, informer, faire savoir
COMMUNICATING TUBES - [phys] vases *mpl* communiquants
~VESSELS - s. communicating tubes
COMMUNICATION - [gen] communication *f*, information *f*, liaison *f*, télécommunication *f*
~BAND - [radio] (the band of frequencies which is occupied by the emission) bande *f*, canal *m*, voie *f*
~CHANNEL - [telecomm] bande *f*, canal *m*, voie *f*
~SATELLITE - [telecomm] satellite *m* de télécommunication
~ZONE INDICATOR - [radio] indicateur *m* de fréquence optimale de trafic
COMMUNICATIONS CENTRE - [aero] (central station for the transmission and reception of signals to and from aircraft in flight) centre *m* de transmission
COMMUNITY ANTENNA s. common aerial
~TELEVISION - [telev] (also called piped television) télévision *f* en circuit fermé
COMMUTATE, to - [el] commuter
COMMUTATING FIELD - [el] (the magnetic field under the compoles of a direct current machine) champ *m* de commutation
~FIELD COIL - [el] bobinage *m* de champ de commutation
~MACHINE - [el] commutatrice *f*
~POLE - [el] (auxiliary pole in a commutation motor, designed to assist commutation) pôle *m* de commutation
~REACTOR - [electron] (interphase transformer causing two anodes to fire simultaneously) transformateur *m* équilibreur de commutation
~WINDING - [el] enroulement *m* de commutation
COMMUTATION - [el] (the reversion of the direction of the current in an electric circuit) commutation *f*
COMMUTATOR - [el] (a device for distributing current to or collecting it from, the armature coil of a d.c. motor or generator) collecteur *m*
[el] (a device for reversing the direction of the current) commutateur *m*, inverseur *m*
~ARMATURE - [el] induit *m*
~BAR - [el] (one of copper bars forming the moving contacts of a commutator) lame *f* de collecteur
~BRUSH - [el] balai *m* de collecteur
~END - [el] (the end of an electric motor at which the commutator is placed) côté *m* collecteur
~FACE - [el] surface *f* du collecteur
~GRINDER - [mech] polisseuse *f* pour collecteur
~HUB - [el] carcasse *f* de collecteur
~LUG - s. commutator riser
~MACHINE - [el] (an electric motor or generator fitted with a commutator) moteur *m* à collecteur
~METER - [instr] compteur *m* à collecteur
~MOTOR - s. commutator machine
~NOISE - [el] (the noise caused by the opening or closing of the circuits) bruit *m* de commutation
COMMUTATOR PITCH - [el] pas *m* au collecteur
~RECTIFIER - [el] (a device for rectifying alternating currents) permutatrice *f*
~RING - [el] bague *f* de collecteur
~RIPPLE - [el] ondulation *f*
~RISER - [el] jonction *f* au collecteur
~SEGMENT - s. commutator bar
~SHELL - s. commutator hub
~SHRINK-RING - [el] bague *f* à ajustage forcé
~SLEEVE - [el] (the sleeve, mounted on the shaft or on the armature spider, which carries the assemblage of bars, insulation and clamping rings) manchon *m* de collecteur
~STRIP - [el] lame *f* de collecteur
~TAG - s. commutator riser
COMMUTED CURRENT - [el] courant *m* redressé
COMPACT, to - [gen] tasser, rendre compact
COMPACT - [gen] compact, solide, dense, aggloméré, ramassé, de faible encombrement, liant
[gen] (pact or agreement) pacte *m*, accord *m*, convention *f*
~IRON ORE - [min] minerai *m* de fer compact
COMPACTED - [hydr] (sludge) consistant
~ROCKFILL - [geol] enrochements *mpl* compactés
COMPACTIBILITY - [soil] compactibilité *f*, aptitude *f* au compactage
COMPACTING CRACK - [metall] crique *f* à la compression
~EFFECT - [met] effet *m* du compactage
~ENERGY - [mech] énergie *f* de compactage
~PRESSURE - [metall] pression *f* de serrage
COMPACTION - [gen] tassement *m*, compacité *f*
[soil] (state of soil due to rolling of heavy machinery) compactage *m*, compaction *f*, damage *m*
~BY ROLLING - [mech] compactage *m* par cylindrage, compactage *m* par roulage
~BY TAMPING - [constr] compactage *m* par damage
~BY VIBRATION - [constr] compactage *m* par vibration, serrage *m* par vibration
~BY WATERING - [soil] compactage *m* par arrosage
~ROLL - [mech] rouleau *m* compresseur
~TEST - [constr] essai *m* de compactage, essai *m* Proctor
COMPACTNESS - [gen] compacité *f*, densité *f*, volume *m* réduit, faible encombrement *m*
[text] compte *m*, contexture *f*, densité *f* du tissu
COMPACTOR - [mech] compacteur *m*
COMPANDOR - [telecomm] (combination of compressor and expandor to reduce contrast in telephone speech) compresseur-expanseur *m*
COMPANION - [naut] capot *m* de descente
~FLANGE - [mech] contrebride *f*
~LADDER - s. companionway
~SCREW - [mech] vis *f* creuse, vis *f* femelle
COMPANIONWAY - [naut] échelle *f* de descente
COMPARATIVE CALCULATION - [math] calcul *m* comparatif
COMPARATOR - [instr] (measuring instrument designed to compare a dimension with a known standard) comparateur *m*
[el] (circuit comparing two signals and supplying an indication of agreement or disagreement) comparateur *m*
~CIRCUIT - s. comparator
COMPARISON - [gen] comparaison *f*

COMPARISON LAMP - [el] lampe *f* témoin
~OSCILLOSCOPE - [instr] (instrument for comparing oscillations) oscilloscope *m*
~PRISM - [phys] (small right-angle prism placed in front of a slit of a spectroscope, so that two spectra may be viewed at the same time) prisme *m* de comparaison
~SPECTROSCOPE - [instr] (instrument comparing spectra) spectroscope *m* comparateur
~SPECTRUM - [phys] (spectrum used as standard for comparison) spectre *m* de comparaison
~SURFACE - [light] surface *f* de comparaison
COMPAROSCOPE - [instr] comparateur *m*
COMPARTMENT - [gen railw etc] compartiment *m*, logement *m*, case *f*
~MILL - [mech] (a ball-mill divided into compartments which contain balls of different diameters) broyeur *m* à boulets à compartiments
COMPASS - [instr] (instrument for indicating the direction of the north) compas *m*, boussole *f*
[draw] compas *m*
[acoust] (the range of notes a voice or instrument can produce) portée *f*, étendue *f*
~BASE - [aero] (device for swinging aircraft for compass adjustment while on the ground) plateforme *f* de compensation des compas
~BEARING - [naut] relèvement *m*
~BOWL - [instr] (the envelope in which the card of a magnetic compass floats) cuvette *f* de compas
~BRICK - [constr] brique *f* circulaire, brique *f* cintrée, brique *f* de puits *m*
~CARD - [instr] (the disc on which the points are marked in a compass) rose *f* (des vents)
~CONDENSATING BASE - [aero] (a concrete base, sometimes a turntable, on which an aircraft is placed for compensating its magnetic compass) plateforme *f* de compensation des compas
~CORRECTOR - [instr] (designed to neutralize the effect of the craft's magnetism) compensateur *m*
~ERROR - [instr] variation *f*
~LENGHTENING BAR - [draw] rallonge *f* de compas
~NEEDLE - [instr] aiguille *f* de boussole
~PLANE - [impl] (plane with a curved sole for round) rabot *m* cintré
~PLATFORM - [naut] passerelle *f* de navigation
~RABBET - [naut] guillaume *m* en navette
~ROSE - [nav] (the graduated circle on a chart for reference purposes) rose *f*
~SAW - [carp] (handsaw with straight tapering blade) scie *f* égoîne, scie *f* à contourner, scie *f* à guichet
~SURVEY -[mining] lévé *m* à la boussole
~SWINGING - [instr] déviation *f*
~TESTING PLATFORM - s. compass base
~TILTING ANGLE - [surv] angle *m* d'inclinaison
~WINDOW - [build] (a type of bay window) fenêtre *f* en saillie ronde
COMPATIBILITY - [gen] compatibilité *f*
[chem] (the tolerance of a dissolved substance towards another dissolved substance) compatibilité *f*
~PROBLEM - [nucl] (the problem of selecting the right chemical substance for nuclear reactors in view of possible mutual reactions) problème *m* du choix des matériaux
COMPATIBLE - [gen & chem] compatible
[telev] (of colour television; of a system of producing colour television which can be received in black and white on ordinary sets) compatible
[bot] (capable of self fertilization) compatible
COMPENDIUM - [gen] abrégé *m*, précis *m*, compendium *m*
COMPENSATE, to - [gen] compenser, balancer, équilibrer
COMPENSATED - [mech & el] compensé, équilibré
~INDUCTION MOTOR - [el] (induction motor with a commutator winding on the rotor) moteur *m* asynchrone compensé
~INSTRUMENT TRANSFORMER - [el] (transformer in which the phase displacement between the primary and secondary quantities is reduced) transformateur *m* de mesure compensé
~-LOOP DIRECTION FINDER - [radar] radiogoniomètre *m* à cadre compensé
~PENDULUM - [phys mech] (a pendulum whose distance between the support and the centre of gravity of the bob is indipendent of temperature, so that this does not alter the time period) pendule *m* composé
~SEMICONDUCTOR - [electron] semi-conducteur *m* équilibré
~SERIES MOTOR - [el] (with a compensated winding to neutralize the effect of armature reaction) moteur *m* série compensé
~SOLENOID - [el] solenoïde *m* compensé
~VOLUME CONTROL - [radio] (device changing the tonal balance of the loudspeaker) réglage *m* de puissance compensé
~WATTMETER - [instr] (electrodynamic wattmeter with an additional reversed current coil) wattmètre *m* compensé
COMPENSATING - [gen mech etc] compensateur, de compensation
~ACTION - [contr] signal *m* de compensation
~BEAM - [constr] balancier *m* de suspension
~CHAMBER - [el] chambre *f* de compensation
~COIL - [el] (coil used on instruments etc to compensate for the effects of friction or other factor causing errors) enroulement *m* de compensation
~COLLAR - [mech] (collar fitted to a shaft) collier *m* de compensation
~COUPLING - [plumb] manchon *m* d'accouplement élastique
~DEVICE - [mech] (to compensate the cylinder pressure in steam locomotives) dispositif *m* de compensation
~DIAPHRAGM - [surv] (fitted to a tacheometer) diaphragme *m* de compensation
~ELEMENT - [mech] élément *m* (ou joint) compensateur, soufflet *m*
~ERRORS - [instr] (accidental errors) erreurs *fpl* qui se compensent
~FIELD - [el] (a term denoting the field produced by a compensating winding) champ *m* de compensation
~FILTER - [photo] filtre *m* de compensation
~GEAR - [mech] engrenage *m* différentiel
~JET -[mech] (auxiliary petrol jet fitted to some carburettors) gicleur *m* de compensation
~LEADS - [el] connections *fpl* compensées
~MACHINE - [rubber ind] (for tyres) machine *f* à équilibrer
~MAGNET - [el] (magnet used with certain galvanometers to compensate for the effect of external

magnetic fields) aimant *m* correcteur
COMPENSATING MEMBER - [mech] s. compensating element
~PENDULUM - s. compensated pendulum
~PIPE - [mech] s. compensating element
[plumb] tuyau *m* compensateur
~REPULSION MOTOR - [el] (having an additional pair of brushes connected in series with the supply circuit) moteur *m* à répulsion compensé
~RESERVOIR - [hydr] résérvoir *m* compensateur, réservoir *m* régulateur du débit
~RESISTANCE - [el] résistance *f* de compensation
~SPRING - [mech] réssort *m* compensateur
~ROD - [aero] tige *f* de compensation
~SIGHT - [aero] viseur *m* de compensation
~TRANSFORMER - [el] transformateur *m* différentiel
~WINDING - [el] (the winding in a compensated induction motor) enroulement *m* de compensation
COMPENSATION - [gen phys etc] compensation *f*
[telecomm] égalisation *f*
[med] (the condition by which no heart failure occurs despite the presence of heart disease) compensation *f*
[instr] (of a compass) compensation *f*
~BALANCE - [horol] balancier *m* compensé
~BAR - [aero] palonnier *m* compensé
~BASIS - [fin] base *f* de réparation
~CHAMBER - [nucl] chambre de compensation
~CIRCUIT - [el] circuit *m* de compensation
~FUND - [fin] fond *m* de compensation
~METHOD - [chem] (a method of measuring the electromotive force) méthode *f* de compensation
~THEOREM - [el] théorème *m* de compensation
~WATER - [hydr] eau *f* de compensation
COMPENSATOR - [gen] compensateur *m*
[mech] balancier *m*
[el] autotransformateur *m*
[opt] (arrangement for the measure of the phase difference between two components of elliptically polarized light) compensateur *m*
[telev] compensateur *m* d'atténuation
[cin] (a speed regulator of the film) régulateur *m* de vitesse
~OSCILLATOR COIL - [el] enroulement *m* de compensation de circuit oscillant
~WINDING - s. compensating winding
COMPENSATORY SURFACES - [aero] plans *mpl* de compensation
COMPETITIVE - [gen] compétitif, concurrent
[chem] (of reactions) concurrent
~BIDDING - [gen] appel *m* d'offres, concurrence *f*
COMPILE, to - [gen] compiler
[comput] (the operation of integrating a number of subroutines into the main routine) compiler
COMPILER - [gen] compilateur *m*
[comput] (a programme making routine) compilateur *m*, autoprogrammeur *m*
COMPLANATE - [bot] (flattened) aplati
COMPLEMENT - [gen] complément *m*
[naut] (the full number, e.g. the number of men in a ship) équipage *m*, effectif *m*
[geom] (of an angle) angle *m* complémentaire
[comput] (a quantity derived from a given quantity) complément *m*
COMPLEMENTARITY PRINCIPLES - [phys] principe *m* de complémentarité
COMPLEMENTARY - [gen] complémentaire, additionnel
COMPLEMENTARY ACCELERATION - [phys] accéleration *f* complémentaire
~AFTER-IMAGE - [opt] (the image which is experienced after visual fatigue caused by colours) image *f* rémanente complementaire
~ANGLE - [geom] angle *m* complémentaire
~CHROMATICITY - [opt] chromaticité *f* complémentaire
~COLOURS - [phys] (of colours having complementary chromaticities) couleurs *fpl* complémentaires
~FACTOR - [biol] (inheritance factor causing the appearance of a character) facteur *m* complémentaire
~FUNCTION - [math] fonction *f* complémentaire
~RECTIFIER - [el] redresseur *m* auxiliaire
COMPLETE, to - [gen] compléter, finir, achever, réaliser
[chem] se terminer, rendre complet
[naut] armer
COMPLETE - [gen] complet, total, intégral
~CARRY - [comput] report *m* final
~CHILL - [metall] trempe *f* totale
~CURE - [rubber ind] vulcanisation *f* complète
~GASIFICATION - [gas ind] gazéification *f* intégrale
~GASIFICATION GAS - [gas ind] gaz *m* intégral
~GELATION - [ind chem] gélification *f* intégrale
~LINKAGE - [genet] liaison *f* absolue
~MINERAL MANURE - [ind chem] engrais *m* chimique complet
~OPERATION - [comput] opération *f* complète
~PURIFICATION - [gas ind] épuration *f* intégrale
~RADIATOR - [phys] corps *m* noir
~REACTION - [chem] (an irreversible reaction which continues until one of the reactants has disappeared) réaction *f* complète
COMPLETED SHELL - [electron] (shell containing the full number of electrons) couche *f* électronique saturée, couche *f* complète
COMPLETELY DIFFUSE SOUND - [acoust] (a sound with uniform energy density in a given region) son *m* diffus
COMPLETION - [gen] achèvement *m*, complétion *f*, conclusion *f*
~GAUGE - [mining] résultat quantitatif pour la mise en production d'un puits
~RADIATION - [phys] rayonnement *m* du corps noir
COMPLEX - [gen] complexe *m*
[adj] complexe, compliqué, difficile
~ADMITTANCE - [el] (a complex quantity having the scalar value of the admittance as its modulus and the phase of displacement between voltage and current as argument) admittance *f* complexe (d'un circuit)
~AMINO-SALTS - [chem] sels *mpl* complexes aminés
~AUTOMATIC CONTROL SYSTEM - [control] système *m* de réglage automatique complexe
~CATHODE - [electron] (its emitting surface contains several emitting components) cathode *f* composée
~COMPOUND - [chem] composé *m* complexe
~DAMPING - [mech] amortissement *m* complexe
~DISPLACEMENT - [phys] déplacement *m* complexe
~DISPLAY - [radar] (display system with data representation given in different displays) présentation *f* combinée
~FRACTION - [math] fraction *f* complexe

COMPLEX FUNCTION - [math] fonction *f* complexe
~IMPEDANCE - [el] (as in admittance, impedance replacing admittance) impédance *f* complexe
~INDEX OF REFRACTION - [opt] indice *m* de réfraction composé
~ION - [phys] (an ion capable of dissociation into simpler ions) ion *m* composé
~LIQUID - [phys] (a liquid in which the rate shear is not merely a linear function of the shearing stress) liquide *m* complexe
~NUMBER - [math] nombre *m* complexe
~POWER - [el] (in this expression active power is the real part and reactive power the imaginary one) puissance *f* complexe
~QUANTITY - [math] quantité *f* complexe
~SOUND - [el acoust] (any sound which is not pure sound) son *m* complexe
~SPECTRUM - [phys] spectre *m* complet
~STEEL - [metall] acier *m* composite
~TISSUE - [bot] (made up of elements of more than one kind) tissu *m* complexe
~TONE - s. combination tone
COMPLEXOMETRY - [instr] complexométrie *f*
COMPLEXUS - [zool] (complicated system of organs) complexe *m*
COMPLIANCE - [gen] conformité *f*, obéissance *f*, soumission *f*
[mech] (of a spring) compressibilité *f*
[med] (of a lung) élasticité *f*
~CONSTANT - [mech] constante *f* d'elasticité
COMPLICATE - [gen] compliqué, complexe
COMPLICATED - [gen] compliqué, complexe
~SHOT - [cin] (picture requiring special technique in shooting) plan *m* difficile
COMPLICATION - [med] complication *f*
COMPO - [build] (a cement mortar) stuc *m*
~BOARD - [build] (narrow strip of wood glued to one another to form a sheet) planche *f* en bois aggloméré
~PIPE - [metall] (alloy made pipe) tube *m* en alliage
COMPOLE - [el] (auxiliary pole in commutator motors) pôle *m* de commutation
COMPONENT - [gen] composant *m*, élément *m*, organe *m*
[mech] pièce *f* détachée
[chem] constituant *m*
[phys] composante *f*
[electron] (general term denoting any circuit element) composant *m*
~FORCES - [phys] forces *f*pl composantes
~OF A SYMMETRICAL SYSTEM - [el] (one of the quantities constituting a symmetrical polyphase system) composante *f* d'un système symétrique
~OF WIND PRESSURE - [phys] composante *f* du vent
~PART - [gen] pièce *f* détachée
~VORTICITY - [phys] (the circulation at a given point in a fluid round an elementary surface at right angle to the direction taken, divided by the area of that surface) tourbillonnement *m*
COMPOSE, to - [gen & print] composer
COMPOSED PIPE - [constr] (lead and tin) tuyau *m* composé plomb et étain
~STAIRS - [constr] escalier composé *m*
COMPOSING - [gen & print] composition *f*
~FRAME - [print] (the structure facing the compositor) casse *f*
~MACHINE - [mech] (monotype, linotype etc) composeuse *f*
COMPOSING ROOM - [print] atelier *m* de composition, composition *f*
~RULE - [print] filet *m* de composition
~STICK - [print] (a container in which the compositor sets the type letter by letter) composteur *m*
COMPOSITAE - [bot] composacées *f*pl, synanthérées *f*pl
COMPOSITE - [gen etc] composé, composite, à structure *f* mixte
~AIRCRAFT - [aero] aéronef *m* de construction mixte
~ANTICLINE - [geol] anticlinorium *m*
~BODY - [auto] carrosserie *f* mixte
~BOX PURIFIER - [gas ind] épurateur *m* monobloc
~CABLE - [el] câble *m* mixte
~CHART - [met] (chart showing the forecast weather conditions at various points along an air route) carte *f* de prévisions météorologiques
~CIRCUIT - [telecomm] (a telephone circuit used as telegraph circuit) circuit *m* approprié
~-COIL WATTMETER - [instr] wattmètre *m* équilibré
~CONDUCTOR - [el] (conductor consisting of more than one material) conducteur *m* composé
~CONTROLLING VOLTAGE - [el] tension *f* de réglage composée
~COURSE - [nav] (a combination of Great Circle and rhumb-line courses, designed to give the shortest praticable path) route *f* mixte
~FUEL - [min] combustible *m* complexe
~GNEISS - [min] magmatite *f*
~JOINT - [metall] (in welding) joint *m* composé
~LENS - [opt] (compound lens) objectif *m* à plusieurs lentilles
~LOUDSPEAKER - [el acoust] (a system of two or more loudspeaker radiating simultaneously in certain frequency bands) haut-parleur *m* à voies multiples
~MOULD - [plast ind] (a mould containing various impressions within a common bolster) moule *m* à empreintes différentes
~PICTURE SIGNAL - s. composite video
~PLATE - [el plating] (coating formed of two or more layers deposited separately) dépôt *m* à plusieurs couches
~PRINT - [cin] (sound and picture on the same film) film *m* sonorisé
~PROPELLANT - [chem] (used in rockets) mélange *m* propulsif hétérogène
~SET - [telecomm] installation *f* terminale de circuit approprié
~SHOT - [photo] prise *f* de vue combinée
~SOIL - [soil] agrégat *m*, mélange *m* de sol
~SYNCHRONIZATION SIGNAL - [telev] signal *m* de synchronisation complexe
~SYNCLINE - [geol] synclinorium *m*
~VESSEL - [shipbuild] bâtiment *m* de construction mixte
~VIDEO SIGNAL - [telev] signal *m* vidéo complexe
~WAVE FILTER - [radio] (selective transducer with two or more filters) filtre *m* de bande composé
COMPOSITION - [gen] composition *f*, mélange *m*, enduit *m*, produit *m*
[print] composition *f*
~OF FABRIC - [text] composition *f* du tissu
~OF FORCES - [phys] (the process of finding the resultant of a number of forces) composition *f* des forces

COMPOSITION OF MILK - [agric] (dairy industry) composition *f* du lait
~OF THE MIXTURE - [rubber ind] composition *f* d'un mélange
~PEDAL - [mus] (in an organ) pédale *m* de composition
~POTENTIOMETER - [el] potentiomètre *m* non bobiné
~RIDER ROLLER - [print] rouleau *m* distributeur
COMPOSITIVE - [gen] synthétique
COMPOSITOR - [print] compositeur *m*
COMPOST, to - [agric] composter, terreauter
COMPOST - [agric] compost *m*, terreau *m*
~HEAP - [agric] tas *m* de terreau
COMPOUND, to - [gen etc] accoupler, composer, raccorder
[ind chem] mélanger, combiner
COMPOUND - [gen] composé, composite, complexe
[chem] (a chemical compound) composé *m*, combinaison *f* chimique, mélange *m*
[chem] (a pharmaceutical compound) préparation *f*
[phys] (a pure substance which can be decomposed into other different pure substances) corps *m* composé
~ACTION - [contr] (also called multiple action; the control action of a controller operating with more than one type of action) action *f* composée
~ARCH - [arch] arc *m* composite
~ARTIFICIAL MANURE - [agric] engrais *m* artificiel composé
~BEAM - [constr] poutre *f* composée
~BEARING - [mech] palier *m* composé
~BRUSH - [el] (a brush collecting current from the commutator of an electric motor) balais *m* composé
~CATENARY CONSTRUCTION - [el] (in electric fraction system) caténaire *f* composée
~COIL - [el] bobine *f* compound
~CONDENSING ENGINE - [mech] machine *f* compound à condensation
~CONTROL ACTION - [contr] réglage *m* convergent
~COUPLING - [mech] couplage *m* mixte
~CURVE - [surv] (curve consisting of two arcs of different radii) courbe *f* polycentrique
[math] courbe *f* composée
~CYCLE - [electron] (thermal cycle using more than one circuit) cycle *m* multiple
~DYEING - [text] teinture *f* combinée, teinture composée
~DYNAMO - [el] dynamo *f* à excitation composée
~ENGINE - [mech] (of a steam engine) machine *f* compound
~EXCITATION - [el] (excitation provided by shunt and series winding) excitation *f* composée
~FILTER - [radiat] filtre *m* composé
~FRUIT - [bot] (formed from various associated flowers) fruit *m* composé
~GENERATOR - [el] (with a shunt and a series field winding) moteur *m* compound
~GIRDER - [build] (rolled strengthened steel joint) poutrelle *f* composée, poutre *f* composée
~LEAF - [bot] feuille *f* composée
~LENS - [opt] objectif *m* composé
~LEVER - [mech] (series of levers to obtain better mechanical results) levier *m* composé
~LOAD - [constr] surcharge *f* composé
~MAGNET - [el] (permanent magnet consisting of several laminations) aimant *m* composé
COMPOUND MICROSCOPE - [instr] microscope *m* (composé)
~MILL FOR DRY PROCESS - [constr] (in cement mixing) moulin *m* combiné pour procédé par voie sêche
~MILL FOR WET PROCESS - [constr] (in cement mixing) moulin *m* combiné pour procédé par voie humide
~MOTION - [mech] mouvement *m* composé
~MOTOR - [el] moteur *m* compound
~NEEDLE - [text] aiguille *f* à rainure, aiguille *f* tubulaire
~NUCLEUS - [mech] (intermediate nucleus: an excited nucleus formed as an intermediate stage in an induced nuclear reaction) noyay *m* composé
~NUMBER - [math] nombre *m* complexe
~OIL - [ind chem] huile *f* composé
~ORDER - [arch] ordre *m* composé
~OVEN - [metall] four *m* compound
~PENDULUM - [mech] (pendulum consisting of a rigid body with no limit of size, shape or composition) pendule *m* composé
~PILLAR - [build] colonne *f* composée
~PROPELLER TURBINE - [aero] turbo-propulseur *m*
~PULLEY - [mech] palan *m*
~PUMP - [mech] (having two stages in series) pompe *f* à deux étages (en série)
~REFLEX - [zool] (combination of various reflexes forming a coordination) réflexe *m* combiné
~RESONATOR - [acoust] résonateur *m* composé
~REST - [mech] chariot *m* à mouvements croisés
~ROTORCRAFT - [aero] (an aircraft combining lifting structures of aeroplane and rotorcraft) giravion *m* composé
~SCREW - [mech] vis *f* differentielle
~SEMI-CONDUCTOR - [electron] semi-conducteur *m* composé
~SLIDE - [mech] chariot *m* supérieur
~SLIDE REST - [mach tools] support *m* à chariot à double coulisse
~SLIDING TABLE - [mech] plateau *m* à mouvements croisés
~SLOTTED WORK-TABLE - [mach tools] table *f* porte-pièce à rainures à double coulisse
~STEAM ENGINE - [mech] machine *f* compound
~STEEL - [metall] acier *m* allié, acier *m* compound
~SWITCH - [railw] traversée-jonction *f*
~TABLE - [mach tools] table *f* à double coulisse
~TIME - [mus] mesure *f* composée
~TOOL-HOLDER - [mach tools] porte-outil *m* à double coulisse
~TRAIN - [mech] (of gear-wheels) train *m* d'engrenages
~TURBINE ENGINE - [aero] (type of gas turbine with compression in separate stages, each being driven by an individual turbine) moteur *m* à turbine compound
~UMBEL - [bot] ombelle *f* composée
~VEIN - [mining] filon *m* composé
~WEDGE - [mining] coin *m* multiple
~WINDING - [el] (a winding in which the field has both shunt and series coils) enroulement *m* compound
~-WOUND - [el] (in electric motors) à excitation composée
~-WOUND DYNAMO - [el] dynamo *f* compound
COMPOUNDED LATEX - [rubber ind] latex *m* conte-

nant des agents de vulcanisation
COMPOUNDER-EXTRUDER - [mech] extrudeuse-mélangeuse *f*
COMPOUNDING - [mech] (the use of the principle of expanding steam in more than one stage) compoundage *m*
[ind chem] mélange *m*, incorporation *f*
~INGREDIENTS - [rubber ind] composants *mpl* d'un mélange, constituants *mpl* d'un composé
~OPERATION - [ind chem] opération *f* de mélange
~PRACTICE - [ind chem] technique *f* de mélange
~ROOM - [rubber ind] atelier *m* de préparation des mélanges
COMPREG - [plast ind] (resin-impregnated compressed wood) bois *m* imprégné densifié
COMPREGNATE, to - [ind chem] (to treat by compression and impregnation at the same time) densifier (par compression et imprégnation)
COMPRESS, to - [gen] comprimer, compresser, presser, refouler
COMPRESS - [text] presse *f*
COMPRESSED - [gen etc] comprimé, compressé
~AIR - [phys] (air under considerable pressure used as a means power transmission) air *m* comprimé
~-AIR CAISSON - [constr] caisson *m* à air comprimé, caisson *m* foncé à l'air comprimé
~AIR DRILL - [oil ind] forage *m* à air comprimé
~AIR DRYING PLANT - [ind chem] installation *f* de séchage à air comprimé
~AIR EJECTION - [plast ind] (the ejection of the moulding from the mould by compressed air) éjection *f* par air comprimé
~AIR HAMMER - [impl] marteau *m* à air comprimé
~AIR HOSE - [rubber ind] (rubber hose with very thick walls) tuyau *m* pour air comprimé
~AIR PUMP - [mech] pompe *f* de compression
~AIR SANDBLASTER - [metall] sableuse *f* à air comprimé
~AIR STARTER - [mech] (arrangement for starting an engine by compressed air) démarreur *m* à air comprimé
~AIR WIND TUNNEL - [aero] (wind tunnel in which highvalues are obtained by using compressed air) soufflerie *f* à densité variable
~ASPHALT - [ind chem] asphalte *m* comprimé
~CHARGE - [mech] (in internal combustion engines) mélange *m* comprimé
~-FIBRE CARD - [text] carton *m* comprimé
~FOLD - [geol] pli *m* serré
~GAS - [gas ind] gaz *m* comprimé
~WOOD - [bot] (wood whose density has been increased at the base of some tree trunks) bois *m* comprimé
~SLAKS - [constr] briquettes *fpl*
COMPRESSIBILITY - [phys] (the relative change of volume per unit of pressure) compressibilité *f*
~BUBBLE - [aero] (shock wave when the speed of the air past an aerofoil approaches the speed of sound) onde *f* de choc de compression
~DRAG - [aero] (the rise in drag value caused by the compressibility of air) résistance *f* (aérodynamique) de compressibilité
~FACTOR - [mech] facteur *m* de compression
COMPRESSIBLE FLOW - [phys] (the flow condition in a fluid when there is appreciable change of density) débit *m* compressible
COMPRESSIBLE SOIL - [soil] sol *m* compressible, terrain *m* compressible
COMPRESSING CHANNEL - [text] canal *m* de compression
~PLATE - [text] tasseur *m*, tôle *f* égalisatrice
~TRAP - [text] clapet *m* de compression
COMPRESSION - [phys mech] (the decrease of a volume of a compressiblesubstancedue to the application of pressure) compression *f*
[int comb] (the stroke during which the working agent is compressed) compression *f*
~CHAMBER - [phys] chambre *f* de compression
~CONE - [radiat] (device designed to exert pressure on the body in radio-therapy) cône *m* de compression
~COUPLING - [mech] (a type of pipe joint in which the seal is compressed) raccord *m* à compression
~CURVE - [el] courbe *f* de compression
~CUTTING TEST - [rubber ind] essai *m* de résistance à l'incision par pression
~DEFLECTION TEST - [rubber ind] essai *m* de déformation par compression
~DIAGONAL - [build] barre *f* oblique
~DISTILLATION - [ind chem] distillation *f* par compression
~FAILURE - [mech] rupture *f* par compression
~GASKET - [mech] joint *m* à écrasement, joint *m* à compression
~GAUGE - [instr] compressiomètre *m*
~HEAT - [phys] chaleur *f* de compression
~JOINT - [geol] piézoclase *f*, fente *f* de compression
~LOAD - [phys] force *f* de compression
~MECHANISM - [mech] dispositif *m* de compression
~MEMBER - [mech] (structural member designed to resist compressive stress) montant *m*
~MOULD - [plast ind] (split mould for pressure forming) moule *m* à compression
~MOULDING - [plast ind] moulage *m* par compression
~OF THE SOIL - [soil] compactage *m* du sol
~PRESS - [mech] presse *f* de compression
~PIMP - [mech] pompe *f* de compression
~RATIO - [mech] (the ratio of the volume of the charge before compression to that when compression is complete) taux *m* de compression
~RIB - [aero] (a structural element running longitudinally in a control surface or wing) nervure *f* de compression, nervure *f* caisson
~RING - [mech] (piston ring in an internal combustion engine whose chief function is to retain the compression) segment *m* d'étanchéité
~ROLLER - [text] cylindre *m* compresseur, rouleau *m* compresseur
~SEAL - [mech] s. compression gasket
~SET - [metall] (a permanent deformation caused by compressive stress) déformation permanente *f* due à la compression
~SHOCK - [phys] choc *m* de compression
~SPACE - [mech] chambre *f* de compression
~STRAIN - [mech] déformation *f* due à la compression
~STRENGTH - [phys] résistance *f* à la compression
~STROKE - [mech] (the piston stroke in a reciprocating internal combustion engine during which the charge air or mixture is compressed) temps *m* de compression

COMPRESSION TEST - [mech] (test for the ductility and malleability of iron and steel bars) essai *m* à la compression, essai *m* à l'écrasement
~VACUUM GAUGE - [mech] manomètre *m* à compression, vacumètre *m* à compression
~ZONE - [plast ind] (the extruder barrel section in which melting is complete) zone *f* de compression
COMPRESSIONAL BAR - [mech] barre *f* travaillant à la compression
~WAVE - [acoust] onde *f* de compression
COMPRESSIVE - [adj] de compression
~STRAIN - [phys] déformation *f* due à la compression
~STRENGTH - [phys] (the ability of a material to resist compressive stress) résistance *f* à la compression, résistance *f* à l'écrasement
~STRESS - [mech] (the force per unit area in a material medium acting at right angle to that area) effort *m* de compression, travail *m* à la compression, contrainte *f* de compression
COMPRESSOR - [mech] (a reciprocating or other pump designed to raise the pressure of a gas) compresseur *m*
[el] (a thermionic amplifier designed to reduce variations in signal amplitude) compresseur *m*, écréteur *m*
[radio] compresseur *m* de volume
~ASSEMBLY - [aero] ensemble *m* compresseur
~BEARING - [aero] appui *m* de compresseur
~CASING - [aero] (the casing which encloses the impeller or compressor) carter *m* compresseur
~DELIVERY DUCTS - [aero] (the passages connecting the compressor discharge to the combustion system) pipes *f*pl de refoulement
~DRUM - [aero] (the central body on which the blades in an axial-flow compressor are mounted) tambour *m* du compresseur
~IMPELLER - [aero] roue *f* de compresseur
~ROTOR - [aero] (in an axial-flow compressor) rotor *m* de compresseur
~SEALING PLATE - [mech] plaque *f* d'étancheité de compresseur
~STATION - [oil ind] station *f* de pompage
COMPRESSURE STRENGTH - [mech] écrasement *m*, résistance *f* à l'écrasement
COMPROMISE BALANCE - [telecomm] équilibreur *m* omnibus
~NETWORK - s. compromise balance
COMPTOMETER - [mech] machine *f* à calculer
COMPTON ABSORPTION - [nucl] (the absorption of an X-ray or gamma-ray photon in the Compton effect) absorption *f* Compton
~COLLISION - [phys] choc *m* Compton
~EFFECT - [nucl] (elastic scattering of photons by electrons) effet *m* Compton
~ELECTRON - [nucl] (set in motion through interaction with a photon in the Compton effect) électron *m* Compton
~METER - [nucl] (ionization chamber for cosmic ray measurements) chambre *f* d'ionisation Compton
~RECOIL PARTICLE - [nucl] (particle finding its momentum in a scattering process) électron *m* de recul Compton
~RULE - [phys] (empirical relationship between the thermal properties of the elements) règle *m* de Compton

COMPTON SCATTERING - [nucl] diffusion *f* Compton, effet *m* Compton
~WAVELENGTH - [nucl] longueur *f* d'onde de Compton
COMPTROLLER - s. controller
COMPULSION - [gen] contrainte *f*, obligation *f*
[med] contrainte *f*
COMPULSION NEUROSIS - [med] (obsessional neurosis) névrose *f* obsessionnelle
COMPULSORY - [gen & leg] obligatoire
COMPUTATION - [gen] calcul *m*, compte *m*, comptage *m*
COMPUTE, to - [gen] calculer, estimer
COMPUTER - [gen] calculateur *m*, machine *f* comptable, calculatrice *f*
~CODE - [comput] code *m* d'ordinateur
~INSTRUCTION - [comput] (instruction understood by the control circuits and effecting an operation) instruction *f*
~OPERATION - [comput] (the operation resulting from an instruction to the computer) opération *f* élémentaire
COMPUTING - [adj] de calcul
~CENTRE - [comput] (the premises where all the computing machinery is concentrated) centre *m* de calcul
~ELEMENT - [contr] (device accepting input signals and whose output is the result of calculation on those signals) élément *m* de calcul
~INTERVAL - [comput] temps *m* de calcul
~MACHINERY - [comput] installation *f* de calcul
~TIME - s. computing interval
CON, to - [naut] gouverner, manoeuvrer
CONCATENATED - [bot] (joined together in a chain-like form) enchaîné, lié, concaténé
~CONNEXION - s. cascade connexion
~MOTOR - s. cascade motor
CONCATENATION - [gen] concaténation *f*, chaîne *f*
CONCAVE - [gen] concave
~CATHODE - [electron] (cathode whose concave surface faces the anode) cathode *f* concave
~-CONVEX LENS - [opt] lentille *f* concavo-convexe
~FILLET WELD - [metall] (fillet weld with concave surface) soudure *f* d'angle concave
~GRATING - [opt] réseau *m* concave
~LENS - [opt] lentille *f* concave
~MILLING CUTTER - [mech mach] fraise *f* concave
~MIRROR - [el] miroir *m* concave
~OF TRESHER - [agric] contre-batteur *m*
~PLANE - [impl] (to work concave surfaces) rabot à moulurer
~REFLECTING GRATING - [opt] réseau *m* de réflexion concave
~TILE - [build] tuile *f* creuse
CONCAVITY - [gen] concavité *f*, creux *m*
CONCAVO-CONCAVE - [gen] biconcave
CONCEAL, to - [gen] cacher, dissimuler, masquer, escamoter
CONCEALABLE - [gen] escamotable
CONCEALED CARCASSING - s. concealed piping
~DAMAGE - [mech etc] avarie *f* occulte
~DOOR HINGE - [auto] charnière *f* encastrée
~EROSION - [geol] érosion *f* occulte
~HEATING - [th eng] (panel heating) chauffage *m* par panneaux radiants
~LIGHTING - [light] (when the source of illumination is not directly to the eye) éclairage *m* indirect
~OUTCROP - [geol] affleurement *m* masqué

.ONCEALED PIPING - [gas & other install] installation *f* encastrée
˜SIGNAL - [railw] signal *m* effacé, signal *m* masqué
ONCENTRATE - [gen etc] concentré *m*
:ONCENTRATED - [gen chem etc] concentré
˜AMMONIACAL LIQUOR - [chem] eau *f* ammoniacale concentrée
˜GAS LATEX - [rubber ind](latex in a concentrated form) latex *m* concentré
˜LOAD - [mech] (load considered as acting through a point) charge *f* concentrée
˜REFLECTOR - [light] réflecteur *m* à concentration
˜WINDING - [el] (winding in which all the conductors of one group are placed in one slot) enroulement *m* concentré
:ONCENTRATING MILL - [min] installation *f* de concentration
˜PERCUSSION TABLE - [min] table *f* à secousses
:ONCENTRATION - [gen] concentration *f*
[chem] concentration *f*, enrichissement *m*
˜BY FLOCCULATION - [rubber ind] concentration *f* par flocculation
˜CELL - [el] (a two-fluid cell in which the same electrolyte is used for both electrodes) pile *f* de concentration
˜COIL - [el] bobine *f* de concentration, bobine *f* de focalisation
˜CUP - [electron] (designed to focus the electron beam) cupule *f* de focalisation
˜FALL - [el] chute *f* de concentration
˜GRADIENT - [el] gradient *m* de concentration
˜OF ELECTRONS - [electron] (the number of free electrons per unit volume) densité *f* électronique
˜OVERVOLTAGE - [el chem] (the portion of the overvoltage due to changes in concentration near the electrode) surtension *f* de concentration
:ONCENTRATOR PANEL - [telecomm] pannneau *m* de commutation
˜PLANT - [min] (all those appliances and structures used in removing the barren elements of mineral material in order to recover the valuable portion) installation *f* de concentration
˜POLARIZATION - [el] polarisation *f* de concentration
:ONCENTRATOR - [min] s. concentration plant
[el] (transformer in which the heating inductor forms a structural part and concentrates the heating effect on a small area) concentrateur *m*
[telecomm] centre *m* de commutation
ONCENTRIC - [gen] concentrique
˜ARCH - [build] (laid in more than one course whose curves have a common centre) arc *m* concentrique
˜CABLE - [el] (cable with two or more concentric conductors) câble *m* coaxial
˜CHUCK - [mech] (a self centring chuck) mandrin *m* auto-centreur
˜CIRCLES - [geom] cercles *mpl* concentriques
˜ELECTRODE SYSTEM - [el] (electrode system geometrically coaxial but electrically asymmetrical) système *m* concentrique d'électrodes
˜LENS - [opt] lentille *f* homocentrique
˜LINE - s. co-axial feeder
˜LOAD - [mech] pression *f* axiale, charge *f* axiale
˜PLUG-AND-SOCKET - [el] (type of plug-and-socket connexion in which a contact is a central pin and the other a concentric ring) prise *f* concentrique
CONCENTRIC ROTARY SCREEN - [impl] (used for grading coal and coke) tambour *m* classificateur
˜TUBE FEEDER - [radio] (type of transmission line with two conductors being concentric tubes) feeder *m* concentrique
˜TUBE TRANSMISSION LINE - [radio] ligne *f* de transmission coaxiale
˜VASCULAR BUNDLE - [bot] faisceau *m* vasculaire concentrique
˜WINDING - [el] (a winding in which each phase comprises a coil group formed of concentric coils) enroulement *m* concentrique
CONCENTRICITY - [gen & mech] concentricité *f*
˜CHECK - [mech] vérification *f* de la concentricité
CONCERT - [gen] accord *m*
[mus] concert *m*
˜FLUTE - [mus] grande flute *f*
˜GRAND - [mus] piano *m* de concert
CONCERTINA - [mus] (accordion without keyboard) concertina *f*
˜FOLDING - [print] plié *m* accordéon
CONCESSION - [gen & comm] concession *f*
CONCESSIONAIRE - [comm] concessionaire *m*
CONCHA - [anat] (the cavity of the outer ear) cornet *m*, conque *f*
CONCHOID - [geom] conchoïde
CONCHOIDAL - [geom etc] (exhibiting concentric circles) chonchoïdal
˜FRACTURE - [mech] fracture *f* conchoïdale
˜STRUCTURE - [min] structure *f* conchoïdale
CONCORDANCE - [gen] concordance *f*
CONCRESCENCE - [bot] (the formation of a single structure through growing together) concrescence *f*
CONCRETE - [gen] concret
[build] (structural material made by mixing cement with sand and crushed stone or gravel and water) béton *m*
˜ADMIXTURE - [constr] adjuvant *m* pour béton
˜-AND-GLASS - [constr] béton *m* translucide
˜-AND-GLASS PAVEMENT LIGHT - [constr] dalle *f* translucide pour chaussée
˜APRON - [constr] aire *f* bétonnée, radier *m* en béton
˜BED - [constr] dalle *f* de béton, socle *m* en béton
˜BLOCK - [constr] parpaing *m*, bloc *m* en béton *m*
˜CASTING - [constr] coulée *f* de béton
˜CONVEYOR - [constr] transporteur *m* à béton
˜DAM - [constr] barrage *m* en béton
˜DEFORMATION PICK-UP - [instr] (an apparatus checking the behaviour of concrete) jauge *f* de déformation
˜FOOTING - [constr] socle *m* en béton
˜FOUNDATIONS - [constr] fondations *fpl* en béton
˜-FRAMED CONSTRUCTION - [constr] construction *f* en ossature en béton
˜GRAVITY SECTION DAM - [constr] barrage-poids *m* en béton
˜IRON - [metall] fer *m* pour béton armé
˜LAYER - [constr] couche *f* de béton
˜-LINED - [constr] à revêtement de béton
˜MIXER - [mech] (machine for mixing the ingredients of concrete) bétonnière *f*, bétonneuse *f*
˜MORTAR - [constr] mortier *m* du béton
˜MUSIC - [mus] (synthetically generated musical sounds) musique *f* concrète

CONCRETE PAINT - [paint] peinture *f* pour béton, peinture *f* pour ciment
~PAVING MACHINE - [constr] bétonneuse *f* pour routes
~PILE - [constr] pilier *m* en béton
~PIPE - [constr] canalisation *f* en béton
~PREPARING EQUIPMENT - [constr] installation *f* de bétonnage
~ROAD - [constr] route *f* en béton
~ROOF - [constr] toit *m* en béton
~RUNWAY - [aero] piste *f* en béton
~SHIELD - [nuch] (a biological shield consisting of concrete) écran *m* en béton
~SLAB - [constr] dalle *f* de béton
~SPREADER - [mech] épandeuse *f* à béton
~VIBRATOR - [mech] vibrateur *m* à béton
CONCRETING - [constr] bétonnage *m*, travaux *mpl* de béton
CONCRETION - [med] (collection of organic matter in organs) concrétion *f*, calcul *m*
[geol] (condensation) concrétion *f*
CONCUR - [gen] concourir, concorder, être d'accord
CONCURRENT CENTRIGUGE - [mech] (used in isotopes separation) centrifuge *f* à écoulement continu
~FLOW APPARATUS - [gas ind] (used in the treatment of gas) appareil *m* à courants parallèles
~FLOW GASIFICATION - [gas ind] gazéification *f* en courants parallèles
~FORCES - [mech] forces *fpl* concourantes
~LINES - [geom] lignes *fpl* concourantes
~OPERATIONS - [comput] opérations *fpl* concourantes
CONCUSSION - [gen] secousse *f*, ébranlement *m*, choc *m*
[med] commotion *f* cérébrale
~OF THE BRAIN - s. concussion [med]
~BURST - [mech] éclatement *m* causé par un choc
~OF THE LABYRINTH - [med] commotion *f* labyrintique
~SPRING - [mech] ressort *m* amortisseur
CONDEMN, to - [gen] condamner, interdire l'usage
[naut] (of ships) radier
CONDENSABLE GASES - [phys] gaz *mpl* condensables
CONDENSATE - [chem] (liquid produced by condensation) condensat *m*, produit *m* de condensation, eau *f* de condensation
~DRAINAGE SYSTEM - [hydr] canalisation *f* d'évacuation des eaux de condensation
~FIELD - [oil ind] gisement *m* de condensat
~OUTLET - [hydr] purge *f* d'eau de condensation
~PUMP - [hydr] pompe *f* de récupération des eaux de condensation
~TANK - [ind chem] récipient *m* collecteur (de condensat), collecteur *m* de condensat
~TRAP - [ind chem] séparateur *m*, piège *m* à condensat
~WELL - [oil ind] (a well with a high ratio of gas-oil) puits *m* à rapport élevé gaz/brut
CONDENSATION - [phys] (change from the vapour to the liquid state) condensation *f*
[chem] (formation of long-chained compounds by the linking of two or more molecules) condensation *f*
~AGENT - [chem] (a substance which initiates or accelerates condensation) agent *m* de condensation
CONDENSATION CLOUD - [nucl] (the water droplets surrounding the fire ball after a nuclear detonation in humid atmosphere) nuage *m* de condensation
~COLLECTOR - [th eng] gouttière *f* de condensation
~COLUMN - [ind chem] colonne *f* de condensation
~GROOVE - s. condensation collector
~GUTTER - S. condensation collector
~LEVEL - [met] (the level at which cooling by ascent causes saturation in rising air) niveau *m* de condensation
~NUCLEUS - [met] (ice crystals on which vapour from supercooled cloud droplets condenses to form raindrops) noyau *m* de condensation
~OF GAS - [gas ind] condensation *f* d'un gas
~PAN - s. condensation collector
~POLYMERIZATION - [chem] (polymerization reaction in which water is eliminated) polymérisation *f* par condensation
~POT - [ind chem] vase *m* de condensation
~PUMP - [mech] pompe *f* à diffusion, pompe *f* à condensation, pompe *f* à entraînement moléculaire
~RESINS - [chem] (resins produced by condensation polymerization) résines *fpl* polymères de condensation
~TRAIL - [aero] traînée *f* de condensation
~WATER - [chem] eau *f* de condensation
CONDENSE, to - [phys chem] condenser
~TANK - [ind chem] réservoir *m* de condensation
CONDENSED - [gen] condensé
~FILM - [phys] pellicule *f* de condensation
~MILK - [agric] lait *m* condensé
~NUCLEUS - s. condensation nucleus
~SYSTEM - [phys] (substance or mixture of substances in the liquid or solid state) système *m* condensé
~WATER SEPARATOR - [ind chem] (a mechanism for the removal of water of condensation from steam) purgeur *m* d'eau de condensation
CONDENSER - [ind chem] (any apparatus designed to convert a substance from the gaseous to the liquid state) condenseur *m*
[chem] réfrigérant *m*
[el] (obsolete for capacitor) condensateur *m*
[opt] (condenser lens; lens or system of lenses used to concentrate light on an object) condensateur *m*
~AERIAL - [radio] antenne *f* capacitive
~ANTENNA - s. condenser aerial
~BELL - [ind chem] boule *f* de condensation
~BOBBIN - [text] bobine *f* de mèches
~BUSHING - [el] douille *f* à condensateur
~CARD - [text] carde *f* finisseuse
~CHAMBER - s. condenser bell
~CIRCULATING PUMP [mech] pompe *f* de circulation d'eau de condensation
~COIL - [th eng] serpentin *m* de refroidissement
~DRUM - [text] tambour *m* dérouleur
~ELECTROSCOPE - [instr] (electroscope fitted with a condenser to increase its sensitivity) électroscope *m* à condensateur
~FOCUSING SLEEVE - [opt] anneau *m* de mise au point (de condensateur)
~IONIZATION CHAMBER - [nucl] chambre *f* d'ionisation à condensateur
~LOUDSPEAKER - [el acoust] (loudspeaker depending for its operation on electrostatic forces) haut-

parleur *m* éléctrostatique
:ONDENSER MERCURY TEMPERATURE - [electron] (the temperature measured on the outside of the envelope where the mercury is condensing) température *f* du mercure condensé
˜METER - [el] électromètre *m* à condensateur
˜MICROPHONE - [el acoust] microphone *m* électrostatique
˜TRANSMITTER - s. condenser microphone
˜TRANSMITTER AMPLIFIER - [el acoust] (amplifier associated with a condenser microphone) amplificateur *m* de microphone électrostatique
˜TRUMPET - [text] entonnoir *m* collecteur
˜TUBES CLEANING BULLETS - [th eng] boucles *fpl* pour le nettoyage des tubes de chaudières
˜WITH DAMPING RESISTANCE - [el] condensateur *m* à résistance d'amortissement
:ONDENSING - [phys] condensation *f*
˜ELECTROSCOPE - s. condenser electroscope
˜FUNNEL - [metall] entonnoir *m*, entonnoir *m* à trop-plein
˜LENS - s. condenser [opt]
˜POWER - [phys] puissance *f* condensatrice, capacité *f* condensatrice, rendement *m* du condenseur
˜STEAM ENGINE - [mech] machine *f* à vapeur à condensation
˜UNIT [th eng] (in a refrigerating unit) ensemble *m* frigorifique
:ONDITION, to - [gen] conditionner, mettre en condition, préparer
[text] (the operation of checking humidity in wool) conditionner
:ONDITION - [gen] condition *f*, état *m*, régime *m*, circonstance *f*
˜OF EQUILIBRIUM - [phys] état *m* d'équilibre
:ONDITIONAL - [gen] conditionnel
[comput] (subject to the result of comparison carried out during computation) conditionnel
˜BREAKPOINT INSTRUCTION - [comput] (conditional instruction which stops the computer, later allowing to continue the same routine or to jump to another routine) instruction *f* d'arrêt conditionnel
˜INSTRUCTION - [comput] (causing a discrimination) instruction *f* conditionnelle
˜INTERLOCK - [el] couplage *m* conditionnel
˜JUMP - [comput] (instruction which, when reacted in the course of a programme, will lead the computer either to deal with the next instruction in the sequence or to jump to another instruction) saut *m* conditionnel
˜STABILITY - [el] (stability conditional to the signal falling within a particular range) stabilité *f* conditionnelle
˜TEST - [gen] essai *m* sous condition
˜TRANSFER OF CONTROL - s. conditional jump
:ONDITIONED - [gen] conditionné
[th eng] (of air] climatisé
˜WEIGHT - [text] poids *m* conditionné
:ONDITIONER - [text] humecteur *m*
[min] conditionneur *m*
˜DRYER - [agric] conditionneur *m*
:ONDITIONING - [gen] conditionnement *m*
[ind] (treatment of a material to bring it to a prescribed standard) conditionnement *m*, traitement *m*
[hydr] (of sludge) amélioration *f*, préparation *f*, traitement *m*
[text] conditionnement *m*
CONDITIONING CABINET - [ind chem] armoire *f* de conditionnement
˜OF SILK - [text] conditionnement *m* de la soie
˜OVEN - [text] four *m* de conditionnement
˜PERIOD - [ind] (the time required for conditioning) durée *f* de conditionnement
˜PROCESS - [ind] procédé *m* de traitement
˜ROOM - [text etc] salle *f* de conditionnement
˜STIMULUS - [el biol] (stimulus of a given configuration applied to a tissue before a test stimulus) stimulus *m* de mise en condition
CONDITIONS OF SEVERITY - [ind] conditions *fpl* d'essai sévère
CONDUCIVE - [gen] favorable à, qui contribue à
CONDUCT, to - [el phys etc] conduire
CONDUCT - [gen] conduite *f*, comportement *m*
˜OF WATER - [hydr] (irrigation) canalisation *f* d'eau
CONDUCTANCE - [el] (the reciprocal of resistance) conductance *f*
˜RATIO - [el] rapport *m* de conductance
CONDUCTIBILITY - [el] (the reciprocal of the specific resistance of a conductor) conductibilité *f*
CONDUCTIMETRIC ANALYSIS - [chem] (a technique utilizing changes in the conductivity of a solution) analyse *f* conductimétrique
CONDUCTING LAYER - [el] couche *f* conductrice
˜MATERIAL - [el] matériau *m* conducteur
˜PERIOD - [electron] (the part of an alternating voltage cycle during which a certain arc path is carrying current) durée *f* de conduction
˜SALTS - [el plating] (salts which are used to increase the conductivity of a plating solution) sels *mpl* conducteurs
˜STRAND - [bot] (vascular bundle) faisceau *m* vasculaire
˜WIRE - [el] fil *m* conducteur
CONDUCTION - [phys] (the transmission of energy) conduction *f*
[mech] (of materials) acheminement *m*
˜BAND - [electron] bande *f* de conduction
˜CURRENT - [electron] (the current produced by the circulation of electrons or ions in a conductive medium) courant *m* de conduction
˜DEAFNESS - [med] surdité *f* de transmission
˜ELECTRON - [electron] (a valence electron with energy within the conduction band) électron *m* de conduction
˜HOLES - [el] trous *mpl* de conduction
˜OF HEAT - [phys] (the transference of heat through a body) conduction *f* de la chaleur
CONDUCTIVE - [phys] (capable of transmitting energy) conducteur
˜BRAID - [el] câble *m* conducteur
˜GLASS - [el] (glass made conductive to allow for deicing etc) verre *m* conducteur
˜LAYER - [el] couche *f* conductrice
˜METER - [instr] (instrument for the measure of conductivity) appareil *m* de mesure de la conductivité
˜RUBBER - [rubber ind] (rubber so compounded that it conducts electricity) caoutchouc *m* conducteur
˜TYRE - [aero] (tyre made conductive by the introduction of carbon black into its structure, so as to discharge static electricity when the aircraft

lands) pneumatique *m* conducteur
CONDUCTIVE WATER - [chem] (very pure water used for conductivity measurements) eau *f* de conductivité
CONDUCTIVITY - [el] (the reciprocal of resistivity) conductivité *f*
[phys] (the coefficient of thermal conduction) conductivité *f*
~ASH DETERMINATOR [metall] détermination *f* conductométrique des cendres
~BRIDGE - [el] pont *m* de conductivité
~MEASURING BRIDGE - [el] pont *m* de mesure de conductivité
~MODULATION - [electron] (the variations of the conductivity of a semiconductor) modulation *f* de conductivité
~MODULATION TRANSISTOR - [electron] transistor *m* à conductivité variable
~TEST - [el] essai *m* de conductivité
~TESTER - [meas] contrôleur *m* de conductivité
CONDUCTOMETER - [instr] (any instrument which measures conductivity) conductomètre *m*
CONDUCTOMETRIC TITRATION - [chem] titration *f* conductométrique
CONDUCTOR - [phys el] conducteur *m*
~ELEMENT - [el](part of a conductor whose assembly is connected in parallel) conducteur *m* élémentaire
~RAIL - [el] (live rail running parallel to the running rails to conduct the current to the train) rail *m* conducteur
~RAIL ANCHOR - [railw] ancrage *m* de rail conducteur
~RAIL INSULATOR - [el] isolateur *m* pour rail conducteur
~STRING - [oil ind] tube-guide *m*
CONDUIT - [el](tube or duct used to protect electric cables) gaine *f*
[hydr] (a pipe or channel for water) canalisation *f*, conduit *m*
~BOX - [el] boîte *f* de connexion
~FITTINGS - [el] (comprising all auxiliary items for the conduit system of wiring) accessoires *mpl* de montage pour canalisations
~SYSTEM - [el] (a system whereby conductors are contained in a steel conduit) câblage *m* électrique sous tubes
~WIRING - s. conduit system
CONDYLE - [zool] (a small protuberance at the end of a bone) condyle *m*
CONDYLOID - [anat] condyloîde
CONDYLOMA - [med] condylome *m*
CONE, to - [text] bobiner, canneter, enrouler
CONE - [geom] cône *m*
[mech] cône *m*
[metall] (of a blast furnace; steel cone suspended in the throat and counterpoised so as to sink when the charge is tipped in, thus preventing escape of gas) cloche *f* (de haut-fourneau)
[text] (a type of bobbin] cône *m*
[bot] pomme *f*, strobile *m*
[electron] (of a cathode-ray) cône *m*, robe *f*
~AERIAL - [radio] (aerial whose radiating element is shaped like a cone) antenne *f* conique
~AND CHEESE WINDER - [text] bobinoir *m* automatique
~ANTENNA - s. cone aerial
CONE BEARING - [mech] palier *m* conique
~BIT - [tool] fraise *f* conique
~CAPACITOR - [el] (capacitor consisting of two conducting cones) condensateur *m* à cônes concentriques
~CELL - [zool] (one of the photosensitive retina cells) rétinier
~CHUCKING RING - [mech] bague *f* de serrage
~CLASSIFIER - [min] appareil *m* classeur à cône
~CLUTCH - [mech] (friction clutch with driving and driven members consisting of conical frustra) embrayage *m* à cônes
~CONNEXION - [plum] jonction *f* conique
~COUPLING - [mech] accouplement *m* à cône
~CRUSHER - [mech] broyeur *m* à cônes
~DIAPHRAGM - [el acoust] (used for radiating sounds in receivers) membrane *f* conique
~DISTANCE - [mech] (in a straight bevel gear) génératrice *f* primitive
~DISTANCE AT FACE ANGLE - [mech] génératrice *f* extérieure
~DISTANCE AT ROOT ANGLE - [mech] génératrice *f* intérieure
~DRAWING - [text] (method of drawing wool whereby the speed of the bobbin decreases with the increase of its diameter) étirage *m* à cônes
~DRIVE - [mech] entraînement *m* par poulie étagée
~DRUMS - [text] (used in cotton spinning) cônes *mpl*
~FIBRES - [anat] fibres *fpl* des cônes rétiniens
~GEAR - s. cone drive
~GEAR WITH FRICTION STRAP - [mech] entraînement *m* à friction par poulie étagée
~HEAD - [mech] tête *f* tronconique
~HEAD RIVET - [mech] rivet *m* à tête tronconique
~INDICATOR - [aero] manche *f* à air
~LENGTH - [mech] (the length of a tapered roller bearing) hauteur *f* du cône de roulement
~LOUDSPEAKER - [el acoust] (a loudspeaker whose main radiating element has the shape of a cone) haut-parleur *m* à cône
~MIXER - [metall] mélangeur *m* à double cône
~OF DEPRESSION - [hydr] entonnoir *m* de dépression
~OF DISPERSION - [astronaut] cône *m* de dispersion
~OF FRICTION - [mech] (conical surface containing the resultant of the force of friction between two surfaces and the force pressing them together) cône *m* de frottement
~OF INTAKE - [hydr] (of a well) entonnoir *m* de prise
~OF LIGHT - [light] cône *m* de lumière
~OF NULLS [radio] (conical surface formed by directions of negligible radiations) cône *m* des zéros
~OF SILENCE - [radio] (conical region immediately above the aerial system of a radio range station, in which range signals are not heard) cône *m* de silence
~OF SILENCE MARKER BEACON - [radar] (radio beacon with emissions radiating in vertical cone-shaped pattern) balise *f* radio à faisceau vertical
~OF SLOPE - [constr] cône *m* de pente
~PULLEY - [mech] (a belt pulley formed with steps of different diameters to obtain changes of speed) poulie *f* étagée, poulie *f* à cône
~PULLEY DRIVE - [mech] (method of belt transmission with a range of speeds) entraînement *m* par poulie à cône

)NE PULLEY LATHE - [mach tools] tour *m* à poulie ›tagée
ʀADIUS - [mech] (of a tapered roller bearing) ·ayon *m* du cône de roulement
·SHAPED - [gen] conique
-SHAPED RUDDER - [aero] gouverne *f* de direcion conique
;HEET - [mining] filon *m* conique
:AP - [hydr] robinet *m* conique, robinet *m* à tourıant
/ARPING MACHINE - [text] ourdissoir *m* à cônes, ›urdissoir *m* sectionnel
/HEEL - [mech] roue *f* à cône
/INDING - [text] bobinage *m*
)NELESS LATHE - [mach tools] tour *m* à une pou.ie
)NES - [opt] (retinal cones) cônes *mpl*
)ISTANCE - [mech] génératrice *f*
)NFERENCE - [gen] conférence *f*, consultation *f*, :onseil *m*
:ALL - [telecomm] appel *m* multiple
:IRCUIT - [telecomm] circuit *m* de conférence
:ONNEXION - s. conference call
;YSTEM - [telecomm] (a telephone system) dis›ositif *m* téléphonique de conférence
)NFETTI - [telev] confetti *mpl*
)NFIGURATION - [gen] configuration *f*
[chem] (the structure of chemical compounds) :onfiguration *f*, structure *f*
:ONTROL - [nucl] (reaction rate control of a :lear reactor by change in the core configuration) réglage *m* de la configuration
)NFINED SPACE - [gen] espace *m* resserré
WATER - [geol] eau *f* captive, nappe *f* captive
)NFIRMATION WELL - [oil ind] (term denoting :he second active well in an oil field) puits *m* de :onfirmation
)NFLICTING - [gen] en contradiction, antagoniste
[chem] en opposition, en désaccord
ROUTE - [railw] itinéraire *m* incompatible
)NFLUENCE - [geogr] confluent *m*
)NFOCAL - [opt] confocal
)NFORM, to - [gen] conformer, se conformer, être d'accord
IN SHAPE, to - [gen el etc] épouser (ou affecter) la forme de
)NFORMABLE FOLD [geol] pli *m* harmonique
)NFORMATION - s. configuration
)NFORMING ANODE - [metall] (in electro-plating; anode shaped to a form similar to that of :he object to be plated) anode *f* préformée
)NFUSION - [photo] diffraction *f*
REGION - [radar] zone *f* de confusion
)NGEAL, to - [gen] congeler, se geler, figer
)NGEALING POINT - [phys] point *m* de congéla.ion
)NGELATION - [phys] (to pass from a liquid to a solid state owing to decrease of temperature, not necessarily to freezing point) congélation *f*
)NGENERIC - [zool] (belonging to the same genus) :ongénère
)NGENITAL - [zool med] congénital
)NGESTION - [med] (a pathological accumulation of blood) congestion *f*
[gen] (of traffic on roads) encombrement *m*, embouteillage *m*
)NGESTUS - [met] (cumulus clouds of considerable vertical extension) congestus *m*
CONGLOMERATE - [geol] (rocks formed of fragments of other formations and cemented together in a matrix) conglomérat *m*
[metall] conglomérat *m*
CONGO CORINTH - [ind chem] (a dye particularly used for cotton) congocorinthe *m*
~RED - [ind chem] (a dye also used in medicine as a diagnostic acid and in analytical chemistry) rouge *m* Congo
CONGRUENCE - [gen & math] congruence *f*, coïncidende *f*
CONGRUENT FIGURES - [geom] figures *fpl* conformes
CONIC - [math] conique
~PROJECTION - [math] projection *f* conique
CONICAL - [gen] conique
~BEARING - s. cone bearing
~BOBBIN - [text] bobine *f* conique
~BORING - [mech] alésage *m* conique
~BRAKE - [mech] frein *m* conique
~BROACH ROOF - [constr] toit *m* conique
~CENTRE - [mech] pointe *f*
~COLLET - [mech] douille *f* de serrage conique
~GEARING - [mech] engrenage *m* conique
~GUIDE - [mech] guide *f* conique
~HORN - [acoust] (a cone-shaped horn with a truncated apex) pavillon *m* conique
~-HORN AERIAL - [radio] antenne *f* en cornet
~-HORN ANTENNA - s. conical-horn aerial
~JOURNAL - [mech] tourillon *m* conique
~LAMP - [el] lampe *f* à pyramide
~MIXER - [mech] (a mixer whose barrel consists of a cylinder terminating in a cone) malaxeur *m* conique
~PENDULUM - [mech] (simple pendulum with bob swinging in a horizontal circle) pendule *m* conique
~PIN - [mech] goupille *f* conique
~PIVOT - [horol] pivot *m* conique
~POINT BOLT - [mech] boulon *m* à pointe conique
~REAMER - [tool] alésoir *m* conique, dégorgeoir *m*
~REFRACTION - [opt] réfraction *f* conique
~ROLL - [metall] cylindre *m* bombé
~RUDDER BAG - s. cone-shaped rudder
~SCANNING - [radar] balayage *m* conique
~SEAT - [mech] (cone-shaped seating for valve) siège *m* conique, portée *f* conique
~SECTIONS - [math] sections *fpl* coniques
~SLEEVE - [mech] (tapering hollow part) manchon *m* conique
~SPINDLE - [text] broche *f* conique
~SPIRAL SPRING - [mech] ressort *m* spiral conique
~SPRING - [mech] ressort *m* conique
~VALVE - [mech] valve *f* conique, obturateur *m* conique
CONICINE - s. coniine
CONICS - [math] coniques *mpl*
CONIFER - [bot] conifère *m*
CONIFERIN - [chem] (a starting material for flavouring agents) coniférine *f*, abiétine *f*, laricine *f*
CONIFEROUS - [bot] conifère
CONIINE - [chem] (an alkaloid with medicinal uses) coniine *f*, conicine *f*
CONING - [text] bobinage *m*, canettage *m*, enroulage *m*
~ANGLE - [aero] (in helicopters; the angle between the longitudinal axis of a blade and the plane in which the path of the blade tip lies) angle *m* de co-

nicité
CONING STOPS - [aero] (in an autogyro, stops limiting the vertical elevation of the flapping angle) arrêts *mpl* limiteurs de l'angle de conicité
CONJUGATE - [gen] conjugué
~ATTENUATION COEFFICIENT - [el] constante *f* d'atténuation conjuguée
~ATTENUATION CONSTANT - s. conjugate attenuation coefficient
~AXIS - [math] axe *m* conjugué
~BRANCHES - [el] dérivations *fpl* conjuguées
~DEVIATION - [med] (deviation of the eyes caused by a brain lesion) déviation *f* conjuguée des yeux
~DIAMETERS - [math] diamètres *mpl* conjugués
~FAULT - [geol] failles *fpl* conjuguées
~FOCI - [opt] foyers *mpl* conjugués
~IMAGE POINTS - [photo] points *mpl* correspondants
~IMAGE RAYS - [photo] rayons *mpl* correspondants
~IMPEDANCES - [el] (having resistance components which are equal and reactance components equal but opposite in sign) impédances *fpl* conjuguées
~LAYERS - [chem] (the two solutions in equilibrium with each other) couches *fpl* conjuguées
~MIRRORS - [opt] miroirs *mpl* conjugués
~PHASE CHANGE COEFFICIENT - [el] (the imaginary part of the conjugate transfer constant) constante *f* conjuguée de déphasage
~POINTS - [geom] points *mpl* conjugués
~SOLUTION - [chem] solution *f* conjuguée
~SYSTEM - [chem] système *m* conjugué
~TRANSFER COEFFICIENT - [el] constante *f* conjuguée de transmission
~TRANSFER CONSTANT - s. conjugate transfer coefficient
~VARIABLES - [phys] variables *fpl* conjuguées
~VEIN - [mining] filon *m* conjugué
CONJUGATED DOUBLE BONDS - [chem] (two or more double bonds separated by a single bond) doubles liaisons *fpl* conjuguées
~HYDROCARBONS - [chem] hydrocarbures *mpl* conjugués
CONJUGATION - [genet] (union of two cells for the development of new ones) conjugaison *f*
CONJUNCTION - [gen] conjonction *f*
[astr] (the position of a celestial body when its celestial longitude and that of the sun are the same and its elongation is zero) conjonction *f*
CONJUNCTIVA - [anat] (the modified epidermis in front of the eyes) conjonctive *f*
CONJUNCTIVE - [gen] conjonctif
CONJUNCTIVITIS - [med] (the inflammation of the conjunctiva) conjonctivite *f*
CONK, to - [mech] (colloquial; said of engines or motors which stop suddenly) caler, avoir des ratées
CONNATE SALT - [chem] (salt trapped in solution in connate water) sel *m* dissous dans l'eau interstitielle
~WATER - [chem] (water existing in the pores of sedimentary deposits, mechanically trapped but not combined) eau *f* interstitielle
CONNECT, to - [gen] connecter, unir, brancher, relier
[el] connecter, brancher
[mech] accoupler, coupler, connecter
~IN PARALLEL - [el] connecter en parallèle, monter en parallèle
CONNECT IN SERIES, to - [el] monter en série
CONNECTED - [gen & el] connecté
~NETWORK - [el] réseau *m* raccordé
CONNECTING - [gen] de connexion, de raccordement, de jonction
~ANGLE - [mech] cornière *f* d'assemblage
~BOX - [el] boîte *f* de jonction
~BRIDGE - [metall] pont *m* de communication
~CABLE - [el] câble *m* de connexion
~CLIP - [mech] pince *f* de raccordement
~COCK - [mech] (a plug cock with tubes on each side) robinet *m* de jonction
~DEVICE - [mech] connexion *f*, organe *m* de connexion, raccordement *m*
~DIAGRAM - [mech] schéma *m* de montage
~DUCT - [el] manchon *m* de raccordement
~FLANGE - [mech] bride *f* de raccordement, bride de connexion
~JACK - [el] jack *m* de raccordement
~LEAD - [el] fil *m* de connexion
~LINE - [el] ligne *f* de raccordement
[railw] voie *f* de raccordement
~LINK - [mech] articulation *f*, fausse maille *f*, manille *f* d'assemblage
~LUG - [el] cosse *f* de connexion, patte *f*
~PIECE - [mech] pièce *f* d'accouplement
~PIPE - [mech] ajutage *m*, tubulure *f* de raccordement, raccord *m* à tubulure
~PLATE - [mech] plaque *f* de liaison
[horol] bielle *f*
~PLUG - [el] fiche *f* de raccordement
~RAIL - [railw] rail *m* de raccordement
~RELAY - [el] relais *m* de connexion
~ROD - [mech] (in a reciprocating engine, the rod linking the piston and the crankshaft) bielle *f*
~ROD BEARING - [mech] coussinet *m* de bielle
~-ROD BIG END - s. connecting-rod head
~-ROD BOLT - [mech] boulon *m* de bielle
~-ROD EYE - [mech] pied *m* de bielle
~-ROD HEAD - [mech] tête *f* de bielle
~-ROD JOURNAL - [mech] tourillon *m* de bielle
~-ROD SHANK - [mech] corps *m* de bielle
~-ROD SMALL END - [mech] pied *m* de bielle
~SCREW - [mech] boulon *m* d'assemblage
~SLEEVE - [mech] manchon *m* de raccordement, manchon *m* de jonction
~STRIP - [el] lame *f* de connexion, barrette *f* de connexion
~TERMINAL - [el] borne *f* de jonction, serre-fil *m*
~TIE ROD - [auto] (the steering track rod) barre *f* d'accouplement
~TISSUE - [zool] tissu *m* connectif, tissu *m* conjonctif
CONNECTOR - [mech] raccord *m*, joint *m*, attelage *m*
[el] pince *f*, serre-fil *m*
[el] (of accumulator) connecteur
[oil ind] ressort *m* de suspension pour bride
[carp] pince *f*
~BAR - [el] barrette *f* de connexion
~BLOCK - [el] bloc *m* de jonction
CONNEXION - [gen] (or connection) connexion *f*, relation *f*
[mech] accouplement *m*, liaison *f*, jonction *f*.
[el] connexion *f*, jonction *f*, branchement *m*
~BOX - [el] boîte *f* de jonction
~CABLE - [el] câble *m* de jonction
~DIAGRAM - [el] (a diagram showing the position

)f the components and connexions) schéma *m* de nontage
)NNEXION ROSE - [el] rosace *f* de raccordement
'O THE MAINS - [el] raccordement *m* au secteur
)NNEXIONS - [plumb] tuyauterie *f*
)NNING TOWER - [naut] kiosque *m*
)NOID - [geom] conoïde
)NOIDAL - s. conoid
)NRADSON CARBON TEST - [ind chem] détermination de la teneur en carbone par volatilisation à 'abri de l'air
)NSCIOUS - [gen] conscient
:RROR - [comput] (in punched cards) erreur *f* remarquée
)NSECUTIVE FIELD - [telev] trame *f* consécutive
;CANNING - [telev] analyse *f* consécutive
)NSEQUENT POLE - [el] (pole in a magnetic substance resulting from the meeting of two magnetizations in opposite directions) pôle *m* conséquent
)NSERVATION - [gen] conservation
[hydr] (of waters) conservation *f*
AGREEMENT - [oil ind] convention *f* de conservation
)F ENERGY - [phys] (the accepted principle that energy is never created or destroyed but only changes in form) conservation *f* de l'énergie
)F ENERGY LAW - [phys] (in any isolated system the amount of energy is constant) loi *f* de la conservation de l'énergie
)F MASS - [phys] (matter is never created or destroyed) conservation *f* de la masse
)F MATTER - s. conservation of mass
)F MOMENTUM - [phys] (the total of the moment in a close system is not affected by processes occurring within that system) conservation *f* de l'impulsion
)NSERVATIVE FLUX - [el] (when it has the same value for all orientable surfaces to the same contour) flux *m* conservatif
'ORCE FIELD - [mech] champ *m* de forces conservatif
'ROPERTY - [met] (any property of an air mass which is not greatly affected by modifying influences) propriété *f* de conservation
;YSTEM - [mech] (system of particles in which the forces acting on any particle can be derived from a potential energy function) système *m* conservatif
)NSERVATOR - [el] (oil container for transformers) éservoir *m* auxiliaire d'huile
)NSERVATORY - [bot] serre *f*
)NSERVE, to - [gen] conserver, mettre en conserve
)NSIGN, to - [gen] consigner, remettre, déposer
comm] expédier, envoyer
)NSIGNEE - [comm] destinataire *m*
)NSIGNMENT - [gen] consignation *f*
comm] envoi *m*, expédition *f*
;ALE - [comm] vente *f* sur commande
)NSIGNOR - [comm] consignateur *m*, expéditeur *m*
)NSISTENCY - [gen] consistance *f*
paint] (the viscosity in coating compositions) consistance *f*, viscosité *f*
chem] consistance *f*, compacité *f*
:HECK - [comput] (check to investigate whether an information conforms to the scope of the prefixed standards) vérification *f* de la cohérence
)F READING - [instr] (of measuring instruments) cohérence *f* des résultats
CONSISTENT - [gen] consistant, visqueux, compatible
CONSISTOMETER - [instr] (instrument designed to measure the consistency of semi-fluid substances) consistomètre *m*
CONSOL BEACON - [radio] système *m* de radionavigation "Consol"
CONSOLE - [gen] console *f*; (for radio receivers, television sets etc) support *m*, meuble *m*
[mus] (the playing control location of an organ) console *f* d'orgue
[radar] (display unit comprising screen etc) meuble *m*
~TABLE - [gen] table *f* console, console *f*
CONSOLIDATE, to - [gen] consolider, renforcer
[phys] comprimer
CONSOLIDATED SAND - [geol] sable *m* compact
CONSOLIDATION - [gen] consolidation *f*
[geol] (the result of the pressure after deposition) consolidation *f*
[soil] (of ground by mechanical means) consolidation *f*
~PRESSURE - [mech] contrainte *f* de consolidation, pression *f* de consolidation
~PROCESS - [soil] processus *m* de consolidation
~STRESS - s. consolidation pressure
CONSOLIDOMETER - [instr] oedomètre *m*, anneau *m* de consolidation
CONSOLUTE - [chem] miscible
~TEMPERATURE - [phys] for two partially miscible liquids, the critical temperature above which they are miscible) température *f* de mélange
CONSONANCE - [acoust] (a combination of two tones with satisfactory effects) consonance *f*
CONSONANT ARTICULATION - [acoust] intelligibilité *f* des consonnes
CONSORTISM- [biol] symbiose *f*
CONSTABLE SKY - [met] (said of a sky with numerous cumulus clouds) ciel *m* couvert de cumulus
CONSTANT - [gen] constante *f*
~AMPLITUDE RECORDING - [el acoust] (mechanical recording in which the resulting recorded amplitude is independent of frequency) enregistrement *m* à amplitude constante
~BOILING - [chem] distillant à point fixe
~BOILING MIXTURE - [ind chem] mélange *m* à point d'ebullition constant
~BOILING POINT - [phys] point *m* d'ébullition constant
~COMPRESSION - [mech] compression *f* constante
~CURRENT CHARGE - [el (mode of charging in which current is kept at a uniform level) charge *f* à courant constant
~CURRENT DYNAMO - [el] (the type of direct current generator in which the current is practically constant at all loads, while the voltage varies to meet load variations) dynamo *f* à courant constant
~CURRENT GENERATOR - [el] génératrice *f* à courant constant
~CURRENT MODULATION - [electron] (method of amplitude modulation in which a common inductor provides the anode supply to a radio-frequency) modulation *f* à courant constant, modulation *f* Heising
~CURRENT TRANSFORMER - [el] transformateur *m* à courant constant

CONSTANT DELIVERY PUMP - [mech] (a pump which gives constant delivery at varying pressure) pompe *f* à débit constant
~DEVIATION PRISM - [opt] prisme *m* à déviation constante
~DIRECTION ERROR - [radar] composante *f* de l'erreur en variation sinusoïdale avec le relèvement
~INDUCTANCE RESISTOR - [el] résistance *f* à inductance constante
~-K FILTER - [electron] (filter circuit in which the product of shunt and series impedances is constant irrespective of frequency) filtre *m* à K constant
~LEVEL CHART - [met] (isobar chart drawn for a given level above mean sea level) carte *f* à niveau constant
~LEVEL REGULATOR - [mech] régulateur *m* de niveau
~LEVEL TUBE - [surv] niveau *m* à réservoir
~LOAD - [mech] charge *f* permanente
~LOSSES - [el] (of a machine or apparatus) pertes *fpl* constantes
~LUMINANCE SYSTEM - [telev] système *m* à luminance constante
~MESH CLUTCH - [mech] embrayage *m* à prise constante
~MESH GEAR BOX - [auto] boîte *f* de vitesse à prise constante
~MESH PINIONS - [auto] engrenages *mpl* à prise constante
~OF A MEASURING INSTRUMENT - [instr] (coefficient by which the reading in divisions must be multiplied to obtain the value of the measured quantity) constante *f* d'un appareil de mesure
~OF A METER - [instr] constante *f* d'un compteur
~OF ELECTROLYTIC DISSOCIATION - [chem] (value obtained by dividing ion activity by molecular activity) constante *f* de dissociation électrolytique
~OF GRAVITATION - [phys] constante *f* de la gravitation
~OF INERTIA - [phys] constante *f* d'inertie
~PITCH - [mech] (of a screw) pas *m* constant
~-PITCH SCREW - [mech] vis *f* à pas constant
~POTENTIAL - [el] tension *f* constante
~-POTENTIAL ACCELERATOR - [el] (to produce high energy or electrons) accélérateur *m* à tension constante
~PRESSURE CHART - [met] (chart of pressure contours) carte *f* des pressions constantes
~PRESSURE COMBUSTION ENGINE - [mech] moteur *m* à combustion à pression constante
~PRESSURE CYCLE - [mech] (in internal combustion engines) cycle *m* de pression constante
~PRESSURE GAS THERMOMETER - [instr] (gas thermometer in which the gas pressure is constant) thermomètre *m* à gaz à pression constante
~PRESSURE TURBINE - [el] turbine *f* à pression constante
~RESISTANCE NETWORK - [el] réseau *m* à resistance constante
~RESISTANCE STRUCTURE - [el] résistance *f* indépendante de la fréquence
[acoust] système *m* acoustique à resistance constante indépendant de la fréquence
~SCREEN-GRID VOLTAGE - [electron] (the direct voltage of constant value applied between the screen-grid and the cathode of a vacuum tube) tension *f* de grille-écran constante
~SPEED PROPELLER - [aero] (propeller fitted with blade-angle control mechanism to provide for co tant engine speed) hélice *f* à vitesse constante
CONSTANT SPEED SCANNING - [telev] balayage *m* à vitesse constante
~-TORQUE RESISTOR - [electron] résistance *f* aut régulatrice
~VELOCITY RECORDING - [el acoust] (mechanica recording in which the resulting recorded ampli de is inversely proportional to the frequency) en registrement *m* à vitesse constante
~VOLTAGE CHARGE - [el] (mode of charging in which voltage is kept at uniform value) charge à tension constante
~VOLTAGE DYNAMO - [el] dynamo *m* à tension constante
~VOLTAGE MODULATION - [radio] modulation *f* à tension constante
~VOLTAGE REGULATION - [contr] régulation *f* à tension constante
~VOLTAGE REGULATOR - [el] stabilisateur *m* de tension
~VOLUME COMBUSTION ENGINE - [mech] moteur à combustion à volume constant
~VOLUME GAS THERMOMETER - [instr] thermomè tre *m* à gaz à volume constant
~VOLUME TURBINE - [el] turbine *f* à volume cons tant
~WHITE - [paint] blanc *m* fixe, blanc *m* de baryte
CONSTANTAN - [metall] (a copper-nickel alloy o high resistivity and a very low temperature coef ficient) constantan *m*
CONSTELLATION - [astr] (an identifiable groupin of stars) constellation *f*
[chem] configuration *f*
CONSTITUENT - [gen] partie *f*, constituante, com sant *m*
~PARTICLE - [phys] particule *f* constituante
CONSTITUTION - [gen] constitution *f*
[phys chem] composition *f*
~DIAGRAM - [phys] (representation of the data o the phase regions of a system) diagramme *m* des phases
CONSTITUTIONAL CHANGE - [metall] (change in solid alloy involving the transformation of one constituent to another) changement *m* de constitu tion
~DIAGRAM - [metall] diagramme *m* d'équilibre
~FORMULA - [phys] (showing the arrangement of atoms in a molecule) formule *f* de constitution
~WATER - [chem] eau *f* de cristallisation
CONSTITUTIVE PROPERTY - [phys] propriété *f* co titutive
CONSTRAIN, to - [gen] forcer, contraindre; (to fasten) lier, attacher
CONSTRAINED - [gen] forcé
~MAGNETIZATION - [phys] magnétisation *f* forcé
~MOTION - [mech] mouvement *m* commandé
~OSCILLATION - [mech] oscillation *f* forcée
CONSTRAINT - [mech] contrainte *f*
CONSTRICTION - [gen] constriction *f*
[chem] (region of reduced cross-section in a fluid passage) étranglement *m*, rétrécissement *m*
CONSTRUCT, to - [gen] construire, fabriquer
CONSTRUCTION - [gen] construction *f*
~ASSEMBLY - [mech etc] montage *m*
~DIAGRAM - [gen] schéma *m* de montage
~OF SCAFFOLDS - [constr] construction *f* d'écha

faudages
ONSTRUCTION PROJECT - [constr] plan *m* de construction
SITE - [constr] chantier *m* de construction
SPECIFICATIONS - [[gen] spécifications *fpl* de construction
THEORY - [constr] science *f* de la construction
ONSTRUCTIONAL IRON - [metall] fer *m* de construction
STEELWORK - [metall] acier *m* profilé pour construction
ONSULTING - [gen] consultant, conseil
ENGINEER - [ind] ingénieur *m* conseil
ONSUME, to - [gen] consommer
ONSUMED PARTICLE - [nucl] (in nuclear reaction, a particle which has completed its work) particule *f* utilisée
ONSUMER - [gen] consommateur *m*
GOODS - [comm] biens *mpl* de consommation
ONSUMPTION - [gen] consommation *f*
[auto] (fuel consumption) consommation *f*
[med] tuberculose *f*, phtisie *f*
BLAST - [metall] consommation *f* de vent
~OF COAL - [th eng] consommation *f* de charbon
TEST - [auto] essai *m* de consommation
ONTACT - [gen el mech etc] contact *m*
~ACTION - [chem] catalyse *f*
~ADHESIVE - [ind chem] (adhesive which requires very little pressure to achieve a bond) colle *f* de contact
~ALTIMETER - [instr] (instrument making or breaking an electrical circuit at a pre-set height) contacteur *m* altimétrique
~ANGLE - [phys] angle *m* de contact
~ARC - [metall] (in welding) arc *m* de contact
~ARC-WELDING - [metall] soudage *m* à l'arc par contact
~AREA OF FLATS - [text] surface *f* de glissement des chapeaux
~BAR - [el] (in arc welder) baguette *f*
~BED - [hydr] lit *m* de contact
~BLADE - [el] lame *f* de contact
~BREAKER - [el] interrupteur *m*
~BREAKER CAM - [el] (the cam actuating the breaker points in an electric ignition system) came *f* d'interrupteur
~BREAKER PLATE - [el] plaque *f* porte-interrupteur
~BREAKER POINT - [el] contact *m*
~BRIDGE-PIECE - [el] pont *m* polaire (d'accumulateur)
~BRUSH - [el] balai *m*
~CONDUCTOR - [el metall] (device to lead electric current into a molten or solid metal or alloy, itself serving as the active electrode in the cell) électrode *f* de contact
~DEPOSIT - [mining] gisement *m* de contact
~DEPTH - [mech] (the depth of a gear tooth) profondeur *f* de dent
~DERMATITIS - [med] dermatite *f* de contact
~ELECTRICITY - [el] effet *m* de contact
~ELECTRO-MOTIVE FORCE - [el] (sometimes occurring when two contactors of different material come into contact) potentiel *m* de contact, effet *m* Volta
~ELECTRODE - s. contact conductor
~FINGER - [el] (a contact pressed by springs against the moving contacts of a controller) doigt *m* de contact, frotteur *m*

CONTACT FLYING - [aero] (flying in which the attitude and the path of an aicraft can always be controlled visually) navigation *f* observée, vol *m* par contact
~FOLLOW - [el] (the further movement of two contacts after connexion) accompagnement *m*
~-FREE MOTION - [mech] frottement *m* de contact
~GETTER - [electron] couche-getter *f*
~GETTERING - [electron] sorption *f* à l'aide de couches-getter (formées par volatilisation du métal getter)
~GLASS - s. contact lens
~GONIOMETER - [instr] (consisting of two flat bars pivoted together like a pair of scissors) radiogoniomètre *m* de contact
~GRILL - [th eng] grilloir *m* par contact
~HEATING - [th eng] échauffement *m* par contact
~INSECTICIDE - [chem] (insecticide killing on contact with the insect surface) insecticide *m* de contact
~JAW - [el] (the clamping mechanism of a resistance welding machine) mâchoire *f* de contact
~LAMINATE - [plast ind] (material laminated without the use of pressure) stratifié *m* à résine de contact
~LAMINATION - [plast ind] (process of manufacture of laminated plastics in which very little mould pressure is required) stratification *f* par contact
~LENS - [opt] verre *m* de contact
~LEVER - [mech] levier *m* de contact
~LODE - [mining] filon *m* de contact
~MAKER - [el] (in the sparking plugs of petrol engines) contact *m* d'allumage
~MASS - [chem] masse *f* de contact
~METHOD - [hydr] (sewage treatment) procédé *m* par contact
~MICROPHONE - [el acoust] microphone *m* de contact
~MODULATED AMPLIFIER - [radio] (amplifier for direct current at very low-frequency signals) amplificateur *m* à interrupteur
~MOULDING - [plast ind] (reinforced plastic material manufactured on a former by hand) moulage *m* à la main
~NOISE - [radio] (a circuit noise) bruit *m* de friture
~PIECE - [el] (of an electrically operated vehicle) plot *m*
~PIN - [el] broche *f* de contact
~PLATING - [metall] (deposition of a coating by immersing the base in contact with another metal in a bath containing the metal which it is intended to deposit) revêtement *m* électrolitique par contact
~POINT - [el] (the mechanical element by means of which an electrical circuit is completed or interrupted) contact *m*, vis *f* platinée
[metall] (the pointed section of a control bar in a welding machine) point *m* de contact
~POINTER - [instr] (pointer with a contact element at its end) aiguille *f* à contact
~POINTS - [el] (mechanical parts) vis *fpl* platinées
~POISON - s. contact insecticide
~POTENTIAL - [el] (the electromotive force due to contact between two bodies in different physical conditions or of a different chemical composition) potentiel *m* de contact, effet *m* Volta
~POTENTIAL BARRIER - [electron] (the potential hill at the contact surface of two bodies arising from

the formation of a barrier layer) barrière *f* de potentiel au contact
CONTACT POTENTIAL DIFFERENCE - [electron] (the difference of potential which arises when two conductors having different work functions are in contact) différence *f* de potentiel au contact
~PRESSURE - [mech] pression *f* par contact
~PRINTING - [photo] (method of copying with the negative in continuous contact with the raw stock for making a positive) tirage *m* par contact
~PROCESS - [ind chem] (a method of producing sulphuric acid using a platinum catalyst for the oxidation of sulphur dioxide to sulphur trioxide) procédé *m* de contact
~RAIL - [el] (the conductor rail) rail *m* de contact
~RAMP - [el] (in railways) contact *m* fixe de voie
~RESINS - [plast ind] synthetic thermosetting resins which cure at comparatively low pressure without water formation) résines *f*pl de contact
~RESISTANCE - [el] résistance *f* de contact
~RING - [el] (of an induction motor) bague *f* de contact
~ROLLER - [metall] (rotating disc used in a seam-welding machine) molette *f*
~SEGMENT - [el] (in motor starters) plot *m* de distribution
~SHOE - s. collector shoe
~SPACE - [chem] chambre *f* de contact
~SPRING - [geol] source *f* de contact
~STRIP - [el] (in a pantograph) lame *f* de contact
~STUD - [el] plot *m*
~SURFACE - [chem] surface *f* de contact
~THERAPY - [radiat] (X-ray therapy by means of a contact tube) radiothérapie *f* superficielle
~THERMOMETER - [instr] (a thermometer with its temperature-sensitive part in contact with the part in question) thermomètre *m* à contact
~TO EARTH - [el] mise *f* à la terre
~TUBE - [radiat] (X-ray tube which can be used at very short distance) tube *m* de contact
~UNIT - [el] contacteur *m*
~VEIN - [mining] (vein between two different rock formations) veine *f* de contact
~VOLTMETER - [instr] voltmètre *m* à contacts
CONTACTING - [chem] agissant catalytiquement
CONTACTOR - [el](power-operated switch which works mechanically, electro-magnetically or electro-pneumatically) conjoncteur *m*, contacteur *m*
~CONTROLLER - [el] combinateur *m* à contact
~STARTER - [el] (electric-motor starter without steps of resistance) démarreur *m* à contacteur
~SWITCHING STARTER - [el] (a switching starter operated by contactors) démarreur *m* à contacteurs
CONTAINER - [gen] récipient *m*, réservoir *m*, cuve *f*, nacelle *f*
[transp] (large box for freight) cadre *m*, emballage *m*
[cin] bobine *f*, chargeur *m*
~CAR - [railw] cadre *m*
~TRANSPORT - [transp] transport in a vessel designed to fit on road or rail vehicles) transport *m* par cadre
CONTAMINANT - [chem] impureté *f*, substance *f* contaminante
CONTAMINATED - [nucl] (made radioactive) contaminé
~ROCKS - [geol] (igneous rocks modified by the incorporation of other rocks) roches *f*pl à inclusion
CONTAMINATION - [gen] contamination *f*
~METER - [instr] (to measure the gamma activity c fission products on clothing etc) contrôleur *m* de contamination
CONTEMPORANEOUS - [gen] contemporain
~EROSION - [geol] (the removal of sediments immediately after deposition) érosion *f* contemporaine
CONTENT - [gen] (the amount of a substance) contenu *m*, volume *m*
[chem] (the amount of a substance in another substance) teneur *f*
~OF FINE ORE - [metall] teneur *f* en fines
CONTIGUOUS - [gen] contigu, voisin, adjacent
CONTINENT - [geogr] continent *m*
CONTINENTAL - [gen & geogr] continental
~CLIMATE - [met] (type of climate characteristic of the interior of a continent) climat *m* continental
~CONDITIONS - [geol] (conditions in large land areas remote from the sea and above sea-level) conditions *f*pl continentales
~DEPOSITION - [geol] (as glacial deposits) dépôts *m*pl continentaux
~FACIES - [geol] faciès *m* continental
~SEGMENT - [geol] plateau *m* continental, socle *m* continental
~SHELL - [geol] (the off-shore zone, sloping down gently) plateform *f* continentale
~SLOPE - [geol] talus *m* continental
CONTINGENT - [gen] contingent *m*
CONTINUAL - [gen] continu, continuel
CONTINUANCE - [gen] continuation *f*, durée *f*
CONTINUE, to -[gen] continuer, prolonger
[leg] ajourner
CONTINUED FRACTION - [math] fraction *f* continue
CONTINUITY - [gen] continuité *f*
[el] (of a circuit] continuité
[cin] (the detailed description of the sequences required in a film) continuité
[telev] déroulement *m* continu, enchaînement *m*
[mining] (of a vein, a bed etc) continuité *f*
~CLERK - [cin] assistant *m*
~CONTROL - [telev] régie *f*
~EQUATION - [phys] (application of the principle of the conservation of matter to fluid motion) équation *f* de continuité
~FITTINGS - [el] (in electric wiring installations to obtain a continuous circuit) joints *m*pl pour tubes de canalisation
~GIRL - [cin] "script-girl", secrétaire *f* de plateau
~OF STATE - [phys] (transition between two states in either direction and without discontinuity) passage *m* continu
~TEST - [el] essai *m* de coupure
~TESTER - [el] ohmmètre *m*
~WRITER - [cin] auteur *m* de scénarios
CONTINUOUS - [gen] continu, ininterrompu, permanent
[bot] (with a smooth surface) lisse
~AGER - [text] vaporiseur *m* continu
~ANNEALING LINE - [metall] four *m* continu à recuire
~ARCHED GIRDER - [constr] poutre *f* à travées continues
~ATTENTION METHOD - [telecomm] méthode *f* de l'attention continue

CONTINUOUS BEAM - [constr] (a beam which is continuous over its supports) filant *m*, poutre *f* continue
~BLEACHING - [text] blanchiment *m* en continu
~BLOWING - [metall] travail *m* continu
~BRAKE - [el mech] (brake system used on trains, by which the operation at one point ensures the application of brakes throughout) frein *m* continu
~CARBONIZATION - [gas ind] distillation *f* continue
~CASTING - [metall] coulée *f* continue
~CASTING PLANT - [metall] installation *f* de coulée continue
~CENTRIFUGE - [mech] turbine *f* continue
~CHARGE-DISCHARGE MECHANISM - [nucl] (in a reactor; mechanism eliminating the need for shutting down during the charging of fuel elements) dispositif *m* de chargement et de déchargement continu
~COMBUSTION TURBINE - [el] (a type of turbine in which air or mixture is taken in at constant or atmospheric pressure, compressed adiabatically, heated and expanded adiabatically) turbine *f* à combustion continue
~CONVEYOR - [mech] carrousel *m*
~COUNTERSHAFT - [mech] arbre *m* intermédiaire de distribution
~CREATION HYPOTHESIS - [phys] hypothèse *f* de la création continue
~CURRENT - [el] (direct current) courant *m* continu
~CUTTER - [mech] machine *f* à tailler continue
~DIFFUSION - [sugar ind] (a beet-sugar extraction system) diffusion *f* continue
~DISTILLATION - [ind chem] (distillation in which the substance to be distilled is fed continuously to the still) distillation *f* continue
~DUTY - [mech] service *m* permanent, régime *m* continu
~DYEING PROCESS [text] processus *m* de teinture en continu
~ELECTRODE - [metall] (a type of electric-arc furnace electrode which can be fed in continuously) électrode *f* prolongée
~EXTRACTION - [chem] (extraction by a solvent which circulates, evaporates and is condensed again to continue the same cycle) extraction *f* continue
~FILTER - [mech] (a percolating filter) filtre *m* continu
~FLOOR - [constr] plancher *m* sans joints
~FLOW CALORIMETER - [instr] (instrument determining the mechanical equivalent of heat) calorimètre *m* à écoulement continu
~FRAMES - [constr] travées *fpl* continues
~FURNACE - [metall] (type of furnace in which treatment is uninterrupted) four *m* continu
~GIRDER - s. continuous beam
~GIRDER BRIDGE - [constr] pont *m* à travées continues
~GROOVE PROFILE - [rubber man] (in tyres) sculpture *f* à rainures longitudinales
~HINGE - [mech] charnière *f* aux dimensions
~HUNTING - [telecomm] recherche *f* continue, sélection *f* continue
~IMPOST - [constr] (not projecting from the surfaces of the pier and arch) sommier *m* d'arc en affleurement
~KILN - [metall] four *m* de séchage à tunnel

CONTINUOUS LOADING - [telecomm] (of submarine cables) krarupisation *f*
~MILL - [metall] (rolling-mill consisting of a series of pairs of rolls) laminoir *m* continu, train *m* à bande
~NETWORK - [radio] (a number of radio stations coupled together) réseau *m* ininterrompu
~OSCILLATIONS - [el] (oscillations generated without interruption, as distinct from damped trains of oscillations) oscillations *fpl* entretenues
~OUTPUT - [el] puissance *f* de sortie continue
~PHASE - [chem] (the liquid portion of a colloidal solution) phase *f* continue
~PICKLING - [metall] décapage *m* continu
~POLYMERIZATION - [chem] (a continuously carried out process) polymérisation *f* continue
~PRINTING- [cin] (contact printing along the length of the film) tirage *m* continu
~PROCESS - [gen] fabrication *f* en continu, processus *m* continu
~PROCESSING - [photo] (of films) développement *m* continu
~RATING - [el] régime *m* permanent
~REHEATING FURNACE - [metall] four *m* poussant
~RINGING BELL - [el] (electric alarm bell which rings continuously when energized) sonnerie *f* continue
~ROTARY CONTACT PRINTER - [cin] (for printing both sound and picture) tireuse *f* continue par contact
~RUNNING - [el] régime *m* continu
~SAND PLANT - [metall] sablerie *f* continue
~SETTLER - [sugar ind] décanteur *m* continu
~SHEETING - [plast ind] (sheet material produced in unbroken bands) feuilles *fpl* continues
~SLOWING DOWN MODEL - [nucl] (treatment of the slowing-down process through a continuous curve replacing the stepwise decrease of energy due to collision) modèle *m* à ralentissement continu
~SPECTRUM - [phys] spectre *m* continu
~STEAMER - [text] s. continuous ager
~TAPPING - [metall] coulée *f* continue (d'un appareil de fusion)
~TEST - [el] (test in which a dry cell or battery is subject to continuous discharge) essai *m* en régime permanent
~TONE - [acoust] (a tone of constant pitch) son *m* continu
~VENT - [plumb] (a waste pipe extension to provide ventilation) évent *m* de ventilation
~VERTICAL RETORT - [ind chem] cornue *f* de distillation continue
~VULCANIZATION - [rubber ind] vulcanisation *f* continue
~WAVE - [radio] (electromagnetic radiation generated from continuous oscillations) onde *f* entretenue (ou non-amortie)
~-WAVE TELEGRAPHY - [telecomm] (radiotelegraphy in which the signal is a train of continuous oscillations) radiotélégraphie *f* à onde entretenue
~WELD - [metall] (weld continuous for its whole length) soudure *f* continue
CONTINUOUSLY ADJUSTABLE INDUCTOR - [el] (variable inductor, the position of whose coils may be changed in relation to one another) inducteur *m* à variation continue
~RUNNING DUTY - [el] (duty with variable loads

and with no intervals of rest) service *m* permanent, régime *m* permanent

CONTINUOUSLY VARIABLE - [mech] à réglage continu, réglable sans graduations

CONTOUR - [gen] contour *m*, silhouette *f*
[surv] (point connecting line) profil *m*

~ACCENTUATION - [telev] (obtained by reducing the rising time of signals) accentuation *f* des contrastes

~GAUGE - [mech] calibre *m* pour profils

~INTERVAL - [surv] (the vertical distance between two adjacent contour lines) intervalle *m* vertical entre les lignes de niveau

~LINE - [surv] (a line following points at the same height) ligne *f* de niveau, courbe *f* de niveau

~MAP - [surv] carte *f* à courbes de niveau
[telev] (a map showing the boundaries of the range of a television transmitter) carte *f* de la zone couverte

~PROJECTOR - [opt] projecteur *m* de profils

~SHAPING - [metall] formation *f* du modèle

~TEST - [mech] (examination of parts for correctness of outline) vérification *f* des profils

CONTOURED CHAPLET - [metall] support *m* de noyau de forme

CONTOURING - [surv] relevé *m* à courbes de niveau

CONTOUROMETER - [instr] vérificateur *m* de profil

CONTRACEPTIVE - [med] anticonceptionnel *m*

CONTRACT, to - [gen] contracter, raccourcir

CONTRACT - [gen] contrat *m*, convention *f*, marché *m*, traité *m*

~NOTE - [comm] bordereau *m*

CONTRACTILE - [gen bot anat] contractile

~TISSUE - [zool] (animal tissue possessing the property of contractility) tissu *m* contractile

CONTRACTILITY - [gen] (the property of becoming reduced in length) contractilité *f*

CONTRACTION - [gen] contraction *f*, compression *f* (of materials) retrait *m*
(hydr] (in water veins) étranglement *m*
[metall] (the shrinkage of a moulding) retrait *m*

~ALLOWANCE - [metall & plast ind] (dimensional allowance applied in the design of a mould for the moulding shrinkage) tolérance *f* de retrait

~COEFFICIENT - [el] (used in the calculations relating to the magnetic circuit of electrical machines) coefficient *m* de contraction

~CRACK - [metall] crique *f* de retrait

CONTRA-FLOW - [mech] (in machines, particularly turbines) à contre-courant, à écoulement inverse

~TURBINE ENGINE - [aero] (type of engine in which the direction of flow in compressor and turbine are opposite) moteur *m* à turbine à écoulement inverse

CONTRAINCISION - [med] (in surgery) contre-ouverture *f*

CONTRAPROPS - [aero] (abbreviation for contra-rotating propellers) hélices *fpl* contra-rotatives

CONTRA-ROTATING - [mech] adjacent mechanical parts revolving in opposite directions) contra-rotatif

~ROTORS - [aero] rotors *mpl* contra-rotatifs

CONTRAROTATION - [mech] (rotation in the opposite sense to an adjacent part) rotation *f* en sens inverse

CONTRARY CROSSING - [text] croisure *f* en directions opposées

CONTRARY DOUBLING - s. contrary crossing

CONTRAST - [gen] contraste *m*
[acoust] (intensity relation between the loudest and the lowest part of a sound) contraste *m*
[telev] (ratio of brightness level) contraste *m*

~AMPLIFICATION - [acoust] amplification *f* à contraste

~BALANCE - [telev] équilibre *m* des contrastes

~CONTROL - [telev] (means of regulating the contrast in a picture) réglage *m* des contrastes

~DYEING - [text] teinture *f* contraste

~EXPANSION - [telev] (increase in white-to-black amplitude range) expansion *f* des contrastes

~GRADIENT - [telev] gradient *m* des contrastes

~LIGHTING - [telev] illumination *f* en contraste

~RANGE - [telev] (the ratio between the whitest and blackest portions of the picture) contraste *m* maximal

~RATIO - [telev] (the ratio between maximum and minimum brightness) rapport *m* des contrastes

~REDUCTION - [telev] (weakness in the contrast caused by internal refraction in the glass covering the screen) perte *f* de contraste

CONTRASTING COLOUR - [text] couleur *f* contrastante

~SHADE - s. contrasting colour

CONTRASTY NEGATIVE - [photo] cliché *m* heurté

~PAPER - [photo] papier *m* contraste

CONTRATE - [mech] (with teeth at right angle to the plane of the wheel) à denture frontale

~WHEEL - [mech] roue *f* à denture frontale
[horol] roue *f* de champ

CONTRIVANCE - [gen] mécanisme *m*, dispositif *m*, appareil *m*

CONTRIVE, to -[gen] inventer, projecter, concerter, trouver le moyen

CONTROL, to [gen] régler, commander, contrôler
[contr] (to regulate the condition or motion of anything) régler, actionner, commander
[aero] piloter
[mech] commander

CONTROL - [gen] réglage *m*, régulation *f*, contrôle *m*
[contr] (the means of regulating the condition or motion of a part or unit) commande *f*
[instr etc] (the operation of adjusting) réglage *m*
[mech] (the device used for adjustments) régulateur *m*, commande *f*
[acoust] (the regulation of the sound) contrôle *m*

~ACCURACY - [contr] précision *f* du réglage, fidélité *f* du réglage

~ACTION [contr] (indicating the controller effect on the control unit) action *f* de réglage

~ADVANCE - [mech] (of rotors etc) avance *f*

~AMPERE-TURNS - [el] ampère-tours *mpl* caractéristiques

~APPARATUS - [el mech etc] appareil *m* de régulation

~AREA - [aero] (the officially defined air-space within which the control is exercised)zone *f* de régulation

~ASSAY [mech etc] essai *m* contradictoire

~BATTERY - [el] batterie *f* d'excitation

~BEVEL GEAR - [mech] couple *m* d'engrenages coniques

~BOARD - [el etc] (a supporting structure for instruments, switches, indicators etc) tableau *m* de commande

CONTROL BOARD - [comput] pupitre *m* de commande
~BOX - [mech] (in jet engines) coffret *m* de commande
[telecomm] boîte *f* de contrôle
~BURNER - [gas ind] brûleur-contrôleur *m*
~BUTTONS - [mach tools] tableau *m* boutons, tableau *m* boutons commandes différentes
~CABIN - [railw] poste *m* de commande
~CABLE - [el] câble *m* de commande
~CAR - [aero] (the car from which the control of an aircraft is operated) poste *m* de pilotage
~CENTRE - [aero] (a station from which civilian air traffic is controlled) station *f* de contrôle
~CHARACTERISTIC - [electron] (of a thyratron; a curve showing the relation between the anode voltage and the critical grid voltage) caractéristique *f* de réglage
~CIRCUIT - [comput] (the circuit for the execution of the instruction in the prescribed sequence) circuit *m* de commande
[contr] (the circuit of a control apparatus or system) circuit *m* de réglage
~CLUTCH - [mach tools] embrayage *m* de commande
~COLUMN - [aero] (a pillar supporting aircraft controls, in particular longitudinal and lateral controls) colonne *f* de commande, "manche *m* à balai"
~CONSOLE - [comput] (assembly of manual controls) pupitre *m* de commande, armoire *f* de commande
~CONSTANT - [el] (intrinsic constant of a galvanometer) constante *f* de rappel
~COUNTER - [comput] (device recording the storage location of the instruction word) compteur *m* d'instructions
~CRANK - [mech] manivelle *f* de commande
~CUBICLE - [radio] cabine *f* de contrôle
~DESK - [contr] pupitre *m* de commande
~DEVICE - [mech etc] dispositif *m* de réglage, dispositif *m* de commande
~ECCENTRIC - [mech] came *f* de commande
~ELECTRODE - [electron] (in a thermionic valve, an electrode regulating the amount of electron current) électrode *f* de commande
~ENGINEERING - [gen] (comprising the science and technology of automatic control) régulation *f* automatique
~FORK - [auto] fourchette *f* de commande
~FREQUENCY - [radio] fréquence *f* pilote
~GEAR - [mech] mécanisme *m* de commande, dispositif *m* de réglage
~GRID - [electron] (control electrode in the form of a grid interposed between the cathode and the other electrodes of a thermionic valve) grille *f* de commande
~GRID BIAS - [electron] tension *f* de grille de commande
~GRID MODULATION - [radio] (modulation of transmission by means of a control grid) modulation *f* par grille de commande
~HEAD - [oil ind] (in natural gas installations) tête *f* de tubage, tête *f* de revêtement
~HOOK-UP - [oil ind] ensemble *m* des "preventers"
~HYSTERESIS - [contr] (double-value control characteristic) hystérésis *f* de commande
~IMPEDANCE - [el] impédance *f* de commande
~INSTRUCTION COUNTER - s. control counter
~KNOB - [mech] bouton *m* de réglage
~LABORATORY - [ind] laboratoire *m* de contrôle

CONTROL LIMIT SWITCH - [el] interrupteur *m* de fin de course
~LINKAGE - [auto] timonerie *f* de direction
~LOOP - [mech] boucle *f* de réglage
~MAGNET - [el] (magnet for the adjustment of the direction of the moving magnet system in some galvanometers) aimant *m* directeur
~MEMORY - [comput] (process controlling data-carrying store) mémoire *f* de commande
~MOTOR - [mech] moteur *m* de commande
~PANEL - s. control board
~PEDESTAL - s. control column
~POINT - [surv] point *m* de relèvement
[contr] (the value of the control variable which is maintained by an automatic controller) valeur *f* prescrite
~PRECISION - [contr] fidélité *f* de la commande
~PULSE - [electron] impulsion *f* de commande
~RANGE - [contr] bande *f* de réglage
~RATE - [contr] vitesse *f* de régulation
~RATIO - [electron] (ratio of firing voltage to critical grid voltage) rapport *m* de commande
~REGISTER - [comput] (register storing current instructions in the control unit of the computer) registre *m* d'instructions
~RELAY - [el] relais *m* de commande
~RESOLUTION - [contr] seuil *m* d'amorçage
~ROCKET - [astronaut] fusée *f* de guidage
~ROD - [nucl] (a rod made of a material which changes the reactivity by motion) barre *f* de commande, barre *f* de réglage
~ROOM - [el] (room used by the technical staff of a power supply system for the supervision of operations) salle *f* de commande, salle *f* de régie
[contr] (in ships, submarines etc) poste *m* central
~SEQUENCE - [comput] (instruction sequence) séquence *f* de commande
~SHUTTER - [aero] volet *m*
~SIGNAL - [radio] signal *m* d'entrée
~SPRING - [mech] ressort *m* régulateur
~STAND - [railw] banc *m* de manoeuvre
~STATION - [surv] point *m* de relèvement
[telecomm] station *f* directrice
~STICK - [aero] (vertical lever controlling the longitudinal and lateral control surfaces) levier *m* de commande, manche *m* à balai
~STORAGE - s. control memory
~STORE - s. control memory
~SURFACE - [aero] (aerofoil used to control an aircraft) gouverne *f*
~SURFACE AXIS - [aero] axe *m* des gouvernes
~SURFACE LOCK - [aero] (locking device preventing unwanted movements of the control surfaces) verrouillage *m* des gouvernes
~SURFACE POSITION INDICATOR - [aero] (a device indicating the true position of the control surface) indicateur *m* de position des gouvernes
~SURVEY - [surv] relevé *m* géométrique
~SYSTEM - [contr] système *m* de régulation
~TOWER - [aero] (elevated part of an airfield building in which traffic control is located) tour *f* de contrôle
~TRACK - [el acoust] (a supplementary sound track, to control some aspects of the reproduction in the main track) piste *f* de contrôle
~WINDING - [el] (winding through which flows a current controlling the output) enroulement *m* de com-

mande
CONTROL WORD - [comput] (a store cell with the information of the address at which more specific information is obtained) mot *m* de commande
~ZONE - [aero] (the zone of one or more aerodromes, where rules additional to those for control area flights are applicable) zone *f* de contrôle
CONTROLLABILITY - [mech etc] manoeuvrabilité *f*, maniabilité *f*
CONTROLLABLE PITCH PROPELLER - [aero] (propeller whose pitch can be modified by the pilot while in flight) hélice *f* à pas commandé
~REACTION - [nucl] (nuclear reaction which can be controlled) réaction *f* en chaîne contrôlable
CONTROLLED ATMOSPHERE - [th eng] (for ovens) atmosphère *f* artificielle
~ATMOSPHERE FURNACE - [metall] four *m* à atmosphère artificielle
~CHAIN REACTION - [nucl] réaction *f* en chaîne contrôlée
~CONDITION - [contr] (the condition of the controlled process) grandeur *f* réglée, grandeur *f* dépendante
~SENDER - [radio] (with frequency kept within specified limits) émetteur *m* pilote
~SYSTEM - [contr] (a regulated process in order to change the physical or chemical state of a material) système *m* réglé
~TAB - [aero] (tab designed to be adjusted in flight) volet *m* réglable en vol
~VARIABLE - [contr] (any quantity or condition that is measured and controlled) grandeur *f* réglée
CONTROLLER - [el] (equipment to control the operation of an electrical apparatus) combinateur *m* (de couplage)
[mech] (regulating device) régulateur *m*, indicateur *m*
[instr] (recording instrument) enregistreur *m*
[comm] contrôleur *m*
~ACTION - [contr] (the modus operandi of an automatic control system) mode *m* de réglage
~RESISTANCE - [el] résistance *f* d'un combinateur
CONTROLLING - [gen] de réglage, de commande, de contrôle
~COUPLE - [el] couple *m* antagoniste
~FACTOR - [gen] facteur *m* principal, facteur *m* déterminant
~GAUGE - [mech] manomètre *m* de réglage
~MAGNET - s. control magnet
~MEANS - [contr] (the automatic controller elements involved in producing a correct action) dispositifs *mpl* de réglage
~STEAM GAUGE - s. controlling gauge
~TORQUE - [mech] (the resultant of the deflecting and restoring torques) couple *m* antagoniste
CONTROLS - [gen mech etc] commandes *fpl*, réglages *mpl*
CONVECTION - [phys] (mechanically or thermally produced upward or downward movement of air currents) convection *f*
[th dyn] (heat transfer by currents flowing from the hotter part of a fluid to a colder region) convection *f*
~CLOUDS - [met] (clouds caused by convection currents carrying moist air up into colder layers) nuages *mpl* de convection
~CURRENT - [phys] courant *m* de convection
CONVECTION DRYING - [text] séchage *m* par contact
~HEATER - s. convector
CONVECTIVE - [gen] (having the property of conveying) de convection, de transmission
~DISCHARGE - [el] (the movement of a stream of particles carrying away charges from a body charged to a sufficient high voltage) effluve *m*
CONVECTOR - [th eng] (heating device in which heat is transferred by convection) convecteur *m*
CONVENIENCE OUTLET - [el] (electric point for domestic uses) prise *f* de courant
CONVENTION - [gen & leg] convention *f*, traité *m*, accord *m*, contrat *m*
CONVENTIONAL - [gen] conventionnel
CONVERGE, to - [gen] converger
CONVERGENCE - [gen] (movement toward the same point) convergence *f*
~CONTROL - [telev] réglage *m* de la convergence
~OF FLUID - [phys] (the accumulation of fluid which accompanies a negative divergence of the flow velocity) convergence *f* d'un fluide
~OF MERIDIANS - [surv] (the convergence of the meridians toward a pole) convergence *f* des méridiens
~PLANE - [telev] plan *m* de convergence
~RECORDER - [mining] appareil *m* enregistreur de l'affaissement du toit
CONVERGENT - [gen] convergent
~CONTROL SYSTEM - [contr] (multiple system in which the output of the controllers is combined) système *m* de réglages convergents
~DIVERGENT NOZZLE - [mech] diffuseur *m* convergent-divergent
CONVERGING LENS - [opt] lentille *f* convergente
~RAYS - [opt] rayons *mpl* convergents
~ROUTES - [railw] itinéraires *mpl* incompatibles
~WAVE - [phys] (spherical wave travelling with a decreasing radius) onde *f* convergente
CONVERSATION TIME - [telecomm] unité *f* de communication
CONVERSE PATTERNS - [text] dessins *mpl* à retour
CONVERSION - [gen] conversion *f*, changement *m*
[el] transformation *f*
[chem] conversion *f*
~ANGLE - [surv] (the correction angle applied to a Great Circle bearing) angle *m* de conversion
~BURNER - [el-mech] brûleur *m* de transformation
~COATING - [metall] (the treatment of a metal with a solution which removes some of the surface and leaves an insoluble film) revêtement *m* par conversion
~COEFFICIENT - [nucl] (ratio between the number of internal conversion electrons and the total number of quanta and conversion electrons emitted in a specified mode of a nucleus de-excitation) fraction *f* de conversion
~CONDUCTANCE - [el] (the relation in a frequency changer of the radio frequency input voltage to the intermediate frequency component of the output) pente *f* de conversion
~DETECTOR - [radio] étage *m* changeur de fréquence
~EFFICIENCY OF A KLYSTRON OSCILLATOR - [electron] (ratio between the high-frequency output power and the direct current power supplied to the beam) rendement *m* de conversion (de klystron oscillateur)
~ELECTRON - [electron] électron *m* de conversion

CONVERSION FACTOR - [phys] (ratio of two measures of the same physical quantities in different units) facteur *m* de conversion
~GAIN - [radio] (of a frequency change stage) gain *m* de conversion
~LOSS OF A FREQUENCY-CHANGER CRYSTAL - [el] perte *f* de conversion d'un changeur de fréquence à cristal
~OF COUNT - [text] conversion *f* du numéro
~RATIO - [nucl] (the number of fissionable atoms produced per fissionable atoms destroyed) taux *m* de conversion
~RESISTANCE - [radio] résistance *f* de conversion
~TABLES - [math] tables *fpl* de conversion
~TRANSCONDUCTANCE - [el] pente *f* de conversion
~TRANSDUCER - [radio] transducteur *m* à conversion
~TRANSITION - [phys] (in the nucleus) transition *f* par conversion
CONVERT, to - [gen] convertir, transformer
[metall] (in a Bessemer converter) fabriquer l'acier dans un convertisseur
~BACK, to - [ind] reconvertir
CONVERTED COUNT - [text] numéro *m* converti
~STEEL - [metall] acier *m* fabriqué par procédé Bessemer
CONVERTER - s. convertor
CONVERTIBLE - [gen] transformable, convertible
[auto] décapotable
~COUPE - [auto] cabriolet *m*
~LENS - [opt] (camera lens made of two separate systems) objectif *m* convertible
~VALVE - [mech] soupape *f* d'inversion
CONVERTING CHEST - [metall] caisse *f* de cémentation
~NETWORK - [el] réseau *m* convertisseur
~PLANT - [metall] installation *f* d'affinage
CONVERTOR - [el] (a rotary unit designed to change the frequency or number of phases of an alternating current, or to convert alternating to direct current) convertisseur *m*
[metall] convertisseur *m*
~BELLY - [metall] calotte *f* du convertisseur
~BOTTOM - [metall] moule *m* pour le fond de cornue
~NOSE - [metall] bec *m* de cornue
~PROCESS - [metall] procédé *m* au convertisseur
~SET - [el] groupe *m* convertisseur
~SLAG - [metall] scories *fpl* de convertisseur
~STEEL - [metall] acier *m* au convertisseur
CONVEX - [gen & geom] convexe
~BOBBIN - [text] bobine *f* convexe
~FILLET WELD - [metall] (fillet weld with a convex face) soudure *f* convexe
~LENS - [opt] lentille *f* convexe
~MILLING CUTTER - [mach tools] fraise *f* convexe
~MIRROR - [opt] (portion of a sphere with a reflecting surface) miroir *m* convexe
~SIDE - [gen] côté *m* convexe
~SURFACE - [gen] surface *f* convexe
~TILE - [constr] tuile *f* ronde
~VEINS - [zool] (veins following the ridges of the wing corrugations) veines *fpl* convexes
CONVEXITY - [gen] convexité *f*
CONVEY, to - [gen] conduire, transporter, transmettre
CONVEYANCE - [gen] transport *m*, moyen *m* de transport, envoi *m*, transmission *f*, communication *f*
[leg] (the transference of property from one person to another) cession *f*
CONVEYING - [gen] transport *m*, transmission *f* de transport
~BAND LINK - [mech] jonction *f* de courroie transporteuse
~CHUTE - [mech etc] descendeur *m*, déversoir *m*, goulotte *f*
~OF CHARGE - [metall] transport *m* au gueulard
~PLANT - [agric] moyens *mpl* de transport agricoles
~TROUGH - [min] gouttière *f* transporteuse
CONVEYOR - [mech] (machine or device for the transport of objects or material by continuous action) transporteur *m*, convoyeur *m*
~BELT - [mech] courroie *f* transporteuse
~BUCKET - [impl] benne *f* de transport, godet *m* transporteur
~CHAIN - [mech] (for the forward movement of trucks) chaîne *f* de transport
~FABRIC - [text] tissu *m* pour courroies transporteuses
~FURNACE - [metall] four *m* à passage
~IDLER - [mech] rouleau *m* de transporteur
~JIB - [mech] bras *m* du convoyeur
~LOADER - [mech] convoyeur-chargeur *m*
~SPAN - s. conveyor belt
~TRUCK - [transp] chariot *m* transporteur
~WORM - [mech] transporteur *m* à vis
CONVOLUTION - [gen] spire *f*, tour *m*, pas *m* de vis
[anat] (an elevation of the brain surface) circonvolution *f*
CONVOY, to - [transp] convoyer
CONVOY - [transp] convoi *m*
COOING - [acoust] (the sound made by pigeons) roucoulement *m*
COOK, to - [gen] cuire
COOKED - [gen] cuit
[photo] (overdeveloped) surdeveloppé
COOKER - [impl] cuisinière *f*, fourneau *m*
COOKING SPACE - [th eng] espace *m* utile
COOL, to - [gen] refroidir, réfrigérer
COOL - [gen] frais, froid
~COLOURS - [opt] couleurs *fpl* froides
COOLANT - [ind chem] réfrigérant *m*, liquide *m* de refroidissement
[mech] (mixture of water, soda, oil and soft soap to cool a cutting tool) liquide *m* réfrigérant
[chem] (any fluid used to convey heat away from a body) réfrigérant *m*
~CLARIFIER - [mech] (an arrangement for the elimination of impurities caused by circulating-oil emulsions) filtre *m* réfrigérant
~ELECTRIC PUMP - [mach tools] électropompe f du réfrigérant
~FLOW REGULATING VALVE - [mach tools] robinet *m* de réglage flux réfrigérant
~LINE - [th eng] (of refrigerating plant) canalisation *f* de refroidissement
~PUMP - [mech] (pump designed to circulate cooling fluid) pompe *f* à fluide de refroidissement, pompe *f* réfrigérante
~RADIATOR - [mech] (a device for cooling a liquid-heat vehicle by the provision of a large heat-exchange surface) radiateur *m* de refroidissement
~TANK - [mech] (cooling liquid reservoir, particularly for machine tools) réservoir *m* du réfrigé-

rant
COOLED - [gen] refroidi, réfrigéré
~ANODE VALVE - [electron] (a type of thermionic valve with special provision for cooling the anode) tube *m* à anode refroidie
~TUYERE - [metall] (blast-furnace tuyere surrounded by a water-jacket) tuyère *f* à refroidissement par air
COOLER - [gen] réfrigérant *m*, refroidisseur *m*
~COIL - [th eng] (of a refrigerating plant) serpentin *m* de refroidissement
~SLUDGE - [th eng] (of a refrigerating plant) dépôt *m*
~SUBSTITUTE - [brew ind] bac *m* refroidisseur profond
~TUN - [brew ind] cuve *f* de refroidissement
COOLIDGE TUBE - [electron] tube *m* de Coolidge
COOLING - [gen] réfrigérant, de refroidissement
~AGENT - [chem] agent *m* réfrigérant
~AIR - [gen] (used as a cooling agent) air *m* de refroidissement
~BATH - [mech] bain *m* de refroidissement
~BED - [metall] refroidisseur *m*
~BRINE - [ind chem] saumure *f* frigorigène
~BY ADIABATIC MAGNETIZATION - [th eng] (technique obtaining very low temperatures) refroidissement *m* par magnétisation adiabatique
~BY EVAPORATION - [phys] refroidissement *m* par évaporation
~BY WATER - [th eng] refroidissement *m* par eau
~CHAMBER - [metall] chambre *f* de refroidissement
~CIRCUIT - [th eng] cycle *m* réfrigérant, circuit *m* de refroidissement
~COIL - [th eng] (a helically-coiled tube conveying cooling fluid) serpentin *m* réfrigérant
~COLUMN - s. cooling tower
~CRACK - [metall] tapure *f* à froid
~CURVES - [metall] courbes *fpl* de refroidissement
~DRAG - [phys] (drag due to the cooling device) retard *m*
~DRUM - [th eng] tambour *m* de refroidissement
~DUCT - [th eng] (in refrigerating plants) canalisation *f* de refroidissement
~EFFECT - [th eng] effet *m* réfrigérant
~FAN - [auto] ventilateur *m*
~FIN - [mech] (a projecting blade to increase the heat-transfer surface) ailette *f* de refroidissement
~FIXTURE - [mech] (a device for supporting a moulded article during cooling) installation *f* pour le refroidissement
[plast ind] jauge *f* conformatrice
~FLANGE - [mech] ailette *f* de refroidissement
~JACKET - [th eng] (chamber or passage surrounding a part or unit to convey heat away from it) chemise *f* de refroidissement
~JIG - [mech] (device for controlling the dimensions of a moulding while cooling) gabarit *m* de contrôle
[plast ind] s. cooling fixture
~POND - [metall] bassin *m* de refroidissement
~RATE - [metall & glass ind] vitesse *f* de refroidissement
~RIB - s. cooling flange
~SERPENTINE - [th eng] serpentin *m* refroidisseur, serpentin *m* frigorifique
~SURFACE - [th eng] surface *f* de refroidissement
~SYSTEM - [th eng] (arrangement for removing heat from any desired part) système *m* de refroidissement
~TOWER - [ind chem] (a structure in which the liquid is cooled by causing it to descend against a current of gas at a lower temperature) tour *f* de réfrigération, réfrigérant *m* à cheminée
~TRAP - s. cold trap
~TRAY - [th eng] bac *m* de refroidissement
~TROUGH - [plast ind] (a small tank through which water is circulated so that extruded pipes passing through it reduce their temperature) cuve *f* de refroidissement
~VANE - s. cooling flange
~VAT - [brew ind] cuve *f* de refroidissement
COOLSHIP - [brew ind] réfrigérant *m*
COOP - [electron] (abbreviation for Mercury Vapour LAMP) (an electric discharge lamp) tube *m* à vapeur de mercure
COOPER - [gen] tonnelier *m*
COOPER'S WOOD - [timber] bois *m* pour tonneaux
COOPERATIVE ASSEMBLY - [mech](assembly in mechanics with appreciable interactions between the assembly composing systems) ensemble *m* à interactions
~PHENOMENON - [phys] phénomène *m* d'ensemble
COOPERITE - [min] sulphide and arsenide of platinum coopérite *f*
COORDINATE - [gen] coordonné
[math] coordonnée *f*
~PHENOMENON - s. cooperative phenomenon
~POTENTIOMETER - [instr] potentiomètre *m* coordonné
~SYSTEM - [phys] (surfaces used for determining the position of a point) système *m* de coordonnées
COORDINATION - [gen] coordination *f*
~NUMBER - [chem] (the number of atoms of a complex salt) indice *m* de coordination
COP, to - [text] (the operation of winding cotton yarn on a cop) caneter, canetter
COP - [text] canette *f*, bobine *f*
~BASE - [text] embase *f* de la bobine, embase *f* de la canette
~BIT - [text] noyau *m* de canette
~BODY - [text] corps *m* de la canette
~BOTTOM - [text] embase *f* du cops, cul *m* de la bobine
~BOX - [text] boîte *f* à cops, boîte *f* à bobines
~BUILDING MOTION - [text] mécanisme *m* de formation de la canette
~CHANGER - [text] porteur *m* de canettes
~CHANGING MOTION - [text] mécanisme *m* changeur
~CREEL - [text] ratelier *m* pour cops
~FEELER - [text] tâteur *m*
~GUARD - [text] tâteur *m*
~GUIDE - [text] dévidoir *m* à guide-fil et guide-canette
~HOLDER - [text] dévidoir *m* à porte-canette et guide fil fixe
~NOSE - [text] pointe *f* du cops
~SHAPE - [text] forme *f* de la canette
~SKEWERING DEVICE - [text] dispositif *m* d'enfilage de la canette
~STEAMING BOX - [text] caisson *m* à vaporisage des bobines
~WINDER - [text] machine *f* à caneter
~WINDING - [text] bobinage *m* de cops
~WINDING MACHINE - s. cop winder
~YARN - [text] fil *m* en canette

COPAL - [chem] (hard resinous exsudations from trees, sometimes found in recent fossil form) copal *m*
~OIL - [chem] huile *f* de copal
~RESIN - s. copal
~VARNISH - [paint] vernis *m* au copal
COPALITE - [min] (pale yellow substance partly mineralized by remaining in the earth) copalite *f*
COPE, to - [constr] couronner, chaperonner
COPE - [constr] s. coping
[metall] dessus *m*, chapeau *m*
~RING - [metall] (in founding) armature *f* du dessus
~STONE - [constr] s. coping
COPING - [constr] (stone or bricks covering the top of a wall) couronnement *m*, chaperon *m*, tablette *f* de mur
~BRICK - [constr] brique *f* de chaperon, brique *f* de tablette
~RAIL - [mech] rail *m* de roulement, barre *f* glissière
~SAW - [impl] scie *f* à découper
~STONE - s. coping
COPLANAR - [geom] coplanaire
~ELECTRODES - [electron] électrodes *f*pl coplanaires
~FORCES - [phys] (when lines of action of all forces lie in a single plane) forces *f*pl coplanaires
~GRID VALVE - [electron] tube *m* à grilles coplanaires
COPOLYCONDENSATION - [chem] (production of polymers by the condensation of two or more molecules) copolycondensation *f*
COPOLYMER - [chem] (compound produced by the reaction together of two or more dissimilar polymers) copolymère *m*
COPOLYMERIZATION - [chem] copolymérisation
~WITH CROSS LINKING - [plast ind] copolymérisation *f* avec réticulation
COPPER, to - [metall] cuivrer
COPPER - [metall] (an element, symbol Cu; reddish tough metal, very ductile and with good electrical conductivity) cuivre *m*
[brew ind] (open vessel for heating liquids) chaudron *m*
~ACCUMULATOR - [el] accumulateur *m* au cuivre
~ACETATE - [chem] (an insecticide, fungicide, pigment for surface coatings and dye mordant) acétate *m* de cuivre
~ACETOARSENITE - [chem] (a wood preservative and insecticide) acéto-arsénite *m* de cuivre
~ARSENATE - [chem] (an insecticide and pigment) arséniate *m* de cuivre
~ASBESTOS - [constr] matériau *m* métalloplastique
~-BEARING - [min] cuprifère
~BEECH - [bot] hêtre *m* rouge
~-BERYLLIUM ALLOYS - [metall] (alloys of copper and beryllium used to produce non-sparking tools for use in mines etc to minimize the danger of explosions) alliages *m*pl cuivre-beryllium
~BIT - [tool] (a tool consisting of a pointed piece of copper at the end of an iron rod) fer *m* à souder
~BIT SOLDERING IRON - s. copper bit
~-BLENDE - [min] kupferblende *f*
~BLUE - [paint] bleu *m* de cuivre
~BOLT - s. copper bit
~BOTTOMS - [metall] fond *m* cuivreux
COPPER BRUSHES - [el] (used in electric commutator machines) balais *m*pl en cuivre
~-CALCIUM ALLOY - [metall] alliage *m* cuivre-calcium
~CARBONATE - [chem] (a fungicide, insecticide and pigment) carbonate *m* de cuivre (déployé)
~CHLORIDE - [chem] (a wood preservative, disinfectant, catalyst and oxidizing agent, and mordant. Copper chloride also has application in petroleum refining and metallurgy) chlorure *m* de cuivre
~CHROMATE - [chem] (a dyeing mordant) chromate *m* de cuivre
~-CLAD STEEL CONDUCTOR - [el] conducteur *m* en acier cuivré
~-COLOURED - [gen] cuivré, cuivreux
~CYANIDE - [chem] (an intermediate in organic synthesis and a component of plating baths) cyanure *m* de cuivre
~DEPOSIT - [min] gîte *m* de cuivre
~DISH - [ind chem] capsule *f* de cuivre
~-DISH TEST - [ind chem] essai *m* à la capsule de cuivre
~DRIFT - [mech] (drift made of copper to avoid damage to the part struck) chasse *f* en cuivre
~FERROCYANIDE - [chem] (a pigment for surface coatings) ferrocyanure *m* de cuivre
~FLUOSILICATE - [chem] [a marble dye) silicofluorure *m* de cuivre
~GASKET - [mech] joint *m* en cuivre
~GLANCE - [min] (natural copper sulphide, crystallizing in the orthorhombic system and commonly associated with the more abundant ores of copper) chalcocite *f*, chalcosine *f*
~GLAZING - [constr] (glazing made with many individual panes separated by copper strips) vitrage *m* à baguettes de cuivre
~GLYCINATE - [chem] (a constituent of plating baths; also has medicinal uses) glycinate *m* de cuivre
~GREEN - [min] vert *m* malachite, vert *m* de montagne
~HAMMER - [impl] (hammer with a copper head) marteau *m* en cuivre
~HYDROXIDE - [chem] (a mordant and pigment; also used in the production of rayon) hydroxyde *m* de cuivre
~INDEX - s. copper number
~JACKET - [metall] chemise *f* en cuivre, revêtement *m* de cuivre
~KETTLE - [metall] chaudière *f* en cuivre, cuve *f* en cuivre
~LACTATE - [chem] (a fungicide) lactate *m* de cuivre
~LEAD - [metall] alliage *m* cuivre-plomb
~LINING - [metall] revêtement *m* de cuivre, garniture *f* de cuivre
~LOSS - [el] (occurring in machinery etc owing to the current flowing in the windings) perte *f* dans le cuivre
~LUSTER - [metall] brillant *m* métallique irisé
~MASHING - [brew ind] mélange *m* en chaudron
~MASTER - [el acoust] (the copper electroplate) matrice *f* en cuivre
~METABORATE - [chem] (insecticide and pigment) métaborate *m* de cuivre
~MILL - [metall] usine *f* de traitement du cuivre
~MINE - [min] mine *f* de cuivre

COPPER MORDANT - [chem] mordant *m* de cuivre
~NAPHTHENATE - [chem] (polymerization agent for cashewnut shell liquid resins; has also uses as a fungicide) naphténate *m* de cuivre
~NICKEL - [min] niccolite *f*
~NITRATE *f* [chem] (has uses as an astringent in medicine, as a catalyst, insecticide, in dyes and in electroplating) nitrate *m* de cuivre
~NITRITE - [chem] nitrite *m* de cuivre
~NUMBER - [chem] (the amount of copper, expressed in milligrams, produced by the reduction of Fehling's solution by one gram of carbohydrate) indice *m* de cuivre
~OLEATE - [chem] (an insecticide and fungicide) oléate *m* de cuivre
~ORE - [min] minerai *m* de cuivre
~OXIDE - [chem] (a catalyst for the reduction of organic compounds) oxyde *m* de cuivre
~OXIDE RECTIFIER - [el] redresseur *m* cuivre-oxyde de cuivre
~OXYCHLORIDE - [chem] (a pesticide) oxychlorure *m* de cuivre
~PIPE - [metall] tube *m* en cuivre
~PLAIT - [el etc] tresse *f* de cuivre
~PLATE - [print] plaque *f* de cuivre, cliché *m* de cuivre
~-PLATE ENGRAVING - [print] chalcographie *f*
~-PLATE PRESS - [print] presse *f* en taille-douce
~PLATING - [metall] (the deposition of a coating of copper of high purity on a metallic electrode) cuivrage *m*
~POWDER - [metall] cuivre *m* en poudre, cuivrée *f*
~PYRITES - [min] (natural sulphite of copper and iron, crystallizing in the tetragonal system; it is an abundant ore of copper and occurs in veins as brassy masses, often iridescent from surface tarnish) pyrite *f* de cuivre, chalcopyrite *f*
~RESINATE - [chem] (insecticide) résinate *m* de cuivre
~RICINOLEATE - [chem] (a component of insecticides and fungicides) ricinoléate *m* de cuivre
~RUST - [metall] (synonym of verdigris) vert-de-gris *m*
~SMITH - [gen] chaudronnier
~-SMITH HAMMER - [tool] (hammer with long ball-pane head) marteau *m* de chaudronnier
~SOAPS - [chem] (in paint manufacture, copper naphthenate and oleate which have preservative, anti-rivelling and polymerizing properties) savons *mpl* de cuivre
~SPRAY - [agric] bouille *f* bordelaise
~STEEL - [metall] acier *m* au cuivre
~STRIKE - [metall] (in electroplating) fond *m* cuivreux
~SULPHATE - [chem] (used as a mordant, astringent in medicine, germicide and insecticide, in ore flotation, in metallurgy, as a wood preservative and in the petroleum and synthetic rubber industries) sulfate *m* de cuivre
~SULPHATE TEST - [ind chem] (made to investigate the quality of galvanized coatings) essai *m* au sulfate de cuivre
~TREATMENT - [agric] application *f* de solution cuivrée
~URANITE - [min] (a hydrous phosphate of uranium and copper) uranite *f*
~VALUE - s. copper number

COPPER VAT - s. copper kettle
~VITRIOL - s. copper sulphate
~WIRE - [metall] fil *m* de cuivre
~-WIRE ARTERY - [med] (in ophthalmology) artère *f* à fil de cuivre
~WIRE GAUZE - [metall] toile *f* en fil de cuivre
COPPERAS - [chem] (a term used for commercial crystalline ferrous sulphate) copperas *m*, couperose *f* verte
COPPERING - [metall] cuivrage *m*
COPPING - [text] cannettage *m*
~RAIL - [text] pièce *f* de mise en forme
COPRA - [bot] (the dried meat of the coconut) copral *m*
COPRECIPITATION - [chem] (the simultaneous formation of two or more precipitates) coprécipitation *f*
COPROIC - [zool] coproïque
COPROLITE - [geol] (fossil faeces of fish and reptile largely consisting of calcium phosphate) coprolithe
COPROSTEROL - [biochem] (a constituent of faeces) coprostérol *m*
COPY, to - [gen] copier
[comput] (the operation of transferring informatio from one register to another) copier
COPY - [gen] copie *f*
~MACHINING - [mech] copiage *m*, reproduction *f*
~-MILLING - [mach tools] fraisage *m* à gabarit, fraisage *m* en copiage
~MILLING MACHINE - [mach tools] fraiseuse *f* à copier
COPYING - [gen] copiage *m*
[photo] épreuve *f*, tirage *m*
[mech etc] reproduction *f*, copie *f*
~INK - [ind chem] encre *m* à copier
~LATHE - [mach tools] tour *m* à reproduire, tour *m* à copier
~MACHINE - [mach tools] (machine producing a number of similar objects by an automatically guided tool) machine *f* à copier
~MILLING MACHINE - [mach tools] fraiseuse *f* à copier
~PRESS - [mech] presse *f* à copier
~PROCESS - [photo] tirage *m*
~ROLLER - [impl] rouleau *m* copieur
CORAL - [zool] corail *m*
~REEF - [geol] (calcareous bank consisting of the skeletons of corals) récif *m* corallien, barrière *f* de corail
CORALLOID - [bot] (having the appearance of a branched coral piece) coralloïde, corallaire
CORB - [mining] (a type of bucket) cuffat *m*
CORBEL - [constr] (bricks or stone projecting from a wall) corbeau *m*
[constr] sous-poutre
CORD - [gen] corde *f*, ficelle *f*, fil *m* souple
[el etc] (flexible cable for telephone etc) conducteur *m* souple, cordon *m*
[text] (rib effect in a fabric) côte *f*
[meas] (a measure of wood, stone or rock) mesure *f* de volume valant 128 pieds cubes (3,62 m^3)
[glass man] corde *f*
[rubber man] (of a tyre) câblé *m*
~ARMOURING - [el etc] blindage *m* de cordon
~CIRCUIT - [el] circuit *m* de fiche
~CIRCUIT REPEATER - [el] répéteur *m* de circuit à fiches
~CUTTING MACHINE - [text] machine *f* à couper le

velours à côtes
CORD CYLINDER - [text] tambour *m* d'entraînement des broches par cordes
~DRIVE - [mech] transmission *f* par corde, commande *f* par cordon
~EDGE - [text] garniture *f* de ganse
~FABRIC - [rubber man] (in tyres) tissu *m* câblé, tissu *m* pour pneumatiques
~HEDDLE - [text] lisse *f* en fil retors, lisse tricotée
~PULLEY - [mech] poulie *f* à gorge
~TRACTION - [mech] traction *f* par corde
~VELVETEEN - [text] velours *m* de coton à côtes
~WEIGHT - [telecomm] (for telephones) contrepoids *m*
~YARN - [text] fil *m* pour cordes
CORDAGE - [gen] cordage *m*, corde *f*, câble *f*
[oil ind] (cable used in oil well sinking) câble *f* d'acier
CORDIERITE - [min] (a silicate of aluminium, iron and magnesium with water) cordiérite *f*
CORDING - [text] (connexion of the treadles of a loom with the leaves of heddles by cords) encordage *m*
~MACHINE - [text] toroneuse *f*, machine *f* à fabriquer les torons
~STRIPES - [text] ligne *f* pour ourlets
~TOOL - [text] pied *m* pour faire l'ourlet
CORDITE - [chem] (a propellant explosive mixture consisting nitroglycerin and nitrocellulose) cordite *f*
CORDLESS SWITCHBOARD - [telecomm] table *f* d'opérateur à clefs
CORDUROY - [text] velours *m* cotelé
CORE, to - [gen] évider, percer, forer
[min] carotter
[metall] (in castings) noyauter; (in founding) renmouler
CORE - [bot] coeur *m*, trognon *m*
[el] (of electric cable) conducteur *m* isolé
[text] (of a rope) âme *f*
[metall] noyau *m*
[nucl] (the fuel or moderator and fuel in a nuclear reactor) coeur *m* de réacteur
[metall] (in casting) noyau *m*
[el] (of a magnet) noyau *m*
[mining] (in well sinking) carotte *f*, stross *m*
[paper man] âme *f*
[el] (in transformers; the substance in the space within the coils, air or ferromagnetic material) noyau *m*
[firearms] (of a bullet; the inner portion of steel) noyau *m* d'acier
[acoust] (in mechanical recording) the central layer of a laminated medium) couche *f* intérieure
[el] (of a carbon arc) mèche *f*
[text] (of a spool) bobine *f*
[electron] (the wire on which the helical grids for electronic tubes are wound) support *m* de cathode
~ANALYSIS - [min] analyse *f* du carottage
~ASSEMBLY - [metall] (in a casting operation) mise *f* en place des noyaux, noyautage *m*
~BACKING PLATE - [metall] (in die casting) contre-plaque *f*
~BAR - [metall] (in casting) armature *f* de noyau
~BARREL - [metall] lanterne *f* du noyau, armature *f* du noyau
[oil ind] tube *m* carottier
CORE BINDER - [metall] (substance used as a binder for cores) lien *m* de noyautage
~BIT - [mech] couronne *f* de sondage, trépan *m* carottier
~BLOWER - [metall] (in casting) machine *f* à souffler les noyaux
~BORING - [min] carottage *m*
~BOX - [metall] (in casting) boîte *f* à noyaux
~CARRIAGE - [metall] chariot *m* d'étuve à noyaux
~CARRIER - [metall] coquille *f* de séchage
~CATCHER - [mining] arrache-carotte *m*
~COMPOUND - [metall] liant *m* de noyautage composé
~CUTTER - s. core bit
~DIAMETER - [mech] (in threading) diamètre *m* du noyau
~DISC - [el] (armature lamination to be built up and form the armature core of an electric machine) tôle *f* d'induit
~DRIER - [metall] (in casting) sécheur *m*
~DRILL - [tool] (used in mining) sonde *f* à tubes, sondeuse *f* à carottes
~-DRILLING - [min] sondage *m* géologique peu profond, carottage *m*
[mech] perçage *m* à tube
~DRYING STOVE - [metall] (casting) étuve *f*
~EJECTOR AND BLOWER - [metall] (in casting) machine *f* à extraire et souffler les noyaux
~FIXTURE - [metall] fixation *f* pour noyau
~FRAME - [metall] (in casting) armature *f* de noyau
~GRID - [metall] armature *f* du noyau
~HARDENED - [metall] trempé à coeur
~HOLE - [mining] trou *m* pour carottage
[metall] trou *m* du noyau
~JIG - [metall] montage *m* de rectification
~LATHE - [mech mach] tour *m* à noyaux
~LIFTER - [mining] extracteur *m* de carottes, arrache-carottes *m*
~LIFTING ROD - [el] (in a transformer) barre *f* de levage du noyau
~LOSSES - [el] (power losses in ferromagnetic core) pertes *f*pl dans le fer
~MACHINE - [metall] machine *f* à noyauter
~MAKER - [metall] noyauteur *m*
~-MAKING - [metall] fabrication *f* des noyaux
~-MAKING MACHINE - [metall] machine *f* à noyauter
~MARK - s. core print
~MATERIAL - [nucl] (material subject to fission in a nuclear reactor) substance *f* fissible
~MOULDING - [metall] moulage *m* des noyaux
~OF A MAGNET - [el] noyau *m* (d'un aimant)
~OF ANTICLINE - [geol] noyau *m* anticlinal
~OF BALLAST - [road constr] noyau *m* de l'empierrement
~OF BOBBIN - [text] bobinot *m* du diamètre intérieur, centre *m* de la bobine
~OF FLOW - s. centre of flow
~OF SYNCLINE - [geol] noyau *m* synclinal
~OF THE ANTICLINE - [geol] noyau *m* d'un anticlinal
~OF THE ATOM - [phys] (the nucleus with its electrons of the alkali metals) noyau *m* de l'atome
~OF THE FRACTURE - [metall] noyau *m* de cassure
~OIL - [ind chem] (used in casting) huile *f* à noyaux
~OVEN - [ovens] étuve *f*
~PIN - [metall] (hard steel pin in a mould to produce

a plain or threaded hole in the moulding) broche *f*
CORE PIN PLATE - [impl] (plate carrying core pins) plaque *f* porte-broches
~PLATE - [metall] (in die casting) plaque *f* porte-noyaux
~PRINT - [metall] (the projection which is attached to a pattern to provide recesses in the mould) portée *f* de modèle, portée *f* de noyaux
~PULLER - [metall] (in casting) démoleur *m* de noyaux
~RACK - [metall] support *m* à noyaux
~RECOVERY - [mining] récupération *f* de carotte
~SAMPLE - [mining] carotte *f*
~SAND - [metall] (for casting) sable *m* à noyaux
~SETTING - [metall] (the complex of cores in the mould) mise *f* en place des noyaux, noyautage *m*; remoulage *m*
~SHEET - [metall] (in lamination; a central sheet of a laminated material) feuille *f* centrale
~SHIFTS - [metall] (defects in casting) déplacement *mpl* des noyaux
~SHOP - [metall] atelier *m* des noyaux
~SPINDLE - s. core bar
~STRAND - [text] (in rope manufacture) toron *m* central
~TANK - [mech] (of nuclear reactor; tank for the uranium compound solution in a light-water moderated reactor) cuve *f* contenant le coeur (du réacteur)
~-TYPE TRANSFORMER - [el] transformateur *m* à noyau
CORED CARBON - [mining] charbon *m* à mèche
~ELECTRODE - [el] électrode *f* creuse
~HOLE - [metall] (in casting; the hole in the core) trou *m* noyauté
~MOULD - [metall] (mould with passages for electric heating elements) moule *m* à voies calorifères
[plast ind] (for fluid circulation) moule *f* à canaux de refroidissement
CORELESS INDUCTION FURNACE - [metall] (type of induction furnace in which heat is developed in the charge itself by means of windings surrounding a crucible) four *m* à creuset à induction
CORF - [min] (wooden or iron tub to carry coal ore) berline *f*
CORING - [nucl] (variation in composition due to lack of equilibrium during solidification) ségrégation *f*
[mining] carottage *m*
[mech] (hot oil core in aviation engines) formation *f* d'un noyau
~PIN - s. core pin
CORINTHIAN - [arch] corinthien
CORIOLI'S EFFECT - [phys] (the deflection to which all objects moving above the ground surface at spatial velocity are subject, due to the earth rotation) effet *m* de Corioli
CORK, to - [gen] boucher
CORK - [bot] liège *m*
[gen] (for bottles etc) bouchon *m*
~BLACK - [chem] charbon *m* de liège
~-BOARD - [acoust] isolant *m* acoustique, plaque *f* de liège, liège *m* aggloméré
~INSULATION - [el] isolant *m* à base de liège, revêtement *m* en liège
~JACKET - [naut] brassière *f* de sauvetage
CORK OAK - [bot] chêne *m* liège
~PIPE COVERING - [constr] revêtement *m* de liège pour canalisation
~SLAB - [gen] plaque *f* de liège
~STOPPER - [gen] bouchon *m* de liège
~TREE - s. cork oak
~WASHER - [mech] (washer made of disintegrated cork) rondelle *f* de liège
CORKSCREW, to - [gen] descendre en spirale, tourner
CORKSCREW - [impl] tire-bouchon *m*
~ANTENNA - [radio] (called helical aerial in G.B. consisting of helix-shaped wire at the top) antenne *f* hélicoïdale
~DIVE - [aero] (in aerobatics) piqué *m* en spirale
~FIELD - [phys] champ *m* hélicoïdale
~SPIN - s. corkscrew dive
~STAIRCASE - [build] escalier *m* en spirale
~TWIST - [text] torsion *f* en tire-bouchon
~WEAVE - [text] armure *f* à cannelé oblique
CORKSCREWING - [mech] vrillage *m*
CORKY - [gen] subéreux
CORLISS VALVE - [mech] distributeur *m* corliss
~VALVE-GEAR - [mech] détente *f* (ou distribution) Corliss
CORN - [bot] (in Great Britain) grain *m*, blé *m*, céréales *f*; (in the USA) maïs *m*
[med] cor *m*, durillon *m*, oignon *m*
~COB - [bot] raffle *f* de l'épi de maïs
~CRIB - [build] (only in the USA) silos *m* à maïs
~DRILL - [agric] semoir *m* à maïs
~MILL - [agric] minoterie *f*, moulin *m*
~OIL - [chem] (pale yellow oil obtained from Indian corn) huile *f* de maïs
~STARCH - [ind chem] amidon *m* de maïs
CORNEA - [anat] (of the eye) cornée *f*
CORNEAL REFLEX - [opt] réflexe *m* cornéen
~TRANSPLANT - [med] greffe *f* de cornée
CORNEL - [bot] cornouiller *m*
CORNEOUS - [gen] corné, de corne
CORNER, to - [gen] mettre dans un angle
[comm in USA] accaparer
CORNER - [gen] angle *m*, coin *m*, arête *f*, recoin *m*
[bookbind] angle *m*
[electron] (waveguide; an abrupt direction change) coude *m*
~AERIAL - [radio] (aerial consisting of primary radiating element and a corner reflector) antenne *f* à réflecteur plat replié
~ANGLE - [mech] (in tools) angle *m* de taille
[arch] cornière *f* d'angle
~ANTENNA - s. corner aerial
~BAND - [metall] équerre *f* d'angle
~BEAD - [build] (round strip of wood to protect corners from damage) bourrelet *m*, baguette *f* d'angle
~BRACE - [constr] jambe *f* de force
[impl] vilebrequin *m* d'angle
~CHISEL - [impl] (with two straight cutting edges) ciseau *m* à angles
~CUTTING - [telev] (the elimination of the corners of the picture) obscurcissement *m* des coins
~DETAIL - [telev] (very clear details at the corner of the screen picture) finesse *f* aux coins
~GUIDING - [mech] (in lifts) guidage *m* par cornière
~HORN - [acoust] (type of loudspeaker fitted to

corners in a room) haut-parleur *m* d'angle
CORNER JOINT - [carp] joint *m* à angle droit
[metall] (weld joint at the junction between two components) joint *m* à l'angle
~KEY - [mech] clavette *f* d'angle
~LAMP - [auto] lampe *f* d'angle
~LOUDSPEAKER - [acoust] (a built-in loudspeaker at the front corners of a set) haut-parleur *m* d'angle
~PIECE - [mech] cornière *f*
~POST - [constr] poteau *m* cornier, montant *m* d'angle
~REFLECTOR - [radar] (type of artificial target) réflecteur *m* métallique
~SEAM - [metall] soudure *f* en angle
~SHALING - [metall] peau *f*, coquille *f* d'oeuf
~SLICK - [metall] lissoir *m* d'équerre
~STONE - [constr] pierre *f* angulaire, pierre *f* d'encoignure *f*, pierre *f* de coin *m*
~TILE - [constr] tuile *f* cornière
~TROVEL - [constr] truelle *f* pour angles
~VALVE - [mech] vanne *f* d'équerre, robinet *m* d'équerre, vanne *f* à passage angulaire
CORNET - [mus] cornet *m*
[naut] pavillon *m*
CORNFLOUR - [bot] (US) farine *f* de maîs
CORNFLOWER - [bot] blenet *m*
CORNICE - [constr] (projecting ornamental mould) corniche *f*; (when on the top of a building) corniche *f*, fronton *m*
~OF THE PEDESTAL - [constr] moulure *f* d'embase
CORNIFICATION - [med] kératinisation *f*
CORNISH BOILER - [metall] (a type of horiozntal boiler) chaudière *f* de Cornouailles
~DIAMOND - [min] (a crystalline quartz) quartz *m* de Cornouailles
~GRANITE - [constr] (coarse)grained granite used for heavy construction work, found in Cornwall) granit *m* de Cornouailles
CORNSTALK DISEASE - [vet] maladie *f* du bétail attribuée à l'ingestion de paille ou de maîs
COROLLARY - [math] corollaire *f*
CORONA - [arch] (section of the cornice) couronne *f*
[astr] (series of coloured rings appearing round the sun or moon) couronne *f* solaire
[el] effet *m* de couronne
[constr] corniche *f* de gouttière
[metall] (the surface round a spot weld) couronne *f*
~DISCHARGE - [el] (luminous discharge occurring round a conductor at high potential, due to the ionization of the surrounding gas) décharge *f* par effet de couronne
~EFFECT - [el] (due to the ionization of the gas surrounding a conductor) effet *m* de couronne
~POWER LOSS - [el] perte *f* par effet de couronne
~VOLTMETER - [instr] voltmètre *m* à effet de couronne
CORONARY - [zool] (shaped like a crown) coronaire
~HEART DISEASE - [med] maladie *f* coronarienne
~THROMBOSIS - [med] thrombose *f* coronaire
CORPORAL - [min] chef *m* d'équipe, contremaître *m*
CORPORATE - [gen] consitué en corps
[comm] social
~BODY - [leg] corporation *f*, société *f* légale
~NAME - [comm] raison *f* sociale
CORPORATION - [gen & comm leg] corporation *f*, société *f* légale; (in the USA) société *f* par action
CORPUSCLE - [gen] corpuscule *m*
[physiol] (of the blood) globule *m* sanguin
CORPUSCULAR - [gen] corpusculaire
CORRADIATION - [radiol] focalisation *f*
CORRASION - [geol] (the cutting, vertical or lateral, made by a river) corrasion *f*, érosion *f* fluviale
CORRECT, to - [gen] corriger
CORRECT - [gen] correct, juste
CORRECTED - [gen] corrigé, compensé, ajusté
[el] remis en phase
~HORSEPOWER - [mech] puissance *f* corrigée
~LENS - [opt] objectif *m* corrigé
CORRECTING COIL - [telev] (coil in amplifier for the increase of the gain in the higher frequencies) bobine *f* correctrice
~CONDITION - [autom] (a control system output constituting an imput to the controlled equipment) grandeur *f* d'influence
~ELEMENT - [autom] (mechanism governing the variations in a control system) organe *m* de réglage
~LENS - [opt] (astigmatism correcting lens) lentille *f* correctrice, verre *m* correcteur
~SIGNAL - [contr] (signal correcting the alignment between the controlling and the controlled elements) signal *m* de correction
~UNIT - [contr] (the correcting element and its servo-motor) organe *m* correcteur
~WHEEL - [instr] roue *f* correctrice
CORRECTION - [gen] correction *f*, rectification *f*, rattrapage *m*
[meas] (the algebraically added quantity to obtain the true value of the quantity to be measured) correction *f*
[aero] (of flight) correction *f* de dérive
~FACTOR - [meas] (the quantity by which a calculated or observed quantity is multiplied to compensate for error) facteur *m* de correction, constante *f*
~OF ANGLES - [surv] correction *f* des angles
~TABLE - [instr] (list of figures for the conversion of the errors in instruments) table *f* de correction
~TIME - [contr] temps *m* de correction
~TO VACUUM - [opt] (the correction of the speed of light measured in air to the value in vacuum) correction *f* par rapport au vide
CORRECTIVE - [gen chem etc] correctif, correcteur *m*
~ACTION - [contr & automat] correction *f*
~FERTILIZER - [ind chem] engrais *m* correctif
~NETWORK - [radio] (inserted in a circuit to improve the transmission) réseau *m* correcteur
CORRECTOR - [horol] correcteur *m*
~OF THE PRESS - [print] (printer's reader, whose main task is to compare the proofs with the author's copy) correcteur *m*
~VANE - [instr] (part of instruments for regulating purposes) plaque *f* de réglage
CORRELATION - [gen] corrélation *f*
~COEFFICIENT - [math] (the ration between a variable and another) coefficient *m* de corrélation
~ENERGY - [phys] (the tendency of electrons to keep apart) énergie *f* de corrélation
~FUNCTION - [math] fonction *f* de corrélation
~TRACKING AND TRIANGULATION - [radar] poursuite *f* et triangulation par corrélation

CORRESPONDENCE - [gen comm etc] correspondance *f*
~PRINCIPLE - [phys] (the principle that in the limit of high quantum numbers, the quantum theory predictions agree with those of classical physics) principe *m* de correspondance
CORRESPONDING PROFILE - [mech] (in gears) profil *m* conjugué
~STATES - [phys] (when pressures and temperatures are equal fractions of the critical values) états *mpl* correspondants
CORRIDOR - [build] corridor *m*, couloir *m*
[aero] (air corridor) corridor *m* (aérien), couloir *m*
[railw] (in a train) couloir *m*
CORRIE - [geol] cirque *m*, entonnoir *m*
CORRODE, to - [gen] corroder, attaquer
CORRODED - [chem] corrodé, erodé, attaqué
CORRODING BRITTLENESS - [metall] fragilité *f* de corrosion
~INHIBITOR - [chem] anti-corrosif, anticorrosion
~LEAD - [metall] (lead of very high purity used for the manufacture of white lead by the corrosion process) plomb *m* pur
~POT - [metall] (glazed stoneware container in which the lead is corroded to form white lead) creuset *m* de corrosion
CORROSION - [chem] (the conversion of iron and other metals into oxides and carbonates by the action of air-and or water) corrosion *f*
~DURING SERVICE - [el] (of dry cells; wearing of the negative electrode while the cell is in service) corrosion *f* en cours de fonctionnement
~DURING STORAGE - [el] (of dry cells; consumption of the negative electrode while in store) corrosion *f* pendant l'emmagasinage
~FATIGUE - [metall] (deterioration of metals due to cyclic stress reversals accompanied by corrosion) fatigue *f* due à la corrosion
~PREVENTING LACQUER - [metall] laque *f* anticorrosive
~PREVENTIVE - [ind chem] inhibiteur *m* de corrosion
~-PROOFING - [ind chem] anticorrosif, anticorrosion
~RESISTANCE - [metall] résistance *f* à la corrosion
~-RESISTANT - [gen] protégé contre la corrosion
~-RESISTANT ALLOY - [metall] alliage *m* résistant à la corrosion
CORROSIVE - [chem] corrosif
~FLUID - [chem] fluide *m* corrosif
~GASES - [chem] gaz *mpl* corrosifs, vapeurs *fpl* corrosives
~SUBLIMATE - [chem] (commercial term for mercuric chloride) sublimé *m* corrosif, chlorure *m* mercurique
CORRUGATED - [gen] (of a surface) ondulé, cannelé, prismatique
~ASBESTOS CEMENT - [constr] feuille *f* ondulée en amiante-ciment
~BOARD - [paper man] carton *m* ondulé
~BOILER TUBE - [metall] (boiler tube whose walls are corrugated transversally to the direction of flow) tube *m* ondulé
~DIAPHRAGM - [acoust] (a conical membrane) membrane *f* striée
~GLASS - [build] verre *m* strié
~IRON - [metall] (sheet iron with parallel corrugations) tôle *f* ondulée
CORRUGATED IRON SHEET PILING - [constr] barrière *f* en tôle ondulée, palissade *m* en tôle ondulée
~ROLL - [metall] (roller formed with undulating surface to obtain similia configuration to the material passed over it) cylindre *m* canelé
~ROOFING SHEET - [constr] tôle *f* ondulée pour toiture
~STITCHER - [rubber man] roulette *f* moletée, molette *f* strié
~WIRE NETTING - [metall] grillage *m* ondulé
CORRUGATOR - [anat] (a muscle which causes wrinkles by contraction) muscle *m* corrugateur
CORSET - [text] corset *m*
~DRILL - [text] coutil *m* pour corsets
CORSITE - [geol] (igneous rock occurring in Corsica) corsite *f*
CORTEX - [bot] écorce *f*
[text] cortex *m*
[anat] cortex *m*, écorce *f*
CORTICAL - [bot] (of bark) cortical
~BUNDLE - [bot] (bundle in the cortex of a steam) faisceau *m* cortical
~-DEPRESSANT - [med] cortico-dépressif
~HORMONES - [physiol] hormones *fpl* corticales
~STIMULATOR - [instr] (electronic instrument delivering an electrical shock, used in medicine) excitateur *m* électronique du cortex
CORTICATE - [bot] cortiqueux
CORTICATION - [bot] (cell-covering around the threads of algae) cortication *f*
CORTICOSTERONE - [chem] (a natural steroid) corticostérone *f*
CORTISONE - [chem] (natural steroid, also prepared synthetically, with a number of applications in medicine) cortisone *f*
CORUBIN - [min] corubin *m*
CORUNDUM - [min] (naturally occurring aluminium oxide) corundum *m*, corindon *m*
CORVE - [mining] (a tram like vehicle for transport underground) berline *f*
CORYDALINE - [chem] (an alkaloid of the isoquinoline group) corydaline *f*
COS - [avvrev for cosine] cosinus *m*
~-LETTUCE - [bot] laitue *f* romaine
COSEC - [math] (abbrev for cosecant) cosécante *f*
COSECANT - [math] cosécante *f*
~-SQUARE BEAM - [radar] (transmission of energy in proportion to the square of the cosecant of the depression angle) faisceau *m* compensé
COSEPARATION - [chem] coprécipitation *f*
COSINE - [math] (triginometrical ratio) cosinus *m*
~EMISSION LAW - [phys] (relating to the emission of radiation from a radiating surface) loi *f* d'émission du cosinus
~LAW - s. cosine emission law
COSLETTIZING - [metall] (corrosion protection of steel by producing a surface coating of phosphate) phosphatation *f*
COSMETIC - [ind chem] cosmétique *m*
~BISMUTH - [chem] oxychlorure *m* de bismuth, chlorure *m* de bismuthyle
COSMIC ABUNDANCE - [phys] abondance *f* cosmique
~NOISE - [radio] bruit *m* cosmique
~RADIO NOISE - [radio] (noise produced in reception by cosmic radiations) ondes *fpl* cosmiques
~RAY - [phys] (high-energy radiation from outer

space, consisting mainly of protons) rayon *m* cosmique
COSMIC-RAY DECAY ELECTRONS - [nucl] (electrons originating from the decay of mesons) électrons *mpl* de désintégration du rayonnement cosmique
~-RAY KNOCK-ON ELECTRONS - [nucl] (electrons originating from the direct impact of fast mesons with the orbital electrons of the oxygen and nitrogen atoms) électrons *mpl* cosmiques de choc
~-RAY SHOWER - [phys] (a simultaneous appearance of cosmic rays) gerbe *f* de rayons cosmiques
COSMICAL GEOLOGY - [geol] géologie *f* cosmique
COSMOGRAPHY - [astr] (science of the consitution of the universe) cosmographie *f*
COSMOLINE - [chem] cosmoline *f*, vaseline *f* blanche
COSMOLOGY - [astr] (theoretical astronomy of the known universe) cosmologie *f*
COSMONAUTICS - [astronaut] navigation *f* spatiale (ou interplanétaire), astronautique, cosmonautique *f*
COSMOS - [astr] cosmos *m*
COSMOTRON - [nucl] (modified synchrotron for the acceleration of protons by frequency modulation) bévatron *m*
COSSETTE - [sugar ind] (slice of beetroot, of a V-shape, for diffusion extraction) cossette *f*
COSSYRITE - [min] cossyrite *f*, aenigmatite *f*
COST, to - [gen & comm] coûter
COST - [gen & comm] coût *m*, prix *m*, frais *m*
~AND FREIGHT - [comm] coût-fret *m*
~CONTROL - [comm] contrôle *m* des prix de revient
~INSURANCE AND FREIGHT - [comm] coût-assurance-fret *m*, C.A.F.
~OF LIVING - [comm] coût *m* de la vie
~OF PRODUCTION - [comm] frais *m* de production
~OF UPKEEP - [comm] frais *mpl* d'entretien
COSTA - [bot] (a rib or a vein in a plant) côte *f*
COSTALGIA - [med] (pains in the rib region) névralgie *f* intercostale
COSTEAN, to - [min] (the operation of removing soil by water to expose rocks) prospecter, sonder
COSTECTOMY - [med] costectomie *f*, résection *f* costale
COSTICARTILAGE - [anat] cartilage *m* costal
COSTING - [comm ind etc] (the operation of determining the cost of an understarking) établissement *m* des prix de revient
COSTMARY OIL - [ind chem] essence *f* de balsamite odorante
COSTOCERVICAL TRUNK - [anat] tronc *m* costocervical
COT - [gen] berceau *m*, lit *m* d'enfant
[naut] couchette *f*, cadre *m*; (a small boat) petite embarcation *f*
[build] cabane *f*, hutte *f*
[text] garniture *f* de cylindre, garniture *f* du rouleau, revêtement *m* du rouleau
[text] tissu *m* pour rouleaux
[math] (abbrev for cotangent) cotangente *f*
~BUFFING ATTACHMENT - [text] appareil *m* à dresser les cylindres, appareil *m* à rectifier les cylindres
~PRESS - [text] dispositif *m* de garnissage des cylindres
COTANGENT - [math] (a trigonometrical ratio) cotangente *f*
COTS - [text] (roller covering) garniture *f* de cylindre
COTTAGE - [build] cottage *m*
~CHEESE - [agric] fromage *m* blanc
~INDUSTRY - [gen] artisanat *m*, travail *m* à domicile
~PIANO - [mus] (small size piano) piano *m* droit
COTTER, to - [mech] (small part, generally a pin, passed through a hole in another part to secure it) goupille *f*, clavette *f*
~BOLT - [mech] (bolt securing a part to a shaft) boulon *m* à clavette
~HOLE - [mech] trou *m* de goupille
~JOINT - [mech] joint *m* à goupille
~MILL - [mech] fraiseuse *f* pour clavettes
~PIN - [mech] (cotter in the form of a pin) goupille *f* fendue
~WAY - [mech] rainure *f* de clavette
COTTERED CHAIN - [oil ind] chaîne *f* à clavette
COTTON - [text] (a downy fibre surrounding the seeds of the cotton plant) coton *m*
~ASBESTOS - [text] (a cotton material used in filters) laine *f* d'amiante
~BATTING - [text] ouatine *f*
~BELT - [agric] zone *f* de culture du coton
~BINDER - [telecomm] fil *m* de couleur (d'identification)
~BLANKET - [text] (heavy cotton fabric with a raised surface) tissu *m* pour couverture, flanelle *f* de coton
~BLEACHING - [text] blanchiment *m* du coton
~BLUSH - [paint] (in lacquer paints) louchissement *m*
~CLASSING - [text] classement *m* du coton
~COMBER - [text] peigneuse *f* pour coton
~COP - [text] cannette *f* de coton
~CORE - [agric] âme *f* de coton
[text] filé *m* d'amiante à âme de coton
~COUNT - [text] numéro *m* de fil
~-COVERED PAPER CORE CABLE - [el] câble *m* isolé au papier à guipage coton
~COVERED WIRE - [el] fil *m* à guipage coton
~DAMASK- [text] damas *m* de coton
~DRILL - [agric] semoir *m* à coton
~DUCK - [text] (used for tropical suitings etc) toile *f* de coton, coutil *m*
~EXCHANGE - [comm] bourse *f* du coton
~FABRIC - [text] tissu *m* de coton, cotonnade *f*
~FIBRE - [text] fibre *f* de coton
~FLANNEL - [text] flanelle *f* de coton
~FLOCK - [text] (filler for plastics and rubber, mainly consisting of finely ground cotton rags) flocon *m* de coton
~GIN - [text] (a cotton-fibre separating machine) égreneuse *f* de coton
~JENNY - [text] carde *f* à coton
~LAP - [text] toile *f* de coton pour carde
~LINTERS - [text] (fragments of cotton fibre, too short for spinning) linters *mpl*
~LOOM - [text] métier *m* à tisser le coton
~LUMP - [text] flocon *m* de coton, motte *f*
~MILL - [text] filature *f* de coton
~MOQUETTE - [text] moquette *f* de coton
~OIL - [ind chem] (specially compounded oil used to lubricate the cotton during spinning) huile *f* d'ensimage
~OPENER - text] ouvreuse *f*
~OPENING - [text] ouvraison *f*

COTTON PARCHMENT - [paper man] papier *m* parchemin de coton
˜PICKER - [mech] machine *f* à cueiller le coton
˜PLANT - [bot] cotonnier
˜PLANTER - [agric] planteuse *f* de coton
˜PLUSH - [text] peluche *f* de coton
˜RAT - [zool] sigmodon *m*
˜SATEEN - [text] satinette *f* de coton
˜-SEED - [bot] graine *f* de coton
˜-SEED CAKE - [fodder) tourteau *m* de graines de coton
˜-SEED MEAL - [agric] (fodder) mouture *f* de graines de coton, farine *f* de graines de coton
˜-SEED OIL - [ind chem] (extracted from the seeds of the cotton plant and used in lubricants and foodstuffs) huile *f* de coton
˜SHUTTLE - [text] navette *f* pour coton
˜SLUB - [text] bouton *m* de coton, bouloche *f*, puce *f*
˜SPINNING - [text] filature *m* du coton
˜SPINNING MILL - s. cotton spinning
˜STAPLE - [text] fibre *f* de coton
˜STRIPPER - s. cotton picker
˜SWAB - [med] tampon *m*
˜THREAD - [text] fil *m* de coton
˜TREE - [bot] cotonnier
˜TWILLS - [text] tissu *m* croisé de coton
˜TWIST - [text] fil *m* retors de coton
˜-VELVET - [text] velours *m* de coton
˜WADDING - s. cotton wool
˜WASTE - [text] déchêts *m* de coton, étoupe *f*
˜WASTE SPINNING - [text] filature *f* des déchets de coton
˜WASTE YARN - [text] (yarn obtained from cotton waste) fil *m* de déchet de coton
˜-WOOL - [text ind] (loose cotton bleached and pressed into sheets) ouate *f*
˜WARP CLOTH - [text] drap *m* à chaîne de coton
˜WARP LINEN - [text] toile *f* mélangée lin-coton
˜WEFT POPLIN - [text] popeline *f*
˜WOOD - [bot] peuplier *m* de la Caroline, peuplier *m* grisard
˜YARN - [text] fil *m* de coton
˜YARN HANK - [text] écheveau *m* de coton
COTTONER - [bot] viorne *f*
COTTONIZATION - [text] (treatment of a flexible fibrous material deriving from the rind of certain trees) cotonisation *f*
COTTONIZED FLAX - [text] lin *m* floconné
COTTRELL HARDENING - [phys] (the migration of impurity atoms in a crystal) durcissement *m* Cottrell
˜LOCKING - [phys] (the mechanism by which an atmosphere of impurity atoms prevents the movement of a dislocation line) phénomène *m* de Cottrell
COTYLEDON - [bot] cotylédon *m*
COUCH, to - [paper man] (the operation of pressing the pulp on the felt) coucher
COUCH - [gen] lit *m*, couche *f*, canapé *m*, divan *m*
[paint] couche *f* de fond
[med] (in radiological examination) table *f* d'examen
[brew ind] (of malt) tas *m*, couche *f*
˜GRASS - [bot] triticum *m*, chiendent *m*
˜ROLL - [paper man] (cylinder to express water from a wet paper web) cylindre *m* de pression
COUGAR - [zool] puma *m*
COUGH - [med] toux *f*
COULIER ARM - [text] bras *m* de cueillage
˜CAM - [text] came *f* de cueillage
COULIERING - [text] cueillage *m*
COULISSE - [carp] (grooved timber, the groove being cut for the sliding in of another piece) coulisse *f*
COULOMB - [el] (unit of electrical quantity, corresponding to a flow of one ampere for one second) coulomb *m*
˜BARRIER - [el] (barrier derived from repulsive electrostatic forces) barrière *f* de Coulomb
˜DEGENERACY - [el] dégénérescence *f* coulombienne
˜ENERGY - [el] force *f* coulombienne
˜FIELD - [el] champ *m* de Coulomb
˜FORCE - s. Coulomb energy
˜METER - s. coulometer
COULOMETER - [instr] (instrument for measuring electric current by the amount of electro-chemical action produced by it in a given time) coulombmètre *m*, voltamètre *m*
COULTER - [agric] (steel disc attached to a plough to cut the ground vertically) coutre *m*
˜STEM - [agric] tige *f* du coutre
COUMARIC ACID - [chem] (hydroxy-cinnamic acid) acide *m* coumarique
COUMARIN - [chem] (flavouring and deodorant for cosmetics, tobacco, rubbers etc) coumarine *f*
COUMARONE - [chem] (coal tar derivative) coumarone *f*
˜-INDENE RESINS - [chem] (thermoplastic polymers) résines *f*pl coumarone-indène
˜RESINS - [chem] (polymers of coumarone and indene, used in rubber compounding and in the modification of other resins) résines *f*pl de coumarone, résines *f*pl coumaroniques
COUNT, to - [gen] compter, calculer
[nucl] faire un comptage
COUNT - [gen] comptage *m*, compte *m*, calcul *m*
[nucl] (the number of ionizing events) comptage *m*
[text] (of the yarn) numéro *m*
˜OF THE FEED SLIVER - [text] numéro *m* de la mèche entrante
˜OF THE REED - [text] compte *m* du peigne, nombre *m* de dents du peigne
˜OF YARN - [text] numéro *m* du fil
˜RATE - [phys] taux *m* de comptage
COUNTDOWN - [gen] compte *m* à rebours
COUNTER - [gen] compteur *m*, calculateur *m*
[phys] (device for counting ionizing events) compteur *m*
[print] (defective alignment of lines) alignement *m* défectueux
[instr] (for electricity etc) compteur *m*
[shipbuild] voûte *f* (arrière)
[shoe man] (stiffener round the heel) protège-talon *m*
˜CELL - [el] (storage-battery; very small capacity cell in opposition to the battery for regulating purposes) élément *m* de force contre-électromotrice
˜CHARACTERISTIC CURVE - [nucl] (the curve of a Geiger-type counter counting rate) caractéristique *f* d'un tube compteur G.M.
˜-CHUTE - [min] cheminée *f* à charbon, cheminée *f* à minerai
˜CONDENSER - [el] condenseur *m* à contre-courant
˜CONTROLLED CLOUD CHAMBER - [nucl] (cloud

chamber expansion triggered by counter) chambre *f* de Wilson commandée par compteur, chambre *f* à nuage declenchée par compteur
COUNTER-COUPLING - [radio] (negative feedback) contre-réaction *f*
~-CURRENT - [gen & hydr] contre-courant *m*
~-CURRENT BOILER - [th eng] chaudière *f* à contre-courant
~-CURRENT BRAKE - [mech] frein *m* à contre-courant
~-CURRENT CENTRIFUGE - [mech] (mechanically or thermically established counter-current circulation in a centrifuge) centrifugeuse *f* à contre-courant
~-CURRENT COOLER - [ind chem] refroidisseur *m* à contre-courant
~-CURRENT DISTILLATION - [ind chem] (distillation method based on counter-current flow) distillation *f* à contre-courant
~-CURRENT EXTRACTION PROCESS - [ind chem] (process in which the streams of immiscible liquids flow counter to one another) extraction *f* à contre-courant
~-CURRENT FLOW - s. counterflow
~DEAD TIME - [instr] (in radiation counters) temps *m* mort (de tube compteur)
~-DRAFT - [plast ind] contre-dépouille *f*
~EFFICIENCY - [instr] effet *m* utile d'un compteur
~-ELECTRODE - [electron] (in dry rectifiers) contre-électrode *f*
~-ELECTROMOTIVE FORCE - [el] (electromotive force which opposes the normal flow of current in a circuit) force *f* contre-électromotrice (de self-induction)
~FILLING SYSTEM - [instr] système *m* de remplissage de compteur
~-FLANGE - [mech] contre-bride *f*
~-FLAP HINGE - [mech] penture *f* à charnière, ferrure *f* à noix
~FLOOR - [constr] sous-plancher *m*, faux-plancher *m*
~FOR YARN - [text] torsiomètre *m* pour filé
~GAS APMLIFICATION - [phys] (the ratio between the collected charge and the liberated charge in the initial ionizing event) amplification *f* due au gaz
~GEAR - [mech] engrenage *m* de renvoi
~-IRRITATION - [med] révulsion *f*, dérivation *f*
~LIFE-TIME - [instr] durée *f* de vie (d'un tube compteur)
~-LIGHT - [photo] (between light source and camera) contre-jour
~OPERATING VOLTAGE - [el] tension *f* de fonctionnement de compteur
~OVERSHOOTING - [el] (when the change in potential of the wire exceeds the counter overvoltage) dépassement *m*, surcharge *f*
~PLATEAU - [instr] (the region of the counter characteristic curve where the rate is independent of voltage) palier *m* d'un compteur
~RANGE - [nucl] (reaction rate range in a reactor whose neutron flux must be measured by counters) plage *f* de démarrage
~-RECOIL - [mech] retour *m*
~RECOVERY TIME - [instr] temps *m* de restitution
~REIGNITION - [nucl] (process by which multiple counts are generated within a counter) réactivation *f* d'un compteur
COUNTER RESOLVING TIME - [instr] (the smallest detectable time interval between counts) temps *m* de résolution
~ROTATING PROPELLERS - [aero] (propellers on concentric shafts, having a common drive and rotating in opposite directions) hélices *fpl* contra-rotatives
~-ROTATION - [mech] contra-rotation *f*
~-SCALES - [impl] balance *f* Roberval
~-SLOPE - [surv] contre-pente *f*
~-SPRING - [mech] ressort *m* antagoniste
~SPURIOUS TUBE COUNTS - [nucl] (in radiation counter tubes) coups *mpl* parasites
~STARTING POTENTIAL - [el] (the voltage required by a Geiger-type counter) tension *f* de démarrage
~-STERN - [shipbuild] voûte *f* arrière
~-TABLE - s. countertable
~TIME LAG - [instr] retard *m*
~TUBE - [instr] (ionisation chamber for the count of electrons or other ionizing particles) tube *f* compteur
~-VAULT - [build] (an inverted arch) arc *m* renversé
COUNTERACTING WINDING - [el] enroulement *m* différentiel
COUNTERBALANCE, to - [gen] contrebalancer
COUNTERBALANCE - [gen] contrepoids *m*, équilibre *m*
COUNTERBALANCED CRANKSHAFT - [auto] vilebrequin *m* équilibré
COUNTERBLADE OF KNIFE-BOX - [agric] contre-plaque *f* du porte-couteaux
COUNTERBORE, to - [mech] (the operation of enlarging the end of a hole) contrepercer, élargir
COUNTERBORER - [tool] (tool for boring the end of a hole to a larger diameter) outil *m* à agrandir
COUNTERBORING - [mech] contrepercage *m*, chambrage *m*
COUNTERBRACE, to - [naut] counterbrasser
COUNTERCLOCKWISE - [gen] dans le sens inverse du sens des aiguilles d'une montre
~POLARIZED WAVE - [phys] (electromagn waves) onde *f* à polarisation elliptique à gauche
COUNTERDRAW, to - [draw] calquer
COUNTERDRAWING - [draw] calque *m*
COUNTERFEIT, to - [gen] contrefaire, falsifier
COUNTERFLOW - [mech etc] (flow of fluid in the opposite direction to that of another fluid) contre-courant *m*
~FURNACE - [metall] four *m* à contre-courant
COUNTERFLUSH DRILLING - [mining] forage *m* avec circulation renversée
~-LODE - [mining] filon *m* croiseur
COUNTERFORT - [build] contrefort *m*
COUNTERGAUGE - [impl] troussequin *m* à deux pointes
COUNTERHATCHING - [constr] hachures *fpl* croisées
COUNTERLODE - [mining] filon *m* transversal
COUNTERMARK, to - [gen] contremarquer
COUNTERPOINT - [mus] contrepoint *m*
COUNTERPOISE - [mech] (weight used in a mechanism to balance another part) contrepoids *m* [phys] équilibre *m*
COUNTERPRESSURE - [phys] contre-pression *f*
~RACKER - [instr] isobaromètre *m*
~STEAM BRAKE - [mech] frein *m* à contre-pression de vapeur
COUNTERSCARP - [constr] contrescarpe *f*
COUNTERSHAFT - [mech] (intermediate shaft between

driving and driven shafts) arbre *m* de renvoi
[auto] (of a gear box) arbre *m* intermédiaire
COUNTERSINK, to - [mech] (the operation of boring the end of hole to a larger diameter) fraiser, noyer (une vis)
COUNTERSINK - [mech] fraise *f* conique, alésoir *m* d'ébauche, fraise *f* angulaire, champignon *m*
COUNTERSINKING BIT - [mech] (rotary cutter designed to form a countersink) foret *m* à centrer
COUNTERSTOP - [mech] contre-butée *f*
COUNTERSUNK - [mech] fraisé, encastré
~HEAD - [mech] (head of screw, bolt or rivet driven into a mating recess) tête *f* fraisée
~HEAD RIVET - [mech] rivet *m* à tête fraisée
~HEAD SCREW - [mech] (screw which can be driven below the surface) vis *f* à tête fraisée
COUNTERTABLE - [mach tools] contre-table *f*
COUNTERTEMPLATE - [print] contre-gabarit *m*
COUNTERWEIGHT, to - [mech] équilibrer
COUNTERWEIGHT - [mech] contrepoids *m*
~CONTROLLED INSTRUMENT - [instr] instrument *m* de mesure à contrepoids
COUNTESS - [constr] (roofing slate) ardoise *f* de couverture
COUNTING - [gen] comptage *m*, compte *m*
[nucl] comptage *m*
~CIRCUIT - [el] (circuit containing elements capable of counting impulses fed to the circuit) circuit *m* de comptage
~FRAME - [impl] boulier *m*
~GLASS - [text] compte-fils *m*
~IONIZATION CHAMBER - [nucl] (used to detect individual ionizing events) chambre *f* d'ionisation de comptage
~LOSS - [instr] correction *f* de temps mort
~MECHANISM OF A METER - [meas] (the register of a meter) intégrateur *m*
~RATE - [nucl] (average rate of occurrence of ionizing events) taux *m* de comptage, débit *m* de coups
~RATE METER - [nucl] (mechanism giving a continuous indication of the average of ionizing events rate) debitmètre *m* de coups
~RELAY - [el] relai *m* compteur
COUNTLESS - [gen] innombrable
COUNTRY HOUSE - [build] maison *f* de campagne
~ROCK - [geol] (rock of no value, forming the walls of a mineral deposit) roche *f* intercalaire, gauge *f*
COUPE' - [auto] coupé *m*
COUPLE, to - [gen] accoupler, coupler, assembler
[railw] atteler
[mech] (on a shaft etc) accoupler, embrayer, engrener
[el] coupler, connecter
[chem] (se) combiner, (se) coupler
COUPLE - [gen] couple *m*, paire *f*, couplage *m*
[mech] (two equal parallel forces of opposite direction) couple *m*
[el] couple *m* voltaïque
[el] (of storage battery; pair of plates of opposite polarity) couple *m*
[instr] (of a meter) couple *m* de freinage
~-CLOSE ROOF - [constr] toit *m* comble à deux égouts, toit *m* à deux versants
COUPLED - [gen] couplé, accouplé, connecté, attelé
COUPLED AXIS - [railw] axe *m* couplé
~AXLE-PIN - [railw] tourillon *m* d'essieu accouplé
~CIRCUIT - [el] circuit *m* couplé
~CIRCUIT EFFECT - [electron] effet *m* de couplage dans les circuits oscillants
~CIRCUIT TRANSFORMER - [el] transformateur *m* à circuit couplé
~CONTROL ELEMENT COMBINATION - [contr] (the combination or actuating signals for the operation of one controlling means) réglage *m* combiné
~-ENGINE POWER UNIT - [aero] (power unit in which two engines are coupled to one propeller or to a pair of counter-rotating propellers) moteurs *mpl* couplés
~FLUTTER - [aero] (also called classical flutter; flutter only due to coupling between two or more degrees of freedom) battement *m* classique
~GATES - [railw] barrières *fpl* accouplées
~PACING MOTIONS - [text] mécanismes *mpl* régulateurs accouplées
~POINTS - [railw] aiguillages *mpl* couplés
~POLES - [telecomm] appui *m* de ligne double
~REFERENCE INPUT COMBINATION - [contr] (multiple system causing one or more controllers to regulate the input of other controllers) réglage *m* en cascade
~SIGNALS - [railw] signaux *mpl* accouplés
~SPACE CHARGE WAVE - [electron] onde *f* de charge d'espace couplé
~SWITCHES - [el] (mechanically linked switches) interrupteurs *mpl* accouplés
~WHEELS - [mech] (locomotive wheels coupled by coupling rods) roues *fpl* accouplées
COUPLER - [gen & mech] (any device for making a connection) accouplement *m*, coupleur *m*, connecteur *m*
[acoust] - coupleur *m*
[mus] (organ device) tirasse *f*
[mech] (on vehicles) attelage *m*
[railw] attelage *m* automatique
[el] (length of tubing) connecteur *m*
[radio] (coil coupling) coupleur *m*
~HEAD - [railw] tête *f* d'attelage
COUPLING - [mech] (the operation of connecting) couplage *m*, embrayage *m*, accouplement *m*, jonction *f*
[plumb] manchon *m*, joint *m*
[railw] attelage *m*
[radio] couplage *m*
[chem] (a combination) combinaison *f*
[genet] attraction *f*, accouplement *m*
[naut] accouplement *m*
~APERTURE - [electron] (waveguides; aperture in a cavity resonator wall) fenêtre *f* de couplage
~-BAR - [railw] barre *f* d'écartement des aiguilles
~BOLT - [mech] boulon *m* d'assemblage
~BOX - [mech] manchon *m* à accouplements
~BUSHING - [mech] manchon *m* d'accouplement
~BY FLEXIBLE HOSE - [ind chem] (temporary connexions between process units) raccordement *m* par tubes souples
~CAPACITOR - s. coupling condenser
~CASES - [phys] (the resultant of the different angular moments in the molecule) possibilités *fpl* de couplage
~COEFFICIENT - [el] (the ratio between the reactance of two circuits and the square root of the pro-

duct of the totals of the reactance in the two circuits) coefficient *m* de couplage
COUPLING COIL - [el] bobine *f* de couplage
~CONDENSER - [radio] (a condenser coupling two circuits) condensateur *m* de couplage
~CONSTANT - [phys] (constant expressing the strength of a coupling) constante *f* de couplage
~DEVICE - [mech] dispositif *m* d'accouplement
~ELEMENT - [el] élément *m* de couplage
~FACTOR - s. coupling coefficient
~FLANGE - [mech] bride *f* de fixation, bride *f* de serrage
~GEAR - [mech] accouplement *m*
~HANDLE - [railw] levier *m* d'attelage
~HEAD FOR AIR BRAKE - [mech] accouplement *m* de frein pneumatique
~IRON - [build] crampon *m*
~KEY FRAME - [text] levier *m* d'embrayage
~LEVER - [text] levier *m* d'accouplement
~LINK - [railw] maillon *m*
~LOOP - [electron] (waveguides; a loop of wire for radiofrequency coupling) boucle *f* de couplage
~NUT - [railw] tendeur *m*
~POSITION - [railw] position *f* d'accouplement
~PROBE - [electron] (metallic rod used to extract or inject energy into the resonant cavity of a klystron) sonde *f* de couplage
~PURLIN - [constr] lierne *f*
~RESISTANCE - [el] (common resistance between two circuits) résistance *f* de circuits couplés
~RESISTOR - [el] (resistor common to two circuits so as to allow transfer of energy) résistance *f* de couplage
~RING - [mech] collier *m* de serrage
~ROD - [railw] bielle *f* d'accouplement
~SCREW - [railw] tendeur *m* à vis
~SCREW LEVER - [railw] levier *m* de tendeur
~SHAFT - [mech] arbre *m* d'accouplement, arbre *m* d'embrayage
~SLEEVE - [mech] manchon *m* d'accouplement
~SPINDLE - [mech] arbre *m* d'accouplement
~SPRING - [mech] ressort *m* d'embrayage, ressort *m* d'accouplement
~SPRING CLICK - [horol] cliquet *m* à ressort d'accouplement
~SPRING FOR PICKING MOTION - [text] ressort *m* de détente de chasse
~STRAP - [text] manchon *m* d'accouplement
~SYSTEM - [radio] (circuit and transmission lines system for energy transfer from transmitter to aerial) système *m* de couplage
~WEDGE - [text] coin *m*
COUPON - [gen] coupon *m*; (sample) échantillon *m*, éprouvette *f*
[metall] lingot-éprouvette *m*
COURSE, to - [min] (the conveyance of air through a mine) ventiler
COURSE - [gen] course *f*, parcours *m*, trajet *m*
[naut] route *f*, cap *m*
[naut] (of sail) basse voile *f*
[build] (length of bricks running throughout the breadth of a wall) assise *f*, couche *f*, lit *m*, rangée *f*
[text] chemin *m*, course *f*, marche *f*, vitesse *f*
[text] (length of stitches) rang *m*
[surv] (the length and bearing of a line) levée *f*
[text] (a repeat of the pattern) répétition *f*

COURSE AND DISTANCE COMPUTER - [aero] (airborne computer dealing with course and distance data) calculateur *m* de route et distance
~AND DRIFT INDICATOR - [aero] indicateur *m* de cap et de dérive
~ANGLE - [aero] angle *m* de cap
~BY DEAD - [naut & aero] estime *f*, route *f* estimée
~-CARD MAGNIFIER - [naut] verre *m* grossissant pour carte marine
~COMPUTER - [radar] calculateur *m* de route
~CONTROL - [aero] contrôleur *m* de cap, conservateur *m* de cap
~CORRECTOR - [naut] correcteur *m* de route
~DEVIATION INDICATOR - [radar] indicateur *m* d'écart
~INDICATING BEACON - [radar] radiophare *m* directionnel
~INDICATOR - [naut & aero] contrôleur *m* de cap
~LIGHT - [naut] feu *m* de route
~LINE - [radio] axe *m* de radio-alignement
~MADE GOOD - [aero] (plotting line of constant direction from the point of departure) route *f* vraie
~MAGNITUDE - [aero] (one of the characteristics of a course) magnitude *f* de la route
~OF A RIVER - [geogr] cours *m* d'une rivière (ou d'un fleuve)
~OF BRICKS - s. course
~OF HEADERS - [constr] appareil *m* en boutisses
~OF STRETCHERS - [constr] appareil *m* en panneresses
~ON GROUND - [naut] route sur le fond
~ON WATER - [naut] route *f* sur l'eau
~VENTILATION - [min] (system of ventilation in mines) aérage *m* en série
COURT - [gen] cour *f*
~-HOUSE - [constr] palais *m* de justice
~-ROOM - [build] salle *f* d'audience
~-YARD - [build] cour *f*
COUVEUSE - [impl] couveuse *f*, incubateur *m*
COVALENCY - [chem] (the union of two atoms sharing a pair of electrons) covalence *f*
COVALENT BOND - [nucl] (a type of linkage between atoms) liaison *f* de covalence
~COMPOUND - [nucl] (compound formed by the sharing of electrons between atoms) composé *m* covalent
COVARIANCE - [math] covariance *f*
COVARIANT - [math] covariant
COVARIATION - [math] (the coincidence of two variations) covariation *f*
COVE - [geogr] baie *f*, crique *f*, anse *f*
[constr] raccord *m* courbe, congé *m*, gorge *f*, cavet *m*
[carp] moulure *f* à gorge
~MOULDING - [carp] moulure *f* à gorge
COVED CEILING - [build] (with a hollow curve between wall and ceiling) plafond *m* à raccord courbe
~TILE - [build] plinthe *f* à gorge
COVELLITE - [min] (natural copper sulphite usually occurring in the form of thin plates) covellite *f*
COVER, to - [gen] couvrir, recouvrir, envelopper
COVER - [gen] couvercle *m*, couverture *f*, capot *m*, enveloppe *f*, housse *f*
[bookbind] couverture *f*
[mech] (of valves) cache-soupape *m*

[instr] couvercle *m*
[fin] couverture *f*, provision *f*
[build] (the covered section of a tile) surface *f* couverte
COVER BY PATENT, to - [leg] protéger par un brevet
~DIE - [casting] demi-moule fixe
~FURNACE - [metall] four *m* de recuit à cloche mobile
~GLASS - [impl] (the strip of glass which is used in microscopy to cover the observed specimen) lamelle *f*
~GLOBE - [photo] globe *m* de recouvrement inactinique
~GRATING - [constr] grille *f* de fermeture
~MOULD - [plast ind] (the fixed half of an injection mould) demi-moule *m* de fermeture
~PLATE - [gen & mech] plaque *f* de fermeture, couvre-joint *m*
~SLAB - [telecomm] dalle *f* de fermeture
~SMUT - [bot] charbon *m* vêtu .
~STRIP - [metall] (in welding; strip of material to overlay and reinforce a butt joint) tôle *f* de recouvrement, couvre-joint *m*
COVERAGE - [radio] (of a broadcasting station) couverture *f*, étendue *f*, zone *f* d'action
[radar] (the actual area covered by the beams of the scanning aerial) couverture *f*, zone *f* de portée utile
~DIAGRAM - s. coverage pattern
~MAP - [radio] carte *f* des zones de réception
~PATTERN - [radar] (surface of space formed by the points at which the target gives a detectable signal) diagramme *f* de couverture
COVERED - [gen] couvert, protégé, recouvert
~BARLEY - [brew ind] orge *m* vêtu
~ELECTRODE - [el] (metal electrode which is covered with a coating of flux) électrode *f* enrobée
~PLATFORM - [railw] quai *m* couvert
~STOP - [mus] tuyau *m* bouché
~WAGON - [railw] wagon *m* de marchandises couvert
~WIRE - [el] fil *m* guipé
~WITH PLUSH - [text] cylindre *m* transporteur recouvert de feutre
COVERING - [gen] couverture *f*, enveloppe *f*
[text] (on a card cylinder) garniture *f*, garnissage *m*, enveloppe *f*
[text] (by spinning) guipage *m*
[aero] (of an aircraft wing) revêtement *m*
~CAM - [mech] came *f* des souteneuses
~FABRICS - [text] tissus *mpl* pour couvertures
~FLANGE - [mech] plaque *f* de recouvrement, couvercle *m*, recouvrement *m*
~HEEL KNIFE - [text] souteneuse *f* de talon
~HEEL NEEDLE - [text] poinçon *m* à talon
~POWER - [paint] (the property of a coating composition to obliterate the colour of a surface in a single coat) pouvoir *m* couvrant
~PRESS - [rubber ind] (a machine for retreading tyres) presse *f* de rechapage
~SLAB - [constr] pierre *f* de couverture *f*
~SPINDLE - [text] broche *f* à guiper
~THREAD - [text] fil *m* de guipage
~VELVET - [text] velours *m* pour boutons
COVERLET - [text] couvre-lit *m*
COVERT - [gen] couvert *m*, abri *m*, gite *m*
~COATING - [text] (a wool, or wool and cotton fabric) toile *f* pour capotes
COVING - [constr] (of a fireplace) voussure *f* (de cheminée)
COVOLUME - [chem] (of the molecules of a gas) covolume *m*
COW - [zool] vache *f*
~-CALF - [zool] génisse *f*
~-CATCHER - [railw] (device in front of locomotives to push aside obstacles) chasse-pierre *m*
~-HERD - [agric] bouvier *m*, vacher *m*, nourisseur *m*
~HIDE - [leather ind] peau *f* de vache tannée
~PARSLEY - [bot] cerfeuil *m* sauvage, cerfeuil *m* des prés
COWBELL - [agric] (used for cloche) cloche *f* à vache, clochette *f*
COWBIRD - [zool] étourneau *m* du bétail
COWHAIR - [text] poil *m* de vache
COWHIDE - [leather ind] cuir *m* de boeuf, vachette *f*
COWL - [gen] capuchon *m*
[constr] manche *f* à vent
[constr] (of chimney) mitre *f*
[auto] auvent *m*
[aero] capot *m*
~LAMP - [auto] lampe *f* de tableau de bord
~SIDE PANEL - [auto] joue *f* d'auvent
~SUPPORT BRACKET - [auto] plaque *f* support de capot
COWLED - [gen] capoté
COWLING - [aero] (removable covering over the engine) capot *m*, capotage *m*
[mech] capot *m* de protection
COWSLIP - [bot] coucou *m*, primevère *f*
COXWAIN - [naut] patron *m* d'embarcation
CRAB, to - [aero] corriger la dérive
[naut] aller "en crabe"
CRAB - [zool] crabe *m*
[bot] pommier *m* sauvage
[mech] treuil *m*, chêvre *f*, chariot *m* de pont-roulant
[mech] (crane fitted with claws) grue *f* à benne
[naut] (of ships) cabestan *m* volant
[mech] (a winch-carrying mechanism) petit treuil *m*
[photo] (in air photography) effet *m* de dérive
CRABBING - [text] (a mechanical process to avoid shrinkage during finishing operations) procédé *m* mécanique évitant le retrait en cours de finissage
[aero] (to land against the wind) atterrissage *m* contre le vent
[cinema] travelling *m* latéral
CRACK, to - [gen] fendre, fêler, casser, se crevasser
[oil ind] (the operation of breaking down a heavy petroleum product into a lighter one by controlled heating processes) fractionner, craquer
[metall] se fêler
[ind chem] fractionner, craquer, soumettre au cracking
CRACK - [gen] (a fissure or a split) fente *f*, fissure *f*, cassure *f*, craquelure *f*, lézarde *f*, garçure *f*
[soil] (a crevice in the soil etc) crevasse *f*
[metall] (in casting) tapure *f*, crique *f*, soufflure *f*, défaut de soudure
[paints] craquelure *f*
[oil ind] fractionnement *m*, craquage *m*, cracking *m*
~DETECTION - [metall] (process for indicating the presence) détection *f* des fissures

CRACK DETECTOR - [instr] détecteur *m* de fissures
~INITIATION - [gen & mech] (the beginning of a crack, as in the mechanical failure of a component) début *m* de fissure
~-LIMITING - [mech] (method for restricting the further development of a crack) limitation *f* des fissures
CRACKED - [gen] crevassé, fêlé, fissuré, fendu
~-CARBON RESISTANCE - [radio] (in radio receivers) résistance *f* au carbone déposé par projection
~EDGE - [metall] fissure *f* d'angle
CRACKER - [rubber ind] broyeur *m* à cylindres
CRACKING - [gen] fissuration *f*
[metall] criquage *m*
[metall] (of a weld) fissuration *f* dans la soudure
[paint] (the occurrence of small fissures) fendillement *m*, craquèlement *m*
[chem] (the breaking of a carbon/carbon bond by heat and a catalyst. Especially important for the thermal decomposition of petroleum) crackage *m*, cracking *m*, destruction *f* pyrogénée
~FUEL - [oil ind] combustible *m* pour craquage
~-OFF RING - [mech] (part of glass-blowing machines) anneau *m* de séparation
~PROCESS - [ind chem] (of carbon black) (preparation of carbon black by cracking natural gas in a preheated chamber) craquage *m*
~PRODUCT - [ind chem] produit *m* de cracking (produit de désintégration thermique)
CRACKLE - [glass man] (system of decorating glass or pottery in which the glaze is covered with small cracks during firing) givrage *m*
[acoust] craquement *m*
~FINISH - [paint] vernis *m* craquelé
CRACKLED - [glass man] givré
CRACKLING - [acoust] (the sound made by wood when burning etc) crépitement *m*, craquement *m*
[telecomm] crachement *m*, friture *f*
~NOISE - [electron] (occurring in electronic tubes) crachement *m*
[acoust] crépitement *m*, craquement *m*
CRACKY - [gen] cassant, fissuré, fragile
CRADLE - [carp] berceau *m*, panier *m*
[mech] (flat support on casters to work under a machine) chariot *m*
[shipbuild] ber *m*, berceau *m*
[mech etc] châssis *m*
[text] support *m* des lanières d'étirage
[mech] (the space into which the engine is fitted) berceau *m*
[metall] affût *m*
[el] (under high-tension cables) filets *m* de garde
[astronaut] (launching platform) tour *f* de lancement
~FOOT - [text] anglaisage *m* de la semelle
~ROCKER - [text] appareil *m* de desserrage
CRADLING - [build] échafaudage *m* volant
CRAFT - [gen] (trade or occupation) métier *m*, art *m*
[naut] bâtiment *m*, navire *m*, embarcation *f*
CRAFTMANSHIP - [gen] habileté *f*, dextérité *f* manuelle
CRAFTSMAN - [gen] artisan *m*, ouvrier *m*, spécialiste
CRAMP - [mech] (a device for holding parts during construction) crampon *m*, serre-joint *m*, presse *f*, crochet *m* d'assemblage, crampe *f*
[constr](locking metal bar binding adjacent stones in a course) agrafe *f*, crampon *m*, bride *f* de serrage
[geol] massif *m*
[med] (a painful contraction of a muscle) crampe *f*
CRAMP IRON - [constr] crampon *m*
~LAPPING - [mech] rodage *m* forcé
CRAMPED CONSTRUCTION - [electron] (to ensure small interelectrode capacities in electronic tubes) construction *f* compacte
CRAMPING - [cinema] contraction *f* d'image
CRAMPON - [mech] (appliance for holding stones or heavy materials while being hoisted by crane) pince *f* de levage
CRAMPOON - s. crampon
CRANAGE - [comm] frais *mpl* de grutage
CRANE - [mech] (machine for lifting and lowering heavy weights) grue *f*
~BEAM - [mech] balancier *m* de grue
~BOOM - [mech] flêche *f* de grue
~CLEARANCE - [mech] dégagement *m* de la grue
~CRAB - [mech] chariot *m* de grue, chariot *m* de pont roulant
~ERECTION - [mech] montage *m* d'une grue
~-FOUNDATIONS - [constr] fondations *fpl* d'une grue
~GIRDER - [mech] pont *m*
~GRAB - [mech] benne *f* preneuse
~HOOK - [metall] (in foundry) croc *m* de suspension
[mech] crochet *m* de grue (ou de levage)
~INSTALLATION - [mech] installation *f* de grue
~JIB - [mech] bras *m*, flêche *f*, volée *f*
~LADLE - [metall] (in foundry) poche *f* de coulée, poche *f* pour grues
~LOCOMOTIVE - [mech] locomotive *f* à grue
~MOTOR - [mech] moteur *m* pour grue
~NECK - [mech] col *m* de cygne
~ON PILED TRESTLES - [mech] grue *f* sur pilotis
~OPERATOR - [gen] grutier *m*, conducteur *m* de grue
~PILLAR - s. crane post
~PONTOON - [shipbuild] ponton-grue *f*
~POST - [mech] colonne *f* de grue
~RATING - [mech] puissance *f* d'un moteur de grue
~SHOVEL - [mech] (earth moving machine) excavatrice *f*
~TRACK - [mech] voie *f* pour grue
~TRUCK - [mech] camion-grue *f*
~WAGON - [railw] wagon-grue *m*
~WINCH - [mech] treuil *m* pour grue
CRANEFLY - [zool] tipule *f*
CRANIAL - [anat] cranien
~FLEXURE - [zool] flexion *f* cranienne
~INDEX - [anat] indice *m* céphalique
CRANIOLOGY - [med] craniologie *f*
CRANIOMETER - [instr] (skull measuring instrument) craniomètre *m*
CRANIOMETRIC POINTS - [med] points *fpl* craniométriques
CRANIOPLASTY - [med] cranioplastie *f*
CRANIUM - [anat] (the brain case) crane *m*
CRANK, to - [mech] tourner la manivelle
[mech] (to revolve an internal combustion engine by external means) démarrer à la manivelle
[mech] (to start an internal combustion motor by using the starter motor) faire démarrer, démarrer au moteur auxiliaire (de démarrage)
CRANK - [mech] (mechanical arrangement to con-

vert reciprocating to rotary movement) manivelle *f*, coude *m*
[mech] (arm attached to a shaft) manivelle *f*, levier *m* coudé
[mech] (of a bicycle) manivelle *f*
CRANK AND CONNECTING ROD SYSTEM - [mech] dispositif *m* à bielle et manivelle
~ARM - [mech] bras *m* de manivelle
~-BEARING - [mech] palier-manivelle *m*
~BLASTING MACHINE - [mech] exploseur *m* à manivelle
~BRACE - [impl] vilebrequin *m* pour forerie, fût *m* pour forerie
~CONNECTING ROD - [mech] bielle *f*
~DISPLACEMENT - [auto] course *f* du vilebrequin
~DRIVE - [mech] commande *f* à manivelle
~DRIVE MOTOR - [mech] moteur *m* à bielle
~DRIVE TABLE - [mech] table *f* à réglage par manivelle
~DRIVEN RIVETING MACHINE - [mech] riveteuse *f* à manivelle
~-END - [mech] (of horizontal stationary engine) avant, AV
[railw] (of a locomotive) arrière
~-END OF CYLINDER - [mech] avant-cylindre *m*, avant *m* du cylindre
[railw] arrière *m* du cylindre
~GEAR - [mech] mécanisme *m* à manivelle
~-HANDLE - [auto] poignée *f* de manivelle
~LEVER - [text] levier *m* à manivelle
~LOOM - [text] métier *m* à manivelle
~MECHANISM - [mech] mécanisme *m* à manivelle
~PICKING MOTION - [text] chasse-navette *m* à manivelle
~PIN - [mech] maneton *m*, tourillon *m*, soie *f* de vilebrequin, bouton *m* de manivelle
~PIN BEARING - [mech] palier *m* de maneton
~PIN THROW - [mech] bras *m* de manivelle
~PRESS - [impl] presse *f* à manivelle
~SHAPING MACHINE - [mach tools] étau-limeur *m* à commande par bielle
~THROW - [mech] coude *m* de vilebrequin
~-WEB - [mech] (one of the two members connecting the crank-pin with the main journal) bras *m* de manivelle, flasque *m* de manivelle
CRANKCASE - [mech] (the enclosed chamber in an internal combustion engine where the crankshaft revolves) carter *m* moteur
~BEARER - [mech] (an arm integral with the crankcase, by which it is supported and anchored) support *m* de carter
~BREATHER - [mech] (a device which allows air to enter or leave the crankcase when pressure in it varies) reniflard *m* de carter
~COMPRESSION - [auto] compression *f* dans le carter
~EXPLOSION - [mech] (in a diesel motor) explosion *f* dans le carter moteur
~FRONT END COVER - [auto] carter *m* de distribution
~SCAVENGING - [auto] vidange *f* du carter
~STIFFENING WEB - [mech] (flange on a crankcase to increase rigidity) nervure *f* de renforcement du carter
~SUMP - [mech] (the part of the crankcase in which oil is collected from the moving parts of the engine to supply the oil pump) puisard *m*, carter *m* inférieur
CRANKED - [gen] coudé
~LEVER - [impl] levier *m* coudé
~LINK - [mech] (of a chain) maillon *m* à crémaillère
~SPANNER - [mech] clé *f* coudée
~TOOL - [tool] (for metal-turning) crochet *m*
~VALVE - [mech] robinet *mm* coudé
CRANKING - [mech] démarrage *m* à la manivelle
~MOTOR - [mech] moteur *m* (auxiliaire) de démarrage
~SPEED - [auto] vitesse *f* de démarrage
~TORQUE - [mech] couple *m* de démarrage
CRANKNESS - [naut] faiblesse *f* d'un navire
CRANKSHAFT - [mech] (the principal shaft of an engine in which cranks are formed to receive the connecting rod ends) vilebrequin *m*, arbre *m* à manivelle, arbre *m* coudé
~BEARING - [mech] (the main bearing on which a crankshaft runs) palier *m* de vilebrequin
~BEARING BUSHING - [auto] bague *f* de palier de vilebrequin
~DAMPER - [mech] (device designed to suppress vibration at a natural frequency) amortisseur *m* pour vilebrequin
~DEFLECTION - [auto] fléchissement *m* du vilebrequin
~GRINDING MACHINE - [mach tools] rectifieuse *f* pour vilebrequin
~JOURNAL - [mech] soie *f* de portée de vilebrequin
~LATHE - [mach tools] tour *m* pour vilebrequin
~THRUST BEARING - [mech] butée *f* de vilebrequin
CRANNY - [gen] crevasse *f*, fente *f*, fissure *f*
CRASH, to - [of cars] (se) heurter, entrer en collision, se tamponner
[aero] (of aircraft) s'écraser au sol, casser du bois
CRASH - [gen] chûte *f*
[auto] collision *f*, accident *m*
[aero] accident *m* d'aviation, écrasement *m*
[text] (coarse texture linen) drap *m* grossier, toile *f* à serviettes
~HELMET - [impl] casque *m* protecteur, serre-tête *m*
~-LAND, to - [aero] faire un atterrissage de fortune, s'écraser à l'atterrissage
~-LANDING - [aero] écrasement *m* à l'atterrissage
~RESCUE EQUIPMENT - [aero] (for prompt aid in case of accident) matérial *m* de sauvetage
CRASHED ENGINE - [aero] (an engine which has been involved in a serious aircraft incident) moteur *m* accidenté
CRASHES - [radio] (very strong atmospherics) parasites *mpl* atmosphériques
CRATE, to - [gen] (to pack into crates) emballer, emballer dans une caisse à claire-voie
[draw] diviser en carreaux, carroyer
CRATE - [gen] caisse *f*) à claire-voie, panier *m* à verrerie, cageot *m*
[packag] caisse *f* d'emballage à claire-voie
~FRAME - [packag] panier *m* à verrerie
CRATER - [geol] (the orifice of a volcano) cratère *m*
[light] cratère *m*
[impl] vase *m*, cratère *m*
[metall] (depression at the end of a bead) cratère *m*
~CRACK - [metall] (in welding) crique *f* à cratères
CRATERIFORM - [bot] cratèriforme
CRATERING - [paint] (formation of small round

holes) cratérisation *f*
CRAUNCH - [mining] massif *m* de protection
CRAWL, to - [gen] ramper, se traîner
CRAWLER - [agric] tracteur *m* à chenilles, chenillard *m*
~GEAR - [mech] ralenti *m*
~TRACK - [mech] chenille *f* de roulement
~TRACTOR CRANE - [mech mach] grue *f* à chenilles
CRAWLING - [el] (a phenomenon occurring in induction motors which sometimes fail to attain more than one seventh of their speed for the presence of a seventh harmonic) rampage *m*
[metall] (creep movement of sheet lead used as tank lining etc) glissement *m*
[telev] filage *m*, formation *f* de rayures
[paint] (formation of wrinkles) retirement *m*
~SPEED - [el] vitesse *f* de rampage
[mech] vitesse *f* au ralenti
CRAYON - [impl] crayon *m* de pastel, craie *f* à dessiner
~CRAZING - [paint] fendillement *m*, craquelage *m*
CRAZE - [build] (hair-like cracks appearing on the surface of pre-cast concrete) fendillement *m*, craquelage *m*
[metall] (minute cracks) fendillement *m*, craquèlement *m*
~HEALING - [metall] élimination *f* du fendillement
CRAZING - [paint] (the appearance of a pattern of fine cracks running in all directions) fendillement *m*, craquelage *m*
[metall] craquelure *f* superficielle
CRAZY PAVING - [build] dallage *m* irrégulier
CREAM, to - [rubber ind] crémer
CREAM - [gen] crème *f*
~-LAID PAPER - [paper man] papier *m* à lettre à filigrane
~OF TARTAR - [chem] crème *f* de tartre
~OFF, to - [gen] (to skim) écrémer
~SEPARATOR - [agric] (in dairy) écrémeuse *f*
CREAMED LATEX - [rubber ind] latex *m* crémé
CREAMING - [gen] (the process of removing floating substances from a liquid) crémage *m*
[paint] (the separation of an emulsion into layers of different concentrations) cassage *m* d'une émulsion
~AGENT - [rubber ind] agent *m* crémant
~TANK - [rubber ind] (open vessel in which creaming is carried out) bac *m* de crémage
CREASABILITY - [text] (the extent to which a fabric retains folds or creases) froissabilité *f*
CREASE, to - [gen & text] plisser, froisser, chiffoner
CREASE - [gen & text] pli *m*, ancrure *f*
[constr] (a course formed of several thicknesses of roofing tiles) rangée *f* faîtière
[paper man] fronce *f*
~-PROOF - s. crease resisting
~-PROOF FINISH - [text] apprêt *m* infroissable
~RECOVERY - [text] défroissement *m*
~-RESISTING - [text] infroissable
~-RESISTING FABRIC - [text] tissu *m* infroissable
CREASER - [tool] suage *m*, dégorgeoir
CREASING - [gen] plissement *m*
[build] s. crease
[paper man] fronce *f*, froncement *m*
~ANGLE - [text] angle *m* de pliage
CREASING SUPPORT - [ind proc] (hinged plate used in folding) plaque *f* de pliage
~TOOL - s. creaser
CREATE, to - [gen] créer, produire
CREATINE - [chem] créatine *f*
CREATININE - [chem] (formed by creatine in acid solution) créatinine *f*
CREATINURIA - [med] (creatinine in the urine) créatinurie *f*
CREATION - [gen] création *f*
~RATE - [electron] (the time rate of creation of electron-hole pairs) vitesse *f* de création des paires
CREDIT, to - [gen & comm] créditer
CREDIT - [gen fin comm] crédit *m*
~LINES - [cinema] générique *m*
CREDITS - s. credit lines
CREEK - [geogr] crique *f*, baie *f*
CREEL - [text] (the frame holding the supply bobbins) râtelier *m*
~BAR - [text] colonne *f* du râtelier
~BOARD - [text] planche *f* à bobines, plaque *f* à bobines
~BOBBIN - [text] bobine *f* à embrocher
~FLYER - [text] ailette *f* de prétorsion
CREELING - [text] garnissage *m*, embrochement *m*, montage *m*
~UP - [text] embrochement *m*, garnissage *m*
CREEP, to - [gen] ramper, progresser, se traîner, grimper
[mech] (the gradual deformation under continuous stress) fluer
CREEP - [metall] (cold flow of a metal under steady continuous stress) fluage *m*, allongement *m* visqueux
[chem] (the formation of crystals of the solute on the sides of a vessel containing a solution above the level of the liquid) grimpement *m* de cristaux
[chem] (the rising of a precipitate up the walls of a vessel) montée *f*, grimpement *m*
[mech] (gradual movement of a mechanical part caused by hydraulic leakage) cheminement *m*, glissement *m*
[metall] (of metals subjected to high temperature) allongement *m* visqueux
[metall] (the gradual deformation due to steady stress) fluage *m*
[mech] (of belts in respect to a pulley) glissement *m*
[mining] (gradual rising of a coal mine floor) gonflement *m*, boursouflement *m*
[instr] (the rotation of a disc of a meter under no load conditions) fonctionnement *m* à vide
~BARRIER - [metall] barrière-écran *f* contre les migrations superficielles
~BUCKLING - [metall] (buckling representing the critical termination of a slow gradual increase in deformation) gauchissement *m* par fluage
~EFFECT - [phys] effet *m* de migration
~LIMIT - [metall] résistance *f* au fluage, limite *f* de viscosité (ou d'allongement), limite *f* de fluage
~OUT, to - [rubber ind] (the operation of rolling out sheets of creep rubber) laminer en feuilles
~PRESSURE - [mech] pression *f* de fluage
~RESISTANCE - [metall] s. creep limit

CREEP-RESISTING ALLOY - [metall] alliage *m* résistant au fluage
~SHIELD - s. creep barrier
~STRAIN - [metall] (plastic deformation caused by constant stress) déformation *f* par fluage
~STRESSES - [metall] contraintes *fpl* de fluage
~TEST - [metall] essai *m* de fluage
CREEPAGE - [el] (the travel of the electrolyte to other parts of the cell above the level of the main body of electrolyte) grimpement *m*
~DISTANCE - [el] ligne *f* de fuite (dans l'air)
~SURFACE - [el] surface *f* de fuite
CREEPER - [bot] plante *f* grimpante
[impl] (flat surface on casters to carry out repairs under a machine) chariot *m*
[impl] (used on ice) crampon *m* à glace
[naut] chatte *f*
~CHAIN - [mining] chaîne *f* traînante
~TRACK - [mech] chenille *f* de roulement
~VALVE - [mech] vanne *f* à ouverture progressive
CREEPIE-PEEPIE - [colloq. for portable television transmitter] récepteur *m* de télévision portable
CREEPING - [mech] (of gears) glissement *m*
[geol] cheminement *m*
[chem] (rising of acid) grimpement *m*, ascension *f* capillaire
[metall] (of railway tracks under load) déformation *f*
[instr] s. creep
[bot] rampement *m*
~LIMIT - [metall] limite *f* de fluage
~MACHINE - [rubber ind] laminoir *m* à crêpe, crêpeuse *f*
~OF SALTS - [chem] (movement of electrolyte along the face of the electrodes) grimpement *m* des sels
~PARALYSIS - [med] paralysie *f* progressive
~STIMULUS - [el] (a current which increases so gradually as to be less effective than it would have been, had the final intensity be attained abruptly) excitation *f* cumulative
~STRENGHT - [metall] s. creep limit
~TITLE - [cinema] titre *m* apparaissant progressivement
CREMATE, to - [gen] incinérer
CREMATION - [gen] incinération *f*, crémation *f*
CREMATORIUM - [build] four *m* crématoire
CREMOMETER - [instr] crémomètre *m*
CREMONE BOLT - [carp] (a window bolt) verrou *m* de crémone
CRENA - [med] (in surgery) échancrure *f*, incision *f*, sillon *m*
CRENATED - [bot] échancré, sillonné, crénelé
CRENEL - [arch] créneau *m*, meurtrière *f*
CRENELLATED - [arch] crénelé
CRENELLATION - [arch] crénelage *m*
CREOSOL - [chem] créosol *m*
CREOSOTE, to - [ind chem] imprégner à l'aide de créosote, créosoter
CREOSOTE - [chem] (coal tar fraction used as a wood preservative and disinfectant) créosote *f*
CREOSOTED POLE - [telecomm] poteau *m* impregné de créosote
CREOSOTING CYLINDER - [timber] (container in which timber is impregnated with creosote under pressure) cylindre *m* de créosotage
CREPE - [text] (worsted, silk or cotton dress material) crêpe *m*
CREPE DE CHINE - [text] (light fabric with crêpe effects) crêpe *m* de Chine
~PAPER - [paper man] papier *m* crêpon, papier *m* crêpé
~RUBBER - [rubber ind] (unvulcanized raw sheet rubber) crêpe *m* de caoutchouc
~SOLE - [shoe man] semelle *f* de crêpe
CREPOLINE - [text] crêpoline *f*
CRESCENT - [gen] croissant *m*
[horol] (a notch in the lever escapement roller) demilune *f*
~COMB - [impl] peigne *m* en demi-lune
~SPANNER - [impl] (spanner of crescent form) clé *f* plate
[impl] (in the USA) clé *f* à molette
~TEAR TEST - [rubber ind] essai *m* de déchirement sur éprouvette en croissant
CRESOL - [chem] (coal tar derivative used in the production of phenolic resins, in disinfectants etc) crésol *m*
~RED - [dye] (used in wood dyeing) rouge *m* de crés
~RESIN - [ind chem] (synthetic resin formed by the condensation of a cresol with an aldehyde) résines *fpl* crésoliques
CRESS - [bot] cresson
CRESSING - [metall] réduction *f* de section
CREST - [gen build etc] crête *f*
[zool] crête *f*
[mech] (of screw threads) sommet *m* du filet
[el] (peak factor) crête *f*
[geogr] (of a muntain) crête *f*
[hydr] (of a weir) crête *f*
[build] (the ridge of a roof) faîte *m*
~FACTOR - [el] (peak factor) facteur *m* de crête
~LINE - [geol] ligne *f* de crête, ligne *f* de faîte
~OF A DAM - [hydr] crête *f* d'une digue
~OF A WAVE - [opt] point *m* haut d'une onde
~OF AN ANTICLINE - [geol] crête *f*, sommet *m* d'ur anticlinal
~TILE - [build] (a V-shaped tile covering the ridge lines of the roof) tuile *f* faîtière
~VALUE - [el] valeur *f* de crête
~VOLTAGE - [el] (peak voltage) tension *f* de crête
CRESYL - [chem] crésyl *m*
~DIPHENYL PHOSPHATE - [chem] (a plasticizer with good low-temperature characteristics) phosphate *m* de diphényle et monocrésyle
CRESYLENE - [chem] crésylène *m*, tolylène *m*
CRESYLIC ACID - [chem] (a coal tar derivative and a starting material for phenolic resins etc) acide *m* crésylique
CRETACEOUS SYSTEM - [geol] (the rocks between the Jurassic and the tertiary system) roches *fpl* crétacées
CRETONNE - [text] (a printed cotton material) cretonne *f*
CREVASSE - [geol] (a fissure in a glacier) crevasse *f*
CREVICE - [gen] fissure *f*, crevasse *f*, fente *f*
~-WATER - [geol] eau *m* de diaclases
CREW - [naut] équipage *m*, personnel *m*
[aero] équipage *m*
CREWEL WORK - [text] tapisserie *f* sur canevas
CREWELS - [med] scrofule *f*
CRIB - [gen] crêche *f*, berceau *m*, couchette *f*
[build] (a type of silos) silos *m*
[mining] (a system of timber support) coffrage *m*

en charpente, boisage *m*
[hydr] (of a water obtaining plant) prise *f*
CRIB-BED - [mining] cadre *m* porteur ou de base
~-WORK - [constr] (timber boxes filled with concrete) coffrage *m*, boisage *m*
CRIBBED - [mining] boisé par cadres jointifs
~CHUTE - [mining] cheminée *f* de fagotage
CRIBBING - [constr] (interior lining of a shaft) revêtement *m* de puits, coffrage *m*
~MACHINE - [road & railw works] épandeuse *f* de ballast
CRICKET - [zool] grillon *m*, criquet *m*
CRIMP, to - [gen] (to cross into folds or corrugations) gaufrer
[mech] (to secure a part to another by deforming it by a special tool) sertir
CRIMP CRÊPE - [text] crêpe *m* ondulé
~YARN - [text] filé *m* crêpé, filé *m* mousse, filé *m* frisé
CRIMPED FABRIC - [text] tissu *m* cloqué
CRIMPING - [metall] sertissage *m*
[plast ind] (process of giving synthetic bristles a wavy form) gauffrage *m*, crêpage *m*
~DEVICE - [text] appareil *m* à friser, appareil *m* à crêper
~MACHINE - [text] s. crimping device
~PLIERS - [impl] (special pliers used for crimping) pince *f* à sertir
~RING - [mech] (metal ring used as an anvil in crimping) bague *f* de sertissage
CRIMSON - [dye] cramoisi *f*
~ANTIMONY - [paint] (pigment, pale orange to deep crimson in colour, with good opacity and staining power) vermillon *m* d'antimoine, cramoisi *m* d'antimoine
~LAKE - [paint] (pigment obtained by precipitating cochineal extract with salts of aluminium and tin) laque *f* carminée
CRINANITE - [geol] (a basic igneous rock) crinanite *f*
CRINKLE FINISH - [mech] gaufrage *m*
CRINKLING - [paint] effritement *m*
CRINUM - [bot] crinole *f*
CRIPPLING STRAIN - [metall] effort *m* de flambage
CRISP - [gen] friable, fragile
CRISPATE - [bot] (with a fizzled appearance) crêpu
CRISPENING - [telev] distorsion *f* d'image, accentuation *f* des contrastes
~CONTOUR ACCENTUATION - s. crispening
CRISS-CROSS - [gen] entrecroisé, enchevêtré, treillissé
~-CROSS GRID - [radio] grille *f* à réticule
~-CROSS INHERITANCE - [genet] hérédité *f* croisée
~-CROSS PATCH - [of tyres] pièce *f* protectrice, emplâtre *m* croisé
~-CROSS REPAIR - [rubber ind] (of tyres) pièce *f* croisée
~-CROSSING - [rubber ind] profil *m* en traits de scie
CRISTATE - [bot etc] crêté, cristé
CRISTOBALITE - [min] (a silica) cristobalite *f*
CRIT - [abbrev for critical mass] masse *f* critique
CRITERION - [gen] critère *m*
~OF DEGENERACY - [phys] (the criterion determining the degeneracy of a system particles) critère *m* de dégénérescence
CRITH - [meas] (weight in grammes of a litre of hydrogen at prescribed temperauture and pressure) crith *m*
CRITICAL - [gen phys etc] critique
~AGEING - [metall] vieillissement *m* total
~ALTITUDE - [aero] (the maximum altitude at which a supercharger can maintain an induction manifold pressure equal to that existing during normal operations at rated power and speed at sea-level) altitude *f* critique, altitude *f* de rétablissement
~ANGLE - [opt] angle *m* limite
~ANGLE OF ATTACK - [aero] (the attack angle at which abrupt changes in the flow round an aerofoil occur, giving rise to rapid changes in lift, drag etc) angle *m* d'attaque critique
~ANODE VOLTAGE - [electron] (of a magnetron) tension *f* anodique d'amorçage
~ASSEMBLY - [nucl] (fissionable material and moderator capable of maintaining a fission chain reaction at low power) ensemble *m* critique, montage *m* critique
~BUILD-UP RESISTANCE - [el] (the limiting resistance of the circuit for the armature, for which the machine builds up under specified conditions) résistance *f* critique d'amorçage
~BUILD-UP SPEED - [el] vitesse *f* critique d'amorçage
~CLOSING SPEED - [aero] (in parachutes, the speed during acceleration at which squidding will occur in a normally inflated parachute) vitesse *f* critique de fermeture
~COEFFICIENT - [phys] (additive property of a substance representing a measure of the space occupied by the molecules) coefficient *m* critique
~COMPOSITION - [phys] composition *f* critique
~CONCENTRATION - [phys] (the point at which two immiscible liquids become consolute) concentration *f* critique
~COUPLING - [el] (the degree of coupling in a transformer producing maximum current in the secundary) couplage *m* critique
~DAMPING - [el] (damping in an oscillatory circuit to prevent oscillations) amortissement *m* critique
[mech] (the minimum amount of viscous damping required in a linear system to prevent a free vibration of any given initial amplitude from passing the equilibrium value) amortissement *m* critique
~DEFINITION - [opt] pouvoir *m* séparateur
~DENSITY - [phys] (density of a substance at its critical temperature and critical pressure) densité *f* critique
~DIMENSION - [electron] (waveguides; dimension determining the critical frequency) dimension *f* critique
~EQUATION - [nucl] (reactor parameter equation which must be satisfied to make the reactor critical) équation *f* critique
~EXPERIMENT - [nucl] (gradual building-up of fissionable material until a chain reaction is attained) expérience *f* de divergence
~FIELD - [phys] (the magnetic field below which the super-conducting transition occurs at a specified temperature) champ *m* critique
~FLOW VELOCITY - [aero] (that above which a qualitative change in the nature of such flow occurs) vitesse *f* critique d'écoulement
~FREQUENCY - [electron] (the minimum frequency at which a travelling wave can be maintained in a wave-guide) fréquence *f* critique

CRITICAL FREQUENCY OF RADIATION - [radiat] (the highest frequency at which vertically-directed radiation is deflected from the ionosphere) fréquence *f* critique de rayonnement

~GRID CURRENT - [electron] (the value of grid current at the moment the anode current begins to flow) courant *m* critique de grille

~GRID VOLTAGE - [electron] (the instantaneous value of the grid voltage at which the anode current starts to flow) tension *f* critique de grille, tension *f* d'amorçage

~HEAT - [metall] chaleur *f* critique

~HUMIDITY - [met] (the level of atmospheric water content at which the water vapour partial pressure is equal to the saturation vapour pressure) humidité *f* critique

~INDUCTANCE - [el] (the minimum inductance required to prevent the current from going to zero) inductance *f* critique

~INDUCTION - [electron] (induction theoretical value in a magnetron) induction *f* critique

~MACH NUMBER - [aero] nombre *m* de Mach critique

~MASS - [nucl] (the mass of fissionable material in a critical reactor) masse *f* critique

~MOISTURE CONTENT - [phys] humidité *f* critique

~OPALESCENCE - [phys] (phenomenon occurring when a homogeneous solution of two liquids at its critical composition is cooled) opalescence *f* critique

~OPENING SPEED - [aero] (the speed during deceleration at which a squidded parachute assumes the normal-inflated form) vitesse *f* critique d'ouverture

~ORGAN - [radiat] (the organ in which radiation has the greatest effect) organe *m* critique

~POINT - [phys] (the point at which two phases become identical and form one phase) point *m* critique
[aero] (point in a flight at which the time to reach destination is equal to that required to return to the starting point) point *m* critique

~POTENTIAL - [electron] (measure of the energy which is required to raise an electron to a higher-level) potentiel *m* d'ionisation

~PRESSURE - [mech] (the pressure of a vapour at its critical point) pression *f* critique

~QUOTA - s. critical altitude

~RADIUS - [nucl] (of a nuclear reactor) rayon *m* critique

~RANGE - [metall] (the range of temperature in which the reversible change from austenite to ferrite etc occurs) intervalle *m* critique

~REACTION - [electron] (the maximum degree of reaction before a thermionic valve falls into self-oscillation) réaction *f* critique

~REACTOR - [nucl] (nuclear reactor at a stage in which the loss of neutrons is balanced by their production) réacteur *m* entrant en divergence

~REGION - [phys] (the region near the critical point) zone *f* critique

~SHEAR STRESS - [mech] effort *m* limite de cisaillement

~SIZE - [phys] (a physical dimension of the core and reflector of a reactor which maintains a critical chain reaction) dimensions *fpl* critiques

~SOLUTION TEMPERATURE - [phys] température *f* critique de solution

~SPEED - [mech] (rotating part speed at which resonant vibrations begin) vitesse *f* critique

~STATE - [nucl] (reactor state at which the loss of neutrons is balanced by their production) état *m* critique

~TEMPERATURE - [phys] (the temperature above which it is impossible to liquefy a gas, however great the pressure) température *f* critique, point *m* de transformation

~THICKNESS OF INFINITE SLAB - [nucl] (in a nuclear reactor, thickness making the geometric buckling equal to the material buckling) épaisseur *f* critique d'une plaque infinie

~VELOCITY - [hydr] (the linear velocity above which the flow will become turbulent) vitesse *f* critique

~-VOLTAGE PARABOLA - [electron] (curve representing in Cartesian coordinates the variation of the critical voltage as a function of the magnetic induction) parabole *f* de coupure

~VOLUME - [phys] (the volume occupied by one gramme of a liquid or gas at its critical temperature and critical pressure) volume *m* critique

~WAVELENGTH - [electron] (waveguides; the free-space wavelength which corresponds to the critical frequency) longueur *f* d'onde critique

~YEAR - [med] année *f* climatérique

CRITICALITY - [nucl] (criticality is reached by a core when the neutron escape rate is equal to the rate at which they are generated by the fission) divergence *f*

CRIZZLING - [gen] (on the surface of frozen water) fendillement *m*
[glass man] (fine looking cracking on the surface of glass, caused by local chilling during manufacture) fendillement *m*

CROAKING - [acoust] (the sound made by frogs) coassement *m*

CROCEIN - [chem] crocéine *f*

~DYES - [dyes] colorants *mpl* à la crocéine

CROCETIN - [chem] crocétine *f*

CROCHET - [gen] (a type of knitting with a crochet hook) travail *m* au crochet

~GAUZE - [text] gaze *f* au crochet

~HOOK - [impl] crochet *m*

~YARN - [text] fil *m* pour travail au crochet

CROCIDOLITE - [min] (commercially important form of asbestos) crocidolite *f*, asbeste *m* bleu

CROCK - [gen] (any open pottery vessel) cruche *f*, pot *m*, terrine *f*

CROCKET - [arch] crochet *m*

CROCODILE - [zool] crocodile *m*

~LEATHER PAPER - [paper man] papier *m* crocodile

~PRESS - [mech] cingleur *m* à levier

~TRUCK - [railw] (railway truck formed by a long very low platform resting on two four-wheel bogies) wagon *m* à plateforme basse

CROCODILING - [paint] (formation of ridges and minute cracks on the surface of paints) craquelage *m* en peau de crocodile

CROCOITE - [min] (also called crocoisite; natural lead chromate, bright red in colour) crocoise *f*, plomb *m* rouge

CROCUS - [bot] crocus *m*, safran *m*

CRONING METHOD - [plast ind] (shell-moulding) moulage *m* à noyau

CROOK - [gen] courbure *f*, coude *m*, croc *m*, cro-

chet *m*
[mus] (curved tube inserted in horns) ton *m* rechange
CROOKED HOLE - [mining] sondage *m* tordu
CROOKES DARK SPACE - [electron] (cathode dark space) espace *m* sombre cathodique
~RADIOMETER - [instr] (for the detection of heat radiation) radiomètre *m* de Crookes
CROOKESITE - [min] (natural copper rhallium selenide, occasionally argentiferous) crookésite *f*
CROP, to - [metall] (to shear off the rough end of a metal bar etc) faire chuter, tronçonner
[mech] (of a propeller blade) couper l'extrémité de pale
CROP - [agric] récolte *f*, culture *f*
[metall] (the sheared off portion of a lingot) chute *f*, chute *f* d'extrémité
~ACREAGE - [agric] surface *f* cultivée, terre *f* labourable
~AREA - s. crop acreage
~DAMAGE - [agric] endommagement *m* des récoltes
~-DESTROYING INSECT - [zool] insecte *m* nuisible aux cultures
~DUSTING - [agric] (the operation of spraying plant disinfectants by helicopter) poudrage *m* des cultures
~-END - [metall] (the rough, sheared-off end of bars, ingots etc) chute *f* d'extrémité
[photo] (US; those sections of photographs or drawings not to be printed) partie *f* à ne pas reproduire
~ESTIMATE - [agric] récolte *f* estimée
~FAILURE - [agric] mauvaise récolte *f*
~FARMING - [agric] culture *f*
~GROWING - [agric] culture *f*
~-HEAD - s. crop-end
~LOSSES - [agric] perte *f* des récoltes
~PRODUCTION - [agric] culture *f*
~PROSPECTS - [agric] prévisions *fpl* de récolte, récolte *f* estimée
~PROTECTION - [agric] protection *f* des cultures
~RESTRICTION - [agric] limitation *f* d'une culture
~ROTATION - [agric] assolement *m*, rotation *f* des cultures
~-SHARE - [agric] métayage *m*
~YEAR - [agric] campagne *f*
~YELD - [mech] rendement *m*
CROPPER - [mech] tronçonneuse *f*
CROPPING - [metall] (the operation of shearing off a piece from one or both end of an ingot to obtain a test specimen or remove defects) tronçonnage *m*, éboutage *m*
[metall] (the operation of cutting wire into specified lengths) tronçonnage *m*
[cin] réduction *f* du plan d'image
[text] (cutting off the excessive pile) affinage *m* de draps
CROSS - [gen] croix *f*
[zool] croisement *m*, fécondation *f* croisée
[adj] en croix, transversal
[plumb] raccord *m* en croix
[mining] travers-bancs *m*
~ADDING - [comput] (a system of checking in which horizontal and vertical sums are interchanged) vérification *f* par totaux croisés
~-ADIT - [mining] s. cross
~ADJUSTMENT - [mech] ajustage *m* transversal
CROSS AIR LEVEL - [opt] croix *f*, niveau *m* en croix
~-AISLE - [constr] transept *m*
~AMPERE TURNS - [el] (the armature ampere-turns components tending to produce a field at right angle to the main field) ampère-tours *mpl* transversaux
~ARM - [telecomm] (the horizontal cross-member attached to a telegraph pole) croisillon *m*, traverse *f*
~AUGER - [oil ind] tarière *f* en croix
~-BANDING - [metall] (in lamination; laying alternate plies at right angle to each other) stratification *f* croisée
~BAR - [mech] (part of frame or structure placed transversally to the longitudinal axis) croisillon *m*, potence *f*, traverse *f*
~BAR - [constr] petit bois *m*, croisillon *m*
[mining] traverse *f*, entretoise *f*
~-BAR MICROMETER - [instr] (in a telescope) micromètre *m* à réticule
~-BAR OF THE FRAME - [text] traverse *f* de cadre
~-BAR SUPPORT - [telecomm] potence *f* d'appui
~-BAR TRANSFORMER - [el] changeur *m* de mode à barres croisées
~-BEAM - [constr] traverse *f*, sommier *m*, chapeau *m*
~-BEAM COUNTERWEIGHT - [mach tools] contrepoids *m* d'équilibrage de la traverse mobile
~-BEAM HEADSTOCK - [mach tools] poupée *f* de traverse
~BEAMS - [build] poutres *fpl* croisées
~-BEARER - [railw] traverse *f* intermédiaire
~-BEARING - [surv] (check bearing between stations) relèvements *mpl* croisés
~BEATER MILL - [ind proc] broyeur *m* à marteaux
~BED - [mining] couche *f* oblique
~-BEDDED - [geol] à stratification *f* entrecroisée
~-BEDDED SAND - [geol] sable *m* à stratification entrecroisée
~-BEDDING - [geol] stratification *f* entrecroisée
~-BELT - [mech] courroie *f* croisée
~-BELT DRIVE - [mech] transmission *f* par courroie croisée
~BIT - [impl] taillant *m* en croix
~-BLAST EXPLOSION POT - [el] (oil switch in which the pressure generated by the arc forces a stream of oil to cut through the arc stream) chambre *f* d'explosion à soufflage transversal
~-BLAST OIL CIRCUIT BREAKER - [el] disjoncteur *m* à huile à soufflage transversal
~BOMBARDMENT - [nucl] (nuclear bombardment using different target materials) bombardement *m* croisé
~-BOND - [el] (a rail-bond connecting the two rails of a track) connexion *f* transversale
~BORDER DOBBY - [text] ratière *f*
~BORDER MOTION - [text] mécanique *f* à armures
~-BOW - [gen] arbalète *f*
~BRACE - [mech] traverse *f*, croisillon *m*
~-BRACKET - [mech] support *m* transversal, traverse *f*
~-BREAKING STRENGTH - [phys] résistance *f* à la flexion, effort *m* transversal
~-CARLING - [naut] barrotin *m*
~-CHECK - [surv] recoupement *m*
~-COIL AERIAL - [radio] (directional aerial system with two coil aerials at right angle to each other)

radiogoniomètre *m* à cadres croisés
CROSS-COMPOUND - [mech] à étages basse et haute pression
~-COMPOUND TURBINE - [el mach] groupe *m* à deux lignes d'arbres
~CONDUCTION - [el] transconductance *f*
~-CONNECTING TERMINAL - [telecomm] boîte *f* de division, sous-répartiteur *m*
~-COUNTRY PROFILE - [rubber ind] (of tyres) profil *m* tous terrains
~-COUNTRY VEHICLE - [mech] véhicule *m* tous terrains
~COUPLING - [el] (measure of unwanted power transferred from one channel to another) puissance *f* de fuite
~-COURSE - [mining] filon *m* croiseur
~-CUT, to - [mining] percer en travers-banc
~-CUT - [mining] (a tunnel or level through the rock to intersect a lode) travers-banc *m*
[mech] coupe *f* en travers, contre-taille *f*, taille *f* croisée
~[telev] raccord *m*
~-CUT CHISEL - [impl] bédane *m*
~-CUT FILE - [impl] (with cutting edges formed by the intersection of the teeth) lime *f* à taille croisée
~-CUT SAW - [impl] (for cutting timber across the grain) scie *f* en travers, passe-partout *m*, scie *f* alternative à tronçonner
~-CUTTER - [mech] (cutting machine of the guillotine type) cisaille
~-CUTTING - [mining] s. cross-cut
[cin] changement *m* rapide de cadre
~-DIPOLE - [radio] (type of non-directional receiving aerial) antenne *f* dipôle en croix
~-DRAINING - [agric] drainage *m* transversal
~DRAUGHT CARBURETOR - [mech] (int.comb.engines) carburateur *m* transversal
~DRIFT - [electron] (drift in electron and ion orbits causing a mass motion with no electric current) mouvement *m* transversal
[mining] recoupe *f*
~DYEING - [text] (system of dyeing in which the warp and the weft are treated by different dyes) tenture *f* en deux couleurs
~ENTRY - [mining] galerie *f* transversale, recoupe *f*
~ESTERIFICATION - [chem] alcoolyse *f*
~-EYE - [med] strabisme *m*
~-FADE - [radio] fondu *m* enchaîné
~-FALL - [constr] bombement *m* de la chaussée
~FAULT - [geol] faille *f* oblique, faille *f* orthogonale, faille *f* de plongement
~-FEED - [mech] avance *f* transversale
~-FEED CONTROL - [mach tools] commande *f* de l'avance transversale
~FERTILIZATION - [agric] fécondation *f* croisée
~FILE - [impl] lime *f* à taille croisée, feuille *f* de sauge
~-FIRE - [el] (interfering current in a signalling channel) induction *f* télégraphique
~FOLD - [geol] pliage *m* transversal
~-FOOTER - [instr] (counting device) compteur *m* à opération horizontale
~FORCE - [mech] effort *m* tranchant, force *f* transversale
~-FRONT - [photo] porte-object *m* à décentrement latéral

CROSS-GARNET HINGE - [mech] penture *f* à T
~GIRDERS - [mech] (acting as a tie between the main girders) poutres *mpl* transversales
~GRAINED - [timber] à fibres irrégulières
~-GRINDER - [mech] défibreuse *f* transversale
~-GRINDING - [paper man] défibrage *m* transversal
~GROOVE - [text] rainure *f* transversale
~HAIR - [surv] (a thread fixed across the diaphragm of a level) réticule *m*
~-HAIR LINES - [opt] réticule *m*
~-HAIRS - [instr] viseur *m*
~HANDS - [mining] traverse *f*
~HATCH - [draw] contre-taille *f*, contre-hachure *f*
~-HEAD - [mech] (the junction piece between the piston-rod and the connecting-rod) pied *m* de bielle
[surv] équerre *f*
[mech] (of engine) crosse *f* de piston, crossette *f*, tête *f* de piston
[plast ind] (extruder head fixed at right angles to the centre-lines of the barrel) filière *f* à tête d'équerre
~-HEAD DIE - [plast ind] filière *f* en croix
~-HEAD GUIDE - [mech] guide *m* de pied de bielle, glissière *f* de crosse
~-HEAD PIN - [mech] goupille *f* à tête en croix, tourillon *m* de crosse (ou de la tête de piston)
~-HEAD SHOE - [cin] (of a projector) patin *m* de crosse
[mech] patin *m* de la crosse, semelle *f* de la crosse
~HEADING - [mining] galerie *f* transversale
~IGNITION - [gas ind] interallumage *m*
~-JACK - [naut] voile *f* de fortune, voile *f* barrée
~-JACK YARD - [naut] vergue *f* barrée
~JOINT - [constr] (the vertical mortar joint for bricks in a wall) joint *m* vertical
[metall] déplacement *m* de moule
[plumb] joint *m* en croix
~JURNAL - [mech] tourillon *m* en croix
~LAMINATION - [plast ind] (of a sheet) stratification *f* croisée
~LATH - [text] pédale *f* auxiliaire transversale
~LEVEL - [instr] (instrument designed to show the lateral inclination of an aircraft) appareil *m* de niveau latéral
~-LINE - [draw] contre-taille *f*
~-LINKED POLYMER - [chem] polymère *m* à liaisons transversales
~-LINKING - [chem] (the formation of additional links between the chains of atoms in polymerized material, a process which may be caused by radiation) liaison *f* transversale, réticulation *f*
~-MATCHING - [chem & med] essai *m* croisé de compatibilité
~MEMBER - [mech] traverse *f*, croisillon *m*, entretoise *f*
~-MODULATION - [radio] (unwanted transfer of a modulation from one signal carrier to another) transmodulation *f*
~-NEUTRALIZATION - [radio] neutralisation *f* symétrique
~-OUT GALLERY - [mining] galerie *f* transversale
~-OVER - [gen] croisement *m*
[railw] voie *f* de croisement
[el] (the point at which the circuit is changed over in transposing) point *m* de transposition
[telev] (position of beam constriction in an electron

optical system) première convergence *f*
[electron] (the first focusing of the beam taking place in the electron gun) première convergence *f*
CROSS-OVER FLUE - [constr] carneau *m* transversal
~-OVER OVEN - [metall] four *m* à carneaux transversaux
~-OVER SPIRAL - [el acoust] (the groove cut between short duration recordings on the same disk) sillon *m* intermédiaire
~-OVER STITCH - [text] point *m* de liaison
~-PANE HAMMER - [impl] marteau *m* à panne croisée
~PATCH - s. criss-cross patch
~-PAWL - [shipbuild] linguet *m*
~-PIECE - [mech] traverse *f*
~-PLATING - [text] vanisage *m* renversé
~-PLATING NEEDLE - [text] aiguillage *f* à vaniser avec bec droit
~POLARIZATION - [el] (the component of the field vector normal to the polarization component) croisement *m* des polarisations
~POLLINATION - [bot] pollinisation *f* croisée, hybridisation *f*
~RAIL OF LOOM - [text] traverse *f* de métier
~REELING - [text] dévidage *m* croisé
~-ROAD - [gen] croisement *m*
~ROLL, to - [metall] laminer par rotation entre cylindres inclinés
~ROLLING - [metall] laminage *m* transversal
~ROW OF KNOTS - [text] ligne *f* transversale de noeuds
~ROW OF SPOTS - [text] ligne *f* transversale de points
~SEAMS - [text] coutures *fpl* transversales
~-SECTION, to - [draw] (a method of representation on which the object drawn is shown as if cut along a line normal to the major axis) représenter en section transversale
~-SECTION - [gen & draw] section *f* transversale section *f* droite
[phys] (the probability of occurrence of a given reaction) section *f* efficace
~-SECTION IRON - [metall] fer *m* en X
~SHAFT - [mech] arbre *m* transversal
~SHAKE - [timber] fente *f* transversale
~SHED - [text] pas *m* croisé
~-SILL - [railw] traverse *f*
~-SLIDE - [mech] (secondary tool-slide of a machine-tool mounted on the main slide and travelling transversely to it) glissière *f* transversale, chariot *m* transversal
~-SPALE - s. cross-pawl
~SPRING - [mech] ressort *m* transversal
~-STAFF - [surv] équerre *f*
[naut] astrolabe *f*
~-STITCH - [text] point *m* de croix
~STICHED GAUZE - [text] gaze *f* au point de croix
~-STRATIFIED - s. cross bedded
~STRIPE PATTERN - [text] dessin *m* à rayures transversales
~STRIPED FABRIC - [text] tissu *m* rayé transversalement
~-TALK - [radio] (interference from another transmission) diaphonie *f*
[acoust] (in recording) effet *m* d'écho, effet *m* d'empreinte magnétique
~-TALK ATTENUATION - [telecomm] atténuation *f* de diaphonie
CROSS-TALK COUPLING [telecomm] (cross-coupling between speech channels) perte *f* par diaphonie
~-TALK LOSS - s. cross-talk coupling
~-TALK METER - [instr] (measuring the attenuation between circuits subject to cross-talk) diaphonomètre *m*
~-TALK VOLUME - [radio & telecomm] puissance *f* perturbatrice
~TAP - [el] prise *f* double
~-TIE - [railw] traverse *f*, tirant *m* transversal
~TRAVEL - [mech] déplacement *m* transversal
~-TRAVERSE HANDWHEEL - [mech] volant *m* pour passe transversale
~TREE - [naut] barre *f* en flêche
~-TUBE BOILER - [metall] (a type of steam generator with transverse water-tubes fitted in the furnace) chaudière *f* à tubes transversaux
~-TWILL - [text] tissu *m* croisé, tissu *m* en diagonale, sergé *m* croisé
~VAULT - [arch] voûte *f* croisée
~VEIN - [mining] filon *m* croiseur
~-WIND, to - [rope man] enrouler à spires croisées
[text] bobiner à fils croisés
~-WIND - [met] (wind blowing transversally to the direction in which a ship or aircraft is travelling) vent *m* de travers, vent *m* contraire
~-WIND AXIS - [phys] (straight line normal to the lift and drag axes, drawn through the centre of gravity) axe *m* perpendiculaire au plan de symétrie
~-WIND FORCE - [met] (the degree of force of a cross-wind) force *f* de dérive
~-WIND LANDING - [aero] atterrissage avec vent de travers
~-WIND LANDING GEAR - [aero] (type of landing gear in which all the wheels or bogies are designed to castor) train *m* orientable
~WIRE - s. cross hair
~WORKING - [mining] méthode *f* en traverse
~WOUND BOBBIN - [text] bobine *f* à fil croisé, bobine *f* croisée
~WOUND COP - [text] machine *f* à cannettes à fil croisé
~ZIGZAG TWILL - [text] tissu *m* croisé à chevrons en largeur
CROSSARM - [photo] (boom) girafe *f*
CROSSBRED - [gen] métis, croisé
CROSSBREED, to - [zool] croiser, métisser
CROSSBREEDING - [genet] croisement *m*, hybridation *f*, métissage *m*
CROSSED AERIAL - s. crossed dipole
~ANTENNA - s. crossed dipole
~GROOVE THREAD - [rubber ind] (of tyres) sculpture *f* à rayures transversales
~HARNESS - [text] lames *fpl* croisées
~-LOOP AERIAL - s. cross-coil aerial
~-LOOP ANTENNA - s. cross-coil aerial
~NICOLS - [opt] (arrangement of two Nicol prisms) nicols *mpl* croisés
~POSITION - [opt] (position of the prisms preventing the passage of light) position *f* croisée
CROSSING - [gen] traversée *f*, passage *m*, croisement *m*
[railw] croisement *m* de voies, passage *m* à niveau
~LAYER - [text] excentrique *m* pour enroulement croisé

CROSSING OF THE SELVEDGE THREADS - [text] croisement *m* des fils de lisière
~OF THREADS - [text] croisement *m* des fils
~OVER - [genet] croisement *m* chromosomique
~POINT - [text] point *m* de croisement
~THE TIES - [text] croisure *f*
~THREAD - [text] fil *m* de croisure
~WARP - [text] chaine *f* de croisure
CROSSLINE SIGHT - [opt] (cross-hair sight) pinnule *f* à fils
CROSSOVER POINT - [genet] (crossing over) point de croisement, d'échange chromosomique
CROSSROADS - [constr] carrefour *m*
CROSSWINDING - [text] bobinage *m* croisé, enroulement *m* en couches croisées
~MACHINE - [text] bobinoir *m* à fil croisé
CROSSWISE - [gen] en travers, en croix, transversalement
~DIRECTION - [paper man] (in resin bonded paper, across the machine direction of the paper) direction *f* transversale
~MOVEMENT - [gen & mech] mouvement *m* transversal
CROTCHET - [mus] noire *f*
CROTONALDEHYDE - [chem] (a solvent and intermediate for vulcanization accelerators etc) crotonaldéhyde *m*, aldéhyde *m* crotonique
CROTONIC ACID - [chem] (intermediate for plasticizers, pharmaceuticals, resins etc) acide *m* crotonique
CROUP - [med] croup *m*
CROW - [zool] corneille *f*
[impl] levier *m*
CROW'S FEET - [anat] pattes *fpl* d'oie
~NEST - [naut] nid *m* de pie
~-FOOT - [oil ind] caracole *f*, crochet *m* de repêchage, harpon *m*
CROWBAR - [impl] (round iron bar used as a lever for heavy objects) levier *m*, pince *f* à panne fendue
~TYRE LEVER - [auto] démonte-pneu *m*
CROWD, to - [gen] serrer, entasser, encombrer, remplir
[naut] (a sail) faire force de voiles
CROWFOOT - [naut] araignée *f*
[cin] (a support preventing the sliding of the camera) support *m* antidérapant
[text] équerres *fpl*, support *m* de porte-bobines
~WRENCH - [impl] clé *f* à pied de biche
CROWING - [acoust] (the sound made by cocks) chant *m* du coq
CROWN, to - [gen] couronner
[mech etc] bomber, donner de la cambrure
CROWN - [gen] couronne *f*
[arch] (on an arch) clef *f* de voûte
[mech] (of a bevel gear) cercle *m* extérieur
[mining] plafond *m*
[horol] couronne *f*
[naut] diamant *m* de l'ancre
[metall] (of furnace) couronne *f*, voûte *f*
[mech] (of a piston) fond *m* de piston
[constr] (of a road) bombement *m*
[paper man] papier *m* couronne
[aero] (top of a parachute) coupole *f*
[dent] couronne *f*
[mining] (a drilling tool) trépan *m* à couronne
[mech] (of a press) tête *f*
~BIT - [mining] fleuret *m* à couronne
CROWN BLOCK - [oil ind] bloc-couronne *m*, moufle *m* fixe
~CORK - [impl] capsule *f*
~CUTTER - [mech] fraise *f* à couronne
~FRAMING - [gas ind] (in gas holders) charpente *f* de la cloche
~GATE - [hydr] (a lock head-gate) porte *f* d'amont
~GEAR - [mech] couronne *f* dentée
~GLASS - [opt] (glass of alkali-lime-silica type) verre *m* crown, crown-glass *m*
~HINGE - [constr] articulation *f* à la clef
~JOINT - [impl] joint *m* de clef
~LENS - [opt] lentille *f* en crown-glass
~OF ABERRATION - [opt] cercle *m* d'aberration
~OF DIAMOND DRILL - [oil ind] couronne *f* à diamants
~PLATE - [metall] tôle *f* du ciel du foyer
~PLATFORM - [metall] plate-forme *f* supérieure
~POST - [carp] (vertical timber tie used in building works) poinçon *m* de faîte
~PULLEY - [oil ind] poulie *f* de forage
~RAIL-BOND - [railw] (a flexible copper cable rail bond) câble *m* souple de connexion
~REST - [gas ind] (in gas holders) charpente *f* intérieure fixe
~SHEET - [metall] (the plate forming the top of the fire-box) plaque *f* de voûte
~SUPPORT - s. crown rest
~TAP - s. crown cork
~TILE - [constr] (a common flat tile) tuile *f* plate
~WHEEL - [mech] couronne *f* dentée
[auto] couronne *f* de différentiel
~WHEEL BRIDGE - [horol] pont *m* de roue à couronne
CROWNED - [gen] couronné
[bot] (with a terminal outgrowth) couronné
~FACE PULLEY - [mech] poulie *f* à bord
CROWNER - [mech] capsuleuse *f*
CROWNING - [mech] (convexity of the surface of a pulley) bombement *m*
~TOOL - [tool] outil *m* à bomber
CROWS-FOOTING - [paint] (defect in a surface coating consisting of small ridges and resembling a bird's foot-print) rides *mpl* en patte d'oie
CROWSFOOT - [aero](in balloons; group of short ropes attached to the end of a single rope, to distribute the pull of the latter on the envelope) araignée *f*
CROY - [constr] (bank erosion protective barrier) digue *f* de protection
CROZE - [carp] (the groove at the end of cask staves) jable *m*
CROZZLE - [constr] (overburnt brick, partially fused and distorted) brique *f* brûlée
CRUCIAL - [gen] crucial, décisif
~TEST - [gen & mech] épreuve *f* décisive
CRUCIBLE - [ind chem, metall etc] (a refractory vessel for heating solids at high temperatures) creuset *m*
~CAST STEEL - s. crucible steel
~GRAPHITE - [min] (graphite suitable for making crucibles) graphite *m* pour creuset
~LIFTER - [metall] brancard *m* de creuset
~MELTING FURNACE - [metall] four *m* à creuset
~STEEL - [metall] (steel made in a crucible; an obsolete system) acier *m* au creuset
~TONGS - [impl] (crucible holding tool) pinces *fpl* pour creusets, tenailles *f*

CRUCIFORM - [gen] cruciforme, croisé, en croix
~GIRDER - [aero] (a structural member with a cross-shaped section) poutre *f* en croix
CRUDE - [gen] brut
~AMMONIACAL LIQUOR - [chem] eau *f* ammoniacale brute
~BOTTOM - [oil ind] résidu *m*, sédiment *m*,
~GAS LIQUOR - s. crude ammoniacal liquor
~METAL - [metall] (a substance containing impurities to make it suitable for specified purposes, or containing some valuable metals to make their recovery justifiable) métal *m* brut
~OIL - [oil ind] (complex mixture of mineral hydrocarbons obtained from geological deposits) pétrole *m* brut
~ORE - [min] (as found in deposits) minerai *m* brut
~PETROLEUM - [min] pétrole *m* brut
~RUBBER - [rubber ind] (rubber not yet subjected to any process) caoutchouc *m* brut
~STEEL - [metall] acier *m* brut
CRUISING - [gen] croisière *f*
~ALTITUDE - s. cruising height
~CEILING - [aero] (the maximum altitude at which it is possible to maintain the cruising threshold) plafond *m* de croisière
~HEIGHT - [aero] (the altitude at which the major part of the flight is to be carried out) altitude *f* de croisière
~POWER - [aero] autonomie *f*, puissance *f* de croisière
~RADIUS - [aero] (the maximum distance from a specified point which aircrafts can cover in cruising conditions) rayon *m* d'action
~RANGE - [aero] autonomie *f* de vol, rayon *m* d'action
~SPEED - [aero] (the speed at normal level flight) vitesse *f* de croisière
~THRESHOLD - [aero] (equivalent airspeed for the lowest acceptable cruising speed) vitesse *f* minimale économique de croisière
CRUMB - [gen] miette *f*
[ind chem] morceau *m*
~-STRUCTURE - [metall] structure *f* granuleuse
CRUMBLE, to - [gen] émietter, tomber en ruines
CRUMBLING - [soil] émiettage *m*
CRUMBLY - [gen] (easily broken up, as between the fingers) friable
CRUMBS - [ind chem] (shredded soda-cellulose in viscose manufacture) flacons *mpl*, déchets *mpl*
CRUNCH SEAL - [metall] scellement *m* métal-céramique, scellement *m* (ou joint) céramique-métal
CRUPPER - [impl] (part of horse harness) croupière *f*
CRUSH, to - [gen] écraser, presser, concasser, briser, casser, rompre
[agric] (e.g. olives) presser
CRUSH - [gen] broyage *m*, concassage *m*
[metall] (a casting defect) écrasement *m*
~GAUGE - [oil ind] dynamomètre *m* à écrasement
~GRINDER - [mech] rectifieuse *f* à meule
~GRINDING - [mech] rectification *f* à la meule
~-RESISTANT - [gen & text] (of a material which is not easily crushed) résistant à l'écrasement
~SEAL - [metall] joint *m* à écrasement, joint *m* à compression, raccord *m* à écrasement
~STRUCTURE - [geol] structure *f* cataclastique
CRUSHED - [gen] broyé, concassé
CRUSHED SPOT - [text] écrasure *f*
~STONE - [constr] pierre *f* concassée, pierre *f* à macadam
CRUSHER - [mech] (machine for reducing the size of coarse material) broyeur *m*
[text] batteur *m*
~BALL - [min] élément *m* broyeur
~ROLL - [min] rouleau *m* à broyer
CRUSHING - [gen] broyage *m*, bocardage *m*, concassage *m*
[text] (of burrs) battage *m*
~LOAD - [mech] charge *f* d'écrasement
~MILL - [mech] broyeur *m*, bocard *m*,
[min] concasseur *m* à cylindres
~PLANT - [min] (for minerals) installation *f* de broyage
~ROLLS - [mech] (hardened steel rollers used to crush materials) broyeur *m* à rouleaux
~STRESS - [constr] effort *m* de compression
~TEST - [constr] essai *m* à la compression, essai *m* d'écrasement
CRUST - [gen] croûte *f*, écorce *f*, dépôt *m*
[constr] (in road constr) enrobement *m*
[geol] écorce *f* (terrestre)
~FRACTURE - [geol] grande *f* faille
~LEATHER - [leather ind] (light skin which has been shaved on the flesh side) croûte *f*
CRUSTACEOUS - [zool] crustacé *m*
CRUTCH - [impl] béquille *f*, support *m*, soutien *m*, étançon *m*
[naut] fourche *f*, chandelier *m* à fourche
~END - [impl] embout *m* de béquille
~HEAD - [impl] garrot *m*
~KEY - [mech] (of a tap) béquille *f*
~PAD - [impl] coussinet *m* d'aisselle pour béquille
~PARALYSIS - [med] paralysie *f* des béquillards
CRUTCHER - [ind chem] (a mixing tank with paddle agitators, used for soap making) malaxeur *m*, mélangeur *m*
CRYESTHESIA - [med] cryesthésie *f*
CRYOBAFFLE - [th eng] baffle *m* cryogénique, baffle *m* refroidi (à basse température)
CRYOGENERATOR - [th eng] machine *f* frigorifique à gaz, cryogénérateur *m*
CRYOGENIC FLUID PUMP - [mech] pompe *f* (de circulation) pour gaz liquéfié
~LINER - [mech] surface *f* cryogénique (du condenseur d'une cryopompe)
~PUMP - [th eng] pompe *f* cryogénique, cryopompe *f*
~SHROUD - [th eng] système *m* des éléments à l'intérieur d'une cryopompe
~SINK - [th eng] condenseur *m* cryogénique, condenseur *m* d'une cryopompe
~SYSTEM - [phys] (a system in which a local temperature is produced which is lower than the surrounding one) système *m* cryogénique
~TRAPPING - [ind chem] cryotrapping *m*, captage *m* cryogénique, cryofixation *f*
CRYOHYDRATE - [chem] (eutectic system consisting of salt and water) cryohydrate *m*
CRYOHYDRIC POINT - [phys] (the eutectic point when there is water) point *m* cryohydratique
CRYOLITE - [min] (aluminium-sodium fluoride, a mineral used in the manufacture of porcelain glass) cryolithe *f*
~GLASS - [glass man] (white porcellaneous glass) verre *m* porcellanique

CRYOMETER - [instr] (low-temperature thermometer) cryomètre *m*

CRYOPLATE - [th eng] surface *f* cryogénique, cryosurface *f*

CRYOPUMP - s. cryogenic pump

~CONDENSER - s. cryogenic sink

CRYOSCOPIC - [phys] cryoscopique

~CONSTANT - [phys] constante *f* cryoscopique

~METHOD - [chem] (measure of the molecular weight of a substance by observing the lowering of the freezing point of a suitable solvent) méthode *f* cryoscopique

CRYOSCOPY - [phys] (study of the lowering of the freezing point of a solution) cryoscopie *f*

CRYOSORB-TRAP - [th eng] piège *m* à froid, piège *m* cryogénique

CRYOSORPTION PUMP - [th eng] pompe *f* cryostatique à sorption

CRYOTHERAPY - [med] cryothérapie *f*

CRYOTRON - [comput] (a computer element, helium cooled) cryotron *m*

CRYPT - [arch] crypte *f*
[anat] follicule *m*, cavité *f* glandulaire, crypte *f*

CRYPTO-STEADY FLOW - [aero] (a non-steady flow which is steady over certain regions in space and time interval is said to be crypto-steady in respect to such domains) écoulement *m* cryptostable

CRYPTOCRYSTALLINE - [phys] (a structure consisting of very minute crystals) cryptocristallin

CRYPTOGAMIC DISEASE - [bot] maladie *f* cryptogamique

CRYPTOMERE - [genet] (unseen genetic factor) cryptomère *m*

CRYPTOMETER - [instr] (measuring the opacity of paints and dyes) cryptomètre *m*

CRYSTAL - [gen] cristal *m*
[min] (a generally solid body whose atoms are in a definite pattern. There are seven systems and thirty different types of crystal) cristal *m*

~ANGLES - [crystall] (the constant angles which are formed by the crystal faces) angles *mpl* d'un cristal

~BLANK - [crystall] (the final cutting result in a crystal slab) lame *f* cristalline

~BOUNDARIES - [metall] (the contact surfaces between adjacent crystals in a metal) surface *f* de contact des cristaux

~CONTROL - [radio] (the maintainance of constant frequency by means of a piezo-electric crystal) pilotage *m* piézoélectrique

~CONTROLLED TRANSMITTER - [radio] (transmitter whose carrier frequency is directly controlled by the electro-mechanical characteristics of a crystal) émetteur *m* piloté par quartz

~COUNTER - [nucl] (counter with a crystal made conducting by ionizing events) compteur *m* à cristal

~CUT - [crystall] (plane section) coupe *f* d'un cristal, face *f* d'un cristal

~CUTTER - [el acoust] (cutter in which the mechanical displacement of the recording stylus are caused by the deformations of a piezo-electric crystal) graveur *m* à cristal

~DETECTOR - s. crystal diode

~DIODE - [radio] (rectifier consisting of a tungsten cat's whisker with a germanium or silicon crystal in point contact) détecteur *m* à cristal

CRYSTAL DRIVE - [radio] (electromechanical drive whose frequency is determined by a crystal) pilotage *m* piézoélectrique

~EFFECTS - [phys] (effect on the crystalline structure of a material by the neutron cross-section of the material) effets *mpl* du réseau

~ELEMENTS - [crystall] éléments *mpl* d'un cristal

~FACE - [phys] face *f* (d'un cristal)

~FIELD - [crystall] (the electrostatic field inside a crystal) champ *m* d'un cristal

~FILTER - [radio] (band-pass filter using piezoelectric crystals for frequency-discrimination) filtre *m* piézoélectrique

~FREQUENCY-CHANGER EFFICIENCY - [radio] (ratio between the output power and the input power of the signal) rendement *m* de changeur de fréquence à cristal

~GLASS - [glass man] (good-quality glass of fine appearance and transparence) cristal *m*, flint *m* flint-glass *m*

~GROWTH - [crystall] croissance *f* des cristaux

~HABIT - [crystall] (the external shape of a crystal) forme *f* d'un cristal

~ICE - [phys] (pure ice) glace *f* cristalline

~LATTICE - [phys] (the spatial arrangement of atoms or molecules in a crystal) réseau *m* cristallin

~LOUDSPEAKER - [radio] (with a diaphragm vibrated by a piezo-electric crystal) haut-parleur *m* piézoélectrique

~MALT - [brew ind] malt *m* caramelisé

~MICROPHONE - [radio] (the diaphragm vibrations are applied to a piezo-electric crystal to generate audio-frequency signals) microphone *m* piézoélectrique

~MIXER - [radio] (crystal receiver which can be at once fed from an oscillator and a signal source for frequency changing) mélangeur *m* à cristal

~OSCILLATOR - [radio] (type of oscillator in which a piezo-electric crystal is used for frequency-control) oscillateur *m* à cristal

~OVEN - [radio] (container used to stabilize the temperature and resonant frequency of a crystal) thermostat *m* pour cristal

~PARAMETER - [crystall] (the length of the cells of the lattice) paramètre *m* cristallin

~PULLING - [crystall] (method of crystal growing) étirage *m*

~RECEIVER - [radio] (waveguide incorporating a crystal detector) récepteur *m* à cristal

~RECTIFIER - s. crystal diode

~SEED - [crystall] germe *m* cristallin, germe *m* de cristallisation

~SET - [radio] récepteur *m* à cristal, récepteur *m* à galène

~SIZE - [sugar ind] grandeur *f* des cristaux

~SLAB - [crystall] (a thick cut across a mother crystal) lame *f* cristalline

~SPECTROGRAPH - [instr] spectrographe *m* à cristal

~SPECTROMETER - [instr] (measuring X-ray or gamma-ray wavelengths) radiodiffractomètre *m*

~SPOTS - [metall] (spots which are caused by the development of crystals of sulphide) taches *fpl*

~STRUCTURE - [phys] (geometric framework and arrangement of atoms and molecules) structure *f* d'un cristal

~SUGAR - [sugar ind] sucre *m* cristallisé

CRYSTAL SYSTEM - [crystall] (a classification of crystals based on the intercepts on the crystallographic axes by certain planes) système *m* cristallin
~TETRODE MIXER - [electron] (tetrode transistor) transistor *m* mélangeur
~VARNISH - [ind chem] (gum varnish for the protection of photographic images) vernis *m* transparent
~VIDEO RECTIFIER - [telev] (crystal rectifier which transforms a high-frequency signal into a video frequency signal) redresseur *m* vidéo à cristal
~WORK - [glass man] cristallerie *f*
CRYSTALLINE - [phys] cristallin *m*
[gen] (clear and transparent) cristallin *m*
~BASEMENT - [geol] soubassement *m* cristallin
~CONE - [zool] cône *m* cristallin
~FORM - [crystall] (the external-geometrical shape of a crystal) forme *f* cristalline
~LENS - [anat] cristallin *m*
~MATERIAL - [phys] (material in crystalline form) matériau *m* cristallin
~OVERGROWTH - [crystall] (the growth of a crystal round another) croissance *m* en macle
~SCHISTS - [geol] (groups of rocks resulted from heat and pressure) schistes *mpl* cristallins
CRYSTALLINITY - [phys] (the extent to which crystal structures exist in a substance) cristallinité *f*
CRYSTALLITES - [phys] (minute crystals of imperfect form) cristallites *fpl*, pseudo-cristaux *mpl*
CRYSTALLIZATION - [chem] (the formation of crystals from solutions) cristallisations *f*
CRYSTALLIZE, to - [phys] cristalliser
CRYSTALLIZED FERROUS SULPHATE - [chem] sulfate *m* ferreux cristallisé
~GREEN - [ind chem] vert *m* cristallisé
~SODA - [chem] cristaux *mpl* de soude
CRYSTALLOBLASTIC TEXTURE - [geol] (metamorphic rocks which recrystallize under pressure and high viscosity) structure *f* cristalloblastique
CRYSTALLOGENESIS - [crystall] cristallogénie *f*
CRYSTALLOGRAPH - [opt] (radiogram of the diffraction pattern produced in a crystal) cristallogramme *m*
CRYSTALLOGRAPHIC AXES - [crystall] (lines of reference intersecting at the centre of a crystal) axes *m* cristallographiques
~NOTATION - [min] notes *fpl* sur la face d'un cristal
CRYSTALLOGRAPHY - [phys] (study of the forms, properties and structure of crystals) cristallographie *f*
CRYSTALLOID - [chem] obsolete term denoting a substance which dissolves to form a true solution) cristalloïde *m*
CRYSTALLOLUMINESCENCE - [chem] (emission of light during crystallization) cristalloluminescence *f*
CRYSTALLOMETER - [instr] (determining the refractive index of crystals) réfractomètre *m* cristallographique
CRYSTALLOMETRY - [phys] cristallométrie *f*
CUBAGE - [gen] (measure of cubic content) cubage *m*
CUBAN EIGHT - [aero] (aerobatic manoeuvre consisting of a combination of two turns) boucle *f* double en forme de huit
CUBATURE - [geom] cubage *m*, cubature *f*
CUBE, to - [math] cuber, élever au cube
[constr] (road works) paver
CUBE - [geom] cube *m*
[constr](road works) petit pavé
~CUTTING MACHINE - [sugar ind] cassoir *m* à sucre
~ICE - [gen] glace *f* en cubes
~ORE - [min] pharmacosidérite *f*
~PRESS - [sugar ind] mouleuse *f* pour morceaux de sucre
~ROOT - [math] racine *f* cubique
~ROOT LAW - [phys] (scaling law relating to blast phenomena) loi *f* de la racine cubique
CUBIC - [geom] cubique
[math] cubique, du troisième degré
~CAPACITY - [mech] cylindrée *f*, volume *m*
~CONTENT - [gen] (the volume contained by a vessel) volume *m*, cubage *m*
~CURVE - [mech] (of motors) courbe *f* cubique
~EQUATION - [math] équation *f* du troisième degré
~FOOT - [meas] pieds *m* cube
~INCH - [meas] pouce *m* cube
~LATTICE - [phys] (the lattice structure of a crystal on the cubic system) réseau *m* cubique, assemblage *m* cubique
~MEASURE - [meas] mesure *f* de volume
~METRE - [meas] mètre *m* cube
~ROOT - s. cube root
~SPACE - s. cubature
~SYSTEM - [crystall] (the crystal system with the highest degree of symmetry) système *m* cubique
~VOLUME - [geom] volume *m*, cubage *m*
CUBICAL - [gen] cubique
~EPITHELIUM - [zool] épithélium *m* à cellules cubiques
~EXPANSION - [phys] dilatation *f* cubique
CUBICLE - [arch] compartiment *m*, cabine *f*, box *m*
[el] for switchboard) cabine *f*, armoire *f*
CUBILOT - s. cupola furnace
CUBING - [constr] (a rough method of assessing building costs) cubage *m*
CUBITUS - [anat] cubitus *m*
[zool] (a primary vein of an insect's wing) cubitus *m*
CUBOID - [geom] cuboïde *m*
CUCUMBER - [bot] concombre *m*
CUCURBIT - [ind chem] cucurbite *f* (d'un alambic), ventouse *f*
CUDBEAR - [dyes] (obtained from certain lichens) orseille *f*
CUDDY - [build] cabinet *m*, armoire *f*, placard *m*
[naut] tille *f*, cabine *f* arrière, rouf *m*
CUE - [impl] (billiard) queue *f*
[telev] signal *m* d'avertissement
[cin] (visual or sound indication for acting) signal *m* de prise de vue
~LIGHTS - [telev] lampes pl de signalisation
~SIGNAL - [cin] (signal by a buzzer or bell) signal *m* de prise de vue
CUFF - [med] manchon *m*
CUL-DE-SAC - [roads] cul-de-sac *m*
CULDOSCOPY - [med] culdoscopie *f*
CULICIFUGE- [med]culicifuge, anti-moustiques
CULL, to - [gen] cueiller, arracher
[min] trier, classer, choisir, cueillir
CULL - [ind] rejet *m*, déchet *m*, rebut *m*

[plast ind] (the material left in the transfer pot after filling the mould) excès *m* de matière
[zool] poule *f* non pondeuse
CULLET - [glass man] (waste glass used to improve the rate of melting. Also generally, a broken scrap glass) verre *m* cassé, verre *m* de rebut
CULLING - [min] triage *m*, classification *f*
CULLIS - s. coulisse
CULM - [bot] tige *f*
[min] (anthracite dust) poussière *f* d'anthracite
CULMEN - [zool] (the edge of the upper beak in birds) culmen *m*
CULMINATION - [astr] culmination *f*, passage *m* au méridien
CULTIVATE, to - [gen] cultiver
CULTIVATION - [gen & agric] culture *f*
~AREA - [agric] zone *f* cultivable
CULTIVATOR - [gen] cultivateur *m*, agriculteur *m*
[agric] cultivateur *m*
[agric] (rotary cultivator) fraise *f* agricole
CULTURE - [gen] culture *f*
[bot] culture *f*
[surv] (details of man-made features on a map) cultures *f*pl
~DISH - [med] plaquette *f* de culture
~MEDIUM - [agric] milieu *m* de culture
CULVERT - [constr] (construction for the free passage of water under a road etc) ponceau *m*, pont *m* dormant, passage *m* sous chaussée, aqueduc *m* souterrain, conduite *f* enterrée
~HEADER - [constr] couteau *m*
~LOCK - [constr] écluse *f* de passage *m* d'eau *f*, écluse *f* à tambour *m*
~STRETCHER - [constr] coin *m*
~SYPHON - [constr] ponceau *m*, siphon *m*, passage *m* d'eau *f*
CUMENE - [chem] (a solvent and an aviation fuel additive) cumène *m*
~HYDROPEROXIDE - [chem] (polymerization catalyst) hydroperoxyde *m* de cumène
CUMIN OIL - [ind chem] (essential oil used in perfumery, medicine etc) essence *f* de cumin
CUMULATIVE - [gen] cumulatif
~COMPOUND EXCITATION - [el] (compound excitation whose two windings supply ampere-turns in the same direction) excitation *f* composée cumulative
~DOSE - [radiat] (the total dose which results from repeated exposure) dose *f* cumulée
~EXCITATION - [phys] (the process raising an atom from one excited state to higher states) excitation *f* cumulative
~GRID RECTIFIER - [radio] (rectifier in which the signal voltage is applied to the grid of a thermionic valve) redresseur *m* à grille
~IONIZATION - [phys] (ionisation caused in a gas or vapour by an electron colliding with a series of atoms or molecules and freing other electrons which continue the process) ionisation *f* cumulative
CUMULIFORM - [met] (with the general shape of a cumulus cloud) cumuliforme
CUMULONIMBUS - [met] (extensive convection cloud supporting an anvil of white cirrus which may reach the tropopause) cumulonimbus *m*
CUMULUS - [met] (vertically developing detached cloud in the form of rising mounds) cumulus *m*
CUMYL PHENOL - [chem] (an intermediate for insecticides and synthetic resins) cumylphénol *m*
CUNEATE - [bot] cunéaire
CUP, to - [med] appliquer des ventouses
CUP - [gen] tasse *f*, coupe *f*, coupelle *f*, cupule *f*, cuvette *f*
[mech] (a cone shaped ring) cuvette *f*, godet *m*
[instr] cuvette *f*
[el] (of an insulator) cloche *f*
[mech] (of ball bearing) alvéole *f*
[ind chem] (of a straying device) réservoir *m*
[bot] (an apothecium) calice *m*, cupule *f*
[med] ventouse *f*
[gas ind] (of gas holders) gorge *f*
[plumb] (of pipes) collet *m* battu
~AND BALL - [mech] (universal join of the ball-and-socket type) joint *m* sphérique
~-AND-BALL STRUCTURE - [metall] structure *f* en bilboquet
~AND CONE FRACTURE - [metall] cassure *f* en cône et coupe
~ANEMOMETER - [instr] (anemometer whose rotating element is moved by hemispherical cups) anémomètre *m* à demi-sphères
~CARRIAGE - [gas ind] (in gas holders) guidage *m* de la partie inférieure des levées
~CHANNEL - [gas ind] (in gas holders) fer *m* en U de gorge
~CHUCK - [mech] (lathe chuck in the form of a cup) mandrin *m* à cuvette
~-FEED DRILL - [agric] semoir *m* à cuillères
~FLOW - [plast ind] (a British Standard Institute test) fluidité *f* au gobelet
~FLOW FIGURE - [plast ind] (the time taken by a mould to close completely) indice *m* de fluidité au gobelet ,
~FLOW TEST - [plast ind] essai *m* de fluidité au gobelet, test *m* de fluidité (ou de plasticité)
~GREASE - [ind chem] graisse *f* consistante
[mech] (for use in cup lubricators) graisse *f* pour boisseau
~-HEAD - [mech] à tête ronde
~-HEADED BOLT - [mech] boulon *m* à tête bombée
~JOINT - [plumb] (joint by made opening out one of the ends which will receive the tapered end of the other) joint *m* à emboîtement
~LEATHER - [mech] (ring of leather used in hydraulic machinery) cuir *m* embouti
~LEATHER PACKING - s. cup leather
~LUBRICATOR - [impl] (cup-shaped device on a hollow stem used to force grease into a bearing) godet *m* graisseur, graisseur *m*, boisseau *m*
~REST - [gas holders] repos *m* de gorge
~SHAKE- [timber] (a crack between concentric layers) crevasse *f* annulaire
~-SHAPED BASE - [electron] (for the coil of a cathode-ray tube) support *m* en cuvette
~SKIRTING PLATE - [gas ind] (in gas holders) tôle *f* de gorge
~SPRING - [mech] ressort *m* Belleville, ressort *m* à rondelle
~TEST - [metall] épreuve *f* de ductilité
~WINDING MACHINE - [text] cannetière *f* à godets
CUPEL - [metall] (small cylinder of compressed bone-ash on which metal specimens are placed in cupellation) coupelle *f*
~FURNACE - [metall] (special tupe of furnace used for cupellation) four *m* de coupellation
CUPELLATION - [metall] (method for purifying noble

metals by treating them with high temperature in the presence of oxygen; the metals forming the impurities are converted to oxides and absorbed by the cupel) coupellation *f*
CUPOLA - [arch] coupole *f*, dôme *m*
[metall] s. cupola furnace
[railw] poste *m* d'observation
~CONTROL EQUIPMENT - [metall] régulateur *m* de débit de vent
~FURNACE - [metall] (shaft furnace used for melting pig-iron to make castings charged from the top and blown from the bottom) cubilot *m*, four *m* à cubilot
~HAND - [metall] cubilotier *m*
~SHELL - [metall] (of cupola furnace) enveloppe *f* de cubilot
~TORCH - [metall] allumeur *m* du cubilot
~VAULT - [arch] coupole *f*, voûte *f* sphérique
~WITH SQUARE SHAFT - [metall] (of cupola furnace) cubilot *m* à cuve carrée
CUPPING - [metall] (in a drawn wire) rupture *f* transversale interne
[ind] emboutissage *m*
[gas ind] accrochage *m*
~-GLASS - [med] ventouse *f*
~MACHINE - [mech mach] machine *f* à emboutir
CUPRAMMONIA - [chem] (cellulose solvent made by the addition of ammonium chloride followed by an excess of caustic soda to a solution of a copper salt) cupro-ammoniaque *f*
CUPRAMMONIUM PROCESS - [text] (method of producing rayon by treating cellulose fibres with an ammoniacal solution of cupric oxide) procédé *m* cupro-ammoniacal
~RAYON - [text] (rayon produced by the cuprammonium process) rayonne *f* cupro-ammoniacale
CUPRATE SILK - [text] soie *f* au cuivre
CUPRIC - [chem] (containing divalent copper) cuivrique
~ACETATE - [chem] acétate *m* de cuivre
CUPRIFEROUS - [min] cuprifère
~PYRITE - s. chalcopyrite
CUPRITE - [min] (natural cupric oxide valuable as an ore of copper) cuprite *f*
CUPRO-ALUMINIUM - [metall] (alloy of copper and aluminium) cupro-aluminium *m*
CUPROLEAD - [metall] (alloy of copper and lead) alliage *m* cuivre-plomb
CUPROSE - [bot] pavot *m*
CUPROURANITE - [min] (also called torbenite, a mineral containing copper and uranium) cuprouranite *f*, torbénite *f*
CUPROUS - [chem] (term for monovalent copper compounds) cuivreux
~VAT - [ind chem] cuve *f* de vitriol
CUPROZIPPEITE - [min] (ore containing uranium) cuprozippéite *f*
CUPULA - [med] cupule *f*, capsule *f*
CUPULAR - [bot] cupulaire
CURAGE - [med] curage *m*, curettage *m*, curetage *m*
CURARE - [chem] (South American Indian arrow poison extracted from various species of Strychnos) curare *m*
CURARINE - [chem] (toxic paralysing alkaloid extracted from curare) curarine *f*
CURARIZATION - [chem] curarisation *f*
CURATIVE - [gen] curatif, thérapeutique
[plast ind] (a resin polymerizing agent) agent *m* de polymérisation
[rubber ind] agent *m* de vulcanisation
CURB, to - [gen] freiner, modérer, restreindre
CURB - [build] (wall plate) corniche *f*, bordure *f*
[constr] (hollow timber cylinder used to line wells in well sinking) cylindre *m* creux
[agric] (type of horse bit) mors *m* à gourmette
~BOX - [gas ind] (a surface box, or vault) bouche *f* à clé
~COCK - [gas ind] (a service valve) robinet *m* sous trottoir
~PINS - [horol] goupilles *fpl* de raquette
~-PLATE - [constr] panne *f* de brisis
~ROOF - [constr] toit *m* mansardé, comble *m* brisé
~SHUT-OFF - s. curb cock
~STONE - [constr] bordure *f* de trottoir
CURBED CHAIN - [mech] chaîne *f* torse
CURBING - [mining] boisage *m* du puits à cadres jointifs
CURCUMIN - [chem] (acid-base indicator and a yellow dye) jaune *m* de curcuma
CURD - [agric] caillé *m*, lait *m* caillé, précipité *m* caillebotté
CURDLE - [chem] (of the milk) (se) figer, (se) couaguler, floculer
CURDLING - [gen] coagulation *f*, floculation *f*
[paint] (the thickening of the paint) épaississement *m*
[agric](dairy ind) caillement *m*
CURDY - [gen] caillebotté
CURE, to - [gen] guérir, soigner
[chem] durcir, prendre
[rubber ind] vulcaniser
[leather ind] sécher
CURE - [gen] cure *f*, guérison *f*
[chem] -term applied to a fermentation agent) vieillissement *m*
[plast ind] (the chemical process occurring while a hot thermoset changes isothermically to the solid condition) durcissement *m*
[rubber ind] vulcanisation *f*
[leather ind] séchage *m*
[constr] (process to eliminate or reduce cracks in cement) cure *f*, séchage *m*
~TIMBER - [rubber ind] (mechanical device controlling the vulcanization time) régulateur *m* de vulcanisation
CURED MALT - [brew ind] malt *m* séché
CURETTAGE - [med] (the scraping of the walls of cavities) curetage *m*, curettage *m*
CURETTE - [med] curette *f*
CURIE - [phys] (the unit of radioactivity defined as 3.7×10^{10} disintegrations per second) curie *m*
~POINT - [phys] (critical temperature, above which a ferromagnetic body is paramagnetic) point *m* de Curie
~TEMPERATURE - s. curie point
CURINE - [chem] (an alkaloid of the quinoline group) curine *f*
CURING CYCLE - [ind chem] (cyclic period of curing) cycle *m* de durcissement
~OVEN - [rubber ind] (heating chamber for rubber vulcanization) étuve *f* de vulcanisation
~PRESS - [rubber ind] (rubber vulcanization press) presse *f* de vulcanisation
~RANGE - [rubber ind] (the temperature range over

which vulcanization occurs) intervalle *m* de vulcanisation
CURING RATE - [rubber ind] (the degree of rubber curing in relation to time) vitesse *f* de vulcanisation
~TEMPERATURE - [ind chem] (the temperature at which curing occurs) température *f* de durcissement
~TIME - [plast ind] (in moulding thermosets) temps *m* de cuisson
CURITE - [min] (a mineral containing lead and uranium) curite *f*
CURIUM - [chem] (synthetic element, symbol Cm; a silvery metal) curium *m*
CURL, to - [gen] courber, plier, friser, recourber, boucler
CURL - [gen] boucle *f*, courbure *f*
[paper man] enroulement *m*
[el] (of a vector) rotation *f*, rotationnel *m*
~FIELD - [el] (a vector field in which the curl is not everywhere zero) champ *m* rotationnel, champ *m* tourbillonnaire
CURLINESS - [text] (of wool) frisure *f*
CURLING - [gen] frisure *f*
[metall] bord *m* roulé
~DIE - [tool] outil *m* à rouler, outil *m* à border
~MACHINE - [metall] machine *f* à border
CURRANT - [bot] groseille *f*
CURRENCY - [gen] cours *m*, circulation *f*
[fin] monnaie *f*, numéraire *m*
CURRENT - [gen] courant *m*
[el] (the passage of electricity through a medium or along a circuit) courant *m*
~ADJUSTING RESISTOR - [el] résistance *f* chutrice
~AMPLIFICATION - [radio] amplification *f* en courant
~AMPLIFICATION FACTOR - [el] facteur *m* d'amplification en courant
~AT THE TERMINALS - [el] courant *m* aux bornes
~ATTENUATION - [radio] (of a transducer) atténuation *f* du courant
~BALANCE - [instr] (instrument measuring the current by balancing the mechanical force between two conductors) balance *f* de courant, électrodynamomètre *m*
~BEDDING - [geol] stratification *f* torrentielle, stratification *f* oblique
~BRANCHING - [el] dérivation *f*
~CARRYING - [of a cable] sous tension
~CARRYING CAPACITY - [el] (the current a cable can carry) courant *m* de régime, charge *f* de régime
~CHANGER - [el] convertisseur *m* statique
~CIRCUIT - [el] (in measuring instruments) circuit *m* de courant, circuit *m* série
~COIL- [el] (a coil in a measuring instrument which is connected in series with the main supply) enroulement *m* série
~COLLECTING RAIL - [railw] (live rail) rail *m* conducteur
~COLLECTOR - [el] (device for passing electric current to a mobile mechanism) collecteur *m*
~CONSUMPTION - [el] consommation *f* de courant
~CONTROL - [el] régulation *f* d'intensité
~DAMPER - [el] (the current flowing per unit cross-sectional area of a conductor) densité *f* de courant
~DISTRIBUTING PLANT - [el] installation *f* de distribution
CURRENT DISTRIBUTOR - [auto] distributeur *m* d'allumage
~DRAINAGE - [el] (to check leaks and corrosion in gas pipes) drainage *m* cathodique, drainage *m* polarisé
~EFFICIENCY - [electron] (relation between the quantity of matter actually produced in electrolysis and the amount calculated in theory) rendement *m* électrochimique
~FEEDBACK - [el] (feedback which is proportional to that flowing in the output circuit) réaction *f* d'intensité
~FENDER - [hydr] (protective work to deflect a current of water from a bank etc) épi *m*
~FIELD - [el] champ *m* d'un courant
~FLOW - [el] flux *m* de courant
~GAIN - [el] (ratio between output and input current in an amplifying circuit) gain *m* en courant
~GALVANOMETER - [instr] galvanomètre *m*
~GRADIENT - [el] (the rate of change in current) gradient *m* de courant
~HARMONICS - [electron] (component of an alternating wave with a frequency which is a multiple of the fundamental one) harmoniques *mpl* de courant
~LIMITER - [el] (relay used to give a warning signal when the current in a circuit reaches a certain value) limiteur *m* de courant
~LOOP - [el] ventre *m* d'intensité
~METER - [el] ampèremètre *m*
[hydr] (instrument measuring the velocity of flow of running water) moulinet *m* hydrométrique
~NETWORK - [el] réseau *m* de courant
~NOISE - [el] (the noise in any conductor caused by the current passing through it) bruit *m*
~OF AIR - [gen] courant *m* d'air
~OF CHARGE - [el] courant *m* de charge
~OF INJURY - s. injury potential
~PATH - [el] trajet *m* du courant
~PEAK - [el] pointe *f* de courant
~RATING - [el] courant *m* de régime
~RATIO - [el] (of a transformer) rapport *m* de transformation
~REGULATOR TUBE - [electron] (a ballast resistor used in receivers) tube *m* régulateur d'intensité
~RELAY - [el] relais *m* d'intensité, relais *m* de courant
~RESONANCE - [el] (the condition in which the positive reactance of a current in a circuit is balanced by the negative reactance) résonance *f* du courant
~REVERSER - [el] commutateur *m* inverseur
~-REVERSING KEY - [el] clé *f* d'inversion de courant
~SENSITIVITY OF A CRYSTAL RECTIFIER - [radio] sensibilité *f* en courant d'un redresseur à cristal
~STRENGTH - [el] intensité *f* de courant
~SUPPLY - [el] alimentation *f* en courant
~TAP - [el] douille *f* voleuse
~TEST - [el] essai *m* de courant
~TRANSFORMER - [el] (transformer in which the secondary current is proportional to the primary current) transformateur *m* d'intensité, transformateur *m* de courant
~TRIANGLE - [el] triangle *m*
~VARIATION - [el] variation *f* d'intensité
~-VOLTAGE CHARACTERISTIC - [el] caractéristique

f intensité-tension
CURRENT WAVELENGTH CHARACTERISTIC - [phys] courbe *f* de sensibilité spectrale
~WEIGHER - [instr] (a current balance in which one coil is movable and suspended from an arm of the balance) balance *f* de courant, balance *f* de Rayleigh
CURRIER - [leather ind] corroyeur *m*
CURRY, to - [leather ind] (the operation of treating leather with greases to make it flexible) corroyer
[agric] (currycombing) étriller
CURRYCOMB - [impl] étrille *f*
CURRYING - [leather ind] corroyage *m*
CURSOR - [instr] (the adjustable part of a drawing instrument) curseur *m*
[el] curseur *m*, contact *m* glissant, index *m*
CURTAIL, to - [gen] réduire, raccourcir, diminuer restreindre
CURTAILMENT - [gen] diminution *f*, réduction *f*
CURTAIN - [gen] rideau *m*, tenture *f*
[auto] (protective covering in front of the radiator) rideau *m*
[nucl] (a thin, generally cadmium screen used to shut off a slow electron flow) écran *m*
~AERIAL ARRAY - [radio] (directional array in which the elements are arranged in two parallel planes) rideau *m* directionnel
~ANTENNA ARRAY - s. curtain aerial array
~ARCH - [metall] (in retorts) autel *m*
~RING - [impl] anneau *m* de rideau
~ROD - [impl] tringle *f* à rideau
~WALL - [build] mur-rideau *m*
CURTAINING - [paint] (the formation of wrinkles also called crawling) coulure *f* en forme de festons
[geol] ridage *m*
CURTATE - [gen] raccourci
[comput] (of a punched card) ligne *f* horizontale
~CYCLOID - [geom] cycloïde *f* raccourcie
CURVATURE - [gen] courbure *f*
[math] (the rate of change of direction of a curve) courbure *f*
~CORRECTION - [surv] correction *f* de courbure
~ERROR - [instr] (an error which is caused by the curvature of the magnetizing curve) erreur *f* de courbure
~GRADIENT - [surv] (the rate of difference in elevation) gradient *m* de courbure
~OF SPECTRUM LINES - [phys] (in a prism produced spectrum the lines are convex towards the red end) courbure *f* des raies spectrales
~OF SURFACE - [mech] (the reciprocal of the radius of curvature) courbure *f* des surfaces
~OF THE FIELD - [telev] (the curved reproduction of a flat object) courbure *f* du champ
~OF TYRES - [rubber ind] courbure *f* d'un pneumatique
CURVE, to - [gen] courber, plier, cintrer, infléchir
CURVE - [gen math etc] courbe *f*, courbure *f*
[impl] curvigraphe *m*
[a bend] courbe *f*, virage *m*
[geom] courbe *f*
~CHART - [gen & statist] diagramme *m*
~IN SPACE - [geom] courbe *f* dans l'espace
~IN THE TRACK - [railw] courbe *f* de la voie
~OF LIGHT DISTRIBUTION - [phys] (a graph showing the relation between the luminous intensity of a light source and the angle of emission) courbe *f* de distribution de la lumière
CURVE PLOTTING - [math etx] tracé *m* des courbes
CURVED AUGER - [ind chem] sonde *f* coudée
~BOTTOM - [gas ind] (in gas holders) fond *m* bombé
~CORE - [metall] noyau *m* coudé
~COURSE - [text] chemin *m* incurvé, chemin *m* de came
~FIELD - [opt] champ *m* courbé
~HEAD - [gen & mech] fond *m* sphérique
~LINE SOURCE - [phys] (source consisting of a large number of point sources of equal strength, vibrating in phase on the arc of a circle) source *f* curviligne
~NEEDLE - [impl] (used in upholstery work) aiguille *f* courbe
~PLATE - [print] plaque *f* courbe pour stéréotypie
~RIM - [auto] jante *f* déformée
~RUNNER PLATE - [cin] (of a film projector) couloir *m* incurvé
~SHUTTLE RACE - [text] battant *m* à cintre
~SURFACE SOURCE - [phys] (portion of a spherical source) source *f* à surface sphérique
~TANK - [gas ind] (in gas holders) cuve *f* bombée
~TRIANGLE - [geom] triangle *m* sphérique
~WELL - [oil ind] sondage *m* infléchi
CURVILINEAR - [gen] curviligne
~CONE - [geom] cône *m* parabolique
~DISTORTION - [photo] (the lack of linearity in the object) distortion *f* curviligne
CURVING - [gen etc] courbure *f*
CURVOMETER - [instr] (to measure accurately the length of curves) curvimètre *m*
CUSH-CUSH - [bot] colocase *f*
CUSHION, to - [mech] amortir, soutenir
[acoust] (using resilient materials isolating microphones etc) isoler
CUSHION - [gen] coussin *m*, coussinet *m*, amortisseur *m*
[constr] (e.g. the capping stone of a pier) coussinet *m*
~COURSE - [constr] (a layer of sand and cement or mortar over the foundations of a road) couche *f* de fondation
~GAS - [gas ind] gaz *m* coussin
~GEAR - [mech etc] (a shock absorber) amortisseur *m*
~SPACE - [phys] tampon *m* d'air
~TYRE - [auto] bandage *m* semi-pneumatique
~UNDERFRAME - [railw] châssis *m* à suspension élastique
CUSHIONED - [mech] amorti
~BLASTING - [mining] tir *m* avec chambre d'expansion
~MOVEMENT - [mech] mouvement *m* élastique
~SOCKET - [radio] support *m* anti-vibratoire
CUSHIONING - [acoust] (the use of elastic material for the isolation of sensitive apparatuses) garniture *f* anti-vibration, rembourrage *m*
~ACTION - [mech] (the action of a part or a substance reducing the shock effect of a movement) action *f* d'amortissement, amortissement *m*
~EFFECT - [mech] (the effect of an elastic or resilient machine element in reducing the shock caused by the change of motion in another part) effet *m* amortisseur
~MAT - [transp] (used in unloading) tapis *m* amortisseur
~SPRING - [mech] (spring modifying the shock action of a moving component) ressort *m* amortis-

seur
CUSP - [arch] (sharp pointed prominence) lobe *m*
[geom] point *m* de rebroussement, sommet *m* d'une courbe
[geol] cap *m*
CUSPIDAL - [gen] cuspidé
CUSTOM - [gen] coûtume *f*, usage *m*, habitude *f*, moeurs *m*
[comm] clientèle *f*, chalandage *m*
~BODY - [auto] carrosserie *f* réalisée sur commande
~BUILT - [gen] réalisé sur commande
~MIXING - [rubber ind] (mixing of rubber compound on the basis of specified specifications) mélange *m* effectué sur commande
CUSTOMER - [gen & comm] client *m*, usager *m*, abonné *m*
CUSTOMS - [comm & leg] douane *f*
~BARRIER - [comm & leg] barrière *f* douanière
~DECLARATION - [comm] déclaration *f* en duane
~DUE - [leg] droit *m* de douane
~DUTY - [leg] droit *m* de douane
~ENTRY - s. customs declaration
~EXAMINATION - [comm] visite *f* de douane
~HOUSE - [gen] douane *f*
~OFFICIER - [gen] douanier *m*, officier *m* des douanes
~TARIFF - [comm] tarif *m* douanier
~UNION - [comm] union *f* douanière
~WAREHOUSE - [comm] entrepôt *m* de douane
CUT, to - [gen] couper, fendre, graver, trancher, inciser, rogner
[cin] (the editing of a film) faire le montage
[el acoust] (of a record) enregistrer
[ind chem] fractionner
CUT - [gen] coupe *f*, taille *f*, fente *f*, incision *f*
[constr] (the material which is removed in making a cutting) déblai *m*, tranchée *f*
[cin] (sudden change from one source of vision to another) changement *m* brusque de séquence
[cin] (the signal for a camera to stop) "coupez"
[cin] (the editing of film) montage *m*
[mech] coupe *f*, passe *f*
[mech] (the result of a cut on a machine) passe *f*
[metall] (thickness of metal cuttings) épaisseur *f* des copeaux
[oil ind] impureté (en % sur la production)
[mining] fouillet *f*, déblai *m*
[text] (length of cloth; also lenght of a linen lea, i.e. 300 yards) coupe *f*
~A SCREW, to - [mech] fileter une vis
~A THREAD, to - [mech] fileter
~ACROSS, to - [gen] couper en travers
~-AND-COVER - [constr work] (a method of recovering cuttings in the construction of railway just under the ground) déblai et recouvrement *m*
~-AND-FILL - [constr] (describing road or railway construction work section, made partly in cutting and partly in filling) déblai-remblai *m*
~-AND-TRY - [mech] vérification *f* point par point
~-AWAY - [draw] en coupe transversal
~-AWAY MODEL - [mech] modèle *m* en coupe
~-AWAY VIEW - [draw] section *f* transversale, vue *f* en coupe
~-BACK - [gen] réduction *f*, diminution *f*
~BACKS - [constr] (blends of asphaltic bitumen with solvents) brai *m* de pétrole fluxé

CUT BY MILLING, to - [mech] fraiser
~BY WELDING TORCH - [mech] coupe *f* au chalumeau
~DOWN, to - [gen] réduire
~EDGE - [book binding] bord *m* rogné
~EDGE OF CLOTH - [text] bord *m* coupé du tissu
~FOR SCALLOPING PATTERNS - [text] machine *f* à couper en festons
~GEAR - [mech] engrenage *m* taillé à la fraise
~GROWTH - [mech etc] (the extension of a cut beyond the point at which it originally ended) agrandissement *m* de la déchirure
~IN, to - [el] mettre en circuit, connecter, se fermer
~-IN PRESSURE - [mech] pression *f* d'amorçage, pression d'enclenchement, pression de mise en marche
~-IN SCENE - [cin] (scene inserted into a film) raccord *m*
~LOOKER - [text] contrôleur *m* des pièces
~LOOP - [text] noeud *m*, boucle *f*
~MARK - [text] marque *f* de pièce
~MARKER - [text] marqueur *m* de pièces
~OFF, to - [gen] couper, interrompre, découper, déconnecter
~OFF - [mech] (condition of the stroke when the steam admission to an engine cylinder is discontinued) fermeture *f* de l'admission
[electron] (condition in a thermionic valve when it becomes non-conductive) blocage *m*
[paper man] rognage *m*
[metall] coupage *m*, tronçonnage *m*
[plast ind] (the land-area part sealing off the moulding) couteau *m*
~-OFF ATTENUATOR - [radio] (used to introduce non-dissipative attenuation) atténuateur *m* réactif
~-OFF BIAS - [electron] (electronic tube grid bias voltage which reduces the anode current to zero) polarisation *f* de coupure
[electron] (in a cathode-ray tube) tension *f* de suppression de faisceau
~-OFF COCK - [hydr] robinet *m* d'arrêt
~-OFF DISTANCE - [nucl] rayons *m* de l'effet d'écran
~-OFF FREQUENCY - [radio] (also called barrier frequency; the frequency below which a radio wave cannot enter a layer at the angle of incidence required for the transmission) fréquence *f* de coupure
~-OFF GRID VOLTAGE - [electron] (the negative voltage which, when applied to the control grid of a thermionic valve, reduces the electron flow to zero) tension *f* de coupure
~-OFF INPUT - [nucl] puissance *f* de coupure
~-OFF JACK - [el] prise *f* de renvoi
~-OFF LEVER - [mech] levier *m* d'arrêt
~-OFF PARABOLE - [electron] parabole *f* de coupure
~-OFF ROD - [nucl] barre *f* de sécurité
~-OFF VOLTAGE - [mech] (of steam engine) tiroir *m* de détente
~-OFF VOLTAGE - s. cut-off grid voltage
~-OFF WAVEGUIDE - [radio] (used at a frequency which is below its critical frequency) guide *m* d'ondes utilisé en dessous de la fréquence critique
~-OFF WAVELENGTH - [radio] longueur *f* d'onde critique
~OIL - [oil ind] (US only) émulsion *f* d'eau et de

pétrole
CUT OUT, to - [gen] échancrer, couper
[mech] (to remove by cutting) découper
[mech] (to stop a machine) arrêter, stopper
[el] mettre hors circuit
~-OUT - [el] (a device for interrupting a circuit automatically under pre-set conditions) coupe-circuit *m*, disjoncteur *m*, fusible *m*, interrupteur *m*
[cin] (discarded sections of a film) chute *f*
[mech] (a valve in the exhaust pipe) vanne *f* de décompression, clapet *m* d'échappement
[gen] ouverture *f*
[print] coupure *f*
~-OUT BOARD - [el] tableau *m* d'interrupteurs
~-OUT HOLE - [mech] trou *m* découpé
~-OUT SWITCH - [el] disjoncteur *m*, interrupteur *m*
~OVER - [acoust] (the cutter on a disc recorder jumping from one groove to another) défaut *m* de gravure
~PARABOLOID REFLECTOR - [radar] (reflector with approximately even sides) réflecteur *m* parabolique asymétrique
~PILE CARPET - [text] tapis *m* de Tournai
~PIPE - [plumb] tube *m* court
~PLUSH - [text] peluche *f* coupée
~PRESSER - [text] roue de presse *f*
~SCREWS, to - [mech] fileter
~STONE - [constr] pierre *f* de taille
~TEMPLE ROLLER - [text] temple *m* à rouleaux
~TO LENGTH, to - [gen] couper à longueur
~VELVET - [text] velours *m* coupé
~-WATER - [constr] (the angular edge of a bridge-pier) bec *m* de pile
~WEFT THREADS - [text] fils *mpl* de trame coupés
~-WIRE - [mech] (wire cut small cylinder) tronçon *m* de fil d'acier
CUTANEOUS - [zool] (of the skin) cutané
CUTICLE - [anat] cuticule *f*, épiderme *m*
CUTIE PIE - [nucl] (colloq. for portable ionization chamber) chambre *f* d'ionisation portative
CUTLASS - [shoe man] clou *m* à chaussures, caboche *f*
CUTTER - [mech] (a rotary cutting tool) fraise *f*, molette *f*, rouleau *m*
[el acoust] (the point of diamond or sapphire, cutting the groove in a record) graveur *m*
[cin] monteur *m*
[naut] cotre *m*, canot *m*
[tool] couteau *m*
[mech] (a guillotine for paper) massicot *m*, coupe-papier *m*
~ARBOR - [mech] mandrin *m* de serrage
~-BAR - [mech] (bar fitted with a cutting tool) porte-outil *m*
~BLOCK - [mech] porte-outil *m*
~-HEAD - [mech] porte-fraise *m*, porte-outil
~-LOADER - [mining] haveuse-chargeuse *f*
~SWITCH - [el] relais-disjoncteur *m*
~WIRE - [text] (wire loom) fer *m* tranchant (métier velours)
~WITH INSERTED TEETH - [mech] fraise *f* à dents rapportées
CUTTING - [gen] coupe *f*, taille *f*, découpage *m*, gravure *f*
[cin] (the operation of editing) montage *m*
[ind chem] séparation *f*
[constr] tranchée *f*
[text] (wool) tonte *f* (du drap), ciselage *m* (du velours)
[metall] rognure *f*, copeau *m*
[wire] tronçonnage *m*
[carp] copeau *m*
[glass man] (diamond scoring and breaking) taille *f*
[comm] (reduction of expenses) réduction *f* des dépenses
CUTTING ABRASION - [rubber ind] (in tyres) abrasion *f* par incisions
~ACTION - [mech] cisaillement *m*, coupe *f*, taille *f*
~AND EDITING - [cin] découpage *m* et montage d'un film
~ANGLE - [of tools & implements] angle *m* de coupe
~BLOWPIPE - [metall] (in welding) chalumeau *m* coupeur
~CHISEL - [impl] ciseau *m* étroit, fermoir *m*, ébauchoir *m*
~COMPOUND - [ind chem] (special fluid usually an emulsion of a cutting oil applied to the work and cutter in machining) liquide *m* de refroidissement
~DOWN - [el-plating] (polishing to remove roughness in a plated surface) polissage *m*
~EDGE - [gen & tools] tranchant *m*, fil *m*, taillant *m*
~EDGE SEAL - [metall] joint *m* à couteau, jonction *f* à pénétration
~FLUID - s. cutting compound
~-GAUGE - [carp] troussequin *m*
~IN - [gen] insertion *f*
[el] mise *f* en circuit
~MACHINE - [mech mach] machine *f* à découper
~MOTION FOR LOOP WARP - [text] mouvement *m* pour couper
~MOTION FOR SELVEDGES - [text] mouvement *m* pour couper les lisières
~OIL - [mech] huile *f* de coupe
~-OUT STOPE - [mining] gradin *m* inférieur (dans l'exploitation par gradins renversés)
~PLANE - [tool] (smoothing plane) rabot *m* à polir
~POINT - [gen & tools] pastille *f*, tranchant *m*
~POWER - [mech] capacité *f* de coupe, puissance *f* de coupe
~RAIL - [text] traverse *f* de coupe
~RATE - [mech] vitesse *f* de coupe, vitesse *f* d'avance
~ROOM - [cin] atelier *m* de montage, salle *f* de montage
~SPEED - s. cutting rate
~SPINDLE FOR THE PILE THREAD - [text] broche *f* à couper le poil
~STEEL - [metall] acier *m* pour outils de découpage
~STRAIN - [metall] contrainte *f* de coupe
~STYLUS - [acoust] (recording stylus which cuts a groove into the recording medium) graveur *m*, burin *m* de gravure
~TABLE - [text] table *f* à couper
~TEST - [ind] essai *m* de coupage
~THE PILE LOOPS - [text] coupe *f* de boucles du poil
~THE PILE TUBES - [text] coupe *f* des tubes de poil
~TIP - [metall] (in welding) bec *m* de chalumeau
~TOOLS - [mech] (used for the machining of metals) outils *mpl* tranchants, outils *mpl* de découpage
~TORCH - s. cutting blowpipe
CUTTINGS - [mining] débris *mpl* de forage
CUTTLER - [text] plieur *m*, plieuse *f* au large
CUTTLING FRAME - [text] dispositif *m* pour dépôt en

plis
CYANAMIDE - [chem] (common name for calcium cyanamide) cyanamide *f*
~PROCESS - [chem] (nitrogen fixation; formation of calcium cyanamide from nitrogen and calcium carbide) procédé *m* de préparation de la cyanamide
CYANEPHIDROSIS - [med] cyanidrose *f*, sueur *f* bleue
CYANHEMOGLOBIN - [med] cyanchémoglobine *f*
CYANIDING - [min] (the extraction of gold and silver from finely-ground ore with a weak solution of sodium cyanide, from which the metal is afterwards precipitated with zinc) cyanuration *f*
CYANINES - [chem] (a group of blue dyes, strongly basic and used in photography as sensitizers) cyanines *f*pl
CYANITE - [chem] (a naturally occurring aluminium silicate used as a refractory material) cyanite *f*, disthène *m*
CYANOACETAMIDE - [chem] (a synthesis intermediate for plastics) cyanoacétamide *m*
CYANOETHYL ACRYLATE - [chem] (an intermediate for copolymers with uses in textile finishing and as adhesives) acrylate *m* de cyanoéthyle
CYANOETHYLATED COTTON - [ind chem] (cotton which has been treated with acrylonitrile to give it increased resistance to mildew, bacteria and heat, and increased receptivity towards dyestuffs) coton *m* cyanoéthylé
CYANOETHYLATION - [chem] (the formation of a cyanoethyl group in an organic compound by reaction of acrylonitrile with a hydroxyl group) introduction *f* du groupe [$-OCH_2-CH_2CN$] dans une molécule organique, cyanoéthilation *f*
CYANOGEN - [chem] (poisonous, colourless gas used as a synthesis intermediate) cyanogène *m*
~BROMIDE - [chem] (poisonous, corrosive, colourless crystals used in ore extraction; as a fumigant and in organic synthesis) bromure *m* de cyanogène
~CHLORIDE - [chem] (colourless, poisonous liquid used in organic synthesis) chlorure *m* de cyanogène
CYANOMETER - [instr] (instrument designed to measure the intensity of blueness) cyanomètre *m*
CYANOPATHY - [med] cyanopathie *f*
CYANOPIA - [med] perversion *f* de la vision qui rend tout bleu
CYANOSIS - [med] cyanose *f*, ictère *m* violet
CYANURIC ACID - [chem] (a synthesis intermediate in the production of dyestuffs, explosives and pharmaceutiicals) acide *m* cyanurique
CYBERNETICS - [autom] (the comparison of the human nerve system and electronic machines) cybernétique *f*
CYBOTACTIC GROUP - [phys] (of the structure of long chain molecule liquids) groupe *m* cybotactique
CYCLAMEN - [bot] cyclamen *m*
CYCLANES - [chem] (term for the cycloparaffins polymethylenes) . (These are hydrocarbons containing saturated carbon rings) cyclanes *m*pl
CYCLE, to [ind chem] (to pass a substance through a repetitive series of processes) cycler, recycler
CYCLE - [gen] cycle *m*, période *f*
[ind chem] (a repetitive series of processes) cycle *m*
CYCLE COUNT - [comput] comptage *m* de cycles
~COUNTER - [comput] (mechanism for the count of cycles) compteur *m* de cycles
~OF EROSION - [geol](development in the evolution of land) cycle *m* d'érosion
~RATE COUNTER - [radar] compteur *m* de cycles
~RESET - [comput] (the return of a cycle index to its initial value) remise *f* à zéro
~SHIFT - s. cyclic shift
~TRACK - [constr] piste *f* cyclable
CYCLES PER SECOND - [C/S or cps] (the number of complete sets of variations in a periodic phenomenon) herz *m*
CYCLETHRIN - [chem] (an insecticide) cycléthrine *f*
CYCLIC - [gen] cyclique, périodique
~ACTIVITY - [biol] activité *f* cyclique
~ADMITTANCE - [el] (of a symmetrical polyphase winding the reciprocal of the cyclic impedance) admittance *f* cyclique
~BATCH - [phys] (molecular distillation) charge *f* intermittente (distillation moléculaire)
~COMPOUNDS - [chem] (closed-chain organic compounds) composés *m*pl cycliques
~DOUBLE HELICAL GEAR - [mech] engrenage *m* à chevrons cyclique
~IMPEDANCE - [el] (of a polyphase symmetrical winding, the potential difference across the terminals of a phase divided by the current flowing in it) impédance *f* cyclique
~IRREGULARITY - [mech] (of a flywheel) irrégularité *f* cyclique
~MEMORY - [comput] (memory which can be found in a periodic manner) mémoire *f* cyclique
~PERMUTATIONS - [math] permutations *f*pl circulaires
~PITCH CHANGE - [aero] (of the rotor blades) changement *m* de pas cyclique
~PITCH CONTROL - [aero] (in helicopters; a control for varying the rotor-blade pitch angle sinusoidally with the position of the blade in azimuth) commande m de pas cyclique
~PITCH STICK - [aero] (in helicopters) levier *m* de commande de pas cyclique
~POLYMERIZATION - [chem] polymerisation *f* cyclique
~REACTANCE - [el] (of a symmetrical poliphase winding; corresponding to the cyclic impedance) réactance *f*
cyclique
~SERIES - [chem] série *f* cyclique
~SHIFT - [comput] (of a digits of a number) décalage *m* circulaire
~TESTING - [mech] essai *m* de fatigue
CYCLICAL BINARY CODE - [comput] code *m* binaire cyclique
CYCLING - [phys] (periodic change of controlled variable) fluctuation *f*
[contr] (the unwanted oscillation of an automatic control) system) pompage *m*
~TIME - [comput] durée *f* du cycle
CYCLIZATION - [chem] cyclisation *f*
CYCLIZE, to - [ind chem] cycliser
CYCLIZINE HYDROCHLORIDE - [chem] (an antihistamine used in medicine as an anti-emetic) chlorhy-

drate *m* de cyclizine
CYCLO- - [chem] (prefix denoting the presence of a closed carbon chain or carbon-ring structure) cyclo-
CYCLOBARBITAL - [chem] (a rapid-action hypnotic used in medicine) cyclobarbital *m*
CYCLOCOUMAROL - [chem] (an anticoagulant used in medicine) cyclocoumarol *m*
CYCLOGENESIS - [met] (the process by which a cyclone is developed) cyclogenèse *f*
CYCLOGRAM - [instr] oscillogramme *m*
CYCLOGRAPH - [instr] (device for recording the revolutions of a wheel) oscillographe *m*[
CYCLOHEXANE - [chem] (a solvent for cellulose derivatives, waxes, resins, rubber and surface coatings and an intermediate in the production of nylon) cyclohexane *m*
CYCLOHEXANOL - [chem] (a raw material in nylon production, solvent for rubber, resins, cellulose derivatives, surface coatings, oils and dyes and a constituent of germicides and plasticizers) cyclohexanol *m*
~ACETATE - [chem] (a solvent for natural and synthetic resins, raw rubber and cellulose derivatives) acétate *m* de cyclohexanol
CYCLOHEXANONE - [chem] (a synthesis intermediate for a number of copolymers, a solvent for rubber, resins, fats, waxes and dyestuff) cyclohexanone *m*
CYCLOHEXENE - [chem] (an intermediate for organic synthesis) cyclohexène *m*
CYCLOHEXYL PHENOL [chem] (an intermediate for synthetic resins) cyclohexylphénol *m*
~STEARATE - [chem] (a plastics plasticizer) stéarate *m* de cyclohexile
CYCLOHEXYLAMINE - [chem] (intermediate for pharmaceuticals, dyestuffs and pesticides. Also has uses as a corrosion inhibitor in boiler feed water treatment) cyclohéxilamine *f*
CYCLOID - [geom] cycloïde *m*
CYCLOIDAL - [geom] cycloïdal
- GEARING SYSTEM [mech] engrenages *mpl* cycloïdaux
~PENDULUM - [phys] (pendulum made independent of the magnitude of the arc) pendule *m* cycloïdal
~TEETH - [mech] (gear-wheel teeth whose profile consists of cycloidal curves) denture *f* cycloidale
CYCLOMETER - [instr] (also called odometer or hofometer) enregistreur *m* de parcours
CYCLONE - [chem] (device for separating solids from gases by inertial and centrifugal effects) séparateur *m* cyclone
[met] (wind system surrounding an area of very low pressure towards which winds blow in a counter-clockwise direction in the northern and clockwise direction in the southern hemisphere) cyclone *m*
[aero] (aeromechanical device) cyclone *m*
~FIRED STEAM GENERATOR - [mech] générateur *m* de vapeur à cyclone
CYCLONIC - [met] cyclonique
~SYSTEM - [met] (system of weather conditions based on a cyclone) système *m* cyclonique
CYCLONIUM - [chem] (symbol Pm; a radioactive element) prométhium *m*, prométhéum *m*
CYCLOPARAFFINS - [chem] (a synonym for cyclanes. Cycloparaffins are saturated hydrocarbons containing one ring and forming a homologous series) cycloparaffines *fpl*
CYCLOPENTANE - [chem] (a solvent for cellulose derivatives) cyclopentane *m*
CYCLOPENTANOL - [chem] (intermediate for pharmaceuticals and dyestuffs) cyclopentanol *m*
CYCLOPENTENE - [chem] (a synthesis intermediate) cyclopentène *m*
CYCLOPENTOLATE HYDROCHLORIDE - [chem] (a mydriatic used in medicine) chlorhydrate *m* de cyclopentolate
CYCLOPENTYL BROMIDE - [chem] (synthesis intermediate for pharmaceuticals) bromure *m* de cyclopentyle
CYCLOPROPANE - [chem] (an anaesthetic and synthetic intermediate) cyclopropane *m*
CYCLOSTROPHIC - [met] (of a wind which blows under the influence of both pressure and centrifugal force) cyclostrophique
~FORCE - [met] (the centrifugal force deflecting an air current) force *f* cyclostrophique
CYCLOSTYLE - [print] (a duplicating machine) appareil *m* à polycopier
CYCLOTHYMIA - [med] cyclothymie *f*
CYCLOTRON - [phys] (device for accelerating charged particles to high energies) cyclotron *m*
~FREQUENCY - [nucl] (the frequency at which an electron orbits in a field) fréquence *f* de cyclotron
CYCRIMINE HYDROCHLORIDE - [chem] (anti-spasmodic used in medicine) chlorhydrate *m* de cycrimine
CYLINDER - [gen] cylindre *m*, tambour *m*
[mech] (of a motor) cylindre *m*
[ind chem] (a gas or fluid container) bouteille *f*
[text] tambour *m*, cylindre *m* rouleau *m*
[auto] (in the brake system) tambour *m*
[print] rouleau *m*
[firearms] (of a revolver) barillet *m*
~BARREL - [mech] (the wall of a cylinder of a motor) fût *m* de cylindre
~BASE - [auto] embase *f* de cylindre
~BATTEN - [text] battant *m* de cylindre
~BIT - [carp] (a steel drill used in carpentry) mêche *f*
~BLANKET - [print] blanchet *m*
~BLOCK - [mech] (of engine) bloc *m* cylindres
~BLOCK STUD - [mech] (for securing the cylinder head) boulon *m* prisonnier
~BOILER - [th eng] chaudière *f* cylindrique
~BORE - [mech] (the internal recess of a compressor cylinder or pump) alésage *m*
~CAPACITY - [mech] (the volume swept out by the piston during one stroke) cylindrée *f*
~CASING - [auto] chemise *f* de cylindre
~CHAIN DRIVE - [mech] commande *f* par chaîne
~CLEARANCE [railw] (of a locomotive cylinder) jeu *m* du cylindre, espace *m* nuisible
~COVER - [mech] (removable part of the cylinder head) plateau *m* de cylindre
~CRUSHER - [mech] broyeur *m* à rouleaux
~-DRIED - [paper man] (paper dried by passing over a heated cylinder) séchage *m* à la machine
~ESCAPEMENT - [horol] échappement *m* à cylindre
~FOR LONG STRIPED FABRICS - [text] cochonnet *m* d'impression de tissus rayés dans la longueur
~GRINDER - [mech] rectifieuse *f* pour cylindres
~GRINDING MACHINE - s. cylinder grinder

CYLINDER HEAD - [mech] (the plate or casting which covers the open end of a working cylinder, e.g. hydraulic ram or an air compressor) culasse *f*
~HEAD BOLT - [mech] boulon *m* de culasse
~HEAD GASKET - [mech] joint *m* de culasse
~HOLDER - [impl] (used for the transport of gas or fluid cylinders) cadre *m* porte-bouteilles
~HONE - [mech] pierre *f* à roder les cylindres
~LINER - [mech] (a cylindrical shell designed to be inserted in the cylinder-block) chemise *f* de cylindre
[metall] (in die casting) cylindre *m* d'injection
~LINER BOTTOM GASKET - [mech] joint *m* inférieur de chemise
~LINER SEAL - [mech] (a sealing ring for a watertight joint between cylinder liner and water-jacket system) garniture *f* de chemise
~MIXER - [mech] (mixer consisting of a plain cylinder revolving about its longitudinal axis) mélangeur *m* à cylindre
~OIL - [mech] huile *f* à cylindres
~OPERATING CATCH - [text] doigt *m* (ou ergot) de commande
~PEG - [text] broche *f*, cheville *f*
~POWER - [opt] puissance *f* d'un cylindre
~PRESS - [print] (a cylinder printing machine) presse rotative *f*
~REBORING - [mech] (process of correcting the effect of cylinder wear) réalésage *m* de cylindres
~SIZING MACHINE - [mech] encolleuse *f*
~SKIRT - [mech] (the lower section of the cylinder of an engine) jupe *f*
~SLEEVE - [mech] chemise *f* de cylindre
[metall] s. cylinder liner [metall]
~WEAR - [mech] (the attrition of the bore of a cylinder) usure *f* des cylindres
CYLINDERS IN LINE - [auto] cylindres *mpl* en ligne
CYLINDRICAL - [gen] cylindrique
~BORING - [mech] forage *m* cylindrique
~CATHODE - [electron] (a cylinder-shaped cathode for electronic tubes) cathode *f* cylindrique
~CAVITY-TYPE MAGNETRON - [electron] magnétron *m* à cavité cylindrique
~COORDINATES - [math] coordonnées *fpl* cylindriques
~COUNTER CHAMBER - [meas] chambre *f* de comptage cylindrique
~GRINDER - [mech] machine *f* à rectifier à table cylindrique
~LENS - [opt] lentille *f* cylindrique
~REACTOR - [nucl] (reactor with a cylindrically shaped tank) réacteur *m* cylindrique
~REAMER - [mech] alésoir *m* cylindrique
CYLINDRICAL REFLECTOR - [radio] (of aerials; reflector consisting of a portion of a cylinder) réflecteur *m* cylindrique
~ROLLER BEARING - [mech] roulement *m* à rouleaux cylindriques
~SPRING - [mech] ressort *m* spiral cylindrique
~WAVE - [phys] (electromagnetic waves) onde *f* cylindrique
~WINDING - [el] enroulement *m* cylindrique
CYLINDRICALITY - [gen] (the quality of being truly cylindrical) cylindricité *f*
CYLINDRITE - [min] (a sulphide of lead, tin and antimony) cylindrite *f*
CYMA - [arch] (moulding showing a reverse curve in profile) cymaise *f*
~RECTA - [arch] cymaise *f* droite
~REVERSA - [arch] cymaise *f* renversée
CYMBALS - [mus] cymbales *fpl*
CYMENE - [chem] (a solvent and an intermediate for synthetic rubber catalysts and synthetic resins) cymène *m*
CYMOGRAPH - [instr] (used to make tracings of profiles) cymographe *m*
CYMOL - [chem] thymène *m*
CYMOMETER - [instr] (to measure electromagnetic waves) cymomètre *m*
CYMOPHANE - [min] (a gem-mineral chrysoberyl) cymophanite *f*, crysobéryl *m*
CYMOSCOPE - [instr] (to detect electromagnetic waves) cymoscope *m*
CYPRESS - [bot] cyprès *m*
CYRTOMETER - [instr] (to determine dimensions and movements of curved surfaces) cyrtomètre *m*
CYST - [med] (non-living membrane) sac *m*, vésicule *f*
CYSTADENOMA - [med] cystadénome *m*
CYSTALGIA - [med] cystalgie *f*, cystodynie *f*
CYSTICERCOSIS - [med] cysticercose *f*
CYSTITIS - [med] cystite *f*
CYSTOSCOPE - [med] cystoscope *m*
CYSTOSCOPY - [med] cystoscopie *f*
CYSTOTOMY - [med] cystotomie *f*
CYSTINE - [chem] (amino acid containing sulphur) cystine *f*
CYTOBLAST - [biol] cytoblast *m*
CYTOCHROMES - [biochem] (generic name for a group of porphyrin/iron proteins important in cell metabolism) cytochromes *mpl*
CYTOLOGY - [biol] (the study of cells, their functions, structure and reproduction) cytologie *f*
CYTOPLASM [genet] (a cell protoplasm without the nucleus) cytoplasme *m*
CYTOTOXIC - [med] cytotoxique

D

D - [chem] (abbreviation for dextro-rotatory) dextrogire
[met] (abbreviation for drizzle) bruine *f*, crachin *m*
D - [chem] (the symbol of Deuterium) - (symbole) Deutérium *m*
D.C. [el] (abbreviation for Direct Current) courant *m* continu
D-LAYER - [phys] (the lowest layer of the ionosphere at an altitude of abour 70 Km, exclusively diurnal) zone *f* D
D LINES - [phys] (group of three Fraunhofer lines caused by sodium in the sun atmosphere) raies *fpl* D
D.M.E. - s. distance measuring equipment
D SLIDE VALVE - [mech] (slide valve resembling in section the letter D) vanne *f* à tiroir en D
D-VALUE - [meas] (difference between the height shown by a radio altimeter and that shown by a pressure altimeter set to standard pressure value) valeur *f* D
DAB, to - [gen] (frapper légèrement) tapoter, tamponner
[constr] (in masonry work) aplanir, polir; (of plaster) crépir; (with a sponge, to mosten or to powder) appliquer à petits coups, humecter
DAB - [gen] petit coup *m*, coup *m* léger, tape *f*
[constr] masse *f* de torchis
[impl] used for markings) poinçon *m*
DABBER - [print] balle *f*, poupée *f*
[metall] (of a loam mould) broche *f*
DABBING - s. daubing
DABREY - [rubber ind] (a small container in which the latex exuding from the tree is collected) pot *m* pour la récolte du latex
DACITE - [geol] (volcanic rock consisting mostly of plagioclases spar with biotite, hornblende etc) dacite *f*
DACRON - [text] (US designation of terylene) dacron *m*
DACTYL - [anat] (a digit] doigt *m*
DACTYLITIS - [med] (inflammation of finger or toe) dactylite *f*
DACTYLOCAMPSODYNIA - [med] flexion *f* douloureuse des doigts
DACTYLOGRYPOSIS - [med] courbure *f* anormale des doigts
DACTYLOID - [bot] (stretching out like fingers) en forme de doigts
DACTYLOPHASIA - [med] langage *m* des sourds-muets
DADO - [arch] (on a pedestal, one of the faces of the block) lambris *m*
[constr] (running border on a wall near the floor) plinthe *f*
DADO CAPPING - [constr] moulure *f*
~RAIL - s. dado capping
DAFFODIL - [bot] asphodèle *m*
DAGGER PIN - [mech] cheville *f* d'arrêt, butoir *m*
DAGGINGS - [text] (clots in the wool caused by dirt or sticking soil) mêches *fpl* de laine coagulées
DAGGLE, to -[gen] salir, souiller, éclabousser
DAGS - s. daggings
DAGUERREOTYPE - [photo] (ancient form of photography) dagguerréotype *m*
DAHLIA - [bot] dahlia *m*
DAILY - [gen] journalier, journellement
[print] (of newspaper) quotidien *m*
DAILYGRAPH - [el acoust] (magnetic recorder to be attached to the telephone) enregistreur *m* magnétique combinable avec un téléphone
DAIRY - [agric] laiterie *f*
[agric] (for the production of cheese) fromagerie *f*
~BRONZE - [agric] bronze *m* de berthes
~BY-PRODUCTS - [agric] sous-produits *mpl* de laiterie, sous-produits *mpl* de fromagerie
~CATTLE - [agric] bétail *m* à lait
~COW - [agric] vache *f* laitière
~EQUIPMENT - [agric] machines *fpl* pour l'industrie laitière (ou fromagère)
~FARM - [agric] laiterie *f*
~PRODUCTS - [agric] produits *mpl* laitiers
~RESIDUES - s. dairy by-products
~WASTES - [agric] (milk-product wastes) eaux *fpl* résiduaires de laiterie
DAIS - [gen] estrade *f*, dais *m*
DAISY - [bot] paquerette *f*
DALE - [geogr] vallée *f*, vallon *m*
DALTON'S LAW - [phys] (the total pressure of a mixture of gases is the sum of the partial pressures of each individual gas) loi *f* des pressions partielles
~LAW OF PARTIAL PRESSURES - s. Dalton's law
DALTONISM - [opt] (colour blindness) daltonisme *m*
DAM, to - [gen & hydr] boucher, endiguer, contenir, construire une digue
DAM - [hydr] digue *f*, écluse *f*, barrage *m*
[mining] (kind of bank to avoid flooding) mur *m* de défense
[metall] (blast furnace equipment) plaque *f* de protection
[metall] (in the sprue basin) seuil *m* dans le bassin de coulée
[plast ind] (slidable metal strip used to control the width of extruded sheet) plaquette *f* de vérification, gabarit *m*
~CREST - [hydr] couronnement *m* de barrage, crête *f* de barrage

DAM FAILURE - [hydr] rupture *f* du barrage
~PLATE - [metall] (of a blast furnace) barrage *m*
~SPRING - [hydr] source *f* de barrage
~STONE - s. dam plate
~-TYPE LIP LADLE - [metall] poche *f* à barrage, poche *f* à dame
DAMAGE, to - [gen & mech] endommager, abimer, gâter
DAMAGE - [gen] dommage *m*, avarie *f*, dégat *m*
[mech] avarie *f*, panne *f*
~CRITERIA - [nucl] (used to assess levels of radiation damage) critères *mpl* de lésion
~REPORT - [comm] (report submitted to the insurance company) certificat *m* d'avarie
~SURVEY - [comm] expertise *f* d'avarie
DAMAGED - [gen & mech] endommagé, gâté, abimé
~GOODS - [comm] marchandises *fpl* avariées
DAMASCENE, to - [gen] damasser, marqueter
DAMASCENING - [metall] damasquinage *m*
DAMASCUS STEEL - [metall] acier *m* damassé
DAMASK - [text] (a rich furnishing fabric; also a linen cloth) damas *m*
~GAUZE - [text] gaze *f* damassée
~HEALD - [text] lisse *f* pour damasser
~LOOM - [text] métier *m* à damasser
~SHUTTLE - [text] navette *f* pour le damas
~WEAVE - [text] armure *f* de damas
~YARN - [text] filé *m* pour damas
DAMASKEEN - [metall] (the inlay of ivory etc on metal) damasquinage *m*
DAMMAR - [chem] (generic name for a number of natural resins used in the manufacture of varnishes) dammar *m*
~GUM - s. dammar resin
~RESIN - [chem] résine *f* de dammar
~VARNISH - [ind chem] vernis *m* dammar
DAMMING UP - [gen] (accumulation) accumulation *f*
[constr] barrage *m*, batardeau *m* digue *f*
[soil] barrer
DAMP, to -[gen] humidifier, baigner, mouiller
[acoust] amortir, atténuer
DAMP - [gen] humide
[gen] humidity, i.e. the presence of moisture) humidité *f*
[mining] [any gas present in a coal mine) grisou *m*
~AIR BLOWER - [mech] humidificateur *m*
~DOWN, to - [metall] suspendre la marche, boucher
[metall] (the furnace] mettre le four hors feu, refroidir
~-PROOF [gen] imperméable
[constr] hydrofuge
~-PROOF COATING - [[constr] revêtement *m* imperméable, imperméabilisation *f*
~-PROOF COURSE - [constr] couche *f* isolante imperméable
~-PROOF SLAB - [constr] chape *f* d'étancheité
~SHEET - [mining] toile *f* d'aérage
DAMPED - [gen] mouillé
[acoust] amorti, atténué
~BALANCE - [instr] (balance fitted with a device to reduce the oscillatory movement of the suspended parts) balance *f* amortie
~BOBBIN - [text] bobine *f* mouillée
~HARMONIC MOTION - [phys] mouvement *m* harmonique amorti
~HARMONIC OSCILLATION - [phys] (oscillation whose motion is subject to resistance) oscillation *f* harmonique amortie
DAMPED LINEAR OSCILLATOR - [mech] (linear oscillator subject to a damping coefficient) oscillateur *m* linéaire amorti
~OSCILLATION - [phys] (an oscillation of progressively decreasing amplitude) ascillation *f* amortie
~PERIODIC ELEMENT - [instr] (moving element attaining its equilibrium position after a number of oscillations) équipage *m* périodique amorti
~PERIODIC INSTRUMENT - [instr] (instrument fitted with a damped periodic element) instrument *m* périodique amorti
~SINUSOIDAL QUANTITY - [phys] (quantity varying according to the product of a sinusoidal function) grandeur *f* sinusoïdal amortie
~WAVES - [phys] (waves originating from trains of oscillations) ondes *fpl* amorties
DAMPEN, to - [gen] humidifier, humecter
[acoust] insonoriser
[phys & chem] amortir, atténuer
DAMPENER - [impl] humidificateur *m*
[mech] amortisseur *m*
DAMPENING - s. damping
~MACHINE - [text] humecteuse *f*, aspergeuse *f*
~ROLLER - [text] rouleau *m* humecteur
~SOLUTION - [print] (used in offset printing) bain *m* affaiblisseur
DAMPER - [gen & impl] humidificateur *m*
[mech] (adjustable plate applied to a flue pipe to control the draught of a boiler, stove etc) registre *m*, régulateur *m* de tirage
[mech] (device for producing mechanical damping) amortisseur *m*
[el] (device to oppose changes in oscillations; also applied to the moving parts of a measuring instrument) amortisseur *m*
[mus] sourdine *f*
~PEDAL - [mus] (pianoforte pedal raising all dampers) pédale *f* forte
~WEIGHT - [mech] (counterweight balancing the damper) contrepoids *m* de régulateur de tirage
~WINDING - [el] (the winding of a short circuit placed in a machine to oppose changes in angular velocities) enroulement *m* amortisseur
DAMPING - [phys] (reduction of amplitude of oscillation, due to energy dissipation) amortissement *m*
[electron] (progressive reduction in the amplitude of electrical oscillations) amortissement *m*
[instr] (the application of a retarding effect to an instrument, so as to ensure that it gives a reading without undue oscillation of the indicating element) amortissement *m*
[mech] (reduction in the amplitude of mechanical vibration) amortissement *m*
[acoust] (vibration absorbing) atténuation *f*, amortissement *m*
[acoust] (removal of echoes etc) insonorisation *f*
~APPARATUS - [paper man] humidificateur *m*
~BASIN - [hydr] bassin *m* d'amortissement, bassin *m* de tranquillisation
~CAGE - [el] cage *f* d'amortissement
~CAPACITY - [mech] pouvoir *m* anti-vibratoire
~COIL - [el] bobine *f* d'amortissement
~CONSTANT - [instr] (intrinsic constant of a galvanometer) constante *f* d'amortissement
[el] s. decay factor

AMPING DIODE - [telev] (discharge tube damping a coupling phenomenon) diode *f* amortisseuse, tube *m* amortisseur
DISK - [mech] disque *m* amortisseur
DOWN - [th eng] (the operation of retarding combustion in a boiler) etc) arrêt *m*, ralentissement *m*
FACTOR - s. decay factor
FACTOR OF A MEASURING INSTRUMENT - [instr] (US expression; In GB, ballistic factor of a measuring instrument; the ratio of the first deflection to the permanent deflection when current impulses are passed through the instrument) facteur *m* balistique d'un appareil de mesure
GRID - s. damper
MAGNET - [instr] (magnet which produces the damping torque in an instrument by acting on a moving part) aimant *m* amortisseur
MICA - [electron] (anti-vibration piece of mica fitted to an electronic tube) mica *m* amortisseur
RATIO - [mech] (the ratio between the actual and the critical damping) rapport *m* d'amortissement
ROLLS- [paper man] (wet cylinders wetting the paper to give it a smoother finish) cylindres *mpl* humidificateurs
SPRING - [mech] ressort *m* amortisseur
STRETCH - [paper man] (stretch due to action of the damping rolls) allongement *m* dû à l'humidité
THE WEFT - [text] mouillage *m* de la trame
TIME - [meas] temps *m* d'arrêt
TORQUE - [instr] (torque tending to prevent oscillations of the moving element) couple *m* de freinage
UP - [text] (the operation of damping skin before stretching) humectation *f*
VALVE - [hydr] (a valve designed to reduce the movement of fluids) vanne *f* régulatrice
VANE - [mech] (vane for damping systems) amortisseur *m*
AMPNESS - [gen] humidité *f*
ANBURITE - [min] (an accessory mineral; a calcium silicate) danburite *f*
ANCER - s. dancer roll
ROLL - [plast ind] (roller used as tension-sensing device in wire coating extruder) cavalier *m*
ANCING - [mech] (the vibration of a valve spring) débattement *m*
STEP - [constr] (a step with its outer end narrower than its inner end) marche *f* balancée
ANDELION - [bot] pissenlit *m*
MARK - [paper man] filigrane *m*
METAL - [metall] alliage *m* d'antimoine de plomb et d'étain
RUBBER - [rubber ind] (rubber produced from the dandelion latex) caoutchouc *m* de pissenlit
ANDER - [physiol] pellicules *fpl*
ANDRUFF - s. dander
ANDY - [mining] brouette *f* à deux roues
[metall] (US for running-out fire) four *m* de premier affinage
BRUSH [impl] brosse *f* de pansage
FINISHER - [text] étirage *m* sans préfinissage
LOOM - [text] (a partly mechanized loom) métier *m* semi-mécanique
MARK - [paper man] vergeure *f*
REDUCING - [text] étirage *m* sans réduction
ROLL - [paper man] (a wire gauze cylinder impressing ribs and watermarks) rouleau *m* vergeur, rouleau *m* à filigraner
DANGER - [gen] danger *m*, péril *m*, risque *m*
~COEFFICIENT - [nucl] (the change in a reactor reactivity caused by a substance) coefficient *m* de danger
~CONE - [aero] pennant or windsock attached to the mooring cable of a captive balloon to warn aircraft of the obstruction) manche *f* à air
~SIGNAL - [gen] signal *m* d'alarme
DANIELL CELL - [el] (a primary cell having a copper positive electrode in copper sulphate solution and a zinc negative in zinc sulphate, the electrolytes being separated by a porous pot) pile *f* Daniell
~HYGROMETER - [instr] (obsolete type of dew-point hygrometer) hygromètre *m* Daniell
DANK - [gen] humidité *f*, humide
DANT - [min] (type of soft coal) charbon *m* gras
[mining] (broken coal) fines *fpl* de charbon
DANTRY - s. dant
DANTZIG OAK - [constr] chêne *f* de choix
DAP, to [carp] (to cut notches for joints) mortaiser
~JOINT - [carp] joint *m* mortaisé
DAPPING - [carp] (when building timber bridges) mortaisage *m*
DARAF - [el] (unit of elastance, the reciprocal of capacitance in farads) daraf *m*
DARBY - [constr] (a trowel, also called derby float) truelle *f*, règle *f* du plafonneur
DARG - [mining] (the production of one day) production *f* journalière
DARIMONT CELL - [el] (a primary cell with carbon and zinc electrodes. The electrolyte is a solution of calcium carbonate and sodium chloride and the depolarizer is ferric chloride) pile *f* Darimont
DARK - [gen] obscur, sombre
~ADAPTATION - [opt] adaptation *f* à l'obscurité, adaptation *f* de la vue à l'obscurité
~AND LIGHT SPOT - [telev] moirage *m*, contrastes *mpl* excessifs
~BEER FOR COLOURING - [brew ind] bière *f* colorante
~BURN - [electron] (decrease of the efficiency of a luminescent material in the course of excitation) fatigue *f* (d'une substance luminescente)
~CONDUCTION - [el] (flow of dark current) courant *m* d'obscurité
~CURRENT - [el] (the current which flows in a photocell when it is not illuminated) courant *m* d'obscurité
~CURRENT CLIPPER - [el] limiteur *m* de courant d'obscurité
~DESATURATION [telev] (in the shadows of the picture) désaturation *f* dans les parties obscures
~DISCHARGE - [el] (an electric discharge in a gas without the emission of light) décharge *f* obscure
~FIELD - [photo] fond *m* noir
~FIELD ILLUMINATION - [opt] éclairage *m* sur fond noir
~FUMES - [metall] fumée *f* de couleur foncée
~GLOWING - [metall] recuit *m* bleu
~GROUND ILLUMINATION - [opt] (technique in microscopy enabling specimens to appear as bright objects against a dark background) éclairage *m* sur fond noir
~INTERVAL - [cin] (the time between two exposures during projection) phase *f* d'obturation
~LANTERN - [light] lanterne *f* sourde

DARK-LINE SPECTRUM - [phys] (spectrum with lines darker than others) spectre *m* de raies sombres
~OIL - [ind chem] (lubricant used for rough work) huile *f* noire
~PERIOD - s. dark interval
~RADIATION - [phys] (an infrared radiation) rayonnement *m* infrarouge
~RED SILVER - [min] pyrargyrite *f*
~RESISTANCE - [el] (the resistance of a photoelectric device in the absence of illumination) résistance *f* obscure
~-ROOM - [photo] chambre *f* noire
~-ROOM LIGHT-FILTER - [photo] lampe *f* inactinique
~-ROOM PACKING - [photo] emballage *m* pour chargement en chambre noire
~SLIDE - [photo] (plate carrier) cadre *m* porte-plaque
~SPACE - [phys] espace *m* sombre
~SPOT - [telev] (result of unequal boundary potential) point *m* sombre
~TIPPED - [agric] à pointes brunes
~-TRACE CATHODE-RAY TUBE - [electron] (a C.R.T. having a screen which gives a magenta trace which under mercury discharge lamp illumonation appears black on a green ground) tube *m* cathodique à trace sombre
~-TRACE SCREEN - [electron] (screen with a spot darker than the remainder of the surface) écran *m* absorbant
~-TUBE - [electron] (type of cathode-ray tube in which the electron beam darkens the fluorescent screen instead of making it brighter) skiatron *m*
DARKEN, to - [gen] assombrir, obscurcir, brunir
[photo] se foncer
DARKENING - [dyes] accentuation *f*, foncement *m*
DARN, to - [gen] repriser
DARN - [gen] reprise
DARNEL - [bot] ivraie *f* enivrante
DARNING COTTON - [text] coton *m* à repriser
~NEEDLE - [text] aiguille *f* à repriser
~WOOL - [text] laine *f* à repriser
D'ARSONVAL CURRENT - [el] (current of intermitting and isolated trains of heavily damped oscillations, of high frequency, high voltage and low amperage) courant *m* d'Arsonvalisation
~GALVANOMETER - [instr] galvanomètre *m* à cadre mobile
D'ARSONVALIZATION - [el biol] (therapeutic use od d'Arsonval current) diathermie *f*, d'Arsonvalisation *f*
DART - [horol] (the pin attached to the end of the lever at the notch) ergot *m* de l'ancre
[text] couture-pinces *f*pl
~CONFIGURATION - [astronaut] configuration *f* à empennage arrière
~MOTH - [zool] papillon *m* (nocturne) des céréales
DASH, to - [gen] jeter, jaillir
DASH - [gen] tiret *m*, élan *m*, trait *m*
[print] tiret *m*
[telecomm] trait *m*
~-BOARD - [auto] tableau *m* de bord
~-BOARD COWL - [auto] auvent *m*
~-DOT MODE - [comput] (a binary digit mode of storage in which 0 is a luminous dash and 1 a luminous dot) méthode *f* trait-point
~DOT RYTHM - [radio] (type of characteristic signal sent out by a marker beacon) signal *m* trait-point
DASH-PANEL - [auto] s. dash-board
~PLATE - [mech] plaque *f* déflectrice
DASHING WATER - [hydr] courant *m* d'eau rapide, eau *f* déferlante
DASHLAMP - [auto] lampe *f* de tableau de bord
DASHPOT - [mech] (a vibration-damping device, consisting of a cylinder containing oil, and fitted with a piston) dash-pot *m*
[mech] (a mechanism to avoid shocks) amortisseur *m*
DASYMETER - [instr] (instrument designed to measure the density of gases) dasymètre *m*
DATA - [gen] caractéristiques *f*pl, données *f*pl, renseignements *m*pl
[comput] (information supplied to or by the computer) données *f*pl, information *f*
~ACQUISITION - [comput] acquisition *f* des données notation *f* des données
~ADDRESS - [comput] (an address which gives the location od data) adresse *f* de données
~BUS LINE - [comput] canal *m* collecteur des données
~CHANNEL - [comput] canal *m* des données
~COLLECTION - [comput] regroupement *m* des données, collationnement *m* des données
~DISPLAY - [comput] affichage *m*
~DUMP - [comput] (loss of information caused by the power supply being interrupted) perte *f* de données
~HANDLING - [comput] manipulation *f* des données, manipulation *f* des informations, traitement *m* de l'information
~INPUT - [comput] entrée *f* des données, introduction *f* des données
~LINK - [radio] (radio link used for sending navigational information) liaison *f* radio pour la transmission d'informations
~LOGGER - [comput] (apparatus storing information and effecting a number of operations) enregistreur *m* (pour l'élaboration de l'information)
~OUTPUT - [comput] sortie *f* des données
~PICK-OFF ELEMENT - [comput] palpeur *m*, organe *m* de prélèvement
~PLATE - [gen] (in machines) plaque *f* de constructeur
~PROCESSING - [comput] (the handling of information in a sequence of operations) traitement *m* des données, traitement *m* de l'information
~PROCESSING MACHINE - [comput] machine *f* pour le traitement de l'information, ordinateur *m*
~PROCESSOR - [comput] installation *f* de traitement des données
~RECORDING - [comput] (on a magnetic tape etc) enregistrement *m* de données
~RECORDING MEDIUM - [comput] milieu *m* mémoire, support *m* d'information
~REDUCTION - [comput] (the conversion of data which have been obtained experimentally into useful information) élaboration *f* de l'information, réduction *f* des données
~SORTER [comput] trieuse *f* des données
~STORAGE POSITION - [comput] position *f* de mémorisation des données
~TRANSDUCER - [comput] convertisseur *m* d'informations
~TRANSFER - [telecomm] (transmission of informa-

ion from ground to air stations and vice-versa) ransmission *f* d'informations
ATA TRANSLATOR - [comput] s. data transducer
TRANSMISSION SYSTEM - [instr] (information which is transferred from an instrument to another) système *m* de transmission de mesures
ATE, to - [gen] dater
ATE - [gen] date *f*
[bot] datte *f*
BAND - [impl] bande *f* pour machine à dater les billets
HAND - [horol] indicateur *m* de date
JUMPER - [horol] détente *f* de date
-LINE - [geogr] (the agreed international line, mainly along the 180th degree of longitude, on which the date is considered to change) ligne *f* de changement de date
NAIL - [mech] clou *m* pour millésime
PALM - [bot] palmier *m* dattier
ATING - [nucl] (the determination of the age of a mineral, a fossil etc, from its radioactive contents and their decay products) détermination *f* de l'âge
ATIVE BOND - [chem] (synonymous of semi-polar bond. A bond in which two atoms share a pair of electrons supplied by one of them. They are thus both charged and form a dipole) liaison *f* semi-polaire
COVALENCE - [phys] liaison *f* de coordination
VALENCE - [phys] valence *f* donneuse
ATOLITE - [min] (hydrated silicate of boron and calcium) datolite *f*
ATUM - [gen] caractéristique *f*, donnée *f*, renseignement *m*
[surv] (a reference surface) point *m* de repère
ERROR - [instr] erreur *f* initiale
HORIZON - [mining] niveau *m* de repère
LEVEL - [surv] (the horizontal reference plane of a chart or map, from which altitudes and the like are measured) plan *m* de niveau, plan *m* de comparaison
MARK - [surv] point *m* de nivellement, point *m* de repère
PLANE - s. datum level
POINT - [surv] point *m* de repère
ATURA - [bot] datura *m*
AUB, to - [gen] barbouiller, colorer
[paint] enduire, couvrir
[constr] (to apply some plaster) plâtrer
AUB - [gen] barbouillage *m*, enduit *m*
[constr] torchis *m*, gobetage *m*
AUBING - [metall] revêtement *m* en matière réfractaire
AUGHTER - [gen] fille *f*
[genet] (first-generation offspring) descendant *m* de première génération
[nucl] (the product formed in the radioactive decay of a nucleus which is called the parent) descendant *m*
ACTIVITIES - [nucl] (radioactivity] activités *fpl* engendrées
ELEMENT - s. daughter [nucl]
NUCLEUS - [phys] noyau *m* engendré
PRODUCT - [nucl] descendant *m*, produit *m* de filiation
AVAINE'S BACILLUS - [med] bacille *m* de Davaine, bactéridie *f* charbonneuse

DAVIDITE - [min] (mixture of ilmenite and carnotite) davidite *f*
~FURNACE - [metall] four *m* à réverbère
DAVIT - [naut] bossoir *m*
~DOOR - [mech] (type of autoclave door so arranged that when opened its weight is supported on a davit or swinging bracket) porte *f* à davier
DAVITS - [naut] bossoirs *mpl* d'embarcation
DAVY - [mining] (the safety lamp designed by Davy) lampe *f* de sécurité, lampe *f* Davy
DAWN - [gen] aube *f*
DAY - [gen] jour *m*, journée *f*
[constr] (of a window) lumière *f*, baie *f*
~BLINDNESS - [med] nyctalopie *f*
~-BOOK - [comm] journal *m*
~-BREAK - [gen] aube *f*, point *m* du jour
~-DRIFT - [mining] fendue *f*, galerie *f*
~EYE - [mining] puits *m* incliné
~FALL - [mining] affaissement *m* du sol (au dessus des travaux)
~RATE - [comm] paie *f* journalière, salaire *m* journalier
DAYLIGHT - [gen] lumière *f* du jour
~COLOUR FILM - [photo] pellicule-couleur *f* lumière du jour
~FACTOR - [constr] (the amount of natural light at any point within a building) facteur *m* d'éclairement par lumière diffuse
~GENERATOR - [light] générateur *m* de lumière du jour
~LAMP - [light] lampe *f* lumière du jour
~LOADING - [photo] chargement *m* au jour
~MANTLE - [light] (gas light) manchon *m* lumière du jour
~MINE - [mining] exploitation *f* au jour, mine *f* à ciel découvert
~OPENING - [mech] (the space between the platens of a press) distance *f* entre plateaux, ouverture *f* de la presse
~PRESS - [mech] presse *f* à étages
DAYMARK - [aero] balise *f* de jour
DAZE - [min] (mica or any glittering stone) mica *f*
DAZZLING - [light] éblouissement *m*
~LIGHT - [auto etc] lumière *f* éblouissante
D.C. BLOCKING - [telev] (for the elimination of unpleasant surges) blocage *m* de la composante continue
D.C. CLAMP DIODE - [electron] diode *f* de niveau
D.C. CONVERGENCE - [telev] (static convergence) convergence *f* statique
D.C. INSERTION - [telev] insertion *f* de la composante continue
D.C. LEVEL - [telev] niveau *m* de la composante continue
DEACCENTUATION - [gen] atténuation *f*
[electron] désaccentuation *f*
DEACCENTUATOR - [el] (deemphasis obtaining network) affaiblisseur *m*, atténuateur *m*
DEACIDIFYING - [chem] désacidification *f*
DEACON'S PROCESS - [chem] (a method for the commercial preparation of chlorine, using hydrogen chloride with air as the oxidizing agent) procédé *m* de Deacon
DEACTIVATE, to - [gen] désactiver
DEACTIVATION - [phys] (the resumption of normal state by an atom, a molecule or a substance) désactivation *f*

DEAD - [gen] mort
[dyes] (of colours) terne
[constr] (of building material) détérioré
[el] sans tension
[acoust] sourd
~ABUTMENT - [constr] faux contrefort *m*
~AHEAD - [naut] droit devant
~AIR [phys] air *m* mort
~AIR REGION - [phys] (a separated region of low air velocity) zone *f* d'air mort
~ANGLE - [mech] (crank angle of a steam engine) angle *m* mort
~AXLE - [mech] (not rotating with the wheels it carries) essieu *m* fixe
~BAND - [phys] (those values through which the variable can be varied without an effective response) zone *f* morte
~BATTERY - [el] batterie *f* déchargée
~-BEAT - [instr] (term used to denote an instrument which is heavily damped, so as to give a reading with the minimum oscillation of the indicating element) apériodique
~-BEAT AMMETER - [el] ampèremètre *m* apériodique
~-BEAT ESCAPEMENT - [horol] (escapement without recoil to the escape wheel) échappement *m* libre
~-BEAT INSTRUMENT - [instr] instrument *m* de mesure apériodique
~-BEAT MOTION - [mech] oscillation amortie
~BLACK - [opt] noir *m* très foncé
~BOLT - [mech] pêne *m* dormant
~CALM - [naut] bonace *f*
~CENTRE - [mech] (that point in the revolution of a crank-connecting rod assembly when the crank-throw, crank centre and rod are in the same straight line) point *m* mort
~-CENTRE LATHE - [mach tools] tour *m* à pointes fixes
~CENTRE POSITION - [gen] point *m* mort
~COIL - [el] bobine *f* court-circuitée
~COLOURING - [paint] couche *f* primaire
~DOOR - [constr] fausse porte *f*
~EARTH - [el] terre *f* parfaite
~END - [plumb] bout *m* mort
[radio] (unused section of inductance coil) spire *f* inactive
[railw] voie *f* de garage
~-END HOLE - [mech] trou *m* borgne
~-END SIDING - [railw] voie *f* de garage
~-END STATION - [railw] gare *f* de tête de ligne, gare *f* de départ
~-END SWITCH - [radio] (multipoint switch short-circuiting the unused end-turns of an inductance coil) commutateur *m* de court-circuit
~-END TIE - [mech] ligature *f* d'arrêt
~-END TRACK - [railw] voie *f* de garage
~-ENDED FEEDER - [el] (an independent feeder) ligne *f* d'alimentation de sous-station
~ENDING - [railw] ancrage *m* fixe
~EYE - [naut] margouillet *m*
~FINISH - [paint] (without any brilliancy of surface) fini *m* mat
~FLOOR - [constr] (a sound-absorbing floor, fitted with deadening material) plancher *m* insonorisé
~FOLD - [metall] (a fold in sheet material which does not unfold of itself) pli *m* permanent
~GROUND [el] terre *f* parfaite
[geol] (also spelt deadground; ground of no value from a mining point of view) roche *f* stérile
DEAD HEAD - [mach tools] (a fixed headstock) pointe *f* fixe
~HOLE - s. dead-end hole
~KNOT - [timber] (knot which is separated from the timber surrounding it) noeud *m* détaché
~LEVEL - [gen] à niveau
[hydr] niveau *m* permanent
~LIMB - [med] engourdissement *m* d'un membre
~LINE - [oil ind] brin *m* mort
~LOAD - [constr] (dead weight; the permanent loading on a structure) charge *f* permanente, charge *f* statique
~LOAD SAFETY VALVE - [mech] soupape *f* de sûreté à contrepoids
~LOCK - [mech] (lock bolt which is key-operated on one side and hand-operated on the other) serrure *f* sans ressort, serrure *f* de sûreté
~-MAN'S CONTROL - [el] (a form of control used in electric vehicles, whereby the current is cut off in case the controller is made incapable) dispositif *m* de sécurité
~-MAN'S HANDLE - [el] (handle used for the dead man's control) manette *f* de sécurité
~MATTER - [print] (used type ready for distribution) matière *f* morte
~MELTING - [metall] chauffage *m* au dessus du point de fusion, fusion *f* surpassée
~-MILLED RUBBER - [rubber ind] (rubber which has been masticated to a point at which it no longer exhibits elastic properties) caoutchouc *m* malaxé à fond
~MILLING - [rubber ind] (process of masticating rubber until it has lost its elastic properties) malaxage *m* à fond
~OIL - [chem] huile *f* de créosote
[oil ind] (crude oil after elimination of gas) huile morte
~RANGE - [el] zone *f* d'insensibilité
~RECKONING - [radar etc] (determination of position) estime *f*, point *m* estimé, navigation *f* à l'estime
~RIPENESS - [agric] maturité *f* avancée
~RISE - [shipbuild] angle *m* de quille, inclinaison *f* des varangues
~ROASTING - [ind chem] (process of roasting a concentrate carried far enough to reduce the sulphur content as much as possible, as distinct from sulphating roasting) grillage *m* à fond
~ROOM - [acoust] chambre *f* sourde
~SECTOR - [radio] secteur *m* mort
~SEGMENT - [el] (non connected commutator segment) lame *f* de collecteur non-connectée
~SHORE - [carp] (vertical timber support) étai *m* provisoire
~SHORT - [el] (colloq.; a very low-resistance short-circuit) court-circuit *m*
~SLOW - [gen & naut] le plus lentement possible
~SMOOTH - [mech] (of files) à taille très fine
~SMOOTH FILE - [mech] (a file having very fine teeth, designed for removing very small amounts of material) lime *f* douce
~SOFT TEMPER - [metall] trempe *f* douce
~SPACE - [mech] espace *m* nuisible, espace *m* mort
[phys] (a chamber under vacuum) espace *m* sous vide, volume *m* évacué
[comput] espace *m* inutilisé, capacité *f* inutilisée

:AD SPACE - [telev] zone *f* de silence
;POT - [radio] (region in which radio reception in
a given frequency band is very poor or non exist-
ent) point *m* mort, zone *f* de silence
;TEAM - [mech] (steam which has been in a machine or process and has given up all its available
heat and pressure) vapeur *f* passive
;TEEL - [metall] acier *m* calmé
;TUDIO - [radio & telev] (studio without resonance) studio *m* insonorisé
'IME - [radar] (the period following the reception
of a signal in which a beacon cannot be triggered)
emps *m* mort
phys & nucl] (the interval after an event, during
which a system does not respond to another) temps
m mort
'IME CORRECTION - [phys] (counting rate correc-
ion in case events occur during dead time) cor-
ection *f* de temps mort
'URNS - s. dead end
VALL - [acoust] paroi *f* à isolement acoustique
VATER - [hydr] petite marée *f*; (non-circulating
vater) eau *f* morte
VEIGHT - s. dead load
shipbuild] port *m* en lourd
-WEIGHT CAPACITY - [shipbuild] portée *f* en lourd
-WEIGHT PRESSURE GAUGE - [mech] (gauge in
which the pressure to be measured acts on a piston
connected to a pointer) balance *f* manométrique
-WEIGHT SAFETY VALVE - s. dead-load safety
valve
-WEIGHT TONNAGE - s. dead-weight
VELL - [constr] (shaft cut through an impermeable
stratum to allow water through) puits *m* stérile
VIND - [naut] vent *m* contraire
VINDOW - [constr] fausse fenêtre *f*
radio] fenêtre *f* sourde
VOOL - [text] laine *f* pelade, laine *f* des peaux
avalies
ZONE - [el acoust] zone *f* morte
contr] (of a control element) zone *f* d'insensibi-
ité
meas] (in a measured variable, the range of
values obtaining no effective response from the
nstrument) zone *f* morte
:ADEN, to - [gen] amortir, atténuer
acoust] insonoriser
:ADENER - [constr] matériau *m* insonorisant
:ADENING - [gen] amortissement *m*
constr] insonorisation *f*
constr] (a special mixture used on the floor) en-
duit *m* insonorisant
:ADLIGHT - [shipbuild] opercule *m* de hublot
:ADLINE - [gen & comm] date *f* limite, dernier de-
ay *m*
:ADLY NIGHTSHADE - [chem] belladone *f*
:ADS - s. dead ground [geol]
:ADSTOCK - [agric] cheptel *m* mort
:ADWOOD - [shipbuild] (the horizontal courses)
bois *m* blanc
shipbuild] (the solid timber at the stern) massif *m*
:AERATE, to - [gen] désaérer, chasser l'air
:AERATING - s. deaeration
:QUIPMENT - [ind] installation *f* de désaération
:AERATION - [phys] évacuation *f* de l'air, désae-
ration, dégazage *m*
:AERATOR - [mech] (apparatus for removing air
from some other substance) désaérateur *m*
DEAF - [med] sourd
~AID - [acoust] (a device used to improve the hear-
ing of external sounds) appareil *m* de prothèse audi-
tive
~-AND DUMB - s. deaf-mute
~-MUTE - [med] sourd-muet *m*
DEAFENING - [constr] (mixture placed between
floor joists to deaden sounds) enduit *m* insonorisant
DEAFNESS - [med] surdité *f*
DEAIRING - [ind chem] (the process of extracting
air from clay in preparing it for ceramic work)
désaération *f*
~MACHINE - [ind chem] (apparatus consisting of
two pug mills in tandem and a vacuum chamber
placed between them, used for the extraction of
air from ceramic clay) désaérateur *m*
DEAL, to - [gen] agir, traiter, commercer, trafiquer
[comm] négocier
DEAL - [gen] quantité *f*
[comm] affaire *f*, marché *m*
[timber] bois *m* blanc, sapin *m*
~FLOOR - [constr] plancher *m* en bois
~FRAME - [mech] scie *f* alternative
~OUT, to - [gen] donner, distribuer
DEALBATION - [chem] blanchiment *m*
DEALUMINIZING - [metall] (elimination of aluminium from other metals) élimination *f* de l'aluminium
DEAN AND STARK APPARATUS - [chem] (a laborato-
ry apparatus used to measure the amount of water
present in an oil) appareil *m* de Dean et Stark
DEARSENIFICATOR - [chem] (apparatus for bringing
sulphuric acid and hydrogen sulphide into contact,
to remove arsenic from the acid) appareil pour
l"élimination de l'arsenic
DEARTH - [gen] disette *f*, manque *m*, cherté *f*
DEATH - [gen] mort *f*, extinction *f*, destruction *f*
~DUTY - [leg] impôt *m* sur les successions
~POINT - [biol] (the limit beyond which an organism
or a cell cannot live) limite *f* de résistance
~RATE - [statist] (the number of persons dying
over a specified period in a given locality) *f*orta-
lité *f*
~RAY - [phys] (radiation which would kill human
beings from a distance) rayon *m* de la mort
~WATCH - [zool] (small wood-boring insect) ver *m*
de bois, horloge *m* de la mort, atropos *m*
~-WATCH BEETLE - s. dead watch
DEATHNIUM - [electron] (electron resuming an
empty place in a crystal rectifier and recombining)
centre *m* de recombinaison
DEBACLE - [gen] ruine *f*, désastre *m*
[met] (the breaking up of ice on the surface of
rivers in spring) dégel *m*, débacle *m*
DEBAGGING OPERATION - [rubber ind] (in tyres)
extraction *f* de la chambre de vulcanisation
DEBAR, to - [gen] exclure, interdire, priver
[leg] suspendre, exclure
DEBARK, to - [gen] écorcer
DEBARKATION - [naut] débarquement *m*
DEBASE, to - dégrader, abaisser, altérer
DEBEADER - [mech] (for tyres) machine *f* à déta-
lonner
DEBEADING - [mech] détalonnage *m*

DEBENTURE - [fin] (instrument, generally a bond, for the repayment of a debt out of some specified security) obligation *f*
[leg] (customs certificate issued for a draw-back) certificat *m* de droit au remboursement des droits de douane
~HOLDER - [fin] obligataire *m*, porteur *m* d'obligations
~STOCK - [fin] capital-obligations *m*
DEBENZOLIZATION OF OIL - [chem] désessenciement *m* de l'huile, dégazolinage *m* de l'huile
~OF THE ABSORBER - [chem] désessenciement *m* de l'absorbeur
DEBENZOLIZING - [chem] débenzolage *m*
~BY REFRIGERATION - [chem] débensolage *m* par le froid
DEBIT, to - [fin] débiter
DEBIT - [fin] débit *m*
DEBITEUSE - [glass man] débiteuse *f*
DEBLOOMING - [oil ind] enlèvement *m* de substances fluorescentes du pétrole, blanchiment *m* du pétrole lampant
~AGENT - [chem] agent *m* de blanchiment
DEBONE, to - [gen] désosser
DEBOOSTER - [mech] équilibreur *m*
DEBREEZING SCREEN - [impl] (for coke) tamis *m*
DEBRIS - [gen] restes *mpl*
[geol] débris *m*
DEBT - [gen and comm] dette *f*, créance *f*
DEBTOR - [gen and comm] débiteur *m*
DEBUG, to - [comput] (the elimination of errors from a programme, or the repairing, generally of circuits, in the computer) corriger un programme
DEBUGGING - [comput] correction *f* d'un programme
DEBUNCHING - [electron] (elimination of electron bunching by space charge effect) dégroupage *m*
DEBURR, to - [mech] ébarber, ébavurer
DEBURRING - [metall] (the action of eliminating burrs) ébarbage *m*, ébavurage *m*
~OF WOOL - [text] échardonnage *m* de la laine, épluchage *m* de la laine
DECABORANE - [chem] (a plastics stabilizer, vulcanization agent, intermediate in the production of polymers and rayon delustrant) décaborane *m*
DECADAL - [math] (in decades) en décades, décimal
DECADE - [comput] (group of ten) échelle *f* décimale
~ATTENUATOR BOX - [el] (box in which the attenuators are divided into groups of ten) boîte *f* d'atténuateurs à décades
~BRIDGE - [el] (a Wheatstone bridge with ten separate coils) pont *m* à décade
~CAPACITANCE BOX - [el] (a box in which the capacitors are divided into groups of ten) boîte *f* de capacités à décades
~CIRCUIT - [el] échelle *f* de dix
~CONDENSER - [el] condensateur *m* à decade
~CONDUCTANCE BOX - [el] (a box in which the conductances are divided into groups called decades) boîte *f* de conductances à décades
~COUNTER - [electron] (a counting device registering in the scale of ten) compteur *m* à decades
~INDUCTANCE BOX - [el] boîte *f* d'inductances à décades
~RESISTANCE BOX - [el] boîte *f* de résistances à décades
DECADE RESISTOR - [el] résistance *f* à décades
~RHEOSTAT - [el] rhéostat *m* à décades
~SCALER - [instr] (scaler with a scaling factor of ten) échelle *f* de comptage décimal
~STAGE - [comput] (of a counter) étage *m* de comptage
~VOLTAGE DIVIDER - [el] réducteur *m* de tension à décades
DECADENCE - [gen] décadence *f*
DECAFFEINATED - [chem] décaféiné
DECAGON - [geom] décagone *m*
DECAGRAM - [meas] décagramme *m*
DECAHEDRON - [geom] sécaèdre *m*
DECAHYDRONAPHTHALENE - [chem] (a solvent for rubber, resins, fats and waxes) décahydronaphtalène *m*
DECALAGE - [aero] (in a biplane, the angle betwe[en] the chords of the upper and lower planes, measur[ed] in a plane parallel to the plane of simmetry) décalage *m*, interinclinaison *f*
DECALCIFICATION - [med] décalcification *f*
DECALCIFYING AGENT - [chem] adoucisseur *m* d'e[au]
DECALCOMANIA - [draw] décalcomanie *f*
~PROCESS - [plast ind] (process of transferring designs from paper backing to plastic products) [dé]calcomanie *f*
DECALESCENCE - [metall] (the phase of heat absorption by an iron or steel melt as its temperature passes the critical point) décalescence *f*
DECALITRE - [meas] décalitre *m*
DECAMETRE - [meas] décamètre *m*
DECANGULAR - [geom] décagonal
DECANNING - [nucl] (the removal of the can from fuel element after irradiation) dégainage *m*
DECANNULATION - [med] enlèvement *m* d'une canule
DECANT, to - [gen] (to pour off liquid from a vessel leaving precipitate or sediment behind) décanter
DECANTATION - [chem] (separation of a liquid an[d] a sediment, or of two immiscible liquids, by pouring, or drawing off through pipes) décantation *f*
~TANK - [ind chem] bac *m* de décantation, cuve *f* [de] décantation
DECANTER - [ind chem] (apparatus for carrying o[ut] decantation operations on an industrial scale) décanteur *m*, verre *m* de décantation
DECANTING - s. decantation
~VESSEL - [ind chem] bac *m* de décantation, décanteur *m*
DECAPSULATION - [mech] décapsulation *f*
DECARBONATE, to - [chem] éliminer l'anhydride carbonique
DECARBONIZATION - s. decarburization
DECARBONIZE, to - [metall] (to remove carbon deposited in a combustion chamber, on a valve-head or the like) décarburer
DECARBONIZED - [metall] décarburé
DECARBOXYLATION - [chem] décarboxylation *f*
DECARBURIZATION - [metall] (the process of abstracting carbon from the surface layer of steel by heating in a suitable atmosphere) décarburisation *f*
DECARBURIZE, to - s. decarbonize, to
DECASTYLE - [arch] (a portico with ten columns) décastyle *m*
DECATIZE, to - [text] (the operation of giving a

finish to fabrics by forcing steam through them) décatir
ECATIZING - [text] décatissage m, délustrage m
CLOTH - [text] doublier m de décatissage
MACHINE - [text] décatisseuse f
ECAY, to - [gen] pourrir, se gâter, croupir, détruire
[timber] se moisir
ECAY - [gen] (decomposition of matter) décomposition f, désintégration f
[nucl] (decrease of radioactivity in a substance) décroissance f, désintégration f
[opt] (of a luminescent screen) extinction f
[electron] (in a damped harmonic oscillator) diminution f
[constr] (of a building or any fabric) dégradation f
CHAIN [nucl] (decay series, so that the produced nuclide forms the next nuclide) série f radioactive
˜CHARACTERISTIC - [electron] (parameter of the decline of luminescence in a C.R.T. screen) caractéristique f de persistance
˜COEFFICIENT - s. decay factor
CONSTANT - [nucl] constante f de désintégration, constante f radioactive
˜CURRENT - [metall] (an annealing current) courant m de recuit
˜CURVE - [nucl] (a curve showing the decrease of activity with the increase of time) courbe f de désintégration, courbe f de décroissance
˜FACTOR - [phys] (the rate of decay of a diminishing oscillation) coefficient m d'amortissement
˜MODULUS - [phys] (in a damped harmonic oscillation) module m d'amortissement
OF LUMINESCENCE - [electron] (the decrease of phosphorescence emission) déclin m de luminescence
˜PARTICLE - [phys] particule f de désintégration
˜PERIOD - [nucl] (half life; the time for the decay of one half of the atoms of a radioactive substance) période f radioactive
˜PROBABILITY - [nucl] probabilité f de désintégration
PRODUCT - [nucl] (radioactive or stable nuclide which results from the disintegration of a radionuclide) produit m de désintégration, descendant m
RATE - [nucl] (the decay rate of a radioactive substance and the transformation rate of a bombarded nuclide) vitesse f de désintégration
˜SEQUENCE - [nucl] série f de désintégration
˜TIME - [metall] (the time taken for the annealing) temps m de recuit
[nucl] temps m de décroissance
[electron] (in charge-storage tubes) période f d'extinction, persistance
ECAYED VEGETABLES - [soil] matières fpl végétables décomposées
ECCA NAVIGATIONAL SYSTEM - [aero] (system for determining the position of an aircraft instantaneously, continuously and very accurately. Continous waves are transmitted from several grouns stations and a special receiver in the aircraft shows the position by automatic measurement of phase differences) système m de radio-navigation Decca
ECELERATE, to - [gen & mech] décélerer, ralentir
ECELERATING ELECTRODE - [electron] (electrode potential providing an electric field which diminishes the velocity of the electrons in a beam) électrode f de ralentissement
DECELERATION - [gen & mech] ralentissement m, freinage m
[nucl] (negative acceleration) décélération
[comput] (e.g. of a magnetic tape) décélération f
˜PARACHUTE - [aero] parachute m de freinage
DECELERATOR - [mech] modérateur m
DECELEROMETER - [instr] (instrument measuring the change of speed during deceleration) décéléromètre m
DECENTRALIZATION - [gen] décentralisation f
DECENTRALIZE, to - [gen] décentraliser
DECEREBRATION - [med] décérébration f
DECHLORINATION - [hydr] déchloration f
DECINEPER - [el] (one tenth of a neper; the unit of voltage and current attenuation) décineper m
DECIBEL - [acoust] (one tenth of a bel) décibel m
˜METER - [instr] (instrument measuring noise level in decibels) décibelmètre m, hypsomètre m
DECIDUOMA - [med] déciduome m
DECIDUOUS - [bot] caduc
˜TEETH - physiol] dents fpl de lait
˜WOOD - [bot] bois m feuillu
DECIGRAM - [meas] décigramme m
DECILE - [statist] décile
DECILOG - [meas] (logarithmic scale division to measure the logarithm of two values of a quantity) décilog m
DECIMAL - [math] décimal
˜BALANCE - [meas] (a balance with arms in the 10:1 ratio) balance f décimale
˜CANDLE - [light] bougie f décimale
˜DIGIT - [comput] chiffre m décimal
˜EQUIVALENT - [math] équivalent m décimal
˜FLOATING POINT - [comput] point m décimal flottant, virgule f flottante
˜FRACTION - [math] fraction f décimale
˜NOTATION - [comput] (writing of quantities in the rate of 10) numération f décimale
˜NUMBER - [math] nombre m décimal
˜POINT - [math & comput] (the point, or comma, separating the integral and the fractional values of ten) virgule f décimale
˜READOUT - [comput] signal m de sortie décimal
˜SCALE - [meas] échelle f décimale
˜SCALER - [comput] compteur m décimal
˜-TO-BINARY CONVERSION - [comput] (the mathematical process whereby a number in the scale of ten is converted into a number in the scale of two) conversion f décimale-binaire
DECIMETRE - [meas] décimètre m
˜WAVES - [radio] (electromagnetic radiations with a wavelength of about ten centimetres) ondes fpl décimétriques
DECINORMAL - [gen] décinormal
DECISION - [gen] décision f
˜ELEMENT - [comput] (circuit performing a logical operation) circuit m logique
˜FEEDBACK - [comput] annonce f en retour
DECK, to - [naut] ponter
DECK - [gen] (a level or a storey) étage m
[shipbuild] pont m
˜BEAM - [shipbuild] (a deck-stiffening member) barrot m de pont
˜BRIDGE - [constr] (a bridge designed to carry the

track) pont *m* à tablier supérieur
DECK CAMBER - [shipbuild] tonture *f*
~CARGO - [naut] chargement *m* en pontée
~CRANE - [naut] grue *f* de pont
~FLOORING - [shipbuild] bordage *m* de pont
~HEAD - [constr] plafond *m*
~HEIGHT - [shipbuild] hauteur *f* d'entrepont
~HOUSE - [shipbuild] roufle *m*, roof *m*
~LANDING AIRCRAFT - [aero] (aircraft designed to land on the deck of an aircraft-carrier) aéronef *m* embarqué
~LIGHT - [shipbuild] écoutille *f*
~LINE - [naut] livet *m* de pont
~LOAD - s. deck cargo
~MACHINERY - [naut] auxiliaires *mpl* de pont
~OF CARDS - [comput] jeu *m* de cartes, pile *f* de cartes
~OF SCREEN - [ind chem] (a single-level or sizing unit of a reciprocating screen) surface *f* criblante
~PLAN - [shipbuild] plan *m* des ponts supérieurs
~PLANKING - [shipbuild] bordé *m* de pont
~PLATING - s. deck planking
~SHEATHING - [shipbuild] soufflage *m*, revêtement *m*
~STRINGER - [shipbuild] (that part of the deck which is adjacent and attached to the shell plates) serre *f*
DECKED - [shipbuild] ponté
DECKING - [constr] platelage *m*, imperméabilisation *f* d'une terrasse
[mining] encagement *m*
[timber] bois *m* de pont
~TIMBER - s. decking [shipbuild]
DECKLE - [paper man] forme *f*
~EDGE - [paper man] (the thin feathery edge of hand-made paper) bavure *f*
~-EDGE TRIMMER - [impl] cisaille *f* dechiqueteuse
~-EDGED PAPER - [paper man] papier *m* à la forme
~PULLEYS - [paper man] poulies *fpl* de guides
~STRAPS - [paper man] (in a paper-making machine, the rubber straps forming a deckle edge on the paper) bandes *fpl* pour bords
DECLARE, to - [gen] déclarer
DECLARED EFFICIENCY - [el] (the efficiency which is guaranteed by the manufacturers for electric machines) rendement *m* garanti
DECLASSIFY, to - [gen] déclasser
DECLENSION - [gen] diminution *f*, inclinaison *f*, déclinaison *f*
DECLINATE, to - [gen] décliner
DECLINATE - [bot] (pointing downwards) décliné
DECLINATION - [astr] (the arc of a celestial meridian, or the angle which it subtends at the centre of the earth, between an observed celestial body and the equinoctial) déclinaison *f*
[surv] (the angular deviation of a magnetic compass) déclinaison *f*
~CIRCLE - [astr] (the great circle passing through the celestial poles and cutting the celestial equator at right angles) cercle *m* de déclinaison
~OF THE COMPASS - [instr] déclinaison *f* du compas
DECLINOMETER - [instr] (designed to measure the angle between the magnetic and the geographic meridians) boussole *f* de déclinaison
DECLIVITY - [gen] déclivité *f*, pente *f*
DECLUTCH, to - [mech] débrayer
DECOATING - [metall] of layers deposited under vacuum) élimination *f* de couches métallisées
DECOBALTER - [ind chem] (apparatus for removing cobalt catalyst in the Oco process) installation *f* pour l'élimination du catalyseur au cobalt
DECOCT, to - [chem] décocter
DECOCTION - [chem] (liquid made by boiling vegetable products in water) décoction *f*, décocté *m*
~METHOD - [brew ind] méthode *f* de la décoction
DECODE, to - [comput] déchiffrer, décoder, interpréter
DECODER - [comput] (circuit in which several inpu are excited simultaneously to produce a single input) décodeur *m*, changeur *m* de code
DECODING - [comput] décodage *m*
~CHECK - [telev] platine *f* de décodage
DECOHERER - [radio] mechanical device for restoring the coherer to its normal condition) décohéreu *m*
DECOILER - [mech] (used for wire in a rolling mill dérouleur *m*
DECOKE, to - [mech] s. decarbonize, to
DECOLLATION - [gen] (decapitation) décapitation *f*
[med] (in obstetrics) décollation *f*
DECOLLEMENT - [med] (in surgery, the separation of an organ from the tissues to which it is attache décollement *m*
DECOLORANT - [chem] décolorant *m*
DECOLORATION - [gen & chem] décoloration *f*
DECOLORIZE, to - [chem] décolorer
DECOLORIZING POWER - [chem] pouvoir *m* décolora
DECOLOUR, to - s. decolorize, to
DECOMETER - [instr] (Decca navigation instrument giving dial indication of position as obtained from the ground stations) décomètre *m*
DECOMPENSATION - [med] (failure of a heart whic though diseased, has maintained its strength) décompensation *f*
DECOMPOSABLE - [gen] décomposable
DECOMPOSE, to - [gen & chem] décomposer
DECOMPOSED ORGANIC MATTER [soil] humus *m*, sol *m* organique
DECOMPOSITION - [chem] (the breaking down of a molecule into simpler molecules or into atoms) décomposition *f*
~BY HEAT - [chem] pyrogénation *f*, décomposition pyrogénée
~HAZARD - [chem] (tendency to decomposition) susceptibilité *f* de décomposition
~OF SLUDGE - [hydr] décomposition *f* des boues
~VOLTAGE - [el chem] (the lowest potential at whi a continuous electrochemical process can continu tension *f* de dissociation
DECOMPOUNDING WINDING - [el] (series winding on a compound-wound generator) enroulement *m* differentiel
DECOMPRESSION - [gen & phys] (the relief of pressure) décompression
~AND RECOMPRESSION LOOP - [soil] boucle *f* de la courbe de décompression et recompression
~CHAMBER - [physiol] (a chamber equipped for the decompression of human bodies subjected to high pressure; also to reduce human pressure in the simulation of high-altitude flying) chambre *f* de décompression
~DEVICE - [mech] (any mechanism designed to facilitate the start of an engine when cold) dispositif *m* de décompression
DECOMPRESSOR - [mech] (in internal-combustion engines) décompresseur *m*

DECONTAMINATE, to - [gen & nucl] décontaminer
DECONTAMINATING AGENT - [chem] agent *m* de décontamination
DECONTAMINATION - [nucl] (the removal of radioactive material, e.g. from clothing) décontamination
~AREA - [nucl] zone *f* de décontamination
~INDEX - [nucl] (logarithm of the ratio between initial specific radioactivity and final specific radioactivity) indice *m* de décontamination
~PLANT - [nucl] (plant where decontamination is effected) installation *f* de décontamination
~SQUAD - [nucl] équipe *f* de décontamination
DECONTROL, to -[gen] (the elimination of controls) débloquer, libérer
DECCOPERIZING - [metall] décuivrage *m*
DECORATIVE BOND - [constr] liaison *f* décorative
~IRON - [metall] fer *m* à dessins
~LAMINATE - [metall] (laminate in which colour and design are integral in the sheet) stratifié *m* à décor incorporé
DECORING - [metall] débourrage *m*
DECORTICATED - [bot] (deprived of bark) décortiqué, écorcé
DECORTICATION - [bot & med] décortication *f*
~OF THE LUNG - [med] (the removal of a thickened pleura from the lungs) décortication *f* du poumon
DECORTICATOR - [mech] (machine for removing the skins or shell of natural products, such as seeds for processing fibres, such as flax, ramie etc) décortiqueur *m*
DECOUPLE, to - [el] (to eliminate or reduce undesired coupling effects between two circuits) découpler
DECOUPLING - [el] (reduction or elimination of any coupling between circuits which is not desired) découplage
~CIRCUIT - [el] circuit *m* de découplage, étage *m* séparateur
~RESISTANCE - [el] résistance *f* de découplage
DECOY - [gen] leurre *m*, appât *m*, piège *m*
DECREASE, to - [gen] diminuir
[naut] (of the wind] tomber
[text] défroisser
DECREASE - [gen] diminution *f*
~OF CURRENT - [el] baisse *f* de courant
DECREASER - [mech] (reducing piece in gas installations) cône-réduction *m*
DECREASING - [text] défroissement *m*
~LOAD - [el] charge *f* décroissante
~MACHINE - [text] machine *f* à diminuer
~PITCH SCREW - [mech] hélice *f* à petit pas
~WORMPITCH - [mech] petit pas *m*
DECREMENT - [math] (abbreviation for logarithmic decrement) décrément *m* logarithmique
~GAUGE - [instr] manomètre *m* à viscosité, jauge *f* à amortissement
~VISCOSITY GAUGE - s. decrement gauge
DECREMETER - [instr] (instrument designed to measure the decrement) décrémètre *m*
DECREPITATION - [crystall] (the sound made by crystals when heated) décrépitation *f*
DECRIMP, to - [text] lisser, défriser
DECRUSTATION - [gen] élimination *f* de la croûte
DECRUSTING OF A FILTER - [hydr etc] décroûtage *m* d'un filtre, ratissage *m* d'un filtre
DECRYPTION - [comput] (of information) déchiffrage *m*, déchiffrement *m*
DECTRA NAVIGATION SYSTEM - [radio] (modified Decca system) système *m* de radio-navigation Dectra
DECUBITUS - [med] (the position of a patient when lying in bed) décubitus *m*
~ULCER - [med] escarre *f* de décubitus
DECUMBENT - [bot] lying flat) étalé
DECURRENT - [bot] with a prolongued base) décurrent
DECURVED - [bot] (bent downwards) courbé vers le bas
DECUSSATE - [bot] (having leaves in pairs at right angle with other pairs) décussé
DECUSSATION - [physiol] (in the central nervous system) décussation, croisement *m* en X
DECYCLIZATION - [chem] decyclisation *f*
DECYL MERCAPTAN - [chem] (a component of synthetic rubber additives) décylmercaptan *m*
~-OCTYL METHACRYLATE - [chem] (polymerizable monomer for plastics) méthacrylate *m* de décyle et d'octyle
DEDENDUM - [mech] (the radial distance between the pitch circle of a gear wheel and base between the teeth) distance *f* entre fond de dent et cercle primitif, creux *m* (d'une dent), profondeur *f* du creux
~ANGLE - [mech] (of gear wheel) angle *m* de pied de dent
~CIRCLE - [mech] cercle *m* de base, cercle *m* de racine
~LINE - [mech] ligne *f* de base, ligne *f* de racine, droite *f* d'évidement
~LINE OF CONTACT - [mech] ligne *f* d'engrènement (de contact)
DEDENT, to - [text] (to beat out the fabric) débosseler
DEDIFFERENTIATION - [biol] (retrogressive changes in differentiated tissues) dédifférentiation *f*
DEDOLOMITIZATION - [geol] (dolomite rock recrystallization) dédolomitisation *f*
DEDUCED POSITION RECKONING - [aero] (method of determining the position of an aircraft from the True Airspeed, True Heading steered and the best possible estimate of wind velocity) point *m* estimé
DEDUCT, to - [gen] déduire, soustraire, défalquer
DEDUCTION - [gen & comm] déduction *f*, réduction *f*
~CARD - [comput] carte *f* retenue
DEDUSTING - [gen] dépoussiérage *m*
~AGENT - [ind chem] (a substance which can be used to reduce or suppress the formation of dust) agent *m* dépoussiérant
~SCREEN - [gen & min] tamis *m*
DEE - [electron] (of a cyclotron; the half hollow cylinder in which a spiral beam of elecyrons is accelerated all the time) dee *m* (chambre d'accélération des particules dans un cyclotron)
~LINE - [electron] (structural element supporting the cyclotron dees) support *m* des dees
DEEM, to - [gen] estimer, juger
DEEMPHASIS - [radio] (reduction of relative strength of the higher audio frequencies) affaiblissement *m*, atténuation *f*, désaccentuation *f*
~NETWORK - [radio] atténuateur *m*
DEENERGIZE, to - [el] désexciter, couper, désamorcer, décrocher
DEEP - [gen] profond, foncé

DEEP - [naut] (of the sea) haute (mer)
~-BAR INDUCTION MOTOR - [el] moteur *m* asynchrone à induit en barres
~BLUE - [dyes] bleu foncé
~BORING - [mining] sondage *m* profond
~CAVITY MOULD - [rubber and plast ind] moule *m* à cavité profonde
~-CHANNEL SCREW - [plast ind] (type of extruder screw in which the channel between the flight turns is deeper than usual) vis *f* à filet profond
~DIMENSION PICTURE - [photo & telev] (picture having an appreciable depth without being stereoscopic) image *f* à profondeur de champ
~-DRAUGHT VESSEL - [naut] navire *m* à grand tirant d'eau
~DRAW - [metall & plast ind] (moulding operation in which the depth of moulding is considerable in relation to its other dimensions) emboutissage *m* profond
~DRAWING - [metall] s. deep draw
~-DRAWING SHEET - [metall] tôle *f* pour emboutissage profond
~-DRAWING STEEL - [metall] acier *m* pour emboutissage
~-DRAWING STRIP - [metall] bande *f* pour emboutissage
~-DRAWING TEST - [metall] essai *m* d'emboutissage, essai *m* de macro-attaque
~DREDGER - [mech] drague *f* profonde
~DRILLING - [mining] forage *f* profond
~ETCH TEST - [metall] essai *m* de macro-attaque
~-FISHING VESSEL - [naut] navire *m* de pêche hauturier
~FLIGHT WORM - [plast ind] (of extruders) vis *f* à filet creux
~FREEZER - [th eng] installation *f* de surgélation
~-FREEZING - [th eng] surcongélation *f*
~-FREEZING PLANT - s. deep freezer
~HOLE - [mining] s. deep drilling
~LEVEL - [mining] voie *f* de fond
~-LEVEL MINING - [mining] exploitation *f* des niveaux inférieurs
~-LOADING TRAILER - [transp] remorque *f* à plateforme basse
~-MINED COAL - [mining] charbon *m* d'exploitation souterraine
~MINING - [mining] exploitation *f* souterraine
~SEA - [geogr] haute mer *f*, eau *f* profonde
~-SEA CABLE - [telecomm] câble *m* sous-marin pour grands fonds
~-SEA DEPOSITS - [geol] (sediments accumulating beyond the reach of ordinary material) dépôts *mpl* pélagiques
~-SEA LOAD - [surv] plomb *m* de grande sonde
~-SEA OOZE - [geol] boue *f* pélagique
~-SEATED BLOWHOLE - [metall] (in welding) soufflure *f* profonde
~STAMPING - [metall] emboutissage *m* profond
~TANK - [shipbuild] (in the hold) cale *f* à eau
~TEST - [mining] sondage *m* profond
~TONE - [acoust] son *m* grave
~-WATER LINE - [naut] flottaison *f* en charge
~WELL - [constr] (shaft sunk into a water containing stratum) puits *m* profond
~-WELL DRAINAGE - [hydr] drainage *m* par pompage dans des puits profonds
~-WELL PUMP - [mech] (a centrifugal pump at the bottom of a deep bore hole) pompe *f* de fond
DEEP X-RAY THERAPY - [med] (effected by hard radiation which passes through the superficial layers) radiothérapie *f* profonde
DEEPEN, to - [gen] creuser, approfondir
[acoust] (of voice) forcer les graves
DEER - [zool] cerf *m*, daim *m*
DEFACE, to - [gen] défigurer
DEFACING OF MARKS - [gen] (on crates, bales etc) effacement *m* des marques
DEFALCATE, to - [gen] défalquer, déduire
DEFAT, to - [agric] dégraisser
DEFAULT, to - [gen & comm] manquer
DEFAULT - [gen] défaut *m*
[leg] non-exécution *f*
DEFAULTER - [leg & comm] défaillant *m*, partie *f* défaillante
DEFECATION - [ind chem] (process of purifying sugar solutions, e.g. by coagulation of albuminoids and neutralization with milk of lime) défécation *f*, épuration *f*
DEFECT - [gen] défaut *m*
[mech] défaut *m*, vice *m* de construction
[railw] (of a rail) courbure *f*
~IN MATERIAL - [gen] défaut *m* dans un matériau
DEFECTIVE - [gen] défectueux
~CASTING - [metall] défaut *m* de la fonte
DEFER, to - [gen] différer
DEFERENT - [astr] (epicycle) déférent
DEFERENTITIS - [med] déférentite *f*, inflammation *f* des canaux déférents
DEFERRED - [gen] différé
~ASSETS - [comm] (expenses paid within a period of business for any item to be accounted for in the next period) actif *m* différé
~CALL - [telecomm] (telephone) communication *f* différée
~SHARES - [fin] (shares whose dividends are paid last) actions *fpl* différées
DEFERRIZE, to - [hydr] déferriser
DEFERVESCENCE - [med] (the period during which temperature falls) défervescence *f*
DEFIBERING - [text] effilochage *m*
DEFIBRATOR - [paper man] défibreur *m*
DEFICIENCY - [gen] déficience *f*, manque *m* défaut
[med] déficience *f*, carence *f*, insuffisance *f*
~AREA - [agric] zone *f* déficitaire
~DISEASE - [med] (disease caused by a deficiency of vitamins etc) maladie *f* de carence
DEFICIENT - [gen] déficient, insuffisant
[med] (of mental deficiency) déficient
~[constr] instable
DEFINABLE - [gen] définissable
DEFINE, to - [gen] définir, délimiter
DEFINITE - [gen] défini, précis, concret, délimité
~INTEGRAL - [math] intégrale *f* definie
~PROPORTIONS LAW - [chem] (the law of definite, or constant, proportions states that pure substances always contain the same elements combined in the same proportions) loi *f* des proportions constantes
~TIME LAG - [el] (fitted to relays to delay their operations) retard *m* (indépendant
~TIME LAG RELAY - [el] relais *m* à retard indépendant
~VARIATION - [bot] variation *f* définie
DEFINITION - [gen] définition

[telev] (a detailed reproduction on a television screen; called "resolution" in the USA) définition *f*
[acoust] (clarity of sounds) fidélité *f*
[photo & cin] profondeur *f* de champ, netteté *f*
DEFINITION CHART - [telev] mire *f*
~CIRCLE - [el] (in impedance matching) cercle *m* de définition
~IN DEPTH - [photo] profondeur *f* de champ
~OF IMAGE - [telev] définition *f* de l'image de l'écran
~TEST CARD - [telev] mire *f* de définition
~WEDGES - [telev] quadrillage *m* de lignes de définition
DEFINITIVE - [gen] définitif
~NUCLEUS - [genet] noyau *m* définitif
DEFLAGRATE, to - [gen and chem] brûler rapidement, enflammer, déflagrer
DEFLAGRATING SPOON - [ind chem] (small metal vessel provided with a long stem to hold small amounts of a substance to be burnt in a vessel of gas) cuillère *f* pour déflagrations
DEFLAGRATION - [chem] (sudden rapid combustion, accompanied by the production of sound, heat and light) déflagration *f*
DEFLAGRATOR - [chem] (of explosives) déflagrateur *m*
DEFLASH, to - [metall] (to remove flash from a product) ébarber, ébavurer
DEFLATE, to - [gen] dégonfler
DEFLATION - [gen] dégonflement *m*
[fin] déflation *f*
~SLEEVE - [aero] manche *f* de dégonflement
DEFLECT, to - [gen] dévier, incliner, détourner, écarter
DEFLECTED JET - [mech] (reversed jet) jet *m* déflecté
DEFLECTING - [gen] déviation *f*
~COIL - s. deflection coil
~ELECTRODE - [electron] (electrode in a C.R.T. used to deflect the elctron beam) électrode *f* de déviation
~FORCE - [phys] (in measuring instruments) force *f* de déviation
~MAGNET - [el] aimant *m* de déviation
~PLATE - s. deflector plate
~ROLLER - [text] rouleau *m* de détour
~TORQUE - [instr] (torque resulting from electrostatic or electromagnetic effects on the moving element) couple *m* actif, couple *m* moteur
~TOOLS - [metall] déviateurs *mpl*
~YOKE - [electron] (mechanism encircling the electron beam tube to produce a deflection of the electron beam) bobine *f* de déviation
DEFLECTION - [gen] déviation *f*, déflexion *f*, déformation *f* par flexion
[phys] flèche *f*, fléchissement *m*
[constr] flexion *f*
[constr] (in bridges) flèche *f*, fléchissement *m*
[electron] (the displacement of the beam caused by the deflecting field) déviation *f*
[instr] (the extent of movement of the indicating element of an instrument when a reading is shown) déviation *f*
[metall] (distorsion of a mould, a press or associated structures under pressure) flexion *f*, déformation *f*
~AMPLIFIER - [radio] amplificateur *m* de déviation
~ANGLE - [surv] angle *m* d'inclinaison
DEFLECTION CIRCUIT - [el] circuit *m* de déviation
~COEFFICIENT - [electron] (of a cathode-ray tube) coefficient *m* de déviation
~COILS - [electron] (coils placed round the neck of a cathode-ray tube and energized to deflect the electron beam) bobines *fpl* de déviation
~CURRENT - [electron] courant *m* de la bobine de balayage
~DEFOCUSING - [electron] (loss of sharpness of the image on a cathode-ray tube screen, caused by the electron beam meeting the screen at an acute angle in the outer part) élargissement *m* du spot, défocalisation *f*
~DISTORTION - [telev] distorsion *f* par déviation
~ELECTRODE CONNEXION - [electron] couplage *m* des électrodes de balayage
~ELECTRODES - [electron] électrodes *fpl* de balayage, électrodes *fpl* de déviation, plaques *fpl* de balayage
~MAGNET - [electron] aimant *m* de balayage
~METHOD - [instr] (method of measurement by reading the deflection of an instrument) méthode *f* de déviation
~OF THE RAIL - [railw] écrasement *m* de la voie
~PLANE - [telev] plan *m* de déviation
~PLATE - [electron] plaque *f* de déviation
~POLARITY - [instr] polarité *f* de déviation
~POTENTIOMETER - [instr] (a measuring instrument replacing the null device) potentiomètre *m* à déviation
~SENSITIVITY - s. deflectional sensitivity
[electron] (of a magnetic deflection cathode-ray tube) sensibilité *f* (de déviation magnétique
[electron] (of an electrostatic-deflection cathode-ray tube) sensibilité *f* de déviation électrostatique
~SPACE - [electron] espace *m* de déviation
~SPEED - [telev] (velocity of the light spot on the screen of the cathode-ray tube under the influence of deviation) vitesse *f* de balayage
~TENSOR - [electron] tenseur *m* de déviation
~TEST - [constr] essai *m* de flexion
~VALVE - [electron] (electronic tube with output current controlled by the deflection of an electron stream) tube *m* à parcours électronique dirigé
~YOKE PULL-BACK - [telev] jeu *m* du bloc de balayage
DEFLECTIONAL SENSITIVITY - [electron] (the linear displacement of a fluorescent spot in a cathode-ray tube) sensibilité *f* de déviation
DEFLECTOR - [mech etc] déflecteur *m*
[el] (arc deflector) passage *m* en chicane
[ind chem] (of fluids) déflecteur *m*
~BRATTICE - [mining] cloison *f* d'aérage
~COIL - [electron] (in a cathode-ray tube) bobine *f* de déviation
~PANE - [auto] déflecteur *m*
~PLATE - [electron] (a deflecting electrode in a cathode-ray tube) plaque *f* de déviation
~ROLL - [metall] rouleau *m* de guide
DEFLOCCULATING AGENT - [ind chem] agent *m* de dispersion, agent *m* défloculant
DEFLOCCULATION - [chem] (redispersal of a sediment) défloculation *f*
DEFLORATION - [bot] défloration *f*
DEFLOREZ PROCESS - [chem] (a liquid-phase cracking process used in petroleum refining) procédé *m* de De Florez

DEFLUXION - [hydr] écoulement *m*
DEFOAM, to - [ind chem] débarasser de la mousse
DEFOAMER - [ind chem] agent *m* antimousse
oil ind] additif *m* antimousse
DEFOCALIZE, to - [opt] défocaliser
DEFOCUS-DASH METHOD - [electron] mode *m* non-focalisé-trait
~-FOCUS MODE - [electron] mode *m* non-focalisé-focalisé
DEFOCUSING - [photo] déréglage *m* de la mise au point
DEFORESTATION - [agric] déboisement *m*, déboisage *m*
DEFORM, to - [mech] (to change shape, e.g. under some stress) déformer
[gen] déformer, défigurer
DEFORMABILITY - [phys] (capacity of substance to undergo change of physical dimensions under certain conditions) déformabilité *f*
DEFORMABLE BODY - [phys] (body undergoing changes in size and shape under the influence of external stresses) corps *m* déformable
DEFORMATION - [phys] (a change in the shape of a body, e.g. bending, caused by forces acting upon it) déformation *f*, distorsion *f*
~BANDS - [crystall] (metal crystal areas which assume different orientations owing to slip) bandes *f*pl de déformation
~BY LOAD - [mech] déformation *f* sous charge
~BY SHEARING STRESS - [mech] déformation *f* au cisaillement
~DUE TO TORSION - [mech] déformation *f* due à la torsion
~ENERGY OF FISSION - [nucl] (the energy required to deform a nucleus) énergie *f* de déformation
~GROOVE - [metall] (for the lamination of deformed bars) cannelure *f* pour la fabrication de barres deformées
~POTENTIAL - [el] (the effective electric potential acting on a free electron in a semiconductor owing to a deformation in the crystal lattice) potentiel *m* de déformation
~STRAIN - [mech] (the change in size and shape accompanying external stresses) déformation *f* par contrainte
~STRENGTH - [mech] résistance *f* à la déformation
DEFORMETER - [instr] (instrument designed to impose a specified distortion on the model of a structure) déformomètre *m*
DEFRIBINATION - [med] (the removal of fibrin) défibrination *f*
DEFROST, to - [gen] dégeler
[th eng] (the elimination of ice on a screen, or window-screen; also in a refrigerator or deep-freezer) dégivrer
[food] (of deep-frozen articles of food) décongéler
DEFROSTER - [gen] dégivreur *m*
DEFROSTING - [gen & th eng] dégel *m*, dégivrage *m*, décongélation *f*
DEFROSTING AGENT - [ind chem] agent *m* de décongélation
DEFUSE, to - [gen] (e.g. a bomb] désamorcer
DEGAS, to - [chem] (to remove gas from a substance) dégazer
DEGASER - [chem] (substance which removes entrained or occluded gases) agent *m* de dégazage, dégazeur *m*
DEGASIFY, to - [gen] dégazer
DEGASSING - [chem & phys] (the process of extracting air or other gases from a substance) dégazage *m*
[metall] (process designed to avoid porosity in castings) dégazage *m*
[electron] (the liberation from all gases in the manufacture of electronic tubes) dégazage *m*
~ANNEAL - [metall] recuit *m* de dégazage
~FLUX - [metall] dégazant *m*
~INSTALLATION - [metall] installation *f* de dégazage, poste *m* de dégazage
~STAGE - [metall] étage *m* de dégazage
DEGATE, to - [metall] (to detach the gate from a moulding) décarotter
DEGAUSSING - [el] (neutralization of magnetization) démagnétisation *f*
DEGEAR, to - [mech] débrayer, désenclencher
DEGENERACY - [electron] (the condition in which several modes have the same resonant frequency) dégénération *f*
DEGENERATE GAS - [electron] (an electronic gas consisting of free electrons in the crystal lattice of a conductor is a degenerate gas) gaz *m* dégénéré
~OSCILLATING SYSTEM - [phys] (vibrating system with several degrees of freedom) système *m* vibrant dégénéré
~STATE - [phys] (so called when different states of motion correspond to the same energy level) état *m* dégénéré
DEGENERATED YEAST - [brew ind] levure *f* dégénérée
DEGENERATION - [gen] dégénération *f*
[med] dégénérescence *f*, appauvrissement *m*, abâtardissement *m*
[radio] (a feedback which decreases the amplification) contre-réaction *f*
[electron] (the decrease of damping in a circuit between the grid and the cathode) dégénération *f*
~RESULTING FROM INBREEDING - [genet] dégénérescence *f* consanguine
DEGENERATIVE AMPLIFIER - [radio] amplificateur *m* à contre-réaction
~FEEDBACK - s. degeneration [radio]
DEGLUTITION - [physiol] (swallowing) déglutition *f*
DEGRADATION - [gen] dégradation *f*
[chem] (process in which a complex alkaloid molecule is reduced to simpler forms) dégradation
[chem] (cleavage of a cellulose molecule to produce hemicelluloses, pentosam etc) dégradation *f*
[nucl] réduction *f* d'énergie
[telev] déterioration *f* de l'image
~ENERGY - [phys] (changes in the form of energy) énergie *f* de dégradation
~OF ENERGY - [phys] dégradation *f* de l'énergie, perte *f* d'énergie
DEGRAS, to - [gen & leather ind] (to remove fats from) dégraisser
DEGRAS - [leather ind] (the fats obtained from sheepskins after the oiling process) degras *m*
DEGREASE, to - [gen] (to remove grease from parts, usually by means of a detergent bath) dégraisser, décaper
DEGREASING - [gen] (treating with suitable solvents to remove grease) dégraissage *m*, décapage *m*
~AGENT - [ind chem] (special preparation for removing grease from mechanical parts) agent *m*

de dégraissage
DEGREASING BATH - [ind chem] bain *m* de dégraissage, bain *m* de décapage
~MACHINE - [mech] (machine designed to remove grease from mechanical parts) machine *f* à dégraisser
~OF THE WOOL - [text] dégraissage *m* de la laine
~OPERATION - s. degreasing
~TANK - [ind chem] (apparatus containing a degreasing agent, in which parts are submerged for cleaning) bac *m* de dégraissage, bac *m* de décapage
DEGREE - [gen] degré *m*, rang *m*, taux *m*, condition *f* qualité *f*
[math] s. degree of arc
[phys] (the unit of temperature difference) degré *m*
~CENTIGRADE - [meas] degré *m* centigrade, degré *m* Celsius
~OF ACCURACY - [meas] (in instruments) degré de précision
~OF ACIDITY - [chem] degré *m* d'acidité
~OF ARC - [math] (one three-hundred-and-sixtieth part of the circumference of a circle) degré
~OF ATTENUATION - [gen] degré *m* d'atténuation
~OF BEATING - [paper man] degré *m* de défibrage
~OF CARBONIZATION - [gas ind] degré de cuisson
~OF COMPACTNESS - [soil] degré *m* de compacité
~OF CONSISTENCY - [soil] degré *m* de consistance
~OF CONSOLIDATION - [soil] degré *m* de consolidation
~OF COOLING - [gen & brew ind] degré *m* de refroidissement
~OF COVERAGE - [electron] (the degree of coverage of the carrier-electrode by the electron-emitting substance) taux *m* de recouvrement
~OF CRUSHING - [min] (of coke) taux *m* de broyage
~OF CURE - [ind chem] (the extent to which curing and hardening has taken place) degré *m* de durcissement
~OF CURRENT RECTIFICATION - [el] rendement *m* d'un redresseur
~OF CURVE - [surv] degré *m* de courbure
~OFDAMPING - [phys] (the measure of the damping in an oscillatory system) taux *m* d'amortissement
~OF DENSITY - s. degree of compactness
~OF DISSOCIATION - [chem] (the relation between the total number of molecules and the number of those dissociated) degré *m* de dissociation
~OF EXCENTRICITY - [mech] excentricité *f*
~OF ELASTICITY - [phys] degré *m* d'élasticité
~OF ELECTROLYTIC DISSOCIATION - s. degree of dissociation
~OF EXPANSION - [phys] taux *m* de dilatation
~OF FILLING - [brew ind] degré *m* de remplissage
~OF FLUIDITY - [chem] (the consistance or "flowability" of a substance) degré *m* de fluidité
~OF FREEDOM - [phys] (the ways in which a system may change in respect of its configuration, e.g. a rigid body can move along three coordinates and also rotate about them, hence it has six degrees of freedom) degré *m* de liberté
~OF GRINDING - [mech] (the fineness to which a substance is ground) taux *m* de broyage
~OF HARDNESS - [phys] degré *m* de dureté
~OF IONICITY - [phys] (relating to excited states) degré *m* d'excitation
~OF IONIZATION - [chem] (the number of ionized molecules in a solute in relation to the total number of molecules in it) taux *m* d'ionisation
DEGREE OF MODULATION - [radio] taux *m* de modulation
~OF PLASTICITY - [rubber ind] degré *m* de plasticité
~OF POLYMERIZATION - [chem] (the extent to which polymerization of two or more monomers has proceeded) degré *m* de polymérisation
~OF PURITY - [electron] (in transistors, the ratio of the number of semiconductor atoms to those of the impurity) degré *m* de pureté
~OF RAMMING - [metall] serrage *m*
~OF REBOUND - [rubber ind] hauteur *f* de rebond
~OF RESET - [radio] niveau *m* de rétalissement
~OF SAFETY - [gen] facteur *m* de securité
~OF SATURATION - [phys & chem] degré *m* de saturation
~OF SHRINKING - [text] degré *m* du retrait
~OF SMEARING - s. degree of fluidity
~OF STEEPING - [brew ind] degré *m* de macération
~OF SUPERCHARGING - [el] taux *m* de suralimentation
~OF SUPERHEAT - [phys] (the extent to which the temperature of superheated steam exceeds that of the vapour phase in the generator) degré *m* de surchauffe
~OF TEMPERATURE - [phys] ((a single unit in a thermometric scale) degré *m* de température
~OF TOLERANCE - [mech] tolérance *f*, limite *f*
~OF TWIST - [text] torsion
~OF VOLTAGE RECTIFICATION - s. degree of current rectification
~OF VULCANIZATION - [rubber ind] (the extent to which crosslinking between sulphur and the double bonds of the rubber molecule has proceeded) degré *m* de vulcanisation
DEGREES K - s. Kelvin temperature scale
DEGRESSIVE - [genet] (step towards degeneration) dégressif
DEGUMMED SILK - [text] soie *f* cuite, soie *f* dégommée
DEGUMMING - [text] dégommage *m*
~AGENT - [text] produit *m* de dégommage, produit *m* de décreusage
~BATH - [text] bain *m* de dégommage
DEHAIRING - [leather ind] dépilage *m*, dépoilage *m*
DEHISCENCE - [bot] (the spontaneous opening of fruits at maturity) déhiscence *f*
DEHORNING - [vet] (the operation of cutting the horns) décornage *m*
DEHUMIDIFICATION - [th eng] (in air conditioning) séchage *m*
DEHUMIDIFIED AIR - [th eng] (in air conditioning) air *m* sec
DEHUMIDIFIER - [ind] (apparatus for removing small amounts of moisture from a material) déshumidificateur *m*, déshydratant *m*
DEHYDRATE, to - [chem] (to remove chemically combined water from a substance) deshydrater
DEHYDRATED - [gen & chem] déshydraté
DEHYDRATING - [gen & chem] déshydratant
~PACKET - [ind chem] (a small parcel of a substance which has the property of taking up moisture, used to preserve parts or small machines) sachet *m* déshydratant
~SACHET - s. dehydrating packet
DEHYDRATION - [chem] (the elimination of water from crystals, oils or other bodies containing it,

as by heating, distillation etc) déshydratation *f*, anhydrisation *f*
DEHYDRATOR - [ind chem] (apparatus for removing substantial amounts of water from a material) déshydrateur *m*
DEHYDROACETIC ACID - [chem] (an intermediate, fungicide and plasticizer) acide *m* déhydro-acétique
DEHYDROGENASE - [biochem] (an enzyme which removes hydrogen from compounds containing it) déshydrogénase *f*
DEHYDROGENATION - [chem] (the removal of hydrogen from a compound by chemical means) déshydrogénation *f*
DEHYDROGENIZE, to - déshydrogéner
DEHYDROTHIOTOLUIDINE - [chem] (an intermediate for direct cotton azo dyestuffs) déhydro-thio-toluidine *f*
DEICE, to -[gen] (to detach ice forming on the external parts of an object, i.e. aircraft etc) dégivrer
DEICER - [aero] (device for detaching ice from the external parts of an aircraft) dégivreur *m*
DEICING - [gen] dégivrage *m*
~AIR - [th eng] (warm air directed upon parts liable to icing) air *m* de dégivrage
~BOOT - [aero] (rubber structure enclosing parts, e.g. leading edges of aerofoils, which are liable to icing, and capable of being expanded by internal pressure, thus breaking the ice layers and detaching it) gaine *f* de dégivrage
~DEVICE - [gen] dispositif *m* antigel
~FLUID - [ind chem] fluide *m* antigel
~SWITCH - [el] interrupteur *m* de commande de dégivrage
DEINHIBIT, to - [mech] (to remove anti-corrosion protection (inhibitor)) désinhiber
DEINKING WASTE - [paper man] (a recycling process) eaux *fpl* de désencrage
DEIONIZATION - [phys] (the process of the return of an ionized medium to a non-ionized condition) désionisation *f*
~POTENTIAL - [electron] (the potential at which a gas discharge tube becomes non-conductive because of cessation of ionization) potentiel *m* de désionisation
~TIME - [electron] (the time required for the control electrode of a gas discharge tube to regain control after anode current interruption) temps *m* de désionisation
DEIONIZE, to - [electron] désioniser
DEIONIZER - [ind chem] désionisant *m*
DEIONIZING GRID - [electron] (electron grid which accelerates deionization in its vicinity) grille *f* de désionisation
~RATE - [electron] taux *m* de désionisation, vitesse *f* spécifique de désionisation
DEIRONING - [metall] déferrisation *f*
DEKAMETER - s. decametre
DEKATRON - [electron] (electronic tube designed to count impulses) compteur *m* décatron
DELAMINATION - [metall etc] (separation of the layers of a laminated material) clivage *m*
DELASTING - [shoe man] démontage *m* de la forme
DELAY, to - [gen] retarder, différer
DELAY - [gen] retard *m*, délai *m*
[telecomm] (in telephony] attente *f*
~-ACTION CIRCUIT BREAKER - [el] disjoncteur *m* retardé
DELAY BREAKER - [el] disjoncteur *m* temporisé
~CABLE - [telecomm] ligne *f* de retard
~CIRCUIT - [comput] circuit *m* à retard
~CORRECTION NETWORK - [comput] correcteur *m* de phase
~DISTORTION - [el] distorsion *f* non-linéaire
~-FREQUENCY DISTORSION - [radio] (occurring when the delay is not constant) distortion *f* retard-fréquence
~LINE - s. delay cable
~-LINE MEMORY- s. delay-line storage
~-LINE REGISTER - [comput] registre *m* à ligne de retard
~-LINE STORAGE - [comput] mémoire *f* à ligne à retard, mémoire *f* dynamique
~MECHANISM - [mech & photo] mécanisme *m* de retardement
~NETWORK - [telecomm] (artificial line providing a predetermined delay in transmission) ligne *f* à retard
~PERIOD-[mech] (in internal combustion engines, the interval between the passing of the spark and the rise in pressure of the petrol) retard *m*
~SCREEN - [telev] écran *m* à retard
~UNIT - [comput] unité *f* à retard
DELAYED - [gen & mech] retardé, temporisé
~ACTION - [el] (causing a delay in the operation of a switch etc) commande *f* temporisée
~-ACTION RELEASE - [photo] déclencheur *m* à action différée
~ALPHA PARTICLES - [nucl] (easily emitted by excited nuclei in a beta-disintegration process) particules *fpl* alpha de long parcours
~AUTOMATIC GAIN CONTROL - s. delayed automatic volume control
~AUTOMATIC VOLUME CONTROL - [radio] antifading *m* retardé
~BLANKING SIGNAL - [telev] signal *m* de suppression retardé
~COINCIDENCE - [nucl] (in a radiation count) coïncidence *f* retardée
~CRITICAL - [nucl] critique *m* différé
~EQUALIZER - [radio] correcteur *m* de phase
~FIRING - [mech] (in internal combustion engines) allumage *m* retardé
~FISSION NEUTRON - [nucl] neutron *m* différé
~FUSE - [chem] (explosives) fusé *f* à retardement
~IMPULSE GENERATOR - [el] générateur *m* d'impulsions retardées
~INHERITANCE - [genet] hérédité *f* retardée, prédétermination
~NEUTRON EMITTER - [nucl] émetteur *m* de neutrons différés
~RECLOSURE - [el] fermeture *f* retardée
~RINGING - [telecomm] (in telephony) sonnerie *f* à retardement
~SCANNING - [telev] (a sweep of the beam whose beginning is delayed after the pulse) analyse *f* différée
~SWEEP - s. delayed scanning
DELEAD, to - [ind chem] éliminer le plomb
DELEGATE, to - [gen] déléguer, députer
DELEGATE - [gen] délégué
DELEGATION - [gen] délégation *f*, députation *f*
DELETE, to - [gen] effacer, rayer
DELETERIOUS - [gen] délétère, nocif, nuisible

DELETION OF ERRORS - [comput] suppression f des erreurs
DELF - [mining] couche *f* de charbon mince
DELFT - [potter] (a close-grained earthenware, made with a tin glaze for painting; after this has been done, a transparent glaze is applied to protect the decoration) faïence *f* de Delft
DELICATE - [gen] délicat, délié, fin
[instr] de précision
[ind chem] (of chemical precision tests) de précision
DELIGNIFICATION - [bot] (the destruction of lignin by a fungus) destruction *f* de la lignine
DELIGNIFY, to [ind chem] (of man-made fibres) éliminer la lignine
DELIME, to - [leather ind] (the operation of removing lime salts before tanning) déchauler
DELIMING - [leather ind] déchaulage *m*
DELIMIT, to - [gen] délimiter, circonscrire
DELIMITATION OF THE IMAGE FIELD - [photo] délimitation *f* du champ de l'image
DELINEATE, to - [gen] ébaucher, esquisser
DELINEATION - [photo] tracé *m*
DELIQUESCE, to - [chem] tomber en déliquescence, devenir déliquescent
DELIQUESCENCE - [chem] (a process resulting from the exposure to humid air of certain substances which, because of the very low vapour pressures of their saturated solutions, can take up enough water to dissolve them under such conditions) déliquescence *f*
[bot] (the liquefaction and gelatinization of cell walls) déliquescence *f*
DELIQUESCENT - [chem] (having the ability to pick up atmospheric moisture) déliquescent
DELIQUIUM - [med] évanouissement *m*, défaillance *f*
DELIRIUM - [med] (a violent disturbance of the consciousness) délire *m*
DELIVER, to - [gen] délivrer, libérer, fournir, livrer, produire
[mech] (discharge from a pump) débiter
[med] (in obstetrics) accoucher, délivrer
[comm] livrer, remettre
~THE WARP, to - [text] fournir la chaîne
DELIVERY - [gen] délivrance *f*, distribution *f*
[mech] (of pump or compressor) débit *m*
[mech] (the amount of fluid discharged by a pump under given conditions) débit *m*, refoulement *m*, décharge *f*
[plumb] (the piping attached to the discharge opening of a pump) tuyau *m* de refoulement
[oil ind] refoulement *m*
[comm] remise *f*, livraison *f*
[med] accouchement *m*, délivrance *f*
~BOBBIN - [text] bobine *f* d'alimentation, bonine *f* fournisseuse
~CANAL - [hydr] conduite *f* de décharge
~CHUTE - [mech] descendeur *m*, déversoir *m*
~CONDUIT - [metall] conduite *f* de refoulement
[hydr] conduite *f* forcée, conduite *f* de refoulement
~COUNTER - [text] compteur *m* de production
~END - [mech] sortie *f*
[mining] pied *m* de taille
~FLAP - [text] (an ejection device) clapet *m* de sortie, clapet *m* d'éjection
~FUNNEL - [text] entonnoir *m* de sortie
DELIVERY GUIDE - [mech] (a device designed to support the metal being delivered by the rolling mill) guide-support *m*
~HEAD - [hydr] (the height to which a pump forces a liquid above its own level) hauteur *f* de refoulement
~INSTRUCTION - [comput] instruction *f* d'extraction, instruction *f* de sortie
~LATTICE - [text] tablier *m* délivreur
~MAIN - s. delivery conduit
~MOTION - [text] dispositif *m* de livraison
~NOTE - [comm] bordereau *m* de livraison
~ORDER - [comm] bon *m* de livraison
~OUTPUT - [mech] débit *m* de pompage
~PIPE - [hydr] tube *m* d'alimentation, tuyau *m* de refoulement
~PIPE OUTLET - [hydr] sortie *f* de tuyau de refoulement
~POINT - [hydr] (in pumps) bouche *f* d'alimentation
~PRESSURE - [mech] pression *f* de refoulement
~PUMP - [hydr] pompe *f* de refoulement
~RAKE - [text] herse *f* de sortie
~RAMP - [metall] rampe *f* de sortie
~RATE - [hydr] (of pumps) débit *m*
~RATE ANALYZER - [meas] analyseur *m* de débit
~RECORD BOOK - [comm] registre *m* de livraison
~ROLLER - [text] cylindre *m* de décharge, cylindre *m* de sortie
~ROLLS - [text] cylindres *mpl* délivreurs
~SPINDLE - [text] broche *f* delivreuse
~TABLE - [print] plateau *m* de sortie
[text] tablier *m* de sortie, table *f* de sortie
~TIME - [comm] delai *m* de livraison
~TRUMPET - s. delivery funnel
~VALVE - [hydr] vanne *f* de décharge, vanne *f* de
~VAN - [transp] camionette *f*, voiture *f* de livraison
~VOUCHER - [comm] bon *m* de livraison
DELL - [geogr] vallon *m*, creux *m*
DELLINGER EFFECT - [phys] (sudden fading of sky-waves signals due to rapid and intense ionization in the ionosphere caused by solar storms) effet *m* Dellinger
DELORENZITE - [min] (ytterbium, uranium, iron and titanium containing mineral) délorencite *f*
DELPHINIC ACID - [chem] acide *m* isovalérique
DELPHININE - [chem] (an intensely toxic alkaloid, having an action similar to that of aconitine) delphinine *f*
DELTA - [gen] (the fourth letter of the Greek alphabet) delta *m*
[geogr] (of a river) delta *m*
[el] - s. delta connexion
~CIRCUIT - [el] montage *m* en delta, montage *m* en triangle
~CONNECTED - [el] monté en triangle
~CONNEXION - [el] (method of connecting the windings of three-phase electrical apparatus, which are in series, current being taken from, or supplied to, the junctions) montage *m* en triangle
~-DELTA CONNEXION - [el] couplage *m* à double triangle
~IRON - [metall] (the polymorphic form of iron which is in a stable state between 1403° C. and the melting point, and has a space lattice identical) fer *m* delta (variété allotropique du fer, stable au dessus de 1400° C)

DELTA-L CORRECTION - [telev] (in colour television) correction *f* delta-L
~-MATCHED IMPEDANCE AERIAL - [radio] (unbroken resonant aerial to which the leads of a transmission line are connected in the shape of a y) antenne *f* à attaque en triangle
~MATCHED IMPEDANCE ANTENNA - s. delta-matched impedance aerial
~-MATCHING - [radio] (a method of connecting a feeder to a point with the correct impedance on a half-wave aerial) attaque *f* en triangle
~METAL - [metall] (a term originally used for an alpha-beta brass with a small addition of iron. It is now applied also to manganese and aluminium brasses) métal *m* delta
~MODULATION - [radio] (a pulse modulation involving quantization) modulation *f* delta
~NETWORK - [el] (formed from three impedances in series) réseau *m* en triangle
~NOISE - [comput] (the core store noise which is caused by the difference in half-select pulse outputs) bruit *m* delta
~RAYS - [phys] (electrons which are ejected by recoil when a charged particle passes through matter) rayons *mpl* delta
~-STAR - [el] étoile-triangle *f*
~-STAR CONNEXION - [el] montage *m* étoile-triangle
~-THREE ANGLE - [el] (the angle between the axis of the flapping-hinge and a line drawn at right angles to the blade axis, in plan view) angle *m* delta trois
~VOLTAGE - [el] (the voltage between the alternate lines of a symmetrical s- and x-phase system) tension *f* en triangle
~WING - [aero] (wing of triangular form) aile *f* delta
DELTAIC DEPOSIT - [geol] (accumulation of sand, clay, animal remains, plant debris etc washed in from land) dépôt *m* deltaique
DELUGE - [gen] déluge *m*
DELUSTRANT - [chem] (agent removing lustre from man-made fibres) agent *m* de délustrage
DELUSTRING - [text] dépolissage *m*
~CALENDER - [text] calandre *f* de matage
DELVE, to - [mining] creuser, approfondir
DEMAGNETIZATION - [phys] (elimination of a magnetic condition in a body, e.g. a steel part which has been machined on a magnetic chuck) démagnétisation *f*, désaimantation *f*
~CURVE - [phys] (curve representing the product of the increasing demagnetizing force and the decreasing flux density of a permanent magnet) courbe *f* de démagnétisation
DEMAGNETIZE, to - [phys] (to restore a magnetized ferromagnetic body to its neutral condition) démagnétiser, désaimanter
DEMAGNETIZER - [phys] appareil *m* de démagnétisation
DEMAGNETIZING - [phys] démagnétisation *f*, désaimantation *f*
~BOBBIN - [el] appareil *m* de démagnétisation
~FACTOR - [phys] (factor used to obtain the value of the demagnetizing field) facteur *m* de démagnétisation
~FIELD - [phys] (the internal field of a magnet) champ *m* démagnétisant
~TURNS - [el] (turns of the armature which produce a magnetomotive force of the field when the machine is running under load) spires *fpl* antagonistes
DEMAL - [meas] (one gram-equivalent of solute per cubic decimeter of solution) unité *f* de mesure de densité
~SOLUTION - [chem] (solution of one demal) solution *f* normale de mesure
DEMAND, to - [gen] demander, exiger
DEMAND - [gen] demande *f*, exigence *f*
~ATTACHMENT - [el] (attachment for a meter, converting it into a maximum demand indicator) indicateur *m* de maximum
~FACTOR - [el] (the ratio between the maximum demand and the total load) coefficient *m* d'utilisation, coefficient *m* de consommation
~INDICATOR - [el] (meter carrying a pointer which indicates the maximum value of the average power utilized during successive equal intervals of time) indicateur *m* de maximum, indicateur *m* de pointe
~METER - [el] (a meter with maximum demand indicator) compteur *m* à indicateur de maximum, enregistreur *m* de pointe
~RECORDER - s. demand meter
~-TYPE MASK - [aero] (oxygen mask which can be regulated according to altitude) masque *m* inhalateur réglable
~WORKING - [telecomm] service *m* rapide, service *m* "à la demande"
DEMANTOID - [min] (a silicate of lime and iron) demantoïde *m*
DEMARCATION CURRENT - s. demarcation potential
~LINE - [gen & surv] ligne *f* de démarcation, délimitation *f*
~POTENTIAL - [el biol] (synonym of injury potential: the difference in the potential between affected and unaffected parts of a living structure) potentiel *m* de lésion
DEMARGINATE, to [ind chem] (the operation of eliminating stearin from vegetable oils) éliminer la stéarine
DEMATERIALIZATION - [nucl] (the process in which a positive and a negatively charged material particle unite and cease to exist as such, the energy which corresponds to their material masses being reproduced as electromagnetic radiation) dématérialisation *f*
DEME - [genet] (local group of closely related organisms) dème *m*
DEMERARA SUGAR - [sugar ind] sucre *m* de second jet, sucre *m* non raffiné
DEMESH, to - [mech] (of gears) s. disengage
DEMI-DITONE - [mus] tierce *f* mineure
~HUNTER - [horol] (a case half metal and half glass) boîtier *m* de montre mi-métal mi-verre
~QUAVER - [mus] double croche *f*
DEMIJOHN - [gen] dame-jeanne *f*
DEMILUSTRE - [text] demi-lustre
DEMINERALIZE, to - [chem] déminéraliser
DEMINERALIZE, to - [chem] (e.g. of water) déminéralisation
DEMISTER - [auto etc] (mechanism designed to demist the windscreens) dispositif *m* anti-buée
DEMISTING - [phys] (elimination of condensed moisture on windscreens or windows, normally by causing evaporation) élimination *f* de la buée
DEMITINT - [paint] demi-teinte *f*
DEMODIFICATION - [metall] (the elimination of

sodium from aluminium alloy scraps) élimination *f* du sodium
:MODULATION - [radio] (the process by which the signal with which a carrier has been modulated is separated from the carrier itself) démodulation *f*
:FFECT - [radio](the apparent reduction in the modulation degree of a received signal) suppression *f* de la modulation
:MODULATOR - [radio] (device used for detection purposes) démodulateur *m*, détecteur *m*, redresseur *m*
:MOGRAPHY - [statist] (the science dealing with population statistics and variations) démographie *f*
:MOLISH, to - [gen & constr] démolir, abattre, détruire
:MOLITION - [constr] démolition *f*
WORK - [constr] travaux *mpl* de démolition
EMONSTRATE, to -[gen & math] démontrer
EMONSTRATION - [gen] démonstration *f*
EYEPIECE - [opt] (of a microscope) dispositif *m* à double oculaire
VAN - [transp] véhicule *m* publicitaire
EMOULDING - [rubber & plast ind] démoulage *m*, démontage *m* d'un moule
EMOUNT, to - [gen & mech] démonter
EMOUNTABLE - [gen & mech] démontable
TUBE - [electron] (transmitting valve which is demountable for inspection purposes) tube *m* démontable
EMPY - [mining] (a sector of a mine where outbursts of gases are likely) secteur *m* dans une mine où des explosions sont susceptibles de se produire
EMULCENT - [chem] (having the property of soothing) émollient *m*, adoucissant *m*
EMULSIBILITY - [chem] (chiefly of oils) résistance *f* à l'émulsification
TEST - [ind chem] essai *m* de résistance à l'émulsification
EMULSIFICATION - [chem] (resistance to emulsification) désémulsification *f*
NUMBER - [ind chem] (an empirical index of the resistance of a lubrificant to emulsification in the presence of water or steam) indice *m* de désémulsification
EMULSIFIER - [chem] (a substance which increases the emulsion resistance of an oil) agent *m* désémulsionnant
EMULSIFY, to - [chem] désémulsionner
EMURRAGE - [naut] surestaries *fpl*, droits *mpl* de magasinage
DAYS - [naut & comm] jours *mpl* de surestaries
DEMYELINATION - [med] démyélinisation *f*
DENARY - [math] dénaire, décimal
DENATURANT - [nucl] (an isotope which is not itself fissionable, and when added to fissile material deprives the latter of its fissile characteristics) dénaturant *m*
DENATURATED ALCOHOL - [chem] (alcohol made unfit for drinking by the addition of noxious substances, e.g. methyl alcohol, pyridine etc) alcool *m* dénaturé
DENATURATION - [nucl] (the addition of an isotope to fissile material to make it unusable in atomic weapons) dénaturation *f*
[chem] (especially of proteins) dénaturation *f*
DENATURE, to - [chem & nucl] dénaturer
DENATURING - [chem] (addition of any product to industrial alcohol, to render it unfit for drinking) dénaturation *f*
DENDRITE - [min] crystal formations with a tree-like shape) dendrite *f*
DENDRITIC - [min] (branching like a tree; arborescent) dendritique
~FIGURE - [biol] (appearing in experimental embryology) figure *f* dendroïde
~MARKINGS - [geol] (branch-like markings appearing on rock faces and fractures) marques *fpl* dendritiques
~ULCER - [med] (branch-like ulcer of the cornea) ulcère *f* dendritique
DENDROGRAPH - [instr] (instrument designed to measure the periodical swelling and shrinking of tree trunks) dendrographe *m*
DENDROMETER - [instr] (instrument designed to measure the diameter and the height of trees) dendromètre *m*
DENERVATED - [med] énervé
DENERVATION - [med] (the deprivation of nerve supply) énervation *f*
DENGUE - [med] (tropical disease transmitted to man by mosquitoes) dengue *m*
DENIAL - [gen] refus *m*, désaveu *m*, dénégation *f*
[telecomm] suspension of service for delayed payment) suspension *f* du service
DENIER - [text] (measure of filament fineness; the weight in grammes of 9000 metres of yarn is its denier) denier *m*
~BALANCE - [instr] (instrument designed to read directly the denier of a yarn, from the weight of a 9-metre length) balance *f* pour la mesure du denier
DENIM - [text] (strong cotton fabric mainly used for overalls) denim *m* (croisé)
DENITRATE, to - [chem] dénitrifier
DENITRIDING - [chem] dénitruration *f*
DENITRIFICATION - [chem] (the liberation of elementary nitrogen from nitrogenous compounds in the soil) dénitrification *f*
~TOWER - [ind chem] (in a sulphuric acid plant) tour *f* de dénitrification
DENITRIFYING - s. denitrification
DENOISE, to - [telev] éliminer les perturbations
DENOMINATION - [gen] dénomination *f*
DENOMINATOR - [math] dénominateur *m*
DENORA CELL - [ind chem] (electrolytic cell for the commercial production of chlorine and sodium hydroxide) pile *f* Denora
~MERCURY CELL - s. Denora cell
DENOTATION - [gen] indication *f*, signe *m*
DENOTE, to - [gen] dénoter, indiquer, signifier
DENSE - [gen] dense, serré, compact, épais
[metall] (in foundry work) compact
[photo] (indicating the lack of good transparency) sombre
~-CRUSH LOAD - [railw] charge *f* d'écrasement
~FLINT - [opt] flint *m* dense
~FOG - [met] brouillard *m* épais
~PACKING - [soil] sol *m* à architecture serrée
~PROPELLANT - [astronaut] propergol *m* lourd
~SINTERING - [metall] frittage *m* à densité maximale
DENSIFIED IMPREGNATED WOOD - [timber] (compressed wood evenly impregnated with a synthetic

resin) bois *m* imprégné comprimé
DENSIFIED LAMINATED WOOD - [timber] bois *m* stratifié comprimé
DENSIFIER - [hydr] (a sludge concentrator) épaississeur *m* de boues
DENSIFYING - [hydr] (sludge thickening) épaississement *m* de boues
DENSIMETER - [instr] (any type of instrument measuring density) aréomètre *m*, densimètre *m*, pèse-acide *m*
DENSIMETRY - [meas] (the measuring of density) densimétrie *f*
DENSITENSIMETER - [instr] (for the determination of the density and the pressure of a vapour) densitensimètre *m*
DENSITOMETER - [instr] (for the measurement of the densities of exposed and developed films) densitomètre *m*
DENSITY - [phys] (mass per unit volume) densité *f*
[el] (current density, i.e. the quantity of current flowing per unit cross-sectional area of a conductor) densité *f* de courant
[opt & photo] densité *f* optique
[photo] (more specifically for negatives) obscurcissement *m*
[metall] (in foundry work) homogénéité *f* du grain
~ALTITUDE - [phys] (altitude corresponding to a given density in a standard atmosphere) altitude *f* densité
~AS SINTERED - [metall] densité *f* après frittage
~BOTTLE - [ind chem] (an accurately calibrated vessel of very thin glass used for the determination of specific gravity) flacon *m* à densité, flacon *m* à mesurer la densité, pycnomètre *m*
~CURRENT - [el] courant *m* spécifique
~CURVE - [opt] courbe *f* de réfringence
~FUNCTION - [astronaut] fonction *f* de répartition
~HEIGHT - [phys] (in the International Standard Atmosphere, the height at which the air density is equal to a given density) altitude *f* densité
~MODULATION - [electron] (of an electron beam; alternative variation in density impressed on the electrons of a beam) modulation *f* de densité
~OF A GAS - [phys] densité *f* d'un gaz
~OF ACID - [chem] concentration *f* d'un acide
~OF AN ELECTRON BEAM - [electron] (the density of the electron current of the beam at any specified point) densité *f* d'un faisceau électronique
~OF AN ION BEAM - [electron] (the density of the ion current of the beam at any specified point) densité *f* d'un faisceau ionique
~OF CHARGE - [el] densité *f* de charge
~OF CLOTH - [text] (according to the number of picks per inch or centimetre) densité *f* du tissu
~DISLOCATION - [crystall] (the density, or concentration lines, i.e.the curves separating displaced and undisplaced portions of a crystal) densité *f* de dislocation
~OF EAR - [agric] densité *f* de l'épi
~OF ELECTRONS - [phys] (the number of free electrons per unit volume) densité *f* d'électrons
~OF KINETIC ENERGY - [acoust] (the total instantaneous energy density of sound per unit volume) densité *f* d'énergie cinétique
~OF SEED - [agric] quantité *f* de semence
~OF SOIL - [soil] densité *f* du sol
~OF SOLARIZATION - [photo] densité *f* de solarisation
DENSITY OF SOUND ENERGY - [acoust] (energy de. sity of sound; the particle vibration energy per ur volume in a reverberating sound field) densité *f* d'énergie acoustique
~OF SURFACE CHARGE - [el] (limiting value of the quotient which is obtained by dividing the charge over the surface of a body by the area of the corresponding element of surface) densité *f* de charg superficielle
~OF THE SWARD - [agric] densité *f* de la surface e herbe
~OF THE TOTAL ELECTROMAGNETIC ENERGY - [el] (the electromagnetic energy divided by the volum densité *f* d'énergie électromagnétique totale
~OF THE WEFT - [text] (the number of filling threa per inch) nombre *m* de fils de trame
~OF TRAFFIC - [railw & road traffic] densité *f* de l circulation, densité *f* du trafic
~OF VOLUME CHARGE - [el] (limiting value of the quotient which is obtained by dividing the charge b the element of volume containg it) charge *f* volumique
~OF WINDING - [text] densité *f* de renvidage
~RANGE - [photo] intervalle *m* des densités extrêmes
~THRESHOLD - [photo] seuil *m* de noircissement
~VALUE - [opt] réfringence *f*
DENSOGRAPHY - [radiat] densographie *f*
DENT, to - [gen] entailler, denteler
[mech] (of a wheel) denter
DENT - [gen] brèche *f*, entaille *f*, dentelure *f*
[auto] (the damage on a car body) coup *m*
[mech] dent *f*
[text] (one of the wires forming a reed) dent *f*
~CORN - [agric] mais *m* à dent de cheval
~MAIZE - s. dent corn
~OF RADDLE - [text] dent *f* de peigne
~OPENER - [text] (reed opener) ouvreuse *f* de peigne
DENTAGRA - s. dentalgia
DENTAL - [anat] (relating to the teeth) dentaire
~ABSCESS - [med] abcès *m* dentaire
~APPLIANCE *f* [dentist] prothèse *f* dentaire
~BURR - [dentist] fraise *f* de dentiste, "roulette" *f*
~DECAY - [med] carie *f* dentaire
~ENGINE - s. dental burr
~FORMULA - [anat] (formula indicating the numbe and arrangement of the teeth in mammals) formule *f* dentaire
~GAS - [chem] acide *m* azoteux
~PLATE - [dentist] dentier *m*
~SURGERY - s. dentistry
~TECHNICIAN - [dentist] mécanicien *m* dentiste
~UNIT - [med] service *m* dentaire
DENTALGIA - [med] mal *m* de dents
DENTAPHONE - [instr] (instrument for hearing sounds by means of vibrations transmitted to the auditory nerve through the teeth) dentiphone *m*, ostéophone *m*
DENTATE - [bot] (with a toothed margin) dentelé, denté
DENTATED SILL - [hydr] seuil *m* denté
DENTATION - [bot] (of a margin) dentelure *f*
DENTEL - s. dentil
DENTELLE - [text] (lace) dentelle *f*
DENTICLE - [anat] denticule *m*

ENTICULAR - [arch] à denticules
ENTICULATE - [bot] (said of a margin with small teeth) denticulé
ENTICULATED - [arch] (said of a moulding decorated with dentils) denticulé
ENTICULATION - [arch] denticulation *f*
ENTIL - [arch] (one of a row of projecting blocks under a cornice
ENTIMETER - [instr] dentimètre *m*
ENTIN - s. dentine
ENTINE - [anat] (the substance of which teeth are mainly composed) dentine *f*, ivoire *m*
ENTING BLADE - [text] passette *f* pour peigne
˜HOOK - [text] crochet *m*, passette *f*
˜KNIFE - [text] crochet *m* de rentrage
˜MACHINE - [text] piqueuse *f* mécanique
˜RESISTANCE - [text] résistance *f* à la déformation
ENTINOGENESIS - [anat] formation *f* de la dentine
ENTIST'S FORCEPS - [med] davier *m*
ENTISTRY - [med] (the treatment of diseased or irregular teeth) chirurgie *f* dentaire, art *m* dentaire
ENTITION - [phusiol] dentition *f*
[zool] (arrangement, number etc of teeth) denture *f*
ENUCLEATED - [genet] privé de noyau
ENUDATED - [gen bot etc] dénudé, dépouillé
ENUDATION - [geol] (chemical and mechanical disintegration of rocks) mise *f* à nu, dénudation *f*, dépouillement *m*
˜PLAIN - [geol] pénéplaine *f*
DENUDER - [ind chem] (apparatus in which the metal is separated from the amalgam formed in a mercury electrolytic cell) séparateur *m*
DENUTRITION - [med] dénutrition *f*
DENVER MUD - [min] bentonite *f*
DEOBSTRUENT - [med] (agent or medicine designed to eliminate obstructions in the passages of the body) désobstruant *m*
DEODORANT - [chem] (a substance having the property of eliminating odours) désodorisant *m*
DEODORIZE, to - [gen] (to eliminate or reduce smells) désodoriser
DEODORIZER - s. deodorant
DE-OIL, to - [chem] (the operation of removing oil from paraffins) déshuiler
DE-OILER - [ind chem] déshuileur *m*
DEOPERCULATE - [bot] (without a cover, thus allowing the escape of spores) désoperculé
DEOXIDATION - [chem] (the removal of oxygen from molten metal before casting by adding suitable elements which form oxides. These float on the melt and can be removed) désoxydation *f*
DEOXIDIZE, to - [metall] désoxyder
DEOXIDIZER - [chem] (substance which removes oxygen from a compound) désoxydant *m*, réducteur *m*
DEOXIDIZING - s. deoxidation
DEOXY - [chem] (or desoxy; a prefix denoting a compound in which hydroxyl has been replaced by hydrogen) désoxy-
DEOXYCORTICOSTERONE - [chem] (an adrenal corticosteroid used in medicine) désoxycorticostérone *m*
DEOXYRIBONUCLEIC ACID -[biochem] (DNA;nucleic acid found in cell nuclei and especially genes, associated with the transmission of genetic information) acide *m* désoxyribonucléique
DEPARTURE - [gen] départ *m*
[naut] partance *f*
[nav etc] (deviation from course) déviation *f*, écart *m*
˜PLATFORM - [railw] quai *m* de départ
˜VELOCITY - [astronaut] of rockets) vitesse *f* de lancement, vitesse *f* initiale
DEPASSIVATION - [chem] (removal of the passive film on a metal by immersion in acid) dépassivation *f*
DEPAUPERATE - [gen & bot] rabougri, chétif
DEPEND, to - [gen] dépendre, compter sur
DEPENDABILITY - [gen] sûreté *f*, confiance *f*
[mech] (particularly of machines) fiabilité *f*, sûreté *f*
DEPENDENT - [gen etc] dépendant, auxiliaire, secondaire
˜SIGNAL - [railw] signal *m* secondaire
˜VARIABLE - [math] (a variable whose value depends upon that of another variable) grandeur *f* dépendante
DEPENTANIZER - [oil ind] tour *f* de séparation du pentane
DEPHASE, to - [el] déphaser, décaler
˜TRANSFORMER - [el] transformateur *m* (de décalage)
DEPHENOLATION - s. dephenolization
DEPHENOLIZATION - [chem] déphénolage *m*
DEPHENOLIZE, to - [chem] déphénoler
DEPHENOLIZER - [ind chem] colonne *f* de déphénolisation
DEPHLEGMATION - [nucl] condensation *f* partielle
DEPHLEGMATOR - [ind chem] (a fractionation column in which fractions of different boiling points can be separated by distillation) déflégmateur *m*, déphlégmateur *m*
˜COLUMN - [ind chem] colonne *f* de déflégmation
DEPHLOGISTICATED AIR - [chem] (Pirestley's name for oxygen) air *m* déphlogistiqué, oxygène *m*
DEPHOSPHORIZATION - [metall] (the removal of phosphorus from iron in the basic steel process) déphosphoration *f*
DEPHOSPHORIZE, to - [metall] déphosphorer
DEPICKLING - [leather ind] (the removal of acid and salt-pickle from sheepskins) rinçage *m*
DEPIGMENTATION -[med] dépigmentation *f*
DEPILATION - [gen & text] (the action of removing hair from a body; also of removing wool from hides) épilation *f*
DEPILATORY - [chem] (reagent used to remove hair, especially from the human body) dépilatoire *m*
DEPITCH, to - [oil ind] éliminer les brais
DEPLANATE - [bot] (flattened) aplati
DEPLETE, to - [gen] diminuer, épuiser; (to empty the contents) vider
DEPLETED AREA - [geol] surface *f* drainée
˜MATERIAL - [nucl] matière *f* épuisée, matière *f* appauvrie
˜SAND - [geol] sable *m* drainé
˜WATER - [chem] (water containing a smaller percentage of deuterium oxide than is normally the case, in consequence of an isotopic exchange treatment) eau *f* appauvrie
DEPLETION - [nucl] (reduction in the fissionable content of nuclear reactor fuel in consequence of the operation of the reactor) appauvrissement *m*

DEPLETION - [oil ind] déplétion *f* de gaz, abaissement *m* de concentration
[med] déplétion *f*
~BARRIER - s. depletion layer
~LAYER - [electron] (the region where the density of donors and acceptors is not neutralized) zone *f* de dépression de charge
~OF BASES - [soil] (removal of axchangeable cations after displacement by H) désaturation *f*
DEPLOY, to - [gen] déployer, étendre
DEPLOYMENT - [gen] déploiement *m*
[aero] (of a parachute canopy) ouverture *f*
~BAG - [aero] (of parachutes) enveloppe *f* de parachute
DEPOLARIZATION - [el chem] (the process of eliminating the hydrogen formed on the positive electrode of a primary cell) dépolarisation *f*
~FIELD - [el] champ *m* de dépolarisation
DEPOLARIZE, to - [el chem] dépolariser
DEPOLARIZER - [el] (chemical, mechanical or other agent which reduces or prevents polarization in primary cells, caused by the accumulation of hydrogen at the positive electrode) dépolariseur *m*
DEPOLARIZING MIX - [el] (for dry cells; mixture comprising a substance to increase conductivity and a depolarizer) dépolarisant *m*
DEPOLYMERIZATE - [chem] produit *m* de dépolymérisation
DEPOLYMERIZATION - [chem] (the breaking-up of a polymer into simpler molecules) dépolymérisation *f*
DEPOLYMERIZE, to - [chem] dépolymériser
DEPOSIT, to - [gen] déposer, poser
DEPOSIT - [gen] dépôt *m*
~[chem] (sediment at the bottom of an accumulator cell etc) dépôt *m*, sédiment *m*
[min] (natural accumulation of ore) gisement *m*
[nucl] gisement *m*, gîte *f*
[comm & fin] dépôt *m*
DEPOSITED CARBON RESISTOR - [electron] résistance à couche de carbone
DEPOSITING-OUT TANK - [metall] cuve *f* de dépôt total, cuve *f* liberatrice
DEPOSITION - [gen & leg] déposition *f*
[nucl] retombées *f*pl
~OF GETTER - [metall] formation *f* du getter, formation *f* de couches-getter
~OF OIL - [ind chem] séparation *f* de l'huile
~OF SPECULAR LAYERS - [metall] dépôt *m* de couches réfléchissantes (ou spéculaires)
DEPOSITOR - [fin] déposant *m*
DEPOT - [gen] dépôt *m*
[railw] dépôt *m*; (in the USA) gare *f*
~SHIP - [naut] bâtiment *m* ravitailleur
~WORKSHOP - [railw] atelier *m* de dépôt
DEPRECIATE, to - [gen] déprécier, rabaisser
[comm & fin] dévaluer
DEPRECIATION - [gen] dépréciation *f*
[comm] amortissement *m*, moins-value *f*
~ALLOWANCE - [comm] indemnité *f* de dépréciation
~FUND - [fin] fond *m* d'amortissement
DEPRESS, to - [gen] abaisser, presser
[mech] (e.g. a lever) abaisser
DEPRESSANT - [chem] (agent or reagent causing the lessening or depressing of some specified property) déprimant *m*, agent *m* abaisseur
DEPRESSED - [gen] déprimé, abaissé
~ARCH [arch] arc *m* déprimé
~CONDUCTOR RAIL - [railw] (conductor rail below the normal level) rail *m* conducteur en contrebas
DEPRESSION - [gen] dépression *f*, enfoncement *m*, abaissement *m*, affaissement *m*
[surv] (either relative to the sea-level or to the surrounding region) dépression *f*
[mech] (condition of pressure in a closed vessel below that of the atmosphere) dépression *f*
[met] (area of low atmospheric pressure) dépression *f*
[firearms] (of gun) site *m* négatif
[bot] (of grain) sillon *m*
~METER - [instr] déprimomètre *m*
~OF FREEZING POINT - [th eng] abaissement *m* du point de congélation
DEPRESSOR - [anat] (muscle supporting the lowering of an organ) muscle *m* abaisseur
[chem] catalyseur *m* négatif
DEPRESSURIZED - [phys] décompressé
DEPROTEINIZED RUBBER - [rubber ind] caoutchouc *m* déproteiné
DEPSIDE - [chem] (a product of hydroxy-aromatic acids) depside *f*
DEPTH - [gen] profondeur *f*, hauteur *f*, enfoncement *m*
[geogr] (of the sea) profondeur *f*, brassiage *m*
[naut] (of a submerged craft) immersion *f*
[meas] (the perpendicular measurement in a vessel) creux *m*
[mech & horol] (the measure by which the teeth of a wheel intersect the teeth of a pinion) hauteur *f* de dent
~AREA - [opt] étendue *f* de la profondeur
~BALLAST - [constr] épaisseur *f* de l'empierrement
~BOMB - s. depth charge
~CHARGE - [firearms] (bomb set to detonate at a specified depth in the sea) grenade *f* anti-sous-marine
~CONTOUR - [surv] (the lines following the points of equal depth in a water basin) courbe *f* isobathe
~DEVELOPMENT - [photo] développement *m* en profondeur
~DOSE - [nucl] (radiation dose at a given depth beneath the surface of a body) dose *f* en profondeur
~EFFECT - [photo] relief *m*
~GAUGE - [instr] (used to measure the depth of a hole) jauge *f* de profondeur
[metall] (in loam moulding) témoin *m* de profondeur
~GEAR - [instr] (instrument designed to measure the depth of the sea) sondeur *m*
~INDICATOR - s. depth gear
~INTERVIEW - [comm] sondage *m* en profondeur
~LINE - [instr] courbe *f* bathymétrique
~LOCALIZATION - [acoust] (auditory perspective; the faculty of appreciating distances through hearing) localisation *f* des sons
~MARK - [surv] repère *m* de profondeur
~OF A TOOTH - [mech] (in a gear wheel) hauteur *f* de dent, profondeur *f* de dent
~OF CASTING - [metall] hauteur *f* de la pièce coulée
~OF CHILL - [metall] profondeur *f* de trempe

EPTH OF CUT - [mech] profondeur *f* de coupe, profondeur *f* de passe
OF DEFINITION - s. depth of focus
OF DRAINING - [hydr] profondeur *f* de drainage
OF ENGAGEMENT - [mech] profondeur *f* d'engrènement
OF FIELD - [photo] (the permissible range of movement of the lens before an object loses its sharpness) profondeur *f* de champ
OF FIELD RING - [photo] bague *f* de profondeur de champ
OF FOCUS - [photo] (the distance between the nearest and the farthest planes in the photographed area) profondeur *f* de champ
OF FOCUS SCALE - [photo] échelle *f* de profondeur de foyer
FOUNDATION - [constr] profondeur *f* de la fondation
OF FREEZING - [soil] profondeur *f* du gel
OF FUSION - [metall] (the depth to which the weld metal penetrates) profondeur *f* de fusion, pénétration *f*
OF GIRDER - [constr] hauteur *f* de poutre
OF HARDENING - [metall] épaisseur *f* de trempe
OF HOLD - [shipbuild] creux *m*
OF LOOM - [text] profondeur *f* de métier
OF MODULATION - [radio] (the extent of the modulation of a wave) taux *m* de modulation
OF PAGE - [print] longueur *f* de la composition
OF PENETRATION - [radio] (in a solid conductor carrying high-frequency current) profondeur *f* de pénétration
[metall] s. depth of fusion
OF PISTON - [mech] hauteur *f* de piston
OF PLOUGHING - [agric] profondeur *f* de labour
OF POLE HOLE - [constr] profondeur *f* de plantation
OF RAIL - [mech] hauteur *f* de rail
OF REED - [text] hauteur *f* du peigne, hauteur *f* des peignes à tisser
OF RIM - [mech] hauteur *f* de couronne
OF ROUND - [mining] avancement *m*
OF SCOUR - [soil] profondeur *f* des affouillements
OF SEAL - [hydr] hauteur *f* de garde
[gas ind] (in the treatment of gas) hauteur *f* de plonge
OF SOIL - [agric] profondeur *f* du sol
OF SOWING - [agric] profondeur *f* de l'ensemencement
OF THE SHED - [text] hauteur *f* du pas
OF THREAD - [mech] (of a screw) profondeur *f* du filet
OF TOP-SOIL - [agric] épaisseur *f* de la terre arable
OF TRENCH - [constr] profondeur *f* de la tranchée
~PERCEPTION - [opt] perception *f* de profondeur, vision *f* tridimensionnelle
~RECORDER - [instr] sondeur-enregistreur *m*
~SWITCH - [el] (in deep tanks etc) interrupteur *m* de niveau
DEPTHING TOOL - [tools] (for the adjustment of the depth of two wheels or a wheel and a pinion) compas *m* pour engrenages
DEPTHOMETER - [instr] (designed to measure the depth of the sea) sondeur *m*
DEPURATION - [gen & chem] dépuration *f*
DEPURATIVE - [chem] dépuratif
DEPURATOR - [ind chem] épurateur *m*
DERAIL, to - [railw] dérailler
DERAILING POINT - [railw] aiguillage *m* de protection
~SWITCH - [railw] aiguillage *m* de sécurité
DERAILMENT - [railw] (abandonment of the rail by a train or part of it) déraillement *m*
~GUARD - [railw] dispositif *m* de protection contre le déraillement
DERATE, to - [mech] (to decrease the power of a motor or engine for expediency reasons) réduire la puissance
DERATED - [mech] (of a motor or engine the power of which has been reduced) à puissance réduite
DERATING CURVE - [comm] (the change in value of an article owing to changing circumstances) courbe *f* de dépréciation
~FACTOR - [mech] facteur *m* de réduction
DERBY FLOAT - [impl] truelle *f*
~RED - [paint] (also called Chinese Red; a basic lead chromate pigment) rouge *m* Derby
DERBYSHIRE NECK - [med] (enlargement of the thyroid gland) goître *m*, bronchocèle *f*
DE-REEL, to - [gen & mech] (to unwind a reel, a bobbin etc) dévider, dérouler
DE-REELING - [mech el etc] dévidage *m*
~DEVICE - [mech] (a mechanism designed to unwind a reel, a bobbin etc) dévideuse *f*
DERESINATED RUBBER - [rubber ind] caoutchouc *m* dérésiné
DERESINIFICATION - [chem] élimination *f* de la résine
DERIVATE - [math] dérivée *f*
DERIVATION - [gen & math] dérivation
DERIVATIVE - [chem] (substance derived from another substance and usually retaining certain structural features of the original one) dérivé *m*
~ACTION - [contr] (action having a continuous linear relation between the rate of change of the controlled variable and the position of the final control element) action *f* par dérivation
~ACTION TIME - [contr] constante *f* de temps de dérivation
~COEFFICIENT - [contr] s. derivative action time
~CONTROL - [contr] (in this control the potential correction is in proportion to the rate of the deviation changes) réglage *m* par dérivation
~HYBRID - [genet] (obtained by the crossing of two hybrids, or of one hybrid with one of its parents) hybride *m* dérivé, hybride *m* double
DERIVE, to - [gen] dériver, évaluer
DERIVED - [gen & el] dérivé
~CIRCUIT - [el] (shunt circuit, voltage circuit, the circuit of an instrument etc) circuit *m* dérivé
~FOSSILS - [geol] (fossils found in a stratum which is younger than the remains themselves) fossiles *mpl* dérivés
~RESISTANCE - [el] résistance *f* dérivée
~UNITS - [el] (derived from the three fundamental units of mass, length and time) unitées *fpl* dérivées
~WEAVE - [text] armure *f* dérivée
~WIRE - [el] fil *m* de dérivation
DERM - [anat] derme *m*
DERMABRASION - [med] abrasion *f* superficielle
DERMAL RESISTANCE - [el] résistance *f* de la peau
DERMALGIA - [med] dermalgie *f*

DERMATERGOSIS - [med] dermatose *f* professionnelle
DERMATITIS - [med] (inflammation of the epidermis) dermatite *f*
DERMATOAUTOPLASTY - [med] dermatoplastie *f* par auto-greffe
DERMATOCYST - [med] kyste *m* de la peau
DERMATOFIBROMA - [med] dermato-fibrome *m*
DERMATOID - [gen & med] dermoïde
DERMATOLOGY - [med] (the branch of science dealing with skin and its diseases) dermatologie *f*
DERMATOMYCOSIS - [med] dermatomycose *f*, dermatophytie *f*
DERMATOPLASTY - [med] dermatoplastie *f*
DERMATORRHEA - [med] dermatorrée *f*
DERMATROPHY - [med] dermatrophie *f*
DERMOID - s. dermatoid
DERMOSTOSIS - [med] ossification *f* du derme
DERRICK - [mech] (machine or arrangement for hoisting materials) mât *m* de charge
[naut] (arrangement for hoisting cargo on to or from a ship) mât *m* de charge, martinet *m*, bigue *f*
[oil ind] derrick *m*, tour *f* de sondage, chevalement *m*
~BRACES - [oil ind] croisillons *mpl* de derrick, entretoises *fpl* de la tour de sondage
~CAR - [railw] wagon-grue *m*
~CELLAR - [mining] avant-puits *m*
~CRANE - [mech] derrick *m*
~FLOOR - [oil ind] (in oil drilling operations) plancher *m* de manoeuvre
~LEG - [oil ind] (the slender support of the derrick) montant *m* de derrick, jambe *f* de la tour
~PLATFORM - [oil ind] plateforme *f* de forage
~POLE - [telecomm] poteau *m* derrick
~-STYLE ANTENNA MAST - [radio] (a lattice mast; aerial mast of lattice-work material) pylone *m* en treillis
DERRICKING JIB CRANE - [mech] (a jib crane with a variable radius of action) grue *f* à volée variable
DERRICKS AND RIGGING - [shipbuild] apparaux *mpl* de levage et agrès
DERUST, to - [metall] dérouiller
DESACTIVATION PROCESS - [chem] procédé *m* de désactivation
DESAERATE, to - [phys etc] désaérer, dégazer
DESAERATING - [phys etc] désaération *f*, dégazage *m*
DESALT, to - [chem] dessaler
DESALTING - [chem] the elimination or reduction of salt contents) dessalement *m*
DESAMINASE - [biochem] (an enzyme inducing the splitting off of the amino group from the amino acids) désaminase *f*
DESAMINATE, to - [text] (the wool) désaminer
DESARTICULATION - [med] désarticulation *f*
DESATURATED COLOURS - [telev] couleurs *fpl* désaturées
DESATURATION - [photo etc] (the extent of grey in colours) désaturation
DESCALE, to - [gen] détartrer, décalaminer
DESCALER - [metall] (for the treatment of laminates after breaking up the scales) désincrustant *m*, décalaminant *m*
DESCALING - [th eng] (removing scale from boilers) détartrage *m*
[metall] (in forging] décalaminage *m*
~FLUID - [ind chem] désincrustant *m*, détartrant *m*

DESCANT - [mus] (the free part which is added ove a melody) chant *m*
DESCARTES LAW - [opt] (the law of refraction) loi de Descartes
DESCEMETITIS - [med] descemétite *f*, kératite *f* ponctuée
DESCEND, to - [gen] descendre, tomber
[mech] abaisser
DESCENDING HORIZONTAL SLICING - [mining] exploitation *f* par tranches horizontales descendante
~LETTER - [print] lettre *f* descendante
~PATH - [astronaut] trajectoire *f* de descente
DESCENSION PIPE - [gas ind] colonne *f* montante, colonne *f* d'ascension
DESCENT - [gen] descente *f*, pente *f* chute *f*
[mech] abaissement *m*
[mining] entrée *f* de mine
[chem] dégradation *f*
[bot] descendance *f*
DESCRIBE, to - [gen] décrire
[draw] tracer
DESCRIPTION - [gen] description *f*, nomenclature *f*
DESCRIPTIVE - [gen] descriptif
~ANATOMY - [anat] anatomie *f* descriptive
~ASTRONOMY - [astr] (only dealing with the description of heavenly bodies) astronomie *f* descriptive
~GEOMETRY - [math] géométrie *f* descriptive
DESCUMMING - [text] enlèvement *f* de l'écume
DESEAM, to - [metall] (to remove superficial blemishes from ingots by means of an oxy-gas flame) décaper au chalumeau
DESEAMING - [metall] décapage *m* au chalumeau
DESEMULSIFICATION - [chem] désémulsification *f*
DEMULSIFY, to - [chem] désémulsifier
DESENSITIZATION - [photo] (the reduction of the reactions caused by the light on photographic materials) désensibilisation *f*
[med] (the elimination of sensitivity to a protein) désensibilisation *f*
DESENSITIZE, to - [photo & med] désensibiliser
DESENSITIZER - [photo] (agent designed to reduce the sensitivity of photographic emulsions) désensibilisant *m*
DESERT, to - déserter, abandonner
DESERT - [geol] désert
~PAVEMENT - [soil] (surface of stones and rocks after the finer material has been blown away) pavés *mpl* du désert
~SOIL - [soil] terrain *m* désertique
~SORE - [med] ulcère *f* du désert
~VARNISH - [soil] (a glossy coating of stones in deserts, due to wind action) patine *f* désertique
DESICCANT - [chem] (a substance which exerts a drying action by absorbing moisture) dessiccant *m*, produit *m* desséchant, produit *m* hygroscopique
~BED - [ind chem] lit *m* dessiccant
DESICCATE, to - [gen] (the action of drying) dessécher, sécher
DESICCATION - [gen & chem] dessiccation *f*
DESICCATOR - [ind chem] (glass vessel with close-fitting lid and a chamber in the base for moisture-absorbing material, in which substances can be placed for the gradual extraction of water) dessiccateur *m*
DESIGN, to - [gen mech etc] projeter, préparer un projet, dessiner (d'original), inventer, construire

établir un plan; (e.g. a circuit) dimensionner
ESIGN - [gen] étude *f*, plan *m* projet *m*, dessin *m*, modèle *m*, type *m*
~FOR INTERLACING - [text] dessin *m* de l'armure
~GROSS WEIGHT - [aero] (the weight at which an aircraft is expected to meet appropriate airworthiness requirements) poids *m* brut de calcul
~IT-YOURSELF SYSTEM - [mech] (assembly by using ready-made parts) construction *f* par éléments d'assemblage
~OFFICE - [gen] bureau *m* d'études
~PAPER - [text] carte *f*, papier *m* de mise en carte, papier *m* quadrillé
~POWER - [nucl] (the planned power of a nuclear reactor) puissance *f* calculée
~SPEED - [gen] (the time within which a design is carried out) rapidité *f* de réalisation
~STRESS - [mech] contrainte *f* calculée
~WEIGHT - [aero] poids *m* de calcul
~WING AREA - [aero] (the area of a wing as designed) surface *f* conventionnelle de la voilure
ESIGNATE, to - [gen] désigner, nommer, élire
ESIGNATION OF CIRCUITS - [telecomm] désignation *f* des circuits
~STRIP - [el & telecomm] (on cables) plaquette *f* distinctive, réglette *f* porte-étiquette
ESIGNER - [gen] projecteur *m*, constructeur *m*, réalisateur *m*
[ind] (of machines, e.g. of cars) dessinateur *m*
[cin] costumier *m*
~ENGINEER - [ind] ingénieur *m* d'études
ESIGNING - [text] mise *f* en carte
ESILICONIZATION - [chem] élimination *f* du silicium
ESILT, to - [min] décrasser, dévaser, décrotter
ESILTING - [hydr] élimination *f* des boues
~DEVICE - [min & hydr] séparateur *m* de schlamm
DESILVERING BATH - [photo] bain *m* de désargenture
DESILVERIZATION - [metall] (the removal of silver from lead) séparation *f* de l'argent du plomb
DESIRED VALUE - [contr] (USA; in a control system: reference input) valeur *f* de référence
DESIZE, to - [text] désencoller
DESIZING - [text] désencollage
~COMPOUND - [text] agent *m* de désencollage
DESK - [gen] pupitre *m* (de commande)
~CALCULATOR - [comput] calculateur *m* de table
~SWITCHBOARD - [el etc] pupitre *m* de commande
DESLAG, to - [metall] (to remove the slag from the surface of a weld) piquer le laitier, décrasser
DESLIME, to - s. desilt, to
DESLIMER - [min] (device or machine to eliminate silt) séparateur *m* de schlamm
[metall] débourbeur *m*
DESLUDGE, to - [hydr] débourber, évacuer les boues
DESLUDGING - [hydr] (sludge removal) évacuer les boues
~OPERATIONS - [hydr] travaux *mpl* d'enlèvement des boues
DESMALGIE - [med] douleur *f* dans les ligaments
DESMAN - [zool] desman *m*, myogale *m*
DESMINE - [min] (silicate of sodium, calcium and aluminium with chemically combined water) stilbite *f*
DESMOID - [anat] desmoïde, fibreux
DESMOSOME - [biochem] desmosome *m*
DESMOTROPISM - [chem] (the change of position of a double bond in which both series of compounds can exist independently) desmotropie *f*
DESMOTROPY - s. desmotropism
DESORPTION - [chem] (the removal of adsorbed material from a substance) désorption *f*
DESPATCH, to - [gen] envoyer, expédier, exécuter rapidement
DESPATCH - [gen] envoi *m*, expédition *f*
~OFFICE - [comm] service *m* d'expédition
DESPATCHER - [aero] (in the air force, the officer in charge of parachutists' launch) dispatcher *m*
DESPUMATION - [med] despumation *f*, purification *f* par enlèvement de l'écume
DESQUAMATION - [med] desquamation *f*
DESTABLIZE, to - [phys mech] (to render unstable) rendre instable
DESTATICIZER - [el] (a device for carrying off static charges from material being processed) dispositif *m* anti-statique
DESTINATION - [gen] destination *f*, désignation *f*, nomination *f*
~BOARD - s. destination sign
~INDICATOR - [railw] indicateur *m*
~PANEL - s. indicator (destination-)
~SIGN - [railw] plaque *f* indicatrice, écriteau *m*
DESTROY, to - [gen] détruire
DESTRUCTION - [gen] destruction *f*
DESTRUCTIVE - [gen] destructif, destructeur
~DISTILLATION - [chem] (distillation of solids accompanied by decomposition, e.g. the treatment of coal to produce coke, tar, town gas etc) distillation *f* sèche
~READING - [comput] (the erasing of the store during reading) lecture *f* destructive
~READOUT - s. destructive reading
~TEST - [mech] essai *m* de rupture
DESTRUCTOR - [ind chem] (apparatus for destroying waste material by burning) incinérateur *m*
DESUBLIMATION - [chem] désublimation *f*
DESUGARIZING - [sugar ind] déssucrage *m*
DESUINT, to - [text] (to remove the grease from wool) désuinter
DESUINTING BATH - [text] (for wool) bain *m* de désuintage
DESULPHURIZATION - [metall] désulfuration *f*, désoufrage *m*
DESULPHURIZE, to - [metall] désulfurer
DESULPHURIZER - [ind chem] (a desulphurizing agent) désulfurant *m*
DESULPHURIZING FURNACE - [metall] four *m* à pyrites
DESUPERHEATER - [ind chem & oil ind] (apparatus for producing saturated or less highly superheated steam by the injection of water) désurchauffeur *m*
DESYL - [chem] désyle *m*
DESYNCHRONIZATION - [el] désynchronisation *f*
DETACH, to - [gen] détacher, séparer, enlever
DETACHABILITY - [gen & mech] amovibilité *f*
DETACHABLE - [gen] amovible, détachable, démontable
~BIT - [mech] taillant *m* à vis, jack-bit *m*
~CAP - [mech] couvercle *m* amovible
~CYLINDER HEAD - [auto] culasse *f* rapportée
~HEAD - [mech] tête *f* amovible
~LANDING GEAR - [aero] (used in early endurance flights and designed to be dropped after take-off)

train *m* d'atterrissage largable
DETACHABLE RIM - [mech] jante *f* démontable
~TYRE LOCKING RING - [auto] jonc *m* de verrouillage fendu
~VALVE SEAT - [mech] siège *m* de soupape démontable
~WAGON FITTINGS - [railw] accessoires *mpl* amovibles pour wagons
DETACHED ARCH CORE - [geol] noyau *m* anticlinal étranglé
DETACHING - [gen] détachement *m*
~COMB - [text] peigne *m* détacheur
~HOOK - [mining] déclic *m* d'attelage, crochet *m* de sûreté
~KNIFE - [text] couteau *m* séparateur
~MOTION - [text] dispositif *m* de déroulement
~OF THE FLOW - [aero] (in aerodynamics, the detaching of the flow from a wing) décollement *m* des filets d'air
~ROLLER - [text] cylindre *m* arracheur
DETACHMENT - [gen] détachement *m*, séparation *f*, éloignement *m*
[mech etc] détachement *m*, décrochement *m*, dételage *m*
[railw] (of a truck) détachement *m*
~DEVICE - [mech] dispositif *m* de dételage
~OF ELECTRONS - [electron] (the liberation of electrons from negative ions by impact) éjection *f* d'électrons
DETACKIFIER - [ind chem] (substance added to another to make the latter less readily adhesive, e.g. to prevent it from sticking to rollers) agent *m* antiadhesif
DETAIL, to - [gen] détailler, spécifier
DETAIL - [gen] détail *m*, particularité *f*, finesse *f*
[mech] détail *m*
[geogr & surv] particularité *f*
~CARD - [comput] carte *f* de détail
~CONTRAST RATIO - [telev] (the contrast determined by a small dark area in the middle of an area of high brilliance) rapport *m* de contraste des détails
~DRAWING - [draw] détail *m* d'exécution, dessin *m* détaillé, dessin *m* d'exécution
~IMAGE - [photo] (the reproduction, mostly enlarged, of the wanted part of an image) détail *m*, plan *m* de détail
~PAPER - [paper man] (used for detail drawing) papier *m* calque
~PICTURE - s. detail image
~PRINTING - [comput] travail *m* en liste
DETAILED BALANCING - [mech] (in statistical mechanism) équilibrage *m* détaillé
DETAILER - [draw] dessinateur *m* de plans de détail
DETANNING - [ind] détannage *m*
DETAR, to - [ind proc] (the elimination of tar) dégoudronner
DETARRER - [oil ind] (a tar extractor) séparateur *m* de goudron, dégoudronneur *m*
~BELL - [oil ind] cloche *f* à double paroi
DETARRING - [oil ind] dégoudronnage *m*
DETEARING - [paint] (formation of droplets) coulage *m*
DETECT, to - [gen] déceler, détecter, découvrir
[radio] (to reproduce an original modulating signal) détecter
[el] (the action of rectifying) redresser
DETECTABLE - [gen] détectable, discernable, décèlable
DETECTING ELEMENT - [contr] (the element directly responding to the value of the controlled condition) détecteur *m*
~GRATING - [electron] (waveguides) réseau *m* détecteur
~INSTRUMENT - [instr] (a device designed to indicate the presence of a phenomenon or occurrence) détecteur *m*, appareil *m* indicateur
DETECTION - [gen] détection *f*, dénonciation *f*, découverte *f*
[el] (rectification) redressement *m*
[nucl] détection *f*
[radio] (the operation of separating the signal from the modulated carrier wave) détection *f*
~CIRCUIT - [telecomm] circuit *m* détecteur
~LIMIT - [phys] limite *f* de détection
~PUNCH - [comput] perforation *f* repère (ou caractéristique)
~RANGE - [contr] étendue *f* de détection
~THRESHOLD - s. detection limit
~UNIT - [comput] détecteur *m*, démodulateur *m*
~UNIT PICTURE - [telev] bloc *m* de détection vidéo
~UNIT SOUND - [telev] bloc *m* détection son
DETECTIVITY - [electron] sensibilité *f*
DETECTOPHONE - [acoust] (microphone with earphones) microphone *m* espion
DETECTOR - [el] (elementary form of galvanometer galvanomètre *m* directionnel
[nucl] (instrument designed to detect the presence and extent of radiation) détecteur *m*
[radio] (device or circuit for reproducing the original modulation from the modulated carrier wave) détecteur *m*
[comput] s. detection unit
[th eng] (in boilers) niveau *m*, indicateur *m* de niveau
~BALANCED BIAS - [radar] (a radar coupling method) polarisation *f* automatique
~BAR - [railw] pédale *f* de sûreté (ou de calage)
~COEFFICIENT - [radio] coefficient *m* de détection
~COHERER - [radio] détecteur *m*
~COIL - [radio] bobinage *m* détecteur
~DIODE - [radio] diode *f* détectrice
~FINGER - [text] aiguille *f* de casse-chaîne
~TUBE - [electron] (electronic tube used to rectify incoming oscillations) tube *m* détecteur
~VALVE - s. detector tube
DETENT - [mech] (a device for retaining a part in a mechanism in a given position for a given period during a cycle of operations, e.g. a dog engaging a toothed wheel) détente *f*, goupille *f* d'arrêt
~ARM - [mech] levier *m* de calage, levier *m* d'arrêt
~ESCAPEMENT - [horol] (escapement using a detent échappement *m* à ancre
~PAWL - [mech] cliquet *m* d'arrêt, cliquet *m* de calage
~PIN - [mech] goupille *f* d'arrêt
~SCREW - [mech] vis *f* d'arrêt
~SPRING - [horol] (flat spring in the escapement) ressort *m* spiral
DETENTION TIME - [chem] (the period during which a liquid is held in a sedimentation vessel) temps *m* de sédimentation
DETERGENCY - [chem] propriété *f* détersive, détergence *f*

DETERGENCY POWER - [ind chem] effet *m* détersif
DETERGENT - [chem] (a cleansing substance, such as soap or the synthetic anionic, nonionic, and cationic detergents) détergent *m*, détersif *m*
~ADDITIVES - [chem] produits *mpl* d'addition détersifs
~OIL - [ind chem] huile *f* détergente
DETERIORATE, to - [gen] détériorer, se détériorer
DETERIORATION - [gen] (any kind of naturally-occurring change which makes an article or substance less suitable for its original purpose) détérioration *f*, dégradation *f*
~OF EMISSION - [electron] (the decrease in electron emission of a tube) appauvrissement *m* de l'émission
DETERMINABLE - [gen] déterminable
DETERMINANT - [math] déterminant
[chem] facteur *m* déterminant
DETERMINATE VARIATION - [genet] variation *f* dans une direction déterminée, orthogénèse *f*
DETERMINATION - [gen] détermination *f*, décision *f*, dosage *m*
~OF CARBON - [metall] dosage *m* du carbone
~OF EARTH THRUST- [geol] calcul *m* de la poussée du sol
~OF EFFICIENCY BY TOTAL LOSSES - [el] (method of indirect measurement to assess the total losses) méthode *f* par détermination des pertes totales
~OF HARDNESS BY INDENTATION - [rubber ind] détermination *f* de la dureté par pénétration
~OF SINGLE SUBSTANCES - [min] dosage *m* d'un seul corps
DETERMINE, to - [gen] déterminer, décider, doser
DETERMINISM - [gen] déterminisme *m*
DETERSIVE - s. detergent
DETIN, to - [metall] désétamer
DETINNING - [metall] désétamage *m*
DETONATE, to - [chem] détoner, faire exploder
DETONATING BULB - [ind chem] bombe-chandelle *f*, larme *f* batavique
~EXPLOSIVE - [chem] explosif *m* détonant
~FUSE - [chem] cordeau *m* détonant
DETONATION - [chem] (chemical decomposition initiated and propagated by intense local shock, which proceeds in wave form through the substance at exceedingly high speed) détonation *f*, explosion *f*
[mech] (in electric-ignition reciprocating engines; sudden spontaneous ignition of the charge throughout its volume, as distinct from the normal advance of a combustion wavefront from the sparking plug) explosion *f*
~METER - [instr] (instrument designed to measure the intensity and frequency of detonation) détonometre *m*
DETONATOR - [ind chem] (a short copper tube, closed at one end and containing mercury fulminate or other suitable compound, used to detonate charges of explosive) détonateur *m*
[railw] (coupled to signals) pétard *m*
[mining] détonateur *m*, amorce *f*
~PLACER - [railw] pose-pétard *m*
DETOUR - [gen] détour *m*
[constr] déviation *f*, voie *f* détournée
DETOXICATE, to - [gen] désintoxiquer
DETOXICATION - [med] désintoxication *f*
DETREADER - s. detreading machine
DETREADING CHIPS - [rubber ind] (fragments of scrap rubber obtained by detreading tyres) déchets *mpl* de déchapage
DETREADING MACHINE - [rubber ind] (machine for stripping off the treads of old tyres) machine *f* à déchaper
DETRIMENT - [gen] détriment *m*, dommage *m*
DETRITAL - [geol] (resulting from detritus) détritique
~DEPOSIT - [geol] dépôt *m* détritique
~FAN - [soil] (cone-shaped deposit where a valley enters a plain) cône *m* de déjection
~MINERALS -[geol] (grains of heavy minerals which are found in sand and other sediments) minéraux *m* détritiques
DETRITUS - [geol] détritus *m*
[soil] (produced by rock weathering) produits *mpl* d'altération des roches
[constr] galets *mpl*, cailloux *mpl* roulés
DETRUSION - [phys] déformation *f* latérale, action *f* de refouler
DETUNE, to - [radio] (the adjusting of a resonant circuit) dérégler, désaccorder
DETUNER - [mech] (a dynamic damper, e.g. in jet engines test beds) silencieux *m*
DETUNING - [radio] déréglage *m*
DEUCE - [gen] (in playing cards) paire *f*
[cin] (lighting device with two incandescent lamps) projecteur *m* à deux lampes
DEUTERANOPIA - [opt] (form of dichromatic vision) deutéranopie *f*
DEUTERATED - [chem] deutérique
DEUTERATION - [chem] deutération *f*
DEUTERIDE - [chem] (a deuterium compound) hydrure *m* lourd, deutérure *m*
DEUTERIUM - [nucl] (the naturally-occurring isotope of hydrogen, A.W. 2.0I6) deutérium *m*
DEUTEROACETYLENE - [chem] deutéro-acétylène *m*
DEUTEROCOMPOUND - [chem] (compound containing heavy hydrogen) composé *m* deutérique
DEUTEROGAMY - [genet] (process replacing normal fertilization) deutérogamie *f*
DEUTERON - [chem] (a heavy hydrogen nucleus, A. M. 2, with unit positive charge) deutéron *m*, deuton *m*
DEUTON - [chem] deuton *m*, deutéron *m*
DEVALUATION - [gen comm & fin] dévaluation *f*, dévalorisation *f*
DEVALUE, to - [gen fin etc] dévaluer
DEVASTATE, to - [gen] dévaster
DEVASTATION - [gen] dévastation *f*, destruction *f*
DEVELOP, to - [gen photo etc] développer, mettre au point, réaliser
[min] (in the extraction of ore, to prove its existence) produire, préparer
DEVELOPED DYE - [chem] (a dye which is produced on the fibre by chemical reaction) colorant *m* développable, colorant *m* développé
DEVELOPER FOG - [photo] voile *m* de développement
DEVELOPER - [chem] (a reducing solution used in photography to reduce the silver salts in the emulsion after exposure to light) révélateur *m*
~IN CARTRIDGE - [photo] révélateur *m* en cartouche
DEVELOPING - [gen] développant
[metall] (the determination of the shape of a blank) détermination *f* de l'ébauche, développement *m*
~AGENT - [photo] développateur *m*, substance *f* révélatrice

DEVELOPING CLIP - [photo] pince *f* à développer
~DRUM - [photo] tambour *m* de développement
~DYESTUFF - [dyes] colorant *m* à développement
~FRAME - [photo] cadre *m* de développement
~HANGER - s. developing frame
~POWDER - [photo] révélateur *m* en poudre
~SOLUTION - [photo] solution *f* de développement
~TANK - [photo] bac *m* de développement, cuve *f* de développement
~TROUGH - s. developing tank
DEVELOPMENT - [gen] développement *m*, évolution *f*, formation *f*
[min] (the amount of exposed ore) préparation *f*
[aero] (of a parachute) ouverture *f*
[ind] préparation *f*, production *f*, mise *f* au point
[mech] (the operation of determining the total length or area of a curved or bent part to find the amount of straight or flat material needed to make it) développement *m*
[photo] développement *m*
~AND DISPLACEMENT CHROMATOGRAPHY - [phys] (in gas analysis) chromatographie *f* par développement et déplacement
~BATH - [photo] bain *m* de développement, bain *m* révélateur
~END - [mining] front *m* de taille, front *m* d'avancement
~OF BRITTLENESS - [metall] (an increase in liability to fracture) accroissement *m* de la fragilité
~REACTOR - [nucl] pile *f* expérimentale
~SCHEME - [gen] plan *m* de développement
[agric] plan *m* de développement foncier
~STOPING - [mining] abattage *m* en traçage
DEVELOPMENTAL STABILITY - [genet] stabilité *f* phénotipique
DEVIATE, to - [gen] dévier
DEVIATING PRISM - [opt] prisme *m* d'inflexion
DEVIATION - [gen] déviation *f*, écart *m*, erreur *m*
[leg] dérogation *f*
[math] déviation *f*
[instr] (the angle between the direction of a compass needle and the true direction of north and south) déclinaison *f*, déviation *f*
[mech] (in a tolerance system) écart *m* (par rapport à une cote)
[phys] (periodic change in the controlled variable) écart *m*, erreur *f*
[electron] excursion *f* de fréquence
~FLAG - [radar] (flight-path deviation indicator) indicateur *m* de dérive
~FROM THE INDEX VALUE - [contr] (instantaneous difference between the value of the controlled condition and the index value) écart *m* de consigne
~GRAPH - [instr] courbe *f* de déviation
~INDICATOR - [radar] (instrument designed to show the extent of the deviation of a vehicle from its course) indicateur *m* de dérive
~OF A DRILL HOLE - [oil ind] (in well sinking) déviation *f* d'un forage
~OF THE TITRE - [text] déviation *f* du titre
~OF THE WIND - [met] (the angle between the direction of the wind and direction of the pressure gradient) déviation *f* du vent
~RATIO - [radio] (the ratio of the maximum change in carrier frequency to the highest frequency of modulation) rapport *m* entre le désaccord maximal et la fréquence maximale de modulation
DEVIATION SENSITIVITY - [radio] (the output power produced by the least frequency deviation) seuil *m* de glissement de fréquence
DEVICE - [gen] appareil *m*, dispositif *m*, mécanisme *m*, système *m*
DEVIL - [metall] chambrière *f*
[impl] (used in metallurgical work) pot *m* à feu
~CLAW - [impl] béquette *f*, bec-de-corbeau *m*
~GAS - [chem] (ammonia by-product) gaz *mpl* méphitiques
~LIQUOR - [chem] (liquid produced by cooling of gases discharged from an ammonium sulphate saturator, smelling of H_2S and pyridine) "devil liquor" *f*
DEVIL'S DICE - [min] limonite *f* en cubes
DEVILLER NEEDLING - [text] garniture *f* de pointes du briseur
DEVISCERATION - [med] éviscération *f*
DEVISE, to - [gen] inventer, découvrir, imaginer, élaborer, concevoir
DEVITRIFICATION - [phys] (change from the glassy state to a microcrystalline condition) dévitrification *f*
DEVITRIFY, to - [phys] dévitrifier, se dévitrifier
DEVOLUTION - [gen & leg] dévolution *f*
DEVULCANIZATION - [chem] (a stage in the reclaiming of rubber. Devulcanization is closely related to depolymerization) dévulcanisation *f*
DEVULCANIZED RUBBER - [rubber ind] (rubber recovered from scrap material) caoutchouc *m* régénéré
DEVULCANIZING - [rubber ind] régénération *f*
~PAN - [impl] (autoclave for regeneration of rubber) chaudière *f* de régénération
DEW - [met] (moisture deposited from the atmosphere upon objects cooler than the ambient air, espec. by night) rosée *f*
~-BOW - [met] (rainbow seen in dewdrops) arc-en-terre *f*
~CLAW - [zool] ergot *m* des chiens
~POINT - [chem] (the temperature at which the moisture in a saturated vapour begins to condense) point *m* de rosée
~-POINT HYGROMETER - [instr] hygromètre *m* à point de rosée
~-POINT MIRROR - [instr] (part of a dew-point meter) miroir *m* d'hygromètre à point de rosée
~POND - [agric] (a shallow pond on elevated pastures) cuvette *f* de condensation de la rosée
~RETTING - [text] rouissage *m* à la rosée
DEWAR FLASK - [ind chem] (container surrounded by an evacuated jacket, for storing liquids at low temperatures) vase *m* Dewar
~VESSEL - s. Dewar flask
DEWATER, to - [chem] (to remove substantial quantities of water) déshydrater, sécher, assécher, dénoyer
DEWATERER - [mech] déshydrateur *m*
[mining] égoutteur *m*
DEWATERING - [chem] (process of removing substantial quantities of water substance without the use of heat) anhydrisation *f*, déshydratation *f*, drainage *m*, séchage *m*
[mining] égouttage *m*, épuisement *m*, exhaure *f*
~CHANNEL - [hydr] canal *m* de vidange
~MACHINE - [hydr] (used in sludges) machine *f* à sécher les boues

DEWATERING OUTLET - [hydr] orifice *m* d'évacuation
~PIT - [hydr] puits *m* d'assèchement
~PLANT - [soil] installation *f* d'épuisement
~PUMP - [hydr] pompe *f* de drainage, pompe *f* d'assèchement
~SCREEN - [min] tamis *m* d'égouttage
~TANK - [min] cuve *f* de décantation
DEWAX, to - [chem] (to eliminate wax) déparaffiner
DEWAXED - [chem] déparaffiné
~SHELLAC - [chem] gomme-laque *f* déparaffinée
DEWAXING - [chem] déparaffinage *m*
~MIX - [chem] mélange *m* déparaffinant
DEWBERRY - [bot] mûre *f* des haies
DEWINDTITE - [min] (lead, uranium and phosphorus containing mineral) dewindtite *f*
DEXTRIN - [chem] (a carbohydrate derived from starches and used as an adhesive, thickening and sizing agent) dextrine *f*
DEXTRINASE - [biochem] (plant enzyme which hydrolyses dextrin) dextrinase *f*
DEXTRINE - s. dextrin
DEXTRINIZE, to - [biochem] convertir *m* en dextrine, transformer en dextrine
DEXTRO-ROTATORY - [phys] (having the property of rotating the plane of polarization of light in the clockwise sense, viewed in the direction of propagation of the light. Indicated by the prefixed symbol "d") dextrogyre
DEXTROCARDIA - [med] dextrocardie *f*
DEXTRORSAL - [gen] dextrorsum, à droite
DEXTRORSE - [gen] dextrorsum
~SUBSTANCE - [chem] substance *f* dextrorsum, substance *f* dextrogyre
DEXTROSE - [chem] (sugar found in plants and in blood. Used in foodstuffs brewing and confectionery) dextrose *m*
DEZINCIFICATION - [chem] (process involving the removal of zinc from metals in the liquid state) dézingage *m*, dézincification *f*
D F - [constr] (abbrev for Drinking Fountain; term used on building sites) fontaine *f*, poste *f* d'eau potable
D/F - s. direction finding
~BEARING - [surv] (bearing obtained by direction-finding equipment) relèvement *m* radiogoniométrique
~NAVIGATION - [aero] navigation *f* par radiogoniométrie
D.H.N. - [initials for Diamond Hardness Number] chiffre *m* de dureté Vickers
DI - [prefix denoting twice or two] di-
di - [diameter] (prefix denoting diameter) diamètre *m*
DIABASE - [chem] diabase *f*
DIABETES - [med] diabète *m*
DIABETIC - [med] diabétique
DIABETOGENIC - [med] (of substances or conditions causing diabetes) diabétogène
DIABETOMETER - [instr] (instrument designed to measure the sugar content in urine) diabétomètre *m*
DIABOLIC - [gen] diabolique, infernal
DIABROSIS - [med] érosion *f*; corrosion *f*
DIACAUSTIC - [opt] (caustic caused by réfraction) diacaustique
DIACETIC ACID - [chem] (acetoacetic acid) acide *m* diacétique
DIACETIN - [chem] (a solvent and plasticizer for cellulose derivatives) diacétine *f*
DIACETONE ALCOHOL - [chem] (a solvent for cellulose derivatives) diacétone alcool *m*, alcool *m* diacétonique
DIACETONYL ALCOHOL - [chem] alcool *m* diacétonique
DIACETYL - [chem] (alpha-diketo-butane, the simplest of the diketones; obtained by reacting nitrous acid with methyl ethyl ketone) diacétyle *m*
DIACETYLENE - [chem] diacétylène *m*
DIACETYLMORPHINE - [chem] (a potent analgesic used in medicine) diacétylmorphine *m*
DIACETYLRESORCIN - [chem] diacétate *m* de résorcine
DIACHRONISM - [geol] (geological formations found in beds of a different age) diachronisme *m*
DIACLASE - [geol] diaclase *f*
DIACRITICAL - [med] diacritique
DIACTINISM - [chem] (the property of transmitting active radiation chemically) diactinisme *m*
DIAERESIS - [print] diérèse *f*
DIAGENESIS - [geol] diagénèse *f*
DIAGENETIC TRAP - [geol] piège *m* diagénétique
DIAGENIC - [genet] diagynique
DIAGEOTROPISM - [bot] (the condition by which a plant grows perpendicularly to the surface of the earth) diagéotropisme *m*
DIAGNOSE, to - [med] diagnostiquer
DIAGNOSIS - [med] (the determination and identification of an illness or complaint) diagnose *f*, diagnostic *m*
DIAGNOSTIC - [med] diagnostique
~CHARACTERS - [zool] (characteristics which enable to differentiate species, groups etc) diagnoses *f*pl
~INSTRUMENT - [med] instrument *m* diagnostique
~ROUTINE - [comput] (routine designed to identify a malfunction or a mistake) programme *m* diagnostique
~TUBE - [electron] (X-ray tube used in diagnostic radiography) tube *m* de diagnostic
DIAGONAL - [geom] diagonale *f*
~ARCH - [constr] arc arêtier, croisée *f* d'ogive
~BOBBIN THREAD - [text] trame *f* oblique
~BOND - [constr] entretoisement *m* en diagonale
~BRACE - [mech] (a member of a structure designed to resist stresses which act in other directions than those of the longitudinal or lateral axes) croisillon *m*, barre *f* oblique
~CLIPPING - [telev] distortion *f* diagonale du signal
~CROSS BRACE - [constr] entretoise *f* diagonale
~CUTTING MACHINE - [mech] (in rubber industry) machine *f* à couper en biais
~FAULT - [geol] rejet *m* oblique, faille *f* oblique
~JOINT - [geol] cassure *f* diagonale
~KNURLING - [metall] moletage *m* croisé
~LINING - [rubber ind] alignement *m* en quinconce
~MEMBER - [aero] traverse *f*, entretoise *f*
~PITCH - [mech] (the distance between the centres of the adjacent rivets) pas *m* diagonal
~RIB - [text] (cloth with a prominent rib) tissu *m* en diagonale
[constr] (of a vault) arête *f* diagonale
~STRUT - [mech] diagonale *f* comprimée
~WINCH - [mech] (steam-winch with inclined engine cylinders, so as to save space) treuil *m* à cylin-

dres inclinés
DIAGONALIZING - [telev] (the radiation of the same programme at different times and on different wavelengths) répétition *f* d'un programme sur longueurs d'onde différentes
DIAGRAM - [geom etc] (outline figure) diagramme *m*, figure *f*, dessin *m*, courbe *f*, schéma *m*
[autom] (the curve showing the sequence of operations) diagramme *m*
~FACTOR - [mech] (the ratio of the average effective pressure of a steam-engine cylinder to the best pressure deduced from the indicator diagram) coefficient *m* d'utilisation
~OF BLOCK FLOW - [mech] (in mills) diagramme *m* de succession des opération
~OF CONNECTIONS - [el] schéma *m* de montage
~OF STRESSES - [phys] polygone *m* des forces
~OF VELOCITIES - [phys] diagramme *m* des vitesses
~PAPER - [print] (paper used for the registration of instruments) papier *m* millimétré
DIAGRAMMATIC - [gen] graphique, schématique
DIAL, to - [telephone] composer le numéro
DIAL - [instr] (the part of the instrument which is observed) cadran *m*, limbe *m*
[telecomm] (calling dial) cadran *m* de numérotation, cadran *m* d'appel
[surv] (large compass for surveying coal-mine workings) boussole *f*, boussole *f* de minière
[radio] (the mechanism designed to adjust the controls) cadran *m*
[mech] (of a machine tool etc) cadran *m*
[astr] (sundial) cadran *m* solaire
~BORE GAUGE - s. dial gauge (to measure cylinder bores etc)
~BRIDGE - [el] pont *m* à contacts multiples
~COMPARATOR - s. dial gauge
~EXCHANGE AREA - [telecomm] central *m* automatique
~FEED PRESS - [mech] (a press fed by means of a revolving turret) presse *f* revolver
~FOOT - [horol] (a circular pin at the back of the dial) pied *m*
~GAUGE - [meas] (very sensitive measuring instrument) montre-comparateur *m*, comparateur *m* à cadran
[instr] (to measure cylinder bores etc) jauge *f* à cadran
~IMPULSES - [telecomm] (the impulses which are received from a dial) impulsions *fpl* du cadran de numérotation
~INDICATOR - [instr] (any device for showing information by means of a pointer moving over a circular scale or the like) indicateur *m* à cadran
~LIGHT - [instr] ampoule *f* de cadran
~MICROMETER - [instr] (an instrument for making fine mechanical measurements, which gives a direct reading on a dial) micromètre *m* à cadran
~NEEDLE - [text] aiguille *f* de plateau
~PLATE - [instr] cadran *m*
~RESISTANCE BOX - [el] boite *f* de résistances à combinaisons multiples
~SNAP GAUGE - [tool] calibre *m* mâchoire avec indicateur de cote
~STITCH - [text] maille *f* formée par les aiguilles du plateau
~SWITCH - [el] (a multi-contact switch, the contacts being arranged in the arc of a circle) commutateur *m* rotatif
DIAL SYSTEM - [telecomm] installation *f* téléphonique automatique
~TELEPHONE - [telecomm] téléphone *m* automatique
~TEST - [meas] (test over a period of time of not less than an hour with constant load at full value) vérification *f*
~TEST INDICATOR - s. dial indicator
~TESTER - [telecomm] appareil *m* d'essai des cadrans d'appel
~THERMOMETER - [instr] thermomètre *m* à cadran
~TONE - s. dialling tone
~TRAIN - [instr] équipage *m* mobile
DIALDEHYDES - [chem] (compounds which contain two aldehyde groups. Glyoxal is the most important of these) dialdéhydes *m*
DIALING - [mining] levé *m* à la boussole
DIALLAGE - [min] (found in basic igneous rock) diallage *f*
DIALLAGITE - [min] (a rock mainly consisting of diallage with small amounts of other minerals) diallagite *f*
DIALLING IN - [telecomm] appel *m* en téléphonie automatique
~OUT - [radio] (signalling by means of a dial) appel *m* en téléphonie automatique (sur un indicatif spécial)
~PULSING - [telecomm] appel *m* par cadran de numérotation
~TONE - [telecomm] tonalité *f*, signal *m* de ligne
DIALLYLBARBITURIC ACID - [chem] (a hypnotic used in medicine) acide *m* diallylbarbiturique
DIALLYL CYANAMIDE - [chem] (an intermediate for polymers) cyanamide *f* de diallyle
~ISOPHTHALATE - [chem] (polymerizable monomer used for moulding compounds) isophtalate *m* de diallyle
~MALEATE - [chem] (colourless or pale yellow readily polymerized liquid) maléate *m* de diallyle
~PHTHALATE - [chem] (plasticizer and polymerizable monomer with uses in laminates and electrical insulants) phtalate *m* de diallyle
DIALOGITE - [min] (a kind of trigonal carbonate of manganese) rhodocrosite *f*, dialogite *f*
DIALOGUE - [gen] dialogue *m*, conversation *f*
~DIRECTOR - [cinema] dialoguiste *m*
DIALYSABLE - [chem] (capable of undergoing dialysis) dialysable
DIALYSATE - [chem] dialysat *m*
DIALYSIS - [chem] (the separation of smaller moleculesfrom larger ones in a solution by means of a semi-permeable membrane which allows the passage of the smaller molecules) dialyse *f*
DIALYTIC - [chem] dialytique
DIALYZE, to - [chem] dialyser
DIALYZER - [ind chem] dialyseur
DIAMAGNETIC - [el] (of a substance having a negative magnetic susceptibility and a permeability less than that of a vacuum) diamagnétique
~SUBSTANCE - [el] (substance which becomes weakly magnetized by an external magnetic field) substance *f* diamagnétique
DIAMAGNETISM - [el] (the properties of a diamagnetic substance under the influence of a magnetizing force) diamagnétisme *m*
DIAMETER - [geom etc] (the length of a straight line joining two points on the circumference of a cir-

cle and passing through its centre) diamètre *m*
DIAMETER INCREMENT - [mech] (of straight bevel gear) différence *m* entre le diamètre du cercle primitif et le diamètre extérieur
~OF THE VOIDS - [soil] diamètre *m* des pores
~OF THE WIRE - [telecomm] diamètre *m* du fil
DIAMETRAL - [geom] diamétral
~PITCH - [mech] (in a gear wheel; the number of teeth per inch of pitch circle diameter) pas *m* diamétral
~VOLTAGE - [el] (the largest voltage between lines of a polyphase system) tension *f* diamétrale
~WINDING - [el] enroulement *m* diamétral
DIAMETRIC - [gen] diamétral
DIAMETRICAL - s. diametric
DIAMINE - [chem] diamine *f*
~BLACK - [chem] noir *m* diamine
DIAMINOPHENOL HYDROCHLORIDE - [chem] (a photographic chemical and analytical reagent) chlorhydrate *m* de (2,4) diaminophénol, amidol *m*
DIAMOND - [min] (naturally occurring, crystalline allotropic form of carbon) diamant *m*
[geom] losange *m*, rhombe *m*
[text] (weave in diamond form) losange *m*
[print] (obsolete term) corps *m* quatre, perle *f*
~AERIAL - [radio] (rhombic aerial) antenne *f* rhombique, antenne *f* en losange
~ANTENNA - s. diamond aerial
~-BACK MOTH - [zool] piéride *f* du chou
~-BEARING - [min] diamantifère
~BLACK - [dyes] (a dye for wools) noir *m* diamant, noir *m* au chrome
~BORING - [mech] forage *m* au diamant
~CHISEL - [tool] grain *m* d'orge, ciseau *m* à pointe de diamant
~CLEAVAGE - [min] boort *m*, clivage *m* du diamant
~CORE-DRILL - [mining] sonde *f* à pointe de diamant (avec tube carottier)
~CROSSING - [railw] coupement *m*, traversée *f* oblique
~CUTTING - [ind] taille *f* du diamant
~DRILL - [tool] (used in mining) foret *m* au diamant, perforatrice *f* à diamants, foreuse *f* à pointes de diamant, sondeuse *f* au diamant
~DUST - [ind] égrisée *f*, poussière *f* de diamant
~EFFECT - [text] quadrillage *m*
~FIELD - [min] champ *m* diamantifère
~FLOOR SHEET - [metall] tôle *f* striée
~FRAME - [mech] (the frame of a bicycle of a diamond pattern) cadre *m* (en losange)
~FRET - [constr] bossage *m* en pointe de diamant
~GREEN - [dyes] (a dye for wools) vert *m* diamant
~GROUND - [text] fond *m* losangé
~KNOT - [naut] pomme *f* de tire-veille
~MAT - [glass man] (loose chopped strands quilted or chemically bound together) tissu *m* en laine de verre
~MATRIX - [geol] roche *f* mère du diamant
~MESH - [metall] (used in the building industry) grillage *m* à mailles carrées, treillis *m* à mailles carrées, tôle *f* étirée avec mailles en forme de losange
~PASS - [text] passage *m* en pointe
[metall] cannelure *f* quadrangulaire
~PENCIL - [tool] diamant *m* de vitrier
~PLATE - [metall] plaque *f* en losange
~POINT - [tool] pointe *f* de diamant
DIAMOND POINT BIT - [tool] trépan *m* pointu
~POINT CHISEL - [tool] s. diamond chisel
~POINT TOOL - [tool] outil *m* à pointe de diamant
~PROFILE - [rubber ind] (of tyres) sculpture *f* à losanges
~PYRAMID HARDNESS TEST - [mech] essai *m* de dureté Vickers
~RIFFLE - [mining] riffle *m* à losanges
~RIVETED JOINT - [mech] rivure *f* convergente
~RUBBISH - [min] fragments *mpl* de diamant
~SAW - [tool] (stone cutting circular saw with diamond dust) scie *f* diamantée
~SHAPED - [geom] en losange, rhomboïdal
~SKIN DISEASE - [vet] rouget *m* du porc
~SPAR - [min] corindon *m*
~SWITCH - [railw] traversée *f* bretelle
~TREAD - s. diamond profile
~WHEEL - [tool] meule *f* diamant
~WINDING - [el] (distributed winding mainly composed of identical coils) enroulement *m* en losange
~WORK - [constr] construction *f* en pointes de diamant
~YARN - [text] fil *m* retors croisé
DIAMONDS - [astronaut] losanges *mpl* formés par des ondes de choc
DIAMYL PHENOL - [chem] (an intermediate for plasticizers, detergents, insecticides, germicides, and rubber chemicals) diamylphénol *m*
~PHTHALATE - [chem] (a plastics plasticizer) phtalate *m* de diamyle
~SULPHIDE - [chem] (intermediate for organic sulphur compounds) sulfure *m* de diamyle
DIANDROUS - [bot] (with two stamens) diandre, diandrique
DIANINE - [chem] (a developer) dianine *f*
DIANISIDINE - [chem] (a dyestuff intermediate) dianisidine *f*
DIANTHINE - [chem] (a dye for cotton) dianthine *f*
DIAPASON - [mus] (fixed standard of musical pitch) diapason *m*
[mus] (in organs; the chief flue stop) principaux jeux *mpl* de fond
DIAPER - [text] couche *f*
[arch] ornementation *f* en losanges, panneau *m* losangé
~WORK - [carp] ornementation *f* en losanges
DIAPHANOMETER - [instr] (instrument for measuring the transparency of air or other substances) diaphanomètre *m*
DIAPHANOUS - [opt] diaphane
DIAPHONE - [mus] (a resonating pipe in an organ) diaphone *m*
DIAPHORESIS - [med] diaphorèse *f*
DIAPHRAGM, to - [opt, photo etc] diaphragmer
DIAPHRAGM - [el chem] (permeable or porous partition in an electrolytic cell dividing anode and cathode regions to prevent mixing of anolyte and catholyte) diaphragme *m*, membrane *f*
[acoust] (a thin plate or sheet designed to vibrate under the action of sound waves, as in a microphone) diaphragme *m*, membrane *f*
[mech] (a thin sheet of material used to separate fluids) diaphragme *m*, membrane *f*
[metall] (in iron structures) plaque *f* de renfort
[ind chem] (a thin membrane used to separate liquids in dialysis or electrolysis) diaphragme *m*, membrane *f*

DIAPHRAGM - [el] (a permeable membrane arranged within an electrolytic cell to prevent mixing of the anolyte and catholyte) diaphragme *m*
[photo] (the movable curved blades which are adjusted to control the aperture) diaphragme *m*
[anat & zool] (a transversal partition in a cavity) cloison *f*, diaphragme *m*, membrane *f*
~APERTURE - [photo] ouverture *f* du diaphragme
~ASSEMBLY - [mech] (in a gas meter) boîte *f* à soufflet
~CELL - [ind chem] (a type of electrolytic cell used in the commercial production of chlorine and sodium hydroxide from a solution of brine) diaphragme *m* électrolytique
~CONTROL - [photo] commande *f* automatique du diaphragme
~CONTROL VALVE - [mech] (type of valve in which the degree of opening is controlled by the deformation of a flexible diaphragm under pressure) vanne *f* à commande barostatique
~DISC - [mech] (in a gas meter) plaque *f* de membrane
~DRIVE RING - [photo] anneau *m* de réglage du diaphragme
~GOVERNOR - [contr] régulateur *m* à membrane
~OPENING - [opt] diaphragme *m*
~PLANE - [photo] plan *m* du diaphragme
~PLATE - [mech] contre-plaque *f* de diaphragme
~POINTER - [photo] index *m* du diaphragme
~PRESETTING - [photo] présélection *f* du diaphragme
~PUMP - [ind chem] (pump in which liquid is displaced by the alternating distortion of a flexible diaphragm) pompe *f* à membrane
~REGULATOR - s. diaphragm governor
~RING - [mech] (in a gas meter, etc; a rim or pan) cadre *m* de fixation de la membrane
~RING MODE FILTER - [electron] (ring-shaped mode filter in a diaphragm) filtre *m* à diaphragme annulaire
~SETTING - [photo] réglage *m* du diaphragme
~TYPE PRESSURE GAUGE - [meas] (type of pressure gauge in which the moving element consists of a deformable diaphragm connected to a pointer) manomètre *m* à membrane, membrane *f* manométrique
~VALVE - [mech] (a pressure controlled valve) soupape *f* pneumatique, soupape *f* à membrane
DIAPHRAGMLESS MICROPHONE - [acoust] (microphone utilizing a discharge modulated directly by the motion of the air particles in a passing sound wave) microphone *m* statique
DIAPHYSIS - [anat] (a limb bone shaft) diaphyse *f*
DIAPOPHYSIS - [anat] diapophyse *f*
DIAPOSITIVE - [photo] (positive transparency on glass or film) diapositif *m*
DIARRHOEA - [med] diarrhée *f*
DIARTHROSIS - [zool] (a joint between two bones subject to frequent movement) diarthrose *f*
DIASCOPE - [photo] (a projection apparatus for diapositives) lanterne *f* de projection pour diapositifs
DIASCOPIC PROJECTION - [med] projection *f* diascopique
DIASPORE - [min] (naturally occurring hydrous aluminium oxide used as an abrasive and refractory material) diaspore *m*
DIASTASE - [chem] (amylotic enzyme with applications in textile and food processing) diastase *f*
DIASTASIS - [med] (in surgery, the separation of an epiphysis from the bone, without any fracture) diastasis *f*
DIASTATIC ENZYMES - [chem] (diastase producing enzymes) enzymes *mpl* diastasiques
~POWER - [chem] (the capacity of malt to produce diastase) pouvoir *m* diastasique
DIASTEMA - [zool] (a toothless gap in a jaw) diastème *m*
[geol] lacune *f* de sédimentation
DIASTIMETER - [instr] (instrument designed to measure distances) diastomètre *m*
DIASTROPHISM - [geol] (periodical changes of the configuration of the earth's surface) diastrophisme *m*
DIASTYLE - [arch] diastyle *m*
DIATHERMACY - [phys] (the transmission of radiant heat) diathermanéité *f*
DIATHERMANOUS - [phys] (capable of transmitting heat radiation) diathermane, diathermique
~SUBSTANCE - [phys] substance *f* diathermane
DIATHERMIC - [heat] diathermique
~COAGULATION - [med] (the destruction of animal tissues by electric sparks) diathermo-coagulation *f*, électrocoagulation *f*
~KNIFE - [med] bistouri *m* électrique
DIATHERMY - [med] (the generation of heat in the tissues by electric current) diathermie *f*
~MACHINE - [electron] (electronic oscillator generating the high-frequency voltages used in diathermy) appareil *m* diathermique
DIATOM - [bot] diatomée *f*
DIATOMACEOUS EARTH - s. diatomite
DIATOMIC -[chem] diatomique, biatomique
DIATOMITE - [min] (hydrated silica formed from the skeletons of diatoms. Diatomite is used as a filler for plastics, rubber, surface coatings, as an absorbant for dynamite, in filtration, and as a thermal insulant) diatomite *f*
~FILTER - [ind chem] (filter in which diatomaceous earth is used to arrest solids) filtre *m* à diatomique
DIATONIC - [mus] diatonique
~INTERVAL - [mus] (musical interval in major scales) intervalle *m* diatonique
DIATREME - [geol] diatrème *f*, cheminée *f* volcanique
DIAZIN - [chem] s. diazine
DIAZINE - [chem] diazine *f*
~BLUE - [chem] (a pigment and dye for cotton) bleu *m* diazine
~GREEN - [chem] (a dye for cotton) vert *m* diazine
DIAZO COMPOUNDS - [chem] (compounds having the general formula R·N:N·R. They are important as dyestuff intermediates) composés *mpl* diazoïques, diazoïques *m*
~COUPLING - [chem] copulation *f* avec diazoique
~REACTION - [chem] diazo-réaction *f*
DIAZOMETHANE - [chem] (an open-chain diazo compound, very reactive and used for the introduction of a methyl group into a molecule) diazométhane *m*
DIAZONIUM SALTS - [chem] (the acid salts of diazobenzene. They are important as dyestuff intermediates) sels *mpl* diazoïques
DIAZOTIZATION - [chem] (the conversion of amino compounds in diazo compounds) diazotation *f*

DIAZOTYPE - [photo] (process for obtaining coloured images on paper or fabrics) diazotypie *f*
DIBASIC - [chem] bibasique, dibasique
˜ACIDS - [chem] (acids which contain two replaceable hydrogen atoms in the molecule) acides *mpl* bibasiques
˜AMMONIUM PHOSPHATE - [chem] (has uses in fertilizers, medicine and fire proofing preparations) phosphate *m* d'ammonium bibasique
˜CALCIUM PHOSPHATE - [chem] (a fertilizer, plastics stabilizer and dietary supplement) phosphate *m* de calcium bibasique
˜MAGNESIUM PHOSPHATE - [chem] (a plastics stabilizer and an antacid used in medicine) phosphate *m* de magnésium bibasique
DIBBLING - [agric] semis *m* en poquets
˜MACHINE - [agric] machine *f* à faire les poquets
DIBENZANTHRONE - [chem] (a blue-violet dye) dibenzanthrone *f*
DIBENZOYL - [chem] dibenzoyle *m*
DIBENZYL ETHER - [chem] (a plasticizer for cellulose derivatives) éther *m* dibenzylique
˜SEBACATE - [chem] (a plasticizer) sébacate *m* de dibenzyle
DIBORANE - [chem] (boron hydride, boro-ethane; colourless evil-smelling gas with applications as a polymerization catalyst, especially for ethylene, intermediate for organic boron compounds, and as a fuel for rockets) diborane *m*
DIBROMODIETHYL SULPHIDE - [chem] (organic synthesis intermediate) sulfure *m* de dibromodiéthyle
DIBROMODIFLUOROMETHANE - [chem] (intermediate for pharmaceuticals and dyestuffs) dibromodifluorométhane *m*
DIBROMOMALONIC ACID - [chem] (intermediate for pharmaceuticals) acide *m* dibromomalonique
DIBROMOPROPANOL - [chem] (intermediate for pharmaceuticals, flame proofing compounds and pesticides) dibromopropanol *m*
DIBUCAINE - [chem] (a local anaesthetic) dibucaïne *f*, butylcaïne *f*
DIBUTOLINE SULPHATE - [chem] (a cholinergic blocking agent and mydriatic used in medicine) sulfate *m* de dibutoline
DIBUTOXYETHOXY ETHYL ADIPATE - [chem] (plasticizer for cellulose nitrate, ethyl cellulose, polyvinyl acetate and polyvinyl butyral) adipate *m* de dibutoxyéthoxyéthyle
DIBUTOXYETHYL ADIPATE - [chem] (a plasticizer for most plastics, giving good resistance to UV light) adipate *m* de dibutoxyéthyle
˜PHTHALATE - [chem] (a plastics plasticizer) phtalate *m* de dibutoxyéthyle
˜SEBACATE - [chem] (plasticizer for vinyl-chloride and vinyl-chloride acetate) sébacate *m* de dibutoxyéthyle
DIBUTOXYTETRAGLYCOL - [chem] (a solvent for insecticides) dibutoxy-tétraglycol *m*
DIBUTYL ADIPATE - [chem] (plasticizer for cellulose acetate butyrate, cellulose nitrate, ethyl cellulose, polyvinyl acetate, vinyl chloride and vinyl chloride acetate) adipate *m* de dibutyle
˜FUMARATE - [chem] (an intermediate for polymers and plasticizers) fumarate *m* de dibutyle
˜MALEATE - [chem] (intermediate for plasticizers and polymers) maléate *m* de dibutyle
˜OXALATE - [chem] (solvent and synthesis intermediate) oxalate *m* de dibutyle
DIBUTYL PHTHALATE - [chem] (plasticizer and solvent for synthetic resins, fixative and solvent for perfumes, and a textile conditioning agent) phtalate *m* de dibutyle, phtalate *m* de butyle
˜SEBACATE - [chem] (plasticizer and component of rubber additives) sébacate *m* de dibutyle
˜SUCCINATE - [chem] (plasticizer for cellulose acetate, cellulose acetate butyrate and cellulose nitrate) succinate *m* de dibutyle
˜TARTRATE - [chem] (solvent for cellulose derivatives) tartrate *m* de dibutyle
DIBUTYLTHIOUREA - [chem] (pickling agent for cleaning castings) dibutyl-thio-urée *f*
DIBUTYLTIN DIACETATE - [chem] (condensation catalyst and stabilizing agent) diacétate *m* de dibutyl-étain
˜DILAUREATE - [chem] (stabilizer and condensation catalyst) dilaurate *m* de dibutyl-étain
˜MALEATE - [chem] (stabilizer and condensation catalyst) maléate *m* de dibutyl-étain
˜OXIDE - [chem] (condensation catalyst) oxyde *m* de dibutyl-étain
˜SULPHIDE - [chem] (antioxidant, lubricant and stabilizer for plastics) sulfure *m* de dibutyl-étain
DICAPRYL ADIPATE - [chem] (plasticizer for cellulose derivatives and vinyl plastics) adipate *m* de dicapryle
˜PHTHALATE - [chem] (plasticizer for cellulose derivatives) phtalate *m* de dicapryle
˜SEBACATE - [chem] (plasticizer for acrylonitrile elastomers) sébacate *m* de dicapryle
DICE - [gen] (the plural of die, cubic structures, cubic fragments) dés *mpl*, cubes *m*
[glass man] (a cubical fracture of tempered glass) fracture *f* du verre trempé
˜PATTERN - [text] quadrillage *m*
˜WEAVE - [text] armure *f* panama
DICER - [mech] (machine to cut cubes from strips or sheets) machine *f* à découper en cubes
DICETYL ETHER - [chem] (intermediate for antistatic compounds, and plastics moulding lubricants) éther *m* dicétylique
DICHLONE - [chem] (insecticide and fungicide) dichlone *f*
DICHLORIDE - [chem] bichlorure *m*
DICHLOROANILINE - [chem] dichloroaniline *f*
DICHLOROBENZENE - [chem] (a solvent for a large number of organic compounds, insecticides and heat-transfer mediums) dichlobenzène *m*
DICHLORODIETHYL SULPHIDE - [chem] (mustard gas, used in warfare and as a synthesis intermediate) sulfure *m* de dichlorodiéthyle
DICHLORODIFLUOROMETHANE - [chem] (aerosol propellant, refrigerant and an intermediate for fluorocarbon plastics) dichlorodifluorométhane *m*
DICHLORODIPHENYLTRICHLOROETHANE - [chem] (an insecticide) dichlorodiphényltrichloroéthane *m*
DICHLOROETHYLETHER - [chem] (solvent for a number of organic compounds, surface coatings, fats and oils) éther *m* dichloroéthylique
DICHLOROETHYL FORMAL - [chem] (intermediate for synthetic rubbers) formal *m* dichloroéthylique
DICHLOROFLUOROMETHANE - [chem] (aerosol propellant refrigerant, and solvent) dichlorofluorométhane *m*
DICHLOROISOPROPYL ETHER - [chem] (solvent for

surface coatings, fats, waxes, and a component of cleaning fluids) éther *m* dichloroisopropylique

DICHLOROMETHANE - [chem] (methylene chloride. Much used as a solvent, especially for cellulose acetate) dichlorométhane *m*

DICHLORPHENAMIDE - [chem] (an oral diuretic used in medicine) dichlorophénamide *f*

DICHLOROPHENE - [chem] (bactericide and fungicide) dichlorophène *m*

DICHLOROPHENYLTRICHLOROSILANE - [chem] (an intermediate for silicones) dichlorophényltrichlorosilane *m*

DICHOGAMY - [genet] (production of elements ensuring cross-fertilization) dichogamie *f*

DICHOTOMY - [bot statist etc] dichotomie *f*

DICHROIC CRYSTAL - [crystall] cristal *m* dichroïque

~FILTER - [opt] miroir *m* dichroïque

~FOG - [photo] (formed by an organic compound of silver) voile *m* dichroïque

~MIRROR - [radar] (almost transparent mirror) miroir *m* dichroïque

DICHROISM - [opt] (the property of some crystals of absorbing a ray to a different extent) dichroïsme *m*

DICHROITE - [min] (cordierite; silicate of aluminium, iron and magnesium) dichroïte *f*, iolite *f*, cordiérite *f*

DICHROMATE CELL - [chem] (primary cell with an electrolyte consisting of a solution of sulphuric acid and potassium dichromate, the electrodes being carbon and zinc) pile *f* au bichromate

DICHROMATISM - [opt] (a kind of colour blindness) dichromatisme *m*

DICING - [aero] (U.S.: flying below 100 feet over enemy territory) vol *m* à très basse altitude

DICKEY - [mech] (in motor-vehicles) siège *m* (du conducteur)

DICKY - s. dickey

~SEAT - [auto] strapontin *m*

DICLINISM - [bot] (with separate sexes) diclinisme *m*

DICLINY - s. diclinism

DICROSCOPIC EYEPIECE - [opt] (in a polarizing microscope) oculaire *m* dicroscopique

DICTAPHONE - [el acoust] (a machine recording dictation) dictaphone *m* (marque déposée)

DICTATING - [gen] dictée *f*

~MACHINE - [mech] machine *f* à dicter

DICTYATE STAGE - [genet] stade *m* dictyotique

DICUMYL PEROXIDE - [chem] (a vulcanizing agent) péroxyde *m* de dicumyle

DICYANDIAMIDE - [chem] (intermediate for pharmaceuticals, dyestuffs, and synthetic resins; stabilizer for cellulose derivatives; fertilizer component; ingredient of detergents and rubber chemicals) dicyandiamide *m*

DICYCLOHEXYLAMINE - [chem] (rubber antioxidant, plasticizer for plastics, and an intermediate for a number of organic compounds) dicyclohexylamine *f*

DICYCLOHEXYL PHTHALATE - [chem] (plasticizer for a number of synthetic resins and cellulose derivatives) phtalate *m* de dicyclohexyle

DICYCLOMINE HYDROCHLORIDE - [chem] (an antispasmodic used in medicine) chlorhydrate *m* de dicyclomine

DICYCLOPENTADIENE - [chem] (an insecticide) dicyclopentadiène *m*

DICYCLOPENTADIENE DIOXIDE - [chem] (intermediate for plasticizers and synthetic resins) bioxyde *m* de dicyclopentadiène

DIDACTYL - [zool] (having two digits) didactyle

DIDECYL ADIPATE - [chem] (a plasticizer) adipate *m* de didécyle

~ETHER - [chem] (antistatic agent and plastics moulding lubricant) éther *m* didécylique

~PHTHALATE - [chem] (plastics plasticizer) phtalate *m* de didécyle

DIDERICHITE - [min] (rare alternative product of uraninite; also called Rutherfordine) diderichite *f*

DIDYMIUM - [chem] (term once used for a supposed element, now known to have been a mixture of praseodymium and heodymium) didyme *m*

DIE - [gen] dé *m*, cube *m*
[mech] (a tool, usually of hardened steel, used to give a specific form to a workpiece by being pressed upon it) matrice *f*, étampe *m*, coussinet *m* de filière
[plast ind] (a steel block recessed in the form of the desired object which receives the injected plastic in injection moulding) coquille *f*; moule *m*
[plast ind] (in extrusion; a steel block pierced with an orifice suitable in size and shape, through which plastic material is forced to produce an extrudate of the required section) filière *f*
[tool] (of forging press) matrice *f* de forgeage
[metall] (in diecasting) coquille *f*, moule *m* de fonderie
[rubber ind] (in extruders) filière *f*
[arch] (of a column) dé *m*, tympan *m*

~ADAPTOR - [mech] (part of an extruder used to hold dies of various types) nez *m* intermédiaire, adapteur *m*

~APPROACH - [mech] (the terminal part of the channel nearest the extrusion die) canal *m* d'injection

~BASE - [mech] (the term generally used in the USA for 'die body', the part of the extrusion head which actually carries the die) corps *m* de filière

~BODY - [metall] (the outer body or barrel of an extrusion die) corps *m* de filière, part *f* fixe du moule

~BOX - s. die head

~CAVITY - [metall] (diecasting) empreinte *f* (du moule)

~CHANNEL - [mech] (the part of the channel in an extrusion machine which is nearest the die) canal *m* d'injection

~CHASER - [mech] chasse-peigne *m*

~CHUCK - [tool] mandrin *m* à mâchoires indépendantes, mandrin *m* à coussinets

~COLLAR - [mech] filière *f* de repêchage

~CONE - [mech] (the tapering element in an extrusion die which guides the material to the webs of the spider) buse *f*, diffuseur *m*

~COOLING - [metall] refroidissement *m* des coquilles (de fonderie), refroidissement *m* des moules

~CUTTING - [metall] usinage *m* des moules

~DRESSING - [metall] (in diecasting) poteyage *m*, lubrifiant *m* pour moules

~FACE CUTTER - [mech] (device which cuts off filaments as they come from the extruder, by means of knives passing across the die face) tronçonneu-

se *f* à l'outil
DIE FORGING - [metall] forgeage *m* par matriçage
~GRINDING - [metall] rectification *f* des matrices
~GUIDE - [mech] guide-coussinets *m*
~HEAD - [mech] (in turret lathes) filière *f* à fileter, filière *f* à peignes circulaires
[constr] tête *f* de pose
~IMPRESSION - [mech] empreinte *f* de la matrice
~LAND - [plast ind] (the final portion of an extrusion die, where the cross-section does not vary) zone *f* à section constante
~LOCK - [mech] s. die locking mechanism
~LOCKING MECHANISM - [mech] dispositif *m* de fermeture du moule, verrouillage *m* du moule
~MARK - [metall] (a marking due to abrasion during drawing) trace *f*, marque *f*, rayure *f*, défaut *m*
~MATCHING - [metall] (in diecasting) centrage *m* des moules
~MILLING MACHINE - [mach tools] machine *f* à molleter
~MOULD - [metall] moule *m* pour moulage sous pression, coquille *f*
~NIPPLE - [oil ind] taraud *m* de repêchage
~NUT - [mech] (a tool of hardened steel in the form of a nut but with clearance slots at intervals, used to correct a damaged thread) écrou *m* taraudeur
~OF EXTRUDER - [mech] filière *f*
~PARTING-FACE - [metall] (in diecasting) face *f* de joint d'un moule
~PLATE - [metall] (in diecasting) (in two-plate injection moulds, the fixed head die plate and moving head die plate are those parts of the mould which are respectively attached to the fixed head and to the moving head of the press) plateau *m* matrice, plateau *m* porte-moule
[tool] (for drawing wire) filière *f*
~PLATEN - [plast ind] plateau *m* porte-moule
~QUENCHING - [metall] refroidissement *m* entre moules
~RESTRICTION - [plast ind] (the degree of convergence in an extrusion die) constriction *f*
~ROLL - [metall] cylindre *m* à matricer
~ROLLING - [metall] laminage *m* entre cylindres matriceurs
~SCORE - s. die mark
~SCRATCH - s. die mark
~SHARPENING - [metall] affûtage *m* des matrices
~SHOE - [metall] porte-matrice *m*
~SINKER - [mech] machine *f* à fraiser les matrices
~SINKING - [metall] (in diecasting ; the process of making dies) usinage *m* des moules, fraisage *m* de matrice
~SINKING MACHINE - s. die sinker
~SLOTTER - [mach tool] machine *f* à mortaiser (les rainures)
~SOCKET - [oil ind] cloche *f* à écrou
~SPACE - [metall] (in diecasting) cavité *f*
~SPINNING NOZZLE - [mech] buse *f* de filière
~SPOTTING - [metall] (in diecasting]) mise *f* en place des moules, adaptation *f* du moule
~SPOTTING PRESS - [mech] presse *f* à essayer les matrices
~STAMP - [bookbind] balancier *m*
[mech] poinçon *m*
~STAMPING - [metall] (the operation of forming parts by impressing specially-shaped blocks of hardened steel upon them) matriçage *m*
DIE STEEL - [metall] acier *m* pour matrices
~STOCK - [mech] porte-filière *m*, fer *m* à tarauder
~STROKE - [metall] (in diecasting) course *f* du moule
~SUPPORT - [metall] porte-moule *m*
~THREADING - [mech] filetage *m*
~TO ROLL GAP - [mech] (distance between the extruder die head and coating rolls) distance *f* filière-rouleaux
~TORPEDO - [mech] s. die cone
DIEBACK - [bot] (a plant disease) nécrose *f*
~OF THE BUDDING - [bot] nécrose *f* des greffés
DIECAST, to - [metall] mouler par injection, couler en coquille, couler sous pression
DIECAST - [metall] coulé *m* sous pression, moulé par injection
DIECASTING - [metall] (production of accurate casting by forcing molten metal into metal moulds under pressure) moulage *m* par injection, moulage *m* sous pression, moulage *m* en coquille
~MACHINE - [metall] machine *f* à mouler sous pression, machine *f* à mouler par injection
~PROCESS - [metall] moulage *m* sous pression, moulage *m* par injection
DIELDRIN - [chem] (an insecticide) dieldrine *f*
DIELECTRIC - [adj] (capable of supporting electric stress) diélectrique
(of a substance which does not conduct electricity) diélectrique
~ABSORPTION - [el] (the persistence of electric polarization occurring in some dielectric substances after the cessation of the polarizing field) absorption *f* diélectrique
~AERIAL - [radio] (an aerial in which the radiation pattern is obtained mainly by the use of a dielectric) antenne *f* diélectrique
~AMPLIFIER - [electron] amplificateur *m* diélectrique
~ANTENNA - [radio] s. dielectric aerial
~BREAKDOWN - [el] rupture *f* diélectrique
~BREAKDOWN TEST - [el] (a test made to determine the voltage at which a given dielectric part will fail) essai *m* de rupture diélectrique
~BREAKDOWN TEST BENCH - [el] (a permanent assembly of apparatus for dielectric testing) banc *m* d'essai de rupture diélectrique
~CIRCUIT - [el] circuit *m* diélectrique
~CONSTANT - [el] (a measure of the ability of a material to conduct electrostatic lines of force) constante *f* diélectrique, permittivité *f*
~DISPLACEMENT - [el] déplacement *m* diélectrique
~DRYING - [el] (of cables) dessiccation *f* diélectrique
~FATIGUE - [el] (the breakdown of a dielectric substance when subjected to repeated stress) fatigue *f* diélectrique
~FURNACE - [metall] four *m* à chauffage diélectrique
~HEATING - [el] (heating by means of dielectric hysteresis losses in the material to be heated) chauffage *m* par hystérésis diélectrique
~HYSTERESIS - [el] (hysteresis occurring in a dielectric) hystérésis *f* diélectrique
~LAYER - [electron] (layer on alkaline metals for sensitizing purposes) couche *f* diélectrique
~LENS - [opt] (lens made of dielectric material) lentille *f* diélectrique

DIELECTRIC LOSS - [el] (power loss in a dielectric due to dielectric heating) perte *f* diélectrique

~MATCHING PLATE - [el] (dielectric plate used as impedance transformer for matching purposes) transformateur *m* diélectrique d'adaptation

~MATERIAL - [el] diélectrique *m*

~MEDIUM - [el] (a substance having dielectric properties) diélectrique *m*

~PHASE ANGLE - [el] (the angular phase difference between the sinusoidal alternating voltage applied to a dielectric and the component of the resulting current of the same frequency as the voltage) angle *m* de pertes

~POLARIZATION - [el] polarisation *f* diélectrique

~PORCELAIN - [el] (special type of porcelain, made for electrical use, made with china clay body, quartz filler and felspar flux) porcelaine *f* diélectrique

~PROPERTIES - [el] propriétés *fpl* diélectriques

~RADIATOR - [ant] antenne *f* diélectrique

~RELAXATION - [el] relaxation *f* diélectrique

~RESISTANCE - [el] résistance *f* diélectrique, résistance *f* d'isolement

~RIGIDITY - s. dielectric strength

~ROD AERIAL - [radio] antenne-pylone *f* diélectrique

~ROD ANTENNA - s. dielectric rod aerial

~STRENGTH - [el] (the electrical pressure which a material can withstand without disruption of its structure) rigidité *f* diélectrique

~SUBSTANCE - s. dielectric medium

~TEST VOLTAGE - [el] (the voltage at which the dielectric test on an instrument is made) tension *f* d'épreuve diélectrique

~VISCOSITY - [el] (the lagging of the polarization of a dielectric behind those of a field producing them) viscosité *f* diélectrique

~WAVEGUIDE - [electron] (waveguide consisting of a dielectric tube) guide *m* d'ondes diélectrique

~WEDGES - [el] (tapering pieces of dielectric material used to match an air-filled waveguide to another) coins *mpl* diélectriques

DIELECTRICITY - [el] diélectricité *f*

DIELS-ALDER REACTION - [chem] (important reaction for the synthesis of six-membered rings. In the reaction a conjugated diene reacts by I,4-addition with the C=C bond of a number of compounds) réaction *f* de Diels-Alder

DIEMAKER - [metall] (diecasting) mouliste *m*

DIENE - [chem] (generic name for aliphatic hydrocarbons containing two double bonds) diène *m*

~SYNTHESIS - [chem] (the production of a cyclic compound from an aldehyde, an acid for ester and a conjugated diene) synthése *f* diénique

DIENOESTROL - [chem] (a synthetic oestrogen used in medicine) diénoestrol *m*

DIES - [mech] coussinets *mpl* de filière, peignes *mpl*

DIESEL CYCLE - [mech] (cycle of operation in a compression-ignition reciprocating engine) cycle *m* diesel

~ELECTRIC - [el] diesel-électrique

~ELECTRIC LOCOMOTIVE - [railw] (locomotive in which the power from a Diesel engine is used to drive an electric generator) locomotive *f* Diesel électrique

~ELECTRIC SET - [el] groupe *m* électrogène à moteur Diesel

~ENGINE - [mech] (type of internal-combustion engine in which the charge is ignited by the heat of compression) moteur *m* diesel

DIESEL ENGINED FLAPPER - [aero] modèle *m* à ailes battantes

~FUEL - [oil ind] carburant *m* diesel

~INDEX - [oil ind] (the measurement of the ignition properties of a diesel oil) indice *m* Diesel

~LOCOMOTIVE - [railw] (locomotive powered by a compression-ignition engine) locomotive *f* diesel

~MOTOR COACH - [railw] automotrice *f* diesel

~NUMBER - s. diesel index

~OIL - [oil ind] (a petroleum product used as a fuel for diesel engines) carburant *m* diesel, huile *f* lourde

~POWERED - [mech] à moteur diesel

~RAILROAD CAR - s. Diesel motor coach

DIESELIZATION - [mech] application *f* du moteur diesel

DIESELIZE, to - [mech] (to convert into diesel motors or engines) convertir en moteur diesel

DIESIS - [mus] dièse *m*
[print] diésis *m*

DIESULFORMING - [oil ind] traitement *m* des huiles diesel par hydrogénation

DIET - [med] diète *f*; régime *m*

DIETETIC - [med] diététique

DIETETICS - [med] diététique *f*

DIETHANOLAMINE - [chem] (an intermediate for plasticizers and synthetic resins, and a component of detergents and shampoos) diéthanolamine *f*

DIETHOXYETHYL ADIPATE - [chem] (plasticizer for cellulose acetate, cellulose acetate butyrate, and cellulose nitrate) adipate *m* de diéthoxyéthyle

~PHTHALATE - [chem] (plasticizer for cellulose derivatives, polyvinyl butyral, vinyl chloride, and vinyl chloride acetate) phtalate *m* de diéthoxyéthyle

DIETHYL ADIPATE - [chem] (a plastics plasticizer) adipate *m* de diéthyle

~CARBONATE - [chem] (synthesis intermediate and solvent for cellulose derivatives and a number of natural and synthetic resins) carbonate *m* de diéthyle

~MALEATE - [chem] (synthesis intermediate) maléate *m* de diéthyle

~PHOSPHITE - [chem] (a synthesis intermediate and solvent for paints) phosphite *m* de diéthyle

~PHTHALATE - [chem] (plasticizer and solvent for cellulose derivatives and a fixative for perfumes) phtalate *m* de diéthyle, salvarome *m*

~SEBACATE - [chem] (plasticizer for cellulose acetate butyrate) sébacate *m* de diéthyle

~SUCCINATE - [chem] (a plasticizer and intermediate) succinate *m* de diéthyle

~SULPHATE - [chem] (an ester of sulphuric acid with applications as an alkylating agent) sulfate *m* diéthyle

~TARTRATE - [chem] (resin plasticizer and solvent) tartrate *m* de diéthyle

DIETHYLAMINE - [chem] (a polymerization inhibitor and intermediate for pharmaceuticals, rubber additives, dyestuffs and insecticides) diéthylamine *f*

DIETHYLALUMINIUM CHLORIDE -[chem] (intermediate for organometallic compounds and a polymerization catalyst) chlorure *m* de diéthylaluminium

DIETHYLAMINOETHANOL - [chem] (an intermediate for textile conditioning agents, pharmaceuticals, and emulsifying agents) diéthylamino-éthanol *m*

DIETHYLAMINOPROPYLAMINE - [chem] (epoxy resins curing agent) diéthylaminopropylamine *f*

DIETHYLBENZENE - [chem] (a solvent and intermediate) diéthylbenzène *m*

DIETHYLCARBAMAZINE CITRATE - [chem] (used in medicine in the treatment of filariasis) citrate *m* de diéthylcarbamazine

DIETHYLDIPHENYLUREA - [chem] (stabilizer for explosives and a component of rubber additives) diéthyldiphénylurée *f*

DIATHYLDITHIOCARBAMIC ACID - [chem] (a reagent in copper identification (characteristic brown colour). The zinc salt of this acid is used as a rubber vulcanization accelerator) acide *m* diéthyldithiocarbamique

DIETHYLENE GLYCOL - [chem] (conditioning and softening agent for textile fibres, paper and vulcanized fibres; solvent for cellulose derivatives, dyestuffs, resins and gums; an intermediate in the production of explosives; and a component of antifreeze mixtures) diéthylèneglycol *m*, diglycol *m*

~GLYCOL DIBENZOATE - [chem] (plasticizer for cellulose acetate butyrate, cellulose nitrate, ethyl cellulose, polymethyl methacrylate, polystyrene, polyvinyl acetate, vinyl chloride and vinyl chloride acetate) dibenzoate *m* de diéthylèneglycol

~GLYCOL DIBUTYL ETHER - [chem] (an inert solvent and diluent) éther *m* dibutylique du diéthylèneglycol

~GLYCOL DIETHYL ETHER - [chem] (solvent for cellulose derivatives and an intermediate for organic synthesis) éther *m* diéthylique du diéthylèneglycol

~GLYCOL DIMETHYL ETHER - [chem] (synthesis reaction medium and solvent) éther *m* diméthylique du diéthylèneglycol

~GLYCOL DIPELARGONATE - [chem] (plasticizer for cellulose acetate butyrate, cellulose nitrate, ethyl cellulose, polyvinyl butyral, vinyl chloride and vinyl chloride acetate) dipélargonate *m* de diéthylèneglycol

~GLYCOL DIPROPIONATE - [chem] (plasticizer for cellulose derivatives) dipropionate *m* de diéthylèneglycol

~GLYCOL MONOBUTYL ETHER - [chem] (intermediate for plasticizers and a solvent for cellulose derivatives, natural and synthetic resins) éther *m* monobutylique du diéthylèneglycol

~GLYCOL MONOBUTYL ETHER ACETATE - [chem](it has application as a plasticizer and as a solvent for synthetic resins and cellulose derivatives) acétate *m* de l'éther monobutylique du diéthylèneglycol

~GLYCOL MONOETHYL ETHER - [chem] (a solvent, component of textile conditioning agents, and a synthesis intermediate) éther *m* monoéthylique du diéthylèneglycol

~GLYCOL MONOETHYL ETHER ACETATE - [chem] (a cellulose ester solvent) acétate *m* de l'éther éthylique du diéthylèneglycol

~GLYCOL MONOMETHYL ETHYL ETHER - [chem] (brake fluid component and synthesis intermediate) éther *m* monométhyléthylique du diéthylèneglycol

~GLYCOL MONOOLEATE - [chem] (plasticizer for ethyl cellulose and cellulose nitrate) mono-oléate *m* de diéthylèneglycol

DIETHYLENETRIAMINE - [chem] (solvent for dyestuffs, resins and sulphur) diéthylènetriamine *f*

DIETHYLMALONATE - [chem] (an intermediate for the barbiturates) diéthylmalonate *m*

DIETHYLSTILBESTROL - [chem] (a synthetic oestrogen used in medicine) diéthylstilbœstrol *m*

~DIPROPIONATE - [chem] (a synthetic oestrogen used in medicine) dipropionate *m* de diéthylstilbœstrol

DIFFER, to - [gen] différer, être en désaccord

DIFFERENCE - [gen] différence *f*, écart *m*

~AMPLIFIER - [electron] (an amplifier having two inputs and an output which is the function of their difference) amplificateur *m* de tension différentielle

~CHECK - [comput] contrôle *m* différentiel

~EQUATION - [math] équation *f* aux différences

~FREQUENCY - [radio] (beat frequency produced by the combination of two different frequencies (equal to the difference between them) fréquence *f* différentielle

~IN COUNT - [text] différence *f* de numéro

~IN HEIGHT - [surv] différence *f* de niveau, dénivellation *f*

~IN TWIST - [text] différence *f* de torsion

~LIMEN - [acoust] (a just noticeable difference) accroissement *m* juste perceptible

~NUMBER - [nucl] (the number by which the neutrons in a nucleus exceed the protons) excès *m* de neutrons

~OF DEPARTURE - [surv] différence *f* de longitude

~OF ELEVATION - [surv] différence *f* d'altitude

~OF LATITUDE - [surv] différence *f* de latitude

~OF LEVEL - s. difference of height

~OF LONGITUDE - [surv] différence *f* de longitude

~OF PHASE - [phys] (between two sinusoidal quantities having the same frequency) déphasage *m*, différence *f* de phase

~OF POTENTIAL - [el] (line integral of the electric intensity between two points) différence *f* de potentiel

~ZONE - [acoust] (combination tone the frequency of which is the difference of the frequencies of the two generating tones) son *m* différentiel

DIFFERENTIABILITY - [gen] (the ability to differentiate) pouvoir *m* de se différencier

DIFFERENTIABLE - [adj] différentiable, que l'on peut différencier

DIFFERENTIAL - [adj] différentiel

[mech] (of a gear, hobbing machine, etc) différentiel *m*

~ABSORPTION RATE - [nucl] (the ratio of concentration of an isotope in a given tissue to the concentration which would be obtained if the isotope were distributed throughout) coefficient *m* d'absorption différentielle

~AILERON LINKAGE - [aero] (method of interconnecting aileron controls such that when the main control is moved the upward motion of one exceeds the downward motion of the other) ailerons *mpl* différentiels

~AMPLIFIER - [radio] amplificateur *m* différentiel

~ANALYSER - [comput] (analog computer designed to solve a variety of differential equations) machine *f* à résoudre les équations différentielles

~ANODE RESISTANCE - [radio] (anode-cathode path resistance in a thermionic valve) résistance *f* interne différentielle

~ARRANGEMENT - [telecomm] montage *m* différentiel

~BEVEL GEARS - [mech] engrenages *mpl* coniques de différentiel

~BEVEL PINION - [auto] satellite *m* du différentiel

DIFFERENTIAL BEVEL WHEEL - [auto] planétaire *m* du différentiel
~BOOSTER - [el] survolteur *m* différentiel
~BRAKE - [mech] frein *m* sur différentiel
~BRIDGE - [el] (containing two symmetric transformers) pont *m* différentiel
~CAGE - [auto] carte *m* de différentiel
~CALCULUS - [math] (the method of obtaining the derivative of a function) calcul *m* différentiel
~CALORIMETER - [instr] (instrument designed to measure a quantity of heat by comparing it with a known quantity of heat at the same temperature) calorimètre *m* différentiel
~CAPACITOR - [electron] condensateur *m* différentiel
~CAR AXLE - [auto] différentiel *m*
~CARRIER - [auto] carter *m* (de pignon d'attaque) de différentiel
~CASE - s. differential carrier
~CASING - [mech] (casing housing the carrier with planetary and satellites) carter *m* de différentiel
~COEFFICIENT - [math] différentielle *f*
~COIL - [el] enroulement *m* différentiel
~CONDENSER - [radio] (used for balancing purposes) condensateur *m* différentiel
~CROSS-SECTION - [nucl] section *f* efficace différentielle
~DIAGNOSIS - [med] diagnostic *m* différentiel
~DUPLEX - [telecomm] (two-way system of telegraphy with differentially wound relays and galvanometers) système *m* duplex différentiel
~ELEMENT - [contr] comparateur *m*, différentiel *m*
~EQUATION - [math] (an equation involving derivatives of an unknown function) équation *f* différentielle
~EXCITATION - [el] (excitation produced by two windings through which flow separate currents whose electromagnetic actions are in opposite directions) excitation *f* différentielle
~FLOTATION - [metall] (process designed to separate metallic sulphides from each other) flottation *f* différentielle
~FLUX - [opt] facteur *m* différentiel de flux du diaphragme
~GAIN CONTROL - [radio] (device for altering the gain of a radio receiver) atténuateur *m* sélectif
~GALVANOMETER - [instr] (designed to measure the difference between two currents) galvanomètre *m* différentiel
~GAP - [contr] (a two-step action with overlap) réglage *m* à deux paliers avec recouvrement
~GEAR - [mech] engrenage *m* différentiel
~HARDENING - [metall] trempe *f* différentielle
~HEAT OF DILUTION - [phys] chaleur *f* différentielle de dilution
~HEAT OF SOLUTION - [phys] chaleur *f* différentielle de dissolution
~HOUSING - s. differential casing
~IONIZATION CHAMBER - [nucl] (a two-section ionization chamber) chambre *f* d'ionisation différentielle
~LEAKAGE FLUX - [el] flux *m* de dispersion dans l'entrefer
~LOCK - [mech] blocage *m* du différentiel
~LOCKING DEVICE - [mech] dispositif *m* de blocage du différentiel
~MEASURING INSTRUMENT - [instr] (instrument designed to measure the difference of two electrical quantities of the same kind) appareil *m* de mesure différentiel
DIFFERENTIAL MICROPHONE - [el acoust] microphone *m* différentiel
~NUT - [mech] écrou *m* différentiel
~PERMEABILITY - [el] perméabilité *f* différentielle
~PINIONS - [mech] (in automobiles) satellites *mpl* de différentiel
~PISTON - [mech] (a piston or plunger having different effective areas on its opposite faces) piston *m* différentiel
~PLANETARY PINION - [auto] satellite *m* de différentiel
~PLUNGER - s. differential piston
~PRESSURE GAUGE - [instr] manomètre *m* différentiel
~PRESSURE TRANSDUCER - [comput] transducteur *m* de pression différentiel
~RELAY - [electron] (type of relay in which the armature movement is due to the difference of current in two actuating coils) relais *m* différentiel
~RESISTANCE - [el] résistance *f* différentielle
~SCATTERING CROSS-SECTION - s. differential cross-section
~SEGMENT - [genet] segment *m* différentiel
~SELF-ACTOR - [text] métier *m* renvideur à filer (à mouvement différentiel)
~SEX GENE - [genet] gène *m* de différenciation sexuelle
~SHAFT - [auto] arbre *m* de roue
~SHRINKAGE - [photo] contraction *f* différentielle
~SIGNALLING - [radio] (signalling method employing a differential relay) signalisation *f* différentielle
~SPIDER - [auto] croisillon *m* du différentiel
~STEAM CALORIMETER - s. differential calorimeter
~SUN GEAR - [auto] planétaire *m* de différentiel
~SUSCEPTIBILITY AND PERMEABILITY - [el] (the rate of change of the magnetization with respect to the magnetic field) susceptibilité *f* et perméabilité *f* différentielles
~TACKLE - [mech] palan *m* différentiel
~THERMOMETER - [instr] (a wet and dry bulb thermometer) thermomètre *m* différentiel
~THREAD - [mech] filet *m* différentiel
~THRESHOLD - s. difference limen
~TRANSFORMER BRIDGE - [el] pont *m* à transformateur différentiel
~WHITE COUNT - [nucl] (the number of every variety of white corpuscles in a count of 100) formule *f* leucocytaire
~WINDING - [el] (a component winding in which two sections are connected in opposition, so that their overall effect is equal to the difference of their magnetomotive forces) enroulement *m* différentiel
~WINDLASS - [mech] treuil *m* différentiel
DIFFERENTIALLY COMPOUND-WOUND MACHINE - [el] (compound-wound direct-current machine in which the magnetomoving forces of the two windings oppose one another) machine *f* à excitation différentielle
~-WOUND MOTOR - [el] moteur *m* à excitation différentielle
DIFFERENTIATE, to - [math] (to obtain a derivative by the differential calculus) différentier
[gen] différencier
DIFFERENTIATED-IMPULSE SIGNALLING - [radio] si-

gnalisation *f* par impulsions différentiées
DIFFERENTIATING CIRCUIT - [radio] circuit *m* linéaire, circuit *m* différentiateur
DIFFERENTIATION - [math] (the operation of finding the derivative of a function with respect to the independent variable) différentiation *f*
[gen] différentiation *f*
DIFFERENTIATOR - [electron] (device whose output signal is in proportion with the derivative of an input signal) différentiateur *m*, dérivateur *m*
DIFFLUENT - [adj] (term used of fluid streams which are following divergent paths) diffluent
DIFFORM - [adj] (of irregular form) difforme
DIFFORMED - s. difform
DIFFRACTED - [phys] diffracté
~WAVE - [acoust] onde *f* déviée
DIFFRACTION - [opt] (spreading of light to a small extent beyond the limits of the geometrical shadow) diffraction *f*
[radio] (property of all magnetic waves to curve round an obstruction in their path) diffraction *f*
~ANALYSIS - [nucl] (the study of atomic arrangement by means of X-rays or material particles) analyse *f* par diffraction
~ANGLE - [opt] (the angle formed by the incident beam of light and the resulting diffracted beam) angle *m* de diffraction
~BANDS - s. diffraction fringes
~EFFECT - [opt] effet *m* diffraction
~ELECTRON MICROSCOPE - [instr] microscope *m* électronique à diffraction
~FRINGES - [opt] franges *fpl* de diffraction
~GRATING - [opt] (optical device for producing spectra by diffusion) réseau *m* de diffraction
~INSTRUMENT - [instr] (used in spectroscopy to study the structure of matter etc) appareil *m* à diffraction
~LINES - [opt] lignes *fpl* de diffraction
~MOTTLING - [metall] taches *fpl* de diffraction
~PATTERN - [opt] diagramme *m* de diffraction
~RING - [opt] anneau *m* de diffraction
~SCATTERING - [nucl] dispersion *f* par diffraction
~SPECTROSCOPE - [instr] spectroscope *m* à réseau
~SPECTRUM - [opt] (the spectrum produced by diffraction) spectre *m* de diffraction
DIFFRACTOGRAPH - [instr] diffractographe *m*
DIFFRACTOMETER - [instr] diffractomètre *m*
DIFFUSATE - [chem] (the material which passes the membrane in dialysis) diffusat *m*
DIFFUSE, to - [gen & phys] diffuser
~DENSITY - [photo] (density measured with diffused light) densité *f* diffuse
~NEBULAE - [astr] (irregularly shaped nebulae) nébuleuses *fpl* diffuses
~NUCLEUS - [cytol] (occurring in non-nucleated cells) noyau *m* diffus
~REFLECTION - [light] (a reflection from a surface, so that the incident light beam is reflected in all directions) réflexion *f* diffuse
~-REFLECTION FACTOR - [light] (the ratio between the diffused luminous flux and the total luminous flux incident on the surface) facteur *m* de réflexion diffuse
~REFRACTION - [opt] (refraction in all directions) réfraction *f* diffuse
~SCATTERING - [nucl] fond *m* continu de diffusion
~SOUND - [acoust] (acoustic oscillation in a sound field in which the sound energy density is uniform) son *m* diffus
DIFFUSE TRANSMISSION - [light] (the transmission of a light beam through a screen) transmission *f* diffuse
~TRANSMISSION DENSITY - [photo] densité *f* de transmission diffuse
DIFFUSED-AIR AERATION - [hydr] aération *f* par diffusion (ou à air comprimé)
~FIELD - [acoust] champ *m* (acoustique) diffus
~FOCUS LENS - [opt] objectif *m* anachromatique
~ILLUMINATION - s. diffused lighting
~JUNCTION - [electron] (junction formed by the diffusion of an impurity in a semiconductor crystal) jonction *f* par diffusion
~JUNCTION TRANSISTOR - [electron] (type of transistor in which the impurity is diffused into the semiconductor) transistor *m* à jonction par diffusion, transistor *m* à jonction diffusé
~LIGHTING - [light] (system of illumination in which the directions of rays are as varied as possible, so as to produce an absence of shadows) éclairage *m* diffusé
~TYPE TRANSISTOR - [electron] transistor *m* diffusé
DIFFUSER - [mech] (a passage so formed as to give rise to a decrease in pressure and an increase in velocity in a fluid passing through it) diffuseur *m*
[light] diffuseur *m*
[photo] (device used to diffuse, or soften the light entering the lens) écran *m* diffusant
[acoust] (in a loudspeaker, to avoid sound focusing) diffuseur *m*
[sugar ind] vase *m* de diffusion
~SCREEN - [photo] écran *m* diffusant
~SCRIM - [photo] (light diffusing material, e.g.gauze) diffuser *m*, tarlatane *f*
~THROAT - [mech] col *m* de diffuseur
DIFFUSIBILITY - [gen] diffusibilité *f*
DIFFUSING ADDITIONAL LENS - [opt] bonnette *f* diffusante
~DISC - [mech] disque *m* diffuseur
~DISK - [photo] disque *m* diffuseur
~SCREEN - [photo] écran *m* diffuseur
~TISSUE - [photo] tissu *m* diffuseur
DIFFUSIOMETER - [instr] (instrument designed to measure diffusion in liquids) diffusiomètre *m*
DIFFUSION - [gen] diffusion *f*
[chem] (the formation of a homogeneous mixture or solution due to the movement of the molecules of each substance) diffusion *f*
[phys] (in a structure, of the transverse stress distribution) diffusion *f*
~AIR PUMP - [mech] pompe *f* à diffusion
~ANALYSIS - [chem] (the determination of the molecular weight, or relative size, of particles by the comparison of their diffusion rate) analyse *f* par diffusion
~APPARATUS - [mech] diffuseur *m*
~BARRIER - [phys] (a partition with submicroscopic holes through which materials are transferred by diffusion) barrière *f* à diffusion
~BEET KNIFE - [sugar ind] couteau *m* de coupe-racines
~BY REFLECTION - [opt] diffusion *f* par réflexion
~CAPACITANCE - [electron] (effect occurring in germanium diodes, requiring a few microseconds for the diodes to acquire a specified inverse resis-

tance): capacité *f* de diffusion
DIFFUSION CASING - [mech] (in pumps) corps *m* de pompe
~CIRCLE - [opt] cercle *m* de diffusion
~CLOUD CHAMBER - [phys] (chamber in which the diffusion of vapour is utilized to produce a supersaturated condition) chambre *f* à nuage
~COEFFICIENT - [phys] coefficient *m* de diffusion
~COLUMN - [phys] (a vertical tube in which the radialtemperature gradient is maintained) colonne *f* de diffusion
~CONSTANT - [phys] (the quotient of diffusion current by the concentration gradient of energy distribution) constante *f* de diffusion
~CROSS-SECTION - [nucl] (the effective target area presented by an atom to an incident particle) section *f* efficace de diffusion
~CURRENT - [phys] (limiting current reached by electrolytic migration of the ions in a solution) courant *m* de diffusion
~EQUATION - [phys] (used in the study of the energy distribution in the motion of ions and electrons) équation *f* de la diffusion
~FACTOR - [light] facteur *m* de diffusion
~IN SOLIDS - [phys] (a very slow yet observable phenomenon) diffusion *f* dans les solides
~INDICATRIX - [opt] (a graph in polar coordinates) diagramme *m* de diffusion
~-JUICE HEATER - [sugar ind] calorisateur *m* du jus de diffusion
~LAYER - [el chem] (the stratum of fluid surrounding an electrode, across which a change in the concentration of the electrolyte is exhibited) couche *f* de diffusion
~LENGTH - [phys] (the average distance covered by a diffusing particle from its formation to its absorption) longueur *f* de diffusion
~LOSSES - [sugar ind] pertes *fpl* à la diffusion
~OF GASES - [phys] (the drift of gas molecules) diffusion *f* des gaz
~OF LIGHT - [light] (transmittance of light by a translucent medium) diffusion *f* de la lumière
~POTENTIAL - [el chem] (liquid junction potential) potentiel *m* de diffusion
~PRESS WATER - [sugar ind] eaux *fpl* de pression de diffusion
~PROCESS - [sugar ind] procédé *m* de diffusion
~RATE - [nucl] (the rate of diffusion or radiations) vitesse *f* de diffusion
~SCREEN - [sugar ind] tôle *f* perforée du diffuseur
~TIME - [sugar ind] durée *f* de la diffusion
~TRANSISTOR - [electron] transistor *m* diffusé
~-TYPE ENLARGER - [photo] agrandisseur *m* à lumière diffuse
~WATER RETURN - [sugar ind] recyclage *m* des eaux de diffusion, reprise *f* des eaux de diffusion
DIFFUSIVE SUBSTANCE - [phys] (a substance easily dialysing through colloidal septa) matière *f* diffusive
DIFFUSIVITY - [phys] (in a heated bar, the quantity determining the rise of temperature at a point which is not directly heated) diffusibilité *f*, diffusivité *f*
DIFFUSOR - s. diffuser
DIFLUOROPHOSPHORIC ACID - [chem] (produced by heating phosphoric acid with ammonium fluoride) acide *m* difluorophosphorique
DIG, to - [gen] creuser, fouiller, piocher, excaver, extraire
DIG - [gen] coup *m* de bêche, coup *m* de pointe, coup *m*
DIGASTRIC - [anat] (said of those muscles which have a fleshy terminal point) digastrique
DIGENESIS - [genet] (the alternation of generations) digénèse *f*, génération *f* alternante
DIGENETIC REPRODUCTION - [genet] (sexual reproduction) reproduction *f* sexuée
DIGESTED SLUDGE - [ind chem] boue *f* digérée
DIGESTER - [ind chem] (for the extraction of solubles) digesteur *m*, fosse *f* de digestion
[paper man] (a receptacle in which raw materials are boiled) lessiveur *m*
[brew ind] autoclave *m*
[rubber ind] digesteur *m* (de dévulcanisation)
~HOUSE - [paper man] salle *f* des lessiveurs
~RECLAIM - [rubber ind] (regeneration by an alkali process) régénération *f* alcaline
DIGESTIBILITY - [physiol] digestibilité *f*
DIGESTING - [gen] digestion *f*
[paper man] lessivage *m*
~COMPARTMENT - s. digestion chamber
DIGESTION - [gen] digestion *f*
[nucl] (a process in uranium ore treatment) schéidage *m*
[photo] (of emulsion) maturation *f* chimique, refonte *f*
~AGENT - [photo] adjuvant *m* de maturation
~CHAMBER - [ind chem] chambre *f* de putréfaction des boues
~FOG - [photo] voile *m* de maturation
~PROCESS - [ind chem] procédé *m* de digestion
~TANK - s. digester
DIGESTIVE - [adj] digestif
~GLAND - [bot] (in carnivorous plants) glande *f* digestive
~ORGAN - [physiol] organe *m* digestif
DIGGER - [gen] terrassier *m*, mineur *m*
[mech] (a mechanical excavator) excavatrice *f*
[agric] (a potato-digger] arracheuse *f*
[constr] terrassier *m*
[mining] chercheur *m*, orpailleur *m*, piqueur *m*
DIGGING - [agric] creusement *m*, terrassement *m*, déblai *m*, fouille *f*, extraction *f*
~FORK - [agric] fourche *f* à bêcher
~MACHINE - s. digger
DIGGINGS - [mining] gisements *mpl*, fouilles *fpl*
DIGIT - [math] (any whole number from I to 9) chiffre *m*, signe *m*
[zool] doigt *m*
~ABSORBER SELECTOR - [telecomm] sélecteur *m* à rappel de chercheur
~ABSORPTION - [telecomm] suppression *f* des impulsions
~DELAY - [comput] retard *m* de l'impulsion digit, retard *m* de l'impulsion de chiffre
~DRIVER - [comput] amplificateur *m* d'écriture
~EMITTER - [comput] émetteur *m* d'impulsions
~IMPULSE - [comput] impulsion *f* de chiffre
~KEY - [comput] touche *f* de chiffre
~PUNCHING - [comput] perforation *f* numérique (ou normale)
~SELECTOR - [comput] délecteur *m* digit
~TRACK - [comput] piste *f* de digits
DIGITAL - [comput] (of a computer using members

expressed in digits to represent all the variables of a problem) à indication numérique, digital
DIGITAL - [mus] (one of the keys of a piano keyboard) touche *f*
~CLOCK - [comput] chronomètre *m* numérique
~CODE - [comput] (a code which consists only of digits) code *m* (de calculatrice) numérique
~CODING - s. digitize, to
~COMPUTER - [comput] (computer calculating by the use of numbers expressed in digits, yes and no for all the variables occurring in a problem) calculatrice *f* digitale, calculatrice *f* numérique
~DIFFERENTIAL ANALYSER - [comput] (machine which solves differential equations) analyseur *m* différentiel digital
~HEAD - [electron] (for magnetic recording) tête *f* magnétique
~MAGNETIC TAPE - [comput] ruban *m* numérique
~PROCESS-COMPUTER - [comput] calculateur *m* numérique industriel
~READOUT - [comput] (from a counter) prélèvement *m* numérique
~SIGNAL - [comput] signal *m* numérique
~SORTING METHOD - [comput] méthode *f* de classification numérique
~TAPE PROGRAMME CONTROL - [contr] (used in mechanical machines etc) commande *f* à programme numérique sur bande
~-TO-ANALOG CONVERTER - [comput] convertisseur *m* digital-analogique
~TRANSDUCER - [comput] convertisseur *m* numérique
DIGITALIFORM - [bot] (shaped like fingers) digitaliforme
DIGITALIN - [chem] (a glycoside obtained from the leaves of Digitalis purpurea and used in medicine) digitaline *f*
DIGITALIS - [chem] (the dried leaves of Digitalis purpurea) digitale *f*
DIGITALISM - [med] intoxication *f* par la digitale
DIGITATE - [bot] (of a hand-shaped compound leaf) digité
DIGITIFORM - [anat] digitiforme
DIGITIGRADE - [zool] (walking on digits) digitigrade
DIGITIZE, to - [comput] (to change an analog measurement into a number) coder numériquement
DIGITOXIN - [chem] (glycoside with medicinal uses obtained from the leaves of Digitalis purpurea) digitoxine *f*, digitoxoside *m*
DIGLYCOL CARBAMATE - [chem] (intermediate for synthetic resins) carbamate *m* de diéthylèneglycol
~CHLOROFORMATE - [chem] (a component of plasticizers) chloroformate *m* de diéthylèneglycol
~LAURATE - [chem] (emulsifying agent with applications in cosmetics, polishes, textiles processing, and papermaking) laurate *m* de diéthylèneglycol
~MONOSTEARATE - [chem] (an emulsifying agent for cosmetics) monostéarate *m* diéthylèneglycol
~OLEATE - [chem] (emulsifying agent) oléate *m* de diéthylèneglycol
~RICINOLEATE - [chem] (plasticizer for synthetic elastomers and resins) ricinoléate *m* de diéthylèneglycol
~STEARATE - [chem] (emulsifying and thickening agent) stéarate *m* de diéthylèneglycol
DIGLYCOLIC ACID - [chem] (intermediate for plasticizers and synthetic resins) acide *m* diglycolique
DIGRESS, to - [gen] digresser
DIHEDRAL - [geom] dièdre *m*
~ANGLE - [geom] (the angle at which the wings of an aeroplane are inclined upward in respect to the plane of reference) angle *m* dièdre
DIHEPTAL BASE - [el] (a 14-pin base for cathode-ray tubes) culot *m* à quatorze broches
DIHEXAEDRON - [geom] dihexaèdre *m*
DIHEXAGONAL - [geom] dihexagonal
DIHEXYL PHTHALATE - [chem] (plasticizer for cellulose derivatives, polymethyl methacrylate, polystyrene, polyvinyl acetate, polyvinyl butyral, vinyl chloride and vinyl chloride acetate) phtalate *m* de di-hexyle
~SEBACATE - [chem] (a plasticizer) sébacate *m* de di-hexyle
DIHYDRID - [genet] (the product of two parents having different heritable characters) dihydride *m*
DIHYDRIC ALCOHOL - [chem] diol *m*, alcool *m* diatomique
DIHYDROABIETYL ALCOHOL - [chem] (a plasticizer) alcool *m* dihydroabiétylique
DIHYDROCHOLESTEROL - [chem] (used in pharmacy to produce hydrophilic ointments) dihydrocholestérol *m*
DIHYDROMORPHINONE HYDROCHLORIDE - [chem] (a potent analgesic used in medicine) chlorhydrate *m* de dihydromorphinone
DIHYDROSTREPTOMYCIN - [chem] (an antibiotic used in medicine) dihydrostreptomycine *f*
DIHYDROXYACETONE - [chem] (an intermediate for plasticizers, insecticides, and pharmaceuticals) oxétone *f*, dihydroxy-acétone *f*
DIIODOHYDROXYQUIN - [chem] (used in medicine for the treatment of amoebiasis) diiodohydroxyquine *f*
DIIODOTHYRONINE - [chem] (intermediate in the synthesis of thyroxine) diiodothyronine *f*
DIISOBUTYL ADIPATE - [chem] (a plasticizer) adipate *m* de di-isobutyle
~ALUMINIUM CHLORIDE - [chem] (a catalyst for polyolefins) chlorure *m* de di-isobutylaluminium
~AZELATE - [chem] (a plasticizer for cellulose acetate butyrate, cellulose nitrate, ethyl cellulose, polymethyl methacrylate, vinyl chloride, and vinyl chloride acetate) azélate *m* de di-isobutyle
~KETONE - [chem] (a solvent for rubber cellulose esters and synthetic resins) di-isobutylcétone *f*
~PHTHALATE - [chem] (a plasticizer) phtalate *m* de di-isobutyl
DIISOBUTYLAMINE - [chem] (a synthesis intermediate) di-isobutylamine *f*
DIISOBUTYLENE - [chem] (plasticizer and intermediate) di-isobutylène *m*
DIISOCYANATES - [chem] (compounds which contain two isocyanate groups) di-isocyanates *mpl*
DIISODECYL ADIPATE - [chem] (plasticizer for synthetic resins) adipate *m* de di-isodécyle
~PHTHALATE - [chem] (a plasticizer) phtalate *m* de di-isodécyle
DIISOOCTYL ADIPATE - [chem] (a plasticizer) adipate *m* de di-isooctyle
~AZELATE - [chem] (plasticizer for cellulose acetate, cellulose acetate butyrate, cellulose nitrate, ethyl cellulose,vinyl chloride and vinyl chloride acetate) azélate *m* de di-isooctyle
~PHTHALATE - [chem] (plasticizer for synthetic resins and elastomers) phtalate *m* de di-isooctyle

DIISOOCTYL SEBACATE - [chem] (a plasticizer) sébacate *m* de di-isooctyle
DIISOPROPANOLAMINE - [chem] (an emulsifying agent) di-isopropanolamine *f*
DIISOPROPYL BENZENE - [chem] (an intermediate and solvent) di-isopropylbenzène *m*
~CRESOL - [chem] (a stabilizer for insecticides) di-isopropylcrésol *m*
~FLUOPHOSPHATE - [chem] (a miotic used in medicine) fluophosphate *m* de di-isopropyle
DIISOPROPYLENE SALICYLATE - [chem](an absorbent for UV light) salicylate *m* de di-isopropylèneglycol
DIKE - s. dyke
~OF A RAILWAY - [constr] (or dyke of a railway) digue *f* d'un chemin de fer
DIKETENE - [chem] (an intermediate) dicétène *m*
DIKETONES - [chem] (compounds containing CO-groups divided into alpha- and beta-types according to their position in the molecule) dicétones *fpl*
DIKTIOMA - [med] tumeur *f* de l'épithélium ciliare
DILACERATION - [med] dilacération *f*
DILAPIDATION - [gen & build] (damage to premises etc) dilapidation *f*, délabrement *m*
DILATABILITY - [phys] (the property of dilating) dilatabilité *f*
DILATABLE SOIL - [soil] sol *m* dilatable
DILATANT - [gen] dilatant
DILATANCY - [phys] (the property of some colloidal solutions of setting under pressure) dilatance *f*
DILATATION - [phys] (the increase of volume) dilatation *f*, expansion *f*, allongement *m*, élargissement *m*
DILATATIONAL STRAIN - [phys] (the increase of volume divided by the original volume) coefficient *m* de dilatation
DILATE, to - [phys] (to increase in volume) dilater, allonger
DILATION - [phys] (the fractional increase in volume caused by deformation) dilatation *f*, expansion *f*
DILATOMETER - [instr] (instrument designed to measure thermal expansion) dilatomètre *m*
DILATOMETRY - [phys] dilatométrie *f*
DILATOR - [med] (surgical instrument) dilatateur *m*
~[zool] (muscle opening an orifice by contracting) dilatateur *m*
DILAURYL ETHER - [chem] (antistatic agent and plastics moulding lubricant) éther *m* dilaurylique
~SULPHUR - [chem] (intermediate for sulphonium compounds) sulfure *m* de dilauryle
~THIODIPROPIONATE - [chem] (plasticizer and antioxidant) thio-di-propionate *m* de dilauryle
DILINOLEIC ACID - [chem] (an emulsifying agent and modifier for synthetic resins) acide *m* dilinoléique
DILL - [chem] (essential oil obtained from Anethum graveolens and used in perfumery and as a flavouring agent) aneth *m*, anet *m*
~OIL - [chem] essence d'aneth
DILLY - [mining] petit plan *m* incliné
DILSH - [min] (band of inferior coal) couche *f* de charbon de mauvaise qualité
DILUENT - [chem] (a substance used to dilute, i.e. to reduce the concentration of another) diluant *m*
~GAS - [gas ind] gaz *m* d'appoint
~STILL - [oil ind] appareil *m* de distillation du diluant
DILUTE, to - [chem] (to reduce the strength of a solution by increasing the quantity of solvent) diluer
DILUTE SOLUTION - [chem] (a solution containing a large amount of water or other solvent) solution *f* diluée
DILUTED RUBBER - [rubber ind] dissolution *f* de caoutchouc
DILUTENESS - s. dilution
DILUTING AGENT - [chem] (a liquid which decreases the concentration of a solution) diluant *m*
DILUTION - [chem] (a decrease in concentration) dilution *f*
~EFFECT - [nucl] (effect resulting from mixing radioactive effluents with sea-water) effet *m* de dilution
~WATER - [hydr] eau *f* de dilution
~ZONE - [metall] (dilution of weld metal with the base metal in welded joints) zone *f* de dilution
DILUVIAL - [geol] diluvial
DILUVIUM - [geol] (obsolete term for accumulation which could not be accounted for) diluvium *m*
DIM, to - [gen] réduire, diminuer, obscurcir, mettre en veilleuse
[auto] passer en codes, se mettre en codes
~LIGHT - [auto] feux *mpl* de croisement, codes *mpl*
~SWITCH - s. dimmer switch
DIMEGALY - [zool] (with spermatozoa of different sizes) dimégalie *f*
DIMENSION - [gen] dimension *f*
[draw] cote *f*
~DRAWING - [gen] plan *m* d'encombrement, croquis *m* d'encombrement
~LINE - [draw] cote *f*
DIMENSIONAL - [gen] dimensionnel
~AND OPERATION TEST AND ACCEPTANCE CONDITIONS - [aero] (in aircraft part inspection, the specification for test made by measurement and functional trial, before acceptance) essai *m* dimensionnel et de fonctionnement et conditions de réception
~CHECK - [comm] (verification of the dimensions of a part by measurement) vérification *f* dimensionnelle
~STABILITY - [phys] (capacity of a material to retain its original dimension under changing conditions) stabilité *f* dimensionnelle
~STABILITY UNDER HEAT - [phys] stabilité *f* dimensionnelle à la chaleur, résistance *f* à la chaleur
~TOLERANCE - [mech] tolérance *f* dimensionnelle
DIMENSIONING - [draw] cotation *f*
DIMENSIONLESS VARIABLE - [math] variable *f* sans dimension
DIMER - [chem] (a molecule formed by the combination of two identical molecules) dimère *m*
~ACID - [chem] (generic name for high molecular weight dibasic acids which will polymerize with alcohols and polyols to form plasticizers etc) acide *m* dimère
DIMERIC - [chem] (capable of forming, or relating to, dimers) dimère
DIMERISM - [chem] dimérie
DIMETHICONE - [chem] (a mixture of silicone oils used in pharmaceuticals) diméthicone *m*
DIMETHOXYETHYL ADIPATE - [chem] (a plasticizer) diméthoxyéthyl adipate *m*
DIMETHYL ANTHRANILATE - [chem] (an intermediate for pharmaceuticals, dyestuffs and flavours) anthranilate *m* de di-méthyle

DIMETHYL CHLOROACETAL - [chem] (a solvent and an intermediate for pharmaceuticals) chloroacétal *m* de di-méthyle
~ETHER - [chem] (a solvent, aerosol propellant, and refrigerant) éther *m* diméthylique
~GLYCOL PHTHALATE - [chem] (cellulose ester plasticizer) phtalate *m* de di-méthylglycol
~ISOPHTHALATE - [chem] (a plasticizer) isophtalate *m* de di-méthyle
~OCTANOL - [chem] (a perfumery agent) diméthyloctanol *m*
~PHOSPHITE - [chem] (synthesis intermediate) phosphite *m* de di-méthyle
~PHTHALATE - [chem] (plasticizer for rubber and synthetic resins) phtalate *m* de di-méthyle
~SEBACATE - [chem] (intermediate for vinyl resins and a solvent and plasticizer for cellulose esters) sébacate *m* de di-méthyle
~SULPHATE - [chem] (an alkylating agent in organic reactions) sulfate *m* de diméthyle, diméthylsulfate *m*
~SULPHIDE - [chem] (a solvent for a number of inorganic compounds) sulfure *m* de di-méthyle
~SULPHOXIDE - [chem] (solvent for synthetic fibres) diméthylsulfinone *m*
~TEREPHTHALATE - [chem] (intermediate for polyester resins) téréphtalate *m* de di-méthyle
DIMETHYLACETAL - [chem] (synthesis intermediate) diméthylacétal *m*
DIMETHYLACETAMIDE - [chem] (intermediate and solvent for natural and synthetic resins) diméthylacétamide *f*
DIMETHYLAMINE - [chem] (a solvent and intermediate for rubber chemicals, pharmaceuticals, and textile chemicals) di-méthylamine *f*
DIMETHYLAMINOPROPYLAMINE - [chem] (curing agent for epoxy resins) diméthylaminopropylamine *f*
DIMETHYLANILINE - [chem] (intermediate for flavourings and dyestuffs) diméthylaniline *f*
DIMETHYLBENZYLCARBINOL - [chem] (a perfumery intermediate) diméthylbenzylcarbinol *m*
DIMETHYLCYCLOHEXANE - [chem] (synthesis intermediate) diméthylcyclohexane *m*
DIMETHYLDICHLOROSILANE - [chem] (intermediate for silicone rubbers, resins, and oils) diméthylchlorosilane *m*
DIMETHYLDIPHENYLUREA - [chem] (stabilizer for explosives) diméthyldiphenylurée *f*
DIMETHYLDITHIOCARBONATE - [chem] (used in rubber industry) diméthyldithiocarbonate *m*
DIMETHYLHYDANTOIN - [chem] (intermediate for synthetic resins) diméthylhydantoine *f*
DIMETHYLISOPROPANOLAMINE - [chem] (synthesis intermediate) diméthylisopropanolamine *f*
DIMETHYLOCTANEDIOL - [chem] (intermediate and surface-active agent) diméthyloctanédiol *m*
DIMETHYLISOBUTYLCARBINOL PHTHALATE - [chem] (plasticizer for cellulose derivatives, polymethyl methacrylate, polystyrene, polyvinyl butyral, vinyl chloride, and vinyl chloride acetate) phtalate *m* de diméthylisobutylcarbinol
DIMETHYLKETON - [chem] diméthylcétone *f*
DIMETHYLOCTYNEDIOL - [chem] (intermediate and surface-active agent) diméthyloctynédiol *m*
DIMETHYLPIPERAZINE - [chem] (lupetazine; an intermediate for synthetic resins, pharmaceuticals, rubber chemicals, and fungicides) diméthylpipérazine *f*

DIMINISH, to - [gen] diminuer, amoindrir
DIMINISHING PIPE - [plumb] (a tapered pipe lenght) raccord *m* conique
DIMINUTION - [gen] diminution *f*, amoindrissement *m*
DIMIXIS - [genet] dimixie *f*
DIMMED LIGHT - [light] lumière *f* atténuée
DIMMER - [light] (device for reducing the brilliance of a lamp) gradateur *m*, variateur *m*
~SWITCH - [auto] (anti-dazzle switch) inverseur *m* phares-codes, commutateur *m* de feux de croisement
DIMMING - [paint] opacifiant
[light] affaiblissement *m*, mise *f* en veilleuse, obscurcissement *m*
~RESISTANCE - [el] (a variable resistance used mainly for stage lighting) résistance *f* chutrice
DIMORPHIC - [chem] (capable of crystallizing in two different forms) dimorphe
DIMORPHISM - [zool] (the condition of having two different forms) dimorphisme *m*
DIMPLE - [mech] (small, approximately spherical depression, e.g. in a plate) logement *m* de tête de rivet
[mech] (initial depression used to guide the drill) trou *m* de centre
[gen] fossette *f*
DIMPLING - [mech] (in sheet metal drawing) emboutissage *m*
DIMYRISTYL ETHER - [chem] (intermediate, antistatic agent, and plastics mould lubricant) éther *m* dimyristique
DIN - [gen] (abbrev for Deutsche Industrie Normen, German Industrial Standards) DIN
[acoust] (loud and confused sound) bruit *m*, tumulte *m*, cliquetis *m*
DINER - s. dining car
DINERIC - [chem] (a solution consisting of two incompatible solvents with a single soluble in each) dinère
~INTERFACE - [chem] (the bounding surface between two immiscible solvents) interface *f* dinère
DINETTE - [constr] (a recess in a room and used as dining room) salle *f* à manger en niche
DINEUTRON - [nucl] (a combination of two neutrons) dineutron *m*, bineutron *m*
DINGEY - s. dinghy
DINGHY - [naut] (a small rowing boat) dinghy *m*, canot *m* pneumatique
[railw] (staff sleeping accommodation) wagon *m* de servitude
DINGING - [constr] crépi *m* grossier
~HAMMER - [impl] marteau *m* de tôlier
DINGLE - [agric] hangar *m*
DINGO - [zool] (a breed of wild Australian dog) dingo *m*
DINGS - [mech] (marks in sheet iron surfaces) bosselures *fpl*
DINGY - [gen] sale, souillé, terne, sombre, foncé
[naut] s. dinghy
DINING - [gen] dîner *m*
~CAR - [railw] wagon-restaurant *m*
~HALL - [constr] réfectoire *m*
DINITROBENZENE - [chem] (intermediate for dyestuffs and a camphor substitute in the production of celluloid) dinitrobenzène *m*
DINITRONAPHTHALENE - [chem] (intermediate for dyestuffs) dinitronaphtalène *m*

DINITROPHENOL - [chem] (intermediate for dyestuffs and photographic chemicals) dinitrophénol *m*
DINITROTOLUENE - [chem] (DNT) (an explosive and an intermediate for dyestuffs) dinitrotoluène *m*
DINKEL - [agric] épeautre *m*
DINKING DIE - [tool] emporte- piéce *m*
DINKY INKY - [cin] (lighting implement with 500-750 W incandescent lamps) projecteur *m* 500 W
DINONYL ADIPATE - [chem] (DNA) (plasticizer with special low-temperature properties) adipate *m* de di-nonyle
~PHENOL - [chem] (a solvent) dinonylphénol *m*
~PHTHALATE - [chem] (plasticizer for vinyl plastics) phtalate *m* de di-nonyle
DIOCTYL ETHER - [chem] (intermediate, antistatic agent, electrical insulator, and plastics mould lubricant) éther *m* di-octylique
~FUMARATE - [chem] (plasticizer for polyvinyl acetate, vinylchloride and vinylchloride acetate) fumarate *m* de di-octyle
~ISOPHTHALATE - [chem] (plasticizer for cellulose acetate butyrate, cellulose nitrate, ethyl cellulose, polystyrene, vinyl chloride and vinyl chloride acetate) isophtalate *m* de di-octyle
~PHOSPHITE - [chem] (intermediate and solvent) phosphite *m* de di-octyle
~SEBACATE - [chem] (plasticizer for cellulose nitrate, ethyl cellulose, polymethyl methacrylate, polystyrene, vinyl chloride and vinyl chloride acetate) sébacate *m* de di-octyle
DIODE - [electron] (an electronic tube having only two electrodes, i.e. cathode and anode) diode *f*
~CHARACTERISTIC - [electron] caractéristique *f* de diode
~CLIPPER - [electron] (clipper circuit using a diode) circuit *m* de seuil à diode
~COUNTER - [instr] (circuit comprising two diodes for counting impulses) compteur *m* à diodes
~DETECTOR - [radio] détecteur *m* à diode
~HEPTODE - [electron] (electronic tube with a diode and a heptode section which are screened from each other and use separate filaments) diode-heptode *f*
~MODULATOR - [radio] modulateur *m* à diodes
~PENTODE - [radio] (electronic tube with separate diode and pentode electrode system and a common filament) diode-penthode *f*
~PROBE-TYPE VOLTMETER - [instr] (voltmeter with a diode in a probing head) voltmètre *m* électronique à sonde
~SQUARE-LAW DETECTION - [radio] détection *f* quadratique par diode
~SWITCH - [telev] commutateur *m* à diode
~TRIODE - [electron] diode-triode *f*
~VALVE - s. diode
DIOECIOUS - [bot] (with male and female organs on separate plants of the same species) diœcie *f*
DIOPTOMETER - [opt] (instrument designed to measure the refraction of the eye) optomètre *m*
DIOPTOMETRY - [opt] optométrie *f*, dioptrique *f* de l'œil
DIOPTRAL - [opt] dioptrique
DIOPTRE - [opt] (the unit of power of a lens) dioptrie *f*
~STEP - [instr] subdivision *f* en dioptries
DIOPTRIC - [opt] dioptrique
~MECHANISM - [opt] (mechanism enabling images to be focused on the retina) mécanisme *m* dioptrique
DIOPTRIC OBJECTIVE - [opt] objectif *m* dioptrique
~SUBSTANCE - [opt] substance *f* dioptrique
~SYSTEM - [opt] (a convergent optical system with both focal lenses positive) système *m* dioptrique
~TELESCOPE - [instr] lunette *f* dioptrique
DIOPTRICAL - [opt] dioptrique
DIOPTRICS - [opt] dioptrique *f*
DIOPTROMETER - [instr] (instrument designed to measure the number of dioptres of a lens) dioptrimètre *m*
DIORAMA - [opt] (picture seen through an opening from a specified distance) diorama *m*, maquette *f*
DIORITE - [min] (a granite rock) diorite *f*
DIOTRON - [electron] (computer circuit using an emission limited diode) diotron *m*
DIOXANE - [chem] (I:4-diethylene oxide; a solvent for a large number of organic compounds) dioxane *m*, dioxyde *m* de diéthylène
DIOXIDE - [chem] bioxyde *m*
DIOXOLANE - [chem] (solvent for cellulose derivat ves, fats, waxes and dyes) dioxolane *m*
DIOXYNAPHTHALENE - [chem] dioxynaphtalène *m*
DIP, to - [gen] plonger, immerger, descendre
[naut] (the flag) marquer (un pavillon), saluer
[metall] (to submerge in a liquid, as part of a pr cess) plonger, tremper
[aero] piquer, s'incliner en avant
[auto] (of lights) passer de phares en codes
DIP - [gen] immersion *f*
[draw] (of a line) inclinaison *f*
[phys] (of the magnetic needle) inclinaison *f* (magnétique)
[aero] piqué *m*
[el] (of a line) flèche *f* (normale)
[geol] (inclination of strata to the horizontal) pen dage *m*, inclinaison *f*, plongement *m*
[geom] (of a curve) chute *f*, descente *f*
[gas ind] (a grip) crochet *m*
[el] (solution designed to produce a chemical rea tion on a metal surface) bain *m* d'immersion
~ANGLE - [phys] (the angle between the magnetic field of the earth and the horizontal) angle *m* d'inclinaison magnetique
~BRAZING - [metall] soudure *f* au pot
~CARRIAGE - [gas ind] (in gasholders) guidage *m* la partie supérieure d'une levée
~CHANNEL - [gas ind] (in gasholders) fer *m* en U crochet
~CIRCLE - [instr] (instrument designed to measur the magnetic dip) inclinomètre *m*
~COAT, to - [metall] (to produce a coating by sub mersions in a liquid) revêtir par immersion
~COAT - [metall] potée *f*
~COATING - [metall] (process of coating a mould by dipping and later removing the solidified coati in one piece) revêtement *m* par immersion
~DYEING - [text] teinture *f* en cuve, teinture *f* pa immersion
~ENAMELLING - [metall] émaillage *m* par immersion, émaillage *m* au trempé
~FAULT - [geol] faille *f* de plongement
~FINISHING - [text] apprêt *m* par immersion
~FOLD - [geol] pli *m* plongeant
~GAUGE - [instr] mire *f* pour vérifier la flèche des lignes
~GETTERING - [electron] (gettering method used the manufacture of electronic tubes) revêtir de

getter par immersion
DIP HEADING - [mining] descenderie *f*
~HEATING - [metall] réchauffement *m* par immersion
~HOT-PROCESS - [metall] galvanisation *f* à chaud
~LOGGING - [geol] mesure *f* de pendage
~NEEDLE - [instr] (compass in which the magnet moves only in a vertical plane and measures the magnetic inclination) boussole *f* d'inclinaison
~OF BED - [geol] s. dip
~OF FLAG - [naut] s. dip
~OF THE NEEDLE - s. dip
~OILER - [mech] graisseur *m* à compte-gouttes
~PIPE - [mech] (in condensation processes) plongeur *m*
~PLANT - [paint] (a plant for painting by dipping) installation *f* de revêtement par immersion
~PLATE - [gas ind] (in gasholders) tôle *f* de garde
~PLATING - [metall] (method of producing a thin coating of a different metal on a metal object by immersion in solution of a salt or of salts of the metal to be deposited) revêtement *m* électrolytique par immersion)
~POLISHING - [metall] (immersion in a hot solution of sodium hypochlorite followed by washing and drying) polissage *m* par immersion
~REST - [mech] (of gasholders) repos *m* de crochet
~ROD - [mech] jauge *f*
~SKIRTING PLATE - s. dip plate
~-SLIP - [geol] rejet *m* en profondeur
~SLOPE - [geol] inclinaison *f* du pendage
~STICK - [impl] (a simple rod which is inserted into a tank to measure the depth of the liquid therein) jauge *f*
~SWITCH - s. dimmer switch
~TANK - [metall] (tank to contain the coating material in dip-coating) bac *m* de trempée, bac de plonge
~WORKING - [mining] exploitation *f* en aval pendage
DIPENTAERYTHRITOL - [chem] (component of protective coatings) dipentaérythritol *m*
DIPENTENE - [chem] (solvent and dispersing agent and an intermediate for synthetic resins) dipentène *m*
~MONOXIDE - [chem] (an intermediate for epoxy resins) oxyde *m* de dipentène
DIPHASE - [el] (term applied to A.C. systems employing two phases) biphasé, diphasé
DIPHENHYDRAMINE HYDROCHLORIDE - [chem] (antihistamine used in medicine) chlorhydrate *m* de diphénhydramine
DIPHENIC ACID - [chem] (an intermediate for pharmaceuticals and dyes) acide *m* diphénique
DIPHENYL - [chem] (a fungicide and heat transfer agent) biphényle *m*, phénylbenzène *m*
~BLACK - [chem] (a dye used for cotton dyeing)noir *m* biphényle
~CARBONATE - [chem] (an intermediate for carbonates and a solvent and plasticizer) carbonate *m* de diphényle
~CRESYL PHOSPHATE - [chem] (flame-retardant plasticizer for PVC, cellulose acetate and nitrocellulose) phosphate *m* de diphényle et monocrésyle
~DECYL PHOSPHITE - [chem] (stabilizer for vinyl resins) phosphite *m* de biphényle et de décyle
~ETHER - [chem] éther *m* biphénylique
~GREEN - [chem] (a dye used for cotton) vert *m* biphényle
~OCTYL PHOSPHATE - [chem] (flame-retardant plasticizer for PVC, cellulose acetate, and nitrocellulose)phosphate *m* de biphényle et d'octyle
DIPHENYL OXIDE - [chem] (a heat transfer medium and synthesis intermediate for perfumes) oxyde *m* de phenyle, diphényloxyde *m*
~PHTHALATE - [chem] (plasticizer for cellulose esters and other synthetic resins) phtalate *m* de biphényle
~ORTHO-XYLENYL PHOSPHATE - [chem] (plasticizer) diphényl-ortho-xylényl-phosphate *m*
DIPHENYLACETONITRILE - [chem] (a herbicide and intermediate for pharmaceuticals) diphénylacétonitrile *m*
DIPHENYLAMINE - [chem] (a stabilizer for plastics and a starting material for pharmaceuticals, pesticides, and dyes) diphénylamine *f*
DIPHENYLBENZIDINE - [chem] (an analytical reagent) diphénylbenzidine *f*
DIPHENYLCARBAZIDE - [chem] (an analytical reagent) diphénylcarbazide *f*
DIPHENYLDICHLOROSILANE - [chem] (intermediate for silicone compounds) diphényldichlorosilane *m*
DIPHENYLENE OXIDE - [chem] oxyde *m* de biphénilène, dibenzofurane *m*
DIPHENYLGUANIDINE - [chem] (vulcanization accelerator) diphénylguanidine *f*
~PHTHALATE - [chem] (vulcanization accelerator) phtalate *m* de diphénylguanidine
DIPHENYLKETONE - [chem] (a plasticizer) diphénylcétone *f*, benzophénone *f*
DIPHENYLMETHANE - [chem] (synthesis intermediate for pharmaceuticals and dyestuffs) diphénylméthane *m*, ditane *m*
~DYES - [chem] (dyestuffs derived from diphenylmethane) colorants *mpl* au diphénylméthane
~GROUP - [chem] (compounds containing two benzene nuclei and a single carbon atom) groupe *m* diphénylméthane
DIPHENYLNAPHTHYLENEDIAMINE - [chem] (synthesis intermediate) diphénylnaphtylènediamine *f*
DIPHENYLPYRALINE HYDROCHLORIDE - [chem] (an antihistamine used in medicine) chlorhydrate *m* de diphénylpyraline
DIPHENYLTHIOUREA - [chem] (plasticizer) diphénylthio-urée *f*, thio-carbanilide *f*
DIPHENYLUREA - [chem] (synonym for carbanilide. Diphenylurea has uses as a synthesis intermediate for pharmaceuticals) diphényl-urée, carbanilide *f*, diphenylcarbamide *f*
DIPHOSGENE - [chem] diphosgène *m*
DIPHTHERIA - [med] diphtérie *f*
DIPHTHEROID - [bact] (term describing a number of bacilli similar to that of diphtheria) diphtéroïde
~BACILLI - [bact] bacilles *mpl* diphtéroïdes
DIPHTHONG - [print] (the union of two vowels pronounced in one syllabe) diphtongue *f*
DIPHYGENIC - [zool] (with two modes of development) diphygénique
DIPHYLETIC - [biol] (belonging to two different ancestral groups) diphylétique
DIPHYODONT - [zool] (with two sets of teeth) diphyodonte
DIPLANETARY - [zool] (having two kinds of zoospores) biplanétaire
DIPLEGIA - [med] (bilateral paralysis) diplégie *f*
DIPLET - [phys] (spectrum line composed of two lines) doublet *m*

DIPLEX - [telecomm] (telegraphy system for the independent transmission of two telegrams over the same circuit) transmission *f* simultanée de deux signaux
~RADIO TRANSMISSION - [radio] (the simultaneous transmission of two signals over a common wave) transmission *f* simultanée de deux signaux sur une même porteuse
~SYSTEM - [telecomm] système *m* permettant la transmission simultanée de deux signaux
DIPLEXER - [telev] (interference preventing equipment) système *m* permettant d'utiliser une même antenne pour deux émission radioélectriques
~FILTER - [telev] filtre *m* diplexeur
DIPLOCOCCUS - [biol] (coccus individuals with a tendency to form pairs) diplocoque *m*
DIPLO-HAPLOID TWINNING - [genet] formation *f* de jumeaux diplo-haploïdes
DIPLOID - [biol] (with the somatic number of chromosomes of the species) diploïde
DIPLOMA - [gen] diplôme *m*, certificat *m*
DIPLOMATIC INK - [chem] encre *f* sympathique
DIPLOPIA - [med] (double vision) diplopie *f*
DIPLOSCOPE - [instr] (measure designed to measure the extent of the double vision of objects) diploscope *m*
DIPLOSIS - [genet] (the restoration of the somatic chromosome number by the fusion of two gametes) diplosie *f*
DIPOLE - [chem](molecule with separate centres of the positive and negative charges) doublet *m*
[radio & telev] s. dipole aerial
[el] (a positive and negative charge at a fixed distance) doublet *m*, dipôle *m*
~AERIAL - [radio & telev] (an aerial comprising two straight conductors in line and end-to-end, connected to the transmitter or receiver at the inner ends) antenne *f* dipôle
~ELECTRIC - [el] (arrangement of two equal quantities of electricity of opposite signs, concentrated at two points infinitely near to one another) dipôle *m* électrique, dipôle *m* magnetique
~LAYER - [electron] (an electric double layer) couche *f* bipolaire
~MICROPHONE - [acoust] microphone *m* bidirectionnel à deux systèmes indépendants
~MOMENT - [chem] (the product of the magnitude of the charges of a dipole and their distance from one another) moment *m* dipolaire
~RADIATOR - s. dipole aerial
~RECORDING METHOD - [comput] méthode *f* d'écriture avec retour à zero
~ZONE - [electron] (a zone in which the density of the hole and of the electron is too small to neutralize the space charges) zone *f* bipolaire
DIPPED - [adj] incliné, abaissé, trempé
~FABRIC - [text] (textile material impregnated with rubber by dipping) tissu *m* gommé
DIPPEL'S OIL - [chem] (another term for bone oil) huile *f* de dippel, huile *f* empyreumatique d'os
DIPPER - [mech] (a dredger with a large bucket at the end of a long arm) drague *f* à cuiller
[impl] (the bucket of the dredger) godet *m* (de drague, de pelle mécanique)
[impl] (a fairly large vessel with a long handle) cuillère *f* a pot
[astr] Grande Ourse *f*
DIPPER ACID - [chem] (colloq) acide *m* sulfurique
~DIG - [impl] pelle *f* de profondeur
~DREDGER - s. dipper
DIPPING - [gen] immersion *f*
[draw] (in a diagram) chute *f* d'une courbe
[ind chem] (any treatment in a liquid bath) traitement *m* par immersion, trempage *m*
[metall] décapage *m*
[paint] vernissage *m* au trempé
[th eng] (fire control) sondage *m* du feu
~ACID - s. dipper acid
~COMPOUND - [rubber ind] (rubber compound for dipping fabrics) mélange *m* de trempage
~DRUM - [text] rouleau *m* immergé
~FILAMENT - [el] (used for dipped beam light) filament *m* pour lampes codes
~FORM - [rubber ind] forme *f* de trempage
~FRAME - [text] abaisseur *m* de cuve
~LACQUER - [paint] vernis *m* au trempé
~LATEX - [rubber ind] latex *m* au trempé
~LIQUOR - s. dipping lye
~LYE - [text] lessive *f* d'immersion
~MICROSCOPE - [opt] microscope *m* à immersion
~NEEDLE - s. dip needle
~PROCESS - [rubber ind] (process for the reclaim of rubber) procédé *m* au trempé
~RACK - [impl] support *m*, ratelier *m* de trempage
~REFRACTOMETER - [instr] (refractometer which is dipped into the liquid to be examined) réfractomètre *m* à immersion
~ROLLER - [text] cylindre *m* de trempage
~SOLUTION - [ind chem] solution *f* au trempé
~TANK - [impl] bac *m* de trempage
~THE YARN INTO THE SIZE - [text] encollage *m* des filés
~VARNISH - [paint] (varnish designed to be applied by dipping the object to be treated into it) vernis *m* au trempé
DIPROPYL KETONE - [chem] (solvent for natural and synthetic resins) dipropylcétone *f*, butyrone *f*
~PHTHALATE - [chem] (a plasticizer) phtalate *m* de dipropyle
DIPROPYLENE GLYCOL - [chem] (a solvent for cellulose derivatives) dipropylèneglycol *m*
~GLYCOL DIBENZOATE - [chem] (plasticizer) dibenzoate *m* de dipropylèneglycol
~GLYCOL MONOMETHYL ETHER - [chem] (a solvent and component of brake fluids) éther *m* monométhylique du dipropylèneglycol
~GLYCOL MONOSALICYLATE - [chem] (plasticizer conferring good UV light degradation resistance) monosalicylate *m* de dipropylèneglycol
DIPS - [chem] liqueurs *fpl* corrosives
DIPSESIS - [med] soif *f* intense
DIPSOMANIA - [med] (the impulse to drink excessively) dipsomanie *f*
DIPSOSIS - s. dipsesis
DIPSOTHERAPY - [med] dipsothérapie *f*
DIPSTICK - s. dip stick
DIPTERA - [zool] (flies, gnats, etc) diptères *mpl*
DIPTERAL - [arch] diptère *m*
DIPYRE - [min] (also called mizzonite; a mixture of the meionite and mariolite molecules) dipyre *m*
DIPYRIDYLETHYLSULPHIDE - [chem] (intermediate for a number of compounds including plasticizers, pharmaceuticals, dyestuffs, weedkillers, and insecticides, rubber and textile chemicals) sulfure *m* de

dipyridyléthyle
DIPYRITE - s. dipyre
DIRECT, to - [gen] diriger, conduire, gouverner, ordonner, renseigner
~ACTING ENGINE - [mech] machine *f* à action directe
~ACTING LOAD - [phys] charge *f* directe
~ACTING PUMP - [mech] (type of steam pump in which the steam piston and the pump plunger are at opposite ends of the same rod, usually without any rotary parts) pompe *f* à action directe
~ACTING RECORDING INSTRUMENT - [instr] (an instrument in which the marking device is connected to the primary detector) enregistreur *m* à action directe
~ADAPTATION - [bot] (an adaptation seemingly independent of natural selection) adaptation *f* directe
~ARC FURNACE - [el] (type of arc furnace in which the arc occurs between an electrode and the charge itself) four *m* à arc direct
~ARC HEATING - [el] (method of arc heating in which the arc current passes through the heated medium) chauffage *m* direct par arc
~-AXIS COMPONENT OF A MAGNETOMOTIVE FORCE- [el] composante *f* longitudinale d'une force magnétomotrice
~-AXIS COMPONENT OF THE ELECTROMOTIVE FORCE [el] composante *f* longitudinale d'une force électromotrice
~-AXIS COMPONENT OF THE VOLTAGE - [el] composante *f* longitudinale d'une tension
~-AXIS SUBTRANSIENT ELECTROMOTIVE FORCE -[el] force *f* électromotrice subtransitoire longitudinale
~-AXIS SUBTRANSIENT IMPEDANCE - [el] impédance *f* subtransitoire longitudinale
~-AXIS SYNCHRONOUS IMPEDANCE - [el] impédance *f* synchrone longitudinale
~-AXIS SYNCHRONOUS REACTANCE - [el] réactance *f* synchrone longitudinale
~-AXIS TRANSIENT ELECTROMOTIVE FORCE - [el] force *f* électromotrice transitoire longitudinale
~-AXIS TRANSIENT IMPEDANCE - [el] impédance *f* transitoire longitudinale
~BLUE - [paint] bleu *m* direct
~CALL - [telecomm] (in telephony:call involving only one international circuit) appel *m* direct
~CASTING - [metall] coulée *f* en moule de fonte de haut-fourneau, coulée *f* directe
~CIRCUIT - [telecomm] (in telegraphy; a circuit from a station to another station without the use of relays at intermediate stations) ligne *f* directe
~CONNEXION - [el] couplage *m* direct
~CONTRACT - [comm] marché *m* par entente directe
~CONTROL - [mech] commande *f* directe
[aero] (system of operating aircraft control surfaces by manual power as distinct from servo-systems) commande *f* directe
~COOLER - [gas ind] condenseur *m* direct
~-COUPLED - [mech] (connected by means of a shaft coupling, as distinct from connexion through belts, gears or the like) à couplage direct, en prise directe
~-COUPLED CIRCUIT - [el] circuit *m* à couplage direct
~-COUPLED EXCITER - [el] (exciter for electric machines mounted on the shaft of the machine it excites) excitatrice *f* à couplage direct
DIRECT-COUPLED GENERATOR - [el] (generator mechanically coupled with the machine it drives) génératrice *f* à couplage direct
~-COUPLED ROLL - [mech] cylindre *m* commandé
~-COUPLED STAGE - [electron] étage *m* à couplage direct
~COUPLING - [radio] (coupling in valve amplifiers in which the anode of one valve is connected to the grid of the next, without the interposition of a capacitor) couplage *m* direct
~CURRENT - [el] (DC) (electric current flowing in one direction only and without any important cyclic variation in magnitude) courant *m* continu
~-CURRENT AMPLIFIER - [el] amplificateur *m* à courant continu
~-CURRENT BALANCER - [el] (a direct-current motor generator which is used to equalize the voltages between the wires of a multiple-wire d.c. system) égalisatrice *f* à courant continu
~-CURRENT INTERRUPTION - [electron] (the opening of the circuit at the rectifier output end) ouverture *f* du circuit de sortie
~-CURRENT MOTOR - [el] (a motor which is only suitable for operations by direct current) moteur *m* à courant continu
~-CURRENT RELAY - [el] (relay for direct current only) relais *m* à courant continu
~-CURRENT RESTORER - [el] régénérateur *m* de composante continue
~DIALLING - [telecomm] (in telephony) appel *m* direct
~-DRAUGHT BOILER - [metall] chaudière *f* à flamme directe
~DRIVE - [mech] prise *f* directe
[radio] (system of transmission with the antenna circuit directly coupled to the oscillator circuit) commande *f* directe
~DRIVE LOCOMOTIVE - [railw] locomotive *f* à transmission directe
~-DRIVEN - [mech] à commande directe
~DYE - [paint] colorant *m* substantif, colorant *m* direct
~FEED (OF THE AERIAL) - [radio] attaque *f* directe de l'antenne
~-FIRE OVEN - [domest gas appliance] (an internally heated oven) four *m* à chauffage direct, four *m* à circulation
~FIRED HEATER - [heat] (a heater in which the heat is brought to the heating surface in contact with the heating fluid) réchaffeur *m* à chauffage direct
~FLYWHEEL DRIVE - [mech] entraînement *m* direct par volant
~FOUNDATION - [constr] fondation *f* directe
~GERMINATION - [bot] (germination of a spore by a filament) germination *f* directe
~HEATED MOULD - [plast ind] moule *m* à chauffage direct
~HEATING - [heat] (heating system by radiation) chauffage *m* direct, chauffage *m* par rayonnement
~HYDRAULIC PISTON - [metall] (in diecasting) piston *m* direct hydraulique
~ILLUMINATION - [opt] (in a microscope, the light which falls on the object under examination from the side from which it is observed) éclairage *m* direct
[light] s. direct lighting
~IMAGING OPTICS - [opt] optique *f* formatrice d'ima-

ges directes
DIRECT IMPULSE - [telecomm] impulsion *f* directe
~INJECTION - [mech] (the injection of liquid fuel direct into the cylinder or induction pipe) injection *f* directe
~INJECTION PUMP - [mech] (pump used to inject fuel in direct injection engines) pompe *f* d'injection directe
~INPUT - [electron comput] (introduction of information into a computer) entrée *f* directe
~INTERELECTRODE CAPACITANCE - [electron] (the direct capacitance between two electrodes) capacité *f* interélectrodes
~LAYING - [telecomm] (of cables) pose *f* directe des câbles
~LEVELLING - [surv] nivellement *m* direct
~LIFT HOIST - [mech] (a hydraulic cylinder assembly connected with the understructure of the body) vérin *m* de levage, vérin *m* à action directe
~LIGHT - [light] lumière *f* directe
~LIGHTING - [light] (light almost totally directed downwards) éclairage *m* direct
~LINE - [telecomm] (in telephony; the line from the subscriber's telephone to the exchange) ligne *f* directe
~MAGNETOSTRICTIVE EFFECT - [el] (mechanical strain in ferromagnetic rods in a magnetic field) effet *m* de magnétostriction directe
~MEASUREMENT OF EFFICIENCY - [el] (in electric machines or transformers; method by which input and output are measured directly) mesure *f* directe du rendement
~NUCLEAR DIVISION - [cytol] (direct division of the nucleus by constriction) amitose *f*, division *f* acinétique, division *f* directe
~OBSERVATION - [gen] observation *f* directe
~OXIDATION - [metall] (the formation of oxide on the surface which occurs at high temperatures) oxydation *f* directe
~PICK-UP - [telev] (transmission of images without intermediate recording) réception *f* directe, prise *f* de vues directe
~PIEZOELECTRIC EFFECT - [el] effet *m* piézoélectrique direct
~POSITIVE - [telev] enregistrement *m* positif de l'image
~PRESSURE CLOSING - [metall] fermeture *f* à pression directe
~-PRESSURE HOT CHAMBER MACHINE - [metall] machine *f* à chambre chaude à pression directe de l'air sur le métal
~PRINTER - [telecomm] (a type of telegraph printer) téléimprimeur *m*
~PRINTING - [telecomm] (in telegraphy; system in which the received signals are automatically printed) impression *f* directe
~PROCESS - [metall] (e.g. the obtaining of wrought iron without first making pig-iron) réduction *f* directe
~PULP DRYING - [sugar ind] séchage *m* des pulpes par gaz de foyer indépendant
~RADIAL TRIANGULATION - [surv] triangulation *f* radiale graphique
~RADIATION - [radiat] rayonnement *m* de fuite
~RADIATOR LOUDSPEAKER - [el acoust] haut-parleur *m* à radiation directe
~RAY - [radio] (the part of the wave directly transmitted to the receiver) parcours *m* direct
DIRECT READING - [instr] à lecture directe
~-READING INSTRUMENT - [instr] (an instrument in which the values are shown directly on the scale, without the need for the application of a multiplying constant) instrument *m* à lecture directe
~RECORDING - [el acoust] (a recording for reproduction without any processing) enregistrement *m* direct
~REFLECTION - [opt] réflexion *f* régulière, réflexion *f* spéculaire
~RESISTANCE FURNACE - [metall] (type of furnace in which heat is developed by passing an electric current through the charge) four *m* à chauffage direct par résistance
~RESISTANCE HEATING - [th eng] (method of heating in which the current used is passed directly through the substance to be heated) chauffage *m* direct par résistance
~ROUTING SYSTEM - [telecomm] système *m* à commande directe (de la sélection)
~SENS - [math] sens *m* direct, sens *m* trigonométrique
~SOUND - [acoust] (sound from the source to the listener) son *m* direct
~STARTING - [el] (method of starting an electric motor by switching it directly into the line, without the use of any apparatus to modify the initial current demand) démarrage *m* direct
~STEAM - [mech] vapeur *f* directe
~STEEL FROM ORE PROCESS - [metall] acier *m* obtenu par le procédé au minerai, acier *m* Siemens
~STROKE - [el] (lightning striking a transmission line) coup *m* de foudre direct
~SULPHATE RECOVERY - [chem] sulfatation *f* directe
~SUSPENSION CONSTRUCTION - [el] (for the overhead contact wires of an electric traction system) ligne *f* de contact à suspension directe
~SWITCHER - [telev] commutateur *m* direct
~SWITCHING STARTER - [el] (a starter in which a motor is started by being connected directly to the line, without the use of any current-limiting device) démarreur *m* direct
~TAKING-UP - [text] (of a loom) enroulement *m* direct
~TOLL CIRCUIT - [telecomm] (in telephony) circuit *m* direct de transit
~TRANSMISSION FACTOR - [opt] facteur *m* de transmission régulière
~TRANSMITTER - [radio] (device transmitting directly to the receiving end) émetteur *m* direct
~VIEW SCREEN - [telev] écran *m* à vision directe
~VIEW TUBE - [telev] (a picture tube without magnification by optical means) tube *m* à vision directe
~-VISION SPECTROSCOPE - [instr] spectroscope *m* à vision directe
~-VISION TUBE - s. direct view tube
~WARPER - [text] ourdissoir *m* direct, ourdissoir *m* au large
~WAVE - [radio] (an electromagnetic wave travelling without reflection or refraction) onde *f* directe
~X-RAY ANALYSIS - [phys] analyse *f* radiocristallographique
DIRECTING COUPLE - [el] couple *m* directeur
~PULSE - [el] impulsion *f* de commande
DIRECTION - [gen] direction *f*, conduite *f*, administration *f*, orientation *f*

[phys] (of a straight line) direction *f*
[cin] (in film production) direction *f*, réalisation *f*
[el] (of current) sens *m*
DIRECTION CONE - [electron] (the cone which is applied to an X-ray tube to direct the rays) cône *m* de centrage
~FINDER - [radio] (directional receiver by which bearing is determined) radiogoniomètre *m*
~-FINDER DEVIATION - [radar] erreur *m* de relèvement
~FINDING - [radio] (D/F) (determination of a bearing by the direction of received radio transmission) radiogoniométrie *f*
~INDICATOR - [instr] indicateur *m* de direction
[aero] (instrument in an aircraft which shows the heading in reference to a scale stabilized by a gyroscope having a horizontal axis) indicateur *m* de direction gyroscopique
~INDICATOR SWITCH - [auto] commutateur *m* d'indicateur de direction
~OF A WALL - [constr] alignement *m* d'un mur
~OF CORDS - [rubber ind] (of tyres) direction *f* des fils
~OF CUTTING - [mech] sens *f* de coupe
~OF FEED - [mech] (in machine tools) direction *f* de l'avance
~OF GRAIN - [timber] sens *m* des fibres, sens *m* du bois
~OF POLARIZATION - [electron] (in waveguides) sens *m* de polarisation
~OF PROPAGATION - [electron] (in waveguides) sens *m* de propagation
~OF THE RIB - [text] sens *m* de la côte
~OF THE TWILL - s. direction of the rib
~OF TWIST (IN THE YARN) - [text] sens *m* de torsion
~-POST - [gen] poteau *m* indicateur, panneau *m* de signalisation
DIRECTIONAL - [gen & radio] directionnel
~AERIAL - [radio] (an aerial which radiates or receives directionally, i.e. more effectively in certain directions than in others) antenne *f* directionnelle
~ANTENNA - s. directional aerial
~CHARACTERISTIC - [radio] caractéristique *f* de directivité
~CONTROL LEVER - [mech mach] levier *m* de commande du sens de rotation
~COUNTER - [instr] (instrument designed to measure radiations) détecteur *m* directif
~COUPLER - [electron] (coupler between a number of waveguides) détecteur *m* directionnel de mesures
~DRILLING - [oil ind] (in oil well sinking) forage *m* dirigé
~FINDER - [radar] radiogoniomètre *m*
~GAIN - [acoust] (of a transducer) indice *m* de directivité
~GYRO - [instr] (a gyroscope giving direct display in azimuth) gyroscope *m* directionnel, indicateur *m* gyroscopique de direction
~HOMING - [radar] (the following of a path so that the objective is kept at a constant bearing) radioguidage *m* directif
~LOUDSPEAKER - [acoust] (loudspeaker fitted with a flared horn) haut-parleur *m* directionnel
~LUMINOUS REFLECTANCE - [opt] facteur *m* de réflexion relative
~LUMINOUS TRANSMITTANCE - [opt] facteur *m* de transmission relative
DIRECTIONAL MICROPHONE - [acoust] (microphone directional in its response) microphone *m* directionnel
~PATTERN - [radio] (a graphic representation of the emission or the reception of the aerial as a function of direction) diagramme *m* de rayonnement
~PHASE CHANGER - [waveguides] déphaseur *m* directionnel
~RECEIVER - [radio] récepteur *m* directionnel
~REFLECTANCE - [light] albédo *m*
~RESPONSE PATTERN - [acoust] (the graphical representation of the response of an electro-acoustic transducer) diagramme *m* directionnel
~SOLIDIFICATION - [metall] solidification *f* dirigée
~STABILITY - [phys] (stability in respect of way, sideslip or any combination of these) stabilité *f* de route
~WELL - [min] puits *m* obtenu par forage dirigé
DIRECTIONALS - [gen] signaux *mpl* de direction
DIRECTIVE - [gen] directive *f*, instruction *f*
~EFFICIENCY - [radio] (of a directional aerial) rendement *m* directionnel
~FORCE - [el] (the couple which causes the turning of a magnetic needle) couple *m* actif
~GAIN - [radio] gain *m* d'antenne directionnelle
~MOVEMENT - [bot] (the movement of orientation) tropisme *m*
DIRECTIVITY - [radio] (the extent of a directional antenna radiation or response in certain directions) directivité *f*
[acoust] (of loudspeakers) directivité *f*
~FACTOR - [acoust] (of a transducer) facteur *m* de directivité
~INDEX - s. directional gain
~PATTERN - [radio] diagramme *m* directionnel de rayonnement
DIRECTLY FED AERIAL - [radio] (aerial receiving power from the transmitter) antenne *f* à attaque directe
~HEATED CATHODE - [electron] (a hot cathode heated directly by current flowing in it) cathode *f* à chauffage direct
DIRECTOR - [gen] directeur *m*, chef *m*, guide *m*, administrateur *m*, gérant *m*
[cin] (the person in charge of the production) réalisateur *m*, metteur *m* en scène
[telev] (a straight conductor placed before a dipole aerial to increase the directional properties of the latter) directeur *m*
[telecomm] (in telephony; automatic telephone) sélecteur *m*
[naut] télépointage *m*
[firearm] (an instrument designed to measure angles) goniomètre *m* boussole
[med] (grooved instrument guiding the surgeon knife) guide *m*
~CIRCLE - [geom] cercle *m* directeur
~OF PHOTOGRAPHY - [cin] chef-opérateur *m*, directeur *m* de la photographie
~TOP - [navy] hune *f* de télépointage
DIRECTORY - [gen] annuaire *m*
[telecomm] annuaire *m* téléphonique
DIRECTRIX - [geom] directrice *f*
DIRESORCINOL - [chem] (synthesis intermediate) dirésorcinol *m*
DIRIGIBLE - [aero] (a lighter-than-air aircraft capa-

ble of being propelled and steered) dirigible *m*
DIRT - [gen] crasse *f*, saleté *f*, impureté *f*
[mech] impuretés *f*pl, boue *f* de polissage, crasse *f*
[min] (broken mineral of no value) stériles *m*pl terreux, déblais *m*pl
[geol] (gravel etc) gravier *m*
[paper man] (a defect) soufflure *f*, grumeau *m*
[metall] (inclusion of non-metallic matter in a metal) inclusion *f* (dans le métal)
[constr] terre *f*, déblais *m*, boue *f*
~COLLECTOR - [mech] (in filters, a container for the impurities) collecteur *m* de poussière
~CONTENT - [gen] (the amount of foreign matter in a substance) teneur *f* en impuretés
~GUARD - [auto] pare-poussière *m*
~INHALATION (COMPRESSORS) - [aero] (the drawing-in of foreign bodies by gas turbine compressors) aspiration *f* de saletés
~PERCENTAGE - [sugar ind] tare *f* de terre, pourcents *m*pl de tare
~TRAP - [hydr] piège *m* à impuretés, collecteur *m* d'impuretés, séparateur *m* d'impuretés
[metall] nid *m* à crasses, piège *m* à crasses
DIRTY - [adj] sale, encrassé
~BOMB - [nucl] (nuclear bomb with an appreciable measure of radioactive fall-out) bombe *f* sale
'DIS' - [radio etc] (slang) (disconnected, on open circuit) debranché, coupé
DISABILITY - [gen] incapacité *f*, impuissance *f*
[med] invalidité *f*, infirmité *f*
DISABLE, to - [gen] rendre incapable, mettre hors d'état
DISABLED - [adj] invalide, infirme
[naut] (of a ship) hors service, hors de combat, désemparé
DISABLEMENT - [gen] invalidité *f*, incapacité *f* de travail
DISACCHARIDES - [chem] disaccharides *m*pl, diholosides *m*pl
DISACCHAROSES - [chem] (a group of carbohydrates) disaccharoses *m*pl
DISADVANTAGE FACTOR - [nucl] (the relation betweenthe mean neutron flux in a cell and the mean neutron flux within the fissionable material) rapport *m* de désadvantage, facteur *m* de désadvantage
DISAFFORESTATION - [agric] déboisement *m*
DISAGGREGATION - [gen chem etc] désagrégation *f*
DISALIGNMENT - [mech] désalignement *m*
DISALLOW, to - [gen] défendre, interdire, rejeter
DISAPPEARING - [gen] escamotable, à disparition
~CARRIAGE - [in artillery)] affût *m* à éclipse
~FILAMENT PYROMETER - [instr] (instrument designed to measure the temperature of a furnace) pyrométre *m* à disparition de filament
DISARM, to - [gen] désarmer
DISARMAMENT - [gen] désarmement *m*
DISARTICULATION - [med] (the amputation of a bone through a joint) désarticulation *f*
DISASSEMBLE, to - [mech] démonter
DISASSEMBLY - [mech] démontage *m*
DISASSOCIATE, to - [gen] dissocier, se dissocier, séparer
DISASSORTATIVE MATING - [genet] accouplement *m* d'individus dissemblables
DISASTER - [gen] désastre *m*, sinistre *m*, catastrophe *f*
DISBARK, to - [gen] débarquer
[bot] (the removing of the bark) écorcer
DISBARKED WOOD - [timber] bois *m* écorcé
DISBUD, to - [agric] ébourgeonner
DISBUDDING - [agric] ébourgeonnage *m*
DISC - s. disk
~SAW - s. circular saw
DISCALER - [mech] machine *f* à désincruster, machine *f* à décalaminer
DISCARD, to - [gen] écarter, rejeter, exclure
[ind] (in overhaul and maintenance work, to discard is to reject and scrap parts which are defective or have exceeded their authorized life) rebuter, mettre à la ferraille
DISCARD - [gen] rebut *m*
[ind] (defective material or useless parts or fragments) rebut *m*, chute *f*, déchet *m*
[metall] (the part which is removed from the upper end of an ingot) chute *f*
DISCARDED - [gen] rejeté, ecarté, exclu
DISCERNIBLE - [gen] perceptible
DISCHARGE, to - [gen] décharger, évacuer
[constr](the building of a structure, e.g. an arch, to protect a space from the weight from above) décharger
[plumb] (of a pipe) sortir, se décharger
[comm etc] (to dismiss) licencier, donner congé, congédier
[leg] libérer, élargir
[med] (of a wound) suppurer
DISCHARGE - [gen] décharge *f*, déversement *m*, évacuation *f*, éjection *f*, écoulement *m*
[gen] (the operation of unloading) déchargement *m*
[firearms] décharge *f*
[hydr] débit *m*, refoulement *m*
[el] (the re-conversion of the chemical energy in a battery to electrical energy) décharge *f*
[leg] élargissement *m*, libération *f*
[med] suppuration *f*
~AIR SHAFT - [mining] puits *m* de ventilation
~AREA - [hydr] zone *f* de résurgence
~BASIN - [hydr] bassin *m* de décharge
~BRANCH - [hydr] conduite *f* de décharge
~BRIDGE - [el] (a device designed to measure the discharge in dielectric material and in cables) pont *m* à décharge
~CAPACITY - [el] pouvoir *m* de décharge (d'un parafoudre)
~CHANNEL - [hydr] (an effluent sewer) canal *m* de décharge, conduite *f* de décharge
~COCK - [hydr] (a small valve used to release fluid from a system) robinet *m* de décharge
~COEFFICIENT - [hydr] (the ratio between tha actual and the theoretical discharge) coefficient *m* de décharge
~CONDUIT - [hydr] conduite *f* d'évacuation
~CORRECTION FACTOR - [mech] coefficient *m* de correction de décharge
~CULVERT - [hydr] buse *f* de décharge
~CURRENT - [el] courant *m* de décharge
~CURVE - [hydr] courve *f* des débits
~DELIVERY SIDE - [hydr] (of a pump) refoulement *m* d'une pompe
~DITCH - [hydr] fossé *m* de décharge
~EFFECT - [text] effet *m* de rongeage
~ELECTRODE - [el] (high potential electrode of an electrical precipitator) électrode *f* d'émission

ISCHARGE GAUGE - [meas] manomètre *m* de refoulement
~GROUND - [text] fond *m* rongé
~HEAD - [phys] (the pressure at which gas leaves a blower or pump) hauteur *f* de refoulement, charge *f* à la sortie
[oil ind] hauteur *f* d'élévation
~HEADER - [aero] (in jet engines) collecteur *m* de décharge
~HOLE - [metall] (or bottom door) porte *f* de vidange
~HOPPER - [hydr] entonnoir *m* à trop-plein
~LAMP - [el] (electric lamp in which light is obtained from a discharge between two electrodes) lampe *f* à décharge
~LINE - [mech] tuyau *m* de refoulement
~NOZZLE - [mech] diffuseur *m*, buse *f* d'alimentation
~OF A CAPACITOR - [el] décharge *f* d'un condensateur
~OF A RIVER - [hydr] débit *m* (d'un fleuve)
~OF A WELL - [hydr] (the yielding capacity of a well) débit *m* d'un puits
~PIPE - [mech] tuyau *m* d'évacuation, tuyau *m* de décharge
[metall] tuyau *m* de déchargeur
~PLUG - [mech] bouchon *m* de vidange
~POTENTIAL - [el] (characteristic value of the electrode potential) potentiel *m* de décharge
~PRESSURE - [el chem] (the pressure at which the chemical energy is converted into energy) tension *f* de décharge
[mech] (in a compressor) pression *f* de refoulement (à la sortie)
~PRINTING - [text] impression *f* par rongeage
~RATE - [el] (the rate of discharge of an accumulator) régime *m* de décharge
~RESISTOR - [el] résistance *f* de décharge
~ROLLER - [text] cylindre *m* délivreur, cylindre *m* de sortie
~STOP VALVE - [hydr] (valve controlling the rate of discharge from a pump) déchargeur *m*
~SWITCH - [el] limiteur *m* de tension
~TECHNIQUE - [text] procédé *m* de rongeage
~THROUGH A VACUUM LOCK, to - [mech] décharger, extraire, faire sortir par un sas à vide
~TUBE - [el] tube *m* de décharge, tube *m* à décharge
~VALVE - [mech] clapet *m* de refoulement, soupape *f* d'échappement, soupape *f* de refoulement, soupape *f* à disque, soupape *f* à bille
~VELOCITY - [hydr] vitesse *f* d'écoulement (ou de débit), vitesse *f* de filtration
~WATER - [hydr] eau *f* de décharge, eau *f* d'écoulement
ISCHARGED - [gen] déchargé
[el] (of a battery) déchargé
ISCHARGER - [el] (device for the discharging of part of an electrical apparatus) éclateur *m*, excitateur *m*
[el] (device for firing explosives in blasting) allumeur *m*
ISCHARGING ARCH - [constr] (arch built to protect an opening from the weight above) arc *m* de décharge, voûte *f* de décharge
~BY WATER-JET - [hydr] déchargement *m* hydraulique par lance orientable
~CRANE - [mech] grue *f* de transbordement
DISCHARGING GANTRY - [gas ind] estacade *f* de déchargement
~HOLE - [metall] ouverture *f* de défournement
~MACHINE - [mech] (for ovens) défourneuse *f*, déchargeuse *f*
~PLOUGH - [sugar ind] charrue *f* de déchargement
~RAKE - [impl] crochet *m* à déluter
~SLUICE - [hydr] écluse *f* de décharge ou d'assèchement
~TONGS - [impl] (pair of metal tongs used for discharging condensers) pince *f* de décharge
~VOLTAGE - [el] tension *f* résiduelle (d'un parafoudre)
DISCLOSE, to - [gen] révéler, découvrir, déclarer
DISCOCARP - [bot] (an open fructification) discocarpe *m*
DISCODACTYLOUS - [zool] (flattened digits forming sucking disks) discodactyle
DISCOID - [bot] (round and flattened) discoïde, discoïdal, discoïdé
DISCOIDAL SEGMENTATION - [genet] (the cleavage of an ovum) segmentation *f* discoïde
DISCOLOUR, to - [gen] décolorer
DISCOLOURATION - [gen] décoloration *f*, changement *m* de couleur
DISCOLOURING - [gen] décolorant
(a paint effect) décoloration *f*, altération *f*
~AGENT - [chem] décolorant *m*
DISCOMMON , to - [leg] désapproprier (des biens communaux)
DISCOMPOSITION - [nucl] (in a crystal lattice, the process by which an atom is dislocated by direct nuclear impact) création *f* de défaut, dislocation *f*
~EFFECT - [nucl] (the change in physical or chemical properties caused by discomposition) effet *m* Wigner
DISCONE - [radio] (biconical aerial in which the vertex angle of cone is 180 degrees) antenne *f* tubulaire
~AERIAL - s. discone
~ANTENNA - s. discone
DISCONFORMITY - [geol] (a break in the rock sequences) discordance *f*, anomalie *f*
DISCONNECT, to - [gen & mech] débrancher, déconnecter, couper, désolidariser, mettre hors circuit
[el] (to interrupt an electric circuit) déconnecter, couper
[telecomm] couper, interrompre
~A CALL, to - [telecomm] interrompre une communication, couper une communication
~SIGNAL - [telecomm] signal *m* de fin de communication
~THE FRICTION, to - [mech] débrayer
DISCONNECTING - [gen & mech] de coupure, de débrayage, de déclenchement
~BOX - [el] boîte *f* de coupure
~CHAMBER - [hydr] puits *m* de raccord à l'égout
~DEVICE - s. disconnecting switch
~LEVER - [mech] levier *m* de débrayage
~LINK - [el] (a link disconnecting a dead circuit) sectionneur *m*
~SWITCH - [el] disjoncteur *m*
DISCONNECTION - s. discontinuity
~FAULT - [telecomm] dérangement *m* dû à une rupture de ligne
DISCONTINUANCE - [gen] interruption *f*, suppression *f*, suspension *f*, cessation *f*

[gen](of gas or electricity supply) désabonnement *m*, effacement *m*
DISCONTINUE, to - [gen] discontinuer, interrompre
DISCONTINUITY - [gen] (any interruption) coupure *f*, discontinuité *f*
~OF THE PORE WATER - [soil] discontinuation *f* de l'eau interstitielle
DISCONTINUOUS - [phys] discontinu
~ACTION - [gen & contr] action *f* intermittente
~-ACTION SERVO MECHANISM - [mech] servomécanisme *m* à action intermittente
~DISTRIBUTION - [biol] (isolated distribution of a species) distribution *f* discontinue
~FUNCTION - [math] fonction *f* discontinue
~PHASE - [chem] phase *f* dispersée
~SPECTRUM - [phys] (a combined band and line spectrum) spectre *m* discontinu
~VARIATION - [biol] (a variation which occurs rarely) variation *f* discontinue
DISCOPLACENTA - [zool] (placenta with a villi arranged on a flat area) placenta *m* discoïde
DISCORDANCE - [gen] discordance *f*
[geol] discordance *f*, anomalie *f*
DISCOUNT, to - [gen & comm] escompter, faire l'escompte
DISCOUNT - [gen & comm] escompte *f*, remise *f*, rabais *m*
~CHARGES - [fin] agio *m*, frais *mpl* d'escompte
~RATE - [fin] taux *m* d'escompte
DISCOURSE - [gen] discours *m*, conversation *f*, entretien *m*, raisonnement *m*
DISCOVER, to - [gen] découvrir, montrer
DISCOVERY - [gen] découverte *f*
~WELL - [min] (in well sinking) puits *m* de découverte
DISCREET - [adj] discret
DISCREPANCY - [gen] différence *f*, désaccord *m*, contradiction *f*
DISCRETE - [adj] (separate, discontinuous) discret, séparé, discontinu, isolé
~DEFECTS - [metall] défauts *mpl* isolés
~MESSAGE - [comput] message *m* distinct, message *m* discret
~RADIO SOURCE - [radio] radio-source *f* discontinue
~SPECTRUM - [phys] spectre *m* discontinu
DISCRETENESS - [phys] (the distribution of allowed values of a physical quantity over a specified interval) distribution *f* discrète
DISCRIMINANT - [math] discriminant *m*
DISCRIMINATE, to - [gen] discriminer, discerner, distinguer
DISCRIMINATING CIRCUIT-BREAKER - [el] (operating only when the current is in given direction) disjoncteur *m* sélecteur
~PROTECTIVE SYSTEM - [el](excess current protective system) systéme *m* de protection sélectif
~SATELLITE EXCHANGE - [telecomm] bureau *m* téléphonique auxiliaire
~SELECTOR - [telecomm] (in telephony; a selector which discriminates calls to be completed through the local exchange and those requiring other exchanges) sélecteur *m* discriminateur
DISCRIMINATION - [gen] discrimination *f*
[comput] (conditional transfer of control) saut *m* conditionnel
[instr] (degree of sensitiveness of an instrument) sensibilité *f*
DISCRIMINATION INDEX - [opt] (the ratio between the luminance and the differential luminance threshold) indice *m* de discrimination
DISCRIMINATOR - [radio] (circuit designed to convert frequency-modulated signals to amplitude-modulated signals) discriminateur *m*, comparateur *m*
~CIRCUIT - [electron] circuit *m* discriminateur
DISCUS - [bot] disque *m*
DISEASE - [med] maladie *f*
~RESISTANCE - [med] résistance *f* à la maladie
DISEMBARK, to - [gen & naut] débarquer
DISEMBARKATION - [gen & naut] débarquement *m*
DISEMBARKMENT - s. disembarkation
DISENERGIZE, to - [el] désexciter, désamorcer, supprimer l'alimentation
DISENGAGE, to - [mech] (to move gears out of mesh) libérer, désolidariser, débrayer, dégager, déclencher
~THE CLUTCH, to - [mech] débrayer
~THE COUPLING, to - [railw] découpler
~THE TAPPET, to - [text] libérer l'excentrique
DISENGAGED - [mech] libéré, débrayé
~LINE - [telecomm] ligne *f* libre
DISENGAGEMENT - [mech] débrayage *m*, découplage *m*, dégagement *m*, déclenchement *m*
~OF THE SPINDLE - [text] dégagement *m* de la broche
DISENGAGING - [gen & mech] de débrayage, de découplage
~ARBOR - [mech] arbre *m* de débrayage
~DRUM - [oil ind] séparateur *m*
~FORK - [mech] fourchette *f* de commande de débrayage
~LEVER - [mech] levier *m* de débrayage, levier *m* de déclenchement
~MECHANISM - [mech] mécanisme *m* de débrayage
~PAWL - [mech] cliquet *m* de déclenchement
~SHAFT - [mech] arbre *m* de débrayage
~YOKE - s. disengaging fork
DISENTANGLE, to - [gen] dégager, débrouiller, débarrasser
DISEQUILIBRIUM - [gen & med] déséquilibre *m*
DISFIGUREMENT - [constr] (in town planning) enlaidissement *m*
DISGREGATE, to - [gen & phys] disperser, séparer
DISGREGATION - [gen] dispersion *f*, désagrégation *f*, séparation *f*
~ENERGY - [phys] (the energy of a body due to the tendency of its particles to repel each other) energie *f* de séparation
DISH, to - [mech] (to give a hollow form, like that of a shallow bowl) emboutir, former à la presse
[carp] (to hollow out a recess in a piece of wood) creuser
[metall] (in forging, to hollow out thus forming a cavity) former à la presse, emboutir
DISH - [gen] plat *m*, capsule *f*, cuvette *f*, coupelle *f*, creuset *m*
[photo] (flat receptacle) bac *m*
[radar] (US term for mirror; reflector resulting by removing pieces of the paraboloid, so that its shape is even-sided) réflecteur *m* paraboloïde
~CLOTH - [text] torchon *m*
~DRAINER - [impl] égouttoir *m* à vaisselle
~PAN - [impl] plonge *f*

DISH ROCKER - [photo] balance-cuvette *m*
~WARMER *f* [impl] chauffe-plat *m*
~WASHER - s. dish-washing machine
~-WASHING MACHINE - [mech] machine *f* à laver la vaisselle
~-WASHING SINK - s. dish pan
DISHED - [adj] (having a hollow form, like a dish) concave, bombé, en creux, en cuvette
~BOTTOM - [metall] fond *m* bombé
~DIAPHRAGM - [mech] diaphragme *m* concave
~PLATE - [metall] tôle *f* emboutie, tôle *f* ondulée
~SPRING WASHER - [mech] rondelle *f* élastique, rondelle Belleville
~TANK - [impl] cuve *f* bombée, réservoir *m* à fond concave
~WHEEL - [mech] roue *f* désaxée
DISHING - [carp] (the recess hollowed out in a piece of wood) cavité *f*, concavité *f*
[metall] (of plates) emboutissage *m*, formage *m* à la presse, cambrage *m*, cambrure *f*
[metall] (in forging) empreinte *f*, cavité *f*
~PRESS - [mech] presse *f* à emboutir
DISINFECT, to - [gen & med] désinfecter, assainir
DISINFECTANT - [gen & med] désinfectant *m*
DISINFECTION - [gen & med] désinfection *f*
DISINFEST, to - [chem] (to destroy insects) désinsectiser
DISINFESTATION - [chem] (the destruction of insects) désinsectisation *f*
DISINTEGRATE, to - [phys] (to break up mechanically into smaller units) désintégrer, se désintégrer
[constr] (the crumbling of masonry or building stones) se désagréger, s'effriter
DISINTEGRATION - [gen] désintégration *f*
[nucl] (the spontaneous change of a nucleus through the emission of a particle) désintégration *f*
[phys & mech] (the loss of form or crumbling in the form of powdering) désintégration *f*, effritement *m*
[constr] (the crumbling away of building stones) détrition *f*, délitation *f*
[chem] (molecular separation of a chemical compound) désintégration *f*
~CONSTANT - [nucl] (the probability of disintegration per unit time) constante *f* de désintégration
~CURVE - [nucl] (curve showing the decrease of activity with the increase of time) courbe *f* de désintégration
~ENERGY - [nucl] (the energy which is released in radio-active disintegration) énergie *f* de désintégration
~PARTICLE - [nucl] (particle undergoing disintegration) particule *f* de désintégration
~PRODUCT - [nucl] (radioactive product of stable nuclide resulting from the disintegration of a radionuclide) produit *m* de désintégration
~RATE - [nucl] (the decay rate of a radioactive substance and the transformation rate of a bombarded nuclide) taux *m* de désintégration
~SCHEME - [nucl] (diagram showing the modes of disintegration of a radioactive nuclide) diagramme *m* de désintégration
~TEMPERATURE - [phys] température *f* de décomposition
DISINTEGRATOR - [mech] (machine for granulating or shredding materials) broyeur *m*, désintégrateur *m*, pulvérisateur *m*
[min] broyeur *m* centrifuge, désintégrateur *m*
[paper man] (machine designed to break into shreds the slices of wood) broyeur *m*, déchiqueteuse *f*, défibreuse *f*
[min] (machine reducing materials to a granular product) broyeur *m*, concasseur *m*
DISJOIN, to - [gen] disjoindre, déjoindre, désunir
DISJUNCT MOTION - [acoust] (movement of a part by leap) mouvement *m* disjoint
DISJUNCTION - [gen] disjonction *f*, séparation *f*
[genet] (the separation of the two members of a pair of homologous chromosomes during meiosis) séparation *f*
DISK, to - [el acoust] enregistrer
[agric] labourer à la charrue à disques
DISK - [gen] disque *m*, obturateur *m*
[med] disque *m*, papille *f*
~AERIAL - [radio] antenne *f* en nappe
~AND DRUM TURBINE - [mech] (an impulse-reaction turbine) turbine *f* à action et à réaction
~ANODE - [electron] (a circular plate with an aperture through which the cathode ray beam passes) anode *f* à disque
~ANTENNA - s.disk aerial
~AREA - [el] (the area of the circle described by the rotor blade tips) surface *f* du cercle balayé
~ATTENUATOR - [electron] (in waveguides; a variable flap attenuator the absorbing material of which is in the shape of an excentrically mounted disk) atténuateur *m* à lame rotative
~BARKER - [paper man] écorceuse *f* à disque
~BIT - [tool] outil *m* de coupe circulaire, trépan *m* à disque
~BRAKE - [mech] (type of brake in which retardation is effected by pads pressed laterally against a disk which revolves with the part to be braked) frein *m* à disque
~CALCULATOR - [instr] disque *m* calculateur
~CAM - [mech] came *f* à disque
~CHANGER - [el acoust] changeur *m* de disques
~CLUTCH - [mech] (friction clutch consisting of disks between which friction is produced to transmit power) embrayage *m* à disque
~COAL CUTTING MACHINE - [mining] haveuse *f* à disque
~CONDENSER - [el] (variable condenser with an axial motion of disks) condensateur *m* plat
~COULTER - [agric] (steel disk on a drill to prepare a trench for the seed) coutre *m* à disque
~COUPLING - [mech] accouplement *m* à disque
~CRANK - [mech] manivelle *f* à plateau, plateau-manivelle *m*
~CRUSHER - [mech] broyeur *m* à plateau
~CULTIVATOR - [agric] cultivateur *m* à disques
~CUTTER - [mech] fraise-disque *f*, couteau *m* circulaire
~DATA STORAGE - [comput] enregistrement *m* des données sur disques, mémoire *f* à disques
~DISCHARGER - [el] (spark gap, in which sparks occur between a fixed contact and studs on a rotating disk) éclateur *m* à disques
~DRILL - [agric] semoir *m* à disques
~DRIVE - [text] entraînement *m* à disques
~DRYER - [mech] (a drying device in which rotating circular shelves are used) séchoir *m* à plateaux
~ENCODER - [comput] traducteur *m* à disque code
~FILE - [comput] mémoire *f* à disques magnétiques
~FILTER - [mech] (rotating disk with sectors with

suitable filters) filtre *m* à disque
DISK GAP - [el] éclateur *m* à disques
~GRINDER - [mech] polisseuse *f* à disque
~HARROW - [agric] (harrow which consists of steel disks on two axles) herse *f* à disques
~HILLERS - [agric] disques *mpl* butteurs
~LOADING - [aero] (quantity obtained by dividing the rotor thrust by the total disk area) charge *f* unitaire sur le cercle balayé
~MILL - [mech] (machine in which grinding is carried out by a pair of iron disks revolving close together but in planes which are at a small angle to each other) moulin *m* à disques
~MIXER - [mech] mélangeur *m* à disque
~OF THE SLICER - [sugar ind] plateau *m* de coupe-racines
~PISTON - [mech] piston *m* à disque
~PLOUGH - [agric] charrue *f* à disques
~PRISM - [telev] (disk scanner consisting of glass so ground that its periphery forms a series of prisms) disque *m* analyseur prismatique
~RECORD - [acoust] (ordinary record of shellac, rotating at constant speed) disque *m*, disque *m* phonographique
~RECORDER - [acoust] (a mechanical recorder in which the recording medium has the form of a disk) enregistreur *m* sur disque
~RECORDING - [el acoust] (a recording on circular wax disks for subsequent disk pressing) enregistrement *m* sur disques
~RIDGERS - [agric] disques *mpl* butteurs
~RING - [mech] anneau *m* plat
~ROLLER - [agric] rouleau *m* à disques
~SAND-PAPERING MACHINE - [mech] ponceuse *f* à disque
~SANDER - [mech] ponceuse *f* à disque
~SANDING - [mech] ponçage *m* au disque
~SAW - [impl] scie *f* circulaire
~SCANNER - [telev] (rotating disk used in mechanical scanning) analyseur *m* à disque, disque *m* analyseur
~SCREEN - [hydr] râteau *m* à disque, grille *f* à disque
~SEAL - [metall] scellement *m* verre-métal, joint *m* verre-métal
~-SEAL VALVE - [electron] (type of valve in which connexion between electrodes and external circuits is made through metal disks sealed into the envelope) mégatron *m*, tube *m* à disques scellés, tube-phare *m*
~SEPARATOR - [mech] (machine designed to separate grain or seeds of various sizes) calibreur *m* à disques
~SHUTTER - [cin] (shutter operating with an obturating disk) obturateur *m* à disque
~SIGNAL - [railw] disque *m* (de signalisation)
~SPINNING MACHINE - [text] métier *m* à filer à plateau
~STORAGE - [comput] mémoire *f* à disques
~STUBBLE CLEANER - [agric] déchaumeuse *f* à disques
~TILLER - s. disk stubble cleaner
~TOP-SOIL PLOUGH - s. disk stubble cleaner
~-TYPE CAPACITOR - [electron] condensateur *m* à disque
~VALVE - [mech] (type of valve used in the section of pumps and air-compressors, in which a flexible steel disk covers ports in a flat seating) vanne *f* à opercule, vanne *f* à tampon d'équerre
DISK WATER METER - [instr] hydromètre *m* à disque scillant
~WHEEL - [mech] (wheel in which the hub and the rim are connected by a disk of metal) roue *f* à disque
~WINDING - [el] (used for medium size and large size transformers) enroulement *m* en disque, bobinage *m* en disque
DISLOCATED DEPOSIT - [geol] gite *m* disloqué, gisement *m* disloqué
DISLOCATION - [med] (abnormal separation of bones at a joint) luxation *f*
[geol] (of strata etc) dislocation *f*
[phys] (imperfection in a crystalline solid) dislocation *f*
~LINE - [phys] (the line, or curve, which separates displaced and undisplaced sections of a crystal) ligne *f* de dislocation
~NETWORK - [phys] (networks of dislocation lines) réseau *m* de dislocation
~OF THE LENS - [med] ectopie *f* du cristallin
DISMANTLE, to - [gen & mech] (to take to pieces; to separate the parts of a mechanism or the like) démonter, déséquiper, dégréer (une grue)
[constr] démanteler
~THE FALSEWORK, to - [constr] démonter un échafaudage, enlever les ouvrages provisoires
DISMANTLING - [mach] (the operation of taking to pieces) démontage *m*
[constr] démontage *m* des ouvrages provisoires
[naut] démantèlement *m*, désarmement *m* d'un navire
DISMAST, to - [naut] démâter
DISMOUNT, to - [gen & mech] démonter, (from a vehicle, e.g. a cycle) descendre
DISMOUNTABLE MAST - [radio] (an aerial mast which can be easily removed or folded together) mât *m* démontable
DISMOUNTING ROD - [impl] (rod used to remove a tyre) démonte-pneu *m*
DISMUTATION - [chem] (a chemical reaction) dismutation *f*
DISORDER - [gen] désordre *m*
[med] désordre *m*, affection *f*, trouble *m*
~PRESSURE - [phys] (the contribution to the pressure arising from molecular disorder) pression *f* d'entropie
DISORGANIZATION - [gen] désorganisation *f*
DISORGANIZE, to - [gen] désorganiser
DISORIENTATION - [med] (disordered mental state) désorientation *f*
DISOXIDATION - [chem] désoxydation *f*
DISPARATE CHIASMA - [genet] chiasma *m* asymétrique
DISPATCH - s. despatch
DISPENSARY - [med] pharmacie *f* (d'un hôpital)
[med] (clinic for treatment of patients) dispensaire *m*
DISPENSE, to - [gen & med] dispenser, distribuer, fournir, débiter
DISPENSER - [gen] distributeur *m* (automatique)
[med] pharmacien *m* (d'un hôpital)
[telecomm] (in radiotelegraphy; a signal transmitting device) émetteur *m* radiotélégraphique
[packag] (a spray device) distributeur *m*
~CATHODE - [electron] (non-coated cathode conti-

nually supplied with suitable emission material from a separate and associated element) cathode *f* à réserve, cathode *f* compensée
DISPERGATOR - s. dispersing agent
DISPERMIC FERTILIZATION - s. dispermy
DISPERMY - [genet] (the penetration of an ovum by two spermatozoa) dispermie *f*
DISPERSAL - [gen] dispersion *f*, diffusion *f*, dissipation *f*
[biol] (the process of the establishment of individuals in a new area) dispersion *f*
~AREA - [aero] aire *f* de dispersion
~EFFECT - [nucl] effet *m* d'évacuation
~GETTER - [electron] getter *m* à dispersion
DISPERSANT - [chem] (a dispersing agent) agent *m* de dispersion, dispersant *m*
~ADDITIVE - [oil ind] (in petroleum refining) additif *m* dispersant
DISPERSE, to - [chem] (to distribute finely-divided material throughout a fluid medium) disperser, émulsionner
DISPERSE DYE - [text] colorant *m* de dispersion
~MEDIUM - s. dispersion medium
~PARTICLE - [phys] (the particles of a colloid in a colloidal system) particule *f* dispersée
~PHASE - s. dispersed phase
DISPERSED CORROSION - [metall] corrosion *f* en plaques
~DEVELOPMENT - [constr] (in town planning) aménagement *m* dispersé
~PHASE - [chem] (the portion of an emulsion suspended as droplets of particles) phase *f* dispersée
DISPERSING AGENT - [chem] (a substance which promotes and maintains the suspension of particles in a colloidal solution) agent *m* de dispersion, dispersant *m*
~SYSTEM - [phys] (colloidal system consisting of two phases, these being the dispersed phase and the dispersion medium) système *m* dispersé
DISPERSION - [gen] dispersion *f*, fuite *f*
[electron] (waveguides; the separation of a wave into its frequencies) décomposition *f*
[phys] (the separation of a radiation in accordance with its frequency, wavelength and energy) dispersion *f*
[paint] dispersion *f*
[light] (change in refractive index with the wavelength of light, as in the formation of a spectrum by a prism) dispersion *f*
[chem] (a system of minute solid, gaseous, or liquid particles suspended in a gaseous, solid or liquid medium) dispersion *f*
~AGENT - [chem] agent *m* de dispersion, adjuvant *m* de dispersion
~AND MASK METHOD - [light] méthode *f* du diaphragme spectral
~COEFFICIENT - [el] (the leakage factor of an induction motor) coefficient *m* de dispersion
~CURVE - [opt] (curve indicating the deviation of light produced by a prism) courbe *f* de dispersion
~EFFECT - [el] (the attraction between electrically neutral molecules the dipoles of which align themselves so as to produce an attractive force) effet *m* de dispersion
~ELECTROSCOPE - [meas] électroscope *m* à dispersion
~FORCE - [el] (the force of the attraction between molecules without permanent dipole) force *f* de dispersion
DISPERSION HARDENING - [metall] (the increase of hardening with the passing of time) durcissement *m* par précipitation, durcissement *m* structural
[metall] (when caused by heat treatment) vieillissement *m*
~INDEX - [math] écart-type *m*
~MEDIUM - [chem] (the medium in which a colloid is dispersed) milieu *m* dispersif, milieu *m* de dispersion
~MIXER - [ind chem] (a machine for the production of dispersions) mélangeur *m*
~OF CONDUCTANCE - [el] (the conductance of an electrolytic solution varies with the frequency) effet *m* Debye-Falkenhagen
~OF ROTATION - [opt] (the variation of the rotation of the plane of polarized light with wavelength for an optically active substance) dispersion *f* rotatoire
~PATTERN - [opt] cône *m* de dispersion
~PHOTOMETER - [instr] (instrument designed to measure the intensity of strong light sources) photomètre *m* à dispersion
~PRISM - [opt] prisme *m* de dispersion
DISPERSITY - [phys & chem] (the degree of dispersion of a colloid) dispersité *f*
DISPERSIVE - [phys & chem] dispersif
~LENS - [opt] lentille *f* divergente
~MEDIUM - s. dispersion medium
~POWER - [opt] pouvoir *m* dispersif, puissance *f* dispersive
~REFLECTOR - [light] réflecteur *m* à dispersion
DISPERSOID - [chem] (a colloidal system with a comparatively important dispersity) système *m* colloïdal à grande dispersion
DISPLACE, to - [gen] déplacer, remplacer
[naut] déplacer
[el] (of a phase) décaler
DISPLACED TERMS - [phys] (anomalous terms) termes *mpl* anomaux
DISPLACEMENT - [gen] déplacement, décalage *m*, dislocation *f*, rejet *m*
[med] ectopie *f*, déplacement *m*
[mech] (the swept volume of a pump cylinder) cylindrée *f*
[hydr] (the weight of the water displaced by a vessel) déplacement *m*
[chem] (the property of chemical elements arranged in the order of their electrode potentials) déplacement *m*
~ADSORPTION - [phys] (the displacement from a surface of an adsorbed by another) adsorption *f* de déplacement
~ANTIRESONANCE - [phys] antirésonance *f* d'amplitude
~CORRECTION - [opt] correction *f* de parallaxe
~CURRENT - [radio](current which is not accompanied by motion of charges in the dielectric) courant *m* de déplacement
~EQUATION - [phys] (in photometry) équation *f* de la parallaxe, équation *f* du déplacement
~FORCE - [chem] force *f* de déplacement
~KERNEL - [nucl] (kernel in which the spatial dependence is only on the distance between the two points) noyau *m* de déplacement
~LAW - [phys] (law relating to the emission of an alpha-particle to the position of the emitting ele-

ment in the periodic table) loi *f* de Wien, loi *f* du déplacement
DISPLACEMENT METER - [instr] (of a pump) compteur *m* volumétrique
~METHOD - [soil] procédé *m* par déplacement
~OF A LIQUID - [hydr] déplacement *m* d'un liquide
~OF BOGIE - [railw] déplacement *m* angulaire du bogie
~OF BRUSHES - [el] décalage *m* des balais
~RESONANCE - [phys] résonance *f* d'amplitude
~TONNAGE - [naut] déplacement *m*
~VOLUME - [mech] cylindrée *f*
DISPLAY, to - [gen] montrer, étaler, exposer, présenter, manifester
DISPLAY - [gen] déploiement *m*, développement *m*, spectacle *m*, parade *f*
[comm] étalage *m*, exposition *f*
[radar] (the visible indication on the screen of a cathode-ray tube when this is used in radar equipment) image *f*, présentation *f*
[telev] reproduction *f*
~DEVICE - [comput] dispositif *m* indicateur, dispositif *m* d'affichage
~FACE - [print] caractères *mpl* pour titres
~INSTRUCTION - [comput] instruction *f* d'affichage
~PANEL - [telecomm] (in telephony) tableau *m* d'annonciateurs
[contr] tableau *m* d'affichage, baie *f* d'indicateurs
~PRIMARIES - [telev] (in colour television) primaires *mpl* à la réception
~ROOM - [comm] salle *f* d'exposition
~TYPE-SETTING - [print] composition *f* pour travaux de ville
~UNIT - [radar] (the display screen unit) récepteur *m*
[comput] baie *f* d'indicateurs
~WORK - [print] travaux *mpl* de ville
DISPOSABLE LOAD - [naut & aero] (the total weight of the crew, fuel, oil, and payload) charge *f* disponible
DISPOSAL - [gen] disposition *f*, arrangement *m*, vente *f*
[nucl] (the elimination of radioactive materials) élimination *f* (de déchets), décharge *f*
~GROUND - [nucl] enfouissement *m*
~WELL - [nucl] puits *m*
DISPOSE, to - [gen] disposer, arranger
DISPOSSESS, to - [gen & leg] déposséder
DISQUALIFICATION - [gen] disqualification *f*, incapacité *f*
DISQUALIFY, to - [gen] frapper d'incapacité, rendre incapable, disqualifier
DISRUPT, to - [gen & phys] rompre, briser, claquer, perforer, faire éclater
DISRUPTION - [gen] rupture *f*, fracture *f*, dislocation *f*, destruction *f*
[nucl] (of molecules) dissociation *f*, rupture *f* (moléculaire)
DISRUPTIVE - [phys] brisant, disruptif
~DISCHARGE - [el] (the result of a breakdown of insulating material caused by electric stress) décharge *f* disruptive
~STRENGTH - [el] (electric strength) rigidité *f* diélectrique
~VOLTAGE - [el] (the voltage which is necessary to produce a disruptive discharge between two conductors) tension *f* disruptive
DISSECT, to - [med] disséquer, exciser
DISSECTING KNIFE - [med] scalpel *m*
DISSECTION - [gen] dissection *f*
[mech] découpage *m*
DISSECTOR MULTIPLIER - [telev] (image dissector multiplier; a unit comprising an image dissector and an electron multiplier) dissector *m*
DISSEMINATE, to - [gen & bot] disséminer, diffuser
~DEPOSIT - [geol] gite *m* porphyrique, gite *m* disséminé
DISSEMINATED ORE - [min] minerai *m* disséminé
~SCLEROSIS - [med] (sclerosis appearing throughout the central system) sclérose *f* en plaques
DISSEMINATION - [genet] (the spreading of species) dissémination *f*, propagation *f*, transmission *f*
DISSEPIMENT - [zool] (a calcareous partition between the septa) cloison *f*, septum *m*
DISSERTATION - [gen] discours *m*, dissertation *f*, mémoire *m*
DISSIMILAR - [gen] dissemblable
DISSIMILARITY - [gen] dissemblance *f*, dissimilitude *f*
DISSIMULATED - [gen] dissimulé, caché
~ENERGY - [el] électricité *f* latente
DISSIMULATION - [gen] dissimulation *f*
DISSIPATE, to - [gen] dissiper, disperser, éparpiller
DISSIPATION - [gen] dissipation *f*
[el] (power losses in condensers and inductances) dissipation *f*
[comput] dissipation *f*, puissance *f* dissipée
~FACTOR - [el] (the reciprocal of the storage factor) facteur *m* de dissipation
~OF ENERGY - [phys] (waste of energy caused by the production of heat in a circuit) dissipation *f* de l'énergie, déperdition *f* d'énergie, dégradation *f* de l'énergie
DISSIPATIONLESS LINE - [el] (ideal transmission line) ligne *f* sans pertes
DISSIPATIVE FORCE - [phys] (any force opposing motion; its action converts mechanical into thermal energy) force *f* dissipative
~NETWORK - [el] (network designed to absorb power) réseau *m* dissipatif
~SYSTEM - [mech] (mechanical system in which dissipation occurs) système *m* dissipatif
DISSOCIATE, to - [gen] désassocier, dissocier
[chem] dissocier, se dissocier
DISSOCIATING SOLVENT - [chem] (solvent in which solutes enter into solution as single molecules) solvant *m* à dissociation
DISSOCIATION - [chem] (a reversible decomposition of a substance into two or more new substances, the dissociating particles recombining when the conditions causing dissociation are reversed) dissociation *f*
~COEFFICIENT - [chem] coefficient *m* de dissociation
~CONSTANT - [chem] (the constant for a chemical reaction breaking up a molecule to obtain simpler molecules) constante *f* de dissociation, constante *f* d'affinité
~CONTINUA - [phys] spectres *mpl* continus de dissociation
~LIMIT - [metall] limite *f* de dissociation
~OF GASES - [chem] (a combustion reaction which takes place at the highest flame temperature, where there is a tendency for water vapour and carbon

dioxide to dissociate into hydrogen and oxygen in the former and into carbon monoxide in the latter case) dissociation *f* des gaz
DISSOCIATION PRODUCT - [electron] (a product which is formed by the dissociation of a cathode structure) produit *m* de dissociation
DISSOCIATIVE CAPTURE - [nucl] (ion recombination) capture *f* dissociative
DISSOLUTION - [gen] dissolution *f*
[chem] (the formation of a homogeneous solution) dissolution *f*
DISSOLVE, to - [gen] dissoudre
[comm] dissoudre, annuler
[chem] (when a liquid takes up a substance) dissoudre, se dissoudre
[cin] (to fade) faire un fondu
DISSOLVE - [cin] (during the transition from one sequence to another) fondu *m*
[telev] fondu *m* enchaîné
~IN - s. fade in
~OUT - s. fade out
DISSOLVED GAS - [phys] gaz *m* dissous
~OXYGEN - [chem] (the atmospheric oxygen dissolved in natural waters. It is an important indicator of the condition of a water supply) oxygène *m* dissous
DISSOLVENT - [chem] dissolvant *m*, solvant *m*
DISSOLVING INTERMEDIARY - [chem] tiers *m* solvant
~SHUTTER - [cin] obturateur *m* pour fondus
~VIEW - s. dissolve
DISSONANCE - [acoust & mus] (the opposite of consonance) dissonance *f*
DISSONANT - [acoust & mus] dissonant
DISSYMMETRIC - s. asymmetric
DISSYMMETRY - s. asymmetry
~FACTOR - [opt] (also called anisotropy factor; quantity expressing the magnitude of circular dichronism) coefficient *m* d'anisotropie
DISTAD - [med] orienté vers l'extrémité
DISTAFF - [text] quenouille *f*
DISTAL - [bot] (with distance in between; widely apart) le plus éloigné, distal, situé vers l'extrémité
DISTANCE, to - [gen] éloigner, reculer
DISTANCE - [gen] distance *f*, éloignement *m*
~BETWEEN CENTRES - [mech] entre-axes *m*
~BETWEEN GIRDERS - [constr] écartement *m* des poutres
~BETWEEN ROLLS - [metall] écartement *m* des cylindres
~BETWEEN RUNNING LINES - [railw] intervoies *m*
~BETWEEN SLEEPERS - [railw] espacement *m* des traverses
~BETWEEN THE COALS - [el] écartement *m* des charbons
~BETWEEN TRACK CENTRES - [railw] entre-axes *m* des voies
~BETWEEN TRAINS - [railw] espacement *m* des trains
~BLOCK - [constr] (wooden block separating two pieces by a specified distance) entretoise *f*
[mech] (in mechanical structures; a metal block separating two pieces by specified distance) pièce *f* d'écartement
~BOLT - [mech] boulon *m* d'écartement
~CONTROL - [contr] (also called remote control) télécommande *f*, commande *f* à distance
~ERROR - [radar] (error which is due to sky-wave transmission during the day and to multiple-path propagation during the night) erreur *f* de distance
DISTANCE FINDING STATION - [radar] station *f* télémétrique
~FOG - [cin] (in long shot exposures) voile *m* lointain
~GAUGE - [railw] calibre *m* d'écartement
~INDICATOR - [instr] télémètre *m*
~-MARKING LIGHTS - [aero] (luminous beacons placed in the approach area to indicate the distance from the threshold lights) feux *mpl* de distance
~MEASURING EQUIPMENT - [radar] (D.M.E.) (instrumentation for observing the distance travelled by an aircraft) télémètre *m* radar, appareillage *m* de mesure de distances
~METER - [meas] télémètre *m*
~ON BEAM - [naut] distance *f* par le travers
~PIECE - [mech] entretoise *f*, rondelle *f* d'èpaisseur, rondelle-entretoise *f*, douille *f* d'écartement
~PROTECTION - [naut] protection *f* de distance
~RECEPTOR - [zool] (capable of perceiving objects at a distance) extérocepteur *m*
~RELAY - [el] (impedance relay) relais *m* télécommandé, télérelais *m* d'impédance
~PIECE - [mech] (a mechanical part used to maintain a given separation between two other parts) entretoise *f*, pièce *f* d'écartement, cale *f* d'épaisseur
~SETTING - [photo] réglage *m* de la distance
~SHOT - [cin] plan *m* d'ensemble
~SIGNAL - [naut] signal *m* à distance
~SPACER - s. distance piece
~TABLE - [railw] indicateur *m* des distances
~TO LANDING - [aero] distance *f* d'atterrissage
~TUBE - [mech] (a tubular distance block) tube *m* entretoise
~VELOCITY LAG - [contr] (in remote control operations) retard *m* de parcours
~WASHER - [mech] rondelle *f* d'épaisseur
DISTANT - [gen] distant, éloigné, lointain, à distance
~COLLISION - [nucl] interaction *f* à grande distance
~CONTROL - [contr] commande *f* à distance, télécommande *f*
~EXCHANGE - [telecomm] (in telephony) bureau *m* correspondant
~OFFICE - s. distance exchange
~READING INSTRUMENT - [instr] (instrument with a reading shown on a scale at a distance from the point of measurement) appareil *m* à téléindicateur
~WARNING DEVICE - [railw] signal *m* avancé, signal *m* à distance
DISTEARYL ETHER - [chem] (an antistatic agent and plastics mould lubricant) éther *m* distéarylique
~THIODIPROPIONATE - [chem] (a plasticizer and oxidation inhibitor) thio-di-propionate *m* de distéaryle
DISTEMPER, to -[paint] (to mix dry pigments with size and water) détremper
(to paint with distemper) peindre en détrempe, peindre à la colle
DISTEMPER - [paint] (a coating composition in which the vehicle is water and the binding agents water-soluble substances, such as glue, casein, etc, a large quantity of pigment being used) peinture *f* en détrempe, peinture *f* à la colle, détrempe *f*

(the process of painting with distemper) détrempe *f*
[vet] (infection attacking particularly dogs and due to a virus) maladie *f* des chiens, maladie *f* de Carré
DISTEND, to - [gen] enfler, dilater, gonfler
DISTENSION - [gen & med] distension *f*, dilatation *f*, gonflement *m*
DISTHENE - [min] (a silicate of aluminium) disthène *m*
DISTICHOUS - [bot] (in two opposite vertical rows) distique
DISTIL, to - [chem] distiller, raffiner
~OFF, to - [chem] éliminer par distillation, chasser par distillation
DISTILLABLE - [chem] distillable
DISTILLATE - [chem] (product of a distillation) distillat *m*, produit *m* de distillation
~FUELS - [chem] (petroleum products having B.P. between 350 and 700 degrees Fahrenheit and flash point of 120 degrees Fahrenheit or higher) fuel oils *mpl* distillés
~GAS - [chem] gas *m* à distillat
DISTILLATION - [chem] (the process of separating liquids by évaporation and re-condensation in relation to their boiling points or boiling ranges) distillation *f*
~APPARATUS - [ind chem] appareil *m* de distillation
~FLASK - [ind chem] (special type of flask having a round bottom and a side outlet in the neck) ballon *m* à distiller
~IN STEAM - [ind chem] distillation *f* à la vapeur d'eau, entraînement *m* à la vapeur
~METHOD - [nucl] (the separation of isotopic species by fractional distillation) méthode *f* de distillation
~OF COAL - [ind chem] distillation *f* de la houille
~OF WOOD - [ind chem] distillation *f* du bois
~OVERLAP - [oil ind] chevauchement *m* de distillation
~POT - [oil ind] vase *m* pour la distillation
~PROCESS - [chem] procédé *m* de distillation
~RANGE - [chem] intervalle *m* de distillation
~RESIDUE - [oil ind] résidu *m* de distillation
~TOWER - s. distilling column
~UNDER HIGH VACUUM - [chem] distillation *f* sous vide poussé, distillation *f* à vide moléculaire, distillation *f* moléculaire
~UNDER VACUUM - [chem] distillation *f* sous vide, distillation *f* dans le vide
DISTILLED - [chem] distillé
~BLUE - [chem] indigo *m* soluble, composition *f* d'indigo
~FUEL - [chem] combustible *m* distillé
~WATER - [chem] eau *f* distillée
DISTILLER - [ind chem] distillateur *m*, appareil *m* à distiller, bouilleur *m*
DISTILLER'S WASH - [ind chem] résidus *mpl* de ditillation, résidus *mpl* de distillerie
DISTILLERY - [brew ind] (plant for the preparation of spirituous liquors) distillerie *f*
DISTILLING - [chem] de distillation, à distiller
~COLUMN - [chem] (the vertical column of a still) colonne *f* de distillation
~FLASK - s. distillation flask
~FURNACE - [ind chem] four *m* à distillation
DISTOMATOSIS - [vet] (infection of the bile glands) distomatose *f*
DISTORT, to - [gen & phys] (to change shape, especially under stress) déformer, se gauchir, se déjeter, se voiler, travailler
DISTORTED - [gen & acoust] déformé
[mech] (deformed) déformé, gauchi
~PILE - [nucl] pile *f* perturbée
~WAVE - [el] (a non-sinusoidal wave-form of curren onde *f* déformée
~WAVEFORM - s. distorted wave
DISTORTING LENS - [opt] lentille *f* anamorphotique
DISTORTION - [mech] (change in the form of a body caused by stress) déformation *f*, gauchissement *m*
[radio] (difference in the waveform of an amplifier output signal from that of the input signal) distorsion *f*
~ANALYZER - [electron] analyseur *m* de distorsion harmonique
~BRIDGE - [instr] (bridge arrangement to determine the degree of distortion of a wave) pont *m* de mesure de distorsion
~CURVE - [opt] courbe *f* de distorsion
~FACTOR - [el] (ratio between the effective value of the harmonic content and the effective value of the non sinusoidal quantity) taux *m* de distorsion
~-FREE - [opt] exempt de distorsion
~METER - [instr] distorsiomètre *m*
~OF A FIELD - [el] (in electric machines; the change in the distribution of flux in the air gap when the machine is put on load) distorsion *f* d'un champ
~OF THE TRACK - [railw] déformation *f* de la voie
~STRAIN - [mech] effort *m* de torsion
~TEMPERATURE - [phys] (the temperature at which a material begins to be subject to deformation) point *m* de déformation thermique
~TOLERANCE - [radio] (the maximum signal distortion permissible in a telegraph receiver) distorsion *f* admissible des signaux
~TRANSMISSION IMPAIRMENT - [radio] réduction *f* de la qualité de transmission
DISTORSIONLESS - s. distortion-free
DISTRAIN, to - [leg] mettre sous séquestre, confisquer, saisir
DISTRESS SIGNAL - [radio] signal *m* de détresse
DISTRIBUTE, to - [gen] distribuer, répartir
[print] (the action of replacing types in their cases after use) distribuer, remettre en casses
DISTRIBUTED - [gen] distribué, réparti
~AMPLIFIER - [electron] (amplifier consisting of two vacuum tubes along two transmission lines) amplificateur *m* de ligne de transmission
~CAPACITANCE - [el] capacitance *f* répartie
~CONSTANTS - [math] constantes *fpl* réparties
~LOAD - [constr] (load spread out over the length of a girder) charge *f* répartie
~NETWORK - [gen] réseau *m* de distribution
~POINT - [el] point *m* d'alimentation
~WINDING - [el] (winding with conductors occupying several slots per pole) enroulement *m* réparti
DISTRIBUTING BOARD - s. distribution board
~BOX - s. distribution box
~CENTRE - [el] (the point at which the power supply is distributed into various feeders) centre *m* de distribution
~FRAME WIRE - [el] (in telecommunication installations) répartiteur *m*
~ROLLER - [print] rouleau *m* distributeur
DISTRIBUTION - [gen] distribution *f*, répartition *f*

[print] distribution *f*
DISTRIBUTION BELL - [impl] (in retorts) cône *m* de chargement
~BOARD - [el] (a panel at which branch electric supplies are connected to the main supply) tableau *m* de distribution
~BOX - [el] (a housing containing connecting devices in electric supply systems) boîte *f* de dérivation, boîte *f* de jonction
[mech] boîte *f* de distribution, boîte *f* de dérivation
~CHAMBER - [mech] s. distribution box
~COCK - [mech] robinet *m* de distribution
~COEFFICIENT - [chem] (the partition coefficient, i. e. the relation between the equilibrium concentrations of a solute in two immiscible solvents) coefficient *m* de distribution
~DIAGRAM - [mech] diagramme *m* de distribution
~ECCENTRIC - [mech] excentrique *m* de distribution
~FACTOR - [el] (the component factor of the winding factor taking into account the coil pitch) facteur *m* de distribution
~FRAME - [el] (a structure designed for the arrangement of circuits) répartiteur *m*
~FUNCTION - [math] fonction *f* de répartition
~GEAR - [mech] pignon *m* de distribution
~HEAD - [mech] tête *f* de distributeur
~LINE - [el] ligne *f* de distribution
~MAIN - [el] ligne *f* de distribution
~MAINS - [hydr] conduite *f* principale de distribution
~NOISE - [telecomm] souffle *m* de distribution
~OF DUNG - [agric] épandage *m* d'engrais
~OF LOAD - [gen] répartition *f* de la charge
~OF SOUND PRESSURE IN SPACE - [phys] répartition *f* spatiale de la pression acoustique
~PANEL - s. distribution board
~PILLAR - [el] (a pillar-like structure with various switches etc for interconnecting the distribution mains) colonne *f* de distribution
~PIPE - [hydr] conduit *m* de distribution
~PUMP - [mech] pompe *f* d'injection
~RATIO - [nucl] rapport *m* de concentrations
~SWITCHBOARD - s. distribution board
~TERMINAL - [el] boîte *f* de dérivation
DISTRIBUTOR - [gen] distributeur *m*
[print] distributeur *m*
[comm] concessionnaire *m*
[constr] (machine evenly distributing the surface compound) répandeuse *f*, épandeuse *f*
[el] (device for passing ignition current at each cylinder of an internal-combustion reciprocating engine at the beginning of the firing stroke) allumeur *m*, distributeur *m*
~ADVANCE - [auto] dispositif *m* d'avance
~CAM - [auto] came *f* de rupteur
~CAP - [el] (in electric motors) chapeau *m* de distributeur, tête *f* de distributeur
~CASING - [auto] boîtier *m* d'allumeur
~DISK - [auto] disque *m* de distributeur
~DUCT - [el] canalisation *f*
~HEAD - s. distributor cap
~HOUSING - s. distributor casing
~PLATE - [auto] platine *f* de rupteur
~POINTS - [auto] vis *fpl* platinées, plots *mpl* de contact
~ROTOR ARM - [el] (in electrical motors) doigt *m* de distribution
DISTRIBUTOR SHAFT - [auto] axe *m* de distributeur
~VACUUM BRAKE - [auto] frein *m* à dépression de distributeur
DISTRICT - [gen] district *m*, quartier *m*, arrondissement *m*, région *f*, province *f*
~HOLDER - [gas ind] (a gasometer serving a district) gazomètre *m* relais
~LINE - [railw] (railway line between a city and outlying districts) ligne *f* de banlieue
DISTRIX - [med] (hair disease) trichoptilose *f*
DISTUNING - s. detuning
DISTURB, to - [gen] agiter, brouiller, déranger, perturber, troubler
DISTURBANCE - [aero] (a change from steady-state conditions) perturbation *f*
[radio] (any signal interfering with the transmission of a station) brouillage *m*, interférence *f*, panne *f*, parasites *mpl*
[contr] (variable which modifies the corresponding condition) grandeur *f* perturbatrice
[nucl] (in a reactor) perturbation *f*
DISTURBED-ONE OUTPUT - [comput] (in static magnetic storage) niveau *m* de sortie positif perturbé
~-ZERO OUTPUT - [comput] (in static magnetic storage) niveau *m* de sortie négatif perturbé
DISTURBER MECHANISM - [text] mécanisme *m* de brouillage
DISTURBING MOMENT - [aero] moment *m* perturbateur
~PULSE - [comput] (pulse caused by a faulty rectangularity of the hysteresis loop) impulsion *f* parasite
~VARIABLE - [contr] grandeur *f* perturbatrice
~VOLTAGE - [el] tension *f* perturbatrice
DISTYLE - [arch] distyle *m*
DISULPHIDE - [chem] bisulfure *m*
~OF CARBON - [chem] (a solvent for sulphur and rubber) sulfure *m* de carbone, anhydride *m* sulfocarbonique
DISULPHITE - [chem] bisulfite *m*
DISUSE ATROPHY - [med] (the wasting of a part caused by lack of activity) atrophie *f* par inaction
DITAN - [chem] diphénylméthane *m*
DITCH, to - [gen] creuser, creuser une tranchée, drainer
[aero] s. ditching
DITCH - [constr] (a channel in the ground for drainage water) fossé *m*, caniveau *m*, tranchée *f*, rigole *f*
[electron] (in waveguides; a groove acting as a choke) gorge *f* du piège
~CLEANING - [agric] nettoyage *m* des fossés
~CLEANING MACHINE - [agric] machine *f* à nettoyer les fossés
~CUTTINGS - [mining] déblais *mpl*, farine *f* de sondage
~DREDGER MACHINE - s. ditcher
~MACHINE - [mech] (earth moving machine) excavatrice *f*, trancheuse *f*
DITCHER - [agric] machine *f* à creuser les tranchées
~PLOUGH - [agric] charrue *f* fossoyeuse
DITCHING - [aero] (a forced descent in the sea) amerrissage *m* forcé
[mining] exécution *f* des tranchées
~OUTLET - [hydr] décharge *f*, canal *m* d'écoulement
DITERPENES - [chem] (unsaturated hydrocarbons for-

med of two terpene molecules) diterpènes *mpl*
DITETRAHYDROFURFURYL ADIPATE - [chem] (a plasticizer for cellulose acetate butyrate) adipate *m* de di-tétrahydrofurfuryle
DITHIOCARBAMIC ACID - [chem] (intermediate in the production of vulcanization accelerators) acide *m* di-thiocarbamique
DITHIONATE - [chem] dithionate *m*
DITHIONE - [chem] (an insecticide) dithione *m*
DITHIONIC ACID - [chem] acide *m* hyposulfurique, acide *m* dithionique
DITRIDECYL PHTHALATE - [chem] (a plasticizer) phtalate *m* de di-tridécyle
~THIODIPROPIONATE - [chem] (a stabilizer and plasticizer) thio-di-propionate *m* de di-tridécyle
DITROITE - [geol] (a type of rock) ditroïte *f*
DITTANY - [bot] dictame *m*, fraxinelle *f*
DITTO - [print] idem, de même
~MARKS - [print] signes *mpl* de répétition
DIURESIS - [med] (excess secretion of urine) diurèse *f*
DIURETIC - [med] diurétique
DIURNAL ARC - [astr] arc *m* diurne
~CHANGES - [met] (changes occurring during the day, especially those which occur at specific times) changements *mpl* diurnes
~PARALLAX - [astr] (geocentric parallax) parallaxe *f* géocentrique
~VARIATION - [phys] (oscillation of a compass needle during the day) variation *f* diurne (du champ magnétique de la terre)
DIVAGATION - [med] (a form of incoherent speech) divagation *f*
DIVALENT - [chem] (capable of combining with two atoms of hydrogen or the equivalent) bivalent, divalent
DIVARICATE, to - [gen] diverger, divariquer, bifurquer
[bot] divariquer, bifurquer
DIVARICATE - [bot] (forked, spreading apart) divariqué
DIVARICATORS - [anat] (set of muscles) muscles *mpl* divaricateurs
DIVE, to - [gen] plonger, pénétrer, enfoncer
[aero] (to cause an aircraft to descend steeply) piquer
DIVE - [gen] plongée *f*, descente *f*
[aero] (steep descent of an aircraft) piqué *m*
~-ANGLE INDICATOR - [instr] (an instrument showing the angle which the flight path of an aircraft makes with the vertical in a dive) indicateur *m* de piqué
~-BOMB, to - [aero] bombarder en piqué
~BRAKE - [aero] (device for increasing the drag) frein *m* de piqué
~FLAP - [aero] frein *m* de piqué
~TOP VELOCITY - [aero] vitesse *f* limite de piqué
DIVER - [naut] plongeur *m*, scaphandrier *m*
DIVERGENCE - [gen] divergence *f*, diffraction *f*
[el] (the scalar quantity equal to the limit of the flux emerging from a closed surface, divided by the volume contained by the surface of indefinitely small dimensions) divergence *f*
[aero] (a disturbance which increases without oscillation) divergence *f*
~LOSS - [acoust] perte *f* par divergence
~OF THE FRONT WHEELS - [auto] défaut *m* de parallélisme des roues avants
DIVERGENT - [gen] (in a condition of divergence) divergent
~DIP - [geol] pendage *m* divergent
~LENS - [opt] (lens designed to increase the divergence of a beam of light) lentille *f* divergente, lentille *f* négative
~NOZZLE - [mech] (nozzle with a cross-section increasing from entry to exit) divergent *m*
~PORTION - s. divergent nozzle
~SEQUENCE - [math] suite *f* divergente
~STRABISMUS - [med] (when the eyes diverge from each other) strabisme *m* divergent
DIVERGING JUNCTION SIGNAL - [railw] signal *m* de bifurcation
~LENS - [opt] lentille *f* divergente
~POINTS - [railw] aiguillage *m* d'embranchement
DIVERS' PARALYSIS - [med] mal *m* des caissons, embolie *f* gazeuse
DIVERSIFIED - [gen] varié
DIVERSION - [gen] (of roads) déviation *f*, déroutement *m*
[gen] déviation *f*, dérivation *f*, détournement *m*
~CHANNEL - [hydr] canal *m* de dérivation
~CUT - [hydr] rigole *f*
~OF PIPES - [plumb] bifurcation *f*
~WEIR - [hydr] barrage *m* de dérivation
DIVERSITY - [gen] diversité *f*
~RECEIVER - [radio] (a receiver which overcomes automatically any fading by selecting the strongest signal impulses from two or more aerials) récepteur *m* diversité
~RECEPTION - [radio] réception *f* simultanée (sur plusieurs antennes ou récepteurs)
~SYSTEM - [radio] (system employing two or more channels) système *m* complexe, système *m* de réception sur plusieurs antennes (ou récepteurs)
DIVERT, to - [gen] dévier, détourner, dérouter, écarter
~THE TRAFFIC, to - [roads & telécomm] détourner le trafic (radio), détourner la circulation (routière)
DIVERTED - [transp] (of a consignement) égaré
DIVERTER - [el] (a low resistance for the diversion of part of the current) résistance *f* de champ
~RELAY - [el] (relay used in excess-current protective systems) relais *m* de dérivation
~VALVE - [mech] (an auxiliary valve providing hydraulic power from the hoist pump) clapet *m* de dérivation
DIVERTICULITIS - [med] (inflammation in the colon) diverticulite *f*
DIVERTICULOSIS - [med] (diverticula in the colon) diverticulose *f*
DIVERTICULUM - [anat] (a protrusion of the mucuous membrane of the colon) diverticule *m*
DIVERTING DISK - [mech] poulie *f* de détour, poulie *f* de déviation
~PULLEY - [mech] galet *m* de détour
~RELAY - s. diverter relay
~ROLLER - [text] galet *m* de détour
DIVI-DIVI - [bot] (used for dyeing and tanning) dividivi *m*, dividiri *m*, libidibi *m*
DIVIDE, to - [gen] diviser, partager, subdiviser, trancher
DIVIDE - [geogr] (US) (line forming a division between two river valleys) ligne *f* de partage des eaux

DIVIDED - [gen] divisé, partagé, gradué, cloisonné
~BEAM - [text] ensouple *f* divisée
~BEARING - [mech] coussinet *m* en deux parties
~COIL - [el] enroulement *m* de déviation
~CREEL - [text] cantre *m* ouvert
~IRON-CORE - [el] noyau *m* divisé, noyau *m* feuilleté
~PISTON - [mech] piston *m* en deux parties
~STOP - [mus] (any of the organ stops which are controlled in two parts) registre *m* de combinaison
~TROUGH KNEADER - [mech] (a mixing machine with the trough divided longitudinally) malaxeur *m* à deux brasseurs
~WINDING - [el] enroulement *m* multiple
DIVIDEND - [math] dividende *m*
DIVIDER -[gen] diviseur *m*
[el] (potential divider; a high resistance with a fixed or adjustable tapping, giving a voltage which is only a fraction of that which is applied across the resistance) diviseur *m* de tension, réducteur *m* de tension
~PLATE - [text] platine *f* de distribution, platine *f* de formage
~STRAP - [text] lanière *f* de division
DIVIDERS - [impl] (instrument consisting of two finely-pointed arms pivoted together, used in measuring distances or in marking-out) compas *m* à pointes sèches, compas *m* à diviser
DIVIDING - [gen] diviseur, de partage, de démarcation
~BOX - [el] (used to separate the cores of a multicore cable) boîte *f* de dérivation
~COMB - [text] (a raddle) peigne *m* séparateur, peigne *m* diviseur
~ENGINE - [mech] (a machine expressly designed for making angular graduations) machine *f* à diviser
~HEAD - [mech] (part of a lathe, by means of which a workpiece can be rotated through accurately measured angular displacements, as in gear-cutting) plateau *m* diviseur
~KNIFE - [agric] diviseur *m*
~NETWORK - [radio] (a frequency selecting network dividing the spectrum to be radiated into two or more parts) montage *m* à deux voies
~PLATE - [mech] plateau *m* diviseur
~ROD - [text] baguette *f* d'envergure
~THE WARP - [text] division *f* de la chaîne
DIVING - [gen] plongée *f*, immersion *f*
[aero] piqué *m*
~APPARATUS - s. diving dress
~BELL - [constr] (watertight chamber, open at the bottom for building works under water) cloche *f* à plongeur
~COMPARTMENT - [naut] (in a submarine) water-ballast *m*
~DOOR - [naut] (in a submarine) clapet *m* de remplissage
~DRESS - [naut] scaphandre *m*
~GEAR - [naut] (in a torpedo) régulateur *m* d'immersion
~HOSE - [impl] tuyau *m* de scaphandrier
~MOMENT - [naut] moment *m* de tangage
~RUDDER - [naut] (in a submarine) gouvernail *m* de profondeur
~SUIT - s. diving dress
~TURN - [aero] virage *m* en piqué
~VELOCITY - [aero] (the speed attained by an aircraft in a dive, with or without power) vitesse *f* de piqué
DIVINING ROD - [impl] (a forked stick by means of which certain people claim they can discover water) baguette *f* de sourcier
DIVINYLBENZENE - [chem] (intermediate for synthetic resins and elastomers) divinylbenzène *m*
DIVINYL SULPHONE - [chem] (intermediate for a number of polymers) divinylsulfone *m*
DIVISIBILITY - [math] divisibilité *f*
DIVISION - [gen] division *f*, partage *m*, répartition *f*, graduation *f*, section *f*
[math] division *f*
[instr] (interval between scale marks) division *f*, graduation *f*
(the application of lines on the scale of a measuring instrument) graduation *f*
~PLATE - [mech] cloison *f*, séparation *f*
~WALL - [constr] cloison *f*, séparation *f*
DIVISIONAL BULKHEAD - [shipbuild] cloison *f* de séparation
DIVISOR - [math] (the quantity by which another quantity is divided) diviseur *m*
DIVORCED ANNEALING - [metall] sphéroïdisation *f*
~PEARLITE - [metall] (pearlite, a microconstituent of steel and cast-iron, in granular form) perlite *f* divisée, perlite *f* sphéroïdale
DIVULSION - [med] divulsion *f*, dilatation *f* forcée, arrachement *m*
DIVULSOR - [med] instrument *m* pour faire une dilatation
DIZZINESS - [gen & med] étourdissement *m*, vertige *m*
DIZZY - [med] pris d'étourdissement, pris de vertige
[gen] (of a height) vertigineux
DJALMAITE - [min] (tantalate of calcium and uranium) djalmaïte *f*
D.M.E. - s. distance measuring equipment
D/N - [comm] (debit note) note *f* de débit
DNA - s.desoxyribonucleic acid
D NICKEL - [metall] (commercial nickel, containing 4,5 p.c. of manganese) nickel *m* commercial
DO, to - [gen] faire, accomplir, exécuter
DOBBIE - s. dobby
DOBBY - [text] (a shedding mechanism on a loom for the manufacture of complex fabrics) ratière *f*
~CARD - [text] carton *m* de mécanique d'armures
~JACK - [text] lame *f* de ratière, levier *m* de ratière
~LOOM - [text] métier *m* à ratière
DOCIMASIA - [med] docimasie *f*
DOCK, to - [naut] faire passer au bassin, faire entrer au bassin, entrer au bassin
DOCK - [naut] (enclosed basin for ships) bassin *m*, cale *f*, forme *f*
[railw] quai *m* d'embarquement
[naut] (pier used for loading and unloading) dock *m*, quai *m*, appontement *m*
~MASTER - [naut] capitaine *m* de port, directeur *m* des docks
~TRIAL - [naut] essai *m* en bassin
DOCKAGE - [comm] droits *mpl* de bassin, droits *mpl* des docks
DOCKER - [comm] déchargeur *m*, docker *m*, débardeur *m*
DOCKET - [gen] étiquette *f*, fiche *f*
[comm] note *f* de remise, récépissé *m* des douanes

DOCKING BLOCKS - [naut] tins *mpl*
~KEEL - [naut] quille *f* d'échouage
~SURVEY - [naut] visite *f* de coque
DOCKSIDE FENDER - [naut] défense *f* de quai
DOCKYARD - [shipbuild] chantier *m* naval, arsenal *m* maritime
DOCTOR, to - [ind] (to spread a coating in a layer of uniform thickness) étaler à la racle
[chem] adultérer, traiter
DOCTOR - [impl] (a blade used to regulate the pulp on a roller, as in papermaking) racle *f*, couteau *m* racleur
[print] (a blade used to scrape ink from a plate) racle *f*, raclette *f*, docteur *m*
[plumb] (colloq for soldering iron) fer *m* à souder
~BAR - [impl] (a scraper used to spread adhesive or the like) racle *f*
~BLADE - [impl] (flat part used to spread paste in an even layer) racle *f*
~KISS-COATER - [impl] (a coating machine using a roller to apply the coating in conjunction with a blade (doctor knife) which spreads it evenly) machine *f* à imprégner à lames docteurs
~KNIFE - [impl] (a blade attached to a machine and used to spread a coating evenly) racle *f*, lame-docteur *f*
~ROLL - [impl] (roller which is given a wiping action by being driven at a different surface speed or in a different direction from that of the material treated) rouleau *m* essuyeur
~RULER - [mech] règle *f* de racle
~SOLUTION - [oil ind] (used to clear petroleum of sulphur) solution *f* au plombite
~TEST - [oil ind] (test for sulphur in petroleum) essai *m* au plombite
~TREATMENT - [chem] (a process for deodorizing petroleum stocks) traitement *m* au plombite, désulfuration *f* au plombite
DOCTOR'S BENCH - [impl] banc *m* d'essai
DOCUMENTARY FILM - [cin] (motion picture film using natural characters and real scenes for cultural purposes or entertainement) film *m* documentaire
~PHOTOGRAPHY - [photo] photographie *f* documentaire
DODDER - [bot] cuscute *f*
DODECAGON - [geom] dodécagone *m*
DODECAHEDRON - [geom & crystall] (of crystals, a form of the cubic system, consisting of twelve faces) dodécaèdre *m*
DODECENE - [chem] (intermediate for pharmaceuticals, dyestuffs, resins and perfumes) dodécène *m*
DODECENYLSUCCINIC ACID - [chem] (synthesis intermediate) acide *m* dodécénylsuccinique
~ANHYDRIDE - [chem] (intermediate for synthetic resins, plasticizers, and corrosion inhibitors) anhydride *m* dodécénylsuccinique
DODECYL ACETATE - [chem] an ingredient of perfumes) acétate *m* de dodécyle
DODECYLBENZENE - [chem] (a starting material in detergent production) dodécylbenzène *m*
DODECYLPHENOL - [chem] (intermediate for pharmaceuticals, bactericides, dyestuffs, resins, detergents, and rubber additives) dodécylphénol *m*
DODGER - [photo] (shading mask) cache *m* mobile
DOE - [zool] daine *f*
DOE-RABBIT - [zool] lapine *f*
DOESKIN - [leather ind] peau *f* de daim
DOFF, to - [text] (to remove full bobbins from a spinning machine) lever, faire la levée
DOFFER - [text] (workman who removes full bobbins from a spinning machine) déchargeur *m*
[text] (carding engine cylinder) peigneur *m*
~ROLLER - [text] rouleau *m* détacheur, rouleau *m* d'appel
~SHAFT - [text] arbre *m* du peigne détacheur
DOFFING - [text] levée *f*
~COMB - [text] peigne *m* détacheur
DOG - [zool] chien *m*
[mech] cliquet *m*, entraîneur *m*, taquet *m*, crabot *m*, griffe *f*, mâchoire *f*
[mech] (in a lock, a pawl, a catch) déclic *m*, cran *m* d'arrêt
[carp] (mechanical device designed to hold timber sections together) crampon *m*
[radar] (colloq for ground beacon) radiophare *m* de sol
[text] butée *f*, taquet *m*
~-AND CHAIN - [mining] arrache-étais *m*
~-APE - [zool] singe *m* mâle
~BENT - [bot] agrostide *f*
~-BOX - [railw] cage *f* pour le transport des chiens
~CLUTCH - [mech] (clutch consisting of opposed flanges with projections and slots) accouplement *m* à crabots
~CLUTCH SHAFT - [mech] arbre *m* d'accouplement à crabots
~-FISH - [zool] chien *m* de mer, roussette *f*
~-FOX - [zool] renard *m* mâle
~HOLE - [mining] passage *m*, recoupe *f*
~-HOUSE - [glass man] (in a furnace) bassin *m* d'enfournage
[mining] abri *m* de sondeur
[astronaut]renflement *m* de l'enveloppe (d'une fusée)
[cin] (US. colloq) cabine *f* de l'opérateur
~IRON - s. dog nail
~KENNEL - [build] chenil *m*, niche *f* à chien
~LEG - [el acoust] (colloq) déraillage *m* au début d'un disque
~-LEG - [oil ind] dogleg *m*
[astronaut] changement *m* de direction (pendant l'envol)
~-LEG CHISEL - [impl] (a carpenter's chisel) butteavant *m*
~-LEGGED STAIR - [build] (stair without a well-hole with flights rising in opposite directions) escalier *m* à limons superposés, escalier *m* sans jour médian
~LEGS - [mech] (in drilling) pattes *fpl* de chien, double cambrure *f*
~MARK - [metall] empreinte *f* de tenailles
~MOVEMENT - [cin] (a system of intermittent movement whereby a film is moved on in a series of beats) entraînement *m* à batteur
~NAIL - [impl] clou *m* à large tête, clou *m* à crochet
[mech] pilier *m*, clou *m* de serrure
~SPIKE - [railw] crampon *m* (de rail)
~-TOOTH COURSE - [constr] assise *f* de briques en dentelure
DOGS - [mech] arrêts *mpl*, cliquets *mpl* d'arrêt
DOG'S EAR - [plumb] (folding joint on a sheet-lead tray) joint *m* à oreilles

DOG'S MERCURY - [bot] (a weed) mercuriale *f*
~TOOTH GRASS - [bot] (a herbage plant) cynodon *m*, chiendent *m*
DOGGERS - [geol] (calcareous or ferruginous concretions) doggers *mpl*
DOGWATCH - [naut] petit quart *m*
DOGWOOD - [bot] cornouiller *m*
DOILY - [text] (US - colloq; any rag to remove oil etc) chiffon *m*
DOLDRUMS - [met] (a belt of light variable winds, extending between latitude I0 to I5 deg N and S of the equator approximately) calmes *mpl* équatoriaux, pot *m* au noir
DOLERITE - [geol] (basic igneous rocks, especially quarried for road works) dolérite *f*
DOLICHOCEPHALIC - [anat] dolichocéphale
DOLICHOCEPHALISM - [med] dolichocéphalie *f*
DOLICHOCOLON - [anat] dolicolon *m*
DOLINE - [geol] doline *f*
DOLL - [gen] (a toy) poupée *f*
~BUGGY - [telev] (colloq for camera dolly) chariot *m* pour caméra
DOLLAR - [nucl] (a unit of reactivity) dollar *m*
DOLLY, to - [cin & telev] (to follow the action on a dolly) tourner en travelling
[mech] (heavy, hammer-like tool for holding the head of a rivet while forming the head at the other end) contre-bouterolle *f*, tas *m* à river
[metall] (in sheet metal work; for straightening sheets) tasseau *m*, tige *f* de bocard
[constr] rallonge *f* de pilotis
[cin & telev] (mobile platform for carrying cameras and personnel) chariot *m*, travelling *m*
[impl] (for the transport of heavy machines etc) chariot *m*, diabolo *m*, support *m* à galets
[mining] (a counterbalance weight) contrepoids *m*
[el] (component of a dry cell, formed of a depolarizing mix moulded round a carbon rod to form the positive electrode) dépolarisant *m*
~BLOCK - [impl] tasseau *m*, tas *m* à river
~IN, to - [cin & telev] tourner en travelling avant
[cin & telev] travelling *m* en avant de la caméra
~OUT, to - [cin & telev] tourner en travelling arrière
[cin & telev] travelling *m* en arrière de la caméra
~SHOT - [cin] (the shooting of a moving object from the same distance) prise *f* de vues en travelling
DOLOMITE - [min] (naturally occurring calcium and magnesium carbonate used as a refractory material, fertilizer, and as a source of magnesium) dolomite *f*, dolomie *f*
~CORNERED STONE - [constr] sable *m* dolomitique
~ROCKS - [geol] roches *fpl* dolomitiques
DOLOMITIC CONGLOMERATE - [geol] (a limestone conglomerate found in South Wales) conglomérat *m* dolomitique
~LIMESTONE - [geol] (calcareous rock which contains calcite in addition to dolomite) calcaire *m* dolomitique
~MARL - [geol] marne *f* dolomitique
DOLOMITIZATION - [geol] (the replacement of calcium carbonate in a limestone by calcium and magnesium carbonate, i.e. dolomite) dolomitisation *f*
DOLORIMETER - [instr] (medical instrument, designed to measure the intensity of pain) dolorimètre *m*
DOLPHIN - [zool] dauphin *m*
[naut] duc *m* d'Albe, bouée *f* de corps mort
[oil ind] pylône *m* d'encrage
DOLPHIN STRIKER - [naut] martingale *f*
DOMAIN - [gen] domaine *m*
[crystall] (a region of spontaneous electric or magnetic polarization in a ferro-electric or a ferromagnetic crystal) domaine *m*
~STRUCTURE - [crystall] structure *f* des domaines
~THEORY - [crystall] théorie *f* des domaines
~WALL - [phys] (e.g. in ferro-magnetic material) paroi *f* de Bloc
DOMAL STRUCTURE - [mining] structure *f* en dôme
DOME - [constr] (a vault built on a circular base) coupole *f*
[mech] (domed cylinder fitted to steam boilers) dôme *m*
[geol] (dome-like roof of an igneous intrusion) dôme *m*
[nucl] (in a nuclear bomb explosion, the water spray thrown upwards) champignon *m*
[packag] (the bottom end) fond *m*, fond *m* bombé
~HEAD - [auto] (of a cylinder) culasse *f* hémisphérique
~-HEAD PISTON - [mech] piston *m* a tête bombée
~LADDER - [impl] (fitted to locomotives) échelle *f* de dôme
~LAMP - [light] plafonnier *m*
~NUT - [mech] écrou *m* borgne à tête ronde
~OF THE COMBUSTION CHAMBER - [mech] dôme *m* de la chambre
DOMED - [gen] à dôme, en forme de dôme, hémisphérique
DOMESTIC - [text] (cotton cloth of plain quality) toile *f* de coton
~TELEPHONE CIRCUIT - [telecomm] circuit *m* téléphonique intérieur
DOMICAL GROIN - [constr] lunette *f* sphérique
~ROOF - [constr] charpente *f* de dôme
~VAULT - [constr] (vault with its centre higher than its sides) voûte *f* en coupole
DOMINANT - [gen] dominant
~CHARACTER - [genet] (the character which emerges in the hybrid results from the cross-breeding of parents) caractère *m* dominant
~MODE - [electron] (in waveguides; with the lowest critical frequency) mode *m* fondamental, mode *m* dominant
~MODE OF PROPAGATION - [electron] (in waveguides) mode *m* fondamental de propagation
~WAVE - [electron] (in waveguides; electromagnetic wave with the lowest cut-off frequency) onde *f* fondamentale, onde *f* dominante
~WAVELENGTH - [phys] (the determining colour in colorimetry) longueur *f* d'onde dominante
DOMING - [geol] formation *f* de dômes
DONKEY - [zool] âne *m*, baudet *m*
~BOILER - [mech] (small-size vertical auxiliary boiler) chaudière *f* auxiliaire
~ENGINE - [mech] petit cheval *m*, moteur auxiliaire *m*
~PUMP - [mech] (small steam reciprocating pump suitable for general duties) pompe *f* alimentaire *f*, petit cheval *m* alimentaire
~WINCH - [mech] treuil *m* à vapeur
DONNAN POTENTIAL - [el biol] (potential difference across an impermeable semi-inert membrane placed between ionic mixtures) potentiel *m* de Donnan

DONOR - [gen] donneur *m*
[electron] s. donor atom
~ATOM - [electron] (those atoms in a semiconductor which furnish electrons) donneur *m*, atome *m* donneur
~ELEMENTS - [chem] éléments mpl donneurs
~IMPURITY - [electron] (any impurity which promotes electronic conduction, especially in transistor elements) impureté *f* donneuse
~LEVEL - [electron] (intermediate level close to the conduction band in the energy diagram of an extrinsic semiconductor) niveau *m* donneur
DONUT - s. doughnut
DONUTRON - [chem] (special type of magnetron, capable of being tuned) magnétron *m* à double cage d'écureuil, magnétron *m* réglable
DOODLE, to - [oil ind] (US colloq) nettoyer une tranchée de pipe-line
DOODLE - [impl] (a diviner's stick used to find oil wells) baguette *f* de sourcier
DOODLEBUG - [min] prospecter
[auto] (US colloq for a small racing car) petite voiture *f* de course
DOOMSDAY BOOK - [surv] (the first survey of the lands of England) le grand cadastre *m* d'Angleterre
DOOR - [gen] porte *f*
[mech] (of a furnace; a hinged barrier) porte *f*
[railw & auto] porte *f*, portière *f*
~APERTURE FRAME - [auto] encadrement *m* de porte
~BELL - [el] sonnette *f*
~BOLT - [mech] verrou *m*
~BUFFER STRIP - [auto etc] bourrelet *m*, garniture *f* de portière
~BY-PASS SWITCH - [el] (in lifts) interrupteur *m* de sécurité (ascenseurs)
~CASE - s. door frame
~CATCH - [mech] loquet *m*
~CHECK - [mech] (mechanism bringing the door to its closed position without slamming) ferme-porte *m*
~CHEEKS - s. door jamb
~CLOSER - s. door check
~CONTACT - [el] (a contact attached to a door, or a gate) contact *m* de porte
~FITTINGS - [auto] accessoires *m*pl de porte
~FRAME - [carp] bâti *m* de porte, dormant *m* de porte, chambranle *m*, huisserie *f*
~FURNITURE - [constr] ferrure *f* d'une porte, garniture *f* d'une porte
~HANDLE - [impl] poignée *f* de porte
~JAMB - [carp] (the jambs of the frame) montant *m* de porte, jambage *m*, poteau *m* d'huisserie
~-KNOB - [gen] bouton *m* de porte, poignée *f* de porte
~LEAF - [constr] vantail *m*, battant *m*
~LINTEL - [constr] linteau *m*
~LOCK - [gen] serrure *f* de porte, fermeture *f* de porte
~-LOCK STRIKER - [auto] gache *f* de porte
~-MAT - [impl] paillasson *m*, essuie-pieds *m*
~PILLAR - [auto] montant *m* de portière
~PLATE - [impl] plaque *f* (de porte)
~POST - s. door jamb
~RAIL - [constr] traverse *f* de porte
~REBATE - [constr] feuillure *f* de porte
~RUNNER RAIL - [mech] glissière *f*
~SCRAPERMAT - [impl] décrottoir *m*
DOOR SEAL - [mech] bourrelet *m*
~SIDE - [build] (the wall into which the door opens) mur *m* lateral
~SIDE BUMPER - [auto] pare-choc *m* latéral
~STONE - [constr] seuil *m*
~STOP - [constr] butée *f* de porte
~STRIP - [carp] (stripping applied to doors to avoid draughts) bourrelet *m*
~UNLOCKED WARNING LIGHT - [aero] (warning lamp showing that an aircraft is not locked) indicateur *m* de 'porte ouverte'
~WEDGE - [impl] coin *m* pour porte
DOORKNOB TRANSFORMER - [electron] (a mode changer designed to convert a coaxial line transmission to a rectilinear waveguide one) transformateur *m* de couplage
~VALVE - [electron] (a short-path low interelectrode-capacity type of valve, resembling a doorknob in shape) tube *m* 'bouton de porte'
DOORWAY - [constr] (any opening into or out of a building) entrée *f* de porte, baie *f* de porte, porte *f* cochère, portail *m*
DOPE, to - [paint] (to apply varnish used as dope) enduire, enrober, revêtir, recouvrir, vernir
[chem] (to treat with dope) doper
DOPE - [chem] (solution of cellulose derivative used to size and also process yarns and fabrics) enduit *m*, enduit *m* d'enrobage, vernis *m* imperméabilisant
[photo] (paste, oil or wax used for retouching) vernis *m* pour retoucher
[mech] (added to petrol) produit *m* d'addition, additif *m*
[oil ind] pâte *f*, graisse *m*
[telev] (in video recording) pâte *f* magnétique
[text] masse *f* d'enduction
[med] (colloq) narcotique *m*, stupéfiant *m*, anesthésique *m*
~DYED - [text] teint *m* dans la masse
DOPED ENVELOPE - [aero] enveloppe *f* enduite
~FUEL - [oil ind] (fuel containing additives to give special characteristics) carburant *m* dopé
~JUNCTION - [electron] (in transistors, a junction formed by adding a controlled 'impurity' to the melt during crystal growth) jonction *f* dopée
DOPING - [electron] (addition of controlled 'impurities' to a semiconductor material in transistor manufacture) dopage *m*
[text] enduisage *m*, imperméabilisation *f*
[astronaut] (with additives) dopage *m*
~COMPENSATION - [electron] compensation *m* des impuretés, compensation *f* par dopage
DOPPLER BROADENING - [phys] (the spreading of frequencies) élargissement *m* Doppler
~DIRECTION FINDER - [radar] (radar direction finder in which differentiation between fixed and moving targets is obtained by the Doppler effect) radiogoniomètre *m* à effet Doppler
~DISPLACEMENT - [phys] déplacement *m* Doppler
~EFFECT - [phys] (modification of the apparent frequency of a wave-train caused by the movement of the source of the waves towards or away from the observer) effet *m* Doppler-Fizeau
~RADAR - [radar] (radar system depending upon Doppler effects in radio transmission) radar *m* à effet Doppler
~SHIFT - [phys] (the magnitude of the change in apparent frequency due to Doppler effects) variation

f de fréquence (due à l'effet Doppler)
DORIC - [arch] dorique
˜ARCHITECTURE - [arch] architecture *f* dorique
DORMANCY - [bot] germination *f*, vie *f* ralentie
DORMANT BOLT - [mech] pêne *m* dormant
˜OIL - [chem] (an insecticide) huile *f* insecticide
˜STATE - [gen] (e.g. in brewing) état *m* de repos
DORMER - [constr] (small window on a roof slope) lucarne *f*, fenêtre *f* en mansarde, fenêtre *f* en chien assis
˜CHECK - [constr] paroi *f* latérale d'une lucarne
˜RAFTER - [constr] chevron *m* de croupe
˜WINDOW - s. dormer
DORMITORY - [constr] (large sleeping room or apartment) dortoir *m*
˜AREA - [constr] (in town planning] zone *f* dortoir
DORMOUSE - [zool] loir *m*
DORNIC - [text] (or dornick; linen fabric used in Scotland for the table) tissue *m* façonné
DORSAL - [anat] dorsal
[carp] (the back of a tall chair) dossier *m*
˜PROJECTION - [phys] (X-ray beam from front to back) projection *f* antéropostérieure
˜TURRET - [aero] tourelle *f* supérieure
DORSALGIA - [med] nevralgie *f* dorsale
DORY - [naut] doris *m*
DOSAGE - [radiat] s. dose
˜PUMP - [packag] (for filling) pompe *f* doseuse
˜RATE - [radiat] (amount of radiation in unit time) débit *m* de dose
DOSAGING VALVE - [ind chem] fuite *f* réglable, robinet *m* de dosage, vanne doseuse
DOSE - [gen] dose *f*
[nucl] (the amount of radiation received by a given area or volume of a body, or by the whole of such body) dose *f*
˜EFFECT CURVE - [nucl] (curve showing the dose and the effect of a radiation) courbe *f* dose-effect
˜FRACTIONATION - [mech] (a system of administering radiations) fractionnement *m* de la dose
˜OF AN ISOTOPE - [med] (the quantity of radiation administered with an isotope) dose *f* administrée avec un isotope
˜PROTRACTION - [med] (a continuous administration of radiation over a comparatively long period) étalement *m* de la dose
˜RATE - [nucl] (dose of radiation per unit of time) débit *m* de dose
˜RATE METER - [instr] (instrument for measuring dosage rate) dosimètre *m*
DOSIMETER - [instr] (instrument for measuring dose of radiation) dosimètre *m*
DOSING FEEDER - [mech] (a device for supplying a machine with weighed quantities of material) doseur *m*
˜MACHINE - [ind chem etc] doseur *m*, machine *f* doseuse
˜PUMP - [ind chem] pompe *f* de dosage, pompe *f* doseuse
DOSOLOGY - [med] dosologie *f*
DOSSERET - [arch] dosseret *m*
DOT, to - [text] tamponner
DOT - [gen & print] point *m*
[telev] (an element of the mosaic) point *m*
˜BAR GENERATOR - [telev] générateur *m* de mire à points et barres
˜-DASH LINE - [telecomm] ligne *f* trait-point
DOT-DASH MODE - [electron] (in a cathode-ray tube storage) mode *m* point-trait
˜ETCHING - [print] retouche *f* par morsure du point
˜FREQUENCY - [telev] (the number of dots of the screen covered per second) fréquence *f* d'analyse
˜GRATING - [opt] treillis *m* à points
˜INTERLACING - [telev] entrelacement *m* à points successifs
˜RECTIFICATION - [telev] augmentation *f* de la luminance des points
˜SEQUENCE - [telev] séquence *f* de points
˜SEQUENTIAL COLOUR TELEVISION - [telev] système *m* à séquence de points
˜SEQUENTIAL SYSTEM - [telev] système *m* à points successifs
DOTTED - [print] pointillé
˜LINE - [print] ligne *f* en pointillé, pointillés *mpl*
˜LINE PROFILE - [rubber ind] (in tyres) sculpture *f* en lignes brisée
˜LINE TREAD - s. dotted line profile
˜MULL - [text] mousseline *f* à pois
˜NET - [text] tulle *m* moucheté, tulle *m* à pois
˜PATTERN - [text] dessin *m* pointillé, armure *f* pointillée
˜PROFILE - [rubber ind] (in tyres) sculpture *f* pointillée
˜TREAD - s. dotted profile
DOTTING - [print draw etc] pointillage *m*
DOUBLE, to - [gen] doubler
[cin] doubler
[naut] doubler (un cap)
DOUBLE - [gen] double, géminé
[constr] (a I6 x I3" slate) ardoise *f* de 40 x 33 cm
[cin] doublure *f*
[mus] (an organ stop) doublette *f*
˜ACTING - [mech] à double effet
˜-ACTING CYLINDER - [mech] cylindre *m* à double effet, vérin *m* à double effet
˜-ACTING FUSE - [chem] fusée *f* à double effet
˜-ACTING HINGE - [mech] charnière *f* à ressort
˜-ACTING HYDRAULIC RAM - [constr] bélier *m* double
˜-ACTING PISTON - [mech] piston *m* à double effet
˜-ACTING PLUNGER PUMP - [mech] pompe *f* à plongeur à double effet
˜-ACTING PUMP - [mech] (reciprocating pump in which both sides of the piston can act alternately) pompe *f* à double effet
˜-ACTING RELAY - [el] (relay operating in two steps) relais *m* à deux temps
˜ACTION - [gen & mech] à double effet
˜ACTION DROP HAMMER - [mech] (used in forging) marteau-pilon *m* à double effet
˜AMPLITUDE - [el] (in an alternating quantity, the sum of the highest values of the positive and negative half-waves) amplitude *f* totale
˜-AMPLITUDE PEAK - [radio] amplitude *f* crête à crête
˜-ANGLE CUTTER - [mach tool] fraise *f* d'angle
˜ANGLE FISH-PLATE - [railw] éclisse *f* à double revers
˜ARC BIT - [mining] taillant *m* double (d'un fleuret)
˜-ARM KNEADER - [mech] (a kneading machine fitted with a pair of arms carrying the paddles) malaxeur *m* à deux brasseurs
˜-BANK ENGINE - [mech] (an internal-combustion engine having two rows of cylinders) moteur *m* à

deux rangées de cylindres
DOUBLE-BANK RADIAL ENGINE - [aero] moteur *m* en double étoile
~-BANKED - [naut] armé à couple
~BAR - [mus] double barre *f*
~-BAR COACH-WRENCH - [impl] clef *f* anglaise double à marteau
~-BARREL - s. double barrelled
~-BARREL PUMP - [mech] pompe *f* à double cylindre
~-BARRELLED - [firearms] à deux canons, à deux coups
~-BASS - [mus] contrebasse *f*
~-BASSOON - [mus] contrebasson *m*
~BATTERY SWITCH - [el] (switch arranged to control independently the number of cells in a battery which are charging or discharging) réducteur *m* double (de batterie)
~-BEAM CATHODE-RAY TUBE - [electron] tube *m* cathodique à deux faisceaux
~BEARING - [mech] palier *m* double
~BEAT UP - [text] coup *m* de battant double
~BELL JAR - [ind chem] cloche *f* double, double enceinte *f*
~BELTING - [mech] (heavy duty, double thickness belting) courroie *f* double
~BEND - [gen] double courbure *f*, à double courbure
~-BETA DECAY - [nucl] (type of radioactivity in which the atomic number increases by two while the mass number is unchanged) désintégration *f* béta double
~BEVEL WHEEL - [mech] roue *f* conique double
~-BEVELLED CHISEL - [tool] fermoir *m*
~BLADE MIXER - [mech mach] (a mixing machine provided with two agitating blades) mélangeur *m* à deux palettes
~-BLAST CIRCULAR BELLOWS - [metall] soufflet *m* cylindrique à double vent
~-BLAST FORGE - [metall] forge *f* à double vent
~BLOCK - [mech] (of a tackle) poulie *f* double
~BOND - [chem] (a covalent linkage in which atoms share two pairs of electrons) liaison *f* double
~-BONDED COMPOUND - [chem] composé *m* à double liaison
~BOTTOM - [gen & naut] double fond *m*
~BOTTOM AFT - [naut] double fond *m* arrière
~BRACED FRAME - [auto] châssis *m* à entretoises (doubles)
~BRANCH GATE - [metall] jet *m* de coulée en fourche
~BREAK - [el] (said of a circuit which is made or broken in two points in each phase) bipolaire, à coupure double
~-BREAK JACK - [el] jack *m* double de rupture
~-BREAK SWITCH - [el] (type of switch in which the circuit is broken at two points in each phase) disjoncteur *m*, bipolaire, interrupteur *m* à double rupture
~BRIDGE - [el] (bridge network with two sets of coils) pont *m* double (de Thomson)
~BRIDGING - [carp] (to connect adjacent floor joists) poutre *f* traversière double
~BURNER - [th eng] brûleur *m* double
~-BUTT NEEDLE - [text] aiguille *f* à deux talons
~CALENDER - [text] calandre *f* double
~CALLIPERS - [impl] compas *m* maître de dan
~CAM - [mech] excentrique *m* double
~CAMERA - [cin & telev] caméra *f* pour image et son
DOUBLE-CARBON ARC LAMP - [el] lampe *f* à arc au carbone double
~CARBURATION - [auto] double carburation *f*
~CARD - [text] carde *f* double
~CASING - [el] (an electrical machine with ventilated frame in which the cooling air circulates in a space between the actual casing of the machine and an outer casing) à double paroi, à double enveloppe
~CASTING - [cin] (the casting of one action in two roles) rôle *m* double
~-CATENARY CONSTRUCTION - [el] (in electric traction systems) suspension *f* caténaire double
~-CATHODE-RAY TUBE - [electron] (a cathode-ray tube containing two sets of beam-forming electrodes operated from the same cathode) tube *m* cathodique à deux faisceaux
~CHAIN STITCH SEWING MACHINE - [text] machine *f* à coudre à point de chaînette double
~-CHAMBERED SLIDE-VALVE - [mech] tiroir *m* à double coquille
~-CLAD VESSEL - [nucl] (a vessel provided with a double layer of protective material) récipient *m* à double enveloppe
~CLAW - [cin] (a claw with two sets of contact points engaging the film perforations) griffe *f* double
~CLOTH - [text] (two separate cloths woven together to obtain a greater thickness) étoffe *f* double, tissu *m* double
~CLOTH WEAVE - [text] armure *f* de tissu double
~CLOTH WITH BINDING WEFT - [text] tissu *m* double à trame de liage
~CLOTH WITH COLOUR EFFECTS - [text] tissu *m* double à effet de couleurs
~CLOTH WITH INNER PACKING WEFT - [text] tissu *m* double à trame de fourrure
~CLOTH WITH WADDING WARP - [text] tissu *m* double à chaîne de fourrure
~-CLUTCH - [mech] (US) (clutch requiring a double depression) double débrayage *m*
~-COATED FILM - [photo] (film with emulsion on both sides) pellicule *f* à double couche
~COCOON - [text] cocon *m* double
~-COLUMN JIG BORING MACHINE - [mach tool] machine *f* à pointer à deux montants
~-COLUMN PLANING MACHINE - [mach tool] (type of large planing machine in which the tool slide is carried on a part of columns between which the bed reciprocates) raboteuse *f* à deux montants
~COMPOUND SWITCH - [railw] traversée-jonction *f* double
~CONCAVE LENS - [opt] lentille *f* biconcave
~CONDENSER SPINNING - [text] filature *f* de coton cardé, filature *f* en gros numéro
~CONE DRUM MIXER - s. double cone mixer
~CONE MIXER - [mech] (mixer consisting of two cones joined at their bases and rotating about a line normal to their common longitudinal axis) mélangeur *m* à tambours coniques
~CONICAL BOBBIN - [text] bobine *f* à bouts coniques
~CONTRAST ENEMA TECHNIQUE - [med] (in a radiological examination, the outlining of the large intestine by injection of air after the evacuation of an opaque enema) technique *f* du double contraste
~CONVEX LENS - [opt] lentille *f* biconvexe
~CONVEYOR WORM WITH INCREASING PITCH - [mech] vis *f* transporteuse double à pas croissant
~CORE BARREL - [mining] carottier *m* double

DOUBLE COUPLING PIECE - [mech] piéce *f* d'accouplement double
~COURSE - [constr] (a double course of bricks) double rangée *f* de briques
~-COVER BUTT JOINT - [mech] assemblage *m* à double couvre-joint
~CRANE-HOOK - [mech] crochet *m* double, crochet *m* à tête de bélier
~-CRANK - [mech] à double manivelle
~-CRANK PRESS - [mech] presse *f* à double manivelle
~CREEL - [text] cantre *m* double
~-CROSS, to - [gen] duper, tromper
~CROSS - [genet] dihybridation *f*
~CROSS-OVER - [railw] traversée *f* double, traversée *f* bretelle
~CROSS-VAULT - [arch] voûte *f* à double croisée
~CROSSING - [text] croisure *f* double
~-CUP GRINDING WHEEL - [mech] meule *f* double boisseau
~CURRENT - [telecomm] (denoting the use of current reversals in telegraphy) bicourant
~-CURRENT FURNACE - [el] (electric furnace using direct and alternating current) four *m* tous courants
~-CURRENT SIGNALLING - [telecomm] (in telegraphy; a method of transmission by means of alternate positive and negative currents) système *m* bicourants
~-CURVE POINTS - [railw] aiguillage *m* en courbe
~-CUT - [mech] taille *f* double, taille *f* croisée
~-CUT BASTARD - [mech] à taille double bâtarde
~-CUT FILE - [impl] lime *f* à taille croisée
~CUT SMOOTH - [mech] à taille double fine
~-CUTTING PLANING MACHINE - [mach tool] machine *f* à raboter dans les deux sens de marche
~-CYLINDER ENGINE - [mech] moteur *m* à deux cylindres
~-CYLINDER STEAM ENGINE - [mech] machine *f* à vapeur à deux cylindres
~-DECK SCREEN - [impl] tamis *m* à deux plateaux
~-DECKER - [autobus, ships, etc] à impériale (autobus), à deux ponts
~DECOMPOSITION - [chem] (bimolecular reaction in which the atoms are redistributed into other molecules) double décomposition *f*
~-DELTA CONNECTION - [el] (connection of windings which can be represented by a diagram consisting of two triangles) couplage *m* double triangle
~DENSITY - [comput] (the doubling of the number of tracks on storage disks, thus reducing the access time) densité *f* double
~DERIVATIVE ACTION - [contr] action *f* par double dérivation
~-DETECTION RECEIVER - [radio] (used for the detection of supersonic waves) récepteur *m* superhétérodyne
~DICHOTOMY - [statis] (double division into two subclasses) dichotomie *f* double
~DIODE - [electron] (type of electron tube in which are two separate diode systems screened from each other) duo-diode *f*, double diode *f*
~DIODE OUTPUT PENTODE - [electron] double diode-penthode *f* de sortie
~DIODE PENTODE - [electron] double diode-penthode *f*
~DIODE TRIODE - [electron] duodiode-triode *f*
~-DISK SAND-PAPERING MACHINE - [mach tool] machine *f* à poncer à deux plateaux
DOUBLE-DISK WINDING - [el] (form of winding used for transformers) enroulement *m* en disque double
~DOFFER CARD - [text] carde *f* à deux peigneurs
~-DOORS - [mining] (for ventilation) portes *fpl* doubles
~-DROP FRAME - [auto] chassis *m* surbaissé
~DRUM-TYPE TAR EXTRACTOR - [mech] (in condensation machinery) dégoudronneur *m* à double cloche
~-DRY SPINNING FRAME - [text] métier *m* à filer sec double
~-EARTH FAULT - [el] (fault occurring when two of the phases go to earth at the same time) contact *m* double à la terre
~ECCENTRIC - s. double cam
~ENCASED GEARING - [mech] double harnais *m* d'engrenages (enfermé dans des gaines protectrices)
~-END CONTROL - [railw] commande *f* réversible
~-ENDED - [gen] à deux fins, à deux façades
~-ENDED BOILER - [th engine] (shell-type marine boiler with a furnace at both ends) chaudière *f* à deux façades
~-ENDED BOLT - [mech] prisonnier *m*
~-ENDED CATCH - [text] cliquet *m* double
~-ENDED CORD - [telecomm] cordon *m* à deux fiches
~-ENDED CRANE - [mech] grue *f* double
~-ENDED CROSS-CUT SAW - [impl] passe-partout *m*, scie *f* en travers
~-ENDED CROWBAR - [impl] pince *f* de carrier, pince *f* à pointe et à tranche
~-ENDED HAND-SAW FILE - [impl] tiers-point *m* double
~-ENDED HAND-WRENCH - [impl] (used in mining) tourne-à-gauche *m* double
~-ENDED MATCH-PLANE - [impl] bouvet *m* à fourchement
~-ENDED NEEDLE - [text] aiguille *f* à deux becs, aiguille *f* à deux clapets
~-ENDED PUNCHING AND SHEARING MACHINE - [mach tool] poinçonneuse-cisaille *f* double
~-ENDED RAIL - [railw] rail *m* à double champignon
~-ENDED SCREW-STOCK - [mech] filière *f* double
~-ENDED SNAP GAUGE - [tool] calibre *m* de tolérance à fourchette
~-ENDED SPANNER - [impl] clé *f* plate double, clé *f* double à fourches
~-ENDED TOOL GRINDING MACHINE - [mech] machine *f* à affûter les outils à deux meules
~-ENDED WRENCH - s. double-ended spanner
~ENTRIES - [mining] galeries *fpl* jumelles doubles
~-EXHAUST FAN - [mech] ventilateur *m* aspirant double
~EXPOSURE - [photo] (the re-exposure of the same film) double impression *f*, surimpression *f*
~-EYE CABLE GRIP - [telecomm] serre-câble double
~FACE TWILL - [text] sergé *m* double face, tissu *m* croisé double face
~-FACED HAMMER - [impl] (hammer with a flat face at both ends) marteau *m* à dresser
~-FACED SLEDGE HAMMER - [impl] marteau *m* à devant
~-FACED WINDING MACHINE - [text] bobinoir *m* double face
~-FEED VALVE - [mech] clapet *m* à double alimentation
~FEEDBACK - [el] (regeneration supplied to two or more stages simultaneously) réaction *f* double

DOUBLE FILAMENT LAMP - [el] lampe *f* à double filament
~FILTRATION - [hydr] double-filtration *f*
~-FLANGED PAPER TUBE - [text] bobine *f* en carbon à plateaux
~FLANGED SEAM - [metall] (in welding) joint *m* bout à bout sur bords relevés
~-FLANGED SHEET METAL TUBE - [text] bobine *f* métallique à plateaux
~FLANGED TRAVELLING WHEEL - [mech] (of a crane) galet *m* à deux joues
~-FLANGED WOODEN BOBBIN - [text] bobine *f* en bois à plateaux
~FLAT - [mus] double bémol *m*
~FLEXIBLE CORD - [el] cordon *m* souple à deux conducteurs
~FLOOR - [constr] plancher *m* sur poutre
~-FLOOR KILN - [brew ind] séchoir *m* à deux plateaux
~-FLOW TURBINE - [mech] (turbine in which the fluid enters the casing in the middle and flows towards both ends) turbine *f* à double flux
~FLOWER - [bot] (a flower with an excess of petals) fleurs *f* double
~FLOWERING - [bot] (plant which abnormally flowers in the spring and in the autumn) floraison *f* double
~-FLUID BATTERY - [el] pile *f* à deux liquides
~-FOCUS TUBE - [electron] tube *m* à double foyer
~FOCUSING MAGNETIC SPECTROMETER - [instr] (instrument designed to focus ions with velocities having a component normal to the median magnetic plane) spectromètre *m* à deux foyers
~-FOLDING DOOR - [gen & auto] porte *f* à deux battants
~FORCE MOULD - [casting] (a mould in which there are two forces) moule *m* à double effet
~-FREQUENCY CHANGING - [el] double changement *m* de fréquence
~-FREQUENCY OSCILLATOR - [radio] (oscillator in which two sets of oscillations are generated at the same time) oscillateur *m* à deux fréquences
~FROG - [railw] (diamond crossing) cœur *m* de traversée
~FURROW PLOUGH - [agric mach] charrue *f* à deux socs, charrue *f* à deux corps
~GAP VULCANIZING PRESS - [rubber ind] presse *f* jumelle de vulcanisation à col de cygne
~GAUGE LINE - [railw] voie *f* à écartement double
~-GAUZE LAMP - [mining] lampe *f* à double toile
~GEARING - [mech] double harnais *m* d'engrenages
~GIB KEY - [text] clavette *f* à double talon, contre-clavette *f*
~GRAFTING - [agric] surgreffage *m*
~GRAVITY-HOIST - [mech] (used in mining) balance *f* sèche à double effet
~-GRID VALVE - [el] tube *m* bigrille, tétrode *f*
~HALF-ROUND FILE - [impl] lime *f* double demironde
~-HALVED JOINT - [carp] tenon *m* croisé
~-HANDED HAND-SHANK - [metall] poche *f* à main à double manche
~HANDED SAW - [impl] scie *f* passe-partout
~HARDENING - [metall] trempe *f* tenace à cœur
~-HEAD PROJECTOR - [cinema] projecteur *m* double poste
~-HEADED - [adj] bicéphale
~-HEADED CHAPLET - [metall] support *m* double

DOUBLE-HEADED RAIL - [railw] rail *m* à double champignon
~-HEADED SHAPING MACHINE - [mach tool] étau-limeur *m* double
~-HEADED TRAIN - [railw] train *m* à deux locomotives en tête
~-HEADED WRENCH - [impl] clef *f* double
~HEADING - [railw] (with two locomotives at the head of the train) double traction *f*
~-HELICAL - [mech] à chevrons
~HELICAL GEAR - [mech] (type of helical gear having two sets of teeth of opposite inclination on the same wheel to eliminate axial thrust) engrenage *m* à chevrons
~HELICAL TOOTH - [mech] dent *f* à chevron
~-HINGED ARM - [mech] bras *m* à double articulation
~HOOK - [mech] crochet *m* double
~HOOK CATCH - [text] platine *f* à bec double
~-HOOK NEEDLE - [text] aiguille *f* à crochet double
~-HOUSING PLANE - [mech] raboteuse *f* à deux montants
~HULL - [naut] coque *f* double (sous-marin)
~-HUNG SASH WINDOW - [build] fenêtre *f* à guillotine
~HUNG WINDOW - [build] fenêtre *f* à guillotine
~IGNITION - [auto] allumage *m* double
~IMAGE - [telev] (undesired images appearing on the screen, thus causing a double picture) image *f* fantôme
~-IMAGE FOCUSING - [photo] mise *f* au point par dédoublement de l'image
~-IMAGE RANGE FINDER - [instr] (a range finder in which two images of the target are obtained) télémètre *m* de Barr et Stroud, télémètre *m* à deux images
~INCLINE - [railw] dos *m* d'âne
~-INLET FAN - [th eng] ventilateur *m* à deux ouies
~INSURANCE - [insur] assurance *f* cumulative
~IRRADIATION EXPERIMENT - [nucl] expérience *f* d'irradiation avec deux radiofréquences
~ISOMORPHISM - [phys] isomorphie *f* double
~-JACK FRAME - [text] banc *m* en surfin double
~-JAW SPANNER - [impl] clef *f* à double machoire
~-JET TURBINE - [hydr] turbine *f* à deux jets
~-JOINTED ARM - s. double-hinged arm
~JUNCTION - [build] (water-pipe with two branches) té *m* (d'embranchement)
~KNIFE EDGE SEAL - [mech] joint *m* à double couteau, joint *m* à arête double
~KNOCK-OVER - [text] abattage *m* double
~LADDER - [impl] échelle *f* double
~-LAYER SCREEN - [telev] écran *m* à deux couches
~-LAYER WINDING - [el] (armature winding in which the coils are arranged in two layers) enroulement *m* à deux couches
~LEASE ROD - [text] baguette *f* d'envergure
~LEATHER BELTING - [mech] courroies *f pl* doubles en cuir
~LEDGER - [comm] grand livre *m* double
~-LENGTH ACCUMULATOR - [comput] accumulateur *m* à double précision, accumulateur *m* à double longueur
~-LENGTH NUMBER - [electron comput] (number having twice as many digits as those which are normally used in a computer) nombre *m* de longueur double

DOUBLE LETTERS - [print] (ligatures and old-face letters) ligature *f*, lettres *fpl* doubles
~-LEVEL BRIDGE - [build] pont *m* à deux niveaux
~LIFT - [mech] came *f* à deux étages
[mining] double travée *f*
~LIFT DOBBY - [text mach] ratière *f* à double lève
~LIFTING ARRANGEMENT - [text mach] double lève *f*
~LIMITER - [el] (two limiters in cascade) limiteur à deux étages
[telev] circuit *m* à déclenchement périodique
~LINKING - [chem] à double liaison
~LOCK - [hydr] écluse *f* à double sas
~LOCKSTITCH - [text] point *m* double
~MODULATION - [modul] (the modulation of a wave by another modulated wave) double modulation *f*
~MORTISE CHISEL - [impl] bédane *m* double
~-MOTION AGITATOR - [mech mach] (a stirring machine with two contra-rotating agitators) agitateur *m* à contre-rotation
~-MOTION PADDLE - [mech] (in mixers) palette *f* à double effet
~NIPPLE - [mech] raccord *m* fileté double
~-NOTCH JOINT - [carp] assemblage *m* à double entailles
~OFFSET EXPANSION U BEND - [mech] joint *m* de dilation à double coude
~OPENER - [text] ouvreuse *f* double
~PARTITION - [build] (a partition with a cavity between two walls) cloison *f* double
~PERSONALITY - [med] dédoublement *m* de la personnalité
~PETTYCOAT INSULATOR - [el] isolateur *m* à double cloche
~PHANTOM CIRCUIT - [telecomm] (in telephony, a circuit using two phantom circuits in parallel) circuit *m* fantôme double, circuit *m* superfantôme
~PICA - [print] corps *m* 24 (points anglais)
~PICK - [impl] pic *m* à deux pointes
[text] duite *f* double
~PITCH - [constr work] à deux versants
~-PITCH ROOF - [constr] comble *m* à deux égouts (ou à deux pentes)
~-PITCH SKYLIGHT - [build] (a skylight with two glazed surfaces) verrière *f* à deux versants
~PLANE-IRON - [impl] fer *m* de rabot double
~-PLATED - [naut] à double bordé
~PLUSH - [text] peluche *f* double
~POINTED PICK - [impl] pioche *f* à deux pointes, pic *m* à deux pointes
~POLE - [el] (acting on two poles) bipolaire, à deux pôles
~POLE BATTERY SWITCH - [telecomm] réducteur *m* double de batterie
~-POLE DOUBLE THROW - [el] bipolaire *m* à deux directions
~-POLE SWITCH - [el] (a switch in which the circuit is made or broken at two poles simultaneously) interrupteur *m* bipolaire
~PRECISION - [electron comput] (the retention of twice the number of digits normally handled by the computer) double précision *f*
~-PRECISION NUMBER - [comput] nombre *m* de precision double, nombre *m* de double longueur
~PRIME - [math] dérivée *f* seconde
~PULLEY - [mech] poulie *f* double, rouleau *m* double
~PUNCH - [comput] (two perforations in the same column) double perforation *f*
DOUBLE-PURCHASE JACK - [mech] cric *m* à double engrenage
~-PURPOSE VALVE - [radio] (a valve which is used for two purposes at the same time) tube *m* à double fonction
~QUOTES - [print] (inverted commas) guillemets *mpl*
~RACK - [mech] crémaillière *f* double
~RAM PRESS - [mech] (a press provided with two rams in separate cylinders) presse *f* à deux coulisseaux
~RANGE - [meas] à deux lectures, à deux sensibilités
~REACTION - [radio] (obtained by combining inductive and capacity coupling) double réaction *f*
~-READING THEODOLITE - [surv] (a theodolite which makes it possible to observe at once the readings of horizontal and vertical circles) théodolite *m* à lecture double
~RECEPTION - [radio] (reception of signals on different wavelengths by two receivers connected to the same aerial) - réception *f* double
~RECORDING SYSTEM - [acoust] enregistrement *m* double piste
~REDUCTION SPEED REDUCER - [mech] réducteur *m* de vitesse à double réduction
~REED - [text] peigne *m* double
~REEL - [text] dévidoir *m* double
~REFLECTION - [opt] double réflexion *f*
~REFRACTION - [phys] (property common to several crystalline substances) biréfringence, double refraction *f*
~REGENERATION - s. double feedback
~RESISTANCE BOX - [el] (box holding two identical sets of resistors, making it possible to introduce simultaneously two resistors of equal value) boîte *f* de résistances double
~RIB - [text] côte *f* anglaise
~RIB FRAME - [text] métier *m* double pointure
~RIB LOOM - [text] métier *m* Rachel
~RIB WARP FRAME - [text] métier *m* Rachel, métier *m* à chaîne de retenue
~RIVETED - [mech] à rivetage double
~-RIVETED JOINT - [mech] rivure *f* double, rivure *f* à deux rangs de rivets
~-RIVETED LAP JOINT - [mech] rivetage *m* à recouvrement à tête affleurée
~RIVETING - [mech] rivetage *m* double
~-ROLLED CATHEDRAL GLASS - [glass man] verre *m* cathédrale double
~-ROTATION RADIAL TURBINE - [mech] turbine *f* à écoulement radial à double rotation
~-ROW BALL BEARING - [mech] (type of bearing in which there are two circles of balls) roulement *m* à deux rangées de billes, roulement *m* double
~-ROW ENGINE - [mech] moteur *m* en double étoile
~SALT - [chem] sel *m* double
~-SCREW CONVEYOR - [mech] transporteur *m* à vis double
~-SCREW EXTRUDER - [mech] boudineuse *f* à deux vis
~SCULL - [naut] nager à couple
~SEAM - [packag] double couture *f*, double agrafage *m*
~-SEAT VALVE - [mech] soupape *f* à double siège
~SERIES - [math] série *f* double
~-SHAFT KNEADER WITH HELICALLY ARRANGED BLADES - [mech] malaxeur *m* à deux brasseurs héli-

coïdaux

DOUBLE SHAPING MACHINE - [mach tool] étau-limeur *m* double

~SHARP - [mus] double dièse *m*

~SHEAR - [mech] (to test the strength of materials) double cisaillement *m*

~SHEAR STEEL - [metall] acier *m* corroyé deux fois

~SHED - [text] pas *m* double

~SHED DOBBY - [text] ratière *f* à double pas

~SHED INSULATOR - s. double pettycoat insulator

~-SHEET BEND - [naut] nœd *m* d'écoute double

~-SHOT MOULDING - [metall & plast ind] (process for making two-colour products by successive moulding operations) moulage *m* par double injection

~SHOVEL PLOUGH - [agric] charrue *f* à deux corps, charrue *f* à deux socs

~-SHUNT FIELD COIL - [el] bobinage *m* de champ à double circuit dérivé

~SHUTTLE - [text] navette *f* double

~SIDE-TIPPING WAGON - [mining] wagonnet *m* à double bascule

~-SIDEBAND TRANSMISSION - [radio] émission *f* sur deux bandes latérales

~-SIDEBAND TRANSMITTER - [radio] émetteur *m* sur deux bandes latérales

~-SIDED GOODS - [text] tricot *m* double face

~-SIDED PLUSH - [text] peluche *f* double face

~-SIDED WINDING FRAME - [text] bobinoir *m* double face

~-SILK COVERED WIRE - [el] (wire insulated with two layers of silk) fil *m* à guipage deux couches soie

~SLIDING DOOR - [auto] (in autobuses) porte *f* coulissante double

~SLIT - [opt] (two parallel openings used in certain diffraction experiments) fente *f* double

~SPINDLE - [text] broche *f* double

~-SPOT TUNING - [radio] (the reception of a specified station by a superheterodyne receiver at two different dial settings which correspond to two local oscillator frequencies) accord *m* sur deux réglages

~SQUIRREL-CAGE MAGNETRON - s. donutron

~SQUIRREL-CAGE MOTOR - [el] (induction motor with a rotor comprising two cages of different resistances) moteur *m* à double cage d'écureuil, moteur *m* à cage d'écureuil de Boucherot

~SQUIRREL-CAGE WINDING - [el] (winding consisting of two concentric squirrel-cages) enroulement *m* à cage d'écureuil de Boucherot, enroulement *m* à double cage d'écureuil

~-STAGED - [gen] à deux étages

~STARS - [astr] (two stars appearing very close together) étoiles *f*pl doubles

~STEP - [text] double pas *m*

~-STEP RELAY - [el] relais *m* à double effet, relais *m* à action échelonnée

~STITCH - [text] maille *f* double, cueillage *m*

~STOPPING - [mus] (the simultaneous sounding of two notes on a string instrument) double corde *f*

~-STRAPPED BLOCK - [mech] (of a tackle) moufle *f* à estrope double

~-STREAM AMPLIFIER - [radio] (travelling-wave amplifier in which the amplification is the result of the interaction of two electron beams with different average velocities) amplificateur *m* à deux faisceaux

~STUD - [metall] support *m* double à platine

DOUBLE SUCTION IMPELLER - [mech] roue *f* à aube à double aspiration

~-SUCTION PUMP - [mech] pompe *f* à double aspiration

~SUPERHETERODYNE RECEPTION - [radio] réception *f* à double changement de fréquence

~SWING-FRONT - [photo] double bascule *f* à l'avant

~TACKLE - [mech] palan *m* double

~TAKE - [cin] prise *f* de vue double

~TEEM - [metall] (a casting defect) repli *m*, goutte *f* froide

~TENON - [constr] double tenon *m*, double dent *f*

~TENSION DISK - [text] disque *m* double de tension

~THREAD - [mech] (of a screw) filet *m* double

~-THREAD HOB - [mach tool] fraise-mère *f* combinée

~-THREADED SCREW - [mech] (term sometimes used for a two-start thread) vis *f* à deux filets, filet *m* double

~-THROAT BURNER - s. double burner

~-THROW SWITCH - [el] (a two-way switch) commutateur *m* à deux directions

~-THRUST BEARING - [mech] (thrust bearing taking axial thrust in both directions) palier *m* à double butée

~TIER NET MACHINE - [text] métier *m* à tulle à deux rangs

~TIMBER - [mining] cadre *m* ordinaire

~TIMBERING - [mining] boisage *m* armé, longrinage *m*

~-TONE INK - [print] encre *f* double-ton

~TRACK - [acoust] double piste *f*

~-TRACK CAGE - [mining] cage *f* à double voie

~-TRACK HAULAGE ROAD - [mining] galerie *f* de roulage à double voie

~TRACK LINE - [railw] ligne *f* à deux voies

~-TRACK TAPE - [el acoust & telev] (magnetic record for video recording etc) enregistrement *m* bipiste

~-TROLLEY SYSTEM - [el] (electric traction) système *m* à double trolley

~-TUNED AMPLIFIER - [radio] amplificateur *m* à accord double

~-TUNED CIRCUIT - [electron] (circuit resonant to two adjacent frequencies) circuit *m* à double syntonisation

~TURN - [plumb] (a length of pipe bent into a loop and inserted in a pipeline to provide for expansion) lyre *f* de dilatation

~-TURN LOCK - [mech] serrure *f* à pêne dormant et demi-retour

~TURNOUT - [railw] aiguillage *m* double, embranchement *m* à trois voies

~-TWIST SPINNING FRAME - [text] continu *m* à retordre, métier *m* à retordre

~-TWIST YARN - [text] fil *m* à double torsion, fil *m* retors

~TWISTED AUGER - [mining] mèche *f* torse

~-TWISTED YARN - s. double-twist yarn

~TYRES - [rubber ind] pneumatiques *m*pl jumelés

~UNIVERSAL PROPELLER SHAFT - [auto] arbre *m* de transmission à double cardan

~UP, to - [naut] doubler (les amarres etc)

~VANDYKE - [text] tricot *m* atlas

~VEE BUTT WELD - [metall] (a weld made by chamfering both upper and lower edges in a butt joint) soudure *f* à double chanfrein

DOUBLE VISION - [med] double vision *f*, diplopie *f*
~-WALL COFFERDAM - [hydr] (cofferdam formed with a pair of parallel walls) bâtardeau *m* à double paroi
~-WALLED - [plumb] (of a pipe) à double enveloppe, à double paroi
~WALLED VACUUM VESSEL - [ind chem] recipient *m* à double paroi pour le vide
~WARP BAGGING - [text] toile *f* à sac
~-WAY REGULATING VALVE - [mech] soupape *f* régulatrice à double action
~-WEFT WEAVE - [text] armure *f* à double trame
~-WEIGHING - [chem] double pesée *f*
~-WHEEL LATHE - [mach tool] tour *m* double à roues montées
~WHEELS - [auto] (in heavy vehicles) roues *fpl* jumelées
~WIDTH LOOM - [text] métier *m* large
~WINDOW - [constr] (system of two windows acting as sound and heat insulator) double fenêtre *f*
~-WING DITCHER - [agric] charrue *f* fossoyeuse
~-WING DOOR - [build] porte *f* à deux battants
~-WIRE - [el] à deux fils, bifilaire
~-WIRE SYSTEM - [el] système *m* bifilaire
~-WOUND - [el] bifilaire
~YARN - [text] fil *m* doublé
DOUBLED SILK - [text] mouliné *m* soie
~WARP - [text] chaîne *f* en fil retors
DOUBLER - [text] assembleuse-retordeuse *f*
[text] (a doubling frame) réunisseuse-doubleuse *f*
[radio] (a frequency multiplier in which the output voltage is twice the input frequency) doubleur *m*
~BOBBIN - [text] bobine *f* réceptrice
DOUBLES - [min] (in coal size classification) morceaux *mpl* de calibre 25-5I mm
DOUBLET - [opt] (a pair of lenses used together) objectif *m* formé de deux lentilles collées, doublet *m*
[radio] (half-wave aerial fed with low impedance transmission line) doublet *m*, dipôle *m*
[light] (pair of associated lines in a spectrum) doublet *m*
~SOURCE - [phys] (two point sources, equal in strength but opposite in phase, which are separated by a disappearing distance) source *f* double
DOUBLETREE - [transp] (in a horse-drawn vehicle) volée *f*
DOUBLING - [text] (combining of two or more threads to form a yarn) doublage *m*
[cin] doublage *m*
[metall] doublage *m* (opération de laminage)
~AND ROLLING MACHINE - [text] dosseuse *f* et enrouleuse
~BOBBIN - [text] bobine *f* de doubleuse, bobine *f* d'assembleuse
~CALENDER - [text] cylindre *m* de doublage
~DRAW FRAME - [text] étireuse *f* à doublage
~FRAME - [text] (machine in which yarns are folded or twisted together) réunisseuse-doubleuse *f*
~IRON OF A PLANE - [mech] couvre-ciseau *m*, contre-ciseau *m* d'un rabot, contre-fer *m*
~TIME - [nucl] (the time required by a breeding reactor to double the amount of fuel contained in it) temps *m* de doublement
~TWISTER - [text] assembleuse-retordeuse *f*
~WINDER - [text] bobinoir *m* assembleur, assembleuse *f* à envidage croisé
~WINDING FRAME - [text] moulineuse *f*

DOUBLY-FED POLYPHASE SHUNT COMMUTATOR MOTOR - [el] moteur *m* polyphasé shunt à collecteur à double alimentation
DOUGH - [gen] pâte *f*
~GUN - [rubber ind] seringue *f* à injection de dissolution
~MILL -[mech] malaxeur *m*
~MIXER - s. dough mill
~MOULDING COMPOUND - [plast ind] (mixture of glass fibre with a synthetic resin of a putty-like consistency) résine *f* coulée renforcée de fibres de verre
~-RAISING OVEN - [th eng] coffre *m* à faire lever la pâte
~-SPREADING MACHINE - [gen] métier *m* à enduire, métier *m* à gommer
DOUGHNUT - [electron] (toroidal-shaped vacuum envelope in which the electrons are accelerated) tore *m*
[nucl] (used in thermal reactors as an assembly of fissionable material, to provide a local increase in the fast neutron flux) chambre *f* torique
~COIL - [el] bobinage *m* toroidal
~TYRE - [rubber ind] pneumatique *m* superballon, pneumatique *m* à très basse pression
DOUGHY - [gen] pâteux
DOUGLAS FIR - [bot] sapin *m* de Douglas
DOUGLASITE - [min] douglasite *f*
DOURINE - [vet] dourine *f*, maladie *f* du coït
DOUSE, to - [cin] (to switch off lights) éteindre (les lumières)
[naut] amener, affaler (une voile)
[metall] (to plunge a hot metal into water to anneal it) tremper
DOUSER - [cin] (an automatic screen between the projector arc and the film) volet *m* pare-feu
DOVE - [zool] colombe *f*
~PRISM - [opt] (prism with the property of inverting a beam of light) prisme *m* de Dove, prisme *m* de Wollaston, prisme *m* basculant
DOVETAIL, to - [mech] assembler à queue d'aronde
[cin] (to synchronize two scenes) synchroniser
DOVETAIL - [carp] (a joint comprising one or more tapering tenons which are wider at the tip than at the root, and corresponding recesses or mortises) queue *f* d'aronde
[mech] assemblage *m* à queue d'aronde
~CUTTER - [mach tool] fraise *f* conique à deux tailles
~DADO - [carp] lambris *m* à queue d'aronde
~GROOVE - [mech] (a groove which is widest at the deepest part and tapers towards the surface) rainure *f* à queue d'aronde
~HALVING - [carp] (a form of jointing with parts cut to a dovetail shape) assemblage *m* à queue d'aronde à mibois
~HINGE - [mech] (hinge with leaves increasing outwards from the joint) charnière *f* à queue d'aronde
~JOINT - [naut] assemblage *m* à queue d'aronde
~KEY - [carp] (used to avoid warping of the timber) clavette *f* à queue d'aronde
~SAW - [impl] (a smaller type of tenon saw) scie *f* à dos, scie *f* à raccourcir
~SLIDE - [mech] glissière *f* prismatique
DOVETAILED - [carp & mech] à queue d'aronde
~GROOVE - [mech] rainure *f* à queue d'aronde
DOVETAILING - s. dovetail

DOVETAILING MACHINE - [mech] machine *f* à faire les queues d'aronde
~PLANE - [tool] rabot *m* à languettes
DOWEL, to - [mech] (to locate by means of cylindrical pins entering in the parts concerned) engoujonner
DOWEL - [mech](a pin,fitted into a mould,which enters a hole in the other part, so that when the mould is closed the two parts become accurately aligned) prisonnier *m*, doigt, goujon *m*, ergot *m* de centrage, goujon *m* d'assemblage
[carp] (a wooden pin used for fastening two sections together) cheville *f*, goujon *m*
~BIT - [carp] (wood-boring bit with the cutting section in a conoidal shape) mèche *f* conique (pour trous de chevilles)
~BUSH - [mech] (a hardened steel bush lining a dowel hole) douille *f*, de goujon de centrage
~BUSHING - s. dowel bush
~HOLE - [mech] (the mating recess which receives a dowel-pin to locate one part in reference to another) logement *m* de goujon de centrage, alésage *m*
~PIN - s. dowel
~SLOT - s. dowel hole
DOWELLING - [carp etc] chevillage *m*, assemblage *m* par chevilles
~JIG - [mech] (used for directing the bits in drilling the holes for the dowels) gabarit *m* de perçage
DOWLAIS MILL - [metall] train *m* de laminage double duo
DOWLAS - [text] (a coarse linen fabric used for towelling etc) toile *f* doulens
DOWN, to - [gen] abattre, terrasser
DOWN - [text] (fine soft hairs) duvet *m*, poil *m* follet
[geogr] dune *f*, colline *f*
[gen] en bas, à bas
~-ACTING TYPE PLATEN PRESS - [mech] presse *f* descendante à plateaux
~BY THE STERN - [naut] sur l'arrière
~CARBURATION - [auto] carburation *f* inversée, carburation *f* à tirage descendant
~-DRAUGHT - [aero] courant *m* d'air descendant
~-DRAUGHT CARBURETTOR - [mech] carburateur *m* inversé
~FAULT - [geol] faille *f* normale
~FLOW - [phys] courant *m* de retour
~FOLD - [geol] pli *m* synclinal
~GRADE - [railw] pente *f*, déclivité *f*, descente *f*
~MILLING - [mech] fraisage *m* en avalant
~PIPE - [hydr] tuyau *m* de descente (des eaux pluviales), descente *f*, chute *f*
~QUILT - [text] édredon *m*
~RANGE - [astronaut] domaine *m* aval
~RULE - [print] filet *m* de séparation
~RUNNER - [metall] canal *m* de coulée vertical, descente *f* de coulée
~SAND - [geol] sable *m* dunaire
~SINKER - [text] platine *f* de décrochage, platine *f* de crochetage
~SPRUE - [casting] jet *m* de coulée
~-STRIKE - s. downward movement
~TIME - [gen] (the period of time during which a machine is not used) période *f* d'arrêt
~WARPING - [mining] tassement *m* (d'une couche)
DOWNBLOW - [phys] (term denoting the backdraught of gas) refoulement *m*
DOWNCAST - [mining](colloq for downcast shaft, through which fresh air enters a mine) puits *m* d'entrée d'air, puits *m* d'aérage
DOWNCAST SHAFT - s. downcast
DOWNCOMER - [plumb] (a pipe or duct in which flow is downward or nearly so) tuyau *m* de descente, déversoir *m*, trop-plein *m*, descente *f*
[metall] (in a blast furnace) descente *f*
[sugar ind] tube *m* (ou puits) central
DOWNCOMING WAVE - [radio] (wave travelling along an ionospheric ray) onde *f* atmosphérique
DOWNDRAUGHT - s. down draught
~TUYERES - [metall] (in a blast furnace) jets *mpl* descendants
DOWNFOLD - [geol] pli *m* synclinal
DOWNGATE - [metall] trou *m* de coulée
DOWNGRADE, to - [gen] déclasser
DOWNHAND WELD - [metall] soudure *f* à plat
DOWNHILL - [gen] déclivité *f*, pente *f*, descente *f*, rampe *f* en aval
~POSITION - [surv] position *f* en pente
DOWNLEAD - [radio] (the almost vertical conductor which connects the uppermost part of the aerial with the receiver or transmitter) descente *f* d'antenne, ligne *f* d'alimentation
DOWNPOUR - [met] averse *f*
DOWNRIGHTS - [text] (a colloquial term denoting the lower part of the sides of the fleece) downrights *fpl*, super *f*, laine *f* courte des environ du cou
DOWNS PROCESS - [el chem] (an electrolytic process for the production of chlorine and metallic sodium from fused sodium chloride) procédé *m* Downs
DOWNSPOUT - [plumb] descente *f* d'eaux pluviales, bec *m* de descente, trop-plein *m*, déversoir *m*
DOWNSTAIRS - [gen] en bas (de l'escalier)
DOWNSTREAM - [hydr] (in a fluid system,the region towards which flow is moving) aval, en aval
DOWNSTROKE - [mech] (the downward motion of a piston) course *f* descendante, coup *m* descendant
~MOULDING PRESS - [mech] presse *f* à mouler descendante
~PRESS - [mech mach] (hydraulic press in which the main ram is above the moving table and operates downward) presse *f* descendante
DOWNSTROKING PRESS - s. downstroke press
DOWNWARD - [gen] descendant, de haut en bas, en bas
~BOREHOLE - [min] sondage *m* descendant
~CONVERSION - [telev] conversion *f* à un système à nombre de lignes plus bas
~ENRICHMENT - [min] enrichissement *m* secondaire
~MODULATION - [radio] (modulation in which the instantaneous amplitude of a carrier is less than the unmodulated amplitude of the carrier) modulation *f* négative
~MOVEMENT - [mech] (of the piston) course *f* descendante
~RUN - [mech] course *f* descendante
~TRAVEL - [mech] mouvement *m* de descente
DOWNWARDS - [gen] en descendant, de haut en bas, vers le bas
DOWNWASH - [aero] (the general downward air flow resulting from the combined effects of the vortex sheet following an aerofoil in motion) déflexion *f* (des filets d'air) vers le bas, déviation *f* verticale, courant *m* descendant
~ANGLE - [aero] (the angle through which an airstream is deflected by the passage of an aerofoil,

measured in the plane of symmetry) angle *m* de déflexion descendante
DOWNWASH EFFECT - [aero] effet *m* de déflexion descendante
~VELOCITY - [aero] vitesse *f* de déflexion descendante
DOWNWIND - [naut] vent arrière
DOWNY - [gen] duveteux, lanugineux
~HAIR - [zool] (fine soft hair) duvet *m*
DOWSE, to - [gen] (to find underground water by the use of a dowser) employer la baguette de sourcier, faire de l'hydroscopie
DOWSER - [impl] (the twig held by the dowser and indicating the existence of underground water by twitching in his hand) sourcier *m*, hydroscope *m*
DOWSING - [gen] (water divining) sourcellerie *f*, hydroscopie *f*
DOZE, to - [mining] remuer la terre avec un bulldozer
DOZEN - [gen] douzaine *f*
DOZER - [mech mach] (earth-moving machine used in pipelaying) bulldozer *m*
D.R. - [naut] (abbrev for dead reckoning) point *m* estimé
DRACONTIASIS - [med] (the presence of the Guinea worm) dracontiase *f*, dracunculose *f*
DRAFF - [bot] drèche *f*
~SEPARATOR - [agric] centrifugeuse *f* à drèches
DRAFT, to - [gen] rédiger, faire un projet; faire un brouillon, dessiner
DRAFT - [gen] (or draught) courant *m* d'air, appel *m* d'air
[gen] (of a written text) projet *m*, brouillon, minute *f*
[draw] (of a drawing) projet *m*, dessin *m*, esquisse *f*
[phys] (or draught) tirage *m*, dépression *f*
[mech] (the mechanism for the regulation of draft) régulateur *m* de tirage
[naut] (or draught) tirant *m* d'eau
[metall] (in wire drawing) passe *f*
[metall] (in diecasting) dépouille *f* d'un modèle
[plast ind] (the amount of taper in a mould necessary to allow the withdrawal of the moulding) dépouille *f*
[hydr] aire *f* (d'une vanne)
[fin] (order of payment) traite *f*, mandat *m*
[text] étirage *m*, remettage *m*, rentrage *m*
~ANGLE - [metall] (in diecasting) angle *m* de dépouille
~ANIMALS - [zool] bêtes *f*pl de trait, bêtes *f*pl de attelage
~ARTICLES - [leg] (of a company) projet *m* de statuts, project *m* de contrat
~CYLINDER - [text] cylindre *m* étireur
~DIVERTER - [th eng] coupe-tirage *m*
~GAUGE - s. draught gauge
~HEAD - [hydr] hauteur *f* d'aspiration
~HORSE - [transp] cheval *m* de trait
~TUBE - [mech] (in a water-turbine, the discharge pipe to the tail race) diffuseur *m*, cône *m* d'aspiration
[hydr] (a downtake pipe) tube *m* de descente, tuyau *m* de chute
~VENTILATION - [th eng] ventilation *f* forcée
DRAFTER SLIVER - [text] ruban *m* d'étirage
DRAFTING - [gen] projet *m*, dessin *m*, étude *f*
[text] (the arrangement of the warp threads) remettage *m*, rentrage *m*
[text] (in spinning) étirage *m*
DRAFTING CYLINDER - [text] rouleau *m* du banc d'étirage
~GEAR - [text] engrenage *m* d'étirage
~MACHINE - [text] étireuse *f*, laminoir *m*
~ROLLER - [text] cylindre *m* étireur
DRAFTSMAN - s. draughtsman
DRAG, to - [gen] entraîner, traîner, tirer, draguer, frotter sur
[naut] chasser (sur le fond), râcler
DRAG - [aero] (the component of air reaction parallel to the direction of mean fluid movement) traînance *f*, traînée *f*, résistance *f* aérodynamique
[naut] (excess of draft at the stern) résistance *f* hydrodynamique
[metall] (in foundry work; the bottom half of a moulding flask) châssis *m* de dessous
[mining] (dragging bucket of an excavator) benne *f* traînante
[mech] (resistance to rolling) résistance *f* au roulement, entraînement *m* dû aux frottements
[soil] (resistance against the flow) résistance *f* à l'écoulement
[agric] herse *f* lourde
~AERIAL - [radio] (mounted in aircraft on a reel which is unreeled in flight) antenne *f* traînante
~ANCHOR - [naut] ancre *f* flottante
~ANTENNA - s. drag aerial
~AXIS - [aero] (the straight line drawn through the centre of gravity and parallel to the direction of undisturbed flow) axe *m* de traînance
~BIT - [mech] trépan *m* à lame
~COEFFICIENT - [aero] coefficient *m* de traînée
[soil] (resistance coefficient) coefficient *m* de résistance
~COMPONENT - [aero] composante *f* de traînée
~CONVEYOR - [mech] (conveyor specially suited for hot material, in which a chain is made to drag along the bottom of trough, carrying the material forward) transporteur *m* à godets, transporteur *m* à raclettes
~DIRECTION - [phys] (in stress analysis, the direction of the relative wind) direction *f* de la résistance aérodynamique
~EXCAVATOR - [mech] excavatrice *f* à pelle traînante
~HEAD - [text] (of a beam) collier *m* de tension
~HINGE - s. lagging hinge
~IN - [el chem] (the quantity of solution conveyed into a plating bath along with the cathodes) solution *f* adhérente
~-LINE - [aero] barre *f* de traction
~LINK - [mech] (a rod moving the steam motion of a steam engine, thus varying the cut-off) bielle *f* d'accouplement, tringle *f* de relevage
[auto] (a steering rod) barre *f* de direction, bielle *f* d'accouplement
~MARK - [mech] rayure *f*
~-ORE - [min] brèche *f* minéralisée, minerai *m* broyé
~OUT - [el chem] (the quantity of solution withdrawn from a plating bath along with the electrodes) solution *f* entraînée
~PARACHUTE - [aero & astronaut] (used to slow down the speed of an aircraft, space capsule etc) parachute *f* de freinage (ou extracteur)

DRAG PER UNIT OF AREA - [aero] traînance *f*
~PIN TENSION - [text] frein *m* à levier
~PIT - [metall] fosse *f* de tension
~RAKE - [agric] rateau *m* traîné
~ROLLER - [metall] rouleau *m* de tension
~ROPE - [aero] guide rope *m* (ballon)
~SCRUBBER CONVEYOR - s. drag conveyor
~SHOCK STRUT - [aero] amortisseur *m* de traction
~SHOE - [railw] sabot *m* d'enrayage
~SHOVEL - [mech] godet *m* rétro
~STRUT - [aero](structural member of an aircraft designed to resist stresses in the direction of the drag) bielle *f* de rétraction, tirant *m*
~TRUSSING - [aero] croisillonnement *m* de recul
~-TYPE TACHOMETER - [instr] tachymètre *m* à courants de Foucault
~WIRE - [aero] (wire used in an aicraft structure to resist forces acting in the drag direction) câble *m* de traînée, hauban *m* traînée
DRAGENDORFF'S SOLUTION - [chem] (analytical reagent consisting of a solution of bismuth-potassium iodide) solution *f* de Dragendorff
DRAGGING - [plast ind] (damage caused by mould faces moving in contact and scoring each other) rayure *f*
DRAGLINE - s. drag line
~BUCKET - [mech] benne *f* traînante
~EXCAVATOR - [mech] (excavating machine fitted with a bucket which is dragged towards the machine itself) excavatrice *f* à benne traînante
~SCRAPER - [mech] drague *f* à câble, scraper *m*
DRAGNET - [fishing ind] (net which is dragged along the bottom of a river) filet *m* de drague, dragant *m* de fond, dranet *m*
DRAGON BALLOON - [aero] ballon *m* anti-aérien
~FLY - [zool] libellule *f*
~PIECE - [constr] entrait *m* retroussé
~TREE - [bot] (a medicinal plant) dragonnier *m*, dracéna *m*
DRAGON'S BLOOD - [chem] (dark red resin obtained from species of Daemonoraps. It is used as a pigment, and as a protective varnish in photo-engraving) sang-dragon *m*
DRAIN, to - [gen] drainer, épuiser, vidanger, évacuer, faire l'exhaure (mines)
[hydr] (to eliminate water or moisture) assécher, essorer, égoutter, exprimer
[gen] (the operation of emptying) purger, vidanger, décharger
DRAIN - [gen] purge *f*, collecteur *m* de purge, évent *m*, drain *m*, conduite *f* de drainage, fossé *m*, rigole *f*
~COCK - [mech] (a plug cock fitted to a vessel or pipe for drawing off liquid) robinet *m* de purge, robinet *m* purgeur, purgeur *m*
~COVER - [plumb] bouchon *m* de vidange, bouchon *m*
~DITCH - s. drainage ditch
DRAIN INSPECTION POINT - s. drain manhole
~MANHOLE - [constr] regard *m* d'égout
~MANIFOLD - [mech] collecteur *m* de purge
~MOUTH - [hydr] (in a drainage system) bouche *f* d'égout
~NEEDLE SCREW - [auto] vis-pointeau *f* de vidange
~OFF, to - [constr] se déverser, s'évacuer, vider, s'écouler
~OIL - [mech] huile *f* de vidange
~OUTLET - [mech] orifice *m* de vidange
DRAIN PIPE - [plumb] tuyau *m* de vidange, tuyau *m* d'évacuation, tuyau *m* d'écoulement
~PLUG - [plumb] (device designed to close the outlet of drain pipes) bouchon *m* de vidange
~SCREW - [mech] vis *f* de purge
~TAP - [auto] robinet *m* de vidange
~-TRAP - [plumb] (the trap preventing foul air from rising out of sinks, drains, sewers, etc) siphon *m*, purgeur *m*
~TRUNK LINE - [constr] collecteur *m*
~VALVE - [mech] (a valve fitted to a vessel or pipe for drawing off liquid) robinet *m* de purge
~WATER - [hydr] eau *f* d'infiltration, eau *f* de drainage
DRAINABLE OIL - [oil ind] huile *f* exploitable
DRAINAGE - [hydr] drainage *m*, épuisement *m*, évacuation *f* des eaux, assainissement *m*, écoulement *m*
[th eng] (from a boiler etc) purge *f*, purgeage *m*
[gen] (the operation of drying) drainage *m*, assèchement *m*
~ADIT - [mining] galerie *f* d'exhaure, galerie *f* d'écoulement
~AREA - s. drainage basin
~AT THE TOE - [soil] drain *m* de pied, caniveau *m* de pied
~BASIN - [geogr] (also called catchment area; the basin from which water runs towards a river valley) bassin *m* hydrographique, aire *f* de drainage, aire *f* d'alimentation
~BLANKET - [soil] tapis *m* drainant, couche *f* drainante
~BY FROST ACTION - [soil] drainage *m* par le gel
~BY PUMPING - [soil] drainage *m* par pompage, épuisement *m* par pompage
~BY SUCTION - [soil] drainage *m* par suction, drainage *m* par aspiration
~CONDUIT - [hydr] conduit *m* d'écoulement
~DITCH - [constr] fossé *m* de drainage, saignée *f* (accotement de route)
~LAYER - s. drainage blanket
~OF SILT - [soil] drainage *m* du silt
~PATTERN - [geogr] réseau *m* hydrographique
~PIT - [constr] (a lined recess in floor or foundation to receive liquid draining off from a machine or process unit) puisard *m*
~PLANT - [soil] (dewatering plant) installation *f* d'épuisement
~PUMP - [mech] pompe *f* d'épuisement, pompe *f* d'exhaure
~SYSTEM - [constr] réseau *m* d'évacuation des eaux, réseau *m* d'égouts
~TERRACE - [soil] (channel for removing surplus water) fossé *m* de colature, terrasse *f* de colature
~TRENCHING - [agric] ouverture *f* de creusement
~WELL - [hydr] puisard *m*, puits *m* de drainage, puits *m* drainant
DRAINER - [paper man] égouttoir *m*, tour *f* d'égouttage
DRAINING - [hydr] drainage *m*, écoulement *m*
[agric] assainissement *m*
~BOARD - [impl] égouttoir *m*
~CANAL - [agric] canal *m* d'assainissement
~DISH - [oil ind] passoire *f*, vase *m* à égoutter
~-ENGINE - [mining] machine *f* d'épuisement, pompe *f* d'épuisement
~SCREEN - [min] tamis *m* d'égouttage
~SHAFT - [mining] puits *m* de drainage

DRAINING TABLE - [ind chem] égouttoir *m*
~TRANSFORMER - [el] (a transformer in which the primary and secondary windings are connected in series with an insulated line and an earthed return for electric traction, thus decreasing the flow of current through the ground) transformateur *m* suceur
DRAINWAY - [mining] galerie *f* de drainage
DRAKE - [zool] canard *m* (mâle)
DRAM - [meas] (equal to 3.888 g) dragme *f* (syst. avoir-dupoids =I,77I84 g, pharmacie = 3,888 g)
DRANK - [bot] (a weed) ivraie *f*, zizanie *f*
DRAPE, to -[gen] draper, tendre
[plast ind] (to lay or hang plastic sheet over a mould, into which it is then blown or sucked) draper
DRAPE - [text] tenture *f*, rideau *m*
~ASSIST FRAME - [plast ind] (a form used in vacuum) drape moulding) chassis *m* à former sous vide
~ASSISTS - [plast ind] (narrow rods or bars used to force the sheet into place in drape forming) baguettes *f*pl de fixation
~FORMING - [plast ind] (a special variant of vacuum forming, in which the sheet material is shaped approximately before entering the mould) drapage *m*, mise *f* en forme par aspiration-emboutissage
~MOULDING - [plast ind] (moulding by draping) drapage *m*
DRAPERY - [text] draperie *f*
~PANEL - [constr] panneau *m* à étoffes pliées
DRAUGHT - s. draft
~BEER - [brew ind] (beer drawn from the cask, as distinct from bottled beer) bière *f* à la pression
~DIVERTER - [mech] coupe-tirage *m*
~EXCLUDER - s. draught tubing
~GAUGE - [instr] (instrument to indicate the draught in a stack or flue) indicateur *m* de tirage, déprimomètre *m*
~SCREEN - [constr] coupe-air *m*
~SLOT - [constr] feuillure *f* coupe-air
~STABILIZER - [mech] stabilisateur *m* de tirage, régulateur *m* de dépression
~TUBING - [rubber ind] (rubber or plastics cord fitted to windows to eliminate draughts) bourrelet *m* isolant, joint *m* étanche, brise-bise *m*
DRAUGHTING UP - [gas ind] admission *f* d'air (pendant la marche en veilleuse)
DRAUGHTSMAN - [gen] dessinateur *m*
DRAUK - s. drank
DRAW, to - [gen] tirer, étirer, emboutir, entraîner aspirer, traîner
[draw] dessiner, tracer
[text] (in cotton spinning) étirer
[metall] (to produce wire or to smooth out the surface of rods) tréfiler
[metall] (in die casting, to remove the pattern from a mould) démouler
[ind chem] extraire, soutirer, prélever
[ceramics] (to take out the dried contents of a kiln) défourner
[oil ind] aspirer
[mining] extraire
[naut] porter (voile)
[naut] caler, s'enfoncer de
[comm] (to cash money) tirer, retirer
DRAW - [gen] extraction *f*, prélèvement *m*, tirage *m*
[constr] tablier *m* mobile
[metall & plast ind] (the amount of taper allowed for the purpose of facilitating the extraction of a moulding from a mould) dépouille *f*
DRAW A DASH LINE, to - [print] hachurer
~A DOTTED LINE, to - [print] pointiller, tracer une ligne pointillée
~A PILE, to - [constr] arracher un poteau
~AFT, to - [naut] adonner (vent)
~AND BUFFING GEAR - [railw] appareil *m* de choc et de traction
~BACK, to - [comm] retirer
[gen] revenir en arrière, se retirer, reculer
~-BACK - [gen] inconvénient *m*, désavantage *m*, obstacle *m*
[metall] (diecasting) pièce *f*, battue, motte *f*
[comm] drawback *m*, détaxe *f*
~-BACK CORE - [metall] pièce *f* rapportée
~-BACK RAM - [mech] (hydraulic ram used to withdraw a press jaw or the like after the working stroke) bélier *m* de retour, vérin *m* de retour
~-BACK SPRING - [mech] ressort *m* de rappel
~-BAR - [mech] (the bar transmitting the traction effort) barre *f* de traction, barre *f* d'attelage
~-BAR CRADLE - [mech] (in tractors) support *m* de barre de traction
~-BAR GUIDE - s. draw-bar cradle
~-BAR HITCH - [mech] (in tractors etc) crochet *m* d'attelage
~-BAR HOOK - s. draw-bar hitch
~-BAR H.P. - [mech] puissance *f* à la barre
~-BAR LOAD - [mech] charge *f* à la barre
~-BAR PULL - [mech] (the effort exerted by a locomotive pulling a train) effort *m* de traction à la barre, traction *f* à la barre
~-BAR SAFETY PIN - [mech] attache *f*, broche *f*
~-BAR SPRING - [mech] (shock absorbing between the draw-bar and the frame of a railway carriage) ressort *m* amortisseur
~BENCH - [mech mach] banc *m* d'étirage
~-BOLT - [mech] tire-fond *m*, boulon *m* de serrage
~BOX - [text] (fluted rollers between the doffer and the coiler of a carding engine) tête *f* d'étirage, boîte *f* d'étirage
~BOY - [text] tireur *m* de lacs
~-BRIDGE - [constr] pont *m* mobile, pont-levis *m*
~CAM SHAFT BEARING - [mech] coussinet *m* de l'arbre de cueillage
~CORE - [metall] (in moulding) noyau *m* rétractable
~-DOOR WEIR - [hydr] (weir fitted with gates which can be raised vertically) barrage *m* à vannes verticales
~DOWN, to - [metall] écrouir
~-DOWN - [hydr] abaissement *m* de la nappe
[plast ind] (reduction in diameter of an extruded section caused by excessive haul-off) étranglement *m* dû au tréfilage
~-DOWN TEST - [mining] essai *m* d'épuisement de puits
~-DOWN WATER - [hydr] eau *f* à écoulement forcée
~FILING - [metall] limage *m* en long
~FORWARD, to - [naut] refuser (vent)
~FULL SIZE, to - [draw] dessiner grandeur nature
~-GATE - [mech] (the valve controlling a sluice) vanne *f* de commande
~-GEAR - [mech] appareil *m* de traction, attelage *m*
~-GEAR HEAD - [mech] tête *f* d'attelage

DRAW-HOOK PIN - [railw] crochet *m* d'attelage
~-HOOK TRACTIVE EFFORT - [mech] effort *m* de traction au crochet, traction *f* au crochet
~IN, to - [gen] aspirer
~IN - [constr] (a lay-by) évitement *m*, parc *m* à voitures (le long d'une route)
~KNIFE - [carp] (a blade with a handle on each side) plane *f*
~-KNOB - [mus] (a knob on the manuals operating the stop mechanism) registre *m*
~-MARKS - [metall] rayures *fpl* d'étirage
~MOULDING MACHINE - [metall] (in diecasting) machine *f* à démouler
~OFF, to - [gen] retirer, soutirer, faire écouler
[chem] extraire, soutirer
~OFF - [text] tirage *m*
~-OFF BRIDGE - [text] glissière *f* de guidage, barre *f* de guidage
~-OFF CARD ROLLER - [text] cylindre *m* de carde détacheur
~-OFF REEL - [text] rouloir *m*
~-OFF ROLL - [mech] (a roller used to draw processed material away from the delivery of the machine, e.g. in extrusion) rouleau *m* de tirage
~-OFF TAP - [hydr] robinet *m* de vidange, purgeur *m*
~-OFF VALVE - [mech] (a domestic type of draw-off tap) robinet *m* de puisage
~ON PINS, to - [metall] démouler sur chandelles
~ON AN APRON, to - [metall] démouler sur cadre
~OUT - [gen] débrochable, amovible
~-OUT METAL CLAD SWITCHGEAR - [el] (a metal-lined switchgear with the switch isolated from the bus-bars) disjoncteur *m* blindé amovible
~PLATE - [mech] (plate for the support of the dies) filière *f*
~RAIL - [text] tirant *m*
~ROLLS - [plast ind] rouleaux *mpl* de tirage
~ROPE - [text] câble *m* tracteur, corde *f* de tirage
~SHUTTER - [constr] volet *m* à coulisse
~SLIDE - [photo] volet *m*, tiroir *m*
~SNAP - [impl] chasse-tôle *m*
~SPEED - [mech] vitesse *f* de tréfilage
~SPINDLE BEAM - [text] travée *f* du fuseau-tirant
~SPRING - [mech] ressort *m* de traction
~-STOP - s. draw-knob
~-STROKE - [metall] (in moulding) course *f* de démoulage
~THE ROVINGS, to - [text] étirer les mèches
~THE WARP THREADS THROUGH THE EYES, to-[text] remettre, rentrer
~THE WARP THREADS TOGETHER, to-[text] resserrer les fils de chaîne
~THREAD - [text] fil *m* de séparation (bonneterie)
~TO SCALE, to - [draw] dessiner à l'échelle
~TONGS - [impl] (in telecommunication work) étau *m* tendeur, étau *m* tenseur, mâchoires *fpl* à tendre
~TUBE - [mech] tube *m* télescopique, tube *m* coulissant
~-TUBE TELESCOPE - [surv] (a type of telescope in which the eye-piece and the lens are fitted in separate tubes) lunette *f* à tubes séparés
~-TWISTER - [text] étireuse-retordeuse *f*
~TWISTING - [text] retordage *m* étiré
~UP, to - [gen] lever, relever, rédiger (un document)
~VICE - s. draw tongs
DRAWBAR - [metall] picot *m*

DRAWEE - [comm] payeur *m*, tiré *m*
DRAWER - [carp] tiroir *m*
[comm] tireur *m*, souscripteur *m*
[draw] dessinateur *m*
[metall] tréfileur *m*, étireur *m*
~PULL - [carp] poignée *f* de tiroir
~SLIP - [carp] tasseau *m* de tiroir
~TUNER - [telev] syntonisateur *m* à tiroir
DRAWING - [gen] dessin *m*, schéma *m*, tracé *m*
[text] (in cotton spinning, the running together of a number of slivers prior to spinning) étirage *m*
[metall] emboutissage *m*
[metall] (in diecasting; the removal of the mould) démoulage *m*
[metall] (forging operation) étirage *m*, laminage *m*
[plast ind] (the operation of stretching plastic material to increase its length or area) étirage *m*
[hydr etc] puisage *m*
~BACK - [mining] abattage *m*
~BENCH - [metall] banc *m* d'étirage
~BOARD - [impl] planche *f* à dessin, planchette *f*
~BOX - s. draw box
~CAGE - [mining] cage *f* d'extraction
~COMPASSES - [impl] compas *m*
~DEPTH - [metall] profondeur *f* d'emboutissage
~DIE - [metall] filière *f* d'étirage, matrice *f* d'emboutissage
~ENGINE - [mining] machine *f* d'extraction
~FRAME - [text] banc *m* d'étirage, laminoir *m*
~FRAME ROLLER - [text] cylindre *m* étireur
~FURNACE - [metall] (for the heating of steel below the critical point) four *m* de revenu
~HEAD - [text] tête *f* d'étirage
~-IN - [text] (in weaving, the placing of the warp threads in the eyes of the healds) rentrage *m*, remettage *m*
~-IN APPLIANCE - [text] mécanisme *m* de rentrage
~-IN DRAFT - [text] graphique *m* de rentrage
~-IN FRAME - [text] chevalet *m* de rentrage, chevalet *m* de remettage
~-IN HOOK - [text] crochet *m*, passette *f*
~-IN MACHINE - [text] rentreuse *f* automatique, machine *f* à rentrer les fils de chaîne, machine *f* à remettre
~MILL - [metall] tréfilerie *f*
~OFF THE ORE - [min] défournement *m* du minerai
~-OUT - [mining] extraction *f*
~PAPER - [draw] papier *m* à dessin
~PASS - [metall] cylindre *m* étireur
~PEN - [impl] tire-ligne *m*
~PENCIL - [impl] crayon *m* à dessin
~PIN - [impl] punaise *f*
~PRESS - [metall] presse *f* à emboutir, presse *f* à étirer
~PUNCH - [metall] matrice *f* de poinçonnage, poinçon *m* à tirer
~RULE - [impl] double décimètre *m*
~SAIL - [naut] voile *f* pleine, voile *f* qui porte bien
~SHAFT - [hydr] puits *m* d'extraction
~TEMPER - [metall] (the tempering of hardened steel by heating and quenching) revenu *m* (traitement thermique)
~THE WARP FORWARD - [text] avance *f* de la chaîne
~TIMBER - [mining] (the removal of props from an area) abattage *m*
~TOOL - [metall] outil *m* à étirer
~TRACER - [draw] papier *m* calque

DRAWN - [gen] tiré, étiré, embouti, extrait, épuisé
[draw] dessiné, tracé
˜BAR - [metall] barre *f* étirée
˜FROM THE WOOD - [brew ind] (term used of beer etc)drawn directly from the cask, not bottled) venant du tonneau
˜HALF - ROUND - [metall] fer *m* étiré demi-rond
˜JUNCTION - [electron] jonction *f* par tirage
˜ORE - [min] minerai *m* défourné
˜-OUT IRON - [metall] fer *m* étiré
˜-OUT PIPE - [metall] tube *m* étiré
˜-OUT WAVE - [radio] onde *f* étalée
˜STEEL - [metall] acier *m* étiré
˜WIRE - [metall] fil *m* étiré
˜-WIRE FILAMENT - [el] filament *m* à fil tiré
DRAWPLATE - s. draw plate
DRAWSLATE - [min] faux toit *m*
DRAWSPRING - s. draw spring
DRAWTONGS - s. draw tongs
DRAY - [transp] camion *m*, fardier *m*, haquet *m*
˜-HORSE - [zool] cheval *m* de charrette, cheval *m* de roulage
DREAM - [gen] rêve *m*
˜STATE - [med] état *m* oniroïde
DREAMY STATE -[med] dreamy state *m*, crise *f* unciforme
DREDGE, to - [constr] (to excavate under water) draguer, dévaser, débourber
DREDGE - [mech] (a machine designed to excavate under water) drague *f*
[min] (a large barge or raft with chain buckets or suction pumps mounted on it to operate in alluvial deposits for gold, tin, platinum, precious stones, etc) drague *f*, drague *f* suceuse, bateau-dragueur *m*
[oceanology] (a type of bag-net to study the fauna of deep sea bottom) drague *f*, dragant *m* de fond
DREDGED EARTH - [constr] terre *f* de draguage *m*
DREDGER - [shipbuild] (special type of vessel used for dredging) drague *f*, bateau-dragueur *m*
˜BELT - [mech] courroie *f* de drague
˜EXCAVATOR - [mech] (a type of bucket-ladder dredger working on land) excavateur *m* à godets
DREDGING - [gen] dragage *m*
˜BUCKET - [mech] godet *m* de drague
˜CHAIN CONVEYOR - [mech] transporteur *m* à chaîne à godets
˜HOSE - [impl] tuyau *m* pour drague
˜MACHINE - [mech] drague *f*
˜PLANT - [hydr] installation *f* de dragage
˜SCOOP - [mech] cuillère *f* de drague
˜SHOVEL - [mech] drague *f* à pelle
˜WORK - [constr] dragage *m*, dragages *mpl*, draguage *m*, draguages *mpl*
DREDGY ORE - [geol] roche *f* avec filonnets de minerai
DREG - [geol] sédiment *m*
DREGS - [gen] lie *f*, sédiment *m*, rebut *m*
[chem] (solid material deposited from suspension, especially in brewing) dépôts *mpl*
DRENCH, to - [gen] tremper, mouiller
DRENCHING - [leather ind] (the removal of the last traces of lime light skins) trempage *m*
DRENCHPROOF - [gen] imperméable (à l'eau)
DREPANOCYTE - [cytol] (a sickle cell) drépanocyte *m*, cellule *f* falciforme

DRESS, to - [gen] habiller, vêtir, se vêtir
(to arrange or prepare) disposer, préparer, traiter, orner
[agric] (a vinyard) tailler, émonder, élaguer
[agric] (to sift flour etc) tamiser
[naut] (to dress a ship) pavoiser
[text] (to prepare the warp for the loom) parer, encoller
[text] (to size the yarn) apprêter
[text] (to size the cloth) apprêter
[text] (of linen) apprêter
[mech] (to grind a wheel) dresser
[mech] (to sharpen and set the teeth of a saw) affûter
[metall] (to smooth out the surface of an ingot) nettoyer, sabler
[metall] (the surface prior to casting) enduire (le moule), poteyer
[leather ind] tanner
[constr] (a stone) tailler, parer
[plumb] (to flatten out sheet lead with a dresser) aplanir
[arch] (to apply ornamental features) orner, décorer
[min] (to treat ore) traiter
[mining] (coal) conditionner, préparer
DRESS - [gen] habillement *m*, habit *m*, vêtement *m*, mise *f*, uniforme *m*, toilette *f*
˜BOOTS - [shoe man] bottines *fpl* vernies
˜-CIRCLE - [theatre] premier balcon *m*, corbeille *f*
˜COAL, to - [min] préparer le charbon, conditionner le charbon
˜FORM - [impl] mannequin *m*
˜GOODS - [text] étoffe *f* pour robes et vêtements
˜-GUARD - [impl] (mounted on bicycles) filet *m* gardejupe
˜JACKET - [gen] habit *m*, frac *m*
˜LINEN - [text] (linen fabric used for curtains and dresses) toile *f* pour vêtements
˜MAKING FORM - s. dress form
˜REHEARSAL - [gen] répétition *f* générale
˜-SHIELD - [impl] (in rubber, nylon, etc) dessous *m* de bras
˜SHIP, to - [naut] pavoiser , arborer le grand pavois
˜TRIMMING - [text] galon *m*, garniture *f*, passementerie *f*
DRESSED - [gen] habillé, vêtu
˜-CEILING BOARDING - [build] plafond *m* de plâtre
˜ORE - [min] minerai *m* préparé
˜STONE - [constr] pierre *f* épincée
˜STUFF - [timber] (timber worked to shape) bois *m* équarri
DRESSING - [gen] habillement *m*, vêtement *m*
[gen] (the completion of arrangements) préparation *f*, décoration *f*
[agric] (the sifting of flour etc) tamisage *m*
[agric] fumage *m*, parage *m*
[agric] (vines, hops) taille *f*, émondage *m*, élagage *m*
[med] pansement *m*
[text] (the preparation of the yarn for the loom) apprêt *m*
(the sizing of cloth, yarn, etc) apprêt *m*, encollage *m*
[mech] (the grinding of a wheel) dressage *m*
[metall] (the smooting out of the surface of an in-

got) parage *m*, sablage *m*
[metall] (in diecasting) poteyage *m*
[leather ind] tannage *m*
[text] (furs) apprêt *m*
[constr] (of stones) taille *f*, épinçage *m*
[min] (the treatment of ore) préparation *f*
DRESSING AGENT - [text] apprêt *m*
~-CASE - [impl] nécessaire *m* de voyage, trousse *f* de voyage
~COAL - [min] traitement *m* du charbon, conditionnement *m* du charbon
~GOWN - [text] robe *f* de chambre, saut de lit *m*, peignoir *m*
~JACKET - [text] peignoir *m* de coiffeur
~MACHINE - [agric] (for barley) machine *f* à nettoyer (le grain)
[agric] (for hops) pulvérisateur *m*
[text] encolleuse *f*
~-OFF -[metall] (in diecasting, the process of removing the runners, cores, and risers from a casting) ébavurage *m*, parachèvement *m*
~POWDER - [chem] (used to protect plants from parasites) poudre *f* antiparasitaire
~STATION - [med] poste *m* de secours
DRIBBLE, to - [gen] dégoutter, égoutter, tomber goutte à goutte, baver
DRIBBLE - [gen] bave *f*
(a leakage) fuite *f*, suintement *m*
DRIED GRAINS - [agric] grains *mpl* séchés
~LATEX RUBBER - [rubber ind] caoutchouc *m* de latex total
~PULP - [sugar ind] pulpes *fpl* séchées
DRIER - [gen] séchoir *m*, dessiccateur *m*, sécheuse *f*
[text] séchoir *m*
[ind chem] (in paints etc) siccatif *m*
[paper man] séchoir *m*, cylindre *m* sécheur
~GLAZER - [mech] sécheuse-glaceuse *f*
DRIERS - [chem] (compounds of cobalt, lead, manganese, cerium, zinc, and iron added to paints, varnishes, etc. to accelerate drying) siccatifs *mpl*
DRIFT, to - [gen & naut] chasser, pousser, entraîner, aller à la dérive, dériver
[aero] dériver
DRIFT - [gen] tendance *f*, impulsion *f*, déviation *f*
[naut & aero] dérive *f*
[mech] (a tapered steel bar) chasse-cône *m*, chasse-clavette *m*, calibre *m* passe-tube
[constr] (of tunnels) avancement *m*
[electron] (gradual slight change in the characteristics of an instrument or circuit element, giving rise to error) dérive *f*
[nucl] variation *f* lente et progressive, dérive *f*
[geol] (superficial formations of the earth's crust) dépôt *m* erratique, dépôt *m* glaciaire
[mining] (in metal mines, underground tunnel for exploration) galerie *f* d'avancement, galerie *f* fendue
[metall] (creep, continuous deformation by heavy load) fluage *m*, allongement *m* visqueux
[el acoust] (a slow variation of the speed of the recording medium as it passes the recording transducers) dérive *f*
[instr] (the deviation of the indication with time) dérive *f*
[aero etc] (the lateral element of the movement of an aircraft or a ship due to a transverse current) dérive *f*
DRIFT ANGLE - [aero] (the angle made by the longitudinal axis of an aircraft and its direction on the ground) angle *m* de dérive
~-BAND - [mining] intercalation *f* stérile
~BOULDER - [geol] bloc *m* erratique
~COMPUTER - [aero] (air-borne computer used to determine drift) calculateur *m* de dérive
~CORRECTION - [instr] (the correction angle applie to a course to obtain heading) correction *f* de dérive
~CURRENT - [met] (current occurring in the oceans and produced by wind) courant *m* de surface
~DEPOSIT - [geol] dépôt *m* glaciaire
~ENERGY - [electron] (the energy of mobility of ions and electrons) énergie *f* de mobilité
~HAMMER - [railw] marteau *m* chasse-clou
~INDICATOR - [instr] (instrument used to show the amount of drift) dérivomètre *m*
~METER - s. drift indicator
~MINE - [mining] mine *f* à puits
~MINING - [mining] exploitation *f* à flanc de coteau
~MOBILITY - [electron] (the average drift velocity of carriers per unit electric field) mobilité *f* moyenne
~PIN - [mech] broche *f* d'assemblage, chasse-goupille *m*
~PLUG - [plumb] bouchon *m* tronconique
~RECORDER - [instr] (DR) (a recording driftmeter) dérivomètre *m* enregistreur
~SIGHT - [instr] (optical device to show drift by observation of points on the groud) dérivomètre *m* optique
~SPACE - [electron] (the space through which the beam passes between the two resonators and in which velocity modulation is changed to density modulation) espace *m* de glissement
~SPEED - s. drift velocity
~STOPING - [mining] abattage *m* en taille chassante
~TERRACE - [geol] terrasse *f* d'accumulation
~TEST - [plumb] essai *m* de mandrinage
~TUNNEL - [electron] (a piece of metal tubing which is held at a fixed potential and forms the drift space) tunnel *m* de glissement
~VELOCITY - [electron] (the mean velocity at which ions move under the influence of an electric field) vitesse *f* moyenne de déplacement
~VELOCITY OF CHARGE - [electron] (the average velocity of the individual charge in a group) vitesse *f* moyenne de déplacement
~WOOD - [timber] bois *m* flottant, bois *m* flotté
DRIFTER - [naut] (a fishing boat using drift-nets for fishing near the surface) drifter *m*
[mining] (a compressed-air rock drill) perforateur *m* d'avancement
DRIFTING - [gen] à la dérive
[mining] percement *m* de galeries, extraction *f*
DRIFTLESS - [adj] sans dérive
[el] à haute stabilité
DRIFTPIN - s. drift pin
DRIFTSAND - [geol] sable *m* mouvant, vent *m* de sable
DRILL, to - [méch] (to pierce by means of a rotating tool) forer, perforer, percer
[agric] semer en lignes
[mining] forer, sonder
[gen] (to train) instruire, faire faire l'exercice
DRILL - [tool] (a revolving tool with its cutting ed-

ges at one end) foret *m*, mèche *f*
[mining] (when relating to mining, a drill is a compressed-air operated rock drill) perforatrice *f*, marteau *m* perforateur
[tool] (cutting tool) mèche *f*, foret *m*
[mech] (the cutting tool of the rock-drill) fleuret *m*
[text] (heavy linen cloth) coutil *m*, treillis *m*, drill *m*
[agric] semoir *m*
DRILL ADAPTER - [mech] (steel sleeve designed to enable taper-shank drills to be used in a mandrel socket of a different taper) douille *f* de réduction, douille *f* intermédiaire
~BIT - [mech] outil *m* de forage, trépan *m*
~BOX - [mech] archet *m*
~BRACE - [mach tool] (a machine for holding and rotating drills, worked by hand) vilebrequin *m*
~BUSH - [mech] guide-mèche *m*, canon *m* de perçage
~BUSHING - s. drill bush
~CASING - [mech] boîte *f* à forets
~CHUCK - [mech] (gripping device fitted to a drilling machine to hold parallel-shank drills) mandrin *m* de perceuse, mandrin *m* porte-forets, porte-mèches *m*
~-CHUCK ARBOR - [mech] queue *f* de montage, arbre *m* de montage
~COLLAR - [oil ind] (in well sinking) masse-tige *f*, maîtresse-tige *f*
~COLUMN - [oil ind] tige *f* de forage
~CORE - [oil ind] carotte *f* de sondage
~COULTER - [agric] coutre *m* de semoir
~CUTTING - [mining] déblais *mpl* de forage
~EJECTOR - [mech] (tapered metal tool used to dislodge taper-shank drills from the mandrel socket) chasse-foret *m*, chasse-cône *m*
~EXTRACTOR - [mech] (drill used for removing a broken or loosened drill from a hole) outil *m* utilisé pour extraire les trépans
~FED BY HAND - [mech] vilebrequin *m*, cliquet *m*
~FEED - [mech] (the mechanism, either hand- or power-operated, feeding the drill into the workpiece) avance *f* du foret
~GAUGE - [meas] calibre *m* d'affûtage pour forets, gabarit *m* d'affûtage pour forets
~GRINDER - [mech] machine *f* à affûter les pointes de forets
~HAMMER - [tool] marteau *m* perforateur
~HOLE - [mining] trou *m* de mine, forage *m*, sondage *m*
[mech] avant-trou *m*
~IN, to - [oil ind] pénétrer (dans une couche pétrolière
~LOCATOR - [mech] centreur *m* de foret
~LOG - [oil ind] coupe *f* de sondage, rapport *m* de sondage
~-OFF TEST - [mining] tests *mpl* sur paramètres de forage
~OUT, to - [mech] (to remove a jammed or broken part, e.g. the broken end of a bolt from its hole by drilling a slightly smaller hole down its longitudinal axis) reforer
~PIPE - [oil ind] tige *f* de forage
~PLATEN PRESS - [mech] presse *f* à plateaux à canaux d'injection
~POINT - [mech] pointe *f* du foret
~POINTER - s. drill grinder
~POST - [oil ind] colonne-support *f* perforatrice
DRILL PRESS - [mech] (a drilling machine with a rapid feed, usually actuated by a hand lever) perceuse *f*, perceuse *f* sensitive
~RIG - [mining] tour *f* de forage, installation *f* de forage
~ROD - [impl] (long rod reaching down into a boring) tige *f* de fleuret
[mining] tige *f* de forage
~SHANK - [mech] queue *f* de foret
~SLEEVE - [mech] douille *f* porte-foret
~SOCKET - [mech] douille *f* porte-foret
~SPINDLE - [mech] porte-foret *m*, broche *f* porte-foret
~STEEL - [metall] acier *m* à fleurets (de mine)
~STEM - [oil ind] maîtresse-tige *f*
~-STEM TEST - [oil ind] (a test designed to measure the productivity of a well) essai *m* aux tiges
~STOCK - [mech] (tool for holding and revolving drills by hand) vilebrequin *m*
~TEMPLATE - s. drilling jig
DRILLED - [gen & mech] perforé, percé, foré
~BURNER - [th eng] brûleur *m* à trous
~-TYPE ROLLER - [mech] (a calender roller having a central bore of the same diameter throughout) cylindre *m* alésé
DRILLING - [gen] perforation *f*
[mech] forage *m*, percage *m*, percement *m*
[mining] foration *f*, sondage *m*
[oil ind] (the action of drilling to find oil or gas) forage *m*, sondage *m*
[agric] semailles *fpl* en lignes
~AND MILLING MACHINE - [mach tool] perceuse-fraiseuse *f*
~AND TAPPING MACHINE - [mach tool] perceuse-taraudeuse *f*
~BARGE - [oil ind] ponton *m* de forage
~BEAM - [oil ind] balancier *m*
~BIT - [mech] taillant *m* de mèche
[mining] trépan *m*
~BY ROTATION - [oil ind] forer par rotation
~CABLE - [oil ind] câble *m* de forage
~CLAY - [oil ind] (a clay used to press down the muds) argile *f* pour boues de forage
~CRAMP - [mech] (frame fixed to a difficult piece of work for the support of a portable drill) support *m* de perçage
~DEPTH - [mech] longueur *f* taillée (du foret)
~DISTANCE - [agric] (the distance between the lines of sowing) distance *f* entre les lignes
~DRUMLINE - [mining] câble *m* de forage
~FLUID - [mech] fluide *m* de forage
~HEAD - [mech] tête *f* de perçage, tête *f* d'alésage
~JAR - [mining] joint *m* hydraulique
~JIG - [mech] (device for holding a workpiece and guiding drills used to work on it) calibre *m* de perçage, gabarit *m* de perçage
~LOCATION - [oil ind] emplacement *m* de forage
~MACHINE - [mech] (a machine for holding, guiding feeding and rotating drills) perceuse *f*, perceuse *f* automatique, perforatrice *f*
[agric] semoir *m*
~MACHINE PILLAR - [mach] colonne *f* de perceuse
~MACHINE STANDARD - s. drilling machine pillar
~MUD - [oil ind] (in well sinking) boue *f* de forage
~PATTERN - [oil ind] (in well sinking; the location and number of wells over a given area) canevas *m* des puits, espacement *m* des puits

DRILLING RIG - s. drill rig
~ROPE - s. drilling cable
~SITE - [mining] emplacement *m* du sondage
~SPINDLE - [mech] (the revolving spindle which holds the drill) porte-forets *m*, broche *f* porte-forets
~TABLE - [mech] table *f* porte-pièce (de perceuse)
~TOOL - [mech] (used with a rock drill) fleuret *m*, outil *m* de dorage
~VESSEL - [oil ind] plate-forme *f* de forage flottante
DRILLOMETER - [instr] (instrument designed to measure the tension of the cable in the drilling operation) drillomètre *m*, indicateur *m* de charge
DRINK OF SYRUP - [sugar ind] (addition of feed liquor) alimentation *f* en sirop, charge *f* de sirop
DRINKABLE - [gen] potable
DRINKING - [gen] potable
~BOWL - [agric] abreuvoir *m*
~WATER - [gen] eau *f* potable
DRIP, to - [gen] dégoutter, tomber goutte à goutte, s'égoutter
DRIP - [gen] égouttement *m*, goutte *f*, purge *f*, égout *m*
[constr] larmier *m* jet *m* d'eau
[oil ind] larmier *m*, tuyau *m* de purge
[ind chem] appareil *m* de séparation des condensats
[med] goutte-à-goutte *m*
~BAND - [aero] s. drip flap
~COCK - [mech] robinet *m* purgeur
~FEED LUBRICATOR - [mech] (type of lubricator in which oil is fed in measured drops) graisseur *m* à compte-gouttes
~FEED OILER - s. drip feed lubricator
~FLAP - [aero] ralingue *f* (ballon)
~MOULDING - [auto] jet *m* d'eau
~NOZZLE - [mech] ajutage *m* à gouttes
~PAN - [ind chem] gouttière *f* de condensation, cuvette *f* à huile, collecteur *m* d'huile
~PIPE - [plumb] tuyau *m* de purge
~-PROOF - [gen] (so constructed as to prevent the entry of dripping liquid, e.g. electric motors)abrité contre les éclaboussures
~-PROOF MOTOR - [el] (an electric motor designed to exclude dripping of condensed water, but not proof against a jet of water or against submersion) moteur *m* abrité, moteur *m* à orifices de ventilation abrités
~PUMP - [mech] pompe *f* de reprise des purges, pompe *f* de purge
~STRIP - [hydr] jet *m* d'eau, rejeteau *m*, bande *f* d'égout
~TRAP - [mech] pot *m* de purge, siphon *m*
~TRAY - [ind chem] égouttoir *m*
DRIPPER - [oil ind] puits *m* produisant goutte à goutte
DRIPPING - [gen] égouttage *m*, égout *m*, égouttement *m*, essorage *m*
[mech] suintement *m*
~BOARD - [mech] égouttoir *m*
~CUP - [mech] cuvette *f* d'huile, godet *m* d'huile
~EAVES - [constr] gouttière *f*, larmier *m*
~PAN - [mech] cuvette *f* à huile
~TUBE - [impl] pipette *f* compte-gouttes, compte-gouttes *m*
DRIPSTICK - [mech] (a tube which can be passed upward into a vessel from below, to determine the amount of liquid contained. When the liquid ceases to drop from the lower end of the dripstick, the upper end of the latter has reached the surface, and the depth of liquid is then read on the stick) indicateur *m* de niveau à pipette
DRIPSTONE - [build] larmier *m*
[chem] stalactite *m*
DRIVE , to - [gen] avancer, actionner, entraîner, faire marcher, asservir, mener, pousser
[auto etc] conduire
[mech] introduire
[mech] (to force an object into another) enfoncer, forcer
[mining] (to drive a tunnel) percer, pousser, creuser
DRIVE - [gen] (a small road) allée *f*, route *f* carrossable
[mech] (a transmission system) transmission *f* commande *f*, entraînement *m*
[auto] conduite *f*, force *f* motrice
[radio] (the signal applied to the control electrode of an electronic tube) attaque *f*, commande *f*
[mining] galerie *f*
~AXLE - [auto] essieu *m* moteur
~CASE - [auto] carter *m* d'engrenages
~CHAIN - [mech] chaîne *f* de transmission
~CLAMP - [oil ind] crampon *m* de fixage
~END - [mech] côté *m* d'entraînement
~FIT - [mech] ajustage *m* à force, emboîtement *m* à force, montage *m* à frottement dur
~GEAR - [mech] engrenage *m* d'entraînement,pignon *m* de commande
~HEAD - [oil ind] tête *f* de tube, manchon *m* de battage, tête *f* de tubage
~HOME, to - [mech] (to screw or hammer a part into another until it has reached the designed position) enfoncer complètement
~IN, to - [mech] enfoncer, serrer (vis), chasser (clou)
~IN - [cin] cinéma *m* (en plein air) pour automobilistes
~IN A PILE, to - [constr] enfoncer un pieu
~JOINT HOUSING - [auto] chape *f* de fusée
~LINE - [auto] arbre *m* de transmission
~MOVEMENT - [railw] (of a locomotive) mécanisme *m*
~OFF, to - s. drive out
~ON THE RIM, to - [auto] (to drive with deflated tyres) rouler sur la jante
~OUT, to - [mech] chasser, décaler, décoincer
~OVER, to - [chem] entraîner par distillation
~PIN - [el acoust] (in disk recording, a pin preventing a disk from slipping) doigt *m* d'entraînement
~-PIN HOLE - [el acoust] trou *m* (de doigt) d'entraînement
~PINION - [auto] pignon *m* d'entraînement, pignon *m* d'attaque
~PIPE - [oil ind] (in well sinking) tube *m* de pilotage, colonne *f* de tubage, tube *m* de fonçage
~PULSE - [comput] (in static magnetic storage system, a pulsed magnetomotive force applied to a magnetic cell) impulsion *f* de commande, impulsion *f* d'attaque, impulsion *f* de sélection, impulsion *f* de lecture-écriture
~PULSE GENERATOR - [comput] générateur *m* d'inscription, générateur *m* d'impulsions de sélection
~RATIO - [mech] rapport *m* d'entraînement
~SHAFT - [mech] arbre *m* de transmission, arbre *m* de commande, arbre *m* menant, arbre *m* moteur

DRIVE-SHAFT HOUSING - [mech] carter *m* de transmission
~WELL - [hydr] puits *m* instantané
~WINDING - [comput] enroulement *m* de commande
DRIVEN - [gen] conduit, guidé
[mech] commandé, entraîné
~ANTENNA - [radio] (a directly fed aerial) antenne *f* à attaque directe
~DISK - [mech] disque *m* mené, disque *m* récepteur
~ELEMENTS - [radio] (of an aerial) éléments *mpl* alimentés
~END - [mech] (the end of an extruder at which the feed screw is driven) extrémité *f* menée, extrémité *f* entraînée
~GEAR - [mech] engrenage *m* mené, engrenage *m* commandé
~IN - [mech] enfoncé, planté, introduit
~MEMBER - [mech] (the part of a transmission assembly which receives power, as distinct from the part which delivers it) organe *m* récepteur
~MULTIVIBRATOR - [radio] multivibrateur *m* pilote
~PULLEY - [mech] (the pulley in a belt drive which receives power from the belt) poulie *f* menée, poulie *f* commandé
~ROLLS - [mech] (rolls rotated by power, as distinct from idler rolls, which are driven by contact with the driven rolls or with the material) rouleaux *mpl* menés
~SENDER - [radio] (a transmitter with a frequency which is determined by a master oscillator) émetteur *m* piloté
~SHAFT - [mech] arbre *m* mené, arbre *m* récepteur
~TRANSMITTER - s. driven sender
~WELL - [oil ind] puits *m* à production forcée
DRIVER - [gen] conducteur *m*
[railw] (locomotive driver) mécanicien *m*
[mech] élément *m* moteur
[mech] pignon *m* d'entraînement, entraîneur *m*, toc *m* d'entraînement, taquet *m*, chasse-clavette *m*, doguin *m*
[electron] (a driver tube; an electronic tube used in the amplifier stage) tube *m* d'attaque, excitateur *m*, driver *m*
[impl] (for a die) poinçon
~-CHUCK - [mach tool] plateau-toc *m*, mandrin *m* à toc
~CIRCUIT - [radio] (circuit designed to produce the signal applied to the control grid of an amplifier) circuit *m* de commande
~PINION - [mech] pignon *m* d'entraînement
~PLATE - [mech] (a disk fixed on the mandrel nose and furnished with a pin to engage a carrier fixed to the workpiece) plateau *m* porte-pièce, plateau *m* à toc
~-SPRINGS - [railw] ressorts *mpl* des roues motrices
DRIVER'S CAB - [railw etc] cabine *f* de conduite
~CAGE - s. driver's cab
~HUT - s. driver's cab
~THIGH - [med] sciatique *f* des automobilistes
DRIVING - [gen] conduite *f*
[gen] menant, motrice, de commande, d'attaque
[mech] (of a screw) serrage *m*
[mining] percement *m*, chassage *m*, avancement *m*
[oil ind] enfoncement *m*
[metall] (of a blast furnace) allure *f* rapide
~AMPLIFIER - [electron] préamplificateur *m*

DRIVING ANCHOR - [naut] ancre *f* flottante
~ARM - [instr] (in electricity meters) bras *m* d'attaque
~ASSEMBLY - [auto] direction *f*
~AXLE - s. drive axle
~BAND - [firearms] (copper hand-pressed round the tail end of a projectile) ceinture *f* (d'un obus)
~BELT - [mech] courroie *f* de transmission
~BLADE - [impl] (of a cutting machine) couteau *m*
~BOX - [railw] boîte *m* d'essieu moteur
~BY FRICTION - [text] entraînement *m* par friction
~CAP - [oil ind] tête *f* de tubage
~CIRCUIT - [el] in an amplifier, a circuit in which flows the signal to be amplified) circuit *m* de commande
~CHAIN - [mech] (endless chain with teeth engaging toothed wheels) chaîne *f* de transmission, chaîne *f* de commande
~CHUCK - [mech] (in a lathe; a driving plate fitted with a device for gripping the workpiece) mandrin *m* de tour, plateau *m* porte-pièce
~CLAW - [mech] crabot *m*, griffe *f* d'entraînement
~CLUTCH - [mech] embrayage *m* à crabots, accouplement *m* à crabots
~COUPLE - [mech] couple *m* moteur
~CYLINDER - [mech] cylindre *m* d'entraînement
~DEPTH - [constr] profondeur *f* de fiche, hauteur *f* fichée
~DISK - [mech] disque *m* d'entraînement
~DOG - [mech] crabot *m*, griffe *f* d'entraînement
~DRUM - [text] cylindre *m* d'entraînement
~ELEMENT (OF INDUCTION METER) - [instr] (a working part of the meter which produces a torque by its action on the moving element) élément *m* moteur (compteur *m* à induction)
~FACE - [naut] face *f* travaillante (d'une hélice)
~FINGER - [text] doigt *m* de commande
~FIT - s. drive fit
~FORCE - [mech] force *f* motrice
~GEAR - [mech] (system of gears, shafts etc for the transmission of power) transmission *f*, commande *f*, mécanisme *m* d'entraînement
~HUB - [mech] moyeu *m* d'entraînement
~-IN TEST - [mech] essai *m* de perforation
~JET - [mech] jet *m* d'entraînement, jet *m* de fluide moteur
~KNOB - [impl] (in electric motors etc) bouton *m* de commande
~MECHANISM - s. driving gear
~MEDIUM - [mech] (of a pump) fluide *m* moteur, fluide *m* actif, fluide *m* d'entraînement
~MEMBER - [mech] (the part of a transmission assembly which delivers power, as distinct from that part which receives it) organe *m* moteur, élément *m* moteur
~MIRROR - [auto] rétroviseur *m*
~MOMENT - [phys] moment *m* moteur, moment *m* des forces actives (cisaillement)
~MOTOR - [mech] moteur *m* d'entraînement, moteur *m* de commande
~PAWL - [mech] cliquet *m* d'entraînement
~PIN - [text] goupille *f* d'entraînement
~PINION - [mech] pignon *m* d'entraînement
~PLATE - s. driver plate
~-POINT ADMITTANCE - [el] (the reciprocal of the driving-point impedance) admittance *f* d'entrée
~-POINT IMPEDANCE - [el] (the ratio between the

voltage of two terminals of a network and the total current caused to flow between the two terminals) impédance *f* d'entrée

DRIVING POTENTIAL - [electron] (in a photoelectric cell, the positive potential applied to the anode) potentiel *m* positif (de l'anode)

~POWER - [mech] force *f* motrice, puissance *f* de commande

~PROPELLER - [aero] hélice *f* propulsive

~PULLEY - [mech] (the pulley in a belt drive from which power is transmitted to the belt) poulie *f* de commande, poulie *f* menante, poulie *f* d'entraînement

~PULSE - [telev] impulsion *f* d'excitation

~ROD - [mech] tringle *f* de commande; barreau *m* d'essai

~SHAFT - [mech] arbre *m* menant, arbre *m* de commande, arbre *m* moteur

~SIDE - [mech] (in a driving belt, the tension side) brin *m* conducteur
(in a gear tooth) face *f* frontale

~SIGNAL - [telev] (signal timing the scanning at the pick-up point) signal *m* d'excitation

~SLEEVE - [auto] manchon *m* d'entraînement

~SPINDLE - [mech] axe *m* moteur, axe *m* d'entraînement

~SPROCKET - [mech] tambour *m* denté d'entraînement, barbotin *m*

~STAGE - [electron] (electronic circuit feeding another electronic circuit) excitateur *m*

~STATION - [railw] (in cable railways and in aerial cable-ways) station *f* de commande

~THE SHUTTLE, to - [text] chasser la navette

~TORQUE - [mech] (the torque on the moving element resulting from electrostatic or other effects) couple *m* moteur

~TRAILER - [transp] motrice *f* (chemins de fer) remorque *f* à cabine de conduite

~WHEEL - [mech] (the first member of a gear train) roue *f* d'entraînement, roue *f* motrice

DRIZZLE - [met] (precipitation in very small droplets, usually from stratus clouds) bruine *f*, crachin *m*

DROGUE - [aero] (sea anchor on seaplanes; shaped like a bottomless bucket to check the way) cône-ancre *m*
[naut] ancre *f* flottante

DROMEDARY - [zool] dromadaire *m*

DROMETER - [instr] (instrument designed to measure the speed of a moving train) dromomètre *m*

DROMOMANIA - [med] dromomanie *f*

DROMOMETER - s. drometer

DRONE, to - [acoust] bourdonner

DRONE - [zool] abeille *f* mâle, faux bourdon *m*
[aero] (pilotless aircraft, controlled by radio) avioncible *m* télécommandé
[naut] bâtiment *m* télécommandé
[acoust] bourdonnement *m*

DROOP, to - [gen] laisser pendre, pendre, pencher, tomber

DROOP - [gen] inclinaison *f*, abattement *m*, fléchissement *m*

DROOPED AILERON - [aero] aileron *m* abaissé (de I0 à I5°)

DROOPING - [gen] chute *f*
[el] (in electric motors or machines) chute *f* de tension

DROP, to - [gen] tomber, laisser tomber, abaisser, tomber goutte à goutte, s'égoutter, lancer
[met] (the easing off of the wind) tomber, mollir
[print] (to withdraw the chase after printing) dégarnir
[naut] culer, laisser culer
[naut] (to drop the anchor) jeter (l'ancre)

DROP - [gen] goutte *f*, abaissement *m*, chute *f*, tombée *f*
[phys] (small quantity of liquid bound by free surfaces) goutte *f*
[gen] (a fall) chute *f*
[comm] (of prices) baisse *f*, chute *f*
[naut] (of a sail) creux *m* (d'une voile)
[metall] (in diecasting) grimace *f*, défaut *m* dû à un manque de serrage
[mech] (of a cam) tombée *f*
[el] (voltage drop) chute *f* de tension

~ANCHOR, to - [naut] jeter l'ancre

~ANNUNCIATOR - [el] (in an electric signalling system, a device indicating the point from which the signal originates) annonciateur *m* à volets

~ARCH - [arch] arc *m* surbaissé

~ARROW - [surv] fiche *f* plombée

~BACK, to - [ind chem] refluer goutte à goutte
[naut] perdre sur l'arrière

~BALL - [impl] (heavy metal ball used to break up castings etc) poire *f* de cassage

~BASEBOARD - [impl] abattant *m*

~BLACK - [paint] (charcoal black pigment) noir *m* d'os, noir *m* de carbone

~BOTTLE - s. dropping bottle

~BOTTOM - [gen] fond *m* mobile, fond *m* ouvrant

~BOTTOM SKIP - [mech] (in earth moving machines) skip *m* à fond ouvrant

~BOX - [text] boîte *f* montante (de battant)

~BOX FLY - [text] battant *m* à boîtes montantes

~BOX LATHE - [text] battant *m* à boîtes montantes

~BOX LEVER - [text] levier *m* de boîtes montantes

~BOX LOOM - [text] métier *m* à boîtes montantes

~BOX MOTION - [text] mouvement *m* des boîtes

~BOX SLAY - [text] battant *m* brocheur

~CENTRE RIM - [rubber ind] (of tyres) jante *f* à base creuse

~CHAINING - [surv] cultellation, mesure *f* par ressauts horizontaux

~COMPASS - [draw] balustre *m* (à pompe pour l'encre)

~CURTAIN - [photo] toile *f* de fond

~DOOR - [gas ind] (of domestic gas cookers) porte *f* abattante

~DROWN, to - [gen] laisser tomber
[chem] (in a reaction) se séparer

~ELBOW - [plumb] (a small elbow fitted with ears) coude *m* à oreilles

~ELECTRODE - [electron] (half element consisting of mercury dropping in a fine stream) électrode *f* à gouttes

~ELL - [plumb] (small ell with ears) coude *m* à oreilles

~FOOT - [med] (due to a muscle paralysis) fléchissement *m* du pied

~FORGE, to - [metall] estamper (à chaud), matricer

~-FORGING - [metall] (forging made with a drop-hammer) matriçage *m*, estampage *m*, forgeage *m* par choc

~FORGINGS - [metall] pièces *fpl* matricées, pièces

*f*pl estampées
DROP FORMATION - [phys] (e.g. liquid flowing drop by drop from the open lower end of a vertical tube of minute diameter) formation *f* de gouttes
~FRAME TRAILER - [transp] remorque *f* à chassis surbaissé
~-FREE - [text] sans gouttes, anti-gouttes
~FRONT - [photo] décentrement *m* vers le bas (du porte-objectif)
~GATE - [metall] (runner leading into the top of a mould) attaque *f* en chute, descente *f* de coulée
~GRATE - [mech] (of a locomotive) jette-feu *m*
~HAMMER - [tool] (mechanical hammer raised by power and falling freely) marteau-pilon *m*, mouton *m*
~HAND - s. drop wrist
~HANGER - [plumb etc] suspension *f*, support *m*
~IN VOLTAGE - [el] chute *f* de tension
~INDICATOR - [railw] annonciateur *m* à volet
~INLET - [hydr] bouche *f* d'égout
~KEEL - [shipbuild] dérive *f*, quille *f* mobile
~LATCH - [constr] (of a window) loquet *m*
~LIFTER - [text] platine *f* tombante
~LUBRICATOR - [mech] graisseur *m* à compte-gouttes
~-MANHOLE - [hydr] regard *m* avec chute
~MESSAGE - [aero] message *m* lesté
~NEEDLE - [text] aiguille *f* tombante, cavalier *m*
~OF POTENTIAL - [el] chute *f* de tension
~OF TEMPERATURE - [met] baisse *f* de température, chute *f* de température
~OFF, to - [gen] diminuer, se séparer
~OUT, to - [gen] se séparer
~-OUT - [el] mise *f* au repos (d'un relais)
[telev] (in reception and video recording) manque *m* de signal
[telecomm] perte *f* à la transcription, intensité *f* de désexcitation
~PIN - [text] lamelle *f* d'arrêt, lamette *f*
~REACTION - [chem] stilliréaction *f*
~ROLLER - [text] rouleau *m* tendeur, rouleau *m* plongeur, rouleau *m* compensateur
~-RUNNER - [metall] chenal *m* de coulée en chute directe
~SHAFT - [oil ind] cuvelage *m* descendant
~-SHAPED - [gen] en forme de goutte
~-SHEET - [mining] toile *f* d'aérage
~SHUTTER - [photo] obturateur *m* à guillotine
~-SIDE BODY - [transp] (a lorry with its sides hinged on the floor level) caisse *f* à ridelles abattantes
~-SONDE - [radio] radiosonde *f* parachutée
~STAMPING - [metall] estampage *m*
~STAMPING HAMMER - [tool] mouton *m*
~STITCH - [text] maille *f* coulée
~TANK - [aero] (tank which is dropped after use) réservoir *m* largable
~TEST - [mech] (for steel tyres, a strength test) essai *m* au choc, essai *m* de rupture au choc
[chem] touche *f*
~TEST MACHINE - [mech] machine *f* pour essai au choc
~TIN - [metall] (granular tin produced by dropping molten tin into water) étain *m* en larmes, étain *m* en grains
~-TYPE SPRAG - [mech] béquille *f* de retenue articulée
DROP VALVE - [mech] (conical-seated valve used sometimes in steam engines) soupape *f* renversée, robinet *m* à siège conique
~-WEIGHT METHOD - [meas] (method of measuring the surface tension of a fluid by weighing the drops falling from the end of capillary tube) méthode *f* par pesée des gouttes
~WINDOW - [auto] glace *f* descendante
~WIRE - [telecomm] (in telephony) branchement *m* d'abonné, entrée *f* de poste
[text] œillet *m* casse-fil, fil *m* d'œillet, cavalier *m*, lamelle *f* d'arrêt, lamette *f*
~WRIST - [med] (a limp flexion of the wrist due to paralysis of the extensor muscles) fléchissement *m* du poignet, paralysie *f* du nerf radial
~ZINC - [metall] (granular zinc produced by dropping molten zinc into water) zinc *m* en grains
DROPHEAD - [auto] décapotable
DROPLET - [phys] (a very small spherule of liquid, bounded by free surfaces and of such dimensions as to float in air under room conditions) gouttelette *f*
DROPLIGHT - [el] baladeuse *f*
DROPOUT - [comput] (on a magnetic tape)"dropout" *f*, lacune *f*
DROPPABLE - [aero] largable
DROPPED BEAT - [med] pouls *m* intermittent
~END - [text] fil *m* tombé
~FRAME - [auto] chassis *m* surbaissé
~POSITION - [telecomm] (in telephone exchange) position *f* inoccupée
DROPPER - [el] (the fitting supporting the contact wire in catenary constructions) biellette *f* de support, pendule *m* de ligne caténaire
[med] (a device to pour a liquid drop by drop) compte-gouttes *m*, stilligoutte *m*
[text] lamelle *f*
~WITH EYE - [text] lamelle *f* à œillet
DROPPING - [gen] chute *f*
[phys] dégouttement *m*, égouttement *m*
[glass man] (heat moulding without pressure) moulage *m* par gravité
~ANGLE - [aero] (of bombs) angle *m* de bombardement
~BOTTLE - [ind chem] (special bottle with grooved glass stopper and spout for pouring liquids drop by drop) flacon *m* doseur, stilligoutte *m*
~FUNNEL - [lab] (glass funnel fitted with a glass stop-cock on the stem) entonnoir *m* à robinet, ampoule *f* à brome
~GLASS - [impl] compte-gouttes *m*
~MERCURY CATHODE - [electron] cathode *f* à gouttes
~METAL - [metall] métal *m* entré en fusion
~MOOR - [naut] affourchage *m*
~OF THE CHARGE - [metall] (in ovens) chute *f* de charge
~OF THE STRIKE - [sugar ind] vidange *m* de la masse cuite
~PART OF THE LATHE - [text] épée *f* (du battant)
~POINT - [chem] (the temperature at which fats liquefy) point *m* de goutte
~RESISTOR - [electron] resistance *f* série, résistance *f* additionnelle
DROPPINGS - [text] (wool droppings) bourres *f*pl
[zool & agric] fientes *f*pl (d'animaux)
DROPS - [arch] (ornamental mouldings) gouttes *f*pl

DROPS ELIMINATOR - [ind chem] séparateur *m* de condensats
DROPSY - [med] hydropisie *f*
DROPWISE - [gen] goutte à goutte
~CONDENSATION - [chem] (deposition of condensate in individual drops rather than as a film) condensation *f* en gouttes
DROPWORT - [bot] (medicinal plant) filipendule *f*, spirée *f*
DROSOMETER - [instr] (instrument designed to measure the quantity of dew on a body) drosomètre *m*
DROSS - [mining] (a small worthless coal) fines *fpl* de charbon
[metall] (like slag, but consisting of oxides) scorie *f*, crasses *fpl*, laitier *m*
[gen] (waste stuff) rebut *m*, déchet *m*, impureté *f*
~HOLE - [metall] trou *m* à laitier
DROSSING - [metall] (method of removing copper and iron from lead by stirring the molten metal from the blast furnace in a drossing kettle. Impurities separate in virtue of their relatively low solubility in the lead) écumage *m*
~KETTLE - [metall] (refractory-lined vessel used in separating impurities from molten lead) écumoire *f*, dispositif *m* retenant les impuretés
DROSSY COAL - [mining] charbon *m* pyriteux
DROUGHT - [met] (prolonged period without precipitation) sécheresse *f*
~RESISTIVITY - [bot] (the fitness of a plant to withstand drought) résistance *f* à la sécheresse
DROVE - [gen] (a number of animals) troupeau *m* (gros bétail)
[hydr] (small channel for irrigation purposes) canal *m* d'irrigation
[impl] (broad-edged chisel for the dressing of stone) ciseau *m* à piquer
~CHISEL - [impl] (used for stone dressing) ciseau *m* à piquer
DROWN, to - [gen] noyer, submerger
[geogr] inonder
[acoust] couvrir (un son)
[constr] (mortar) noyer le mortier
DROWNED - [gen] noyé, inondé, submergé
~DAM - [hydr] barrage *m* submergé
~PIPE - [plumb] (an inlet pipe to a tank, the end of which is below the surface of the liquid) tuyau *m* d'arrivée immergé
~PISTON - [mech] piston *m* plongeur
~SPRING - [geol] source *f* sous-marine
DROWNING - [gen] noyage *m*, noyade *f*, submersion *f*
DRUG, to - [gen] droguer, administrer un narcotique, administrer des stupéfiants
DRUG - [chem] drogue *f*, médicament *m*, produit *m* pharmaceutique
~ADDICT - [med] toxicomane *m*
~ALLERGY - [med] allergie *f* médicamenteuse
~DISEASE - [med] maladie *f* médicamenteuse
~-FAST - [med] résistant à la thérapeutique
~HABIT - [med] toxicomanie *f*
DRUGGED - [gen] drogué
DRUGGIST - [gen] pharmacien *m*
DRUGSTORE - [gen] pharmacie *f* (USA)
DRUM - [gen & mus] tambour *m*, caisse *f*
[anat] tympan *m*
[mech] (hollow cylindrical barrel) fût *m*, cylindre *m*
[mech] (the rotor of a reaction turbine) rotor *m*
[constr] (of a cement mixer) tambour *m*
[transp] (for storing and transport of gas, petrol, etc) tonnelet *m*, fût *m*, réservoir *m* sous pression, cylindre *m*
[th eng] (of a boiler) corps *m* cylindrique
[comput] tambour *m* (d'un traducteur analogique-numérique etc)
[impl] (for embroidery, lace etc) tambour *m*
DRUM ARMATURE - [el] (in electric machines, an armature with a drum winding) induit *m* en tambour
~BARKER - [paper man] (a machine intended to remove the bark from the logs) écourceuse *f* à tambour
~BARROW - [impl] dévidoir *m* (pour câbles)
~BRAKE - [aero] (type of brake in which friction elements, such as pads or bands, are applied to a cylindrical rotating element) frein *m* à tambour
~CARRIER - [transp] (fitted to fork-lift lorries) transporteur *m* pour fûts
~COLOURING - [ind chem] coloration *f* au tonneau
~CONTROLLER - [el] (fixed contact-finger operated controller; the fingers bear on contact strips mounted in the form of a rotating drum) combinateur *m* cylindrique, combinateur *m* à tambour
~CURB - [build] (hollow cylinder used to line the shaft of a well) cuve *f* (d'un puits enfoncé)
~DRIVE - [text] commande *f* du tambour
~DRYER - [ind chem] (apparatus consisting of heated cylinders onto which a paste or sludge is fed for drying) séchoir *m* à tambour, sécheur *m* rotatif, tambour *m* sécheur
~FRICTION TEST - [ind proc] essai *m* de résistance à l'usure au tambour
~HAY MAKER - [agric] faneuse *f* à tambour
~-HEAD - [anat] membrane *f* du tympan
[gen] peau *f* de tambour
~LAYOUT CHART - [comput] diagramme *m* d'attribution (du tambour)
~MEMORY - [comput] mémoire *f* tambour, mémoire *f* à tambour magnétique
~MIXER - [mech] (type of cement mixer) bétonnière *f* à tambour
~POLISHING - [plast ind] polissage *m* au tonneau
~PUMP - [mech] (type of rotary pump) pompe *f* (rotative à tambour)
~SCANNER - [telev] rotating drum fitted with mirrors at its periphery) analyseur *m* à tambour
~SETTING - [text] réglage *m* du tambour
~SHAFT - [mech] arbre *m* du tambour
~SHUTTER - [cin] obturateur *m* à barillet
~SLICER - [sugar ind] coupe-racines *m* à tambour
~SPEED - [radio] (the rotation speed of the message drum) vitesse *f* d'exploration
~STARTER - [el] démarreur *m* à cylindre
~SWITCH - [el] interrupteur *m* à tambour
~TEST - [packag] essai *m* au tambour
~-TYPE CONTROLLER - s. drum controller
~-TYPE MAGNETIC SEPARATOR - [min] (magnetic separator consisting of a drum with internal electromagnets, the material treated being shot onto the revolving drum) séparateur *m* magnétique à tambour
~WASHER - [paper man] (in a hollander, a gauze-covered cylinder for the washing of the pulp) cylindre *m* laveur
~WINCH - [mech] treuil *m* à tambour
~WINDER - [text] bobinoir *m*
~WINDING - [text] (the winding of yarn onto a flan-

ged bobbin by means of a driving drum) enroulement *m* en tambour

DRUM WINDING WITH DIAMETRICAL PITCH - [el] enroulement *m* en tambour à pas diamétral

~WINDING WITH SHORTENED PITCH - [el] enroulement *m* en tambour à pas raccourci

DRUMLIN - [geol] (a small hog-backed hill) drumlin *m*

DRUMMING - [leather ind] (treatment of skins in a revolving drum) rinçage *m* au tonneau

DRUMSKIN ACTION - [acoust] (the vibration of the whole of a wall under the action of an incident wave) vibration *f* de la paroi entière

DRUNKEN - [phys] imprégné, trempé

DRUPE - [bot] fruit *m* charnu, fruit *m* à noyau (USA)

DRUSE - [bot] (in some plant cells, a globose mass of crystals of calcium oxalate) druse *f*

DRY, to - [gen] sécher, dessécher, faire sécher, mettre à sec

DRY - [gen] sec, tari, à sec, déshydraté, desséché, altéré
[metall] fragile, cassant, impur

~-ADIABATIC LAPSE RATE - [met] gradient *m* vertical de température adiabatique sèche

~AIR CURE - [rubber ind] vulcanisation *f* à l'air chaud

~ASH-FREE COAL - [min] charbon *m* pur

~ASH-FREE COKE - [min] coke *m* pur

~ASSAY - [min] (the determination of the valuable contents of an ore specimen by high temperature treatments as distinct from those involving solution techniques) essai *m* par voie sèche

~BATTERY - [el] (a battery of dry cells) pile *f* sèche

~BINDER - [ind chem] (powdered material, e.g. certain resins, which is mixed with chopped glass rovings in preforming) résine *f* (en poudre) pour poudres à mouler

~-BLEACHING - [text] blanchiment *m* à sec

~BLEND EXTRUSION - [plast ind] extrusion *f* de poudre à mouler

~BLENDING - [plast ind] (the operation of blending different materials in the extruder barrel itself) mélange *m* à sec

~BOND - [metall] corps *m* de sable à sec

~BONE ORE - [min] (also called smithsonite; a carbonate of zinc occurring in calcareous rocks) smithsonite *f*, calamine *f*

~BORING - [oil ind] sondage *m* à sec

~BOX - [nucl] (a box with hand holes on which rubber gloves are fitted for manipulation within it) boîte *f* à gants

~-BULB THERMOMETER - [met] (an ordinary thermometer of the fluid expansion type, in distinction from a wet-bulb thermometer) thermomètre *m* à boule sèche, thermomètre *m* sec

~CELL - [el chem] (a primary cell, usually of leclanché type, having a paste electrolyte and therefore containing no liquid) pile *f* sèche

~CLEANING - [gen] nettoyage *m* à sec

~CLUTCH - [mech] embrayage *m* sec

~COLOUR - [dyes)] pigment *m*

~COOLING - [metall] extinction *f* à sec, refroidissement *m* à sec

~-CORE CABLE - [el] (used in telephone or telegraph lines) câble *m* à circulation d'air

~CRUSHING - [min] broyage *m* à sec

~DASH - [constr] (plastering of an inferior quality) enduit *m* grossier

DRY DECATIZING - [text] décatissage *m* à sec

~DEFECATION - [sugar ind] chaulage *m* roche

~DEPOSITION - [nucl] (the deposition of small particles from a nuclear explosion, without any relevant rate) retombées *fpl* négligeables

~DISTILLATION - [chem] distillation *f* sèche

~-DOCK, to - [naut] mettre en cale sèche

~-DOCK - [naut] (dock used for the repairing of ships) cale *f* sèche, bassin *m* de radoub

~DREDGER - [mech] drague *f* sèche, excavatrice *f*

~DRILLING - [oil ind] forage *m* à sec

~ELECTROLYTE CONDENSER - [el chem] (electrolytic condenser with a negative pole in the form of a sticky paste) condensateur *m* électrolytique sec

~ENAMELLING - [metall] émaillage *m* au poudré

~END - [paper man] sécherie *f*

~FARMING - [agric] culture *f* sèche

~FILLER - [ind chem] charge *f* sèche

~FLASHOVER VOLTAGE - [el] (the voltage at which air surrounding a dry insulator breaks down, thus causing the insulator to flash over) tension *f* d'arc à sec dans l'air

~FLEX-STRENGTH - [text] résistance *f* à la flexion à sec

~FOG - [met] (haze caused by the presence of solid particles, e.g. dust or smoke, as distinct from that due to water droplets) nuages *mpl* de poussière

~FRUIT - [bot] (a dry fruit with a pericarp which does not become fleshy at maturity) fruit *m* sec

~GASHOLDER - [gas ind] (a waterless gasholder) gazomètre *m* sec, gazomètre *m* à piston

~GRAZING - [agric] pâturage *m* non irrigué

~GRINDING - [mech] broyage *m* à sec

~HARBOUR - [naut] port *m* asséchant à marée basse

~HEAT CURE - s. dry air cure

~HOLDER - s. dry gasholder

~HOLE - [oil ind] puits *m* non productif, puits *m* sec, forage *m* improductif

~HOPPING - [brew ind] (the addition of a small quantity of raw hops to certain types of beer) houblonnage *m* en cuve

~ICE - [chem] (solidified carbon dioxide; it is used as a refrigerant, source of carbon dioxide and as a coolant) neige *f* carbonique

~JOINT - [constr] joint *m* sec
[mech] soudure *f* défecteuse, soudure *f* sèche

~LUTE - [ind chem] joint *m* sec

~MATTER - [gen & chem] substance *f* sèche, matière *f* solide

~METER - [instr] (gas meter) compteur *m* sec

~MIXING - [ind chem] mélange *m* à sec

~MOULDING - [metall] (the preparation of moulds in dry sand) moulage *m* en sable étuvé

~NATURAL GAS - [min] gaz *m* naturel 'sec', gaz *m* pauvre

~PEAT - [geol] tourbe *f* sèche

~PERIOD - [agric] période *f* d'arrêt de lactation

~PILE - s. dry cell

~PLATE - [photo] (the glass plate supporting the emulsion used for photographic purposes) plaque *f* sèche

~-PLATE RECTIFIER - [radio] (a metal rectifier) redresseur *m* sec à oxyde métallique

~POINT - [chem] point *m* sec

~POTASH - [chem] potasse *f* caustique

~PROCESS - [constr] (method of making Portland ce-

ment in which the raw materials are crushed and ground in the dry state) procédé *m* par voie sèche
[nucl] traitement *m* par voie sèche

DRY RECTIFIER - [electron] (device consisting of one positive and one negative electrode and a rectifying junction) redresseur *m* sec

~-REED RELAY - [electron] relais *m* hermétique à gaz inerte, contacteur *m* à languettes magnétiques

~RIVER - [geol] oued *m*

~ROOT AND COLLAR ROT - [bot](a plant disease) maladie *f* sèche des racines et pourriture du collet

~ROT - [constr] (timber decay caused by fungi) carie *f* sèche, pourriture *f* sèche

~RUBBER - [rubber ind] caoutchouc *m* sec

~-RUBBER CONTENT - [rubber ind] teneur *f* en caoutchouc sec

~RUN - [telev] (a pre-transmission show of a television programme) avant-première *f*

~SAND - [metall] (moulding sand with good cohesion and strength when dry) sable *m* étuvé

~SANDING - [ind proc] sablage *m* à sec

~SCREENING - [min] criblage *m* à sec

~SEAL - [gas ind] grille *f* sèche, joint *m* sec

~-SEAL JOINT - [metall] joint *m* sec

~SOLIDS CONTENT - [phys] (the amount of solids present in a substance, in terms of such solids in the dry state) teneur *f* en substances sèches, teneur *f* en matières solides

~SPINNING - [text & plast ind] (extrusion of synthetic fibres into air, as distinct from wet spinning) filage *m* à sec

~SPOT - [plast ind] (in laminates, an area which has not been fully impregnated) vide *m* en surface (ou entre strates)

~SPRAY - [packag] pulvérisation *f* sèche

~STAMPING - [min] bocardage *m* à sec

~STEAM - [phys] (steam free from water in the liquid state, but not super-heated) vapeur *f* sèche

~STEAM-COAL - [min] houille *f* maigre, charbon *m* à courte flamme, charbon *m* maigre, charbon *m* anthraciteux

~STONE - [constr] pierre *f* sèche

~STORE - [gen] entrepôt *m* climatisé, entrepôt *m* réfrigéré

~STRENGTH - [phys] (the strength of a material when dry) résistance *f* mécanique à sec
[soil] résistance *f* à sec

~SUMP - s. dry-sump lubrication

~-SUMP LUBRICATION - [mech] (system of lubrication in internal-combustion reciprocating engines, compressors and the like, in which there is no body of lubricating oil in the sump) graissage *m* à carter sec

~SWEATING - [metall] (a process of copper purification, whereby copper is exposed to oxidizing temperature under its melting point) ressuage *m*

~TACKING - [ind chem] pouvoir *m* collant

~TUNNEL - [packag] tunnel *m* de séchage

~UP, to - [gen] dessécher
[mech] désamorcer (une pompe)

~WALL - [constr] mur *m* en pierres sèches

~WASHING - [min] lavage *m* à sec

~WAY PROCESS - [chem] traitement *m* par voie sèche

~WEIGHT - [mech] (the weight of an engine with all accessories indispensable for operation) poids *m* à vide
[astronaut] (weight of a rocket vehicle without fuel and payload) poids *m* à sec, poids *m* de la fusée vide

DRY WELL - s. dry hole

~WOOD - [timber] (seasoned, sap-free timber)bois *m* séché

DRYABILITY - [chem] (degree or extent to which a substance can be readily dried) aptitude *f* au séchage

DRYER - [paint] siccatif *m*, dessicatif *m*
[mech] sécheur *m*, séchoir *m*, déshydrateur *m*

DRYING - [gen] siccatif
[paint] séchage *m*
[chem] (the removal of small quantities of water from liquids, gases or solids) dessèchement *m*

~AGENT - [ind chem] agent *m* déshydratant, produit *m* dessiccant, dessiccateur *m*

~AIR - [gen] air *m* desséchant

~APPARATUS - [gen] séchoir *m*, sécheuse *f*, déshydrateur *m*

~CABINET - [photo] séchoir *m*

~CHAMBER - [gen] (enclosed space in which drying is carried out) armoire *f* de séchage, séchoir *m*

~CLIP - [impl] pince *f* à linge

~CUPBOARD - [gen] étuve *f*, chambre *f* chaude

~CYLINDER - [paper man] (hollow, steam-heated cylinder over which the web of paper is passed for drying) cylindre *m* sécheur

~DRUM - [gen] tambour *m* sécheur
[photo] tambour *m* de séchage

~FRAME - [photo] égouttoir *m*

~FURNACE - [th eng] four *m* de séchage

~MARKS - [photo] traces *f*pl de séchage, marques *f*pl de séchage

~OIL - [chem] (vegetable, animal or synthetic oils which readily absorb atmospheric oxygen with formation of a tough film. Drying oils have major applications in the production of surface coatings) huile *f* siccative

~OVEN - [ind chem](closed heated chamber fitted with a door for drying substances in the laboratory) étuve *f*, four *m* de séchage

~PLANT - [gen] installation *f* de séchage

~PROCESS - [gen] procédé *m* de séchage, procédé *m* de dessiccation

~RACK - [impl] chassis *m* de séchage, étagère *f* de séchage

~RATE - [phys] vitesse *f* de séchage

~ROLLER - s. drying cylinder

~ROOM - [gen & paper man] étuve *f*, séchoir *m*, chambre *f* chaude

~STENTER - [text] rame *f* sécheuse

~TIME - [gen] durée *f* de séchage

~TUBE - [ind chem] (tube packed with moisture-absorbent substances for extracting water from gases) tube *m* de dessiccation

~TUNNEL - [gen] (long narrow chamber in which material is dried by air currents as it passes through) tunnel *m* de séchage

~UP - [gen] dessèchement *m*, assèchement *m*
[mech] (in engines or motors) manque *m* d'huile

DRYNESS - [gen] siccité *f*, sécheresse *f*

DRYWAY PROCESS - s. dry process

D SCOPE - [radar] indicateur *m* type D

D.T.I. - [telecomm] (abbrev for distortion transmission impairment) réduction *f* de la qualité de la transmission due à la limitation de la bande des fréquences transmises effectivement

DUAL - [gen] double

DUAL AMPLIFICATION CIRCUIT - [radio] (a circuit with one or more valves used for simultaneous high- and low-frequency amplification) circuit *m* réflex
~CARD - [comput] (punch card containing the normal combination and a document) carte *f* à double usage
~CARRIAGE-WAY - [constr] route *f* à deux voies
~CHANNEL SOUND - [telev] (technique used in television receivers) système *m* sonore à deux canaux
~COLLECTION ANALYSIS - [phys] analyse *f* en double collection
~COMBUSTION CYCLE - [mech] (in internal combustion engines) à double combustion
~COMPLETION - [min] complètement *m* double, complétion *f* double
~COMPONENT - [comput] (component which can serve two separate mechanisms) composante *f* à double fonction
~-CONTROL - [mech] à double commande
~-CONTROL COLUMNS - [aero] doubles commandes *fpl*
~-CONTROL HEPTODE - [electron] (an indirectly heated heptode in which grid number one and grid number three have a common base) heptode *f* à double commande
~CONTROLS - [aero] (arrangement of aircraft controls in duplicate, so that either of two pilots can operate them at need, as for instruction or in emergency) double commande *f*
~FUEL ENGINE - [mech] (in internal combustion engines) moteur *m* polycarburant
~HEADLIGHTS - [auto] phares *mpl* jumelés
~HECKLING MACHINE - [text] (for flax) machine *f* (double) à peigner le lin
~HIGHWAY - s. dual carriage-way
~IGNITION - [mech] (duplicated system of ignition with battery ignition and magneto ignition having separate sparking plugs) double allumage *m*
~INSTRUCTION - [aero] (instruction of a pilot by the use of dual controls) instruction *f* sur appareil à double commande
~INSTRUCTION AIRCRAFT - [aero] (training aircraft fitted with dual controls) avion *m* à double commande
~INTAKE AND DUAL EXHAUST PORTS - [mech] (in internal combustion engines) orifices *mpl* doubles d'admission et d'échappement
~METER - [meas] appareil *m* de mesure à deux lectures
~MODULATION - [radio] (the modulation of a single carrier by two simultaneous methods) modulation *f* double
~NETWORK - [el] (a network in which all the elements can give opposite signs) réseau *m* réciproque
~OPERATION - [mech] à double effet
~PRINTING - [comput] impression *f* en double
~PRODUCER - [oil ind] puits *m* à double production
~PROPELLERS - [aero] (counter-rotating propellers) hélices *fpl* contra-rotatives
~PURPOSE - [gen] à double usage, à double fonction
~-SEAL TUBE - [rubber ind] (of tyres, type of air-tube in aircraft tyres, in which there are two air chambers, one inside the other, the inner being capable of maintaining rolling radius should the outer fall) chambre *f* à double paroi
~SENSITIVITY ELECTRONIC INDICATOR - [instr] (tuning indicator) indicateur *m* d'accord à deux sensibilités
DUAL TRACK RECORDER - [el acoust] (tape recorder with a head covering only half of the tape, so that a second recording can be made) enregistreur *m* magnétique à double piste
~-TYRED - [auto] à pneus jumelés
~WHEELS - [auto] roues *fpl* jumelées
~-ZONE WELL - [oil ind] s. dual producer
DUALITY - [gen] dualité *f*
DUARC - s. durarc light
DUB, to - [el acoust] (the production of a copy of a tape recording made by recording on one machine what another machine is playing; also the combining of sound and picture, taken separately, into one tape) doubler, monter
~COPY - [el acoust] (copy of a tape recording made by recording on one machine what is played by another machine) réenregistrement *m*
DUBBED - [cin] (of a film) doublé
DUBBER - [el acoust] console *f* de mixage, doubleur *m*, mélangeur *m*
DUBBING - [el acoust] (the combining of two or more recordings into a single composite recording) montage *m*
[cin] doublage *m*
[telev] (the combination of several sound tracks into a single track)
[leather ind] préparation *f* du cuir avec dégrat
~MIXER - [el acoust] mélangeur *m* doubleur
DUBBS PROCESS - [oil ind] procédé *m* Dubbs
DUCK - [zool] canard *m*, cane *f*
[text] (a heavy cotton fabric used for tropical suiting; also a heavy coarse linen cloth) coutil *m*, toile *f* à voile
[transp] véhicule *m* amphibie
~-BILL - [mech] (a power shovel) pelle *f* mécanique
~-BILL PLIERS - [impl] pinces *fpl* à bec canard
~BOARD - [naut] caillebotis *m*
~CLOTH - [text] coutil *m*
~EGG - [zool] œuf *m* de cane
~FOOT BEND - [mech] coude *m* à patin
DUCKLING - [zool] canneton *m*
DUCT - [gen & hydr] canal *m*, conduit *m*, conduite *f*
[mech] (a large thin-walled tube or casing to carry air, gases, etc) conduit *m*, conduite *f*, drain *m*
[el] (a pipe or conduit for cables, especially underground or in a wall) alvéole *m*, gaine *f*, canalisation *f*
[print] (also called ductor) encrier *m*
[med] canal *m*, conduit *m*
~CLEANER - [impl] écouvillon *m*, brosse-écouvillon *m*
~CLEANING TOOL - s. duct cleaner
~-KEEL - [shipbuild] quille *f* tubulaire, quille *f* en caisson
~LINING - [gen] revêtement *m* intérieur de conduit
~PROPULSION - [astronaut] propulsion *f* par moteur à double flux
~ROD - [telecomm] aiguille *f* de tirage
~ROLLER - [print] (for the distribution of ink) rouleau *m* preneur
~WINDING - [mech] (the manufacture of large tubes by winding glass-reinforced tape on a mandrel) fabrication *f* de tubes par enroulement
DUCTED - [gen mech etc] sous gaine, sous tunnel
~FAN - [mech] (a fan working in a closed annular chamber, as in a ducted-fan engine) turbine *f* sous

carrossage
[aero] hélice *f* carenée
DUCTED FAN ENGINE - [aero] réacteur *m* à hélice carenée
~-FAN TURBINE ENGINE - [aero] (a type of turbojet engine in which a fan working in an annular duct delivers air both to the high-pressure compressor and to the exhaust discharge) moteur *m* à ventilation auxiliaire
~GAS SYSTEM - [aero] (method of driving helicopter rotors by means of gas turbine exhaust ducted through the blades and discharged through jets at the tips) propulsion *f* par réaction
~IMPELLER - [aero] turbine *f* sous carrossage
~OIL COOLER - [aero] radiateur *m* d'huile caréné
~RADIATOR - [aero] radiateur *m* caréné
DUCTILE - [gen] ductile, malléable, souple, flexible
~CAST-IRON - [metall] fonte *f* malléable
~IRON - [metall] fer *m* doux
DUCTILIMETER - [instr] (instrument designed to measure the ductility of metals) ductilimètre *m*
DUCTILITY - [metall] (the capacity of metals for cold flow, which is accompanied by progressively increasing resistance to such flow ('work-hardening'). It is this property which makes possible the drawing of wire, cold pressing and the like operations) ductilité *f*
DUCTLESS - [gen & anat] sans conduit
~GLAND - [biol] (ductless glandular tissues which discharge their product direct into the blood) glande *f* à sécrétion interne, glande *f* close
DUCTOR - [print] (ink-holding reservoir in a printing machine) encrier *m*
~ROLLER - s. duct roller
DUCTULE - [anat] canalicule *m*
DUCTUS - [zool] (also called cut; a tube formed of cells) conduit *m*, canal *m*
DUD - [gen] (old clothes or rags) frusques *fpl*, vieilles hardes *fpl*, nippes *fpl*
[oil ind] puits *m* improductif, puits *m* qui s'arrête de produire
DUE - [gen] dû, juste, mérité, convenable, propre
[comm] dû, exigible
[fin] payable
DUES - [gen & fin] droits *mpl*, frais *mpl*, honoraires *mpl*, redevances *fpl*
DUFF - [min] (fine coal which is obtained by screening) poussier *m*, fines *fpl* lavées
[soil] (US for raw humus) matte *f*.
DUFFEL - [text] drap *m* molletonné, molleton *m*
DUFFING OUT PROCESS - [photo] séparation *f* des couleurs par grattage, bouchage *m* d'un cliché
DUFFLE - s. duffel
DUG - [zool] mamelle *f* (d'un animal), pis *m* de vache
~-OUT - [gen] abri *m*, cagna *f*, abri-caverne *m*
~-OUT SHELTER - [gen] abri *m* enterré
DULCIMER - [mus] tympanon *m*
DULCIN - [chem] (a synthetic sweetening agent, having about 200 times the sweetening power of sugar) dulcine *f*
DULL, to - [paint light etc] amortir, assourdir, ternir, dépolir
DULL - [gen] sourd, assourdi, triste, sombre, terne
[met] (of the sky) lourd, gris, nuageux, couvert
[mech] (of a blade) émoussé, usé
[paint] mat, terne, trouble
[light] pâle, sans éclat
DULL BIT - [tool] outil *m* émoussé
~COAL - [min] charbon *m* mat, houille *f* mate
~COATED PAPER - [paper man] papier *m* couché mat
~EMITTER - s. dull emitting cathode
~EMITTING CATHODE - [electron] (cathode in which the core filament is covered with alkaline earth oxides) cathode *f* à oxydes
~FINISH - [paper man] fini *m* mat
~-FINISH PAPER - [paper man] papier *m* mat
~FRACTURE - [min] cassure *f* terne
~RED - [metall] rouge *m* sombre
~SURFACE - [light] (a rough or poorly-reflecting surface) surface *f* terne, surface *f* mate
DULLING - [gen] dépolissement *m*, ternissement *m*
[paint] ternissement *m*, matage *m*
~AGENT - [ind chem] (substance in powder form incorporated in surface coatings to reduce reflection) agent *m* de matité
DULLNESS - [gen] matité *f*, ternissure *f*
~OF HEARING - [acoust & med] hypoacousie *f*, ouïe *f* peu sensible
DULONG AND PETIT'S LAW - [chem] (the product of the specific heat and the atomic weight of an element are constant and approximately equal to 6.4) loi *f* de Dulong et Petit
DUMB - [gen] muet, aphasique
[naut] (deprived of locomotion means) désemparé
~AERIAL - [radio] (aerial which does not resonate) antenne *f* non résonnante
~ANTENNA - s. dumb aerial
~BARGE - [naut] chalaud *m* remorqué
~-BELL SLOT - [electron] (a dumb-bell shaped hole in a diaphragm of a waveguide thus obtaining the required resonant properties) fenêtre *f* résonnante en forme d'haltère
~BELL TEST-PIECE - [rubber ind] éprouvette *f* en forme d'haltère
~-BELL TEST-STRIP - s.dumb-bell test-piece
~-BELL WAVEGUIDE - [electron] guide *m* d'onde en forme d'haltère
~COMPASS - [surv] (also called pelorus; a circular plate graduated in degrees and used to take the bearing of some object) taximètre *m*
~SCAB - [metall] adhérence *f* de sable vitrifié dartre *m*
~SWITCH - [railw] aiguillage *m* à deux aiguilles fixes
~WAITER - [constr] monte-plat *m*, dressoir *m*
DUMBNESS - [med] mutisme *m*, muétisme *m*, aphasie *f*
DUMBO - [radar] (colloq) for air-to-surface vessel; a radar system designed to detect targets on the sea surface) radar *m* aéroporté pour la détection d'objets à la surface de la mer
DUMET SEAL - [mech] scellement *m* verre-métal, joint *m* verre-métal
DUMMY - [gen] mannequin *m*, pantin *m*, prête-nom *m*
[print] (unprinted volume, made up for the assessment of costs) maquette *f*
[comput] (artificial unit of information which is inserted but does not affect operations) donnée *f* fictive
[railw] (steam locomotive fitted with a condenser) locomotive *f* à condenseur
[el] section *f* morte (d'un enroulement)
[astronaut] étage *m* mannequin, symbole *m* muet

DUMMY AERIAL - [radio] (a network designed to have the characteristics of an aerial but without radiation, for testing purposes, and also in some navigation systems) antenne *f* fictive
~BOLT - [mech] boulon *m* bouche-trou
~CARTRIDGE - [firearm] cartouche *f* d'exercice
~COIL - [el] (coil on an armature to keep the mechanical balance, but not connected to the winding) section *f* morte
~DIFFUSER - [nucl] diffuseur *m* fictif
~INSTRUCTION - [comput] (instruction having no influence on the operation) instruction *f* fictive
~LOAD - [el] (dissipative, non-radiating substitute device) charge *f* fictive
~PIPES - [mus] (the ornamental organ pipes) tuyaux *mpl* de facade
~PISTON - [mech] (in a reaction turbine; steam pressure is applied to one side of it thus balancing the end thrust) piston *m* d'équilibrage
~PLUG - [mech] fausse bougie *f*
[el] fiche *f* isolante
~RISER - [metall] masselotte *f* borgne
~SECTION - [telecomm] meuble *m* pour l'entrée des câbles
~SLOT - [el] encoche *f* borgne, encoche *f* vide
DUMMYING - [metall] (preliminary rough shaping of heated metal) dégrossissage *m*
DUMONTITE - [min] (lead, uranium and phosphorus containing mineral) dumontite *f*
DUMORTIERITE - [min] (natural basic aluminium silicate used as a refractory material) dumortiérite *f*
DUMP, to - [gen] basculer, culbuter; décharger en vrac, déverser
[comput] (to transfer the contents of one section of the storage into another section) transférer
DUMP - [gen] tas *m*, amas *m*, crassier *m*, dépotoir *m*
[constr] dépôt *m*, chantier *m* de dépôt
[mech] (railway trucks, lorries, etc) culbuteur *m*, mécanisme *m* de basculement
[comput] (the intentional or accidental withdrawal of power) coupure *f* d'alimentation
[comm] pratiquer le dumping
~ANGLE - [mech] angle *m* de basculement
~BAILER - [oil ind] (in well sinking operations, for the condensation of the bottom of the well) cuillier *m* de cimentation
~BODY - [mech] (of trucks etc) benne *f* basculante
~BOX - [mech] benne *f* basculante
~BUCKET - [metall] (a special vessel capable of being inverted, used in shell moulding) benne *f* basculante
~CAR - [railw] wagon *m* basculant, berline *f* basculante
~CHECK - [comput] (the operation of checking all digits during dumping and verifying the sum when retransferring) vérification *f* de transfert
~CRADLE - [mech] culbuteur *m*
~GAS - [oil ind] (US term for low-grade petrol) essence *f* de mauvaise qualité
~OIL - [oil ind] (US - colloq) huile *f* transportée en barils
~PIT - [min] terril *m*
~RAKE - [agric mach] rateau *m* mécanique
~TRAILER - [transp] remorque *f* à benne basculante
~TRUCK - [transp] camion *m* à benne basculante, wagon *m* basculant
DUMP VALVE - [aero] (special valve for discharging fuel overboard in emergency) vide-vite *m*, clapet *m* de décharge
DUMPED MORAINE - [geol] moraine *f* rempart
DUMPER - [transp] véhicule *m* basculant, wagon *m* basculant, culbuteur *m*
DUMPING - [gen] déchargement *m* en vrac, versage *m*, dépôt *m*
[transp] déversement *m*, culbutage *m*, renversement *m*
[comput] (the operation of withdrawing the power) coupure *f* de l'alimentation
[comput] (the operation of transferring the contents of one section of the store into another section) transfert *m*
[metall] chute *f* de la charge
[comm] dumping *m*
~GROUND - [gen] dépôt *m*, chantier *m* de dépôt, décharge *f*
~MECHANISM - [auto] (the tipping gear) mécanisme *m* de basculement
~STONES - [constr] cailloux *mpl* et pierres
~(TYPE) TANK - [aero] réservoir *m* vide-vite
~WAGON - [railw] wagon *m* basculant
DUMPY LEVEL - [surv] (a type of level with a rigid connexion between telescope and vertical spindle) niveau *m* à lunette (fixe)
DUN - [gen] brun foncé
DUNDER - [brew ind] (lees or dregs of cane-juice, used in the West Indies for the fermentation of rum) écumes *fpl* de jus de canne
DUNE - [geol] (mount of loose sand piled up by the wind) dune *f*
~RANGE - [geol] cordon *m* de dunes
DUNG - [agric] fumier *m*, engrais *m* naturel
[gen] fiente *f*, bouse *f*, crottin *m*
~CUTTER - [agric] hache-fumier *m*
~FORK - [agric] fourche *f* à fumier
~HEAP - [agric] fumier *m*
~HILL - s. dung heap
DUNGAREE - [text] (cotton cloth, mostly coloure, used for overalls) treillis *m*
[gen] bleu *m*, combinaison *f*, salopette *f*
DUNGING CHANNEL - [agric] couloir *m* à purin
~SALT - [chem] silicate *m* de sodium
DUNITE - [geol] (coarse-grained igneous rock occurring in many parts of the world) dunite *f*
DUNKING - [cin] (the process of dipping the film into a chemical bath) immersion *f* de la pellicule
~PULSE - [radio] (colloq for recycling pulse) impulsion *f* de blocage
DUNN BASS - [min] schiste *m* argileux
DUNNAGE - [naut] fardage *m*, calage *m*, bardis *m*
DUNNING - [cin] (a system of double exposure) double impression *f*
DUNNITE - [min] dunnite *f*, nitrate *m* d'ammonium
DUO-CONE LOUDSPEAKER - [acoust] haut-parleur *m* à double cône mixte
DUODECIMAL - [math] duodécimal
~NOTATION - [math] (the scale notation with 12 as base) notation *f* duodécimale
DUODENAL - [anat] duodénal
DUODENECTOMY - [med] (the surgical excision of the duodenum) duodénectomie *f*
DUODENUM - [anat] (the small intestine following the pylorus) duodénum *m*
DUODIODE-TRIODE - [electron] duodiode-triode *f*

DUOLATERAL COIL - [radio] (coil in which the turns are wound in criss-cross fashion) bobine *f* duolatérale
DUOSOL PLANT - [oil ind] installation *f* duosol
DUOSPRUNG FORK - [mech] (in motorbicycles) fourche *f* avant à double ressort
DUOTONE - [print] à deux teintes, bicolore
~PRINTING - [print] tirage *m* en deux couleurs
DUOTRICENARY NOTATION - [math] (scale of notation with 32 as base) notation *f* duotrigésimale
DUPE - [cin] (colloq for duplicate) contre-type *m*
~NEGATIVE - [photo] (negative made from a positive) négatif *m* contre-type
~POSITIVE - [photo] (positive made from a positive) positif *m* contre-type
DUPING - [cin] (making a duplicate negative or positive) contretypage *m*
~PROCESS - s. duping
DUPION - [text] doupion *m*, duppion *m*, cocon *m* double
DUPLET - [chem] (the structure in which two atoms share a pair of electrons, which form a single bond) doublet *m*
DUPLEX - [gen] double, duplex
~ARTIFICAL CIRCUIT - [radio] (a balancing network) réseau *m* équivalent
~BALANCE - s. duplex artificial circuit
~BURNER - s. duplex fuel burner
~CARBURETTOR - [auto] carburateur *m* double
~COMMUNICATION - [telecomm] liaison *f* duplex
~DIODE - [radio] double diode *f*
~FUEL BURNER - [th eng] (type of burner in gas turbines in which low- and high-pressure sprays are combined, so as to provide a wide range of power demand) brûleur *m* double
~-HEAD VERTICAL MILLING MACHINE - [mach tool] fraiseuse *f* Duplex verticale
~MILLER - [mach tool] fraiseuse *f* Duplex (à deux fraises opposées)
~OPERATION - [radio] (the concurrent operation of transmission and reception) fonctionnement *m* en duplex
~PROCESS - [metall] (the combination of two methods in performing one operation) procédé *m* duplex
~PUMP - [mech] (pump working with two cylinders placed side by side) pompe *f* duplex
PURIFIER - [ind chem] (a tower purifier) tour *f* de épuration
~SYSTEM - [radio] (circuit fitted with balancing devices, so that the simultaneous both-way transmission of telegraph signals by the modulation of a continuous current is possible) système *m* duplex, télégraphe *m* duplex
DUPLEXED LINE - [telecomm] liaison *f* duplex
DUPLEXER - [radar] (single aerial used in radar systems for receiving and transmitting) antenne *f* unique pour l'émission et la réception
DUPLEXING - s. duplex process
DUPLICATE, to - [gen] reproduire (en double), faire un double, polycopier
DUPLICATE - [gen] calque *m*, copie *f*, double *m*, duplicate *m*
~CIRCUITRY - [comput] (the circuitry which makes a duplication check possible) circuit *m* double, circuit *m* à deux voies
MOULDING - [metall] surmoulage *m*
~PARTS - [gen] pièces *fpl* de rechange
DUPLICATE PLATE - [plast ind] plateau *m* de duplication
DUPLICATED - [gen] tiré au duplicateur, polycopié
~NEGATIVE - s. dupe negative
DUPLICATING - [gen] double *m*, reproduction *f*, duplication *f*
~LATHE - [mach tool] tour *m* à reproduire, tour *m* à copier
~MACHINE - [print] duplicateur *m*
~PAPER - [paper man] papier *m* pour duplicateur
DUPLICATION - [gen] duplication *f*, reproduction *f*, doublage *m*
[cin] contretypage *m*
~CHECK - [comput] (check by obtaining the identical result from two independent performances of the same operation) vérification *f* par duplication
DUPLICATOR - s. duplicating machine
DUPLICATURE - [med] duplicature *f*, repli *m*
DURA - [anat] (the outermost of the three membranes surrounding the brain and spinal cord) dure-mère *f*
DURABILITY - [gen] durabilité *f*, durée *f*, endurance *f*, résistance *f* (des matériaux)
DURABLE - [gen] durable, résistant
DURAIN - [min] (a constituent of some types of coal, of a compact granular texture, probably formed chiefly from seeds and spores) durain *m*
DURALUMIN - [metall] (aluminium alloy containing copper, manganese and magnesium, used in construction) duralumin *m*
DURAMEN - [timber] (the inner part of a trunk, hard and often dark-coloured) duramen *m*, cœur *m* du bois
DURAN - [metall] (contraction for duralumin) duralumin *m*
DURAPLASTY - [med] (a graft of the dura) opération *f* de la dure-mère
DURARC LIGHT - [cin] lampe *f* à arc à deux paires de charbons, duarc *m*
DURATION - [gen] durée *f*
~OF THE FREE VIBRATIONS - [electron] durée *f* de l'amortissement
DUREMATOMA - [med] hématome *m* dural, hématome *m* de la dure-mère
DURENE - [chem] (intermediate for polymers and synthetic fibres) durène *m*
DURESS - [mech] force *f*, pression *f*, contrainte *f*
DURN - [mining] cadre *m* de galerie
DUROMETER - [instr] (an instrument designed to measure hardness, usually a type of drill working under a specified pressure) duromètre *m*
DUROSARCOMA - [med] tumeur *f* des méninges
DURRA - [bot] (a type of Indian millet) doura *m*, sorgho *m*
DUSCHEZIA - [med] (a form of constipation) dyschésie *f*, dyschézie *f*
DUST, to - [gen] saupoudrer, épousseter, poudrer (traitement insecticide)
[min] pulvériser, poudrer
DUST - [gen] poussière *f*, poudre *f*
~ARRESTER - [mech] collecteur *m* de poussières
~BIN - [impl] poubelle *f*, boîte *f* à ordures
~-BRUSH - [impl] balayette *f*
~CAP - [mech] bouchon *m* de protection, cache-poussière *m*
~CAR - [transp] camion *m* pour l'enlèvement des ordures

DUST CART - [transp] tombereau *m* à ordures
~CATCHER - [metall] (a chamber for the extraction of gases from metal furnaces) capteur *m* de poussières, séparateur *m* de poussières, dépoussiéreur *m*
~CATCHING APPARATUS - s. dust catcher
~CLOUD - [met] (a cloud composed of fine particles of dust, met with in arid regions) nuage *m* de poussière
~COAL - [min] charbon *m* pulvérulent, poussier *m*
~COLLECTOR - [mech] (an apparatus designed to separate dust from a gaseous medium and collect it) collecteur *m* de poussières, aspirateur *m* de poussières
~CONTENT - [soil] teneur *f* en poussière
~CORE - [el] noyau *m* à poudre de fer, noyau *m* de fer pulvérulent
~COUNTER - [instr] (an instrument designed to measure the number of dust particles in a unit volume of air or other gas) détecteur *m* de poussières
~COVER - [gen & mech] cache-poussière *m*, pare-poussière *f*
~-DEVIL - [met] (a rotating thermal, often hollow, wind which lifts a column of dust in arid regions) tourbillon *m* de sable, tornade *f* de sable
~EXHAUSTER - [mech] (apparatus for withdrawing dust-laden air) aspirateur *m* de poussières
~EXPLOSION - [mech] (explosion caused by concentration of inflammable dust) explosion *f* de poussières, coup *m* de poussier
~EXTRACTING PLANT - [mech] installation *f* de dépoussièrage
~FILTER - [mech] filtre *m* à poussières
~GOLD - [min] poudre *f* d'or
~HOOD - [auto] housse *f* pour automobiles
~LAVING - [min] abattage *m* des poussières
~LAYING - [mining] abattage *m* des poussières
~LINEN - [text] (strong cotton cloth) toile *f* à stores
~MASK - [impl] masque *m* anti-poussières
~PLUG - [impl] bouchon *m* de protection
~POCKET - [metall] sac *m* à poussières
~PRECIPITATOR - [mech] (apparatus for extracting dust from air or other gases) précipitateur *m*
~-PROOF - [mach] (of electrical motors and machines) protégé contre les poussières
~-PROTECTION CAP - [mech] s. dust cap
~-REMOVAL INSTALLATION - [mech] installation *f* de dépoussièrage
~SEPARATOR - [mech] capteur *m* de poussière, collecteur *m* de poussière, appareil *m* de dépoussièrage ·
~SHIELD - [mech] écran *m* pare-poussières
~STORM - [met] (a large cloud of dust lifted and carried by a windstorm in arid regions) tempête *f* de poussières
~-TIGHT - [mech] protégé contre les poussières
~TRAP - [mech] collecteur *m* de poussières
~TRUNK - [text] caisse *f* à poussières
~TUFF - [geol] tuf *m* volcanique
DUSTER - [gen] chiffon *m* à poussière, torchon *m*
[agric] poudreuse *f*
[impl] (a small brush used to remove dust from the surface before painting) brosse *f*
[paper man] (used for dusting rags etc) dépoussiéreur *m*
[oil ind] (colloq for dry hole) puits *m* improductif
DUSTING - [gen] dépoussièrage *m*, poudrage *m*, époussetage *m*
[tyres etc] poudrage *m*, talquage *m*
[agric] poudrage *m* (des cultures)
DUSTING AIRCRAFT - [aero] (aircraft specially fitted for spreading powder over crops, usually for insecticidal purposes) avion *m* pour le poudrage des cultures
~APPARATUS - [mech] poudreuse *f*, machine *f* à talquer
~BRUSH - [gen] pinceau *f* à épousseter
~CROPS - [agric] (the operation of spreading insecticide powder over crops from aircraft) poudrage *m* des cultures
~SPACE IN THE EXTRUDING NOZZLE - [mech] circuit *m* de talquage de tête de boudineuse
DUSTY - [gen] poussièreux, pulvérulent
DUTCH ARCH - [constr] (also called French arch) arc *m* plate-bande
~BOND - [constr] appareil *m* croisé
~CLOVER - [bot] trèfle *m* blanc
~GOLD - [metall] clinquant *m*(d'or)
~LINEN - [text] hollande *f*, toile *f* de Hollande
~LIQUID - [chem] (synonym for ethylene chloride) liqueur *f* des Hollandais, chlorure *m* d'éthylène
~MARBLE PAPER - [paper man] (strong paper with marble patterns) papier *m* marbré
~OIL - s. Dutch liquid
~PROCESS - [chem] (a method of manufacturing white lead by attacking lead with dilute acetic acid) procédé *m* hollandais
~ROLL - [aero] (lateral roll round the vertical axis) boucle *f* latérale
DUTCHMAN - [gen] hollandais *m*
[carp] (piece or strip of wood inserted in a gap left in a joint) flipot *m*, morceau *m* rapporté
DUTCHMAN'S PIPE - [bot] aristoloche *f* en arbre
DUTY - [gen] devoir *m*, fonction *f*, service *m*
[leg] droit *m*, taxe *f*, impôt *m*, redevance *f*
[mech] (performance of a machine in proportion to the energy required) rendement *m*, débit *m*
[mech] (e.g. heavy-duty, light-duty, etc, to indicate the conditions of service of a product or machine) rendement *m*, service *m*
~CHART - [gen] tableau *m* de service
~CYCLE - [el] (the cycle of operations which must be performed by an electrical apparatus) coefficient *m* d'utilisation, période *f* de fonctionnement
~FACTOR - [electron] (the ratio between the operating period and the total period during which an electronic tube is operating) coefficient *m* d'utilisation, facteur *m* d'utilisation
~FREE - [comm & leg] exempt de droit, en franchise
~PAID ENTRY - [comm & leg] déclaration *f* d'acquittement des droits
~PAID SALE - [comm & leg] vente *f* à l'acquitté
D.W. - s. dead weight
DWALE - [bot] belladonne *f*
DWARF - [zool] nain *m*
~BEAN - [bot] haricot *m* nain
~FAN PALM - [bot] palmette *f*
~STAR - [astr] (a low-luminosity star) étoile *f* naine
DWARFISHNESS - [med] nanisme *m*
DWARFISM - [bot] nanisme *m*
DWARFNESS - s. dwarfism
DWELL, to - [gen] demeurer, habiter, résider, rester
[mech] (to remain for a fixed time in the same po-

sition, as of a reciprocating part) s'arrêter (momentanément)
[plast ind] (to allow the escape of gas from the moulding material) dégazer
DWELL - [gen] arrêt *m*, pause *f*, repos *m*
[metall & plast ind] (a pause in the application of pressure to a mould, made just before the mould is completely closed, to allow the escape of gas from the moulding material) pause *f* de fermeture du moule
~OF THE SLAY - [text] arrêt *m* du battant
~TIME - [mech] temps *m* d'arrêt
DWELLING - [gen] habitation *f*, demeure *f*, résidence *f*
~HOUSE - [gen] maison *f* d'habitation
DWINDLE, to - [constr] (e.g. of cement) se rétrécir, se contracter
DWINDLING - [constr] (of cement) retrait *m*
DYAD - [math] dyade *f*
[chem] radical *m* divalent
DYAKISDODECAHEDRON - [phys] (of crystals) diploèdre *m*
DYCLONINE HYDROCHLORIDE - [chem] (a local anaesthetic) chlorhydrate *m* de cyclonine
DYE, to - [gen] (to colour with a dye) teindre, colorer
DYE - [gen] teinte *f*, teinture *f*, matière *f* colorante
[chem] (soluble pigment having the property of becoming attached to fibres) colorant *m*
[text] teinture *f*
~AFFINITY - [dyes] pouvoir *m* colorant
[text] affinité *f* tinctoriale, substantivité *f*
~BACKING - [photo] couche *f* dorsale teintée
~BATH - [photo] bain *m* de teinture
~BEAM - [text] ensouple *f* pour teinture
~BECK - [text] cuve *f* de teinture
~BOBBIN - [text] tube *m* de teinture, douille *f* de teinture
~COUPLER - [photo] copulant *m*
~INTERMEDIATE - [chem] intermédiaire *m* pour colorants
~PLANT - [bot] plante *f* tinctoriale
~REDUCER - [photo] faiblisseur *m* de couleur
~RETARDER - [ind chem] retardateur *m* pour colorants
~SPOT - [gen & paper man] tache *f* de couleur
~SPRAYER - [mech] (a device for applying a dye by means of a blast of compressed air or by a spray of liquid dye-stuff) pulvérisateur *m* de colorant
~TONING - [photo] virage *m*
~WOOD - [constr] bois *m* colorant
~-WORKS - [text etc] teinturerie *f*, atelier *m* de teinture
DYEABILITY - [text] teintabilité *f*, affinité *f* à la teinture
DYED - [gen] teinté, teint, coloré
~CLOTH - [text] tissu *m* teint
~IN THE PIECE - [text] teint en pièce
~LAYER - [photo] couche *f* colorée
~YARN - [text] fil *m* teint
DYEING - [gen text etc] teinture *f*, teintage *m*
~APPARATUS - [gen] appareil *m* de teinture
~IN THE YARN - [text] teinture *f* en fil
~MACHINE - [text] machine *f* à teindre
~VAT - [impl] cuve *f* de teinture
~WASTES - [ind chem] eaux *fpl* résiduaires de teinture
DYER - [gen] teinturier *m*
DYER'S OAK - [bot] (oak from which dyestuffs can be extracted) chêne *m* des teinturiers
~OIL - [ind chem] huiles *fpl* tournantes
DYESTUFF - [ind chem] (compounds used as dyes) matière *f* colorante
DYEWOOD - [timber] (quality of timber used for the extraction of dyestuffs) bois *m* tinctorial, bois *m* de teinture
DYING OUT - [radio] évanouissement *m*
DYKE - [hydr] (a wall, embankment, etc, used to confine the flow of a river) digue *f*
[geol] (minor intrusive vein into the crust) dyke *m*, filon *m* d'injection
~ROCK -[geol] roche *f* filonienne
DYNAFLOW - [mech] (proprietary name for a type of fluid drive) transmission *f* hydraulique 'Dynaflow'
DYNAMIA - [med] dynamie *f*
DYNAMETER - [instr] (instrument designed to measure the magnifying power of telescopes) dynamètre *m*
DYNAMIC - [gen] dynamique
~ACCENT - [acoust] (the stressing of a sound by relatively greater volume) accent *m* dynamique
~ALLOTROPY - [phys] (allotropic phenomena in which the transition from one form to another is reversible) allotropie *f* dynamique
~BALANCE - [mech] (condition of a rotating part or assembly, such that no cycle stresses are set up when rotating at the speed for which such balance exists) équilibre *m* dynamique
~BALANCING - [mech] (adjustment of rotating parts so that there is no cyclic stress variation due to rotation) équilibrage *m* dynamique
~BRAKING - [el] (rheostatic braking; braking an electromechanical system with an electric motor) freinage *m* rhéostatique
~BRECCIA - [geol] brèche *f* tectonique
~CHARACTERISTIC - [electron] (the characteristic under operating conditions) caractéristique *f* dynamique
~CHARACTERISTICS OF A WELDING SET - [metall] caractéristique *f* dynamique d'un appareil d'alimentation pour soudage à l'arc
~CONDENSER ELECTROMETER - [instr] (instrument designed to move the impressed charge in an electrostatic field by mechanical energy) électromètre *m* à condensateurs
~CONTROL SYSTEM - [contr] réglage *m* dynamique
~CONVERGENCE - [telev] (in colour television) convergence *f* dynamique
~DEMOSTRATOR - [telev] (a working diagram of a radio or television receiver with components mounted on the diagram) schéma *m* imagé
~EFFECT - [chem] effet *m* dynamique
~ELECTRICITY - [el] (electric charges in motion) électricité *f* dynamique
~ELECTRODE POTENTIAL - [el chem] (the electrode potential which exists when current flows between electrode and electrolyte) potential *m* dynamique d'une électrode
~EQUILIBRIUM - s. dynamic balance
~EQUILIBRIUM RATIO - [phys] rapport *m* dynamique d'équilibre
~FACTOR - [aero] (the ratio between the load on a

structural element of an aircraft under acceleration and the basic load) coefficient *m* dynamique
DYNAMIC FLOW METER - [meas] débitmètre *m* dynamique
~FOCUSING - [telev] (the automatic process of varying the voltage on the focusing electrode of a TV picture tube) focalisation *f* dynamique
~GEOLOGY - [geol] géologie *f* dynamique
~HEAD - [mech] (in the mechanics of fluids) charge *f* dynamique, pression *f* dynamique
~HEATING - [aero] (the heat caused by the motion through the air) échauffement *m* dynamique
~HYSTERESIS LOOP - [phys] (the curve of magnetization as opposed to applied magnetomotive force per unit length which is obtained with the magnetic material cyclically magnetized at some specified rate) courbe *f* d'hystérésis dynamique
~IMPEDANCE - [radio] (the impedance of a rejector at its resonant frequency) impédance *f* dynamique
~ISOMERISM - [chem] (a synonym for tautomerism; a property by which a substance may exist as an equilibrium mixture of two forms which are inconvertible, thus giving rise to two distinct series of derivatives) tautomérie *f*
~LIFT - [aero] (the component of the total aerodynamic forces on an aerofoil which is normal to the wind direction) force *f* ascensionnelle dynamique
~LOAD - [aero] (load on aircraft structure due to acceleration) charge *f* dynamique
~LOUDSPEAKER - [el acoust] haut-parleur *m* électrodynamique
~LUMINOUS SENSITIVITY - [electron] (of a photoelectric device) sensibilité *f* lumineuse dynamique
~MEMORY - [comput] mémoire *f* dynamique
~METAMORPHISM - [geol] (metamorphism due principally to greatly increased pressure) dynamométamorphisme *m*
~METEOROLOGY - [met] (the part of the science of meteorology which deals with movement in the atmosphere) météorologie *f* dynamique
~MICROPHONE - [el acoust] microphone *m* électrodynamique
~MODEL - [mech] (a model having mass distribution and linear dimensions, such as to ensure that it will behave in the same way as the prototype) modèle *m* dynamique
~MODULUS - [phys & acoust] (the ratio between stress and strain under vibratory conditions) module *m* dynamique
~PRESSURE - [phys] (the difference between the static and total pressure) pression *f* dynamique
~QUALITY - [mech] facteur *m* de qualité dynamique
~RANGE - [el acoust] (the range of sounds which a tape recorder can produce without distortion) gamme *f* dynamique
~RUNOUT - [comput] excentrage *m* dynamique, faux rond *m* dynamique
~SENSITIVITY - [electron] (of a photo-electric device) sensibilité *f* dynamique
~STABILITY - [phys] (complete stability of movement, making allowance for all aerodynamic forces, and for inertia and gravity) stabilité *f* dynamique
~STORAGE - s. dynamic memory
~STRESS - [phys] contrainte *f*, dynamique, effort *m* dynamique
~VISCOSITY - s. absolute viscosity
DYNAMIC WAVEMETER - [meas] ondemètre *m* électrodynamique
DYNAMICAL ANALOGIES - [phys] (the formal similarities among differential equations of electrical, mechanical, and acoustic systems) analogies *f*pl dynamiques
~UNIT - [phys] unité *f* dynamique
DYNAMICIZER - [comput] (circuitry converting a time sequence of states representing digits into a corresponding time sequence) convertisseur *m* série-parallèle
DYNAMICS - [phys] (the science of the motion of bodies affected by forces) dynamique *f*
DYNAMISM - [gen] dynamisme *m*
DYNAMITE - [chem] (commercial explosive consisting of nitroglycerine absorbed in kieselguhr or other suitable substance) dynamite *f*
DYNAMITING - [gen] dynamitage *m*
DYNAMO - [el] (commonly denoting a direct-current generator) dynamo *f*
~-BATTERY IGNITION UNIT - [auto] allumage *m* par batterie
~COUPLING - [el] entraînement *m* de la dynamo
~-ELECTRIC - [el] dynamo-électrique, électrodynamique
~-ELECTRIC AMPLIFIER - [radio] amplificateur *m* électrodynamique
~GOVERNOR - [el] régulateur *m* de dynamo
~FIELD COIL - [el] enroulement *m* d'excitation
DYNAMOGENY - [med] dynamogénie *f*
DYNAMOMETAMORPHISM - [geol] dynamométamorphisme *m*
DYNAMOMETER - [meas] (apparatus for measuring mechanical power, as of an engine) dynamomètre *m*
~AMMETER - [el] (ammeter operating on the dynamometer principle) ampèremètre *m* dynamométrique
~CAR - [railw] wagon *m* dynamométrique
DYNAMOTOR - [el] (electric machine with two armature windings, acting respectively as a generator and a motor) dynamoteur *m*
DYNAMOMETRIC - [mech] dynamométrique
~WATTMETER - [el] wattmètre *m* dynamométrique
DYNAMOMETRY - [mech] dynamométrie *f*
DYNATRON - [electron] (a multi-electrode thermionic valve) dynatron *m*
~CHARACTERISTIC - s. dynatron effect
~EFFECT - [electron] (effect equivalent to a negative resistance when the electrode characteristic has a negative slope) effet *m* dynatron
~OSCILLATOR - [radio] oscillateur *m* dynatron
DYNE - [mech] (a unit of force in the c.g.s. system) dyne *f*
DYNETRIC BALANCING - [contr] (electronic method of balancing rotating parts) équilibrage *m* électronique
DYNODE - [electron] (of an electronic tube; electrode supplying a secondary electron emission) cathode *f* secondaire
DYOTRON - [electron] (single-cavity microwave oscillator tube with three electrodes) dyotron *m*
DYSCRASITE - [min] (natural silver antimonide; one of the less common sources of silver) dyscrase *f*, dyscrasite *f*
DYSCRYSTALLINE - [geol] mal cristallisé
DYSENTERY - [med] dysenterie *f*
DYSFUNCTION - [med] dérèglement *m*
DYSGENESIA - [med] (a tendency towards degenera-

tion) homogénie *f* dysgénésique, dysgénésie *f*
DYSLEXIA - [med] dyslexie *f*
DYSMENORRHEA - [med] (painful and irregular menstruation) dysménorrhée *f*
DYSMIGRATION - [min] dysmigration *f*
DYSOPSIA - [med] dysopie *f*, dysopsie *f*
DYSOSTOSIS - [med] dysostose *f*
DYSPEPSIA - [med] (any form of indigestion) dyspepsie *f*
DYSPHAGIA - [med] (a difficulty in swallowing) dysphagie *f*
DYSPHOTIC ZONE - [geol] région *f* dysphotique
DYSPHRENIA - [med] psychose *f* fonctionnelle
DYSPLASIA - [med] (abnormal development) dysplasie *f*
DYSPNEA - [med] dyspnée *f*
DYSPROSIUM - [chem] (a rare earth element, A.N. 66, A.W. 162.5, symbol Dy. It is the most basic of the erbium sub-group of the yttrium family) dysprosium *m*
DYSTECTIC MIXTURE - [chem] (a mixture in which the proportions of the constituents are such that it has the highest melting point attainable for such a combination) mélange *m* dystectique
DYSTOCIA - [med] (difficult and painful childbirth) dystocie *f*
DYSTROPHIC - [med] dystrophique
DYSURIA - [med] (the painful passage of urine) dysurie *f*
DZUS FASTENER - [mech] (type of hollow rivet) attache *f* 'dzus'

E

e - [phys] (coefficient of restitution) coefficient *m* de restitution
[math] (base of hyperbolic or natural logarithms) base *f* des logarithmes naturels (ou népériens)
[met] (symbol for wet air) symbole *m* de l'air humide

E - [phys] (symbol for potential difference) symbole *m* de différence de potentiel
[el] (symbol for elasticity modulus) symbole *m* du module d'élasticité ou module d'Young
[el] (symbol for effective voltage) symbole *m* de la tension efficace

EAGLE - [zool] aigle *m*
[cin] (insect flying across the view during shooting) insecte *m* dans le champ
[photo] (a perfect take) prise *f* de vue excellente
~STONE - [min] aélite *f*, pierre *f* d'aigle
~-WOOD - [bot] calambac *m*, calamboc *m*, calambar *m*

EAGLET - [zool] aiglon *m*

EAM - [comput] (Electric Accounting Machine) machine *f* comptable

EAR - [gen] oreille *f*, anse *f*, œillet *m*
[anat] oreille *f*
[bot] épi *m*
[plumb] (on pipes) oreille *f*
~-ACHE - [med] (a pain in the drum of the ear) mal *m* d'oreille, douleur *f* d'oreille, otalgie *f*
~BIT - [mech] trépan *m* à oreilles
~BLOCK - [med] obstruction *f* de la trompe d'Eustache
~BONES - [anat] osselets *mpl* de l'oreille
~-DEAFENING - [acoust] assourdissant
~DEFENDER - [impl] (any device protecting the ear drums against too loud sounds) protège-tympan *m*, protège-oreille *m*, protecteur *m* auriculaire
~DROPS - [med] gouttes *fpl* auriculaires
~DRUM - [anat] tympan *m*
~FLAP - [anat] (the exterior part of the ear which collects air vibrations) lobe *m* de l'oreille
~-PIECE - [acoust] (a shaped piece of plastic material inserted into the ear when used in connection with a hearing aid) embout *m*, écouteur *m* interne
~PLUG - [impl] antiphone *m*, tampon *m* auriculaire
~SCOOP - [impl & med] cure-oreilles *m*
~-SHOT - [acoust] portée *f* de la voix, distance *f* d'écoute
~TAG - [agric] (a marking of cattle) bouton *m* de marquage (auriculaire)
~TINKLING - [acoust] tintement *m* d'oreilles
~-TRUMPET - [impl] (a straight or curved tube used by partially deaf people to intensify sounds) cornet *m* acoustique

EAR WAX - [med] cérumen *m*

EARED SCREW - [mech] vis *f* à oreilles

EARING - [agric] (the growing of the ears) épiage *m*, épiaison *m*, épiation *f*
[naut] (small rope fastening the upper corner of a sail to the yard) empointure *f* de voile, raban *m*
~END - [agric] fin *m* de l'épiage

EARLY - [gen] matinal, précoce, hâtif, prématuré
~FLOWERING - [agric] floraison *f* hâtive
~RIPENING - [agric] hâtif
~SETTING CEMENT - [constr] ciment *m* à prise rapide, ciment *m* prompt
~SPARK - [mech] avance *f* à l'allumage, étincelle *f* anticipée
~TIMING - [text] (in knitting) réglage *m* des cames de chute
~WARNING - [radar] veille *f* avancée, couverture *f* radar avancée
~WHEAT - [agric] blé *m* hâtif

EARMARK, to - [gen] marquer, assigner, faire une marque, affecter

EARMARK - [gen] marque *f*, affectation *f*

EARN, to - [gen] gagner, mériter

EARNEST MONEY - [comm] arrhes *fpl*, dépôt *m* de garantie

EARNINGS - [comm] salaire *m*, gain *m*, gages *mpl*, profit *m*, recette *f*, récompense *f*
~CARD - [comput] carte *f* de salaire

EARPHONE - [acoust] (transducer closely coupled acoustically to the ear) écouteur *m*
~COUPLER - [acoust] (cavity used for testing earphones in conjunction with a calibrated microphone) coupleur *m* pour écouteur

EARTH, to - [el] (to connect to earth) mettre à la terre, mettre à la masse
[constr] terrasser, enfouir, enterrer

EARTH - [astr] (the planet) terre *f*
[el] (connexion to the mass of the earth by a very-low impedance conductor) terre *f*, masse *f*
[radio] (system of plates or wires in the ground, to which the connection is made) terre *f*
~AERIAL - [radio] (an aerial buried in the ground) antenne *f* enterrée
~ALMOND - [bot] amande *f* de terre, souchet *m* comestible
~ANCHOR - [agric] bêche *f* d'ancrage
~ARCH - [constr] arc *m* établi sur terre coupée en cintre
~AUGER - [mining] tarière *f* à large spire
~BASIN - [hydr] bassin *m* en terre
~BORER - [constr] sonde *f*, tarière *f*, tire-sable *m*, fer *m* à sonder
~BORING MACHINE - [mining] sondeuse *f*

EARTH CEMENT MORTAR - [constr] mortier *m* ciment-terre
~CIRCUIT - [radio] (the transmitter or receiver circuit which includes the earth lead) circuit *m* de terre
~CLOSET - [hydr] cabinet *m* sec
~COIL - [el] (pivoted, large coil designed to measure the strength of the earth magnetism) inducteur *m* de terre
~COLOURS - [paint] (pigments made from earths, generally silicates, oxides and hydrated oxides of such elements as aluminium, iron, manganese, silicon, etc) terres *fpl* colorées
~COMPACTION - [soil] (soil compaction) compactage *m* du sol, serrage *m* du sol
~CONNEXION - [el] prise *f* de terre
~CONTACT - s. earth connexion
~-CREEP - [geol] glissement *m* de terrain
~CURRENT - [el] (the current which is flowing in the earth lead) courant *m* à la terre, courant *m* de fuite, courant *m* de retour
~CURRENT PROSPECTING - [geol] (study of the subsoil by the observation of earth currents) prospection *f* électrique
~CURRENTS - [el] (the currents in the earth which affect submarine cables; also the currents in the earth which cause corrosion in the cable sheathing) courants *mpl* telluriques, courants *mpl* telluriens
~CURVATURE - [geol] courbure *f* de la terre
~DAM - [hydr] barrage *m* en terre, digue *f* en terre
~DAM EMBANKMENT - [constr] levée *f* de terre
~DETECTOR - [el] (a leakage indicator) indicateur *m* de terre
~DIKE - [hydr] digue *f* en terre
~DRILL - [impl] sonde *f* à tarière
~ELECTRODE - [el] (a conductor, or a combination of conductors, buried in the earth to make a connexion with it) prise *f* de terre, électrode *f* de masse
~ENBANKMENT - s. earth dam
~FALL - [geol] éboulement *m*
~FAULT - [el] (accidental connexion) mise *f* à la terre accidentelle
~GRAB - [mech] benne-drague *f*
~INDUCTOR - s. earth coil
~-INDUCTOR COMPASS - [aero] (apparatus for determining the angle between the direction of the earth's field and the heading of an aircraft by means of a conductor system, the E.M.F. generated in the latter being read in a scale of degrees) compas *m* à induction terrestre, boussole *f* d'induction
~LEAD - [el] (a wire or cable used to make connection to earth) câble *m* de masse
~LEAKAGE - [el] contact *m* à la terre, perte *f* à la terre, fuite *f* à la terre
~LEAKAGE CURRENT - [el] (the current which is flowing to the earth owing to a faulty insulation) courant *m* de perte à la terre
~LEAKAGE PROTECTION - [el] (dispositif *m* de protection contre les défauts à la terre
~LIGHT - s. earth shine
~LOAD - [constr] poussée *f* des terres
~MAGNETISM - [el] (also called terrestrial magnetism; the magnetic properties within, on and outside the earth's surface) magnétisme *m* terrestre
~MASS - [soil] masse *f* de sol
~MOVEMENTS - [geol] (the change of attitude of the strata) mouvements *mpl* de l'écorce terrestre
EARTH MOVER - [mech] engin *m* de terrassement
~MOVING - [constr] terrassement *m*
~-MOVING EQUIPMENT - [mech] engins *mpl* de terrassement
~MOVING MACHINERY - [mech] matériel *m* de terrassement
~-NUT - [bot] terre-noix *f*, truffle *f*, arachide *f*
~-NUT CAKE - [agric] (foodstuff for animals) tourteau *m* d'arachides
~-NUT OIL - [agric] huile *f* d'arachide
~OIL - [min] huile *f* minérale, huile *f* de roche, huile *f* de pierre
~PEA - [bot] pois *m* des champs
~-PHANTOM CIRCUIT - [radio] (earth-return circuit derived from pairs with wires effectively in parallel) circuit *m* fantôme à retour par la terre
~PITCH - [min] pissasphalte *m*
~PLATE - [el] (metal plate in the earth to provide a connexion between an electrical system and the earth) prise *f* de terre, plaque *f* de masse, plaque *f* de terre
~POTENTIAL - [el] (the electric potential of the earth; as this is regarded as zero, all potentials are referred to the earth potential) potentiel *m* de terre
~PRESSURE - [constr] (the pressure exerted by the earth when it is laterally supported by a wall) poussée *f* des terres, pression *f* des terres
~PRESSURE AT REST - [soil] pression *f* naturelle des terres, pression *f* des terres au repos
~RAMMER - [impl] pilon *m*, dame *f*
~REFINING - [ind chem] raffinage *m* à la terre
~RESERVOIR - [hydr] réservoir *m* enterré
~RESISTANCE - [el] (the resistance provided by the earth between two points of connexion) résistance *f* de terre
~RESISTANCE METER - [instr] (instrument designed to measure the resistance between an earthing plate and the surrounding ground) tellurohmmètre *m*
~RETAINING WALL - [constr] mur *m* de soutènement
~RETURN - [el] (a return circuit through electrically bonded parts) retour *m* par la masse
~RETURN SYSTEM - [el] circuit *m* avec retour par la masse
~RIPPER - [mech] excavateur *m* de tranchée
~ROD - [radio] (rod driven into the ground for a good earth connexion) piquet *m* de terre
~RUBBER - [rubber ind] (the rubber coagulating on the ground) caoutchouc *m* de terre, scrap *m* de terre
~SATELLITE - [astronaut] (a man made object orbiting the earth) satellite *m* artificiel
[astr] satellite *m* de la Terre
~SCOOP - [mech] pelle *f* niveleuse
~SCREEN - [radio] (group of conductors above the earth and supplementary to an earth system) contrepoids *m*
~SCULPTURE - [geol] glyptogenèse *f*
~SENSOR - [astronaut] (light-sensitive diode operating a photo-multiplier and advising the attitude control system when it sees the earth) détecteur *m* d'orientation terrestre
~SHINE - [astr] lumière *f* cendrée
~STAY - [telecomm] tirant *m*
~SYSTEM - [el] (system of electrically bonding metal parts together) mise *f* à la masse
~TANK - s.earth reservoir

EARTH THERMOMETER - [instr] (instrument designed to measure the temperature of the earth at a given depth) thermomètre *m* servant à mesurer la température du sol
˜UP, to - [agric] butter, terrer, chausser
˜WAVE) [geol] onde *f* sismique, onde *f* séismique
˜WAX - [min] (mineral paraffin-wax) cire *f* minérale, ozokérite *f*
˜WIRE - [el] (wire electrically connected with earth) fil *m* de terre
˜-WORK ENGINEERING - [constr] technique *f* des ouvrages en terre
EARTH'S MAGNETIC FIELD - [phys] champ *m* magnetique terrestre
˜RATE CORRECTION - [astronaut] compensation *f* de mouvement diurne
EARTHCLOD PLANTING - [agric] plantation *f* en mottes
EARTHED - [el] mis à la terre, mis à la masse
˜CATHODE OPERATION - [electron] (mode of operating an amplifier valve, the cathode being maintained at earth potential) méthode *f* de fonctionnement avec la cathode à la masse
˜CIRCUIT - [el] (assembly of conductors with one or more points permanently connected to earth) circuit *m* mis à la terre
˜GRID - [el] grille *f* d'arrêt, grille *f* de freinage
˜GRID OPERATION - [electron] (mode of operating an amplifying valve, in which the grid is maintained at earth potential) méthode *f* de fonctionnement avec la grille à la masse
˜NEUTRAL - [el] (neutral point of a polyphase sys tem connected to the earth) point *m* neutre à la terre
˜SWITCH - [el] (switch making it possible to earth all the exposed metal parts) commutateur *m* de mise à la terre
˜SYSTEM - [el] (electric supply system in which one pole is earthed) système *m* mis à la terre
EARTHEN - [gen] de terre
EARTHENWARE - [pottery] (ordinary, soft-body pottery, with or without glaze; a product inferior to chinaware) poterie *f*, faience *f*, grès *m*, terre-cuite *f*
˜DRAIN - [hydr] drain *m* en poterie
˜DUCT - [el] (earthenware conduit carrying underground cables) conduit *m* en terre cuite, tuyau *m* en grès
˜INDUSTRY - [pottery] faïencerie *f*
˜PIPE - [plumb] tuyau *m* en poterie, tuyau *m* en terre cuite
˜SLAB - [constr] carreau *m* de terre cuite
˜TILE - [constr] tuile *f* de terre cuite
EARTHFAST - [gen] fixé au sol
EARTHING - [el] (the operation of effecting electrical connection between an electrical apparatus or an electric circuit and earth) mise *f* à la terre, mise *f* à la masse
˜CABLE BOND - [el] connexion *f* de gaine métallique
˜CLAMP - [el] collier *m* de mise à la terre
˜CONTACT - [el] contact *m* de mise à la terre
˜PLATE - [el] (in transformers and electrical machines) plaque *f* de terre
˜RESISTOR - [el] (resistance for the earthing of a neutral point in a supply system) résistance *f* de mise à la terre, court-circuiteur *m*
˜SWITCH - [el] (a switch used to connect some part of a circuit to earth) commutateur *m* de mise à la terre
EARTHING TERMINAL - [el] (terminal used to make a connexion to earth) borne *f* de terre
EARTHQUAKE - [geol] (a shaking of the crust of the earth) tremblement *m* de terre, séisme *m*
˜RECORD - [geol] séismogramme *m*
˜RESISTING - [constr] résistant aux tremblements de terre
˜WAVE - [geol] onde *f* séismique
EARTHQUAKING RESISTING BUILDING - [constr] construction *f* résistant aux tremblements de terre
EARTHWORK - [constr] (earth-moving work) terrassememnt *m*
[constr] (banking or cutting) ouvrage *m* en terre
EARTHWORM - [zool] lombric *m*
EARTHY - [gen] terreux, terrestre
[el] (of a circuit connected to earth) mis à la terre
˜COAL - [min] charbon *m* terreux
˜COBALT - [min] cobalt *m* oxydé noir
˜MARL - [soil] terre *f* marneuse, sol *m* marneux
˜WATER - [hydr] eau *f* dure
˜WAX - [min] ozokérite *f*
EARWIG - [zool] (an insect) forficule *f*, perce-oreilles *m*
E.A.S. - s. equivalent airspeed
[electron] s. electronic automatic switch
EASE, to - [gen] alléger, adoucir, calmer, desserrer, décharger, modérer
[mech] desserrer, donner du jeu, réduire la vitesse
[mech] (to ease the pressure) soulager la pression
[naut] redresser (la barre), choquer, mollir
EASE - [gen] facilité *f*, repos *m*, aisance *f*
[gen] (in handling) manœuvrabilité *f*, manipulation *f* simple, maniabilité *f*
˜OF BREATHING - [railw & road constr] respirabilité *f*
˜OF PROCESSING - [mech] bonne usinabilité *f*, facilité *f* de traitement
˜ON THE GAS, to - [auto] (colloq US: to reduce speed) réduire la vitesse, ralentir
EASEL - [gen] chevalet *m*
EASEMENT - [leg] (the right which one might obtain at law over a piece of land belonging to others) servitude *f*, droit *m* d'usage, droit *m* de pas
[gen] commodité *f*, soulagement *m*, décharge *f*
˜CURVE - [road & railw] (transition curve, i.e. one connecting a straight to a circular arc) courbe *f* de confort, courbe *f* de raccordement, courbe *f* de transition
[phys] courbe *f* de transition
EASER - [text] (in a loom) barre *f* de tension
[mining] mine *f* de dégraissage
EASING - [constr] (shaping a curve with no abrupt change in it) arrondi *m* d'une courbe
˜BAR - s. easer
˜FISH-PLATE - [railw] éclisse *f* de jonction
˜OF CURVES - [railw] raccordement *m* de courbes
˜THE BRAKE - [mech] relâchement *m* du frein, déclenchement *m* du frein
˜THE CENTRE - [constr] abaissement *m* du cintre
EAST - [gen] (a cardinal point) est *m*, levant *m*, orient *m*
˜-BOUND - [gen] allant vers l'est, venant de l'est
˜VARIATION - [[instr] déclinaison *f* positive

EAST-WEST EFFECT - [phys] (the east-west asymmetry in the cosmic-ray particle number which is caused by the deflection of the cosmic rays due to the earth's magnetic field and the rotation of the earth from west to east) effet *m* azimuthal, asymétrie *f* Est-Ouest (des rayons cosmiques)
EASTERN - [gen] oriental, à l'est, au levant, de le est
~COTTON WOOD - [bot] peuplier *m* grisard
EASTING - [surv] (east departure) longitude *f* est, abscisse *f*
EASY AXIS OF MAGNETIZATION - [comput] axe *m* aisé (de magnétisation)
~CLEAVAGE - [geol] clivage *m* distinet
~CONTROL - [contr] (for radio, television sets etc) commande *f* commode
~FLOWING ELECTRODE - [electron] électrode *f* douce
~GRADIENT - [surv] pente *f* légère
~RUNNING - [mech] (of an engine) bon fonctionnement *m*, bonne marche *f*
EAT, to - [gen] manger
[gen] (to make a way by snawing at something) ronger, éroder, attaquer, saper
EAVE - [constr] (the lower part of a roof projecting from the face of the wall) bord *m* de toit, gouttière *f*, retombée *f*
~TROUGH - [constr] gouttière *f* pendante, chéneau *m*
EAVES - [constr] larmier *m*
~DROP - [constr] égout *m* du toit
~GUTTET - [constr] gouttière *f* creusée dans un madrier
~OVERHANG - [constr] avant-toit *m*
EBB, to - [gen] refluer, baisser, descendre
EBB - [gen] reflux *m*, déclin *m*
[hydr] courant *m* de reflux, courant *m* de jusant
[naut] (of the sea) jusant *m*, marée *f* descendante
EBBA TIDE - [naut] marée *f* descendante, courant *m* de jusant
EBBING AND FLOWING SPRING - [hydr] (an intermittent spring) source *f* intermittente
E BEND - [electron] (a waveguide so bent that all along the length of the bend the longitudinal axis of guide is in a plane parallel to the direction of polarization) coude *m* 'E'
EBIGITE - [min] (a calcium uranyl carbonate) uranothallite *f*
EBONITE - [rubber ind] (hard, vulcanized rubber used in the manufacture of small articles and as an electrical insulator) ébonite *f*, vulcanite *f*, caoutchouc *m* durci
~DUST - [rubber ind] poudre *f* d'ébonite
~LINING - [rubber ind] revêtement *m* d'ébonite
EBONIZE, to [gen] (to make look like ebony) ébéner, noircir
EBONIZED - [gen] ébéné
EBONY - [bot] (very hard, black wood obtained in Ceylon, Madagascar and the Mauritius island) ébène *f*, ébénier *m*
~BLACK - [paint] noir *m* de charbon
EBRANLEMENT - [gen & med] ébranlement *m*, commotion *f*
EBULLIOMETER - s. ebullioscope
EBULLIOSCOPE - [instr] (instrument designed to measure the molecular weight of a substance by a deviation from a known boiling point) ébullioscope *m*
EBULLIOSCOPIC CONSTANT - [phys] (quantity representing the molal elevation of the boiling point of a solution) constante *f* ébullioscopique
EBULLIOSCOPY - [chem] (the determination of the molecular weight of a substance by the deviation from their known boiling points of suitable solvents, ébullioscopie *f*
EBULLITION - [phys] ébullition *f*
EBURNATION - [med] éburnation *f*
ECAD - [genet] (organism modified by environment; acquired character) écade *f*
ECCENTRIC - [mech] (term often used collectively for the whole assembly of eccentric sheave strap and rod) excentrique *m*
[gen] (not concentric with some associated part) excentrique, excentré
~ACTION - [mech] commande *f* par excentrique
~ANGLE - [astr] (of orbital motion) latitude *f* réduite
~ARBOR - [mech] arbre *m* excentré
~ARM - [mech] tige *f* de levée
~BASE RIM - [mech] jante *f* à base excentrique
~BIT - [mech] (in the lathe) trépan *m* excentrique
~BOLT - [constr] targette *f*
~BREAKER - [metall] presse *f* à excentrique
~BUSH - [mech] (an annular part of which the outer and inner diameters are not concentric) douille *f* excentrique, bague *f* excentrée
~BUSHING - s. eccentric bush
~CAM - [mech] came *f* excentrique
~CARD - [text] carton *m* des excentriques
~CIRCLE - s. eccentric groove
~CRANK - [railw] essieu *m* coudé
~DISENGAGEMENT - [mech] débrayage *m* à excentrique
~DRIVE - [mech] commande *f* par excentrique
~GAUGE - [mech] (measuring device to show the degree of core eccentricity in a cable during extrusion) calibre *m* d'excentricité
~GRIP - [mech] (in lifts or hoisting machines) excentrique *m* d'arrêt
~GROOVE - [el acoust] (a locked groove with a centre other than that of the disk record and used in connection with mechanical control) sillon *m* central
~LATHE - [mach tool] tour *m* à tourner les excentriques
~LOAD - [constr] (non-axial load carried by a structural member) charge *f* décentrée
~MOTION - [mech] mouvement *m* d'excentrique, mouvement *m* par excentrique
~MOVEMENT - [mech] mouvement *m* excentrique
~PIN - [mech] goupille *f* de l'excentrique
~REDUCTION - [geol] réduction *f* au centre
PRESS - [mech] (a press in which the moving element is actuated by eccentries) presse *f* à excentrique
~ROD - [mech] (the rod or bar connecting an eccentric to the assembly to which reciprocating motion is to be given) bielle *f* de commande à excentrique
~ROLLER - [mech] galet *m* excentré, galet *m* à rattrapage de jeu
~SHAFT - [mech] arbre *m* à excentrique
~SHAFT PRESS - [mech] presse *f* à excentrique
~SHEAVE - [mech] (a disk mounted on a shaft so as to revolve with it but not to be concentric with it, used to convert rotary to reciprocating motion) plateau *m* d'excentrique

ECCENTRIC SPUR WHEEL - [text] (in weaving looms) roue *f* dentée excentrique
~STIRRUP - [text] étrier *m* d'excentrique, collier *m* d'excentrique
~STRAP - [mech] (a metal band encircling the disk of an eccentric to connect it to the rod which it actuates) collier *m* d'excentrique
~STUD - [mech] goujon *m* excentré, téton *m* excentré
~THREAD GUIDE - [text] guide-fil *m* excentrique
~THROW-OUT - [mech] (mechanism for the engagement of the back-gear of a lathe) débrayage *m* à excentrique
~VALVE GEAR - [mech] distribution *f* à commande par excentrique
~WHEEL - [mech] roue *f* excentrique
ECCENTRICITY - [mech] excentricité *f*, décentrement *m*, désaxage *m*, décentrage *m*
[geom] (of a curve; deviation from the circular form) excentricité *f*
[constr] (the perpendicular distance from the centre of application of the load to the centroid of that part of the structural member supporting it) excentricité *f*
ECCHONDROMA - [med] (a tumour on the surface of a bone) ecchondrome *m*
ECCHONDROSIS - [med] (in chronic arthritis, a growth of joint cartilage) ecchondrose *f*
ECCHYMOSIS - [med] (discoloured patch caused by extravasation of blood under the skin) ecchymose *f*
ECDYSIS - [zool] (the casting off of the outer layer of the integument by snakes etc) ecdysis *f*, exuviation *f*
ECHELLE GRATING - s. echelon grating
ECHELON, to - [gen] échelonner
ECHELON - [gen] échelon *m*
[phys] (a specialized form of diffraction grating) réseau *m* de Michelson, échelon *m*
~AERIAL - [radio] (an aerial consisting of long wires arranged in echelon formation) antenne *f* directionnelle à fils échelonnés
~ANTENNA - s. echelon aerial
~FAULT - [geol] faille *f* en échelons
~GRATING - [spectr] (a greatly specialized form of diffraction grating) réseau *m* à échelon
~LENS ANTENNA - s. echelon lens aerial
~STRAPPING - [electron] (the strapping of a magnetron with overlapping of the strap sections) jumelage *m* échelonné
ECHELONED - [gen] en forme d'escalier, échelonné
ECHINODERM - [zool] échinoderme *m*
ECHINUS - [arch] (egg-shaped ornament carved on a moulding) échine *f*, ove *f*
ECHO, to - [gen & acoust] faire écho, répéter
ECHO - [acoust] (repetition of sound due to the reflection of the sound waves) écho *m*
[radar] (a signal arriving after the main signal because of reflection of part of the original transmission through a longer route) écho *m*
[telev] (unwanted double images) image *f* fantôme
~ALTIMETER - [aero] altimètre *m* à écho
~ATTENUATION - [telecomm] affaiblissement *m* des courants d'écho
[telev] affaiblissement *m* d'écho
~BOX - [radar] (a cavity resonator) cavité *f* résonnante
~BOX ATTENUATOR - [radar] commutateur *m* de cavité résonnante
ECHO CANCELLATION - [telev] (elimination of the echo through the aerial) suppression *f* des échos, élimination *f* des échos
~CHAMBER - [acoust] chambre *f* d'échos
~CHECKING - [radio] (check obtained by reflecting the transmitted information back to the transmitter and comparing the reflected information with that which is transmitted) essai *m* d'écho
~DETECTION SWEEP - [radar] (radar system used in the navy) balayage *m* circulaire
~EFFECT - [telecomm] effet *m* d'écho
~EQUALIZER - [telev] piège *m* d'écho, correcteur *m* d'écho
~IMAGE - [telev] image *f* fantôme
~INTENSITY - [radar] intensité *f* d'écho
~METER - [instr] (instrument designed to measure and record the intervals between a sound and its echo) échomètre *m*
~KILLER - [telecomm] suppresseur *m* d'écho
~ORGAN - [mus] (church organ consisting of quietly-voiced pipes) orgue *m* d'écho
~PATH - [acoust] parcours *m* d'écho
~POWER - [radar] puissance *f* d'écho
~SIGNAL - [radar] signal *m* objectif, signal *m* d'écho
~SOUNDER - [naut] (sounding apparatus designed to measure the depth of the sea beneath a vessel) sondeur *m* à écho, sondeur *m* ultrasonore, sondeur *m* à ultrasons
~SOUNDING - [meas] (depth determining method by the use of ultrasonic waves, used in geology, metallurgy, oceanography, etc) sondage *m* ultrasonique
~SOUNDING GEAR - s. echo sounder
~SUPPRESSOR - [el acoust] (in a telephone channel, an arrangement designed to eliminate the echo of the speaker's voice through a reflexion at the opposite termination) suppresseur *m* d'écho
~TEST SET - s. echometer
~TRANSMISSION TIME - [telecomm] temps *m* d'écho
~TRAP - [telev] (generally a circuit designed to eliminate unwanted echoes) suppresseur *m* d'échos
~TROUBLE - [telecomm] perturbation *f* due aux effets d'écho
ECHOER - [acoust] répétiteur *m*
ECHOSPHERE - [ecology] (the sphere round the Earth which can be lived in by living organisms) biosphère *f*, écosphère *f*
ECLIMETER - [instr] (instrument designed to measure the difference of level between two points) éclimètre *m*
ECLIPSE - [astr] (intervention of one celestial body between the observer and another such body) éclipse *f*
~BLINDNESS - [med] rétinite *f* solaire
ECLIPTIC - [astr] (the Great Circle on the celestial sphere describing the apparent path of the Sun in the course of the year) écliptique *m*
~SYSTEM - [astr] système *m* de coordonnées écliptiques
ECLOGITE - [min] (a coarse-grained ultramafic rock) éclogite *f*
ECLOSION - [genet] (the emergence from an egg) éclosion *f*
ECOBIOTIC ADAPTATION - [genet] adaptation *f* écobiotique
ECOCLINE - [genet] (series of intergrading forms between two ecological niches) écocline *f*
ECOGRAPHY - [med] (the ability to copy associated

with the inability to express, due to a brain lesion) échographie *f*
ECOLOGICAL - [biol] (that which pertains to the environment of an organism) écologique
ECOLOGY - [biol] (the study of organisms in relation to their environment) écologie *f*
ECONOMETER - [instr] (instrument designed to measure and record the carbon dioxide percentage in flue gas) éconornètre *m*
ECONOMETRICS - [instr] économétrie *f*
ECONOMIC - [gen] économique
~DECAY - [gen] dévaluation *f*, dépréciation *f*
~GEOLOGY - [geol] géologie *f* économique
~SPEED - [auto aero etc] vitesse *f* économique
ECONOMICS - [gen] économie *f*
ECONOMIZE, to - [gen & comm] économiser, épargner
ECONOMIZER - [instr] (apparatus for transferring heat from flue gases to boiler feed water) économiseur *m*
[light] (a hood made of refractory material and placed over the tip of an arc-lamp carbon) économiseur *m*
[text] mécanisme *m* économiseur, dispositif *m* réducteur, économiseur *m*
[astronaut] économiseur *m* d'oxygène
~JET - [auto] gicleur *m* d'économie
ECONOMIZING TUBE - [metall] tuyau *m* économiseur
ECONOMY - [gen] économie *f*
~COIL - [el] bobine *f* de réactance
~RESISTANCE - [el] (the resistance inserted into the circuit of an electromagnetic device, so as to reduce the current) résistance *f* chutrice
~TEST - [gen] essai *m* de consommation
ECOPHENE - s. ecad
ECORTICATE - [bot] (without the cortex) décortiqué, sans écorce
ECOSPECIES - [genet] (subdivision of cenospecies capable of free gene interchange between its members) écospecies *f*
ECOTYPE - [genet] (type based on genetical behaviour) écotype *m*
ECTASIS - [med] (distention of a structure due to pathological reasons) ectasie *f*
ECTOCORNEA - [anat] conjonctive *f* de la cornée
ECTOGENESIS - [bot] (a variation due to conditions outside the plant) ectogenèse *f*
ECTOGENOUS - [zool] (self-supporting) ectogène
ECTOPIA - [med] (displacement from normal position) ectopie *f*
ECTOPLASM - [cytol] (at the periphery of a cell) ectoplasme *m*
ECTOSPHERE - [genet] (the cortical zone of the attraction sphere) ectosphère *f*
ECTROMELIA - [med] (congenital absence of limbs) ectromélie *f*
ECZEMA - [med] (an inflammatory condition of the skin) eczéma *m*
EDAPHON - [biol] (the entire living community of the soil) édaphon *m*
EDDY - [mech] (interruption in the flow of fluids due to some obstacle in the line of flow) tourbillon *m*
[hydr] (water which runs contrary to the direction of the current) remous *m*, contre-courant *m*
[met] (vortex in the atmosphere, due to local wind irregularity) tourbillon *m*, revolin *m*
~CHAMBER - [hydr] chambre *f* à tourbillons
EDDY CURRENT BRAKE - [el] (brake for the loading of motors during testing; also a type of brakes used in tramways) frein *m* (électro)magnétique, frein *m* à courants de Foucault
~CURRENT DAMPING - [instr] (type of damping for measuring instruments) amortissement *m* par courants de Foucault
~CURRENT ENERGY - s. eddy current loss
~CURRENT ERRORS - [instr] (occurring when large pieces of metal are near the measuring bridge) erreurs *fpl* dues aux courants de Foucault
~CURRENT HEATING - [el] (heating by inducing eddy currents in a conducting body) chauffage *m* par courants de Foucault
~CURRENT LOSS - [el] (in electric machines, the loss caused by eddy currents in any part of a machine or a transformer) pertes *fpl* par courants parasites, pertes *fpl* de Foucault
~CURRENTS - [el] (currents induced in a conductor by the variation of a magnetic field surrounding it, e.g. in a transformer core) courants *mpl* de Foucault, courants *mpl* parasites
~FLOW - [hydr] (unsteady motion of particles in a fluid flow) mouvement *m* turbulent
~FORMATION - [gen] tourbillonnement *m*
~LOSS - [hydr] porte tourbillons (turbine)
~MOTION - [aero] écoulement *m* tourbillonnaire
~VELOCITY - [phys] fluctuation *f* de vitesse
~WATER - [naut] (the turbulent water left behind by a moving craft) remous *m* (de sillage)
~WIND - [met] (a whirlwind) revolin *m*, renvoi *m* de vent
EDELEANU PROCESS - [ind chem] (commercial method for removal of aromatics from lubricating oils) procédé *m* Edeleanu (d'extraction au solvant)
EDELWEISS - [bot] edelweiss *m*
EDEMA - [med] (also oedema, the presence of abnormally large amounts of fluids in a part of the body) œdème *m*
EDENTATE - [med] édenté
EDGE, to - [gen] affiler, aiguiser, affûter, tomber (un bord), border, ébarber
[text] border, ourler
EDGE - [gen] fil *m*, can *m*, tranchant *m*, taillant *m*, bord *m*, marge *f*, tranche *f*, arête *f*, périphérie *f*
[mech] (of a knife, blade, etc) fil *m*, tranchant *m*
[phys] (of a pulse) front *m* (d'un impulsion rectangulaire)
[naut] accore *f* '(d'un banc)
[text] bordure *f*, ourlet *m*, lisière *f*
~ACTION - [soil] découpage *m*, effet *m* de coin
~ANGLE BAR - [metall] cornière *f* d'angle
~BREAK - [metall] fissure *f* du bord
~BUILD-UP - [telev] (in video recording) bord *m* redressé
~CAPTION - [telev] (caption with an electronically produced border on the vertical edges of each character) titre *m* à effet de profondeur
~CARRIAGE - [text] chariot *m* de l'extrémité
~COAL - [mining] (a coal seam on a steeply inclined bank) dressant *m* de houille
~CONTROLLER - [text] égalisateur *m* de bordure, guide-lisière *m*
~CORRECTION - [telev] (in video recording) correction *f* de contour
~COURSE - [constr] (of a brick course) assise *f* de champ

EDGE CRACK - [metall] (occurring in rolling) fissure *f* de bord
~CUTTER - [text] découpeur *m* de lisières
~DAMPING - [acoust] (in a loudspeaker, an additional damping at the edge of the membrane) amortissement *m* périphérique
~DISLOCATION - [cryst] (a dislocation in which the vector showing the displacement of the material of the lattice is normal to the line of dislocation) dislocation *f* en gradins
~EFFECT - [el] (effect introducing a correction to the capacitance) effet *m* de bord
[photo] effet *m* Eberhard
~FLARE - [telev & cin] (unwanted flaring at the edge of the picture) distortion *f* de contraste sur les bords
~FOG - [telev & cin] (fog at the edges of the picture) voile *m* aux bords, voile *m* marginal
~-FORMING - [metall] (another term for beading, the operation of bending over the edge of a sheet) bordage *m*, dudgeonnage *m*
~GRADIENT - [photo] gradient *m* de bord
~-HOLDING PROPERTY - [metall] résistance *f* de tranchant, dureté *f* de tranchant
~INSULATOR - [el] (insulator to prevent contact between plates and the side wall of the cell) isolateur *m* latéral
~JOINT - [carp] assemblage *m* sur l'arrête
[metall] (welded joint connecting the edges of two or more parallel components) joint *m* de bord
~-JOINTING ADHESIVE - [ind chem] (specific adhesive for the edge-to-edge joining of materials) adhésif *m* pour joints
~-JOINTING CEMENT - [ind chem] (an adhesive used in plywood making to bond strips of veneer by the edges to form larger sheets) colle *f* (à froid) pour joints
~LIST - [text] lisière *f*
~MILL - [mech] (grinding mill in which rollers are made to follow a circular path in a pan in which the material treated is placed) moulin *m* chilien, broyeur *m* à meules verticales
~NAILING - [carp] (invisible nailing on boarded surfaces) clouage *m* caché
~NOTCHES - [photo] (on flat films) encoches *fpl* marginales d'orientation et d'identification
~NUMBERING MACHINE - [cin] machine *f* à piéter
~OF A PULSE - [electron] flanc *m* d'impulsion
~OF A WEIR - [hydr] paroi *f* de déversoir
~OF CLOTH - [text] lisière *f*, bordure *f*
~OF RAIL - [railw] bord *m* interne de rail
~OF RIM FLANGE - [rubber ind] (of tyres) bord *m* de la jante
~OF THE LENS - [opt] bord *m* de la lentille
~PIPING - [text] passepoil *m* en bordure
~PLANING MACHINE - [print] planeuse *f*
~PREPARATION - [metall] (prepared contour on the edge of a component part) préparation *f* de bord
~PRESSER - [text] machine *f* à repasser les lisières
~-PUNCHED CARD - [comput] (a marginally punched card) carte *f* à perforation marginale
~RAY - [opt] rayon *m* marginal
~ROLL - [metall] cylindre *m* de refoulement
~-ROLLER - [text] broyeur *m*, concasseur *m*
~RUNNER - [mech] (a disk or roller in an edge-runner mill) meule *f* (verticale)
~-RUNNER MILL - [mech] (mill consisting of a pan in which two or more disks run like wheels in a circular path) broyeur *m* à meules (verticales), concasseur *m*
EDGE SEAM - [min] dressant *m*
~-SHOT - [carp] (of a board with a planed edge) planche *f* à bord raboté
~STEEPNESS - [radar] (the degree of steepness of a pulse) raideur *f* de front
~STONE - [constr] garde-pavé *m*
~STRESSES - [soil] contraintes *fpl* au contour, tensions *fpl* aux limites
~STRIP - [carp] couvre-joint *m* à franc-bord
~STRIP COIL - [el] enroulement *m* à ruban
~TRIMMER - [mech] (device for cutting away the irregular edge of sheet or film) machine *f* à ébarber, ébarbeuse *f*
~WATER - [hydr] (in the extraction of natural gas) eau *f* de bordure, eau *f* de gisement
~WELL - [min] puits *m* à la bordure d'un gisement
~WELD - [metall] (weld along the edges of sheets placed at right angles to each other) cordon *m* d'angle
~WELDING - [metall] soudage *m* à bord relevés
EDGED - [gen] acéré, affilé, tranchant, anguleux
EDGER - [carp] (a small circular saw) scie *f* circulaire
[metall] distributeur *m*
[photo] rogneuse *f*, cisaille *f*
EDGEWISE - [gen] de côte, de champ, sur champ
~BEND - [electron] (of a rectangular waveguide) coude *m* conservant le champ magnétique
EDGING - [gen] ourlet *m*; bordure *f*, lisière *f*
[metall] (in rolling, a rectangular section) ébavurage *m*, ébarbage *m*, finissage *m* des bords
[glass man] (the operation of grinding the edge of a flat pane) débordage *m*
[shoe man] (of a shoe) bord *m*
~MACHINE - [text] machine *f* à border, machine *f* à ourler
MILL - [metall] (a rolling mill) train *m* finisseur de bords
~STAND - [metall] cage *f* finisseuse des bords
~TRAIN - [metall] train *m* finisseur des bords, cage *f* de refoulement
EDIBLE - [gen] comestible, mangeable
~OILS - [gen] huiles *fpl* comestibles
EDIFICE - [constr] (generally a large building) édifice *m*
EDISON ACCUMULATOR - [el] (a nickel-iron-alkaline accumulator) accumulateur *m* Edison, accumulateur *m* au fer-nickel
~EFFECT - [el] (the escape of negative electricity from hot filaments) effet *m* Edison
~SCREW-CAP - [light] (screw-shaped lamp cap in which the outer wall forms one of the contact) culot *m* à vis
~SOCKET - [el] (American type of lamholder in which the lamp is fixed by a screw thread) douille *f* à vis
E DISPLAY - [radar] (radar display in which the signal appears like a bright spot with range and elevation as the horizontal and vertical coordinate respectively) indicateur *m* type 'E'
EDIT, to - [print] éditer, publier, préparer pour la publication, annoter, rédiger (un journal)
[comput] (Error Detection by Iterative Transmission) suppression *f* des erreurs par répétition de

la transmission
[telev] (in video recording) point *m* de montage
EDITING - [print] préparation *f*, annotation *f*
[comput] (the arrangement of the information to be printed by the output unit) préparation *f*, mise *f* en page
[cin] (the selection of parts of a recording, and the slicing them together in the wanted sequence) montage *m*
~TABLE - [cin] table *f* de montage
EDITION - [gen] édition *f*
EDITOR - [print] rédacteur *m* en chef, éditeur *m*, directeur *m* (d'une revue), correcteur *m*
[cin] chef-monteur *m*
EDITORIAL NEWSROOM - [gen] centre *m* de rédaction
~PROCESS - [cin] (US) montage *m*
EDPM - [comput] (Electronic Data Processing Machine) installation *f* de traitement électronique des données
EDUCTION - [mech] (exhaust) décharge *f*, évacuation *f*, échappement *m*, dégagement *m*, émission *f*, sortie *f*
~OF STEAM - [mech] échappement *m* de vapeur
~PORT - [mech] (the exhaust port of an engine) orifice *m* d'échappement
EDUCTOR - [mech] éjecteur *m*
EDULCORATE - [chem] édulcorer, purifier (par lavage)
EEL - [zool] anguille *f*
~-GRASS - [bot] (a sea-plant with grass-like leaves which are used for sound insulation) vallisnérie *f* spirale
EELWORM - [zool] (harmful to plants) anguillule *f*
EFFACE, to - [gen] effacer, oblitérer
EFFECT - [gen] effet *m*, action *f*, résultat *m*, influence *f*, conséquence *f*
~AMPLIFIER - [el acoust] correcteur *m* pour pièces radiophoniques
~CARBON - [cin] charbon *m* panchromatique
~COLOUR - [text] couleur *f* à effets
~FILTER - [opt] filtre *m* pour effets speciaux
~LACQUER - [paint] (vacuum coating) laque *f* transparente (métallisation sous vide)
~LIGHT - [cin & telev] éclairage *m* d'effet
~LIGHTING - [cin] (special lighting effects) effets *mpl* lumineux
~LOUDSPEAKER - [acoust] (in cinema installations) haut-parleur *m* pour effets sonores
~OF IMPACT - [phys] effet *m* de choc
~OF LATERAL FRICTION - [soil] effet *m* de frottement latéral
~OF LIGHT - [text] effet *m* de lumière
~OF SHEARS - [phys] effet *m* de cisaillement
~PROJECTOR - [cin] (a special projector for colour and animated scenic effects) projecteur *m* d'effets, projecteur *m* pour effets lumineux
~YARN - [text] filé *m* fantaisie, filé *m* d'effets
EFFECTIVE - [gen] efficace, effectif, réel, énergique
[leg] en vigueur
~ABSORPTION COEFFICIENT - [phys] coefficient *m* d'absorption
~ACOUSTIC CENTRE - [el acoust] (for a sound-emitting transducer, a point from which the spherical waves appear to diverge) centre *m* acoustique effectif
~ADDRESS [comput] (the address as changed by the modification of an instruction) adresse *f* effective
EFFECTIVE ANGULAR VELOCITY - [mech] (at one given point, the root square value of the instantaneous angular velocity over a complete cycle at that point) vitesse *f* angulaire efficace
~APERTURE - [photo] ouverture *f* utile, ouverture *f* (photométrique)
~AREA - [radio] (the effective area of an aerial) portée *f* efficace
~ASPECT RATIO - [aero] (the aspect ratio of an aerofoil of elliptical plan projection which has the same induced drag coefficient, for the same lift coefficient, as the aerofoil in question) allongement *m* efficace (d'une aile)
~ATOMIC CHARGE - [nucl] charge *f* efficace
~ATOMIC NUMBER - [nucl] (number calculated from composition and the atomic numbers of a compound) numéro *m* atomique effectif
~BANDWIDTH - [radio] (that range of frequencies in which performance falls within given limits in respect to some characteristics) largeur *f* de bande utile
~BUNCHING ANGLE - [electron] (in a reflex klystron) angle *m* de groupement équivalent
~CAPTURE CROSS-SECTION - [nucl] (the effective cross-section of radioactive capture) section *f* efficace de capture
~CHARGE - [nucl] charge *f* efficace
~COLLISION CROSS-SECTION - [nucl] (value obtained by dividing the probability of collision by the concentration of the gas in question) section *f* efficace de choc
~CROSS-SECTION FOR RESONANCE - [nucl] (the effective cross-section for the capture of an incident particle into a resonant level of the resultant compound nucleus) section *f* efficace de résonance
~CURRENT RANGE OF A METER - [instr] (the range of current values through which the meter has the highest degree of accuracy) domaine *m* de précision des courants d'un compteur
~CUT-OFF FREQUENCY - [radio etc] (the frequency at which power transmission falls below a specified value) fréquence *f* de coupure efficace
~ELECTROMOTIVE FORCE - [el] (the root-mean-square electromotive force) force *f* électromotrice efficace
~FALL-OUT WIND - [nucl] vent *m* effectif de retombée
~FIELD - [photo] champ *m* utile
~GAP CAPACITANCE - [electron] (measured at the gap for frequences which are near resonance) capacité *f* équivalente de l'espace d'interaction
~GAP LENGTH - [el acoust] (a value of the gap length used in certain experimental tests) largeur *f* équivalente d'entrefer
~HALF-LIFE - [nucl] (of a radioactive isotope) période *f* effective
~HEAD - [phys] chute *f* effective, chute *f* réelle
~HEIGHT - [radio] (of an aerial) hauteur *f* effective
~HELIX ANGLE - [aero] (the angle of the helix described by a point on a propeller blade moving through motionless air) angle *m* hélicoidal effectif
~IMAGE FIELD - [photo] champ *m* d'image utile
~INPUT ADMITTANCE - [electron] (of an electronic tube) admittance *f* effective d'entrée
~INPUT CAPACITANCE - [electron] (of an electronic tube) capacité *f* équivalente d'entrée
~INPUT IMPEDANCE - [electron] (of an electronic

tube) impédance *f* effective d'entrée
EFFECTIVE LENGTH - [nucl] longueur *f* utile (d'un compteur)
~LENGTH OF WELD - [metall] longueur *f* effective de soudure
~MASS - [mech] (parameter of the dimensions of a mass used in the band theory of solids) masse *f* effective
~OUTPUT - [gen] production *f* effective
[mech] (of engines) rendement *m* utile
~OUTPUT ADMITTANCE - [electron] (of an electronic tube) admittance *f* effective de sortie
~OUTPUT CAPACITANCE - [electron] (of an electronic tube) capacité *f* effective de sortie
~OUTPUT IMPEDANCE - [electron] (of an electronic tube) impédance *f* effective de sortie
~PAY - [mining] partie *f* effectivement productive d'un gisement
~PERCENTAGE MODULATION - [radio] taux *m* de modulation effectif
~PICTURE SIGNAL - [telev] signal *m* d'image effectif
~PITCH [aero] (the distance through which a propeller advances during one revolution under standard conditions) pas *m* effectif
~POWER - [mech] puissance *f* effective, puissance *f* au frein
~PRECIPITATION - [soil] (the part of the precipitation which is useful to plants; also the part of the precipitation which infiltrates the soil) pluie *f* efficace
~PROPELLER THRUST - [aero] traction *f* efficace de d'hélice, poussée efficace de l'hélice
~PULSE AMPLITUDE - [radio] (taken over the pulse duration) amplitude *f* efficace
~RADIATED POWER - [radio] (of an aerial) puissance *f* émise efficace, puissance *f* rayonnée efficace
~RADIATING POWER - [radio & telev] (of an aerial) puissance *f* effective émise
~RADIUS OF THE EARTH - [phys] (el magn waves; radius of a hypothetical earth) rayon *m* efficace
~RANGE - [instr] (the part of the scale where measurements can be made with the specified degree of accuracy) étendue *f* de mesure (utile)
[nucl] rayon *m* d'action efficace, portée *f* efficace
[photo] champ *m* d'action efficace
~REACTANCE - [el] (the component of the voltage in quadrature with the current divided by the current) réactance *f* effective
~RESISTANCE - [el] (resistance to alternating current) résistance *f* effective, résistance *f* en courant alternatif
~RUNWAY LENGTH - [aero] longueur *f* de piste utile
~SHEAR - [mech] cisaillement *m* effectif
~SIMPLE PROCESS FACTOR - [nucl] (the separation factor which is obtained by a real separative element) facteur *m* de séparation unitaire
~SIZE - [gen] dimension *f* réelle
~SLIT WIDTH - [cin] (the best width in the reproduction of sound from optical recordings) largeur *f* de fente utile
~SPAN - [aero] (the span of a wing after deducting tip loss) envergure *f* effective
[constr] (horizontal distance between the centres of the two bearings at the ends of a beam) portée *f* effective
~TARGET AREA - [nucl] (the area which is reached by the incident particles) aire *f* efficace de cible
EFFECTIVE TEMPERATURE - [th eng] (the temperature resulting from a gas flame) température *f* résultante
~THRUST - [aero] poussée *f* efficace, traction *f* efficace
~TRANSMISSION - [radio] (the transmission performance of a telephone circuit assessed by calculations based on the result of the repetition-rate tests) transmission *f* normalisée
~TRANSMISSION EQUIVALENT - s. effective transmission
~UNIT WEIGHT - [soil] (unit weight of dry soil) densité *f* sèche; poids *m* volumètrique sec
~VALUE - [el] (of an electrical quantity) valeur *f* efficace
~VELOCITY - [mech] (the effective velocity at one point is the root mean square of the instantaneous velocity over a complete cycle at that point) vitesse *f* efficace
~VOLUME - [nucl] volume *m* utile (d'une chambre d'ionisation)
~WAVELENGTH - [opt] (the hue of the colour determining the match in colorimetry) longueur *f* d'onde effective
EFFECTIVENESS - [gen] efficacité *f*, rendement *m*
EFFECTOR - [zool] (a muscle, gland, etc) effecteur *m*
EFFECTS BANK - [telev] unité *f* de commutation pour effets sonores
~MAN - [cin] bruiteur *m*
~MIXER - [telev] (in colour television) mélangeur *m* d'effets
~SPOTLIGHT - [cin] projecteur *m* d'effets
EFFERENT - [zool] (carrying away from, e.g. efferent nerves) efférent
~NERVE - [zool] (nerve which carries impulses away from the central nervous system) nerf *m* efférent
EFFERVESCE, to - [chem] faire effervescence, entrer en effervescence, mousser
EFFERVESCENCE - [gen] effervescence *f*
[phys] (the escape of small gas bubbles from a liquid as a result of physical or chemical action) effervescence *f*
EFFERVESCENT - [gen] effervescent
~LIQUID - [chem] liquide *m* effervescent
~STEEL - [metall] (unkilled steel) acier *m* effervescent
EFFERVESCING CLAY - [min] argile *f* sujette au gonflement
EFFICIENCY - [gen] efficacité *f*
[mech etc] rendement *m*, effet *m* utile
[el] (in storage batteries, the relation between the electrical energy obtained on discharge to that required for charging to the initial level under the same physical conditions) rendement *m*
~BREEDING - [agric] élevage *m* commercial
~BY INPUT-OUTPUT TEST - [el] (in testing an electrical machine) essai *m* de rendement utile
~BY SUMMATION OF LOSSES - [el] (in an electrical machine) rendement *m* conventionnel
~CURVE - [mach] courbe *f* de rendement
~DIODE - [telev] diode *f* économisatrice shunt
~OF CATCH - [aero] (the relation between the total number of water droplets in the path of an aircraft to the number which actually come in contact with

it) effet *m* de la pluviosité
EFFICIENCY OF RECTIFICATION - [el] rendement *m* d'un redresseur à contact
~RATIO - [mech] (of thermal efficiency of heat engines) rapport *m* de rendement
~WAGES - [comm] salaire *m* proportionné à la production
EFFICIENT - [adj] efficient, capable, compétent, habile, de rendement élevé, à bon rendement
EFFLEURAGE - [med] (a very light massage) effleurage *m*
EFFLORESCE, to - [chem] s'effleurir, tomber en efflorescence
EFFLORESCENCE - [chem] (ability of certain salts to lose their water of crystallization with formation of a powdery incrustation) efflorescence *f*, délitescence
[constr] salpêtre *m*, efflorescence *f*; carie *f*
EFFLUENCE - [chem] dégagement *m*, effluence *f*, émanation *f*
EFFLUENT - [gen] effluent, sortant (liquide)
[hydr] (in sewage) effluent *m*, eau *f* usée traitée
[chem] (stream leaving the process equipment) effluent *m*, eaux *fpl* résiduaires
~CHANNEL - s. effluent sewer
~CONDUIT - s. effluent sewer
~DRAIN - [hydr] canalisation *f* de sortie (égouts)
~SEEPAGE - [hydr] (of the ground water) affleurement *m* de l'eau souterraine
~SEWER - [hydr] canal *m* de décharge, conduite *f* de décharge
~TROUGH - [hydr] rigole *f* de décharge
~WATER - [hydr] eau *f* usée
EFFLUX - [gen] flux *m*, écoulement *m*, émanation *f*
~VELOCITY - [astronaut] (velocity of the flow of gases from a rocket, particularly when on the launching pad) vitesse *f* d'éjection
EFFLUXION - s. efflux
EFFORT - [gen] effort *m*, tentative *f*
~SYNDROME - [med] symptoms of circulatory inefficiency in the absence of heart disease) asthénie *f* neurocirculatoire
EFFRACTION - [med] effraction *f*
EFFUSER - [mech] (device which converts the pressure energy of a fluid into kinetic energy) diffuseur *m*
EFFUSIOMETER - [instr] (instrument designed to measure the rate of effusion from an orifice) effusiomètre *m*
EFFUSION - [phys] (the molecular flow of gases through small orifices) effusion *f*
[med] épanchement *m*
~COOLING - [th eng] (method of cooling gas turbine blades by passing air through a permeable blade material into the laminar flow region on the blade surface) refroidissement *m* par effusion
EFFUSIVE ROCKS - [geol] roches *fpl* effusives, roches *fpl* d'épanchement
EGADS BUTTON - [astronaut] bouton *m* de déstruction automatique
EGEST, to - [zool] (to throw out, to excrete) évacuer
EGG, to - [ind chem] (to pump by means of an egg fed with compressed air) élever par monte-jus
EGG - [zool] œuf *m*, ovule *m*
[ind chem] (a pressure vessel used for transferring acid from one container to another by compressed air. The egg is filled, e.g. by gravity, and the contents forced out again through a pipeline by air pressure) monte-jus *m*
EGG ALBUMEN - [chem] (a simple protein, water-soluble and coagulable by heat) albumine *f* d'œuf
~-AND-ANCHOR - [arch] (carving resembling eggs and vertical anchors) godrons *mpl*
~-AND-DART - [arch] oves et dards *mpl*
~-BEATER ROTORS - [el] (helicopter rotor system comprising two rotors revolving in opposite directions)rotors *mpl* contra-rotatifs
~-SOUND - [vet] (the oviduct of birds obstructed by an egg) obstrué par un œuf
~-CELL - [biol] (the ovum itself) œuf *m*, ovule *m*
~COAL - [min] boulet *m* d'anthracite
~GRADER - s. egg weigher and grader
~LAYING - [zool] ponte *f*
~NUCLEUS - [zool] (the female pronucleus) pronucléus *m* (de l'ovule)
~-SHAPED SEWER - [constr] (used in the case of a fluctuating flow) égout *m* ovoide, égout *m* à cunette
~-SHELL - [zool] coquille *f*, coque *f*
~-SHELL GLOSS - [paint] fini *m* coquille d'œuf
~STONE - [geol] oolithe *f*, calcaire *m* oolithique
~WEIGHER AND GRADER - [agric] machine *f* à calibrer (les œufs)
~YOLK - [zool] jaune *m* d'œuf
EGOPHONY - [med] égophonie *f*
EGRESS - [gen] sortie *f*, échappement *m*
EHT - s. extra high tension
EICHORN'S HYDROMETER - [instr] (special type of hydrometer showing specific gravities on a scale marked on its stem) aéromètre *m* de Eichorn
EICHWALDITE - [min] eichwaldite *f*, jéréméjévite *f*
EICOSANE - [chem] (a solid hydrocarbon, insoluble in water but soluble in ether. A synthesis intermediate for plasticizers) eicosane *m*
EIDERDOWN - [text] édredon *m*, couverture *f* piquée
~QUILT - [text] édredon *m*
EIDOGRAPH - [instr] (instrument designed to enlarge or reduce plans) pantographe *m*
EIGENFUNCTION - [math] (the solution of a differential equation which satisfies specified conditions) fonction *f* propre
EIGENPERIOD - [phys] (frequencies at which resonance occurs in rectangular chambers) période *f* propre
EIGENVALUE - [math] (the value of an eigenfunction parameter) valeur *f* propre
EIGHT-ANGLE - [math] octogonal
~BALL - [radio] microphone *m* pomme
[astronaut] indicateur *m* d'altitude
~-DAY MOVEMENT - [horol] huitaine *f*
~-HEADED - [gen] à huit pans, octogonal
~-PLY - [rubber ind] (in tyres) à huit plis
~-SHAPED PATTERN - [comput] cardioïde *m*
~-TO-PICA LEADS - [print] (strips of metal to space out lines of typres) interligne *m* (de I.I/2 points anglais)
~-TRACK TAPE - [el acoust] bande *f* à huit pistes
EIGHTH - [met] (the extent to which the sky is covered by clouds is estimated in eighths (sometimes called oktas), i.e. from 0 to 8/8) unité *f* de degré de nébulosité (en France, elle se compte de I à I0)
~NOTE - [mus] (US term for quaver) croche *f*
EIKONOMETER - [instr] (scale fitted to the eyepiece of a microscope and used to measure the size of the object viewed) iconomètre *m*

EILUROPHOBIA - [med] phobie *f* des chats
E INDICATOR - [radar] (radar indicator showing range and elevation on its screen) indicateur *m* type 'E'
EINSCHLEICHENDER STIMULUS - [el biol] (creeping stimulus: a current increasing so gradually in intensity that it is less effective than it would have been, had the final intensity be abruptly reached) stimulus *m* cumulatif
EINSTEIN EQUATION - [phys] (the formula for mass-energy equivalence) équation *f* d'Einstein
~EQUATION FOR HEAT CAPACITY - [phys] équation *f* d'Einstein pour la chaleur spécifique
EINSTEINIUM - [chem] (a synthetic radioactive element, symbol Es. Chemical properties are similar to those of holmium. A.N. 99) einsteinium *m*
EINTHOVEN GALVANOMETER - [meas] galvanomètre *m* à crode (ou d'Einthoven)
EITHER DIRECTION WORKING - [railw] circulation *f* banalisée
~-OR - [comput] opération *f* de non-équivalence
EJACULATE, to - [gen] éjaculer
EJACULATION - [gen] éjaculation *f*
EJECT, to - [gen] jeter, émettre, éjecter, expulser, extraire, chasser
[metall] (in diecasting) éjecter
EJECTA - [geol] projections *fpl* volcaniques, éjections *fpl*
EJECTED PARTICLE - [phys] (particle emitted from a solid or liquid) particule *f* émise
EJECTING FLAP - [text] clapet *m* d'expulsion
~LATCH - [text] loquet *m* d'expulsion, loquet *m* de sortie
~PAWL - [text] loquet *m* d'expulsion, loquet *m* de sortie
EJECTION - [gen] expulsion *f*, éjection *f*, projection *f*, jet *m*
[metall] (the process of removing a moulding from a mould) éjection *f*
~BAR - s. ejection plate
~BOX - [metall] (in diecasting) sommier *m*
~CAPSULE - [astronaut] compartiment *m* éjectable
~CYLINDER - [metall] (in diecasting) cylindre *m* d'extraction
~DEVICE - [text] dispositif *m* d'éjecteur
~FLAP - [text] clapet *m* de sortie, clapet *m* d'éjection
~HEEL - [metall] talon *m* d'éjection
~MECHANISM - [aero] mécanisme *m* d'éjection
[comput] mécanisme *m* d'éjection, éjecteur *m* (de cartes)
~PLATE - [metall] (a metal plate used to operate ejector pins by the application of uniform pressure to them) plaque *f* d'éjection, plaque *m* supérieure d'éjection
~RAM - [mech] (a small hydraulic ram fitted to a press to operate the ejector pins) piston *m* d'éjection, piston *m* éjecteur
~SEAT - [aero] (type of pilot's seat which can be ejected bodily with the pilot, for abandonment in emergency) siège *m* éjectable
~TIE BAR - [mech] (a steel bar connecting the ejection plate to the ejector frame) barre *f* d'éjecteur
EJECTOR - [mech] (an attachment to a hydraulic press, to operate the ejector pins. Hydrodynamic device using a jet of fluid to expel or suck out fluid from a vessel or the like) éjecteur *m*, exhausteur *m* à jet
[metall] éjecteur *m*, extracteur *m*
[firearms]éjecteur *m*
EJECTOR BALER - [agric] presse *f* à fourrage à expulseur
~BAND - [text] bande *f* d'expulsion
~BOOSTER PUMP - [mech] éjecteur *m*, pompe *f* à éjecteur, éjecteur *m* à vapeur
~CHOPPER - [agric] ramasseuse-hacheuse *f* à expulseur
~DIE - [metall] (in diecasting) bloc *m* mobile
~DIE HALF - [metall] (in diecasting) bloc *m* mobile
~FRAME - [mech] (a metal frame attached to the ejection ram of a press, to locate and secure the ejection tie bars) cadre *m* d'éjecteur
~FRAME GUIDE - [mech] (bar or plate attached to the ejector frame, to guide it when in movement) guide *m* du cadre d'éjecteur
~HALF-DIE - [metall] demi-moule *m* mobile
~MARK - [metall] (in diecasting) trace *f* d'éjecteur
~NOZZLE - [mech] tuyère *f* à éjecteur
~PAD - [metall] (a movable part of a mould impression used in the process of ejection to push the moulding from the impression) butée *f* d'éjecteur
~PIN - [metall] (a movable pin fitted to a mould and used in the process of ejection to push the moulding from the mould impression) broche *f* d'éjecteur, éjecteur *m*
~PIN BAR - s. ejection plate
~PIN PLATE - s. ejection plate
~PIPE - [aero] (a pipe of such form or position that it can produce an appreciable thrust in a forward direction) pipe *f* réactive
~PLATE - [metall] (in diecasting, to actuate the ejector pins) plaque *f* d'éjection
~PLATE RETURN PIN - [mech] (a pin in a mould which pushes back the ejector assembly as the mould closes) butée *f* de renvoi d'éjecteur
~PUMP - [mech] pompe *f* à éjecteur, éjecteur *m*, éjecteur *m* à vapeur
~ROD - s. pull rod
~STAGE - [mech] étage *m* d'éjection de vapeur, étage *m* d'éjecteur
EKING - [shipbuild] allonge *f*
ELABORATE, to - [gen] élaborer, développer, préparer
ELAEOLITE - [min] (rough form of the mineral nepheline) éléolite *f*
ELAEOMETER - [instr] (instrument designed to determine the grade of purity of oil) oléomètre *m*, éléomètre *m*
ELAPSE, to - [gen] s'écouler, se passer
ELAPSED TIME INDICATOR - [instr] indicateur *m* de temps passé, indicateur *m* chronométrique
~-TIME METER - [comput] compteur *m* horaire
ELASTANCE - [el] (the reciprocal of capacitance) élastance *f*
ELASTIC - [gen] élastique
~AFTER-EFFECT - [phys] (the time delay occurring in some substances before returning to the original shape after being stretched within their elastic limits) effet *m* élastique rémanent
~BAND - [text] élastique *m*, bande *f* élastique
~BANDAGE - [gen] bandage *m* élastique
~BEARING - [mech] palier *m* élastique
~BITUMEN - s. elaterite
~COEFFICIENT - [phys] coefficient *m* d'élasticité
~COLLISION - [phys] (a collision without a change

occurring in the internal energies of the participating systems) choc *m* élastique
ELASTIC COUPLING - [mech] (a flexible coupling) joint *m* élastique, accouplement *m* élastique
~CURVE - [phys] (the curve of the neutral surface of a structural member subjected to loads causing bending) courbe *f* d'élasticité
~DEFLECTION - [phys] flexion *f* élastique
~DEFORMATION - [phys] (change in the shape of material when subjected to a stress, reverting to the original shape upon removal of the stress) déformation *f* élastique
~ELONGATION - [phys] allongement *m* élastique
~ENERGY - [phys] énergie *f* élastique
~FABRIC - [text] tissu *m* élastique
~FAILURE - [phys] (the permanent deformation caused by load) déformation *f* permanente
~FATIGUE - [mech] (increase in the damping coefficient of an elastic solid occurring after a large number of oscillations) fatigue *f* élastique
~FLUID - [phys] (a fluid with large elastic stresses if compared with viscous stresses) fluide *m* élastique
~HARNESS - [rubber ind] (in a gas-mask) brides *fpl* élastiques (de masque à gaz)
~HOSE - [text] bas *m* élastique
~HYSTERESIS - [mech] hystérésis *f* élastique
~IMPACT - [mech] (the collision between two objects which regain their original shape immediately after the collision itself) choc *m* élastique
~LIMIT - [mech] (that value of the deforming force beyond which the stressed body does not return to its original dimensions after release from load) limite *f* élastique, limite *f* d'élasticité
~MEDIUM - [phys] milieu *m* élastique
~MEMORY - [plast ind] (certain plastics, if shaped at a suitable temperature and cooled in that shape, will return to their original form on re-heating. This is sometimes called 'elastic memory') mémoire *f* élastique
~MODEL - [aero] (a model so constructed in respect of stiffness distribution and general dimensions as to behave aeroelastically in the same way as the prototype) maquette *f* pour essais aéro-élastiques
~MODULI - [phys] (stiffness constants) modules *mpl* d'élasticité
~NYLON - [text] (a modified form of nylon 610) nylon *m* élastifié
~PACKING - [mech] garniture *f* élastique
~RANGE - [phys] (the stress range in which a body recovers its original shape once the load is removed) domaine *m* élastique, zone *f* d'élasticité
~REBOUND - [soil] restitution *f* élastique, décompression *f* élastique
~RECOVERY - [phys] (measure of the return of a body towards its original dimensions after release from stress) degré *m* d'élasticité
~RIB - [text] bord *m* élastique
~SCATTERING - [nucl] (occurring through the agency of elastic collisions) diffusion *f* élastique
~STOCKING - [text] s. elastic hose
~STRAIN - [phys] (strain occurring in a material in the elastic state) déformation *f* élastique, déformation *f* réversible, limite *f* des déformations élastiques
~STRENGTH - [phys] limite *f* d'élasticité, limite *f* élastique
ELASTIC SUPPORTING SURFACE - [mech] surface *f* d'appui élastique
~TISSUE - [zool] (connective tissue with a predominance of elastic fibres) tissu *m* élastique
~TOW ROPE - [auto] câble *m* de remorque élastique
~WAVE - [phys] onde *f* élastique
~YARN - [text] fil *m* élastique
ELASTICATOR - [ind chem] (substance added to another to give it elasticity) agent *m* élastifiant
ELASTICITY - [gen & phys] (the ability of a material to deform under stress and to return to its original shape upon removal of the stress) élasticité *f*, flexibilité *f*, souplesse *f*
[med] élasticité *f*, tonicité *f* (des muscles)
~MODULUS - [phys] coefficient *m* d'élasticité, module *m* d'élasticité
~OF BULK - [phys] (the elasticity for the changes of volume of a body due to changes in the pressure) élasticité *f* de volume, élasticité *f* volumétrique
~OF COMPRESSION - s. elasticity of bulk
~OF ELONGATION - [phys] (the stretching force) élasticité *f* de traction
~OF FLEXURE - [phys] (the elasticity of a bent object tending to straighten it) élasticité *f* de flexion
~OF GASES - [phys] (the volume of gas changes with the changing of pressure) élasticité *f* des gaz
~OF RIGIDITY - s. elasticity of shear
~OF SHEAR - [phys] (elasticity of a body pulled out of shape by a shearing force) élasticité *f* de cisaillement
~OF TORSION - [phys] (the elasticity of a body which has been twisted) élasticité *f* de torsion
~TO SHEAR STRESS - [mech] élasticité *f* à l'effort de cisaillement
~TO TORSION STRESS - [mech] élasticité *f* à l'effort de torsion
ELASTIN - [zool] élastine *f*
ELASTOMER - [chem] (synthetic polymer which can be vulcanized and has rubber-like properties) élastomère *m*
ELASTOMERIC - [chem] (possessing the characteristics of an elastomer) élastomère *m*
~PLASTICS - [plast ind] (plastics which have the elastic properties of an elastomer) élastomères *mpl*
ELASTOMERS - [chem] (high polymeric materials which exhibit elasticity when subjected to stress. Usually applied to natural and synthetic rubbers) élastomères *mpl*
ELASTOMETER - [instr] (an instrument for measuring elasticity in elastomers) élastomètre *m*
ELASTO-PLASTIC - [rubber ind] élastoplastique
ELATER - [bot] (a weed) élatère *m*
ELATERITE - [min] (a dark-brown solid natural bitumen with elastic properties) élatérite *f*, caoutchouc *m* minéral
ELATEROMETER - [instr] (instrument designed to measure the pressures of gases) élatéromètre *m*
ELATROMETER - s. elaterometer
E LAYER - [radio] (layer of ionized gas in the ionosphere, almost wholly reflecting medium-frequency transmissions) région *f* 'E'
ELBOW - [anat] coude *m*
[mech] (a fitting to join pipes at right angles, where a curved element is not necessary; it is shaped angularly, not smoothly curved, as in the case of a bend) coude *m*, raccord *m* coudé
[el] (sharp bend) conduit *m* coudé, raccord *m*

[mining] dressant *m*
ELBOW BOARD - [carp] (the window-board in the interior) appui *m* intérieur
˜DIMENSION - [plumb] dimension *f* angulaire, dimension *f* du coude
˜JOINT - [mech] raccord *m* coudé
[anat] articulation *f* du coude
˜PIPE (TUBE) - [plumb] coude *f*, tube *m* coudé, raccord *m* coudé, raccord *m* angulaire
ELBOWED - [gen & mech] coudé
˜SPANNER - [tool] clé *f* coudée, clé *f* à pipe
˜WRENCH - s. elbowed spanner
ELDER - [bot] sureau *m*
ELDERBERRY OIL - [ind chem] huile *f* de sureau
ELDERWOOD - [bot] bois *m* de sureau
ELECTRET - [el] (a permanently polarized dielectric, analogous to a permanent magnet) électret *m*
ELECTRIC - [el] électrique
˜ACCUMULATOR - [el] (a battery of storage cells) accumulateur *m*, batterie *f* d'accumulateurs
˜ALARM - [el] sonnerie *f* électrique
˜ARC - [el] (in a gas; discharge characterized by a comparatively small cathode drop) arc *m* électrique
˜ARC WELDING - [metall] (welding in which the necessary heat is produced by one or more arcs formed between one or more electrodes and the workpiece) soudure *f* à l'arc
˜AXIS - [phys] (that axis of a crystal which offers minimum resistance to the passage of current) axe *m* électrique
˜BAKING OVEN - [th eng] four *m* électrique de boulangerie
˜BELL - [el acoust] sonnerie *f* électrique
˜BLANKET - [el] (blanket containing a wire heating conductor threaded through the heating portion) couverture *f* chauffante (électrique)
˜BLAST FURNACE - [metall] haut-fourneau *m* électrique
˜BLENDER - [el] (an electro-domestic appliance) mélangeur *m* électrique, mixeur *m* électrique
˜BOILER - [el] (vessel with a fitted heating element) bouilloire *f* électrique
˜BOILING PAN - [el] (electro-domestic appliance) marmite *f* électrique
˜BOILING PLATE - [el] (the heating element of a hot plate) foyer *m* de cuisson électrique
˜BRAKE - [el] (a special brake for electrically driven vehicles) frein *m* électrique
˜BRAKING - [el] (for electrically driven vehicles) freinage *m* électrique
˜BURNER - [el] brûleur *m* électrique
˜CABLE - [el] câble *m* électrique
˜CALAMINE - [min] (carbonate of zinc, also called smithsonite) smithsonite *f*, calamine *f*
˜CALCULATING MACHINE - [comput] calculatrice *f* électrique
˜CAPACITANCE OF A CONDUCTOR - [el] capacité *f* électrique d'un conducteur
˜CARBON - [el] charbon *m* électrique
˜CASH-REGISTER - [el] (appliance used in commerce etc) caisse-enregistreuse *f* électrique
˜CATFISH - [zool] silure *m* électrique
˜CAUTERY - [med] (the burning of parts of the body by means of an electrically heated instrument) cautère *m* électrique, électro-cautère *m*
˜CELL - [el] pile *f* électrique
ELECTRIC CEMENT - [constr] ('ciment fondu', i.e. a rapidly hardening cement made by heating lime and alumina in an electric furnace) ciment *m* fondu, ciment *m* alumineux
˜CHARGE - [el] (a synonym for quantity of electricity) charge *f* électrique
˜CHART DRIVE - [meas] (a clock-like device to which the chart is fastened) entraînement *m* électrique (de bande enregistreuse)
˜CIRCUIT - [el] (arrangement of media through which current can flow) circuit *m* électrique
˜CLEANER - [el] (electro-domestic appliance) aspirateur *m* électrique
˜CLOCK - [el] (a clock with an electrically operated movement) horloge *f* électrique
˜COAL CUTTER - [el] (used in coal mines; a coal-cutter operated by an electric motor) haveuse *f* électrique
˜COMPONENT - [radio] (the electrostatic component of an electromagnetic wave) composante *f* électrique
˜CONDUCTION - [el] (process in which the atoms are stationary; also a migration of ionized atoms) conduction *f* électrique
˜CONDUIT - [el] conduit *m*, gaine *f*, tube *m*
˜CONSTANT - [el] constante *f* électrique
˜-CONTACT - [el] contact *m* électrique
˜-CONTACT MINE - [el] mine *f* à contact électrique
˜CONTROLLER - [contr] (mechanism governing the electric power delivered to a machine or apparatus) régulateur *m* électrique
˜CONVEYOR - [el] (electric space heater emitting heat mainly by convection) radiateur *m* électrique à convection
˜COOKER - [el] (electro-domestic appliance) cuisinière *f* électrique
˜COOKING OVEN - [el] (electro-domestic appliance) four *m* électrique de cuisine
˜CORING - [mining] carottage *m* électrique
˜CRANE - [mech] grue *f* électrique
˜CURRENT - [el] (the movement of electricity along a circuit) courant *m* électrique
˜DEEP FRIER - [el] (electro-domestic appliance) friteuse *f* électrique
˜DEFLECTION - [electron] (deflection of the beam caused by an electric field across its path) balayage *m* électrique, déviation *f* électrique
˜DEFLECTION SENSITIVITY - [electron] (of a cathode ray tube) sensibilité *f* de la déviation électrique
˜DEHYDRATOR - [el] (container with low-powered heating elements and ventilation openings intended for the dehydration of fruit, vegetables, etc) dessiccateur *m* électrique
˜DELAY LINE - [comput] ligne *f* électrique à retard
˜DETONATOR - [el] détonateur *m* électrique
˜DIPOLE - [el] (a pair of equal and opposite charges divided by an infinitesimal distance) dipôle *m* électrique, doublet *m* électrique
˜DIPOLE RADIATION - [el] rayonnement *m* d'un dipôle électrique
˜DISCHARGE - [el] (the passage of electricity through a gas as a result of the ionization of the gas) décharge *f* électrique
˜DISCHARGE LAMP - [el] (electric lamp in which light is obtained from an electric discharge) lampe *f* électrique à lueur, lampe *f* électrique à luminescenze

ELECTRIC DISH-WASHER - [el] (electro-domestic appliance) machine *f* à laver la vaisselle électrique
~DISPLACEMENT - [el] (in a dielectric, the charge dispersed over a unit area) déplacement *m* électrique, flux *m* électrostatique
~DISSIPATION - [el] dissipation *f* électrique
~DOUBLE LAYER - [el] (hypothetical distribution of charge) couche *f* bipolaire
~DOUBLET - s. electric dipole
~DRILL - [tool] perforatrice *f* électrique, perceuse *f* électrique
~DRIVE FOR STEERING GEAR - [naut] commande *f* électrique de barre
~DUSTER - [el] (electro-domestic appliance) aspirateur *m* électrique
~DYNAMOMETER - [instr] (type of dynamometer in which the energy from the unit under examination is converted into electric power for measurement) dynamomètre *m* électrique
~EEL - [zool] gymnote *m*, anguille *f* électrique
~ELECTRON LENS - [opt] (device using electric fields to focus an electron beam) lentille *f* électronique électrique
~ENERGY - [el] énergie *f* électrique
~EQUIVALENT - s. electrochemical equivalent
~ETCHER - [mach] machine *f* à graver électrique
~EVAPORATOR - [el] industrial appliance) évaporateur *m* électrique
~EYE - [el] (a photoelectric cell) cellule *f* photoélectrique
~FALL - [el] baisse *f* électrique, chute *f* électrique
~FAN - [el] (electrically operated fan) ventilateur *m* électrique
~FIELD - [el] (space in which a potential gradient exists, so that force is exerted on electric charges within it, e.g. electrons) champ *m* électrique
~FIELD INTENSITY - [el] intensité *f* de champ (électrique, champ *m* électrique
~FIELD VECTOR - [el] (at one point in an electric field, the force on a stationary positive charge per unit charge) vecteur *m* de champ électrique
~FLUX - [el] (the differential coefficient of the electric flux density in a dielectric with respect to time) flux *m* électrique
~FLUX DENSITY - [el] induction *f* électrostatique, déplacement *m* électrique
~FRIER - s. electric deep frier
~FURNACE - [th eng] (a furnace which is heated electrically) four *m* électrique
~FURNACE PIG-IRON - [metall] fonte *f* au four électrique
~GAS-LIGHTER - [el] (spark producing device to ignite the gas from a gas burner) allumeur *m* électrique (pour le gaz)
~GENERATOR - [el] (a machine designed to convert mechanical energy into electrical energy) génératrice *f*
~GENERATOR SET - [el] groupe *m* générateur, groupe *m* électrogène
~GRID - [el] (system for the distribution of electricity on a national or regional basis) réseau *m* d'interconnexion
~GYRO - [el] (a gyroscope driven by electric power) gyroscope *m* électrique
~HARMONIC ANALYZER - [el] (electrical device designed to measure the magnitudes of the harmonics in the wave shape of an alternating-current or voltage) analyseur *m* électrique d'harmoniques
ELECTRIC HEATED SOLDERING IRON - s. electric soldering iron
~HEATING - [el] (generation of heat by electrical means) chauffage *m* électrique
~HOIST - [mech] (electrically operated hoist) treuil *m* électrique
~HORN - [el acoust] avertisseur *m* électrique, sirène *f* électrique
~HOT PLATE - [el] (domestic appliance for table use to keep food hot) chauffe-plat *m* électrique
~HOT TABLE - [el] (table designed to keep vessels which contain food hot) table *f* chauffante électrique
~IGNITION - [el] allumage *m* électrique
~IMAGE - [telev] image *f* électrique de charge
~INDUCTION - [el] (the product of the electric intensity at a point in a dielectric and the dielectric constant) induction *f* électrique
~INSULATION - [el] isolement *m* électrique
~IRON - [el] (electro-domestic appliance) fer *m* à repasser électrique
~IRONING MACHINE - [el] (machine with a motor-driven drum which applies the material to be ironed against a plate) machine *f* à repasser électrique
~IRONING PRESS - [el] (appliance comprising two electrically heated for ironing work) presse *f* à repasser électrique
~KITCHEN - [el] (electro-domestic appliance) cuisine *f* électrique
~LAP WELDING - [metall] (also called resistance lap welding; resistance welding of two or more overlapping pieces) soudage *m* électrique par recouvrement
~LEAKAGE - [el] perte *f* électrique
~LENGTH - [el] (the physical length expressed in wavelengths or degrees) longueur *f* effective
~LIGHT - [el] (light which is produced by electric power) lumière *f* électrique
~LIGHTING - [el] éclairage *m* à l'électricité, éclairage *m* électrique
~LOCOMOTIVE - [railw] locomotive *f* électrique
~LOG - [naut] loch *m* électrique
~LOGGING - [mining] diagraphie *f* électrique
~MASKING - [telev] masquage *m* électrique
~MOMENT - [el] moment *m* électrique
~MOTOR - [el] (a machine which converts electrical into mechanical energy) moteur *m* électrique
~ORGAN - [zool] (tissue suitable for the production and storage of electric energy) appareil *m* électrique
~OSCILLATIONS - [el] (electric currents periodically reversing their direction of flow) oscillations *fpl* électriques
~OUTLET - [el] prise *f* de courant
~OVEN - [el] (electro-domestic appliance) four *m* électrique
~PANEL HEATER - [el] (electro-heating appliance) panneau *m* chauffant électrique
~POTENTIAL - [el] (the potential difference between a point and some equipotential surface) potentiel *m* électrique
~POWER - [el] (the product of electric current and electromotive force) puissance *f* électrique, énergie *f* électrique
~POWER PLANT - [el] centrale *f* électrique
~POWER SUPPLY - [el] alimentation *f* en énergie électrique
~PRECIPITATION - s. electrial precipitation

ELECTRIC PRINTING PUNCH - [comput] perforatrice *f* imprimante
~PROSPECTING - [mining] prospection *f* électrique
~PUMP - [mech] pompe *f* électrique, électropompe *f*
~PUMPING PLANT - [mech] installation *f* de pompage électrique, station *f* de pompage électrique
~PUNCH - [comput] perforatrice *f* à relais, poinçonneuse *f* à relais
~RADIATOR - [el] (electric heating appliance consisting principally of a radiant source of energy) radiateur *m* électrique (à rayonnement)
~RANGE - s. electric cooker
~RAZOR - s. electric shaver
~RESISTANCE WELDING - [metall] soudage *m* par résistance
~RESISTIVITY - [el] résistivité *f* électrique
~RING - [el] (electro-domestic appliance) foyer *m* de cuisson électrique
~RIVETING - [mech] rivetage *m* électrique
~RIVETING MACHINE - [mech] riveuse *f* électrique
~ROCK DRILL - [tool] perforatrice *f* électrique
~RUNABOUT TRUCK - s. electric truck
~SALINOMETER - [el chem] (an instrument for determining the salt content of water by measuring its conductivity) salinomètre *m* électrique
~SCREEN - [el] écran *m* électrostatique
~SCREEN-WIPER - [auto] essuie-glace *m* électrique
~SCREENING - [el] (the isolation of apparatus etc from electrostatic fields by enclosure in a metal container) blindage *m* électrostatique
~SHOCK - [el] (the sudden convulsion caused by the passage of an electric current through the body) secousse *f* électrique, commotion *f* électrique
~SHOCK THERAPY - [med] électrochoc *m*, méthode *f* de Cerletti et Bini
~SHOVEL - [mech] pelle *f* mécanique électrique, drague *f* électrique
~SLEWING CRANE - [mech] grue *f* pivotante électrique
~SOLDERING IRON - [tool] (a tool for soldering metallic pieces consisting of a heating element, a heat-conducting part and a handle) fer *m* à souder électrique
~SPARK - [el] étincelle *f* électrique
~SPARK MACHINING - [metall] usinage *m* par électroérosion, usinage *m* par étincelage
~STARTER - [el] (an electric motor with a suitable drive for revolving internal-combustion engines in order to start them) démarreur *m* électrique
~STEAM BOILER - [el] (electrically heated boiler to produce steam) chaudière *f* électrique à vapeur
~STEAM GENERATOR - [el] (electrically heated apparatus for producing steam) générateur *m* de vapeur électrique, chaudière *f* électrique à vapeur
~STEAM RAISER - [el] (closed container with electric heating for the production of steam) chaudière *f* électrique à vapeur
~STEEL - [metall] acier *m* électrique
~STORM - [met] (condition of high electric fields in a cloud) orage *m* à éclairs
~STRAIN GAUGE - [instr] (instrument designed to determine the stress in materials) jauge *f* de contrainte électrique
~STRENGTH - [el] (the property of a dielectric opposing a disruptive discharge) rigidité *f* diélectrique

ELECTRIC STRESS - s. electric strength
~SYSTEM - [el] circuit *m* électrique
~TABLE COOKER - [el] (electro-domestic appliance) réchaud *m* électrique
~TERMINALS - [el] pôles *mpl* électriques (d'une machine), bornes *fpl* (d'une machine)
~THERMOMETER - [instr] (instrument designed to measure temperatures electrically) thermomètre *m* électrique, pyromètre *m* électrique
~TILTING FURNACE - [metall] four *m* basculant électrique
~TISSUE - [zool] (modified tissue capable of generating electricity) tissu *m* électrique
~TOASTER - [el] (electro-domestic appliance) grille-pain *m* électrique
~TOOL TIPPER - [mech] affûteuse *f* électrique d'outils
~TORCH - [el] (battery operated torch) torche *f* électrique, lampe *f* électrique
~TRACTION - [railw] (vehicle operated by electric motor)traction *f* électrique
~TRAIN - [railw] train *m* électrique
~TRANSDUCER - [el] (a transucer in which all waves are electric) transducteur *m* électrique, capteur *m* électrique
~TRUCK - [transp] chariot *m* électrique
~TYPEWRITER - [el] (a typewriter fitted with an electric motor) machine *f* à écrire électrique
~VACUUM CLEANER - [el] (electro-domestic appliance) aspirateur *m* électrique
~VANE - [el] électro-vanne *f*
~WATER-HEATER - [el] (device for heating water by means of electric current) chaudière *f* électrique à eau chaude
[el] (electric water-heater in which the heat is produced in a resistor and transmitted to water) chaudière *f* électrique à résistance
~WAVE FILTER - [electron] (a frequency-discriminating filter) filtre *m* de bande
~WELDING - [metall] (method of welding in which the necessary heat is produced by an electric arc or by electrical resistance) soudage *m* électrique
~WIND - [el biol] (stream of particles carrying away charges from a body which has been charged to a sufficiently high voltage) effluve *f* électrique
~WIRE-BREAK ALARM - [el] avertisseur *m* de rupture de fil

ELECTRICAL - [el] électrique
~ABSORPTION - [el] (in the dielectric) absorption *f* électrique
~ACTUATOR - [el] moteur *m* électrique d'asservissement
~ANALOGY - [el] (analogy for the evaluation of stresses) analogie *f* électrique
~ANESTHESIA - [el biol] (suspension of sensibility produced by electrical means) électro-anesthesie *f*
~ANGLE - [el] (in a multipolar machine) angle *m* électrique (machine multipolaire)
~BALANCE - [el] (of a bridge circuit) équilibre *m* électrique
~BEHAVIOUR - [phys] comportement *m* électrique
~BIAS - [el] (polarized winding on a relay core to adjust the sensitivity of the relay to signal currents) enroulement *m* de retenue (de relais polarisé)
~BIREFRINGENCE - [opt] biréfringence *f* électrique, phénomène *m* de Kerr
~BORE-HOLE PROSPECTING - [mining] (the identici-

cation of strata by electrical methods) carottage *m* électrique

ELECTRICAL BREAKDOWN TENSION - [el] tension *f* disruptive

~CENTRING - [telev] (the regulating of the picture on the screen by electrical means) centrage *m* par commande électrique

~CHAIN - [el] (a number of coupled circuits) chaîne *f* de circuits électriques

~CLEANUP - [electron] pompage *m* ionique

~COMMUNICATION - [el] (the conveying of information by electrical means) communication *f* électrique, transmission *f* électrique

~CONDUCTIVITY - [el] (the conductance of a cubic centimetre) conductivité *f* électrique

~CONDUCTOR - [el] conducteur *m* électrique

~CONTINUITY - [el] (the existence of an unbroken electrical path through mechanically separated objects) continuité *f* électrique

~CORING - [mining] carottage *m* électrique

~DEGREE - [el] angle *m* de calage, angle *m* électrique

~DEPOSITION - s. electrodeposition

~DETARRER - [ind chem] (in condensation) dégoudronneur *m* électrostatique

~DISTANCE - [el] (distance measured in units based on the velocity of light) distance *f* parcourue par une onde radio pendant une unité de temps

~ELEMENT - [el] (a cell) pile *f* électrique

~ENGINEERING - [gen] électrotechnique *f*

~EQUIPMENT - [el] équipment *m* électrique

~EQUIPMENT TESTING BENCH - [ind proc] banc *m* d'essai pour appareils électriques

~FITTINGS - [el] accessoires *mpl* électriques

~FOOTWEAR - [shoe man] chaussures *fpl* pour électriciens

~FORMING - [electron] (the process of applying electric energy to a semi-conductor) électro-formage *m*

[el chem] galvanoplastie *f*

~FUEL GAUGE - [instr] indicateur *m* de niveau électrique

~IMAGE - s. electric image

~IMPEDANCE - [el] (the ratio between voltage and current in an a.c. circuit) impédance *f* électrique

~INERTIA - [el] inductance *f*

~INSTALLATION - [el] installation *f* électrique

~INSTRUMENT - [instr] (an instrument which indicates electrical quantities or conditions) appareil *m* de mesure électrique

~INTERCONNEXION - [el] (the connecting of electrical machines in order to combine their properties or effects) accouplement *m* électrique

~INTERFERENCE - [el] (undesired effects on equipment caused by other equipment, materials or atmospheric conditions) perturbation *f* d'origine électrique

~INTERLOCK BOARD - [railw] (signalling system) table *f* d'enclenchements électriques

~INTERLOCKING POST - [railw] poste *m* d'interconnexion

~LENGTH - s. electric length

~MEASURING INSTRUMENT - [instr] appareil *m* de mesure électrique

~NOISE - [el] (unwanted electrical energy present in transmission systems or in any measuring device) bruit *m* d'origine électrique

ELECTRICAL OUTPUT - [el] puissance *f* fournie en électricité, puissance *f* de sortie

~PORCELAIN - [el] (porcelain of special type used for electrical insulation) porcelaine *f* pour isolateurs

~PRECIPITATION - [el] (extraction of very fine particles from a stream of gas by means of electrodes carrying a high voltage) précipitation *f* électrostatique

~RECORDING - [el acoust] (sound recording by electrical means) enregistrement *m* électrique

~REPRODUCTION - [el acoust] (the reproduction from records by means of electromagnetic devices) reproduction *f* électrique

~RESISTANCE - [el] (property of a substance whereby the flow of electric current through it is resisted) résistance *f* électrique

~SHIELDING - [el] (the shielding of an apparatus from interferences caused by electrical disturbances) blindage *m* électrique

~STANDARDS - [el] normes *fpl* électriques

~SURFACE PROSPECTING - [mining] (in the study of the subsoil) prospection *f* électrique

~TECHNOLOGY - [el] (the science relating to the practical application of electricity) électrotechnique *f*

~TIME DISTRIBUTION SYSTEM - [el] (in factories etc) télécommande *f* électrique d'horloges

~UNIT - [el] unité *f* électrique

~ZERO - [contr] zéro *m* électrique

ELECTRICALLY HEATED PAD - [el] (electric heating appliance consisting of a fabric in which a heating conductor is woven, or otherwise fixed, especially for therapeutical purposes) thermoplasme *m*

~HEATED THERMOCOUPLE - [el] (device comprising a thermocouple and a conductor heated by the current to be measured) thermocouple *m*

~OPERATED VALVE - [el] électro-vanne *f*

ELECTRICIAN'S PLIER - [tool] pince *f* d'électricien

~SCREWDRIVER - [impl] (special type of screwdriver fitted with a handle of high insulating power) tournevis *m* d'électricien

~SIDE CUTTER - [tool] pince *f* coupante d'électricien

ELECTRICITY - [el] (the manifestation of a form of electric energy) électricité *f*

~METER - [meas] (an instrument designed to measure the quantity of power supply used in a given time) compteur *m* électrique

ELECTRIFICATION - [el] (the production of electric charges on a body; also the conversation to electric operations) électrification *f*, électrisation *f*

~OF A GAS - [electron] (the state of a gas made conductive by the introduction of electrons, without ionizing the gas) électrisation *f* d'un gaz

ELECTRIFIED - [gen] électrifié, électrisé

ELECTRIFY, to - [gen] électrifier, électriser

ELECTRO - [el] (a prefix) électro-

[print] (abbrev for electrotype) galvano *m*, galvanotype *m*

ELECTROACOUSTIC - [el acoust] électroacoustique

~COUPLING IMPEDANCE - s. electroacoustic force factor

~FORCE FACTOR - [el acoust] (the value of the complex quotient of the resulting sound pressure in the blocked acoustical system and the corresponding current in the electrical system) coefficient *m* de couplage électroacoustique

ELECTROACOUSTIC TRANSDUCER - [el acoust] (transducer which supplies energy to an acoustical system) transducteur *m* électroacoustique

ELECTROACOUSTICS - [el acoust] (the practical application of acoustics by electrical means) électroacoustique *f*

ELECTROANALYSIS - [el chem] (the determination of the quantity of an element in a solution by electrodeposition) analyse *f* électrolytique

ELECTROBIOLOGY - [el biol] (the study of electrical phenomena in relation to biological systems) électrobiologie *f*

ELECTROBRIGHTENING - [el] brillantage *m* électrolytique

ELECTROBRONZING - [el metall] bronzage *m* électrolytique

ELECTROCAPILLARITY - [el] (the science dealing with electrocapillary phenomena) électrocapillarité *f*

ELECTROCAPILLARY CURVE - [el] (the curve is obtained by plotting the interfacial tension of a mercury surface in contact with the aqueous solution of an electrolyte) courbe *f* électrocapillaire

~PHENOMENA - [el] (phenomena which depend on the variation of surface tension at the boundary layer of two liquids when a potential difference is established between these two liquids) phénomènes *mpl* électrocapillaires

ELECTROCARDIOGRAM - [med] (a graphic record of the variation with time of the voltage associated with cardiac activity) électrocardiogramme *m*

ELECTROCARDIOGRAPH - [med] électrocardiographe *m*

ELECTROCAST BRICK - [metall] brique *f* réfractaire alumineuse

ELECTROCAUTERY - s. electric cautery

ELECTROCHEMICAL - [el chem] électrochimique

~CLEANING - [el chem] dégraissage *m* électrolytique

~DIFFUSION - [el chem] (movement of a charged constituent from one region to another of the same phase) diffusion *f* électrochimique

~DISSOCIATION - [el chem] (decomposition of some parts of the molecules concerned, with the formation of ions) dissociation *f* électrolytique

~EQUIVALENT - [el chem] (the weight of a substance or ion associated with a specific electrochemical reaction due to the passage of a given quantity of electricity) équivalent *m* électrochimique

~FORCE SERIES - s. electrochemical series

~INDUSTRY - [gen] industrie *f* électrochimique

~MIGRATION - [el chem] (movement of a charged element of a phase from one part of the latter to another, due to electrochemical potential differences between the constituents of the regions of the phase concerned) migration *f* électrochimique

~OXIDATION - [el chem] (anodic process consisting of the removal of electrons from atoms or ions or the addition of positive charges to them) oxydation *f* anodique

~PASSIVATION - [el chem] (production of approximately perfect passivity in a metal by electrochemical means) passivation *f* électrochimique

~PASSIVITY - [el chem] (the occurrence of a polarization which is so great in certain metals used as anodes in electrolytic cells, that it must be ascribed to irreversibility in the ionization process) passivité *f* électrochimique

ELECTROCHEMICAL PICKLING - [metall] décapage *m* électrochimique

~RECORDING - [el acoust] (recording by means of a chemical reaction) enregistrement *m* électrochimique

~REDUCTION - [el chem] (cathodic process consisting of the addition of electrons to atoms or ions, or the removal of positive charges from them) réduction *f* électrochimique

~SERIES - [el chem] (a tabular statement of the metals in order of magnitudes of the potential differences between each metal and a normal solution of ions of its salts) classification *f* électrochimique, série *f* des éléments d'après les potentiels électrochimiques

~VALVE - [el chem] (apparatus comprising a metal element in contact with a solution or compound so arranged as to have an electrical rectifying action) soupape *f* électrolytique, redresseur *m* électrochimique

~VALVE METAL - [el chem] (metal or alloy suitable for use in an electrochemical valve) métal *m* pour soupape électrolytique

ELECTROCHEMISTRY - [el chem] (science and technology dealing with interrelated transformations of chemical and electric energy) électrochimie *f*

ELECTROCOAGULATION - [el biol] (the clotting of tissue by heat generated by impressed electric current) électrocoagulation *f*

ELECTROCOAGULATOR - [el biol] électrocoagulateur *m*

ELECTROCOPPERING - [el metall] cuivrage *m* électrolytique

ELECTROCORTICOGRAM - [el biol] (graphic record of the voltage variations taken from exposed cerebral cortex) électrocorticogramme *m*

ELECTROCREMAGE - [el chem] (electric decantation) décantation *f* électrique

ELECTROCULTURE - [el biol] (stimulation of growth seeding, etc by means of electricity) électroculture *f*

ELECTROCUTED - [gen] électrocuté

ELECTROCUTION - [el biol] (destruction of life by means of electricity) électrocution *f*

ELECTRODE - [el] (conductor leading an electric current into a liquid or a gas) électrode *f*

~ACTIVE SURFACE - [el chem] (the part of the surface of an electrode at which reaction is proceeding) surface *f* active d'une électrode

~ADMITTANCE - [electron] (the reciprocal of electrode impedance) admittance *f* d'électrode

~BIAS - [electron] (the voltage at which an electrode is stabilized under operating conditions) tension *f* de repos

~BOILER - [el] (boiler in which heat is produced by the passage of electrodes through the liquid to be heated) chaudière *f* à électrodes

~-CARBON - [metall] charbon *m* à électrodes

~CHARACTERISTIC - [electron] (the relation between an electrode current and voltage) caractéristique *f* d'électrode

~CONDUCTANCE - [electron] (the reciprocal of the electrode resistance) conductance *f* d'électrode

~CONTROL - [arc furnaces] (device for advancing or retracting electrodes during operation) commande *f* des électrodes

~CURRENT - [electron] (of an electronic tube) courant *m* d'électrode

ELECTRODE-CURRENT AVERAGING TIME - [electron] (the time interval over which the current is averaged) temps *m* d'intégration

~-CURRENT DENSITY - [el] (current per unit area of electrode active surface) densité *f* de courant de électrode

~DISSIPATION - [electron] the power which is dissipated in the form of heat by an electrode when bombarded by electrons or ions) dissipation *f* d'électrode

~FORCE - [mech] effort *m* sur l'électrode

~GAP - [electron] (the region between two electrodes) distance *f* interélectrode

~HOLDER - [el] (device for holding the electrode and leading the current to it) porte-électrode *m*, pince *f* porte-électrode

~IMPEDANCE - [electron] (the quotient of the sinusoidal component of the electrode voltage by the corresponding sinusoidal component of the electrode current) impédance *f* d'électrode

~LEAD-IN - [el] traversée *f* d'électrode

~PASSAGE - [el] passage *f* de l'électrode

~POTENTIAL - [el biol] (the potential between a metallic electrode and a biological material) potentiel *m* d'électrode

~RADIATOR— [electron] (used to facilitate the dissipation of the heat generated in the electrode) radiateur *m* d'électrode

~REACTANCE - [electron] (imaginary component of the electrode impedance) réactance *f* d'électrode

~RESISTANCE - [electron] (the real component of the electrode impedance) résistance *f* d'électrode

~SALT BATH - [el chem] (type of furnace in which a bath of salt is maintained at the required temperature by passing current between electrodes immersed in it) four *m* à bain de sel à électrodes

~SKID - [mech] dérapage *m* de l'électrode

~STEAM GENERATOR - [el] (type of apparatus in which steam is generated by passing current between electrodes immersed in water) chaudière *f* à vapeur à électrodes

~STEAM RAISER - s. electrode steam generator

~SUPPORT - [electron] (the support in an electronic tube which holds the electrode in place) support *m* d'électrode

~SUSCEPTANCE - [electron] (the reciprocal of the electrode reactance) susceptance *f* d'électrode

~TIP - [el] tête *f* d'électrode

~VOLTAGE - [electron] tension *f* d'électrode

~WATER-HEATER - [el] chaudière *f* à eau chaude à électrodes

ELECTRODEPOSIT, to - [el chem] déposer par électrolyse

ELECTRODEPOSITED METAL - [el metall] métal *m* déposé électrolytiquement, métal *m* déposé par voie galvanique

ELECTRODEPOSITION - [el chem] (the process of deposition on an electrode by electrolysis or electrophoresis) électrodéposition *f*, galvanostégie *f*

ELECTRODESICCATION - [el biol] fulguration *f*, étincelage *m*

ELECTRODIAGNOSTICS - [med] (examination and diagnosis with the aid of electrical means) électrodiagnostic *m*

ELECTRODIALYSIS - [el chem] (dialysis accelerated by the use of electrodes placed on the sides of the semipermeable membrane) électrodialyse *f*

ELECTRODIALYSIS PURIFICATION PROCESS - [ind chem] (method of purifying water by mignation through an electrically charged membrane) épuration *f* par électrodialyse

ELECTRODIALYZER - [phys] électrodialyseur *m*

ELECTRODISSOLUTION - [el chem] (the operation of dissolving the material of an electrode by electrolysis) dissolution *f* électrolytique, électrodissolution *f*

ELECTRODELESS DISCHARGE - [el] (discharge in a gas tube when the tube is placed in intense high-frequency electromagnetic fields) décharge *f* lumineuse dans un tube sans électrode

ELECTRODYNAMIC - [el] électrodynamique

~BALANCE - [instr] (an electrodynamometer in which electrodynamic forces are balanced by weights) balance *f* électrodynamique

~INSTRUMENT - [instr] (instrument making use of the force between fixed and moving coils carrying currents) appareil *m* électrodynamique

~LOUDSPEAKER - [el acoust] (loudspeaker operating on the motion of a conductor) haut-parleur *m* à conducteur mobile, haut-parleur *m* électrodynamique, haut-parleur *m* à bobine mobile

~MICROPHONE - [el acoust] (microphone depending for its operation on the generation of an electromotive force in a conductor moving in a magnetic field) microphone *m* à conducteur mobile, microphone *m* électrodynamique

~VIBRATION PICK-UP - [meas] (an apparatus designed for the exact measurement of amplitude, velocity and acceleration of mechanical vibrations) capteur *m* de vibration électrodynamique

~WATTMETER - [instr] wattmètre *m* électrodynamique

ELECTRODYNAMICS - [el] (the science of the mutual influence of electric currents) électrodynamique *f*

ELECTRODYNAMOMETER - [instr] (a meter with a fixed coil and a movable coil) électrodynamomètre *m*

ELECTROENCEPHALOGRAM - [el biol] (graphic record of the voltage variation taken from the cerebral skin) électroencéphalogramme *m*

ELECTROENDOSMOSIS - [el chem] (the movements of fluids through diaphragms produced by an electric potential) électroendosmose *f*

ELECTROEROSION - [el] électroérosion *f*, étincelage *m*

~METAL WORKING PROCESS - [el metall] (a spark metal working process) usinage *m* par électroérosion, usinage *m* par étincelage

ELECTROETCHING - [metall] (etching on metals etc) gravure *f* électrolytique

ELECTROEXTRACTION - [el chem] (extraction of metals from compounds or ores by electrochemical means) extraction *f* électrochimique

ELECTROFACING - [el chem] (the process of coating by electrodeposition) revêtement *m* électrolytique

ELECTROFORMING - [el metall] (the production or copying of objects by electrodeposition) électroformage *m*, galvanoplastie *f*

ELECTROGALVANIZING - [el metall] (the electrodeposition of a coating of zinc) zingage *m* électrolytique

ELECTROGILDING - [el metall] (the deposition of a coating of gold on metal objects) dorure *f* électrolytique

ELECTROGRAPH - [instr] (recording electrometer) électromètre *m* enregistreur
ELECTROGRAPHIC INK - [comput] encre *f* conductrice
~PEN - [comput] crayon *m* à mine conductrice, crayon *m* de graphitage
ELECTROGRAPHY - [telecomm] (phototelegraphy) phototélégraphie *f*
ELECTROHYDRAULIC - [el] électrohydraulique
~SERVO VALVE - [mech] valve *f* d'asservissement électrohydraulique
~STEERING GEAR - [naut] appareil *m* à gouverner électrohydraulique
ELECTROJET - [astr] (current in the ionosphere) électrojet *m*
ELECTROKINETIC EFFECTS - [el] (the movement of particles under the influence of an electric field) effets *mpl* électrocinétiques
~POTENTIAL - [el] (set of four velocity potentials accompanying relative motion between solids and liquids) potentiel *m* électrocinétique
ELECTROKINETICS - [el] (the science dealing with the phenomena of electricity in motion) électrocinétique *f*
ELECTROKYMOGRAPH - [instr] (instrument designed to record the time rate of motion of the shadow on a fluoroscopic screen) électrocymographe *m*
ELECTROLIER - [el] (hanging ornamental cluster of electric lamps) lustre *m* électrique, plafonnier *m*, suspension *f*
ELECTROLOGY - [med] (synonym for electrotherapy) électrothérapie *f*
ELECTROLUMINESCENCE - [phys] électroluminescence *f*
ELECTROLYSIS - [el chem] (chemical change in an electrolyte caused by the passage of a current through electrode reactions and ionic migration) électrolyse *f*
~ARRESTER - [el] (a lightning arrester consisting of a number of aluminium trays) parafoudre *m* à cuvettes d'aluminium
ELECTROLYTIC - [el chem] électrolytique
~ANALYSIS - [el chem] analyse *f* électrolytique
~BATH - [el chem] bain *m* électrolytique
~CAPACITOR - [el] (capacitor in which the dielectric between the plates is an electrolyte) condensateur *m* électrolytique
~CELL - [el chem] (a vessel in which chemical reactions are induced by the passage of an electric current) bac *m* d'électrolyse
~CLEANING - [el plating] (removal of grease in an alkaline solution through which a current is passed) dégraissage *m* électrolytique, nettoyage *m* alcalin électrolytique
~CONDENSER - s. electrolytic capacitor
~CONDUCTOR - [el] conducteur *m* électrolytique, électrolyte *m*
~COPPER - [metall] (copper refined by electrolysis) cuivre *m* électrolytique
~DETECTOR - [radio] (a platinum wire immersed in an electrolyte and polarized by a small steady voltage) détecteur *m* électrolytique
~DISSOCIATION - [el chem] (dissociation of a substance into ions) dissociation *f* électrolytique
~DISSOCIATION CONSTANT - [el chem] (value obtained by dividing the product of the ion activities by the molecular activity) constante *f* de dissociation électrolytique
ELECTROLYTIC GRINDING - [el metall] rectification *f* électrolytique
~IRON - [metall] fer *m* électrolytique
~LIGHTNING ARRESTER - [el] (lightning arrester consisting of a number of electrolytic cells in series) parafoudre *m* à cuvettes d'aluminium
~METER - [instr] (instrument designed to record the quantity of electricity by electrolytic means) compteur *m* électrolytique
~MUD - [el chem] boue *f* électrolytic
~PARTING - [el chem] (separation of alloyed metals by electrolysis, one being deposited on the cathode and the other in the anode slime) séparation *f* électrolytique
~PICKLING - [metall] (method of pickling in which current is passed through the solution) décapage *m* électrolytique
~POLARIZATION - [el chem] polarisation *f* électrolytique
~POLISHING - [metall] (by making a metal anodic in an electrolytic solution, a smooth shining surface is obtained) polissage *m* électrolytique
~RECTIFIER - [el] (device in which an alternating current can be rectified by electrolytic action) redresseur *m* électrolytique, soupape *f* électrolytique
~REFINING - [metall] (the production of pure metals by making the impure metal the anode in an electrolytic cell) affinage *m* électrolytique
~SLIME - s. electrolytic mud
~SOLUTION PRESSURE - [metall] (the tendency of a metal to dissolve in solution with the formation of ions) tension *f* d'ionisation, tension *f* de dissolution électrolytique
~STRIPPING - [metall] (removal of metal from an object by making it the anodic in an electrolytic bath) décapage *m* électrolytique
~TANK - [electron] (used to obtain diagrams of the equipotential curves in a plane of symmetry of an electronoptical system) cuve *f* rhéographique
~TIN PLATE - [metall] fer *m* blanc électrolytique
~VALVE - s. electrolytic rectifier
~VALVE RATIO - [el] (the relation between the impedance offered by an electrolytic valve to a current flowing in a given direction and that which it offers to a current flowing in the opposite direction) pourcentage *m* de redressement
~ZINC PROCESS - [metall] électrolyse *f* du zinc
ELECTROLYZE, to - [el chem] électrolyser
ELECTROLYZER - [el chem] (an industrial electrolytic cell) cuve *f* électrolytique
[med] (used in urethral disturbances) électrolyseur *m*
ELECTROMAGNET - [el] (a magnet which is energized by electric current and loses the greater part of its field when the electric supply ceases) électroaimant *m*
ELECTROMAGNETIC - [el] électromagnétique
~BRAKE - [el] (brake in which the force is produced by the friction between two surfaces pressed by magnetic attraction) frein *m* électromagnétique
~CONTACTOR - [el] contacteur *m* électromagnétique
~DAMPING - [el] (the checking of the oscillations of a moving system by eddy currents in a magnetic field) amortissement *m* électromagnétique
~DEFLECTION - [electron] (deflection of the electron beam in cathode-ray tubes by an electromagnetic

field produced by a coil at the neck of the tube) déviation *f* électromagnétique, balayage *m* électromahnétique
ELECTROMAGNETIC FIELD - [el] champ *m* électromagnétique
~FLOWMETER - [instr] fluxmètre *m* électromagnétique
~FOCUSING - [electron] focalisation *f* électromagnétique
~IGNITION - [auto] allumage *m* magnétoélectrique
~INDUCTION - [el] (the production of an electromotive force by relative motion of a conductor and magnetic field in which it lies) induction *f* électromagnétique
~INSTRUMENT - [instr] (an instrument depending for its operations on electromagnetic forces) appareil *m* électromagnétique
~ISOTOPE SEPARATION UNIT - [nucl] (for the separation of isotopes by electromagnetic methods) appareil *m* de séparation électromagnétique des isotopes
~LENS - [electron] lentille *f* électromagnétique
~LOUDSPEAKER - [el acoust] (loudspeaker depending on variations of the reluctance of a magnetic circuit) haut-parleur *m* électromagnétique
~MASS - [phys] masse *f* électromagnétique
~MASS SEPARATOR - [nucl] spectrographe *m* de masse électromagnétique
~METHOD - [nucl] (for the separation of isotopes) méthode *f* électromagnétique
~MICROPHONE - [el acoust] (microphone operating on variations of the reluctance of a magnetic circuit) microphone *m* électromagnétique
~PERCUSSION WELDING - [metall] soudage *m* électromagnétique par percussion
~PUMP - [el] pompe *f* électromagnétique
~RADIATION - [phys] (the propagation of energy in the form of electromagnetic waves) rayonnement *m* électromagnétique
~RAIL BRAKE - [railw] frein *m* électrique à patin
~REACTION - [electron] (reaction between the anode and grid circuits of a thermionic valve obtained by electromagnetic coupling) réaction *f* électromagnétique
~REPULSION - [el] (the force of repulsion between two circuits) répulsion *f* électromagnétique
~SCREEN - [instr] (conducting shield protecting the operating part of a measuring instrument) écran *m* électromagnétique
~SCREENING - [el] blindage *m* électromagnétique
~SEPARATION - [el] (the removal of ferrous objects from refuse by setting up a magnetic field while they travel along a conveyor) triage *m* magnétique
~SEPARATOR - [metall] séparateur *m* électromagnétique
~SLIPPER BRAKE - [el mech] frein *m* électromagnétique à patin
~SPECTRUM - [el] (the whole range of frequency over which energy is radiated in wave form) spectre *m* électromagnétique
~STRESS - [phys] force *f* électromagnétique
~VALVE - [el] vanne *f* électro-magnétique, vanne à commande électro-magnétique, électrovanne *f*
~WAVE - [el] (wave characterized variations of the electric and magnetic fields) onde *f* électromagnétique
ELECTROMAGNETICALLY CONTROLLED VALVE - s. electromagnetic valve
ELECTROMAGNETICS - s. electromagnetism
ELECTROMAGNETISM - [el] (science dealing with the relations between electricity and magnetism) électromagnétisme *m*
ELECTROMECHANICAL - [el] électromécanique
~CONVERTER - [el] transducteur *m* électromécanique
~DRIVE - [radio] (a master oscillator) commande *f* électromécanique
~FORCE FACTOR - [el] coefficient *m* de couplage électromécanique
~RECORDING - [el acoust] (recording by means of a mechanical device) enregistrement *m* électromécanique
~TRANSDUCER - [el] (transducer supplying energy to a mechanical system) transducteur *m* électromécanique, capteur *m* électromécanique
~TRANSMISSION - [mech] transmission *f* électromécanique
ELECTROMECHANICS - [el] électromécanique *f*
ELECTROMERISM - [chem] (a form of tauterism caused by a redistribution of electrons) électromérie *f*
ELECTROMETALLURGICAL PLANT - [el metall] installation *f* électrométallurgique
ELECTROMETALLURGY - [metall] (the science of the application of electrochemistry and electrothermics to metallurgy) électrométallurgie *f*
ELECTROMETER - [instr] (instrument designed to measure a charge) électromètre *m*
~TUBE - [electron] (vacuum tube used to amplify very weak voltages) tube *m* électromètre
~VALVE - s. electrometer tube
ELECTROMETRIC - [el] électrométrique
~TITRATION - [chem] (titration with the use of an electrode immersed in the titrated solution) titration *f* électrométrique
ELECTROMOBIL - [transp] véhicule *m* électrique
ELECTROMOTIVE - [el] électromoteur, électromotrice
~FORCE - [el] (that force which gives rise to the motion of electric charges) force *f* électromotrice
~SERIES - [el chem] (tabular arrangement of the metals in the order of the magnitude of potential difference which exists between the metal and a normal solution of one of its salts) série *f* des éléments d'après les potentiels électrochimiques, classification *f* électrochimique
ELECTROMOTOR - [el] électromoteur *m*, moteur *m* électrique, machine *f* électromotrice
ELECTRON - [phys] (fundamental, negatively charged atomic particle having a mass I/I840th of a hydrogen atom) électron *m*
~AFFINITY - [electron] (the minimum energy necessary for an electron to pass through the potential barrier) travail *m* de sortie, travail *m* d'extraction
~AVALANCHE - [electron] (a group of electrons freed by cumulative ionization) avalanche *f* électronique
~BEAM - [electron] (a stream of electrons convergently focused and moving at high velocity, as in a cathode-ray tube) faisceau *m* électronique, faisceau *m* d'électrons
~BEAM DEFLECTION FACTOR - [electron] coefficient *m* de déviation
~-BEAM FILM SCANNING - [telev] analyse *f* de films par faisceau électronique
~-BEAM FOCUSING - [electron] focalisation *f* de

faisceau électronique
ELECTRON-BEAM FURNACE - [metall] four *m* à bombardement électronique
~BEAM MAGNETOMETER - [instr] (magnetometer measuring the field by means of an electron beam immersed in it) magnétomètre *m* à faisceau électronique
~-BEAM MELTING - [metall] fusion *f* par bombardement électronique
~-BEAM TRANSMISSION EFFICIENCY - [electron] rendement *m* de transmission du faisceau électronique
~BEAM TUBE - [electron] tube *m* à faisceau électronique, tube *m* cathodique, tube *m* à parcours électronique commandé
~BEAM VALVE - s. electron beam tube
~-BEAM WELDING - [metall] soudage *m* par bombardement électronique
~BOMBARDMENT - [phys] bombardement *m* électronique
~BOMBARDMENT FURNACE - s. electron beam furnace
~CAMERA - [telev] (any device which converts an optical image into an electric current by electronic means) caméra *f* électronique
~CAPTURE - [phys] capture *f* d'électrons
~-CHARGE MASS RATIO - [electron] (ratio between charge and electron mass) rapport *m* charge-masse de l'électron
~CLOUD - [electron] (a concentration of electrons in a comparatively stationary condition in the space between two electrodes of an electron tube) nuage *m* électronique, cortège *m* électronique planétaire, orbite *f* électronique, enveloppe *f* électronique
~COLLECTION - [electron] (used in ion chamber measurements) collection *f* électronique
~COLLECTOR - [electron] (in a microwave tube, the electrode which receives the electron beam at the end of its path) collecteur *m*
~CONCENTRATION - [phys] (the ratio between the number of valence electrons and the number of atoms in a molecule) concentration *f* électronique
~-COUPLED OSCILLATOR - [ratio] oscillateur *m* à couplage cathodique
~COUPLING - [electron] (the coupling of two circuits in a vacuum tube) couplage *m* électronique
~CURRENT - [electron] (the current produced by the movement of free electrons) courant *m* électronique
~DIFFRACTION - [phys] (diffraction phenomena which are analogous to those obtained with light) diffraction *f* électronique
~DISCHARGE - [electron] décharge *f* électronique
~DISCHARGE TUBE - [electron] tube *m* à décharge électronique
~DRIFT - s. electron flow
~-ELECTRON SCATTERING - [electron] diffusion *f* électron-électron
~EMISSION - [electron] (the liberation of electrons into the surrounding space) émission *f* électronique
~-EMITTING AREA - [electron] surface *f* d'émission électronique
~FLOW - [electron] (the movement of electrons in a definite direction) flux *m* électronique, flux *m* d'électrons
~GAS - [electron] gaz *m* électronique
~GUN - [electron] (an assembly of electrodes in a cathoderay tube designed to produce an electron beam) canon *m* à électrons, canon *m* d'électrons, canon *m* électronique
ELECTRON HOLE - [electron] (the space left in a crystal lattice by an electron passing to another position in the lattice) électron-trou *m*, lacune-trou *f*, trou *m*
~INJECTOR - [electron] (the electron gun of a betatron) injecteur *m* d'électrons
~IMAGE - [telev] (virtual image formed in a beam) image *f* électronique
~IMAGE TUBE - [telev] tube *m* à image électronique
~IMPACT - [electron] (the impact of the beam on the screen of the cathode-ray tube) impact *m* d'électrons
~IMPINGEMENT - s. electron impact
~-ION WALL RECOMBINATION - [electron] recombinaison *f* ion-électron à la paroi
~JUMP - [electron] transition *f* électronique
~LENS - [electron] (a system of electromagnetic and/or electrostatic fields which has an electron beam analogous to that of an optical lens on a light ray)lentille *f* électronique
~MEAN FREE PATH - s. mean free path
~MICROSCOPE - [instr] (thermionic tube in which the electrons are so focused as to form an enlarged image of the cathode on a fluorescent screen) microscope *m* électronique
~MIRROR - [electron] (device causing the total reflection of an electron beam) miroir *m* électronique
~MULTIPLIER - [electron] multiplicateur *m* électronique
~OPTICS - [electron] (the electronic optical means in an electronic tube, e.g. electronic lens etc) optique *f* électronique
~ORBIT - [phys] orbite *f* électronique
~PAIR - [electron] (feature of the form of molecular structure) paire *m* d'électrons
~RANGE-FINDER - [instr] télémètre *m* électronique
~-RAY INDICATOR TUBE - [electron] (elementary form of cathode-ray tube used to indicate a change of voltage) indicateur *m* cathodique
~RELAY - [electron] (an electron discharge tube in which the discharge is controlled) relais *m* électronique
~SCANNING BEAM - [telev] faisceau *m* électronique analyseur
~SHEATH - [electron] (film of electrons on a surface) gaine *f* d'électrons
~SHELL - [phys] (a group of electrons round the atomic nucleus, which have adjacent energy levels) couche *f* électronique
~SPIN - [comput] spin *m*, spin *m* des électrons, précession *f* des électrons
~-STREAM POTENTIAL - [electron] potentiel *m* du faisceau électronique
~-STREAM TRANSMISSION EFFICIENCY - [electron] rendement *m* de transmission du faisceau électronique
~TELESCOPE - [instr] télescope *m* électronique
~TRAJECTORY - [electron] (the path of an electron in a beam) trajectoire *f* électronique
~TRANSIT TIME - [electron] temps *m* de transition, durée *f* de parcours des électrons
~TRANSITION - [electron] (the passage of an electron from one level of energy to another) transition *f* électronique
~TUBE - [electron] (a device in which the movement of electrons or other charged particles in a vacuum

or a highly attenuated gas can be controlled in a manner and to a degree which would not be possible in a solid conductor) tube *m* électronique
ELECTRON-TUBE BASE GAUGE - [electron] calibre *m* pour culots de tubes
~-TUBE TESTER - [electron] appareil *m* d'essai pour tubes électroniques
~VOLT - [electron] électron-volt *m*
~-WAVE TUBE - [electron] (electron tube in which interacting electron streams with different velocities cause a signal modulation to change along their length) tube *m* électronique à onde
ELECTRONEGATIVE ELEMENT - [el] (an element with a tendency to attract electrons) élément *m* électronégatif
ELECTRONEGATIVITY - [el] (the extent to which a given atom tends to attract valence electrons) électronégativité *f*
ELECTRONIC - [electron] (relating to operations or apparatus involving the control of electron movement) électronique
~ADMITTANCE - [electron] (of a vacuum tube) admittance *f* électronique
~ALTIMETER SET - [instr] altimètre *m* électronique
~ANALOGUE COMPUTER - [comput] (computer using physical analogues of the variables by electronic means) calculateur *m* analogique électronique
~AUTOMATIC SWITCH - [electron] commutateur *m* électronique automatique
~AUTOPILOT - [electron] (arrangement whereby flight deviations of an aircraft are detected and corrected) pilote *m* automatique à commande électronique
~BEAM - [electron] (stream of electrons emanating from a cathode-ray tube) faisceau *m* d'électrons
~BRAIN - [electron] calculateur *m* électronique
~BUG - [electron] (a transmitting key and circuit which automatically give correct dot and dash length and spacing) manipulateur *m* électronique
~BUILDING BRICK - [comput & contr] (a sub-assembly for the construction of data processing equipment etc) sous-ensemble *m* électronique
~CALCULATING PUNCH - [comput] perforatrice-calculatrice *f* électronique
~CALCULATOR - [comput] (computer which can undertake simple operations by means of electronic circuits) calculateur *m* électronique
~CHARACTER SENSING - [comput] lecture *f* électronique des caractères
~CHIMES - [electron] (chimes system the sound of which can be picked up) carillon *m* électronique
~COMMUTATOR - [comput] commutateur *m* électronique
~COMPUTER - [comput] (computer operating by means of electronic devices) calculatrice *f* électronique, ordinateur *m* électronique
~CONDUCTIVITY CONTROL - [electron] commande *f* électronique de la conductivité
~CONTROL - [contr] (control by electronic means) commande *f* électronique
~COUNTING INSTRUMENT - [meas] appareil *m* de comptage électronique
~DATA PROCESSING MACHINE - [comput] installation *f* de traitement électronique des données
~DEWPOINT RECORDER - [instr] enregistreur *m* électronique de point de rosée
~DIGITAL COMPUTER - [comput] calculatrice *f* numérique électronique
ELECTRONIC EFFICIENCY - [electron] rendement *m* du faisceau
~EMISSION - [electron] émission *f* électronique
~FIELD FREQUENCY CONVERTER - [electron] convertisseur *m* électronique de la fréquence de trame
~FLIGHT IMITATOR - s. electronic flight simulator
~FLIGHT SIMULATOR - [aero] (flight simulator operating electronically) simulateur *m* de vol électronique
~FLYING-SPOT SCANNER - [telev] analyseur *m* indirect électronique à spot mobile
~GAP ADMITTANCE - [electron] admittance *f* électronique de l'espace d'interaction
~GRID - [electron] grille *f* électronique
~IGNITION - [electron] (based on a transistor) allumage *m* électronique
~KEYING - [radio] (method of keying with control by electronic means) manipulation *f* électronique
~LENS - [opt] lentille *f* électronique
~MAGNIFICATION - [telev] (the magnifying of a television picture by electronic means) agrandissement *m* électronique
~MASKING - [telev] (in colour television) masquage *m* électronique
~MISSILE ACQUISITION - [radar] système *m* électronique de repérage
~MULTIPLICATION CIRCUIT - [comput] circuit *m* multiplicateur électronique
~MUSIC - [mus] (music generated by synthetic means) musique *f* électronique
~PEAK-READING VOLTMETER - [instr] voltmètre électronique, indicateur *m* de tension crête
~PRINT READER - [comput] (an apparatus which can read printed texts, including numbers and punctuation marks) lectrice *f* électronique d'imprimés
~PROFILOMETER - [instr] (instrument designed to measure surface roughness) profilomètre *m* électronique
~PUNCH - [comput] perforatrice *f* électronique
~PUNCHING MACHINE - [comput] (punch-card machine with electronic control) perforatrice *f* électronique
~RASTER SCANNING - [telev] analyse *f* électronique canevas
~RAY INDICATOR - [electron] (cathode-ray tube used to indicate the tuning of a circuit) indicateur *m* d'accord à rayons cathodiques, indicateur *m* cathodique
~RECTIFIER - [electron] redresseur *m* électronique
~ROUGHNESS TESTER - [instr] (instrument designed to determine roughness in machines etc) rugomètre *m* électronique
~SCANNING - [telev] (dissection of the picture by means of a cathode-ray tube) analyse *f* électronique
~SEWING - [electron] (high-frequency heating used to join plastic objects) soudage *m* à haute fréquence, soudage *m* électronique
~SWITCH - [radio] commutateur *m* électronique, interrupteur *m* électronique
~TEMPERATURE RECORDER - [electron] enregistreur *m* électronique de température
~TIMER - [electron] (time-switch for control purposes) minuterie *f* électronique
~TUBE - [electron] (electronic device in which conduction occurs through a vacuum or a gaseous medium inside a gas-tight envelope) tube *m* électroni-

que
ELECTRONIC TUBE SHIELD - [electron] écran *m* pour tubes électroniques
~TUNING - [electron] (the process of changing the operating frequency of a system) accord *m* électronique
~TUNING RANGE - [electron] plage *m* d'accord électronique
~VALVE - s. electronic tube
~VIEW FINDER - [telev] (view finder operated by means of electrons) viseur *m* électronique
~WORK FUNCTION - [phys] potentiel *m* d'extraction (ou de Helmholtz)
ELECTRONICS - [electron] (the science of controlled electron movement and its applications) électronique *f*
ELECTROOPTICS - [opt] électro-optique *f*
ELECTROOSMOSIS - [el chem] (migration of a fluid through a diaphragm, due to an electric field) électro-osmose *f*
ELECTROOSMOTIC DRAINAGE - [soil] drainage *m* par électro-osmose, drainage *m* électro-osmotique
~POTENTIAL - [el biol] (the electrokinetic potential which causes unit liquid flow velocity through a porous body) potentiel *m* électro-osmotique
~STABILIZATION OF SOILS - [soil] stabilisation *f* des sols par électro-osmose
ELECTROPHONIC EFFECT - [el acoust] (the sensation of hearing when an alternating current of adequate frequency and magnitude is passed through an animal) effet *m* électrophonique
ELECTROPHORE - [el] électrophore *m*
ELECTROPHORESIS - [el chem] (the migration of colloidal particles in a liquid under the influence of an electric field) électrophorèse *f*
ELECTROPHORETIC POTENTIAL - [el chem] (the electrokinetic potential gradient which will give rise to the flow of a suspension or colloid through a liquid electrolyte at unit velocity) potentiel *m* d'électrophorèse
ELECTROPHOROUS - [el] (device consisting of a sheet of ebonite attached to a metal plate) électrophore *m*
ELECTROPLATE, to - [el metall] déposer par électrolyse, plaquer
ELECTROPLATING - [metall] (process of forming an adherent coating of metal on an object by electrochemical means) dépôt *m* électrolytique, revêtement *m* électrolytique, métallisation *f* électrolytique
~GENERATOR - [metall] (a special type of electric generator designed to produce current for electroplating. Also called plating generator) génératrice *f* pour galvanoplastie
ELECTROPLEXY - [med] électroconvulsion *f*, électrochoc *m*
ELECTROPNEUMATIC - [el] (operated by compressed air and electrically controlled) électropneumatique
~CONTACTOR - [el] contacteur *m* électropneumatique
~CONTROL - [el] commande *f* électro-pneumatique
~CONTROLLER - [contr] régulateur *m* électropneumatique
~VALVE - [contr] (electrically operated valve controlling a compressed-air line) soupape *f* électropneumatique, vanne *f* électropneumatique
ELECTROPOLISHING - [metall] (the operation of making a metal surface brighter and smoother by making it anodic in a suitable electrolytic solution) polissage *m* électrolytique
ELECTROPOSITIVE - [chem] (carrying a positive charge of electricity) électropositif
ELECTROPULT - [aero] (electrically operated take-off catapult) catapulte *f* électrique
ELECTROREFINING - [el metall] (the process of removing impurities from metals by making the crude metal the anode in an electrodeposition bath, the required metal being deposited in a purified form on the cathode) affinage *m* électrolytique
ELECTRORETINOGRAM - [med] électroretinogramme *m*
ELECTROSCOPE - [instr] (electrostatic device designed to indicate an electric charge) électroscope *m*
ELECTROSHOCK THERAPY - [med] (production of a reaction in the central nervous system by the application of electric current to the cranium) électrochoc *m*, méthode *f* de Cerletti et Bini, électroconvulsion *f*
ELECTROSTATIC - [el] (relating to the behaviour of electric charges and electric potentials) électrostatique
~ACTUATOR - [el acoust] (a device making it possible to apply a given electrostatic force to the diaphragm of a microphone) excitateur *m* électrostatique
~COUPLING - [el] (coupling between two circuits effected by a condenser also called capacity coupling) couplage *m* électrostatique
~DEFLECTION - [electron] (the bending of an electron beam by electrically charged plates, as distinct from electromagnetic fields) déviation *f* électrostatique
~FIELD - [el] (a field produced by an electric charge, as distinct from a magnetic field) champ *m* électrostatique
~FILTER - [nucl] (a screen used in nuclear energy plants to collect charged dust particles) filtre *m* électrostatique
~FOCUSING - [electron] (the focusing of an electron beam by means of an electrostatic electron lens) concentration *f* électrostatique
~GENERATOR - [el] (generator depending for its action on electrostatic processes producing charges at high potential) machine *f* électrostatique, générateur *m* électrostatique
~INDUCTION - [el] influence *f* électrostatique, électrisation *f* par influence
~INSTRUMENT - [instr] (instrument acting on electrostatic forces) appareil *m* électrostatique
~LENS - [electron] (electronic lens in which the result is obtained by an electrostatic field) lentille *f* électrostatique
~LOUDSPEAKER - [el acoust] (loudspeaker depending for its operation on the electrostatic forces applied to the plates of a capacitor) haut-parleur *m* électrostatique
~MEMORY - [comput] mémoire *f* électrostatique
~MEMORY TUBE - [electron] (cathode-ray tube storing information in the form of the presence or the absence of spots bearing electrostatic charges) tube *m* à mémoire électrostatique
~MICROPHONE - [electron] (microphone depending for its operation on variations of electric capacitance) microphone *m* électrostatique
~PERCUSSION WELDING - [metall] soudure *f* électrostatique par percussion

ELECTROSTATIC POTENTIAL - [el] potentiel *m* électrostatique
~PRESSURE - [el] pression *f* électrostatique
~SCREEN - [el] (the Faraday cage; an enclosure protecting from the effects of electrified bodies outside it) écran *m* électrostatique, cage *f* de Faraday
~SEPARATOR - [el metall] (machine designed to separate dry materials of different electrostatic characteristics) séparateur *m* électrostatique
~STORAGE TUBE - s. électrostatique memory tube
ELECTROSTATICS - [el] (science dealing with phenomena associated with electricity at rest) électrostatique *f*
ELECTROSTENOLYSIS - [el chem] (precipitation of metals in the pores of a membrane by electrolysis) électrosténolyse *f*
ELECTROSTRICTION - [el] (strain produced by an electric field in certain materials, analogous to magnetostriction) électrostriction *f*
ELECTROTAXIS - [el biol] (the response of an organism to an electric stimulus)électrotaxie *f*
ELECTROTECHNIC - [el] électrotechnique
ELECTROTECHNICS - [el] (electrical technology) électrotechnique *f*
ELECTROTELECLINOMETER - [instr] téléclinomètre *m* électromagnétique
ELECTROTHERAPEUTICS - s. electrotherapy
ELECTROTHERAPY - [med] (treatment of diseases by electricity) électrothérapie *f*
ELECTROTHERMAL RECORDING - [el acoust] (electrochemical recording in which the change is produced mainly by thermal action) enregistrement *m* électrothermique
~RELAY - [el] relais *m* électrothermique
ELECTROTHERMIC INSTRUMENT - [instr] (thermal instrument; an instrument acting on the heating effect of a current) appareil *m* thermique
ELECTROTHERMICS - [phys] (the science dealing with the direct conversion of electrical energy into heat or viceversa) électrothermiė *f*
ELECTROTONIC WAVE - [el biol] (brief change of potential on an excitable membrane near an applied stimulus) onde *f* électrotonique
ELECTROTONUS - [el biol] (of a nerve) électrotonus *m*
ELECTROTROPISM - [med] électrotropisme *m*
ELECTROTYPE - [print] (printing plate) galvanotype *m*, galvano *m*
ELECTROTYPING - [print] (formation of copying and printing plates and similar objects by electrodeposition) galvanotypie *f*
ELECTROTYPY - s. electrotyping
ELECTRO-ULTRAFILTRATION - [el chem] électrodialyse *f*
ELECTROVALENCE - [el chem] (a chemical bond due to the transfer of electrons from one atom to another) électrovalence *f*
ELECTROVALENT COMPOUND - [el chem] (compound in which a transfer of electrons occurs) complexe *m* électrovalent
ELECTROVALVE - [el metall] (a device for controlling fluid flow by means of a mechanical valve actuated by a solenoid) électrovanne *f*
ELECTROWINNING - [el metall] (process of obtaining metals from their ores by dissolving the latter and electrolysing the solution) extraction *f* par voie électrolytique

ELECTRUM - [min] (argentiferous gold, occurring in nature and containing up to 26 percent of silver) électrum *m*, électre *m*
ELEKTRON - [metall] (magnesium based alloys) élektron *m*
ELEMENT - [phys] (a substance the atoms of which all have the same atomic number) élément *m*
[el] (storage battery; complete assembly of positive and negative plates and separators) élément *m*, bloc *m* de plaques
[chem] (substance which cannot be decomposed into simpler substances) corps *m* simple, élément *m*
[biol] cellule *f*
[mech] élément *m*, membre *m*
[gen] élément *m*
~FORMER - [th eng] (refractory material on which the heating resistor is wound in a heating element) support *m* d'élément chauffant
~OF A WINDING - [el] spire *f* (d'un enroulement)
ELEMENTAL AREA - [telev] point *m* d'image, élément *m* d'image
ELEMENTARY - [gen] élémentaire
~AERIAL - [radio] (radiating constituent of an aerial array) élément *m* d'antenne
~ANALYSIS - [chem] (the quantitative and qualitative analysis of the elements present in an organic compound) analyse *f* élémentaire
~ANTENNA - s. elementary aerial
~BODIES - [med] (in infections) corps *mpl* élémentaires
~CELL - [phys] (the simplest geometric figure including all the characteristics of the lattice structure of a crystal) cellule *f* élémentaire
~CHARGE - [el] (elementary quantity of negative electricity associated with the electron) charge *f* élémentaire
~COLOURS - [paint] (primary colours) couleurs *fpl* fondamentales
~DIAGRAM - [gen] diagramme *m* schématique
~DIFFUSION EQUATION - [phys] équation *f* de la diffusion
~FIBRE - [text] fibre *f* élémentaire
~FILAMENT - [text] filament *m* élémentaire
~PARTICLE - [phys] (a general term for electrons, protons, etc) particule *f* élémentaire
~THREAD - [text] fil *m* élémentaire
ELEMI GUM - [plast ind] résine *f* élémi
ELEPHANT - [zool] éléphant *m*
~GRASS - [bot] herbe *f* à éléphants
ELEPHANTIASIS - [med] (appreciable enlargement of limbs or scrotum) éléphantiasis *f*
ELEVATE, to - [gen] élever, soulever, hausser
[aero] cabrer
ELEVATED - [gen] élevé, soulevé
(of a railway) aérien
~AERIAL - [radio] (an aerial, placed sufficiently high to be free of obstacles) antenne *f* extérieure
~ANTENNA - s. elevated aerial
~BASIN - [phys] réservoir *m* en élévation (ou surélevé)
~CABLEWAY CRANE - [mech] grue *f* à câble, blondin *m*
~PEDESTRIAN CROSSING - [constr] passage *m* pour piétons surélevé, passerelle *f* pour piétons
~RAILWAY - [railw] (overhead railway running on girders) chemin *m* de fer aérien
~SOURCE - s. elevated spring

ELEVATED SPRING - [hydr] source *f* haute (ou de montagnes)
ELEVATING - [aero] cabrage *m*
˜ARC - [firearms] crémaillère *f* de pointage
˜DIAL - [firearms] secteur *m* denté de pointage en hauteur
˜GATE - [transp] (in lorries, an endgate used in conjunction with a hoisting mechanism to allow the gate to come down to ground level) hayon *m* élévateur
˜GEAR - [firearms] mécanisme *m* de pointage en hauteur
˜HANDWHEEL - [mech] volant *m* à main de levage
ELEVATION - [gen] élévation *f*
[constr] (the front of a building) facade *f*, élévation *f*
[astr] élévation *f*
[geogr] (the height of a point on the earth's surface above a given datum, e.g. M.S.L.) altitude *f*, cote *f*, hauteur *f*
[text] (of warper) hauteur *f* du cône
[firearms] angle *m* de hausse
[mech] déplacement *m* vertical
[draw] élévation *f*
˜ANGLE - [radar] angle *m* d'élévation
˜HEAD - [hydr] (the energy per unit weight of a fluid due to its elevation above a given datum) hauteur *f* de chute
˜MARKER - [surv] point *m* trigonométrique
˜OF BOILING POINT - [chem] (the raising of the boiling point by substances in the solution) élévation *f* du point d'ébullition
˜OF THE WELL - [mining] côte *f* du sondage
˜POSITION INDICATOR - [radar] (radar display showing simultaneously angular elevations and distances of objects) radar *m* de position (indiquant la distance de l'altitude)
˜VIEW - [draw] élévation *f*, projection *f* verticale
ELEVATOR - [aero] (a control surface designed to produce a pitching movement in an aeroplane) gouverne *f* de profondeur
[mech] (a type of conveyor) élévateur *m*, élévateur-transporteur *m*
[mech] (US for lift) ascenseur *m*, monte-charge *m*
[agric] élévateur *m*, silo *m* à grains
[med] (surgical instrument) instrument *m* pour relever un organe
[auto] (carlifter) élevateur *m*
˜ANGLE - [aero] (the angle which the chord of an elevator makes with that of the corresponding fixed surface) braquage *m* de la gouverne de profondeur
˜BELT - [mech] élévateur-transporteur *m*
˜BUCKET - [mech] auget *m* de noria, godet *m* d'élévateur
˜CAGE - [mining] cage *f*
˜CAR - [mech] cabine *f* d'ascenseur
˜CASING - [mech] cage *f* d'ascenseur, puits *m* d'ascenseur
˜CONTROL - [aero] commande *f* de profondeur
˜DREDGER - [mech] (earth moving machine) drague *f* à godets
˜LATTICE - [text] tablier *m* sans fin montant
˜LINK - [mining] anse *f* de l'élévateur
˜LOADER - [agric mach] élévateur-chargeur *m*
˜PLATFORM - [auto] pont *m* de levage, plateforme *f* de levage
˜SCOOP - s. elevator bucket
ELEVATOR SHAFT - [constr] cage *f* d'ascenseur
˜TRIM TAB - [aero] (a trim tab applied to an elevator) volet *m* compensateur de gouverne
ELEVON - [aero] (a control surface which performs the function of both an elevator and an aileron) élevon *m*, gouverne *f* combinée aileron-profondeur
˜ANGLE - [aero] (the angle which the chord of an elevon makes with that of the corresponding fixed surface) angle *m* d'élevon
ELIASITE - [min] (impure gummite) éliasite *f*
ELIMINATE, to - [gen] éliminer
ELIMINATION - [gen] élimination *f*
˜FILTER - [radio] (a filter with a single attenuation band) filtre *m* coupe-bande
˜OF THE WEIGHTING - [text] envelage *m* de la charge, délestage *m* (soie)
ELIQUATION - [metall] liquation *f*
ELIXATION - [chem] lessivage *m*
ELIXIR - [med] élixir *m*
ELK - [zool] élan *m*
ELL - [plumb] coude *m* (de tuyau), raccord *m* coudé en 'L'
ELLIPSE - [geom] ellipse *f*
˜OF ESSENTIAL INFORMATION - [telev] (this is obtained by laying an oval celluloid sheet across a view finder screen) cône *m* d'observation
ELLIPSOID - [geom] (solid body formed by the revolution of an ellipse about one of its axes) ellipsoïde
˜CORE AERIAL - [radio] (antenne *f* à noyau magnétique ellipsoïdal
˜CORE ANTENNA - s. ellipsoid core aerial
˜OF REVOLUTION - s. ellipsoid
˜OF ROTATION - [geom] (rotation generated by the revolution of an allipse about one of its axes) ellipsoïde *m* de révolution
ELLIPSOIDAL - [geom] ellipsoïdal
˜BASALT - [geol] basalte *m* ellipsoïdal
˜REFLECTOR - [light] réflecteur *m* ellipsoïdal
ELLIPTIC - [geom] elliptique
˜COMPASS - [impl] compas *m* à ellipse
˜MIRROR - [opt] miroir *m* elliptique
˜POLARIZATION - [radio] polarisation *f* elliptique
˜SPRING - [mech] ressort *m* elliptique
˜TRAMMEL - [draw] (instrument for drawing ellipses) compas *m* à ellipse, compas *m* elliptique, ellipsographe *m*
ELLIPTICAL - [geom] elliptique
˜ARCH - [arch] (arch formed to an elliptical curve) arc *m* elliptique
˜COMPASS - s. elliptic compass
˜FIELD - [el] (a field of which the representative vector rotates at the same time as it sweeps the area of an ellipse) champ *m* elliptique
˜GEARS - [mech] engranages *mpl* elliptiques
ELLIPTICALLY POLARIZED LIGHT - [opt] (light produced by two mutually perpendicular plane-polarized components which are not in phase) lumière *f* polarisée elliptiquement
˜POLARIZED WAVE - [electron] (in waveguides) onde *f* à polarisation elliptique, onde *f* polarisée elliptiquement
ELLIPTICITY - [geom] ellipsité *f*
ELM - [bot] orme *m*
ELONGATING STAGE - [bot] (the time during which an organ increases in length) période *f* d'allongement
ELONGATION - [gen] allongement *m*, élongation *f*

[astr] (the difference between the celestial longitude of a heavenly body and that of the sun) élongation *f*, disgression *f*
[phys] allongement *m*
ELONGATION AT BREAK - [phys] (the amount , expressed as a percentage of the length of the specimen, by which a yarn can be stretched before it breaks) allongement *m* de rupture
~AT RUPTURE - [phys] allongement *m* de rupture
~CABLE - [el] câble *m* de rallonge, câble *m* de prolongement
~DUE TO PULL - s. elongation due to tension
~DUE TO TENSION - [phys] allongement *m* à la traction
~INDEX - [text] indice *m* d'allongement
~LIMIT - [text] limite *f* d'allongement
~OF THE IMAGE - [photo] élongation *f* de l'image
~PIECE - [el] rallonge *f*, allonge *f*, pièce *f* de rallonge
~PROOF LIMIT - [metall] limite *f* élastique conventionnelle
~STRAIN - [phys] déformation *f* d'allongement
[metall] limite *f* apparente d'extension
~STRENGTH - [text] pouvoir *m* d'allongement, élasticité *f*
~STRESS - [phys] tension *f* d'allongement
~TEST - [mech] essai *m* d'allongement
~TUBE - [plumb] tube *m* de rallonge
ELUATE GRADE - [chem] titre *m* d'éluat
ELUENT - [chem] éluant
ELUTE, to - [chem] éluer
ELUTION - [chem] (process used in chromatography analysis) élution *f*
ELUTRIATE, to - [chem] éluer, séparer par décantation
[min] épurer par lavage et filtrage
ELUTRIATION - [chem] (separation of solids in particle form in a rising current of fluid) élution *f*, séparation *f* par décantation
ELUTRIATOR - [chem etc] (device used to wash or size very fine powders in a rising current of water) séparateur *m*
ELUVIAL DEPOSIT - [geol] éluvion *m*
~GRAVEL - [geol] (gravel formed by the local disintegration of rocks) éluvium *m*, éluvion *m*
~HORIZON - [soil] (layer from which material has been removed in solution or in water suspension) horizon *m* éluvial
ELUVIUM - s. eluvial gravel
e/m - [chem] (the symbol for emanation) émanation *f*
E/m - [phys] (the ratio of the electric charge to the mass for electrons etc) $\frac{e}{m}$ (charge spécifique électronique)
EM - [print] (the square of the body of any size of type, used as a unit of measurement) mesure *f* typographique
~QUAD - [print] cadratin *m*
~SCALE - [print] typomètre *m*
EMANATION - [phys] (general term for the radioactive, chemically-inert gases derived from the disintegration of radium, thorium and actinium, viz, radon and its isotopes) émanation *f*
[gen] émanation *f*
~PROSPECTING - [min] prospection *f* émanométrique
EMANOMETER - [instr] (instrument designed to measure small emanations) émanomètre *m*
EMANON - [chem] (a term sometimes used for radon, radium emanation) émanon *m*, radon *m*
EMASCULATE, to - [gen & med] émasculer, châtrer
EMASCULATION - [med] (the removal of testes) émasculation *f*
[bot] (the removal of the stamens) castration *f*
EMASCULATOR - [vet] (an instrument designed to castrate bulls and horses by crushing the spermatic cord) instrument *m* pour la castration
EMBALM, to - [gen] embaumer
EMBALMENT - [gen] embaumement *m*
EMBANK, to - [constr] encaisser, endiguer
EMBANKMENT - [constr] remblais *m*, talus *m*, levée *f* de terre, banquette *f*, digue *f*
~SLOPE - [constr] talus *m* de remblai
~WALL - [constr work] (a retaining wall) mur *m* de soutènement
EMBARGO, to - [comm & leg] mettre l'embargo sur (e.g. imports and exports) interdire
EMBARGO - [leg naut etc] embargo *m*, confiscation *f*, mise *f* sous séquestre
EMBARK, to - [gen & naut] embarquer
[fig] s'embarquer
EMBARKATION - [naut] embarquement *m*
EMBATTLED - [build] crénélé
EMBATTLEMENT - [build] (part of a building indented like a battlement) parapet *m* crénelé, crénelage *m*
EMBAYMENT - [geogr] baie *f*
EMBED, to - [gen] enrober, enchasser, enfouir, sceller, planter, enfoncer, noyer dans
[mach] (to build into a machine) encastrer, incorporer
[agric & bot] planter
EMBEDDED COLUMN - [build] (column which is partly built into a wall) colonne *f* adossée, colonne *f* engagée
~RECTIFIER - [el] (a dry rectifier embedded in a suitable material) redresseur *m* encastré
~TEMPERATURE DETECTOR - [el] (resistance thermocouple which is built into a piece of equipment during its construction) détecteur *m* interne de température
EMBEDDING - [mech] incrustation *f*, enrobage *m*, enrobement, enfoncement *m*, encastrement *m*, noyage *m*
~MATERIAL - [mech] matériau *m* de revêtement
EMBELLISH, to - [gen] agrémenter, embellir, orner
EMBELLISHMENT - [gen] embellissement *m*, enjolivure *f*
[arch] (a general term denoting ornamentation) ornement *m*
EMBER - [gen] braise *f*, charbons *mpl* ardents
EMBODY, to - [gen] contenir, renfermer, incarner, incorporer, réunir
EMBOLECTOMY - [med] (the excision of an embolus) embolectomie *f*
EMBOLIC ABSCESS - [med] abcès *m* métastatique
EMBOLISM - [med] (blocking of a blood vessel) embolisme *m*
EMBOLITE - [min] (important ore of silver occurring in Chilean mines) embolite *f*
EMBOLUS - [med] (a mass which forms in the circulation of the blood) embolus *m*, embole *m*
EMBOLY - [zool] (invagination) embolie *f*, invagination *f*
EMBOSS, to - [gen] repousser (le métal), estamper, emboutir

[leather ind] gaufrer
[text] gaufrer, frapper
EMBOSSED - [metall text] gaufré
~CHARACTER - [print] caractère *m* en relief
~CLOTH - [text] tissu *m* gaufré
~EFFECT - [text] effet *m* de gaufrage
~FINISH - [text] gaufrage *m*
~PRINTING - [text] impression *f* en relief
~WORK - [ind proc] emboutissage *m*, repoussage *m*, grainage *m*, gaufrage *m*
EMBOSSER - s. embossing machine
EMBOSSING - [gen & metall] bosselage *m*, repoussage *m*, estampage *m*, gaufrage *m*
[print] gaufrage *m*, estampage *m*, bombage *m*
[text] gaufré *m*
[photo] filigranage *m*
~ANVIL - [impl] table *f* de gaufrage
~BOWL - [text] rouleau *m* à gaufrer
~CALENDER - [paper man] gaufreuse *f*, calandre *f* gaufreuse
~CONE - [text] cône *m* gaufré
~DIE - [metall] matrice *f* de gaufrage
~MACHINE - [mech] (machine for embossing by means of heated plates and rollers) gaufreuse *f*
~OF PRINTS - [photo] repoussage *m* de photos
~PRESS - [mech] presse *f* à matricer
~ROLLER - [print] cylindre *m* en relief
[text] rouleau *m* d'impression
~TOOL - [metall] outil *m* à matricer
EMBOUCHURE - [mus] (the mode of application of the lips to the mouthpiece of a brass instrument) embouchure *f*
EMBRASURE - [constr] (splayed opening of a window) embrasure *f*
EMBRITTLE, to - [gen] rendre fragile, rendre cassant
EMBRITTLEMENT - [phys] (the increase in the tendency of a metal to fracture under stress) fragilisation *f*, accroissement *m* de fragilité *f* caustique
EMBRITTLING GAS - [phys] gaz *m* causant la fragilisation
EMBROCATION - [med] (the application of a liquid into an injured part) embrocation *f*, fomentation *f*
EMBROIDER, to - [gen & text] broder
EMBROIDERING - [text] broderie *f*
~NEEDLE - [text] aiguille *f* à broder
EMBROIDERY - [gen & text] broderie *f*
~FABRIC - [text] tissu *m* pour broderie
~MACHINE - [text] métier *m* à broder, machine *f* à broder
~YARN - [text] fil *m* à broder
EMBRYO - [biol] (organism in the early stages of development) embryon *m*
[bot] gemmule *f*
~SAC - [biol] (cavity in the ovule) sac *m* embryonnaire
EMBRYOCARDIA - [med] (the heart condition, in which the sounds of the heart are similar to those heard in the foetus) embryocardie *f*
EMBRYOGENETIC - [biol] embryogénique
EMBRYOGENY - [biol] (the process which leads to the formation of the embryo) embryogénie *f*
EMBRYOLOGY - [biol] embryologie *f*
EMBRYOMA - [med] (tumour consisting of foetal elements) embryome *m*
EMBRYONIC - [biol] embryonnaire
~TISSUE - [biol] tissu *m* embryonnaire
EMBRYOTOMY - [med] (mutilation of a foetus for removal) embryotomie *f*
EMEDULLATE, to - [med] extraire la moelle
EMERALD - [min] (brilliant green gemstone) émeraude *f*
~COPPER - [min] dioptase *f*
~GREEN - [paint] (a complex copper acetate-arsenite pigment, now little used) vert *m* émeraude
~NICKEL - [min] (hydrated basic nickel carbonate; also called zaratite) zaratite *f*, texasite *f*
EMERGE, to - [gen] émerger, surgir, se montrer
EMERGED BOG - [geol] tourbe *f* émergée
EMERGENCE OF LAND - [geol] (the emergence of land from the sea) émersion *f*, exondation *f*
~PLANE - [opt] émergence *f*, plan *m* d'émergence
EMERGENCY - [gen] (a situation developing suddenly and calling for immediate action) urgence *f*, circostance *f* critique
[gen] (used of any device or equipment designed for use in an emergency, e.g. reserve apparatus to take the place of the normal unit in case of failure of the latter) de secours, de réserve
~AIR - [aero] (compressed air provided for emergency operation of hydraulic or compressed air equipment) alimentation *f* pneumatique de secours
~BATTERY - [el] (electric battery for use in case of failure of other source of current) batterie *f* de secours
~BRAKE - [auto] (the hand brake) frein *m* de secours
~BYPASS - [telev] dérivation *f* de secours
~CABLE - [mech] câble *m* de secours
[telecomm] câble *m* provisoire
~CALL - [gen & telecomm] sonnerie *f* d'alarme
~CAR - [auto] dépanneuse *f*, voiture *f* de dépannage
~CARTRIDGE - [mech] (a cartridge to provide gases of combustion under pressure for the operation of a hydraulic or compressed air system in emergency) cartouche *f* de secours
~CIRCUIT - [telecomm] circuit *m* de secours
~COOLING - [nucl] refroidissemment *m* de sécurité
~DOSE - [nucl] (dose absorbed in excess) dose *f* d'urgence
~EXIT - [gen] sortie *f* de secours
~EXPOSURE TO EXTERNAL RADIATIONS - [nucl] exposition *f* externe exceptionnelle concertée
~HIGH EXPOSURE - [nucl] (to radioactive materials) contamination *f* interne exceptionnelle concertée
~LANDING - [aero] atterrissage *m* forcé
~LANDING FIELD - [aero] terrain *m* de secours
~LANDING GROUND - s. emergency landing field
~LIGHT - [gen] (in factories, cinemas etc) éclairage *m* de sécurité, éclairage *m* de secours
~MACHINE - [mech] machine *f* de réserve
~PARACHUTE - [aero] (a parachute intended to be used for escape from an aircraft in case of disaster) parachute *m* de secours
~PLAN - [gen] plan *m* de secours, plan *m* d'urgence
~REPAIR - [gen & mech] (a temporary repair made to rectify unexpected breakdown) réparation *f* de urgence, réparation *f* provisoire
~ROUTE - [telecomm] ligne *f* de réserve
~SHUT-DOWN - [nucl] arrêt *m* d'urgence
~SHUT-OFF ROD - [nucl] barre *f* de sécurité
~STOP - [el] (switch, easily accessible, for cutting off supply to an electrical machine in case of emergency) interrupteur *m* de secours
~SWITCH - s. emergency stop
~TANK - [auto] réserve *f*

EMERGENCY TRANSMITTER BEACON - [radar] (a very small beacon transmitter fitted to life-belts) radiobalise *f* de sauvetage
~TRIP - [nucl] arrêt *m* brusque, interruption *f* d'urgence
EMERGENT - [phys] émergent
~RAY POINT - [radiat] (the centre of the area through which a beam of radiation leaves the body) point *m* d'émergence, point *m* de sortie
EMERIZING - [text] rectification *f*, aiguisage, polissage *m*
~MACHINE - [text] machine *f* pour émeriser
EMERSION - [gen] émersion *f*
[astr] (the appearance of a celestial body from a shadow) émersion *f*
EMERY - [min] (a natural abrasive consisting of corundum mixed with other oxides) émeri *m*
~BUFF - s. emery cloth
~CLOTH - [mech] (cloth coated with finely-ground emery) toile *f* d'émeri
~DISC - [mech] meule *f* d'émeri
~DUST - [mech] poudre *f* d'émeri, potée *f* d'émeri
~FILLET - [mech] ruban *m* d'émeri, ruban *m* émerisé
~FLOUR - s. emery dust
~GRINDING - [mech] émerisage *m*
[metall] polissage à l'émeri
~MACHINE - [mech] machine *f* pour émeriser
~PAPER - [mech] (paper coated with finely-ground emery, used for finishing or cleaning etc) papier *m* émerisé, papier *m* d'émeri
~PASTE - [ind chem] (finely- ground emery mixed to a paste with oil, used for grinding-in valves etc) pâte *f* émerisée, potée *f* d'émeri
~POWDER - s. emery dust
~ROLLER - [mech] cylindre *m* d'émeri, cylindre *m* émerisé
~RUBBING - [mech] rodage *m*, émerisage *m*
~STICK - [impl] rodoir *m*, polissoir *m*
~STONE - [mech] pierre *f* d'émeri, meule *f* d'émeri
~WHEEL - [impl] meule *f* d'émeri
EMESIS - [med] (vomiting) vomissement *m*
EMETIC - [med] (causing vomiting) émétique, émétisant
e.m.f. - [el] (initials meaning electromotive force; also written emf or E.M.F.) f.e.m. (force électromotrice)
~AMPLIFICATION FACTOR - [radio] coefficient *m* de amplification
EMICTORY - [med] (a drug facilitating the excretion of urine) diurétique *m*
EMINENTLY HYDRAULIC LIME - [chem] (lime obtained by burning a limestone rich in clay) chaux *f* hydraulique
EMISSARIUM - [hydr] (a flood-gate) émissaire *m*
[anat] émissaire *m*, canal *m* veineux du crane
EMISSARY - [gen] émissaire
[anat] émissaire *m*
EMISSION - [gen] émission *f*, dégagement *m*, lancement *m*
[electron] (the rate of emission of electrons) émission *f*
[radio] émission *f*
~BANDS - [radio] bandes *f*pl d'émission
~CHARACTERISTIC - [electron] (curve showing the relation between the saturation current from a cathode and one of the factors controlling the emission) caractéristique *m* d'émission
EMISSION CELL - [electron] cellule *f* photo-émissive
~CURRENT - [electron] (the current resulting from electron emission) courant *m* d'émission, courant *m* cathodique
~EFFICIENCY - [electron] (the quotient of the saturation current divided by the heating power which the cathode absorbs) rendement *m* d'émission
~MEASURING DEVICE - [electron] (instrument designed to check the quality of electronic tubes) appareil *m* de mesure de l'émission
~OF GAS - [phys] désorption *f* de gaz, dégagement *m* gazeux
~OF HEAT - [phys] cession *f* de chaleur, émission *f* de chaleur
~SOURCE - [phys] source *f* d'émission
~SPECTRUM - [phys] (spectrum produced by radiation from any emitting source) spectre *m* d'émission
~STABILIZATION - [mech] stabilisation *f* d'émission
EMISSIVE POWER - [phys] pouvoir *m* émissif, pouvoir *m* d'émission
EMISSIVITY - [phys] (the ratio of the quantity of energy emitted by a body to that emitted by a theoretical black body) pouvoir *m* émissif, pouvoir *m* rayonnant
EMIT, to - [gen] émettre, dégager, lancer, jeter (of a smell) exhaler
EMITTANCE - [phys] pouvoir *m* émissif, émittance *f*
EMITTED PARTICLE - [phys] (the particle emitted from a solid or a liquid) particule *f* émise
EMITTER - [electron] distributeur *m* d'impulsions, émetteur *m* d'impulsions
~CURRENT - [electron] (of a transistor) courant *m* émetteur
~ELECTRODE - [electron] électrode *f* émetteur
~FOLLOWER - [comput] (of a transistorized circuit) émetteur *m* cathodyne, émetteur *m* follower
~HUB - [comput] plot *m* "Distributeur", jack *m* "Emetteur"
~JUNCTION - [electron] jonction *f* émetteur
~TERMINAL - s. emitter electrode
EMMETROPIC EYE - [opt] (an eye in which light from a distant source is focused at the retina) œil *m* émmétrope
EMOLLIENT - [chem] émollient *m*
EMPENNAGE - [aero] (the whole of the tail structure of an aircraft) empennage *m*
EMPHASIZE, to - [gen] accentuer
EMPHASIZER - [radio] (filter used to emphasize a portion of the frequency spectrum) accentuateur *m*
EMPHYSEMA - [med] (presence of air in connective tissues) emphysème *m*
EMPIRIC TEST - [gen] essai *m* empirique
EMPIRICAL - [gen] (induced only from observation) empirique
~FORMULA - [chem] (a formula which expresses the simplest numerical relation between the atoms of a compound) formule *f* empirique
~MASS FORMULA - s. empirical formula
EMPIRICISM - [gen] (a procedure whereby scientific laws are set out by reasoning from observation) empirisme *m*
EMPLACEMENT - [gen] emplacement *m*
EMPORIUM - [gen] entrepôt *m*, centre *m* commercial

EMPOWER, to - [gen] autoriser,
EMPTIES - [gen] (containers emptied out of contents) caisses fpl vides, emballages mpl vides
EMPTY, to - [gen] vider
EMPTY - [gen] vide
~BAND - [electron] (in a specified substance, a band of possible energy levels none of which includes electrons in a specified state) bande f vide
~FIELD VISION - [opt] vision f sans repères
~OF AIR - [phys] vide d'air, évacué
~OUT, to - [gen] vider complètement
~WEIGHT - [aero] (the weight of the structure, propelling plant and fixed equipment of an aircraft, exclusive of equipment which is not indispensable under regulations, and load) poids m à vide
[gen] poids m à vide
EMPYEMA - [med] empyème m, pleurésie f purulente
EMPYREUMATIC TASTE - [brew ind] goût m empyreumatique
EMU - [zool] émeu m
EMULSIBLE OIL - [oil ind] huile f émulsionable
EMULSIFIABILITY - [chem] (the quality of being emulsifiable) émulsibilité f
EMULSIFIABLE - [chem] émulsionnable
EMULSIFICATION - [chem] (the preparation of an emulsion) émulsification f, émulsionnement m
EMULSIFIED OIL - [oil ind] huile f emulsionnée
EMULSIFIER - [chem] émulseur m, émulsionneur m
[oil ind] émulsifiant m
EMULSIFY, to - [chem] (to make an emulsion) émulsionner
EMULSIFYING AGENT - [chem] émulsifiant m, agent m émulsionnant
EMULSIN - [chem] (an enzyme facilitating emulsion) émulsine f
EMULSION - [chem] (a suspension in one fluid of another immiscible one, e.g. milk) émulsion f
[photo] (the light-sensitive material coated on a support, e.g. a film) émulsion
~ADHESIVE - [ind chem] colle f en émulsion
~FREEZE STABILIZER - [chem] stabilisant m antigel pour émulsions
~MUD - [oil ind] boue f à l'émulsion
~PAINT - [paint] (a surface coating in which the vehicle is an emulsion of a binding agent, commonly latex, in water) peinture-émulsion f
~POLYMERIZATION - [chem] (polymerization of a monomer in minute particles which are dispersed in an (aqueous medium with the aid of an emulsifying agent) polymérisation f par émulsion
~SIDE - [photo] côté m de l'émulsion
~SPEED - [photo] sensibilité f de l'émulsion
~SUPPORT - [photo] émulsion, support m de l'émulsion
~TROUGH - [photo] auge f à émulsion
EMULSIVE - [chem] émulsif
EMULSOID - [chem] (a lyphilic colloid, i.e. a colloid which can be readily dispersed in an appropriate medium, and can be re-dispersed after coagulation) émulsoïde m
EN-QUAD - [print] (type space corresponding to half an em) demi-cadratin m
ENABLE, to - [gen] rendre capable de, permettre à, rendre possible
ENABLING PULSE - [electron] (a pulse which opens a normally closed electric gate) impulsion f de déclenchement, impulsion f d'ouverture
ENALITE - [min] (zirconium and uranium containing mineral) énalite f
ENAMEL, to - [ceram] (to apply enamel) émailler
[paint] peindre à la peinture émail
ENAMEL - [metall] (a hard ceramic coating fired on to a metal for decorative or protective purposes) émail m
[paint] (an oil-based paint that dries to a hard smooth surface) peinture f émail
~COATING - [metall] couche f d'émail
~FIRING - [metall] cuisson f de l'émail
~FRIT - [metall] fritte f d'émail
~INSULATED WIRE - [el] (used for the winding of small magnetic coils) fil fm émaillé
~KILN - [metall] (a kiln for firing painting ware) four m de recuit pour l'émail
~PAINT - [paint] (a special, ready for use, oil paint) peinture f émail
ENAMELLED BRICK - [constr] brique f vernissée
~CLOTH - [text] (used in upholstery) toile f cirée
~HARDBOARD - [constr] panneau m (de fibres) laqué
~IRON - [metall] fer m émaillé
~LEATHER - [leather ind] (leather coated with a varnish) cuir m verni
~PAPER - [paper man] (highly finished paper coated with china clay) papier m couché, papier m glacé
~SHEET IRON - [metall] tôle f émaillée
~STEEL TUB - [brew ind] cuve f en acier émaillé
~STEEL VESSEL - s. enamelled steel tub
~WIRE - [mech] (wire insulated with a coating of a special flexible enamel) fil m émaillé
ENAMELLING - [metall] émaillage m, émaillerie f
[photo] (high glaze on a print) glaçage m
ENANTIOMERIDE - [phys] énantiomorphe m
ENANTIOMERISM - [opt] (the difference in optical activity between the isomers of compounds with asymmetric bonds) éniantomorphie f
ENANTIOMORPHIC ALLOTROPY - [phys] allotropie f éniantomorphe
ENANTIOMORPHOUS - [phys] (term used of two forms, one of which is related to the other as an object to its mirror image) énantiomorphe
ENANTIOMORPHS - [phys] (crystals without a plane or a centre of symmetry, so that they cannot be brought into coincidence with their reflected image) cristaux mpl éniantiomorphes
ENANTIOTROPIC - [phys] (term used of a substance which can exist in two crystalline forms, one of which is stable below a certain temperature and the other above it) énantiotropique
ENANTIOTROPY - [phys] (the property of a substance to exist in two crystal forms) énantiotropie f
ENARGITE - [min] (natural sulpharsenate of copper, sometimes occurring with antimony; black; orthorhombic) énargite f
ENBLOC CYLINDER - [mech] cylindre m monobloc
ENCAPSULATE, to - [gen] encapsuler, protéger
ENCAPSULATED FUEL UNIT - [nucl] (fuel element enclosed in a jacket) élément m de combustible sous gaine étanche
~MODULE - [comput] module m encapsulé
ENCAPSULATION - [nucl] (the covering of neutron sources to avoid leakage under normal circumstances) scellement m
ENCASE, to - [gen] encaisser, enfermer, blinder
ENCAUSTIC - [metall] (a term, often loosely used,

for ceramic products with impressed designs, which are afterwards filled with tinted slip and fired) encaustique
ENCAUSTIC PASTE - [photo] encaustique *f*, cérotine *f*
~TILE - [constr] carreau *m* vernissé
ENCEPHALITIS - [med] encéphalite *f*
ENCEPHALOGRAPHY - [med] encéphalographie *f*
ENCEPHALOMYELITIS - [vet] (a porcine infection) encéphalomyélite *f*
ENCEPHALOTROPHY - [med] atrophie *f* cérébrale
ENCIRCLE, to - [gen] encercler, ceindre
ENCLOSE, to - [gen] enclore, clôturer, envelopper, enfermer, inclure
[mech] blinder, enfermer dans un carter, protéger
ENCLOSED - [gen] clos, clôturé, enfermé, inclu, ci-joint
[mech] (fitted in an enclosure) blindé, protégé, en boîtier, hermétique
~ARC LAMP - [el] lampe *f* à arc en vase clos
~FLAME ARC-LAMP - [el] (enclosed arc-lamp with flame carbons) lampe *f* à arc à flamme en vase clos
~FUSE - [el] fusible *m* protégé
~MOTOR - [mech] moteur *m* protégé
~SELF-COOLED MACHINE - [el] (so built that circulation of air between the inside and the outside of the machine is eliminated) moteur *m* à autoventilation protégé
~SLAG - [metall] scorie *f* interposée
~-VENTILATED - [el] (of electrical machines) à ventilation protégé
ENCLOSING BEDS - [geol] couches *fpl* encaissantes
~ROCK - [geol] roche *f* encaissante
ENCLOSURE - [gen] enceinte *f*, clôture *f*, enveloppe *f*, coffret *m*
[constr] (a kind of framing) clôture *f*, enceinte *f*
[aero] capot *m*
[geol] inclusion *f*, enclave *f*
~WALL - [constr] mur *m* d'enceinte, mur *m* de clôture
ENCODE, to - [comput] (also to code; to express information in a language which is acceptable to a computer) coder, chiffrer
ENCODER - [comput] codeur *m*, chiffreur *m*
~MATRIX - [comput] matrice *f* de codage
ENCOMPASS, to - [gen] entourer, environner, investir, cerner
ENCOUNTER - [nucl] (a random event) choc *m* aléatoire
ENCROACH, to - [gen] empiéter sur, avancer, envahir
ENCRYPTION - [comput] chiffrage *m* (de messages), chiffrement *m* (de messages)
ENCUMBER, to - [gen] encombrer, embarrasser, entraver, surcharger
ENCYSTED - [med] enkysté
END, to - [gen] achever, cesser, finir, terminer, clôturer
END - [gen] fin *f*, aboutissement *m*, extrémité *f*, bout *m*
[gen] (of a cable, a rope, etc) bout *m*, extrémité *f*, tête *f*
[mech] extrémité *f*, fond *m*, plateau *m*
[text] (of a warp) fil *m* de chaîne
~-AND-END - [text] (in weaving) envergure *f* fil à fil
~-ARCH BRICK - [constr] coin *m*
END-AROUND CARRY - [comput] transfert *m* circulaire, report *m* circulaire
~ARTERY - [anat] (a terminal artery) artère *f* terminale
~BAY - [constr] (of a bridge) travée *f* d'extrémité
~-BEARING CAPACITY OF PILES - [constr] capacité *f* portante en pointe des pieux
~BELL - [el] (a robust metal covering placed over the end-winding of a rotor) tête *f* de câble
~BINDER - [text] agrafe *f* reliant les extrémités (courroie)
~BLOCK STATION - [railw] poste *m* de blocage terminal
~-BOARD - [auto] (the rear board in a truck) hayon *m*
~BRACKET - [mech] (open structure at the end of an electrical machine) palier *m* flasque, palier *m* console
~BRAKE - [railw] frein *m* de queue
~CAP - [plumb] protecteur *m*
~CELL - [el] (one cell of a battery arranged to be inserted into or cut out of the circuit for voltage regulation) élément *m* de réduction, élément *m* de régulation
~CLOTH - [text] doublier *m*, toile *f* de tête
~CONNEXION - [el] borne *f*
~CUTTER - [mach tool] fraise *f* à deux tailles (en bout), fraise *f* (cylindrique) à queue
~CUTTING EDGE - [mach tool] arête *f* secondaire, arête *f* de coupe latérale
~DELIVERY THRESHER - [agric] batteuse *f* en long
~DEVICE - [meas] (the final system element responding quantitatively to the measure and performing the actual operation) élément *m* final
~DUMP - [auto] benne *f* basculant vers l'arrière
~ECHO PATH - [telecomm] parcours *m* des courants d'écho produits à l'extrémité du circuit
~EFFECT - [el] effet *m* d'extrémité
~ERROR - [contr] erreur *f* d'extrémité
~FADE - [cin] fermeture *f* en fondu
~-FED VERTICAL AERIAL - [radio] (vertical aerial insulated from earth and energized at the base) antenne *f* verticale alimentée en série
~-FED VERTICAL ANTENNA - s. end-fed vertical aerial
~FEED BOX - [el] boîte *f* d'alimentation
~-FIRE AERIAL ARRAY - [radio] (a type of linear array) réseau *m* d'antennes à rayonnement longitudinal
~-FIRE ARRAY - s. end-fire aerial array
~FLAKE - [med] plaque *f* nerveuse terminale
~FLOAT - [mech] jeu *m* axial, flottement *m*
~FRAME - [mech] paroi *f* extérieure du bâti de la machine
~GABLE - [constr] facade *f* (latérale) à pignon
~-GATE - s. end-board
~GRAIN WOOD - [constr] bois *m* de bout
~LAP - [photo] recouvrement *m* longitudinal (de deux photos)
~LAP-JOINT - [carp] (joint between the ends of two pieces of timber) assemblage *m* à mi-bois
~LEAKAGE - [el] (leakage associated with the end connexions of an electric motor) flux *m* de dispersion aux bornes
~LINK - [mech] (the link at one end of a chain) maillon *m* terminal, anneau *m* de fermeture
~LOADING - [railw] chargement *m* en bout

END-LOADING PLATFORM - [railw] quai *m* de chargement en bout
~LOBE - [anat] lobe *m* occipital
~MILL - s. end cutter
~MILLING - [mech] fraisage *m* en bout
~MORAINE - [geol] moraine *f* terminale
~MOUNTING - [mech] dispositif *m* de fixation à deux côtés
~-OF-HEADING SIGNAL - [comput] signal *m* de fin d'en-tête
~-OF-LINE - [comput] fin *f* de ligne
~OF PULSING SIGNAL - [radio] (signal conveying the information that no more digit signals will follow) signal *m* de fin de transmission
~OF-RECORD GAP - [comput] blanc *m*, intervalle *m*
~OF STROKE - [mech] (of a piston) fin *f* de course
~OF TAPE - [comput] fin *f* de bande (perforée), fin *f* de ruban (magnétique)
~OF WAGON - [railw] extrémité *f* de wagon
~-OF-WORD CHARACTER - [comput] caractère *m* de fin de mot
~ON - [naut] debout (vent)
~-ON ARMATURE - [el] armature *f* frontale
~-ON COLLISION - [railw] tamponnement *m*
~-ON DIRECTIONAL ARRAY - [radio] antenne *f* directionnelle multiple, antenne *f* Yagi
~-ON WELDING - [metall] soudure *f* en bout
~ORGAN - [anat] organe *m* terminal
~PAPER - [bookbind] feuillet *m* de garde, feuille *f* de garde, garde *f*
~-PIECE - [gen] embout *m*, bout *m*, extrémité *f*
~PLATE - [mech] plaque *f* d'extrémité, flasque *m* latéral
[metall] plaque *f* de fond, plaque *f* tubulaire
~PLATE BEARING - [mech] support *m* de plaque d'extrémité
~PLAY - [mech] jeu *m* axial, jeu *m* en bout, jeu latéral
~POINT - [chem] (the point, indicated by a marked colour change, which indicates that a titration is completed) point *m* de virage, point *m* final
[contr] position *f* extrême
~-POINT VOLTAGE - [el] (the specified voltage at which the discharge is considered complete) tension *f* finale
~PRINTING MACHINE - [comput] imprimante *f* marginale (pour cartes perforées)
~PRODUCT - [chem] queue *f* de distillation
[gen] produit *m* fini, produit *m* final
~PULLEY - [mech] poulie *f* de retour
~QUENCHING - [metall] trempe *f* des extrémités
~RELIEF ANGLE - [mach tool] angle *m* de dépouille en bout, angle *m* de dégagement avant
~RESISTANCE - [meas] résistance *f* résiduelle
~RIDGE TILE - [constr] faîtière *f* d'about
~SCALE - [instr] (on an indicating instrument) pleine déviation *f*, déviation *f* à pleine échelle
~SEAL - [metall] joint *m* de tête
~SETTING - s. end resistance
~SHACKLE - [naut] manille *f* d'extrémité
~SHEET - [el] (a sheet of insulated material between the end section of accumulators and their containers) plaque *f* isolante
~SHRINK - [timber] retrait *m* dans la longueur
~SIZING - [metall] calibrage *m* de tête de tube
~SPEED - [aero] (in catapult launching, the special of the aircraft in respect to the ship at the instant when the former is released from the catapult) vitesse *f* finale
END SPRING - [mech] (in lead-acid accumulators, small spring of hard lead) ressort *m* d'extrémité
~SUPPORT - [mach tool] lunette *f*
~TEETH - [mach tool] (in milling cutters) dents *mpl* en bout, denture *f* en bout
~TERMINAL - [el] borne *f*
~THRUST - [mech] (longitudinal or axial thrust, as of a shaft) poussée *f* axiale
~TIPPER - [mech] benne *f* basculant vers l'arrière, véhicule *m* à benne basculant vers l'arrière
~TO END SIGNALLING - [radio] transmission *f* sans répetition de signaux
~VACUUM - [phys] vide *m* final, vide *m* limite
~WALL - [ovens] pilier *m* d'extrémité
~WALL OF WAGON - [railw] paroi *f* d'extrémité de wagon
~WINDING - [el] (the part of the coil which is included between two slots, outside the core) connexion *f* frontale, tête *f* de bobine
~-WINDOW COUNTER - [mech] (counter which is irradiated from one end) détecteur *m* à fenêtre (terminale), compteur *m* à fenêtre en bout
~ZONE - [aero] aire *f* de dégagement
ENDAORTITIS - [med] (inflammation of the aorta) inflammation *f* de la tunique interne de l'aorta
ENDELIONITE - [min] (bournonite; wheel ore; natural copper-lead-antimony sulphide, crystallizing in the orthorhombic system) bournonite *f*
ENDIVE - [bot] endive *f*, chicorée *f*
ENDLESS - [gen] infini, sans fin
[mech] (of a screw etc) sans fin
~BAND ELEVATOR - [mech] élévateur *m* à bande continue
~BELT - [mech] bande *f* continue, courroie *f* sans fin
~KNIFE-BLADE CUTTING MACHINE - [mech] découpeuse *f* à ruban
~PAPER - [paper man] (used on rotary printing machines) bobine *f* de papier (pour rotatives)
~ROPE - [aerial railw] câble *m* sans fin
~ROPE HAULAGE - [transp] (the hauling of trucks by means of long loop of rope) traction *f* par câble sans fin
~SAW - [impl] (a band saw) scie *f* à ruban
~SCREW - [mech] vis *f* sans fin
~TAPE - [el acoust] (lengths of magnatic tape joined together to form an endless loop) bande *f* sans fin, boucle *f*
[comput] bande *f* sans fin, bande *f* ininterrompue
~TRACK VEHICLE - [auto] véhicule *f* à chenilles
~VULCANIZATION - [rubber ind] vulcanisation *f* continue
ENDOBIOTIC - [bot] (growing into another plant) endobiotique
ENDOCARDIAC - [anat] (within the heart) endocardiaque
ENDOCARDITIS - [med] (inflammation of the lining of the heart membranes) endocardite *f*
ENDOCRINE - [physiol] (said of glands pouring their secretion into the blood) endocrine
~GLANDS - [anat] glandes *fpl* endocrines
ENDODYNE - [radio] (also autodyne ; a device generating local oscillations) autodyne
ENDOENZYME - [biochem] (endocellular enzyme) enzyme *f* intracellulaire, endoenzyme *f*

ENDOERGIC - [phys](synonym for endothermic) endothermique
ENDOGAMY - [bot] (pollination between two flowers on one plant) endogamie *f*
[genet] (breeding between related animals) endogamie *f*
ENDOGENOUS - [genet] (formed inside another organ) endogène
ENDOMITOSIS - [genet] (division of chromosomes in a nucleus without nuclear division causing duplication of the chromosome complex) endomitose *f*
ENDOMIXIS - [genet] (periodic nuclear reorganization) endomixie *f*
ENDOPOLYPLOIDY - [genet] (state of cells, tissues etc in which the chromosomes divide without subsequent mitosis) endoployploïdie *f*
ENDOSCOPE - [instr] (instrument designed to examine cavities) endoscope *m*
ENDOSCOPIC CAMERA - [photo] chambre *f* endoscopique
ENDOSMOMETER - [instr] endosmomètre *m*
ENDOSMOSIS - [chem] (osmosis in which the solvent dialyzes into the system) endosmose *f*
ENDOSOME - [genet] (of a vescicular nucleus) endosome *m*
ENDOSPERM - [bot] (multicellular tissue inside a developing seed) endosperme *m*
ENDOSTEUM - [anat] (the tissue which lines the cavities of bones) endoste *m*
ENDOTHELIUM - [anat] (layer of cell lining the inner surface of circulatory organs) endothélium *m*
ENDOTHERMAL - s. endothermic
ENDOTHERMIC - [chem] (term used of a reaction or process in which heat is absorbed) endothermique
~COMPOUND - [chem] (compound the formation of which is endothermic) composé *m* endothermique
~PROCESS - [chem] procédé *m* endothermique
~REACTION - [chem] réaction *f* endothermique
ENDOVASCULITIS - [med] inflammation *f* de la tunique interne d'un vaisseau
ENDOVENOUS - [med] intraveineux
ENDS DOWN - [text] (in weaving, the warp threads which have broken) fils *mpl* cassés
~PER DENT - [text] densité *f* des fils dans le peigne
ENDURANCE - [gen] résistance *f*, endurance *f*
[gen] (of objects etc) durée *f*
[aero] autonomie *f*
~FAILURE - [mech] rupture *f* par fatigue
~FLIGHT - [aero] (the maximum period of time an aircraft can continue to fly without refuelling) vol *m* d'endurance
~LIMIT - [phys] (limiting range of stress) limite *f* d'endurance, limite *f* de fatigue, limite *f* de résistance à la fatigue
~RATIO - [phys] (the endurance limit divided by the endurance strength) rapport *m* de fatigue
~STRENGTH - [phys] résistance *f* à la fatigue
~TEST - [mech] essai *m* de fatigue, essai *m* de durée
ENDWAY - [gen] bout à bout, de champ, dans les deux sens
ENERGETICS - [phys] (the study of the energy relation of changes) énergétique *f*
ENERGIC NUCLEUS - [genet] noyau *m* quiescent
~STAGE - [genet] phase *f* métabolique
ENERGIZE, to - [el] alimenter, activer, exciter, mettre sous tension, stimuler, amorcer
ENERGIZING CIRCUIT - [el] circuit *m* d'excitation
ENERGIZING POWER - [el] puissance *f* d'excitation
ENERGY - [phys] (the capacity to do work) énergie *f*, travail *m*
~ABSORPTION - [mech] absorption *f* de l'énergie (amortissement)
~BALANCE - [phys] (the amount of energy which is released in each individual reaction) bilan *m* énergétique
~BAND - [nucl] (the complex of the discrete but closely adjacent energy levels) bande *f* d'énergie
~BAND STRUCTURE - [nucl] structure *f* de bandes d'énergie
~BARRIER - [nucl] barrière *f* d'énergie
~BATTERY - [auto] batterie *f* d'accumulateurs
~BUILD-UP FACTOR - [nucl] facteur *m* d'accumulation en énergie
~COMPONENT - [el] composante *f* active
~CONSUMPTION - [gen] consommation *f* d'énergie
~CONVERSION EFFICIENCY - [mech] rendement *m* cinétique
~CONVERSION EFFICIENCY OF A SCINTILLATOR - [electron] rendement *m* énergétique de conversion d'un scintillateur
~CONVERSION FACTOR - [phys] (the ratio of two measures of the same quantity of energy) facteur *m* de conversion d'énergie
~DECREMENT - [phys] (in a system) décrément *m* de énergie
~DENSITY - [phys] (the ratio between the total amount of energy in a volume and the volume itself) densité *f* d'énergie
~DENSITY OF SOUND - [acoust] (the particle vibration energy per unit volume) densité *f* d'énergie acoustique
~DEPENDANCE - [mech] dépendance *f* de l'énergie
~DEPOSITION - [nucl] (absorption of energy by interaction with the incident radiation when a substance is irradiated) acquisition *f* d'énergie
~DEPOT - [nucl] dépôt *m* d'énergie
~DISTRIBUTION - [electron] (distribution of the electrons emitted from a surface) distribution *f* de l'énergie
~EFFICIENCY - [phys] (the ratio between the energy radiated as luminescence and the energy absorbed) rendement *m* énergétique
~EQUIVALENCE - [phys] (one mass unit equals 930 million electron volts) équivalence *f* de la masse et de l'énergie
~FLUENCE - [phys] fluence *f* d'énergie
~GAP BETWEEN TWO BANDS - [electron] (the difference of energy between the lowest level of the higher band and the highest level of the lower band)écart *m* énergétique entre deux bandes
~LEVEL - [mech] (the stationary state of energy of any physical system) niveau *m* d'énergie
~LEVEL DIAGRAM - [phys] (diagram representing the energy levels of the particles of a quantized system by horizontal lines, with ordinates representing the energy of these particles) diagramme *m* énergétique
~LEVEL OF A PARTICLE - [phys] (in a quantized system, the value of the energy level of a particle) niveau *m* d'énergie d'une particule
~LEVEL WIDTH - [phys] largeur *m* du niveau d'énergie
~LOSS - [phys] perte *f* d'énergie
~LOSS PER ATOM - [nucl] (the measure of the energy

lost by an atom per unit area normal to the motion of the particle) perte *f* d'énergie par atome
ENERGY METER - [instr] compteur *m* d'énergie active
~OF ACTIVATION -[phys] énergie *f* d'activation
~OF DISLOCATION - [phys] énergie *f* de dislocation
~OF DISTORSION - [mech] énergie *f* de distorsion
~OF FLOW - [phys] énergie *f* du flux, énergie *f* cinétique
~OF LIGHT - [phys] (light being a form of energy, one watt of radiant power gives approximately 685 lumens of light flux at a wavelength of 555 millimicrons) énergie *f* lumineuse
~OF PILE DRIVING - [constr] énergie *f* de battage
~OF RESONANCE ABSORPTION - [nucl] (in a nuclear reactor) énergie *f* d'absorption par résonance
~QUANTUM - [phys] quantum *m* d'énergie
~RANGE - [phys] (the distance a particle will penetrate a substance before its kinetic value falls below that of ionization) portée *f*
~REGION - s. energy range
~RELEASE - [phys] libération *f* d'énergie
~SENSITIVITY - [electron] sensibilité *f* statique
~SPECTRUM - [phys] spectre *m* énergétique
~STORAGE BRAKING - [el] freinage *m* à accumulation
~TRANSFER - [phys] (from one system to another) transfert *m* d'énergie
~TRANSMISSION - [mech] transmission *f* d'énergie, transport *m* d'énergie, transmission *f* de force motrice
~UNIT - [phys] unité *f* d'énergie
~YIELD - [nucl] (the total of the effective energy which is released in a nuclear explosion) rendement *m* énergétique, coefficient *m* G
ENGAGE, to - [gen] engager, enclencher, actionner
[mech] (to bring into action as of gears or clutches or the like) mettre en prise, accrocher
~GEARS, to - [mech] mettre en prise, engrener
~THE CLUTCH, to - [mech] embrayer
ENGAGED - [adj] en prise, engagé, occupé, employé
~COLUMN - [constr] (a column which is embedded by less than half) colonne *f* semi-encastrée
~SIGNAL - [telecomm] signal *m* d'occupation
ENGAGEMENT - [mech] engrènement *m*, accrochement *m*, prise *f*, enclenchement *m*
~DEPTH OF THREAD - [mech] hauteur *f* du filet, profondeur *f* du filet
~LEVER - [mech] levier *m* d'embrayage
~SPEED - [aero] (in alighting on a flight deck) vitesse *f* d'appontage
ENGINE - [mech] (in general, a machine in which heat is converted into work) moteur *m*, machine *f*, appareil *m*
[railw] locomotive *f*
~AIR INTAKE - [aero] admission *f* d'air, diffuseur *m* d'entrée d'air
~BASE OIL SUMP - [auto] fond *m* de carter
~BEARER - [mech] bâti-moteur *m*, berceau *m*
~BLOCK - [auto] bloc *m* moteur
~BONNET - [auto] capot *m*
~BONNET ANTICHAFING STRIP - [auto] tresse *f* repos de capot
~BRACKET - [mech] (part formed on or attached to an engine to fix it in place) support *m* de moteur
~CAB - [railw] poste *m* de conduite, cabine *f* de conduite
~CASE - [mech] carter *m* de moteur
~COMPRESSOR - [mech] groupe *m* motocompresseur
ENGINE CONTROLS - [mech] commandes *fpl* du moteur
~COVER - [auto] capot *m*
~COWLING - [aero] (removable covering over an aircraft engine) capot *m*, capotage *m*
~DISPLACEMENT - [mech] (the total volume swept out by all the working pistons during a single stroke of each) cylindrée *f*, déplacement *m*
~-DRIVEN WELDING SET - [metall] groupe *m* électrogène de soudage à l'arc
~EFFICIENCY - [mech] rendement *m* du moteur
~EXHAUST - [mech] échappement *m*
~FAILURE - [mech] panne *f* de moteur
~FRAME - [mech] (the crankcase) châssis *m*, bâti *m*, carcasse *f*
~FRONT COVER - [auto] carter *m* de distribution
~FRONT MOULDING - [auto] patin *m* de support avant de moteur, support *m* avant de moteur
~GEAR-BOX UNIT - [auto] boîte *f* de vitesses, groupe *m* moteur-boîte de vitesses
~GENERATOR UNIT - [el] groupe *m* convertisseur
~GROUND TROLLEY - [aero] (a light carriage for carrying engines between aircraft and repair shops etc) chariot *m* porte-moteur
~HATCHWAY - [naut] écoutille *f* de la machine
~HOLDING-DOWN BOLT - [mech] (bolt used to secure an engine in place) boulon *m* de retenue, boulon *m* de fixation
~HOOD - s. engine bonnet
~INDUCTION MANIFOLD - [auto] collecteur *m* d'aspiration de moteur
~INTAKE MANIFOLD - s. engine induction manifold
~LATHE - [mach tool] tour *m* parallèle (à filter)
~LOG - [naut] journal *m* de la machine
~LOG-BOOK - [aero] livre *m* de vol
~LUG - [auto] patte *f* de fixation de moteur
~MOUNTING - [mech] bâti-moteur *m*
~NACELLE - [aero] fuseau *m* moteur
~OIL - [mech] huile *f* pour moteurs
~OVERHAUL - [mech] révision *f* du moteur
~OVERHAUL BENCH - [mech] banc *m* de révision, banc *m* de réparation
~PIT - [constr] (the hole in the floor of a garage which enables a mechanic to work from below) fosse *f* à réparations
[auto] carter *m* de moteur
~POWER RATING - [mech] puissance *f* indiquée
~PRIMER - [aero] enrichisseur *m* de mélange, injecteur *m* de départ
~PROTECTING PLATE - [auto] plaque *f* de protection de moteur
~ROOM - [mech] (housing of the torpedo motor) compartiment *m* moteur
[naut] compartiment *m* des machines
~ROOM TELEGRAPH - [naut] transmetteur *m* d'ordres
~SHED - [railw] dépôt *m* de locomotives
~SHOP - [gen] atelier *m* machines
~SIDE SHEET METAL - [auto] tôle *f* de protection latérale de moteur
~-SIZED PAPER - [paper man] (a paper which is hardened by the addition of rosin and alum to the pulp) papier *m* collé à la résine
~-SIZING - [paper man] collage *m* du papier
~SPEED - [mech] (the speed at which an engine runs under the given conditions) régime *m* du moteur
~-SPEED INDICATOR - [instr] compte-tours *m*
~-SPEED SYNCHRONIZATION - [mech] (the operation

of bringing two or more engines to the same speed) synchronisation *f* de moteurs
ENGINE SPRAY - [astronaut] (spray of cooling water during lauching) aspersion *f* des propulseurs
~STARTER - [mech] démarreur *m*
~STARTER DOG - [mech] dent *f* de loup, griffe *f* de démarreur
~TEST - [mech] essai *m* de moteur
~TESTER - [mech] banc *m* d'essai (pour moteurs)
~TIMING - [mech] réglage *m*
~TORQUE - [mech] couple *m* moteur
~TRIAL - s. engine test
~TROUBLE - [mech] panne *f* de moteur
~WARM-UP - [mech] réchauffage *m* du moteur
ENGINEER, to - [gen] inventer, combiner, arranger
ENGINEER CHAIN - [surv] (a I 00 ft long chain consisting of I 00 links) chaîne *f* d'arpenteur
ENGINEERING - [gen] science *f* de l'ingénieur, génie *m*
[gen] (the preparation of technical details) usinage *m*, étude *f* (technique), construction *f* mécanique
~BRICK - [constr] (machine-made and pressed bricks used when strength and good appearance are required) brique *f* recuite à maconner, brique *f* hollandaise
~TIME - [comput] temps *m* d'entretien
~WORKS - [gen] ateliers *mpl* de constructions mécaniques
ENGLER DISTILLATION - [chem] (a process for determining the boiling range of petroleum distillates) distillation *f* Engler
~FLASK - [ind chem] (a specially dimensioned I 00 cc. flask used in the Engler distillation) ballon *m* pour distillation Engler
ENGLISH - [print] corps *m* I2-I3, Saint-Augustin *m* (environ)
~BOND - [constr] appareil *m* à assises alternées de paneresses et de boutisses
~COUNT - [text] numéro *m* anglais
~CUT - [bookbind] découpe *f* anglaise
~HORN - [mus] cor *m* anglais
~MELTING POINT - [oil ind] point *m* de goutte anglais
~PITCH - [text] (in Jacquard looms) division *f* anglaise
~PLASTER - [rubber ind] taffetas *m* gommé, taffetas *m* anglais
~RED - [paint] colcotar *m*
~ROOF TRUSS - [constr] (truss for large span roofs) ferme *f* de type anglais
~SEAM - [text] couture *f* à l'anglaise
~VERMILION - [chem] (a mercury sulphide pigment) vermillon *m* anglais
~WHITE - [chem] carbonate *m* de calcium
ENGOBE - [ind chem] (a fine paste coating) engobe *m*
ENGORGEMENT - [med] (of the breast) engorgement *m*
ENGRAILED-[carp] (of an edge indented with curvilinear notches) engrêlé
ENGRAVE, to - [gen] graver
ENGRAVING - [gen] gravure *f*, estampe *f*
~CYLINDER - [print] cylindre *m* d'impression
~FOR EMBOSSING - [text] gravure *f* à embossage
~MACHINE - [mech] machine *f* à graver, pantographe *m*
~NEEDLE - [impl] pointe *f* pour taille douce
~RUBBER - [print] caoutchouc *m* pour clichés (plastotypie)
ENGRAVING TOOL - [text] poinçon *m* d'incision, poinçon *m* de gravure
ENHANCE, to - [gen] accroître, augmenter, exalter, rehausser
ENHANCED CARRIER DEMODULATION - [radio] démodulation *f* à porteuse renforcée
~LINE - [phys] (spectral line from a very hot source) raie *f* renforcée
ENLARGE, to - [gen] agrandir, accroître, étendre, aléser
ENLARGEMENT - [gen] agrandissement *m*, alésage *m*, extension *f*
~AFTER CROPPING - [photo] agrandissement *m* après élangage
~MAGNIFICATION - [photo] rapport *m* d'agrandissement
~TO SCALE - [photo] agrandissement *m* à l'échelle
ENLARGER - [photo] agrandisseur *m*
~HEAD - [photo] tête *f* de l'agrandisseur
~PILLAR - [photo] colonne *f* d'agrandisseur
ENLARGING LENS - [photo] objectif *m* pour agrandissement
~MACHINE - s. enlarger
~MASKING FRAME - [photo] margeur *m*, châssis-margeur *m*
~NEGATIVE HOLDER - [photo] porte-négatif *m*, passe-vues *m*
~NEGHOLD-(US) - s. enlarging negative holder
~THE WEAVE - [text] augmentation *f* de l'armure
ENMESHING - [mech] engrènement *m*
ENNEANDROUS - [bot] (with nine stamens) ennéandr
ENNEODE - [electron] (electronic tube including a cathode, seven grids and an anode) ennéode *m*
ENOLIZATION - [chem] énolisation *f*
ENOSTOSIS - [med] (growth on internal surface of bone) énostose *f*
ENOUNCE, to - [gen] annoncer, proclamer, énoncer
ENRICH, to - [gen] enrichir
[agric] fertiliser, amender
ENRICHED - [gen] enrichi
[nucl] (containing a higher percentage than the normal concentration) enrichi
~MATERIAL - [nucl] (material in which the isotope content has been increased) matière *f* enrichie
~MIXTURE - [mech] (in internal combustion engines mélange *m* enrichi
~REACTOR - [nucl] (nuclear reactor operation with enriched uranium) réacteur *m* enrichi
~URANIUM - [nucl] (uranium containing a higher percentage of U^{235} than the normal value) uranium *m* enrichi
ENRICHING SECTION - [chem] (that part of a still column in which the vapour enrichment occurs) colonne *f* d'enrichissement
ENRICHMENT - [gen] enrichissement *m*
[arch] (ornamentation) ornement *m*
[nucl] (process changing the isotopic percentage) enrichissement *m*
[agric] (of the soil) fertilisation *f*, amendement *m*
~FACTOR - [nucl] (the ratio between isotopic ratios after enrichment and before enrichment) taux d'enrichissement
~JET - [mech] (in internal combustion engines) jet *m* d'enrichissement
~PLANT - [nucl] installation *f* d'enrichissement, usine *f* d'enrichissement
ENROCKMENT - [hydr] (for the protection against

the impact of water) enrochement *m*
ENSEMBLE - [gen] ensemble *m*
[acoust] (the mixing of individual sounds in an orchestra) ensemble *m*
[arch] (a number of units within an edifice or a block) ensemble *m*
ENSILAGE - [agric] (cattle food in portable or fixed silos) ensilage *m*, silotage *m*
[agric] (the process of ensiling) ensilage *m*, ensilotage *m*, silotage *m*
~HARVESTER - [agric] ensileuse *f*
ENSILATION - s. ensilage
ENSILE, to - [agric] ensiler, ensiloter
ENSTATITE - [min] (a silicate of megnesium) enstatite *f*
ENTABLATURE - [constr] (all that is supported on columns) entablement *m*
ENTAIL, to - [gen] entailler, sculpter
[leg] (to settle an estate on a number of persons) substituer (un héritier), léguer
ENTAIL - [arch etc] sculpture *f* (sur bois)
ENTANGLED HANK - [text] écheveau *m* vrillé
ENTANGLEMENT - [gen] enchevêtrement *m*, réseau *m*
[text] entrelacement, enchevêtrement *m*
~OF THREADS - [text] enchevêtrement *m* des fils
ENTASIS - [arch] (slight swelling in the middle of a column) entasis *f*, renflement *m* imperceptible (de une colonne)
ENTEQUE - [vet] (cattle, horse and sheep disease occurring in South America) septicémie *f* hémorragique des bestiaux
ENTER, to - [gen] entrer, pénétrer, embouquer (un canal)
[leg] (e.g. a petition) intenter (un procès)
[comput] (data into memory) introduire, inscrire
[comm etc] inscrire, enregistrer
ENTERAL - [med] (within the intestine) intestinal
ENTERCLOSE - [constr] (a corridor between two rooms) corridor *m* entre deux pièces
ENTERECTOMY - [med] entérectomie *f*
ENTERIC - [med] (relating to the intestines) entérique
~-COATED - [chem] kératinisé
ENTERICOID FEVER - [med] fièvre *f* d'aspect typhoïdique
ENTERING CATCH - [mech] loquet *m* encastré
~CHANNEL - [text] canal *m* d'introduction, canal *m* d'alimentation
~EDGE - [el] (the brush edge of an electrical machine) bord *m* d'attaque
[aero] (leading edge) bord *m* d'attaque
~GRID - [text] grille *f* d'entrée
~GUIDE - [text] guide *m* d'introduction, guide *f* d'entrée
~TABLE - [text] table *f* d'alimentation, table *f* d'introduction
ENTERITIS - [med](inflammation of the small intestine) entérite *f*
ENTEROBIASIS - [med] (an infection) entérobiase *f*, oxyurose *f*
ENTEROBROSIA - [med] perforation *f* intestinale
ENTEROCINERIA - [med] substance *f* grise interne de l'encéphale
ENTEROCINESIA - [med] péristaltisme *m*
ENTEROIDEA - [med] fièvre *f* intestinale
ENTERPRISE - [gen] entreprise *f*
ENTHALPY - [phys] (heat content of a substance per unit mass) enthalpie *f*
ENTHALPY CHART - [phys] diagramme *m* enthalpique
ENTIRIS - [anat] uvée *f*
ENTITY - [gen] entité *f*
ENTOMOLOGY - [zool] (zoology branch dealing with insects) entomologie *f*
ENTOMOPHAGOUS - [zool] (feeding on insects) entomophage
ENTRAILS - [zool] entrailles *f*pl, viscères *f*pl
ENTRAINER - [chem] (an additive used in distillation to form an azeotropic mixture with one of the components and thus assisting in their separation) entraîneur *m*, agent *m* de déplacement, agent *m* d'entraînement
ENTRAINMENT - [chem] (the mist of fine droplets transported by the vapour of a boiling liquid) entraînement *m*
~FILTER - [chem] (type of filter used in distillation) filtre *m* par entraînement
~SEPARATOR - [ind chem] séparateur *m* par entraînement
ENTRANCE - [gen] entrée *f*, accès *m*, admission *f*, pénétration *f*, commencement *m*
[gen] (of a channel) embouchure *f*, goulet *m*
[shipbuild] formes *f*pl d'avant
[telecomm] entrée *f* de poste
~CHANNEL SPIN - [nucl] (in a nuclear reaction) spin *m* de voie d'éntrée
~FEE - [comm] droit *m* d'entrée, droit *m* d'inscription
~HALL - [constr] salle *f* d'entrée, salle *f* des pas perdus
~LOCK - [naut] (the lock through which vessels must pass when entering or leaving a dock) écluse *f* d'entrée
~LOSS - [hydr] perte *f* de charge à l'entrée
~PORT - [opt] (the image of the aperture stop viewed from the object) pupille *f* d'entrée
~PUPIL - [opt] pupille *f* d'incidence
~SLIT - [opt] (the slit through which light enters a spectrometer) fente *f* d'entrée
ENTRAPPED AIR - [metall] air *m* occlus, bulle *f* d'air, air *m* entrainé
~COLD SHOT - [oil ind] goutte *f* froide
ENTREPOT - [gen] entrepôt *m*, dépôt *m*
ENTROPY - [phys] (unavoidable energy of a substance which is due to the internal irregular and compensating motion of the molecules) entropie *f*
~DIAGRAM - [phys] diagramme *m* entropique
~OF DISORDER - [phys] (the part of the entropy of a substance due to lack of order in the arrangement of the molecules) entropie *f* de désordre
~OF FUSION - [phys] (the unavailable energy during a fusion) entropie *f* de fusion
~OF MIXING - [phys] (the unavailable energy of a mixture) entropie *f* de mélange
~OF SOLUTION - [phys] (heat of solution less free energy of solution divided by the absolute temperature) entropie *f* de solution
~OF VAPORIZATION - [phys] (when changing from the liquid to the vapour state, the entropy increases) entropie *f* de vaporisation
ENTRUST, to - [gen & comm] confier, remettre à
ENTRY - [gen] entrée *f*, accès *m*
[comm] inscription *f*, enregistrement *m*, article *m*
[leg] (customs) déclaration *f* d'entrée
~CORRIDOR - [astronaut] (the method and conditions for continuous flight in the earth's atmosphere,

bounded by a curve over which the weight is greater than the lift and by a curve below which the vehicle becomes too hot) couloir *m* d'entrée

ENTRY LINE - [railw] voie *f* d'accès

~PLATE - [text] tôle *f* de guidage

~PORTAL - [phys] (the field through which a beam of radiation enters a body) champ *m* d'irridiation

~RAMP - [metall] rampe *f* d'alimentation

~ROLLER - [text] cylindre *m* introducteur, cylindre *m* d'alimentation

~SIDE - [metall] ouverture *f* d'entrée

~SIGNAL - [railw] signal *m* d'entrée

~STAND - [text] chevalet *m* d'entrée, râtelier *m* d'alimentation

~STUMP - [mining] pilier *m* de galerie

~TIMBERING - [mining] boisage *m*, soutènement *m* d'une galerie

ENTWINE, to - [text] enlacer, guiper

ENUMERATE, to - [gen] énumérer, détailler, dénombrer

ENUMERATION - [gen] énumération *f*

ENURESIS - [med] (incontinence of urine) énurésie *f*, énurèse *f*

ENVELOPE - [gen] enveloppe *f*, pli *m*
[aero] (of an airship) enveloppe *f* (de dirigeable)
[telev & radar] (graph indicating the variations in amplitude of successive oscillations in an amplitude-modulated wave) enveloppe *f*
[electron] (the vessel of glass or other material which contains the electrode assembly in an electron tube) ampoule *f*
[geom](curve tangent to each of a family of curves) enveloppante *f*
[nucl] (of a reactor) couche *f* fertile de la pile

~DELAY - [radio] (the propagation time between two points on the envelope of a wave) retard *m* de groupe

~DELAY/FREQUENCY CHARACTERISTIC - [radio] caractéristique *f* phase-fréquence

~DELAY/FREQUENCY DISTORTION - [radio] distortion *f* phase-fréquence

~DELAY METER - [instr] (instrument designed to measure group delay in television receivers) appareil *m* de mesure de retard de groupe

~DEMODULATION - [telev] détection *f* de phase, redressement *m* d'enveloppante

~OF A BATTERY - [el] (the battery case) bac *m* d'accumulateur

~VELOCITY - [radio] (the velocity of propagation of a feature of the wave envelope) vitesse *f* de groupe

ENVIRONMENT - [gen] milieu *m* ambiant, environnement *m*

ENVIRONMENTAL - [gen] ambiant, du milieu

~CHAMBER - [astronaut] (a type of test chamber) chambre *f* d'environnement, simulateur *m* d'espace

~CONDITIONS - [gen] (the total of external and internal conditions determining growth, development, activity, mechanical and other operations etc) conditions *fpl* du milieu, conditions *fpl* ambiantes

~CONTAMINATION - [nucl] (the contamination of the surroundings of a nuclear reaction) contamination *f* de l'ambience

~FACTOR - [gen] facteur *m* ambiant

~LAPSE RATE - [phys] gradient *m* de température atmospherique

~POLLUTANTS - [ecology] polluants *mpl* des milieux

~PRESSURE - [phys] pression *f* environnante, pression *f* ambiante

ENVIRONMENTAL TEST - [mech] essai *m* d'ambiance

~TEST CHAMBER - [astronaut] chambre *f* d'environnement, chambre *f* de simulation

ENZYMATIC - [chem] enzymatique

~CLARIFICATION - [chem] clarification *f* enzymatique

~COAGULATION - [chem] coagulation *f* enzymatiqu

ENZYME - [biochem] (an organic catalyst produced by the living cells. Enzymes have a specific action and are classified accordingly as hydrolytic enzymes, lipases, proteases, peptidases, oxidizing, fermenting and deaminizing enzymes) enzyme *m* (*f*

~INACTIVATOR - [chem] agent *m* de désactivation pour enzymes

ENZYMIC ACTIVITY - [chem] activité *f* enzymatiqu

ENZYMOLOGY - [chem] (the study of enzymes) enzymologie *f*

EOCENE - [geol] (the oldest of the tertiary systems éocène *m*

EOCRETACEOUS - [geol] (layers of unfossiliferous slates occurring in Mexico) éocrétacé

EOLIENNE - [text] (a silk fabric) éolienne *f*

EOLITH - [geol] (of the oldest stone implements so far known) éolithe *m*

EONISM - [med] travestisme *m*, éonisme *m*, transvetisme *m*

EOSIN - [chem] (a red dye for textiles) éosine *f*

EOSINOPHIL - [biochem] (a cell with a protoplasmic granule readily stained by eosin) éosinophile

EOSINOPHILIA - [med] (excess of eosinophils in the blood) éosinophilie *f*

EOTVOS TORSION BALANCE - [meas] balance *f* de torsion d'Eotvos

EOZOON STRUCTURE - [geol] structures *fpl* énigmatiques observées dans l'Antécambrien

EPACTAL - [anat] intercalé

EPEIRIC SEA - [geol] mer *f* épicontinentale

EPEIROGENESIS - [geol] épirogénie *f*

EPEIROGENIC EARTH MOVEMENTS - [geol] (movements of the earth crust forming very large areas) mouvements *mpl* épirogéniques de la terre

EPENDYMA - [anat] (epithelium lining the spinal co: central canal and the brain centricles) épendyme *m*

EPENDYMITIS - [med] (inflammation of the ependyma) épendymite *f*

EPENDYMOMA - [med] épendymome *m*

EPHANTINITE - [min] (a rare, ianthinite containing uranium) éphantinite *f*

EPHEBIC - [zool] (of the period of maximum strengt éphébique

EPHEDRINE - [chem] (an alkaloid deriving from an aliphatic amine) éphédrine *f*

EPHEMERAL - [gen &bot] éphémère

~FEVER - [vet] fièvre *f* éphémère

EPHEMERIS - [astr] (an almanac showing the daily positions of celestial bodies) éphémérides *mpl*, almanach *m*

EPIBOLY - [zool] (overgrowth) épibolie *f*, épibole *f*

EPICADMIUM - [phys] (energies exceeding the cadmium cut-off) épicadmique

EPICARDIUM - [zool] (the serous membrane coveri the heart) épicarde *m*, feuillet *m* viscéral de la séreuse péricardique

EPICAUMA - [med] épicaume *m*

EPICENTRAL - [zool] (attached to the vertebral centra) épicentrique

EPICENTRE - [geol] (the point on the surface of the earth above the focus of an earthquake) épicentre *m*
EPICHLOROHYDRIN - [chem] (a solvent for paints, varnishes and resins) épichlorhydrine *f* (de la glycérine)
EPICLASTIC - [geol] épiclastique
EPICYCLIC - [geom] épicycloidal
~GEAR TRAIN - [mech] (a type of gear train in which a central gear meshes with others placed round it on a common pitch circle, and these in turn with an internally-toothed ring gear encircling them) train *m* d'engrenages épicycloïdaux
~TRAIN - s. epicyclic gear train
EPICYCLOID - [mech] (the path of a point on the circumference of a circle as it rolls round the circumference of another circle) épicycloïde *f*
~CURVE - [geom] courbe *f* épicycloïdale
EPIDEMIC - [med] (an infectious disease spreading widely) épidémique
EPIDEMICS - [med] épidémie *f*
EPIDERMIS - [zool] (the epithelium which covers the body) épiderme *m*
EPIDERMAL - [zool] épidermique
EPIDIASCOPE - [opt] (a projection lantern) épidiascope *m*
EPIDOTE - [min] (a minor gemstone) épidote *m*
EPIDROME - [med] congestion *f*
EPIFOLLICULITIS - [med] folliculite *f* décalvante, folliculite *f* du cuir chevelu
EPIGASTRIUM - [anat] (abdominal region) épigastre *m*
EPIGENESIS - [genet] (development involving gradual diversification) épigénèse *f*
EPIGENETIC - [min] (class of ores which have developed later than the enclosing rocks) épigénétique, épigène
EPILATION - [med] (the removal of hair) épilation *f*
EPILEPSY - [med] épilepsie *f*
EPILEPTIC - [med] épileptique
~FIT - [med] crise *f* d'épilepsie, accès *m* épileptique
~SEIZURE - s. epileptic fit
~STATE - [med] (the condition of one who suffers from epilepsy) état *m* épileptique
EPILEPTIFORM - [med] (which resembles epilepsy) épileptiforme
EPILEPTOGENIC - [med] (provoking an attack of epilepsy) épileptogène
EPIMENORRHEA - [med] (the frequent occurrence of menstruation) menstruation *f* trop fréquente
EPINEPHRINE - [chem] (a synonym for adrenalin, the hormone produced by the adrenal glands. Epinephrine has uses in medicine) adrénaline *f*
EPIPHYSIS - [zool] (a separately ossified bone) épiphyse *f*
EPIPLOCELE - [med] (a type of hernia) épiplocèle *f*
EPIPOLE - [opt] point *m* nucléal
EPIPOLAR PLANE - [surv] plan *m* nucléal
EPIROCK - [geol] épi-roche *f*, roche *f* superficielle
EPISCOPE - [opt] (projection lantern showing on a screen an enlarged image of a brilliantly illuminated object) épiscope *m*
EPISCOPIC ATTACHMENT - [photo] dispositif *m* épiscopique additionnel
EPITAXIS - [med] (bleeding from the nose) épitaxis *f*
EPITASIMETER - [instr] (an instrument designed to control processes by continuous measurement of the surface tension) épitasimètre *m*
EPITAXY - [cryst] (intergrowth between two solid phases) épitaxie *f*
EPITENON - [anat] gaine *f* fibreuse entourant le tendon
EPITHELIAL - [biol] (pertaining to the epithelium) épithélial
~CANCER - [med] épithélioma *m*
~CELL - [biol] cellule *f* épithéliale
~LAYER - [bot] (layer of elongated cells in a grain) couche *f* épithéliale
~TISSUE - [biol] tissu *m* épithélial
EPITHELIOMA - [med] (malignant growth derived from epithelium) épithelioma *m*
EPITHELIUM - [anat] (non-vascular tissue which forms the outer layer of the mucous membrane) épithélium *m*
[bot] (epidermis consisting of young cells) épithélium *m*
EPITHERMAL NEUTRONS - [nucl] (neutrons of which the energies are greater than those of thermal neutrons, i.e. to I000 eV) neutrons *mpl* epithermiques
~REACTOR - [nucl] (nuclear reactor in which some fissions are induced by epithermal neutrons) réacteur *m* épithermique
~VEIN - [geol] filon *m* épithermal
EPITUBERCULOSIS - [med] (inflammation of the region surrounding a tuberculous focus) épituberculose *f*
EPIZOON - [zool] (an animal living on the skin of another animal) épizoaire *m*
EPIZOOTIC - [zool] (pertaining to epizoon) épizootique
~APHTHA - [vet] (foot-and-mouth disease) aphte *f* épizootique
~LYMPHANGITIS - [vet] lymphangite *f* épizootique
E PLANE BEND - [waveguides] coude *m* 'E'
EPOXIDE - [chem] époxyde *m*
EPOXY - [chem] (prefix denoting an oxygen atom bonded to two other atoms which are already united by other bonds) époxy
~PAINT - [paint] peinture *f* époxyde
~RESINS - [chem] (a large class of synthetic resins with uses as adhesive and casting resins, in surface coatings and electrical encapsulation) résines *fpl* époxydes, éthoxylines *fpl*
EPOXYETHANE - [chem] époxyéthane *m*
EPOXYPROPANE - [chem] oxyde *m* de propylène
EPSOM SALTS - [chem] (common name for hydrated magnesium sulphate) sulfate *m* de magnésie, sels *mpl* anglais, sels *mpl* d'Epsom, sels *mpl* d'Angleterre
EPSOMITE - [min] (naturally occurring hydrated magnesium sulphate with application in medicine, textile processing, and fertilizer manufacture) epsomite *f*, sel *m* amer
EPULIS - [med] (a tumour of the gums) épulie *f*
EQUAL - [gen] égal
~AND OPPOSITE FORCES - [phys] forces *fpl* égales et opposées
~AREA MAP PROJECTION - [geol] projection *f* équivalente
~CARDS - [comput] cartes *f* identiques
~ENERGY SOURCE - [opt] (light source with a rate of energy emission per unit of wavelength which is constant throughout the spectrum) source *f* équiénergétique
[telev] (in colour television) source *f* d'énergie

égale
EQUAL ENERGY WHITE - [telev] (radiance uniform on all wavelengths) blanc *m* idéal, blanc *m* d'énergie égale
~FLUTE SPACING - [mach tool] (of a reamer) denture *f* régulière, pas *m* régulier
~LEG ANGLE - [mech] cornière *f* à ailes égales
~LENGTH MULTI-UNIT CODE - [radio] code *m* à moments
~RESISTANCE BRANCH CIRCUIT - [el] circuit *m* de branchement à résistance identique
~SIGNAL WHITE - [telev] (total radiance with the signals having the same value in all channels for colour television) blanc *m* moyen
~TEMPERED SCALE - [mus] (the musical scale of instruments and voices) gamme *f* tempérée
EQUALITY - [gen & math] égalité *f*, équipollence *f* (de deux vecteurs)
[leg] concours *m*, concurrence *f*
~BRIGHTNESS PHOTOMETER - [instr] (visual photometer in which the parts of the comparison field are viewed simultaneously) photomètre *m* à plages juxtaposées à égalité de brillance
~OF CONTRAST PHOTOMETER - [instr] photomètre *m* à plages juxtaposées à égalité de contraste
~OF LUMINOSITY PHOTOMETER - s. equality of brightness photometer
~OF SETS - [contr] égalité *f* de quantité
EQUALIZATION - [gen] égalisation *f*, régularisation *f*, compensation *f*, équilibrage *m*
[electron] (the reduction of distortion through introduction of networks compensating for the particular type of distortion over the required frequency band) correction *f* de distorsion
[telecomm] (the improvement of sound signal-to-noise ratio) amélioration *f* du rapport signal sonore-bruit
~OF LEVELS - [telecomm] égalisation *f* des niveaux
~OF LOAD - [constr] répartition *f* uniforme de la charge
~OF STRAIN - [text] (in weaving) compensation *f* de la tension
EQUALIZE, to - [gen] égaliser, compenser
(of load) répartir uniformément
~THE SLIVER, to - [text] égaliser le ruban (de carde)
EQUALIZER - [el] (connection between points on a winding to equalize the potential on the points) égaliseur *m* de potentiel, connexion *f* équipotentielle
[mech] compensateur *m*, égalisateur *m*, équilibreur *m*
[railw] (of electric locomotives) balancier *m* compensateur
[radio] (a network correcting the frequency response of a radio system) correcteur *m*, compensateur *m*
[acoust] correcteur *m*
~SPRING - [mech] ressort *m* compensateur
EQUALIZING - [surv] égalisation *f*, compensation *f*, péréquation *f*
~CHARGE - [el] (storage battery; long-period charge to ensure that all active material is fully restored) charge *f* d'égalisation
~CIRCUIT - [radio] circuit *m* correcteur
~COIL - [el] bobine *f* égalisatrice
~CURRENT - [el] (the current which flows in an equalizer ring or bar) courant *m* compensateur, courant *m* de compensation
EQUALIZING DEVELOPER - [photo] révélateur *m* à action compensatrice
~MAIN - [el] câble *m* d'équilibrage
~NETWORK - [el] (network introduced into a circuit to alter its response in a specified manner) réseau *m* de compensation
~RESERVOIR - [hydr] reservoir *m* compensateur, reservoir *m* régulateur du débit
~TANK - [hydr] bassin *m* de compensation
EQUALLY TEMPERED SCALE - s. equal tempered scale
EQUALS - [math] quantités *fpl* égales
EQUATE, to - [math] mettre en équation
~TO ZERO, to - [comput] rendre égal à zéro
EQUATION - [math] équation *f*
~OF A CURVE - [math] équation *f* d'une courbe
~OF CONDITIONS - [math] équation *f* de condition
~OF CONTINUITY - [phys] (equation relating to the inflow into a volume to the rate of increase of the amount of the quantity within the volume) équation *f* de continuité
~OF LIGHT - [phys] (the travel time of a beam of light from a celestial body to the earth) équation *f* de la lumière
~OF STATE - [phys] (an expression relating to volume, pressure and temperature for a given system) équation *f* d'état
~OF THE FIRST DEGREE - [math] équation *f* du premier degré
~OF TIME - [astr] (the difference between the apparent and mean sun) équation *f* du temps
~SOLVER - [comput] (computing device designed to solve systems of equations etc) machine *f* électronique à résoudre les équations
EQUATIONS OF MOTION - [mech] équations *fpl* du mouvement
EQUATOR - [geogr] (the Great Circle which is equidistant from the poles) équateur *m*
[el] (of a magnet) équateur *m*
~OF HEAT - [met] équateur *m* thermique
~OF THE HEAVENS - [astr] (synonym for celestial equator) équateur *m* céleste
EQUATORIAL - [geogr] équatorial
[instr] (an astronomical telescope) équatorial *m*
~BULGE - [astr] renflement *m* équatorial
~COORDINATES - [astr] coordonnées *fpl* équatoriales
~RADIUS OF THE EARTH - [astr] rayon *m* équatorial de la terre
EQUIANGULAR - [geom] équiangle
~SPIRAL - [geol] loxodromie *f*
EQUIAXED - [phys] équiaxe
~CRYSTALS - [metall] (formed in the interior of a mass of metal in a mould) cristaux *mpl* équiaxes
EQUI-BAND CODER - [telev] (for colour television) codeur *m* à largeurs de bande égales
~-BAND COLOUR TELEVISION RECEIVER - [telev] téléviseur *m* couleur à largeurs de bande égales
EQUIDISTANCE - [gen & geom] équidistance *f*
EQUIENERGY SPECTRUM - [light] spectre *m* équiénergétique
EQUILATERAL - [geom] équilatéral, équilatère
~ARCH - [arch] ogive *f* en tiers-point
~POINTED ARCH - [constr] ogive *f* équilatérale, arc *m* en tiers points
~TRIANGLE - [geom] triangle *m* équilatéral
EQUILIBRATE, to - [gen & mech] équilibrer, contre-

balancer
EQUILIBRATING OPERATION - [med] cure *f* du strabisme paralytique
EQUILIBRATION - [mech] (the production of balance) équilibration *f*, mise *f* en équilibre, équilibrage *m*
EQUILIBRIUM - [gen] équilibre *m*, aplomb *m*
[phys] (the exact counterbalancing or neutralizing of the forces of tendencies) équilibre *m*
[phys] (in a reversible reaction, the condition when the reaction velocities in the two opposite directions are equal and there is therefore no tendency to further change) équilibre *m*
~CONCENTRATION - [chem] concentration *f* équilibrée
~CONDITION - [phys] (the condition in which all forces present are balanced or neutralized) condition *f* d'équilibre
~CONSTANT - [chem] (the ratio between the products of the active masses of the molecules on each side of a reversible reaction at equilibrium) constante *f* d'équilibre
~DIAGRAM - [metall] (constitutional diagram) diagramme *m* d'équilibre
~ELECTRODE POTENTIAL - [el chem] (the potential difference between an electrode and an electrolyte when both are in equilibrium for the reaction determining the electrode potential) potentiel *m* d'équilibre d'électrode
~ENERGY - [electron] énergie *f* d'équilibre
~ENRICHMENT FACTOR - [nucl] (the relation between stable isotopic ratios before and after enrichment) facteur *m* d'enrichissement équilibré
~FLOW - [phys] écoulement *m* en équilibre
~FLIGHT - [astronaut] vol *m* plané en équilibre
~HEIGHT - [aero] (the height at which the lift and weight of a free balloon are equal) altitude *f* d'équilibre
~METHOD - [electron] (method of calculating the coefficient of recombination) méthode *f* d'équilibre
~OF A PARTICLE - [phys] (condition of a particle when the vector sum of all the forces through it is zero) particule *f* en équilibre
~OF FLOATING BODIES - [phys] (due to the weight of the body being equal to the weight of the fluid it displaces) équilibre *m* des corps flottants
~OF FORCES - [phys] équilibre *m* des forces
~OF STRESSES - [phys] équilibre *m* des tensions (ou contraintes)
~ORBIT - [phys] (the circle of constant radius which is described by accelerated particles) orbite *f* de équilibre, orbite *f* stable
~POINT - [mech] (the external conditions at which a system is in equilibrium) point *m* d'équilibre
~POSITION - [phys] position *f* d'équilibre
~PRESSURE - [phys] pression *f* d'équilibre
~REACTION POTENTIAL - [el] (the lowest value of voltage at which an electrochemical reaction can occur) potentiel *m* d'équilibre d'une réaction
~RING - [mech] (a ring placed between the back of a slidevalve and the cover in a steam engine) bague *f* compensatrice, rondelle *f* d'équilibrage
~SLIDE-VALVE - [mech] (slide-valve balanced by means of an equilibrium ring) tiroir *m*
~STATE - [phys] état *m* d'équilibre
~TIME - [nucl] (of a reactor) temps *m* de démarrage
~WATER - [chem] (the water content of a solid remaining unchanged by exposure to air) teneur *f* normale en eau, humidité *f* d'équilibre
EQUIMOLAR - [chem] équimoléculaire
EQUINE - [zool] (pertaining to horses) équin, de cheval
EQUINIA - [vet] morve *f*, farcin *m*
EQUINOCTIAL - [astr & met] équinoxial
~DAY - [astr] jour *m* sidéral
~GALES - [met] (strong wind storms occurring about the time of the equinoxes) vents *mpl* d'équinoxe
~POINT - [astr] (one of the two points in which the celestial equator is cut by the ecliptic) équinoxe *f*
~STORMS - s. equinoctial gales
~YEAR - [astr] année *f* équinoxale
EQUINOX - [astr] (the instant in which the sun crosses the celestial equator) équinoxe *m*
EQUIP, to - [gen] armer, équiper, monter, outiller
[naut] (of a ship) armer
[mech] (of a factory) outiller
EQUIPARTITION OF ENERGY - [chem] (the kinetic energy possessed by a large number of particles which are rapidly moving in an enclosure and mingle together, becomes distributed according to the Boltzmann's law of the principle of equipartition of energy) équipartition *f* de l'énergie
EQUIPHASE ZONE - [radar] (that region in space where the phase difference of radio signals cannot be distinguished) zone *f* neutre
EQUIPMENT - [gen] équipement *m*, matériel *m*
[naut] (of a ship) armement *m*
[mech] (all that is required by a factory for its production) outillage *m*, appareillage *m*
[el] (radio etc) appareillage *m*
~CAPACITY - [telecomm] capacité *f* d'un équipement
EQUIPOISE - [phys] équilibre *m*
EQUIPOTENTIAL - [gen] équipotentiel
~CATHODE - [electron] (emitting-cathode with a surface all at the same potential) cathode *f* équipotentielle
~CONNEXION - [el] (an equalizer bar or an equalizer ring) connexion *f* équipotentielle
~LINE - [el] (line with all its points of the same potential) ligne *f* équipotentielle
~REGION - [el] (a field-free region) région *f* équipotentielle
~SURFACE - [el] (a surface on which no two points have a different potential) surface *f* équipotentielle
EQUISIGNAL RADIO RANGE BEACON - [radar] radiophare *m* à signaux équilibrés
~SECTOR - [radar] (the sector of the equisignal line) droite *f* de balisage
~ZONE - [radar] (the region in which two signals are of equal strength) zone *f* neutre
EQUISPACED - [gen] équidistant
EQUITEMPERED SCALE - s. equal tempered scale
EQUIVALENCE - [gen] équivalence *f*
~FACTOR - [metall] (copper-base alloys) titre *m* fictif en cuivre
~PRINCIPLE - [mech] principe *m* d'équivalence
~RELATION - [phys & math] relation *f* d'équivalence
EQUIVALENT - [gen] équivalent *m*
[radio] (the total gain between the ends of a line) équivalent *m*
[radio] (the total loss between the ends of a line) équivalent *m*
~ABSORPTION - [acoust] (of a room or of any object,

or objects in a room) absorption *f* équivalente
EQUIVALENT ACOUSTICS - [acoust] (sound effects which give a better copy of the sound to be heard) effets *mpl* sonores équivalents
~AIR SPEED - [aero] (E.A.S.) (indicated airspeed multiplied by the square root of the relative density) vitesse *f* équivalente au sol, équivalent *m* de vitesse
~BEAM METHOD - [constr] méthode *f* de la poutre équivalente
~BUILDING-UP TIME - [radio] (of a receiver or an amplifier) temps *m* de montée équivalent
~CIRCUIT - [el] circuit *m* équivalent
~CIRCUIT OF A CONTACT RECTIFIER - [el] circuit *m* équivalent d'un redresseur à contact
~CONCENTRATION - [phys] (the ion concentration divided by the valency of the ion considered) concentration *f* équivalente, concentration *f* moléculaire
~CONDUCTANCE - [el] (the conductance of that quantity of a solution containing one gramme equivalent of solute measured between parallel electrodes placed one centimetre apart) conductance *f* équivalente
~CONDUCTIVITY - [electron] conductivité *f* équivalente, conductivité *f* moléculaire
~CONSTANT POTENTIAL - [el] (the constant potential to be applied to an X-ray tube to produce a given radiation) potentiel *m* constant équivalent
~CONTROL CIRCUIT RESISTANCE - [radio] résistance *f* équivalente du circuit de commande
~DIODE OF A TRIODE - [electron] (imaginary diode in which the cathode current in the diode is the same as in the tube considered) diode *f* équivalente d'une triode
~DIODE VOLTAGE - [electron] (the anode voltage of the equivalent diode) tension *f* de la diode équivalente
~DISTURBING CURRENT - [radio] (in transmission, the inductive influence of an electric supply circuit) courant *m* perturbateur équivalent
~DISTURBING VOLTAGE - [radio] (in transmission) tension *f* perturbatrice équivalente
~ELECTRONS - [phys] (electrons in the same orbit) électrons *mpl* équivalents
~FIELD LUMINANCE - [opt] luminance *f* équivalente du champ visuel
~FOCAL LENGTH - [opt] longueur *f* focale équivalente, distance *f* focale équivalente
~FOCUS - [opt] foyer *m* équivalent
~HEIGHT OF AERIAL - [radio](the height of a perfect aerial, i.e. an aerial which, when carrying a uniformly distributed current equal to the maximum current in the actual aerial, radiates the same amount of power) hauteur *f* d'antenne équivalente
~HEIGHT OF ANTENNA - s. equivalent height of aerial
~HEIGHT OF THE SURCHARGE - [constr] hauteur *f* équivalente de la surcharge
~LENS - [opt] (lens equivalent to a system of lenses) lentille *f* équivalente
~MAP PROJECTION - [surv] projection *f* équivalente
~MONOPLANE - [aero] (a monoplane équivalent in aerodynamic properties to some given combination of wings, e.g. a biplane) monoplan *m* équivalent
~NETWORK - [el] (network which can replace another network without any alteration of the performance) réseau *m* équivalent
EQUIVALENT NUCLEI - [phys] (those nuclei in a molecule which can be transformed into one another) noyaux *mpl* équivalents
~PRESSURE - [soil] pression *f* de compactage équivalente
~REACTANCE - [el] réactance *f* équivalente
~RESISTANCE - [el] resistance *f* équivalente
~RESISTANCE BEAM - [constr] poutre *f* isostatique
~RESPONSE-PULSE DECAY-TIME - [radio] temps *m* d'affaiblissement de l'impulsion équivalente
~RESPONSE-PULSE GROWTH-TIME - [radio] temps *m* de montée de l'impulsion équivalente
~SIMPLE PENDULUM - [phys] (centre of oscillation) pendule *m* simple équivalent
~SINE WAVE - [el] (sine wave with the same frequency as a specified wave) onde *f* sinusoidale équivalente
~SOLUTION - [chem] solution *f* équivalente
~SOUND INTENSITY LEVEL - [acoust] niveau *m* d'isosonie
~STILL AIR RANGE - [aero] (theoretical horizontal distance which can be flown by an aircraft at given values for altitude, true airspeed, power and fuel weight) autonomie *f* équivalente en air calme
~THICKNESS - [nucl] épaisseur *f* équivalente (de plomb)
~VEILING LUMINANCE - [opt] luminance *f* équivalente de voile
~VOLT - s. electron volt
~WEIGHT - [phys] (the number of grams of an element which will combine with or replace one gram of hydrogen) équivalent *m* chimique
EQUIVALVE - [zool] (a bivalve with equal halves of the shell) équivalve
EQUIVISCOUS TEMPERATURE - [phys] température *f* d'équiviscosité
Er - [chem] (symbol for erbium) Er (erbium)
ERA - [gen & geol] ère *f*
ERADICATE, to - [gen] déraciner, extirper
ERADICATION - [gen] déracinement *m*, extirpation *f*
ERASABILITY - [paper man] (resistance to erasing) résistance *f* au grattage
ERASABLE MEMORY - [comput] mémoire *f* effaçable
ERASE, to - [gen] effacer, gommer, raturer, rayer, gratter
~-BIT - [comput] bit *m* d'effacement
ERASER - [impl] gomme *f*, grattoir *m*
ERASING - [electron] (in charge-storage tubes) effacement *m*
~HEAD - [el acoust] tête *f* d'effacement
~SPEED - [comput] (the rate of erasing storage elements in charge-storage tubes) vitesse *f* d'effacement
ERASION - [med] (the removal of diseased parts of a joint by scraping) curettage *m*
ERASURE - [gen] rature *f*, effaçure *f*, effacement *m*, [comput] effacement *m*(de données enregistrées sur un ruban magnétique)
~SIGNAL - [telecomm] signal *m* d'erreur
ERBIUM - [chem] (a rare earth element, symbol Er. A.N. 68, A.W. 167.7, with uses in nuclear technology) erbium *m*
ERECT, to - [gen] dresser, ériger, élever
[constr] construire
[mech] monter, installer
[opt] droite (image)

ERECT-IMAGE VIEWFINDER - [photo] viseur *m* à image redressée
~PILE - [text] (of carpets) poil *m* droit
~THE FALSEWORK, to - [constr] (in concrete work) mettre en place le coffrage
ERECTILE TISSUE - [anat] tissu *m* érectile
ERECTING - [gen] construction *f*
[mech] montage *m*, installation *f*
~BAY - [mech] atelier *m* de montage
~FRAME - [mech & constr] échafaudage *m*
~LENS - [opt] lentille *f* redresseuse
~PRISM - [opt] (right-angle prism for erecting the image formed by an inverting projection system) prisme *m* redresseur
~SCAFFOLD - s. erecting frame
~SHOP - [mech] (in an engineering works, the part where assembly takes place) atelier *m* de montage
~STAGE - [mech] plateforme *f* de montage
~STAND - [mech] banc *m* de montage
~WEDGE - [photo] coin *m* de redressement
~YARD - [constr] chantier *m* de construction
ERECTION - [gen] érection *f*, construction *f*, édifice *m*
[mech] montage *m*, assemblage *m*
(of a crane) installation *f*
~CUT-OUT - [contr] interruption *f* du signal d'érection
~DRAWING - [mech] plan *m* de montage
~INSTRUCTIONS - [mach] instructions *fpl* pour le montage
~MAST - [mech] mât *m* de levage
~RATE - [constr] divergence *f* d'érection
ERECTOR - [zool] (muscle raising a part or an organ by its contraction) muscle *m* érecteur, érecteur *m*
[opt] inverseur *m*
[astronaut] (the equipment used to position a rocket for vertical launching) érecteur *m*
~-LAUNCHER - [astronaut] véhicule *m* d'érection et de lancement
E REGION - [phys] (el magn waves) région *f* E, zone *f* E
EREPSIN - [biochem] (an enzyme of the intestinal juices) érepsine *f*
ERETHISM - [med] (abnormal irritation of a tissue) éréthisme *m*
ERG - [mech] (the unit of work or energy) erg *m*
ERGATE - [zool] (a worker ant) ouvrière *f*
ERGATOPLASM - [biol] (a functional protoplasm) ergatoplasme *m*
ERGMETER - [instr] (or ergometer; instrument designed to measure in ergs the energy expended) ergmètre *m*
ERGOGRAM - [instr] ergogramme *m*
ERGOGRAPH - [instr] ergographe *m*, ergomètre *m* enregistreur
ERGOMETER - [instr] s. ergmeter
ERGONOMICS - [ind] (the distribution of work in accordance with the operators' ability) ergonomie *f*
ERGOSTEROL - [chem] (a vegetable sterol of importance as the precursor of vitamin D) ergostérol *m*, ergostérine *f*
ERGOT - [bot] (a plant disease) ergot *m*
ERGOTINE - [chem] ergotine *f*
ERINOID - [plast ind] (proprietary name for a casein thermoplastic) érinoid *f*, galalithe *f*
ERIOMETER - [instr] (instrument designed to measure small diameters, e.g. those of fibres) ériomètre *m*
ERLENMEYER FLASK - [chem] (thin glass flask of conical form, with flat bottom) erlenmeyer *m*
ERODE, to - [gen] éroder, ronger, user, corroder
[geol] éroder, affouiller
ERODENT - [adj] corrosif
EROSION - [gen] érosion *f*, usure *f*
[geol] (the lowering of the land surface by weathering etc) érosion *f*
(of the coastline by the action of the sea) affouillement *m*, érosion *f*
~CONTROL - [hydr] prévention *f* de l'érosion
~LEVEL - [geol] niveau *m* d'érosion
~THRUST - [geol] charriage *m* d'érosion
EROSIONAL GAP - [geol] lacune *f* d'érosion
~UNCONFORMITY - [geol] discordance *f* d'érosion
EROSIVE - [gen] érosif
~ACTION - [soil] action *f* érosive, effet *m* d'attrition
~CAPACITY - [soil] capacité *f* d'érosion, puissance *f* d'affouillement
EROTICISM - s. erotism
ERRANT SPACECRAFT - [astronaut] véhicule *m* spatial sans contrôle
ERRATIC - [gen] erratique, variable, faux
[mech] irrégulier, intermittent
~BLOCKS - s.erratics
~BOULDERS - s. erratics
~ERROR - [instr] erreur *f* aléatoire
~FIRING - [auto] (of the ignition) allumage *m* irrégulier
ERRATICS - [geol] (small or large stones transported by ice far from their original location) blocs *mpl* erratiques
ERRATUM - [print] erratum *m*
ERROR - [gen] erreur *f*
[meas] (the difference between the indicated and the true value) erreur *f*, écart *m*
[instr] (the measure of the loss of precision) erreur *f*
~AND OMISSIONS EXCEPTED - [comm] (on invoices) sauf erreur ou omission
~CALCULATION - [math] calcul *m* d'erreur
~CARD - [comput] carte *f* erronée, carte *f* défectueuse
~-CHECKING CHARACTER - [comput] caractère *m* de contrôle d'erreur, signal *m* de contrôle d'erreur
~-CORRECTING CHARACTER - [comput] caractère *m* de correction d'erreur, signal *m* de correction de erreur
~-CORRECTING CODE - [telecomm] code *m* correcteur d'erreurs, code *m* de sécurité
~-CORRECTING PROGRAM - [comput] programme *m* de correction des erreurs
~-DETECTING CODE - [comput] code *m* détecteur de erreur
~DETECTION - [comput] recherche *f* des erreurs, détection *f* des erreurs, identification *f* des erreurs
~FUNCTION - [contr] fonction *f* d'erreur
~OF APPROXIMATION - [comput] (in electronic computers, the error caused by the mathematical operation when approaching a process numerically) erreur *f* d'approximation
~OF CLOSURE - [surv] (the distance between the final and the starting points in a closed transverse) erreur *f* de fermeture
~OF CURVATURE - [meas] (the error caused by the

curvature of the magnetizing curve) erreur *f* de courbure
ERROR OF ESTIMATION - [gen] erreur *f* de mesure
~OF READING - [instr] erreur *f* de lecture
~RATE - [comput] fréquence *f* d'erreurs, taux *m* d'erreurs
~RATIO - [contr] rapport *m* de dérive
~SENSING DEVICE - [contr] détecteur *m* d'écart
~SIGNAL - [telecomm] signal *m* d'erreur
~TRIANGLE - [math] triangle *m* d'erreur
ERTOR - [astronaut] température *f* de rayonnement de la couche d'ozone terrestre
ERUBESCITE - [min] (bournite; variegated copper ore; a natural sulphide of copper and iron, of variable composition, generally about five times as much copper as iron. Cubic; red to brown; a valuable ore of copper) érubescite *f*, bornite *f*, phillipsite *f*
ERUCIC ACID - [chem] acide *m* érucique, acide *m* brassique
ERUCTATION - [med] (belching of gas from the stomach) éructation *f*
ERUPTION - [geol] éruption *f*
[med] (the breaking out of a rash on the skin) éruption *f*, poussée *f*
~CLOUD - [geol] nuée *f* volcanique
ERUPTIVE - [gen & geol] éruptif
~ROCK - [geol] (general term for igneous rocks) roche *f* éruptive
ERYSIPELAS - [med] (an infection of the skin) érysipèle *m*, érésipèle *m*
ERYTHEMA - [med] (redness of the skin) erythème *m*
ERYTHRAEMIA - [med] (blood disease affecting the spleen) érythrémie *f*, maladie *f* de Vaquez
ERYTHRENE - [chem] (colourless gas with principal applications in the production of synthetic rubbers and nylon) érythrène *m*
ERYTHRITE - [min] (natural cobalt arsenate, crystallizing in the monoclinic system. Used especially for colouring glazes in ceramics) érythrite *f*, érythrol *m*
ERYTHROCYTE - [biol] (haemoglobin-containing blood corpuscle) érythrocyte *m*, hématie *f*
ERYTHRODEXTRIN - [chem] (the most complex of the dextrins; it gives a wine-red colour with iodine) érythrodestrine *f*
ERYTHROID - [med] érythroide
ERYTHROSINE - [chem] (a red colourant) érythrosine *f*
ESCALATOR - [transp] (conveyor for passenger transport). escalier *m* roulant
~BANISTER RAIL - [transp] main-courante *f* pour escalier roulant
ESCAPE, to - [gen] fuir, échapper
[mech] (of gas etc) se dégager, s'échapper
ESCAPE - [gen] fuite *f*, échappement *m*, dégagement *m*,
[constr] (a fire escape) sortie *f* de secours
~APPARATUS - [naut] (generic term for life-saving devices) dispositif *m* de sauvetage
~CHANNEL - [mech] conduit *m* d'échappement
~CLAUSE - [leg] clause *f* échappatoire, clause *f* dérogatoire
~DOOR - [constr] porte *f* de secours
~HARCH - [astronaut] trappe *f* de sortie de secours
~MAGNET - [comput] aimant *m* de transport, aimant *m* d'avance
ESCAPE OF GAS - [phys] échappée *f* de gaz, dégagement *m* spontané de gaz
~OF NEUTRONS - [phys] (due to leaks etc) fuite *f* de neutrons
~OF VOLATILE CONSTITUENTS - [chem] dégagement *m* des constituants volatils
~PIPE - [mech] tuyau *m* d'échappement
~PROBABILITY - [nucl] facteur *m* antitrappe, probabilité *f* d'échapper à la capture par resonance
~ROCKET - [astronaut] fusée *f* de séparation de la capsule
~STEAM - [gen] vapeur *f* d'échappement
~TOWER - [astronaut] système *m* de séparation de la capsule
~VALVE - [mech] soupape *f* d'échappement
~VELOCITY - [astronautics] (the speed which is required to leave the gravitational pull of the earth) vitesse *f* de libération
~WHEEL - [horol] roue *f* d'ancre
ESCAPEMENT - [gen] échappement *m*
[horol] (a device for coverting circular into reciprocating motion)échappement *m*, assortiment *m* ancre
[photo] (of a shutter) échappement *m*
~BEARING PLATE - [horol] platine *f* porte-échappement
~CONTACT - [comput] contact *m* de cliquet de transport, contact *m* de cliquet d'avance
~WHEEL - [comput] roue *f* de transport, roue *f* d'avance
[photo] roue *f* de rencontre
ESCARPMENT - [geol] escarpement *m*, talus *m*
ESCHAR - [med] (dry slough) escarre *f*
ESCHAROSIS - [med] escarrification *f*
ESCHEAT - [leg] déshérence *f*
ESCHYNITE - [min] (thorium containing ore) eschynite *f*
ESCORT, to - [gen] escorter, accompagner, protéger
ESCULIN - [chem] (a constituent of cosmetic lotions) esculine *f*
ESCUTCHEON - [carp] (perforated plate around an opening) écusson *m*, entrée *f* de serrure
[gen] écusson *m*
[auto] écusson *m*, cache *m*
[naut] (the letters forming the name of the ship on the stern) écusson *m*
~PLATE - [carp] entrée *f* de serrure
ESERINE - [chem] (alkaloid obtained from the seeds of calabar beans) ésérine *f*, physostigmine *f*
ESOPHAGESTASIS - [med] dilatation *f* de l'œsophage
ESOPHAGUS - [anat] œsophage *m*
ESOPHORIA - [med] ésophorie *f*
ESOTERIC - [adj] (intellegible to the initiated) ésotérique
ESPARCET - [bot] (herbage plant) esparcet *m*, sainfoin *m*
ESPARTO - [bot] (a cultivated grass used in the manufacture of paper cordage etc) spart *m*, sparte *m*, alfa *m*
[paper man] Alfa *m*, papier *m* d'alfa
~GRASS - [bot] sparte *m*, alfa *m*
~WAX - [ind chem] (a wax extracted from esparto grass and used in the production of polishes) cire *f* de fibre, cire *f* de sparte
ESPLANADE - [gen] esplanade *f*
ESQUILLECTOMY - [med] esquillectomie *f*
ESSENCE - [gen] essence *f*, nature *f*

[chem] essence *f*, extrait *m*, huile *f* essentielle
ESSENTIAL - [gen] essentiel
~MINERALS - [min] (components present in a rock) minéraux *mpl* essentiels
~OIL - [chem] (water-insoluble oil present in small amounts in a large variety of plants and having application in the production of perfumes and flavourings) huile *f* essentielle
ESTABLISH, to - [gen] affermir, établir, ériger
ESTABLISHING SHOT - [cin] (shot reproducing the whole scene) plan *m* de situation, plan *m* de présentation
ESTABLISHMENT - [gen] établissement *m*, affermissement *m*, fondation *f*
(the total number of employees etc in an organization) effectif *m*
~CHARGE - [telecomm] (of telephone) frais *mpl* d'installation
ESTATE - [gen] propriété *f*, bien *m*, domaine *m*
~CAR - [auto] limousine *f* commerciale, break *m*
~RUBBER - [rubber ind] caoutchouc *m* de plantation
ESTER - [chem] (organic compounds formed by the reaction of an acid with an alcohol and corresponding to inorganic salts) ester *m*, éther-sel *m*
~GUMS - [chem] (synthetic resins derived by the esterification of natural resins) gomme-ester *f*
ESTERIFICATION - [chem] (the direct formation of esters by the action of an acid on an alcohol) estérification *f*
ESTERIFY, to - [chem] estérifier
ESTHESIOMETER - [instr] (instrument designed to determine the distance by which two points must be separated in order that they feel as separate on the skin) esthésiomètre *m*
ESTIMATE, to - [gen] estimer, évaluer, apprécier
[comm] (to calculate costs, expenses, quantities etc) estimer, faire un devis
ESTIMATE - [gen] appréciation *f*, évaluation *f*, estimation *f*
[comm] budget *m*, devis *m*, prévision *f*
~OF QUANTITIES AND COSTS - [comm] devis *m* estimatif
ESTIMATED CHARGES - [comm] imputations *fpl* estimatives
~ERROR - [meas] erreur *f* approximative
ESTIMATING DEPARTMENT - [comm] service *m* d'études
ESTIMATION - s. estimate
ESTIMATOR - [comm] estimateur *m*, priseur *m*, expert *m*
ESTRAGOLE - [chem] (a component of certain essential oils used in perfumery) estragol *m*
ESTRIOL - [med] œstriol *m*, hydrate *m* de folliculine
ESTRONE - [med] folliculine *f*
ESTUARY - [geogr] estuaire *m*, embouchure *f*
ETAMINE - [text] étamine *f*
ETCH, to - [chem] décaper, attaquer (par voie chimique), corroder, ronger
[metall] (to print on metal) graver
[litho] préparer
~FIGURE - s. etching figure
~MACHINING - [mech] (the removal of material from a workpiece by chemical means) usinage *m* chimique
ETCHANT - [chem] décapant *m*
ETCHED-OUT PATTERN - [text] déssin *m* à corrosion
ETCHING - [chem] (the operation whereby the structure of metals is revealed, by attacking its surface with a reagent) décapage *m*, attaque *f* à l'acide, gravure *f*, corrosion *f*
[gen] eau-forte *f*, gravure *f* à l'eau forte
[print] morsure *f*, mordançage *m*, préparation *f*
ETCHING ACID - [chem] acide *m* corrosif, acide *m* fluorhydrique
~BASIN - [print] cuvette *f* à morsure
~FIGURE - [metall] figure *f* de corrosion
~GROUND - [print] (asphalt and wax paint) enduit *m* à graver
~INK - [print] vernis *m* de rebouchage
~MACHINE - [print] machine *f* à graver
[text] machine *f* à ronger
~PITS - [metall] pores *mpl* d'attaque à l'acide
~REAGENT - [chem] décapant *m*, mordant *m*
ETHANAL - [chem] éthanal *m*, aldéhyde *m* acétique
ETHANE - [chem] (a colourless odourless gas, the second member of the paraffin series. Obtained chiefly from the fractionation of natural gas. Used as a fuel, a refrigerant and in organic synthesis) éthane *m*
ETHANOL - [chem] (ethyl alcohol) éthanol *m*, alcool *m* éthylique
ETHANOLAMINE - [chem] (an intermediate for pharmaceuticals, insecticides, and textile chemicals) éthanol *m*
ETHENE - s. ethylene
ETHENOID PLASTICS - s. ethenoid resins
~POLYMERS - [chem] polymères *mpl* éthénoides
~RESINS - [chem] (the acrylic, vinyl and styrene groups; made from compounds in which there is a double bond between two carbon atoms) résines *fpl* éthénoides
ETHER - [chem] (diethyl ether; colourless, volatile, flammable liquid with uses as an anaesthetic, solvent, intermediate and refrigerant) éther *m*, éther *m* officinal, éther *m* éthylique, éther *m* diéthylique, éther *m* sulfurique
~EXTRACT - [chem] extrait *m* éthéré
~PRIMING SYSTEM - [mech] (used in cold starting) démarrage *m* à l'éther
~SOLUBILITY - [chem] solubilité *f* dans l'éther
ETHEREAL - [chem & gen] éthéré, volatif
EHTEREALIZE, to - [chem] estérifier
ETHERIFICATION - [chem] éthérification *f*
ETHERIFY, to - [chem] éthérifier
ETHERIZE, to - [med] anesthésier par l'éther
ETHERS - [chem] (organic compounds having two hydrocarbon radicals joined through an oxygen atom. This term is sometimes, but erroneously, used to mean esters) éthers *mpl*
ETHIDE - [chem] éthylure *m*
ETHINE - [chem] (an alternative term for acetylene) acétylène *m*
ETHIODIDE - [chem] iodoéthylate *m*
ETHMOID - [med] ethmoïde *m*
ETHNOGRAPHY - [gen] ethnographie *f*
ETHNOLOGY - [gen] ethnologie *f*
ETHOXIDE - [chem] éthylate *m*
ETHRIOSCOPE - [instr] (instrument designed to measure changes of temperature due to different conditions of the air) éthrioscope *m*
ETHYL - [chem] éthyle *m*
~ACETANILIDE - [chem] (a camphor substitute in the manufacture of cellulose plastics) éthylacétanilide *f*

ETHYL ACETATE - [chem] (a solvent for plastics and an intermediate in the manufacture of perfumes and flavourants) acétate *m* d'éthyle, éther *m* acétique
˜ACETOACETATE - [chem] (synthesis intermediate for pharmaceuticals and dyestuffs) éther *m* acétylacétique, acétylacétate *m* d'éthyle
˜ACETYLENE - [chem] (a synthesis intermediate) éthylacétylène *m*
˜ACRYLATE - [chem] (intermediate for polymers) acrylate *m* d'éthyle
˜ALCOHOL - [chem] (solvent, intermediate for plastics, dyestuffs, pharmaceuticals, elastomers, cosmetics and explosives, and a constituent of a large number of beverages) alcool *m* éthylique, alcool *m* (ordinaire), alcool *m* de vin, éthanol *m*, méthylcarbinol *m*
˜ALDEHYDE - [chem] aldéhyde *m* éthylique
˜AMYL KETONE - [chem] (a solvent used in perfumery) éthyl-amyl-cétone *f*
˜ANTHRANILATE - [chem] (a perfumery intermediate) anthranilate *m* d'éthyle
˜BENZOATE - [chem] (a solvent for cellulose derivatives and a component of perfumes and flavours) benzoate *m* d'éthyle
˜BORATE - [chem] (a fireproofing agent and disinfectant) borate *m* d'éthyle
˜BROMIDE - [chem] (an anaesthetic, solvent, insecticide, and refrigerant) bromure *m* d'éthyle
˜BUTYL CARBONATE - [chem] (solvent for natural and synthetic resins) éthylbutylcarbonate *m*, carbonate *m* d'éthyle et butyle
˜BUTYL KETONE - [chem] (solvent for surface coatings) éthyl-butyl-cétone *f*
˜BUTYRATE - [chem] (solvent for natural and synthetic resins, flavouring agent and perfumery component) butyrate *m* d'éthyle
˜CAFFEATE - [chem] (preservative used in food processing) caféate *m* d'éthyle
˜CAPRATE - [chem] (intermediate for the production of flavourants) caprate *m* d'éthyle
˜CAPROATE - [chem] (intermediate for flavouring essences) caproate *m* d'éthyle
˜CAPRYLATE - [chem] (intermediate for flavouring essences) caprylate *m* d'éthyle
˜CARBAMATE - [chem] (synonym for urethane) carbamate *m* d'éthyle, uréthane *m*
˜CELLULOSE - [chem] (white, thermoplastic solid with uses in moulding and coating materials, plastics additives, and in textile finishing) éthylcellulose *f*
˜CHLORIDE - [chem] (an anaesthetic and synthesis intermediate) chlorure *m* d'éthyle, chloréthane *m*, éther *m* éthylchlorhydrique
˜CHLOROCARBONATE - [chem] (synthesis intermediate) chlorocarbonate *m* d'éthyle, éther *m* chlorocarbonique
˜CHLOROFORMATE - [chem] chloroformiate *m* d'éthyle, éther *m* chlorocarbonique, éther *m* chloroformique
˜CINNAMATE - [chem] (a flavouring and perfumery agent) cinnamate *m* d'éthyle, éther *m* éthylcinnamique
˜CROTONATE - [chem] (a solvent and synthesis intermediate) crotonate *m* d'éthyle
˜CYANIDE - [chem] (synthesis intermediate and solvent) cyanure *m* d'éthyle
ETHYL DIIODOBRASSIDATE - [chem] di-iodo-brassidate *m* d'éthyle, lipoidine *f*
˜FORMATE - [chem] (an intermediate for flavourants and synthetic resins) formiate *m* d'éthyle, éther *m* formique
˜FUEL - [chem] (anti-detonating fuel for electric ignition i.c. reciprocating engines, containing ethylene dibromide and tetraethyl lead) essence *f* éthylée, carburant *m* éthylé
˜GLYCOL ACETATE - [chem] acétate *m* d'éthylglycol
˜IODIDE - [chem] (an adjuvant to oral iodides in medicine) iodure *m* d'éthyle, iodo-éthane *m*
˜ISOBUTYRATE - [chem] (an intermediate for the production of flavours) isobutyrate *m* d'éthyle
˜ISOCYANATE-[chem] (intermediate for pharmaceuticals) isocyanate *m* d'éthyle
˜LACTATE - [chem] (solvent for resins and cellulose derivatives) lactate *m* d'éthyle, oxy-propionate *m* d'éthyle
˜LAURATE - [chem] laurate *m* d'éthyle, dodécylate *m* d'éthyle
˜MAGNESIUM BROMIDE - [chem] (a Grignard reagent) bromure *m* de magnésium et d'éthyle
˜MAGNESIUM CHLORIDE - [chem] (a Grignard reagent) chlorure *m* de magnésium et d'éthyle
˜MALONATE - [chem] (intermediate for dyestuffs and pharmaceuticals) malonate *m* d'éthyle, éther *m* malonique
˜MERCAPTAN - [chem] mercaptan *m* éthylique, mercaptan *m*
˜METHACRYLATE - [chem] (an intermediate for polymers) métacrylate *m* d'éthyle
˜NITRATE - [chem] (intermediate for dyestuffs and perfumes) nitrate *m* d'éthyle, éther *m* nitrique
˜NITRITE - [chem] (synthesis intermediate) nitrite *m* d'éthyle, éther *m* nitreux
˜OENANTHATE - [chem] (synthetic flavouring agent) œnanthate *m* d'éthyle
˜OLEATE - [chem] (a plasticizer and solvent) oléate *m* d'éthyle
˜OXALATE - [chem] (intermediate for drugs, dyes, and perfumes, and a solvent for natural and synthetic resins) oxalate *m* d'éthyle
˜PELARGONATE - [chem] (a synthetic flavouring agent) pélargonate *m* d'éthyle
˜PHENYLACETATE - [chem] (a flavouring and perfumery agent) phénylacétate *m* d'éthyle
˜PROPIONATE - [chem] (solvent for synthetic resins and cellulose derivatives) propionate *m* d'éthyle
˜PROPYL KETONE - [chem] éthyl-propyl-cétone *f*, hexanone 3 *f*
˜SALICYLATE - [chem] (ingredient of flavours and perfumes) salicylate *m* d'éthyle
˜SILICATE - [chem] (a preservative for stones and brickwork) silicate *m* d'éthyle
˜SULPHIDE - [chem] (a solvent and synthesis intermediate) sulfure *m* d'éthyle
ETHYLAMINE - [chem] (dyestuffs intermediate and a component of latex stabilizers) éthylamine *f*
ETHYLATE, to - [chem] éthyler
ETHYLATION - [chem] éthylation *f*
ETHYLBENZENE - [chem] (an intermediate for styrene and a solvent) éthylbenzène *m*
ETHYLBENZYL CHLORIDE - [chem] (synthesis intermediate) chlorure *m* d'éthyle et de benzyle
ETHYLBENZYLANILINE - [chem] (intermediate for dyestuffs) éthylbenzylaniline *f*

THYLBUTYL ALCOHOL - [chem] (perfumery intermediate and solvent for dyestuffs, resins and waxes) alcool *m* éthylbutylique

'SILICATE - [chem] (heat transfer medium) silicate *m* d'éthyle et de butyle

THYLBUTYRALDEHYDE - [chem] (intermediate for resins, drugs, and rubber chemicals) éthylbutyraldéhyde *m*

THYLBUTYRIC ACID - [chem] (an intermediate for dyestuffs and pharmaceuticals) acide *m* éthylbutyrique

THYLCYCLOHEXANE-[chem]synthesis intermediate) éthylcyclohexane *m*

THYLDICHLOROARSINE - [chem] (a war gas) éthyldichloroarsine *f*

THYLDIETHANOLAMINE - [chem] (solvent) éthyldiéthanolamine *f*

THYLENE - [chem] (olefiant gas; a major synthesis intermediate for plastics and a number of organic compounds. Ethylene also has uses as an anaesthetic) gaz *m* oléifiant, éthylène *m*, éthène *m*

'CARBONATE - [chem] (intermediate for textile chemicals, pharmaceuticals and rubber additives, and a solvent for a number of synthetic resins) carbonate *m* d'éthylène

'CHLORIDE - [chem] chlorure *m* d'éthylène, liqueur *f* des Hollandais

'CHLOROHYDRIN - [chem] (synthesis intermediate and a solvent for cellulose derivatives) éthylène-chloro-hydrine *f*, monochlorohydrine *f* du glycol

'CYANIDE - [chem] (synthesis intermediate) cyanure *m* d'éthylène

'CYANOHYDRIN - [chem] (synthesis intermediate and a solvent for certain cellulose derivatives) cyanhydrine *f* d'éthylène

DIBROMIDE - [chem] (a solvent, fumigant, and insecticide) dibromure *m* d'éthylène

DICHLORIDE - [chem] (solvent and intermediate for synthetic resins, surface coatings, and cellulose derivatives) dichlorure *m* d'éthylène

'GLYCOL - [chem] (solvent for dyestuffs, pharmaceuticals, cellulose esters, waxes and resins, and a component of anti-freeze coolant mixtures) éthylène glycol *m*, glycol *m*, éthanédiol *m*, monoéthylène glycol *m*

GLYCOL DIACETATE - [chem] (a perfume fixative and solvent and plasticizer for cellulose derivatives) diacétate *m* d'éthylène glycol

'GLYCOL DIMETHYL ETHER - [chem] (a solvent) éther *m* diméthylique de l'éthylène glycol

GLYCOL DIPROPIONATE - [chem] (a plastics plasticizer) dipropionate *m* d'éthylène glycol

'GLYCOL MONOACETATE - [chem] (a solvent for cellulose derivatives) monoacétate *m* de glycol

'GLYCOL MONOBENZYL ETHER - [chem] (a perfumery fixative, synthesis intermediate, and a solvent for resins and dyestuffs) éther *m* monobenzylique du glycol

GLYCOL MONOBUTYL ETHER - [chem] (a textile processing agent and a solvent for cellulose derivatives) éther *m* (mono)butylique du glycol, butylglycol *m*, butylcellosolve *m*

GLYCOL MONOBUTYL ETHER ACETATE - [chem] (a solvent for epoxy resins and cellulose derivatives) acétate *m* de butylglycol

GLYCOL MONOBUTYL ETHER LAURATE - [chem] (a plasticizer) laurate *m* de butylglycol

ETHYLENE GLYCOL MONOBUTYL OLEATE - [chem] (a plasticizer) oléate *m* de butylglycol

~GLYCOL MONOETHYL ETHER - [chem] (textile processing agent and a solvent for natural and synthetic resins) éther *m* (mono)éthylique du glycol, éthylglycol *m*, éther *m* éthylglycolique, cellosolve *m*

~GLYCOL MONOETHYL ETHER ACETATE - [chem] (a solvent used in paint manufacture) acétate *m* d'éthylglycol

~GLYCOL MONOETHYL ETHER LAURATE - [chem] laurate *m* d'éthylglycol

~GLYCOL MONOETHYL RICINOLEATE - [chem] (plasticizer) ricinoléate *m* d'éthylglycol

~GLYCOL MONOMETHYL ETHER - [chem] (a solvent for natural and synthetic resins, cellulose derivatives and dyestuffs) éther *m* monométhylique du glycol

~GLYCOL MONOMETHYL ETHER ACETATE - [chem] (a solvent for resins and cellulose derivatives, and a textile printing agent) acétate *m* de méthylglycol

~GLYCOL MONOPHENYL ETHER - [chem] (a perfumery fixative, intermediate for plasticizers, pharmaceuticals and disinfectants, and a solvent for resins and dyes) éther *m* phénylique de l'éthylèneglycol

~GLYCOL MONORICINOLEATE - [chem] (a plasticizer) monoricinoléate *m* d'éthylèneglycol

~GLYCOL SILICATE - [chem] (a weather-proofing agent for surface coatings) silicate *m* d'éthylèneglycol

~HYDROCARBONS - [chem] carbures *mpl* éthyléniques, oléfines *fpl*

~OXIDE - [chem] (intermediate for polymers, lubricants, solvents, ethylene glycol and a number of other organic compounds) oxyde *m* d'éthylène, époxy-éthane *m*

~SERIES - [chem] série *f* éthylénique

ETHYLENEDIAMINE - [chem] (a solvent, emulsifier and stabilizing agent for a number of organic compounds. Also used as an intermediate for rubber and textile chemicals) éthylènediamine *f*

ETHYLETHANOLAMINE - [chem] (synthesis intermediate) éthyléthanolamine *f*

ETHYLHEXYL ACETATE - [chem] (solvent for cellulose derivatives) acétate *m* de 2-éthylhexyle, acétate *m* de capryle, acétate *m* d'octyle

~ACRYLATE - [chem] (an intermediate for synthetic resins) acrylate *m* de capryle

~ALCOHOL - [chem] (a surface-active agent, an intermediate for plasticizers and a solvent for resins, dyestuffs, fats, and waxes) alcool *m* éthylhexylique

ETHYLHEXYLAMINE - [chem] (intermediate for insecticides, vulcanization agents and detergents) éthylhexylamine *f*

ETHYLIC - [chem] éthylique

ETHYLIDENE - [chem] ethylidène *m*

ETHYLISM - [med] éthylisme *m*

ETHYLMORPHINE HYDROCHLORIDE - [chem] (sedative and analgesic used in medicine) chlorhydrate *m* d'éthylmorphine, "chlorhydrate" *m*

ETHYLSULPHURIC ACID - [chem] (synthesis intermediate) acide *m* sulfovinique, acide *m* éthylsulfurique

ETIOLATE, to - [gen] étioler, s'étioler

ETIOLATED PLANT - [bot] plante *f* étiolée

ETIOLATION - [bot] (the condition of plants which have suffered under lack of light) étiolement *m*, chlorose *f*

ETIOLOGY - [med] (the study of the causation of di-

sease) étiologie *f*
EUCAINE LACTATE - [chem] (a local anaesthetic) lactate *m* d'eucaine
EUCALYPTOL - [chem] (a component of eucalyptus oil used in perfumery and pharmacy) eucalyptol *m*, cinéol *m*, oxyde *m* de terpilène
EUCALYPTUS - [bot] eucalyptus *m*
~GUM - [chem] gomme *f* eucalyptus
~OIL - [chem] (essential oil obtained from species of Eucalyptus and used as a flotation in ore extraction and in perfumery) essence *f* d'eucalyptus
EUCHLORINE - [chem] (a mixture of chlorine and chlorine peroxide) euchlorine *f*
EUCHROMATIN - [genet] (the part of the chromatin which is active) euchromatine *f*
EUCHROMATOPSY - [med] chromatopsie *f* normale
EUCLASE - [min] (a hydrated silicate of beryllium and aluminium) euclase *f*
EUCLIDEAN SPACE - [geom] espace *m* euclidien
~GEOMETRY - [geom] géométrie *f* euclidienne
EUCRITE - [geol] (a basic igneous rock) eucrite *f*
EUDIOMETER - [chem] (a long glass tube, closed at one end and graduated in volumetric units, into which platinum wires are fused, so that a spark can be passed through the contained gas) eudiomètre *m*
EUDIOMETRY - [chem] (the determination of the composition of gases) eudiométrie *f*
EUGENIC - [genet] (valid for the production of good offsprings) eugénique
EUGENICS - [med] eugénie *f*, eugénique *f*, callipédie *f*
EUGENOL - [chem] (flavouring agent, germicide, intermediate for vanillin and perfumery agent) eugénol *m*, acide *m* eugénique, acide *m* caryophyllique
~ACETATE - [chem] (a perfumery agent) acétate *m* d'eugénol
EUMERISM - [zool] (aggregation of similar parts) eumérisme *m*
EUMEROGENESIS - [zool] (a form of segmentation) eumérogénèse *f*
EUNATROL - [chem] oléate *m* de sodium
EUPHORIA - [med] (a feeling of well-being)euphorie *f*
EUPLOIDY - [genet] (chromosome number which is a multiple of the monoploid number) euploidie *f*
EUROPIUM - [chem] (a rare element, symbol Eu.A.N.63, A.W. I52. It behaves both as a divalent and as a trivalent element) europium *m*
EURYBATHIC - [ecology] (capable of living in a wide range of depths) eurybathique
EUSTACHIAN MUSCLE - [anat] muscle *m* acoustico-malléen
~TUBE - [anat] (the duct connecting the tympanic cavity and the pharynx) trompe *f* d' Eustache
EUSTATIC MOVEMENT - [geol] (a change of sea-level over a vast area) mouvement *m* eustatique
EUSTON PROCESS - [ind chem] (process for the manufacture of white lead in which the lead is reproduced to a spongy condition, oxydized by atmospheric air and treated with normal lead acetate solution, which is then carbonated with carbon dioxide) procédé *m* de Clichy
EUSTYLE - [build] (a type of colonnade) eustyle *m*
EUTECTIC - [chem] (a mixture of two substances or more such that its melting point is the lowest of any mixture of these. Eutectics may behave in some ways like a true compound) eutectique *m*
EUTECTIC ALLOY - [metall] alliage *m* eutectique
~CHANGE - [chem] (the change of a eutectic from the liquid to the solid state) transformation *f* eutetique
~EXUDATION - [metall] ressuage *m* eutectique
~HALT - s. eutectic point
~MIXTURE - [metall] mélange *m* eutectique
~POINT - [chem] (the point in a constitutional diagram of an alloy which indicates the composition of the eutectic and its freezing point) point *m* eutectique
~STRUCTURE - [metall] (the arrangement of the constituents of a eutectic alloy) structure *f* eutectique
~SYSTEM - [metall] (alloy system in which one alloy solidifies at a temperature lower than all other alloys) système *m* eutectique
EUTECTOID - [metall] (with the same properties as eutectic but occurring only in completely solid regions) eutectoïde *m*
~STEEL - [metall] acier *m* eutectoïde (constitué par la perlite)
~TRANSFORMATION - [metall] transformation *f* eutectoïde
EUTROPIC SERIES - [phys] (a serial arrangement of substances in which the crystalline forms and other properties show a regular variation) série *f* eutropique
EUTROPY - [phys] (regular variation in crystalline form in a series of compounds in accordance with the atomic number of the element concerned) eutropie *f*
EUXENITE - [min] (naturally-occurring mixture of niobates and titanates of yttrium, erbium, cerium and uranium; usually massive and brownish-black sometimes occurring as orthorhombic crystalline forms) euxénite *f*
eV - [el] (or ev) eV (électron-volt)
EVACUATE, to - [gen] évacuer, abandonner, faire le vide, pomper
EVACUATED SPACE - [phys] espace *m* sous vide, volume *m* évacué
~VOLUME - s. evacuated space
EVACUATION - [gen] évacuation *f*
[electron] (the removal of gases from the envelope of an electric tube) pompage *m*
~LINE - [mech] conduite *f* de désaération, tube d'évacuation
~PIPE - [mech] ajutage *m* d'évacuation, raccord *m* de pompage
~PORT - [mech] ajutage *m* de vide, raccord *m* pour le vide
~TIME - [phys] temps *m* de pompage
EVALUATE, to - [gen] évaluer, estimer
EVALUATION - [gen] évaluation *f*
EVANESCENT MODE - [waveguides] (mode of oscillation in a cut-off waveguide) mode *m* évanescent
~WAVE - [phys] onde *f* évanescente
EVAPORATE, to - [phys] (to convert liquid into vapour) évaporer
~DOWN, to - [chem] concentrer, réduire
~OFF, to - [chem] chasser par évaporation
EVAPORATED LATEX - [rubber ind] latex *m* concentré par évaporation
~MILK - [dairy ind] lait *m* condensé
EVAPORATING - [phys] d'évaporation
~BASIN - [metall] capsule *f* à évaporation
~DISH - [ind chem] (shallow open vessel with spou

of heat-resisting porcelain, used for evaporating small quantities of liquid) vase *m* évaporateur, capsule *f* d'évaporation
EVAPORATING FLASK - [metall] matras *m* à évaporation
~LOSS - [phys] perte *f* par évaporation
~METAL - [metall] matériau *m* (ou matière) d'évaporation, métal *m* d'évaporation
~PAN - [metall] vase *m* d'évaporation
~PRESSURE - [phys] pression *f* d'évaporation
~SURFACE - [phys] surface *f* évaporatrice, surface d'évaporation
~TEMPERATURE - [phys] température *f* d'évaporation, température de vaporisation
EVAPORATION - [phys] (the change of a liquid to a vapour, at either normal or elevated temperatures, by the escape of molecules at the surface) évaporation *f*, vaporisation *f*
~APPARATUS - s. evaporator
~AT THE SURFACE - [phys] évaporation *f* de la surface, évaporation *f* superficielle
~CATHODE - [electron] (cathode in which the tungsten filament is covered with barium by the evaporation of the latter in a vacuum) cathode *f* formée par évaporation
~NUCLEON - [nucl] nucléon *m* d'évaporation
~OF GETTER - [electron] évaporation *f* du getter, vaporisation *f* du getter
~PLANT - [sugar ind] installation *f* d'évaporation
~POINT - [phys] point *m* de vaporisation
~POND - [ind chem] (body of sea-water impounded for natural evaporation) vase *m* évaporateur
~PSYCHROMETER - [instr] (instrument designed to measure humidity) psychromètre *m* (à évaporation)
~SHIELD - [mech] écran *m* (protecteur) contre l'évaporation
EVAPORATIVE CENTRIFUGE - [ind chem] centrifugeur *m* à évaporation
~COOLING - [phys] (reduction of temperature brought about by inducing evaporation by other means than by heating, the latent heat being furnished by the body itself, which is thus cooled) refroidissement *m* par évaporation
~ICE - [phys] (ice formed on or near an engine induction system by the abstraction of heat caused by the evaporation of the fuel) glace *f* d'évaporation
EVAPORATOR - [naut] bouilleur *m*
[instr] (apparatus for increasing the concentration of a solution) évaporateur *m*, distillateur *m*, vaporisateur *m*
~BOAT - [metall] creuset *m* d'évaporation (métallisation sous vide)
~BODY - [sugar ind] caisse *f* d'évaporation
~BOILER - [ind chem] évaporateur *m*, élément *m* évaporateur
~COIL - [ind chem] serpentin de détente
~CRUCIBLE - [metall] creuset *m* d'évaporation (métallisation sous vide)
~FOR CONCENTRATION - [ind chem] évaporateur *m* pour concentration
~FOR DISTILLATION - [ind chem] évaporateur *m* pour distillation
~GETTER PUMP - [electron] pompe à évaporation, pompe getter à évaporation
~POTENTIAL - [ind chem] potentiel *m* de l'évaporateur
EVAPORATOR TOWER - [oil ind] tour *f* d'évaporation
~VESSEL - s. evaporator body
~WITH FORCED JUICE CIRCULATION - [sugar ind] évaporateur *m* avec circulation de jus forcée
EVAPORIMETER - [instr] (instrument for measuring the rate of evaporation of water into the atmosphere) évaporimètre *m*, atmomètre *m*
EVASION - [gen] évasion *f*
~COEFFICIENT - [phys] coefficient *m* d'évaporation
EVEN, to - [gen] aplanir, égaliser, niveler
EVEN - [gen] uni, plan, égal, régulier
[math] pair
~DISTRIBUTION - [gen & chem] distribution *f* uniforme
~-EVEN NUCLEUS - [nucl] (a nucleus which contains an even number of neutrons and protons) noyau *m* pair-pair
~FOLIO - [print] feuillet *m* à nombre pair de pages
~FRACTURE - [geol] cassure *f* plane
~FUNCTION - [comput] fonction *f* paire
~-GRAINED - [gen] à grain régulier
~HARMONIC - [phys] harmonique *m* d'ordre pair
~LAP - [text] nappe *f* régulière
~LINE - [print] ligne *f* sans renfoncement
~-LINE INTERLACE - [telev] analyse *f* entrelacée à nombre pair de lignes
~NUMBER - [math] nombre *m* pair
~-ODD NUCLEUS - [nucl] (a nucleus which contains an even number of protons and an odd number of neutrons) noyau *m* pair-impair
~PAGE - [print] page *f* de gauche, page *f* portant un numéro pair
~RUNNING - [mech etc] marche *f* uniforme, stabilité *f* de marche
~SPEED - [mech] vitesse *f* uniforme
~THE SOIL, to - [constr] égaliser un terrain
EVENER - [text] cylindre *m* égalisateur
~COMB - [text] peigne *m* détacheur, peigne *m* égaliseur
~ROLLER - [text] cylindre *m* égalisateur, rouleau *m* d'égalisation, cylindre *m* de réglage
EVENNES - [gen] égalité *f*, régularité *f*
~TESTER - [text] (a dark polished table on which wool fibres and yarns are examined) vérificateur *m* de régularité du fil
EVENT - [gen] évènement *m*, cas *m*
EVERGREEN - [bot] à feuilles persistentes
EVERSET SHUTTER - [photo] (self-setting shutter) obturateur *m* toujours armé
EVERSION - [gen] éversion *f*
EVIDENCE - [gen & leg] évidence *f*, marque *f*, preuve *f*, signe *m*, témoignage *m*
EVOLUTE - [gen] évolué
[geom] développée *f*
EVOLUTION - [gen] évolution *f*, déroulement *m*, développement
[biol] (the gradual development of complex organs) évolution *f*
[math] extraction *f* de la racine
[chem] (of gases) dégagement *m*
~OF GAS - [chem] dégagement *m*
EVOLUTIONARY SUCCESS - [genet] succès *m* évolutif
EVOLVE, to - [gen] dégager, développer, évoluer
EVOLVENT - [math] développante *f*
EWE - [zool] brebis *f*
~-LAMB - [zool] agnelle *f*
EWE'S MILK - [agric] lait *m* de brebis

EWE'S MILK CHEESE - [agric] fromage *m* de brebis
EXACT - [gen] exact, précis
EXACTNESS - [gen] exactitude *f*, précision *f*
EXALTATION - [chem] (abnormal increase in the molecular refractivity of a compound) exaltation *f*
EXALTED CARRIER RECEPTION - [radio] réception *f* à amplification constante de l'onde porteuse
EXAMINATION - [gen] examen *m*, inspection *f*
~PULSE - [comput] impulsion *f* d'interrogation, impulsion *f* de consultation
EXAMINE, to - [gen] examiner
[naut] visiter, arraisonner
EXAMINER - [gen] inspecteur *m*, examinateur *m*
EXAMINING BOARD - [gen] commission *f* d'examen
EXAMPLE - [gen] exemple *m*, modèle *m*
EXANTHEMA - [med] (a type of eruption on the surface of the body) exanthème *m*
EXANTHEMATIC FEVER - [vet] (infectious disease affecting dogs) fièvre *f* exanthémateuse
EXCAVATE, to - [gen & constr work] excaver, creuser, fouiller
EXCAVATED SHAFT WITH STEEL SHELL - [constr] pile-colonne *f* havée dans le sol
EXCAVATION - [gen] excavation *f*, fouille *f*, creux *m*, fosse *f*
[constr] déblai *m*, excavation *f*
~BY DREDGING - [constr] havage *m*
~IN THE DRY - [constr] excavation *f* exécutée à sec
EXCAVATOR - [mech mach] (earth-moving machine) excavateur *m*, excavatrice *f*
EXCEED, to - [gen] excéder, dépasser
EXCELSIOR - [paper man] copeaux *mpl* d'emballage (US), laine *f* de bois, vrillons *mpl*
EXCENTRIC - [mech] (not concentric) excentrique
[mech] (a disk so arranged as to rotate about a point other than its geometrical centre, and thus giving reciprocating motion to a rod by way band embracing it) excentrique *m*
~BIT - [mech] trépan *m* décentré
EXCEPT - [gen] excepté
[comput] (also called 'AND NOT'; a logical operator) ET-NON
~GATE - [comput] (AND NOT-gate; a gate where a pulse is present on one line and a pulse is absent on the other line) circuit *m* ET-NON, circuit *m* INHIBITEUR
EXCESS - [gen] excès *m*, excédent *m*
~ACTIVATED SLUDGE - [hydr] boue *f* activée en excès
~AIR FACTOR - [phys] (the relation of the actual volume of air used in combustion to the theoretical value) excès *m* d'air, coefficient *m* d'excès d'air
~AND TOTAL METER - [instr] (meter registering the excess energy consumed when the power exceeds a certain value) compteur *m* à dépassement totalisateur
~CURRENT - [el] surintensité *f*
~HEAD - [hydr] surcharge *f* hydraulique
~LIME PROCESS - [ind chem] désacidification *f* par chaux en excès
~LOAD - [soil] accroissement *m* de charge, augmentation *f* de charge
~METER - [instr] meter registering the excess energy in a circuit when the power exceeds a specified value) compteur *m* à dépassement
~MODIFIED REFRACTIVE INDEX - [phys] (refractive modulus) module *m* de réfraction
EXCESS POWER - [mech] (in engines) réserve *f* de puissance
~RAPPING - [metall] (in fondry) excès *m* d'ébranlage
~REACTIVITY - [nucl] excès *m* de réactivité
~SLUDGE - [hydr] boue *f* en excès
~THREE CODE - [comput] (a coded decimal notation) code *m* plus trois
~VOLTAGE - [el] surtension *f*
~WATER PRESSURE - [soil] surpression *f* interstitielle, surpression *f* hydrostatique
~WEIGHT - [gen] poids *m* excédentaire, excédent *m* de poids
EXCESSIVE - [adj] excessif
~PRODUCTION - [gen] production *f* excédentaire
~REVERBERATION - [acoust] (echo effect in an auditorium) réverbération *f* excessive
EXCHANGE, to - [gen] échanger, changer, intervertir, permuter, remplacer
EXCHANGE - [gen] échange *m*, change *m*, troc *m*
[telecomm] central *m*
~ACIDITY - [soil] (acidity produced by the treatment of soil with a solution of a neutral salt) acidité *f* de échange
~AREA - [telecomm] (the area which is covered by an exchange) circonscription *f* téléphonique
~CABLE - [telecomm] câble *m* régional
~CAPACITY - [soil] (milliequivalents of ions which can be absorbed by I00 g of material at a specific pH) capacité *f* d'échange
~DATA, to- [comput] échanger des informations *f*, transférer des informations *f*
~ENERGY - [phys] (a quantum-mechanical effect) énergie *f* d'échange
~ENGINE - [mech] moteur *m* de remplacement
~FIELD - [comput] champ *m* d'échange
~FORCES - [phys] (the forces between two particles which exist because of the presence of a third particle) forces *fpl* d'échange
~OF BASE - [chem] (in a reaction) échange *m* de base
~OF HEAT - [phys] échange *m* de chaleur
~OF KINETIC ENERGY - [phys] (the exchange of energy in a body caused by the motion of the latter) échange *m* d'énergie cinétique
~REACTION - [chem] (process in which atoms of the same element in two different molecules or atoms in two different positions in the same molecule change places) réaction *f* de substitution
~TESTING POSITION - [telecomm] table *f* d'observation
EXCHANGEABLE - [soil] (of ions capable of replacement in the absorbing complex) échangeable
EXCHANGER - [chem] échangeur *m*
[mech] échangeur *m* de chaleur
EXCIPIENT - [med] (in a medicine, the inert ingredient holding the other ingredients together) excipient *m*
EXCISE DUTY - [comm] droit *m* de régie
EXCISION - [gen & med] excision *f*, coupure *f*
EXCITABILITY - [gen] excitabilité *f*
~CURVE - [el biol] courbe *f* d'excitabilité
EXCITABLE TISSUE - [zool] (a tissue responding to stimulation) tissu *m* excitable
EXCITANT - [el] (the electrolyte in a primary cell) électrolyte *m* d'une pile
EXCITATION - [gen] excitation *f*, stimulation *f*
[phys] (the addition of energy to a system) excita-

tion *f*
[el] (the production of magnetic flux through a magnetic circuit by an electrical current) excitation *f*, amorçage *m*
[opt] (e.g. of the retina) excitation *f*
[electron] (the input signal of any stage in electron tube circuits) excitation *f*
XCITATION ANODE - [electron] (anode used to maintain a cathode spot on a pool cathode in the absence of output through the main anode) anode *f* d'amorçage
(converters) anode *f* d'entretien
BAND - [electron] (range of values to which the energies of the electrons of an atom can be raised by excitation) bande *f* d'excitation
COIL - [el] enroulement *m* d'excitation, bobine *f* d'excitation
DRIVE - [radio] tension *f* d'excitation
ENERGY - [nucl] (the minimum energy which is necessary to carry a non-excited atom to a degree of excitation) énergie *f* d'excitation
FREQUENCY - [phys] fréquence *f* d'excitation
LAMP - [cin] lampe *f* excitatrice, lampe *f* phonique
LEVEL - [nucl] (the energy which characterizes a given degree of excitation) niveau *m* d'excitation
OF A GAS - [phys] (the change of structure of the atoms or molecules of a gas by the passage of an electron from one energy level to a higher one) excitation *f* d'un gaz
OF AN AERIAL - [radio] (by means of an oscillation generator) excitation *f* d'une antenne
OF AN ANTENNA - s. excitation of an aerial
POTENTIAL - [phys] (the difference of potential which is required to give an electron starting from rest the required energy so that it can excite by collision an atom of a molecule) potentiel *m* d'excitation
PROBABILITY - [phys] (probability which depends on the difference between the energy of the electron and the critical energy) probabilité *f* d'excitation
PURITY - [phys] (of light and colour) facteur *m* de pureté d'excitation
STATE - s.excitation level
WINDING - [el] (of an electrical machine) enroulement *m* inducteur, enroulement *m* d'excitation
XCITE, to - [gen etc] provoquer, exciter
XCITED ATOM - [nucl] (an atom which possesses more energy than is normally the case with an atom of that kind) atome *m* excité
ATOM DENSITY - [phys] (the number of excited atoms in a gas per unit volume) densité *f* d'atomes excités
FIELD LOUDSPEAKER - [el acoust] (loudspeaker in which the steady magnetic field is produced by an electromagnet) haut-parleur *m* à excitation indépendante
STATE - [phys] (condition of a particle which the energy level is higher than the normal) état *m* excité
XCITER - [radio] (also called active aerial; that portion of a directive aerial which is connected with the source of power) excitateur *m*
[el] (small current-producing machine) excitatrice *f*
CURRENT - s. excitation current
LAMP - [cin] (a lamp which makes it possible to reproduce optically recorded sounds) lampe *f* d'excitation, lampe *f* phonique
EXCITER SET - [el] (assembly of one or more exciters with an electric motor) groupe *m* d'excitation
EXCITING - [gen] passionnant
[el] d'excitation
~CIRCUIT - [el] (the circuit through which the current flows to excite an electric machine) circuit *m* d'excitation
~COIL - [el] (a coil on a field magnet) bobine *f* inductrice, bobine *f* de champ
~DYNAMO - [el] dynamo *f* d'excitation, dynamo *f* excitatrice
~VOLTAGE - [el] tension *f* d'excitation
~WINDING - [el] (the winding producing the magnetomotive force to set up the flux in an electric machine) enroulement *m* d'excitation
EXCITOMOTOR - [med] excito-moteur
EXCITON - [electron] (combination of an electron and a hole in a semiconductor in an excited state) exciton *m*
EXCITOR - [med] excitateur *m*
EXCITRON - [electron] (a single-anode pool rectifier) excitron *m*
EXCLUSION - [gen] exclusion *f*
~AREA - [nucl] (the area surrounding a nuclear plant) zone *f* interdite
~-ADDRESS CODE - [comput] code *m* d'adresse exclusif
EXCOCHLEATION - [med] curettage *m* d'une cavité
EXCORIATION - [med] (a superficial loss of skin) excoriation *f*, écorchure *f*
EXCREMENT - [physiol] excrément *m*
EXCRESCENCE - [gen] aspérité *f*, excroissance *f*
[med] (an outgrowth of tissue) excroissance *f*
EXCRETA - [zool] (waste substances eliminated by an organism) excréta *f*pl, excrétions *f*pl
EXCRETE, to - [physiol] excréter
EXCRETION MONITORING - [nucl] contrôle *m* de l'activité de l'excrétion
EXCURSION - [mech] (the distance covered) course *f*
[comput] excursion *f*
[astr] (of the sun) excursion *f*
~POWER - [nucl] accès *m* de surpuissance
EXECUTE, to - [gen] exécuter, accomplir
EXECUTION - [gen etc] exécution *f*
EXECUTIVE - [gen] exécutif, directeur
~OFFICER - [naut] commandant *m* en second
~PRODUCER - [cin] directeur *m* de production
~PROGRAM - [comput] (routine intended to control other routines) programme *m* directeur
~SIGNAL - [naut] signal *m* d'exécution
EXEDRA - [arch] exèdre *f*
EXEIRESIS - [med] (the evulsion of a part) exérèse *f*
EXEMPT, to - [gen] exempter, exonérer, affranchir
EXEMPTION - [gen] exemption *f*, exonération *f*, dispense *f*, affranchissement *m*
EXENTERATION - [med] exentération *f*, éviscération *f*
EXERCISE, to - [gen] exercer
EXERESIS - s. exeiresis
EXERT, to - [gen] employer, faire usage
[phys etc] (to apply) exercer (une action)
EXERTION - [gen] effort *m*, emploi *m*, usage *m*
EXFETATION - [med] grossesse *f* extra-utérine
EXFOLIATE, to - [gen] s'exfolier, s'effeuiller
EXFOLIATION - [bot] (the process of falling away in

flakes etc) exfoliation *f*
[constr] ébréchures *fpl*, effritements *mpl*
[metall] affouillement *m*, écaillement *m*
[med] desquamation *f*, exfoliation *f*
[geol] (the splitting of thin sheets of rock from exposed surfaces) exfoliation *f*, effeuillage *m*
EXFOLIATION JOINT - [geol] fissure *f* d'exfoliation
EXHALATION - [gen] exhalation *f*
EXHALE, to - [gen] exhaler
EXHALING VALVE - [impl] (in gas masks) clapet *m* d'expiration
EXHAUST, to - [gen] aspirer, épuiser, faire le vide, vider
EXHAUST - [mech] (the products of combustion in an internal-combustion engine) échappement *m*, évacuation *f*
~AIR - [mech] gaz *m* d'échappement, gaz *m* perdu
~ALARM - [auto] avertisseur *m* de fuite de gaz
~ANALYSER - [auto] analyseur *m* de gaz d'échappement
~BACK PRESSURE - [mech] (in internal-combustion engines) contre-pression *f* à l'échappement
~BOX - [auto] pôt *m* d'échappement, silencieux *m*
~COLLECTOR RING - [mech] (circular duct running round the cylinder heads of a radial engine to receive the exhaust gases) anneau *m* du collecteur de échappement, collecteur *m* annulaire d'échappement
~CONE - [mech] (in a jet engine) cône *m* de fuite
~DEFLECTING RING - [mech] anneau *m* déviateur de échappement
~DEFLECTOR - [auto] déflecteur *m* d'échappement
~DRAIN TAP - [mech] robinet *m* de purge
~FAN - [mech] (used in artificial draught systems) ventilateur *m* aspirant, exhausteur *m*
~FILTER - [mech] filtre *m* d'échappement
~FLAME DAMPER - [mech] anti-retour *m* de flammes
~GAS - s. exhaust air
~GAS ANALYSER - [mech] (apparatus for determining the composition of exhaust gases from an engine) analyseur *m* de gaz d'échappement
~GAS OUTLET - [auto] sortie *f* des gaz brûlés
~GASES - [auto] (the gaseous products of combustion of engines) gaz *mpl* d'échappement, gaz *mpl* brûlés
~HOSE - [mech] tuyau *m* d'échappement
~JACKETED CARBURETTOR - [auto] carburateur *m* à réchauffage par gaz d'échappement
~LINE - [mech] conduite *f* d'échappement, tuyau *m* d'échappement
[contr] conduite *f* de retour
~MANIFOLD - [mech] (a branched pipe designed to collect the exhaust from a multicylinder internal-combustion engine) tubulure *f* d'échappement, collecteur *m* d'échappement, pipe *f* d'échappement
~NOISE - [mech] (sound produced by the escape of exhaust gases from an internal-combustion engine) bruit *m* d'échappement
~NOZZLE - [mech] orifice *m* d'échappement, tuyère *f* de sortie
~PIPE - [mech] (pipe which receives and carries off the exhaust gases) tuyau d'échappement, tuyau *m* d'évacuation
~PIPING - [auto] tuyau *m* d'échappement
~PORT - [mech] (in a two-stroke engine) orifice *m* d'échappement, canal *m* d'échappement
~PRESSURE - [phys] état *m* du vide (dans un appareil etc)
EXHAUST PRESSURE - [mech] pression *f* de décharge
~PUMP - [min] pompe *f* d'épuisement
~REHEATER - [mech] (a device for burning additional fuel after the last stage of a gas turbine to give extra thrust) dispositif *m* de post-combustion
~RETARDER - [auto] ralentisseur *m* sur échappement
~RING - [mech] (in radial engines) collecteur *m* annulaire d'échappement, anneau *m* du collecteur de échappement
~SIDE - [mech] (in internal-combustion engines) côté *m* échappement
~SIDE TRAP - [mech] séparateur *m* côté d'échappement, séparateur *m* côté pression
~SILENCER - [auto] silencieux *m*, pôt *m* d'échappement
~STACK - [aero] collecteur *m* d'échappement
~STATOR-BLADES - [mech] (a set of fixed blades placed behind the last stage of a gas turbine to straighten out the exhaust flow) aubages *mpl* fixes, guides d'échappement
~STEAM - [mech] (steam discharged after use especially in a heat engine) vapeur *f* d'échappement
~STEAM PIPE - [mech] tube *m* d'échappement de vapeur
~STEM - [mech] tubulure *f* de pompage
~STREAM - [aero] courant *m* d'échappement
~STROKE - [mech] (in an Otto cycle engine, the piston stroke during which the exhaust gases are allowed to escape) course *f* d'échappement
~SYSTEM - [mech] système *m* d'échappement
[aero] circuit *m* des gaz de combustion (d'un réacteur)
~TEMPERATURE - [mech] (the temperature level of the exhaust gases escaping from an internal-combustion engine) température *f* des gaz d'échappement
~TRAIL - [aero & astronaut] trainée *f* d'échappement
~VALVE - [mech] (the valve used to release the products of combustion from the cylinder of an internal-combustion engine) soupape *f* d'échappement
~VALVE GUIDE - [mech] guide *m* de soupape d'échappement
~VAPOUR - [phys] vapeur *f* d'échappement
~VELOCITY - [aero & astronaut] vitesse *f* d'échappement
~WHISTLE - s. exhaust alarm
EXHAUSTED - [gen] vidé d'air, épuisé
~WELL - [min] puits *m* épuisé
EXHAUSTER - [mech] aspirateur *m*, exhauster *m*, ventilateur *m* aspirant
[mech] (a pump) pompe *f* à vide
~DEVICE - [mech] aspirateur *m*, dispositif *m* d'aspiration
EXHAUSTING FAN - [mech] (fan used to eliminate fumes etc) ventilateur *m* aspirateur
~PERIOD - s. exhaust stroke
EXHAUSTION - [mech] aspiration *f*, exhaustion *f*, pompage *m*, évacuation *f*, ventilation *f*
[gen] (of fuel, of a well, a photoelectric cell etc) épuisement *m*
~MACHINE - s. evacuating machine
EXHIBIT, to - [gen] exhiber, montrer, faire voir, produire
EXHIBIT - [gen] object *m* exposé, pièce *f* à conviction
EXIT - [gen] sortie *f*
[comput] sortie *f* (p.ex. d'une ligne)
~DOSE - [nucl] (the radiation dose on the surface of

a body by which a beam leaves it) dose *f* de sortie
EXIT HUB - [comput] jack *m* de sortie
~INSTRUCTION - [comput] (a device enabling the machine to carry on with the main program after carrying out reruns) instruction *f* de sortie
~LOSS - [hydr] perte *f* de charge à la sortie
~PORT - [opt] (the image of the aperture stop as viewed from the image) pupille *f* de sortie
~PORTAL - [radiat] (the exit through which the beam of a radiation leaves a patient's body) champ *m* de sortie
~PUPIL - s. exit port
~WINDOW - [opt] lucarne *f* de sortie
EXOCARDIAC - [med] (of the region outside the heart) exocardiaque
EXOCRINE - [zool] (of glands with secretion into a cavity of the body) exocrine
EXOERGIC - [phys] (associated with the evolution of energy) exoénergétique
EXOGAMY - [genet] (union of not closely related gametes) exogamie *f*
EXOGENOUS - [genet] exogène
~INCLUSION - [metall] inclusion *f* métallique étrangère
[geol] enclave *f* exogène
EXOMORPHISM - [geol] exomorphisme *m*
EXOPHTHALMOMETER - [instr] (medical instrument designed to measure the extent of a protrusion particularly of the eyeball) exophtalmomètre *m*
EXOPLASM - s. ectoplasm
EXOREIC DRAINAGE - [hydr] drainage *m* exoréique
EXOSMOSIS - [chem] (osmosis in an outward direction) exosmose *f*
EXOSPHERE - [astr] (a space region) exophère *f*
EXOTHERMAL REACTION - [metall] réaction *f* exothermique
EXOTHERMIC - [therm] (term used of a reaction or process in which heat is evolved) exothermique
~CHANGE - [phys] (physical or chemical change involving the development of heat) transformation *f* exothermique
~COMPOUND - [chem] (compound with an exothermic formation) composé *m* exothermique
~NUCLEAR DISINTEGRATION - [nucl] désintégration *f* nucléaire exothermique
~PROCESS - [chem] (a process in which heat is evolved) procédé *m* exothermique
~REACTION - [chem & phys] réaction *f* exothermique
EXOTIC - [metall] (of metals which are difficult to machine) difficile à travailler
~FUELS - [astronaut] propergols *mpl*
EXPAND, to - [gen] dilater, gonfler, développer, étaler, élargir
[mech] (by means of a machine tool) mandriner, dudgeonner
[math] développer
EXPANDED - [gen] allongé, dilaté, développé, expansé, déplyé
[plast ind] (of a material having a cellular structure and thus light in weight) expansé
~CENTRE PLAN DISPLAY - [radar] indicateur *m* à repère annulaire de zéro
~CONTRAST - [telev] (method of improving the contrast in the pictures) expansion *f* des contrastes
~METAL - [metall] (a metal network formed by stretching it to form a lattice) métal *m* déployé
EXPANDED PIPE - [plumb] tube *m* expansé
~RUBBER - [rubber ind] caoutchouc *m* mousse, caoutchouc *m* cellulaire
~RUBBER PACKING - [rubber ind] garniture *f* en caoutchouc mousse
~SWEEP - [electron] (of the electron beam or a cathode-ray tube) balayage *m* compensé
EXPANDER - [radio] étaleur *m* de bande
[mech] (for tubes etc) expanseur *m*, dudgeon *m*, mandrin *m*
[gas ind] bloc *m* mono-détendeur, détendeur *m*
~FOR DRILL PIPE - [oil ind] appareil *m* pour monter les protecteurs sur les tiges
EXPANDING - [gen] expansible, extensible
[mech] (of tools) de détente
~ANCHOR - [oil ind] ancre *f* extensible
~BIT - [mech] trépan *m* réglable
~BRAKE - [mech] frein *m* à expansion
~CLAY - [constr] argile *f* gonflante
~CONE - [mech] expandeur *m*
~DIE - [mach tool] outil *m* à expansion
~MANDREL - [mech] (a partially split turned shaft which can be expanded by a tapered plug) bouchon *m* expandeur, mandrin *m* extensible
~OIL RING - [mech] segment *m* racleur à expansion
~PIN - [mach tool] (of a reamer) pointeau *m* central, vis *f* centrale
~PULLEY - [mech] poulie *f* à diamètre variable
~REAMER - [mach tool] (a partially slit reamer which can be adjusted in its diameter by means of an internal plug) alésoir *m* expansible
~REED - [text] peigne *m* extensible
~REEL - [text] dévidoir *m* extensible
~STOPPER - [impl] obturateur *m* par expansion
EXPANSIBLE - [gen] expansible, dilatable
EXPANSION - [gen] dilatation *f*, accroissement *m*, expansion *f*, gonflement *m*
[math] développement *m*
[phys] (the process by which a mass of a substance undergoes an increase in size) dilatation *f*
[mech] (of a pipe) mandrinage *m*, dudgeonnage *m*
[telev] (contrast expansion) expansion *f* (des contrastes)
~BOLT - [mech] boulon *m* de scellement
~CHAMBER - [phys] (a vessel in which gases are allowed to expand) chambre *f* à détente, chambre *f* de Wilson
[packag] (in pressurized packing) chambre *f* d'expansion
~CIRCUIT BREAKER - [el] (circuit breaker operated by the cooling produced by the expansion of steam or gases) disjoncteur *m* à expansion
~COIL - [mech] serpentin *m* de détente
~COMB - [text] peigne *m* extensible
~COUPLING - [mech] embrayage *m* à segments extensibles
~CURVE - [phys] (on a diagram, the line showing the pressure of the working fluid during the expansion stroke) courbe *f* de détente
~DUE TO SHEAR - [constr] gonflement *m* provoqué par cisaillement
~ENGINE - [mech] (turbine or reciprocating engine in which compressed gas is expanded in refrigeration, e.g. in the Claude process) machine *f* à détente
~GAP - [railw] (rail joints) espace *m* de dilatation au joint

EXPANSION JOINT - [mech] (a joint between two parts to allow these parts to expand) joint *m* de dilatation
~LINE - s. expansion curve
~LOOP - [constr] boucle *f*, lyre *f* de dilatation
~ORBIT - [electron] (the last part of the electron path which terminates at the target and is outside the equilibrium orbit) orbite *f* d'expansion
~PORT - [mech] (of a hydraulic brake) orifice *m* de dilatation
~RATIO - [mech] (in internal combustion engines) taux *m* de dilatation, taux *m* de détente
[astronaut] (of a rocket; ratio between the nozzle outlet area and the nozzle throat area) taux *m* de détente
~ROLLER - [constr] (a roller supporting a girder or a bridge to allow for any movement resulting from expansion) rouleau *m* de dilatation
~STROKE - [mech] (in internal combusion engines) détente *f*, course *f* de détente
~TANK - [build] (the tank above the hot water system for the expansion of the water when heated) vase *m* d'expansion
[mech] (in cooling systems of engines) conservateur *m*, réservoir *m* auxiliaire d'huile
~WAVE - [phys] onde *f* de détente
EXPECTED VALUE - [phys] valeur *f* probable
EXPECT, to - [gen] attendre, prévoir
EXPEDIENT - [gen] expédient *m*, moyen *m*
[gen] expédient, convenable, utile, pratique
EXPEL, to - [gen] expulser, chasser, dégager
EXPELLED MOISTURE - [chem] (in carbonization) buée *mf*
EXPELLER - [mech] expulseur *m*
EXPENDABLE - [gen] consommable
EXPENDITURE - [gen] dépense *f*, coût *m*
EXPENSE - [gen] dépense *f*, charges *fpl*, frais mpl,
~ITEM - [comm] chef *m* de dépense
~RATIO - [comm] pourcentage *m* des frais généraux
EXPENSES OF COLLECTION - [comm] frais *mpl* d'encaissement
EXPERIENCE, to - [gen] expérimenter, essayer, éprouver, rencontrer (une difficulté), observer, constater
EXPERIENCE - [gen] expérience *f*, pratique *f*
EXPERIMENT, to - [gen] expérimenter, faire une expérience
EXPERIMENT - [gen] expérience *f*, épreuve *f*, essai *m*
EXPERIMENTAL - [gen] (of machines, structures, etc) expérimental
~ERROR - [meas] erreur *f* de mesure
~HOLE - [nucl] (hole through a reactor shield) canal *m* expérimental
~LOOP - [nucl] boucle *f*, circuit *m* expérimental
~MEAN PITCH - [aero] (the axial travel of a propeller during one revolution when it is not developing thrust) pas *m* efficace (d'une hélice)
~MEAN PITCH DIAMETER RATIO - [aero] pas *m* efficace relatif
~MODEL - [aero etc] modèle *m* expérimental, prototype *m*
~REACTOR - [nucl] (a reactor used for experiments) réacteur *m* expérimental
~TANK - [naut] bassin *m* des carènes
EXPERIMENTATION - [gen] expérimentation *f*
EXPERT - [gen] spécialiste *m*, expert *m*
EXPIRATION - [gen & med] expiration *f*
EXPIRE, to - [gen] expirer
[comm] expirer, cesser
EXPIRED - [gen & comm] expiré, périmé
EXPLETIVE - [constr] (stone used as a filling for a cavity) remplissage *m*, matériau *m* de remplissage
EXPLICIT ADDRESS - [comput] adresse *f* explicite, adresse *f* vraie
~FUNCTION - [comput] fonction *f* explicite
EXPLODE, to - [gen & explos] exploser, faire explosion, éclater, faire éclater, faire exploser
EXPLODED - [gen] explosé, éclaté
[draw] (of a drawing etc) éclaté
~VIEW - [draw] (in mechanical drawing, a method of showing all the parts of an assembly individually, in perspective and in such relative positions that they present a general appearance of having been ejected from the assembly in the sequence in which they are arranged in it when fitted) vue *f* éclatée
EXPLODER - [el] (appliance used to fire electrically the charges of explosives) exploseur *m*
EXPLOIT, to - [gen] exploiter, mettre en exploitation, exécuter
EXPLOITABLE - [gen] exploitable, utilisable
EXPLOITATION - [gen] exploitation *f*
EXPLORATION - [gen] exploration *f*, reconnaissance *f*, recherche *f*
[telev & radar] balayage *m*, exploration *f*
~WORK - [gen & geol] travaux *mpl* de recherche
EXPLORATIVE - [gen] exploratif
~DRILLING - [oil ind] sondage *m* d'exploration, sondage *m* de recherche
EXPLORATORY - [adj] préliminaire
~WELL - [oil & gas ind] sondage *m* d'exploration
EXPLORER - [gen] explorateur *m*, instrument *m* de recherche
EXPLORING COIL - [el] (small coil which is used to measure the flux in a magnetic field) bobine *f* exploratrice, bobine *f* d'exploration
EXPLOSIMETER - [instr] explosimètre *m*
EXPLOSION - [chem] explosion *f*, déflagration *f*, éclatement *m*
~BOMB - [phys] (used in measuring the specific heat at constant volume for gas at very low temperature) bombe *f* calorimétrique
~BURETTE - [instr] (an alternative term for eudiometer; a long glass tube closed at one end and graduated volumetrically, having platinum wires fused through it, so that a spark can be passed through the contained gas or gases) eudiomètre *m*
~DOOR - [metall] porte *f* de sécurité, porte *f* d'explosion
~ENGINE - [mech] (obsolete for internal-combustion engine) moteur *m* à explosion
~LIMIT - [phys] (inflammability limit) limite *f* d'inflammabilité
~METHOD - [meas] méthode *f* de la bombe calorimétrique
~MOTOR - s. explosion engine
~POINT - [chem] point *m* d'explosion, stade *m* explosif
~POT - [el] (metal container for the contacts of an oil circuit breaker) chambre *f* d'explosion
~-PROOF - [el] (also called flame-proof; an electrical machine so designed that explosion of gas inside will not ignite inflammable gas outside) antidéflagrant
~STROKE - [mech] (in an Otto cycle engine, the

piston stroke in which the charge is burnt and its energy transferred in part to the crankshaft) détente *f*, course *f* de détente
EXPLOSION TURBINE - [mech] turbine *f* à explosion
˜VENT - [mech] évent *m* d'explosion
˜WAVE - [phys] onde *f* explosive
EXPLOSIVE - [gen] explosif détonant
[chem] (a substance which undergoes a rapid chemical change on heating or detonation, with evolution of great heat and a large volume of gases) explosif
˜BOLT - [mech] (explosive charge or loaded spring, set off electrically and providing release of one part from another) boulon *m* explosif
˜CHAIN REACTION - [nucl] réaction *f* explosive en chaîne
˜DECOMPRESSION - [astronaut] décompression *f* explosive
˜FISSION - [nucl] (the type of fission which occurs in nuclear weapons) fission *f* explosive
˜FISSION REACTION - [nucl] réaction *f* de fission explosive
˜FORMING - [metall] formage *m* par explosion
˜GAS MIXTURE - [chem] mélange *m* de gaz explosif, mélange *m* tonnant
˜MATERIAL - [chem] (any material which can be detonated) matière *f* explosive
˜MIXTURE - [mech] (the proportional mixture of air and fuel vapour in an internal-combustion engine) mélange *m* détonant
˜OIL - [chem] nitroglycérine *f*
˜REACTION - [nucl] (a reaction occurring in a nuclear weapon) réaction *f* explosive
˜WAVE - s. explosion wave
EXPLOSIVENESS - [chem] explosibilité *f*
EXPONENT - [gen & math] exposant *m*
EXPONENTIAL - [gen & math] exponentiel
[math] exponentielle *f*
˜ABSORPTION - [phys] (the removal of particles from a beam according to a specified exponential relationship)
˜BAFFLE - [el acoust] haut-parleur *m* exponentiel
˜CURVE - [math] courbe *f* exponentielle
˜DAMPING FACTOR - [contr] facteur *m* exponentiel d'amortissement
˜DEATH PHASE - [genet] phase *f* de mort exponentielle
˜DECAY - [nucl] décroissance *f* exponentielle
˜DISTRIBUTION - [comput] distribution *f* exponentielle
˜EQUATION - [math] équation *f* exponentielle
˜EXPERIMENT - [nucl] (experiment carried out with subcritical assembly of fissionable material and moderator) expérience *f* exponentielle
˜FUNCTION - [math] fonction *f* exponentielle
˜HORN - [acoust] (horn with a sectional area doubled for every 45 cm of its length) pavillon *m* exponentiel
˜LAG - [contr] retard *m* exponentiel
˜METHOD - [nucl] (system of construction of a reactor-moderator assembly using a quantity of fissile material too small to sustain a chain reaction) méthode *f* exponentielle
˜REACTOR - [nucl] (assembly of fissionable material and moderator incapable of sustaining a chain reaction) réacteur *m* exponentiel
˜SERIES - [math] série *f* exponentielle
EXPONENTIAL THEOREM - [math] théorème *m* exponentiel
˜TRANSMISSION LINE - [electron] (a two-conductor transmission line with impedances varying exponentially with electrical length over the line) ligne *f* de transmission exponentielle
˜TUBE - [electron] tube *m* à pente variable
˜WELL - [nucl] puits *m* exponentiel
EXPORT GOLD POINT - [fin] point *m* de sortie de l'or
EXPOSE, to - [gen] exposer
[photo] (to let light into a sensitive emulsion) exposer
˜BY STRIPS, to - [photo] exposer par bandes *fpl* (successives)
EXPOSED - [gen] exposé
[photo] exposé
˜FILM - [photo] film *m* impressionné, pellicule *f* exposée
˜PIPING - [plumb] installation *f* visible
˜PROPELLER SHAFT - [auto] arbre *m* de transmission sans protection
˜STEM CORRECTION - [instr] (correction applied to a thermometer reading to compensate for errors introduced by the exposed part of the thermometer column) correction *f* de la colonne émergente
EXPOSING CHART - [photo] exposition, table *f* de exposition
EXPOSITION - [gen] exposition *f*, interprétation *f*
EXPOSURE - [gen] exposition *f*, étalage *m*, mise *f* à nu, dévoilement *m*
[photo] (the exposure of the film to the action of light) exposition *f*
[met] exposition *f* aux agents atmosphériques
[nucl] exposition *f*, radioexposition *f*, irradiation *f*
˜CHART - [nucl] (chart showing the exposure which is suitable for different thicknesses of a given material) table *f* de temps d'exposition
[photo] table *f* des temps de pose
˜COUNTER - [photo] compteur *m* de poses
˜CRACKING - [gen] craquelure *f* due aux agents atmosphériques
˜DOSE - [nucl] dose *f* d'exposition
˜DOSE RATE - [nucl] débit *m* d'exposition
˜INDEX - [photo] indice *m* de pose
˜INTERVAL - [photo] intervalle *m* des luminations
˜LOCK - [photo] verrouillage *m* du déclencheur
˜METER - [photo] (instrument designed to determine the photographic exposure to be given to a sensitive emulsion) posemètre *m*, photomètre *m*, actinomètre *m*
˜RATE - [nucl] débit *m* d'exposition
˜RATE METER - [nucl] débitmètre *m* d'exposition
˜SUIT - [nucl] vêtement *m* protecteur
˜TIME - [photo] temps *m* de pose, durée d'insolation, durée *f* d'exposition
˜TIME RANGE - [nucl] plage *f* de temps d'exposition
˜TIMER - [photo] (instrument designed to determine the right time of exposure) posemètre *m*
˜TO RADIATION - [radiat] irradiation *f*
EXPRESS, to - [gen] exprimer, énoncer
EXPRESS TRAIN - [railw] train *m* express, express *m*
EXPRESSION - [gen] expression *f*
[math] expression *f*
˜OF LENS - [med] expression *f* du cristallin
EXPROPRIATE, to - [gen & leg] exproprier
EXPROPRIATION - [gen] expropriation *f*
EXPULSION - [gen] expulsion *f*

EXPULSION ARRESTER - [el] parafoudre *m* à expulsion
~FUSE - [el] (a fuse, the fusible part of which can be expelled through an aperture in the container) coupe-circuit *m* à expulsion dirigée
~GAP - [el] (expulsion fuse connected in series with a gap) parafoudre *m* à expulsion
~PROTECTIVE GAP - s. expulsion gap
EXPUNGE, to - [gen] effacer, rayer, biffer
EXSICCATION - [geol] (the draining away of water) assèchement *m*, dessèchement *m*
EXSICCATOR - [ind chem] dessiccateur *m*, séchoir *m*
EXSICCOSIS - [med] déshydratation *f*
EXSTROPHY - [med] (congenital malformation) exstrophie *f*, extroversion *f*
EXTEND, to - [gen] étendre, allonger, développer, prolonger
[comm] proroger, reporter
[paint] (to add an ingredient to a mix in order to increase its covering power) diluer
EXTEND PORT - [contr] lumière *f* d'extension
EXTENDED DOUBLE ZEPPELIN AERIAL - [radio] antenne *f* dipôle prolongée
~DOUBLE ZEPPELIN ANTENNA - s. extended double zeppelin aerial
~PLAY RECORD - [el acoust] disque *m* de long durée (45 t/m)
EXTENDER - [paint] (an additive to a coating composition as an adulterant or to give some special quality) extendeur *m*, allongeant *m*, charge *f*, diluant *m*
EXTENDING WALL LAMP - [el] applique *f* extensible
EXTENSIBILITY - [gen] extensibilité *f*
EXTENSIBLE - [gen] extensible, allongeable
~MULTIPLE SWITCHBOARD - [telecomm] multiple *m* téléphonique
EXTENSION - [gen] élargissement *m*, développement *m*, extension *f*, allongement *m*, dilatation *f*, traction *f*
[comm] prorogation *f*
[mech] (of a pipe etc) rallonge *f*, raccord *m*
[text] (the amount, expressed as a percentage of the length of the specimen, by which a yarn can be stretched before breakage occurs) extension *f*
[photo] (a device for lengthening the distance between the lens and the sensitive emulsion) tirage *m*
[el] cordon *m* prolongateur, prolongateur *m*
~ADAPTER - [photo] adapteur *m* permettant d'augmenter le tirage
~ARM - [impl] bras *m* extensible, traverse *f* en porte-à-faux (téléphone)
~BAR - [mech] allonge *f*, embout *m*, rallonge *f*, rallonge *f* (de compas)
~BELLOWS - [photo] (attachment) soufflet *m* de rallonge
~CABLE - [el] câble *m* de rallonge, câble *m* prolongateur
~CARRIAGE - [text] chariot *m* renvideur
~CORD - [el & telecomm] câble prolongateur *m*, prolongateur *m*
~COVER - [print] couverture *f* débordante
~FLAP - [aero] (a flap which can be moved to increase the effective chord of an aerofoil) volet *m* type Fowler
~FLASH - [photo] éclair *m* simultané
~FORK - [text] tendeur *m*
~LADDER - [impl] échelle *f* extensible
~LEAD - s. extension cable
~LIGHT - [illum] baladeuse *f*
EXTENSION MAST - [radio] mât *m* télescopique (pour antenne)
~OF THE PAPER CARDS - [text] (in Jacquard looms) allongement *m* des cartons
~ORGAN - [mus] multiplex *m*
~ROD - [mech] rallonge *f*
~ROLLER - [agric] rouleau *m* articulé
~SPEED - [mech] vitesse *f* d'allongement
~TELEPHONE - [telecomm] poste *m* supplémentaire
~TUBE - [instr & photo] tube *m* télescopique
[chem] réfrigérant *m* à air
EXTENSIONAL MODE OF VIBRATION - [radio] (a mode of vibration) vibration *f* longitudinale
EXTENSIVE - [adj] étendu, vaste, ample
~SHOWER - [nucl] (a vast shower of cosmic particles) gerbe *f* géante
EXTENSOGRAPH - [instr] extensographe *m*
EXTENSOMETER - [instr] (instrument designed to measure small length changes of a piece undergoing strain) extensomètre *m*
EXTENSOR - [zool] (muscle straightening a part of the body by its contraction) extenseur *m*
EXTENT - [gen] étendue *f*, degré *m*, mesure *f*, importance *f*, grandeur *f*
EXTERIOR - [adj] extérieur, externe
~AERIAL - [radio] (an outdoor aerial) antenne *f* extérieure
~BALLISTICS - [astronaut] balistique *f* extérieure
~BRICKWORK - [th eng] maçonnerie *f* externe
~CAMERA - [cin] (camera for use in the open air) caméra *f* pour extérieurs
~FINISHES - [paint] (coatings designed to resist the effects of weather in outdoor locations) peintures *fpl* pour extérieurs
~SHOOTING - [cin] (shooting in the open air and without artificial lighting) prise *f* de vues en extérieur
EXTERNAL - [gen] externe, extérieur
~AILERON - [aero] (a separate aileron mounted clear of the wing surfaces) aileron *m* externe
~ANGLE - [geom] angle *m* externe
~BRAKE - [auto] frein *m* à serrage externe
~BREEDING RATIO - [nucl] rapport *m* de régénération externe
~BROACHING MACHINE - [mach tool] machine *f* à brocher extérieure
~BURNING - [astronaut] combustion *f* extérieure
~CHARACTERISTIC - [el] (of a generator; a curve showing the voltage at the terminals on load as a function of the current supplied by the machine under given conditions) caractéristique *f* externe ((d'une génératrice)
~CIRCUIT - [el] (circuit with current supplied by a generator etc) circuit *m* extérieur
[hydr] circuit *m* extérieur
~COMBUSTION - [mech] combustion *f* externe
~COMBUSTION ENGINE - [mech] moteur *m* à combustion externe
~CONTACT BASE - [electron] (electronic base with electrode contacts at the outside of the base) pied *m* à contacts extérieurs, culot *m* à contacts extérieurs
~DIAMETER - [gen & mech] diamètre *m* extérieur
~DRAIN - [hydr] drainage *m* externe
~FIELD - [el] (externally applied electric field) champ *m* extérieur
~FIRING - [th eng] (the heating of a boiler by a furna-

ce which is outside the shell of the boiler itself) chauffage *m* extérieur
EXTERNAL FOCUSSING - [opt] mise *f* au point externe
~FURNACE - [th eng] foyer *m* extérieur
~FURNACE BOILER - [th eng] chaudière *f* à foyer extérieur
~GRINDER - [mach tool] rectifieuse *f* d'extérieur
~GRINDING - [mech] rectification *f* extérieure
~HOT TEARS - [metall] (colloq) criques *fpl* (de retrait)
~IDLE TIME - [comput] (that period of time in which a machine could be used but is not operating) temps *m* mort
~LEAKAGE - [hydr] fuite *f* externe
~LOAD CIRCUIT - [el] (circuit exterior to an amplifier) circuit *m* de charge extérieur
~MASONRY - s. exterior brickwork
~MEMORY - [comput] (a store which is separated from the computer) mémoire *f* externe, mémoire *f* périphérique
~OPTICAL DENSITY - [opt] densité *f* optique externe
~PARASITE - [zool] ectoparasite *m*, parasite *m* externe
~PROGRAMME - [comput] (any programme which is introduced step by step into the machine) programme *m* externe
~QUENCHING - [nucl] (quenching by a momentary reduction of the applied potential difference) coupure *m* externe
~RESISTANCE - [el] (the resistance of the section of a circuit lying outside the source of current) résistance *f* externe
~SCREW VALVE - [mech] robinet *m* à vis extérieure
~STORAGE - s. external memory
~TEETH - [mech] denture *f* extérieure
~UPSET DRILL PIPE - [oil ind] tige *f* de forage à refoulement externe
EXTERNALIA - [med] organes *mpl* génitaux externes
EXTERNALLY HEATED ARC - [electron] (electric arc with a thermionic cathode heated by the arc current itself) arc *m* à chauffage externe
~HEATED OVEN - [th eng] four *m* à parois chauffantes
~HEATED RETORT - s. externally heated oven
EXTERPOLATE, to - [math] extrapoler
EXTINCT - [gen] éteint, détruit, effacé
EXTINCTION - [gen] extinction *f*, atténuation *f*
[phys] (of radiations) atténuation *f*
~COEFFICIENT - [chem] (measure of the absorption of light by a dissolved substance) coefficient *m* d'absorption
~OF LIGHT - [phys] (the absorption of light) absorption *f* de la lumière
~PHOTOMETER - [instr] photomètre *m* à disparition
~TIME - [el] durée *f* d'extinction
~VOLTAGE - [electron] (the anode voltage at which the discharge ceases when the supply voltage is decreasing) tension *f* d'extinction
EXTINGUISH, to - [gen] éteindre
EXTINGUISHER - [impl] (device designed to extinguish a fire) extincteur *m*
EXTIRPATE, to - [gen] extirper
EXTIRPATION - [gen] extirpation *f*, éradication *f*
EXTIRPATOR - [agric] extirpateur *m*
EXTRA - [gen] en sus, en plus, extra
[cin] (actor engaged for small parts) figurant *m*
EXTRA BOLD - [print] gras (caractère)
~CURRENT - [el] extra-courant *m*
~FINE - [mech] de precision
~FOVEAL - [opt] vision *f* indirecte
~-GALACTIC - [astr] (outside the galaxy) extragalactique
~-GALACTIC NEBULA - [astr] nébuleuse *f* extragalactique
~-HARD STEEL - [metall] acier *m* extra dur, acier *m* diamant
~-HEAVY - [gen] extra-lourd
~-HIGH TENSION - [el] très haute tension *f*
~-LENGTH DRILL - [mach tool] foret *m* série extralongue
~-LONG STAPLE - [text] mèche *f* longue
~-LOW VOLTAGE LIGHTING - [light] éclairage *m* à très basse tension
~-NUCLEAR - [nucl] (outside the nucleus) extranucléaire
~-NUCLEAR ELECTRON - [electron] (electron of the outer shells) électron *m* extranucléaire
~-NUCLEAR STRUCTURE - [electron] (the structure of an atom outside the nucleus) structure *f* extranucléaire
~-SLIM TAPER-SAW FILE - [impl] lime *f* tiers-point extra-mince
EXTRACT, to - [gen] extraire, épuiser
EXTRACT - [gen] extrait *m*, condensé *m*
[text] (recovered wool) extract *m*
[oil ind] (the portion of an unrefined petroleum product resulting from a solvent extraction process) extrait *m*
~FAN - s. extract ventilator
~INSTRUCTION - [comput] instruction *f* d'extraction, instruction *f* de substitution
~VENTILATION - [phys] (ventilation by means of an appliance on the top of a shaft in a building, so as to provoke an up-draught) ventilation *f* à aspiration
~VENTILATOR - [constr] ventilateur *m* à aspiration
EXTRACTA - [chem] extraits *mpl* officinaux
EXTRACTABLE - [gen] extractible
EXTRACTANT - [chem] (liquid used in an extraction process) solvant *m* pour extraction, extracteur *m*, agent *m* d'extraction
EXTRACTION - [chem] (the operation of dissolving some given constituent or constituents from a mixture by the use of a solvent which can dissolve such constituents only) extraction *f*
~APPARATUS - [ind chem] (glass device for extraction operations on a laboratory scale) appareil *m* d'extraction
~DRUM - [sugar ind] tambour *m* d'extraction
~FLASK - [ind chem] matras *m* d'extraction
~HEATER - [th eng] réchaffeur *m* par soutirage
~LIQUOR - [chem] (hydro-metallurgical solution used in concentration of mineral values) solvant *m* pour lixiviation
~OF A CATARACT - [med] enlèvement *m* du cristallin
~OF A ROOT - [math] extraction *f* d'une racine
~OF WATER - [hydr] (drainage) drainage *m*, extraction *f* de l'eau
~PLANT - [ind chem] (chemical plant for extraction) installation *f* d'extraction
~PUMP - [impl] pompe *f* d'extraction
~THIMBLE - [ind chem] (thimble-shaped vessel for extraction operations) cartouche *f* d'extraction
~TOWER - [ind chem] (tower with packing baffle pla-

tes for extraction processes) colonne *f* d'extraction
[sugar ind] tour *f* d'extraction
EXTRACTION TURBINE - [mech] (a turbine from which steam is tapped at a suitable stage) turbine *f* à soutirages, turbine *f* à condensation et prise de vapeur, turbine d'extraction
EXTRACTIVE AGENT - [chem] extracteur *m*, agent *m* d'extraction
EXTRACTOR - [gen & mech] extracteur *m*
[ind chem] (implement for extraction operations) extracteur *m*
[med] extracteur *m*
[mech] essoreuse *f*
~CHUCK - [mech] mandrin *m* extracteur
EXTRADOS - [constr] (the top surface of an arch) extrados, ligne *f* d'extrados *m*
EXTRADURAL - [anat] (outside the dura mater) en dehors de la dure mère
EXTRAGALACTIC - [astr] (outside the Galaxy) extragalactique
EXTRAMAGMATIC - [geol] extramagmatique
EXTRANEOUS - [gen] étranger, en surplus
~CUSHION GAS - [gas ind] (in gasholders) gaz-coussin *m* injecté
EXTRAORDINARY - [gen] extraordinaire, rare
~INDEX - [opt] (the index of refraction for the extraordinary ray in a double refracting crystal measured perpendicularly to the optic axis) indice *m* de réfraction pour le rayon extraordinaire
~RAY - [opt] rayon *m* extraordinaire
~WAVE - [phys] (el magn waves) onde *f* extraordinaire
EXTRAPOLATE, to - [math] extrapoler
EXTRAPOLATED BOUNDARY - [phys] limite *f* extrapolée
~VISCOSITY - [chem] viscosité *f* extrapolée
EXTRAPOLATION - [math] (the method of finding by calculation based on the known terms of a series) extrapolation *f*
~CHAMBER - s. extrapolation ionization chamber
~DISTANCE - [nucl] (the distance between the extrapolated boundary and the true boundary of a reactor) longueur *f* extrapolée
~IONIZATION CHAMBER - [nucl] (used to make a series of measurements, in which one factor is varied in suitable steps) chambre *f* d'ionisation à extrapolation
EXTRASYSTOLE - [med] extrasystole *f*
EXTRAUTERINE - [med] (occurring outside the uterus) extra-utérin
EXTRAVASATION - [med] (an abnormal escape of fluids) extravasion *f*, extravasation *f*
EXTREME - [gen] extrême
~BREADTH - [shipbuild] largeur *f* extrême
~LONG SHOT - [photo & telev] plan *m* de grand ensemble
~PRESSURE - [oil ind] extrême pression *f*
EXTREMELY-HIGH FREQUENCY - [radio] (term used for frequencies between 30,000 and 300,000 Mc/s (millimetric waves) fréquences *fpl* extrêmement élevées, ondes *fpl* millimétriques
EXTREMITY - [gen] extrémité *f*
EXTRICATION - [chem] libération (d'un gaz)
EXTRINSIC - [zool] (muscles running from the trunk to the girdle) extrinsèque
~PROPERTY - [electron] (the property of a semi-conductor in relation to imperfection in the crystal) propriété *f* extrinsèque
EXTRINSIC SEMI-CONDUCTOR - [electron] semi-conducteur *m* extrinsèque
EXTRORSE - [gen & zool] (bent outwards) extrorse
EXTROVERSION - [med] extroversion *f*, exstrophie *f*
EXTRUDE, to - [metall & plast ind] (to form bars etc by forcing material in plastic state through a die) extruder, boudiner, tréfiler, expulser, chasser, refouler
EXTRUDED - [metall & plast ind] extrudé, boudiné, refoulé
~BAR - [metall] (bar or rod of special profile, produced by forcing metal in a plastic state through a die of the required form) barre *f* extrudée
~ELECTRODE - [electron] électrode *f* boudinée
~FLANGE - [mech] bride *f* extrudée
~GOODS - [metall & plast ind] articles *mpl* boudinés, articles *mpl* extrudés
~HOSE - [rubber ind] tube *m* extrudé
~RUBBER SECTION - [rubber ind] profilé *m* de caoutchouc boudiné
~RUBBER THREAD - [rubber ind] fil *m* de caoutchouc boudiné
~SECTION - s. extruded bar
~SHAPE - [metall] profil *m* extrudé
~SHEET - [plast ind] feuille *f* extrudée, plaque *f* extrudée
~TUBE - [plast ind] tube *m* extrudé
EXTRUDER - [mech mach] extrudeuse *f*, boudineuse *f*
~HEAD - [mech mach] tête *f* de boudineuse
EXTRUDING MACHINE - s. extruder
~PRESS - [mech mach] presse *f* à refouler (métal), presse *f* à extrusion (plastiques)
EXTRUSION - [metall] (process of forming bars and the like by forcing metal in a plastic state through a die which determines the section of the finished product) extrusion *f*, refoulement *m*
[plast ind] boudinage *m*, extrusion *f*
~BILLET - [metall] ébauche *m* pour presse à filer
~DIE - [plast ind] filière *f* de boudineuse
[metall] filière *f* de refoulement, tuyère *f* sous pression
~HEAD - [plast] tête *f* d'extrusion, filière *f*
~MOULDING - [plast ind] moulage *m* par extrusion
~PRESS - s.extruding press
~SPINNERET - [plast ind] filière *f* multiple
EXTRUSIVE - [gen] extrusif
~ROCKS - [geol] (rocks which form on the surface of the ground) roches *fpl* extrusives
EXTUMESCENCE - [med] tuméfaction *f*, enflure *f*
EXUDATE - [gen] exsudat *m*
EXUDATION - [gen] exsudation *f*
[med] (the exuding of fluid from blood vessels into the tissue, caused by inflammation) exsudation *f*, suintement *m*
EXUDE, to - [gen] exsuder, se dégager, suinter
EX WORKS - [comm] départ *m* usine
EYE - [anat] œil *m*
[mech] (a loop at the end of a wire) œillet *m*, boucle *f*
[mech] anneau *m*
[arch] œil *m*
[bot] œil *m*
[print] œil *m* (d'un caractère)
[text] chas *m*, œillet *m* de lisse, œilleton *m* de navette
~BASE - [opt] distance *f* interpupillaire, écartement

m des yeux
EYE-BATH - [med] œillière *f*, bain *m* d'œil, gondole *f*
~-BOLT - [mech] (a bolt having a ring or loop instead of an ordinary head for lifting purposes or for the attachment of another part, e.g. a clevis) boulon *m* à œil, piton *m*
[naut] piton *m* à boucle
~CAP - [opt] lunettes *fpl* de sécurité, lunettes *fpl* de protection
~DOCTOR - [med] oculiste *m*
~DROPPER - [impl] compte-gouttes *m*
~-END - s. clevis
~GLASSES - [opt] lunettes *fpl*
~GNEISS - [geol] gneiss *m* œillé
~GUARD - s.eye cap
~IRRITANT - [gen] lacrymogène
~JOINT - [mech] biellette *f* à œil, articulation *f*
~JOINT LINK - [mech] articulation *f* annulaire
~-LASH - [anat] cil *m*
~LENS - [opt] loupe *f*, verre *f* d'œil (d'un oculaire), lentille *f* supérieure (d'un oculaire)
~LOTION - s.eyewash
~MEMORY - s. eye mindedness
~MINDEDNESS - [gen] mémoire *f* visuelle
~NEEDLE - [text] passette *f* de guidage
~OF STORM - [met] (small calm area in the centre of a tropical cyclone) œil *m* de la tempête
~OF THE WIND - [naut] lit *m* du vent
~PROTECTORS - [opt] (special spectacles for the protection of the eyes) lunettes *fpl* de protection, lunettes *fpl* de sécurité
~-SHAPED - [gen & med] oculiforme
~SHIEL - [nucl] (shield of protective material against ionizing radiation) œuilleton *m* protecteur
~-SPLICE - [naut etc] (a splice in a rope forming a loop at the end of such rope) épissure *f* œil
~STRUCTURE - [geol] structure *f* œillée
EYEBALL - [anat] globe *m* oculaire, globe *m* de l'œil
EYEBRIGHT - [bot] (a medicinal plant) eufraise *f*, euphrasie *f*, luminet *m*
EYEBROW - [anat] sourcil *m*
[naut] gouttière *f* (de hublot)
[auto] visière *f* (de phare)
EYECUP - [opt] (of eyepiece) bonnette *f* oculaire
EYEGLASS - [opt] monocle *m*, longue-vue *f*, loupe *f* *f* de visée
[opt] (a synonym for eyepiece) oculaire *m*, verre *m* d'œil (d'un oculaire)
EYEGROUND - [anat] fond *m* de l'œil
EYEHOLE - [anat] cavité *f* de l'œil, orbite *f* de l'œil
EYELET - [mech] (a hollow circular rivet-like part, set into flexible sheet material) œillet *m*, cosse *f*; (for cloth etc, a metal ring round a hole) œillet *m*
~CIRCULAR KNITTING MACHINE - [text] tricoteuse *f* circulaire eyelet, métier *m* à pélerines
~FACING - [shoe man] garant *m*
~FORMING MACHINE - [mech mach] machine *f* à faire les œillets
~MACHINE - [text] métier *m* pour tricot à jour
~STRAP - [mech etc] sous-œillet *m*, renfort *m*
EYELETTED PLATE - [text] plaque *f* à œillets
EYELETTING MACHINE - [shoe man] machine *f* à poser les œillets
EYELID - [anat] paupière *f*
[aero] (a curved pivoted structure designed to deflect the discharge from a jet engine) paupière *f*, déflecteur *m*
~REVERSER - [aero] (a pair of eyelids arranged to deflect the jet discharge in the reverse direction, so as to give a retarding thrust) déflecteurs *mpl* inverseurs
EYEPIECE - [opt] (the lens to which the eye is applied) oculaire *m*
[mech] regard *m*
[mech] (a part formed with an eye for attachment to another) articulation *f*
EYEPIECE HOLDER - [opt] porte-oculaire *m*
~MICROMETER - [opt] (in the focal plane of a micrometer eyepiece) micromètre *m* oculaire
EYESIGHT - [gen & opt] vue *f*, portée *f* de la vue
EYESPOT - [bot] (plant disease occurring in cereals) ocelle *m*
EYESTRAIN - [med] fatigue *f* oculaire
EYETOOTH - [dent] dent *f* œillère, dent *f* canine (supérieure)
EYEWASH - [med] collyre *m* liquide
EYEWATER - [med] humeur *f* vitrée, humeur *f* aqueuse

F

f - [symbol for activity coefficient] symbole *m* du coefficient de correction
[met] (symbol for fog in the Beaufort symbols) symbole *m* du brouillard
F - [phys] (symbol for Fahrenheit) symbole *m* de Fahrenheit
[chem] (symbol of Fluorine) symbole de Fluorine
[phys] (symbol for Faraday's constant) symbole de la constante de Faraday
FABRIC -[gen] (whatever is fabricated) édifice *m*, structure *f*
[text] (textile material] tissu *m*, étoffe *f*, texture *f*
[constr] (the walls, floors and roof of a building) structure *f*, fabrique *f*
[aero] (in small aircraft, the textile material which is used to cover wings and control surfaces) entoilage *m*
~BATCH - [text] balle *f*, ballot *m* de tissu fini
~BREAK - [rubber ind] (a break in the textile material of tyres) rupture *f* des plis (des cables), cassure *f* des plis
~COATING - [text] (the coating of a continuous film of synthetic resin on textile fabrics) revêtement *m* sur textile
~CUTTINGS - [text] déchets *mpl* de tissu
~ENVELOPE - [aero] enveloppe *f* de tissu
~EXPANDER ROLL - [text] cylindre *m* étireur (pour tissu)
~FACE - [text] endroit *m*, beau côté *m* d'une étoffe
~-FILLED MOULDING COMPOUND - [text] (a moulding compound in which disintegrated textile fabric is used as a filler) matière *f* chargée de fragments de tissu
~IMPRESSION - [rubber ind] (on tyres etc) empreinte *f* de textile
~IMPRINT - s. fabric impression
~INFEED - [text] entrée *f* du tissu, introduction *f* du tissu
~INSERT - [rubber ind] (in tyres etc) couche *f* de textile intermédiaire
~LAYER - [mech] armature *f* textile, couche *f* de tissu
~RESISTANCE - [text] résistance *f* du tissu
~SLEEVE - [ind chem] manchon *m* de toile (pour filtre)
~SLITTING MACHINE - [rubber ind] machine *f* à refendre le tissu
~SPEED - [text] vitesse *f* de passage des tissus
~SPOOLING MACHINE - [rubber ind] machine *f* à enrouler la toile
~SURFACE BELT - [rubber ind] courroie *f* avec toile superficielle
~TAKE-UP - [text] enroulement *m* du tissu
FABRIC WINDING - [text] enroulement *m* du tricot, enroulement *m* du tissu
FABRICATE, to - [gen] (to build, to manufacture, to construct) fabriquer, construire
(the operation of assembling various parts) monter
FABRICATED STEEL - [metall] acier *m* marchand
FABRICATION - [gen] fabrication *f*, construction *f*
FABRICATOR - [gen] constructeur *m* fabricateur *m*, faiseur *m*
FACADE - [constr] (the front of a building) façade *f*
~WALL - [constr] mur *m* de façade *f*, mur *m* de pignon *m*
FACE, to - [gen] faire face à, se présenter de face
[metall] dresser, surfacer, usiner, revêtir
[mach tool] (on a lathe) surfacer
FACE - [gen] (the outer and principal surface of an object) face *f*, pan *m*
[anat] face *f*, visage *m*
[geom] face *f*
[impl] (of a hammer) tête *f*, aire *f*
[aero] (of a propeller blade) face *f*
[mech] (of a ball bearing) face *f*
[constr] (of a building stone) pan *m*, parement *m*, façade
[mech] (the section of a gear tooth beyond the pitch line) face *f* d'une dent d'engrenage
[horol] (the dial) cadran *m*
[phys] (crystal face) face *f*, facette *f*
[mining] (the exposed surface of a mineral deposit) front *f* de taille, front *f*, front *f* d'abattage
[text] endroit *m*
[print] (of a type) œil *m* (du caractère)
[print] (printing surface) côté *m* image
[mech] (in a gear cutting machine) plateau *m*
[bookbind] découpure *f*
[metall] (in welding) surface *f*
[electron] (of a cathode-ray tube) fenêtre *f*, fond *m*
~ANGLE - [mech] (of a bevel gear) inclinaison *f* de la surface de contact
~ARCH - [constr] arc *m* de front, voûte *f* de front
~ASHLAR - [constr] (accurately squared masonry work for facades) parement *m* frontal en pierres de taille, revêtement *m* frontal en pierre appareillée
~BAR - [auto] barre *f* frontale
~BEND TEST - [metall] essai *m* de flexion de la face de soudure
~BRICK - [constr] (specially treated bricks used for facades) brique *f* de parement
~CAM - [mech] came *f* plane
~-CENTRED CUBIC LATTICE - [phys] (crystal structure consisting of a cubic cell having an atom at each corner and one in the centre of each face) réseau *m* cubique à faces centrées

FACE CHUCK - s. face plate
~CLOTH - [text] drap *m* fin
~CONVEYOR - [mining] convoyeur *m* de taille
~CUTTER - s. face lathe
~FLANGE - [mech] bride *f* plane
~GEAR - [mech] roue *f* à couronne dentée
~GRINDER - [mech] machine *f* à meule travaillant sur face, machine *f* à rectifier les surfaces, planes extérieures
~GRINDING WHEEL - [mech] meule *f* à affûter
~HAMMER - s. facing hammer
~-HARDEN, to - [metall] tremper, endurcir superficiellement, cémenter
~-HARDENING - [metall] (through heat treatment) durcissement *m* superficiel
~LATHE - [mach tool] (lathe for work on large disks, wheels etc) tour *m* en l'air (à plateau vertical), tour *m* à facer
~MARK - [carp] (a mark on the face of a piece of timber which has been used as basis for truing other surfaces) repère *m*, marque *f* de repère
~MASK - [gen] (a protective mask) masque *m* facial
~MILLING CUTTER - [mech] fraise *f* de face, fraise *f* à lames rapportées
~MIX - [constr] (cement and stone dust mixture giving a good imitation of stone) mélange *m* pour parement
~MOULD - [constr] (used for shaping the face of stone, wood etc) moule *m*
~OF A PLANE - [mech] semelle *f* d'un rabot
~OF A RIVET HOLE - [mech] paroi *f* du trou (de rivet)
~OF A WELL - [mining] fond *m* du sondage
~OF AN ARCH - [constr] front *f* d'arc
~OF COAL - [mining] stratification *f* de charbon, limet *m*
~OF STOPE - [mining] front *m* de taille
~OF TRANSFER HAMMER - [text] (in the weft supply) tête *f* de marteau
~OF WORKING - [mining] front *m* d'exploitation
~ON - [mining] front *m* d'attaque parallèle aux limets
~-PIECE - [rubber ind] (of protective masks) couvre-face *f* (de masque à gaz)
~PLATE - [mach tool] (in a lathe, a circular plate screwed to fit the mandrel nose, on which a workpiece can be fixed for machining) plateau *m* (de tour etc)
[electron] (of a cathode-ray tube; the large transparent end of the envelope) fenêtre *f*, plaque *f* de couverture
~-PLATE BREAKER CONTROLLER - [el] (face-plate starter with a separate contactor for breaking the circuit) combinateur *m* à interrupteur et plaque à contacts
~-PLATE CONTROLLER - s. face plate starter
~-PLATE JAW - [mech] (in a machine tool) poupée *f* à griffes
~-PLATE STARTER - [el] (starter for electric motor with a contact lever moving over a number of contacts) combinateur *m* à plaque à contacts
~PROP - [mining] étai *m* posé à front
~SEAL - [mech] joint *m* à anneaux glissants
~SHIELD - [metall] masque *m* de soudeur
~SHOVEL - [metall] drague *f* à godet poussant
~SIDE - [carp] (that face of a piece of timber which bears the face mark) côté *m* du repère

FACE SPANNER - [impl] clef *f* à griffes sur le côté
~STITCH - [text] maille *f* unie endroit
~VALUE - [gen & comm] valeur *f* nominale
~WALL - [constr] (front wall) mur *m* de revêtement
[a retaining wall] mur *m* de talus, mur *m* de soutènement
~WEFT - [text] trame *f* d'endroit, fil *m* de dessus
~WEFT SHUTTLE - [text] navette *f* de trame d'endroit
~WHEEL - [mech] roue *f* de champ
[mach tools] meule *f* travaillant sur face, meule *f* lapidaire
~WIDTH - [mech] (the width of a gear tooth) largeur *f* de la dent
~-WORK - [constr] façade *f*
~YARN - [text] fil *m* de fond (pour molletonné)
FACER - [tool] outil *m* à facer
FACET, to - [gen] (the cutting of precious stones) facetter
FACET - [gen] (the flat surface of a precious stone) facette *f*
[arch] (of a column) côte *f* (de colonne), listel *m*
~REFLECTOR - [light] réflecteur *m* à facettes (planes)
FACETTE - [arch] (projecting flat surface) listel *m*
FACIA - [arch] (projection from the face of a member) fasce *f*, bandelette *f*
FACIAL - [gen] facial
~ANGLE - [anat] angle *m* facial
~PANEL - s. fascia
FACIES - [geol] (the lithological and faunal characters of a sediment) facies *f*
FACILITATION - [gen] facilitation *f*
[el biol] (brief rise of excitability after a response or after a series of sub-treshold stimuli) facilitation *f*
FACILITIES - [gen] facilités *f*pl
FACILITY - [gen] facilité *f*
FACING - [gen] mouvement *m* de face, placage *m*, surfaçage *m*
[text] revers *m*, parement *m*
[constr] parement *m*, revêtement *m*, perré *m* (de tranchée)
[mech] (turning a flat face) fraisage *m* de face, surfaçage *m*
[metall] (the facing of a core box etc) poncif *m*
[auto] (of a clutch) garniture *f*
[metall] (in foundry work) noir *m* de fonderie
~ARM - [mech] porte-outil *m* à facer
~BAR - [text] (in plain net looms) barre *f* de support, lisière *f* de support
~BOND - [constr] (bond consisting of stretchers) assise *f* de panneresse
~BRICK - [constr] brique *f* de parement
~GAUGE - [instr] (instrument designed to measure the total head of a stream of fluid) tube *m* piézométrique
~HAMMER - [impl] (with a flat peen) massette *f*
~HEAD - [mach tool] poupée *f* porte-outil
~LATHE - [mach tool] tour *m* en l'air, tour *m* à facer, tour *m* en l'air à plateau vertical
~LOOP - [text] point *m* de diminution, maille reportée
~MACHINE - [mach tool] fraise *f* de face
~POINT BAR - [railw] pédale *f* mécanique d'aiguille en pointe
~POINT LOCK - [railw] verrou *m* d'aiguille
~POINTS - [railw] aiguille *f* prise en pointe
~SAND - [metall] (moulding sand with some finely ground coal dust round the faces of a pattern) sa-

ble *m* à la houille
FACING THE ENGINE - [railw] (of a seat, direction of traffic) sens *m* de la marche
~TOOL - [tool] outil *m* à dresser
~-TOOL BLOCK - [mach tool] porte-outil *m* à dresser
~WORK - [constr] maçonnerie *f* en briques brutes (ou de parement)
FACIOPLASTY - [med] (facial plastic surgery) plastie *f* faciale
FACSIMILE - [gen] fac-similé *m*, copie figurée
~BROADCASTING STATION - [radio] (station transmitting pictures by radio link) poste *m* d'émission de phototélégraphie
~EQUIPMENT - [telecomm] appareillage *m* de phototélégraphie
~POSTING MACHINE - [comput] reporteuse *f*, machine *f* de report
~RECEIVER - [radio] récepteur *m* facsimilé
~SIGNAL - [telecomm] signal *m* d'image
~TRANSMITTER - s. facsimile broadcasting station
FACTIVE - [rubber ind] (an ingredient in rubber manufacture, produced by vulcanizing vegetable oils with sulphur or sulphur dioxide formerly used as a substitute for rubber itself) factice *m*
FACTIS - s. factice
FACTITIOUS - [gen] (artificial) artificiel, contrefait
FACTOR - [math] (a number by which a quantity is multiplied to obtain a certain modification of such a quantity) facteur *m*, diviseur *m*, coefficient *m*
[comm] agent *m*, courtier *m* de marchandises
~OF MERIT - [el] (the millimetric deflection on the scale of reflecting galvanometers) coefficient *m* de qualité
[radio] s. factor of quality
~OF QUALITY - [radio] (measure of the field strength at a specified zenith angle at one kilometre for I KW input) coefficient *m* de qualité
~OF SAFETY - [mech] (the figure by which a limit load is multiplied to give the working load) coefficient *m* de sécurité
~OF TEN - [math] puissance *f* de dix
~-PAIR - [genet] couple *f* de facteurs héréditaires
FACTORABLE - [math] factorisable
FACTORIAL - [math] factorielle *f*
~NOTATION - [math] notation *f* factorielle, représentation *f* factorielle
FACTORIZATION - [math] factorisation *f*
FACTORIZE, to - [math] factoriser, décomposer en facteurs
FACTORY - [gen] fabrique *f*, usine *f*, manufacture *f*
~ACCEPTANCE GAUGE - [mech] calibre *m* d'essai de réception
~CALIBRATION - [instr] (the first application of a scale division of manufactured instruments) calibrage *m* préliminaire
~CHIMNEY - [constr] cheminée *f* d'usine
~COST - [comm] prix *m* de fabrication
~FITTING - [el] (electric light fitting with lamps in strong glass globes) appareillage *m* d'éclairage (pour usines)
~LENGTH OF CABLES - [metall] longueur *f* de fabrication de câbles
~LIMIT - [gen] limite *f* (ou tolérance) de fabrication
~MADE - [gen] fait à la machine
~OVERHEADS - [comm] frais *mpl* généraux de fabrication
~PRICE - [comm] prix *m* de fabrication
FACTORY TESTED - [gen] essayé en usine
FACTUAL - [gen] effectif, réel
FACULA - [astr] (large bright area of the sun photosphere) facule *f*
FACULTATIVE - [gen] facultatif
FACULTY - [gen] faculté *f*
FADE, to - [gen] se faner, se flétrir
[cin & telev] (the gradual appearance or disappearance of an image) operer une fusion, enchaîner
FADE - [cin & telev] fusion *f*
[radio] (of sound) évanouissement *m*, fading *m*
~-DOWN - [radio] (to reduce to a lower level the output of a channel) abaissement *m* graduel du niveau
[telev] (gradual disappearance of an image towards the bottom of the screen) disparition *f* graduelle en bas
~-IN AND -OUT - [telev] apparition et disparition graduelle
~MARGIN - [comput] plage *f* de régulation
~-OVER - s. fade-in or fade-out
~-UP - [radio] (gradual increase to a higher level of the output of a channel) relèvement *m* graduel du niveau
[telev] (gradual appearance of an image from bottom to top) apparition *f* graduelle en haut
FADEOMETER - [instr] (apparatus to determine the fastness of colours to light) fadéomètre *m*
FADER - [el acoust] (a potentiometer designed to control the volume of sound) régleur *m* de niveau acoustique, atténuateur *m* variable
FADGE - [text] (a bale of wool of no specified weight) balle *f* de laine
FADING - [cin & telev] fondu *m*
[photo] (the weakening of the contrasts in a print) altération *f*
[paint] (gradual loss of intensity) décoloration *f*
[auto] (of the brakes) fading *m*
[radio] (diminution of a received signal due to changes in ionosphere reflection) évanoussement *m* des signaux, fading *m*
[text] décoloration *f*, fading *m*, perte *f* de couleur
~AMPLIFIER - [telev] amplificateur *m* régleur du niveau de sortie
~BACKGROUND - [photo] fond *m* perdu
~BY INTERFERENCE - [telecomm] évanouissement *m* par interférence
~HEXODE - [radio] hexode *f* antifading
~TEST - [paint] essai *m* fadéométrique
FADOMETER - s. fadeometer
FAECES - [physiol] (fecal substances) matières *fpl* fécales, excreta *mpl*
FAGGOT - [gen] fagot *m*, bourrée *f*, fascine *f*
[metall] (of scraps etc held together by a boxlike container made of four flat bars) paquet *m* de fer à souder
FAGGOTED IRON - [metall] (a wrought-iron bar obtained through heating a faggot to melting point) fer *m* corroyé
~IRON FURNACE - [metall] fourneau *m* d'affinage de ferraille
~METALS - [metall] métaux *mpl* misés
FAGGOTING - [metall] paquetage *m*
FAHLBAND - [min] (sulphide containing layer) fahlbande *f*
FAHLERZ - [min] (grey copper ore) fahlerz *m*
FAHRENHEIT - [phys] Fahrenheit

FAHRENHEIT DEGREE - [phys] (degree of the Fahrenheit temperature scale) degré *m* F
~SCALE - [phys] (a temperature scale in which 212° is the boiling point and 32° the freezing point of water at standard pressure) échelle *f* Fahrenheit
~TEMPERATURE - [phys] (temperature on the Fahrenheit°) température *f* Fahrenheit
~TEMPERATURE SCALE - [phys] (temperature scale in which the freezing point of water is defined as 32 degrees and the boiling point as 212 degrees) échelle *f* de température Fahrenheit
~THERMOMETER - [instr] thermomètre *m* Fahrenheit
FAIENCE - [pottery] (good quality glazed earthenware) Faience *f*, poterie *f* vernissée
FAIL, to - [gen] manquer, faire défaut
[comm] faire faillite
[mech] (the failing of an engine etc) rester en panne, flancher
[mech] (of a pump) se désamorcer
[soil] céder, s'affaisser, se déplacer
FAIL SAFE - [mech] (a term used of a structure in which an alternative path is provided for the load, so that failure does not cause a general chain reaction of collapse) dispositif *m* de protection, à autoprotection, à sûreté absolue
~-SAFE BRACE - [mech] (a brace designed to have fail-safe characteristics) renforcement *m* à l'épreuve d'avaries
~-SAFE CONTROL - [mech] (positive safety) sécurité *f* positive
~-SAFE WING - [aero] (a wing designed to have fail-safe characteristics) aile *f* à l'épreuve d'avaries
FAILED ELEMENT DETECTION - [nucl] équipement *m* d'avertissement et de localisation des ruptures de gaines
~ELEMENT INDICATOR - [nucl] signaleur *m* de ruptures de gaine
FAILING - [gen] défaillant, baissant
~LOAD - [soil] charge *f* de rupture, charge *f* ultime, charge *f* limite
FAILURE - [gen] manque *m*, défaut *m*
[comm] faillite *f*
[mech] defaillance *f*, avarie *f*
[rubber ind] (of tyres) crevaison *f*
[soil] rupture *f*, cassure *f*, bris *m*, rupture *f* d'équilibre
[mining] éboulement *m*
~BY HEAVE - [soil] rupture *f* par soulèvement
~BY PIPING - [soil] rupture *f* par renard
~BY SHEAR - [soil] rupture *f* par glissement, rupture *f* par cisaillement
~BY SINKING - [soil] rupture *f* par enfoncement
~BY TILTING - [soil] rupture *f* par rotation
~CRACK - [metall] fente *f*
~LINE - [soil] ligne *f* de rupture
~LOAD - s. failing load
~RATE - [comput] taux *m* de defaillance, taux *m* de détérioration, taux *m* d'avarie
~STRAIN - [phys] déformation *f* de rupture
~STRESS - [phys] tension *f* de rupture, contrainte *f* de rupture
~WEDGE - [soil] prisme *m* de glissement
FAINT, to - [med] s'évanouir
FAINT - [gen] faible
~RED - [gen] rouge *m* pâle
~-RED HEAT - [metall] (approximately 500 centigrades) chaleur *f* rouge (500°)
FAINT-RUN - [metall] (colloq; e.g. of a casting) mal-coulé
FAIR, to - [shipbuild] (the operation of making sure that the intersection lines of all planes are fair) caréner
[auto etc] profiler, caréner (la carrosserie)
FAIR - [comm] foire *f*, exposition *f*
[met] beau temps *m*
~-LEAD - [naut & aero] passage *m* libre (pour cordage)
~-WEATHER CUMULUS - [met] (small cumulus humilis clouds occurring during fine weather conditions) cumulus *m* de beau temps
~WIND - [naut] vent *m* favorable
FAIRED CURVE - [math] courbe *f* moyenne, courbe *f* médiane
FAIRING - [shipbuild & aero] (any structure used purely to reduce drag) carénage *m*, profilage *m*
FAIRWAY - [naut] chenal *m*, passage *m*
FAITHFUL REPRODUCTION - [acoust] (high quality sound reproduction) reproduction *f* fidèle
FAKE, to - [gen] truquer, altérer, adultérer
[naut] lover (le cordage)
FAKE - [gen] article *m* faux, falsification *f*
[naut] plet *m*, glène *f*
[metall] mastique *m* de fonte
~CEILING - [constr] plafond *m* à glissière
FAKES - [mining] havrits *mpl*
[geol] grès *m* micacé
FALCATE - [gen & anat] falqué, falciforme
FALCULA - [anat] faux *f* de cervelet
FALL, to - [gen] tomber, descendre, crouler, s'éffondrer
[light] se projeter
[met] (of temperature etc) baisser
[comm] (of a date) expirer
[instr] (of an instrument) descendre
[met] (of the wind) tomber, s'apaiser
[agric] (a tree) abattre
[geom] (of a curve) décroître
[astr] (of a star) filer
FALL - [gen] tombée *f*, chute *f*, descente *f*
[hydr] chute *f*, chute *f* d'eau, hauteur *f* de chute,
[comm & fin] baisse *f*
[geogr] (a waterfall) cascade *f*, chute *f*, saut *m*
[instr] abaissement *m*, baisse *f*
[agric] (of trees) abattage
[naut] (of a rope) garant *m*, courant *m*
[met] (of the temperature) abaissement *m*
[constr] (of a rammer) hauteur *f* de chute d'un mouton
[geol] pente *f*
[gen] (US for Autumn) automne *m*
[plumb] pente *f* (d'un tuyau)
[soil] éboulement *m* de terre
[mining] écrasée *f* (du plafond)
~APART, to - [soil] désintégrer, se désagréger
~-AWAY SECTION - [astronaut] partie *f* larguée
~BACK, to - [gen] tomber en arrière
~BACK CIRCUIT - [telecomm] (telephone circuit carrying a voice frequency telegraph system if the normal circuit fails) circuit *m* de secours
~BACK POSSIBILITY - [comput] (the possibility of proceeding with an operation on reserve machines) possibilité *f* de continuer l'operation
~CALM, to - [naut] calmir
~DANDELION - [bot] (a weed) pissenlit *m* d'automne

FALL HEAD - [hydr] hauteur *f* de chute, chute *f*
~IN, to - [constr] s'écrouler
~IN SPIN, to - [aero] descendre en vrille, tomber en vrille
~INTO DISUSE, to - [gen] tomber en désuétude
~OF CAGE - [mining] chute *f* de la cage
~OF ROOF - [mining] éboulement *m* du toit
~OFF, to - [gen] diminuer, ralentir
[naut] tomber sur le vent
~-OUT - [nucl] (radioactive dust falling back to the earth's after a nuclear explosion) retombées *f*pl radioactives
~OF POTENTIAL - [el] (a fault in an insulated conductor) chute *f* de potentiel (ou de tension)
~OF POTENTIAL TEST - [el] essai *m* de chute de potentiel
~PIPE - [constr] drain *m*, canal *m* de drainage
~ROPE - [mech] (cable ways) corde *f* de levage, câble *m* de levage
~WEBWORM - [zool] chenille *f* blanche
~-WIND - [met] (wind of foehn type blowing down a mountainside) vent *m* descendant
FALLEN-IN SHAFT - [mining] puits *m* éboulé
~TREE - [timber] arbre *m* chablis
~WOOL - [text] laine *f* morte, moraine *f*
FALLER - [text] (of a spinning machine) baguette *f*
~PIN - [text] aiguille *f* de barrette
~ROLLER - [text] rouleau *m* de compensation
~SHAFT - [text] arbre *m* des baguettes
~SPINNING FRAME - [text] métier *m* à filer à barrettes de gill's
FALLERS - [mining] taquets *m*pl à abaissement
FALLING - [gen] tombant, en baisse
~BACK - [chem] (decline in fermentative activity) ralentissement *m*
~BALL TESTING MACHINE - [metall] machine *f* à chute de bille
~BALL VISCOSIMETER - [instr] (instrument designed to measure viscosity on the principle of the falling velocity of a ball in a fluid) viscosimètre *m* à chute de bille
~-DART TEST - [metall] (impact resistance test in which a steel point or ball is allowed to fall on the sheet from a known height) essai *m* à bille tombante
~FILM - [chem] film *m* tombant
~-FLAP RELEASE - [photo] déclencheur *m* de volet
~FRONT - [photo] décentrement *m* vers le bas
~GRADIENT - [geol] pente *f*, inclinaison *f*
~LEAF - [aero] (an aerobatic manoeuvre consisting in a series of stalls and side-slips, resembling the descending motion of a dead leaf falling from a tree) feuille *f* morte
~OFF - [gen] (a deterioration of the value) baisse *f* de valeur
[photo] (in definition) diminution *f*, affaiblissement *m*
[photo] (in speed) atténuation *f* (de la sensibilité)
~-OFF BURNER - [electron] (used in the manufacture of electronic tubes) chalumeau *m* de coupage
~OUT OF STEP - [el] (when synchronism is destroyed) décrochage *m*
~RATE DRYING PERIOD - [phys] (that period in the later stages of drying in which the rate of moisture loss is decreasing) intervalle *m* de temps de séchage à vitesse décroissante
~SLUICE - [hydr] vanne *f* plongeante
~TIDE - [naut] marée *f* descendante

FALLOPIAN TUBE - [anat] oviducte *m*, trompe *f* utérine (ou de Fallope)
FALLOW, to - [agric] écroûter, jachérer
FALLOW - [agric] jachère *f*, friche *f*
~LAND - [agric] terre *f* en jachère, terre *f* en repos
FALLSTREAKS - [met] (descending trails of precipitation from the underside of a cloud, which do not reach the earth's surface) queues *f*pl de pluie
FALSE - [gen] faux, inexact
~ACACIA - [bot] faux acacia *m*, robinier *m*
~BARS - [th eng] faux barreaux *m*pl
~BODY - [chem] (any substance with an appaerently high viscosity which disappears under mechanical agitation) corps *m* apparent
~BOTTOM - [mech] (a removable bottom to facilitate cleaning operations) double fond *m*, faux fond *m*
[mining] faux bedrock *m*
~BRINELLING - [metall] détachement *m* de la couche
~CAP - [mining] chapeau *m* provisoire
~CIRRUS - [met] (cirrus-like cloud at the top of storm cloud) faux cirrus *m*
~CLEAVAGE - [geol] pseudo-clivage *m*
~CURVATURE - [electron] trace *f* fictive
~DECK - [constr] faux comble *m*
~DOOR - [constr] porte *f* fausse
~DOVETAIL - [mech] contre-queue *f* d'aronde
~DRAFT - [text] mauvais rentrage *m*
~ECHO - [radar] (false echo caused by objects which near the horizontal path of the beam) faux echo *m*
~ELLIPSE - [constr] (approximate ellipse consisting of circular arcs) fausse ellipse *f*
~EQUILIBRIUM - [mech] (a condition of apparent equilibrium in a system) équilibre *m* apparent
~FRONT - [cin] (background facades representing buildings etc) décor *m* de fond
~GRAIN - [sugar ind] (condition in which there are too few nuclei for rapis crystal formation in evaporated juice) faux grains *m*pl
~GRATE - [th eng] fausse grille *f*
~KEEL - [naut] fausse quille *f*
~LINE LOCK - [telev] décalage *m* de phase du signal de synchronisation de ligne
~LINING - [metall] faux garnissage *m*, fausse-chemise *f*
~RELATION - [mus] fausse relation *f*
~RIB - [aero] fausse nervure *f*
~ROOF - [constr] faux comble *m*
[mining] faux toit *m*
~SET - [mining] cadre *m* provisoire
~STULL - [mining] plateforme *f* mobile
~WORK - s. falsework
FALSET VOICE - [mus] (a man's voice imitating a woman's) fausset *m*
FALSEWORK - [constr] (any temporary work, like scaffolding etc) étaiment *m*, échafaudages *m*pl, coffrage *m*, caisson *m* de coulée, ferme *f* à contre-fiches
FALSIFICATION - [gen] falsification *f*, altération *f*
FALSIFY, to - [gen] falsifier, altérer
FALX - [anat] (sickle-like structure) faux *f*
FAMATINITE - [min] (natural copper antimony sulphide; orthorhombic) famatinite *f*
FAMILY - [gen] famille *f*
[chem] (a group characterized by similar chemical properties, e.g. valence, solubility of salts, etc) groupe *m*

FAMILY OF CHARACTERISTICS - s. family of curves
~OF CURVES - [mech] (a series of curves showing the relation between a pair of variables as modified by a set of values of a third, e.g. the output of a centrifugal pump against varying heads for a series of increasing rotor speeds) famille *f* de courbes
~OF DEPRESSIONS - [met] (series of depressions arising successively from the tail-ends of cold fronts) série *f* de dépressions
FAN, to - [gen] ventiler, souffler (le feu)
[agric] (the grain) vanner
[mining] ventiler
FAN - [mech] (rotary mechanical device for imparting motion to gases) ventilateur *m*
[auto] ventilateur *m*
[mech] aile *f*, pale *f*, hélice *f*
[el] éventail *m*
[agric] van *m*, cribleur *m*
~AERIAL - [radio] (unidirectional aerial with approximately elliptical cross sections of the major lobe) antenne *f* à éventail
~ANTENNA - s. fan aerial
~BAFFLE - [auto] déflecteur *m* d'air pour le ventilateur
~BARLEY - [agric] orge *f* éventail
~BELT - [auto] courroie *f* de ventilateur
~-BELT IDLER - [auto] tendeur *m* de ventilateur
~BLOWER - [mech] ventilateur *m*
~BRAKE - [mech] (a testing device) frein *m* à palettes, frein *m* dynamométrique à brassage d'air
~COOLED - [metall] refroidi par l'air
~COOLING - [gen & auto] refroidissement *m* par ventilateur
~-COWLING - s. fan shroud
~CUPOLA - [metall] cubilot *m* à aspiration
~DELTA - [geol] cône *m* de déjection
~DRIFT - [mining] (the intake duct of a ventilation fan) œillard *m*, galerie *f* de ventilateur
~FLYWHEEL - [auto] volant *m* formant ventilateur
~-FOLD - [geol] pli en éventail
~GLOMERATE - [geol] conglomérant *m* alluvionnaire
~HUB - [mech] moyeu *m* de ventilateur
~-IN UNIT - [nucl] (part of the 400 channel pulse height analyzer) aiguilleur-mélangeur *m*
~LIGHT - [constr] fenêtre *f* à soufflet *m*
~MARKER-BEACON - [radio] (radio marker beacon producing a vertical fan-pattern) radiobalise *f* en éventail
~OF SIDINGS - [railw] faisceau *m* de voies
~OUT, to - [gen] étaler en éventail
~-OUT SCREW - [mining] sondage *m* à la corde
[oil ind] forage *m* à la corde, forage *m* au câble
~PALM - [bot] (a fibre for textiles) talipot *m* de l'Inde
~ROTOR - [auto] rotor *m* de ventilateur
~SHAFT - [mech] arbre *m* du ventilateur
~SHAPED BAFFLE - [mech] baffle *m* à éventail, écran *m* à éventail
~-SHAPED FOLDS - [geol] (structure consisting of folds with axial planes converging like the ribs of a fan) plis *mpl* en éventail
~-SHAPED MARKINGS - [photo] (caused by static electrical charges on the negative) marques *fpl* en forme d'éventail
~-SHROUD - [aero] (casing surrounding a fan to restrict air flow to the blade disk area) déflecteur *m* d'air de ventilateur

FAN SPINDLE - [mech] arbre *m* du ventilateur
~STRAIGHTENER - [aero] (in a wind-tunnel, radial vanes designed to correct the flow from a fan by eliminating a tendency to rotary flow) redresseur *m* de ventilateur
~-TAIL DIE - [plast ind] (an extrusion die of divergent form) filière *f* plate
~TAILED BURNER - [metall] brûleur *m* circulaire à jet éventail
~-TALUS - [geol] cône *m* d'éboulis
~TRAINING - [agric] taille *f* en palmette
~TURBINE - [el] turbo-ventilateur *m*
~TYPE BAFFLE - s. fan shaped baffle
~-TYPE EXHAUSTER - [oil ind] (in gas extraction) extracteur *m* à palettes
~VAULTING - [arch] voûte *f* en éventail, voûte *f* normande
~WHEEL - [mech] (of a blower) tourniquet *m* de volant, volant *m* ventilateur
~WINDOW - s. fanlight
FANCY - [text] (card) volant *m*
~CLEANER ROLLER - [text] cylindre *m* nettoyeur
~CLOTH - [text] étoffe *f* façonnée
~HEALDS - [text] lames *fpl* de dessin, harnais *m* de dessin
~LIGHTING - [light & cin] (special lighting effects obtained by a special arrangement of lamps) effets *mpl* lumineux
~PASS - [text] remettage *m* faconné
~ROLLER - [text] volant *m* (carde)
~RULE - [print] filet *m* ondulé
~TWILL - [text] (a weave of various colours and designs) croisé *m* façonné, croisé *m* fantaisie
~TWIST - [text] fil *m* retors fantaisie
~TYPE - [print] caractères *mpl* ornés
~WEAVING - [text] tissage *m* des étoffes façonnées, tissage *m* au Jacquard
~YARN - [text] (a yarn produced for fancy, or decorative purposes) fils *mpl* fantaisie
FANFOLD PAPER - [comput] papier *m* replié en accordéon, papier *m* replié en zigzag
FANG - [zool] croc *m*
[mech] soie *f*
[mining] canard *m* d'aérage
~-BOLT - [mech] boulon *m* de scellement
FANGOTHERAPY - [med] fangothérapie *f*
FANION - [surv] fanon *m*
FANLIGHT - [constr] (sash, mostly semicircular, over a door) éventail *m*, imposte *f*, vasistas *m*
FANNED-BEAM AERIAL - [radar] (used for height finding) antenne *f* à faisceau à éventail
~-BEAM ANTENNA - s. fanned-beam aerial
FANNING - [metall] (the operation of easing the draught in a furnace) diminution *f* du vent
~BEAM - [radio] faisceau *m* en éventail
FANNY BOARD - [cin] (a flat board on rollers) chariot *m*, plancher *m* roulant
FANTAIL - [carp] queue *f* d'aronde, queue *f* d'hironde
[impl] (of a windmill) gouvernail *m*
~VAULT - s. fan vaulting
FAR EAST - [geogr] extrême orient *m*
~-END CROSSTALK - [radio] (occurring in a disturbed channel) télédiaphonie *f*
~-END CROSSTALK ATTENUATION - [radio] (the crosstalk attenuation corresponding to the far-end crosstalk) affaiblissement *m* télédiaphonique

FAR POINT OF THE EYE - [opt] (the point on which the eye is focussed when relaxed) point *m* éloigné
FARAD - [el] (unit of electrical capacity: the capacity which exhibits a potential difference of one volt between its plates when charged with one coulomb) farad *m*
FARADAY - [el] (the charge carried by an univalent gramme ion) faraday *m*
~CAGE - [el] (an electric screen) cage *f* de Faraday, écran *m* électrostatique
~DARK SPACE - [electron] (a non luminous region following the negative glow and in which the electrons acquire enough energy to produce a new luminous region) espace *m* sombre de Faraday
~DISK MACHINE - [el] (a rotating copper disk which cuts the flux between the poles of an electromagnet) machine *f* à plateau de Faraday
~SCREEN - s. Faraday cage
~SHIELD - s. Faraday cage
FARADAY'S CONSTANT - s. Faraday
~LAW - [el] (the fundamental law of electromagnetic induction) loi *f* de Faraday
~LAW OF ELECTROLYSIS - [chem] (the amounts of different substances liberated or deposited by a given quantity of electricity are proportional to the chemical equivalent weights of such substances) lois *fpl* de Faraday d'électrolyse
~LAW OF ELECTROMAGNETIC INDUCTION - [phys] loi *f* de Faraday d'induction électromagnetique
~LAW OF INDUCTION - s. Faraday's law
FARADIC - [el] faradique
~BRUSH - [med] pinceau *m* faradique
~CURRENT - [el biol] (currents used for medical purposes) courant *m* faradique
FARADISM - s. faradization
FARADIZATION - [med] (treatment by means of currents from an induction coil) faradisation *f*, faradisme *m*
FARADIZE, to - [med] (the operation of stimulating living tissues with faradic currents) faradiser
FARADOTHERAPY - s. faradization
FARE - [transp] (the price of transport) prix *m* de voyage
FAREWELL ROCK - [mining] roche *f* d'adieu
FARINACEOUS - [gen] farineux, farinacé
FARINOSE - [bot] (covered with white, easily removable short hair) farinaceous
FARM, to - [agric] cultiver, prendre à ferme
FARM - [agric] ferme *f*
~-BUILDINGS - [agric] dépendances *fpl* de la ferme
~-HOUSE - [agric] maison *f* de ferme, ferme *f*
~IMPLEMENT - [agric] outil *m* agricole
~-LAND - [agric] terrain *m* agricole
~TRACTOR - [agric] tracteur *m* agricole
~-YARD - [agric] cour *f* de ferme, bâtiments *mpl* d'exploitation
~-YARD MANURE GAS - [agric] gaz *m* de fumier
FARMING - [gen] agriculture *f*, exploitation *f* agricole
FARNESOL - [chem] (a perfumery agent) farnésol *m*
FARRIER - [gen] maréchal *m*
FARROW, to - [zool] (of a sow) cochonner, mettre bas
FARROW - [zool] cochonnée *f*, portée *f* de cochons
FASCIA - [arch] (broad flat surface) fasce *f*, bandelette *f*
[auto] (the instrument board) tableau *m* de bord, planche *f* de bord
[med] ligament *m*
FASCICLE - [bot] (a tuft of leaves) fascicule *m*, brassée *f*
FASCINE - [gen] (bundle of brushwood) fascine *f*
[hydr & constr] fascine *f*
~HEDGE - [constr] tune *f*
~-WORK - [constr] fascinage *m*
FASH - [metall] plaque *f* d'écoulement
FASHIONING MACHINE - [text] machine *f* à former
FAST - [gen] rapide
[gen] (of objects etc) ferme, fixe, solide
[dyes] (term used to indicate power of resistance of colours to the fading action of light or other influences) solide, résistant
[photo] (relating to the sensitivity of a film emulsion) à sensibilité élevée
[horol] en avance
~ACCELERATOR - [rubber ind] accélérateur *m* rapide
~ACTING RELAY - [telecomm] relais *m* à action rapide
~ACTIVATION CROSS SECTION - [nucl] section *f* efficace d'activation rapide
~AND LOOSE PULLEY - [mech] poulies *fpl* folle et fixe
~BREEDER - [nucl] réacteur *m* surrégénérateur à neutrons rapides
~CHARGED PARTICLE - [nucl] particule *f* chargée rapide
~CHOPPER - [electron] (high-speed operating chopper) interrupteur *m* de faisceau à action rapide
~COMB - [text] peigne *m* fixe
~COUPLING - [mech] (coupling permanently connecting two shafts) accouplement *m* rigide
~DIP - [geol] pente *f* rapide
~-DYED - [dyes] à grand teint
~DYESTUFF - [text] colorant *m* solide
~EFFECT - s. fast fission effect
~FERMENTATION - [brew ind] fermentation *f* rapide
~FINISH - [text] finissage *m* solide
~FISSION - [nucl] (fission caused by fast neutrons) fission *f* rapide
~FISSION EFFECT - [nucl] (change in reactivity in a thermal reactor due to fission caused by fast neutrons) facteur *m* de fission rapide
~FLUX - [nucl] flux *m* rapide
~FRAGMENT - [nucl] (in a thermal reactor, a fragment due to fast neutrons) fragment *m* rapide
~HEAD - [mech tool] (in a lathe etc, the fixed headstock) poupée *f* fixe
~LINE - [mining] câble *m* d'enroulement
~MILLING PIGMENT - [paint] (a pigment which can be rapidly dispersed in milling the paint mixture) pigment *m* à dispersion rapide
~MODE - [comput] (of magnetic tape) grande vitesse *f* de défilement
~MORDANT - [chem] mordant *m* rapide
~MORDANT DYESTUFF - [text] colorant *m* solide
~MOTION EFFECT - [cin] effet *m* de ralentissement de la camera
~MULTIPLICATION EFFECT - s. fast fission effect
~NEUTRON - [nucl] (neutron with an energy greater than 0.I Mev) neutron *m* rapide
~NEUTRON BREEDER REACTOR - s. fast neutron reactor
~NEUTRON FISSION - [nucl] fission *f* par neutrons rapides

FAST NEUTRON RANGE - [nucl] (the distance which a fast neutron will penetrate into a given substance before its kinetic energy falls below the level at which it can cause ionization) portée *f* de neutrons rapides
~NEUTRON REACTOR - [nucl] (a nuclear reactor in which moderation is small and fission is primarily caused by fast neutrons which have lost little of their original energy) réacteur *m* rapide
~NEUTRON SHIELD - [nucl] blindage *m* pour neutrons rapides
~-OPERATE RELAY - [contr] relais *m* à vitesse élevée
~PIGMENT - [chem] pigment *m* solide
~PLUTONIUM BREEDER - [nucl] réacteur *m* surrégénérateur au plutonium
~PULLEY - [mech] (pulley fixed to a shaft) poulie *f* fixe
~PULP - [paper man] pâte *f* maigre
~REACTOR - s. fast neutron reactor
~REGISTER - [comput] mémoire *f* rapide
~REGISTRATION - [meas] (excess registration in meters) avance *f*
~SHUTTER SPEED - [photo] grande vitesse *f* d'obturation
~-THERMAL COUPLED REACTOR - [nucl] réacteur *m* thermique-rapide
~TIME GAIN CONTROL - [contr] (used to reduce the gain) réglage *m* differentiel de puissance
~TOOL - [tool] (firmly fixed tool) outil *m* fixe
FASTEN, to - [gen] attacher, fixer, ancrer, s'agrafer, cheviller
[mech] (to lock) fermer
[naut] cheviller
~SEAT BELTS, to - [aero] (order given to occupants of an aircraft to secure safety belts in anticipation of a possible bad landing) attacher les ceintures
~WITH PEGS, to - [mech] cheviller
FASTENER - [gen] attache *f*, fermeture *f*, agrafe *f*,
[mech etc] (any device designed to secure parts together) bride *f*, collier *m* de serrage, dispositif *m* de fixation, fermoir *m*
[mech] (of a belt) agrafe *f* (pour courroie)
FASTENING - [gen] attache *f*, attachement *m*, liaison *f*, fixation *f*, armature *f* de charpente
[mech etc] dispositif *m* de fixation, fermoir *m*
~BELT - [mech] courroie *f* de fixation
~BOLT - [mech] boulon *m* de fixation
~CLIP - [auto] douille *f* d'accouplement
~HOOP - [text] arc *m* de fixation
~NAIL - [constr] clou d'attache *f*
~ROD - [mech & text] barre *f* de fixation
~SCREW - [mech] vis *f* de fixation
~SPIDER - [mech] croisillon-support *m* (d'une plaque chauffante)
FASTNESS - [dyes] solidité
~TO BLEACHING - [text] solidité *f* au blanchiment
~TO DETERGENTS - [text] solidité *f* aux détergents
~TO LIGHT - [dyes] résistance *f* à l'action de la lumière
[text] solidité *f* à la lumière
~TO PLEATING - [text] solidité *f* au plissage
~TO RAIN - [dyes] résistance *f* à l'action de la pluie
~TO SOLVENTS - [text] solidité *f* aux solvants
~TO WASHING - [dyes] résistance *f* au lavage
~TO WEAR - [text] solidité *f* au porter
FAT - [chem] graisse *f*, gras *m* (de viande)
[gen] gras, gros, adipeux
[soil] riche, fertile, gras
[zool] suif *m* (de mouton)
FAT CLAY - [geol] argile *f* grasse
~COAL - [min] (volatile matter containing coal) houille *f* grasse, charbon *m* bitumineux
~EDGE - [paint] (term applied to the collection of a disproportionately large amount of a heavy coating at the lowest edge or point of the object to which it is applied) surépaisseur *f* en bordure
~LIME - [constr] (lime obtained by burning pure limestone) chaux *f* grasse
~-LIQUORING - [leather ind] (the impregnation of leather with fats) graissage *m* (du cuir)
~OIL - s. fat turpentine
~SOIL - [agric] terre *f* grasse
~TRAP - [ind chem] dégraisseur *m*
~TURPENTINE - [paint] (thickened product obtained by blowing turpentine with air at high temperature, or by exposing it to the atmosphere for long periods, and used in decorating pottery) térébentine *f* soufflée
FATAL - [gen] fatal, mortel
~ACCIDENT - [gen] accident *m* mortel
FATALITY - s. fatal accident
FATHOM, to - [naut] sonder
FATHOM - [meas] (measure, usually of depth of water, equal to six feet) brasse *f* (1,828 m)
~TALE - [mining] paie *f* suivant le volume extrait
FATHOMETER - [instr] (electronic instrument designed to measure the depth of water) sonde *f* électronique
FATIGUE - [mech] (the deterioration of a material caused by repeated stresses) fatigue *f*
[electron] (the decrease of efficiency of a luminescent material during excitation) fatigue *f* d'une substance luminescente
~ALLOWANCE - [metall] effort *m* admissible
~CRACK - [mech] (a crack due to fatigue) fissure *f* par fatigue, rupture *f* due à la fatigue
~FRACTURE - [metall] rupture *f* par fatigue
~LIFE - [phys] vie *f*, longévité *f*, durée *f* de vie, durée de service
~LIMIT - [mech] (the point above which a material cannot withstand indefinite stress) limite *f* de fatigue
~LOAD METER - [instr] (a type of recording accelerometer) appareil *m* mesureur de la fatigue
~METER - s. fatigue load meter
~OF METALS - [mech] (the occurrence of the failure of metals under the stress) fatigue *f* des métaux
~OF THE CASING - [rubber ind] (in tyres) fatigue *f* de la carcasse
~OF THE MATERIAL - [mech] (the failure of a material under the renewed application of stress) fatigue *f* des matériaux
~POISON - [med] toxine *f* de la fatigue
~-PRONE - [gen] (liable to fatigue) sujet à la fatigue
~RATIO - [mech] (the ratio of the fatigue limit to the ultimate tensile stress) rapport *m* limite de fatigue
~STRENGTH - [phys] résistance *f* à la fatigue
~STRESS - [metall] sollicitation *f* de fatigue
~TEST - [mech] (the test which is made on a material to determine the range of stress to which it can be subjected) essai *m* de résistance à la fatigue
~TEST TANK - [aero] (a water tank in which a cabin or other parts of an aircraft structure can be sub-

merged and subjected to variations of pressure, to investigate fatigue phenomena) bassin *m* d'essai d'endurance
FATIGUE TESTING MACHINE - [rubber ind] machine *f* de fatigue
~UNDER FLEXING - [rubber ind] fatigue *f* de foulage, fatigue *f* par flexion
FATTEN, to - [gen] engraisser, gaver
[agric] (the soil) engraisser, enrichir
FATTENING - [gen & agric] engraissement *m*, engraissage
[paint] épaississement *m*
FATTY - [gen] graisseux, gras, adipeux
[chem] gras
~ACID - [chem] (naturally occurring monobasic organic acids) acide *m* gras
~ACID PITCH - [chem] (pitch-like residual material from the distillation of animal or vegetable fats and oils, or of fatty acids obtained from these) poix *f* d'acides gras
~ALCOHOLS - [chem] (high molecular weight alcohols from C_8 to C_{20} with applications as solvents) alcools *mpl* gras
~AMINES - [chem] (aliphatic amines in which the alkyl group of a monohydric alcohol replaces active hyxrogen) amines *fpl* alphatiques
~COMPOUND - [chem] composé *m* gras (ou aliphatique)
~DEGENERATION - [med] (degeneration of cell substance which is revealed by the appearance of droplets of fat) dégénérescence *f* graisseuse
~ESTERS - [chem] (fatty acids in which the alkyl group of a monohydric alcohol replaces active hydrogen) esters *mpl* d'acides gras
~HEART - [med] (fatty degeneration of the heart muscles) dégénérescence *f* graisseuse du coeur
~MATTERS - [ind chem] matières *fpl* grasses
~NITRILES - [chem] (organic cyanides with uses as plasticizers) nitriles *mpl* aliphatiques
FAUCES - [anat] gosier *m*, fosse *f* gutturale
FAUCET - [mech] (US term for a small valve) robinet *m*
FAULT - [gen] défaut *m*, imperfection *f*
[mech & el] défaut *m*, dérangement *m*, défaillance *f*
[geol] (fracture in the rock which has caused displacement) faille *f*, dislocation *f*
[min] (fracture causing displacements and movements of mineral veins, coal seams etc) paille *f*, crapaud
[metall] (in a casting) défaut *m* de coulage, grains *mpl*
~BENCH - [geol] gradin *m* de faille
~BRECCIA - [geol] (breccia type fragments of rock produced by faults) brèche *f* de faille
~CLIFF - [geol] escarpement *m* (ou rocher) de faille, côte *f*, ressaut *m* de faille
~COMPLAINT SERVICE - [telecomm] service *m* de dérangements
~CURRENT - [el] (current flowing from a conductor to earth owing to a fault) courant *m* de défaut
~DIP - [geol] inclinaison *f* de faille
~ELECTRODE CURRENT - [electron] (the peak current flowing through an electrode under fault conditions) courant *m* anormal d'électrode
~FINDING - [el etc] (location and diagnosis of defects, espec. of an electrical character) dépannage *m*, décèlement *m* (p. ex defuites)
FAULT FISSURE - [geol] paraclase *f*
~-LIABLE - [comput] sujet à des dérangements
~LINE - [geol] ligne *f* de faille
~-LINE SCARP - [geol] escarpement *m* de faille
~LOCALIZING BRIDGE - [instr] (bridge arrangement for the determination of the distance of a cable fault) pont *m* déceleur de défauts (ou de pertes)
~LOCATION COIL - [el] (a coil used to find a fault) chercheur *m*
~ON A LINE - [telecomm] dérangement *m* sur une ligne
~OUTCROP - [geol] affleurement *m* de faille
~OUTLIER - [geol] lambeau *m* de charriage
~-PIT - [geol] effondrement *m* circulaire
~PLANE - [geol] plan *m* de faille
~-POLISH - [geol] miroir *m* de glissement
~RESISTANCE - [el] (the resistance of a fault) résistance *f* de défaut
~SCARP - s. fault cliff
~STRIKE - [geol] direction *f* d'une faille
~SURFACE - s. fault plane
~TRACE - s. fault line
~WALL - [geol] lèvre *f* de faille
FAULTAGE - [geol] formation *f* des failles
FAULTED LINE - [comput] ligne *f* en dérangement
~STRUCTURE - [geol] terrain faillé
FAULTING - [geol] (the occurrence of faults in rocks dislocation *f* des couches
~SWITCHING - [telecomm] faux montage *m*
FAULTLESS - [gen] parfait, sans défaut
FAULTY - [gen] défectueux
[mech etc] (with a defect in a mechanism) défectueux, imparfait, mauvais
~EXPOSURE - [photo] exposition *f* ratée, raté *m*
~IGNITION - [mech] allumage *m* défectueux
~LINE - [telecomm] ligne *f* en dérangement
~LOADING - [transp] vice *m* de chargement
~MIXTURE - [mech] (in internal combustion engines) mélange *m* défectueux
~PACKING - [transp] vice *m* d'emballage, défectuosité *f* d'emballage
~SELECTION - [telecomm] fausse sélection, sélection *f* déformée
FAUNA - [zool] (the animals living in a particular region) faune *f*
FAURE PLATE - [el] (a storage-battery plate coated with a paste of lead oxides or salts which are subsequently converted into active material) plaque *f* faure
FAUSER AMMONIA PROCESS - [chem] (variant of the Casal process, using lower pressure and separation of ammonia by cooling through expansion) procédé *m* ammoniacal de Fauser
~PROCESSES - [chem] (processes for the production of nitric acid and the synthesis of ammonia, using high pressure and an catalyst) procédés *mpl* Fauser
FAVOURABLE - [gen] favorable, de faveur
~WIND - [naut] vent *m* favorable
FAVUS - [med] (an infectious skin disease) teigne *f* faveuse
FAWN - [zool] faon *m*
~-COLOURED - [text] brun-fauve
FAY, to - [gen] (to prepare part of a metal or wood surface so that it fits an adjoining part) affleurer
FAY - [gen] bouvet *m*, rabot
FAYALITE - [min] (a silicate of iron) fayalite *f*
FAYENCE - s. faience

FAYING FACE - s. faying surface
~SURFACE - [mech etc] (the surface of a structural member which is permanently held in contact with another such surface) surface *f* d'affleurement (ou de contact)
F-BAND - [opt] (the optical absorption bandcharacteristic of F-centres in alkali halide crystals) bande *f* F
F-CENTRE - [phys] (the simplest type of ocular centre in an alkali halide crystal) centre *m* F
fe - [met] (Beaufort symbol for wet fog) brouillard *m* humide
Fe - [chem] (the symbol for Iron) symbol du fer
FEASABILITY - [gen] praticabilité *f*, possibilité *f*
FEASIBLE - [gen] faisable, possible, réalisable
FEATHER, to - [aero] (to set the blades of a propeller to the minimum- drag position in case of engine failure) mettre en drapeau
[aero] (the operation of altering the angle of incidence) changer l'angle d'incidence (de pas)
FEATHER - [gen] plume *f*
[mech] (a long key let into a shaft to allow a wheel to move axially while being driven) languette *f*, clavette *f*, plat-coin *m*
[carp] (a think tongue on the edge of a board)coin *m* demi-rond, languette *f*
~-ALUM - [chem] alun *m* de plume
~-EDGE - [carp] (thin cresting on chairs etc)biseau *m*, chanfrein *m*
[metall] bavure *f*
[tools] (of the cutting edge) bavure *f*, morfil *m*
~-EDGE BRICK - [constr] (a brick resembling a compass brick, mainly used for arches) brique *f* chanfreinée
~-EDGE FILE - [impl] (type of file in which one edge has a thin flattened firm) lime *f* en losange
~-EDGED - [constr] (boarding) à recouvre-joints (voligeage)
~-EDGED FLAT - [metall] fers *mpl* à arêtes vives
~EDGED HALFROUND - [metall] demi-rond *m* irrégulier
~FILE - s. feather-edge file
~JOINT - [carp] assemblage *m* à languette rapportée (ou à rainure et languette)
~KEY - [mech] languette *f*, clavette *f* linguiforme
~ORE - [min] (a variety of Jamesonite, a natural sulphide of lead and antimony, which takes a plumose (feather-like) or acicular form) antimoine *m* sulfuré capillaire, hétéromorphite *f*
~PROOF - [text] étanche aux plumes, clos
~RULE - [nucl] (relation between energy and range of continuous beta radiation, determined empirically) loi *f* de Feather
~SHOT - [metall] cuivre *m* granulaire
~SPLINE - [mech] clavette *f*
~SPRING - [mech] ressort *m* de gâchette
~TWILL - [text] sergé *m* croise
~VALVE - [mech] soupape *f* d'échappement
FEATHERED - [gen] emplumé
[metall] granulaire
[mech] nervuré, à nervures, cannelé
~RING - [mech] bague *f* nervurée
~TIN - [metall] étain *m* en larmes
FEATHERING - [aero] s. feather, to
[arch] foliation *f*, lobes *mpl* (d'arc)
[mech] (roughness of edges) nervurée
[mech] (the changing of the angle of incidence) variation *f* de pas (périodique)
FEATHERING HINGE - [aero] (the pivot on which a rotor or propeller blade turns to alter the blade angle) axe *m* de variation de pas
~PITCH - [aero] (the minimum-drag position of a blade when the engine is not supplying power) pas *m* de mise en drapeau
~PROPELLER - [aero] hélice *f* à ailes articulées
~PUMP - [aero] (hydraulic pump to actuate feathering mechanism in a propeller) pompe *f* de mise en drapeau
FEATURE, to - [gen] caracteriser, représenter, exposer
FEATURE - [gen] trait *m*, caractéristique *f*
~FILM - [cin] (the main film in a programme)grand film *m* du programme, long métrage *m*
FEBRICIDE - [med] antithermique, fébrifuge
FEBRICITY - [med] fébrilité *f*
FEBRICULA - [med] fébricule *f*, fièvre *f* éphémère
FEBRIFUGE - [med] fébrifuge *m*
FEBRILE - [med] fébrile
FECALOMA - [med] tumeur *f* stercorale, copromé *m*
FECULA - [chem] fécule *f*
FECULENT - [chem] féculent, fétide
FECUNDATE, to - [gen etc] féconder
FECUNDATION - [gen zool etc] fécondation *f*
FECUNDITY - [gen] fécondité *f*, fertilité *f*
FEE - [gen] honoraires *mpl*
~-TELEVISION - [telev] (pay as you see television set) télévision *f* à prépaiement
FEEBLE - [gen] faible, débile
[med] (of the pulse) déprimé, rare
~CURRENT - [el] courant *m* faible
~CURRENT CABLE - [telecomm] câble *m* à courant faible
~-MINDEDNESS - [med] déficience *f* mentale
FEED, to - [gen] alimenter, nourrir
[gen] (to load or supply with something) alimenter
[mech tool] (to bring a work against a tool) avancer
FEED - [gen] alimentation *f*, pâture *f*, nourriture *f*, fourrage *m*
[mach tool] (the rate at which work is brought against the tool) avancement *m*
[comput] alimentation *f* (p.ex. en cartes perforées)
[hydr] dosage *m*
~ALLEY - [agric] (for cattle) couloir *m* d'alimentation, passage *m* d'affourragement
~AND DISCHARGE LOCK - [mech] sas *m* à vide, écluse *f* à gaz
~APRON - [text] tablier *m* d'alimentation
~-BACK - [mech el etc] (the operation of returning some energy from the output to the input end of a circuit, e.g. for amplification) réaction *f*, alimentation *f* en retour
~-BACK ADMITTANCE - [el] (supplementary admittance in a circuit caused by the feed-back effect of the anode of an electronic tube) admittance *f* de réaction
~-BACK AMPLIFIER - [radio] (an amplifier with a positive feed-back) amplificateur *m* à contre-réaction
~-BACK CHANNEL - [comput] (in error correction equipment) canal *m* de retour, voie *f* de retour
~-BACK CIRCUIT - [radio] (circuit by which a programme can be heard at a programme source)

contrôle *m* d'écoute, retour *m* d'écoute
[el] circuit *m* de réaction inverse, montage *m* à contre-réaction
FEED-BACK COIL - [radio] bobine *f* de réaction
[comput] (of a magnetic coil) enroulement *m* de réaction
~-BACK CONTROL - [contr] (method of control using a feed-back loop) réglage *m* à contre-réaction
~-BACK CONTROL LOOP - [electron] boucle *f* de commande à réaction
~-BACK CONTROL SYSTEM - [contr] (control system maintaining a specified relation ship between one system variable and another) système *m* de réglage à contre-réaction
~-BACK CONTROLLER - [contr] régulateur *m* de réaction
~-BACK COUPLING - [el] couplage *m* par réaction
~-BACK ELEMENTS - [contr] (that part of the feed-back control which establishes the relationship between the primary feed-back and the controlled variable) organes *mpl* de réaction
~-BACK EQUALIZER - [el] (in an audio amplifier) circuit *m* d'égalisation à réaction négative
~-BACK GEAR - [mech] appareil *m* d'alimentation en retour
~-BACK LIMITER - [contr] limiteur *m* à contre-réaction
~-BACK LOOP - [comput] boucle *f* de réaction, boucle *f* d'asservissement
~-BACK OSCILLATOR - [el] (circuit in which oscillation is set up by feeding back energy from the output to the input) oscillateur *m* à réaction
~-BACK PATH - [el] parcours *m* de réaction
~-BACK RATIO - [radio] taux *m* de contre-réaction
~-BACK SIGNAL - [contr] signal *m* de réaction
~-BACK SUSCEPTANCE - s. feed-back admittance
~-BACK TRANSDUCER - [contr] transducteur *m* à réaction
~-BACK TRANSFER FUNCTION - [contr] fonction *f* d'alimentation en retour
~-BACK VOLTAGE - [el] tension *f* de réaction
~-BACK VOLTAGE TRANSFER RATIO - [el] réaction *f* de tension à vide
~-BACK WINDINGS - [radio] (in a saturable reactor, those windings to which are connected the feed-back connections) enroulements *mpl* de réaction
~BARLEY - [agric] orge *f* fourragère, escourgeon *m*
~BELT - [mech] tablier *m* d'alimentation, courroie *f* d'alimentation
~-BOARD - [print] table *f* à pile, table *f* de marge
[text] table *f* d'alimentation
~-BOX - [mech] boîte *f* d'alimentation
[agric] mangeoire *f*
[mach tool] boîte *f* du méchanisme d'avancement, boîte *f* des avances
[text] boîte *f* de remplissage
~BUSH - [plast ind] (hardened steel ring forming a seal between the heating cylinder nozzle and the mould) cheminée *f* du moule, douille *f* de carotte
~CASSETTE - [photo] cassette *f* débitrice
~CHAIN - [mech] chaîne *f* d'alimentation
~CHANGE - [mach tool] changement *m* de vitesse d'avancement
~CHANGE GEAR - [mach tool] engrenage *m* de rechange pour l'avancement
~CHANGE KNOB - [mech] poignée *f* de changement des avances
FEED-CHANGE LEVER - [mech] levier *m* sélecteur vitesse d'alimentation
~-CHECK VALVE - [mech] (a non-return valve in a delivery pipe) soupape *f* de retenue
~CHUTE - [mech] puits *m* d'alimentation
~CISTERN - [constr] (water tank connected to the main) reservoir *m* d'alimentation
~CLEANING MACHINE - [agric] machine *f* à nettoyer les fourrages
~COCK - [gen] robinet *m* d'alimentation, robinet *m* de remplissage
~COLLAR - [mech] collier *m* d'accés, cylindre-support *m* de traversées
~CONE - [mech] cône *m* de renvoi
~-CONE PULLEY - [mech] poulie *f* à gradins
~CONTROL - [mech] régulateur *m* d'alimentation
~CONTROL UNIT - [comput] unité *f* de sélection
~CURRENT - [radio] (in a thermionic valve, the direct-current component of the anode current) courant *m* d'alimentation
~ELEVATOR - s. feed lift
~ENGAGE LEVER - [mech] levier *m* d'engrenage d'alimentation
~FILTER - [mech] filtre *m* d'alimentation
~-FORWARD - [contr] réaction *f* positive
~FREQUENCY - [el] fréquence *f* d'alimentation
~GEAR - [mach tool] boîte *f* de changement de vitesse pour l'avance
~GRID - [text] grille *f* alimentaire
~GRINDER - [agric] hache-fourrages *m*
~GUIDES - [print] côte *m* de la marge
~HEAD - [metall] masselotte *f*
~HEATER - [mech] (in a locomotive) économiseur *m*
~HOLES - [telecomm] trous *mpl* de transport
[comput] trous *mpl* d'entraînement
~HOLES PER MINUTE - [telecomm] nombre *m* de trous de transport par minute
~HOPPER - [mech] (a conical receptacle through which material is fed to a machine) trémie *f* d'alimentation, trémie *f* de chargement
[comput] magasin *m* d'alimentation en cartes, chargeur *m* d'alimentation en cartes
~IN, to - [el acoust] insérer
~INLET - [text] ouverture *f* d'introduction, ouverture *f* d'alimentation
~INTERLOCK - [comput] blocage *m* de l'alimentation en cartes
~LATTICE - [text] toile *f* d'alimentation
~LAUNDER FOR ORE - [min] goulotte *f* alimenteuse pour minerais
~LIFT - [cin] (used in film processing machines) élévateur *m* d'alimentation
~LINE - [el] (line of large current-carrying capacity) ligne *f* d'alimentation
[hydr] conduite *f* nourricière, conduite *f* d'alimentation
~LOCK - s. feed and discharge lock
~MECHANISM - [mech] (any device for feeding a machine) dispositif *m* d'alimentation
[el] (the mechanism causing the carbons of an arc lamp to move towards the arc) mécanisme *m* d'avancement
~MIXER - [agric] machine *f* à mélanger les fourrages
~MOTION - [mach tool] mouvement *m* d'avancement mouvement *m* de pression
~OPENING - [plast ind] (port in the barrel of an ex-

truder through which pellets pass from the hopper into the barrel) orifice *m* de remplissage
FEED ORIFICE - [plast ind] (short restricted passage at the entrance to the die cavity in an injection- or transfer-mould) point *m* d'injection, entrée *f* d'empreinte
~PAWL - [text] cliquet *m* d'alimentation, loquet *m* d'alimentation
~PHASE - [contr] phase *f* de façonnage
~PIPE - [mech] (pipe used to supply fluid to an engine or unit) tuyau *m* d'alimentation, tuyau *m* de prise d'eau
~PIPING - [hydr] tuyautage *m* d'alimentation
~PISTON - [mech] plangeur *m* (ou piston) d'alimentation
~PLATE - [ind chem] (in a distillation column) plateau *m* d'alimentation
~PRESSURE - [mech] pression *f* d'alimentation
~PUMP - [mech] (a pump specifically used to supply a boiler or other device) pompe *f* d'alimentation
~RACK - [mech] rampe *f* de distribution, rampe *f* de graissage
~REEL - [cin] (the reel of film not yet passed through the aperture) bobine *f* débitrice
~-REEL SPROCKET - [cin] tambour *m* débiteur
~REGULATOR - [mech] régulateur *m* de l'avancement
~RELEASE - [mech] déclanchement *m* du mouvement de pression
~REVERSE LEVER - [mech] levier *m* inverseur du mouvement de pression
~RING - [mech] collier *m* d'accès, cylindre-support *m* de traversées
~ROD - [mech] barre *f* de chariotage
~ROLL - [metall] rouleau *m* entraîneur
[mech] nourisseur *m*
[photo] rouleau *m* débiteur
~ROLLER - [sugar ind] (the roller which nips the cane against the top roller and passes it over the trash-plate to the bagasse roller) rouleau *m* alimenteur
[text] cylindre *m* d'alimentation
~SCREW - [mech] (the screw which forces the plastic material through the die) vis-mère *f*, vis *f* de commande de l'avance, vis *f* sans fin d'alimentation
[oil ind] vis *f* de rallonge
~SHAFT - [mach tool] barre *f* de chariotage, barre *f* de commande de chariotage
~SLIDE - [plast ind] (a valve or shutter in an extruder to control the feed) vanne *f* doseuse
~SPROCKET - [cin] (the sprocket used to feed unexposed films into the camera) tambour *m* débiteur
~STEAMER - [agric] étuveuse *f*
~-STOCK - [mech] (material used to feed a machine) matière *f* utilisée
[metall] charge *f*
[oil ind] stockage *m*
~TABLE - [text] table *f* d'alimentation
~TANK - [gen] bâche *f* d'alimentation
~-THROUGH - [mech] passage *m* (étanche) pour le chargement
~-THROUGH CAPACITOR - [electron] condensateur *m* de traversée
~TIMER - [contr] régulateur *m* d'alimentation
~TRUMPET - [text] entonnoir *m* d'alimentation

FEED TRUNK - [text] puits *m* d'alimentation
~VALVE - [hydr] vanne *f* de remplissage
~WATER - [th eng] (water supplied to a boiler for evaporation) eau *f* d'alimentation
~WATER HEATER - [th eng] (device for pre-heating water supplied to a boiler) réchauffeur *m* d'eau de alimentation
~-WATER PUMP - [mech] pompe *f* d'alimentation
~-WATER SYSTEM - [nucl] circuit *m* d'alimentation en eau
~WIRE - [el] fil *m* d'amenée, artère *f*
~WORKS - [mech] mécanisme *m* d'avancement
~ZONE - [plast ind] (the part of the extruder barrel into which the hopper discharges, terminating at the beginning of the compression zone) zone *f* de alimentation
FEEDER - [el] (cable interconnecting generating stations) artère *f*, feeder *m*
[metall] (the riser hole of a mould) masselotte *f*
[radio] (radio frequency transmission line used to convey energy from the generator to the work electrodes) feeder *m*
[mech] (in motors) alimentateur *m*
[text] mécanisme *m* d'alimentation, alimentateur *m*, cylindre *m* alimentaire
[hydr] (water supplying channel) conduite *f* d'alimentation, canal *m* d'alimentation, conduite *f* nourricière
[geogr] (a tributary) affluent *m*
[ind chem] appareil *m* doseur
[metall] conduite *f* d'alimentation
[mining] filon *m* nourricier
[railw] embranchement *m*
[print] (a device for passing paper into the machine) margeur *m*, dispositif *m* de marge
[firearms] (of a gun) transporteur *m*
[mining] (mechanical device for the transport of crushed ore to a crusher) chargeur *m* mécanique
~BAND - [paper man] (of a guillotine) bande *f* d'alimentation
~BOARD - [paper man] table *f* de marge
~BOX - [el] (a junction box] boîte *f* d'artère, boîte *f* de dérivation
~BUS-BAR - [el] (one of the bus-bars to which the outgoing feeder are connected in generating stations) barre *f* collectrice
~CABLE - [el] (large capacity conductor used in electrical supply to substations and principal points on a power network, without intermediate branches) câble *m* d'alimentation
~CAGE - [text] coulisse *f* de guide-fil
~EAR - [el] (used for attaching an overhead contact wire of a tramway system) griffe *f* d'alimentation
~HEAD - [metall] (refractory insulation at the top of an ingot mould) masselotte *f* chaude, masselotte *f* à étranglement
~LINE - [railw] embranchement *m*
~MAIN - [el] feeder *m* primaire
~OX - [zool] bœuf *m* de boucherie
~PANEL - [el] (switchboard panel controlling feeders) tableau *f* de distribution
~PILLAR - [el] (a pillar with switches, fuses etc) colonne *f* à câble
~PIPE - [hydr] (of a horizontal well) tuyau *m* d'amenée
~REGULATOR - [el] régulateur *m* d'alimentation
~RING - [text] bague *f* intérieure sans épaulement,

anneau *m* porte-guide-fil
FEEDER WHEEL - [text] fournisseur *m* de fil
FEEDING - [gen] alimentation *f*
[mech] alimentation *f*, pompage *m*
[ind chem] (a fault in pigmented coating compositions characterized by rise in viscosity value) épaississement *m*
[metall] nourissage *m*
~APPARATUS - [mech] mécanisme *m* d'alimentation, mécanisme *m* d'avancement
~BAR - [text] tringle *f* d'avance, tringle *f* de connexion
~CABLE - [el] artère *f* d'alimentation
~CAKE - [agric] tourteau *m*
~CAKE MEAL - [agric] farine *f* de tourteau
~CAM - [text] excentrique *m* d'alimentation
~CHANNEL - [hydr] canal *m* d'amenée
~COMB - [text] peigne *f* alimentaire
~CYCLE - [comput] cycle *m* d'alimentation
~DEVICE - s. feed mechanism
~FUNNEL - [text] entonnoir *m* de remplissage
~GATE OF A FURNACE - [metall] gueulard *m* de chargement
~HOPPER - [metall] trémie *f* de chargement
[text] boîte *f* d'alimentation
~LATTICE - [text] tablier *m* sans fin d'alimentation
~LINE - [hydr] conduite *f* d'amenée
~MOUTH - [metall] ouverture *f* d'entrée
~NECK - [metall] col *m* de liaison de la masselotte
~NIPPERS - [text] pince *f* d'alimentation
~PLUNGER - [mech] plangeur *m* de pompe alimentaire
~POINT - [el] (junction point between feeder and system) nœud *m* de réseau
~RAKE - [text] herse *f* de sortie, extracteur *m*
~RATE - [mech] (the rate at which a work is bought against the tool) vitesse *f* d'avancement
~SCREW - s. feed screw
~SUCKER - [book bind] ventouse *f*
~SUGAR - [agric] sucre *m* fourrager
~SYRUP - s. feeding sugar
~TROUGH - [agric] mangeoire *f*, auge *f*
~-UP - s. feeding
FEEL, to - [gen] toucher, sentir
[text] être au toucher *m*
FEEL - [gen] toucher *m*, tact *m*
[astronaut] sensation *f* des commandes
~OF CLOTH - [text] toucher *m* du tissu
~SIMULATOR SYSTEM - [aero] (an arrangement for imparting to controls a degree of resistance to movement by the pilot which is proportional to the actual resistance of control surface to movement by the servo-mechanism which actuates them) dispositif *m* de sensation musculaire
FEELER - [gen] antenne *f*, palpe *f*
[mech] repère *m* d'aile
[text] (used to determine when replenishment is required) tâteur *m*
[mech] jauge *f* d'épaisseur, calibre *m* à lames
~ARM - [text] bras *m* de tâteur
~BOW - [text] arc *m* de tâteur
~FINGER - [text] doigt *m* tâteur
~GAUGE - [instr] (thin slips of steel ground accurately to thickness and used to determine the interval between two parts by progressive selection) calibre *m* à lames, cale *f* d'épaisseur
~NEEDLE - [text] aiguille *f* de tâteur
FEELER STOCK - s. feeler gauge
~WHEEL - [text] roulette *f* de tâteur
FEELING LEVER - [metall] levier-palpeur *m*
FEET - [mech] (of an accumulator plate etc) base *f* d'appui
~SWITCH - [el] interrupteur *m* hydrofuge
FEHLING'S SOLUTION - [chem] (a solution of cupric sulphate and potassium sodium tartrate in alkali, used as an oxidizing agent in the analysis of sugars, which reduce it to cuprous oxide) liqueur *f* de Fehling
FEINTS - [ind chem] residu *m* de distillation alcoolique
FELDSPAR - [min] (a common mineral consisting of a mixture of calcium, aluminium, potassium, sodium and barium silicates. It has applications in ceramics, fertilizers, and abrasive compounds) feldspath *m*
FELDSPATHIC - [geol] feldspathique
~PORPHYRY - [geol] (an igneous rock) porphyre *m* feldspathique
~SANDSTONE - [geol] grès *m* feldspathique
FELDSPATHOIDS - [min] (rock-forming minerals) feldspathoïdes *mpl*
FELL, to - [gen] abattre
FELL - [gen] peau *f*, fourrure *f*
[text] (edge) rabattage *m*
[text] (of the cloth) ligne *f* de serrage de la duite
[timber] (the quantity of timber felled in a given period of time) abattis *m* (nombres des arbres abattus)
[geogr] colline *f* rocheuse
[ind chem] (of a sieve) matières *fpl* traversant le crible
[min] minerai *m* de plomb
FELLING - [timber] abattage *m*, coupe *f*
[mining] abattage *m*
~AXE - [impl] hache *f* de bûcheron
~SAW - [impl] scie *f* passe-partout
FELLOE - [mech] (the outer part of a framing of a wheel) jante *f* (d'une roue)
[carp] (segment of the rim of wooden wheel) section *f* de la jante
~-BAND - [rubber man] vulcanisateur *m* des bourrelets (ou de flaps)
FELLY - s. felloe
FELON - [med] panaris *m*
FELSITE - [min] (fine grained igneous rocks) felsite *f*
FELSPAR - s. feldspar
FELSTONE - s. felsite
FELT, to - [ind proc] (to convert material into felt) feutrer
[constr] (e.g. a roof) couvrir de carton bitumé
[text] (of wool etc) se feutrer, s'agglutiner
FELT - [text] (dense matted fabric of fibrous material) feutre *m*
[acoust] (material used for underlining for soundproofing purposes) feutre *m* insonorisant
[paper man] (blanket carrying the web of paper and squeezing the moisture from it) feutre *m*
~BOARD - [print] tendeur *m* de feutre
~CALENDER - [text] calandre *f* à feutre
~COVERING - [gen] feutrage *m*
~DIRECTION MARK - [paper man] flèche *f* de direction du feutre
~DRYING CYLINDER - [print] sécheur *m* de feutre

FELT JOINT - [mech] (a gasket or packing made of felt) joint *m* de feutre, garniture *f* de feutre
~NUMBER - [meas] (the aweight of a ream of paper cut I2"xI2") marque *f* du feutre
~OIL SEAL - [mech] (felt packing to prevent the passage of oil) joint *m* d'huile en feutre
~PACKING - [mech] garniture *f* de feutre
~PAPERS - [constr] feutre *m* de couverture
~RETAINER - [mech] cage *f* de retenue du feutre
~ROLL - [paper man] rouleau-guide *m* du feutre
~WASHER - [mech] (packing ring made of felt) anneau *m* de feutre
FELTED YARN - [text] fil feutré
FELTING - [text] (the matting together of wool fibres) feutrage *m*
[paper man] (the binding of fibres) feutrage *m*
~MACHINE - [text] foulon *m*, machine *f* à feutrer
FELTWORK - [anat] réseau *m* élémentaire diffus, neuropile
FELUCCA - [naut] felouque *f*
FEMALE - [gen] femme *f*
[gen] féminin
[mech] (of screw threads etc) femelle *f*
~CLUTCH CONE - [mech] cuvette *f* d'embrayage
~CONE - [mech] cône *m* femelle
~DIE - [metall etc] matrice *f*, moule *m* femelle
~FRICTION CONE - [mech] contre-cône *m*
~SCREW - [mech] vis *f* femelle, vis *f* creuse, écrou *m*
~TAP - [mining] cloche *f* de repêchage
~THREAD - [mech] (a screw thread cut on the inside of a hole or bore) filetage *m* femelle, filetage *m* intérieur
FEMORAL - [anat] fémoral
~ARTERY - [anat] artère *f* fémorale
~NERVE - [anat] nerf *m* fémoral
FEMOROCELE - [med] hernie *f* crurale
FEMUR - [anat] (a supporting bone) fémur *m*
FEN - [geogr] marais *m*, marécage *m*
FENCE, to - [gen] clôturer, enclore
FENCE - [gen] clôture *f*, barrière *f*, mur *m* de clôture
[mech] (guide for material) guide *f*, mentonnet *m*, ergot *m* (de pêne)
[aero] (for air threads) aube *f* directrice
[astronaut] réseau *m* de détection des satellites
~BAR - [metall] fer *m* à grilles
~WALL - [constr] mur *m* de clôture
FENCHYL ALCOHOL - [chem] (a synthesis intermediate and solvent) alcool *m* fenchylique
FENCING - [gen] enceinte *f*, barrière *f*
~WIRE - [metall] fil *m* de fer pour clôture
FEND-OFF - [mech] balancier *m*
FENDER - [gen] garde-feu *m*
[carp] (timber baulk used as temporary kerb) bouteroue *f*
[auto] (only US) pare-choc *m*
[mech] amortisseur *m*
[impl] (as a protection against splashing) aile *f*, garde-boue *m*
[naut] pare-battage *m*, défense *f*
~BRACKET - [auto] support *m* d'aile
~FLAP - [auto] (mud flap) bavette *f*, rabat-eau *m*
~LAMP - [auto] feu *m* d'aile
~MARKER - [mech] (width indicator) indicateur *m* de gabarit
~POST - [road] bord *m*, bordure *f*
FENDER SKIRT - [auto] jupe *f* d'aile
FENESTRATION - [build] (the arrangement of windows in a building) fenêtrage *m*
FENNEL - [bot] (the dried ripe fruits of varieties of Foeniculum vulgare used in medicine as a carminative) fenouil *m*, fenouil *m* officinal
FENNY - [soil] marécageux, tourbeux
~SOIL - [soil] sol *m* marécageux, sol *m* tourbeux
FENT - [text] (a damaged piece of cloth) coupon *m* de drap
FERAL - [gen] (wild, savage) sauvage
FERBAM - [chem] (a fungicide) ferbam *m*
FERBERITE - [min] (essentially natural iron tungstate, but some part of the iron is often replaced by manganese) ferbérite *f*
FERETORY - [arch] châsse *f*
FERGHANITE - [min] (uranium containing secondary mineral) ferghanite *f*
FERGUSONITE - [min] (natural columbate and tantalate of yttrium, though this is often partially replaced by cerium, iron or calcium) fergusonite *f*
FERMAT PRINCIPLE - [opt] (stating that a ray of light follows the path which can be covered in the minimum time) principe *m* de Fermat
FERMENT, to - [chem] (the cause decomposition of organic substances by means of micro-organisms or enzymes) fermenter
FERMENT - [chem] (a substance which induces fermentation) ferment *m*, fermentation *f*
FERMENTATION - [chem] (decomposition of organic material by enzymes or micro-organisms) fermentation *f*
~AMYL ALCOHOL - [chem] alcool isoamilique
~FLASK - [ind chem] flacon *m* pour fermentation
~GAS - [gas] gaz *m* biologique, gaz *m* de fermentation
~-INHIBITING - [chem] (term used of a substance or operation which reduces or stops fermentation) freinant la fermentation
~-SEPTIZATION PROCESS - [ind]chem] procédé *m* de fermentation-digestion
~TUBE - [ind chem] tube *m* de fermentation
FERMENTED - [chem] fermenté
FERMENTER - [chem] (vessel in which fermentation is carried on) cuve *f* de fermentation
FERMENTING CELLAR - [brew ind] (underground room used for fermenting operations) cave *f* de fermentation
~TANK - [ind chem] (open vessel in which fermentation is carried on) cuve *f* de fermentation, tank *m* de fermentation
~TUB - [brew ind] cuve *f* de fermentation
FERMI AGE - [phys] (of neutrons) équation *f* de Fermi, équation *f* de l'âge
~-AGE MODEL - [nucl] (model for the study of the slowing down of neutrons by elastic collisions) modèle *m* de l'âge de Fermi
~CHARACTERISTIC-ENERGY LEVEL - [electron] (the inner work function in the case of a metal) niveau *m* caractéristique de Fermi
~CONSTANT - [nucl] (universal constant appearing in beta-disintegration theory) constante *f* de Fermi
~-DIRAC DISTRIBUTION FUNCTION - [phys] fonction *f* de distribution de Fermi-Dirac
~-DIRAC GAS - [electron] (assembly of independent particles following the Fermi statistics) gaz *m* de Fermi-Dirac

FERMI-DIRAC-SOMMERFIELD VELOCITY-DISTRIBUTION LAW - [phys] (FDS law; the algebraic of a quantized system) loi *f* de la distribution de vitesse de Fermi-Dirac-Sommerfield
~DISTRIBUTION - [phys] (the energy distribution of the electrons in a metal) distribution *f* de Fermi
~ENERGY - [phys] (the energy of the Fermi-Dirac gas) énergie *f* de Fermi
~LEVEL - [nucl] (the value of the electron energy at which the Fermy distribution function has the value one half) niveau *m* de Fermi
~PLOT - [nucl] diagramme *m* de Fermi, droite *f* de Fermi
~RESONANCE - [nucl] (in polyatomic molecules) résonance *f* de Fermi
~STATISTICS - [phys] (the study of the probability of the macroscopic states of quantized systems of particles) statistique *f* de Fermi
~SURFACE - [nucl] (the constant energy surface corresponding to the Fermi level) surface *f* de Fermi
~TEMPERATURE - [nucl] (the degeneracy temperature of a Fermi-Dirac gas) température *f* de Fermi
~THEORY - [nucl] (Fermi's 1934 theory on beta-decay) théorie *f* de Fermi (sur la désintégration bêta
FERMION - [nucl] (a particle) fermion *m*
FERMIUM - [chem] (a synthetic radioactive element, A.N. 100, symbol Fm, with properties similar to those of erbium) fermium *m*
FERN - [bot] fougère *f*
~OIL - [ind chem] essence *f* de fougère
FERRET - [zool] furet *m*
[text] padou *m*, filoselle *f*
[glass man] ferret *m*
FERRIC - [chem] ferrique
~ACETYLACETONATE - [chem] (a curing accelerator and catalyst) acétylacétonate *m* ferrique
~AMMONIUM CITRATE - [chem] (a therapeutic source of iron) citrate *m* de fer ammoniacal
~AMMONIUM OXALATE - [chem] (photo-sensitive salt used in blue-printing) oxalate *m* de fer ammoniacal
~AMMONIUM SULPHATE - [chem] (an astringent used in medicine and as a dyeing mordant) alun *m* ferrique-ammoniacal
~ARSENATE - [chem] (an insecticide) arséniate *m* ferrique
~CACODYLATE - [chem] (used in medicine in the treatment of skin complaints) cacodylate *m* ferrique
~CHLORIDE - [chem] (an astringent used in medicine; dyeing mordant; feed additive and disinfectant) chlorure *m* ferrique
~CHROMATE - [chem] (a pigment for surface coatings and ceramics) chromate *m* ferrique
~FERROCYANIDE - [chem] (prussian blue. A blue pigment) ferrocyanure *m* ferrique
~FLUORIDE - [chem] (a glaze for ceramics) fluorure *m* ferrique
~GLYCEROPHOSPHATE - [chem] (a source of iron in medicine) glycérophosphate *m* ferrique
~HYDROXIDE - [chem] (component of rubber pigments) hydrate *m* ferrique
~HYPOPHOSPHITE - [chem] (a source of iron in medicine) hypophosphite *m* ferrique
~NAPHTHENATE - [chem] (a fungicide and a drier for paints and varnishes) naphténate *m* ferrique
~NITRATE - [chem] (analytical reagent and dyeing mordant) nitrate *m* de fer
FERRIC OXALATE - [chem] (a photographic chemical) ocalate *m* ferrique
~OXIDE - [chem] (a catalyst, analytical reagent, dyeing assistant, and rubber pigment) oxyde *m* ferrique
~PHOSPHATE - [chem] (a medicinal source of iron) phosphate *m* ferrique
~POTASSIUM CITRATE - [chem] (medicinal source of iron) citrate *m* ferrique de potassium
~POTASSIUM TARTRATE - [chem] (a source of nutritional iron) tartrate *m* ferrique de potassium
~RESINATE - [chem] (a drier for paints and varnishes) résinate *m* ferrique
~SODIUM OXALATE - [chem] (a photographic chemical) oxalate *m* ferrique de sodium
~STEARATE - [chem] (a drier for surface coatings) stéarate *m* de fer
~SULPHATE - [chem] (textile dyeing assistant, analytical reagent, source of nutritional iron, and disinfectant) sulphate *m* ferrique
FERRICYANIDE - [chem] cyanure *m* de fer
FERRIERITE - [min] ferriérite *f*
FERRITE - [metall] (any solid solution which is basically alpha-iron; also special mixture containing trivalent iron oxide and a divalent metal, e.g. manganese, zinc, nickel, cobalt etc., giving unusual magnetic hysteresis properties) ferrite *f*
~CORE - [comput] noyau *m* de ferrite, tore *m* de ferrite
~CORE MEMORY - [comput] mémoire *f* à tores de ferrite
~FILLED ROD RADIATOR - [electron] (a ferromagnetic dielectric rod) élément *m* rayonnant à ferrite
~FILLED ROD AERIAL - [radio] (a coil aerial in which a rod of ferrite acts as conductor) antenne *f* à ferrite
~FILLED ROD ANTENNA - s. ferrite-rod aerial
~-FILLED WAVEGUIDE - [electron] guide *m* d'ondes à ferrite
~PHASE-DIFFERENTIAL CIRCULATOR - [electron] circulateur *m* à ferrite et décalage de phase
~ROD - [comput] bâtonnet *m* de ferrite
~STORE - [comput] mémoire *f* à noyaux de ferrite, mémoire *f* à ferrites
~SWITCH - [electron] (ferrite device blocking the flow of energy through a waveguide) commutateur *m* de guide d'ondes à ferrite
~TRANSCHARGER - [electron] transporteur *m* de charge en ferrite
~YELLOWS - [chem] (a group of iron pigments made by precipitating ferric hydroxide) jaunes *mpl* de ferrite (oxydes de fer jaunes)
FERRITIC CAST IRON - [metall] fonte *f* ferritique
~STEEL - [metall] acier *m* ferritique
FERRITIZATION - [metall] ferritization *f*
FERRO-ALLOY - [metall] ferro-alliage *m*
~-ALUMINIUM - [metall] ferro-aluminium *m*
~-BORON - [metall] ferrobore *m*
~-CHROME - [metall] (alloy of iron with 60-72 p.c. Cr used in making additions of chromium in iron and steel manufacture) ferrochrôme *m*
~-CONCRETE - [constr] béton *m* armé
~-CYANIDE - [chem] ferrocyanure *m*
~-ELECTRIC - [el] (dielectric materials which exhibit spontaneous polarization and hysteresis) ferroélectrique

FERRO-ELECTRIC EFFECT - [el] effet *m* ferroélectrique
~-ELECTRIC MATERIAL - [el] matières *f*pl ferroélectriques
~-MAGNETIC - [el] (material with a permeability greater than unity) ferromagnétique
~-MAGNETIC RESONANCE - [el] (resonance occurring in a magnetic material) résonance *f* ferromagnétique
~-MAGNETIC SUBSTANCE - [metall] substance *f* ferromagnétique
~-MAGNETISM - [el] ferromagnétisme *m*
~-MANGANESE - [metall] (alloy of iron with 80 p.c. manganese, used in making additions of manganese in steel and iron manufacture) ferro-manganèse *m*
~-MOLYBDENUM - [metall] (alloy of iron with 55 to 65 p.c. molybdenum used in making additions of Mo to steel and iron) ferro-molybdène *m*
~-NICKEL - [metall] (alloy of iron with more than 30 p.c. Ni) ferro-nickel *m*
~-SILICON - [metall] (alloy of iron with I5 p.c. of silicon, used for making additions of silicon in steel and iron manufacture) ferro-silicium *m*
~-TITANIUM - [metall] (an alloy of iron and titanium) ferro-titane *m*
~-TUNGSTEN - [metall] (alloy of iron with a large percentage of tungsten, used in making tool steels) ferro-tungstène *m*
FERROCENE - [chem] (derivative of cyclopentadiene and iron with uses as a curing agent for rubber, fuel additive, and intermediate for polymers) ferrocène *m*
FERRODYNAMIC INSTRUMENT - [instr] (instrument in which the electrodynamic effect is increased by the presence of ferromagnetic material) appareil *m* ferrodynamique
~RELAY - [el] relais *m* ferrodynamique
FERROMAGNETIC MEMORY - [comput] mémoire *f* ferromagnétique
FERROMETER - [instr] (instrument designed to measure magnetic losses) ferromètre *m*
FERRORESONANCE - [telecomm] ferrorésonance *f*
FERRORESONANT CIRCUIT - [el] circuit *m* ferrorésonant
FERROSPINEL - [radio] (ceramic-like material used in aerial-loops etc) ferrospinel *m*
FERROSTATICAL PRESSURE - [metall] pression *f* ferrostatique
FERROTYPE - [photo] (a wet collodion process) ferrotype *m*
FERROUS - [chem] ferreux
~ACETATE - [chem] (a mordant in dyeing) acétate *m* ferreux
~AMMONIUM SULPHATE - [chem] (analytical reagent and a source of nutritional iron) sulfate *m* ferreux ammoniacal
~ARSENATE - [chem] (an insecticide) arséniate *m* de fer
~CHLORIDE - [chem] (intermediate for ferric chloride, dyeing mordant, and source of nutritional iron) chlorure *m* ferreux
~FLUORIDE - [chem] (a component of glazes for ceramics) fluorure *m* ferreux
~FUMARATE - [chem] (a source of nutritional iron) fumarate *m* ferreux
~GLUCONATE - [chem] (a source of iron in medicine) gluconate *m* ferreux

FERROUS IODIDE - [chem] (a source of nutritional iron) iodure *m* ferreux
~LACTATE - [chem] (a source of nutritional iron) lactate *m* ferreux
~OXIDE - [chem] (a starting material for ferrous salts) oxyde *m* ferreux
~PHOSPHATE - [chem] (used in medicine) phosphate *m* ferreux
~PHOSPHIDE - [chem] (water-insoluble greyish powder used in metallurgy) phosphure *m* ferreux
~QUININE CITRATE - [chem] (a source of dietary iron) citrate *m* ferreux de quinine
~SULPHATE - [chem] (a fertilizer, feedstuffs additive, weedkiller, analytical reagent, and dietary source of iron) sulfate *m* ferreux
~SULPHIDE - [chem] (a source of hydrogen sulphide) sulfure *m* ferreux
FERROVANADIUM - [chem] (an alloy of iron and vanadium) ferrovanadium *m*
FERROXYL INDICATOR - [instr] (a special indicator used to show which parts of an iron body are positive or negative. It consists of potassium ferricyanide and phenolphthalein in a corrosive solution, such as an aqueous solution of sodium chloride, the whole made up with agar to form a jelly) indicateur *m* à ferroxyl
FERRUGINOUS - [bot] (the colour of rusty iron) ferrugineux
~CEMENTING MATERIAL - [constr] ciment *m* ferrugineux
~CLAY - [geol] (impure type of clay) argile *f* ferrugineuse
~DEPOSIT - [geol] (iron containing sedimentary rocks) gisement *m* ferrugineux
FERRULE, to - [mech] viroler, baguer, ferrer
FERRULE - [mech] (a short length of tube) virole *f*, bague *f*, frette, bout *m*, ferré, coupelle *f*
[mech] (a small bushing) manche *f*, virole *f* (de manche d'outil)
FERRY, to - [gen] passer l'eau, passer en bac
[naut] (the operation of transporting trains cars etc) passer en bac, transborder
[aero] (the operation of transporting troops and/or armament by air) transporter par voie aérienne
FERRY - [gen] passage *m*
[naut] bac *m*, bateau *m* de passage
~ACROSS, to - [naut] transborder
~-BOAT - [naut] bac *m*, bateau *m* de passage
~ROPE - [rope man] cordage *m* en six, cordage *m* à six torons, filin *m* en six
~SLIP - [naut] dock *m* de bac
FERRYMAN - [naut] passeur *m*
FERTILE - [nucl] (term used of material which can be transformed into a fissionable substance by the capture of a neutron) fertile
~ELEMENT - [nucl] (an element which can be transformed into fissile material in a reactor) élément *m* fertile
~MATERIAL - [nucl] (material which is not fissionable in itself, but can be transformed into fissionable material in a reactor) matière *f* fertile
FERTILITY - [gen] fertilité *f*, fécondité *f*
FERTILIZATION - [agric & gen] fertilisation *f*, fécondation *f*
[bot] pollinisation *f*
FERTILIZE, to - [gen & agric] fertiliser, féconder (le sol), amender

[biol] inséminer
FERTILIZER - [chem] (a specific chemical compound or mixture of compounds prepared for application to the soil to provide plant foods and to adjust pH values: e.g. ammonium and sodium sulphates and nitrates, phosphates of calcium etc) engrais *m*, fertilisant *m*, engrais *m* chimique
~DISTRIBUTOR - [agric] (for industrial fertilizers) distributeur *m* d'engrais
~DRILL - [agric] distributeur *m* d'engrais en lignes
~-WATER IRRIGATION - [agric] irrigation *f* fertilisante
FERTILIZING CAPACITY - [genet] pouvoir *m* fécondant
~TEST - [agric] essai *m* de fumure
~TRIAL - s. fertilizing test
~VALUE - [agric] valeur *m* engrais, valeur *f* fertilisante
FERTIRRIGATION - s. fertilizer-water irrigation
FESCOLING - s. fescolizing
FESCOLIZING - [metall] (special process of electrodeposition of metal, espec. on wearing surfaces, usually Ni, Co or Cr) fescolisation *f* (galvanoplastie)
FESCUE - [bot] fétuque *f*
FESSLER COMPOUND - [chem] (a flocculating agent used in the production of wine) composé *m* Fessler
FESTOON - [gen & arch] (ornamental feature in the shape of a garland of flowers) feston *m*, guirlande *f*
[plast ind] (a sheet of plastic material draped over a mould in thermo-forming) guirlande *f*
~ARRANGEMENT - [genet] disposition *f* en festons
~CLOUD - s. mammatocumulus
~CUTTING MACHINE - [paper man] rogneuse *f* en festons
~DRYER - [paper man etc] (a drying chamber in which the material is hung up in loops or festoons) séchoir *m* à guirlandes
~TRIMMING - [text] ruban *m* festonné
~WORK - [text] broderie *f* de feston
FESTOONING - [paint] décoration *f* en festons
~THE YARN - [text man] suspension *f* des écheveaux (sur perches)
FETCH, to - [comput] (locating and loading data from storage) ramasser, placer, mettre en place
FETID - [gen] (also foetid) fétide
FETTLE, to - [gen] (to preparare, to complete) ajuster, ranger, mettre en ordre
[metall] (dressing off) ébarber, décaper, épiler, débourrer
[metall] (the preparation of an open-hearth furnace) retorcher
FETTLE - [gen] condition *f*
[metall] garniture *f*
FETTLER - [metall] ébarbeur *m*
FETTLING - [metall] (dressing off) ébarbage *m*
[metall] (the preparation of an open hearth furnace) garnissage *m*, revêtement *m*
[text] (the cleaning of fibrous material) débourrage *m* des garnitures de carde
~-HAMMER - [metall] marteau *m* à ébarber
~-MACHINE - [metall] ébarbeuse *f*
~SHOP - [metall] atelier *m* d'ébarbage
~TOOL - [metall] outil *m* d'ébarbage
FETUS - [anat] foetus *m*, embryon *m*
FEVER - [med] fièvre *f*
~BLISTER - [med] bouton *m* de fièvre
F-HEAD - [mech] (type of i.c. engine cylinder head in which one valve is at the side of the cylinder and the other in the head, opening above the side valve and in the opposite direction) culasse *f* à soupapes opposées
f HOLES - [mus] (the sound holes in a violin) ouies *f*pl
F H P - [mech] (initials for Friction Horse Power) énergie *f* absorbée par le frottement
FIBER - s. fibre
FIBERING - [min] structure *f* fibreuse
FIBRE - [bot text plast etc] (a filament or thread-like body, such as a hair) fibre *f*, filament *m*
~BLEND - [text] mélange *m* de fibres
~-BOARD - [paper man] carton *m* de fibre
~BUILDING BOARD - [constr] board *m* panneaux de fibre
~BUNCHING - [plast ind] (segregation of glass fibres from the resin/filler matrix) rebroussement *m* des fibres
~CUP - [text] (in a cup winding machine) entonnoir *m* en fibre vulcanisée
~CUTTING MACHINE - [text] machine *f* à couper les fibres
~DEGRADATION - [text] dégradation *f* de la fibre
~DRAFT TESTER - [text] appareil *m* de contrôle de l'étirage des fibres
~DUST - [text] poussière *f* de fibres
~DUSTING PLANT - [rubber ind] installation *f* de flocage
~EXTRACTING MACHINE - [text] défibreuse *f*, racleuse *f*, décortiqueuse *f*
~-FILLED MOULDING MATERIAL - [plast ind] matière *f* fibreuse de moulage
~FINENESS - [text] titre *m* de la fibre
~FLY - [text] duvet *m* de fibres, tontisse *f*
~GAUGE - [instr] (instrument designed to measure pressure by gauging the rate of damping of the vibrations of a suitable material) manomètre *m* à fil
~-GLASS - [plast ind] laine *f* de verre
~HAMMER - [impl] (hammer of which the head is made of compacted fibrous material to avoid damage to objects struck with it) marteau *m* de fibre
~LAYER - [text] couche *f* de fibres
~MIXTURE - [text] mélange *m* de matières filables
~ORIENTATION - [text] disposition *f* des fibres, orientation *f* des fibres
~PINION - [mech] pignon *m* en fibre
~SLURRY - [text] pâte *f* de fibres
~STAPLE - [text] appareil *m* pour établir le diagramme des fibres
~STRENGTH TESTER - [instr] dynamomètre *m* pour fibres
~STRESS - [metall] effort *m* dans la fibre
~STRUCTURE - [gen] structure *f* de fibre
[text] structure *f* de la fibre
~TEXTURE - [metall] texture *f* fibreuse
~THICKNESS - [text] grosseur *f* des fibres
~TIP - [text] pointe *f* de la fibre, extrémité *f* de la fibre
~TUFT - [text] barbe *f* de fibres, mèche *f* de fibres
~YIELD - [text] rendement *m* en fibres
FIBREMIA - [med] fibrinémie *f*
FIBRIL - [bot etc] (a small fibre) fibrille *f*
FIBRILLAR - [bot & anat] fibrillaire
FIBRILLATION - [physiol] (contraction of individual muscle fibres) fibrillation *f*

[paper man] défibrillation *f*
FIBRIN - [chem] (insoluble substance precipitating in the form of fibres when the blood coagulates) fibrine *f*
˜CLOT - [med] caillot *m* fibrineux
˜GLUE - [ind chem] colle *f* de fibrine
FIBRINOGEN - [chem] (protein in the blood plasma) fibrinogène
FIBRO-ADENOMA - [med] (adenoma with an overgrowth of fibrous tissue) adénofibrome *m*
FIBROBLAST - [anat] (irregularly shaped connective-tissue cell) fibroblaste *m*, cellule *f* fibreuse
FIBROCARTILAGE - [anat] (cartilage with fibres in the matrix) fibrocartilage *m*, cartilage *m* fibreux
FIBROCYST - [med] kyste *m* fibreux
FIBROCEMENT - [constr] fibro-ciment *m*
˜DISEASE - [med] (disease caused by a cyst in the bones) ostéite *f* fibro-kystique
FIBROFASCITIS - [med] rhumatisme *m* musculaire
FIBROID - [med] fibroïde
˜TUMOR - [med] tumeur *f* fibroïde
FIBROIN - [chem] (a component of raw silk) fibroïne *f*
FIBRONEUROMA - [med] neurofibrome *m*
FIBROTUBERCOLOSIS - [med] tuberculose *f* fibreuse
FIBROUS - [gen] fibreux
[metall] (of iron) nerveux
˜GRAIN - [metall] structure *f* fibreuse
˜FRACTURE - [min] cassure *f* fibreuse
˜IRON - [metall] fer *m* nerveux
˜IRON ORE - [metall] minerai *m* de fer fibreux
˜PEAT - [text] fibre *f* de tourbe
˜RED IRON ORE - [min] fer *m* oligiste concrétionné
˜ROOT - [bot] racine *f* fasciculée (ou fibreuse)
˜ROOTED GRASS - [bot] froment *m* des chiens
˜TALC - [min] saponite *f*, stéatite *f*, talc *m*
˜TISSUE - [anat] (connective tissue consisting of bundles of fibres) tissu *m* fibreux
FIBROVASCULAR - [bot] fibrovasculaire
FIBULA - [anat] péroné *m*
[constr work] (iron bar used to fasten together adjacent stones) fibule *f*
FICIN - [biochem] (a proteolytic enzyme) ficine *f*
FICTILE - [gen] (a general term applied to products made from clay by the use of its plastic properties) céramique, plastique
FICTITIOUS PRIMARIES - [telev] couleurs *fpl* primaires imaginaires
˜YEAR - [astr] (conventionally assumed) année *f* fictive
FID - [naut] (of the mast) clef *f* de mât
(a splicing wood or iron pin) coin *m*, cale *f*
FIDDLE - [mus] violon *m*
[naut] fiche *f* de roulis
[mech] arcon *m*
[ceramics] égouttoir *m*
˜BLOCK - [naut] poulie *f* à violon
˜BOW - [mus] archet *m*
˜DRILL - [tool] drille *f* à arcon
˜STICK - s. fiddle bow
FIDDLEY - [naut] (in the boiler room) partie *f* supérieure de la chambre de chauffe
FIDELITY - [gen] fidélité *f*
[el] (the measure with which a system reproduces the main characteristics of the signal) fidélité *f*
˜OF FREQUENCY RESPONSE - [el acoust] précision *f* en fréquence
FIDUCIAL - [surv] (of a line or a point taken as fixed line of reference) fiduciel
˜AXIS - [instr] axe *m* de foi
˜LINE - [surv] ligne *f* de foi
˜MARKS - [instr] (lines on the scale of an instrument which are obtained by direct calibration, as distinct from graduations interpolated between them) lignes *fpl*
˜TEMPERATURE - [met] (the temperature at which a barometer reads correctly) température *f* fiducielle
FIELD - [gen] champ *m*, étendue *f*, espace *m*
[el] (a vectorial field, or intensity of field) champ *m*
[phys] (region in which the forces under examination are appreciable) champ *m*
[telev] trame *f*
[of a punch card] zone *f*
[radiat] (entry portal) champ *m* d'incidence, porte *f* d'entrée
˜ADAPTER - [photo etc] adaptateur *m* de champ
˜AMMONIATION - [rubber ind] addition *f* de l'ammoniaque au latex dans la plantation
˜AMPERE-TURNS - [el] (the ampere-turns which produce the magnetic field) ampère-tour *m* de champ
˜AMPLITUDE - [radio] amplitude *f* de champ
[telev] amplitude *f* verticale
˜APPARATUS - [cin & telev] appareillage *m* pour extérieurs
˜BALANCE - [el] balance *f* de champ
˜BEAN - [bot] féverole *f*
˜BEND - [telev] (component of the video signal) signal *m* parabolique de correction de trame
˜BLANKING - [telev] suppression *f* de trame
˜BLOCK - [med] infiltration *f* anesthésique du champ opératoire
˜BOOK - [surv] (records of field measurements in a chain-survey) carnet *m* d'opérations
˜-BREAKING RESISTANCE - s. field discharge resistance
˜-BREAKING SWITCH - [el] s. field-discharge switch
˜BROME-GRASS - [bot] brome *m* des champs
˜CALIBRATION - [instr] (the calibration following the factory calibration; also called Checking the calibration) étalonnage *m* de contrôle, étalonnage *m* de champ
˜CAMOMILE - [bot] fausse camomille, œil *m* de vache
˜CAPACITY - [agric] (irrigation) capacité *f* absolue en eau
˜CENTRING CONTROL - [telev] réglage *m* de centrage de trame
˜CHOPPER - [agric] ramasseuse-hacheuse *f*
˜CIRCUIT - [el] circuit *m* de champ
˜COIL - [el] (an element of the winding of a generator, motor or electromagnet, which produces a magnetic field) bobine *f* de champ, bobine *f* inductrice
˜CONDITIONS - [constr] conditions *fpl* au chantier, état *m* du terrain
˜CONSOLIDATION LINE - [soil] courbe *f* de consolidation en place
˜CONTROL - [el] (system for controlling the speed of direct-current motor by varying the exciting flux) régulation *f* du champ
˜CONVERGENCE - [telev] convergence *f* de trame
˜CONVERGENCE SHAPE CONTROL - [telev] réglage *m*

de la configuration de la convergence de trame
FIELD CONVERGENCE TILT CONTROL - [telev] réglage *m* linéaire de la convergence de trame
~CONVERGENCE YOKE - [telev] bobine *f* de convergence de trame
~COPPER - [el] (term denoting the total quantity of copper in a field winding) cuivre *m* de l'enroulement d'induction
~CORE - [el] noyau *m* du champ
~COVERAGE - [opt] champ *m* de vue
~CURING - [agric] fanage *m* sur le sol, séchage *m* en champ
~CURRENT - [el] (the current in the field winding) courant *m* de champ
~DEFECT - [opt] défaut *m* de champ de miroir plan
~DEFINITION - [comput] définition *f* de la zone, délimitation *f* de la zone
~DEFLECTION - [telev] (the downward movement of the electron beam in a picture tube while scanning every line) déviation *f* de la trame
~DENSITY - [el] densité *f* de champ
~DESORPTION - [electron] (a form of field emission of positive ions) désorption *f* de champ
~DIRECTION - [telecomm] direction *f* du champ
~-DISCHARGE SWITCH - [el] (switch used to control the field circuit of a generator) interrupteur *m* court-circuitant le champ
~DISTORTION - [el] distorsion *f* du champ
~DISTRIBUTION - [comput] distribution *f* de champ
~DISTRIBUTION OF THE APERTURE - [radio] (in aerials) distribution *f* de champ de l'ouverture
~DIVERTER RHEOSTAT - [el] (rheostat used to control the magnetomotive force produced by the winding) rhéostat *m* de dérivation de champ
~DIVIDER - [telev] (an apparatus for the reproduction of signals at field frequency) diviseur *m* de fréquence de trame
~DOSE - [radiat] (the dose of radiation which is given to an irradiated surface area) dose *f* en surface
~DRAG - [agric] herse-trainoir *f* pour champs
~DRIVE SIGNAL - [telev] signal *m* synchroniseur de trame
~DRYING - s. field curing
~DURATION - [telev] durée *f* de trame
~DYNAMIC FOCUS - [telev] foyer *m* dynamique vertical
~EFFECT TRANSISTOR - [electron] transistor *m* à effet de champ
~ELIMINATION GATE - [telev] porte *f* de trame
~ELM - [bot] orme *f*
~EMISSION - [nucl] (emission of electrons from an unheated metals surface, induced by a strong electric field) émission *f* de champ, émission *f* froide
~EMISSION COUNTER - [nucl] (radiation counter in which an ionic charge releases an electron from the cathode surface of the counter) compteur *m* à émission froide
~-EMISSION MICROSCOPE - [electron] microscope *m* ionique à émission par champ électrique
~-ENHANCED PHOTOELECTRIC EMISSION - [electron] (in light-sensitive devices) photoémission *f* augmentée par champ électrique
~EXCITATION - [el] excitation *f* de champ
~EXCITATION CURRENT - [el] courant *m* d'excitation de champ
~EXPERIMENT - [auto] (for tyres) essai *m* sur route
FIELD EXTENSION WELL - [oil ind] puits *m* de délimitation du gisement
~FLYBACK - [telev] retour *m* de trame
~FORCING - [contr] réglage *m* par excitation du champ
~FORM - [el] (curve indicating the value of the flux density in an air gap) courbe *f* de champ
~-FORM FACTOR - [el] (the ratio of the average value of the flux density) coefficient *m* de courbe de champ
~FRAME - [el] (in a transformer etc) culasse *f* magnétique
~FRAME MASK - [photo] cache *m* délimitant le champ de l'image
~-FREE EMISSION - [electron] émission *f* à champ nul
~-FREE EMISSION CURRENT - [electron] (the emission current from an emitter when the surface gradient is zero) courant *m* d'émission à champ nul
~FREQUENCY - [telev] (the number of pictures per second; also called Frame Frequency) fréquence *f* de balayage vertical
~FREQUENCY CONTROL - [telev] réglage *m* de la fréquence de trame
~GATING CIRCUIT - [telev] (circuit for extracting or inserting a signal) circuit *m* déclencheur de trame
~GLASS - [opt] verre *m* de champ (de jumelles ou microscope)
~HARROW - [agric] herse *f* brise-mottes
~HAY DRYING - [agric] fanage *m* sur le sol
~HOIST - [agric] grue *f* agricole
~HOLD CONTROL - [telev] régleur *m* de la synchronisation de la trame
~HORSETAIL - [bot] (a weed) prêle *f* des champs
~HOSPITAL - [med] hôpital *m* de campagne
~-INTENSIFIED GAS DISCHARGE - [el] (conduction current in a gas due to ionization of the gas from an external source) décharge *f* Townsed
~INTENSITY - s. field strength
~INTENSITY METER - [instr] (a calibrated radio receiver used to measure field strength) mesureur *m* d'intensité de champ
~ION EMISSION - [electron] émission *f* ionique par champ électrique
~KEYSTONE WAVEFORM - [telev] signal *m* de correction de trapèze-trames
~LATEX - [rubber ind] latex *m* de plantation
~LENS - [opt] (auxiliary lens used to focus light which has already passed through stages of an optical system) lentille *f* de champ
~LINE - [el] ligne *f* de champ, ligne *f* de force
~LINEARITY CONTROL - [telev] réglage *m* de linéarité de la trame
~MAGNET - [el] (the electro-magnet which produces the magnetic field in a motor or generator) aimant *m* de champ, inducteur *m*, électroaimant *m* d'excitation
~MAINS - [oil ind] (in natural gas installations) réseau *m* de collecte, canalisations *fpl* de gaz brut
~MASKING - [cin & telev] limitation *f* de champ
~MELILOT - [bot] mélilot *m* officinal
~MINT - [bot] (a weed) baume *m* des champs
~MONITORING TUBE - [telev] tube *m* moniteur de trame
~NEUTRALIZING COIL - [telev] bobine *f* de blindage du champ magnétique

FIELD NEUTRALIZING MAGNET - [telev] (in colour television) aimant *m* de blindage
~OF APPLICATION - [ind proc] domaine *m* d'emploi
~OF FIXATION - [opt] champ *m* de vue
~OF FORCE - [el] (the force field set up by a source) champ *m* de force, champ *f* d'intensité
~OF PLANE MIRROR - [opt] champ *m* de miroir plan
~OF SELECTION - [telecomm] couronne *f* de contacts, arc *m* de broches, champ *m* de sélection
~OF VIEW - [opt] (the area which is visible through an optical instrument) champ *m* de vision, champ *m* de vue, champ *m* de la lunette
~-OF-VIEW ERROR - [photo] erreur *m* de champ de vision
~OF VIEW LIMITATION - [cin] (the reduction of the field of view which is obtained by optical means) limitation *f* du champ visuel
~OF VISION - s. field of view
~OF VISION CAMERA COVERAGE - [cin] champ *m* visuel
~OPERATOR - [nucl] (operator representing the creation or destruction of a particle in quantized field theory) opérateur *m* de champ
~OVERHAUL - [mech etc] remise *f* au état sur le terrain
~PATTERN - [agric] distribution *f* des champs
~PHASING - [telev] mise *f* en phase de la trame
~PICK-UP - [telev] (outside broadcast) prise *f* de vues en extérieur
~POINT - [el] (the point at which the field strength is measured) point *m* de mesure du champ
~POLE - [el] (the structure carrying the field coil) pièce *f* polaire
~PULSE - [telev] impulsion *f* de trame
~QUANTUM - [phys] quantum *m* du champ
~RATE CONVERTER - [telev] convertisseur *m* du nombre de trames
~RATE FLICKER - [telev] papillotement *m* en trames
~REGULATOR - [el] régulateur *m* du champ, rhéostat *m* automatique
~RHEOSTAT - [el] (variable resistance connected in series with a field winding to vary the current in the latter) rhéostat *m* de champ
~RIVET - [constr] (rivet put in when the work is on the site) rivure *f* de chantier
~SALAD - [bot] doucette *f*, mâche *f*
~SCAN GENERATOR - [telev] base *f* de temps de trame
~SELECTION - [comput] sélection *f* de la zone
~SENSITIVITY - [acoust] (of a microphone) efficacité *f* dans le champ acoustique libre
~SEQUENCE - [telev] séquence *f* d'entrelacement
~SEQUENTIAL SYSTEM - [telev] (the colour system in which a succession of individual coloured fields are sent and received) système *m* à fréquence de trames
~SHUNTING - [el] (in motors) shuntage *m* par dérivation
~SIMULTANEOUS SYSTEM - [telev] (a colour television system in which a complete colour field is sent and received as a unit) système *m* simultané (de télévision)
~SPIDER - [el] croisillon *m* de champ
~-SPLITTING SWITCH - [el] (switch used to prevent excessive voltage during the starting period) sectionneur *m* de champ
~SPOOL - [el] (bobbin carrying a field coil) carcasse *f* de bobine
FIELD STOP - [opt] diaphragme *m* du champ de vision
~STORE CONVERTER - [telev] (a converter which can change the field rate electronically) convertisseur *m* électronique du nombre des trames par seconde
~STRENGTH - [el] (electric field strength) intensité *f* de champ
~SUPPLY - [el] alimentation *f* du champ
~SUPPRESSOR - [el] (in a generator, a device, to reduce the field current) éliminateur *m* de champ
~SWEEP - [telev] balayage *m* de la trame
~SWITCH - [el] interrupteur *m* de champ
~SYNC PULSE - [telev] signal *m* de synchronisation de trame
~SYNCHRONIZING SIGNAL - s. field sync pulse
~SYNCHRONIZING SYSTEM - [telev] système *m* de synchronisation de trames
~TEST - [ind proc] essai *m* sur place
[comput] essai(s) *m* pratique(s)
~TESTED APPARATUS - [mech] appareil *m* fiable
~THEORY - [el] (the theory of the dynamic motion of a set of electromagnetic fields) théorie *f* de champ
~TILT - [telev] signal *m* de correction linéaire de trame
~TUBE - [nucl] (devise used in a proportional counter) tube *m* limiteur du champ
~USE - [gen] (of equipment) mise *f* en œuvre pratique, exploitation *f* pratique
~VOLE - [zool] (a pest) campagnol *m* agreste
~VOLTAGE - [el] tension *f* du champ
~WEAKENING - [el] shuntage *m* des inducteurs
~WINDING - [el] (the whole of the winding in an electrical device, designed to produce the electric field) enroulement *m* d'excitation, enroulement *m* inducteur
~WOOD-RUSH - [bot] luzule *f* des champs
FIELDED - [carp] (of a panel; divided into smaller panels) à panneaux
FIELDWORK - [gen] travail *m* sur le terrain
FIERCE CLUTCH - [mech] (a friction clutch which has the defect of engaging in a series of snatches rather than smoothly) embrayage *m* brutal
FIERY - [mining] grisouteux
FIFE - [mus] fifre *m*
FIFTH - [math] cinquième
[mus] (the interval covering five degrees of the scale) quinte *f*
~ORDER THEORY - [opt] (theory relating to corrections to lenses) théorie *f* du cinquième ordre
~WHEEL - [mech] cercle *m* horizontal
FIG - [bot] figue *f*
~TREE - [bot] figuier *m*
~-WORT - [bot] (a weed) ficaire *f*
FIGHTER - [aero] avion *m* de combat, avion *m* de chasse
FIGHTING BULL - [zool] toreau *m* d'arène
FIGURATIVE CONSTANT - [comput] constante *f* prévue
FIGURE - [gen] figure *f*, forme *f* extérieure
[math] chiffre *m*
[geom] figure *f* (géométrique)
[mus] figure *f*, motif *m*
~COLOUR - [text man] couleur *f* du dessin
~HEAD - [naut] figure *f* de proue
~MASK - [electron] (component of counter tubes) masque *f* de chiffres

FIGURE-OF-EIGHT BANDAGE - [med] bandage *m* en huit de chiffre
~-OF-EIGHT CALIPERS - [impl] huit *m* de chiffre, compas *m* dit huit de chiffre
~OF LINEARITY - [telev] facteur *m* de linéarité
~OF LOSS - [el] (in transformers, the energy loss per pound of material) indice *m* de pertes
~OF MERIT - [nucl] (numerical quantity based on one or more characteristics under specified conditions, used to indicate comparative effectiveness) coefficient *m* de qualité
~SHADING - [text] ombrage *m* faconné
~SHIFT - [comput] (e.g. of a pageprinter) inversion *f* "Chiffres"
~SHIFT - [print] (in a teletypewriter) commutation *f* du clavier
~WITH WARP EFFECT - [text] dessin *m* à effet de chaîne
~WITH WEFT EFFECT - [text] dessin *m* à effet de trame
FIGURED - [gen] faconné, ouvré, broché, à dessins
~GAUZE - [text] gaze *f* façonnée
~HONEYCOMB WEAVE - [text] armure *f* nid d'abeille au Jacquard
~MELODY - [mus] mélodie *f* figurée
~PATTERN - [text] dessin *m* de figures
~TILE - [constr] carreau *m* figuré
~WEAVING - [text man] (on a Jacquard loom) tissage *m* d'étoffe façonnée
FIGURES CASE - [telecomm] série *f* chiffres
~SHIFT - [telecomm] (a case shift due to the translation of received signals as secondary functions) inversion *f* chiffres
~-SHIFT SIGNAL - [telecomm] signal *m* d'inversion chiffres
FIGURING - [text] (on looms for figured goods) brochage *m*
~HARNESS - [text] harnais *m* de figures, harnais de dessin
~MACHINE - [text] mécanisme *m* pour le dessin
~WARP - [text] chaîne *f* brochée, chaîne de brochage
FILAMENT - [gen] filament *m*, fil *m*, filet *m*
[el] (fine, high-resistance wire) filament *m*
[text] (the fibre produced by the silkworm) filament *m*
[electron] (the metallic wire which is heated by the passage of current in a thermionic tube) filament *m*, cathode à chauffage
[plast ind] (material in a threadlike form produced by the coagulation of a continuous jet of liquid, by extrusion or cold drawing) filament *m*
[bot] filet *m*, fibre *f*
~ACTIVITY TEST - [electron] essai *m* pour le rebut de tubes à vide
~BATTERY - [el] batterie *f* de chauffage
~BREAK- [el] rupture *f* de filament
~CIRCUIT - [radio] circuit *m* de filament
~CURRENT - [electron] (the current used to heat the filament in a thermionic tube) courant *m* de filament
~CURRENT CONSUMPTION - [electron] consommation *f* du courant de chauffage
~CURRENT POWER - [el] puissance *f* de chauffage
~EFFICIENCY - [electron] (in thermionic tubes, the ratio between the current emitted from a filament and that used to heat it) rendement *m* du filament

FILAMENT GETTER - [electron] (for a ligh vacuum) getter *m* filiforme
~LAMP - [el] lampe *f* à filament, lampe *f* à incandescence
~LIMITATION - [electron] limitation *f* du courant anodique
~NOISE - [electron] (in semiconductors) bruit *m* de résistance
~POWER SUPPLY - [electron] (for electron tubes) alimentation *f* filament
~RAYON - [text ind] fil *m* de viscose
~REACTIVATION FLASHING - [electron] (of thoriated-tungsten filaments) réactivation *f* du filament
~RESISTANCE - [electron] (in a thermionic tube) résistance *f* du filament
~SATURATION - [electron] (the condition of a thermionic tube when all the electrons elitted from the filament are drawn to the electrode) saturation *f* cathodique
~TRANSFORMER - [el] transformateur *m* de chauffage
~VOLTAGE - [electron] (the voltage applied to the terminals of the filament) tension *f* de chauffage
~WINDING - [electron] (separate winding on a power transformer for tube filaments) enroulement *m* de chauffage
FILAMENTARY - [gen] filiforme
~CATHODE - [electron] (the filament of an electronic tube; a hot cathode heated directly by the current flowing in it) cathode *f* filiforme
~TRANSISTOR - [electron] transistor *m* à une jonction, transistor *m* filiforme
FILAMENTOUS - [gen] (thread-like) filamenteux
FILATURE - [text ind] (the production of silk filaments from coccons) filature *f*, dévidage *m*
FILBERT - [bot] aveline *f*
[aero] (colloq) pilote *m* automatique
FILE, to - [gen & mech] (to remove metal by means of a file) limer, enveler à la lime
[comm etc] (of papers, documents etc) classer, enfiler
[leg] (an application) déposer
[leg] (a signature) enregistrer
[comput] (of cards) classer
FILE - [tool] (a tool for removing metal, consisting of a flat bar of hardened steel cut into chisel-like teeth) lime *f*
[gen] (a row) file *f*
[comm] classeur *m*, fichier *m*
~CARD - [impl] carde *f* à limes
~CARRIER - [mech] arbalète *f*, porte-lime *m*
~CONDENSATION AND CONSOLIDATION - [comput] (concentration of files in a specified order system fusion *f* de fichiers
~-CUTTING MACHINE - [mech] tailleur *m* de limes
~DOWN, to - [mech] adoucir à la lime
~DUST - [mech] limaille *f*, râpure *f*, sable *f* de limage
~HANDLE - [impl] (handle, usually of wood, for fitting to a file) manche *m* de lime
~-HARD - [gen] résistant à la limure
~HARDEN, to - [metall] tremper les limes
~HOLDER - [office impl] porte-lime *m*, porte-limes *m*
~MARK - [comput] (on magnetic tape) caractère *m* "Fin de bloc" marque *f* "Fin de bloc"
~MEMORY - [comput] mémoire *f* de fichier, mémoire *f* de grande capacité

FILE STEEL - [metall] acier *m* à limes
~TEMPERING CRUCIBLE - [metall] creuset *m* pour la trempe des limes
~UPDATING - [comput] mise *f* à jour d'un fichier
FILEGREE - [gen] (ornamental work) filigrane *m*
FILIAL GENERATION - [genet] génération *f* des descendants
FILIFORM - [gen] filiforme
FILING - [mech] limage *m*, limure *f*
~BLOCK - [carp] taquet *m* pour limage
FILINGS - s. file dust
FILIOMA - [med] fibrome *m* de la sclérotique
FILL, to - [gen] remplir, s'emplir, se combler
[transp] (to load into) remplir, charger
[print] plomber
FILL - [gen] content *m*, suffisance *f*
[constr] remblayage *m*, charge *f*
[text] duite *f*, trame *f* (d'un tissu)
~AND DRAW CONTACT BED - [hydr] (a type of contact filter) lit *m* de contact
~BODIES - [ind chem] corps *mpl* de remplissage
~BODIES COLUMN - [ind chem] colonne *f* à corps de remplissage
~FROM THE BOTTOM, to - [metall] (in foundry work) couler en source
~IN, to - [gen] combler, remblayer, boucher
[constr] (e.g. a door) condamner
[oil ind] tamponner
~-IN LIGHT - [cin] lumière *f* d'appoint (pour adoucir les ombres)
[light] éclairage *m* d'appoint
~MATERIAL - [soil] remblai *m* matériau *m* de remblai
~IN WITH EATH, to - [constr] remblayer, couler (les joints de ciment)
~THE BRAKES, to - [railw] alimenter le frein
~THE DENT, to - [text] (reeding operation) garnir la dent
~THE WARP, to - [text man] garnir la trame
~UP, to - [gen] (with fuel) faire le plein
[soil] remblayer, mettre en remblai, surélever
[oil ind] mastiquer
~-UP - [gen] remplissage *m*
[cin] (a short film inserted in the programme) film *m* complément du programme
~WITH ZEROS, to - [comput] inscrire des zéros
FILLED BAND - [nucl] (an energy band in which each level is occupied by an electron) bande *f* remplie
~BEAM - [text] ensouple *f* garnie
[constr] poutre *f* pleine
~FLAT-BACK STOPING - [mining] exploitation *f* à échelon remblayée
~RILL STOPE - [mining] gradin *m* incliné avec remblayage
~STOPE - [mining] taille *f* remblayée
~-UP GROUND - [constr] remblai *m*
~WELD - [metall] (welding) cordon *m* de soudure
FILLER - [mech] (of a fuel tank) boyau *m* de remplissage
[paint etc] (a reinforcing material introduced into a moulding composition to give bulk and strength) mastic *m*, blanc *m* de charge
[plast ind] (material added to a moulding composition to improve its working qualities) charge *f*
[acoust] (the inert material of the compound of the record) charge *f*
[constr] (material used to fill in holes, pores etc) stuc *m*, plâte *m* de moulage
[rubber ind] (a loading material) charge *f*
[print] bouche-trou *m*, bouchon *m*
[in welding] apport *m*, métal *m* d'apport
[metall] (in plate work) coquille *f*
[paper man] (material added to the composition to improve specified properties) charge *f*
(a mineral powder) remblai *m*
[gas ind] (blending agent) amaigrissant *m*, dégraissant *m*
[railw] entretoise *f*
FILLER CAP - [mech etc] (the device which closes an opening through which a vessel, e.g. a fuel tank, is filled) bouchon *m* de remplissage
~COAT - [paint] couche *f* de blanc de charge
~INGREDIENT - [plast & rubber man] agent *m* de charge
~LOAD - [plast & rubber ind] quantité *f* de la charge
~MATERIAL - [constr] matériau *m* de remplissage
~METAL - [metall] métal *m* d'apport
~NECK - [mech] boyau *m*
~PLATE - [mech] (a plate in a compression moulding machine which can be removed for charging) plaque *f* de remplissage
[metall] semelle *f* avec le dispositif à tourner en cône
~PLUG - [mech] (a part designed to close a filling opening) bouchon *m* de remplissage
[oil ind] bouchon *m* de remplissage
~ROD - [metall] baguette *f* de soudure
~SAND - [metall] sable *m* de remplissage
~SPECKS - [rubber ind] taches *fpl* de charge
FILLET, to - [mech etc] arrondir (un bord etc)
FILLET - [gen] filet *m*, bandelette *f*
[metall & plast ind] (the rounding of a mould angle) arrondi *m*, congé *m*
[arch] fasce *f*, bande *f*, congé *m*
[constr] embrasure *f*, pièce *f* d'ébrasement
[mech] (narrow strip of metal above the surface level) raccordement *m*
[mech] (at the intersection of two surfaces) congé *m* de raccordement
[aero etc] (straightening the junction of two surfaces) carénage *m* de raccordement
[text] ruban *m*, bande *f* d'étoffe
[carp] languette *f*, languette *f* rapportée
[mech] (of a screw) filet *m*
~GUTTER - [constr] solin *m*
~IRON - [metall] fer *m* à congé
~-LIKE ARCHITRAVE - [constr] baguette *f*, filet *m*, flipot *m*
~OF A SCREW - [mech] filet *m* d'une vis
~-SLICK - [metall] (in welding) lissoir *m* à congé
~WELD - [metall] (a weld run along the angle between two sheets) soudure *f* d'angle, joint *m* d'angle, cordon *m* d'étanchéité
FILLETED - [mech etc] raccordé, rapporté
FILLING - [gen] remplissage *m*, chargement *m*
[paint] blanc *m* de charge
[text] (in the USA, the weft) trame *f*, duite *f*
[text] remplissage *m*
[constr work] remblayage *m*, comblement *m*, matière *f* inerte, blocage *m*
[min] (the filling up of areas in a metal mine) remplissage *m*, comblement
[shipbuild] mailletage *m* de la carène

[med] (in dentistry) plombage *m* dentaire, obturation *f*
FILLING BOBBIN - [text] fuseau *m* de trame
~BOX - [text] entonnoir *m* de remplissage, hotte *f* de chargement
~BUGGY - [gas ind] (coal charging car) chariot *m* à charbon
~BURR APPARATUS - [text] chaîneuse *f*
~FORK - [text] fourchette *f* du casse-trame
~FRAME - [mining] cadre *m* supérieur
~FUNNEL - [impl] entonnoir *m* de remplissage
~HOLE - [metall] fouille *f* de chargement
~-IN - [constr] comblement *m*, remblaiment *m*
[metall] tamponnage *m*
~IN LAND - [hydr] (in sludge disposal) remblayage *m* du terrain
~IN WITH GROUND - [constr work] maçonnerie *f* de remplage
~LAYER - [rubber ind] couche *f* de rembourrage
~MARK - [of tanks, tuns, vats etc] repère *m* de remplissage
~MATERIAL - [mining] matériaux *mpl* de remblayage
[constr] matière *f* inerte, blocage *m*, remblai *m*
~NECK - [aero] (for a balloon) raccordement *m* à manche
~OPENING - [mech] (an orifice through which a tank or other vessel is filled) orifice *m* de remplissage
~PLUG - [mech] bouchon *m* de remplissage
~PORT - [metall] (in diecasting) trou *m* de remplissage (du cylindre sur machines à chambre chaude à piston)
~PRESSURE - [liquid gas] pression *f* de remplissage
~RAISE - [mining] descenderie *f* des remblais
~SLEEVE - [aero] raccordement *m* à manche
~STATION - [auto] poste *m* d'essence
~STRAINER - [auto] tamis *m* de remplissage
~STROKE - [railw] (of brakes) à-coup *m* de remplissage
~THREAD - [text] fil *m* de trame
~TWIST - [text] torsion *f* de la trame
~UP - [soil] (land fill) remblaiment *m*, surélévation *f*
~VALVE - [mech] (a feeding valve) vanne *f* de remplissage
~WEFT - [text] trame *f* de remplissage, trame *f* de fourrure
~WINDER - [text] canneteuse *f*
~WITH EARTH - [constr] remblayage *m*
FILLISTER - [carp] (a rabbeting plane with a movable stop) feuilleret *m*
[a groove, to receive glass, putty] feuillure *f*
~HEAD - [mech] tête *f* cylindrique
~HEAD SCREW - [mech] vis *f* à tête cylindrique
~SLOTTED HEAD - [mech] tête *f* cylindrique fendue
FILM, to - [cin] filmer, cinématographier
FILM - [phys chem] (a very thin layer, which may have a thickness substance from that of which it consists) pellicule *f*, film *m*, couche *f*
[photo & cin] film *m*, pellicule *f*
[nucl] (in a boundary layer, where the motion approximates streamline flows) couche *f* limite laminaire
~ADHESIVE - [ind chem] colle *f* en feuille
~BADGE - [nucl] (personal radiation-exposure monitoring device, consisting of a packet of suitable photographic film) film-témoin *f*, plaquette *f* de film
FILM BASE - [photo & cin] (the flexible base, generally of celluloid carrying the emulsion) support *m*
~BENDING - [cin] (the curvature of the film when placed in the projector gate) courbure *f* du film
~BUILDING PROPERTIES - [chem] (the properties in a coating composition vehicle which make it possible to produce applied films of the desired thickness) capacité *f* de former une pellicule
~CAMERA - [photo] chambre *f* à pellicules
~CAMERA CHAIN - [telev] chaîne *f* de caméra
~CAR - [cin] (the vehicle which carries the equipment) voiture *f* de reportage
~CASE - [cin] boîte *f* pour films
~CASTING MACHINE - [mech] appareil *m* pour la fabrication par coulée des pellicules
~CEMENT - [chem] (a glue used to join two sections of a film) colle *f* à film
~CHAMBER - [photo] logement *m* de la pellicule
~CHANNEL - s. film track
~CLIP - [photo] fichoir *m*
~COATING - [plast ind] couchage *m*
~COEFFICIENT - [phys] coefficient *m* de transmission superficielle
~COEFFICIENT OF HEAT TRANSFER - [phys] (the rate of transfer of heat through unit area of gas or liquid film under unit driving force) coefficient *m* de transmission superficielle de chaleur
~CONCEPT - [phys & chem] hypothèse *f* du voile fluide intermédiaire
~CONTRAST - [photo] contraste *m*
~COOLING - [astronaut] refroidissement *m* pelliculaire, refroidissement *m* par film fluide
~COPYING MACHINE - [cin] (a machine used to make positive copies) tireuse *f* cinématographique
~CUTTER LAMP - [cin] lampe *f* pour visionneuse
~DRYING DRUM - [cin] tambour *m* de séchage
~EDITOR - [cin] monteur *m*
~FEEDING - [cin] transport *m* de la pellicule
~FORMING YEAST - [brew ind] mycoderme *m*
~GATE - [cin] presse-film *m*
~GLUE - [chem] (a synthetic resin adhesive, usually termosetting, in the form of a thin dry film of resin with or without a paper carrier) colle *f* en feuille
~GRADER - [cin] (the technician who assesses the exposure time required) étalonneur *m*
~GRAIN - [ind chem] (the particles composing the light-sensitive layer) grain *m* d'émulsion
~GUIDE - [photo] rail *m* de guidage
~GUILDING - [photo] guidage *m* de la pellicule
~HOLDER - [photo] cadre *m* porte-film
~ILLUMINATOR - [radiat] (equipment with a good source of illumination for examining radiographs) négatoscope *m*
~KEY - [photo] clef *f* d'entraînement de la pellicule
~LENGHT - [cin] métrage *m*
~LIBRARY - [cin] (any collection of special films) cinémathèque
~LOCK - [photo] verrouillage *m* de la pellicule
~LOOP - [cin] (stretch of film) boucle *f* de film
~MAGAZINE - [photo] (the container for virgin films) chargeur *m*
~MARKER - [radiat] marqueur *m* de cliché
~MOVEMENT - [cin] avancement *m* du film
~OF GETTER METAL - [electron] miroir *m* de getter
~OF MOISTURE - [chem] film *m* de liquide, film *m* d'humidité

FILM OF PAINT - [paint] couche *f* de peinture
~OF RUBBER - [rubber ind] pellicule *f* de caoutchouc
~OF WATER - [th eng] (in a boiler) lame *f* d'eau, cloison *f* d'eau
[chem] pellicule *f* d'eau, couche *f* d'eau (absorbée sur les parois d'un récipient)
~PERFORATION - [cin] perforation *f* du film
~PICK-UP - [telev] prise *f* de vues
~POLISHING MACHINE - [mech] machine *f* à polir les films
~PROCESSING - [cin] (the operations of developing, fixing etc) traitement *m* du film
~PROCESSING APPARATUS - [cin] installation *f* de traitement du film
~RACK - s. film holder
~-REVIND INDICATOR - [photo] indicateur *m* de réenroulement
~RING - [nucl] (ring-shaped film badge) dosimètre *m* photographique annulaire
~ROLL - [photo & cin] bobine *f*
~RUPTURE - [cin] (any breakdown of the film) rupture *f* du film
~SCANNER - [telev] (television film scanner) analyseur *m* de film, télécinéma *m*
~SCANNING - [telev] (motion picture pickup) analyse *f* de film
~SCRAP - [photo] pellicule *f* sans émulsion
~SHOOTING TECHNIQUE - [cin & telev] technique *f* de prise de vues
~SLIDE - [photo] couloir *m* passe-film
~SOUND RECORDER - [el acoust] enregistreur *m* magnétique de son
~SPLICER - [cin] (a device used to join films together) colleuse *f*, presse *f* à coller
~SPOOL - [photo] bobine *f*
~STOCK - [cin] film *m* vierge
~STUDIO - [cin] studio *m* cinématographique
~SUPPORT - [photo] support *m* pelliculaire
~TAKE-UP FILM - [cin] bobine *f* d'enroulement (du film)
~TELEVISION - [telev] télécinématographie *f*
~TENSION DEVICE - [photo] dispositif *m* de tension de la pellicule
~TENSION FRAME - [photo] cadre *m* tendeur pour pellicule
~THREADING - [cin] (starting the film through the projector) mise *f* en place du film
~TRACK - [cin] (a component of camera and projector) couloir-film *m*, guide-film *m*
~TRAP - [cin] (in the film projector) fenêtre-image *f*
~-TYPE RESISTOR - [electron] résistance *f* à couche
~VIEWER LAMP - s.film cutter lamp
~WASTE - [cin] déchet *m* de film
~WINDING KEY - [photo] clef *f* d'enroulement de la pellicule
~WINDOW - [cin] (a movable element holding the film in its proper position against the aperture plate) presse-film *m*
FILMPACK HOLDER - [photo] châssis *m* film-pack
FILTER, to - [gen] filtrer, épurer, tamiser, suinter
FILTER - [impl] (in general, a device used to separate solids from liquids) filtre *m*
[photo] (in general, a glass plate used to correct the relative intensity of the beam of light) filtre *m*, écran *m*
[el] (frequency discriminating filter, or electric wave filter) filtre *m*, éliminateur *m*, réseau *m*
[metall] (in general a canvass cloth through which a solution of finely crushed ore is forced) filtrant, étouffeur *m* d'harmoniques, filtre *m*
FILTER AID - [ind chem] (finely divided solid material) adjuvant *m* de filtration
~AREA - [impl] surface *f* filtrante
~ATTACHMENT - [photo] raccord-filtre *m*
~ATTENUATION - [radio] atténuation *f* d'un filtre, affaiblissement *m* de filtre
~ATTENUATION BAND - [radio] (transducer transmitting energy at frequencies within bands and attenuates all other frequencies) bande *f* d'atténuation d'un filtre, bande *f* eliminée par un filtre
~BAG - [brew ind] trub-sac *m*
[photo] filtre-housse *m*
~BEAKER - [ind chem] vase *m* à filtrations chaudes
~BED - [hydr] (in sewage, a bed used for filtering purposes) couche *f* filtrante, bassin *m* de filtration
[mining] couche *f* filtrante
~BODY - [hydr] corps *m* filtrant
~BOTTOM - [hydr] (drainage) fond *m* d'un filtre, faux-fond *m* de filtre
~BOX - [hydr] bassin *m* filtrant
[mech] (in autos) boîtier *m* de filtre
~BY SUCTION, to - [chem] essorer
~CAKE - [chem] (the aggregate of solids on the filtering medium after filtration) gâteau *m* de masse filtrante
[brew ind] gâteau *m*
~CAKE PRESS - [mech] (in brewing industry) presse *f* à gâteaux de masse filtrante
~CARTRIDGE - s. filter element
~CASING - [mech] corps *m* du filtre
~CASING GASKET - [mech] garniture *f* du corps du filtre
~CHAMBER - [brew ind] chambre *f* de filtration
~CHOKE - [radio] (iron-core coil allowing the passing of direct current and opposing the passing of other currents) self *m* de filtrage
~CIRCUIT - [radio] (resonant circuit connected to a transmitting aerial) circuit *m* d'aspiration
~CLOTH - [text] toile *f* filtrante, étamine *f* à filtrer
~COIL - [radio] bobine *f* de filtre
~CONE - [impl] cône *m* à filtrer
~CORE - [metall] noyau-filtre *m*
~CRUCIBLE - [impl] creuset *m* de filtration
~DENSITY - [photo] densité *f* du filtre
~DISC - [metall] plaque *f* à filtrer
~DISCRIMINATION - [telecomm] sélectivité *f* d'un filtre
~DRAINAGE - [hydr] fond *m* d'un filtre
~DRUM - [hydr] tambour *m* filtrant
~DYE - [photo] (screening dye) colorant-filtre *m*
~ELEMENT - [mech] (the active portion of a filter, espec. if changeable) élément *m* filtrant
~FACTOR - [photo] (the increase of the exposure time due to the presence of a filter) coefficient *m* du filtre
~FLASK - [ind chem] (in a filter pump) flacon *m* à filtrer
~FRAME - [mech] cadre *m* de filtre
~FUNNEL - [impl] entonnoir *m* à filtre
[metall] entonnoir *m* à filtrer
~GALLERY - [ind chem] galerie *f* filtrante
~GATE - [metall] attaque *f* annulaire
~-GAUZE - [text & mech] filtrant *m*, toile *f* filtrante

FILTER GOVERNOR - [hydr] régulateur *m* automatique du débit à la sortie des filtres
~GRAVEL - [hydr] gravier *m* de filtre, gravier *m* filtrant
~HUMUS - [hydr] boue *f* secondaire
~IMPEDANCE COMPENSATOR - [radio] compensateur *m* d'impédance de filtre
~LAYER - [soil] (filter bed) couche *f* filtrante, tapis *m* filtrant
~LOADING - [biochem] charge *f* d'un lit bactérien
~MASS - [brew ind] masse *f* filtrante
~MEDIUM - [chem] (the material on which the solids removed are collected) matériau *m* filtrant
~NOZZLE - [mech] tuyère *f* d'un filtre
~OFF, to - [gen & ind chem] séparer par filtration
~PAPER - [chem] (pure cellulose paper used for filtration) papier *m* filtre, papier *m* à filtrer
~PASSBAND - [radio] bande *f* passante d'un filtre
~PASSER - [bacteriol] (an ultramicroscopic through a filter) filtrant *m*, bacille *m* filtrant
~-PASSING BACTERIUM - s. filter passer
~PIPE - [hydr] tube *m* filtrant
~PLANT - [hydr] installation *f* de filtration
~-PLATE - [chem] plaque *f* filtrante
~PRESS - [ind chem] (apparatus in which a filtering medium is supported in frames, through which the liquid under treatment is pumped) filtre-presse *m*
~-PRESS CLOTH - [ind chem] toile *f* de filtre-presse
~PRESS FOR DREGS - [brew ind] filtre-presse *m* à trouble
~PRESS FOR SLUDGE - s. filter press for dregs
~-PRESS WITH CENTRAL FEED - [sugar ind] filtre-presse *m* à plateaux avec un canal central d'alimentation
~PULP - [paper man] (cakes of rag fibres) gâteau *m* de pâte de papier
[brew ind] masse *f* filtrante
~PUMP - [ind chem] (glass air-ejector device used to exert suction on a filter when connected to a water-tap) pompe *f* à filtrer
~RANGE - [photo] zone *f* de transmission
~REGULATOR - s. filter governor
~RING - [metall] support *m* de filtre
[impl] anneau *m* de filtre
~RUFFLE - [impl] manchon *m* de filtre
~RUN - [hydr] durée *f* d'activité d'un filtre
~SAFETY DEVICE - [el] dispositif *m* de sécurité à filtres
~SAND - [hydr] sable *m* d'un filtre
~SCREEN - [photo] écran *m* filtre
~SECTION - [el] section *f* de filtre
~SHEET - [brew ind] couche *f* clarifiante
~SKIN - [soil] matériau *m* de remplissage, membrane *f* filtrante
~SLAB - [hydr] (a diffuser plate) plaque *f* filtrante, diffuseur *m*
~SPONGE - [impl] éponge *f* à filtrer
~STONE - [min] pierre *f* de liais
~SURFACE - [hydr] surface *f* du filtre
~TANK - [oil ind] réservoir *m* de filtration
~TOE - [soil] tapis *m* filtrant, tapis *m* drainant, filtre *m* de pied
~TRANSMISSION BAND - [radio] bande *f* passante d'un filtre
~VAT - [brew ind] cuve-filtre *f*
~WELL - [soil] puits *m* filtrant
FILTERABLE - [gen] (also filtrable) filtrant
FILTERABLE VIRUS - s. filter passer
FILTERED - [gen] filtré
~LIGHT - [light] lumière *f* filtrée
FILTERING - [gen chem etc] filtrage *m*, filtration *f*
(the material with which a filter is made) matériau *m* à filtrer
~BASIN - [hydr] (a tank in which passing liquid is subjected to filtering) bassin *m* de filtration
~CRUCIBLE - [ind chem] (crucible in which filtering can be carried out, the collected precipitate being afterwards ignited in the crucible itself) creuset *m* de filtration
~FLASK - [ind chem] (special flask to take a filter, with tubular neck for connexion to a filter-pump) flacon *m* à filtrer, fiole *f* conique (pour filtrer à la pompe)
~ROOM - [brew ind] chambre *f* de filtration
~STONE - [ind chem] pierre *f* à filtrer
FILTERPLEXER - [telev] combinaison *f* diplexeur-filtre de bande latérale
FILTHY WATER - [hydr] purin *m*
FILTRATE, to - s. to filter
FILTRATE - [chem etc] (the clear liquid obtained by filtering) produit *m* filtré, filtrat *m*
FILTRATION - [chem etc] (the separation of solids from a liquid by passage of the latter through a porous substance) filtration *f*, filtrage *m*
[nucl] (the removal of some components of a beam of radiation) filtration *f*
~PERCOLATION - [oil ind] filtrage *m* à percolation
~PROCESS - [chem & ind proc] procédé *m* de filtration
FIMBRIA - [anat] frange *f*, villosité *f*
FIMBRIATE - [med] frangé
FIN - [zool] nageoire *f*
[mech] (a projecting strip of metal) ailette *f*
[metall] (thin projection of excess material forming at the parting line of a mould) bavure *f*, ébarbure *f*, balêtre *f*, baleine *f*
[aero] (a fixed vertical surface designed to give lateral stability of motion) plan *m* fixe vertical, dérive *f*
[carp] (on the edge of a board) languette *f*
[metall] (a casting defect) ailette *f* de soudure
[aero] (the tail fin) dérive *f*
[astronaut] (of a rocket) nervure *f*, dérive *f*
[aero] (of an airship) empennage *m*
~AREA - [aero] surface *f* de dérive
~-BLOCK - [mech] (in water-heating apparatuses) corps *m* de chauffe
~CRACK - [metall] crique *f* de lamination
~FIXING POINT - [aero] point *m* de fixation de la dérive
~KEEL - [naut] aileron *m*
~NECK BOLT - [mech] boulon *m* à clavette
~POST - [aero] (the vertical structural member of a fin) longeron *m* vertical de la dérive
FINAL - [gen] final, dernier
~AMPLIFIER - [radio] (the last stage in an amplifying system) amplificateur *m* de sortie
~ANODE - [electron] (in a cathode-ray tube) anode *f* finale
~APPROACH - [aero] (the last part of the approach procedure, from the time when the aircraft passes the airfield boundary to the beginning of the final flare-out) approche *f* finale
~APPROACH ALTITUDE - [aero] (the height at which

the final approach is begun) hauteur *f* d'approche finale

FINAL BALANCE OF PROPELLER - [aero] équilibrage *m* de l'hélice

~BLANKING - [telev] postsuppression *f*

~CLARIFICATION - [hydr] décantation *f* secondaire

~CLARIFICATION TANK - [hydr] bassin *m* de décantation finale, décanteur *m* secondaire

~CONNECTOR - [telecomm] sélecteur *m* de ligne

~CONTROL ELEMENT - [control] (in a controlling means, the portion which changes the value of the manipulated variable) organe *m* de réglage final

~CONTROLLED CONDITION - [control] (the value which the control system is aiming to achieve) grandeur *f* réglée finale

~CONTROLLER - [aero] (radar controller transmitting landing information during final approach) contrôleur *m* radar de precision

~CURVES - [contr] courbes *fpl* finales

~DEGREE OF ATTENUATION - [brew ind] atténuation *f* limite

~DRIVE - [mech] propulsion *f* par les roues arrière

~DRIVE BOX - [mech] boîte *f* des engrenages de réduction

~DRIVE MECHANISM - [mech] engrenages *mpl* de réduction des roues motrices

~DRIVE MOUNT - [mech] carter *m* du réducteur

~DRIVE SPUR GEAR - [mech] transmission *f* à engrenages cylindriques des roues motrices

~FILTRATION - [hydr] filtration *f* terminale

~HARDENING - [metall] trempe *f* finale

~IMAGE - [telev] image *f* finale

~LIMIT SWITCH - [el] (limit switch used in electric lifts) interrupteur *m* de fin de course

~MASHING TEMPERATURE - [brew ind] température *f* finale de la trempe

~PASS - [metall] cylindre *m* finisseur

~PRINT - [cin] (the release print) copie *f* de distribution, copie *f* d'exploitation

~PROCEDURE TURN - [aero] (the turn made by an aircraft from a course reciprocal to the direction of the runway to begin the final approach) virage *m* conventionnel final

~PRODUCT - [gen] produit *m* final
[nucl] (the stable nuclide which is the final member of a radioactive series) produit *m* terminal, produit final

~PROOF - [print] morasse *f*

~PURIFICATION - [hydr] épuration *f* secondaire

~SCREEN - [radar] écran *m* final

~SELECTOR - [telecomm] (in automatic exchange telephones) sélecteur *m* de ligne, connecteur *m* final

~SETTLING TANK - s. final clarification tank

~SHOT - [cin & telev] prise *f* définitive

~STAGE - [electron] étage *m* final

~STAGE VALVE - [telecomm] lampe *f* de sortie, tube *m* de sortie

~SULFITATION OF THIN JUICE - [sugar ind] sulfitation *f* finale du jus épuré

~TEMPERATURE - [sugar ind] température *f* d'évacuation
[el] température *f* de régime

~TEST - [gen] essai *m* final

~TRIAL COMPOSITE - s. final print

~TWIST - [text] torsion *f* finale

~UP-RUN PURGE - [water gas install] balayage *m* par vapeur montante

FINAL VALUE - [contr] valeur *f* finale, valeur *f* réglée asymptotique

~VALUE OF AMPLIFICATION - [radio] valeur *f* finale de l'amplification

~VOLTAGE - [el] tension *f* d'arrêt, tension *f* finale

~WRITING AMPLIFIER - [comput] (of a drum memory) amplificateur *m* final d'écriture

FINANCIAL YEAR - [fin] année *f* budgétaire

FIND, to - [gen] trouver, découvrir
[a course] s'orienter
[math] (a value) chercher

FIND - [gen] trouvaille *f*, découverte *f*

~A BEARING, to - [naut aero etc] s'orienter

FINDER - [opt] lunette *f* de repère
[photo] (view finder, for viewing the image field of the camera) viseur *m*
[instr] (small auxiliary telescope parallel to the tube of a large telescope) lunette *f* de repère, chercheur *m*

~FRAME SELECTOR - [photo] sélecteur *m* de champ

~SWITCH - [telecomm] chercheur *m* d'appel

FINE, to - [ind chem] (to clarify a fermented liquor) clarifier

FINE - [gen] fin, pur
[instr] (of adjustment) précis
[mech] (in tolerance systems) fin
[text] (of wool, indicating good quality) fin

~ADJUSTMENT - [instr etc] (a device for making a secondary and more accurate setting) réglage *m* précis

~ADJUSTMENT SCREW - [instr] vis *f* de fin calage

~BORE, to - [metall] reforer

~BULLION - [metall] lingot *m* d'or (ou d'argent pur)

~CHROMINANCE PRIMARY - [telev] (in colour television) primaire *f* principale de chrominance

~CONTROL - [mech] réglage *m* de précision
[contr] (to correct low amplitude reactivity variations) pilotage *m*

~-CONTROL CHANNEL - [nucl] chaîne *f* de pilotage

~-CONTROL MEMBER - [nucl] élément *m* de pilotage

~-CONTROL ROD - [nucl] (a nuclear reactor control-rod designed for exact regulation) barre *f* de pilotage, barre *f* de réglage fin

~COUNT - [text] (of yarns) numéro *m* fin

~COUNT SPINNING - [text] filage *m* en fin

~-COUNT YARN - [text ind] fil *m* à numéro fin

~CRUSHING - [min] broyage *m* fin

~DRAWING - [metall] étirage *m* fin, tréfilage *m* fin

~ENGRAVING - [rubber man] (in tyres) profil *m* en traits de scie (antidérapant)

~FEED - [mech] avancement *m* micrométrique

~FIBROUS - [rubber man] (of tyres) à fibres fines

~-FINISH - [gen & mech] fini *m* de précision

~-FINISH MACHINING - [mech mach] usinage *m* de précision

~FOCUS - [phys] (in X-rays) petit foyer *m*

~-FOCUSING SCREW - [photo] vis *f* de mise au point précise

~GAUGE SHEET SCRAP - [metall] chutes *fpl* fines de tôle

~GOLD - [metall] or *m* au titre

~-GRAINED - [geol] (in rock classification) à grains fins
[metall] à grain fin

~-GRAINED EMULSION - [chem] émulsion *f* à grains fins

~-GRAINED SAND - [geol] sable *m* à grains fins

FINE-GRAINED WHEEL - [mech] meule *f* à grain fin
~GRAVEL - [gen] menu gravier *m*
~GRITS - [agric] gruau *m* fin
~HACKLE - [text ind] peigne *m* fin, peigne *m* à repasser
~-HAND FEED - [mech mach] avancement *m* micrométrique
~LAPPING - [metall] fleur *f* (sur les pieces coulée)
~MESHED SIEVE - [min] tamis *m*
~MILL - [min] broyeur *m* fin
~ORE - [min] minerai *m* fin
~PARTICLE - [ind chem etc] (as obtained through filtering) particule *f* fine, grain *m* fin
~PITCH - [aero] pas *m* minimum
~PITCH GEAR HOB - [mach tool] fraise-mère *f* à pas fin
~PITCH STOP - [aero] arrêt *m* du pas minimum
~-PITCH THREAD - [mech] pas *m* fin, filetage *m* fin
~-REGULATING ROD - s. fine-control rod
~RUBBLE - [constr] blocaille *f* fine
~SCANNING - [telev] (scanning using a very small light beam and a large number of lines) analyse *f* à haute définition
~SCREEN - [impl] tamis *m*
~-SCREENING - [gen] criblage *m* fin, tamisage *m* fin
~SIDE - [phys] côté *m* vide poussé, côté *m* haut vide
~SIEVE - s. fine screen
~-SIEVING - s. fine-screening
~SILVER - [metall] argent *m* de coupelle
~SOLDER - [metall] soudure *f* fine
~SPECTRUM - [spectrology] (the resolution of a line into two or more fine and close lines in atomic emission spectra) spectre *m* à structure fine
~STRUCTURE - [phys] (of an alpha-particle spectrum when several groups are present) structure *f* fine
~-STRUCTURE CONSTANT - [phys] constante *f* de structure fine
~TOOTHED COMB - [text] peigne *m* aux dents fines
[impl] décrassoir *m*
~TROMMEL - [min] trommel *m* des fins
~TUNING - [radio] (accurate adjustment of the amplitude value) réglage *m* précis
~VACUUM - [phys] vide *m* moyen, vide intermédiaire
~WEATHER - [met] beau temps *m*
FINELY DIVIDED CARBON - [chem] poussière *f* de charbon
~DIVIDED CONCENTRATES - [min] concentrés *mpl* très fins
~-GRAINED - s. fine-grained
~-GRANULAR - s. fine-grained
~-GROUND - [gen] fin
~GROUND CONCENTRATE - [min] (concentrate of ore which is pulverized by mechanical means) concentré *m* pulverisé
FINENESS - [gen] finesse *f*, pureté *f*
[chem] (the state of subdivision) finesse *f*
[metall] (the gold or silver content of an alloy consisting predominantly of one of these metals, stated as parts per thousands) titre *m*, aloi *m*
~MODULUS - [constr] (numeral indicating the fineness of an aggregate) module *m* de finesse
~OF SCANNING - s. fine scanning
~OF SCANSION - s. fineness of scanning
~OF SHAPE - [gen] finesse *f* de forme
~OF WOOD - [timber ind] finesse *f* du bois
~RATIO - [shipbuild] allongement *m*, finesse *f*

FINERY - [metall] (furnace for high-quality bar-iron) four *m* d'affinage
~FIRE - [metall] four *m* (ou foyer) d'affinerie
~PROCESS - [metall] travail *m* au bas-foyer
FINES - [min] (in size classification; less than one sixteenth of an inch) fines *fpl*
FINGER, to - [gen] manier, tâter, doigter
[mus] (a guitar) agacer
FINGER - [anat] doigt *m*
[mech] doigt *m*, touche *f* guidée
[die-casting] noyau *m*, plongeur *m*
~-AND-TOE DISEASE - [bot] (plant disease; clubroot) gros pied *m*, hernie *f* du chou
~BAR - [mech] doigt *m* de retenue, doigt *m* de relevage
~-BOARD - [mus] touche *f*
[mining] râtelier *m* à tiges
~CHUTE - [mining] cheminée *f* étroite
~DAMPENER - [impl] mouilleur-doigts *m*
~DAMPER - s. finger dampener
~DIE - [metall] étampe *f* à noyau glissant
~DISK - [telecomm] (of telephones) disque *m* d'appel, envoyeur *m*
~GAUGE - [mech] guide-tôle *m*
~GRASS- [bot] digitaria *m*
~GRIP - [mech] dispositif *m* mécanique de préhension
~GUIDE - [mech] guide *f* de tige
~KNIFE - [impl] couteau *m* à doigts
~MILKING - [agric] traite *f* sur le pouce
~POST - [gen] pateau *m* indicateur
~RELEASE - [photo] déclenchement *m* au doigt
~SCREW - [mech] vis *f* à oreilles
~SLOT - [metall] fente *f* de levage
~STOP OF THE DIAL - [telecomm] (of telephones) butée *f* du cadran d'appel
~-TIP CONTROL - [mech] réglage *m* simple
~WEEDER - [agric] désherbuse-sarcleuse *f*
FINGERHOLE - [mus] (in wind-instruments) trou *m*
FINGERING - [oil ind] digitation *f*
FINGERNAIL TEST - [brew ind etc] (of grains etc) essai *m* à l'ongle
FINGERPRINT - [gen & leg] empreinte *f* digitale
FINGERSTALL - [impl] doigtier *m*
FINGERY COKE - [fuel] (in size classification) coke *m* en morceaux allongés
FINIAL - [constr] (an ornament at the top of a pillar etc) fleuron *m*, faîteau *m*
[el] capuchon *m* en fonte
FINING - [glass man] affinage *m*
[metall] affinage *m*, fabrication *f* de fonte malléable
[brew ind] collage *m*
~-OFF - [build] (the operation of applying the setting coat) application *f* du voile
~PERIOD - [metall] période *f* des fumées
~PROCESS - [metall] méthode *f* d'affinage
FININGS - [brew ind] (isinglass used to clarify beer) colle *f* poisson
FINISH, to - [gen] finir, achever, cesser, terminer se terminer
[metall] planer, finir
FINISH - [gen] fin *f*
[constr] deuxième couche *f*
[text ind] (the final operation on woven material) apprêt *m*
[metall] (the stock which is removed) laminage *m*
[glass man] calage *m*

[paint] laque *f*, peinture *f*, vernis *m*
FINISH BORING - [mech] alésage *m* de précision
~BRIGHT, to - [metall] polir finement
~CARD - [text] (also called finisher) carde *f* en fin
~-GRINDING - [mech] meulage *m* de précision
[metall] finissage *m* à la meule, finissage *m* de meulage
~HARDWARE - [metall] quincaillerie *f*
~MACHINING - [metall] finissage *m*
~MILLING - [metall] fraisage *m* de précision
~TURNING - [mech] finissage *m* au tour
FINISHED - [gen] fini, achevé, définitif, apprêté
~ARTICLE - [ind proc] (the final product of a manufacturing process) article *m* fini (ou apprêté)
~BEER - [brew ind] (beer ready for drinking) bière *f* prête au débit
~PART - [mech] pièce *f* finie
~PRODUCT - [metall] produits *m* de laminage
~SHEET - [metall] tôle *f* finie, tôle *f* nettoyée
~STEEL - [metall] acier *m* fini
FINISHER - [text] carde *f* en fin
[die-casting] filière *f* à finir
~BOX - [text] (machine designed to straighten the fibres before drawing and spinning) finisseur *m*, frotteur *m* en fin
~CARD - [text] carde *f* en fin
FINISHING - [gen] achèvement *m*, finissage *m*, apprêtage *m*
[text] apprêtage *m*
[paper man] apprêtage *m*
~AGENT - [text] adjuvant *m* de finissage
~BIT - [tool] alésoir *m*
~BOARD - [text] forme *f* pour bas
~CALENDER - [text] calandre *f* d'apprêt
~COAT - [paint] (the last layer of paint or varnish or the like, applied to a surface) deuxième couche *f*
~CUT - [mech] entaille *f* de précision
~CUTTER - [mech mach] fraise *f* finisseuse, taraud *m* finisseur
~DOPE - [text] pâte *f* d'apprêtage
~DRAFT - [text] étirage *m* finisseur
~DRUM - [min] trommel *m* finisseur
~GILL BOX - [text] gill-box *m* finisseur
~GROOVE - [metall] cannelure *f* (de laminoir) finisseuse, finisseur *m*
~HOB - [mach tool] fraise-mère *f* à finir
~IMPRESSION - [metall] (in diecasting) empreinte *f* de finissage
~MACHINE - [text] machine *f* à apprêter
[metall] machine *f* de finissage
~METAL - [metall] apport *m*, fonte *f* d'addition
~MILL - [metall] laminoir *m* finisseur
~NAIL - [metall] pointe *f* ronde à tête "homme"
~OF PAPER - [paper] apprêtage *m* du papier
~PASS - [metall] (in the body of the roll) cannelure *f* finisseuse
~PROCESS - [text] procédé *m* d'apprêt
~RATE - [el] (in a storage battery, the reduced rate at which charging is completed) régime *m* de fin de charge
~REFINER - [rubber ind] cylindres *mpl* affineurs
~ROASTING - [min] grillage *m* définitif
~ROLL - [mech mach] (of a rolling mill) cylindre *m* finisseur (de laminage)
~ROLLING MILL - [metall] laminoir *m* finisseur
~ROLLS - s. finishing rolling mill
~RUBBING PORCUPINE BOX - [text ind] frottoir *m* finisseur
FINISHING SCUTCHER - [text] batteur *m* finisseur
~SLOT MILL - [metall] fraise *f* pour finir les rainures
~STAND - [metall] cage *f* finisseuse, finisseuse *f*
~TAP - [mech] taraud *m* finisseur, troisième taraud *m*
~TEMPERATURE - [metall] température *f* de finissage
~TOOL - [tool] achevoir *m*
~WAX - [carp] cire *f* pour polissage
~WINDING FRAME - [text] bobinoir *m* finisseur
FINISHINGS - [metall] corrections *fpl*
FINITE - [gen] fini, limité
[math] fini
~CYLINDRICAL REACTOR - [nucl] (cylindrical theoretical reactor having finite dimensions) réacteur *m* cylindrique fini
~LINE - [geom] segment *m*
~PROGRESSION - [math] progression *f* finie, série *f* finie
~SERIES - [math] série *f* finie, progression *f* finie
~SQUARE WELL - [nucl] puits *m* de potentiel carré de dimensions finies
FINNED - [gen & mech] à ailettes
~CYLINDER - [auto] cylindre *m* muni d'ailettes
~CYLINDER HEAD - [auto] culasse *f* munie d'ailettes
~JACKET - [mech] enveloppe *f* à ailettes
~PIPE - [plumb] tuyau *m* à ailettes
~TUBE - [plumb] tube *m* à ailettes
~TUBULAR RADIATOR - [auto] radiateur *m* en tubes à ailettes
FINNING - [mech] ailette *f*, application *f* d'ailettes
FIORD - [geogr] (narrow and winding sea inlets bounded by steep slopes) fiord *m*, fjord *m*
FIPPLE FLUTES - [mus] (any small flute of the whistlemouthpiece type) flûtes *fpl* à biseau
F.I.R. - s. flight information region
[nucl] (initials of Food Irradiation Reactor) réacteur *m* d'irradiation de denrées
FIR - [bot] sapin *m*
~CONE - [bot] cône *m* de sapin, pigne *f*
~OIL - [chem] essence *f* de sapin
FIRE, to - [gen] mettre feu, incendier
[firearm] (of guns etc) décharger, faire feu, allumer (un fourneau de mine)
[astronaut] (a rocket) lancer
[mech] chauffer, charger
[pottery] cuire
[photo] (the flash) déclencher (l'éclair)
FIRE - [gen] feu *m*, incendie *m*
[firearms] tir *m*, feu *m*, coups *mpl* de feu
[auto] allumage *m*
~ALARM - [instr] (devices of various types, both detecting and warning) avertisseur *m* d'incendie
~APPLIANCE - [gen] (any device for use in extinguishing fire) matériel *m* d'incendie
~ASSAY - [metall] (quantitative analysis of an ore by high temperature "dry" methods e.g. cupellation) coupellation *f*
~-BACK BOILER - [th eng] (boiler which is fitted to a kitchen range) brûleur *m* à retour de flamme
~BALL - [astr] aérolithe *m*
~BAR - [th eng] barreau *m* de grille
~BAR BOTTOM GAS-PRODUCER - [th eng] gazogène *m* à grille
~BARRIER - s. fire wall
~BED - [th eng] grille *f*

FIRE BLENDE - [min] pyrostilpnite *f*
~BOARD - [mining] écriteau *m* pour quartier grisouteux
~BOX - [th eng] (the structure containing the furnace of a boiler or the like) foyer *m*
~BOX STEEL - [metall] acier *m* à refouler, acier *m* pour foyers
~BRICK - [constr] brique *f* réfractaire
~BRIDGE - [metall] autel *m*, pont *m*
~CEMENT - [build] ciment *m* réfractaire
~CHAMBER - [metall] foyer *m*, chauffe *f*
~CHUTE - [el] boîte *f* pare-feu
~CLAY - [metall] argile *f* réfractaire, terre *f* à pipe
[geol] (used widely as refractories, also in construction work and metallurgy) argile *f* réfractaire, terre *f* réfractaire
~CLAY BOX - [metall] caisse *f* en matière réfractaire
~CLAY CRUCIBLE - [metall] creuset *m* en argile
~CLAY PLUG - [oil ind] bouchon *m* d'argile réfractaire
~-COAT - [metall] revêtement *m* réfractaire
~COCK - [constr] bouche *f* incendie
~CONTROL - [contr] conduite *f* de tir
~-CONTROL COMPUTER - [comput] calculateur *m* pour la conduite de tir
~-CONTROL ROOM - [navy] salle *f* de direction du tir
~COPPER - [brew ind] chaudière *f* à feu nu
~CRACK - [metall] (a crack occurred during firing) crique *f*, criqûre *f* de recuit, criqûre *f* de recristallisation
~CRACKING - [metall] criqûre *f* de recristallisation
~-DAMP - [chem] (mixture of methane and other hydrocarbon gases evolved from coal and occurring in some coalmines) grisou *m*
~DETECTOR - [instr] (thermostatically controlled device operating an alarm on a sudden rise of temperature) détecteur *m* d'incendie
~DEVIL - [impl] brasier *m*
~DOOR - [th eng] (sliding or hinged cover to close the main opening to a furnace) porte *f* de foyer
~DRENCHER - [impl] extincteur *m*
~EARTH - [geol] terre *f* réfractaire
~ENGINE - [mech] (a self-propelled vehicle fitted pumps for fire fighting) pompe *f* à incendie
~ESCAPE - [constr] (special exit used in case of fire) échelle *f* de sauvetage
~ESCAPE TRAILER - [transp] remorque *f* pour engins de sauvetage
~EXTINGUISHER - [ind appliance] (generally portable antifire devices) extincteur *m*
~EXTINGUISHING TANKER - [mech] (a self-propelled vehicle fitted with a tank to contain a fire extinguishing medium) camion-citerne *m* pour service d'incendie
~FIGHTING - [gen] (the operation of extinguishing a fire) service *m* d'incendie
~FLAVOUR - [brew ind] goût *m* de fumée
~-FLY - [zool] luciole *f*
~GRATE - [th eng] grille *f*
~HEATING - [brew ind] cuisson *f* à feu nu
~-HOOK - [impl] crochet *m* à incendie
~HOSE - [impl] manche *f* à feu, tuyau *m* de pompe à incendie

FIRE-HOSE BOX - [impl] porte-manche *m* à feu
~-HOSE CONNECTION - [impl] manchon *m* pour tuyau à incendie
~HYDRANT - [impl] bouche *f* d'incendie, prise *f* de incendie
~IRONS - [impl] garniture *f* de foyer
~KILN - [min] four *m* de grillage
[brew ind] touraille *f* à feu nu
~OPAL - [min] (brilliant opal of an orange-flame colour) opale *f* de feu, opale *f* à flammes
~PLUG - s. fire hydrant
[gen] poste *m* d'incendie, prise *f* d'incendie
~POINT - [phys] (the lowest temperature at which a liquid will vapourize sufficiently to burn continuosly) point *m* d'inflammation
[of materials] température *f* de combustion spontanée, seuil *m* d'inflammabilité en phase vapeur
~POT - [metall etc] creuset *m*
~PROOF - [gen] ignifuge, incombustible, résistant au feu
~PROOFING OF CABLES - [el] ignification *f* de câbles
~PROTECTION - [gen] protection *f* contre l'incendie
~PUMP - [fire extinguishing] (a power drive pump mounted on a carriage (usually a trailer) and used for fire fighting) pompe *f* à incendie, pompe *f* à feu
~RAKE - [impl] pique-feu *m*, râble *m*, attisoir *m*
[metall] râble *m*
~RESISTANT - s. fire resisting
~RESISTING - [gen] résistant au feu
~-RESISTING COVERING - [el] (for electric cables) enveloppe *f* ignifuge
~-RESISTING PAINT - [paint] peinture *f* resistante au feu
~-RESISTING PROPERTY - [gen & chem] résistance *f* au feu
~RETARDANT - [chem] ignifuge
~RING - [mech] (the piston ring which is nearest to the combustion chamber of a reciprocating i.c. engine) segment *m* de piston
~SAND - s. fire earth
[metall] sable *m* de moulage
~SCALE - [metall] couche *f* d'oxyde de cuivre
~-SCREEN - [impl] écran *m* ignifuge
~SHIP - s. fire boat
~-SHUTTER CONTROL - [cin] (mechanism releasing the safety fire-shutter to isolate the danger) dispositif *m* de déclenchement du volet de securité
~-STONE - [min] (a type of sandstone with uses in the construction of glass furnaces) pierre *f* à feu, silex *m*
~STOP - [metall] autel *m*, pont *m*
~STORM - [nucl] (stationary mass of fire) tempête *f* de feu
~STRIP - [cin] (a stretch of fireproof film) bande *f* anti-feu
~TRUCK - s. fire engine
~TUBE - [th eng] tube *m* à fumée, tube *m* de chaudière
~-TUBE BOILER - [th eng] (type of boiler in which the products of combustion pass through tubes surrounded by water) chaudière *f* à tube-foyer
~WALL - [constr] facade *f* coupe-feu
[aero & naut] (fireproof bulkhead interposed between the engine compartment and other parts of the structure in some aircraft) cloison *m*, pare-feu
~WASTE - [metall] pertef au feu
FIREARM - [gen] arme *f* à feu
FIREBALL - [astr] bolide *m*, éclair *m* en boule

IREBARS - [mech] (cast-iron bars used to form the grate of a furnacen on which the burning fuel rests) fers mpl à barreaux de grille
IREBOX - s. fire box
~CROWN - [th eng] ciel m de foyer
IREBRICK - [build etc] (refractory brick) brique f réfractaire
~ARCH - [th eng] (built at the end of a boiler furnace) voûte f en briques réfractaires
~LINING - [th eng] revêtement m de briques réfractaires
IRECLAY - s. fire clay
IREDAMP - s. fire-damp
~DRAINAGE - [mining] drainage m du grisou
~FEEDER - [mining] soufflard m de grisou
~POCKET - [mining] poche f à grisou, ballon m
~PROOF MACHINE - [el] (flame-proof machine used in mines) appareil m électrique pour atmosphère grisouteuse
IREDOG - [impl] chenet m, landier m
IREGRATE - [ovens] autel m du foyer
IRELIGHT - [light] lumière f du feu
IREMAN - [th eng] (a stoker) chauffeur m, cuiseur m
IREPLACE - [constr] cheminée f
~INSERT - [th eng] (inset fire) radiateur m encastré
IREPROOF, to - [gen] (to render fireproof) rendre ininflammable, ignifuger
IREPROOF - [gen] (not liable to combustion, e.g. of a coating or a structure) résistant au feu, ignifuge, ininflammable
~BULKHEAD - [aero] (a partition in an aircraft designed to prevent a fire from spreading beyond it) cloison m pare-feu
~CEMENT - [constr] ciment m réfractaire
~FABRIC - [text] tissu m ignifuge
~MATERIAL - [ind chem] matériaux mpl incombustibles
IREWOOD - [timber] bois m à brûler
IREWORKS - [gen] feu m d'artifice
IRING - [firearms] tir m, feu m
[astronaut] lancement m, mise f à feu
[ceram] (the process of baking clay objects to obtain hardness etc) cuisson f, cuite f
[electron] (in a magnetic amplifier) activation f, amorçage m
[mining etc] (the ignition of an explosive device) allumage m, mise f à feu
[metall etc] (the process of feeding a furnace) chauffage m, chargement m
[mus] carillonnement m
[mech] (in internal combustion engines) allumage m
[mining] (of a mine) tirage m
~A BOILER - [metall] allumage m du foyer d'une chaudière
~ANGLE - [radio] (the angle at which the gate impedance changes from high to a low angle) angle m de saturation
~CHARGE - charge f d'amorçage
~CURRENT - [el] courant m d'allumage
~DOOR - [th eng] tisard m
~ELECTRODE - [radio] électrode f d'amorçage
~GLASS - [glass man] verre m à fond renforcé
~KEY - [el] (to fire a charge of explosive) détente f
~ORDER - [mech] (the ignition sequence in an i.c. reciprocating engine) rythme m d'allumage
~PIN - [firearms] percuteur m
~POINT - [phys] point m d'inflammation
[radio] point m d'activation
~POTENTIAL - [electron] (the grid-cathode tension necessary to make a gaseous triode conductive) potentiel m d'ionisation, tension f d'amorçage
~ROD - [el] (in a rectifier) barre f d'amorçage
~SEQUENCE - s. firing order
~STROKE - [mech] (in internal combustion engines) course f d'explosion, course f de détente
~TOOL - [impl] tisonnier m
[metall] attirail m de chauffe
FIRKIN - [meas] (cask holding 9 gallons) quartaut m (approx.)
[meas] (for liquids, butter etc) tonnelet m, tinette f
FIRM - [gen] ferme, compact, consistant
[comm] (commercial undertaking) firme f, maison f, société f commerciale
FIRMAMENT - [astr] firmament m
FIRMER - s. firmer chisel
~CHISEL - [carp] (a wood-cutting chisel) ciseau m ordinaire, ciseau m à panne
FIRMNESS - [gen] fermeté f, solidité f, consistance f
FIRN - [geol] (a fairly compacted mixture of snow and ice occurring above the snow-line) névé m
FIRST - [gen] premier
~AID - [med] (immediate attention given to an injured person before medical treatment is available) premiers secours mpl
~AID KIT - s. first aid outfit
~AID OUTFIT - [med] (compact set of equipment for rendering first-aid) infirmerie f portative
~ANODE - [electron] (in a cathode-ray tube, the positive electrode which is nearest to the cathode) première anode f
~BREAKER - [text] carde f briseuse
~CLASS LEVER - [phys] levier m du premier genre
~COAT - [paint] première couche f
[build] couche f de fond
~DETECTOR - [radio] (the modulating portion of a frequency changer) convertisseur m
~DRAFTING - [text] étirage m préparatoire, étirage m en gros, premier étirage
~DRAWING FRAME - [text] étirage m préparatoire, étirage en gros
~FIXING - [carp] (the basis for the support of joinery) accessoires mpl de menuiserie
~FLOOR - [constr] (in GB) premier étage m
[constr] (in the USA) rez-de-chaussée m
~GENERATION TAPE - [el acoust] (the original recording) bande f de première génération, enregistrement m original
~GROUP SELECTOR - [telecomm] premier sélecteur m de groupe
~-HAND - [gen] (adj) de première main
~HARMONIC - [radio] (the fundamental wave) onde f dominante
~INVERSION OF A CHORD - [mus] première inversion f d'un accord
~IONIZATION POTENTIAL - [nucl] premier potentiel m d'ionisation
~ITEM LIST - [comput] tabulation f de groupe
~LINE FINDER - [telecomm] chercheur m primaire (ou de lignes)

FIRST LOADING COIL SECTION - [telecomm] longeur *f* d'amenée, complément *m* d'une section de pupinisation
~MATE - [naut] second *m*
~MEAN CHORD - [aero] (the value is obtained by dividing the gross wing area by the span) corde *f* moyenne géométrique, profondeur *f* moyenne
~MERIDIAN - [astr] premier méridien *m*, méridien *m* d'origine
~MOTION SHAFT - [mech] arbre *m* principal
~MOVEMENT FORM - [mus] forme *f* sonate
~ORDER EQUATION - [math] équation *f* du premier ordre
~ORDER THEORY - [opt] théorie *f* du premier ordre
~OVERTONE BAND - [phys] (a spectral band) bande *f* de la deuxième harmonique
~PASS - [metall] (first weld) premier cordon *m* de soudure
~PHASE - [cin] verrouillage *m* principal
~PLY - [rubber ind] (of tyres) premier pli *m* de câblé
~POINT OF ARIES - [astr] (the intersection of the celestial equator and the ecliptic: the vernal equinox. The origin of the celestial co-ordinates) point *m* vernal, point *m* gamma
~PRINTING - [print] impression *f* en blanc
~QUANTUM NUMBER - [phys] nombre *m* quantique principal
~QUARTER - [astr] premier quartier *m*
~RADIATION CONSTANTE - [nucl] première constante *f* de rayonnement
~REFLECTION - [telev] première réflexion *f*
~REMOVE SUBROUTINE - [comput] sous-programme *m* du premier étage
~ROUTE - [telecomm] voie *f* normale
~RUN - [cin] première *f*
~RUNNINGS - [chem] (the first fraction in fractional distillation, often chiefly impurities) têtes *fpl* (de distillation)
~SPEED - [auto] première vitesse *f*
~SPEED WHEEL - [auto] couple *m* d'engrenages de première
~SPINNING - [text] premier tors *m*
~STRING - [mus] (of a violin) chanterelle *f*, mi *m* du violon
~TEST - [gen] examen *m* préalable
~-SURFACE DECORATION - [plast ind] (application of decoration to the side of the material which will be visible in use) décoration *f* sur la surface visible
~WELD - s. first pass
~WORT - [brew ind] premier moût *m*, moût *m* trouble
FIRTH - [geogr] estuaire *m*, bras *m* de mer
FISH, to - [gen] pêcher
[naut] (the anchor) traverser
[min] (the operation of retrieving objects from a borehole, such as parts of drills or tools) repêcher, saisir
[oil ind] repêcher
FISH - [zool] poisson *m*
~-BAR - s. fish-plate
~BASKET - [fixhing] bourriche *f* à poissons
~-BELLIED GIRDER - [constr] poutre *f* à ventre de poisson
~-BOLT - [mech] boulon *m* d'éclisse
~BONE - [zool] arête *f*
~-EYE LENS - [photo] grand-angulaire *m* extrême
FISH-EYES - [plast ind] (blemishes in plastic films, especially polythene) yeux *mpl* de poisson
-GLUE - [ind chem] (isinglass. A strong glue prepared from skins and offal of fish) colle *f* de poisson, isinglass *m*
~-HOOK - [fishing] hamecon *m*
[naut] croc *m* de traversière
~JOINT - [metall] assemblage *m* à couvre-joints
[railw] (a rail-joint) joint *m* à éclisses
~MANURE - [agric] engrais *m* de poisson
~MEAL - [ind chem] (used as a fertilizer) farine *f* de poisson
~OIL - [ind chem] huile *f* de poisson
~-PLATE - [mech] (cover plate used to unite two metal parts, e.g. rails) éclisse *f*, couvre-joints *m*, fourrure *f*
PLATING - [railw] éclissage *m*
~SCALE - [zool] écaille *f* de poisson
[paint] effritement *m*
~-TAIL - [gen] à queue de poisson
[metall] (a defect) cavité *f* en V
~-TAIL BIT - [tool] (used in mines) trépan *m* (ou bit) à queue de poisson
~-TAIL BURNER - [gas ind] bec *m* à queue de poisson
~-TAIL KNEADER - [mech] (a kneader fitted with fan-shaped blades) mélangeur *m* à ailerons de requin
~-TAIL MANIFOLD - [plast ind] (type of manifold in which the distributing channels are at an obtuse angle to the main supply) tubulure *f* à aileron de requin
~TRACKS - [nucl] (in a cloud chamber) traces *fpl* de poisson
FISHBONE AERIAL - [radio] antenne *f* en sapin
FISHED JOINT - [mech] (butt joint between beams) assemblage *m* à couvre-joints
FISHER - [gen] pêcheur *m*
[naut] bateau *m* de pêche
FISHER'S REAGENT - [chem] (mixture of three parts of sodium acetate, two parts of phenyl-hydrazine hydrochlorate and 20 parts of water: used in sugar analysis) réactif *m* de Fischer
~TROPSCH PROCESS - [chem] (a method of synthesizing hydrocarbons from carbon monoxide and hydrogen) procédé *m* Ficher- Tropsch
FISHERMAN - s. fisher
FISHERMAN'S BEND - [naut] nœud *m* de grappin, nœud *m* anglais
FISHERY - [gen] pêche *f*, pêcherie *f*
FISHING - [gen] pêche *f*
[railw] éclissage *m*
[min] (retrieving objects from boreholes) pêchage *m*; repêchage *m*
~BOAT - [naut] bateau *m* de pêche, pêcheur *m*
~GROUND - [fishing] pêcherie *f*
~HOOK - [mining] caracole *f*
[oil ind] crochet *m* de repêchage
~JAR - [mining] coulisse *f* de pêchage
~LINE - [fishing] ligne *f* de pêche
~MAGNET - [oil ind] aimant *m* de repêchage
~NET - [fishing] filet *m* de pêche, échiquier *m*
~REEL - [fishing] rouet *m* de canne à pêche
~ROD - [fishing] canne *f* à pêche
~SMACK - [naut] s. fishing boat
~SOCKET - [oil ind] sourcière *f*
~TAP - [oil ind] taraud *m* de repêchage
~TOOL - [impl] (used for 'fishing' in boreholes)

outil *m* de repêchage, outil *m* de secours
FISHING VESSEL - [naut] s. fishing boat
FISHPOLE - [el acoust] girafe *f* de microphone
FISHTAILING - [aero] (manoeuvre consisting in swinging the tail of an aircraft from side to side to reduce speed during landing) embardée *f*
FISHY - [gen] goût *m* de poisson
FISSILE - [nucl] (capable of undergoing nuclear fission) fissile
~ATOM - [nucl] (atom which can undergo fission) atome *m* fissile
~DERIVATIVE - [nucl] (a fission product which can undergo fission again) dérivé *m* fissile
~MATERIAL - [nucl] (the material of a nuclear reactor) matière *f* fissile
~NUCLEUS - [nucl] (nucleus used in fission process) noyau *m* fissile
FISSION - [gen] fission *f*
[nucl] (the process of splitting a nucleus into two parts) fission *f*
[genet] (reproduction by spontaneous division of the body) fissiparité *f*
~BOMB - [nucl] bombe *f* atomique, bombe *f* nucléaire
~CAPTURE - [nucl] (the capture of a particle by the target nucleus) capture *f* utile
~CHAIN - [nucl] (the successive beta-decays beginning with a fission product and ending in a stable nucleus) chaîne *f* de fission
~CHAIN REACTION - [nucl] (a reaction in which the fission of an atom causes the successive fission of other nuclei) réaction *f* en chaîne de fission
~CHAMBER - [nucl] (ionization chamber lined with uranium, used for the detection of slow neutrons) chambre *f* d'ionisation à fission
~COUNTER - [nucl] tube *m* compteur à fission
~CROSS-SECTION - [nucl] (cross-section for fission) section *f* efficace de fission
~DECAY CHAIN - s. fission chain
~ENERGY - [nucl] (energy released by fission processes) énergie *f* de fission
~ENERGY RELEASE - s. fission energy
~ENERGY SPECTRUM - [nucl] spectre *m* ébergétique de fission
~FACTOR - [nucl] coefficient *m* de fission
~FRAGMENT - [nucl] (the nuclear species produced when an atom undergos fission) fragment *m* de fission
GASES - [nucl] (gases which are freed from uranium fuels) gaz *m* de fission
~HEAT - [nucl] (the heat produced by fission) chaleur *f* de fission
~IONIZATION CHAMBER - s. fission chamber
~NEUTRON - [nucl] (neutrons emitted from nuclear fission) neutron *m* de fission
POISON - [nucl] (fission products which absorb neutrons unproductively) poison *m* de fission
~PROBABILITY - [nucl] (the probability that fission will happen) probabilité *f* de fission
~PROCESS - [nucl] (reaction in which fission occurs) procédé *m* de fission
~PRODUCT - [nucl] (nucleus produced in the fission process) produit *m* de fission
~PRODUCT BETA ACTIVITY - [nucl] activité bêta du produit de fission
~-PRODUCT-RETAINING COATED FUEL PARTICLE - [nucl] particule *f* combustible à retention de produits de fission
~PRODUCT SEPARATION - [nucl] séparation *f* des produits de fission
~PRODUCT SEPARATOR - [nucl] moniteur *m* de rupture de gaine à séparation des produits de fission
~RATE - [nucl] débit *m* de fission
~RECOIL - [nucl] (fission fragment at the separation) particule *f* de recul à la fission
~SPECIES - [nucl] (one of various kinds of fission) espèces *f*pl de fission
~SPECTRUM - [nucl] (the wide range of elements and isotopes formed in fission) spectre *m* de fission
~SPIKE - [nucl] pic *m* de déplacement de fragments de fission
~THRESHOLD - [nucl] énergie *f* de seuil, seuil *m* de énergie
~YIELD - [nucl] (that fraction of fission which originates a specified nuclide) rendement *m* de fission
FISSIONABLE - [in the USA] s. fissile
[nucl] (capable of undergoing nuclear fission, i.e. the splitting of the atomic nucleus) fissile
~ATOM - s. fissile atom
~DERIVATIVE - s. fissile derivative
~MATERIAL - s. fissile material
~NUCLEUS - s. fissile nucleus
FISSIONING DISTRIBUTION - [nucl] distribution *f* de l'énergie des neutrons dans le réacteur
FISSIPED - [zool] (with free digits) fissipède
FISSURATION - [min] (coke treatment) fissuration
FISSURE, to - [gen] fissurer, fendiller, se fissurer
FISSURE - [gen] fissure *f*, fente *f*
[geol] (a cleft caused by a facture) crevasse *f*
[hydr] (in a river weir) renard *m*
[th eng] (in a boiler plate) renard *m*
~SPRING - [geol] source *f* vauclusienne, source *f* de faille
~VEIN - [mining] filon *m* de fracture, fente *f* de remplissage
FISTULA - [med] (an infected canal in the body) fistule *f*
[zool] (of a whale) évent *m*
FISTULAR - [med] fistulaire, fistuleux
FISTULIZATION - [med] (formation of a fistula) formation *f* d'une fistule
FIT, to - [gen] adapter, ajuster, accommoder
[mech] ajuster, s'agencer, monter
[mech] (e.g. a pin to a nut) garnir, munir, cheviller
[plumb] adapter (p.ex. un ajutage)
[ind chem] (in soap making, to bring the fluid soap to the condition in which it will separate into two strata, the upper one purer than the lower) traiter le savon (pour obtenir le produit final)
FIT - [med] (a sudden attack due to a disturbed function) accès *m*, attaque *f*
[mech] montage *m*, ajustage *m*, frottement *m*
[mech] (a fitting of two parts) accouplement *m*
[gen] convenable, juste, propre
~A TYRE, to - [auto] monter un pneu
~IN, to - [gen & mech] emboîter, enclaver, s'emboiter
~OUT, to - [gen] équiper, outiller
[naut] équiper, armer
~UP, to - [gen & mech] monter, ajuster
FITCH - [zool] (the dress skin of the polecat) putois *m*
FITCHEW - [zool] s. fitch

FITMENT - [mech] garniture *f*, support *m*, montage *m*
FITMENTS - [carp] (in furniture etc) meubles *mpl*
FITNESS - [gen] aptitude *f*
FITS AND CLEARANCES - [mech] (the specified allowances in size between parts which are inserted one into another) tolérance *f* et jeu
FITTED WAGON - [railw] wagon *m* à conduite blanche
FITTER - [gen etc] monteur *m*, ajusteur *m*, mécanicien *m* mécanicien
FITTIG'S SYNTHESIS - [chem] (synthesis of benzene hydrocarbon homologues by reacting metallic sodium with a brominated benzene hydrocarbon and alkyl bromide or iodide in a dry ether solution) synthèse *f* de Fittig
FITTING - [gen] ajustage *m*, adaptation *f*
[mech] agencement *m*, montage *m*
[mech etc] (generic term for any small part) monture *f*
[el] (any device used to support or contain a lamp etc)plafonnier *m*, applique *f*
~DEPARTMENT - [gen] département *m* de montage
~PIPE - [plumb] pièce *f* de raccordement, pièce *f* de forme
~SHOP - [mech] (of machines etc) usine *f* d'ajustage
~SLEEVE - [mech] manchon *m* de raccordement
FITTINGS - [mech] (small auxiliary parts of an engine etc, or of a boiler) accessoires *mpl*
FIVE BEER WARP - [text] chaîne *f* a cinq portées
~-BOWL CALENDER - [mech] calandre *f* à cinq cylindres
~-ELECTRODE VALVE - [electron] (pentode valve) pentode *f*
~-HOLE PUNCHED TAPE - [comput] bande *f* perforée à cinq canaux
~-IMPULSE - [comput] cinq *m* d'arrondi
~LEVEL CODE - [telecomm] alphabet *m* à cinq unités
~-ROLLER MILL - [mech mach] broyeur *m* à cinq cylindres
-SHAFT SATIN - [text] satin *m* turc
~-SIDED BROACH - [mech] alésoir *m* à cinq pans
~SPOT - [oil ind] puits *m* en quinconce
~-UNIT CODE - [telecomm] (code used for machine transmission of telegraphic signals consisting of five-unit elements of the same duration) code *m* à cinq moments
~-WIRE CABLE - [el] câble *m* à cinq conducteurs
-WIRE SYSTEM - [el] (system used for electrical distribution) système *m* à cinq fils
FIX, to - [mech & carp] fixer, monter, caler, arrêter, poser, assujettir
[chem] fixer
[photo] fixer
[gen US] remettre en état, réparer
FIX - [aero] (the ground position of an aircraft as determined by observation) point *m*, point *m* observé
[naut] détermination *f* du point
[mech] attache *f*
~A TOOL, to - [mech] fixer en outil
~IN, to - [mech & carp] encastrer
FIXATION - [chem] fixation *f*
[opt] (the concentration of the eye on a fixed point) fixation
[text] fixage *m*
~ABSCESS - [med] abcès *m* de fixation
~BATH - [text & photo] bain *m* de fixage, bain de fixation
FIXATION MACHINE - [text] machine *f* à fixer
~OF NITROGEN - [chem] (the conversion of atmospheric nitrogen into a combined form) fixation *f* de l'azote
FIXATIVE - [paint] (in perfumery, substances which prevent the too rapid volatilization of the components) fixatif *m*
FIXED - [gen mech etc] fixe, arrêté, stationnaire, prisonnier, à demeure
[chem] fixé
[photo] fixé
~AERIAL - [radio] (in an aeroplan, an aerial fixed to the fuselage) antenne *f* fixe
~AMMONIA - [chem] (ammonia in the combined state, especially in ammonia liquor) ammoniac *m* combiné
~ANTENNA - s. fixed aerial
~-AREA EXHAUST NOZZLE - [mech] tuyère *f* d'échappement de section invariable
~ARM - [mech] bras *m* fixe
~ARMATURE - [el] induit *m* fixe
~AXIS OF ROTATION - [mech] axe *m* fixe de rotation
~BACK REST - [text] (in the loom) traverse *f* arrière fixe
~BARRAGE - [hydr] barrage *m* fixe
~-BASE REPRESENTATION - [math] représentation *f* à base fixe
~BEAM - [constr] (beam with fixed ends) poutre *f* encastrée
~-BEAM CATHODE-RAY TUBE - [electron] tube *m* cathodique à cinq faisceaux
~BEARING - [constr] appui *m* encastré
~BIAS - [electron] polarisation *f* constante de grille
~CALANDRIA - [sugar ind] faisceau *m* fixe
~CARBON - [chem] carbone *m* fixe, carbone *m* combiné
~CENTRE - [mech] pointe *f* fixe
~CHARGE - [gen] (the amount to be inserted into the slots of gas meters etc) taxe *f* fixe
~-COMB HIVE - [agric] (in bee-keeping) ruche *f* à rayons fixes
~CONE - [auto] cône *m* fixe
~CONTACT - [el] (a contact which is fixed to the circuit terminal) plot *m*
~CORE - [metall] (die casting) noyau *m* fixe
~CYCLE OPERATION - [comput] (the fixed time allocated to an operation by a computer, independently of the actual time it actually takes) opération *f* à cycle fixe,fonctionnement *m* à cycle fixe
~DELIVERY PUMP - [mech] pompe *f* à déplacement constant
~DIE - [metall] bloc *m* fixe
~DIE HALF - [die casting] bloc *m* fixe
~DIE HEAD - [plast ind] filière *f* fixe
~DIE PLATE - [plast ind] (the stationary die-plate of an injection mould) plaque *f* fixe du moule
~DISPLACEMENT COMPRESSOR - [mech] compresseur *m* à déplacement constant
~DISPLACEMENT MOTOR - [mech] machine *f* à déplacement constant
~DRUM MIXER - [mech] malaxeur *m* à tambour fixe (ou claveté)
~ECCENTRIC - [mech] (eccentric permanently fixed to a shaft) excentrique *m* fixe
~ECHO - [radar] (permanent echo) écho *m* fixe, écho *m* permanent
~

FIXED ELECTRODE - [el] électrode *f* fixe
~EMITTING STATION - s. fixed transmitter
~END - [constr] (the fixed end of a beam) extrémité *f* encastrée
~-END CYLINDER - [mech] cylindre *m* fixe
~-END MOMENT - [constr] moment *m* d'encastrement
~-END PLATE - [mech] sommier *m* fixe
~EXPANSION - [mech] (of a steam engine working with a constant expansion ratio) expansion *f* constante
~FIELD ALTERNATING GRADIENT ACCELERATOR - [nucl] accélérateur *m* à focalisation par champ magnétique alternant en espace
~FIN - [aero] plan *m* fixe vertical
~FLANGE - [constr] aile *f* encastrée
~FOCUS CAMERA - [photo] chambre *f* à foyer fixe
~FOCUS LENS - [opt] lentille *f* à mise au point fixe
[photo] objectif *m* à mise au point fixe
~FORMAT - [comput] disposition *f* fixe
~-FREQUENCY TRANSMITTER - [radio] (transmitter operating on a single fixed carrier-frequency) émetteur *m* à frequence constante
~GAUGE - [impl] calibre *m* fixe
~GRID - [text] (in Jacquard looms) grille *f* fixe
~HANDLE - [auto] poignée *f* fixe
~HANDLE CIRCUIT BREAKER - [el] (a circuit breaker with a tripping mechanism which cannot operate while the breaker is closed) déclenchement *m* bloqué
~HEAD - [comput] (in magnetic drum memory) tête *f* magnétique fixe
~HINGE - [mech] charnière *f* fixe
~INSTALLATION - [railw etc] matériel *m* fixe
~-JIB CRANE - [mech mach] grue *f* fixe
~KEY - [mech] clavette *f* encastrée
~LADLE - [metall] poche *f* fixe
~LANDING GEAR - [aero] (type of landing gear used in low speed aircraft, which cannot be retracted for flight) train *m* d'atterrissage fixe
~-LENGTH RECORD - [comput] enregistrement *m* à longueur fixe
~LIGHT - [opt] (a light of which the luminous intensity as seen from a fixed point, is of constant value) feu *m* fixe
~LOOP AERIAL - [radio] (a permanently fixed loop aerial, used for homing) cadre-fixe *m*
~MOUNTING - [mech] dispositif *m* de montage fixe
~NIP ROLLERS - [text] (in warping machines) cylindres *mpl* tournant continuellement
~OIL - [chem] (vegetable oils unchanged by heat or distillation) huile *f* fixe, huile *f* grasse
~OPEN-HEARTH FURNACE - [metall] four *m* Martin fixe
~PAN MILL - [min] broyeur *m* à cuve fixe
~PITCH - [aero] pas *m* fixe
~PITCH PROPELLER - [aero] (propeller which is not fitted with arrangements for varying the pitch) hélice *f* à pas fixe
~PIVOT - [mech] cheville *f* fixe
~PLATE - [metall & plast ind] (a plate which is fixed permanently in a press and forms part of the construction of a two or three plate mould) matrice *f* fixe, plateau *m* fixe
~POINT - [math] (system of arithmetic in which all numerical quantities are expressed by a specified number of digits with the point located at a given position) virgule *f* fixe
[railw] aiguille *f* fixe
FIXED-POINT ADDITION - [comput] addition *f* en virgule fixe
~POINT CALCULATION - [comput] opération *f* en virgule fixe, calcul *m* en virgule fixe
~-POINT DIVISION - [comput] division *f* en virgule fixe
~-POINT OPERATION - [comput] opération *f* en virgule fixe
~-POINT PART - [math] mantisse *f*, partie *f* à virgule fixe
~-POINT REPRESENTATION - [math] notation *f* à virgule fixe
~-POINT SUBTRACTION - [comput] soustraction *f* en virgule fixe
~POINTS - [instr] (the standard temperatures defining a thermometer scale) points *mpl* de repère
~PRESCALING - [comput] prédétermination *f* du facteur d'échelle
~PULLEY - [mech] (pulley fixed to the shaft) poulie *f* fixe
~RADIAL ENGINE - [mech] moteur *f* radiale fixe
~-RADIX NOTATION - [comput] notation *f* à base fixe
~RAIL - [cin] (component part of a film perforacting machine) couloir *m* fixe
RANGE MARK GENERATOR - [radar] (used for calibrating the scale having a frequency stabilized by crystal control) marqueur *m* d'étalonnage
~RECORD LENGTH - [comput] longueur *f* de bloc (d'information) fixe
~RESISTOR - [el] (resistor of which the value is not variable at will) résistance *f* non variable
~REST - [mach tools] support *m* fixe, lunette *f* fixe
~ROOF TANK - [oil ind] réservoir *m* à toit fixe
~SALTS - [chem] sels *mpl* fixes
~SASH - [constr] (sash fixed to the frame) chassis *m* (de fenêtre) fixe
~SATELLITE - [astronaut] satellite *m* stationnaire
~SEQUENTIAL FORMAT - [comput] disposition *f* à séquence fixe
~SHEAVE - [mech] moufle *f* fixe
~-SIEVE JIG - [impl] bac *m* à piston, crible *m* hydraulique à piston
~SIGNAL - [railw] signal *m* fixe
~SONG - [mus] (a melody made suitable for contrapuntal treatment) canto *m* fermo
~SPARK GAP - [radar] éclateur *m* fixe
~-SPINDLE CIRCULAR-SAW BENCH - [mech] scierie *f* circulaire à axe fixe
~SPOTLIGHT - [photo] spot *m* à foyer fixe
~STAR - [astr] étoile *f* fixe
~STORE - [comput] mémoire *f* morte
~TARGET REJECTION FILTER - [radar] filtre *m* de suppression d'écho fixe
~TIME-LAG - [el] (definite time-lag) appareil *m* à action differée
~TRANSMITTER - [radio] (transmitter operated in a fixed location) station *f* émettrice fixe
~TRIP - s. fixed handle circuit breaker
~-TUNED CIRCUIT - [radio] (radio reception circuit of which the tuning is pre-set and cannot be adjusted by the operator) circuit *m* accordé non variable
~-TYPE METAL-CLAD SWITCHGEAR - [el] (with all parts permanently fixed) interrupteur *m* blindé fixe
~UNDERCARRIAGE - s. fixed landing gear
~VECTOR - [el] vecteur *m* lié

FIXED WEIR - s. fixed barrage
~-WORD LENGTH - [comput] longueur *f* de mot fixe
FIXER - [chem] fixatif *m*
FIXING - [gen] fixage *m*, mise *f* en place, calage *m*, ancrage *m*, réparation *f*
[chem] fixation *f*
[photo] fixage *m*
[radar] détermination *f* du point
~AGENT - [chem] adjuvant *m* de fixage
~AIDS - [aero] (systems for determining the geographical position of an aircraft) dispositifs *mpl* de repérage de position
~BATH - [photo] bain *m* de fixage, bain *m* fixateur
~BLOCK - [carp] (brick-shaped piece of material on which window frames etc can be fixed) taquet *m* de fixage
~BRACE - [photo] (fixing clip) étrier *m* de fixation
~CLIP - [mech] griffe *f* de serrage, barrette *f* de fixation
~DEPTH - [constr] profondeur *f* d'encastrement
~PARTS - [mech etc] (any parts used for securing others, e.g. bolts, rivets, nuts or the like) accessoires *mpl* de raccordement
~POINT - [aero] point *m*
~ROD BRACKET - [text] (in winding machines) support *m* de la tige de maintien
~SCREW - [mech] (screw used in assembling a machine) vis *f* de fixation
~SOLUTION - [chem] (in photography) solution *f* de fixage
~STRIP - [mech] ruban *m* de fixation
~TANK - [photo] cuve *f* de fixage
FIXTURE - [constr] (an attachement to a building) agencement *m*
(any device designed to support the workpiece) pièce *f* fixe, partie *f* fixe
FIZEAU TOOTHED WHEEL - [opt] (designed to measure the velocity of light) roue *f* dentée de Fizeau
FJORD - s. fiord
F.L. - [initials for Fligh Level] altitude *f*
FLABELLATE - s. flabelliform
FLABELLIFORM - [bot & zool] (fan shaped) flabellé, flabelliforme
FLACCID - [gen] (flabby, limp) mou, flasque
FLAG, to - [gen] (of a ship) pavoiser
[constr] (road work) daller
FLAG - [gen] drapeau *m*
[a flagstone] dalle *f*
[surv] (any cloth tied to a pole at a survey station for visibility purposes) drapeau *m*, pige *f*
[print] (as inserted between types) signe *m* indiquant un bourdon
[geol] dalle *f*
[brew ind] (material used to caulk a barrel) ganse *f*
[cin] (large sheet of metal used to shield camera lenses from light) écran *m* pare-lumière
[electron] ailette *f* de getter
[telev] coupe-flux *m*, nègre *m*
~AT HALF MAST - [gen] pavillon *m* en berne
~LEAF - [agric] dernière feuille *f*
~ORE - [min] hématite *f* stratifiée
~ROD - [mech] (in meters) arbre *m* vertical
~-STAFF - [impl] hampe *f* de drapeau
~STANDARD - s. flag-staff
~-STONE - [constr] (road work) dalle
FLAGELLATE - [bot] flagellé
FLAGELLUM - [zool] (filiform extension of a cell protoplasm) flagellum *m*
[bot] stolon *m*
FLAGEOLET - [mus] flageolet *m*
FLAGGING - [constr] (road work) dallage *m*
[naut] (of a sail) fouettement *m*
FLAGMAN - [gen] starter *m*
FLAGPOLE - [impl] hampe *f* de drapeau
[telev] (black line on the screen, used for testing and occasionally appearing as interference) barre *f*
~AERIAL - [radio] (aerial which consists of a bar of metal) antenne *f* à barres
~ANTENNA - s. flagpole aerial
FLAGS - [cin] (intercepting screens) petits écrans *mpl*
FLAIL, to - [gen] flageller, fouetter
[agric] battre
FLAIL - [agric] (used to thresh by hand) fléau *m*
FLAKE, to - [gen] s'écailler, se diviser en paillettes, s'épraufer
[phys] se diviser en lamelles
FLAKE - [gen] flocon *m*, écaille *f*, lamelle *f*
[metall] (a defect; internal fissures which appear as bright scales on the fractured surface of forgings) craquelure *f*, gerçure *f*
~GRAPHITE - [min] (lamellar form of graphite occurring in metamorphic rocks) graphite *m* lamellaire
~OFF, to - [gen] s'écailler
~-SHAPED PARTICLES - [soil] particules *fpl* en forme de paillettes (ou d'écailles)
~WHITE - [paint] (paint base consisting of pure white lead, prepared in the form of scales) blanc *m* de céruse
FLAKES - [metall] gerçures *fpl*
FLAKING - [paint] (the detachment of a coating from the surface to which it is applied, caused by loss of adhesion for a variety of reasons) boursoufflure *f*
[geol, etc] écaillement *m*
FLAKY - [gen etc] écailleux, lamellé, lamelliforme
~ASHES - [gen] suie *f*
~LIGHT - [oil ind] (natural gas) torche *f*
FLAMBOYANT - [arch] (French architectural style in the I5th and I6th century) flamboyant
~QUARTZ - [min] quartz *m* aventuriné
~TRACERY - [arch] réseau *m* flamboyant
[constr] tracé *m* flamboyant
~TREFOIL - [arch] trèfle *m* flamboyant
FLAME, to - [gen] flamber, flamboyer
FLAME - [gen] flamme, feu *m*, éclat *m* (de lumière)
~ANNEALING - [metall] recuit *m* au chalumeau
~ARC - [el] (electric arc between carbons containing metallic salts) arc *m* à flamme
~ARC LAMP - [el] lampe *f* à arc flamme
~ARRESTER - [gen] (structure or device to limit the expansion of a flame) pare-feu
[mining] coupe-feu *m*
~BAFFLING - [metall] conduite *m* des flammes
~BLOW-OFF FACTOR - [th eng] (in industrial burners; the rate of flame propagation varies considerably with the nature of the gas burnt, and in the relation between this rate and the velocity imparted to the fuel-air mixture is termed the Flame Blow-Off Factor) facteur *m* de propagation de la flamme
~BRIDGE - [metall] autel *m*
~BUCKET - [astronaut] auge *f* à flamme
~CARBON - [light] (carbon electrodes with metallic

salts) charbon *m* mineralisé, charbon *m* pour arc à flamme
FLAME CLEANING - [metall] décalaminage *m* au chalumeau
~CUT - [mech] coupé à l'autogène
~COLOURING - [metall] essai *m* à la flamme
~DAMPER - [aero] (a device preventing the detection of exhaust flames) cache-flamme *m*
~DEFLECTOR - [astronaut] déviateur *m* de flamme
~DESCALING - s. flame-cleaning
~-FAILURE DEVICE - [gas ind] (safety pilot) veilleuse *f* de sécurité
~FRONT - [phys] (in combustion) front *m* de flamme
~GOUGING - [metall] rainurage *m* à la flamme, gougeage *m* à la flamme
~GUN - [impl] (hand-pistol type gas burner adapter for flame-spraying) pistolet *m* de projection à la flamme
~-HARDEN, to - [metall] tremper au chalumeau
~HARDENER - [metall] appareillage *m* pour trempe au chalumeau
~HARDENING - [metall] trempe *f* au chalumeau
~HOLDER - [th eng] (device for maintaining a flame in a fuel burner, e.g. a hot element placed downstream of a burner) brûleur *m*
~-OUT - [mech] (in jet engines) arrêt *m* de la combustion
~PHOTOMETER - [electron] (used for a rapid determination of certain chemicals in liquids) photomètre *m* à flamme
~PLANING - [metall] rabotage *m* dans la flamme
~PLATING - [metall] placage *m* dans la flamme
~PORT - [th eng] orifice *m* de formation de la flamme
~-PROOF, to - [gen] rendre ininflammable
~-PROOF - [gen] ininflammable, ignifuge
[el] antidéflagrant
~-PROOF MACHINE - [el] (totally enclosed electrical machine) appareil *m* antidéflagrant
~-PROOF WIRE - [el] fil *m* ignifugé
~PROPAGATION - [phys] propagation *f* de la flamme
~RADIATION - [phys] radiation *f* de la flamme
~REACTION - [chem] essai *m* par coloration de la flamme
~SCARFING - [metall] (the removal of surface defects by flame) chalumeautage *m*, décriquage *m* à la flamme
~SPECTRUM - [phys] (spectrum of a substance obtained at the temperature of flame) spectre *m* de flamme
~SPEED - [phys] (the rate of flame propagation) vitesse *f* de propagation de la flamme
~SPRAYING - [paint] (coating with metal or plastic projected through a flame gun) plastication *f* à chaud
~TEAT - [th eng] bec *m*
~TEMPERATURE - [phys] température *f* de combustion
~TEST - [chem] (a test for the identification of an element by the colour which it imparts to a flame) essai *m* de coloration
~-THROWER - [impl] lance-flammes *m*
~-TRAP - [int comb engines] (a device fitted to the indication system of an engine to check a return flame, e.g. from a backfire) anti-retour *m*
~TREATMENT - [plast ind] (process of bathing thermoplastic products in an open flame to render them receptive to inks, colours etc) traitement *m* à la flamme
FLAME TUBE - [mech] (the tubular part of a combustion chamber in which combustion takes place) tube *m* à la flamme
[metall] (of a brazing lamp) tube *m* brûleur
~VELOCITY - s. flame speed
~WELDING - [metall] soudure *f* autogène
FLAMING - [gen] en flammes, flambant
~ARC - [el] arc *m* à la flamme
~COAL - [min] charbon *m* bitumineux
FLAMINGO - [zool] flamant *m*
FLAMMABILITY - [phys] (liability to take fire) inflammabilité *f*
~LIMITS - [combustion] limites *fpl* d'inflammabilité
FLAMMABLE - [gen] inflammable, combustible
FLAMME DYEING - [text] teinture *f* flammée, teinture jaspé
~EFFECT - [text] effet *m* flammé
FLANG - [mining] pic *m* à deux pointes
FLANGE, to - [mech] (to forma a flange) border, rabattre le bord de
[metall] (of sheet metal) border, refouler
FLANGE - [mech] (a flat portion formed on a part, usually for attachment) bride *f*, rebord *m*, boudin *m*, collet *m*, membrure *f*, embase *f*
[an external rib] bord *m*, cornière
[plumb] (disc-shaped rim on the end of pipes) bride *f* collerette *f*
[constr] (the top or bottom members of an I-beam) aile *f*
[horol] rondelle *f*
[railw] (of a rail) patin *m*, bourrelet *m*
[mech] (of a groove pulley) mâchoire *f*
[mech] (in cranes) joue *f* (d'un galet)
[constr] (of a road) mentonnet *m*
[metall] taquet *m* (de chassis de moulage)
~BEAM - [constr] (a beam of I-section) poutre *f* en I
~CAT - [mech] bride *f* avec tubulure, bride *f* avec ajutage
~COUPLING - [mech] (shaft coupling) manchon *m* à plateaux, accouplement *m* à plateaux
~DOWN, to - [mech] tomber le bord
~FITTING - [mech] raccord *m* à bride
~GASKET - [mech] joint *m* de bride
~HUB - [mech] moyeu *m* à bride, moyeu *m* à flasque
~JOINT - [plumb] assemblage *m* à brides, raccord *m* à bride
~MEMBER - [constr] barre *f* de membrure
~-MOUNTED - [mech] accouplé à bride
~-MOUNTING TYPE - [mech] (designed to be attached by means of a flange, espec. of electric motors or generators, the carcase of which is provided with a drilled flange to bolt to another assembly) accouplement *m* à brides
~NUT - [mech] écrou *m* à bride, écrou *m* à collet
~OF A WHEEL - [mech] boudin *m*, bourrelet *m*, mentonnet *m*
~OF DRUM - [mech] joue *f* d'un touret pour câble
~OF RAIL - [railw] patin *m* d'un rail
~PACKING - [mech] garniture *f* de brides
~PIPE - [plumb] tube *m* à bride
~PLATE - [metall] plaque *f* à rebord
~RAIL - [railw] rail *m* à T
~SEAL - [mech] joint *m* de bride
~SLEEVE - [mech] manchon *m* à plateau
~STEEL - [metall] acier *m* à refouler, acier *m* malléable

FLANGE TYPE MOTOR - [mech] moteur *m* couplé par bride *f*
~UNION - [mech] raccord *m* à bride
~WHEEL - [mech] roue *f* à boudin, roue *f* à bourrelet
~WITH GROOVE - [mech] bride *f* à rainure, bride *f* rainurée, bride *f* avec gorge
FLANGED - [mech etc] à bride, à brides, à rebord (fitted with external ribs) à plateaux, à joues
~BEAM - [mech] (joist of I-section in rolled-steel) poutre *f* en I
~BOBBIN - [text] bobine *f* à rebords, bobine *f* à disques
~BOLT - [mech] (a bolt having an integral flange under the head) boulon *m* à bride
~CHUCK - [mech] (a face-plate) plateau *m*
~COUPLING - s. flange coupling
~FITTING - [mech] accouplement *m* à brides
~HOLE - [mech] trou *m* à rebords
~NIPPLE - [mech] bride *f* avec tubulure, bride *f* avec ajutage
~NUT - s. flange nut
~PIPE - [plumb] tuyau *m* à bride, conduite *f* à joints à brides
~PLATE - [metall] tôle *f* à bord rabattu
~PROFILE - [metall] profil *m* biseauté, profil *m* chanfreiné
~PULLEY - [mech] poulie *f* à rebords, poulie *f* à joues
~RAIL - [railw etc] (rail section of inverted T-shape) rail à patin et à champignon unique, rail *m* Vignole, rail *m* à T
~SLEEVE - [mech] (a tubular part provided with a flange) manchon *m* à plateau
~SOCKET - [mech] manchon *m* épaulé
~SPIGOT - [mech] tourillon *m* à patin, tourillon *m* à flasque
~SPINDLE - [mech] mandrin *m* à bride
~TUBULURE - [mech] bride *f* avec tubulure, bride *f* avec ajutage
FLANGER - [railw] (a scraping device for ice and snow used to provide a clear run for the wheel flanges) taille-neige *m*
[mech] machine *f* à border
FLANGING - [mech] bordage *m*, rabattement *m* des brides, tombage *m* des bords
[of sheet metal] bordage *m*, façonnage *m*
~MACHINE - [mech] machine *f* à border
~OF ENDS - [plumb] rabattement *m* des bords
~PRESS - [mech] presse *f* à border
FLANK - [gen] flanc *m*
[mech] (of a gear tooth) flanc *m* (d'une dent d'engrenage)
[of a tool] flanc *m* (d'outil)
[constr] (of a building) côté *m*
[arch] (of an arch) rein *m* (d'une voûte)
~OF TAPPET - [mech] côté *m* de l'excentrique
~WELL - [oil ind] puits *m* de flanc
FLANKING CHANNEL - [comput] voie *f* adjacente, canal *m* adjacent
FLANNEL - [text] (all-wool material made from soft wool) flanelle *f*
~YARN - [text] fil *m* pour flanelles
FLAP, to - [gen] battre
[phys] frapper
[naut] (of a sail) battre, fouetter
FLAP - [gen] battement *m*, trappe *f*, abattant *m*, rabat *m*
[aero] (any surface designed to increase lift, and to be adjustable in flight) volet *m*
[carp & mech] (a hinged leaf or plate) abattant *m*
[text] (in looms for carpets or tapestries) clapet *m*
[med] manchette *f*, lambeau *m*
[rubber ind] bande *f* de fond de jante, flap *m*
FLAP ANGLE - [aero] (the angle which the chord of a flap makes with that of the corresponding fixed surface) angle *m* de braquage
~ATTENUATOR - [electron] (attenuator with a strip of absorbing material introduced in the waveguide) affaiblisseur *m* à cloison longitudinal
~CONTROL - [aero] commande *f* volets
~EXTRATION - [med] (in ophtalmology) extraction *f* (de la cataracte) avec un lambeau conjonctival
~HINGE - [mech] briquet *m*, charnière *f* de battant
~LOCK - [auto] serrure *f* de battant
~MIRROR - [photo] miroir *m* escamotable
~MOULD - [rubber ind] vulcanisateur *m* de flaps
~RING - [mech] anneau *m* à clapet
~SEAT - [gen] (a chair in a cinema or theatre etc fitted with a seat which can be tilted) strapontin *m*
~SHUTTER - [photo] obturateur *m* à double volet
~TILE - [constr] (made to fit over a hip line etc) tuile *f* flamande, plate *f* recourbée
~-TRAP - [mech] (an anti-flood valve used in sanitary engineering) puits *m* étanche à l'eau
~VALVE - [mech] (a non-return valve in which the obturating element consists of a pivoted shutter, moved by the fluid passing it) clapet *m* à charnière, soupape *f* à clapet, clapet *m*
~VULCANIZER - [rubber ind] (in tyres) vulcanisateur *m* de flaps
FLAPPER - [mech] (component part of valve) palette *f*
FLAPPING - [gen] battement *m*
[aero] (angular movement of a rotor blade up on the flapping hinge) levée *f* de pale, battement *m*
[mining] (of a winding rope) fouttement *m*, coup *m* de molette
[metall] rupture *f* de la couche de scories
[mech] coup *m* de fouet
~ANGLE - [aero] (the angle through which a rotor blade moves on the flapping hinge) angle *m* de basculement du disque balayé
~HINGE - [el] (the pivot on which a rotor blade can move in a zenithal sense) axe *m* de levée de pale, axe *m* de levée de battement
~OF A BELT - [mech] battement *m* de la courroie
FLARE, to - [gen] flamboyer, brûler
[metall] (to shape out the end of a pipe) évaser
[metall] (of a metal plate etc) s'évaser, se dilater
[mining] vaciller
FLARE - [metall] (divergent form, usually at the end of a tube or the like, of a trumpet) évasement *m*, renflement *m*, bavure *f*
[signalling] (pyrotechnic device to show a bright light, e.g. as a warning) signal *m* pyrotechnique
[constr] mitre *f* de cheminée
[photo] (extraneous image of a light source caused by faulty focusing) tache *f* par réflexion
[metall] plaque *f* d'écoulement, ailette *f* de soudure
[oil ind] (natural gas wells) torche *f*
[radio] (elementary type of aerial) cornet *m*
[telev] diffusion *f* parasite, tache *f* hyperlumineuse

[med] flambeé *f*
[radar] écho *m* deformé, filage *m* horizontal
[astr] éruption *f* solaire
FLARE ANGLE - [radar] (the orifice of a horn-shaped radiator) angle *m* d'ouverture
~BLEEDER - s. flare [oil ind]
~FITTING - [mech] accouplement *m* à évasement
~GHOST - [photo] tache *f* hyperlumineuse
~HEADER - [build] (brick burnt to a darker shade at one end) boutisse *f* brulée
~NIMBUS - [met] nuage *m* corpusculaire (associé à une éruption solaire)
~PISTOL - [signal] pistolet *m* éclairant
~-SHAPED - s. flared
~SIGNAL - [signal] signal *m* pyrotechnique
~SPOT - [telev] (bright spot of light due to internal reflections within the lens) tache *f* hyperlumineuse
~UP, to - [gen] s'enflammer, lancer des flammes, augmenter d'intensité
~WALL - [constr] mur *m* de tête *f* (ou en aile)
FLARED - [gen] (trumpet-shaped) évasé, épanoui
~BOTTOM - [naut] fond *m* évasé, carène *f* evasée, bisauté
~FITTING - [mech] raccord *m* embouti
~RADIATING GUIDE - [electron] (flares replacing part of the sheath in a waveguide for radiation direction) guide *m* d'ondes à évasements rayonnants
~THROAT - [acoust] (of a loudspeaker) embouchure *f* profilée
FLARING - [mech] (said of the end of flared pipe) évasement *m*
[gen] éclatant, ébouissant
[shipbuild] (of the timbers) dévoiement *m*
~CUP GRINDING WHEEL - s. flaring grinding wheel
~GRINDING WHEEL - [tool] meule *f* à noyau rentrant
~MACHINE - [mech] (used in particular for flaring glass tubes for electronic tubes) machine *f* à évaser
~TOOL - [tool] outil *m* à évaser
FLASER-GNEISS - [geol] flasergneiss *m*
FLASH, to - [gen] flamboyer, lancer des étincelles, faire flamboyer, projeter, scintiller
[telecomm] (of signals) rappeler sur lampe de supervision
[glass man] (to expand into a sheet) plaquer
[hydr] (e.g. a river) chasser
[el] carburer, nourrir
[metall] brûler
[mining] abattre à l'explosif
FLASH - [phys] éclair *m*, éclat *m* de flamme
[hydr] chasse *f* d'eau
[metall] (thin fragments of metal at the sides of a forging) barbe *f*, bavure *f*, ailette *f* de soudure, soufflure *f*
[metall] (in welding) étincellage *m*
[print] dernière nouvelle *f*
[plast ind] (moulding material escaping from a mould at a space between its moving parts) bavure *f*, toile *f*
[el] débordement *m*
[telev] surimpression *f*
~A SIGNAL, to - [signal] clignoter un signal
~ARC - [electron] (rocky-point effect; sudden increase in space current caused by irregularities in the surface of the electrode) effet *m* Rocky-Point
~-BACK - [mech] (in internal combustion engines) retour *m* de flamme

FLASH BARRIER - [el] (insulating, fire-proof screen fitted to electrical apparatus to prevent the formation of an arc) écran *m* antiarc
~BOILER - [th eng] (a boiler consisting of a coil of steel tube, which is heated while the feed water is pumped through it) chaudière *f* à vaporisation instantanée
~-BULB - [photo] lampe-éclair
~BURN - [nucl] (caused by thermal radiation) lucite *f* par rayonnement thermique
~BUTT WELDING - [metall] soudage *m* en bout par étincellage
~CASTING - [metall] coulée *f* en châssis
~CURRENT - [el] (initial value of current when the cell or battery is connected to a circuit of negligible resistance) courant *m* de court-circuit
~DISTILLATION - [chem] (a method of distillation in which the final liquid and vapour are in equilibrium) distillation *f* flash
~DRUM - [oil ind] chambre *f* de vaporisation
~DRYER - [mech] (apparatus for very rapid drying) séchoir-éclair *m*
~DRYING - [paint] (process of rapid drying of a coated surface by exposure for a short time to a relatively high temperature) séchage *m* istantané
~EVAPORATION - [chem] (a method of very rapid evaporation of saline solutions, used in the Alberger process for obtaining sodium chloride of high purity) évaporation *f* instantanée
~EXPOSURE - [photo] demi-pose *f*, prise *f* de vue à la lumière d'un éclair, flash *m*
~FIRING SWITCH - [photo] déclencheur *m* de l'éclair
~GAMMA RADIATION - [nucl] (radiation accompanying the fission process without measurable delay) rayonnement *m* gamma instantané
~GENERATOR - s. flash boiler
~GROOVE - [metall & plast ind] gorge *f* d'échappement, rainure *f* d'échappement
~-GUARD - [el] (the refractory material which directs any arc arising on the operation of a circuit breaker) écran *m* protecteur contre le contournement
[metall] (in diecasting) protecteur *m* contre les éclaboussures
~GUN - [photo] raccord *m* pour prise flash
[photo] éclair *m*, appareil-éclair *m*
~GUTTER - [metall] rainure *f* d'échappement, chenal *m* à barbe
~HOLE - [metall] s. flash gutter
~LAMP - [el] (a filament lamp used in connection with a dry battery) lanterne *f* de clignotant, lampe *f* de poche électrique
~LAND - [metall & plast ind] (the total area of the mould surfaces which are in contact when the mould is closed) surface *f* de contact
~-LIFT MOULDING MACHINE - [metall] machine *f* à démouler
~-LIGHT - [photo] (artificial light giving a bright illumination for a short period) lumière-éclair *f*
[signalling] clignotement *m*
[el] (small battery torch) lampe *f* de poche électrique
[naut] feu *m* à éclats
[telecomm] feu *m* à éclipses
-LIGHT BATTERY - [el] (battery used to light an electric hand lamp) pile *f* de lampe portative
~-LIGHT POWDER - [chem] photopoudre *f*

FLASH LINE - [plast ind] (a raised line on the surface of a moulding, formed at the joint between the members of a mould) jointure *f*
~MOULD - [rubber ind] moule *m* à couteau, moule *m* à déversage
~OF LIGHTNING - [met] éclair *m*
~-OFF - [paint] (process of promoting very rapid drying of a coating containing a volatile solvent) séchage *m* instantané
~-OVER - [el] (electric discharge across a dielectric, resulting from excessive voltage) contournement *m*
~-OVER PROTECTION - [el] protection *f* contre les detournements
~OVERFLOW MOULD - [plast ind] (type of mould provided with a space to receive flash) moule *m* à échappement
~PACK - [photo] générateur *m* portatif pour éclair électronique
~PERIOD - [paint] (the time taken by the evaporation of the volatile solvents in paints) temps *m* de vaporisation
~PIN - [mech] goujon *m* de châssis
~PLATE - [el plating] (a thin deposit produced rapidly) voile *f*
~PLATING - s. flash plate
~POINT - [phys] (the minimum temperature at which the vapour given off by a liquid forms an explosive mixture with the atmosphere) point *m* d'inflammabilité
~RADIOGRAPHY - [radiat] (radiography with an extremely short exposure) milligraphie *f*
~RIDGE - [metall & plast ind] bord *m* d'appui
~RING - [metall] (the annular flash groove) gorge *f* circulaire
~ROASTING - [metall] grillage *m* de concentrés pulvérulents
[min] grillage *m* superficiel
~SETTING - [oil ind] prise *f* éclair
~SHOE - [photo] griffe *f* pour lampe-éclair
~SUPPRESSOR - [el] (mechanism for preventing flashovers) suppresseur *m* de contournement
~SYNCHRONIZER - [photo] synchronisateur *m* pour éclair
~TEST - [el] (text applied to electrical equipment for insulation strength) test *m* lumineux scintillant
[chem] essai *m* du point d'inflammabilité
~TRIMMING PRESS - [metall] ébavureuse *f*, rogneuse *f*
TUBE - [electron] (gas discharge tube) lampe-éclair *f* électronique
~-TYPE MOULD - s. flash mould
~WEIGHT - [metall] poids *m* de la bavure
~WELDER - [metall] soudeuse *f* flashing
~WELDING - [el metall] soudage *m* par étincellage
[metall] soudure *f* flashing
FLASHER - s. flash boiler
[el] clignoteur *m*
[radar] (colloq; a reflecting object consisting of mutually intersecting surfaces) réflecteur *m* angulaire
~BULB - [auto] ampoule *f* clignotante
~LAMP - [auto] lanterne *f* de clignotant
FLASHING - [gen] flamboiement *m*
[auto] (rapid succession of warning flashes with headlights) signal *m* clignotant, clignotement *m*
[build] (the burning of bricks with a periodical stoppage of the air supply, so as to obtain an irregular colouring) cuisson *f* des briques (colorées)
[hydr] (the construction of a convergent passage from a high to a low level in a river) ouverture *f* des écluses
[plumb] (a strip of lead or zinc around the junction of two surfaces, e.g. on a roof) pose *f* du chaperon
[el] (the sparks on a commutator etc) scintillement *m*
[paint] (glossy patches on a surface) flamboiement *m*
[electron] (a system of coil testing) passage *m* à la bobine
[glass man] (the operation of expanding into a thin sheet) placage *m*
[glass man] (the reheating of glass) réchauffage
[photo] (flash exposure) demi-pose *f*
[plumb] (a joint to render a junction watertight) noquet *m*, noue *f*
[telecomm] rappel *m* sur supervision
FLASHING - [metall] plaque *f* de protection
FLASHING BEACON - [telecomm] balise *f* à éclats
~BOARD - [build] (the board to which flashing are fixed) reverseau *m*, rejéteau *m*
~LIGHT - [telecomm] (a cyclically-intermittent light, in which the periods of light are perceptibly shorter than those of darkness) feu *m* à éclats
~OVER - [el] contournement
~POINT - s. flash point
~RELAY - [el] relais *m* clignotant
~SIGNAL - [telecomm] (in telecommunications, a luminous signal between operators) rappel *m* sur supervision
[telecomm] test *m* lumineux scintillant
~TIME - [metall] (in flash welding) durée *f* d'etincellage, temps *m* d'étincellage
FLASHLESS POWDER - [explos] explosif *m* sans flamme
FLASHLINE - [metall] témoin *m* d'ébarbure, ligne *f* d'ébarbure
FLASK - [gen] flacon *m*
[ind chem] (laboratory vessel in a variety of shapes, in general having a body with a flattened bottom and a neck) ballon *m*, flacon *m*
[chem] (a vessel containing 76 1/2 lbs of mercury) flasque *f*
[metall] (a moulding box holding the sand mould in which the casting is made) châssis *m*
~BAND - [metall] bande *f* de châssis de moulage
~CLAMP - [ind chem] (device for holding a flask, for attachment to a stand) pince *f* de support
~MOULDING BOX - [metall] châssis *m*
~PIN - [metall] goupille *f* de châssis
FLAT, to - [gen] aplatir
[paint] (to smooth a coating by abrasive action) mater, amatir
[acoust] insonoriser
FLAT - [gen] plat, horizontal, plan
[paint] mat
[constr] appartement *m*
[photo] sans relief
[acoust] sourd
[mus] bémol
[cin] (a tormentor; a large portable wall covered with sound-absorbing material) décor *m* insonorisé
[arch] (on top of a building) toit *m* en terrasse
[text] chapeau
[metall] (small metal bar of a thickness not exceed-

ing 1/4") barre *f* plate
FLAT - [naut] plate *f*, bateau *m* plat
[auto] (colloq for flat tyre) dégonflé, à plat
[surv] horizontal
[med] plat
[brew ind] (of beer, wine etc) fade, insipide
[shipbuild] varangue *f* de fond
~ARCH - [arch] (arch with intrados without curvature) voûte *f* plate, arc *m* déprimé
~BACK - [book bind] (of a volume) dos *m* plat
~-BACK METHOD - [mining] exploitation *f* par tailles en échelon
~-BACK OVERHAND STOPE - [mining] taille *f* montante en échelon
~BACK STOPE - [mining] taille *f* en échelon
~BAND - [constr] (a square impost stone) plate-bande *f*
~-BAND CONFECTION - [rubber tyres] confection *f* sur tambour plat
~BAR - s. flat bar iron
~BAR IRON - [metall] fer *m* plat
~BASE - [rubber ind] support *m* lisse
~BASE RIM - [auto] jante *f* à base plate
~BASTARD FILE - [impl] lime *f* bâtarde plate
~BED - [print] marbre *m* plan
~BELT - [mech] (a flat endless band used to transmit power between cylindrical pulleys) courroie *f* plate
~BILLET - [metall] billette *f* plate, large-plat *m*
~BLOOM - [metall] bloom *m* plat
~-BOAT - [naut] bateau *m* plat
~BOBBIN - [text] (of a weft winding machine) bobine *f* plate
~-BOTTOM BOAT - [naut] bateau *m* plat, plate *f*, bateau *m* ras
~BOTTOM TAPPET - [mech] (in auto) poussoir *m* à fond plat
~-BOTTOMED - [gen] (of flasks, containers, boats etc) à fond plat
~BOTTOMED CENTRIFUGE BASKET - [sugar ind] panier *m* de turbine à fond plat
BOTTOMED FLASK - [ind chem] (thin glass flask with round body and flat bottom) ballon *m* à fond plat
~-BOTTOMED RAIL - [railw] rail *m* à patin, rail *m* Vignole
~-BRICK PAVING - [constr] pavage *m* en briques de champ
[constr] carrelage *m* en briques à plat, carrelage *m* en damier
~BULB IRON - [metall] fer *m* plat à boudin
~CALM - [met] calme *m* plat
~-CAR - [railw] (or flatcar) wagon *m* plat
~CARD RESOLVER - [instr] (a type of sine-cosine potentiometer) résolveur *m* potentiométrique à carte carrée
[contr] sin-cos-potentiomètre *m* plat
~CARDBOARD ROOF - [constr] toit *m* en carton plat
~CARVED NAIL - [metall] pointe *f* plate rayée
~CEILING - [naut] vaigrage *m* plat
~CHISEL - [tool] (cold chisel with a broad cutting edge) trépan *m* plat (au simple), burin *m* plat
~COAT - [paint] (a type of undercoating) couche *f* de fond
~COAXIAL TRANSMISSION LINE - [electron] guide *m* d'ondes à rubans
~COIL MEASURING INSTRUMENT - [instr] (moving-coil instrument with a flat coil) instrument *m* de mesure à bobine plate
FLAT-COMPOUNDED - [el] (of a compound-wound generator with a series winding so designed that the voltage remains constant) à compoundage à tension constante
~CONICAL CHEESE - [text] bobine *f* soleil
~CONVEYOR-BELT - [mech] courroie *f* transporteuse plate
~COUNTER TUBE - [nucl] tube *m* compteur plan
~COUNTERSUNK HEAD RIVET - [mech] rivet *m* à tête noyée
COURSE OF BRICKS - [constr] aire *f* de briques à plat
~DIE - [metall & plast ind] (a slot-die: one with a wide narrow orifice) filière *f* plate
~DOME - [arch] voûte *f* plate, dôme *m* plat
~DRILL - [tool] mèche *f* plate, foret *m* à langue d'aspic
~ENAMEL - [paint] peinture *f* mate au vernis
~-ENDED TUBE - [electron] (cathode-ray tube with a flat surface at the top) tube *m* à fond plat
~FILAMENT LAMP - [el] lampe *f* à filament plat
~FILE - [impl] lime *f* plate pointue
~FILISTER SCREW - [mech] vis *f* à tête cylindrique noyée
~FINISH - [paint] (non-glossy finish) surface *f* mate
~FISH-PLATE - [railw] éclisse *f* plate
~FIXED CALANDRIA - [sugar ind] faisceau *m* tubulaire fixe à faces planes
~-FOLDED SEAM - [mech] agrafage *m* plat
~FOOT - [med] pied *m* plat
~FOUNDRY NAIL - [metall] pointe *f* à tête plate pour mouleurs
~GASKET - [mech] joint *m* plat, rondelle *f* plate, platine *f* d'étanchéité
~GATE - [metall] attaque *f* plate
~GAUGE - [impl] calibre *m* plat
~GOLD WIRE - [text] (in tapestries) lame *f* d'or
~GOUGE - [tool] (gouge with a cutting edge shaped to a large radius of curvature) ciseau *m* à gouge
~GRID - [mech] grille *f* plate
~-GRID TYPE - [mech] (in a gauge, in which the wire is wound in a zig-zag shape) à grille plate
~GROOVE RAIL - [railw] rail *m* à gorge plate
~GUIDE - [mech] guide *m* plat
~HAIRSPRING - [horol] spiral *m* plat
~HALF-ROUND - [metall] demi-rond *m* plat, demi-rond *m* irrégulier
~HAMMER - [impl] marteau *m* plat
~HEAD - [mech] à tête plate
~-HEAD BOLT - [mech] boulon *m* à tête plate
~HEAD COUNTERSUNK RIVET - [mech] rivet *m* à tête plate noyée
~HEAD NAIL - [mech] pointe *f* à tête plate
~-HEAD RIVET - [mech] rivet *m* à tête plate
~-HEAD SCREW - [mech] vis *f* à tête plate
~-HEAD TRIPOD - [photo] trépied *m* à tête plate
~-HEADED RAIL - [railw] rail *m* à champignon plat
~-HEADED SMOOTHING TOOL - [tool] plat *m* carré, plat *m* à navette
~-IRON - [impl] fer *m* à repasser
[metall] fer *m* méplat
~JOINT - [constr] (type of mortar joint) joint *m* plein
[metall] diaclase *f* horizontale
~KEEL - [naut] (the lower horizontal member of the

ship's backbone) quille *f* plate
FLAT-KEELED - [naut] à quille plate
~KEY - [mech] clavette *f* plate
~KNITTING FRAME - [text] tricoteuse *f* rectiligne, métier *m* à tricoter rectiligne
~KNITTING LOOM - [text] tricoteuse *f* rectiligne
~LAND - [geol] (the flat regions on the sides of a river) plaine *f*, marécage *m*
~LAYER - [constr] (of bricks) assise *f* couchée
~LEAD - [metall] feuille *f* de plomb, plomb *m* laminé
~LEAD FLOAT FILE - [tool] lime *f* à simple taille pour plomb
~LINEAR INDUCTION PUMP - [mech] (used in vacuum technique) pompe *f* à induction linéaire à disposition plate des enroulements
LINK - [mech] (of chain) tige *f* de chaîne
~LINK CHAIN - [mech] chaîne *f* à articulations, chaîne *f* à fuseaux
~LOUDSPEAKER - [acoust] (a small-depth loudspeaker) haut-parleur *m* plat
~MARSHALLING YARD - [railw] gare *f* de triage en palier
~NEGATIVE - [photo] negatif *m* à faible contraste
~-NOSED SCREW - [plast ind] (obsolete type of extruder screw in which the nose end is flat, not tapered or pointed) vis *f* à nez plat
~NUT - [mech] écrou *m* plat
~OIL-PAINT - [paint] peinture à l'huile
~OPEN CUT SMOOTH FILE - [tool] lime *f* douce plate
~OPEN SECOND CUT FILE - [tool] lime *f* demi-douce
~OUT - [gen] en vitesse
~PARACHUTE - [aero] (a parachute of which the gores are triangular, so that it is flat when laid out on a plane surface) parachute *m* plat
~PICTURE - [photo & cin] (lacking contrast and depth) image *f* à faible contraste, image *f* douce
~PILE - [constr] palplanche *f*
~PIN - [el] broche *f* plate
~-PIN PLUG - [el] prise *f* à broches plates
~PISTON - [mech] piston *m* plat, piston *m* à disque
~PLATE - [print] cliché *m* plan
[railw] plaque *f* de manoeuvre
~-PLATE KEEL - [naut] quille *f* plate
~PLIERS - [impl] pince *f* plate
~-POINT BOLT - [mech] boulon *m* à pointe plate
~PROOF - [print] épreuve *f* en première
~RAMMER - [impl] pilette *f*, batte-plate *f*
~RANDOM NOISE - [el acoust] bruit *m* blanc
~RASP - [impl] râpe *f* plate-à-main
~RATE - [comm] tarif *m* à forfait, tarif *m* de base
~-RATE TARIFF - [comm] tarif *m* forfaitaire
~RESPONSE - [acoust] réponse *f* uniforme
~-RING ARMATURE - [el] induit *m* en anneau plat
~ROD - [mech] tige *f* de transmission, barre *f* de renvoi
~ROLLING - [metall] laminage *m* plat
~ROOF - [constr] toit *m* plat
~ROOF AERIAL - s. flat-top aerial
~-ROOF ANTENNA - s. flat-top aerial
~RUBBER BELT - [mech] courroie *f* plate
~SEALING RING - [mech] anneau *m* plat, joint *m* plat, rondelle *f* plate
~SEAM - [mining] couche *f* plate
~SEAT - [mech] (of a valve) siège *m* plan
~SECOND CUT FILE - [tool] lime *f* demi-douce plate

FLAT SLIDE VALVE - [mech] coulisse *f*, palette *f* coulissante (d'une pompe à piston tournant)
~SLOTTED HEAD - [mech] tête *f* plate fendue
~SPACE - [astr] (the Euclidean space) espace *m* euclidien
~SPIN - [aero] (a continuous spiral descent at a large mean angle of incidence, the longitudinal axis of the aircraft being nearer to the horizontal than to the vertical) vrille *f* à plat
SPRING - [mech] ressort *m* plat
~STITCH - [text] point *m* plat
~-STITCH EMBROIDERY - [text] tissu *m* à point brodé plat
~STRIPS - [text] bandes *fpl* plates
~SURFACE - [geom] surface *f* plane
[paint] couche *f* mate
~-SURFACE GOBELIN - [text] gobelin *m* ras
~-SURFACE GRINDING - [mach tool] rectification *f* des surfaces planes
~-SURFACE GRINDING MACHINE - [mach tool] rectifieuse *f* pour surfaces planes
~-SURFACE LAPPING - [mech] machine *f* à roder les surfaces planes
~TILE - [constr] tuile *f* plate
~TINT - [photo] (uniformity in brightness and colour) couleur *f* mate
~TIRE - s. flat tyre
~TONGS - [impl] pinces *fpl* plates
~-TOP AERIAL - [radio] antenne *f* en nappe, traversier *m*
~-TOP ANTENNA - s. flat-top aerial
~-TOP BOGIE TROLLEY - [mech] trolley *m* à plate-forme
~TOP PISTON - [mech] piston *m* à tête plate
~TORSION SPIN - [mech] ressort *m* spiral
~TRAILER - [of lorry] remorque *f* plate-forme
~TROWEL - [impl] truelle *f*
~TRUCK - [railw] wagon *m* plat, wagon *m* plate-forme
~TUNING - [telev] syntonisation *f* mauvaise
[radio] accord *m* flou
~TURN - [aero] (turn made without banking) virage *m* à plat
~TWIN - [mech] (type of engine having two horizontal cylinders placed on opposite sides of the crankshaft) moteur *m* à deux cylindres opposés horizontaux
~TWIN-CABLE - [el etc] câble *m* à conducteurs parallèles
~-TYPE ARMATURE - [el] armature *f* latérale
~-TYPE RELAY - [el] relais *m* plat
~TYRE - [tyres] (a deflated pneumatic tyre) pneu *m* à plat
~-VARNISH - [paint] (type of varnish in which the natural glossiness of the coating has been diminished by the addition of wax, metal soaps, pigments, fillers etc - not to be confused with flatting varnish q.v.) vernis *m* mat d'apprêt
~-VAULTED CEILING - [constr] voûte *f* bohémienne plate, voûte *f* bohémienne à calotte
~WAGON - [railw] wagon *m* plat
~WALL - [mining] mur *m*
~WASHER - [mech] (a flat ring used as packing or placed under a nut) rondelle *f* plate
~WELD - [metall] (weld laid down in the downhand position) soudure *f* à plat
~-WIRE BRAID - [text] galon *m* lamé
~WOOD FILE - [impl] lime *f* plate à main

FLAT WOOD RASP - [impl] râpe *f* plate-à-main
~WOUND - [el] bobiné à plat, enroulé à plat
~-WOUND TENSION SPRING - [mech] ressort *m* de traction à spirale plate
FLATNESS - [gen] nature *f* plate, égalité *f* de surface
[photo] (a picture lacking contrasts) absence *f* de distorsion du champ
[paint] (lack of gloss) embu *m*
[med] matité *f*
[astronaut] tension *f* de la trajectoire
FLATS - [paper man] in-plano *m*
[mech] (the width across the flats of a nut) plats *m*pl, acier *m* plat, fers *m*pl plats
FLATTEN, to - [gen] aplatir, aplanir, redresser, s'affadir
[aero] (to bring the aircraft to horizontal flight) se redresser, allonger le vol
[naut] (of sails) border plat
[chem] (of wine) s'éventer
[metall] laminer, écacher
~A SLOPE, to - [constr work] aplatir une pente
~OUT, to - [gen] aplatir
FLATTENED - [gen] aplati, aplani, écaché
[constr] (of an arch or vault) déprimé
~PATTERN - [text] dessin *m* aplati
~REGION - [nucl] (in a nuclear reactor) zone *f* de aplatissement
~ROD - [mech] barre *f* brunie
~SHEET - [metall] tôle *f* étirée
~SILK - [text] soie *f* laminée
THREAD - [text] fil *m* aplati
FLATTENER - [tool] chasse à parer
[metall] (roller leveller) dresseuse *f* à galets
FLATTENING - [gen] aplatissement *m*, aplatissage *m*
[metall] laminage *m*, écachement *m*, brunissage *m*, lissage *m*
[med] (use of filtering medium to obtain flat isodose curves in a body submitted to irradiation) égalisation *f* des isodoses
~FILTER - [radiat] filtre *m* de nivellement
~MACHINE - [mech mach] aplatissoire *f*
[metall] planeuse *f*
~MATERIAL - [nucl] (in a reactor) substance *f* de aplatissement
~-OUT - [aero] (the change from approach the horizontal flight before alighting) arrondi *m*
~RADIUS - [nucl] (in a cylindrical reactor core) rayon *m* d'aplatissement
~TOOL - [tool] (tool resembling a flat-face hammer) chasse *f* à parer
FLATTER - [mech] (draw-plate designed to produce flat wire) machine *f* à dresser à rouleaux
[metall] banc *m* à étirer les bandes
FLATTING - [paint] (in general, paint used for undercoating) vernis *m* mat
[metall] laminage *m*, aplatissage *m*
[constr] couche *f* de mortier
[metall] (kinking) rupture *f* par déformation
~AGENTS - [chem] (substances in powder form incorporated in surface coatings to reduce reflection from the surface of the dried film) agents *m*pl de matité
~KNIFE - [constr] couteau *m* à lisser le lut
~MILL - [metall] aplatissoir *m*
~PUTTY - [constr] glacis *m* de mastic pour la peinture, lut *m*

FLATTING VARNISH - [paint] (special varnish suitable for under-coatings, which are to be flatted, not to be confused with Flat Varnish q.v.) couche *f* de vernis mat
FLATULENCE - [med](accumulation of air or gas in the intestines) flatulence *f*, ventosité *f*
FLATULENT - [med] flatulent, flatueux
FLATUS - [med] (the air or gas accumulated in the intestines) flatuosité *f*
FLATWISE - [gen & metall] (perpendicular to the plane of lamination) à plat
~BEND - [electron] (of waveguides) coude *m* à plat
FLAUNCHING - [constr] (the slope at the top surface of a chimney) inclinaison *f* de la souche de cheminée
FLAVEDO - [med] coloration *f* jaune de la peau
FLAVESCENT - [bot] (with yellow spots on the green) flavescent
[gen] jaunâtre
~SUBSTANCE - [chem] (a dyeing substance) matière *f* flavescente
FLAVIN - [chem] (synonym for acriflavine; used as an antiseptic) flavine *f*
FLAVONE - [chem] (the phenyl derivative of chromone, the basis of several natural vegetable dyes) flavone *f*
FLAVOPHENINE - [chem] (yellow dyestuff) flavophénine *f*
FLAVOPROTEINS - [chem] (enzymes which act as dehydrogenases in the cell) flavoprotéines *f*pl
FLAVOUR, to - [gen] assaisonner, parfumer
FLAVOUR - [gen] saveur *f*, goût *m*
[gen] (of wines etc) bouquet *m*
[gen] (of tea etc) arome *m*
FLAW - [gen] défaut *m*, imperfection *f*, vice *m* de forme
[metall & plast ind] craquelure *f*, soufflure *f*
[mech] (a crack, e.g. on a wall etc) crique *f*, paille *f*, fissure *f*, fente *f*
[metall] fêlure *f*, fissure *f*
[gen] (error or imperfection) erreur *f*, imperfection
~FAULT - [geol] faille *f* à rejet horizontal
~SENSITIVITY - [radiat] (in a radiograph) sensibilité *f* au défaut d'épaisseur
~IRON - [metall] fer *m* paillieux, fer cendreux
FLAWLESS - [gen] parfait, impeccable, sans défaut
FLAX - [bot] lin *m*
~COMB - [text] sérançoir *m*, peigne *m* pour lin
~DODDER - [bot] (a weed) cuscute *f* du lin
~DRESSING - [text] préparation *f* du lin
~FLEA BEETLE - [zool] altise *f* du lin
~GROWING - [agric] culture *f* linière
~HACKLING - [text] peignage *m* du lin, sérançage *m* du lin
~HACKLING MACHINE - [text ind] peigneuse *f* du lin
~LINE YARN - [text] (linen yarn) fil *m* de lin
~MILL - [text] filature *f* de lin
~OAKUM - s.flax tow
~OIL - [chem] (linseed oil; oil obtained from the seeds of flax) huile *f* de lin
~PULLER - [agric] arracheuse *f* de lin
~RETTERY - [text] rouissage *m* du lin
~RUST - [bot] (a plant disease) rouille *f* du lin
~-SEED - [bot] (linseed) grain *m* de lin, linette *f*
~SPINNING - [text] filature *f* du lin

FLAX SPREADER- [text] étendeuse *f* pour lin
~STRAW - [text] lin *m* en paille, lin *m* en tige
~TOW - [text] étoupe *f* de lin
~-TOW PREPARATION - [text] préparation *f* des étoupes de lin
~WEAVING - [text] tissage *m* du lin
~WILT - [bot] (a plant disease) flétrissement *m* du lin
~YARN - [text] fil *m* de lin
FLAXEN HOSE - [rubber ind] tuyau *m* en lin
FLAXSEED COAL - [mining] culm *m*, ménu *m*
~ORE - [min] minerai *m* oolithique
FLAY, to - [gen] écorcher, dépouiller
F LAYER - [astr] (the upper layer of the ionosphere) couche *f* en F
FLAYING - [gen] (stripping of the skin) écorchement *m*, dépouillement *m*
FLEA - [zool] puce *f*
FLEAM - [vet] (a type of lancet for phlebotomy) flamme *f* pour saignées
FLECK - [metall] flocon *m*
FLEECE, to - [gen] tondre
FLEECE - [text] (the woolly covering of a sheep etc) toison *f*
[plast ind] (mat or blanket of glass fibre) toison *f*
[text] (the tin layer of cotton or wool fibre obtained from the breaking-cards) couche *f* de voile
~FABRIC - [text] étoffe *f* imitant la fourrure
~FORMER - [text] nappeur *m*
FLEECY CLOUDS - [met] nuages *mpl* moutonnés, nuages *mpl* cotonneux
~FABRIC [text] (a fabric for hosiery with a pile raised at the back) étoffe *f* moutonnée
FLEET ANGLE - [mining] angle *m* de déviation de câble
FLEETING ANGLE - [mech] angle *m* de courroie en V
FLEMING VALVE - [electron] (the original form of thermionic diode used as a detector) tube *m* de Fleming
FLEMISH BOND - [constr] (a bond which consists of stretchers and headers laid alternatively in each course) appareil *m* polonais
FLESH - [gen] chair *f*, viande *f*
[zool] chair *mf*
~SIDE - [leather ind] (the internal surface of the hide, used in driving belts) côté *m* chair
~SPLIT - [leather ind] croûte *f*
~-TINT - [photo] carnation *f*
FLESHING - [leather ind] (the removal of flesh from hides and skins) écharnage *m*
~BEAM - [leather ind] chevalet *m* de rivière
FLESHY - [gen & bot] charnu
FLEX, to -[gen] (to bend) fléchir
[geol] plier, se plier
FLEX - [el] (colloq for flexible cable) câble *m* souple, conducteur *m* flexible
[math] point *m* d'inflexion
[gen] (short for flexion) flexion *f*
~BREAK - [rubber ind] rupture *f* de carcasse par flexion
~CRACKING - [rubber ind] (the appearance of cracks due to continuous stresses) craquelures *fpl* de flexion
[mech] craquelure *f* due à la flexion
~CRACKING RESISTANCE - [rubber ind] résistance *f* aux craquelures par flexion
~LEAD TORQUE - [mech] moment *m* de torsion du conducteur flexible
FLEX LIFE - [phys] (period during which a material will endure repeated bending) résistance *f* à la flexibilité
~STIFFNESS - s. flexural rigidity
~TEST - s. flexing test
FLEXED - [gen] fléchi, plié
FLEXIBILITY - [mech] flexibilité *f*
[paint] (capable of withstanding flexural stresses without cracking) flexibilité *f*
~TEST - s. flexing test
FLEXIBLE - [gen] flexible
~BETA-RAY SHEETING - [radiat] feuille *f* flexible à émission bêta
~BRAIDED METAL LINER - [auto] revêtement *m* souple en métal tressé
~BRAKE HOSE - [auto] tuyau *m* flexible de freins
~BRAKE PIPE - [mech] conduite *f* souple
~CABLE - [el] (a cable of sufficiently fine stranding to make it flexible) câble *m* souple, cordon *m* souple
~COLLAR - [plumb] collet *m* élastique
~CONDUIT - [gen] conduite *f* flexible, conduite *f* souple
~CONNECTOR - [el] connecteur *m* flexible
~CORD - [el] (flexible cable of small cross-section) cordon *m* flexible, conducteur *m* souple de raccordement
COUPLING - [mech] (a shaft coupling used when a perfectly rigid alignment is impossible) manchon *m* élastique, accouplement *m* élastique
~CROSS-SPAN SUSPENSION BY MEANS OF CABLE - [el] portique *m* souple, suspension *f* transversale souple par câble
~FASTENING - [el] fixation *f* flexible
~FIBRE - [text] fibre *f* flexible
~GANGWAY BELLOWS - [railw] soufflet *m* de communication entre deux wagons
~HARROW - [agric] herse *f* flexible
~HEALD - [text] lisse *f* flexible
~HYDROCARBONS - [chem] hydrocarbures *mpl* à structure déformable
INTERLOCK - [mech] enclenchement *m* flexible
~LEAD - [el] conducteur *m* flexible
~-LEAD CONNECTOR - [el] prolongateur *m*
~LINE - [mech] (used to feed oil or fuel) conduite *f* souple
~METAL TUBING - [mech] tuyau *m* flexible en métal, tuyau *m* métallique flexible
~PACKING - [mech] garniture *f* flexible
~PIPE - s. flexible line
~PLASTICS - [plast ind] plastique *m* flexible
~RISING PIPE - [ind chem etc] tuyau *m* ascendant de refoulement
~SEAL - [gas ind] (in gasholders) jupe *f* de joint
~SEPARATOR ACCUMULATOR - [el] accumulateur *m* à membrane flexible
~SHAFT - [mech] (device for transmitting small amounts of power through a flexible spiral of metal) flexible *m*, arbre *m* flexible, transmission *f* flexible
~SHAFT MACHINE - [mech mach] (used in filing, grinding etc) moteur *m* à transmission flexible
~SHAFT RASP - [impl] meule *f* sur flexible
~SHAFT RUBBER - [rubber ind] caoutchouc *m* souple pour tiges de bottes
~SHEETING - [metall & plast ind] (non-rigid sheeting: sheeting which can be flexed or folded) feuille *f* souple

FLEXIBLE SUPPORT - [el] (for overhead lines) support *m* flexible
~SUSPENSION - [el] (a method of suspending the contact wire of a traction system, so as to allow for a measure of lateral and vertical movement) suspension *f* flexible
[acoust] (the suspension of a loudspeaker cone in such way that a bass reproduction is excellently obtained) suspension *f* souple
~THERMOCOUPLE HARNESS - [el] support *m* protecteur flexible de couple thermoélectrique
~WAVEGUIDE - [electron] guide *m* d'ondes souple
~WIRING - [el] (the use of flexible cables for installations) câblage *m* à conducteurs souples
FLEXING - [gen etc] flexion *f*
[rubber ind] (of tyres) aire *f* de flexion, zone *f* de flexion
~LIFE - [phys] (the durability of an object in regard to frequent flexure) résistance *f* à la flexion alternée
~MACHINE - [mech] (for tests) appareil *m* pour essai de flexion
~RESISTANCE - s. flexural strength
~TEST - [mech] essai *m* de flexibilité, essai *m* de flexion
FLEXION - [gen & mech] flexion *f*
~LOAD - [phys] charge *f* de flexion
FLEXOGRAPHIC PRESS - [print] machine *f* flexographique
~PRINTING - [print] flexographie *f*, impression *f* flexographique, impression *f* à l'aniline
FLEXOR - [zool] (muscle which bends a limb by its contraction) fléchisseur *m*
FLEXURAL - [gen] fléchissant, flexionnel
~CENTRE - [phys & aero] (also called shear centre; the point at which a shearing forcecauses only bending but no twist) centre *m* de cisaillement
~LOAD - [mech] charge *f* de flexion
~MODE OF VIBRATION - [radio] (common mode of vibration) mode *m* de vibration flexionnel
~MOMENT - [phys] moment *m* de flexion
~RIGIDITY - [rubber ind] rigidité *f* de flexion, raideur *f* de flexion
~STRENGTH - [phys] (measure of the resistance to fracture, on transverse loading, as with a load centrally placed on a simple beam of the material. Known also as cross-breaking or transverse strength) résistance *f* à la flexion
~STRESS - [phys] résistance *f* au cisaillement
~VIBRATION - [mech & aero] vibration *f* de flexion
FLEXURE - [mech etc] (the bending of a member) fléchissement *m*
[geol] (folding) flexure *f*, pli *m*, courbure *f*
[math] courbure *mf*
~TEST - s. flexing test
FLICK - [gen] petit coup, petit bruit, claquement *m*
[cin] (colloq for film) ciné *m*
[mech etc] impulsion *f*
~ROLL - [aero] (a quick roll in which the rudder is used at high angles of incidence to aid the rolling action) tonneau *m* déclenché
FLICKER - [gen] tremblotement *m*, battement *m*
[opt] (the sensation which is produced by a fluctuation in brightness) papillotement *m*
[cin] (effect caused by lack of complete persistence of vision) scintillement *m*
~CONTROL - [contr] commande *f* par tout ou rien

FLICKER EFFECT - [radio] (interference in amplifiers caused by impurities in the cathode of the electronic tubes) bruit *m* de scintillation, effet *m* de grêle
~-FREE REPRODUCTION - [telev] reproduction *f* sans papillotement
~FREQUENCY - [opt] (value of the image frequency at which flickering occurs) fréquence *f* de papillotement
~FUSION THRESHOLD - s. flicker photometry
~NOISE - [electron] (noise due to variations in the activity of cathode emitting surfaces in electronic tubes) bruit *m* de scintillation
~PHOTOMETER - [instr] (photometer designed to reveal any difference between two illuminations by an observable flicker) photomètre *m* à papillotement
~PHOTOMETRY - [phys] vitesse *f* critique de la fusion (des excitations lumineuses)
~SHUTTER - [cin] (a rotary shutter) obturateur *m* antipapillotant
FLICKERER - [el] (in electric lighting) limiteur *m*
FLICKERING - [gen] tremblotant, vacillant
[cin] scintillation *f*
LIGHT - [light] lumière *f* intermittente
FLIER - [constr] (a rectangular step forming part of a stair) marche *f* carrée (d'escalier)
[agric] (of a windmill) moulinet *m*
FLIERS - [constr] escalier *m* à rampe droite
FLIGHT - [gen] vol *m*, envol *m*
[constr] (of stairs) volée *f* d'escalier
[mech] (the thread of a feed screw) filet *m*
[firearms] trajectoire *f*
[astronaut] vol *m*
[mech] (of a conveyor) palette *f*
[zool] (of birds) volée *f*
[agric] balle *f* (d'avoine)
[impl] raclette *f*, racloir *m*
~ALTITUDE - [aero] altitude *f*
~ANALYZER - s. flight recorder
~CONTROL - [contr] conduite *f* de vol
~-CONTROL COMPONENT - [contr] organe *m* conducteur de vol
~CONTROLS - [aero] commandes *fpl* de vol
~CONVEYOR - [mech] (conveying device in which transverse plates attached to a chain scrape the material along a trough) transporteur *m* à palettes, convoyeur *m* à palettes
~CORRIDOR - [aero] couloir *m* de vol
~-COURSE COMPUTER - [aero] calculateur *m* de route
~DECK - [aero] (the deck of an aircraft carrier on which aircraft land or from which they take off) pont *m* d'envol
~DIRECTION - [aero] direction *f* de vol
~ENGINEER - [aero] (member of an aircraft crew occupied with engineering duties) mécanicien *m* navigant
~EQUATION - [math] équation *f* de vol
~IN AUTOROTATION - [aero] (gliding flight, in which the airflow through the rotor due to the forward motion maintains its rotation) vol *m* en autorotation
~INDICATOR - [instr] indicateur *m* de la position de vol
~INFORMATION REGION - [aero] (defined airspace within which an air traffic control centre is responsible for giving information and other services) région *f* d'information de vol

FLIGHT LEVEL - s. flight altitude
~LOG - [aero] (a recording instrument indicating the position of the plane on a map) enregistreur *m* de route
~MACH NUMBER - [aero] (the relation between the flying speed of an aircraft and the speed of sound under the same conditions) nombre *m* de Mach de vol
~OF MOLECULES - [phys] trajectoire *f* moléculaire, parcours *m* des molécules
~OF STAIRS - [constr] volée *f* d'escalier
~OF STEPS - [constr] escalier *m*
~PATH - [aero] trajectoire *f* de vol
~-PATH ANGLE - [aero] (the angle between the flight of an aircraft and the horizontal) angle *m* de trajectoire de vol
~-PATH COMPUTER - [comput] (a machine calculating the course and height of an aircraft) calculateur *m* de route
~-PATH DEVIATION INDICATOR - [aero] (instrument designed to indicate any deviation from the flight path) indicateur *m* de dérive
~-PATH RECORDER - [aero] (recording instrument showing the inclination of the flight-path to the horizontal) enregistreur *m* de trajectoire de vol
~PLAN - [aero] (detailed instruction for a proposed flight) plan *m* de vol
~PROFILE - [astronaut] profil *m* du vol
~RECORDER - [instr] (instrument designed to record simultaneously the characteristic value of a flight) enregistreur *m* de route
~REFUELLING - [aero] (the operation of re-filling the fuel-tank of an aircraft during flight) ravitaillement *m* en combustible en vol
~SENSING - [comput] lecture *f* cinématique (pendant le déplacement de la carte)
~SIMULATOR - [aero] (apparatus used in training etc to simulate on the ground conditions produced by actual flight) simulateur *m* de vol
~STRIP - [aero] bande *f* d'atterrissage
[air photography] bande *f* photographique
~TEST - [aero] essai *m* de vol
~TIME - [aero] temps *m* de vol (de cale à cale)
~TRACK - [aero] (the track in space of an aircraft) route *f*
FLIMSY - [gen] faible, sans solidité
[paper man] papier *m* pelure
F. LINE - [phys] (a Fraunhofer line) ligne *f* F
FLINT - [geol] (crystallized form of naturally occurring silica. Used as a filler in powder form) silex *m*
[gen] (as used for lighters) pierre *f* à briquet
[paper man] (coated paper with a polished surface) papier *m* satiné, papier *m* lissé
~CLAY - [min] argile *f* à silex, argile *f* flint
~GLASS - [glass man] (a type of glass having a higher refractive index than that of soda glass: high-quality silica is needed for its manufacture, hence the name, which has no significance as to composition) flint *m*, verre *m* de plomb
~GLAZE - [paper man] (the polished surface on flint paper) satinage *m*
~GRAVEL - [geol] (gravel deposit predominantly consisting of flints) gravier *m* siliceux
~HIDE - [leather ind] peau *f* séchée à fond
~LIGHTER - [gen] briquet *m*
~PAPER - [paper man] papier *m* satiné, papier *m* lissé
FLINT STONE - [geol] silex *m*
~WALL - [constr] (a wall built with fragments of flint) mur *m* de mortier siliceux
~WARE - [gen] (a synonym of stoneware) grès *m* fin, faïence *f* fine
~WHEAT - [agric] blé *m* dur
FLINTY FRACTURE - [metall] cassure *f* conchoïdale, fracture *f* conchoïdale
~GROUND - [geol] sol *m* siliceux
~SLATE - [geol] schiste *m* siliceux
FLIP AND FLOP GENERATOR - [el] (used in television circuits; a relaxation generator) générateur *m* flip-flop, générateur *m* bistable
~-FLOP - [electron] (two vacuum, or gas-filled valves, so arranged that only of them is conducting) flip-flop *m*, bascule *f*
~-FLOP CIRCUIT - [electron] (an electronic circuit with two stable states) bascule *f*, circuit *m* flip-flop, montage *m* en flip-flop, multivibrateur *m* bistable, basculeur *m*
~-FLOP GENERATOR - [telev] (a relaxation generator which is changed from a stable to a quasistable state by means of a pulse of short duration) générateur *m* flip-flop
~-FLOP OSCILLATOR - [radio] (trigger circuit with two conditions of permanent stability) oscillateur *m* flip-flop, univibrateur
~-OVER - [nucl] (in nuclear physics) fustigation *f*
~-OVER PROCESS - [nucl] (a collision between photons, or photons and electrons, in which crystal momentum is not conserved) processus *m* de fustigation
FLIPPER - [zool] nageoire *f*
[swimming aid] nageoire *f*
[billiard-like play] billard *m* électrique
[aero] (colloq) gouvernail *m* de profondeur
[rubber ind] (in tyres) tringle *f*, bandelette *f*, support *m* de talon
~MACHINE - [mech mach] (in rubber industry) machine *f* à enrober les tringles
FLIT-PLUG - [el] (a connecting box which can be detached, used for coupling cables) boîte *f* de dérivation portative
FLITCH - [timber] (a fairly large piece of timber used for reconversion) dosse *f* de bois
[constr] plaque *f* rapportée
[constr] (veneer sheets) feuilles *f pl* de placage
~BEAM - [build] (a beam constructed in layers of timber pinned together) poutre *f* en tôles, poutre à âme pleine, poutre *f* à âme composée
FLIVVER, to - [auto] (colloq. USA; to drive a very old and dilapidated car) se balader en bagnole (ou en tacot)
FLOAT, to - [gen & naut] flotter, nager, mettre à flot, renflouer, surnager
[comm] (the operation of forming a company) lancer, fonder
[constr] (the operation of plastering) dresser, aplanir
[gen] (in the air) entraîner
[text] (said of picks at the back of the cloth) flotter
[metall] (in diecasting) se déplacer
FLOAT - [gen] flot *m*, masse *f* flottante, ras *m*, brelle *f*, radeau *m*
[auto] (of a carburettor) flotteur *m*
[instr] (in a mercury measuring instrument) pointe *f*

FLOAT - [mech] jeu *m* axial
[hydr] (a buoyant structure used in devices for controlling the level of a liquid) flotteur *m*
[aero] (a buoyant structure used to support a seaplane when on the water) flotteur *m*
[aero] (the distance travelled horizontally by an aircraft after flattening-out before landing) palier *m* d'atterrissage
[impl] (a plasterer's trowel) aplanissoire *f* de plâtrier
[transp] wagon *m* à plate-forme
[agric] (an irrigation channel) canal *m* d'irrigation
[agric] (used to flatten clods) herse-trainoir *f* pour champs
[chem] engrais *mpl* phosphatés
[tool] lime *f* à simple taille
[cin] (swinging scenery) rampe *f*
[text] (a weaving defect) brèche *f*, forlançure *f*, nid *m*
[min] minerai *m* pulvérisé flottant
[telev] instabilité *f* d'image
[ind chem] auge *f* de sedimentation
[hydr] aube *f*, palette *f*
~BOWL - s. float chamber
~CARBURETTOR - [mech] (type of carburettor in which a constant level of fuel to supply the jets is maintained by means of a float operating a needle-valve) carburateur *m* à flotteur
~CASE - [hydr] (water-tight air receiver used to float a sunken vessel) caisson *m*
~CHAMBER - [mech] (a vessel forming part of a carburettor in which the float operates) cuve *f* à niveau constant, chambre *f* du flotteur
~CHAMBER COVER - [auto] couvercle *m* de cuve à niveau constant
~COLLAR - [oil ind] manchon *m* à soupape
~COPPER - [min] (fine copper powder) cuivre *m* pulvérisé
~CUT FILE - [tool] lime *f* à taille simple
~FEED - [auto] alimentation *f* par cuve à niveau constant
~FOR CEILINGS - [constr] taloche *f* pour le plafond
~GAUGE - [meas] indicateur *m* de niveau
~GEAR - s. float landing gear
~INDICATOR - [oil ind] indicateur de niveau à flotteur
~LANDING GEAR - [aero] (landing gear incorporating buoyant elements, to support a helicopter when alighting on water) train *m* à flotteurs
~LEVEL - [mech] (in autos) niveau *m* du flotteur
~MINERAL - [min] paillettes *fpl* flottantes
~NEEDLE - [mech] (a needle valve operated by a float to maintain a uniform fuel level in a carburettor) pointeau *m* du flotteur
~NEEDLE GUIDE - [auto] guide *m* du pointeau du flotteur
~-PLANE - s. float seaplane
~-RAIL - [th eng] (a burner rail) rampe *f*, rampe *f* d'alimentation
~SEAPLANE - [aero] (a seaplane which is carried by floats when on the water) hydravion *m* à flotteurs
~SHOE - [mining] sabot *m* à soupape
~SPINDLE - [auto] axe *m* de flotteur
~SPINDLE GUIDE - [auto] guide *m* de l'axe de flotteur
~STONE - [constr] (iron block used to remove marks on brickwork) aplanissoire *f* en fer
[min] (a porous variety of silica) quartz *m* spongieux, spongite *f*
FLOAT STOP VALVE - [mech] soupape *f* d'arrêt à flotteur
~SWITCH - [el] (switch operated by a float in a tank etc) automate *m* à flotteur, interrupteur *m* à flotteur
~TEST - [oil ind] (viscosity test) essai *m* de flottage
~TRAP - [th eng] soupape *f* à flotteur
~-TYPE CARBURETTOR - s. float carburettor
~-TYPE PRESSURE GAUGE - [instr] manomètre *m* à flotteur
~UNDERCARRIAGE - [aero] (the undercarriage of a float seaplane) atterrisseur *m* d'hydravion à flotteurs
~VALVE - [mech] soupape *f* à flotteur
[in big lorries] (a double acting valve with a fourth position which is called 'float' to allow the return by gravity of a double-acting cylinder) robinet *m* à quatre voies dont une de retour libre
FLOATED - [constr] aplani
~COAT - [constr] (plaster smoothed with a float) enduit *m* aplani
FLOATER - [brew ind] flotteur *m*, nageur *m*
FLOATING - [gen] flottant, à flot
[mech] libre, mobile
[naut] flottant
[constr] (the second of the three coats which are applied in plastering) aplanissage *m*, aplanissement *m*
[chem] en suspension
[auto] (of the valves) shimmy *m*
[el] charge *f* à courant constant
[auto] (of a car) flottement *m*
[opt] flottaison *f* sur bain de mercure
~ACTION - [phys] (the action in which a predetermined relation is found between and the rate of motion of a final control element) action *f* flottante
~ADDRESS - [comput] (a label identifying a word or any other information in a routine which is indipendent of the information within the routine itself) adresse *f* symbolique
~ANCHOR - [naut] (a sea anchor) ancre *f* flottante
~ARM - [hydr] bras *m* pivotant
~AXLE - [auto] pont *m* arrière flottant
~BATTERY - [el] (alternative term for "Buffer battery". A battery connected in parallel with a D.C. generator to smooth out variations in voltage and current) batterie *f* équilibrée, batterie *f* de transfert, batterie *f* flottante, batterie *f* tampon
~BEACON - [signal] (a beacon designed to float on a surface of water) balise *f* flottante
~BEARING - [text] (of a cloth roller) support *m* mobile (d'ensoupleau)
~BODY - s. floater
~BRIDGE - [constr] (bridge supported on floating pontoons) pont *m* flottant
~BUSH - [mech] (a bush which is not fixed to the body of the assembly) douille *f* mobile
~CAISSON - [hydr] caisson *m* flottant
[shipbuild] caisson *m* creux
~CALANDRIA - [sugar ind] faisceau *m* suspendu
~CAPITAL - [comm] capital *m* circulant
~CARD COMPASS - [naut] (compass with a circular card floating in a bowl on a mixture of water and alcohol, or oil) boussole *f* à rose mobile
~CARRIER MODULATION - [radio] (system of amplitude modulation used in transmitters in which the carrier level is varied in proportion to the ampli-

tude of the modulating signal) modulation *f* d'amplitude à taux de modulation constante
FLOATING CARRIER SYSTEM - s. floating carrier modulation
~CELLAR - [constr] cave *f* flottante
~CHASE - [mech] (a movable frame) châssis *m* mobile
~CHUCK - [machine tool] mandrin *m* flottant
~CONTROLLER ACTION - [contr] (a control method with a rate of variation of the controlled condition related to the controller error of the set value error) commande *f* à régulateur flottant
~COUPLING - [mech] joint *m* flottant
~CRANE - [mech] (a crane carried on a pontoon) grue *f* flottante
~DAM - [hydr] (a floating structure placed across the entry to a dock etc) barrage *m* mobile
~DEVICE - [mech] système *m* flottant
~DIAL - [instr] rose *f* mobile
~DIE - [rubber and plast ind] matrice *f* avec fixation élastique
~DOCK - [naut] (floating structure open at the ends) dock *m* flottant, chantier *m* à flot
~DREDGER - [mech] bateau-drague *m*
~DRY DOCK - s. floating dock
~ELEVATOR - [mech] élévateur *m* flottant
~FLOOR - [el] plancher *m* de cabine mobile
~GATE - [hydr] (of a floating dock) porte *f* de bassin flottant
~GOLD - [min] or *m* flottant, paillettes *fpl* d'or
~GRID - [electron] (an insulated grid with a potential which is not fixed) grille *f* flottante
~GUDGEON PIN - [mech] (a gudgeon-pin which is not fixed to either the connecting rod or to the piston, and hence is free to move in respect to either) axe *m* de piston flottant
~HARBOUR - [hydr] (a system of booms so fastened together that they form a break-water) digue *f* à flot
~HEAD CONDENSER - [el] condensateur *m* à tête flottante
~HOOK - [mech] crochet *m* libre
~ICE - [gen] glaces *fpl* flottantes
~KIDNEY - [med] (nephroptosis) rein *m* flottant
~KNIFE - [mech] (a doctor knife which is free to move within limits while spreading a coating) racle *f* en l'air
~LEVER - [mech] levier *m* différentiel
~LOAD - [el] charge *f* flottante
NOZZLE - [mech] buse *f* élastique
~PISTON PIN - [mech] axe *m* de piston flottant
~PLATE - [mech] (a plate which is free to move within limits) plateau *m* libre
~PLATEN - [mech] (a press platen which is not fixed to either the upper or the lower platen and is free to move between them) plaque *f* intermédiaire mobile
POINT - [comput] (a notation) virgule *f* flottante
~-POINT CALCULATION - [comput] (a calculation taking into account the varying location of the decimal or binary point) calcul *m* à virgule flottante
~-POINT REPRESENTATION - [comput] notation *f* à virgule flottante
~-POINT ROUTINE - [comput] (routine of coded instructions in proper sequence directing the computer to perform a calculation with floating-point operation) programme *m* à virgule flottante

FLOATING POWER ENGINE MOUNTING - [auto] suspension *f* flottante du moteur
~POWER SUPPLY - [el] alimentation *f* flottante
~PUNCH - [mech] (a punch attached to the head of a press in such a manner that it is free to align itself in the lower portion of the mould during the operation of closing the mould) poinçon *m* flottant
RAIL - [cin] (of a film perforating machine) couloir *m* mobile
~RATE - [contr] (in proportional-speed floating controller) allure *f* flottante, vitesse *f* flottante
[phys] allure *f* flottante, vitesse *f* flottante
~RING - [el] anneau *m* dansant
~ROOF TANK - [oil ind] réservoir *m* à toit flottant
~SADDLE - [mech] (in autos) selle *f* flottante
~SCUM-BOARD - [hydr] pare-écumes *m* flottant, déflecteur *m* de surface
~SLUDGE - [hydr] boue *f* flottante, chapeau *m* d'écume
~SPEED - [contr] étendue *f* de la grandeur de correction
~SPEED - s. floating rate
~SUPPORT - [constr] appui *m* flottant, support *m* flottant
~THREAD - [text] flotté *m*
~-TRICKLE - [el] (battery permanently connected to a discharge circuit which it supplies, and to a charge circuit so regulated, that the mean charging current balances the quantity of electricity discharged and the loss of current due to local action) batterie *f* équilibrée
~ZERO - [contr] zéro *m* flottant
FLOATS - [cin] (footlights) rampe *f*
FLOC - [chem] (abbrev. for flocculent mass) flocon *m*
FLOCCILATION - [med] carphologie *f*, crocidisme *m*
FLOCCITATION - s. floccilation
FLOCCOSE - [bot] (with a tangled covering of wool-like hairs) floconneux
FLOCCULATE, to - [chem] (the settling from a suspension) floculer
FLOCCULATED LATEX - [rubber ind] latex *m* floculé
~SEDIMENT - [soil] sédiment *m* floculé
FLOCCULATING AGENT - [constr] agent *m* de floculation (ou de coagulation), coagulant *m*
~YEAST - [brew ind] levure *f* floculente, levure *f* caséeuse
FLOCCULATION - [chem] (settling from a suspension by the agglomeration of particles into groups or flocs) floculation *f*
[oil ind] floculation *f*
FLOCCULENT - [chem] (in the form of tufts resembling clouds) floculeux, floconneux
FLOCCUS - [met] (fleecy, wool-like cloud) nuage *m* floconneux
FLOCK, to [paper man] (to coat with cotton flock) velouter
FLOCK - [gen] troupe *f*, troupeau *m*
[gen] (of soft material) flocon *m*
[text] (small fibres obtained by the disintegration of cotton, silks etc) déchets *mpl*
~FILLER - [rubber and plast ind] charge *f* floconneuse
~PAPER - [paper man] (paper coated with cotton flock) papier *m* velouté, papier *m* soufflé
~SPRAYING - [paint] (the production of a felt- or suede-like effect by blowing flock with a jet of compres-

sed air into a surface previously coated with an adhesive) floconnage *m*
FLOCKED FABRIC- [text] tissu *m* velouté
FLOCKING - [text] (the operation of coating with flock) flocage *m*
~INSTINCT - [zool] instinct *m* grégaire
FLOCKY - [gen] à flocons
~GRAPHITE - [min] graphite *m* colloïdal
FLOCS - [chem] (coagulated masses of material formed in a suspension) flocons *mpl*, précipité *m*
FLOG, to - [gen] fustiger, flageller
[carp] (the operation of rough-dressing a piece of timber to a shape) dégrossir
[metall] dessabler
[die casting] (the operation of eliminating loose sand from a casting) décocher
[naut] (of sails) fouetter
FLOGGING - [carp] dégrossissage *m*
[die casting] décochage *m*
[naut] (of sails) fouettement *m*
FLONG - [print] (sheets of papier maché used for moulds from which stereo plates are cast) flan *m*, matrice *f*, écrou *m*
FLOOD, to - [gen] inonder, submerger, déborder
[hydr] (of a river) être en crue
FLOOD - [gen] inondation *f*
[hydr] inondation *f*
[hydr] (of water course) débordement *m*, crue *f*
[cin & telev] (in a studio) lampe *f* à faisceau élargi, lampe *f* pour éclairage égal
[naut] marée *f* montante, flot *m*
~ARCH - [hydr] arche *f* de décharge pour les hautes eaux
~BED - [geogr] lit *m* majeur
~BRIDGE - [constr] pont *m* audessus des terrains de inondations
~COCK - [hydr] robinet *m* d'arrosage
~DAM - [constr] digue *f* contre les crues
~DEPOSIT - [geol] gîte *f* alluvionnaire
~FENCING - [hydr] (fencing designed to withstand the pressure of flood waters) protection *f* contre les crues
~FLANKING - [hydr] (the erection of an embankment by means of stiff moist clay in separate small loads) digue *f* d'argile et de terre
~-GATE - [hydr] vanne *f* de passe, porte *f* d'écluse, vanne *f* plongeante
~-LAMPS -[light] lampes *fpl* pour éclairage par projection
~LEVEL - [hydr] niveau *m* de crue
~-LIGHT - [el] (system of illumination) lumière *f* à grands flots
~-LIGHT PROJECTOR - [light] projecteur *m* à flots de lumière
~-LIGHT SCANNING - [telev] analyse *f* à éclairage par projection
~-LIGHT SYSTEM - [light] (arrangement for producing general illumination of an area or the face of a building) système *m* d'éclairage à flots de lumière
~-LIGHTING - [light] (illumination of an area by means of lights projected from some distance from the area itself) éclairage *m* à grands flots, illumination *f* diffusée, éclairage *m* par diffusion
~PLAIN - [geol] plaine *f* d'inondation
[geogr] (of a river) lit *m* majeur
~-PLAIN DEPOSITS - [geol] dépôts *mpl* de plaine alluviale
FLOOD TIDE - [met] flux *m*, marée *f* montante
FLOODBANK - [hydr] dyke *m*
FLOODED - [gen] inondé, noyé, submergé
[mech] (of a carburettor) engargé
~LIGHT - [light] (a light illuminating a large surface by means of projectors placed at some distance from the surface itself) lumière *f* diffuse
~SUCTION - [oil ind] aspiration *f* immergée
FLOODING - [gen] inondation *f*, arrosage *m*, noyage *m*
[mech] (of a carburettor) noyé
[electron] (the screen of a camera) arrosage *m*
~OF DISTILLATION COLUMN - [oil ind] (condition when the spaces between the trays become filled with liquid) noyage *m*
~CATHODE - [electron] cathode *f* d'arrosage
~PATTERN - [mining] patron *m* d'injection
~POINT - [chem] point *m* d'engorgement
~POINT-RATE - [chem] vitesse *f* au point d'engorgement
FLOODLIT - [gen] illuminé par projecteurs
FLOOR, to [gen] terrasser, aplatir
[constr] planchéier, parqueter
FLOOR - [gen & build] plancher *m*, parquet *m*, carrelage *m*, pavement
[build] étage *m*
[naut] frodage *m*, parquet *m* de chargement
[of a dock] plafond *m*
[min] (the upper surface of a stratum) éponte *f* inférieure
[shipbuild] varangue *f*
[auto] plancher *m*
[constr] (the actual division of one floor from another) plancher *m*
[meas] (of a scale) chevalet *m*
[metall] (foundry; the bed of sand forming the floor of a foundry) lit *m*
[hydr] (of a barrage) radier *m*
[mech] fond *m*, aire *f*, tablier *m*
[agric] airée *f*
~ARMOUR - [auto] renfort *m* de plancher
~BEDDING - [metall] mise *f* en chantier
~BLOCK - [mining] moufle *m* de plancher
~BOARD - [constr] planche *f* de plancher, lame *f* de plancher
[shipbuild] varangue *f*
~BOARDING - [constr] voligeage *m*
~BRICK - [constr] brique *f* à paver
~CARPET - [text] tapis *m* de pied
~CHISEL - [carp] (chisel used to lift floor boards) ciseau *m*
~CONTACT - [el] (switch fitted to the floor of an automatic lift) interrupteur *m* à pédale
~ENAMEL - [chem] émail *m* pour planchers
~FOUNDATION - [constr] soubassement *m*
~FURNACE - [space heating] (underfloor warm air heater) calorifère *m* sous plancher
~HANGER - [constr] chaise *f* de sol
~HEIGHT - [railw] (the distance between rails and the coach floor) hauteur *f* du plancher
~INDICATOR - [el] (in automatic lifts) indicateur *m* d'étage
~JOIST - [build] (timber beam supporting the floorboards) solive *f* de plancher
~LAMP - [light] lampadaire *m*
~LEVEL - [mining] plaque *f* intérmédiaire d'entre-

toisement
FLOOR LINE - [carp] (mark on a door or window post to indicate the level of the floor) repère *m* de niveau du plancher
~MALT - [brew ind] malt *m* sur aire
~MALTING - [brew ind] malterie *f* sur aire
~MAT - [gen & auto] tapis *m* de plancher
~OF A LOCK - [constr] plancher *m* de l'écluse *f*
~OF A WEIR - [hydr] radier *m* d'un barrage
~OF FOUNDRY - [metall] sol *m* de la fonderie
~OF SEAM - [mining] mur *m* d'une couche
~ON COUCH, to - [brew ind] étendre en couche
~PAN - [auto] dessous *m* de caisse, plancher *m*
~PAN FRONT - [auto] avant *m* du plancher
~PLAN - [build] (special detailed plan relating to one floor of a building) plan *m* d'étage
[auto] plan *m* du plancher
[comput] (of a computer system) plan *m*, plan *m* d'aménagement
~PLATE - [shipbuild] tôle-varangue *f*
[railw] (of the gangway between coaches) passerelle *f*
[metall] tôle *f* pour couvertures de planchers
~PLOUGH - [agric] retourneur *m*
~ROUGH - [constr] soubassement *m*
~SHOOTING - [cin & telev] prise *f* en studio
~SLAB - [constr] damme *f*
~SPACE - [gen] (the actual surface occupied by a machine) encombrement *m*
[railw] (of a wagon) surface *f* du plancher
[constr] (of a building) surface *f* des étages
~STANDARD - [light] (a portable standard lamp) lampadaire *m*
~SWITCH - [el] (in a lift) commutateur *m* à pédale
~TIMBER - [shipbuild] varangue *f*
~TRAP - [gen] trappe *f*, abattant *m*
~-TRAP FASTENER - [auto] fixation *f* de la trappe du plancher
~WAX - [ind chem] cire *f* à parquet
~WORK - [metall] moulage *m* à découvert
FLOORING - [gen & constr] planchéiage *m*, carrelage *m*, dallage *m*, parquetage *m*, platelage *m*
[agric] (particularly in connexion with brewing industry) conduite *f* de la germination
[gas ind] tablier *m*
~CEMENT - [constr] colle *f* à tapis
~OF RUBBER TILES - [rubber ind] carrelage *m* en caoutchouc
~ON JOISTS - [build] planchéiage *m* sur soliveaux
~PLATE - [rubber ind] dalle *f* en caoutchouc, carreau *m* en caoutchouc
~PRESS - [mech] presse *f* à revêtement de plancher
FLOP VALVE - [mech] vanne *f* à clapet, robinet *m* à clapet articulé
FLORA - [bot] (the plant population and distribution in a specified area) flore *f*
FLORAL AXIS - [bot] (the enlarged end of a flower stem) axe *m* floral
~DESIGN - [text] dessin *m* à ramages
~ENVELOPE - [bot] (calyx and corolla) tunique *f* florale
~LEAF - [bot] (a petal) feuille *f* florale
FLORENCE - [text] (a silk fabric) florence *f*
FLORENTINE ARCH - [arch] (arch with a semicircular intrados and a pointed extrados) arc *m* florentin
~BLIND - [constr] (roller blinds with outside pieces) store *f* à l'italienne, banne *f*, store-banne *f*
FLORENTINE DRILL - [text] florentine *f*
~LAKE - [paint] (a pigment obtained by precipitating cochineal extract with salts of aluminium and tin. It is also called Crimson Lake) laque *f* florentine, laque *f* carminée
FLORESCENCE - [bot] fleuraison *f*, floraison *f*
FLORET - [bot] (individual flower in a massed inflorescence) fleuron *m*
SILK - [text] filoselle *f*
~SPINNING - [text] filature *f* de schappe, fil *m* de poil
FLORID - [gen & med] fleuri, floride
FLOSS - [gen] frison *m*, filoselle *f*, strasse *f*
[metall] floss *m*
[text] (of silk) bourre *f* de soie, fleuret *m*
~HOLE - [foundry] (an orifice through which the floss is removed) trou *m* à laitier
~-SILK - [text] bourre *f* de soie
FLOTAGE - [naut & gen] flottage *m*
FLOTATION - [gen & naut] flottaison *f*
[min] (concentration of ores by the production of minute bubbles which attach themselves selectively to mineral particles and carry them to the surface of a tank in the form of a froth) flottage *m*, flottation *f*
[comm] (of a company) lancement *m*
~BLANKET - [min] (in the tank used for separation) couverture *f*
~GEAR - [aero] (emergency float equipment for land aircraft) dispositif *m* de flottaison, appareil *m* d'amerrissage
~PROCESS - [min] s. flotation
FLOTILLA - [naut] flottille *f*
FLOTSAM - [naut] épave *f* flottante, corps *m* charrié par un fleuve
~AND JETSAM - [naut] choses *f*pl de fot et de mer
FLOUR, to - [chem] (of mercury) réduire en farine
FLOUR - [agric] (ground wheat) farine *f*
[constr] (fine dust which is formed when materials are crushed) farine *f*
~BEETLE - [zool] ténébrion *m*
~DRESSING - [ind proc] criblage *m*
~EMERY - [metall] potée *f* d'émeri
~GOLD - [min] or *m* flottant
~-MILL - [mech mach] moulin *m* à farine, minoterie *f*
~MITE - [zool] (a pest) pyrale *f* de la farine
~MOTH - [zool] (a pest) mite *f* de la farine
FLOURISH, to - [gen & bot] être florissant, croître
[chem] (of bacteria) se développer
FLOURISH - [mus] (trumpet music) fanfare *f* (de trompettes)
FLOW, to - [gen] couler, s'écouler
[hydr] couler
[chem] (of a fluid) écouler
[oil ind] jaillir, érupter
FLOW - [gen] coulement *m*, écoulement *m*
[met] flot *m*, flux *m* de la marée
[chem] (of a fluid) écoulement *m*, cours *m*
[hydr] (a current) courant *m*
[hydr] (quantity of fluid passing through a pipe etc in a specified period of time) débit *m*, portée *f*, affluence *f*, cours *m*
[paint] (the property in a coating of levelling out to a smooth surface when applied) fluidité *f*, fluage *m*
[mech] (the pipe through which the water leaves a boiler) conduite *f* montante

FLOW - [metall] glissement *m*
[el] passage *m* (de courant)
[geol] (e.g. of lava) coulée *f*
~ALIGNMENT - [plast ind] (the extent to which glass fibres lies in the direction of moulding flow) centrage *m* de la coulée
~AND EBB - [met] (the rising and the return of the tide water) flux *m* et reflux
~ASSEMBLY - [instr] (a mechanism designed to measure continuously the pH-value of liquids in closed-flow circuits) mesureur *m* de pH dans une conduite
~BACK, to - [gen] refluer, regorger
~BEAN - [oil ind] (colloq. for flow nipple) tuyère *f*
[oil ind] (natural gas) duse *f*
~BY HEADS, to - [oil ind] jaillir par intermittence
~CALORIMETER - [instr] calorimètre *m* à flux continu
~CASTING - [rubber ind] moulage *m* par absorption, procédé *m* flow-casting
~CHART - [constr] (a diagram showing the series of operations, machines, processes, etc. through which materials or objects pass during manufacture) organigramme *m*, schéma *m* de fluence
~CONTROL VALVE - [mech] soupape *f* réglante de la vitesse d'écoulement
~CONTROLLER - [contr] régulateur *m* de débit
~CRACKS - [metall] craquelures *fpl* d'écoulement
~CURVE - [mech] (the graphical representation of flow) diagramme *m* d'écoulement
~DIAGRAM - s. flow chart
~DIVIDER - [mech] distributeur *m* d'écoulement
~-DIVIDING VALVE - [mech] soupape *f* distributrice d'écoulement
~EQUALIZER - [contr] égalisateur *m* d'écoulement
~GAUGE - [hydr] débitmètre *m*, jauge *f* de passage
~GRADE - [plast ind] (arbitrary grading of moulding powders according to flow temperatures) fluidité *f*
~HEAD - [oil ind] tête *f* d'éruption
~HOLE - [glass man] (the passage between the melter and the refiner in a tank) barrage *m*
~INDICATOR - [instr] indicateur *m* de débit
~LIMIT - [metall] limite *f* de glissement
~LINE - [metall] bec *m* de déversement
~LINES - [metall] (visible lines in a moulding, due to premature freezing) lignes *fpl* de coulée
~MARK - [metall & plast ind] (a mark on a moulding due to the flow of partially hardened moulding material within the mould) ligne *f* d'écoulement, trace *f* d'écoulement
[metall] (in diecasting) fleur *f*, trace *f* d'écoulement
~METER - [instr] (instrument designed to measure the rate of flow) fluxmètre *m*
~NIPPLE - [oil ind] tuyère *f*
~NOZZLE - [gas ind] (in flow pipes) tuyère *f*
~OBSTRUCTING SPOT - [hydr] étranglement *m*, point *m* d'étranglement
~OF A FLUID - [phys] (the movement of a fluid with a continual change of place) écoulement *m* d'un fluide
~OF INFORMATION - [comput] flux *m* d'informations
~OF MATERIALS - [nucl] (in a nuclear reactor) circulation *f* du matériel
~OF THE MASSECUITE - [sugar ind] écoulement *m* de la masse cuite
~OF WATER - [hydr] courant *m* d'eau, écoulement *m* de l'eau

FLOW-OFF, to - [gen &hydr] sortir, s'écouler
~-OFF - [metall & plast ind] (a channel cut from a riser to allow metal to escape once it has reached a certain height) décharge *f*
~-OFF CASTING - [metall] dégorgement *m* de la coulée
~-OFF GATE - [metall & plast ind] déversoir *m*
~PATTERN - [contr] diagramme *m* d'écoulement
~PIPE - [hydr] tuyau *m* adducteur
~PROCESS - [glass man] (or gob process; the process of delivering glass to a forming unit in a gob form) procédé *m* à goutte
~PROPORTIONER - [contr] dispositif *m* de dosage
~RESISTANCE - [acoust] (the quotient of pressure difference between the two surfaces of sound absorbing material) résistance *f* acoustique CC
[el] résistance *f* au flux
~SET - [plast ind] essai *m* de durcissement complet
~SLIDE - [soil] glissement *m* par écoulement
~SPOOL - [oil ind] bride *f* d'écoulement
~STRESS - [metall] effort *m* de fluage
~STRING - [oil ind] (natural gas install) tube *m* de pompage, tube *m* de production
~STRUCTURE - [geol] (a frequently contorted banding caused by flow movement in a viscous molten fluid) structure *f* fluidale
[metall] structure *f* due à la déformation plastique
~TANK - [el chem] cuve *f* électrolytique
[oil ind] réservoir d'écoulement
~TEMPERATURE - [metall & plast ind] (the temperature at which a material softens to a degree that enables it to flow in a specified manner, commonly under pressure) température *f* d'écoulement
~TEST - [ind chem] essai *m* de fluidité
~TEXTURE - s. flow structure
~-THROUGH CENTRIFUGE - [nucl] (centrifugal cream separator applied to isotope separator) centrifugeur *m* à courant parallèle
~VALUE - [soil] coefficient *m* de glissement, coefficient *m* de glissance
FLOWABILITY - [gen, metall etc] fluidité *f*, pouvoir *m* de s'écouler, aptitude *f* au moulage, coulabilité *f*
FLOWED WAX - [acoust] disque *m* à cire coulée sur support métallique
FLOWER, to - [gen & bot] fleurir, être en fleur
FLOWER - [bot] fleur *f*
~BED - [bot] parterre *f*, plate-bande *f*
~BUD - [bot] bourgeon *m* à fleur
~BULB - [bot] bulbe *m*, oignon *m*
~FORCING - [bot] forçage *m* de fleurs
~FORMATION - [bot] formation *f* de fleurs
~GROWER - [bot] fleuriste *m*, jardinier *m* fleuriste
~GROWING - [bot] floriculture *f*, culture *f* florale
FLOWERING TIME - [agric] fleuraison *f*
~OIL - [bot] semences *fpl* de fleurs
~SEED - [bot] semences *fpl* de fleurs
~SEEDLING - s. flower seed
FLOWERS OF ANTIMONY - [chem] trioxyde *m* d'antimoine
~OF CAMPHOR - [chem] fleurs *fpl* de camphrées
~OF SULPHUR - [chem] (commercial purified sulphur in fine powder form, obtained by sublimation) fleur *f* de soufre
~OF TIN - [chem] fleur *f* d'étain
~OF ZINC - [chem] blanc *m* de Chine
FLOWERING TIME - [agric] fleuraison *f*
FLOWING - [gen] coulant, flottant

FLOWING BY HEADS - [oil ind] production *f* intermittente
~FURNACE - [metall] four *m* réverbère
~LIFE - [oil ind] durée *f* d'éruption
~OF SLAG - [metall] (slag floating on the surface of a weld) écoulement *m* de laitier
~PRESSURE - [oil ind] pression *f* d'écoulement
~THROUGH - [phys] (of gas or liquid) écoulement *m*, courant *m*, flux *m*, passage *m* (de gaz ou liquide)
~WATER - [hydr] eau *f* courante
~WELL - [oil ind] puits *m* jaillissant, puits *m* en éruption naturelle, sonde *f* éruptive
FLOWINGNESS - [gen] fluidité *f*
FLOWRATOR - [instr] (used only in the USA) débitmètre *m* à flotteur, rotamètre *m*, gyromètre *m*
FLOX, to - [astronaut] ravitailler en fluor-oxygène
FLOX - [astronaut] fluor-oxygène *m* liquide
FLUCTUATE, to - [gen] fluctuer, varier, osciller
FLUCTUATING POWER - [el] (vector quantity, the vector representing the alternating part of the power and rotating at speed double the angular velocity of the current) puissance *f* fluctuante
~VOLTAGE - [el] variation *f* de tension, inversion *f* de tension
ELUCTUATION - [gen] oscillation *f*, variation *f*
[electron] (in thermionic tubes) fluctuation *f*
[comm] (of prices, costs etc) variation *f*
~NOISE - [radio] (in an amplifier, the noise caused by shot and flicker effect) effet *m* de grenaille
~OF AMPLITUDE - [electron] fluctuation *f* d'amplitude, variation *f* d'amplitude
~OF DEMAND - [comm] variation *f* des besoins
~OF PRESSURE - [mech] variation *f* de pression, fluctuation *f* de pression
~STRUCTURE - [geol] structure *f* fluidale
~TIME - [phys] (a physical value) temps *m* de fluctuation
~VOLTAGE - [electron] (in thermionic tubes) tension *f* irrégulière
FLUE - [constr] (part of a chimney) conduite *f* de fumée, tuyau *m* de cheminée
[th eng] (the channel through which the combustion products of a boiler are led to the chimney) carneau *m* intérieur, tube *m* foyer
[railw] (in locomotives) conduite *f* de fumée
[metall] carneau *m*, rampant *m*
~ASH RETAINER - [gen] boîte *f* à recueillir les cendres
~BLOCK - [build] boisseau *m*
~BOILER - [boilers] chaudière *f* à carneaux intérieurs
~BRIDGE - [constr] (synonym of fire-brick arch) arc *m* en briques réfractaires
~COLLAR - s. flue nozzle
~DAMPER - [th eng] registre *m*
[metall] registre *m* de tirage, soupape *f* de tirage
~DUST - [th eng] cendres *fpl* volantes
[metall] escarbilles *fpl*
~GAS - [th eng] (the gaseous products of combustion discharged to a furnace flue) fumée *f*, gaz *m* de fumée
~GROUPING - [constr] (the bringing together of flues within a single stack) groupement *m* de cheminées
~LINING - [constr] (used for the protection of the walls) revêtement *m* intérieur de cheminée
~LOSS - [th eng] (in chimneys) pertes *fpl* à la cheminée

FLUE NOZZLE - [gas ind] buse *f* d'évacuation, buselot *m*
~PIPE - [gas ind] conduit *m* de raccordement
~SPIGOT - s. flue nozzle
~SWEAT - [th eng] (water containing soot, tar, etc which condenses on the sides of a flue when a solid-fuel is first lighted) bistre *m*
~ROOF - [metall] voûte *f* du rampant
~TERMINAL - [constr] (in chimney) aspirateur *m* statique
~TUBE - [metall] tube-foyer *m*
~WALL - [th eng] (side of flue structure) piédroit *m* de chauffage
FLUELESS HEATER - [th eng] (space heating) appareil *m* de chauffage sans dégagement, radiateur *m* sans dégagement
~SPACE HEATING APPLIANCE - s. flueless heater
FLUFF, to - [text] (of the pile in furniture plush etc) réduire en peluches, lainer
[leather ind] poncer
[zool] (the feathers) hérisser
FLUFFINESS - [text] état *m* de ce qui est duveteux
FLUFFY MUD - [oil ind] boue *f* gazéifiée
FLUID - [gen] (a flowing substance) fluide *m*, liquide *m*
[phys & chem] (a state of matter, e.g. a gas or liquid) fluide *m*
~AGGREGATE STATE - [chem] état *m* liquide d'agrégat
~ASSETS - [fin & comm] fonds *mpl* de roulement
~BED - [plast ind] lit *m* fluide, lit *m* suspendu
~-BED FURNACE - [metall] four *m* à lit facilement fusible
~CIRCUIT - [auto] circuit *m* hydraulique
~-COMPRESSED - [metall & plast ind] fluide
~CONDITIONER - [contr] dispositif *m* de contrôle pour fluides
~CONVERTER - [auto] convertisseur *m* hydraulique (de couple)
~COUPLING - [auto] embrayage *m* hydraulique
~CYLINDER - [auto] cylindre *m* hydraulique
~DENSITY - [phys] (the mass per unit volume of fluid) densité *f* de fluide
~DRIVE - [mech] (a method of power transmission using hydraulic means) transmission *f* hydraulique
~DYNAMICS - [mech] (the study of the motion of fluids) dynamique *f* des fluides
~EQUIPOTENTIAL SURFACE - [phys] (surface on which the velocity potential of the motion has a constant value) surface *f* équipotentielle de fluide
~FLOW - [phys] (boundary layer; the motion of a low-viscosity fluid, e.g. air, around a stationary body) couche *f* limite, couche *f* sous-jacente
~FLOW STAGNATION POINT - [phys] (in non-viscid fluids, a point on a bounding surface on which the fluid velocity is zero in relation to the boundary) point *m* de stagnation du courant d'un fluide
~FLYWHEEL - [mech] (system of transmission through the medium of the change in momentum of a fluid) embrayage *m* hydraulique
~FORCE - [phys] force *f* fluide
~FRICTION - [phys] (the stresses in a distorted fluid) friction *f* de fluide, frottement *m* de fluide
~FRICTION DAMPING - [instr] (used in electrical measuring instruments) amortissement *m* par frottement de fluide
~GUN - [impl] pistolet *m* à huile

FLUID HYDROCARBON - [chem] hydrocarbure *m* fluide
~JET PUMP - [mech] trompe *f* à vide, pompe *f* à jet de liquide
~LEVEL GAUGE - [instr] indicateur *m* de niveau de liquide
~LUBRIFICATION - [mech] (a state of perfect lubrification whereby a fluid or viscous oil film separates completely the bearing surfaces) graissage *m* fluide
~MEASURE - [meas] mesure *f* pour les liquides
~MECHANICS - [mech] mécanique *f* des fluides
~MIXING TANK - [nucl] (vessel in which the solution of uranium compound is prepared, for transfer to the core tank) cuve *f* mélangeuse
~OBTAINED BY DISTILLING - [chem] distillat *m*
~PACKED PUMP - [oil ind] pompe *f* à garniture liquide
~PARCEL - [phys] élément *m* de fluide
~PARTICLE - [phys] hypothetical particle moving with the flow velocity of the fluid) particule *f* de fluide
~PORT - [mech] lumière *f* de fluide
~PRESSURE - [phys] pression *f* du fluide, pression *f* hydraulique
[mech] pression *f* du liquide
~RUBBER - [rubber ind] caoutchouc *m* liquide
~SOLID- [phys] solide *m* fluidisé
~TIGHT PACKING - [mech] garniture *f* étanche
~TON - [meas] (in metallurgy, equals 32 cubic feet) poids *m* de 32 pieds cubes
~TORQUE CONVERTER - [mech] convertisseur *m* de couple hydraulique
~VELOCITY POTENTIAL - [mech] (function of position of the flow) potentiel *m* de vitesse d'un fluide
FLUIDAL - [gen] (of a fluid) fluidal
~STRUCTURE - [geol] structure *f* fluidale
FLUIDIMETER - [instr] (instrument designed to measure fluidity) fluidimètre *m*
FLUIDITY - [gen & phys] fluidité *f*
FLUIDIZATION - [gen] fluidisation *f*
~DIP COATING - [plast ind] (the position of a plastic coating on a heated object by immersion in finely powdered plastic material agitated by an air blast) recouvrement *m* par immersion en lit fluide
FLUIDIZE, to - [gen] fluidifier
FLUIDIZED - [gen] fluidisé
~BED - [chem] lit *m* fluidisé
~BED GASIFICATION - [gas ind] (in gasification under pressure) gazéification *f* en lit fluidisé
~GAS GASIFIER -[gas ind] (in gasification under pressure) gazogène *m* à combustible fluidisé
~PURIFICATION - [chem] épuration *f* par fluidisation
~REACTOR - [nucl] (type of reactor in which the fuel has quasi-fluid properties) réacteur *m* à combustible fluidisé
FLUIDIZER - [metall] fondant *m*
FLUKE - [naut] (one of the arms of the anchor) patte *f*, aile d'ancre
[med] trématode *m*, douve *f*
~DISEASE - [med] distomatose *f*
FLUME - [hydr] (artificial channel for industrial uses) canal *m* d'amenée, flume *f*, auge *f*, canalisation *f* sur chevalets
[mech] (turbine chamber) chambre *f* d'eau
~WATER - [sugar ind] eau *f* de lavage des betteraves
FLUMES - [sugar ind] caniveaux *mpl*
FLUOBORIC ACID - [chem] (acid formed by the combination of hydrogen fluoride and boron trifluoride) acide *m* fluoborique
FLUOCERITE - [min] fluocérite *f*, fluocérine *f*
FLUOPHOSPHORIC ACID - [chem] (intermediate for fungicides and disinfectants) acide *m* fluophosphorique
FLUOR - [min] (generic term for minerals resembling gems but readily fusible) fluorine *f*, fluorite *f*, chaux *f* fluatée
~CROWN - [opt] fluocrown *m*
FLUORACETOPHENONE - [chem] (reagent used in organic synthesis) fluoroacétophénone *f*
FLUORESCEIN - [chem] (resorcinol-phthalein. A yellowish-green dye; soluble in alkalies with red colour and green fluorescence) fluoroscéine *f*
~PAPER - [paper man] papier *m* à la fluoroscéine
FLUORESCENCE - [phys] (the property of certain substances of emitting visible light when excited by radiations in the invisible range) fluorescence *f*
~BAND - [electron] bande *f* de fluorescence
~COATING - [photo] couche *f* fluorescente
~EFFICIENCY - s. fluorescence yield
~LIGHT - [light] lumière *f* fluorescente
~YIELD - [nucl] rendement *m* de fluorescence
FLUORESCENT - [phys] (proceeding from fluorescence) fluorescent
~DYE - [nucl] (used for the measure of dispersal and dilution effects when discharging radioactive waste into water) colorant *m* fluorescent
~LAMP - [el] (mercury-vapour electric discharge lamp) lampe *f* fluorescente
~LIGHTING - [light] éclairage *m* à fluorescence
~PIGMENTS - [chem] (pigments which show luminosity after being subjects to irradiation by light) pigments *mpl* fluorescents
~RADIATION - [radiat] (characteristic X-radiations emitted when X-rays of higher frequency are absorbed) rayonnement *m* de fluorescence, rayonnement *m* caractéristique
~SCANNING - [telev] analyse *f* de l'écran fluorescent
~SCREEN - [radiat] (surface layer of fluorescent material emitting visible light when subjected to ionizing radiation) écran *m* fluorescent
~SUBSTANCE - [chem] (substance possessing the property to absorb radiant energy of a given wavelength and to emit it as waves of a different length) matière *f* fluorescente
FLUORESCING DYE - [dyes] colorant *m* fluorescent
FLUORIDATE, to - [chem] (the operation of treating with fluoride) fluorurer
FLUORIDATION - [chem] fluoruration *f*
FLUORIDE - [chem] fluorure *m*
~RESISTANT - [chem] résistant aux fluorures
FLUORIDIZATION - s. fluoridation
FLUORIDIZE, to - s. fluoridate
FLUORIMETER - [instr] (instrument designed to measure the contents of fluorine in a substance) fluorimètre *m*
FLUORIMETRIC ANALYSIS - [chem] analyse *f* fluorométrique
FLUORINATED ETHYLENE PROPYLENE- [chem] (copolymer of hexafluoropropylene and tetrafluoroethylene which can be applied by fluidized bed techniques or used in conventional moulding and extrusion

equipment) éthylène-propylène *m* fluoré
FLUORINATED PARAFFIN - [chem] (a heat transfer medium and a lubricant) paraffine *f* fluorinée
FLUORINE - [chem] (a gaseous element, symbol F., with uses as an oxidizing agent, and as a starting material for the production of fluorides) fluor *m*
FLUORITE - s. fluorspar
FLUOROANILINE - [chem] (an intermediate for dyestuffs) fluoroaniline *f*
FLUOROBENZENE - [chem] (intermediate for insecticides and an analytical reagent) fluorobenzène *m*
FLUOROBORIC ACID - [chem] (a starting material for fluoroborates) acide *m* fluoroborique
FLUOROCARBON RESINS - [chem] (a series of polymers which have a low coefficient of friction, and high dielectric strength: they retain their properties over a wide range of temperatures) résines *fpl* fluorocarbonées
FLUOROCARBONS - [chem] (compounds of carbon and fluorine with uses as refrigerants, aerosol propellants, lubricants, and fire extinguishing agents) fluorocarbones *mpl*
FLUOROCHEMICAL ANALYSIS - [chem] analyse *f* par fluorescence
FLUOROMETER - [instr] (instrument designed to measure fluorescence) fluoromètre *m*
FLUOROMETRY - [opt] (the measurement of intensity and colour of fluorescent radiations)fluorométrie *f*
FLUOROPHOTOMETER - s. fluorometer
FLUOROSCOPE - [radiat] (a fluorescent screen so mounted, that X-ray shadows of objects the tube and the screen become visible) appareil *m* radioscopie, röntgenoscope *m*
FLUOROSCOPY - [radiat] (X-ray examination by means of a fluoroscope) radioscopie *f*, röntgenoscopie *f*
FLUOROSIS - [med] fluorosis *f*
FLUOROTHENE - [chem] (a polymer with uses as a plastics material) fluorothène *m*
FLUOROTITRIMETRY - [chem] (the continuous observation of potential changes through changes in fluorescence) analyse *f* volumétrique à fluorescence
FLUOROUS - [chem] fluoré
FLUORSPAR - [min] (naturally occurring calcium fluoride with uses the manufacture of paints, in metallurgy, ceramics, and the production of hydrofluoric acid) fluorine *f*, spath *m* fluor
FLUOSILIC ACID - [chem] (a disinfectant, wood preservative , and component of ceramics glazes) acide *m* fluosilicique
FLUOSULPHONIC ACID - [chem] (a catalyst in organic synthesis) acide *m* fluosulphonique
FLUSH, to - [gen] (to clean out by flushing) donner une chasse, curer
[mech] affleurer, affronter
[auto] (the radiator) rincer
[mining] (wet refuse) embouer
[mining] procéder au remplissage hydraulique
[min] débourber
FLUSH - [gen] (of a surface) très plein, débordant, ras
[gen] (at the same level as a given surface) de niveau, affleurant, dans le même plan
[instr] (of an instrument) encastré, noyé
[med] bouffées *fpl* de chaleur
[hydr] chasse *f* d'eau, canal *m* de fuite
[light] éclat *m* de lumière
[min] lavage *m*
FLUSH BINDING - [bookbind] (binding in which the boards are trimmed with the book) cartonnage *m* rogné à ras de page
~BOLT - [mech] (sliding board flush with a surface) boulon *m* à tête noyée, boulon *m* à tête perdue
~BORING - [hydr] forage *m* par pression hydraulique, sondage *m* à injection d'eau
~DECK - [naut] pont *m* ras, pont *m* de bout à bout, pont entier
~DRILLING - [mining] forage *m* avec circulation
~EAVES - [constr] égout *m* encaissé
~FITTING SLIDING ROOF - [auto] toit *m* ouvrant sans aucune saillie
~GATE - [hydr] vanne *f* de chasse
~HARDWARE - [auto] ferrures *fpl* encastrées
~-HEAD - [mech] (of rivets) à tête noyée, à tête perdue
~-JOINT CASING - [mining] tube *m* fileté à mi-épaisseur
~MOUNTED - [gen] affleuré, encastré, monté dans le même plan
~-MOUNTED INSTRUMENT - [instr] (an instrument mounted so that its front is level or flush with the instrument panel) appareil *m* de mesure encastré
~MOUNTING - [gen] montage *m* affleuré
~MOUNTING TYPE - [mech] (designed to be mounted flush with a panel or other surface, as of an instrument) à montage affleuré
~PANEL - [carp] (with a surface which is in line with the faces of the stiles) panneau *m* ras
~PIPE - [plumb] (in a lavatory system) tuyau *m* du réservoir de chasse
~PLATE - [el] (switch plate) plaque *f* d'interrupteur encastré
~PLATING - [shipbuild] bordé *m* à ras
~PRODUCTION - [oil ind] production *f* éruptive
~RIVET - [mech] (a rivet let into a surface so that its head is in the same plane as that surface) rivet *m* à tête noyée
~RIVETING - [mech] rivure *f* à couvre-joint, rivure *f* à franc-bord
~SCREW - [mech] vis *f* noyée
~SOFFIT - [constr] (continuous surface under a stair) intrados *m* de niveau
~-SWITCH - [el] (switch which is mounted flush with the wall) interrupteur *m* encastré
~TANK - s. flushing tank
~-TANK SIGN - [med] décharge *f* urinaire dans l'hydronéphrose
~-TYPE DIRECTION INDICATOR - [auto] flèche *f* de direction encastrée
~VALVE - [mech] (the valve regulating the flushing system) robinet *m* de chasse
FLUSHER - [gen] appareil *m* de chasse, dispositif *m* de chasse
FLUSHING - [hydr] (the process of cleaning) lavage (the actual flow of water caused by flushing) chasse *f*, canal *m* de fuite
(of a lavatory) chasse *f*
[metall] (ovens and retorts) écartement *m* du laitier
[mining] (the clearing off the accumulation) remblayage *m* hydraulique par embouage
[mining] (if obtained hydraulically) embouage *m*
[constr] (the crushing of the edges of stones at a

hollow bed) broyage *m* (des pierres)
FLUSHING - [ind chem] décrassage *m*
[min] lavage *m* (d'or)
~AUGER - [mining] tarière *f* à rincer
~CHANNEL - [hydr] goulotte *f* de rinçage
~CISTERN - [hydr] réservoir *m* de chasse
~DEVICE - [hydr] appareil *m* de chasse
~DREDGER - [constr] drague *f* à courant *m* d'eau
[mining] drague *f* aspirante
~MANHOLE - [hydr] regard *m* de chasse
~OIL - [mech] (oil used for washing out engine crank-cases and the like) huile *f* de rinçage
~SHAFT - [min] puits *m* d'embouage
[mining] puits *m* de remblayage
~SHIELD - [hydr] vanne *f* roulante
~TANK - [hydr] (the tank which is used to collect water for flushing) bassin *m* de chasse
~WITH AIR - [ind chem] rinçage *m* d'air, balayage *m* d'air, rinçage à air
FLUTE, to - [arch] (the operation of fluting a column) canneler, strier
[mech] rainer, rainurer
[tool] évider
FLUTE - [gen] rainure *f*, cannelure *f*
[constr] (vertical, generally circular groove in the surface of a column etc) cannelure *f*
[tool] outil *m* à évider
[glass man] entaille *f*
[paper man] rainurage *m*
[mus] (instrument) flûte *f*
[mus] (an organ stop of flute-like quality) flûte *f*
[metall] outil *m* à rainurer
~FLUE-STOPS - [mus] (series of flue-stops including harmonic or non-harmonic pipes) jeux *mpl* de flûte
FLUTED - [gen] cannelé, strié
[of glass] à cannelures
[acoust] (of voices etc) flûté
~BOWL - [mech] (a roller formed with lengthwise corrugations) cylindre *m* rainuré
~COLUMN - [arch] colonne *f* striée
~DISK - [mech] disque *m* denté
~MAT - [rubber ind] tapis *m* cannelé
~ROLLER - [mech] cylindre *m* à cannelures
~SHAFT - [mech] arbre *m* cannelé
~TUBE FOR CHEESES - [text] tube *m* cannelé pour bobines croisées
FLUTES - [mech] (of a hob) cannelures *fpl* longitudinales
[glass man] (parallel depression cut in the glass for decorative purposes) cannelures *fpl*
FLUTING - [gen etc] cannelure *f*, évidage *m*, rainurage *m*
[metall] rupture *f* par flexion
~CUTTER - [tool] outil *m* à évider
~PLANE - [carp] (plane for cutting grooves) rabot *m* à évider
FLUTTER, to - [gen] flotter, s'agiter, secouer
[mech] (of a mechanical structure) battre
(of a sail) flotter
[med] (of the pulse) battre irregulièrement
FLUTTER - [gen] volètement *m*, flottement *m*
[mech] battement *m*, vibration *f*
[el acoust] (undesired vibration of frequency modulation introduced in the signal by an irregular motion of the recording medium) battement *m*
[aero] (unstable continuous mechanical oscillation in any part of an aircraft initiated by a transient disturbance and maintained by the interaction of inertial and aerodynamic forces and the elasticity of the material) flutter *m*, flottement *m*
[of piston rings] (a movement of expansion and contraction in piston rings, as the piston travels, caused by lack of cylindricality in the cylinder bore) battement *m*
[acoust] (pulsation of intensity in reproduced sound) battement *m*, oscillation *f* du son
[med] flutter *m* auriculaire
[radio] variation *f* de perte de transmission
[radio & telev] (spurious vibration of the dipole) vibration *f* parasite
[el acoust] (in tape recording) pleurage *m*, sautillement *m*
FLUTTER ECHO - [acoust] (a multiple echo) échos *mpl* multiples
~EFFECT - [radio] effet *m* de variation de perte de transmission
[acoust] effet *m* vibratoire, effet *m* de battement
~SPEED - [aero] (the lowest equivalent airspeed which gives rise to flutter) vitesse *f* de flutter
FLUVIAL - [gen] (of river) fluvial
FLUVIATILE - [bot etc] (which occurs in streams etc) fluviatile
~DEPOSIT - [geol] (the sand and gravel which are deposited on the bed of a river) dépôt *m* fluviatile,
~DEPOSITS - [geol] dépôts *mpl* fluviaux
FLUVIOGRAPH - [instr] (instrument designed to measure and record the rise and fall of a river) fluviographe *m*
FLUVIOMARINE - [zool] (living in rivers and in the sea, e.g. the salmon) fluvio-marin
FLUVIOTERRESTRIAL - [zool] (can be found in rivers and on river-banks) fluvio-terrestre
FLUX, to - [gen & phys] (to become fluid, to melt) fondre, mettre en fusion, devenir liquide
(to flow abundantly) ruisseler, jaillir
[metall] (to treat with a flux) additionner de fondant, ajouter le fondant
[metall] (to sprinkle with borax) rocher
FLUX - [gen] (a copious flowing) flux *m*
[el] (the integral of the product of each element of a surface on which a positive and a negative face can be distinguished, by the component of the vector) flux *m* (d'un vecteur)
[ind chem] (a material of low melting point, such as borax, used in mixing enamels or glazes) fondant *m*, flux *m*
[metall] (a substance added, e.g. to a furnace charge, to form fusible slags with impurities or gangue) fondant *m*
[metall] (a substance, used in soldering, brazing or welding, to clean the surfaces to be joined and form a slag with the impurities removed) fondant *m* de brasage, décapant *m*
[electron] (in the preparation of phosphors, ingredient promoting the formation of a luminescent solid) fondant *m*
[met] flot *m*
~ADDITIONS - [metall] additions *fpl*
~-AMPERE-TURN LOOP - [electron] (dynamic hysteresis loop) boucle *f* d'hystérésis dynamique
~-COVERING - [metall] flux *m*
~DENSITY - [el] (the quantity of flux passing through the unit area) densité *f* de courant
[nucl] débit *m* de fluence de neutrons

FLUX DISTRIBUTION - [phys] (the number of electric lines of force per unit area at right angles to the lines) distribution *f* des lignes de force
[phys] (in nuclear physics) distribution *f* de flux de neutrons
~LINE - [comput] ligne *f* de flux
~LINKAGE - [el] couplage *m* inductif
~LINKING A COIL - [el] (the total of the fluxes linking the turns of the coil) flux *m* à travers une bobine
~LINKING A TURN - [el] (the flux across any rotating surface of which the turn forms the circumference) flux *m* à travers une spire
~LOSS - [ind chem] (of ingredients) perte *f* par volatilisation
~METER - [instr] (instrument designed to measure variations in magnetic flux) fluxmètre *m*
~OF A VECTOR - [el] flux *m* d'un vecteur
~OIL - [ind chem] huile *f* fluidifiante
~REFRACTION - [el magnet] réfraction *f* des lignes de force
~REVERSAL - [comput] réversibilité *f* du flux
~TRAP - [nucl] piège *m* à flux
~TRAP REACTOR - [nucl] réacteur *m* à flux concentré
~TUBE - [phys] ligne *f* de flux, ligne *f* de force
FLUXGATE - [instr] (a detector designed to produce a signal proportional to an external magnetic field acting along its axis) sonde *f* magnétométrique
~COMPASS - [instr] (instrument using a fluxgate detector to give indication of the magnetic meridian direction) compas *m* à sonde
~MAGNETOMETER - [instr] magnétomètre *m* à sonde
FLUXING - [metall] abaissement *m* du point de fusion
[metall] addition *f* de fondant
[plast ind] (the conversion into a plastic mix) plastification *f*
~ORE - [metall] minerai *m* fondant
~POWDER - [metall] (fusible material for dissolving and preventing formation of oxides) poudre *f* à souder
~POWER - [metall] pouvoir *m* scorifiant
FLUXIONAL - [geol] (of a texture) fluidal
FLUXVALVE - s. fluxgate
FLY, to - [gen] voler
[aero] voler
[aero] (the action of piloting an aircraft) piloter
[gen] (a flag) battre
FLY - [zool] (an insect) mouche *f*
[fishing] (the insect attached to the hook) mouche *f*
[mech] moulinet *m*
[naut] (of a patent log) sillomètre *m*, loch *m* à hélice
[text] (in combing, the waste discarded) blousse *f*, freinte *f*
[text] (in drawing) déchets *mpl* d'étirage
[text] duvet *m*
~ASH - [gen] (ash form furnaces, etc. of such small particle size that it is normally separated electrostatically) cendres *fpl* volantes
~-BOAT - [naut] (type of fast vessel generally used on canals) canot *m* ailé
[naut] (a type of flat-bottom vessel) bateau *m* à fond plat
~CATCHING INSTRUMENT - [text] instrument *m* pour attraper les duvets
~COLLECTOR - [text] ramasseur *m* de duvet
~COMB SPINDLE - [text] arbre *m* du peigne détacheur

FLY CORD - [text] corde *f* de chasse-navette, cordon *m* de tirage
~CRANK - [mech] contre-manivelle *f*
~CUTTER - [mach tool] (used for cutting slots) fraise *f* à une taille
~FIBRE - [text] fibre *f* volante, fibre *f* étrangère
~FRAME - [text] (machine used to attenuate roving) banc *m* à broches
~GOVERNOR - [mech] (speed-regulating device, generally consisting of vanes on a rotating shaft) volant *m*
~LEAF - [book bind] (the blank leaf at the beginning and at the end of a volume) allonge *f*, feuille *f* attachée
~NUT - [mech] (wing nut) écrou *m* à oreilles, papillon *m*
~OVER - [railw] saut-de-mouton *m*
~-OVER CROSSING - s. fly-over
~PRESS - [mech] (press actuated by a screw provided with revolving fly-weights) balancier *m*, découpoir *m* à la main, presse *f* à vis
~-PROOF - [text] étanche au duvet
~ROLL - [paper man] rouleau *m* compensateur
[text] volant *m*
~SHUNTING - [railw] manœuvre *f* au lancer
~-SHUNTING LINE - [railw] voie *f* de lancement
~SHUTTLE - [text mach] navette *f* volante
~SHUTTLE LATHE - [text mach] battant *m* à navette volante
~SWITCHING - s. fly shunting
FLYBACK - [electron] (the rapid return of the electron beam to its original position after each scan) retour *m*
[telev] (the rapid return of the scanning beam to the starting point) retour *m* du spot
~ACTION - [horol] (part of the action in a cronometer which causes the hands to return to zero) mécanisme *m* de remise à 0
~BLANKING - [telev] suppression *f* du retour
~EXTRA-HIGH TENSION SUPPLY - [telev] alimentation *f* en très haute tension par retour du spot
~LEVER - [horol] bascule *f* de remise à 0
FLYBALL ARM - [mech] bras *m* du volant
~GOVERNOR - [mech] régulateur *m* à boules, régulateur *m* à force centrifuge
FLYER - [gen] aviateur *m*
[text] (of a spindle) ailette *f* (de banc à broches)
[constr] marche *f* carrée (d'escalier)
~BOBBIN - [text] bobine *f* de banc à broches
~DRAG - [text] frein *m* à ailettes
~FRAME - s. fly frame
~LATHE - [text] battant *m* à ailettes
~SPINDLE - [text] broche *f* à ailette
~SPINNING - [text] filature *f* continue à ailettes
~SPINNING FRAME - [text] métier *m* continu à ailette
~THREAD GUIDE - [text] guide-fil *m* à ailette
~THROSTLE - [text] métier *m* continu à ailettes
~YARN - [text] méche *f* de banc à broches
FLYING - [gen] volant, vol *m*
[constr] (screed) saillie *f*, porte-à-faux *m*, encorbellement *m*
~BEDSTEAD - [aero] (term applied to an experimental VTOL craft, using vertical jets for lifting) "flying bedstead" *m*
~BOAT - [aero] (a type of aircraft in which the main body rests on the water and forms the support when not flying) hydravion *m* à coque
~

FLYING BOMB - [milit] (small pilotless aeroplane with a warhead) bombe *f* volante
~BRIDGE - [naut] passerelle *f* volante
[constr] (used during road-works) pont *m* volant
~BULKEAD - [shipbuild] cloison *f* volante
~BUTTRESS - [constr] (arch buttress supporting the base of another arch) arc-rampant *m*
~CHARACTERISTICS - [aero] caractéristiques de vol
~CONTROLS - [aero] (the means by which an aircraft pilot operates the control surfaces) commande *f* de vol
~CRANE - [aero] (a helicopter used for lifting, conveying and depositing loads in structural work) grue *f* volante
~DISK - s. flying saucer
~FISH - [zool] poisson *m* volant
~FLOCK - [text] déchets *mpl* volants
~HEAD - [comput] (of magnetic drum memory) tête *f* magnétique mobile
~HEIGHT - [aero] altitude *f*
~JIB - [naut] contre-foc *m*
~-JIB BOOM - [naut] bâton *m* de contre-foc
~LANE - [aero] direction *f* du vol
~LIFT - [gas ind] (in gasholders) levée *f* intérieure d'un gazomètre (guidée seulement à sa partie inférieure)
~MICROMETER - [instr] (instrument designed to measure the thickness of metals while in movement) micromètre *m* à action rapide
~-OFF DECK - [aero] (aircraft-carrier deck for the take-off of aircraft) pont *m* de décollage
~-ON DECK - [aero] (aircraft-carrier deck for the landing of aircraft) pont *m* d'atterrissage
~PERFORMANCE - s. flying characteristics
~RANGE - [aero] distance *f* franchissable
~RIGGING - [aero] (the rigging connecting a kite balloon to its winch cable) haubannage *m* de vol
~SAUCER - [aero] (flying object of unknown origin said to have seen in the sky) soucoupe *f* volante
~SCAFFOLD - [constr] (a suspended scaffold) échafaudage *m* volant, échafaud *m* à bascule
~SCHOOL - [aero] école *f* de pilotage
~SHEARS - [metall] cisailles *fpl* à porte-à-faux
~SHORE - [constr] (horizontal baulk of timber temporarily supporting two walls) entroisement *m* provisoire
~SHUTTLE - [text] navette *f* volante, navette *f* droite
~SPOR SCANNER - [telev] analyseur *m* à spot lumineux
~-SPOT STORE - [comput] mémoire *f* à tube de Williams
~SPOT SYSTEM - [telev] analyse *f* à spot lumineux
~-SPOT TUBE SCANNING - [telev] tube *m* analyseur à spot lumineux (ou mobile)
~START - [auto etc] départ *m* lancé
~SURFACE - [aero] voilure *f*
~TEST BED - [aero] (an aircraft specially designed or used for testing engines or other features) banc *m* d'essai volant
~TIME - [aero] (the period between departure and arrival times) durée *f* de vol
~TRIANGLE - s. flying wheel
~WEIGHT - [aero] poids *m* total (en vol)
~WHEEL - [aero] volant *m*
~WIRES - [aero] (lift wires) haubans *mpl* porteurs
LYWHEEL - [mech] (a relatively heavy wheel attached to a mechanism to increase the uniformity of rotary motion) volant *m*, volant *m* de commande, volant *m* d'entraînement
~ALTERNATOR - [el] (alternator with a heavy spider and therefore without a separate flywheel) alternateur *m* à volant
~ARM - [mech] bras *m* du volant
~CASING - [mech] carter *m* de volant
~CIRCUIT - [el] circuit *m* à effet du volant
~EFFECT - [radio] (the ability of a resonant circuit to operate continuosly from short poulses of energy of constant frequency and phase) effet *m* de volant
[mech] effet *m* volant
~ENGINE - [mech] machine *f* à volant
~HOUSING - s. flywheel casing
~HUB - [mech] moyeu *m* du volant
~MAGNETO - [mech] (used in the ignition system of two-stroke engines) volant *m* magnétique
~MOMENT - [phys] moment *m* d'inertie
~PIT - [gen] fouille *f* du volant
~PUMP - [mech] pompe *f* à volant
~RACE - s. flywheel pit
~RING GEAR - [mech] (a toothed ring formed on or bolted to the fly wheel of an i.c. reciprocating engine, to engage the starter motor pinion) couronne *f* dentée du volant moteur
~STARTER GEAR - [mech] couronne *f* dentée du volant
~SYNCHRONIZATION - [radio] (synchronization of a sinusoidal oscillator using pulses of constant frequency and phase) synchronisation *f* par effet de volant
~TEETH - [auto] couronne *f* dentée du volant
~TIMEBASE - [telev] base *f* des temps à effet de volant
~WITH FRICTION PLATE - [mech] volant *m* avec disque d'embrayage
F.M. - (initials for Frequency Modulation) s. frequency modulation
FM DETECTOR - [radio] détecteur *m* de modulation de fréquence
~RADAR - [radar] (a system in which the emission of waves is on the whole continuous while the frequency of the waves is modulated) radar *m* avec modulation de fréquence
F-NUMBER - [opt] (relative aperture) ouverture *f* relative
FOAM, to - [gen] écumer, mousser
[hydr] (of a rapid water course) écumer
FOAM - [gen] écume *f*, mousse *f*
~BATH - [med] bain *m* de mousse
~BEATER - [rubber ind] batteuse *f* à mousse, fouetteuse *f*
~COLLAPSE - [plast ind] (condition resulting from insufficient expansion of a plastic) moussage *m* insuffisant
~CONDITION - [oil ind] condition *f* de mousse
~CONNEXION - [el] connexion *f* en matière plastique mousse
~EQUIPMENT - [fire-fighting equip] (apparatus for producing artificial foam for fire-fighting) appareillage *m* pour la production de mousse
~EXTINGUISHER - [fire-fighting equip] extincteur *m* à mousse
~GLUE - [ind chem] (glue which has been foamed to increase the area covered by a specific quantity of glue) colle *f* battue en mousse
~RUBBER - [rubber ind] (rubber expanded into a spon-

gy form) caoutchouc *m* mousse, mousse *f* de latex, écume *f* latex
FOAM RUBBER SLAB - [rubber ind] plaque *f* de latex mousse, plaque *f* de mousse de latex
~RUBBER SOLE - [shoe man] semelle *f* en caoutchouc mousse
~SANDWICH MOULDING - [plast ind] (the process of moulding sandwiched foam materials) moulage *m* sandwich avec mousse
~SHAMPOO - [detergent ind] détergent *m* à pouvoir moussant
~WIPING CUP - [impl] écumoire *f*
FOAMED IN THE MOULD - [plast ind] mousse *f* moulée in situ
~PLASTICS - [plast ind] (plastic materials which have been given a cellular structure in which the voids are interconnected) mousses *fpl* plastiques, plastiques *fpl* cellulaires
FOAMING - [phys] écumage *m*, moussage *m*
~AGENT - [chem] (any substance or combination of substance used to produce foaming) générateur *m* d'écume
[oil ind] produit *m* moussant
~MACHINE - [rubber ind] (machine for producing sponge or foam rubber by agitation) batteuse *f* à mousse
~PROCESS - [rubber ind] (the operation of expanding a material) méthode *f* de fabrication de mousse
FOAMY - [gen & chem] écumeux, mousseux
F.O.B. - [comm] (Free On Board) franco à bord
FOCAL - [opt] focal
~APERTURE - [photo] (the diameter of the section of the lens which is used to take a photograph) ouverture *f* focale
~COLLIMATOR - [opt] (collimator consisting of an objective lens at one end of a tube and a pair of cross haris accurately placed in its focal plane at the other end) collimateur *m* de mesure
~DEPTH RANGE - [opt] profondeur *f* de foyer
~DISTANCE - [opt] (the distance between the principal focus and the second principal point as measured along the principal axis) distance *f* focale
~LENGTH - s. focal distance
~LENGTH OF A LENS - s. focal length
~LINE - [opt] (one of the two short lines in the principal sections of a narrow-astigmatic bundle of light rays) ligne *f* focale
~PLANE - [opt] (plane normal to the optical axis of a lens) plan *m* focal
~PLANE CAMERA - [photo] chambre *f* à obturateur de plaque
~PLANE OF THE OBJECTIVE - [opt] plan *m* focal de l'objectif
~POINT - [opt] foyer *m*
~SPOT - [electron] (the target area which is struck by the main electron stream) foyer *m*
~TIME - [opt](in an aperture electron lens, the equivalent of the focal length in a lens) distance *f* focale de lentille électronique, temps *m* focal
FOCALIZE, to - s. focus, to
FOCIMETER - [instr] (instrument designed to measure the focal length of an optical system) focimètre *m*
FOCIMETRY - [instr] focimètrie *f*
FOCOMETER - s. focimeter
FOCOMETRY - s. focimetry
FOCUS, to - [opt] (to adjust an optical instrument to give a sharp image) faire converger, mettre au point
FOCUS - [opt] (the point at which rays converge after passing through an optical system) foyer *m*
[geol] (of an earthquake) centre *m* (d'un tremblement de terre)
[math] foyer *m*
[med] foyer *m*
~CONTROL - [telev] réglage *m* du spot lumineux
[electron] commande *f* de la concentration
~EYE GLASS - [opt] (used by clockmakers) monocle *m* d'horloger
~LAMP - [opt] lampe *f* focale
~MODULATION - [electron] modulation *f* de la convergence
~OF A CURVE - [math] foyer *m* d'une courbe
~POINT - [opt] (the principal focus) foyer *m*
~POWER - [opt] puissance *f* du foyer
~PULLER - [cin] (the person responsible for the adjustment of the focus) assistant *m* de caméra
~-TO-FILM DISTANCE - [radiat] (the distance between the focus of a radiation source and the film used for the radiographic camera) distance *f* foyer-film
~-TO-SKIN DISTANCE - [radiat] (the distance between the focus of an X-ray tube and the area of incidence of the patient) distance *f* foyer-peau
~UNIT - [telev] (a control unit) appareil *m* de télécommande de la focalisation
FOCUSED BEAM - [phys] faisceau *m* de rayons
~SPOTLIGHT - [photo] spot *m* à mise au point
FOCUSING - [opt] (the process of controlling and regulating a beam of light) focalisation *f*
[telev] (the process of controlling the electron beam) focalisation *f*
~AND SWITCHING GRILLE - [telev] (in colour television) grille *f* de Lawrence, sélecteur *m* de couleurs commutateur
~ANODE - [electron] anode *f* de concentration, anode *f* de focalisation
~ARC LAMP - [light] (arc lamp having a feed mechanism so arranged that the position of the crater does not alter) lampe *f* à arc à réglage automatique
~ARRANGEMENT - [opt] dispositif *m* de focalisation
~BY SCALE - [photo] mise *f* au point à l'aide de l'échelle
~CAM- [photo] came *f* (de commande) de la mise au point
~CLOTH - [photo] voile *m* noir pour la mise au point
~COIL - [electron] (coil producing a magnetic field for focusing beam) bobine *f* de focalisation
~CONTROL - [opt] commande *f* de la mise au point
~CUP - [electron] (a metal device used to focus the electron beam) cupule *f* de concentration, cylindre *m* de Wehnelt
~DEVICE - [opt] dispositif *m* de focalisation
~DISTANCE - [photo] mise *f* au point
~ELECTRODE - [electron] (electrode with a potential adjusted to focus an electron beam) électrode *f* de concentration, électrode *f* de focalisation
~FIELD - [telev] (the electrostatic or electromagnetic field produced by the focusing electrodes) champ *m* de concentration, champ *m* de focalisation
~GLASS - [photo] (small eyepiece used to examine the sharpness of focus of an image) loupe *f* de mise au point
~HOOD - [photo] capuchon *m* de mise au point

FOCUSING LENS - [opt] lentille *f* de concentration
~MAGNET - [electron] (an electro-magnet producing a magnetic field for focusing an electron beam) aimant *m* de concentration
~MICROSCOPE - [opt] (used in rack-over viewfinding) viseur *m* à microscope focalisateur
~MOUNT - [photo] monture *f* de mise au point
~RACK - [photo] coulisse *f* à crémaillère pour la mise au point
~RANGE - [photo] latitude *f* de mise au point
~RING - [photo] bague *f* de mise au point
~SCALE - [photo] échelle *f* de mise au point
~SCREEN - [photo] glace *f* dépolie, verre *m* dépoli
~SLIDE - [photo] diapositive *f* de mise au point
~VOLTAGE - [electron] tension *f* de focalisation
~WHEEL - [photo] volant *m* pour la mise au point, molette *f* de mise au point
~WORM - [photo] vis sans *f* fin de mise au point
FOCUSSING - s. focusing
FODDER, to - [agric] affourager, affenager
FODDER - [agric] fourrage *m*, affouragement *m*
~BARLEY - [agric] orge *f* fourragère
~BEET - [agric] betterave *f* fourragère
~CARROT - [agric] carotte *f* fourragère
~CONCENTRATE - [agric] fourrages *mpl* concentrés
~CROP GROWING - [agric] cultures *fpl* fourragères de plein champ
~CROPPING - [agric] culture *f* fourragère
~GRAIN - [agric] céréales *fpl* fourragères
~KALE - [agric] chou *m* fourrager
~OATS - s. fodder grain
~STEAMER - [agric] cuiseur *m*, étuveuse *f*
~THROUGH - [agric] auge *f*, mangeoire *f*
~YEAST - [agric] levure *f* alimentaire
FOEHN - (föhn, fön) [met] (dry, mainly downward wind occurring in mountainous country) foehn *m*
FOETAL - [zool] fœtal
FOETUS - [zool] fœtus *m*, embryon *m*
FOG, to - [phys] (the action of obscuring a transparent surface by condensation) voiler, embrouiller
[photo] se voiler
FOG - [met] (obscurity caused by condensed moisture or smoke in the lower levels of the atmosphere, defined as reducing visibility to less than 1100 yards) brouillard *m*
[photo] voile *m*
[telev] (the level difference between the darkest part of the image signal and the "black") voile *m*
~BELL - [signal] cloche *f* de brume
~-BOUND - [gen] arrêté par le brouillard, pris dans la brume
~-BOW - [met] arc-en-ciel dans le brouillard
~DENSITY - [photo] densité *f* du voile
~-DOG - [met] (a clearing spot, indicating the imminent lifting of a dog) éclaircie *f*
~-FOAM - [chem] écume *f* pulverisée
~-FOAM FIRE PROTECTION - [fire-fighting] protection *f* contre l'incendie avec écume pulverisée
~GRAIN - [photo] grain *m* de voile
~HORN - [naut] corne *m* de brume, sirène *f*
~LAMP - s. fog light
~LIGHT - [auto] phare *m* antibrouillard
~NOZZLE - [impl] (in fire-fighting) pulvérisateur *m*, atomiseur *m*
~QUENCHING - [metall] refroidissement *m* rapide en vapeur d'eau
~SICKNESS - [vet] tympanite *f*
FOG SIGNAL - [signal] signal *m* de brume
[railw] pétard *m*, détonateur *m*
~STREAKS - [photo] stries *fpl* de voile
~TRACKS - [phys] (linear condensation regions in air or other gases supersaturated with water vapour, produced by the passage of electrified particles) lignes *fpl* de brouillard, traces *fpl* de brouillard
~-TYPE INSULATOR - [el] (overhead insulator designed for areas fog occurs frequently) isolateur *m* anti-brouillard
FOGGED - [photo] voilé
FOGGER - [cin] (small incandescent lamp to record a fog mark) lampe *f* pour voiler
FOGGING - [phys] (obscuring of a transparent surface by condensation) embouage *m*, ternissement *m*
[railw] voile *m*, signalisation *f* par temps de brouillard
~AGENT - [photo] adjuvant *m* de voile
~MACHINE - [agric] atomiseur *m*, nébuliseur *m*
FOGGY - [met] brumeux
~WEATHER - [met] temps *m* de brouillard
[naut] temps *m* gras
FOHN - s. foehn
FOIL - [gen & arch] lobe *m*, feuille *f*
[metall] (very thin metal sheet) lame *f*, feuille *f*, clinquant *m*
[paper man] feuille *f* (doublée)
[impl] (in fencing) fleuret *m*
FOLD, to - [gen] plier, replier, rabattre, envelopper
[mech] (the action of clasping) agrafer, replier
[agric] (of sheep) parquer (les moutons)
FOLD - [gen] pli *m*, repli *m*
[agric] (an enclosure for sheep etc) parc *m* (à moutons), bergerie *f*
[glass man] (a defect) cassure *f*
[math] (as a suffix to cardinal number it serves as multiplicative) -uple, -iple
[geol] (of strata) flexure *f*, plissement *m*
[bookbind] pliure *f*
~-BACK LOUDSPEAKER - [el acoust] hautparleur *m* à membrane repliée
~CARPET - [geol] pli-nappe *m*
~-FAULT - [geol] pli-faille *m*
~OVER - [telev] (colloq. USA) double image *f*, image *f* fantôme
~-OVER MULTIPATH EFFECT - [telev] (double image; unwanted images appearing on the screen and causing a double picture) image *f* multiple (ou fantôme)
~TESTER - [instr] (device for measuring folding endurance) pliagraphe *m*, appareil *m* d'essai au pliage
FOLDED CAVITY - [electron] (in klystrons, to allow the action of incoming waves on the electron stream) cavité *f* repliée
~DIPOLE AERIAL - [radio] antenne *f* dipôle *m* replié
~DIPOLE ANTENNA - s. folded dipole aerial
~FILTER - [impl] (used in the brewing industry) filtre *m* plissé
~LIGHT BEAM - [opt] faisceau *m* de lumière réfléchi
~SELVEDGE - [text] lisière *f* rentrée, lisière *f* couchée
~YARN - [text] retors *m*
FOLDER - [gen] chemise *f*
[text] plieuse *f*
[print] plioir *m*
[pubblicity] prospectus *m*, dépliant *m*
FOLDING - [adj] pliant, repliant, rebattable

FOLDING - [geol] (the bending of strata) plissement *m*
[bookbind] (in a book sheet) pliure *f*
[agric & of sheep] parcage *m*
~AND ROLLING MACHINE - [text] dosseuse-enrouleuse *f*
~ANGLE - [gen] (the angle through which sheet is folded) angle *m* de pli
~BENCH - [gen] banquette *f* pliante
~BOBBIN - [text] bobine *f* de doubleuse, bobine *f* de assembleuse
~BULWARKS - [naut] pavois *mpl* rabattables
~CAMERA - [photo] chambre *f* pliante, folding *m*
~DOORS - [constr] portes *fpl* battantes
~ENDURANCE - [phys] (the ability of a material to resist repeated folding without cracking) résistance *f* au pliage
~FRAME - [text] dispositif *m* de pliage
~LADDER - [impl] échelle *f* brisée
~MACHINE - [text] machine *f* à plier
~MAGNIFYING GLASS - [opt] (used to examine fabrics) loupe *f* à coulisse
~MEASURE - [meas] mètre *m* pliante
~NUMBER - [text] numéro *m* du retors
~OF THE BEDS - [geol] plissement *m* des couches
~PHASE - [geol] phase *f* de plissement
~PLATE - [text] plaque *f* de pliage
~PRESS - [mech] presse *f* pliante
~RESISTANCE - s. folding endurance
~RULE - s. folding measure
~SEAT - [gen & auto] (a seat hinged to fall into a vertical position when out of use) strapontin *m*, siège *m* rabattable
~SHUTTER - [carp] (boxing shutter) volet *m* pliant
~SIGHT - [opt] pinnule *f* à charnière
~SLIDE DRUM - [rubber ind] (for giant tyres) tambour *m* à charnières coulissantes (pour pneus géants)
~STRIPE - [text] rayure *f* de plissage
~SUPPORT - [impl] (a device for folding sheet material) tablette *f* pliante à charnière
~TABLE - [carp] table *f* pliante
~TEST - [metall] essai *m* de pliage
~TOP - [auto] voiture à capote
~WINGS - [aero] (wings of aircraft designed to fold back along the fuselage to reduce space in storage or transport) ailes *fpl* repliables
FOLIAGE - [bot] feuillage *m*, frondaison *f*
[arch] (ornamental work) rinceau *m*
~PLANT - [bot] plante *f* à feuilles
FOLIATE, to - [gen] se diviser en lames, s'écailler, étamer
[min] se diviser en lamelles
FOLIATED - [gen] folié, lamellé, lamelleux
[geol] (of a structure) lamellaire, schisteux
[metall] laminé
~STRUCTURE - s. foliation
~TELLURIUM - [min] tellure *m* auroplombifère
FOLIATION - [geol] (minerals arranged in folia or leaves) schistosité *f*
~STRUCTURE - [geol] structure *f* feuilletée
FOLIC ACID - [chem] (a member of the vitamin B complex used in medicine) acide *m* folique
FOLIO - [print] (a sheet of paper folded in half) folio *m*, feuillet *m*
[print] (the number of a page) numéro *m* d'une page
[print] (book made up of sheets folded only once, thus obtaining four pages per sheet) in-folio
FOLIUM - [math] courbe *f* foliacée
FOLLICLE - [bot] follicule *m*
[zool] cocon *m*
FOLLICULITIS - [med] (inflammation of a follicle, particularly of the ovary) folliculite *f*
FOLLOW, to - [gen] suivre
[gen] (a profession etc) exercer
[naut] s'accrocher (à un navire)
FOLLOW BOARD - [metall] plaque *f* porte-modèle
~-FOCUS DEVICE - [cin] (a view finder with the focusing mechanism mechanically coupled to the focusing ring on the lens) viseur *m* couplé
~-FOCUS SHOT - [cin & telev] prise *f* de vues poursuite
~-FOCUS VIEWER - [telev] viseur *m* suivant le sujet
~REST - [mach] lunette *f* à suivre
~SHOT - [cin] (the shooting of a moving object by keeping at the same distance from it) plan *m* pris en mouvement, travelling *m*
~-UP, to - [gen] poursuivre, donner suite immédiate
~-UP - [gen & comm] complémentaire
~-UP CONTROL - [contr] (method of control in which the control quantity may vary rapidly and frequently) régulation *f* de correspondance
[contr] allongement *m* d'une barre de commande
~-UP SLEEVE - [mech] manchon *m* à glissière
FOLLOWER - [constr] (intermediate length of timber transmitting the blow from the monkey to the pile) rallonge *f* de sonnette
[horol] (the driven wheel of a pair of wheels) roue-menée
[mech] chapeau *m* d'un presse-étoupe
[mech] (roller following a cam profile) roulette *f*
[auto] boîtier *m* du poussoir
[text] (a driven friction disk) disque *m* de friction entraîné
[mech] (a driven pulley) poulie *f* menée
[mech] (to hold piston-rings) couvercle *m* du piston
[photo] galet *m* de guidage
[oil ind] presse- étoupe *m* de cémentation
~DRIVE - [mech] entraînement *m* auxiliaire
~OPERATION - [contr] opération *f* séquentielle
FOLLOWING CAM - [text] came *f* succédante
FOMENTATION - [med] fomentation *f*
FOMES - [med] (any infected object with the exception of food) vecteur *m* de contagion
FONT - [print] (or fount; complete set of tyres) fonte *f*, assortiment *m* complet (d'un œil)
FOOD - [gen] nourriture *f*, aliments *mpl*, denrées *fpl*
~CANNING - [food ind] mise *f* en conserve
~CANNING FACTORY - [food ind] fabrique *f* de conserves, conserverie *f*
~COLOURS - [food] colorants *mpl* pour produits alimentaires
~CONTAMINATION MONITOR - [food ind] moniteur *m* de denrées
~FLAVOUR ADDITIVE - [chem] additif *m* alimentaire
~INDUSTRY - [food ind] industrie *f* alimentaire
~JAR - [food ind] pot *m*, récipient *m*
~-JAR RING - [mech] joint *m* pour bocal à conserves
~POISONING - [med] intoxication *f* alimentaire
~PRESERVATIVES - [chem] (substances preventing the fermentation of foodstuffs) préservatifs *mpl* pour produits alimentaires
~PROCESSING - [ind proc] transformation *f* des pro-

duits alimentaires
FOOD-STUFF - [gen] vivres *m*pl, denrées *f*pl
~YEAST - s. fodder yeast
FOOL-PROOF - [mech] (term used of anything which cannot be damaged or cause damage when improperly operated) à l'épreuve de maladresses, à toute épreuve, de sûreté
FOOL'S GOLD - [min] (iron pyrite) pyrite *f* (de fer ou cuivre)
FOOLFOOT - [bot] (a weed) tussilage *m*, pied *m* de poulain
FOOLSCAP - [paper man] (standed size of paper) papier *m* ministre, tellière *f*
FOOT - [anat] pied *m*
[meas] (30.48 cm) pied *m* (0,3048 *m*)
[ind chem] fond *m*, sédiment *m*
[print] (the margin at the bottom) bas *m* de page
[mus] (the lowest part of an organ pipe) pied *m*
[naut] (of a sail) bordure *f* (de voile)
[railw] patin *m*, semelle *f*, appui *m* d'un rail
[geom] point *m* d'intersection (ou de recoupement)
[mech] (of anvil) patin *m*
[constr] (of a stair) départ *m*
~-AND-MOUTH DISEASE - [vet] (acute febrile condition of cloven-footed animals) fièvre *f* aphteuse
~-BAND - [naut] bordure *f* de fond
~BELLOWS - [impl] soufflerie *f* à pédale
~-BOARD - [railw & gen of vehicles) marchepied *m*
[auto] plancher *m*
[railw] (of a locomotive) plate-forme *f*
~-BOARD BRACKET - [mech] marche-pied *m*
~BRAKE - [auto etc] frein *m* à pied
~-BRAKE PEDAL - [auto] (the pedal operating the brake shoes) pédale *f* de frein à pied
~-BRIDGE - [constr] railw etc] (narrow bridge for the use of pedestrians) passerelle *f*, pont *m* dormant
[gen] pont *m* pour piétons
~CANDLE - [light] (a unit of illumination) unité *f* de éclairement
~CONTROL - [auto] commande *f* des gaz, pédale *f* d'accélérateur
~CONTROL LEVER - [auto & motorcycles] pédale *f* de commande
~COURSE COUNTER - [text] compteur *m* de rangées du pied
~CROSSING - [railw] (between platforms) passage *m* de quais à niveau
~FEED - [auto] (colloq USA) accélérateur *m*
~-HOLE - [mus] (the wind-admitting opening at the foot of an organ pipe) trou *m*
[mining] mine *f* de relevage
~LATHE - [mech] tour *m* à pédale, tour *m* marchant au pied
~LEASE - [text] (in a warping mill) envergeure *f* du bas
~-LIGHTS - [light] (in a theatre etc) rampe *f*
~LINE - [print] ligne *f* de pied, ligne *f* au bas d'une page
~-NOTE - [print etc] apostille *f*, renvoi *m* (au bas de la page)
~-OPERATED AIR-PUMP - s. foot pump
~OF A DIKE - [constr] bas *m* (ou base) d'une digue
~OF A SLOPE - [constr] pied *m* de talus
~-PATH - [gen] sentier *m* pour piétons, banquette *f*
[constr] (in a street) trottoir *m*
~-PATH BRIDLEWAY - [roads] route *f* muletière
FOOT-PATH PLATFORM - [constr] passerelle *f*
~PEDAL - [mech etc] pédale *f*
~PEDAL PAD - [auto] patin *m* de pédale
~PEG - [text] (in a warping mill) cheville *f* inférieure
~PIECE - [mining] semelle *f*, sole *f*
~-PLATE - [constr] saille *f*
[railw] (the platform for the driver of a locomotive) tablier *m*, plate-forme *f*
~-POUND - [meas] (unit of work in the British system of units) pied-livre *m*
~PUMP - [mech] (pump operated by foot) pompe *f* à pied
~RAIL - [railw] rail *m* à patin
~-REST - [gen] repose-pied *m*
~-ROPE - [naut] marchepied *m*
(section of boltrope) ralingue *f* de fond
~ROT - [vet] (the ulceration of the skin of the coronary band of sheep) piétin *m*, fourchet *m*
~RULE - [impl] verge *f* de charpentier, règle *f*
~RUN - [constr] (foot of length) longueur *f* d'un pied
~SCREW - [mech] (plate screw) vis *f* de calage, vis *f* de bride, vis *f* arrêtoir
~-STALL - [build] (the base of a pillar) piédestal *m*, socle *m*
~STARTER - [auto] démarreur *m* au pied
~STEP - [mech] palier *m* de pied, butée *f*
~-STEP BEARING - [mech] (thrust bearing supporting the lower end of a vertical shaft) palier *m* de butée
~-STEP LEVER - [mech] levier *m* à gradins
~-STOCK - [mach tool] contre-pointe *f*, poupée *f* mobile, poupée *f* courante
~SWITCH - [el] (switch operated by the foot) commutateur *m* à pédale
~THROTTLE - [auto] pédale *f* d'accélérateur
~TON - [meas] (2240 pounds) pied-tonne *m*
~-VALVE - [mech] (a non-return valve placed at the lower end of a pump suction line to retain the fluid the pump is not working) clapet *m* de pied, soupape *f* de pied
~-WALL - [mining] sol *m*, chevet *m*, mur *m*
[geol] lèvre *f* inférieure
~WINCH - [mech] treuil *m* à pédale
FOOTAGE - [meas] longueur *f* (en pieds), métrage
[mus] s. foot
[cin] (the length of a film) métrage *m*
[mining] avancement *m* du forage au mètre
~INDICATOR - [cin] compteur *m* de film, métreuse *f*
FOOTFLEXER - [rubber ind] boule *f* pour massage des plantes des pieds
FOOTING - [constr] (foundation or a wall or other masonry structure) embase *f*, socle *m*, empattement *m*, scellement *m*
[el] (foundation supporting the tower of an overhead line) fondation *f*
[hydr] base *f*
[sugar ind] pied *m* de cuit
~BLOCK - [constr] couchis *m*, semelle *f*, patin *m*
~MACHINE - [text] machine *f* à pieds de bas
~OF A COLUMN - [constr] construction *f* de la base d'une colonne
~OF A WALL - [constr] base *f*, fondation *f* d'un mur
FOOTS - [ind chem] (insoluble sludge deposited at the bottom of tanks containing varnish, oil etc. when undisturbed) sédiments *m*pl, tourteau *m* (p.ex. d'huile)
FOOTSTALL - [constr] piédestal *m*, socle *m*, sca-

bellon *m*
FOOTWALL - [geol] (the lower wall of rock in contact with a vein) sol *m*
~SEAM - [mining] gîte *m* en aval pendage
FOOTWAY - [min] (a colliery shaft with ladders) échelle *f* de puits
[constr] trottoir *m*
~JOINTING CHAMBER - [telecomm] chambre *f* sous trottoir
~JOINTING MANHOLE - s. footway jointing
FOOTWEAR - [shoe man] chaussures *fpl*
~FABRIC - [text] tissu *m* à chaussures
FORAGE - [agric] fourrage *m*, affouragement *m*
~DRYER - [agric] séchoir *m* pour fourrage vert
~HARVESTER - [agric] faucheuse-hacheuse *f*
~PIT - [agric] meule *f* à fourrage
~STRAW - [agric] paille *f* fourragère
FORAGING BEE - [beekeeping] butineuse *f*
FORAMEN - [anat] foramen *m*, orifice *m*
FORBIDDEN BAND - [nucl] (a range of energies in which there are no electronic levels) bande *f* interdite
~-CODE COMBINATION - [comput] combinaison *f* de code inadmissible
~-CODE COMBINATION CHECK - [comput] contrôle *m* de combinaisons prohibées
~COMBINATION - [comput] combinaison *f* prohibée
~COMBINATION CHECK - [comput] contrôle *m* des combinaison de code inadmissibles
~NUCLEAR TRANSITION - [nucl] (nucleus transformation which result in a change of more than one unit of angular momentum) transition *f* interdite
~TRANSITION - [nucl] (transition between two states of a quantum-mechanical system) transition *f* interdite
FORCE, to - [comput] intervenir
FORCE - [phys] (in dynamics & statics) force *f*, épuissance *f*
[mech] (any operating agency) force *f*, énergie *f*
[gen] personnel *m*, effectif *m* de la main-d'œuvre
~BALANCE - [contr] bobine *f* des forces
~CONSTANTS OF LINKAGES - [nucl] (expressions of the forces which act between nuclei to restrain displacement) constantes *fpl* de forces de liaison nucléaire
~COUPLE - [mech] couple *f*
~CUP - [impl] débouche-évier *m*
~DIAGRAM - [mech] (diagram of the internal forces in a framed structure shown to scale) diagramme *m* des forces réciproques
~-DISPLACEMENT CURVES - [contr] caractéristiques *fpl* force-déplacement
~FACTOR - [acoust] (of an electro-mechanical transducer) facteur *m* de force, facteur *m* de couplage
~FEED - [mech] (the lubrication of an engine by forcing the oil through) graissage *m* sous pression
~FIT - [mech] (a fit between two parts such that a hydraulic or other press is required to bring them into the required position in respect to each other) montage *m* à force
~FUNCTION - [mech] (the negative of the potential energy of a mechanical system) fonction *f* des forces
~IN, to - [gen] faire entrer à force, enforcer
~LIMITING DEVICE - [el] limiteur *m* d'effort
~-LINE - [phys] ligne *f* de force
~OF ATTRACTION - [phys] (the force acting on a particle to direct it towards the agency responsable for the force) force *f* d'attraction
~OF EXPLOSION - [auto] force *f* d'explosion
~OF FRICTION - [phys] force *f* de frottement
~OF GRAVITATION - s. force of gravity
~OF GRAVITY - [phys] force *f* de gravité, pesanteur *f*
~OF IMPACT - [phys] puissance *f* d'impact
~OF INERTIA - [phys] force *f* d'inertie
~OF REPULSION - [phys] force *f* repoussante, force *f* répulsive, force *f* de répulsion
~ON, to - [mech] emmancher par effort
~PLATE - [metall] (plate which carries force plug or plugs in a compression mould) plaque *f* porte-poinçon
~PLUG - (projection on the top force of a positive mould) poinçon *m*
~POLYGON - [phys] (polygon of forces) polygone *m* des forces
~PUMP - [mech] (any pump delivering at a pressure which is greater than its suction pressure) refouleur *m*, pompe *f* foulante
[plumb & gas install] (used to clean out gas pipes etc) pompe *f* à air comprimé, pompe *f* à piston plongeur
~SIDE PART - [metall] (term used in the U.S.A. for the movable part of a mould) partie *f* mobile du moule
FORCED - [gen & mech etc] forcé, sous pression
~-AIR BURNER - [gas ind] brûleur *m* à air soufflé
~-AIR COOLING - [el] refroidissement *m* à air soufflé
~-AIR HEATING - [space heating] chauffage *m* par air pulsé
~BIREFRINGENCE - s. forced double refraction
~CIRCULATION BOILER - [th eng] (steam boilers in which steam and water are circulated by pumps) chaudière *f* à circulation forcée
~CODING - [comput] (programming so made that a very short time is required to obtain the information) programmation *f* absolue
~CONVECTION - [th eng] (the forced transfer of heating) convection *f* forcée
~DOUBLE REFRACTION - [opt] (produced as a result of strain) biréfringence *f* mécanique
~DRAUGHT - [mech] (an air supply to a furnace artificially produced by fans or the like, as distinct from the natural draught caused by the combustion of the fuel) tirage *m* forcé, courant *m* d'air forcé
[el] (electrical apparatus cooled by ventilation) à ventilation forcée
~-DRAUGHT BLOWER - [mech] compresseur *m*
~-DRAUGHT FURNACE - [metall] four *m* à tuyères et courant d'air forcée
~DRYING TEMPERATURE - [plast ind] (temperature between room temperature and 65.6 C. maintained to accelerate drying) température *f* de séchage accéléré
~-FLOW BOILER - s. forced circulation boiler
~-FUEL PIPING - [mech] (in autos) alimentation *f* sous pression
~LANDING - [aero] (landing made unavoidable by weather or defect in an aircraft) atterrissage *m* forcé
~LUBRICATION - [mech] (lubrication in which the lubricant is supplied to the moving parts under pressure) graissage *m* sous pression

FORCED MAGNETIZATION - [radio] (impeded harmonic operation) magnétisation *f* forcée
~MALTING - [brew ind] malter à sueur chaude
~OSCILLATION - [phys] (oscillation resulting from an external periodic driving force when this is applied to a system capable of free oscillations) oscillation *f* forcée
~PUMP - [oil ind] pompe *f* foulante
~VENTILATION - [mech] ventilation *f* mécanique par insufflation
~VIBRATION - [phys] (vibration resulting from the application of a periodic force to a body capable of vibrating) vibration *f* forcée
~WATER COOLING - [auto] refroidissement *m* par eau à circulation forcée
FORCEPS - [impl] (tool for picking up small beads of metal or the like) pince *f*
[med] pince *f*, bec *m* de corbeau
[dent] tire-racines *m*
~DELIVERY - [med] accouchement *m* aux fers
FORCES OF REPULSION - [phys] (forces between particles tending to keep them apart) forces *f*pl de répulsion
FORCING - [mech] forcement *m*
[contr] intensification *f* de la valeur d'ajustage
[agric] culture *f* forcée
~FRAME - [agric] châssis *m* pour culture forcée
~HOUSE - [agric] forcerie *f*, serre *f* à forcer
~MACHINE - [rubber ind] boudineuse *f*, extrudeur *m*
~PUMP - s. force pump
~RESISTANCE - [el] (resistance placed in series with a magnetic amplifier to obtain a higher speed of response) résistance *f* accélératrice, résistance *f* activatrice
FORD, to - [gen] guéer, traverser à gué
FORD - [gen] gué *m*
FORDING - [gen] gué *m*, passage *m* du gué
~HEIGHT - [gen] profondeur *f* du gué
FORE - [gen] antérieur, de devant
[naut] avant
[naut] (the formast) mât *m* de misaine
~AND AFT - [naut] de l'avant à l'arrière
~-AND-AFT LEVEL - [instr] (pitch-indicating instrument, operated by gravity) niveau *m* longitudinal
~-AND-AFT RIGGED SHIP - [naut] navire *m* à voiles auriques
~-AND-AFT SAIL - [naut] voile *f* aurique
~-AND-AFT STABILITY - [naut] (stability in respect to pitch) stabilité *f* longitudinale
~ARM - [mech] raccord *m* (in vacuum technology)
~-BRAIN - [zool] (the anterior brain-vesicle) prosencéphale *m*, encéphale *m* antérieur
~HEARTH - [metall] chaufferie *f*, sole *f* à réchauffer
~-PLANE - [tool] (a type of bench plane) riflard *m*
~PUMP - [mech] pompe *f* primaire, pompe à prévide, pompe préparatoire, pompe pour vide primaire
~-PUMP SYSTEM - [mech] groupe *m* de pompes à vide primaire
~-PUMP VALVE - [mech] vanne *f* de vide primaire
-RUNNINGS - [chem] (head-products from a contінous spirit still, containing about 97 p.c. acetaldehyde and only about 3 p.c. of alcohol) produit *m* de tête
~-STAGE - [mech] étage *m* préliminaire, premier dégré *m*
~VACUUM - [hydr] (depression maintained downstream of a diffusion pump for efficient operation of the latter) vide *m* préliminaire, vide *m* préalable à basse pression
FORE-VACUUM COOLER - [mech] refroidisseur *m* pour le vide primaire
~-VACUUM PORT - [mech] raccord *m* au vide primaire
FOREARM - [anat] avant-bras *m*
FOREBAY - [hydr] (reservoir at the head of a pipeline) chambre *f* d'équilibre
FOREBODY - [shipbuild] (the section from the ship from midship to bow) avant *m*
[auto] avant *m* de carrosserie
FOREBREAST - [mining] front *m* de taille
FOREBRIDGE - [naut] pont *m* avant
FORECABIN - [naut] cabine *f* d'avant
FORECARRIAGE - [of vehicles] avant-train
FORECAST, to - [gen] prévoir, calculer
FORECAST - [gen] prévision *f*
[met] (prediction of future weather conditions from existing data) prévisions *f*pl (du temps)
FORECASTING - [gen] prévision *f*
[met] prévisions *f*pl météorologiques
FORECASTLE - [naut] gaillard *m*
FORECLOSE, to - [comm] saisir (ce qui est hypothequé)
FORECLOSURE - [leg] forclusion *f*, saisie *f*
FORECOOLER - [mech] (apparatus for reducing the temperature of a gas before the first stage of compression) pré-réfrigérant *m*
FOREDECK - [naut] pont-avant *m*
FOREFLIPPER - [zool] (of a whale) nageoire *f* pectorale
FOREFOOT - [shipbuild] brion *m*, étrave *f*
FOREGOING - [gen] précedent
FOREGROUND - [gen photo etc] premier plan *m*, avant-plan *m*
[cin] devant *m*, avant-plan *m*
FOREHAND WELDING - [metall] soudure *f* à gaz à gauche
FOREHEAD - [anat] front *m*
[mining] front *m* d'une galerie
FOREHEARTH - [metall] (of a furnace) avant-creuset *m*
FOREIGN ATOM - [chem] (atom in a crystal which is foreign to the crystal) atome *m* d'impurité
~BODY - [gen & chem] (alien or undesired object unintentionally introduced) corps *m* étranger
~CUSHION GAS - [gas ind] (in gasholders) gaz-coussin *m* injecté
~GAS - [gas ind] gaz *m* injecté
~MATTER - s. foreign body
~-MATTER CONTENT - [ind chem etc] (dirt content) teneur *m* en impuretés
~SERVICE POSITION - [telecomm] (telephone service) position *f* internationale
~SUBSTANCE - s. foreign body
FORELAND - [geogr] (salient point of coastline) promontoire *m*, cap *m*
FORELINE - [mech] (vacuum technology) conduite *f* (ou canalisation) de vide primaire
~TANK - [mech] réserve *f* de vide, réservoir *m* tampon (entre pompe à diffusion et pompe primaire)
~TRAP - [mech] piège *m* à vide primaire, piège *m* pour le vide primaire
~VALVE - [mech] vanne *f* de vide primaire

FORELOCK - [mech] clavette *f*, goupille *f*
FOREMAN - [gen] chef *m* d'équipe, contremaître *m*
[print] prote *m*
FOREMARKER - [radar] (the signal which is received by an aircraft from a transmitter at a distance of about two miles) signal *m* de radiobalise extérieure
FOREMAST - s. fore-mast
FORENSIC - [gen & leg] légal
~CHEMISTRY - [leg] chimie *f* légale
~MEDICINE - [leg] médicine *f* légale
FOREOBSERVATION - s. foresight (surv)
FOREPART - [gen] partie *f* antérieure, avant *m*, devant *m*, avant-corps *m*
FOREPAW - [zool] patte *f* antérieure
FOREPEAK - [shipbuild] avant-bec *m*
~BULKHEAD - [shipbuild] cloison *f* de choc
FOREPRESSURE - [phys] pression *f* de vide primaire
~GAUGE - [mech] tube *m* de mesure pour vide primaire
FORESAIL - [naut] misaine *f*, voile *f* de misaine
FORESEE, to - [gen] prévoir
FORESHAFT - [mining] avant-puite *m*
FORESHORE - [geogr] (the part of the shore uncovered by ebb-tide) plage *m f*
FORESHORTENING - [photo] (exaggerated perspective) raccourci *m*, effet *m* de raccourci
FORESHORTING - [draw] raccourci *m*
FORESHOTS - s. fore-runnings
FORESIGHT - [gen] prévision *f*, prévoyance
[firearms] fronteau *m* de mire, bouton *m* de mire
[surv] (the levelling staff reading, taken forward to a station which has not been passed by the instrument) coup *m* avant
FORESKIN - [anat] prépuce *m*
FOREST - [bot à geogr] forêt *f*
~MEADOW - [agric] prairie *f* forestière
~SOIL - [agric] sol *m* forestier
FORESTATION - [agric] boisement *m*
FORESTRY - [agric] économie *f* forestière
FORETOP - [naut] hune *f* de misaine
FORETOPSAIL - [naut] petit hunier *m*
FOREWINNING - [mining] exploitations *f* des chambres, tracage *m*
FOREWORD - [gen] avant-propos *m*
FOREYARD - [naut] vergue *f* de misaine
FORFEIT, to - [gen] perdre par confiscation
[leg] laisser périmer, être déchu (d'un droit)
[leg] (to subject to forfeiture) confisquer
FORFEITURE - [leg] perte *f* par confiscation, forfaiture *f*
FORGE, to - [metall etc] (to shape while hot by hammering or pressing) forger, corroyer
[gen] (of money) contrefaire
FORGE - [gen] forge *f*
[metall] (the plant used forging) forge *f*
[metall] (the action of forging metals etc) forgeage *m*
[metall] (the plant used for the production of wrought iron from ore) usine *f*, atelier *m* de forge
~BLAST - s. forge blower
~BLOWER - [mech] soufflerie *f* de forge, soufflet *m* de forge
~CHIMNEY - [metall] cheminée *f* d'une forge
~CINDER - [metall] laitier *m* de forge
~COAL - [metall] charbon *m* de forge, houille *f* maréchale
~COLD, to - [metall] forger à froid
FORGE HAMMER - [tool] marteau *m* à forger, pilon *m*
~HEAT - [metall] température *f* de forgeage
~HOT, to - [metall] forger à chaud
~IRON - [metall] fer *m* malléable
~MACHINE - [metall] machine *f* à refouler
~OUT, to - [metall] étirer sous le marteau
~PIGS - [metall] (pig-iron suitable for the production of wrought-iron) fonte *f* blanche
~PRESS - [metall] presse *f* à forger
~RANGE - [metall] température *f* idéale de forgeage
~SCALE - [metall] (iron oxide coating on iron and steel) scories *fpl* de forge, paille *f* de fer, battitures *fpl* de forge, mâchefer *m*
~SHOP - [metall] forge *f*, usine *f* métallurgique
~STEEL - [metall] (steel suitable for forging) acier *m* de forge
~TESTS - [metall] (made to check the ductility and malleability) essais *mpl* de forgeage
~TONGS - [tool] pince *f* à marteler
~WELDING - [metall] soudage *m* à la forge
FORGEABILITY - [metall] forgeabilité *f*
FORGEABLE - [gen] forceable
FORGED - [gen] forgé
~CHROMIUM STEEL - [metall] acier *m* au chrome forgé
~IRON - [metall] fer *m* forgé
PIECE - [metall] pièce *f* forgée
~PISTON - [auto] piston *m* forgé
~STEEL - [metall] acier *m* forgé
FORGER - [gen] forgeur *m*, forgeron *m*
[mech] forgeuse *f*, machine *f* à refouler
FORGING - [metall] (the operation of shaping hot metals) forgeage *m*
[metal part] (shaped while hot by hammering, or by means of a hydraulic press) pièce *f* forgée
~BILLET - [metall] billette *f* pour forge
~BRASS - [metall] laiton *m* à forger
~HAMMER - [impl] marteau *m* de forge, maillot *m*
~INDUCTION HEATER - [metall] four *m* à induction pour forgeage
~INGOT - [metall] lingot *m* à forger
~MACHINE - [mech mach] (system of power hammers and presses used for forging) forgeuse *f*, machine *f* à refouler
~PLANT - [metall] forge *f*
~PRESS - [mech mach] (press used for forging and drop forging) presse *f* à forger
~PRESSURE - [metall] pression *f* de forgeage
~QUALITY STEEL - s. forging steel
~ROLLS - [metall] laminoir *m* à forger, machine *f* à forger à rouleaux
~SCALE - [metall] déchets *mpl* de forge, scorie *f* de forge
~SHOP - [metall] atelier *m* de forge
~STOCK SIZE - [metall] dimension *f* des billettes
~STRAIN - [metall] effort *m* dû au forgeage
FORK, to - [gen] fourcher, bifurquer, se bifurquer
FORK - [gen] fourche *f*, fourchette *f*
[agric] fourche *f*, fourchet *m*
[mech] (bicycle) fourchette *f*
[horol] (the end of the lever receiving the impulse pin) fourchette *f*
[gen] (on a road) bifurcation *f*, jonction *f*, fourche *f* de routes
[mech] (of a fork-truck) fourchette *f*
[auto] fourchette *f*
[mining] fond *m* du puisard; (in haulage) fourchette

f d'accrochage
FORK - [mech] (of a knuckle joint) chape *f*
[metall] fourche *f* de retenue
~A MINE - [mining] épuiser une mine, assécher une mine
~BLADE - [auto] branche *f* de fourchette
~BRIDGE - [auto] tête *f* de fourchette
~CENTRE - [mach tool] pointe *f* à trois dents
~CHUCK - [mach tool] mandrin *m* à tulipe, griffe *f*
~COMB - [text] peigne *m* à fourchettes
~END - [mech] extremité *f* de fourchette
~EXTENSION - [mech] (in a fork-truck) barre *f* de rallonge d'une fourche
~KEY - [mech] clavette *f* à fourche
~LIFT - [impl] élevateur *m* à mâchoires
~LIFT STACKER - s. fork truck
~LIFT TRUCK - s. fork truck
~LINK - [auto] articulation *f* à chape
~SIDE-SHIFT DEVICE - [mech] (in a fork-truck) mécanisme *m* pour le déplacement latéral des fourches
~SPANNER - [impl] clef *f* à fourche
~SPRING - [mech] soupape *f* de la fourchette
~TEDDER - [agric] faneuse *f* à fourches
~TONE MODULATION - [telev] (modulation of the picture carrier frequency by the synchronizing tone) modulation *f* par la fréquence du diapason
~TRUCK - [mech] gerbeuse *f* à fourche
[mech] (a powered vehicle fitted with a fork or platform raised by power, used to handle, transport and stack heavy objects) élévateur-camion *m* à fourche, transporteur *m* de levage
~WRENCH - s. fork spanner
FORKED - [gen] fourchu, bifurqué
~ANGLE LEVER - [text] (in a dobby) levier *m* coudé à fourche
~ASHLAR - [constr] ancre *f* à fourchette
~ASSEMBLY - [mech] (connecting-rod assembly of a radial engine, in which articulated-rods are pivoted upon a single master rod) embiellage *m* à fourche
~AXLE - [auto] axe *m* à fourchette
~CHANGE LEVER - [text] (in weft changing) levier *m* de changement de vitesse à fourchette
~CLAMP - [mech] crampon *m* à fourchette
~CONNECTING ROD - [auto] bielle *f* fourchue
~LEVER - [mech] levier *m* à fourchette
~PIN - [mech] cheville *f* à fourchette
~ROD - [auto] tige *f* forchue
~ROOT - [bot] racine *f* forchue
~SCREWDRIVER - [impl] tournevis *m* à double lame
~SLOT - [text] (of the hook) (in twilling jacquards) fente *f* en forme de fourchette de la platine
~STAND - [photo] pied *m* à fourche
~TENON - [carp] (joint astride a tenon) tenon *m* fourchu
FORKING - [biol] bifurcation *f*
~PIECE - [brew ind] couche *f* jeune, tas *m* jeune
FORM, to - [gen] (to give shape to) former, façonner
FORM - [gen] forme *f*, conformation *f*, figure *f*, relief *m*, modelé *m*
[gen] (in schools) classe *f*
[gen] (a blank form) bulletin *m*, formulaire *m*
[metall] (a mould) empreinte *f* d'un moule, forme *f*, moule *m*
[print] forme *f* d'impression
[crystall] forme *f*

FORM - [constr] (for concrete work) coffrage *m*
[carp etc] banc *m*, banquette *f*
~BOARD - [constr] planche *f* de coffrage
~CUTTER - [mach tool] fraise *f* de forme
~DRAG - [aero] (a value obtained by deducting trailing-vortex drag from normal-pressure drag) traînée *f* de forme
~FACTOR - [el] (the ratio between the root mean square value of an alternating wave and its mean value taken over half a cycle) facteur *m* de forme
~FEEDING DEVICE - [comput] introducteur *m* de formulaires
~GRINDING - [mach tool] (profile grinding) meulage *m* de forme
~LUMBER - [timber] bois *m* pour coffrage
~OIL - [metall] huile *f* à moules
~TOOL - s. form cutter
FORMABLE SHEET - [plast ind] feuille *f* postformable
FORMAL LOGIC - [contr] logique *f* formale
FORMALDEHYDE - [chem](colourless poisonous gas with applications in the production of synthetic resins and plastics, dyestuffs, disinfectants and tissue preservatives) formaldéhyde *f*, formal *m*, méthanal *m*
~ANILINE - [chem] (a synthesis intermediate and vulcanization accellerator) aniline *f* de formaldéhyde, formaniline *f*
~SOLUTION - [chem] solution *f* aqueuse de formaldéhyde
FORMALIN - [chem] (a commercial solution of 40 p. c. of formaldehyde in water) formaline *f*
FORMALITY - [gen] formalité *f*
FORMAMIDE - [chem] (synthesis intermediate and solvent) formamide *f*, formiamide *f*
FORMANILIDE - [chem] formanilide *f*
FORMANILINE - [chem] formaniline *f*
FORMANT - [acoust] (of a sound) son *m* caractéristique
FORMANTS - [acoust] fréquences *f*pl principales d'un son
FORMAT - [print] (the appearance of the book) format *m* du livre
[bookbind] format *m*
[comput] (of an instruction word) structure *f*
~CONTROL - [comput] (the selection of the position of each character in a punch-card system) commande *f* de format
~EFFECTOR - [comput] caractère *m* de mise en page
~TAPE - [comput] bande *f* pilote
FORMATE - [chem] (salt of formic acid) formiate *m*
FORMATION - [gen] formation *f*
[geol] (generally a series of strata) formation *f*
[constr] (in earthmoving works) plate-forme *f*, niveau *m* des remblais
[el & electron] formation *f*
[aero] (the arrangement in flight of more than one aircraft) formation *f*
~BOUNDARY - [geol] limites *f*pl d'une formation géologique
~FLYING - [aero] vol *m* en formation
~LEVEL - [railw] (road-bed) plate-forme *f* des terrassements
~OF AGGLOMERATES - [rubber ind] agglutination *f*
~OF BUBBLES - [rubber ind] formation *f* de bulles d'air
~OF BUMPS - [rubber ind] (in tyres) formation *f* de bosses

FORMATION OF CONE - [text] formation *f* du cône
~OF EDDIES - [hydr] formation *f* de remous
~OF GETTER FILM - [electron] formation *f* du getter
~OF HEAT - [phys] (the heat increase in a system when one gramme-molecule of a substance is formed from its elements) chaleur *f* de formation
~OF PELLETS - [agric] enrobage *m* des semences
~OF VACUUM - [phys] formation *f* du vide
~PACKER - [oil ind] (to obtain formation samples) essayeur *m* de couches
~PRESSURE - [oil ind] (of natural gas) pression *f* de gisement
~SAMPLE - [oil ind] (sample obtained from a reservoir formation) échantillon *m* de couche mineralisée
~TIME LAG - [electron] (in the formation of photoions with certain elements) retard *m* de formation
~VOLTAGE - [el metall] (the final impressed voltage at which the film is formed on the valve in an electrochemical valve) tension *f* de formation
~WATER - [oil ind] (salt water underlying oil and gas in a formation) eau *f* de couches mineralisées
~YARD - [railw] gare *f* de formation
FORME - [print] (the type matter assembled and locked up in a chase) forme *f*
~INKING ROLLERS - [print] toucheur *m*, rouleau *m* de imprimeur
~RACK - [print] rang *m*
FORMED - [gen] formé
~CUTTER - [metall] fraise *f* profilée
~HEAD - [mech] tête *f* de forme
~PLATE - [el] (plate used in lead-acid accumulators) plaque *f* à grande surface
~SHED - [text] (in shedding) pas *m* formé
FORMER - [mech] (a device used to give shape to a built-up structure, e.g. a coil, during construction) calibre *m* de forme, calibre *m* reproducteur, gabarit *m*
[labour] patron *m*
[aero] cadre *m* fermoir
[el] gabarit *m* de bobinage
~OF COIL - [el] carcasse *f* de bobine
~RIB - [constr work] faux cintre *m*
~SLEEVE - [plast ind] (a hollow part used to limit the external size of extruded pipe) manchon *m* calibré
~-WOUND COIL - [el] (armature coil correctly shaped by means of a former) bobine *f* sur gabarit
FORMIC ACID - [chem] (colourless, caustic liquid with uses as a synthesis intermediate, food preservative, latex coagulant, and in medicine) acide *m* formique
FORMICATION - [med] formication *f*
FORMICIN - [chem] (a disinfectant) formicine *f*
FORMING - [gen] formation *f*, développement *m*
[mech] (of sheet metal) profilage *m*, formage *m*, emboutissage *m*
[glass man] formage *m*
[el] (of the plate of lead-acid accumulator) formation *f*
~A TRAIN - [railw] (the making up of a train) formation *f* d'un train
~ATTACHMENT - [mach tool] appareil *m* à calibrer
~BOARD - [el] gabarit *m* pour peignes
~CUTTER - s. form cutter
~DEPARTMENT - [metall] usine *f* emboutissage
~DIE - [metall] estampe *f* d'emboutissage
FORMING LATHE - [mach tool] tour *m* à singer, tour *m* à copier
~PRESS - [mech mach] presse *f* à formage
~PROCESS - [el chem] (the process resulting in a change of impedance at the interface of a valve metal to the passage of current from metal to electrolyte, when the voltage is applied) formation *f*
~THE KNOTS - [text] (in the weaving of tapestry and carpets) formation *f* des nœuds
~TOOL - [tool] outil *m* de forme, outil *m* à calibrer
FORMOL - [chem] formol *m*
~TITRATION - [chem] (the volumetric assessment of amino acids present in a solution) titrage *m* à formol
FORMULA - [gen] formule *f*
[chem] (symbolic expression of the composition, construction, and proportions of the elements present in a chemical compound) formule *f*
~EVALUATION - [math] évaluation *f* d'une formule, calcul *m* d'une formule
~TRANSLATOR - [comput] fortran *m*
~WEIGHT - [chem] (the sum of the atomic weights of the elements present in a compound) poids *m* de formule
FORMULATE, to - [gen] formuler, élaborer
FORMULATION - [chem] formulation *f*, élaboration *f*
FORMWORK - [constr] (a woodform for concrete) coffrage *m*
FORMYL FLUORIDE - [chem] (a synthesis intermediate) fluorure *m* de formyle
FORNIX - [anat] fornix *m*, trigone *m* cérébral
FORSTERITE - [min] (silicate of magnesium) forstérite *f*
FORT - [gen] fort *m*
FORTIFICATION - [gen] fortification *f*, renforcement *m*
FORTIFIED - [gen] fortifié
[build] (reinforced, strengthened) fortifié, renforcé
FORTIFY, to - [gen] fortifier, renforcer
[chem] (to strengthen by increasing the proportion of some essential constituent, e.g. by adding alcohol to wine, or oleum to sulphuric acid) renforcer, fortifier
FORTIN'S BAROMETER - [instr] (mercury barometer giving accurate readings) baromètre *m* de Fortin
FORTNIGHT - [of time] quinzaine *f* (deux semaines)
FORTNIGHTLY - [gen] bimensuel
FORTRESS - [gen] forteresse *f*
FORTUITOUS - [gen] fortuit, imprévu
~DISTORTION - [telecomm] distorsion *f* irrégulière
FORWARD, to - [gen] expédier, avancer
[bot] forcer, pousser, hâter
[print] (in bookbinding) coller et endosser
FORWARD - [gen] de devant, situé en avant
[naut & aero] (near, or nearer to the front or leading part of a structure) avant, sur l'avant
[bot] (of plants) precoce
[comm] à terme
~-ACTING REGULATOR - [radio] (a transmission regulator) régulateur *m* à action
~AND BACKWARD - [mech] d'avance et de recul
~AND REVERSE - [mech] avance *m* et recul
~AXLE - [auto] essieu *m* avant
~BEARING - [surv] orientation *f* conséquente
~BIAS - [electron] polarisation *f* en sens direct
~BOX - [railw] poste *m* de block aval

FORWARD CHANNEL - [comput] voie *f* d'aller, canal *m* d'aller
~COMBUSTION - [metall] combustion *f* en avance
~CONDUCTANCE - [el] (the reciprocal of forward resistance) conductance *f* en sens direct
~CONE - [aero] (the tapering part of the fuselage at the leading end) cône *m* avant
~CREEP - [metall] fluage *m*
~CURRENT - [el] (current resulting from a forward voltage) courant *m* direct, courant *m* en sens direct
[electron] (of a semiconductor diode) courant *m* direct
~CURRENT DENSITY - [el] (the forward current per unit of active rectifying area) densité *f* du courant direct
~-CURRENT TRANSFER RATIO - [electron] amplification *f* de courant en court-circuit
~DEAD-CENTRE - [mech] point *m* mort avant
~DELIVERY - [comm] livraison *f* à terme
~DIRECTION - [el] (the direction of lesser resistance to current flow through the cell) sens *m* conducteur, sens *m* direct
~ECCENTRIC - [mech] excentrique *m* de marche en avant
~ELEMENTS - [contr] organes *mpl* d'action
~END OF CYLINDER - [mech] partie *f* inférieure du cylindre
~END OF CYLINDERBLOCK - [mech] partie *f* inférieure du bloc-cylindre
~ENGINE MOUNTING - [mech] montage *m* du moteur en avant
~FEED - [mech] avance *f*
~FEED OF MATERIAL - [mech mach] avance *f* de la pièce
~-INVERSE ELECTRONIC RESOLVER - [comput] machine *f* électronique avant-arrière à résoudre des équations
~LEAD - s. forward shift
~MOTION - [mech mach] mouvement *m* en avant
~PATH - [mech] parcours *m* en avant
~RECALL SIGNAL - [radio] signal *m* de rappel en avant
~RESISTANCE - [el] (the contact resistance opposing the forward current) résistance *f* en sens direct
~ROLL - [mech] (a roller running in the same direction as the band of material passing over or round it) cylindre *m* à marche avant
~SCATTER - [nucl] (scattered radiation with a scattering angle of less than 90 degrees) diffusion *f* en avant
~SHIFT - [el] (movement of the brushes around the commutator in the same direction as that of rotation) décalage *m* dans le sens du mouvement
~SIGNAL - [contr] (order transmission through a circuit) signal *m* d'action
~SIGNALLING - [telecomm] signalisation *f* vers l'avant
~SLIP, to - [aero] glisser sur l'aile
~SPEED - [mech auto etc] marche *f* avant
~STROKE - [mech] (of a piston etc) course *f* avant
~-STROKE INTERVAL - [telev] durée *f* de la ligne
~TILT - [aero] (the angle between the plane of rotation and a line passing through the centroids of the blade sections) déport *m* de la pale dans le plan de avancement de l'avion
~-TRANSFER ADMITTANCE - [el] transadmittance *f* grille-anode
FORWARD TRANSFER FUNCTION - [contr] fonction *f* de transfert en avant
~TRANSFER SIGNAL - [telecomm] signal *m* d'intervention
~VOLTAGE - [el] (the voltage applied between the two terminals of a contact rectifier in the sense which corresponds to the direction of the highest current for this voltage) tension *f* en sens direct
~VOLTAGE DROP - [el] (in the cell) chute *f* de tension en sens direct
~WAVE - [electron] (in a travelling-wave tube) onde *f* directe
~WELDING - [metall] soudage *m* en avance
~WIND KEY - [el acoust] touche *f* défilement rapide
FORWARDER - [gen] expéditeur *m*
FORWARDING - [gen] expédition *f*, envoi *m*
[bot] forçage *m*
[print] (in bookbinding) collage *m* et endossage
~AGENT - [comm] agent *m* expéditeur, commissaire *m* de transport, entreprise *f* de groupage
~DEPARTMENT - [comm] bureau *m* d'expédition
~RAILWAY - [railw] chemin de fer *m* expéditeur
~ROUTE - [railw] itinéraire *m* d'acheminement
~STATION - [railw] gare *f* expéditrice
FOSSE - [anat] (also called foss or fossa; a pit-like depression) fosse *f*, fossette *f*
FOSSETTE - [anat] fossette *f*
FOSSICKING - [mining] maraudage *m*
FOSSIL - [gen] fossile *m*
~-BEARING - [geol] fossilifère
~FLOUR - [min] (used in the rubber industry) farine *f* fossile
~MEAL - s. fossil flour
~OIL - [min] huile *f* de roche, pétrole *m*, naphte *m* minéral
~PAPER - [min] papier *m* fossile, papier *m* de montagne
~PLANT - [geol] (fossilized plant) plante *f* fossile
~WAX - [min] cire *f* fossile, cérésine
~WOOD - [geol] bois *m* fossile
FOSSILIFEROUS - s. fossil-bearing
FOSSILIZATION - [geol] fossilisation *f*
FOSSILIZE, to - [geol] (the process of fossilization) fossiliser, se fossiliser
FOSTER-SEALEY DETECTOR - [el] (type of F.M. demodulator using two diodes) détecteur *m* Foster-Sealey
FOT - [Optimum Traffic Frequency] (radio) fréquence *f* optimum de trafic
FOTHER, to [naut] (to stop a leak with ropeyarn etc) aveugler
FOUCAULT CURRENT - [el] (eddy current) courant *m* de Foucault
~KNIFE-EDGE TEST - [opt] (method of testing concave lenses) méthode *f* de Foucault
~ROTATING MIRROR - [opt] (a device demonstrating that the velocity of light in water was less than in air) miroir *m* rotatif de Foucault
~PENDULUM - [instr] (instrument designed by Foucault to demonstrate the rotation of the earth) pendule *m* de Foucault
FOUL, to - [gen] salir, souiller
[naut] (of the anchor) surjaler, engager
[naut] (to collide) entrer en collision
[naut] (of the keel) encrasser
[mech] (of piping) entartrer
[firearms] encrasser

FOUL, to - [auto] (of the sparking) encrasser
[hydr] entartrer (les tubes d'une chaudière)
[mech] fausser, se fausser
FOUL - [gen] sale, infect, fétide
[hydr etc] (of a pipe) obstrué
[naut etc] (of a cable) engagé, embarassé
[naut] (of the keel) sale
[naut] (collision) collision *f*
~AIR - [gen] air *m* vicié
~-AIR FLUE - [constr] conduite *f* de ventilation
~BOTTOM - [naut] carène *f* sale
~CLAY - [constr] (brick earth consisting of silica and alumina with a small percentage of lime, soda or other salts) glaise *f*
~ELECTROLYTE - [chem] (an electrolyte containing enough impurities to impede the proper action of the cell) électrolyte *m* imprope
~MAIN - [gas ind] (gas collecting main) collecteur *m* de gaz
FOULARD - [text] (very light dress fabric made of silk or fine cotton) foulard *m*
FOULBROOD - [vet] loque *f*, pourriture *f* du couvain
FOULING - [gen] (naut mech firearms etc) encrassement *m*, engorgement *m*
[naut] abordage *m*
[chem] (in chemical purification) engorgement *m*
FOULNESS - [gen] saleté *f*, impureté *f*
[min] grisou *m*
FOUND, to - [gen] fonder, poser, établir
[constr] fonder, poser les fondations
[metall] mouler
FOUNDATION - [constr] (the formation on which a construction is built) fondation *f*, massif *m* de base, fondement *m*, soubassement *m*
[text] base *f*, fondement *m*
~BLOCK - [constr] massif *m* de base, massif de fondation
~BOLT - [mech] (in machines etc) tire-fond *m*, boulon *m* de fondation
~FABRIC - [text] tissu *m* de fond
~FRAME - [mach] châssis *m*, assiette *f*
~GRILL - [constr] grillage *m* en pieux
~MASS - [constr] soubassement *m*
~MUSLIN - [text] (particularly used for hats) mousseline *f* pour chapeaux
~OF A PILLAR - [constr] massif *m* d'encrage
~PILE - [constr] (the pile driven into the ground to supply a rigid support for a structure) pieu *m* de fondation, pilotis *m* de fondation
~PIT - [constr] fouille *f* de construction, fouille *f* de fondation
~-PLATE - [mech] plaque-semelle *f*, plaque *f* d'assise
[text] plaque *f* de fondation, plaque *f* de fondement
~PLINTH - [constr] plinthe *m* de fondation
~RING - [mech] (a locomotive fire-box) cadre *m* du foyer
~SLAB - [constr] semelle *f*
~SOIL - [constr] terrain *m* à bâtir, sol *m* de fondation
~STONE - [constr] pierre *f* fondamentale
~STRATUM - [constr] couche *f* de fondation
~TESTER - [instr] (used to measure the strength of a foundation work) vérificateur *m* du sol de fondation
~TEXTURE - [text] tricot *m* de base, tricot *m* de soutien
~THREAD - [text] fil *m* de fond, fil *m* de support, âme *f*
FOUNDATION TIE ROD - [th eng] ancre *f* à clavette
~TRENCH - [constr] fouille *f* de construction, fouille *f* en rigole, tranchée *f* de fondation, fosse *f* de fondation
~WALL - [constr] jambage *m*, substruction *f*
~WARP - [text] chaîne *f* de fond
~WASHER - [mech] plaque *f* d'encrage
~WORK - [constr] travaux *mpl* de fondation
[mus] fonds *mpl*
FOUNDATIONS - [constr] fondations *fpl*
FOUNDED - [constr] (of a caisson sunk to a firm level) fixé sur fondation
FOUNDER, to - [gen] s'effondrer, s'écrouler
FOUNDER - [gen] fondateur *m*
[metall] fondeur *m*
[vet] fourbure *f*, sole *f* battue
[geol] éboulement *m*
FOUNDER'S FLANGE - [impl] lissoir *m* de soudeur
~SAND - [metall] sable *m* de fonderie
~TYPE - [print] caractère *m* fondeur
FOUNDING - [gen] fonte *f*, moulage *m*
~FURNACE - [metall] four *m* de fusion
~METAL - [metall] métal *m* de coulée
~OF ROLLERS - [metall] fonte *f* de rouleaux
FOUNDRY - [mech] (establishment in which founding of metal or glass is carried out) fonderie *f*
[print] fonderie *f* (de caractères)
~AIR FURNACE - [metall] four *m* à réverbère de fonderie
~BLACKING - [metall] noir *m* de fonderie
~BRUSH - [metall] brosse *f* pour fontes brutes
~BUCKET - [metall] poche *f* de coulée
~CAR - [metall] chariot *m* de fonderie, chariot *m* de pont de coulée
~CASTING WORK - [metall] coulée *f*, fusion *f*
~COKE - [fuel] coke *m* métallurgique, coke *m* de cubilot
~CRANE - [mech] grue *f* de fonderie
~CRANE WITH FIXED LADLE GUIDE - [mech mach] pont *m* de coulée
~CUPOLA - [metall] cubilot *m* de fonderie
~DRESSING SHOP - [metall] atelier *m* d'ébardage
~FLASK - [metall] châssis *m* de moulage, châssis *m* de fonderie
~HOUSE - [metall] halle *f* de fonderie
~IRON - [metall] (foundry pig) fonte *f* de moulage
~LADLE - [metall] (fireclay-lined steel ladle) cuiller *m* à couler, poche *f* à fonte, poche *f*, poche *f* de coulée
~LOSS - [metall] (any casting which is rejected from the foundry) refus *m* de fonderie
~MOULDING MACHINE - [mech mach] machine *f* à mouler
~PATTERN - [metall] modèle *m* de fonderie
~PIG-IRON - [metall] (the cast-iron bars which are bought by the foundry) fonte *f* grise, fonte *f* de moulage
~PIT - [metall] (a hole in the floor of a foundry used as a moulding box for large or very deep castings) fosse *f* de moulage
~PROOF - [print] (the final proof before stereo-or electrotyping) dernière épreuve *f*, épreuve *f* corrigée
~RATTLER - [metall] tonneau *m* dessableur
~RETURNS - [metall] refus *m* de fonderie
~REVERBERATORY FURNACE - [metall] four *m* à réverbère de fonderie

FOUNDRY-RIDDLE - [metall] tamis *m* de mouleur
~RUN - [metall] coulée *f* de fonderie
~SAND - [metall] sable *f* de fonderie, sable *f* de moulage
~SLEWING CRANE - [mech mach] grue *f* de fonderie tournante
[metall] grue *f* de fonderie pivotante
~STOVE - [metall] (large stove used to dry moulds and cores) étufe *f* de fonderie
~TRAVELLER - [metall] pont *m* roulant de fonderie
FOUNT - [gen] source *f*
[print] fonte *f*, lettres *fpl*, types *mpl*
[impl] (e.g. of a lamp) réservoir *m*
FOUNTAIN - [gen] fontaine *f*, source *f* d'eau, jet *m* d'eau
[ind chem] (liquid container) cuvette
[metall] (US only; a spout in GB) canal *m* de coulée, chenal *m* de coulée
[oil ind] puits *m* tubulaire américain
~HEAD - [gen] source *f* (de rivière)
~-PEN - [impl] stylographe *m*
FOUR-ADDRESS - [comput] (each instruction sets out the operation and the addresses of four registers) à quatre adresses
~-ADDRESS INSTRUCTION - [comput] instruction *f* à quatre adresses
~-ARMED LEVER - [mech] levier *m* à croix
~-BLADE FAN - [el] ventilateur *m* à quatre aubes
~-CENTRED ARCH- [arch] (a pointed arch struck from four centres) voûte *f* à quatre centres, arc *m* aplati
~-CHANNEL AUDITORIUM CONTROL - [cin] régulateur *m* de volume à quatre voies
~-CHANNEL MAGNETIC HEAD - [acoust] (magnetic head used for the reproduction of four-channel stereophonic films) lecteur *m* de son à quatre voies
~-CHANNEL POWER AMPLIFIER - [acoust] amplificateur *m* de puissance à quatre voies
~-CHANNEL PREAMPLIFIER - [acoust] (used for the reproduction of stereophonic films) préamplificateur *m* à quatre voies
~-CHANNEL STEREOPHONIC FILM - [cin] (film with four magnetic sound-tracks) film stéréophonique à quatre voies
~-COLOUR PRINTING - [print] quadrichromie *f*, impression *f* en quatre couleurs
~-COLUMN PRESS - [mech] presse *f* à quatre colonnes
~-CORE CABLE - [el] câble *m* quadripolaire
~-COURSE - [agric] assolement *m* quadriennal
~COURSE BEACON - [radio] (a facility supplying radial equisignal zones) radiophare *m* d'alignement (4 axes)
~-CYCLE ENGINE - s. four-stroke engine
~-CYLINDER COMPOUND LOCOMOTIVE - [mech] locomotive *f* compound à 4 cylindres
~-DECK - [naut] à quatre ponts
~-DIMENSIONAL - [math] à quatre dimensions
~-ELECTRODE TUBE - s. four-electrode valve
~-ELECTRODE VALVE - [electron] (thermionic valve with a cathode and three other electrodes) tétrode *f*
~-END TWILL - [text] croisé *m* à quatre lames, sergé *m* de quatre
~-ENGINE - [aero] quadrimoteur *m*
~FACTOR FORMULA - [nucl] (multiplication factor in a reactor with no leakage) formule *f* des quatre facteurs

FOUR-FLUTE TWIST HAND-REAMER - [tool] alésoir *m* à quatre tranchants en hélice
~-FOLD - [gen] quadruple
~-FOLD BLOCK - [mech] moufle *f* à quatre poulies
~-FOOT WAY - [railw] (a side path) entre-rail *m* de la voie normale
~-FORCE - [mech] (the four-vector describing the rate of change of four-momentum along the space-time path of a particle) résultante *f* quadrivectorielle
~-FREQUENCY DIALLING - [telecomm] sélection *f* à distance à quatre fréquences
~-FREQUENCY KEY SENDING - s. four-frequency dialling
~-HABNESSED TWILL - [text] étoffe *f* à quatre lames
~-HIGH FOUR-STAND CONTINUOUS STRIP MILL - [metall] laminoir *m* continu à feuillards à quatre cages
~-HIGH MILL - [metall] laminoir *m* à quatre cylindres
~-HIGH ROLL STAND - [metall] cage *f* quatre
~-HIGH UNIVERSAL REVERSING MILL - [metall] laminoir *m* universal réversible à quatre cylindres
~-JAW CHUCK - [mech] plateau *m* à quatre griffes
~-LINE BINARY CODE - [comput] code *m* tétradique, code *m* binaire 4 fils
~-MASTER - [naut] quatre-mâts *m*
~-MOMENTUM - [phys] (of a classical particle) moment *m* quadratique
~-PART FLASK - [oil ind] bride *f* double
~-PHASE SYSTEM - [el] (in practice, a two-phase system in which the mid-points are connected to form a neutral point) système *m* à quatre phases
~-POINT PROBE TECHNIQUE - [electron] (for semiconductors) méthode *f* des sondes
~-POLE - [el] tétrapolaire, à quatre pôles
~-ROLL CALENDER - [mech] (a finishing machine for sheet plastics in which the desired texture is imparted by means of four rollers) calandre *f* à quatre cylindres
~-ROLLER MILL - [metall] laminoir *m* à quatre cylindres
~-ROWED - [agric] (of barley etc) à quatre rangs
~-SCREW BELL-CHUCK - [mach tool] mandrin *m* à quatre vis
~-SEATER - [auto] voiture *f* à quatre places
~-SHAFT TWILL - s. four-end twill
~-SLIDED BED - [mach tool] banc *m* d'un tour à quatre guides
~-SPINDLE DRILLING-MACHINE - [mach tool] machine *f* à percer à quatre forets
~-STAGE COMPRESSOR - [mech mach] (air or gas compressor in which the pressure of the fluid is raised in four successive steps) compresseur *m* à quatre étages
~-STAND TANDEM MILL - [metall] laminoir *m* continu à quatre cages en tandem
~-START SCREW - [mech] vis *f* à quatre filets
~-STROKE CYCLE - [i.c. engines] (in reciprocating i.c. engines, the system of operations in which induction, compression, explosion and exhaust occurring during four successive piston strikes) cycle *m* à quatre temps
~-STROKE ENGINE - [mech] (i.c. engine working on the Otto cycle, i.e. one explosion in 2 revolutions) moteur *m* à quatre temps
~-TERMINAL NETWORK - [el] (any group of impedan-

ces which can exchange electric power with external systems by their input and output connections, both comprising two terminals) quadripôle *m*
FOUR-TERMINAL RESISTANCE - [el] résistance *f* à quatre bornes
~-TERMINAL TRANSMISSION NETWORK - [telecomm] quadripôle *m*
~-THREAD SCREW - s. four-start screw
~-TRACK SOUND RECORDING - [el acoust] enregistrement *m* sonore à quatre pistes
~-VELOCITY - [phys] (the four quantities describing the velocity of a classical particle in relativity theory) vélocité *f* quadratique
~-WAY - [plumb etc] à quatre voies
~-WAY BIT - [mining] trépan *m* à quatre ailettes
~-WAY COCK - [plumb] robinet *m* à quatre voies
~-WAY RIM WRENCH - [auto] clef *m* en croix démonte-roue
~-WAY SWITCH - [el] interrupteur *m* à quatre voies
~-WAY TOOL BLOCK - [mach tool] porte-outil *m* à quatre faces
~WEFT BINDING - [text] liage *m* par quadruple duite
~-WHEEL - [gen] à quatre roues
~-WHEEL BOGIE - [aero] (landing gear bogie having four wheels grouped together) double diabolo *m*
~-WHEEL BRAKE - [mech] frein *m* sur les quatre roues
~-WHEEL DRIVE - [mech] à quatre roues motrices
~-WHEEL STEERING - [auto] à quatre roues directrices
~WINGS BIT - [oil ind] burin *m* à quatre ailettes
~-WIRE CIRCUIT - [el] (used in telecommunications) circuit *m* à quatre fils
~-WIRE SIDE CIRCUIT - [el] circuit *m* réel à quatre fils
~-WIRE SYSTEM - [el] (system of electric power distribution requiring four wires) réseau *m* à quatre fils
~-WIRE TERMINATING SET - [telecomm] termineur *m*
FOURBLE - [oil ind] jeu *m* de quatre
FOURCHET - [anat] fourchette *f* vulvaire
FOURIER INTEGRAL - [math] (applied in electric communications) intégrale *f* de Fourier
~SERIES - [math] série *f* de Fourier
~SYNTHESIS - [math] (applied in diffraction analysis) synthèse *f* de Fourier
FOURMARIERITE - [min] (uranium and lead containing mineral) fourmarierite *f*
FOURNIER'S DISEASE - [med] gangrène *f* foudroyante des organes
FOURTH - [num] quatrième
[mus] quarte *f*
~CANAL - [cin] (used in sound projection) piste *f* d'ambiance
~DIMENSION - [math] quatrième dimension *f*
~PINION - [horol] (the pinion on which the fourth wheel is mounted) pignon *m* de la quatrième roue
~POWER LAW - [opt] loi *f* de Stefan-Boltzmann
~RAIL - [el] (the conductor rail on a traction system) rail *m* conducteur
~VENTRICLE - [anat] (the cavity of the hind-brain) quatrième ventricule *m*
~WHEEL - [horol] aiguille *f* des secondes
~WIRE - [el] (the neutral wire in a three- or four-phase system) neutre *m*
FOWLER FLAP - [aero] (a flap designed to increase the chord length of an aerofoil) volet *m* Fowler
FOWLER'S SOLUTION - [chem] (a solution of potassium arsenite used in medicine) liqueur *f* de Fowler
FOX - [zool] renard *m*
[naut] (tarred and twisted rope yarns) tresse *f*, commande *f*
~BRUSH - [zool] queue *f* de renard
~-GLOVE - [bot] digitale *f*
~-HOUND - [zool] (a hunting dog) chien *m* courant
~WEDGE - [mech] contre-clavette *f*
[carp] coin *m* en bois
FOXED - [paper man] piqué
[agric] (of timber) pouilleux
[brew ind] (of beer) aigre, piquée
FOXHOLE ARTHRITIS - [med] polyarthrite *f* tropical aigue
FOXTAIL SAW - [impl] (dovetail saw) scie *f* à queue d'aronde
~WEDGING - [carp] assemblage *m* à contre-clavette
FOXY - [paper man] piqué
FOYER - [constr] foyer *m*
[metall] (a crucible for molten metal) creuset *m*
F.P. - [Freezing Point] température *f* de congélation
[phys] point *m* de consolidation
FRACTION, to - [gen] fractionner
FRACTION - [gen & math] fraction *f*
~EXCHANGE - [nucl] (number denoting the progress of an isotopic exchange reaction) facteur *m* d'interchange
~LINE - [math] trait *m* de fraction
~OF SATURATION - [met] humidité *f* relative
FRACTIONAL - [gen & chem] fractionnaire
~CARD - [comput] (of punched-cards) carte *f* ventilée
~COAGULATION - [ind chem] (incomplete coagulation of rubber) coagulation *f* fractionnée
~COMBUSTION - [chem] combustion *f* fractionée
~CRYSTALLIZATION - [chem] (the process of separating substances by reiterated partial crystallization from solution) cristallisation *f* fractionnée
~CURRENT - [el] courant *m* partiel
~DISTILLATION - [chem] (a process of selective distillation using special columns, according to the different boiling points of the constituents of a mixture of liquids) distillation *f* fractionnée
~HORSEPOWER - [mech] de puissance inférieure à un cheval
~IONIZATION - [nucl] degré *m* d'ionisation
~ISOTOPIC ABUNDANCE - [nucl] (relative isotopic abundance) abondance *f* isotopique relative
~LIMING - [sugar ind] chaulage *m* fractionné
~MASHING OF THE GRIST - [brew ind] procédé *m* de empâtage par séparation de moutures
~OXIDATION - [chem] oxydation *f* fractionnée
~PITCH - [mech] (of a screw thread) pas *m* partiel
~PITCH WINDING - [el] enroulement *m* à pas partiel
~PRECIPITATION - [chem] précipitation *f* fractionnée
~-SLOT WINDING - [metall] enroulement *m* à nombre fractionnel d'encoches par pôle et par phase
~YIELD - [chem] (intermediate yield in a step-by-step process) rendement *m* par étage
FRACTIONATE, to - [chem] (the operation of selective distillation) fractionner
FRACTIONATED TREATMENT - [radiat] (course of tratment in doses spread over a period of time)

fractionnement *m*
FRACTIONATING - [ind chem] fractionnant
~BRUSH - [ind chem] brosse *f* fractionnate (distillation), brosse tournante
~BRUSH STILL - [ind chem] installation *f* de distillation à brosse fractionnante
~COLUMN - [ind chem] (column attached to a still) colonne *f* de fractionnement
~DIFFUSION PUMP - [ind chem] pompe *f* (à diffusion) fractionnante
~TOWER - [oil ind] tour *f* de distillation
FRACTIONATION - s. fractional distillation
FRACTIONATOR - [ind chem] (used in cracking processes) fractionneur *m*
FRACTO - [met] (prefix before a cloud-type, e.g. "fracto-cumulus", indicating that clouds consist of separate masses, not in a continuous body) fracto-
FRACTOCUMULUS - [met] (cloud formation consisting of isolated masses of cumulus) fracto-cumulus *m*
FRACTOSTRATUS - [met] (cloud formation consisting of isolated masses of stratus) fract-stratus *m*
FRACTURE, to - [gen] casser, briser
[med] fracturer
FRACTURE - [gen] (breakage) fracture *f*, rupture *f*
[med] fracture *f*
[text] (on woven materials) surface *f* de rupture
[metall] grain *m* d'une cassure, crique *f*
[geol] cassure *f*, fracture *f*
~LINE - [min] (the line of the broken surface of a mineral) ligne *f* de fracture
~ON DEFORMATION - [metall] cassure *f* ductile
~STRESS - [metall] effort *m* de cassure
~STRUCTURE - [metall] structure *f* de la cassure
~SURFACE - [constr] surface *f* de rupture, plan *m* de rupture
FRACTURED - [geol] crevassé
FRACTURING - [geol] formation *f* des fissures
FRAGILE - [gen] fragile
FRAGILITY - [gen] fragilité *f*
[metall] frangibilité *f*
FRAGMENT - [gen] fragment *m* morceau *m*, brin *m*
FRAGMENTAL DEPOSIT - [geol] (deposits consisting of fragments of rocks or minerals) dépôts *mpl* détritiques
~TEXTURE - [geol] structure *f* clastique
FRAGMENTATION - [gen bot etc] fragmentation *f*, séparation *f*
~OF NUCLEUS - [nucl] (nuclear explosion) fragmentation *f* du noyau
FRAME, to - [gen] encadrer
[cin & photo] encadrer
[telev] (the operation of adjusting the frequency of the sawtooth voltage generator of the local frame to that of the transmitter) encadrer
[min] concentrer sur la table dormante
FRAME - [gen] cadre *m*, encadrement *m*
[constr] (framework) structure *f*, treillis *m*, charpente *f*, pan *m*
[el mach] carcasse *f*
[el] (of storage battery; structure in an alkaline battery equivalent to the grid in a lead-acid one) cadre *m*
[text] cadre *m*, bâti *m*, bâti-support *m*, métier *m*
[mech] châssis *m*
[constr] (of a wondow, a door etc) dormant *m*, bâti *m*, chambranle *m*

FRAME - [shipbuild] membrure *f*, carcasse *f*, cadre *m* de l'hélice
[opt] (of eyeglasses) monture *f*
[aero] cadre *m* (couple)
[cin] (a single picture on a picture film) image *f*
[telev] (subdivision of the complete picture) panneau *m*
[anat] forme *f*, structure *f*
[mining] châssis *m* de mine
[hydr] (of a sluice-gate) tableau *m* (d'une vanne)
[min] table *f* dormante
[comput] (on a analog display unit) image *f*
~AERIAL - [radio] cadre *m*, antenne *f* fermée
~AERIAL RECEPTION - [telev] réception *f* à cadre
~AMPLITUDE CONTROL - [telev] (control of the vertical amplitude of the scanning) réglage *m* de la hauteur d'image
~ANTENNA - s. frame aerial
~BAR - [telev] barrette *f*
~BATTEN - [text] listeau *m*
~BLANKING - [telev] (suppression of the image signal at the end of the frame) suppression *f* du cadre
~BORDER - [telev] (US only; picture edge in GB) bord *m* de l'image
~BRACKET - [auto] patte *f* de châssis
~BY FRAME DISPLAY - [telev] marche *f* image par image
~CENTRE SECTION - [auto] partie *f* centrale du châssis
~CLAMP - [carp] crampon *m*
~COIL - [telev] (picture control coil) bobine *f* d'image
~CONTROL - [telev] réglage *m* du cadrage
~COUNTER - [photo] compteur *m* d'images
~CROSS MEMBER - [mech] traverse *f* de châssis
~DEFLECTION - [telev] (the movement of the electron beam downwards as each line is scanned) déviation *f* de l'image
~DISTORTION TILT - [telev] (undesired component of the signal causing the upward tilting of the horizontal scan) distorsion *f* de l'image
~DRAWING - [text] étirage *m* préparatoire, étirage *m* en gros
~END - [text] montant *m*
~EXTENSION - [auto] prolonge *f* de châssis
~FITTING - [auto] accessoires *mpl* de châssis
~FREQUENCY - [telev] (the number of pictures per second) fréquence *f* d'image
~GRID - [electron] grille *f* à cadran
~HOUSE - [constr] maison *f* en pans de bois
~LEVEL - [instr] (mason's level) règle *f* pour nivellement
~LINE - [cin] (the line which separates two frames) ligne *f* de séparation de deux images
[telev] (the registration of a frame) cadrage *m*
~LINE NOISE - [cin] (noise caused by incorrect adjustment of the light source for the sound track) bruit *m* des lignes de séparation
~LINEARITY CONTROL - [telev] réglage *m* de la linéarité
~NEEDLE BAR - [text] barre *f* à aiguilles, fonture *f*
~NOISE - [cin] (caused by the film when it is pulled too far to the right in the projector) bruit *m* de cadre
~NUMBER - [auto] numéro *m* de châssis
~OF HEALD SHAFT - [text] cadre *m* à lisses
~OF PLATE - [el] (storage battery) cadre *m* de la

plaque
FRAME OF REFERENCE - [phys] système *m* de référence
~PLATE - [storage battery] (a plate consisting of a frame on which the active material is fixed) plaque *f* à cadre
~POINTER - [photo] aiguille *f* du cadre
~PRESS - [mech] presse *f* à châssis
~RATE - [telev] durée *f* de l'image
~REPETITION RATE - s. frame frequency
~RETRACE TIME - [telev] temps *m* de retour d'image
~SAW - [impl] (thin-blade saw held taut in a special frame) scie *f* ordinaire, scie *f* à châssis
~SCAN - [telev] (the downward movement of the cathode-spot ray in a tube after the scanning of each line) analyse *f* du cadre
~SEQUENTIAL SYSTEM - [telev] système *m* à séquence d'images
~SET - [mining] châssis *m* de mine
~SIDE - [mach] longeron *m* du châssis
~SIMULTANEOUS SYSTEM - [telev] système *m* additif de télévision couleur
~SLIP - [telev] (lack of exact synchronization of vertical scanning and incoming signal) décalage *m* vertical
~SPINNING - [text] filature *f* à anneaux
~STRAIGHTENER - [mech] appareil *m* à redresser les châssis
~SUPPRESSION - s. frame blanking
~SWEEP UNIT - [telev] générateur *m* de déviation de image
~SYNC PULSE - [telev] impulsion *f* de synchronisation d'image
~SYNCHRONIZATION - [telev] (effect obtained by the insertion of the synchronizing pulses in the circuit) synchronisation *f* d'image
SYNCHRONIZING PULSE - [constr] impulsion *f* de synchronisation de base
~SYNCHRONIZING SIGNAL - [telev] signal *m* de synchronisation verticale
~TIE - [mech] tirant *m* de châssis
~TILT - [telev] correction *f* de la distorsion d'image
~TIME BASE - [telev] (the time base for deflecting the electron beam vertically downwards and to bring it back to the starting point of the frame) base *f* des temps d'image
~VIEW-FINDER - [photo] viseur *m* iconométrique, viseur *m* à cadre
~WEIR - [hydr] barrage *m* mobile
FRAMED BUILDING - [constr] construction *f* en treillis, charpente *f* triangulée
~CORE BOX - [metall] boîte-caisson *f*
~GROUNDS - [constr] huisserie *f* accouplée
~WALL - [constr] charpente *f* en bois, colombage *m*, ossature *f*
~PARTITION WALL - [constr] coison *f* en charpente
FRAMELESS - [auto] véhicule *m* sans châssis
FRAMER - [telev] appareil *m* de cadrage
FRAMEWORK - [constr] (the supporting skeleton of a structure) charpente *f*, bâti *m*, ossature *f*
[naut aero & mech] châssis *m*, charpente *f*
~ARCH - [constr] arc *m* en treillis
~BRIDGE - [constr] point *m* triangulé, point *m* en treillis
~STAY - [aero] entretoise *f* diagonale
FRAMING - [gen] encadrement *m*
[constr] charpente *f*, bâti *m*
[cin] cadrage *m*
[telev] cadrage *m*
~DEVICE - [cin] appareil *m* de cadrage
~KNOB - [cin] (knob of the framing lever) bouton *m* de cadrage
~LEVER - [cin] levier *m* de cadrage
~MASK - [telev] (the mask used in sets to frame the picture) masque *m* de cadrage
~OF THE PICTURE AREA - [photo] découpage *m* du champ de l'image
~STRUCTURE - [build shipbuild etc] ossature *f*, carcasse *f*
~TIMBER - [constr] bois *m* pour châssis
FRANCHISE - [gen & leg] droit *m* de vote, privilège *m*
[comm] concession *f*
FRANCIS WATER TURBINE - [mech] (reaction turbine in which water flows radially inwards) turbine *f* Francis
FRANCIUM - [chem] (symbol Fr. The name assigned to element number 87, existing only as a radioactive isotope) francium *m*
FRANKFORT BLACK - [dyes] (a pigment of similar character and origin to vine black obtained by the destructive distillation vine twigs: an artist's pigment) noir *m* d'Allemagne
FRANKINCENSE - [bot] (aromatic gum resin used for burning as incense) encens *m*, oliban *m*
FRANKLIN AERIAL - [radio] (a type of directive aerial) antenne *f* directive de Franklin
~ANTENNA - s. Franklin aerial
FRANKLIN'S GLASS - [opt] lentille *f* bifocale
FRANKLINISM - [el biology] (frictional electricity) franklinisation *f*
FRANKLINITE - [min] (natural zinc-manganese spinel of uncommon occurrence) franklinite *f*
FRAP, to - [naut] brider, ceintrer
FRASCH PROCESS - [min] (a process for winning underground deposits of sulphur: a double concentric tube is driven into the deposit and steam or hot air forced down the outer tube: the melted sulphur is thus expelled through the inner tube) procédé *m* Frasch
FRASS - [zool] (excrement) chiures *f*pl
FRATER - [constr] (eating room of a monastery) réfectoire *m*
FRAUNHOFER LINES - [spectroscopy] (the dark line which form the absorption spectrum exhibited by sunlight) raies *f*pl de Fraunhofer
~REGION - [radio] (of aerials) zone *f* de Fraunhofer
FRAY, to - [gen] érailler, effilocher, s'effranger, s'étriper
~OUT, to - [mining] se coincer en biseau
FRAYING - [text] effilochage *m*
~BAND - [text] lisière *f* de pantalon
~BRAID - s. fraying band
FRAZE, to - [metall] (the operation of removing edges from forgings) ébarber, ébavurer
FRAZE - [metall] (ragged edge, or other defect on a forges piece) bavure *f* de découpage
FRAZING - [metall] bavure *f*
FRECKLE - [physiol] (on the skin) éphélide *f*
[metall] (dull spots on metal plates) décoloration *f*
FREE, to - [gen] libérer, delivrer, exempter
[constr] (a passage) ouvrir

FREE, to - [mech] (to release from a seized or jammed condition) (mechanism) désobstruer, dégorger, franchir, dégager
[naut] (of the wind) fraîchir
FREE - [gen] libre, exempt, disponible
[gen] (of money) gratuit
[comm] franco
[naut] (of a boat) largue
[mech] (of materials) peu résistant
[chem] non-combiné, libre
~AIR DISPLACEMENT - [phys] débit *m* à la pression atmosphérique
~AIR DOSE - [nucl] (an alternative term for Air Dose: a dose of radiation, measured in air, excluding secondary radiation) dose *f* dans l'air
~AIR IONIZATION CHAMBER - [nucl] chambre *f* de ionisation à air libre
~AIR OVERPRESSURE - [phys] (as created in the air by the blast wave from an explosion) surpression *f* en air libre
~AIR PRESSURE - [phys] pression *f* en air libre
~ALONGSIDE SHIP - [F.A.S.] franco le long du bord (FAS)
~AMMONIA - [chem] (ammonia in the free or uncombinated state, especially in ammonia liquor) ammoniac *m* libre
~BAGGAGE - [aero] (passenger's baggage carried in civil aircraft without extra charge) franchise *f* de bagages
~BALLOON - [aero] (a balloon floating freely and not attached to the earth) ballon *m* libre
~-BALLOON SET - [aero] (the net enclosing the envelope of a free balloon) filet *m* de ballon libre
~BEAM - [constr] poutre *f* en porte à faux
~BEARING - [mech] palier *m* libre
~-BODY DIAGRAM - [mech] (diagram showing the forces acting on a body) diagramme *m* de corps libre, diagramme *m* de forces
~BULKHEAD - [constr] rideau *m* de palplanches sans ancrage
~CALL - [telecomm] appel *m* non taxé
~CARBON - [chem] carbone *m* libre
~CEMENTITE - [metall] (iron-carbide in cast-iron or steel) cémentite *f* libre
~CHARGE - [el] (electric charges on the surface of a conductor) charge *f* libre
~-CUTTING BRASS - [metall] (brass containing some lead to improve machining) laiton *m* de décolletage
~-CUTTING PROPERTY - [metall] usinabilité *f* facile
~-CUTTING STEEL - [metall] (steel to which a certain degree of brittleness is induced to facilitate speedy machining) acier *m* de décolletage
~CYANIDE - [el plating] (excess of alkaline cyanide beyond the minimum necessary for a clear solution. Excess of cyanide beyond the minimum needed to form the desired cyanides) cyanure *m* libre
~-DISCHARGE VALVE - [mech] clapet *m* de décharge libre
~ELECTRON - [electron] (electron which has escaped the attraction of a nucleus and can move freely in matter or in a vacuum) électron *m* libre
~END - [constr] (cantilever end which is not built in) poutre *f* encastrée à une extremité (et libre à l'autre)
[mech] (of a pulley-block chain) brin *m* libre
~-END BEARING - [mech] palier *m* frontal libre
FREE -END OF THE ROPE - [text] (of the rope brake of the warp and cloth movement) extrémité libre de la bande
~ENERGY - [chem] (the capacity of a system to carry out work) énergie *f* libre
~-ENERGY CHANGE - [phys] échange *m* d'énergie libre
~-ENERGY FUNCTION - [phys] fonction *f* de l'énergie libre
~ENGINE CLUTCH - [auto] roue *f* libre
~FALL - [space & rocketry] chute *f* libre
~-FALL BORING - [mining] forage *m* à chute libre
~FALLING - [text] (negative box movement) à chute libre
~-FALLING SKID - [aero] (landing skid designed to drop into the alighting position by its own weight) patin *m* d'atterrissage à ouverture libre
~-FALLING SLAY - [text] battant *m* à chute libre
~FERRITE - [metall] (ferrite in steel or cast-iron) ferrite *f* libre
~FIELD - [phys] (a field in an isotropic medium free from boundaries) champ *m* libre
~-FIELD CURRENT RESPONSE - [el acoust] (of a microphone) réponse *f* en champ libre, intensité *f* en champ libre
~-FIELD VOLTAGE RESPONSE - [el acoust] (of a microphone; ratio of the output electromotive force to the sound pressure in the undisturbed free field) tension *f* en champ libre
~FIT - [mech] (used only in the USA) montage *m* libre
~-FLIGHT WIND-TUNNEL - [aero] (a wind tunnel designed for observation of a model in free light) soufflerie *f* de vol libre
~-FLOATING CAMERA MOUNT - [photo & telev] montage *m* flottant de la caméra
~FLOWING OIL - [ind chem] huile *f* fluide, huile *f* à haute fluidité
~FROM KNOTS - [text] (of a warp) sans nœuds (chaîne)
~GAS - [oil ind] gaz *m* libre
~GRID - [electron] grille *f* flottante
~GYRO - [mech] (a gyroscope not constrained in movement) gyroscope *m* libre
~-HAND - [draw] à main levée
~-HAND DRAWING - [draw] dessin *m* à main levée
~-HAND SKETCHING - s. free-hand drawing
~-HEARTH ELECTRIC FURNACE - [metall] (direct-arc furnace with one electrode forming part of the bottom of the hearth) four *m* à arc avec électrode de sole
~IMPEDANCE - [el acoust] (of an electrochemical transducer) impédance *f* de court-circuit
~IONIZATION CHAMBER - [nucl] (ionization chamber in which radiation passes through the electrodes without striking them) chambre *f* d'ionisation libre
~ISOTOPE HOLD-UP - [nucl] charge *f* ouvrable libre
~LENGTH - [mech] (of a spring) hauteur *f* libre (de ressort)
~LINE SIGNAL - [telecomm] lampe *f* d'inoccupation
~LIQUID KNOCKOUT - [oil ind] (natural gas) séparateur *m* de liquides
~-MACHINING STEEL - [metall] acier *m* de coupe
~MAGNETIZATION - [el](natural magnetization) magnétisation *f* propre
~MILLINGORE - [min] minérais *m* contenant métal à

l'état libre
FREE MILLING QUARTZ - [min] quartz *m* à or libre
~MOISTURE - [min] (in coal testing) humidité *f* brute
[metall] teneur *f* libre d'humidité
~-MOLECULE DIFFUSION - [phys] (the flow of a gas through a long tube at such pressures that the mean free path is greater than the tube radius) diffusion *f* de molécules libres, courant *m* thermique moléculaire
~MOTIONAL IMPEDANCE - [el acoust] impédance *f* cinétique libre
~NEUTRON - [nucl] (netron liberated from the nucleus) neutron *m* libre
~NUCLEAR DIVISION - [genet] (division which is not accompained by the formation of cell walls) division *f* nucléaire libre
~OF BLOWHOLES - [metall] exempt de soufflures, sans soufflures
~OF BUBBLES - [metall] exempt de soufflures, sans soufflures
~OF DUTY - [comm] exempt de droits d'entrée
~-OF-GROUND - [el] exempt de terre, sans mise à la terre
~OF HUM - [telecomm] exempt de ronflement
~OF RIPPLES - [telecomm] exempt d'ondulations
~ON BOARD (FOB) - [comm] franco à bord
~OSCILLATION - [phys] (natural oscillation) oscillation *f* libre, vibration *f* libre, oscillation *f* propre
~PARACHUTE - [aero] (a parachute which is opened by the parachutist, as opposed to one opened automatically on jumping) parachute *m* à ouverture commandée
~PATH - [phys] (the distance travelled by a particle before it collides with another. s. also "mean free path") libre parcours *m* moyen
~-PATH DISTILLATION - [chem] distillation *f* (moléculaire) à libre parcours *m*
~PERIOD OF CIRCUIT - [el] (natural period; the reciprocal of the natural frequency of the circuit) période *f* propre
~PHENOL - [chem] (uncombined phenols present in a phenol-formaldehyde based resin or plastic) phénol *m* propre
~PISTON ENGINE - [mech] (type of piston engine having no connecting-rods or crankshaft, used as gasifiers for gas turbines)moteur *m* à pistons libres
~PISTON GAUGE - [meas] (designed to measure high fluid pressures) manomètre *m* à piston libre
~POINT - [oil ind] point *m* libre
~POINT TESTER - [meas] (electron tube tester) lampemètre *m*
~POLE - [el] (in theory, magnetic pole supposed to be separate from its opposite pole) pôle *m* libre
~PROGRESSIVE WAVE - [el acoust] (wave propagated under conditions which are equivalent to those in an infinite isotropic medium) onde *f* progressive libre
~PROPELLER - [aero] hélice *f* libre
~RADICAL - [phys] (one or a group of atoms containing an unpaired electron) radical *m* libre, radical *m* non saturé
~REED - [mus] (reed which vibrates through an air slot) anche *f* libre
~RING - [mech] bague *f* libre
~ROUTING - [comput] acheminement *m* libre
~-RUNNING - [telev] (system in which the frame frequency is indipendent of the main frequency) système *m* libre
FREE-RUNNING CIRCUIT - [electron] auto-oscillateur *m*, circuit *m* auto-oscillateur
~-RUNNING FREQUENCY - [radio] fréquence *f* propre
~-RUNNING MULTIVIBRATOR - [electron] multivibrateur *m* auto-oscillant
~-RUNNING SPEED - [el] (the speed reached by a vehicle when propelled by a constant tractive effort) vitesse *f* de régime, vitesse *f* d'équilibre
[contr] (balancing speed) vitesse *f* d'équilibre
~-RUNNING SWEEP - [telev] balayage *m* non-synchronisé, déviation *f* non-synchronisée
~SOUND FIELD - [el acoust] (sound field in which the effect of the boundary is negligible) champ *m* acoustique libre
~-SPACE CHARGE WAVE - [electron] (space charge wave with constant energy in the direction of the flux) onde *f* de charge spatiale libre
~-SPACE DIAGRAM - [radio] diagramme *m* de rayonnement direct
~-SPACE FIELD INTENSITY - [el] (the field intensity which in the theory would exist in the absence of waves reflected from the earth or other reflecting bodies) intensité *f* de champ en espace libre
~-SPACE PATTERN - s. free-space diagram
~-SPACE PROPAGATION - [telev] rayonnement *m* direct
~-SPINNING WIND TUNNEL - [aero] (a wind-tunnel provided with a vertical air stream for testing spinning characteristics) soufflerie *f* à courant de air vertical
~STANDING CASING - [oil ind] tubage *m* libre
~STREAM - [phys] écoulement *m* non perturbé
~-STREAM CAPTURE AREA - [phys] surface *f* de captation de l'écoulement non perturbé
~STREAMLINE - [astronaut] frontière de jet libre, ligne *f* de jet
~STUFF - [timber] (free from knots and other blemishes) sans nœuds, net de nœuds
~SULPHUR - [chem] (sulphur in an uncombined state) soufre *m* libre
~SURFACE - [phys] (phase boundary between two fluids) surface *f* libre
~-SURFACE ENERGY - [phys] énergie *f* de surface libre
~SUSPENSION - [mech] suspension *f* libre
~SWINGING LOUDSPEAKER - [acoust] hautparleur *m* à membrane libre
~SWINGING PLANSIFTER - [min] cribleuse *f* à oscillation libre
~-TRADE AREA - [comm] zone *f* de libre-échange
~TRAVEL - [mech] jeu *m*
~TRAVEL OF PEDAL - [auto] course *f* libre de la pédale, garde *f* à la pédale
~TRIP - [el] (a switch in which the tripping mechanism is indipendent of the closing mechanism) à déclenchement libre
~TURBINE - [mech] (type of turbine in which the compressor and the output shaft are driven by separate turbines without mechanical interconnection) turbine *f* libre
~TYING UP - [text] (in a wire heald twisting machine) encordage *m* libre
~VALENCE - [phys] (e.g. the valence of a free radical) valence *f* libre

FREE VECTOR - [contr] vecteur *m* libre
~VIBRATION - [phys] (natural vibration) vibration *f* libre
~VOLUME - [phys & chem] (used in theories of the liquid state) volume *m* libre
~VORTEX PATTERN OF FLOW - [hydr] schéma *m* d'écoulement à tourbillon libre
~WALL - [constr] mur *m* isolé, mur *m* dégagé
~WATER - [phys] (the amount of water which is removed in drying a solid to its equilibrium water content) eau *f* libre
~-WHEEL - [mech] roue *f* libre
[auto] (a one-way cluth in the transmission line) roue *f* libre
~-WHEEL BICYCLE - [mech] bicyclette *f* à roue libre
~-WHEEL CONTROL - [mech] commande *f* de roue libre
~-WHEEL HUB - [mech] moyeu *m* de roue libre
~-WHEEL MECHANISM - [mech] mécanisme *m* à roue libre
~-WHEEL TEST - [auto] essai *m* de la roue libre
~ZING SHAFT - [metall] puits *m* frigorifique
FREEBOARD - [shipbuild] (set out by law to prevent overloading) franc-bord *m*
[auto] garde *f* au sol
FREEDOM - [gen] liberté *f*, franchise *f*
FREEHOLD - [leg] propriété *f* foncière (perpétuelle et libre)
~ESTATE - s. freehold
FREEING PORT - [naut] sabord *m*
FREESTONE - [constr] (building stone which can be worked with a chisel) pierre *f* de taille, grès *m* à bâtir
FREEZE, to [phys] (to change from the liquid to the solid state) geler, congeler, prendre se glacer
[phys] (to change into ice) geler
[food ind] (food storage) congeler
[mech] (to seize) se coller, se cramponner
[oil ind] se coincer
FREEZE DRIER - [ind chem] installation *f* de lyophilisation, poste *m* de lyophilisation, appareil *m* de lyophilisation
~-DRY, to - [ind chem] lyophiliser
~DRYING - [chem] séchage *m* à froid
~-DRYING PROCESS - [ind chem] procédé *m* de lyophilisation
~FRAME - [telev] arrêt *m* d'image, tirage *m* en image fixe
~OUT - [telecomm] (brief refusal of a telephone circuit by a system of speech interpolation) interruption *f* de ligne en tiers
~POINT - [oil ind] point *m* neutre d'une colonne en puits
~-THAW STABLE - [phys] (resistant to repeated freezing and thawing) résistant au gel et à la rosée
FREEZER - [th eng] (electrically or other-wise operated machine producing a permanent temperature lower than freezing point) congélateur *m*, appareil *m* frigorifique
[food ind] compartiment *m* congélateur
[railw] wagon *m* réfrigérant
FREEZING - [gen] congélation *f*
[phys] (the conversion of a liquid into a solid form) congélation *f*, solidification *f*
[metall] (in a furnace cooling continued until the metal is solidified) solidification
[food ind] (cold storage) réfrigération *f*, congélation *f*
FREEZING - [oil ind] calage *m*
~CHEST - [ind chem] armoire *f* frigorifique, congélateur *m* à basse température
~COMPARTMENT - [th eng] compartiment *m* congélateur
~CURVE - [th dyn] (graph of temperature against time) courbe *f* de congélation, courbe *f* de solification
[metall] courbe *f* de fusibilité
~MIXTURE - [chem] (generally ice and salt, producing a temperature below freezing point) mélange *m* réfrigérant
~NUCLEI - [phys] (permanently-solid particles in the atmosphere, e.g. mineral dust, on which water is precipitated by freezing) nucléi *mpl* de congélation
~OUT THE ORE - [min] préparation *f* des minerais par le froid
~POINT - [phys] (the temperature at which a liquid passes into the solid state under given conditions) point *m* de congélation, point *m* de solidification, température *f* de congélation
~PROCESS - [mining] (the freezing of unstable materials in mining) procédé *m* de congélation
~RANGE - [metall] intervalle *m* de solidification
~TEMPERATURE - [phys] température *f* de congélation, température *f* de solidification
F REGION - [astr] (ionospheric region higher than 160 km above the earth) zone *f* F
FREIBERGITE - [min] (obsolete term, formerly used for argentiferous copper-antimony sulphide) freibergite *f*
FREIGHT, to - [naut & comm] fréter, affréter, noliser
FREIGHT - [gen] fret *m*, transport *m*, nolage *m*
[comm] fret *m*, cargaison *f*
[comm] (payment for transport) nolage *m*, nolis *m*, prix *m* du louage
~CAR - [railw] (goods wagon; goods carrying truck) wagon *m* à marchandises
~TARIFF - [comm] tarif *m* marchandises
~TRAIN - [railw] train *m* de marchandises
FREIGHTAGE - [comm] frètement *m*
[comm] (transport of goods) transport *m* des marchandises
FREIGHTER - [aero] avion *m* de marchandises
[naut] navire *m* de charge
[railw] wagon *m* de marchandises
FRENCH BLUE - [paint] (an alternative term for Ultramarine) bleu *m* d'outremer
~BOILER - [metall] chaudière *f* à bouilleurs
[th eng] chaudière à bouilleurs
~CASEMENT - s. French window
~CHALK - [min] (finely-ground talc) talc *m*, craie *f* de Meudon, stéatite
~COLUMN - [ind chem] appareil *m* pour distillation fractionnée
~COMB - [text] (french wool-combing system) peigneuse *f* Heilmann
~CROSS-HEAD - [surv] (a form of cross-staff) équerre *f* d'arpenteur
~DOOR - s. French window
~DRAIN - [build] (a drain formed by partially filling the trench with rubble) puisard *m* à absorption, fosse *f* à absorption
~DRAW LOOM - [text] métier *m* Jacquard

FRENCH GROUND - [text] (in bobbin net fabrics) réseau *m* francais
~GUTTER TILE - [contr] tuile *f* à coulisse à la francaise
~HARP - [mus] harmonica *m*
~LAWN - [text] linon *m* de coton
~OCHRE - [paint] (pale yellow ochre obtained from France) ocre *m*
~PITCH - [text] (in the Jacquard loom) division *f* francaise
~-POLISH, to - [paint] vernir au tampon
~POLISH - [paint] (solution of shellac applied to wood surfaces) vernissage *m* au tampon
~ROOF - [constr] (mansard roof) comble *m* brisé, comble *m* en mansarde
~TRUSS - [constr] (symmetrical roof truss for large spans) ferme *f* à la Mansarde
~WINDOW - [build] (a glazed casement, at once door and window)porte-croisée *f*, porte-fenêtre *f*,
[constr] porte *f* à deux vantaux à gueule-de-loup
FRENCHMAN - [metall] polissoir *m*
FRENOTOMY - [med] section *f* du frein
FRENULUM - [anat] frein *m*, filet *m*
FRENZY - [gen & med] frénésie *f*, délire *m* furieux
FREQUENCY - [phys etc] (the number of vibrations, cycles or waves of any periodic phenomenon, usually per second) fréquence *f*
[gen] fréquence *f*
[el] (in an alternating current, the number of cycles of alternation in a given time, usually one second) fréquence *f*
~ALLOCATION - [el] allocation *f* des fréquences, attribution *f* des fréquences
~ANALYSIS - [el acoust] analyse *f* harmonique
~BAND - [radio] (frequencies lying between certain given limits and allotted for specific radio transmissions) bande *f* de fréquences
~BAND OF EMISSION - [radio] (the band of frequencies which is occupied by the emission) bande *f* de communication
~CALIBRATION - [radio & telecomm] étalonnage *m* de la fréquence
~CHANGE OSCILLATOR - [radio] (used to produce a change of carrier frequency in super-heterodyne reception) oscillateur *m* local
~CHANGE SIGNALLING - [radio] formation *f* de signaux par modulation de fréquence
~CHANGER - [radio] (circuit including a local oscillator and a mixer) changeur *m* de fréquence, mélangeur *m*, tube *m* mélangeur
~-CHANGER CRYSTAL - [electron] (crystal rectifier) cristal *m* convertisseur
~CHARACTERISTIC - [radio] (the relation to frequency of the ratio of the excitation under given conditions) caractéristique *f* de fréquence
~CLOCK - s. frequency meter
~CONTROL - [radio] régulation *f* par la fréquence, correcteur *m* automatique de fréquence
~CONVERSION - [radio] conversion *f* de fréquence
~CONVERTER - [el] (a set for converting alternating current at one frequency to alternating current at another frequency)changeur *m* de fréquence, convertisseur *m* de fréquence, transformateur *m* de fréquence
~CORRECTION - [telecomm] correction *f* de fréquence

FREQUENCY COUNTER - [meas] compteur *m* de fréquence
[instr] compteur-fréquencemètre *m*
~COVERAGE - [comput] (of a receiver) gamme *f* de fréquences, plage *f* de fréquences
~DEMODULATOR - [radio] (mechanism producing an output which is proportional to the variation of the instantaneous frequency of the input voltage) démodulateur *m* de fréquence
~DEMULTIPLICATION - s.frequency division
~DEMULTIPLICATOR - s. frequency divider
~DEPARTURE - [radio] (the variation of a carrier frequencyfrom its given value) excursion *f* de fréquence
~DETECTOR - [instr] détecteur *m* de fréquence, discriminateur *m*
~DEVIATION - [radio] (the peak difference between the instantaneous frequency of the modulated wave and the carrier frequency) déviation *f* de fréquence, désaccord *m* maximum
~-DISCRIMINATING FILTER - [radio] (electric wave-filter capable of discrimination among component frequencies of electric signals) filtre *m* discriminateur de fréquence
~DISCRIMINATION - [radio] (the selection of a desired frequency) sélection *f* d'une fréquence
~DISTORTION - [radio] (signal distortion due to the presence of more than one frequency such frequencies being amplified to different extents) distorsion *f* d'amplitude/fréquence
~DISTORTION CHARACTERISTIC - [radio] caractéristique *f* de distorsion d'amplitude/fréquence
~DISTORTION FACTOR - [radio] facteur *m* de distorsion amplitude/fréquence
~DISTORTION TRANSMISSION IMPAIRMENT - [radio] réduction *f* de la qualité de transmission
~DISTRIBUTION CURVE - [math] courbe *f* des distributions des fréquences
~DIVIDER - [radio] (electrical apparatus with an output frequency which is a submultiple of the frequency at the input) diviseur *m* de fréquence
~DIVISION - [radio] (the production of a current the output frequency of which is an exact submultiple of the frequency at the input) répartition *f* en fréquence
~-DIVISION MULTIPLEX - [radio] (used for the transmission of two or more signals over the same path and using a different frequency band for each signal) multiplexage *m* par répartition en fréquence
~DOMAIN - [phys] domaine *m* de fréquences
~DOUBLER - [radio] (a type of frequency multiplier) doubleur *m* de fréquence
~DOUBLING - [radio] doublement *m* de fréquence
~DRIFT - [radio] (variation in natural frequency, caused by external influences such as temperature changes) glissement *m* de fréquence
~-EXCHANGE KEYING - s. frequency shift keying
~-EXCHANGE SIGNALLING - [radio] formation *f* de signaux par modulation avec mutation de fréquence
~FILTER - s. frequency-discriminating filter
~FLUTTER - [acoust] effet *m* de flottement, effet *m* ondulatoire
~FOR COMPARISON - [telecomm] fréquence *f* de référence
~HALVER - [electron] dédoubleur *m* de fréquence, circuit *m* dédoubleur de fréquence
~HYSTERESIS - [radio] (effect produced by coupling

an oscillator to a load through a transmission line which is very long is compared with a wavelength) effet *m* de ligne longue

FREQUENCY INDICATOR - [meas] (showing when two alternating currents have the same frequency) indicateur *m* de fréquence

~INTERLACE - [telev] (of interfering signal frequencies) entrelacement *m* de fréquences, entrelacement *m* spectral

~INVERTER - [radio] inverseur *m* de fréquence

~JUMPING - [radar] (of a magnetron during the existence of a pulse) saut *m* de fréquence

~LIMIT OF EQUALIZATION - [telecomm] limite *f* de la contre-distorsion

~LIMITATIONS - [telev] (in television recording) limitations *fpl* de fréquence

~LOCK - [electron] verrouillage *m* de fréquence

~LOCUS - [phys] lieu *m* géométrique des fréquences

~MEMORY - [electron] (memory consisting of an oscillator with a number of possible oscillation frequencies) mémoire *f* de fréquences

~METER - [instr] (instrument designed to measure radio frequencies) fréquencemètre *m*

~MODULATED CYCLOTRON - [electron] (electronic device for accelerating charged particles to high energies) cyclotron *m* à modulation de fréquence

~-MODULATED TRANSMITTER - [radio] (transmitter for frequency-modulated wave) émetteur *m* à modulation de fréquence

~MODULATION - [FM - radio] (method of radio transmission in which the signal modifies the frequency of the carrier-wave) modulation *f* de fréquence

~MODULATION DETECTOR - [instr] détecteur *m* de modulation de fréquence, détecteur *m* MF

~-MODULATION STATION MONITOR - [radio] moniteur *m* d'émetteur

~MODULATOR - [radio] (device for the production of frequency modulation) modulateur *m* de fréquence

~MONITOR - [radio] (indicating the departure of an oscillation frequency from its assigned value) moniteur *m* de fréquence

~MULTIPLIER - [radio] (electrical apparatus with an output frequency which is an exact integral multiple of the frequency at the input) multiplicateur *m* de fréquence

~-MULTIPLIER KLYSTRON - [electron] klystron *m* multiplicateur de fréquence

~OF A SOUND WAVE - [phys] fréquence *f* d'une onde acoustique

~OF OSCILLATIONS - [phys] (the number of complete oscillations of a given system, usually per second) fréquence *f* oscillante

~OFFSET TRANSPONDER - [radio] répondeur *m* à décalage de fréquence

~OVERLAP - [telev] (in colour television) bande *f* commune

~PARAMETER - [aero] (in a structure which is subject to mechanical oscillation, the relation to the airspeed of the value obtained by multiplying the frequency of oscillation by the representative length) paramètre *m* de fréquence

~PULL-IN RANGE - [radio] plage *f* de synchronisation

PULLING - [radio] (the displacement of the frequency of a self-maintained oscillation towards the frequency of an applied oscillation) entraînement *m* de fréquence

[electron] glissement *m* aval

FREQUENCY PULSING - [radio] (in oscillators with two likely frequencies of oscillations) variation *f* rythmique de fréquence due au battement

~PUSHING - [electron] glissement *m* à mont

~PUSHING FIGURE - [electron] indice *m* de glissement amont

~RANGE - [phys] (of the frequency spectrum) gamme *f* de fréquences

~RANGE OF A TRANSMISSION SYSTEM - [radio] bande *f* de fréquences transmises par un système de transmission

~RANGE OF EQUALIZATION - [radio] intervalle *m* de contre-distorsion

~RATE - [gen & statist] taux *m* de fréquence

~RECORD - [radio] tableau *m* de fréquences

~RELAY - [el] relais *m* de fréquence

~RESPONSE - [radio] courbe *f* de fréquence, réponse *f* de fréquence

[contr] réponse *f* harmonique

~RESPONSE DISTORTION - [electron] distorsion *f* de la réponse de fréquence

~SCANNING - [comput] balayage *f* de fréquence

~SELECTIVE SIGNALLING - [radio] signalling transmitted to any selected station independently of the others) signalisation *f* sélective en fréquence

~SENSITIVITY - s. frequency hysteresis

~SEPARATING CIRCUIT - [radio] circuit *m* tampon

~-SHIFT KEYING - [radio] (method of telegraphic signalling in which keying is transmitted as change in frequency of the signal) formation *f* de signaux par modulation avec déplacement de fréquence, modulation *f* par déplacement de fréquence

~-SHIFT SIGNALLING - s. frequency-shift keying

~-SHIFT DIPLEX OPERATION - [radio] (twinplex) twinplex *m*

~SLIDING - [radar] (of a magnetron) glissement *m* de fréquence

~SLIP - [electron] glissement *m* de fréquence

~SPACING - [radio] interval *m* de fréquence

~SPECTROGRAPH - [phys] (spectrograph which scans continuosly part of the frequency spectrum) spectrographe *m* de fréquence

~SPECTRUM - [phys] spectre *m* de fréquence

~STABILITY - [electron] (the quality of electronic equipment whereby its output frequency is maintained constant) stabilité *f* de fréquence

~STABILIZATION - [radio] (the prevention of changes in the frequency of oscillation of a self-oscillating circuit) stabilisation *f* de fréquence

~STANDARD - [radio] étalon *m* de fréquence

~SWEEP - s. frequency deviation

~SWING - [radio] balayage *m* de fréquence, swing *m* de fréquence

~TOLERANCE - [radio] (of a radio transmitter) tolérance *f* de fréquence

~TRANSFER UNITS - [radio] (consisting of a heterodyne frequency meter and a heterodyne detector) dispositif *m* de transposition de fréquence

~TRANSFORMER - [el] (transformer arranged for the transformation of frequency of supply) changeur *m* de fréquence, transformateur *m* de fréquence

~TRANSLATION - [radio] transposition *f* de fréquence

~TRANSLATOR - [electron] transpositeur *m* de fréquence

FREQUENCY TRIPLER - [radio] (device delivering output voltage at a frequency three times the input frequency) tripleur *m* de fréquence
~WOW - [acoust] (modulation effect caused by periodic variations of the tape speed) huhulement *m*
FRESCO, to - [paint] (the painting on plastered walls) peindre à fresque
FRESCO - [paint] (method of painting on plastered walls) fresque *m*
FRESH - [gen] frais, nouveau
[dyes] (of colours) frais, fleuri
[hydr] (of water) (eau) fraîche, douce
~AIR INLET - [plumb] (inlet admitting fresh air into a drainage) conduite *f* d'air fraîche
~BREEZE - [met] (wind of a velocity of about 20 m.p.h. Beaufort force 5) bonne brise *f*, brise *f* fraîche
~GALE - [met] (wind of a velocity of about 35 m.p., Beaufort force 8) vent *m* frais, coup *m* de vent
~PAINT - [paint] (recently applied paint, not yet dry) peinture *f* fraîche
~WATER - [chem] eau *f* douce
FRESHEN, to - [gen] rafraîchir
[chem] (the sea water) dessaler
[naut] rafraîchir
[met] (of the wind) fraîchir
[zool] (USA only; to calve) vêler
FRESHENER - [ind chem] (a solvent spread over unvulcanized sheets of rubber before building them up into plies) refraîchissant *m*
FRESNEL - [meas] (of frequencies) fresnel *m*
~BIPRISM - [opt] (very flat prism with two very acute and one very obtuse angles) biprisme *m* de Fresnel
~COEFFICIENT OF DRAG - [mech] coefficient *m* de résistance de Fresnel
~DIFFRACTION - [opt] diffraction *f* de Fresnel
~MIRRORS - [opt] (two plane mirrors, yet not quite in the same plane) miroirs *mpl* de Fresnel
~REGION - [phys] (the region between an aerial and the Fraunhofer region) zone *f* de Fresnel
~RHOMB - [opt] (glass rhomb with about 55 degree angles) parallélépipède *m* de Fresnel
FRET, to - [gen] ronger
[brew ind] fermenter
[metall] (of metals) ronger, corroder, éroder
FRET - [arch] fret *m*, frette *f*, grecque *f*
~-SAW - [impl] scie *f* à découper, bocfil *m*
~-SAWING MACHINE - [mech mach] scie *f* à chantourner, sauteuse *f*
[metall] scie *f* alternative à découper
~-WORK - [build] vitrail *m* historié
[carp] découpage *m*, découpure *f*; ouvrage *m* à claire-voie
FRETS - [mus] (the raised line across the fingerboard of string instruments) touchettes *fpl*
FRETTAGE - [metall] rongement *m*
FRETTED - [gen] orné, sculpté
~LEAD - [metall] ruban *m* en plomb
FRETTING - [gen] érosion *f*, corrosion *f*
[of road surface] rupture *f*
[brew ind] fermentation *f* secondaire infectieuse
~CORROSION - [mech] (corrosion occurring on the surface of coupled elements) corrosion *f* par friction
~FATIGUE - [metall] fatigue *f* de friction, fatigue *f* de frottement

FREYALITE - [min] (aluminium, iron, manganese and sodium containing mineral) fréyalite *f*
FRIABLE - [gen] (liable to crumble) friable
FRIABILITY - [gen] (liability to crumble) friabilité *f*
~TEST - [constr] (test for determining the suitability of a stone in asphalt work) essai *m* de friabilité
FRICATIVE - [acoust] (a consonant sound produced by the friction of the breath between two of the mouth-organs through a narrow aperture) fricative *f*, sifflante *f*
FRICTION, to - [rubber ind] frictionner
FRICTION - [mech] (the resistance set up when one substance slides in contact with another) friction *f*, frottement *m*, glissement *m*
~AND WINDAGE LOSS - [el] (losses in electrical machines caused by the friction of sliding parts) pertes *fpl* de frottement et ventilation
~BEARING - [mech] (a plain bearing) palier *m* à glissement, palier *m* ordinaire, palier *m* lisse
~BRAKE - [mech] dispositif *m* de blocage par frottement
~BRUSH - [rubber ind] brosse *f* à frictionner
~CALENDER - [paper man] supercalandre *f*, calandre *f* à friction
CHECK - s. friction resistance
CLUTCH - [mech] (mechanism for connecting and disconnecting two co-axial shafts under any condition of rotation) embrayage *m* à friction
~COEFFICIENT - [mech] coefficient *m* de frottement
~COMPOUND - [rubber ind] mélange *m* de friction
~CONE - [mech] (conical surface always containing the resultant of the friction between two surfaces and the normal force pressing the two surfaces together) cône *m* de friction
~CONE DRIVE - [mech] commande *f* par cônes à friction
~COUNTERSHAFT - [mech] arbre *m* de renvoi à friction
~COUPLING - [mech] manchon *m* à friction
DISK - [mech] (in autos etc) disque *m* de friction
DISK BRAKE - [mech] frein *m* à friction
~DISK CLUTCH - [mech] embrayage *m* à friction à disques
~DISK SPRING - [mech] ressort *m* à friction
~DRAW-GEAR - [railw] appareil *m* de traction à friction
DRIVE - s. friction gear
~FABRIC - [rubber ind] tissu *m* fractionné
~FACTOR - s. friction coefficient
~FORCE - [phys] force *f* de frottement
~GEAR - [mech] (gear wherby the power is transmitted from one shaft to another the tangential friction between a pair of wheels) transmission *f* par frottement
~GEARING - s. friction gear
~GLAZING - [paper man] (method of glazing with one or more cylinders of the calender revolving at greater speed than others) satinage *m* par frottement
~HEAD - [mech] (loss of head in a pump system due to skin friction in pipes and fittings) hauteur *f* correspondant à la perte par frottement
~-HEADSTOCK - [mach tool] (of a lathe) poupée *f* à frictions
~HORSE-POWER - [mech] (the power which is absorbed in frictional losses) perte *f* par frottement
~LAYER - [met] (the lowest level of the atmosphere,

below I5000 - 3000 ft. in which air movement is affected by friction with the surface of the earth) couche *f* de friction
FRICTION LINING- [mech] garniture *f* de friction
~LOSS - [el] (the power absorbed in any sliding contact of an electrical machine) perte *f* par frottement
~OF MOTION - [phys] frottement *m* en mouvement, frottement *m* pendant le mouvement
~OF REST - [mech] frottement *m* au départ
~PAWL - [mech] cliquet *m* à friction, mâchoire *f*
[of a lock] mordache *f* de serrage
~PAWL GEAR - [mech] blocage *m* par frottement
~PLATE - [mech] (US for friction disk) disque *m* de friction
~PRESS - [mech] presse *f* à frottement
~PULLEY - [mech] poulie *f* de friction
~RING - [mech] bague *f* extensible
~ROLL - [metall] (top roll) cylindre *m* libre
~ROLLER - [mech] (anti-friction bearing) galet *m* de friction, rouleau *m* de friction
~RUNNING LIGHT - [mech] frottement *m* à vide
~SAW - [mech] scie *f* à friction
~SCREW PRESS - [mech] poulie *f* de friction à vis
~SHOCK-ABSORBER - [mech] (type of shock-absorber in which energy is absorbed by friction) amortisseur *m* à friction
~SPRING - [mech] ressort *m* de friction
~SURFACE BELT - [mech] courroie *f* avec surface de frottement
~TAPE - [el] ruban *m* isolant, chatterton *m*
~TEST - [ind chem] essai *m* de frottement
~TORQUE - [mech] couple *m* de frottement
~-TYPE BEARING - [mech] coussinet *m* en métal antifriction
~WELDING - [metall] soudage *m* par friction
~WHEEL - [mech] roue *f* de friction, cylindre *m* de friction
FRICTIONAL - [gen] de frottement, à friction
~BLASTING MACHINE - [mech] exploseur *m* électrostatique
~DAMPER - [mech] amortisseur *m* à friction
~DAMPING - [mech] amortissement *m* à friction
~ELECTRIC MACHINE - [el] (an electrostatic generator) générateur *m* électrostatique
~ELECTRICITY - [el] (term denoting the electric charges which are produced by rubbing together two insulating materials) électricité *f* électrostatique
~RESISTANCE - [mech] résistance *f* de frottement
[text] résistance *f* frictionale, force *f* de frottement
FRICTIONING - [rubber ind] (the operation of impregnating a fabric with rubber) frictionnage *m*
FRICTIONLESS - [phys] sans friction
FRIEDEL-CRAFTS REACTION - [nucl] (a reaction in which nuclear hydrogen atoms are replaced by alkyl radicals) réaction *f* Friedel-Crafts
FRIEZE, to - [text] ratiner
FRIEZE - [arch] (the centre of an entablature) frise *f*
[text] (heavy, rough woollen fabric) ratine *f*, frise *f*
~VELVET - [text] velours *m* frisé
FRIEZED CLOTH - [text] drap *m* frisé, montagnac *m*
FRIEZING - [text ind] ratinage *m*
FRIGORIFIC - [gen] frigorifique
~MIXTURE - [chem] mélange *m* frigorifique
FRILLING - [photo] (on a film, the crinking of the emulsion) décollement *m*, plissement *m*
[text] froncage, plissage *m*
FRINGE, to - [gen & text] franger
[text] (of fabrics) s'effiler, s'effilocher
FRINGE - [gen] frange *f*, bord *m*
[photo] (overlapping of colours) frange *f* (chromatique)
[text] frange *f*, crépine *f*
~AREA - [telev] (the region in which reception of television broadcast is not always adequate) zone *f* limite de propagation
~KNOTTER - [text] noueur *m* de franges
FRINGED - [text] (of a fabric) frangé, effilé
~TRIMMING - [text] ruban *m* frangé
FRINGING - [gen & text] frangeage *m*, effilochage *m*
[el] (in a magnetic field, the spreading of the lines of force at the edges of an airgrap) déformation *f* de champ
[telev] effet *m* de défaut de calage
[electron] (the bulginf of the electrostatic field at the edges of the deflection plates of a cathode-ray tube) déformation *f* de champ
~COEFFICIENT - [el] (in calculation relating to magnetic circuits, it allows for the effect of fringing of the flux) coefficient *m* de dispersion du flux
~MACHINE - [mech] machine *f* à franger
FRISE AILERON - [aero] (aileron fitted to trailing edge of wing, but having its leading edge forward of its hinge axis) aileron *m* de Frise
FRIT, to - [ceramics, glass etc] (to heat a ground or finely-crushed material until individual particles begin to melt at corners and angles and so join together. The temperature is not raised to the general fusion point) fritter
[glass man] (to agglomerate) fritter
FRIT - [ceramics] (flint, glass or sand, calcined and finely ground. It is mixed with glazes or body pastes to prevent the constituents of these from dissolving in the mixing water) fritte *f*
[glass man] (used for filetering) fritte *f*
~FLY - [zool] (a pest) mouche *f* de frit, oscinie *f* ravageuse
FRITTED GLASS FILTERING CRUCIBLE - [ind chem] (type of filtering crucible in which disks of fritted glass powder are used as filter elements) creuset *m* de filtrage à verre fritté
~METAL - [metall] métal *m* fritté, métal *m* sintérisé
FRITTING - [metall] frittage *m*
~FURNACE - [metall] four *m* de frittage
FRITZSCHEITE - [min] (variety of autunite containing manganese) fritzschéite *f*
FROG - [zool] grenouille *f*
[el] (a trolly-frog) aiguillage *m* de trolley
[railw] (intersection point of inner rails) pointe *f* de cœur
[text] (in the weft supply motion) épaulement *m* du butoir, bloc *m* d'arrêt
[text] (in the loom brake) butoir *m*
[text] (of the warp protector) plan *m* incliné
[med] aphte *f*
~AND SWITCH PLANER - [mech] raboteuse *f* pour cœurs d'aiguille
~BELLY - [med] ventre *m* de batracien
~BOX - s. frog bracket
~BRACKET - [text mach] (in a cotton loom) support *m* du butoir
~CLAMP - [railw] serrure *f* d'aiguille

FROG-LEG WINDING - [el] (composite winding consisting of one lap winding and one wave winding connected to the same commutator) enroulement *m* à patte de grenouille
~LINE - [el] (trolley-frog) croisement *m* aérien
~RAMMER - [impl] sauterelle *f*, grenouille *f*, dame *f* sauteuse
FROGMAN - [naut] "homme *m* grenouille"
~SUIT - [rubber ind] costume *m* pour "homme-grenouille"
FROGSUIT - [nucl] (protective garments) vêtements *mpl* de sûreté
FRONT - [gen] front *m*, devant *m*, partie *f* antérieure, façade *f*
[text] (of a shirt) à plastron *m*
[metall] taille *f*
[met] (surface of discontinuity between two adjacent air-currents of different densities) front *m*
[arch] (the frontal part of a building) façade *f*, élevation *f*
[mus] (of an organ) montre *f*
~ANGLE - [tools] angle *m* de résistance
ARMATURE RELAY - [el] relais *m* à armature frontale
~AXLE - [auto] essieu *m* avant
~AXLE RADIUS ROD - [auto] bielle *f* de train avant
~BEARING - [mech] palier *m* antérieur *f*
[in machine tools] portée *f* antérieure (de mandrin)
~BEARING SPRING - [auto] ressort *m* de suspension avant
~BEVEL GEAR - [mech] engrenage *m* conique frontal
~BODY - [print] (of a character) corps *m* (de la lettre) antérieur
~BRAKE - [auto etc] frein *m* avant
~CARRIAGE - [mech] (of vehicles) avant-train *m*
~CLEARANCE - [mech] angle *m* de détalonnage antérieur
~CONTACT - [el] contact *m* frontal, contact *m* de travail
~COVER - [mech] (of engine etc) couvercle *m* antérieur
~DELIVERY - [print] sortie *f* frontale
~DOOR - [constr] porte *f* d'accès principale, porte *f* d'entrée
DRIVE - [mech] traction *f* avant
~ELEVATION - [draw] (drawing showing an object as seen from the front) élévation *f*, vue *f* de face
~END - [railw] boîte *f* à fumée
~END COVER - [auto] carter *m* de distribution
~ENGINE-TRUCK WHEEL - [mech] roue *f* porteuse d'avant
~FENDER - [auto] pare-choc *m* avant
~FOCAL LENGTH - [opt] première distance *f* focale
~FORK - [mech] (of bicycles, autos) fourche *f* avant
~FORK SPRING - [auto] ressort *m* de fourche avant
~FRAME - [auto] châssis *m* avant
~IDLER - [mech] rouleau-tendeur *m* avant
~-LENS FOCUSING - [photo] mise *f* au point par lentille frontale
~LIGHTING - [photo] éclairage *m* de face
~LOADER - [agric] chargeur *m* frontal
~MILL TABLE - [metall] ligne *f* d'amenée
~OF A VAULT - [constr] front *m* d'une voûte
~PANEL - [mech] platine *f* frontale
~PANEL GRILLE - [el acoust] grille *f* de hautparleur
~PART - [gen & vehicles] partie *f* antérieure, partie *f* frontale
FRONT PLATE - [th eng] (of boiler furnace) plaque *f* de devanture
~PORCH - [telev] (part of the video signal at black level) palier *m* avant
~POWER TAKE-OFF SHAFT - [mech] prise *f* de force avant
~PROJECTION - [telev] projection *f* directe
~RELEASE - [mech] débrayage *m* sur le devant
~-RING FOCUSING - [photo] mise *f* au point par anneau frontal
~SCANNING - [cin] (a sound reproduction system) balayage *m* antérieur
~SCREEN - [auto] pare-brise *m*
~SEAT - [gen & auto] siège *m* avant
[gen] (theatres etc) siège *m* aux premières loges
~SHOE - [plast ind] (the upper clamp plate of a mould) plaque *f* supérieure
~SHUTTER - [photo] obturateur *m* frontal
~SIGHT - [firearms] guidon *m* de mire, bouton *m* de mire
~SOARING - [aero] (soaring flight in a glider, obtained by making use of upward air currents before a front) vol *m* plané frontal
~SPAR - [mech] longeron *m* d'avant
~SPRING - [auto] ressort *m* avant
~TAP - [med] réflexe *m* tibial
~TIPPER - [transp] basculeur *m* en avant
~-TO-BACK RATIO - [radio] (in a directional aerial, the ratio of the effectiveness toward the front and the rear) rapport *m* d'atténuation (avant-arrière)
~-TO-REAR RATIO - s. front-to-back ratio
~TOOL - [tool] outil *m* de face
~TOP RAKE - [mech] angle *m* de dégagement supérieur
~TOW HOOK - [auto] crochet *m* de remorquage avant
~VIEW - [gen] vue *f* de face
[arch] élévation *f*
~WALL - [gen & in ovens] mur *m* de face
[constr] (of a vault) mur *m* de front *m*
~WHEEL - [mech] roue *f* avant
~WHEEL DRIVE - s. front drive
~WHEEL SUSPENSION - [auto] suspension *f* avant
FRONTAGE LINE - [constr] alignement *m*
FRONTAL - [arch] fronteau *m*, bandeau *m*
[gen] frontal
~ANALYSIS - [chem] (in gas analysis) chromatographie *f* frontale
~BONE - [anat] os *m* frontal, os *m* coronal; frontal *m*
~MOULDING AREA - [metall] (in die casting) surface *f* de moulage
FRONTGENESIS - [met] (the process which gives rise to a front) frontogénèse *f*
FRONTIER - [gen & geogr] frontière *f*
~STATION - [railw] gare *f* frontière
FRONTISPIECE - [print] (illustration facing the title page of a book) frontispice *m*
[arch] frontispice *m*, fronton *m*
FRONTOLYSIS - [met] (the process which destroys a front) frontolyse *f*
FROST, to - [gen] geler, givrer
[glass men] dépolir
[metall] mater
FROST - [phys] (condition prevailing when the atmospheric temperature is below the freezing point of water) gelée *f*, gel *m*
[met] gelée *f*
[gen] (on glass panes, screens, aircraft wings etc)

glace *m*
FROST HEAVE - [soil] soulèvement *m* dû au gel, gonflement *m* dû au gel
~LINE - [plast ind] (term used for the crystallization point in the extrusion of polyethylene, polypropylene and nylon) givrage *m*
~NAIL - [agric] (for horses) crampon *m*
~PENETRATION - [soil] pénétration *f* du gel
~PRECAUTIONS - [gen] (measures taken against damage by freezing) précautions *fpl* contre le gel
~-PROOF - [gen] résistant au gel
~SHIELD - [auto] (a winter cover for the radiator grill) cache-radiateur *m*, rideau *m* de radiateur
~SMOKE - [met] (fog produced over the sea by the presence of air at a temperature considerably below freezing-point- brouillard *m* maritime
~UPHEAVEL - s. frost heave
FROSTBITE - [med] gelure *f*
[agric] brûlure *f* par la gelée
FROSTED - [gen] givré
[glass man] (treated so as to break up direct rays of light) dépoli
[metall] (of gold) mat
~GELATINE - [photo] gélatine *f* cristallisée
~GLASS - [glass man] verre *m* dépoli
~LAMP - [el] (filament lamp with sand-blasted bulb in order to break up direct light rays) lampe *f* dépolie
FROSTING - [paint] (misty appearance in coatings which have dried in foul atmosphere) matage *m*
FROSTS - [lighting] (frosted glass screens to diffuse light) écrans *mpl* de verre dépoli
FROTH, to - [phys & chem] faire mousser, écumer
[el] (of electric battery) bouillonner
FROTH - [chem] (foam) écume *f*, mousse *f*
FROTH, to OVER - [brew ind] faire déborder la mousse
~SUPPRESSOR - [brew ind] destructeur *m* de mousse
FROTHER - [metall] (substance producing foam in the flotation process) agitateur *m* d'écume
FROTHING - [ind chem] formation *f* de mousse
~AGENT - [chem] générateur *m* d'écume
~MACHINE - [rubber ind] batteuse *f* à mousse, fouetteuse *f*
~PERCENTAGE - [rubber ind] pourcentage *m* de battage en mousse
~UP - [rubber ind] battage *m*
FROTHY SHEET - [rubber ind] (soft sheet containing too many bubbles due to excess of fermentation) "frothy sheet" (feuilles molles contenant des bulles)
FROUDE BRAKE - [instr] (an absorption dynamomèter consisting of a rotor inside a casing) dynamomètre *m* de Froude
~NUMBER - [phys] (a number expressing the relation of the gravitational to inertial forces) nombre *m* de Froude
FROZEN - [gen] gelé, glacé
~BEARING - [mech] grippage *m*
~EQUILIBRIUM - [phys] équilibre *m* figé
~FLOW - [phys] écoulement *m* figé
~FOOD - [food] produits *mpl* alimentaires congelés
~RUBBER - [rubber ind] caoutchouc *m* gelé
FRUCTIFICATION - [bot] (development after fertilization) fructification *f*
FRUCTOSE - [chem] (a natural sugar occurring in a large number of fruits) fructose *f*, sucre *m* de fruit
FRUIT - [gen & bot] fruit *m*
~GRADER - s. fruit grading machine
~GRADING MACHINE - [agric mach] calibreur *m* de fruits
~-GROWING - [agric] fructiculture *f*
-JUICE - [agric] jus *m* de fruits
~PEEL - [agric] pelures *fpl* de fruits
~SHRUB - [bot] arbuste *m* fruitier
~SUGAR - s. fructose
~SYRUP - [agric] sirop *m* de fruits
~TREE - [bot] arbre *m* fruitier
~VINEGAR - [agric] vinaigre *m* de fruits
~WINE - [agric] vin *m* de fruits, cidre *m*
FRUITERER - [comm] fruitier *m*
FRUSTUM - [geom] (that part of a solid, usually a cone or a pyramid, which lies between two parallel planes cutting it) tronc *m*
~OF CONE - [geom] tronc *m* de cône
~OF PYRAMID - [geom] tronc *m* de pyramide
FRUTESCENT - [bot] (having the appearance of a shrub) frutescent
FRY - [acoust] (undesired noises which are accidentally recorded) bruit *m* de microphone, friture *f*
FRYING - [acoust] (undesired noise accidentally recorded, or the noise caused by excessive current passing through the carbon in a telephone transmitter) bruit *m* de microphone, crépitements *mpl*
F SCOPE - [radar] (F display in GB; single signal appearing as a bright spot) indicateur *m* type F
F-SYSTEM - [opt] (method of designating relative apertures) système *m* type F
FUCHSIN - [chem] (a synthetic red dye which also has uses in medicine as an antiseptic) fuchsine *f*
FUCHSITE - [min] (variety or white mica) fuchsite *f*
FUDGE, to - [gen] faire cadrer (des comptes)
FUDGE - [print] (in a newspaper, space reserved for late news) manchette *f* des dernières nouvelles
FUEL, to - [gen] alimenter en combustible, s'approvisionner en combustible
FUEL - [gen] combustible *m*
[for a petrol engine] carburant *m*
[nucl] (fissile material for use in a nuclear reactor core) combustible *m* nucléaire
~ACCUMULATOR - [jet engines] (a device which temporarily stores fuel during the starting phase, so as to give an additional supply when a pre-set fuel pressure is attained) accumulateur *m* de départ
~ADDITIVE - [chem] additif *m* de combustible
-AIR RATIO - [mech] (the ratio between fuel and air in the mixture fed to an engine) rapport *m* combustible-air
~ASSEMBLY - [nucl] (structure of fuel and other material used in some types of reactors) assemblage *m* combustible
~BREEDING CYCLE - [nucl] (process in which the core material is regenerated and brought into operation again) cycle *m* du combustible
~CAN - [auto] bidon *m* à carburant, nourrice *f* à carburant
~CAN NOZZLE - [auto] bec *m* verseur de nourrice
~CANISTER - s. fuel can
~CAPACITY - [railw] (of a locomotive) capacité *f* de la soute à combustible
~CHANNEL - [nucl] (a hole through which fuel is introduced into a nuclear reactor) canal *m* de combustible

FUEL CHARGE TUBE - [nucl] tuyau *m* de charge de combustible
~CLEANER - [chem] épurateur *m* du carburant
~COCK - [mech] (a valve used to interrupt flow of fuel) robinet *m* de carburant
~CONSUMPTION - [mech] (the amount of fuel consumed, e.g. for unit time, distance or work) consommation *f* de carburant
~CONTROL - [nucl] commande *f* par combustible
~CORRECTION - [metall] clause *f* du charbon
~CUT-OFF - [mech] (a device for cutting off the supply of metered fuel) clapet *m* de ralenti
~CYCLE - s. fuel breeding cycle
~DAMAGE - [nucl] déterioration *f* du combustible
~DELIVERY - [auto] alimentation *f* en essence
~DELIVERY PIPE - [auto] tuyau *m* d'alimentation en carburant, tuyau *m* d'alimentation en essence
~DEPLETION - [nucl] appauvrissement *m* du combustible
~DISTANCE - [aero] (the distance which can be flown by an aircraft without further fuelling) distance *f* franchissable sans ravitaillement en combustible
~DISTRIBUTOR - [th eng] distributeur *m* de combustible, répartiteur *m* de combustible
~DOPE - [chem] anti-détonant *m*
~ELEMENT - [nucl] (a portion of the reactor core) élément *m* combustible
~ELEMENT CORE - [nucl] noyau *m* d'un élément combustible
~ELEMENT NOZZLE - [nucl] distributeur *m* du réfrigerant
~FEED PIPE - [auto] tuyau *m* d'alimentation en essence
~-FEEDING PUMP - [mech] (for internal combustion engines) pompe *f* à essence
~FILTER - [mech] filtre *m* à carburant
~FLOW TEST - [aero] essai *m* de débit de combustible
~FLOWMETER - [instr] (instrument measuring the rate of flow in a fuel system) debitmètre *m* de combustible
~GAS - [th eng] gaz *m* combustible
~GAUGE - [auto] jauge *f* de niveau d'essence
~GRADE - [chem] (the "knock-rating" quality of a fuel) indice *m* d'octane
~HEATER - [auto] appareil *m* de réchauffage du carburant
~HOSE - [auto] tuyau *m* d'alimentation en essence, tuyau *m* à essence
~INJECTION - [int comb eng] injection *f* de carburant
~INJECTION EQUIPMENT - [auto] dispositif *m* d'injection
~INJECTION PUMP - [mech] (pump used to spray fuel into the combustion chambers of a Diesel engine) pompe *f* d'injection
~INJECTION VALVE - [mech] soupape *f* d'injection
~-JETTISON GEAR - [aero] (equipment for discharging fuel overboard rapidly in emergency) vide-vite *m* réservoirs
~LEVEL GAUGE - [instr] jauge *f* de niveau d'essence
~LEVEL INDICATOR - s. fuel level gauge
~LIFETIME - [nucl] vie *f* de combustion
~LINE - [auto] tuyau *m* d'alimentation en essence
~MAKE-UP - [nucl]recharge *f* de combustible
~MANIFOLD - [mech] (branching pipe supplying fuel to turbine burners) rampe *f* d'alimentation de combustible
FUEL MIXTURE - [fuel] (for small engines) mélange *m*
~NOZZLE - [mech] pulvérisateur *m* du carburant
~-OIL - [th eng] (petroleum fraction used in furnace burners) mazout *m*
~-OIL SETTLING TANK - [mech] bac *m* de décantation du mazout
~-OIL STRAINER - [mech] filtre *m* mazout
~PELLET - [nucl] (pellet-shaped fuel) comprimé *m* de combustible
~PENCIL - [nucl] aiguille *f* de combustible
~PRESSURE GAUGE - [instr] (instrument measuring the pressure in a fuel system) manomètre *m* de combustible
~PRESSURE PIPE - [mech] alimentation *f* en essence sous pression
~PRESSURE SWITCH - [el] (electrical device designed to prevent the starter motor from receiving full current before the fuel pressure has reached pre-set level) monocontact *m* de basse pression de alimentation en combustible
~PROCESS CELL - [nucl] casematte *f*, cellule *f* chaude
~PROCESSING - [nucl] traitement *m* de combustible
~PUMP - [mech] (in autos etc) pompe *f* d'alimentation, pompe *f* à essence
~RATING - [nucl] puissance *f* spécifique
~RECONDITIONING - s. fuel regeneration
~REGENERATION - [nucl] (the treatment of used core material in order to use it again) traitement *m* de combustible irradié
~REPROCESSING - s. fuel regeneration
~ROD - [nucl] (long slender fuel element in a nuclear reactor) barre *f* de combustible
~ROD COATING - [nucl] barrière *f* de diffusion, couche *f* protectrice de barre de combustible
~SHUT-OFF - [mech] (system designed to stop the flow of fuel to an engine in emergency, e.g. fire) coupe-feu *m*
~SHUT-OFF COCK - [mech] (the valve which cuts off the fuel in a shut-off system) robinet *m* coupe-feu
~SLUG - [nucl] (lump of nuclear fuel inserted into holes or channels in the active lattice of a reactor) barreau *m* de combustible
~SOLUTION - [nucl] (a solution containing fuel in a homogeneous reactor) solution *f* de combustible
~SPRAY - [mech] (in internal combustion engines) jet *m* de carburant
~SYSTEM - [aero] (the group of units supplying fuel to an aircraft engine) circuit *m* carburant
~TANK - [gen & aero] (the vessel which contains the fuel in an aircraft) réservoir *m* à combustible
~TANKER - [transp] (a self-propelled tank vehicle for supplying fuel to aircraft on the ground) camion-citerne *m* à carburant
~TAP - s. fuel cock
~TESTER - [chem] analyseur *m* du carburant
~TRANSFER POND - [nucl] chambre *f* de transfert de combustible
~TUBE - [nucl] tube *m* de combustible
FUELLED LOOP - [nucl] circuit *m* actif
FUGACITY - [phys] (quantity measuring the true escaping tendency of a gas) fugacité *f*
FUGACIOUS - [bot] (which lasts for a very short

time) caduc, éphémère
FUGATIVE COLOURS - [paint] couleurs *fpl* fugatives
~PIGMENTS - [paint] (pigments which lose their colour under given conditions) pigments *mpl* fugatifs
FUGITOMETER - [paint] (apparatus for testing the resistance of colours to light, using ultra-violet radiation) fugitomètre *m*
FUGUE - [mus] (a musical style) fugue *f*
FULCHRONOGRAPH - [instr] (instrument designed to register the characteristics of lightning currents) fulchronographe *m*
FULCRUM - [mech] (the pivot about which a lever turns) palier *m*, support *m*
[constr] point *m* d'appui
~PIN - [mech] (a pin used as a fulcrum in a mechanism) axe *m* de pivotement d'un levier
FULGURATION - [el biology] (obsolete for electrodesiccation) fulguration *f*
FULGURATOR - [instr] (used in spectroscopy) pulvérisateur *m*
FULGURITE - [min] (a tubular bodyproducedby lightning in loose sand and caused by the vitrification of sand grains) fulgorite *m*
FULIGINOUS - [gen] fuligineux
FULL - [gen] plein, comble, rempli
[gen] (of a period of time, e.g. one year) complet, entier
~ADVANCE - [mech] (the ignition setting at which the spark occurs at the earliest point in the range) pleine avance *f*
~AGEING - [metall] vieillissement *m* complet
~ANNEAL, to - [metall] recuire
~ANNEALING - [metall] (of steel) recuit *m* complet
~-AUTOMATIC ELECTRO-PLATING - [el metall] (method of plating in which the cathodes move automatically through a series of cleaning and plating baths) galvanoplastie *f* automatique
~-AUTOMATIC WORKING - [telecomm] (in telephony) exploitation *f* en automatique intégrale
~BLEACH - [text] blanchiment *m* complet
~BLOSSOM - [bot & agric] en pleine fleur
~-BORE VALVE - [plumb] robinet *m* à écoulement libre
~-CENTRE ARCH - [constr] (semicircular) arc *m* en plein ceintre
~CIRCLE MOULD - [rubber ind] (for tyres) moule *m* circulaire entier
~CIRCLE SPEAR - [mining] harpon *m* de repêchage
~CURE - [rubber ind] vulcanisation *f* complète, cuisson *f* totale
~-CUT HEADING MACHINE - [mining] excavateur *m* à section entière
~DEFLECTION - [mech] déviation *f* maxima (instrument de mesure)
~DEPTH - [mech] (of a gear tooth) hauteur *f* totale
~DIP - [geol] (of a bed) plongée *f*
~DROP CENTRE RIM - [auto] jante *f* à base creuse, jante *f* à dépression médiane
~-FASHIONED - s. fully fashioned
~-FAT CHEESE - [dairy ind] fromage *m* à pâte grasse
~-FILLED WELDING - [metall] soudure *f* à cordon complet
~FINISH - [text] apprêt *m* complet
~-FLIGHTED SCREW - [plast ind] (an extruder screw in which the flight extends over the whole length of the screw) vis *f* à pas sur toute la longueur

FULL FLOATING AXLE - [auto] axe *m* entièrement flottant
~-FORCE FEEDING - [mach] graissage *m* forcé
~-HOLE TOOL JOINT - [oil ind] joint *m* à passage total
~HARD TEMPER - [metall] trempe *f* dure
~HARDENING - [metall] trempe *f* à coeur
~LENGTH SOLE - s. full sole
~LINER - [oil ind] chemise *f* d'une pièce
~LOAD - [transp & aero] charge *f* totale
[el] charge *f* nominale, pleine charge *f*
~LOAD METER ADJUSTMENT - [el] (device for the regulation of the moving element speed at the normal voltage) dispositif *m* de réglage aux grands débits d'un compteur
~-MAGNETIC CONTROLLER - [contr] (electric controller operated by electromagnets) combinateur *m* tout-électro-magnétique
~MATURITY - [bot agric and brew ind] pleine maturité *f*
~MOON - [astr] (when the moon is directly opposite the sun) pleine lune *f*
~MOULDED TYPE RUBBER BEARING - [rubber ind] monobloc *m* rainuré, coussinet *m* en caoutchouc
~-ON POSITION - [el] position *f* de fonctionnement total
~ORGAN - [mus] orgue *m* à plein jeu
~PIPE - [hydr] conduite *f* forcée
~-PITCH WINDING - [el] (armature winding in which the span of the coil is equal to a pole pitch) enroulement *m* à pas diamétral
~RADIATOR - [opt] (black body) corps *m* noir, radiateur *m* intégral
~RATE - [gen] plein tarif *m*
~RETARD - [mech] (ignition setting at which the spark occurs at the latest point in the range) plein retard *m*
RETREADING - [rubber man] (of tyres) rechapage *m* de talon, rechapage *m* total
~-RIGGED - [naut] gréé en trois-mâts carré
~ROTATION OF A SELECTOR - [telecomm] (in telephony) rotation *f* à bout de course d'un sélecteur
~-SCALE - [draw] de grandeur naturelle
~SCALE DEFLECTION - [instr] pleine déviation *f*, lecture *f* totale, déviation totale
~-SCALE VALUE - [contr] (the largest value of the actuating quantity which can be indicated on the scale) valeur *f* maximale nominale d'une grandeur de calculatrice
~-SIZE - [draw] grandeur *f* naturelle
~SOLE - [shoe man] semelle *f* complète
~-SPEED - [gen] à toute vitesse
SPIRAL TAPPING - [rubber ind] (of tyres) saignée *f* en spirale entière
~STEEPED - [brew ind] assez mouillé
THROTTLE - [mech] (setting at which gas admission to an engine is at maximum value) gaz *mpl* ouverts en grand
~-THRUST DURATION - [astronaut] durée *f* de la pleine poussée
~TIMBERING - [mining] boisage *m* complet
~WAVE - [phys] onde *f* pleine
~-WAVE RECTIFICATION - [radio] redressement *m* biphasé
~-WAVE RECTIFIER - [radio] (device for rectifying an alternating current so as to obtain unidirectional flow from both halves of the cycle) redresseur

m biphasé
FULL-WAY VALVE - s. full-bore valve
FULLCAPPING - [rubber ind] (of tyres) rechapage *m* à mi-flanc
FULLED CLOTH - [text] tissu *m* foulé
FULLER - [tool] dégorgeoir *m*
[metall] dégorgeoir *m*
FULLER'S CELL - [el] (a primary cell of the same composition as the bichromate cell, but in which the electrolyte and depolarizer are separated by a porous pot and the zinc electrode stands in a pool of mercury to maintain amalgamation) cellule *f* de Fuller
~EARTH - [min] (naturally-occurring highly absorbent clay-like material with uses as a catalyst in the production of dehydrated castor oil and as emulsifying and thixotropic agent in surface coatings) terre *f* à foulon, argile *f* smectique
~HERB - [bot] (medicinal plant) savonnière *f*
~THISTLE - [bot] (a weed) cirse *m* lancéolé
FULLERING - [mech] (caulking a riveted joint) matage *m*, refoulement *m*
~TOOL - [mech] (tool used to produce circumferential grooves on a circular work) matoir *m*, dégorgeoir *m*, matrice *f*
FULLING - [text] (milling; a preliminary process in the finishing of woollen fabrics) foulage *m*, foulement *m*
[rubber ind] (of a tyre) déformations *fpl* sous charge
~AGENT - [chem] (used for milling the fabric) adjuvant *m* de foulement
~DEFLECTION - [of a tyre] déformation *f* sous charge
~FELT - [text] feutre *m* foulé
~HAIR - [text] bourre *f* de foulage
~MACHINE - [text] foulon *m*, machine *f* à fouler, fouleuse *f*
~STRAIN - s. fulling deflection
FULLNESS - [gen] (also spelt fulness) plénitude *f*, état *m* plein, ampleur *f*
[of a colour, a smell etc] intensité *f*
FULLY AUTOMATIC REPERFORATOR SWITCHING - [radio] (reperforator switching system whereby the automatic transmitter is automatically switched to a selected out-going channel) commutation *f* automatique avec retransmission par bande perforée
~CONTAINED EXPLOSION - [nucl] explosion *f* complètement enfermée
~FACTORED LOAD - [phys] (the greatest load which a structure must withstand) charge *f* maximum
~INTERMESHED NETWORK - [telecomm] réseau *m* maillé
~LOADED - [gen] à pleine charge
~PROTECTIVE TUBE HOUSING - [nucl] gaine *f* de tube à protection totale
FULMINATE OF MERCURY - [chem] fulminate *m* de mercure
FULMINATES - [chem] (explosive salts of isocyanic acid employed as detonators) fulminates *mpl*
FULMINATING -CAP - [light] amorce *f* fulminante
~GOLD - [chem] (an explosive; obtained from a solution of gold chloride treated with ammonia) or *m* fulminant
~MERCURY - s. fulminate of mercury
~OIL - [chem] nitroglycérine *f*
FULMINIC ACID - [chem] acide *m* fulminique
FULVOUS - [zool & bot] fauve
FUMARIC ACID - [chem] (intermediate for synthetic resins for paints and plastics, in food processing, and as a dyeing mordant) acide *m* fumarique
FUMAROLE - [geol] (small vent on the flank of volcanic cones) crevasse *f* de sortie de fumerolles, fumerolle *f*
~ACID - [chem] acide *m* borique
FUMARYL CHLORIDE - [chem] (intermediate for drugs, insecticides and dyes) chlorure *m* de fumaryle
FUME - [gen] exhalaison *f*, fumée *f*, vapeur *f*
~CUPBOARD - [chem] (enclosed glazed chamber with sliding door and connexion to flue or fan, in which apparatus emitting fumes is placed during laboratory operations) canal *m* d'aspiration, sorbonne *f*
~DISPOSAL - [gen] élimination *f* des exhalaisons
~HOOD - [ind chem] (enclosing structure designed to catch and draw off fumes from an operation) hotte *f*
~-TIGHT OPERATION - [rubber ind] procédé *m* en système étanche au gaz
FUMELESS DISSOLVING - [nucl] solution *f* sans fumée
FUMES - [gen] fumée *f*, exhalaison *f*, vapeurs *fpl*
~OF SULPHUR - [chem] vapeurs *fpl* de soufre
FUMIGANTS - [chem] (substances which used in the vapour or gaseous state can destroy pests and moulds or act as disinfectants) produits *mpl* fumigatoires
FUMIGATION - [chem] fumigation *f*
[med] fumigation *f*
[chem] désinfection *f*, méchage *m*
FUMIGATOR - [med] appareil *m* fumigatoire
FUMING - [chem] (of liquids which give off vapour) fumant, émettant de la vapeur
~LIQUIDS - [chem] (liquids giving off vapour) liquides *mpl* fumants
~SULPHURIC ACID - [chem] (solution of sulphur trioxide in concentrated sulphuric acid) acide *m* vitriolique
"FUMULUS" - [met] (term sometimes used for dense fog due in part to the accumulation of smoke over cities. Also "smog") fumulus *m*
FUNCTION, to - [gen] fonctionner, agir
FUNCTION - [gen] fonction *f*
[math] fonction *f*
[biol] fonction *f*
~CHART - [comput] table *f* de fonctions, diagramme *m* de fonctions
~CODE - [comput] (the instruction item designating the operation to be performed) code *m* fonctionnel, commande *f* de fonctions
~DIGITS - [comput] (digits determining the arithmetical or logical operation to be performed) digits *mpl* de fonctionnement
~GENERATOR - [comput] (mechanism producing the value of a specified function as the indipendent variable increases) générateur *m* de fonctions
~KEY - [comput] touche *f* à fonction
~MULTIPLIER - [comput] multiplicateur *m* de fonctions, multiplicateur *m* fonctionnel
~NUMBER - s. function code
~TABLE - [math] (tabulation of the values of a mathematical function for a set of values of the indipendent variable) tabulation *f* de fonctions

FUNCTION TRANSLATOR - [comput] traducteur *m* de fonction
FUNCTIONAL - [gen] fonctionnel
~ARCHITECTURE - [arch] architecture *f* fonctionnelle
~DIAGRAM - [el] diagramme *m* fonctionnel d'un circuit
[comput] (diagram consisting of functional symbols and the connexion between them) diagramme *m* fonctionnel
~MODULE - [comput] module *m* fonctionnel
~SYMBOL - [comput] (logical symbol) symbole *m* fonctionnel
FUNCTIONING - [gen & mech] fonctionnement *m*
~TEST - [ind proc] (working test of an item of equipment to ensure that it will perform correctly in service) essai *m* de fonctionnement
~VALUE - [el] données *fpl* de fonctionnement
FUND - [gen] fonds *mpl*, caisse *f*
FUNDAMENTAL - [gen] fondamental
~BAND - [phys] (the spectral band produced when the vibration energy of a molecule changes from an initial to a higher level) niveau *m* de vibration originale
~CIRCUIT - [radio] (principal circuit) schéma *m* de principe
~COLOURS - [paint] (red, yellow and blue pigments) couleurs *fpl* fondamentales
~COMPONENT - [el] (of an alternating wave, the harmonic component with the lowest frequency, usually representing the major portion of the wave) composante *f* fondamentale
~COMPONENT DISTORTION - [radio] distorsion *f* de amplitude
~ELECTRICAL UNIT - [el] unité *f* électrique fondamentale
~FIELD PARTICLE - [phys] (generic term denoting an electron, a proton, a positron etc) particule *f* élémentaire
~FREQUENCY - [el] (the frequency of the fundamental component of an alternating wave) fréquence *f* fondamentale
~INTERVAL - [meas] (used to define a temperature scale) intervalle *m* fondamental
~MODE - [radio] (of a waveguide, the mode with the lowest critical frequency) mode *m* dominant, mode *m* fondamental
~PARTICLE - [phys] (any particle which cannot be demostrated to contain simpler units) particule *f* fondamentale
~TISSUE - [bot] (the ground tissue) tissu *m* parenchymateux
~TONE - [acoust] note *f* fondamentale
~UNIT - [phys] (the unit-of length, time etc- on which a system of unit is based) unité *f* fondamentale
~WAVELENGTH - [radio] (corresponding to the fundamental frequency of an aerial)longueur *f* d'onde fondamentale
FUNGICIDAL - [bot agric chem] (of a substance which destroys fungi) fungicide *m*
~PAINTS - [paint] (coatings containing substances which prevent the growth of fungi, e.g. pentachlorphenol) peintures *fpl* fungicides
FUNGICIDES - [chem] (substances which destroy fungal growths) fungicides *mpl*
FUNGINERT - [chem] résistant aux moisissures
FUNGISTATS - [chem] (substances which inhibit the growth of fungi) produits *mpl* fungicides
FUNGOID DISEASE - [bot] fongosité *f*
FUNGUS OF THE BRAIN - [med] hernie *f* cérébrale, encéphalocèle *f*
~-RESISTANT - s. fungicidal
~-RESISTANT VARNISH - s. fungicidal paints
FUNICLE - [bot & zool] funicule *m*, cordon *m*
FUNICULAR - [gen transp] (type of cable-railway) funiculaire *f*
~POLYGON - [constr] polygone *m* funiculaire
~RAILWAY - [railw] funiculaire *f*
~WORKING IN TWO DIRECTIONS - [transp] funiculaire *f* à va-et-vient
FUNIFORM - [gen & bot] (rope-like) funiforme
FUNNEL - [gen & ind chem] (device in the form of a hollow cone, used for pouring liquids from one vessel to another) entonnoir *m*
[a shaft for ventilating purposes] cheminée *f*, tuyau *m* d'aérage
[metall] (of a mould) trou *m* de coulée, jet *m* de coulée
~-BLOCK - [mech] poulie *f* à chape simple
~BREAST - [anat] thorax *m* en entonnoir, thorax *m* de cordonnier
~CASING - [constr] enveloppe *f* de cheminée
~CHEST - s. funnel breast
~CLOUD - [impl] (device in the form of a hollow cone, used for pouring liquids from one vessel to another) entonnoir *m*
~HOLDER - [ind chem] (stand to support a funnel, e. g. for filtering) support *m* de filtre
~-SHAPED - [gen] en entonnoir
~-SHAPED AERIAL - [radio] (shaped as an inverted pyramid) antenne *f* à entonnoir
~-SHAPED ANTENNA - s. funnel-shaped aerial
~STAND - [photo] porte-entonnoir *m*
~-TYPE AERIAL - s. funnel-shaped aerial
~-TYPE ANTENNA - s. funnel-shaped aerial
~WITH FILTER - [impl] entonnoir *m* avec filtre
FUR - [gen] fourrure *f*, pelleterie *f*, poil *m*
[plumb] (the incrustation in pipes) incrustations *fpl*, entartrage *m*, calcin *m*
[metall] (on metal surfaces) fourrure *f*
FURAL - s. furfural
FURAN - [chem] (a synthesis intermediate; it contains a ring of four carbon atoms and one oxygen atom) furanne *m*
~RESINS - [chem] (polymers of furfuryl alcohol and also phenol-furfuryl resins. They have good chemical resistance and are also used as modifiers for other synthetic resins) résines *fpl* furanniques
FURBISH, to - [gen] polir, fourbir
FURBISHING - [gen & carp] polissage *m*, fourbissage *m*
FURCAL - [gen & anat] fourché
FURCATE - [gen & bot] bifurqué
FURFUR - [med] (dandruff) furfure *f*
FURFURACEOUS - [med] furfuracé
FURFURAL - [chem] (furfuraldehyde. A solvent for cellulose derivatives intermediate for nylon and synthetic resins, fungicide and weed-killer) furfural *m*
FURFUROL - s. furfural
FURFURYL ACETATE - [chem] (a synthetic flavourant) acétate *m* de furfuryle
~ALCOHOL - [chem] (a solvent for resins and dyes

and an intermediate for synthetic resins) alcool *m* furfurylique
FURL, to - [naut] serrer, ferler
FURLONG - [meas] (one eighth of a mile) furlong (201 *m*)
FURNACE - [gen] four *m*, fourneau *m*
[th eng] (of a boiler) foyer *m*
[metall] four *m*, pot *m*
~BLACK - [ind chem] (a form of carbon black used as a filler and reinforcing agent in synthetic rubbers) noir *m* fourneau, noir *m* de carbone
~BOTTOM - [metall] sole *f* de four
~BRAZING - [metall] brasage *m* au four
~BRIDGE - [metall] autel *m*, pont *m*
~CHARGE - [ind proc] (the whole amount of material treated in a furnace in a single operation) charge *f*, charge *f* d'un four
~CLINKER - [th eng] (the residue of the combustion of coke etc) mâchefer *m*
~COAL - [metall] charbon *m* pour fours métallurgiques
~CROWN - [ovens] couronne *f* d'un fourneau
[metall] ciel *m* du foyer
~DELIVERY TABLE - [metall] tablier *m* à rouleaux
~DOOR - [metall] porte *f* du foyer
~END - [metall] extrémité *f* du four
~FLUE - [metall] carneau *m* du four
GAS - [metall] combustible *m* de gueulard
~HOIST - [mech] monte-charge *m*
~LINING - [metall] (made of heat-, abrasion and chemical acid resistant materials) revêtement *m* du foyer
~OIL - [oil ind] (distillate fuel principally used for domestic heating) combustible *m* pour chaudières
~PLATE - [metall] tôles *f*pl pour foyers
~SCALE - [metall] calamine *f* de four
~SHAFT - [metall] cuve *f* de four
~SHELL - [metall] revêtement du four
~TOP - [th eng] orifice *m* de chargeur (de haut fourneau)
~-TOP BELL - [metall] cloche *f* de haut fourneau
~-TOP HOPPER - [metall] trémie *f* de chargement de haut fourneau
~WITH MOVABLE FLOOR - [th eng] four *m* à sole mobile
~WITH MOVABLE HEARTH - [metall] four *m* sole mobile
~WITH MULTISTAGE GRATE - [metall] foyer *m* avec grille à étages
~WITH STEPPED GRATE - [metall] foyer *m* avec grille à gradins
FURNISH, to - [gen] fournir, pourvoir, garnir
[comm] (to supply) fournir, aprovisionner
FURNISH - [paper man] (the raw material treated in the beater) charge *f*
FURNISHING FABRIC - [text] étoffe *f* d'ameublement
FURNITURE - [gen] mobilier *m*, meubles *m*pl
[constr] (all fittings for doors, windows etc) ferrures *f*pl
[print] (lengths of wood or metals for margins) garniture *f*
~AND SPACING MATERIALS - [print] blancs *m*pl
~CABINET - [print] lingotier *m*, lingotière *f*
~LACQUER - [paint] encaustique *f* pour les meubles
~LEG-PAD - [rubber ind] embout *m* pour pied de meubles
~REST-PAD - s. furniture leg-pad
FURRED - [plumb] (of a pipe the inner surface of which is affected by hard lime deposit) incrusté, entartré
[arch] fourru
FURRING - [plumb] (of pipes) calcin *m*, tartre *m*
[arch] fourrure *f*
[shuipbuild] soufflage *m*
FURROW, to - [gen & agric] labourer, rayonner, creuser des sillons
[carp etc] canneler, rainer
~-SLICE - [agric] motte *f*, glèbe *f*
~WHEEL - [agric] roue-support *f*
FURROWING - [metall] formation *f* des cannelures
FURS - [gen] fourrures *f*pl, pelleterie *f*
FURUNCLE - [med] furoncle *m*
FURUNCOLOSIS - [med] furonculose *f*
FURYLACRYLIC ACID - [chem] (an intermediate in the production of perfumery compounds) acide *m* furylacrylique
FURZE - [bot] ajonc *m*
FUSAIN - [chem] (the constituent of coal, commonly alternating with durain etc., derived from plant remains from which the volatile elements have been naturally eliminated) fusain *m*
FUSE, to - [gen] fondre, mettre en fusion, fusionner
FUSE - [el] (current-limiting device generally consisting of a length of wire of low melting-point and suitable resistance) coupe-circuit *m*, coupe-circuit à fusibles
[mining] fusée *f*, amorce *f*
[min] (small waterproof canvas tube containing gunpowder) étoupille *f*, mèche *f*
~BLOCK - [el] porte-fusible *m*
~-BOARD - [el] (distribution fuse-board) panneau *m* des fusibles
~-BOARD STRIP - [el] conducteur *m* fusible
~BOX - [el] (a box in which fuses are mounted) boîte des fusibles
~CARRIER - s. fuse block
~CLIPS - [el] douilles *f*pl de fusible
~ELEMENT - [el] fusible *m*, conducteur *m* fusible
~HOLDER - [el] (device in which a fuse is mounted) porte-fusible *m*
~LINK - [el](fuse element; the principal part of the cut-out) élément *m* de remplacement
~PANEL - s. fuse board
~PANEL BLOCK - [el] (in electrical machines, autos etc) boîte *f* des fusibles
~PLUG - [el] fusible *m* à fiche
~STRIP - [el] conducteur *m* fusible
~SWITCH - [el] commutateur *m* à fusible
~TONGS - [impl] pince *f* à fusible
~WIRE - [el] fil *m* pour fusibles
FUSED - [gen] fondu
~ALUMINA - [chem] alumine *f* fondue
~BORAX - [chem] borax *m* fondu
~DRIERS - [paint] (driers obtained by heating the ingredients together, as distinct from those made by precipitation) siccatifs *m*pl fondus
~ELECTROLYTE - [el;chem] (electrolyte consisting of a fused anhydrous compound) électrolyte *m* fondu
~ELECTROLYTE CONTAINER - [el chem] cuve *f*
~ELECTROLYTE CELL - [el chem] (primary cell having a molten electrolyte) pile *f* à électrolyte fondu
~JUNCTION - [electron] (formed by the recrystallization) jonction *f* par fusion recristallisée

FUSED JUNCTION TRANSISTOR - [electron] transistor *m* à jonction à alliage
~MASS - [min] masse *f* fondue, fonte *f*
~ORE - [min] masse *f* fondue, fonte *f*
~SALT - [el chem] (molten anhydrous electrolyte) électrolyte *m*, sel *m* fondu
~-SALT REACTOR - [nucl] réacteur *m* à sels fondus
FUSEE - [mech] (a cable tensioner for a railway berth) tambour *m*, poulie *f* conique
[horol] (a spirally grooved pulley) fusée *f*
[ind chem] (wind-resting match) allumette-tison *f*, tison *m*
[signal] signal *m* lumineux
[astronaut] amorce *f* d'allumage, allumeur *m*
~ARBOR - [horol] arbre *m* de la fusée
~BARREL - [horol] (barrel used for a fusee) tambour *m* de fusée
~ENGINE - s. fusee lathe
~LATHE - [mach tool] (special lathe for cutting fusees) petit tour *m* pour fusées
FUSEL OIL - [min] (a naturally occurring mixture of amyl alcohols with uses as an intermediate, raw material, and solvent) huile *f* de fusel
FUSELAGE - [aero] (the main body of an aircraft except in the case of an amphibian or flying boat, in which case it is called the hull) fuselage *m*
~BOX - [aero] (box-structure forming part of fuselage) caisson *m* de fuselage
~FRAME - [aero] bâti *m* du fuselage
~STRUT - [aero] étrésillon *m* de fuselage
FUSES AND PROTECTOR BLOCKS - [el] groupe *m* de protection
FUSIBILITY - [gen phys metall] fusibilité *f*
FUSIBLE ALLOYS - [metall] (alloys which fuse at temperatures between 180° C. and 60° C. They usually contain lead, tin and bismuth, sometimes with cadmium or mercury, and are used for safety devices and the like) alliages *mpl* fusibles
~CLAY - [geol] argile *f* fusible
~CONES - s. fusion cones
~CUT-OUT - [el] coupe-circuit *m*
~METAL - [metall] métal *m* fusible
~PLUG - [metall] (screwed into the crown of a furnace to prevent excessive over-heating) rondelle *f* fusible
~WIRE - [el] fil *m* fusible
FUSIFORM - [gen] fusiforme
FUSILLADE - [firearms] fusillade *f*
FUSING AGENT - [metall] agent de fusion
~BURNER - [metall] brûleur *m* à couper les métaux
~FACTOR - [el] (the minimum current necessary to blow a fuse) coefficient *m* de fusion
~OVEN - [metall] four *m* de fusion
~POINT - s. fusion point
~STRIP - [el] conducteur *m* fusible
FUSION - [phys] (the change from the solid to the liquid state) fusion *f*
[metall] (die casting) vitrification *f*
[metall] (a defect on a casting; having a vitrified surface) pièce *f* de fonderie vitrifiée
FUSION BOMB - [nucl] bombe *f* à hydrogène, bombe *f* H
~CONES - [meas] (also called Seger cones; small cones made of clay and oxide mixtures used in determining temperature in ceramic furnaces) cônes *mpl* de fusion, cônes *mpl* de Seger
~CUTTING - [metall] sciage *m* par friction
~ENERGY - [nucl] (fusion power) énergie *f* de fusion
~FACULTY - [opt] pouvoir *m* de fusion
~FREQUENCY - [opt] fusion *f* de sensations de couleurs
[phys] fréquence *f* de fusion rétinienne
~FUEL - [nucl] combustible *m* à fusion
~HEAT - [phys] chaleur *f* de fusion
~POINT - [phys] (melting point) point *m* de fusion
~PROCESS - [metall] (method of obtaining metallic aluminium from bauxite by fusing with sodium carbonate precipitation with carbon dioxide and electrolytic refining) procédé *m* de fonte, procédé *m* de fusion
~REACTION - [nucl] (a nuclear reaction in which energy is produced by the fusion of nuclei) réaction *f* de fusion nucléaire
~REACTOR - [nucl] réacteur *m* à fusion
~WELDING - [metall] (welding made without the application of heat) soudage *m* par fusion
~WELDING WITH PRESSURE - [el metall] (fusion welding with static or dynamic pressure to complete the union) soudage *m* à pression par fusion
FUST - [arch] (the shaft of a column) fût *m*
FUSTET - [bot] (plant used for curing) sumac *m*, arbre *m* à perruque
FUSTIAN - [text] (generic term for heavily wefted cotton fabrics) futaine *f*
FUSTIC - [paint] (yellow pigments obtained from natural sources) fustet *m*, bois jaune
FUSTY - [brew ind] moisi
FUTTOCK - [shipbuild] (a middle timber of a ship frame) genou-allonge *m*
~SHROUD - [naut] (small shroud securing the lower dead-eyes and futtock plates of topmast rigging to a band on a lower mast) jambe *f* de hune
~STAFF - [naut] bastet *m*
FUTURE - [gen] futur, à venir
[noun] avenir *m*
FUTURES - [comm] cotations *fpl* à terme, livraisons *fpl* à terme
FUZZY - [photo & gen]flou
~NOSE OF COP - [text] pointe *f* de cannette éboulée
~PICTURE - [photo & cin] image *f* floue
FWWMR - [text](denotes textile fabrics which are Fire, Water, Weather and Mildew resistant) résistant au feu, à l'eau, aux intempéries et à la moisissure

G

g - [meas] (an abbreviation for gramme) gramme *m*
[chem] (symbol for osmotic coefficient) symbole du coefficient d'osmose
[phys] (symbol of acceleration due to gravity at the surface of the earth) "g"
G - [phys] (symbol for constant of gravitation) symbole de constante d'attraction universelle
[chem] (in dyestuff names) jaune
[chem] (symbol for thermodynamic potential) symbole de potentiel thermodynamique
[met] (for whole gale) vent *m* volant
Ga - s. gallium
GAB - [tool] (pointed tool used for hard stone) pied-de-biche *m*
[mech] (of an eccentric rod) encoche *f*
[constr] (grosse) pointe *f*
[constr] (a knobbling iron) ciseau *m* pointu
GABARDINE - [text] (a twill fabric) gabardine *f*
GABARIT - [draw] gabarit *m*
GABBART - [constr] bois *m* équarri, bois *m* carré
~SCAFFOLDING - [build] échafaudage *m* à bois carré
GABBLE - [acoust] (the sound made by geese) criaillement *m*, bredouillement *m*
GABBRO - [geol] (Italian name for a rock composed mainly of felspar and diallage) gabbro *m*
GABBROIDAL - [geol] (of rock) gabbroïdal
GABERDINE - s. gabardine
GABERS SCAFFOLD - s. gabbart scaffolding
GABION - [gen] (wicker basket filled with earth for fortification purposes) gabion *m*, corbeille *f*
[el] bobine *f* en lattis, gabion *m* en flanc, en flanc de panier
GABLE - [constr] (triangular section of outside wall) pignon *m*, gable *m*
[constr] (of a gothic cathedral) pignon *m* ajouré
~-BOARD - [constr] (ornamental board under the gable of a roof) gable *m*, gâble *m*
~END - [constr] pignon *m*
~MOULDING - [constr] moulure *f* de gable
~ROOF - [constr] (ridge roof ending in a gable-end) comble *m* sur pignons, comble *m* à dos d'âne
~TILE - [constr] (roofs covering the intersection between gable and roof) tuile *f* de pignon, tuile *f* de rive
GABLET - [constr] (small decorated gable) gablet *m*
GABLING - [electron] (a form of aperture illumination) répartition *f* du champ en cloche, répartition *f* triangulaire du champ
GABLÖCK - [impl] (more correctly spelt: gavelock; an iron crowbar) pied-de biche *m* en fer
GAD - [min] (short chisel used to break ore) coin *m*
~PICKER - [mining] coin *m*, perforatrice *f* à main
~TONGS - [impl] pince *f* de mineur
GADFLY - [zool] (an insect) taon *m*
GADGET - [gen] (any small mechanical device) accessoire *m*, machin *m*
GADOLINITE - [min] (a naturally-occurring silicate of beryllium, iron and yttrium, sometimes with cerium: found in pegmatites as greenish or brownish black crystals) gadolinite *f*
GADOLINIUM - [chem] (a rare earth element, symbol Gd. A. N. 64, A. W. I57.3I, trivalent. Used as an oxygen and nitrogen scavenger in metallurgy. It is only known in combination, and is obtained from the same sources as Europium) gadolinium *m*
~OXIDE - [chem] (a catalyst and ingredient of special glasses and ceramics) oxyde *m* de gadolinium
GAFF - [impl] (a staff armed with an iron hook) gaffeau *m*
[naut] (a type of spar) corne *f*
~SAIL - [naut] voile *f* à corne
GAFFER - [gen] (a foreman) conducteur *m* des travaux, chef *m* d'équipe
[cin] (in a sound-film studio, the head electrician) chef-électricien *m*
[glass man] (a blower or a foreman) verrier *m*, souffleur *m*
GAG, to - [mech] (to stop up a valve) obturer, obstruer
[railw] (the straightening of rails by means of a gag) dresser (les rails)
GAG - [gen] bâillon *m*
[railw impl] barre *f* pour dresser les rails
[med] (a gadget thrust into the mouth to keep it open) ouvre-bouche *m*
[mech] (of a valve) engorgement *m*, obstruction *f*
[min] (obstruction in the bucket of a pumping set) engorgement *m*
GAGATE - [min] (jet) jais *m*
GAGE - [U.S.] s. gauge
GAGGER - [metall] (a lifter, or dabber) crochet *m* de fonderie, tirette *f*
[metall] (used to hold the mould) clou *m* de mouleur
GAGGING - [metall] dressage *m*, redressage *m*
GAGGLE - s. gabble
GAGING - [US] s. gauging
GAHNITE - [min] mineral of the spinel group) gahnite *f*
GAIN, to - [gen] gagner
[carp] faire une jonction à serrage
GAIN - [gen] gain *m*
[el] (increase of power provided by the insertion of an amplifier) gain *m*
[radio] (the relation of the input of a circuit, e.g. an amplifier, to its output) gain *m*, coefficient *m* de amplification

GAIN - [radio & acoust] (the increase, in decibels, in transmission) gain *m* de transmission
[carp] (in a wall, or timber) mortaise *f*
[contr] (the steady-change of output for a unit of input variation at a specified working point) coefficient *m* d'amplification, gain *m*
[mining] recoupe *f*
~ADJUSTING AMPLIFIER - [radio] amplificateur *m* à limitation automatique de volume
~-BAND - [radio] bande *f* pour amplification
~-BAND MERIT - [radio] largeur *f* de bande pour amplification
~-BANWIDTH PRODUCT - [radio] (the product of amplification of an amplifier stage at midband by the bandwidth of the amplifier in megacycles) produit *m* gain/largeur de bande passante
[telev] produit *m* amplification/largeur de bande
~CHARACTERISTIC - [electron] caractéristique *f* de gain, courbe *f* de gain
~CONSTANT - [contr] coefficient *m* d'amplification
~CONTROL - [radio] (the control which adjusts the degree of amplification) réglage *m* de l'amplification
~CONTROL CHARACTERISTIC - [radio] (in reception) caractéristique *f* de réglage de gain
~CROSSOVER FREQUENCY - [contr] fréquence *f* de recouvrement d'amplification
~EQUATION - [phys] (equation for the study of ion and gas motions) équation *f* de gain
~MARGIN - [radio] (stability criterion in feed-back systems) plage *f* de sécurité dans l'amplification
[electron] marge *f* de gain
~MEASURING SET - [radio] (a device for the measure of the gains of repeaters) kerdomètre *m*
~OF AN AMPLIFIER - [electron] (the ratio of input in a given system, expressed in the same unit) coefficient *m* d'amplification
~-SENSITIVITY CONTROL - [radio] atténuateur *m* sélectif
~SPEED, to - [mech] (particularly in electrical machines) se mettre en vitesse
GAIT - [gen] (the manner of moving, walking, particularly of horses) allure *f*, démarche *f*
~UP LAP - [text] (for joining the warp threads) tissu *m* à rappondre
GAITER - [text & leather ind] (a covering of the ankle) guêtre *f*
[mech] (protective covering of a leaf spring in autos) gaine *f* de ressort
GAITERS - [auto] (on half swinging axles) gaine *f* souple d'étanchéité
GAITING - [text] (preparing a loom for weaving) montage *m* (du métier)
~UP THE WARP - [text] montage *m* de la chaîne
GAL - [meas] (acceleration unit) gal *m*
GALACTANS - [chem] anhydrides *mpl* du galactose
GALACTIC - [astr] (relating to the galaxy) galactique
~CIRCLE - [astr] (the great circle of the celestial sphere) cercle *m* galactique
~CONCENTRATION - [astr] (the crowding of the stars towards the galactic plane) concentration *f* galactique
~COORDINATES - [astr] (the two spherical coordinates referred to the galactic plane) coordonnées *fpl* galactiques
~PLANE - [astr] (the plane passing through the centre of the galaxy) plan *m* galactique
GALACTOCELE - [med] (a cystic swelling) galactocèle *f*
GALACTOMETER - [meas] (instrument designed to measure the density of milk) galactomètre *m*, lactodensimètre *m*
GALACTOPHLEBITIS - [med] œdème *m* blanc douloureux
GALACTOSE - [chem] (a naturally occurring sugar with applications in medicine) galactose *f*
GALACTOSTASIS - [med] retention *f* lactée
GALALITH - [chem] galalithe *f*
GALATEA - [text] (white cotton material striped in blue, used for children's sailor suits) coutil *m* blanc à raies
GALAXY - [astr] (the milky way) Galaxie *f*, voie *f* lactée
[astr] (the star dust, gases etc within which the sun moves) galaxie *f*
GALBANUM - [chem] (a resin) galbanum *m*
GALBULUS - [bot] galbule *m*
GALE - [met] (a wind of force between 7 and I0 in the Beaufort Scale (5I to I00 kph) vent *m* fort, grand vent *m*, coup *m* de vent, vent *m* volant
~WARNING - [met] avis *m* de tempête, signal *m* de tempête
GALEATE - [bot etc] (shaped like a hood) casqué
GALEIFORM - s. galeate
GALENA - [min] (a natural lead sulphide occurring, often with zinc blende, in grey cubic crystals in mineralized veins. It often contains up to 1 p.c. of argentite (natural silver sulphide) which is isomorphous with it. It is the commonest ore of lead) galène, plomb *m* sulfuré
~DETECTOR - [radio] (crystal detector with an active material consisting of galena) détecteur *m* à galène
~RECEIVER - [radio] récepteur *m* à galène
GALILEAN TELESCOPE - [opt] (a telescope with a divergent lens as an ocular) lunette *f* de Galilée
GALILEE - [arch] galilée *f*, porche *m* (d'église)
[constr] portail *m* (ou porche) d'église
GALIPOT - [paint] (alternative name for Bordeaux turpentine) galipot *m*, térébenthine *f* de Bordeaux
GALL, to - [gen] écorcher, excorier
[mech] écorcher
GALL - [bot] (abnormal growth on a plant) galle *f*
[physiol] (bile) fiel *m*
[vet] (injury caused by harness pressure) blessure *f*, excoriation *f*
[glass man] (layer of molten sulphates) fiel *m* de verre, suint *m*
[text] défaut *m*, éraillure *f*
[anat] vésicule *f* biliaire
~BLADDER - [anat] vésicule *f* du fiel, vesicule *f* biliaire
~EXTRACT - [chem] extrait *m* de noix de galle
~NUT - [bot] noix *f* de galle
~SICKNESS - [vet] anaplasmose *f*
~-STONE - [med] calcul *m* biliaire
GALLED - [mech] (worn through friction) écorché, enlevé par frottement
[agric] dénudé
[vet] frayé
GALLERY - [gen] galerie *f*, tunnel
[constr] galerie *f*, balcon *m*, véranda *f*
[min] (tunnel in a coal mine) galerie *f* en rocher
[el] galerie *f*

GALLERY - [theatre] poulailler *m*
[arch] (a large hall) estrade *f*
[constr] (long balcony projecting beyond a wall) galerie *f*
[instr] (in meters; US for top chamber) compartiment *m* supérieur, tête *f*
[hydr] (a filter gallery) galerie *f* filtrante
[light] (used to support a lamp-globe) griffe *f*
~FRAMEWORK - [mining] boisage *m* en porte
~FURNACE - [metall] four *m* de galère, galère *f*, four *m* à tunnel
~OF A CHURCH - [constr] déambulatoire *m*
GALLET - [constr] (a splinter of stone) éclat *m*
GALLETING - [constr] (also called garreting-insertion of gallets into joints) calage *m*
GALLEY - [shipbuild] (a type of vessel) galère *f*
[naut] (in a ship) cuisine *f*, coquerie *f*
[aero] (compartment in aircraft fitted and used for cooking) cuisine *f*
[print] (steel tray holding the type matter after setting) galée *f*, violon *m*
~PRESS - [print] presse *f* à bras
~PROOF - [print] épreuve *f* en placard, épreuve *f* en première
GALLIC ACID - [chem] (an intermediate for pharmaceuticals,dyeing agent, analytical reagent, and intermediate for photographic chemicals and tanning agents) acide *m* gallique
GALLING - [mech] écorchure *f*
[metall] (fretting corrosion due to natural seizure of two metal surfaces during sliding) enlèvement *m* par frottement, écorchure
[mech] (fretting due to faulty lubrication) grippage *m*
[oil ind] éraillure *f*
GALLIOLINO - [paint] (alternative term for Naples yellow; lead antimoniate) jaune *m* de Naples
GALLIUM - [chem] (a metallic element, A.N. 3I, A. W. 69.9, symbol Ga) gallium *m*
GALLOCYANINE - [chem] (a biological stain) gallocyanine *f*
GALLON - [meas] (4,54 lit. in G.B. and 3,78 lit. in the USA) gallon *m*
GALLONAGE - [meas] quantité *f* en gallons
GALLOON - [text] galon *m*
GALLOPING - [mech] (US term denoting irregular running of an engine due to very rich mixture) fonctionnement *m* irrégulier
GALLOTANNIN - [chem] acide *m* tannique
GALLOWS - [gen] potence *f*, gibet *m*
[min] (a timber framework to support the roof) chevalement *m*
[print] support *m* de feuille de couverture
[naut] (spare mast supports) potence *f*
[mining] cadre *m* incomplet, plancher *m*
~ARM - [telev] potence *f* pour projecteur
~FRAME - [min] (timber set supporting the roof) chevalement *m*
[mining] (head frame) chevalement *m*
~TIMBER - s. gallows frame
GALOSH - [shoe man] galoche *f*, couvre-chaussure *m*
[shoe man] (in boots) claque *f*
GALVANI'S EXPERIMENT - [el biol] (the production of a muscular contraction by closing a circuit formed with a muscle in simultaneous contact with two dissimilar metals) expérience *f* de Galvani

GALVANIC - [el] galvanique
~ACTION - [el] action *f* galvanique
~BATH - [el chem] bain *m* galvanique
~CELL - [el] (obsolete term for a primary cell) élément *m* galvanique
~CORROSION - [el metall] corrosion *f* galvanique
~COUPLE - [el] couple *m* galvanique
~CURRENT - [el] (obsolete, denoting an essentially steady unidirectional current) courant *m* galvanique, courant *m* voltaïque
~ELECTRICITY - [el] (obsolete term for electric current) courant *m* électrique
~POLARIZATION - [el] polarisation *f* galvanique
GALVANICALLY ISOLATED - [electron] à isolement *m* galvanique, isolé galvaniquement
GALVANISM - s. galvanization
GALVANIZATION - [el biol] (the use of a galvanic current for biological effects) galvanisation *f*, galvanisme *m*
[metall] galvanisage *m*, galvanisation *f*, zingage *m* au trempé
GALVANIZE, to - [metall] galvaniser, zinguer, plaquer par galvanoplastie
GALVANIZED COPPER WIRE - [metall] fil *m* de cuivre galvanisé (ou zingué)
~SIGNAL STRAND - [metall] câble *m* à torons (7 ou I9 fils)
~STEEL DECKING - [metall] caillebotis *mpl* galvanisés
~STRIP - [metall] bande *f* galvanisée
GALVANIZING - [el metall] (the process of coating steel or iron with zinc, either by immersion in a bath of molten zinc, or by deposition from a solution of zinc sulphate, to give protection against corrosion) galvanisation *f*, étamage *m*, zingage *m*
~PROCESS - [metall] procédé *m* de galvanisation
GALVANNEALING - [metall] recuit *m* après galvanisation
GALVANOCAUTERY - [med] galvanocautère *m*
GALVANOIONIZATION - [med] ionothérapie *f*
GALVANOMETER - [instr] (instrument designed to measure small currents) galvanomètre *m*
~CONSTANT - [el] (used to give a reading of current in amperes) constante *f* galvanométrique
~RECORDER - [instr] (instrument consisting of mirror and coil suspended in a magnetic field) enregistreur *m* de galvanomètre
~SHUNT - [el] (used to reduce the galvanometer sensitivity) shunt *m* du galvanomètre
~WITH MOVING MAGNET - [instr] galvanomètre *m* à aimant mobile
GALVANOMETRIC - [el] galvanométrique
~RELAY - [el] relais *m* galvanométrique
GALVANOMETRY - [el] (the process of measuring the strength of an electric current) galvanométrie *f*
GALVANOPLASTIC - [el chem] galvanoplastique
GALVANOPLASTICS - [el] (the process of coating a substance with metal by galvanism) galvanoplastie *f*
GALVANOPLASTY - s. galvanoplastics
GALVANOSCOPE - [instr] (instrument designed to indicate the passage of an electric current) galvanoscope *m*
GALVANOTAXIS - s. galvanotropism
GALVANOTHERAPY - s. galvanization
GALVANOTHERMY - [el] (the process of producing heat by electricity) galvanothérmie *f*
GALVANOTROPISM - [el biol] (the response of an

organism to an electrical stimulus) galvanotropisme *m*

GAM - [fishing] (a school of whales) troupe *f* de baleines
[naut] (meeting of whalers at sea) réunion *f* en mer, soirée *f* des pêcheurs de baleine (de différents bateaux)

GAMBOGE - [chem] (naturally occurring gum with uses as a yellow pigment) gomme-gutte *f*
~GUM - s. gamboge

GAMBREL - s. gambrel roof
~ROOF - [build] (a mansard roof) toit *m* en croupe

GAME - [zool] gibier *m*
~COCK - [zool] (a cock trained to flight) coq *m* de combat

GAMEKEEPER - [gen] garde-chasse *m*

GAMETE - [biol] gamète *m*

GAMETIC NUMBER - [genet] (the number of chromosomes in the nucleus of a gamete) nombre *m* gamétique

GAMETOBLAST - [genet] gamétoblaste *m*

GAMETOCIDE - [genet] gamétocide *m*

GAMETOGAMY - [genet] gamétogamie *f*

GAMETROPIC - [genet] (relating to the movement of organs before or after fertilization) gamétropique

GAMMA - [gen] (the third letter of the Greek alphabet) gamma *m*
[photo] (for a film emulsion, the intensity of the contrast) gamma *m*
~ACID - [chem] acide *m* gamma
~ANTIMONY - [chem] antimoine *m* commun, antimoine *m* gamma
~ARSENIC - [chem] arsenic *m* ordinaire
~BACKSCATTER THICKNESS METER - [nucl] épaisseurmètre *m* à rétrodiffusion gamma
~BRASS - [metall] (an alloy of zinc and copper, stable at N.T.P. when containing between 60 p.c. and 68 p.c. of zinc) laiton *m* gamma
~COMPENSATION - [nucl] compensation *f* du rayonnement gamma
~CONTAMINATION INDICATOR - [nucl] signaleur *m* de contamination gamma
~CORRECTION - [photo telev etc] (the restoration of the correct gradation) correction *f* du gamma
~CROSS-SECTION - [phys] section *f* efficace pour rayons gamma
~EMITTER - [nucl] (atom with a radioactive decay process associated with the emission of gamma rays) émetteur *m* gamma
~ERROR - [telev] (gradation distortion) distorsion *f* des demiteintes, erreur *f* de gradation
~FLUX - [nucl] flux *m* gamma
~HEATING - [nucl] (thermal effect of gamma radiation in a nuclear reactor) chauffage *m* par rayonnement gamma
~IRON - [metall] (polymorphic form of iron existing between 906° and I406° C. Non-magnetic. It has a face-centred cubic lattice and is the basis of austenitic solid solutions) fer *m* gamma
~LEAKAGE PEAK - [nucl] pic *m* d'échappement, pic *m* de fuite
~MILKER - [nucl] séparateur *m* répétiteur d'isotope
~QUANTUM - [nucl] (quantum of electro-magnetic radiation) quantum-gamma *m*
~RADIATION - [nucl] (radiation consisting of high-energy photons) rayonnement *m* gamma
~RADIOGRAPHY - [radiat] (radiography by means of gamma rays) gammaradiographie *f*

GAMMA RAY - [phys] (electro-magnetic rays emitted by radioactive substances; their frequencies are generally higher than those of X-rays) rayonnement *m* gamma
~-RAY ACTIVITY - [nucl] (the emission of gamma rays) rayonnement *m* gamma
~-RAY CAPSULE - [radiat] capsule *f* à rayonnement gamma
~-RAY CONTAINER - [radiat] (thick container allowing a safe handling) bombe *f* gamma
~-RAY LOGGING - [oil ind] (form of radioactivity logging which is based on the variations of natural gamma radiation resulting from the different minute amounts of radioactive material in a formation. The log thus obtained distinguishes the different layers of rock) diagraphie *f* de rayons gamma
~-RAY PHOTON - [phys] (gamma-rays are composed of high energy photons) photon *m* gamma
~-RAY SOURCE - [radiat] (used in radiology) source *f* de rayonnement gamma
~-RAY SOURCE STRENGTH - [radiat] intensité *f* de source gamma
~-RAY SPECTROMETER - [instr] (instrument designed to measure the energy distribution of gamma rays) spectromètre *m* du rayonnement gamma
~RAY SPECTRUM - [phys] (sharp lines corresponding to intensity and energy characteristics of the source) spectre *m* du rayonnement gamma
~SPACE - [phys] (an Euclidean hyperspace) espace *m* de phase, extension *f* en phase
~SPECTROSCOPY - [phys] (spectroscopy using gamma rays as a source of radiation) spectroscopie *f* gamma
~URANIUM - [nucl] (allotropic modification of uranium metal becoming stable over 770 centigrades) uranium *m* gamma
~VALUE - [photo] valeur *f* de gamma

GAMMAGRAPHY - s. gamma radiography

GAMMASCOPE - [instr] gammascope *m*

GAMMON, to - [food ind] (to cure bacon by salting and smoking) saler et fumer
[naut] faire la liure du beaupré

GAMMON - [food ind] lard *m* fumé, quartier *m* de lard fumé, jambon *m* fumé
~THE BOWSPRIT, to - [naut] faire la liure du beaupré

GAMMONING - [naut] (the lashing of ropes fastening the bowsprit to the cutwater) liure *f*

GAMOGENESIS - [biol] (reproduction by the union of sexual elements) gamogénèse *f*

GAMOPETALOUS - [bot] (of petals united by their edges) gamopétale

GAMOTROPISM - [biol] (the tendency of gametes to attract one another) gamotropisme *m*

GAMUT - [acoust] (the range of frequencies associated with a given type of reproduction) gamme *f* diatonique
[mus] clavier *m* (de la clarinette)

GANDER - [zool] jars *m*

GANG - [gen] groupe *m*, troupe *f*
[mech etc] (group of set of appliances arranged to operate as one) série *f*
[ind] (group of workers performing the same task) équipe *f*
~CONDENSERS - s. ganged capacitors
~CULTIVATOR - [agric] cultivateur *m* polysoc
~CUTTERS - [mach] (set of cutters arranged on the

same spindle) fraise *f* multiple
GANG DIE - [metall] matrice *f* multiple
˜DRILL - [mach tool] machine *f* à percer multiple, perceuse *f* à broches multiples
˜JOB CARD - [comput] carte *f* de travail-équipe
˜MACHINING - [mach tool] usinage *m* en série
˜MILLING - [mach tool] (the use of several cutters on one spindle) fraisage *m* des pièces en série
[metall] fraisage *m* multiple
˜MOULD - [metall] (mould in which several units can be cast simultaneously) estampe *f* multiple
˜PLOUGH - [agric mach] charrue *f* polysoc
˜PRESS - [mech] presse *f* à matrices multiples
˜PUNCH, to - [comput] perforer en série
˜PUNCH - [tool] poinçonneuse *f* multiple
[comput] perforatrice *f* récapitulative, perforatrice *f* bloc
˜SAW - [timber] (a number of parallel saws secured in one frame) scie *f* multiple, scie *f* à plusieurs lames
SLITTING MACHINE - [mech] découpeuse *f* en feuillards
˜SLITTING MACHINE - [metall] laminoir à bandes d'acier
˜SUMMARY PUNCH - [comput] (a type of punching machine) perforateur *m* totalisateur, reperforatrice *f* récapitulative
˜SWITCHES - [el] (a number of electrical switches connected together) commutateurs *mpl* à commande unique
˜TOOL - [mach tool] (holder with a number of cutters each one of which cuts deeper than the one ahead) porte-outil *m* à broches multiples
˜-TUNING CAPACITOR - s. ganged capacitors
GANGED CAPACITORS - [radio] (two or more variable tuning capacitors mounted on the same shaft. Each capacitor is destined to tune a different circuit) condensateurs *mpl* en ligne, condensateurs *mpl* jumelés
˜CIRCUITS - [radio] (two or more tuned circuits coupled mechanically so that their resonance frequencies are adjusted by one control) circuits *mpl* à commande unique
˜CONDITION - [electron] régime *m* de synchronisme
˜CONTROL - [contr] commande *f* mécanique bloquée
GANGING - [radio] (mechanical coupling of the tuning controls) accouplement *m* mécanique
[el & mech] couplage *m* mécanique, couplage *m* jumelé
[mining] roulage *m*
˜OSCILLATOR - [radio] (oscillator giving a constant output used for testing purposes) oscillateur *m* décalé
GANGLIECTOMY - [med] gangliectomie *f*
GANGLIOMA - [med] gangliome *m*
GANGLION - [anat] (aggregation of nerve cells) ganglion *m*
[med] (localized cystic swelling) kyste *m* synovial
GANGLIONOSTOMY - [med] kystostomie *f*
GANGPLANK - [gen & naut] passerelle *f*
GANGRENE - [med] (virtual death of part of the body) gangrène *f*
˜IN WOOD - [constr] carie *f* du bois, gangrène *f*, échauffures *fpl* dans le bois
GANGUE - [min] (a worthless mineral found in lodes and veins) gangue *f*, roche *f* mère
GANGWAY - [gen] passage *m*, couloir *m* central
GANGWAY - [naut] passerelle *f*
[naut] (in the bulwark of a vessel) coupée *f*, passavant *m*
[min] (an elevated roadway) galerie *f* maîtresse, voie *f* principale
˜BETWEEN COACHES - [railw] organes *mpl* d'intercirculation, passerelle *f*
˜LADDER - [naut] échelle *f* de coupée
˜PORT - [naut] coupée *f*
GANISTER - [geol] (also "gannister") ganister *m*
[min] (a siliceous sedimentary rock used as a refractory material) ganister *m*
(the mixture used to line furnaces etc) coulis *m*, matériau *m* réfractaire
GANTRIES - [cin] (or cat-walk; elevated platform for the men in charge of lighting) passerelle *f*
GANTRY - [gen] chevalet *m* de levage, portique *f*
[constr] châssis *m*, lattis *m*
[mech] (an erection for the support of cranes etc) portique *m*, beffroi *m*, charpente *f* en forme de portique, pont *m* roulant
[astronaut] portique *m*
[radar] (colloq for a radar aerial array) ensemble *m* d'antennes radar
˜CRANE - [mech] grue *f* à portique
GAP - [gen] trou *m*, ouverture *f*, vide *m*, intervalle *m*, interstice *m*
[el] (between the electrodes) distance *f*, intervalle *m*
[el] (the short gap between the ferromagnetic parts of a magnetic circuit; preferably called 'air gap') entrefer *m*, espace *m* d'interaction, trajet *m* de décharge
[mech] (between rolls; the space between the rolls of a machine, e.g. a calender) espace *m* entre les rouleaux
[mech] (between the ends of a piston ring) jeu *m*, ouverture *f*, écartement *m*
[aero] (the interval between the leading edge of the upper plane of a multiplane and its projection on the chord of the plane below it) jeu *m* à la coupe (d'un segment)
[comput] (between two informations) intervalle *m*
[gen] (in fortification or lines) brèche *f*
[dent] (between teeth) écartement *m*
[med] brèche *f*, lacune *f*
[mach tool] coupure *f*
[metall etc] renard *m*
˜ARRESTER - [el] (a type of lightning arrester) parasurtension *f* à entrefer
˜AT JOINT - [railw] ouverture *f* du joint de rail
˜BED - [mech] (lathe bed with a gap near the headstock) banc *m* rompu
˜BRIDGE - [mech] (bridge casting of the same cross-section as the bed in a gap-bed lathe, used to close the gap) pont *m*
˜-CHORD RATIO - [aero] (in an aircraft having more than one plane, the ratio between the distance separating the leading edges of upper and lower planes, and the chord) rapport *m* de l'entreplan à la profondeur
˜CRANK PRESS - [mech] presse *f* à excentrique
˜EXTENSION - [railw] jeu *m* dû à l'expansion de rail
˜FAULT - [geol] faille *f* ouverte
˜FILLING ADHESIVE - [chem] (an adhesive suitable for joining surfaces which cannot be brought into intimate contact) mastic *m* bouche-pores

GAP GRADATION - [soil] (discontinuous grading) granulométrie *f* discontinue
~LATHE - [mach tool] (lathe with a gap-bed) tour *m* à banc rompu
~LENGTH - [el acoust] (the distance between the pole pieces of a magnetic head used for longitudinal recording) largeur *f* d'entrefer
[el] longeur *f* d'une coupure
~LOADING - [electron] admittance *f* électronique de l'espace d'interaction
~LOSS - [el acoust] perte *f* d'entrefer
~SECTION - [el] section *f* de séparation
~-TYPE PRESS - [mech] presse *f* à col de cygne
GAPFILLER - [radar] (auxiliary radar aerial to compensate for possible gaps in the main diagram) antenne *f* de complétement du diagramme
GAPING - [gen] ouverture *f*
[geol] bâillement *m*
GAPLESS-TYPE ICE GUARD - [mech] (an ice-guard fitted inside the mouth of the air intake) grille *f* anti-givre sans passage
GAPPED-TYPE ICE GUARD - [aero] grille *f* anti-givre à passage devié
GAPPER - [agric] démarieuse *f*
GARAGE, to - [auto] garer, remiser
GARAGE - [constr] garage *m*
[railw] garage *m*
~A LOCOMOTIVE, to - [railw] remiser une locomotive
GARBAGE - [gen] immondices *fpl*, ordures *fpl*
[food ind] tripaille *f*, issues *fpl*
[astronaut] débris *m* (en orbite)
~DISPOSAL - [gen] destruction *f* des immondices
~REMOVAL TRUCK - [transp] camion *m* à ordures
GARBLED INFORMATION - [comput] information *f* mutilée
GARBOARD - [shipbuild] (the first range of planks near the keel) gabord *m*
GARBOARD STRAKE - [shipbuild] virure *f* de gabord
GARDEN - [gen] jardin *m*
~BED - [agric] (flower bed) parterre *m*, corbeille *f*
~CITY - [town planning] cité-jardin *f*
~HOSE - [impl] tuyau *m* d'arrosage
~KNIFE - [impl] serpette *f*
~SHEARS - [impl] sécateur *m*
~TROWEL - [impl] houlette *f*
~-WALL BOND - [build] (used for low-boundary walls) appareil *m* à joint emboîté
GARDENING - [gen] jardinage
~SHEARS - s. garden shears
GARDEROBE - [constr] garde-robe *f*
GARDNER-HOLT VISCOSITY TUBES - [ind chem] (viscosity measurement apparatus, consisting of glass tubes of small constant standard bore, which are filled with the liquid for measurement, through which air-bubbles are allowed to rise, the time of passage being a measure of the viscosity) tubes *mpl* Gardner-Holt de mesure de viscosité
GARE - [text] (a coarse type of wool) pattelettes *fpl*
GARGET - [vet] mastite *f*, mammite *f*
GARGLE - [med] gargarisme *m*
GARGOYLE - [arch] (a spout, grotesquely shaped, projecting from the upper part of a building) gargouille *f*
GARLIC - [bot] ail *m*
~OIL - [chem] (a natural oil obtained from Allium sativum and used as a flavourant) essence *f* d'ail
GARMENT - [gen] vêtement *m*
~INDUSTRY - [text] industrie *f* du vêtement, industrie *f* d'habillement
~LENGTH MACHINE - [text] métier *m* circulaire pour panneaux
~MACHINE - [text] machine *f* garnett
GARNET - [min] (silicate minerals used as gem-stones and abrasives) grenat *m*
[naut] (a tackle used to hoist light goods) palan *m* de charge
~HINGE - [constr] penture *f* à gond
~LAC - [paint] (specially refined shellac, a solution of which in alcohol has a deep red colour) laque *f* grenat
~PAPER - [carp] papier *m* verré
GARNETTING - [text] (the treatment of waste material) traitement *m* des déchets
~MACHINE - [text] machine *f* garnett
GARNIERITE - [min] (essentially a hydrated nickel magnesium silicate, of varying composition, green to white in colour: an important ore of nickel) garniérite *f*
GARNISH, to - [gen] garnir, orner
GARNISH - [gen] garniture *f*
~MOULDING - [arch] moulure *f* garnie
~STRIP - [auto] bande *f* de garniture
GARRET - [constr] mansarde *f*, galetas *m*
~WINDOW - [constr] fenêtre *f* en mansarde
GARRETING - s. galleting
GARROT - [med] (used in surgery; a tourniquet) garrot *m*, tourniquet *m*
GAS, to - [chem] passer au gaz
[text] (the operation of passing yarns through flame thus enhancing their appearance) gazer, flamber
[el] (of accumulators) bouillonner
[chem] (of liquids etc) dégager des gaz
GAS - [phys] (matter in a fluid state, having neither definite shape nor volume, and consisting of molecules moving freely in space) gaz *m*
[only USA; abbrev for gasoline] essence *f* (de pétrole)
[mining] grisou *m*
~ABSORPTION PLANT - [oil ind] installation *f* d'extraction de gasoline du gaz
~ACTIVITY METER - [instr] activimètre *m* de gaz, radioactivimètre *m* de gaz
~ALARM - [gen] (warning of gas escape) avertisseur *m* de fuite de gaz
~AMPLIFICATION - [electron] (the increase in sensitivity in a gas-filled photo-electric cell) amplification *f* d'un gaz
~AMPLIFICATION FACTOR - [phys chem] (a factor showing the increase in sensitivity of a gas-filled tube caused by ionisation of the gas filling) facteur *m* d'amplification
~ANALYSIS APPARATUS - s. gas analyzer
~ANALYZER - [ind chem] (apparatus for determining the constituents of a gas and their proportions) analyseur *m* de gaz
~ANCHOR - [oil ind] (pipe extension under a well pump piston to check gas escape into the pump itself) ancrage *m* de gaz
~-BAG - [aero] (an airship unit filled with gas) ballonnet *m*
[gas ind] (gas mains) ballon *m* obturateur
[gen] (for oxygen) ballon *m* à gaz

GAS BAILER - [oil ind] pompe *f* mammouth, pompe *f* pneumatique
~BALANCE - [ind chem] (special balance to weight gases in determination of specific gravity) balance *f* à gaz
~BALLAST - [mech] lest *m* d'air, injection *f* d'air
~BALLAST PORT - [mech] orifice *m* d'entrée du lest d'air
~BALLAST PRINCIPLE - [phys] principe *m* de lest d'air
~BALLAST PUMP - [mech] pompe *f* à lest d'air, pompe *f* rotative à lest d'air, pompe *f* à injection d'air
~BALLASTING - [mech] admission *f* du lest d'air
~BALLOON - [gas ind] (gasholders) ballon *m* gazométrique
~BARREL - [mech] (a wrought iron tube, so called because of its original connotation; a tube for conducting gas from mains into buildings) tuyau *m* en fer
~-BEARING STRATUM - s. gas stratum
~BLACK - [chem] (a variety of carbon black used in rubber compounding; made by burning natural gas with a restricted air supply) noir *m* de gaz
~-BLAST CIRCUIT BREAKER - [el] (extinction of the arc by a blast of gas across the contacts) interrupteur *m* à gaz
~BLEED - [mech] rinçage *m* à gaz, rinçage *m* par gaz, balayage *m* avec un gaz
~BLEED FLANGE - [mech] bride *f* de rinçage à gaz
~BLOW-OUT - [mining] éruption *f* due au grisou
~BLOWING ENGINE - [metall] soufflerie *f* à gaz
~BOTTLE - [gas ind] (colloq for gas cylinder) bouteille *f* à gaz, bombonne *f* à gaz
~BRAZING - [mech] brasage *m* par gaz
~BREAKDOWN - [phys] rupture *f* dans un gaz
~BUBBLE - [phys] bulle *f* de gaz, bulle *f* gazeuse
~BUOY - [naut] (a marine floating beacon exhibiting a light by means of compressed or liquefied gas stored in it) bouée *f* lumineuse, bouée-balise *f*
~BURETTE - [ind chem] (special burette used to measure gas) burette *f* à gaz
~BURNER - [burners] brûleur *m* à gaz
[light] bec *m* de gaz
~BURST - [phys] échappée *f* de gaz
~BY-PASS VALVE - [mech] régulateur *m* de retour (du gaz)
~CABLE - [telecomm] (gas impregnated cable, in order to minimize ionization) câble *m* sous pression de gaz
~CALORIMETER - [instr] (instrument for measuring the calorific value of a gas in fuel analysis measurements) calorimètre *m* analyseur de gaz
~CAP - [oil ind] (free gas found in the highest part of a reservoir rock) cape *f* de gaz, calotte *f* de gaz
[mining] (a safety lamp) auréole *f*
~CAP DRIVE - [oil ind] expansion *f* de gaz, poussée *f* de cape
~CARBON - [chem] (deposit of almost pure carbon found in coal gas retorts) charbon *m* de cornue
~CARBURIZING - [metall] (for mild steel) cémentation *f* par gaz
~CASE HARDENING - [metall] cémentation *f* gazeuse
~CELL - [el chem] (a primary cell in which one or more of the reacting substances in a gas. s. also fuel cell) pile *f* à gaz
[aero] (in an airship) compartiment *m* du gaz
~CHROMATOGRAPHY - [ind chem] (a chromatographic technique employing a carrier gas stream to volatilize the sample being analyzed) chromatographie *f* des gaz
GAS CIRCULATION - [nucl] (the movement of the coolant gas in a nuclear reactor) circulation *f* du gaz
~CIRCULATION RETORT - [metall] (used in low-temperature carbonization) four *m* à chauffage interne par gaz de balayage
~COAL - [min] (gas which contains an appreciable percentage of volatile hydrocarbons) charbon *m* à gaz, houille *f* à gaz
~COCK - [gas ind] robinet *m* à gaz, robinet *m* de conduite de gaz
~COKE - [gas ind] (coke obtained in the manufacture of gas (as distinct from coke made expressly for metallurgical purposes) coke *m* de gaz, coke *m* d'usine à gaz
~COLLECTING MAIN - [metall] cuve *f* de carbonisation à basse température
~COMPONENTS - [chem] constituants *mpl* du gaz
~COMPRESSION - [phys] compression *f* du gaz
~CONCRETE - [constr] (a porous type of concrete) béton *m* poreux
~CONDITIONING - [gas ind] traitement *m* du gaz
~CONNETION - [gas install] arrivée *f* du gaz
~CONSTANT - [phys] (the constant of proportionally in the equation of state of a perfect gas) constante *f* de gaz
~CONSTITUENTS - s.gas components
~CONTAINER - [gas ind] (gasholders) cloche *f*
~CONTENT COEFFICIENT - [electron] (ratio of an ionizing current to the resulting ion current) coefficient *m* de gaz résiduel
~COOKER - [gas ind] réchaud *m* à gaz
~-COOLED REACTOR - [nucl] (reactor cooled by a gaseous medium) réacteur *m* à refroidissement au gaz
~COOLING - [gas ind] réfrigération *f* du gaz
~COULOMETER - [instr] (coulometer measuring the quantity of electricity by determining the volume of the gas evolved) coulombmètre *m* à gaz
~COUNTER - [nucl] (counter in which the sample is in the form of gas and introduced into the counter tube) compteur *m* à gaz
~CURE - [rubber ind] vulcanisation *f* au gaz
~CURRENT - [radio] courant *m* d'ionisation (dans un gaz)
~CURTAIN - [phys] gaine *f* gazeuse
~CUSHION CABLE - [telecomm] (used to minimize the effect of ionization) câble *m* sous pression de gaz
~-CUT - [gen] coupé à l'autogène
[oil ind] venue *f* de gaz dans la boue
~-CUT MUD - [oil ind] boue *f* gazéifiée
~CUTTING - [mech] coupage *m* à l'autogène, oxycoupage *m*
~CUTTING AND PROFILING MACHINE - [mach tool] (machine for cutting profiles in metal sheets automatically by means of a cutting flame the movement of which is controlled by a model profile) machine *f* à tailler et profiler à gaz
~DENSITY - [phys] densité *f* de gaz
~DENSITY APPARATUS - [meas] densimètre *m*
~DENSITY METER - [meas] (apparatus for measuring the density of gases) densimètre *m*
~DEPOSIT - [oil ind] gisement *m* de gaz, formation *f*

productive
GAS DESORPTION - [phys] (from surfaces) désorption *f* de gaz
~DETECTOR - [minig] détecteur *m* de grisou
~DIFFUSION - [phys] diffusion *f* gazeuse
~DISCHARGE GAUGE - [instr] (a pressure-gauge in which values are determined from an electric discharge in the gas under examination) manomètre *m* à décharge lumineuse
~DISCHARGE LAMP - [light] (electric discharge lamp) lampe *f* à luminescence
~DISPOSER - [gas ind] (Canadian term for incenerator) incinérateur *m*
~DISTRIBUTION - [gas ind] distribution *f* du gaz
~DRAIN - [min] (tunnel built to eliminate gas from workings) galerie *f* d'aérage
~DRIVE - [oil ind] pompage *m* par injection de gaz
~DRIVE RESERVOIR - [oil ind] (a reservoir rock from which the oil is produced by energy supplied by the expansion of free gas present in the reservoir) gisement *m* exploité par pression du gaz libre
~DUCT - [metall] trajet *m* de gaz
~EJECTION - [phys] expulsion *f* de gaz (par chauffage)
~-ELECTRIC SET - [el] groupe *m* électrogène à moteur
~ELECTRODE - [el] (electrode which contains a gas presenting a gaseous surface to a solution in contact with the electrode) électrode *f* à surface gazeuse
~ENGINE - [mech] (type of internal combustion engine in which gaseous fuel is mixed with air) moteur *m* à gaz
~ENTRAINMENT - [nucl] entraînement *m* de gaz
~ENVELOPE - [metall] enveloppe *f* de gaz
~ESCAPE - [geol] émanation *f* de gaz
~ESCAPE PIPE - [gasholders] évent *m*
~EVOLUTION - [oil ind] expansion *f* de gaz
~EVOLUTION RATE - [phys] quantité *f* de gaz désorbée
~EVOLVING - [phys] dégageant du gaz, émettrice *f* de gaz
~EXHAUSTER - [mech] (in gas-works, a large blower for the elimination of gas from the retorts), extracteur *m* de gaz
~-EXPANSION INSTRUMENT - [instr] appareil *m* à dilatation de gaz
~-EXPANSION THERMOMETER - [instr] (thermometer indicating by means of the expansion of a body of gas) thermomètre *m* à dilatation de gaz
~FACTOR - [oil ind] rapport *m* gaz-pétrole
~FADING - [text] gaz-fading *m*, sensibilité *f* aux gaz
~FEED - s. gas connexion
~FIELD - s. gas deposit
~FILLED BULB - [el] (an electric lamp containing an inert gas) ampoule *f* à remplissage gazeux
~FILLED CABLE - s. gas-cushion cable
-FILLED COUNTER - [meas] (a ionization chamber used as a counter) compteur *m* à remplissage gazeux
~-FILLED FILAMENT LAMP - [light] (bulb filled with an inert gas) lampe *f* à filament à atmosphère gazeuse
~-FILLED PHOTOCELL - [electron] (photocell with anode and photo-cathode in an atmosphere of gas at low pressure to increase its sensitivity) cellule *f* photoélectrique à gaz
~-FILLED RECTIFIER - [electron] (tube gaving unidirectional properties) tube *m* redresseur à gaz
GAS-FILLED RELAY - [electron] (grid-controlled thermionic tube) relais *m* à gaz (ionisé)
~-FILLED TUBE - [electron] (electronic tube evacuated to such a degree that its electrical characteristics are practically unaffected by ionization of residual gas) tube *m* à gaz
~-FILLED VALVE - s. gas-filled tube
~FIRED BOILER - [th eng] bouilleur *m* chauffé au gaz
~FIRED FURNACE - [metall] four *m* chauffé au gaz
~FIRED MUFFLE FURNACE - [metall] four *m* à moufle au gaz
~FLOAT - [gas ind] rampe *f* d'alimentation
~FLOW - [phys] courant *m* de gaz, flux *m* gazeux, flux *m* de gaz
~FLOW ADJUSTING FLANGE - [mech] bride *f* à conduite de gaz
~FLOW COUNTER - [meas] compteur *m* à courant de gaz
~-FLOW COUNTER TUBE - [instr] tube *m* compteur à balayage gazeux
~-FLOW IONIZATION CHAMBER - [nucl] chambre *f* d'ionisation à courant gazeux
~FOCUSING - [electron] (focusing of electron beam by the action of ionized gas) concentration *f* du faisceau
~-FURNACE - [metall] four *m* à gaz, four *m* chauffé au gaz
~GANGRENE - [med] (wound infection spreading with gas-forming anaerobic bacteria) gangrène *f* gazeuse, œdème *m* malin
~GAP - [el] éclateur *m* à gaz
~GENERATOR - [gas ind] gazogène *m*
~-GRAPHITE REACTOR - [nucl] (reactor in which the coolant is a gas and the moderator consists of graphite) réacteur *m* à gaz-graphite
~GROOVES - [electro-chemistry] (grooves in deposited metal caused by streams of gas rising along the cathode during deposition) ondulations *fpl* au gaz
~GUN - [gas ind] (in fles and chimneys) canon *m*, canal *m* de gaz riche
~-HEATED CONVECTOR - s. gas radiator
~-HOLDER - s. gasholder
~HOLE - [metall] soufflure *f*
~HOOD - [aero] (in airship) capuchon *m* du gaz
~HYDRATES - [chem] (synonym for the clathrates formed by certain gases with water) hydrates *mpl* de gaz
~IMPLOSION - [phys] implosion *f* (due à la pression de gaz)
~-IMPREGNATED CABLE - s. gas cable
~INJECTION - [gas ind] (gasholders) injection *f*, mise *f* au stock
~INLET - s. gas connection
~INPUT - s. gas injection
~INTERLOCK - [mech] sas *m* pour des gaz, écluse *f* à gaz
~IONIZATION - [phys] ionisation *f* des gaz
~JET - [mech] jet *m* de gaz, brûleur *m* à gaz,
~JET PUMP - [mech] éjecteur *m* à gaz, éjecteur *m* à air
~LASER - [phys] laser *m* à gaz
~-LEADING TUBE - [ind chem] tube *m* abducteur de gaz
~LEAK - [gen] fuite *f* de gaz
~LEAK VALVE - [mech] vanne *f* d'admission d'air,

robinet *m* d'entrée d'air
GAS LIBERATION - [metall] dégagement *m* de gaz
~LIFT - [oil ind] (extraction by compressed gas) gaz lift *m*, air lift *m*, puisage *m* au gaz
~LIFT WELL - [oil ind] puits *m* d'extraction à injection de gaz
~-LIGHT - [light] lumière *f* du gaz
~-LIGHT PAPER - [photo] papier *m* au gélatino-chlorure, papier *m* pour épreuves à la lumière
~LIGHTER - [impl] allume-gaz *m*
~LIME - [chem] (spent lime which has been used to absorb hydrogen sulphide and carbon dioxide in purifying coal gas) chaux *f* de gaz
~LINE - [gas ind] conduite *f*, gazoduc *m*
~LIQUOR - [chem] (ammoniacal liquor obtained in the production of coal gas) eau *f* de gazomètre, eau *f* ammoniacale
~LOG - [th eng] radiateur *m* à bûches réfractaires
~MAGNIFICATION - s. gas amplification
~MAIN - [gas ind] conduite *f* maîtresse de gaz
[aero] (for the distribution of gas in an airship) conduite *f* du gaz
~MANTLE - [gas ind] manchon *m* à incandescence
[oil ind] chemise *f* de gaz
~MASER - [phys] (maser in which the microwave radiation interacts with the molecules of a gas) maser *m* à gaz
~MASK - [gen] masque *m* respiratoire
~METER - [meas] compteur *m* de gaz
~MIXER - [ind chem] malaxeuse *f* de gaz
~MIXTURE - [metall] mélange *m* gazeux
~-MUD - [oil ind] boue *f* gazéifiée
~MULTIPLICATION - [nucl] multiplication *f* due au gaz
~NITRIDING - [metall] nitrurer au gaz
~NOISE - [electron] (the noise which is caused by the random production of ions in gas-filled tubes) bruit *m* d'ionisation
~NOZZLE - [gas ind] brûleur *m*
~OFF-TAKE - [gas ind] tubulure *f*
OFF-TAKE PIPE - [gas ind] (in the treatment of gas) colonne *f* montante, colonne *f* d'ascension
~OFF-TAKE VALVE - [gas ind] (used in condensation process) vanne *f* de barillet
~OIL - [petr ind] (a petroleum distillate of flash point 168° F. similar to diesel oil) gasoil *m*, gazole *m*
~-OIL INTERFACE - [oil ind] surface *f* de contact gas-huile
~OUTLET - [metall] (in a blast furnace) prise *f* de vent
~OUTPUT - [gasholders] reprise *f*, soutirage *m*
~OVEN - [gas ind] four *m* à gaz
~PASSAGES - [nucl] tuyauterie *f* de gaz
~PERMEABILITY - (the extent to which a specified gas can pass through the pores of a specimen of material) perméabilité *f* au gaz
~PICKLING - [metall] décapage *m* au gaz
~PIPE - [plumb] tuyau *m* de gaz
~PIPE LINE - [gas ind] canalisation *f* de gaz
~PIPE TONGS - [impl] pinces *fpl* de gazier
~PLANT - s. gas-works
~PLIERS - [impl] pinces *fpl* à gaz
~PLUG - [gas ind] prise *f* de gaz (mâle)
~POCKET - [gas ind] soufflure *f*
[metall] (foundry) inclusion *f* de gaz, poche *f* de gaz
[oil ind] poche *f* de gaz
GAS POISONING - [chem] (poisoning by the inhalation of a toxic gas) asphyxie *f*
~POOL - s. gas deposit
~PORT - [metall] parcours *m* du gaz
~PRESSURE - [gas ind] pression *f* d'alimentation
~PROCESSING - s. gas conditioning
~PRODUCER - [gas ind] gazogène *m*
~PRODUCER PLANT - s. gas producer
~-PROOF - [paint] (term used for a coating composition which does not exhibit defects (e.g. Gas Checking) when exposed to an atmosphere containing the combustion products of coal gas) à l'épreuve des gaz
[gen] imperméable aux gaz
[el] (of electrical machines etc) étanche au gaz
~PURIFIERS - [gas ind] (layers of hydrated oxides of iron to eliminate hydrogen sulphide) épurateurs *mpl* de gaz
~PURIFICATION - [gen] épuration *f* du gaz
~RADIATOR - [heat] (gas-heated convector) poêle *m* à gaz
~RANGE - s. gas cooker
~RATION- [electron] (the quotient of ion grid current and anode current) degré *m* de vide
~REFRIGERATING SYSTEM - [th eng] machine *f* frigorifique à gaz, cryogénérateur *m*
~REGULATOR - [gas ind] (valve maintaining a steady pressure) régulateur *m* de pression du gaz
[mech] (the throttle valve of a petrol engine) papillon *m* des gaz
~RELAY - [electron] (gastriode; thermionic three-electrode tube) tube *m* relais électronique
~RESERVOIR - s. gas deposit
~RETORT - [ind chem] cornue *f* à gaz
~-REVERSING VALVES - [gas ind] (water gas) vanne *f* d'inversion
~RING - [impl] réchaud *m* à gaz (à un feu)
~ROCK - [geol] roche *f* gazifère
~SAND - [min] sable *m* à gaz
[geol] sable *m* gazifère
~SCATTERING - [nucl] (in a counter tube) diffusion *f* dans le gaz
~SCINTILLATOR - [nucl] scintillateur *m* gazeux
~SCRUBBER - [ind chem] (apparatus for removing impurities, especially dust, from a gas) laveur *m* de gaz
~SCRUBBING - [metall] épuration *f* de gaz
~SEPARATOR - [oil ind] (installation to separate natural gas from oil) séparateur *m* de gaz
~SHOW - [geol] (surface indication of the escape of natural gas) émanation *f* de gaz
~SINGEING MACHINE - [text] flambeuse *f* à gaz, grilleuse *f* à gaz
~SOCKET - s. gas plug
~SPARGING - [nucl] (in a reaction vessel) purgation *f* à gaz
~STARTER - [aero] (type of engine starter) démarreur *m* à gaz
~STATION - [auto] (only USA) (petrol station) station *f* service
~STOCKS AND DIES - [mech] (stocks and dies used to cut threads on gas pipes) filière *f* à gaz
~STOPPER - [gas ind] obturateur *m*
~STORAGE - [gas ind] (gasholders) stockage *m* du gaz
~STORAGE WELL - [gas ind] puits *m* de stockage sou-

terrain
GAS STOVE - s. gas cooker
~STRATUM - [min] (natural gas) couche *f* à gaz
~SUPPLY - s. gas distribution
~SWITCH - s. gas-filled tube
~SWITCHING TUBE - [electron] (in radar practice) tube *m* commutateur à gaz
~TAR - [ind chem] (coal-tar condensed from coal-gas) goudron *m* de gaz, coaltar *m*
~TARGET - [nucl] (layer or jet of gas) cible *f* en gaz
~TEE PIECE - [metall] tube *m* de séparation des gaz
~THERMOMETER - [instr] (type of thermometer in which the expanding and contracting element is gaseous) thermomètre *m* à gaz
~THREAD - [plumb] pas *m* du gaz
~-TIGHT - [gen] étanche
~-TIGHT CASING - [chem] enveloppe *f* étanche aux gaz
~TRAP - [gas ind] (relating to network equipment) siphon *m*, pot *m* de purge
~TRAPPING SURFACE - [phys] surface de sorption (par getter)
~TREATMENT - [gas ind] traitement *m* du gaz
~TRUNK - [aero] (in an airship) tuyau *m* d'évacuation du gaz
~-TUBE RING COUNTER - [electron] compteur *m* en anneau avec tubes à gaz, compteur *m* annulaire avec tubes à gaz
~TURBINE - s. gas turbine engine
~TURBINE ENGINE - [mech & aero] (an engine in which gas heated by combustion within the engine itself is expanded through a turbine) turbine *f* à gaz
~VALVE - [oil ind] vanne *f* à gaz
~VANE - [astronaut] volet *m* placé dans le jet des gaz d'échappement
~VENT - [casting] (of a mould) évent *m*
~VOLUME FACTOR - [gas ind] (of calorific value) coefficient *m* de correction
~WASHING - [natural gas] lavage *m* du gaz
~WASHING BOTTLE - [ind chem] flacon *m* laveur du gaz
~WELDING - [metall] (joining of parts by fusing them or making them plastic by means of heat supplied by a gas flame) soudure *f* à flamme de gaz, soudure *f* au gaz
~WELL - [oil ind] (natural gas) puits *m* de gaz naturel
~WITHDRAWAL - s. gas output
~-WORKS - [gas ind] usine *f* à gaz
~YIELD - [gas ind] rendement *m* en gaz, rendement *m* volumétrique
GASEOUS - [phys] gazeux
~COMPOUND - [gas ind] mélange *m* gazeux, mélange *m* de gaz
~DIFFUSION - [phys] diffusion *f* gazeuse
~DIFFUSION METHOD - [nucl] (used to separate isotopes) méthode *f* à diffusion gazeuse
~DISCHARGE - [electron] (in many electronic tubes, the basis of operation) décharge *f* lumineuse
~FILM - [phys] voile *m* gazeux
~FLOW - [phys] écoulement *m* gazeux, courant *m* gazeux
~FUEL - gaz *m* combustible
GAP - [electron] espace *m* gazeux (dans un tube électronique)
~IONIZATION - [phys] (the process by which molecules of gas form charged particles) ionisation *f* d'un gaz
GASEOUS MINE - [mining] mine *f* grisouteuse
~PHASE - [phys] (the condition in which a substance is a gas) phase *f* gazeuse
~PATH - s. gaseous gap
~PROPELLANT - [astronaut] laser *m* à gaz
~THERMAL YIELD - [phys] rendement *m* thermique d'un gaz
~YIELD - s. gas yield
GASH, to - [gen] couper, entailler
[mech] (a rough milling) fraiser brut
GASH - [gen] entaille *f*, incision *f*
[med] blessure *f* saignante
GASHING - [metall] fraisage *m* brut
GASHOLDER - [gas ind] gazomètre *m*
~WITH CROWN FRAMING - s. gasholder with
~WITH TANK FRAMING - s. gasholder with untrussed crown
~WITH TRUSSED CROWN - [gas ind] gazomètre *m* à charpente intérieure mobile
~WITH UNTRUSSED CROWN - [gas ind] gazomètre *m* à charpente intérieure fixe
GASIFICATION - [gen] gazéification *f*
GASIFORM - [phys] gazéiforme, gazeux
GASKET - [mech] (specially made packing for sealing joints) joint *m* d'étanchéité, garniture *f* pour joint
[naut] (used to secure a sail) garcette *f*
[oil ind] garniture *f* ronde
~CEMENT - [metall] matière *f* de bourrage
[oil ind] matière *f* de bourrage
~GROOVE - [mech] (a groove formed in a part to receive a gasket) gorge *f* pour le joint
~RING - [mech] (an annular gasket) anneau *m* joint
~SEAL - [mech] scellement *m* rigide
[mech] joint *m* à écrasement, joint à compression
~SEAL PASTE - [ind chem] (special paste for use with gaskets) pâte *f* à joint, pâte *f* d'étanchéité
~-SEALED RELAY - [contr] relais *m* scellé à étoupage
GASOLINE - [fuel] (U.S. synonym for petrol) essence *f*, gazoline *f*
~CAN - [impl] bidon *m* d'essence
~LEVEL GAGE - [meas] (USA only; a fuel level gauge) jauge *f* de niveau d'essence
~PRESSURE GAGE - [meas] (USA only-fuel pressure gauge) manomètre *m* de l'essence
~PUMP - [auto] pompe *f* de distribution d'essence
~STOVE - [heat] poêle *m* à gazoline
~TORCH - [U.S.] s. blow-lamp
~TRAP - [oil ind] appareil *m* pour l'extraction de la gazoline
GASOMETER - [gas ind] (a gasholder) gazomètre *m*
GASPING - [acoust] (sound made when breathing is difficult) halètement *m*
GASSED - [chem] gazé
GASSER - [min] (gas producing well) puits *m* de gaz
GASSING - [el] (the evolution of gas in an accumulator at the end of its charging period) ébullition *f*
[text] (passing yarns through a flame to remove outstanding fibres and improve appearance) flambage *m*, grillage *m*
[aero] (the operation of filling a balloon with gas) gonflement *m*
[ind chem] passage *m* au gaz
GASSY IRON - [metall] fonte *f* soufflée
GASTRALGIA - [med] gastralagie *f*
GASTRIODE - s. gas relay

GASTRITIS - [med] gastrite *f*
GASTROENTERITIS - [med] gastro-entérite *f*
GASTROSCOPE - [med] (instrument used to examine the interior of the stomach) gastroscope *m*
GASTUNITE - [min] (rare ore containing uranium) gastunite *f*
GATE, to - [el] (the operation or the activation of a gate circuit) déclencher
[metall] (foundry) jeter
[radar] (USA only; strobing in G/B; the operation of selecting a signal) commander le passage d'un signal
[hydr] vanner
GATE - [gen] porte *f*
[gen] (barrier) barrière *f*
[hydr] vanne *f*, vannelle *f*
[electron] (electronic circuit having two or more input lines and one output, the essential characteristic being that an output pulse will only occur as the result of a given combination of pulses on the input lines) circuit *m* de porte, circuit *m* de gâchette, porte *f*
[metall] (die casting) (the opening through which the material enters the die cavity) attaque *f* de coulée, trou *m* de coulée
[plast ind] (of a mould) entrée *f*
[aero] (on the extension of a runway) portail *m* de approche
[metall] (the metal which solidifes in the gate) carotte *f*, jet *m* de coulée
[mech] (on a lever) encoche *f*
[radar] (only USA) (step or strobe in G/B; the spot which is generated by the gate generator under the control of a variable delay pulse) sélectionneur *m* d'un signal
[telev & radar] (square voltage wave which operates a circuit electronically) circuit *m* à impulsions périodiques
[radio] (the part of a saturable reaction which exhibits gate action) porte *f*
[cin] (the channel in which the film is held during projection) fenêtre *f*, porte *f*
[nucl] (movable shielding material used for filling a hole) porte *f* de blindage
~AGITATOR-TYPE MIXER - [mech] (in rubber industry) mélangeur *m* agitateur à grille tournante
~AND CULL - [plast ind etc] (in a press moulding, the material left in the gate and other openings) culot *m*
~ANGLE - [radio] (the angle at which the gate impedance changes) angle *m* de saturation
~AREA - [metall] section *f* d'attaque
~BY-PASS SWITCH - [el] (fitted in an electric lift) interrupteur *m* de sécurité automatique
~CHAMBER - [hydr] (in a lock, side wall recesses for the open gates) enclave *f* d'écluse
~CIRCUIT - [electron] (circuit passing a signal only when the gate is opened by an appropriate gating signal) circuit *m* de gâchette, circuit *m* de porte
~CLOSER - [el] (in lifts) dispositif *m* automatique de fermeture de la porte
~-CONTROLLED TURN-OFF TIME - [electron] temps *m* d'ouverture à commande de gâchette
~-CONTROLLED TURN-ON TIME - [electron] temps *m* de fermeture à commande de gâchette
~CURRENT - [radio] (in a magnetic amplifier) courant *m* de porte

GATE CUTTER - [mech] (device for shearing off gate from a moulding) coupe-coulées *m*
[metall] outil *m* à trancher la coulée
~-END BOX - [el] (used in coal mines) plaque *f* à bornes pour câble flottant
~GENERATOR - [electron] (generator of gate pulses) générateur *m* d'impulsion de déclenchement
~-HOUSE - [constr] loge *f* de garde
~IMPEDANCE - [radio] (impedance of a gate winding) impédance *f* de porte
~INLET - [metall] attaque *f* de coulée
~INTERLOCK - [el] (door switch and door lock combination) enclenchement *m* de porte
~KNIFE - [metall] coupe-coulées *m*
~KNIFE - s. gate cutter
~MARK - [plast ind etc] (mark left on a moulding where the gate has been removed) trace *f* de la carotte
~MONEY - [comm] recette *f*
~NON-TRIGGER CURRENT - [electron] courant *m* de gâchette non-commutateur
~NON-TRIGGER VOLTAGE - [electron] tension *f* de gâchette non-commutatrice
~PADDLE AGITATOR - [mech] (paddle mixer in which the blades are arranged in a form resembling a verticalbar gate) agitateur *m* à treillis, agitateur *m* à grille
~PIER - [constr] pied-droit *m* de porte
~PIN - [metall] modèle *m* de jet de coulée
~POST - [gen] pilier *m*
[constr] (of a door) montant *m* de porte
~PULSE - [comput] impulsion *f* de porte, impulsion *f* de coincidence
~RESISTANCE - [radio] (resistance on a gate winding) résistance *f* de porte
~RING - [metall etc] (a ring-shaped opening to allow material to enter a mould) entrée *f* annulaire
~SEATING ACTION VALVE - [mech] robinet-vanne *m*
~STICK - [metall] (a runner stick) bâton *m* de coulée
~SWITCH - [el] (a safety device designed to isolate an electrical circuit and consisting of a switch attached to a gate, door or removable panel) interrupteur *m* de porte
~TENSION - [text] frein *m* à chaînes
~TERMINAL - [electron] borne *f* de gâchette (ou de commande)
~TRIGGER CURRENT - [electron] courant *m* de gâchette commutateur
~TRIGGER DIODE - [electron] (the vacuum tube releasing the gate pulse) diode *f* de gâchette
~TRIGGER VOLTAGE - [electron] tension *f* de gâchette commutatrice
~TURN-OFF CURRENT - [electron] courant *m* de gâchette d'ouverture
~TURN-OFF VOLTAGE - [electron] tension *f* de gâchette d'ouverture
~VALVE - [mech] (a device for controlling fluids, in which a vertically-sliding plate is moved by a screw and handwheel) robinet-vanne *m*
[oil ind] vanne *f* à coin
~VOLTAGE - [radio] tension *f* de commande, tension *f* de gâchette
~WASH - [metall] nettoyage *m* du trou de coulée
GATED AUTOMATIC GAIN CONTROL - [telev] antifading *m* à déclenchement périodique
~-BEAM TUBE - [electron] (X-ray tube depending on the presence of a small quantity of residual gas for

the supply of electrons) tube *m* à commande du faisceau
GATED-BEAM TUBE DISCRIMINATOR - [electron] (a frequency discriminator circuit) discriminateur *m* à tube à contre-huis
~PATTERN - [metall] modèle *m* à entonnoirs
GATES - [metall] trous *mpl* de coulée
GATEWAY - [constr] porte *f* cochère
GATHER, to - [gen] assembler, recueillir, récolter
[text] (of fabrics) froncer
[bookbind] (the operation of collecting and arranging the folded sections) rassembler
[glass man] (to collect molten glass of a blow-pipe of gathering iron) cueillir le verre
[med] (of wounds) abcéder
[comput] obtenir, acquérir
GATHERED PARASHEET - [aero] (a parasheet having a hem cord round the periphery) parachute *m* à bord froncé
GATHERER - [glass man] (a workman who collects molten glass on a blow- pipe or a gathering-iron for forming an article or feeding a machine) cueilleur *m*
GATHERING - [gen] rassemblement, accumulation *f*
[med] abcédation *f*
[bookbind] assemblage *m*
[metall] (defect on the surface of plates) surface *f* irregulière
~GROUND - [hydr] (catchment area; the area from which the water runs forth) aire *f* d'alimentation, bassin *m* hydrographique
~IRON - [glass man] (iron used to collect molten glass) tube *m* à cueillir
~LINE - [oil ind] conduite *f* d'amenée
~LINES - [gas ind] (natural gas install) canalisations *fpl* de gaz brut, réseau *m* de collecte
~LOCOMOTIVE - [min] (small locomotive used in mines) locomotive *f* de manoeuvre
~MACHINE - [bookbind] machine *f* à assembler, assembleuse *f*
~PUMP - [oil ind] pompe *f* auxiliaire
~ROD - s. gathering iron
~SYSTEM - s. gathering lines
~TANK - [oil ind] réservoir *m* collecteur
GATING - [radar] (only USA) sélection *f* du signal
[metall] système *f* d'alimentation
[electron] séparation *f* d'impulsions
~ADJUSTMENT - [text] réglage *m* des aiguilles du plateau par rapport aux aiguilles du cylindre
~AND RISERS - s. gate and cull
[metall] jets *mpl* de coulée
~PULSE - [electron] (pulse operating on a circuit for the duration of the pulse itself) impulsion *f* de déclenchement
[radar] (only USA; strobing pulse in G.B.; pulse used to scrutinize a particular epoch) impulsion *f* de sélection
~PULSE GENERATOR - [electron] (USA only; strobing pulse generator in G.B.; generator of gating pulses) générateur *m* d'impulsion de séparation
~UNIT - [electron] élément *m* porte
GATISM - [med] gâtisme *m*
GAUFRED PLATE - [metall] (a chequered plate) tôle *f* striée, tôle *f* rainurée
GAUGE, to - [gen] mesurer, gabarier
[mech] calibrer ,jauger, gabarier, cuber
GAUGE - [meas] (an object or an instrument used to measure) calibre *m*, jauge *f*
GAUGE - [instr] (the device applied to the point of measurement) indicateur *m*, calibre *m*, appareil *m* vérificateur
[of a vehicle] (the distance between the wheels) voie *f*
[railw] (the distance between the rails) écartement *m*, écartement *m* de voie
[metall] (diameter of wire, or thickness) diamètre *m*, épaisseur *f*
[text] compte *m*
[constr] (proportion of gypsum in a lime plaster) mortier *m* d'enduit
[carp] (a marking tool) trusquin *m*
[mech] (accurately dimensioned metal piece used for testing purposes) trait *m* utile
[firearms] calibre *m*
[naut] tirant *m* d'eau d'un navire
[constr] pureau *m* (d'une tuile etc)
[instr] (pressure-gauge) manomètre *m*
[metall] (of a blast furnace) tige *f* de jaugage, bécasse *f*
[oil ind] niveau *m* (productif)
~BLOCK - [meas] (used in mechanical measurements) gabarit *m*
~BOARD - [metall] fond *m* de bois gabarié
~CIRCUITRY - [electron] coffret *m* d'alimentation, appareil *m* de contrôle
~CLEARANCE - [railw] (on the track) surécartement, jeu *m* de la voie
~COCK - [hydr] robinet *m* de jauge, robinet *m* de hauteur d'eau
~CONNECTION - [instr] raccord *m* pour tube de mesure, ajutage *m* pour une jauge
~CONTROL UNIT - s. gauge circuitry
~DOOR - [min] (door for the control of air supply) porte *f* à guichet
~FACTOR - [meas] facteur *m* d'étalonnage
~FOR BUFFET HEADS - [railw] gabarit *m* pour plateaux de tampons
~GLASS - [instr] (a glass tube for a water gauge q.v.) tube *m* de niveau d'eau, niveau *m*
~-GLASS RING - [mech] (the annular sealing element fitted at each end of the glass tube of a water-level gauge) anneau *m* de tube de niveau
~HEAD - [instr] tête *f* manométrique, tube *m* de mesure
~HOLE - [mech] trou *m* de repère
~LATHE - [mach tool] tour *m* à gabarit
~NUMBER - [metall] numéro *m* de jauge, titre *m* de jauge
~OF A WIRE - [meas] diamètre *m* d'un fil
~OF SHEET - [metall etc] (the thickness of sheet material) épaisseur *f* de feuille
~OF THE CLOTH - [text] compte *m* (densité du tissu)
~PIN - [metall] (in sheet metal) goupille *f* de repère
[text] tampon-calibre *m*, calibre *m* mâle
~PRESSURE - [mech] (pressure as shown on an ordinary pressure gauge, i.e. pressure above the atmospherical level, as distinct from absolute pressure) pression *f* manométrique
~ROD - [meas] (dipstick) jauge *f*
~SAW - [carp] sciotte *f*
~SETTING DEVICE - [railw] (a device designed to check the rail gauge) dispositif *m* pour le contrôle de l'écartement

GAUGE SIDE - [railw] (of rails) face *f* intérieure des rails
~STICK - [metall] (in foundry work) bâton *m* de contrôle
~TANK - [mech] cuve *f* de niveau
~TUBE - [meas] tube *m* jaugeur
~WIDENING - s. gauge clearance
GAUGED MORTAR - [build] (mortar consisting of cement, lime and sand) mortier *m* dosé
~ORIFICE - [mech] trou *m* calibré
GAUGING - [gen] calibrage *m* étalonnage *m*, mesurage *m*
[constr] (the mixture of gypsum and lime plaster) dosage *m*
[mech etc] standardisation *f*, vérification *f*, gabariage *m*
~LINE - [meas] échelle *f*
~ROD - [instr] jauge *f*
~ROLLS - [plast ind etc] rouleaux *mpl* de calibration
~RULE - [meas] jauge *f*, velte *f*
~SURFACE - [mech] (of a tool) surface *f* ouvrière
~TANK - [oil ind] reservoir *m* jaugeur
GAUNTLET - [impl] (protective hand covering) crispin *m*
[med] gantelet *m*
~BANDAGE - [med] bandage *m* à gantelet
GAUNTRY - s. gantry
GAUSS - [el] (electromagnetic unit of magnetic flux density) gauss *m*
~EYEPIECE - [opt] (eyepiece with a thin glass plate at 45 degrees to the optical axis between the two lenses) oculaire *m* de Gauss
GAUSSIAN IMAGE POINT - [opt] (the image point to which the paraxial rays converge) point *m* image de Gauss
~PLANE - [phys] plan *m* complexe
~WELL - [nucl] (well in a nuclear potential) puits *m* de Gauss
GAUZE - [text] (very thin fabric) gaze *f*
[mech] (fine mesh of metal wire used in filters etc) toile *f* métallique, tamis *m* métallique, crépine *f*
[photo] voile *m*
[metall] toile *f* métallique
~BAND - [plast ind] flanelle *f* à polir, crêpe *m* à polir
~CLOTH - [text] tissu *m* de gaze
~DRAFT - [text ind] (in a looming appliance) remettage *m* gaze, remettage *m* sineux
~FILTER - [impl] filtre *m* de gaze
~HARNESS - [text ind] (in looms for gauze weaves) harnais *m* de la gaze
~REED - [text] peigne *m* gaze
~SHAFT - [text] lame *f* gaze
~WEAVE - [text] armure *f* pour gazes
~WIRE CLOTH - [metall] toile *f* métallique
GAVAGE - [med] gavage *m*
GAVEL - [impl] (mallet used for setting stone) marteau *m*
[agr] javelle *f*
GAVELKIND - [leg] (equal division of a tenant's land among his sons) partage *m* égal (entre les fils du défunt)
GAVELOCK - [impl] (iron crowbar) pied-de-biche *m*, levier *m*
GAY-LUSSAC ACID - [chem] (a mixture of nitrogen oxides and sulphuric acid obtained at one stage of the production of sulphuric acid by the chamber process) acide *m* de Gay-Lussac
GAY-LUSSAC LAW OF VOLUMES - [chem] (law relating to the volumes of reacting gases) loi *f* de Gay-Lussac
~-LUSSAC TOWER - [chem] (tower used in sulphuric acid manufacture for absorbing nitrous gases from the air leaving the chambers) tour *f* de Gay-Lussac
GAYLUSSITE - [min] (a natural hydrated double carbonate of sodium and calcium, crystallizing with five molecules of water in the monoclinic system. It occurs as a saline residue) gaylussite *f*
GAZEBO - [constr] (type of summerhouse) belvédère *m*
[constr] (a type of balcony) fenêtre *f* en saillie
G.C.A. - [aero] s. ground controlled approach
g-cal - [phys] calorie-gramme *f*
G.C.A.S. - [aero] s. ground controlled approach system
GDF - [telecomm] (Group Distribution Frame; in carrier telephony) répartiteur *m* de groupe
G DISPLAY - [radar] (single signal only which appears as a bright spot) indicateur *m* type G
GEAN - [bot] cerise *f* sauvage, merise *f*
GEAR, to - [gen] engrener, gréer, s'enclencher
[mech] engrener, embrayer
[ind] développer
[a mechanical instrument] régler
GEAR - [gen] appareil *m*, engins *mpl*, mécanisme *m*
[mech] (relating to transmission) engrenage *m*, roue *f* dentée
[auto] vitesse *f*, couronne *f* de pont, grande couronne *f*
[mech] (bicycle) développement
[mech] (a toothed wheel) roue *f* d'engrenage, roue *f* dentée
[naut] apparaux *mpl*
[mech] (of a pump) armature *f*
~BLANK - [metall] lopin *m* d'engrenage
~-BOX - [mech] (auto etc) boîte *f* de changement de vitesse
[railw] (of electric locomotive) boîte *f* à engrenages
[aero] carter *m* de transmission
~BOX AND AUXILIARY GEAR BOX - [mech] boîte *f* de vitesses et engrenages de réduction
~BURNISHER - [mech] brunissoir *m* pour engrenages
~BURRING - [mech] barbure *f* des engrenages
~BURRING MACHINE - [mech mach] ébavureuse *f* pour engrenages
~CASE - [mech] carter *m*, couvre-engrenages *m*
[auto] carter *m* des engrenages
~CASING CORE - [metall] (foundry) noyau *m* de la boîte des engrenages
~CHAMFERING MACHINE - [mech mach] machine *f* à chanfreiner les engrenages
~CHANGING - [auto] changement *m* de vitesse
~CHECKER - [mech] dispositif *m* pour le contrôle des engrenages
~CONTROL - [mech] changement *m* de vitesse
~CONTROL LEVER - s. gear-shift lever
~-CONTROLLED SHUTTER - [photo] obturateur *m* commandé par engrenage
~CONVEYOR - [agric] vis *f* élévatrice, transporteur *m* à vis sans fin
~COUPLING - [mech] embrayage *m* par engrenage
~CUTTER - [mach tool] machine *f* à tailler les engre-

nages
[mach tool] (a cutting machine for teeth on gear-wheels) machine à tailler par génération les engrenages
GEAR CUTTING - [mech] (the cutting of teeth on a gear-wheel by milling cutters, hobs etc) taille *f* (d'engrenages)
~-CUTTING MACHINE - [mach tool] tailleuse *f* d'engrenages
~DOWN AND LOCKED - [aero] (indication that landing gear is in the alighting position and is locked there) train *m* verrouillé sorti
~-DRIVEN PRESS - s. geared press
~FLANK - [mech] flanc *m* d'une dent d'engrenage
~GRINDING MACHINE - [mach tool] machine *f* à rectifier les engrenages
~HOBBER - [mach tool] machine *f* à tailler par génération les engrenages
~HOUSING - s. gear-box
~IN NEUTRAL - [auto] position *f* neutre
~LAPPING - [mech] rodage *m* d'engrenages
~LEVER - [auto] levier *m* de changement de vitesse
~LUBRICANT - [mech] graisse *f* pour engrenages
~OIL - s. gear lubricant
~PLANER - [mach tool] raboteuse *f* mécanique pour engrenages
~PUMP - [mech] (type of pump consisting of a pair of meshing gears in a losely fitting casing) pompe *f* à engrenage
~-QUENCHING MACHINE - [mech] machine *f* à tremper les engrenages
~RATIO - [mech] (the relation between the speeds of a pair or of a train of gears) rapport *m* d'engrenage, multiplication *f*, rapport *m* de démultiplication
~RETRACTION SAFETY LOCK SOLENOID - [aero] (solenoid designed to actuate the landing-gear safety lock) solénoide *m* d'interdiction de relevage du train
~RING - [auto] couronne *f* dentée
~ROLLER - [mach] brunissoir *m*
~SELECTOR ROD - [auto] levier *m* de changement de vitesse
~-SET - [mech] jeu *m* de roues d'engrenage
~SHAFT - [auto] arbre *m* intermédiaire
[mech] train *m* de pignons
~SHAPER - [mach tool] (cutting and slotting machine) mortaiseuse *f* pour roues d'engranages
~SHAVING - [mech] ébardage *f* d'engrenages
~-SHIFT - [auto] changement *m* de vitesse
-SHIFT ATTACHED TO STEERING COLUMN - [auto] levier *m* de changement de vitesse sur le volant
~-SHIFT BAR - s. gear-shift lever
~-SHIFT FORK - [auto] fourchette *f* de changement de vitesse
~-SHIFT LEVER - s. gear lever
~-SHIFT LEVER BALL - [auto] rotule *f* du levier de changement de vitesse
~SOUND TESTER - [instr] (instrument designed to measure the degree of gear noise) appareil *m* contrôleur du bourdonnement des engrenages
~SPEEDER - s. gear sound tester
~SPINNING PUMP - [text] pompe *f* de filature par engrenage
~-TESTING MACHINE - [mech mach] machine *f* pour le contrôle des engrenages
~TOOTH - [mech] dent *f* d'engrenage
~TOOTH CUTTER - [mech] fraise *f* à tailler les engrenages
GEAR-TOOTH ROUNDING MACHINE - [mach tool] machine *f* à arrondir les dents
~TRAIN - s. gear set
~TRANSMISSION - [auto] transmission *f* à engrenages
~-TYPE INJECTION PUMP - [mech] pompe *f* d'injection à engrenages
~-TYPE PUMP - s. gear pump
~UP, to - [gen & mech] accélérer
~UP AND LOCKED - [aero] (indication that the landing-gear is in the flying position and is locked there) train *m* verrouillé rentré
~-WHEEL - [mech] roue *f* d'engrenage, roue *f* dentée
[mech] (a cogwheel) roue *f* dentée
~-WHEEL MOULDING MACHINE - [mach tool] machine *f* à tailler les engrenages (par génération)
WHINE - [auto] bourdonnement *m* des engrenages
~WITHDRAWER - [auto] extracteur *m* de pignon
GEARBOX - [mech] (housing or casing enclosing a gear train or change speed device) boîte *f* de vitesses
~CASING - [auto] carter *m* de boîte de vitesses
~LEVER - s. gear lever
GEARED - [mech] à engrenages
~AIRSCREW - s. geared propeller
~BENDER - [mech] (used for pipes) cintreuse *f* à engrenages
~BRAKE - [photo] (of a shutter) frein *m* à engrenage
~CHUCK - [mach tool] mandrin *m* universel
~DOWN - [mech] demultiplié
~HEAD - [mach tool] (of a lathe) poupée *f* fixe à harnais d'engrenages
~LADLE - [metall] poche *f* à engrenages
~LATHE - [mach tool] (lathe with a multi-speed gearbox between the driving motor and the head) tour *m* à engrenages
~MOTOR - [mech] machine *f* (électrique) à engrenages
~PRESS - [mech] presse *f* à engrenages
~PROPELLER - [aero] (a propeller connected to the engine through gearing) hélice *f* démultipliée
~PUNCHING AND SHEARING MACHINE - [mach tool] poinçonneuse-cisaille *f* à engrenages
~-QUILL DRIVE - [mech] (form of quill-drive used in electric locomotives) arbre *m* creux à engrenages
~RATCHET-BRACE - [mech] vilebrequin *m* à engrenages
~RING - [mech] anneau *m* denté
~SHAPING MACHINE - [mach tool] étau-limeur *m* à engrenages
~TURBINE - [el] (electric generator driven through a reduction gear) turbo-réducteur *m*
~TURBO-GENERATOR - s. geared turbine
~WINCH - [mech] (a lifting winch) treuil *m* à engrenages
GEARING - [mech] (set of gear-wheel transmitting motion) engrenages *mpl*, harnais *m* d'engrenages, rouage *m*
~-DOWN - [mech] (reduction in speed between driving and driven wheel) démultiplication *f*
~OF TEETH - [mech] engrènement *m* des dents
~SWITCH - [mech] commande *f* par engrenage
~-UP - [mech] (increasing the speed of the driven unit beyond that of the driving unit) multiplication *f*
GEARLESS - [mech] (of motor armature mounted di-

rectly on the driving axle) sans engrenages

GEARLESS LOCOMOTIVE - [railw] (electric locomotive with the motor armatures are mounted on the driving axle) locomotive *f* à entraînement direct

GEE - [radar] (system of radar navigation in which position is found from interrelated pulses transmitted by pairs of ground stations) GEE *m*

~H - [radar] (system of radar navigation in which the aircraft's position is determined from two beacons) GEE H

~H SYSTEM - s. Gee H

GEIGER COUNTER - s. geiger-Muller counter

~FORMULA - [nucl] (relation between the initial velocity and range of alpha particles) formule *f* de Geiger

~-MULLER COUNTER - [nucl] (apparatus for counting particles by means of ionization produced by them) tube-compteur *m* Geiger-Müller

~-NUTALL RELATION - [nucl] relation *f* de Geiger-Nutall

~REGION - [nucl] (voltage interval for a counter tube in which the charge transferred per isolated discharge does not depend on the number of primary ions in the initials ionizing event) région *f* de Geiger-Müller

~THRESHOLD - [nucl] (the lowest voltage which, when applied to a Geiger counter, will produce pulses which are in each case of substantially the same magnitude irrespective of the number or primary ions produced) seuil *m* de Geiger-Müller

GEISSLER PUMP - [ind chem] (an exhaustion device, made of glass and operated by the ordinary water supply) pompe *f* Geissler

~TUBE - [electron] (vacuum tube in which the luminous effect of discharges is shown) tube *m* Geissler

GEL, to - [chem] se coaguler, se prendre en gelée

GEL - [chem] (the material formed when a colloidal solution is allowed to stand; often of jelly-like appearance) gel *m*, colloïde *m* coagulé, gèle *m*

~CEMENT - [geol] gel-ciment *m*

~GAS - [oil ind] (oil in a synthetic resin solution) pétrole *m* en solution gélatineuse

~STATE - [chem] état *m* de gèle

~STRENGHT - [geol] effet *m* tissotropique de la boue

~TIME - [chem] (the time required for a colloid to form a gel) temps *m* de gélification

GELATIN - [min] (mineral protein used in medicine biology, papermaking, textile processing, food processing, and in the production of films and adhesives) gélatine *f*

~FOIL - [photo] feuille *f* de gélatine

~RELIEF IMAGE - [photo] image *f* de gélatine en relief

GELATINE - s. gelatin

~PAPER - [paper man] papier *m* gélatiné

GELATING FILTER - [photo] filtre *m* gélatine

GELATINIZE, to - [photo] (to coat with a film of gelatine) gélatiniser

GELATINOUS - [chem] gélatineux

GELATION - [chem] gélification *f*, congélation *f*

GELLING - [chem] (the solidification of colloidal solutions) gélification *f*

[paint] (defect) gélatinisation *f*

~AGENT - [chem] adjuvant *m* de gélatisation

~POINT - [phys & chem] (the temperature at which colloidal solutions become solid) température *f* de formation de gels

GELSEMINE - [chem] (an alkaloid obtained from Gelsemium sempervirens and used in medicine) gelsémine *f*

GEM - [geol] gemme *f*, pierre *f* précieuse

GEMEL WINDOW - [constr] fenêtres *fpl* jumelées

GEMINATE - [bot] (with twin branches arising from one node) géminé

GEMINATION - [bot & crystall] gémination *f*

GEMINI - [astr] Les Gémeaux

[genet] (bivalent chromosomes) couples *mpl* bivalents

GEMMA - [bot] gemme *f*, cellule *f*, bourgeon *m*

GEMMATION - [bot] (the formation of gemma) gemmation *f*, reproduction *f* par bourgeonnement

GEMMULE - [bot] gemmule *f*

GEMSTONE - [min] pierre *f* précieuse

GENE ARRANGEMENT - [genet] disposition *f* des gènes

~INTERACTION - [genet] interaction *f* des gènes

~-MUTATION - [genet] (heritable variation) mutation *f* des gènes

~STRING - [genet] (hypothetical component of a chromosome) chromatide *f*, cordon *m* de gènes

~TAGGED CHROMOSOMES - [genet] chromosomes *m* pl couverts de gènes

GENEALOGY - [biol etc] (the study and investigation of hereditary characters) généalogie *f*, ascendance *f*

[gen] (a pedigree) arbre *m* généalogique

GENERAL - [gen] général

~ACCOUNTS - [comm] comptes *m* généraux

~ARRANGEMENT DRAWING - [draw] dessin *m* de disposition générale

~CHART - [auto] plan *m* d'ensemble, diagramme *m* d'ensemble

~COLOUR RENDERING INDEX - [telev] indice *m* général de reproduction de la couleur

~DIFFUSION EQUATION - [nucl] (in nuclear theory) équation *f* générale de diffusion

~INFERENCE - [met] (the meteorological situation at a stated time, as deduced from existing conditions, including references to changes currently occurring and a general forecast) prévision *f* générale

~INTERPRETATIVE PROGRAMME - [comput] programme *m* interprétatif général

~MONITOR CHECKING ROUTINE - [comput] programme *m* d'analyse général

~OUTLINE - [draw] apercu *m* d'un projet, grandes lignes *fpl*

~PARESIS - [med] paralysie *f* générale

~POST-MORTEM PROGRAMME - [comput] (post-mortem programme giving all instructions to write out the contents of all stores and registers) programme *m* autopsie général

~PRACTITIONER - [med] médecin *m* ordinaire

~-PURPOSE - (of tools) universel, à toutes fins

~-PURPOSE COMPUTER - [comput] (computer for various problems) calculateur *m* universel

~-PURPOSE CONVEYOR - [agric] transporteur *m* universel

~PURPOSE DIODE - [electron] (diode which can be used in various couplings) diode *f* universelle

~-PURPOSE MANIPULATOR - [nucl] (remotely-controlled device for handling experimental material etc. for general purposes in a nuclear reactor installation) télémanipulateur *m* universel

~-PURPOSE PLOUGH - [agric mach] charrue *f* universelle

GENERAL-PURPOSE RADAR - [radar] (general warning set capable of detecting surface targets within a 100,000 yards) radar *m* universel
~-PURPOSE RELAY - [contr] relais *m* universel
~-PURPOSE TRACTOR - [agric] tracteur *m* universel, polyculteur
-PURPOSE WRENCH - [impl] clef *f* à toutes fins
~RELATIVITY THEORY - [phys] théorie *f* générale de la relativité
~ROUTINE - [comput] programme *m* universel
~STAIN - [chem] (one which imparts the same depth of colour to all the structures treated with it, i.e. has no selective qualities) coloration *f* générale
~STORAGE - [comput] mémoire *f* générale, mémoire *f* principale
~TARIFF - [comm] tarif *m* normal
~TREND - [gen] tendence *f* générale
[el] direction *f* générale
GENERALIZED VELOCITIES - [phys] (in particle mechanics) vitesses *fpl* généralisées
GENERATE, to - [gen] engendrer, produire
[comput] (the operation of producing coding by assemblying primitive elements) générer
~A CODE, to - [comput] engendrer un code *m*, produire un code *m*, former un code *m*
GENERATING - [gen] générateur, producteur
[mech] (of a hobbing machine) mouvement *m* de génération
[el] électrogène
~CIRCLE - [mech] (of gears) cercle *m* générateur, groupe *m* électrogène, groupe *m* générateur
~PLANT - s. generating station
~PROGRAMME - [comput] (programme designed to construct other programmes) générateur *m*
~ROLL - [mech] (of a gear generation) cylindre *m* de base
~ROLLING CIRCLE - [mech] (of a gear) cercle *m* générateur de base
~ROLLING CURVE - [mech] (of a gear) courbe *f* génératrice
~ROLLING LINE - s. generating rolling
~SET - [el] (one or more electric generators driven by a prime mover) groupe *m* électrogène
~-STATION - [el] (the building in which the equipment for generating electric energy is found) installation *f* de production, centrale *f* électrique
~STATION CAPACITY - [el] puissance *f* installée
GENERATION - [gen & biol] génération *f*
[geom] engendrement *m*, génération *f*
~DATA GROUP - [comput] famille *f* d'ensemble de données
~OF GAS - [phys] production *f* de gaz
~OF GEARS - [mech] taille *f* par génération des engrenages
~OF HEAT - [phys] production *f* de chaleur
~OF NUCLEI - (the successive number of nuclei produced in a nuclear chain reaction) génération *f* des noyaux
~RATE - [electron] (the speed of creation of electron-hole pairs) formation *f* d'une paire
~TIME - [nucl] (the mean time necessary for neutrons produced by a fission to produce neutrons on their own account) temps *m* de génération
GENERATOR - [el] (an electrical machine) générateur *m*, génératrice *f*
[el] (direct current machine) dynamo *f*
~ARMATURE - [el] induit *m*
GENERATION-BATTERY IGNITION - [auto] allumage *m* par batterie
~BODY - [gas ind] cuve *f* de gazogène
~BOILER - [metall] (annular boiler) chaudière *f* annulaire
~BRUSH - [el] balai *m*, charbon *m*
~BUS-BARS - [el] (the bus-bars in a power station to which generators are connected) barres *fpl* collectrices
~CASING - [el] boîtier *m* du générateur
~COIL - [el] bobine *f* de génératrice
~COUPLING - [el] (dynamo coupling) entraînement *m* de la dynamo
~EXCITER - [el] excitatrice *f* pour génératrice
~FIELD COIL - [el] bobinage *m* inducteur de la dynamo, bobinage *m* d'excitation de la dynamo
~FIELD CONTROL - [el] (variable voltage control) régulation *f* par variation dans le champ de la génératrice
~FRAME - [el] carcasse *f* de la dynamo
~PANEL - [el] (switchboard panel for the equipment necessary to control a generator) panneau *m* de tableau de distribution
~REGULATOR - [auto] régulateur *m* de dynamo
~SHELL - s. generator body
~SIGNALLING - [telecomm] appel *m* sur les circuits par clef
~UNIT - [el] groupe *m* électrogène
~VOLTAGE REGULATOR - [el] régulateur *m* de tension de la génératrice
~WARNING LIGHT - [auto] lampe-témoin *f* de charge
~WITH AUTOMATIC GRATE - [gas ind] gazogène *m* à grille tournante
~WITH ROTARY GRATE - s. generator with automatic grate
GENERATRIX - [geom] génératrice *f*
GENERIC HYBRID - [genet] hybride *m* générique
GENES - [biol] (hypothetical units concerned with the development of hereditary character) gènes *mpl*, facteurs *mpl* d'évolution
GENESIS - [biol] (the origin and development of a species) genèse *f*
GENET - [text] (dressed skin of the genet) genette *f*
GENETIC ASSORTATIVE MATING - [genet] accouplement *m* assortatif
~BACKGROUND - [genet] milieu *m* génétique
~COMPLEX - [genet] (the total of the hereditary factors) complexe *m* génétique
~EFFECT OF RADIATION - [nucl] (of radiation; those changes in the reproductive cells of living matter which are caused by the absorption of ionizing radiation) effet *m* génétique du rayonnement
~LOAD - [genet] charge *f* génétique
~SPIRAL - [genet] (hypothetical line drawn on a stem for research purposes) spirale *f* génétique
~VARIATION - [genet] variation *f* génotypique
GENETICS - [biol] (the study of variation and heredity) génétique *f*
GENEVA LOCKING DEVICE - [mech] dispositif *m* de verrouillage par mouvement de Genève
~MOVEMENT - s. Geneva wheel mechanism
~WHEEL MECHANISM - [mech] (a mechanism in which the driven member receives intermittent motion from the driving member and is locked in the intervals: in some forms, the whole device locks after a certain number of revolutions) croix de Malte

GENIC BALANCE - [genet] (the principle that the characters of any organism are determined by the interaction of genes) balance *f* génique
˜STERILITY - [genet] stérilité *f* génique
GENICULATE - [zool etc] géniculé
GENIOPLASTY - [med] (plastic surgery of the chin) génioplastie *f*
GENITALS - [zool] (reproductive organs) organes *mpl* génitaux
GENITO-URINARY - [med] (relating to the genital and urinary organs) génito-urinaire
GENLOCKING - [telev] système *m* à générateur, système *m* à générateur verrouillé
GENOM - [genet] (the chromosome content of the nucleus of a gamete) génome *m*, garniture *f* chromosomique
GENOTYPE - [genet] (group of individuals possessing the same factorial constitution) génotype *m*
˜-ENVIRONMENT INTERACTION - [genet] interaction *f* entre hérédité et environnement
GENTIAN - [bot] gentiane *f*
[chem] (a bitter flavourant used in medicine) gentiane *f*
˜BLUE - [dyes] bleu *m* de gentiane
GENTLE - [fishing] (a maggot, used as a bait by fishermen) asticot *m* achée *f*
BREEZE - [met] légère brise *f*
GENTLEMAN - [metall] (a rotating soldering table) table *f* rotative pour soudures
GENUINE - [gen] pur
˜TISSUE - [genet] (tissue derived from a division of related cells) tissu *m* pur
˜WHITE LEAD - [chem] blanc *m* de céruse
GEOBIOTIC - [zool] (living on land) géobiotique
GEOCARPY - [bot] (the ripening of fruit underground) géocarpie *f*
GEOCENTRIC - [astr] (having the centre of the earth as a point of reference) géocentrique
˜PARALLAX - [astr] (the apparent change in the position of a heavenly body due to the shift of the observer caused by the rotation of the earth) parallaxe *f* géocentrique
GEOCHEMICAL OIL EXPLORATION - [min] (method of exploration by analysing the gaseous content of surface soils to ascertain the presence of microseepages of hydrocarbon gases) recherche *f* géochimique des gisements pétrolifères
GEOCHEMISTRY - [chem] (study of the chemical composition and properties of the earth's crust) géochimie *f*
GEOCRATIC PERIOD - [geol] période *m* géocratique
GEODE - [med] géode *f*, caverne *f*
[geol] s. geodes
GEODES - [geol] (cavities in rocks, lined with crystals) géodes *fpl*, poches *fpl* à cristaux
GEODESY - [surv] (surveying of large areas) géodésie *f*
GEODETIC CONSTRUCTION - [aero] (system of airframe construction in which structural members follow geodetic lines on the external surface and are all in compression or tension) construction *f* géodesique
˜DATUM - [astronaut] donnée *f* géodésique
˜FRAME - s. geodetic construction
˜LINE - [geom] ligne *f* géodésique
˜STRUCTURE - s. geodetic construction
˜SURVEYING - s. geodesy
GEODETIC SYSTEM - [surv] réseau *m* géodésique
˜VALUE - [surv] valeur *f* géodésique
GEODETICAL - [surv] géodésique
GEODYNAMICS - [geol] géodynamique
GEOGNOSY - [geol] (obsolete; complete knowledge of the earth) géognosie *f*
GEOGRAPHIC - [geogr] géographique
˜COORDINATES - [geogr] (latitude and longitude) coordonnées *fpl* géographiques
˜GRID - [geogr] (the grid formed by parallels and meridians) graticule *m* géographique
˜POLES - [geogr] (the two points on the earth's surface through which its axis of rotation passes) pôles *mpl* géographiques
˜POSITION (OF HEAVENLY BODY) - [astr] (the point at which a line drawn from the centre of the body to that of the earth cuts the surface of the latter) position *f* géographique
˜TONGUE - [med] langue *f* géographique, glossite *f* exfoliatrice marginée
GEOGRAPHICAL - [geogr] géographique
˜MILE - [meas] (the length of one minute of latitude, i.e. I,852 m in G.B.; the length of one minute of latitude at the equator, i.e. approx I,855 in the USA) mille *m* marin, mille *m* nautique
˜PANEL - [el] tableau *m* géographique de commande et de contrôle
˜WIND DIRECTION - [met] direction *f* du vent (géographique)
GEOGRAPHY - [geogr] géographie *f*
GEOID - [surv] (geometrical solid almost identical with the terrestrial spheroid and with a surface which is at every point perpendicular to the direction of gravity) géoide *m*
GEOISOTHERMS - [geol] (surfaces having equal temperature in the outer rocky shells surrounding the centrosphere) géoisothermes *fpl*
GEOLOGICAL - [geol] (of geology) géologique
˜EPOCH - [geol] époque *f* géologique
˜ERA - [geol] âge *m* géologique
˜GROUND SURVEY - [surv] (survey of an area by geological means) levé *m* géologique terrestre
˜LAYER - [geol] couche *f* géologique
˜PROFILE - [geol] coupe *f* géologique, structure *f* géologique
˜SUBSURFACE MAP - [geol] carte *f* géologique du sous-sol
˜TIME - [geol] (the time between the end of the formative period of the history of the earth and the beginning of the historical-period) époque *f* géologique
˜WINDOW - [geol] fenêtre *f* tectonique
GEOLOGIZE, to - [min] (only USA, to prospect) étudier la géologie, prospecter
GEOLOGRAPH - [instr] (instrument designed to record the speed of boring) géolographe *m*
GEOLOGY - [geol] (study of the history of the earth's crust) géologie *f*
GEOMAGNETIC EFFECT - [phys] (effect due to the earth's magnetic field) effet *m* de latitude, effet *m* géomagnétique
˜POLE - [phys] pôle *m* magnétique
GEOMETRIC - [geom] géométrique
˜BUCKLING - [nucl] (in reactor theory) laplacien *m* géométrique
˜CAPACITANCE - [el] (the capacitance of an isolated conductor in a vacuum) capacité *f* géométrique

GEOMETRIC DILUTION OF PRECISION - [astronaut] perte *f* de précision due à la géométrie du système
˜DISTORTION - [opt] (aberration causing the picture to be geometrically dissimilar to the projection of the original) distorsion *f* géométrique
˜ISOMERISM - [chem] (the phenomenon of compounds having the same molecular formulae but different properties due to dissimilar arrangement of the atoms within the molecule) isomérie *f* géométrique
˜ISOMERS - [chem] (s. geometric isomerism) isomères *mpl* géométriques
˜LOCUS - [geom] lieu *m* géométrique
˜MEAN - [math] moyenne *f* géométrique
˜PATTERN - [gen & text] dessin *m* géométrique
˜PITCH - [aero] (the distance through which a given element of a propeller whould move in one revolution along a helix such that the line of the blade angle would form a tangent to it) pas *m* géométrique
˜PROGRESSION - [math] progression *f* géométrique
˜TWIST - [aero] (the spanwise variation of the chord of an aerofoil in respect to a given datum) vrillage *m* géométrique
GEOMETRICAL ATTENUATION - [acoust] (the reduction of sound-wave intensity in relation to the distribution of energy in space) atténuation *f* géométrique
˜CONFIGURATION - [nucl] configuration *f* géométrique
˜FIGURE - [geom] figure *f* géométrique
˜OPTICS - [opt] optique *f* géométrique
˜SIMILARITY OF FLUID FLOW - [phys] (two fluid flows are geometrically similar if one flow is a model of the other) similitude *f* géométrique des courants fluides
˜STAIR - [constr] escalier *m* à noyau
˜STAIRS - [constr] (staircase with a well) escalier *m* à jour *m*, escalier *m* à œil
GEOMETRY - [geom] géométrie *f*
[mech etc] (the shape and arrangement of parts) géométrie *f*, forme *f*
[nucl] (the spatial locations, symmetry, and dimensions of elements of an assembly, especially of a nuclear reactor) géométrie
˜FACTOR - [nucl] sensibilité *f* du compteur
˜HUM - [telev] (positional hum) ronflement *m* de géometrie
GEOMORPHOLOGY - [geol] géomorphologie *f*
GEON - [phys] (an entity consisting of an electromagnetic field held together by the gravitational attraction produced by the energy of the field) géon *m*
GEOPHILIC - [bot] (growing in the soil) géophilique
GEOPHONE - [instr] (instrument for seismic surveys used to record vibrations) géophone *m*
GEOPHYSICAL EXPLORATION - [soil] reconnaissance *f* géophysique, exploration *f* géophysique
˜EXPLORATION - s. geophysical prospecting
˜PROSPECTING - [min] (method of prospecting by studying the various properties of the earth's crust) prospection *f* géophysique
GEOPHYSICS - [geol] (study of the physical characteristics of the earth) géophysique
GEOPOTENTIAL - [astronaut] géophysique *f*
˜HEIGHT - [astronaut] altitude *f* géopotentielle
˜SURFACE - [astronaut] surface *f* équipotentielle terrestre
GEORDIE - [mining] mineur *m* de charbonnage

"GEORGE" - [aero] (colloq for automatic pilot) pilote *m* automatique
[radar] (colloq for a type of anti-jamming device for radar receivers) dispositif *m* anti-brouillage
GEOSCOPY - [geol] géoscopie *f*
GEOSPHERE - [geol] (the solid body of the earth) géosphère *f*
GEOSTROPHIC - [met] (relating to geostrophy) géostrophique
˜FORCE - [met] (force producing deflection in bodies relatively to the surface of the rotating earth) force *f* géostrophique
˜WIND SPEED - [met] (the speed of a wind as derived from the rotational speed of the earth, the pressure gradient, air density and latitude, the curvature of its path being disregarded) vitesse *f* du vent (géostrophique)
GEOSTROPHY - [met] (the study of the forces due to the rotation of the earth) géostrophie *f*
GEOSYNCLINE - [geol] (long narrow area of marine sedimentation) synclinal *m*, géosynclinal *m*
GEOTAXY - [biol] (the response to the stimulus of gravity) géotropisme *m*
GEOTECHNICAL PROPERTIES - [soil] propriétés *fpl* géotechniques
GEOTECHNICS - [geol] (the various systems used for the study of the earth's crust) géotechnique *f*
GEOTHERMAL - [geol] (of the temperature of the earth's crust) géothermique
˜GRADIENT - [geol] (the rate of the increase with depth of the earth's temperature) gradient *m* géothermique
GEOTHERMOMETER - [instr] (a thermometer designed to record the temperatures below the earth's surface) géothermomètre *m*
GEOTROPISM - s. geotaxis
GERANIOL - [chem] (a constituent of geranium-oil used in perfumery) géraniol *m*
˜BUTYRATE - [chem] (a perfumery agent) butyrate *m* de géraniol
˜FORMATE - [chem] (a component of perfumes) formiate *m* de géraniol
GERANIUM - [bot] géranium *m*, bec-de-grue *m*
˜OIL - [chem] essence *f* de géranium
GERANYL ACETATE - [chem] (a perfumery agent) acétate *m* de géranyle
GEREOLOGY - s. geriatrics
GERIATRICS - [med] gériatrie *f*
GERM - [zool] (primitive rudiment destined to develop into an individual) germe *m*, microbe *m*
˜CELL - [phys] (those cells of an organism with the function of reproduction) cellule *f*, spermatozoide *m*, ovule *m*
˜NUCLEUS - [biol] (nucleus of a germ cell) noyau *m* de l'œuf, pronucléus *m*
˜PLASM - [biol] plasma *m* germinatif
˜TRACT - [biol] tractus *m* germinal
GERMAN - [gen] allemand
[print] gothique
˜MEASLES - [med] rubéole *f*
˜NOZZLE - [mech] (a parabolic nozzle) tuyère *f* à section parabolique
˜SILVER - [metall] (term sometimes used for nickel silver, i.e. a series of copper-zinc-nickel alloys) maillechort *m*, métal *m* blanc
GERMANIUM - [chem] (a metallic element, symbol Ge, A.N. 32, A.W. 72.5, used in electronics)

germanium *m*
GERMANIUM DIODE - [electron] (semiconductor device) diode *f* au germanium
DIOXIDE - [chem] (compound used for germanium diodes) bioxyde *m* de germanium
~MELT - [electron] (semiconductors; the molten material from which germanium crystals are obtained) coulée *f* de germanium
~OXIDE - [chem] (a constituent of special glasses) oxyde *m* de germanium
~PELLET - [electron] (used for the manufacture of semiconductors) pastille *f* au germanium
~WAFER - [electron] (used for the manufacture of semiconductors) gaufrette *f* de germanium
GERMICIDAL - [chem] antiseptique, germicide
GERMICIDE - [chem] (a substance which destroys bacteria) germicide *m*
GERMINAL APERTURE - [bot] (the aperture through the pollen tube emerges) ouverture *f* germinale
~VESICLE - [anat] vésicule *f* germinative
GERMINATING APPARATUS - [brew ind] germoir *m*
~BOX - s. germinating apparatus
~CAPACITY - [brew ind] pouvoir *m* germinatif
~DRUM - [brew ind] tambour-germeur *m*
~TEST - [brew ind] essai *m* de germination
~TIME - [brew ind] durée *f* de germination
GERMINATION - [bot] (the beginning of growth in a seed) germination *f*
~BED - [agric] pépinière *f*
~ENERGY - [agric] (of seeds) énergie *f* germinative
GERMINATOR - [brew ind & agric] germoir *m*
GERMPLASM - [chem] plasma *m* germinatif
GERONTAL - [med] sénile
GERONTOLOGY - s. geriatrics
GESSO - [ind chem] (used for a basis for painting) plâtre *m* de Paris, gypse *m*
GESTATION - [zool] (pregnancy) gestation *f*
GESTOSIS - [med] gestose *f*, intoxication *f* gravidique
GET, to - [astronaut] dégazer par sorption
GET - [min] production *f*, rendement *m*
~A BONE, to - [oil ind] (the operation of perforating slaty material while boring for oil) perforer des roches stratifiées pendant le forage
[mining] perforer une roche dure
~-AWAY SPEED - [aero] (the speed at which a seaplane or flying-boat becomes fully airborne) vitesse *f* de décollage
~CHALKY, to - [constr] (said of paint) s'épuiser
GETAWAY - [cin] (movable scene) coulisse *f* de studio
GETTER - [electron] (substance used to increase the decree of vacuum in an electronic tube) getter *m*, dégazeur *m*
[mining] (colloq for a worker engaging in earth-moving work) piqueur *m* à la veine
~CAPACITY - [for a specified gas] capacité *f* de getterisation
~EFFECT - [electron] effet *m* getter
~EVAPORATION - [electron] evaporation *f* du getter
~FILM - [electron] couche-getter *f*, couche *f* absorbante
~FLASH - s. getter evaporation
~FORMATION - [electron] formation *f* du getter
~LAYER - s. getter film
~METAL - [electron] métal *m* getter
~MIRROR - [electron] miroir *m* de getter
GETTER PATCH - s. getter mirror
~PLATE - [electron] palette *f* du getter
~PUMP - [electron] pompe *f* à getter
~SURFACE - [electron] surface *f* de sorption (par getter)
GETTERING - [electron] (the use of a gas-binding substance to eliminate residual gases in an electronic tube) évaporation *f* du getter
~PUMP - s. getter pump
GETTING - [mining] abattage *m*, exploitation *f*, souscavage *m*
GEYSER - [geol] (intermittent hot spring) geyser *m*
[gas ind] (water-heating appliance) chauffe-bain *m* à gaz
~WATERHEATER - s. geyser [gas ind]
GEYSERITE - [min] (siliceous sinter) aggloméré *m* silicieux, geysérite *f*
G-FORCE - [astronaut] force *f* G, force *f* d'inertie
G.H.A. - s. greenwhich hour angle
GHERKIN - [bot] cornichon *m*
GHOST - [gen] fantôme *m*, spectre *m*
[opt] (appearing in optical instruments) double image *f*, image *f* fantôme
[metall] (band of steel with insufficient carbon content) bande *f* de ferrite libre
[telev] (unwanted images causing double and sometimes treble pictures) double image *f*
[photo] tache *f* centrale
~CIRCUIT - [telecomm] (in telegraphy) circuit *m* combiné double
~LINE - s. ghost [metall]
~MODE - [electron] (waveguides) mode *m* fantôme
GHOSTS - [phys] fantômes *mpl*
GIANT - [gen] géant *m*
[adj] gigantesque
~AIR SHOWER - [nucl] (shower of a large number of cosmic-ray particles over a large area) gerbe *f* extensive
~CELL - [zool] (exceptionally large cells, like those of the bone-marrow, cerebrum etc) cellule *f* géante
~FIBRES - [zool] (enlarged nerve-fibres of the ventral nerve cord) fibres *fpl* géantes
~RESONANCE - [phys] crête *f* de résonance
~STAR - [astr] (more luminous than other stars of the same spectral class) étoile *f* géante
~TYRE - [rubber ind] (of tyres) pneu *m* géant
GIANTISM - [med] (abnormal growth of the body) géantisme *m*, gigantisme *m*
GIB - [mech] (metal piece transmitting the thrust of a wedge or a cotter) patin *m*
[mech] (of steam engine) contre-clavette *f*
~AND COTTER - [mech] (of steam engine) clavette *f* et contre-clavette
~GUIDE - [mech] guide *m* à clavette
~-HEADED KEY - [mech] (used for securing wheels) clavette *f* à mentonnet, clavette *f* à tête
~KEY - [mech] clavette *f* à nez, clavette *f* à talon
~STEEL - [metall] acier *m* pour clavettes
GIBBERELLIC ACID - [chem] (a naturally occurring plant hormone with uses as a growth promoter) acide *m* gibbérellique
GIBBERISH - [acoust] (a confused hardly understandable way of talking) baragouinage *m*
~TOTAL - [comput] total *m* mêlé
GIBBET - [mech] flèche *f* de grue
GIBBOSITY - [gen] gibbosité *f*, bosse *f*

[med] gibbosité *f*
GIBBOUS - [gen] (of convex form, as of the moon between half and full) gibbeux, convexe
GIBBSITE - [min] (hydrated oxide of aluminium) gibbsite *f*
GIG, to - [fishing] foéner
[text] gratter, lainer
GIG - [min] (a skip) cage *f*
[transp] (a vehicle) tapecul *m*
[naut] petit canot *m*, yole *f*, guigue *f*
[fishing] foène *f*
[text] échardonneuse *f*
[mining] machine *f* d'extraction, cage *f* d'extraction à deux étages, benne *f*
GIGANTISM - [bot] (abnormal increase in size) gigantisme *m*
GIGANTOCYTE - s. giant cell
GIGGING - [text] lainage *m*
GILBERT - [el] (the c.g.s. unit of magnetomotive force) gilbert *m*
GILD, to - [gen] (to apply gold-leaf) dorer
GILDED - [gen] doré
GILDING - [gen] dorure *f*
[bookbind] (the application of lettering or design to the back of a book) dorage *m*, dorure *f*
~BY AMALGAMATION - [metall] dorure *f* au mercure
~METAL - [metall] (a copper-zinc alloy) alliage *m* cuivre-zinc
~PRESS - [metall] presse *f* à dorer
GILL - [zool] (of acquatic animals) branchie *f*
[th eng] (of a heat radiator) ailette *f*
[text] peigne *m*
[meas] (measure for liquids, i.e. one quarter of a pint) gill *m*, canon *m*
[comput] (unit of milliseconds required by the machine for a given operation) gill *m*
~ARCH - [zool] (in fish) arc *m* branchial
~BAR - [text] barrette *f* de gills
~BOOK - [zool] (book-like respiratory gills) branchies *f*pl
~BOX - [text] étirage *m* à barrettes, gill-box *m*
~FRAME - [text] (hemp) barrette *f* de gills, peigne *m* pour fibres libériennes
~NEEDLE - [text] aiguille *f* de barrette
~NET - [fishing] (vertical net with the head rope fixed to a buoy) araignée *f*, sanglon *m*
~PIN - [text] (of a gill box) dent *f* du peigne
GILLED RADIATOR - [heat] (a radiator consisting of tubes fitted with fins or gills to increase radiation surface) radiateur *m* à ailettes
[auto] radiateur *m* à ailettes
~TUBE WITH FINS - [metall] tube *m* à ailerons
GILLING - [text] étirage *m*
GILLS - [th eng] (thin flat plates fixed to a pipe at right angles to its axis, to increase heat transfer surface) ailettes *f*pl
[mech] ailettes *f*pl
GILPINITE - [min] (copper and uranium containing mineral) gilpinite *f*
GILSONITE - [min] (a naturally occurring asphalt with application in the production of protective coatings, dielectric compounds, and waterproof coatings) gilsonite *f*
GILT - [gen] dorure *f*, dorage *m*
[zool] (a young sow) jeune truie *f*
[metall] métal *m* doré
GILT WIRE - [text] fil *m* doré
GIMBAL - [mech] balancier *m*
~AXIS - [mech] axe *m* de cardan
~FREEDOM - [mech] degré *m* de liberté d'un balancier
[mech] degré *m* de liberté d'un cardan
~GAIN - [contr] amplification *f* de balancier
~LOCK - [mech] blocage *m* de cardan
[mech] verrouillage *m* du balancier
~MOMENT OF INERTIA - [contr] moment *m* d'inertie d'un balancier
~-RING - [mech] (one of the rings forming part of a set of gimbals) anneau *m* du balancier
GIMBALS - [mech] (a "universal" mounting, consisting of two rings pivotes to each other at points 90 deg. apart, used for supporting a compass, gyroscope etc) suspension *f* à la Cardan
GIMLET - [impl] (hand tool designed to bore holes in wood) vrille *f*, avant-clou *m*, perçoir *m*
GIMP, to - [text] tresser-recouvrir
GIMP - [text] (a fancy yarn) gros fil *m* pour contours
~MACHINE - [text] machine *f* à faire les ganses, guimpeuse *f*, moulin *m* guimpier
GIMPING MACHINE - [text] tresseuse-recouvreuse *f*
GIN, to - [text] (of cotton) égrener
GIN - [bot] genièvre *m*
[mech] (a hand hoist) chèvre *f*, palan *m* de chèvre
[mech] (type of tripod for considerable weights) chèvre *f*, sonnette *f*
[text mach] machine *f* à égrener, égreneuse *f*
[constr] (a trestle) chevalet *m*, tréteau *m*, grue *f* à chevalet, chèvre *f*, baudet *m*
~BLOCK - [naut] poulie *f* à chape croisée
~PIT - [mining] (gin operated shallow shaft) puits *m* à cabestan
~POLE - [mech] (a hoisting device) bigue *f*
[oil ind] bigue *f*
~PULLEY - [mech] poulie *f* de chèvre
GINGER - [bot] (the dried rhizome of Zingiber officinale with uses in foodstuffs and medicine) gingembre *m*
~BEER - [brew ind] bière *f* de gingembre
GINGERBREAD WORK - [arch] (a rather pretentious decoration) architecture *f* prétentieuse
GINGHAM - [text] (cotton or linen cloth, generally in checks) guingan *m*
GINGILI - [bot] sésame *m*
~OIL - [chem] huile *f* de sésame
GINGIVAL - [zool] (of the gums) gingival
GINGIVITIS - [med] (inflammation of the gums) gingivite *f*
GINNERY - [text] (a ginning mill) filature *f* de coton
GINNING - [text] égrenage *m* (du coton)
~MACHINE - [text] machine *f* à égrener, gin *m*, égrenoir *m*
~MILL - s. ginning plant
~OUTTURN - [text] rendement *m* d'égrenage
~PLANT - [agric] installation *f* d'égrenage
GIORGI SYSTEM - [el] (a system in which the principal units are the metre, the kilogramme, the second and the ampere) système *m* Giorgi, système *m* MKSA
~UNIT - [meas] (metre-kilogram-second system) unité *f* Giorgi
GIRAFFE - [zool] girafe *f*
GIRANDOLE - [gen] (tall stand for a lamp) girandole *f*, lustre *m*

GIRBOTOL PROCESS - [chem] (an absorption process for the removal of carbon dioxide or hydrogen sulphide from a mixture of gases) procédé *m* Girbotol
GIRDER - [gen and constr] (built-up structure designed to act as beam) solive *f*, longrine *f*, poutre *f*, poutrelle *f*, ferme *f* métallique
[shipbuild] carlingue *f*
[aero] (supporting structure in an airship) poutre *f*
[mech] (of a machine) ferme *f* de support
~BRIDGE - [constr] (bridge made of beams resting on the bridge supports) pont *m* en poutres, pont *m* à poutre
~CONNEXION - [constr] attache *f* de poutres
~DEPTH - [constr] hauteur *f* d'une poutre
~FOR CRANES - [mech] pont *m* pour grues
~FRAMEWORK - [constr] treillis *m* à longerons
~IRON - [metall] fer *m* pour poutres
~PASS - [metall] cannelure *f* à poutrelles
~POLE - [constr] colonne *f* triangulé, poteau *m* triangulé
~RAIL - [railw] rail-poutre *m*
~-TYPE INCLINED ENLARGER - [photo] agrandisseur *m* à colonne inclinée
GIRDLE, to - [gen] ceinturer, encercler
GIRDLE - [gen] ceinture *f*, gaine *f*
[geol] (thin layer of stone) couche *f* de pierre
GIRDS - [oil ind] (derrick) entretoisage *m*
GIRT - [gen] ceinture *f*
[constr] (a small girder) poutrelle *f*
[timber] (circumference of a round timber) circonférence *f*
[constr] (horizontal supporting element) entretoise *f*
GIRTH, to - [gen] (to surround, to encompass) ceindre, entourer, encercler
[mech] (to measure, e.g. by means of a string closed round the surface) mesurer la circonférence
GIRTH - [gen] tour *m*, contour *m*, circonférence *f*
[naut] cartahu *m*
[constr] (in mines) poutre *f* horizontale
[agric] (of a saddle) sous-ventrière *f*
[oil ind] (horizontal brace supporting the frame of the derrick) poutre *f*
~SHEET - [metall] tôle *f* latérale
GIT - s. gate (foundry)
GITOXIN - [chem] (a glycoside or digitals used in medicine) gitoxine *f*
GIVE - [constr] tassement *m*
[mech] élasticité *f*
[mining] affaissement *m* du toit
GIVEN DOSE - [nucl] (the dose of radiation which is given to an irradiated surface) dose *f* en surface
GIZZARD - [zool] (part of bird stomach) gésier *m*
GLABROUS - [zool] (with a smooth surface) glabre
GLACE ICE - [met] (glass-like ice-layer forming on leading edges and extending afterward, due to freezing raindrops) verglas *m*
~KID - [leather ind] (glossy finished goat skin) chèvre *f* au chrome, peau *f* de chevreau glacée
GLACIAL ACETIC ACID - [chem] (term used for pure acetic because it forms crystals resembling ice, its M.P. being I6.6° C.) acide *m* acétique cristallisable
~ACTION - [geol] (the effect of the ice mixed with rock fragments) action *f* glaciaire
~ALLUVION - [geol] alluvions *fpl* glaciaires
GLACIAL DENUDATION - [gen] (the disintegration of rocks due to glacial conditions) dénudation *f* glaciaire
~DEPOSITS - [geol] dépôts *mpl* glaciaires
~DRIFT - [geol] alluvion *m* glaciaire
~EROSION - [geol] érosion *f* glaciaire
~FACIES - [geol] faciès *m* glacial
~MUD - [geol] boue *f* glaciaire
~PHOSPHORIC ACID - [chem] acide *m* métaphosphorique
~PLOUGHING - [geol] labourage *m* glaciaire
~SANDS - [geol] (the outwash from vast ice-sheets) sables *fpl* glaciaires
~SCOUR - [geol] strie *f* glaciaire
~TILL - [geol] moraine *f*, argile *f* à blocaux
~TROUGH - [geol] auge *f* glaciaire
~VALLEY - [geol] vallée *f* glaciaire
GLACIATION - [met] (the production of ice- crystals) glaciation *f*
GLACIER - [geol] (a river of ice in mountain regions) glacier *m*
~BREEZE - [met] (a katabatic wind flowing down a glacier, and due to cooling by the ice surface) brise *f* de glacier
GLACIERET - [geol] glacier *m* de second ordre
GLACIOLOGY - [geol] (the study of the developpemen and movements of glaciers) glaciologie *f*
GLACIS - [constr](naturally of artificially inclined bank) glacis *m*
[agric] (of a forest) glacis *m*
GLADE - [mech] couvercle *m* du presse-étoupe
[constr] clairière *f*
GLADIATE - [bot] (of a sword-blade shape) en glaive
GLAIR - [bookbind] (glue made of white of egg and vinegar used as an adhesive for gold-leaf in bookbinding) glaire *f*, blanc *m* d'œuf
GLAIRY - [med] glaireux, muqueux
GLANCE - [mech etc] (a swift movement, or impact) ricochet *m*, coup *m* en biais
[min] (a variety of ore with a superficial lustre revealing a metallic nature) minerai *m* lustré
~COAL - [min] anthracite *f*
GLANCING ANGLE - [opt] (the angle between a ray and the tangent plane to a surface) angle *m* de Bragg
GLAND - [anat] (cells which are specialized for secretions required by an organism) glande *f*
[mech] (a complete stuffing box) presse-étoupe *m*, bague *f* d'emboîtement
[mech] (the follower which compresses the packing in a stuffing box) gland *m* de presse-étoupe
[bot] glande *f*, cellule *f*
[mining] (for a guide-rope) pince *f* pour câble-guide
~BOLT - [mech] (bolt designed to hold down a gland) boulon *m* du presse-étoupe
~BONNET - [mech] boîte *f* de presse-étoupe
~COCK - [plumb] (a taper plug cock) robinet *m* à tournant, robinet *m* à boisseau
~-FOLLOWER - [mech] (that part of a gland assembly usually annular, which is forced against the packing to form a seal) chapeau *m* d'un presse-étoupe
~HOUSING - s. gland bonnet
~-NUT - [mech] (annular threaded part acting as a follower in a stuffing-box) écrou *m* de presse-étoupe
~OIL - [ind chem] huile *f* pour presse-étoupe
~PACKING - [mech] garniture *f*, bourrage *m*, joint *m*

de garniture
GLAND RING - [mech] (a loose flange) contre-bride *f*, bride *f* de serrage
GLANDERS - [vet] (contagious disease affecting horses, mules and asses) morve *f*, farcin *m*
GLANDLESS - [mech] sans-presse-étoupe, sans garniture, sans bourrage
~VALVE - [mech] robinet *m* sans garniture, vanne *f* sans presse-étoupe
GLANDULAR - [gen & anat] glandulaire
~ABSCESS - [med] adénite *f* suppurée
~HAIR - [bot] (of hops) poil *m* secréteur
GLANS - [anat] gland *m* (du pénis)
[bot] gland *m*
GLARE, to - [gen] éblouir, briller
GLARE - [gen] éclat *m*, lumière *f* éblouissante
[opt] éblouissement *m*, aveuglement *m*
~INDEX - [telev] échelle *f* d'éblouissements
~PROTECTION - [gen] (protection against intense or dazzling light) protection *f* contre l'éblouissement
~SHIELD - [aero, auto etc] (device to prevent a pilot being dazzled by light from exhaust flames) écran *m* antiéblouissement
GLASERITE - [min] (mineral consisting of potassium sodium sulphate) glasérite *f*
GLASS, to - [gen] glacer, lustrer
[gen] (to enclose in glass) vitrer, garnir de vitres
[text] (a cloth) satiner
GLASS - [glass] (a hard amorphous mixture made by fusing oxides of silicon, boron or phosphorus followed by rapid cooling. This is a general term including many types of mixtures having the same typical composition and physical characteristics) verre *m*
[opt] (a magnifying glass) loupe *f*
~AND CONCRETE - [constr work]) ciment *m* vitreux
~ARM - [med] épaule *f* douloureuse (des joueurs de golf)
~-BACKED MICA TAPE - [el] ruban *m* isolant en lamelle de mica
~BAR - [constr] petit bois *m*, croisillon *m*, fer *m* à vitrage
~BATTS - [glass man] (glass fibre fleeces) toisons *mpl* en fibre de verre
~BEAD SEAL - [metall] perles *fpl* de verre scellées par fusion, scellement *m* à perles de verre
~BEAD TUBE - [metall] tube *m* à perles
BLOCK - [constr] (for construction work) brique *f* en verre
~BLOWER - [glass man] souffleur *m* de verre, verrier *m*
~-BLOWER'S DISEASE - [med] maladie *f* des souffleurs de verre
~-BLOWER'S MOUTH - [med] hypertrophie *f* parotidienne des souffleurs de verre
~BLOWING - [glass man] soufflage *m* du verre
~BONDED DIODE - [electron] diode *f* encastrée en verre
~-BONDED MICA - [glass man] (combination of natural mica and a glass of electrical type, with a low melting point) verre-mica *m*
~BULB - [glass man] ampoule *f* de verre
~-BULB RECTIFIER - [el] (mercury-arc rectifier with an arc occurring in a glass bulb) soupape *f* à ampoule de verre
~CEILING-BOWL - [el] (lighting device) plafonnier *m*
GLASS CLOTH - [el] (material formed by weaving glass fibre) toile *f* verrée
~CONCRETE - s. glass and concrete
~-CUTTER - [tool] (tool for cutting glass)diamant *m* de vitrier, coupe-verre *m*
~CUTTING-SHAPE - [photo] calibre *m* à découper en glace forte
~DROPPER - [impl] compte-gouttes *m* en verre
~ELECTRODE - [electron] (a half cell having a glass membrane through which potential measurements are made)électrode *f* de verre, demi-cellule *f* en verre
~-ENAMELLED - [constr] vitrifié
~FABRIC - s. glass cloth
~FIBRE - [glass man] (a fibre made from glass and used as a reinforcing agent in laminated plastics, as a dielectric, and as a filtration medium for corrosive fluids) fibre *f* de verre
~-FIBRE LAMINATE - [plast ind] (laminated material containing glass fibre reinforcement) stratifié-verre *m*
~FIBRE MAT - [plast ind] mat *m* en fibre de verre, feutre *m* en fibre de verre
~FILAMENT - [glass man] (a single thread of glass) silionne *f*
~FLAKES - [glass man] (very thin flakes of glass made by extrusion) flocons *mpl* de verre
~-FURNACE - [glass man] four *m* de verrerie
~GAUGE - [instr] (in boilers etc) tube *m* jaugeur
~GRINDING - [glass man] façonnage *m* de glaces
~-GROUND JOINT - [mech] rodage *m* en verre
~HALF CELL - [el chem] (half-cell with glass diaphragm through which measurements can be made) demi-cellule *f* en verre
~HARDNESS - [metall] trempe *f* parfaite
~HOLDER - [glass man] (ring supporting a lamp-globe) griffe *f*
~IN FIBRE - s. glass fibre
~LINED - [gen] verré
~LIQUOR - [chem] silicate *m* de sodium
~LUBRICATOR - [impl] graisseur *m* à vase verre
~MAIL - [text] (of a Jacquard loom) maillon *m* en verre
~PANE - [glass man] vitre *m*, carreau *m*
~PANEL - [constr] (sheet of glass) panneau *m* de verre
~-PAPER - [paper man] papier *m* de verre, papier *m* verré
~PLIERS - [impl] tenaille *f* vitrier
~POWDER - [glass man] poudre *f* de verre
~REBATE - [constr] (groove) feilleure *f*
~-REINFORCED PLASTICS - [plast ind] (plastics material strengthened by the incorporation of glass fibre) plastique *m* renforcé à la fibre de verre
~RIM - [gen] (of a mirror, a screen, etc) encadrement *m* de glace
~ROD - [impl] (solid round rod of glass, used for stirring and the like) baguette *f* en verre, agitateur *m* en verre
~ROOFING TILE - s. glass tile
~SHEARS - [impl] cisailles *fpl* de vitrier
~SIDE - [photo] (of a plate) côté *m* vitré
~SILK - [acoust] (absorbing quilt made of layers of glass-silk threads) flocons *mpl* de verre
~SPINNING - [glass man] filature *f* du verre
~STAPLE FIBRE - [glass man] (glass fibre in short pieces) fibre *f* discontinue

GLASS-STARVED AREA - [plast ind] (region in a preform moulding in which there is insufficient chopped glass roving) zone *f* à insuffisance de roving
~TILE - [constr] (in a roof, a small glass pane bonded in with tiles) brique *f* en verre
~-TO-METAL SEAL - [electron] (airtight seal between the component parts of an electronic tube) scellement *m* verre-métal
[mech] scellement *m* verre-métal, joint *m* verre-métal
~TUBE - [impl] (tube of chemical glass, used for laboratory work) tube *m* en verre, tube *m* de verre
~VESSEL - [impl] vase *m* de verre
~WIRE - [glass man] fil *m* de verre
~WOOL - [glass man] (used in air-filters and as an insulant) coton *m* de verre, laine *f* de verre
GLASSED - [constr] (of marble and granite, highly polished) vitré, vitrifié, poli
GLASSES - [opt] lunettes *fpl*
GLASSHOUSE - [constr] serre *f*
[glass man] verrerie *f*
~PLANT - [bot] plante *f* de serre
GLASSINE - [paper man] (transparent glazed paper, used for wrapping etc) papier *m* cristal
GLASSPAPER - [carp etc] (strong paper coated with finely-crushed glass, used as abrasive agent) papier *m* verré
GLASSWARE - [glass man] articles *mpl* de verre, verrerie *f*, cristaux *mpl*
GLASSWORK - [glass man] vitrage *m*, vitraux *mpl*
GLASSWORKS - [glass man] verrerie *f*
GLASSY - [gen] vitreux
[of minerals] hyalin
~FELDSPAR - [min] (potash feldspar occurring as transparent crystals) feldspath *m* vitreux, sanidine *f*
GLAUBERITE - [min] (natural sodium calcium sulphate; monoclinic: occurs with rock salt in saline deposits) glaubérite *f*
GLAUBER'S SALT - [min] (natural hydrated sodium sulphate, crystallizing in the monoclinic system with ten molecules of water) sel *m* de Glauber
GLAUCODOT - [min] (iron cobalt sulpharsenide mineral, occurring with cobaltite, Orthorhombic) glaucodot *m*
GLAUCOMA - [med] (eye condition) glaucome *m*
GLAUCONITE - [min] (natural hydrated silicate of iron and potassium, green in colour and used to some extent in distempers) glauconie *f*, sable *m* glauconifère
GLAUCONITIC SAND - [geol] sable *m* glauconieux (ou glauconifère)
~SANDSTONE - [min] (a moulding sand rich in organic matter) sable *m* glauconifère
GLAUCOUS - [bot] (coated with a greenish bloom) pruiné, pruineux
GLAZE, to - [gen] (the operation of fitting glass-panes) vitrer, garnir de vitres
[gen] (to become glazed) se glacer
[gen] (to overlay or coat with a lustrous substance) vitrifier, lustrer, vernisser
[paper man] (to render smooth and shiny) satiner
[gen] (to make smooth and lustrous) lisser, glacer, surglacer
[text] briller
GLAZE - [ind chem] (colourless glass used to form a lustrous impermeable surface coating on pottery) glaçure *f*, émail *m*
[ind chem] (any coating a lustrous surface) vernis *m*, enduit *m*, émail *m*
[met] (only USA; a coating of ice) verglas *m*
GLAZED - [gen] (of a window etc) vitré
[gen] (glossy) lustré, vernissé
[gen] (of ceramics) émaillé
[paper man] satiné
~BOARD - [paper man] carton *m* glacé, carton *m* lissé
~BRICK - [constr] (brick with a glossy finish) brique *f* vernie, brique *f* émaillée, brique *f* vitrifiée
~DOOR - [constr] (door fitted with glass panels) porte *f* vitrée
~FINISH - [text] glaçage *m*
~FRAME - [constr] encadrement *m* sous verre
~FROST - [met] (layer of smooth ice formed by rain freezing on a surface) verglas *m*
~MECHANISM - [mech] mécanisme *m* sous gaine de verre
~SILK - [text] soie *f* glacée
~YARN - [text] fil *m* glacé
GLAZIER - [gen] vitrier *m*
GLAZIER'S PUTTY - [build] (mixture of whiting and linseed oil) mastic *m* de vitrier
GLAZING - [ceramics] (process to give a smooth lustrous surface to pottery ware by means of a colourless glass fired on to the surface) lustrage *m*, glaçage *m*
[constr] (the fitting of glass panes) pose *f* des vitres, vitrage *m*
[photo] (the operation of applying a shiny surface to prints) émaillage *m*
[paint] (to cover with a thin layer of transparent colour) glaçage *m*
[paper man] satinage *m*
[metall] polissage *m* préliminaire
~BEAD - [constr] tasseau *m* pour vitres
~CALENDER - [paper man] (a set of rollers used to impart a very smooth surface to sheet material or fabrics) calandre *f*, calandre *f* à satiner
[text] calandre *f* à glacer, calandre *f* à lisser
~HEAD - [constr] (loose mould) parclose *f*
~MACHINE - s. glazing calender
[photo] glaceuse *f*, machine *f* à glacer
[text] machine *f* à brillanter, machine *f* à glacer, machine à lustrer
~MILL - [electron] (used for finishing the glass tubes) machine *f* à rebrûler les queusots
~PRESS - [text] presse *f* à glacer, presse *f* à satiner
[photo] presse *f* plate *f* à glacer
~PUTTY - s. glazier's putty
~ROLLS - [mech] (system of rolls arranged to impart a high gloss to material passed between them) cylindres *mpl* à satiner
GLEAM, to - [gen] luire, rayonner
GLEAM - [gen] lueur *f*, rayon *m* (de lumière)
GLEAN, to - [agric] glaner
GLEANER - [agric] glaneur *m*
GLEANING - [agric] glanage *m*, glane *f*
GLIDE, to - [aero] (the operation of descending at normal incidence with very little or no power) planer
GLIDE - [aero] (controlled descent at normal incidence, with very little or no power) vol *m* plané
~ANGLE INDICATOR - [instr] indicateur *m* d'angle

de planement
GLIDE LANDING - [aero] (landing in an unpowered glide) atterrissage *m* moteur réduit
~PATH - [astronaut] trajectoire *f* de descente planée
~PATH BEACON - [radio] (radio-directional beacon giving altitude information to an aircraft during approach and landing) radiophare *m* d'alignement de descente
~-PATH TRANSMITTER - [radar] (the transmitter which sends out signals while the aircraft is on the glide path) émetteur *m* d'atterrissage
~PLANES - [crystall] (symmetry elements of a space lattice) plan *m* de glissement
~PLATE - [text] tôle *f* incliné, glissoir *m*
~ROLL - [text] galet *m* à glissement
~SLOPE - s. glide path
[astronaut] guidage *m* d'atterrissage, angle *m* de pente, pente *f*
GLIDER - [aero] (an aircraft which flies without power by itself, or is towed by a powered aircraft) planeur *m*
[naut] (a power boat with a flat heel) hydroglisseur *m*
[text] curseur *m*, cavalier *m*, lamelle *f*
~PILOT - [aero] pilote *m* de planeur
GLIDING - [aero] plané *m*, planement *m*
~ANGLE - [aero] (the angle which the flight path of an aircraft in a glide makes with the horizontal) angle *m* de plané
~BAND - [metall] ligne *f* de glissement
~BLOCK - [text] coulisseau *m* des glissoires
~BOAT - [naut] hydroglisseur *m*
~CERTIFICATE - [aero] certificat *m* de vol plané, certificat *m* de vol à voile
~MACHINE - s. glider
~PLANES - s. glide planes
~PLATE - [el] plaque *f* de glissement
~RIDER - [text] cavalier *m* glissant
~TUBE AND SHEATH - [photo] (of a tripod) tube *m* coulissant et fourreau
GLIOMA - [med] (a type of tumour) gliôme *m*
GLIOMATOSIS - [med] (in the brains or the spinal cord, a diffuse overgrowth of neuroglia) gliose *f*, gliomatose *f*
GLIOSIS - s. gliomatosis
GLITTER, to - [gen] (to emit a sparkling light) scintiller, étinceler, papilloter
GLITTER - [metall etc] (special decorative material in flakes large enough to produce individual reflection, and thus to give a sparkling effect) scintillement *m*, brillant *m*
GLOBE - [gen] globe *m*, sphère *f*
[anat] globe *m* de l'œil
~JOINT - [mech] joint *m* à rotule sphérique, joint *m* à genou
~LIGHTNING - s. globular lightning
~MILL - [min] moulin *m* à boulets
~PATTERN VALVE - s. globe valve
~VALVE - [mech] (a control device for fluids, in which the body is globular, and the obturating element is a disk moved into or off its seat by a screw and handwheel) soupape *f* à boulet
GLOBERINA OOZE - [geol] (deep-sea deposit covering large sections of the ocean floor) boue *f* à globigérines
GLOBOMYELOMA - [med] sarcome *m* à cellules rondes
GLOBULAR - [gen] globulaire, sphérique, globuleux
~CEMENTITE - [metall] (cementite occurring in the form of globules in steel) cémentite *f* globulaire
~FLASK - [ind chem] ballon *m*
~LIGHTNING - [met] (slowly moving luminous ball sometimes seen during thunderstorms) éclair *m* en boule
~PEARLITE - [metall] (a granular pearlite) perlite *f* globulaire
~SHRINKING CAVITY - [foundry] retassure *f* sphérique
GLOBULE - [gen] (a small, approximately spherical body) globule *f*, goutelette *f*
[photo] (in colour photo) granule *m*
GLOBULINS - [chem] (proteins, such as globulin, fibrin, myosin, fibrinogen and others belong to this group. They are insoluble in water but soluble in dilute saline solution) globulines *fpl*
GLOBULUS - [anat] noyau *m* globuleux, noyau *m* sphérique
GLORY - [opt] (concentric coloured rings seen round a shadow cast on a cloud or fog bank (caused by diffraction) gloire *f*, spectre *m* de Brocken
~-HOLE - [glass man] (a subsidiary furnace for reheating) four *m* de réchauffage
[metall] (a peep-hole of a furnace) regard *m*; (in a reactor) canal *m* expérimental
~-HOLE SYSTEM - [mining] exploitation *f* par entonnoir souterrains
GLOSS, to - [gen] lustrer, glacer
GLOSS - [paint] (a bright smooth surface on a coating) lustre *m*, brillant *m*
[gen] luisance *f*, lustre *m*
[text] cati *m*
~OIL - [paint] (an inexpensive coating medium, made by dissolving resin in a suitable solvent) huile *f* vernis-polie
~WHITE - [paint] (composite filler, made by co-precipitating barium sulphate and alumina) blanc *m* brillant
GLOSSIMETER - [instr] (an instrument for measuring the ratio of the light reflected regularly or specularly from a surface to the total amount reflected) brilliancemètre *m*
GLOSSINESS - [gen & paint] glacé *m*, lustre *m*, aspect *m* brillant
GLOSSING - [text] lustrage *m*
~MACHINE - [text] machine *f* à cheviller
GLOSSITIS - [med] (inflammation of the tongue) glossite *f*
GLOSSOCELE - [med] glossocèle *f*
GLOSSOPHOBIA - [med] lalophobie *f*
GLOSSOPYROSIS - [med] langue *f* rôtie
GLOSSY - [gen] lustré, glacé, brillant
[paint] brillant
~COLOUR - [gen] couleur *f* brillante
~FINISH - [text] apprêt *m* glacé
~PRINT - [photo] épreuve *f* glacée
~SKIN - [med] liodermie *f*
GLOST FIRING - [ceram] (operation in which pottery ware is strongly heated to vitrify the glaze) émaillage *m* au fer
~KILN - [ceram] (kiln used for glazing pottery ware) four *m* à émailler
GLOTTIS - [anat] (the opening from the pharynx to the trachea) glotte *f*
GLOVE, to - [gen] ganter

GLOVE - [gen] gant *m*
~BOX - [auto] boîte *f* à gants, vide-poche *m*
~COMPARTMENT - s. glove box
~FORMER - [rubber ind] moule *m* de gant, forme *f* de gant
~LEATHER - [leather ind] (a tanned skin) cuir *m* pour ganterie, peaux *f*pl pour ganterie
GLOVER TOWER - [chem] (tower used in sulphuric acid manufacture, in which the sulphur dioxide is nitrated) glover *m*, tour de Glover
GLOW, to - [gen] luire (rouge), rayonner
[metall] rougir
GLOW - [gen] lueur *f* rouge, luminescence *f*
~DISCHARGE - [el] (silent luminous discharge of electricity through a gas in a vacuum tube) décharge *f* à lueur, effluve *m*, décharge *f* luminescente
~DISCHARGE LAMP - [electron] tube *m* à décharge
~FILAMENT - [light] filament *m* à incandescence
~LAMP - [light] (a gas-discharge lamp) lampe *f* à incandescence
~PLUG - [mech] bougie *f* à incandescence, bougie *f* de préchauffage
~PLUG FILAMENT - [el] filament *m* de bougie à incandescence
~POTENTIAL - [el] (the potential across a glow discharge) tension *f* d'amorçage
~SCREEN - [metall] (device to cut off glow from heated surfaces) écran *m* de protection
~TUBE - [electron] (cold-cathode gas-filled diode) tube *m* à décharge luminescente
~-WORM - [zool] ver *m* luisant, luciole *f*
GLOWING - [gen & in combustion] incandescence *f*, embrasement *m*, rayonnement *m*
GLOWPROOF - [metall] (capable of retaining given properties at red-heat) stabile à l'incandescence
GLUCINIUM - [chem] (a term formerly used for the element beryllium q.v.) glucinium *m*, beryllium *m*
GLUCOMETER - [instr] (instrument designed to measure the amount of glucose) glucomètre *m*
GLUCONIC ACID - [chem] (an oxidation product of dextrose used in the preparation of foodstuffs and pharmaceuticals) acide gluconique
GLUCOSE - [chem] (dextrose obtained from starch by the action of sulphuric acid) glucose *f*
GLUCURONOLACTONE - [chem] (an intermediate for pharmaceuticals) glucuronolactone *f*
GLUE, to - [gen] coller
GLUE - [chem] (adhesive based on impure gelatine or synthetic resins) colle *f*
~HEATING APPARATUS - [carp] chauffe-colle *m*
~LAYER - s. glue spread
~LINE - [ind chem] (the layer of adhesive uniting a glued joint) collure *f*
~PENETRATION - [chem] (the degree to which glue is absorbed by a surface) pénétration *f* de la colle
~POT - [carp etc] (vessel for heating glue) pot *m* à colle
~SOLUTION - [chem] solution *f* de colle
~SPREAD - [chem] (the weight of glue in a glue line per unit area of surface) consommation *f* de colle
GLUEING - [ind proc] (the joining together of two articles with glue) collage *m*
[bookbind] collure *f*
~MACHINE - [mech mach] machine *f* à coller, encolleuse *f*
GLUEPOT - s. glue pot
GLUING - s. glueing
GLUME - [bot] (a dry membrane) glume *f*
GLUMIFEROUS - [bot] (with flowers enclosed by glumes) glumifère
GLUT, to - [gen] encombrer, rassasier
[comm] (a market) surcharger (le marché)
GLUT - [gen] encombrement *m*, assouvissement *m*
[mech] point *m* d'appui
[carp] coin *m*
[metall] (of a kiln) ouverture *f* (d'un four de grillage)
[constr] (a closer) clausoir *m*, closoir *m*
GLUTAMIC ACID - [chem] (an amino acid present in a number of proteins and having medicinal uses) acide *m* glutamique
GLUTAMINE - [chem] (a biological culture medium) glutamine *f*
GLUTARALDEHYDE - [chem] (a component of water-repellent treatments for textiles) aldéhyde *f* glutarique
GLUTARIC ACID - [chem] (a synthesis intermediate) acide *m* glutarique
~ANHYDRIDE - [chem] (an intermediate for drugs, plasticizers, and dyes) anhydride *m* glutarique
GLUTARONITRILE - [chem] (a synthesis intermediate) glutaronitrile *m*
GLUTEAL - [anat] fessier *m*
GLUTELINS - [chem] (simple proteins soluble in dilute acids and alkalies, but not in water or neutral salt solutions) glutélines *f*pl
GLUTEN - [chem] (a mixture of vegetable proteins with uses in food processing) gluten *m*
~TURBIDITY - [chem] turbidité *f* glutinique
GLUTINOUS - [gen & chem] (with a clammy exudation) glutineux
GLYCERALDEHYDE - [chem] (an intermediate for polyesters) glicéraldéhyde *m*
GLYCERIDE - [chem] (generic name for esters of glycerol with organic acids) glycéride *m*
GLYCERIN, GLYCERINE - s. a synonym for Glycerol
~CARBONATE - [chem] (a synthesis intermediate and solvent) carbonate *m* de glycérine
~LITHARGE CEMENT - [chem] (mixture of lead oxide and glycerin setting to a very hard mass) ciment *m* à glycérine et litharge
GLYCEROL - [chem] (an intermediate for synthetic resins, solvent, plasticizer, component of anti-freeze mixtures, and an ingredient of cosmetics, pharmaceuticals, and foodstuffs) glycérol *m*, glycérine *f*
~BORIBORATE - [chem] (a textile processing agent and an adhesive) boriborate *m* de glycérine
~DIACETATE - [chem] (plasticizer for cellulose nitrate and acetate) diacétate *m* de glycérine
~ETHER ACETATE - [chem] (plasticizer for cellulose derivatives and polyvinyl acetate) acétate *m* de l'éther glycérique
~MONOACETATE - [chem] (plasticizer for cellulose acetate and nitrate) monoacétate *m* de glycérine
~MONOLAURATE - [chem] (plasticizer for cellulose nitrate, ethyl cellulose, polystyrene, polyvinyl butyral, vinyl chloride and vinyl chloride acetate) monolaurate *m* de glycérine
~MONORICINOLEATE - [chem] (lubricant, solvent and plasticizer) monoricinoléate *m* de glycérine
~MONOSTEARATE - [chem] (thickening agent for cosmetics and foodstuffs) monostéarate *m* de glycérine

GLYCEROL TRIACETATE - [chem] (plasticizer for cellulose derivatives, polymethyl methacrylate, and polyvinyl butyral) triacétate *m* de glycérine
GLYCEROPHOSPHATE - [chem] (having medicinal uses) glycérophosphate *m*
GLYCEROPHOSPHORIC ACID - [chem] (starting material in the production of glycerophosphates) acide *m* glycérophosphorique
GLYCERYL - [chem] glycéryle *m*
~ABIETATE - [chem] (an ingredient of synthetic flavourants) abiétate *m* de glycéryle
~PHTHALATE - [chem] (a synthetic resin used in the production of protective coatings) phtalate *m* de glycéryle
~TRIACETOXYSTEARATE - [chem] (a plasticizing for PVC and cellulose derivatives) tri-acétoxi-stéarate *m* de glycéryle
~TRIACETYLRICINOLEATE - [chem] (a plasticizer for cellulose derivatives and PVC) triacétylricinoléate *m* de glycéryle
~TRIBUTYRATE - [chem] (a plasticizer) tributyrate *m* de glycéryle
~TRIHYDROXYSTEARATE - [chem] (an intermediate and a component of cosmetics, lubricants, and plasticizer) trihydroxystéarate *m* de glycéryle
~TRIPROPIONATE - [chem] (a plasticizer) tripropionate *m* de glycéryle
~TRIRICINOLEATE - [chem] (an emulsifying agent) triricinoléate *m* de glycéryle
GLYCIDOL - [chem] (a plastics plasticizer, textile dyeing agent, and demulsificant) glycidol *m*
GLYCINE - [chem] (a non-essential amino acid with uses in medicine and as a synthesis intermediate) glycine *f*
~ETHYL ESTER HYDROCHLORIDE - [chem] (a synthesis intermediate) chlorhydrate *m* éthylestérique de glycine
GLYCOGEN - [chem] (a polysaccharide found in animal tissue) glycogène *m*
GLYCOL - [chem] (dihydric alcohol) glycol *m*
GLYCOL - DIMERCAPTOACETATE - [chem] (curing agent for synthetic resins and a crosslinking agent for rubbers) dimercaptoacétate *m* de glycol
~SALICYLATE - [chem] (an external analgesic used in the treatment of rheumatism) salicylate *m* de glycol
GLYCOLIC ACID - [chem] (an intermediate for plasticizers) acide *m* glycolique
GLYCOLONITRILE - [chem] (intermediate and solvent) glycolonitrile *m*
GLYCOSIDES - [chem] (glucose derivatives formed by alkylation of the -OH group on C_1) glucosides *mpl*
GLYCOSURIA - [med] (the present of sugar in the urine) glycosurie *f*, méliturie *f*
GLYCYRRHIZA - [chem] (liquorice. The dried extract of varieties of Glycyrrhiza used in confectionery and medicine) glycyrrhizine *f*
GLYOXAL - [chem] (an insolubilizing agent for compounds containing polyhydroxyl groups and a stabilizer for rayon) glyoxal *m*
GLYPH - [arch] (short vertical flute) glyphe *m*
G-METER - [instr] accéleromètre *m*
GMELIN TEST - [chem] (used for the presence of bile pigments) essai *m* de Gmelin
GMELINITE - [min] (hydrated silicate of aluminium, sodium and calcium) gmélinite *f*
G.M.T. - s. greenwich mean time

GNASHING - [acoust] (the sound made teeth) crissement *m*, grincement *m*
GNAT - [zool] moucheron *m*, poustique *m*
GNATHALGIA - [med] névralgie *f* maxillaire
GNATHITIS - [med] gnatite *f*, inflammation de la mâchoire
GNAW, to - [gen] ronger
GNEISS - [geol] (coarse metamorphic rock) gneiss *m*
GNOME - [gen] gnome *m*
~CALF - [med] pseudohypertrophie des mollets
GNOMON - [surv] (primitive instrument for the determination of time and latitude) gnomon *m*
GNOMONIC CHART - [surv] carte *f* gnomonique
~PROJECTION - [surv] (method of map projection in which the centre of projection is that of the earth, the plane of projection tangential to the latter) projection *f* gnomonique
GNOTOBIOTICS - [astronaut] étude *f* de la vie *f* aseptique
GO-DEVIL - [petr ind] (round steel bar dropped down the drill pipe to open the top valve of the tester, so as to bring tester and empty drill pipe into communication) passe-diable *m*
[oil ind] (device used for cleaning out a pipeline) écouvillon *m* pneumatique
[oil ind] (device for the separation of two liquids pumped one after the other through the same pipeline) séparateur *m* de pétrole
[oil ind] ramoneur *m*
[mining] installation *f* transportable du plan incliné
[mech] (a gravity plane) plan *m* incliné
[railw] (used for the transport of personnel) bogie *m* de service
[agric] (a sled cultivator) cultivateur *m* à traîneau
~-NO-GO-TEST - [autom] (in quality control) test *m* oui-non
GOAF - [min] (the empty space in a coal seam after extraction) remblai *m*
~STOWAGE - [mining] remblayage *m*
GOAT - [zool] chèvre *f*
~GRASS - [bot] égilope *m*
~-SKIN - [leather ind] peau *f* de chèvre
GOAT'S HAIR - [text] poil *m* de chèvre
~HAIR YARN - [text] fil *m* de poil de chèvre
~MILK CHEESE - [dairy] fromage *m* de chèvre
~RUE - [bot] (herbage plant) rue *f* des chèvres
GOB - [glass man] (small lump of glass taken as sample during processing) prélèvement *m*
~FIRE EXPLOSION - [mining] explosion *f* de grisou
~PROCESS - [glass man] (the delivery to a forming unit in a gob form) procédé *m* à goutte
~ROAD - [mining] galerie *f* dans les remblais
~STUFF - [mining] remblai *m*
GOBBING - [mining] remblayage *m*, remblai *m*
GOBELIN - [text] (type of smooth coloured tapestry) gobelin *m*
GOBO - [cin] (a panel covered with sound-absorbing material) écran *m* insonorisé portable, (US) écran *m* pare-lumière
GODET - [text] (rayon spinning) galette *f*
~ROLLS - [mech] (pairs of rollers running at different speeds, around which filaments are passed to give orientation tension) cylindres *mpl* à godets
~WHEELS - s. Godet rolls
GOES-OVER - [cin] (colloq for lens shield) parasoleil *m*

GOETHITE - [min] (an ore of iron, resembling limonite, consisting of ferric oxide crystallizing in the orthorhombic system with one molecule of water) gœthite *f*
GOFFER, to - [text etc] gaufrer
GOFFER - [text etc] gaufreur *m*
[mech] presse *f* à gaufrer
GOFFERED CLOTH - [text] (embossed cloth) tissu *m* gaufré
~PLUSH - [text] peluche *f* gaufrée
~TINSEL - [text] lame *f* gaufrée
GOFFERING - [text etc] gaufrage *m*, plissage *m*, tuyautage *m*
[metall] froncement *m*, plissotement *m*
~IRON - [text etc] fer *m* à tuyauter, godron *m*
GOGGLES - [impl] (device for protecting the eyes from radiation, gases, splashing or other dangers) lunettes *fpl* protectrices
GOING - [build] (in a stair) giron *m*, foulée *f*, emmarchement *m*
~PART - [text ind] (colloq) semelle *f* du battant
GOITRE - [med] (enlargement of the thyroid gland) goitre *m*
GOITROGENIC - [med] (with a tendency to produce goiter) strumigène
GOITROGENOUS - s. goitrogenic
GOITROUS - [med] (pertaining to goiter) goitreux
GOLD - [metall] (metallic element, symbol Au, A. N. 79, A.W. I97.2, one of the precious metals) or *m*
~AMALGAM - [min] (variety of gold containing about 60 p.c. pf mercury) amalgame *m* d'or
~BEAD - [text] perle *f* dorée
~BEATER'S SKIN - [impl] (membrane obtained from animal intestines) baudruche *f*
~BEATING - [metall] battage *m* d'or
~BLOCKING - [bookbind] (the process of pressing a design on gold leaf) dorure *f* en feuilles, impression *f* en or
~-BONDED DIODE - [electron] diode *f* à connexion en or, diode *f* à pointe or
~BRAID - [text] galon *m* d'or
~BROCADE - [text] brocart *m* en broché or
~BRONZE - [paint] (gold-coloured metallic powder, used in coatings, obtained from copper and its alloys) bronze *m* doré, bronze *m* d'or
~BULLION - [metall] or *m* en lingots
~CHLORIDE - [chem] (a ceramics pigment, photographic chemical, and starting material for certain dyes) chlorure *m* d'or, aurichlorure *m*
~COATING - [gen] dorure *f*
[electron] (with the aim of making the grid a heat reflector, the control-grid of some oxide-cathode tubes is coated with gold) couche *f* en or
~CONTENTS - [min] teneur *f* en or
~CUSHION - [bookbind] coussin *m* pour impression en or
~DIGGING - [min] orpaillage *m*, exploitation *f* des alluvions aurifères, exploitation *f* aurifère
~DUST - [metall] poussière *f* d'or
~ELECTRODE - [electron] électrode *f* en or
~-FILLED - [metall] doublé
~-FILM GLASS - [glass man] (special window-glass, especially for aircraft, in which a thin film of metallic gold, which can be heated by passing an electric current through it, is sandwiched between two sheets of special glass. The whole sheet can thus be kept at a controlled temperature, to prevent ice-or mist - formation) verre *m* à pellicule d'or
GOLD-FILM WINDOW PANEL - [glass man] (type of laminar glass panel which is heated internally by means of a thin film of gold through which an electric current is passed) vitre *f* à pellicule d'or
~FOIL - [metall] feuille *f* d'or
~HYDROXIDE - [chem] (a component of gold plating baths, ceramic glazes, and rubber pigments) hydroxyde *m* d'or
~IN BARS - [metall] or *m* en lingots
~LEAF - [metall] (gold beaten out into very thin sheets and used for gilding) feuille *f* d'or, or *m* en feuille
~-LEAF ELECTROSCOPE - [el] (electroscope which consists of a glass jar containing two pieces of gold leaf attached to a metal rod passing through the top of the jar) électroscope *m* à feuilles d'or
~MINE - [min] mine *f* d'or, exploitation *f* aurifère
~NUMBER - [chem] (the weight in milligrams of a protective colloid just sufficient to prevent the change from red to blue in I0 cubic centimetres of a standard gold sol after the addition of one cubic cm. of a I0 p.c. sodium chloride solution) nombre *m* d'or
~OXIDE - [chem] (a rubber pigment and component of ceramic glazes) oxyde *m* d'or
~NUGGET - [min] pépite *f* d'or
~PAN - [mining] batée *f*
~PLATED - [metall] doublé d'or
~-PLATED CONTACTS - [electron] (contacts coated with gold to prevent corrosion) contacts *mpl* dorés
~PLATING - [metall] dorage *m*
~POINT - [fin] gold-point *m*, point *m* d'or
~POTASSIUM CHLORIDE - [chem] (component of photographic chemicals and ceramic glazes) chlorure *m* d'or et de potassium
~POWDER - [metall] poussière *f* d'or
~RIMMED - [metall] cerclé d'or
~SIZE - [chem] (a mixture of linseed oil, turpentine and copal varnish used as an adhesive and protective coating) or *m* couleur, vernis *m* d'apprêt, vernis *m* à sceller
~SODIUM CHLORIDE - [chem] (a colourant for glassware) chlorure *m* d'or et de sodium
~SODIUM THIO-MALONATE - [chem] (a gold salt having application in medicine) thiomalonate *m* d'or et de sodium
~SPRING - [horol] (in the detent escapement) ressort *m* spiral
~STOVING VARNISH - [paint] (transparent varnish, treated by heat after application, to give a gold - or brass-like appearance to tinplate or the like) vernis *m* doré à l'étuvage
~TELLURIDES - [min] (these occur naturally as Sylvanite, Petzite, and Calaverite, Nagyagite is a sulpho-telluride of lead and gold) tellulures *mpl* d'or
~THREAD - [text] fil *m* d'or
~-TIN PURPLE - [chem] (purple of Cassius. A colourant for glass and ceramics) pourpre *m* de Cassius
~TONING - [photo] (addition of a gold film to the surface of the silver image of a print) virage *m* à l'or
~TRIBROMIDE - [chem] (an analytical reagent) tribromure *m* d'or
~VEIN - [mining] veine *f* aurifère
GOLDEN - [gen] d'or

GOLDEN BERYL - [min] beryl m jaune

~ NUMBER - [astr] (the place of any year in the Metonic cycle of nineteen years) nombre m d'or

GOLDSMITH - [gen] orfèvre m

GOLDSMITHING - [metall] orfèvrerie f

GOLDSTONE - [min] aventurine f

GOLF HOSE - [text] bas m de sport

GOLFERS - [text] (type of rag) laine f renaissance

GOLIATH CRANE - [mech] grue-chevalet f (sur roues), portique m

GOLOSH - [shoe man] (also galosh, overshoe) galoche f

GONAD - [zool] (a sex gland) gonade f

GONDOLA - [railw] (only USA) wagon m découvert

[aero] (of airship) nacelle f, gondole f

[naut] (only USA) bachot m pour navigation fluviale

~ CAR - [railw] s. gondola

GONEITIS - [med] (joint inflammation) gonite f, gonarthrite f

GONG - [mus] (bronze instrument in the form of a disk) gong m

GONIOMETER - [instr] (instrument designed to measure the angles between two planes) goniomètre m

GONIOMETRY - [meas] goniométrie f

GONIOSCOPE - [instr] gonioscope m

GONITIS - s. goneitis

GONOCOCCUS - [med] (the causative agent of gonorrhea) gonocoque m

GONYCAMPSIS - [med] flexion f permanente du genou

GONYOCOELE - [med] tuméfaction f du genou

GOOCH CRUCIBLE - [ind chem] (special type of porcelain crucible containing an asbestos filter) creuset m de Gooch

GOOD BOTTOM - [mining] bon mur m

~ EARTH - [el] (an earth connexion of negligible resistance, hence virtually at zero potential) bonne terre f

GEOMETRY - [phys] (in nuclear physics measurements) bonne géométrie f

~ PACK - [mining] bon remblai m

GOODS - [gen] biens mpl, effets mpl, marchandises fpl

~ SHED - [railw] halle f à marchandises

~ STATION - [railw] gare f à marchandises

~ TRUCK - [railw] wagon m à marchandises

GOOSE-NECK - [mech] col m de cygne

[metall] (the shape of a pipe) coude m porte-vent

[metall] (diecasting) col m de cygne

~-NECK BEND - [plumb] col m de cygne, bec m d'oie

~-NECK DYE-CASTING MACHINE - [metall] machine f à fondre à air comprimé à chambre mobile

~-NECK MACHINE - s. goose-neck press

~-NECK NOZZLE - [metall] (diecasting) buse f de injection

~-NECK PRESS - [metall] (in diecasting; type of press in which the vertical structure connecting the head and the base is curved somewhat in the shape of the letter "C", in stead of consisting of straight colums) presse f col de cygne

GOOSEBERRY - [bot] groseille f verte

~-STONE - [min] grossularite f, grenat m calcifère

GOPHER HOLE - [mining] fourneau m de mine

GOPHERING - [mining] exploitation f irrationnelle

GORE - [aero] (segment of a balloon envelope) fuseau m

[aero] (a shaped section of a parachute canopy usually between rigging lines) fuseau m

GORE - [text] coin m, clavette, rainure f

~ HEEL - [text] talon m américain

GORGE - [geol] gorge f, défilé m

[constr] (also called drip or throat; groove in the undersurface of a stone) mouchette f

[mech] (a groove) gorge f, rainure f

GORGERIN - [arch] (of a column) gorgerin m

GORILLA - [zool] gorille m

GORSE - [bot] genêt m épineux

[impl] (a kind of barrel for carrying water underground) tonnelet m

GOSLARITE - [min] (zinc sulphate) goslarite f

GOSSAGE'S PROCESS - [chem] (a method for the manufacture of sodium hydrate by boiling sodium carbonate in solution with lime) procédé m Gossage

GOSSAN - [min] (the oxidized upper part of a vein of sulphide minerals) chapeau m de fer

GOSSYPOL - [min] (a naturally occurring polyphenol with uses as an antioxidant for rubbers and as a stabilizer for certain powders) gossypol m

GOTH - [min] (sudden bursting of the roof) rupture f soudaine du plafond

GOTHIC - [arch] gothique

[print] gothique m, caractères mpl gothiques

~ ARCH - [arch] arc m gothique

~ PASS - [metall] (of a rolling mill) cannelure f ogive

~ ROOF - [arch] toit m gothique

~ SECTION - [métall] profil méplat

GOTTEN - [min] (minerai) extrait

[min] (the coal which is ready to be carried away) charbon m extrait, houille f extraite

GOUACHE - [gen] (used in impasto style) gouache f

GOUFING - [constr] (the strengthening of wall foundations) renforcement m des fondations

GOUGE - [impl] (tool used to form curved lines) gouge f, ciseau à gouge

[carp] gouge f à ébaucher

[mining] (salvage) salbande f argileuse, lisière f

[med] (a hollow chisel) gouge f

~-BIT - [carp] (wooden boring bit) ciseau m à gouge

~ SLIP - [carp etc] pierre f à aiguiser une gouge

~-TAPPING - [carp etc] taraudage m à gouge

GOUGING - [gen] (removal of a part by sudden cutting) travail m à la gouge

GOUNDOU - [med] goundou m

GOURD - [bot] courge f, gourde f

GOUT - [med] (disease due to excess of uric acid in the blood) goutte f, podagre f

~ FLY - [zool] (a pest) chlorops m, mouche f jaune des chaumes

GOUTY - [med] goutteux

GOVERN, to - [gen] gouverner

[mech] régler

GOVERNING - [mech] réglage m; (adj) modérateur

~ FACTOR - [el] facteur m dominant

~ THE WARP - [text] livraison f de la chaîne

GOVERNOR - [mech] (a mechanism to provide automatic control of speed) régulateur m, régulateur m de vitesse

[el] monostat m régulateur

~ BOARD - [mech] tableau m du régulateur

~ CASING - [mech] carter m du régulateur

~ HEAD - [mech] pendule m du régulateur

~ HOUSE - [gas ind] (gas distribution) salle f d'émission, émission f

GOVERNOR HOUSING - s. governor casing
~LEVER - [mech] balancier *m* du régulateur
~MOVEMENT - [cin] (the mechanism which controls the automatic shutter) mécanisme *m* régulateur
~SLEEVE - [mech] douille *f* du régulateur
~SPINDLE - [mech] arbre *m* du régulateur
~WEIGHT - [mech] (a mass of metal used in a governor to provide control by centrifugal action) masselotte *f* centrifuge du régulateur
~WITH AUTOMATIC LOADING - [gas ind] (gas distribution) régulateur *m* à surcharge automatique
~WITH OVERRIDING TIME CONTROL - [gas ind] (gas distribution) régulateur *m* à surcharge horaire
GOX - [astronaut] oxygène *m* gazeux
G.P. - (of heavenly body) s. geographic position
G.P.I. - s. ground position indicator
GRAB, to - [gen] saisir, empoigner
[mech] agripper
(the action of collecting material by means of a grab) rassembler (des matériaux) par grue à grappin
GRAB - [mech] (device attached to the fall end of a crane rope, which automatically picks up a load when dropped upon loose material) grappin *m*, benne *f* automatique, benne *f* piocheuse
~BUCKET - s. grab
~CHAIN - [mech] (of earthmoving machines) chaîne *f* de bennes
~CRANE - s. grabbing crane
~DREDGE - [mech mach] drague *f* à mâchoires
~-HANDLE - [mech] manche *m*, poignée *f*
~-HOOK - [impl] grappin *m*, croc *m*
~IRON - [mining] accrocheur *m*
~JAWS - [mech] (of a crane) mâchoire *f* de benne
~RAIL - [auto] rampe *f*
~SAMPLE - [mining] échantillon *m* pris au grappin
GRABBING CLUTCH - [US for fierce clutch] embrayage *m* brutal
~CRANE - [mech mach] (excavator which consists of a crane and a large grab) grue *f* à grappin
~OF BRAKES - [mech] (fierce action in brakes) broutement *m* des freins
GRABEN - [geol] (geological structure caused by subsidence) graben *m*, fossé *m*
GRACE NOTES - [mus] broderies *fpl*, fioritures *fpl*
GRACES - s. grace notes
GRADATION - [gen] gradation *f*
[opt] (the slope of the tangent to the brightness reproduction characteristic at each point) gradation *f*
GRADE, to - [gen] classer, trier, échelonner, nuancer
[photo] étalonner
GRADE - [gen] (quality) grade *m*, rang *m*, degré
[gen] (gradient) pente *f*, montée *f*, descente *f*
[chem] (the rating) indice *m* d'octane
[geol] profil *m* d'équilibre, pente-limite *f*
~CROSSING - [railw] (USA only; level crossing) passage *m* à niveau
~OF A COAL - [min] calibre *m* d'un charbon
GRADED COKE - [fuel] (classified) coke *m* classé
[min] (screened) coke *m* classé
[min] (divided in sizes) coke *m* calibré
~JUNCTION - [electron] (semiconductors) jonction *f* à variation de la vitesse de croissance
~LIGHT FILTER - [photo] filtre *m* dégradé
~MULTIPLE - [telecomm] multiplage *m* échelonné
~NEUTRAL DENSITY FILTER - [telev] filtre *m* neutre à graduation
GRADED POTENTIOMETER - [meas] (potential divider in which the output voltage is a specified non-linear function of the displacement of a sliding contact from a arbitrary zero) potentiomètre *m* non-linéaire
GRADEFINDER - s. gradiometer
GRADER - [mech] (in road constr) nivellement *m*
[impl] (for sorting) classeur *m*, trieur *m*
[paint] (type of brush used for graining) veinette *f*, spalter *m*
~ROLLER - [constr] rouleau *m* compresseur de nivellement
GRADIENT - [gen] inclinaison *f*, rampe *f*
[surv] (inclination) pente *f*, déclivité, inclinaison *f*
[phys] (scale of variation, e.g. heat) gradient *m*, échelle *f*
[el] gradient *m*
~OF POTENTIAL - [el] gradient *m* du potentiel
~OF SLOPE - [surv] angle *m* de déclivité
~PEGS - [surv] jalonnettes *fpl*
~POST - [railw] indicateur *m* de pentes et de rampes
~TINTS - [surv] (in cartography) teintes *fpl* hypsométriques
~WIND - [met] (a wind of sufficient velocity to balance the pressure gradient) vent *m* isobarique
~WIND SPEED - [met] (geostrophic wind speed after taking account of the curvature of the wind path) gradient *m* du vent
GRADIENTER - [surv] (micrometer head attachment) mesureur *m* d'angle de déclivité
GRADING - [gen] classement *m*, gradation *f*
[constr] (levelling) nivellement *m*, ménagement *m*
[min] (the operation of selecting) triage *m*
[telecomm] (scheme for a group of selectors for their access to individual trunks) échelonnement *m* du multiplage
[text] classement *m*, triage *m*
[min] (of coke) calibrage *m*
[min] granulométrie *f*
~INSTRUMENT - s. gradiometer
~MACHINE - [mech] trieur *m*
~OF TRAFFIC - [telecomm] brassage *m* du trafic
~PLANT - [gas ind] (for coke) atelier *m* de triage (de coke)
[min] installation *f* de triage
~SCREEN - [min] crible *m* classeur
~TABLE - [photo] table *f* de sortage
GRADIOMETER - [instr] (instrument designed to measure slopes) indicateur *m* de déclivité, indicateur *m* de pente
GRADOMETER - s. gradiometer
GRADUAL - [gen] graduel, progressif
~FILTER - [telev] (in colour television) filtre *m* dégradé
GRADUATE, to - [instr] graduer
GRADUATED - [mech & of instr] gradué
~BAR - [surv] mire *f* graduée
~BEAM - [mech] (of a balance) réglette *f*
~CIRCLE - [surv] (a marked circular plate used in surveying instrument) cercle *m* gradué, cercle *m* divisé
~COLLAR - [mech] bague *f* graduée
~CYLINDER - s. graduated pipette
~DIAL - [instr] cadran *m* gradué
~GLASS - [ind chem etc] tube *m* de verre gradué
~HYDROMETER - [auto] densimètre *m* gradué
~MEASURE - [ind chem] mesure *f* graduée
~PIPETTE - [ind chem] (pipette with marks to show

quantity of liquid drawn up) pipette *f* graduée
GRADUATED SCALE - [surv] (of a compass) échelle *f* graduée (de boussole)
~SHADING - [photo] dégradé *m*
~SLIDE-WIRE - [meas] (component of potentiometers with a graduation ranging from 0 to I0 ohms) contact *m* glissant
~VESSEL - [ind chem] (vessels used to measure liquids) bac *m* gradué
GRADUATING CONTROLLER - [meas] (in telemetering) régulateur *m* à action continue
GRADUATION - [gen] graduation *f*
[meas] (the application of graduation lines on a scale of instrument) graduation *f*
[instr] (mark on an instrument indicating degree or value) échelle *f*, graduation *f*
[chem] concentration *f*
~OF THE LIMB - [constr] graduation *f* du limbe
GRAFT, to - [gen] greffer, enter
[agric] enter
[med] (in surgery) greffer, implanter
GRAFT - [gen etc] greffe *f*, greffon *m* ente *f*
~COPOLYMER - [chem] (a copolymer having side chains of different chemical composition attached to the main chain) copolymère *m* à greffe
~HYBRID - [agric] hybride *m* de greffe
GRAFTING - [agric] (the insertion of part of a plant into the part of another plant) greffage *m*, greffe *f*
[med] greffe *f*, implantation *f*
[carp] entement *m*
~CLAY - [agric] argile *f* à greffer
~KNIFE - [agric] greffoir *m*, écussonnoir *m*
~WAX - [agric] mastic *m* à greffer
GRAHAM LAW - [phys] (the rate of diffusion of a gas is inversely proportional to the square root of its density) loi *f* de Graham
GRAHAMITE - [min] (mineral asphaltum of varying composition) grahamite *f*
GRAIN, to - [paint] veiner (une surface), crépir
[ind chem] (the salting out in soap manufacture) grener, saler
[leather ind] grainer
GRAIN - [gen & bot] grain *m*
[agric] (the seed of a cereal) céréale *f*, grain *m*
[geol] (e.g. a small hard particle of sand) grain *m*
[meas] (unit of weight, i.e. 0.0648 g) grain *m* (0,0648 g)
[leather] (the granular texture) graine *f*, grenure *f*
[timber] (the arrangement and size of the constituent fibres) fil *m*
[metall] (the crystal) grain *m*
[metall] (of a grinding wheel) graine *f*
[metall] (the texture of the metal) grain *m*, texture *f*
[ind chem] (a particle) particule *f*
[photo] (the element in the sensitive emulsion) graine *f*
[paper man] grain *m*
[el acoust] (a small particle of plastic material used for the manufacture of disks) granule *m*
~ALCOHOL - [chem] alcool *m* éthylique
~BINDER - [agric] lieur *m* de céréales
~BLOWER - [agric mach] pelleteuse *f* pneumatique
~BOUNDARY - [crystall] (the surface which separates two regions of a solid in which the crystal axes are differntly oriented) joint *m* de grains
GRAIN-BOUNDARY RELAXATION - [crystall] (source of internal friction in solids caused by the motion of grain boundaries under stress) relaxation *f* des cristaux externes
~CLEANING MACHINE - [agric] machine *f* à nettoyer le blé, nettoyeur *m* de grains
~COUTING - [photo] (in photographic emulsion technique) comptage *m* des grains
~CRUSHER - [agric] aplatisseur *m*
~DRILL - [agric] semoir *m* à céréales
~DUBBIN - [leather ind] nourriture *f* de fleur
~ELEVATOR - [agric] élévateur *m* à grains
~FARMING - [agric] céréaliculture *f*, culture *f* céréalière
~GRADE - [soil] (grain size) grosseur *f* de grains, dimension *f* des grains
~GROWTH - [metall] croissance *f* des grains
~HARVEST - [agric] moisson *f*
~HARVESTER - [agric] moissonneuse *f*
~IMPLANTATION GUN - [radiat] (appliance for implanting radioactive grains into the body) pistole *f* d'injection de grains radioactifs
~LEATHER - [leather ind] (used for heavy sport shoes) chagrin *m*, peaux *fpl* glacées
~LIFTER - [agric] releveur *m* d'épis
~-MAIZE - [agric] mais-grain *m*
~NOISE - [nucl] trouble *m* de trace
~NUMBER - [mech etc] indice *m* du grain
~REFINING - [metall] affinage *m* du grain
~SIEVE - [agric impl] crible *m* à grains
~SILO - [agric build] silo *m* à grains
~SIZE - [metall & geol] (the average size of the grain or crystal) grosseur *f* du grain
~-SIZE COMPOSITION - [soil] composition *f* granulomètrique
~-SIZE CONTROL - [metall] (the control of the rate at which the grains grow when steel is heated) réglage *m* de la grosseur des grains
~-SIZE FREQUENCY CURVE - [soil] courbe *f* de fréquence granulométrique
~-SIZE NUMBER - [metall] nombre *m* de grains par unité de surface
~SKELETON - [soil] squelette *m* des grains, ossature *f* des grains
~SORTER - [agric] trieur *m* de grains
~SPLIT - [leather ind] fleur *f* de cuir fendu
~-STORE - [agric] grenier *m* à céréales
~STRIPS - [zool] thrips *m* des céréales
~TESTER - [agric] farinatome *m*
~TIN - [metall] étail *m* en larmes
~-TO-GRAIN STRESS - [soil] tension *f* effective, pression *f* intergranulaire
~WEEVIL - [zool] (a pest) charançon *m* du blé
~WETTING MACHINE - [agric] (used in flour mills) mouilleur *m*
~WOOD - [constr] (long cut wood) bois *m* de long *m*, bois *m* de fil
GRAINED - [gen] granulé, granuleux, grenu
~IRON - [metall] fer *m* à grains
~ROCK - [geol] roche *f* grenue, roche *f* plutonienne
~SHEET METAL - [metall] tôle *f* grenue
GRAINENESS - [photo] (the visible coarseness due to silber grains in a developed film) granulation *f*
GRAINING - [paint] (process of simulating the grain of wood by working on a wet coating with combs, rags, brushes etc) veinage *m*, décor *m* (en bois, en marbre)

GRAINING - [leather ind] (the process of bringing up the natural grain of a skin) crépissage *m*, grenure *f*
[ind chem] (in soap making, the salting out) relargage *m*
[metall] grenage *m*
~ POINT - [ind chem] (in sugar manufacture, stage of concentration of juice at which crystals are beginning to form) point *m* de cristallisation, point *m* de granulation
GRAINS DRYER - [agric] séchoir *m* à grains
~ SETTLING - [agric] couche *f* de drêches
~ TANK - [agric] silo *m* à drêches
GRAM - [meas] (or gramme; the unit of mass in the metric system) gramme *m*
~-ATOM - [phys] (the mass, in grams, of an element which is equal to its atomic weight) atome-gramme *m*
-ATOMIC WEIGHT - s. gram-atom
~ CALORIE - [phys] petite calorie *f*
~-ELEMENT SPECIFIC ACTIVITY - [nucl] (the total radioactivity of a given isotope per gram of element) activité *f* spécifique par élément-gramme
~-EQUIVALENT - [phys] (the equivalent weight of a substance in grams) gramme-équivalent *m*
~ ION - [phys] (the mass of an ion with a value in grammes expressed by a number equal to the sum of the atomic weights of which the ion is composed) ion-gramme *m*
~ MOLE - [chem] mole *m*
~-MOLECULAR VOLUME - [phys] (the volume of one gram-molecule of a gas at normal temperature and pressure) volume *m* d'un gramme-molécule
~ MOLECULAR WEIGHT - s. gram molecule
~ MOLECULE - [chem] (also called mole; the weight of a pure substance which gives the same number as its molecular weight when expressed in grammes) molécule-gramme *m*, mole *m*
~-NEGATIVE - [chem] (term used of bacteria which do not stain when treated successively with methyl violet, iodine and acetone or ethyl alcohol. s. Gram-positive) Gram-négatif *f*
~-POSITIVE - [chem] (term used of bacteria which stain when treated with methyl violet, followed by iodine and finally with acetone or ethyl alcohol. NOTE: This term has no connexion with the metric unit of mass: it is derived from the name of the Danish biologist, H. Gram) Gram-positif
~-RAD - [radiat] (unit of integral absorbed dose) gramme-rad *m*
~-ROENGTEN - [radiat] (unit regarded as the real energy conversion when a dose of one roengten is delivered to one gram of air) gramme-roengten *m*
GRAMICIDIN - [chem] (a strong bactericide produced by the soil bacterium B. brevis. It is a polypeptide, and has therapeutic uses as an antibiotic) gramicidine *f*
GRAMINACEOUS - [bot] (relating to grasses) graminé
GRAMMATITE - [min] (rarely used synonym for tremolite) grammatite *f*, trémolite *f*
GRAMOPHONE - [el acoust] phonographe *m*, gramophone *m*
~ AUDIOMETER - [instr] (instrument designed to test the hearing of a number of subjects) audiomètre *m* phonographique
~ DISK - s. gramophone record
GRAMOPHONE RECORD - [el acoust] (disk with grooves modulated by sound vibrations) disque *m* de phonographe
GRAMPUS - [metall] tenaille *f* à billettes
GRANARY - [agric] (a grain store) grenier *m*
GRANDFATHER-CLOCK - [horol] horloge *f* de parquet, pendule *m* à gaine
GRANITE - [geol] (coarse-grained igneous rock) granit *m*, granite *m*
~ PEGMATITE - [min] (granite consisting of quarz and feldspath) pegmatite *f* granitique
~ PORPHIRY - [geol] (rock of granitic composition including large crystals) porphyre *m* granitique
~ SAND - [geol] sable *m* granitique
~ SLAB - [constr] tranche *f* de granite
~ WEAVE - [text] armure *f* granitée, armure *f* à grai de lime
[text man] armure *f* petit granité
GRANITIC FABRIC - [text ind] granité *m*
~ SUBSOIL - [geol] soubassement *m* granitique
~ TEXTURE - s. granitoid texture
GRANITOID - [geol] granitoïde
~ TEXTURE - [geol] (a rock fabric) structure *f* granitoïde
GRANOBLASTIC TEXTURE - [geol] (arrangement of grains in a rock of an origin similar to that of granite) structure *f* granoblastique
GRANODIZING - [metall] (protective coating of phosphate) phosphatage *m* (électrolytique) préventif
GRANOLITHIC - [constr] (cement and fine granite chippings) en ciment à parement de granit concassé
GRANOPHYRE - [geol] (igneous rock) granophyre *m*
GRANT, to - [gen] accorder, concéder, conférer
GRANT - [gen] concession *f*, cession *f*, octroi *m*
~ A PATENT, to - [leg] délivrer un brevet
GRANULAR - [gen] granulaire, grenu, en grains
~ ASH - [chem] (coarse-grained dense soda ash, used in glass making etc) cendre *f* granulaire, soude *f* granulaire
~ CARBON - [min] charbon *m* granulaire
~ CARD - [text] carde *f* granulaire
~ FRACTURE - [metall] cassure *f* granulaire
~ IRON - [metall] (globular pearlite) perlite *f* granulaire
~ SNOW - [met] (small opaque grains of snow) neige *f* granulaire
~ STRUCTURE - [phys] (composed of grain-like particles) structure *f* granulaire, structure *f* grenue
GRANULATE, to - [constr] boucharder
GRANULATED - [phys] (in particles or crystals of a grain-like appearance) granulé, granulaire
~ BORIC ACID - [chem] acide *m* borique à grains
~ CARBON - [min] grenaille *f* de charbon
~ CORK - (cork reduced to small particles) liège *m* granulé
~ SLAG - [metall] laitier *m* granulé
~ SUGAR - [sugar man] (loose sugar) sucre *m* cristallisé
GRANULATES - [phys] (substances having a granular structure) matières *f*pl à structure granulaire
GRANULATING - [gen] granulation *f*
~ CRUSHER - [mech] concasseur-granulateur *m*
~ HAMMER - [impl] marteau *m* à granuler
~ MACHINE - [print] machine *f* à granuler
~ PLANT - [min] installation *f* de granulation
GRANULATION - [med] (new formation of vascular connecting tissue) granulation *f*

GRANULE - [gen] (small individual bodies or grains) granule *m*, granule *m* de poudre
GRANULITE - [min] (granular-textured metamorphic rock) granulite *f*
GRANULITIC TEXTURE - [geol] (arrangement of interlocking mineral grains which resemble a granitic texture but are developed in metamorphic rocks) structure *f* granulitique
GRANULO-FATTY - [med] granulo-adipeux
GRANULOCYTE - [biol] (any cell which contains conspicuous granules) granulocyte *m*
GRANULOMETER - [instr] (an apparatus for measuring grain or granule size) granulomètre *m*
GRAPE - [bot] grain *m* de raisin, raisin *m*
~ MILK - [brew ind] moût *m* non-fermenté
~ SEED - [bot] pépin *m*
~-SEED OIL - s. grapestone oil
~ SUGAR - [chem] (d-glucose) sucre *m* de raisin, glucose *f*
GRAPEFRUIT - [bot] pamplemousse *f*
GRAPES - [vet] tuberculose *f* bovine
GRAPESTONE - [agric] pépin *m* de raisin
~ OIL - [chem] huile *f* de pépin
GRAPEVINE - [bot] vigne *f*
GRAPH, to - [gen] graphiquer, tracer graphiquement
GRAPH - [gen] graphique *m*, diagramme *m*
~ PAPER - [paper man] (used for drawing) papier *m* quadrillé
~ PLOTTER - [comput] traceur *m* de courbes; enregistreur *m* graphique
GRAPHER - s. graphic instrument
GRAPHIC - [gen] graphique
~ ARTS - [print] industrie *f* graphique, arts *mpl* graphiques
~ FORMULA - [chem] (diagram showing the relative positions of atoms and groups in two dimensions) formule *f* de constitution
~ GRANITE - [geol] (a granite in which the intergrown quarz simulates runic characters) granite *m* graphique
~ INDICATOR - s. graphic instrument
~ INSTRUMENT - [instr] (electrical instrument with a pointer consisting of a pen and moving ovar a paper) appareil *m* de mesure enregistreur
~ SIGHT-FINDER - [photo] viseur *m* à réticule
~ TELLURIUM - [min] (rarely used synonym of sylvanite) or *m* graphique, sylvanite *f*
GRAPHICAL CALCULATION - [gen] calcul *m* graphique
~ DETERMINATION OF THE DOMINANT WAVELENGTH - [radio] détermination *f* graphique de la longueur d'onde dominante
GRAPHITE - [chem] (a naturally occurring allotropic form of carbon with uses as a refractory material, in electrodes, electrical equipment, and in the production of pencils) graphite *m*, mine *f* de plomb, carbone *m* graphitique
~ BRICK - [min] brique *f* graphitique
~ BRONZE - [metall] bronze *m* à graphite
~ BRUSH - [el] (a current-collecting brush made of graphite) balais *m* de graphite
~ CRUCIBLE - [impl] creuset *m* de graphite
~ ELECTRODE - [el heating] (for arc furnaces; arc-furnace electrode of graphite) électrode *f* de graphite
~ METAL - [metall] alliage *m* de plomb, étain et antimoine
GRAPHITE-MODERATED REACTOR - [nucl] (nuclear reactor with a moderator consisting of graphite) réacteur *m* modéré au graphite
~ MODERATOR - [nucl] modérateur *m* au graphite
~ PAINT - [paint] (paint made of silica-graphite ground in oil) peinture *f* au graphite
~ PEBBLES - [nucl] (small balls of graphite used as a reflecting agent in heterogeneous enriched-fuel reactors) billes *fpl* de graphite
~ REFLECTOR - [nucl] réflecteur *m* en graphite
~ RESISTANCE - [el] (resistance consisting of a rod of graphite) résistance *f* de graphite
~ ROSETTE - [metall] rosette *f* de graphite
~ SOFTENING - [metall] corrosion *f* graphitique
~ TREATED OIL - s. graphite lubricant
~-URANIUM LATTICE - [nucl] (a lattice in which the fissionable material is uranium and the non-fissionable material graphite) réseau *m* d'éléments combustibles modéré au graphite
GRAPHITIC - [metall] (of cast iron, when carbon occurs as graphite and not as cementite) graphitique
~ ACID - [chem] (graphite treated with nitric acid and potassium chlorate) acide *m* graphitique
~ CARBON - [metall] (carbon occurring as graphite instead of cementite in cast-iron) carbone *m* graphitique
~ CAST IRON - [metall] fonte *f* graphiteuse
~ LUBRICANT - [ind chem] (in natural flake form or suspended in oil) huile *f* graphitée
~ OIL - [oil ind] huile *f* graphitée
~ STEEL - [metall] (steel containing minute quantities of free carbon) acier *m* à graphite
GRAPHITIZATION - [gen] (the process of producing a superficial coating of graphite) graphitisation *f*, revêtement *m* avec une couche mince en graphite
[chem] (the process of transforming combined carbon into graphitic carbon) graphitisation *f*
GRAPHITIZED FILAMENT - [el] filament *m* graphité
GRAPHITIZING - [metall] graphitisation *f*, recuit *m* de graphitisation
GRAPHOMETER - [instr] (instrument for measuring angles) graphomètre *m*, demi-cercle *m*
GRAPNEL - [tool] (instrument for grasping or clutching) grappin *m*, crochet *m*
[naut] (small anchor with three or more flukes) grappin *m*
[naut] (instrument fitted with iron claws) grappin *m*, harpeau *m*
GRAPPLE, to - [gen] (to seize or hold) accrocher, agripper
[gen] (to seize and hold with a grapnel) grappiner
[naut] jeter le grappin, aborder
GRAPPLE - [oil ind] harpion (pour câbles)
GRAPPLING IRON - [impl] grappin *m*, crochet *m*
~ OF AN ARCH - [constr] ancrage *m* de l'arc, chaînage *m* de l'arc
GRASS - [bot] herbe *f*
[radar] (random vertical deflections of a C.R.T. electron beam caused by noise currents in the receiver) signaux *mpl* parasites sur l'écran du tube à rayons cathodiques, herbe *f*
~ BLEACHING - [text] (bleaching of tissues of a very white colour; the whiteness resembles that obtained in air-bleached linen) blanchiment *m* sur l'herbe, blanchiment *m* au pré
~ CHOPPER - [agric] hache-herbe *m*

GRASS DISEASE - s. grass sickness
~HUSBANDRY - [agric] exploitation *f* des herbages
~MEAL - [agric] farine *f* d'herbe séchée
~MOWER - [agric] faucheuse *f* à herbe
~PEA - [bot] gesse *f*, lentille *f* d'Espagne
~ROOTS - [min] (at surface) surface *f*
~SICKNESS - [vet] spirochétose *f* gastrique
GRASSED FLAX - [text] lin *m* roui au pré
GRASSHOPPER - [zool] sauterelle *f*
~PLANE - [aero] (colloq for light aircraft) avionnette *f*
GRASSING - [text] (bleaching of linen in the open air) blanchiment *m* au pré
GRASSLAND - [agric] prairie *f*, pré *m*
GRATE, to - [gen] râper, crisser, froisser
[mech] (production of a grating noise) grincer
GRATE - [gen] (of a boiler, a stove etc) plancher *m*, grille *f*
[phys] réseau *m*, trame *f*, treillis *m*
[constr] grillage *m*, treillis *m*, lattis *m*
[hydr] gril *m*, pommelle *f*
~AREA - [th eng] (the area of a grate in a furnace) palier *m* de grille
~BARS - [ovens etc] (fire bars sufficiently spaced to admit air) barreaux *mpl* de grille *f*
~DRIVE - [th eng] mécanisme *m* d'entrainement de la grille
~DRIVING GEAR - s. grate drive
~DRUM - [th eng] tambour *m* de grille
~FIRING - [th eng] chauffage *m* sur grille
~GAS PRODUCER - [gas ind] gazogène *m* à grille
GRATER - [tool] râpe *f*
GRATICULE - [opt] (reticle composed of lines marked on a transparent plate) réseau *m* à traits
[electron] réticule *m*, trame *f*
GRATING - [gen] frottement *m*, râpage *m*
[light] réseau *m*
[hydr] (perforated cover over a drain etc) pommelle *f*, grille *f*
[naut] (the wood-work covering the hatchways) plancher *m*
[acoust] (periodic spatial variation of the index of refraction caused by the presence of acoustic waves within the medium) variation *f* de l'indice de réfraction
~CONVERTER - [radio] (wave converter consisting of a double grating placed ahead of a coaxial sheet grating in a circular waveguide) convertisseur *m* à quadrillage
~OF THE GEARS - [mech] frottement *m* des engrenages
~REFLECTOR - [radio] (open-work metal structure providing a good reflecting surface) réflecteur *m* à quadrillage
~SPECTROSCOPE - [instr] spectroscope *m* à réseau
GRAUPEL - [met] (a soft hail) neige *f* fondue (à moitié)
GRAVE, to - [gen] (to dig) creuser
[gen] (to engrave) graver, ciseler
[naut] (to clean a ship's bottom by burning off all accretions) gratter le fond, radouber un navire
GRAVEL - [geol] (mostly pebbles and sand) gravier *m*
[constr] (mixture of sand, flints and loam) gravier *m*, gravillon *m*
[med] calcul *m*, concrétion *f*
~BACKFILL - [constr] remblai *m* en gravier
~BALLAST - [constr] ballast *m* en gravier
GRAVEL BLANKET - [soil] (gravel deposit) couche *f* de gravier, banc *m* de gravier
~CONCRETE - [constr] béton *m* de gravier
~DRAIN - [hydr] drain *m* en gravier
~FACE - [mining] (in hydraulic mining) front *m* de gravier
~FILL - [constr] remblai *m* de gravier
~FLAG - [constr] carreau gravuleux *m*
~MINE - [mining] exploitation *f* alluviale
~PACKING - [oil ind] filtre à gravier
~PIT - [mining] gravière *f* cailloutière *f*
~WIRE NETTING - [impl] treillis *m* métallique pour gravier
GRAVER - [tool] (a burin) burin *m*, échoppe *f*, gravoir *m*, ciselet *m*
GRAVESTONE - [arch] pierre *f* tumulaire, pierre *f* tombale
GRAVEYARD - [nucl] (area of ground set apart for burying waste radio-active material) cimetière *m*
GRAVID - [zool & med] (pregnant) gravide
GRAVIMETER - [instr] (measuring instrument for gravimetric analysis) gravimètre *m*, aréomètre *m*
GRAVIMETRIC ANALYSIS - [chem] (analysis of a solid mixture by separation and weighing of the constituents) analyse *f* gravimétrique
~METHOD - [oil ind] (method of exploration based on measurements of variations in gravity of the earth's surface) méthode *f* de recherches gravimétrique
GRAVIRECEPTOR - [astronaut] récepteur *m* de pesanteur
GRAVITATE, to - [gen & phys] graviter
GRAVITATION - [phys] (the mutual attraction between masses) gravitation *f*, attraction *f* universelle
~CONSTANT - [phys & astr] constante *f* de gravitation
~MARSHALLING YARD - [railw] gare *f* de triage en pente continue
GRAVITATIONAL - [phys] (subject to gravitation) de gravitation, gravitationnel
~ASTRONOMY - [astr] (in astronomy, the study of the motion of heavenly bodies under the forces of gravitation) astronomie *f* gravitationnelle
~ENERGY - [phys] (the energy which is inherent in the gravitational force) énergie *f* de la gravitation
~FIELD - [phys] (any region in which a particle is subject to gravitational force) champ *m* de gravitation
~FLUX - [phys] (the surface integral of the normal component of the gravitational field over a given area of the product of that integral by the inverse of the gravitation constant) flux *m* de gravitation
~INDUCTION - [bot] (development of structure from the under side of a plant member) induction *f* gravitationnelle
~MASS - [phys] (the mass of an object which is considered the generator of a gravitational field) masse *f* gravitationnelle
~POTENTIAL - [phys] potentiel *m* de la gravitation
~PULL - [phys] gravitation *f*
~RADIUS - [phys] rayon *m* gravitationnel
~RED SHIFT - [phys] (in spectrology) déplacement *m* gravitationnel vers le rouge, décalage *m* gravitationnel vers le rouge
~TIDE - [astr] marée *f* gravitationnelle
~WATER - [geogr] eau *f* hydrostatique

GRAVITOMETER - [instr] (instrument designed to measure changes in the specific gravity of oil flowing in a pipeline) gravimètre *m*
GRAVITON - [phys] (the elementary quantum of the gravitational field) graviton *m*
GRAVITY - [phys] (the attractive force by which all bodies tend to move towards the centre of the earth) gravité *f*, pesanteur *f*
~ABSCESS - [med] abcès *m* migrateur, abcès *m* par congestion
~ACCELERATION - [mech] accélération *f* de gravité
~BALANCE - [meas] balance *f* de torsion gravimétrique
~CABLE - [mech] câble *m* de traction
~CASTING - [metall] coulée *f* sans pression
~CELL - [el] (two-fluid cell with horizontal electrodes and two electrolytes in separate layers because of their difference in specific gravity) pile *f* à différence de poids spécifique
~CENTRE - [mech] (centre of gravity) centre *m* de gravité
~CLOSING - [mech] (a method of closing a downstroke press in which the closing motion is actuated only by the weight of the ram and associated parts) fermeture *f* sans pression
~CONTROLLED INSTRUMENT - [instr] (instrument in which the controlling torque is provided by the action of gravity) instrument *m* à gravité
~CONSTANT - [phys] accélération *f* de la pesanteur
~CONVEYOR - [mech mach] tapis *m* roulant en pente
~DAM - [hydr] (dam which cannot overturn to its weight) barrage-poids *m*
~DIE-CASTING - [metall] (castings of alloys in steel moulds into which the molten metal is poured by hand) moulage *m* en coquille
~DRAINAGE - [hydr] drainage *m* par gravité
~DROP ANNUNCIATOR - [el] tableau *m* commutateur à annonciateur
~ESCAPEMENT - [horol] (escapement in which the impulse is given to the pendulum by a weight continually falling through a constant distance) échappement *m* à gravité
~FEED - [mech] (supply of material by allowing it to move under the action of gravity; also feeding of the engine by gravites) alimentation *f* par pesanteur
~FEED SYSTEM - [mech] système *m* d'alimentation en charge
~FILTER - [ind chem] (filter in which the liquid passes through under the action of gravity and is not forced) filtre *m* à gravité
~FORCE - [phys] force *f* de gravitation
~-FREE - [phys] sans pesanteur, sans gravité
~-FREE FLIGHT - [astr] vol *m* à pesanteur nulle,
~FAULT - [geol] faille *f* normale
~FLOW - [hydr] écoulement *m* par gravité
~GRADIENT - [phys] gradient *m* de gravité
~HOIST - [mech] balance *f* sèche, balance *f* à gravité
~HOT-AIR HEATING - [th eng] aérotherme *m*
~INCLINE - [railw] plan *m* incliné automoteur
~LOADING INCLINE - [railw] plan *m* incliné de chargement
~LUBRICATION SYSTEM - [mech] graissage *m* à huile par gravité
~METER - s. gravimeter
~PLANE - [min] (inclined plane where the full trucks which descend pull up the ascending trucks) plan *m* incliné
GRAVITY PURIFIER - [mech] (of flour mill) purificateur *m* à gravitation
~SEGREGATION - [metall] (a variation in the composition of casting caused by the settling of heavier constituents) ségrégation *f* par gravité
~SEPARATION - [mech] (the separation of materials by means of their different gravities) séparation *f* par gravité
~SHUNTING - [railw] manœuvre *f* par gravité
~SLIDE - [soil] glissement *m* par gravité
~SOLUTION - [chem] liqueur *f* dense
~STAMP - [constr] pilon *m* à chute libre
[min] bocard *m*
~TANK - [aero] (type of tank from which fuel is fed by gravity only) réservoir *m* en charge
~WATER - [hydr] eau *f* gravitationnelle
~WATER SYSTEM - [hydr] (system in which the flow occurs under the natural pressure caused by gravity) alimentation *f* par gravité
GRAVURE - [print] photogravure *f*
~PRINTING - [print] (process of printing on film or sheet from an engraved cylinder) impression *f* en héliogravure
~ROLLERS - [print mach] cylindres *mpl* d'impression hélio
GRAY - s. grey
GRAZE, to - [agric] pâturer, paître
GRAZING - [gen] pâturage *m*
GREASE, to - [mech] (to apply grease, usually under pressure) graisser, lubrifier
GREASE - [mech] (semi-solid lubricant, consisting of emulsified mineral oil and soda or lime soap) graisse *f*
[text] laine *f* en suint, laine *f* surge
[vet] crapaud *m*
~BOX - s. grease cup
~CONNEXION - [auto] raccord *m* de graissage
~CUP - [mech] (a cylindrical internally-threaded device to be filled with grease and screwed down on a hollow, threaded stem, to force the lubricant into a bearing) godet *m* graisseur
~FORMING - [mech] (deep drawing by air- or liquid-pressure combined with the use of grease as a lubricant) emboutissage *m* avec lubrifiant
~GATE - [mech] entrée *f* de graisse
~GUN - [mech] (device for forcing grease into lubricators under pressure) pistolet *m* graisseur, injecteur *m* de lubrifiant
~HOLE - [oil ind] nouveau puits *m*
~NIPPLE - [mech] (a special fitting screwed into the housing of a bearing for connexion to a grease gun) nipple *f* de graissage
~PAINT - [chem] (used for making-up) fard *m*
~-PROOF PAPER - [paper man] (imitation parchment) papier *m* imperméable à la graisse, papier *m* parcheminé
~PUMP - [mech] pompe *f* à graisse
~REMOVER - [gen] dégraisseur *m*
~RESISTANT - [gen] (capable of withstanding the deteriorating effect of grease) à l'épreuve de la graisse
~RETAINER - [mech] (seal to prevent grease escaping from a bearing) déflecteur *m* à graisse
~SOLVENT - [chem] dissolvant *m* de la graisse
~-SPOT PHOTOMETER - [light] (means of comparting

the intensity of two light sources) photomètre *m* à tache d'huile
GREASE TRAP - [plumb] siphon *m* de dépôt de graisse
~TRAP-BAND - [agric] (used for pest control) bande-piège *f*, ceinture *f* gluante
GREASINESS - [paint] (defect in a coating, taking the form of a greasy exudation, caused by incompatibility of ingredients) onctuosité *f*
[text] (of wool fibres) caractère *m* gras
GREASING - [gen & mech] graissage *m*, lubrification *f*
GREASY - [gen] graisseux, onctueux
~PULP - [paper man] (of the wood pulp) pâte *f* graisse
~ROAD - [gen] chausser *f* glissante
~SHEETS - [rubber ind] feuilles *f*pl échauffées, feuilles *f*pl collantes
~WOOL - [text] laine *f* en suint, laine *f* graisse
GREAT - [gen] grand
~CALORY - [chem] grande calorie *f*, kilocalorie *f*
~CIRCLE - [astr] (the shortest distance between any two points on a sphere) grand cercle *m*
~CIRCLE CHART - [geogr] carte *f* en projection gnomonique
~CIRCLE COURSE - [aero] route *f* orthodromique
~CIRCLE FLYING - [aero] navigation *f* orthodromique
~COATING - [text] étoffe *f* pour paletot
~MAPLE - [bot] érable *m* sycomore, érable *m* faux-platane
~PRIMER - [print] (previous name of the type now called I8-point) corps *m* I8
~TOE - [anat] gros orteil *m*
GRECIAN TYPE AERIAL - [radio] (an aerial shaped like an inverted V) antenne *f* en grècque
GREEN - [colour] vert
~ACIDS - [chem] (mixed sulphonic acids, produced by the reaction of sulphuric acid with petroleum. When reacted with open-chain organic acids form detergent soaps) acides *m*pl verdiques
~ADDER - [telev] circuit *m* mélangeur pour le vert
~APEX - [telev] point *m* de couleur de la primaire verte
~BEAM - [telev] faisceau *m* pour le vert
~-BEAM MAGNET - [telev] aimant *m* du faisceau pour le vert
~BLACK LEVEL - [telev] niveau *m* minimal pour le signal vert
~BLINDNESS - [opt] achloropsie *f*, deuteranopie *f*
~BRICK - [constr] brique *f* moulée
~BRICKS - [constr] (the clay moulds which will become bricks after burning) brique *f* brute, brique *f* verte
~CARBONATE OF COPPER - [min] malachite *f*, cendre *f* verte
~CAST - [photo] dominante *f* verte
~CASTING - [metall] moulage *m* à vert
~CLAY - [geol] argile *f* maigre
~COAL - [min] charbon *m* frais, charbon vierge
~COMPOUND - [rubber ind] mélange *m* non-vulcanisé
~COPPER - [min] malachite *f*
~COPPERAS - [chem] couperose *f* verte
~CROP DRYER - [agric] séchoir *m* à herbe
~DENSITY - [metall] densité *f* du comprimé
~EARTH - [paint] (a green earthy mineral, primarily silicate of iron, mixed with other compounds) terre *f* verte, glauconie *f*
~ENGINE - [aero] (colloq. USA; a new and unused engine) moteur *m* nouveau (non-essayé)
GREEN FILM - [photo] (film just out of the processing room) copie *f* neuve
~FILTER - [photo] filtre *m* vert
~FLASH - [astr] (seen when the upper rim of the sun disappears below the horizon) rayon *m* vert
~GLASS - [glass man] verre *m* ordinaire
~GOLD - [metall] (consisting of 75 p.c. gold and 25 p.c. silver) or *m* vert
~HIGHS - [telev] hautes fréquences *f*pl pour le vert
~LEAD ORE - [min] mimétite *f*
~LIGHT - [signal] lumière *f* verte
[cin & broadcasting studios] (the signal to start the recording) lumière *f* verte
~LIZARD - [zool] margouillate *m*
~LOWS - [telev] basses fréquences *f*pl pour le vert
~LUMBER - [timber] bois *m* frais, bois *m* vert
~MALT - [ind chem] (malt which has not been kiln-dried) malt *m* vert
~MANURE - [agric] engrais *m* vert
~MORTAR - [constr] (colloq for mortar in the hardening stage) mortier *m* frais
~OVERCAST - [telev] (green shading) teinte *f* de vert
~OXIDE - [chem] oxyde *m* de nickel
~PEAK LEVEL - [telev] niveau *m* maximal pour le signal vert
~PRIMARY SIGNAL - [telev] signal *m* du vert primaire
~RUN - [aero] (colloq. USA; engine test) vol *m* d'essai
~SAND - [metall] (moulding sand rich in organic matter) sable *m* vert, sable *m* à vert
~SAND CASTING - [metall] moulage *m* en sable vert
~SAND CORE - [metall] noyau *m* en sable vert
~SAND MOULDING - [metall] moulage *m* à vert
~STRENGTH - [metall] résistance *f* du comprimé, cohésion *f* à vert
~VITRIOL - [min] (popular term for melanterite) vitriol *m* vert, sulfate *m* de fer
~WOOD - s. green lumber
GREENFINCH - [zool] verdier *m*
GREENGAUGE - [agric] reine-Claude *f*
GREENHEART - [timber] (a South American very strong green-yellow timber) greenheart *m*, ébène *m* vert
GREENHOUSE - [agric] serre *f*
[pottery] chambre *f* de séchage
~EFFECT - [astronaut] effet *m* de serre
GREENLAND SPAR - [min] (an alternative name for cryolite, a natural sodium-aluminium fluoride, q.v.) cryolite *f*
GREENOCKITE - [min] (a naturally occurring cadmium sulphide of no importance as a source of cadmium) greenockite *f*
GREENWICH APPARENT TIME - [geom] (time referred to the passage of the apparent sun over the meridian or anti-meridian of Greenwich) temps *m* apparent de Greenwich
~HOUR ANGLE - [geogr] (the distance of a given celestial body from the meridian of Greenwich, measured in terms of arc along the celestial equator) angle *m* horaire origine
~MEAN TIME - [G.M.T.] (the hour angle measured in terms of time of the mean sun W. of Greenwich ; I2 hours) temps *m* universel (T.U.)
GREGE - [text] (soie) grège, grège *f*
GREGORIAN TELESCOPE - [opt] télescope *m* grégorien

GREGORIAN TONES - [mus] chant *m* grégorien
GRENADE - [gen] grenade *f*
GRENADINE - [text] grenadine *f*
~ SILK - [text] soie *f* grenadine
GRENZ-RAYS - [radiat] (X-rays produced at voltages of 5 to 20 kv) rayons *mpl* limite, rayons *mpl* mous
~ TUBE - [radiat] (X-ray with a low-absorption window which allows the transmission of X-rays at low voltages) tube *m* à rayons X mous
GREY - [colour] gris
[text] écru
~ ACETATE - [chem] acétate *m* de calcium
~ BODY - [opt] (radiator with a spectral emissivity which remains constant through the spectrum) corps *m* gris
~ CAST IRON - [metall] fonte *f* grise
~ CLOTH - [text] tissu *m* écru
~ COPPER ORE - [min] (a sulphide of copper and antimony) tétrahédrite *f*
~ COTTON CLOTH - [text] tissu *m* de coton écru
~ FILTER - [photo] (neutral-density filter) filtre *m* gris
~ FORGE PIG - [metall] fer *m* gris malléable
~ IRON FOUNDRY - [metall] fonderie *f* à fonte grise
~ MATTER - [anat] (cell bodies forming an area of the central nervous system) substance *f* grise
~ METAL - [min] (a greyish shale) schiste *m* gris
~ MOULD - [chem] rouille *f* grise
~ PIG-IRON - [metall] fer *m* gris malléable
~ SCALE - [opt] (series of achromatic tones from black to white) échelle *f* des gris
~ SCALE SIGNAL - [telev] signal *m* de gradation
~ SCALE TRACKING - [telev] reproductions *fpl* de l'échelle de gris
~ SCALE VALUE - [telev] luminosité *f* équivalente
~ SLAG - [metall] scorie *f* verte de plomb
~ STONE LIME - [constr] chaux *f* grise
~ WARP - [text] chaîne *f* écrue
GRID - [gen] grille *f*
[el] (network) réseau *m*
[gas ind] (gas mains network) réseau *m*
[light] grille *f*
[photo] (wooden rack) claie *f*
[el] (of a battery-metallic framework) grille *f*
[electron] (electrode with openings) grille *f*
[constr] (perforated cover) grillage *m*
[surv] grille *f*, quadrillage *m*, graticule *m*
[radio] grille *f*
[text] griffe *f*
~ -ANODE CAPACITANCE - [electron] (in a thermionic tube, the capacity which exists between grid and anode) capacité *f* grille-anode
~ BAR - [metall] fer *m* à barreaux
~ BASE - [electron] (the bias voltage which is required to reduce the anode current to a negligible value) polarisation *f* de coupure
~ BATTERY - [electron] pile *f* de polarisation
~ BIAS - [electron] (the negative potential which, when applied to the control grid of a thermionic tube, will cause Cut-off, i.e. reduce the electron flow to zero) polarisation *f* de grille
~ BIAS BATTERY - [electron] (battery used to furnish grid bias potential) pile *f* de polarisation de grille
~ BIAS RESISTANCE - [radio] résistance *f* de tension de polarisation de grille
~ BLOCKING - [electron] (paralysis of capacitance-coupled stages in an amplifier) blocage *m* de grille
GRID BLOCKING CAPACITOR - [electron] condensateur *m* de grille
~ CAP - [radio] (top cap terminal for the grid) téton *m* de grille
~ CAPACITOR - [electron] condensateur *m* de grille
~ CATHODE SPACE - [electron] espace *m* grille-cathode
~ CIRCUIT - [radio] (in a thermionic valve) circuit *m* de grille
~ CLIP - [electron] (metal part fastening a conductor to the grid cap) capuchon *m* de grille
~ CONDENSER - [radio] condensateur *m* de grille
~ CONNECTING RING - [el] (in electronic tubes) anneau *m* de connexion de la grille
~ CONTROLL - [telecomm] commande *f* par la grille
~ CONTROL RECTIFIER - [electron] redresseur *m* à grille commandée
~ CURRENT - [electron] (the current flowing from the grid to the cathode) courant *m* de grille
~ CUT-OFF VOLTAGE - [electron] (the cut-off voltage when the electrode is a control grid) tension *f* de coupure de grille
~ -DIP METER - [instr] (instrument designed to measure resonance frequencies) ondemètre *m* à absorption
~ -DIP OSCILLATOR - [electron] (vacuum tube oscillator) oscillateur *m* grid-dip
~ DRIVING POWER - [electron] (average product of instantaneous values of the alternating grid over a complete cycle) puissance *f* d'attaque
~ EMISSION - [electron] (negative grid current) émission *f* de grille
~ GAS - [gas ind] (gas distribution) (long distance gas) gaz *m* de réseau de transport
~ GLOW TUBE - [electron] (a cold-cathode triode) tube *m* à décharge à grille et cathode froide
~ HUM - [electron] (hum voltage between grid and earth) ronflement *m* de grille
~ KEYING - [radio] (keying by changing the grid bias) manipulation *f* par la grille
~ LAYOUT - [gen] quadrillage *m*
~ LEAK - [radio] (a high resistance to provide a d.c. path from grid to cathode) résistance *f* de fuire de grille
~ -LEAK RESISTANCE - s. grid leak
~ LOCKING - [radio] (a defect of the electronic tube operation) effet *m* d'émission excessive de grille
~ MODULATION - [radio] modulation *f* par la grille
~ NEUTRALIZATION - [radio] neutralisation *f* par grille
~ NOISE RESISTANCE - [electron] (imaginary resistance) résistance *f* fictive du bruit de grille
~ OF GLASS RODS - [text] (in a Jacquard loom) grille *f* avec barrettes en verre
~ OF MIRRORS, MIRROR GRID - [telev] grille *f* à miroirs
~ PIN - [electron] (the contact making element for the grid) broche *f* de grille
~ PITCH - [electron] (the pitch of the helix of a helical grid) pas *m* de grille
~ POWER SUPPLY - [electron] source *f* de la tension de polarisation de grille
~ PRIMING VOLTAGE - [electron] tension *f* de grille à l'amorçage
~ PULSE MODULATION - [radio] (modulation of a carrier-wave operating the grid of a valve in any stage of an amplifier) modulation *f* de grille par

impulsions
GRID RATIO - [radiography] (the ratio of the total area of holes to the total area of the grid) rapport *m* de grille
~ RESISTANCE - [electron] (bias resistor) résistance *f* de grille, résistance *f* de fuite
~ RING - [instr] (component part of aircraft compass) anneau *m* indicateur de route
~ RUNNER - [gas ind] (an oven shelf support) gradin *m* de four
~ SHIELDING CAN - [electron] (metallic cover for the grid of an electric tube) blindage *m* de grille
~ STOPPER - [radio] (parasitic stopper connected to the grid of a tube) impédance *f* d'étouffement de grille
~ STRETCHER - [electron] (wire lengthening a helical grid) mandrin *m* de grille
~ SWEEP - [electron] (total cariation) amplitude *f* de tension grille-cathode
~ SWING - s. grid sweep
~ THERAPY - [med] (irradiation of the body surface through a protective perforated sheet) traitement *m* par grille
~ TRANSPARENCY - [electron] (the reciprocal of the amplification factor) transparence *f* de grille
~ TUBE - [electron] (an X-ray tube) tube *m* à rayons X à grille
~ -TYPE WHEEL EXTENSION - [mech] (in agricultural machines) barillets *mpl* de jumelage
~ VARIATION - [astr] déclinaison *f* magnétique de la carte
~ WINDING MACHINE - [impl] bobineuse *f* de grille
~ WIRE - [instr] (component part of aircraft compasses) fil *m* de grille
GRIDDLE - [impl] crible *m*
GRIDIRON - [impl] gril *m*
~ TRACK - [railw] faisceau *m* de triage
GRIESS REAGENT - [chem] (an analytical reagent for nitrous acid) réactif *m* de Griess
GRIFFE - [text] (of a dobby, a knife box) griffe *f*
~ BAR - [text] levier *m* de la griffe
~ BOX - [text] griffe *f*
~ LIFT ECCENTRIC - [text] tapette *f* de la griffe
GRIGNARD REAGENTS - [chem] (organic reagents of considerable importance, consisting of an ethersoluble organo-magnesium halide, having the general formula R. Mg X.) réactifs *mpl* de Grignard
GRILL - [gen] gril *m*, grille *f*,
[auto] grille *f*, calandre *f*
GRILLAGE - [constr] (a system of beams crossing at right angles, used to support a heavy structure) grillage *m*, treillis *m* en sous-oeuvre
~ BEAM , GRATING - [constr] radier *m* en traverses
~ FOUNDATIONS BUILD - [constr] grillage *m* des fondements d'un édifice
GRILLE - [constr] grille *f*
GRILLO-SCHROEDER PROCESS - [chem] (a process for the synthesis of sulphuric acid) procédé *m* Grillo-Schroeder
GRIME - [gen] saleté *f*, poussière *f* (de charbon etc)
GRIND, to - [gen] moudre, concasser, réduire en poudre, meuler, broyer
[mech] (the action of sharpening) auguiser, affûter, roder, aiguiser
[glass man] (the action of smoothing) polir, meuler, doucir
[mach tool] affûter, meuler
[auto] (of the gears) grincer, crisser
GRIND, to - [opt] (a lens) meuler (une lentille), tailler (une lentille)
GRIND A FLAT, to - [mech] roder une surface
~ A TOOL - [mach tool] affûter un outil
~ DOWN, to - [mech] émeuler, aiguiser, roder
~ DRY , to - [mech] affûter à sec
~ FINISH - [mech] rectification *f* de la meule
~ -IN, to - [mech] (to ensure good seating of a part by moving it in contact with its mating part while a paste abrasive is applied between them, espec. of i.c. engine valves) rectifier, meuler, fraiser
~ THE SEAT, to - [mech] rectifier un siège
~ TO POWDER, to - [ind proc] réduire en poudre
GRINDELIA - [chem] (an extract of Grindelia camporum with uses in medicine) grindélia *f*
GRINDER - [a tool] meuleuse *f*, rodoir *m*, rectifieuse *f*
[mach tool] (a grinding machine) rectifieuse *f*, machine *f* à rectifier
[mach tool] (a sharpening machine) machine *f* à affûter
[mach tool] (for smoothing out imperfections) machine *f* à meuler, machine *f* à ébarber
[radio] (a rather long atmospheric disturbance) décharge *f* atmosphérique
[ind chem] machine *f* broyeuse
[paper man] (a large grit-stone circular stone used in the manufacture of mechanical wood-pulp) defibreur *m*
GRINDER'S PHTHISIS - s. grinder's rot
~ ROT - [med] (a lung disease which is caused by the inhalation of metallic particles) pneumoconiose *f* des rémouleurs
GRINDERY - [metall] atelier *m* de meulage
GRINDING - [mech] meulage *m*, émeulage *m*
(the operation of removing fins etc by means of a grinding wheel) ébarbage *m*
[glass] meulage *m*
[min] (ore treatment to facilitate the recovery of the metal) meulage *m*
[acoust] (the sound made by the teeth) grincement *m*, crissement *m*
[opt] (of lenses) taille *f* (de lentilles)
~ ALLOWANCE - [mech] surépaisseur *f* de rectification
~ BALLS - [mech] (of a grinder) boulets *mpl*
~ BENCH - [metall] machine *f* à meuler
~ BLOCK - [impl] bloc *m* à affûter
~ BOARD - [metall] planchette *f* à polir
~ CHARCOAL - [metall] charbon *m* à polir
~ CLOTH - [text] toile *f* d'émeri, toile *f* émerisée
~ COMPOUND - [ind chem] pâte *f* à roder
~ CRACKS - [metall] criques *fpl* de meulage
~ CYLINDER - [mech] cylindre *m* broyeur
~ DISK - [impl] disque *m* à rectifier, disque *m* affûter
~ DOWN - [mech] rectification *f*, action *f* de frotter
~ ELEMENTS - [mech] éléments *mpl* broyeurs
~ EMERY - [metall] émeri *m*
~ FINENESS - [mech] finesse *f* de broyage
~ GAUGE - [mech] calibre *m* de rectifieuse
~ HEAD - [mach tool] (a spindle carried in bearings, with a pulley or motor for the drive, and carrying abrasive wheels at its ends) porte-meule *m*, machine *f* simple à meuler
~ -IN - [mech] (the operation of adjusting two mating parts (espec.a valve and its seat))to a good fit by

applying abrasive paste between them and moving them in contact and under pressure) rectification *f*, rodage *m*
GRINDING MACHINE - [mach tool] (machine tool for finishing surfaces by the abrasive action of a high-speed grinding wheel) meuleuse *f* rectifieuse, machine *f* broyeuse, machine *f* à surfacer
~ MILL - [mech mach] broyeur *m*, moulin *m*, concasseur *m*, machine *f* à morceler
[min] moulin *m* chilien
~ OF A CYLINDER - [mech] rectification *f* d'un cylindre
~ OUT GROOVES - [mech] rectification *f* des rainures
~ PAN - [min] amalgamateur *m*
~ PAPER - [paper man] papier *m* abrasif
~ PASTE - [mech] pâte *f* à roder
~ PATH - [mech] chemin *m* de rectification
[metall] plateau *m* des moules, cuvette *f*
~ PLATE - [mech] plaque *f* à affûter
~ ROLLER - [text] cylindre *m* aiguiseur
~ SLIP - [carp] (piece of oil-stone used for sharpening purposes) meule *f* à aiguiser
~ STOCK - [mech] abrasif *m*
~ STONE - [tool] (also called grindstone) meule *f*
~ TEETH - [zool] (molar and premolar teeth) molaires *f*pl
~ UNIT - [constr] (in a cement factory) groupe *m* ponçage
~ WHEEL - [mech] (disk of cemented abrasive material rotated rapidly and used to remove material from a workpiece) roue *f* à meuler, meule *f* de rectification, disque *m* abrasif
~ WHEEL CENTRE - [mech] disque *m* porte-meule
~-WHEEL DRESSING - [mech] rhabillage *m* de meules, redressage *m* et mise au rond des meules, rhabillage *m*
GRINDSTONE - s. grinding stone
GRINNING THROUGH - [paint] (the appearance of the colour of a previous coat through a subsequent one, caused by lack of opacity in the latter) se voyant au travers
GRIP, to - [gen and mech] saisir, serrer, pincer
[rubber ind] (of tyres) avoir une bonne tenue de route
GRIP - [gen] prise *f*, serrage *m*
[mech] (any device which must be gripped to perform an operation) griffe *f*, prise *f*, douille *f* de serrage, pince *f* poignée
[constr] (small channel for rain-water during the construction of the foundations) fossé *m*, rigole *f*
(a safety mechanism in machines) dispositif *m* de blocage
[mech] (the distance between the centres of the rivets) distance *f* entre les rivets
[agric] (in a cowhouse) fossé *m*
[transp] (a haulage clip) tenaille *f* d'attelage
[firearms] (of a pistol) crosse *f*
[rubber ind] (of tyres) adhérence *f* au sol
[telecomm] manchon *m* à mailles, amarrage *m* de câble
[brew ind] essai *m* à la main
~ CHEEK - [mech] (in a hoist etc) mâchoire *f* de serrage
~ DIE - [metall] matrice *f* de machine à forger
~ ECCENTRIC - [mech] (in hoists etc) excentrique *m* de blocage
GRIP GEAR - [mining] parachute *m*
~ GOVERNOR - [mech] (in hoists etc) régulateur *m* de contact
~ IRON - [mech] (in hoists etc) mâchoire *f*
~ LENGTH - (the minimum length of the reinforcing bar to be embedded in concrete) longueur *f* de barre (minimum)
~ PAWL - [mech] (in hoists etc) cliquet *m* de serrage
~ REGULATOR - s. grip governor
~ REST - [gas ind] (in gasholders) repos *m* de crochet, repos *m* de la tôle de garde
~ SHOT - [mining] coup *m* de mine incliné
~ SOCKET - [mech] manchon *m* de serrage d'un tour
~ TYRE - [rubber ind] pneu *m* à bonne adhérence au sol
~ WEDGE - [mech] coin *m* d'arrêt
GRIPE - [shipbuild] brion *m*, étrave *f*
GRIPES - [vet] colique *f*
[naut] saisines *f*pl
GRIPING - [naut and aero] (having a tendency to go to the windward) venant au vent
~ ANGLE - [metall] angle *m* de contact
GRIPPE - [med] grippe *m*, influenza *f*
GRIPPED - [gen and mech] saisi, serré, empoigné
GRIPPER - [gen] pince *f*, griffe *f*
[text] barre *f* de pinçage
[text] (in a large loom) pince *f*
~ ARM - [text] bras *m* de pinçage
~ EDGE - [print] (the edge of the paper which is gripped by the grippers) bord *m* des pinces
~ LOOM - [text] machine *f* à tisser à griffes, machine à tisser à pinces
~ SHUTTLE - [text] navette *f* à griffe, navette *f* à pinces
GRIPPERS - [print mach] (attachments gripping the edges of paper when it is fed into the machine) pinces *f*pl
GRIPPING - [gen and mech] serrage *m*, blocage *m*, étreinte *f*
~ ARM - [text] bras *m* de pinçage
~ CLUTCH - [auto] embrayage *m* brutal
~ DEVICE - [mech] (in hoists etc) grip *m*, frein *m* de sûreté, pince *f* d'accrochage
~ EFFECT - [text] effet *m* de pinçage
~ FLANGE - [mech] (a split flange) bride *f* de serrage
~ OF BRAKES - [auto] broutement *m* des freins
~ SHUTTLE - [text] navette *f* à griffe
GRIST - [brew ind] (the mixture of malts which is sufficient for one brewing) mouture *f*, versement *m*
GRISTLE - [anat] cartilage *m*
GRIT - [geol] (a substance consisting of sharp-edged hard particles) sable *m*, graviers *m*pl, grès *m*
[mech] (the general size of abrasive particles in grinding wheels) boue *f* de meule
[paper man] (a defect) toucher *m* graveleux
~ BASIN - [hydr] bassin *m* de dessablement, chambre *f* à sable
~ BLASTING - [mech] grenaillage *m*
~ BLASTING-MACHINE - [mech mach] machine *f* à grenailler
~ CATCHER - [hydr] collecteur *m* de sable
~ CHAMBER - [hydr] (tank through which sewage is first passed to allow the heaviest suspended matters to fall to the bottom) bassin *m* de dessablement, chambre *f* à sable
~ CLEANING - [metall] grenaillage *m*

GRIT COLLECTOR - [hydr] collecteur *m* de sable
~NUMBER - [mech] (to indicate the consistency of the grit) indice *m* du grain
~WASHING - [hydr] rinçage *m* des sables
GRITS - [agric] gruau *m* d'avoine
GRITSTONE - [metall] grès *m* à gros grains
GRITTER - [mech mach] (in road construction work) machine *f* à gravillonner
GRITTY - [gen] graveleux, cendreux
GRIVATION - [navig] (the horizontal angle made between a grid,datum and the magnetic meridian) décligrille *f*
GRIZZLE - [min] (coal intermixed with iron pyrites) charbon *m* pyritifère, charbon *m* pyriteux
~BRICKS - [build] (underburnt and badly shaped bricks)briques *fpl* de qualité inférieure
GRIZZLY - [mining] (strong grating a heavy parallel steel bars, usually placed under a coarse-crushing machine to separate different sized products)grille *f* à barreaux, cribon *m* à barres parallèles
~ELEVATOR - [mining] élévateur *m* à grille,grille *f*
~FEED - [metall] alimentation *f* à grille
~LEVEL - [metall] niveau *m* de grilles
GROAN, to - [gen] gémir
GROANING - [acoust] (sound expressing pain) geignement *m*, gémissement *m*
GROATS - [agric] (hulled and crushed grain) gruau *m* d'avoine, gruau *m* de froment
GROCER'S ITCH - [med] eczéma *m* des épiciers
GROG - [ceram] (material made by grinding fired clay, used as an additive to clay to reduce shrinkage) chamotte *f*, argile *f* cuite pulvérisée
GROGGINESS - [vet] (arthritis affecting horses) faiblesse *f* de jambes
GROIN - [anat] aine *f*
(the line of junction of the two arches in a groined arch) arête *f* de voûte, nervure *f*
~BOOT - [shoe man] bottes *fpl* de marin
GROINED - [arch] fourni d'arêtes
~ARCH - [arch] (arch intersected by other arches) voûte *f* à arêtes, voûte *f* en arc-doubleaux
~VAULT - s. groined arch
~VAULTING - s. groined arch
GROMMET - [mech] (a ring of soft material inserted in a hole in sheet material to allow a cable to pass without damage to the insulation) anneau *m*, passe-fil *m*
(a metal eyelet) rondelle *f* métallique
[naut] erseau *m*, anneau *m* de corde
[plumb] (mixture of lead and putty used as a jointing material) bague *f* d'étoupe
[oil ind] attache *f* continue du câble sans bout
~BELT - [oil ind] courroie *f* trapézoïdale à deux câbles
GROOVE, to - [gen] rainer, canneler, entailler, silloner, écrancher
GROOVE - [gen] rainure *f*, cannelure *f*, entaille *f*
[mech] gorge *f*
[mining - galerie *f*
[rubber ind] (of a tyre) rainure *f* , sculpture *f*
[el acoust] (the track in mechanical recording) sillon *m*
[firearms] rayure *f*
[print] (the groove of a type) jet *m*, saumon *m*
[metall] (of a rolling mill) cannelure *f*
[mech] (of a screw) écuelle *f*, creux *m*
[th eng] (in a boiler-plate) sillon *m*
[mining] creux *m*, tranchée *f*
GROOVE AND TONGUE - [carp] gueule-de-loup *f*
~ANGLE - [el acoust] (the angle between the two walls of an unmodulated groove of a disk) angle *m* du sillon
~CORD PULLEY - [mech] poulie *f* à gorge
~CRACKING - [rubber ind] (in tyres) craquelure *f* à fond de sculptures
~JUMPING - [el acoust] (the jumping of the needle one groove to another) déraillage *m*
~OF INSULATOR - [el] gorge *f* de l'isolateur, cou *m* de l'isolateur
~SHAPE - [el acoust] (in disk recording, the contour of the groove in a radial plane perpendicular to the surface of the record) profil *m* du sillon
~SHEAVE - [mech] poulie *f* à gorge
~SPACING - [el acoust] taux *m* de gravure
~SPEED - [el acoust] (the linear speed of the groove in relation to the stylus) vitesse *f* linéaire du sillon
~TURNING - [mech] rainurage *m*
~WALL - [el acoust] paroi *f* de sillon
~WALL STIFFNESS - [el acoust] rigidité *f* de la paroi du sillon
~WELD - [metall] cordon *m* frontal
GROOVED - [gen mech etc] (formed with longitudinal recesses) cannelé, rayé, à rainures, à gorge, raine
~AND FEATHERED JOINT - [carp] assemblage *m* à fausse languette
~BELT PULLEY - [mech] poulie *f* pour courroie à gorge
~BIT - [tool] mèche *f*
~COLLAR BEARING - [mech] roulement *m* à gorges annulaires
~DRUM - [metall] tambour *m* cannelé
~FLANGE - [mech] bride *f* à rainure, bride *f* rainurée, bride *f* avec gorge
~INSULATOR - [el] isolateur *m* à gorge
~PROFILE - [rubber ind] (of tyres) sculpture *f* rayurée
~PULLEY - [mech] poulie *f* à gorge
~RAIL - [mech] rail *m* à gorge
~RAIL PAD - [mech] semelle *f* cannelée
~RUBBER BEARING - [mech] coussinet *m* cannelé de caoutchouc
~SPRING STEEL - [metall] acier *m* cannelé pour ressorts
~TREAD - s. grooved profile
~TOOL - [tool] (for moulding test specimen) moule *m* d'eprouvettes normalisées
~WIRE - [el] (used for overhead contact-wire of electric traction system) caténaire *f* polygonale
[text] tringle *f* à rainures, fer *m* coupant
GROOVER - [tool] (in forging) outil *m* à rainurer
GROOVING - [mech] (in steam boilers, the cracking of the plates due to wear) sillon *m*
[mining] excavation *f*
[carp etc] entaillage *m*, rainurage *m*, bouvetage *m*
GROOVING AND TONGUING - [mech] rainure *f* et languette
~CUTTER - [tool] fraise *f* à rainurer
~PLANE - [tool] (plane specially designed for cutting grooves) bouvet *m* à rainure, bouvet *m* femelle
~SAW - [tool] (circular saw used for cutting grooves) scie *f* circulaire à bouveter
GROS - [text] (union fabric) gros *m*

GROSS - [meas] (twelve dozens) grosse f, douze douzaines fpl
[adj] gros, gras
~-ALMERODE CLAY - [glass man] (refractory clay used for glass-melting pots) argile f Gross-Almerode
~ANATOMY - [med] anatomie f macroscopique
~BLOWHOLES - [metall] bouillonnement m
~CALORIFIC VALUE - [th eng] (the quantity of heat produced by the combustion of the unit mass of a substance with oxigen under pressure) puissance f calorifique supérieure
~ERROR - [comput] erreur f d'accumulation, erreur f cumulative
~HEATING VALUE - s. gross calorific value
~LIFT - [aero] force f ascensionnelle totale
~THRUST - [astronaut] poussée f totale
~TONNAGE - [naut] jauge f brute
~WEIGHT - [meas] (the total weight of an aircraft fully loaded) poids m brut
~WING AREA - [aero] (the area of the wings as limited by the wing tips and leading and trailing edges, and the centre line or straight lines joining the intersections of the edges and the fuselage and nacelles(fillets neglected) surface f alaire totale
GROSS' DISEASE - [med] kyste m mouqueux du rectum
GROSSULARITE - [min] (a green carnet) glossuralite f, grenat m calcifère
GROTTO - [geol] grotte f
GROUND, to - [gen] fonder, baser, appuyer,
[el] mettre à la terre, mettre à la masse
[naut] échouer
[aero] (of a balloon) atterrir
GROUND - [gen] sol m, terre f, fond m, champ m, base f
[agric] sol m, terrain m
[el] (earth in GB) terre f, masse f
[naut] fond m
[surv] terrain m
[min etc] (passed through a mill) moulu, broyé
[paint] (the first coat of a paint) fond m, première couche f de peinture
[text] (the cloth on which a design is printed etc) fond m, tissu m de fond
[min] (the mineralized deposit) couche f
~A LINE, to - [el] mettre une ligne à la terre
~ABSORPTION - [radio] (the loss of energy in the transmission of radio-waves caused by the dissipation in the ground) absorption f du sol
~AIR - [phys] (the air which is contained in the upper layer of the subsoil, frequently harmful) air m au sol
~ANGLE - [aero] (landing angle) angle m d'atterrissage
~ANGLE SHOT - [photo] (shot taken under a certain angle from the earth's surface, the floor etc) contre-plongée f
~ANTENNA - [radio] (only US; earth aerial in GB; aerial buried into the earth) antenne f enterrée
~AUGER - [tool] (used for boring holes in the ground) tarière f
~BAR - [el] collecteur m de terre
~-BASED DUCT - [radio] (a layer in the atmosphere possessing the quality of conducting radio-frequency waves in the same way as a true wave-guide) conduite f de surface

GROUND BASS - [mus] basse f contrainte
~BEAM - [build] poutre f de rive
[text] (in weaving) ensouple f de chaîne principale
~BEARING CAPACITY - [constr] résistance f du terrain
~BED - [el] (earth terminal) prise f de masse
~BOARD - [constr] (of a fence) planche f de fond
~BUS - [el] barre f omnibus de mise à la terre
~CABLE - [el] conducteur m de terre
~CIRCUIT - [el] (earth circuit) circuit m de connexion à la masse
~CLAMP - [el] borne f de terre, collier m de terre
[metall] (in welding, a clamp used to hold parts firmly in position) crampon m de base
~CLEARANCE - [auto] garde f au sol, hauteur f libre au dessous de la voiture
~CLEARANCE INDICATOR - [aero] indicateur m position sol
~COAL - [mining] charbon m de la partie inférieure de la couche
~COLOUR - [text] couleur f de fond
~CONDUIT - [el] tube m protecteur de mise à la terre
~CONE - [mech] rodage m conique, cône m rodé
~CONSOLIDATION - [constr] consolidation f du sol
~CONTACT - [el] contact m de mise à la terre
~-CONTROLLED APPROACH (G.C.A.) - [radar] (device using primary radar to determine the position of an aircraft during its approach; also called colloquially "talking down") système m d'approche contrôlé du sol GCA
~-CONTROLLED INTERCEPTION - [radar] interception f contrôlée du sol
~CREW - [aero] (staff occupied in the servicing etc. of aircraft on the ground) personnel m non navigant
~CURRENT - [el] courant m à la terre
~CUSHION - [aero] (air cushion caused by 'ground effect') coussin m pneumatique par effet de sol
~DETECTOR - [el] (instrument designed to indicate the value of a leakage current to earth) indicateur m de terre
~DIRECTION FINDING - [radar] (a system of radiolocation in which a transmitter tries to locate a moving object) radiogoniometrie f terrestre
~DISTANCE - [radar] (the horizontal distance from one object to another) distance f terrestre
~EFFECT - [aero] (of helicopters) (the phenomenon of the development of an air-cushion under a helicopter hovering close to the ground) effet m de sol
~EFFECT VEHICLE - [mech] (vehicle which is supported by an air-cushion built up under it by the discharge of air, e.g.from ducted fans. Loosely called "hovercraft") hovercraft m, véhicule m à coussin d'air
~ELECTRODE - [el] (earth electrode) prise f de terre
~ENGINEER - [aero] (member of ground staff concerned with aircraft engine care and maintenance) ingénieur m d'aérodrome
~EQUALIZER CONDUCTOR - [el] (coils of low inductance placed in the circuit connected to points of an aerial to distribute the current in a specified manner) élément m d'égalisation
~FAULT - [el] dérangement m dû à une mise accidentelle à la terre
~FLOOR - [constr] rez-de-chaussée m
~FOG - [met] (fog due to condensation in a very shallow layer on the ground) brume f de sol

GROUND GEAR - [mech] (synonym for landing gear) train *m* d'atterrissage
~GLASS - [glass man] (used in cameras etc) verre *m* dépoli, verre *m* biseauté
~-GLASS FOCUSING - [photo] mise *f* au point sur verre dépoli
~-GLASS IMAGE - [photo] image *f* projetée sur le verre dépoli
~-GLASS SCREEN CIRCLE - [photo] cercle *m* dépoli
~-GLASS VIEWFINDER - [photo] viseur *m* à chambre noire
~GRADING - [constr] aplanissement *m* du terrain
~HANDLING EQUIPMENT -[astronaut] équipement *m* de manutention au sol
~HARNESS - [text] harnais *m* de fond
~-HOG - [mining] contre-poids *m* d'un plan incliné
~ICE - [met] glace *f* de fond
~INDICATOR - s. ground detector
~JOINT - [mech] jointure *f* rodée
~LEAD - [radio] prise *f* de terre, fil *m* de terre
~LEAK - [el] contact *m* à la terre, perte *f* à la terre
~LEVEL - [soil] terrain *m* naturel, niveau *m* du sol naturel, cote *f* du sol naturel
[mining] niveau *m* du sol
~LEVELLING - [constr] aplanissement *m*, nivellement *m*
~LIMESTONE - [constr] calcaire *m* broyé
~LINE - [math] ligne *f* d'intersection
[el] conduite *f* à la terre
~LOOPING - [aero] (uncontrolled sudden turn by an aircraft while moving under power on the ground) cheval *m* de bois
~MASS - [min] masse *f* microcristalline
~MERISTEM - [bot] (giving rise to ground tissue) méristème *m* des racines
~MOULD - [constr] (timber frame used to bring an embankment etc to the required shape)bâti *m* de support
~NOISE - [radio] bruit *m* de fond
~OF THE CLOTH - [text] fond *m* du tissu
~-OIL PAINT - [paint] peinture *f* à huile
~PAINT - [paint] couche *f* de fond
~PICK - [text] trame *f* de fond, trame *f* d'envers
~PLAN - [surv] projection *f* horizontale
[constr] plan *m* de fondation
~PLANE - [geom] (any horizontal plane) plan *m* géométrique
~-PLANE AERIAL - [radio] antenne *f* à polarisation horizontale
~-PLANE ANTENNA - s. ground-plane aerial
~PLATE - [el] plaque *f* de terre
[constr] (the horizontal timber to which the frame is secured) sole *f*, semelle *f*, seuil *m* de dormant
~PLOT - [constr] terrain *m* à bâtir
~POSITION - [aero] (the point on the earth's surface vertically beneath an aircraft) position *f* sol
~POSITION INDICATOR - (G.P.I.) - [aero] (automatic computing and displaying position indicator, usually employing air position and pre-set wind data) totalisateur *m* d'estime, indicateur *m* de position
~-PRESSURE PICK-UP - [instr] (instrument designed to measure ground pressures at any depth) indicateur *m* de pression du sol
~PROBE - [geol] (a probe to obtain a sample of the subsoil) lance *f* de sonde
~-REFLECTED WAVE - [radio] (the component of the ground wave which is reflected from the ground) onde *f* réfléchie par le sol
GROUND REFLECTION - [radar] (the reflection of the beam by the ground) réflexion *f* terrestre
~RESONANCE - [aero] (the shaking caused by the tyre flexibility upon the landing of a helicopter) résonance *f* au sol
~RETURN - [radar] (the echo which is received from the ground by air-borne radar sets) écho *m* de sol
~RETURN CIRCUIT - [telecomm] circuit *m* unifilaire
RIG - [mining] sonde *f*
~ROD - [el] (earth rod ; rod driven into the earth to ensure a good ground connexion) plaque *f* de mise à la terre
~ROPE - [text] palombe *f*, palonne *f*
~RUNNING - [aero] fonctionnement *m* au sol du moteur
~RUNNING TIME - [aero] (that part of the total engine running time during which an aircraft is motionless or taxying on the ground) temps *m* de fonctionnement au sol du moteur
~SCRAP - [rubber ind] déchets *mpl* broyés de caoutchouc
~SCREEN - [radio] (earth screen) écran *m*
~SEA RETURNS - [radar] (radar echoes from earth or sea) échos *mpl* de mer ou de sol
~SETTLING - [constr] affaissement *m*, tassement *m* du terrain
~SILL - [carp] sablière *f* basse (de cadre)
~SILLS - [hydr] (underwater walls preventing excessive scour of the bed) heurtoir *m*, radier *m* d'écluse
~SLIPPING - [gen] glissement *m* de terre
~SLUICE - [constr] canal *m* à sluices
~SPACE - [constr] aire *f*
~SPEED - [aero] (speed of an aircraft in respect to the surface of the earth) vitesse *f* sol
~SPRING - [rubber ind] (any kind of ground scrap rubber free from fabric material) déchet *m* broyé sans fibres textiles
~STAFF - [aero] (persons employed in connexion with aircraft on the ground) personnel *m* non navigant
~START - [aero & astronaut] démarrage *m* au sol
~STARTER - [aero] (any device for starting an aircraft engine which is not carried in the aircraft) démarreur *m* de piste
~STATE - [phys] (the state of a quantized system which is that of lowest energy) état *m* normal, niveau *m* normal
~STATE DISINTEGRATION ENERGY - [phys] énergie *f* de désintégration dans l'état fondamental
~STRAP - [el] (earth strap) lame *f* de contact de terre
~SUBSIDENCE - [soil] affaissement *m* du terrain
~SURVEY - [surv] (the general examination of an area) levé *m* du terrain, levé *m* terrestre
~SWINGING PLATFORM - [aero] aire *f* de régulation des compas
~SYSTEM OF AN ANTENNA - [radio] système *m* antenne-terre
~TEMPERATURE - [met] température *f* au sol
~TERMINAL - [el] (terminal used to make a connexion to earth) borne *f* de terre, borne *f* de mise à la terre
~TEST - [el] essai *m* au sol
~TEST COUPLINGS - [aero] (hydraulic or compressed air connexions for testing aircraft equipment

from ground supplies) connexions *f*pl d'essais au sol

GROUND TESTS - [aero] (the testing of the engines on the ground) essais *m*pl au sol

~TEXTURE - [text] (the back of the cloth) fond *m*, tissu *m* d'envers

~THREAD - [text] (in fancy yarns) fil *m* de fond, âme *f*, fil *m* de support

~TIMBER - [constr] (ground sleeper) longrine *f*

~TINT - [print] sous-impression *f* en à-plat

~-TO-AIR COMMUNICATION - [radio] (undirectional communication from ground to aircraft) communication *f* dans le sens sol-air

~TRACK POINT LOCK - [railw] circuit *m* de voie d'immobilisation d'aiguille

~TRAFFIC - [aero] (movement of wheeled vehicules on the ground) circulation *f* au sol

~-TRAFFIC SIGNAL LIGHT - [aero] (signalling light designed for the control of traffic on the ground) feu *m* de signalisation de circulation au sol

~TWILL - [text] sergé *m* de fond

~TYRE - [rubber ind] déchets *m*pl de pneu broyés

~WASTE - [rubber ind] déchets *m*pl broyés de caoutchouc

~WATER - [hydr] (water which is naturally contained in the subsoil) nappe *f* superficielle

~-WATER ARTERY - [hydr] veine *f* d'eau souterraine

~-WATER BASIN - [hydr] bassin *m* d'eau souterraine

~-WATER DISCHARGE - [hydr] affleurement *m* de l'eau souterraine

~-WATER DIVIDE - [hydr] ligne *f* de partage des eaux souterraines

~-WATER DRAINING - [agric] captage *m* des eaux souterraines

~-WATER FLOW - [hydr] écoulement *m* de l'eau souterraine

~-WATER LEVEL - s. ground-water table

~-WATER LOWERING - [soil] abaissement *m* de la nappe phréatique

~-WATER SUPPLY - [hydr] approvisionnement *m* en eau souterraine

~-WATER TABLE - [geol] nappe *f* phréatique

~-WATER VEIN - s. ground-water artery

~WAVE - [radio] (radiowave reaching the receiver by propagation along the earth's surface) onde *f* de sol

~WHEEL - [mech] roue *f* porteuse

~WIRE - [el] (earth wire; conductor leading to an electric conduction with the ground) fil *m* de terre

~WOOD BOARD - [paper man] carton *m* blanc volumineux

~WOOD PULP - [paper man] pâte *f* de bois

~-ZERO - [phys] (the region vertically below or above the centre of a burst of a nuclear explosion) hypocentre *m*

GROUNDED ANODE AMPLIFIER - [radio] amplificateur *m* à anode à la masse, amplificateur *m* à charge cathodique

~BASE - [electron] base *f* commune

~CATHODE AMPLIFIER - [radio] amplificateur *m* à cathode à la masse

~CIRCUIT - s. ground circuit

~CONDUCTOR - s. ground wire

~EMITTER - [electron] (semiconductors) émetteur *m* commun

~GRID AMPLIFIER - [radio] amplificateur *m* à grille à la masse

GROUNDED-GRID CIRCUIT - [electron] montage *m* avec grille à la masse

~-GRID TRIODE - [electron] triode *f* à grille à la masse

GROUNDING - [el] (U.S. term for "earthing") mise *f* à la terre

[naut] échouage *m*

~CHOKE COIL - [el] bobine *f* de réactance de mise à la terre

~OF A PHASE - [el] mise *f* à la terre d'une phase

~REACTOR - [el] réactance *f* de mise à la terre

~TERMINAL - s. ground terminal

GROUNDMASS - [geol] (the finer grained portion of an igneous rock which has crystallized in to stages) masse *f* microcristalline

GROUNDNUT - [bot] carvi *m*, gland *m* de terre, arachide *f*

~CAKE - [agric] tourteau *m* d'arachide

~OIL - [chem] huile *f* d'arachide

GROUNDS AND LAGS - [metall] modèle *m* en douves

GROUNDWALL - [constr] mur *m* de fondation

GROUNDWOOD - [paper man] pâte *f* de bois

GROUNDWORK - [constr] (the work which is carried out for preparing a site for a foundation) travaux *m*pl de fondation

GROUNDS - [ind chem] (solids deposited from a suspension on standing, especially in brewing) dépôt *m*

[el] pertes *f*pl à la terre

GROUP, to - [gen] grouper, disposer en groupes

GROUP - [gen] groupe *m*

[el] (in storage batteries, the assembly of plates of the same polarity connected by a strap) faisceau *m*, groupe *m*

[chem] (metallic radicals which are precipitated together during the initial separation in quantitative analysis) radical *m*

[phys] (number of atoms occurring together in several compounds) groupe *m* d'atomes

[geol] faisceau *m*

[min] genre *m*

~ALLOCATION -[telecomm] (allocation of groups on a coaxial cable) répartition *f* de groupes

~BAND FILTER - [telecomm] filtre *m* de bande de groupe

~BUSY SIGNAL - [telecomm] (in telephony) signal *m* d'occupation interurbain

~CENTRE - [telecomm] centre *m* de distribution

~CODE - [comput] code *m* de groupe

~CONTROL CHANGE - [comput] changement *m* de groupe

~DELAY - [radio] (the time taken by an individual signal to travel from the transmitter to the receiver) temps *m* de propagation de groupe

~DELAY-FREQUENCY - [radio] caractéristique *f* phase-fréquence

~DELAY FREQUENCY-DISTORTION - [radio]distorsion *f* phase-fréquence

~-DELAY METER - [instr] (instrument designed to measure group-delays in networks, particularly in television receivers) appareil *m* de mesure du temps de retard de groupe

~-DELAY VARIATION - [radio] variation *f* de temps de retard de groupe

~DIFFUSION METHOD - [nucl] (theoretical treatment of nuclear reactors) méthode *f* de groupes

~DISTRIBUTION FRAME - [radio] (frame producing

flexibility of interconnexion of apparatuses used for the transmission of frequencies in the basic group range) répartiteur *m* de groupe
GROUP DRIVE - [mech] (method of electric motor drive, in which a single motor drives a number of machines) commande *f* à groupes
~FREQUENCY - [radio] (the frequency of repetition of the trains of waves emitted from a spark transmitter) fréquence *f* de groupe
~INDICATION CYCLE - [comput] cycle *m* d'indication contrôle de groupe
~LIGHTING SYSTEM - [el] (of a train) éclairage *m* collectif
~LINK - [radio] liaison *f* en groupe primaire
~OF CONDUCTORS - [el] (conductors placed in the same slot, belonging to the sale coil) faisceau *m* de conducteurs
~OF CONTACTS - [el] couronne *f* de contacts
~OF FAULTS - [geol] groupement *m* de failles
~OF LINES - [telecomm] faisceau *m* de lignes, faisceau *m* de jonctions
[railw] faisceau *m* de voies
~PILOT - [radio] (group reference pilot) onde *f* pilote de groupe primaire
~PULLEY BLOCKS - [mech] moufle *f* à plusieurs poulies
~REACTION - [chem] (the reaction by which the metallic radicals are precipitated) réaction *f* de groupe
~SECTION - [radio] (part of a group link between two adjacent group distribution frames) section *f* de groupe primaire
~SORTING DEVICE - [comput] trieuse *f* par cartes maîtresses
~SWITCH - [telecomm] sélecteur *m* de groupe
~TRANSFER POINT - [radio] point *m* de transfert de groupe primaire
~TRANSLATING EQUIPMENT - [radio] installation *f* de modulation de groupe
~VELOCITY - [radio] (the velocity of displacement of beats produced by the superposition of two or more plane sinusoidal waves of slightly different frequencies propagated in the same direction) vitesse *f* de groupe
GROUPAGE - [transp] groupage *m*
~TRAFFIC - s. groupage
GROUPED POSITIONS - [telecomm] positions *f*pl groupées
GROUPING - [gen] groupement *m*
[telecomm] groupement *m*, répartition *f* de groupes
[chem] structure *f*
[el acoust] (non uniform spacinq between the grooves of a disk) répartition *f* des sillons
[med] détermination du groupe sanguin
~KEY - [telecomm] clé *f* de groupement, clé *f* de concentration
GROUSE - [zool] tétras *m*, faisan *m* bruyant
GROUSER - [mech] (the gripping section of a track) ailette *f*, dent *f*
~PLATE - [mech] (of a tractor etc) semelle *f* de chenillard
~TRACK SHOE - s. grouser plate
GROUT, to - [constr] (the operation of injecting cement grout into the foundations) sceller au ciment
[soil] colmater
GROUT - [constr] (cement grout) mortier *m* liquide, lait *m* de ciment
(a plaster consisting of mortar and small stones) enduit *m* bretté
GROUTING - [constr] jointoiement *m* au mortier liquide
[constr] (the operation of applying a mortar mixed with small stones) coulis *m*
[mech] (cementation for the base of a machine) coulée *f* de ciment
[mining] fonçage *m* de puits par cimentation, cimentation *f* des fissures aquifères
GROVE - [mining] travers-bancs *m*, puits *m*, ouvrage *m* souterrain
~CELL - [el] (a primary cell, with platinum positive electrode, zinc negative electrode, dilute sulphuric acid electrolyte and concentrated nitric acid depolarizer) élément *m* de Grove
GROW, to - [gen] croître, augmenter, grandir
[agric] croître, pousser, revenir
GROWER WASHER - [mech] rondelle *f* élastique, rondelle *f* Grower, rondelle *f* à ressort
GROWING - [agric] élevage *m*
~ABILITY - [agric] faculté *f* de croissance
~PIECE - [bot] (of barley) couche *f* jeune, tas *m* jeune
~POINT - [bot] (the apical meristem of a growing axis) point *m* de croissance
~STRENGTH - [bot] énergie *f* végétative
GROWL, to - [acoust] grogner, gronder
GROWL - [acoust] grognement *m*, grondement *m*
[min] (noise due to a great pressure on strata) crissement *m*
GROWLER - [met] glaçon *m*
[el] (device designed to test an armature) vibreur *m* pour la vérification des enduits
GROWTH - [biol] croissance *f*, néoplasme *m*
[bot & agric] arriver à maturité
[metall] (the tendency of cast-iron to increase in volume when repeatedly heated and cooled) gonflement *m*
[gen] croissance *f*, développement, venue *f*
[med] tumeur *f*, grosseur *f*
~CHECK - [bot] arrêt *m* de végétation
~CURVATURE - [bot] courbure *f* de croissance
~CURVE - [nucl] (curve showing the activity increases with time) courbe *f* de grandissement
~OF MILDEW - [chem] formation *f* de rouille
~SPIRAL - [phys] (structure seen on the surfaces of crystals after growth) spirale *f* de croissance
~STEP - [phys] (on the surface of crystals) échelon *m* de croissance
GROYNES - [hydr] (barrier walls) estacade *f*
GR-S RUBBER - [ind chem] (government rubber-styrene, a general-purpose rubber if the styrene group) caoutchouc *m* styrolène-butadiène
GRUB, to - [gen & agric] fouir, défricher, fouiller
GRUB - [zool] larve *f*, ver *m* blanc
~BREAKER PLOUGH - [agric] charrue *f* défricheuse
~SAW - [tool] (hand saw for cutting marble) scie *f* à marbre
~SCREW - [mech] tourillon *m* fileté, goujon *m* fileté
GRUBBER - [agric] extirpateur *m*, machine *f* à essarter
GRUBBING - [agric] essouchement *m*, extirpation *f* des racines, défrichage, essartement *m*
GRUBSCREW - [mech] (a screw which has no head and can therefore be driven below the surface of the material which it enters) vis *f* sans tête

GRUMBLING - [acoust] grommellement *m*, grognement *m*
GRUMMET - s. grommet
GRUMOUS - [bot] (with flesh consisting of small grains) grumeleux
GRUNTING - [acoust] grognement *m*
G SCOPE - [radar] indicateur *m* type G
G - SUIT - [astronaut] vêtement *m* anti-g
G - TOLERANCE - [astronaut] tolérance *f* à l'accélération
GUAIAC - [chem] (a gum obtained from species of Guaiacum and used in medicine and in food processing) résine *f* de guaiac
GUAIACOL - [chem] (an antiseptic and deodorant used in medicine) guaiacol *m*
GUANIDINE - [chem] (a synthesis intermediate) guanidine *f*
~ CARBONATE - [chem] (a synthesis intermediate) carbonate *m* de guanidine
~ NITRATE - [chem] (a component of explosives) nitrate *m* de guanidine
GUANINE - [min] (a naturally occurring purine with uses in medical research) guanine *f*
GUANO - [agric] (deposits of bird excrement used as a fertilixer) guano *m*
GUANOSINE - [chem] (a nucleoside with in medical research) guanosine *f*
GUANYL ACID - [chem] (a nucleotide with uses in research) acide *m* guanilique
~ NITROSOAMINOGUANYLIDENE HYDRAZINE - [chem] (a detonator for explosives) guanylnitrosoamino guanyldènehydrazine *f*
~ NITROSOAMINOGUANYL TETRACENE - [chem] (an initiating explosive) guanylnitrosoamino guanyltetracene
~ UREA SULPHATE - [chem] (an intermediate for dyestuffs and an analytical reagent) sulphate *m* de guanylurée
GUARANTEE, to - [gen & comm] garantir
GUARANTEE - [gen & comm] garantie *f*, caution *f*
GUARANTOR - [gen & comm] garant *m*, avaliste *m*
GUARANTY - s. guarantee
GUARD, to - [gen] garder, proteger
[bookbind] fournir d'onglets
GUARD - [gen etc] (any device or structure to give protection, e.g. against moving parts or electrical apparatus) protecteur *m*, gaine *f*, volet *m*
[constr] parapet *m*
[bookbind] onglet *m*
~ A CIRCUIT, to - [telecomm] garder un circuit
~ BAND - [radio] (narrow frequency band which is vacant between two channels, so as to give a margin of safety against interference) espace *m* libre entre deux canaux
~ BOARD - [text] planchette *f* de protection
[constr] planche *f* de rive *f* d'un échafaudage, planche *f* de côtè *m*
~ CAM - [text] came *f* de guidage
~ CIRCLE - [el acoust] (of records) sillon *m* fermé, sillon *m* final
~ CIRCUIT - [radio] (in Voice Frequency receivers) circuit *m* de garde
~ CIRCUIT COEFFICIENT - [radio] sensibilité *f* relative du circuit de garde et du circuit de signalisation
~ IRON - [railw] (a rail guard) chasse-pierre *m*
~ LOCK - [hydr] (the lock which separates tidal water from the waters in a basin) écluse *f* intermédiaire
[naut] sas *m* de dûreté
GUARD MEMORY - [comput] mémoire *f* de surveillance
~ POSITION - [comput] (digit position associated with the accumulator to guard against the loss of a digit during an overflow) position *f* de réserve
~ -RAIL - [railw] (a check-rail) contre-rail *m*
~ RELAY - [el] relais *m* de garde
~ RING - [el] (in insulators) anneau *m* de garde
[nucl] (in a counter tube) électrode *f* auxiliaire
~ VACUUM - [ind chem] vide *m* de protection, espace *m* vide intermédiaire
~ WIRE - [instr] (used in measuring methods to eliminate errors caused by surface leakage) fil *m* de protection
GUDGEON - [mech] goujon *m*, tourillon *m*, arbre *m*
[constr] (metal pin holding together adjacent stones) goujon *m*
[mech] (in a motor) axe *m*, tourillon *m* de la tête de piston
[mech] (hook and hinge) penture *f* de gond
~ HOLE HONE - [tool] alésoir *m* pour logement d'axe de piston
~ PIN - [mech] (pin on which the connecting-rod of a reciprocating engine is pivoted in the piston) tourillon *m* de crosse, axe *m* de piston
GUESS-ROPE - [naut] (a line from a ship to a boat on tow) faux-bras *m* de ceinture
GUGGENHEIM PROCESS - [chem] (a method for the production of sodium nitrate from Chilean caliche) procédé *m* Guggenheim
GUIDANCE - [gen] direction *f*, conduite *f*
[astronaut] guidage *m*
GUIDE, to - [gen] guider, diriger, conduire
[mech] conduire
GUIDE - [gen] guide *m*
[mech] guide *m*
[railw] (a check rail) contre-rail *m*
[mech] (in a motor, the guide of a valve stem) guide *m*, glissière *f* (de la crosse de piston)
[radio] (a waveguide) guide *m* d'ondes
[mech] (in hoists, elevators etc) guidonnage *m*, guide *m*
[mech] (of a turbine) directrice *f*
~ BAR - [mech] (bars guiding the crosshead of a steam engine) glissière *f* de crosse, guide *m*
~ BEAM - [text] rayon *m* conducteur, rayon *m* vecteur
~ BEARINGS - [mech] support *m* de glissière
~ BELT - [text] (in bobbin net machines) courroie-guide *f*
~ BLADE - [mech] (of a turbine) aube *f* directrice, aube *f* fixe
~ BLOCK - [mech] tasseau *m* de fixation des glissières
~ BOLT - [text] cheville *f* de guidage
~ BOWL - [mech] galet-guide *m*
[text] galet-guide *m*, roulette *f*
~ BRAKE SHOE - [auto] guide *m* de mâchoire de frein
~ BUSHING - [mech] douille *f*
~ CAM - [mech] (of hoists etc) came *f* de guide
~ CARD - [comput] (strong index card not to be passed through the machine) carte *f* indicatrice
~ CARRIAGE - [gas ind] (gasholders) dispositif *m* de guidage à col de cygne
~ CHARACTERISTIC WAVE IMPEDANCE - [electron] (of a travelling wave in a waveguide) impédance *f*

caractéristique d'un guide d'ondes
GUIDE COLUMN - [gas ind] (gasholders) colonne *f* de guidage, montant *m*
~COMB - [text] peigne *m* guide-fil
~DISK - [mech] disque *m* de guidage
~EDGE - [mech] bord *m* de guidage, rail *m* de guidage
~EYE - [text] œillet *m* de guidage
~FIELD - [phys] (the magnetic flux used to keep particles in a stable circular orbit in a particle accelerator) champ *m* magnétique guide
~FLAT - [gas ind] (gasholders) plaque *f* de raidissement et de guidage
~FOSSIL - [geol] fossile *m* caractéristique
~FRAME - [gas ind] (gasholders) charpente *f* de guidage
~FRAMING - s. guide frame
~GROOVE - [mech] rainure *f* de guidage
[text] gorge *f* de guidage, rainure *f*
~HAMMER - [text] marteau *m* de guidage
~HEAD - [mech] tête *f* de guidage
~LAMP - [auto] phare *m* orientable
~LEVER - [text] levier-guide *m*
~MARK - [metall] (in rolling, a defect due to abrasion) trace *f* de laminage, strie *f* de laminage, marque *f* de guide
~MILL - [metall] (rolling-mill with guides to make sure that the stock enters the mill at the right point and at the right angle) laminoir *m* à guides
~NUMBER - [photo] nombre-guide *m*
~PILES - [constr] (timber piles at the sides of on excavation for additional support) pieux *mpl* d'appui
~PIN - [metall] (diecasting) goujon *m*, colonnette *f*
[metall] tige *f* de démouleuse
[mech] broche-guide *f*
~PLATE - [text] (of spinning machine) plaque-guide *f*
[auto] plaque-guide *f*
~-POLES - [constr] (of a pile-driver) jumelles *fpl* d'une sonnette
~PULLEY - [mech] (loose pulley guiding a driving belt over an obstruction) galet-guide *m*, poulie-guide *f*
[telecomm] (of aerial controls) poulie *f* de guidage
~RAIL - [cin] (component part of a film perforating machine) couloir *m*
[text] (in an automatic loom for pile weaving) glissière-guide *f*, rail-guide *m*
[railw] contre-rail *m*
~RING - [mech] (of a turbine) couronne *f* directrice, distributeur *m*
[photo] bague *f* de réglage
~ROD - [mech] tringle *f* de guide
[cin] (part of film guiding mechanism) tige *f* de guidage
~ROLL - [mech] (a roller used to direct a continuous band of material) rouleau-guide *m*
~ROLLER - s. guide roll
~ROPE - [aero] (a cable used to control a balloon) câble-guide *m*
~SCORE - s. guide mark
~SCRATCH - s. guide mark
~SOUND - [el acoust] piste *f* de contrôle, piste *f* de ordres
~SPRING - [auto] ressort-guide *m*
~STANDARD - s. guide column
~SURFACE - [photo] surface *f* de guidage, glissière *f*
~TABLE - [text] table *f* de guidage
~TRACK - [metall] rampe *f*
[el acoust] s. guide sound
GUIDE VANE - [mech] (flat element used to direct the flow of a fluid) aube *f* directrice
~WAVELENGTH - [electron] longueur *f* d'onde dans un guide d'ondes
~WHEEL - [cin] (a toothed wheel guiding the film to the projector gate) galet *m* de guidage, pouile *f* de guidage
~WINCH - [text] tourniquet *m* conducteur
~WIRE - [telecomm] (a single wire transmission line) ligne *f* de transmission unifilaire
~WORM - [text] vis *f* sans fin, vis de commande
~-YOKE - [mech] support *m* des glissières
GUIDED DISTRIBUTION - [mech] distribution *f* commandé
~MISSILE - [astronaut] (missiles guided to a target by remote control) missile *m* guidé
~MISSILE COUNTERMEASURE - [radar] (the measures which are taken on the ground to counteract the effect of guided missiles) action *f* contre les missiles radioguidés
~WAVES - [electron] (wave the energy of which is concentrated within the boundaries between materials of different properties and is propagated along the path thus defined) onde *f* guidée
GUIDEPOST - [gen] poteau *m* indicateur (de route)
[constr] montant *m* de sonnette
GUIDES - [paper man] (of printing machines) guides *mpl* du papier
GUIDING - [gen] guidage *m*, conduite *f*, direction *f*, guidonage *m*
[gas ind] (gasholders) guidage *m*
~EDGE - [cin] (the side of the film which lays against the rail) flanc *m* de guidage
~PULLEY - [mech] rouleau *m* conducteur, galet-guide *m*
~RAFTER - [constr] chevron *m* de guide
~RIBS - [rubber ind] (in tyres) nervures *fpl* concentriques
~SHOE - [mech] (in elevators) main *f* courante, main *f* de guidage
GUIGNET GREEN - [paint] (a chrome green used as a paint pigment. Also called Viridian Green q.v.) vert *m* Guignet, oxyde *m* de chrome
GUILD - [gen] corporation *f*
~COLORIMETER - [instr] (type of colorimeter in which beams of light of the three primary colours are mixed and adjusted until a match is obtained with the colour under examination) colorimètre *m* de Guild
GUILDHALL - [gen] hôtel *m* de ville
GUILLEMINE LINE - [radar] (in high-level pulse modulation, the network which is used in generating an almost square pulse with steep rise and fall) ligne *f* du Guillemine
GUILLOCHE - [arch] (ornament in the shape of interlaced bands) guillochis *m*
GUILLOTINE, to - [gen & mech] guillotiner
[bookbind] massicoter
GUILLOTINE - [mech mach] (machine for cutting by shearing action) guillotine *f*, cisaille *f*
[med] (instrument used to cut tonsils) amygdalotome *m*
[bookbind] presse *f* à rogner, massicot *m*
[metall] coupe-jet *m*
~CUTTER - [mech] guillotine *f*
~CUTTING MACHINE - [mech] massicot *m*, machine

f à couper
GUILLOTINE SHEARS - [impl] cisailles *fpl* à guillotine
~TRIMMER - [impl] guillotine *f*, cisaille *f* à guillotine, massicot *m*
GUINEA GREEN - [paint] (term sometimes used for Viridian Green, a chrome green pigment) vert *m* de Guinée
~FOWL - [zool] pintade *f*
~PIG - [zool] cochon *m* d'Inde, cobaye *m*
GUITAR - [mus] guitare *f*
GULF - [geogr] golfe *m*
~STREAM - [geogr] courant *m* du golfe
GULL - [zool] mouette *f*, goéland *m*
GULLET - [constr] (a narrow trench) goulet *m*, chenal *m*
[mech] (that part of the space between the teeth of a saw which is farthest from the cutting edge) dent-de-loup *m*, écranchure *f*
~TOOTH - s. gullet [mech]
GULLETING - [constr] (the process of preparing road cuttings in a series of simultaneously worked steps) attaque *f* à gradins
GULLEY - [hydr] (in a sanitary engineering, fitting at the upper end of a drain) siphon *m* de décantation
[geogr] (a channel or ravine in the earth) ravine *f*, ravin *m*, goulet *m*
~EMPTIER - [transp] véhicule *m* de vidange
~GRATING - [hydr] (perforated cover for a gulley trap) grille *f* de puisard
~HOLE - [hydr] (of a drain) puisard *m*, fosse *f* d'aisances
~TRAP - [hydr] (in sanitary engineering, a device for the elimination of foul air in the drain pipe) puisard *m* à siphon
GULLY - s. gulley
~TRAP - [constr] puisard *m*, chambre *f* d'écluse
GUM, to - [gen] gommer, engommer
[mech] gommer, encrasser, se gommer
[bookbind] encoller
GUM - [chem] (substances occurring in plant tissues, colloidal and not volatile. They can by hydrolyzed to give complex organ acids and pentoses and heroxoses) gomme *f*, colle *f*
(only USA; photo) copie *f* caoutchouc
[oil ind] (the substance produced by oxidation of petroleum) gomme *f*
[anat] gencive *f*
~ABSCESS - [med] abcès *m* gingival
~ACACIA - s. gum arabic
~ACAROIDES - [chem] résine *f* acaroïde
~ARABIC - [chem] (synonym for acacia gum. Used as a stabilizer for colloids and as a thickening agent) gomme *f* arabique
~BENZOIN - [chem] benjoin *m*
~BYCHROMATE PROCESS - [photo] (the use of gum as a vehicle for pigments and bichromate on printing papers) procédé *m* aux sels de chrome
~BOIL - [med] parulie *f*, abcès *m* pyorrhéique
~CAMPHOR - [chem] camphre *m*
~COPAL - [chem] copal *m*
~ELASTIC - [chem] gomme *f* élastique
~FLUX - [bot] gommose *f*
~FORMATION - [oil ind] (the oxidation of petroleum sometimes produces a sticky substance known as gum) formation *f* de gomme
~LAC - s. gum shellac
GUM LINE - [anat] liséré *m* gingival
~RESINS - [chem] (naturally-occurring complex polysaccharides usually obtained as vegetable secretions. True gums are water-soluble) résines *fpl* naturelles
~SHELLAC - [chem] gomme *f* laque
~SPIRIT OF TURPENTINE - [chem] (a distillate obtained from the oleoresin of certain species of pine trees) essence *f* de térébenthine
~SUGAR - [chem] arabinose *m*
~TRAGACANTH - [chem] (naturally-occurring gum with uses as an emulsifying agent) gomme *f* adragante
~TREE - [bot] kaori *m*
~TURPENTINE - [chem] essence *f* de térébenthine végétable
~UP, to - [print] (of lithographic plates) gommer
GUMBO - [geol] gumbo *m*
GUMMED PAPER - [paper man] papier *m* gommé
GUN, to - [auto] (colloq) ouvrir les gaz en grand
GUN - [firearms] canon *m*, fusil *m*
[paint etc] (a spray-gun) pistolet *m*, injecteur *m*
[electron] (the assemblage of electrodes from which the electron beam is emitted) canon *m* électronique
[metall] canon *m* à boucher le trou de coulée
[min] canon *m* perforateur
~BARREL - [firearms] canon *m* de fusil
[oil ind] réservoir *m* de décantation, séparateur *m* d'eau
~BOAT - [mining] skip *m*
~CARRIAGE - [firearms] (the support for an artillery weapon) affût *m* de canon
~COTTON - [chem] coton *m* fulminant, nitrocoton *m*
~CURRENT - [electron] (the total electronic current flowing to the anode) courant *m* du faisceau électronique
~-DRILL - [tool] canon *m* perforateur
~DRILLING - [oil ind] (holes and perforations made by a gun or a casing perforator through oil-well casing into the formation, so as to admit oil into the borehole) perforation *f* à canon
~EFFICIENCY - [electron] rendement *m* de canon
~FLUE - [ind chem] (in the heating of retorts) canal *m* de gaz riche, canon *m*
~LATHE - [mech] tour *m* à conons
~-METAL - [metall] (an alloy of copper, tin and zinc, sometimes with the addition of lead and nickel. Has good resistance to corrosion and wear) bronze *m* à canon, bronze *m* industriel
~-METAL BEARING - [mech] coussinet *m* en bronze
~MIKE - [radio] (rifle mike) microphone *m* canon
~PERFORATION - s. gun drilling
~PERFORATOR - [oil ind] canon *m* perforateur
~-POWDER - [chem] (an explosive consisting of a mixture of potassium or sodium nitrate, charcoal and sulphur in varying proportions. The first explosive made by man; it is still much used in pyrotechnics) poudre *f*
~RIFLING MACHINE - [mech] rifloir *m*
~-RIVET, to - [mech] riveter avec un marteau pneumatique
~SIGHT - [firearms] guidon *m*, bouton *m* de mire
~WELDER - [metall] (for spot welding) pince *f* à souder
GUNCOTTON - [chem] (a nitrocellulose made by nitrating cotton. It is a very safe and convenient ex-

plosive, especially for purposes of military engineering) coton-poudre *m*, fulmicoton *m*
GUNK - [chem] (pre-fixed moulding compound of polyester and glass fibre) composé *m* polyester-fibre de verre
[nucl] (colloq.; an undesirable, semi-solid material) crasse *f*
GUNNEL - s. gunwale
GUNNY - [text] toile *f* de jute, toile *f* à sac
~CLOTH - s. gunny
GUNSMITH - [gen] armurier *m*
GUNSTOCK MOUNT - [photo] support *m* (de la chambre) en fût de fusil
GUNWALE - [naut] plat-bord *m*, fargue *f*
GURLEY DENSIMETER - [instr] (apparatus for measuring the porosity of paper, in which the sample is made to close one end of a weighted cylinder, which sinks in a tank of water as air escapes through the specimen) densimètre *m* de Gurley
GUSH - [oil ind] jaillissement *m*
GUSHER - s. geyser
[oil ind] puits *m* jaillissant
GUSSET, to - (to fix by means of gussets) renforcer de goussets
GUSSET - [mech] (a plate or bracket attached to two or more parts to maintain the desired angle between them) gousset *m*, plaque *f* de jonction, éclisse *f*
~PLATE - [metall](a plate used as a gusset) plaque *f* de jonction
GUSSETTED - [constr] (provided or strengthened with gussets) renforcé de goussets, éclissé
GUSSETTING - [mech] (fixing of parts by means of gussets) ancrage *m* par goussets, assemblage *m* par goussets
GUST - [met] (sudden change in wind velocity or direction) rafale *f*
~DETECTOR - [instr] détecteur *m* de rafales
~ENVELOPE - [met] diagramme *m* des rafales
~GRADIENT DISTANCE - [met] (the distance over which gust velocity changes from zero to maximum measured horizontally) gradient *m* de rafale
~LOADING - [aero] chargement *m* des rafales de vent
~LOCK HANDLE - [aero] (locking device used to prevent undesired movements of a control surface under suddenly imposed load, as a wind-gust) verrouillage *m* du gouverne
~TUNNEL - [aero] soufflerie *f* à rafales
~VELOCITY - [met] (velocity of wind in a gust) vélocité *f* de rafale
GUSTINESS - [met] (condition under which gusts of wind occur frequently) tendance *f* aux rafales
GUT - [zool] boyau *m*
~BAND - [text] (used in looms for the weaving of tapestry or carpets) corde *f* à boyau
~LINE - [min] (in sulphur mines; a small pipe for the steam) veine *f*
~STRING - [mus] (string made with the guts of sheep) corde *f* de boyau
GUTS - [zool] intestin *m*
GUTTA - s. gutta-percha
~-PERCHA - [bot chem] (a naturally occurring rubber-like material) gutta-percha *f*
GUTTATE - [gen and med] guttiforme, tacheté
GUTTER, to - [gen] rainer, silloner
[gen] (of a candle) couler
[gen] (of water) couler en ruisseaux
GUTTER - [constr] (channel alongside a road) cunette *f*, caniveau *m*, ruisseau *m*
[constr] (channel along the eaves of a building used to dispose of the rain water) gouttière *f*, chéneau *m*
[hydr] (trench alongside a canal) canal *m* de décharge, conduite *f* de décharge
[bookbind] petits fonds *mpl*
[metall] (diecasting) rainure *f* pour l'ébarbage
[mining] galerie *f* d'aérage
[mining] (dry river bed with traces of gold) lit *m* d'ancien cours d'eau avec traces de minerai d'or
~BEARER - [carp] (timber carrying gutter boarding) appui *m* de gouttière
~BOARD - [constr] corniche *f*
~BRACKET - [constr] (a type of support for the drain pipes) crochet *m* de gouttière
~CHANNEL - [hydr] conduite *f*
~HOOK - s. gutter bracket
~LEAD - [constr] noue *f*, noquet *m*
~SUPPORT - [constr] crochet *m* de gouttière
~WOUND - [med] (a flesh wound) plaie *f* cavitaire, plaie *f* en seton
GUTTERING - [mech] (on a valve) rainure *f*
GUY - [gen] (a rope or chain) hauban *m*, gui *m*, verboquet *m*
[naut] hauban *m*, câble *m* de retenue
~ATTACHMENT - [mech] console *f* d'ancrage pour haubans
~HOOK - [mech] crochet *m* de hauban à scellement
~-ROPE - [constr] (rope used to hold a structure in a given position) câble *m* de haubanage, corde *f* de manoeuvre
[rope man] cordage *m* à sept torons
~STRAND - [rope man] corde *f* à torons de sept fils
~THIMBLE - [mech] casse *f* pour haubans
~WIRE - [telecomm] fil *m* d'attache
GUYED POLE - [telecomm] appui *m* haubanné
GYM-SHOE - [shoe man] (shoes for gymnastics) chausson *m* pour gymnastes
GYMNASTICS - [gen] gymnastique *f*
GYMNOCOLON - [med] lavement *m* colique
GYMNOSOPHY - [gen] (nudism) gymnosophie *f*
GYNAECOLOGIST - [med] gynécologiste *m*
GYNAECOLOGY - [med] (also spelt gynecology; the branch of medicine dealing with women's functions and diseases) gynécologie *f*
GYNATRESIA - [med] atrésie *f* vaginale
GYPSITE - [min] (an impure form of gypsum mixed with sand and the like) gypsite *f*
GYPSOMETER - [instr] (instrument designed to assess the contents of potassium sulphide in chemical substances) gypsomètre *m*
GYPSUM - [min] (naturally occurring hydrated calcium sulphate with uses in the production of plasters, paint and paper fillers, metallurgy, fertilizers, and in the production of sulphuric acid) gypse *m*, plâtre *m*
~CEMENT - [metall] moulage *m* en plâtre
~CEMENT PATTERN - [metall] modèle *m* en plâtre
~QUARRY - [min] carrière *f* de gypse
GYRATE - [gen and anat] sinueux, tortueux
GYRATOR - [el] (circuit element which variates the reciprocity theorem like the gyroscope) gyrateur *m*
GYRATORY CRUSHER - [mech] (type of crusher in which the moving jaw is a gyrating cone) broyeur *m* giratoire, concasseur *m* giratoire
~PADDLE - [mech] (blade in a mixing machine which

is made to move with a gyrating motion) palette *f* gyroscopique
GYRO - [aero] (autogyro) autogyre *m*
[instr] (a gyrocompass) boussole *f* gyrostatique
~FREQUENCY - [el magnetic waves] gyrofréquence *f*
~-HORIZON - [aero etc] (gyroscope-controlled artificial horizon) horizon *m* gyroscopique
~NORTH - [instr] (north as indicated by a gyrocompass) nord *m* gyroscopique
~PICKOFF - [instr] détecteur *m* d'angle
~REPEATER - [aero] répéteur *m* gyroscopique
~SYSTEM - [radar] système *m* gyroscopique
GYROCAR - [railw] wagon *m* de monorail gyroscopique
GYROCOMPASS - [instr] (compasso built round an electrically rotated gyro wheel) boussole *f* gyrostatique
GYROGRAPH - [instr] (instrumenti indicating rotary speeds) compte-tours *m*
GYROMAGNETIC - [phys] gyromagnétique
~COMPASS - [instr] (a gyrostabilized magnetic compass) ćompas *m* gyromagnétique
~EFFECT - [phys] (the amount of angular momentum which is acquired by a magnetized substance) effet *m* gyromagnetique
~RATIO - [phys] (the ratio between the magnetic moment of a system and its angular momentum) rapport *m* gyromagnétique, taux *m* gyromagnétique
GYROMETER - [instr] (instrument designed to measure the rotary speed) gyromètre *m*, compte-tours *m*
GYROPILOT - [aero] (automatic control device to keep aircraft on set course and in level flight) pilote *m* automatique
GYROPLANE - [aero] (a rotor-craft in which the rotors are not power-driven) autogyre *m*
GYROSCOPE - [instr] (a system comprising a rapidly-rotating rotor mounted in gimbals to give one or more additional degrees of freedom) gyroscope *m*
GYROSCOPIC - [instr] (relating to gyroscope devices) gyroscopique
~COMPASS - [instr] (a compass which depends for its indications on the special properties of a gyroscope) compas *m* gyroscopique
~SEXTANT - [instr] (an air-sextant having a gyroscopic artificial horizon) sextant *m* gyroscopique
~STABILIZER - [aero] plan-fixe *m* gyroscopique
GYROSTABILIZER - s. gyroscopic stabilizer
GYROSTAT - s. gyroscope
GYROSYNE COMPASS - [instr] (compass which operated on the principles of the magnetic modulator) compas *m* à induction terrestre
GYRUS - [zool] (convolution of the surface of the cerebrum) circonvolution *f*

H

h - [phys] (Planck constant) symbole de la constante de Planck
[met] (the Beaufort letter for hail) grêle *f*
H - [chem] (the symbol for Hydrogen) symbole de l'hydrogène
H-AERIAL - [radio] (dipole primary radiator with a dipole reflector) antenne *f* en H
H-ARMATURE - [el] armature *f* à double T
H-BEND - [electron] coude *m* progressif H
H-BOMB - [nucl] bombe *f* H, bombe *f* à hydrogène
H-LINES - [phys] (contour along which the electromagnetic field is constant in relation to a reference plane) lignes *f*pl H
H-PARAMETER - [electron] paramètre *m* H, paramètre *m* hybride
HAAR - [met] (cold foggy conditions) brume *f* (glaciale)
HABER AMMONIA PROCESS - [chem] (ammonia synthesis process in which nitrogen and hydrogen are passed through a catalyst bed under high pressure, the uncombined gases being treated for ammonia extraction and then recycled) procédé *m* de l'ammoniac de Haber
HABERDASHERY - [text] mercerie *f*
HABIT - [gen] habitude *f*, coutume *m*
[text] (a dress) habit *m*
[crystall] (the development of crystal forms by a mineral) facies *m*
[zool & bot] (the characteristic mode of development) habitus *m*, manière *f* de croître
HABITAT - [biol] habitat *m*
HABITATION - [gen] habitation *f*, demeure *f*
HABITUATION - [med] adaptation *f* progressive
HABROMANIA - [med] hédonisme *m*
HACHEMENT - [med] (hacking) hachure *f* (massage), tapotement
HACHURE - [draw] (a form of shading) hachure *f*
HACK, to - [gen] hacher, tailler
HACK - [zool] cheval *m* de louage
[mech] (a notch) entaille *f*
[constr] (long bank on which bricks are laid to dry them in the open) claie *f* de séchage
~-SAWING MACHINE - [mech] scie *f* à métaux
HACKBERRY TREE - [bot] micocoulier *m*
HACKING - [constr] (the roughening of a surface prior to plastering) boucharder
[of bricks] (the piling-up of green bricks on the hacks) pose *f* des briques sur claie de séchage
[carp] entaille *f*
~HAMMER - [impl] boucharde *f*
HACKLE, to - [text] (the operation of combing linen in the hackling machine) peigner, sérancer, échanvrer, émoucheter
HACKLE - [text] peigne *m*, sérançoir *m*
[zool] plume *f* de cou
~COMB - [text] barrette *f* de gills, peigne *m* pour fibres libériennes
~DRAWING FRAME - [text] gills *m*, étirage *m* à barettes
~PIN - [text] dent *f* du peigne
HACKLED FLAX - [text] lin *m* peigné
~HEMP - [text] chanvre *m* peigne
~JUTE - [text] jute *m* peigné
~JUTE YARN - [text] fil *m* de jute peigné
HACKLING - [text] (the process of combing flax or linen in the hackling machine, so as to place the long fibres along parallel lines and eliminate the short ones) peignage *m*, sérançage *m*
~BENCH - [text] banc *m* de peigne
~MACHINE - [text] machine *f* à peigner
~-THRESHING MACHINE - [agric] batteuse *f* en long
HACKLY FRACTURE - [geol] cassure *f* hachée
HACKNEY TAXI - [auto] (inGB only) taxi *m*
HACKSAW - [saw for cutting metal] scie *f* à métaux
~FRAME - [mech] porte-scie *m* à métaux, porte-scie *m* emmanche
HADDOCK - [zool] aiglefin *m*, aigrefin *m*
HADE - [geol] (angle of inclination of a fault-plane) inclinaison *f* de plan de faille
~-SLIP FAULT - [geol] faille *f* normale
HAEMAL - [zool] (relating to the blood) hémal
HAEMATACHOMETER - [instr] (instrument designed to measure the velocity of the blood circulation in the veins) hématachymètre *m*
HAEMATAL - s. haemal
HAEMATIN - [chem] (the pigment group of haemoglobin) hématine *f*
HAEMATINE - [min] (a very important ore of iron chiefly ferric oxide) hématite *f*
HAEMATOMETER - [instr] (instrument designed to determine the number of corpuscles in the blood) hématomètre *m*
HAEMATOPORPHYRIN - [chem] (a component of haemoglobin) hématoporphyrine *f*
HAEMATOXYLIN - [chem] (a logwood extract important in microscopy for staining preparations) hématoxyline *f*
HAEMIC - s. haemal
HAEMOCYANIN - [biochem] (a blue copper containing respiratory pigment in the blood of crustacea) hémocyanine *f*
HAEMOGLOBIN - [chem] (the complex oxygen-carrying protein of the red blood corpuscles) hémoglobine *f*
~METER - [instr] hémoglobinemètre *m*
HAEMOGLOBINOMETER - s. haemoglobin meter

HAEMORRHAGE - [med] (bleeding) hémorragie *f*
HAFNIUM - [chem] (a metallic element, symbol Hf, A.N. 72, A.W. 178.6, with application as a control element in atomic piles. It occurs naturally in zirconium minerals) hafnium *m*
HAFT - [tool] (a tool handle) poignée *f*, manche *m*
HAGBERRY TREE - [bot] cerisier *m* à grappes
HAIL - [met] (precipitation taking the form of hard ice pellets) grêle *f*
~DISEASE - [agric] (disease affecting grapes) rot *m* blanc, coitre *m*, maladie *f* de la grêle
HAILSTONE - [met] (pellet of ice formed by precipitation at freezing temperatures) grêlon *m*
HAILSTORM - [met] abat *m* de grêle
HAIR - [zool] poil *m*, cheveu *m*
[text] (generally a mixture of coarse hair and cotton yarn) poil *m*, crin *m*, soie *f*
~BELT - [impl] courroie *f* de crin
~BLOWER - [text] séchoir *m* pour cheveux
~BULB - [text] bulbe *m* pileux
~CARPET - [text] tapis *m* à poil
~CLIPPER - [tool] tondeuse *f*
~CLOTH - [text] (also written haircloth; material generally consisting of coarse hair and cotton yarn) tissu *m* en crin
~COMPASSES - [instr] (a drawing instrument) compas *m* à cheveu, compas *m* de précision
~COPPER - [min] chalcotrichite *f*
~CRACK - [metall] (very fine crack, e.g. in metall beginning to fail by fatigue) gerçure *f*
~CRACKS - [soil] fissures *fpl* capillaires, craquelures *fpl*
~-CROSS - [metall] réticule *m*
~DRIER - [impl] séche-cheveux *m*
~FELT - [text] (used for heat insulation purposes) feutre *m* en crin
~FOLLICLE - [anat] follicule *m* pileux
~GLAND - [anat] glande *f* sebacée
~HYGROMETER - [instr] (hygrometer which is controlled by the varying length of human hair with humidity) hygromètre *m* à cheveu
~LINES - [opt] fils *mpl* croisés
~MORDANT - [text] mordant *m* pour poil
~PENCIL - [paint] (a very fine brush made of camel or sable hair) pinceau *m*
~-PIN TUNING BAR - [electron] barre *f* de syntonisation en U
~PYRITE - [min] millérite *f*
~SALT - [min] halotrichite *f*
~SEAM - [metall] repliure *f* de laminage
~SIEVE - [impl] crible *m* de crin
~TUFTS - [text] bouclettes *fpl* de poils
~YARN - [text] filé *m* de poil
HAIRBRUSH - [impl] brosse *f* à cheveux
~PAD - [rubber ind] tampon *m* de brosse à cheveux
HAIRCURLER - [impl] frisoir *m*, bigoudi *m*
HAIRINESS - [gen and bot] aspect *m* hirsute
HAIRLESS - [gen etc] chauve, sans cheveux
HAIRLINE - [print] délié
~CRACKS - [metall] craquelures *fpl*, gerçures *fpl*
~LETTER - [print] (a very fine-faced type) capillaire *f*
HAIRPIN - [gen] épingle *f* à cheveux
~BEND - [constr] lacet *m*
~CATHODE - [electron] cathode *f* en U
~FILAMENT - [el] filament *m* en U
~FLUE - [ovens] (of retorts) conduits *mpl* jumeaux

HAIRPIN SHAPED WIRE - [mech] fil *m* métallique en épingle à cheveux
~SPRING - [mech] (spring formed of strip or wire bent through 180 deg) ressort *m* en épingle de cheveux
HAIRSPRING - [horol] (the balance spring) ressort *m* spiral, spiral *m*
~COLLET - [horol] virole *f* pour les spirals
HAIRY - [gen] poilu, velu, hirsute
HAKE - [zool] merluche *f*
~BOX - [text] (of a warping mill) plot *m*
HALATION - [photo] (the fogging of an emulsion due to light reflection) halo *m*, auréole *f*
[gen opt] halo *m*
[electron] (in a cathode-ray tube, the annular area surrounding a spot) halo *m*
[med] éblouissement *m*
HALAZONE - [chem] (a disinfectant for water supplies) halazone *f*
HALF - [gen] moitié *f*
[adj] demi
~-A-BRICK - demi-brique *f*, deux quartiers *mpl*
~-ADDER - [comput] (circuit with two output channels for binary signals) demi-additionneur *m*, demi-totalisateur *m*
~-ANCHOR EAR - [el] griffe *f* d'ancrage pour câble
~AND HALF SOLDER - [metall] soudure *f* de plombier
~-BACK LIGHTING - [cin] éclairage *m* latéral d'ambiance
~-BALL VALVE - [mech] (valve in which the obturating element is hemispherical) soupape *f* à semi-sphère
~BAT - [constr] demi-brique
~BATTEN - [constr] latte (trés menue, pigeonnière)
~BEARING - [mech] demi coussinet *m*
~BINDING - [bookbind] demi-reliure *f*, demi-reliure *f* à petits coins
~BLEACH - [text] blanchir prélablement
~-BOIL BATH - [text] bain *m* d'assouplissage
~-BOILED SILK - [text] soie *f* souple, soie *f* mi-cuite
~-BOILING - [text] assouplissage *m*, demi-cuite *f*
~-BOUND - [bookbind] en demi-reliure à coins
~BOX - [mech] coquille *f*
~-BREADTH - [shipbuild] plan *m* horizontal
~-BREED - [zool] demi-sang, métis
~-BRICK WALL - [constr] (wall built completely of stretchers) mur *m* de l'épaisseur d'une demie brique, cloison *f* en briques à plat, mur *m* d'une demi-brique
~-BROTHER - [gen] demi-frère *m*
~-CELL - [el] (a cell comprising one electrode immersed in an electrolyte) demi-cellule *f*, hémicellule *f*
~CIRCUMFERENCE TAPPING - [rubber ind] saignée *f* en demi-spirale
~CLIP - [mech] bride *f* semi-circulaire
~-CLOSE - [mus] (harmonic progression) cadence *f* imparfaite
~-CLOSED POINTS - [railw] aiguille *f* entrebâillée
~CLOTH - [bookbind] mi-toile *f* à coins
~-COILED WINDING - [el] (form of single layer winding) enroulement *m* à couche unique
~COLUMN - [constr] colonne *f* encastrée
~-COMPRESSION CAM - [mech] (relief cam) came *f* de décompression

HALF-COMPRESSION GEAR - [mech] (device for reducing compression in i.c. engine when starting) décompresseur *m*
~-CONVERGENCY - [astr] (conversion angle) angle *m* de conversion
~CORE - [metall] demi-noyau *m*
~CORNS - [agric] (broken barley) demi-grains *mpl*
~-CROSS GAUZE - [text] gaze *f* à demi-tour
~-CROSS LENO - [text] gaze *f* à demi-tour
~CYCLE - [el] demi-cycle *m*
~-CURRENT - [comput] (for core selection) demi-courant *m*
~DIPOLE WITH COAXIAL STUB - [radio] (aerial consisting of half of a sleeve-dipole aerial projecting from the metal surface) dipôle *m* à adaptateur coaxial
~-DUPLEX OPERATION - [telecomm] (circuit designed for duplex operations, but operating in one direction only at a time) semi-duplex *m*
~-ELLIPTIC SPRING - [mech] ressort *m* à demi-pincette
~-FINISHED GOODS - [gen] produits *mpl* semi-finis
~-FINISHED RUBBER - [rubber ind] (partly-manufactured rubber, before vulcanization) caoutchouc *m* semi-ouvré
~-HARD STEEL - [metall] acier *m* mi-dur
~HARNESS - [text] (in a Jacquard loom) mi-harnais *m*
~HEADER - [constr] (half-brick used for corners etc) demi-boutisse *f*
~HERRINGBONE TAPPING - [rubber ind] saignée *f* en demi-arête de poisson
~HITCH - [naut] (type of knot) demi-clé
~-INTENSITY PERIOD - [acoust] (the time in which the energy of vibration decays to one half of the initial value) demi-période d'extinction
~INTERMEDIATE CROSS-MEMBER - [auto] demi-traverse *f* intermédiaire
~LAP - [text] peigne *m* circulaire
~-LAP JOINT - [carp] assemblage *m* à mi-bois
[mech] assemblage *m* à mi-ter
~-LATTICE GIRDER - [constr] poutre *f* en treillis
~-LIFE - [nucl] (the time required for the decay of one-half of the atoms of a given radioactive substance) période *f* de demi-vie, période *f* radioactive
~LIGHT - [light] pénombre *f*, demi-jour *m*
~-LINEN CLOTH - [text] toile *f* mixte, tissu *m* mi-lin mi-coton
~-MAST - [gen] à mi-mât
[naut] en berne, à mi-mât
~MOON - [astr] (the moon as seen when half its visible surface is illuminated) demi-lune *f*
~-MOON PLATE - [mech] disque *m* semi-circulaire
~-NAKED BARLEY - [agric] orge *f* demi-nue
~-NOTE - [mus] (only USA; minim in GB) blanche *f*
~NUT - [mech] semi-écrou *m*
~-PACE - [constr] (landing at the top of a flight of stairs) demi-palier *m*
[in a window bay] palier *m* de repos
~-PEAK DELAY - [photo] délai *m* d'allumage
~-PERIOD ZONE - [phys] zone *f* de Fresnel
~-POWER POINT - [radio] (of an aerial) point *m* à demi-puissance
~-POWER WIDTH - [radio] largeur *f* de bande à demi-puissance
~ROLL - [aero] (a manoeuvre in which the aircraft makes half a revolution about the longitudinal axis) demi-tonneau *m*
HALF ROUND - [arch] demi-rond
[metall] (of a bar) demi-rond *m*, demi-rondin *m*
~-ROUND BAR - [metall] barre *f* demi-ronde
~-ROUND BASTARD FILE - [impl] lime *f* bâtarde
~-ROUND CHISEL - [impl] (cold chisel with a small half-round cutting edge) ciseau *m* demi-rond
~-ROUND DRILL - [mech] foret *m* demi-rond
~-ROUND FILE - [mech] (file with a cross-section having one flat and one convex face) lime *f* demi-ronde
~-ROUND SCREW - [mech] (a button-headed screw) vis *f* à tête demi-ronde
~-ROUND SLEEPER - [railw] traverse *f* demi-ronde
~-ROUND SPLICING - [text] dispositif *m* pour le renfort de la pointe et de la semelle
~SECTION - [el] (the section of an electric wave filter divided in the centre) demi-section *f*
~-SELECT CURRENT - [comput] (in a coincident-current memory) demi-courant *m*
~-SHADE PLATE - [opt] (semicircular plate of quart between the polarizer and analyzer) lame *f* à pénombre
~SHROUD - [mech] (of a gear wheel) disque *m* de pro tection
~SHUT - [gen] entre-clos
~-SILVERED - [opt] à argenture semi-transparente, demi-argenté
~SIZE - [constr] (of a paving stone) retendu
~-SIZED - [paper man] demi-collé
~-SOLE - [shoe man] demi-semelle *f*
~SPEED - [naut] demi-vitesse
~SPEED SHAFT - s. half time shaft
~STEEL - [metall] acier *m* ferreux
~STEELY - [gen] semi-vitreux
~STEP - [mus] (semitone) blanche *f*, demi-ton *m*
~STICK - [print] (half-column setting) demi-reglette *f*
~STORY - [constr] mezzanine *f*, entresol *m*
~STUFF - [paper man] (the raw material which has been converted into pulp by a breaker) défilé *m*
~SUBTRACTER - [comput] unité *f* de demi-soustraction
~THICKNESS - [nucl] s. Half Value Layer
~-TIMBER - [timber] demi-bois *m*
~-TIMBERING - [constr] (method of building in which foundations and principal members were made of strong timber and the walls formed by filling the spaces between members with plaster) demi-boisage *m*
~-TIME EMITTER - [comput] émetteur *m* demi-point
~-TIME OF EXCHANGE - [chem] demi-période *f* de échange
~-TIME SHAFT - [mech] (in a reciprocating i.c. engine, a shaft running at half the speed of the crankshaft, e.g. for valve operation) arbre *m* tournant à demi-vitesse, arbre *m* à cames
~TONE - [mus] seconde *f* mineure
[photo] demi-teinte *f*, simili *m*
~-TONE DISTORTION - [telev] distorsion *f* de demi-teintes
~-TONE IMAGE - [photo etc] image *f* à demi-teinte
~-TONE SCREEN - [photo] trame *f*, réseau *m*
~TRACK - [mech] (a heavy vehicles with tracks in the rear section) véhicule *m* semi-chenillé
~TRACK RECORDER - [el acoust] magnétophone *m* double-piste

HALF TRUSS - [constr] demi-ferme *f*
~TURN OF A STAIRS - [constr] tournant *m* de 90° de escalier *m*
~TWIST - [text] demi-tour *m*
~-VALUE LAYER - [nucl] (the thickness of a given material which will reduce the dose rate of a beam of radiation to one-half of its original value) couche *f* de demi-atténuation
~WARP - [text] demi-chaîne *f*
~-WAVE - [radio] demi-onde *f*
~-WAVE AERIAL - [radio] (aerial with an electrical length which is approximately half the wavelength of the signal to be received) antenne *f* demi-onde
~-WAVE ANTENNA - s. half-wave aerial
~-WAVE DIPOLE - [radio] dipôle *m* demi-onde
~-WAVE RECTIFICATION - [electron] (method of rectifying A.C., in which only half of the wave is delivered in undirectional form) redressement *m* monophasé
~-WAVE RECTIFIER - [el] (rectifier circuit operating only during alternating half-cycles) redresseur *m* biphasé
~-WAVE RECTIFIER CIRCUIT - [el] circuit *m* de redresseur biphase
~-WET SPINNING - [text] filature *f* en demi-mouillé
~WIDTH - [phys] largeur *f* de raie à demi-puissance
~-WOOLLEN CLOTH - [text] tissu *m* mi-laine
~WORD - [comput] demi-mot *m*
~-WORSTED YARN - [text] fil *m* de sayette, peigne *m* mixte
HALFWAY - [gen] à mi-chemin
HALFWIDTH - [phys] (of a spectral line) largeur *f* de valeur moyenne
HALIBUT LIVER OIL - [chem] (oil obtained from the liver of the halibut: a rich source of vitamins A and D) huile *f* de foie de fletan
HALIDE - [chem] (generic name for binary halogen compounds) halogenure *m*, haloïde *m*
~ABSORBER - [chem] absorbeur *m* d'halogènure
~LEAK DETECTOR - [electron] détecteur *m* de fuites aux halogènes
~SALT - [chem] sel *m* halogène
HALITE - [min] (natural sodium chloride, rock salt) halite *f*
HALL - [gen] vestibule *m*, hall *m*, salle *f*
[arch] manoir *m*, château *m*
~ANGLE - [el] (the ratio of the electrical field developed across the current and the field generating the current) angle *m* de Hall
~COEFFICIENT - [el] constante *f* de Hall
~CONSTANT - s. Hall coefficient
~EFFECT - [el] (change in the distribution of current in a strip of metal, caused by a magnetic field) effet *m* Hall
~MOBILITY - [electron] (the carrier mobility) mobilité *f* du porteur
~PROCESS - [el chem] (electrolytic process for the production of aluminium from fused cryolite. The working temperature is about I000C. and the aluminium produced is drawn off in a molten state) procéde *m* Hall
HÄLLEFLINTA - [geol] (obsolete name for tough, fine-grained volcanic ashes) cendres *fpl* volcaniques compactes
HALLMARK - [gen] (on silver etc) cachet *m*, empreinte *f*
HALLOYSITE - [min] (an amorphous mineral resembling kaolin, but containing a larger proportion of the hydroxyl radical. When immersed in water it disintegrate with audible crackling) halloysite *f*
HALLUCINATION - [med] (sensation with no objective reality) hallucination *f*
HALLUCINOSIS - [med] hallucinose *f*
HALLUX - [zool] gros orteil *m*
HALO, to - [gen etc] auréoler
HALO - [met] (a series of rings seen round sun or moon, caused by diffraction through ice crystals) auréole *f*
[opt] (faint coloured ring surrounding a source of light) halo *m*, cercle *m* lumineux
[photo] (the ring surrounding a photographic image of a bright source, due to a scattering the light) halo *m*
~-COUPLED LOOP - [electron] (coupling loop in a multi-cavity magnetron) boucle *f* de couplage
HALOCARBON PLASTICS - [chem] (plastics based on monomers containing only carbon and a halogen) plastiques *mpl* halogénés
HALOCHROMISM - [chem] halochromie *f*
HALOGEN GAS - [phys] gaz *m* halogène
HALOGENATED DERIVATIVE - [chem] (a compound produced by causing the combination of the original substance with a halogen) dérivé *m* halogéné
~RUBBER - [ind chem] caoutchouc *m* halogéné
HALOGENATION - [chem] (the introduction of a halogen molecule into a chemical compound) halogénation *f*
HALOGENS - [chem] (the elements chlorine, fluorine, bromine, and iodine) halogénes *mpl*
HALOGENYL - [chem] halogényle *m*
HALOHYDRIN - [chem] halohydrine *f*
HALOID - [chem] haloïde *m*
~ACIDS - [chem] (the acids formed by the combination of the halides with hydrogen, viz. hydrochloric, hydrofluoric, hydrobromic and hydroiodic acids) hydracides *mpl* halogéniques
~SALTS - [chem] sels *mpl* halogènes
HALT, to - [gen] s'arrêter, stationner
HALT - [gen] halte *f*, arrêt *m*, stationnement *m*
HALTED - [gen] arrêté
HALTER - [of the harness] longe *f*
HALVE, to - [gen] diviser en deux, partager en deux
HALVED JOINT - [carp] s. halving
~PATTERN - [diecasting etc] modèle *m* en deux sections
HALVING - [gen] partage *m* en deux
[carp] (method of jointing) assemblage *m* à mi-bois, coupe *f* à mi-bois
HALYARD - [naut] (also Hallyard) (rope for raising or lowering a sail) drisse *f*
HAM - [gen] jambon *m*
[anat] jarret *m*
HAMMER, to - [gen] marteler
HAMMER - [impl] marteau *m*, masse *f*, tetu *m*
[firearms] chien *m*, percuteur *m*
[metall] (forging machine) marteau *m* à refouler
[anat] (an ossicle of the ear) osselet *m*, marteau *m*
[mus] (the mechanism beating on the strings of a piano) marteau *m*
[text] (in a Jacquard loom) presseur *m*, marteau *m*
~AXE - [impl] malebete *f*
~BEAM - [constr] (short cantilever beam projecting into a room) saillie *f*
~-BEAM ROOF - [carp] (type of timber roof) comble

m polygonal
HAMMER BLOCK - [gen & mech] coup *m* de marteau
[metall] (in forging) martelage *m*
[railw] (the alternating force between the driving wheels of a locomotive and the rails) choc *m* (d'une roue contre le rail)
~BREAK - [el] (electromagnetic trembler device for electric bells) interrupteur *m* à marteau
~BREAKER - [metall] broyeur *m* à marteaux
~COGGING - [metall] laminage *m* à marteaux
~CONTACT - [el] contact *m* à marteau
~CRUSHER - [mech mach] concasseur *m* à marteaux
~CYLINDER - [mech] cylindre *m* de massette
[metall] cylindre *m* du mouton
~DOWN, to - [metall] refouler
~-DRESSED - [build] (of stone surfaces with a rough finish) dressé au marteau
~DRILL - [tool] (compressed-air rock drill) marteau *m* perforateur, perforatrice *f* à percussion
~DRILL STOPER - [mining] foratrice *f* télescopique
~DRILLING - [mining] forage par battage
~EYE - [tool] œil *m* du marteau
~FACE - [tool] tête *f* de marteau
~-FINISH PAPER - [paper man] papier *m* martelé
~FORGING - [metall] martelage *m*
~GLAZING - [paper man] satinage *m* à marteau
~HARDENING - [metall] écrouissage *m*, martelage *m* à froid
~HEAD - [mech] tête *f* de marteau
[mech] (of a power-hammer) pilon *m*, mouton *m*
~-MILL - [mech mach] (type of mill in which free-swinging hammers are used to reduce the material treated) broyeur *m* à marteaux
~PALSY - [med] maladie *f* professionnelle des forgerons
~PEEN - [mech] panne *f*
~PISTON - [mech] piston *m* du marteau
~RIVETING - [mech] rivetage *m* au marteau
~RIVETING MACHINE - [mech mach] rivoire *f*
~ROD - [horol] (one of the rods connecting the lifting cams of a turret clock to the hammers) tige *f* du martelet
~SCALE - [mech] (iron oxide scale formed on a workpiece when heated for forging) battitures *fpl* de martelage
~SLAG - [metall] laitier *m* de puddlage
~STALK - [horol] (the rod on which the hammer is fastened) tige *f* de fixation du marteau
~TRACKS - [nucl] (the tracks which are produced in a nuclear emulsion) traces *fpl* dues au noyau Li-8
~UNION - [mech] union *f* à marteau
~VALVE GEAR - [metall] distribution *f* du marteau
~WELDING - [metall] soudure *f* au marteau
~WITH RUBBERBALL - [impl] (for mill) marteau *m* à boule en caoutchouc
HAMMERED - [gen] martelé, battu
~PLATE - [metall] tôle *f* martelée
HAMMERING - [gen] martelage *m*, battage *m*
[mech] martelage *m*
(the operation of forging by hammer) usinage *m* à martelage
~MACHINE - [mech] (a machine which is used specifically to reduce the diameter of tungsten rods for manufacturing filaments) machine *f* à marteler
~TEST - [metall] essai *m* d'élargissement
HAMMERLESS - [firearms] hammerless, sans chien
HAMMERWELDING - [metall] soudure *f* au marteau

HAMMOCK - [gen] hamac *m*
HAMPER, to - [gen] empêcher, brouiller
HAMPER - [impl] panier *m*, calais *m*, banne *f*
[naut] accessoires *mpl* lourds
HAMSTRING - [anat] tendon *m* du jarret
HANCOCK JIG - [min] tamis *m* hydraulique à étage mobile
HAND - [anat] main *f*
[instr] (the pointer in a dial) aiguille *f*, indicateur *m*
[horol] (of a watch) aiguille *f*
[carp] (the method of hanging a door) emplacement *m* des gonds
[meas] paume *f*
[constr] relais *m* (transport à la brouette), jetée *f* (transport à la bêche)
~ACCELERATOR - s. hand throttle
~ADVANCE - [auto] avance *f* à main
~AIR-PUMP - [impl] (manually-operated device for exhausting air) pompe *f* à air
~AUGER - [tool] tarière *f* à main
~-AXE - [tool] (light hand-operated axe) hache *f*
~-BARROW - [impl] brouette *f*, civière *f*, brancard *m*
~BEAMER - [text] enroulement *m* à la main
~BELLOWS - [metall] soufflet *m*
~BRACE - [tool] (a brace) vilebrequin *m* à main
~BRAKE - [auto etc] frein *m* à main
~CAMERA - [photo] chambre *f* à main, appareil *m* photographique à main
~CAPACITANCE - [el] (specific type of body resistance) effet *m* de capacité de la main
~CAPACITY EFFECT - s. hand capacity
~CAPSTAN - [mech] cabestan *m* à bras
~CARD - [text] carde *f* à main
~CART - [gen] charrette *f* à bras
~CAULKING - [gen] calfatage *m* à la main
~CLUTCH CONTROL - [mech] commande *f* à main de l'embrayage
~CRANE - [constr] guindole *f*, gérance *f*
~CRANK - [mech] manivelle *f* de mise en marche
~DOG - [mining] tourne-à-gauche
[oil ind] clef *f* à tige
~DOLLY - [mech] (used for riveting) étampe *f*
~DRAWING - [draw] dessin *m* à main levée
~DRILL - [tool] drille *f* à main, sondeuse *f* à bras, perforatrice *f* à main, porte-foret *m* à main, chignole *f*
DRIVE - [mech] commande *f* à main
~-DRIVEN - [mech] commandé à main
~-DRIVEN COMPRESSOR - [mech] compresseur *m* à main
~EJECTION - [metall & plast ind] (removal of a moulding from the mould by the use of the hand alone) éjection *f* à la main, éjection *f* manuelle
~FEED - [mech] (operation of a feed mechanism by hand) pression *f* à main
~-FEED PUNCH - [comput] (manually actuated punch perforateur *m* manuel
~FERTILIZER - [agric] distributeur *m* d'engrais à main
~FILE - [impl] lime *f* plate-à-main
~FILLER - [auto] (hand-controlled device for the adjustement of pressure in a tyre) controleur *m* de pression à main
~FORCE PUMP - [constr] pompe *f* à main, pompe *f* foulante à bras *m*
~FORGING - [metall] forgeage *m* à la main

HAND GAUGE - [impl] aiguille *f*, comparateur *m*
~GRENADE - [firearms] grenade *f* à main
~GRINDER - [mach tool] machine *f* à rectifier à main
~GRIP - [gen] poignée *f*
~HAMMERING - [metall] (in forging work) martelage *m* à la main
~-HELD VIEW - [photo] prise *f* de vue à la main
~HOE - [agric] houe *f* à bras, binette *f* à bras
~-HOLE - [mech] (small hole with a removable cover at the side of a pressure vessel, a tank etc) trou *m* de bras, regard *m* de lavage
~JACK - [auto] cric *m* à main
~JIG - [min] bac *m* à piston
~KNITTING LOOM - [text] métier *m* à cueillir manuel, métier *m* à cueillement à main (tricotage)
~LADLE - [impl] (in foundry work) poche *f* à main
~LAMP - [el] (portable electric lamp, used for inspection etc) lampe *f* à main, lampe *f* portative
~LATHE - [tool] tour *m* simple avec support à main
~LAY-UP MOULDING - [plast ind] moulage *m* au contact
~LEAD - [surv] petite sonde *f*
~LEVEL - [instr] niveau *m* à main
~LEVER - [mech] manette *f*, levier *m* de manœuvre
~LIFTING - [agric] arrachage *m* à la main
~LIMITED INTEGRATOR - [math] intégrateur *m* à limite tranchée
~LINE - [fishing] ligne *f*
~LOOM - [text] métier *m* à tisser à la main
~LUBRICATOR - [auto etc] graisseur *m*
~-MADE - [gen] fait à la main, fabriqué à la main
~-MADE PAPER - [paper man] (made by dipping a mould into the pulp and distributing it into a sheet by shaking) papier *m* à la main
~MICROTELEPHONE SET - s. hand set
~MILKING - [agric] traite *f* à la main
~MILL - [impl] moulin *m* à bras
~MILLING MACHINE - [tool] fraiseuse *f* à la main
~-MINING - [mining] abattage *m* à la main
~MOULD - [plast ind] (a small mould designed to be stripped and loaded by hand away from the press) moule *m* à main
[metall] moule à main
~MOULDING - [metall] moulage *m* à main
~MOULDING MACHINE WITH TURNOVER PATTERN PLATE - [metall] machine *f* à mouler à la main avec plaque à modèle renversable
~OF SPIRAL - [mech] sens *m* de la spirale
~-OPERATED - [gen] (worked or driven by human strength) commandé à main
~OPERATION - [gen] commande *f* à main
~PICKING - [min] triage *m* à la main
~PIT - [mining] excavation *f* à la main
~PLACED STONE - [constr] hérisson *m* (fondation de route)
~-PLIERED NEEDLE - [text] aiguille *f* dressée à la main
~-PLIERING OF NEEDLES - [text] dressage *m* à la main des aiguilles et poinçons
~PNEUMATIC RAMMER - [mech] damoir *m* à main et à air comprimé
~POLLINATION - [bot] (artificial pollination) pollinisation *f* artificielle
~POWER - [gen] force *f* de bras, à bras, à main
~PRESS - [mech] (press in which the mechanical effort is applied by the operator's muscles, with amplification by mechanical or hydraulic devices) press *f* à main
HAND PUMP - [auto etc] pompe *f* à main
~-RADAR - [radar] radar *m* portatif
~-RAIL - [gen] main *f* courante, lisse *f*, garde-fou *m*
~RAIL OF A BALUSTRADE - [constr] filet *m* de recouvrement *m* liteau *m*
~-RAILING IRON - [metall] fer *m* main-courante
~RAKE - [brew ind] fourquet *m*
~-RAMMER - [impl] fouloir *m* à main
[constr] (for pile driving) sonnette *f* à main
~-RAMMING MACHINE - [metall] machine *f* à mouler à main
~RATCHET ADJUSTMENT - [mech] adjustage *m* à la main du relâchement par cliquet
~REAMER - [tool] alésoir *m* à main
~REGULATION - [gen & mech] réglage *m* à la main
~REST - [of a machine tool] support *m* à main, support *m* à éventail
~ROLLER - [print] rouleau *m* à main (d'imprimeur)
~-SAW - [impl] scie *f* à main, égoïne *f*
~SCREW - [tool] (a clamp) presse *f* à main, presse *f* à serrer
~SERIAL TAP - [metall] taraud *m*
~-SET - [print] (type-matter composed by hand) composé à la main
[telecomm] (rigid combination of telephone microphone and receiver which can be held simultaneously to the mouth and the ear) appareil *m* à microtéléphone combiné
~-SETTING - [print] (composition by hand) composition *f* à la main
~SHANK LADLE - [metall] poche *f* à main
~SHEARS - [impl] cisailles *f*pl à main
~-SIZING - [paper man] encollage *m* à main
~SORTING - [min] triage *m* à la main
~-SPIKE - [impl] broche *f*
~SPINNING - [text] filage *m* à la main
~-STAMP - [metall] étampeuse *f* à main
~STARTER - [mech] (built-in manual engine starting device, in which it is necessary to turn the propeller by hand) démarreur *m* à main
~STEERING - [naut] direction *f* par levier à main
~-STOPING - [mining] abattage *m* à la main
~STRIPPER - [text] débourreuse *f* à main
~STRIPPING BOARD - [text] débourreuse *f* à main
~TAMP - [constr] (hand rammer) demoisele *f*, dame *f*, hie *f*, pilon *m*
~TAP - [tool] robinet *m* à main
~TEARING TEST - [text] essai *m* d'arrachage à la main
~TEST - [gen] (also called 'grip') essai *m* à la main
[telev] bout *m* d'essai
~THROTTLE - [mech] accélérateur *m* à main
~TOGGLE PRESS - [mech] presse *f* à genouillère à main
~TOOL - [tool] outil *m* à main
~TRAMMING - [mining] roulage *m* à bras
~-TRUCK - [gen] chariot *m* à main, diable *m*
~TURNING - [mech] tournage *m* à la main
~-VICE - [mech] (small vice designed to grip a workpiece and to be held in the hand) étau *m* à main, étau *m* à vis
~WEAVING - [text] tissage *m* à la main
~WELR TURNING - [text] transfert *m* à la main, report *m* à la main
~-WHEEL - [mech] (a wheel designed to be turned by the hand for control purposes, e.g. the operation

of a gatevalve) volant *m*, volant *m* à main
HAND WINCH - [mech] treuil *m* à bras
HANDAUFWINDE - [met] (anabatic winds flowing up the sides of a valley after sunrise) vents *mpl* anabatiques
HANDBAG - [gen] sac *m* à main
HANDBOOK - [gen] manuel *m*
(a booklet containing instructions) manuel *m* d'instructions
HANDCART - [gen] charrette *f* à bras
HANDCUFF - [impl] menottes *fpl*
HANDED - [mech etc] (term used of a part which is similar to another, but is laterally inverted, i.e. is a mirror image of the other, or of a pair of such parts)à symetrie spéculaire
HANDHOLD - [impl] poignée *f*
HANDICRAFT - [gen] travail *m* manuel
[ind proc] métier *m* manuel
HANDIE-TALKIE - [radio] (small portable two-way radio unit) radiotéléphone *m* portatif
HANDINESS - [gen and naut] maniabilité *f*
HANDKERCHIEF - [text] mouchoir *m*
~MOTION - [text] (a card saving motion) mécanisme *m* réducteur de cartons
HANDLE, to - [gen] tâter des mains, manier
[mech] (to regulate etc) contrôler
(a machine etc) manoeuvrer
[mech] (to provide with a handle) emmencher
HANDLE - [text etc] (term used of yarns, fibres and woven materials, referring to the sensation produced by feeling them in the hand) toucher *m*
[gen] poignée *f*, manche *m*
[agric] (of a plough) mancheron *m*
(of an oar) giron *m*
[mech] manivelle *f*, poignée *f*
[constr] (shaft) tige *f*, manchon *m*, manche *m*, hampe *f*
~KNOB - [gen and auto] bouton *m*, olive *f*
~LATCH - [mech] loquet *m* de poignée
HANDLEBAR - [mech] (of bicycle etc) guidon *m*
~GRIP - [mech] poignée *f* du guidon
HANDLERS - [leather ind] (pits where the hides are laid flat after the first stage of tanning) fosse *f* à tanner
HANDLING - [gen] maniement *m*, manipulation *f*
[mech etc] (of a machine etc) manoeuvre *f*
HANDPICK , to - [agric] ramasser à la main
HANDRAIL - [gen] (along train,corridors, and public vehicles) garde-fou *m*, main *f* courante, barre *f* d'appui
HANDS - [horol] aiguilles *fpl*
HANDSAW - [constr] scie *f* à main ou à manche, scie *f* égoine
HANDSEL - [comm] première vente *f*, arrhes *fpl*
[gen] étrenne *f*
HANDSHAKING - [comput] (exchange of predetermined signals) affirmation *f* de connexion
HANDWORK - [gen] travail *m* à la main, travail *m* manuel
HANDY - [gen] maniable, commode, facile à manier
HANG, to - [gen] pendre, accrocher, suspendre
[gen] (of an object) pendre, étre suspendu
HANG UP - [oil ind] arrêt de la colonne pendant le tubage
HANGABWINDE - [met] (katabatic winds, flowing down mountain slopes) vents *mpl* catabatiques
HANGAR - [aero] (building designed to accomodate aircraft) hangar *m*
[gen] (sometimes used for vehicles) remise *f*
HANGENDES - [mining] toit *m* de tunnel
HANGER - [mech] crochet *m*, crôchet *m* de suspension
[constr] (type of bracket bolted to a wall to support overhead shafting) sabot *m*
[constr] (steel strip bracket supporting a pipe from a roof) étrier *m* de suspension
[el] (for the overhead contact wire) console *f*
[impl] (used in foundry work) crochet *m* de fonderie
[gen] (a weapon, short sword) coutelas *m*
[mining] (the overhanging rock over a vein or lode) palier *m*, échafaudage *m* volant
[th eng] (ovens and retorts) (only USA; of a charge) accrochage *m*
[mech] chaise *f*, suspenseur *m*
[photo] tringle *f*
[metall] (foundry) chariot *m*, porte-poche *f*
~CRACK - [metall] crique *f* transversale
~IRON - s. hanger
HANGFIRE - [mining] long feu *m*
[astronaut] arrêt *m* de la mise à feu
HANGING - [gen] suspendu, pendant
[th eng] (ovens and retorts) (of a charge) accrochage *m*
[carp] accrochage *m*
[metall] suspension *f*
[metall] accrochage *m* du cubilot, accrochage *m* des charges
~-BAR EJECTION - [plast ind] (system in which the ejector mechanism is actuated by tie bars hung from the head of the press) éjection *f* par barre de tête
~BATTENS - [light] (suspended stage lighting) herses *fpl* d'éclairage suspendues
~BEARING - [mech] support *m* suspendu
~BOLT - [mech] boulon *m* libre
~BUTTRESS - [constr] (buttress based on a corbel) arc-boutant *m*
~COMPASS - [instr] (in vessels) boussole *f* pendante
~CORE - [metall] (diecasting) noyau *m* suspendu
~GLACIER - [geol] glacier *m* suspendu
~GUTTER - gouttière *f*
~LAYER - [mining] toit *m*
~-ON - [mining] recette *f*
~POST - [constr] (post on which the door is hung) chardonnet *m*, poteau *m* tourillon
~ROD - [mining] tringle *f* de suspension
~SASH - [carp] (sash arranged to slide in vertical grooves) vantail *m* de fenêtre
~SCAFFOLD - s. hanging stage
[constr] échafaudage *m* suspendu, échafaud *m* volant
~STAGE - [constr] pont *m* volant, échafaudage *m* volant
~STEPS - [constr] (steps which are fixed into a wall on one side only, and are unsupported on the other side) escalier *m* suspendu, escalier *m* en encorbellement
~STILE - [constr] (the door stile on which the hinges are fixed) montant *m* vertical
~TIE - [constr] (a tie beam supported by a bar to prevent sagging) traverse *f* de suspension
~TRUSS - [constr] comble *m* à plancher suspendu
~VALLEY - [geol] (tributary valley higher than the

main one) valleuse *f*, vallée *f* suspendu
HANGING WALL - [mining] (the overhanging wall of a load or vein) toit *m*
HANGINGS - [gen] tapisserie *f*, tenture *f*, rideaux *mpl*
HANGOVER - [telev] (in colour television) traînage *m*
~TIME - [radar] (of an echo suppressor) temps *m* de blocage
HANK - [text] (general term indicating a reeled length of yarn) écheveau *m* (laine), matteau *m* (soie)
[naut] (a hoop or ring of rope or wood) anneau *m*, cosse *f*
[gen] (a circular loop of anything which is flexible) anneau *m*
~BRUSHING MACHINE - [text] machine *f* à brocher les écheveaux
~CLOCK - [text] compteur *m* d'écheveaux
~COUNTER - [text] compteur *m* d'écheveaux
~DRYING MACHINE - [text] sécheuse *f* pour écheveaux
~DYEING - [text] teinture *f* en écheveaux
~HOOK - [text] crochet *m* à battre
~OF YARN - [text] écheveau *m* de filé
~SIZING AND WRINGING MACHINE - [text] machine *f* à encoller les écheveaux
~TIE - [text] centaine *f*, ligature *f*, attache *f*
~WINDING MACHINE - [text] bobinoir *m* pour écheveaux
~YARN - [text] fil *m* en écheveaux
HANSA YELLOW - [dyes] (generic name for a group of organic azo pigments having good heat and light resistance) jaune *m* Hansa
HANSGIRG PROCESS - [metall] (a metallurgical process for the production of magnesium, giving a product of high purity) procédé *m* Hansgirg
HAPLITE - [geol] (preferably 'Aplite'; fine, light-coloured igneous rock) aplite *f*
HAPLOID - [genet] (having a single complete set of chromosomes) haploide
~APOGAMY - [genet] apogamie *f* haploïde
HAPLOPIA - [med] vision *f* simple
HAPTOTAXIS - [med] stéréotropisme *m*
HARBOUR, to - [gen] héberger
HARBOUR - [gen] abri *m*, refuge *m*
[geogr] port *m*, rade *f*
[glass man] (type of trough for the mixed ingredients which must be brought to the pot for fusion) creuset *m*
HARD - [gen] dur
[electron] (of a thermionic tube evacuated to a high degree) (vide) élevé
[radiat] (to indicate a good penetrating quality) dur
[acoust] (indicating dullness of hearing) dur (d'oreille)
[phys] (of sounds, colours etc) dur, à contrastes
[chem] (of a water which contains dissolved mineral salts) (eau) crue
[comm] (indicating a very stable condition of the market) tendu, soutenu
[met] (of frost etc) dur, sévère, fort
[fin] (of currency) (monnaie) forte
~ANTIREFLECTION COATING - [photo] couche *f* antiréfléchissante dure ou stable
~APORT - [naut] tribord toute
~BAKED SLAB - [constr] carreau *m* biscuit
~BAST - [bot] écorce *f* dure
HARD BOBBIN - [text] bobine *f* dure
~BRASS - [metall] laiton *m* laminé
~BRAZING - [metall] brasure *f* dure
~BROACHING - [mech] alésage *m* à l'état dur
~CAST-IRON - [metall] fonte *f* dure
~CAST STEEL - [metall] moulage *m* d'acier dur
~-CENTRE - [mach tool] (supporting point in the tailstock of a lathe, hardened to resist wear by workpieces revolving on it) contre-pointe *f*
[metall] (of a bar etc) noyau *m* dur
~CHROME - s. hard plating
~CLAY - [geol] argile *f* ferme
~COAL - [fuel] (only USA; anthracite in GB) anthracite *f*
~COMPONENT OF COSMIC RAYS - [radiat] (that part of the cosmic radiation which can penetrate a moderate thickness of an absorber) composante *f* dure des rayons cosmiques
~CORE - [constr] (the basis of a foundation consisting of hard stone, broken bricks etc) plate-forme *f* des terrassements
[constr] clinkers *mpl* concassés
~CORE LAYER - [constr] couche *f* de débris *m*, couche *f* de briquaillons
~CORN - [agric] blé *m* dur
~CURRENCY - [fin] monnaie *f* forte
~-DRAWN - [metall] trempe *f* produite par étirage au froid
~DRY STAGE - [paint] (term used for a coating which has practically completed its drying, is free of tack and has dried throughout its thickness, not merely on the surface) tout à fait sec
~END - [agric] (of the corn) pointe *f* dure
~-FACING - [metall] (the process of welding on the surface of soft steel a layer of a harder material) rechargement *m* dur
~FIBRE - [text] fibre *f* dure, fibre libérienne
[rubber ind] fibre *f* vulcanisée
~FINISH - [build] (fine coat of plaster applied with a trowel) dernière couche *f*
~FLOW - [phys] (flow under conditions of high viscosity) fluidité *f* grande dureté *f*
~GLASS - [chem] (borosilicate glass which is resistant to heat and to chemical action) verre *m* dur
~GLOSS PAINT - [paint] (a type of paint containing resin and having a specially hard surface) peinture *f* de lustre durcie
~GRADE - [metall] degré *m* de dureté
~GRAIN - [metall] grain *m* d'acier dur
~HEAD - [metall] métal *m* blanc dur
[geol] concrétion *f* dure dans le grés
~HEADING - [min] (hard rock met in making tunnels) avancement *m* difficile
~HOLE - [metall] carotte *f*
~IMAGE - [telev] (caused by excess voltage of the signal) image *f* dure
~IRON - [metall] fonte *f* dure
~LANDING - [astronaut] atterrissage *m* dur
~LEAD - [metall] (lead in which the malleability is eliminated by the presence of impurities, in particular antimony) plomb *m* dur, plomb *m* antimonié
~LIGHT - [cin] source *f* lumineuse (avec ombres) à contours nets
~LIGHTING - [photo] (strong contrast between light and shade) éclairage *m* donnant des ombres à contours nets
~MATERIAL - [text] tissu *m* dur, tissu *m* rêche

HARD NEGATIVE - [photo] cliché *m* dur
~PAN - [metall] calcin *m*, carapace *f* calcaire
~PAPER - [paper man] papier *m* compact
~-PASTE PORCELAIN - [pottery] (porcelain composed of china stone and china clay) porcelaine *f* dure
~PLASTER - [constr] (hard-setting plaster of Paris) enduit *m* dur
~PLATING - [metall] (a thich chromium plating directly deposited on the base metal) chromage *m* dur
~RADIATION - [radiat] (the quality of penetration) rayonnement *m* dur
~RAY - [radiat] (an X-ray with a high penetrating power) rayon X *m* dur
~RUBBER - [rubber ind] (ebonite, or vulcanite) caoutchouc *m* durci, ébonite
~SETTING - [metall] posage *m* de plaquettes de coupe métal dur
~SIZE - [text] colle *f* dure
[text] fils *mpl* collés, marque *f* d'encollage
~SOAPS - [chem] (sodium salt containing soaps) savons *mpl* durs
~SOLDER - [metall] soudure *f* au quart, soudure *f* au cuivre, brasure *f*
~SOLDERING - [metall] brasage *m*, brasure *f*
~SPOT - [metall] zone *f* dure
~-SPUN - [text] (strongly twisted yarn) très tordu
~STOCKS - [build] (slightly defective and overburnt bricks) brique *f* recuite
~TIN - [metall] alliage *m* d'étain dur
~TUBE - s. hard valve
~TWIST - [text] (yarn with more than the usual measure of twist) torsion *f* forte
~VACUUM - [phys] vide *m* très poussé
~VALVE - [electron] (term used of a tube exhausted to a high vacuum) tube *m* à vide élevé
~WASTE - [text] (the waste of the final manufacturing process) déchet *m* dur
~WASTE BREAKER - [text] loup-briseur pour déchet dur
~WATER - [chem] (water containing dissolved minerals salts) eau *f* crue, eau *f* calcaire
~WHEAT - s. hard corn
~WINDER - [text] doigt *m* du secteur, doigt *m* pour serrer les pointes des couches
~WINDING - [text] renvidage *m* très compact
~WHEEL - [mech] meule *f* dure
~WIRE TELEMETRY - [meas] télémesure *f* par fil
HARDBOARD - [paper man] (compressed fibre-board) carton *m* compact
HARDEN, to - [gen] durcir, endurcir
[metall] (the operation of making steel hard) tremper, durcir
[constr] (to make a surface hard etc) durcir, s'affermir
[photo] (the hardening of gelatine by the use of formalin) durcir, aluner
[chem] durcir
[constr] (of plaster etc) s'endurcir
[leather ind] se racornir
HARDENABILITY - [metall] pouvoir *m* trempant
[metall] aptitude *f* à la trempe
HARDENABLE STEEL - [metall] acier *m* trempable
HARDENED - [gen] durci, endurci
[metall] (heat-treated to give hardness) trempé
~ALLOY STEEL - [metall] acier *m* trempé
~AND TEMPERED - [metall] (acier) trempé
~CASE - [metall] couche *f* de cémentation
HARDENED FATS - [chem] (natural oils or fats the melting points of which have been raised by the conversion of the unsaturated fatty acid radicals to saturated types) huiles *fpl* durcies
~OILS - s. hardened fats
~ROSIN - [paint] (rosin which has been treated with lime to give a product which is chiefly calcium resinate) colophane *f* durcie
~ZONE - [metall] (the zone of metal near the weld hardened by the welding process) zone *f* trempée
HARDENER - [ind chem] (a material used to promote the setting of synthetic thermosetting resins) durcisseur *m*
[constr] (accelerator) produit *m* durcissant
HARDENERS - [paint] (catalyst which promote polymerization or resinification) durcisseurs *mpl*
HARDENING - [gen] durcissement *m*
[metall] (the process of making steel hard) trempe *f*, trempe *f* à l'air
[photo] (the use of formalin to harden gelatine emulsion which thus becomes insoluble and more durable) durcissement *m*
[ind chem] (a material used for hardening) durcissement *m*
[radiat] (the increase of the penetrating value) durcissement *m*
[text] (rope man) tordage *m*
~AGENT - [ind chem] agent *m* durcissant, siccatif *m*, durcisseur *m*
~ALLOY - [metall] alliage-mère *m*
~AND TEMPERING - [metall] trempe *f*
~BATH - [photo] (formalin bath) bain *m* de fixage, bain *m* de durcissement
~BY NITRIDATION - [metall] trempe *f* par nitruration
~COMPOUND - [metall] composé *m* de cémentation
~CRACK - [metall] procédé *m* de trempe
~DEPTH - [metall] profondeur *f* de trempe
~EFFECT - [rubber ind] effet *m* durcissant
~FIXING BATH - [photo] fixateur *m* tannant
~FURNACE - [metall] four *m* à tremper
~IN OIL - [metall] trempe *f* à huile
~MEDIUM - [metall] (the liquid used for the hardening of steel) bain *m* de trempe
~OF CONCRETE - [constr] faire durcir le béton, durcissement *m* du béton
~OF OILS - [chem] (the hydrogenation of oils in the presence of a catalyst) durcissement *m* des huiles
~POWDER - [metall] poudre *f* pour cémentation
~RESINS - [chem] (a term for thermosetting resins) résines *fpl* thermodurcissables
~SOLUTION - [photo] solution *f* de durcissement
~TEMPERATURE - [metall] température *f* de trempe
HARDENITE - [metall] (martensite; the hard constituent in hardened steel. Obsolete) hardénite *f*
HARDFACING ALLOY - [metall] métal *m* d'apport, alliage d'apport
HARDGLASS - [glass man] (the glass which is used for the envelope of X-ray tubes) verre *m* dur
HARDHEAD - [metall] alliage *m* fer-étain de scories
HARDIE - [tool] tranchet *m* d'eclume, casse-fer *m*
HARDINESS - [gen] robustesse *f*
HARDNESS - [gen metall phys etc] dureté *f*
[min] dureté *f*
[metall] (general resistance to deformation) trempe *f*
[radiat] (term used to describe the penetrating quality of a radiation) dureté *f*

HARDNESS - [chem] (presence of dissolved mineral salts in water, which affect its soap consuming power, causes scaling in boilers etc) dureté *f*
[electron] (of a thermionic tube; the degree of vacuum) degré *m* du vide
[photo] dureté *f*, tons *m*pl heurtés
~GRADIENT - [metall] (the ratio of hardness between the centre and the surface of a casting) rapport *m* de dureté
~NUMBER - [metall] indice *m* de dureté
~OF HEARING - [acoust] amblyacousie *f*, dureté *f* d'oreille
~OF RADIATION - [radiat] dureté *f* de rayonnement
~OF THE BOBBIN - [text] dureté de la canette, dureté de la bobine
~OF THE PIRN - [text] dureté de la canette, dureté de la bobine
~OF TWIST - [text] degré *m* de torsion
~OF WATER - [chem] crudité *f* de l'eau
~PENETRATION - [metall] profondeur *f* de trempe
~TEST - [mech] essai *m* de dureté
~TESTER - [instr] duromètre *m*
HARDPAN - [geol] (rock layer just under soft soil) couche *f* d'argile, carapace *f* calcaire, calcin *m*
[agric] (a hardened subsoil) sol *m* résistant
HARDTOP - [auto] toit *m* rigide
HARDWARE - [gen] quincaillerie *f*, ferronnerie *f*
[mech] (all the parts pertaining to endgate attachments in a dump truck) ferrures *f*pl
[comput] (all the mechanical, electrical and electronic devices of a computer) matérial *m* de traitement
[comput] (electronic components and finished pieces of equipment) éléments *m*pl de construction
~FINISH - [metall] (in diecasting) belle peau *f* (permettant la décoration directe)
HARDWOOD - [timber] (a dense, close-grained wood, e.g. oak, teak, ash etc) bois *m* dur
HARDY - [gen] robuste, endurci
[tool] (a cutting chisel for anvils) tranchet *m* d'enclume, tasseau *m*, tranche *f*
~HOLE - [mech] trou *m* carré de tasseau
HARE - [zool] lièvre *m*
~-LIP - [med] bec-de-lièvre *m*
HARE'S EYE - [med] lagophtalmie *f*
HAREWOOD - [bot] érable *m*
HARGREAVES PROCESS - [chem] (a method for the commercial production of sodium sulphate) procédé *m* Hargreaves
HARICOT BEAN - [bot] haricot *m*
HARL - [text] (filament or fibres of hemp or flax) tille *f*, teille *f*
HARM, to - [gen] faire du mal, léser
HARM - [gen] mal, tort *m*
HARMATTAN - [met] (hot dry wind, carrying extremely fine dust, blowing off the western Sahara at certain season) harmattan *m*
HARMFUL - [gen] nocif, nuisible
HARMLESS - [gen] inoffensif
HARMONIC - [acoust] (of a fundamental tone, any overtone which is an exact multiple of the pitch frequency) harmonique *m*
[math] (of quantities whose reciprocals are in arithmetic progression) harmonique *m*
[el] (component of an alternating wave, its frequency being a multiple of the fundamental frequency) harmonique *m*
[radio] (sinusoidal oscillation with a frequency which is a multiple of a fundamental frequency) harmonique
HARMONIC AERIAL - [radio] (standing wave aerial designed to operate at a frequency which corresponds to a multiple of the frequency of the aerial) antenne *f* harmonique
~ANALYSIS - [math] analyse *f* harmonique
~ANALYZER - [el] (instrument designed to determine the magnitude and phase angle of the constituent harmonics of an alternating wave form) analyseur *m* des harmoniques
~ANTENNA - s. harmonic aerial
~BALANCER - [mech] (vibration dampener) amortisseur *m* de vibrations, damper *m*
~CAN MOVEMENT - [cin] (colloq. for intermittent movement) mouvement *m* intermittent, mouvement *m* saccade
~COMPLEX - [geom] complexe *m* harmonique
~COMPONENT - [el] (development terms of a periodic function) composante *f* harmonique
~CONJUGATE - [math] conjuguée *f* harmonique
~CONTENT - [el] (function obtained by subtracting the fundamental wave from a non-sinusoidal periodic function) résidu *m*
~CONVERSION TRANSDUCER - [radio] transducteur *m* à conversion d'harmoniques
~CURRENT - [el] courant *m* harmonique
~CURVE - [math] sinusoïde *f*
~DISTORTION - [acoust] (the percentage of harmonics in the output voltage of the amplifier when this is fully modulated by a sinuosoidal signal) distorsion *f* non-linéaire
~DISTORTION ATTENUATION - [acoust] atténuation *f* de distorsion non-linéaire
~DISTORTION CHARACTERISTIC - [radio] caractéristique *f* de distorsion non-linéaire
~DISTORTION FACTOR - [radio] (in reception) coefficient *m* de distorsion non-linéaire
~DIVISION - [math] division *f* harmonique
~EXCITATION - [radio] (the excitation of a transmitter from a harmonic of the oscillator, or the excitation of an aerial on one of its harmonic modes) excitation *f* d'harmonique
~FILTER - [radio] étouffeur *m* d'harmoniques, réseau *m* filtrant
~FLUTE - [mus] flûte *f* harmonique
~FREQUENCY - [el] (frequency which is multiple of its fundamental frequency) fréquence *f* harmonique
~FUNCTION - [math] fonction *f* harmonique
~GENERATOR - [radio] (frequency multiplier) générateur *m* d'harmoniques
[electron] générateur *m* d'harmoniques
~INTEGRATOR - [math] intégrateur *m* harmonique
~INTERFERENCE - [radio] (interference caused by the harmonic radiation) interférence *f* d'harmonique
~LEAKAGE POWER - [electron] puissance *f* transmise à fréquences harmoniques
~MEAN - [math] moyenne *f* harmonique
~MINOR - [mus] (that form of the minor scale in which the leading note is sharpened) gamme *f* mineure
~MOTION - [phys] (a type of periodic motion which is characteristic of elastic bodies) mouvement *m* harmonique
~OF FUNDAMENTAL SOUND - [acoust] harmonique *f* du son générateur

HARMONIC OSCILLATOR - [mech] (a mechanical system with one degree of freedom) oscillateur *m* harmonique
~OSCILLATOR INTERFERENCE - [radio] interférence *f* harmonique de l'oscillateur local
~PIPE - [mus] tuyau *m* harmonique
~PROGRESSION - [math] progression *f* harmonique
~PROPORTION - [math] proportion *f* harmonique
~RANGE - s. harmonic proportion
~RESONANCE - [radio] résonance *f* sur harmonique
~RESPONSE - [control] (the change in steady-state output produced by a sinusoidal input signal) réponse *f* harmonique
~SELECTIVE RINGING - s. harmonic selective signalling
~SELECTIVE SIGNALLING - [telecomm] (the ringing on a party line by alternating currents of selected frequencies) appel *m* sélectif harmonique
~SERIES - [math] série *f* harmonique
[acoust & mus] (succession of sounds with frequencies that are in proportion to the whole members) série *f* d'harmoniques
~SUPPRESSOR - [radio] (generally a filter used to suppress the radiation of harmonics from a transmitter) étouffeur *m* d'harmoniques
~TELEGRAPHY - [telecomm] télégraphie *f* à fréquences acoustiques, télégraphie *f* harmonique
~UNMODULATED OUTPUT POWER - [radio] (of a transmitter) puissance *f* de sortie non-modulée de l'harmonique
~VOLTAGE - [el] tension *f* harmonique
~WAVE - [radio] onde *f* harmonique
HARMONICA - [mus instr] harmonica *m*
HARMONIUM - [mus] harmonium *m*
HARMONIZE, to - [gen etc] harmoniser, s'harmoniser
HARMONY - [mus] harmonie *f*
HARMOTOME - [min] (hydrated silicate of aluminium and barium) harmotome *m*
HARNESS, to - [gen] harnacher
[hydr etc] aménager, mettre en valeur
HARNESS - [gen] harnais *m*, harnachement *m*
[el] (set of pre-formed electric leads) fils *mpl* couplés
[aero] (a system of screened ignition leads to prevent electromagnetic radiations affecting the ratio equipment) blindage *m*
[aero] (of a parachute) harnais *m*
[text] (of a drawing appliance) harnais *m*, lisses *fpl*
[text] harnais *m*, remise *f*, lisse *f*
[comput] peigne *m* de câbles, forme *f* de câbles
~BOARD - [text] planche *f* d'arcades, planche *f* à trous, planche d'empoutage
~CHAIN - [text] chaîne *f* de harnais
~CLEANER - [text] brosseur *m* de lames
~CORD - [text] (strong linen twine) fil *m* d'arcade
~CORDING - [text] empoutage, encordage *m*
~FOR STATIONARY WARP - [text] corps *m* de fils fixes
~FRAME - [text] porte-lisse *m*
~HOOK - [text] crochet *m* de harnais
~LEATHER - [leather ind] cuir *m* à harnais, cuir *m* à bourrellerie
~MOTION - [text] mouvement *m* des lames
~MOUNTING - [text] empoutage *m*, encordage *m*
~THREADING - [text] empoutage, encordage *m*
~TYING - [text] colletage *m* du harnais
~WIRE - [text] fil *m* métallique de suspension des lames
HARNESS WITH SLIDING HEALDS - [text] harnais *m* à lisses mobiles
HARP - [mus] harpe *f*
[min] claie *f* à sable
~AERIAL - [radio] (undirectional aerial with almost elliptical transverse cross sections of the major lobe) antenne *f* à éventail
~ANTENNA - s. harp aerial
~PENDANT - [light] (gas pendant with a burner at the centre of a suspended loop) lampe *f* à gaz en forme de harpe
HARPOON, to - [gen] harponner
HARPOON - [fishing] harpon *m*, lance *f*
[mining] harpon *m*
HARPSICHORD - [mus] clavecin *m*
HARRIS PROCESS - [chem] (a method for the purification of lead, in which arsenic, antimony and tin are oxidized and the resulting oxides are converted to arsenates, antimonates and stannates by reactions using sodium hydroxide and chloride) procédé *m* Harris
HARROW, to - [agric] herser
HARROW - [agric] herse *f*
~MACHINE - [text] (used to wash wool) léviathan *m*
HARROWING - [gen] navrant, poignant
[med] hersage *m* des nerfs
HARSH - [gen] rêche, dur, âpre
[brew ind] (the taste of beer) âcre
~MATERIAL - [text] tissu *m* grossier
HARTREE VOLTAGE - [electron] (treshold voltage) tension *f* de seuil de Hartree
HARTSHORN - [chem] (contraction for Spirits of Hartshorn, an old term for an aqueous solution of ammonia) corne *f* de cerf
HARVARD CLASSIFICATION - [astr] (method of classifying stellar spectra) classification *f* de Harvard
HARVEST, to - [gen & fig] moissonner, récolter
HARVEST - [agric & gen] moisson *f*, récolte *f*, fenaison *f*, vendange *f*
~MITE - [vet] (larval forms penetrating the skin) aoûtat, rouget *m*
~MOON - [astr] (full moon) lune *f* de moisson
~MOUSE - [zool] souris *m* des moissons
HARVESTER - [agric] moissonneuse *f*, moissonneuse-lieuse *f*
~-THRESHER - [agric] (a combined binder and thresher) moissenneuse-batteuse *f*
HARVEYIZE, to - [metall] (the use of producer gas to harden a metal) tremper, durcir
HARZ JIG - [min] crible *m* hydraulique à étage fixe
HASH TOTAL - [comput] total *m* de vérification
HASHISHISM - [med] cannabisme *m*
HASLOCK - [text] (a type of hard wool shorn from the throat) laine *f* du cou inférieur
H.A. SOLVENT - [chem] (alternative name for cyclohexyl acetate) acétate *m* de cyclohexyle
HASP - [gen & mech] (a fastening mechanism) moraillon *m*, fermoir *m*
HASTATE - [bot] hasté, hastiforme
HAT - [gen] chapeau *m*
[cin] (generally a tripod for the camera) support *m*
~BLOCKER - [text] machine *f* à former les chapeaux
~-LEATHER - [mech] cuir *m* pour chapellerie
HATCH, to - [agric] faire éclore, sortir de la coquille
[zool] incuber, faire couver

[draw] hacher, hachurer
HATCH - [gen] demi-porte *f*
[naut] (a hatchway) descente *f*, écoutille *f*
[naut] (a close hatch) panneau *m*
[constr] (door closing only the lower half) partie *f* basse d'une porte coupée
[railw] ouverture *f* de charge
[draw] hachure *f*
[hydr] vanne *f* d'écluse
~BEAM - [shipbuild) barreau *m* de tasseau
~BOARD - s. hatch cover
~COVER - [naut] panneau *m* de descente
HATCHEL, to - [text] (to dress flax with a hatchel) peigner, sérancer, échanvrer
HATCHEL - [text impl] peigne *m*, séran *m*
HATCHERY - [food ind] (hatching establishment, in particular for the ova of fish by artificial means) établissement *m* de pisciculture
HATCHET - [impl] hachette *f*
~IRON - [plumb] (a form of copper bit) fer *m* coudé
~STAKE - [tool] (a smith's tool) enclumette *f*
HATCHETTITE - [min] (a mineral which contains a decomposed pyrochlorine compound) hatchettine *f*, suif *m* minéral
HATCHING - [gen] (of eggs) éclosion *f* d'une couvée
[agric] couvée *f*
[draw] hachure *f*
~OUT OF SPORES - [bot] dissémination *f* des spores
HATCHWAY - [naut] descente *f*, écoutille *f*
~LIFT - [mech mach] monte-charge *m* de lucarne
H.A.T.S. - s. Hour Angle of True Sun
HATTER'S SHAKES - [med] hydrargisme *m* des chapeliers
HAUL, to - [gen] traîner, remorquer, haler
[mech etc] traîner, transporter
[naut] (with a rope) haler
[naut] (in sailing boats) serrer (le vent)
HAUL - [gen] amenée *f*, effort (pour tirer etc)
[transp] (the total of the products of each load by its haul distance) distance *f* de transport
[telecomm] (the relative length of a trunk line) longueur *f* d'une ligne
~CLIP - [mining] tenaille *f* d'attelage
~DISTANCE - [constr] (the distance excavated material must be carried before depositing it) distance *f* de transport (du remblai)
~DOWN, to - [naut] haler bas
~DOWN A ROPE, to - [naut] haler sur une manœuvre
~DRIFT - [mining] galerie *f* de roulage
~IN, to - [naut] haler en dedans
~-OFF - [mech] (mechanism used to draw extruded cable, rod or tube away from the head as it is extruded) dispositif *m* de réception
~-OFF ROLL - [mech] (roller used to pull material away from the machine producing it) rouleau *m* dévideur
~OUT OF LINE, to - [naut] quitter la ligne
~PLANE - [mining] plan *m* incliné
~UP, to - [naut] remonter dans le vent
HAULAGE - [gen] transport *m* par roulage, traction *f*, camionage *m*, charriage *m*
[comm] frais *mpl* de roulage
[min] (the pulling on the level by animals or other means) roulage *m*, herschage *m*
~PLANT - [gen] matériel *m* de roulage
~ROPE - [impl] câble *m* tracteur
~TRACK - [mech] voie *f* de roulage
HAULER - [mech] treuil *m*
HAULIER - [mining] hercheur *m*, traîneur *m* de wagonnets
HAULING - s. haul, to
[gas ind] (in the operation of mainlaying) transport *m* des tubes
~CABLE - [mech] câble *m* tracteur
~JOIST - [constr] poutre à lever
~ROPE - s. hauling cable
~TACKLE - [naut] palan *m*, moufle *f*
HAUNCH - [anat] hanche *f*
[arch] aisselle *f* d'une voûte
[constr] épaulement *m* coupé d'équerre et parallèlement aux épaulements du tenon
~BONE - [anat] os *m* iliaque, os *mpl* coxaux
~OF A BEAM - [constr] console *f*
OF A VAULT - [arch] aisselle *f* d'une voûte
HAUNCHED TENON - [carp] (tenon with a haunch near its root) tenon *m* à mordâne
HAUSMANNITE - [min] (a manganese oxide) hausmannite *f*
HAVEN - [geogr] havre *m*, port *m*
HAVERSACK - [gen] havresac *m*, musette *f*
HAVERSIN - [math] (half of the versine, i.e. the function I -cos alfa) la moitié d'un sinus verse
HAWAIIAN GUITAR - [mus] (ukulele) guitare *f* hawaienne
HAWK - [zool] faucon *m*
[build] (square board with a handle underneath used to carry mortar) taloche *f* de plâtrier, taloche *f*
HAWSE - [naut] (that section of the ship's bows in which the hawse-holes are cut for cables to pass through) écubiers *mpl*
HAWSEHOLE - s. hawse
HAWSEPIECE - [shipbuild] tape *f* d'écubier
HAWSEPIPE - [shipbuild] (the tubular casting through which the cables run) manchon *m* d'écubier
HAWSER - [naut] (large rope or small cable) aussière *f*, grelin *m*
~BEND - [naut] nœud *m* d'écoute
~-LAID - [naut] cordage *m* commis en grelin
~PLUG - [naut] tape *f* d'écubier
HAWTHORN - [bot] aubépine *f*
HAY - [agric] foin *m*
~AND STRAW PRESS - [agric] presse *f* à foin et à paille
~BACILLUS - [agric] bacille *m* du foin
~BALER - [agric] presse *f* à foin
~BALING PRESS - s. hay baler
~-BAND - [gen] (a straw rope) corde *f* en foin, torche *f* de foin
~-BARN - [agric] fenil *m*
~BRIDGE - [instr] (instrument designed to measure high-Q inductors) pont *m* de Hay
~CART - [agric] chariot *m* à foin, fourragère *f*
~CHOPPER - [agric] ramasseuse-hacheuse *f*
~COCK - [agric] tas *m* de foin, moyette *f*
~DRYING - [agric] séchage *m* du foin
~ELEVATOR - [agric] monte-foin *m*
~EQUIVALENT - [agric] unité *f* de foin
~FEVER - [med] fièvre *f* des foins, asthme *f* d'été, rhume *m* des foins
~FIELD - [agric] pré *m* de fauche
~FORK - [agric] fourche *f* à foin
~GRAB - [agric] déchargeuse *f* à griffe
~LAND - [agric] prairie *f* de fauche
~LOADER - [agric] chargeur *m* à foin

HAY MAKING - [agric] fenaison *f*, fanage *m*
~MEADOW - [agric] pré *m* de fauche
~PASTURE - [agric] pâturage *m* et fauche alternés
~RACK - [agric] siccateur *m* à foin
~RAKE - [agric] râtelier *m* à foin
~STACKER - [agric] râteau *m* ameulonneur
~SWEEP - [agric] râteau-ramasseur *m*
~TEDDER - [agric] faneuse *f*
TOWER - [agric] tour *f* à foin
~TRAP - [agric] trappe *f* à foin, abat-foin *m*
~YIELD - [agric] rendement *m* en foin
HAYING - [agric] fanage *m*, fenaison *f*
HAYLOFT - [agric] grenier *m* à foin
HAYMAKING - [agric] fenaison *f*
~MACHINE - [mach agric] machine *f* de fenaison
HAZARD BEACON - [naut & aero] (luminous beacon giving warning of a navigational hazard) phare *m* de danger
HAZE - [met] (obscurity caused primarily by dust particles in unsaturated air) brume *f* sèche
[photo] voile *m*
~OF GAS - [phys] gaine *f* gazeuse, voile *m* gazeux
HAZEL - [bot] noisetier *m*
~-NUT - [bot] noisette *f*
HAZELNUT - [bot] balanin *m* des noisettes
~BUTTER - [food] beurre *m* de noisette
HAZINESS - [telev] manque *m* de netteté
HAZLE - [geol] grès *m* schisteux compact
HAZY PICTURE - [photo] image *f* vaporeuse, image *f* enveloppée
HEAD, to - [gen] conduire, mener, commander
[gen] (to be at the head of) être à la tête, diriger
[print] (a page) intituler
[gen] (in a competition) être à la tête, venir en tête
[naut] mettre le cap
[metall] (in forging) refouler
[mining] avancer
HEAD - [gen] tête *f*
[gen] (of a school) directeur *m*, principal *m*
[gen] (of a firm) principal *m*, chef *m*, directeur
[gen] (of a newpaper) haut *m* de page, titre *m* courant
[geogr] (of a river) source *f*
[mech] (of a nail) tête *f*
[naut] (of a ship) avant *m*
[mech] (of a shell) ogive *f*
[gen] (of a arrow) pointe *f*
[agric] (of an asparagus) pointe *f*
[arch] (of a column) chapiteau *m*
[phys] (pressure) hauteur *f* de pression, hauteur *f* manométrique
[naut] (of a jetty) musoir *m*
[mech] (of a pump) hauteur *f* d'élévation, hauteur *f*
[geogr] (a head land) cap *m*, promontoire *m*
[naut] (of a sail) têtière *f*, envergure *f*
[print] (the top edge of a volume) blanc *m* de prise
[firearms] (of a cartridge) fond *m*; (de fusée) chapiteau *m*
[transp] (of a lift-truck) plate-forme *f*
[hydr] (of water) colonne *f* (d'eau), charge *f*, chute *f*
[mining] galerie *f* d'avancement, avancée *f*
[mining] (of a lode or vein) bouche *f*, carrière *f* de mine
[el] (small electromagnet used for recording or reading polarized spots on a magnetic surface) tête *f*
HEAD - [agric] (of a cabbage) tête *f*, pomme *f*
[brew ind] (of a beer) mousse *f*, faux-col *m*
[geogr] (of a bay) fond *m*
[constr] (of a bridge) tête *f* (de pont)
[impl] (of an axe) fer *m*
[mach tool] poupée *f*, poupée *f* fixe, porte outil *m*
[mech] (of an engine) tête *f*
[metall] (foundry work) masselotte *f*
[constr] (a lintel) linteau *m*, poitrail *m*
[metall] (of an iron beam) champignon *m*
[mech] (of a valve) tête *f* (de soupape)
[mech] (of a draw-gear) tête *f* (d'appareil de traction)
[railw] (of a rail) champignon *m*
[ind chem] (circular wooden plate, chamfered at the edge to enter grooves in the staves and forming the end of the vessel) fond *m*
[mech] (of the spokes in a bicycle) tête *f*
[mus] (the end of the bow of stringed instruments) pointe *f*
[constr] (at the top of a rain-water downpipe) hotte *f*, cuvette *f*
[oil ind] charge *f* d'eau, hauteur *f* d'élévation
[el acoust] tête *f* magnétique
~ADJUSTMENT - [el acoust] équilibrage *m* de l'entrefer
~AMPLIFIER - [telev] (the amplifier which is contained in the television camera) préamplificateur *m* vidéo
(in a motion picture projector) amplificateur *m* de projecteur
~ASSEMBLY - [el acoust] porte-têtes *m*
~-BAND - [telecomm] ressort *m* du récepteur serre-tête
~BANDING - [el acoust] effet *m* de bande
~BAY - [hydr] (the section of a canal lock immediately above the head-gates) bief *m* d'amont
~BEAM - [constr] solive *f* de tête, traverse *f*
[text] traverse *f* supérieure, cintre *m*
~BIRTH - [med] présentation *f* de la tête
~BOARD - [text] planche *f* supérieure
[shipbuild] herpe *f* de guibre, pavois *m* de poulaine
[impl] cale *f*, écoin *m*
~BOLT - [auto] (colloq. USA; cylinder head-bolt) boulon *m* de culasse
~BOX - [hydr] (of sluices) boîte *f* de tête
~BRACE - [auto] (reinforcing pieces strengthening the body head panels) renfort *m* de pavillon, renfort *m* avant
~CAPACITY CURVE - [hydr] (of a pump) courbe *f* caractéristique hauteur/débit
~CLAMP - [print] pressoir *m* de blanc de prise
~CLEARANCE - [constr] (under a bridge etc) hauteur *f* libre
~CLIP - [auto] crochet *m* de blocage
~CLOGGING - [el acoust] colmatage *m* de tête
~CYLINDER GASKET - [oil ind] joint *m* de culasse
~DROP - [med] chute *f* de la tête
~DRUM - [el acoust] tambour *m* à têtes magnétiques
~EFFICIENCY - [el acoust] rendement *m* de tête
~-END - [mech] (of engine) arrière
~ [text] tête *f* du métier, têtière de machine
~FLASHING - [constr] gouttière-solin *f*
~FLOW - [oil ind] débit *m* intermittent, refoulement *m* intermittent
~FRAME - [mining] chevalement *m*, belle-fleur *f*

HEAD GASKET - [mech] (of a motor or engine) garniture *f* de la tête

~-GATE - [hydr] (in a lock, the gates at the high-level end) porte *f* d'amont
[hydr] (admitting water to a water-wheel) vanne *f* lançoire
[mining] cadre *m* de superficie

~GEAR - [mining] chevalement *m*

~-GEAR RECEIVER - [acoust] casque *m* téléphonique, écouteur *m* serre-tête

~GUARD - [impl] volet *m*

~-HOUSE - [mining] chevalement-abri *m*

~IRRIGATION - [agric] irrigation *f* en surface par ruissellement

~-LAMP - [light] (lamp and reflector projecting a beam of light) phare *m*, projecteur *m*
[impl] (lamp strapped to the forehead) lampe *f* frontale
[railw] (of a locomotive) feu *m* d'avant

~-LAMP BRACKET - [mech] support *m* de phare

~-LAMP INSERT - [auto] optique *f* de phare

~LEADER - [telev] amorce *f* de départ

~LEASE - [text] envergeure *f* du haut

~-LIGHT DIPPING - [auto] mouvement *m* de bascule des phares

~-LIGHT SUPPORT TIE ROD - [auto] hauban *m* de fixation des supports de phare

~LIMIT - [hydr] point *m* d'eau

~-LOSS - [hydr] perte *f* de charge, perte *f* de pression

~METAL - [metall] métal *m* de masselotte

~MOTION - [text] ratière *f* d'armures, mécanique *f* d'armures

~OF A PILE - [constr] tête *f* de pieu *m* ou de pilotis *m*

~OF A RAIL - [railw] champignon *m* du rail

~OF A SAIL - [naut] têtière *f*, envergure *f*

~OF CATTLE - [agric] têtes *fpl* de bétail

~OF COLUMN - [arch] chapiteau *m*

~OF CONNECTING ROD - [mech] tête *f* de la tringle de connexion

~OF DOCK - [naut] musoir *m*

~OF DRAIN - [constr] tête *f* du tuyau d'écoulement

~OF FEELER - [text] tête *f* du tâteur

~OF INGOT MOULD - [metall] tête *f* de la lingotière

~OF MUSCLE - [anat] chef *m* d'un muscle

~OF STAMP - [metall] sabot *m* de bocard

~OF THE BOBBIN - [text] embase *f* de la bobine

~OF THE MAST - [naut] tête *f* de mât, haut *m* du mât

~OF THE PUNCH - [text] tête *f* du poinçon

~-ON - [gen] de front

~-ON COLLISION - [gen] collision *f* frontale
[nucl] (a collision between two particles moving in opposite directions along the same line) collision *f* frontale

~PEG - [text] cheville *f* supérieure

~-PIECE - [mining] (of timbering) chapeau *m*

~PIECE - [text] morceau *m* de la tête, têtière *f*

~PLATE - [text] plaque *f* supérieure
[constr] sablière *f* haute

~PRODUCTS - [chem] (in fractional distillation) produits *mpl* de tête

~RACE - [hydr] (channel converying water to a machine which is operated hydraulically) canal *m* d'amenée, bief *m* d'amont

~-RACE TUNNEL - [hydr] galerie *f* d'amenée

~RECEIVER - s. head-gear receiver

~RESISTANCE - [aero] résistance *f* aérodynamique

HEAD REST - [gen] appui-tête *m*

~RETENTION - [brew ind] tenue *f* de la mousse

~ROLL - [gen] support *m* de la tête

~SLAB CORE - [metall] noyau *m* détachable

~STICK - [print] lingot *m* de tête

~STOCK - [mach tool] poupée *f*
[mining] chevalement *m*

~SWITCH - [el] commutateur *m* de tête

~SWORD - [mining] eau *f* de galerie

~TAPPING CUT - [rubber ind] incision *f* principale

~TIPS - [el acoust] protubérances *fpl* de la tête magnétique

~-TO-DRUM SEPARATION - [comput] écart *m* tête-tambour

~TOOL - s. heading punch

~TRAVERSE MICROMETER DIAL - [mach tool] cadran *m* déplacement micrométrique poupée de traverse

~TREE - [carp] traverse *f* de cadre

~TWISTING - [text] torsion *f* supplémentaire, surfilage *m*

~VALVE - [mech] (the delivery valve of a pump) soupape *f* d'amenée

~VOICE - [mus] (the highest register) voix *f* de tête

~WHEEL - [mining] molette *f*
[el acoust] disque *m* porte-tête

~WIND - [naut] vent *m* contraire

~WINDING - [el acoust] bobinage *m* de tête

~WORK - [mining] chevalement *m*

~YOKE - [agric] joug *m* frontal, joug *m* simple

HEADBAND - [bookbind] (decorative band at the head of a book) tranchefile *f*

HEADED - [gen] dirigé (vers)
[mech mach] muni de tête
[agric] décolleté

HEADER - [th eng] (in a water-tube boiler, a transverse tubular member into which a number of pipes are connected) collecteur *m*, distributeur *m*
[build] (brick with its length at right angle to the face of the wall) boutisse *f*
[metall] jet *m* de coulée
[of a roof] chevron *m*
[mech] (manifold supplying fluid to a number of pipes) collecteur *m*
[carp] (of a framing) frise *f*
[metall] (forging mach) machine *f* à refouler
[agric] écimeuse *f*, coupeuse *f* d'épis
[comput] (a message header) titre *m* de message
[mining] collecteur *m*

~AND THRESHER - [agric] moissonneuse-batteuse *f*

~BRACE - s. head brace

~CLIP - s. head clip

~DIE - [metall] matrice *f* de refoulage

~LABEL - [comput] label *m* de tête

~PIPE - [hydr] collecteur *m*

~POINT - [mech] (for machine screws) arrondissement *m* de l'extrémité (pour vis à métaux)

~TANK - [auto] boîte *f* à eau

HEADFRAME - [min] (the frame at the top of a shaft) chevalet *m* d'extraction

HEADGEAR - s. headframe

HEADGUT - [anat] intestin *m* céphalique, intestin *m* antérieur

HEADING - [aero] (the angle made by the longitudinal axis of an aircraft or strip with a given meridian) cap *m* géographique, gisement *m* d'un point observé
[metall] façonnement *m* des têtes

HEADING - [metall] (in forging) refoulage *m*
[min] (passage-way in a coalmine, through solid coal) galerie *f* d'avancement
[print] intitulé *m*, tête *f*
[gen] (of a letter etc) en-tête *m*
[print] (of a newspaper) titre *m* (du journal), tête *f*
[build] (heading course) appareil *m* en boutisses
[oil ind] produit *m* de tête
[min] concentré *m*
[constr] (in mainlaying) galerie *f* provisoire
[agric] écimage *m*
[metall] façonnement *m* des têtes
~ CARD - [comput] carte *f* d'un-tête
~ FACE - [mining] front *m* d'avancement
~ FLASH - [radar] feu *m* de route
~ INDICATOR - [radar] indicateur *m* de route
~ LINE - [naut & aero] route *f*
~ MACHINE - [mech mach] (in forging) machine *f* à refouler
[mining] coupeuse *f*, excavateur *m*
~ MARKER - s. heading indicator
~ PUNCH - [tool] (used in head forging) poinçon *m* à façonner
~ STATEMENT - [comput] titre *m* informatif
~ STOPE - [mining] chantier *m* d'avancement
~ TOOL - s. heading punch
~ UP - [mech] (operation of closing rivets) rivetage *m*, rivure *f*
HEADLAND - [geogr] promontoire *m*, cap *m*
[agric] tournière *f*
HEADLEDGE - [shipbuild] fronteau *m* d'écoutille
HEADLESS - [gen] sans tête
~ BOLT - [mech] boulon *m* sans tête
~ SCREW - [mech] vis *f* sans tête
HEADLIGHT - [auto etc] phare *m*, projecteur *m*
[radio] (type of radar aerial which is small enough to be housed in the wing of an aircraft like a car headlight) antenne *f* à phare
~ BEAM CONTROL - [auto] commande *f* de l'éclairage de route
HEADLINE - [print] (line at the top of a page giving the title) titre *m*
HEADPHONE - [acoust radio etc] écouteur *m*
(of a telephonist) casque *m* téléphonique, serre-tête *m*
HEADPIECE - [gen] casque *m*, armure *f* de tête
HEADPLATE - [metall] plaque *f* frontale, gendarme *m*
HEADQUARTERS - [gen] quartier *m* général
[comm] siège *m* social
HEADRACE - s. head race
HEADROOM - [constr] (the free passage under a bridge etc) échappée *f*
[mech] hauteur *f* libre (pour les pistons etc)
HEADROPE - [naut] (of a sail) ralingue *f* de têtière, amarre *f* de bout
HEADSET - [telecomm] casque *m* (téléphonique)
HEADSTOCK - [mech] (generally, a mechanism supporting the head of a part; in particular, that part of a lathe which carries the spindle) poupée *f*, poupée *f* fixe, poupée *f* mobile, contre-pointe *f*
[railw] (of the underframe) traverse *f* de châssis
[mining] chevalement *m*
[text] chapeau *m* de métier
HEADTWIST - [text] prétorsion *f*
HEADWATER - [hydr] eau *f* d'amont
~ CHANNEL - [hydr] canal *m* d'amenée
~ OF A RIVER - [hydr] bassin *m* de la source

HEADWAY - [constr] s. headroom
[gen] (a motion or movement) progrès *m*
[gen] (the distance between a vehicle and a following one) intervalle *m*
[railw] intervalle *m* (entre deux trains)
~ OF A BRIDGE - [constr] échappée *f* d'un pont
HEADWORK - [hydr] barrage-réservoir *m*
HEAL, to - [gen] guérir
[of wounds] cicatriser
HEALD - [text] lisse *f*, lice *f*
~ BRAIDING MACHINE - [text] tresseuse *f* mécanique pour lisses
~ CALCULATION - [text] calcul *m* des lisses
~ EYE - [text] maillon *m*
~ FRAME - [text] lame *f*, cadre *m* lisses, cadre de lame
~ FRAME MOTION - [text] mouvement *m* des lames
~ FRAME RIDER - [text] curseur *m* de lisse, crochet *m* porte-tringle glissant, cavalier *m*
~ FRAME SLIDER - [text] curseur *m* de lisse, crochet *m* porte-tringle glissant, cavalier *m*
~ FREE FROM KNOTS - [text] lisse *f* sans nœuds
~ HOOK - [text] passette *f*
~ KNITTER - [text] lamier *m*, lissier *m*
~ KNITTING FRAME - [text] banc *m* pour tricoter les lisses
~ KNITTING MACHINE - [text] tricoteuse *f* pour lisses
~ LATH - [text] lamette *f*, liseron *m*, verge *f*
~ LEVELLING MOTION - [text] niveleur *m* des lames régleur *m* des lames
~ LIFT - [text] contrôle *m* du remettage
~ LOOP - [text] boucle *m* des lisses
~ MACHINE - [text] ratière *f*
~ MOTION - [text] changement *m* des lames
~ SHAFT - [text] lame *f*
~ SHAFT CORD - [text] corde *f* des lames
~ TINNED AT THE EYE - [text] lisse *f* étamée au maillon
~ TWINE - [text] retors *m* pour lisses, câblé *m* pour lisses
~ WIRE - [text] fil *m* de fer pour lisses
~ WIRE KEY - [text] clef *f* pour lisses métalliques
~ WITH GLASS EYE - [text] lisse *f* avec œillet en verre
~ WITH LONG EYE - [text] lisse *f* avec maillon long
~ WITH LONG LOOPS - [text] maille *f* à grande coulisse
~ WITH MAILED EYES - [text] lisse *f* à maillon
~ WITH TWINE EYE - [text] lisse *f* avec œillet en fil retors
~ WITHOUT EYE - [text] lisse *f* sans œillet
HEALDER - [text] remetteuse *f* de chaînes
HEALDS FOR OPEN SHEDS - [text] lames *f*pl immobiles
HEALED PRUNED OFF SURFACE - [bot] greffage *m* cicatrisé
HEALING - [gen & med] guérison *f*, cicatrisation *f*
[adj] curatif, cicatrisant, sanatoire
[constr] (the covering of a roof with tiles etc) couverture *f*
~ BY GRANULATION - [med] cicatrisation *f* par granulation
~ STONE - [geol] ardoise *f*
HEALTH - [gen] santé *f*
~ CERTIFICATE - [gen] certificat *m* médical
~ HAZARDS - [nucl] (the danger of radioactivity and

radiations) risque *m* d'irradiation
HEALTH MONITOR - [nucl] moniteur *m* de radioprotection (ou de rayonnement)
~MONITORING - [nucl] (the control of radioactivity and other radiations for their danger to living beings) contrôle *m* radiologique
~OFFICER - [gen] médicin *m* sanitaire, inspecteur *m* de l'hygiène
~PHYSICS - [nucl] (the branch of radiological physics relating to the protection of living beings from the dangers of ionizing radiations) science *f* de la protection contre les rayonnements ionisants
HEAP, to - [gen] entasser, mettre en pile
HEAP - [gen] tas *m*, amas *m*, amassage *m*
[min] halde *f*
~CLOUDS - [met] (clouds which have the appearance of being piled up vertically) nauges *mpl* en monceaux
~LEACHING - [min] lixiviation *f* en tas
HEAPED CONCRETE - [constr] béton *m* coulé
HEAPSTEAD - [min] (the surface works around a colliery shaft) plâtre *m* du puits
HEAR, to - [gen] ouir, entendre
HEARING - [gen] ouie *f*, audition *f*
~AID - [acoust] (apparatus used by deaf people) appareil *m* de prothèse auditive
~DEFECT - [med] défaut *m* de l'ouie
~DISTANCE - [acoust] (the maximum distance at which a sound of normal intensity can be heard by a person with a normal hearing) distance *f* maximum d'audibilité
~LIMIT - s. hearing distance
~LOSS - [acoust] (the diminution of audibility) surdité *f*
LOSS FOR SPEECH - [acoust] taux *m* de surdité tonale
HEARSE - [transp] char *m* funèbre
HEART - [anat] cœur *m*
[text] (of a rope) âme *f*, mèche *f*
[bot] cœur *m*
[naut] (a kind of dead-eye, with one large hole in the middle) moque *f*
[mech] (a heart-shaped cam) came *f* en forme de cœur, excentrique *m* en forme de cœur
~AND DART MOTH - [zool] noctuelle *f* point d'exclamation
~ARREST - [med] arrêt *m* du cœur
~BLOCK - [med] bloc *m* pariétal, bloc *m* intraventriculaire
~-BURN - [med] pysoris *f*; aigreurs *fpl*
~-CAM - [mech] (a heart-shaped cam) excentrique *m* en forme de cœur
~CHERRY - [food] guigne *f*
~CHERRY TREE - [bot] guignier *m*
~CUT - [metall] fraction *f* de cœur
~DISEASE - [med] cardiopathie *f*
~FAILURE - [med] décompensation *f* cardique
~MURMUR - [med] souffle *m* cardique
~OF THE TREE - [bot] cœur *m* du bois
~OF THE WOOD - [constr] moelle *f* du bois
~ROT OF BEETS - [plant disease] pied *m* noir de la betterave, pouriture *f* du cœur
~-SHAPED CAM - [text] excentrique *m* en forme de cœur came *f* en cœur
~-SHAPED DIAGRAM - [el] diagramme *m* en cardioïde
~-SHAPED PILE WARP GRIPPER - [text] pince *f* en forme de cœur de la chaîne en poil
HEART SOUND - [med] bruit *m* du cœur
~WHEEL - [text] came *f* en cœur, plateau *m* en forme de cœur
HEARTH - [gen] (floor of the fireplace) foyer *m*
[metall] (in a reverberating furnace) aire *f*, foyer *m*, sole *f*, laboratory (d'un four à réverbère)
[glass man] (the part of the furnace where heat is developed for the melting of glass) lit *m* de fusion
~BOTTOM - [metall] fond *m* de creuset
~FINING - [metall] affinage *m* sur sole
~FURNACE - [metall] four à sole
~HOLE - [metall] ouverture *f* de la poitrine
~JACKET - [metall] blindage *m* du creuset
~LEVEL - [metall] fond *m* du creuset, sole *f* du creuset
~LINING - [metall] garnissage *m* de la sole
~MOULDING - [metall] moulage *m* en creuset
~PLATE - [metall] plaque *f* du foyer
~STEEL - [metall] acier *m* Martin
HEARTING - [constr] remplissage *m* d'un espace (entre deux parements), maçonnerie *f* de remplissage
HEAT, to - [gen] chauffer, échauffer, s'échauffer
HEAT - [phys] chaleur *f*, ardeur *f*
[phys] (the energy possessed by a substance in the form of kinetic energy of molecular translation, rotation and vibration) chaleur *f*, énergie *f* thermique
[metall] (the process of heating) chaleur *f*
[metall] (in foundry works) coulée *f*
[metall] (the forging period) piquée *f*
[metall] (the product of one furnace charge) coulée *f*
[zool] (the period of mating) rut *m*, chaleur *f*
~-ABSORBING FILTER - [cin] filtre *m* catathermique, filtre *m* anticalorique
~-AFFECTED ZONE - [metall] zone *f* attaquée par la chaleur
~AGEING INHIBITOR - [chem] (agent used to combat the ageing effects of heat, espec. in rubber) anti-vieillisseur *m* contre hautes températures
~APOLEXY - [med] coup *m* de chaleur
~BALANCE - [mech] (the heat-energy prepared for a boiler and an engine trial) bilan *m* calorifique
~BODIES OILS - [paint] (oils which are polymerized only by the action of heat, as distinct from those which can be polymerized by catalyst at normal temperature, or by hot or cold air blowing) huiles *fpl* cuites
~BOOSTER - [rubber ind] (for retreading) élément *m* de chauffage pour réparations
~BUILD-UP - [phys] (internal accumulation of heat in a body or substance) échauffement *m* accumulatif interne
~-BUMP - [med] boule *f* d'œdème, papule *f* ortiée
~CAPACITY - [phys] (the quantity of heat which is required to raise the temperature of a body by a specific amount) capacité *f* calorifique
~CHAMBER - [brew ind] chambre *f* de chaleur, cochon *m*
~CHECKING - [metall] (in die-casting) formation *f* de criques due au chaleur
~COIL - [el] (small protective resistance coil) fusible *m* thermique
~CONDUCTION - [phys] conduction *f* de chaleur
~CONDUCTIVITY - [phys] conductivité *f* thermique
~CONSTANT - [phys] constante *f* calorifique

HEAT CONSUMPTION - [phys] consommation *f* de chaleur
~CONTENT - [phys] (thermodynamic property; the total heat of a substance or a system) teneur *f* en chaleur
~CONTROL - [instr] thermostat *m*
~CONVECTION - [phys] convection *f*
~CRACKING - [metall] crique *f* due à la chaleur
~DISSIPATION - [phys] dissipation *f* de la chaleur [electron] dissipation *f*, puissance *f* dissipée
~DISTORTION POINT - s. heat distortion temperature
~DISTORTION TEMPERATURE - [phys] (the temperature at which a material begins to distort) point *m* de déformation à chaud
~DROP - [phys] (adiabatic heat drop) chaleur *f* de la détente adiabatique
~DRYING MACHINE - [photo] sécheuse *f* à chaud
~EFFECT - [phys] effet *m* thermique
~EMISSION - [phys] émission *f* de chaleur
~ENGINE - [mech mach] moteur *m* à air chaud
~ENGINE SET - [el] groupe *m* thermique
~EQUIVALENT OF WORK - [phys] équivalent *m* calorifique du travail
~EXCHANGER - [chem] (device for transferring heat from one fluid to another) échangeur *m* de chaleur
~EXHAUSTION - [med] coup *m* de chaleur avec collapsus
~FASTNESS - [dyes] (the property of resisting colour-change caused by heat) résistance *f* à la chaleur, stabilité *f* sous l'effet de la chaleur
~FLUSH - [chem] (a method for the separation of He^3 and He^4 by means of a flow or heat in super-helium) coup *m* de chaleur
~FLUX - [phys] flux *m* de chaleur
~INDICATOR - [instr] thermomètre *m*
~INPUT - [th eng] (of burners) débit *m* calorifique
~INSULATING JACKET - [heat] enveloppe *f* calorifuge, manteau *m* d'isolation thermique
~INSULATING MATERIAL - [th eng] calorifuge *m*, isolant *m* thermique
~INSULATING SLEEVE - [metall] manchon *m* isolant
~INSULATION - [gen] (provision for preventing or retarding the passage of heat) isolation *f* calorifuge
~INSULATOR - s. heat insulating material
~JACKET - [mech] gaîne *f* chauffante
~-JACKETED DRUM - [mech] tambour *m* à double enveloppe chauffante
~LOSS - [radio] (part of the transmission which is lost due to the conversion of electric energy into heat) pertes *f*pl par effet Joule [metall] perte *f* de chaleur
~MOULDING PROCESS - [rubber ind] moulage *m* à chaud, moulage *m* au latex thermosensible
~OF ABLATION - [astronaut] chaleur *f* d'ablation
~OF ACTIVATION - [phys] (the increase of heat content upon the transformation of a substance from a less active to a more reactive form) chaleur *f* d'activation
~OF ADSORPTION - [chem] (the increase of heat content in the adsorption process) chaleur *f* d'adsorption
~OF AGGREGATION - [phys] (the increase of heat content on the formation of various aggregates of matters) chaleur *f* d'agrégation
~OF ASSOCIATION - [phys] (the increase of heat content when one mole of a compound is formed from its constituent molecules) chaleur *f* d'association

HEAT OF COMBUSTION - [phys] (the increase in the heat content) chaleur *f* de combustion
~OF COMPRESSION - [chem] (heat generated by the compression of gas) chaleur *f* de compression
~OF CONDENSATION - [phys & chem] (the increase of heat content when a unit mass is converted into liquid at its boiling point) chaleur *f* de condensation
~OF COOLING - [phys] (the increase of heat content at certain temperatures on its cooling curve) chaleur *f* de refroidissement
~OF CRYSTALLIZATION - [phys] (the increase of heat content when one mole of a substance is transformed into its crystalline state) chaleur *f* de cristallisation
~OF DECOMPOSITION - [phys] (increase of heat content when one mole of a compound is decomposed into its elements) chaleur *f* de décomposition
~OF EMISSION - [electron] (the additional energy supplied to an electron emitting surface to keep it at a constant temperature) chaleur *f* d'émission
~OF EVAPORATION - s. heat of vaporization
~OF FORMATION - [chem] (the heat evolved when one grammolecule of a compound is formed from its constituent elements) chaleur *f* de formation
~OF FUSION - [chem] (the amount of heat necessary to change a solid to the liquid state) chaleur *f* de fusion
~OF HYDRATION - [phys] (the increase of heat content when one mole of a hydrate is formed from the anhydrous form of the compound) chaleur *f* d'hydratation
~OF IONIZATION - [phys] (the increase of heat content upon the complete ionization of one mole of a substance) chaleur *f* d'ionisation
~OF LINKAGE - [phys] (bond energy of a valence linkage between atoms) énergie *f* de liaison
~OF NEUTRALIZATION - [phys] (the increase of heat content in a system undergoing neutralization reaction) chaleur *f* de neutralisation
~OF RADIOACTIVITY - [phys] (the quantity of heat which is produced by radioactive decay) chaleur *f* de décroissance, chaleur *f* de désintegration
~OF REACTION - [chem] chaleur *f* de réaction
~OF SOLIDIFICATION - [phys] (the increase of heat content accompanying the solidification of a substance in the liquid state) chaleur *f* de solidification
~OF SOLUTION - [chem] (the amount of heat taken in or given out when a substance is dissolved in a large volume of solvent) chaleur *f* de dissolution
~OF SUBLIMATION - [phys] (the increase of heat content on the transformation of a solid direct to the gaseous condition) chaleur *f* de sublimation
~OF TRANSITION - [phys] (the increase of heat content upon the transformation of a substance into an allotropic form) chaleur *f* de transformation
~OF VAPORIZATION - [phys] (the increase of heat content upon the change of a substance from the liquid or solid state to gaseous conditions) chaleur *f* d'évaporation, chaleur *f* de vaporisation
~OUTPUT - [mech] puissance *f* thermique
~PERIOD - [zool] période *f* du rut, période *f* des chaleurs
~PERMEABILITY - [phys] transmission *f* du chaleur [el heat] diathermie *f*
~-PROOF - [gen] (capable of resisting a considera-

ble degree of heat without damage) résistant à la chaleur
HEAT PROTECTION SYSTEMS - [astronaut] systèmes *mpl* de protection thermique
~ PUMP - [mech] (a device designed to convert mechanical into thermal energy) pompe *f* thermique
~ RADIATION - [electron] (emission of heat in the form of electro-magnetic rays) rayonnement *m* thermique
RADIATOR - [electron] (a radiator used with power transistors to dissipate heat) élément *m* de rayonnement thermique
~ RAISING - [th eng] réchauffage *m*
~ RATING - [el] degré *m* thermique
~ RECOVERY - [chem] (the transfer of heat from a substance which has been heated during a process to the sustance about to be treated, to avoid waste of energy) récupération *f* thermique
~-REDUCING FILTER - [telev] (to control the colour temperature) filtre *m* absorbant la chaleur
~-REFLECTING - [phys] réfléchissant la chaleur
~ REQUIREMENT FOR HEATING - [thermo] chaleur *f* d'échauffement
~-RESISTANT - [gen] (not readily affected by heat) résistant à la chaleur
~-RESISTANT ALLOY - [metall] alliage *m* résistant à la chaleur
~ RESISTING CAST IRON - [metall] fonte *f* réfractaire
~-SEAL, to - [metall] (to weld easily fusible materials such as plastics by the application of heat) souder à chaud, sceller à chaud
~ SEALABLE - [metall] (suitable for being closed by welding) scellable à chaud
~-SEALING - [metall] (bonding of a material by heat alone, without adhesives) scellage *m* à chaud
~-SEALING MACHINE - [plast ind] (machine for welding plastics and the like by means of heat) machine *f* à souder à chaud
~-SENSITIVE - [gen] (easily affected by heat) thermosensible
~-SENSITIVE LATEX - [rubber ind] latex *m* thermosensible
~-SENSITIVE PAINT - [paint] (special paint formulated to change colour at a given temperature) peinture *f* sensible à la chaleur
~ SENSITIVITY - [phys] thermosensibilité *f*
~ SENSITIZATION - [rubber ind] thermosensibilisation *f*
~-SET PLEAT - [text] pli *m* fixé à chaud
~-SHIELD - [electron] (metallic surface surrounding a hot cathode to reduce the radiation losses) écran *m* thermique
[astronaut] (device protecting from heat, e.g. a shield in front of a re-entry capsule) écran *m* thermique
~ SHOCK TEST - [plast ind] (test to determine tendence of a plastic coating to crack under heat) résistance *f* au choc à chaud
~ SINK - [aero] (a heavy layer of heat absorbing material placed within the outer skin of aircraft designed to reach high supersonic speeds) couche *f* de absorption de chaleur, puits *m* thermique
[mech] (any objects, such as a solid mass of heat-conducting material, placed of work to prevent local overheating during an operation such as soldering) source *f* froide, dispositif *m* disperseur de chaleur
HEAT SPONGE - [aero] (in the combustion chamber) éponge *f* de chaleur
~ STABILIZER - [rubber ind] stabilisant *m* contre la coagulation thermique
~-STERILIZED - [chem] stérilisé thermiquement
~ STORAGE CAPACITY - [electron] (the amount of heat which can be absorbed by the anode of an X-ray tube) capacité thermique
~ STORAGE VESSEL - [metall] chaudière *f* accumulatrice de chaleur
~ TIME - [metall] (the time during which the current flows in a welding process) temps *m* de chauffe
~ TRANSFER - [phys] (movement of heat from one body to another) transmission *f* de chaleur, échange *m* de chaleur, échange *m* calorifique
~ TRANSFER BY CONDUCTION - s. heat conduction
~-TRANSFER CYCLE - [phys] cycle *m* de transmission de chaleur
~ TRANSFER EQUIPMENT - [th eng] échangeur *m* de chaleur
~ TRANSFER FLUID - [chem] (used in high-temperature processes) fluide *m* évacuateur de chaleur
~ TRANSMISSION - [phys] transmission *f* de chaleur
~ TRANSMISSION COEFFICIENT OF A WALL - [th eng] coefficient *m* de transmission d'une paroi
~ TREATABLE - [metall] améliorable par trempe et revenu
~-TREATED RAIL - [railw] rail *m* traité thermiquement
~-TREATING INDUCTION HEATER - [metall] appareil *m* à induction pour le traitement thermique
~ TREATMENT - [chem] (the process of changing the properties of a substance by the controlled application of heat) traitement *m* thermique
~-TREATMENT BATH - [chem] (furnace in which heat treatment is effected by immersing the object to be treated in a medium which is liquid at the temperature required) four *m* à bain
~-TREATMENT FURNACE - [metall] (furnace expressly designed for heat treatment) four *m* pour le traitement thermique
~-TREATMENT SALTS - [chem] (salts used for heat treatment baths, in fusion or solutions, to produce accurate temperature control) sels *mpl* pour le traitement thermique
~ UNIT - [phys] calorie *f*
~ VALUE - [phys] pouvoir *m* calorifique, puissance *f* calorifique
~ WAVE - [met] (a periodic of abnormally high atmospheric temperature. It has been defined as three or more consecutive days on each of which the temperature is 90 deg. **F**. or more) vague *f* de chaleur
HEATABLE - [heat] chauffable
HEATED BARLEY - [bot] orge *f* échaudée
~ DIGESTION CHAMBER - [hydr] digesteur *m* chauffé
~ GROOVED ROLLER - [text] cylindre *m* cannelé chauffé
~ TOOL WELDING - [plast ind] (welding of plastic sheet material using a hot metal tool) soudage *m* à chaud par conduction
~ WEDGE WELDING - [plast ind] (alternative term for Heated Tool Welding, q.v.) soudage *m* à panne chauffante
HEATER - [gen] réchauffeur *m*, radiateur *m*, bouilleur *m*
[electron] (a separate element used to heat the cathode) fil *m* chauffant

HEATER - [el] radiateur *m* électrique
[auto] appareil *m* de chauffage
~BATTERY - [electron] (the power supply battery for the filaments of electronic tubes) batterie *f* de chauffage, batterie *f* A
~CIRCUIT - [electron] (the circuit designed to heat the filament of an electronic tube) circuit *m* de chauffage
~CURRENT - [electron] (current used for heating cathodes of electronic tubes) courant *m* de chauffage
~CURRENT CONSUMPTION - [electron] consommation *f* du fil de chauffage
~ELEMENT - [th eng] élément *m* de chauffage, chafferette *f*, résistance *f* de chauffage
~HEAD - [plast ind] (threaded steel plug which forms the end of the heating cylinder and carries the injection nozzle) tête *f* chauffante
~FILAMENT - [electron] filament *m* chauffé
~INPUT - [th eng] consommation *f* de courant de chauffage
~OUTPUT - [th eng] puissance *f* de chauffage
~PLUG - [mech] (of a Diesel engine) bougie *f* à incandescence, bougie *f* de préchauffage
~POWER - s. heater output
~RECLAIMING PROCESS - [rubber ind] procédé *m* de régénération par la vapeur et l'huile
~TAPE - [electron] ruban *m* de chauffe, ruban *m* chauffant
~TEMPERATURE - [electron] température *f* du fil chauffant
~TUNNEL - [rubber ind] four *m* tunnel
~VOLTAGE - [electron] (the voltage applied to the terminals of the filament or heater) tension *f* de chauffage
~WINDING - [el] bobine *f* de chauffage
HEATH - [geogr] bruyère *f*, lande *f*
[bot] bruyère *f*, brande *f*
HEATHER - s. heath [bot]
HEATING - [adj] échauffant
[gen] chauffage *m*
[metall] chauffe *f*
~AND COOLING CHART - [electron] (for X-ray tubes) diagramme *m* de chauffage et de refroidissement
~BATTERY - s. heater battery
~BODY - [th eng] (in water heaters) corps *m* de chauffe
~BY INFRARED RADIATION - [el heat] (heating method in which the heat energy is transferred by infrared radiation) chauffage *m* par rayonnement infrarouge
~CABLE - [el] (special type of low-temperature heating element designed to be used in contact with the material to be heated) câble *m* chauffant
~CHAMBER - [th eng] chambre *f* de chauffage
~CHANNEL - [plast ind] (spaces within a mould or similar through which a heating medium such as steam may be circulated or in which electrical heating elements) canal *m* de chauffage
~CIRCUIT - [el] circuit *m* de chauffage
~COIL - [th eng] (spiral coiled tube for supplying heat, especially of liquid in a vessel) serpentin *m* de chauffage
~COUPLING - [railw] accouplement *m* de chauffage
~CURRENT - s. heater current
~CURRENT SOURCE - [electron] source *f* de courant de chauffage
HEATING CYLINDER - [plast ind] (the chamber in an injection moulding machine in which the material is heated before injection) pot *m* de chauffage
~EFFECTS - [electron] (effects borne of the resistive heating of the cathode) effets *mpl* de chauffage
~ELEMENT - [el] (electrical assembly forming part of a heating unit) élément *m* chauffant
~FILAMENT - [el] filament *m* de chauffage
~FLUE - [th eng] (ovens and retorts) carneau *m* de chauffage
~GRID - [th eng] grille *f* de chauffage
~INDUCTOR - [el] (winding designed to induce heating currents in the charge of a furnace) inducteur *m* de chauffage
~JACKET - [plast ind] (chamber or passage surrounding a part or unit to give heat to it) chemise *f* de réchauffage
~[metall] chemise *f* de réchauffage
~KETTLE - [impl] (used in mainlaying) fondoir *m*
~MANURE - [agric] fumier *m* chaud
~MEDIUM - [gen] agent *m* chauffant
~MUFF - [aero & auto] (jacket placed round an exhaust pipe to provide heated air) manchon *m* à air chaud
~PATTERN - [chem] (the distribution of temperature in a load) marche *f* de la température
~PIN - [plast ind] (a projection, often forming part of an ejector pin, designed to reduce the thick parts of a moulding and to improve the rate of conduction of heat into such parts) mandrin *m* chauffant
~PLANT - [gen] installation *f* de chauffage
~PLATE - [el] plaque *f* chauffante, plateau *m* chauffant
~[metall] (in welding) flamme *f* de chauffe
~RESISTOR - [el] (conductor producting heat from electricity by the Joule effect) résistance *f* chauffante, conducteur *m* chauffant
~ROD - [el] calentador en forme de barre
~SERPENTINE - [th eng] serpentin *m* de chauffage
~SOURCE - [phys] source *f* calorifique
~SPACE - [th eng] chambre *f* de réchauffage
~SPIRAL - [th eng] serpentin *m* de chauffage
[el] spirale *f* de chauffage
~STOVE - [metall] four *m* à recuire
~STRIP - [electron] ruban *m* de chauffe, ruban *m* chauffant
~SURFACE - [th eng] surface *f* de chauffe
~SYSTEM - s. heating plant
~TIME - [electron] (the time required by a cathode to reach its best working conditions) temps *m* de chauffe
~TO WHITENESS - [metall] chauffe *f* à blanc
~TUBE - [nucl] (a duct or conduit in a reactor filled with a fluid at high temperature) tube *m* de chauffe, tuyau *m* de chauffe
~TUNNEL - s. heater tunnel
~UP - [th eng] chauffage *m*, préchauffage *m*, réchauffage *m*
~-UP PERIOD - [th eng] (the time taken to raise a furnace to its normal operating temperature from the ambient level) durée *f* de mise en température
~VALUE - s. heat value
~VAN - [railw] fourgon-chaudière *m*
~VOLTAGE - s. heater voltage
~WALL - [th egn] piédroit *m* de chauffage
~WIRE - [electron] s. heater

HEATING WORM - [th eng] serpentin *m* de chauffage
HEAVE, to - [gen] lever, soulever, lancer
[mech etc] (the action of lifting) soulever
[naut] jeter
[naut] (the careening of a ship) caréner
[naut] (to heave the log) sonder, jeter la sonde
[mining] (of the soil) se boursoufler
HEAVE - [gen and mech] soulèvement *m*, effort *m* pour soulever
[geol] (along a fault) déplacement *m* latéral
[mining] boursouflement *m*, gonflement *m* (dans une galerie)
[naut] (of the sea) poussée *f*
[naut] (of a ship) paraître à l'horizon, poindre
[geol] recouvrement *m* horizontal
~ OF THE SOIL - [soil] soulèvement *m* du fond, remontée *f* du fond
~ TO, to - [naut] se mettre en panne
HEAVEN - [gen] ciel *m*
HEAVENLY BODY - [astr] (celestial body) astre *m*
HEAVER - [impl] levier *m* de manœuvre
HEAVIER-THAN-AIRCRAFT - [aero] aéroplane *m*, avion *m*
HEAVINESS - [gen] pesanteur *f*, poids *m*
[aero] (of the nose, or of the tail) pesanteur *f*
[soil] nature *f* grasse
HEAVING BOTTOM - [mining] sol *m* qui gonfle
~ FAILURE - [soil] rupture *f* par soulèvement
~ SANDS - [geol] sable *m* coulant
~ SHALE - [geol] schiste *m* gonflant, argile *f* gonflante
HEAVISIDE-KENNELLY LAYER - [radio] (the lower or E-layer of the ionosphere) couche *f* de Heaviside Kennelly
~ LAYER - s. Heaviside-Kennelly layer
HEAVY - [gen] lours, pesant
~ AGGREGATE CONCRETE - [constr] béton *m* lourd
~ AGGREGATE CONCRETE SHIELD - [nucl] bouclier *m* en béton lourd
~ ALLOY - [metall] alliage *m* lourd, alliage *m* de tungstène
~ ANODE - [electron] (anode consisting of a solid piece of metal) anode *f* massive
~ ARMATURE RELAY - [el] relais *m* à armature lourde
~ ATOM - [nucl] (atom of an element having a high atomic weight) atome *m* lourd
~ ATOM METHOD - [phys] (direct x-ray analysis) analyse *f* par radiocristallographie directe
~ BEDDED - [mining] en couches épaisses
~ BURR WOOL - [text] (wool which includes up to I6 ‰ of burrs) laine *f* chardonneuse
~ CASTINGS - [metall] grosses pièces *fpl*
~ CLOTH - [text] tissu *m* lourd
~ CONCRETE - [constr] (concrete also used for nuclear reactors in which the aggregate used is of dense material such as iron ore or steel scrap) béton *m* lourd
~ COSMIC-RAY TRACKS - [nucl] (tracks which are found in emulsions which are exposed to cosmic radiation at very high altitudes) traces *fpl* de particules cosmiques lourdes
~ CRUDE OIL - [min] pétrole *m* brut
~ CUT - [mach tool] forte passe *f*
~-DUTY - [gen] (term often applied to any appliance specially intensed for exacting service) de grande puissance, à grand rendement
HEAVY-DUTY LATHE - [mach tool] tour *m* parallèle très renforcé
~ DUTY MACHINE - [mech mach] machine *f* à grand rendement
~-DUTY TRAILER - [auto] remorque *f* pour service sévère
~-DUTY TRUCK - [auto] camion *m* pour service sévère, camion *m* type travaux publiques
~ EARTH - [chem] terre *f* pesante, baryte *f*
~ ELEMENT - [phys] (relating to actinide elements, in particular thorium and uranium) élément *m* lourd
~ ELEMENT CHEMISTRY - [chem] chimie *f* des éléments lourds
~ ENDS - [oil ind] fractions *fpl* lourdes
~ FLUID SEPARATION - [min] séparation *f* par liquides denses
~ FUEL - [fuel] carburant *m* lourd, huile *f* lourde
~-GAUGE WIRE - [mech] fil *m* métallique à grande épaisseur
~ GLASS - [glass man] (a flint glass) flint *m*
~ HYDROCARBON - [chem] hydrocarbure *m* lourd
~ HYDROGEN - [nucl] (term used for deuterium, the hydrogen isotope of atomic mass 2) deutérium *m*, hydrogène *m* lourd
~ ICE - [nucl] (frozen heavy water) glace *f* lourde
~ TON - [nucl] ion *m* lourd
~ IRONWORK - [metall] grosse serrurerie *f*
~ ISOTOPE - [nucl] (isotope of an element having an atomic weight greater than of another isotope of the same element) isotope *m* lourd
~ LOOM - [text] métier *m* lourd
~ MAKE OF WATER - [mining] grosse venue *f* d'eau
~ MESON - [nucl] (present in cosmic radiations) méson *m* lourd, kaon *m*, méson *m* K
~ METAL - [metall] métal *m* lourd
~ MILLING - [mach tool] fraisage *m* fort
~ MINERALS - [geol] (small grains of mineral having a specific gravity greater than that of bromoform) minerais *mpl* lourds
~ NAPHTHA - [chem] (a coal tar derivative with uses as a solvent) essence *f* lourde
~ NUCLEUS - [phys] (the nucleus of heavy elements) noyau *m* lourd
~ OIL - [min] huile *f* lourde
~ OIL ENGINE - [mech] moteur *m* à essence lourde
~ OVERCOAT MATERIALS - [text] étoffe *f* pour vêtements d'hiver
~ OXYGEN - [chem] (term used for isotopes of oxygen) oxygène *m* lourd
~ PICK - [text] chasse *f* dure, coup *m* dur
~ PLATE MILL - [metall] laminoir *m* à grosses tôles
~ PLATES - [metall] tôles *fpl* fortes, grosses tôles *fpl*
~ PUMPING COSTS - [oil ind] frais *mpl* élevés d'épuisement
~ RAINS - [met] grandes pluies *fpl*
~ RAPPING HAMMER - [metall] mailloche *f*
~ SCRAP - [metall] débris *mpl* de fonte de bâtiment
~ SEAS - [met] grosse mer *f*
~-SECTION BULL-HEADED RAIL - [metall] rail *m* dissymétrique (type renforcé)
~-SECTION ROLLS - [metall] train *m* à gros profilés
~ SECTIONS - [metall] gros profilés *mpl*
~ SEED WOOL - [text] (wool containing up to I6 ‰ of seeds) laine *f* à pourcentage élevé de semence
~-SERVICE - [ind chem] (in the oil index) "heavy service" *m*

HEAVY SHADOWS - [photo] ombres *fpl* profondes
~SPAR - [min] (another name for barytes, natural sulphate of barium, used as a filler to give weight to materials) spath *m* pesant, barytine *f*
~STOCK - [railw] matériel *m* lourd
~STRATIFICATION - [geol] stratification *f* puissante
~THREAD - [text] fil *m* lourd
~TREADLE - [text] pas *m* dur
~-TYPE BALL BEARINGS - [metall] roulements *mpl* à billes (pour forte charge)
~WATER - [chem] (deuterium oxide. Water composed of oxygen and heavy hydrogen) eau *f* lourde
~WATER CONTENT METER - [instr] teneurmètre *m* d'eau lourde
~-WATER REACTOR - [nucl] réacteur *m* à eau lourde
HEBEPHRENIA - [med] démence *f* précoce
HEBETOMY - [med] pubiotomie *f*
HEBETUDE - [med] hébétude *f*
HECK BOX - [text] plot *m*
~CORD - [text] cordon *m* du plot
HECKLE, to - [text] s. hackle, to
HECKLE - s. hackle
HECTARE - [meas] hectare *m*
HECTOGRAM - [meas] hectogramme *m*
HECTOLITRE - [meas] hectolitre *m*
HECTOMETRE - [meas] hectomètre *m*
HECTOMETRIC WAVES - s. medium frequency
HEDDLE - s. heald
~HOOK - [text] passette *f*
HEDDLING - [text] remettage *m* des fils de chaîne
~THE WARP THREADS - s. heddling
HEDGE - [gen] haie *f*
~-BILL - [impl] serpe *f*, vouge *m*
~CLIPPING SHEARS - [agric] taille-buissons *m*, cisaille *f* à haies
~MUSTARD - [med & agric] roquette *f*
~TRIMMER - [agric] s. hedge clipping shears
HEDGEHOG - [zool] hérisson *m*
HEDGEHOP, to - [aero] (colloq. for very low flying) voler à ras de terre
HEDGEROW - [gen] bordure *f* de haies
HEDONIC GLANDS - [zool] (skin glands secreting a pleasant-smelling substance in reptiles) glandes *fpl* hédonistiques
HEDONISM - [gen] hédonisme *m*
HEDONOPHOBIA - [med] phobie *f* du plaisir
HEEL, to - [naut] (to lean on one side) donner de la bande, pencher sur le côte
HEEL - [anat] talon *m*
[shoe man] talon *m*
[zool] (of a horse) derrière *m* du sabot
(of a tree) pied *m* (d'un arbre)
[naut] (of a mast) pied *m*, de carène
[shipbuild] (of the keel) bande *f*, gîte *f*, inclinaison *f*
[railw] (of points) talon *m* (d'aiguille)
[railw] (of blades) talon *m*
[constr] (the lower end of a timber) pied *m*, gros bout *m*
[hydr] inclinaison *f*
[metall] (in an induction furnace) réserve *f* (liquide de) de métal
[rubber ind] (of tyres) talon *m*
[chem] (the residue of a completed batch distillation) résidu *m*
[mus] (the end of the bow of string instrument, at which it is held) talon *m*
[bookbind] queue *f*
[zool] (of a bird) éperon *m*
HEEL BEARING - [mech] (of gears) contact *m* de fond
~BINDING - [shoe man] bordure *f* de crêpe pour protection du talon
~COUNTER - [shoe man] contrefort *m*
~CUSHION - [shoe man] coussin *m* contre excoriation du talon
~EFFECT - [radiat] (the alteration of intensity at the anode side of the X-ray beam) effet *m* de talon
~FABRIC - [rubber man] (of tyres) toile *f* de talon, tissu *m* talon
~FIXING DISK - [rubber ind] (of tyres) cercle *m* pose-talon
~LIFT - [shoe man] bloc *m* de remplissage du talon
~OF A DAM - [hydr] pied *m* d'une digue
~OF BLADE - [railw] talon *m* d'aiguille
~OF POINTS - [railw] talon *m* d'aiguille
~-PIECE - [text] talonnette *f*
[el] (of a magnet) culasse *f*
~PLATE - [firearms] plaque *f* de couche
[shoe man] (of a boot) fer *m* de botte
~POST - [hydr] (the vertical post at one side of the lock-gate) poteau *m* tourillon
[constr] montant *m*
~REST - [shoe man] rebut *m* de talon
~RIM - [rubber ind] jante *f* à rebord de talon
~STRAP - [build] ferrure *f* de suspension
~-TAP - [shoe man] rondelle *f* en cuir
~TIP - [shoe man] (of rubber) bon-bout *m* en caoutchouc
HEELING - [gen & naut] gîte *f*, inclinaison *f*
[shoe man] pose *f* du talon
HEIFER - [zool] génisse *f*
~CALF - [zool] velle*f*, vêle *f*
HEIGHT - [gen] hauteur *f*
[aero] altitude *f*
[constr] (the rise of an arch) flèche *f*
[constr] (of a bridge) flèche *f*
[gen] (of man) stature *f*
[geogr] hauteur *f*, éminence *f*, colline *f*
~ABOVE BASE - [surv] (vertical height above an arbitrary datum) hauteur *f* (d'instrument) en station
~BETWEEN FLOORS - [constr] hauteur *f* d'étage
~CLEARANCE - [constr] hauteur *f* de libre passage
~FINDER - s. height indicator
~GAIN - [phys] gain *m* obtenu par effet de miroir, gain *m* de hauteur
~INDICATOR - [instr] (instrument designed to measure the distance between an aircraft and a point on the ground directly below it) sondeur *m*
~LOSS - [aero] perte *f* d'altitude
~OF BREAST BEAM - [text] hauteur *f* de poitrinière
~OF BUFFERS - [railw] hauteur *f* de tamponnement
~OF BURST - [nucl] (the height at which a nuclear bomb is detonated over the earth's surface) altitude *f* d'explosion
~OF CENTRES - [mach tool] hauteur *f* des pointes
~OF DAMMING - [hydr] hauteur *f* de refoulement
~OF EQUILIBRIUM - [aero] altitude *f* d'équilibre
~OF FLANGE - [mech] (the dimension of a flange measured from the edge radially to the body of the part, e.g. a pipe, on which it is formed) hauteur *f* de rebord de jante
~OF HEAD OF A RAIL - [railw] hauteur *f* du champignon
~OF INSTRUMENT - [surv] (the vertical distance of

the horizontal axis of the instrument from the ground level) hauteur *f* de l'instrument en station
HEIGHT OF LOAD - [transp] gabarit *m* de chargement
~OF THE PROFILE - [rubber ind] (of tyres) hauteur *f* de profil
~OF TYRES - [tyres] hauteur *f* du pneu
~OF WATER - [hydr] hauteur *f* de la colonne d'eau
~POWER FACTOR - [aero] (the relation between the power developed by an engine at a specified height and that which it would produce at M.S.L.) coefficient *m* de puissance en altitude
HEIGHTEN, to - [gen] surélever, surhausser, augmenter
HEILIGENSCHEIN - [opt] (a diffraction effect) nimbe *m*
HEISING MODULATION - [telev] (a system of amplitude modulation by absorption) modulation *f* à courant constant
HELBERGER FURNACE - [metall] (a type of electric crucicle furnace) four *m* Helberger
HELCOMA - [med] ulcère *f* de la cornée
HELIARC WELDING - [metall] soudure *f* en atmosphère de hélium
HELICAL - [gen & mech] helicoïdal, en hélice, en spirale
~AERIAL - [radio] (aerial which consists of a wire in the form of a helix open at the top) antenne *f* helicoïdale
~CUT - [mech] taille *f* hélicoïdale
~ELEMENT - [el heating] (wire resistor wound in a helix) élément *m* en hélice
~-FINNED PIPE - [mech] tuyau *m* à nervures hélicoïdales
~GEAR - [mech] (type of gear in which the teeth are not parallel with the axis of the gear, but aligned with a helix described about it) pignon *m* à taille hélicoïdale, roue *f* hélicoïdale
~GEAR SHAPER CUTTER - [tool] fraise *f* à tailler les engrenages hélicoïdaux
~GROOVE - [mech] rainure *f* hélicoïdale
~PLATE - [el] (storage battery) (plate formed of helices of ribbed strips of soft lead fixed in pockets in cells of hard lead) plaque *f* à rosettes
~-SCAN VIDEO RECORDER - [telev] magnétoscope *m* à analyse hélicoïdale
~SCANNING - [radar] (scanning motion roughly describing a distorted helix) balayage *m* hélicoïdal, exploration *f* hélicoïdale
~SLOT - [text] fente *f* (du tambour) courant obliquement
~SPRING - [mech] (a spring in the form of a helix) ressort *m* hélicoïdal, ressort *m* spiral
~-TOOTHED - [mech] (of gears) denté en spirale
~WORM GEAR - [mech] vis *f* sans fin hélicoïdale
HELICOID - [geom] hélicoïde *m*
[gen & geom] hélicoïdal
~TOOTHING - [mech] (of gears) engrenages *mpl* hélicoïdaux
HELICOPTER - [aero] (a havier-than-aircraft supported by power-driven rotors operating on approximately vertical axes) hélicoptère *m*
~BUS SERVICE - [aero] (colloq. USA) transport *m* par hélicoptère
HELIOCENTRIC - [astr] héliocentrique
HELIOCHROME - [photo] héliochromique
HELIOGRAPH - [instr] (instrument used for photoengraving) héliographe *m*, héliostat *m*
(type of heliostat fitted with a spring device by which it can flash long or short flashes) héliographe *m* de signalisation
HELIOGRAVURE - [photo] héliogravure *f*, photogravure *f*
HELIOMETER - [instr] (instrument originally designed to measure the diameter of the sun) héliomètre *m*
HELION FILAMENT - [el] (filament consisting of carbon and an outer coating of silicon) filament *m* de carbone et silicium
HELIOSCOPE - [instr] hélioscope *m*
HELIOSENSITIVITY - [med] héliosensibilité *f*
HELIOTHERAPY - [med] (sun treatment) héliothérapie *f*
HELIOTHERMOMETER - [instr] (instrument designed to measure the intensity of solar radiation) héliothermomètre *m*
HELIOTROPE - [min] (a bloodstone; frequently used in signet rings) héliotrope *m*, jaspe *m* sanguin
[bot] héliotrope *m*
~EXTRACT - [chem] essence *f* d'héliotrope
HELIOTROPIN - [chem] (a component of perfumes) héliotropine *f*
HELIOTYPY - [photo] héliotypie *f*
HELIPORT - [aero] (a landing stage or field designed for the use of helicopters) héliport *m*
HELIUM - [chem] (a gaseous element; symbol He, A.N. 2, A.W. 4.002 chemically inert and the second lightest gas. Helium is used for the inflation of lighter- than aircraft, as an oxygen diluent in the atmosphere supplied to those working under pressure, as a filler for fluorescence light tubes, and as an inert atmosphere for reactions. It occurs in the atmosphere and in natural gases, from which it is derived for use) hélium *m*
~COOLING - [nucl] (cooling method of a reactor in which helium is employed as a coolant) refroidissement *m* à l'hélium
~DIVING-BELL - [naut] (a diving-bell in which the nitrogen in compressed air is replaced by helium) cloche *f* de plongeurs à l'hélium
~LEAK DETECTOR - [nucl] détecteur *m* de fuites à l'hélium
~LIQUEFACTION - [chem] liquéfaction *f* de l'hélium
~LIQUEFIERS - [chem] appareillage *m* pour la liquéfaction de l'hélium
~MASS SPECTROGRAPHIC METHOD - [nucl] méthode *f* spectrographique de masse pour essais de fuites de l'hélium
~NUCLEUS - [phys] (the nucleus of the helium atom) noyau *m* d'hélium
~PERMEATION - [nucl] pénétration *f* d'hélium
~PRESSURE TANK - [nucl] (tank used to receive the fission gases from the accumulator tank) récipient *m* d'hélium sous pression
~SOLIDIFICATION - [chem] (by subjecting it to external pressure) solidification *f* de l'hélium
HELIX - [geom] (a spiral line) hélice *f*
[arch] (of a column) volute *f*, spirale *f*
[anat] ourlet *m* de l'oreille
~ANGLE - [mech] angle *m* d'inclinaison de l'hélice
~ANGLE MICROMETER - [instr] micromètre *m* pour la mesure de l'angle hélicoïdal
~FLASH TUBE - [electron] (masers and lasers) dispositif *m* de concentration hélicoïdal de lumières
~OF A COIL - [mech] direction *f* de la spirale (d'un

ressort)
HELIX WAVEGUIDE - [electron] guide *m* d'ondes hélicoïdal
HELL - [telecomm] (a telegraph code, in which each character is represented by a fixed number of unit elements) code *m* Hell
~BOX - [print] boîte *f* à défets, caisse *f* pour les caractères de rebut
~PRINTER - [telecomm] imprimeur *m* Hell
~RECEIVER - s. Hell printer
~SENDER - [radio] émetteur *m* Hell
~SYSTEM - s. Hellschreiber system
HELLEBORE - [chem] (the dried rhizome of varieties of Helleborus. It is used in medicine) ellébore *m* varaire *f*
HELLEBOREIN - [chem] (a glucoside, the active constituent of hellebore) elléboreine *f*
HELLIGE COMPARATOR - [paint] (instrument for colour determination of liquids, by comparison with a series of disks of coloured glass) comparateur *m* Hellige
HELLSCHREIBER SYSTEM - [telecomm] système *m* Hell
HELM - [naut] timon *m*, gouvernail *m*
[naut] (the handle or tiller of the rubber) barre *f* (du gouvernail)
~DOWN - [naut] barre *f* dessous
~UP - [naut] barre *f* au vent
~INDICATOR - [instr] axiomètre *m*
HELMET - [gen] casque *m*
[naut] (of the diver) casque *m* de plongeurs
[metall] (in welding) casque *m*
~HEADACHE - [med] casque *m* neurasthénique
~SHAPED - [bot] (corolle) en forme de casque
HELMINTH INFESTATION - [vet] helminthiase *f* (ver intestinal)
HELMINTHICIDE - [chem] helminthicide *m*
HELMSMAN - [naut] timonier *m*
HELOMA - [med] durillon *m*
HELVE - [tool] (the handle of an axe, a hatchet etc) manche *m*
~HAMMER - [impl] (obsolete; a trip hammer) martinet *m*, marteau *m* à bascule
~-RING - [impl] bague *f*, jurasse *m*
HELVITE - [min] (an ore of beryllium) helvite *f*
HEM, to - [gen & text] border, mettre un bord
HEM - [text] bord *m*, ourlet *m*
~WAVE - [electron] (electromagnetic wave with transverse and longitudinal components of displacement) onde *f* électromagnétique hybride
HEMACHROMATOSIS - [med] hémochromatose *f*
HEMACYTOMETRY - [med] numération *f* globulaire
HEMAL NODES - [med] glandes *fpl* hémales
HEMALEXIN - [med] alexine *f* du sang
HEMANALYSIS - [med] analyse *f* du sang
HEMANGIECTASIS - [med] hémangiectasie *f*
HEMANGIOBLAST - [med] hémangioblaste *m*
HEMAPHOBIA - [med] hématophobie *f*
HEMATIC ABCESS - [med] hématome *m* suppuré
HEMATIDROSIS - [med] hématidrose *f*, sueur *f* de sang
HEMATIMETER - [instr] hémocytomètre *m*, hématimètre *m*
HEMATITE - s. haematite
HEMATOMA - [med] hématome *m*
HEMATOMOLE - [med] hématome *m* sous-chorial
HEMATOMYELIA - [med] hémorragie *f* médullaire, hématomyélie *f*
HEMATOMYELITIS - [med] hématomyélite *f*
HEMATOPLANIA - [med] mentration *f* vicariante
HEMATOSPERMIA - [med] hémospermie *f*
HEMATURIA - hématurie *f*
HEMIANOPIA - [med] hémipsie *f*, hémiablepsie *f*
HEMIATAXIA - [med] hémiataxie *f*
HEMIATHETOSIS - [med] hémiathétose *f*
HEMIATROPHY - [med] atrophie *f* unilatérale
HEMICELLULOSE - [chem] (a component of vegetable tissues) hémicellulose *f*
HEMICOLLOID - [phys] (colloid composed of small-size particles) semicolloïde *m*
HEMICRYSTALLINE ROCKS - [geol] (rocks of igneous origin containing some interstitial glass) roches *fpl* semicristallines
HEMIDEMISEMIQUAVER - [mus] quadruple-croche *f*
HEMIDROSIS - [med] hémidrose *f*
HEMIHEDRAL CRYSTAL - [crystall] (crystal with half of its faces developed) cristal *m* hémière
HEMIHEDRITY - [crystall] hémiédrie *f*
HEMIHYPESTHESIA - [med] hypoesthésie *f* unilatérale
HEMIHYPOTONIA - [med] hémihypotomie *f*
HEMIOMORPHITE - [min] (sometimes called Electric Calamine. A natural hydrated zinc silicate, occurring with zinc sulphide ores) hémiomorphite *f*, calamine *f*
HEMIPARESIS - [med] hémiparesie *f*
HEMIPLEGIA - [med] hémiplégie *f*
HEMISACRALIZATION - [med] sacralisation *f* unilatérale
HEMISPASM - [med] spasme *m* unilatéral
HEMISPHERE - [gen] hémisphère
HEMITHORAX-[anat] hémithorax *m*
HEMITROPIC - [crystal] hémitrope, maclé
HEMLOCK - [bot] cigue *f*, cicutaire *f*
~BARK - [bot] (the bark of Tsuga canadensis with uses in tanning and the preparation of pharmaceuticals) écorce *f* de sapin-ciguë
~SPRUCE - [bot] sapin-ciguë, sapin *m* du Canada
HEMMING - [gen] ourlet *m*
[metall] (in sheet metal working) bordure *f*
~MACHINE - [text] machine *f* à border
HEMOBLASTOSIS - [med] hémoblastose *f*
HEMOCLASIA - [med] crise *f* hémoclasique
HEMODYNAMICS - [med] hémodynamique *f*
HEMOGLOBIN - s. haemoglobin
HEMOLYTIC DISEASE OF NEWBORN - [med] érythroblastose *f* fœtale
HEMOLYTIC JAUNDICE - [med] ictère *m* hémolitique
HEMOPATHY - [med] maladie *f* du sang
HEMOPEXIS - [med] hémopexie *f*
HEMOPHILIA - [med] hémophilie *f*
HEMOPHTALMIE - [med] hémophtalmie *f*
HEMOPTYSIS - [med] hémoptysie *f*
HEMORRHAGE - s. haemorrhage
HEMOSTASIA - [med] hémostase *f*
HEMP - [text] (fibres obtained from Cannabis sativa and used for the production of ropes and heavy fabrics) chanvre *m*
~BELT - [impl] courroie *f* de chanvre
~BRAIDING - s. hemp coiling
~BRAKE - [text] broie *f*
~CLOTH - [text] tissu *m* de chanvre
~COILING-[text] (rope man) tresse *f* de chanvre, enveloppe *f* de chanvre
~COMB - [text] échanvroir *m*

HEMP FLOCK - [text] flocon *m* de chanvre
~JOINTING - s. hemp packing
~LINEN - [text] toile *f* de lin
~MEAL - [agric] chanvre *m* moulu
~-NETTLE - [bot] (a weed) ortie *f* rouge
~OIL - s. hempseed oil
~PACKING - [mech] garniture *f* de chanvre
~ROPE - [text] (rope man) corde *f* en chanvre
~SERVING - s. hemp coiling
~STRAP - [text] sangle *f* en chanvre
~TOW YARN - [text] fil *m* d'étoupe de chanvre
~TWILL-CLOTH HOSE - [text] tuyau *m* en tissu de chanvre entrecroisé
~YARNS - [text] fils *mpl* de chanvre
HEMPSEED - [bot] chènevis *m*
~CAKE - [agric] (animal feeding) tourteau *m* de chènevis
~OIL - [chem] (oil obtained from Cannabis sativa and used in the production of varnishes and paints) huile *f* de chènevis
HEMSTITCH - [text] ourlet *m* à jour
HEN - [zool] poule *f*
~HOUSE - [agric] poulailler *m*
~HUT - s. hen house
~-ROOST - [agric] juchoir *m*, perchoir *m*
~-RUN - [agric] parcours *m*, courette *f*
HEN'S EGG - [zool] œuf *m* de poule
HENBANE - [bot] (med plants) jusquiame *f*, hanebane *f*
H-ENGINE - [mech] (type of reciprocating i.c.; engine having two crankshafts and four banks of cylinders vertically-opposed, the end view of the arrangement having the form of the letter "H") moteur *m* en H
HENNA - [dy] henné *m*
HENNERY - [agric] poulailler *m*, basse-cour *f*
HENRY - [el] (unit of inductance, equivalent to the induction of one volt by a current change rate of one ampere per second) henry *m*
HENRY'S LAW - [chem] (the amount of a gas dissolved by a given amount at a given temperature is proportional to the temperature) loi *f* de Henry
HEPARIN - [chem] (an anticoagulant used in medicine) héparine *f*
HEPATATROPHY - [med] atrophie *f* du foie
HEPATECTOMY - [med] hépatectomie *f*
HEPATITIS - [med] hépatite *f*
HEPATOCELE - [med] hernie *f* du foie
HEPATOLITHECTOMY - [med] lithectomie *f* (calcul biliaire)
HEPATOMEGALIA - [med] hépatomégalie *f*
HEPTABARBITAL - [chem] (a sedative, hypnotic and anaesthetic used in medicine) heptabarbital *m*
HEPTACHLOR - [chem] (an insecticide) heptachlore *m*
HEPTADECANOL - [chem] (a plasticizer and synthesis intermediate) heptadécanol *m*
HEPTADECYLGLYOXALIDINE - [chem] (a fungicide) heptadécylglyooxalidine *f*
HEPTAFLUOROBUTYRIC ACID - [chem] (a surface active agent and synthesis intermediate) acide *m* heptafluorobutyrique
HEPTANAL - [chem] (intermediate for rubber chemicals) heptanal *m*
HEPTANE - [chem] (one of the paraffin series; there are seven heptanes having the same formula and containing seven hydrogen atoms in the molecule. Normal heptane is one of the constituents of petrol and resembles hexane in chemical behaviour) heptane *m*
HEPTANOL - [chem] (solvent and synthesis intermediate) heptanol *m*
HEPTAVALENT - [chem] (capable of combining with seven hydrogen atoms, or their equivalent) heptavalent, septivalent
HEPTENE - [chem] (intermediate for synthetic resins, flavours, perfumes, dyestuffs and pharmaceuticals) heptène *m*
HEPTODE - [electron] (a seven-electrode thermionic tube) heptode *f*
HEPTOSES - [chem] (mono-saccharoses, containing seven atoms of oxygen) heptoses *mpl*
HEPTYL FORMATE - [chem] (intermediate for flavourants) formiate *m* d'heptyle
HERB - [bot] herbe *f*
~PATIENCE - [agric] patience *f*, oseille-épinard *f*
HERBACEOUS - [bot] herbacé
~CUTTING - [bot] (of hops) bouture *f* herbacée
HERBAGE CROP - [agric] fourrage *m* herbacé
HERBBANE - [bot] (a weed) orobanche *f* rameuse
HERBARIUM - [bot] (collection of dried plants) herbier *m*, jardin *m* sec
HERBICIDES - [chem] (compounds which destroy plant life) herbicides *mpl*
HERCOGAMY - [bot] hercogamie *f*
HERCYNITE - [min] (iron spinel) hercynite *f*
HERD, to - [agric] assembler en troupeaux
HERD - [agric] troupeau *m*, troupe *f*, bande *f*
HERDBOOK - [agric] livre *m* généalogique des races bovines
HERDING INSTINCT - [zool] instinct *m* grégaire
HEREDITARY - [gen & anat] héréditaire
~DEFECT - [genet] (defect in the human body which is caused by mutation of the reproductive cells) lésion *f* héréditaire
~MECHANISM - [mech] (that field of mechanism which is related to boundary conditions extending over continuous intervals of space and time and demanding integrals for their representation) mécanique *f* à conditions limites
HEREDITY - [gen] hérédité
[metall] (the characteristics which remain unchanged in a metal after successive melting) hérédité
HERING FURNACE - [metall] (a special type of electric melting furnace, in which the Pinch Effect is made use of in circulating the melt) four *m* Hering
HERKOGAMY - s. hercogamy
HERMAPHRODITE - [genet] (having both reproductive organs) hermaphrodite *m*
HERMAPHRODITISM - [genet] hermaphrodisme *m*
HERMETIC - [gen] hermétique
~SEAL - [mech etc] scellement *m* hermétique
[electron] scellement *m* hermétique
HERNIA - [med] hernie *f*
HERNIARY BANDAGE - [med] bande *f* herniaire
HERNIOPLASTY - [med] hernioplastie *f*
HERNIOTOMY - [med] herniotomie *f*, kélotomie *f*
HEROIN - [chem] (diacetylmorphine, obtained by the acetylation of morphine. This alkaloid is a narcotic, with action resembling that of morphine, and is classes as a dangerous drug of addition) héroïne *f*
HEROULT FURNACE - [metall] (direct arc furnace in which there is no bottom electrode, two electrodes passing through a refractory roof and forming arcs

in series to and from the charge) four *m* Héroult
HERPES - [med] herpès *m*, bouton *m* de fièvre
HERRINGBONE - [gen] à chevrons, en arête de hareng
[constr] en épi, en feuille de fougère
~ASHLAR - [constr] (block of stone in grooves of a herringbone design) appareil *m* en épi
~BOND - [constr] appareil *m* en épi
~GEAR - [mech] (double helical gear) engrenage *m* à chevrons, engrenage *m* hélicoïdal double
~PATTERN - [telev] (interference pattern occurring on a TV screen) diagramme *m* de perturbation en boucles
~STRUTTING - [carp] (small struts which are fixed diagonally in pairs) latte *f* en croix, croisillon *m*
~TOOTH - [mech] (of gear) dent *f* chevronnée
~TWILL - [text] coutil *m* en jute
HERRINGER-HULSTER EFFECT - [electron] focalisation *f* de phase à modulation de vitesse
HERTZ - [el] (unit of frequency = one cycle per sec.) hertz *m*, cycles *mpl* par seconde
~DIPOLE - [radio] (the double pole) dipôle *m* hertzien
HERTZIAN - [el] hertzien
~OSCILLATOR - [el] (system producing electro-magnetic waves) oscillateur *m* hertzien
~WAVES - [el] (electromagnetic waves produced by a Hertzian oscillator) ondes *fpl* hertziennes
HESPERIDIN - [chem] (a glycoside obtained from the rind of certain citrus fruits and used in medicine for the control of capillary bleeding) hespéridine *f*
HESS'S LAW - [phys] (the heat absorbed or generated in a chemical process is constant irrespective of wheter the process is carried out in one or several steps) loi *f* de Hess
HESSIAN - [text] (coarse material woven loosely from jute yarn) étoffe *f* grossière de chanvre
~FLY - [zool] mouche *f* de Hesse, cécidomyie *f*
HETERO - [gen] (prefix denoting "different") hétér(o)
HETEROCHROMATIC - [phys] (of colours of different hues) hétérochromatique
~X-RAYS - [radiat] (X-rays of a considerable number of frequencies) rayons *mpl* X hétérogènes
HETEROCYCLIC COMPOUNDS - [chem] (cyclic compounds, in which other atoms, e.g. of nitrogen, oxygen, sulphur, form part of the ring) composés *mpl* hétérocycliques
HETERODYNE - [radio] (the method of combining oscillations of different frequencies, espc. to produce an oscillation of frequency equal to the difference between the original two) hétérodyne *m*
~CONVERSION TRANSDUCER - [radio] transducteur *m* à conversion hétérodyne
~FREQUENCY - [radio] (frequency resulting from a combination of two signals of different frequency) fréquence *f* de battement
~FREQUENCY METER - [instr] fréquencemètre *m* hétérodyne, ondemètre *m* hétérodyne
~INTERFERENCE - [radio] interférence *f* par hétérodynation
~OSCILLATOR - [radio] (in a superheterodyne receiver) oscillateur *m* hétérodyne
~PRINCIPLE - [radio] principe *m* de l'hétérodyne
~RECEPTION - [radio] (beat reception) réception *f* par battement, réception *f* hétérodyne
~WARBLER OSCILLATOR - s. heterodyne oscillator
~WAVEMETER - [instr] ondemètre *m* hétérodyne
HETERODYNE WHISTLE - [radio] (the steady tone which is heard in the output of an amplitude-modulation receiver caused by the heating of two carriers having a frequency difference) sifflement *m* d'hétérodyne
HETERODYNING - [radio] hétérodynation
HETEROGAMY - [genet] (condition whereby gametes of more than one type are produced) hétérogamie *f*
HETEROGENEITY - [gen] (the opposite of homogeneity) hétérogénéité *f*
HETEROGENOUS - [gen] hétérogène
~EQUILIBRIUM - [phys] (equilibrium between phases) équilibre *m* hétérogène
LATTICE - [nucl] (in a heterogeneous reactor) réseau *m* hétérogène
~MISTURE - [nucl] (mixture of fuel and moderator in a heterogeneous reactor) mélange *m* hétérogène
~PROPELLANT - [chem] propergol *m* hétérogène, propergol *m* solide composite
~REACTOR - [nucl] (type of nuclear reactor in which the fissile material and the moderator are arranged as separate bodies of dimensions such that a non-homogeneous medium is presented to the neutron) réacteur *m* hétérogène
~SYSTEM - [phys] (system with more than one phase) système *m* hétérogène
HETEROION - [phys] (complex ion, i.e. simpler ion absorbed on a molecule) complex *m* ion-molécule
HETEROKINESIS - [med] hétérokinèse *f* (mitose)
HETEROMOLYBDATES - [chem] (generic term for a group of complex molybdenum compounds having application in analytical chemistry and in thedyeing of textiles) hétéromolybdates
HETEROPHORIA - [med] (latent strabismus) hétérophorie
HETEROPOLAR - [el] (having an unequal charge distribution like that in a semipolar bond) hétéropolaire
~BOND [nucl] (valence linkage between two atoms) liaison *f* hétéropolaire
~COLLOID - [phys] (colloidal system in which the dispersed particles are polar compounds) colloïde *m* hétéropolaire
~FIELD MAGNET - [el] (field magnet in which the poles are of opposite polarity) inducteur *m* hétéropolaire
HETEROPOLYMER - [chem] (a polymer produced from two or more monomers) hétéropolymère *m*
HETEROPSIA - [med] hétéropsie *f*
HETEROSIS - [genet] hétérosis *f*, vigueur *f* hybride
HETEROSPHERE - [astr] hétérosphère *f*
HETEROSTATIC CIRCUIT - s. heterostatic method
~INSTRUMENT - [instr] (instrument with quadrant electrometer having the needle maintained at a high potential independently of the quadrants) instrument *m* de mesure hétérostatique
~METHOD - [instr] (method of using an electrometer requiring an external source giving a constant difference of potential) méthode *f* hétérostatique
HETEROTONIA - [med] tension *f* variable
HETEROTOPIA - [med] hétérotopie *f*
HETEROTOPIC - [nucl] (having a different atomic number or nuclear charge; the anonym of isotopic) hétérotopique
HETEROTRICHOSIS - [med] hétérotrichose *f*
HETEROZYGOUS - [genet] (derived from germ cells which are genetically dissimilar) hétérozygote

HEULANDITE - [min] (a natural zeolite) heulandite *f*
HEURISTIC - [gen] (valuable for empirical research) heuristique
~APPROACH - [phys] approximation *f* heuristique
HEURTELOUP - [med] sangsue *f* artificielle
HEVEA - [bot] hévéa *f*
HEW, to - [gen] couper, tailler avec une hache [constr] (of stones) tailler, dégrossir
~DOWN, to - [gen] abattre
HEX - [chem] (abbrev. for Uranium hexafluoride) hexafluorure *m* d'uranium
HEXABROMOETHANE - [chem] (a synthesis intermediate) hexabromoéthane *m*
HEXACHLOROBENZENE - [chem] (an agricultural fumigant and a synthesis intermediate) hexachlorobenzène *m*
HEXACHLOROBUTADIENE - [chem] (colourless, non-flammable liquid with uses as a solvent for natural and synthetic rubbers and other polymers) hexachlorobutadiène *m*
HEXACHLOROCYCLOHEXANE - [chem] (an insecticide) hexachlorocyclohexane *m*
HEXACHLOROCYCLOPENTADIENE - [chem] (an intermediate for non-flammable resins) hexachlorocyclopentadiène *m*
HEXACHLORODIPHENYL OXIDE - [chem] (a solvent and intermediate) oxyde *m* d'hexachlorodiphényle
HEXACHLOROETHANE - [chem] (a rubber accelerator and camphor substitute in celluloid manufacture) hexachloroéthane *m*
HEXACHLOROPROPYLENE - [chem] (a plasticizer and solvent) hexachloropropylène *m*
HEXACHORD - [mus] (a group of six sounds) hexacorde *f*
HEXADECANE - [chem] (an intermediate and solvent) hexadécane *m*
HEXADECENE - [chem] (intermediate for pharmaceuticals,dyestuffs, flavouring agents, perfumes, and resins) hexadécène *m*
HEXADECIMAL NOTATION - [comput] système *m* hexadécimal
HEXADECYL MERCAPTAN - [chem] (a polymerization modifier) mercaptan *m* d'hexadécyle
HEXADECYLTRICHLOROSILANE - [chem] (a silicone intermediate) hexadécyltrichosilane *m*
HEXAETHYL TETRAPHOSPHATE - [chem] (an insecticide) tétraphosphate *m* d'hexaéthyle
HEXAGON - [geom] hexagone *m*
~BOLT - [mech] (a bolt having a six-sided head) boulon *m* à six pans
~BROACH - [mech] équarissoir *m* pour trous hexagonaux
~HEAD - [mech] tête *f* à six pans
~HEAD SCREW - [mech] vis *f* à tête à six pans
~TURRET - [mach tool] tourelle *f* hexagonale
~TURRET LATHE - [mach tool] tour *f* à tourelle hexagonale
~VOLTAGE - [el] (the voltage between two lines of a six-phase system) tension *f* hexagonale (à six phases)
HEXAGONAL - [chem] hexagonal
~BARREL MIXER - [mech mach] (mixer consisting of a hexagonal drum revolving about its longitudinal axis) mélangeur *m* à tonneau hexagonal
~CLOSE-PACKED - [crystall] (a form of crystal structure) hexagonal compact
~CLOSE-PACKED STRUCTURE - [crystall] structure *f* hexagonale compacte
HEXAGONAL MESH - [text] maille *f* hexagonale
~MESH NET - [text] tissu *m* à mailles hexagonales
~-NUT ANGLE-GAUGE - [impl] équerre *f* à six pans
HEXAHYDROBENZOIC ACID - [chem] (rubber stabilizer) acide *m* hexahydrobenzoïque
HEXAHYDROPHTHALIC ACID - [chem] (intermediate for resins and plasticizers) acide *m* hexahydrophtalique
~ANHYDRIDE - [chem] (an intermediate for plasticizers and alkyd resins) anhydride *f* hexahydrophtalique
HEXALDEHYDE - [chem] (intermediate for plasticizers, insecticides, dyestuffs,and synthetic resins) hexaldéhyde *f*
HEXALIN - [chem] (a term for cyclohexanol) cyclohexanol *m*, hexaline *f*
HEXAMETHYLENEDIAMINE - [chem] (colourless leaflets which give a corrosive solution in ether or ethyl alcohol. A basic material in a number of polymers, specifically (with adipic acid) of nylon) hexaméthylènediamine *f*
HEXAMETHYLENETETRAMINE - [chem] (a rubber accelerator and curing agent for plastics. Its addition to cellulosic fibres gives increased elasticity) hexaméthylène tétramine *f*
HEXAMETHYLPHORIC TRIAMIDE - [chem] (a light stabilizer for PVC) triamide *f* hexaméthylphosphorique
HEXANE - [chem] (one of the paraffin series, having six atoms of hydrogen in the molecule. There are five hexanes, of which normal hexane is an important constituent of petrol and a valuable solvent) hexane *m*
HEXANETRIOL - [chem] (a solvent and intermediate for synthetic resins) hexanetriol *m*
HEXANITRODIPHENYL AMINE - [chem] (an analytical reagent and a component of certain explosives) hexanitrodiphénylamine *f*
HEXANOL - [chem] (an intermediate for pharmaceuticals, dyestuffs, perfumery materials, and flotation agents) hexanol *m*
HEXAVALENT - [chem] (capable of combining with six atoms of hydrogen or the equivalent) hexavalent
HEXENE - [chem] (synthesis intermediate for synthetic resins, pharmaceuticals, flavourants, and perfumes) hexène *m*
HEXENOL - [chem] (a perfumery compound) hexénol *m*
HEXOBARBITAL - [chem] (a hypnotic used in medicine) hexobarbital *m*
HEXODE - [electron] (six-electrode electronic valve) hexode *f*
HEXOKINASE - [chem] (an enzyme with applications in research) hexokinase *f*
HEXOSES - [chem] (mono-saccharoses containing six atoms of oxygen) hexoses *mpl*
HEXYL ACETATE - [chem] acétate *m* hexilique
~BROMIDE - [chem] (synthesis intermediate) bromure *m* hexilique
~ETHER - [chem] (a solvent) éther *m* hexilique
~MERCAPTAN - [chem] (applications in the processing of synthetic rubbers) mercaptan *m* d'hexyle
~METHACRYLATE - [chem] méthacrylate *m* d'hexyle
HEXYLENE GLYCOL- [chem] (textile processing agent, brake fluid, and emulsificant) hexylèneglycol*m*

HEXYLPHENOL - [chem] (intermediate for resins) hexylphénol *m*
HEXYLRESORCINOL - [chem] (an anthelmintic used in medicine) hexylrésorcine *f*
HEZASTYLE - [arch] (a six column portico) hexastyle *m*
H.F. - s. High Frequency
Hg - s. Mercury Column Pressure
HIATAL TEXTURE - [geol] structure *f* poreuse
HIATUS - [med] hiatus *m*, orifice *m*
HIBBERT CELL - [el chem] (a standard primary cell, differing from the Clark cell only in having a zinc chloride electrolyte in place of zinc sulphate) élément *m* Hibbert
HIBERANTING SPACECRAFT - [astronaut] spationef *f* sur orbite d'attente
HIBERNATION - [gen] hibernation *f*
HICCOUGH - [acoust] hoquet *m*
HICKORY - [bot] hickory *m*, noyer *m* d'Amérique
~WOOD - [bot] noyer *m* américain
HICKS HYDROMETER - [instr] (a special hydrometric device designed for checking the specific gravity of accumulator electrolyte, consisting of a glass tube containing a number of differently coloured beads weighted to float according to the Sp. gr. of the liquid) hydromètre *m* Hicks
HIDDEN BIT - [constr] (assemblage à) paume *f*
~WADDING THREADS - [text] fils *mpl* de fourrure recouverts
HIDE, to - [gen] cacher, se cacher
[leather ind] tanner (le cuir)
HIDE - [gen] peau *f*, cuir *m*
[plast ind] (thick rough sheet of plastic material produced by rolling) feuille *f* brute
~GLUE - [ind chem] colle *f* animale
HIDEBOUND SKIN - [med] sclérodermie *f*
HIDES - [gen] cuirs *mpl*
~AND SKINS - [gen] cuirs et peaux
HIDING POWER - [paint] (the capacity of a paint or the like to conceal the surface to which it is applied) pouvoir *m* couvrant
HIDRADENOMA - [med] lymphangiome *m* tubéreux multiple
HIDRORRHEA - [med] hydrorrhée *f*
HIDROSIS - [med] hidrose *f*
HIERALGIA - [med] sacralgie *f*
HIERARCHY - [gen etc] hiérarchie *f*
HI-FI - s. high fidelity
HIGH - [gen] haut
[metall] (steel with large chromium content) à haute teneur
[met] (a high pressure area) anticyclone *m*
[acoust] (gen) fort
[acoust] (referring to the pitch of a note) aigu
~-ACTIVITY WASTE - [nucl] (highly active radioactive waste materials from nuclear power plants) déchets *mpl* fortement radioactifs
~ALLOY - [metall] alliage *m* à haute teneur
~ALLOY STEEL - [metall] acier *m* afiné
~ALTITUDE - [geogr aero etc] haute altitude *f*
~ALTITUDE FLIGHT - [aero] vol *m* à haute altitude
~ALTITUDE GENERATOR - [el] (electric generator specially constructed for use at high altitudes) génératrice *f* haute altitude
~AMPERAGE - s. high current
~ANGLE HOLE - [mining] puits *m* à grande déviation
~ANGLE SHOT - [photo & cin] (shot taken from an elevated spot) plan *m* en plongée
HIGH ANGLED TWILL - [text] croisé *m* à angle aigu
~ATTENUATING YEAST - [brew ind] levure *f* à forte atténuation
~BAND - [telev] bande *f* des plus hautes fréquences
~-BEAM HEADLIGHTS - [auto] phares *mpl* à longue portée
~-BEAM VELOCITY VIDICON - [electron] vidicon *m* à haute vitesse d'électrons d'analyse
~BEAMS - s. high-beam headlights
~-BOILING - [paper man] (the intensive boiling of pulp) ébullition *f* élevée
~BOILING PHENOLS - [chem] (mixtures composed mainly of meta-substituted alkyl phenols. They have applications in phenolic resins, solvents and rubber chemicals) phenols *mpl* de point d'ébullition élevé
~BOOST - [telev] (only USA; high frequency compensation in GB) accentuation *f* des fréquences élevées
~-BOW CARRIAGE - [text] chariot *m* à grand étrier
~BRASS - [metall] laiton *m* en feuilles (65% cuivre et 35% zinc)
~BULK YARN - [text] fil *m* gonflant, filé *m* mousse
~BUTT - [text] haut talon *m*
~-CAPACITY WAGON - [railw] wagon *m* de grande capacité
~CARBON STEEL - [metall] acier *m* à haute teneur de carbone
~CHROMIUM ALLOY - [metall] alliage *m* à haute teneur de chrome
~CLOUDS - [met] (clouds, usually consisting of ice-crystals, and of an average height exceeding 20,000 feet) nuages *mpl* supérieurs
~COMPRESSION - [auto] compression *f* élevée
~COMPRESSION ENGINE - [mech] (an i.c. reciprocating engine working at specially high compression) moteur *m* à compression élevée
~-COMPRESSION-RATIO PUMP - [mech] pompe *f* à rapport volumétrique élevé
~CONDUCTIVITY - [el] haute conductivité *f*
~CONDUCTIVITY COPPER - [metall] (a metal of high purity with a very high electrical conductivity) cuivre *m* à haute conductivité
~CONTENT ALLOY - [metall] alliage *m* à haute teneur
~CONTRAST IMAGE - [telev] image *f* contrastée
~COUNT YARN - [text] filé *m* fin
~CURRENT - [el] courant *m* de haute intensité
~-DEFINITION - [photo] haute définition *f*
~-DEFINITION PICTURE - [photo & cin] image *f* à haute définition
~-DEFINITION TELEVISION - [telev] télévision *f* à grand nombre de lignes
~-DENSITY ALLOY - [metall] alliage *m* à haute densité
~DRAFT - [text] grand étirage *m*
~-DUMPER - [transp] basculeur *m* à double effet et à grand angle de levée
~DUTY CAST-IRON - [metall] (cast iron with a very high tensile strength) fonte *f* résistante
~-ELECTRON-VELOCITY CAMERA TUBE - [electron] tube *m* de prise de vues stabilisateur au potentiel de l'anode
~-ENERGY ALPHA PARTICLE - [nucl] (an alpha particle from a nuclear reaction in the energy scale up to about 50 Mev) particule *f* alpha à grande énergie
~-ENERGY DEUTERON - [nucl] (a deuteron resulting from a nuclear reaction in the energy region up to about 30 to 50 Mev) deutéron *m* à grande énergie

HIGH-ENERGY ELECTRON - [nucl] (an electron resulting from a nuclear reaction in the energy region up to about 30 to 50 Mev) électron *m* à grande énergie
~-ENERGY FISSION - [nucl] (fission of Uranium 238 by high-velocity neutrons) fission *f* rapide
~-ENERGY GAMMA RADIATION - [nucl] (in the region of 30 to 50 Mev) rayonnement *m* gamma à grande énergie
~-ENERGY LEVEL REACTOR - [nucl] (a high-power reactor with an energy level in the region of 30 to 50 Mev) réacteur *m* à grande puissance
~-ENERGY NEUTRON - [nucl] (neutron from a nuclear reaction, in the region of 30 to 50 Mev) neutron *m* à grande énergie
~ENERGY RATE FORMING - [metall] formage *m* à haute énergie
~EXPLOSIVE - [explos] grand explosif *m*
~FIDELITY - [acoust] (the quality of a sound reproducing system) haute fidélité *f*
~-FIELD EMISSION ARC - [electron] (electric arc in which the electron emission is due to the effect of a high electric field in the immediate neighbourhood of the cathode) arc *m* à effet de champ
~FLASH SOLVENT - [paint] (a solvent of which the flash point is considerably above normal air temperatures) solvent *m* à degré élevé d'inflammation
~FLASHING POINT - [oil ind] point *m* d'inflammabilité élevé
~FLOW - [ind chem] haute fluidité *f*
~-FLUX REACTOR - [nucl] (reactor designed to operate with a high neutron flux) réacteur à flux élevé
~-FREQUENCY - [el] (having a frequency between 3 and 30 Mc/s) haute fréquence *f*
~-FREQUENCY ABSORPTION - [el acoust] absorption *f* des aigus
~-FREQUENCY ALTERNATOR - [el] (alternator with a frequency above I0 kc/s) alternateur *m* H.F.
~-FREQUENCY BAND FILTER - [el] filtre *m* de bande à haute fréquence
~-FREQUENCY BRIDGE - [el] pont *m* à haute fréquence
~-FREQUENCY CABLE - [telecomm] câble *m* à haute fréquence
~-FREQUENCY CALIBRATION - [el] étalonnage *m* à haute fréquence
~-FREQUENCY CARRIER CURRENT TELEPHONE CHANNEL - [telecomm] voie *f* de communication par courants porteurs de haute fréquence
~-FREQUENCY CHOKE - [radio] (inductance coil with a high impedance at high frequencies) bobine *f* de arrêt à haute fréquence, bobine *f* de choc à haute tension
~-FREQUENCY COAGULATION - [chem] (coagulation induced by means of ultra-sonic, or near ultra-sonic, vibrations) coagulation *f* par courant à haute fréquence
~-FREQUENCY COMPENSATION - s. high boost
~-FREQUENCY CONDENSER MICROPHONE - [acoust] (condenser microphone in which the polarizing voltage is alternating at high radio frequency) microphone *m* électrostatique à H.F.
~-FREQUENCY DIELECTRIC SEALING - [el metall] (sealing by the use of heat produced by dielectric effects) soudage *m* électronique à H.F.
~-FREQUENCY FEEDTHROUGH - [el] traversé *f* de haute fréquence
~-FREQUENCY FLASH UNIT - [photo] éclair, appareil éclair *m* de haute fréquence
HIGH-FREQUENCY HEAT-SEALING DEVICE - [mech] (apparatus for welding plastics and the like by means of dielectric heating using high-frequency current) soudeuse *f* par haute fréquence
~-FREQUENCY INDUCTION FURNACE - [metall] (a type of air transformer in which the primary consists of a water-cooled spiral of copper pipes and the secondary of the metal which is melted) four *m* à induction à haute fréquence
~-FREQUENCY MOULDING - [rubber ind] radiovulcanisation *f*
~-FREQUENCY OSCILLATION - [el] oscillation *f* haute fréquence
~-FREQUENCY PREHEATER - [el] (preheater by dielectric losses) appareil *m* à préchauffage par pertes diélectriques
~-FREQUENCY PROSPECTING - [surv] (used to locate mineral deposits) prospection *f* à haute fréquence
~-FREQUENCY SEWING - [electron] (for plastic objects) soudure *f* électronique, soudure *f* haute fréquence
~-FREQUENCY TRANSFORMER - [el] transformateur *m* à haute fréquence
~-FREQUENCY TURBINE GENERATOR - [el] turbo-alternateur *m* à haute fréquence
~-FREQUENCY VACUUM FURNACE - [metall] four *m* à vide à haute fréquence
~-FREQUENCY VULCANIZATION - [rubber ind] (rubber vulcanization carried out with the aid of high-frequency heating) vulcanisation *f* par irradiation, radiovulcanisation
~-FREQUENCY WELDING - [el metall] (welding in which the heating is produced by high frequency current)soudage *m* à haute fréquence
~FURNACE - [metall] four *m* à cuve, haut fourneau *m*
~-GAIN AERIAL - [radio & telev] antenne *f* à gain élevé
~GAMMA TUBE - [telev] (a television camera tube) tube *m* à gamme élevé
~GEAR - [auto] prise *f* directe
~GEAR CLUTCH - [auto] crabot *m* de prise directe
~-GLASS CALENDER - [text] calandre *f* pour effets haut brillant
~-GRADE NUCLEAR FUEL - [nucl] combustible *m* fortement enrichi
~-GRADE ORE - [min] minerai *m* riche
~-GRADE URANIUM ORE - [min] (ore containing a large amount of uranium) minerai *m* à haute teneur d'uranium
~HAT - [cin] (colloq for a very low stand) pied *m* de table
~-HAT TRIPOD - s. high hat
~IMPEDANCE TRIODE - [electron] tube *m* à impédance élevée
~-INTENSITY TAPPING - [rubber ind] méthode *f* de saignée intensive
~INVERSE-VOLTAGE RECTIFIER - [electron] (crystal rectifier whose inverse characteristic has a high-voltage maximum followed by a negative slope) redresseur *m* à crête de tension inverse
~-KEY IMAGE - [telev] image *f* lumineuse
~-LEAD BRONZE - [metall] bronze *m* à haute teneur de plomb
~LEVEL - [gen] au niveau supérieur
~-LEVEL MODULATION - [radio] (high-power modu-

lation *f* dans l'anode de l'étage final
HIGH-LEVEL RADIATION - [nucl] (radiation having a high energy level) rayonnement *m* à niveau d'énergie élevé
~-LIFT HOIST - [mech] élévateur *m* de grande amplitude, élévateur *m* à ciseaux
~-LIGHT AREAS - [photo] plages *f*pl correspondant aux hautes lumières
~-LIGHT BRIGHTNESS - [telev] luminosité *f* des blancs
~-LIGHT READING - [photo] temps de pose, détermination *f* du temps de pose
~LIGHTS - [el plating] (the region of a plated article which are most affected by polishing operations e. g. because of their position or contour) points *m*pl brillants
[cin] blancs *m*pl, plages *f*pl lumineuses
~LINE - [oil ind] câble *m* d'extraction
~-LOW CONTROL - [contr] action *f* par tout ou peu
~-LUSTRE CALENDER - s. high-glass calender
~MELTING - [metall] à point de fusion élevé
~MOORLAND - [geol] plateau *m* marécageux
~MU TUBE - [electron] (vacuum tube with high amplification factor) tube *m* à forte pente
~PASS FILTER - [telecomm] (a network designed to allow the passage of frequencies above a specific value, and to attenuate all others) filtre *m* passe-haut
~-PEAKER TEST - [telev] examen *m* des contrastes
~PERCENTAGE ALLOY - [metall] alliage *m* à haute teneur
~-PILED - [text] à long poil, "high pile"
~-PITCHED - [acoust] (of a note) aigu
~-PITCHED ROOF - [constr] comble *m* à forte pente
~PLATFORM - [railw] quai *m* haut
~POLYMER - [chem] (a polymer containing a very large number of monomer molecules) haut polymère *m*
~POTENTIAL GRINDER - [mech] broyeur *m* de grande puissance
~POWER FEEDTHROUGH - [el] sortie *f* de courant à haute intensité
~POWER LEAD-IN - s. high power feedthrough
~-POWER MAGNIFIER - [opt] loupe *f* puissante
~-POWER MODULATION - s. high-level modulation
~PRECISION - [gen] de haute précision
~PRESSURE - [phys] haute pression *f*
~-PRESSURE BOILER - [th eng] chaudière *f* à haute pression
~-PRESSURE BURNER - [th eng] brûleur *m* à mélange surpressé
~-PRESSURE CLOUD CHAMBER - [phys] (cloud chamber in which the gas is maintained at high pressure to reduce the range of high-energy particles) chambre *f* à détente à haute pression
~PRESSURE COMPRESSOR - [mech] compresseur *m* à haute pression
~-PRESSURE GAS BURNER - [burners] brûleur *m* à gaz surpressé
~-PRESSURE GAS LIGHT SOURCE - [light] foyer *m* à gaz surpressé
~-PRESSURE GASHOLDER - [gas ind] (gasholders) réservoir *m* sous pression
~-PRESSURE HEATING - [heat] chauffage *m* à haute pression
~-PRESSURE HYDRO-PNEUMATIC ACCUMULATOR - WITH DIFFERENTIAL PISTON - [el mach] accumulateur *m* hydro-pneumatique à haute pression avec piston differentiel
HIGH-PRESSURE HYDRO-PNEUMATIC ACCUMULATOR WITH FLOAT - [el mach] accumulateur *m* hydro-pneumatique à haute pression avec flotteur
~-PRESSURE INSPIRATOR BURNER - s. high-pressure
~-PRESSURE LUBRICANT - [ind chem] lubrifiant *m* pour hautes pressions
~-PRESSURE MERCURY LAMP - [el] lampe *f* à vapeur de mercure sous haute pression
~-PRESSURE MOULDING - [plast ind] (moulding at pressures exceeding 500 psi) moulage *m* à haute pression
~-PRESSURE PACKING - [mech] garniture *f* pour haute pression
~-PRESSURE RELIEF VALVE - [mech] soupape *f* de surpression
~-PRESSURE STEAM ENGINE - [mech] machine *f* à vapeur à haute pression
~-PRESSURE STEAM INLET - [text] tuyau *m* d'arrivée pour vapeur à haute pression
~-PRESSURE STEAM PROCESS REGENERATION - [rubber ind] régéneration *f* à vapeur sous haute pression
~-PRESSURE STEAM RECLAIMING PROCESS - [rubber ind] procédé *m* de régéneration à la vapeur à haute pression
~-PRESSURE TYRE - [auto] pneumatique *m* à haute pression
~-PRESSURE UNIT - [mech] (any assembly in which pressures are used) ensemble *m* pour haute pression
~-PRESSURE VALVE - [mech] robinet-vanne *m* à haute pression
~QUALITY MILK - [food] lait *m* de haute qualité
~-RADIATION FLUX - [nucl] (the time rate of flow of radiant energy of high intensity) flux *m* de rayonnement intensif
~RANGE - [mech] grande étendue *f*
~RATIO PULSING - [radar] (the ratio between the various pulse durations on the radar set) taux *m* élevé *m* d'impulsion
~RECOMBINATION RATE CONTACT - [electron] contact *m* à vitesse de recombinaison élevée
~RESISTANCE-CEMENT - [constr] ciment *m* à haute résistance
~ROAD - [gen] (in a town or village) rue *f* principale
[gen] (a highway) grande route *f*
~-SIDED OPEN WAGON - [railw] wagon *m* tombereau
~SPEED - [gen] grande vitesse *f*, marche *f* rapide
[adj] à grande vitesse, à allure rapide
~-SPEED AUXILIARY JET - [auto] gicleur *m* auxiliaire de puissance
~-SPEED CENTRIFUGE - [mech] ultracentrifugeur *m*
~-SPEED CHARGED PARTICLE - [nucl] (travelling at high velocity) particule *f* chargée rapide
~-SPEED COMPRESSOR - [mech] compresseur *m* à grande vitesse
~-SPEED COMPUTER - [comput] calculateur *m* rapide
~-SPEED DRILLING MACHINE - [mech] machine *f* à percer à grande vitesse
~-SPEED ELECTRON - [electron] (electron travelling at high velocity) électron *m* rapide
~-SPEED EMULSION - [photo] (specially prepared for short exposures) émulsion *f* rapide
~-SPEED EXTRUDER - [plast ind] (extruder in which the screw speed is 300 rpm or more) extrudeuse *f*

à grande vitesse
HIGH-SPEED FISSION NEUTRON - [nucl] neutron *m* rapide de fission
~-SPEED FLASH - [photo] éclair *m* rapide
~-SPEED ION - [nucl] ion *m* rapide
~-SPEED LENS - [photo] objectif *m* à très grande ouverture
~-SPEED LOOM - [text] métier *m* à grande vitesse, machine *f* à tisser à grande vitesse
~-SPEED NEEDLE - [auto] aiguille *f* de puissance
~-SPEED NEUTRON - [nucl] neutron *m* rapide
~-SPEED NEUTRON FISSION CROSS-SECTION - [nucl] (the cross-section for high-speed neutrons) section *f* efficace de fission pour neutrons rapides
~-SPEED OIL ENGINE - [mech] moteur *m* diesel à régime rapide
~-SPEED PARTICLE - [phys] particule *f* rapide
~-SPEED PLUNGER MOULDING - [plast ind] moulage *m* par transfert avec deux pistons
~-SPEED RELAY - [el] relais *m* ultrarapide
~-SPEED SELECTOR - [telecomm] sélecteur *m* rapide
~-SPEED STEEL - [metall] (special tool steel which does not lose its hardness at high cutting speeds) acier *m* rapide, acier *m* à coupe rapide
~-SPEED STEEL TOOL - [tool] outil *m* d'acier rapide
~-SPEED SWITCHGEAR - [el] disjoncteur *m* ultrarapide
~-SPEED TOOL - [tool] outil *m* rapide
~-SPEED TOOL-STEEL - [metall] acier *m* pour outils rapides
~-SPEED TURRET LATHE STEEL - [metall] acier *m* rapide pour tour-révolver
~-SPEED WIND TUNNEL - [aero] (subsonic wind-tunnel, in which the stream velocity is high enough to allow of observation of the compressibility of the fluid) soufflerie *f* à grande vitesse
~SPOT - [nucl] (a point where the radiation dose is appreciably above the general dose level) point *m* chaud
~-STABILITY UNLOCKED OPERATION - [telev] opération *f* à déclenchement de haute stabilité
~-STRENGTH BRASS - [metall] (brass composed of 60% copper and 40% zinc with additions of iron, manganese and aluminium to increase its strength) laiton *m* à grande ténacité
~STRESS GRINDING ABRASION - [metall] usure *f* par abrasion sous pression de contact élevée
~TAP - [el] prise *f* à haute tension
~TAR ACIDS - [ind chem] phénols *mpl* supérieurs
~TEMPERATURE - [ind chem] haute température *f*
~-TEMPERATURE BULB - [ind chem] (for gas analysis) tube *m* à haute température
~-TEMPERATURE CARBONIZATION - [gas ind] distillation *f* à haute température
~-TEMPERATURE CERAMIC MATERIAL - [nucl] (used in building nuclear reactors) matière *f* céramique résistante aux températures élevée
~-TEMPERATURE DISTILLATION - s. high-temperature
~-TEMPERATURE GAS-COOLED REACTOR - [nucl] réacteur *m* à gaz à température élevée
~-TEMPERATURE PROCESSING - [ind chem] traitement *m* à haute température
~-TEMPERATURE PYROMETER - [instr] pyromètre *m* pour hautes températures
~-TEMPERATURE REACTOR - [nucl] (reactor designed to operate with the active section at high temperature) réacteur *m* à température élevée
HIGH-TEMPERATURE RESISTING STEEL - [metall] acier *m* résistant aux hautes températures
~TENSILE BRASS - [metall] (high tension brass) acier *m* à haute résistance
~-TENSILE STEEL - [metall] (steel of specially high tensile strength) acier *m* à haute résistance
~TENSION - [el] (voltages such as are used on the anodes of electron tubes, in contradistinction to those used for cathode heating) haute tension *f*, tension *f* anodique
~-TENSION AMMETER - [instr] ampèremètre *m* à haute tension
~-TENSION BUS - [el] barre *f* collectrice de haute tension
~-TENSION CIRCUIT - [el] circuit *m* de haute tension
~-TENSION DETONATOR - [el] (obsolete form of detonator in which the charge was fired by an electric spark) exploseur *m* électrique à haute tension
~-TENSION IGNITION - [el] (for internal combustion engines) allumage *m* à haute tension
~-TENSION KEYING - [radio] (keying of a radio transmitter by interrupting the anode-supply circuit) manipulation *f* anode, manipulation *f* dans la haute tension
~-TENSION MAGNETO - [el] (the magneto used for internal combustion engines) magnéto *f* à haute tension
~-TENSION PROBE - [electron] (a component part of electronic volt-ohmmA-meter, increasing its measuring possibilities) sonde *f* à grande vitesse
~-TENSION WINDING - [el] bobinage *m* de haute tension
~TIDE - [geogr] marée *f* haute
~-VACUUM - [phys] (term used for a vessel or system which has been exhausted to a high degree) vide *m* élevé, vide *m* poussé
~-VACUUM CATHODE-RAY TUBE - [electron] tube *m* cathodique à vide élevé
~-VACUUM COATING - [metall] évaporation *f* (thermique) sous vide poussé
~-VACUUM COUPLING - [mech] raccord *m* à vide poussé
~-VACUUM CUT-OFF - [phys] fermeture *f* de vide élevé
~-VACUUM DISTILLATION - [phys] distillation *f* sous vide poussé
~-VACUUM EVAPORATOR - [metall] installation *f* de métallisation sous vide
~-VACUUM FLANGE - [mech] bride *f* pour vide poussé
~VACUUM FURNACE - [metall] four *m* à vide poussé
~VACUUM PLANT - [phys] installation *f* de vide poussé
~VACUUM PUMPING PLANT - [phys] poste *m* de pompage à vide poussé
~VACUUM SINTERED - [metall] fritté (ou sintérisé) sous vide poussé
~VACUUM SPACE - [phys] espace *m* sous vide moléculaire (ou sous vide poussé)
~VACUUM TECHNOLOGY - [phys] technique *f* du vide poussé
~-VELOCITY SCANNING - [telev] analyse *f* par électrons de haute vitesse
~VISCOUS - [chem] (exhibiting a high degree of viscosity) à viscosité élevée
~-VOLATILITY - [phys] (the property of evaporating

readily, e.g. of engine fuel) volatilité *f* élevé
HIGH VOLTAGE - [el] haute tension *f*
~-VOLTAGE COIL - [el] bobine *f* à haute tension
~-VOLTAGE CURRENT - [el] courant *m* à haute tension
~-VOLTAGE FUSE - [el] coupe-circuit *m* à haute tension
~-VOLTAGE LINE - [el] ligne *f* à haute tension
~-VOLTAGE TEST - [el] (insultation test on electrical equipment) essai *m* haute tension
~-WATER BASIN - [hydr] bassin *m* de retenue de la crue
~-WATER FLOODGATE - [hydr] déversoir *m*, évacuateur *m* de crues
~-WATER MARK - [hydr] repère *m* de crue
[of the tide] laisse *f* de haute mer
~-WATER OVERFLOW - s. high-water floodgate
~WEBBED TEE-IRON - [metall] fer *m* à T à âme allongée
~WIND - [met] grand vent *m*, vent *m* fort
~-WING - [aero] à aile haute
~-WING AIRPLANE - [aero] (aircraft in which the plane is at or near the top of the fuselage) avion *m* à aile haute
~-WING MONOPLANE - [aero] (a type of monoplane in which the plane is placed at or near the top of the fuselage) monoplane *m* à aile haute
~-YIELDING - [gen] (adj) à haut rendement, à rendement élevé
HIGHER-ORDER EQUATION - [math] équation *f* d'ordre élevé
~-ORDER DIGIT - [of a number] (comput) digit *m* de rang immédiatement supérieur
HIGHLIGHT - [telev] aire *f* de grande intensité, plage *f* lumineuse
~LUMINOSITY - [telev] luminosité *f* maximale
HIGHLY ACTIVE - [nucl] (highly radioactive) chaud, fortement radioactif
~RADIOACTIVE - s. highly active
~RADIATED PIG - [metall] fonte *f* blanche à texture franchement rayonnée
HIGHMOOR - [soil] (peaty accumulation mainly of sphagnum, which produces a mound effect) tourbière *f* haute
HIGHPASS FILTER - [comput] filtre *m* passe-haut
HIGHWAY - [roads] (main route on land or sea) route *f*, chemin *m* de grande communication, autoroute *f*
[comput] (path over which information is transferred) bus *m*
[el] (electrical conductor capable of carrying a large amount of current) barre *f* collectrice
~MAP - [topogr] carte *f* routière
HILL - [geol & geogr] colline *f*, coteau *m*
[geogr] (on a topographic map) cote *f*
[constr] (on a road) côte *f*, montée *f*, descente *f*
~AND DALE RECORDING - [el acoust] (a mechanical recording in which the modulation is in a direction perpendicular to the surface of the medium) enregistrement *m* en profondeur
~CATTLE - [agric] bétail *m* de montagne
~CLIMBING - [math] (the adjustment of a self-regulating adaptive control system to obtain an optimum result) recherche *f* de l'extrême
~-CLIMBING ABILITY - [auto] capacité *f* de gravissement des côtes
~-CLIMBING GEAR - [auto] rapport *m* de vitesse pour le gravissement des côtes
HILL FARMING - [agric] agriculture *f* montagnarde
~FOG - [met] (fog consisting of clouds lying on a hillside) cumulus-brouillard *m* de montée
~MEADOW - [agric] prairie *f* alpestre, pré *m* de montagne
~MOOR - [geol] tourbière *f* haute, fagne *f*
~PASTURE - [agric] pâturage *m* de montagne
HILLOCK-[geogr] butte *f*, petite colline *f*, tertre *m*
HILT - [gen] poignée *f*
HIND - [zool] biche *f*
[adv] postérieur, de derrière
~CARRIAGE - [mech] arrière-train *m*
~WHEEL - [mech] roue *f* arrière
HINDER, to - [gen] retarder, entravers, susciter des obstacles
HINDERED SETTLING - [chem] (method of particle separation in which natural setting is modified by an upward current of liquid) décantation *f* retardée
HINDMOST - [gen] dernier
HINDRANCE - [gen] obstacle *m*, empêchement *m*
HINGE, to - [gen] mettre les charnières, monter sur des gonds
[mech] tourner, pivoter
HINGE - [mech & carp] (mechanical device in which two parts are joined by a pin or pivot so as to be capable of relative movement) charnière *f*, fiche *f*, penture *f*
[constr]. paumelle *f*, briquet *m*
~AIR GAP - [el] entrefer *m* à charnière
~AND PIN - [mech] charnière *f* et gond
~-BLOCK - [mech] charnière *f*
~BOLT - [mech] boulon *m* de charnière
~HOLE - [carp] œil *m* de la charnière
~IRON - [constr] fer *m* à gonds
~LEVER - [mech] levier *m* articulé
~LIGAMENT - [zool] (the tough membrane connecting the two valves of a shell) ligament *m* de la charnière
~LINE - [mech] ligne *f* de charnière
~MOMENT - [aero] moment *m* de charnière
~PIN - [mech] broche *f* de charnière, cheville *f*, tourillon *m* de pivotement
~[constr] (of a band hinge) gond *m*
~SHAFT - [auto] (across the rear of a hoist or truck, a round bar on which the body rotates) arbre *m* de pivotement
~-TYPE DOOR - [ind chem] (in pressure vessels) porte *f* à charnières, porte *f* à clapet à charnière
HINGED - [gen] à charnière, articulé
~ARM - [mech] bras *m* articulé
~BACK - [photo] (of printing frame) volet *m* à charnière
~BAR - [mech & in text] tringle *f* oscillante
~BENDER - [mech] angle *m* de pliage
~BOBBIN PEG - s. hinged spindle peg
~CLOISING TRAP - [carp] trappe *f* à charnière
~CORE BOX - [metall] boîte *f* à noyaux (ou à charnière)
~DROP PIN - [text] (in a beam warping machine) aiguille-tâteur *f*
~FOLLOWER MOULD - [plast ind] moule *f* à charnière
~GUARD - [text] écran *m* protecteur mobile
~JOINT - [mech & carp] assemblage *m* à charnière
~LEVER - [mech] levier *m* à charnière
~MOULDING BOX - [metall] châssis *m* à démotter
~NOSE PIECE FOR EXTRUDER - [plast ind] tête *f* arti-

culée d'une boubineuse
HINGED PEG - [mech] cheville *f* articulée
~ REED - [text] peigne *m* suspendu librement
~ SHUTTLE PEG - [text] broche *f* de navette à clapet
~ SIDE - [auto] (of the body) ridelle *f*
~ SPINDLE PEG - [text] porte-bobine *m* à charnière
~ TYPE OF CONNECTING ROD - [mech] bielle *f* articulé
HINGING POST - [constr] (the postform which a gate is hung) montant *m* de côtière
HINTERLAND - [geogr] hinterland *m*, arrière-pays *m*
HIP - [anat] hanche *f*
[arch] (the outer angle between two intersecting roof slopes) chevron *m* d'arête
[constr] (hip rafter) arêtier *m*, chanlatte *f*
~ AND VALLEY ROOF - [constr] combles *mpl* s'intersectant
~ -BATH - [impl] demi-bain *m*
~ BONE - [anat] os *m* coxal
~ BOOT - [shoe man] botte *f* cuissarde
~ JACK RAFTER - [constr] empanon *m* d'arête
~ KNOB - [constr] (on a gable or hipped roof) boule *f* d'arête
~ OF A ROOF - [constr] pan *m* de toit, pan *m* de comble
~ RAFTER - [carp] (the rafter at the hip of a roof) chevron *m* d'arêtier, arêtier *m*
~ ROOF - s. hipped roof
~ SUPPORT - [constr] planche *f* sur l'arêtier
~ TILE - [constr] (arris-tile across the hip of the roof) tuile *f* arêtière
~ TRUSS - [constr] ferme *f* en croupe
HIPPED GABLE - [constr] pan *m* coupé
~ ROOF - [build] (plitched roof with sloping ends) comble *m* en croupe
HIPPURIC ACID - [chem] (synthesis intermediate) acide *m* hippurique
HIRE, to - [gen] louer, engager
[gen] (to hire out for a consideration) louer pour la saison, donner en location
HIRE - [gen] louage *m*, location *f*
~ CAR - [auto] voiture *f* de location
~ PURCHASE - [comm] location-vente *f*, vente *f* à tempérament
HIRST - [geol] banc *m* de sable
HISS - [radio] (a valve hiss) sifflement *m*
HISSING - [acoust] (high-pitched sound) sifflement *m*
[radio etc] crachement *m* friture *f*, crépitements
~ ARC - [el] (arc producing a hiss owing to the incorrect adjustment of the carbon electrodes) arc *m* sibilant
HISTAMINE - [chem] (a decarboxylation product of Histidine, occurring in plants and animals and used in medicine) histamine *f*
~ PHOSPHATE - [chem] (a diagnostic agent used in medicine) phosphate *m* d'histamine
HISTIDINE - [chem] (an amino acid occurring in protein) histidine *f*
HISTOCHEMISTRY - [chem] histochimie *f*
HISTOGEN - [bot] (in a plant, the region where tissues are subject to differentiation) histogène
HISTOGENESIS - [biol] (the formation of new tissues) histogénèse *f*
HISTOLOGY - [biol] (the study of the structure of tissues and organs) histologie *f*
HIT, to - [gen] frapper, donner un coup
[gen] (a target) atteindre
HIT - [gen] coup *m*
HITCH, to - [gen] accrocher, empoter, remuer par saccades
[mech] s'enfoncer (dans la pièce)
HITCH - [gen] nœud *m*, saccade *f*
[naut] nœud *m*, clef *f*
[mech] (for power take-off) prise *f* de force, raccord *m*
[mech] (of agric machines etc) accrochage *m*
[mech] (a hook) attache *f*
[geol] légère faille *f*
[mining] potelle *f*
~ BACK - [text] défaut *m* de tension
~ POINT - [mech] attache *f*
~ RAIL - [mech] (a drawbar) barre *f* d'attelage, barre *f* d'accrochage
~ ROLL - [paper man] tendeur *m* de feutre
HITCHER - [mining] encageur *m*
HITTORF DARK SPACE - [electron] (cathode dark space) espace *m* sombre de Hittorf
HIVE - [agric] ruche *f*
~ ENTRANCE - [agric] trou *m* de vol
~ STAND - [agric] rucher *m*
HJELMITE - [min] (mineral which contains stannium, tantalium, niobium yttrium, uranium, iron, calcium and manganese) hjelmite *f*
HLOPINITE - [min] (mineral which contains cobalt, titanium, radium and uranium) chlopinite *f*
H NETWORK - [radio] (network composed of five impedances) réseau *m* en H
HOARDING - [constr] palissade *f*, clôture *f* en planches
HOARFROST - [met] (opaque semi-crystalline ice formed on an aircraft when the air temperature is below the hoarfrost point) givre *m*, gelée *f* blanche
~ POINT - [met] (dew point, when the temperature is below freezing) point *m* de gelée blanche
HOARHOUND - [bot] (a medicinal plant) marrube *m*
HOARSE - [acoust] rauque, enroué
HOARSENESS - [gen] raucité *f*, enrouement *m*
HOARY - [gen] blanchi, chenu
HOB, to - [mech] (the operation of cutting with a knob) fraiser, tailler, tailler par génération (les engrenages)
HOB - [mech] (the hardened steel master tool used for hobbing) vis *f* fraise, fraise-mère *f*, taraud *m* mère
[mech] maîtresse-matrice *f*
[metall] (diecast) poinçon *m*
[impl] plaque *f* de côté
[metall] (of furnace) plateau *m* de four
~ SHARPENING MACHINE - [mech mach] affûteuse *f* pour couteaux à denter
~ SPINDLE - [mech] mandrin *m* porte-fraise
~ SWIVEL - [mach] support *m* pivotant porte-fraise
~ TAP - [tool] taraud *m* matrice, matrice *f*
HOBBED - [mech] fraisé, taillé
HOBBER - s. hobbing machine
HOBBING - [mech] (the cutting of teeth on gear blanks by means of a hob) taille *f* d'engrenages, fraisage [of a grooved shaft] taille *f* par génération
[metall] (with a hob tap) enfonçage *m*, noyage *m*
[metall] impression *f*
~ CUTTER - s. hob
~ MACHINE - [mach tool] (machine for cutting teeth on gear blanks; also for the production of worm, spur and helical gears) machine *f* à tailler les en-

grenages
HOBBING OF GEAR TEETH - [mech] taille *f* des engrenages par génération
~PRESS - [mech] (press used for forming dies with hobs) presse *f* à frapper
~STEEL - [metall] acier *m* pour frappe, acier *m* à poinçons
HOBBY - [zool] hobereau *m*
HOBNAIL - [shoe man] caboche *f*, becquet *m*
[shoe man] (for the soles) clou *m* à ferrer
~LIVER - [med] foie *m* atrophié
HOD - [impl] (open receptacle used to carry mortar; bricks etc) hotte *f*
HODOGRAPH - [mech] (curve used to measure the acceleration of a particle along a curved path) odographe *m*
HODOMETER - [instr] odomètre *m*, compteur *m* enregistreur
HODOSCOPE - [instr] (apparatus used to trace the path of a charged particle in a magnetic field) odoscope *m*
HOE, to - [agric] gouer, biner, sarcler
HOE - [impl] houe *f*, binette *f*
[mech] (a type of drag shovel) drague *f* à cuiller
[mining] sape *f*
COULTER - [agric] (coulter in the shape of a small hoe) coutre *m* à houe
~FOR CEREALS - [agric] bineuse *f* pour céréales
HOEING - [agric] houement *m*, binage *m*, sarclage *m*
~PLOUGH - [agric] charrue *f* à biner
HOFMANN'S REACTION - [chem] (a process for the preparation of primary amines by means of bromine and caustic soda) réaction *f* de Hofmann
HOG, to - [gen] prendre de l'arc, s'arquer, cintrer
[naut] (of a ship, the arching upward in the centre, due to strain) s'arquer
[naut] (the cleaning of a ship's sides with a hog) nettoyer la carène
HOG - [text] (said of wool of a year-old sheep) brebis *m* d'un an
[aero] (a hobbing girder) courbure *f* de l'axe longitudinal
[naut] (type of brush used to clean the sides of a ship) goret *m*
[shipbuild] arqûre *f*
[paper man] (used to stir the pulp) agitateur *m*
[zool] porc *m* châtré, porc *m*, cochon *m*
~BACK - [constr] (road constr etc) dos *m* d'âne, ligne *f* de crête
[geol] crête *f* isoclinale
~-BACK GIRDER - s. hogging girders
~-BACKED - [gen] en dos d'âne
~-BACKED CURVE - [photo] courbe *f* en dos d'âne
~-BACKED CUTTING - [railw] coupage *m* de terrain ondulé
~FENNEL - [bot] (a medicinal plant) peucédan *m* officinal
~-FRAME - [constr] (form of truss bulding on the upper side) ferme *f* à dos d'âne
~-NOSE DRILL - [tool] foret *m* à canon, mèche *f* demi-ronde
~-PEN - [agric] porcherie *f*
~WIREX - [metall] (barbed wire weighing approx 400 lbs per mile) fil *m* barbelé
HOGGED - [gen] arqué, cassé
[constr] (road) fortement bombée, en dos d'âne
HOGGER - [plumb] (a short piece of pipe used as connexion) tuyau *m* de raccordement
[mach tool] fraise *f* pour aciers rapides
HOGGER PUMP - [mech] (used in deep mines) pompe *f* supérieure
HOGGET - [zool] (yearling sheep) agneau *m* antenais, antenais *m*
[a year-old colt] poulain *m* d'un an
HOGGIN - s. hogging
HOGGING - [constr] (mixture of gravel and clay) gravier *m* criblé
[shipbuild] (the tendency of the keel to hog towards the bow or the stem) auqûre *f*
[mech] ébarbage *m* à la flamme
~CUTS - [mech] pièces *f*pl d'ébarbage
~FRAME - s. hog frame
~GIRDERS - [constr] (girders bulging along their top edges) poutres *f*pl en arc
HOGHORN - [radio] (elementary aerial with a smooth transition from a waveguide to a cheese) cornet *m* parabolique
HOGSHEAD - [impl] (large cask for liquids) tonneau *m*, barrique *f*
[meas] (cask containing 63 gallons) fût *m* de 240 litres
[brew ind] barrique
HOGWASH - [agric] eaux *f*pl grasses, lavasse *f*
HOGWEED - [bot] (herbs eaten by hogs) berce *f* commune
HOHLRAUM - [phys] corps *m* noir, radiateur *m* intégral
HOIST, to - [gen] (the operation of lifting materials by mechanical means) monter, remonter, lever
[naut] hisser
[naut] (to lift by means of tackle) embarquer (un canot)
[mining] (coal from the mine) extraire
HOIST - [mech] (mechanical lifting device; in particular one using a rope wound on a drum) appareil *m* de levage, treuil *m*
[naut] ralingue *f*, guindant *m*
~A SAIL, to - [naut] guinder une voile, mettre au vent
~BLOCK - [mech] moufle *f* mobile, poulie *f* mobile
~-BRIDGE - [mech] pont-grue *m*
~BUCKET - [mech] benne *f* d'extraction
~DRUM - [mech] tambour *m* du treuil
~FRAME - [mech] (the stationary structural part of a hoistresting on the chassis frame) treillis *m* du treuil
~FRAME SIDE RAIL - [mech] (the main channel enclosing the sides of a hoist mechanism) châssis *m* de charpente du treuil
~PIPE - [min] (in well sinking) câble *m* d'extraction
HOISTING - [gen & mech] levage *m*, remontée *f*, remonte *f*, hissage *m*
~APPARATUS - s. hoisting equipment
~APPLIANCE - [mech] appareil *m* de levage
~BLOCK - [mech] moufle *f* mobile
~CABLE - [mining] câble *m* d'extraction
~CHAIN - [mech] chaîne *f* de levage
~EQUIPMENT - [mech] (a crab, a jack, a hydraulic lift etc) engins *m*pl de levage, appareils *m*pl d'extraction
~MOTOR - [mech] (a lift motor) moteur *m* d'ascenseur
~PIT - s. hoisting shaft
~ROPE - [mining] câble *m* d'extraction

HOISTING SHAFT - [min] puits *m* d'extraction
~SHEAVE - [mech] molette *f*
~SLING - [of a rope] élingue *f* de levage
~STRING-UP - [mech] ensemble *m* d'élévation
~SYSTEM - [mech] système *m* de levage
~TACKLE - [mech] appareil *m* de hissage
~WINCH - [mech] treuil *m*
HOISTAWAY - [mech] (only USA; mechanical lift or elevator) monte-charge *m*
HOLD, to - [gen] tenir, tenir ferme, contenir, maintenir
[mech] serrer
[electron] charger
HOLD - [gen] prise *f*, soutien *m*, point *m* d'appui
[naut] cale *f*
[mech] blocage *m*
[comput] (information in a storage device which is retained by copying it into a second storage device) lecture *f* non-destructive
[contr] valeur *f* de maintien nominale
~A CALL, to - [telecomm] retenir an appel
~A CIRCUIT, to - [telecomm] bloquer un circuit
~BACK, to - [gen] retenir, rester en arrière
~-BACK AGENT - [nucl] (the inactive isotope of a radioactive agent) inhibiteur *m* d'entraînement
~-BACK CARRIER - s. hold-back agent
~-BACK SPROCKET - [cin] (a sprocket lodged between the film reels and the intermittent sprocket) tambour *m* inférieur
~BEAM - [shipbuild] barre *f* sèche
~CONTROL - [telev] (to adjust the synchronization of the electron-beam deflection) commande *f* de synchronisme
~CURRENT - [el] (the minimum current which is required to maintain a relay in an operating position) courant *m* de maintien
~-DOWN BOLT - [mech] (for machines etc) boulon *m* di fixation, boulon *m* d'assujetissement
~-DOWN CLAMP - [mech] (metal jaw which descends on the work in advance of the blade to hold it during shearing) serre-flan *m*
~-DOWN GROOVE - [mech] cannelure *f* de retenue
~-DOWN PLATE - [mech] plaque *f* de serrage
~-DOWN ROLL - [paper man] rouleau *m* compensateur
~-DOWN TEST - [astronaut] essai *m* au banc
~FAST, to - [gen] tenir ferme, tenir bon
~-FAST NAIL - [mech] détente *f*, doigt *m* d'encliquetage
~IN RANGE - [telev] zone *f* de l'enclenchement
~-OFF VOLTAGE - [electron] tension *f* maximale de blocage
~ON COIL - [el] (electromagnet holding the moving arm of a motor starter) relais *m* de démarrage, relais *m* de mise en marche
~RANGE - [telev] (the frequency range over which the oscillator will remain synchronized when the synchronization frequency is altered) bande *f* de synchronisation
~TAKE - [cin] (a possibly useful shot) prise *f* de vue de réserve
~THE RAVEL, to - [text] tenir le vautoir
~THE WIRE WITH THE CLAMP, to - [text] maintenir le fil métallique par des mâchoires
~TIME - [metall] (in a welding operation) durée *f* de la pression
~-UP - [nucl] (the quantity of material retained in a separation plant: commonly, the amount of the desired isotope) charge *f* ouvrable
HOLDER - [gen] teneur *m*, détenteur *m*
[gen] (of object) poignée *f*
[gen] (of a licence etc) titulaire *m*, concessionnaire *m*
[comm] (a lamp holder) porte-lampe *m*
[fin] (of check etc) accrédité *m*
[gas ind] (abbrev for gasholder) gazomètre *m*
[gen] (a vessel) récipient *m*
~CAPACITY - [gas ind] (gasholder) capacité *f* gazométrique
~FOR END OF WEFT YARN - [text] support *m* de bout du fil de trame
~FOR TOP RADDLE PLATE - [text] goupille *f* pour le chapeau du vautoir
~HEATERS - s. holder heating system
~HEATING SYSTEM - [gas ind] (gasholder) dispositif *m* de chauffage des gorges
~TANK SHELL - [gas ind] (gasholder) robe *f* de la cuve, manteau *m*
~UP - [mech] contre-bouterolle *f*
HOLDERBAT - [impl] (metal collar in two half-round parts, used for pipes) collier *m* d'attache
HOLDFAST - [carp] (wrought-iron spike) valet *m* de établi, valet *m* de menuisier
[impl] (a type of nail) clou *m* à patte
~COUPLING - [text] accouplement *m* par manchon à frettes
HOLDING ALTITUDE - [aero] altitude *f* d'attente
~ANODE - [electron] électrode *f* d'accumulation
~BEAM - [electron] (diffuse beam of electrons for regenerating the changes of an electrostatic tube) faisceau *m* d'accumulation
~BRAKE - [el] freinage *m* de maintien
~CIRCUIT - [telecomm] circuit *m* de maintien, circuit *m* de collage
[electron] (in a relay network) circuit *m* de maintien
~COIL - [el] enroulement *m* de blocage
~CONTROL - s. hold control
~CORD - [text] cordon *m* de retenue
~CYCLE - [el] phase *f* de maintien
~DETENT - [telecomm] (of telephone exchange) cliquet *m* d'arrêt
~DEVICE - [text] support *m*, dispositif *m* de soutien
~DOG - s. holding detent
~-DOWN BOLT - [mech] boulon *m* de fixation
~-DOWN SPRING - [horol] ressort *m* de fixation
~INTERLOCK - [mech] blocage *m* de maintien, verrouillage *m* de maintien
~KEY - [telecomm] clé *f* de garde
~MAGNET - [el] électro-aimant *m* de retenue
~POINT - [aero] (a point which is readily identified, in the neighbourhood of which an aircraft is instructed to remain) point *m* d'attente
~PROCEDURE - [aero] (the specified flight track of an aircraft near a holding point) circuit *m* d'attente
~PUMP - [mech] (to maintain operating pressure) pompe *f* d'entretien
~ROD - [text] tringle *f* de fixation
~SCREW - [mech] vis *f* de fixation
~TANK - [hydr] bac *m* intercepteur
~TIME - [telecomm] (the total time a trunk line is in use for one call) durée *f* d'occupation
[electron] temps *m* de maintien
~UP - [mech] (the action of holding a hammer against the head of a rivet) suspension *f*

HOLDING VOLTAGE - [el] tension *f* de retenue
~VACUUM - [mech] vide *m* d'entretien
~WINDING - [el] eroulement *m* de maintien
HOLE, to - [gen] trouer, percer
[mining] haver, sous-caver, souchever
HOLE - [gen] trou *m*, œillard *m*, orifice, ouverture *f*, creux *m*
[mining] (a bore-hole) trou *m* de sonde
[constr] (any depression made for a blasting charge) fosse *f*, sondage *m*
[gen] (on a road surface) trou *m*, cavité *f*
[mining] (a pit) puits *m*, fouille *f*
[opt] (of a sight) lumière *f*
[electron] (space left in a crystal lattice by an electron which has gone on to another position) lacune *f*, trou *m*
[radar] (part of the scan from which no reflection is obtained) zone *f* morte
[oil ind] puits *m* sterile
~AND SLOT MAGNETRON - [electron] magnétron *m* à rainures et trous
~BOARD - [text] planche *f* à trous, planche *f* d'arcades
~-BORING CUTTER - [mach tool] alésoir *m* à fraise
~CALIPER - [mech] diamétreur
~CAPACITOR - [electron] condensateur *m* à semiconducteur
~CAPTURE - [electron] (the capture of a hole by an impurity other than the foreign atom) capture *f* de trous
~CONCENTRATION GRADIENT - [telecomm] gradient *m* de concentration de lacunes
~CONDUCTION - [electron] conduction *f* par trous
~CURRENT - [electron] (in a semiconductor) courant *m* de trous
~GAUGE - [mech] calibre *m* pour trous
~IN THE CARD - [text ind] trou *m* dans le carton
~INJECTION - [electron] (the emission of holes at the surface of a semiconductor when a sharp metallic point is applied) injection *f* de trous
~MAKER - [tool] (a hollow punch) découpoir *m* pour trous
~MOBILITY - [electron] (in a semiconducteur) migration *f* de trous
~PLANTING - [agric] plantation *f* en trous
~STORAGE EFFECT - [electron] (in germanium diodes, which may require some microseconds before attaining a specified inverse resistance) capacité *f* de diffusion
~THEORY OF LIQUIDS - [phys] théorie *f* des cavités des liquides
~THROUGH SPINDLE - [mech] largeur *f* du mandrin
~TRAP - [electron] (in semiconductors) piège *m* de trous
HOLES IN SPACED ROWS - [text] trous *mpl* en quinconce
~IN THE TRAP BOARD - [text ind] trous *mpl* dans la planche à cordons
HOLIDAY - [min] (colloq. USA only; section of pipe on which paint is missing) section *f* de tuyau sans couche de peinture
HOLING - [gen & constr] percement *m*
[agric] préparation *f* des trous pour plantation
[mining] havage *m*, souchevage *m*, sous-cave *f*, percement *m* (du massif)
HOLLANDER - [paper man] (a beating engine) pile *f* défileuse

HOLLOW - [gen] creux *m*, cavité *f*, bas-fond *m*
[adj] creux, évidé
[acoust] sourd
[carp] cannelure *f*
[build] excavation *f*
[metall] tube *m* ébauché
[geol] creux *m* de terre, anfractuosité *f*, dénivellation *f*
~ABUTMENT - [build] culée *f* décomposée, culée *f* creuse
~ANODE - [electron] (of tubular shape) anode *f* cylindrique
~ARM - [text] support *m* creux
~-BACK - [med] lordose *f*
~BAND - [text] ruban *m* creux
~BAR - [hydr] barreau *m* creux
~BLOCK - [constr] (hollow blocks or tiles used especially for floors) brique *f* creuse
~BRICK - s. hollow block
~-CAST FURNITURE - [metall] lingots *mpl* en fonte, garniture *f* creuse
~CAST-IRON PILLAR - [constr] colonne *f* creuse en fonte
~CASTING - [metall] fonte *f* creuse
~CATHODE - [electron] (of tubular shape) cathode *f* creuse, cathode *f* cylindrique
~CLAY TILE - [constr] (hollow burnt-clay tile) brique *f* perforée
~COLUMN - [build] colonne *f* creuse
~CONCRETE FLOOR - [constr] hourdis *m* en béton
~COP - [text] cocon *m*, cannette *f* sans tube,
~-CORE SCREW - [plast ind] (type of extruder screw in which there is a passage along the longitudinal axis for temperature control) vis *f* à corps creux
~DRILL STEEL - [metall] acier *m* pour forets
~FIBRE - [text] fibre *f* creuse
~FORGING - [metall] forgeage *m* creux, forgeage à mandrin
~FUSEE - [horol] (fusee with the top pivot sunk into the body, thus reducing the height of the movement) fusée *f* creuse
~GAUGED BRICK - [constr] brique *f* de hourdis
~GRAVITY DAM - [hydr] barrage *m* évidé
~-GROUND - [mech] (term used of a saw in which the blade is ground so as the thinner than the width of cut, to avoid binding) creusé, évidé à la meule
~IRON - [metall] tube *m* en fer
~MANDREL LATHE - [mach tool] (used for repetition work) tour *m* mandrin creux
~MILL - [mach tool] fraise *f* creuse
~MOULD - [plast ind] moule *f* à couler
~MOULDING - [constr] (latte à) cavet *m*
~NEWEL - [constr] (of a winding stair) noyau *m* creux
~OUT, to - [gen] creuser, évider, creuser
~PART OF THE SHUTTLE - [text] fosse *f* de la navette
~PIECE - [gen] pièce *f* creuse, corps *m* creux
~PIN - [mech] goujon *m* tubulaire
~PINION - [mech] (drilled throughout its length) pignon *m* creux
~PISTON - [mech] piston *m* creux
~PROFILE - s. hollow tread
~PUNCH - [tool] emporte-pièce *m*, découpoir *m*
~RAM PUMP - [hydr] (an abyssinian driven well) puits *m* instantané, puits *m* abyssin
~ROD - [mech] tige *f* creuse
~SHAFT - [mech] arbre *m* creux
~SCREW - [mech] (a screw pierced axially with a

hole) vis *f* creuse
HOLLOW SPACE - [constr] espace *m* creux, cavité *f*
~SPAR - [min] andalousite *f*
~SPINDLE - [tool] mandrin *m* percé de part en part
~-SPUN - [constr] (of concrete) centrifugé
~STEEL SPOKE WHEEL - [auto] roue *f* à rayons d'acier creux
~TILE - s. hollow clay-tile
~TREAD - [rubber ind] (tyres) sculpture *f* creuse
~TRUNNION - [mech] tourillon *m* creux
~WALL - [constr] (a cavity wall) mur *m* creux
~WARP BEAM - [text] ensouple *f* à chaîne creuse
HOLLOWS - [min] (abandoned mine) vieux travaux *mpl*
HOLLOWED OUT - [gen carp etc] creusé, cavé
HOLLY - [bot] houx *m*
HOLLYHOCK - [bot] rose *f* trémière
HOLM OAK - [bot] yeuse *f*, chêne *f* verte
HOLMIUM - [chem] (a rare earth element, symbol Ho, A.N. 67, A.W. 163.5) holmium *m*
~OXIDE - [chem] (a refractory material) oxyde *m* de holmium
HOLOAXIAL - [crystal] (of those crystals which are characterized by axes of symmetry) holoaxe
HOLOCARDIUS - [med] holocarde *m*
HOLOCRINE - [med] glande *f* holocrine
HOLOCRYSTALLINE ROCK - [geol] (igneous rock in which all the components are crystalline) roche *f* holocristalline
HOLOGAMY - [genet] (fusion between two mature cells) hologamie *f*
HOLOHEDRAL - [geom] holoèdre *m*
~CRYSTAL - [crystall] (crystal in which all faces are developed) cristal *m* holoédrique
HOLOHEDRIC - s. holohedral
HOLOMETER - [instr] (instrument designed to measure the elevation angle of a spot above the horizon) holomètre *m*
HOLOPHYTIC - [zool & bot] (able to manufacture food from the simplest beginnings) holophytique
HOLOSCHISIS - [med] amitose *f*
HOLOSTERIC - [phys] holostérique
HOLOTETANUS - [med] tétanos *m* généralisé
HOLOTONIA - [med] hypertonie *f* musculaire généralisée
HOLSTER - [gen] fonte *f*
[firearms] étuis *m* de revolver
[mech] (of mill rolls) cage *f*, colonne *f*
HOLYSTONE - [naut] (soft sandstone used for scouring the decks of a ship) brique *f*
HOMATROPINE - [chem] (a mydriatic employed in medicine) homatropine *f*
HOME CONTACT - [mech] contact *m* de repos
~DEPOT - [railw] dépôt *m* d'attache
~-GROWN - [agric] indigène, du pays
~LOT - [cin] (colloq. for film studio) studio *m* cinématographique
~ON, to - [aero] se diriger
~POSITION - [instr] (the zero position) position *f* de repos
[comput] position *f* initiale
~-PRODUCED - [gen] du pays
~SIGNAL - [railw] (stop signal) signal *m* d'arrêt
~STATION - [railw] gare *f* d'attache
~SWEET HOME - [radar] (colloq. for H_2S system) radar *m* du type H_2S
~WEAVING - [text] tissage *m* à domicile
HOMENERGIC FLOW - [phys] (type of flow in which the total of the potential and kinetic energies, and the enthalpy, are respectively the same throughout the fluid, and continue at a constant level) flux *m* homénergétique
HOMENTROPIC FLOW - [phys] (flow in which entropy is at the same level throughout the fluid, and remains constant) flux *m* homentropique
HOMEOGRAFT - [med] homéogreffe *f*, homoplastie *f*
HOMESPUN - [text] (spun at home) de fabrication domestique
~LINEN - [text] toile *f* de ménage
HOMESTEAD - [agric] ferme *f*
HOMEWARD BOUND - [aero, naut etc] retournant au port
~JOURNEY - [gen & railw] (return journey) voyage *m* de retour
~RUN - [railw] parcours *m* d'entrée
HOMING - [radar] (procedure for bringing in an aircraft by directional radio methode) radioralliement *m*, homing *m*
[contr] retour *m* au repos
~ACTION - [telecomm] (in telephony) retour au repos d'un sélecteur
~AIDS - [radar] (systems designed for the guidance of aircraft to an airfield or aircraft carrier) dispositifs *mpl* radioralliement
~DEVICE - [radar] appareillage *m* de radioralliement
~GUIDANCE - [radar & radio] radioguidange *m*
~INDICATOR - [radar] indicateur *m* de homing
~RECEIVER - [radio] (radio apparatus designed to indicate wheter an aircraft is holding the homing flight path) récepteur *m* de radioralliement
~RELAY - [el] relais *m* retournant
HOMOCENTRIC RAYS - [opt] (with the same focal point) rayons *mpl* à foyer commun
HOMOCHROMATIC - [opt] homochromatique
~RADIATION - [radiat] radiation *f* homochromatique
HOMOCYCLIC - [chem] (term used to describe cyclic compounds which contain a ring consisting wholly of atoms of the same element) homocyclique
~COMPOUNDS - [chem] composés *mpl* homocycliques
HOMODROMOUS - [med] homodrome
HOMODYNAMIC - [biol] homodynamique
~HYBRID - [genet] (hybrid with equal grouping of characters derived from both parents) hybride *f* homodyname
HOMODYNE - [radio] homodyne
~RECEPTION - [radio] (reception with an oscillating valve adjusted to the same frequency as the carrier of the incoming signal) réception *f* homodyne
HOMOEOPATHY - [med] (a system of medicine) homéopathie *f*
HOMOEOSIS - [genet] (type of variation) homéose *f*
HOMOEROTIC - [med] homosexuel
HOMOGAMY - [genet] (inbreeding, due to a variety of causes but principally to isolation) homogamie *f*
HOMOGENEITY - [gen & chem] (the quality of being homogeneous) homogénéité *f*
HOMOGENEOUS - [gen] homogène
~BODY - [phys] corps *m* homogène
~EQUILIBRIUM - [phys] (conditions of equilibrium in a system constituting a single phase) équilibre *m* homogène
~FLUID - [phys] (a fluid having properties which are the same at all points) fluide *m* homogène
~FLUIDIZATION - [chem] (for chemical purification

purposes) fluidisation *f* homogène
HOMOGENEOUS LIGHT - [opt] lumière *f* monochromatique
~PROPELLANT - [chem] propergol *m* homogène
~RADIATION - [nucl] (radiation with a very narrow band of frequencies) rayonnement *m* homogène
~REACTOR - [nucl] (a reactor in which the fissile material and moderator are combined in a mixture in such a way that a homogeneous medium is presented to the neutrons) réacteur *m* homogène
~SOLUTION REACTOR - [nucl] (a reactor in which uranium is uniformly dispersed in the aqueous solution) réacteur *m* type solution homogène
~STEEL - [metall] acier *m* homogène
~SYSTEM - [phys] (with only one phase) système *m* homogène
~X-RAYS - [radiat] (X-rays of a single frequency or of a very narrow band of frequencies) rayons *mpl* X homogènes
HOMOGENIZER - [ind chem] homogénisateur
HOMOGENIZING - [ind chem] (producing or tending to produce a homogeneous mixture) homogénisation *f*
[metall] (through special annealing) recuit *m* d'homogénisation
HOMOGRAPHY - [geom] (the same anharmonic ratio) homographie *f*
HOMOKERATOPLASTY - [med] homokératoplastie *f*
HOMOLOG - s. homologous
HOMOLOGOUS - [gen & math] (having the same ratio or relative value) homologue
[chem] (said of a series of compounds which differ in composition by a specified amount of certain constituents) homologue
~CHROMOSOMES - [biol] (chromosomes in which the same gene loci occur in the same sequence) chromosomes *mpl* homologues
~SERIES - [chem] (a series of organic compounds having a similar structure, each succeeding compound having one more CH_2 group than the previous one) série *f* homologue
HOMOLOGUES - [chem] (a series of compounds having a similar structure, each successive member differing in composition byCH_2 and any two members differing by a multiple of CH_2) homologues *mpl*
HOMOLOGY - [gen] (homologous quality or condition) homologie *f*
HOMOMETRIC PAIRS - [crystall] (a pair of crystal structures with the same X-ray diffraction pattern) paires *fpl* homométriques
HOMOMORPHS - [chem] (molecules of similar size and shape) homomorphes *mpl*
HOMONUCLEAR MOLECULE - [phys] (a molecule whose atoms have identical nuclei) molécule *f* homonucléaire
HOMOPHONE - [mus] (in the harp, two strings which are tuned to produce the same note) homophone
HOMOPHONY - [acoust] (music performed in unison; also, the opposite of poliphony) homophonie *f*
HOMOPLASY - [genet] (the similarity between two different organisms) homoplasie *f*
HOMOPOLAR - [el chem] (term used to denote an equal distribution of charge, as in the case of a covalent bond) homopolaire
[el] (unipolar) homopolaire, unipolaire
~BOND - [phys] (a type of linkage between atoms, each atom contributing one electron to a shared pair which forms a chemical bond) liaison *f* covalente, liaison *f* homopolaire
HOMOPOLAR COLLOID - [chem] (colloidal system in which the dispersed particles are non-polar compounds) colloïde *m* homopolaire
~DYNAMO - [el] dynamo *f* homopolaire
~FIELD MAGNET - [el] (field magnet in which the poles passing successively before a point in the armature are of like polarity) inducteur *m* homopolaire
~GENERATOR - [el] génératrice *f* unipolaire
~INDUCTION - [el] induction *f* unipolaire
~MACHINE - [el] (unipolar machine) machine *f* acyclique, machine *f* unipolaire
~WINDING - [el] enroulement *m* unipolaire
HOMOPOLYMER - [chem] (a polymer produced from a single monomeric material) homopolymère *m*
HOMOSEXUAL - [med] homosexuel
HOMOTAXIS - [geol] (of two strata in different areas sharing the same faunal characters) homotaxie *f*
HOMOTHETIC - [phys] (having a symmetrical direction) homothétique
HOMOZYGOSIS - [genet] (the condition whereby a specified genetic factor is inherited from both parents) homozygotie *f*
HOMOZYGOUS - [genet] (which is derived from germ cells which are genetically identical) homozygote
HONE, to - [mech] (to produce a very fine dinish by an abrasive process) affiler, aiguiser
[mech] (the action of sharpening a tool or a blade) rectifier
HONE - [impl] (smooth stone, designed to give a sharp edge to a cutting tool) pierre *f* à aiguiser, pierre *f* à huile, affiloire *f*
HONESTONE - s. hone
HONEY - [agric] miel *m*
~BEE - [zool] abeille *f* domestique
~CLOVER - [bot] mélilot *m* blanc
~-COMB - [zool] rayon *m* de miel, gaufre *f* de miel
~COMBED ROCK - [geol] roche *f* alvéolaire
~COMBING - [metall] formation *f* de soufflures
~DEW - [bot] (plant disease) miellat *m*
~EXTRACTOR - [agric] mello-extracteur *m*
~FLOW - [agric] miellée *f*
~HARVEST - [agric] récolte *f* de miel
~PLANT - [bot] plante *f* mellifère
~SEPARATOR - [agric] extracteur *m* de miel
~-STONE - [min] mellite *f*
~YIELD - [agric] rendement *m* en miel
HONEYCOMB - [gen & text] nid d'abeilles, alvéolaire
[aero] (a system of intersecting surfaces resembling the cell-walls in a comb of honey; used e.g. for the guidance of an airstream) filtre *m* en nid d'abeilles
~CASTING - [metall] fonte *f* alvéolaire
~COIL - [el] (a type of inductance coil in which the wire is wound in a crisscross fashion) bobine *f* en nid d'abeilles
~CORE - [phys] matériau *m* en nid d'abeilles
~LENS - [photo] lentille *f* à cloisonnement cellulaire
~RADIATOR - [mech] (i.c. engine radiator consisting of tubes arranged in honeycomb) radiateur *m* alvéolaire
~RINGWORM - [vet] teigne *f* faveuse
~SANDWICH - [plast ind] (structure in which resin-impregnated sheet is formed into cells and enclosed in external layers of like material) panneau *m* sandwich en nid d'abeilles

HONEYCOMB STRUCTURE - [plast ind] (formed of numerous small cells of equal size side by side as in a honeycomb) sandwich *m* nid d'abeilles
[geol] structure *f* alvéolée
HONEYSUCKLE - [bot] chèvrefeuille *f*
HONING - [mech] (the process of finishing cylinder bores) affilage *m*, repassage, rectification *f*, rectification *f* intérieure
~DEVICE - s. honing fixture
~FIXTURE - [tool] outil *m* à repasser, affiloire *f*
~MACHINE - [mech] (machine used to hone to bores of cylinders) polisseuse, machine *f* à pierrer
~STONE - s. hone
~TOOL - [tool] affiloire *f*
HOOD - [gen] capuchon *m*, capot *m*
[constr] (a chimney cowl) mitre *f*, capote *f* de cheminée
[el] (for cables) cloche *f* d'isolateur
[auto] capote *f*, soufflet *m*
[metall] hotte *f*, auvent *m* de forge
[auto] (USA only) panneau *m* ouvrant de capot
[auto] (a roof made of canvas) soufflet *m*
[naut] (canvas covering) capot *m* (d'habitable etc)
[radar] (protection againstlight when looking at the screen of the cathode-ray tube) abat-jour *m*
[nucl] (stainless steel structure used for low-level energy work with radioactive liquids) coupole *f*, dôme *m*
[ind chem] auvent *m* de laboratoire
~CATCH - [auto] (only USA) attache-capot *m*, verrou *m* de capot
~FABRIC - [text] étoffe *f* pour capote, tissu *m* pour capote
~FASTENER - [auto] (only USA) serrure *f* de capot
~FOR LENS - [opt] parasoleil *m* pour objectif
~LATCH - s. hood fastener
~LOUVRE - [auto] fente *f* d'aération de capot
~REST STRIP - [auto] sangle *f* d'appui de capot
~SAFETY CATCH - [auto] attache-capot *m*
~STRAP - [auto] sangle *f* de capot
HOODED ANODE - [electron] (type of anode in which the target is placed in a metal hood which intercepts electrons from the focus) anode *f* à cible inserée
~BARLEY - [agric] orge *f* trifurquée
HOODING DUCK - [text] toile *f* de capote
HOOF - [zool] sabot *m*
~-AND-MOUTH DISEASE - s. foot-and mouth disease
~-BOUND - [vet] encastelé
HOOK, to - [gen] accrocher, suspendre, agrafer
[fishing] accrocher, prendre à l'hameçon
[naut] gaffer, crocher
HOOK - [mech] croc *m*, crochet *m*, agrafe *f*
[knitting implement] crochet *m*
[fishing impl] hameçon *m*
[agric] faucille *f*, échardonnet *m*
[constr] (the pin on which a door is hung) crochet *m* de suspension
[geogr] cap *m*, coude *m* d'une rivière
~AND EYE - [constr] crochet *m* à pitons
~AND HINGE - [mech] gond *m* et penture
~BAR - [text] barre *f* à crochets
~BLOCK - [mech] poulie *f* à croc
~BLOCK ASSEMBLY - [metall] moufle à crochet combiné
~BOLT - [mech] (used for fixing corrugated sheets) boulon *m* à crochet, boulon *m* à croc
~CLIP - [mech] pince *f* à crochet

HOOK COMB - [text] peigne *m* à crochets
~END - [text] tête *f* de la platine
~EXTENSION - [text] pièce *f* d'allongement de la platine
~FOR TWILLING BARS - [text] platine *f* des tringles
~FOR WITHDRAWING THE WIRES - [text] crochet *m* de retrait de fers
~FOUNDRY NAIL - [metall] crochet *m* pour mouleurs
~GAGGER - [metall] armature *f* à crochet
~GAGGERS - [metall] tirette *f* de la hotte
~-GEAR - [mech] (valve motion) bec-de-cane *m*
~GUIDE - [text] guide-crochet *m*
~HEAD - [text] tête *f* de la platine
~KEY - [text] clef *f* à crochet
~LEVER - [text] levier *m* des crochets, levier *m* à crochet
~LOOM FOR PLUSH FABRICS - [text] métier *m* à crochets pour tissus pelucheux
~NEEDLE - [text] aiguille *f* à crochet
~ON, to - [gen] accrocher, accrocher à, suspendre à
~-ON METER - [instr] (instrument for electrical tests) pince *f* ampérométrique
~ROD - [text] tringle *f* d'arrêt; tringle *f* à crochet
~SPANNER - [impl] griffe *f*
~STICK - [el] perche *f* isolante
~SWITCH - [el] crochet *m* commutateur
~TRANSISTOR - [electron] (junction transistor using an extra p-n junction to act as trap for holes) transistor *m* à dépression, transistor *m* à jonction en piège
~-UP, to - [radio etc] conjuguer
[el] mettre dans le circuit
~-UP - [el] (temporary connexion) montage *m* provisoire
[radio] schéma *m* de montage
[gen el] connexion *f*
~-UP ATTACHMENT - [text] dispositif *m* d'accrochage des griffes à revers
~-WALL PACKER - [oil ind] remblayeuse *f* pour trou non tubé
~WRENCH - s. hook spanner
HOOKE'S COUPLING - [chem] (universal joint consisting of two forks connected 90 degree a part by a cross-shaped member on which the forks can pivot) joint *m* universel
~JOINT - s. hook's coupling
~LAW - [phys] (within the elastic limit of an elastic body the strain is proportional to the stress producing it) loi *f* de Hooke
HOOKED NEEDLE - [text] crochet *m*
~PEG - [text] broche *f* à crochet
~PIN - [text] aiguille *f* à crochet, cheville *f* à crochet
~ROD - [text] tige *f* à crochet
~WIRE - [text] crochet *m*
HOOKER - [naut] (two-masted coasting or fishing vessel) hourque *f*
~CELL - [el chem] (an electrolytic cell for the production of sodium hydroxide and chlorine from sodium chloride) élément *m* d'Hooker
HOOP, to - [gen] cercler, relier, fretter
HOOP - [gen] cercle *m*, cerceau *m*
[mech] anneau *m*, bague *f*
[mech] (of a wheel) jante *f*, bandage *m*
[meas] (ring round a pot, to indicate a measure) anneau *m*
[text] (ring of whalebone or steel, used to expand

a woman's dress) cercle *m* de baleine, vertugadin *m*
HOOP - [firearms] (of a gun] frettage *m*
[firearms] (on a gun jacket) embrassure *f*
[aero] (load ring) cerce *m*
[gen] (of a barrel) sommier *m*
~BINDING - [agric] cerclage *m*
~DAMPER - [aero] (of landing gear; damping device to prevent excessively rapid vertical movement of a bogie) amortisseur *m* d'atterrissage
~-DROP RELAY - [electron] régulateur *m* à étrier mobile
~IRON - [mech] (strip iron used to secure barrels etc) fer *m* feuillard, feuillard *m* de fer
~OF A PILE - [constr] frette *f* d'un pieu, frette *f* d'une palplanche
~STRESS - [mech] (tensile stress in the shell of a pressure vessel) tension *f* tangentielle
HOOPES PROCESS - [el chem] (an electrolytic process for the purification of aluminium) procédé *m* de Hoopes
HOOPING - [gen & mech] cerclage *m*
[constr] (reingorcing bars for ferroconcrete) bandage *m*
HOOSIER POLE - [plumb] (used in mainlaying) balancier *m*
HOOTER - [auto] trompe *f*, klaxon *m*
[gen] (of factories etc) sirène *f*
HOP, to - [gen] sauter, sautiller
[brew ind] houblonner, cueillir le houblon
HOP - [gen] sautillement *m*
[aero] (colloq. USA only) vol *m*
[bot] s. hops
[radio] (transmission path embodying a single reflection from the ionosphere to the earth) trajet *m* de réflexion
[telecomm] (of radio link) section *f* radioélectrique
~AROMA - [brew ind] arôme *m* de houblon
~BACK - [brew ind] (a hop strainer) panier *m* à houblon
~CAPSID - [zool] (a pest) punaise *f*
~CLOVER - [agric] petit trèfle *m* d'or
~CONE - [bot] cloche *f* de houblon, cône *m* de houblon
~CULTURE - [agric] culture *f* de houblon
~DOWN, to - [brew ind] houblonner
~DOWNY MILDEW - [bot] mildiou *m*
~EXTRACT - [brew ind] extrait *m* de houblon
~EXTRACTING APPARATUS - [brew ind] laveur *m* de houblon
~FLEA BEETLE - [zool] (a pest) altise *f*
~GARDEN - [agric] houblonnière *f*
~GROWER - [agric] houblonnier *m*
~GROWING - [agric] culture *f* de houblon
~MILDEW - [bot] (plant disease) oïdium *m* du houblon
~MILL - [brew ind] effeuilleuse *f* à houblon
~NOSE - s. hop aroma
~OIL - [brew ind] huile *f* de houblon
~PICKING - [agric] cueillette *f* du houblon
~POCKETING - [text] toile *f* pour sac à houblon
~POLE - [agric] perche *f* à houblon
~RATE - [brew ind] houblonnage *m*
~RESIN - [agric] résine *f* du houblon
~SACKING - [text ind] s. hop pocketing
~STORAGE - [brew ind] magasin *m* de houblon
~STRAINER - s. hop back
~STRIG - [agric] rachis *m* de houblon
HOP TANNIN - [chem] tanin *m* de houblon
~TREFOIL - [bot] petit trèfle *m* d'or
~VINE - [agric] tige *f* de houblon
~YARD - s. hop garden
HOPPED WORT - [brew ind] moût *m* hoblonné
HOPPER - [impl] (hollow conical structure used to feed loose material into an opening) trémie *f*, hotte *f* (d'un moulin)
[hydr] (a water tank in sanitary fitments) chasse *f* d'eau
[min] (a bin for broken ore) trémie *f*
[metall] (of blast furnace) cup *m*
[agric] semoir *m*
[mus] (of a piano) échappement *m* du marteau
[comput] magasin *m* d'alimentation (en cartes), chargeur *m* de cartes
~BALE BREAKEN - [text] ouvreuse *f* brise-balles
~BARGE - [transp] (vessel divided into compartments to convey dredged material) marie-salope *f*, bac *m* à vase
~BODY - [auto] (a non-dumping type of load carrier) tombereau *m* à déchargement par gravité
~-BOTTOM CAR - s. hopper car
~-BOTTOM TANK - [hydr] bassin *m* à fond pyramidal
~CAR - [railw] wagon-trémie *m*
~CHAIN - [mech] (in a conveyor, a dredger etc) chaîne *f* de trémies
~CLOSET - [hydr] (sanitary equipment) bassin *m* de décantation pyramidal
~DRYER - [mech] (a feed hopper supplied with an upward current of warm air for pre-drying of the material fed) trémie *f* de séchage préliminaire
~FEED - [text] boîte *f* d'alimentation
~FEEDER - [text] chargeuse *f* alimenteuse *f*
~GONDOLA CAR - s. hopper car
~HEAD - [constr] cuvette *f*, hotte *f*
~LOADER - [mech] (a duct for pneumatic conveyance of moulding powder from shipping drums to moulding press feed hoppers) chargeur *m* à trémie
~VIBRATOR - [mech] (device for imparting vibration to a feed hopper, to facilitate the delivery of material from it) vibrateur *m* à trémie
~WAGON - s. hopper car
HOPS - [bot] (the ripe cones borne by the female hop plant, used for giving bitterness to beer) houblon *m*
HOPSACK WEAVE - [text] armure *f* panama
HORDENINE - [chem] (an alkaloid having a similar action to that of adrenaline and used in medicine) hordénine *f*
HORIZON - [astr] horizon *m*
[opt] (as experienced subjectively by the blind) horizon *m*
[surv] (the plane which is perpendicular to the direction of gravity at the point of observation) horizon *m*
~DISTANCE - [surv] rayon *m* d'horizon
~GLASS - [instr] (of a sextant) miroir *m* du sextant (the sextant itself) sextant *m*
~LIGHTS - [signal] (luminous beacons showing reference points on the ground for the information of air pilots) feux *mpl* d'horizon
HORIZONTAL - [gen & astr] (in the plane of the horizon) horizontal
~ANCHORAGE - [telecomm] ancrage *m* horizontal
~ANGLE - [astr] angle *m* horizontal
~AXIS - [surv] (the horizontal axis about which a telescope is rotated) axe *m* horizontal

HORIZONTAL BEARING - [gen] palier *m* horizontal
~BELT CONVEYOR - [mech] transporteur *m* horizontal à courroie
~BED MACHINE - [text] métier rectiligne *m* à mailles retournées à main
~BLACK-OUT PERIOD - [telev] (USA only) (Frame Suppression Period in GB) (the interval of time during which the frame synchronizing signals and the frame suppression signals are transmitted) intervalle *m* de suppression d'image
~BLANKING - [telev] (USA only) (Frame Suppression in GB) (the suppression of the image signal at the end of the frame) suppression *f* de ligne
~BLANKING LEVEL - [telev] niveau *m* de suppression de ligne
~BOILER - [th eng] chaudière *f* horizontale
~BORING AND MILLING MACHINE - [mach tool] machine *f* à aléser et à fraiser horizontale
~BORING MACHINE - [mech mach] machine *f* à aléser horizontale
~BRACING - [mech] entretoisement *m* horizontal
~BREAK SWITCH - [el] interrupteur *m* à couteau horizontal
~CASTING - [metall] coulée *f* horizontale
~CENTRING - [telev] (manual control) dispositif *m* de réglage du centrage horizontal
~CHAMBER - [th eng] chambre *f* horizontale
~CIRCLE - [surv] (graduated circular plate for the measurement of horizontal angles) cercle *m* horizontal
~COMPONENT - [el] (the component of the earth's magnetic field acting in a horizontal direction) composante *f* horizontale
~CONTACT PRESSURE - [soil] contrainte *f* effective horizontale, pression *f* de contact horizontale
~CREEL - [text] (a spinning frame) contre *m* horizontal
~CUT-AND-FILL STOPE - [mining] taille *f* en échelon avec remblayage
~CYLINDRICAL PADDLE MIXER - [mech] (mixer with horizontal cylindrical drum containing paddles on a horizontal shaft) mélangeur *m* à tonneau cylindrique horizontal à pales
~DEFINITION - [telev] (the number of dots in a horizontal line) définition *f* horizontale
~DEFLEXION - [telev] déviation *f* horizontale
~DEFLECTION - [electron] déviation *f* horizontale
~DISPLACEMENT - [geol] rejet *m* horizontal
~DRIVE - [mech] commande *f* horizontale
~ENGINE - [mech] (engine in which the cylinder is horizontal) moteur *m* à cylindre horizontal
~ESCAPMENT - [horol] (cylinder escapment) échappement *m* horizontal
~EXTRUDER HEAD - [rubber & plast ind] tête *f* d'extrusion horizontale, tête *f* droite de boudineuse
~FLUKES - [zool] (the rear fins of a whale) queue *f* horizontale
~FLYBACK - [electron] retour *m* horizontal
[telev] retour *m* du spot de ligne
~FRAME SAW - [impl] scie *f* à châssis horizontal
~GENERATOR - [el] génératrice *f* à axe horizontal
~GRATE - [mech] grille *f* horizontale
~HOLD CONTROL - [telev] (manual control) réglage *m* de la fréquence de lignes
~HUNTING - [telev] (the horizontal movement of the image which is caused by defective synchronization) instabilité *f* horizontale de l'image, écart *m* horizontal
HORIZONTAL INJECTION PRESS - [plast ind] (injection moulding machine in which the material is injected in a horizontal direction) presse *f* d'injection horizontale
~KILN - [brew ind] touraille *f* horizontale
~LATHE - [mach tool] tour *m* parallèle
~LINE - [geom] ligne *f* horizontal
[surv] ligne *f* d'horizon
~LINING - s. horizontal sheeting
~MILL - [mach tool] fraiseuse *f* horizontale
~MOTION - [mech] mouvement *m* horizontal
~MOVEMENT - [text] mouvement *m* horizontal
~OVERLAP - [geol] recouvrement *m* horizontal
~PARALLAX - [surv] (the value of the geocentric parallax for a celestial body, when this is on the observer's horizon) parallaxe *f* horizontale
~PARITY CHECK - [comput] contrôle *m* de parité horizontal
~PASS-BELT TYPE OVEN - [rubber ind] étuve *f* à ruban transporteur horizontal
~PENDULUM - [mech] (a compound pendulum with a rotation axis almost perpendicular) pendule *m* horizontal
~PLANER - [mach tool] rabot *m* horizontal
~POLARIZATION - [radio] (a transmission in which the electrostatic field is in a horizontal plane) polarisation *f* horizontale
~POWER - [telev] (USA only; Line Frequency in GB; the number of lines scanned per second) fréquence *f* d'analyse de ligne
~RESOLUTION - s. horizontal definition
~RETORT - [th eng] cornue *f* horizontale
~RETORT FURNACE - [th eng] four *m* à cornue horizontale
~RETRACE - s. horizontal flyback
~RETURN TUBE BOILER - [boilers] chaudière *f* tubulaire à retour de fumée
~RIBBON MIXER - [mech] melangeur *m* horizontal à bandes
~RUDDER - [aero] gouvernail *m* de profondeur
~SCANNING - [telev] analyse *f* à lignes, analyse *f* horizontale
~SCREW PRESS - [mech] presse *f* à vis horizontale
~SECTION - [mech] coupe *f* horizontale
~SEISMOGRAPH - [instr] (a seismograph wich registers the horizontal components of the earth's vibrations) sismographe *m* horizontal
~SHAFT - [mech] arbre *m* de transmission horizontal
~SHEETING - [constr] (long horizontal boards on each side of a trench cut in bad ground) coffrage *m* horizontal
~SHORE - [constr] étrésillon *m*
~SIZE CONTROL - [telev] (USA only; Line Amplitude Control in GB; mechanism for the control of the horizontal amplitude of the scanning) dispositif *m* de réglage de la largeur de ligne
~SPINDLE GRINDER - [mach tool] rectifieuse *f* pour surfaces planes
~STEAM PUMP - [mech] pompe *f* à vapeur horizontale
~STRESS - [phys] effort *m* horizontal
~SWEEP - [telev] aller et retour horizontal
~SYNCHRONIZATION PULSE - [telev] impulsion *f* de synchronisation de ligne
~THROW - [geol] rejet *m* horizontal transversal
~THRUST - [mech] poussée *f* horizontale

HORIZONTAL VULCUNIZER - [rubber ind] vulcanisateur *m* horizontal
~WATER WHEEL - [mech] roue *f* à cuiller
HORIZONTALLY-OPPOSED ENGINE - [mech] (type of reciprocating i.c. engine in which the cylinders are placed in pairs on opposite sides of the crank-shaft) moteur *m* à cylindres horizontaux opposées
~POLARIZED WAVE - [electron] (wave with a horizontal direction of polarization) onde *f* à polarisation horizontal
HORMONES - [chem] (complex substance produced by the endocrine glands which regulate and initiate a number of important physiological functions) hormones *fpl*
HORMONOTHERAPY - [med] hormonothérapie *f*
HORN - [gen] corne *f*
[radio] (elementary aerial consisting of a wave guide with dimensions increasing towards the aperture) cornet *m*
[acoust] (a warning device emitting a continuous note) avertisseur *m*, klaxon *m*
[el acoust] (od a loudspeaker) pavillon *m*
[el] corne *f*
[auto] cornet *m* avertisseur, trompe *f*
[mech] (of a machine) bras *m*
[mus] (wind instrument) cor *m*, cornet *m*
[zool] antenne *f*, aigrette *f*
[impl] (of an anvil) bigorne *f*
~ARRESTER - [el] (lightning arrester consisting of a horn gap) parasurtension *f* à cornets
~BALANCE - [aero] (local balance area at tip of control surface) corne *f* de compensation
~BAR - [mech] traverse *f* de voiture
~BLOCK - [railw] (of the axle blox) plaque *f* de garde
~BULB - [impl] poire *f* de trompe
~BUTTON - [auto] bouton *m* d'avertisseur
~BUTTON RELAY - s. horn relay
~CASTING - [metall] coulée *f* en cornichon
~FEED - [radio] (the surface of a wave guide of lare facing into a parabolic reflector) alimentation *f* du cornet
~FLINT - [min] silex *m* corné
~FUSE - [el] coupe-circuit *m* à antennes
~GAP - [el] (spark gap of gradually increasing length) parafoudre *m* à cornes
~GAP ARRESTER - s. horn arrester
~GATE - [metall] (horn-shaped in-gates supplying several small moulds made in the same moulding box) attaque *f* en cornichon, chenal *m* de coulée en source
~LEAD - [min] (a nautral mineral of lead) plomb *m* corné
~LOUDSPEAKER - [acoust] hautparleur *m* à pavillon
~MEAL - [agric] farine *f* de corne torréfiée
~MERCURY - [chem] calomel *m*
~MOUTH - [mus] embouchure *f*
~PLAQUE STAY - [mech] entretoise *f* de plaque de garde
~-PLATE - [railw] plaque *f* de garde
[mech] para-essieux *m*
~PRESS - [mach tool] presse *f* à bras
~RADIATOR - [radio] (radiating element shaped like a horn) cornet *m*
~REED - [acoust] anche *f*
RELAY - [auto] relais *m* de commande d'avertisseur
~RING - [zool] bourrelet *m* de corne, anneau *m* de corne
HORN SILVER - [min] (cerargyrite, chlorargyrite. Natural silver chloride, an ore of silver) argent *m* corné, cérargyrite *f*
~SOCKET - [mining] douille *f* de secours, cloche *f* de repêchage
[mech] tube *m* de repêchage
~SPACING - [mech] (of a welding machine) écartement *m* des bras
~THROAT - [mus] partie *f* antérieure d'un pavillon
~-TYPE ANTENNA - [radio] (funnel-shaped aerial) antenne *f* à cornet
~-TYPE RADIATOR - s. horn radiator
~WELDING - [metall] soudure *f* par martelage
HORNBEAM - [bot] (a tree of hard, closed grained wood) hêtre *m* blanc
HORNBLENDE - [min] (rock forming mineral, essentially a silicate of calcium, magnesium and iron) hornblende *f*
HORNBLENDITE - [min] hornblendite *f*
HORNING PRESS - s. horn press
HORNS - [mining] rainures *fpl* du tambour d'extraction
~OF A VALVE - [el] cornes *fpl* d'une lampe
HORNSTONE - [min] silex *m* noir, pierre *f* de corne
HOROLOGY - [horol] (the science of time-measuring and the production of watches) horlogerie *f*
HOROPTER - [opt] (in the field of binocular vision, the locus of the point which are seen single) horoptère *m*, cercle *m* de Müller
HORRIPILATION - [med] (the erection of the hairs on the skin) horripilation *f*
HORSE - [zool] cheval *m*
[impl] (a trestle supporting a piece of timber while it is sawn) chevalet *m*
[tool] chevalet *m*
[min] (mass of barren rock found in veins or lodes) nerf *m*, cheval *m*
[constr] (mason impl] trépied *m*
[gen] (a kind of stand or trestle) chevalet *m* de montage
[naut] marchepied *m*, cheval *m* de vergue
[mining] inclusion *f* de stériles dans une veine
~-BEAN - [bot] féverole *f*
~BEEDING - [zool] élevage *m* des chevaux
~BRUSH - [impl] brosse *f* de pansage
~-CAPSTAN - [constr] manège *m*
~CARD - [agric mach] brosse *f* à cheval
~CHESTNUT TREE - [bot] maronnier *m* d'Inde
~CLOTH - [text] couverture *f* pour chevaux
~CLOTH STRAP - [text] sangle *f* pour écurie
~DAISY - [bot] matricaire *f* inodore
~DUNG - [agric] fumier *m* de cheval, crottin *m* de cheval
~-FLASH ORE - [min] (name given to bornite by cornish miners) bornite *f*, cuivre *m* panaché
~FLY - [zool] taon *m*
~GRAM - [agric] dolic *m* asperge
~-HAIR - [zool & text] crin *m* (de cheval)
~HEAD - [oil ind] secteur *m* de suspension
~HOE - [agric] bineuse *f* à cheval
~HUSBANDRY - [zool] élevage *m* des chevaux
~LATITUDES - [met] (region of calms and variable winds coincident with the subtropical high-pressure belt on the poleward side of the trade-winds (usually applied to the Northern Hemisphere only) calmes *mpl* subtropicaux

HORSE MANURE s. horse dung
~OPERA - [cin] (a western picture) western *m*, film *m* de cowboy
~PLOUGH - [agric] charrue *f* à attelage
~-POWER - [H.P.] [meas] (unit of power, equal to 33.000 ft./lb. per minute) puissance *f* en cheveaux, force *f* motrice, cheval *m* vapeur
~-POWER FORMULA - [mech] formule *f* de la puissance
~-POWER HOUR - [meas] cheval-heure *m*
~-POWER RATING - [mech] puissance *f* nominale
~-POWER YEAR - [meas] cheval-an *m*
~POX - [vet] (contagious virus infection affecting equines) variole *f* équine
~RADISH - [bot] cran *m*, raifort *m* sauvage, moutarde *f* des capucins
~RAKE - [agric] râteau *m* à cheval
~RASP - [impl] râpe *f* maréchal
~-SHOE - [impl] fer *m* à cheval
~-SHOE CURVE - [surv] virage *m* en fer de cheval
~-SHOE DRAIN - [plumb] (drain pipe with a U-section) canal *m* de vidange à fer de cheval
~-SHOE FILAMENT - [light] (filament having the shape of a single half-turn) filament *m* en fer à cheval
~-SHOE MAGNET - [el] aimant *m* en fer à cheval
~-SHOE MIXER - [mech] (type of mixer having a hemispheric bowl containing a U-shaped rotating blade) agitateur *m* à palette en U
~-SHOE RUBBERS - [rubber ind] garnitures *fpl* en caoutchouc de fers à cheval
~-SHOE SHAPED - [gen] (having a crescentic form) en fer à cheval
~-SHOE VETCH - [bot] hyppocrépide *f* à toupet
HORSEFLESH - [food] viande *f* de cheval
HORSEHAIR CLOTH - [text] tissu *m* en crin
~FABRIC - s. horsehair cloth
~HEALD - [text] lisse *f* en crin de cheval
~SEAT COVERING - [text] tissu *m* en crin pour siège
~YARN - [text] fil *m* de crin
HORSEHIDES - [leather ind] peaux *fpl* de cheval
HORSEHOE FISTULA - [med] fistule *f* en fer à cheval
~KIDNEY - [med] rein *m* en fer à cheval
~WEIGHT - [text] poids *m* en forme d'étrier, poids *m* en forme d'U
~WHEEL PATH - [text] anneau *m* en fer à cheval
HORSEHOER - [labour] maréchal *m* ferrant
HORSESHOEING - [agric] ferrage *m*, ferrure *f* des chevaux
HORSING - [zool] monte *f*, saillie *f*
HORST - [geol] (the structure resulting from two parallel normal faults forming buttresses against which surrounding tracts have been pressed) horst *m*, môle *m*, butoir *m*
HORTICULTURAL PRODUCT - [agric] produit *m* horticole
~SEED - [agric] semences *fpl* horticoles
~TRACTOR - [agric] tracteur *m* horticole
HORTICULTURE - [agric] horticulture *f*
HOSE - [impl] (flexible pipe of rubber, plastic, textile material or the like) tuyau *m* souple, gaine *f*
~ADAPTER - [impl] (used in the installation of gas appliances) raccord *m* de durite
~ARMOURED WITH TARRED ROPE - [rubber ind] tuyau *m* enroulé de corde goudronnée
~BANDAGING MACHINE - [mech] machine *f* à bandeler les tuyaux
~BUILDING MACHINE - [mech] machine *f* à faire les tuyaux
HOSE CLAMP - s. hose clip
~CLIP - [mech] (device for securing a hose to a rigid spigot) collier *m* de durite
~CONNEXION - s. hose coupling
~COUPLING - [plumb etc] (joints for connecting lengths of hose) raccord *m* de tuyau, accouplement *m* de tuyau
~COVER - [rubber ind] couverture *f* de tuyau
~DUCK - [text] toile *f* pour tuyaux
~FITTING - s. hose coupling
~INSERT - [rubber man] armature *f* de tuyau
~LINE - [brew ind] tuyau *m* flexible
~MACHINE - [text] métier *m* circulaire automatique à chaussettes
~NIPPLE - s. hose adapter
~NOZZLE - [impl] (used in the installation of gas appliances) mamelon *m* porte-tuyau souple
~-PROOF - [el] (of an enclosure preventing water to reach the apparatus when washed with a hose) protégé contre les jets d'eau
~UNION - s. hose coupling
~WITH BRAIDED INSERTS - [rubber ind] tuyau *m* avec armature tressée
~WITH BRAIDED WIRE INSERT - [rubber ind] tuyau *m* avec armature en fil de métal tressé
~WITH FABRIC COATING - [rubber ind] tuyau *m* à revêtement de tissu
~WITH REINFORCED SOCKET - [rubber ind] tuyau *m* avec manchon renforcé
~WRAPPING MACHINE - [mech] machine *f* à bandeler les tuyaux
~YARN - [rubber ind] fil *m* de renfort pour tuyaux
HOSIERY - [text] (a variety of knitted fabrics) bonneterie *f*
~FRAME - [text] métier *m* à cueillage
~YARN - [text] fil *m* de bonneterie
HOSPITAL - [gen] hôpital *m*
~BUS-BARS - [el] (set of bus-bars for emergency purposes) barre *f* omnibus de secours
~GANGRENE - [med] gangrène *f* nosocomiale
~SWITCH - [el] (switch used for changing a circuit over to an emergency supply) interrupteur *m* d'émergence
[el] (switch in trains or trams to cut out a faulty motor) bouton *m* d'arrêt d'émergence
HOST - [gen] hôte *m*
[electron] s. host crystal
~CRYSTAL - [electron] (the predominant crystalline constituent of a luminescent material) base *f*
HOT - [gen] chaud
[nucl] (of a space, a body etc in which a high level of radioactivity exists) chaud, fortement radioactif
[el] (charged to a very high potential) sous tension
[ind chem & metall] chaud
~ADJUSTMENT - [radio] régulation *f* sous tension
~-AIR - [adj] à air chaud
~-AIR AGEING - [ind proc] vieillissement *m* à l'air chaud
~-AIR BLOWER - [photo] ventilateur *m* à air chaud
~-AIR CIRCULATOR - [th eng] (for space heating) aérotherme *m*
~-AIR CURE - [rubber ind] vulcanisation *f* à l'air chaud
~-AIR DRIER - [ind proc] séchoir *m* à air chaud
~-AIR ENGINE - [mech] moteur *m* à air chaud

[th eng] moteur *m* à air chaud

HOT-AIR FURNACE -[ovens] (supplying warm air through gratings etc) appareil *m* de chauffage à air chaud

~-AIR HEATING SYSTEM - [th eng] (a system of heating by a circulation of hot air) chauffage *m* à air chaud

~-AIR HORN - [auto] prise *f* d'air chaud

~-AIR INTAKE - [mech] (admission port for hot air) entrée *f* d'air chaud

~-AIR KILN - [brew ind] touraille *f* à air chaud

~-AIR MAIN - [metall] conduit *m* annulaire de vent chaud

~-AIR SEASONING - [timber] (a process of dessication) séchage *m* à air chaud

~-AIR SIZING MACHINE - [text] encolleuse *f* à air chaud, pareuse *f* à air chaud

~-AIR TURBINE - [mech] (a set consisting of a turbine, an air-compressor, a regenerator, a water cooler and a starting motor) turbine *f* à air chaud

~ATOM - [nucl] (atom with high interval energy) atome *m* excité

~ATOM CHEMISTRY - [nucl] chimie *f* des atomes fortement excités

~-BANBURY RECLAIMING PROCESS - [rubber ind] (thermo-mechanical reclaiming process) procédé *m* de régénération thermomécanique

~BED - [agric] couche *f* chaude, forcerie *f*

~BENDING TEST - [metall] essai *m* de flexion à chaud

~BLAST - [metall] (hot air supplied to a furnace for combustion) courant *m* d'air chaud

~-BLAST CUPOLA - [metall] cubilot *m* à vent chaud

~-BLAST FURNACE - [metall] four *m* soufflé au vent chaud

~-BLAST HEATING SYSTEM - [th eng] chauffage *m* à air chaud à circulation sous pression

~-BLAST MAIN - [metall] conduite *f* de vent chaud

~-BLAST PIG - [metall] fonte *f* à l'air chaud

~-BLAST STOVE - [metall] (a large chamber containing brickwork columns used for preheating air for blowing a blast furnace) appareil *m* régénérateur de vent chaud, appareil *m* de Cowper

~-BLAST VALVE - [mech] vanne *f* à vent chaud

~BOX COREMAKING - [metall] manufacture *f* de noyaux en boîte chaude

~-BULB - [mech] (internal combustion engines) chapeau *m* incandescent

~BULB MOTOR - [mech] moteur *m* semi-diesel

~CATHODE - [electron] (a thermionic cathode; a cathode functioning primarily by the process of thermionic emission) cathode *f* incandescente, cathode *f* thermionique

~-CATHODE ELECTRON TUBE - [electron] tube *m* à cathode incandescente

~-CATHODE GAS-FILLED TUBE - s. hot-cathode gas-filled valve

~-CATHODE GAS-FILLED VALVE - [electron] (or thyratron; a hot-cathode gas-filled valve in which control electrodes initiate, but do not control, the anode current) thyratron *m*

~-CATHODE IONIZATION GAUGE - [electron] jauge *f* à ionisation, manomètre *m* à ionisation à cathode chaude

~-CATHODE RECTIFIER - [electron] redresseur *m* thermionique

~-CATHODE VACUUM GAUGE - [electron] jauge *f* à cathode chaude (ou incandescente)

HOT CAVE - s. hot cell

~CELL - [nucl] (shielded storage space of a hot laboratory) cellule *f* chaude, casematte *f*

~CHAMBER MACHINE - [metall] (diecasting) machine *f* à chambre chaude

~CHANNEL FACTOR - [nucl] facteur *m* de canal chaud

~CHISEL - [tool] burin *m* à chaud

~CIRCUIT - [el] (charged to a dangerously high potential) circuit *m* sous tension

~CLEAN REACTOR - [nucl] réacteur *m* chaud et propre

~COINING - [metall] frappe *f* à chaud

~COMPACTING - [metall] compression *f* à chaud

~CRACK - [metall] (crack appearing on a casting before it cools) rupture *f* à chaud

~CURE - [rubber ind] cuisson *f* à chaud, vulcanisation *f* à chaud

~CUTTER - [tool] tranche *f* à chaud

~DEFROSTER - [auto] dégivreur *m* chauffant

~DESEAMER - [metall] décriqueuse *f* à chaud

~DIE SEALING - [metall] (sealing by electrically-heated tools of special shape) soudage *m* à chaud, thermosoudure *f*

~-DIE STEEL - [metall] acier *m* pour étampage à chaud

~-DIP COMPOUND - [plast ind] mélange *m* anticorrosif de trempage

~-DIP GALVANIZING BATH - [metall] (tank of liquid zinc in which objects are dipped to coat them with metal) bain *m* chaud de galvanisation

~-DIP TINNING BATH - [metall] (tank of liquid tin in which objects are dipped to coat them with metal) bain *m* chaud d'étamage

~DIPPING - [metall] revêtement *m* par immersion en bain chaud

~-DRAWN - [metall] étiré à chaud

~DRYING - [gen] séchage *m* à chaud

~EMBOSSING - [text] gaufrage *m* à chaud

~END - [glass man] (any operation with hot glass) opération *f* sur verre chaud

~-EXTRUDED - [plast ind] extrudé à chaud

~EXTRUSION - [plast ind] extrusion *f* à chaud

~FACE TEMPERATURE - [metall] température *f* superficielle maximale

~FINISHING - [metall] finissage *m* à chaud

~FLAT IRON - [metall] fer *m* plat à planer à chaud

~GALVANIZING - [el chem] galvanisation *f* en bain chaud

~GAS WELDING - [plast ind] (a method of welding thermoplastic materials in which the material is heated by a jet of hot air or inert gas directed from a welding torch on to the area of contact of the surfaces which are being welded) soudage *m* au gaz chaud

~GLUE - [ind chem] colle *f* à chaud

~-GRINDING PROCESS - [text & paper man] procédé *m* de ponçage à chaud

~HARDNESS - [metall] maintien *m* de la dureté à température élevée

~HEADING - [mech] façonnement *m* des têtes à chaud

~-HOUSE (or Hothouse) - [agric] (an artificially heated structure for growing plants) serre *f* [agric] (a drying chamber) séchoir *m*

~-INGOT SHEAR - [metall] cisaille *f* à chaud de lingots

~IRON - [metall] (obtained in the melting zone of a blast furnace) fonte *f* chaude

~JUNCTION - [mech] (of a thermocouple) soudure *f*

chaude
HOT LABORATORY - [nucl] (laboratory for research with radioactive materials) laboratoire *m* chaud
~ LACQUER PROCESS - [paint] vernissage *m* à chaud
~ LEAD - [el] câble *m* sous tension
~ LIGHT - [telev & cin] (essential lamps in a studio) éclairage *m* principal
~ METAL - [metall] métal *m* en état de fusion
~ METAL CAR - [metall] wagon *m* mélangeur
~ METAL CRANE - [metall] pont *m* de coulée
~ MIX - [constr] (bituminous surfacing consisting of a mixture of crushed stone, sand and filler coated with bitumen) mélange *m* chaud
~-MIX COURSE - [constr] couche *f* deposée à chaud
~ NOSE ICE PROTECTION - [el mach] (prevention of icing on the nose of a turbine nacelle effected by heating the nose) protection *f* antigivre par réchauffement du nez
~ OIL-PRODUCTION - [colloq. USA] production *f* de pétrole illegal
~ PEENING - [mech] matage *m* à chaud
[metall] trempe *f* chaude
~ PENETRATION TEST - [plast ind] (test in which the specimen is heated and pierced by a wire under specified conditions) test *m* de pénétration à la chaleur
~ PLASTER SEALER - [paint] mastic *m* pour plâtre
~ PLATE - [impl] (metal slab heated by electricity, gas or otherwise, for heating glass vessels in laboratory work) plaque *f* de chauffage
[el] chauffe-plat *m* électrique
[gas ind] réchaud-plat *m*
~-PLATE BURNER - [gas ind] brûleur *m* de dessus, brûleur *m* de taque
~-PLATE PRESS - [mech] presse *f* à plateaux chauds
~ PLUG - [auto] bougie *f* surchauffée
~-PRESS, to - [metall] estamper au pilon, forger à l'estampe
[paper man] (glazing by hot plates) calandrer à chaud
~ PRESS - [mech] presse *f* à chauffage des plateaux
[paper man] calandre *f* à cylindre réchauffés
~ QUENCHING - [metall] trempe *f* chaude
~ RIVETED - [mech] rivé à chaud
~ RIVETING - [mech] rivure *f* à chaud
~ ROD - [auto] (colloq. USA) hot rod *m*
~-ROLL, to - [metall] laminer à chaud, cylindrer à chaud
[paper man] (glazing by means of steam-heated cylinders) calandrer à chaud
~-ROLLED - [metall] laminé à chaud
[paper man] calandré à chaud
~ ROLLING - [metall] laminage *m* à chaud
~ ROLLING MILL - [metall] laminoir *m* à chaud
~-RUNNER MOULD - [plast ind] (type of mould in which the runners are kept at a relatively much higher temperature than the cavity) moule *m* à canal chauffé
~ SAW - [mech] (metal-cutting circular saw used to cut off the ands of heated steel forgings) scie *f* à chaud
~ SEALING - [metall & plast ind] soudure *f* à chaud
~ SETT - [tool] tranche *f* à chaud
~-SETTING - [rubber & plast ind] (thermosetting) durcissable à chaud
~-SETTING ADHESIVE - [chem] (an adhesive forming a permanent bond when subjected to heat) colle *f* durcissable à chaud
HOT-SETTING GLUE - [ind chem] colle *f* à chaud
~-SETTING RESIN - [plast ind] (thermosetting resin) résine *f* durcissable par chaleur
~-SHORT - [metall] (brittle when hot) cassant à chaud
~ SHORTNESS - [metall] fragilité *f* à chaud
~ SHRINK FIT - [mech] (shrink-fit in which the external part is designed to be heated) emmanchement *m* à chaud
~ SPOT - [mech] (in internal combustion engines) dispositif *m* de réchauffage de l'admission
(of cables; a location at which thermal generation is high) point *m* chaud
[th eng] (point of local excessive temperature in a heating device) endroit *m* chaud
[cin] (bright area of light on an image projected by a lens system) superintensité *f* lumineuse brusque
~-SPOT CHAMBER - [mech] (in internal combustion engines) chambre *f* de réchauffage
~ SPRAYING - [paint] vaporisation *f* à chaud
~ SPRING - [geol] source *f* thermale
~ STAMPING - [metall] étampage *m* à chaud
~ STRIP MILL - [metall] laminoir *m* à bandes larges à chaud
~ TAP - [gas ind] (gas network equipment) (USA only; under-pressure connexion in GB) piquage *m* sur conduite en charge
~ TEAR - s. hot crack
~ TEARING - [metall] criqûre *f* de retrait
~ TEST - [gen] essai *m* à chaud
[text] (shrinking test) essai *m* de rétrécissement à chaud
~ TIN PLATE - [metall] fer-blanc *m* étamé à chaud
~ TINNING - [metall] étamage *m* à chaud
~ TOP - [metall] (refractory insulation at the top of a lingot mould) masselotte *f* chaude
~ TRAP - [nucl] piège *m* chaud
~-TRIMMING - [metall] ébarbage *m* à chaud
~ WASTE - [nucl] déchets *mpl* radioactifs
~-WATER - [adj] à eau chaude
~-WATER BOILER - [th eng] chaudière *f* à eau chaude
~-WATER FIT - s. hot shrink fit
~-WATER HEATING - [th eng] chauffage *m* à eau chaude
~ -WATER RADIATOR - [th eng] (for space heating) radiateur *m* à eau chaude
~-WATER RINSING - [ind proc] rinçage *m* à eau chaude
~ WAVE - [U.S.] s. Heat Wave
~ WELL - [mech] (tank into which the condensate from a steam-engine is pumped) bâche *f*, citerne *f*, réservoir d'alimentation
~ WIRE - [el] (of an instrument whose operation depends on the thermal expression) fil *m* chaud
~-WIRE AMMETER - [el instr] ampèremètre *m* à fil chaud
~-WIRE ANEMOMETER - [instr] (instrument in which a gas flow is measured by exposing a heated wire to it and measuring the change of resistance due to cooling) anéomètre *m* à fil chaud
~-WIRE DETECTOR - [radio] détecteur *m* à fil chaud
~-WIRE GAUGE - [instr] (instrument designed to measure pressures by determining the heat losses from an electrically heated wire) manomètre *m* à fil chaud
~-WIRE INSTRUMENT - [instr] (a measuring device actuated by the expansion of a wire when heated by

the passage of a current) instrument *m* à fil chaud, appareil *m* à dilatation, appareil *m* à chaud
HOT-WIRE MICROPHONE - [acoust] microphone *m* à fil chaud, thermophone *m*
~-WORKING - [metall] (the process of working metals by rolling, forging, extruding etc at high temperatures) travail *m* à chaud
~WORT RECEIVER - [brew ind] bac *m* fermé
HOTCHING - [min] criblage *m* par dépôt
~MACHINE - [mining] crible *m* hydraulique, hydrotamis *m*
HOTHOUSE FRUIT - [agric] fruits *mpl* de forcerie
~GARDENING - [agric] culture *f* maraîchère sous serre
HOUDRIFORMING - s. houdry process
HOUDRY PROCESS - [ind chem] (a catalyst process for the upgrading of petroleum) procédé *m* Houdry
HOUND - [zool] chien *m* de meute, chien *m* chasseur
[mech] (a wooden bar connecting the fore-carriage of a wagon to the shaft) renforcement *m* latéral
[naut] capelage *m* de mât, noix *fpl*
HOUNDFISH - [zool] chien *m* de mer, squale *m*
HOUR - [gen] heure *f*
~ANGLE - [astr] (the distance, measured westwards in terms of arc, from a given celestial meridian to that of a celestial body) angle *m* horaire
~ANGLE OF MEAN SUN - (H.A.M.S.) [astr] (the angle, at the celestial pole, of the mean sun measured westwards from the meridian of the observer) angle *m* horaire de soleil moyen
~ANGLE OF TRUE SUN - (H.A.T.S.) [astr] (the angle, at the celestial pole, of the true sun, measured westwards from the meridian of the observer) angle *m* horaire du soleil vrai
~CIRCLE - [astr] (the great circle passing through the celestial poles and a heavenly body, cutting the celestial equator at 90°) cercle *m* horaire
[instr] (the graduated circle of an equatorial telescope) cercle *m* horaire
~COUNTER - s. hour meter
~-GLASS - [instr] (generally a glass vessel with ends connected by a constricted neck) sablier *m*
~-GLASS CALIPERS - [instr] compas *m* dit 8 de chiffre
~-GLASS HEAD - [med] crâne *m* en sablier
~-GLASS SCREW - [mech] vis *f* globique, vis *f* à filets convergents
~-GLASS STOMACH - [mech] estomac *m* en sablier
~HAND - [horol] aiguille *f* des heures, petite aiguille *f*
~METER - [instr] (instrument designed to measure the time during which electric energy is measured) compteur *m* horaire
~WHEEL - [horol] (the wheel in the motion carrying the hour hand) roue *f* des heures
HOUSE, to - [gen] loger, héberger
[mech] encastrer, emboiter
[agric] mettre à l'abri
HOUSE - [gen] maison *f*, demeure *f*, bâtiment *m*
[mech] (of a crane) cabine *f*, guérite *f*
~-BOAT - [naut] (boat used as a dwelling) bateau-maison *m*
~-KEEPING - [comput] (colloq. USA only; those operations required by a programme, which do not directly contribute to the solution of the problem) aménagement *m*
~SEWER - [constr] drainage *m* domestique
HOUSEBREAKER - [gen] (employed in the business of demolishing a house) démolisseur *m*
HOUSEKEEPER SEAL - [metall] scellement *m* verre-métal, joint *m* verre-métal
HOUSING - [gen] longement *m*, mise *f* à l'abri
[mech] (a hollow structure enclosing a mechanical or electrical assembly) longement *m*, cage *f*, colonne *f* (de laminoir)
[auto] (a box) boîte *f*, boîtier *m*
[mech] (of a machine) jumelle *f*, coquille *f*
[mech] (of a bearing) siège *m*
[naut] (part of the lower mast) lusin *m*, merlin *m*
[el & mech] capsule *f*
~DEVELOPMENT - [town planning] projet *m* d'aménagement
~ESTATE - [town planning] groupe *m* d'habitations, quartier *m* d'habitation
~JOINT - [constr] assemblage *m* à fourche
HOVEN - [vet] tympanite *f*
HOVER, to - [phys & aero] (to remain motionless while airborne and supported by the aircraft's own mechanical power, as of a helicopter) être en vol stationnaire, planer
HOVERCRAFT - [transp] (a vehicle which is supported by an air-cushion built up under it by the discharge of air) hovercraft *m*
HOVERING CEILING - [aero] plafond *m* absolu
~FLIGHT - [aero] vol *m* stationnaire
HOWITZER - [firearms] obusier *m*, canon *m* court
HOWLBACK - [acoust] (undesired effect arising from acoustic or mechanical feed-back from a loudspeaker to a microphone) hurlement *m*
HOWLING - [acoust] (the sound made by some animals) hurlement *m*
[radio] (the sound due to back-coupling) réaction *f* dans l'antenne
h.p. - s. Horse Power
H.P. - s. High Pressure
H.P.I. - s. Height Position Indicator
H PLANE BEND - [electron] (wave guides) coude *m* à plan H
H PLANE T JUNCTION - [electron] (wave guides) jonction *f* à plan H
HT - s. High Tension
H2S - [radar] (type of radar giving a display of the earth's surface below the aircraft) H2S
HUB - [mech] (the central portion of a rotary structure, e.g. a propeller) moyeu *m*
[mining] (of a reel for a flat winding rope) estomac
~BOLT - [mech] boulon *m* de moyeu
~CAP - [mech] chapeau *m* de moyeu
~COVER PLATE - [aero] (the plate which covers the hub mechanism of a helicopter rotor) plaque *f* de recouvrement du moyeu
~FLANGE - [mech] (of bicycle) boudin *m* du moyeu
~PULLER - [impl] arrache-moyeu *m*
~SLEEVE - [mech] manchon *m* de moyeu
~SPRING - [mech] ressort *m* de moyeu
~TIP RATIO - [mech] (the relation between the o/d of the hub and the diameter of the tip path of the blades) rapport *m* moyeu/tête d'aube
~TYRE - [rubber ind] (tyres) pneu *m* monté sur moyeu
HUBBUB - [acoust] (an undefined mixture of loud sounds) tapage *m*, vacarme *m*
HUBER'S REAGENT - [chem] (an analytical reagent consisting of an aqueous solution of ammonium molybdate and potassium ferr-cyanide, used in the

detection of the free mineral acids) réactif *m* d'Hüber
HUBERNITE - [min] (or huebnerite; tungstate of manganese) hübnérite *f*
HUCKABACK - [text] (a cotton or linen cloth used for towels etc) tissu *m* granité
~WEAVE - [text ind] armure *f* granité
HUE - [gen] (the characteristic of a colour which is determined by the wave-length of line reflected by it, e.g. blue, red etc.) teinte *f*, couleur *f*, tonalité *f* chromatique
HUG TO THE WIND, to- [naut] pincer le vent, tâter le vent
HULK - [gen] carcasse *f*
[naut] ponton *m*, vaisseau *m* rasé
HULL - [naut aero] (the main body of a boat or a flying boat, which rests on and in the water) coque *f*
[bot] (the shell, pod etc of peas, beans etc., or seeds) cosse *f*, gousse *f*
[aero] (of an airship) coque-fuselage *f*
~DISPLACEMENT - [aero] (the volume of water displaced by a hull when waterborne under normal conditions) déplacement *m* de la coque
HULLING - [agric] décorticage *m*
~MACHINE - [agric] dépouilleuse *f*
HUM - [acoust] bourdonnement *m*, ronflement *m*
[radio] (continuous noise of low frequency in receiver sound output, due to interference byA.C. supplies) bruit *m*, ronflement *m*, bruit *m* d'alimentation
[aero] vrombissement *m*
~-BUCKING COIL - [radio] (coil of wire designed to reduce the interference caused by the mains) bobine *f* anti-ronflement
~-DINGER - [radio](small potentiometer in series across the heater supply secondary winding of the main transformer of an A.C. receiver) potentiomètre *m* anti-ronflement
~MODULATION - [radio] (modulation of a radio-frequency by hum) ronflement *m* d'onde porteuse
~-NEUTRALIZING COIL - s. hum-bucking coil
~NOISE - [acoust & radio] bruit *m*, ronflement *m*
~VOLTAGE - [radio] tension *f* de ronflement
HUMECTANT - [chem] (a substance controlling the water-content of a product) humectant *m*
HUMERAL - [anat] huméral
HUMERUS - [anat] humérus *m*
HUMIC ACID - [chem] acide *m* humique, humine *f*
HUMID - [phys] humide
[physiol] (of the skin) moite
HUMIDIFICATION - [chem] (process for the increase of water content of air or other gases) humidification *f*
HUMIDIFIER - [th eng] (an apparatus designed to maintain specified conditions of humidity in a building) humidificateur *m*
[text] appareil *m* à humecter
HUMIDITY - [phys] (water-vapour content of the atmosphere) humidité *f*
~CELL - [instr] (an apparatus designed for accurate dewpoint measurements) appareil *m* hygrométrique
~CONTROLLER - [gen] (device or arrangement for regulating humidity) régulateur *m* d'humidité
~LOSSES - [soil] déperdition *f* d'humidité
~RATIO - [phys] rapport *m* hygrométrique
~RESISTANCE - [phys] résistance *f* à l'humidité
HUMIDOSTAT - [th eng] humidistat *m*
HUMILIS - [met] (form of cumulus cloud, consisting of separate rounded masses) humilis *m*
HUMITE - [geol] (coals derived from humic materials) humite *f*
HUMMELER - [agric] (colloq. GB) ébardeur *m*
HUMMER-SCREEN - [min] tamis *m* à vibration magnétique
HUMMING - [acoust] (a sibilant sound, as emitted by bees and other insects) bourdonnement *m*
~OF GEARS - [auto]bourdonnement *m* des engrenages
~TOP - [acoust] (a small top which emits musical sounds when rotating) toupie *f* d'Allemagne
HUMOUR - [biol] (a fluid) humeur *f*
HUMP - [gen] bosse *f*
[aero] (on an air route) chaîne *f* de montagnes sur la route
[railw] rampe *f* de débranchement
~-AVOIDING LINE - [railw] voie *f* évite-bosse
~CABIN - [railw] poste *m* de bosse, poste *m* de butte
~LOCOMOTIVE - [railw] locomotive *f* de butte
~SHUNTING - [railw] manœuvre *f* par gravité
~SPEED - [aero] (the speed of a flying-boat or amphibian at which hull resistance in the water is highest) vitesse *f* critique de passage sur le redan, vitesse *f* au déjauger
HUMPING SIGNAL - [railw] signal *m* de débranchement
HUMUS - [chem] (decomposed organic material in colloidal form having no trace of its original structure) humus *m*, terre *f* végétale
~CARBONATE SOIL - [soil] rendzine *f*
~CONTENT - [soil] teneur *f* en humus
HUNDREDWEIGHT - [meas] (weight corresponding to II2 lbs in GB and I00 lbs in the USA) poids *m* de II2 livres) (I00 livres U.S.)
HUNG GUTTER - [constr] gouttière *f*
~TRAVELLING ROLLER - [constr] penture *f* à galets, galet *m* haut d'une porte coulissante
HUNGARIAN BROME-GRASS - [bot] brome *m* inerme
~MILLET - [bot] moha *m*
~VETCH - [bot] vesce *f* de Hongrie
HUNT, to - [gen] chasser
[mech] s. hunting [mech]
HUNTER - [gen] chausseur *m*
[zool] cheval *m* de chasse, hunter *m*
[horol] savonnette *f*
HUNTING - [mech] (in a self-controlled or governed mechanism, successive over-correction resulting in repeated approximation to the required conditions without the attainment of a steady level) oscillation *f*, irrégularité *f* de marche, instabilité *f*
[aero] (the rytmic variation in the speed of the engine) oscillation *f*
[railw] galop *m*
[acoust] fluctuation *f*
[cin] (regular up-and-down movement on the picture) glissement *m* vertical, instabilité *f* verticale faible
[aero] (of helicopter rotor blade) oscillation *f*
[el] (periodic variation of the speed of two or more synchronous machines) oscillation *f* pendulaire
[contr] (the continuous attempt on the part of an automatically controlled system to attain an equilibrium condition) pompage *m*, recherche *f* continue d'équilibre
[electron] rattrapage *m*, poursuite *f*
~HORN - [mus] cor *m* de chasse

HUNTING MOVEMENT - [railw] (of a vehicle) mouvement *m* de galop
~OSCILLATION - [electron] oscillation *f* pendulaire
~PROBE - [meas] (the up and down movement of the electrode of a liquid level indicator for a very accurate measurement) sonde *f* à mouvement de montée et de descente
HURDY-GURDY - [mus] (stringed instrument) vielle *f*
HURRICANE - [met] (wind of force 12 in the Beaufort scale) vent *m* d'ouragan, houragan *m*
[naut] tempête *f*
~LAMP - [impl] lanterne-tempête *f*
HURTER - [railw] heurtoir *m*
[mech] (of a vehicle) heurtequin *m*
[impl] chasse-roues *m*
HUSBANDRY - [agric] agronomie *f*, agriculture *f*
HUSH, to - [gen] calmer, imposer silence
[min] (the operation of washing away the surface soil from the rock formation for prospecting purposes) enlever les terres de couverture par un fort courant d'eau
HUSK, to - [gen] décortiquer, écosser, ébrouer, perler, monder
[agric] ébrouer, cerner, écorcer, monder
[agric] (of corn) éplucher
HUSK - [gen & bot] cosse *f*, gousse *f*, brou *m*, pelure *f*, bogue *f*
~FINENESS - [bot] (particularly of barley) finesse *f* des enveloppes
~PERCENTAGE - [brew ind] (in barley) pourcentage *m* d'enveloppes
HUSKER - [agric] dépouilleure *f*
HUT - [gen] hutte *f*, cassine *f*
HUTCH - [gen] (king of chest used to store flour etc) huche *f*
[agric] clapier *m*, lapinière *f*
[min] (small wagon) wagonnet *m*
[impl] (basket for coal) benne *f*, wagonnet à minerai
[min] concentré *m* de minerai
HUTMENTS - [gen] baraques
HUTTONITE - [min] (rare ore containing about 72 percent of thorium) huttonite *f*
H.V. - [High Voltage] haute tension *f*
HVL - [nucl] (abbrev. for Half Value Layer q.v.) CDA, couche *f* de demi-atténuation
HYACINTH BEAN - [agric] dolic *m* d'Egypte
HYALINE - [gen] hyalin, transparent
~LEUCOCYTE - [biol] monocyte *m*
HYALINOSIS - [med] hyalinose *f*, dégénerescence *f* fibrinoïde
HYALITE - [min] (transparent variety of opal, colourless) hyalite *f*
HYALITIS - [med] hyalite *f*
HYALOGENESIS - [biol] (secretory process in a cell) hyalogénèse *f*
HYALOGENS - [biol] (the particles which are formed by the secretory process in the cell) hyalogènes *mpl*
HYALOGRAPHY - [photo] (a process for transferring photographic images to glass) hyalographie *f*
HYALOID - [anat] (almost transparent) haloïde, haloïdien
~MEMBRANE - [anat] (e.g. of the eye) membrane *f* haloïde, membrane *f* du corps vitré
HYALOPHANE - [min] (a rare feldspar) hyalophane *f*
HYALOPLASM - [biol] (a clear, non granular protoplasm) hyaloplasme *m*
HYALURONIC ACID - [chem] (a component of animal tissues) acide
HYALURONIDASE - [chem] (an enzyme having uses in medicine) hyaluronidase *f*
HYBRID - [genet] (of the product of two different species, or races) hybride
~BALANCE - [radio] (degree of balance between two impedances) facteur *m* différentiel
~CLOVER - [bot] trèfle *m* hybride
~COIL - [el] (single transformer with three windings designed to be connected to four branches of a circuit, so as to make these branches conjugate in pairs) transformateur *m* différentiel
~COMPUTER - [comput] calculateur *m* hybride
~CORN - [bot] mais *m* hybride
~COUPLER - [electron] (hybrid junction in the form of a directional coupler) coupleur *m* à 3dB
~ELECTROMAGNETIC WAVE - [electron] onde *f* électromagnétique hybride
~JUNCTION - [electron] jonction *f* hybride
~MAIZE - [bot] s. hybrid corn
~MATRIX - [electron] (the matrix expression of a set of equations of a quadripole expressed in tensions and currents) matrice *f* hybride
~PLANT - [bot] plante *f* hybride
~ROCKS - [geol] (type of contaminated rocks) roches *fpl* hybrides
~SET - [radio] réseau *m* differentiel
~TELEVISION RECEIVER - [telev] téléviseur *m* à tubes électroniques et transisteurs
~TRANSFORMER - s. hybrid coil
HYBRIDATION - [biol] hybridation *f*
HYBRIDISM - [biol] hybridisme *m*
HYBRIDIZATION OF EIGENFUNCTIONS - [nucl] (any linear combination of the eigenfunctions of one problem used to represent an eigenfunction of another problem) hybridisation *f* des fonctions propres
HYDANTOIN - [chem] (glycolurea; intermediate for synthetic resins) hydantoïne *f*
HYDATID DISEASE - [med] échinococcose *f*
HYDNOCARPIC ACID - [chem] (a fatty acid obtained from the ripe seeds of Hydnocarpus wightiana and used in the treatment of leprosy) acide *m* hydnocarpique
HYDRACRYLIC ACID - [chem] acide *m* hydracrylique
HYDRANT - [plumb] (connexion to a water main, so as to make it possible for a hose to be attached) bouche *f* d'eau, bouche *f* hydrante
HYDRARGILLITE - [min] (hydrated oxide of aluminium; gibbsite) gibbsite *f*, hydrargyllite *f*
HYDRARGYRISM - [med] hydragyrisme *m*
HYDRARGYRIUM - [chem] (quicksilver) mercure *m*
HYDRASTINE HYDROCHLORIDE - [chem] (the hydrochloride of an alkaloid obtained from Hydrastis and used in medicine) clorhydrate *m* d'hydrastine
HYDRATE - [chem] (a loosely combined compound of water and another substance, in which the water is assumed to retain its molecular state) hydrate *m*
~OF LIME - [chem] chaux *f* hydratée
HYDRATED - [chem] hydraté
~ALUMINA - [chem] hydrate *m* d'aluminium
~CELLULOSE - [ind chem] (used in paper manufacture) cellulose *f* hydratée
~ION - [chem] (complex particle consisting of an ion combined with one or more molecules of water)

ion *m* hydraté
HYDRATES - [chem] (salts which contain water of crystallization, i.e. water retained on crystallization from aqueous solution) hydrates *mpl*
HYDRATION - [chem] (I) the absorption or combination of water with a substance.
2) in paper manufacture a process conferring water-resistance to the finished product) hydratation *f*
HYDRAULIC - [adj] (actuated by liquid under pressure) hydraulique
~ACCUMULATOR - [el] (energy store comprising a cylinder and piston, the latter being loaded with weights, a spring or a pneumatic cushion) accumulateur *m* hydraulique
~ACTUATOR - [mech] servomoteur *m* hydraulique
~AIR VESSEL - [plumb] réservoir *m* d'air
~AMPLIFIER - [hydr] (device used in hydraulic systems, which draws additional power from a supply in proportion to requirements) amplificateur *m* hydraulique
~BEDPLATE - [paper man] plaque *f* de fondation hydraulique
~BENDING MACHINE - [mech] cintreuse *f* hydraulique, rouleuse *f* hydraulique
~BORING - [mining & hydr] forage *m* par pression hydraulique, sondage *m* à injection d'eau
~BOUNDARY CONDITION - [soil] conditions *fpl* hydraulique aux limites
~BRAKE - [mech] (a brake in which the shoes are expanded by small pistons operated by oil pressure) frein *m* à commande hydraulique
~BRAKE DRAIN TUBE - [auto] tube *m* de purge pour freins hydrauliques
~BRONZE - [metall] brone *m* de robinetterie
~BURSTING TEST - [tyres] (for aeroplane tyres) essai *m* de pression hydraulique d'éclament
~CABLE PRESS - [mech] presse *f* hydraulique pour câbles
~CEMENT - [constr] (cement which hardens under water) ciment *m* hydraulique
~CIRCUIT - [mech] (a system or part of a system of hydraulic mechanism and connexions) circuit *m* hydraulique
~CLEANING - [metall] ébarbage *m* par jet d'eau
~CLUTCH - [mech] (transmission device in which the degree of mechanical coupling can be varied by hydraulic means) couplement *m* hydraulique
~CONDUCTIVITY - [soil] coefficient *m* de perméabilité
~CONTROL - [mech] régulateur *m* hydraulique
~CONVERTER - [mech] convertisseur *m* hydrodynamique
~COUNTERPOISE - [mech] contrepoids *m* hydraulique
~COUPLING - [mech] accouplement *m* hydraulique, embrayage *m* hydraulique
~COUPLING DRIVE - [mech] (in a fuel drive) convertisseur *m* de couple
~CRANE - [mech] grue *f* hydraulique
~CUSHION - [mech] amortisseur *m* hydraulique
~DAMPER - [mech] amortisseur *m* hydraulique
~DAMPING - [mech] (damping effected by means of restricted flow in a hydraulic cylinder) amortissement *m* hydraulique
~DREDGER - [mech] (type of suction dredger) drague *f* hydraulique, pompe *f* de dragage
~DRIVE - [contr] commande *f* hydraulique
~DUMP HOIST - [transp] vérin *m* hydraulique

HYDRAULIC ELEVATOR - [mech] élévateur *m* hydraulique
~ENGINEERING - [gen] construction *f* hydraulique, technique *f* hydraulique
~EXTRUDER - [mech] (an extrusion machine using a hydraulic ram to produce the necessary pressure) boudineuse *f* hydraulique
~FEED - [mech] avancement *m* hydraulique
~FILL - [soil] remblai *m* hydraulique, remblai *m* mis en place hydrauliquement
~FILL DAM - [hydr] barrage *m* en remblai hydraulique
~FILL METHOD - [hydr] procédé *m* de remblai hydraulique
~FIT - [mech] accouplement *m* bloqué
~FLATTENING - [metall] aplatissement *m* hydraulique
~FLUID - [ind chem] (specially prepared oil, used as the working medium in hydraulic devices) fluide *m* hydraulique, fluide *m* pour presses hydrauliques
~GEAR - [mech] dispositif *m* hydraulique
~GLUE - [chem] (glue which is partially moisture resisting) colle *f* résistante à l'humidité
~GOLD MINING - [mining] exploitation *f* hydraulique de l'or
~GOVERNOR - s. hydraulic control
~GRADIENT - [hydr] (of a system of fluid flow) niveau *m* piézométrique
~HAMMER - [mech] marteau *m* hydraulique
~HEAD - [hydr] pression *f* hydraulique, poussée *f* de l'eau
~HOIST - [mech] treuil *m* hydraulique
~HORSEPOWER - [mech] puissance *f* en cheval hydraulique
~HOSE - [rubber ind etc] (specially thick-walled strong hose used for hydraulic connexions) tuyau *m* souple à haute pression, tuyau *m* armé
~INTENSIFIER - [mech] (device for augmenting hydraulic pressure in a system) élévateur *m* de pression hydraulique
~JACK - [impl] (hydraulic lifting device (also sometimes used for a hydraulic actuator or ram) vérin *m* hydraulique
~JUMP - [hydr] ressant *m* hydraulique
~LEAKAGE TESTBENCH - [mech] (assembly of testing equipment to determine existence and amount of leakage in hydraulic equipment) banque *m* d'essai d'hydraulique
~LIFT - [mech] ascenseur *m* hydraulique, monte-charge *m* hydraulique
~LIME - [chem] (type of lime which will harden under water) chaux *f* hydraulique
~LIME MORTAR - [constr] mortier *m* de chaux hydraulique, mortier *m* hydraulique
~LOCK - [mech] (immobilisation of a hydraulic device by preventing the flow of fluid in it) verrouillage *m* hydraulique
~MACHINE - [mech mach] machine *f* hydraulique
~MAIN - [gas ind] (large horizontal steel pipe into which the products of distillation are discharged) barillet *m* humide
[metall] (collecting main in coke ovens) barillet
~MAIN AGITATOR - [gas ind] (a scraper chain) chaîne *f* à raclettes
~MEAN DEPTH - [hydr] (the ratio of the sectional area of flow through a pipe to the wetted perimeter) rayon *m* moyen, rayon *m* hydraulique
~MINING - [min] (also called 'hydraulicking; the

operation of breaking down and working alluvial deposits by jets of water under high pressure) abattage *m* à l'eau, abattage *m* hydraulique, hydrauliquage *m*, exploitation *f* hydraulique

HYDRAULIC MORTAR - [constr] (mortar under water) chaux *f* hydraulique

~ MOTOR - [mech] (multi-cylinder reciprocating engine driven by water under pressure) moteur *m* hydraulique

~ OIL - [mech] (used in hydraulic hoidts) liquide *m* hydraulique

~ ORGAN - [mus] (organ-like instrument built in the second century B/C) orgue *m* hydraulique

~ PACKING - [mech] (rings providing self-tightening packing under fluid pressure) garniture *f* hydraulique

~ PERCUSSION METHOD - [hydr] procédé *m* de percussion à injection d'eau

~ POWER - [mech] force *f* hydraulique

~ PRESS - [mech mach] (press operated by hydraulic pressure) presse *f* hydraulique

~ PRESSURE - [phys] pression *f* hydraulique, poussé *f* de l'eau

~ -PRESSURE EXTRUDER - [rubber ind] expulseur *m* à pression hydraulique

~ PULLING TOOL - [metall] vérin *m* hydraulique

~ RADIUS - s. hydraulic mean depth

~ RAM - [mech] (the plunger of a hydraulic press) piston *m* hydraulique
(for the delivery of water under pressure) bélier *m* hydraulique

~ RIVETER - [mech] (machine used to close rivets by hydraulic power) riveuse *f* hydraulique

~ SEAL - [gas ind] (of a oven) fermeture *f* hydraulique

~ SHOCK ABSORBER - [mech] (type of shock absorber in which energy is absorbed by hydraulic means) amortisseur *m* hydraulique

~ SQUEEZER - [mach] (moulding machine operated by hydraulic power) cingleur *m* rotateur hydraulique

~ SYSTEM - [gen] (arrangement of hydraulic mechanism and connexions) système *m* hydraulique
[mech] (a type of transmission) transmission *f* hydraulique

~ TACHOMETER - [instr] (a tachometer based on the fact that pumps produce a velocity which can be converted into static pressure) tachymètre *m* hydraulique

~ TEST - [mech] (test for pressure tightness and strength applied to boilers) épreuve *f* à l'eau
[metall] (of follow metal parts) essai *m* hydraulique

~ TIPPING GEAR - [auto] dispositif *m* hydraulique de basculement

~ TORQUE CONVERTER - [auto] convertisseur *m* hydraulique de couple

~ TUBING - [mech] (soft thick-walled steel tube used for hydraulic circuits) tuyau *m* métallique pour transmission hydraulique

~ VULCANIZING PRESS - [mech mach] (used in the rubber industry) presse *f* hydraulique à vulcaniser

~ WATER - [mech] eau *f* motrice

~ WHEEL PRESS - [mech] presse *f* hydraulique pour le calage des roues

~ WORKING - [mech] manœuvre *f* hydraulique

HYDRAULICALLY OPERATED LIFT - [agric] élevateur *m* hydraulique

HYDRAULICITY - [chem] (the property of lime, cement etc to set under water) hydraulicité *f*

HYDRAULICKING - s. hydraulic mining

HYDRAZIDES - [chem] (mono-acyl derivatives of hydrazine) hydrazides *fpl*

HYDRAZINE - [chem] (compound of hydrogen and nitrogen ($H_2N - NH_2$), used as a high-energy rocket fuel) hydrazine *f*

~ HYDRATE - [chem] (a strong diacid base, which forms salts with acids: it is a very powerful reducing agent and attacks glass, rubber and cork) hydrate *m* d'hydrazine

~ SULPHATE - [chem] (diamine sulphate; diamidogen sulphate, a powerful reducing agent. Also has applications as a catalyst in acetate fibre production, and in condensation reactions) sulphate *m* d'hydrazine

HYDRAZO COMPUNDS - [chem] (symmetrical hydrazine derivatives, obtained by reducing azo compounds) hydrazoïques *mpl*

HYDRAZOBENZENE - [chem] (a synthesis intermediate) hydrazobenzène *m*

HYDRAZONES - [chem] (compounds obtained by condensing aldehydes and ketones with hydrazine) hydrazones *fpl*

HYDREMIA - [med] hydrémie *f*

HYDRIATRICS - [med] hydrothérapie *f*

HYDRIDES - [chem] (compounds of hydrogen in which only other elements is present) hydrures *mpl*

HYDRIOTIC ACID - [chem] (a disinfectant and analytical reagent) acide *m* iodhydrique

HYDRO-FINISHING - [mech] (finishing by jets of a mixture of water and emergy powder) finissage *m* à jets d'eau

~ -RUBBER - [rubber ind] (rubber to which hydrogen has been added in combined form by catalytic action) caoutchouc *m* hidrogéné

HYDROAIRPLANE - s. hydroplane

HYDROBLAST - [mining] dessablage *m* hydraulique

HYDROBORONS - [chem] hydrures *mpl* de bore

HYDROBROMIC ACID - [chem] (an analytical reagent and a component of pharmaceuticals) acide *m* hydrobromique

HYDROCARBON - [chem] (a compound containing only carbon and hydrogen) hydrocarbure *m*, carbure *m*

~ POLYMER - [chem] (a polymer based on a hydrocarbon) hydrocarbure *m* polymérique

~ SOLVENT - [chem] (a liquid fuel) solvant *m* blanc

HYDROCELE - [med] hydrocèle *f*

HYDROCELLULOSES - [chem] (compounds formed by treating cellulose with cold concentrated acids. They retain the fibrous structure of cellulose) hydrocelluloses *fpl*

HYDROCEPHALUS - [med] hydrocéphalie *f*

HYDROCHLORIC ACID - [chem] (aqueous solution of hydrogen chloride. Strong, caustic acid with wide industrial uses) acide *m* chlorhydrique

HYDROCHLORINATED RUBBER - [rubber ind] caoutchouc *m* hydrochloré

HYDROCINNAMIC ACID - [chem] (a perfumery fixative) acide *m* hydrocinnamique

HYDROCONDENSER - [el] (used in cinema projectors) hydrocondensateur *m*

HYDROCOPYING - [mech] reproduction *f* hydraulique

HYDROCOPYING MACHINE - [mach tool] tour *m* hydraulique à copier
HYDROCORTISONE - [chem] (a steroid hormone with uses in medicine) hydrocortisone *f*
HYDROCRACKING - [ind chem] (a refining process for petroleum, in which cracking is carried out in the presence of hydrogen and a catalyst) hydrocraquage *m*
HYDROCYANIC ACID - [chem] (intensely poisonous substance used in the manufacture of rubber and plastics (acrylonitrile, acrylates, etc.) acide *m* cyanydrique
HYDROCYCLONE - [hydr] épaississeur-cyclone *m*, hydrocyclone *m*
HYDRODESULPHURIZATION - [chem] (the removal of sulphur compounds by the elimination of the sulphur as hydrogen sulphide by reaction with hydrogen in the presence of a catalyst) hydrodésoufrage *m*
HYDRODYNAMIC - [phys] hydrodynamique
~ PRESSURE - [soil] pression *f* hydrodynamique
HYDRODYNAMICS - [phys] (the science treating of the flow of liquids) hydrodynamique *f*
HYDRODYNAMOMETER - [instr] (instrument designed to measure the speed of the flow of a fluid) hydrodynamomètre *m*
HYDROELECTRIC - [el] (relating to a water-turbine) hydro-électrique
~ GENERATING SET - [el] (a generating set with a prime mover using hydraulic energy) groupe *m* hydro-électrique
HYDROEXTRACTOR - [mech] (machine consisting of a perforated cylinder, in which water is removed from loose material by flinging it outwards) hydroextracteur *m*, essoreuse *f* centrifuge
HYDROFLUORIC ACID - [chem] (an intensely corrosive solution of hydrogen fluoride in water used in etching glass, metal processing, and cleaning stonework) acide *m* hydrofluorique
HYDROFOIL - [aero & naut] (a hydrodynamic surface, espec. one placed below a float in a seaplane to give additional lift at take-off) surface *f* auxiliaire d'hydroplanage
HYDROFORMING - [chem] formage *m* hydraulique, hydroforming *m*
~ PROCESS - [chem] (catalytic process for the production of cyclic and aromatic hydrocarbons) procédé *m* d'hydroformage
HYDROFUGE - [gen] hydrofuge
HYDROFURAMIDE - [chem] (furfuramide; hardening agent and accelerator for resins) hydrofuramide *f*
HYDROGEL - [chem] (an aqueous gel, i.e. one in which water forms the liquid constituent) hydrogel *m*
HYDROGEN - [chem] (a gaseous element, symbol H, the lighest element. A starting material for the synthesis of ammonia, hydrochloric acid, methanol, the hydrogenation of coal, in the atomic hydrogen and oxy-hydrogen blowpipe to produce high temperatures, and in the hydrogenation of oils) hydrogène *m*
~ BOMB - [nucl] (nuclear bomb operating on the transmutation of hydrogen and lithium isotopes into helium) bombe *f* à hydrogène, bombe *f* H
~ BOND - [chem] (valence linkage joining two electro-negative atoms through a hydrogen atom) liaison *f* d'hydrogène
HYDROGEN BROMIDE - [chem] (a starting material and intermediate for pharmaceutical compounds, and a catalyst in petroleum processing) acide *m* bromhydrique
~ CHAMBER - [phys] chambre *f* d'ionisation à hydrogène
~ CHLORIDE - [chem] (colourless, poisonous gas with a strong affinity for water and with application in the production of a number of polymers) acide *m* chlorhydrique
~ COOLING - [chem] (cooling system, espec. for electrical machinery, in which gaseous hydrogen is circulated as a heat transfer medium) refroidissement *m* par hydrogène
~ CYANIDE - [chem] (intensely poisonous gas with uses as a fumigant and as a starting material for nylon and other polymers, pharmaceuticals, and dyestuffs) acide *m* cyanhydrique
~ DIOXIDE - [chem] (H_2O_2) eau *f* oxygénée
~ ELECTRODE - [electron] (a platinized platinum electrode surrounded by hydrogen and immersed in an aqueous solution) électrode *f* à hydrogène
~ EMBRITTLEMENT - [metall] (increase on the brittleness of a metal caused by hydrogen absorption during any plating operation) fragilité *f* par l'hydrogène
~ FLUORIDE - [chem] (colourless, corrosive gas with application as a catalyst and starting material for fluorine compounds) acide *m* hydrofluorique
~ FUSION BOMB - s. hydrogen bomb
~ GENERATING PLANT - [ind chem] installation *f* pour la production d'hydrogène
~ IODIDE - [chem] (a heavy colourless gas, which fumes strongly in contact with air) acide *m* iodhydrique
~ ION - [chem] (an atom of hydrogen carrying a positive charge) ion *m* hydrogène
~ -ION CONCENTRATION - [chem] (a measure of the acidity of a solution) concentration *f* en ions d'hydrogène
~ ISOTOPES - [nucl] (protium, deuterium and tritium) isotopes *mpl* d'hydrogène
~ -LIKE ATOM - [nucl] (an atom or ion resembling the shell structure of a hydrogen atom) atome *m* hydrogénoïde
~ OXIDE - [chem] oxyde *m* d'hydrogène
~ PEROXIDE - [chem] (colourless, corrosive liquid with oxidising properties with applications in foam rubber production and epoxidation processes) eau *f* oxygénée
~ PHOSPHIDE - [chem] phosphure *m* d'hydrogène
~ SCALE - [el] (used for the measurement of electrode potentials) échelle *f* à hydrogène
~ SULPHIDE - [chem] (poisonous, evil-smelling gas with uses as a reagent) hydrogène *m* sulfuré
~ SULPHIDE DUAL TEMPERATURE EXCHANGE PROCESS - [chem] procédé *m* d'échange à deux températures à l'hydrogène sulfure
~ THYRATRON - [electron] (thyratron with a gas-filling of hydrogen) thyratron *m* à hydrogène
HYDROGENATE, to - [chem] hydrogéner
HYDROGENATED - [chem] hydrogéné
~ OILS - [chem] (oils which have been hydrogenated by the catalytic addition of hydrogen to give products having a higher melting point than the original substance) huiles *fpl* hydrogénées
~ RUBBER - [chem] (rubber to which hydrogen has

been added in combination, by a catalytic process) caoutchouc *m* hydrogéné
HYDROGENATION - [chem] (the addition of hydrogen to an unsaturated compound) hydrogénation *f*
[of coal] (treatment of coal with hydrogen by a direct process, to obtain liquid and gaseous fuels and industrial chemical products) hydrogénation *f*
HYDROGENERATOR - [ind chem] (closed vessel fitted with temperature control arrangements, sprays etc. for hydrogenating oils) appareil *m* d'hydrogénation
HYDROGENOUS - [chem] (containing a high percentage of hydrogen) hydrogénique
~ MODERATOR - [nucl] modérateur *m* hydrogéné
HYDROGLIDER - [aero] hydroglisseur *m*
HYDROGRAPHER - [gen] hydrographe *m*
HYDROGRAPHIC - [gen] hydrographique
HYDROGRAPHY - [science] (the study of the conditions of navigable waters) hydrographie *f*
HYDROHALOGENATED PRODUCT - [chem] (product treated with hydrochloric acid) produit *m* hydrohalogéné
HYDROKINETIC - [phys] se rapportant à la cinétiques des liquides
HYDROLAPPING - [metall] rodage *m* hydraulique
HYDROLASE - [chem] (an enzyme) hydrolase *f*
HYDROLIZING TANK - [hydr] (a septic tank) fosse *f* septique
HYDROLOGY - [science] hydrologie *f*
HYDROLYSIS - [chem] (the breaking down of a molecule by water together with the decomposition of the water) hydrolyse *f*, électrolyse *f* de l'eau
HYDROLYTIC - [chem] hydrolitique, qui agit par hydrolyse
~ STABILITY - [mech] stabilité *f* hydrolytique
HYDROLYZE, to - [chem] hydroliser
HYDROMA - [med] boursite *f* rotulienne chronique
HYDROMASSAGE - [med] massage *m* sous l'eau
HYDROMATIC - [mech] hydromatique
~ BED - [mach] banc *m* hydromatique
~ PROPELLER - [aero] propulseur *m* hydromatique
HYDROMECHANICS - [mech] hydromécanique
HYDROMEL - [zool] hydromel *m*
HYDROMETALLURGY - [metall] (the science of obtaining metals from ores or compounds by solution in water) hydrométallurgie *f*
HYDROMETEOR - [met] (bodies of precipitated liquid or solid water (e.g. rain or hail) hydrométéore *m*
HYDROMETER - [instr] (instrument for determining the approximate specific gravity of a liquid, consisting of a glass float with graduations showing the depth to which it will sink in different liquids) hydromètre *m*, aréomètre *m*, pèse-acide *m*
HYDROMETRY - [chem] hydrométrie *f*
HYDROMOTOR - [naut] (a propulsion device actuated by the reaction of a water jet) hydromoteur
HYDROMYOMA - [med] fibrome *m* kystique
HYDRONEPHROSIS - [med] uronéphrose *f*
HYDROPAROTITIS - [med] parotidite *f* séreuse
HYDROPENIA - [med] hydropénie *f*
HYDROPEXIA - [med] hydropexie *f*
HYDROPHILIC - [chem] (having affinity for water) hydrophile
HYDROPHOBIA - [vet] hydrophobie *f*
HYDROPHOBIC - [chem] (substances which are immiscible with water) hydrophobe
HYDROPHONE - [instr] (electroacoustic transducer transforming waterborne acoustic signals into electrical form) hydrophone *m*
HYDROPHYSOMETRA - [med] physohydrométrie *f*
HYDROPLANE - [aero] hydravion *m*
[naut] hydroglisseur *m*
[naut] (of a submarine; planing surface enabling it to submerge) barre *f* de plongée (de sous-marin)
HYDROPLANING - [naut] hydroplanage *m*
HYDROPNEUMATIC - [mech] (actuated by air and water) hydropneumatique
~ ACCUMULATOR - [el] accumulateur *m* hydropneumatique
HYDROPNEUMOPERICARDIUM - [med] hydropneumopéricarde *m*
HYDROPNEUMOPERITONEUM - [med] hydropneumopéritoine *m*
HYDROPNEUMOTHORAX - [med] hydropneumothorax *m*
HYDROPOWER - [mech] force *f* hydraulique
HYDROQUINONE - [chem] (polymerization inhibitor and a component of rubber coating compounds) hydroquinone *f*
~ BENZYL ETHER - [chem] (organic intermediate stabilizer and polymerization inhibitor) benzyléther de l'hydroquinone
~ DIMETHYL ETHER - [chem] (a perfumery fixative and intermediate for resins and plastics additives) diméthyléther *m* de l'hydroquinone
~ HYDROCHLORIDE - [chem] (starting material for the production of quinine) chlorhydrate *m* d'hydroquinone
~ MANOMETHYL ETHER - [chem] (organic intermediate with uses in plastics plasticizers, antioxidants, inhibitors, stabilizers and UV stabilizers) monométhyléther *m* de l'hydroquinone
HYDRORRHEA - [med] hydrorrhée *f*
HYDRORUBBER - [rubber ind] caoutchouc *m* hydrogéné
HYDROSALPINX - [med] hydrosalpinx *m*, sactosalpinx *m*
HYDROSARCOCELE - [med] hydrocèle *f* avec sarcocèle
HYDROSTAT - [el] (controlling device regulating the water level) hydrostat *m*
HYDROSTATIC - [phys] hydrostatique
~ BALANCE - [instr] (balance designed to ascertain the specific gravity of substances) balance *f* hydrostatique
~ CENTRE OF PRESSURE - [mech] centre *m* de pression hydrostatique
~ COMPRESSION - [soil] compression *f* hydrostatique
~ HEAD - [hydr] (the pressure exerted by a column of fluid, equal to the height of the column by the fluid density by the acceleration of gravity) charge *f* hydrostatique
~ PRESSURE - [phys] pression *f* hydrostatique
~ PRESSURE GRADIENT - [hydr] gradient *m* de la pression hydrostatique
~ PULL - [hydr] traction *f* hydrostatique
~ TEST - [mech] (test to find leakages in a drain) épreuve *f* hydrostatique
[th eng] (of boilers) s. hydraulic test
~ UPLIFT - [hydr] sous-pression *f* hydrostatique
~ VALVE - [mech] (apparatus tending to maintain an underwater body at the desired depth) plongeur *m* hydrostatique
HYDROSTATICS - [phys] (the science of the equilibrium of liquids and of fluid pressure) hydrostatique *f*

HYDROSUCTION MACHINE - [mech] extracteur *m* hydraulique
HYDROTHERAPY - [med] hydrothérapie *f*
HYDROTHORAX - [med] hydrothorax *m*
HYDROTROPES - [chem] (substances which can increase the solubility in water of certain organic compounds) hydrotropes *mpl*
HYDROTYMPANUM - [med] hydrotympan *m*
HYDROUS - [phys] (the presence in a substance of an unspecific amount of water) hydrique, hydraté, aqueux
HYDROVANE - s. hydrofoil
HYDROXIDES - [chem] (compounds of the basic oxides with water) hydroxydes *mpl*
HYDROXYACETIC ACID - [chem] (emulsificant, textile dyeing agent, and cleaning agent) acide *m* hydroacétique
HYDROXYBENZALDEHYDE - [chem] (an intermediate for pharmaceuticals, plastics, and dyestuffs) hydrobenzaldéhyde *f*
HYDROXYBUTYRIC ACID - [chem] (a synthesis intermediate) acide *m* hydroxibutyrique
HYDROXYCITRONELLAL - [chem] (a perfumery fixative) hydroxycitronellal *m*
HYDROXYDIBENZOFURAN - [chem] (a synthesis intermediate) hydroxydibenzofurane *m*
HYDROXYDIPHENYLAMINE - [chem] (a starting material for the production of sulphur dyes) hydroxydiphénylamine *f*
HYDROXYETHYLCELLULOSE - [chem] (vinyl polymerization stabilizer) hydroxyéthylcellulose *f*
HYDROXYETHYLENEDIAMINE - [chem] (starting material for synthetic rubbers and resins) hydroxyéthylènediamine *f*
HYDROXYETHYLHYDRAZINE - [chem] (a synthesis intermediate) hydroxyéthyldrazine *f*
HYDROXYETHYL PIPERAZINE - [chem] (intermediate for pharmaceuticals) hydroxyéthylpiperazine *f*
HYDROXYETHYLTRIMETHYLAMMONIUM BICARBONATE - [chem] (synthesis intermediate and catalyst) bicarbonate *m* d'hydroxyéthyltriméthylammonium
HYDROXYL GROUP - [chem] (monovalent group consisting of a hydrogen atom and an oxygen atom linked together) hydroxyles *mpl*
HYDROXYLAMINE - [chem] (synthesis intermediate) hydroxylamine *f*
~ACID SULPHATE - [chem] (intermediate for rubber chemicals) sulphate acide d'hydroxylamine
~HYDROCHLORIDE - [chem] (photographic chemical and synthesis intermediate) chlorhydrate *m* d'hydroxylamine
~SULPHATE - [chem] (synthesis intermediate, textile processing agent, depilatory for hides and skins, photographic chemical, and catalyst) sulphate *m* d'hydroxylamine
HYDROXYMERCURICHLOROPHENOL - [chem] (an agricultural disinfectant) hydroxymercurichlorophénol
HYDROXYMERCURICRESOL - [chem] (an insecticide) hydromercuricrésol
HYDROXYMERCURINITROPHENOL - [chem] (an insecticide) hydroxymercurinitrophénol
HYDROXYMETHYLBUTANONE - [chem] (synthesis intermediate) hydroxyméthylbutanone
HYDROXYNAPHTHOIC ACID - [chem] (an intermediate for dyestuffs) acide *m* hydroxynaphtoïque
~ANILIDE - [chem] (dyestuffs intermediate) anilide *f* hydroxynaphtoïque
HYDROXYNAPHTHOQUINONE-[chem] (intermediate for antiseptics, drugs and dyestuffs) hydroxynaphtoquinone *f*
HYDROXYPHENYLMERCURIC CHLORIDE - [chem] (an antiseptic and fungicide) chlorure *m* hydroxyphénylmercurique
HYDROXYPROLINE - [chem] (amino acid forming colourless, water-soluble crystals) hydroxyproline *m*
HYDROXYPROPYL TOLUIDINE - [chem] (intermediate for dyestuffs) hydroxypropyltoluidine *f*
HYDROXYPROPYLGLYCERIN - [chem] (plastics plasticizer, soluble in methanol and water) hydroxypropylglycérine *f*
HYDROXYPYRIDINE OXIDE - [chem] (a bactericide) oxyde *m* d'hydroxypyridine
HYDROXYQUINOLINE - [chem] (an intermediate for fungicides) hydroxyquinoléine *f*
~BENZOATE - [chem] (an agricultural fungicide) benzoate *m* d'hydroxyquinoléine
~SULPHATE - [chem] (an antiseptic) sulphate *m* d'hydroxyquinoléine
HYDROXYSTEARYL ALCHOL - [chem] (intermediate for plastics, resins and pharmaceuticals) alcool *m* hydroxystéarylique
HYDROXYTITANIUM STEARATE - [chem] (a crosslinking agent) stéarate *m* d'hydroxytitane
HYDROZINCITE - [chem] (a zinc ore, consisting essentially of natural basic zinc carbonate, also called bloom) hydrozincite *f*
HYETOGRAM - [met] pluviogramme *m*
HYETOGRAPH - [instr] (instrument designed to collect and to measure the fall of rain) pluviographe *m*
HYETOMETER - [instr] (instrument designed to measure the fall of rain over a specified period) pluviomètre *m*
HYGIENE - [gen & med] (practices relating to the maintainance of health) hygiène *f*
HYGIENIC - [gen] hygiénique
HYGRODEIK - [instr] (type of hygrometer with wet and dry bulb thermometer) hygromètre *m* à condensation
HYGROGRAPH - [instr] (instrument making a continuous record of barometric pressure, temperature and humidity on a single chart) hygrographe *m*
HYGROMETER - [instr] (instrument to measure humidity) hygromètre *m*
HYGROMETRIC STATE - [phys] (relative humidity) humidité *f* relative
HYGROMETRY - [met] (the measurement of relative humidity) hygrométrie *f*
HYGROSCOPE - [instr] (instrument designed to show variations in the moisture of the atmosphere) hygroscope *m*
HYGROSCOPIC - [phys] (having the property of absorbing moisture from the air) hygroscopique
~PAPER - [paper man] papier *m* hygroscopique
HYGROSCOPICITY - [phys] (the property of absorbing moisture from the air) hygroscopicité *f*
HYGROSTAT - [th eng] (in air conditioning) hygrostat *m*
HYLA - [med] partie *f* latérale de l'aqueduc de Sylvius
HYLOMA - [med] hylome *m*
HYMEN - [anat] hymen *m*
HYOID - [med] hyoïde *m*
HYP - [meas] (one tenth of a neper) hyp *m*

HYPABYSSAL - [geol] hypabyssal
HYPADRENIA - [med] insuffisance *f* surrénalienne
HYPAETHRAL - [arch] (of a building without a roof) hypètre
HYPALGESIA - [med] hypoalgésie *f*
HYPASTHENIA - [med] hypoasthénie *f*
HYPERACID - [chem] hyperacide
HYPERACIDITY - [med] hyperacidité
HYPERACUSIS - [med] hyperacousie *f*
HYPERADRENALEMIA - [med] hypersurrénalisme *m*, hyperepinéphrie *f*
HYPERAFFECTIVITY - [med] hyperaffectivité *f*
HYPERALGESIA - [med] hyperalgie *f*
HYPERAPHIA - [med] hyperesthésie *f* tactile
HYPERBOLA - [math] (the section of a right circular cone by a plane intersecting the cone on both sides of the plane) hyperbole *f*
HYPERBOLIC - [math] hyperbolique
~CONTINUOUS RADAR SYSTEM - [radar] système *m* hyperbolique continu de radionavigation
~GEAR WHEEL - [mech] engrenage *m* hyperboloïde
~NAVIGATION - [radar] (radar navigation by synchronized slave stations, as in Gee, Lorac and similar systems) navigation *f* hyperbolique
~PULSED RADAR SYSTEM - [radar] système *m* hyperbolique de radionavigation à impulsions
~RADAR SYSTEM - [radar] (system in which the radio waves are emitted from two basis points and received in a third one, the position of which is wanted) système *m* hyperbolique de radionavigation
HYPERCALCEMIA - [med] hypercalcémie *f*
HYPERCALCINURA - [med] hypercalciurie *f*
HYPERCHOLIA - [med] hypercholie *f*
HYPERCHROMASIA - [med] hyperchromie *f*
HYPERCINESIA - [med] hypercinèse *f*
HYPERCONDUCTIVITY - [el] hyperconductivité *f*
HYPERCONJUGATION - [phys] (the description of the properties of a molecule in terms of resonance structures) hyperconjugaison *f*
HYPERCRYALGESIA - [med] hypersensibilié *f* au froid
HYPERDYNAMIA - [med] hyperactivité *f* musculaire
HYPEREMESIS - [med] hyperémèse *f*, vomissement *m*
HYPEREMIA - [med] engorgement *m*, hyperémie *f*
HYPEREMOTIVITY - [med] hyperémotivité *f*
HYPEREPHIDROSIS - [med] hyperhydrose *f*
HYPERERGASIA - [med] hyperfonctionnement *m*
HYPERERGIA - [med] hyperallergie *f*
HYPERESTHESIA - [med] hyperesthésie *f*
HYPEREUTECTIC - [metall] hypereutectique
~IRON - [metall] fonte *f* hypereutectique
HYPEREUTECTOID - [metall] (of steel; with more carbon than is contained in pearlite) hypereutectoïde
~CAST IRON - [metall] fonte *f* à matrice hypereutectoïde
~STEEL - [metall] acier *m* hypereutectoïde
HYPEREXTENSION - [med] hyperextension *f*
HYPERFINE SPECTRUM - [phys] (spectrology) spectre *m* hyperfin
~STRUCTURE - [phys] (set of a very closely spaced lines forming a spectral line) structure *f* hyperfine
HYPERFOCAL - [photo] hyperfocal
~DISTANCE - [photo] (the distance beyond which all objects are substantially in focus) distance *f* hyperfocale
HYPERFOLLICULINISM - [med] hyperfolliculinisme
HYPERFORMING - [ind chem] hyperforming *m*
HYPERGALACTIA - [med] polygalactie *f*
HYPERGASIA - [med] hypofonctionnement *m*
HYPERGENESIS - [med] hyperplasie
HYPERGEUSIA - [med] hypergueusie *f*, oxyhuneusie *f*
HYPERHIDROSIS - [med] diaphorèse *f*, transpiration *f* excessive
HYPERINOCEMIA - [med] hyperfibrinémie *f*
HYPERINVOLUTION - [med] superinvolution *f*
HYPERKALEMIA - [med] hyperkaliémie *f*
HYPERLORDOSIS - [med] hyperlordose *f*
HYPERMASTIA - [med] hypermastie *f*, hypertrophie *f* mammaire
HYPERMETROPIA - [opt] hypermètropie *f*
HYPERMNESIA - [med] hypermnésie *f*
HYPERMYOTONIA - [med] hypertonie *f* musculaire
HYPERNATREMIA - [med] hypernatrémie *f*
HYPERON - [nucl] (a particle of mass intermediate between that of a neutron and a deutron) hypéron *m*
HYPEROVARIA - [med] hyperovarie *f*
HYPEROXEMIA - [med] hyperoxie *f*
HYPERPHASIA - [med] hyperphasie *f*
HYPERPNEA - [med] hyperpnée *f*
HYPERPYRETIC - [med] hyperpyrétique
HYPERREFLEXIA - [med] surréflectivité *f*
HYPERSCOPIC VIEW - [opt] (a type of stereoscopic view) projection *f* hyperscopique
HYPERSECRETION - [med] hyperchlorhydrie *f*
HYPERSENSITIZATION - [photo] (treatment of a sensitive emulsion to increase its speed) hypersensibilisation *f* (d'une pellicule)
HYPERSOMNIA - [med] hypersomnie *f*
HYPERSONIC - [phys] hypersonique
~SPEED - [aero] (high supersonic speed, generally taken as - Mach 5.) vitesse *f* hypersonique
HYPERSTATIC - [phys] hyperstatique
HYPERTENSION - [med] hypertension *f*
HYPERTHERMIA - [med] hyperthermie *f*
HYPERTHYMIA - [med] hyperthymie *f*
HYPERTONIC - [med] hypertonique
HYPERTROPHIA - [med] hypertrophie *f*
HYPERVISCOSITY - [med] hyperviscosité *f*
HYPHEDONIA - [med] hyphédonie *f*
HYPHEMIA - [med] hypéma *m*, hémorragie *f* dans la chambre antérieure de l'oeil
HYPINOSIS - [med] hypofibrinémie *f*
HYPNAGOGUE - [med] somnifère *m*
HYPNESTHESIA - [med] somnolence *f*
HYPNOLEPSY - [med] hypnolepsie *f*
HYPNONARCOSI - [med] hypnonarcose *f*
HYPNOSIA - [med] hypnosie *f*
HYPNOSIS - [med] hypnose *f*
HYPNOTISM - [med] hypnotisme *m*, mesmérisme *m*
HYPO BATH - [photo] bain *m* d'hyposulfite
HYPOADRENALEMIA - [med] hyposurrénalisme *m*, insuffisance *f* surrénalienne
HYPOADRENOCORTICISM - [med] hypocorticalisme *m*
HYPOBULIA - [med] déficience *f* de la volonté
HYPOCALCEMIA - [med] hypocalcémie *f*
HYPOCENTRE - [nucl] (the region vertically below or above the centre of an explosion of a nuclear weapon) hypocentre *m*
HYPOCHLOREMIA - [med] chloropénie *f*, hypochlorémie *f*
HYPOCHLORITE - [chem] (hypochlorous acid) hypochlorite *m*
HYPOCHLOROUS ACID - [chem] (a weak unstable

acid, which only exists in solution. It is formed when chlorine dissolves in water and thus acts as a bleach) acide *m* hypochloreux
HYPOCHOLIA - [med] oligocholie *f*
HYPOCHONDRIA - [med] hypocondrie *f*
HYPOCHROMASIA - [med] hypochromie *f*
HYPOCINESIA - [med] hypokinésie *f*
HYPOCOLASIA - [med] déficience *f* des mécanismes inhibiteurs
HYPOCYCLOID - [geom] hypocycloïde *f*, épicycloïde *f* intérieure
HYPOCYCLOSIS - [med] déficience *f* de l'accomodation de l'oeil
HYPOCYSTOTOMY - [med] cystotomie *f* périnéale
HYPODERM - [med] tissu *m* cellulaire sous-cutané, hypodermie *f*
HYPODYNAMIA - [med] hypodynamie *f*
HYPOERGIA - [med] hypoergia *f*
HYPOEUTECTIC - [metall] hypoeutectique
HYPOEUTECTOID - [metall] (containing less than 0,9 % carbon) hypoeutoctoïde
~CAST IRON - [metall] fonte *f* à matrice hypoeutectoïde
HYPOFUNCTION - [med] hypofonction *f*
HYPOGALACTIA - [med] hypogalactie *f*
HYPOGASTRIUM - [med] hypogastre *m*
HYPOGENESIS - [med] hypogénésie *f*
HYPOGENITALISM - [med] hypogénitalisme *m*, hypogonadisme *m*
HYPOGEUM - [arch] hypogée *m*
HYPOID - [mech] (of a gear in which the axes of the driving and driven shafts are at right angle but not in the same plane) hypoïde
~GEAR - [mech] engrenages *mpl* à taille hypoïde
~OIL - [chem] huile *f* hypoïde
HYPOIDROSIS - [med] hypohidrose *f*
HYPOLIMNION - [hydr] (in limnology) hypolimnion *m*
HYPOMENORRHEA - [med] hypoménorrhée *f*
HYPOMNESIS - [med] déficience *f* de la mémoire
HYPOMYOTONIA - [med] hypotonie *f* musculaire
HYPONITROUS - [chem] hypoazotique
HYPOPEPSIA - [med] hyposthénie *f* gastrique
HYPOPHOBIA - [med] hypophobie *f*
HYPOPHORIA - [med] hypotropie *f*, hypophorie *f*
HYPOPHOSPHITE - [chem] (hypophosphorous acid) hypophosphite *m*
HYPOPHOSPHOROUS ACID - [chem] (a starting material for the production of hypophosphites for medicinal use) acide *m* hypophosphoreux
HYPOPHYSECTOMY - [med] hypophysectomie *f*
HYPOPHYSIS - [med] hypophyse *f*, glande *f* pituitaire
HYPOPHYSOMA - [med] tumeur *f* hypophysaire
HYPOPLASIA - [med] hypoplasie *f*
HYPOPLAST(US) - [genet] hypoblaste *m*, feuillet *m* interne
HYPOPLOIDY - [biol] hypoploïdie *f*
HYPOPRAXIA - [med] inactivité *f*
HYPOSCOPIC VIEW - [opt] projection *f* hyposcopique
HYPOSENSITIVE - [med] hyposensitif
HYPOSIALADENITIS - [med] inflammation *f* de la glande sous-maxillaire
HYPOSPHAGMA - [med] ecchymose *f* sous-conjonctivale
HYPOSPHRESIA - [med] hyposmie *f*
HYPOSTHENIA - [med] hyposthénie *f*
HYPOSTATIC CONTROL - [contr] régulation *f* hypostatique
HYPOSTYLE - [arch] hypostyle
HYPOSULPHITE - [chem] (hyposulphurous acid) hyposulfite *m*, thiosulfate *m*
HYPOSULPHUROUS ACID - s. hyposulphite
HYPOSYNERGIA - [med] troubles *mpl* de la coordination
HYPOSYSTOLE - [med] hyposystolie *f*
HYPOTAXIA - [med] hypotaxie *f*
HYPOTENSION - [med] hypotension *f*
HYPOTENSOR - [med] hypotenseur *m*
HYPOTENUSE - [geom] (that side of a right-angled triangle which subtends the right angle) hypoténuse *f*
HYPOTERMIA - [med] hypothermie *f*
HYPOTESIS - [gen] hypothèse *f*
HYPOTHERMAL VEIN - [mining] filon *m* hypothermal
HYPOTHETICAL - [gen] hypothétique
~EXCHANGE - [telecomm] bureau *m* fictif
~REFERENCE CIRCUIT - [radio] (hypothetical circuit with a specified length and a specified amount of terminal and intermediate equipment) circuit *m* fictif de référence
HYPOTHYREA - [med] hypothyroïdie *f*
HYPOTONIA - [med] hypotonie *f*
HYPOTROPHY - [med] hypotrophie *f*
HYPOVOLEMIA - [med] hypovolhémie *f*
HYPOXEMIA - [med] hypoxémie *f*
HYPSOGRAPH - [instr] (instrument recording the transmission levels on a circuit) hypsographe *m*
HYPSOMETER - [instr] (instrument designed to determine heights by a boiling point thermometer) hypsomètre *m*, thermomètre *m* à ébullition, thermobaromètre *m*
HYPSOMETRIC TINTS - [cartography] teintes *fpl* hypsométriques
HYPSOMETRY - [meas] (the determination of heights by means of a boiling-point thermometer) hypsométrie *f*
HYSSOP - [agric] hysope *f*
HYSTERATRESIA - [med] atrésie *f* utérine
HYSTERESIS - [phys] (the difference in the quantitative effect of a force according to wheter it is increasing or diminishing) hystérésis, hystérèse *f*
~COEFFICIENT - s. hysteresis factor
~CONSTANT - [el] constante *f* d'hystérésis
~CURVE RECORDER - [instr] hystérésigraphe *m*
~DISTORTION - [el] (the distortion of voltage waveforms in circuits containing magnetic components) distortion *f* due à l'effet d'hystérésis
~ENERGY - [el] (the energy used per cycle of operation to overcome the effect of hysteresis) travail *m* d'hystérésis
~ERROR - [instr] (an error due to hysteresis in the instrument) erreur *f* d'hystérésis
~FACTOR - [el] (the increase in the effective resistance of a coil produced by hysteresis of its magnetic core when a current of one ampere at a given frequency is flowing) coefficient *m* d'hystérésis
~LOOP - [el] (the closed figure obtained by plotting the hysteresis curves for the same substance under identical conditions, first with ascending and then with descending values of stress) boucle *f* d'hystérésis
~LOSS - [el] (loss of energy due to hysteresis) pertes *fpl* par hystérésis
~METER - [instr] (instrument designed to measure

hysteresis torque) histérésimètre *m*
HYSTERESIS MOTOR - [el] moteur *m* à hystérésis
~TESTER - s. hysteresis meter
HYSTEREURYSIS - [med] dilatation *f* intrumentale de l'utérus
HYSTERIA - [med] pithiatisme *m*, hystérie *f*
HYSTERIC - [med] hysterique *f*
HYSTEROCELE - [med] hernie *f* de l'utérus
HYSTEROGRAM - [med] hystérogramme *m*
HYSTEROLITH - [med] calcul *m* utérin
HYSTEROMETRY - [med] hystérométrie *f*
HYSTEROMYOTOMY-[med] hystéromyotomie *f*
HYSTEROSALPINGOGRAPHY - [radiat] (the radiological examination of the uterus and the Fallopian tubes) hystérosalpingographie *f*
HYSTEROSCOPY - [med] métroscopie *f*
HYSTEROSPASM - [med] spasme *m* utérin
HYSTEROTOMY - [med] métrotomie *f*, hystérotomie *f*
HYSTEROTRACHELOPLASTY - [med] plastie *f* du col utérin
HZ - [el] (abbrev. for Hertz) hertz *m*

I

I - [chem] (the symbol of iodine) symbole *m* de l'iode
[mech] (the symbol for Moment of Inertia) symbole du moment d'inertie
I.A. - [phys] (International Angstrom) unité internationale d'Angstrom
I.A.S. - [aero] (Indicated Airspeed) vitesse *f* indiquée
I.A.T.A. - [aero] (initials of International Air Traffic Association) Association *f* Internationale Trafic Aérien
I - BAR - [metall] fer *m* double T
I - BEAM - [mech & constr] (also called H-beam) poutre *f* en double T
I - BEAM AXLE - [auto] essieu *m* rigide à section en I
I - RAIL - [railw] rail *m* à double champignon
IANTHINITE - [min] (rare alternation product of uraninite, containing approximately 70% of uranium) ianthinite *f*
IBEX - [zool] bouquetin *m*, ibex *m*
I/C - [mech] (initials of Internal Combustion) combustion *f* interne
I.C.A.O. - (initials of International Civil Aviation Organisation) Organisation *f* Internationale d'Aviation Civile
ICAO STANDARD ATMOSPHERE - [met] (standard set of atmospheric conditions) (pressure, temperature, density and altitude laid down by ICA) atmosphère *f* standard OACI
ICE, to - [gen] congeler, geler
[gen] (to convert into ice) geler
[gen] (the operation of chilling) refroidir, congeler
[gen] (of objects, roads etc) prendre dans les glaces
ICE - [gen phys & met] glace *f*
~ACCRETION - [aero] (accumulation of ice on the outside surfaces of an aircraft, caused by flight through moisture-laden air at temperatures below 34 deg. F) accumulation *f* de glace
~APRON - [constr] (construction covering the upstream side of a bridge pier) brise-glace *m*
~-BAG - [med] vessie *f* à glace, poche *f* à glace
~BLINDNESS - [med] cécité *f* des neiges
~BLOCK - [gen] bloc *m* de glace
~-BOX - [gen] glacière *f*, compartiment *m* congélateur
~BREAK - s. ice-breaker
~-BREAKER - [naut] (vessel used to break floating ice) brise-glace *m*
[constr] s. ice apron
~CALORIMETER - [instr] (instrument designed to measure the specific heat of a solid or liquid of which only a small quantity is available) calorimètre *m* à glace
ICE CAN - [gen] forme *f* pour glace
~CAP - [geol] calotte *f* glaciaire
~CAPACITY - [gen] production *f* de glace
~CAR - [railw] wagon *m* frigorifique
~CELL - [gen] compartiment *m* frigorifique
~COLOURS - [chem] (term used for dyestuffs formed directly on the cotton fibre, using a second component interacted with an ice-cooled solution of a diazo salt) teintures *fpl* azoïques
~CONDENSER - [chem] condenseur *m* à glace
~CREAM - [food] glace *f*
~CRYSTAL CLOUDS - [met] (clouds, e.g. cirrus, composed of fine ice particles) nuages *mpl* contenant des aiguilles de glace
~CUBE - [gen] morceau *m* cubique de glace, cube *m* de glace
~FALL - [geol] cascade *f* de séracs
~FLOE - [naut] (floating ice) glaçon
~FORMATION - [aero & auto] givrage *m*
~GUARD - [aero] (in a ram air-intake, a screen of wire mesh on which ice forms rather than within the system) grille *f* anti-givre
~HOOK - [impl] marteau *m* d'escalade
~-HOUSE - s. ice-box
~INDICATOR - [aero] indicateur *m* anti-givre
~MOULD - s. ice can
~NEEDLES - [met] (very small slender bodies of ice, usually suspended in the air, as in circus cloud) aiguilles *fpl* de glace
~NUCLEUS - [phys] noyau *m* de glace
~PACK - [met] banquise *f*
~POINT - [phys] température *f* de la glace fondante
~QUAKE - [met] fracas *m* des glaces
~RAIN - [met] (a shower of fine particles of precipitated ice) pluie *f* de glace, grésillon *m*
~RIVER - [met] fleuve *m* de glace
~SPAR - s. Iceland spar
~-STORM - s. ice rain
~STRENGTHENING - [shipbuild] renforcement *m* pour la navigation
~TONGUE - [geogr] glacier *m* d'écoulement entre les glaces
~TRAY - [th eng] tiroir *m* à glace, bac *m* à glace
ICEBERG - [met] (a mass of floating ice detached from the end of a glacier flowing into the sea) iceberg *m*
ICEBOAT - [transp] bateau *m* à patins
ICEBOUND - [gen] (made impassable by ice) pris dans les glaces, bloqué par les glaces
ICELAND MOSS - [bot] lichen *m* d'Islande, mousse

f d'Islande
ICELAND SPAR - [min] (natural crystalline calcium carbonate in a pure state, transparent and having perfect cleavage and double refraction. Used in the construction of Nicol prisms) spath *m* d'Islande, cristal *m* d'Islande
ICELANDIC LOW - [met] (low-pressure system typically occurring in the neighbourhood of Iceland) minimum *m* islandais
ICHNOGRAM - [gen] ichnogramme *m*
ICHNOGRAPHY - [draw] (preparation of the plan of a building) ichnographie *f*, plan *m* géométral
ICHOR - [med] (watery discharge) ichor *m*
ICHTHYOL - [chem] (a disinfectant) ichtyol *m*
~OIL - [chem] huile *f* d'ichtyol
ICHTHYOPHOBIA - [med] ichtiophobie *f*
ICHTHYOSIS - [med] (dryness and scaliness of the skin) ichtyose *f*
ICING - [gen] (on surfaces) congélation *f*
[aero] (accumulation of ice on an aircraft in flight) givrage
[food] surglaçage
~FORMATION - [aero] (accumulation of ice on an aircraft in flight) givrage *m*
~HEART - [med] péricardite *f* séreuse
~INDEX - [aero] (an estimate of ice-formation probability at specific points and times) index *m* de givrage
~LIVER - [med] cirrhose *f* du foie
~UP - s. icing
ICONOMETER - [instr] (instrument determining the distance of an object of known size, or the size of an object at a known distance) iconomètre *m*
ICONOSCOPE - [telev] (camera tube in which the image of a scene is projected on to a photoelectric cathode) iconoscope *m*
~MOSAIC - [telev] mosaique *m*
ICONOSTASIS - [arch] iconostase *f*
ICOSAHEDRON - [geom] icosaèdre *m*
ICTERUS - [med] jaumisse *f*, éctère *m*
ICTUS - [med] ictus *m*, coup *m*
I.C.W. - [radio] (abbrev. for Interrupted Continuous Waves) s. Interrupted continuous waves
IDEA OF REFERENCE - [med] délire *m* de signification
IDEAL ASSEMBLY - [mech] (in statistical mechanics, an assembly in which the interactions between the systems composing the assembly can be neglected) composition *f* idéale
~BLACK BODY - [phys] corps *m* noir idéal, radiateur *m* intégral
~BUNCHING - [electron] (the theoretical condition in which the bunching of the electrons in a velocity-modulation tube would give a single infinitely large current peak during each cycle) groupement *m* idéal
~CASCADE - [electron] (a cascade in which the number of elements in parallel and the flow of each stage vary continuosly, so that a minimum number of separative elements are used for the production of material of a given concentration) cascade *f* idéale
~ENGINE - [mech] moteur *m* à cycle théorique
~GAS - [phys] (a gas the molecules of which have mass but no finite size, and do not exert any force on each other) gaz *m* idéal
~MAGNETIC MEDIUM - [el acoust] bande *f* magnétique standard
~MAGNETIZATION - [el] (magnetization remaining at constant magnetizing force) magnétisation *f* idéale
IDEAL PERMEABILITY - [phys] (in magnetism) perméabilité *f* idéale
~RECTIFIER - [electron] (a rectifier in which the back conductance, forward resistance and capacity are zero) redresseur *m* idéal
~SATURABLE CORE - [radio] noyau *m* saturable idéal
~SATURABLE REACTOR - [radio] réacteur *m* saturable idéal
~SEPARATION FACTOR - [nucl] (the ideal ratio of the abundance of two isotopes) facteur *m* de séparation idéal
~SIMPLE PROCESS FACTOR - [nucl] facteur *m* de séparation idéal d'un étage
~SOLENOID - [el] (a cylindrical coil in which all the turns are assumed to be in planes normal to the axis and equally spaced) solénoïde *m* d'Ampère
~SOLUTION - [phys] (a solution following Raoult's Law despite variations in concentration and temperature and with which, on mixing there is no change of internal energy) solution *f* idéale
~TRANSDUCER - [el acoust] (hypothetical passive transducer transferring the maximum possible power from the source to the load) transducteur *m* idéal
~TRANSFORMER - [radio] (transformer employing an ideal core operating in the unsaturated region of the core characteristic, with perfect coupling between windings) transformateur *m* idéal
~VALUE - [contr] valeur *f* desirée
IDEALIZED SYSTEM - [contr] (a system the performance of which is supposed to define the relationship between the ideal value and the command) système *m* idéal
IDEALLY IMPERFECT CRYSTAL - [phys] cristal *m* mosaique parfait
IDENTIFICATION - [gen] identification *f*
~BEACON - [signal] (a luminous beacon emitting code signals, for the identification of a point on the ground) phare *m* d'identification
~HEAD LEADER - [cin] (length of pictures; 24 frames with reel number and title) bande *f* amorce
~LAMP - s. identification light
~LEADER - s. identification head leader
~LIGHT - [signal] (light shown by an aircraft to enable an observer to identify it) lumière *f* d'identification de l'avion
~MARKER - [radiography] marqueur *m* de cliché
~PLATE - [mach] (plate bearing information, e.g. voltage, marker's name, type and the like, attached to a unit or part) plaquette *f* d'identification
~SIGN - [aero] (a sign giving the name of a point, visible from the air) signe *m* d'identification
~TAG - [gen] marque *f* d'identification
[photo] feuille *f* d'identification
~TAIL LEADER - [cin] (24 frames with reel number and title at the end of film) amorce *m* en fin de bobine
~TALLY - s. identification tag
~TAPE - [gen] (tape, usually adhesive, attached to an electric cable or lead for identification) ruban *m* d'identification
~TRAILER - s. identification tail leader
IDENTIFYING INFORMATION - [comput] indicatif *m* critère *m* d'identification
IDENTITY - [gen] identité
~PERIOD - [phys] (the distance between identical

atomic groupings in the chain molecule of an associated substance, or in a crystal lattice) distance *f* d'identité
IDEOPHRENIA - [med] démence *f* avec perversion mentale
IDIOCHROMATIC - [gen] idiochromatique
~CRYSTAL - [crystall] (crystal having photoelectric properties which are not due to foreign matter) cristal *m* idiochromatique
IDIOCHROMATIN - [biol] (substance controlling the reproduction of the cell) idiochromatine *f*
IDIOCRASY - [med] idiosyncrasie *f*
IDIOCY - [med] idiotie *f*
IDIOELECTRIC - [phys] idioélectrique
IDIOGLOSSIA - [med] (a child's wrong use for consonants) idioglossie *f*
IDIOMORPHIC CRYSTALS - [min] (rock minerals bounded by the crystal faces peculiar to the species) cristals *mpl* idiomorphes
IDIOMORPHOUS - [crystall] (which appears in distinct crystals) idiomorphe
IDIOPATHETIC - [med] essentiel, idiopathique
IDIOPLASM - [med] idioplasma *m*
[zool] idioplasma *m*
IDIOSTATIC CIRCUIT - s. idiostatic method
~INSTRUMENT - [instr] (a measuring instrument in which the voltage is applied between the needle and one pair of quadrants) appareil *m* de mesure idiostatique
~METHOD - [el] (method of using an electrometer which does not require an external supply source of current) méthode *m* idiostatique
IDIOTISM - [med] idiotie *f*
I DISPLAY - [radar] indicateur *m* type I
IDLE, to - [mech] (of an i.c. engine, to run slowly without supplying power, in readiness for work) tourner au ralenti
[mech] (of a machine) marcher au ralenti, marcher à vide
IDLE - [el] déwatté
[mech] décalé, (roue) folle, parasite
~CAPACITY - [autom] capacité *f* inutilisée
~COIL - [el] (winding element which does not belong to the electric circuit of a winding, but has the only purpose of filling in a certain space) section *f* morte
~COMPONENT - [el] composante *f* réactive du courant
~COURSE ATTACHMENT - [mech] dispositif *m* pour rangée à vide, dispositif d'abattage
~COURSE LEVER - [mech] levier *m* de commande pour passage à vide, levier de commande de la came
~CURRENT - [el] courant *m* déwatté
~CURRENT WATTMETER - [el] (electrical measuring instrument) wattmètre *m* à courant déwatté
~GEAR - [mech] roue *f* folle d'engrenage
~PERIOD - [el] (the part of an alternating-voltage cycle during which a certain arc path is not carrying current) temps *m* de repos
~RACKING - [mech] pousser à vide, commander à vide
~ROLLER - [mech] galet *m* libre
~SHOT - [photo] prise *f* de vue à vide
~STROKE - [mech] course *f* à vide
~TIME - [comput] temps *m* mort, temps *m* inactif externe
~WHEEL - [mech] roue *f* parasite
[horol] (intermediate wheel) engrenage *m* intermédiaire
IDLER - [mech] (roller, e.g. in a belt conveyor, which serves to guide or tension the belt and does not transmit power to it) poulie *f* de tension, galet *m* tendeur, galet *m* de renvoi, pignon *m* fou
(a wheel or gear) s. idler gear
(an idle pully) s. idler pulley
~GEAR - [mech] (a gear used to connect two other gears, but not driving anything itself) roue *f* parasite, roue *f* intermédiaire
~LEVER - [mech] levier *m* du rouleau de tension, levier-relais *m*
~NOZZLE - [auto] gicleur *m* de ralenti
~PULLEY - [mech] (a pulley, espec. in a belt conveyor, which supports or changes the direction of a belt without transmitting or receiving power) poulie *f* de tension
~SPINDLE - [mech] (a fixed spindle carrying an idler gear) arbre *m* d'engrenage de renvoi
IDLERS - [bot] (of barley) grains *mpl* légers
IDLING - [mech] (term used of the operation of an i.c. engine which is not delivering power, but is running slowly and ready for use) marche *f* au ralenti
~ADJUSTMENT - [auto] régulation *f* du ralenti
~CONTROL - [mech] (minimum burner pressure valve; a device to maintain burner fuel pressure above a minimum level) valve *f* commande ralenti
~CYCLE - [comput] cycle *m* blanc
~GEAR - [mech] marche *f* à vide
~JET - [mech] (carburettor jet designed to supply fuel when the engine is idling) gicleur *m* de ralenti
~POSITION - [mech] position *f* de débrayage
~ROLLER - [mech] galet *m* à glissement
~SPEED - [mech] (the speed at which an engine is operated when it is not supplying power but is running slowly and ready for use) régime *m* du ralenti
IDP - [comput] s. Integrated Data Processing
I.F. REJECTION FACTOR - [telev] (Intermediate Frequency rejection factor) pénétration *f* de la moyenne fréquence
I.F.R. - [aero] (Instrument Flight Rules) I.F.R., règles *fpl* pour le vol aux instruments
I.F.R. FLIGHT - [aero] (a flight carried out in accordance with Instrument Flight Rules) vol *m* I.F.R.
IGEWESKY'S SOLUTION - [chem] (an etching agent, consisting of a 5 p.c. solution of picric acid in absolute alcohol: used in steel microscopic) liqueur *f* d'Igewesky
IGNEOUS FUSION - [metall] fusion *f* ignée
~INTRUSIONS - [geol] (the many types of emplacement of igneous rocks) intrusions *fpl* ignées
~MAGMA - [geol] (molten fluids generated within the earth) magma *m* igné
~METALLURGY - [metall] métallurgie *f* par le voie ignée, pyrométallurgie *f*
~ROCKS - [geol] (rocks formed from magmatic material injected into, or extruded upon the earth's crust) roches *fpl* éruptives, roches *fpl* ignées
IGNIFEROUS - [gen] ignifère
IGNITABLE - [adj] (capable of being put into combustion by merely being heated to a certain temperature in air) inflammable, allumable
IGNITE, to - [gen] enflammer, allumer
[chem] (the heat a gaseous mixture to the temperature at which combustion occurs, in particular

by means of an electric spark) s'enflammer
[ind chem] (the operation of calcining) calciner, fritter
IGNITER - [aero] (device for initiating combustion in a gas turbine) allumeur *m*, interrupteur *m*
[el] (in i.c; engines) bougie *f* de démarrage
[firearms] allumeur *m*, déflagrateur *m*, boute-feu *m*
[electron] igniteur *m*, électrode *f* d'amorçage
~PILOT - s. ignition pilot
~PLUG - [el] (in a gas turbine, a special type of electric ignition plug for initiating combustion) bougie *f* d'allumage
~WIRE - [el] fil *m* explosif, fusible *m* d'allumage, amorce *f*
IGNITION - [gen] allumage *m*, ignition *f*, inflammation *f*
[mech] (the process of initiating combustion in the cylinder of an i.c. engine) allumage *m*
[chem] ignition *f*
~ACCUMULATOR - [el] accumulateur *m* d'allumage
~ADVANCE - [mech] (the crank angle at which the spark is timed to pass) avance *m* d'allumage
~ALLOY - [metall] alliage *m* pyrophore
~ANALYZER - [contr] (ignition-controlling device operating by means of a cathode-ray oscillograph) oscillographe *m* analyseur de l'allumage
~ARCH - [gas ind] (water gas) voûte *f* d'allumage
~BY INCANDESCENCE - [phys] ignition *f* à incandescence
~CABLE - [auto] fil *m* d'allumage
~CHECK - [mech] contrôle *m* de l'allumage
~COIL - [auto] (induction coil converting the battery low-tension current into the high-tension current required by the sparking plug) bobine *f* d'allumage
~CONTROL - [auto] commande *f* de l'allumage, contrôle *m* de l'avance à l'allumage
~CURRENT - [el] courant *m* d'allumage
~DELAY - [mech] retard *m* d'amorçage, amorçage *m* retardé
~DEVICE - [auto] dispositif *m* d'allumage
~DISTRIBUTOR - [auto] allumeur *m*, distributeur *m* d'allumage
~ELECTRODE - [el] électrode *f* d'amorçage, électrode d'allumage
~FILAMENT - s. igniter wire
~HARNESS - [auto] rampe *f* d'allumage
[aero] (of a screened ignition engine) rampe *f* de allumage
~HEAT - [phys] chaleur *f* d'ignition
~INTERFERENCE - [radio] (effects in radio or radar receivers caused by radiation from the ignition system of i.c. engines) brouillage *m* dû à l'allumage
~KEY - [auto] clé *f* de contact
~LAG - [mech] (the interval between the passage of the spark and the pressure rise due by combustion) retard *m* d'allumage
~LOCK - [auto] serrure *f* de contact d'allumage
~LOCK SWITCH - [auto] commutateur *m* d'allumage
~LOSS - [chem] perte *f* au feu
~PILOT - [gas ind] veilleuse *f* d'allumage
~PLUG - [el] bougie *f* d'allumage
~POINT - [phys] (the lowest temperature at which a substance will support continuous combustion) point *m* d'ignition
~QUALITY - [mech] (a measure of the ignition delay of a fuel in a Diesel engine) qualité *f* d'allumage
IGNITION RATING - [el] (of accumulators employed for supplying ignition systems) régime *m* d'allumage
~RECTIFIER - [el] (a mercury-arc rectifier in which the cathode spot is initiated by a voltage impulse applied to a special electrode) redresseur *m* d'allumage
~SPARK - [auto & el motors] étincelle *f* d'allumage
~SYSTEM - [mech] (the whole complex of arrangements for initiating combustion in an i.c. engine) système *m* d'allumage
~TEMPERATURE - [mech] (flash point) point *m* d'inflammabilité
~TIMING - [auto etc] calage *m* de l'allumage
~VELOCITY - [phys] (flame velocity) vitesse *f* de combustion
~VOLTAGE - [el] (the voltage required to start the discharge in a cold-cathode electron tube) tension *f* d'amorçage
~WIRE - s. ignition cable
IGNITOR - [electron] (a stationary starting electrode which is partly immersed in a cathode pool) ignitor *m*, tige *f* d'amorçage
~CURRENT TEMPERATURE DRIFT - [electron] variation *f* du courant d'amorçage en fonction de la température
~FRING TIME - [electron] intervalle *m* d'amorcage
~LEAKAGE RESISTANCE - [electron] résistance *f* de fuite de l'ignitor
IGNITRON - [electron] (a single-anode pool rectifier in which the arc-discharge is initiated by an ignitor) ignitron *m*
~CONTROL SYSTEM - [el] système *m* de commande à ignitrons
IGNORE - [comput] caractère *m* non-valable
~INSTRUCTION - [comput] instruction *f* non valable, instruction *f* "ignorer"
IGNORED CONDUCTOR - [el] conducteur *m* fantôme
I.H.P. - (initials for Indicated Horse Power) puissance *f* indiquée en chevaux
I-HEAD ENGINE - [mech] (term sometimes used for i.c. reciprocating engines of which both valves are located in the cylinder head) moteur *m* à soupapes verticals en tête
IHRIGIZING - [metall] imprégnation *f* au silicium
ILEECTOMY - [med] résection *f* de l'iléon
ILEITIS - [med] iléite *f*
ILEOCOLITIS - [med] iléocolite *f*
ILEOCOLOSTOMY - [med] iléo-colostomie *f*
ILEOCOLOTOMY - [med] iléo-colotomie *f*
ILEOILEOSTOMY - [med] iléo-iléostomie *f*
ILEORRHAPHY - [med] iléorraphie *f*
ILEOSIGMOIDOSTOMY - [med] iléo-sigmoïdostomie *f*
ILEOSTOMY - [med] iléostomie *f*
ILEOTOMY - [med] iléotomie *f*
ILEOTRANSVERSOSTOMY - [med] iléo-transversostomie *f*
ILEUM - [med] iléon *m*
ILL - [gen & med] malade, souffrant
ILLEGIBLE - [gen] illisible
ILLEGITIMACY - [gen & leg] illégitimité *f*
ILLEGITIMATE - [gen & leg] illégitime
ILLICIT - [gen] illicite
ILLINIUM - [chem] (also called cyclonium or promethium; radioactive element; symbol Pm, atomic number 61) prométhium *m*
ILLITE - [min] (a refractory material) illite *f*

ILLITERACY - [gen] analphabétisme *m*
ILLITERATE - [gen] analphabète *m*
ILLNESS - [med] maladie *f*
ILLUMINANCE - [phys] (the quotient of the luminous flux incident on an infinitesimal element of surface containing the wanted point by the area of that element) éclairement *m*
ILLUMINANT METAMERISM - [electron] erreur *f* chromatique dû à illuminant erroné
ILLUMINATE, to - [gen] éclairer, illuminer
[print] (of manuscripts etc) enluminer, colorier
ILLUMINATED - [gen] éclairé
[print] (of manuscripts) enluminé
~DIAGRAM - [el] (track diagram in a railway signal box) diagramme *m* lumineux, tableau *m* de contrôle optique
~DIAL INSTRUMENT - [instr] instrument *m* à tableau optique
~TRACK DIAGRAM - s. illuminated diagram
~WIND CONE - [aero] manche *f* à vent éclairée
ILLUMINATING GAS - [light] (obsolete; town gas preferable) gaz *m* de ville
~OIL - [chem] (paraffin oil) pétrole *m* lampant, kérosène *m*
~POWER - s. illumination intensity
~SCALE BULB - [el] lampe *f* pour l'éclairage du cadran
~WINDOW - [photo] fenêtre *f* d'éclairage
ILLUMINATION - [gen] éclairage *m*, illumination *f*
[of manuscripts] enliminure *f*
[opt] (of a lens) éclat *m*
~CONTROL - [telev] réglage *m* de l'éclairement
~INTENSITY - [opt] (the flux density of light incident on a surface) intensité *f* d'éclairage
~METER - [instr] (instrument for the measurement of illumination) photomètre *m*
~PHOTOMETER - s. illumination meter
ILLUMINATOR - [impl] dispositif *m* d'éclairage
ILLUMINOMETER - [instr] (type of photometer) illuminomètre *m*
ILMENITE - [min] (a natural oxide of iron and titanium; a widely-distributed secondary mineral and an important source of the latter metal) ilménite *f*, fer *m* titané
~BLACK - [paint] (a black pigment of good opacity, made from ilmenite) noir *m* d'ilménite
ILMENORUTILE - [min] (titanium oxide, a black variety) ilménorutile *f*
I.L.S. - s. Instrument Landing System
ILVAITE - [min] (silicate of iron and calcium) ilvaïte *f*
IMAGE, to - [gen] représenter par une image
IMAGE - [gen & opt] image *f*
~ABERRATION - [telecomm] défaut *m* de l'image
~ANALYSIS - [telev] analyse *f* de l'image
~AREA - [cin & telev] (the picture area) cadrage *m*, champ *m* de l'image
~ATTENUATION COEFFICIENT - [telecomm] (the real part of the image transfer coefficient of a network) facteur *m* d'affaiblissement sur images
~ATTENUATION CONSTANT - s. image attenuation coefficient
~BAND - [comput] bande *f* de fréquence image, bande *f* image
~BLACK - [telev] noir *m* d'une image
~CAMERA TUBE - [telev] tube *m* analyseur à transfert d'image

IMAGE CARRIER - [telev] (a carrier wave modulated by the video signal) porteuse *f* vidéo
~CHANNEL - [telev] canal *m* d'image
~COIL - [telecomm] bobine *f* représentation
~CONTRACTION - [cin & telev] (distortion of the reproduced sound track) resserrement *m* de l'image
~CONTRAST - [photo] contraste *m* de l'image
~CONTROL COIL - [telev] bobine *f* de cadrage
~-CONVERTER TUBE - [electron] (electronic tube in which an optical image applied to a photo-emissive surface produces a corresponding image on a luminescent surface) convertisseur *m* d'image
~DETAIL - [telev] (the clearness of the details of the picture) détail *m* d'image
~DISPLACEMENT - [photo] déplacement *m* de l'image
~DISSECTOR - [telev] (a camera tube in which the image is projected on to a photo-cathode) tube *m* dissecteur
~DISTANCE - [photo] distance *f* ultranodale postérieure
~-DRIFT - [telev] oscillation *f* de l'image
~FIELD - [telecomm & televl] champ *m* d'image
~FIELD CURVATURE - [telecomm] courbure *f* de champ d'image
~FLYBACK - [telev] retour *m* du spot analyseur
~FREQUENCY - [radio] (interfering frequency) fréquence-image *f*
~-FREQUENCY REJECTION RATIO - [radio] taux *m* de réjection fréquence-image
~-FREQUENCY RESPONSE - [radio] (unwanted response to an excitation at an image frequency) réponse *f* de fréquence-image
~GROWTH - [cin & telev] extension *f* de l'image
~ICONOSCOPE - [telev] image-iconoscope *m*
~IMPEDANCES - [el acoust] (of a transducer) impédances *f*pl images
~INTENSIFIER - [opt] (X-ray tube used to obtain a very high brilliance) tube *m* intensificateur d'image
~INTERFERENCE - [telev] (interference caused by the video frequency) interférence *f* d'image par la fréquence vidéo
~LOCK - [telev] calage *m* de l'image
~MONITOR - [telev] écran *m* de contrôle
~ORTHICON - [telev] (a camera tube in which the image to be televized is projected on the a photo-cathode, the photo-electrons being focussed on a mosaic) image-orthicon *m*
~OUTPUT - [telev] (the image base taking care of the horizontal scanning of the picture) base *f* de temps vidéo
~OUTPUT TRANSFORMER - [telev] (the time-base which is provided for the scanning of the frame) transformateur *m* de la base de temps vidéo
~PATTERN - [electron] (the total of the charged particles of the luminescent substance on an insulating surface in a cathode-ray tube) image *f* de potentiel, image *f* électronique de charge
~PHASE CONSTANT - [telecomm] (the imaginary part of the transfer constant) partie *f* imaginaire du coefficient de transfert
~PHASE FACTOR - [telecomm] facteur *m* de dephasage sur images
~PLANE - [telev] plan *m* image
~PROPAGATION CONSTANT - s. image propagation factor
~PROPAGATION FACTOR - [telecomm] constante *f* de propagation, facteur *m* de transfert

IMAGE RATIO - [radio] rapport *m* signal/image
~REACTOR - [nucl] (virtual reactor; a reactor in which the method of images is used to solve the critically equations) réacteur *m* virtuel
~REGISTRATION - [telev] superposition *f* de l'image
~REJECTION - [telev] (partial suppression of the image carrier frequency) intensité *f* relative du signal reproduit
~REPRODUCTION - [cin] reproduction *f* de l'image
~RESPONSE - [radio] réponse *f* image
[comput] réponse *f* image
~RETENTION - [telev] (in a cathode-ray tube, the image is occasionally retained by the screen and thus appears again in subsequent recordings) rémanence *f* d'image
~SCALE - échelle *f* de reproduction
~SHELL - [photo] caustique *f*
~SHIFT - [telev] déplacement *m* de l'image
~SOURCE - [nucl] (virtual source; theoretical source using infinite kernels to solve the reactor equation) source *f* virtuelle
~SPACE - [telev] (the space in which the image from the object in the object space is reproduced) espace *m* d'image
~SPREAD - [telev] dispersion *f* de l'image
~STORAGE TUBE - [electron] (an electron tube containing a mosaic electrode on which an optical image is focussed and subsequently scanned by the electron beam) tube *m* analyseur à accumulation
~STRIP - [telev] ligne *f* d'analyse
~SURFACE - [telev] surface *f* de l'image
~SWEEP FREQUENCY - [telev] fréquence *f* d'analyse vidéo
~TEST - [telev] mére *f*
~TRANSFER COEFFICIENT - [radio] exposant *m* de transfert sur images
~TRANSFER CONVERTER - [telev] convertisseur *m* d'image par transfert
~TRANSFER CONSTANT - s. image transfer coefficient
~TRANSMISSION - [telecomm] (radiotelegraphic transmission of pictures etc) transmission *f* d'images
~TRUMBLING - [photo] renversement *m* des images
~VIEWING TUBE - s. image converter
IMAGINARY - [gen] imaginaire
[math] imaginaire
~QUANTITY - [math] quantité *f* imaginaire, imaginaire *f*
IMBALANCE - [med] déséquilibre *m*
IMBECILE - [gen & med] imbécile, faible d'esprit
IMBECILITY - [med] imbécillité *f*
IMBED, to - s. embed
IMBIBITION - [chem] (the absorption or adsorption of a liquid by a solid) imbibition *f*, trempage *m*
[soil] imbibition *f*
~PRINTING - [photo] tirage *m* par imbibition
IMBOWMENT - [arch] voûte *f*
IMBRICATE - [bot] (or imbricated; of overlapping leaves) imbriqué
~STRUCTURE - [geol] (mountain structures produced by intense pressure) structure *f* à écailles
IMBRICATED PLATE - [metall] plaque *f* en forme de écaille
IMBRICATION - [med] imbrication *f*, chevauchement *m*
IMBUE, to - [gen] imbibern imprégner
I.M.C. - s. Instrument Meteorological Conditions
IMIDAZOLE - [chem] (a synthesis intermediate) imidazole *m*, glyxaline *f*
IMIDES - [chem] (organic compounds which contain the group -CO.NH.CO.-) imides *mpl*
IMINOBISPROPYLAMINE - [chem] (an intermediate for drugs, dyest, insecticides, and rubber additives) iminobispropylamine *f*
IMITATE, to - [gen] imiter, copier
IMITATION - [gen] imitation
(used as an adjective) factice, simili
~FUR - [text] étoffe *f* imitant la fourrure
~LEATHER - [text] imitation *f* de cuir, tissu *m* enduit plastique
[text] (upholstery) cuir *m* artificiel, similicuir *m*
~LINEN - [text] imitation *f* de lin
~LOG FIRE - [th eng] radiateur *m* à bûches réfractaires
~PARCHMENT - [paper man] (wood-pulp paper to which certain qualities have been given by a prolongued beating of the pulp) parchemin *m* artificiel
~PIPES - [mus] (ornamental dummy organ pipes) tuyaux *mpl* de façade
~SILVER THREAD - [text] fil *m* similargent
~VELVET - [text] épinglé *m*
~WORSTED - [text] semipeigné *m*
IMMANENT - [gen] immanent
IMMATERIAL - [gen] peu important, sans conséquence
IMMATURE - [gen & med] pas mûr, immature
~CRYSTAL - [crystal] cristal *m* embryonnaire
IMMEASURABLE - [gen] immesurable, immense
IMMEDIATE - [gen] immédiat
~ACCESS - [comput] accès *m* instantané
~-ACCESS STORE - [comput] mémoire *f* rapide, mémoire *f* à court temps d'accès, mémoire *f* à accès immédiat
~ACTION ALARM - [telecomm] signalisation *f* immédiate
~APPRECIATION PERCENTAGE - [telecomm] taux *m* de compréhension immédiate
~SET - [rubber ind] (of tyres) déformation *f* initiale, rémanence *f* instantanée
IMMELMANN TURN - [aero] (manoeuvre consisting of a half loop followed by a half-roll, which leaves the aircraft on a reciprocal course at a greater altitude) virage *m* d'Immelmann, virage *m* à la verticale et en épingle à cheveux
IMMERGE, to - [gen] immerger, plonger, submerger
IMMERSE, to - s. immerge, to
IMMERSED BOG - [soil] tourbière *f* immergée
~DENSITY - [soil] poids *m* volumétrique sous l'eau
IMMERSIBLE APPARATUS - [el] (an electrical apparatus operating under water) appareil *m* électrique immersible
IMMERSION - [gen] immersion *f*
~BATH - [ind chem] bain *m* à immersion
~BOBBIN - [text] bobine *f* d'immersion
~BURNER - [th eng] bruleur *m* immergé
~CELL - [instr] (instrument for the continuous measurement of the pH-value of a liquid in a tank or in an open container) sonde *f* à immersion
~COUNTER - [nucl] (a counter designed to be immersed in a liquid, to measure its radioactivity) compteur *m* à immersion
~HARDENING - [metall] trempe *f* par immersion
~HEATER - [th eng] (an appliance chiefly consisting

of a heating element, which can be placed in a liquid or plastic material to heat it) chauffe-liquides *m*

IMMERSION HEATER COIL - [el] serpentin *m* thermoplongeur

~ LENS - s. immersion objective

~ LUBRICATION - [mech] (method of lubrication in which the parts concerned dip into or run in an oil bath) graissage *m* par trempage

~ OBJECTIVE - [opt] (in some high-power microscopes, the objective lenses have the space between these and the object filled with an oil in order to reduce refraction losses) objectif *m* à immersion

~ PLATING - [metall] (deposition of a thin coating by immersing a metal object in a solution of the metal intended to form the coating) dépôt *m* par immersion

~ ROLLER - [text] cylindre *m* de trempage, cylindre *m* d'immersion

~ WASHING - [ind chem] lavage *m* par immersion

IMMIGRANT - [gen] immigrant *m*

IMMIGRATE, to - [gen] immigrer

IMMIGRATION - [gen] immigration *f*

IMMINENT - [gen] imminent

IMMISCIBLE - [chem] (term used for liquids which form more than one phase when brought together, i.e. which do not mingle) non-miscible, immiscible

~ SOLUTIONS - [chem] (solutions which, when brought together, do not form a single phase but remain each in its own separate phase) solutions *fpl* immiscibles

~ SOLVENT - [chem] dissolvant *m* immiscible

IMMOBILITY - [gen] immobilité *f*

IMMOBILIZATION - [gen] immobilisation *f*
[railw etc] (of a vehicle) mise *f* en hors de service

IMMOVABLE - [gen] immobile
[gen] (of machines etc) fixe

IMMUNE - [gen & med] immun, immunisé

IMMUNITY - [gen] immunité *f*, exemption *f*
[el chem] (the condition of an unprotected metal surface in which electro-chemical corrosion is theoretically impossible) immunité *f*

IMMUNIZATION - [med] (the operation of rendering immune) immunisation *f*

IMPACT - [gen] impact *m*, choc *m*, collision
[mech] (the action of two bodies in collision, whereby their velocity is changed) choc *m*
[hydr] (the relative velocity of a liquid before and after impact) choc *m*, coup *m*, percussion *f*

~ ACCELEROMETER - [instr] (instrument for measuring the deceleration of an aircraft when landing) accéléromètre *m* d'impact

~ ANGLE - [phys] (of a falling body) angle *m* d'impact

~ BAR - [metall] (a specimen used for shock fracture tests) barre *f* de choc

~ BEHAVIOUR - [phys] fonctionnement *m* au choc

~ BENDING RESISTANCE - [phys] résistance *f* de flexion au choc

~ EXTRUSION - [plast ind] extrusion *f* par choc, filage *m* à la presse par choc

~ FLUORESCENCE - [opt] (fluorescence of a material due to bombardment by high-energy molecules of a different material) fluorescence *f* par choc

~ IONIZATION - [phys] ionisation *f* par choc

~ MIXER - [mech] (type of mixer in which a rotating disk throws out solids centrifugally) broyeur *m* centrifuge

IMPACT MOULDING - [plast ind] moulage *m* par choc

~ NOISE ABATEMENT - [constr] insonorisation *f* contre le bruit de contact

~ PARAMETER - [mech] paramètre *m* de choc

~ PENDULUM - [impl] mouton *m* pendule, appareil *m* à marteau-pendule

~ PRESSURE - [mech] (the difference between the static and the pitot tube pressures) pression *f* d'arrêt, pression *f* de choc

~ RESILIENCE - [phys] (elasticity on impact) résilience *f*

~ RESISTANCE - [constr] résistance *f* au choc, résilience *f*

~ SCREEN - [min] tamis *m* vibreur, tamis *m* à vibrations

~ STRENGTH - [constr etc] (a measure of the brittleness of a material, usually obtained by the Izod technique) résilience *f*

~ SWITCH - [el] (device designed to switch off the high-voltage circuits of an aircraft through the effect of an impact, e.g. on a crash landing) interrupteur *m* de choc

~ TAR EXTRACTOR - [mech] séparateur *m* de goudron par choc

~ TEST - [metall & constr work] essai *m* au choc
[paint] (a test for the adhesion and flexibility of coatings, in which a weighted plunger is made to strike the back of a panel coated with the composition under test) essai *m* de flexion par choc

~ TESTING MACHINE - [mech] appareillage *m* pour essais au choc

~ TOUGHNESS - [metall] résilience *f* au choc

~ WHEEL MIXER - [mech] (a machine in which rotary impellers disintegrate the material treated by impact caused by centrifugal action) broyeur *m* centrifuge

~ WRENCH - [impl] clé *f* pneumatique

IMPACTED FRACTURE - [metall] fracture *f* pénétrante

IMPACTION - [med] impaction *f*, enclavement *m*

~ TOOL - [tool] (only USA; a crowbar) pince *f* à levier

IMPAIR, to - [gen] affaiblir, diminuer
[gen] (the quality of something) détériorer
[gen] (of machines etc) endommager

IMPAIR THE CARBURATION, to - [mech] (in internal combustion engines) fausser la carburation

IMPAIREMENT - [radio] (transmission performance rating) indice *m* de qualité de transmission

IMPALEMENT - [med] empalement *m*

IMPARIPINNATE - [bot] (of a pinnate leaf with a terminal small leaf) imparipenné

IMPART, to - [gen]communiquer, annoncer, faire connaître

IMPARTIAL - [gen] impartial

IMPASSABLE - [gen] infranchissable, impracticable
[gen] (of roads) impracticable

IMPEACH, to - [gen] (to discredit etc) attaquer, récuser, accuser
[leg] (to accuse of a crime against the strate) accuser d'un crime

IMPEACHMENT - [gen] dénigrement *m*, récusation *f*
[leg] mise *f* en accusation (d'un ministre etc)

IMPECCABLE - [gen] irréprochable, parfait

IMPEDANCE - [phys] (the ratio between a force-like quantity and a related velocity-like quantity) impé-

dance *f*
[el] (the total opposition to a current in a circuit, consisting of non-inductive resistance, inductive reactance and capacitative reactance) impédance *f*
IMPEDANCE ANGLE - [el] décalage *m*, argument *m* d'une impédance
~BRIDGE - [el] (device in which a bridge arrangement is used to measure impedances) pont *m* d'impédance
~COIL - [el] bobine *f* de réactance
~COMPARATOR - [instr] comparateur *m* d'impédances
~COMPENSATOR - [radio] (electric network associated with a line in order to give the impedance of the combination a specified characteristic over a specified frequency range) réseau *m* compensateur, correcteur *m* d'impédance
~CORRECTOR - s. impedance compensator
~COUPLING - [electron] couplage *m* par impédance
~LEVEL - [el] (value of circuit impedance) niveau *m* d'impédance
~MAGNETOMETER - [instr] (instrument measuring local variations of the magnetic field of the earth) magnétomètre *m* à variation d'impédance
~MATCHING - [el] (the adjustment of the load impedance to the impedance of the source of supply, in order to ensure maximum transfer of power) adaptation *f* des impédances, équilibrage *m* des impédances
~MATRIX - [electron] matrice *f* d'impédance
~OF THE GRID CIRCUIT - [el] impédance *f* du circuit de grille
~PROTECTION - [el] dispositif *m* de protection à impédance
~SCREW - [mech] vis *f* d'étranglement
~SIMILATING NETWORK - [el] (a balancing network) équilibreur *m*, réseau *m* d'équilibrage
~STARTER - [el] démarreur *m* à impédance
~TRANSFORMER - [electron] transformateur *m* d'impédance
~UNBALANCE MEASURING SET - [el] (apparatus designed to measure the return loss) équilibromètre *m*
~VOLTAGE OF A TRANSFORMER - [el] tension *f* de court-circuit d'un transformateur
IMPEDE, to - [gen] entraver, empêcher
IMPEDED HARMONIC OPERATION - [radio] (in a magnetic amplifier) magnétisation *f* forcée
IMPEDIMENT - [gen] obstacle *m*, entrave *f*
IMPEDOMETER - [instr] appareil *m* de mesure de l'impédance
IMPEDOR - [el] (a circuit element having impedance) impédance *f*
IMPELLER - [mech] (the rotary element in a centrifugal pump or rotary compressor) roue *f* mobile, couronne *f* mobile, turbine *f*, roue *f* à aubes
~PUMP - [oil ind] pompe *f* centrifuge
~RAMMING - [metall] serrage *m* par projection
~-TYPE PUMP - [mech] (a pump in which a rotating element imparts movement to the fluid dealt with) pompe *f* à turbine, pompe *f* à couronne mobile
IMPELLING RATIO - [phys] taux *m* de poussée
IMPENDING - [gen] imminent, prochain
IMPENETRABILITY - [gen & phys] impénétrabilité *f*
IMPENETRABLE - [gen] impénétrable
IMPERCEPTIBLE - [gen] imperceptible
IMPERCEPTIBILITY - [gen] imperceptibilité *f*
IMPERFECT - [gen] imparfait, défectueur
~CADENCE - [mus] cadence *f* imparfaite
~CRYSTAL - [crystall] cristal *m* imparfait
~HYBRIDIZATION - [bot] hybridisation *f* imparfaite
~TAPE - [comput] (magnetic tape containing such imperfections as to make it impossible to use them for storage purposes) bande *f* magnétique imparfaite
IMPERFECTION - [phys] (any structural deviation from the structure of an ideal crystal) imperfection *f*
[gen] défaut *m*, imperfection *f*, défectuosité *f*
[print] sortes *fpl* pour parer aux remplacements
IMPERFORATE - [med] (abnormally closed) imperforé
IMPERIAL - [gen] impérial
[constr] (a pointed dome-shaped roof) comble *m* en dôme
[constr] (a type of slate 33 x 24 in) feuille *f* d'ardoise de 33 x 24 pouces
[paper] (a standard size of paper) papier *m* grand jésus
[text] (heavy cotton fustian fabric) futaine *f*
~CAP - [paper man] (standard size of brown paper) papier *m* d'emballage (22 x 29 pouces)
~GALLON - [meas] (4,54 l.) gallon *m* (dans le Royaume-Uni)
~ROOF - [constr] comble *m* en dôme
IMPERMEABILITY - [gen chem geol etc] imperméabilité *f*
IMPERMEABILIZATION - [gen] (waterproofing) étanchement *m*, aveuglement *m*, imperméabilisation *f*
IMPERMEABLE - [gen etc] (which does not permit the passage of liquids or gas) imperméable
~COATING - [rubber ind] couche *f* imperméable
~TO AIR - [text] imperméable à l'air
~TO OIL - [text] étanche à l'huile
IMPERMEATOR - [mech] (a lubrificating device) dispositif *m* de graissage
IMPERVIOUS - [gen etc] (having the property of preventing the passage of water) imperméable
(of an electric machine) étanche
~LAYER - [geol] couche *f* imperméable
~TO MOISTURE - [phys] imperméable à l'humidité
IMPERVIOUSNESS - [gen & build] (the quality of being resistant to the passage of water) imperméabilité *f*, impénétrabilité *f*
IMPETIGINIZATION - [med] impétiginisation *f*
IMPETIGO - [med] (contagious skin disease) impétigo *f*
IMPETUS - [gen & phys] élan *m*, force *f* de jet, impulsion *f*, force *f* acquise
IMPILATION - [med] impilement *m*, empilage *m*
IMPINGE, to - [gen] (to come into violent contact) venir en contact, entrer en collision, se heurter contre
IMPINGEMENT - [gen] heurt *m*, collision *f*
~RATE - [phys] taux *m* d'incidence
IMPLANT, to - [gen] implanter
IMPLANT - [gen] implant *m*
IMPLANTATION - [gen] implantation *f*
IMPLEMENT, to - [gen] rendre effectif, mettre en œuvre
IMPLEMENT - [gen] outil *m*, ustensile *m*, instrument *m*
[agric] instrument *m*, instrument *m* à bras, outil *m*
~BAR - [agric] barre *f* d'attelage, barre *f* d'accrochage

IMPLEMENT CARRIER - [agric] porte-outils *m*
~DRAW BAR - [agric] s. implement bar
~PORTER - s. implement carrier
IMPLEMENTS - [agric] outils *mpl*, instruments *mpl* à bras
IMPLICATE, to - [gen] impliquer, intéresser
IMPLODE, to - [phys] (of a hollow structure under external pressure) imploder
IMPLOSION - [phys] (sudden and violent failure in an inward direction of a hollow structure under external pressure) implosion *f*
~GUARD - [electron] (of a cathode-ray tube) glace *f* de protection, pareimplosion *m*
~WEAPON - [nucl] (device in which a quantity of fissionable material which is less than a critical mass is so compressed that it becomes supercritical and an explosion can occur) arme *f* à implosion
IMPLUVIUM - [arch] (of Roman houses) impluvium *m*
IMPLY, to - [gen] impliquer
IMPONDERABILITY - [phys] impondérabilité *f*
IMPONDERABLE - [gen] impondérable
IMPORT, to - [comm] importer
IMPORT - [gen] sens *m*, signification *f*, importance *f*
[comm] importation *f*
~CERTIFICATE - [comm] certificat *m* d'importation
~DUTY - [leg] droits *mpl* d'importation, droit *m* de entrée
~FREEZE - [comm] suspension *f* des importations
~LICENCE - [leg] permis *m* d'importation
~PERMIT - s. import licence
~TRADE - [comm] commerce *m* d'importation
IMPORTANCE - [gen] importance *f*
~FUNCTION - [nucl] (of a neutron in a given position and with a given velocity in a nuclear reactor) fonction *f* importance
IMPORTATION - [comm] importation *f*
IMPOSE, to - [gen] imposer
[print] (the operation of assembling type pages on the imposing stone, to prepare the forme) mettre en pages
IMPOSING STONE - [print] (heavy, iron-top table on which the type matter is locked up) marbre *m*
~SURFACE - s. imposing stone
IMPOSITION - [gen] imposition
[print] mise *f* en pages
IMPOST - [leg] impôt *m*, taxe *f*
[constr] (the pillar from which an arch springs) imposte *f*, sommier *m*
IMPOTENCE - [gen] impuissance *f*
IMPOUND, to - [leg] (to take possession by legal right) confisquer, saisir
IMPOVERISHED MATERIAL - [nucl] (reactor fuel which has lost its effective content of fissile material) matière *f* appauvrie
IMPRACTICABILITY - [gen] impracticabilité *f*
IMPREG - [wood] (impregnated with resin) imprégné (par la créosote)
IMPREGNATE, to - [ind proc] (to saturate with another substance) imprégner, saturer
[biol] féconder, imprégner
IMPREGNATED - [gen] imprégné, saturé
~CLOTH - [text] tissu *m* imprégné
~FABRIC - [ind chem] (fabric impregnated with synthetic resin) tissu *m* imprégné
~PAPER - [paper man] (paper impregnated with resin) papier *m* imprégné
~POLE - [constr] poteau *m* injecté

IMPREGNATED SLEEPER - [railw] traverse *f* imprégnée
~TAPE - [el acoust] ruban *m* homogène
~WITH RUBBER - [rubber ind] imprégné au caoutchouc
~WOOD SHUTTLE - [text] navette *f* en bois imprégné
IMPREGNATING AGENT - [ind chem] agent *m* d'imprégnation
~BATH - [text] bain *m* de piétage ,bain *m* de mordançage
~MACHINE - [mech] (machine designed for impregnating materials with resin) machine *f* à imprégner, machine *f* à tremper
~VESSEL - [ind chem] cuve *f* d'imprégnation
~WAX - [ind chem] cire *f* d'imprégnation
IMPREGNATION - [gen] (saturation with another substance) imprégnation *f*, saturation *f*
[ind chem] (the controlled penetration of a material or an object with synthetic resin) imprégnation
[timber] (the saturation of timber with creosote in order to preserve it) imprégnation *f* par la créosote
~BATH - [text] bain *m* d'imprégnation
~BY DIPPING - [ind proc] imprégnation *f* au trempé
~OF SLEEPERS - [railw] imprégnation *f* des traverses
~ORE - [min] minerai d'imprégnation
~TANK - s. impregnating vessel
~UNDER VACUUM - [ind chem] imprégnation *f* sous vide
~VEIN - [mining] filon *m* d'imprégnation
~VESSEL - [impl] cuve *f* d'imprégnation
~YARD - [railw] chantier *m* d'imprégnation des traverses
IMPRESS, to - [gen] imprimer, empreindre
[gen] (to stamp) faire une impression
[el] (in a conductor) appliquer (p.e. une tension)
IMPRESSED CURRENT DRAINAGE - [gas ind] soutirage *m* sur rail, drainage *m* forcé
~CURRENT PROTECTION - [gas ind] (in mainlaying) soutirage *m* de courant
~ELECTROMOTIVE FORCE - [el] (the open-circuit electromotive force of a source connected into a network) force *f* d'électromotrice imprimée
~FIELD - [el] champ *m* imprimé
~FORCE - [mech] (external force acting on a particle in a dynamic system) force *f* imprimée
IMPRESSION - [gen] impression *f*
[mech] empreinte *f*
[metall] (the space in a mould in which the moulding hardens and takes its form) empreinte *f*
[metall] (diecasting) empreinte *f* d'un moule
[print] (all the copies of a book printed from the same type) impression *f*, empreinte *f* des caractères
~BLOCK - [metall] (the part of the die used to carry the impression pad, insert or block) plaque *f* de portée (dans un moule)
~CYLINDER - [print] cylindre *m* d'imprimerie, rouleau *m* imprimeur
~MOULDING - [plast ind] moulage *m* au contact
IMPRINT, to - [gen] imprimer, fixer
[comput] (of punched cards) imprimer (à clavier)
IMPRINT - [print] (the name of the publishiers etc on the frontispice of a book) firme *f* de l'éditeur
[photo] empreinte *f*
IMPROPER - [gen] impropre, incorrect
~FRACTION - [math] expression *f* fractionnaire
~INTEGRAL - [math] intégrale *f* indéfinie

IMPROPER ROUTING CHARACTER - [comput] critère *m* d'acheminement incorrect
IMPROVE, to - [gen] améliorer, s'améliorer, perfectionner, bonifier
~THE POWER FACTOR, to - [el] augmenter le facteur de puissance
IMPROVED WOOD - [timber] (wood processed, usually under pressure and at high temperatures with or without a synthetic resin, to give improved physical properties) bois *m* amélioré
IMPROVEMENT - [gen] amélioration *f*
[gen] (of land or town planning) embelissement *m*
[leg] (of a patent) perfectionnement
~TRESHOLD - [radio] seuil *m* d'amélioration de la caractéristique signal/bruit
IMPSONITE - [min] (a mineral asphaltum, occurring in Oklahoma, U.S.A.) impsonite *f*
IMPUBERAL - [med] impubère
IMPUGN, to - [gen] contester, attaquer
[leg] récuser, attaquer
IMPULSE - [gen] impulsion *f*, poussé *f*, poussée motrice
[el] (a undirectional flow of current of very short duration) impulsion *f*
[mech] (the force imparted to a movable object) impulsion *f*, choc *m* propulsif
[horol] (the force imparted to the pendulum) impulsion *f*
~BLADING - [mech] (of a gas turbine) aubage *m* à force impulsive
~CAM - [telecomm] (in automatic telephone exchanges) came *f* d'impulsions
~CHARGE - [chem] charge *f* de lancement
~CIRCUIT - [telecomm] (of an automatic telephone exchange) circuit *m* d'impulsions
~CIRCUIT BREAKER - [el] (circuit breaker in which the arc is extinguished by a mechanically produced flow of oil) interrupteur *m* à impulsions
~CONTACT - [el] contact *m* d'impulsion
~CONTACT OF THE DIALS - [telecomm] (in telephony) contact *m* du cadran d'appel
~COUNTER - [instr] (instrument designed to measure the thermal neutron flux in the counter region) compteur *m* d'impulsions
~-DRIVEN CLOCK - [el] (a clock in which the hands are driven by current impulses) horloge *m* électrique à impulsion
~EXCITATION - [radio] (method of exciting the grid of a thermionic tube) excitation *f* par choc
~EXCITER - [telecomm] (in telephony) contacteur *m*, générateur *m* d'impulsions
~FLASHOVER - [el] (of an insulator or other apparatus) contournement *m* d'impulsion
~FLASHOVER VOLTAGE - [el] (the value of the impulse voltage causing the flashover) tension *f* de contournement d'impulsion
~FREQUENCY - [telecomm] (in dialling etc) fréquence *f* de répétition des impulsions, fréquence *f* d'impulsions
~GENERATOR - [el] (device producing very short pulses of high voltage) génératrice *f* de choc
~INERTIA - [el] rigidité *f* diélectrique dynamique
~MACHINE - [telecomm] (in automatic telephone exchange) émetteur *m* d'impulsions, envoyeur *m* de impulsions
~METER - [instr] (instrument registering the number of current impulses passing through a circuit) compteur *m* d'impulsions
IMPULSE MODULATION - [radio] (modulation of a carrier by a pulse train) modulation *f* par impulsions
~MOTOR - [el] moteur *m* type contact
~NOISE - [radio] (noise die to transient disturbances) parasite *m* de courte durée
~OF CURRENT - [el] émission *f* de courant
~OF RECOIL - [phys] (the impulse obtained by a body in a collision opposite to the colliding mass) impulsion *f* de recul
~PERIOD - [telecomm] (in telephony) période *f* de impulsion
~PIN - [horol] (in the roller of the lever escapement) cheville *f* de l'échappement à ancre
~PLANE - [horol] (part of the pellet on which a tooth of the escape wheel acts when giving an impulse) plan *m* d'impulsion
~PULLEY - [text] galet *m* d'impulsions
~RATIO - [telecomm] (in telephony) rapport *m* d'impulsions
~REACTION TURBINE - [mech] (a disk and drum turbine) turbine *f* à action et réaction
~RECORDER - [instr] enregistreur *m* d'impulsions
~RELAY - [el] relais *m* récepteur d'impulsions
~REPEATER - [telecomm] (in an automatic telephone exchange) répétiteur *m* d'impulsions
~SEALING - [metall] (sealing by heating for short periods followed by periods of cooling under pressure) soudure *f* par impulsion
~SELECTOR - [el] sélecteur *m* d'impulsions
~SENDER - s. impulse exciter
~SEPARATOR - [radio] séparateur *m* à impulsion
~SERIES RELAY - [el] relais *m* de pontage
~STARTER - [mech] (device for starting reciprocating i.c. engines by giving a series of mechanical impulses to the magneto) impulseur *m*, lanceur *m*
~TACHOMETER - [instr] (tachometer of the capacitor) tachymètre *m* à impulsions
~TIMER - [horol] (synchronous timer used in radiation) chronomètre *m* à impulsions
~TRACK CIRCUIT CURRENT - [el] courant *m* de voie pulsé
~TURBINE - [mech] (steam turbine operating without any change in pressure as the steam passes the blade ring) turbine *f* à action
~VOLTAGE - [el] (transient voltage lasting for only a few microseconds) tension *f* de choc
~WAVE - [phys] onde *f* de choc
~WITHSTAND VOLTAGE - [el] tension *f* de tenue au choc
IMPULSING - [el & telecomm] émission *f* des impulsions
~RELAY - [el] relais *m* batteur
IMPULSIVE MOMENT - [phys] (the time integral of a torque) impulsion *f* angulaire, moment *m* d'inertie géométrique
~NOISE - [radio] bruit à intervalles
~SOUND - [acoust] (sound consisting of short bursts) son *m* à impulsions
I.M.S. - abbrev. for Industrial Methylated Spirit q.v.
IMPURITIES - [electron] (in a semi-conductors) impureté *f*pl
IMPURITY - [gen] impureté *f*
~ACTIVATION ENERGY - [electron] (the energy gap between an intermediate level (due to an impurity) and the adjacent energy band) énergie *f* d'activité

en fonction d'impuretés
IMPURITY ATOM - [phys] (an atom within a crystal which is alien to the crystal) élément *m* d'impurité
~COMPENSATION - [electron] compensation *f* par dopage
~DIFFUSION TECHNIQUE - [electron] technique *f* par diffusion des impuretés
~ELEMENT - s. impurity atom
~LEVEL - [electron] niveau *m* d'impuretés
~SEMICONDUCTOR - [electron] semiconducteur *m* par impuretés
~SPOT - [electron] (in germanium or silicon layers) spot *m* d'impureté
in. - [meas] pouce *m*
IN - [gen] en, dans, à
~-AIR DOSE - [radiat] (dose of radiation measured in the air) dose *f* dans l'air
~BULK - [gen & comm] en masse, en bloc
~EXCESS - [gen & chem] (term used of a substance added to another for a chemical reaction, and in excess of the stoichiometric proportion) en excès
~KIND - [gen] en nature
~LINE - [mech] en ligne
~-LINE HEADS - [el acoust] têtes *fpl* magnétiques alignées verticalement
~-LINE COMPRESSOR - [mech] compresseur *m* à cylindres en ligne
~-LINE ENGINE - [mech] (a reciprocating i.c. engine in which the cylinders are arranged in lines or banks) moteur *m* à cylindres en ligne
~MESH - [mech] (term used of gears which are engaged, i.e. acting on each other) en prise
~-PHASE - [el] en phase
~-PHASE REJECTION - [el biol] (common-made rejection) réjection *f* commune
~-PHASE REJECTION QUOTIENT - [el biol] (common-mode rejection quotient) quotient *m* de réjection en phase
~-PHASE SIGNAL - [el] (a signal applied equally to the inputs of a balanced amplifier stage or other differential device) signal *m* commun, signal *m* en phase
~-PILE TEST - [nucl] (an irradiation test in a reactor) essai *m* en réacteur
~-PROCESS MATERIAL - [chem] (product which is formed during a chemical process) produit *m* intermédiaire
~RUNNING CONDITION - [mech] (term used of a machine which is fit for immediate use) en état de marche
~RUNNING ORDER - [mech] (term used of a unit or machine which is ready for service) en ordre de marche
~SITU COMBUSTION - [oil ind] combustion in situ
~VIVO - [chem] (term used concerning a reaction or process occurring in the living cell) en-vivo
~WINDING - [mech] (term used of a flat object which is distorted by torsion, so that its extremities are not co-planar) ondulé
INABILITY - [gen] incapacité *f*
INACCESSIBILITY - [gen] inaccessibilité *f*
INACCESSIBLE - [gen] inaccessible
INACCURACY - [gen] inexactitude *f*, imprécision *f*, infidélité *f*
INACCURATE - [gen] inexact, incorrect
INACTINIC LIGHT - [phys] lumière *f* inactinique
INACTION - [gen] inaction *f*, inertie *f*
INACTIVATION - [chem] (the distruction of the activity, e.g. of a catalyst) inactivation *f*
INACTIVE - [gen] inactif, inert
~AREA - [nucl] zone *f* à conditions de travail non-réglementées
~FILLER COMPOUND - [rubber ind] mélange *m* de charges inertes
~GASES - [chem] (rare gases) gaz *mpl* inerts
~REFERENCE ELECTRODE - [el] électrode *f* de référence inactive
INANITION - [med] (exhaustion due to lack of food) inanition *f*
INARCHING - [agric] greffage *m* en approche
INAUDIBLE - [acoust] inoui
INBAND - [build] (a header stone) boutisse *f*
INBOARD - [aero] (towards the inside of an aircraft e.g. "the inboard end of a wing" is that nearest to the centre-line of the fuselage) intérieur *m*
[naut] en abord
~ENGINE - [naut] moteur *m* en abord
INBORN - [gen] inné
INBRED LINES - [genet] lignes *fpl* consanguines
INBREEDING - [genet] (breeding by related animals) accouplement *m* entre consanguins
INBUILT - [gen] incorporé, encastré
~COMPARTMENT FOR LUGGAGE - [auto] coffre *m* à bagages incorporé
INBYE - [min] (direction from a haulage way to a working face) vers le front d'attaque
INC - [met] s. incus
INCALCULABLE - [gen] incalculable
INCANDESCE, to - [phys] rendre incandescent, entrer en incandescence, mettre en incandescence
INCANDESCENCE - [phys] (the emission of light by a substance due to its high temperature) incandescence
INCANDESCENT - [light] incandescent
~CATHODE - [electron] cathode *f* incandescente
~LAMP - [el] (lamp in which light is produced by heating a substance to a very high temperature) lampe *f* à incandescence
~MANTLE - [light] (of gas light) manchon *m* (à incandescence)
INCARCERATION - [gen & med] incarcération *f*
~OF A HERNIA - [med] incarcération *f* d'une hernie
INCASEMENT - [med] préformation *f*
INCENDIARY - [gen] (with a tendency to cause combustion) incendiaire
INCENTIVE - [gen] stimulant *m*, aiguillon *m*, ressort *m*, encouragement *m*
INCEPTION - [gen] commencement *m*, début *m*
INCEST - [med] inceste *m*
INCH, to - [gen & mech] (to move by small increments) avancer peu à peu, reculer peu à peu
INCH - [meas] (25.4 mm) pouce *m*
INCHING - [gen] avancement pouce par pouce
[plast ind] (a technique of application of pressure to a mould in which a reduction in rate of application of pressure is made just before the mould is completely closed) marche *f* par mouvements saccadés
[el] commande *f* par fermetures successives rapides d'un circuit
~CONTROL - [mech] (mechanism for producing movement by small increments) commande *f* d'avancement à mouvements saccadés
~SWITCH - [el] (of a machine) interrupteur *m* d'in-

termittence
INCIDENCE - [gen phys opt] incidence *f*
~ANGLE - [gen] (angle of incidence) angle *m* d'incidence
~INDICATOR - [instr] (instrument for measuring the angle of incidence of an aircraft) indicateur *m* d'incidence
~OF TRAFFIC - [telecomm] allure *f* de trafic
~PLANE - [opt] plan *m* d'incidence
~WIRES - [aero] (wires used to brace a main plane structure in the plane of a pair of struts) haubans *mpl* d'incidence
INCIDENT - [gen phys opt etc] incident
~BEAM - [radiat] (beam falling on a target under an angle) faisceau *m* incident
~LIGHT - [light] lumière *f* incidente
~PARTICLE - [nucl] (a bombarding particle; a particle coming into collision with another particle) particule *f* incidente
~SIDE - [phys] plan *m* d'incidence
~SOUND - [acoust] (any sound falling on the microphone at an angle) son *m* incident
~WAVE - [phys] onde *f* incidente
INCIDENTAL - [gen] incidentel, fortuit
~CONSTITUENT - [metall] partie *f* accessoire (de la texture)
~MUSIC - [mus] (music played during the action of a play) musique *f* scénique
~SOUNDS - [cin] (all those sounds which are incorporated in the final sound track, e.g.foot-steps, bells, etc) sons *mpl* complémentaires spéciaux
INCINERATE, to - [gen] incinérer, carboniser, crémer
INCINERATOR - [impl] (any appliance designed to burn waste material) incinérateur *m*
[metall] four *m* à gadoues
INCIPIENT - [gen] naissant, qui commence
~CRACK - [constr work, metall etc] (a crack which is just beginning to develop) amorce *f* de fissure
~FAILURE - [geol] début *m* de rupture (ou du glissement)
INCISE, to - [gen] inciser
INCISION - [gen] incision *f*, entaille *f*
[med] (a cut by a surgical knife) incision *f*
[bot] découpure *f*
INCISOR - [med] dent *f* incisive
INCISURA - [med] incisure *f*, échancrure *f*
INCLINABLE - [gen] inclinable
~POWER PRESS - s. inclinable press
~PRESS - [mech] (a tilting-head press; one in which the head can be set at an angle) presse *f* inclinable
INCLINATION - [gen] inclination *f*, inclinaison *f*, pente *f*
[geom] inclinaison *f*
[el] (dip) flèche *f* normale
~INDICATOR - [photo] indicateur *m* d'inclinaison
~METER - [constr] éclimètre *m*
~OF STEERING KNUCKLE - [auto] inclinaison *f* des pivots de furée
~OF THE IMAGE - [photo] inclinaison *f* de l'image
~OF THE TWILL LINE - [text] inclinaison *f* du sillon
~OF THE WIND - [met] (the angle between the wind and the isobar at the point of observation) inclinaison *f* du vent
INCLINE, to - [gen] incliner, pencher
INCLINE - [gen] inclinaison *f*, plan *m* incliné, déclivité *f*, pente *f*
[min] (a sloping tunnel) rampe *f*, pente *f*
[railw] (hump) rampe *f* de débranchement
[mech] plan *m* automoteur
[mining] plan *m* inclin, descenderie *f*, puits *m* incliné
INCLINE CUT AND FILL - [mining] exploitation *f* par gradin incliné avec remblayage
INCLINED - [gen] incliné
~ARCH - [arch] arc *m* incliné
~BALANCE - [impl] bascule *f* à cadran
~BUCKET-HOIST - [mech] élévateur *m* à godets incliné
~-CARBON ARC LAMP - [light] (arc lamp with carbons set at an angle) lampe *f* à arc à électrodes de charbon inclinées
~CATENARY CONSTRUCTION - [el] (for overhead contact wires of an electric traction system) construction *f* à caténaire gauche
~CHORD CATENARY SUSPENSION - s. inclined overhead contact
~COKE BENCH - [metall] (in ovens & retorts) rampe *f* d'extinction
~CYLINDER ENGINE - [mech] moteur *m* à cylindres inclinés
~-DRUM TYPE MIXER - [mech] pétrisseur *m* à tambour
~FAULT - [geol] faille *f* inclinée
~FOLD - [geol] pli *m* oblique
~HOIST - [mech] élevateur *m* incliné
~LIFT - [mining] monte-charges *m* incliné
~MIRROR - [photo] miroir *m* incliné
~OVERHEAD CONTACT LINE - [el] caténaire *f* gauche, caténaire *f* inclinée
~PIT - [mining] puits *m* incliné
~PLANE - [mech] (simple machine in which the force required to lift an object is appreciably reduced) plan *m* incliné
~POSITION OF THE BLADES - [text] position *f* oblique des couteaux
~RETORT - [ind chem] cornue *f* inclinée
~SHEAR PLANE - [geol] plan *m* de cisaillement incliné
~TOP-SLICING - [mining] exploitation *f* par tranches inclinées descendantes avec foudroyage
~TUBE BOILER - [th eng] chaudière *f* tubulaire inclinée
~TUBE MANOMETER - [instr] (a form of liquid manometer) manomètre *m* à tube incliné
~VALVE - [mech] (in autos) soupape *f* incliné
INCLINOMETER - [instr] (instrument for measuring the inclination of an aircraft to the horizontal) clinomètre *m*, inclinomètre *m*
[instr] (a dip circle); an instrument consisting of a magnetic needle pivoted on a horizontal axis and designed to measure the magnetic dip) boussole *f* d'inclinaison
INCLUDE, to - [gen] comprendre, embrasser
INCLUDED WATER - [geol] eau *f* d'interposition
INCLUSION - [metall etc] (a foreign body enclosed within the mass of a solid, e.g. a fragment of refractory in furnace pig, or of some other mineral in a mineral) inclusion *f*
[metall] (a welding defect) inclusion *f* de scorie
~BODIES - [med] inclusions *fpl* cellulaires
~DISEASE - [med] maladie *f* des inclusions cytomégaliques
~OF PEAT - [geol] inclusion *f* de tourbe

INCLUSION STRINGER - [metall] inclusion *f* allongée
INCLUSIONS - [nucl] (particles of impurities formed within solid metal during freezing or by a subsequent reaction) inclusions *fpl*
INCLUSIVE - [gen] qui comprend, global
~-OR CIRCUIT - [comput] circuit *m* ou inclusif, circuit *m* réunion, mélangeur *m*
INCOMBUSTIBLE - [gen] incombustible
INCOMING CALL - [telecomm] (telephone) communication *f* d'arrivée
~FEEDER - [el] (in a substation, the feeder through which power is received) feeder *m* d'arrivée
~JUNCTION - [telecomm] ligne *f* auxiliaire entrante
~LINE - [telev] ligne *f* d'arrivée
~MESSAGE HOLDING UNIT - [comput] mémoire *f* tampon d'entrée, mémoire *f* de transit d'entrée
~POSITION - [telecomm] (of telephone exchange) groupe *m* d'arrivée, position *f* d'arrivée
~SELECTOR - [telecomm] (telephone) sélecteur *m* de arrivée, sélecteur *m* entrant
~SIGNAL - [telev] signal *m* d'entrée
~TRUNK - s. incoming junction
INCOMMENSURABILITY - [gen & math] incommensurabilité *f*
INCOMPATIBLE - [gen] incompatible
~ROUTES - [railw] itinéraires *mpl* incompatibles, itinéraires *mpl* antagonistes
INCOMPENSATION - [med] décompensation *f*
INCOMPETENCE - [med] insuffisance *f*
INCOMPLETE - [gen] incomplet, imparfait
~COMBUSTION - [fuel] (partial combustion) combustion *f* incomplete
~DIALLED CALL - [telecomm] (telephone) appel *m* incomplet
~OVERFALL - [hydr] déversoir *m* à seuil submergé
~REACTION - [chem] (reversible reaction which is allowed to reach equilibrium, so that a mixture of reactants and reaction products are obtained) réaction *f* incomplète
INCOMPREHENSIBLE - [gen] incompréhensible
INCOMPRESSIBILITY - [phys] (of liquids, gases etc) incompressibilité *f*
INCOMPRESSIBLE - [phys] incompressible
~FLOW - [phys] (type of flow in which changes of density in the medium are negligible) courant *m* incompressible
~FLUID - [mech] (a fluid with a density which is substantially unaffected by change of pressure) fluide *m* incompressible
~VOLUME - [phys] (the volume which the molecules of a gas actually occupy) volume *m* incompressible
INCONCEIVABLE - [gen] inconcevable
INCONGRUENT MELTING POINT - [chem] point *m* de fusion incongruent
~SATURATED SOLUTION - [chem] solution *f* saturée incongruente
INCONSISTENCY - [gen] (lack of accordance, of harmony) inconsistance *f*
[gen] (lack of agreement; incongruity) contradiction *f*, incompatibilité *f*
INCONTINENCE-[med] incontinence *f*, acathexie *f*
INCONTROVERTIBLE - [gen] incontroversable, incontestable
INCONVENIENCE - [gen] inconvénient *m*, dérangement *m*
INCONVERTIBLE - [gen] inconvertible
INCOORDINATION - [med] incoordination *f*

INCORPORATE, to - [gen] incorporer, mêler, unir
INCORPORATED - [gen] incorporé, uni
[leg] (constituent into a legal body) (société) constituée, autorisée
~REDUCTION GEAR - [mech] engrenages *mpl* réducteurs incorporés
INCORPORATING TIME - [rubber ind] temps *m* d'incorporation
INCORPORATION - [gen] incorporation *f*
[leg] (the action of forming a legal body, a company, a society) constitution *f* municipalité
[town planning] érection *f* en municipalité
~OF BEAD WIRE - [rubber ind] (tyres) fixation *f* de la tringle dans le talon
INCORRECT - [gen] incorrect, inexact
[gen] (of a procedure) irrégulier
[mech] (of a part or component) incorrect
INCREASE, to - [gen] augmenter, croître, grandir
[comm] (of prices) augmenter
[phys etc] (e.g. temperature) augmenter, élever
INCREASE - [gen] augmentation *f*, accroissement *m*
~IN EFFICIENCY - [gen] augmentation *f* du rendement
~IN EXPOSURE - [photo] augmentation *f* d'exposition, majoration *f* d'exposition
~IN POPULATION - [gen] taux *m* d'accroissement démographique
~IN PRESSURE - [geol] accroissement *m* de pression, incrément *m* de pression
~IN VALUE - [gen] plus-value *f*, augmentation *f* de valeur
~OF DENSITY - [photo] accroissement *m* de noircissement
~OF POTENTIAL - [el & mech] accroissement *m* de potentiel
~OF SENSITIVITY - [photo] accroissement *m* de la sensibilité
~OF STRAIN - [geol] accroissement *m* de la déformation
[mech] augmentation *f* de la tension
~OF TENSION - s. increase of strain
~OF TORQUE - [mech] augmentation *f* de la torsion
INCREASER - [plumb] (coupling piece between two pipes of different size) cône *m* réducteur
[oil ind] raccord *m* de réduction
INCREASING LENGTH - [text] longeur *f* croissante
~THROW - [text] course *f* accélérée
INCREMENT - [gen] augmentation *f*, accroissement *m*
[math] incrément
~-CUT FILE - [tool] lime *f* à dents irrégulières
~OF GROWTH - [gen] taux *m* d'accroissement
~OF SETTLEMENT - [soil] augmentation *f* du tassement
~RATE - s. increment of growth
INCREMENTAL - [gen] additionnel
~DUPLEX - [telecomm] duplex *m* par addition
~INDUCTANCE - [el] (the induction an iron-cored coil offers to A.C. when it is superimposed on D.C. through the coil) induction *f* additionnelle, induction *f* différentielle
~IRON LOSSES - [el] (iron losses in an a.c. machine caused by frequencies which are higher than the fundamental ones) pertes *fpl* dites dans le fer additionnelles
~MAGNETIZING FORCE - [el] force *f* magnétisante différentielle
~PERMEABILITY - [el] (of magnetic material) per-

méabilité *f* différentielle
INCRETION - [med] sécrétion *f* interne
INCRIMINATE, to - [gen & leg] incriminer, impliquer
INCRUST, to - [gen & of boilers] incruster, entartrer
INCRUSTATED IRON ORE - [min] minerai *m* de fer veiné d'autres matières
INCRUSTATION - [th eng] (the deposition of dissolved salts from water) incrustation *f*, entartrage *m*
[bot] (a coating of calcium carbonate) incrustation *f*
[constr work] (a wall facing) revêtement *m*
[mech] tartre *m*, dépôt *m* calcaire
INCRUSTED BALLAST - [railw] ballast *m* colmaté
INCUBATE, to - [gen etc] incuber
[zool] couver
[med] (of a disease) couver
INCUBATING ROOM - [agric etc] chambre *f* couveuse
INCUBATION - [zool] (the process of causing eggs to hatch) incubation *f*, accouvage *m*
[med] (the period between the actual infection and the appearance of symptoms) incubation *f*
~PERIOD - [med] période *f* d'incubation
INCUBATOR - [biol & agric] incubateur *m*, couveuse *f* artificielle
[med] (apparatus for rearing prematurely born children) incubateur *m*
[chem] (apparatus for the development of bacteria) étuve *f* à incubation
INCUMBRANCE - [agric] grèvement *m* de bien-fonds
INCUNABULUM - [print] (fifteenth century book) incunable *m*
INCUR, to - [gen] courir, encourir
[gen] (e.g. debts) contracter (des dettes)
INCURABLE - [gen & med] incurable
INCURVATION - [opt] incurvation *f*
INCURVED - [gen] incurvé, courbé
INCUS - [met] (the upper part of a cloud base, extended in anvil-like form and smoothly fibrous or striated in appearance) incus *m*
[anat] (the middle one of the three ear-ossicles) enclume *f*
INDAMINES - [chem] (the derivatives of the phenylated p-quinone-dimines. They are important intermediates in azime and sulphide dyes) indamines *fpl*
INDANTHRENE - [chem] (a blue dyestuff) indanthrène *m*
~DYES - [dyes] teintures *f* à l'indanthrène
INDEFENSIBLE - [gen] (which admits of no defence) indéfensible, indéfendable
INDEFINABLE - [gen] indéfinissable
INDEHISCENT FRUIT - [agric] fruit *m* indéhiscent, akène *m*
INDELIBLE - [gen] indélébile, ineffaçable
~COLOUR - [dyes] couleur *f* ineffaçable
~INK - [ind chem] encre *m* indélébile
[text] encre *f* à marquer le linge
INDEMNITY, to - [gen & comm] (to compensate for a loss etc) indemniser, garantir
INDEMNITY - [gen & leg] indemnité *f*, garantie *f* (payment for damage or loss suffered) indemnisation *f*, compensation *f*
INDENE - [chem] (an intermediate for synthetic resins) indène *m*
INDENT, to - [gen] denteler, entailler (un bord)
[carp] endenter, adenter
[comm] (to send an order for goods from abroad) passer une commande
INDENT - [comm] (an order for goods to be sent from abroad) ordre *m* d'achat
[leg] (official requisition of stores) ordre *m* de réquisition
[gen] empreinte *f* creuse
[print] (the blank space in front of the first line of a paragraph) renforcement *m*
[carp] (a notch) endent *m*, adent *m*
[metall] (by the tool) brouture *f*
INDENTATION - [gen] endentement *m*, découpage *m*, dentelure *f*
[carp & mech] indentation *f*, endentement *m*
[metall] empreinte *f*
[metall] (in welding, a depression in the outer surface) incision *f*
[geogr] indentation *f*, écranchure *f*
[mech] (a dent) pénétration *f*
[mech] (a notch) empreinte *f* creuse
~HARDNESS - [metall] dureté *f* Vickers
[rubber ind] dureté *f* à la pression de bille, dureté *f* par pénétration
~HARDNESS TEST - [mech] essai *m* de dureté à pression de bille
~TEST - [metall] essai *m* d'emboutissage
[constr] essai *m* de pression *f* à la bille
INDENTED - [arch] bossué, bossolé
~BOLT - [mech] boulon *m* denté
~V-BELT - [rubber ind] courroie *f* trapézoïdale dentée
INDENTER - [mech] (a roller fitted with projections used on newly laid road surfaces to roughen it in order to avoid skidding) rouleau *m* à bossuer
INDENTING BALL - [impl] (used for hardness test) bille *f* de pénétration, bille *f* d'enforcement
INDENTION - s. indent [print]
INDENTOR - s. indenter
INDEPENDENCE OF A MOTOR-DRIVEN VEHICLE - [gen] rayon *m* d'action d'un véhicule à moteur
INDEPENDENT - [gen] indépendant, autonome
[mech] indépendant, libre
~AXLE DRIVE - [el traction] traction *f* à moteurs indépendants
~CHUCK - [mech] (lathe chuck in which each of the jaws can be moved independently) plateau *m* à griffes indépendantes
~CONTACT - [el] contact *m* de circuit unique
~FISSION YIELD - [nucl] (primary fission yield) rendement *m* de fission primaire
~FRONT SUSPENSION - s. independent front wheel suspension
~FRONT WHEEL SUSPENSION - [mech] (of autos) suspension *f* avant indépendante
~GUIDE BARS - [text] barres *fpl* indépendantes
~HEARTH FIRE - [th eng] appareil *m* de chauffage indépendant raccordé
~HEATING - [th eng] (space heating) chauffage *m* divisé, chauffage *m* indépendant
~-JAW CHUCK - s. independent chuck
~MANUAL OPERATION - [contr] (an operation by hand to complete a cycle of operations) manœuvre *f* indépendante
~MOTION - [mech] mouvement *m* indépendant
~PARTICLE MODEL OF NUCLEUS - [nucl] (nuclear model based on the principle that each nucleon moves independently in the field corresponding to the average position of the rest of the nucleons) modè-

le *m* du noyau à particules indépendantes
INDEPENDENT SIDEBAND TRANSMISSION - [radio] émetteur *m* sur bande latérale indépendante
~TIME-LAG - [contr] retard constant
~TIME-LAG RELAY - [el] relais *m* à retard constant
~TRACKS - [el] indépendance *f* des voies
~TRIP - [el] (tripping device in which the current operating it is independent of the current flowing in the circuit) déclenchement *m* à courant indépendant
~VARIABLE - [math] variable *f* indépendante
INDEPENDENTLY MOUNTED NIP ROLLER - [text] cylindre *m* d'arrêt
INDESTRUCTIBILITY - [gen] indestructibilité *f*
INDETERMINANCY PRINCIPLE-[phys](a postulate of quantum mechanics) principe *m* d'incertitude, principe *m* d'indétermination
INDETERMINATE - [gen] indéterminé, imprécis
[math] indéterminé
~ANALYSIS - [math] analyse *f* indéterminée
INDEX, to - [gen] faire l'index, dresser l'index
[instr] graduer
[contr] positionner
INDEX - [gen] index *m*
[instr] indice *m*
[math] exposant *m*
[horol] (the regulating lever for the adjustment of the rate of a watch) raquette *f*
(of an instrument dial) aiguille *f*
[surv] (simple plane table alidade) alidade *f*
[meas] (meter index) totalisateur *m*
[anat] index *m*, premier doigt *m*
~ACCUMULATOR - [comput] registre *m* d'index, registre-index *m*, registre *m* d'indices
~BAR - [instr] (of a sextant) alidade *m*
~BOARD - [paper man] (standard board size, 25 1/2 x 30 1/2 in) carton *m* pour fichiers
~BOOK - [comm] (a gas or electricity meter index card) carnet *m* de relevé
~BURNER - [gas ind] (a test burner) brûleur *m* d'essai
~CENTRES - [mach tool] appareil *m* diviseur (pour fraises)
~CHANGE GEARS - [mach tool] roues *fpl* d'appareil diviseur
~CONTOUR - [surv] courbe *f* de direction
~CRANK - [mach tool] manivelle *f* du diviseur
~DIAL - [mech tool] cadran *m* de division
~DRUM - [text] (in knitting) tambour *m* compteur
~FINGER - [med] index *m*
~GLASS - [instr] (the movable glass of the sextant) miroir *m* du sextant
~GRADUATIONS - [meas] (the longest division marks adjacent to the scale figures) lignes *fpl* de graduation
~HEAD - [mach tool] (attachment for rotating the work through any angle) diviseur *m*
~LINE - [instr] ligne *f* de référence
~MACHINE - [text] ratière *f*, mécanique *f* armure
~NUMBER - [gen] chiffre *m* indicateur
~OF COOPERATION - [radio] module *m* de coopération, module *m* d'hélice d'exploration
~OF REFRACTION - [light] (the ratio of the velocity of radiation in a vacuum to the velocity of the same radiation in a medium) indice *m* de réfraction
[el] (the square root of the relative dielectric constant of a medium) indice *m* de réfraction
OF THE ROOT - [math] indice *m* de la racine
INDEX PATH - [meas] trajectoire *f* de l'aiguille
~PAWL - [comput] cliquet *m* d'index
~PIN - [horol] cheville *f* de la raquette
~PLATE - [instr] plaque *f* graduée
[mach tool] (of the index head) plateau *m* diviseur, plateau *m* de division
~POINT - [contr] position *f* de référence
~SPINDLE - [mech] (of a meter) arbre *m* de commande de la minuterie
~STRIPS - [telev] (in colour television) filets *mpl* colorés
~TRACK - [comput] piste *f* d'index
~VALUE - [contr] (set point; the value of the controlled condition to which the mechanism is set) valeur *f* mesurée
~WITH POINTERS - [meter] minuterie *f* à aiguilles, dispositif indicateur à aiguilles
~WORD - [comput] modificateur *m*
INDEXING - [gen] (the operation of preparing an index) indexation *f*
[mech] (of machining, drilling etc) division *f*
~CAM - [text] came *f* d'avancement, came *f* de poussée de l'appareil de marque
~ELECTRON BEAM - [telev] faisceau-pilote *m*
~HEAD - s. index head
~INSTRUCTION - [comput] instruction *f* d'indexage
~LEVER - [text] levier *m* d'avancement, levier *m* de commande
~MACHINE - [mach tool] diviseur *m*
~PLATE - s. index plate
~REGISTER - [comput] (machine register in which a variable is put in to change another variable) registre *m* index
~SIGNAL - [telev] signal *m* de tube à index
INDIA GUM - [ind chem] (a synonym of gum Arabic) gomme *f* arabique
~INK - [chem] encre *m* de Chine, noir *m* de Chine
~MILLET - s. Indian millet
~PAPER - [paper man] papier *m* bible, pelure *f* d'oignon
~RUBBER - [chem] (rubber) caoutchouc *m*
~RUBBER CABLE - [el] câble *m* isolé au caoutchouc
~RUBBER CORE CABLE - [telecomm] câble *m* de raccordement aérosouterrain sous caoutchouc
~RUBBER WIRE - [el] fil *m* sous caoutchouc
~THREAD - [text] ourdissage *m* des fils de caoutchouc
~WARP - [text] chaîne *f* de caoutchouc
INDIAN CALICO - [text] percale *f*
~CARPET - [text] tapis *m* brodé
~CORN - [agric] mais *m*
~FIG - [agric] figue *f* d'Inde
~FIG CACTUS - s. indian fig
~HEMP - [text] chanvre *m* indien
~INK - s. india ink
~MALLOW - [text] urène *f*, paka *m*
~MEAL - [agric] farine *f* de mais
~MILLET - [bot] millet *m*, mil *m*
~OCHRE - [dye] ocre *f* rouge
~RED - [paint] (term now used for an artificial iron oxide pigment; ormerly an alternative name for Gulf Red) rouge *m* des Indes
~RUBBER TREE - [bot] figuier *m* élastique
~SHIRTING - [text] shirting *m* indien
~SUMMER - [met] (period of mild calm hazy weather in autumn or early winter in N. America not a regular seasonal phenomenon, though popularly believed to be one) été *m* de la Saint-Martin

INDIAN YELLOW - [dye] (a cobalt potassium nitrate pigment) jaune *m* de cobalt, jaune *m* des Indes
INDICATE, to - [gen] indiquer, montrer
INDICATED - [gen etc] indiqué
~AIRSPEED - [aero] (the reading of an airspeed indicator, corrected only for calibration errors) vitesse *f* indiquée
~ALTITUDE - [aero] (altimeter reading with barometric setting, corrected for instrumental and installation errors) altitude *f* indiquée
~ANGLE - [contr] (the angle shown on a pointer of an apparatus) angle *m* affiché (d'un synchro-récepteur)
~DYNAMIC PRESSURE - s. impact pressure
~HORSE-POWER (I.H.P.) - [met] (value for horsepower obtained by the use of an indicator) puissance *f* indiquée en C.V.
~MEAN EFFECTIVE PRESSURE - [mech] (the mean pressure exerted by the fluid in a working cylinder throughout the cycle) pression *f* effective moyenne indiquée
~STALLING SPEED - [aero] (the airspeed shown by the indicator at the point of stall) vitesse *f* de décrochage indiquée
~TERRAIN CLEARANCE ALTITUDE - [aero] (reading on an altimeter set to read altitude above a given point on the earth's surface, after application of instrumental and installation) hauteur *m* minimum autorisée
~THERMAL EFFICIENCY - [mech] (of a reciprocating engine) rendement *m* thermique indiqué
INDICATING ACCELEROMETER - s. accelerometer
~INSTRUMENT - [instr] (a device designed to show the instantaneous value of a quantity) appareil *m* de mesure indicateur
~LEAD - [el] fil *m* d'amenée de courant à l'indicateur
~SCALE - [meas] (of a recording instrument) échelle *f* d'indicateur
~TUBE - [comput] tube *m* afficheur, tube *m* d'affichage
INDICATION - [gen] indication *f*
[med] indication *f*
~ERROR - [meas] erreur *f* d'indication
INDICATIVE - [gen] indicatif
INDICATOR - [mech] (apparatus designed to be temporarily attached to a working reciprocating engine, which automatically produces a curve of pressure in the cylinder against piston movement) indicateur *m*
[instr] (in indicating device) appareil *m* de mesure indicateur, voyant *m*
[chem] (substance used in reactions to indicate the stage reached by a chang of colour) indicateur *m*
[a sign on roads] signalisateur *m*
[telecomm] annonciateur *m*
~CARD - s. indicator diagram
~DIAGRAM - [mech] (curve obtained from a reciprocating engine indicator) diagramme *m* d'indicateur
~DISC - [text] disque *m* diviseur, disque indicateur, disque *m* à divisions
~DRUM - [text] plateau *m* à doigt
~ELEMENT - [nucl] (radioactive element used as an indicator) traceur *m* isotopique
~EXPONENT - [chem] (the pH-value at which the most rapid change of the colour of an indicator occurs) exposant *m* d'indicateur
INDICATOR FINGER - [text] aiguille *f* indicatrice
~GATE - [electron] (only USA; sensitizing pulse in GB) impulsion *f* de sensibilisation
~GRID - [electron] (the grid which is used in a magic eye) grille *f* de tube indicateur
~INSTRUMENT - s. indicating instrument
~LAMP - [el] (small electric lamp arranged to give notice of some event or state e.g. excessive pressure, or position of landing gear) voyant *m*
~LIGHT - [light] (pilot light) lampe-témoin *f*, lampe-pilote *f*
~ORGANISM - [biol] organisme *m* indicateur, espèce *f* représentative
~PIN - [text] cheville *f* de déclenchement
~PLANTS - [agric] plantes *fpl* indicatrices
~RANGE - [chem] (the range of the pH value within which an indicator changes colour) champ *m* de virage de l'indicateur
~ROD - [text] tringle *f* de déclenchement, barre *f* de déclenchement
~STAND - [mining] indicateur *m* d'extraction
~WHEEL - [text] roue *f* à aiguille, roue *f* indicatrice
~WINDOW - [photo] fenêtre *f* de lecture
~WITH ALARM CONTACT - [telecomm] voyant *m* à signalisation
INDICATRIX - [math] indicatrice *f*
INDICES OF CRYSTAL FACES - [crystall] (symbols used to indicate the spatial location of crystal faces as referred to a given set of axes) indices *mpl* des formes cristallines
INDICIAL ADMITTANCE - [el] (of a network) admittance *f* indicielle
~RESPONSE - [contr] réponse *f* indicielle
INDICOLITE - [min] (or indigolite; blue variety of tourmaline) indigolite *f*
INDICT, to - [leg] accuser, inculper
INDICTABLE - [leg] traduisible en justice
~OFFENCE - [leg] délit *m*
INDICTEMENT - [leg] incrimination *f*, accusation *f*
INDIFFERENT - [med] indifférent, impartial
INDIGENOUS - [gen] (native) indigène
~GAS - [gas ind] gaz *m* d'origine, gaz *m* indigène
~VARIETY - [brew ind] (home-grown barley) orge *f* indigène
INDIGESTIBLE - [med] indigeste
INDIGESTION - [med] indigestion *f*, dyspepsie *f*
INDIGO - [chem] (a blue dye derived from indole and occurring in several plants. It can be synthesized and is an important vat dyestuff) indigo *m*, inde *m*
~CARMINE - [dye] (a dyestuff and a diagnostic aid in medicine) carmin *m* d'indigo
~COPPER - [min] (copper sulphide) covelline *f*
~EXTRACT - [chem] extrait *m* d'indigo
~PLANT - [bot] indigotier *m*, anil *m*
~PRINT - [text] impression *f* à l'indigo
~PURPURIN - [dye] purpurine *f* d'indigo
~RED - [dye] rouge *m* d'indigo
~VAT DYEING - [text] teinture *f* à l'indigo
INDIRECT - [gen] indirect
~-ACTING RECORDING INSTRUMENT - [instr] instrument *m* enregistreur à action indirecte
~ADDRESSING - [comput] (non-systematic addressing) adressage *m* indirect
~ARC FURNACE - [metall] (type of electric furnace in which an arc is maintained above the charge, but does not pass through it) four *m* à arc indirect

INDIRECT ARC HEATING - [heat] (method of arc heating in which the arc current is not passed through the medium to be heated) chauffage *m* indirect par arc
~CALL - [telecomm] communication *f* de transit
~COLORIMETRY - [opt] colorimétrie *f* indirecte
~CONTROL - [contr] (requiring energy from an external source) réglage *m* indirect
~COUPLING - [el] couplage *m* indirect
~-CYCLE NUCLEAR ENGINE - [mech] (type of nuclear propulsion unit in which heat is transferred from the reactor to the turbojet air by way of two successive closed circuit systems) réacteur *m* à cycle indirect
~DEMOSTRATION - [gen & math] demonstration *f* par l'absurde
~DIPOLE MOMENT - [el] moment *m* dipôle induit
~ECHOES - [radar] (false echoes) échos *mpl* faux
~FITTINGS - [light] (fittings for indirect lighting) installation *f* pour éclairage indirect
HEATING - [heat] (heating by convection) chauffage *m* indirect
~HEATING SURFACE - [heat] (in heating by convection) surface *f* de chauffe indirect
~ILLUMINATION - [opt] (in microscopic examination, the light which strikes the object at right angles to the direction of the microscope axis) éclairage *m* indirect
[light] s. indirect lighting
~INPUT - [comput] (the introduction of information into a machine by means of a separate machine) entrée *f* indirecte
~JACQUARD PATTERN - [text] dessin *m* Jacquard indirect
~LIGHT - [light] lumière *f* indirecte
~LIGHTING - [light] (lighting by radiations which undergo one or more reflections in its journey from the source to the object) éclairage *m* indirect
~LOAD - [phys] charge *f* indirecte
~MEASUREMENT - [surv] mensuration *f* indirecte
~MEASUREMENT OF EFFICIENCY - [el] (the method by which the efficiency is calculated from the measurement of the losses) méthode *f* de mesure indirecte du rendement
~POSITIVE TAKING UP MOTION - [text] régulateur *m* d'ensouple positif et indirect
~RESISTANCE FURNACE - [metall] (type of furnace in which heat is generated by a resistance, but not by passing current through the charge) four *m* électrique à chauffage indirect par résistance
~RESISTANCE HEATING - [heat] (method of electrical heating in which the heat is developed by the Joule effect and is then transferred to be object to be heated) chauffage *m* indirect par résistance
~ROUTE - [gen & railw] itinéraire *m* détourné
~ROUTING - [telecomm] utilisation *f* des voies indirectes
~SCANNING - [telev] analyse *f* indirecte
~SULPHATE RECOVERY - [chem] sulfatage *m* indirect
~TAKING UP - [text] enroulement *m* indirect
~TRANSMISSION - [telev] (taken from a photographic image) émission *f* indirecte
~WAVE - [phys] (electromagnetic waves) onde *f* réfléchie
~WEIGHTING - [mech] charge *f* indirecte
INDIRECTLY CONTROLLED SYSTEM - [contr] système *m* de réglage indirect
INDIRECTLY FED AERIAL - [radio] (that portion of a directive aerial which is not directly connected to the transmitter or the receiver) antenne *f* à éléments parasites
~HEATED CATHODE - [electron] (a cathode heated by an element, i.e. the heater, which may be independent of it) cathode *f* à chauffage indirecte
INDISSOLUBILITY - [gen] indissolubilité *f*
INDISSOLUBLE - [gen] indissoluble, insoluble
INDISTINCT INTERLACING EFFECT - [text] effet *m* indistinct du liage
~PATTERN - [text] dessin *m* caprice
INDIUM - [chem] (metallic element, symbol In, A.N. 49, A.W. II4.8, with applications in jewellery and the production of low-melting point alloys) indium *m*
~ANTIMONIDE - [chem] antimoniure *m* d'indium
~ARSENIDE - [chem] arséniure *m* d'indium
INDIVIDUAL - [gen] individuel, particulier
~DISTORTION - [radio] (in modulation) distorsion *f* individuelle
~DRIVE - [el] (industrial system whereby each machine in a factory is driven by an individual motor) moteur *m* indépendant
~FLUE - [plumb etc] conduit *m* individuel, départ *m* individuel
~FOCUSSING - [opt] (of binoculars) mise *f* au point individuelle
~PICTURE - [cin] (a single picture on a film) image *f*
~SHOT IDENTIFICATION - [cin] marque *f* de synchronisme
~TRUNK GROUP - [telecomm] faisceau *m* des jonctions individuelles, sectionnement *m* particulier
~VULCANIZER - [rubber ind] (device for vulcanizing single pieces, one at a time) vulcanisateur *m* individuel
INDIVISIBILITY - [gen] indivisibilité *f*
INDIVISIBLE - [gen] indivisible
~TRAIN SET - [railw] unité *f* indéformable, rame *f* indéformable
~TRAIN UNIT - s. indivisible train set
INDOIN BLUE - [dye] (used in the textile industry, in particular for wool) bleu *m* d'indoïne
INDOLE - [chem] (the basis of the indigo molecule. It is used as a component of perfumes and as an analytical reagent) indol *m*
INDOLEBUTYRIC ACID - [chem] (a plant hormone with uses in horticulture) acide *m* indolbutyrique
INDOLO-ACETIC ACID - [chem] acide *m* indolacétique
INDOOR - [gen] (within doors) d'intérieur
~AERIAL - [radio] antenne *f* intérieure
~ANTENNA - s. indoor aerial
~FARM EQUIPMENT - [agric] matériel *m* de ferme
~FLYING MODEL - [aero] modèle *m* en appartement
~RECORDING - [el acoust] enregistrement *m* en intérieur
INDOORS - [gen] à la maison
INDOPHENOLS - [chem] (compounds similar in constitution to the indamines: used in cotton and wool dyeing) indophénols *mpl*
INDOXYL - [chem] (an indole derivative) indoxyle *m*
INDOXYLEMIA - [med] indoxylémie *f*
INDOXYLURIA - [med] indoxylurie *f*
INDUCE, to - [gen] causer, produire, amener
INDUCED - [gen] induit
[el etc] induit

INDUCED ANGLE OF ATTACK - [aero] (the difference between the actual angle of attack and the angle of attack for a wing of the same lift coefficient but of infinite aspect ratio) angle *m* d'induit
~CHARGE - [el] (electric charge produced on a conductor as a result of a charge on a neigh-bouring conductor) charge *f* induite
~CIRCUIT - [el] circuit *m* induit
~CURRENT - [el] (current produced by an induced electromotive force in the circuit) courant *m* induit
~DIPOLE MOMENT - [el] (a dipole moment induced in a system because it is brought into a magnetic field) moment *m* de dipôle induit
~DRAFT - s. induced draught
~DRAG - [aero] (term formerly used for trailing-vortex drag (subsonic flow) or for any type of drag dependent on lift) traînée *f* induite
~DRAUGHT - [mech] (forced draught system for boiler furnaces) aérage *m* mécanique par aspiration, aérage *m* négatif
~-DRAUGHT FAN - [mech] ventilateur *m* aspirant
~ELECTRICITY - [el] électricité *f* induite
~e.m.f. - [el] (electro-magnetic force occurring in a circuit as a result of a change in the amount of magnetic flux linked with the circuit) force *f* électromagnétique induite
~FISSION - [nucl] (fission induced under controlled conditions) fission *f* induite
~GAS - [gas ind] (burners) fluide *m* induit
~MAGNETISM - [el] magnétisme *m* induit
~NATURAL RADIONUCLIDES - [nucl] (natural radionuclides having a short time-life geologically) radionucléides *mpl* naturals induits
~NOISE - [radio] (circuit noise produced by electrostatic or electromagnetic induction from near-by power lines) bruit *m* induit
~NUCLEAR DISINTEGRATION - [nucl] (occurring when a nucleus comes into contact with another nucleus) désintégration *f* nucléaire induite
~POLARIZATION - [el] (due to the action of an electric field on a dielectric not containing permanent dipoles) polarisation *f* induite
~POTENTIAL - [el] potentiel *m* induit
~POWER LOSS - [aero] (the power consumed by imparting velocity to the air passing through the propeller in respect to its centre) puissance *f* absorbée induite
~RADIO-ACTIVITY - [nucl] (radio-activity produced under artificial conditions) radioactivité *f* artificielle, radioactivité *f* induite
~REACTION - [chem] (one which can be accelerated by the occurrence in the same system and at the same time, of another and rapid reaction) réaction *f* induite
~VOLTAGE - [el] tension *f* induite
INDUCING CURRENT - [el] courant *m* inducteur
INDUCTANCE - [el] (the property of a circuit which produces E.M.F. as result of a change in the magnetic flux through the circuit) inductance *f*
~BOX - [el] (assembly of inductors of known nominal value mounted in a box) boîte *f* d'inductances
~BOX WITH PLUGS - [el] boîte *f* d'inductances à fiches
~-CAPACITANCE COUPLING - [el] couplage *m* par impédance
~COIL - [el] bobine *f* d'induction, bobine *f* inductrice
INDUCTANCE METER - [instr] (instrument for measuring inductance in henrys, millihenrys etc) inductomètre *m*
~MULTIPLIER - [electron] multiplicateur *m* d'inductance
~PER UNIT LENGTH - [el] inductance *f* linéique
~STRAIN GAUGE - [meas] (strain gauge using the variations of the inductance) extensomètre *m* à induction
~UNBALANCE - [el] déséquilibre *m* de l'inductance
INDUCTION - [el] (the process which gives rise to an E.M.F. in a circuit through which the magnetic flux is changing) induction *f*
[mech] (a process of drawing in the air or mixture by piston suction in i.c. reciprocating engines) induction *f*, aspiration *f*, admission *f*
~ACCELERATOR - [nucl] (e.g. the betatron or rheotron) bêtatron *m*
~BRAZING - [el metall] soudure *f* forte à induction
~COIL - s. inductor
~CHANNEL FURNACE - [ovens] (type of furnace in which the charge is contained in an annular channel which acts as the secondary of a transformer, in which heavy currents are induced by a primary winding) four *m* à induction à canal
~COMPASS - [instr] ("eart-inductor compass") boussole *f* d'induction
~CURRENT - [el] courant *m* induit
~EFFECT - [el] effet *m* d'induction
~FIELD - [el] (the magnetic field around a conductor by the current in the conductor) champ *m* d'induction
~FLOWMETER - [instr] fluxmètre *m* à induction
FORGING - [metall] (forging by induction heating) forgeage *m* par induction
~FURNACE - [el] '(metal melting electric furnace in which heat is produced by current induced in the charge) four *m* à induction
~FURNACE WITH SUBMERGED CHANNEL - [metall] four *m* à induction à canal submergé
~GENERATOR - [el] (asynchronous machine working as an alternating-current generator) alternateur *m* asynchrone
~HARDENING - [metall] trempe *f* par induction
~HEAT-TREATMENT - [metall] traitement *m* thermique par induction
~HEATER - s. induction furnace
~HEATING - [el] (heating by inducing currents in the material to be heated) chauffage *m* par induction
HEATING APPARATUS - [el heat] réchauffeur *m* à induction
~-HEATING EQUIPMENT - [el] appareillage *m* pour le chauffage par induction
~INSTRUMENT - [instr] (electrical measuring instrument in which the pointer depends on the interaction between an alternating flux produced by the measured quantity and the currents which are induced by this flux in a disk) appareil *m* à induction
~INTERFERENCE - s. induced noise
~LOG - [mining] carottage *m* à induction
~MACHINE - [el] (electrostatic generator) machine *f* électrostatique
~MANIFOLD - [mech] (branched pipe through which the mixture is distributed to individual cylinders in a reciprocating i.c. engine) collecteur *m* d'admission
~MELTING FURNACE - [el heat] four *m* de fusion par

induction

INDUCTION METER - [instr] (meter with fixed coils acting on a conducting moving element in which currents induced by the coils flow) compteur *m* à induction

~ METHOD - [instr] méthode *f* inductif

~ MOTOR - [el] (an alternating-current motor without a commutator, in which only one part is connected to the rotor, and the other part works by induction) moteur *m* d'induction

~ MOTOR-GENERATOR - [el] (motor-generator set driven by an induction motor) génératrice *f* commandée par moteur d'induction

~ NOISE - s. induced noise

~ PERIOD - [chem] (the time which elapses between the initiation of a reaction and the actual occurrence of it) période *f* d'induction

~ PIPE - [mech] (in a reciprocating engine) tuyau *m* d'admission

~ REGULATOR - [contr] survolteur *m* d'induction, régulateur *m* d'induction

~ RELAY - [el] relais *m* à induction

~ RESISTANCE WELDING - [metall] soudage *m* par induction

~ STROKE - [mech] (in a four-cycle reciprocating i. c. engine, the outward piston movement during which the charge is introduced into the cylinder) course *f* d'aspiration

TACHO-GENERATOR - [el] (tacho-generator with two stator-windings excited by a single-phase current and with a short-circuited rotor) générateur *m* tachymétrique asynchrone

~ TYPE MAGNETIC SEPARATOR - [el] (separator provided with rotors magnetized by induction from suitable windings, the material leaving the rotors at different points according to its magnetic quality and being divided by knife edges) séparateur *m* magnétique à induction

~ VALVE - s. inlet valve

~ WELDING - [metall] soudage *m* par induction

~ VELOCITY - [mech] (velocity of the induced air or mixture in a reciprocating i.c. engine) vitesse *f* de l'aspiration, vitesse *f* des gaz aspirés

~ VOLTAGE REGULATOR - [el] (transforming apparatus) régulateur *m* à induction

~ WATTHOURMETER - [instr] compteur *m* d'énergie active à induction

~ ZONE - [radio] (of a transmitting aerial) zone *f* d'induction

INDUCTIVE - [el] (possessing self or mutual inductance) inductif

~ CAPACITY - [el] (dielectric constant) constante *f* diélectrique

~ CIRCUIT - [el] (circuit of which the inductance is not negligible in the conditions under consideration) circuit *m* inductif

~ COUPLING - [el] (coupling between two circuits by means of mutual inductance) couplage *m* inductif

~ DISPLACEMENT PICK-UP - [meas] (instrument designed to measure static and dynamic moments up to I mm)détecteur *m* inductif de déplacement

~ DROP - [el] (the voltage drop which is produced in an a.c. circuit by its inductance) chute *f* de potentiel inductive

~ FEEDBACK - [radio] réaction *f* inductive

~ LOAD - [el] (electric load whose impedance has a positive imaginary component) charge *f* inductive

INDUCTIVE NEUTRALIZATION - [radio] (method of neutralizing an amplifier) neutralisation *f* inductive

~ -OUTPUT TUBE - [electron] tube *m* à sortie inductive

~ POTENTIAL DIVIDER - [contr] (precision-type of auto-transformer) potentiomètre à induction

RAIL CONNEXION - [el] raccord *m* inductif des rails

~ REACTANCE - [el] (the opposition which a circuit offers to the passage of a current as a consequence of the inductive effects in it) réactance *f* inductive

~ REACTION - [radio] (electro-magnetic reaction) réaction *f* électro-magnétique, réaction *f* inductive

~ RESISTOR - [el] (resistor with considerable inductance) résistance *f* inductive

~ SAWTOOTH GENERATOR - [el] (generator which produces a saw-tooth current by means of a self-inductance) générateur *m* inductif de courants en dents de scie

~ SHUNT - [el] shunt *m* inductif

~ TUNING - [radio] accord *m* inductif

~ VOLTAGE DIVIDER - [el] diviseur *m* de tension inductif

~ WINDING - [el] enroulement *m* inductif

~ WINDOW - [electron] (conducting diaphragms producing the effect of arc-inductive susceptance) fenêtre *f* inductive

INDUCTIVITY - [el] inductivité *f*

INDUCTOMETER - [instr] inductomètre *m*

INDUCTOPYREXIA - [med] électropyrexie *f*

INDUCTOR - [el] (a component used to give inductance to a circuit) inducteur *m*, inductance *f*

~ ALTERNATOR WITH MOVING IRON - s. inductor generator

~ COIL - s. inductance coil

FLAME DAMPER - [mech] (a device fitted to the induction system of an engine to check a return flame) anti-retour *m*

~ GENERATOR - [el] (alternating-current generator in which all the coild are fixed, so that the variation of the flux linking them is produced by the movement of ferromagnetic masses) alternateur *m* à fer tournant

~ LOUDSPEAKER - [el acoust] (cone loudspeaker with an electro-magnetic drive) haut-parleur *m* à inducteur

INDULINES - [chem] (azine dyestuffs containing three or four amino groups) indulines *f*pl

INDURATED - [med] induré, sclérosé

INDUSTRIAL - [gen] industriel

~ ALCOHOL - [chem] (alcohol made for industrial operations, as distinct from that made for drinking) alcool *m* dénaturé

~ CHEMISTRY - [chem] (the science of chemical processes carried out on a commercial scale) chimie *f* industrielle

~ CLOTHING - [text] vêtement *m* de travail

~ CONTROL - [contr] (generally, the methods and means of regulating the performance of equipment) conduite *f* industrielle

~ DATA PROCESSING - [comput] traitement *m* industriel de données

~ DISEASE - [med] maladie *f* professionnelle

~ FABRIC - [text] tissu *m* technique, tissu *m* industriel

~ FREQUENCY - [el] (the frequency of the a.c. used for ordinary industrial purposes) fréquence *f* industrielle

~ INSTRUMENT - [instr] (instrument of a robust construction and easy to read at a distance) instrument *m* industriel

INDUSTRIAL LABELLING - [gen] désignation *f* de la qualité

~ LOCOMOTIVE - [railw] locomotive *f* d'usine, locomotive *f* industrielle

~ METHYLATED SPYRIT - [chem] (specially denatured ethyl alcohol, used in making varnish etc) alcool *m* dénaturé industriel

~ PLANT - [gen] industrie *f*

~ PROPERTY - [leg] (the site, buildings etc belonging to an industrial concern) propriété *f* industrielle

~ RADIOGRAPHY - [metall] radiographie *f* industrielle

~ RAILWAY - [railw] chemin *m* de fer industriel

~ REACTOR - [nucl] (reactor for industrial purposes) réacteur *m* industriel

~ TELEVISION - [telev] (used for remote viewing of industrial processes or operations) télévision *f* industrielle

~ TRACK - [railw] (a private siding) voie *f* d'usine, voie *f* industrielle

~ TRACTOR - [mech] tracteur *m* industriel

~ TUBE - [electron] (electronic tube used in apparatus for industrial uses) tube *m* industriel

~ WASTE - [plumb] eaux *fpl* résiduaires industrielles

INDUSTRIALIZATION - [gen] industrialisation *f*

INDUSTRIALIZE, to - [gen] industrialiser

INDUSTRY - [gen] industrie *f*

INEBRIANT - [chem] inébriant *m*

INEBRIETY - [med] ivresse *f*, ébriété *f*

INEFFECTIVE - [gen] inefficace, ineffectif

~ CALL - [telecomm] appel *m* perdu, appel *m* inefficace

INEFFICACIOUS - [gen] inefficace

INEFFICIENCY - [gen] inefficacité *f*, insuffisance *f*

INEFFICIENT - [gen] ineffectif, inefficace, incapable

INELASTIC - [phys] inélastique, raide

~ COLLISION - [mech] (collision in which changes occur both in the internal energy of the partecipant systems and in the sums of the kinetic energies of translation) collision *f* inélastique

~ CROSS-SECTION - [nucl] (inelastic scattering cross-section) section *f* efficace inélastique

~ IMPACT - [mech] (plastic impact) choc *m* inélastique

~ SCATTERING - [nucl] (scattering effected by inelastic collision) diffusion *f* inélastique

~ THERMAL STRESS - [metall] effort *m* thermique inélastique

INEMIA - [med] hyperfibrinémie *f*

INEPT - [gen] inepte, déplacé
[leg] nul

INEPTITUDE - [gen] inaptitude *f*, inepsie *f*

INEQUALITY - [gen etc] inégalité *f*

INERADICABLE - [gen] indéracinable, inextirpable

INERASABLE - [gen] (of ink etc) indélébile

INERT - [chem] (not readily, or not at all, subject to change by chemical action) inerte, inactif

~ CARRIER - [ind chem] élément *m* porteur inerte

~ CELL - [el] (dry cell with ingredients forming an electrolyte when water is added) pile *f* non amorcée

~ FILLER - [paint] (substance added to a paint to give it more body) charge *f* inerte

~ GAS - [chem] (one of the six gases in the first group of the periodic table, which form no compounds (helium, argon, krypton, neon, xenon and radon) gaz *m* inerte

INERT-GAS ARC WELDING - [el heat] (welding process in which the arc operates in an inert atmosphere protecting the molten metal from the action of the air) soudage *m* à l'arc en atmosphère inerte

~ GAS PURIFICATION - [chem] épuration *f* de gaz inertes

~ GASES - [chem] (the gases argon, krypton, neon, xenon, helium and radon, which occupy group O in the periodic table, and show practically no tendency to chemical activity) gaz *mpl* inertes

~ SOLID - s. inert carrier

INERTIA - [phys] (the property of matter of resisting change in state of motion) inertie *f*

~ CONSTANT - [el] (acceleration constant of a machine) constante *f* d'accélération d'une machine

~ -FREE SWITCH - [el] interrupteur *m* exempt d'inertie

~ GOVERNOR - [mech] (shaft type of centrifugal governor using an eccentrically pivoted weighted arm) régulateur *m* à inertie

~ OF THE SPRING - [mech] inertie *f* du ressort

~ PINION - [auto] pignon *m* à inertie

~ SHOCK - [phys] choc *m* d'inertie

~ STARTER - [mech] (device for starting an i.c. engine by means of energy stored in a rapidly-rotating flywheel, which can be connected to the crankshaft through a friction clutch. The flywheel is usually given its rotary speed by hand) démarreur *m* à inertie

~ TO DEFORMATION - [phys] inertie *f* à la déformation

INERTIAL - [phys] d'inertie, inertiel

~ AXIS - [phys] axe *m* d'inertie

~ FORCE - [phys] (the resistance of any force against change of movement due to its inertia) force *f* d'inertie

~ GUIDANCE - [astronaut] (of a space vehicle) guidage *m* inertiel

~ MASS - [phys] (any mass opposed to movement) masse *f* d'inertie

~ MODEL - [mech] (a model with such mass distribution and linear dimensions, that it will behave in the same way as the prototype) modèle *m* dynamique

~ NAVIGATION - [astronaut] navigation *f* inertielle

~ ORBIT - [astronaut] orbite *f* naturelle

~ SPACE - [astr] espace *m* inertiel

~ VELOCITY - [astronaut] vitesse *f* dans un référentiel inertiel

INERTS - [chem] (mineral constituents) inertes *mpl*

"I"NEUTRONS - [nucl] (neutrons the energy of which is such that they are readily absorbed by iodine) neutrons *mpl* I

INEXTENSIBLE - [gen] inextensible

INFANCY - [med] première enfance *f*

INFANT - [med] enfant *m* du premier âge

~ MORTALITY - [med] mortalité *f* infantile

INFANT'S MILK - [chem] lait *m* maternisé

INFANTILE - [gen] (maladie) infantile

~ PARALYSIS - [med] (poliomyelitis) paralysie *f* spinale infantile

INFANTILISM - [med] infantilisme *m*, arrêt *m* de croissance

INFARCT - [med] (organ left without blood supply)

infarctus *m*
INFARCTION - [med] infarcissement *m*, processus *m* infarctogène
INFECTION - [med] (a diseased condition caused by living micro-organism invading the body tissue) infection *f*
[radar] (the dropping of metallized paper tape from bombers in order to jam the enemy radar system) lancement *m* de papier metallisé
INFECTIOUS - [gen & med] infectieux, infect
~DISEASE - [med bot vet] maladie *f* infectieuse
INFECTIVE GERM - [med] germe *m* infectieux
INFEED - [mach tool] avancement *m* en profondeur
INFERIOR - [gen] inférieur
~CONJUNCTION - [astr] (the same apparent geocentric longitude of two heavenly bodies) conjonction *f* inférieure
~FIGURES - [print] (figures or letters, smaller than the type used, set below the level of the line, e.g. in chemical formulae) petites lettres *f*pl inférieures
INFERRED ZERO - [instr] (suppressed-zero instrument in which the zero position is below the limit of travel of the indicating means) compteur *m* à zéro déduit
INFERTILE - [gen] infertile, stérile
INFERTILITY - [gen agric etc] infertilité *f*, stérilité *f*, infécondité *f*
INFEST, to - [gen] infester
INFESTED WITH WEEDS - [bot] envahi par les mauvaises herbes
INFIBULATION - [med] (the fastening of the genital organs) infibulation *f*
INFILLING DRILLING - [oil ind] forage *m* intercalaire
~WELL - [oil ind] puits *m* intermédiaire
INFILTRATE, to - [gen & phys] (to spread gradually) infiltrer, s'infiltrer, imprégner, pénétrer
INFILTRATION - [gen] infiltration *f*
[med] (the gradual spread of infection in an organ) infiltration *f*
~ANALGESIA - [med] analgésie *f* par infiltration
~GALLERY - [hydr] galerie *f* drainante, saignée *f*
~SLOT - [hydr] fente *f* d'infiltration
~VEIN - [min] filon *m* d'incrustation
~WATER - [geol] eau *f* d'infiltration
INFINITE - [gen & math] infini
~LATTICE - [nucl] (a theoretical reactor lattice, the dimensions of which are of infinite magnitude) réseau *m* infini
~MULTIPLICATION CONSTANT - [math] (the multiplication constant in an infinite medium) constante *f* de multiplication infinie
~PLAN SOURCE OF NEUTRONS - [nucl] (theoretical plane source of neutrons of infinite dimensions) source *f* de neutrons plane infinie
~SERIES - [math] série *f* infinie
~SET - [contr] ensemble *m* infini
~SLAB - [nucl] (of infinite dimensions) plaque *f* infinie
~SLAB REACTOR - [nucl] (theoretical reactor with a slab of infinite dimensions) réacteur *m* à plaque infinie
INFINITELY DENSE MEDIUM - [acoust] (medium acting acoustically like a rigid wall) milieu *m* idéalement dense, milieu *m* sans absorption
~RARE MEDIUM - [mech] (medium which does not support any pressure changes) milieu *m* à raréfaction infinie
INFINITELY SAFE GEOMETRY - [nucl] (of a reactor) géométrie *f* toujours sûre
~THICK TARGET - [nucl] (used for the complete absorption of the incident particles) cible *f* infiniment épaisse
~VARIABLE - [mech] (term used of a change-speed gear or control which operates continuously and not in discrete steps or stages) infiniment variable
~-VARIABLE GEAR - [mech] (type of transmission in which the speed ratio is continuously variable) engrenage *m* infiniment variable
~VARIABLE GEAR TRANSMISSION - [mech] variateur *m* de vitesse
INFINITESIMAL - [math] infinitésimal
INFINITY - [gen & math] infini *m*, infinité *f*
[photo] infini *m*
[meas] valeur *f* infinie
[comput] (any number which is larger than the maximum number the computer can store in any register) nombre *m* dépassant la capacité
~FOCUSSING - [photo] mise *f* au point sur l'infini
INFIRM - [gen] infirme, faible
INFIRMARY - [med] infirmerie *f*
INFIRMITY - [gen] infirmité *f*, débilité
INFLAMMABILITY - [phys & chem] (the property of taking easily) inflammabilité *f*
~LIMITS - [phys] limites *f*pl d'inflammabilité
INFLAMMABLE - [gen] inflammable
~AIR - [chem] (general term denoting combustible gases) gaz *m* inflammable
~MIXTURE - [chem] mélange *m* inflammable
INFLAMMATION - [med] (the reaction of living tissue to infection) inflammation *f*
[gen] (catching fire) inflammation *f*
INFLATABLE - [gen] pneumatique
~CUSHION - [rubber ind] coussin *m* gonflable, coussin *m* pneumatique
~EXPOSURE SUIT - s. inflatable suit
~PACKING - [rubber ind] joint *m* gonflable
~STRIP SEAL - s. inflatable packing
~SUIT - [aero] (inflatable suit for airmen) combinaison *f* pneumatique pour aviateur
~TOY - [rubber ind] jouet *m* gonflable, jouet *m* pneumatique
INFLATE, to - [gen] gonfler, souffler
[aero] (a balloon) insuffler
INFLATING AGENT - [chem] (a substance used to cause foaming) agent *m* gonflant
INFLATION - [fin] (increase beyond proper limits) inflation *f*
[aero] (the process of filling a balloon etc) gonflement *m*
[auto etc] (of a tyre) gonflement *m*
[med] insufflation *f*
[gas ind] (gasholders) remplissage *m*, gonflage *m*
~NET - [aero] filet *m* pour gonflement
~PRESSURE - [auto] (of tyres) pression *f* de gonflage
~SLEEVE - [auto] (tyres) manche *f* de gonflement
INFLATOR - [med] pompe *f*
[impl] (a tyre pump) pompe *f* pour pneus
INFLECTED ARCH - [arch] arc *m* renversé
INFLECTOR - [nucl] (type of deflector used in particle accelerators) inflecteur *m*
INFLEXION - [phys] (curving or bending inwards) inflexion *f*, fléchissement *m*

[math] (point of inflection) inflexion *f* (d'une courbe)
INFLEXION POINT - [phys] (the point at which a curve takes a definite change in direction) point *m* d'inflexion
~POINT EMISSION CURRENT - [radio] courant *m* de émission au point d'inflexion
INFLICT, to - [gen] appliquer, imposer
INFLORESCENTE - [bot] (the part of the shoot which bears flowers) inflorescence *f*
INFLOW - [hydr] entrée *f*, affluence *f*
[aero] (the flow of air into a propeller) courant *m* d'air en avant de l'hélice
~NOZZLE - [mech] tuyère *f* d'admission
~PERFORMANCE - [oil ind] caractéristiques *f*pl de productivité
~RATIO - [el] (the relation between the total axial flow velocity through the rotor disk and the blade tip speed) rapport *m* d'affluence
INFLUENCE, to - [gen] influencer
[phys] influer (sur)
INFLUENCE - [gen] influence *f*
[el] (induction) influence *f*
[telecomm] (interference) interférence *f*, perturbation *f*
[astr] influence *f*
~FUNCTION - [contr] fonction *f* d'influence
~LINE - [constr] (curve with an ordinate which at any point represents the value of a variable at another point in the structure) ligne *f* d'influence
~MACHINE - [el] (electrostatic generator) machine *f* électrostatique à influence
INFLUENT - [hydr] adduction *f*, introduction *f*
~CONDUIT - [hydr] canal *m* d'amenée
INFLUENZA - [med](infectious disease of the respiratory tract) grippe *f*, influenza *f*
INFLUX - [hydr] (inward flow) entrée *f*, affluence *f*
[geogr] (of a river) affluence *f*
[oil ind] afflux *m*, venue *f*
INFORMATION - [gen] informations *f*pl, renseignements *m*pl
[gen] (news) informations *f*pl
[telev] (signal complex) ensemble *m* des signaux
~BIT - [comput] bit *m*, unité *f* d'information binaire
~CALL - [telecomm] (telephony) demande *f* de renseignement
~CAPACITY - [telev] capacité *f* de l'information
~CHANNEL - [comput] canal *m*
~CIRCUIT - [comput] (circuit for the transmission of measured values) circuit *m* d'information
~COMMUNICATION SYSTEM - [comput] système *m* de communication d'information
~CONTROL - [comput] contenu *m* en informations
~FEEDBACK SYSTEM - [comput] système *m* de réinjection d'information
~FLOW - [comput] débit *m* des informations
~FLOW ANALYSIS - [comput] analyse *f* du flux d'information
~FORMAT - [comput] structure *f* d'une information, format *m* d'une information
~-READ-WIRE - [comput] fil *m* de lecture d'information
~RECTIEVAL - [comput] (the recovery of information from any document) dépistage *m* de l'information, recherche *f* des informations
~STORAGE - [comput] mémorisation *f* des informations
INFORMATION THEORY - [gen] (the study and understanding of human communications) théorie *f* de l'information
~TRANSFERAL - [comput] transfert *m* d'informations
~TRUNK - [telecomm] (telephony) ligne *f* de renseignements
~WIRE - [comput] (of the magnetic core) fil *m* de lecture
~-WRITE-WIRE - [comput] fil *m* de lecture/écriture
INFRAACOUSTIC TELEGRAPHY - [telecomm] télégraphie *f* infra-acoustique
INFRACLAVICULAR - [anat] sous-claviculaire
INFRACOMMISSURE - [anat] commissure *f* inférieure
INFRACOSMIC RAY - [phys] rayon *m* infracosmique
INFRACTION - [gen & leg] infraction *f*, transgression *f*
INFRADIAPHRAGMATIC - [anat] sous-phrénique
INFRAGLOTTIC - [anat] sous-glottique
INFRAHUMAN - [med etc] cobaye *m*
INFRAHYOID - [anat] sous-hyoïdien
INFRAMAMMARY - [anat] sous-mammaire
INFRAMANDIBULAR - [anat] sous-maxillaire
INFRAMICROBE - [med] virus *m* filtrant
INFRANGIBLE - [gen] infrangible
~ATOM - [phys] atome *m* non-fissile
INFRAORBITAL - [anat] sous-orbitaire
INFRAPATELLAR - [anat] sous-rotulien
INFRAPROTEINS - [chem] (protein derivatives which are formed by hydrolisis) infraprotéines *f*pl
INFRARED - [phys] (electromagnetic radiations having wavelenghts between 8,000 to 4,000,000 A.U.) infrarouge
~ABSORPTION - [crystall & opt] (absorption of infrared radiation by crystal due to the excitation of lattice vibrations, in which ions of opposite charge move relative to one another) absorption *f* de l'infrarouge
~ABSORPTION SPECTRUM - [spectrology] spectre *m* d'absorption infrarouge
~DRYING - [radiat] (process of drying materials by infrared radiation) séchage *m* à l'infrarouge
~DRYING APPARATUS - [mach] séchoir *m* à rayons infrarouges
~EMULSION - [photo] (photographic emulsion sensitive to infrared rays) émulsion *f* sensible aux rayons infrarouges
~FILTER - [photo] filtre *m* pour infrarouge
~FLASH - [photo] éclair *m* infrarouge
~HEATING - [th eng] (heating by infra-red radiation) chauffage *m* à rayons infrarouges
~IMAGE CONVERTER - [telev] tube *m* convertisseur d'image à éclairement par rayons infrarouges
~LINK - [telev] (in colour television) liaison *f* par rayons infrarouges
~MASER - [opt] laser *m*
~MICROSCOPE - [instr] (microscope provided with an infrared light-source) microscope *m* infrarouge
~PHOTOGRAPHY - [photo] photographie *f* à l'infrarouge
~PLATE - [photo] plaque *f* sensible aux rayons infrarouges
~RADIATION - [phys] (electro-magnetic radiation of wave-lengths exceeding 76000 A.U.) rayonnement *m* infrarouge
~RADIATION HEATING - [th eng] (heating by the generation of infra-red radiation, which is transmit-

ted to be object to be heated) chauffage *m* à rayons infrarouges
INFRARED RANGE AND DIRECTION DETECTION EQUIPMENT - [radar] (a method of location) radar *m* à rayons infrarouges
~RAYS - [phys] (radiation in the band between the limit of the visible spectrum (about 750 millicrons) and the shortest microwaves (about 1000 millicrons) rayons *mpl* infrarouges
~SENSITIZER - [photo] sensibilisateur *m* à l'infrarouge
~SPECTRUM - [spectrology] spectre *m* infrarouge
INFRASCAPULAR - [anat] sous-scapulaire
INFRASIZER - [min] classeur-trieur *m* intermédiaire
INFRASONIC - [acoust] infrasonique
~FREQUENCY - [acoust] fréquence *f* infraacoustique, fréquence *f* infrasonore
INFRASOUND - [acoust] (acoustic oscillation with a frequency which is too low to affect the sense of hearing) infrason *m*
INFRASPINOUS - [anat] sous-épineux, infraspinal
INFRASTERNAL - [anat] sous-sternal
INFRATEMPORAL - [anat] sous-temporal
INFRATONSILLAR - [anat] sous-amygdalien
INFRATROCHLEAR - [anat] sous-trochléen
INFRA-UMBILICAL - [anat] sous-ombilical
INFRINGE, to - [gen & leg] transgresser, violer
~A PATENT, to - [leg] empiéter sur un brevet
INFRINGEMENT - [gen & leg] infraction *f*, violation *f*
INFRUCTESCENCE - [agric] infrutescence *f*
INFUNDIBULAR - [gen] (shaped like a funnel) infundibuliforme, en forme d'entonnoir
INFUNDIBULUM - [anat] infundibulum *m*
INFUSE, to - [chem] (to immerse in a liquid, in order to extract some constituent) infuser
INFUSIBILITY - [phys] (the property of a substance which cannot be melted at a specified temperature) infusibilité *f*
INFUSIBLE - [adj] (term used of a substance which cannot be melted at the temperature in question) infusible, non fusible
INFUSION - [chem] (extraction of the active constituent of a crude by treatment with hot or cold water) infusion *f*
~METHOD - [brew ind] méthode *f* d'infusion
INFUSORIAL EARTH - [min] farine *f* fossile, terre *f* à infusoires, tripoli *m* silicieux
INGATE - [metall] amorce *f* de coulée, attaque *f* dirigée
~-PLOT - [mining] accrochage *m*, recette *f* du fond
INGESTION - [med] ingestion *f*
INGLENOOK - [constr] (fire-side corner) coin *m* du feu
INGOT, to - [metall] lingoter
[die casting] mouler en gueusets
INGOT - [metall] lingot *m*, saumon *m*
~BLEEDING - [metall] perforation *f* de la croûte du lingot
~-BLOOM - [metall] lingot-bloom *m*, lingot *m* prelaminé
~BUGGY - [impl] chariot *m* à lingots, gaillot *m* à lingots
~CASTING - [metall] coulée *f* de lingots
~CASTING UNDER VACUUM - [metall] coulée *f* de lingots sous vide
~CHARGING CARRIAGE - [metall] chariot *m* enfourneur des lingots
INGOT COPPER - [metall] cuivre *m* en lingots
~CORNER SEGREGATION - [metall] ségrégation *f* de coin du lingot
~CRANE - [metall] pont *m* roulant pour le transport des lingots, grue *f* avec pince à lingots
~CROP END - [metall] tête *f* du lingot
~DOGS - s. ingot tongs
~IRON - [metall] (iron of fairly high purity) fer *m* doux
~LATHE - [mach tool] tour *m* pour lingots
~MOULD - [metall] (the mould into which the molten metal is cast and left to solidity to form an ingot) lingotière *f*
~PIT - [metall] four *m* à réchauffer
~PUSHER - [metall] dispositif *m* pousse-lingots, pousseuse *f*, pousseuse *f* des lingots
~SLICING MACHINE - [metall] machine *f* à découper des lingots
~STEEL - [metall] acier *m* de lingot
~STOOL - [metall] base *f* de lingotière, fromage *m*
~STRIPPER - [metall] stripeur *m*
~STRIPPING - [metall] démoulage *m* du lingot
~STRIPPING CRANE - [metall] grue *f* stripeuse
~STRUCTURE - [metall] (the general arrangement of the crystals in an ingot) structure *f* cristalline du lingot
~TILTING DEVICE - [metall] culbuteur *m* de lingots
~TONGS - [impl] pinces *fpl* à lingots
[metall] tenaille *f* à lingots
~TOP END - [metall] tête *f* de lingot
~UPSETTING - [metall] (the operation of increasing the cross-section of an ingot during forging) refoulage *m* du lingot
INGOTISM - [metall] structure *f* dendritique des lingots
INGRAIN - [paper man] (wood pulp wallpaper dyed during the process of manufacture) papier *m* teint en fibre
~CARPETS - [text] tapis *m* double pli
~DYE - [dye] colorant *m* pour la teinture en fibre
INGRAVESCENT - [med] (gradually increasing in gravity) qui s'aggrave
INGREDIENT- [gen] ingrédient *m*
[chem] partie *f* constituante
INGRESS - [gen & chem] entrée *f*, admission *f*
[astr] ingression *f*
INGROWN - [med] incarné
~TOE-NAIL - [med] ongle *m* incarné
INGUEN - [anat] aine *f*
INGUINAL - [anat] inguinal, inguinaire
~CANAL - [anat] canal *m* inguinal
~HERNIA - [med] hernie *f* inguinale
~REGION - [anat] région *f* inguinale
INGUINODYNIA - [med] (a pain in the groin) douleur *f* inguinale
INHALATION - [gen] (breathing in) aspiration *f*, inhalation *f*
~MASK - [gen] masque *m* pour inhalation
INHALATOR - s. inhaler
INHALER - [med] (appliance used for inhaling) inhalateur *m*
INHAULER - [naut] (a rope) hale-à-bord *m*
INHERENT - [gen] (term used for a quality or property which is inseparable from the concept in question) inhérent, naturel
[med] inné, intrinsèque
~CHARACTERISTIC - [contr] (any formula showing

the relationship between two values) caractéristiques *f*pl naturelles
INHERENT FEEDBACK - [contr] (feedback of a signal along the main forward path) auto-réaction *f*, réaction *f* propre
~FILM NOISE - [cin] (the noise in the projection system) bruit *m* propre du système
~FILTER - [electron] filtre *m* inhérent
~FILTRATION - [electron] (of an X-ray beam) filtration *f* inhérente
~LARGE-SIGNAL FORWARD CURRENT TRANSFER RATIO - [electron] taux *m* inhérent de courant pour signaux forts
~MOISTURE - [min] (in coal testing) humidité *f* constitutionnelle
~REGULATION - [el] (the change in voltage at the output terminals of an electric machine when the load is removed) auto-régulation
[contr] (the inherent characteristic of a system which will return to its normal position after a disturbance) auto-régulation
~STABILITY - [aero] (the property in an aircraft which induces it to return to its original attitude of flight after disturbance) stabilité *f* propre, stabilité *f* de la forme
INHERITABLE - [gen] transmissible, apte à hériter
INHERITANCE - [leg] héritage *m*, succession *f*
INHERITED ERROR - [comput] (the error in the error in the initial values) erreur *f* entraînée, erreur *f* héritée
~MEMORY - [zool] instinct *m*
INHIBIT, to - [gen] inhiber, empêcher
[med] (a secretion) paralyser
INHIBIT DRIVER - [comput] (in a core memory) amplificateur *m* inhibiteur
~GATE - [comput] circuit *m* inhibiteur, circuit *m* d'inhibition
~LINE - [comput] (of a logic circuit) ligne *f* d'inhibition
~PULSE - [comput] (in static magnetic storage) impulsion *f* d'inhibition
~WINDING - [comput] (a driving winding of magnetic core) enroulement *m* d'inhibition
~WIRE - [comput] (of the magnetic core storage) fil *m* inhibiteur
INHIBITED BURNING - [phys] combustion *f* inhibée
~OIL - [ind chem] huiles *f*pl contenant des inhibiteurs
INHIBITING - [chem] (preventing or retarding undesirable changes in a product) inhibition *f*
~CIRCUIT - [comput] circuit *m* inhibiteur, circuit *m* d'inhibition
~FACTOR - [chem] élément *m* inhibiteur
~INPUT - [comput] signal *m* inhibiteur, entrée *f* de inhibition
[electron] entrée *f* de blocage
~PIGMENT - [paint] colorant *m* inhibiteur
INHIBITION - [gen] (the effect produced by inhibitors) inhibition *f*
[zool] (the stopping or retarding of a metabolic process) inhibition *f*
~OF REACTION - [chem] (the retarding of a reaction) inhibition *f* d'une réaction
INHIBITIVE CAPACITY - [chem] pouvoir *m* de rétention, pouvoir *m* inhibiteur
INHIBITOR - [chem] (a substance the presence of which in a product prevents undesiderable changes in its qualities, or in the conditions of the equipment in which the product is used) inhibiteur *m*
INHIBITOR - [metall] (substance with which metals are treated to prevent corrosion) inhibiteur *m*, anti-corrosif *m*
[ind chem] (substance the presence of which in a pickling bath reduces the attack on the descaled areas of metals) inhibiteur *m*
INHIBITORY - [zool] (of a nerve the stimulation of which determines the activities of a muscle or gland on a decreasing curve) inhibitoire, inhibiteur
~-GATE - [comput] (and not gate) circuit *m* d'inhibition
~PHASE - [chem] (protective colloid in a lyophobic colloidal solution) phase *f* inhibitoire
INHOMOGENEOUS - [phys] (non homogeneous) hétérogène
INHOUR - [nucl] (inverted hour; a measure of reactivity) inhour *f*
INION - [anat] (protuberance at the back of the skull) protubérance *f* occipitale externe
INITIAL, to - [gen] initialer, parafer
INITIAL - [gen] initial
~ACTUATION TIME - [contr] durée *f* d'excitation initiale
~ADDRESS - [comput] adresse *f* initiale
~ADJUSTEMENT - [instr] (zero setting) mise *f* au point initiale
~APPROACH - [metall] (approach phase extending from point of entry to the gate) approche *f* initiale
~BODY RETENTION - [radiat] (the fraction of radiation which is initially retained in a critical organ) coefficient *m* de rétention
~BOILING POINT - [phys] (the temperature at which ebullition begins) point *m* d'ébullition initial
~CAPACITANCE - [el] capacité *f* résiduelle
~CARD - [comput] carte *f* initiale
~CELL - [genet] cellule *f* germinale, gamète *m*, cellule *f* sexuelle
~CHARGE - [el chem] (the first charge given to a new storage battery) charge *f* initiale
~CHLORINE DEMAND - [chem] besoin *m* en chlore, dose *f* nécessaire de chlore
~CREEP - [metall] (the initial creep of metals when subjected to a constant load) fluage *m* primaire
~ENERGY OF NEUTRONS - [nucl] énergie *f* initiale de neutrons
~FAILURES - [comput] défaillances *f* prématurées, déchets *m* infantiles
~INVERSE VOLTAGE - [electron] (the value of inverse anode voltage immediately following the conduction period) tension *f* initiale inverse
~IONIZING EVENT - [phys] (initiating a count in a radiation counter tube) événement *m* d'ionisation primaire
~LOAD - [phys] charge *f* initiale
~OUTPUT TEST - [el chem] (test of the rate of sicharge shortly after the manufacture of the battery) essai *m* initial
~PARTICLE - [phys] particule *f* primordiale
~PERMEABILITY - [el] perméabilité *f* initiale
~RADIATION - [nucl] rayonnement *m* initial
~RECOMBINATION - [electron] recombinaison *f* initiale
~SET - [constr] (of cement) début *m* de prise
~SPEED - [mech] vitesse *f* initiale
~START UP - [nucl] (in reactor technology) démar-

rage *m* initial
INITIAL STRESS - [phys] tension *f* initiale, prétension
~SUSCEPTIBILITY AND PERMEABILITY - [el] (limiting value of susceptibility and permeability of a ferromagnetic body at the origin of the curve of first magnetization) susceptibilité *f* et perméabilité initiales
~SYMMETRICAL SHORT CIRCUIT CURRENT - [el] courant *m* initial symétrique de court-circuit
~TENSION - [mech] (of a spring) tension *f* initiale
~TEST TEMPERATURE - [el chem] (the average temperature of the electrolyte in all cells at the beginning of discharge) température *f* initiale
~VALUE - [contr] valeur *f* initiale
[math] valeur *f* initiale
~VELOCITY - [gen phys missil] vitesse *f* initiale
~VOLTAGE - [el] (closed circuit voltage of a cell or battery when current begins to flow) tension *f* initiale, potentiel *m* d'étincelle
~VOLTAGE RESPONSE OF AN EXCITER - [el] rapidité *f* de réponse initiale d'une excitatrice
INITIALLY - [gen] initialement
INITIATING ELECTRON - [electron] (electron initiating an avalanche) électron *m* germe
INITIATOR - [chem] (a substance which causes a reaction to begin) initiateur *m*
INJECT, to - [gen] injecter
[plast ind] (to force plastic material into a mould through a nozzle) mouler par injection
[comput] injecter
[astronaut] (of a satellite) injecter sur orbite
INJECTABLE - [med] injectable
INJECTABILITY - [gas ind] (gasholders) débit *m* injectable
INJECTED BODY - [metall] massif *m* d'injection
~FUEL - [oil ind] carburant *m* injecté
~GAS - [gas ind] (gasholders) gaz *m* injecté
INJECTING FLUID - [th eng] (burners) fluide *m* inducteur
~STREAM - s. injecting fluid
INJECTION - [gen] injection *f*
[mech] (the operation of spraying fuel in an i.c. engine) injection *f*
[metall] (diecasting) injection *f*
[geol] (emplacement of fluid rock matter in crevices etc) intrusion *f*
~ADVANCE - [mech] avance *f* à l'injection
~ADVANCE DEVICE - [mech] (for i.c. engines) dispositif *m* d'avance à l'injection
~CAPACITY - [plast ind] (the volume of material which can be injected into an injection mould) capacité *f* d'injection
~CARBURETTOR - [mech] carburateur *m* à injection
~COCK - [mech] prise *f* de vapeur de l'injecteur
~CONDENSER - [mech] (condenser in which exhaust steam is condensed by a jet of cold water) condenseur *m* à jet, condenseur *m* à injection
~FORCE - [metall] (diecasting) force *f* d'injection
~GRID - [electron] (injection introducing the oscillator signal into the mixer stage) grille *f* d'attaque, grille *f* d'injection
~LAG - [mech] (in i.c. engines) retard *m* de l'injection
~LOCKING - [el] blocage *m* à injection
~MOULD - [plast ind] (type of mould into which material is injected under pressure the form the moulding) moule *m* à injection
INJECTION MOULDED THREAD - [plast ind] (a screw thread formed from plastic material by injection moulding) filet *m* injecté
~MOULDING - [plast ind] (process of forcing softened thermoplastics into a cool closed mould) moulage *m* par injection
~MOULDING MACHINE - [plast ind] (a machine for forming objects by injecting plastic material into a special mould under pressure) machine *f* pour le moulage à injection
~NOZZLE - [mech] (the jet through which fuel is sprayed in an i.c. engine) injecteur *m*
~PISTON - s. injection plunger
~PLUNGER - [rubber & plast ind] piston *m* d'injection
~PRESSURE - [plast ind & metall] (the pressure at which moulding material is injected into the mould) pression *f* d'injection
[soil] (grouting pressure) pression *f* d'injection
~PRESSURE REDUCER - [plast ind & metall] régulateur *m* de pression d'injection
~PUMP - [mech] pompe *f* à injection
~RAM - [mech] (the moving plunger in a ram-type injection machine) piston *m* d'injection
~RINSING MACHINE - [text] rinceuse *f* à injection
~SHOT - [plast ind] (a single complete injection moulding operation) injection
~SYRINGE - [med etc] seringue *f*
~TIMING - [mech] (the operation of setting injection so as to occur at a given point in the operating cycle of an i.c. engine) calage *m* de l'injection
~WELL - [hydr] puits *m* absorbant
[mining] puits *m* d'injection
INJECTIVITY - [oil ind] injectivité *f*
~PROFILE - [oil ind] courbe *f* d'injectivité
INJECTOR - [mech] (a device or mechanism for feeding) injecteur *m*
[mech] (of i.c. engines) carburateur *m* à injection
[mech] (of a Diesel engine) injecteur
~CHANNEL - [metall] buse *f* d'injection
~NEEDLE - [th eng] (burners) pointeau *m*
~NIPPLE - s. injection nozzle
~NOZZLE - s. injection nozzle
~PIN - s. injector needle
~THROTTLE - [mech] prise *f* de vapeur de l'injecteur
~TUBE - [th eng] (only USA; mixing tube in GB) mélangeur *m*, chambre *f* de mélange
~VALVE - [mech] (of a fuel pump) soupape *f* de l'injecteur
INJURE, to - [gen] blesser, nuire à
[mech etc] déformer, s'avarier
INJURIOUS - [gen] nuisible, pernicieux
~EFFECTS OF WEEDS - [bot] dégâts *mpl* dus aux mauvaises herbes
INJURY - [gen] tort *m*, préjudice *m*
[gen] (a wound) blessure *f*, lésion *f*
[gen] (damage) dommage *m*, dégât *m*
~POTENTIAL - [el biol] (demarcation potential; the difference in potential between the injured and uninjured parts of a living structure) potentiel *m* de démarcation, potentiel *m* de lésion
INK - [gen] encre *m*
~BALL - [print impl] tampon *m* encreur
~FEED ROLLER - s. inker
~FORMULATION - [ind chem] formulation *f* des encres
~FOUNTAIN - [print] (fitted to a printing machine)

encrier *m*
INK KNIFE - [print] couteau *m* d'encrier
~-PLANT - [bot] sumac *m* des corroyeurs
~-STAND - [impl] encrier
~-VAPOUR RECORDING - [el mech] (electro-mechanical recording in which vapourized ink particles are deposited on the record sheet) enregistreur *m* à jet d'encre
~WELL - s. ink-stand
INKER - [print] (the roller which applies ink to the type surface) rouleau *m* encreur, toucheur *m*
INKING - [print & photo] encrage *m*
~PAD - [impl] tampon *m* encreur
~-RIBBON - [print] (e.g. of a typewriter) ruban *m* encreur
~ROLLER - s. inker
INKY DINKY - [cin] (lighting implement) petit projecteur *m*, projecteur de 500 W
~LIGHT - [light] (tungsten light) lumière *f* au tungstène
INLAID - [gen] incrusté, parqueté, marqueté
[bookbind] encarté
~CAPTION - [telev] titre *m* inseré
~FLOOR - [constr] plancher *m* parqueté
~PANEL MOULDING - [constr] embrevure *f*, grand cadre *m*
~WORK - [gen] marqueterie *f*
INLAND - [geogr] intérieur *m* (d'un pays)
INLAY, to - [decor] (the operation of sinking shaped sections of wood, ivory, metal etc in a counterpart design on wood or other material) incruster, marqueter
INLAY - [decor] incrustation *f*, marqueterie *f*
[dent] obturation *f* d'une dent, plombage *m*
[telev] procédé *m* des caches électroniques, système *m* électronique d'insertion
INLET - [geogr] petit bras *m* de mer, anse *f*
[mech] entrée *f*, orifice *m* d'admission
[mech] (i.c. engines) aspiration *f*, admission *f*
[mech] (of a pump etc) ouïe *f*, œillard *m*
[metall] rorifice *m* de coulée
~AIR - [gen] air *m* frais, air *m* d'admission
~AND OUTLET PIPE PIT - [gas ind] (gasholders) fosse *f* des tuyaux
~AND OUTLET WELL - s. inlet and outlet pipe pit
~ANGLE - [mech] (in a turbine) angle *m* d'admission
~AREA - [gen] section *f* d'admission
~BEND - [hydr] coude *m* d'une conduite d'amenée
~CHAMBER - [hydr] chambre *f* d'arrivée
~CONDUIT - s. influent conduit
~CONNEXION - [plumb] raccord *m* d'alimentation
~-CONTROL WATER-HEATER - [th eng] chauffe-eau *m* à écoulement libre
~-CURRENT TRANSFORMER - [el] transformateur *m* de courant de traversée
~EDGE - [mech] (of a turbine) arête *f* d'admission
~FLANGE - [mech] (in a plumbing works) bride *f* d'admission
~FRAME - [text] râtelier *m* d'alimentation
~JOINT - [carp] assemblage *m* tête à tête
~MANIFOLD - [mech] collecteur *m* d'admission, pipe *f* d'admission, tuyauterie *f* d'aspiration
~NOZZLE - [mech] tuyère *f* d'admission
~-OVER-EXHAUST - [mech] (arrangement of valves in an i.c. engine, the inlet valve being placed in the upper part of the combustion chamber and opening downward, while the exhaust valve is directly below and opens upward) soupape *f* d'admission au dessus de la soupape d'échappement
INLET PIPE - [mech] (in an i.c. engine) pipe *f* d'admission, tuyau *m* d'admission
~PLUG - [el] fiche *f* de prise de courant d'un appareil
~PLUG AND SOCKET - [el] connecteur *m*
~PORT - [mech] (auto) orifice *m* d'admission, raccord *m* d'admission
~PRESSURE - [mech] pression *f* d'aspiration
~RILL - [agric] rigole *f* d'amenée, rigole *f* d'adduction
~SCREEN - [mech] (wire mesh or other arrangement placed over the air intake of a gas turbine compressor to protect it against foreign bodies) filtre *m* de admission
~SIDE - [mech] (of a pump) admission *f*, côté *m* vide
~SIDE PRESSURE - [mech] pression *f* d'aspiration
~SIDE TRAP - [mech] séparateur *m* côté aspiration (ou côté vide)
~SILENCER - [auto] silencieux *m* d'aspiration, silencieux *m* d'admission
~SOCKET - [el] douille *f* de prise de courant d'un appareil
~SOCKET PIECE - [text] bouche *f* d'entrée
~STRAINER - [hydr] crépine *f* d'entrée
~STROKE - [mech] (induction stroke) course *f* d'admission, course *f* d'aspiration
~VALVE - [mech] (the valve which admits the charge to the cylinder of the i.c. reciprocating engine) soupape *f* d'admission
~VELOCITY - [auto] vitesse *f* d'aspiration
~VENTILATION - [mech] ventilation *f* par compresseur
INLETS - [text] toile *f* à matelas, coutil *m* pour literie
INLIER - [geol] fenêtre *f* (géologique)
IN-LINE DATA PROCESSING - [comput] traitement *m* simultané des données
~ENGINE - [mech] moteur *m* à cylindres en ligne
~SUBROUTINE - [comput] sous-programme *m* ouvert
INN - [gen] auberge *f*, hôtellerie *f*
INNER - [gen] intérieur, interne
~BAND BRAKE - s. inner brake
~BASEBOARD - [photo] glissière *f* (de la base)
~BRAKE - [auto] frein *m* à friction interne
~BREMSSTRAHLUNG - [radiat] rayonnement *m* de freinage interne
~CAPSULE - [nucl] (capsule inside another capsule) capsule *f* intérieure
~CONDUCTOR - [el] (the central conductor of a concentric cable) conducteur *m* intérieur
[el] (in a three-wire system, the neutral conductor) conducteur *m* neutre
~CONE - [mech] (of a jet engine) cône *m* bleu
~CORE - [plast & rubber ind] noyau *m* intérieur
~CORE BARREL - [oil ind] tube *m* carottier intérieur
~COURT - [constr] cour *f* intérieure
~COVER - [metall] coiffe *f* de protection
~COVER OF MANHOLE - [build] couvercle *m* intérieur de regard
~COVERING - [mech etc] revêtement *m* intérieur
~CROWN - [gas ind] (of a gas cooker) plafond *m*
~DEAD-CENTRE - [mech] (the piston position at the beginning of the outstroke in a reciprocating engine) point *m* mort supérieur
~DRIVE JOINT - [mech] joint *m* intérieur d'entraîne-

ment
INNER ECCENTRIC SHEAVE - [text] excentrique *m* intérieur
~END PAPER - [print] (of book) papier *m* de garde
~ENVELOPE - [mech] revêtement *m* intérieur
~GRID - [electron] (the grid which is nearest to the cathode in a multi-grid electronic tube) grille *f* intérieure
~HEAT - [phys] (heat stored or accumulated within a body) chaleur *f* emmagasinée
~LIFT - [gas ind] (gasholders) cloche *f*, première levée *f*
~LINER - [mech] chemise *f* interne
~MARKER BEACON - [aero] (a beacom indicating the boundary of the airfield) radiobalise *f* intérieure, radioborne *m* de bordure
~PACK - [aero] (the bag in which the parachute is packed) sac *m* à parachute
~PACKING WEFT - [text] trame *f* de remplissage
~PART - [gen] partie *f* interne
~PARTITION - [constr] paroi *f* de séparation
~QUANTUM NUMBER - [quantum theory] nombre *m* quantique intérieur
~RADIUS - [railw] petit rayon *m*
~RACE - [mech] chemin *m* de roulement
~RIDGE GIRDER - [aero] (element of the inner ring of a transvers frame in an aircraft) poutre *f* transversale intérieure
~RING - [mech] (of ball bearings) cercle *m* interne
~RING BALL BEARING - [mech] coussinet *m* à billes à cercle interne
~SEAL - [ind chem] (retorts) garde *f* intérieure
~SEAM - [rubber ind] couture *f* intérieure, joint *m* intérieur
~SHELL - [mech] (of a turbine) cuirasse *f*
~SHELL ELECTRON - [electron] (an electron belonging to an electron shell other than the outer shell) électron *m* interne
~STRING BOARD - [constr] (of stairs) limon *m*
~SYNCHROMESH DISK - [auto] disque *m* intérieur de synchronisation
~TOOTH-CROWN WHEEL - [mech] roue *f* de champ à dents intérieures
~TRIANGLE - [mech] (of turbines) triangle *m* d'admission
~TUBE - [rubber ind] (the toroidal rubber chamber containing air under pressure in a pneumatic tyre) chambre *f* à air
[opt] (of a telescope) tube *m* porte-réticule
~-TUBE REPAIR PLATE - [rubber ind] table *f* pour réparation de chambres à air
-TUBE REPAIR PRESS - [rubber ind] presse *f* pour réparation des chambres à air
~-TUBE VALVE - [mech] valve *f* de chambre à air
~-TUBE VULCANIZER - [rubber ind] vulcanisateur *m* à chambres à air
~WIDTH - [gen] largeur *f* interne
~WORK FUNCTION - [electron] (the maximum of the energy of the electrons at absolute zero temperature) travail *m* interne
INNERVATE, to - [zool] innerver
INNERVATION - [zool] (the distribution of nerves to an organ) innervation *f*
INNS OF COURT - [build] (four sets of buildings of the four legal societies which have the exclusive right of admitting lawyers to the bar) écoles *fpl* de droit

INNUMERABLE - [gen] innombrable
INOCHONDRITIS - [med] inochondrite *f*
INOCHONDROMA - [med] fibrochondrome *m*
INOCULANT - [gen] inoculant
[metall] inoculant *m*
INOCULATE, to - [biol] (to introduce a culture of a micro-organisme into material in which it is intended to multiply) inoculer
[med] (the introduction of a vaccine for protection against infection) vacciner
[agric] greffer
INOCULATION - [chem] inoculation *f*, greffage *m*
[metall] (the introduction of specified elements into liquid cast iron) inoculation *f*
INOCULUM - [med] (the inoculated material) inoculum *m*
INODOROUS - [gen] inodore
INOPERATIVE - [gen] inopérant
INORDINATE - [gen] démesuré, excessif
INORFIL - [chem] (term used for inorganic fibre) fibre *f* inorganique
INORGANIC - [chem] (compounds of mineral origin) inorganique
~CHEMISTRY - [chem] (originally used for the chemistry of substances belonging to the mineral, as distinct from the animal and the vegetable kingdom, this term is now used for the study of substances in which carbon is either absent or of minor importance only) chimie *f* minérale
~COMPOUND - [chem] composé *m* inorganique
~FERTILIZER - [agric] engrais *mpl* chimiques, engrais *mpl* artificiels
~PIGMENT - [chem] pigment *m* minéral
INOSINE - [chem] (a riboside used in medical research) inosine *f*
INOSITE - s. inositol
INOSITOL - [chem] (a component of latex) inositol *m*
INOTROPIC - [med] inotrope
INOXIDABLE - s. inoxidizable
INOXIDIZABLE - [chem] (not oxidizable; incapable of rusting) inoxydable
INOXIDIZE, to - [metall] (to render inoxidizable) rendre inoxydable
INPUT - [mech] énergie *f* absorbée, puissance *f* absorbée, prise *f* de vapeur
[radio] (the energy which is fed into a circuit) entrée *f*, énergie *f* d'entrée
[electron] (the signal fed into a device) signal *m* d'entrée
[electron] (the terminals of an electronic device) bornes *fpl* d'entrée
~AMPLIFIER - [el] préamplificateur *m*, amplificateur *m* préliminaire
~APERTURE - [electron] ouverture *f* d'entrée
~AXIS - [mech] axe *m* d'entrée
~BLOCK - s. input buffer
~BUFFER - [comput] (section of internal storage) zone *f* d'entrée
~BUFFER STORAGE - [comput] mémoire *f* tampon de entrée, mémoire *f* de transit d'entrée
~CAPACITANCE - [electron] (of an electron tube) capacité *f* d'entrée
~CARD - [comput] carte *f* d'entrée
~CAVITY BUNCHER - [electron] (in a klystron tube) cavité *f* d'entrée
~CIRCUIT - [electron] (circuit connecting the driving circuit and the input electrode) circuit *m* d'en-

trée
INPUT DAMPING - [radio] amortissement *m* d'entrée
~DATA - [comput] données *f*pl d'entrée
~DATA TRANSLATOR - [comput] traducteur *m* d'entrée, convertisseur *m* d'entrée
~DEVICES - s. input equipment
~DOCUMENT - [comput] document *m* de base
~ELECTRODE - [electron] (the electrode to which is applied the voltage to be amplified, modulated etc) électrode *f* d'entrée
~ELEMENT - [contr] élément *m* d'entrée, organe *m* d'entrée
~ENERGY - [mech] énergie *f* absorbée
~EQUIPMENT - [electron] (the equipment designed to introduce the information to a computer) appareillage *m* d'entrée
~FILE - [comput] fichier *m* d'entrée
~GAP - [electron] (interaction gap to start a variation in an electron stream) espace *m* d'interaction
~IMPEDANCE - [radio] (the impedance existing between the control grid of a thermionic valve and its cathode) impédance *f* d'entrée
~IMPEDANCE OF A TRANSMISSION LINE - s. input impedance
~JOB STREAM - [comput] flot *m* des travaux en entrée
~OUTPUT - [comput] entrée/sortie *f*
~/OUTPUT BUFFER STORE - [comput] mémoire *f* d'entrée-sortie
~OUTPUT CONTROL - [comput] commande *f* d'entrée/sortie, contrôle *m* entrée/sortie
~OUTPUT LIMITED SYSTEM - [comput] limiteur *m* d'entrée
~POWER WINDING - [contr] enroulement *m* d'alimentation
~QUANTITY - [contr] (the signal set by an agency which is independent of the control system) valeur *f* d'entrée
~RANGE - [contr] étendue *f* réglante
~RESISTANCE - [el] résistance *f* d'entrée
~RESISTOR - [el] résistance *f* d'entrée
~RESOLUTION - [contr] résolution *f* de l'entrée
~RESONATOR - [electron] (a resonant cavity producing velocity modulation of the electron beam) résonateur *m* d'entrée
~ROUTINE - [comput] (programme converting instruction in any notation into a notation which is acceptable to the machine) programme *m* d'entrée
~SIGNAL - [contr] (the signal supplied to a circuit) signal *m* d'entrée
~SPEED - [mech] nombre *m* de tours d'entraînement
[comput] vitesse *f* d'entrée
~STAGE - [el] étage *m* d'entrée
~STAND - [mech] (device to carry a coil of cable to feed a cable extruder) socle *m* d'introduction
~STATE - [comput] état *m* d'entrée
~STORE - s. input buffer
~TAPE - [comput] bande *f* d'entrée, bande *f* perforée d'entrée
~TRANSLATOR - [comput] traducteur *m* d'entrée, transcodeur *m* d'entrée
~TUBE - [electron] (the first electronic tube of an amplifier circuit) tube *m* d'entrée
~UNIT - [comput] (the unit introducing outside information into the computer) unité *f* d'entrée
~VOLTAGE - [electron] tension *f* d'entrée

INPUT WELL - [oil ind] soudage *m* d'injection, puits *m* d'alimentation
~WINDINGS - [electron] (windings of a saturable reactor to which independent variables are applied) enroulement *m* d'alimentation
INQUEST - [gen & leg] enquête *f*
INQUIRY - [gen] s. enquiry
INRUSH - [oil ind] irruption *f*, dégagement *m* instantané
~CURRENT - [electron] courant *m* de démarrage, courant *m* d'enclenchement
~OF AIR - [phys] entrée *f* d'air accidentelle
INSALUBRIOUS - [med] insalubre, malsain
INSALUTARY - [gen] malsain
INSANE - [med] aliéné , dément, fou
INSANITY - [med] (a mental disorder) aliénation *f* mentale, folie *f*
INSCRIBE, to - [gen] (to draw a figure enclosed within another) inscrire, graver
[comput] (to read the data recorded and rewrite them for the application of automatic reading) inscribir
INSCRIBER - [comput] (for punching purposes) dispositif *m* de codage
INSCRIPTION - [gen] inscription *f*
[gen] (on a coin) légende *f*
INSECT - [zool] insecte *m*
~CONTROL - [bot] désinsectisation *f*, lutte *f* contre les parasites
~PEST - [bot] parasite *m*, organisme *m* nuisible
[agric] ravageur *m*
~PESTS - [zool] insectes *m*pl nuisibles
~POLLINATION - [bot] (pollination caused by insects) pollinisation *f* par les insectes
~-POWDER - [chem] insecticide *m*
~-POWDER BLOWER - [mech] pulvérisateur *m* de poudre insecticide
~TRANSMISSION - [bot] (plant disease) transmission *f* par insectes
INSECTICIDE - [chem] (a substance which destroys insects) insecticide *m*
INSECTIFUGE - [chem] insectifuge *m*
INSECTIVOROUS PLANT - [bot] plante *f* insectivore
INSECURE - [gen] peu sûr, peu solide
[constr] (of a build etc) dangereux
INSEMINATE, to - [zool] inséminer
~ARTIFICIALLY, to - [zool] inséminer artificiellement
INSEMINATION - [zool] insémination *f*
[med] insémination *f*
INSENSIBILITY - [gen] insensibilité *f*
[med] insensibilité *f*
INSENSITIVE TO GAS - [chem] insensible aux gaz
INSERT , to-[gen] insérer, introduire, intercaler
INSERT - [gen] insertion *f*, introduction *f*
[mech] pièce *f* rapportée
[plast ind] (part moulded in position in a product) prisonnier *m* , insertion *f*
[cin] (simple shot inserted into a sequence) scene-raccord *f*
[draw] (enlarged drawing of a detail)agrandissement *m* d'un détail
[metall] (die casting) insertion *f*, partie *f* rapportée
[rubber ind] couche *f* intermédiaire
[metall] insertion *f*
[telev] insertion *f* statique

INSERT A SLEEVE, to - [mech] insérer une chemise
~CAMERA - [telev] caméra *f* d'insertions
~DIE - [metall] (in die casting) matrice *f* échangeable
~EARPHONES - [el acoust] (small earphone fitting inside the ear) embout *m* téléphonique
~OF FABRIC - [rubber ind] pièce *f* d'insertion tissu
~OF METAL - [rubber ind] pièce *f* d'insertion métal
~PIN - [die casting] (a pin used to locate and maintain an insert in position during moulding) goupille *f* d'insertion
~RETAINING RING - [text] anneau *m* de fixation des cloisons
~SPINDLE - [text] broche *f* à mortaise
~TREAD - [rubber ind] (a spacer ring) anneau *m* intermédiaire, anneau *m* de raccord
~TUBE - [electron] (X-ray tube inserted in the shield of an X-ray apparatus) tube *m* d'insertion
INSERTED - [gen] inséré
[mech] rapporté
~BLADE CUTTER - [mach tool] fraise *f* à lames rapportées
~JOINT - [mech] joint *m* à insertion
~JOINT CASING - [oil ind] tube *m* à renflement, tubage *m* à emboîtement
~PEG - [text] broche *f* à mortaise
~RIB - [metall] nervure *f* rapportée
~SOCKET - [metall & plast ind] raccord *m* mandriné emmanché
~SPIKE - [text] poignée *f* insérée
~TOOTH - [mech] dent *f* rapportée
~VALVE SEAT - [mech] siège *m* de soupape rapporté
~WELD JOINT - [plumb] (a sleeve joint) joint *m* soudé à insertion
INSERTING - [gen] insertion *f*, introduction *f*
[mech] emmanchement *m*
INSERTION - [gen] insertion *f*, introduction *f*
~ATTENUATION - [radio] atténuation *f* par insertion
~CLOTH - [rubber ind] (in tyres) toile *f* d'insertion
~GAIN - [radio] gain *m* d'insertion
~JOINT - [mech] (used to make watertight joints) joint *m* à insertion
~OF THE WEFT THREADS - [text] insertion *f* des fils de trame
~OF WEFT WITHOUT SHUTTLES - [text] insertion *f* de la trame sans navettes
~PARAMETER FILTER - [radio] filtre *m* à perte d'insertion en fonction déterminée de la fréquence
~PEG - [text] broche *f* à mortaise
~PHASE CHANGE - [radio] déphasage *m* par insertion d'un réseau
~PHASE SHIFT - [radio] déphasage *m* par insertion
~TRANSFER FUNCTION - [radio] fonction *f* de transfert par insertion
INSET - [cin] (single shot in a sequence) scène-raccord *f*
[bookbind] encart *m*
[print] hors-texte *m*
[mining] recette *f* inférieure
~CORE - [metall] pièce *f* battue
~FIRE - [th eng] (in space heating) radiateur *m* encastré
~MAP - [print] (in an atlas etc) cartouche *m*, papillon *m*
INSHEATED - [med] enkysté, engainé
INSIDE - [gen] dedans, intérieur
~AERIAL - [radio] (indoor aerial) antenne *f* interne
INSIDE AND OUTSIDE CALIPERS - [impl] maître *m* à danser, compas *m* maître de danse
~ANTENNA - s. inside aerial
~CALIPERS - [mech] (tool for measuring inside diameters) compas *m* d'intérieur
~CLEARANCE - [mech] (of a slide-valve) découvert *m* intérieur
~-COLOUR-SPRAYED LAMP - [light] (filament lamp having a bulb sprayed on the inside with a white or coloured material) lampe *f* colorée intérieurement
~CYLINDERS - [mech] (in a locomotive) cylindres *mpl* intérieurs
~DIAMETER - [mech] diamètre *m* intérieur, diamètre *m* dans œuvre
~DIE - [plast ind] (in extruders) matrice *f* intérieure
~DOOR-HANDLE - [auto] poignée *f* intérieure de porte
~DRIFT - [oil ind] calibre *m* pour manchon
~EDGE OF A RAIL - [railw] bord *m* intérieur du rail
~-FIRED BOILER - [th eng] chaudière *f* à foyer intérieur
~-FROSTED LAMP - [light] (lamp with a bulb sandblasted on the inside) lampe *f* dépolie intérieurement
~GAUGE - [impl] calibre *m* tampon
~GEAR - [mech] enregane *m* intérieur
~GROOVE - [mech] rainure *f* interne
~LAP - [mech] recouvrement *m* à l'échappement
~LEAD - [mech] (of slide-valve) avance *f* à l'émission
~LOOP - [aero] looping *m*
~MICROMETER CALIPERS - [meas] micromètre *m* de intérieur
~PANEL - [auto] panneau *m* de revêtement intérieur
~PLANT - [telecomm] outillage *m* téléphonique
~SHEDDING - [text] formation *f* intérieure du pas
~TAPPET - [mech] excentrique *m* à l'intérieur
~THREAD - [mech] (a screw thread cut on the inside of a hole or bore) filet *m* intérieur, pas *m* d'écrou
~TREADING MOTION - [text] mouvement *m* de marches intérieures
INSIDES - [anat] entrailles *fpl*
INSIDIOUS - [gen] insidieux
[med] insidieux
IN SITE CONCRETE - [constr] béton *m* coulé
INSOLATION - [met etc] (solar radiation) insolation *f*, ensoleillement *m*
[med] (a sunstroke) coup *m* de soleil
INSOLE - [shoe man] première semelle *f*
INSOLUBILITY - [chem] (incapability of being dissolved) insolubilité *f*
INSOLUBILIZATION - [chem] insolubilisation *f*
INSOLUBILIZE, to - [chem] (to render insoluble (in a given medium) insolubiliser
INSOLUBLE, to - [chem] insoluble, indissoluble
~COLORANT - [chem] (colorant in the form of a fine dispersion) colorant *m* insoluble
~IN WATER - [chem] insoluble dans l'eau
INSOLVENCY - [fin] insolvabilité *f*
INSOLVENT - [comm] insolvable, en faillite
INSOMNIA - [med] insomnie *f*
INSONOROUS - [acoust] insonore
~MATERIAL - [acoust] matériau *m* insonore
INSPECTING MACHINE - [text] machine *f* à visiter, visiteuse *f*
INSPECTION - [gen] inspection *f*, vérification *f*

[mech etc] (of machines etc) révision *f*, contrôle *m*, essai *m*
INSPECTION BENCH - [impl] banc *m* d'essai
~BY ATTRIBUTES - [autom] inspection *f* de qualité non mesurable
~BY VARIABLES - [autom] inspection *f* des variables
~CHAMBER - [build] (pit along a drain or sewer giving access for inspection) regard *m* pour de visite
~COVER BAND - [mech] (a sliding cover) regard *m* de visite
~DOOR - [gen & aero] porte *f* de visite
~GAUGE - [mech] (used for testing accuracy) calibre *m* de contrôle
~GLASS - [gen] regard *m*, fenêtre *f* d'observation
~HOLE - [mech] regard *m*
[auto] (timing pointer hole) orifice *m* de visite
~JIG - [mech] calibre *m* de contrôle
~JUNCTION - [mech] (in sanitary engineering, a length of drain pipe into which a vertical pipe is fitted to provide an access for inspection) joint *m* d'inspection
~LAMP - [light] (a portable lamp fed from the supply by a flexible cable) baladeuse *f*
~MACHINE - s. inspecting machine
~PANEL - [ind chem etc] (a sheet of transparent material fitted to a vessel to permit a view of the contents) panneau *m* de visite
~PEEP HOLE - s. inspection hole
~PIT - [railw & auto] fosse *f* à visiter
~PLUG - [el] (in the cover of an accumulator) bouchon *m* pour l'insertion de l'électrolyte
~PORT - [gen] regard *m*, fenêtre *f* d'observation
~SHAFT - [mech] fosse *f* à visiter
~STAMP - [ind proc & leg] (special mark made on a part to certify it has been inspected) cachet *m* de vérification
~TEST - [autom] contrôle *m* de recette, contrôle *m* d'acceptation
~TRAP - [mach] porte *f* d'inspection
~WINDOW - [mach] regard *m*
INSPECTIONISM - [med] voyeurisme *m*, scotophilie *f*
INSPECTOR - [gen] (a person appointed to examine aircraft, equipment, parts etc. to ensure that they possess specified dimensions, qualities or the like) inspecteur *m*, contrôleur *m*
INSPIRATOR BURNER - [th eng] brûleur *m* à induction
INSPISSATE, to - [gen] (to thicken) épaissir, se condenser
INSPISSATE A LIQUID, to - [chem] épaissir par évaporation
INSPISSATION - [gen chem med] épaississement
INSTABILITY - [gen] instabilité *f*
[met] (atmospheric condition favouring vertical flow of air currents) instabilité *f*
[aero] (the quality which makes it necessary for a pilot to use the controls to return an aircraft to its original attitude after disturbance) instabilité *f*
[radio] (the trip action of a magnetic amplifier) instabilité *f*
INSTABLE - [gen] instable
INSTALL, to - [gen] installer, monter
INSTALLATION - [gen] installation *f*
[mech] (of engines of a plant, e.g. a space-heating system) montage *m*, pose *f*
[telecomm] poste *m*, appareil *m*
[comput] aménagement *m*, implantation *f*
~PIPES - [plumb] installation *f* intérieure
INSTALLED CAPACITY - [gas & el] puissance *f* installée
INSTALMENT - [fin] fraction *f*, acompte *m*, versement *m* partiel
[print] fascicule *m*
~PAYMENT - [fin] payement *m* fractionné
INSTANCE - [gen] example *m*
INSTANT - [gen] instant
~COFFEE - [food] café *m* soluble
~-RETURN DIAPHRAGM - [photo] diaphragme *m* à retour instantané
~VALUE - [gen] valeur *f* instantanée, valeur *f* momentanée
INSTANTANEOUS - [gen] instantané
~ACOUSTIC KINETIC ENERGY PER UNIT VOLUME - [el acoust] densité *f* d'énergie cinétique
~ACOUSTIC POTENTIAL ENERGY PER UNIT VOLUME - [el acoust] densité *f* d'énergie acoustique potentielle
~ACOUSTIC POWER ACROSS A SURFACE ELEMENT - [el acoust] (sound energy flux) flux *m* d'énergie sonore
~ACOUSTIC POWER PER UNIT AREA - [el acoust] (the quotient of the instantaneous acoustic power transkitted across a surface element and the area of the surface element) puissance *f* acoustique instantanée unitaire
~ASSEMBLY - [nucl] (in an atomic bomb, the contact which is necessary to initiate the explosive reaction) ensemble *m* instantané
~AUTOMATIC GAIN CONTROL - [radar] régulateur *m* automatique de niveau
~AXIS - [mech] (the axis which is perpendicular to the plane of the motion passing through the point of a body which is instantaneously at rest) axe *m* instantané
~CARRYING-CURRENT - [el] (the maximum value of current carried instantaneously) courant *m* instantané maximal de charge
~CENTRE - [mech] centre *m* instantané
~CENTRE OF ROTATION - s. instantaneous centre
~COMPANDING - [radio] compression/expansion *f* immédiate
EXPOSURE - [photo] pose *f* instantanée
~FLOW - s. instantaneous rate of flow
~FORCE - [mech] (the total instantaneous force at a point) force *f* instantanée
~FREQUENCY - [radio] (of an oscillation) fréquence *f* instantanée
~MECHANOMOTIVE FORCE - s. instantaneous force
~RATE OF FLOW - [gas ind] débit *m* instantané
~RECORDING - [el acoust] (system of recording and reproducing requiring practically no delay in processing) enregistrement *m* instantané
~RELAY - [el] relais *m* instantané
~RELEASE - [el] action *f* instantanée
~SAFETY GEAR - [el] excentrique *m* arrêteur
~SAMPLING - [radio] (the process to obtain a sequence of instantaneous values of a wave) analyse *f* instantanée (d'une onde)
~SOUND ENERGY DENSITY - [el acoust] densité *f* de énergie acoustique
~SOUND PRESSURE - [el acoust] pression *f* acoustique instantanée
~SPEECH POWER - [acoust] (the rate a sound energy is radiated by a speech source at any given moment) radiation *f* sonore instantanée

INSTANTANEOUS VALUE - [el] (the value of the quantity at a particular point) valeur *f* instantanée
~WATER-HEATER - [el or gas] (a geyser) chauffe-eau *m* instantané
INSTEAD - [gen] au lieu de
INSTEP - [anat] cou-de-pied *m*
[text] dessus, cou-de-pied *m*
~BAR - [text] barre *f* de dessus de pied, barre *f* de cou-de-pied
~CUSHION - [shoe man] coussin *m* de cou-de-pied
INSTIGATE, to - [gen & leg] inciter, instiguer
INSTALLATION - [med] instillation *f*
INSTINCT - [zool] (complex coordination of relex actions resulting insome specified achievement without any previous experience; inherited memory) instinct *m*
INSTINCTIVE - [gen] instinctif
INSTITUTE, to - [gen & leg] instituer, constituer
[leg] (a legal action) intenter (un procès)
INSTITUTE - [gen] institut *m*, institution *f*
INSTITUTION - [gen] institution *f*
INSTROKE - [mech] (of an i.c. engine piston) course *f* retour, course *f* arrière
INSTRUCTION - [gen] instruction *f*
~ADDRESS - [comput] adresse *f* instruction
~ADDRESS REGISTER - [comput] registre *m* de contrôle de séquence
~BOOK - [gen] (a booklet issued by a manufacturer for guidance concerning a product) manuel *m* d'instruction
~CARD - [comput] carte *f* instruction
~CLASSIFICATION - [comput] classification *f* des instructions
~CODE - [comput] code *m* d'instructions
~COUNTER - [comput] compteur *m* ordinal, compteur *m* d'instructions, compteur-instructions *m*
~COUNTING REGISTER - [comput] registre-compteur *m* d'instructions
~DECODER - [comput] décodeur *m* d'instructions
~DISTRIBUTION CHANNEL - [comput] canal *m* distributeur d'instructions
ELEMENT - [comput] (part of an instruction) éléménť *m* d'instruction
~FOR USE - [gen] (instructions issued by a manufacture for guidance in the use of a product) instructions *fpl* pour s'en servir
~MODIFICATION - [comput] modification *f* d'instruction
~REGISTER - [comput] (a register which contains the instructions) registre *m* d'instructions
~SEQUENCE - [comput] (the normal sequence of selection of instructions for execution) séquence *f* d'instructions
~WORD - [electron comput] mot *m* d'instruction
INSTRUCTOR - [gen] instructeur *m*, professeur (de conduite)
INSTRUMENT, to - [gen] instrumenter
INSTRUMENT - [gen & instr] instrument *m*
[leg] (a document) acte *m* juridique
~ALIGHTING CHANNEL - [aero] (in landing by instruments) canal *m* pour l'atterrissage aux instruments
~AUTOTRANSFORMER - [instr] (instrument transformer with primary and secondary windings having common parts) autotransformateur *m* de mesure
~BOARD - s. instrument panel
~CARRIER PANEL - s. instrument panel
INSTRUMENT CORD - [telecomm] (for a telephone) cordon *m* de poste d'abonné
~CURRENT TRANSFORMER - [el] (transformer used in conjunction with measuring instruments) transformateur *m* de courant
~ERROR - [instr] erreur *f* instrumentale
~FLIGHT RULES (I.F.R.) - [instr] (official regulations governing instrument-flying) règles *fpl* pour le vol aux instruments
~FLIGHT TRAINER - [aero] (an aircraft specially fitted for training pilots in instrument flying) avion *m* d'entraînement de vol aux instruments
~FLYING - [aero] (flying solely by the use of instruments, as in clouds or darkness) vol *m* aux instruments
~FLYING HOOD - [aero] (collapsible hood over the cockpit of an aircraft, to similate zero visibility in instrument flying training) capote *f* pour vol sans visibilité
~FOR ABSOLUTE MEASUREMENTS - [instr] (an instruments the constants of which are determined by measurements depending on fundamental quantities) instrument *m* pour mesure absolues
~LAMP - [auto] lampe *f* d'éclairage du tableau de bord
~LANDING - [aero] (a landing effected solely by the use of instruments, without external visual observation) atterrissage *m* sans visibilité
~LANDING SYSTEM (I.L.S.) - [aero] (system designed to give lateral and vertical guidance and beacon indications to an aircraft during approach and landing) système *m* d'atterrissage aux instruments, I. L.S.
~LEADS - [el] (instrument connecting conductors) conducteur *m* de mesure, cordon *m*
~METEOROLOGICAL CONDITIONS - [met] (atmospheric conditions such as to make instrument flying necessary) conditions *fpl* météorologiques pour vol sans visibilité
~MOVEMENT - [instr] (the active part of a measuring instrument) élément *m* de mesure d'un appareil
~PANEL - [auto, aero etc] (the flat structure on which instruments are mounted) tableau *m* de bord, tableau *m* de commande, planche *f* de bord
~PLUG - [instr] fiche *f* d'instrument, fiche *f* de prise de courant d'un appareil
~RANGE - [instr] (the range of values covered by an instrument) domaine *m* de fonctionnement
~RELAY - [contr] relais *m* compteur
~RUNWAY - [aero] (runway provided with non-visual aids landing and take-off) piste *f* aux instrument, chenal *m* aux instruments
~SUPPORT - [gen] support *m*
~SWITCH - [el] interrupteur *m* d'appareil de mesure
~TABLE - [instr] (a test table) banc *m* d'épreuve
~TRANSFORMER - [instr] (a transformer expressly designed to supply an electrical instrument) transformateur *m* de mesure
~WITH CONTACTS - [instr] (instrument with moving element capable of opening or closing contacts at specified positions) appareil *m* à contacts
WITH ELECTROMAGNETIC SCREENING - [instr] (instrument with conducting screen protecting it from the influence of external electromagnetic fields) appareil à écran électromagnétique
~WITH LOCKING DEVICE - [instr] appareil *m* à blocage d'équipage

INSTRUMENT WITH MAGNETIC SCREENING - [instr] (instrument protected from the influence of external magnetic fields by ferromagnetic material) appareil *m* à protection magnétique, appareil *m* cuirassé
~WITH OPTICAL POINTER - [instr] appareil *m* à index lumineux
INSTRUMENTAL - [gen] instrumental, contributif
~STRAGGLING - [instr] (due to noise, poor geometry etc) dispersion *f* instrumentale
INSTRUMENTATION - [instr] (the application and arrangement of apparatus to measure, record and control industrial processes) emploi *m* des instruments, instrumentation *f*
~TAPE - [comput] ruban *m* numérique
INSUFFICIENCY - [gen] insuffisance *f*
[med] insuffisance *f*
INSUFFLATE, to - [gen] insuffler
[med] insuffler
INSUFFLATION - [med] (the blowing of air, gas etc into a cavity of the body) insufflation *f*
INSUFFLATOR - [med instr] insufflateur *m*
INSULANT - s. insulating material
INSULAR - [geogr] insulaire
~CLIMATE - [met] (climate characteristic of islands, in general more variable and less extreme than that prevailing over large land masses) climat *m* insulaire
INSULATE, to - [el] (to provide means to prevent the passage of electricity along unwanted paths) isoler
[acoust, heat etc] isoler, insonoriser
INSULATED - [el etc] isolé
~BOLT - [el] boulon *m* isolé
~CAR - [railw] wagon *m* isotherme
~CLIP - [el] pince *f* à œillet isolé
~CONTAINER - [railw] container-isotherme *m*
~CONTAINER WITH HEATING APPLIANCE - [railw] container *m* calorifique
~JOINT - [el] joint *m* isolant
~REFRIGERATOR CONTAINER - s. insulated container
~SCREWDRIVER - [impl] (electrician's screwdriver; screwdriver fitted with a handle of good insulating power) tournevis *m* d'électricien
~TONGS - [el] pince *f* isolante
~WAGON - s. insulated car
~WIRE - [el] cable *m* isolé
INSULATING - [gen el acoust etc] isolant
~AIR CUSHION - [el] matelas *m* d'air isolant
~AND SHEATHING COMPOUND FOR ELECTRIC CABLES - [rubber & plast ind] composition *f* pour l'isolation et le revêtement des câbles
~BARRIER - [el] écran *m* protecteur contre le contournement
~BASE - [el & radio] (the base to which wirings are applied) plaque *f* isolante, sole *f* isolante
~BEADS - [el] (generally glass beads over a bare conductor to provide an insulating covering) perles *fpl* isolantes
BOARD - [rubber & plast ind] (thick sheet material formed under pressure and used for heat insulation) panneau *m* isolant
~BRICK - [constr] brique *f* isolante
~CLAMP - [el] pince *f* isolante
~COMPOUND - [ind chem] (a mixture of substances used for electrical insulation) composition *f* isolante
~FERRULE - [el] bague *f* isolante
INSULATING FLANGE - [el] plateau *m* isolant
~GASKET - [mech] garniture *f* isolante
~GLOVE - [rubber ind] gant *m* isolant
~LAGGING - [th eng] enveloppe *f* calorifuge
~LAYER - [el] couche *f* isolante
~MAT - [rubber ind] (a flat sheet of insulating material placed on the floor, e.g. in front of high-tension equipment, as a safety measure) tapis *m* isolant
~MATERIAL - [el etc] (a substance or a body, the conductivity of which is zero or very small) isolant *m*, matière *f* isolante
~OIL - [el] (type of oil with good insulating properties and used for oilimmersed transformers etc) huile *f* isolante
~PAPER - [paper man] papier *m* isolant
~PLATE - [mech] s. insulation plate
~PROPERTY - [el] (general term denoting the suitability of a substance to act as an insulant) pouvoir *m* isolant
~REFRACTORIES - [metall] (refractory bricks having good heat insulating qualities, used to reduce heat-losses from furnaces) briques *fpl* réfractaires isolantes
~ROD - [el] (in knife switches) perche *f* isolante
~SHEATH - [el] revêtement *m* isolant, enveloppe *f* isolante, gaine *f* isolante
~SHEET - [rubber & plast ind] (plastics sheet material used as insulation) feuille *f* isolante
~SIDE PLATE - [el] plaque *f* isolante
~SLEEVE - [electron] (in an electronic tube) manchon *m* isolant
~SPINDLE - [el] (in telecommunications work) console *f*
~STRIP - [el] ruban *m* isolant
~TAPE - [el] (strong textile tape impregnated with adhesive insulating composition, chiefly used for temporary repairs) ruban *m* isolant, ruban *m* isolateur
~VARNISH - [paint] (varnish compounded for use as insulation) vernis *m* isolant
INSULATION - [el chem] (material used to inhibit the passage of heat or electricity) isolation *f*, isolement *m*
~BARRIER - [el] écran *m* protecteur contre le contournement
~COVERING - [gen] revêtement *m* isolant
~FAULT - [el] (a decrease in insulation resistance which can be considered abnormal) défaut *m* d'isolement
~INDICATOR - [instr] (an instrument measuring insulation resistance) indicateur *m* d'isolement
~MEASURING BENCH - [el] (used for measuring the isolation between the various electrodes in electronic tubes) banc *m* de mesure d'isolement
~PLATE - [mech] (shielding device between the rear face of a turbine disk and exhaust gas heat radiation) plaque *f* d'isolement
~RATING - [el] dimensionnement *m* de l'isolation
~RESISTANCE - [el] (the electrical resistance of an insulating material) résistance *f* d'isolement
~-STRIPPER - [tool] (tool for removing insulation from electric wires when making connection etc) pince *f* écorchante
~-STRIPPER PLIERS- [tool] (special type of pliers provided with a device for stripping insulation from wires) pince *f* à denuder

INSULATION TAPE - [el] chatterton *m*, ruban *m* isolant
~TEST - [el] (test to determine the insulation qualities of a substance, part or unit) essai *m* d'isolation
~TEST BOX - [el] (complete assembly of circuitry and instruments fitted in a portable cabinet, for marking insulation tests) appareillage *m* pour le contrôle de l'isolation
~TESTER - [instr] (instrument designed to measure insulation resistance) appareil *m* contrôleur d'isolement
INSULATOR - [el] (a substance offering resistance to an electric current) isolateur *m*, isolant *m*
~ARC OVER - [el] décharge *f* superficielle d'un isolateur
~ARCING HORN - [el] (metal projection deflecting an arc from the insulator surface) corne *f* de protection pour isolateur
~CAP - [el] (metal cap used in a string of insulators) calot *m*, capot *m*
~CHAIN - [el] chaîne *f* d'isolateurs
~CLIP - [el] agrafe *f* d'isolateur
~CUP - [el] isolateur *m* à calotte
~FRAMEWORK - [el] pont *m* des isolateurs
~GRADING SHIELD - [el] anneau *m* de garde perfectionnée
~PIN - [el] (the central support of a pintype insulator) ferrure *f* d'isolateur
~STRING - [el] chaîne *f* d'isolateurs
~-TYPE TRANSFORMER - [instr] (instrument transformer with a casing of insulating material) transformateur *m* type isolateur
INSULIN - [chem] (an important hormone secreted by the pancreas and used in the treatment of diabetes) insuline *f*
INSWEPT - [mach] étranglé, rétréci à l'avant
INTAGLIO - [print] intaille *f*
~PRINTING - [print] gravure *f* en creux
INTAKE, to - [mech] aspirer
INTAKE - [mech] (of a pump etc) prise *f*, admission *f*, appel *m*
[phys] énergie *f* absorbée, quantité absorbée
[min] (the channel through which the air is led into the workingplace) galerie *f* d'appel d'air
[mech] (of a centrifugal pump) ouïe *f*, œillard *m*
[med] apport *m*, absorption *f*
[hydr] aire *f* d'alimentation
[hydr] (of water) prise *f* d'eau, galérie *f* de captage d'eau
[metall] entré *f*, admission *f*, alimentation *f*
[mining] chantier *m* d'aérage
~AIR HEATER - [th eng] (arrangement for heating air before it is fed into an engine) réchauffeur *m* d'air d'admission
~CHAMBER - [mech] chambre *f* d'aspiration
~CONSTRUCTION - [hydr] ouvrage *m* de prise d'eau, bâtiment *m* pour la prise d'eau
~GATE - [hydr] vanne *f* de prise d'eau
~HEADER - [mech] collecteur *m* d'admission
~MANIFOLD - [mech] pipe *f* d'admission, collecteur *m* d'admission
~MANIFOLD VACUUM - [mech] dépression *f* de l'admission
~OF AIR - [mech] entrée *f* d'air, admission *f* d'air
~OF PUMP - [mech] admission *f* d'une pompe
~OF WATER - [hydr] prise *f* d'eau
INTAKE OPENING - [text] ouverture *f* de remplissage
~OPPOSITE EXHAUST - [auto] admission *f* opposée à l'échappement
~PORT - [auto] (inlet port) orifice *m* d'admission
[mech] raccord *m* d'admission
[mech] (pump) admission *f* d'une pompe
~PRESSURE - [mech] pression *f* d'aspiration
[oil ind] pression *f* d'introduction
[mech] (side) pression *f* côté vide (ou d'aspiration)
~SCREEN - [auto] crépine *f* d'aspiration, tamis *m* d'aspiration
~SHAFT - [mining] puits *m* d'entrée d'air
~SIDE - [mech] admission *f*, côté *m* vide
~SIDE TRAP - [mech] séparateur *m* côté aspiration (ou côté vide)
~SILENCER - s. inlet silencer
~SILENCER WITH WET AIR CLEANER - [auto] silencieux *m* d'aspiration avec filtre à air humide
~STROKE - [mech] (in i.c. engines) course *f* d'admission
~TUBE - [mech] (of a pump) tuyau *m* d'aspiration
~VALVE - [mech] soupape *f* d'admission
~WELL - [mining] sondage *m* d'injection
~WORK - [mech] (in i.c. engines) travail *m* d'aspiration
INTANGIBILITY - [gen] intangibilité *f*
INTANGIBLE - [gen] intangible
INTEGER-SLOT WINDING - [el] (distributed winding with the number of slots pole and phase integral and the same for all poles) enroulement *m* à nombre entier d'encoches par pôle et par phase
INTEGRAL - [gen] intégrant
[math] intégral
[mech] (a single piece) solidaire
~ABSORBED DOSE - [radiat] dose *f* absorbée intégrale
~ACTION - [contr] action *f* I, action *f* par intégration
~ACTION COEFFICIENT - [contr] coefficient *m* d'action par intégration
~ACTION TIME - [contr] temps *m* de compensation
~BUMPERS - [auto] pare-choc *m* incorporé
~CALCULUS - [math] calcul *m* intégral
~CONSTRUCTION - [mech etc] (construction of a part or member as a single piece i.e. not built up from a number of components) construction *f* intégrale
~CONTROL - [contr] (control in which the proportional correction changes in proportion to the deviation) régulation *f* I, régulation *f* par intégration
~CONTROL FACTOR - [contr] coefficient *m* d'action par intégration
~ELEMENT - [contr] organe *m* d'action intégrale
~EQUATION - [math] équation *f* intégrale
~FUNCTION - [math] fonction *f* intégrale
~HEAT OF DILUTION - [chem] (the increase of heat content which takes place when a given amount of solvent is added to a solution) chaleur *f* intégrale de dilution
~HEAT OF SOLUTION - [chem] (the difference between the heat-content of the solution and that of its components) chaleur *f* intégrale de solution
~INTAKE MANIFOLD - [auto] pipe *f* d'admission incorporée
~NUMBER - [math] nombre *m* intégral
~PLOUGH - [agric] charrue *f* portée

INTEGRAL REACTOR - [nucl] réacteur *m* à échangeur intégré
~REAR TRUNK - [auto] coffre *m* arrière incorporé
INTEGRATE, to - [gen, mech, chem] (to perform the summation of a series of differentials, i.e. the inverse of differentiation) intégrer, completer
[math] intégrer
INTEGRATED CIRCUIT - [comput] circuit *m* intégré
~DATA PROCESSING (IDP) - [comput] (the execution of operation in one cycle without human intervention) traitement *m* intégré des données, traitement *m* intégré des informations
~NEUTRON FLUX - [phys] (the total of the neutrons per square cm striking the irradiated material) flux *m* neutronique intégré
~PATH - [nucl] (of an ionizing particle; the total path) parcours *m* total
~REFLEXION - [radiat] réflexion *f* intégrée
~X-RAY REFLEXION - [radiat] réflexion *f* integrée de rayons X
INTEGRATING CIRCUIT - [el] s. integrator
~DIVIDER - [radio] diviseur *m* intégrateur
[telev] diviseur *m* de fréquence à intégration
~DOSE METER - [instr] (instrument designed to measure total radiation during an exposure) débitmètre *m* à intégration
~EXPOSURE METER - [photo] intégrateur *m* de luminations
~FREQUENCY METER - [instr] fréquencemètre *m* intégrateur
~FUNCTION - [math] fonction *f* intégrale
~GALVANOMETER - [instr] galvanomètre *m* intégrateur
~INDICATOR - [instr] indicateur *m* intégrateur
~INSTRUMENT - [instr] (a device designed to indicate and/or to record, the integral of the quantity measured from a given time datum to the time at which the measurement is made) instrument *m* intégrateur, appareil *m* intégrateur, compteur *m*
~IONIZATION CHAMBER - [nucl] chambre *f* d'ionisation à intégration
~METER - s. integrating instrument
~NETWORK - [el] s. integrator
~PHOTOMETER - [instr] (instrument designed to determine the luminous flux by a single measurement) lumenmètre *m*
~SPHERE - [photo] collecteur *m* hémisphérique
INTEGRATION - [math etc] (the summation of a series of differentials) intégration *f*
[med] intégration *f*
[comput] (the repeated use of the same machinable information carrier or carriers obtained from the first carrier) intégration *f*
~CONSTANT - [math] constante *f* d'intégration
~INTERVAL - [instr] temps *m* d'intégration
INTEGRATOR - [el] (a circuit whose output voltage is approximately proportional to the time integral of the input voltage) circuit *m* intégrateur
[comput] (a device with a varying output which is proportional to the integral of a varying input) intégrateur *m*
[chem] (in photography, for photografic baths) intégrateur
INTEGROMETER - [instr] (instrument designed to assess the moment of inertia) intégromètre *m*
INTEGUMENT - [zool & bot] (a cell layer) tégument *m*
[anat] entégument *m*, enveloppe *f*.

INTELLIGENCE - [telecomm] information *f* du signal, message *m*
[comput] intelligence *f*, contenu *m* d'une information
~BANDWIDTH - [radio & telev] (the sum of the audio or video frequency bandwidths) bande *f* de fréquences utile
INTELLIGIBILITY - [gen] intelligibilité *f*
INTELLIGIBLE CROSSTALK - [radio] (crosstalk resulting in intelligible sounds) diaphonie *f* intelligible
INTENSE NEGATIVE - [photo] négatif *m* intense, négatif *m* dense
INTENSIFICATION - [gen] intensification *f*
[photo] (the operation of increasing the contrast by a further deposit on the exposed parts) renforcement *m*, renforçage *m*
~BY OPTICAL PRINTING - [photo] renforcement *m* par tirage optique
~BY SULPHURING - [photo] renforcement *m* par sulfuration
INTENSITY MODULATION - [radar] (the control of the luminosity of the trace on a cathode-ray screen according to the magnitude of a signal) modulation *f* de luminosité
INTENSIFIER - [mech] (device used to increase pressure in a hydraulic line) multiplicateur *m* de pression
~ELECTRODE - [electron] (electrode which increases the velocity of electrons at the end of their trajectory) électrode *f* post-accélératrice
~RING - [electron] anneau *m* postaccélérateur
INTENSIFY, to - [gen] intensifier, augmenter
[mech] (the pressure) augmenter, renforcer
[photo] (to increase the contrast) renforcer
INTENSIFYING BATH - [photo] bain *m* de renforcement
~FACTOR - [radiat] facteur *m* d'intensification
~LEAD SCREEN - [photo] écran *m* renforçateur au plomb
~SCREEN - [radiat] écran *m* renforçateur
[photo] écran *m* renforçateur
INTENSITOMETER - [instr] (instrument designed to measure relative X-ray intensities during a radiography) intensitomètre *m*
INTENSITY - [gen] intensité *f*
[photo] densité *f*
[acoust] (of a reaction) énergie *f*
~MODULATION - [telev] modulation *f* de luminosité
~OF ACTIVATION - [nucl] (activation energy) énergie *f* d'activation
~OF COLOUR - [dye] profondeur *f* des couleurs, intensité *f* des couleurs
~OF CURRENT - [el] puissance *f* de courant
~OF FIELD - [el] (the vector quantity by which an electric field at a point is measured) intensité *f* de champ
~OF ILLUMINATION - [light] intensité *f* lumineuse
~OF INFESTATION - [bot] extension *f* de l'invasion de parasites
~OF MAGNETIZATION - [el] (the magnetic moment per cubic centimeter) aimantation *f*
~OF PRESSURE - [hydr] (the pressure which is exerted by a fluid on a unit area) charge *f* de pression
~OF RADIATION - [radiat] (the quantity of a radiation) intensité *f* de rayonnement
~OF RADIOACTIVITY - [nucl] (the number of atoms

disintegrating per unit time) intensité *f* de radioactivité
INTENSITY OF SOUND - [acoust] intensité *f* du son
~OF STRESS - [metall] (stress) sollicitation *f*
~SCALE - [photo] échelle *f* de gris à temps constant et lumination variable
INTENSIVE - [gen] intensif
~CULTIVATION - [agric] exploitation *f* intensive du sol
~GRAZING - [agric] pâturage *m* intensif
~PASTURE - [agric] pâture *f* intensive
~PROPERTY - [phys & chem] (the property of a system which does not depend on the quantity of material in the system itself) propiété *f* intensive
~REFLECTOR - [el] (reflector for incandescent lamps producing intense illumination at the required point) réflecteur *m* à faisceau concentré
INTERACTION - [gen] (action or influence of persons, objects or forces on each other) interaction *f*
~CIRCUIT PHASE VELOCITY - [electron] (of a travelling-wave-tube) vitesse *f* de phase dans le circuit d'interaction
~CROSSTALK COUPLING - [radio] couplage *m* diaphonique total
~FACTOR - [radio] (for a transducer) coefficient *m* d'interaction
~GAP - [electron] (interaction space between electrodes) espace *m* d'interaction entre électrodes
~IMPEDANCE - [electron] (of a travelling-wave-tube) impédance *f* de l'espace d'interaction
~LOSS - [radio] affaiblissement *m* d'interaction
~MEAN FREE PATH - [electron] parcours *m* libre moyen
~REPRESENTATION - [phys] (in quantum theory; representation of the equations of motion) représentation *f* de l'interaction
~SPACE - [electron] (in an electronic tube, the region where electrons interact with an alternating electromagnetic field) espace *m* d'interaction
~TIME - [electron] temps *m* d'interaction
INTERANNEALED WIRE - [metall] fil *m* recuit entre deux étapes d'étirage
INTERATOMIC FORCE - [nucl] (the force inside an atom) force *f* interatomaire
INTERBASE CURRECT - [electron] courant *m* transversal de base
INTERBEDDED - [geol] interstratifié
INTERBEDDING - [geol] interstratification *f*
INTERBLOCK SPACE - [comput] intervalle *m* interblock
INTERBREED, to - [zool & bot] croiser, entre-croiser
INTERBREEDING - [zool & bot] (hybridization of different species of animals or plants) croisement *m*, entrecroisement *m*
INTERCALARY - [bot] (between other bodies in a row) intercalaire
~CELL - [bot] cellule *f* intercalaire
INTERCALATE, to - [gen] (to insert, to add) intercaler
INTERCALATED BED - [geol] intercalation *f*
~FABRIC - [rubber ind] (tyres) toile *f* d'intercalaire
INTERCARRIER BEAT - [telev] interférence *f* de l'interporteuse
~SOUND SYSTEM - [telev] (system for amplification of combined vision and sound signals) réception *f* par battements
INTERCELLULAR - [zool] intercellulaire
~SPACE - [gen] espace *m* intercellulaire
INTERCEPT, to - [gen] intercepter
[telecomm] capter
INTERCEPT - [astr] (the difference between the observed and calculated positions of a celestial body) différence *f* entre les altitudes calculée et observée
[radio] appel *m* capté
[math] point *m* d'intersection
~DATA STORAGE POSITION - [comput] position *f* de interception
~METHOD - [astronaut] (method of celestial navigation, based on the determination of the difference between C.Z.D. and T.Z.D. (the intercept) méthode *f* d'interception
INTERCEPTER - s. interceptor
INTERCEPTING GRID - [telecomm] grille *f* d'arrêt, électrode *f* de freinage
~SCREENS - [cin] (sound or light intercepting screens) petits écrans *mpl*
~TRUNK - [telecomm] circuit *m* de renvoi
INTERCEPTION - [gen] interception *f*
[telecomm] captation *f*
~CIRCUIT - [telecomm] (in telephony) dispositif *m* de blocage d'une communication
INTERCEPTOR - [build] (in sanitary engineering, a trai providing a water seal against foul gases) siphon *m* d'égout
[mech] (a spoiler so arranged as to intercept the airflow through a slot) intercepteur *m*
INTERCHANGE - [gen] échange *m*
[genet] (the mutual transfer of portions between two chromosomes) interchange *m*, translocation *f* réciproque
[chem] (in a chemical reaction) échange *m*
[el] interversion *f*
[phys] (of electrons) tendance *f* d'échange
~POINT - [railw] station *f* de jonction
~STATION - s. interchange point
~TRACK - [railw] voie *f* d'échange entre faisceaux
~TRISOMIC - [genet] trisomique *f* par interchange
INTERCHANGEABILITY - [gen] (the quality of being interchangeable) interchangeabilité *f*, permutabilité *f*
[mech] (of machine components etc) interchangeabilité *f*
~OF GASES - [th eng] (burners) interchangeabilité *f* des gaz
INTERCHANGEABLE - [gen & mech] interchangeable, permutable
[instr] (said of the accessories of measuring instruments when they can be used in place of one another) interchangeable
~LENS PANEL - [photo] planchette *f* d'objectif interchangeable
~LIP - [metall] bec *m* amovible
~PART - [mech] (a part which can be substituted for an identical one without change of dimensions or impairment of functions) pièce *f* interchangeable
~RIM - [rubber ind] jante *f* amovible
INTERCHANGEABLENESS - s. interchangeability
INTERCHANGING MECHANISM - [mech] dispositif *m* d'échange
INTERCHANNEL INTERFERENCE - [radio] (the interference which is caused in one channel by the radiation in other neighbouring channels) brouillage *m* entre les canaux

INTERCILIUM - [anat] glabelle *f*
INTERCOLUMNIATION - [constr] (in a colonnade, the distance between the columns) entre-colonne *f*, entre colonnement *m*
INTERCOM - [telecomm] (slang abbreviation for internal telephone system) interphone *m*
INTERCOMMUNICATING COACH - [railw] voiture *f* à intercirculation
~POROSITY - [metall] porosité *f* à communication interne
INTERCOMMUNICATION - [telecomm] (telephonic communication system between members of the crew of an aircraft) intercommunication *f*
~SYSTEM - [telecomm] (system, usually telephonic, to allow of members of the crew of an aircraft speaking to each other) interphone *m*, système *m* de transmission d'ordres
INTERCONDENSER - [ind chem] condenseur *m* intermédiaire
INTERCONNECTING- [telecomm] (of telephones) interconnexion *f*
~FEEDER - [telecomm] artère *f* d'interconnexion
INTERCONNECTOR - [mech] (in gas, turbines, a connecting pipe between neighbouring combustion chambers) interconnecteur *m*
[telecomm] feeder *m* d'interconnexion
INTERCOOLER - [mech] (device for extracting heat from a gas between successive stages of compression) réservoir *m* réfrigérant intermédiaire
~RADIATOR - [th eng] (device used to dissipate the heat collected by an intercooler) radiateur *m* intermédiaire
INTERCOOLING - [th eng] (the process of abstracting heat from air between stages of compression) réfrigération *f* intermédiaire
INTERCOSTAL - [zool] (between the ribs) intercostal
~GIRDER - [shipbuild] longis *m*
INTERCOUPLE, to - [electron] interconnecter, interrelier
INTERCRYSTALLINE - [metall] (of a substance having adjacent crystal face) intercristallin
~CORROSION - [metall] (corrosion occurring between adjacent crystal faces in a substance) corrosion *f* fissurante, corrosion *f* intercristalline
~FAILURE - [metall] (metal fractures along the crystal boundaries) rupture *f* intercristalline
~FRACTURE - [metall] cassure *f* intercristalline
INTERDENDRITIC - [metall] interdendritique
~GRAPHITE - [min] graphite *f* interdendritique
~POROSITY - [metall] porosité *f* interdendritique
~SEGREGATION - [metall] microségrégation *f*
INTERDENTIUM - [anat] espace *m* interdentaire
INTERDEPENDENCE - [gen] interdépendance *f*
INTERDIFFUSE, to - [phys] diffuser l'un dans l'autre
INTERDIGIT HUNTING TIME - [telecomm] (in telephony) temps *m* de recherche libre
INTERDIGITAL - [zool] (between the fingers or the toes) interdigital
~MAGNETRON - [electron] (magnetron having the axial anode segments shaped like intersecting combs) magnétron *m* à segments entrelacés, magnétron *m* interdigital
INTERDOT FLICKER - [telev] (interference in colour-television in the dot-interlacing system) papillotement *m* multiple entre les points
INTERELECTRODE CAPACITANCE - [electron] capacité *f* interélectrode
INTEREST, to - [gen] intéresser
INTEREST - [gen] intérêt *m*
[comm] participation *f*
INTERFACE - [phys] (the boundary surface between two different media) surface *f* de séparation, interface *f*, distance *f* réticulaire cristalline
[chem] (the surface common to two adjacent phases) interface *f*, surface *f* de contact
[metall] surface *f* de séparation
~ACTIVITY - [phys] activité *f* interfaciale
~EFFECT - [electron] effet *m* de couche de limite
~REGION - s. interfacial zone
INTERFACIAL - [gen] (relating to interface) de surface de séparation
~ANGLE - [crystall] (the dihedral angle between adjacent crystal faces) angle *m* de surface de séparation
~FILM - [ind chem] (skin like structure developing between two emulsion phases) couche *f* entre deux faces
~SURFACE ENERGY - [phys] (the tension between phase interfaces) énergie *f* de surface de séparation
~SURFACE TENSION - s. interfacial tension
~TENSIOMETER - [instr] (instrument designed to measure facial tension in oxydation and polymerization processes) appareil *m* de mesure de la tension de surface de séparation
~TENSION - [chem] (the surface tension at the boundary between two immiscible liquids) tension *f* de surface de séparation
[phys] tension *f* superficielle limite, tension *f* interfaciale
~ZONE - [phys & chem] (the boundary area between two media, or two phases) zone *f* de surface de séparation
INTERFERE, to - [gen] s'ingérer, intervenir
[opt] interférer
INTERFERENCE - [gen] intervention *f*, ingérence *f*
[aero] (aerodynamic influence of one body on another) interaction *f* aérodynamique
[radio] (simultaneous reception of wanted and unwanted signals, e.g. atmospherics, ignition radiation or other stations) brouillage *m*, troubles *mpl*
[mech] arc-boutement *m* (des engrenages)
[el] (the introduction of electromotive force) interférence *f*
[phys] (the effect of superimposing two or more trains of waves of equal wavelength) interférence *f*
[med] interférence
[vet] entretaillure *f*
~BAND - [opt] (light and dark areas apparent on a surface and due to phase displacement of light) bande *f* d'interférence
~BLANKER - [electron] (for electronic sets) suppresseur *m* de brouillage
~COLOURS - [light] (e.g. colours on soap-bubbles, or oil on water etc) couleurs *fpl* interférentielles
~DRAG - [aero] s. interference
~ELIMINATOR - [telev] filtre *m* antiparasite
~FIELD - [radio] (the radiation field on an interfering transmitter) champ *m* brouilleur
~FILTER - [photo] filtre *m* interférentiel
~FIT - [mech] (a fit between two parts such that the entering part is larger than the receiving part, re-

quiring pressure or shrinkage to unite therm) ajustage *m* à tolérance negative
INTERFERENCE FRINGE - [opt] frange *f* d'interférence
~GUARD BANDS - [radio] (the two bands of frequency which are sometimes provided to minimize interferences) bandes *fpl* de protection
~INVERTER - [radio] (a special diode for the suppression of short duration interference in a television receiver) diode *f* antibrouilleuse
~LAYER - [opt] couche *f* d'intérference
~LIMITER - [radio & telecomm] limiteur *m* de parasites
~MICROSCOPE - [instr] (an optical instrument determining the degree of smootheness of a surface) microscope *m* à interférence
~PATTERN - [telev] moirage *m*, barres *fpl* d'interférences
~PEAK - [radio & telecomm] crête *f* de perturbation
~PULSE - [telev] (caused by the ignition of motocars) impulsion *f* parasite
~SPECTROSCOPE - s. inferometer
~SUPPRESSOR - [el] (device to prevent radio-frequency apparatus from emitting radiations capable of interfering with radio transmissions) suppresseur *m* de parasites
~THRESHOLD - [radio] seuil *m* d'interférence, seuil *m* de perturbation
INTERFEROMETER - [instr] (instrument based on the interaction of two beams of light into which a single beam is broken) interféromètre *m*
INTERFORMATIONAL SHEET - [mining] filon-couche *m*
INTERGLACIAL PERIOD - [geol] (period of comparatively mild temperatures between two glacial periods) période *f* interglaciaire
INTERGRAFTING - [agric] surgreffage *m*
INTERGRANULAR - [metall] intergranulaire
~CORROSION - [metall] (corrosion preferably occurring at the boundaries between the crystal grains) corrosion *f* intergranulaire
~FRACTURE - [metall] cassure *f* intergranulaire
~FRICTION - [geol] frottement *m* intergranulaire
~PRESSURE - [geol] tension *f* effective, pression *f* intergranulaire
~TEXTURE - [geol] (the texture of holocrystalline basalts and doleritic rocks) texture *f* intergranulaire
INTERHEATER - [mech] (a combustion chamber assembly interposed between stages of a gas turbine) échauffeur *m* intermédiaire
INTERIOR - [gen] intérieur
[gen] (of a building etc) intérieur *m*
~ANGLE - [math] angle *m* interne
~LANDING PLATFORM - [gas ind] (gasholders) (interior landing frame) charpente *f* intérieure fixe
~LINING - [th eng] (of domestic ovens etc) flasque *m*
~PHASE - [chem] phase *f* interne
~WOOD FINISHING - [constr] ménuiserie *f* intérieure
INTERJOIST - [constr] distance *f* de poutre *f* en poutre
INTERLACE, to - [gen & text] entrelacer, entrecroiser
INTERLACE CONTROL - [telev] ajustage *m* de l'analyse entrelacée
INTERLACED RECORDING - [comput] enregistrement *m* entrelacé
INTERLACED SCANNING - [telev] (a system whereby each picture is scanned in two or more successive fields) analyse *f* à intercalage
[mining] exploration *f* intercalaire
~SPOT SCANNING - [telev] système *m* à entrelacement de points
~STORAGE ASSIGNMENT - [comput] affectation *f* optimum des informations aux cellules d'une mémoire
~THREADS - [text] fils *mpl* entrelacés, fils *mpl* tressés ensemble
INTERLACEMENT - [text] (interweaving) entrecroisement *m*
INTERLACING - [gen] entrelacement *m*
[text] (the interweaving of wool yarns) croisement *m*, entrelacement *m*
[telev] s. interlaced scanning
[comput] (only USA, interleave in GB; the assigning of successive storage location numbers to separated locations on a magnetic drum) enchevêtrement *m*
[comput] imbriquage *m*, entrelacement *m*
~EFFECT - [text] effet *m* de liage
~INOPPOSITE DIRECTION - [text] liage *m* en direction opposée
~OF THE THREADS - [text] croisement *m* des fils, entrecroisement *m*
~PLAN - [text] mise *f* en carte
~POINT - [text] point *m* de liage
~THE WARP AND WEFT THREADS - [text] entrelacement *m* des fils de chaîne et de trame
INTERLAID - [print] (of a paper inserted between the plate and its mount so as to raise the plate) mis entre cuir et chair
INTERLAMINAR BONDING - [metall & plast ind] (bonding between adjacent sheets of a laminated material) assemblage *m* de laminés
~STRENGTH - [plast ind] (the strength of the adhesion between adjacent layers of a laminar material) résistance *f* au clivage
INTERLAY, to - [gen] insérer
[print] (the insertion of a paper between the printing plate and the mount to raise the plate to type height) mettre en train des planches
INTERLAY - [gen & ind proc] couche *f* intermédiaire
[print] taquet *m*
INTERLAYER - [plast ind] couche *f* de liaison, intercouche *f*
INTERLAYING - [gen] insertion *f* d'une couche
[print] mise *f* en train des planches
INTERLEAVE, to - [comput] imbriquer, entrelacer
INTERLEAVED - [gen] inséré, interfolié, intercalé
~JOINTS - [mech] joints *mpl* encastrés
~TRANSMISSION SIGNAL - [telev] signal *m* de transmission cocanalisé
INTERLEAVING - [telev] (a technique of colour TV transmission) cocanalisation *f*
INTERLINE FLICKER - [telev] (in interlaced scanning, the alternating of the brightness) papillotement *m* interligne
INTERLINING - [text] doublure *f*, triplure *f*
~FABRIC - [text] lin *m* pour doublures
INTERLOBITIS - [med] pleurésie *f* interlobaire
INTERLOCK, to - [el mech etc] verrouiller, enclencher, engrener
INTERLOCK - [el & mech] (any electrical or mecha-

nical arrangement making one operation dependent upon another) verrouillage *m*, enclenchement *m*, blocage *m*
INTERLOCK - [metall] (diecasting) blocage *m*
[cin] (the synchronization of camera and sound registering devices) synchronisation *f*
[cin] (the actual mechanism for the above) dispositif *m* de synchronisation
[railw] enclenchement *m*
[contr] (a device which is actuated by the operation of another device with which it is directly associated) dispositif *m* de verrouillage
[comput] (mechanism controlling the synchronization of concurrent operations) verrouillage *m* mutuel
[text] blocage *m* de commande
~BOARD - [railw] châssis *m* d'enclenchement
~MACHINE - [text] machine *f* interlock, métier *m* circulaire interlock
~RELAY - [el] relais *m* de blocage
~SYSTEM - [cin] (the arrangement for the synchronous device of cameras and recorders) système *m* de verrouillage
~VALVE - [mech] vanne *f* d'arrêt
INTERLOCKED - [gen & el & mech] enclanché, solidaire
~CONTROLS - [mech] commandes *fpl* accouplées (ou couplées)
~OPERATION - [comput] opération *f* à verrouillage
~POINTS - [railw] aiguille *f* enclenchée
~SIGNAL - [railw] signal *m* enclenché
INTERLOCKING - [gen el & mech] enclenchement *m*, engrènement *m*
[contr] (the action of interlocking) enclenchement *m*, asservissement *m*
[constr] emboîtement *m* latéral
~CIRCUIT - [el] (a type of circuit in which the operation of one or more units included in it may be arranged to be inhibited by the existance of certain specified conditions in other units) circuit *m* asservi
~FEED SELECTING LEVER - [mach tool] levier *m* de avancement avec dispositif de blocage
~FRAME - [railw] châssis *m* d'enclenchement
~GEAR - [mech] mécanisme *m* à action solidarisée
~MECHANISM - [mech] mécanisme *m* de verrouillage
~MILLING CUTTER - [mach tool] fraise *f* composée
~POST - [el] poste *m* d'enclenchements électriques
~TAP - [gas ind] robinetterie *f* à enclenchement
INTERLOCUTORY - [leg] interlocutoire
~JUDGEMENT - [leg] jugement *m* interlocutoire
INTERLOOP, to - [text] entrecroiser, entrelacer
INTERLUDE - [mus] interlude *m*
INTERMAXILLARY - [anat] intermaxillaire
INTERMEDIARY LANGUAGE - [comput] (artificial language suggested by Russian scientists to be used in machine translation) idiome *m* intermédiaire
INTERMEDIATE - [gen] intermédiaire
[chem] s. intermediates
~BASE-STRUT - [aero] mât *m* inférieur intermédiaire
~BEARING - [mech] (in motors and engines) coussinet *m* intermédiaire
~BLOCK - [mech] bloc *m* intermédiaire
~BOX - [text] foule *f* intermédiaire
~BRAKE CONTROL - [mech] (in autos) renvoi *m* des freins

INTERMEDIATE CIRCUIT - [radio] (used for coupling an aerial to a transmitter or receiver) circuit *m* intermédiaire
~COMPOUND - [chem] composé *m* intermédiaire
~CONDENSER - [ind chem] condenseur *m* intermédiaire
~CONTOUR - [cartography] distances *fpl* intermédiaires
~COOLER - [ind chem] refroidisseur *m* intermédiaire
~COOLING - [ind chem] refroidissement *m* intermédiaire
~DIE PLATE - [die casting & plast ind] (an intermediate die plate is that part of a three piece injection mould which floats between the moving head die plate and the fixed head die plate) plaque *f* matrice intermédiaire
DRAW FRAME - [text] banc *m* d'étirage intermédiaire
~ECHO SUPPRESSOR - [radio] (echo suppressor at an intermediate point on the circuit) suppresseur *m* d'écho intermédiaire
~ENERGY REGION - [phys] (region of average energies) zone *f* d'énergie moyenne
~FILM SYSTEM - [telev] (the system of recording on a photographic film which is subsequently televized) télévision *f* par film intermédiaire
~FLANGE - [mech] bride *f* intermédiaire
~FORGING - [metall] dégrossissage *m* à chaud
~FRAME - [photo] cadre *m* intermédiaire
[text] banc *m* à broches intermédiaire
~FREQUENCY - [radio] (frequency produced in a superheterodyne receiver by combining the carrier of the incoming signal with a local oscillation) fréquence *f* intermédiaire
~FREQUENCY AMPLIFIER - [radio] amplificateur *m* moyenne fréquence
FREQUENCY HARMONIC RESPONSE - [radio] réponse *f* aux harmoniques M.F.
~FREQUENCY REJECTION RATIO - [radio] taux *m* de rejection fréquence/image
~-FREQUENCY RESPONSE - [radio] (the unwanted response to an excitation at an intermediate frequency) réponse *f* M.F.
~-FREQUENCY RESPONSE RATIO - [radio] gain *m* M.F.
~GEAR - [mech] engrenage *m* de transmission
~GLASS - [instr] verre *m* intermédiaire
~IGNEOUS ROCKS - [geol] (igneous rocks containing approximately 6% silica) roches *fpl* ignées intermédiaires
~LANDING - [aero] champ *m* d'atterrisage intermédiaire
~LANDING-STATION - [mining] recette *f* intermédiaire
~LAYER - [electron] (the layer between the counter-electrode and selenium layer) couche *f* intermédiaire
~LOOP - s. intermediate circuit
~MEANS - [meas] (those system elements which are used for distinct operations in a measurement sequence) éléments *mpl* intermédiaires
~MULTIPLE - [telev] (split image) suroscillation *f*, franges *fpl*
~NEGATIVE - [photo] négatif *m* intermédiaire
~NEUTRONS - [nucl] (resonance neutrons) neutrons *mpl* intermédiaires
~NUCLEUS - [nucl] (compound nucleus) noyau *m* com-

pound nucleus) noyau *m* composé, noyau *m* intermédiaire

INTERMEDIATE OVERHAUL - [railw] (of locomotives, coaches etc) révision *f* intermédiaire

~OXIDES - [chem] oxydes *mpl* intermédiaires

~PROCEDURE - [aero] (landing approach procedure) procédure *f* intermédiaire

~PRODUCT - s. intermediates

~PUMP - [mech] booster *m*, pompe *f* intermédiaire

~RADIAL STRUT - [aero] traverse *f* radiale intermédiaire

~REACTOR - [nucl] (nuclear reactor in which the fission is induced maily by neutrons having energies greater than thermal but smaller than that of fission neutrons) réacteur *m* à neutrons intermédiaires

~REPEATER - [telecomm] (for use in a trunk at any point other than an end) répéteur *m* intermédiaire

~RIB - [constr] arc *m* doubleau

~RING - [photo] bague *f* intermédiaire

~SHAFT - [auto] arbre *m* secondaire, axe *m* secondaire

~SHAFT TUBE - [auto] tunnel *m* d'arbre intermédiaire

~SIGNAL BOX - [railw] poste *m* de block intermédiaire

~SPEED OF NEUTRONS - [nucl] vitesse *f* intermédiaire de neutrons

~STATE - [el] (the state of a super conducting material in an external magnetic field which approaches the critical field) état *m* intermédiaire

~STORAGE - [comput] mémoire *f* intermédiaire, mémoire *f* tampon

~SUBCARRIER - [radio] sous-porteuse *f* intermédiaire

~STRUT - [constr] entretoise *f*, arbalétrier *m* intermédiaire

~TRAIN DISTANCING POINT - [railw] poste *m* d'espacement

~TRANSVERSE FRAME - [aero] cadre *m* intermédiaire de liaison voilure-fuselage

~VALVE - [mech] vanne *f* intermédiaire, robinet *m* intermédiaire

~WHEEL - [mech] roue *f* intermédiaire, roue *f* parasite

INTERMEDIATES - [chem] (manufactured chemicals which are used in further chemical operations to produce end-products such as dyes, drugs, etc) produits *mpl* intermédiaires

INTERMESHING FEED SCREWS - [mech] (extruder screws formed with right-hand and left-hand threads, and arranged to engage together) vis *fpl* engrenantes

~FLIGHTS - [plast ind] (arrangement of two screws such that the flights of each enter the channels of the other) pas *mpl* enchevêtrés

~PADDLE MIXER - [mech] (paddle mixer in which the operating planes of the paddles intersect) mélangeur *m* à pales enchevêtrées

~PADDLES - [mech] (blades of a mixer so arranged that their circular paths intersect) pales *fpl* entremêlées

INTERMETALLIC COMPOUND - [metall] (intermediate constituent; a constituent of alloys formed when atoms of two metals combine in certain proportions, thus forming crystals which have a structure different from that of either metal) composé *m* intermétallique

INTERMINGLE, to - [gen] mélanger, entremêler

INTERMISSION - [gen] interruption *f*
[med] (a temporary cessation of fever etc) intermission *f*, intermittence *f*
[cin] (the interval between two parts of a film, or between two films in a programme) entr'acte *m*

INTERMITTENCY EFFECT - [opt] effet *m* d'intermittence

INTERMITTENT - [gen] intermittent

~CARBONIZATION - [ind chem] distillation *f* discontinue

~CLAUDICATION - [med] (intermittent laminess) claudication *f* intermittente

~CONTACT PRINTER - [cin] (step by step contact printer) tireuse *f* simple par contact

~CURRENT - [el] courant *m* intermittent

~DUTY - [el] (sequence of periods of working at constant load followed by intervals of rest) service *m* intermittent

~-DUTY RATING - [el] rendement *m* en service intermittent

~EARTH - [el] (accidental earth connexion occurring intermittently and thus difficult to locate) mise *f* à la terre accidentelle

~FORWARD MOVEMENT - [mech] mouvement *m* intermittent, mouvement *m* saccadé

~GAS LIFT - [oil ind] gaz *m* lift intermittent

~IN SITU REVIVIFICATION - [ind chem] (in purification processes) revivification *f* intermittente in situ

~INTEGRATION - [meas] (used in electricity meters) intégration *f* intermittente

~JET - [mech] (a type of jet engine in which combustion occurs in discrete pulses) pulsoréacteur *m*

~LET OFF - [text] déroulement *m* intermittent

~LOADING - [el] (cables with conductors which are loaded continuosly for sections of their length only) charge *f* intermittente

~MECHANISM - [cin] (of a projector) mécanisme *m* intermittent

~MOVEMENT - [cin] (intermittent movement for motion picture cameras) mouvement *m* intermittent, mouvement *m* saccadé

~OPTICAL PRINTER - [cin] tireuse *f* optique simple

~PERIODIC DUTY - [el] (a sequence of identical cycles of intermittent duty) service *m* intermittent périodique

~RATING - [el] régime *m* nominal pour service intermittent

~SPINNER - [text] renvideur *m* selfacting

~SPINNING PROCESS - [text] filature *m* intermittente

~SPROCKET - [mech] (a sprocket behind the gate in a projector) tambour *m* denté de croix de Malte

~TAKING UP MOTION - [text] régulateur *m* d'enroulement intermittent

~TENSILE STRESS - [mech] effort *m* de traction intermittent

~TEST - [el] (test in which a cell or battery is subjected to periods of discharge and recuperating) essai *m* intermittent

~VERTICAL RETORT - [ind chem] (retorts) four *m* à distillation discontinue, cornue *f* verticale discontinue

INTERMITTENTLY - [gen] par intervalles, par intermittence

INTERMODULATION - [radio] (the modulation of the components of a complex wave by each other) inter-

modulation *f*, transmodulation *f*
INTERMODULATION DISTORTION - [radio] distorsion *f* d'intermodulation
~DISTORTION CHARACTERISTIC - [radio] taux *m* de distorsion d'intermodulation
~NOISE - [radio] bruit *m* d'intermodulation
~PRODUCTS - [radio] (parasitic signals in an amplifier) produits *mpl* d'intermodulation
INTERMOUNT BASIN - [geol] bassin *m* structural, cuvette *m f*
INTERNAL - [gen] intérieur, interne
~BAND BRAKE - [auto] frein *m* à friction interne
~BASE RESISTANCE - [electron] (the internal resistance of the base of a transistor) résistance *f* interne de base
~BLACKENING - [electron] (on the interior of the glass bulb of vacuum tubes due to gases or carbon deposit) noircissement *m* intérieur
~BULKHEAD - [naut] cloison *f* interne
~CHILL - [metall] refroidisseur *m* interne
~COMBUSTION ENGINE - [mech] (a heat engine in which the fuel is burnt in the engine itself and not externally, as in a steamengine) moteur *m* à combustion interne
~COMBUSTION ENGINED RAILCAR - [railw] autorail *m* à moteur à combustion interne
~COMPENSATION - [chem] (term used for the effect of the combination of two enetiomorphous groups in neutraling optical activity within a molecule) compensation *f* intérieure
~CONDUCTOR - [el] (the inner conductor of a cable) conducteur *m* interne
~CONVERSION - [nucl] conversion *f* interne
~CONVERSION COEFFICIENT - [nucl] coefficient *m* de conversion interne
~CONVERSION ELECTRON - [electron] électron *m* de conversion interne
~COOLING - [ind chem] refroidissement *m* interne
~CORRECTION VOLTAGE - [electron] tension *f* de correction interne
~CRACK - [metall] fissure *f* interne
~CYLINDRICAL GAUGE - [meas] (used to measure the diameter of holes) calibre *m* à bouchon, tampon *m*
~DAMPING - [phys] (damping inherent in the elements of a structure) amortissement *m* interne
~DETECTOR OF TEMPERATURE - [meas] (device registering the temperature in an electric machine) détecteur *m* intérieur de température
~DIAMETER - [gen] diamètre *m* intérieur
~DRAG WIRE - [aero] hauban *m* de traînée interne
~DUSTING - [rubber ind] (the application of powdered talc to the inside of a rubber article) talquage *m* intérieur
~e.m.f. - [el] (the electromotive force generated in an electric machine) force *f* électromotrice interne
~EXPANDING BRAKE - [mech] (wheel brake consisting of a drum against which two or more shoes are expanded by a cam) frein *m* à expansion interne
~-EXTERNAL GEAR CLUTCH - [auto] accouplement *m* à dentures intérieure et extérieure
~FLUSH TOOL JOINT - [oil ind] joint *m* lisse interne
~FOCUSING TELESCOPE - [instr] (surveying telescope which is focused by the movement of an internal lens) télescope *m* à focalisation interne
~FRICTION - [phys] (lack of elasticity) frottement *m* intérieur
INTERNAL GAUGE - [meas] calibre *m* pour trous
~GEAR - [mech] engrenage *m* intérieur
~GRINDER - [mach tool] machine *f* à rectifier les surfaces intérieures
~GRINDING WHEEL - [tool] meule *f* aléseuse
~HEATING - [el chem] (the method of maintaining the electrolyte in a molten condition by Joule effect) chauffage *m* interne
~IDLE TIME - [comput] (time during which a machine part cannot be used) temps *m* inactif interne
~IGNITER - [light] (for safety lamps) allumeur *m* intérieur
~INDICATOR - [ind chem] (an indicator dissolved in the solution in which the reaction which it is to indicate takes place) indicateur *m* intérieur
~INDUCTION HEAT-TREATMENT - [mech] traitement *m* à chauffage par induction
~LEAKAGE - [gas ind] fuite *f* interne, débit *m* de fuite
~LUBRICANT - [ind chem] (substance added to a plastic to provide internal lubrication) lubrifiant
~MASONRY - [constr] (the setting brickwork) maçonnerie *f* interne
~MEMORY - [comput] (storage which is automatically accessible to the computer) mémoire *f* interne
~MICROMETER - [instr] calibre *m* à tige coulissante et vis micrométrique
~MILLING CUTTER - [mach tool] machine *f* à fraiser les surfaces intérieures
~MIXER - [rubber ind] mélangeur *m* fermé, mélangeur *m* interne
~NAVIGATION - [naut] (in canals, rivers etc) navigation *f* interne
~OPERATING RATIO - [comput] (measure of the efficiency of the machine) rendement *m* interne
~PIPING - [constr] conduit *m* intérieur, canalisation *f* intérieure, réseau *m* intérieur
~PLANT - [telecomm] ensemble *m* des installations téléphoniques
~-POLE DYNAMO - [el] dynamo *f* à pôles internes
~PRESSURE - [metall] pression *f* intérieure
~QUENCHING - [nucl] (terminating internally a pulse of ionization current in a Geiger-Müllercounter) autocoupure *f*, étouffement *m*
~RACK AND PINION VALVE - [plumb] vanne *f* à pignon et crémaillère internes
RELATIONSHIP - [electron] (in an electronic tube) relation *f* interne
~RESISTANCE - [el] (the resistance which the cell itself offers, internally, to the current flowing through it on closed circuit) résistance *f* interne
~SCREW THREAD - s. internal thread
~SCREW VALVE - [plumb] robinet *m* à vis intérieure
~SECRETION - [zool] (secretion poured into the blood vessels) sécrétion *f* interne
~SHRINKAGE - [metall] retassure *f* interne
~SILL - [constr] (of window) rebord *m* interne de fenêtre
~STANDARD - [spectrology] (sample used for reference in spectral measurements) étalon *m* interne
~STANDARD LINE - [phys] (in spectrology) ligne *f* de référence
~STORAGE - s. internal memory
~STORAGE CAPACITY - [comput] capacité *f* de mémoire interne
~STRESS - [metall] (residual stress between diffe-

rent parts of metal products) tension *f* interne
INTERNAL TEETH - [mech] (teeth formed on the inner side of a hollow gear) denture *f* intérieure
~THREAD - [mech] filetage *m* intérieur
~YIELD - [metall] limite *f* élastique intérieure
INTERNALLY FIRED BOILER - [th eng] (a boiler with the fire box inside it and surrounded by water) chaudière *f* à foyer intérieur
~HEATED IRON - [impl] fer *m* à chauffage interne
~HEATED OVEN - [th eng] four *m* à chauffage direct, four *m* à circulation
~-TOOTHED GEAR - [mech] (an annular gear having teeth cut on the inner circumference) engrenage *m* à denture intérieure
INTERNATIONAL - [gen] international
~AMPERE - [el] (unit of current in common use) ampère *f* internationale
~CALL - [telecomm] conversation *f* internationale
~CANDLE - [light] (unit of luminous intensity) bougie *f* internationale
~CIRCUIT - [telecomm] (circuit connecting trunk lines in different countries) circuit *m* international
~CIVIL AVIATION ORGANISATION - ICAO - [aero] (international body concerned with the regulation of civil flying) Organisation *f* Internationale d'Aviation Civile
~DATE-LINE - [geogr] (an internationally-agreed line approximating the 18th degree, on which the calendar is adjusted by adding or dropping a day as the line is crossed) limite *f* de date internationale
~DISTRESS FREQUENCY - [radio] (the frequency used by radiomarine stations asking for help) fréquence *f* de détresse internationale
~GAUGE - [railw] écartement *m* international
~METRIC COUNT - [text] (unit measurement for yarns) numéro *m* international
~OHM - [el] (unit of resistance in common use) ohm *m* international
~TEMPERATURE SCALE - [meas] échelle *f* de température internationale
~UNITS - [el] (internationally adopted system of units, based on the values of the international ampere and the international ohm) unités *f*pl internationales
~VOLT - [el] (common unit of potential difference) volt *m* international
INTERNODE - [bot] entre-nœud *m*, mérithalle
[anat] phalange *f*
INTEROCULAR ADJUSTMENT - [photo] réglage *m* interloculaire
~DISTANCE - [opt] distance *f* interloculaire
[photo] écart *m* des centres de rotation des yeux
INTEROFFICE TRUNK - [telecomm] (in telephony) ligne *f* de jonction, ligne *f* auxiliaire
INTEROSSEUS - [anat] interosseux
INTERPARIETAL - [anat] (membrane bone of the skull) interpariétal
INTER-PASS ANNEALING - [metall] recuits *m*pl entre passages
INTERPENETRATION TWINS - [min] (to crystals united in accordance with a fixed plan) macles *m*pl d'interpénétration
INTERPHONE - [telecomm] (office, house, aircraft, etc. intercommunication telephone) téléphone *m* domestique
INTERPLANAR SPACING - s. interface
INTERPLANE AILERON - [aero] (an aileron placed between the planes of a multiplane) aileron *m* de plan central
INTERPLANE STRUTS - [aero] (in a multiplane, struts placed between upper and lower planes) mâts *m*pl de cellule
INTERPLANETARY - [astr] interplanétaire
~FLIGHT - [astronaut] vol *m* interplanétaire
~JOURNEY - [astronaut] (a space journey) voyage *m* interplanétaire
~PROBE - [astronaut] sonde *f* interplanétaire
~SPACE - [astronaut] espace *m* interplanétaire
INTERPOLATE, to - [gen & math] interpoler, intercaler
[comput] (the operation of a combining two sequences of items of information) interpoler
INTERPOLATION - [math] interpolation *f*
[radio] (signal interpolation in submarine cable telegraphy) interpolation *f* de signal
~OF CONTOURS - [surv] interpolation *f* du profil
INTERPOLE - [el] (compole; a supplementary magnetic pole designed to improve commutation) pôle *m* auxiliaire, pôle *m* de commutation
INTERPOSE, to - [gen] interposer
INTERPOSITION CIRCUIT - [telecomm] ligne *f* d'appel des tables interurbaines
~TRUNK - [telecomm] ligne *f* de renvoi, circuit *m* de renvoi
INTERPRET, to - [gen] interpréter
[comput] interpréter
INTERPRETATION - [gen] interprétation
~ROUTINE - [comput] programme *m* interprétatif
~SUBROUTINE - [comput] sous-programme *m* interprétatif
INTERPRETER - [gen] interprète *m*
[comput] programme *m* interprétatif, traductrice *f* d'écriture perforée
[comput] (of punch cards) traductrice *f*
~CODE - [comput] code *m* interprétatif
~INTERPRETIVE COSE - s. interpreter code
INTERPUPILLARY - [opt] interpupillaire
~DISTANCE GAUGE - [meas] appareil *m* de mesure de distance interpupillaire
INTERROGATE, to - [gen] interroger, questionner
INTERROGATION - [gen] interrogation *f*
[radar] (transmission of a radio-frequency pulse to trigger a transponder) interrogation *f*
INTERROGATOR - [radar] (combined transmitter and receiver used to interrogate a transponder and display the replies) interrogateur *m*
INTERRUPT, to - [gen] interrompre, suspendre
INTERRUPTED CADENCE - [mus] cadence *f* feinte
~CONTINUOUS WAVE TELEGRAPHY (ICWT) - [telecomm] (system of telegraphic transmission in which a continuous wave modulated at audio frequency is keyed in telegraph code) télégraphie *f* à ondes entretenues modulées
~CONTINUOUS WAVES - [radio] ondes *f*pl entretenues modulées
~RINGING - [telecomm] (of a telephone) courant *m* d'appel cadencé
~SPOT WELDING - [metall] soudage *m* par points à impressions
~THREAD - [mech] (screw thread having axial grooves in male and female elements, to allow them to be engaged without screwing them together for the whole travel) vis *f* à filet interrompu
~TOOTH TAP - [mech] taraud *m* à denture interrompue

INTERRUPTED TWILL - [text] sergé *m* interrompu
INTERRUPTER GEAR - [aero] (mechanism preventing the firing of a gun when any part of the airscrew is the line of fire) mécanisme *m* d'interruption du tir
INTERRUPTING RATING - [electron] capacité *f* de coupure, puissance *f* de coupure
INTERRUPTION - [gen] interruption *f*
INTERRUPTOR - [el etc] interrupteur *m*, rupteur *m* de courant
~CONTACT - [el] contact *m* du coupe-circuit
INTERSECT, to - [gen] intersecter, entrecouper
INTERSECTING COMB - [text] barrette *f* d'intersecting
~GILL-BOX - [text] intersecting *m* à double champ de barrettes
~PLANE - [geol] plan *m* d'intersection
INTERSECTION - [gen] intersection *f*
[surv] (a method of plane table surveying) recoupement
[gen] (of roads) carrefour *m*, croisement *m* de chemins
~ANGLE - [surv] (angle of deflexion) angle *m* d'intersection
~POINT - [surv] (the point at which the straights of a rail- or highway curve would meet) point *m* d'intersection, point *m* de recoupement
INTERSPACE - [gen] intervalle *m*
[mech] espacement *m*
INTERSPECIFIC CROSSING - [bot] croisement *m* interspécifique
~HYBRID - [bot] hybride *m* interspécifique
~HYBRIDIZATION - [bot] hybridation *f* interspécifique
INTERSPERSE, to - [gen] entremêler, parsemer
INTERSTAGE CONDENSER - [el] condenseur *m* auxiliaire, condenseur *m* intermédiaire
~COOLING - [ind chem] refroidissement *m* intermédiaire
~COUPLING - [radio] circuit *m* de couplage entre étages
~RESERVOIR - [ind chem] réservoir *m* intermédiaire
INTERSTATION NOISE SUPPRESSION - [radio] (silent tuning) réglage *m* silencieux
INTERSTELLAR - [astr] interstellaire
~SPACE - [astr] espace *m* interstellaire
INTERSTICE - [gen] (a small space or cavity within the structure of a substance) interstice *m*
INTERSTICES - [geol] (pores) espaces *mpl* interstitiels, vides *mpl*
INTERSTITIAL - [gen] interstitiel
~ATOM - [nucl] (atom in an interstitial position) atome *m* interstitiel
~COMPOUNDS - [nucl] (compounds in which the metalloid atoms occupy the interstices between the atom and the metal lattice) composés *mpl* interstitiels
~POSITION - [cristall] position *f* interstitielle
~RADIATION - [radiat] (irradiation by sources introduced into the tissues) irradiation *f* interstitielle
~TISSUE - [zool] tissu *m* interstitiel
~WATER - [oil ind] (natural gas) eau *f* interstitielle
INTERSTRATIFIED TUFF - [geol] tuf *m* interstratifié
INTERTRACK BOND - [railw] (conductor electrically connecting the rails of separate tracks) jonction *f* électrique de voies
~TIME DISPLACEMENT - [comput] décalage *m* dans le temps
INTERTRACTION - [chem] (the increase in density of a colloidal solution) contraction *f* interne
INTERTRIPPING - [el] déclenchement *m* interdépendant
INTERTROPICAL - [geogr] intertropical
~FRONT - [met] (the border line between the trade-wind belts of the Northern and Southern hemispheres. See also "Doldrums") front *m* intertropical
INTERTWINE, to - [mech etc] (to twist together) entrelacer
INTERURBAN - [gen] interurbain
INTERVAL - [gen] intervalle *m*
[el] (in a winding) intervalle *m*
[acoust] (between two values of the same variable) intervalle *m*
~BETWEEN KNOPS - [text] écartement *m* des boutons
~ERROR - [contr] (the difference obtained by subtracting the indication error at one of two scale marks from that at the other) erreur *f* d'intervalle
~-SELECTOR CIRCUIT - [radio] (a coincidence circuit) circuit *m* discriminateur de temps
~SIGNAL - [radio] (a signal, usually a musical signal, transmitted in the intervals) signal *m* de repos
INTERVALVE COUPLING - [electron] (undesired coupling between two adjacent electronic tubes in a circuit) couplage *m* entre tubes
~TRANSFORMER - [el] (transformer operating between the anode circuit of a valve and a grid of another valve) transformateur *m* intermédiaire
INTERVENE, to - [gen] intervenir, s'interposer
INTERWEAVE, to - [text] enlacer, entrelacer, enchevêtrer, tresser
INTERWEAVE THREAD - [text] fil *m* d'entrelacement, fil à entrelacer
INTESTINAL - [anat] intestinal
~OCCLUSION - [med] occlusion *f* intestinale
~TRACT - [anat] voies *fpl* intestinales
INTESTINE - [anat] intestin *m*
INTOLERANCE - [gen & biol] intolérance *f*
INTONATION - [mus] intonation *f*, cadence *f*
INTOXICANT - [gen] boisson *f* enivrante
INTRA - [a prefix meaning within] intra
INTRAANNULAR TAUTOMERISM - [chem] (the redistribution of bonds in a ring of atoms) tautomérie *f* intraannulaire
INTRACAVITARY APPLICATOR - [radiat] (applicator introduced into a cavity of the human body) applicateur *m* intercavitaire
~IRRADIATION - [radiat] irradiation *f* intercavitaire
INTRACAVITY X-RAY THERAPY - [radiat] röntgenthérapie *f* intercavitaire
INTRACELLULAR PARTICLE - [biol] particule *f* intercellulaire
INTRACERVICAL - [anat] intracervical
INTRADOS - [constr] (the lower surface of an arch) intrados *m*, ligne *f* d'intrados *m*
INTRAGRANULAR FRACTURE - [metall] cassure *f* intragranulaire
INTRAMOLECULAR - [phys] intramoléculaire
~FORCES - [phys] (forces within the molecules) forces *fpl* intramoléculaires
INTRAMUSCULAR - [anat] intramusculaire
INTRANUCLEAR FORCE - [nucl] (the force between proton and proton, neutron and neutron, or proton and neutron) force *f* intranucléaire
INTRASYSTEM INTERFERENCE - [comput] perturba-

tions *f*pl à l'intérieur
INTRINSIC - [gen] intrinsèque
~ANGULAR MOMENTUM - [phys] (the angular momentum which is associated with axial rotation) moment *m* cinétique intrinsèque
~BRILLIANCE - [light] (brightness) luminosité *f*
~COERCIVE FORCE - [el] (magnetizing force) champ *m* coercitif intrinsèque
~COUNTER EFFICIENCY - [radiat] (the proportion of particles reaching the sensitive part of a counter) sensibilité *f* intrinsèque de compteur
~CRYSTAL - [phys] cristal *m* idiochromatique
~ENERGY - [chem] (the store of energy of a material system) énergie *f* intrinsèque
~INDUCTION - [el] (magnetic polarization) polarisation *f* magnétique
[comput] aimantation *f* intrinsèque
~PROPERTIES - [gen] proprietés *f*pl intrinsèques
[electron] (the properties of a semiconductor which are characteristic of the ideal crystal) proprietés *f*pl intrinsèques
~REACTANCE - [el] (the imaginary part of intrinsic impedance) réactance *f* intrinsèque
~RESISTANCE - [el] (the real part of intrinsic impedance) résistance *f* intrinsèque
~SEMICONDUCTOR - [electron] (a substance in the energy diagram of which the gap between the normal gap and the adjacent excitation band is narrow (approx. one electron-volt) semiconducteur *m* intrinsèque
~SPEED - [mech] débit *m* intrinsèque
~TEMPERATURE RANGE - [electron] (semiconductors) région *f* intrinsèque de température
~THROUGHPUT - [mech] débit *m* massique intrinsèque (ou théorique)
~TRANSCONDUCTANCE - [electron] (in a transistor) transconductance *f* intrinsèque
~VISCOSITY - [phys] (the limiting value at infinite dilution of the ratio of the specific viscosity of a polymer solution to its concentration in primary moles per litre) viscosité *f* intrinsèque
INTRODUCE, to - [gen] introduire
INTRODUCTION - [gen] introduction *f*, avant-propos *m*
INTROITUS - [anat] (the entry into a cavity of the body) entrée *f*
INTRORSE - [gen & bot] (towards the centre) introrse
INTROVERSION - [med] (of an organ) invagination *f*
INTRUSION - [geol] intrusion *f*, injection *f*
INTRUSIVE - [geol] (said of igneous rocks intruded into preexisting rocks of the earth's crust) intrusif
~SHEET - [geol] nappe *f* intrusive
INTUBATION - [med] (the introduction of tube through the larynx to convey air into the lungs) tubage *m*
[metall] tubage *m*
INTUBATOR - [med instr] dispositif *m* de tubage (du larynx)
INTUMESCENCE - [gen & med] intumescence *f*, enflure *f*
[plast ind] (synonym for swelling; applied particularly to plastics when subjected to high temperatures) gonflement *m*
INTUMESCENT - [med] intumescent, tuméfié
~COATING - [plast ind] (type of coating which swells and forms bubbles when exposed to intense heat, used to make the coated body selfextinguishing) revêtement *m* intumescent
INULIN - [chem] (a polysaccharide occurring in the tubers of dahlias and Jerusalem production of foodstuffs for diabetics) inuline *f*
INULINASE - [chem] (enzyme hydrolizing inulin) inulinase *f*
INUNCTION - [med] onction *f*, onguent *m*
INUNDATE, to - [gen] inonder
INUNDATION - [gen] inondation *f*, débordement *m*
INVADE, to - [gen] envahir
INVAGINATION - [zool] (insertion into a sheath) invagination *f*
INVALID - [gen & med] invalide *m*, malade *m*
INVALIDATE, to - [gen & leg] invalider, rendre nul
INVAR - [metall] (iron-nickel alloy) invar *m*
INVARIABLE PLANE - [astr] (plane completely unchanged by any mutual action between the planets) plan *m* invariable
INVARIANT - [phys] (without degree of freedom; the ways in which a system may change in respect to its configuration) invariant *m*
INVASION - [gen] invasion *f*
[agric] (of insects, weeds etc) envahissement *m*
[bot] (the movement of plants from a area to another) envahissement *m*
~BY WEEDS - [bot] envahissement *m* par les mauvaises herbes
~COEFFICIENT - [chem] (factor used to indicate the number of millilitres of a gas absorbed by one square cm. of surface in one minute) facteur *m* de absorption spécifique
INVENT, to - [gen] inventer
INVENTION - [gen] invention *f*, découverte *f*
INVENTORY, to - [gen & comm & leg] inventorier
INVENTORY - [gen & comm] inventaire *m*
~ASSETS - [comm] stocks *m* comptables
~CONTROL - [comm] contrôle *m* d'inventaire
~STORE - [comput] mémoire *f* d'inventaire
INVERSE - [gen] inverse
~-AGITATION TANK - [photo] boîte *f* de développement à agitation par renversement
~ANNEALING - [metall] recuit *m* inverse
~CHILL - [metall] trempe *f* inverse
~CURRENT - [electron] (forward current; resulting from a forward voltage) courant *m* inverse
~DIRECTION - [electron] (the direction of greater resistance to current flow through the cell) direction *f* inverse, sens *m* non conducteur
~ELECTRODE CURRENT - [electron] (the current flowing through an electrode in the direction opposite to that for which the tube is designed) courant *m* inverse d'électrode
~FUNCTION - [math] fonction *f* inverse
~GATE - [comput] circuit *m* inverseur, circuit *m* NE PAS
~HARDENING - [metall] trempe *f* negative
IMPEDANCES - [el] impédances *f*pl inverses
~LEAKAGE CURRENT - [electron] courant *m* de fuite en sens inverse
~NETWORK - [el] (two terminal networks are called inverse when the product of their impedance is independent of frequency within the range wanted) réseau *m* inverse
~PHOTOELECTRIC EFFECT - [electron] (the emission of photons which is due to the impact of electrons) effet *m* photoélectrique inverse
~PIEZOELECTRIC EFFECT - [el] (the mechanical strain in certain asymmetric crystals in an electric

field) effet *m* piézoélectrique inverse
INVERSE RELATION TELEMETER - [instr] appareil *m* de télémesure de proportionalité à inversion
~SQUARE LAW - [math] loi *f* de l'inverse des carrés
~TIME LAG RELAY - [el] relais *m* à retard inverse
~VOLTAGE - [el] (voltage developed between anode and cathode in a thermionic valve in a direction opposite to that of normal anode flow) tension *f* inverse
INVERSELY - [gen] inversement
~PROPORTIONAL - [math] inversement proportionnel
INVERSION - [gen] inversion *f*, renversement *m*
[chem] (the hydrolisis of a dextrorotatory solution of sucrose to produce a laevorotatory solution of fructose and glucose) inversion *f*
[met] (of the usual temperature gradient) inversion *f* atmosphérique
[telecomm] (the inversion of the order of speech frequencies) inversion *f*
~LAYER - [met] couche *f* d'inversion
~OF A COMMON CHORD - [mus] accord *m* parfait inversé
INVERT, to - [gen] renverser, invertir
[gen] (to turn in the opposite direction) retourner, mettre à l'envers
INVERT - [chem] inverti
[plumb] (of a non-vertical sewer) fond *m* de l'égout, radier *m* de l'égout
~SUGAR - [ind chem] (an equimolecular mixture of fructose and glucose obtained by the hydrolysis of sucrose) sucre *m* inverti, sucrase *f*
INVERTASE - [chem] (an enzyme with uses as an analytical reagent and in the production of invert sugars) invertase *f*
INVERTED - [gen chem etc] inverti
~AMPLIFICATION FACTOR - [electron] (the ratio of differential change in grid voltage) facteur *m* d'amplification de seuil à anode négative
~AMPLIFIER - [radio] amplificateur *m* de seuil à anode négative
~ARC LAMP - [light] (with the positive carbon above the negative carbon) lampe *f* à arc à électrodes de charbon inverties
~ARCH - [constr] (like the floor of a tunnel) arc *m* renversé, radier *m*
~BRUSH CONTACT - [el] (type of laminated switch contact) contact *m* à balais invertis
~BURNER - [gas ind] (gas lighting) bec *m* renversé
~CROSSTALK - [telecomm] diaphonie *f* inintelligible
~CYLINDER ENGINE - s. enverted engine
~ENGINE - [mech] (type of engine in which the cylinders are placed below the crankshaft) moteur *m* inversé
~FIELD PULSES - [telev] impulsions *f*pl renversées de trame
~FLIGHT - [aero] (flight in which the aircraft is upside down) vol *m* sur le dos, vol *m* renversé
[aero] (wing loading condition which simulates that prevailing when an aircraft is flying upside down) cas *m* de vol sur le dos
~FOLD - [geol] pli *m* renversé
~FRAME PULSES - s. inverted field pulses
~HOUR - [nucl] (inhour; measure of reactivity) inhour *f*
~IMAGE - [opt] image *f* renversée
~-IMAGE RANGE FINDER - [instr] télémètre *m* à images renversées
INVERTED JET - [mech] jet *m* inversé
~L AERIAL - [radio] antenne *f* à L renversé
~L ANTENNA - s. inverted L aerial
~LOOP - [aero] (a loop executed with the upper surface of the aircraft outward) looping *m* à l'envers
~MORDENT - [mus] battement *m*
~NORMAL LOOP - [aero] (a manoeuvre beginning with inverted flight, followed by dive, normal flight, climb and finally inverted flight again) looping *m* normal à l'envers
~OUTSIDE LOOP - [aero] (a manoeuvre beginning with inverted flight, normal flight and dive, ending in inverted flight again) looping *m* extérieur à l'envers
~PLEAT - [text] pli *m* creux
~POSITION OF A COMMON CHORD - s. inversion of a common chord
~RECTIFIER - [el] (rectifier so arranged that it converts d.c. to a.c.) onduleur *m*
~ROTARY CONVERTER - [el] (rotary converter used to convert d.c. to a.c.) convertisseur *m*
~SADDLE - [geol] selle *f* inverse
~SPEECH - [telecomm] parole *f* démodulée
~SPIN - [aero] (a spin in which the mean angle of incidence is negative) vrille *f* inversée
~VAULT - [constr] voûte *f* renversée
INVERTER - [el] (device for converting direct current to alternating current) onduleur *m*
INVERTOR - s. inverter
INVESTIGATE - [gen] examiner, étudier, enquêter
INVESTIGATION - [gen] investigation *f*, recherche *f*
INVESTMENT - [gen] investissement *m*
[fin] (of a capital etc) placement *m*, mise *f* de fonds
~CASTING - [metall] moulage *m* à cire perdue
INVISCID - [phys] non visqueux
INVISIBILITY - [gen] invisibilité *f*
INVISIBLE - [gen] invisible
~MENDING - [text] stoppage *m* invisible
INVITATION TO BID - s. invitation to tender
~TO TENDER - [comm] sollitation *f* des soumissions
INVOICE, to - [comm] facturer
INVOICE - [comm] facture *f*
INVOLUCRUM - [zool] enveloppe *f* membraneuse
INVOLUNTARY - [gen] involontaire
INVOLUTE - [gen & bot] involuté, involutive
[el mech etc] à développante, de développante de cercle
~CONNEXION - [el] (curved end connexion for the winding of an electric motor)connexion *f* à développante
~CURVE - [math] développante *f*
~GEAR TEETH - [mech] (wheel teeth with a flank profile consisting of an involute curve) engrenages *m*pl à développante
INVOLUTION - [math] (the evaluation of an exponential quantity) involution *f*, élevation *f*
~FORM - [med] forme *f* régressive
INVOLVE, to - [gen] envelopper, impliquer, engager
[math] élever (à une puissance)
INVULNERABILITY - [gen] invulnérabilité *f*
INVULNERABLE - [gen] invulnérable
INWARD - [gen] intérieur, interne
~BOARD - [telecomm] (of telephone exchange) positions *f*pl d'arrivée
~-FLOW TURBINE - [mech] (turbine in which the gases flow radially inward) turbine *f* centripète

INWARD SETTING - [mech] (in autos, of the front wheels) pincement *m* des roues avant
INWARDS - [gen] en dedans, vers l'intérieur
IODARGYRITE - [min] (a mineral, essentially silver iodide) iodargyrite *f*
IODATE - [chem] (salt of iodic acid) iodate *m*
IODESIN - [chem] (an analytical reagent) iodésine *f*
IODIC ACID - [chem] (an analytical reagent and a starting material in the preparation of iodates) acide *m* iodique
~ ANHYDRIDE - [chem] (forming iodic acid when dissolved in water) anhydride *m* iodique
IODIDE - [chem] iodure *m*
IODINATED - [chem] iodé
IODINE - [chem] (a poisonous, non-metallic element, A.N. 53, A.W. 126.92, symbol I, with uses as a catalyst, as an intermediate for pharmaceuticals, dyes, and photographic chemicals) iode *m*
~ ABSORPTION - s. iodine number
~ ACNE - [med] acné *m* iodique
~ INDEX - s. iodine number
~ MONOBROMIDE - [chem] (a synthesis intermediate) monobromure *m* d'iode
~ MONOCHLORIDE - [chem] (an analytical reagent) monochlorure *m* d'iode
~ NUMBER - [chem] (the weight in grams of iodine which will combine with an usaturated compound under standard conditions. It is a measure of the degree of saturation of a compound) indice *m* d'iode
~ PENTAFLUORIDE - [chem] (formed by the direct combination of fluorine and iodine) pentafluorure *m* d'iode
~ SOLUTION - [chem] solution *f* iodique
~ TEST - [chem] essai *m* à l'iode
~ TINCTURE - [chem] (a solution of iodine and potassium iodide in alcohol, used as an antiseptic) teinture *f* d'iode
~ TRICHLORIDE - [chem] (a source of iodine and chlorine in organic synthesis, and a strong disinfectant) trichlorure *m* d'iode
~ VALUE - s. iodine number
IODISM - [med] (condition resulting from overdosage of iodine) iodism *m*
IODIZED OIL - [chem] (an X-ray contrast medium) huile *f* iodée
IODODERMA - [med] iodide *f*
IODOFORM - [chem] (a weak antiseptic, analogous to chloroform in composition) iodoforme *m*
IODOL - [chem] iodole *m*
IODOMETRY - [chem] iodométrie *f*
IODOSUCCINIMIDE - [chem] (a source of iodine in organic synthesis) iodosuccinimide *f*
IOLITE - [min] (silicate of aluminium, iron and magnesium) iolite *f*
ION - [phys] (an electrically charge atom) ion *m*
~ ACCELERATION - [nucl] accélération *f* d'ions
~ ACCELERATOR - [nucl] (apparatus for accelerating ions) accélérateur *m* d'ions
~ ACCEPTOR - [nucl] (substance which accepts an ion) accepteur *m* d'ions
~ ACTIVITY - [phys] (the thermodynamic concentration of a specific type of ion) activité *f* ionique
~ AVALANCHE - [electron] (group of ions freed by cumulative ionization) avalanche *f* ionique
~ BEAM - [electron] (beam of ions emitted from a single source) faisceau *m* d'ions
~-BEAM SCANNING - [nucl](the analysis of a mass spectrum of an ion beam) balayage *m* d'un faisceau ionique
ION BURN - [electron] (the dark area caused by ion burning appearing in cathode-ray tubes with magnetic deflection) tache *f* ionique
~ BURNING - [electron] (the destruction of the active material of the luminescent screen under the bombardment of its surface by the negative ions formed in the tube) brûlure *f* ionique
~ CHAMBER - s. ionization chamber
~ CLUSTER - [nucl] (the ions which are close together at the end of the path of an ionizing particle) essai *m* d'ions
~ COLLECTION CHAMBER - [nucl] chambre *f* à collection ionique
~ CONCENTRATION - [phys] (the number of gramme ions (of a given type) per unit volume of electrolyte) concentration *f* ionique
~ COUNTER - [nucl] (cylindrical ionization chamber used for the measurement of ionization of the air) compteur *m* d'ions
~ DENSITY - [nucl] (the number of ion pairs per unit volume) densité *f* ionique, nombre *m* volumique d'ions
~ DETECTOR - [phys] détecteur *m* déparateur d'ions
~-DIPOLE INTERACTION - [chem] réaction *f* réciproque ion-dipôle
~ ENGINE - [astronaut] moteur *m* ionique
~ EXCHANGE - [chem] (reversible reaction involving an ion exchanger which can be regenerated by a low-cost solution) échange *m* ionique, échange *m* d'ions
~ EXCHANGE RESINS - [chem] (synthetic resins having active groups enabling the resin to combine with or exchange ions with a solution) résines *f*pl échangeuses d'ions
~ EXCHANGER - [chem] échangeur *m* d'ions
~ EXCLUSION - [chem] (ability of certain synthetic resins to absorb non-ionised solutes) exclusion *f* d'ions
~ FLOW - [nucl] (the flow of groups of ions) flux *m* ionique
~ FLUX - [phys] courant *m* d'ions
~ GUN - [electron] (device similar to an electron gun in which the charged particles are ions) source *f* d'ions
~-ION RECOMBINATION - [nucl] (a basic model for recombination) recombinaison *f* ion-ion
~ LIMIT - [nucl] (in a cloud chamber) limite *f* de parcours d'ionisation
~ MOBILITY - [el chem] (the speed reached by an ion under the action of an electric field equal to one volt per centimetre) mobilité *f* ionique
~ MOBILITY ISOTOPE SEPARATION - [nucl] séparation *f* d'isotopes par mobilité ionique
~-MOBILITY METHOD - [nucl] (process of separation of isotopes based on the difference of mobility in different ions) méthode *m* de la mobilité ionique
~ PAIR - [phys] (a positive ion and a negative ion having charges of the same magnitude) paire *f* de ions
~-PAIR YIELD - [nucl] rapport *m* M/N, rendement *m* de paires d'ions
~ PUMP - [el mech] pompe *f* ionique
~ SHEATH - [electron] (film of ions on a surface) gaine *f* d'ions

ION SORPTION PUMP - [el mech] pompe *f* ionique à sorption
~SOURCE - [phys] (device for generating ions to be accelerated) source *f* d'ions
~SPOT - [telev] (localized deterioration of the screen of a cathode-ray tube caused the bombardment of heavy ions) tache *f* ionique
~TARGET - [phys] collecteur *m* d'ions
~THERAPY - [el biol] (the forcing of ions through biological interfaces by means of an electric field) ionothérapie *f*
~TRAJECTORY - [phys] (the path of an ion in a beam) parcours *m* d'un ion
~TRANSFER - s. ion therapy
~TRANSIT TIME - [phys] temps *m* de transit d'un ion
~TRAP - [electron] (a device preventing ion burns by removing the ions from the beam) piège *m* d'ions
~TUBE - [electron] tête *f* de manomètre à ionisation
~VELOCITY - s. ionic velocity
IONIC - [chem] ionique
[arch] (of a column) ionique
~BOND - [nucl] (valence linkage in which two atoms are held together by electrostatic forces) liaison *f* ionique
~CENTRIFUGE - [nucl] (an instrument used for the electromagnetic separation of isotopes) centrifugeur *m* ionique
~CHARGE - [chem] charge *f* ionique
~COMPOUND - [chem] (a readily ionizing compound) composé *m* ionique
~CONDUCTANCE - [el] conductance *f* ionique
~CONDUCTION - [electron] (the continuous movement of charges within a substance caused by the displacement of ions in a crystal lattice) conduction *f* ionique
~CRYSTAL - [nucl] (crystal effectively consisting of ions bound together by their electrostatic attraction) cristal *m* ionique
~CURRENT - [el] (current which is produced by the movement of ions) courant *m* ionique
~EQUILIBRIUM - [chem] équilibre *m* ionique
~FORMULA - [phys] (the formula of an ion) formule *f* d'un ion
~GAUGE - [instr] manomètre *m* à ionisation
~HEATED CATHODE - [electron] cathode *f* à chauffage ionique
~LOUDSPEAKER - [el acoust] (loudspeaker depending for its operation on the interaction between an ionic plasma and the surrounding air) haut-parleur *m* ionique
~MICELLES - [phys] (aggregates of ions with characteristic properties) micelles *fpl* ioniques
~MICROPHONE - [el acoust] (microphone depending for its operations on the interaction between an ionic plasma and the surrounding air) microphone *m* ionique
~MIGRATION - [el] (the movement of charged particles of an electrolyte toward the electrodes under the influence of electric current) migration *f* des ions
~MOBILITY - [phys] (the ratio between the average drift velocity of an ion in a solution and the electric field) mobilité *f* d'un ion
~POLARIZABILITY - [el] polarisabilité *f* ionique
~POTENTIAL - [el] rapport *m* charge-rayon d'un ion
~SEMICONDUCTOR - [electron] semiconducteur *m* ionique

IONIC STRENGTH - [chem] (a measure of the intensity of the electric field in a solution; obtained by multiplying the square of the valency of each ion in a solution by its activity and taking half the sum of the terms thus obtained) force *f* ionique
~VALENCE - [electron] (electrostatic valence) électrovalence
~VALVE - [electron] tube *m* ionique
~VELOCITY - [el chem] (the speed reached by an ion in an electric field) vitesse *f* d'un ion
~WIND VOLTMETER- [instr] voltmètre *m* à vent ionique
IONICITY - [phys] (term denoting a fractional ionic character of the molecule in its grouns state) probabilité *f* d'excitation
IONIUM - [nucl] (an isotope of thorium; it is radioactive and has a half-life of 8 x 10^4 years. Symbol Io) ionium *m*
~AGE - [nucl] âge *m* de radium
IONIZATION - [nucl] (the dissociation of a molecule to form ions) ionisation *f*
~BY COLLISION - [nucl] ionisation *f* par choc, ionisation *f* par collision
~CHAMBER - [nucl] (instrument designed to measure quantities of ionizing radiation) chambre *f* d'ionisation
~COUNTER - [nucl] compteur *m* à ionisation
~CROSS-SECTION - [nucl] section *f* d'ionisation
~DOSEMETER - [instr] (a dosemeter depending for its action on the ionization of a gas) dosimètre *m* à ionisation
~ELECTROMETER - [instr] électromètre *m* d'ionisation
~ENERGY - [nucl] (the minimum energy sufficient to ionize an atom or molecule) énergie *f* d'ionisation
~EVENT - [nucl] (the occurrence of a process in which an ion or a group of ions is produced) evenement *m* d'ionisation
~FOAMING - [plast ind] (foaming of polyethylene by subjecting it to irradiation which evolves hydrogen from the material itself) production *f* de mousse par ionisation
~GAGE - s. ionization gauge
~GAUGE - [instr] (instrument measuring the residual gas pressure in the tube by using the rate of collection of positive ions on one electrode of a triode vacuum tube) manomètre *m* à ionisation
~MEAN FREE PATH - [radiat] libre parcours *m* moyen d'ionisation
~PATH - [nucl] (the trail of ion pairs produced by an ionizing particle passing through matter) parcours *m* d'ionisation
~POTENTIAL - [nucl] (the electrical energy suffecient to give an electron the minimum energy necessary to ionize by impact another atom or molecule) potentiel *m* d'ionisation
~RATE - [nucl] vitesse *f* d'ionisation
~SPECTROMETER - [instr] spectromètre *m* à cristal
~TIME - [nucl] (the time interval between the beginning and the establishment of conduction of a stated value of tube-storage drop) temps *m* d'ionisation
~TRACK - s. ionization path
IONIZE, to - [phys] (to produce ions) ioniser
IONIZED ATOM - [nucl] atome *m* ionisé
IONIZING ENERGY - [nucl] (the average energy which is lost by an ionizing particle in producing an ion pair in a gas) énergie *f* d'ionisation
~MEDIUM - [chem] (an agent used to promote ioni-

zation) agent *m* ionisant
IONIZING PARTICLE - [nucl] particle producing ion pairs in its passage through matter) particule *f* ionisante
~RADIATION - [nucl] (high energy particles from radioactive sources with the ability, on collision, to remove electrons from other atoms) rayonnement *m* ionisant
IONOGENIC - [phys] ionogène
IONOMETER - [instr] (instrument designed to measure intensity of X-rays) ionomètre *m*
IONONE - [chem] (a synthesis intermediate for Vitamin A and a component of perfumes) ionone *f*
IONOSPHERE - [phys] (region in the atmosphere extending from approximately 50 km to 600 km from the earth's surface and consisting of layers of ionized gas) ionosphère *f*
IONOSPHERIC - [phys] ionosphérique
~CROSS-MODULATION - [phys] (el magnet waves; the Luxembourg effect) effet *m* Luxembourg, effet *m* Tellegen
~DISTURBANCE - [radio] perturbation *f* ionosphérique
~STORM - [radio] (prolongued type of ionospheric disturbance causing a general deterioration of communication conditions) tempête *f* ionosphérique
~WAVE - [phys] (wave propagated by way of the ionosphere) onde *f* ionosphérique
IONOTHERAPY - s. ion therapy
IONTOPHORESIS - s. ion therapy
IPECACUANHA - [chem] (the dried root of Cephaelis ipecacuanha with uses in medicine and in the manufacture of confectionery) ipécacuanha *m*
I.R. DROP - [chem] (value obtained by multiplying the current passing through an electrolytic call by the resistance of the latter) chute *f* de tension ohmique
IR-WIRE - [comput] (information-read-wire) fil *m* de lecture d'information
IRIDAUXESIS - [med] hypertrophie *f* de l'iris
IRIDECTOMY - [med] iridectomie *f*
~STENOPEIC - [med] iridectomie *f* sténopéique
IRIDEMIA - [med] hémorragie *f* de l'iris
IRIDESCENCE - [phys] (the production of doclours on a surface due to the interference of light) iridescence *f*
IRIDESCENT - [phys] iridescent, irisé
~CLOUD - [met] (high clouds showing delicate colours) nuage *m* irisé
~COLOUR - [dyes] couleur *f* changeante
IRIDIC CHLORIDE - [chem] (an analytical reagent) chlorure *m* iridique
IRIDIUM - [chem] (a metallic element, symbol Ir, A.N. 77, A.W. 193.1 used in jewellery, electronics, etc) iridium *m*
~POTASSIUM CHLORIDE - [chem] (a ceramic glaze) chlorure *m* d'iridium et de potassium
~SEQUIOXIDE - [chem] (a ceramic glaze) sesquioxyde *m* d'iridium
IRIDOAVULSION - [med] excision *f* de l'iris
IRIDOCERATITIS - [med] iridokératite *f*
IRIDOCINESIA - [med] mouvement *m* de l'iris
IRIDOCOLOBOMA - [med] colobome *m* de l'iris
IRIDOCYCLITIS - [med] irite *f*, iridocyclite *f*
IRIDODESIS - [med] iridodésis *f*
IRIDODIALYSIS - [med] iridodialyse *f*
IRIDOMALACIA - [med] ramollissement *m* de l'iris
IRIDOMEDIALYSIS - [med] iridomédialyse *f*
IRIDOPARALYSIS - [med] iridoplégie *f*
IRIDORHEXIS - [med] iridorrhexie *f*
IRIDOSCLEROTOMY - [med] iridosclérotomie *f*
IRIDOSMINE - [metall] (a native alloy of platinum and osmium with uses as a bearing material in precision instruments and as a hardening agent for platinum) iridosmine *f*
IRIDOTOMY - [med] irodotomie *f*
IRINITE - [min] (variety of loparite containing approximately II% of thorium) irinite *f*
IRIS - [anat] (of the eye) iris *m*
[min] (a form of quarz) pierre *f* d'iris
[radio] (of a waveguide) diaphragme *m* iris
~DIAPHRAGM - [opt] (an optical device consisting of overlapping thin plates guided by pins moving in slots so as to give a nearly circular opening of adjustable diameter) diaphragme *m* iris
~DIASTASIS - [med] iridodiastase *f*
~IN - [photo] (in a camera) fermeture *f* graduelle du diaphragme
~LENS DIAPHRAGM - [anat] diaphragme *m* iridocristallinien
~OIL - [chem] huile *f* d'iris
~OUT - [photo] (in a camera) ouverture *f* graduelle du diaphragme
~PAIN - [med] douleur *f* irienne
~SHADOW - [med] ombre *f* portée sur l'iris
IRISED PRINT - [text] impression *f* irisée, impression iris
IRISH HARP - [mus] guimbarde *f*
~MOSS - [chem] (dried and powdered seaweed (Chondrus) with uses as a thickening agent) mousse *f* perlée, mousse *f* d'Irlande
IRISOPSIA - [med] irisopsie *f*
IRON, to - [gen] ferrer
IRON - [metall] (a metallic element, symbol Fe, A.N. 26, A.W. 55.84, important as a constructional material, both as cast iron and in the form of steel) fer *m*
[impl] fer *m* (à repasser)
[metall] (made of cast iron) en fonte, métallique
~ACETATE LIQUOR - [chem] (a dyeing mordant) pyrolignite *m* de fer
~ALLOYS - [metall] ferr-alliages *mpl*
~ALUM - [min] (halotrichite) alun *m* de fer, halotrichite *f*
~-ALUMINIUM GARNET - [min] grenat *m* aluminoferreux
~ARC - [opt] (between iron electrodes used for spectrometer and spectrograph calibrations) arc *m* à fer
~ARCHED BRIDGE - [constr] pont *m* à arc en fer
~BAR - [metall] barre *f* de fer
~BARK WOOD - [bot] bois *m* de gommier
~BLOCK - [mech] poulie *f* ferrée
~-BOUND - [gen] cerclé de fer
~BRACE - [mech] entretoise *f* en fer
~BRACKET - [metall] console *f* en fonte
~BUFF - [dye] (a yellow dye for cotton materials) jaune *m* de fer
~CARBIDE - [chem] carbure *m* de fer
~-CARBON ALLOY - [metall] alliage *m* fer-carbone
~-CARBON DIAGRAM - [metall] diagramme *m* fer-carbone
~CARBONATE - [chem] carbonate *m* de fer
~CASTING - [metall] pièce *f* en fonte, moulage *m* de

fonte
IRON CASTINGS - [metall] pièces *fpl* en fonte
~CEMENT - [metall] mastic *m* de fer
~CHAIN - [mech] tirant *m*
~CHAMBER - [metall] (of a reverberatory furnace) laboratoire *m*
~CHILL - [metall] lingotière *f*
~CIRCUIT - [el] circuit *m* fer
~COMPOUNDS - s. ferrous & ferric
~CONTENT - [chem] teneur *f* en fer
~CORE - [metall] noyau *m* de fer
~CORE CHOKE - [el] bobine *f* à noyau magnétique
~CORE MAGNETIC CIRCUIT - [el] circuit *m* magnétique à noyau en fer
~CROSS TWIN - [crystall] macle *f* de la croix de fer
~DUST CORE COIL - [el] bobine *f* à noyau de fer comprimé
~FIDDLE - [mus] (nail violin) violon *m* de fer
FILINGS - [metall] limaille *f* de fer
~FITTINGS - [railw] attachés *fpl* des rails
~FOUNDRY - [metall] fonderie *f* de fer
~FRAMEWORK - [constr] treillis *m* en fer
~FROM WASHERY - [metall] fonte *f* provenant des lavoirs
~GLANCE - [min] (iron ore) eisenglimmer *m*, hématite *f*, spécularite
~GOSSAN - [mining] chapeau *m* de fer
~GRATING - [constr] grillage *m*
~GREY - [dye] gris *m* de fer
~LOSS - [el] (energy loss caused by alternating flux in the iron of the magnetic circuit) pertes *fpl* dites dans le fer
~LUNG - [med] poumon *m* d'acier
~METALLURGY - [metall] sidérurgie *f*
~METEORITES - [geol] (common name for meteorites consisting essentially of nickel-iron) météorites *fpl* ferreuses
~MIKE - [aero] (US colloq. for automatic pilot) pilote *m* automatique
~MINE - [min] mine *f* de fer
~MONOXIDE - [chem] protoxyde *m* de fer
~-MOULD - [metall] lingotière *f*
~MUFFLE - [metall] moufle *m* en fer
~-NICKEL STORAGE BATTERY - [el chem] (alkaline storage battery having electrodes in which nickel hydroxide predominates in the positive and iron alloy in the negative) accumulateur *m* au fer-nickel
~NITRATE - [chem] nitrate *m* de fer
~NOTCH - [metall] trou *m* de coulée
~OCHRE - [paint] ocre *m* de fer
~ORES - [min] (deposits containing iron-rich compounds) minerais *mpl* de fer
~OXIDE PIGMENTS - [paint] (generic term for many pigments which consist chiefly of oxides or iron, e.g. umber, sienna, ochre, vandyke brown, Persian Gulf red and others) pigments *mpl* d'oxyde de fer
~OXIDE PROCESS - [chem] (a method for the removal of sulphides from a gas) procédé *m* à l'oxyde de fer
~OXIDE REDS - [paint] (pigments for rubber, anticorrosion and priming paints) oxydes *mpl* de fer rouges
~OXIDE YELLOWS - [paint] (a series of pigments based on hydrated ferric oxide and having good lightfastness) oxydes *mpl* de fer jaunes
~PAN - [min] (hard layer frequently occurring in sands gravels) carapace *f* de fer
IRON PATTERN - [metall] (pattern made of cast-iron and used when large numbers of castings are required) modèle *m* en fer
~PAVING - [constr] (type of road surface) platelage *m* en fer, pavé *m* en fer
~PENTACARBONYL - [chem] (a syntesis intermediate) fer *m* pentacarbonyle
~PIG - [metall] gueuse *f* de fonte, saumon *m* de fer
~PIGLETS - [metall] petites gueuses *fpl* de fonte
~-PLACER - [min] minière *f* de minerais de fer de alluvion
~-PLATING - [metall] tôlerie *f*
~POTASSIUM TARTRATE - [chem] (a source of physiological iron) tartrate *m* ferrico-potassique
~PUDDLING - [metall] puddlage *m* du fer
~PYRITEX - [min] (brassy-yellow natural sulphide of iron, crystallizing in the cubic system) pyrite *f* de fer, pyrite *f* jaune
~RAILING - [constr] grille *f*, clôture *f* en fer
~RED - [paint] oxyde *m* de fer
~RING - [mech] douille *f* en fer
~ROD - [metall] (used for construction work) tige *f* de fer, tringle *f* de fer
~ROLLING MILL - [metall] laminoir *m*
~RUNNER - [metall] rigole *f* de coulée
~SALTS - [chem] sels *mpl* de fer
~SANDSTONE - [min] grès *m* ferrugineux
~SCALE - [metall] batitures *fpl* de fer
~SCRAP - [gen] ferraille *f*
~SHEET - [metall] tôle *f* de fer
~SHEEL - [metall] manteau *m* en tôle, enveloppe *f* en tôle
~SINTER - [min] pharmacosidérite *f*
~SMELTING FURNACE - [metall] four *m* de fusion du fer
~SPAR - [min] (spathic iron) fer *m* oxydé carbonaté
~SPINEL - [min] fer *m* spinelle, hercynite *f*, pléonaste *m*
STANCHION - [constr] chandelier *m*
~STOPPER - [ind chem] mastic *m* pour fer
~STRUT - [constr] contre-fiche *f* en fer
~SULPHATE - [chem] sulphate *m* de fer
~SULPHIDE - [chem] sulfure *f* de fer
~TASTE - [brew ind] goût *m* de fer
~TIE - [constr] traverse *f* en fer
~TOOLING - [metall] usinage *m* de fer
~TUBING - [metall] (wrought iron or steel tubing used for boiler-tubes) tuyaux *mpl* pour chaudières
~TURBIDITY - [brew ind] trouble *m* de fer
~WIRE - [metall] fil *m* de fer
~WIRE MESH - [metall] treillage *m* en fil de fer
~WORKS - [ind] usine *f* sidérurgique
IRONCLAD, to - [metall] cuirasser
IRONCLAD - [metall] cuirassé, à enveloppe de fer
~ELECTROMAGNET - [el] (electromagnet in which the return path for the flux is formed by an iron covering which surrounds the winding) électroaimant *m* cuirassé
~FURNACE - [metall] (used for mercury ore) four *m* pour minerai de mercure
~HEADSTOCK - [mach tool] poupée *f* blindée
~MOTOR - [el] moteur *m* électrique cuirassé
~SHAFT - [metall] four *m* à minerais de mercure
~SWITCHGEAR - [el] (metalclad switchgear) interrupteur *m* à enveloppe de fer
IRONER - s. ironing machine

IRONING - [gen] repassage *m*
~DUMMY - [text] mannequin *m* de repassage
~MACHINE - [gen] machine *f* à repasser, calandre *f*
~PRESS - [text] presse *f* à repasser
IRONMONGERY - [gen] quincaillerie *f*, ferronerie *f*
IRONSTONE - [min] (carbonate of iron, clay and carbonaceous matter) minerai *m* de fer argileux
IRONWARE - [gen] s. ironmongery
IRONWORK - [ind proc] construction *f* en fer, charpente *f* en fer
IRRADIANCE - [opt] (radiant flux density) éclairement *m* énergétique
IRRADIATE, to - [phys opt etc] (to subject to radiation) irradier, éclairer
IRRADIATED - [gen] rayonnant
[nucl] (of a material which has been exposed to radiation, or bombarded) irradié
~PLASTIC - [plast ind] (plastic substance which has been subkected to cathode-ray bombardment) plastique *m* irradié
IRRADIATION - [phys & nucl] radioexposition *f*, irradiation *f*
~CREEP - [metall etc] fluage *m* d'irradiation
~HARDENING - [metall] durcissement *m* par irradiation
~REACTOR - [nucl] réacteur *m* d'irradiation
IRRATIONAL - [gen] irraisonnable, irrationnel
[math] (which cannot be measured by means of ordinary quantities) irrationnel
~DISPERSION - [opt] dispersion *f* irrationnelle
~NUMBER - [math] nombre *m* irrationnel
IRREGULAR - [gen etc] irrégulier
~BEAMING - [text] ensouplage *m* irrégulier
~COUNT OF REED - [text] compte *m* du peigne irrégulier
~CURVE - [draw] courbe *f* irrégulière
~DISPOSITION OF WEFT - [text] tramage *m* irrégulier
~DISTORTION - [telecomm] distorsion *f* irrégulière
~FLOATS - [text] flottés *mpl* irréguliers
~OCCURRENCE OF THE OIL - [oil ind] venue *f* irrégulière du pétrole
~ORE FORMATION - [min] gîte *m* de substitution
~ROUGHNESS - [plumb] rugosité *f* d'asperité
~RUNNING - [text] marche *f* irrégulière (du métier)
~SATEEN - [text] satin *m* brisé
~WEFT - [text] trame *f* irrégulière
~YARN - [text] fil *m* irrégulier
~YARN TENSION - [text] tension *f* irrégulière du fil
IRREGULARITY - [gen] irrégularité *f*, inégalité *f*
IRREGULARLY PITCHED REED - [text] peigne *m* à écartement de dents irrégulier
~SPUN YARN - [text] fil *m* irrégulier
IRREPARABLE - [gen] irréparable
IRRESPECTIVE - [gen] indépendant
IRRESPECTIVELY - [gen] indépendamment
IRREVERSIBLE - [gen] irrévocable, irréversible
[mech] (any mechanism which transmits power in one direction only) irréversible
~ACTION - [chem] action *f* irreversible
~COAGULATION - [chem] (in rubber technology) coagulation *f* irréversible
~CONTROLS - [contr] (type of control system in which forces imposed on the control surfaces cannot be transmitted back to the pilot) commandes *fpl* irréversibles
~GEL - [chem] (a gel which cannot be liqufied) gel *m* irréversible
IRREVERSIBLE MECHANISM - [mech] (any mechanism which operates in one direction or sense only) mécanisme *m* unidirectionnelle
~PROCESS - [el chem] (an electromechanical process during which electrode overvoltage occurs) processus *m* irreversible
~REACTION - [chem] (on which can take place in one direction only) réaction *f* irréversible
~STEERING - [auto] (steering mechanism by which road shocks cause no motion on the steering wheel) direction *f* irréversible
IRRIGATE, to - [gen & agric] irriguer
IRRIGATED FARMING - [agric] culture *f* irriguée
~LAND - [agric] terrains *mpl* irrigués
~SOIL - [soil] sol *m* irrigué
IRRIGATION - [agric] irrigation *f*
~BY FLOODING - [agric] irrigation *f* par déversement
~BY OVERDAMMING - [agric] irrigation *f* par submersion
~CANAL - [hydr] canal *m* d'irrigation
~DITCH - [agric] fossé *m* d'irrigation
~FROM THE SEPTIC TANK - [agric] irrigation *f* fertilisante
~FURROW - [agric] fossé *m* d'irrigation
~GALLERY - [agric] galerie *f* d'irrigation
~PLANT - [agric] appareils *mpl* d'arrosage
~PUMP - [agric] pompe *f* d'arrosage
~RESERVOIR - [agric] réservoir *m* d'irrigation
~RILL - [agric] rigole *f* d'écoulement
~STORAGE - [agric] stockage *m* d'eau pour l'irrigation
~SYSTEM - [agric] système *m* d'irrigation
~WATER - [hydr] eau *f* d'irrigation
~WORKS - [agric] installations *fpl* d'irrigation
IRRIGATOR - [impl] arroseuse *f*, machine *f* à arroser
[med appl] irrigateur *m*, seringue *f* à injection
IRRITABILITY - [gen] irritabilité *f*
[el biol] (electrical excitability; the inherent ability of a tissue to start its specific reaction in response to an electrical current) irritabilité *f* électrique
IRRITANT - [med] irritant *m*
IRRITATION - [gen] irritation *f*
[biol] stimulation *f*
~THERAPY - [el biol] thérapeutique *f* de choc, thérapeutique *f* stimulante
IRROTATIONAL - [phys etc] (said of a fluid motion) irrotationnel
~FIELD - [el] (field in which the circulation is everywhere zero) champ *m* irrotationnel
~FIELD MOTION - [phys] movement *m* irrotationnel
~FLOW - [phys] (flow in a region of zero vorticity) écoulement *m* irrotationnel
~FLUID MOTION - s. irrotational flow
IRRUMATION - [med] coït *m* buccal
IRRUPTIVE - [geol] intrusif
ISABEL - [zool] isabelle *m*
ISALLOBAR - [met] (a line passing through all points at which changes of pressure have been the same in a given period) isallobare *f*
ISALLOBARIC FIELD - [met] (an area over which a complex of isallobars extends) aire *f* isallobarique
ISANOMAL - [phys] (an isogram of anomaly, i.e. departure from the local mean value of a characteristic from the mean local norm for it) isanomale

f, ligne *f* d'égale anomalie
ISANOMALOUS LINE - s. isanomal
ISATIN - [chem] (an intermediate for pharmaceuticals, and an analytical reagent) isatine *f*
ISCHEMIA - [med] ischémie *f*
ISCHIDROSIS - [med] s. ischemia
ISCHIORECTAL ABSCESS - [med] phlegmon *m* de la fosse ischiorectable
ISCHURIA - [med] ischurie *f*, rétention *f* d'urine
I SCOPE - [radar] (used to indicate range and direction) indicateur *m* type I
ISENERGIC FLOW - [phys] (flow in which the sum of potential and kinetic energies, and the enthalpy is always constant) écoulement *m* isenergique
ISENTROPIC - [phys] (taking place without change in entropy) isentrope
~CHANGE - s. isentropic transformation
~FLOW - [phys] (flow in which the entropy of all the fluid elements is constant) écoulement *m* isentrope
~TRANSFORMATION - [phys] (transformation which takes place without change of entropy) transformation *f* isentrope
ISETHIONIC ACID - [chem] (an intermediate for detergents) acide *m* iséthionique
ISINGLASS - [min] isinglass *m*
[chem] (a gelatin obtained from the swim bladders of fish and used as a clarifying agent, especially for beer) colle *f* de poisson, ichtyocolle *f*
ISLAND - [geogr] île *f*
[of a road roundabout] île *f* de sécurité, refuge *m*
~EFFECT - [electron] (the restriction of the emission from the cathode to certain small areas of it (islands) in the case of the grid voltage being lower than a certain value) effet *m* d'îlot
~FLAP - [med] greffe *f* en îlots
ISLANDS OF ISOMERISM - [chem] îles *f*pl d'isomérie
ISLET - [geogr] îlot *m*
I S O - [International Standard Organization] Organisation *f* Internationale pour l'Unification
ISO - [chem] (prefix signifying the isomer of a compound) iso
ISOACTINIC CURVE - [opt] courbe *f* isoactinique
ISOALLOXAZINE - [chem] (intermediate for dyestuffs and pharmaceuticals) isoalloxazine *f*
ISOAMYL ACETATE - [chem] (a component of perfumes and synthetic flavourants) acétate *m* d'isoamyle
~ALCOHOL - [chem] (intermediate for pharmaceuticals, a photographic chemical, and a solvent) alcool *m* d'isoamyle
~BENZOATE - [chem] (a component of perfumes and flavouring agents) benzoate *m* d'isoamyle
~BENZYL ETHER - [chem] (a perfumery agent) isoamylbenzyléther *m*
~BUTYRATE - [chem] (a solvent and plasticizer for cellulose derivatives) butyrate *m* d'isoamyle
~CHLORIDE - [chem] (a solvent and synthesis intermediate) chlorure *m* d'isoamyle
~SALICYLATE - [chem] (a component of perfumes) salicylate *m* d'isoamyle
~VALERATE - [chem] (a component of artificial flavourants) valérianate *m* d'isoamyle
ISOAMYLENES - [chem] (starting materials in the production of isoprene) iso-amylènes *m*pl
ISOBAR - [chem] (one of several nuclides having the same mass number) isobare *f*
[met] (line drawn on a map through places with the same atmospheric pressure) ligne *f* isobare
ISOBARIC - [met] isobarique
~NUCLEUS - [nucl] (nuclide with the same number of nucleus in their nuclei) noyau *m* isobarique
~CHART - [met] carte *f* isobarique, carte *f* isobarométrique
~SPACE - [nucl] (isotopic space; symbolic space in which certain orientations occur) espace *m* isobarique, espace *m* isotopique
~SPIN - [nucl] (the spin which is characteristic of the two quantum states) spin *m* isotopique
~SPIN QUANTUM NUMBER - [quantum theory] nombre *m* quantique de spin isotopique
~TRIAD - [nucl] triplet *m* isobarique
ISOBARS - [nucl] nucléides *m*pl isobares
ISOBATH - [impl] (a type of inkstand, in which the ink in the dipping well is kept at constant level) courbe *f* isobathe
ISOBORNEOL - [chem] (a perfumery agent) isoborneol *m*
ISOBORNYL ACETATE - [chem] (a perfumery agent) acétate *m* d'isobornyle
~SALICYLATE - [chem] (a fixative for perfumes) salicylate *m* d'isobornyle
~THIOCYANOACETATE - [chem] (a peduculicide used in medicine) thiocyanoacétate *m* d'isobornyle
ISOBUTANE - [chem] (an important synthesis intermediate; also has uses as a refrigerant) isobutane *m*
ISOBUTENE - [chem] (intermediate for synthetic resins and rubbers) isobutène *m*
ISOBUTYL ACETATE - [chem] (a perfumery agent and a solvent for cellulose derivatives) acétate *m* d'isobutyle
~ALCOHOL - [chem] (a solvent and an intermediate for flavours and perfumes) alcool *m* isobutylique
~AMINOBENZOATE - [chem] (a local anaesthetic) aminobenzoate *m* d'isobutyle
~BENZOATE - [chem] (a component of synthetic flavourants) benzoate *m* d'isobutyle
~CINNAMATE - [chem] (an ingredient of perfumes) cinnamate *m* d'isobutyle
~PROPIONATE - [chem] (a solvent for paints and varnishes) propionate *m* d'isobutyle
~SALICYLATE - [chem] (a perfumery agent) salicylate *m* d'isobutyle
ISOBUTYLAMINE - [chem] (a synthesis intermediate) isobutylamine *f*
ISOBUTYLENE - [chem] s. isobutene
ISOBUTYLUNDECYLENAMIDE - [chem] (a component of insecticides) isobutylundécylénamide *f*
ISOBUTYRALDEHYDE - [chem] (intermediate for rubber accelerators and antioxidants) isobutyraldéhyde *m* aldéhyde *m* isobutyrique
ISOBUTYRIC ACID - [chem] (intermediate in perfumery, the production of flavourants, and solvents) acide *m* isobutyrique
~ANHYDRIDE - [chem] (a synthesis intermediate) anhydride *m* isobutyrique
ISOBUTYRONITRILE - [chem] (a synthesis intermediate) isobutyronitrile *m*
ISOCETYL LAURATE - [chem] (a plasticizer, mould release agent and solvent) laurate *m* d'isocétyle
~MYRISTATE - [chem] (a perfumery fixative and a lubricant and softening agent for pharmaceuticals) myristate *m* d'isocétyle

ISOCETYL STEARATE - [chem] (a component of cosmetics and pharmaceuticals) stéarate *m* d'isocétyle
ISOCHROMATIC - [opt] (of the same colour) isochromatique
~LINE - [opt] ligne *f* isochromatique
~STIMULI - [telev] stimuli *mpl* isochromatiques
ISOCHRONAL - s. isochronous
ISOCHRONE - [radio] (line joining points associated with a constant time difference in the reception of radio aid signals) isochrone *f*
~DETERMINATION - [radio] détermination *f* isochrone
ISOCHRONISM - [phys] (regular periodicity) isochronisme *m*
ISOCHRONOUS - [phys] isochrone, isochronique
~MODULATION - [radio] modulation *f* isochrone
~SCANNING - [telev] (scanning system in which the synthesis of an image is at the same rate as the dissection) analyse *f* isochrone
ISOCLINAL FOLD - [geol] (partially overturned fold, in which both limbs dip in the same direction) pli *m* isoclinal
ISOCLINE - [geol] isoclinal *m*
ISOCLINIC LINES - [geol] lignes *fpl* isoclines
ISOCORIA - [med] isocorie *f*
ISOCYANATES - [chem] (organic compounds containing an -NCO radical) isocyanates *mpl*
ISOCYANURIC ACID - [chem] (intermediate for the production of disinfectants and bleaching agents) acide *m* isocyanurique
ISOCYCLIC COMPOUNDS - [chem] (closed chain (ring) compounds in which the ring consists entirely of carbon atoms) composée *mpl* isocycliques
ISODECALDEHYDE - [chem] (intermediate for synthetic resins,dyes and pharmaceuticals) isodécaldéhyde *m*
ISODECANOIC ACID - [chem] (intermediate for plasticizers) acide *m* isocaproïque
ISODECANOL - [chem] (a textile processing agent) isodécanol *m*
ISODECYL CHLORIDE - [chem] (resin solvent and intermediate for resins, plasticizers and synthetic rubbers) chlorure *m* isodécylique
~OCTYL ADIPATE - [chem] (a plasticizer for vinyl polymers) adipate *m* isodécyl-octylique
ISODESMIC STRUCTURE - [phys] (ionic crystal structure) structure *f* isodesmique
ISODIAPHERE - [nucl] (one of several nuclides in which the difference between the numbers of neutrons and protons in the nucleus is the same) isodiaphère *f*
ISODIMORPHISM - [crystall] (double isomorphism) isidomorphie *f*
ISODOMON - s. isodomum
ISODOMUM - [arch] (masonry with facing consisting of square stones) isodôme *m*
ISODOSE CHART - [nucl] (diagram on which lines are drawn passing through all points which receive the same dose of radiation) table *f* isodosique
~CONTOUR - [nucl] (a curve formed by the intersection of an isodose surface with a given plane) courbe *f* isodosique
~SURFACE - [nucl] (a surface on which all points receive equal doses of radiation under given conditions) surface *f* isodosique
ISODURENE - [chem] (a synthesis intermediate) isodurène *m*
ISOELECTRIC POINT - [el chem] (when the charge of a colloid is zero, the pH value is termed the Isoelectric Point) point *m* isoélectrique
ISOELECTRONIC - [nucl] (relating to similar electronic arrangements) isoélectronique
~SEQUENCE - [nucl] (a series of atoms of which the electronic configuration outside the nucleus in the same) atomes *mpl* isoélectroniques
ISOEUGENOL - [chem] (intermediate for perfumes and flavourants) isoeugénol *m*
~ETHYL ETHER - [chem] (a perfumery fixative) éther *m* éthylique de l'isoeugénol
ISOFORMING - s. isomerization
ISOGONAL - [el magn] (a line passing through all points at which magnetic variation is the same) zone *f* isogone
ISOGONIC LINE - [el] (an isogram of magnetic variation) ligne *f* isogone
ISOGRAM - [phys] (a line drawn through all points at which equal values of a given characteristic prevail) isogramme *m*
ISOHEXANE - [chem] (2-methylpentane. Used as a solvent) isohexane *m*
ISOLATE, to - [gen] isoler
[chem] isoler
[comput] (to obtain a digit from a machine word; also, to remove from a set of items all those items which meet some arbitrary condition) extraire
ISOLATED CHIMNEY - [metall] cheminée *f* indépendante
~FARM - [agric] domaine *m* isolé
ISOLATING AMPLIFIER - [electron] amplificateur *m* séparateur
~CASING - [metall] gaine *f* isolante
~LINK - [el] barrette *f* de sectionnement
~SPARK GAP - [el] (a spark-gap provided in an ignition booster cable to eliminate feedback from the magneto) éclateur *m*
~SWITCH - s. isolator
~TRANSFORMER - [el] transformateur *m* de découplage
ISOLATION - [gen] isolement *m*
~SYNDROME - [astronaut] syndrome *m* d'isolement
~SWITCH - [el] (in electric railway) sectionneur *m*, clef *f* de séparation
ISOLATOR - [electron] (special type of switch designed to interrupt of close a circuit under noload conditions only) affaiblisseur *m* non-réciproque
[el] (as above, but only when magnetizing or charging current are flowing) sectionneur *m*
[telecomm] (a buffer stage) étage *m* tampon, étage *m* intermédiaire
ISOLEUCINE - [chem] (an essential amino acid with uses in medicine and research) isoleucine *f*
ISOLOGUES - [chem] (compounds of which the molecular structure is similar, but which contain atoms of different nature and the same valency) isologues *mpl*
ISOLUX LINE - [opt] courbe *f* isophote d'éclairement
ISOMAGNETIC - [el magnet] (of lines connecting points at which a property of the earth's magnetic field is constant) isomagnétique
ISOMER - [chem] (a nuclide which has the same number of neutrons and protons as another, but can exsist for a measurable period in different quantum states with different energies and radioactive properties) isomère *m*

ISOMER SEPARATION - [chem] séparation *f* d'isomères
ISOMERIC - [chem] isomérique
~NUCLEAR LEVEL - [nucl] niveau *m* nucléaire isomérique
~STATE - [nucl] état *m* isomérique
~TRANSITION - [nucl] (a radioactive change from one nuclear isomer to another of lower energy) transition *f* isomérique
ISOMERIZATION - [chem] (the rearrangement of organic molecules to give substances with different chemical properties but which have the same number and type of molecules as the original substance) isomérisation *f*
ISOMERS - [chem] (different compounds made up to the same number and type of molecules) isomères *mpl*
ISOMETRIC - [meas] isométrique
~CURVE - [phys] courbe *mf* isométrique
~DRAWING - [draw] (method of drawing) projection *f* isométrique
ISOMORPHIC - [phys] (having similar crystalline form) isomorphe
ISOMORPHISM - [chem] (the existence of different substances possessing the same chemical composition and the same crystal form) isomorphie *f*
ISOMORPHOUS - [chem] isomorphe
~CRYSTAL - [crystall] cristal *m* isomorphe
ISONICOTINIC ACID - [chem] (an intermediate for pharmaceuticals) acide *m* isonicotinique
ISONORMOCYSTOSIS - [med] formule *f* normale (globules blancs)
ISOOCTANE - [chem] (a synthesis intermediate and a fuel for internal combustion engines) isooctane *m*
ISOOCTENE - [chem] (a synthesis intermediate) isooctène *m*
ISOOCTYL ADIPATE - [chem] (a plasticizer) adipate *m* iso-octylique
~ALCOHOL - [chem] (a solvent for resins, a synthesis intermediate, and a component of a number of plasticizers) alcool *m* iso-octylique
~ISODECYL PHTHALATE - [chem] (plasticizer) phtalate *m* d'iso-octyl-isodécyle
~PALMITATE - [chem] (rubber and secondary plasticizer for synthetic resins; extrusion lubricant) palmitate *m* isotylique
~THIOGLYCOLATE - [chem] (used as stabilizer, plasticizer and polymerization additive) thioglycolate *m* isooctylique
ISOPENTALDEHYDE - [chem] (intermediate for synthetic resins) isopentaldéhyde *m*
ISOPENTANE - [chem] (a solvent) isopentane *m*
ISOPENTANOIC ACID - [chem] (suggested as an intermediate for plasticizers and vinyl stabilizers) acide *m* isopentanoïque
ISOPHORONE - [chem] (a solvent for a number of synthetic resins) isophorone *m*
ISOPHOT CURVE - [opt] (curve of equal light intensity) courbe *f* isophote d'éclairement
ISOPHTHALIC ACID - [chem] (plasticizer and starting material for poly-urethanes and polysters) acide *m* isophtalique
ISOPHTHALOYL CHLORIDE - [chem] (intermediate for synthetic resins and fibres) bichlorure *m* de metaphtaloyle
ISOPIESTIC SOLUTIONS - [chem] (solutions exerting equal vapour pressures) solutions *fpl* à pression de vapeur égale
ISOPOLYMORPHISM - [phys] (the occurrence of two forms of a polymorphic substance being isomorphous with two forms of another polymorphic substance) isopolymorphisme *m*
ISOPRAL - [chem] (a hypnotic used in medicine) isopral *m*
ISOPRENE - [chem] (a constituent of natural rubber and a starting material for a number of synthetic resins) isoprène *m*
ISOPROPANOLAMINE - [chem] (emulsifying agent and plasticizer) isopropanolamine *f*
ISOPROPENYLACETYLENE - [chem] (a synthesis intermediate) isopropénylacétylène *m*
ISOPROPYL ACETATE - [chem] (solvent for natural and synthetic resins, nitrocellulose; other uses are in plastics and organic synthesis) acétate *m* d'isopropyle
~ALCOHOL - [chem] (solvent and raw material for acetone) alcool *m* isopropylique
~ANTIMONITE - [chem] (a cross-linking and fireproofing agent for textiles) antimonite *f* d'isopropyle
~BROMIDE - [chem] (intermediate for dyestuffs and drugs) bromure *m* d'isopropyle
~BUTYRATE - [chem] (a solvent for cellulose derivatives) butyrate *m* d'isopropyle
~CHLORIDE - [chem] (an intermediate and solvent) chlorure *m* d'isopropyle
~ETHER - [chem] (resin and rubber solvent) éther *m* isopropylique
~IODIDE - [chem] (an intermediate for pharmaceuticals) iodure *m* d'isopropyle
~MERCAPTAN - [chem] (a standard in analysis) isopropylmercaptan *m*
~METHYL PYRAZOLYL DIMETHYL CARBAMATE - [chem] (an insecticide) isopropylméthylpyrazolyldiméthylcarbamate
~MYRISTATE - [chem] (plasticizer for cellulose derivatives) myristate *m* d'isopropyle
~NITRATE STARTER - [mech] (gas turbine starter using a small turbine fed with gas from the decomposition of isopropyl nitrate) démarreur *m* à nitrate d'isopropyle
~OLEATE - [chem] (plasticizer for ethyl cellulose, cellulose nitrate and polystyrene) oléate *m* d'isopropyle
~PALMITATE - [chem] (plasticizer for cellulose nitrate and ethyl cellulose) palmitate *m* d'isopropyle
~PERCARBONATE - [chem] (a catalyst for the production of polythenes) percarbonate *m* d'isopropyle
~PEROXYDICARBONATE - [chem] (a catalyst for the production of medium density polythene) peroxydicarbonate *m* d'isopropyle
ISOPROPYLAMINE - [chem] (intermediate for rubber accelerators) isopropylamine *f*
ISOPROPYLAMINODIPHENYLAMINE - [chem] (a rubber antioxidant) isopropylaminodiphénylamine *f*
ISOPROPYLAMINOETHANOL - [chem] (a synthesis intermediate) isopropylaminoéthanol *m*
ISOPROPYLPHENOL - [chem] (an intermediate for perfumes, synthetic resins, and plasticizers) isopropylphénol *m*
ISOPROPYL PHENYLCARBAMATE - [chem] (a weedkiller) phénylcarbamate *m* d'isopropyle
ISOPULEGOL - [chem] (a perfumery component) isopulégol *m*

ISOQUINOLINE - [chem] (intermediate for dyestuffs, rubber chemicals, pesticides, and pharmaceuticals) isoquinoléine *f*
ISOSAFROLE - [chem] (intermediate for flavourants and perfumes) isosafrol *m*
ISOSCELES - [geom] isocèle, isoscèle
~TRIANGLE - [geom] triangle *m* isoscèle
ISOSEISMAL - s. isosismic
ISOSISMIC - [geol] isoséiste, isoséismique
ISOSTASY - [geol] (a condition whereby equal earth masses underlie equal areas down to an assumed level of compensation) isostasie *f*
ISOSTATIC - [phys] isostatique
ISOSTERES - [chem] (pair of compounds showing a measure of agreement in physical properties) isostères *mpl*
ISOSTERIC MOLECULE - [nucl] (a molecule which has essentially the same valence configuration as one or more others) molécule *f* isostérique
ISOSTERISM - [phys] (physical similarity of certain compounds due to their similar or identical molecular arrangement) isosterisme *m*
ISOTACTIC - [chem] (a polymer having all those atoms not structurally part of the main chain arranged either all above or all below it) isotactique
ISOTEMPERATURE LOCI - [opt] lieux *mpl* des températures égales de couleur
ISOTENSOID STRUCTURE - [phys] structure *f* à filaments à contrainte uniforme
ISOTHERE - [met] (the curve across point having the same mean summer temperature) isothère *f*
ISOTHERM - [met] (isogram of temperatures) isothèrme *f*
[phys] (process occurring at constant temperatures) processus *m* isotherme
~ADSORPTION - [thermo] adsorption *f* isotherme
ISOTHERMAL - [met] (taking place under unchanging conditions of temperature) isothermique
~ATMOSPHERE - [met] (region of the atmosphere in which temperature does not change with altitude, i. e. neither a normal lapse rate nor inversion prevails) atmosphère *f* isothermique
~COMPRESSION - [phys] (compression during which the temperature remains normal) compression *f* isotherme
~EQUILIBRIUM - [phys] équilibre *m* isothermique
~EXPANSION - [phys] (expansion during which the temperature remains constant) expansion *f* isotherme
~LINE - [phys] (a curve showing the relation between quantities measured at constant temperature) ligne *f* isotherme
ISOTONE - [nucl] (a nuclide which has the same number of neutrons in its nucleus as another, of which it is said to be an isotone) isotone *m*
ISOTONIA - [med] isotonie *f*
ISOTONIC - [phys] (having the same osmotic pressure as some other substance in question) isotonique
ISOTOPE - s. isotopes
~CASK - [nucl] (isotope container) château *m* de transport d'isotopes
~CHART - [nucl] (a chart in which the properties of atomic metal nuclei are summarized) diagramme *m* d'isotopes, table *f* d'isotopes
~CONTAINER - [nucl] (a lead-lined flask) château *m* de transport d'isotopes
~DILUTION ANALYSIS - [nucl] analyse *f* par dilution isotopique
ISOTOPE EFFECT - [nucl] effet *m* isotopique
~PRODUCTION REACTOR - [nucl] réacteur *m* de production d'isotopes
~SEPARATION - [nucl] (the technology relating to the changing of the relative abundance of isotopes) séparation *f* d'isotopes
~SEPARATION FACTOR - [nucl] facteur *m* de séparation
~SHIFT - [nucl] (the displacement of the special line of one isotope as compared with another) déplacement *m* isotopique
~SPECIFIC ACTIVITY - [nucl] activité *f* spécifique de isotope
~THERAPY - [nucl] (radiotherapy in which isotopes are used) thérapie *f* isotopique
ISOTOPES - [nucl] (varieties of an element having the same atomic number but different atomic weights) isotopes *mpl*
ISOTOPIC ABUNDANCE - [nucl] (the relative number of atoms of a given isotope in a given quantity of an element) abondance *f* isotopique, teneur *f* isotopique
~ATOMIC WEIGHT - [nucl] (the atomic weight of an isotope in comparison with the A.W. of I6.00 for the lighter isotope of oxygen) poids *m* atomique isotopique
~CARRIER - [nucl] porteur *m* d'isotope
~COMPOSITION - [nucl] (relative abundance) composition *f* isotopique
~DATING - [nucl] (age determination of a substance by the use of counters to measure the isotope age) datation *f* isotopique
~DILUTION - [nucl] (the mixing of a particular nuclide with its isotopes) dilution *f* isotopique
~EQUILIBRIUM - [nucl] (the relative abundance of the isotopes of an element in natural material) équilibre *m* isotopique
~EXCHANGE PROCESS - [nucl] (process used in the separation of heavy water from natural water) processus d'échange isotopique
~INDICATOR - s. isotopic tracer
~LEVEL GAUGE - [instr] limnimètre *m* isotopique
~NUMBER - [nucl] (the number by which the numerical quantity of neutrons in a nucleus exceeds that of the protons) excès *m* de neutrons
~RATIO - [nucl] (the ratio between the number of atoms of two isotopes of the same element) rapport *m* des teneurs isotopiques
~SPACE - s. isobaric space
~SPIN - s. isobaric spin
~SPIN QUANTUM NUMBER - s. isobaric spin
~TRACER - [nucl] (a radionuclide or an allobar used as a chemical tracer for the element of which it is an isotope) traceur *m* isotopique
ISOTRON - [nucl] (a device for separating isotopes by electrical methods) isotron *m*
ISOTROPE - s. isotropic
ISOTROPIC - [phys] (of a substance possessing the same properties in all directions) isotropique
~AERIAL - [radio] antenne *f* sphérique
~ANTENNA - s. isotropic aerial
s. isotropic radiator
~BODY - s. isotropic medium
~DIELECTRIC - [el] diélectrique *m* isotrope
~MEDIUM - [nucl] (one of which the properties are the same in any direction) substance *f* isotropique

ISOTROPIC RADIATOR - [phys] radiateur *m* sphérique
[radio] antenne *f* isotrope, antenne *f* sphérique
~SCATTERING - [nucl] (scattering in which there is no preferential direction) diffusion *f* isotropique
~SOIL - [soil] sol *m* isotrope, terrain *m* isotrope
~SOURCE OF RADIATION - [nucl] (emitting equally in all directions) source *f* isotropique de rayonnement
~STRESS - [geol] tension *f* isotrope
ISOTROPY - [phys] (the lacking of any predetermined axes) isotropie *f*
ISOVALERALDEHYDE - [chem] (applications in rubber accelerators and synthetic resins) isovaléraldéhyde *m*
ISOVALERIC ACID - [chem] (a component of synthetic flavours and perfumes) acide *m* iso-valérianique
ISSUE, to - [gen] sortir, découler
[print] (to publish) publier
[hydr] (of water)jaillir, découler
[gen] (a permit) émettre, fournir
ISSUE - [gen] sortie *f*, décharge *f*, issue *f*
[hydr] déversoir *m* d'un barrage
[print] publication *f*, édition *f*
[comm] émission *f*
[fin] (delivery) délivrance *f*
[genet] progeniture *f*, descendance *f*
[geogr] (of a river) embouchure *f*
ISTHMUS - [geogr] isthme *m*
ITACONIC ACID - [chem] (copolymerization modifier and a plasticizer component) acide *m* itaconique
ITALIAN CLOTH - [text] drap *m* italien, satin *m* de Chine
~CLOWER - [agric] trèfle *m* incarnant, trèfle *m* anglais
~RYE-GRASS - [agric] ray-grass *m* d'Italie
ITALIC - [print] italique
ITCH - [med] démangeaison *f*, prurit *m*
ITEM - [gen] article *m*
[comm] écriture *f*
~COUNTER - [comput] (a machine part) compteur *m* d'operations
ITEMIZE, to - [gen] détailler
ITERATED FISSION EXPECTATION - s. iterated fission probability
~FISSION PROBABILITY - [nucl] (the value, after a long period, of the number of fissions per generation time occurring in a critical assembly from the daughter neutrons of a given neutron)espérance *f* de descendance, espérance *f* de fission itérée
ITERATION INDEX - [comput] index *m* d'itération
~LOOP - [comput] boucle *f* d'itération
ITERATIVE ATTENUATION COEFFICIENT - [radio] affaiblissement *m* itératif
~ATTENUATION CONSTANT - s. iterative attenuation coefficient
~IMPEDANCE - [rel] (of a two terminal pair network) impédance *f* itérative
~INSTRUCTIONS - [comput] (instructions which is repeated) instruction *f* réitérée
~PHASE CHANGE COEFFICIENT - [radio] (the imaginary part of the iterative transfer coefficient) coefficient *m* de déphasafe itératif
~PHASE-CHANGE CONSTANT - s. iterative phase-change coefficient
~TRANSFER COEFFICIENT - [el] constante *f* de transfert itérative
~TRANSFER CONSTANT - s. iterative transfer coefficient
ITINERANT INSTRUCTOR - [agric] moniteur *m* itinérant
~TEACHER - [gen] s. itinerant instructor
ITINERARY - [gen] itinéraire *m*
IVORY - [zool] (the dentine of teeth, in particular the dentine of elephant tusks) ivoire *m*
~BLACK - [paint] (pigment formerly made from scrap ivory, now from bone black) noir *m* d'ivoire
[chem] noir *m* d'ivoire
~BOARD - [paper man] carton *m* Bristol, carte *f* Bristol
IW-WIRE - [comput] (part of a magnetic-core storage) fil *m* d'inscription
IXODIASIS - [med] fièvre *f* à tique
IZOD TEST - [metall] (impact or notched-bar test) essai *m* d'Izod

J

J - [chem] (symbol for yellow) (symbole de) jaune
[meas] (symbol for 'gram equivalent weight') symbole de gramme-équivalent
[mech] (symbol denoting the polar moment of inertia of a shaft) symbole du moment d'inertie polaire
[heat] (symbol for Joule's equivalent) symbole de l'équivalent mécanique de la chaleur
[el] Joule
J ACID - [chem] (an intermediate for dyestuffs) acide *m* j
J AERIAL - [radio] (half-wave aerial having the form of a J) antenne *f* à j
J ANTENNA - s. Jaerial
J DISPLAY - [radar] (a modification of the range amplitude display) indicateur *m* type j
J SCOPE - s. J display
JACINTH - [min] (or hyacinth; red variety of transparent zircon) jacinthe *f*
JACK, to - [mech] soulever avec un cric, soulever avec un vérin
JACK - [mech] (a lifting device acting beneath the object raised) cric *m*, vérin
[telecomm] (the unit of the vertical face of a telephone switchboard for the insertion of plugs) jack *m*, fiche *f*
[text] (cotton spinning) retordoir *m*
[text] (cotton machine) onde *f*
(of a knitting machine) tringle *f* à crochet
[constr] demi-brique *f*, deux quartiers *mpl*
[impl] chèvre *f*, chevalet *m*
[naut] barre *f* de cacatois
[mus] (part of the harpsichord) sautereau *m*
[min] fausse galène *f*
[mining] récipient *m* métallique
~AND CIRCLE - [mech] cric *m* à crémaillère
~ARCH - [arch] (a flat arch, i.e. an arch having an intrados with no curvature) arc *m* en platebande
~BAR - [text] barre *f* à ondes
~BED - [text] fonture *f* à ondes
~BIT - [impl] taillant *m* amovible
~BOARDS - [min] (supporting boards in well sinking operations) piliers *mpl*
~BOX - [aero] boîte *f* des jacks
~CLUTCH - [mech] embrayage *m* à dents
~DOOR - [text] portillon *m* pour l'échange d'aiguilles et de platines
~ENGINE - [impl] petit cheval *m*
~FIELD - [telecomm] (a field of jacks) panneau *m* de jacks
~FISHING - [fish] (fishing at night by means of a jack-light) pêche *f* au brocheton
~FRAME - [text] (in cotton spinning, the last of a set of four fly frames, used to prepare rovings for the spinning of various counts) banc *m* à broches en surfin
JACK GEAR - [mech] engrenage *m* de renvoi
~GUIDE PLATE - [text] tôle *f* de guidage, barre *f* de guidage
~HAMMER - [impl] marteau *m* brise-béton
[mining] marteau-perforateur *m*
~HANDLE - s. jack lever
~-IN-THE-BOX - [mech] (also 'jack-in-a-box') (a lifting jack) vérin *m*
[instr] (an instrument with a small powerful screw) vérin *m* à vis
[mech] (a differential gear) différentiel *m*
[mech] (sun and planet motion) engrenage *m* épicycloidal
~-KNIFING - [mining] écrasement *m* de boisage
~LAMP - [mining] lampe *f* Davy
~LEG - [ind chem] (only USA; overflow pipe in G.B. a draw-off at the bottom of a process vessel so as to maintain a specified liquid level) tuyau *m* de trop-plein
~LEVER - [mech] levier *m* de manoeuvre du cric
~LUG - [auto] ergot *m* d'accrochage du cric
~-O'-LANTERN - [met] (ignis fatuus, or will-o'-the-wisp) feu *m* follet
~PIT - [min] (small secundary pit) petit puits *m* auxiliaire
~PLANE - [impl] riflard *m*, galère *f*, demi-varlope *f*
~PULLEY - [text] poulie *f*, moufle *f*
~RAFTER - [constr] empannon *m*, chevron *m* de croupe
~ROVING FRAME - [text] banc *m* pour la torsion des mèches
~-SCREW - [mech] vérin *m*
~SELECTOR - [text] clavette *f*, poussoir *m*, jack *m*
~-SHAFT - [mech] (term often applied to an intermediate shaft) arbre *m* secondaire, arbre *m* de renvoi
[mech] (of a stamp mill) arbre *m* de relevage
~-SINKER - [text] platine *f* cueillante standard
~SPRING - [text] ressort *m* oscillant, ressort *m* de ondes
~STRIP - [telecomm] réglette *f* de jacks
~TRUSS - [arch] (of the roof) ferme *f* secondaire
~UP, to - [mech] (to lift by means of a jack) soulever au cric
JACKAL - [zool] chacal *m*
JACKANAPES - [mech] (used for mining machinery; guiding pulleys) galets *mpl* de renvoi
JACKBIT - [impl] jackbit *m*, couronne *f* de forage
JACKET - [gen] jaquette *f*, veston *m*
[mech] (hollow structure surrounding another part, so as to provide an enclosed space round it) chemise *f*, tôle *f* d'enveloppe
[el] (a coating of non-conductive material) envelop-

pe *f*, garniture *f*
JACKET - [mech] (outer casing surrounding a fluid container in order to maintain a temperature constant) chemise *f*
[gen & print] couverture *f*
[firearms] manchon *m*
[med impl] corset *m*, jaquette *f*
[auto] (steering column tube) jupe *f* de colonne de direction
[nucl] (thin container for a fuel slug to prevent the escape of fission products) gaine *f*
~-BAND - [railw] (of a locomotive boiler) cercle *m* de fixation de l'enveloppe
~-COOLING - [mech] (a method of cooling an enclosed space by means of another closed space surrounding it wholly or partially, a cooling medium being circulated through the outer space) refroidissement *m* par chemise réfrigérante
~ CORE - [metall] (in foundry work) noyau *m* tubulaire, noyau *m* tubulaire cylindrique
~ SPACE - [auto] chambre *f* de réchauffage
JACKETED - [gen] (furnished with a closed space surrounding another) garni de chemise, à double chemise
~ CASTING - [metall] jet *m* enveloppé
~ FEED TUBE - [mech] tube *m* d'ascension à double paroi
~ HOT AIR PRESS - [rubber ind] (a vulcanizing machine consisting of a press heated by a jacket in which hot air is circulated)presse *f* à chemise de vulcanisation à l'air chaud
~ LAMP - [light] lampe *f* à chemise
~ SYPHON - s. jacketed feed tube
~ TROUGH - [plast ind] cuve *f* à double enveloppe
~ VACUUM TANK - [ind chem] recipient *m* à double paroi pour le vide
JACKETING - [el] (covering of electric wires with plastic for mechanical protection, as distinct from electrical insulation) gainage *m*
JACKHAMMER - [mining] (compressed-air hammer drill for rock-drilling) marteau *m* pneumatique, marteau-piqueur *m*
JACKING MOTION - [text mach] étirage *m* supplementaire par le chariot
~ PAD - [mech] plaque *f* de soulèvement
~ POINT - [mech] (position at which jacks can be applied under an aircraft) point *m* d'application du cric, point *m* de levage
~ UP - [mech] soulèvement *m* sur cric
JACKNIFE RIG - [oil ind] installation *f* à antenne
JACKRABBIT - [oil ind] calibre *m* "entre"
JACKSTRAW - [glass man] (unorientated chopped glass fibre) silionne *f* coupée
JACOB'S LADDER - s. jack ladder
JACOBSITE - [min] (oxide of magnesium, iron and manganese) jacobsite *m f*
JACONET - [text] jaconas *m*
JACQUARD - [text] (textile machine used in conjunction with a loom to operate the shedding and to control the figuring) jacquard *m*
~ CARD - [text] carton *m* jacquard
~ CARD CUTTER - [text] perceuse *f* de cartes à dentelles, perceuse *f* de cartons jacquard
~ CARD PUNCHER - [text] piqueur *m* de cartons jacquard
~ CYLINDER - [text] prisme *m* jacquard, cylindre *m* à cartons
JACQUARD DESIGN - [text] dessin *m* jacquard
~ DRUM - [text] cylindre *m* jacquard
~ FABRIC - [text] tissu *m* à la jacquard
~ FRAME - [text] bâti *m* du jacquard
~ GAUZE - [text] gaze *f* au jacquard
~ GAUZE HEALD - [text] lisse *f* gaze à la jacquard
~ HEALD - [text] lissette *f*, maille *f* de corps
~ LOOM - [text ind] métier *m* jacqurd
~ MACHINE - [text] machine *f* jacquard
~ MAIL - [text] s. jacquard heald
~ PATTERN - [text] dessin *m* jacquard
~ TRESTLE - [text] s. jacquard frame
~ WEAVE - [text] tissage *m* au jacquard
~ WITH EXTRA FINE PITCH - [text ind] mécanique *f* Lacasse
~ WITH X HOOKS - [text] mécanique *f* jacquard de x crochets
JACTATION - [med] jactation *f*, anxiété *f*
JACTITATION - s. jactation
JACUPIRANGITE - [min] (a type of nepheline-gabbro consisting of various minerals, including iron-ores) jacupirangite *f*
JAD, to - [mining] haver, souscaver
JAD - [mining] havée *f*
JADE - [min] (generic name for natural metalsilicates of sodium, iron, aluminium, calcium and magnesium used as a semi-precious stone in jewellery) jade *m*
JADEITE - [min] (metasilicate of sodium and aluminium) jadéite *f*
JAG, to - [gen] denteler, ébrécher
[mech] mater
~-BOLT - [mech] (rag-bolt; foundation bolt with a long tapered head) boulon *m* de scellement à crans
~ RESISTANCE - [rubber ind] (of tyres) ténacité *f* de entaille
JAGGED - [gen] dentelé, entaillé, déchiqueté
JAGUAR - [zool] jaguar *m*
JALAP - [bot] jalap *m*
~ RESIN - [chem] (a mixture of glycosidal resins obtained from jalap and used as a purgative) résine *f* de jalap
JALOUSIE - [build] (hanging or sliding shutters) jalousie *f*
JAM, to - [gen] serrer, presser, coincer
[mech] coincer, caler, bloquer, se gommer
[radio] (to interfere with a transmission) déranger brouiller
[gen] (road traffic) obstruer, causer un encombrement
[of yarn, films out of a box etc] bourrer
[naut] (of a rope) étriver
[mech] (of a machine parts) se coincer, s'engager, se gommer
[mech] (the operation of forcing in, wedging in) coincer
[comput] (of punched cards) bourrage *m*
[food] confiture *f*
[road traffic] encombrement *m*, embouteillage *m*
[mech] collage *m*, serrement *m*, coincement *m*
~ CIRCUIT - [comput] circuit *m* d'antibourrage
~ NUT - [mech] contre-écrou *m*
~-PROOF - [mech] à l'abri du blocage
~ THE BRAKES, to - [mech] serrer les freins à bloc
~ UP, to - [mech] (of a machine) s'arrêter, se bloquer, s'engager
~ UP - [mech] blocage *m*, coincement *m*

JAM WELDING - [metall] (butt welding) soudure *f* par encollage
JAMB - [constr] (the side of an opening) montant *m* (de fenêtre), jambage *m*
[constr] (of a door) chambranle *f*, dosseret *m*
[constr] (the vertical section of a door frame) montant *m*
[constr] tableau *m* d'une baie, ébrasement *m* poteau *m* de croisée hollandaise, montant *m*
~ LINING - [constr] ébrasement *m* en bois
~ LININGS - [carp] (the panelling at the sides of a window recess) chambranle *m*
~ OF A CHIMNEY - [constr] paroi *f* de cheminée, jambage *m* de cheminée *f*
~ POST - [carp] (an upright member on the side of a door opening) poteau *m* d'huisserie
~ STONE - [constr] (the stone forming the upright side of an opening in a wall) jambage *m* en pierre
JAMESONITE - [min] (a natural sulphide of lead and antimony, crystallizing in the orthorhombic system) jamesonite *f*
JAMMER - [metall] (a spring chaplet) support *m* de noyau de forme
JAMMING - [mech] blocage *m*
[of brakes] (undesired sudden gripping of brakes) freinage *m* brusque, blocage *m*
[radio] (interference with a transmission) brouillage *m*, parasite *m*, interference *f*, perturbation *f* radiophonique
[of a machine, a firearms etc] enrayage *m*, enrayement *m*
[mech] (of a valve) coincement *m*
~ ROLLER - [mech] (a roller designed to lock two other parts together) cylindre *m* de blocage
~ TRANSMITTER - [radio] perturbateur *m*
JAPAN, to - [paint] (the operation of covering metal slate etc with a copal-oil varnish which is then stoved) vernir
JAPAN - [paint] (a glossy black enamel; based on asphaltum) vernis *m* du Japon
~ CAMPHOR - [chem] camphre *m*
~ WAX - [min] (a naturally-occurring wax containing a large percentage of palmitin. It is obtained from sumach) cire *f* du Japon
JAPANESE - [gen] japonais
~ CURTAIN - [text] rideau *m* japonais
~ LAQUER - s. Japan
~ PAPER - [paper man] (japanese hand-made paper prepared from the mulberry-bark) papier *m* Japon
JAPANNED LEATHER - [leather ind] cuir *m* verni
JAPANNING - [paint] vernissage *m*
JAR, to - [mining] secouer, vibrer
JAR - [gen] récipient *m*, verre *m*, vase *m*
[el] (vessel containing plates and electrolyte) bac *m*
[mining] coulisse
[ind chem] bac *m*
[mech] choc *m*, secousse *f*, contre-coup *m*
[acoust] son *m* discordant
~ -DOWN SPEAR - [oil ind] arrache-coulisse *m*
~ KNOCKER - [oil ind] récupérateur *m*
~ MOULDING MACHINE - [metall] machine *f* de moulage à secousses
~ RAMMING - [mech] serrage *m* par secousses
~ RING - [rubber ind] anneau *m* pour bocal de conserves
~ SAFETY JOINT - [oil ind] outil *m* de repêchage
~ SOCKET - [oil ind] arrache-coulisse *m*
JAR STEM - [oil ind] corps *m* de la coulisse de forage
~ WAGON - [railw] (carboy wagon) wagon-jarres *m*
JARGON - [gen] jargon *m*, argot *m*
[acoust] baragouin *m*
[min] (translucent variety of mineral zircon found in Ceylon) jargon *m*, jacinthe *f* citrine
~ APHASIA - [med] (unintelligible speech due to an injury in the brain) jargonaphasie *f*, paraphasie *f* littérale
JAROSITE - [min] (hydrated sulphate of iron and potassium) jarosite *f*
JARRAH - [bot] (dense, deep-red Australian wood) jarrah *m*
JARRING - [mech] ébranlement *m*, trépidation *f*
JASMINE - [bot] jasmin *m*
~ OIL - [chem] (an essential oil obtained from the flowers of Jasmium grandiflorum and used in perfumery) essence *f* de jasmin
JASPER - [min] (a kind of precious stone) jaspe *m*
JATRORRHIZINE - [chem] (alkaloid of the isoquinoline group) jatrorrhizine *f*
JAUNDICE - [med] (yellow coloration of the skin due to an excess of bile pigments) jaunisse *f*, ictère *m*
JAUNE BRILLIANT - [paint] (an artists 'pigment consisting of a mixture of cadmium yellow' vermilion and white lead) jaune *m* brillant
~ D'OR - [paint] (alternative name for Martius yellow, the calcium derivative of naphthalene yellow) jaune *m* d'or
JAVELIN-SHAPED FUEL ROD - [nucl] (fuel rods used in fast breeder reactors) barre *f* combustible en forme de javelin
JAVELLE WATER - [chem] (eau de Javelle, an aqueous solution of potassium hypochlorite and potassium chloride used as a disinfectant and bleaching agent) eau *f* de Javel
JAW - [anat] mâchoire *f*
[mech] mâchoire *f*, mors *m*, bec *m*, gorge *f*
[impl] (of a vice) (pair of members between which an object is held) mâchoire *f*, branles *f*pl
[mach tool] (of a chuck) griffe *f*
[mech] (of engine or motor) chape *f*, griffe *f*
~ BOLT - [mech] boulon *m* à becs
~ BREAKER - s. jawbreaker
~ CHUCK - [mech] poupée *f* à mâchoires, plateau *m* à griffes
~ CLUTCH - [mech] accouplement *m* à crabots
~ CRUSHER - [mech mach] (coarse-crushing machine in which a massive jaw swings back and forth, nipping the material treated against a stationary jaw) broyeur *m* à mâchoires, concasseur *m* à mâchoires
~ DOG - [mech] toc *m* à coussinets
~ JERK - [med] réflexe *m* massétérin
~ OF A CHUCK - [mech] mâchoire *f* de mandrin
~ OF A CONNCTING ROD - [mech] bras *m* de bielle
~ OF THE TONGS - [mech] mâchoire *f* des tenailles
~ ROPE - [naut] (the rope fastening the two prongs of the boom or gaff) cavité *f* des gorges
~ -SOCKET - [mech] mordache *f*
JAWBREAKER - [gen] (a word difficult to pronounce) mot *m* difficile à prononcer
[mech] concasseur *m* à mâchoires
JAWS - [of a vice) (the parts of a vice which grip the work) mâchoires *f*pl, mors *m*, mords *m*
~ OF A CLAMP - [mech] mâchoires *f*pl d'une pince

JAY - [zool] geai *m*
JEAN - [text] (heavy cotton-twill material) coutil *m*, treillis *m*
~BACK VELVETEEN - [text] velours *m* de coton à fond croisé
JEEP - [auto] (compact vehicle with a powerful traction) jeep *m*
JEJUNECTOMY - [med] jéjunectomie *f*
JEJUNITIS - [med] jéjunite *f*
JEJUNOCECOSTOMY - [med] jéjunocaecostomie *f*
JEJUNOCOLOSTOMY - [med] jéjunocolostomie *f*
JEJUNOILEITIS - [med] jéjuno-iléite *f*
JEJUNOILEOSTOMY - [med] jéjuno-iléostomie *f*
JEJUNOSTOMY - [med] jéjunostomie *f*
JEJUNOTOMY - [med] jéjunotomie *f*
JEJUNUM - [anat] jéjunum *m*
JELLY, to - [chem] faire prendre en gelée
JELLY - [chem] (synonym for gel) gel *m*, gelée *f*
[cin] (coloures gelatine screen) écran *m* diffuseur en gélatine
JEMMY - [tool] (a small crowbar) broche-levier *f*
JENA GLASS - [glass man] verre *m* d'Iéna
JENNET - [zool] genet *m*
JENNY - [text] machine *f* à filer
[cin] (small portable generator) générateur *m* portatif
[mech] (a frame on winch carrying wheels) chariot *m* de roulement
JEOPARDIZE, to - [gen] exposer au danger, hasarder
JERK - [gen] (sudden violent movement) secousse *f*, saccade *f*, soubresaut *m*
[med] réflexe *m*
[astronaut] suraccélération *f*
~DAMPER - [mech] amortisseur *m* de secousses
~-IN - [text] duite *f* rentrée
~LINE - [oil] (in well sinking) câble *m* auxiliaire
JERKED WARP - [text] chaîne *f* tiraillée
JERKING - s. jarring
JERKMETER - [astronaut] suraccéléromètre *m*
JERKY START - [mech] départ *m* irrégulier
JERRY-BUILD, to - [constr] (to build with cheap material and badly) bâtir avec du matériau de camelote
~-BUILDING - [build] construction *f* de maisons de pacotille
JERSEY - [text] jersey *m*, tricot *m* de laine
JERUSALEM ARTICHOKE - [bot] topinambour *m*
JERVINE - [chem] (alkaloid depressing the circulation) elléborine *f*
JET, to - [gen] émettre un jet, gicler, s'élancer en jet
JET - [min] (a hard, black, lustrous form of coal with ornamental uses) jais *m*, jayet *m*
[hydr] (fluid stream from an orifice or nozzle) jet *m*, ajutage *m*
[mech] (of a carburettor) gicleur *m*, buse *f*
[metall] (pouring gate) jet *m* de coulée, trou *m* de coulée
[aero] (motor actuated by atmospheric oxygen) réacteur *m*
[metall] (foundry) (the solidified metal attached to a casting) coulée *f*
[phys] veine *f*, veine *f* fluide
~AGER - [text] vaporisateur *m* à jets
~AGITATOR - [mech] (type of mixing device in which jets of liquid are used for stirring) agitateur *m* à jet
JET ASSISTED TAKE-OFF - [gen] décollage *m* assisté
~-BLACK - [gen] noir comme du jais
~BOOSTER - [mech] compresseur *m* à jet, éjecteur *m*
~CARRIER - [auto] porte-gicleur *m*
~CHAMBER - [auto] chambre *f* du gicleur
~CHARGE - [metall] charge *f* creuse
~CHIMNEY - [mech] tuyau *m* de vapeur
~COAL - [min] cannel *m*, candle-coal *m*
~CONDENSER - [mech] (a condenser in which the exhaust steam is condensed by jets of water) condenseur *m* à jet, condenseur *m* par injection
~COOLING - [mech] refroidissement *m* par tuyères
~COWL - [mech] capot *m* de tuyère, capot *m* de diffuseur
~DEVELOPMENT - [photo] développement *m* sous jet
~DRYING - [text] séchage *m* à tuyères
~ELEVATOR - [mech] palette *f* de commande
~ENGINE - [mech] (an i.c. propulsion unit in which the thrust is obtained solely from the reaction of the escaping products of combustion) réacteur, moteur *m* à réaction
~FUEL - [aero] combustible *m* pour réacteurs
~GOVERNOR - [mech] régulateur *m* du jet
~INTERRUPTER - [el] interrupteur *m* à jet de mercure
~MILL - [mech] (grinding machine in which the material is ground by impelling the particles against each other by jets of air or steam) broyeur *m* à jet
~MIXER - [mech] (machine in which mixing is carried out by means or liquid jets) agitateur *m* à jet
~MOULDING - [rubber & plast ind] boudinage *m*
~NOZZLE - [mech] tuyère *f* de moteur à réaction
~-NOZZLED ROCK BIT - [min tool] perforatrice *f* à jet
[mining] sondage *m* à jet
~PERFORATOR - [mining] perforateur *m* à charge creuse
~PIPE - [mech] (the terminal duct of a jet engine, through which the stream of gas is discharged) réaction d'échappement d'un moteur à réaction
[mech] tuyère *f* à jet, lance *f* de soufflage à air comprimé
[soil] lance *f* d'injection
~PIPE TEMPERATURE INDICATOR - [instr] (instrument designed to show temperature prevailing in the jet pipe) indicateur *m* de température de la tuyère de échappement
~PLANE - [aero] (aircraft fitted with jet engines) avion *m* à réaction
~POWER UNIT - [aero] groupe *m* propulseur
~-PROPELLED - [mech] à moteur à réaction
~PROPULSION - [mech] (propulsion by jet engine) propulsion *f* par réaction
~PUMP - [mech] (a device in which a jet of high velocity fluid is used to accelerate another fluid) pompe *f* à jet
~PURIFICATION - [ind chem] dépuration *f* à jet
~RINSING - [photo] lavage *m* sous jet
~SKIRT - [mech] socle *m*, partie *f* inférieure
~SPEED - [mech] (the velocity of the gases ejected from the jet pipe of a turbojet engine) vitesse *f* de échappement du jet
~SPIDER - [mech] croisillon *m* de centrage
~SPINNING - [plast ind] (process for production of fine fibres from molten polymer by a blast of hot gas) filage *m* par jet
~SPRAY NOZZLE - [mech] diffuseur *m*

JET STEM - [auto] économiseur *m*
~STREAM - [met] (a horizontal, approximately cylindrical body of air, of great extent in distance along the direction of wind, and moving at much higher velocity than the latter) jet *m*
[aero] courant-jet *m*
~TAPPING - [metall] couverture *f* explosive du trou de coulée
~THRUST - [astronaut] poussée *f* de réaction
~TRIMMING - [text] galon *m* en jais
~TURBINE ENGINE - [mech] (a gas turbine engine in which the whole useful power output takes the form of a jet) turboréacteur *m*
~VANE - [astronaut] gouverne *f* de jet
~WAVE RECTIFIER - [el] (a form of commutator rectifier in which a jet of mercury is used) redresseur *m* à jet de mercure
JETAVATOR - [astronaut] déviateur *m* de jet
JETSAM - [naut] marchandise *f* jetée à la mer
JETTING - [min] (earth-boring processes) forage *m* à jet
[plast ind] (turbulent flow in a mould, due to small a gate, or sudden change of mould section) turbulence *f*
~DOWN PROCESS - [hydr] procédé *m* de percussion à injection d'eau
~OF PILES - [constr] enfoncement *m* de pieux par injection d'eau, lancage *m* de pieux
JETTISON , to - [aero & naut] (to throw something out of an aircraft or ship with the object of disposing of it, e.g. jettisoning fuel before a crash landing) jeter à la mer, se délester de la cargaison
[astronaut] (the jettisoning of the booster rocket sections) abandonner
[aero] alléger, vider en vol
JETTISON - [gen] jet *m* à la mer
(to discharge fuel from a tank in the air) vidage *m* en vol
[naut] jet *m* de la marchandise à la mer
~GEAR - [aero] mécanisme *m* pour le vidage en vol
~VALVE - [mech] (special valve for releasing fuel which is to be jettisoned) vide-vite *m*
JETTISONABLE TANK - [aero] (a fuel tank designed to be detached and dropped bodily from an aircraft in emergency, or after its contents have been consumed) réservoir *m* largable
JETTY - [naut] jetée *f*, digue *f*
JEWEL - [gen] bijou *m*, joyau *m*
[horol] (rubis or sapphires, sometimes synthetic stones used for pivot bearings) rubis *m*
[mech] (rubis, sapphires or synthetic stones used for mechanical purposes) pierre *f* gemme
~BEARING - [mech] (part with a small jewel at the end as a support for the pivot of a moving element) rubis *m*, vis *f* à pierre
~HOLE - [horol] trou *m* du rubin
~LIGHT - [el] (a small indicator light, e.g. on a control panel, the transparent cover of which is shaped with facets to make it more conspicuous) ampoule *f* indicatrice
~SETTING - [jewel & horol] montage *m* des rubis
JEWELLED - [horol] monté sur rubis
JEWELLER'S FLUSH NIPPER - [impl] pinces *fpl* de bijoutier
JEWELLING - s. jewel setting
JIB, to - [naut] (to pull a sail or a rope from one side of a vessel to the other) coiffer
JIB - [mech] (the boom of a crane) volée *f*, flèche *f*
[naut] (a stay sail) foc *m*
[mech] crochet *m*
~BOOM - [naut] bâton *m* de foc
~CRANE - [mech mach] grue *f* à volée, grue *f* à flèche
~-POST - [mech] fût *m* d'une grue
~SLEWING CRANE - s. jib crane
JIG, to - [min] cribler, laver au bac à piston, laver au crible
JIG - [mech] (a device for holding a workpiece and guiding the tool operating on it) gabarit *m*, calibre *m*
[min] plan *m* automoteur
[min] (a fixed screen through which water is caused to pulsate by means of a plunger, so as to separate heavy minerals from waste) crible *m* hydraulique, hydrotamis *m*
[min] (a gravity separator) crible *m* vibrant
[min] lavoir *m* à secousses
[mech] (for the assembly of structural parts) dispositif *m* de serrage
[metall] gabarit *m* de remoulage
[text] jigger *m*
~-BACK - [transp] (an aerial to- and fro-funicolar) funiculaire *f* va-et-vient
~BORER - [mach tool] fraiseuse *f* d'après calibre
~BORING MACHINE - s. jig borer
~BUSHING - [mech] bague *f* de guide
~FOR FLATS - [text] gabarit *m* pour le réglage des chapeaux
~GRINDING MACHINE - [mach tool] rectifieuse *f* d'après calibre
~LOCATORS - [mech] repères *mpl*
~SAW - [impl] scief sauteuse, scie *f* à chantourner
~STENTER - [text] rame *f* à égaliser
~TABLE - [min] (apparatus for sorting and grading ores) crible *m* vibrant
~TANK - [min] caisse *f* du jig
~WELDING - [welding] (welding or thermoplastic material between suitably shaped jigs) soudage *f* au gabarit
JIGGED ORE - [min] minerai *m* lavé
JIGGER - [mech] (a hydraulic elevator) élévateur *m* hydraulique
[min] plan *m* incliné automoteur à simple effet
[naut] (a tackle) palan *m* à fouet
[telecomm] transformateur *m* d'oscillations
[mining] tenaille *f* d'attelage, tenaille *f* d'accrochage
JIGGING - [min] (separation of heavy minerals from gangue or waste by means of a jig) criblage *m*, séparation *f* à secousses
~CONVEYOR - [min] gouttière *f* transporteuse
~POINT - [metall] repère *m* de départ d'usinage
~SCREEN - [min] cricle *m* vibrant
[min] crible *m* à secousses
JIGGLING - [cin] (a rapid intermittent lateral movement caused by faulty machinery) mouvement *m* latéral rapide
JIGSAW - [impl] (a narrow-bladed saw reciprocated by hand or power, used for cutting sharply-curved forms from sheet) scie *f* alternative à découper
JIM-CROW - [mech] (manually or hydralically operated device, used to bend rails) presse *f* à cintrer les rails
[impl] (crowbar fitted with a claw) pince *f* à pied

de biche
JIM-CROW - [mach tool] (swivelling tool-head on a planing machine, cutting during each stroke of the table) tête *f* porte-outil pivotante
JIMMY - [mech] broche-levier *f*, pince *f* monseigneur
JINGLES - [mus] (the disk of metal which are fastened round the sides of the tambourine) grelots *mpl*
JINGLING - [acoust] cliquetis *m*, tintement *m*
~JOHNNY - [mus] chapeau *m* chinois
JINNY - [min] (a slope) plan *m* incliné automoteur à simple effet
[mech] (colloq) chariot *m* de roulement
[mech] treuil *m* de traction
JIPIJAPA - [text] carludovique *f*
JITTER - [radio] (a departure from temporal regularity) instabilité *f*, vacillement *m*
[telev & radar] (an unsteady movement of the image at high frequency, due to faulty synchronization) instabilité *f* d'image
[telev] (only USA; horizontal hunting in GB; the horizontal movement of the image caused by faulty synchronization) sautillement *m* horizontal de l'image
J-J COUPLING - [nucl] (the interaction between particles, each of which shows spin-orbit coupling) couplage *m* j-j
JOB - [gen] travail *m*, ouvrage *m*, tâche
[gen] (of an employee) emploi *m*, profession *f*
[mech] (a work piece) pièce *f*
~ANALYSIS - [autom] analyse *f* des travaux
~CONTROL INFORMATION - [comput] ordre *m* de contrôle des travaux
~LIBRARY - [autom] bibliothèque *f* de travaux
~LOGGING - [autom] (registre *m* chronologique de travaux réalisés
~MANAGEMENT - [autom] gestion *f* des travaux
~NUMBER - [automat] numéro *m* de travail-commande
~ORDER SHEET - [automat] bon *m* de travail
~SCHEDULER- [autom] programmateur *m* de travaux
~TICKET - [automat] bordereau *m* de travail
JOBBER - [fin] marchand *m* de titres
JOBBING FOUNT - [print] caractères *mpl* travaux de ville
~MACHINES - [print] (machines used for jobbing works) presse *f* pour travaux de ville
~PLATE - [print] (handbillis, letter headings, cards etc) travaux *mpl* de ville, bibelots *mpl*
JOCKEY - [mech] (device used to tension a belt) galet *m* de tension, pignon *m* tendeur
[mech] dispositif *m* d'attelage
~CHUTE - [mining] cheminée *f* auxiliaire à minerai
~GEAR - [mech] (a group of jochey pulleys) système *m* de rouleaux de tension
~POT - [glass man] (small-size pot placed on top of another pot for the melting of special glasses) pot *m* pour verre spécial
~PULLEY - [mech] (in a belt transmission, a pulley movably pounted and controlled by a spring or weight, to provide belt tension) poulie *f* de tension
~ROLLER - [mech] (a roller acting as belt tensioner) rouleau *m* de tension
~WALL - [cin] (easily removable stage decorations) panneau *m* transportable
~WEIGHT - [mech] poids *m* mobile
~WHEEL - s. jockey pulley
JODEL - [acoust] (sudden change of the voice from chest to falsetto) ioulement *m*
JOG, to - [gen] pousser d'un coup
[mech etc] avancer à secousses
[mach tool] (to feed intermittently) avancer à mouvement intermittent
[print] (the operations shaking sheets of paper lightly and intermittently to align the margins) dresser, égaliser
JOG - [gen] coup *m*, secousse *f*
[mech] mouvement *m* intermittent
[crystall] (step in a dislocation line) discontinuité *f*
~MOULDING - [casting & plast ind] moulage *m* par secousses
JOGGING - [mech] mouvement *m* intermittent
JOGGLE, to - [gen] secouer légèrement
[constr] (to prevent slipping by the insertion of joggles) goujonner, embrever
[mech] (to prevent lateral movements by fitting small projections on a piece of metal) assembler à crémaillère
JOGGLE - [constr] (a shoulder on a structural member taking the thrust from another member) goujon *m*, joint *m* à goujon
[constr] (a kind of pin binding together adjacent stones in a course) goujon *m*
[mech] (a small projection on a piece of metal fitting into a corresponding hole in another piece, to prevend slipping) adent *m*
~JOINT - [carp] (assemblage *m* à embrément
~-PIECE - [carp] (a king-post) poinçon *m*
~PLATE - [comput] plaque *f* antibourrage
~-POST - s. joggle-piece
~WORK - [constr] emboîtement *m* des pierres
JOGGLED BUTT JOINT - [constr] joint *m* emboîte, joint *m* en crémaillère
JOHNSON'S NOISE - [electron] (interference due to the movement of the electrodes) souffle *m* d'origine thermique, effet *m* thermique
JOIN, to - [gen] joindre, unir, adjouter, rapprocher, s'assembler
[carp] raboutir
[plumb] raccorder
[med] (a fractured bone) souder
JOINER - [carp] menuisier *m*
~BENCH - [impl] établi *m* de menuisier
JOINER'S CHISEL - [impl] (a long chisel used to finish work by hand) ciseau *m* de menuisier
~CRAMP - [carp] serre-joint *m*
~DOGS - [carp] clameaux *mpl*, clampe *f*
~GAUGE - [meas] (a marking gauge) trusquin *m*, troussequin *m*
~PENCIL GAUGE - [impl] équerre *f* épaulée
JOINERY - [gen] menuiserie *f*
JOINING BALK - [constr] poutre *f* de liaison
~PIN - [constr] broche *f*, goupille *f*
~PIPE - [plumb] tuyau *m* de raccord
JOINT, to - [gen] joindre, assembler
[mech] articuler
[carp] assembler, amboîter
[plumb] emmancher
[constr] jointoyer
[carp] varloper
JOINT - [gen] (connexion between two pieces) joint *m*, jointure *f*
[mech] (the region at which two parts are connec-

ted together, so as at admit motion but prevent the passage of fluids) assemblage *m*, articulation *f*
JOINT - [anat] articulation *f*, joint *m*
[carp] assemblage *m*, empatture *f*
[geol] (divisional planes occurring in most rocks) cassure *f*, diaclase *f*, joint *m*
[bookbind] (lateral projections) mors *m*
[food] pièce *f* de viande, quartier *m*
~ACCOUNT - [fin] compte *m* conjoint
~AGEING TIME - [plast ind] (the time necessary for a joint made with adhesive to reach full strength) durée *f* de durcissement naturel
~-AID - [oil ind] support *m* à ressort
~ANGLE - [build] cornière *f* d'assemblage
~-BOARD - [metall] plaque-modèle *f* en bois
~BOX - [telecomm] point *m* de raccordement
~CHAIR - [railw] (a support for the end of rails) coussinet *m* de joint, selle *f* de joint
~FACE - [metall] plan *m* de joint
~FASTENING - [railw] (device for fastening together the adjacent lengths of rails) mâchoires *fpl* de joints
~FILLER - [rubber ind] (compressible material used for gaskets) matériau *m* d'étanchéité, mélange *m* mastic
~FILLER CASTING MATERIAL - [mech] masse *f* coulée à jointoyer
~FLANGE - [mech] flasque *m* d'accouplement
~FLASH - [metall] bavure *f* de joint
~HINGE - [mech] charnière *f* d'assemblage
~HOLE - [plumb] niche *f*
~LINE - [metall] ligne *f* de joint
~OF PISTON RING - [auto] coupe *f* d'un segment de piston
~PACKING - [mech] garniture *f* de joints, garniture *f* de brides
~PARTY CONTROL - [telecomm] (in telephony) libération *f* au raccrochage des deux correspondants
~PIN - [mech] cheville *f* d'articulation
~PLANE - [geol] plan *m* de séparation, plan *m* de cassure
~RING - [mech] bague *f* de garniture
~SCRAPER - [impl] grattoir *m* du maçon
~SEALING - [metall] cordon *m* d'étanchéité, trainée *f* d'étanchéité
~SECTION - [railw] tronc *m* commun (à plusieurs lignes)
~SETTING - [railw] réglage *m* des joints
~SLEEVE - [plumb] manchon *m* d'accouplement
~STAVES, to - [brew ind] dresser les douves
~STRENGTH - [oil ind] capacité *f* de joint
~STRIP - [mech] (compressible material in strip form, used for sealing longitudinal joints) bande *f* de joint
~TRANSMISSION - [mech] transmission *f* à articulation
~TRANSMITTERS - [radio] émetteurs *mpl* relayeurs
~TRUNK WORKING - [telecomm] établissement *m* des communications interurbaines à la commande de l'opératrice locale
~WELDING - [welding] soudure *f* au galet, soudage *m* à la molette
JOINTED - [gen] articulé
[mech] articulé, jointé
[carp] jointé, encastré
~CHARLOCK - [bot] ravenelle *f*, raifort *m* sauvage
~COUPLING - [mech] manchon *m* articulé universel
JOINT ECCENTRIC ROD - [mech] tirant *m* d'excentrique articulé
JOINTER - [gen] assembleur *m*
[carp] varlope *f*
[impl] mirette *f* (de maçon)
[oil ind] tronçon *m* court de tubage
JOINTING - [gen] jointement *m*
[mech] (compressible material, usually in sheet form, from which flat gaskets can be cut for pipe flanges) mélange *m* mastic, matériau *m* d'étanchéité
[brew ind] (the staves) dressage *m* des douves
[constr] jointoiement *m*
[carp] varlopage *m*
~CEMENT - [ind chem] coulis *m*
~COMPOUND - s. joint filler
~HOLE - s. joint hole
~IRON - [constr] fer *m* à rejointoyer
~MATERIAL - [th eng] (ovens & retorts) lut *m*
~NAIL - [metall] polissoir *m*
~JOINTLESS FLOORING - [build] plancher *m* sans joints, badigeon *m*
JOIST - [constr] (horizontal beam) solive *f*, poutrelle *f*, soliveau *m*
~CEILING - [constr] plancher *m* en poutres
JOLLY BALANCE - [meas] (used for the determination of the specific gravity of a substance by weighing it in air and in liquid) balance *f* de Jolly
JOLLY'S APPARATUS - [instr] (an apparatus designed to make volumetric analyses of air) appareil *m* de Jolly
JOLT, to - [gen] secourer, cahoter
JOLT - [gen] cahot *m*, choc *m*, secousse *f*
[auto] (of a vehicle on a bad road) cahot *m*, secousse *f*, coup *m* de raquette
[metall] (the drop by gravity of the box, pattern and sand of a jolt-ramming machine) cahotage *m*
[metall] (in foundry) secousse *f*
[mech] à-coup
~MOULDING - [metall] moulage *m* par secousses
~PIN-LIFT MACHINE - [metall] machine *f* à mouler, moulage *m* à secousses avec démoulage sur chandelles
~RAMMING - [metall] serrage *m* par secousses
~ROLL-OVER PATTERN-DRAW MACHINE - [metall] machine *f* à mouler à secousses avec démoulage par culbuteur
~SQUEEZE - [metall] plaque *f* de serrage
~-SQUEEZE MACHINE - [mech mach] (moulding machine for deep patterns) machine *f* à mouler par secousses et pression
JOLTER - [metall] (a moulding machine) machine *f* à mouler par secousses
JOLTING - s. jarring
~MACHINE - s. jolter
[min] crible *m* laveur à secousses
~PLATE - [metall] plaque *f* de serrage
JONQUIL - [bot] jonquille *f*
JORDAN COSMOLOGICAL THEORY - [phys] (the theory stating that the total energy of the universe is zero) théorie *f* cosmologique de Jordan
JOSHI EFFECT - [el] (an electric current through a gas is increased or descreased by irradiating the gas with light) effet *m* Joshi
JOSTLE, to - [gen] bousculer, presser
JOULE - [phys] (unit of energy equal to 10^7 ergs) joule *m*

JOULE CYCLE - [mech] cycle *m* de Joule
~EXPERIMENT - [phys] (for the detection of the presence of intermolecular attraction in a gas) expérience *f* de Joule
~HEATING - [phys] (a method of plasma heating) chauffage *m* ohmique, chauffage *m* par effet Joule
~-KELVIN EFFECT - s. Joule-Thomson effect
~LOSSES - [el] pertes *fpl* par effet Joule
~MAGNETOSTRICTION - [el] (change in length parallel to an applied magnetic field and produced by the field itself) magnétostriction *f* de Joule
~-THOMSON EFFECT - [el] (the fall in the temperature of a gas when it expands without doing work externally to itself. This is caused by the absorption of energy in overcoming intermolecular cohesion. The effect is used in the Lindé Process q.v.) effet *m* Joule-Thomson
~-THOMSON INVERSION TEMPERATURE - [phys] température *f* d'inversion de Joule Thomson
JOULE'S EFFECT - s. Joule-Thomson effect
~LAW - [phys] (this states that: I. The intrinsic energy of any mass of gas in a function of the temperature only, and is not dependent on pressure or volume. 2. The molecular heat of a solid compound is the sum of the atomic heats of its component elements in the solid state) loi *f* de Joule
JOULEMETER - [instr] (an integrating meter using the Joule effect) joulemètre *m*
JOURNAL, to - [mech] tourner les tourillons d'un arbre
JOURNAL - [gen] journal *m*, livre *m* de loch
[mech] (the section of a shaft which is in contact with a bearing) tourillon *m*, fusée *f*
[naut] (a logbook) journal *m* de bord
[oil ind] (a detailed list of the drilled strata) liste *f* des couches perforées
~BEARING - [mech] coussinet *m* de fusée, palier *m*
~BOX - [mech] boîte *f* des coussinets, boîte *f* d'essieu
~BOX JACK - [mech] vérin *m* pour paliers
~BRASS - s. journal bearing
~BUSH - [mech] bague *f* de palier
~FRICTION - [mech] frottement *m* du tourillon
~JACK - s. journal box jack
~TURBINE - [mech] turbine *f* axiale, turbine *f* parallèle
JOURNEY - [gen] voyage *m*, parcours *m*
[min] cordée *f*, trait *m*, rame *f* de wagons
[mech] cycle *m* de travail
[glass man] cycle *m* de transformation
[mining] train *m* de berlines
JOWLING - [mining] sonder le toit en le frappant
JOYSTICK - [aero] (colloq. for control column) manche *f* pilote
JUDAS TREE - [bot] arbre *m* de Judée, gainier *m*, cercis *m*
JUDD - [mining] gros bloc *m* de charbon
JUDDER - [mech] (irregular shuddering action, as of a clutch of which the bearings are worn) ébranlement *m* trépidation *f*
[radio] (irregular rotation or motion of the transmitter or receiver mechanism) soubresaut *m*
JUGGLER - [mining] contre-fiche *f*
JUGULAR - [anat] (of the throat or nech region) jugulaire
~VEIN - [anat] veine *f* jugulaire
JUGULATION - [med] jugulation *f*
JUGULUM - [anat] clavicule *f*
JUGUM - [anat] joug *m*
JUICE - [gen] jus *m*, suc *m*
[food] (of fruit) pressis *m* de fruit, jus *m* de fruits
[el] (colloq. for electric current) courant *m* électrique
[auto] (colloq. for petrol) essence *f*, jus *m*
~CONCENTRATE - [agric] jus *m* de fruits concentré
~CONTENT - [agric] teneur *f* en jus
~PUMP - [mech] monte-jus *m*
JUJUBE TREE - [bot] jujubier *m*, gingeolier *m*
JUMBO - [metall] manchon *m* de refroidissement
~RING - [mining] installation *f* de sondage autotransportée
JUMP, to - [gen] sauter
[mech] (the operation of jointing by welding) souder par encollage
[cin] (irregular up-and-down movement) sauter verticalement
[metall] (the operation of upsetting in forging) refouler
[mining] forer au fleuret
JUMP - [gen] saut *m*, bond *m*
[aero] (from an aircraft, by parachute) saut *m*
[comput] (instruction specifying the location of the next instruction) saut *m*, instruction *f* de saut
[mech] chevauchement *m*
[geol] accident, cran *m*, rejet *m*
~DRILLING - [mining] sondage *m* par battage
~FLAP - [med] greffe *f* par transplantation successive
~IN BRIGHTNESS - [telev] (sharp transition from black to white in the picture) contraste *m* brusque
~INSTRUCTION - [comput] instruction *f* de transfert, ordre *m* de transfert, instruction *f* d'aiguillage
~SCANNING - [telev] analyse *f* double de cadre
~SPARK - [el] (between electrodes) décharge *f*
~WELDING - [welding] soudure *f* par encollage, soudure *f* par rapprochement
JUMPER - [el] (a conductor, e.g. of copper braid or tape, used to provide electrical continuity between metal parts) bretelle *f*, jarretière *f*
[min] (the bit of a compressed-air rock drill) barre *f* de mineur, fleuret *m*
[el traction] câble *m* de pontage
[text] jumper *m*, vareuse *f*, casque *f*
[metall] refouloir *m*
[electron] pont *m*
~-CABLE - [el] (for electrical traction connexions) câblot *m* d'accouplement, câble *m* de pontage
~CONNEXION - [el traction] s. jumper
~SOCKET - [el] (in railways, for the electrical coupling of the coaches) prise *f* de courant pour câble de pontage
~WIRE - [el] (a wire used to ensure electrical connexion between parts, to avoid a difference of potential between them) câble *m* d'accouplement
JUMPING - [gen] saut *m*
[metall] (upsetting) refoulement *m*, aplatissement *m*, écrasement *m*
[mech] (vibration by flexibility) broutage *m*, broutement *m*
[telev] (vertical hunting) instabilité *f* verticale de l'image
[instr] cahotage *m*
JUNCTION - [gen] jonction *f*, embranchement *m*

JUNCTION - [railw] bifurcation *f*, embranchement *m*
[mech etc] raccordement *m*, abouchement *m*
[electron] (in a semiconductor) jonction *f*
[el] connexion *f*, prise *f*
~BLOCK - [el] plot *m* de contact
~BOX - [el] (a housing covering connexions between electrical cables) boîte *f* de jonction, rosace *f* de raccordement
~CENTRE - [telecomm] centre *m* de secteur
~DIODE - [electron] diode *f* à jonctions
~DRAIN - [hydr] collecteur *m* d'entrée
~GROUP - [telecomm] faisceau *m* de jonctions individuelles
~LINE - [el] ligne *f* de jonction
[railw] ligne *f* affluente, voie *f* de raccordement
~LINE NETWORK - [telecomm] réseau *m* urbain de câbles auxiliaires
~LOSS - [el] perte *f* de jonction
~MANHOLE - [telecomm] chambre *f* de répartition
~PARTICLE DETECTOR - [nucl] détecteur *m* de particules à jonction semiconductrice
~PLATE - [constr] bande *f* de jonction
~POINT - [surv] point *m* de collimation
[railw] (branching-off point) point *m* de bifurcation
~POINTS - [railw] aiguille *f* de raccordement
~POLE - [telecomm] appui *m* de bifurcation
~RAIL - [railw] rail *m* de raccord
~RAILWAY - [railw] chemin de fer de jonction
~SERVICE - [telecomm] service *m* entre réseau voisins
~STATION - [railw] gare *f* de bifurcation, gare *f* de jonction
~SWITCHING POSITION - [telecomm] groupe *m* interurbain, position *f* intermédiaire
~TEMPERATURE - [electron] (the maximum temperature to which the junction in a semiconductor may be subjected) température *f* critique de la jonction
~TRANSISTOR - [electron] transistor *m* à jonctions
JUNCTURE - [gen] conjoncture *f*
[mech etc] jointure *f*, jonction *f*
JUNE BERRY - [bot] amélanchier *m* à grappes
~-BUG - [zool] petit hanneton *m* de la St. Jean
~GRASS - [agric] pâturin *m* des prés
JUNGLE - [gen] jungle *f*, brousse *f*
~FEVER - [med] fièvre *f* des jungles
JUNIPER - [bot] genévrier *m*, genièvre *m*
~BERRY - [bot] (a medicinal plant) baie *f* de genièvre
~OIL - [chem] (an essential oil used as a flavourant) essence *f* de genièvre
JUNK - [gen] rebut *m*, déchet *m*
[naut] jonque *f*
[naut] étoupe *f*, vieux filin *m*
~BASKET - [oil ind] panier *m* de sédimentation
~CATCHER - [oil ind] collier *m* de repêchage à lames
~PACKING - [mech] garniture *f* d'étoupe
~PUSHER - [oil ind] outil *m* pour pousser les détritus au fond du puits
~RING - [mech] (a ring designed to retain packing) anneau *m* de joint
[mech] (of a piston) couvercle *m* du piston
[mech] (a ring between the cylinder head and the sleeve valve) garniture *f* de piston
JUNKET - [food] jonchée *f*, lait *m* caillé
JURASSIC - [geol] jurassique
~LIME - [geol] calcaire *m* du Jura
~SYSTEM - [geol] (middle division of the Mesozoic era) système *m* jurassique
JURIN LAW - [mech] (the law dealing with the rise of a liquid in a capillary tube) loi *f* de Jurin
JURY - [leg] jury *m*
[naut] de fortune, improvisé
~BOX - [leg] banc *m* des jurés
~MAST - [naut] mât *m* de fortune
~RUDDER - [naut] gouvernail *m* de fortune
~STRUT - [aero] mât *m* provisoire
JUST GAP - [el] (only USA; air gap in GB) entrefer *m*
~NOTICEABLE DIFFERENCE - [acoust] seuil *m* différentiel d'amplitude
~SCALE - [acoust] (natural scale) gamme *f* tempérée
JUSTIFICATION - [gen] justification *f*
[print] (the spacing of the words in the lines to obtain an even margin) justification *f*, longueur *f* de ligne
JUSTIFY, to - [gen] justifier
[print] espacer régulièrement, égaliser, justifier, ajuster
JUT, to - [gen] être en saillie, avancer
JUT - [gen] saillie *f*, projection *f*, avancement *m*
JUTE - [text] (coarse vegetable fibre used for making hessian and the like) jute *m*
~BAG - [gen] sac *m* de jute
~BOARD - [paper man] carton *m* de jute
~BREAKING MACHINE - [text] broyeuse *f* pour jute
~CARPET - [text] tapis *m* de jute
~CLOTH - [text] tissu *m* de jute
~COVERING MACHINE - [mech] (a machine for applying a covering of jute to electric cables) machine *f* à recouvrir de jute
~FABRIC - [text] tissu *m* de jute
~FIBER - [text] fibre *f* de jute
~FILLING WARP - [text] chaîne *f* de remplissage en jute
~LONG LINE YARN - [text] fil *m* de jute peigné
~PAPER - [paper man] papier *m* jute
~PLUSH - [text] peluche *f* de jute
~SERVING - [of cables] enroulement *m* de jute
~TOW YARN - [text ind] fil *m* d'étoupe de jute
~YARNS - [text] fils *mpl* de jute
JUTTING - [gen] saillant, débordant
~OUT - s. jutting
~WINDOW - [constr] fenêtre *f* en saillie
JUVENILE WATER - [hydr] eau *f* juvénile
JUXTAPOSE, to - [gen] juxtaposer
JUXTAPOSITION - [gen] juxtaposition *f*

K

k - [phys] (the symbol for Boltzmann constant) constante *f* de Boltzmann
[chem] (symbol for velocity constant of a chemical reaction) symbole du coefficient de vitesse d'une réaction chimique
K - [chem] (the symbol for potassium) symbole de potassium
[light] (symbol forK line) ligne *f* K
K AERIAL - [radio & telev] (a modified form of H aerial) antenne *f* en K
K-CAPTURE - s. K-electron capture
K DEGREE - [thermo] degrés *mpl* Kelvin
K DISPLAY - [radar] indicateur *m* type K
K-ELECTRON - [nucl] (an electron forming part of the first shell surrounding the nucleus) électron *m* K
K-ELECTRON CAPTURE - [electron] (electron capture from the K-shell by the nucleus of the atom) capture *f* K
K-LINE - [spectrology] ligne *f* K
K SCOPE - s. K display
K-SERIES - [phys] (the shortest wavelengths in the characteristic X-ray spectra of the elements) série *f* K
K-SHELL - [nucl] couche *f* K
K-SPACE - [crystall] (abbrev. for momentum space) espace *m* K
KACHCHAN WIND - [met] (foehn wind occurring in Ceylon) vent *m* Kachchan
KAFFIR PIANO - [mus] (instrument resembling a xylophone) marimba *m*
KAHLERITE - [min] (rare mineral containing approximately 47% of uranium) kahlérite *f*
KAINIT - s. kainite
KAINITE - [min] (a natural double salt of potassium and magnesium used as a source of potassium salts and as a fertiliser) kaïnite *f*
KALA-AZAR - [med] kala-azar *m*, fièvre *f* doumdoum
KALE - [bot] chou *m* frisé
KALEIDOPHON - [acoust] (instrument producing tones by vibrations) caleidophone *m*
KALIEMIA - [med] kaliémie *f*, potassiémie *f*
KALIOPHILLITE - [min] (a silicate of potassium and aluminium) caliophyllite *f*, phacellite *f*
KAMPOMETER - [instr] (instrument designed to measure radiant energy) appareil *m* de mesure de l'énergie de rayonnement
KANG CANCER - [med] cancer *m* des chaufferettes
KANGAROO - [zool] kangourou *m*
KAOLIN - [min] (china clay; white bole; bolus alba, argilla, porcelain clay, white clay, terra alba. Inorganic, naturally occurring clay used as a filler for rubber) kaolin *m*, terre *f* à porcelaine
KAOLINITE - [min] (a natural aluminium silicate, occurring as very small flaky monoclinic crystals; derived from altered felspar) kaolinite *f*
KAPLAN WATER TURBINE - [mech] (propeller-type water turbine) turbine *f* Kaplan
KAPOK - [text] (fibres obtained from species of Bombax and Ceiba and used in upholstery and as an insulant) capoc *m*, kapok *m*
~ TREE - [text] capoquier *m*
KAPOSI'S SARCOMA - [med] sarcome *m* multiple hémorragique
KAPP COEFFICIENT - [el] facteur *m* de Kapp
~ LINE - [el] (unit of magnetic flux) ligne *f* Kapp
KARAKUL - [text] (a breed of sheep) caracul *m*
KARAT - [meas] (preferably 'carat') carat *m*
KARAYA GUM - [chem] (a gum obtained from varieties of Sterculia and used in medicine and as a thickenning and suspending agent in food processing) gomme *f* de karaya
KARMAN STREET OF VORTICES - [phys] rue *f* de vortices de Karman
KARYASTER - [biol] (group of chromosomes arranged like the spokes of a wheel) astéroïde *m*
KARYOGAMY - [biol] (union of two nuclei) caryogamie *f*
KARYOKINESIS - [biol] caryocinèse *f*, division *f* mitotique
KARYOLYSIS - [biol] (dissolution of the nucleus) caryolyse *f*
KARYOMICROSOME - [biol] (nuclear granule) nucléomicrosome *m*
KARYON - [biol] (nucleus) nucléole *m*
KARYOPLASM - [biol] cytoplasma *m*, protoplasma *m*
KARYOPLASTIN - [chem] parachromatine *f*
KARYOSOME - [med] caryosome *m*, nucléole *m* nucléinien
KARYOTIN - [biol] (the substance of the nuclear reticulum) caryotine *f*
KASHMIR GOAT - [zool] chèvre *f* de Cachemire
KATA-POSITIVE - [photo] (of a positive on an opaque surface) catapositif
KATABATIC WIND - [met] (a wind which blows down hillside and slopes, and is due to convective action) vent *m* catabatique
KATABOLISM - [biol] (the total of the disruptive metabolic processes) catabolisme *m*
KATACLASTIC - [geol] cataclastique
~ STRUCTURES - [geol] (produced in a rock by the action of mechanical stress) structures *fpl* cataclastiques
KATAFLOW - [met] (an airflow on a mountain-side giving rise to katabatic winds) écoulement *m* cata-

batique
KATAGENESIS - [zool] (retrogressive evolution) evolution *f* regressive
KATAMORPHISM - [geol] catamorphisme *m*
KATATHERMOMETER - [instr] (instrument measuring the cooling power of air on the human body) katathermomètre *m*
KATHODE - [el] (preferably cathode) cathode *f*
KATION - [phys] (preferably cation) cation *m*
KATISALLOBAR - [met] (an isogram of barometric pressure fall in a given period) catisallobare *f*
KATOPTRIC - [opt] catoptrique
~SYSTEM - [opt] (of a convergent optical system) système *m* catoptrique
KAURI - [bot] dammara *m* austral
~COPAL - [chem] dammar *m* kauri
~RESIN - [chem] (fossil gum found in New Zealand formerly much used in varnish making) résine *f* fossile de kauri
kc/s - [el] (abbrev. for kilocycles per second) ks/s
KECKLING - [naut] (old ropes wound about a cable to protect it from chafing) fourrure *f*
KEDGE, to - [naut] (the operation of moving a ship by winding in a hawser attached to a small anchor) touer sur une ancre à jet
KEDGE - [naut] (small anchor used in mooring or warping) ancre *f* toueuse, ancre *f* de touée
~ANCHOR - s. kedge
~ROPE - [naut] câble *m* de touée
KEDLOCK - [bot] moutarde *f* des champs, sanve *f*
KEEL - [shipbuild] quille *f*
[aero] (of a rigid airship) quille *f*
[bot] carène *f* (de feuille etc)
~AREA - [aero] surface *f* de quille
~BLOCK - [metall] (used in foundry work) bloc *m* à quille
~-BLOCKS - [shipbuild] tins *mpl* de cale sèche
~LAYING - [shipbuild] pose *f* de la quille
KEELBLOCK - [naut] (short piece of timber on which the keel rests during building) tin *m*
KEELING THE WARP - [text] marquage *m* de la chaîne
KEELSON - [aero] (in a flying-boat hull, a longitudinal member running inside in correspondence with the keel on the outside) carlingue *f* centrale
[shipbuild] (longitudinal strength member) carlingue *f*, contre-quille *f*
KEEN - [mech] (of a cutting tool) affilé, aiguisé
[mech] (of a point) aigu, acéré
KEENE'S CEMENT - [constr] (quick-setting hard plaster) mortier *m* d'enduit
KEEP-ALIVE ARC - [electron] courant *m* d'entretien
~-ALIVE ELECTRODE - [electron] (excitation anode) électrode *f* de maintien
~MATTER STANDING, to - [print] conserver la composition
KEEPER - [gen] garde *m*, gardien *m*
[el] armature *f* d'aimant
[mech] cliquet *m*, gâche *f*, détente *f*, contre-écrou *m*
KEEPING QUALITY - [chem] degré *m* de conservation
KEEPS - [mining] taquets *mpl*
KEG - [gen] barillet *m*, tonnelet *m*
[brew ind] tonneau *m*, fût *m*
[light] spot *m* nain
~ROLLING MACHINE - [brew ind] rouleuse *f* de fûts
~WASHER - [brew ind] laveuse *f* pour fûts
KEIROSPASM - [med] crampe *m* des barbiers

KELIM CARPET - [text ind] tapis *m* Kilim
~CURTAIN - [text] rideau *m* Kilim
KELLER FURNACE - [metall] (a type of electric furnace used in iron smelting, in which the heat is produced partly by the resistance of the charge to the current passed through it and partly by arcs formed between the electrodes and the charge) four *m* Keller
KELLOGG EQUATION - [mech] (equation of state relating the pressure, the absolute temperature and the density of a gas) équation *f* de Kellogg
KELLY - [oil ind] (a hollow 40 ft long square pipe attached to the top of the drilling string and turned by the rotary table during drilling) tige *f* carrée
[mech] (auger stem) tige *f* carrée d'entraînement
~BUSHING - [oil ind] carré *m* d'entraînement
~STOPCOCK - [oil ind] robinet *m* de la tige carrée
KELOID - [med] (excessive scar formation) chéloïde *m*
~GROWTH - s. keloid
KELOIDOSIS - [med] chéloïdose *f*
KELOMA - [med] s. keloid
KELOPLASTY - [med] opération *f* réparatrice d'une cicatrice
KELP - [chem] (general term for the larger seaweeds (laminaria and fucus) used as a source of iodine and potassium) varech *m*, soude *f* brute
KELVIN - [el] (rarely used term denoting kilowatthour) kilowatt-heure *m*
~BALANCE - [el] balance *f* Thomson
~CIRCULATION THEOREM - [mech] (in the motion of a non-viscid fluid) théorème *m* de circulation de Kelvin
~COMPASS - [instr] (ship's compass with a number of special features) compas *m* de Kelvin
~DEGREE - [meas] (the unit of the Kelvin's absolute scale of temperature) degré *m* absolu, degré *m* Kelvin
~DOUBLE BRIDGE - [el] (double bridge) pont *m* double de Thomson
~EFFECT - [el] (the depth of penetration of electric currents into a conductor decreases with the increase of frequency) effet *m* de peau, effet *m* Kelvin
~EQUATION FOR SURFACE TENSION - [mech] équation *f* de Kelvin pour la tension superficielle
~SCALE - s. Kelvin's absolute scale of temperature
~SOUNDER - [instr] (instrument indicating the pressure and hence the depth of water) sonde *f* Kelvin
KELVIN'S ABSOLUTE SCALE OF TEMPERATURE - - [meas] (temperature scale based on thermodynamical considerations) échelle *f* Kelvin
KEMP - [text] jarre *m*
KENDALL EQUATION FOR VISCOSITY - [phys] équation *f* de viscosité de Kendall
KENNEL - [gen] chenil *m*, loge *f* (de chien)
[constr] (the surface drain of a street) ruisseau *m* de rue
~COAL - [min] charbon *m* à longues flammes, cannel-coal *m*
KENOTRON - [electron] (industrial hot-cathode diode) kénotron *m*
KENTALLENITE - [geol] (coarse igneous rock) kentallenite *f*
KENTLEDGE - [constr] (heavy iron scraps, stones etc used as loading on a structure) gueuse *f*, surcharge *f* en maconnerie
KERATINIZATION - [zool] (the formation of horns)

kératinisation *f*
KERATITIS - [med] (inflammation of the cornea) kératite *f*
KERATOCELE - [med] (protrusion of the inner membrane of the cornea through an ulcer) descémétocèle *f*, kératocèle *f*
KERATOSE - [anat] kératique
KERATOTOMY - [med] kératotomie *f*
KERB - [gen] (the edge of a pavement, a raised path etc) bordure *f* du trottoir
~WEIGHT - [auto] (the weight of a motovehicle ready to travel but without passengers or load) poids *m* en ordre de marche
KERBSTONE - [constr] pierre *f* de parement du trottoir
[hydr] (of a well) margelle *f*
[constr] (a post) garde-pavé *m*
KERCHIEF - [text] fichu *m*, fanchon *m*, mouchoir *m* de tête
KERF - [mech] (groove made with a saw) trait *m* de scie, voie *f* de scie
[mech] (a notch) entaille *f*, trait *m* de chalumeau
[mining] sous-cave *f*, havage *m*
[timber] surface *f* de coupe
KERFING - [carp] trait *m* transversal de scie
KERMES - [min] (the dried of the female of Coccus ilicis used as a source of a red dye) kermès *m*
KERMESITE - [min] (natural oxy-sulphide of antimony (also called red antimony), an alteration product of stibnite (natural sulphide of antimony), orthorhombic or monoclinic, occurring as red acicular crystals) kermésite *f*
KERN - [print] (projection of type letters which rests on the body of the preceding or following letter) crénage *m* (de caractère)
~BUTT - [geol] gradin *m* de faille
KERNED LETTER - [print] lettre *f* crénée
KERNEL - [bot] amande *f*, pignon *m*, grain *m*
[metall] (a core) noyau *m*
KERNITE - [min] (natural sodium borate. It is an important source of borax) kernite *f*
KEROGEN - [min] (the source of oil in oil-bearing shales) kérogène *m*
KEROSENE - [chem] (a hydrocarbon boiling in the range of I50-300°C., and used as a fuel and illuminant. Equivalent to the English term paraffin (in the common, not in the chemical, sense) kérosène *m*
[gen] pétrole *m* lampant
~ENGINE - [mech] moteur *m* à pétrole
~OIL - [ind chem] huile *f* de kérosène
KEROSINE - s. kerosene
KERSEY - [text] (woollen or worsted fabric made with coarse crossbred wool) créseau *m*
KETCH - [naut] (robust, two-masted vessel) ketch *m*, caiche *f*
KETENE - [chem] (the first of a series of compounds having the general formula $R_2C:C:O$, and homologous with carbon monoxide: unstable and readily polymerized) cétène *m*
KETO-ENOLIC TAUTOMERISM - [chem] (certain compounds form two series of derivatives based on their ketonic or enolic constitution) tautomérie *f* céto-énolique
~-FORM - [chem] (form of a substance showing keto-enolic tautomerism with the properties of a ketone) cétoforme
KETOBENZOTRIAZINE - [chem] (a synthesis intermediate) kétobenzotriazine
KETOGENESIS - [med] cétogenèse *f*
KETOGENIC - [med] (which produces ketone bodies) cétogène
KETONES - [chem] (compounds in which a carbonyl group is attached to two hydrocarbon radicals, the simplest being acetone) cétones *fpl*
KETONIC - [chem] cétonique
~ACID - [chem] acide *m* cétonique
~CLEAVAGE - [chem] clivage *m* cétonique
~SOLVENT - [chem] solvant *m* cétonique
KETONIMINE DYESTUFFS - [dyes] (a class of dyes containing an -NH=C=group) teintures *fpl* de kétonimine
KETOSES - [chem] (monosaccharoses of ketonic constitution) cétoses *mpl*
KETOSURIA - [med] cétosurie *f*
KETOXIMES - [chem] (compounds formed by reacting ketones with hydroxylamine) cétoximes *fpl*
KETTLE - [chem] (general term for a vessel in which a moderate degree of heat is applied to the contents) marmite *f*, bouilleur *m*
[metall] (open vessel used in metallurgical operations on metals having a relatively low melting point, e.g. the drossing of lead) poêle *f* à flamber
[geol] bassin *m*, cirque *m* glaciaire
~HOLE - [geol] pli *m* synclinal
KETTLEDRUM - [mus] (basin-shaped drum) timbale *f*
KeV - [electron] (kilo electron Volt) KeV
KEVEL - [impl] (hammer pointed at one and edged at the other) bitton *m*
[naut] (a peg or cleat) oreille *f* d'âne
KEY, to - [mech] (to secure a part to a shaft by means of a key) claveter, caler, coincer
[carp] adenter
[constr] bander (une voûte)
KEY - [gen] clef *f*, clé *f*
[mech] (small metal part designed to engage grooves or slots in a shaft) clavette *f*, cale *f*, coin *m*
[mech] (small metal part designed to engage grooves or slots in a rotating part fixed to a shaft, to transmit power from one to the other) clavette
[mech] (of stopcock) tournant *m*, clé *f* (de robinet)
[constr] (a wedge shaped hardwood block) adent *m*
[telecomm] (a multiple switch) clé *f*, manipulateur *m*
[telecomm] (of a code) clef *f*
[telecomm] (Morse telegraphy) manette *f*, manipulateur
[gen] (of a diagram etc) légende *f*
[cin etc] (the slope of the tangent to the brightness reproduction) caractéristique *f* de luminosité
[mus] (of piano etc) touche *f*
[mus] (the scale and keynote of a composition) clé *f*
[oil ind] coin *m* de fixation
[instr] (of a watch or clock) clavette
[el] fiche *f*
~A PULLEY, to - [mech] claveter une poulie (sur l'arbre)
~ANIMATION - [cin & telev] intermédiaire *m*
~ARCH BRICK - [th eng] clef *f* de voûte
~BED - [mining] couche *f* guide
~-BIT - [mech] panneton *m* de clef
~BLOCK - [arch] (keystone) clef *f* de voûte
~BOLT - [mech] boulon *m* à clavette
~BOSS - [mech] (the thickening of boss where the

keyway is cut) bossage *m* pour clavette
KEY BRICK - s. key arch brick
~BUGLE - [mus] (keyed modification of the natural bugle) bugle *m* à clés
~BUTTON - [gen] (of a keyboard) touche *f* de clavier
~BY WEDGE, to - [mech] claveter avec un coin de calage
~CARD PUNCH - [comput] (card punch) perforateur *m* de cartes
~CARD PUNCHING MACHINE - [text] machine *f* à piquer à clavier
~CLICK - [telecomm] claquement *m* de manipulation
~-COURSE - [build] (the course of stones which correspond to the keystone in an arch) assise *f* de clef de voûte
~CUT HOLES - [mining] coups *mpl* de bouchon
~DRIFT - [mech] chasse-clavette *m*, chasse-clef *m*
[text] mandrin *m* rond
~DROP - [mech] cache-entrée *m*
~GROOVE - s. keyway
~GUIDE - [mech] guide *m*, canon *m* d'entrée
~HOLDER - [el] douille *f* à clef
~HORIZON - [geol] horizon *m* repère
~LEVER - [mech] levier *m* de touche
~LIGHT - [telev] (only GB; hot light in the USA; the main lamp in a studio during broadcasting) éclairage *m* principal
~MOTION - [text] mécanisme *m* des touches, clavier *m*
~PLAN - [gen] (small-scale plan indicating the arrangement of some items in a scheme) table *f*, tableau *m*
~PLATE - [mech] écusson *m*, entrée *f* de serrure
~PULSING - [telecomm] sélection *f* à distance par clavier
~-RING - [impl] anneau *m* brisé *m*, porte-clefs *m*
~SEAT - [oil ind] encoche *f* de la clavette
~SEAT WIPER - [oil ind] dispositif *m* pour nettoyer l'encoche de clavette
~SEATING - s. keyseat
~-SEATING MACHINE - [mech mach] (machine used for milling keyways in shafts) machine *f* à fraiser les rainures de cales
~-SENDING POSITION - [telecomm] position *f* à clavier
~SET - s. keyboard
~SIGNATURE - [mus] armure *f* de la clé
~SOCKET - [el] douille *f* à interrupteur
~STATION - [contr] station *f* d'enregistrement
~SUBSTANCE - [chem] (the basic substance in a chemical process) matière *f* de base
~SWITCH - [el] interrupteur *m* à culbuteur
~WELL - [mining] puits *m* d'alimentation
KEYBALL - [mech] bille *f* clavette
KEYBOARD, to - [print] composer à clavier
KEYBOARD - [gen] clavier *m*
[telecomm] clavier *m*, tablette *f* à clés
~-ACTUATED TAPE PUNCH - [comput] perforatrice *f* de bande à commande par boutons-poussoirs
~PERFORATOR - [telecomm] (perforator provided with a bank of keys) perforateur *m* morse à clavier
~SELECTION - [telecomm] numérotation *f* à clavier
~SEND/RECEIVE - [telecomm] clavier *m* transmission/réception
~SENDER - [telecomm] manipulateur *m* dactylographique
~TRANSMITTER - [telecomm] (keyboard-controlled device for the automatic transmission of coded electric signals) transmetteur *m* à clavier
KEYED - [mech] à clavette
[telecomm] manipulé
[gen] à touches
~AUTOMATIC GAIN CONTROL - [electron] commande *f* automatique de gain par impulsions
~BUSH - [mech] bague *f* à clavette
~CONTINUOUS WAVES - [telecomm] (continuous waves keyes in accordance with a telegraphic code) ondes *fpl* A1, ondes *fpl* entretenues manipulées
~MODULATED WAVES - [telecomm] (a carrier wave) ondes *fpl* A2, ondes *fpl* continues modulées manipulées
~PROCESS - [chem] (a method of distillation for obtaining absolute alcohol) procédé *m* Keyes
~SIGNAL - [telecomm] signal *m* commutateur
KEYHOLE - [mech] entrée *f* de serrure
[mech] (on a shaft, or a boss, or a hub) trou *m* de clef
~BUSH - [mech] canon *m* de l'entrée de clé
~HEALD - [text] lisse *f* à trou de serrure
~MASK - [photo] (mat over the camera lens, which has an aperture having the shape of a keyhole) masque *m* en forme de trou de serrure
~PLATE - [mech] cache-entré *m*, écusson *m*
~SAW - [impl] (a compass saw) scie *f* à guichet, scie *f* d'entrée
~-STRIP - [rubber ind] bande *f* profilée en trou de serrure
KEYING - [telecomm] manipulation *f*
[carp] coinçage *m*
[mech] clavetage *m*, calage *m*
[mus] accordage *m*
~FILTER - [radio] (network in a circuit reducing the higher-frequency components in the keying signal waveshape, so as to reduce sideband spread) préfiltre *m*
~WAVE - [telecomm] (marking wave) onde *f* de manipulation
KEYLESS RINGING - [telecomm] appel *m* automatique
~WATCH - [horol] montre *f* à remontoir
KEYMAT - [comput] (in programming) plaque *f* à touches
KEYNOTE - [mus] note *f* de clé
KEYSEAT - [mech] (groove or slot to receive a key) rainure *f* de clavette, rainure *f* de cale
KEYSEND, to - [telecomm] taper un numéro sur le clavier
KEYSENDER - [telecomm] (strip of plunger keys for the transmission of marginal currents representing dialled impulse trains) manipulateur *m* à touches, émetteur *m* d'impulsions
KEYSTONE - [arch] clef *f* de voûte, voussoir *m* de clé
~DISTORTION - [telev] (keystone-shaped raster produced by scanning in a rectilinear manner a plane target area which is not normal to the average direction of the beam) distorsion *f* trapezoïdale
~EFFECT - s. keystone distortion
KEYSTONING - s. keystone distortion
KEYSTROKE - [comput] frappe *f*
KEYWAY, to - [mech] mortaiser les rainures de cales
KEYWAY - [mech] (a groove formed in a shaft or other part to receive a key) entrée *f* de serrure, rainure *f* de clavette

KEYWAY OF BLOCK - [text] mortaise *f* du bloc
~SLOTTING MACHINE - [mach tool] mortaiseuse *f* des rainures de cales
KEYWORD - [comput] mot-clé *m*
kg. - [abbrev. for kilogramme] kg.
kg-cal - [phys] (kilogram-calorie) kg/cal, calorie/kilogramme *f*
KHAKI - [text] (dull brown colour) kaki *m*
KHAMSIN - [met] (a warm wind blowing from the Sahara in advance of a low pressure system) khamsin *m*
KHLOPINITE - [min] (cobalt, titanium, radium and uranium containing mineral) chlopinite *f*
KIBBLE - [mining] (large buclet used in well sinking) benne *f*, cuffat *m*, tonne *f*
~FILLER - [mining] chargeur *m* de bennes
KIBBLER - [mech] (a machine for breaking seeds, glue etc. into smaller fragments) moulin *m* à égruger
[min] broyeur *m*
KICK, to - [gen] donner un coup de pied, ruer
[of firearms] reculer
KICK - [gen] coup *m* de pied
[of firearms] recul *m*, gifle *f*
[mech] (of a motor) cahot *m*, secousse *f*
~BACK, to - [mech] (of an i.c. engine) donner des retours en arrière
~BACK DUMP - [mech] calbuteur *m* de tête
~OF A WHEEL (ON A RAIL JOINT) - [railw] saut *m* de la roue sur le joint
~OFF - [oil ind] mise *f* en marche de puits
~-OFF MECHANISM - [astronaut] mécanisme *m* de séparation
~-OFF POINT - [oil ind] point *m* de déviation
~-OFF VALVE - [oil ind] (flow valve) soupape *f* de mise en marche
~SORTER - [electron] (colloq. for pulse height analyzer) analyseur *m* d'amplitude
~STARTER - [mech] (motorbicycle) kick *m*, pédale *f* de mise en marche
~STARTER RUBBER - [rubber ind] manchon *m* de démarreur à pied, garniture *f* de démarreur
KICK'S LAW - [phys] (the energy required a given amount of a substance to a specified fraction of its original size remain constant irrespective of the original particle size) loi *f* de Kick
KICKBACK - [telev] (the return of the scanning beam to the starting point after completion of a frame) retour *m* du spot
[radio] haute tension *f* de retour
~POWER SUPPLY - [telev] alimentation *f* en très haute tension par retour du spot
KICKER - [metall] (mechanism designed to move the sheet metal work from a press) extracteur *m* à secousses
[mech] éjecteur *m* à déclenchement
~LIGHT - [cin] (booster light) éclairage *m* principal
KICKING - [oil ind] puits *m* actif
KICKSTARTER - [mech] (of motorbicycle or scooter) lanceur *m*, pédale *f* de mise en marche
KID - [zool] chevreau *m*
[leather ind] peau *f* de chevreau
[naut] (small wooden tub) gamelle *f*, écuelle *f*
[hydr] (bundle of brushes which serve as a groyne) épi *m*
~FINISH - [paper man] papier *m* genre peau de chèvre
KID GLOVE - [leather ind] gant *m* de chevreau
~LEATHER - [leather ind] peau *f* de chevreau
~-SKINS - [agric] peaux *f*pl de chevreau
KIDNEY - [anat] rein *m*
[food] rognon *m*
[geol] rognon *m*
~BEAN - [agric] haricot *m* nain
~ORE - [min] (a type of haematite occurring in reniform masses) hématite *f* rouge en rognons
~-SHAPED COCOON - [zool] cocon *m* réniforme
KIER - [text] (a boiler in which yarn and cloth are treated) cuve *f* à débouillir
[paper man] autoclave *m* à blanchiment
[gen] bouilleur *m*
~-BOILING - [text] débouillissage *m* sous pression
~DECATIZING - [text] décatissage *m* sous vide
KIERING - [text] débouillissage *m* sous pression
KIESELGUHR - [min] (synonym for diatomite. A natural siliceous powder consisting primarily of the frustules of diatoms. Used in insulating compounds, fireproof cements, as an absorbent in explosive manufacture, and as a filler for rubber and plastics) kieselguhr *m*
KIESERITE - [min] (mineral consisting of hydrous magnesium sulphate crystallizing in the monoclinic system) kiesérite *f*
KILDERKIN - [meas] (cask holding I8 gallons) baril *m* de 72 à 80 litres)
KILK - [bot] moutarde *f* des champs
KILL - [metall] caler
~A WELL, to - [oil ind] (the operation of overcoming the tendency of a well to flow by filling the well-bore with drilling fluid of suitable density) tuer un puits
~LIME, to - [chem] étendre de la chaux
KILLED RUBBER - [rubber ind] (with the nerve gone permanently) gomme *f* malaxée à mort
~STEEL - [metall] (steel which has been completely freed from oxygen before it is cast, by the addition of manganese etc) acier *m* calmé
KILLER - [electron] (the constituent of a luminescent material which affects luminescence efficiency) poison *m*
KILLICK - [naut] (a small anchor, or a heavy stone used as an anchor) petite ancre *f*
KILLING - [metall] (the deoxidation before casting) passage *m* à froid
[soap man] saponification *f*
~LINE - [oil ind] conduit *m* pour tuer le puits
~OF RUBBER - [rubber ind] mastication *f* à mort du caoutchouc
KILN, to - [gen] (to dry in a kiln) cuire, étuver
KILN - [gen] (a special type of furnace, of various forms, in which ceramic ware is fired) four *m*
[brew ind] (a large brick oven in which malt etc. is dried) touraille *f*, séchoir *m*
[gen] séchoir *m*, sécherie *f*
[metall] four *m* de grillage
~-DRIED - [gen] séché au four
~-DRY, to - [gen] (to subject to moderate heat in a kiln or oven) sécher au four
[brew ind] donner le coup de feu
~FAN - [mech] ventilateur *m* de touraille
~FLOOR - [th eng] (the flat platform in a kiln on which material is laid out for drying) plateau *m* de touraillage
~-LINING - [th eng] revêtement *m* de four

KILN MALT - [brew ind] (malt which has been dried in the kiln) malt *m* touraillé
~ SURFACE - [brew ind] surface *f* de la touraille
~ TEMPERATURE - [ovens] (the temperature maintained in a kiln during drying) température *f* de four
~ TURNER - [brew ind] retourneur *m* de touraille
KILNING - [gen] cuisson *f*, étuvage *m*
KILOAMPERE - [el] kiloampère *m*
KILOCALORIE - [heat] kilocalorie *f*
KILOCURIE - [radiat] (KC, 1000 curies) kilocuire *m*
KILOCYCLE - [el] (unit of electrical frequency, equal to 1000 cycles per second) kilocycle *m*
KILOGRAM - [meas] kilogramme *m*
KILOGRAMMETER - [meas] kilogrammètre *m*
KILOLITRE - [meas] mètre *m* cube
KILOLUMEN - [light] kilolumen *m*
KILOMETRE - [meas] (1,000 metres) kilomètre *m*
KILOMETRIC WAVES - [radio] (radiowaves with wavelength 1000 and 10000metres) ondes *fpl* kilométriques
KILOTON - [nucl] (atom bomb; the equivalent of 1,000 tons of dynamite) kilotonne *f*
KILOVOLTAMPERE - [el] (1,000 voltamperes) kilovoltampère *m*
KILOWATT - [el] kilowatt *m*
KILOWATT HOUR - [el] kilowatt-heure *m*
~-HOUR METER - [meas] kilowattheuremètre *m*
KIMBERLITE - [geol] (type of mica peridotite) kimberlite *f*
KIN - [gen] souche *f* (d'une famille), parents *mpl*
KINANESTHESIA - [med] perte *f* du sens de position
KIND - [gen] genre *m*, espèce *f*, sorte *f*, nature *f*
KINDLE, to - [gen] allumer, enflammer, embraser
KINDLING - [wood] bois *m* d'allumage
~ TEMPERATURE - [chem] température *f* d'allumage
KINDRED - [gen] parenté, du même genre
KINEMATIC - [mech] (relating to pure motion) cinématique
~ CHAIN - s. kinematic linkage
~ LINKAGE - [mech] (a sequence of mechanical parts through which mechanical movement is transmitted) raccordement *m* cinématique
~ MOTION - [mech] (pure motion; the theory of pure motion) cinématique *f*
~ RELATIVITY - [phys] (a relativity theory) relativité *f* cinématique
~ VISCOSITY - [phys] (the fluid viscosity divided by the fluid density) viscosité *f* cinématique
KINEMATICS - [mech] (the geometry of abstract motion) cinématique *f*
KINEMOMETER - [instr] (tachometer of high sensitivity) cinémomètre *m*
KINEOPHONE - [cin] (the talking picture machine designed by Edison) cinéophone *m*
KINESCOPE - [telev] (cathode-ray tube) cinescope *m*
~ RECORDING - [telev] (film recording of a television programme as it appears on a cathode-ray tube) enregistrement *m* d'images de télévision d'un cinescope sur film
KINESIONEUROSIS - [med] cinésinévrose *f*
KINESIOTHERAPY - [med] cinésithérapie *f*
KINESOPHOBIA - [med] phobie *f* du mouvement
KINETIC - [phys] (causing, or relating to, motion) cinétique
~ BLURRING - [radiography] flou *m* cinétique
~ BODY - [cytol] corps *m* cinétique
~ CONSTRUCTION - [cytol] (of a chromosome) construction *f* cinétique
KINETIC-CONTROL SYSTEM - [contr] (dynamic, or mechanical control system) système *m* de réglage cinétique
~ ENERGY - [phys] (the sum of the kinetic energies of all particles at a given instant) énergie *f* cinétique
~ ENERGY HEAD - [mech] (the pressure equal to the kinetic energy of the fluid flow per unit volume) pression *f* de vitesse
~ FRICTION - [mech] friction *f* cinétique
[soil] frottement *m* de glissement
~ HEAD - s. kinetic energy head
~ HEATING - [heat] (heating of a body caused by movement through a gas) chauffage *m* aérodynamique
~ INSTABILITY - [phys] micro-instabilité *f*, instabilité *f* cinétique
~ MOMENTUM - [phys] (of a charged particle in an electromagnetic field) moment *m* cinétique
~ PRESSURE - [phys] (the difference between the total and the static pressure) pression *f* cinétique
~ REACTION - [phys] (the physical agent causing a change of momentum) force *f*
[phys] (the physical agent causing an elastic strain in a body) force *f*
~ THEORY - [phys] (theory explaining the phenomena of heat due to the kinetic motion of atoms and molecules) théorie *f* cinétique de la chaleur
KINETICS - [phys] (the study of the effect of forces on the motion of material bodies) cinétique *f*
KINETOMERES - [biol] (molecules of protoplasm) cinétomères *mpl*
KING-BOLT - s. king-rod
~-CLOSER - [constr] clausoir *m*
~-PIECE - s. king-post
~-PIN - [mech] cheville *f* ouvrière, pivot *m* de fusée, pivot *m* d'attelage
~-PIN ANGLE - [auto] inclinaison *f* des pivots de fusée
~-PIN GEOMETRY - [auto] orientation *f* du pivot de fusée
~-PIN INCLINATION - [auto] inclinaison *f* du pivot de fusée
~-PIN SUPPORT - [auto] porte-fusée *m*
~-POST - [carp] poinçon *m*, aiguille *f*
[mech] (of a metal structure) poinçon *m* central
~-POST TRUSS - [carp] (timber roof-truss) ferme *f* à un poinçon, ferme *f* simple
~-ROD - [constr] (vertical rod connecting the ridge and tie-beam of a couple-close roof) poinçon *m* de ferme
~ TRUSS - s. king-post truss
KING'S BLUE - [paint] (alternative name for cobalt blue, a bright blue pigment consisting of a complex mixture of oxides of cobalt and aluminium) bleu *m* royal
~ HOOK - [mech] barre *f* d'attelage
~ YELLOW - [paint] (alternative name for Orpiment, arsenic trisulphide, with some uses a pigment) jaune *m* royal
KINGFISHER - [zool] martin-pêcheur *m*
KINGSTON'S VALVE - [mech] (sea-valve fitted to a ship's side) robinet *m* de prise d'eau à la mer
KINK, to - [gen] nouer, se tortiller
[text] (of yarn etc) se tortiller, nouer
[railw] (to buckle) tordre
KINK - [gen] tortillement *m*

KINK - [naut] coque *f*
[text] (a short twist in a thread etc) vrille *f*, boucle *f*
[mech] (of a pin, a wire etc) jarret *m*, faux pli *m*, coude *m*
[gen] (on a rope) tortillement *m*
[railw] (of a rail) jarret *m*
[med] coudure *f*
~INSTABILITY - [phys] instabilité *f* de flexion
~TEST - [rubber ind] (for fire hose) essai *m* de tortillement
KINKING - [text] (of a yarn) vrille *f*, boucle *f*
[text] cueillage *m*
[metall] (of wire) rupture *f* par deformation
KINOPLASM - [cytol] (protoplasm composed of fibrils) kinoplasma *m*
KIOSK - [gen] kiosque *m*
KIOTOMY - [med] uvolotomie *f*, incision *f* de la luette
KIP - [meas] (unit of force corresponding to 1,000 lb) unité *f* de charge (mille livres)
[mining] accrochage *m*
~HIDE - [leather ind] peau *f* de veau, peau *f* d'agneau
KIPP'S APPARATUS - [ind chem] (self-regulating appliance for generating hydrogen sulphide in the laboratory) appareil *m* de Kipp
~GENERATOR - s. Kipp's apparatus
KIRCHHOFF LAWS - s. Kirchhoff laws of networks
~LAWS OF NETWORK - [e] (laws relating to electric networks carrying steady currents) lois *f*pl de Kirchhoff
KIRVE, to - [mining] (to undercut coal) sous-caver, haver
KISH - [metall] (solid graphite floating on the top of a molten bath of cast-iron or pig-iron rich in carbon) flocons *m*pl de graphite
[min] écume *f* de graphite
~GRAPHITE SPOTS - [metall] piqûre *f* de graphite
KISS-ROLL COATING - [plast ind] (application of a metered film of coating to a web, the film being split at the contact line and half of it retained on the roll) revêtement *m* par enduction sur rouleaux
KISSER - [metall] (colloq. for a scale on a metal sheet) mâchefer *m* de contact local
KIT - [gen] trousseau *m*, trousse *f*
[gen] (the container holding working implements) trousse *f* d'outils
~BAG - [gen] sac *m* de voyage
KITCHEN - [gen] cuisine *f*
~CLOTH - [text] linge *m* de cuisine
~GARDEN - [agric] jardin *m* potager
~HERBS - [agric] fines herbes *f*pl
~LINEN - [text] linge *f* de cuisine
~RANGE - [gen] fourneau *m* de cuisine
KITCHENER - [impl] cuisinière *f*
KITE - [aero] (unpowered heavier-than-aircraft anchored to the ground or a moving vehicle on land or water, and supported by the wind or by motion imparted to it by towing) cerf-volant *m*
[air balloon] ballon *m* cerf-volant, ballon *m* captif
~BALLOON - [aero] (a captive balloon designed to derive stability and some measure of lift, from the wind) ballon *m* cerf-volant, ballon *m* captif
~WINDER - [build] marche *f* d'escalier en limaçon
KJELDAHL FLASK - [ind chem] (long-necked distillation flask for digestion followed by distilling) flacon *m* de Kjeldahl
KJEDAHL'S METHOD - [ind chem] (a process for determination of nitrogen in organic compounds) méthode *f* de Kjeldahl
KJELLIN FURNACE - [metall] (a type of induction furnace with an iron core, used for ferrous metals) four Kjellin
KLAXON - [auto] klaxon *m*
KLEPTOMANIA - [med] kleptomanie *f*
KLIEG EYES - [opt] (effect of the strain caused by the brilliance of lights) fatigue *f* oculaire
~LIGHT - [cin] (type of incandescent lamp) projecteur *m* grand angle
[light] (US) soleil *m*
KLINKER BRICK - [constr] (very hard type of brick, used mainly for pavings) brique *f* recuite, carreau *m*
~PAVING - [constr] pavé *m* en briques recuites
KLINOTROPIC - [bot] (growing at a slant) clinotrope
KLIPPE - [geol] (a term denoting thrust masses of strata mapping out like outliers) couche *f* massive de recouvrement
KLIRRFACTOR - [el] (non-linear distortion factor) facteur *m* de distorsion nonlineaire
KLYSTRON - [electron] (velocity-modulated tube) klystron *m*
KNAGGY - [gen] (knotty) noueux
~WOOD - [timber] bois *m* noueux
KNAPPING HAMMER - [impl] (hammer used to break and shape stone) casse-pierre *m*
[min] (implement or machine used to break rock and produce the minimum possible quantity of fine material) casse-pierres *m*
KNAPSACK - [impl] havresac *m*, sac *m* alpin
~DUSTER - [agric] poudreuse *f* à dos
~SPRAYER - [agric] pulvérisateur *m* à dos
KNEAD, to - [gen] (to mix a pasty material by alternately compressing and folding it) pétrir, malaxer
KNEADER - [mech] (a machine consisting of a chamber contining rotating blades to give an action like that of kneading dough, for mixing compound materials) pétrin *m* mécanique, malaxeur *m*
~BLADE - [mech] (a moving element in a mixing or kneading machine) palette *f*
KNEADING - [gen] pétrissage *m*, malaxage *m*
~HEAD - [mech] (a device which kneads the plastic material in a screw extruder before it reaches the feed unit) tête *f* du malaxeur
~MACHINE - s. kneader
~MILL - [text] moulin *m* à pétrir
~RUBBER - [rubber ind] gomme *f* à pétrir, gomme *f* à modeler
~TROUGH - [impl] pétrin *m*
KNEE - [anat] genou *m*
[plumb] (elbow pipe) coude *m*
[mech] genou *m*, équerre *f* genouillère *f*
[naut] courbe *f* de consolidation
[math] changement *m* brusque de direction
[electron] (point of inflection on the characteristic curve) coude *m*
~ACTION - [auto] suspension *f* indépendante par quadrilatères déformables transversaux
~ACTION SUSPENSION - [auto] s. knee action
~BEND - [plumb] coude *m* d'équerre
~BOARD - [text] planche *f* à genoux
~-BOOT - [shoe man] botte *f* de travail
~BRACKET - [gen] console-équerre *f*

KNEE CAP - [anat] rotule *f*
[gen] genouillère *f*
~-DEEP - [gen] à hauteur du genou
~GRIP - [mech] (of motorcycle) genouillère *f* pour moto
~JERK - [med] réflexe *m* rotulien
~JOINT - [med] articulation *f* du genou
[mech] joint *m* articulé, jarret *m*
~RAFTER - [constr] arêtier *m*
~ROOF - [build] toit *m* à la Mansard
~TIMBER - [carp] bois *m* coudé, bois *m* courbant
KNELL - [acoust] (bell tolled for the dead) glas *m*
KNIB - [print] (the projection of a setting rule) talon *m*
KNIFE - [impl] couteau *m*
[mech] (of a machine) couteau *m*, lame *f*
~BAR - [impl] racle *f*, racloir *m*
~BARKING MACHINE - [timber ind] machine *f* à écorcer à lames
~BLADE - [mech] lame *f*
~CARRIER - [text] support *m* du couteau
~COATER - [mech] (spreading a fluid material over a surface by the passage of a wide flat blade over it) machine *f* à enduire avec racle sur rouleau
~DISC - [text] disque *m* à couteau, couteau *m* circulaire
~DRUM - [paper man] tambour *m* porte-lames
~-EDGES - [mech] (a sharp edge formed on any part) lame *f* de couteau
[opt] méthode *m* de Foucault
[mech] (of a scale) couteau *m* d'une balance
[th eng] (in ovens and retorts) couteau *m* d'étanchéité
~-EDGE LIGHTNING PROTECTOR - [el] paratonnerre *m* à couteau
~-EDGE RELAY - [el] relais *m* à couteau
~EDGE SEAL - [metall] joint *m* à couteau, joint *m* à arête vive
~FILE - [impl] lime *f* à couteau, lime-couteau *f*
~FOR CUTTING DOUBLE PLUSH - [text] couteau *m* pour découpage de la peluche double
~FRAME - s. knife carrier
~GRINDER - [gen] remouleur *m*
~GRINDING - [mech] remoulage *m*, aiguisage *m*
~GRINDING MACHINE - [mech] machine *f* à affûter les lames de couteaux
~GUARD - [text] garde-couteau *m*
~HOLDER - s. knife carrier
~LEVER - [text] levier *m* des couteaux
~PIVOT - [text] arbre *m* des couteaux
~PLUG - [el] fiche *f* à couteau
~SHAFT - [mech] arbre *m* porte-couteaux
~SHARPENING MOTION - [mech] support *m* d'aiguisage du couteau
~SWITCH - [el] (switch in which the moving element consists of a flat blade) interrupteur *m* à couteau
~-TAPPING - [agric] incision *f* à couteau
~TOOL - [tool] (lathe tool with a straight lateral edge) outil *m* à couteau
~WITH HOOK - [impl] couteau *m* avec crochet
KNIGHTHEADS - [naut] apôtres *mpl*
KNIT, to - [text] tricoter
KNITTED - [text] tricoté, en tricot
~FABRIC - [text] tissu *m* par mailles
KNITTER - s. knitting machine
KNITTING - [med] soudure *f*
~FRAME - [text] banc *m* pour tricoter
KNITTING HEAD - [text] tête *f* de tricotage
~LOOM - s. knitting frame
~MACHINE - [text] tricoteuse *f*
~NEEDLE - [impl] aiguille *f* à tricoter
KNOB - [gen] bouton *m*, pomme *m*
[mech] (for manual regulation) bouton *m* de réglage
[instr] bouton *m*
[constr] (a handle of a door, window etc) bouton *m*, olive *f*
[firearms] pommeau *m*
[text] poignée *f* de la tringle de marque
KNOBBING - [constr] (the breaking off of projecting pieces of stone) bosselage *m*
KNOBBLE, to - [metall] cingler, tringler
KNOCK, to - [gen] frapper, cogner
[mech] (to emit a periodic percussive sound caused by excessive play in mechanical parts) cliqueter, taper, pilonner
[auto] (the sound caused by advanced ignition timing) cogner
[auto] (to detonate) détoner
KNOCK - [gen] coup *m*, heurt *m*, choc *m*
[auto] (of a motor) détonation *f*
[mech] (the sound due to excessive play in mechanical parts) cognement *m*
[auto] (the sound caused by advanced ignition timing) cognement *m* du moteur
~DOWN, to - [gen] renverser, abattre
[mech] (dismantling of machines) démonter
[constr] enfoncer (les pieux)
~DOWN TEST - [metall] essai *m* d'aplatissement
~OFF - [text] nez *m* d'arrêt
~-OFF - [mech] (any disconnecting mechanism) butée *f*, déclenchement *m*
[oil ind] décrocheur *m*
~-ON - [phys] (in nuclear physics) communication *f* d'énergie, percussion *f*
~-ON ATOM - [nucl] atome *m* percuté
~-OUT, to - [gen] faire sortir, chasser, repousser
[diecasting & plast ind] décocher
~-OUT - [metall] (ejecting a product from a mould) décochage *m*
~-OUT BAR - [impl] (a bar designed to strike a number of knock-out pins so as to actuate them) barre *f* de piquage
~OUT CORE - [metall] extracteur *m* de carottes
~-OUT GRID - [metall] grille *f* de décochage
~-OUT PIN - [mech] (a sliding part designed to expel a moulding from the mould by pressing on or striking it) broche *f* d'éjecteur
~-OUT PIN PLATE - [mech] (plate or block in which knock-out pins slide) plaque *f* d'éjection
~-OUT STATION - [transp] station *f* de déchargement
~-OUT WINDOW - [build] fenêtre *f* à battants
~UP, to - [bookbind] tapoter pour faire coincider les bords
~-UP - [mech] (of a press die) éjecteur *m*, extracteur *m*
KNOCKER - [impl] marteau *m* pour signaux
KNOCKING - [gen] tapage *m*, cognement *m*, détonation *f*
[min] minerai *m* en morceaux
~-FINGER - [text] bec *m* sur tringle d'arrêt
~HANDLE - [text] manivelle *f* de débrayage
~MOTION - [text] mécanisme *m* de débrayage
~OUT - [nucl] déplacement *m* d'un atome

KNOCKING-OVER BAR - [text] barre *f* d'abattage
~-OVER CAM - [text] came *f* d'abattage
~ROD - [text] boulon *m* de débrayage
KNOLL - [geogr] butte *f*, tertre *m*, monticule *m*
KNOP - [text] (a knot or a lump in a yarn) bouton *m*
~BOX - [text] réservoir *m* à boutons
~YARN - [text] (fancy yarn with lumps placed at intervals) fil *m* mélangé boutonneux
KNOT, to - [gen] nouer, faire un nœud
[naut] (ropes) abouter (deux cordages)
KNOT - [gen] nœud *m*
[meas] (unit of speed equal to one nautical mile per hour; not a unit of distance) nœud *m*
[in glass] druse *f*
~-GRASS - [bot] renouée *f* des oiseaux, herbe *f* à cochon
~SEALER - s. knotting
~STRENTH - [mech] (a tensile strength test on a filament with a knot in it) résistance *f* sur nœud
KNOTTED - [gen] à nœuds
[geol] noduleux
KNOTTER - [text] noueur *m*
[paper man] (device for the elimination of unbeaten particles in the pulp) épurateur *m* de pâte
KNOTTING - [paint] (a strong solution of shellac in industrial alcohol, used to seal knots in new timber, i.e. to prevent exudation of oleoresin into a coating applied later) vernis *m* à masquer les nœuds
KNOTTY - [gen] noueux, plein de nœuds
[gen] (of wood) noueux, raboteux
KNOTWEED - s. knot-grass
KNOW-HOW - [gen] (colloq., but in general use) know-how *m*, technique *f* opérationelle, savoir-faire *m*
KNOWLES CELL - [el chem] (an electrolytic cell for the production of hydrogen and oxygen from an alkaline electrolyte) pile *f* de Knowles
KNUCKLE - [carp] (the part of the hinge on which the pin is fitted) charnon *m*, articulation *f*
[mech] articulation *f*, joint *m* en charnière
[anat] jointure *f* du doigt
[mech] genouillère *f*
[el] (overhead junction) aiguillage *m* tangentiel
~GEAR - [mech] engrenage *m* à dents arrondies
~JOINT - [mech] (hinged joint between two rods) articulation *f* à genouillère, joint *m* en charnière
[mech] fermeture *f* à genouillère
~PADS - [anat] coussinets *mpl* des phalanges
~PIN - [auto] axe *m* de fusée, pivot *m* de fusée
[mech] (in a radial engine, the pin which serves to couple an articulated- to a master- connecting-rod) axe *m* de tête de biellette
[railw] axe *m* d'attelage
~PRESS - [metall] presse *m* matricer
~THREAD - [mech] (type of thread having a rounded cross-section) filetage *m* articulé
~TOOTH - [mech] dent *f* à profil semi-rond
KNUDSEN FLOW - [phys] (free molecule diffusion) diffusion *f* de molécules libres
~NUMBER - [phys] (the ratio between the mean free path of the molecules and a typical linear dimension in the field of flow) nombre *m* de Knudsen
KNURL, to - [mech] (to form single or intersecting serrations on a part intended to be turned with the fingers) moletter, godronner
KNURL - [mech] moletage *m*, godronnoir *m*
[timber] nœud *m*
KNURLED - [mech] (formed with serrations to afford a grip to the fingers, e.g. the head of an adjusting screw) moleté, godronné
~HEAD - [mech] (head of a screw or the like which is knurled to give for the fingers) tête *f* moletée
~NUT - [mech] (nut formed with serration to provide a grip for the fingers) molette *f*, écrou *m* moleté
KNURLING - [mech] moletage *m*, godronnage *m*, molettage *m*
~MACHINE - [mech] machine *f* à moleter
~ROLLER - [mech] godronnoir *m*
~TOOL - [tool] (small and very hard steel roller which are pressed against circular work in the lathe) molette *f*, godronnoir *m*
K/O - s. knock-out
KOH-NUMBER - [chem] indice *m* de potasse
KO-KNEADER - [plast ind] ko-kneter *m*
KOHLRABI - [agric] chou-rave *m*
KOJIC ACID - [chem] (a bactericide and synthesis intermediate) acide *m* cojique
KOLA - [agric] cola *m*
~NUT - [agric] noix *f* de cola
KOLLERGANG - [paper man] meuleton *m*
KOLLSMAN BAROMETER - [instr] baromètre *m* de Kollsman
KOPHEMIA - [med] logocophose *f*, surdité *f* verbale
KOSSEL-SOMMERFELD DISPLACEMENT LAW - [phys] loi *f* de déplacement spectroscopique
KRAFT - [paper man] (strong brown paper made from a good quality of sulphate wood pulp and used as a dielectric) papier *m* Kraft
~PAPER - s. kraft
KRARUP CABLE - [el] câble *m* à charge continue, câble *m* krarupisé
KRAUROSIS - [med] kraurosis *f*
KROLL PROCESS - [ind chem] (a commercial method for the production of titanium) procédé *m* Kroll
KRUPP KRANKHEIT - [metall] (German expression equivalent to temper brittleness; a tendency to break easily due to tempering) fragilité *f* de revenu
KRYPTON - [chem] (gaseous element, symbol kr, A.N. 36, A.W. 83.7. Zerovalent, monoatomic. It occurs in the air, from which it is obtained by liquefaction and fractionation and is used for filling electric lamps) krypton *m*
KRYPTOSCOPE - [radiat] (a small portable fluoriscopic screen for use in undarkened areas) kryptoscope *m*
K-SHELL - [phys] (the innermost electron shell surrounding the nucleus, which can contain only two electrons. The k-shell electrons are involved in the production of k-rays) couche *f* k
KULM - [min] poussière *f* d'anthracite
Kv, KV - [el] (abbrev. for kilovolt (=1000 volts) kilovolt *m*
Kv-T PRODUCT - [radio] (equivalent disturbing voltage) tension *f* perturbatrice équivalente
KYANITE - [min] (a silicate of aluminium) cyanite *f*
KYANIZING - [timber] (a process for the preservation of timber by impregnation with corrosive sublimate solution) cyanisation *f*, kianisation *f*
KYMOGRAPH - [instr] (instrument designed to record the variations of fluid pressure) kymographe *m*
KYMOGRAPHY - [radiol] (method of recording in one radiograph the executions of moving organs in

the body) kymographie *f*
KYNOPHOBIA - [med] cynophobie *f*
KYPHOSCOLIOSIS - [med] (a type of deformity of the spine) cypho-scoliose *f*
KYPHOSIS - [med] gibbosité *f*, cyphose *f*

L

l - [meas] (abbrev. for litre) l. (litre)
[met] (the letter for lightning in the Beaufort notation) éclairs *mpl*
[chem] (abbrev. for laevorotatory) à rotation à gauche (du plan de polarisation)
L - [chem] (symbol for latent heat per mole) symbole de la chaleur latente (molaire)
L ACID - [chem] (abbrev. for laurent acid) (an intermediate for dyestuffs) acide *m* de Laurent
L AERIAL - [radio] antenne *f* en L
L ANTENNA - s. L aerial
L-BAR - [metall] fer *m* en L, fer *m* d'angle
L CAPTURE - [nucl] (a type of decay in which an electron from an L-shell is captured by a nucleus) capture *f* L
L DISPLAY - [radar] (the A-type display with signals from two aerials placed back to back) indicateur *m* type L
L ELECTRON - [phys] électron *m* L
L-HEAD CYLINDER - [mech] cylindre *m* à soupapes latérales
L-LINE - [phys] (one of the lines in the L-series of X-rays produced by excitation of the electrons of the L-shell) ligne *f* L
L MESONS - [nucl] mésons *mpl* légers, mesons *mpl* L
L-RADIATION - [el] (one of a series of X-ray characteristics of each element) rayonnement *m* L
L SCOPE - s. L display
L-SECTION - [el] (elementary section of a network) élément *m* de filtre en L
L-SHELL - [phys] (the second layer of electrons round the nucleus of an atom) couche *f* L
LAB - [abbrev. for laboratory] laboratoire *m*
LABARRAQUE'S SOLUTION - [chem] (eau de Javelle, Javel water. A solution of sodium hyppochlorite, sodium hydroxide, and sodium chloride with uses as a bleaching agent and disinfectant) eau *f* de Labarraque
LABDACISM - [med] lambdacisme *m*
LABDANUM - [chem] (gum resin exuding from plants) labdanum *m*
~OIL - [chem] (essential oil obtained from Cistus ladaniferus and used in perfumery) essence *f* de labdanum
LABEL, to - [gen] étiqueter, attacher une étiquette
[gen] (to place in a category or assign a name to) désigner
LABEL - [gen] étiquette *f*, désignation *f*
[constr] (a projecting moulding above an opening) corniche *f* principale
[constr] (a dripstone) capucine *f*, larmier
[comput] (unit of information which can be used as a marker) étiquette *f*, label *m*
LABEL CORBEL TABLE - [build] (dripstone supported by a corbel) larmier *m* en console
LABELLED ATOM - [nucl] (the atomic position in a molecule distinguished by an isotope tracer) atome *m* marqué
~COMPOUND - [nucl] (compound partly consisting of labelled molecules) composé *m* marqué
~MOLECULE - [nucl] (molecule with one or more atoms disintegrated by non-natural isotopic composition) molécule *f* marquée
LABIAL - [zool] labial
LABIALISM - [med] labialisation *f*
LABIATE - [bot] labié
LABICHOREA - [med] tremblement *m* des lèvres
LABILE - [gen] labile, instable
[chem] (adjective descriptive of unstable compounds which readily undergo change) labile
~OSCILLATOR - [radio] (oscillator with a frequency controlled from a remote location) oscillateur *m* télécommandé
LABILITY - [gen] (lack of stability) instabilité *f*, labilité *f*
[med] (week state of equilibrium) labilité *f*, instabilité *f*
LABIOMANCY - [med] labio-lecture *f*, lecture *f* sur les lèvres
LABIOPLASTY - [med] chéiloplastie *f*
LABLAB - [bot] dolic *m* Lablab
LABORATORY - [gen] (room or building in which experimental work, tests, estimations and the like are carried out) laboratoire *m*
[metall] (of a reverbatory furnace) laboratoire *m*
[metall] sole *f* du four
~BENCH - [ind chem] (specially designed table for chemical operations in the laboratory) table *f* de laboratoire
~ROLLER MILL - [mech] (a small crusher consisting of a pair of rollers, for use in the laboratory) laminoir *m* de laboratoire
~SYSTEM - [chem etc] (system of reference related to the observer's laboratory) système *m* de référence au laboratoire
~TEST - [chem] (a test undertaken in a laboratory to ascertain the properties or qualities of a substance) essai *m* de laboratoire
~YIELD - [ind chem] rendement *m* de laboratoire
LABOUR - [gen] travail *m*, labeur *m*, façon *f*
[med] travail *m*, couches *fpl*
[gen] (workers) main-d'œuvre, travailleurs *mpl*
~COST - [fin] (the financial expenditure necessary to maintain a labour force. That portion of the total cost of an artifact or manufactured substance which must be set against wages) prix *m* de main-

d'œuvre
LABOUR EXCHANGE - [gen] bureau *m* de placement
~FORCE - [gen] effectif *m* de la main-d'œuvre
~OFFICE - s. labour exchange
~PAINS - [med] douleurs *fpl* de l'enfantement
~ROOM - [med] salle *f* de travail
LABRADOR FELDSPAR - s. labradorite
LABRADORESCENCE - [min] (the bright change of colour in some stones, caused by the presence of minute crystalline plates) chatoiement *m* de labradorite
LABRADORITE - [min] (a feldspar occurring in basic igneous rocks) labradorite *f*
LABROCYTE - [med] labrocyte *m*, mastocyte *m*
LABURNUM - [bot] aubour *m*, cytise *m* à grappes
~WOOD - [timber] faux ébénier *m*
LABYRINTH - [gen] labyrinthe *m*
[mech] (a sealing arrangement) labyrinthe *m*
[anat] labyrinthe *m* de l'oreille
~OF THE ETHMOID - [anat] masse *f* latérale de l'ethmoide
~PACKING - [mech] (for turbines) garniture *f* à labyrinthe
~RING - s. labyrinth seal
~SEAL - [mech] (pressure sealing arrangement in hydraulic mechanism, comprising a series of annular grooves and/or packings, in which the leakage pressure is progressively attenuated) scellement *m* au labyrinthe
LAC - [paint] (the product of the lac insect, Coccus lacca, and the source of shellac) laque *f*
LACCOLITE - s. laccolith
LACCOLITH - [geol] (a form of intrusion) laccolithe *m*, laccolite *f*
LACE - [text] (a fabric) dentelle *f*, lanière *f*
[gen] (a shoe string) lacet *m*, cordon *m*
~BOBBIN - [text] fuseau *m*
~EDGING - [text] lisière *f* de la dentelle
~GROUND - [text ind] réseau *m* de dentelle
~MACHINE - [text ind] métier *m* à dentelle
~PAPER - [paper man] papier *m* dentelle
~PUNCHING - [comput] perforation *f* intégrale de toutes les colonnes d'une carte perforée
~TRIMMING - [text] galon *m* dentelle
~UP, to - [mining] garnir de planches
LACED BELT - [mech] courroie *f* lacée
~CARDS - [text] cartons *mpl* lacés
~VALLEY - [constr] (in a tiled roof, a valley formed by interlacing tile-and-a-half tiles across a valley board) noue *f* de toit en tuiles entrecroisées
LACELIKE MATERIAL - [text] tissu *m* en forme de dentelle
LACERATE, to - [gen] lacérer, déchirer
LACERATION - [gen] lacération *f*, déchirement *m*
[med] déchirure *f*
LACHRYMATOR - [chem] (tear gas) gaz *m* lacrymogène
LACHRYMATORY - [chem] (stimulating the secretion of tears) lacrymogène
LACING - [mech] (joining of the ends of belts) attache *f*
[mining] chapeau *m*
~BAR - [constr] barre *f* de triangulation
~FLAT BAR - [metall] plat *m* de triangulation
~FRAME - [text] établi *m* pour laçage
~HOLE - [text] trou *m* de laçage
[naut] oeillet *m*
LACING THREAD - [text] fil *m* de séparation
~TWINE - [text] lacet *m*, cordon *m* de laçage
LACK, to - [gen] manquer (de quelque chose)
LACK - [gen] manque *m*, défaut *m*, pénurie *f*
~-LUSTRE - [gen] (lacking in brightness) terne, éteint
~OF AIR - [gen] manque *m* d'air
[chem] (in combustion) défaut *m* d'air
~OF CONSOLIDATION - [rubber & plast ind] (matrix out of register) matrice *f* mal ajustée dans son centrage, sculpture *f* pas en rapport
~OF DEFINITION - [photo] manque *m* de définition
LACKER - s. lacquer
LACMOID - [chem] (an analytical indicator) lacmoïde *m*
LACQUER, to - [paint] vernir, laquer
[gen] (coating by a lacquer obtained from natural sources) émailler
LACQUER - s. lacquers
~DISK - [acoust] (lacquer-coated mechanical recording disk) disque *m* laqué
~MASTER - [acoust] (original recording on a lacquer surface) enregistrement *m* sur disque laqué
~ORIGINAL - s. lacquer master
~PRESERVATIVE - [paint] produit *m* protecteur de la peinture
LACQUERING - [paint] laquage *m*, vernissage
LACQUERS - [paint] (coatings originally obtained from natural sources, but non applying equally to those based on synthetic materials. The essential criterion of a lacquer lies in the volatilization of the solvent leaving a tough residual film) vernis *m*, laque *f*
LACRIMAL - [zool] (near the tear gland) lacrymal
~DUCT - [anat] canal *m* lacrymal
~GLAND - [anat] (gland at the outer angle of the eye, secreting a watery substance) glande *f* lacrymale
LACTACIDURIA - [med] lactacidurie *f*
LACTALBUMINS - [chem] (albumins found in milk) lactalbumines *fpl*
LACTAM - [chem] (generic name for cyclic acid amides obtained from alpha and delta-amino-acids) lactame *m*
LACTASE - [chem] (an enzyme concerned with the production of glucose from lactose) lactase *f*
LACTATE - [chem] (a salt of lactic acid) lactate *m*
LACTATION - [med] allaitement *m*
LACTEAL - [gen] lacté
LACTEOUS - [chem] (milky, milk-like) laiteux, lactaire
LACTESCENCE - [med] lactescence *f*
LACTESCENT PLANT - [bot] plante *f* lactescente
LACTIC - [chem] lactique, caséique
~ACID - [chem] (a monocarboxilic acid with applications in food processing, the production of plastics and pharmaceuticals and in textile processing) acide *m* lactique
~ACID BACTERIA - [chem] bacteries *fpl* lactiques
~ACID FERMENTATION - [chem] fermentation *f* du lait
~COAGULATION - [chem] coagulation *f* lactique
~FERMENTATION - s. lactic acid fermentation
LACTIFEROUS - [zool] (producing milk) lactifère
[bot] (yielding a milky fluid) lactifère
[rubber ind] (yielding latex) lactifère
~CELL - [bot] (vessel containing latex) cellule *f*

lactifère
LACTIFUGE - [med] lactifuge
LACTIOBOSE - [med] lactose *m*
[chem] s. lactose
LACTOBUTYROMETER - [instr] (apparatus for determining the butter-fat content of milk) lactobutyromètre *m*
LACTODENSITOMETER - [instr] (instrument designed to measure the density of milk) lactodensimètre *m*
LACTOFLAVIN - [chem] (riboflavin, vitamin B2. A thermostable, water-soluble vitamin essential to heath) lactoflavine *f*
LACTOMETER - [instr] (instrument designed to determine the quality of milk) pèse-lait *m*, lactodensimètre *m*
LACTONES - [chem] (intramolecular anhydrides of hydroxy-carboxylic acids) lactones *mpl*
LACTONITRILE - [chem] (a solvent) lactonitrile *m*
LACTONIZATION - [chem] (formation of lactones) lactonisation *f*
LACTOPHENINE - [chem] (a febrifuge and analgesic used in medicine) lactophénine *f*
LACTOSCOPE - s. lactodensitometer
LACTOSE - [chem] (milk sugar. The least soluble and the least sweet of the common sugars. Lactose is used in medicine, food processing and bacteriology) lactose *m*, sucre *m* de lait
LACTOSURIA - [med] lactosurie *f*
LACTOTHERAPY - [med] régime *m* lacté
LACUNA - [gen] (a missing part, a hyatus) lacune *f*
LACUNAR - [constr] (a panel formed in a ceiling) caisson *m* de plafond
[arch] (ceiling consisting of coffers or panels) plafond *m* à caissons
LACUSTRINE - [gen] (of animals and plants in habiting lakes) lacustre
~DEPOSITS - [geol] (of strate originated by deposition at the bottom of lakes) dépôts *mpl* lacustres
LACY - [gen] de dentelle
[rubber ind] (of rubber crepe) (crêpe) ajouré
LADANUM - s. labdanum
LADD-FRANKLIN THEORY - [opt] (theory assuming the existence of molecules in the retina, which are affected by the action of light) théorie *f* de Ladd-Franklin
LADDER - [impl] échelle *f*
[mech] (the continuous line of mud-buckets in a dredger) élinde *f*, échelle *f* d'une drague
[text] (the breaking of a stitch) maille *f* lâchée
[mining] échelle *f* d'excavateur
~BACK - [carp] (of a chair) dossier *m* à barres horizontales
~DITCHER - s. ladder dredger
~DREDGER - [mech mach] drague *f* à élinde, drague *f* à échelle
~EXCAVATOR - s. ladder dredger
~NETWORK - [telecomm] (network consisting of a series of shunt impedances in alternate succession) réseau *m* en échelles
[electron] circuit *m* itératif
~RACK - [mech] crémaillère *f*
~SCAFFOLD - [constr] (light scaffold consisting of ladders which are braced together) échafaudage *m* de montage à échelles
~SHAFT - [mining] puits *m* aux échelles
~SOLLAR - [mining] palier *m* de repos
~TRUCK - [mech] (truck fitted with an extensible ladder used for fire extinguishing) camion *m* porte-échelle
LADDER-TYPE DITCHER - [mech] excavateur *m* à chaîne à godets
LADDERVEIN - [mining] filon à échelle
LADE, to - [gen & naut] charger
LADE - [metall] embouchure *f*
LADEN - [gen] chargé
[naut] chargé
LADING - [gen] chargement *m*, embarquement *m*
LADLE - [metall] (large steel vessel lined with refractory material used to receive and transport molten metal, espec. iton) poche *f* de fonderie, cuillère *f*
[gen] (kitchen impl) cuiller *m* à pot
[metall] (long cup-shaped spoon used in foundries) poche *f*, poche *f* de coulée
[gen] (for industrial uses) puisoir *m*, puchet *m*
[hydr] (the paddle of a water-wheel) palette *f*
~ANALYSIS - [metall] analyse *f* de coulée
~BAIL - [metall] étrier *m* de la poche de coulée
~CAR - [metall] chariot *m* porte-poche
~CARRIER - [metall] griffe *f* à poche de coulée
~CRANE - [metall] (a special crane for handling ladles of molten metal) pont *m* de coulée
~DRIER - [metall] sécheur *m* de poche
~GUIDE - [metall] guidage *m* de la poche de coulée
~HANDLE - [metall] étrier *m* de la poche de coulée
~LIP - [mech] lèvre *f* de poche
~POURING APPLIANCE - [metall] orifice *m* de coulée
~SAMPLE - [metall] prise *f* d'essai de métal liquide
~SHANK - [metall] étrier *m* de la poche de coulée, brancard
~SKULL - [metall] croûte *f* de poche de coulée
~SLAG - [metall] laitier *m* de poche
LADY'S BEDSTRAW - [bot] gaillet *m* jaune
~MANTLE - [bot] pied *m* de lion, manteau *m* de Notre-Dame
~SMOCK - [bot] cardamine *f* des prés
~THUMB - [bot] pied *m* rouge, renouée *f* persicaire
LAEVOROTATORY - [phys chem etc] (a compound having the power to rotate the plane of polarized light to the left. Such compounds have the prefix to distinguish them from dextro-rotatory compounds) lévogyre
LAEVULIC ACID - [chem] (an intermediate for pharmaceuticals and other organic compounds) acide *m* lévulique
LAEVULOSE - [min] (naturally occurring laevorotatory sugar) lévulose
LAG, to - [gen] rester en arrière, traîner
[gen] (to coat or surround with insulating material) revêtir, envelopper, garnir
[el] être déphasé en arrière
LAG - [gen] (of movement etc) retard *m*, décalage *m*
[mining] (timber lining protecting a shaft) garnissage *m*
[mech] (in machines) revêtement *m*
[el] (of a lagging current or lagging load) retard *m*, déphasage *m* en arrière
[a stave] douve *f*
[mech] (stave-shaped form round a cylinder) enrobage *m*
[med] phase *f* de latence, décalage *m*
[instr] (the delay in the change of reading of an altimeter during a sudden change of altitude e.g. as

the result of encountering a rapid vertical air current) retard *m*
LAG - [telecomm] (the period of time elapsing between one event and a succeeding one) retard *m*
[text] carton *m* à chevilles
[telev] (persistence of the electrical-charge image for a small number of fields) effet *m* de rémanence
~BOLT - [mech] boulon *m* à tête carrée
~COMPENSATION - [telev] compensation *f* de rémanence
~SCREW - [carp] tire-fond *m*, vis *f* à bois à tête carrée
LAGER BEER - [brew ind] (light beer of low alcohol content and a special flavour) bière *f* blonde
LAGGING - [th eng] (heat insulating material round a part or unit) enrobage *m* d'isolation thermique
[th eng] (of ovens, boilers etc) enveloppement *m*, garnissage *m*
[mining] (the timber lining protecting a shaft) garnissage *m*
[constr] (wooden boards across the framework of a centre forming the supporting surface of the arch) bois *m* de couchis, lattage *m*
[el] (of current or load) déphasage *m* en arrière
[mining] (behind wedging-curb) lambourde *f*
~COIL - [el] (a part of the watthourmeter) bobine *f* de retard
~CURRENT - [el] (a.c. current reaching its maximum value in the cycle later than the voltage producing it) courant *m* déphasé en arrière
~HINGE - [aero] (the pivot which allows azimuthal displacement of rotor blades) pivot *m* de mouvement azimutal des pales
~LOAD - [el] (reactive load in which the current kags behind the voltage) charge *f* inductive
~OF TIDES - [astr] retard *m* des marées
~PHASE - [el] (in three-phase circuits, a phase the voltage of which is lagging behind that of the other phases) phase *f* en retard
LAGOON - [geogr] lagune *f*
[geogr] (of an atoll) lagon *m*
LAGOONAL DEPOSITION - [geol] (accumulation of sediments in a shallow arm of the sea) dépôts *mpl* lagunaires
LAGUNE - s. lagoon
LAID DRY - [build] (of bricks laid without mortar) rangé à sec
~LINES - [paper man] (ribbed watermarks) verges *fpl*
~PAPER - [paper man] (paper with a ribbed watermark derived from the mould) papier *m* vergé
~PILE - [text] tissu *m* à poil couché
~WOOL - [text] laine *f* en suint, surge *f*
LAITANCE - [constr] (milky scum formed on wet concrete from various causes) croûte *f*
LAKE - [geogr] lac *m*
[chem] (organic pigment combined with inorganic base) laque *f*
[dye] (red pigment) colorant *m* rouge
~ASPHALT - [min] (natural asphaltic material occurring is surface deposits) asphalte *m* naturel
~BLOOM - [hydr] prolifération *f* d'algues
~COPPER - [min] (copper obtained from minerals in the region of Lake Superior) cuivre *m* du lac supérieur
~MARL - [geol] travertin *m*, marme *f* de lac, craie *f* lacustre
LAKE ORANGE - [dye] (an orange pigment) orangé *m* pour laque
~TONERS - [dye] (nearly pure heavy-metal salts of dyes, insoluble in water) colorants *mpl* organiques de laque
LAKES - [min] masse *f* de ségrégation
LAKY - [dye] (of the colour of lake) laqué, (sang) laqué
LALANDE CELL - [el chem] (a primary cell; electrodes, zinc and iron; electrolyte, sodium hydroxide solution; depolarizer, copper oxide) élément *m* de Lalande
LALLATION - [med] lallation *f*
LALOGNOSIS - [med] compréhension *f* de la parole
LAMB - [zool] agneau *m*
~-SKINS - [leather ind] peaux *fpl* d'agneau
~SUCKING - [zool] agneau *m* de lai
LAMB'S LETTUCE - [agric] mâche *f*, doucette *f*
~SHIFT - [spectrology] déplacement *m* de Lamb
~WOOL - [text] (fleece wool) laine *f* d'agneau, laine *f* agneline
LAMBDA LIMITING PROCESS - [nucl] (the definition of a point electron as the limiting case in which a timelike vector tends to zero) limitation *f* lambda
~PARTICLE - [nucl] (hyperon with an extremely short life and a mass between that of neutrons and deuterons) particule *f* lambda
LAMBENT - [gen] blafard
LAMBERT - [light] (unit of brightness) lambert *m*
LAMBERTITE - [min] (uranium containing mineral, very rare) lambertite *f*, uranotile *f*
LAMBING - [zool] agnelage *m*
LAMBSKIN - [text] peau *f* d'agneau
~WITH DOUBLE PILE - [text] peau *f* d'agneau double poil
LAME - [gen & med] boiteux, estropié
LAMELLA - [gen] (a thin plate of film) lamelle *f*
~PROFILE - [rubber ind] (of tyres) gravure *f* à lamelles
~TREAD - s. lamella profile
LAMELLAR - [gen] (constructed from lamellae) lamellaire, lamelliforme
[metall] (of a thin plate) lamellaire
~PEARLITE - [min] perlite *f* lamellaire
~SERPENTINE - [min] antigorite *f*
LAMENESS - [vet] claudication *f*, boiterie *f*
LAMINA - [gen] (a thin plate or film) lamelle *f*, feuillet *m*
[geol] structure *f* laminaire
[bot] limbe *m*, écaille *f*
LAMINAR - [gen] (formed from lamina) laminaire
~AIR NAVIGATION ANTI-COLLISION - [radar] (LANAC; air and ground radar system and beacon equipment with height coding of the aircraft transmitter pulses) système *m* de radar LANAC
~BOUNDARY LAYER - [phys] (boundary layer in which flow is laminar) couche *f* limite laminaire
~FLOW - [phys] (fluid motion which is steady and free from periodic variation and turbulence) écoulement *m* laminaire
~FLOW BURNER - [th eng] brûleur *m* à flamme laminaire
~LAYER - [gen] couche *f* laminaire
~TEXTURE - [geol] texture *f* lamelleuse
LAMINATE, to - [build] (to build up from layers of material cemented together with resin under heat

and pressure) laminer
LAMINATE, to - [metall] (to coat with lamina) laminer, séparer en couches
[geol] (of shales etc; a very fine stratifying) lamelliforme
LAMINATE - [gen] laminé
[plast ind] (sheet materials made by treating paper or textile bases with plastics, e.g. in solution) laminé *m*
LAMINATED - [gen] laminé, stratifié
~AERIAL - [radio] (ultra-short wave aerial) antenne *f* laminée
~ANTENNA - s. laminated aerial
~ARMATURE - [el] induit *m* à lames
~BEAM - [build] (a beam formed of timbers and plates) poutre *f* en tôles et cornières, poutre *f* composée
~BRUSH - [el] (brush made from a number of layer which are insulated from one another) balai *m* à lames
~-BRUSH SWITCH - [el] (switch laminated contacts) interrupteur *m* à contacts laminés
~CHANNEL SECTION - [metall] profilé *m* stratifié en U
~CLAY - [geol] argile *f* feuilletée
~CLOTH - [gen] (cloth formed by cementing successive layers of fabric) stratifié-tissu *m*
~CONDUCTOR - [el] (conductor made up of a number of thin strips and used for armature winding of large machines) conducteur *m* à contacts laminés
~CONTACT - [el] (switch contact made up of a number of laminations) contact *m* laminé
~CORE - [el] (core built up of laminations) noyau *m* de fer divisé, noyau *m* en fer feuilleté
[el] tore *m* à structure laminée
~FABRIC - s. laminated cloth
~FRACTURE - [geol] cassure *f* lamellaire
~GLASS - [glass] (a safety glass) glace *f* de sécurité
~IRON CORE - [el] noyau *m* de fer divisé
~MAGNET - [el] (a magnet up of magnetized strips) électroaimant *m* à noyau feuilleté
~MOULDING - [plast ind] pièce *f* moulée en stratifié
~PANEL - [metall] (a thick sheet of laminate) panneau *m* en stratifié
~PAPER - [paper man] (sheet material prepared by cementing layers of paper together) feuille *f* de papier
~PLASTICS - [plast ind] (materials consisting of superimposed layer of synthetic resin coated or impegnated filler bonded together, usually , by heat and pressure to form a single piece) stratifié *m*, matières *fpl* plastiques laminées
~POLE - [el] (pole with a core made of laminations) pôle *m* à noyau de fer divisé
~POLE-SHOE - [el] (field-magnet pole with a pole-shoe built up of laminations) épanouissement *m* polaire laminé
~RECORD - [acoust] (a record with a surface material different from the core material) disque *m* feuilleté
~SECTION - [metall] profilé *m* stratifié
~SHEET - [plast ind] (plastic sheet material built up from superimposed layers of thinner sheet bonded together) plaque *f* stratifiée
~SOIL - [soil] sol *m* feuilleté; sol *m* lamellé

LAMINATED SPRING - [mech] ressort *m* à lames
~TUBE - [plumb] tuyau *m* stratifié
~VENEERS - [plast ind] (sheet material built up from thin layers of wood (veneer) bonded together) strates *mpl* à mouler
~WIRE-CORE - s. laminated iron core
~WOOD - [timber] (sheet material formed of thin layers of wood bonded with plastic material) bois *m* stratifié
~YOKE - [el] culasse *f* à lames
LAMINATES - [plast ind] (materials produced by bonding together reinforcing sheets of paper or fabric, impregnated with a synthetic thermosetting resin, by the application of heat and pressure) stratifiés *mpl*, produits *mpl* laminés
LAMINATING - [plast ind] (the process of building and consolidating the layers of a laminated plastics material) stratification *f*, laminage *m*
[metall] calandrage *m*
~RESIN - [chem] (synthetic resin used to cement the sheets of laminated materials) résine *f* pour stratifiés
~SHEETS - [plast ind] strates *mpl* imprégnés pour fabriquer des stratifiés
~WEB - [plast ind] strate *m* continu
LAMINATION - [plast ind] (the process of building up material from sheets cemented together) lamination *f*, strate *m*
[metall] laminage *m*, séparation *f* en couches
[el] (shaped piece of special iron alloy used in building up the field cores of electro-magnetic machines or devices, to reduce hysteresis) tôle *f* de noyau, feuilletage *m*
[geol] schistosité *f* de stratification
~LAYER - [el] couche *f* de tôle
LAMINATOR - [mech] (a combiner) laminoir *m* pour combiner des feuilles
LAMINITIS - [vet] fourbure *f*
LAMINOGRAPHY - [radiat] (body section radiography) radiographie *f* en coupe, tomographie *f*
LAMP - [gen] (a vessel, suitably equipped with a wick and filled with oil or a similar combustible, for giving light) lampe *f*, lanterne *f*
[el] lampe *f*, ampoule *f*
[auto] lampe *f*
~ADJUSTMENT - [photo] réglage *m* de la lampe
~BANK - [cin] banc *m* d'éclairage, herse *f*
~BASE - [light] (lamp cap) douille *f* de lampe
~BLACK - [dye] (a variety of carbon black used as a pigment) noir *m* de lampe
~BRACKET - [gen] support *m* de lampe
~BULB - [el] ampoule *f* d'éclairage
~CALL - [telecomm] appel *m* par signalisation lumineuse
~CAP - [el] (the cap at the base of a filament containing the terminals) capuchon *m* pour lampe, calotte *f* pour lampe
~-CAPPING CEMENT - [ind chem] (a heat-hardening cement used for fixing the cap of an electric lamp to the envelope) ciment *m* de scellement pour lampe
~CENTERING - [photo] centrage *m* de la lampe
~COLUMN - s. lamp standard
~CORD - [el] (flexible cord) cordon *m* flexible
~HOUSE - [cin] (a structure containing the projector lamp) lanterne *f*
~INDICATOR - [telecomm] signal *m* lumineux

LAMP JACK STRIP - [el] réglette *f* de lampes
~ METHOD - [chem] (apparatus designed to determine the sulphur contents in petrol) methode *m* à la lampe
~ OIL - [chem] (kerosene) kérosène *m*, pétrole *m* lampant
~ PANEL - [telecomm] panneau *m* de lampes
~ POST - [constr] mât *m* d'éclairage, montant *m* de réverbère
~ POWER - [el] puissance *f* d'une lampe
~ REGULATOR - [el] régulateur *m* de la tension de la dynamo d'éclairage
~ RESISTANCE - [el] (resistance consisting of one or more filament lamps) résistance *f*
~ ROOM - [railw] lampisterie *f*
~ SHADE - [impl] abat-jour *m*
~ SIGNAL CALL - s. lamp call
~ SOCKET - s. lampholder
~ STANDARD - [impl] torchère *f*
LAMPAS - [med] lampas *m*
[vet] (swelling of the mucous membrane behind the incisor teeth of a horse) fève *f*, lampas *m*
LAMPHOLDER - [el] (device for supporting an electric lamp) porte-lampe *m*, douille *f*
~ PLUG - [el] (device for connecting an electric cord to an ordinary lampholder) douille *f* à prise de courant
LAMPREY - [zool] lamproie *f*
LAMPROPHYRES - [geol] (igneous rocks related to intrusive bodies) lamprophyres *mpl*
LANARKITE - [min] (a very rare sulphate of lead occurring in Lanarkshire, Scotland) lanarkite *f*
LANATOSIDE C - [chem] (a glucoside obtained from Digitalis lanata and used in medicine) lanatoside C *m*
LANCASHIRE BOILER - [metall] (cylindrical steam-boiler with two longitudinal furnace tubes) chaudière *f* à deux tubes-foyers, chaudière *f* de Lancashire
LANCE, to - [med] percer, inciser
LANCE - [gen etc] lance *f*
[tool] (of a cutting tool) lance *f*
LANCEOLATE - [gen & bot] (spear-shaped, tapering to each end) lancéolé, en fer de lance
LANCET - [med] lancette *f*, bistouri *m*
[arch] (sharply pointed arch) à lancette
~ ARCH - [arch] (sharply pointed arch) arc *m* en lancette, arc *m* en ogive
LANCINATING - [gen & med] lancinant
LANCING - [metall] (in sheet-metal work) oxycoupage *m*
LAND, to - [aero] (to bring an aircraft down into the surface of the ground) atterrir
[of a seaplane] amerrir
[naut] débarquer, placer, amener
[oil ind] placer
[oil ind] poser une colonne de tubes
LAND - [gen] terre *f*, terrain *m*, pays *m*
[mech] (surface between two grooves) plat *m*, intervalle *m*
[mech] (the surface between the grooves of a piston) cordon *m*
[metall] (of die or mould) surface *f* d'appui
[firearms] cloison *f*
[acoust] (of a record surface, between two adjacent grooves of a mechanical recording) pas *m*
~ AREA - [plast ind] (the whole of the area of contact, perpendicular to the direction of application of the pressure, of the seating faces of a mould, i.e. those faces which come into contact with one another when the mould is closed) surface *f* d'appui
LAND-BREEZE - [met] (wind blowing off-shore during the night in fair weather, caused by unequal cooling over land and sea) brise *f* de terre
~ FILL - [constr] remblaiement *m*, surélévation *f*, opération *f* de remblayage
~ FILTRATION - [hydr] filtration *f* dans le sous-sol
~ IMPROVEMENT - [agric] ameliorations *fpl* foncières
~ LENGTH - [mech] (of an extruder head) guidage *m* de la vis
~ LEVELLER - [agric] niveleuse *f*
~ PACKER - [agric] rouleau *m* compacteur
~ PRESSER - s. land packer
~ RECLAMATION - [agric] défrichement *m*, récupération *f* de terres incultes
~ SURVEY - [surv] arpentage *m*, levé *m* de terrains
~ -SURVEYOR - [agric] arpenteur *m*
~ TENURE - [agric] structure *f* agricole
~ UNDER CROP - [agric] surface *f* cultivée
~ UTILIZATION - [agric] exploitation *f* du sol
~ WHEEL - [agric] roue *f* de terre
~ WIDTH - [mech] (of the bore of a mould) largeur *f* du filet
LANDED - [naut] débarqué
[aero] atterri
[leg] foncier, prédial
~ FORCE - [die cast & plast ind] (a die provided with lands) poinçon *m* avec portée
LANDFALL - [naut] atterrissage *m*
[geol] (a landslide) glissement *m* de terrain
LANDING - [aero] (the operation or act of brinding an aircraft from the airborne condition to that of being fully supported by the ground of sea) atterrissage *m*
[of seaplanes] amérissage *m*
[constr] (platform at the head of a stair) palier *m*, repos *m*, carré *m*
[constr] (pier or quayside) pilier *m*
~ ANGLE - [aero] angle *m* d'atterrissage
~ AREA - [aero] (that part of an aircraft expressly designed for take-off and landing) aire *f* d'atterrissage
~ ATTITUDE - [aero] attitude *f* d'atterrissage
~ BEACON - [aero] (beacon transmitting the radio beam) balise *f* d'atterrissage
~ BEAM - [radio] (a radio beam transmitted from a beacon to show an aircraft pilot his altitude and position in respect to the centre-line of the runway) faisceau-guide *m* d'atterrissage
~ CABLE - [aero] câble *m* d'atterrissage
~ COMPASS - [instr] (a standard magnetic compass used for calibrating aircraft instruments) compas *m* étalon
~ CONFIGURATION - [aero] (outline of an aircraft ready for landing, i.e. with landing gear and flaps etc., extended) configuration *f* d'atterrissage
~ CRAFT - [naut] petit bâtiment *m*
~ DIRECTION INDICATOR - [aero] (device to show direction for take-off or landing) indicateur *m* de direction d'atterrissage
~ DISTANCE - [aero] parcours *m* d'atterrissage
~ DOGS - [mining] taquets *mpl*, clichage *m* pour cages
~ FIELD - [aero] champ *m* d'atterrissage

LANDING FLAP - [aero] volet *m* auxiliaire d'atterrissage
~GEAR - [aero] (a carriage fixed under a land aircraft to support it when not in flight) train *m* d'atterrissage
[of a seaplane] flotteurs *mpl* d'amérissage
~GEAR BAY - [aero] (recess into which landing-gear is retracted) travée *f* du train d'atterrissage, baie *f* du train d'atterrissage
~-GEAR BRAKES - [aero] (brakes operating on landing wheels, for controlling movement of aircraft on the ground) freins *mpl* du train d'atterrissage
~-GEAR EXTENDED SPEED - [aero] (speed of aircraft with landing gear in the "down" or extended position) vitesse *f* maximum train d'atterrissage sorti
~-GEAR LOCK SWITCH - [aero] interrupteur *m* d'arrêt du train d'atterrissage
~-GEAR OPTICAL INSPECTION SYSTEM - [aero] (visual device for checking down-locking of landing gear) indicateur *m* optique d'arrêt du train d'atterrissage sorti
~-GEAR RETRACTION LOCK - [aero] (device for securing landing gear in the "up" or retracted position) cliquet *m* d'arrêt du train d'atterrissage rétracté
~-GEAR UP LOCK SWITCH - [aero] interrupteur *m* de verrouillage du train d'atterrissage rétracté
~-GROUND - [aero] (an area specifically used for aircraft take-off and landing) champ *m* d'atterrissage
~LANE - [aero] (a straight portion of a landing strip, specifically used for take-off and landing) piste *f* d'atterrissage
~LIGHT - [aero] (luminous beacon for use while landing) éclairage *m* d'atterrissage de l'avion
~NET - [fishing] épuisette *f*
~PATH - [aero] (the portion of the flight path in the immediate vicinity of the landing area) trajectoire *f* d'atterrissage
~PROCEDURE - [aero] (the part of the approach extending from the time when the aircraft reaches the centre-line of the runway until landing is effected or given up as missed-approach) procédure *f* d'approche finale
~RUN - [aero] (the distance which an aircraft travels during alighting from the first contact with the ground until it ceases to move) longueur *f* d'atterrissage
~SLIP - [oil ind] coin *m* d'ancrage
~SPEED - [aero] (the normal alighting speed of an aircraft) vitesse *f* d'atterrissage
~SPOOL - [oil ind] bobines *fpl* d'ancrage de la colonne
~STAGE - [naut] débarcadère *m*, ponton *m* de débarquement
~STRIP - [aero] (an elementary form of landing ground, consisting merely of a track for take-off and landing) terrain *m* d'atterrissage
~WIRES - [aero] (wires designed to resist forces acting oppositely to lift) haubans *mpl* de soutien au sol
LANDMARK - [surv] (any prominent object of which the position is shown on a chart or otherwise known, e.g. a church spire or outstanding natural feature) borne *f*, repère *m* topographique
~BEACON - [aero] radiophare *m* de navigation
LANDPLANE - [aero] (an aircraft designed to operate from the land, as opposed to one operating from water surfaces) avion *m* terrestre
LANDSBERGER APPARATUS - [instr] (a boiling point measurement apparatus, in which the vapour of the solvent is used to heat the solution under examination) appareil *m* de Landsberger
LANDSCAPE - [gen] paysage *m*
~LENS - [photo] (panoramic lens) objectif *m* simple
LANDSIDE - [agric] (flat plate attached to the body of a plough) contre-sep *m*
LANDSLIDE - s. landslip
LANDSLIP - [geol] (sudden of masses or rocks and soil from higher to lower levels) éboulement *m*, glissement *m*
LANE - [gen] chemin *m* rural, chemin *m* creux, ruelle *f*
[naut & nav] route *f* de navigation
[road traffic] (subdivision of road) voie *f*
LANGBEINITE - [min] (mineral consisting of magnesium potassium sulphate) langbeinite *f*
LANGITE - [min] (rare ore of copper) langite *f*
LANGUAGE - [gen] langue *f*, langage *m*
[mus] (thin block; the metal part in a flue pipe separating the resonator from the foot) biseau *m*
[comput] langue *f*, langage *m*
~TRANSLATION COMPUTER - [comput] machine *f* à traduire
LANOLIN - [chem] (wool fat. A base of cosmetics and pharmaceuticals) lanoline *f*, lanoléine *f*
LANOSTEROL - [chem] (a sterol present in lanolin and having surface-active properties) lanostérol
LANTERN - [constr] (erection on the toop of a roof to admit light or for ventilation purposes) lanterne *f*, lanterneau *m*
[light] lanterne *f*, fanal *m*
~CHUCK - [mech] mandrin *m* de serrage à lanterne
~LIGHT - [constr] lanterneau *m*
~OF THE REVOLVING BOX - [text] lanterne *f* de la boîte revolver
~PICTURE - [cin] (the image projected by a lantern slide) image *f* d'une diapositive
~PINION - [mech] lanterne *f*
~RING - [mech] anneau *m* à lanterne
~SIDE OF CYLINDER - [text] côté *m* de la lanterne de cylindre
~SLIDE - [cin] (transparent picture used for projection) diapositive *f* de projection
LANTHANIDE CONTRACTION - [nucl] (the decreasing sequence of radio of the tripositive rare-earth ions with increasing atomic number in the group lanthanum through lutecium) contraction *f* de lanthanides
~SERIES - [chem] (generic name for the rare earth elements) lanthanides *fpl*
LANTHANIDES - s. lanthanide series
LANTHANUM - [chem] (a rare earth element, symbol La, A.N. 57, A.W. 139.0. It has uses as a reducing agent and in electronics) lanthane *m*
~ACETATE - [chem] acétate *m* de lanthane
~BROMATE - [chem] bromate *m* de lanthane
~CHLORIDE - [chem] chlorure *m* de lanthane
~NITRATE - [chem] (a germicide) nitrate *m* de lanthane
~OXIDE - [chem] (a component of special glasses and ceramics) oxyde *m* de lanthane
~SULPHATE - [chem] sulfate *m* de lanthane

LANTHANUM TUNGSTATE - [chem] tungstate *m* de lanthane
LANTHIONINE - [chem] (a non-essential amino acid used in research) lanthionine *f*
LANUGINOUS - [zool] (with a woolly coating) lanugineux
LANUGO - [zool] (pre-natal hair) poil *m* de duvet
LANYARD - [naut] aiguillette *f*, ride *f*
[fire] tire-feu *m*, cordon *m* tire-feu
LAP, to - [mech] (to produce a fine finish on metals etc. by means of a rotating cylinder or disk of soft material charged with fine abrasive) roder
[carp] faire un joint à recouvrement
[el] (to lag cover a cable) guiper
[constr] enchevaucher
LAP - [impl] (a disk cylinder of soft material used in lapping) rodoir *m*
[mech] (a soft metal body, normally a disk or cylinder, which is fed with fine abrasive and rotated in close contact with a part to receive a fine finish) disque *m* à abrasif
[tool] (a tool designed to cut glass or precious stones) meule *f* polissoire
[metall] (a surface defect on forged or rolled steel) repliure *f* de laminage
[glass] (a fold on a surface) pli *m*
[text] (rolled sheet of fibres) nappe *f*
[el] guipage *m*
[mech] (in internal combustion engines) chevauchement *m*
[constr] (the length of the overlap of one slate over the next one) recouvrement *m*, chevauchement *m*
[paint] (a defect caused by the overlapping of a fresh coat) chevauchure *f*
[mech] (of a steam engine) métal *m* de repliure
[metall] (in welding) cordon *m* (de soudure)
~BELT - [astronaut] ceinture *f* de sécurité
~DISSOLVE - [cin] (a fade over) fondu *m* enchaîné
~DISSOLVE SHUTTER - [cin] obturateur *m* à rideau pour fondu
~DOVETAIL - [carp] (a kind of engle joint) assemblage *m* (par recouvrement) à queue d'aronde
~-ENDED PISTON RING - [mech] (type of piston ring having diagonally cut overlapping ends) segment *m* de piston à extrémités chevauchantes
~JOINT - [mech] (method of joining two parts in which one is superimposed on the other for some part of their areas) assemblage *m* à recouvrement, joint *m* à recouvrement, assemblage *m* à clin
~MACHINE - [text] plieuse *f*, enrouleur *m*
~-RIVETED JOINT - [mech] rivure *f* à recouvrement
~ROLLER - [text] (of a lap machine) cylindre *m* dérouleur
~SCALES - [text] balance *f* pour rouleaux
~SEAM WELDING - [metall] (in welding) soudure *f* continue à recouvrement
~THE WARP TIGHTLY, to - [text ind] enrouler la chaîne
~WELD - [metall] (a weld made with a lap joint) soudure *f* à recouvrement, soudure *f* par amorces, soudure *f* en écharpe
~-WELDING - [metall] (method of welding in which part of the material at the joint lies upon the other part to be joined to it) joint *m* à recouvrement
~WINDING - [el] (a form of two-layer winding) enroulement *m* imbriqué
LAPARECTOMY - [med] laparectomie *f*
LAPAROCOLECTOMY - [med] colectomie *f*
LAPAROCOLOSTOMY - [med] colostomie *f* latérale
LAPAROCOLOTOMY - [med] colotomie *f* iliaque
LAPARORRHAPHY - [med] laparorraphie *f*
LAPAROSCOPY - [med] laparoscopie *f*, péritonéoscopie *f*
LAPAROTRACHELOTOMY - [med] césarienne *f* basse
LAPEL - [gen] revers *m*
[text] revers *m* d'un habit
~MICROPHONE - [acoust] (microphone fitted in the clothing) microphone-boutonnière *m*
LAPILLI - [geol] (small round pieces of lava) lapilli *mpl*
LAPIS LAZULI - [min] (a semi-precious stone, (the sapphire of the ancients), consisting of calcite stained deep blue by sodalite, lazurite and hauyne and used in making ultramarine) lazulite *m*, pierre *f* d'azur, lapis *m*
LAPLACE'S LAW - [el] (the law giving the force exerted on an element carrying a current placed in a magnetic field) loi *f* de Laplace
LAPLACIEN - [math] laplacien
[nucl] (of a critical reactor) laplacien
LAPPAGE - [gen mech etc] (the length of the overlap) (longueur du) recouvrement
LAPPED ARMOURING - [el] armure *f* à recouvrement
LAPPER - s. lapping machine
LAPPET - [gen] (a fold, an overlapping part of a garment) love *m*, cache-entrée *m*
[text] (fabric resembling an embroidered cloth) basque *f* de vêtement, pan *m*, plumetis *m*
~LOOM - [text] métier *m* brodeur, métier *m* pour plumetis
~THREAD - [text] fil *m* brodeur
~WEAVING - [text] tissage *m* à points brodés
LAPPING - [mech] rodage *m*, polissage *m*
[text ind] (the arrangement of cloth in folds) pliage *m*
[gen] (lagging) recouvrement *m*
[gen] (the operation of wrapping an object) enroulement *m*, guipage *m*
[paper man] (the arrangement of paper infolds) pli *m* en portefeuille
[gen] (a finishing process) polissage *m*
[shoe man] chevauchure *f*, recouvrement *m*
~MACHINE - [mach tool] (machine tool for finishing the bores of cylinders etc) machine *f* à rader
[rubber ind] (for cables) rubaneuse *f* pour câbles
~WHEEL - [impl] meule *f* à polir
LAPPIT LOOM - [text ind] s. lappet loom
LAPSE, to - [gen] déchoir, manquer
[hydr] (of a stream) s'écouler
LAPSE - [gen] déchéance *f*, écoulement *m*
~RATE - [met] (the rate at which atmospheric temperature decreases as altitude increases) gradient *m* thermique vertical
LAPWING - [zool] vanneau *m*
LARCH - [bot] mélèze *m*
~BLACK - [bot] mélèze *m* occidental
~-TREE - [bot] mélèze *m*
LARD - [food] saindoux *m*, graisse *f* de porc
~HOG - [zool] porc *m* gras
~OIL - [food etc] (oil obtained from the fát of pigs, formerly in use as a cutting lubricant) huile *f* de saindoux
~PIG - s. lard hog
~STONE - [min] (soapstone, or steatite ; a coarse

variety of talc which is greasy to the touch) stéatite *f*
LARDER - [gen] garde-manger *m*
~BEETLE - [weeds] dermeste *m* du lard
LARGE - [gen] grand, gros
[met] (of the wind) largue
~-ANGLE SCATTERING - [nucl] diffusion *f* sous grand angle
~AREA CONTRAST - [telev] contraste *m* gros
~CAPACITY - [hydr etc] grande capacité *f*, grand débit *m*
[adj] à grande capacité, à grand débit
~-CAPACITY CABLE - [el & telecomm] câble *m* à grande capacité (à grand nombre de fils)
~-DIAMETER PIPE - [gas ind] (gas network) tube *m* de gros diamètre
~END OF CONNECTING ROD - [mech] grosse tête de bielle
~ORIFICE - [hydr] orifice *m* de grandes dimensions
~SCALE - [gen] à grande échelle
~SCREEN PICTURE TECHNIQUE - [cin] (picture projected in very large dimensions on a special large-size screen) technique *f* du grand écran
~-SCREEN TELEVISION PROJECTION - [telev] projection *f* de télévision à grand écran
~SIZED CABLE - s. large-capacity cable
LARMIER - [constr] (a corona over a door or window, designed as a dripstone) larmier *m*, bordure *f* de pignon
LARNITE - [min] (orthosilicate of calcium) larnite *f*
LARRY - [impl] (tool with a curved steel blade at the end of a long handle, used for mixing mortar) broyon *m* à mortier
[constr] (a kind of grout) mortier *m* liquide
[mining] plate-forme *f*
~CAR - [min] (coal charging car) chariot *m* à charbon, enfourneuse *f*
LARRYING - [constr] (the operation of pouring liquid mortar) coulage *m* de mortier
LARVA - [zool] larve *f*
LARVAL - [zool] larvaire, en forme de larve
[med] latent, larvé
LARVIPAROUS - [zool] larvipare
LARYNGEAL - [anat] laryngé
~EPILEPSY - [med] ictus *m* laryngé, vertige *m* laryngé
LARYNGECTOMY - [med] laryngectomie *f*
LARYNGISMUS - [med] cornage *m*
LARYNGITIS - [med] laryngite *f*
LARYNGOCELE - [med] laryngocèle *f*
LASER - [electron] (optical maser using an infrared frequency as pumping frequency) laser *m*, maser *m* infrarouge
LASH, to - [gen] fouailler, fouetter, cingler
[gen] (to bind together) lier, attacher
LASH - [gen] coup *m* de fouet, cinglon *m*
[gen] (a piece of string used for a whip) lanière *f* de fouet
[mech] (of valves) jeu *m*
[anat] cil *m*
[text] lacet *m*
LASHING - [gen] liage *m*, ligature *f*
[naut] (anything used to bind together) amarrage *m*, amarre *f*
[min] (South African expression meaning the removal of broken rock) enlèvement *m* des matières détritiques
[min] enlévement *m* de minerai
LASHING - [constr] corde *f* d'echafaud, chablot *m*
[constr] drisse *f* (d'échafaudage), drisse *f* d'échafaudage, vingtaine *f* d'échafaudage
[mech] fouettement *m*
~-POINT - [mech] (a securing fixture e.g. an eye-bolt, to which lashings may be made fast in stowing cargo or the like) point *m* d'attache
LAST - [gen] dernier
[shoe man] (wooden or metal model of the foot) forme *f*
~CARD - [comput] dernière carte *f*, carte *f* de fermeture
~CARD INDICATION - [comput] indication *f* contrôle de groupe
~HOOK - [impl] tire-forme *m*
~QUARTER - [astr] (of the moon) dernier quartier *m*
~RUNNINGS - [brew ind] dernier lavage *m*
LASTIC - [plast ind] (synthetic material exhibiting high elastic properties) lastic *m*
LASTING - [gen] durable, permanent
~YARN - [text] fil *m* lasting
L.A.T. - s. Local Apparent Time
LAT - [abbrev. for latitude] latitude *f*
LATCH, to - [gen] fermer à la clenche
[naut] (to secure) ancrer
LATCH - [mech] (a locking device in which the moving portion turns on a pivot) loquet *m*, clenche *f*
[mech] (the movable piece of a locking device) cliquet *m*, verrou *m*, valet *m* d'arrêt
[auto] loquet *m*
[mining] levé *m* de plan de mine
~BAR - [impl] bras *m* de calage
~HOOK - [impl] crochet *m* de calage, oreille *f* de verrouillage
~-IN RELAY - [electron] relais *m* à adhérence
~JACK - [oil ind] fourche *f* à barette
~LEVER - [auto] levier *m* de loquet
~-ON CENTRALIZER - [oil ind] centralisateur *m* de pression
~PLATE - [mech] (a plate used for retaining a removable core of relatively large diameter or for holding insert carrying pins on the upper part of a mould) plateau *m* support de prisonniers
~-TIGHTENING SCREW - [mech] vis *f* de calage, vis *f* de blocage
LATCHED HOOK ROD - [text] tige *f* à crochet enclenché
LATCHET - [shoe man] cordon *m* de soulier
LATCHING - [mech] encliquetage *m*
[mech] blocage *m*, verrouillage *m*
~RELAY - [el] relais *m* à verrouillage
LATE - [gen] tard, en retard, tardif
~CROP - [agric] moisson *f* tardive
~EFFECT - [nucl] effet *m* tardif
~IGNITION - [auto] retard *m* à l'allumage
~TIMING - [auto] calage *m* en retard
LATEEN SAIL - [naut] voil *f* latine
LATENCY - [el biol] (the condition in an excitable tissue in the interval between the application of a stimulus and the first indication of a response) latence *f*
[comput] (the delay while waiting for the information called from the store) temps *m* d'attente
~PERIOD - s. latency
~TIME - [el biol] (the interval between radiation and the appearance of the effect) période *f* latente
LATENSIFICATION - [photo] (intensification of the

latent image) renforcement *m* de l'image latente
LATENT - [gen] latent, caché
[zool] (in a state of arrested development but capable of resuming activity and undergoing further development) latent, dormant
[phys chem etc] (in a condition of rest but capable of undergoing further development) latent
~ELECTRONIC IMAGE - [electron] (the electrical image of an optical image in the form of elementary charges distributed over a surface in certain camera tubes) image *f* électronique latente
~HEAT - [phys chem] (the heat expended in changing the state of a body without increasing its temperature) chaleur *f* latente, calorique *m* latent
~HEAT OF FUSION - [phys] (the increase of heat content when one mole of a solid is converted into liquid at its melting point) chaleur *f* latente du fusion
~HEAT OF SUBLIMATION - [phys] (the increase of heat content when one mole of a solid is converted into a vapour) chaleur *f* latente de sublimation
~IMAGE - [photo] (the image existing in chemical form on an exposed but undeveloped photographic plate) image *f* latente
~PERIOD - s. latency
~PHOTOGRAPHIC IMAGE - s. latent image
~SOLVENT - [chem] solvant *m* indirect
~TISSUE INJURY - [radiat] (injury appearing only some time after the irradiation) lésion *f* latente de tissu
LATERAL - [gen etc] (relating to a side, or situated on a side) latéral
[oil ind] dérivation *f* de conduite
~ACCELEROMETER - [instr] (an instrument designed to show lateral acceleration) accéléromètre *m* latéral
~AXIS - [aero] (a line drawn through the C.G. of an aircraft at right angles to the plane of symmetry) axe *m* de tangage, axe *m* latéral
~BLOCK PROFILE - [rubber ind] (of tyres) sculpture *f* à blocs latéraux
~BRANCH - [bot] (of hops) limbe *m*, pousse *f* latérale
~BUCKLING - [railw] (of the track) déjettement *m* horizontal (de la voie)
~BULGING - [soil] sol *m* se dérobant latéralement
~CHROMATISM - [opt] chromatisme *m* latéral
~CONE - [geol] (volcanic) cône *m* adventif
~CONFINEMENT - [soil] étreinte *f* latérale
~-CORRECTION MAGNET - [electron] aimant *m* de correction latérale
~DISPLACEMENT - [soil] refoulement *m* latéral, déplacement *m* latéral
~DIVERGENCE - [aero] (any divergence in the directions of roll, side-slip or yaw) engagement *m* spiral
~EXPANSION - [soil] dilatation *f* transversale ou latérale
~FORCE - [phys] (the component which acts along the lateral axis (including the resolved gravity component) force *f* transversale
~GUIDE - [mech] (in lifts) guidage *m* latéral
~GUIDE FORCE - [rubber ind] (of tyres; the deflecting force) force *f* déviante, force *f* de guidage latéral
~INCLINATION - [gen & auto] inclinaison *f* latérale
~INSTABILITY - [aero] (the tendency of the aircraft to develop an increasing oscillatory movement after a disturbance in the sense of roll) instabilité *f* de roulis, instabilité *f* latérale
LATERAL INVERSION - [photo] (reversal of the image) renversement *m* de l'image
~KEELSON - [shipbuild] carlingue *f* latérale
~MAGNIFICATION - [opt] agrandissement *m* latéral
~MAGNIFYING POWER - [telev] grandissement *m* frontal, grossissement *m* latéral
~MIRAGE - [astronaut] mirage *m* latéral
~OSCILLATION - [aero] (periodic variation in the sense of roll, side-slip or yaw) oscillation *f* transversale
~PARITY CHECK - [comput] contrôle *m* de parité vertical
~PRESSURE - [phys] pression *f* latérale
~RECORDING - [el acoust] (mechanical recording in which the groove modulation is perpendicular to the motion of the recording medium) enregistrement *m* latéral
~SEPARATION - [aero] (in air traffic regulation, the spacing-out in a horizontal plane of aircraft flying at the same height) espacement *m* latéral
~SHIFT - [min] (displacement of outcrops horizontally) mouvement *m* latéral
~SPHERICAL ABERRATION - [opt] aberration *f* sphérique latérale
~SPIKELET - [bot] épillet *m* latéral
~STABILITY - [aero] (stability in respect of roll, side-slip and yaw) stabilité *f* latérale
[railw] stabilité *f* transversale
~STRAIN - [phys] déformation *f* latérale
~THRUST - [phys] poussée *f* latérale
[soil] pression *f* transversale
~TRAVERSE - [mech] (the longitudinal play given to a locomotive trailing axles) jeu *m* axial
~VELOCITY - [aero] (the resolved velocity component of an aircraft in respect to the undisturbed air, along the direction of the lateral axis) vitesse *f* latérale
~YIELD - [soil] gonflement *m* latéral
LATERALLY - [gen] latéralement
LATERITE - [geol] (type of clay formed under tropical climatic conditions) latérite *f*
~SOIL - [geol] sol *m* latéritique
LATERIZATION - [geol] (the process whereby rocks are converted into laterite) latérisation
LATEROFLEXION - [med] latéroflexion *f*
LATEROPOSITION - [med] latéroposition *f*
LATEROPULSION - [med] latéropulsion *f*
LATEROTORSION - [med] latérotorsion *f*
LATEX - [rubber ind] (plural latices or latexes: the former preferred. A water-based milky fluid containing globules of natural or synthetic rubber or plastic in suspension) latex *m*
~-ADHESIVE - [rubber ind] (an adhesive prepared directly from latex) adhésif *m* au latex
~-BITUMEN EMULSION - [rubber ind] émulsion *f* latex-bitume
~-CEMENT - [rubber ind] latex-ciment *m*
~-CEMENT COMPOSITION - [rubber ind] composition *f* latex-ciment
~CHANNEL - [rubber ind] rigole *f* d'écoulement du latex
~CLARIFYING CENTRIFUGE - [rubber ind] centrifuge *f* à clarifier le latex
~CONCENTRATED BY CREAMING - [rubber ind] latex *m* concentré par crémage

LATEX DEPOSIT - [rubber ind] dépôt *m* de latex
~FOAM - [rubber ind] mousse *f* de latex, caoutchouc *m* spongieux
~FOAM CUSHIONING - [rubber ind] rembourrage *m* de mousse de latex
~GELLING - [rubber ind] gélification *f* du latex
~MOULDING - [rubber ind] moulage *m* au latex
~PAINT - [paint] vernis *m* au latex
~RUG-BACKING MIX - [rubber ind] mélange *m* antidérapant de latex pour tapis
~SKIM - [rubber ind] (skimmed latex) skim *m* latex (résidu de centrifugation)
~SPINNING - [rubber ind] fabrication *f* de fils en latex par filage
~THREAD - [rubber ind] fil *m* de latex
~TUBE - [bot] tube *m* laticifère, canal *m* de latex
~WALL PAINT - s. latex paint
LATH, to - [constr] (to cover with laths for plastering) latter, voliger
LATH - [gen] (a thin strip of wood) latte *f*
[constr] (the covering material before plastering) latte *f*, palançon *m*, échalas *m*, volige *f*
[constr] (a metal lathwork for ceilings) échalas *m* de treillis
[mining] palplanche *f*
~HAMMER - [impl] hachotte *f*
~NAIL - [constr] clou *m* à latter
~-SEAT - [constr] siège *m* à claire voie
~WITH PINS - [text] latte *f* avec chevilles
~-WOOD - [constr] bois *m* de fente
LATHE - [mach tool] (a machine tool for producing cylindrical forms, facing, boring and cutting screws) tour *m*
[text] (of a loom) battant *m*, chasse *f*
~APRON - [mach tool] tablier *m* d'un tour, traînard *m* d'un tour
~ATTACHMENTS - [mach tool] accessoires *mpl* du tour
~BED - [mach tool] banc *m* de tour
~BORE, to - [mech] aléser au tour
~-BORING - [mach tool] alésage *m* au tour
~CARRIAGE - [mech] chariot *m* de tour
~CARRIER - [mach tool] (the clamp attached to the work between the centres) toc *m* de tour, doguin *m*
~CENTRE - [mach tool] pointe *f* de tour
~CHUCK - [mach tool] mandrin *m* d'un tour
~DOG - [mach tool] toc *m* pour tours
~FEED - [mach tool] avance *f*
[mach tool] (the controlled movement of the working tool) pression *f*
~HEADSTOCK - [mach tool] poupée *f* de tour
~JAW - [metall] mâchoire *f* de tour
~SADDLE - [mach tool] traînard *m* du tour
~SPINDLE - [mach tool] mandrin *m* de tour
~TAILSTOCK - [mach tool] contre-pointe *f* de tour
~TOOLS - [mach tool] (turning tools with edges of various shapes) outils *mpl* de tour
~WITH CIRCULAR MOVING SHUTTLES - [text ind] battant *m* à navettes circulaires
LATHER, to - [chem] (to cover with froth, usually of soap and water) mousser
LATHER - [chem] (the foam produced by a detergent in a washing operation) mousse *f*
[zool] (of horses) écume *f*
~BOOSTER - [chem] (a substance used to increase the foaming properties of a detergent) exalteur *m* de mousse
LATHER COLLAPSE - [chem] (subsidence or destruction of the foam produced by a detergent) chute *f* de la mousse
~VALUE - [chem] (a numerical index of the foaming properties of a detergent) indice *m* de mousse
LATHING - [constr] (system of laths) lattage *m*
(the operation intended to cover a surface with laths prior to plastering) voligeage *m*
LATHWORK - s. lathing
LATICIFEROUS - [rubber ind] (carrying latex) laticifère
~CELL - s. laticiferous duct
~DUCT - [bot] (cavity from which latex is secreted) canal *m* laticifère
LATITUDE - [astr] latitude *f*
~AND LONGITUDE NETWORK - [surv] méthode *f* des coordonnées
LATRINE - [gen] latrine *f*
LATTEN - [metall] feuille *f* de laiton
~BRASS - s. latten
LATTICE - [gen] (any structure consisting of diagonal crossing bars) treillis *m*, treillage *m*
[nucl] (the arrangement of fissile and non-fissile material in a nuclear reactor) réseau *m*
[text] (of a carding machine) claie *f*
[cryst] (regular array of points in space) chaîne *f*
[chem] (space lattice) réseau *m*
~BARS - [constr] (diagonal gracing of struts) barres *fpl* en treillis
~BEAM - [constr] poutre *f* en treillis, poutre *f* évidée, poutrelle *f* à croisillons, poutre *f* à jour
~BRIDGE - [constr] pont *m* en treillis
~CALCULATION - [nucl] calcul *m* des réseaux
~CELL - [phys] (the simplest geometric figure including all the characteristics of a crystal lattice) cellule *f* élémentaire
[nucl] (of a reactor) cellule *f* de réacteur
~COIL - [el] (for armature windings) bobine *f* en lattis, gabion *m*
~COMPOUNDS - [phys] composés *mpl* à valence angulaire
~CONSTANT - [phys] (length representing the size of the unit cell in a crystal lattice) constante *f* de réseau
~DEGENERATION OF CORNEA - [med] kératite *f* en grillage
~DESIGN - [nucl] (the planning of the assembly of material in a reactor) géométrie *f* du réseau
~DIMENSIONS - [phys] dimensions *fpl* du réseau
~DRIERS - [text ind] séchoirs *mpl* à tablier
~DYNAMICS - [phys] (branch of the theory of the solid state relating to the properties of the thermal vibrations of crystal lattices) dynamique *f* du réseau
~ENERGY - [phys] (the potential energy of a crystal lattice) énergie *f* du réseau
~FILTER - [telecomm] filtre *m* à treillis
~GIRDER - [constr] (a built-up girder with diagonal bracing in both directions) poutre *f* en treillis, poutre *f* à/en treillis *m* éventuellement panne *f* en trellis *m*
~IMPERFECTIONS - [nucl] (any deviation from a perfect homogeneous crystal lattice) défauts *mpl* du réseau
~MAST - [radio] (aerial mast consisting of latticework material) mât *m* en lattis, pylône *m* en treillis
~NETWORK - [el] (filter network consisting of two

pairs of identical arms on opposite sides of a square) réseau *m* maillé
[electron] circuit *m* en pont, montage *m* en pont
LATTICE PITCH - [nucl] pas *m* du réseau
~PLANE - [nucl] (atomic plane) plan *m* réticulaire
~REACTOR - [nucl] (nuclear reactor in which the fuel elements are arranged in a tallice pattern) réacteur *m* à réseau
~SPAN - [constr] portique de fer profilé à ferme
~STRUCTURE - [nucl] (the form of a lattice in a reactor) structure *f* du réseau
~SUM - [cryst] total *m* d'un réseau
~THEORY - [cryst] (the theory of the lattice principle in crystallography) théorie *f* du réseau
~TOWER - [el constr etc] pylône *m* en treillis
~TRUSS - [constr] ferme *f* à treillis
~UNITS - [phys] corps *mpl* élémentaires du réseau
~VIBRATION - [cryst] (thermal vibration of a crystal lattice) vibration *f* du réseau
~WINDING - [el] (winding consisting of lattice coils) enroulement *m* réticulaire
~WINDOW - [constr] (window with diamond-shaped panes supported in a leaden frame) fenêtre *f* treillagée, fenêtre *f* à vitraux sertis de plomb
LATTICEWORK - [gen] treillage *m*, treillis *m*, grillage *m*
LAUDANINE - [chem] (a opiate used in medicine) laudanine *f*
LAUDANOSINE - [chem] (an opiate derived from opium) laudanosine *f*
LAUDANUM - [chem] (tincture of opium. A narcotic and analgesic) laudanum *m*
LAUE PHOTOGRAPHIC METHOD - [opt] (a method of x-ray analysis of crystal structures) méthode *f* photographique de Laue
LAUGHING GAS - [chem] (old name for nitrous oxide) gaz *m* hilarant
LAUMONITE - [min] (a zeolite essentially consisting of silicate of calcium and aluminium) laumonite *f*, laumontite *f*
LAUNCH, to - [gen] lancer
[naut] (of ships) lancer, mettre à l'eau
[aero] (from an aircraft carrier etc) lancer
LAUNCH - [gen] chaloupe *f*
[naut] chaloupe *f*, vedette *f*
[naut] (of ships) lancement *m*, mise *f* à l'eau
[astronaut] lancement *m*, moment *m* du lancement
~COMPLEX - [astronaut] base *f* de lancement
~CONTROL - [astronaut] commande *f* et réglage lors du lancement
~EMPLACEMENT - [astronaut] emplacement *m* de lancement, aire *f* de lancement
~PAD - [astronaut] table *f* de lancement, plate-forme *f* de lancement
~SITE - [astronaut] aire *f* de lancement
~STAND - [astronaut] installation *f* de lancement
~VEHICLE - [astronaut] lanceur *m*
LAUNCHER - [astronaut] lanceur *m*
LAUNCHING - [gen] lancement *m*
[naut] (of ships) lancement *m*, mise *f* à l'eau
[electron] (the act of feeding energy from a coaxial cable into a waveguide) attaque *f*
~ANGLE - [astronaut] angle *m* de lancement
~BASE - [astronaut] base *f* de lancement
~FRAME - [astronaut] rampe *f* de lancement
~PAD - [astronaut] plate-forme *f* de lancement, table *f* de lancement
LAUNCHING PLATFORM - [rockets] s. launching pad
~RAIL - [astronaut] rampe *f* de lancement, rail *m* de lancement
~SITE - [astronaut] aire *f* de lancement
~STAND - s. launching frame
~TOWER - [astronaut] portique *m* de lancement
LAUNDER, to - [gen] (of linen etc) blanchir
LAUNDER - [min] (wooden through for conveying water or crushed ore) auge *f*, caniveau *m*
[impl] rigole *f*
[metall] (foundry) canal *m* de coulée, chenal *m* de coulée
[metall] rigole de coulée
LAUNDERING - [gen] blanchissage *m*
LAUNDRY - [gen] blanchisserie *f*, lessive *f*
~BLUE - [chem] (a mixture of Chinese blue and oxalic acid, or ultramarine and sodium carbonate, used to impart whiteness to linen) bleu *m* fixe
LAUREL - [bot] (Laurus nobilis, the berries of which are reputed to have some slight therapeutic effect) laurier *m*
~OIL - [chem] (volatile, yellow liquid obtained by distillation of leaves and fruits of the laurel. A starting material for the production of lauric acid) huile *f* de graines de laurier
~TREE - [bot] laurier *m*
LAURENT HALF-SHADE PLATE - [opt] (semicircular half-wave plate of a crystal between the polarizer and analyzer, with its optic axis at a small angle with the principal section of the polarizer) lame *f* à pénombre de Laurent
LAURIC ACID - [chem] (a starting material for alkyd resins) acide *m* laurique
~ALCOHOL - [chem] alcool *m* laurique
LAUROYL CHLORIDE - [chem] (a synthesis intermediate) chlorure *m* de lauroyle
~PEROXIDE - [chem] (polymerization catalyst) peroxyde *m* de lauroyle
LAURYLALCOHOL - [chem] (derivative of coconut oil fatty acids with uses in rubber compounding) alcool *m* laurylique
~ALDEHYDE - [chem] (a perfumery agent) aldéhyde *m* laurylique
~CHLORIDE - [chem] (used in styrene-butadiene and in the synthesis of esters) chlorure *m* de lauryle
~MERCAPTAN - [chem] (a mixture of isomeric compounds with applications in the production of plastics and synthetic rubbers) lauryl mercaptan *m*
~METHACRYLATE - [chem] (a polymerizable monomer) métacrylate *m* de lauryle
~PYRIDINIUM CHLORIDE - [chem] (a component of insecticides and a dispersant and surface active agent) chlorure *m* de lauryle et pyridium
LAUTARITE - [min] (monoclinic iodate of calcium) lautarite *f*
LAUTER-MASH - [brew ind] trempe *f* claire
LAUTERTUB - [brew ind] (filter vat) cuve-filtre *f*
LAVA - [geol] (molten rock material) lave *f*
~FLOWS - [geol] coulées *fpl* laviques
~SLAG - [geol] scorie *f* de lave
~STONE - [geol] roche *f* lavique
LAVATORY - [gen] cabinet *m* de toilette
[impl] lavabo *m*
LAVENDER - [bot] (Lavendula officinalis; the source of oil of lavender) lavande *f*
~OIL - [chem] (an essential oil obtained from species of Lavandula and used in perfumery) essence

f de lavande
LAVENDER PRINT - [photo] copie *f* lavande
~WATER - [chem] (a mixture of oil of lavender and alcohol used as a perfume) eau *f* de lavande
LAW - [gen] loi *f*
~OF ALTERNATION - [phys] (in spectrology) loi *f* de l'alternance
~OF APPEARANCE OF UNSTABLE FORMS - [phys] (the unstable forms of monotropic substances are obtained from a liquid state before the stable form appears) loi *f* d'apparition de formes instables
~OF AREAS - [phys] (a particle in motion under the influence of any central force moves so, that the radius vector covers equal areas in equal intervals of time) loi *f* des aires
~OF CAILLETET AND MATHIAS - [phys] (law of the rectilinear diameter) loi *f* de Cailletet et Mathias
~OF CONSERVATION OF ANGULAR MOMENTUM - [phys] loi *f* de la conservation du moment cinétique
~OF CONSERVATION OF MECHANICAL ENERGY - [phys] loi *f* de la conservation de l'énergie mécanique
~OF CONSTANT HEAT SUMMATION - [phys] (the energy change involved in a chemical reaction) loi *f* de Hess
~OF COSINES - [comput] loi *f* des cosinus
~OF DEGRADATION OF ENERGY - [phys] loi *f* de diminution de l'énergie
~OF DIMINISHING RETURNS - [agric] loi *f* du rendement décroissant du sol
~OF DISTRIBUTION - [phys] loi *f* de répartition de Nernst
~OF FORMATION - [math] loi *f* de formation
~OF MUTUALLY OF PHASES - [phys] loi *f* de réciprocité des phases
~OF OCTAVES - [phys] (the name given by Newman to his hypotesis of the periodic system) loi *f* des octaves
~OF SEGREGATION - [genet] loi *f* de la ségrégation
~OF SINES - [mech] loi *f* des sinus
~OF SPECTROSCOPIC DISPLACEMENT - [phys] loi *f* de déplacement spectroscopique
~OF SUCCESSION - [leg] droit *m* de succession
~OF SUPPLY AND DEMAND - [comm] loi *f* de l'offre et de la demande
~OF THE MEAN - [math] loi *f* de la moyenne
~OF THE TRANSMISSIBILITY OF PRESSURE - [mech] loi *f* de l'accroissement de la pression, loi *f* de Pascal
LAWN - [bot] pelouse *f*, gazon *m*
[text] (fine plain cloth) batiste *f*
~BROOM - [agric] balai *m* à gazon
~MOWER - [agric] tondeuse *f* à gazon
~RAKE - [agric] râteau *m* à herbe
~SPRINKLER - [agric] arroseur *m*
~WEAVER - [text] tisseur *m* de linon, tisseur *m* en batiste
LAWRENCIUM - [chem] (suggested name for the synthetic element number 103) lawrencium *m*
LAWS OF MAGNETISM - [phys] (laws of electromagnetism under conditions of steady-state) lois *f*pl du magnétisme
~OF VORTEX MOTION - [mech] lois *f*pl du mouvement tourbillonnaire
LAXATIVE - [med] laxatif *m*
LAXATOR - [med] relâcheur *m*

LAY, to - [gen] poser, mettre, placer
[ropemaking] (the strands of a rope) commettre
[constr] (the concrete) verser en place
[shipbuild] (the keel of a vessel) poser la quille
[el etc] (cables) poser
[agric] (the crop) abattre
LAY - [el etc] (the axial length of a turn of the helix formed by the strand of a conductor) pas *m* de l'hélice de câblage
[print] (the position of the print on a sheet of paper) mise *f* en pages
[gen] (the position of direction in which something is laid) position *f*
[text] (of a rope) (the measure of the twist) commettage *m*
[text] (of a loom) battant *m*
[agric] soc *m*
[mech] (the position and direction of tool markings on a workpiece) direction *f* des rainures
[naut] (the proportion of the proceeds of a voyage paid to the sailors) part *f* des bénéfices de la pêche
~BRICKS OVERHAND, to - [constr] maçonner un mur du côté opposé au parement
~-BY - [gen] terre-plein *m* de stationnement
~CONCRETE, to - [constr work] couler le béton
~EGGS, to - [zool] pondre
~-FLAT TUBING - [mech] (large-bore tubing extruded and flattened out to form sheet, without trimming or slitting at the edges) feuille *f* aplatie
~-ON ROLLER - [telev] (for video recording) galet *m* presseur
~OUT, to - [mech] (the operation of marking out material for cutting etc) tracer, marquer
[surv] faire le tracé
[mining] (a mine) aménager (une mine)
~-OUT - [gen] étude *m*, dessin *m*
[mech] (in a shop or factory) disposition *f*, projet *m*
[autom] (of instructions and rules) schéma *m*
[draw] dessin *m*, tracé *m*
[metall] (the marking-out of material full size for further work) comparaison *f* des dimension
~-OUT OF A LINE - [railw] tracé *m* de la ligne
~SHAFT - [mech] (auxiliary gear shaft) arbre *m* intermédiaire
~-STOOL - [paper man] table *f* élévatrice
~THE SLEPPERS, to - [railw] poser les traverses
~THE WARP THREADS ON THE FOOT PEGS - [text] disposer les fils de chaîne sur les chevilles d'envergeure
~UP, to - [gen] accumuler, mettre en réserve
[comm etc] amasser
[naut] (a ship) désarmer, mettre en rade
~-UP MACHINE - [rubber ind] machine *f* à confectionner les pneus
LAYER, to - [agric] provigner, marcotter
LAYER - [gen] couche *f*
[constr] (bed of mortar between courses) assise *f*, lit *m*
[zool] (of a hen) pondeuse *f*
[bot] marcotte *f*
[bot] (of a vine) provin *m*
[photo] (on a film) couche *f*
[ropemaking] machine *f* à commettre
[print] (a worker removing the sheets off the felts) receveur *m* des feuilles

LAYER - [plast ind] (a single sheet as used in making laminated material) couche *f*
~HEIGHT - [phys] (el magnetic waves) altitude *f* de la couche
~LATTICE - [phys] réseau *m* à couches
~OF A DISTRIBUTED WINDING - [el] (all the conductors situated at the same distance from the bottom of the slot) couche *f* d'un enroulement réparti
~OF BITUMEN - [constr] couche *f* de bitume
~OF CHARGE - [el] (sheet of charge of one sign; e. g., the surface charge on a charged conductor) couche *f* de charge
~OF HUMUS - [soil] couche *f* de terre végétable, couche *f* humique
~OF MATERIAL - [ind chem] (layer of oxide, bed) lit *m*
~OF OXIDE - s. layer of material
~OF PEAT - [soil] couche *f* de tourbe
~OF SHEETS - [el] couche *f* de feuilles
~OF SILT - [soil] couche *f* de vase, couche *f* de limon, couche *f* de silt
~OF WARP - [text] couche *f* de fils de chaîne
~-ON - [mech] (in machine tools etc) alimentateur *m* automatique
~POROSITY - [metall] porosité *f* à couches
LAYERED MAP - [cartography] carte *f* avec teintes (pour indiquer les différentes altitudes)
LAYING - [gen] pose *f*, posage *m*
[el telecomm etc) (of cables) pose *f*
[constr] (the first of two coats) première couche *f* d'enduit
[ropemaking] commettage *m*
[railw] (of the track) pose *f* (de la voie)
~A BELT ON - [mech] embrayage *m* d'une courroie
~CAT - [mech mach] (colloq. US only; a sideboom tractor, or pipe layer) grue *f* pour la pose des tuyaux
~HEN - [zool] pondeuse *f*
~HOUSE - [agric] poulailler *m* de ponte
~MACHINE - [print] machine *f* de transmission
~NEST - [agric] nid *m* pondoir
~OF CABLES - [el & telecomm] pose *f* des câbles
~OF PIPES - [plumb] montage *m* des tuyauteries
~OF A SECOND TRACK - [railw] doublement *m* de la voie
~ON THE CARDS - [text] pose *f* des cartons
~OUT LAND - [agric] (for irrigation) préparation *f* du sol
~OUT MACHINE - [el] dérouleuse *f*, chariot *m* dérouleur
~PERIOD - [agric] saison *f* de ponte
~SEASON - s. laying period
~STRAIN - [zool] race *f* pondeuse
~TROWEL - [constr] truelle *f* à maçonner
~UP - [plast ind] (the operation of placing resin-impregnated material in or on a mould for post-forming) disposition *f*
LAYOUT - [mech] plan *m* de montage
LAZARET - s. lazaretto
LAZARETTO - [med] (house for the reception of the sick) lazaret *m*, maladrerie *f*
LAZULITE - [min] (natural basic aluminium phosphate used as a semi-precious stone) lazulite *f*
LAZY EIGHT - [aero] (an aerobatic manoeuvre consisting of two successive loops forming a figure eight lying in a vertical plane, altitude, heading and attitude being approximately the same at the end of the evolution as at the beginning) double looping *m*, vertical *m* en huit
LAZY H - [radio] (dipole aerial consisting of two collinear dipoles) série *f* de deux dipôles
~PULLEY - [mech] poulie *f* de renvoi à chape souble articulée
~TONGS - [impl] ciseaux *mpl*
lb - [abbrev. of pound] livre *f*
L.B. - [telecomm] (abbrev. for Local Battery) batterie *f* locale
L.C. - [print] (or l.c.; abbrev. for lower case) bas-de-casse *m*, d.d.c.
L-C COUPLING - [radio] (form of intervalve coupling) couplage *m* par bobine d'arrêt
LC DRIVE - [radio] oscillateur *m* pilote LC
LC MACHINE - [mech mach] (linear-contact H.F. sealing machine) oscillateur LC
L/D RATIO - [plast ind] (the relation between the length of an extrusion screw its diameter) rapport *m* L/d
LEA - [text] (length of yarn) échevette *f* de fil, écheveau *m*
[agric] (of land) en jachère
~TESTER - [text] dynamomètre *m* pour écheveaux
LEACH, to - [gen] filtrer, s'infiltrer, suinter
[chem] lessiver
[hydr] (of a cesspool) filtrer
~LIQUOR - [chem] (solution used to extract metals from ores and compounds by chemical solution) solution *f* de lixivation
LEACHING - [ind chem] (the extraction of a soluble substance in a mixture by solution, usually in water, and subsequent separation) lessivage *m*
~CESSPOOL - [hydr] (cesspool which is not watertight) puits *m* perdu
~PLANT - [ind chem] (installation for the extraction of material by solution, especially of minerals) installation *f* de lessivage
~TRENCH - [of a sewer] drain *m* à ciel ouvert, tranchée *f* d'infiltration
LEAD, to - [print] (to insert a thin strip of lead between of print) interligner
[el] avancer
[gen] plomber, garnir de plomb
LEAD - [metall] (symbol Pb; metallic element, the heaviest and softest of the common metals; strongly resistant to corrosion) plomb *m*
[print] (a thin strip of lead inserted between lines of type) interligne *f*, entre-ligne *f*
[electron] ligne *f* d'amenée
~ACCUMULATOR - [chem] (a deying mordant, intermediate in the production of driers for paints, and an analytical reagent) acétate *m* de plomb
~-ACID ACCUMULATOR - s. lead-acid battery
~-ACID BATTERY - [el chem] (secondary battery using lead and lead oxide plates in a sulphuric acid electrolyte) accumulateur *m* au plomb
~ANNEALING - [metall] recuit *m* en plomb fondu
~ANTIMONATE - [chem] (a paint pigment and a pigment for ceramic glazes) antimoniate *m* de plomb
~APRON - [constr] garnissage *m* de plomb
~ARSENATE - [chem] (a weedkiller and insecticide) arséniate *m* de plomb
~ARSENITE - [chem] (an insecticide) arsénite *m* de plomb
~ASH - [chem] monoxyde *m* de plomb
~ATOMIZER - [mech] (device for reducing liquid

lead to fine particles by means of an air blast, used in white lead manufacture by the Carter process). pulvérisateur *m* de plomb

LEAD AZIDE - [chem] (a detonation initiator) azide *f* de plomb, azothydrure *m* de plomb

~BATH - [ind chem] (bath containing molten lead and intended to keep objects at a constant temperature) bain *m* de plomb

~BATH FURNACE - [metall] (type of furnace in which material to be heat treated is immersed in a tank of liquid lead) four *m* à bain de plomb

~-BATH QUENCH - [metall] trempe *f* au plomb

~BEARING - [metall] plombifère

~BORATE - [chem] (an ingredient of special glasses and a drier for paints and varnishes) borate *m* de plomb

~BOROSILICATE - [chem] (a constituent of certain glasses) borosilicate *m* de plomb

~BRICK - [nucl] (special brick containing a large percentage of lead, used in building protective walls in nuclear reactors) brique *f* plombifère

~BRONZE - [metall] (used for bearings) bronze *m* au plomb

~BUCKLE - [mech] (perforated disk of lead used in the Dutch white lead process) disque *m* perforé de plomb

~BURN, to - [plumb] (the operation of welding together two pieces of lead and forming a joint without using a solder) souder le plomb par fusion

~-BURNING - [metall] (autogenous welding of lead) soudure *f* autogène du plomb

~-BURNT - [plumb] soudé par fusion

~CARBONATE - [chem] (basic lead carbonate; white lead. A pigment for paints and a consistent of putty and ceramic glazes) carbonate *m* basique de plomb

~CASTLE - [nucl] (enclosure of lead designed to shield a radiation detecting device) château *m* de plomb

~CHLORIDE - [chem] (an analytical reagent and a starting material for lead pigments) chlorure *m* de plomb

~CHROMATE - [chem] (poisonous, yellow crystals used in the production of chrome yellow pigments) chromate *m* de plomb

~CHROMES - [paint] (a group of yellow and orange pigments, based on chromates of lead) chromes *mpl* au plomb

~-CLAD - [constr] garni de plomb

~COATING - [metall] gaine *f* de plomb, garnissage *m* de plomb

~COPING - [constr] soline *f* en plomb

~COVERED CABLE - [el] câble *m* sous plomb

~COVERED SHEET - [metall] tôle *f* sous plomb

~CYANATE - [chem] cyanate *m* de plomb

~CYANIDE - [chem] cyanure *m* de plomb

~DIOXIDE - [chem] (a powerful oxidizing agent with application in the manufacture of rubber substitutes) peroxyde *m* de plomb

~DISILICATE - [chem] (used as frit to incorporate lead oxide in ceramic glazes) bisilicate *m* de plomb

~DOORS - [nucl] (doors made of outer sheets of stainless steel plate and poured lead between them) portes *fpl* à couche intérieure de plomb

~DRIERS - [paint] (driers based on lead componds, e.g. lead acetate, lead linoleate, lead resinate, lead naphthenate, litharge etc) siccatifs *mpl* au plomb

LEAD ENCEPHALOPATHY - [med] encéphalopathie *f* saturnine

~EQUIVALENT - [nucl] (the thickness of metallic lead which would give the same protection for a given irradiation as the material in question) équivalent *m* de plomb

~ERASER - [rubber ind] gomme *f* à crayon

~EXTRUSION PRESS - [rubber ind] presse *f* boudineuse à plomb

~FERRICYANIDE - [chem] ferricyanure *m* de plomb

~FERRITE - [chem] ferrite *f* de plomb

~FERROCYANIDE - [chem] ferrocyanure *m* de plomb

~FLUORIDE - [chem] fluorure *m* de plomb

~FOIL - [metall] feuille *f* de plomb, plomb *m* en feuilles

~FORMATE - [chem] (an analytical reagent) formiate *m* de plomb

~FREE - [paint] (of paints to be used for special purposes) sans plomb

~FRIT - s. lead disilicate

~GLANCE - [glass man] (glass containing lead oxide) verre *m* de plomb

~GLAZE - [ceramics] glaçure *f* plombifère

~GOUT - [med] goutte *f* saturnine

~GUTTER - [constr] gouttière *f* garnie de plomb

~HAMMER - [impl] (a hammer of which the head is made of lead, to avoid damage to the object struck with it) marteau *m* de plomb

~HARDEN, to - [metall] tremper dans un bain de plomb

~HYDROXIDE - [chem] (a starting material in the production of a number of lead salts) hydroxyde *m* de plomb

~-IN INSULATOR - [el] isolateur *m* de traversée

~IODIDE - [chem] (a photographic chemical. Lead iodide also has applications in medicine) iodure *m* de plomb

~JOINT - [plumb] (a joint made between cast-iron spigot-and-socket pipes, by pouring molten lead into the collar and setting it up with caulking irons when cold) joint *m* de plomb

~LACTATE - [chem] lactate *m* de plomb

~LADLE - [impl] (used in plumbing) cuiller *m* de plombier

~LIGHTS - [constr] (or leaded lights) fenêtres *fpl* plombées

~LINE - [surv] (line to take soundings) ligne *f* de sonde

[metall] plombure *f*, aile *f* d'un vitrail

~LINING - [gen] chemise *f* de plomb, revêtement *m* de plomb

~LINOLEATE - [chem] (a drier for paints and varnishes) linoléate *m* de plomb

~-LOADED INSULATING SLEEVE - [electron] (cylindrical sleeve surrounding the neck of an X-ray tube of plastic material loaded with lead) manchon *m* en résine synthétique au plomb

~MELTING KETTLE - [metall] four *m* de fusion du plomb

~METAL - [metall] plomb métal *m*

~MOLYBDATE - [chem] (a component of certain pigments) molybdate *m* de plomb

~MONONITRORESORCINATE - [chem] (a detonator) mononitrorésorcinate *m* de plomb

~MONOXIDE - [chem] (a chemical term for litharge q.v.) litharge *f*, monoxyde *m* de plomb

~MOULD - [metall] matrice *f* en plomb

~NAIL - [gen] clou *m* pour plomb

LEAD NAPHTHALENESULPHONATE - [chem] (a synthesis intermediate) naphtalène-sulfonate de plomb
~NAPHTHENATE - [chem] (an insecticide and paint and varnish drier) naphténate *m* de plomb
~NITRATE - [chem] (a photographic chemical, dyeing mordant, pigment for paints, and component of explosives) nitrate *m* de plomb
~OLEATE - [chem] (a drier for paints and varnishes) oléate *m* de plomb
~OXYCHLORIDE - [chem] (also called Cassel's yellow it is prepared by heating lead oxide with ammonium chloride) oxychlorure *m* de plomb
~PAINT - [paint] (paint made with a basis of white lead) peinture *f* à base de plomb
~PAPER - [paper man] papier *m* au plomb
~PATENTING - [metall] trempe *f* au plomb
~PENCIL - [impl] crayon *m* à mine de plomb
~PEROXIDE - [chem] (a strong oxidizing agent; formed on the positive plates of lead acid cells during some parts of the cycle of operation) peroxyde *m* de plomb
~PHOSPHITE - [chem] (an ultra violet stabilizer for vinyl plastics) phosphite *m* de plomb
~PIPE - [gen] tuyau *m* de plomb
~PIPE COLON - [anat] corde *f* colique
~PIPE FRACTURE - [med] fracture *f* en tuyau de plomb
~PLASTER - [med] emplâtre *m* à l'oléate de plomb
~POISONING - [med] intoxication *f* par le plomb, saturnisme *m*
~PRESS - [mech] presse *f* à plomb
~PRESS CURE - [rubber ind] vulcanisation *f* sous plomb
~PROTECTION - [radiat] protection *f* en plomb
~-REFINING FURNACE - [metall] fourneau *m* d'affinage à plomb
~RESINATE - [chem] (a drier for paints and varnishes) résinate *m* de plomb
~RESTRICTED - [paint] (containing a specified amount of lead) à teneur limite en plomb
~RUBBER - [rubber] (rubber containing a large proportion of lead components) caoutchouc *m* plombeux
~-RUBBER APRON - [radiat] tablier *m* en caoutchouc au plomb
~-RUBBER GLOVES - [radiat] (protection gloves) gants *mpl* protecteurs
~SALT - [chem] sel *m* de plomb
~SALT-ETHER METHOD - [paint] (a method of quantitative estimate of saturated and unsaturated fatty acids in drying oils etc., based on the greater solubility of lead salts of the unsaturated fatty acids in ether) méthode *f* de sels de plomb-éther
~SCREEN - [radiat] (protective screen consisting of lead) écran *m* en plomb
~SESQUIOXIDE - [chem] (a paint pigment and component of ceramic glazes) sesquioxyde *m* de plomb
~SETT - [impl] (a caulking tool) matoir *m* du plomb
~SHEATH - [el & telecomm] (of a cable) enveloppe *f* de plomb, gaine *f* de plomb
~SHEATHING MACHINE - [mech] presse *f* à plomb
~SHOT - [mech] grenaille *f* de plomb
~SHOT LOADER - [mech] dispositif *m* pour charger la grenaille de plomb
~SILICATES - [chem] (rubber fillers with confer resistance to degradation by sunlight) silicates *mpl* de plomb
~SLEEVE - [radiat] (cylindrical sleeve surrounding the neck of X-ray tubes) manchon *m* en plomb

LEAD SPAR - [min] cérusite *f*, céruse *f*
~STANNATE - [chem] (has applications in electronics) stannate *m* de plomb
~STEARATE - [chem] (applications as a drier in varnishes, stabilizer for vinyls, and as an extrusion lubricant) stéarate *m* de plomb
~-STORAGE CELL - s. lead-acid battery
~SULPHATE - [chem] (a pigment for paints) sulfate *m* de plomb
[chem] (basic lead sulphate; a pigment and a component of ceramic glazes) sulfate *m* basique de plomb
~SULPHIDE - [chem] (a source of lead and a constituent of ceramic glazes) sulfure *m* de plomb
~TELLURIDE - [electron] (a semi-conductor with uses in electronics) tellulure *m* de plomb
~TETRACETATE - [chem] (an oxidizing agent in organic synthesis) tétracétate *m* de plomb
~TETRAETHYL - [chem] (an anti-knock additive for motor fuels) tétraéthyle *m* de plomb
~THIOCYANATE - [chem] (a constituent of detonating compounds) thiocyanate *m* de plomb
~-THROUGH CAPACITOR - [electron] condensateur *m* de traversée
~TIN ALLOY - [metall] alliage *m* d'étain et plomb
~TITANATE - [chem] (a white pigment having the formula $PbTiO_3$, made by heating lead monoxide and titanium dioxide together) titanate *m* de plomb
~TRINITRORESORCINATE - [expl] (a detonating explosive) trinitrorésorcinate *m* de plomb
~TUNGSTATE - [chem] (a paint pigment) tungstate *m* de plomb
~VANADATE - [chem] (a pigment) vanadate *m* de plomb
~VINEGAR - [chem] acétate *m* de plomb
~VITRIOL - [min] anglésite *f*
~WASH - s. lead water
~WATER - [chem] solution *f* d'acétate de plomb
~WOOL - [metall] laine *f* de plomb
LEAD, to - [gen] conduire, guider
[el] (of an alternating current reaching its peak value earlier in the cycle than the voltage producing it) déphaser en avant
[naut] (a rope) faire passer
[hydr] amener
LEAD - [el] (wire or cable used for connection) câble *m*, fil *m*, conducteur *m* principal
[el] (of a current) calage *m* en avant, avance *f*
[mech] (of a screw) hauteur *f* du pas
[mech] (of steam engine) avance *f*
[min] filon *m*
[hydr] canal *m* d'amenée
~-ANGLE - [mech] (of gears) angle *m* d'inclinaison (de l'hélice)
~APERTURE - [el] ouverture *f* de passage
~AWAY, to - [chem] éliminer
~-IN - [radio] (the conductor which connects an aerial to transmitting or receiving equipment within a building or aircraft) descente *f*, entrée *f* de poste
~-IN GROOVE - [acoust] (in a record) sillon *m* initial
~SPIRAL - s. lead-in groove
~-IN WIRE - [el] fil *m* d'amenée de courant
~OF A PROOF - [print] en-tête *m*
~OF SPIRAL - [mech] pas *m* de l'hélice
~-OUT GROOVE - [acoust] sillon *m* de sortie
~-OVER GROOVE - [acoust] sillon *m* intermédiaire
~SCREW - [mech] (running alongside the bed of a la-

the) vis *f* mère
LEAD SCREW NUT - [mech] (of a machine tool) écrou *m* de vis mère
~TERMINAL - [el] oreille *f* de plaque
LEADED - [metall] plombé, sous plomb, au plomb
[print] interligné
~BRONZE - [metall] (an alloy bearing metal consisting of copper 5-10 p.c. of tin and up to 30 p.c. of lead) bronze *m* au plomb
~FUEL - [fuel] (fuel containing tetraethyl lead as an additive) carburant *m* au plomb
~LIGHTS - s. lead lights
~LUBRICANTS - [ind chem] (lubricants containing tetraethyl lead) lubrifiants *mpl* au plomb
~STRIP - [metall] feuillard *m* plombé
~ZINC OXIDE - [paint] (a mixed pigment containing zinc oxide and basic lead sulphate) oxyde *m* de zinc plombifère
LEADER - [gen] conducteur *m*, guide *m*
[journalism] article *m* de fond
[cin] (blank strip of film used to facilitate threading) bande *f* amorce
[min] (thin mineralized vein) filon *m* guide
[plumb] (conductor) tuyau *m* d'amenée
[constr] descente *f* d'eau, tuyau *m* de descente
[mech] (of a machine) roue *f* maîtresse
[print] (series of dots printed at intervals to guide the eyes) point *m* de conduite
[mus] premier violon *m*, solo *m*
[surv] jalonneur *m*
[agric] conduit *m*
[bot] bourgeon *m* terminal
[metall] laminoir *m* polisseur
[telev] amorce *f*
[oil ind] conducteur *m*
~AND TRAILER - [cin] amorces *fpl* avant et arrière
~PIN - [mech] (a pin which guides the movement of another part) goupille *f* de guidage
~STROKE - [el] prédécharge *f*
~TAPE - [cin] ruban *m* de guidage
LEADHILLITE - [min] (a naturally-occurring mixture of carbonate and sulphate of lead) leadhillite *f*
LEADING - [gen] plombage *m*
[print] interlignage *m*
~AXLE - [mech] (front-axle of locomotive) essieu *m* porteur d'avant
~BEARING - [aero] relèvement *m* de guide
~BOGIE - [railw] avant-train *m* de locomotive
~CAR - s. leading coach
~COACH - [railw] voiture *f* de tête
~CURRENT - [el] (alternating current reaching its peak value earlier in the cycle than the voltage producing it) courant *m* dephasé en avant
~EDGE - [aero] (the edge of an aerofoil or strut which is foremost in the direction of motion) bord *m* d'attaque
[cin] flanc *m* de guidage
[electron] (of a pulse) front *m*, flanc *m* avant
~EDGE PULSE TIME - [radio] durée *f* d'établissement d'impulsion
~-EDGE RADIATOR - [aero] (type of radiator located in the leading-edge of a wing) radiateur *m* de bord d'attaque
~GHOST - [telev] (ghost to the left of the image on a TV screen) image *f* fantôme à gauche
~-IN CABLE - [el] câble *m* d'entrée
~-IN INSULATOR - [radio] (tubular form of insulator through which a lead-in enters a building) isolateur *m* de raccordement, isolateur *m* d'entrée, pipe *f*
LEADING-IN LINE - [telecomm] entrée *f* de poste
[railw] voie *f* d'accès
~-IN POINT - [telecomm] entrée *f* de poste d'abonné
~-IN TUBE - [el] douille *f* de pénétration
~-IN WIRE - [el & telecomm] descente *f* d'antenne
~LOAD - [el] (reactive load on a.c. circuit) charge *f* capacitive
~MARK - [surv] repère *m* de jalonnage
~NOTE - [acoust] note *f* sensible
~ON PULLEY - [mech] poulie-guide *f*, galet *m* de renvoi
~-OUT TERMINAL - [el] borne *f* de sortie
~-OUT WIRE - [el] (flexible wire attached to the wire used for windings of a transformer and used for connecting terminals) câble *m* de sortie
~PHASE - [el] (in measuring equipment) phase *f* en avance
~POLE EDGE - [el] arête *f* d'entrée de la pièce polaire
~POLE HORN - [el] (the portion of the pole-shoe of an electric machine first met by a point on the armature as the machine revolves) corne *m* polaire d'entrée
~POLE-TIP - s. leading pole horn
~SWEEP - [aero] (in a propeller blade, an angular deviation towards the leading edge) déport *m* dans le plan de rotation, compensation *f* dans le plan de rotation
~ZERO - [comput] zéro *m* à gauche
LEADY - [gen] (resembling plomb) plombeux
LEAF - [bot] feuille *f*
[gen] (term loosely applied to an object having a large area in relation to its thickness) feuille *f*
[mech] (of a spring) feuille *f*, lame *f* de ressort, feuillet *m*
[text] lame *f*
[carp] battant *m*, vantail *m*
[constr] tablier *m*
[tool] truelle *f* à feuille de laurier
~AXIL - [bot] aisselle *f* des feuilles
~BASE - [bot] base *f* de la feuille
~BEET - [agric] bette *f*, poirée *f*
~-BLADE - [mech] lame *f* (fine)
[bot] limbe *m* de feuille
~BLIGHT - [bot] rouille *f* des feuilles
~BUD - [bot] bourgeon *m* à feuille
~CUSHION - [bot] gonflement *m*
~ELECTROSCOPE - [instr] (electroscope in which the detecting element is made of gold or aluminium leaf) électroscope *m* à feuilles d'or
~FILTER - [mech] filtre *m* à disques
~FOR A SPRING - [mech] feuille *f* pour ressort
~HAY - [agric] herbages *mpl* déshydratés
~MOULD - [agric] terreau *m* de feuille, terre *f* d'engrais
~OF A DOOR - [carp] battant *m* d'une porte
~OF A PINION - [horol] aile *f* d'un pignon
~SHEATH - [agric] gaine *f*
~SHEDDING - [bot] (wintering) chute *f* des feuilles
~-SPINE - [bot] épine *f* foliaire
~SPOT OF BEET - [bot] (plant disease) cercosporiose *f* de la betterave
~SPRING - [mech] ressort *m* à lames
~-SPRING SHOCK DAMPER - [mech] amortisseur *m*

à ressort
LEAF-STALK - [bot] pétiole *m*
~VEIN - [bot] nervure *f*
LEAFAGE - [bot] feuillage *m*
LEAFING - [paint] foliation *f*
LEAFLESS - [bot] effeuillé, sans feuilles
LEAFLET - [gen] feuillet *m*, imprimé *m*
[bot] foliole *f*
LEAK, to - [gen] couler, suinter, fuir
[naut] faire eau
LEAK - [gen] fuite *f*, écoulement *m*, infiltration *f*
[mech] (an escape of fluid from a system) fuite *f*
[naut] (of a ship) voie *f* d'eau
[el] dispersion *f*, fuite *f*
[hydr] voie *f* d'eau
[radio] (high resistance, sometimes used as a discharging path for a condenser) fuite *f*
~CLAMP - [plumb] serre-joint *m*
~DETECTION - [gen] indication *f* des fuites, indication *f* des pertes
~DETECTOR - [instr] indicateur *m* de terre
~FINDER - [instr] (in an airship) cherche-pertes *m*
~LOCATION ASSEMBLY - [plumb] (for service pipes etc) déceleur *m* de fuites
~-PROOF TUBE - [rubber ind] chambre *f* à air increvable
~RATE - [plumb] taux *m* de fuites
~TEST - [mech] (of machines, autos etc) essai *m* à l'eau
~TRANSFORMER - [el] transformateur *m* à fuites
LEAKAGE - [gen] fuite *f*, défaut *m* d'étanchéité, perte *f*
[gen] (of fluids) voie *f* d'eau, fuite *f*
[el] dispersion *f*
[med] déperdition *f*, fuite *f*
[radio] atténuation *f*, décrement *m*, affaiblissement *m*
~COEFFICIENT - [el] (of a magnetic circuit of an electric machine) coefficient *m* de dispersion
~CONDUCTANCE - s. leakance
~CURRENT - [el] (conduction current flowing between electrodes by any path other than across the vacuous space between the electrodes) courant *m* de dispersion
[med] courant *m* d'excitation
~DISTANCE - [el] distance *f* de fuite
~ERROR - [meas] (in bridge measurement methode) erreur *f* de fuite
~FACTOR - s. leakage coefficient
~FLUX - [el] (in a magnetic circuit) flux *m* magnétique de dispersion
~INDICATOR - [instr] (instrument designed to detect a leakage of current from an electric system to earth) indicateur *m* de terre
~PATH - [el] (the shortest distance across the surface of a piece of insulating material between two points at different potentials) ligne *f* de fuite
~PIPE - [mech] (in diesel engines) tuyau *m* de retour au réservoir
~POWER - [electron] puissance *f* transmise
~PROTECTIVE SYSTEM - [el] (protective system based on the leakage of current from electrical machines to earth) protection *f* contre les défauts à la terre
~RADIATION - [radiat] (direct radiation) rayonnement *m* de fuite
~RATE - s. leak rate
LEAKAGE REACTANCE - [el] (the reactance of a circuit resulting from the interlinkages it and leakage flux) réactance *f* de dispersion
~RESONANCE - [el] résonance *f* de dispersion
~SPECTRUM - [nucl] (the distribution of energy of the neutrons which leave the reactor) spectre *m* de fuite de neutrons
~SURVEY - [plumb] recherche *f* des fuites
~TEST - [ind proc] essai *m* d'étanchéité
~WATER - [hydr] (water derived from leaks, especially in hydraulic equipment) eau *f* de fuites
LEAKANCE - [el] (the reciprocal of insulation resistance) conductance *f* en dérivation
LEAKER - [metall] (any casting which is defective owing to leakage at the pressure test) fonte *f* à soufflures
[metall] caffut *m*
LEAKPROOF - [gen] étanche, à l'épreuve des fuites
LEAKY - [gen etc] qui cole, qui perd
~GRID RECTIFICATION - [radio] (method of detection of a radio signal, in which the signal voltage developed across the aerial tuning circuit is applied to the grid of a thermionic valve through a capacitor, which a resistor connected between the grid and cathode) détection *f* à fuites de grille
~TUBE - [electron] tube *m* fuyant
LEAN, to - [gen] s'appuyer, incliner
[gen] (the action of bending etc) s'incliner, se pencher
LEAN - [gen] maigre
[ind chem] (of a mixture) pauvre
[min] (of minerals generally) pauvre
~CLAY - [min] argile *f* pauvre
~COAL - [min] houille *f* maigre, charbon *m* maigre
~CONCRETE - [constr] béton *m* maigre
~EARTH - [agric] terrain *m* pauvre
~GAS - [min] gaz *m* de gazogène, gaz *m* à l'air, gaz *m* pauvre
~GLASSHOUSE - [agric] serre *f* adossée, serre *f* à un seul versant
~IRON ORE - [min] minerai *m* de fer pauvre
~LIME - [constr] (another term for hydraulic lime, capable of setting under water) chaux *f* hydraulique
~MIXTURE - [mech] (in i.c. engines; mixture containing a low proportion of fuel to air) mélange *m* pauvre
~-OUT - [mech] (of a mixture) appauvrissement *m*
~SOLVENT - [oil ind] (uncharged solvent entering a solvent extraction unit) solvant *m* maigre
~-TO - s. lean-to roof
~-TO ROOF - [constr] (roof with only one slope covering the distance between two walls) toit *m* en appentis, toit *m* à un seul égout, toit *m* à une pente
LEANNESS - [gen] maigreur *f*
LEAP, to - [gen] sauter, bondir
LEAP - [gen] saut *m*, bond *m*
~YEAR - [astr] année *f* bissextile
LEAPFROG - [gas ind] (mechanical rammer) fouloir *m*, pilonneuse *f*, dame *f* mécanique
~TEST - [comput] essai *m* saute-mouton
LEAPFROGGING - [gen] avancement *m* à saut-mouton
LEAR - s. Lehr
LEASE - [text] envergeure *f*, croisure *f*, encroix *m*
~BAND - [text] ficelle *f* d'encroix
~BAR - s. lease rod
~CORD - s. lease band
~PEG - [text] cheville *f* d'encroix

LEASE REED - [text] peigne *m* d'envergeure
~ROD - [text] (the rods across the warp which are used to separate the threads and keep them in the right position) traverse *f* d'envergeure
~STAND - [text] bâti *m* du peigne d'envergeure
~WARPING - [text] ourdissage *m* à la main
LEASED CIRCUIT - [telecomm] circuit *m* de location
~LINE - [railw] ligne *f* affermée
[telecomm] circuit *m* de location
~-LINE NETWORK - [comput] liaison *f* spécialisée
LEASEHOLD DEED - [agric] contrat *m* de bail, contrat *m* de fermage
LEASEHOLDER - [agric] cultivateur *m* à bail
LEASH - [gen] laisse *f*, attache *f*
[text] cordon *m* transversal de lame
[text mach] lacet *m*
LEASING - [text] piquage *m* en peigne
~HECK - [text ind] grillette *f*
LEAST - [gen] le moindre, minimal
~CIRCLE OF ABERRATION - [opt] cercle *m* de confusion minimale
~CIRCLE OF CONFUSION - s. least circle of aberration
~COMMON MULTIPLE - [math] le plus petit commun multiple
~ENERGY PRINCIPLE - [mech] principe *m* d'énergie minimale
~MEAN SQUARE ERROR - [math] (root mean square error) erreur *f* quadratique moyenne
~SIGNIFICANT POSITIONS - [comput] positions *f*pl de valeurs minimales
~WORK - [phys] travail *m* minimal
LEAT - [hydr] (small open water course for mining works) canal *m* d'amenée, bief *m*
LEATHER, to - [gen] garnir de cuir, tanner
LEATHER - [gen] (tanned and dressed animal skin or hide) cuir *m*
~BELT - [impl] courroie *f* de transmission
~BLACK - [dye] (pigment for leather) noir *m* pour cuir
~BOARD - [paper man] carton-cuir *m*
~-BOTTLE - [impl] outre *f*
~BRAKE BAND - [mech] bande *f* de frein en cuir
~CLOTH - [text] (fabric coated with a natural or synthetic plastic material and given a leather-like finish) toile *f* cuir, simili-cuir *m*
~COLLAR - [mech] bride *f* en cuir
~COVERED ROLLER - [text] cylindre *m* recouvert de cuir
~DIAPHRAGM - [mus] (leather joint in an organ wind-chest) boursette *f*
~-FIBRIN - [rubber ind] cuir *m* factice
~FILLET - [metall] congé *m* de cuir
~HOLLOWS - [mech] (strips of leather for the fillets in wood patterning) bandes *f*pl de cuir
~JOINT - [mech] joint *m* en cuir
~LINED FRICTION BLOCK - [text] coquille *f* recouverte de cuir
~LINK BELTING - [mech] courroies *f*pl en cuir articulé
~PACKING - [mech] (sealing arrangement using leather, especially in hydraulic equipment) garniture *f* en cuir
~PAPER - [paper man] papier *m* maroquin
~PIECE - [text] morceau *m* de cuir
~PURSE - s. leather diaphragm
~ROLLER - [text] rouleau *m* de cuir
LEATHER SEAL - s. leather joint
~SQUEEZER - [impl] rouleau *m* essoreur pour peaux
~STROP - [impl] cuir-lanière *m*
~UPHOLSTERY - [gen] garniture *f* en cuir
~YELLOW - [chem] phosphine *f*
LEATHERBOARD - [paper man] carton-cuir *m*
LEATHERETTE - [plast ind] (artificial leather) similicuir *m*, tissu *m* cuir
LEATHERJACKET - [zool] tipule *f*
[bot] leatherwood *m*
LEATHEROID - [dielectric] (proprietary name for vulcanized fibre) cuir *m* d'oeuvre artificiel
LEATHERY - [gen] coriace
LEAVE, to - [gen] laisser, abandonner, quitter
[leg] léguer
LEAVE - [metall] angle *m* de retrait
LEAVEN, to - [chem] faire lever
LEAVEN - [chem] levain *m*
LEAVING EDGE - [el] (the edge of a brush of an electric machine which is last met during the revolution by a point on the commutator) arête *f* de sortie
LEBLANC PROCESS - [chem] (process in which salt is treated with sulphuric acid to obtain sodium sulphate and hydrochloric acid, the sodium sulphate being then heated with limestone and coal to obtain sodium carbonate) procédé *m* Leblanc
LE CHATELIER-BRAUN PRINCIPLE - [phys] (this lays down that when a system is in equilibrium and some change of conditions is imposed upon it, the system will alter so as to counteract such a change. The principle can be applied very widely) principe *m* de Le Chatelier
LECHOSOS OPAL - [min] (deep-green variety of precious opal) opale *f* vert foncé
LECITHIN - [chem] (a group of phosphatides; higher carboxylic acid glycerides with uses in rubber and plastics manufacture) lécithine *f*
LECITHOL - [chem] (a phosphatide with applications in rubber and plastics manufacture) lécithine *f*
LECLANCHE CELL - [el chem] (a primary cell with a carbon positive and zinc negative electrode, ammonium chloride electrolyte and manganese dioxide depolarizer: the basic form of the ordinary dry cell) pile *f* de Leclanché
LECTERN - [gen] aigle *m*, lutrin *m*
LED HORSE - [agric] cheval *m* de main
LEDEBURITE - [min] lédeburite *f*
LEDGE - [gen] rebord *m*, saillie *f*
[carp] corniche *f*
[mech] (ridge on a surface) rebord *m*
[geogr] banc *m* de récifs, chaîne *f* de rochers
[min] (a vein) filon *m*, veine *f*
[electromagnetic waves] (a stratum in the ionosphere within which the gradient of ionization density in relation to height decreases and then increases without becoming negative) banc *m* d'ionisation
[constr] berme *f*, ressaut *m*
~OF A DOOR - [constr] traverse *f*
~ROCK - [soil] assise *f* rocheuse, fond *m* rocheux
LEDGED BATTENS DOOR - [constr] porte *f* à claire-voie, porte *f* en lattis
LEDGER - [gen & comm] grand livre *m*
[constr] (horizontal pole or member) moise *f*, filière *f*
[constr] (flat stone covering a grave) pierre *f* tombale
[mining] chevet *m* de filon

LEDGER BOARD - [carp] (a ribbon strip) rampe *f* (d'escalier)
[constr] (of a scaffolding) planche *f*
~ CARD - [autom] carte *f* compte
LEE - [naut] (region of protection from the wind e.g. "in the lee of a building") côté *m* sous le vent
~ SHORE - [naut] terre *f* sous le vent
~ SIDE - [naut] côté *m* sous le vent
~ WAVE - [met] (a standing wave occurring in the atmosphere on the lee side of a mountain) onde *f* stationnaire du côté de pente sous le vent
~ WAVE CLOUDS - [met] (smooth bars of cloud, characteristic of lee waves) nuages *f*pl caractéristiques d'ondes stationnaires du côté de pente sous le vent
LEEBOARD - [naut] (strong plank fixed to the side of a flat-bottomed vessel) aile *f* de dérive
LEECH - [zool] sangsue *f*
[naut] chute *f* arrière
~ LINE - [naut] cargue-bouline *f*
LEEK - [agric] poireau *m*
LEER - s. lehr
LEES - [brew ind] lie *f*
LEEWARD - [naut] (in the direction more remote from the wind) sous le vent, côté *m* sous le vent
LEEWAY - [naut & aero] (undesired movement of ship or aircraft in the general direction of, and due to, the wind) dérive *f*
LEFT - [gen] gauche
~ CROSSING - [text] tour *m* à gauche
~ GAUZE - [text] gaze *f* à croisement à gauche
~-HAND - [gen] de gauche
[mech etc] à gauche, renversé, sinistrorsum
~-HAND DRIVE - [auto] conduite *f* à gauche
~-HAND HELICAL GEARS - [mech] engrenages *m*pl hélicoidaux avec hélice à gauche
~-HAND HELIX - [mech & el] hélice *f* à pas à gauche
~-HAND ROPE - [rope making] corde *f* à torsion à gauche
~-HAND RULE - [el] (a simple rule for relating the direction of the flux, motion and electromotive force in an electric machine) règle *f* de la main gauche
~-HAND STEERING - [auto] conduite *f* à gauche
~-HAND THREAD - s. left-handed thread
~-HAND TWIST - [text] tors *m* de droite à gauche
~-HAND WORM - [mech] vis *f* sans fin à pas à gauche
~ HANDED ENGINE - [mech] (an engine in which the sense of rotation is counter-clock wise when seen from the end furthest from the propeller and looking towards the latter) moteur *m* tournant à gauche
~-HANDED POLARIZED WAVE - [electromagnetic waves] onde *f* à polarisation elliptique à gauche
~-HANDED PROPELLER - [aero] (a propeller which rotates counter clockwise as seen from the after part of the aircraft) hélice *f* à gauche
~-HANDED THREAD - [mech] (a screw thread so formed that when viewed axially, and turned anti-clockwise, it moves away from the observer) filet *m* à gauche, pas *m* à gauche
~ JUSTIFIED - [comput] aligné à gauche
~ SHIFT - [comput] décalage *m* à gauche
LEG - [anat] jambe *f*, patte *f* (de chien etc)
[el] partie *f* sous tension maximale
[instr] (of compasses) branche *f*
[hydr] branchement *m* des tuyaux

LEG - [geom] (of a triangle) côté *m*
[metall] (of an angle iron) côté *m*
[mech] (of a derrick) montant *m*, pied *m*, jambage *m*
[el] (of a transformer) montant *m*
[instr] (of a U- shaped magnet) branche *f*
[electron] (part of a V-shaped filament in an electronic tube) branche *f*, patte *f*
[th eng] (boilers) culotte *f*, naissance *f*
[electron] (of a circuit) branche *f* de circuit
[metall] aile *f*
~-ROOM - [gen] (in a vehicle etc) emplacement *m* pour les jambes
LEGACY - [leg] legs *m*, héritage *m*
LEGAL - [leg] légal
~ CAP - [paper man] papier *m* pour documents
~ OHM - [el] (unit of resistance adopted by the International Congress of Electricians in Paris in 1884; legal recognition was never granted) ohm *m* légal
LEGEND - [gen] légende *f*
[print] explication *f*, légende *f*
LEGER LINES - [mus] (the short lines which are added above or below the staff) lignes *f*pl supplémentaires
LEGGINGS - [gen] jambières *f*pl, guêtres *f*pl
LEGIBILITY - [gen] lisibilité *f*
LEGISLATIVE - [leg] législatif
LEGISLATURE - [leg] législature *f*
LEGITIMACY - [leg] légitimité *f*
LEGUME - [bot] gousse *f*, cosse *f*, légume *m*
LEGUMEN - [med] légumineux
LEGUMES - [agric] légumineuses *f*pl
LEGUMINOUS - [bot] légumineux
~ PLANT - [bot] plante *f* légumineuse
LEHR - [metall] (continuous annealing oven, provided with a steel wire conveyor passing through it from end to end) arche *f* à recuire
LEIASTHENIA - [med] liasthénie *f*
LEIOMYOSARCOMA - [med] liomyosarcome *m*
LEIPZIG YELLOW - [paint] (pigment consisting of lead chromate) jaune *m* de chrome, jaune *m* de Leipzig
LEISHMANIASIS - [med] leishmaniose *f*
LEITMOTIV - [mus] (the guiding theme in a composition) leitmotiv *m*, thème *m* principal
LEMMA - [math] lemme *m*
LEMMOBLAST - [biol] lemnoblaste *m*
LEMMOCYTE - [biol] lemnocyte *m*
LEMNISCATE - [math] lemniscate *f*
LEMNISCUS - [zool] (pair of bag-like organs at the base of the proboscis) lémnisque *m*
LEMON - [bot] citron *m*, limon *m*
~ BALM - [agric] mélisse *f* officinale, citronnelle *f*
~ CHROME - [dye] (a pale yellow pigment consisting of a mixture of lead sulphate and lead chromate) jaune *m* citron
~ COLOURED - [dye] jaune citron
~ GRASS - [bot] jonc *m* odorant
~-GRASS OIL - [chem] (an essential oil obtained from varieties of Cymbopogon and used in perfumery and the production of flavourants) essence *f* de lemongrass
~ JUCE - [agric] jus *m* de citron
~ OIL - [chem] (an essential oil obtained from lemon peel and used as a flavourant) essence *f* de citron
~ PEEL OIL - [chem] huile *f* de zeste de citron
~ PLANT - [bot] verveine *f* citronnelle
~ TREE - [bot] limonier *m*, citronnier *m*
~-WOOD - [bot] pittospore *m*
LEMOPARALYSIS - [med] paralysie *f* de l'œsophage

LEMURS - [zool] lemurs *mpl*
LEND, to - [gen] prêter
[fin] (money) prêter, placer
LENGTH - [gen] longueur *f*
[gen] (of time) durée *f*
[gen] (distance) distance *f*
[mech] (of a cable) (a lay) pas *m* de l'hélice
[plumb] (of a pipe) tronçon *m*
[timber] morceau *m*
[text] (of a cloth) pièce *f*, coupon *m*
~AT CREST - [hydr] longueur *f* au sommet
~BETWEEN BUFFERS - [railw] longueur *f* hors tampons
~BETWEEN COUPLINGS - [railw] longueur *f* entre attelages
~CREEP LINE - [soil] distance *f* d'infiltration
~OF A BLOCK SECTION - [railw] longueur *f* d'un canton de block
~OF A SPAN - [el] longueur *f* de la portée
~OF BLOCK - [text] largeur *f* du bloc
~OF BRAKE ARM - [text] longueur *f* du bras de levier
~OF BREAK - [el] longueur *f* totale de coupure
~OF CONE - [text ind] hauteur *f* du cône
~OF CONTACT - [mech] (of the gear teeth) longueur *f* de contact
~OF CUT - [carp & mech] (in machine tool work) longueur *f* de coupe
~OF DRUM - [text] longueur *f* du tambour
~OF IRON - [el] longueur *f* du fer
~OF PLATEAU - [radiat] longueur *f* du palier
~OF SCREW - [mech] longueur *f* de la vis
~OF STEP - [constr] emmarchement *m*, largeur *f* de l'escalier
~OF STROKE - [text] course *f* de la levée
[mech] (of a tool) course *f* de l'outil
~OF TUFTS - [text] longueur *f* des brins
~OF WARP - [text] longueur *f* de la chaîne
~ON THE WATERLINE - [naut] longueur *f* de la ligne d'eau
~OVER ALL - [mech] (of screws etc) longueur *f* tête comprise
LENGTHEN, to - [gen] allonger, rallonger
LENGTHENING BAR - [mech] clef *f* de rallonge
~PIECE - [gen] allonge *f*, rallonge *f*
LENGTHWISE - [gen] en longueur, longitudinal
~DIRECTION - [paper man] (resin bonded paper; in the same direction as the machine direction of the paper) direction *f* en longueur
~MOVABLE SIDE COMB - [text] peigne *m* de côté déplaçable dans le sens de la longueur
~SECTION - [mech] coupe *f* en long, profil *m* longitudinal
~TRAVEL - [mach tool] course *f* longitudinale
LENIENCE - [gen] clémence *f*
LENIENCY - s. lenience
LENITIVE - [med] lénitif *m*
LENO - [text] (fabric with an openwork effect) toile *f* à patron
~BROCADE - [text] gaze *f* avec effet de broché par chaîne
~CLOTH - [text] armure *f* pour gazes
~EFFECT - [text] effet *m* de gaze
~FABRICS - [text] effet *m* de gaze sur tissu serré
~THREAD - [text] fil *m* de tour
~TREADLE - [text] pas *m* dur
~WEAVE - [text] (open fabric formed by warp threads passing alternately over and under the fillers and crossing over each other between each of the latter) métier *m* pour la gaze
LENO WEAVING - [text ind] tissage *m* à chaîne croisée
LENS - [opt] (a transparent substance, usually glass, bounded by spherical surfaces and capable of converging or diverging light rays) lentille *f*
[photo] (of a camera etc) objectif
[radio] (of a lens aerial) lentille *f*
[opt] (magnifying glass) loupe *f*
[min] (of mineral) lentille *f*
~ADAPTER - [photo] anneau *m* intermédiaire
~AERIAL - [radio] antenne *f* à lentille
~ANTENNA - s. lens aerial
~APERTURE - [photo] (the largest circle of light used by the lens in forming the image) ouverture *f* de lentille
~BARREL - [opt] barillet *m* d'objectif
~BOARD - [photo] planchette *f* d'objectif
~CAP - [photo] bouchon *m* d'objectif
~COLLAR - [photo] embase *f* d'un objectif
~COVERAGE - [photo] (the field of vision) champ *m* visuel
~DISK - [telev] (disk used for mechanical scanning purposes, in which the holes are provided with optical lenses) disque *m* à lentilles
~DISTORTION - [opt] distorsion *f* de lentille
~DRUM - [telev] (an early mechanical system of scanning) tambour *m* à lentilles
~DRUM SCANNER - [telev] analyseur *m* à tambour à lentilles
~FIELD - [photo] champ *m* de l'objectif
~GRINDER - [tool] rodoir *m* pour lentilles
~GRINDING - [mech] rodage *m* des lentilles
~HOOD - [photo] (a shield to keep stray light out) parasoleil *m* d'objectif
~MOUNT - [photo] monture *f* d'objectif
~OF THE EYE - [anat] cristallin *m*
~OPENING - s. lens aperture
~PORT - [cin] (opening in the projection room for letting through the beam) fenêtre *f* de projection
~POWER - [opt] puissance *f* d'une lentille
~SCREEN - s. lens hood
~SHIELD - s. lens hood
~TURRET - [cin] (part of a camera which makes it possible to exchange the lenses mechanically) tourelle *f* à objectifs
~VESICLE - [anat] vésicule *f* cristallinienne
LENSING - [geol] stratification *f* lenticulaire
LENSMETER - [instr] (instrument designed to measure the optical properties of lenses) focimètre *m*, appareil *m* pour mesurer les lentilles
LENSOMETER - s. lensmeter
LENTICULAR - [gen bot etc] (shaped like a double convex lens) lenticulaire, lentiforme
~CLOUD - [med] (cloud of the general form of a lens, i.e. convex above and below) nuage *m* lenticulaire
~GIRDER BRIDGE - [constr] pont *m* à treillis
~PROCESS - [opt] (process of colour photography) système *m* additif à éléments lenticulaires
~SAND - [geol] sable *m* lenticulaire
~SCREEN - [cin] (a tissue screen covered with minute bead-like reflecting particles) écran *m* gauffré
LENTICULARIS - (abbrev. LENT) - [met] (term applied to a cloud having convex upper and lower surfaces) nuage *m* lenticulaire

LENTICULATION - [photo] (for additive colour photography) réseau *m* lenticulaire
LENTIGO - [med] lentigo *m*
LENTIL - [bot] lentille *f*
~VETCH - [bot] vesce *f* blanche
~WEEVIL - [zool] bruche *m* des lentilles
LENTOPTOSIS - [med] phacocèle *f*
LEONITE - [min] (mineral consisting of hydrous magnesium and potassium sulphates) léonite *f*
LEOPARD - [zool] léopard *m*
LEPER - [med] lépreux *m*
LEPIDINE - [chem] (a synthesis intermediate) lépidine *f*
LEPIDOLITE - [min] (a mica with uses in ceramics and as a source of lithium) lépidolite *f*
LEPIDOSIS - [med] dermatose *f* squameuse
LEPRID - [med] lépride
LEPROMA - [med] léprome *m*
LEPROSARIUM - [med] léproserie *f*
LEPROSY - [med] lèpre *f*, ladrerie *f*
LEPTOCHROA - [med] fragilité *f* de la peau
LEPTOCLASE - [geol] leptoclase *f*
LEPTOCYTE - [biol] leptocyte *m*, cellule-cible *f*
LEPTOMENINGITIS - [med] piemérite *f*
LEPTOMETER - [instr] (instrument for comparing oil viscosity with a standard value) leptomètre *m*
LEPTOMONAS - [med] stade *m* primitif des trypanosomes
LEPTON - [nucl] (a particle of small mass, i.e. an electron, positron, neutrino or anti-neutrino) lepton *m*
LEPTONEMA - [med] stade *m* leptoténique de la mitose
LEPTOPELVIC - [med] femme *f* atteinte de bassin rétréci
LEPTOTRICHOSIS - [med] leptotrichose *f*
LESION - [gen & med] lésion *f*
LESSEE - [leg] locataire *m*
LESSEN, to - [gen] (to diminuish or decrease) diminuer, amoindrir, s'atténuer
LESSOR - [leg] bailleur *m*
LESTE WIND - [met] (a warm wind occurring in Madeira) vent *m* leste
LET, to - [gen] laisser
[leg] (to lease) louer
LET (LINEAR ENERGY TRANSFER) - [radiat] transfert *m* linéique d'énergie, TLE
~DOWN, to - [gen] abaisser
[metall] (to soften by tempering) recuire, recuire après trempe, faire revenir
~GO, to - [gen] lâcher prise
[naut] (a rope) lâcher
~-GO - [metall & plast ind] (area in a lamination over which adhesion has failed) vide *m* entre strates
~IN, to - [gen] laisser entrer
[mech] encastrer, empatter
~IN THE CLUTCH, to - [auto] (to engage the clutch) embrayer
~INTO, to - [gen] laisser entrer
[constr] (to introduce something into the concrete) encastrer
~-OFF MOTION - [mech] mouvement *m* d'avance
~-OFF REEL - [mech] dérouleuse *f*
~-OFF SPINDLE - [mech] (the shaft on which a reel of continuous material is mounted for feeding to an operation) broche *f* d'écoulement
LET OUT, to - [gen] laisser sortir
[mech] débrayer
[mech] (to loosen) lâcher
~SLIP THE CLUTCH, to - [auto] faire patiner l'embrayage
LETHAL DOSE - [nucl] (the amount of radiation which will cause the death of a given organism) dose létale
~MUTATION - [genet] mutation *f* létale
LETHALITY - [med] mortalité *f*
~RATE - [med] mortalité *f* (d'une maladie)
LETHARGIC - [med] léthargique
~ENCEPHALITIS - [med] encéphalite *f* épidémique, encéphalite *f* léthargique
LETHARGUS - [med] maladie *f* du sommeil, trypanosomiase *f* africaine
LETHARGY - [gen & med] léthargie *f*
LETHEOMANIA - [med] narcomanie *f*
LETTER - [gen] lettre *f*
[print] lettre *f*, caractère *m*
~CASE - [print] casse *f*
~-HEAD - [gen & comm] en-tête *m* de lettre
~LOCK - [mech] serrure *f* à combinaison
~NAIL - [impl] marque *f* distinctive
~NOTATION - [mus] (the indication of the musical notes by means of letters) notation *f* alphabétique
~PAPER - [paper man] papier *m* à lettres
~PRESS - [impl] presse *f* à copier
~SHIFT - [telecomm] inversion *f* lettres
~-SHIFT SIGNAL - [telecomm] signal *m* d'inversion lettres
LETTERING - [print] repoussage *m*
[comm] (markings) estampillage *m*
LETTERPRESS - [print] (method of printing from type or block surfaces) impression *f* typographique
[adj] typographique
[print] (the actual reading matter) texte *m*
~PAPER - [paper man] (used for letterpress printing) papier *m* pour impression typographique
~PRINTING - s. letterpress [print]
LETTERS CASE - [telecomm] (case into which the characters and functions of a telegraph code are grouped) série *f* lettres
~PATENT - [leg] lettres *f*pl patentes
~SHIFT - [electron] (in a teletypewriter) commutation *f* de lettres
LETTING DOWN - [aero] (the part of the approach and landing process extending from the end of the initial approach to be beginning of the actual landing) descente *f*
[metall] revenu *m*, adoucissage *m*, recuit *m*
LETTUCE - [bot] laitue *f*
LEUCEMIA - [med] leucémie *f*
LEUCINE - [chem] (an amino acid with uses in nutritional research) leucine *f*
LEUCITE - [min] (natural aluminium potassium silicate) leucite *f*
LEUCITIC - [geol] leucitique
LEUCO-BASE - [chem] (in dyeing, a colourless base readily oxidizable to a dyestuff) leucobase *f*
~-COMPOUNDS - [chem] (colourless compounds capable of reconversion to dyestuffs by oxidation) leucodérivés *m*pl
LEUCOBLAST - [biol] promyélocyte *m*
LEUCOLINE - [chem] (quinoline. An intermediate for pharmaceuticals) quinoléine *f*
LEUCOPENIA - [med] leucopénie *f*

LEUCOSAPPHIRE - [min] (white sapphire) saphir *m* blanc
LEUCOTOMY - [med] lobotomie *f* frontale
LEUKENCEPHALITIS - [med] leuco-encéphalite *f*
LEUKIN - [biol] alexine *f* leucocytaire
LEUKOAGGLUTININ - [biol] leuco-agglutinine *f*
LEUKOCYTE - [biol] leucocyte *m*, globule *m* blanc
~INCLUSIONS - [med] inclusions *fpl* de Dohle
LEUKOCYTOLYSIS - [med] leucolyse *f*
LEUKOCYTOSIS - [med] leucocytose *f*
LEUKODERMA - [med] leucodermie *f*
LEUKOGRAM - [med] formule *f* leucocytaire
LEUKOSARCOMA - [med] leucosarcome *m*
LEUKOSIS - [med] leucose *f*
LEUKOTAXIS - [med] leucotaxis *f*
LEUKOTOXICITY - [med] leucotoxicité *f*
LEUKOTOXIN - [biochem] leucotoxine *f*
LEVANT - [geogr] Levant *m*
LEVANTER - [met] vent *m* aigre de l'est
LEVECHE WIND - [met] (a warm easterly wind occurring in Spain) vent *m* leveche
LEVEE - [hydr] (strong bank of earth used to control river flows) levée *f*, digue *f*
LEVEL, to - [gen] niveler
[dyes] (in dyeing, the production of a uniform shade throughout the material) égaliser
[surv] déniveler
LEVEL - [gen] niveau *m*
[meas] niveau *m*
[acoust] (interval between two magnitudes) niveau *m*
[instr] (for surveying, to establish the difference in height between two points) niveau *m*
[instr] (for obtaining a horizontal line) niveau *m*
[min] (a roughly horizontal tunnel) étage *m*, niveau *m*
[chem] titre *m*
[electron] (a charge value) niveau *m* de charge
[railw etc] (a level stretch) palier *m*
~ABOVE TRESHOLD - [acoust] (the sensation level) niveau *m* de sensation auditive
~ADJUSTEMENT - [surv] ajustement *m* du niveau
~BREAKDOWN - [el] perte *f* de niveau
~CANAL - [hydr] (a canal level throughout) canal *m* à niveau
~CHANGE VALUE - [contr] valeur *f* de commutation
~COMPENSATOR - [radio] (automatic transmission regulating mechanism) équilibreur *m* de niveau
~-COMPOUND EXCITATION - [el] (of a generator; cumulative compound excitation) excitation *f* compound à tension constante
~COMPOUNDED - [el] à compoundage à tension constante
~CONTROL - [hydr] régulation *f* de niveau
[telev] contrôle *m* du niveau
~COURSE - [mining] ouvrage *m* en direction
~CROSSING - [railw] passage *m* à niveau, croisement *m* à niveau
~-CROSSING GATE MOTOR - [el] moteur *m* de barrière
~CROSSING SIDE GATE - [railw] portillon *m* de passage à niveau
~CROSSING SIGNAL - [signal] signal *m* de passage à niveau
~CUTTING - [constr] recoupement *m* en palier
~DEMAND SIGNAL - [contr] signal *m* de niveau de puissance de consigne
LEVEL DENSITY - [nucl] (nuclear level density) densité *f* de niveau nucléaire
~DEVIATION - [telev] (of audio or video signals) variation *f* de niveau
~DIAGRAM - [telecomm] (of a telephone circuit) hypsogramme *m*
~DISTORTION - [telev] distorsion *f* de niveau
~DISTRIBUTION - [phys] (the distribution of the energy levels) distribution *f* des niveaux d'énergie
~FLIGHT - [aero] vol *m* horizontal
~GAUGE - [instr] indicateur *m* de niveau
~GROUND - [surv] terrain *m* de niveau, terrain *m* en palier
~HAULAGEWAY - [mining] galerie *f* de roulage en palier
~-HEARTH REVERBERATORY FURNACE - [metall] four *m* à réverbère à sole plate
~HEIGHT - [hydr] hauteur *f* du niveau
~HUNTING - [telecomm] sélection *f* sur plusieurs niveaux
~INDICATION - [railw] indication *f* d'achèvement
~INDICATOR - [acoust] indicateur *m* de niveau
[acoust] décibelmètre *m*, indicateur *m* de la profondeur de la modulation
~LANDING - [aero] (in stress analysis, that loading condition for fuselage and landing gear which corresponds to a two-point landing with the fuselage horizontal) redressement *m* correspondant à un atterrissage sur deux points
~LINE - [surv] (line which is at all points perpendicular to the direction of gravity) ligne *f* de niveau
~-LUFFING CRANE - [mech] (jib crane in which the load is moved radially in a horizontal path) grue *f* à volée basculante (radiale)
~MEASURING SET - [radio] hypsomètre *m*
~OF ENERGY - [phys] (the value of an energy measured vertically above a fixed origin) niveau *m* d'énergie
~OF THE REED - [text] affleurement *m* du peigne
~OF THE WATER TABLE - [soil] niveau *m* de la nappe aquifère
~OF WATER - [hydr] niveau *m* de l'eau, surface *f* de l'eau
~OFF, to - [aero] (to bring an aircraft into an attitude in which the longitudinal axis is horizontal especially in landing) redresser
[constr] prendre une allure horizontale
~PATH - [gen] chemin *m* égal
~REGULATION - [comput] régulation *f* du niveau
~SETTING - [telev] alignement *m* du niveau
~SPACING - [nucl] écartement *m* de niveau
~STAFF - s. levelling staff
~SURFACE - [surv] (surface at all points perpendicular to the direction of gravity) surface *f* de niveau
[phys] surface *f* équipotentielle
~TANGENT TRACK - [railw] voie *f* en alignement droit
~THE SOIL, to - [agric] niveler le sol, égaliser le sol
~TRIER - [instr] (apparatus designed to measure the angular value of a division of a level tube) appareil *m* pour essai des niveaux
~WIDTH - [phys] (measure of the spread in excitation energy of an unstable state of a quantized system) largeur *f* de niveau
~-WIND DEVICE - [mech] (a mechanism designed to ensure that cable, pipe or the like is coiled evenly

on a drum, without gaps or overriding) régulateur *m* d'enroulement
LEVELLED - [gen] nivelé, égalisé
LEVELLER - [metall] dresseuse *f* à galets, machine *f* à dresser à rouleaux
~DOOR - s. levelling door
LEVELLING - [surv] (the operation of finding the difference of elevation between two points) nivellement *m*
(the operation of levelling a ground) nivellement *m*, aplanissement *m*
[electron] filtrage *m*
~AGENT - [chem] (to obtain uniformity in dyeing solutions) agent *m* nivelateur
~BAR - [metall] barre *f* de repalage
~BLOCK - [mech] (large flat plate on which iron and steel plates are laid for flattening) plaque *f* à dresser
~BOLT - [mech] boulon *m* de nivellement
~BOOK - [surv] carnet *m* de nivellement
~BOTTLE - [ind chem] (used in gas analysis) flacon *m* niveleur
~DOOR - [th eng] (in ovens) portillon *m* de rapalage
~FILLER - [ind chem etc] (in powder form) matériau *m* de nivellement
~INSTRUMENT - [instr] (used in survey work) niveau *m* à lunette
~LAYER - [constr] couche *f* de nivellement
~NET - [surv] réseau *m* de nivellement
~OF MOLE-HILLS - [agric] étaupinage *m*
~OF THE HEALDS - [text] nivellement *m* des lames
~POLE - s. levelling staff
~ROD - s. levelling staff
~ROLLS - [metall] rouleaux *mpl* à dresser
~SCREW - [mech] vis *f* de calage, vis *f* de bride
~SOLUTION - [ind chem] solution *f* de nivellement
~STAFF - [surv] (the graduated wooden rod giving the vertical distance between the line of sight of the level and the point on which the staff is held) mire *f* de nivellement, jalon *m* d'arpentage
LEVELNESS - [dyes & paint] (uniform shade throughout) teinte *f* iniforme
LEVER - [mech] levier *m*
[horol] (the pivoted arm carrying the pellets in the lever escapement) ancre *f*
[mech] (on a machine etc) levier *m*
[mus] (of a piano) levier *m*
[mech] (of a lock) bascule *f*
~ARM - [mech] bras *m* de levier
~ARRANGEMENT - [mech] système *m* de leviers
~BALANCE - [impl] peson *m* à contrepoids
~BOX - [railw] boîte *f* de manoeuvre des aiguilles
~BRAKE - [mech] frein *m* à levier
~CONTROL - [contr] commande *f* à levier
~DRAFT MACHINE - [metall] machine *f* à démouler à levier
~ESCAPEMENT - [horol] échappement *m* à ancre
~FRAME - [text] bâti *m* de leviers
~FULCRUM - [mech] pivot *m* du levier
~FUSEE MOVEMENT - [horol] mouvement *m* à fusée
~GAUGE - [instr] jauge *f* de précision à levier
~-GRIP TONGS - [metall] écrevisse *f*
~HANDLE - [mech] bec *m* de cane
~JACK - [mech] cric *m* à levier
~KNIFE - [text] couteau *m* à levier
~LOCK - [constr] bénarde *f*, serrure *f* à gorges
~NOTCH - [text] cran *m* de la bascule

LEVER OF THE FIRST KIND - [mech] levier *m* du premier genre
~PICK LOOM - [text] métier *m* à chasse-navette audessous
~PRESS - [mech] presse *f* à levier
~PUNCHING AND SHEARING MACHINE - [mech] poinçonneuse-cisaille *f* à levier
~SAFETY-VALVE - [mech] (a type of automatic pressure-relieving valve for pressure vessels, in which a weighted lever holds the valve on its seating) soupape *f* de sûreté à levier
~SCALES - [impl] romaine *f*, balance *f* romaine
~SHEARS - [mech] cisailles *fpl* à levier
~SWITCH - [el] commutateur *m* à manette
~TACKLE - [text] mouvement *m* à leviers pour manoeuvrer les lames
~TUMBLER LOCK - [mech] serrure *f* de sûreté à gorges mobiles
~-TYPE BRUSH-HOLDER - [el] (type of brush-holder in which the brush is held at the end of arm which is pivoted the brush spindle) porte-balais *m* à levier
~-TYPE STARTER - [el] (electric motor starter with a contact lever moving over a number of contacts) démarreur *m* à contacts
~WATCH - [horol] (watch fitted with a lever excapement) montre *f* à ancre
LEVERAGE - [gen & mech] force *f* de levier, rapport *m* des bras de levier, abattage *m*
LEVIGATE, to - [gen] (to make smooth) léviger
[chem etc] (to reduce a substance to fine particles) réduire en poudre
LEVIGATION - [chem] (the reduction of a substance to fine particles by grinding in water and selective settlement) lévigation *f*
LEVOCLINATION - [med] lévotorsion *f*
LEVODUCTION - [med] lévoversion *f*
LEVOROTATORY - s. laevorotatory
LEVOTORSION - s. levoclination
LEVOVERSION - s. levoduction
LEVULOSE - s. laevulose
LEVY - [gen] levée *f*
[leg] (tax) impôt *f*, contribution *f*, cotisation *f*
LEWIS - [impl] (used to lift heavy stones) louve *f*, louve *f* à pierres
~BOLT - [constr] (foundation bolt secured in the anchoring masonry by molten lead round it) vis *f* à scellement
~PROCESS - [ind chem] (a commercial method for the production of carbon black from natural gas) procédé *m* Lewis
LEWISITE - [min] (chlorovinyl dichlorarsine. A vesicant war gas) léwisite *f*
LEXEME - [gen] (the written word or particle which gives the meaning) lexème *m*
LEY - [agric] pature *f* temporaire, pâturage *m* à base de trèfle
~FARMING - [agric] assolement *m* avec pâtures temporaires, ley-farming *m*
LEYDEN JAR - [el] (high-voltage, low-capacitance capacitor using a glass jar as the dielectric) bouteille *f* de Leyde
L.F. - s. low frequency
~ACCELERATION PICK-UP - [instr] (an instrument designed to measure static acceleration in vehicles, lifts, aeroplanes etc) indicateur *m* d'accélération à basse fréquence
~BEAT OSCILLATOR - [instr] (instrument used for

acoustical measurements) oscillateur *m* à battements à basse fréquence
L.H.A. - s. Local Hour Angle
L-HEAD CYLINDER - [mech] cylindre *m* à soupapes latérales
L-HEAD ENGINE - [mech] moteur *m* à soupapes latérales
LIABILITY - [gen & leg] responsabilité *f*
[comm] passif *m*
~INSURANCE - [insur] assurance *f* responsabilité
LIABLE - [gen] responsable
[leg] sujet, exposé
~TO DUTY - [comm] passible de droits de douane
LIANA - s. liane
LIANE - [bot] liane *f*
LIAS - s. liassic system
LIASSIC SYSTEM - [gen] (thick series of marine strata along the English south-eastern and eastern coastline) lias *m*
LIBERATE, to - [gen] libérer
[chem] (to set free by chemical action) libérer, dégager
LIBERATED HEAT - [phys] (latent heat) chaleur *f* latente
LIBERATOR TANK - [el chem] (electrolytic cell with insoluble anodes used to extract metal from the electrolyte by deposition on cathodes) cuve *f* libératrice, cuve *f* de dépôt total
LIBETHENITE - [min] (hydrous phosphate of copper) libéthénite *f*
LIBIDO - [med] libido *f*
LIBRA - [astr] (a constellation) la Balance *f*
LIBRARY ROUTINE - [comput] programme *m* de bibliothèque
~SUBROUTINE - [comput] sous-programme *m* de bibliothèque
LIBRATION - [astr] (periodic oscillation in a celestial body, especially the moon) libration *f*
[mech] (generally a quivering or swaying motion) oscillation *f*, balancement *m*
LICENCE, to - [gen] accorder un permis, patenter
LICENCE - [gen] autorisation *f*, permission *f*
[auto] (a driving licence) permis *m* de conduire
~PLATE - [auto] plaque *f* d'immatriculation
LICENSE - s. licence
LICENSED AIRCRAFT ENGINEER - [aero] (an engineer licensed by an authority to certify inspections) mécanicien *m* responsable
LICENSEE - [leg] patenté *m*, titulaire *m*, concessionnaire *m*
LICENSER - [leg] concesseur *m*, octroyeur *m*
LICHEN - [bot] (a group of plants, some of which have application in the manufacture of dyestuffs) lichen *m*
LICHENIFICATION - [med] lichénification *f*
LICHENIN - [chem] (a derivative of Iceland moss with demulcent properties) lichénine *f*
LICHENIZATION - s. lichenification
LICKER-IN - [text] (revolving cylinder in a carding engine, transferring the lap to the swift) tambour *m* de réunisseuse
~-IN DEVICE - [text] dispositif *m* briseur, installation *f* briseur
~-IN DRUM - [text] tambour *m* briseur
~-IN ROLLER - [text] cylindre *m* briseur
LICORICE - s. licorice root
~ROOT - [chem] (the root of Glycyrrhiza glabra, an extract of which is used in medicine and confectionery) réglisse *f*
LID - [gen] couvercle *m*
[carp] cale *f* de bois
~OF A VALVE - [mech] chapeau *m* d'une soupape
~WITH BAYONET CATCH - [mech] blocage *m* à baionette
LIE, to - [gen] être couché
[to tell lies] mentir
LIE - [gen] mensonge *m*
[gen] (a position) disposition *f*, configuration *f*
~AT ANCHOR, to - [naut] être à l'ancre, être mouillé
~DETECTOR - [instr] (apparatus used in cross examination to detect whether the interrogated person is lying) détecteur *m* de mensonges, polygraphe *m*
~FALLOW, to - [agric] être en jachère, être en friche
~KEY - [oil ind] clef *m* de retenue
~OF THE LAND - [geogr] configuration *f* du terrain
LIEBIG CONDENSER - [ind chem] (laboratory condenser consisting of a helical glass tube enclosed in a glass jacket tube through which cooling water is circulated) condensateur *m* Liebig
LIEBIGITE - [min] (mineral containing compounds urano-hallite) liebigite *f*
LIEBMANN EFFECT - [opt] effet *m* Liebmann
LIEN - [gen] privilège *m*
[leg] (the right to retain possession of a property until a debt is paid) droit *m* de rétention, privilège *m*
LIENOGRAPHY - [med] splénographie *f*
LIENOMALACIA - [med] ramollissement *m* de la rate
LIENOPATHY - [med] splénopathie *f*
LIENTERY - [med] lientérie *f*
LIFE - [gen] vie *f*
[adj] à vie
[gen](of objects) durée *f*
[mech] (of machines) durée *f*
[nucl] s. life time
[el] (of a lamp) durée *f*
[electron] (the maximum number of hours an electronic tube can be used for a perfect operation) durée *f* de débit
~BELT - [naut] ceinture *f* de sauvetage, nautile *m*
~-BOAT - [naut] canot *m* de sauvetage, baleinière *f* de sauvetage
~-BUOY - [naut] (buoyant device designed to support a person in water in emergency) bouée de sauvetage, couronne *f* de sauvetage
~CYCLE - [biol] (the various stages of an organism) cycle *m* vital
~EXPENTANCY - [mech etc] durée *f* de vie probable
~-JACKET - [naut] corset *m* de sauvetage, brassière *f* de sauvetage
~-LINE - [naut] ligne *f* de sauvetage
[aero] (of a balloon) câble *m* de retention
[nucl] (curve in space-time to represent the position of a particle as a function of time) parcours *m* espace-temps d'une particule classique
[mining] câble *m* de sûreté
~OF FURNACE - [metall] campagne *f* d'un four
~PRESERVER - [naut] (device designed to keep a float) appareil *m* de sauvetage
~RAFT - [naut] (inflatable device to support persons on water in emergency) radeau *m* de sauvetage
~SAVING WAISTCOAT - [naut] (a life-saving device) corset *m* de sauvetage
~-SIZE - [gen] de grandeur naturelle

LIFE TEST - [mech etc] essai *m* de durée
~TIME - [nucl] (the time which is required for the decay of one half of the atoms of a specimen of a radioactive substance) période *f* radioactive
~-VEST - [naut] (inflatable garment to support the wearer in water in emergency) gilet *m* de sauvetage
LIFELESS SHADOWS - [photo] ombres *fpl* creuses
LIFELONG - [gen] de toute la vie
LIFETIME - s. life time
LIFT, to - [gen] lever, soulever, hausser, enlever
LIFT - [gen] élévateur *m*, monte-charge *m*
[el] (in GB) ascenseur *m*
[aero] (the aerodynamic lift) portance *f*
[mech] (of a valve etc) levée *f*, hauteur *f* de levage
[mech] (of a pump) hauteur *f* d'élévation, hauteur *f*
[mech] (of a cam) course *f* de came
[plast ind] (a complete set of mouldings made at a single press operation) moulée *f*
[naut] balancine *f*
[mining] jeu *m*
[metall] (foundry) démoulage *m*
[constr] course *f* de pilons
[mining] (of the rope in hoisting) cordée *f*, trait *m*
[mining] (in a quarry) (the plane along which the rock is split) gradin *m*, foncée *f*
[aero] (of a lighter than aircraft) force *f* ascensionnelle
[railw] rame *f* (de wagons)
~ARM - [mech] (of an arm type hoist) bras *m* élévateur
~AXIS - [aero] (a line drawn through the C.G. of an aircraft at right angles to the direction of the corresponding airflow in the plane of symmetry) axe *m* de sustentation
~BRIDGE - [constr] (a movable bridge which can be lifted) pont *m* levant, pont-levis *m*
~CAR - [el] cabine *f* d'ascenseur
~CAR ANNUNCIATOR - [el] panneau *m* de signaux
~COEFFICIENT - s. lift/drag ratio
~COMPONENT - [phys] (any component force acting in the direction of lift) composante *f* de portance
~CONTROLLER - [el] appareillage *m* de commande d'un ascenseur
~DIRECTION - [phys] (the direction in the plane of symmetry which is normal to the relative wind direction) direction *f* de la portance, direction *f* de la sustentation
~/DRAG RATIO - [aero] (the relation between lift and drag) résistance *f* relative, coefficient *m* de planement
~DUMP - [transp] véhicule *m* à benne basculante
~FACTOR - s. lift/drag ratio
~FRAME - [mech] châssis *m* d'ascenseur
~GATE - [constr] (gate opening by bodily vertical movement) barrière *f* à bascule, barrière *f* oscillante
~-LATCH - [mech] loquet *m* à bouton
~LINKS - [mech] (connecting attachments between the lift arms and the body floor connexions) articulations *fpl* d'élévateur
~LOSS - [aero] perte *f* de portance
~MEMBERS - [mech] (structural components of the body understructure with the connecting holes for attachment to the hoist) longerons *mpl* de berceau de benne
~OF THE BOX - [text] course *f* de la boîte
LIFT OF THE SHUTTLE BOX - [text] lève *f* de la boîte à navette
~OFF - [astronaut] départ *m* vertical de la fusée
~PIN - [mech] (short shaft used to attach the lifting mechanism to the lift members) axe *m* de fixation de longeron de berceau au dispositif élévateur
~PIPE - [mech] tuyau *m* élévatoire
~POTENTIAL SWITCH - [el] interrupteur *m* du réseau d'alimentation
~PUMP - [mech] pompe *f* élévatoire
~SHAFT - [constr] puits *m* de l'ascenseur
~STRUT - [aero] (the structural part of an aircraft which serves the purpose of a strut but is so aerodynamically shaped as to contribute an element of lift) montant *m* porteur
~THE FLOOR, to - [mining] couper le mur
~TRUCK - [mech] chariot *m* élévateur
~-VALVE - [mech] soupape *f*, clapet *m*
~-WALL - [hydr] (of a canal lock) mur *m* de chute
~WEB - [aero] (in a parachute, a webbing element connecting rigging lines to harness) sangle *f* de suspension
~WELL - [constr] gaine *f* d'ascenseur
~WIRES - [aero] (wires designed to transmit the lift of the wings to the main structure) haubans *mpl* porteurs
LIFTED FLAME - [th eng] (burners) flamme *f* aérienne, flamme *f* flottante
~SIDE - [geol] lèvre *f* soulevée
~WARP - [text] chaîne *f* levée
LIFTER - [mech](a lifting cam) came *f*, virgule *f*
[gen] élévateur *m*, monte-charge *m*
[metall] (foundry work) (L-shaped bar used for supporting the sand in a cope) crochet *m* de fonderie
[plast ind] (of a mould) crochet *m*
[text] platine *f*, crochet *m*
[mining] mine *f* de revelage
[mining] tirant *m* d'extraction
[el] contact *m*, armature *f*
~BLADE - [text] lame *f*, couteau *m*
~HEAD - [text] tête *f* de la platine
~JACK - [text] arcade *f*
~MOTION - [text] mouvement *m* de montée et descente, mouvement *m* de levée
~ROLLER - [text] cylindre *m* de levée
~WHEEL - [text] roue *f* de levée
LIFTING - [gen] soulèvement *m*, levage *m*, remonte *f*
[paint] (a surface defect) soulèvement *m*
[railw] (in a major overhaul of locomotives) levage *m*
[agric] (e.g. of potatoes) arrachage *m*, récolte *f*
[mining] remontée *f* (du minerai), élevation *f*, remonte *f*
~AND FORCING PUMP - [mech] pompe *f* aspirante et foulante
~APPARATUS - [mech] appareil *m* de levage
~BALL - [mining] étrier *m* d'élévation
~-BEAM - [metall] balancier, palonnier *m*
~BLADE - [text] couteau *m* des platines, couteau *m* des crochets
~BLOCKS - [mech] porte-palan *m*
~BOLT - [mech] boulon *m* à œil
~-BOW - [mech] (of a hoisting bucket) anse *f* d'une benne
~CAM - [mech] came *f* de levage
~CAPACITY - [mech] force *f* de levage
[phys] (of a magnet) force *f* portante(d'un aimant)

LIFTING CORD - [text] corde *f*, cordon *m*, collet *m*
~DESCENT - [astronaut] descente *f* en vol plané
~FLAME - s. lifted flame
~FORK - [mech] râteau *m*, homme *m* de fer
~GEAR - [mech] (of lifting apparatus) appareil *m* de levage
~HANDLE - [metall] poignée *f* pour enlever le modèle
~HOOK - [mech] crochet *m* de levage, croc *m* de hissage
~INJECTOR - [mech] injecteur *m* aspirant
~IRONS - [metall] (shaped rods used to lift the patterns) barres *fpl* d'extraction
~JACK - [mech] cric *m* de levage, vérin *m*
~KNIFE - [text] couteau *m* de levée
~LUG - [mech] oreille *f* de soulèvement
~MAGNET - [el] (a powerful electro-magnet suspended from a crane, for handling magnetic materials, espec. scrap iron) électro-aimant *m* porteur, aimant *m* de suspension
~MAGNET WITH FIXED POLES - [el] électro-aimant *m* de levage à pôles indéformables
~MAGNET WITH MOVABLE POLES SHOES - [el] électro-aimant *m* de levage à pôles déformables
~OF A TRACK - [railw] relevage *m* d'une voie
~OF FLAME - [th eng] décollement *m* de la flamme
~OF PATTERNS - [metall] extraction *f* des modèles
~PIN - [text] goupille *f* de levée
~PLATES - [metall] (iron plates led into a pattern into which a lifting screw is inserted for taking it out of the mould) plaques *fpl* de levage
~PLUG - [oil ind] bouchon *m* de levage
~POWER - [mech] (of a crane) puissance *f* de levage, force *f* de levage
~ROD - [metall] noyau *m* mobile
[text] tige *f* de levée
~ROPE - [constr] câble *m* de sonnette
~SCREW - [metall] (iron rod screwed into a pattern to take it out of the mould) vérin *m* à vis
~SHAFT - [mech] arbre *m* de relevage, arbre *m* de changement de marche
~SPEED - [mech] (of a lifting apparatus) vitesse *f* de levage
~TACKLE - [mech] engins *mpl* de levage
~TRUCK - s. lift truck
~WHEEL - [mech] roue *f* élévatoire
~WINCH - [mech] treuil *m* de levage
~WIRE - [text] platine *f*
LIGAMENT - [anat] ligament *m*
LIGASOID - [chem] (colloidal system in which the dispersed particles are liquid and the dispersion medium is gaseous) dispersion *f* d'un liquide dans un gaz
LIGATION - s. ligature
LIGATURE - [gen] ligature *f*
[med] (any material for tying blood-vessels) ligature *f*
[print] (a monogram) entrelacement *m*
LIGHT, to - [gen] allumer, éclairer, illuminer
[naut] mettre les feux
LIGHT - [gen] (the radiation which induces visual sensation) lumière *f*
[constr] (any opening which let in light) fenêtre *f*, lucarne *f*
[constr] (of a window) vitre *f*
[naut] feux *mpl*
[naut] (of a lighhouse) phare *m*
LIGHT - [gen] (adj) léger
[print] éclairé, maigre
[met] (of wind) léger
~ACCENT - [light] accent *m* de lumière
~ADAPTATION - [opt] adaptation *f* (des yeux) à la lumière
~AGEING - [phys] vieillissement *m* à la lumière
~AIR - [met] vent *m* léger
~ALLOY - [metall] (special alloy of aluminium, magnesium or other metal or metals of low specific gravity) alliage *m* léger
~APPLICATION BAR - [telev] barre *f* noire flottante
~APPLICATION RATIO - [telev] rapport *m* de temps d'exposition
~ATOM - [nucl] (atom of an element the atomic weight of which is below a certain value) atome *m* léger
~-BACK - [th eng] (back-fire) retour *m* de flamme, prise *f* de feu à l'injecteur
~BEAM - [opt] rayon *m* de lumière
~-BEAM BUNDLE - [opt] faisceau *m* lumineux
~-BEAM INSTRUMENT - [instr] (any instrument utilizing the position of a light beam on a scale) instrument *m* à faisceau lumineux
~-BEAM LOCALIZING DEVICE - [radiat] (device used to direct the incident beam of X-rays) localisateur *m* lumineux
~-BEAM PICK-UP - [acoust] (a pick-up in which a beam of light is a coupling element of the transducer) phonocapteur *m* électro-optique
~BLOW - [text] coup *m* léger (du battant)
~BOX - [telev] boîte *f* à lumière
~-BOX FOR REFLEX AND CONTACT PRINTING - [photo] appareil *m* de copie par contact
~BREEZE - [met] légère brise *f*
~BUOY - [naut] bouée *f* lumineuse, photophore *m*
~BURR WOOL - [text] laine *f* légèrement chardonneuse
~CAR - [auto] voiture *f* légère
~-CENTRE LENGTH - [light] distance *f* du point lumineux
~CIRCUIT - [el] circuit *m* d'éclairage
~CLOTH - [text] tissu *m* léger
~CONSTANT - [instr] constante *f* lumineuse
~CORRECTION FILTER - [opt] filtre *m* correcteur de lumière
~-CRACKING - [rubber ind] (surface fissuring in rubber due to the action of light) craquelage *m* causé par la lumière
~CRUDE OIL - [min] (a source of organic compounds) pétrole *m* léger
~CURRENT - [electron] (photo-current) courant *m* photoélectrique
~CUT-OFF - [cin] (part of film track to localize fire danger) volet *m* de sécurité
~CUTTING - [agric] (of hops) taille *f* longue
~DENSITY - [opt] densité *f* lumineuse
~-DIFFUSING MEDIUM - [photo] dispositif *m* de diffusion de la lumière
~DISPERSION - [photo] dispersion *f* de la lumière
~DISTILLATE - [ind chem] produit *m* de distillation léger
~EFFECT - [text] effet *m* de lumière
~ELEMENT - [nucl] (an element the atomic weight of which is below a specified value) élément *m* léger
~EMISSION - [light] émission *f* de lumière
~ENDS - [oil ind] fractions *fpl* légères

LIGHT ENERGY DISTIBUTION CURVE - [phys] caractéristique *f* de la fonction de distribution pour l'énergie lumineuse
~ENGINE - [railw] locomotive *f* haut-le-pied
~ENGINE RUNNING - [railw] circulation *f* de locomotive isolée
~-FAST - [dyes] résistant à la lumière
~FASTNESS - [paint] (the property of resisting fading due to light) résistance *f* à la lumière
~FIELD ILLUMINATION - [photo] éclairage *m* fond clair
~FILTER - [photo] écran *m* orthochromatique
[opt] filtre *m* optique
~FILTER FACTOR - [opt] facteur *m* de filtre
~-FLOODED - [photo] inondé de lumière
~FLUX - [phys] (the quantity of light passing through an area) flux *m* lumineux
~-FOG - [photo] (fog caused by leakage of light into the camera, or access of light during development) voile *m* lumineux
~GATHERING POWER - [light] réceptivité *f* lumineuse
~GLOBE - [el] diffuseur *m* à globe
~GUIDE - [light] conduit *m* de lumière
~HOURS - [gen] heures *f*pl creuses
[telecomm] période *f* de faible trafic
~HYDROCARBON - [chem] hydrocarbure *m* léger
~HYDROGEN - [chem] hydrogène *m* léger
~INTENSITY - [light] (the lighting power) intensité *f* lumineuse
~LEAKAGE - [photo] infiltration *f* de lumière
~LEVEL - [telev & cin] intensité *f* d'éclairage
~LOADING - [el] charge *f* légère
~METAL - [metall] (a metal of low specific gravity) métal *m* léger
~METER - [photo] (exposure timer) compte-pose *m*
~MICROSECOND - [meas] microseconde-lumière *f*
~MINIMUM - [opt] seuil *m* lumineux
~MODULATION - [opt] (the control of the intensity of light by electrical means) modulation *f* de lumière
~MODULATOR - [opt] modulateur *m* de lumière
~-NEGATIVE - [photo] (synonym for photographic negative) photonégatif *m*
[opt] (having negative conductivity) photorésistant
~NUCLEUS - [nucl] (nucleus of light element) noyau *m* léger
~OIL SEPARATION - [ind chem] (debenzolization of oil) dégazolinage *m* de l'huile, désessenciement *m*
~OILS - [ind chem] (coal-tar distillates with boiling points in the range of 110-210°C. They are the source of a number of organic compounds) huiles *f*pl légères
~PIPE - [electron] (solid transparent plastic rod transmitting light from one end to the other) barre *f* transparente
~-PIPING - [light] (the conveyance of light to a distance by means of internally reflecting tubes, e.g. polymethyl methacrylate) amenée *f* de la lumière
~POSITIVE - [opt] (having increasing conductivity as a result of increased radiation) photoconducteur, photopositif
~-PROOF - [photo] résistant à la lumière
~QUANTUM - [phys] (single train of waves emitted by an atom or a molecule without change of phase) photon *m*, quantum *m* de lumière
~RAIL - [railw] (a rail weighing less than 60 lbs per yard) rail *m* léger
LIGHT RAIL MOTOR TRACTOR - [railw] locotracteur *m*
~RAILWAY - [railw] chemin *m* de fer à voie étroite
~RANGE - [light] portée *f* de la lumière
~RATIO - [astr] (the scale by which the brightness of the stars can be measured) rapport *m* de brillance
~REACTION - [biol] réaction *f* à la lumière
~-RELAY - [el] (a relay operated by a photo-electric cell) relais *m* photoélectrique
~RESISTANCE - [el] (of a photoelectric cell) résistance *f* à la lumière
~RUNNING - [mech] marche *f* en machine isolée, marche *f* haut-le-pied
~SATIN - [text] satin *m* léger
~SCREENS - [cin] (movable screens used to control the light rays) écrans *m*pl ajustables, écrans *m*pl opaques
~SECTIONS - [metall] profils *m*pl légeres
~SELECTOR DISC - [photo] disque *m* sélecteur de lumination
~SENSATION - [biol] sensation *f* lumineuse
~SENSITIVE - [electron] (photo electric cell; capable of excitation by light rays) photosensible
~-SENSITIVE CELL - [el] cellule *f* photoélectrique
~-SENSITIVE LAYER - [photo] (the light-sensitive coating on photographic films) émulsion *f* photographique
~-SENSITIVE MATERIAL - [photo] surface *f* photosensible
~SHAFT - [mining] puits *m* au jour, prise *f* de lumière
~SIGNAL - [light] panneau *m*, signal *m* lumineux
~SOLVENT - [paint] (a solvent with a relatively low boiling range) solvant *m* à bas point d'ébullition
~SOURCE - [light] (any device for producing illumination) source *f* lumineuse
~-SPOT GALVANOMETER - [instr] galvanomètre *m* à indice lumineux
~-SPOT SCANNER - [telev] analyse *f* à spot lumineux
~STABILITY AGENT - [chem] (a substance used to render a dye or pigment more resistant to the fading action of light) agent *m* de stabilité à la lumière
~STANDARD - [phys] étalon *m* de lumière
~-STRING - [oil ind] équipement *m* électrique dans la tour
~THREAD - [text] fil *m* léger
~-TIGHT - [photo] étanche à la lumière
~TRANSMISSION - [photo] transparence *f*
[light] transmission *f* de la lumière
~-TRAP - [cin] incerpteur *m* de lumière
~TREADLE - [text] pas *m* doux, pas de gaze
~TRUCK - [cin] (small car with additional lights energized from batteries in it) chariot *m* porte-lampe
~UP, to - [gen] éclairer
~VALVE - [el] (mechanism with a light transmission that can be made to vary according to an externally applied electrical quantity) relais *m* optique, valve *f* de lumière
~WARP - [text] chaîne *f* légère
~WAVE - [phys] onde *f* lumineuse
~-WEIGHT - [gen] léger
~-WEIGHT CONCRETE - [constr] (special concrete used when lightness rather than strength is required, made with aggregates containing pumice, cork and the like, and in cellular form) béton *m* léger

LIGHT-WEIGHT PISTON - [mech] piston *m* de métal léger
~-YEAR - [astr] (spatial unit expressing distances in the universe) année-lumière *f*
LIGHTEN, to - [gen] alléger, reduire le poids
[met] éclairer, éclaircir, s'illuminer
~A WAGON, to - [railw] soulager un wagon
LIGHTENING - [gen] allégement *m*, allégeage *m*
~OF THE SOIL - [agric] ameublissement *m* du terrain
LIGHTER - [impl] (lighting device) allumeur *m*, allumoir *m*
[impl] (for cigars and cigarettes) allume-cigare *m*
[naut] (flat-bottomed barge used for lightening or unloading ships) chaland *m*, péniche *f*, bette *f*
~PILOT - [gas ind] (of gas cooker) veilleuse *f* d'allumage
~-THAN-AIR AIRCRAFT - [aero] (an aircraft which is principally supported by its own buoyancy) aérostat *m*
LIGHTERAGE - [comm] frais *mpl* de chaland
[naut] déchargement *m* par allèges
LIGHTERING - [naut] transport *m* par chalands
LIGHTFACE - [print] caractère *m* léger
LIGHTHOUSE - [naut] (a structure on rock at sea, or at a point of a coastline, equipped with powerful lights to warn and guide ships) phare *m*
~TUBE - [electron] (electronic tube for radar purposes) tube *m* à disques scellés
LIGHTING - [gen] éclairage *m*, illumination *f*
[light] système *m* d'éclairage
[el] allumage *m*
[photo] (the illumination of objects better to define the objects to be photographed) éclairage *m*
~APPLIANCE - [light] appareillage *m* d'éclairage
~BRANCH CIRCUIT - [el] distribution *f* lumière
~FITTINGS - [el] appareils *mpl* d'éclairage, luminaire *m*
~GAS - [light] (a combustible gas used as an illuminant) gas *m* d'éclairage
~GRID - [telev] gril *m* d'éclairage
~HOLE - [gas ind] (of gas cooker) orifice *m* d'allumage
[metall] porte *f* d'allumage
~INSTALLATION - [light] installation *f* d'éclairage
~OUTLET - [el] prise *f* lumière
~PANEL - [el] panneau *m* interrupteurs
~PORT - s. lighting hole
~POWER - [light] (light intensity) intensité *f* lumineuse
~SET - s. lighting installation
~TECHNIQUE - [light & photo] technique *f* de l'éclairage
~TUBE - [gas ind] (of gas cooker) tube *m* d'onde
~UP - [gen] allumage *m*
~WIRE - s. light circuit
LIGHTLY COATED ELECTRODE - [el] électrode *f* à enrobage pelliculaire
LIGHTNESS - [gen] légèreté *f*, facilité *f* (d'une tâche)
~OF A COLOUR - [opt] clarté *f* d'une couleur
LIGHTNING - [met] (a spark discharge of atmospheric electricity between cloud and cloud or cloud and earth) éclairs *mpl*, foudre *f*
~ARRESTER - [el] (over-voltage protection device against lightning) parafoudre *m*
~CONDUCTOR - [el] (metal conductors connected between the highest point of a building and earth, thus providing and easy passage for a lightning discharge) paratonnerre *m*
LIGHTNING DISCHARGER - [el] s. lightning arrester
~-FLASH - [met] (spark discharge of atmospheric electricity between two clouds or between clouds and earth) éclair *m*
~GROUNDING SWITCH - [radio] commutateur *m* antenne/terre
~LINE - s. lightning conductor
~PROTECTOR - s. lightning arrester
~ROD - [el] tige *f* du paratonnerre
LIGHTSHIP - [naut] (a floating lighthouse) bateau-feu *m*, bateau-phare *m*
LIGNE - [horol] (unit in the measurement of wath movements, equal to 2.256 mm) ligne *f*
LIGNEOUS - [gen] (woody) ligneux
~PLANTS - [agric] plantes *fpl* ligneuses
~TISSUE - [bot] tissu *m* ligneux
LIGNIFICATION - [bot] (the process by which lignin is deposited in the cell walls of plants) lignification *f*
LIGNIFIED - [bot] lignifié
LIGNIFY, to - [agric] lignifier
LIGNIN - [chem] (the non-carbohydrate constituent of plants, with uses as an extender for plastics and rubber) lignine *f*
LIGNITE - [min] (a variety of coal, in which the process of conversion from the original vegetable material has not proceeded as far as with true coals. It is dull brown in colour and considerably lighter than the latter) lignite *f*, houille *f* brune
~COKE - [min] (solid residue from the carbonization of lignite) coke *m* de lignite
~OIL - [chem] (a constituent of some water-proofing paints, obtained by countercurrent extraction of lignite) huile *f* de lignite
~WAX - [chem] cire *f* montaine
LIGNIVOROUS - [zool] (who eats wood) lignivore
LIGNOCELLULOSES - [chem] (lignin-cellulose compounds occurring in plant fibres, espec. wood) lignocelluloses *fpl*
LIGNOCERIC ACID - [chem] (a fatty acid occurring in natural fats and having application in research) acide *m* lignocérique
LIGNOSE - s. lignin
LIGNUM VITAE - [bot] gaiac *m*, bois *m* saint
LIGROIN - [chem] (a petroleum with a boiling range of from 90° C, to I20° C. approximately; used as a solvent) ligroine *f*
LIGULA - [zool] (in insects, between the labial palps) ligule *f*
LIGULATE - [bot] (long, flat and narrow) ligulé
LIKE SIGN - [math] même signe *m*
LILAC - [bot] lilas *m*
LILY - [bot] lis *m*
LIMB - [anat] membre *m*
(of optical instrument) bord *m* gradué
[instr] (of a sextant) limbe *m*
[geol] flanc *m* d'un pli
[el] (of an electromagnet) branche *f*
[astr] (of heavenly body) (the upper of lower extremity of a celestial body, as seen in navigational observations) limbe *m*
~CENTRE - [el biol] (in electrocardiography, a reference lead; the junction of three equal resistors to the limb leads) centre *m* des membres, potentiel *m* V, centre *m* de Wilson

LIMB OF A MAGNET - [el] branche *f* d'un électro-aimant
LIMBER - [firearms] (the connecting link between the gun and its tractive power) avant-train *m*
LIMBERNECK - [vet] (bird botulism) botulisme *m* des oiseaux
LIMBERS - [shipbuild] (the holes which are cut through the floor timbers on the sides of the keelson to form a channel for water to the pump-well) canaux *mpl* des anguillers
~BOARD - [shipbuild] axe *m* des anguillers
LIME, to - [build] gluer
[leather ind] (the operation of soaking hides and skins in milk of lime) plainer, pelainer
[agric] chaleur
LIME - [chem] (calcium oxide, obtained by heating natural calcium carbonate, e.g. limestone or chalk, to a high temperature in a special kiln) chaux *f*
[bot] lime *f*
~BIN - [metall] rigole *f* à chaux
~BLAST - [leather ind] (dark patches appearing on limed skins, caused by carbonate of lime) ombres *fpl* de chaux, taches *fpl* de chaux
~BLUE - [paint] (pigment consisting of a mixture of ultramarine and terra alba) bleu *m* de Brême
~-BURNING - [constr] cuisson *f* de la chaux, chaufournerie *f*
~-CEMENT MORTAR - [constr] mortier bâtard *m*
~CHLOROSIS - [bot] (occurring in plants which grow in a soil too rich in calcium carbonate) chlorose *f*
~CONCRETE - [constr] béton *m* de chaux
~CONTENT - [chem] indice *m* de chaux
~DEFECATION - [sugar ind] (clarification of juice with milk of lime) défécation *f* au lait de chaux
~FLUX - [min] castine *f*
~HYDRATED - [chem] (slaked lime. Calcium hydroxide, a white powder formed by the action of water on calcium oxide) chaux *f* délitée
~IN CLODS - [constr] chaux *f* en mottes
~IN POWDER - [constr] chaux *f* en poudre
~-KILN - [ind chem] (simple type of brick furnace in which limestone or chalk is calcined to produce quickline) four *m* à chaux
~LIQUOR - [chem] bouille *f* de chaux
~MARL - [geol] calcaire *m* marneux
~MORTAR - [constr] (mortar consisting of lime and sand) mortier *m* de chaux
~OIL - [chem] (an essential oil obtained from varieties of Citrus, and used in the preparation of perfumes and flavourants) essence *f* de limon
~PASTE - [chem] (slaked lime) chaux *f* délitée
~PIT - [constr] fosse *f* à garder la chaux
~PLASTERING - [constr] plâtrage *m* de chaux
~PRECIPITATION - [chem] précipitation *f* de chaux
~PROOF - [gen] solide à la chaux
~PUTTY - [constr] mastic *m* à la chaux
~RED - [dyes] (a lake pigment made by adsorption of magenta on an earth) rouge *m* de chaux
~-SAND BRICK - [constr] brique *f* silico-calcaire
~SOAP - [ind chem] (a soap of palmitic, oleic or stearic acid) savon *m* de chaux
~SODA PROCESS - [ind chem] (a commercial method of softening hard water) procédé *m* chaux-soude
~TREATMENT - s. lime defecation
~TREE - [bot] limettier *m*
~WASHING - [ind chem] bain *m* au lait de chaux
~-WATER - [chem] (a solution of calcium hydroxide in water, with excess in suspension: used in medicine, and tin the laboratory to detect the presence of carbon dioxide) eau *f* de chaux
LIME WOOL - [text] laine *f* de peaux
~YELLOW - [paint] (a lake pigment made by adsorption of auramine or other yellow dye on an earth) jaune *m* de chaux
LIMED - [leather ind] plainé
~ROSIN - [paint] (commercial calcium resinate) résine *f* durcie
LIMEKILN - s. lime-kiln
LIMELIGHT, to - [gen] diriger le projecteurs (sur)
LIMELIGHT - [light] feux *mpl* de la rampe
[constr] (of a theatre stage) rampe *f*
LIMEN - [el biol] (electrical treshold) seuil *m*
LIMESODA FELDSPAR - [geol] gabbro *m*
LIMESTONE - [geol] (a rock consisting mainly of calcium carbonate with uses in metallurgy, agriculture as a building material and in the production of lime) pierre *f* à chaux, calcaire *m*
[constr] pierre *f* à chaux
[metall] (the limestone used in metallurgical processes) castine *f*
~FILLER - [constr] charge *f* calcaire
~FLUX - [metall] castine *f*, calcaire *m*
~FORMATION - [geol] formation *f* calcaire
~QUARRY - [mining] carrière *f* de pierre à chaux
~SOIL - [geol] (soil containing limestone) terrain *m* calcaire
LIMEWASH - [chem] (a mixture of quickline, tallow and water, used as a cheap coating for walls and the like) lait *m* de chaux, blanc *m* de chaux
LIMEWATER - s. lime-water
LIMEY - [soil] (calcareous) calcaire
LIMINAL - [physiol] (relating to the treshold value of perception) liminaire, du seuil
~CONTRAST - [phys] seuil *m* différentiel de l'excitation lumineuse
~VALUE - [opt] (the treshold of the value of a given colour, below which there is no visual appreciation) valeur *f* liminaire
LIMING - [leather ind] (process to facilitate removal of hair from hides, in which they are soaked in a suspension of calcium hydroxide in water) chaulage *m*
~COLUMN - [ind chem] colonne *f* à chaux, colonne *f* à sels fixes
~STILL - s. liming column
LIMIT, to - [gen] limiter, borner
LIMIT - [gen] limite *f*, borne *f*
[math] limite *f*
~CONTROL - [mech] (thermal cut-off) pyrostat *m*
~GAUGE - [meas] (a set gauge used to determine whether a part is within given tolerances) calibre *m* de tolerance
~GAUGING - [meas] calibrage *m* de tolerance
~LOAD - [aero] (the maximum expected load in any given part of an aircraft) charge *f* limite
~OF AUDITION - [acoust] limite *f* de l'ouïe
~OF BACKWATER - [hydr] limite *f* de la retenue
~OF CAPACITY - [gen etc] limite *f* de capacité
~OF COMPRESSION - [mech] limite *f* de compression
~OF ELASTICITY - [phys] limite *f* d'élasticité, charge *f* limite d'élasticité
~OF ERROR - [instr] (the maximum error throughout the scale under given conditions) limite *f* d'erreur
~OF PROPORTIONALITY - [mech] (the point on a

stress-strain curve at which the strain is no longer proportional to the stress) limite *f* de proportionnalité
LIMIT OF RESOLUTION - [opt] limite *f* de résolution
~OF SHUNT SIGN - [railw] tableau *m* d'arrêt pour les rames de manoeuvre
~OF SHUNT SIGNAL - [railw] pancarte *f* de limite de manoeuvre, signal *m* de limite de manoeuvre
~OF STRETCHING STRAIN - [metall] (the yield point) limite *f* de fluage
~OF TENSION - [metall] tension *f* limite
~POINT - [math] point *m* limite
~PRESSURE - [mech] pression *f* limite
~PRIORITY - [comput] priorité *f* limite
~STOP - [mach tool] arrêt *m* de fin de course
~SWITCH - [el] (switch fitted to electric lifts, cranes etc. to cut off power supplies in case the moving carriage travels beyond a given speed) contacteur *m* de fin de course
LIMITATION - [gen] limitation *f*
[leg] prescription *f*
~OF INTERFERENCE IN THE VISION RECEIVER - [telev] limitation *f* d'interférence vidéo
~OF MOBILITY - [crystall] (the limited mobility of electrons in crystals) limite *f* de mobilité
LIMITED - [gen] limité, borné
[leg] (of a company) à responsabilité limitée
~ACCESS SATELLITE - [telev] satellite *m* à accès limité
~COMPANY - [leg] (incorporated company with limited responsability) société *f* à responsabilité limitée
~CONVERTIBILITY - [agric] convertibilité *f* limitée
~FLAME PROOF - [el] à sécurité interne
~PARTNERSHIP - [leg] commandite *f* par actions
~ROTARY MOTOR - [mech] machine *f* à angle de rotation limité
~STABILITY - [el] (conditional stability in electrical telecommunication systems) stabilité *f* limitée
~SAFE - [nucl] de sûreté limitée
~SOLUBILITY - [chem] solubilité *f* limitée
~STROKE - [text] course *f* limitée
LIMITER - [el] (any transducer setting a boundary value on a signal) limiteur *m*
~CHARACTERISTIC - [radio] (of a limiter stage) caractéristique *f* de limiteur
~CIRCUIT - [electron] circuit *m* limiteur
~DIODE - [telev] (diode used as limiter) diode *f* limitatrice
~VALVE - [el] soupape *f* limitatrice
LIMITING - [gen] limitatif, limitateur
[el] (of electric devices) limitateur *m* de charge
[radio] (in radio communications, the action performed on a signal by a limiter) limitation *f*
[electron] (of pulse amplitudes) limitation *f*, écrêtage *m*
~AMPLIFIER - [telev] amplificateur *m* limiteur de crêtes
~COIL - [el] bobine *f* limiteuse
~CONDUCTIVITY - [chem] (the equivalent conductivity of a substance when completely ionized) conductivité *f* limite
~CURVES - [phys] (any line on a graphical representation of the conditions of a system, at which two phases are coexistent) courbes *fpl* limite
~DEFINITION - [telev] limite *f* de définition
~DENSITY - [chem] (of a gas) densité *f* limite
LIMITING DEVICE - [el] limiteur *m* (de charge)
[el] (in telecommunication circuits) limiteur
~DITCH - [hydr] fossé *m* de clôture *f*
~FACTOR - [physiol] (the slowest acting factor of a group of factors which affect simultaneously a physiological process) facteur *m* limite
~FEEDBACK - [contr] réaction *f* limitatrice
~FRAME - [photo] cadre *m* de delimitation
~FREQUENCY - [el] (the frequency at which there is a percepible change in response) fréquence *f* limite
~FRICTION - [mech] frottement *m* limite
~GRADIENT - [constr] pente *f* aux limites
~PIN - [text] piton *m* d'arrêt, boulon *m* de limitation
~QUANTITY - [meas] (quantity affecting the indications of a measuring instrument, but is not the quantity measured by the instrument) gradeur *f* d'influence
~RANGE OF STRESS - [metall] (the greatest range of stress a metal can withstand without failure for an indefinite number of cycles) gamme *f* limite des efforts
~RESOLUTION - s. limiting definition
~STAGE - [telev] (circuit for limiting the output signal amplitude of a receiver or transmitter to a predetermined value) étage *m* limitateur
~STATE OF STRESS - [mech] état *m* limite de tension
~STRESS - [mech] (critical pressure) contrainte *f* limite, pression *f* limite
~THERMORESISTOR - [el] (current limiting device automatically controlled by temperature, used in de-icing gear) thermorésistance *f* limitatrice
~VALUES - [gen] valeurs *fpl* limites
[electron] (the limits of values which should not be exceeded during the operation of standard electronic tubes) valeurs *fpl* limites
~VELOCITY - [aero] (the steady speed of an aircraft on a straight flight path under given conditions) vitesse *f* d'équilibre
LIMITLESS - [gen] illimité, sans bornes
LIMITROPHES - [anat] ganglions *mpl* du sympathique
LIMITS OF EFFECTIVE CURRENT RANGE - [el] (the values of current corresponding to the upper and lower limits of the effective range) puissance maximum (or minimum) d'un compteur
~OF POWER RANGE FOR ACCURACY OF A METER - [el] puissance *f* maximum (or minimum) de précision d'un compteur
LIMNEMIA - [med] cachexie *f* palustre
LIMNIC - [geol] limnique
LIMNIMETER - [instr] (sensitive instrument designed to measure the level of lakes) limnimètre *m*
LIMNOBIOTIC - [zool] (living in fresh water) limnobiotique
LIMNOLOGY - [biol] (the study of biological conditions in fresh waters) limnologie *f*
LIMNOMETER - s. limnimeter
LIMONIN - [chem] limonie *f*
LIMONITE - [min] (an amorphous iron ore, chiefly ferric oxide with water of crystallization. It is often pseudomorphic after magnetite and haematite, and also occurs as Bog Iron Ore) limonite *f*, hématite *f* brune, fer *m* oxydé hydraté
LIMOSIS - [med] faim *f* exagérée
LIMPID - [gen] limpide, clair, transparent
LIMPIDITY - [gen] (clarity of liquids and atmospheres) limpidité *f*
LIMY - [chem] (containing lime) calcaire, calcique

LIMY SOIL - [geol] sol *m* calcaire
~WOOL - [text] laine *f* pelade, avalies *f*pl de peaux
LINAC - s. linear accelerator
LINAGA - [ind chem] (Manila hemp bagasse) bagasse *f* de chanvre de Manille
LINALOE OIL - [chem] (an essential oil derived from various species of Bursera and having uses in perfumery) essence *f* de linaloe
LINALOOL - [chem] (a component of a number of essential oils, also prepared synthetically and used in perfumery) linalol *m*
LINALYL ACETATE - [chem] (a perfumery component) acétate *m* de linalyle
LINCHPIN - [mech] (a pin fitted to retain a wheel) cheville *f* d'essieu, esse *f*, clavette *f* d'essieu
~HUB - [mech] moyeu *m* claveté
LINDE PROCESS - [chem] (industrial process used to obtain nitrogen and oxygen from the atmosphere by liquefaction and fractionation) procédé *m* Linde
LINDEMANN ELECTROMETER - [instr] (portable electrometer which is insensitive to changes of level) électromètre *m* de Lindemann
LINE, to - [gen] ligner, rayer, border
[mech] tuber, doubler, recouvrir, baguer
[mech] (to fit linings to brakes) garnir
[shipbuild] revêtir intérieurement
[railw] (a road) dresser (une voie), riper
LINE - [gen] ligne *f*, raie *f*
[gen] corde *f*, cordage *m*
[surv] cordeau *m*
[el] ligne *f*, canalisation *f* électrique
[geom] ligne *f*
[constr] (used for rough measurement) cordeau
[railw] ligne *f* (de chemin de fer)
[naut] corde *f*, amarre *f*, passeresse *f*
[fishing] ligne *f* (de pêche)
[text] (yarn spun from flax) fil *m* de lin
[spectrology] (the internal standard line) ligne *f*, raie *f*
[gen] (a cloths line) étendoir *m*, corde *f* (à linge)
[naut] (rope for small boats) passeresse *f*
[geogr] (sometimes used as 'equator line') équateur *m*
[radio] (that part of the circuit which is esternal to the apparatus) ligne *f*
[electron] (the path traced by a moving spot) ligne *f*, trace *f*
[telev] (a single trace of the electron beam from left to right across the screen) ligne *f*
[comput] (on a punch card) ligne *f*
[telecomm] (of telephone) ligne *f* de transmission
[telecomm] (a circuit) circuit *m*
[gas ind] (only USA, main or pipeline in GB) pipeline *f*, conduite *f*, canalisation *f*
[photo] trait *m*
[horol] s. ligne
[autom] série *f* de fabrication
~AT-A-TIME PRINTING - s. line printing
~ACTION OF FORCE - [phys] (the direction along which a force acts) ligne *f* de force
~ADVANCE - [radio] (scanning pitch) distance *f* de ligne
~AMPLIFIER - [radio] (the amplifier which feeds the transmission lines with modulated paper) amplificateur *m* de ligne
~AMPLITUDE CONTROL - [telev] (device for the control of the horizontal amplitude of the scanning) dispositif *m* de réglage de la largeur de ligne
LINE AND PLUMMET - [naut] sonde *f*
~AND TACK - [plumb] (only USA; liming-up in GB; expressions used in mainlaying) mise *f* bout pour soudage, alignement *m* et pointage *m*
~BALANCE - [el] impédance *f* balancée
~BANK - [telecomm] banc *m* de contacts de ligne
~BEND - [telev] (parabolic waveform at line frequency consituting a component of the video signal distortion in the camera output) signal *m* parabolique de correction de ligne
~BLANKING - [telev] (suppression of the vision signal at the end of each lines) impulsion *f* de suppression horizontale
~BLOCK - [print] (printing block consisting of black and white parts only) cliché *m* au trait, gravure *f* au trait
~-BORE, to - [mech] aléser en ligne
~-BORING - [mech] alésage *m* en ligne
~BREAK - [telecomm] renversement *m* de ligne
~BREAKER - [el] coupeur *m* de ligne
~BROADCASTING - [radio & telev] télédiffusion *f*
~BRUSH - [telecomm] bras *m* porte-contacts
~BUILDING-OUT NETWORK - [el] (type of matching network) réseau *m* supplémentaire
~BUMPER - [mining] amortisseur *m* du câble
~BURNER - [th eng] brûleur *m* rectiligne
~BUSHING - [el] traversée *f* isolée de ligne
~-BY-LINE SCANNING - [telev] analyse *f* ligne par ligne
~CARRIED ON BRACKETS - [telecomm] ligne *f* en consoles
~CLAMP AMPLIFIER - [telev] amplificateur *m* stabilisateur du signal
~CORD - [electron] câble *m* secteur
~CRAWL - [telev] (in colour television) déformation *f* de la structure des lignes
s. line flickering
~DIFFUSEUR - [telev] circuit *m* d'estompage de ligne
~DIFFUSION - [telev] estompage *m* de ligne
~DIVIDER - [telev] diviseur *m* de fréquence de lignes
~DRAWING - [draw] dessin *m* linéaire
~DRIVE - [telev] circuit *m* de déclenchement de ligne
~DROP - [el] chute *f* de tension de ligne
~DROP COMPENSATION - [el] compensation *f* de chute de tension de ligne
~EQUALIZER - [el] filtre *m* correcteur
~FEED - [print] (feed of the paper on a page printing machine) avance *m*
~FEED CHARACTER - [comput] caractère *m* de change de ligne
~-FEEDING AMPLIFIER - [telev] amplificateur *m* de sortie
~FINDER - [telecomm] chercheur *m* d'appel
~FLAX - [text] lin *m* à longs brins
~FLICKER - [telev] papillotement *m* de lignes
~FLICKERING - [telev] (undesirable effect in colour television, whereby a vertical crawling of a colour is seen on the screen) effet *m* de papillotement
~FLYBACK PULSE - [telev] retour *m* du spot de ligne
~FOCUS - [electron] (elongated rectangular focus placed in such way that an effectively square source of X-ray is achieved through foreshortening) foyer *m* linéaire
~-FOCUS TUBE - [electron] tube *m* à rayons X à foyer linéaire
~FREE - [telecomm] déblocage *m*

LINE FREE CIRCUIT - [telecomm] circuit *m* de déblocage
~FREQUENCY - [telev] (the number of lines scanned per second) fréquence *f* d'analyse de ligne
~FREQUENCY TO FIELD FREQUENCY RATIO - [telev] rapport *m* fréquence lignes/fréquence trame
~FUSE - [el] fusible *m*, relais *m* thermique
~GAUGE - [print] typomètre *m*
~HEIGHT - [telev] hauteur *f* de ligne
~HEMP - [text] chanvre *m* à longs brins
~HYDROPHONE - [instr] (directional hydrophone consisting of one straightline element) hydrophone *m* à ligne
~INCLUSIONS - [metall] inclusions *fpl* alignées
~INCREMENT - [telev] (in video recording) accroissement *m* de ligne
~INTEGRAL - [el] (of a vector) intégrale *f* de ligne
~JACK - [el] prise *f* monopolaire
~KEYSTONE WAVEFORM - [telev] signal *m* de correction de trapèze-lignes
~LENGTHENER - [electron] extenseur *m* de ligne à phase variable
~LOAD - [soil] charge *f* linéaire, ligne *f* de charge
~LOCK - [telev] verrouillage *m* de fréquences trame et réseau
~LOOP - [telecomm] fonctionnement *m* en connecté
~LOSS - [comput] affaiblissement *m* de ligne
~MICROPHONE - [el acoust] (type of directional microphone) microphone *m* à ligne
~MONITOR - [electron] moniteur *m* de câblage
~MONITORING TUBE - [telev] tube *m* moniteur de ligne
~NOISE - [el] (noise occurring in an electrical circuit) bruit *m* de circuit, bruit *m* de ligne
~OCCUPANCY - [comput] occupation *f* d'une ligne
~OF ACTION - [mech] (the line along which a force acts) ligne *f* de poussée, ligne *f* de pression
[mech] (of gears) normale *f* commune des profils
~OF AIM - [gen] ligne *f* de mire
~OF BEARING - [surv] ligne *f* de repère
~OF BUCKETS - [mech] chapelet *m* de godets
~OF CENTRES OF A WAVEGUIDE - [electron] ligne *f* des centres d'un guide d'ondes
~OF CREEP - [soil] (path of percolation) ligne *f* de cheminement
~OF DIP - [geol] ligne *f* de plongement
[el] inclinaison *f*
~OF ELECTRIC FLUX - [el] (in a graphical representation of an electric field, a line so drawn that its direction at any point is the direction of the electric flux at that point) ligne *f* de flux électrique
~OF EQUAL MAGNETIC DIP - [geol] ligne *f* isoclinale
~OF FLIGHT - [aero] ligne *f* de vol
~OF FLUX - s. line of electric flux
~OF FORCE - [el] (a graphic representation of the direction of either an electric or a magnetic field) ligne *f* de force
~OF INDUCTION - [el] ligne *f* d'induction
~OF INFLUENCE - [phys] ligne *f* d'influence
~OF INTERSECTION - [soil] ligne *f* d'intersection
~OF INTRADOS - [constr] profil *m* d'intrados d'une voûte
~OF LEAST RESISTANCE - [mech] ligne *f* de moindre résistance
~OF LODE - [mining] direction *f* du filon
~OF MAGNETIC FLUX - [el] ligne *f* d'induction magnétique
LINE OF MAGNETIC FORCE - [el] ligne *f* de force magnétique
~OF OUTCROP - [geol] ligne *f* d'affleurement
~OF POSITION - [radar] (line on which the observer is computed to be at the time of the observation) ligne *f* de position
~OF RAILS - [railw] file *f* de rails
~OF RUPTURE - [soil] ligne *f* de rupture
~OF SEGREGATION - [metall] limite *f* de ségrégation
~OF SHAFTING - [mech] ligne *f* de transmission, ligne *f* d'arbres
~OF SIGHT - [surv] (line of collimation) ligne *f* de collimation
[astr] (the line along which a heavenly body is approaching, or receding from the observer) ligne *f* de visée
[opt] rayon *m* visuel
~OF SIGHT VELOCITY - [phys] vitesse *f* radiale
~OF SIGHTING - [firearms] ligne *f* de mire
~OF STRIKE - [mining] alignement *m* de la direction
[geol] direction *f* de stratification
~OF SUPPORT - [constr] (line of pressure) ligne *f* de poussé
~OF TANGENCY - [surv] ligne *f* de tangence
~OF TWILL - [text] ligne *f* du sergé
~OUTPUT - [telev] (the time basis of the scanning circuit of the lines) base *f* de temps de ligne
~OUTPUT PENTODE - [electrode] pentode *f* de sortie de ligne
~OUTPUT TRANSFORMER - [telev] (transformer belonging to the scanning circuit) transformateur *m* de retour du spot
~PAIR - [phys] (an analytical line and the internal standard line with which it is compared) paire *f* de lignes
~PERIOD - [telev] (the internal between the beginning of one pulse and the beginning of the next one) periode *f* de ligne
~PIPE - [gas ind] tube *m* de canalisation
~POSITIVE - [photo] positif *m* à traits
~-PRESSURE - [mech] (the normal working pressure in a compressed air or hydraulic supply system) pression *f* de fluide
~PRINTER - [comput] imprimante *f* en lignes
~PRINTING - [comput] impression *f* en lignes
~PRODUCTION - [ind proc] travail *m* à la chaîne
~PROFILE - [railw] profil *m* d'une ligne
~RATE CONVERTER - [telev] convertisseur *m* de normes de télévision à fréquences différentes de lignes et à fréquences identiques de trames
~-REAMING - [mech] alésage *m* en ligne
~RECORDS - [telecomm] cahier *m* de distribution
~REGISTER - [comput] registre *m* de lignes
~RELAY - [el] relais *m* de ligne, relais *m* d'appel
~REPEATER - [telecomm] répéteur *m* de ligne
~REPEATING COIL - [telecomm] translateur *m* de la ligne
~RESIDUAL CURRENT - [el] (ground return current) courant *m* de retour à la terre
~RESIDUAL EQUALIZER - [el] correction *f* de distorsion restante
~-REVERSAL PYROMETER - [instr] pyromètre *m* à inversion de raie
~RINGING - [telev] barres *fpl* verticales à gauche
~SAWTOOTH - [electron] forme *f* d'onde en dents de scie à fréquence de déviation de ligne

LINE SCAN CIRCUIT - [telev] circuit *m* de déviation de ligne
~ SCRATCHES - [telecomm] bruits *mpl* de friture, crachements *mpl*
~ SELECTOR - [telecomm] chercheur *m*
~ SEQUENCE - [telev] séquence *f* de lignes
~ SEQUENTIAL - [telev] à séquence de lignes
~ SEQUENTIAL SYSTEM - [telev] système *m* de séquence de lignes
~ SHAFT - [mech] arbre *m* de renvoi
~ SHAFT DRIVE - [mech] commande *f* par arbres de transmission
~ SHAFTING - [mech] (an overhead shafting used in factories to transmit power to individual machines) ligne *f* de transmission
~ SHORT CIRCUIT - [el] court-circuit *m* de ligne
~ SIMULATOR - [el] (network designed to simulate the characteristics of a line) ligne *f* artificielle
~ SKEW - [print] biais *m* de ligne
~ SLIP - [telev] défilement *m* horizontal, glissement *m* horizontal
~ SOURCE - [nucl] (a linear source of neutrons) source *f* linéaire
~ SPACE - [print] entre-ligne *m*, interligne *m*
~ SPECTRUM - [spectro] (a non continuous spectrum) spectre *m* de lignes
[acoust] (sound spectrum with components being confined to a number of discrete frequencies) spectre *m* de lignes
~ SPLITTER - [comput] distributeur *m* de voies
~ SQUALL - [met] (a sudden moving forward on a wide front, caused by a body of colder air taking the place of a warmer one) ligne *f* de grains
~ -STABILIZED OSCILLATOR DRIVE - [radio] oscillateur *m* à pilotage par ligne de transmission
~ STRETCHER - [electron] (in waveguides) extenseur *m* de ligne à longueur variable
~ STROBE - [telev] (a waveform monitor) moniteur *m* de forme d'onde
~ STRUCTURE - [telev] structure *f* du canevas
~ SYNCHRONIZING PULSE - [telev] impulsion *f* de synchronisation de ligne
~ SYNCHRONIZING SIGNAL - s. line synchronizing pulse
~ TELEPHONY - [telecomm] téléphonie *f* par fil
~ TEMPERATURE-COMPENSATING EQUALIZER - [radio] réseau *m* correcteur des différences de température
~ TERMINALS - [el] (of a polyphase machine) bornes *fpl* de phase
~ TESTER - [instr] (instrument designed to determine leaks and corrosion of the coating on pipes) balai *m* électrique
[print] compte-fils *m*
~ THE BEARING, to - [mech] recouvrir de métal antifriction un palier
~ TILT - s. line bend
~ -TO-LINE SPACING - [comput] interligne *m*
~ -TO-STORE TRANSFER - [comput] stockage *m*
~ TRANSFORMER - [telev] (an apparatus used for transforming the number of lines of a transmitter to another number of lines of another transmitter) transformateur *m* du nombre de lignes
~ TRANSLATION - [telev] (the operation of transforming the number of lines of a transmitter to another number of lines of another) transformation *f* du nombre de lignes

LINE TRANSMISSION - [el] (the transmission of energy over a line) transmission *f* par ligne
[comput] (the transmission of information over a transmission line) transmission *f*
~ TRAVERSING - [telev] décrochage *m* de lignes
~ -UP - [gen] alignement *m*
[mech] mise *f* au point
[comput] mise *f* en circuit
~ VOLTAGE FLUCTUATIONS - [electron] variations *fpl* de la tension du secteur
~ VOLTAGE OF A POLYPHASE SYSTEM - [el] tension *f* composée d'un système poliphasé
~ VOLTAGE REGULATOR - [electron] régulateur *m* de la tension secteur
~ VORTEX - [phys] (a line such that its direction is at all points the same as that of the local vorticity) ligne-tourbillon *f*
~ WELDING - [mech] (seam welding) soudure *f* continue
~ WIDTH - [telev] (of a scanning line) largeur *f* de ligne
[phys] (in a spectre) largeur *f* de ligne
~ WINDING - [telecomm] enroulement *m* de ligne
~ WIPER - [telecomm] balai *m*, frotteur *m* de ligne, bras *m* porte-contacts
~ WIRE - [el] fil *m* de ligne
LINEAL - [gen] linéal
LINEAR - [gen] linéaire, linéique
~ ABSORPTION COEFFICIENT - [nucl] (the fractional decrease in intensity per unit distance crossed) coefficient *m* d'absorption linéaire
~ ACCELERATOR - [nucl] (an apparatus for accelerating particles in a straight line) accélérateur *m* linéaire
~ ACTIVITY - [nucl] (the activity per unit length of an elongated gamma-ray source) activité *f* linéique
~ AMPLIFICATION - [radio] amplification *f* de ligne
~ ARRAY - [radio] (collinear array; a number of dipoles placed end to end, thus forming a single line) série *f* de dipôles
~ ATTENUATION - [telecomm] affaiblissement *m* linéique
~ ATTENUATION COEFFICIENT - [math] coefficient *m* d'atténuation linéique
~ CALIBRATOR - [instr] calibreur *m* linéaire
~ CHARGE DENSITY - [el] (the quantity of electric charge per unit length of a linear charged body) densité *f* de charge linéaire
~ CIRCUIT NETWORK - [electron] circuit *m* linéaire
~ CODE - [comput] code *m* linéaire
~ CONTROL ELECTROMECHANISM - [contr] électromécanisme *m* de commande linéaire
~ DAMPING - [el] affaiblissement *m* linéique
~ DELAY UNIT - [contr] élément *m* à retard linéaire
~ DETECTION - [electron] (rectification in which the application of a sinusoidal input gives rise to an output which is proportional to the input) détection *f* linéaire, redressement *m* linéaire
~ DIRECT CURRENT AMPLIFIER - [el] amplificateur *m* linéaire pour courant continu
~ DISPERSION - [phys] dispersion *f* linéaire
~ DISPLACEMENT - [phys] déplacement *m* linéaire
~ DISTORTION - [el] (form of distortion independent of the amplitude of the signal) distorsion *f* linéaire
~ ELECTRICAL CONSTANTS - [radio] (of a uniform line) paramètres *mpl* de ligne
~ ELECTRON ACCELERATOR - [electron] accélérateur

m linéaire d'électrons
LINEAR ELEMENT - [contr] organe *m* linéaire
~ENERGY TRANSFER - [electron] transfert *m* d'énergie par unité de longueur
~EQUATION - [math] équation *f* du premier degré, équation *f* linéaire
~EXPANSION - [phys] (increase in unit length of a body per centigrade rise in temperature) expansion *f* linéaire, dilatation *f* linéaire
~EXPANSION COEFFICIENT - [phys] coefficient *m* de expansion linéaire
~EXTRAPOLATION DISTANCE - [nucl](the value of the range in matter) longueur *f* linéaire d'extrapolation
~FIELD - [math] fonction *f* linéaire
~GAMMA ABSORPTIOMETER - [instr] absorptiomètre *m* linéaire à rayonnement gamma
~HARMONIC OSCILLATOR - [electron] oscillateur *m* linéaire
~INDEXING MACHINE - [comput] appareil *m* à indexation linéaire
~INDUCTION MOTOR - [el] moteur *m* d'induction à mouvement linéaire
~INDUCTIVE COUPLING STORE - [comput] mémoire *f* à couplage inductif linéaire
~INSULATION - [el] (the insulation of an insulated conductor) isolement *m* linéaire
~ION DENSITY - [nucl] nombre *m* linéique d'ions
~IONIZATION - [nucl] ionisation *f* linéique
~MACROMOLECULE - [chem] (a molecule containing a large number of atoms in linear configuration i. e., a polymer molecule) macromolécule *f* linéaire
~MAGNIFICATION - [opt] agrandissement *m* linéaire
~MATRIX - [telev] (in colour television) matrice *f* linéaire
~MEASURE - [meas] mesure *f* de longueur
~MEASURING ASSEMBLY - [meas] ensemble *m* de mesure linéaire
~MEMORY - [comput] mémoire *f* linéaire
~MOLECULAR CHAIN - [chem] (chain-like structure formed by monomeric the end-to-end attachment of molecules) chaîne *f* moléculaire linéaire
~MOVEMENT - [comput] translation *f*
~NETWORK - [el] (network in which the impedances of the elements are independent of the magnitudes of the current) réseau *m* linéaire
~OSCILLATOR - [radio] (oscillator of the simple harmonic type) oscillateur *m* linéaire
~PASSIVE ELECTRIC NETWORK - [el] réseau *m* électrique passif linéaire
~POLYMER - [chem] (a polymer in which the molecules are linked in the form of long chains) polymère *m* linéaire
~POWER AMPLIFIER - [radio] amplificateur *m* linéaire de puissance
~PROGRAMMING - [comput] programmation *f* linéaire, optimisation *f* linéaire
~RANGE - [phys] (range expressed in units of length) portée *f* linéaire
~RATEMETER - [electron] ictomètre *m* linéaire
~RECTIFICATION - [radio] (rectification in which the unidirectional output current is directly proportional to the instantaneous peak amplitude of the applied alternating voltage) redressement *m* linéaire
~RECTIFIER - [radio] (for modulation) redresseur linéaire
~SCALE - [instr] (an equally divided scale; each division represents the same value) échelle *f* linéaire
LINEAR SCALE LENGTH - [meas] (the distance between the two end of the scale) longueur *f* d'échelle
~SELECTION - [comput] (in core memory) sélection *f* directe
~SPEED - [phys] (vector quantity denoting the time rate and the direction of a linear motion) vitesse *f* linéaire
~STOPPING POWER - [phys] (the energy loss per unit distance) pouvoir *m* d'arrêt linéique
~SUPERPOLYMER - [chem] (a polymer in which the molecules are essentially in long chains with an average molecular weight greater than I0,000) superpolymère *m* linéaire
~TAPER - [instr] (a resistance with a uniform change over the range of a potentiometer) résistance *f* à variation linéaire
~TRANSDUCER - [el acoust] transducteur *m* linéaire
~UNIT - [comput] organe *m* de calcul linéaire
~VARYING PARAMETER NETWORK - [el] réseau *m* à paramètre à variation linéaire
~VELOCITY - s. linear speed
LINEARITY - [gen] linéarité *f*
[instr] (the maximum deviation from the curve formed by plotting potentiometer electrical output versus mechanical travel, and the best straight line through this curve) linéarité
~CONTROL - [telev] réglage *m* de la linéarité
~CORRECTION - [telev] ajustage *m* de la linéarité
~IN PHASE - [telecomm] linéarité *f* de phase
~SLEEVE - [telev] bande *f* d'ajustage de la linéarité
LINEARLY POLARIZED SOUND WAVE - [acoust] onde *f* sonore polarisée dans un plan
LINED - [gen] ligné, ripé
[mech] garni, chemisé
[text] doublé, fourré
[mech] (of brakes) garni
[constr] (with tiles) en tuiles, carrelé, à carreaux
~BOARD - [paper man] carton *m* doublé
LINEN - [text] (cloth woven from the fibres of the flax) toile *f* de lin
[text] (domest linen) lingerie *f*, linge *m*
~BORDER - [text] lacet *m*
~DAMASK - [text] damassé *m* lin
~HEALD - [text] lisse *f* en câblé lin
~HOSE - [rubber ind] tuyau *m* en lin
~INTERLINING - [text] toile *f* doublure
~PAPER - [paper man] papier *m* de lin
~PLUSH - [text] peluche *f* de lin
~WEAVING - [text] tissage *m* de la toile
~YARN - [text] fil *m* de lin (long)
LINER - [naut] paquebot *m* de ligne
[metall] (of a furnace) contre-porte *f* de chaudière
[impl] fileteur *m*
[mech] cale *f* d'épaisseur
[rubber ind etc] revêtement *m*, fourrure *f*, doublure *f*
[mech] (a cylindrical sleeve, not forming an integral part of the block, which takes the place of a bore machined in the latter, as in reciprocating engines, pumps and the like) chemise *f*, manchon *m* intérieur
[el] (in a dry cell, a paper or card divider between the inner face of the negative electrode and the depolarizing mix) séparateur
[metall] bande *f* métallique de garnissage
[oil ind] tubage *m* perdu
~CLEANER - [oil ind] laveur *m* du tube perdu

LINER HANGER - [oil ind] support *m* du tube perdu
~SETTER - [oil ind] joint *m* du packer
LINGEL - [shoe man] (a thread rubber with beeswax) ficelle *f* de cordonnier
LINGERING - [gen] retard *m*, attardement *m*
~PERIOD - [nucl] (the time during which an electron remains in the orbit of highest excitation before jumping to a lower orbit) période *f* d'attardement
LINGUALIS - [anat] lingual
LINGULA - [anat] lingula *f* du cervelet
~OF THE SPHENOID - [anat] lingula *f* du sphénoide
LINEMENT - [med] (a fluid, usually containing oil and medicaments, intended to be applied to the body by inunction) liniment *m*
LINING - [gen] revêtement *m*, garnissage *m*, revêtement *m* intérieur
[text] doublure *f*
[mech] revêtement *m* intérieur
[mech] (of a cylinder) garnissage *m*
[el] revêtement *m* isolant
[auto etc] (of brake) garniture *f*, fourrure *f*
[hydr] (layer of clay puddle covering the sides of a canal to make it watertight) revêtement *m*
[constr] (boarding covering an interior surface) coffrage *m*, couchis *m*
[railw] (of a road) dressage *m*
[metall] brasque *f*
~BAR - [impl] pince *f* à riper
~FABRIC - [text] tissu *m* de doublure
~FILM - [text] feuille *f* de revêtement
~OF SHAFT - [metall] chemine *f* réfractaire
~-UP - [gen] alignement *m*
[plumb] s. line and tack
~UP THE CAMERA - [cin telev] réglages *mpl* généraux de camera
LINITIS - [med] linite *f*
LINK, to - [gen] lier, attacher, enchaîner
[mech] articuler, s'articuler
LINK - [gen] anneau *f*, maillon *m*, maille *f*
[mech] (a connecting piece) articulation *f*, coulisse *f*
[el] (the fuse link) élément *m* de remplacement
[auto] biellette *f*, jumelle *f*
[phys] liaison *f*
[of a chain] maille *f*, maillon *m*
[comput] instruction *f* d'enchaînement, lien
[comput] (the part of a subroutine which connects it with the rest of a programme) liaison *f*
[telecomm] (a two-way system forming part of a telecommunication circuit) liaison *f* radioélectrique, câble *m* hertzien
[radio telev telecomm] voie *f* de transmission
~BELT - [auto] chaîne *f* de transmission
~BLOCK - [mech] (sliding block pivoted to the end of the valve rod; it operates in the slotted link of a link motion) coulisseau *m*
~BOARD - [el] tableau *m* de liaison
~-BY-LINK SIGNALLING - [telecomm] signalisation *f* à répétition des signaux dans la voie de transmission
~CHAIN - [mech] chaîne *f* à maillons
~CIRCUIT - [comput] ligne *f* de jonction
~COUPLING - [el] accouplement *m* de liaison
~-HANGER - [mech] bielle *f* de suspension de la coulisse
~HEALD - [text] lisse *f* coulante, maille *f* à noeud simple

LINK MECHANISM - [mech] mécanisme *m* articulé
~MOTION - [mech] (valve motion for controlling the cut-off of a steam engine) distribution *f* par coulisse
~NEUTRALIZATION - [radio] neutralisation *f* par coupleurs
~OF PATTERN CHAIN - [text] chaînon *m* de la chaîne d'armure
~PIN - [mech] tourillon *m*, fuseau *m*, goujon *m*, axe *m* de maillon de chaîne
~PLATE - [mech] flasque *m* de coulisse
~ROD - [mech] (the connecting rods of a radial aero-engines) biellette *f*
~STUD - [mech] (of a chain) goujon *m* de chaîne
~TRANSMITTER - [telev] émetteur *m* à couplage link, émetteur *m* à relais
LINKAGE - [chem] (a bond, espec. a co-valent bond in a molecule of an organic compound) liaison *f*
[el] (linkage coefficient; the ratio of the mean flux per turn of an inducing winding to the means flux per turn of another winding linked with it) enchaînement *m*, facteur *m* d'enchaînement
[mech] mécanisme *m* de liaison
[med] liaison *f*, enchaînement *m*
[genet] linkage *m*, liaison *f*
~BALL JOINT - [auto] rotule *f* de timonerie
~COEFFICIENT - s. linkage [el]
~FRAME FOR ATTACHEMENTS - [agric] cadre *m* d'attelage pour les outils de culture
~MULTIPLIER - [comput] multiplicateur *m* mécanique
~RODS - [text] tringlage *m*
~SWITCHES - [el] (switches which are linked mechanically, thus operating in a definite sequence) commutateurs *mpl* à couplage
~SUBROUTINE - [comput] sous-programme *m* fermé
LINKED VEINS - [geol] veines *fpl* réticulées
LINKING COURSE - [text] rangée *f* lâche, rangée *f* à remailler
~POINT - [text] point *m* d'attache
~SEQUENCE - [comput] séquence *f* d'instructions de raccordement
~THE WARP YARN INTO CHAIN - [text] mise *f* en mailles de la chaîne ourdie
LINKS AND LINKS FABRIC - [text] envers *m*, tricot *m* uni, tricot *m*, Jersey uni
LINKWORK - [mech] s. linkage
LINN - [geogr] (a torrent running on rocks, a waterfall) petite cataracte *f*
LINNAEITE - [min] (natural cobalt sulphide, crystallizing in the orthorhombic system) linnéite *f*
LINOLEATE DRIERS - [paint] (driers derived from linseed oil) siccatifs *mpl* au linoléate
LINOLEIC ACID - [chem] (an unsaturated fatty acid used in food processing) acide *m* linoléique
LINOLEIN - [chem] (a component of linoleic acid) linoléine *f*
LINOLENIC ACID - [chem] (an intermediate in the production of driers for paints and varnishes) acide *m* linoléique
LINOLENYL ALCOHOL - [chem] (an intermediate for synthetic resins, plastics, and surface coatings) alcool *m* linolenylique
LINOLEUM - [gen] (a mixture of granulated cork, pigments and drying oil spread on a fabric backing to form a floor-covering material) linoléum *m*
LINOLEYL ALCOHOL - [chem] (an intermediate for synthetic resins and surface coatings) alcool *m* li-

noléilique
LINOLEYLTRIMETHYLAMMONIUM BROMIDE-[chem] (a disinfectant) bromure *m* de linoléyltriméthylammonium
LINOTYPE - [print] linotype *f*
~METAL - [metall] métal *m* linotype
LINOXYN - [chem] (a substance produced by the oxidation of linseed oil. It has elastic properties and is the basis of linoleum) linoxyne *f*
LINSEED - [bot] linette *f*, graine *f* de lin
~CAKE - [agric] (animal feeding) tourteau *m* de lin
~MEAL - [agric] (animal feeding) farine *f* de tourteau de lin
~OIL - [chem] (an oil obtained from the seeds of Linum usitatissimum and used in the manufacture of paints and varnishes, pharmaceuticals, plastics and linoleum) huile *f* de lin
LINSEY - [text] breluche *f*
~WOOLSEY - [text] tiretaine *f*
LINT - [text] (cloth made from cotton or linen, with a soft surface, generally used for dressings) tissu *m* charpie
~DOCTOR - [impl] contre-racle *f*
~-FREE CLOTH - [text] (special textile material which does not allow of the detachment of fluff or lint when cleaning mechanical parts) tissu *m* sans filasse
LINTEL - [constr] (beam across the top of an opening) linteau *m*, sommier *m*
~GIRDER - [metall] marâtre *m*
LINTER - [text] (machine designed to eliminate the lint) machine *f* à éliminer les bourres
LINTERS - [text] (fibres obtained as a by-product of cotton ginning and used as a raw material for rayon and plastics) bourres *fpl* de coton, bas-cotons *mpl*
LINTLESS COTTON - [text] coton *m* à long fibre
LION - [zool] lion *m*
LIONESS - [zool] lionne *f*
LIP - [anat] lèvre *f*
[gen] bord *m*, rebord *m*
[mech] (of a drill) lèvre *f*
[mech] (of a tool) tranchant *m*, lèvre *f*
[metall] (of a ladle; the edge of the spout, so formed as to direct the contents in a narrow stream when poured) bec *m*, coulée *f*
[metall] (of a furnace) rive *f* (d'un four)
~-AND-LEG ULCERATION - [vet] (an infection affecting animals) nécrobacillose *f*
~ANGLE - [mech] angle *m* de coupe
~APPLICATOR - [radiat] (applicator which can be applied to the lip of a patient) localisateur *m* labial
~-BOLT - [mech] boulon *m* à ergot
~GUIDE RAIL - [text] rail-guide *m* à rainure
~MICROPHONE - [el acoust] (microphone used in contact with the lips of the speaker) microphone *m* de bouche
~OF DRILL - [mining] taillant *m* de fleuret
~-POUR LADLE - [metall] poche *f* de coulée à bec
~RING - [oil ind] garniture *f* à bec
~-SYNCHRONIZED SPOT - [telev] prise *f* à haute qualité de synchronisation
~UNION - [plumb] (union with an inner annular projection) joint *m* avec bords
LIPARITE - [geol] (the granitic rocks which occur as lava flows, as those in the Lipari islands) liparite *f*, rhyolite *f*
LIPAROID - [med] gras, graisseux
LIPAROMPHALUS - [med] lipome *m* ombilical
LIPASE - [biol] lipase *f*
LIPASES - [chem] (enzymers which hydrolyse lipids; they have application in food processing) lipases *fpl*
LIPECTOMY - [med] lipectomie *f*
LIPEMIA - [med] lipémie *f*
LIPID - [chem] (or lipide; biochemycally important natural compound insoluble in water but soluble in a number of organic solvents) lipide *m*
LIPOCHROMES - [chem] (the substances which pigment butter-fat) lipochromes *mpl*
LIPOCLASIS - [med] lipolyse *f*
LIPOCYTE - [cytol] lipocyte *f*
LIPOFRIBOMA - [med] fibrolipome *m*
LIPOGENOUS - [zool] (which produces fat) lipogène
LIPOID - [chem] (generic term for naturally-occurring fats) lipoide *m*
LIPOIDOSIS - [med] s. lipidosis
LIPOLYSIS - [chem] (the decomposition of a fat by a lipase) lipolyse *f*
LIPOMA - [med] lipome *m*
LIPOTROPIC AGENT - [chem] (an agent which assists in the metabolism of fat) agent *m* lipotropique
LIPOXIDASE - [chem] (an enzyme) lipoxydase *f*
LIPPING - [med] formation *f* pathologique d'un rebord osseux
LIPURIA - [med] lipurie *f*, adiposurie *f*
LIQUATE, to - [gen etc] (to render liquid) rendre liquide
[metall] (copper, lead etc) liquater
~OUT - [metall] séparer par liquation
LIQUATION - [metall] (a process of separating the components of an alloy or mixture of metals, by gradual cooling and selective freezing-out) liquation *f*
~LEAD - [metall] plomb *m* de ressuage
~PAN - [metall] chaudière *f* de liquation
~SLAG - [metall] scorie *f* de ressuage
LIQUEFACTION - [phys] (change from a gasseous or solid state to a liquid condition) liquéfaction *f*
LIQUEFIED - [gen phys & metall] liquefié
~PETROLEUM GAS - [gas ind] gaz *mpl* de pétrole liquéfiés
LIQUEFIER - [gas ind] appareil *m* pour la liquéfaction
LIQUEFY, to - [gen] liquéfier
[phys] (of solids and of gases) se liquéfier, se fluidifier
[gen oils etc] se défiger
LIQUID - [gen] liquide *m*
[adj] liquide
~AGGREGATE STATE - [chem] état *m* liquide d'agrégat
~AIR - [phys] air *m* liquide
~AMMONIA - [chem] ammoniac *m* liquide
~CHARGE - [metall] (hot metal charge) charge *f* liquide
~CHLORINE - [chem] chlore *m* liquide
~CLINKER - [metall] (in foundry work) scorie *f* liquide
~CONTROLLER - [el] (liquid rheostat arranged for control of an electric motor) démarreur *m* à résistance liquide
~-COOLED - [mech] (term applied to machines e.g. engines, which are cooled by the circulation of liquid through channels formed in the cylinder block or the like) refroidi à liquide
~COOLING - [mech] (system of cooling turbine blades by circulation of liquid through them) refroidis-

sement *m* à liquide
LIQUID COOLING - [nucl] (cooling of a reactor by means of liquid metal or gas) refroidissement *m* à liquide
~COUNTER - [nucl] (counter for the measure of radioactivity of a liquid) compteur *m* de la radioactivité de liquides
~CRYSTALS - [chem] (pure liquids which are no longer clear and become, like crystals, anisotropic over a range of temperature above their freezing point) cristaux *mpl* liquides
~DAMPING - [el] amortissement *m* à liquide
~DETERGENT - [ind chem] (a detergent in liquid form) détergent *m* liquide, lessive *f* liquide
~DIMMER - [el] (dimmer using liquid resistance) affaiblisseur *m* à résistance liquide
~DRIES - [paint] (driers made by dissolving organic driers in suitable hydrocarbon solvents) siccatifs *mpl* liquides
~-DROP MODEL - [nucl] (model in which the atomic nucleus is supposed to behave like a drop of liquid) modèle *m* de la goutte liquide
~-DROP NUCLEAR MODEL - s. liquid-drop model
~END - [oil ind] devant *m* d'une pompe
~EXPANDED FILM - [phys] (state of film between gaseous and condensed films) voile *m* fluide intermédiaire
~-EXPANDING INSTRUMENT - [instr] instrument *m* à expansion de liquide
~-EXPANSION THERMOMETER - [instr] thermomètre *m* à dilatation de liquide
~-FILLED SHIELDING WINDOW - [nucl] fenêtre *f* de blindage à remplissage liquide
~FILTER - [photo] filtre *m* liquide
~-FLOW COUNTER - [nucl] (counter measuring the radioactivity of a flowing liquid) compteur *m* à fluide traversant
~FUEL - [fuel] combustible *m* liquide
~FUSE UNIT - [el] ensemble *m* de fusible à extinction en liquide
~HONING - [metall] honage *m* au jet de vapeur
~JOINT SEAL - [mech] joint *m* d'étanchéité aux liquides
~JUNCTION - [el] jonction *f* liquide
~JUNCTION POTENTIAL - [el] (diffusion potential) potentiel *m* de diffusion
~LEVEL INDICATOR - [instr] (an instrument for showing the level of the liquid in a tank or other vessel) limnimètre *m*
~LEVEL MANOMETER - [instr] (differential manometer using a liquid as the movable partition) manomètre *m* à flotteur
~LEVEL RECORDER - [instr] (recording instrument measuring the level of a liquid) limnigraphe *m*
~-LEVEL SWITCH - [el] automate *m* à flotteur, interrupteur *m* à flotteur
~LIMIT - [soil] (the limit content of water in the soil) limite *f* de liquidité
~-LIQUID EXTRACTION - [chem] (a method of separation in which one or more components in a liquid mixture are removed by solution in a liquid insoluble in the component) extraction *f* par partage liquide-liquide
~-LIQUID INTERFACE - [phys] (the common surface between two liquid phases) surface *f* de séparation liquide-liquide
~MANOMETER - [instr] (a manometer employing communicating vessels containing a liquid) manomètre *m* à liquide
LIQUID MANURE - [agric] purin *m*
~MANURE BARREL - [agric mach] tonne *f* à purin
~MANURE DRILL - [agric] distributeur *m* de purin en lignes
~MANURE PUMP - [agric] pompe *f* à purin
~MANURE SPREADER - [agric] distributeur *m* de purin
~MANURE TANK - [agric] s. liquid manure barrel
~-METAL COOLANT - [nucl] réfrigérant *m* métallique liquide
~-METAL FUEL - [nucl] combustible *m* métallique liquide
~-METAL FUEL REACTOR - [nucl] réacteur *m* à combustible métallique liquide
~MODERATOR LEVEL METER - [nucl] limnimètre *m* d'un modérateur liquide
~PARAFFIN - [chem] huile *f* paraffinée
~PETROLEUM - [oil ind] gaz *m* liquide de pétrole
~-PHASE SINTERING - [metall] frittage *m* en présence de la phase liquide
~PRESSURE - [soil] pression *f* exercée par un liquide
~-PRESSURE PICK-UP - [instr] (an instrument designed to measure the pressure of ground water) indicateur *m* de pression de liquides
~PUDLING - [metall] puddlage *m* du métal liquide
~PURIFICATION - [chem] épuration *f* liquide
~-QUENCHED FUSE - [el] (fuse in which a liquid is used for quenching the arc) fusible *m* à extinction en liquide
~RESISTANCE - [el] (resistance consisting of a liquid of low conductivity) résistance *f* liquide
~RHEOSTAT - [el] (liquid resistance the value of which can be varied continuously) rhéostat *m* liquide
~ROSIN - [paint] (alternative name for Tall Oil; a mixture of fatty and rosin acids, the former being chiefly linoleic and oleic acids. It is used in paint manufacture) colophane *f* liquide
~SCINTILLATOR - [nucl] (in counters) scintillateur *m* liquide
~SHUT-DOWN SYSTEM - [nucl] système *m* d'arrêt à liquide
~SOAP - [ind chem] savon *m* liquide
~STARTER - [el] (liquid rheostat acting as a motor starter) rhéostat *m* liquide de démarrage
~STATE - [phys] état *m* liquide
~SURFACER - [paint] (used to smooth out surfaces) mastic *m* liquide
~TACHOMETER - [instr] tachymètre *m* à liquide
~TARGET - [nucl] cible *f* métallique liquide
~THERMOMETER - [instr] (type of thermometer in which the expanding and contracting element is liquid) thermomètre *m* à liquide
~-WALL IONIZATION CHAMBER - [nucl] chambre *f* d'ionisation à paroi liquide
~WASTE - [nucl] (radioactive waste material in liquid form) déchets *mpl* liquides
LIQUIDATE, to - [gen & comm] liquider
LIQUIDATION - [comm] liquidation *f*
LIQUIDITY - [gen] liquidité *f*
[phys] état *m* liquide
[agric] (water content of the soil) liquidité *f*
~INDEX - [phys] indice *m* de liquidité
LIQUOR - [brew ind] (this term is always used in English breweries for the water actually employed directly in brewing processes. The name 'water' is only used for water used for accessory purposes,

e.g. washing, boiler feed etc.) eau *f* de brassage
LIQUOR - [chem] solution *f*
~TANK - [brew ind] (a cistern) bâche *f* à eau
LIQUORICE - s. licorice
LISSAJOUS FIGURES - [electron] (traces shown on the screen of a cathode ray tube having specific form due to the character of the deflecting voltages) figures *f*pl de Lissajous
LIST, to - [gen] cataloguer, enregistrer
[naut] (to incline to one side) incliner, gîter
[text] faire une lisière
LIST - [gen] liste *f*, tableau *m*
[comm] inventaire *m*
[carp] (the narrow edge of a board) lisière *f*, bourrelet *m*
[carp] règle *f*
[text] (a selvage) lisière *f*
[constr] latte *f*
[naut] (the inclination of a ship to one side) bande *f*, gîte *f*, faux bord *m*
[metall] (tin plating) rebord *m*
LISTEL - [constr] (the projecting flat surface between two flutes in a column) listel *m*
LISTEN, to - [gen] écouter
~IN, to - [radio telephone etc] écouter, se porter en écoute
LISTENER - [gen] auditeur *m*
[radio etc] auditeur *m*
~ECHO - [telecomm] écho *m* (pour la personne qui écoute)
LISTENING AND SPEAKING KEY - [telecomm] clé *f* combinée d'écoute et de conversation
~COIL - [el acoust] (built-in coil in listening aids) bobine *f* d'écoute
~JACK - [telecomm] jack *m* d'écoute
~KEY - [telecomm] clé *f* de surveillance, clé *f* de écoute
~POSITION - [telecomm] position *f* d'écoute
~THROUGH - [telecomm] écoute *f* intersigne
LISTERELLOSIS - s. listeria infection
LISTERIA INFECTION - [vet] listerellose *f*
LISTING - [naut] bande *f*, gîte *f*
[text] lisière *f*
[comput] impression *f* en liste
LITE - [photo] (diffuser filter) trame *f* diffusante
LITER - s. litre
LITERALS - [print] (errors of composition) coquilles *f*pl
LITHARGE - [chem] (lead monoxide. An oxide of lead existing in red and yellow forms and used as a rubber pigment, in ceramic glazes, in match manufacture, in the production of pigments, as a drier for paints and varnishes and in the production of special glasses) litharge *f*
LITHEMIA - [med] uricémie *f*
LITHIA - [chem] (lithium monoxide) lithine *f*, oxyde *m* de lithium
~MICA - [min] (a member of the mica group of minerals) mica *m* lithinifère
~WATER - [chem] eau *f* lithinique
LITHIASIS - [med] lithiase *f*
LITHIFICATION - [geol] pétrification *f*
LITHIUM - [chem] (the lightest of the alakli metals; has uses as a catalyst in the production of synthetic rubber and plastics, in the preparation of Grignard reagents, in metallurgy, and as lithium salts, in medicine. It occurs combined in several minerals, mineral waters and plant ashes) lithium *m*
LITHIUM ALCOHOLATES - [chem] (compounds used as catalyst in Claisen condensation reactions) alcoholates *m*pl de lithium
~ALUMINATE - [chem] (a component of ceramic glazes) aluminate *m* de lithium
~ALUMINIUM HYDRIDE - [chem] (a polymerization catalyst) hydrure *m* de lithium et aluminium
~AMIDE - [chem] (a synthesis intermediate in the production of pharmaceuticals) lithiumamide *f*
~BOMB - [nucl] bombe *f* à lithium
~BOROHYDRIDE - [chem] (a reducing agent) borohydrure *m* de lithium
~BROMIDE - [chem] (a sedative used in medicine) bromure *m* de lithium
~CARBONATE - [chem] (a catalyst and a component of ceramic glazes and special glasses) carbonate *m* de lithium
~CHLORATE - [chem] (has application in air conditioning as a means of controlling humidity) chlorate *m* de lithium
~CHLORIDE - [chem] (a humectant in air-conditioning, a heat transfer agent, and a source of metallic lithium) chlorure *m* de lithium
~CHROMATE - [chem] (an oxiding agent and corrosion inhibitor) chromate *m* de lithium
~CITRATE - [chem] (a constituent of effervescent beverages) citrate *m* de lithium
~COBALTITE - [min] (a constituent of ceramic glazes) cobaltite *f* de lithium
~DEUTERIDE - [chem] (compound of lithium and deuterium) deutérure *m* de lithium
~FLUOPHOSPHATE - [chem] (a constituent of ceramic glazes) fluophosphate *m* de lithium
~FLUORIDE - [chem] (a welding flux, ceramic glaze, and heat exchange medium) fluorure *m* de lithium
~HYDRIDE - [chem] (a catalyst in organic reactions, desiccant and source of hydrogen) hydrure *m* de lithium
~HYDROXIDE - [chem] (an absorbent for carbon dioxide and a consituent of ceramics glazes) hydroxyde *m* de lithium, lithine *f* caustique
~HYPOCHLORITE - [chem] (a bleaching agent) hypochlorite *m* de lithium
~IODIDE - [chem] (a photographic chemical) iodure *m* de lithium
~MANGANITE - [chem] (a constituent of frits for ceramic glazes) manganite *f* de lithium
~METASILICATE - [chem] (a flux for ceramic glazes) métasilicate *m* de lithium
~MOLYBDATE - [chem] (a catalyst in petroleum refining) molybdate *m* de lithium
~NITRATE - [chem] (a constituent of ceramics glazes) nitrate *m* de lithium
~PEROXIDE - [chem] (a source of active oxygen) peroxyde *m* de lithium
~RICINOLEATE - [chem] (a catalyst in organic reactions) ricinoléate *m* de lithium
~SALYCYLATE - [chem] (an alcohol-soluble salt with uses in the treatment of rheumatism) salicylate *m* de lithium
~STEARATE - [chem] (a lubricant additive for plastics) stéarate *m* de lithium
~TETRABORATE - [chem] (a constituent of ceramics glazes) tétraborate *m* de lithium
~TITANATE - [chem] (a constituent of ceramics glazes) titanate *m* de lithium

LITHIUM ZIRCONATE - [chem] (a flux for ceramics and special glasses) zirconate *m* de lithium
LITHO - [abbrev. for lithography] lithographie *f*
~-OFFSET - [print] (offset printing) impression *f* offset, impression *f* planographique
~OILS - [paint] s. lithographic oils
~PAPER - [paper man] (lithographyc paper) papier *m* pour offset
LITHOCHOLIC ACID - [chem] (a bile acid) acide *m* lithocholique
LITHOCLASE - [geol] lithoclase *f*
LITHOCYSTOTOMY - [med] taille *f* vésicale
LITHOGENESIS - [med] lithogenèse *f*
LITHOGRAPH, to - [print] (the operation of lithographic printing) litographier
LITHOGRAPHIC - [print] lithographique
~INK - [print] encre *f* lithographique
~OILS - [print] (heat-bodied oils used in printing) huiles *fpl* lithographiques
~PAPER - [paper man] (supercalendered or plate-glazed paper for lithographic printing) papier *m* pour impression offset
~STONE - [geol] (porous, fine-grained limestone used in lithography) calcaire *m* lithographique
~VARNISHES - [paint] vernis *mpl* lithographiques
LITHOGRAPHY - [print] (process of printing based on the reciprocal repulsion of water and greasy ink) lithographie *f*
LITHOLAPAXY - [med] (surgical operation designed to crush a stone in the bladder) litholapaxie *f*
LITHOLOGY - [geol] (the characters of a rock in terms of their mineral composition) lithologie *f*
LITHOMARGE - [geol] (a term denoting various clays) lithomarge *f*
LITHOPEDION - [med] (or lithopaedion; calcified dead foetus) lithopédion *m*
LITHOPONE - [chem] (white pigment used for inside paintwork, composed of barium sulphate and zinc sulphide) lithopone *m*
LITHOTRITE - [med] (instrument fitted with special blades to crush stones in the bladder) lithotriteur *m*
LITHOTRITY - [med] (the operation of crushing stones in the bladder) lithotritie *f*
LITMUS - [chem] (an analytical indicator obtained from various lichens. It is red in the presence of acids and blue in that of alkalies) tournesol *m*
~PAPER - [chem] (paper impregnated with litmus for use as a chemical indicator) papier *m* de tournesol
LITRE - [meas] (the volume occupied by one kilogram of water at 4°C. and standard atmospheric pressure; 35.196 fluid ounces) litre *f*
LITTER - [med] litière *f*
[gen] immondices *fpl*
[agric] litière *f*
[zool] cochonnée *f*, portée *f*
~CARRIER PLANT - [agric] installation *f* d'évacuation du fumier
~MEADOW - [agric] prairie *f* à litière
LITTLE - [gen] petit, peu
~BEAR - [astr] la petite Ourse *f*
~DIPPER - [astr] (in the USA) s. little bear
~END OF CONNECTING ROD - [mech] petite tête de bielle, pied *m* de bielle
LITTLE'S DISEASE - [med] (a spastic paralysis of the lower half of the body) diplégie *f* cérébrale infantile, maladie de Little
LITTORAL - [geogr] littoral *m*
LITTORAL DEPOSITS - [geol] (deposits of the shore area between high and low-water marks) dépôts *mpl* littoraux
~ZONE - [bot] (the part of the shore which is below average low water level and on which plants are living) zone *f* du littoral
LITTRE'S GLANDS - [anat] (in males, small mucous glands in the membrane of the urethra) glandes *fpl* de Littré
LIVE - [gen] vivant, vif
[el] (of an electric conductor when there is a potential difference between it and earth) sous tension
[acoust] (with normal reverberation) réverbérant, réfléchissant
[mech] roulant
[firearms] (of a projectile etc) chargé
[radio & telev] en direct
[chem] (of coals etc) ardent
~AXLE - [mech] essieu *m* tournant
~CABLE TEST CAP - [el] capuchon *m* isolant d'extrémité de câble
~CENTRE - [mach tool] pointe *f* de la poupée fixe
~COAL - [gen] charbon *m* ardent
~END - [acoust] (reflecting wall) paroi *f* réfléchissante
~-FRONT SWITCHBOARD - [el] panneau *m* sous tension d'un tableau de distribution
~HEAD - [mach tools] poupée *f* fixe
~HOLE OF A KILN - [ovens] carneau *m*
~INJECT - [telev] insertion *f* actuelle
~LOAD - [phys] (the load imposed by forces not due to the weight of a structure itself) poids *m* roulant
(of a cargo on a vehicle) charge *f* utile
~MAIN - [gas ind] (charged main) conduite *f* en charge
~MATTER - [print] composition *f* permanente, marbre *m*, conserve *f*
~OAK - [bot] chêne *m* vert
~OIL - [gas ind] (crude oil containing gas) huile *f* contenant du gaz
~PARTS - [el] parties *fpl* sous tension
~PIPE - s. live main
~RAIL - [railw] (the conductor rail) rail *m* conducteur, troisième rail *m*
~ROLLER - [metall] rouleau *m* automoteur
~ROLLING MILL - [metall] plan *m* à rouleaux commandé
~ROOM - [acoust] (room having a very small measure of sound absorption) chambre *f* à grande réverbération
~STEAM - [mech] (the steam which is supplied directly from a boiler) vapeur *f* vive
~STEAM RECLAIMING PROCESS - [rubber ind] (method of recovering rubber from scrap, using steam at boiler temperature and pressure) procédé *m* de régénération à la vapeur libre
~-STOCK IMPROVEMENT - [genet] amélioration *f* des races
~STUDIO - [cin telev] studio *m* réverbérant
[acoust] (a studio in which reverberation is increased to produce special effects) studio *m* à grande réverbération
~-TANK OIL CIRCUIT-BREAKER - [el] disjoncteur *m* à faible volume d'huile
~-WEIGHT - [agric] poids *m* vif
~WIRE - [el] fil *m* sous tension
LIVER - [anat] foie *m*

LIVER DAMAGE - [med] lésion *f* hépatique
~FLUKE - [vet] douve *f* hépatique
~-FLUKE DISEASE - [vet] distomatose *f* hépatique
~OF SULPHUR - [chem] (a mixture of polysulphides and thiosulphate of potassium, obtained by heating sulphur with potassium carbonate) foie *m* de soufre
~PARENCHYMA - [med] parenchyme *m* hépatique
~ROT - [vet] (fluke disease; distomiasis) distomatose *f*
LIVERING - [paint] (alternative term for feeding, a defect in paint oils characterised by an increase in viscosity) épaississage *m*
LIVERWORT - [bot] (a medical plant) hépatique *f* trilobée, trinitaire *f*
LIVERY CLOTH - [text] tissu *m* pour livrées
LIVERSTOCK - [agric] cheptel *m* vif, bétail *m*
~FARM - [agric] ferme *f* d'élevage
~FEEDING - [agric] affourragement *m* du bétail
~HUSBANDRY - [zool] exploitation *f* du bétail, production *f* animale
~LOADING DOCK - [railw] quai *m* à bestiaux
~MARKET - [comm] marché *m* aux bestiaux
LIVID - [gen & med] livide, plombé
LIVIDITY - [med] lividité *f*
LIXIVIATE, to - [ind chem] (to extract a substance from a mixture containing it by means of a solvent) lixivier, lessiver
LIXIVIATION - [ind chem] (extraction of soluble substances from insoluble material by means of water or other liquid) lixiviation *f*, lessivage *m*
~PLANT - [ind chem] (installation for extracting material by solution) installation *f* à lixiviation
~RESIDUE - [chem] résidu *m* de lessivage
~VAT - [ind chem] cuve *f* à lessiver
LIZARD - [zool] lézard *m*
~-STONE - [geol] marbre *m* serpentin
L.M.T. - [astr] temps *m* civil local, T.C.G.
LOAD, to - [gen] charger
[firearms] (a gun) charger
[naut] (a ship) prendre en charge
[mech] (the operation of gumming) s'engorger, se noyer
[photo] (a camera) charger
[telecomm] (in telephony) pupiniser
[mech] (a spring) serrer, bander
LOAD - [el] (the part of an electric circuit in which the output is developed or to which it is applied) charge *f*
[mech] (any force which is supported by a body) charge *f*
[metall & plastic ind] (the material used to load a mould cavity) charge *f*
[metall] (the material to be heated in high-frequency furnaces) charge *f*
[gen] fardeau *m*
[oil ind] charge *f* d'amorçage
~A SPRING, to - [mech] serrer un ressort, bander un ressort
~ADJUSTING RHEOSTAT - [el] rhéostat *m* de réglage de charge
~AT CENTRE - [aero] charge *f* au centre
~AT ELASTIC LIMIT - [phys] charge *f* à la limite de élasticité
~-BEARING - [constr] portant
~-BEARING WALL - [constr] mur *m* de soutènement
~BINDER - [mech] tendeur *m* à chaîne
~BRAKE - [mech] frein *m* de monte-charge
LOAD-CARD - [comput] carte *f* de charge
~CARRYING CAPACITY - [el] capacité *f* de charge
[rubber ind] (of tyres) force *f* portante
~CARRYING STRUCTURE - [constr] structure *f* portante
~CHAIN - [mech] chaîne *f* de charge, chaîne *f* de levage
~CHARACTERISTIC - [el] (the curve representing a relation between two of the quantities characterizing the functioning the machine on load under given conditions) caractéristique *f* en charge
~CIRCUIT - [radio] (that part of the radio frequency generating circuit associated with the electrodes used for preheating) circuit *m* de charge, circuit *m* d'utilisation
~CIRCUIT EFFICIENCY - [el] rendement *m* du circuit de charge
~-CIRCUIT POWER INPUT - [el] (the power which is delivered to the load circuit) puissance *f* d'entrée du circuit de charge
~-CONSOLIDATION CURVE - [soil] courbe *f* de charge-tassement
~-CONVEYOR - [mech] chargeuse-convoyeur *f*
~CURRENT - [electron] (the current output of a tube utilized as an external load circuit) courant *m* de charge
~CURVE - [mech] (the influence line for a structure) diagramme *m* de charge
[el] courbe *f* de consommation
~DEFLECTION CURVE - [rubber ind] (of tyres etc) courbe *f* charge/flèche, courbe *f* charge-déformation
~DESPATCHER - [el] (the man responsible for the distribution of load and for the control of the system) répartiteur *m* de charge
~DIAGRAM - [el] diagramme *m* de charge
~DISPLACEMENT - [naut] (ship's displacement at load draught) déplacement *m* à pleine charge
~DRAUGHT - [shipbuild] tirant *m* d'eau en charge
~DURATION - [soil] durée *f* du chargement
~EXTENSION CURVE - [metall] (curve showing the relations between the applied load and the extension produced) diagramme *m* de charge et allongements
~FACTOR - [aero] (the relation between a given external load and the weight of an aircraft) facteur *m* de charge
[el] (the ratio between the average load on an electric circuit during a specified period and the maximum load during that period) facteur *m* d'utilisation d'une charge
~INCREMENT - [soil] accroissement *m* de charge
~INDEX - [radio] (of an oscillator) taux *m* de charge
~-INDICATING RESISTOR - [el] résistance *f* de mesure de charge
~-LEVELLING RELAY - [el] (a relay which automatically switches off an apparatus when the demand exceeds a specified value) relais *m* de disjonction automatique
~LIFE - [instr] (the ability of an instrument to withstand its full power rating over a long period of time) durée *f* à charge totale
~LIMIT - [gen] (the load known (or assumed) to be safe, but which must not be exceeded in operation) charge *f* limite
~LIMIT GAUGE - [railw & transp] profil *m* limite de chargement
~LINE - [naut] (the line marked on the outside of a

ship to indicate the depth it may reach when loaded) ligne *f* de flottaison en charge
LOAD LINE - [electron] (the line which represents the variations with load current of the voltage at the terminals of a load impedance) droite *f* de charge
~OIL - [oil ind] huile *f* d'amorçage
~ON SECTION - [railw] charge *f* de la ligne
~ON THE WHEEL - [mech] charge *f* sur la roue
~ON YARN - [text] charge *f* du fil
~PEAK - [el] pointe *f* de charge
~PER AXLE - [mech] charge *f* par essieu
~RESISTANCE - [el] résistance *f* ballast, résistance *f* de charge
~REVERSAL - [mech] renversement *m* de charge (ou d'effort)
~RHEOSTAT - [el] rhéostat *m* d'absorption
~RING - [aero] (in a balloon, the ring to which the net and the load are attached) cercle *m* de suspension
~ROPE - [mining] câble *m* d'équilibre
~-SETTLEMENT CURVE - [mech] courbe *f* de charge-tassement
~-SHIFTING RESISTOR - [el] résistance *f* commutatrice
~STAGE - [mech] palier *m* de chargement, phase *f* de chargement
~-SWELLING CURVE - [soil] courbe *f* charge-gonflement
~SWITCH - [el] interrupteur *m* de charge
~TEST - [gen & el] essai *m* en charge
~TRANSFER - [mech] transmission *f* de la charge
~VARIATION - [el] variation *f* de la charge (d'un moteur électrique)
~VOLTAGE - [el] tension *f* de sortie
~WATER - [oil ind] eau *f* d'amorcage
~WATER LINE - s. load line
LOADED - [gen] chargé
[mech] (of a wheel etc) sous charge
~AERIAL - [radio] (aerial with added series inductance to increase its wavelength) antenne *f* en charge
~ANTENNA - s. loaded aerial
~CABLE - [el] (in telecommunications) câble *m* krarupisé, câble *m* à charge continue
~CIRCUIT - [el] (circuit with loaded cable) circuit *m* pupinisé
~IMPEDANCE - [el acoust] (the impedance which is measured at the input of a transducer when the output is connected to its normal lad) impédance *f* en charge
~MIXTURE - [rubber ind] mélange *m* chargé
~-POTENTIOMETER FUNCTION GENERATOR - [comput] générateur *m* de fonction à potentiomètre chargé
~RADIUS - [mech] (of a wheel) rayon *m* sous charge
~REEL - [cin] bobine *f* chargée
~SILK - [text] soie *f* chargée
~SOIL - [soil] sol *m* chargé
~YARN - [text] fil *m* chargé
LOADER - [mech] chargeuse *f*
[agric] (kind of mechanical shovel) élévateur *m*
[min] (a device for loading trucks underground) drague *f* de chargement
[mining] drague *f* chargeuse
LOADING - [gen] chargement *m*
[mech] (the maximum quantity of material which the equipment can handle) charge *f* limite
[aero] charge *f*
[paper man] (the addition of some material to the pulp to produce solidity) charge *f*
LOADING - [radio] (addition of inductance to an aerial) charge *f*, accumulation *f*
[text] (size or metallic compounds used to weight a fabric) charge *f*
[comput] (the process of feeding information into the shore) alimentation *f*
[auto] engorgement *m*
~AGENT - [ind chem] substance *f* pour la charge
~BRIDGE - [railw] pont *m* de chargement, pont *m* mobile
~CHAMBER - [plast ind] (the space into which the charge is loaded in compression moulding) chambre *f* de compression
[firearms] (of gun) chambre *f* de chargement
~COIL - [el] (the inductance coil placed in transmission lines to unify the attenuation with frequency) bobine *f* de charge, bobine *f* pupin
~COIL SPACING - [el] (in telecomm) pas *m* de pupinisation
~DOCK - [railw] quai *m* de chargement
~ERROR - [comput] erreur *f* de stockage
~FUNNEL - [impl] (a truncated cone through which substances are introduced into a vessel or other receptacle) trémie *f* de remplissage
~GANGWAY - s. loading bridge
~GAUGE - [railw] gabarit *m* de chargement, profil *m* d'encombrement
~HEIGHT - [gen] hauteur *f* de chargement
~HOPPER - [mining] trémie *f* de chargement
~IN BULK - [comm] chargement *m* en vrac
~MACHINE - [mech] chargeuse *f*
[nucl] (the machine which is used to load a nuclear reactor wi+h fuel) appareil *m* de chargement
~MATERIAL - [gen] charge *f*
~OUT STAGE - [transp] quai *m* de chargement
~PLATE - [plast ind] (an intermediate plate in which the transfer chamber is formed) plaque *f* intermédiaire de remplissage
~PLATFORM - [transp] plateforme *f* de chargement
~PLUNGER - [mech] piston *m* de remplissage
~POINT - [el] point *m* de charge, point *m* de pupinisation
~RAMP - [railw] rampe *f* de chargement
~RANGE - [metall] limites *f pl* de charge
~RESISTOR - [contr] (a rheostat constituting a load) rhéostat *m* d'absorption, rhéostat *m* de charge
~RHEOSTAT - s. loading resistor
~ROUTINE - [comput] programme *m* de charge
~SHOE - [plast ind] plaque *f* de remplissage
~SHOVEL - [mech] pelle *f* mécanique
~SIDING - [railw] voie *f* de chargement
~STATION - [railw & transp] station *f* de chargement
~TEST - [constr] essai *m* de chargement
~TRAY - [plast ind] (a device in the form of a specially designed tray which is used to load the charge simultaneously into each cavity of a multi-impression mould by the withdrawal of a sliding bottom from the tray) chargeur *m* d'empreinte
~WELL - [plast ind] chambre *f* de transfert
~WHARF - [naut] quai *m* de chargement
LOADSTONE - s. lodestone
LOAF - [food] pain *m*, miche *f* de pain
LOAM, to - [agric] recouvrir de limon
[constr] torcher, glaiser
[constr] enduire d'argile *f*, bousilier, glaiser
LOAM - [min] (a paste of clay, water, sand and chop-

ped straw, used in brickmaking and foundry work) terre *f* grasse, leh *m*, terreau *m*
LOAM - [gen] terreau *m*
[metall] terre *f* glaise
~BRICKS - [constr] briques *fpl* en terre glaise
~CASTING - [metall] moulage *m* en argile
~-CORE - [metall] (a core of loam used in foundry work) noyau *m* de terre glaise
~MILL - [metall] malaxeur *m* pour terre glaise
~MOULD - [metall] moule *m* en terre glaise
~MOULDING - [metall] moulage *m* en terre glaise
~PLATES - [metall] plaques *fpl* pour le moulage en terre glaise
~ROCK - [geol] marne *f*
~SAND - [metall] sable *m* argileux
~SANDSTONE - [geol] grès *m* glaiseux, grès *m* argileux
LOAMY - [geol] argileux
~PASTE - [ind chem] pâte *f* grasse collante
~SAND - [geol] sable *m* argileux, sable *m* gras
LOB, to - [gen] lancer
[min] scheider, trier
LOB - [mining] (a step in a mine) filon *m* en gradins, filon *m* à échelle
[min] (in ore dressing) scheider *m*
LOBBY - [build] antichambre *f*, vestibule *m*, vestiaire *m*
(of a theatre) entrée *f*
LOBE - [gen & bot] lobe *m*
[anat] (of the ear) lobe *m* de l'oreille
[mech] (rounded projection, a cam) oreille *f*
[radio] (the boundary of the volume in which the field strength of a radio wave is everywhere greater than a chosen value) lobe *m*, pétale *f*
~OF RADIATION - [radio] (in the aerial) lobe *m* de rayonnement
~SWITCHING - [radar] (the direction of a beam caused by a dipole array can be shifted by adjusting the relative phases of the currents in the various dipoles) commutation *f* de lobes
~WIDTH - [radio] (of an aerial; the half-power width of a radiation lobe) largeur *f* du lobe
LOBECTOMY - [med] lobectomie *f*
LOBED LEAVES - [bot] feuilles *f* lobées
LOBELIA - [chem] (the dried leaves of Lobelia inflata and nicotianifolia used in the treatment of asthma and similar conditions) lobélie *f*
LOBELINE - [chem] (the active principal of lobelia; having uses in medicine. When heated with water it gives acetophenone) lobéline *f*
LOBENGULISM - [med] obésité *f* hypogénitale
LOBING - s. lobe switching
LOBSTER - [zool] homard *m*
~BACK - [plumb] (an angle connexion formed of three segments of pipe welded together) coude *m* en segments rapportés
LOBULAR - [gen] lobulaire
LOBULATED - [anat] lobulé
LOCAL - [gen] local, régional
~ACTION - [chem] (the corrosion of a piece of metal caused by the formation of galvanic cells between different parts of its surface) action *f* locale
~ANAESTHESIA - [med] (or local anesthesia) anesthésie *f* locale
~ANNEALING - [metall] recuit *m* sélectif
~APPARENT TIME (L.A.T.) - [astr] (the tume interval which has elapsed since the passage of the true sun across the anti-meridian of the observer) temps *m* apparent local
LOCAL ATTRACTION - [surv] (effect due to mineral deposits in the ground and giving some trouble in compass surveying) déclinaison *f* due à attraction locale
~AUTOMATIC EXCHANGE - [telecomm] bureau *m* central urbain automatique
~BATTERY - [el] (in telegraphy) batterie *f* locale
~BATTERY AREA - [telecomm] réseau *m* à batterie locale
~CALL - [telecomm] conversation *f* locale, appel *m* urbain
~-CARRIER FREQUENCY ERROR - [radio] (in a single-side-band receiver) déviation *f* de fréquence de la porteuse locale
~CONTROL - [telecomm] réglage *m* local
~CORRECTION - [telecomm] (in submarine cable telegraphy) correction *f* de signal au récepteur
~DISEASE - [med] maladie *f* locale
~-DISTANT CONTROL - [telev] réglage *m* local à distance
~EXCHANGE - [telecomm] (in telephony) bureau *m* local, bureau *m* urbain
~EXTENSION - [metall] (the extension produced by a tensile test) allongement *m* local
~HEATING - [electron] (effect of kinetic energy) chauffage *m* local
[metall] chauffage *m* sélectif
~HOUR ANGLE (L.H.A.) - [astr] (the angle between a given celestial meridian and that of a celestial body, measured westward along the celestial equator) angle *m* horaire origine
~MEAN TIME (L.M.T.) - [astr] (the hour angle of the mean sun measured westward from the meridian of the observer (+ I2 hrs.) temps *m* civil local, T.C.G.
~OSCILLATOR - [radio] (oscillator used to produce a change of carrier frequency in superheterodyne reception) oscillateur *m* local
~PREHEATING - [mech] préchauffage *m* local
~RECORD - [telecomm] contrôle *m* local, transcription *f*
~SENSITIVITY - [telecomm] (of an echo suppressor) sensibilité *f* locale
~SIDEREAL (L.S.T.) - [astr] (the hour angle (in terms of time) measured westward from the meridian of the observer to that of the First Point of Aries) temps *m* sidéral local, T.S.G.
~STRESS RELIEVING - [heat] relaxation *f* locale (de contrainte)
[metall] stabilisation *f* locale
~TRAIN - [railw] chemin *m* de fer vicinal
~VENT - [plumb] tube *m* aérateur
LOCALIZATION - [gen] localisation *f*, repérage *m*
[acoust] (the faculty of the human ear to appreciate relative distances of sources of sounds) localisation *f* d'une source sonore
~OF A FAULT - [gen] détermination *f* d'un dérangement
LOCALIZE, to - [gen] localiser
[acoust] (to gauge the distance of the source of sound) localiser
LOCALIZED REACTION PINHOLES - [metall] piqûres *fpl*, hétérogènes bleuâtres
LOCALIZER - [radar] (a radio facility providing signals for lateral guidance of aircrafts) radio *m* de

localisation
LOCALIZER - [radiat] (an applicator, a treatment cone) localisateur *m*, cône *m*
LOCATE, to - [gen] localiser, repérer, déterminer
[gen] (to find etc) établir, reconnaître
[comput] localiser
LOCATING - [gen] localisation *f*, repérage *m*
[surv] (to locate a point) détermination *f*
~ ARRANGEMENT - [el etc] dispositif *m* de marquage de crans
~ CONE - [metall] repère *m* de remoulage,
~ CONES - [metall] bicône *m*
~ HOLE - [mech] trou *m* de repère
~ PIN - [mech] (a pin used to fix the position of another part) goujon *m* de guidage
[mech] goupille *f* d'arrêt
~ POINTS - [mech] points *mpl* de montage
~ RING - [mech] anneau *m* de centrage
~ SPIGOT - [mech] broche *f* de repère
~ SPOT - [gen & mech] point *m* de repère
~ STRIP - [mech] bande *f* de repère
LOCATION - [gen] emplacement *m*, situation *f*
[gen] (the determination of the position of an engineering project) emplacement *m*
[leg] location *f*
[cin] (a site outside the studio used to make a motion-picture) aire *f* de tournage à l'extérieur
[mining] (only USA) concession *f* minière
~ OF A CORE - [metall] position *f* d'un noyau
~ OF INSTRUCTION - [comput] emplacement *m* de l'instruction
~ PLAN - [draw] planimétrie *f*
LOCATOR - [telecomm] (of beacons etc) indicateur *m*, jalon *m*
~ BEACON - [radar] (a directional radar beacon designed to give an aircraft its lateral position in respect to the runway during approach and landing) radiophare *m* de jalonnement
LOCH - [geogr] (a galeic word; a lake or an arm of the sea, narrow and almost totally land-locked) lac *m*, bras *m* de mer
LOCHIA - [med] (discharge from the vagina after childbirth) lochies *fpl*, suites *fpl* de couches
LOCK, to - [gen] fermer
[mech] bloquer, caler, enclencher
[el] (of a relay) verrouiller
[hydr] écluser, pourvoir d'écluses
LOCK - [gen] serrure *f*, fermeture *f*
[hydr] (communicating channel with gates at both ends) écluse *f*
[mech] enclenchement *m*, crabotage *m*
[text] presse *f*, loquet *m*
[railw] (of a point) serrure *f* d'aiguille, verrou *m* d'aiguille
[firearms] (of a gun) platine *f*
[auto] crabotage *m*, angle *m* de braquage
[mech] (of a wheel) enrayure *f*
[metall] irregularité *f*
~-AND-BLOCK SYSTEM - [railw] système *m* à bloc enclenché
~ AND INLAND CANAL - [hydr] canal *m* à écluses et à lac intérieur
~ BLOCK - [text] obturateur *m*
~ BOLT - [mech] pêne *m*
~ CARD - [text] carton *m* de serrure
~ CHAMBER - [hydr] (the space between the gates) sas *m* d'écluse, chambre *f* d'écluse
[hydr] (of a canal) neptune *m*
LOCK CIRCUIT CONTROLLER - [el] commutateur *m* de verrou d'aiguille
~ GATE - [hydr] porte *f* d'écluse
~-IN - [el] (the synchronization of one oscillator with another) enclenchement *m*
~-IN AMPLIFIER - [instr] (instrument giving an accurate detection method) détecteur *m* synchrone, détecteur *m* à sensibilité de phase
[electron] amplificateur *m* synchronisé
~-IN BASE - [electron] culot *m* local
~-IN RANGE - [telev] plage *f* de rattrapage
[contr] (range indicating the point at which a circuit for frequency control of a generator locks in at the frequency of the control pulse) marge *f* de synchronisation
~ KEEPER - [hydr] éclusier *m*
~ KNITTING - [text] tricotage *m* à piqué
~-NUT - [mech] (a second nut, usually of less thickness, used to prevent a fixing nut from becoming loose, as by vibration, by being screwed up against it after tightening) écrou *m* indésserrable, contre-écrou *m*
~ OUT - [telecomm] (in telephone circuits) blocage *m* de ligne
[comput] barrage *m*
~ PIT - [hydr] fouille *f* de l'écluse
~ PLATE - [mech] retient *m*
[metall] (corp de) platine *f*
~ ROD - [mech] tige *f* de verrouillage
~ SCREW - [mech] vis *f* de fixation
~ SLOT - [mech] encoche *f* de blocage
~ STITCH - [shoe man] point *m* de navette
~ WASHER - [mech] (a special washer used to prevent accidental rotation of a nut) rondelle *f* frein
~ WHEEL - [text] roue *f* à cames, roue *f* à gradins
~-WOVEN MESH - [constr] (steel-wire fabric used as reinforcement) toile *f* métallique pour béton
LOCKAGE - [hydr] éclusage *m*
LOCKED CLAMP - [mech] agrafe *f* filée
~-COVER SWITCH - [el] (switch can be operated after unlocking the cover) interrupteur *m* fermé à clef
~ GROOVE - [el acoust] (a concentric groove) sillon *m* final
~ OSCILLATOR - [el] oscillateur *m* bloqué
~ POINTS - [railw] aiguille *f* verrouillée
~ PUNCH - [text] poinçon *m* en prise
~ ROTOR TORQUE - [el] (of a motor; the lock which is developed on the shaft of the motor at the moment of starting) couple *m* initial de démarrage
LOCKER - [gen] coffre *m*, armoire *f*
[naut] caisson *m*, coffre *m*
~ BAR - [text] arbre *m* à ailettes
~ ROOM - [gen] (in factories etc) vestiaire *m*
LOCKING - [mech] verrouillage *m*, enclenchement *m*, blocage *m*
[railw] (of points) blocage *m* (d'aiguille)
[el] (a mechanical interlock) enclenchement *m*, verrouillage *m*
[plast ind] (the holding together of the parts of the closed mould) serrage *m*
[hydr] éclusage *m*, sassement *m*
[telev] réglage *m* de fréquence par un signal à fréquence constante
~ BAR - [mech] barrette-verrou *f*
[text] tringle *f* de pinçage
[gen] verrou *m*

LOCKING BAR - [railw] pédale *f* de calage
~BOLT - [mech] verrou *m* de fermeture
~BY TRACK CIRCUIT - [el] enclenchement *m* par circuit de voie
~CAM - [mech] (part of a maltese cross) roue *f* à loquet
~CAP - [mech] chapeau *m* de sûreté
~CATCH - [text] loquet *m*, encliquetage *m* d'arrêt
~CLAW - [mech] griffe *f* de fixation
~COG - [mech] (in electric time meters) dent *f* de arrêt
~CONE - s. locking catch
~CROW - [text] levier *m* de blocage
~DEVICE - [mech] (any arrangement for holding parts in a given position) verrouillage *m*, blocage *m*
~DISC - [text] disque *m* d'arrêt
~FINGER - [text] cheville *f* d'arrêt
~FORCE - [metall] (in diecasting) force *f* de verrouillage
~FRAME WITH WIRE GEAR - [railw] appareil *m* de manoeuvre à fil de transmission
~-IN - [radio] (the shifting and automatic holding of the frequencies of two oscillating systems) verrouillage *m*, synchronisme *m*
~KEY - [telecomm] bouton *m* à enclenchement, clé *f* à enclenchement
~KNOB - [mech & auto] bouton *m* de verrouillage
~LEVER - [mech] levier *m* d'accrochage, levier *m*
[railw] (of a point) levier *m* de verrouillage d'aiguille
~NUT - [metall] (in diecasting) écrou *m* de blocage
~OF POINTS - [railw] blocage *m* d'aiguille
~OF THE SWITCH BLADES - [railw] calage *m* des lames d'aiguille
~OF WHEELS - [mech] enrayement *m* des roues
~PIN - [text] cheville *f* d'arrêt, goupille *f* d'arrêt
~PLATE - [mech] clapet *m* de fermeture
~PLUNGER - [mech] bonhomme *m* de verrouillage
~PRESSURE - [plast ind] (pressure used to keep a mould closed, as distinct from the moulding pressure itself) pression *f* de serrage
~RELAY - [el] relais *m* à verrouillage
~RING - [mech] (a slotted plate in an injection or transfer mould which locks the parts of the mould together and prevents it from opening while the material is being injected) bague *f* de fermeture, collier *m* de serrage
~ROD - s. locking knob
~SCREW - [mech] vis *f* d'arrêt, vis *f* de blocage
~SLEEVE - [mech] manchon *m* de blocage
~SPRING - [mech] ressort *m* du cliquet
~STILE - [constr] montant *m* pour serrure
~STRIP - [mech] bande *f* de blocage
~UP - [print] (the locking of the types in the chase) serrage *m*
[gen] fermeture *f* (à clef)
~WHEEL - [mech] (in electric time timers) roue *f* à loquet
~WIRE - [mech] (a wire passed through holes formed in bolts, muts or screws to prevent undesired rotation) câble *m* de blocage
LOCKJAW - [med] tétanos *m*
LOCKNUT - [mech] (a second nut, usually thinner, used to secure another nut against accidental rotation) contre-écrou *m*
LOCKSMITH - [mech] serrurier *m*
LOCKSTITCH - [text] point *m* piqué
LOCKSTITCH BAR - [text] barre *f* à poinçons, plaque *f* à poinçons, porte-poinçons *m*
~MACHINE - [text] piqueuse *f*, machine *f* à point de navette
LOCOMOTION - [gen] locomotion *f*
LOCOMOTIVE - [railw] (vehicle driven by various forms of power, hauling carriages on rails) locomotif *m*, machine *f* locomobile, locomotive *f*
~CAR - [railw] voiture *f* automotrice
~CHANGING POINT - [railw] gare *f* de relais de locomotives
~COAL - [min] charbon *m* de traction
~CRANE - [mech] grue *f* locomotive
~FAIRING - [mech] carénage *m* de la locomotive
~PULSE - [med] pouls *m* de Corrigan
~TESTING PLANT - [railw] banc *m* d'essai pour locomotives
LOCOMOTOR - [gen] locomoteur *m*
LOCUS - [gen] lieu *m*, situation *f*
[chem] (of a reaction) centre *m*
[math] lieu *m*
[genet] (the position of a gene in a chromosome) locus *m*
LOCUST - [zool] locuste *f*
~TREE - [bot] caroubier *m*, robinier *m*, faux acacia *m*
LODE - [min] (a vein of mineral ore) filon *m*, veine *f*
[constr] (an artificial dyke) dyke *m*
~CHANNEL - [geol] chenal *m* filonien
~-MINING - [mining] exploitation *f* filonienne
~PLOT - [mining] filon *m* horizontal
~ROCK - [geol] gangue *f*
LODESTAR - [astr] étoile *f* polaire
LODESTONE - [min] (a form of magnetite) pierre *f* d'aimant, aimant *m* naturel, minerai *m* de filon
LODESTUFF - [min] (the mineral constituents of a lode) matière *f* filonienne
LODGE - [arch] loge *f*
LODGEMENT - [gen] logement *m*
[min] recette *f* à eau
[mining] albraque *m*
LOFT - [constr] grenier *m*, soupente *f*
[gen] (a large room in which the shapes of large parts can be set out on the floor for cutting out, bending, etc) atelier *m*
~YARN - [text] fil *m* mousse
LOFTING - [mining] boisage *m* du toit
LOFTY - [gen] haut, élevé
[text] (of wool and woollen fabrics with good springy quality) élastique
LOG - [gen] bloc *m* de sciage
[bot] tronçon *m* de bois
[instr] (instrument designed to measure a ship's speed through the water) loch *m*, sillomètre *m*
[naut aero etc] s. log-book
[plast ind] (a roughly cylindrical mass of cellular plastic material for further processing) bloc *m*
[gen] enregistrement *m*, diagramme *m*
~-BOOK - [naut, aero] (record of navigational data and events aboard an aircraft or a ship) livre *m* de loch
[gen] (for the machines) journal *m* de travail, registre *m*
[auto] carnet *m* de route
~-CABIN - s. log-house
~CONVEYOR - [mech] convoyeur *m* de grumes
~DECODER - [electron] décodeur *m* logarithmique des contenus des canaux

LOG-EXPOSURE - [photo] (exposure as a function of the logarithm of the illumination) exposition *f* en fonction du logarithme de l'éclairement
~FRAME - [mech] scie *f* alternative à tronçonner
~-HOUSE - [constr] cabane *f* de bois
~LINE - [naut] ligne *f* de loch
~REEL - [instr] tour *m* de loch, tambour *m* de loch
~RUNNING - [timber ind] flottage *m* du bois
~SHIP - [instr] bateau *m* de loch
LOGADITIS - [med] sclérite *f*
LOGARITHM - [math] logarithme *m*
~FUNCTION - [math] fonction *f* logarithmique
LOGARITHMIC - [math] logarithmique
~AMPLIFIER - [radio] (amplifier having an output signal which is a logarithmic function of the input signal) amplificateur *m* logarithmique
~DECREMENT - [phys] (the natural logarithm of the ratio of two successive swings) décrément *m* logarithmique
~DECREMENT OF CIRCUIT - [el] décrément *m* logarithmique de circuit
~DECREMENT OF VIBRATION - [acoust] (the natural logarithm of the ratio of successive amplitudes of vibrations) décrément *m* logarithmique de vibration (ou d'oscillation)
~ENERGY DECREMENT - [nucl] (the mean value of the increase in lethargy per collision) décrément *m* logarithmique d'énergie
~HORN - [acoust] (exponential horn) pavillon *m* exponentiel
LOGASTHENIA - [med] logasthénie *f*
LOGATOM - [telecomm] (isolated meaningless syllable) logatome *m*
~ARTICULATION - [telecomm] articulation *f* du logatome
LOGBOOK - s. log-book
LOGGER - [instr] (device which automatically records physical processes) enregistreur *m* automatique
LOGGIA - [arch] (covered gallery, a portico) loge *f*, loggia *f*
LOGGING - [timber ind] exploitation *f* des bois et forêts
[telecomm] (of stations) répérage *m* des stations
~LINE - [mining] câble *m* de carottage électrique
LOGICAL-AND CIRCUIT - [comput] conditionneur *m*
~CIRCUIT - [contr comput etc] (circuit with at least two inputs and one output, so that a switching function can be obtained) circuit *m* logique
~COMPARISON - [comput] comparaison *f* logique
~DIAGRAM - [comput] diagramme *m* logique
~ELEMENT - [comput] (in a computer or in a data-processing system) organe *m* logique
~-OR CIRCUIT - [comput] circuit *m* OU
~PROGRAMME - [comput] programme *m* logique
~SHIFT - [comput] décalage *m* logique
~SYMBOL - [comput] (a functional symbol) symbole *m* logique
LOGOCLONIA - [med] logoclonie *f*
LOGOMANIA - [med] loquacité *f*
LOGONEUROSIS - [med] logonévrose *f*
LOGOSPASM - [med] élocution *f* saccadée
LOGWOOD - [chem] (haemotoxylon. The unfermented heartwood of Haemotoxylon campechianum with uses as a dye and in medicine as a mild astringent) bois *m* de campêche, extrait *m* de campêche
LOIN - [anat] lombes *mpl*
LONE - [gen] seul, solitaire
~ELECTRON - [nucl] (an electron occupying an energy level which does not contain any other electrons) électron *m* célibataire
LONG - [gen] long
~-BASE RANGE-FINDER - [instr] (a range finder with home-base and two points of observation) télémètre *m* à longue base chaloupe *f*
~BOAT - [naut] chaloupe *f*
~BORER - [mining] barre *f* double
~-BRISTLE BRUSH - [impl] queue-de-morue *f*, brosse *f* à poils longs
~COUNTER - [nucl] compteur *m* à enveloppe cylindrique en cire
~-DATED - [gen] à longue échéance
~-DAY PLANT - [agric] plante *f* de longue journée
~-DISTANCE CALL - [telecomm] communication *f* à grande distance
~-DISTANCE LIGHT - [auto] éclairage *m* à longue portée
~-DISTANCE LINE - [telecomm] (a line between toll centres for long-distance calls) ligne *f* interurbaine
~-DISTANCE SCATTER - [phys] dispersion *f* à longue distance
~-DISTANCE TRANSPORT - [transp] transport *m* sur longues distances
~DURATION - [gen] longue durée
[adj] de longue durée
~-DURATION ECHO - [acoust] (artificial echo of long duration used for measuring purposes) echo *m* de longue durée
~EYE - [text] maillon *m* long
~FEED - [mach] avance *f* longitudinale
~-FLAMING COAL - [min] charbon *m* flambant
~FLAX - [text] lin *m* long, iin *m* à long brins
~-FOCUS LENS - [photo] objectif *m* à long foyer, objectif *m* à grande focale
~GRINDING - [paper man] broyage *m* longitudinal
~-HAUL LINE - [electron] ligne *f* de télécommunication
~HOB - [impl] taraud *m* mère
~HOOK - [text] crochet *m* long
~-LINE EFFECT - [electron] effet *m* de ligne de transmission
~-LIVED FISSION PRODUCT - [nucl] (fission product having a comparatively long life) produit *m* de fission à longue période
~-LIVED RADIATION - [radiat] (radiations having a comparatively long duration) rayonnement *m* à longue période
~MACHINE-TAP - [mech] taraud *m* long pour machines
~MALT - [brew ind] (malt which has been allowed to germinate for about twenty days, for distilling) malt *m* vert
~PILED FUSTIAN - [text] velours *m* par trame long-poil
~PILLAR WORK - [mining] exploitation *f* par panneaux
~-PITCH SCREW - [mech] vis *f* de pas allongé, vis *f* à pas rapide
~-PITCH WINDING - [el] enroulement *m* à pas allongé
~PLANE - [impl] rabot *m* de menuisier
~-PLAYING RECORD - [el acoust] disque *m* microsillon, disque *m* de longue durée
~-RANGE ACCURACY RADAR SYSTEM - s. lorac
~-RANGE ALPHA PARTICLES - [nucl] (particles with

spectrum lines due to groups with very low energies) particules *fpl* alpha à portée longue
LONG-RANGE COMMUNICATION - [telecomm] trafic *m* de télécommunication
~-RANGE NAVIGATION SYSTEM - s. lorac
~-RANGE ORDER - [crystall] ordre *m* étalé
~-RANGE STATIC ELIMINATOR - [el] (device for removing static charges from material by a stream of ionized air) éliminateur *m* d'électrostatique
~RESIDUE - [oil ind] (the residue which is resulting from the atmospheric distillation of crude oil) résidu *m* de la distillation du pétrole
~RUNNING - [ind proc] plan *m* lointain, prise *f* de vue à distance
~SHUNT - [el] longue dérivation *f*
~-SIGHTED - [opt] presbyte
~SLOT BURNER - [mining] brûleur *m* à embouchure oblongue
~-STAPLE - [text] coton *m* long-soie
~-STAPLE WOOL - [text] à laine longue
~STRING - [oil ind] colonne *f* de production
~STRIPES - [text] rayures *fpl* dans le sens de la longueur du tissu
~STROKE ENGINE - [mech] moteur *m* à longue course
~-STROKE PRESS - [mech] (type of press having a longer platen travel than usual) presse *f* à longue course
~STROKE PUMP - [oil ind] pompe *f* à longues coups
~-TAILED PAIR - [radio] circuit *m* pseudo-symétrique à résistance cathodique commune
~-TERM - [gen & comm] à longue échéance
~THREADS - [oil ind] filetage *m* type long
~THRESHER - [agric] batteuse *f* en long
~-WALL WORKING - [mining] exploitation *f* par tailles, chassantes *fpl* à front continu
~WAVES - [radio] (waves longer than 1000 m) ondes *fpl* longues
~-WIRE AERIAL - [radio] (linear aerial providing a directional diagram) antenne *f* longue
~-WIRE ANTENNA - s. long-wire aerial
~-WOOL BREED - [zool] race *f* à laine longue
~ZIGZAG TWILL - [text] chevron *m* en long
LONGERON - [aero] (a principal longitudinal airframe element) longeron *m* de fuselage
LONGEVITY OF SEED - [agric] longévité *f* des semences
LONGILINEAL - [anat] longiligne
LONGITUDE - [geogr] (the angular distance East or West of a given meridian) longitude *f*
[surv] longitude *f*
~EFFECT - [phys] (the variation of cosmic ray intensity along the equator as a function of longitude) effet *m* de longitude
~IN ARC - [geogr] longitude *f* en degrés
~IN TIME - [geogr] longitude *f* en heures et minutes
LONGITUDINAL - [gen, geogr & surv] longitudinal
[aero] (of a hull) longeron *m*
~AXIS - [aero] (a line drawn fore and aft through the centre of gravity and lying wholly in the plane of symmetry) axe *m* de roulis, axe *m* longitudinal
~CARRIER-CABLE - [telecomm] câble *m* de décharge, câble *m* porteur (du câble aérien)
~CHROMATIC ABERRATION - [opt] aberration *f* chromatique longitudinale
~CLINOMETER - [instr] (instrument indicating the angle between the longitudinal axis of an aircraft and the horizontal) clinomètre *f* longitudinal
LONGITUDINAL CRACK - [soil] fissure *f* longitudinale
~CRYSTAL - [phys] (piezoelectric crystal oscillating in the longitudinal mode of motion) cristal *m* chromatique longitudinal
~CURL - [metall] cambrure *f* longitudinale
~DIHEDRAL ANGLE - [aero] angle *m* d'attaque longitudinal
~DIRECTION - [metall] sens *m* longitudinal
~DIRECTIONAL ARRAY - [radio] rideau *m* à rayonnement longitudinal
~DIVERGENCE - [aero] (any periodic divergence taking place in the plane of symmetry) engagement *m* longitudinal
~DRAINING - [agric] drainage *m* longitudinal
~DUCTILITY - [metall] ductilité *f* en sens longitudinal
~EXTENSION - [metall] allongement *m* longitudinal
~FEED - [mach] avance *f* longitudinale
~-FINNED PIPE - [plumb] tuyau *m* à nervures longitudinales
~FINNING - [nucl] (cooling fins used to improve the heat transfer of nuclear reactors) à ailettes longitudinales
~FLANGE - [mech] bride *f* longitudinale
~FORCE - [aero] (any force acting in the direction of the longitudinal axis) force *f* longitudinale
~FRAME - [shipbuild] (a stiffening member of a ship's hull) membrure *f* longitudinale
~GROOVE - [mech] rainure *f* longitudinale, gorge *f* longitudinale
~GROOVED PROFILE - s. longitudinal grooved tread
~GROOVED TREAD - [rubber ind] (of tyres) (a parallel grooved tread) profil *m* à sillons longitudinaux
~INSTABILITY - [aero] (tendency in an aircraft towards the increase of any disturbance in the plane of symmetry) instabilité *f* longitudinale
~MAGNETIZATION - [el acoust] (magnetization of the recording medium in a direction which is essentially parallel to the line of travel) aimantation *f*, magnétisation *f* longitudinale
~MAGNIFICATION - [opt] (linear magnification measured to the optical axis) grandissement *m* axial, grossissement *m* longitudinal
~MEMBER - [aero] (longitudinal wing-frame element) longeron *m*
~MILLING - [mach tool] fraisage *m* longitudinal
~OSCILLATION - [aero] (periodic variation in speed, altitude and pitch) oscillation *f* longitudinale
~PHONON - [crystall acoust] (one of the acoustic modes of crystals) phonon *m* longitudinal
~PITCH - [mech] pas *m* longitudinal
~PROFILE - [geol] profil *m* en long
~SECTION - [draw] (a view in which the subject is seen as if cut in two longitudinally) profil *m* en long, coupe *f* longitudinale
~SEPARATION - [aero] (in air traffic regulation, the spacing-out of aircraft in terms of time) espacement *m* longitudinal
~SPHERICAL ABERRATION - [opt] aberration *f* sphérique longitudinale
~STABILITY - [aero] (stability of vertical and forward movement and of pitch in the plane of symmetry) stabilité *f* longitudinale
~STAY - [telecomm] hauban *m* dans le sens de la ligne
~STIFFENER - [aero] (a wing-frame stiffening ele-

ment) raidisseur *m*
LONGITUDINAL STIFFENER - [aero] (a longitudinal airframe element) renfort *m* longitudinal
~ TRAVEL - [mech] course *f* longitudinale
~ TRAVERSE - [mach tool] (of a milling machine) chariotage *m* longitudinal
~ TRAVERSE OF THE WHEEL - [mach tool] chariotage *m* longitudinal de la meule
~ TRICOT - [text] drap *m* tricot à rayures en long
~ VELOCITY - [aero] (velocity in the direction of the longitudinal axis, with respect to the undisturbed) vitesse *f* longitudinale
~ WAVE - [electron] (or transverse wave; a wave characterized by a vector parallel with the direction of propagation) onde *f* longitudinale
LONGITUDINALLY SPLIT RIM - [rubber ind] (of tyres) jante *f* divisée longitudinalement
LONGSHOREMAN - [labour] arrimeur *m*
LONGSOLE - [shoe man] (a full length sole) semelle *f* complète
LONGWALL - [mining] grand front *m* aligné
~ COAL CUTTER - [mining] haveuse *f* pour grands fronts
~ FACE - [mining] grand front *m* aligné
~ MINING - [mining] exploitation *f* par longwall
~ STOPING - [mining] exploitation *f* par longwall
LOOK-THROUGH - [paper man] transparence *f*
LOOKING-GLASS - [gen] miroir *m*, glace *f*
LOOM, to - [gen] apparaître indistinctement
LOOM - [text] métier *m* à tisser
[aero] (a pre-assembled set of electric cables) câblage *m* préassemblé
[naut] (the shaft of an car) silhouette *f* estompée
~ AND SPINDLE OIL - [fuel] (oils of low viscosity used in the textile industry) huile *f* à faible viscosité
~ BATTEN - [text] battant *m*, châsse *f*
~ BEAM - [text] ensouple *f*
~ BLOCK - [text] taquet *m* pour métier
~ BRAKE - [text] frein *m* du métier
~ DRIVE - [text] commande *f* du métier
~ FLY - [text] duvet *m* de métier
~ FOOTPLATE - [text] plaque *f* de support du métier
~ FOR BEADED MATERIAL - [text] métier *m* pour tissu perlé
~ FOR BRAIDINGS - [text] métier *m* pour galons
~ FOR EMBROIDERED GOODS - [text] métier *m* pour brochés-poils-traînants
~ FOR FIGURED GAUZE - [text] métier *m* pour gaze damassée
~ FOR FRINGED MATERIAL - [text] métier *m* pour articles à franges
~ FOR STRAPS BELTS - [text] métier *m* pour sangles
~ FOR TAPES - s. loom for strap belts
~ FOR TRIMMINGS - [text] métier *m* pour soutaches
~ FOR WEAVING PLUSH - [text] métier *m* à tisser la peluche
~ FOR WOVEN BELTING - [text] métier *m* pour tissage de courroie
~ FRAMING - [text] bâti *m* du métier à tisser
~ PICKER - [text] chasse-navette *m*
~ SEATING - [text] bâti *m* du métier
~ SHUTTLE - [text] navette *f* de tissage
~ STOP - [text] arrêt *m* du métier à tisser
~ WASTE - [text] déchets *mpl* de fil
~ WITH ONE SHAFT - [text] métier *m* à arbre unique
~ WITH ROCKING SHAFTS - [text] métier *m* à balancier
LOOM WITH STOCK AND BOWLS - [text] métier *m* à poulies
LOOMING - [met] (the indistinct appearance of objects or, in the nautical sense, of land, through poor visibility) apparence *f* indistincte, mirage *m*
[text] (the drawing of the warp threads through the eyes of the heald shaft before beginning weaving operations) remettage *m*, rentrage *m*, passage *m*
[opt] (type of mirage) effet *m* de mirage verse le haut
~ APPLIANCES - [text] outils *mpl* de rentrage
~ FRAME - [text] banc *m* à rentrer les chaînes, chevalet *m* de rentrage, chevalet *m* de remettage
~ LINE - [text] chemin *m* de remettage
LOOMSTATE - [text] tombant *m* du métier
LOOP, to - [gen] boucler, faire une boucle
[text] (rope man) boucler
[aero] boucler la boucle
[el] faire un circuit bouclé
LOOP - [gen] boucle *f*, anse *f*, ganse *f*
[aero] (aerobatic manoeuvre in which the aircraft performs one complete revolution in flight, the upper surface being on the inside of the circle described) boucle *f*, looping *m* normal
[geogr] (of river etc) anse *f*, sinuosité *f*
[el] (an electrical circuit) circuit *m* bouclé
[electron] (in electronic control equipment, the cyclic repetition of a set of instructions) boucle *f*
[phys] (one of the points or lines in a standing wave system where a characteristic of the wave field has maximum amplitude) ventre *m* d'une oscillation
[cin] (piece of film which is not under tension) boucle *f*
[text] attache *f*
[mech] (of a coil) spire *f*, tour *m*
[railw] voie *f* de raccordement, boucle *f* d'évitement
[text] (in the weave for looped fabric) bride *f*, arcade *f*
[text] (of woven lace) picot *m*
[text] (of the reed) noeud *m*
[metall] (a mass of iron not yet solidified) lingot *m* incandescent, loupe *f*, balle *f*
[nucl] (closed circuit of conduits in a nuclear reactor) boucle *f*, circuit *m* expérimental
~ A LINE, to - [telecomm] boucler une ligne
~ AERIAL - [radio] (an aerial, usually employed for directional purposes, in which the conductor forms one or more complete turns) antenne *f* fermée, cadre *m*
~ ANTENNA - s. loop aerial
~ BACK ONE CIRCUIT WITH ANOTHER, to - [telecomm] renvoyer en retour un circuit sur un autre
~ BAR - [text] barre *f* de cueillage
~ CABLE - [el] (twin cable) câble *m* à deux conducteurs
~ CHECKING - [comput] contrôle *m* d'erreurs à transmission rétrograde
~ CIRCUIT - [el] circuit *m* bifilaire, ligne *f* double, circuit *m* métallique, circuit *m* sans retour par la terre
~ CLOTH - [text] tricot *m* peluche, tricot *m* bouclette
~ CONVEYOR - [metall] (in foundry work) convoyeur *m* de coulée à circuit fermé
~ CURRENT - [el] (in electrical telecommunications, the normal current flowing round a loop) courant *m* de circuit
~ DIALLING - [telecomm] (the use of break pulses in

a loop circuit) sélection *f* à boucle
LOOP DIALLING SYSTEM - s. loop dialling
~ELEMENT - [contr] organe *m* de boucle
~EXPANSION - [metall] courbe *f* de dilatation
~FABRIC - [text] tricot *m* peluche, tricot *m* bouclette
~FEEDER - [el] bouclage *m*, circuit *m* bouclé
~FORMATION - [text] formation *f* de frisures
~GAIN - [electron] gain *m* de boucle
[telecomm radio] (the gain around the feedback loop) gain *m* de boucle de réaction
~GALVANOMETER - [el] (sensitive instrument in which the moving element is a U-shaped loop of aluminium coil) galvanomètre *m* à boucle
~GROUND - [text] réseau *m* anglais
~IMPEDANCE - [el] impédance *f* de boucle
~INPUT RESOLUTION - [contr] résolution *f* de l'entrée de la boucle
~LINE - [oil ind] canalisation *f* parallèle
~LUBRICATION - [mech] lubrification *f* en circuit fermé
~MEASUREMENT - [telecomm] mesure *f* en boucle, mesure *f* en retour
~OPERATION - [comput] boucle *f* à auto-restauration
~OUTPUT RESOLUTION - [contr] résolution *f* de la sortie de la boucle
~PHASE ANGLE - [contr] déphasage *m* en boucle ouverte
~PILE - [text] (type of surface covering) tissu *m* éponge
~RESISTANCE - [el] (the total resistance of a circuit) résistance *f* de circuit
[telecomm] (of an external line) résistance *f* de couple
~RESISTANCE MEASUREMENT - s. loop measurement
~RESISTANCE TEST - s. loop measurement
~ROAD - [constr] route *f*
~SERVICE - [el] alimentation *f* duplex
~STOP - [comput] boucle *f* sans fin
~TEST - [el] (used for locating faults in electric cables) méthode *f* des boucles
~THE WARP, to - [text] mettre la chaîne en mailles
~TRANSFER FUNCTION - [contr] transmittance *f* en boucle ouverte
~TUNING ERROR - [radio] (the error in the bearings given by a direction finder whenever the loop is not rightly tuned) erreur *f* d'accord décalé
~TWO CIRCUITS, to - s. loop one circuit with another
~-TYPE CABLE CLAMP - [el] (a cable clamp consisting of a strip of metal curved round a cable and pierced with a hole for attachment through both the superimposed ends) collier *m* de câble à deux conducteurs
~-TYPE DIRECTIONAL COUPLER - [electron] coupleur *m* directif du type cadre
~TYPE DRYER - [plast ind] séchoir *m* à guirlandes
~WARP - [text] chaîne *f* de poil, chaîne *f* de boucles
~YARN - [text] (fancy type of yarn, with small loops) retors *m* bouclé
LOOPED BRAID - [text] galon *m* bouclé
~CARPET - [text] moquette *f* bouclée, moquette *f* brisée, tapis *m* bouclé
~FABRIC - [text] tissu *m* bouclé
~FILAMENT - [light] (a filament shaped like a large helix generally used for carbon filament lamps) filament *m* à spires
~YARN - [text] filé *m* bouclé
LOOPER - [zool] arpenteuse *f*
LOOPHOLE - [gen] rayère *f*
LOOPING - [comput] déroulement *m*
~PIT - [mining] emmagasineur *m*
LOOSE, to - [gen] délier, détacher
[naut] larguer
[mech] desserrer
LOOSE - [gen] dégagé, délié, détendu, lâche
[gen] (of a mixture) inconsistant
(of a mechanical part) mobile, dégagé, branlé
[mech] (of moving parts) desserré, (roue) folle
[gen] (of a rope) mou, détendu
~BOTTOM - [naut] fond *m* de mauvaise tenue
~COILS - [text] spires *fpl* enroulées lâchement
~COUPLING - [radio] (any degree of coupling which is less than the critical coupling) couplage *m* lâche
[mech] (shaft coupling which can be easily and instantly disconnected) accouplement *m* libre
~EARTH - [constr] terre *f* meuble, deblai *m*
~FIT - s. loose running fit
~FLANGE - [mech] contre-bride, bride *f* mobile
~FRAMING - [telev] (term denoting excessive smallness of the actors in the frame) décadrage *m*
~HEAD - [mach tool] poupée *f* mobile, poupée *f* courante
~INSERTED PEG - [text] broche *f* à mortaise
~-JOINTED - [mech] à joint lâche
~KIBBLE - [mining] benne *f* flottante
~LAID WALL - [constr] mur *m* à sec
~MOULD - [metall & plast ind] (a mould which can be removed from the press for cleaning etc) moule *m* mobile
~PART - [mech] partie *f* dégagée
~PATTERN - [metall] (pattern which can be removed) modèle *m* mobile
~PIECE - [metall] (in diecasting) noyau *m* détachable
~PIN - [metall] broche *f* de remoulage
~PULLEY - [mech] (pulley which is mounted freely on a shaft) poulie *f* folle
~PUNCH - [metall & plast ind] (the male portion of a mould when this is so constructed that it remains attached to the moulding when the press is opened and is removed from the mould with the moulding for the purpose of extraction) poinçon *m* amovible
~REED MECHANISM - [text ind] mécanisme *m* du peigne à échappement
~ROLL - [plast ind] (synonym for "Dancer"; a roller which can move up and down with the material running over it) rouleau *m* fou
~ROLLER - s. loose roll
~RUNNING FIT - [mech] montage *m* à glissement libre
~SMUT - [bot] (a plant disease affecting oats) charbon *m* nu
~STRIKER CAM - [text] excentrique *m* pour lisière
~TOOL - [tool] outil *m* à main
~TREAD - [rubber ind] (of tyres) bande *f* de roulement décollé
~VEINS - [geol] (in limestone) limé *m*
~WARP - [text] chaîne *f* tendue lâche
~WOVEN FABRIC - s. loose woven goods
~WOVEN GOODS - [text] tissus *mpl* lâches
LOOSEN, to - [gen] délier, desserrer, défaire, décoller
[gen] (a rope) détendre, relâcher
[mech] (a screw) dégager, donner du jeu à (un ressort), dégripper
[naut] se délier, se défaire
~BY VIBRATION, to - [mech] (of screws, nuts, cou-

plings etc) se défaire par vibration
LOOSENESS - [gen] état *m* branlant
[mech] desserrage *m*, jeu *m*
[gen] (of a rope) relâchement *m*
[soil] (of the ground) inconsistance *f* (du terrain)
LOOSENING - [mech] (of a screw etc) dégagement *m*, dégrippage *m*, desserrage *m*
[agric] (of the ground) ameublissement *m* (du sol)
[med] (of cough) dégagement *m* (de la toux)
~OF THE SLEEPER BED - [railw] repiquage *m* du moule de traverse
~OF TREAD PLUG - [rubber ind] (in tyres) décollement *m* d'une obturation de pneu
LOOSING - [mining] descente *f* de la cage d'extraction
LOP, to - [gen] (to cut off branches from a tree) ébrancher, tailler, égayer
~-EAR - [zool] oreille *f* pendante
~-EARED PIG - [zool] race *f* à oreilles longues
LOPPER - [el] limiteur *m*
LOPPING - [agric] (controlled cutting off of branches from trees) élagage *m*
~CHISEL - [agric] ciseau *m* à ébrancher
LOPSIDED - [gen] penchant d'un côté, déjété
[gen] (of structures etc) manquant de symétrie
LORAC - [radar] (long-range accuracy radar system similar to Lorac) système *m* radar Lorac
LORAN - [radio & radar] (a radionavigation system using medium or low frequency, similar to "Gee" but of greater range) Loran, système *m* radar Loran
LORDOSIS - [med] lordose *f*
LORRY - [transp] camion *m*
[railw] (long flat wagon without sides) lorry *m*
[mining] (a running bridge over a pit) char *m* à bennes
LOSCHMIDT NUMBER - [math] (synonym for Avogadro's Number) nombre *m* de Loschmidt
LOSE, to - [gen] perdre
LOSS - [gen] perte *f*
[el] (in a transmission system) perte *f*
[meas] (of electric meters) perte *f*
[of gas] perte *f*, fuite *f*
~ANGLE - [el] (the difference between 90° and the angle of lead of current over voltage in a capacitor) angle *m* de perte
[el biol] (complement of the phase angle) angle *m* de perte biologique
~BY EVAPORATION - [phys] perte *f* par évaporation
~BY FRICTION - [mech] perte *f* par frottement
~DUE TO DRESSING - [min] perte *f* au triage
~DUE TO LEAKAGE - [mech] perte *f* par défaut d'étanchéité
~DUE TO ROASTING - [min] perte *f* au grillage
~DUE TO SCREENING - [min] perte *f* au triage
~FACTOR - [el] (the product of the dielectric constant and the power factor) facteur *m* de pertes
~-FREE CONDENSER - [el] condensateur *m* sans pertes
~IN EFFICIENCY - [gen] diminution *f* du rendement
~IN MELTING - [metall] perte *f* de fusion
~IN SUPPRESSED BAND OF FREQUENCIES - [telecomm] affaiblissement *m* pour la bande des fréquencies non transmises
~IN TREATMENT - [min] perte *f* au traitement
~IN WEIGHT - [chem etc] (reduction in the weight of a specimen of a substance in consequence of some operation performed upon it) perte *f* de poids
LOSS METER - [instr] (meter installed on the load side of a transformer to measure the iron and copper losses) compteur *m* de pertes
~OF A TOOL - [mining] perte *f* d'un outil
~OF ACCURACY - [comput] (the difference between the number of significant digits of the input data and the results of an operation) perte *f* d'approximation
~OF CHARGE - [el] perte *f* de charge
~OF CHARGE METHOD - [el] (a method of measuring very high resistances) méthode *f* de la perte de charge
~OF CIRCULATION - [oil ind] (in oil wells) perte *f* de circulation
~OF COMPRESSION - [mech] (leakage in i.c. reciprocating engine resulting in low compression pressure) perte *f* de compression
~OF COOLANT ACCIDENT - [nucl] incident *m* dangereux de perte de réfrigérant
~OF CURRENT - [el] déperdition *f* de courant
~OF FLOW ACCIDENT - [nucl] incident *m* dangereux de réduction de flux
~OF HEAD - [hydr] perte *f* de charge
~OF HEAT - [phys] déperdition *f* de chaleur
~OF HEIGHT - [aero] perte *f* d'altitude
~OF NUTRIENTS - [agric] perte *f* d'éléments fertilisants
~OF PICTURE LOCK - [telev] décrochage *m* vertical de l'image
~OF POWER - [mech] déperdition *f*, travail *m* nuisible
~OF PRESSURE - [mech] perte *f* de charge
~OF TEMPER - [metall] perte *f* de trempe
~OF YIELD - [agric etc] perte *f* de rendement
~ON WASHING - [rubber ind] perte *f* au délavage
~-SUMMATION METHOD - [el] (the method of indirect efficiency measurement in which the losses are determined separately) méthode *f* des pertes séparées
LOSSES - [el] (of an electrical machine; the power which is lost during the operations and generally appears in the form of heat) pertes *fpl*
[gen mech] (the difference between input and useful output of a system) pertes *fpl*, déperdition *f*
~FROM WEEDS - [bot] dégâts *mpl* dus aux mauvaises herbes
LOST CIRCULATION - [oil ind] perte *f* de circulation
~FORCE - [phys] (the lost force on the particle of a dynamical system is equal to the difference between the external force applied and the effective force) force *f* perdue
~MOTION - [mech] (in a linkage or gear train, movement which is not transmitted, e.g. because of looseness in the parts) perte *f* de travail
~-MOULDING - [metall] moulage *m* à la cire perdue
~SAND CORE - [metall] noyau *m* perdu
~-WAX CASTING - [metall] coulée *f* en cire perdue
~WAX MOULDING - [metall] coulée *f* en cire perdue
~-WAX PROCESS - [metall] (process used in statuary founding) procédé *m* à la cire perdue
LOT - [gen] lot *m*, part *m*, partage *m*
[comm] lot *m*, partie *f*
[cin] (the total area of a studio) terrain *m* du studio
[agric] (of land) lot *m* (de terrain)
~-SORTING CONVEYOR GRADING EQUIPMENT - [min] équipement *m* d'estimation et de triage sur bande par lot

LOTION - [med] lotion *f*
LOTUS - [bot] lotus *m*, trèfle *m* cornu, nélombo *m*
~WOOD - [bot] micocoulier *m*, fabrecoulier *m*
LOUD - [acoust] retentissant, sonore
~PEDAL - [mus] (on a pianoforte) pédale *f* forte
~TRAILER - [el acoust] haut-parleur *m* directionnel de grande puissance
LOUDNESS - [acoust] (the subjective measure of the intensity of sound or of noise) intensité *f* sonore
~CONTOURS - [acoust] (the curves showing the related values of sound pressure level and frequency required to produce a specified loudness sensation for the average listener) lignes *fpl* de force sonore
~LEVEL - [acoust] niveau *m* d'intensité sonore
~VOLUME EQUIVALENT - [telecomm] équivalent *m* relatif, équivalent *m* de référence
LOUDSPEAKER - [el acoust] (device for converting audio-frequency currents into corresponding sound waves) haut-parleur *m*
~CAPACITOR - [el acoust] (the capacitor over the supply leads of a loudspeaker) condensateur *m* de haut-parleur
~CONDENSER - s. loudspeaker capacitor
~DIVIDING NETWORK - [el acoust] montage *m* à deux voies
~RESPONSE - [el acoust] (the response measured over its operating frequency range and direction) réponse *f* de haut-parleur
~TELEPHONE SET - [telecomm] appareil *m* téléphonique à réception amplifiée
~VOICE COIL - [acoust] bobine *f* mobile de haut-parleur
LOUISINE - [text] louisine *f*
LOUPE - [opt] loupe *f*
LOUSE - [zool] (pl lice) pou *m*
LOUSEWORT - [bot] (a weed) pédiculaire *f*, herbe *f* aux poux
LOUVER - s. louvre
LOUVRE - [constr] (or louver; window opening fitted with horizontal slats with spaces for ventilation purposes) lucarne *f*, abat-vent *m*
[auto] persienne *f*, fente *f*
[gas ind] (gas installation) (a deflector) déflecteur *m*
~BOARD - [constr] (one of the sloping slats fitted in the window space) abat-vent *m*
~STATOR GUARD - [el] (louvre-like cover for the protection of the stator) dispositif *m* de protection à persienne pour le stator
LOVAGE - [bot] livèche *f*, ache *f* de montagne
LOVE GRASS - [bot] éragrostide *f*
~VINE - [bot] cuscute *f*, barbe *f* de moine
LOVIBOND TINTOMETER - [paint] (instrument used to measure colour, based on the principle of matching the specimen with that of light seen through standard coloured glasses) colorimètre *m* de Lovibond
LOW - [gen] bas, peu élevé
[astr] bas
[met] (a region of low pressure) dépression *f*
[acoust] bas, profond
~-ALLOY STEEL - [metall] acier *m* d'alliage pauvre, acier *m* à faible alliage
~-ANGLE SPOT - [opt] contreplongée *f*
~BEAMS - [auto] (dim, or antidazzle light) faisceau *m* code (antiéblouissant)
~-BED TRAILER - [transp] (truck trailer for a low-loading height) remorque *f* surbaissés
LOW-BUILT CHASSIS - [auto] châssis *m* surbaissé
~-CAPACITANCE WINDING - [el] enroulement *m* à faible capacité
~-CARBON STEEL - [metall] (steel containing only a small proportion of carbon) acier *m* doux
~CHERRY END - [metall] (temperature of steel round 730 centigrades) chaleur *f* rouge-cerise
~CLOUDS - [met] (clouds of which the mean height is less than 8000ft.) nuages *mpl* bas
~-COMPRESSION ENGINE - [auto] moteur *m* à rapport volumétrique peu élevé
~-CONTRAST PICTURE - [telev] image *f* trop peu contrastée
~CURLINGS - [brew ind] pics *mpl* bas, mousse *f* basse
~DEFINITION TELEVISION - [telev] (any system operating on less than 200 lines per picture) télévision *f* à basse définition
~-DRAG ANTENNA - s. low-drag trailing aerial
~-DRAG TRAILING AERIAL - [radio] (an aircraft trailing aerial offering but little resistance to the air) antenne *f* à faible résistance
~-ELECTRON VELOCITY CAMERA TUBE - [telev] tube *m* de prise de vues stabilisateur au potentiel de la cathode
~-ENERGY ARC - [el] arc *m* à faible énergie
~-ENERGY NEUTRON - [nucl] (a neutron resulting from a reaction having an energy level below 30 Mev.) neutron *m* à faible énergie
~-ENERGY PARTICLE - [nucl] (particle resulting from a reaction in which the energy level is less than 30 Mev.) particule *f* à faible énergie
~-ENRICHED FUEL CYCLE - [nucl] cycle *m* de combustible légèrement enrichi
~-EXPANSION - [metall] à bas coefficient de dilatation
~FLOW - [plast ind] basse fluidité *f*
[hydr] débit *m* d'étiage
~-FLUX REACTOR - [nucl] (nuclear reactor with a comparatively low neutron flux) réacteur *m* à flux bas
~FOAMER - [chem] (a detergent which forms a restricted amount of suds) détergent *m* peu moussant
~-FREQUENCY - [radio] (the "Long Wave" range, i. e. between 30 and 300 kc/s.) basse fréquence *f*
~-FREQUENCY AMPLIFIER - [radio] amplificateur *m* à basse fréquence
~-FREQUENCY DIALLING - [telecomm] (in telephony) sélection *f* à distance avec impulsions de courant à fréquence infra-acoustique
~-FREQUENCY IMPEDANCE CORRECTOR - [telecomm] réseau *m* correcteur d'impédance B.F.
~-FREQUENCY MAGNETIC FIELD - [el] champ *m* magnétique à basse fréquence
~GEAR - [auto] première vitesse *f*
~-GRADE - [gen] basse teneur *f*
[min] (of an ore, containing a considerable proportion of worthless matter) à basse teneur
~-GRADE ORE - [min] (ore which contains a small quantity of mineral) minerai *m* pauvre, minerai *m* à bas titre, minerai *m* à faible teneur
~-GRADE STOCK - [rubber ind] mélange *m* pauvre en gomme
~-GRADE SUGAR - [ind chem] sucre *m* roux
~-GRAVITY CRUDE OIL - [oil ind] pétrole *m* brut lourd
~-IMPEDANCE TRIODE - [electron] (triode the internal impedance of which has a low value) triode *f* à

basse impédance
LOW IN FAT - [agric] pauvre en matières grasses
~INSULATION - [el] mauvais isolement *m*
~-LEVEL AMPLIFIER - [electron] amplificateur *m* de faibles signaux
~-LEVEL MODULATION - s. low-power modulation
~-LEVEL SWITCH - [electron] commutateur *m* à faible puissance de coupure
~-LEVEL WASTE - [nucl] déchets *mpl* de faible radioactivité
~-LIFT PUMP - [hydr] pompe *f* à faible hauteur
~-LOAD ADJUSTEMENT - [instr] (device for varying the auxiliary torque which is intended to improve the shape of the meter error curve for small currents) dispositif *m* de réglage aux petits débits
~-LOADING CHASSIS - [auto] châssis *m* surbaissé
~-LOADING TYRE - [rubber ind] pneumatique *m* pour faible charge
~-LOSS LINE - [el] ligne *f* à faibles pertes
~MELTING POINT - [metall] point *m* de fusion bas
~MOUNT - [auto] (of bodies which are mounted so low, that housings are necessary for the wheels) caisse *f* surbaissée
~-ORDER DIGIT - [math] chiffre *m* de poids faible
~PASS FILTER - [el] (filter which passes frequencies lower than a nominal cut-off frequency and attenuates those above) filtre *m* passe-bas
~POINT - [gas ind] (pipe lines) point *m* bas
~POWER DRAIN - [el] faible consommation *f* de courant
~POWER LOSS MATERIAL - [el] (one which has good electrical properties at high frequencies) produit *m* à faible perte de puissance
~-POWER MODULATION - [radio] (used in radio-telephone translitters) modulation *f* dans un étage intermédiaire
~-POWER RANGE - [nucl] (source range) domaine *m* des sources
~-POWER TRANSMITTER - [radio] émetteur *m* de faible puissance
~PRESSURE (L.P.) - [mech phys etc] (where more than one pressure value exists in the same unit or apparatus, the term "low pressure" is often applied to the lesser or least of such values) basse *f* pression, basse tension
~-PRESSURE BURNER - [th eng] brûleur *m* à basse pression
~-PRESSURE CLOUD CHAMBER - [nucl] (cloud chamber in which the gas in kept at low pressure to decrease the scattering of particle tracks) chambre *f* à détente à basse pression
~-PRESSURE CYLINDER - [mech] (the largest cylinder on a multi-expansion steam-engine, in which the steam is expanded) cylindre *m* à basse pression
~-PRESSURE ENGINE - [mech] moteur *m* à basse pression
~-PRESSURE FUEL FILTER - [mech] (in a jet engine) filtre *m* à essence à basse pression
~-PRESSURE GAUGE - [instr] manomètre *m* pour basses pressions
~-PRESSURE HEATING - [th eng] chauffage *m* à basse pression
~-PRESSURE LAMINATE - [metall plast ind] (term, formerly general, now in decreasing use, for laminates made under pressures of 400 psi or less) stratifié *m* à basse pression
~-PRESSURE LIGHT SOURCE - [light] (of a gas light) foyer *m* à basse pression
LOW-PRESSURE MOULDING - [plast ind] procédé *m* de moulage à basse pression
~-PRESSURE PLASTICS - [plast ind] (obsolete term for reinforced plastics) plastiques *mpl* renforcés
~-PRESSURE STEAM - [phys] vapeur *f* à basse pression
~-PRESSURE STEAM ENGINE - [mech] machine *f* à vapeur/à basse pression
~-PRESSURE TURBINE - [mech] turbine *f* à basse pression
~-PRESSURE VACUUM PUMP - [el] pompe *f* à vide élevé
~RANGE - [mach & mach tool] à avance limitée
~-RANGE PRESSURE GAUGE - s. low-pressure gauge
~RED HEAT - [metall] (temperature of steel between 550 and 700 centigrades) chaleur *f* au rouge foncé
~-REFLECTION COATING - [photo] (antireflection coating) couche *f* antiréfléchissante
~-RESISTANCE TRAP - [chem] (trap having a low resistance to molecular flow) fixateur *m* à faible résistance
~SHAFT - [metall] cuve *f* basse
~SIDE - [geol] lèvre *f* affaissée, lèvre *f* abaissée
~SOLIDS MUD - [oil ind] boue *f* à faible teneur en solides
~SPEED - s. low gear
~-SPEED EMULSION - [photo] émulsion *f* lente
~-SPEED LENS - [photo] objectif *m* de faible luminosité
~-SPEED NEEDLE - [auto] (of the carburettor) aiguille *f* de ralenti
~-SPEED NOZZLE - [auto] (of the carburettor) gicleur *m* de ralenti
~-SPEED WHEEL - [hydr] roue *f* à allure lente
~STEEL - [metall] acier *m* à faible teneur en carbone
~-STOP FILTER - [el] (high-pass filter; electric wave filter in which currents having frequencies higher than a nominal cut-off frequency are passed) filtre *m* passe-haut
~SUDSER - s. "Low Foamer"
~-TEMPERATURE ACCELERATOR - [rubber ind] (an agent which increases the rate of vulcanization at relatively low temperatures) accélérateur *m* agissant à basse température
~-TEMPERATURE BULB - [gas ind] (in gas analysis) tube *m* à basse température
~-TEMPERATURE CARBONIZATION - [ind chem] (the destructive distillation of coal at temperatures below 500° C., thus producing more liquid products and relatively little gas) carbonisation *f* à basse température, semi-distillation *f*
~-TEMPERATURE CARBONIZATION GAS - s. low-temperature gas
~-TEMPERATURE CARBONIZATION VESSEL - [gas ind] cuve *f* de carbonisation à basse température
~-TEMPERATURE COKE - [gas ind] (coke produced by the incomplete destructive distillation of coal, in which the maximum temperature is lower than in the ordinary method) semi-coke *m*, coke *m* de distillation à basse température
~-TEMPERATURE CRUDE LIGHT OIL - [ind chem] essence *f* de distillation à basse température
~-TEMPERATURE DISTILLATION - s. low-temperature carbonization
~-TEMPERATURE GAS - [gas ind] gaz *m* primaire, gaz *m* produit par la distillation à basse température

LOW-TEMPERATURE OVEN - [gas ind] four *m* à semi-coke
~-TEMPERATURE OXIDATION - [ind chem] (treatment of coal with alkalies and oxygen at about 500 deg. F. to produce carboxylic acids) oxydation *f* à basse température
~-TEMPERATURE REACTOR - [nucl] (nuclear reactor operating at comparatively low temperatures) réacteur *m* à température basse
~-TEMPERATURE RESISTANT INK - [ind chem] (special ink for marking parts to be exposed to low temperatures) encre *f* résistante aux basses températures
~-TEMPERATURE RETORT - s. low-temperature oven
~-TEMPERATURE TAR - [ind chem] goudron *m* primaire
~-TENSION BATTERY - [el] (of currents and voltages associated with the heater-circuits of a thermionic tube) batterie *f* de chauffage
~-TENSION CABLE - [el] câble *m* basse tension
~-TENSION DETONATOR - [el] (detonator in which a charge is fired by a heating a wire by means of an electric current) détonateur *m* à basse tension
~-TENSION IGNITION - [el] (in the cylinder of an internal combustion engine) allumage *m* à basse tension
~-TENSION LINE - [el] ligne *f* à basse tension
~-TENSION MAGNETO - [el] (for a low-tension ignition system) magnéto *m* à basse tension
~-TENSION TRANSFORMER - [el] transformateur *m* à basse tension
~-TENSION WINDING - [el] enroulement *m* basse tension
~TIDE - [astr] marée *f* basse, basse mer *f*
~-TIDE LEVEL - [naut] niveau *m* de basse mer
~VACUUM - [electron] (a degree of vacuum in an electronic tube which makes it possible for ionization to occur) vide *m* peu poussée
~VELOCITY SCANNING - [telev] analyse *f* avec électrons de basse vitesse
~VISCOSITY - [rubber ind etc] faible viscosité
~-VOLATILE BITUMINOUS COAL - s. low-volatile steam coal
~-VOLATILE STEAM COAL - [min] charbon *m* demi-gras, houille *f* demi-grasse
~-VOLATILITY - [chem] (term applied to a liquid which is not readily volatile at normal temperatures) faible volatilité *f*
~-VOLT RELEASE - [el] (an undervoltage release) déclenchement *m* à tension minimale
~VOLTAGE - [el] (any voltage which does not exceed 250 volts) basse tension *f*
~-VOLTAGE CURRENT - [el] courant *m* à basse tension
~-VOLTAGE INDUCTION FURNACE - [metall] four *m* à induction à basse tension
~-VOLTAGE RELEASE - s. low-volt release
~-VOLUME MIST BLOWER - [agric] pulvérisateur-atomiseur *m*
~WARP - [text] (of a woven fabric, a small number of threads in the warp) basse-lisse *f*
~WARP LOOM - [text ind] métier *m* à basse-lisse
~WATER - s. low tide
~-WATER ALARM - [mech] (in a boiler, an arrangement to warn that the water-level is too low) indicateur *m* à sifflet d'alarme
~-WATER VALVE - [mech] (boiler safety valve) soupape *f* de sûreté pour niveau d'eau trop bas
LOW-WING - [aero] à aile basse
~-WING MONOPLANE - [aero] (type of monoplane having the wings at or near the lower part of the fuselage) monoplan *m* à aile basse
LOWER, to - [gen] baisser, abaisser, s'abaisser
[naut] (the sails) amener, caler
[gen] (a load) descendre
[auto] (e.g. the chassis) surbaisser
LOWER AQUIFER - [hydr] nappe *f* inférieure, deuxième nappe *f*
~BEAM - [constr] membrure *f* inférieure
~BOX - [metall] châssis *m* inférieur
~BRACKET - [mech] (of a machine) support *m* inférieur
~BRUSH - [comput] (metal brush reading information from punched cards) balai *m* de lecture
~CAMBER - [aero] (the amount by which the lower section of the curve of an aerofoil rised along the span) ventre *m* de la courbure
~CASE - [print] (the type case in which the small letters are kept) bas-de-casse *m*
~-CASE LETTER - s. lower letter
~COAL-MEASURES - [mining] terrain *m* carbonifère inférieur
~COURSE OF STONES - [constr] empierrement *m*
~COVERING WARP - [text] chaîne *f* de recouvrement inférieure
~CRANKCASE - [auto] (the oil sump) carter *m* inférieur
~CRITICAL VELOCITY - [hydr] (the critical velocity of change from eddy to viscous flow) valeur *f* minimum de la vitesse limite
~CULMINATION - [astr] (the lowest altitude of a heavenly body as it crosses the meridian) culmination *f* inférieure, passage *m* inférieur
~CURTATE - [comput] (of a punch card) zone *f* inférieure
~DEAD CENTRE - [mech] (of bottom dead centre; the dead centre at which the piston is furthest from the cylinder head) point *m* mort bas
~DECK - [naut] (the deck below the weather deck) pont *m* inférieur, premier pont *m*
~-DECK BEAM - [shipbuild] barrot *m* de pont inférieur
~GATES - [hydr] tête *f* d'aval
~LETTER - [print] miniscule *f*
~LIMB - [mining] flanc *m* inférieur
~LIMIT - [math] limite *f* inférieure
~MEAN HEMISPHERICAL CANDLE-POWER - [light] valeur *f* limite de l'intensité hemisphérique moyenne
~PILE THREAD - [text] fil *m* de bouclage inférieur
~PITCH LIMIT - [acoust] (the minimum frequency producing a sensation of pitch) limite *f* inférieure de tonalité
~PLATEN - [mech] (the bottom platen of a press) plateau *m* inférieur
~PUNCH - [metall] estampe *f* inférieure
~RADDLE PLATE - [text] jumelle *f* inférieure du vautoir
~REACHES - [geogr] (of a rivet) cours *m* inférieur
~SHED - [text] pas *m* inférieur, pas *m* de rabat
~TRANSIT - s. lower culmination
~WARP FRAME - [text] support *m* de chaîne inférieur
~WEFT - [text] trame *f* inférieure
LOWERED SIDE - [geol] lèvre *f* inférieure
LOWERING - [gen] descente *f*, abaissement *m*, diminution *f*

LOWERING - [photo] dépression *f* (de la sensibilité)
~HEALD - [text] maille *f* de rabat
~-IN - [gas & oil ind] (pipe lines) mise *f* en fouille, descente *f* en fouille
~OF CONE - [metall] descente *f* du cône
~OF PRESSURE - [mech etc] réduction *f* de la pression
~OF PRICES - [comm] rabais *m*, diminution *f* des prix
~OF THE BEAM - [text] abaissement *m* de l'ensouple
~OF THE BINDING WARP - [text] abaissement *m* de la chaîne de liage
~OF THE WATER TABLE - [geol] rabattement *m* de la nappe aquifére
~OF TRACKS - [railw] (the levelling of railway tracks with the road surface) abaissement *m* des voies
~SHAFT - [text] lame *f* de rabat
~STAGE - [metall] chargeur *m* descendant
~THE REED - [text] baisse *f* du peigne
~THE VOLTAGE - [el] dévoltage *m*
~TWILLING BAR - [text] tringle *f* de rabat
LOWEST BOBBIN - [text] (la) bobine *f* plus basse
~CHARGE - [comm etc] (minimum charge) minimum *m* de perception
~COMMON MULTIPLE - [math] plus petit commun multiple *m*
~TERMS - [math] plus petits termes *mpl*
~USEFUL HIGH FREQUENCY - [radio] (LUF; the lowest frequency between 3 and 30 Mc/s which can be used for a specified service at a specified time) fréquence *f* minimum d'utilisation
LOWLAND - [geogr] plaine *f* basse, terre *f* en contre-bas
~BOG - [geol] tourbière *f* basse
~BREEDS - [agric] race *f* de plaine
~CATTLE - [agric] bétail *m* de plaine
LOX - [astronaut] (slang abbreviation for liquid oxygen) oxygène *m* liquide
LOXIC - [med] tordu, oblique
LOXING - [astronaut] remplissage *m* en oxygène liquide
LOXODROME - [nav] (aline which cuts successive meridians at the same angle) ligne *f* loxodromique
LOXODROMIC - [nav] loxodromique
~AIRWAY MAP - [aero] carte *f* aéronautique loxodromique
LOXOTOMY - [med] amputation *f* par section oblique
LOXYGEN - [astronaut] oxygène *m* liquide
LOZ - [astronaut] ozone *m* liquide
LOZENGE - [gen] (rhombus shape) losange *m*
[med] (medicated sweetmeat) pastille *f*, tablette *f*
[arch] (an ornament) losange *m*
L.P. - s. Low Pressure
LPG - [oil ind] gaz de pétrole liquéfié
L-SHELL - [nucl] (the electron shell second from the nucleus and surrounding the K-shell) couche *f* L
LST - s. local sidereal time
LUBBER'S HOLE - [naut] trou *m* du chat
~LINE - [aero] (reference line in a compass bowl to show the heading of the aircraft) ligne *f* foi
~POINT - s. lubber's line
LUBOIL - [oil ind] (lubricating oil) huile *f* de graissage, huile *f* de lubrification
LUBRICANT - [mech] (any substance used to reduce friction in moving parts) lubrifiant *m*
~-BLOOM - [plast ind] (cloudy exudation on the surface of plastic material) exsudat *m*
LUBRICANT EXUDATION - [plast ind] (migration of lubricant to the surface of a formed article) exsudation *f* de lubrifiant
LUBRICATE, to - [mech] lubrifier, graisser
LUBRICATING - [mech] lubrifiant
~AGENT - [mech] lubrifiant *m*
~COMPRESSOR - [mech] compresseur *m* de graissage
~GREASES - [ind chem] (mixtures of metallic soaps with mineral and/or vegetable oils) graisses *fpl* lubrifiantes
~OIL - [mech] (oil used to reduce friction and dissipate heat) huile *f* lubrifiante, huile *f* de graissage
~OIL COOLER - [mech] (for machines) huile *f* lubrifiante réfrigérante
~OIL DELIVERY - [mech] alimentation *f* de l'huile de graissage
~OIL PURIFIER - [ind chem] purificateur *m* de l'huile lubrifiante
~POWER - [mech] pouvoir *m* lubrifiant
~PROPERTY - [ind chem] qualité *f* lubrifiante
~QUALITY - s. lubricating power
LUBRICATION - [mech] lubrification *f*, graissage *m*
~CHART - [mech] (table of instructions for lubrication) tableau *m* de graissage
~DIAGRAM - [mech] (diagram showing points which require lubrication) schéma *m* de graissage
~NIPPLE - [mech] graisseur *m*
~SYSTEM - [mech] (the whole of the arrangements for supplying lubricant to the moving parts of the machine) système *m* de graissage
LUBRICATOR - [mech] graisseur *m*
[oil ind] chambre-écluse *f*
LUBRICITY - [chem] onctuosité *f*, lubricité *f*
LUCANTHROPY - [med] lycanthropie *f*
LUCERN - [bot] luzerne *f*
LUCID - [gen] brillant, lumineux
[bot] luisant
[astr] visible à l'oeil nu
LUCIDIFICATION - [physiol] éclaircissement *m*
LUCIDITY - [gen] luminosité *f*, transparence *f*, lucidité *f*
LUCIFER - [astr] (the morning star; Venus when she appears before sunrise) Lucifer *m*, Vénus *f*
LUCIFERIN - [zool] (protein-like substance appearing in the luminous organs of some animals) luciférine *f*
LUES - [med] (any kind of plague) infection *f* luétique
LUETIC - [med] luétique
LUFF, to - [mech] (the jib) transborder par la volée
[naut] (to steer nearer the wind) lofer
LUFF - [naut] (the part of the ship towards the wind) lof *m*, ralingue *f*
[naut] (the manoeuvre bringing the ship nearer to the wind) chute *f* avant
LUFFER BOARD - [acoust] (a device designed to direct the sound emitted by bells) abat-son *m*
LUFFING-JIB CRANE - [mech mach] (crane in which the jib is so hinged as to allow for alteration in its radius of action) grue *f* à volée variable
LUG - [mech] (a projection formed on a mechanical part for the attachment of another part or unit) oreille *f*, ergot *m*, languette *f*
[el] (on an accumulator plate) cosse *f*, attache *f* de conducteur
[plumb] (a type of connexion for tubes) saillie *f*,

point *m* d'attache
LUG - [metall] (in foundry work) (ear) tasseau *m* d'une pièce venue de fonte
[metall] (welding) (small piece of metal for soldering purposes in capacitors etc) serre-fil *m*, étrier *m*
~BASE TYRE - [rubber ind] (off-the-road tyre, cross-country tyre) pneu *m* tous terrains
~OF INSULATOR - [el] patte *f* de la console
~-PLATE - [el] queue *f* conductrice
~STRAP - [text] lanière *f* à taquet, bride *f* de chasse-navette
LUGGAGE - [gen] bagage *m*
~BOOT - [auto] coffre *m* à bagages
~CARRIER - s. luggage boot
~GRID - [auto] grille *f* porte-bagage
~NET - [railw] filet *m* à bagages
~RACK - [auto] (rack or net for light parcels and the like) porte-bagage *m*
[railw] filet *m* à bagages
~VAN - [railw] fourgon *m* à bagages
LUGGER - [naut] (vessel carrying a lugsail) lougre *m*
LUGS - [mining] molettes *f*pl, taquets *m*pl
LUGSAIL - [naut] (four-cornered sail) voile *f* à bourcet
LUKEWARM - [gen] (at about the temperature of the human body) tiède
LULL - [gen] moment *m* de calme
[met] (in the wind etc) bonace *f*
LUMBAGO - [med] (rheumatic affection affecting the lower part of the back) lumbago *m*
LUMBAR - [anat] (near the lower or posterior part of the back) lombaire
~ARTERIES - [anat] artères *f*pl lombaires
~REGION - [anat] région *f* lombaire
LUMBARIZATION - [med] lombalisation *f*, sacrum *m* lombalisé
LUMBER - [timber] (sawn wood) bois *m* de charpente, bois *m* de construction
LUMBERJACK - [gen] bûcheron *m*
LUMBERMAN - s. lumberjack
LUMBOSACRAL - [anat] lombosacré
LUMEN - [light] (the unit of luminosity) lumen *m*
[med] (in surgery) passage *m*, ouverture *f*
~-HOUR - [meas] (unit of quantity of light) lumen-heure *m*
LUMENMETER - [instr] (instrument measuring the intensity of light) lumenmètre *m*
LUMINAIRE - [light] (US term denoting light fittings) luminaire *m*, corps *m* lumineux
LUMINANCE - [opt] (the quantitative attribute of light) luminance *f*
[phys] (photometric brightness) luminance *f*
~BAND - [opt] bande *f* de signal de luminance
~CHANNEL - [telev] (in colour television) canal *m* de luminance
~DISTORTION - [telev] distorsion *f* de luminance
~FLICKER - [telev] (in colour television) papillotement *m* de luminance
~NOTCH - [telev] encoche *f* de luminance
~PRIMARY - [telev] (in colour television) primaire *f* prépondérante
~SCALE - [opt] échelle *f* de luminance
~TEMPERATURE - [opt] (for a source of radiation) température *f* de luminance monochromatique
~VALUE - [opt] valeur *f* de luminance
LUMINESCENCE - [phys] (the emission of light by a substance without appreciable rise in temperature, e.g. as the effect of ultra-violet radiation) luminescence *f*
LUMINESCENCE THRESHOLD - [electron] (the minimum frequency of radiation which can excite a luminescent material) seuil *m* de luminescence
LUMINESCENT - [phys etc] luminescent
~CENTRE - [nucl] (an atom or a group of atoms which can produce luminescence when suitably excited) centre *m* luminogène
~DIGITAL INDICATOR - [comput] (99 dot electroluminescent indicator block) indicateur *m* numérique luminescent
~PIGMENTS - [dyes] (pigments which are activated by ultra-violet radiation and produce strong luminescence) pigments *m*pl luminescents
~POINTER - [instr] aiguille *f* luminescente
~SCREEN - [electron] (screen giving a luminous spot) écran *m* luminescent
~-SCREEN TUBE - [electron] (cathode-ray tube in which the image on the screen is more luminous than the background) tube *m* à écran luminescent
LUMINOPHORE - [phys] (molecule emitting electrons when illuminated) électronogène
LUMINOSITY - [light] (brightness) brillance *f*, luminosité *f*
[astr] (the amount of light emitted by a star; absolute magnitude) intensité *f* lumineuse, luminosité *f*
~COEFFICIENTS - [opt] (the constant multipliers for the respective tristimulus values of a colour) facteurs *m*pl d'efficacité lumineuse
~CURVE - [opt] (graphical relationship between wavelength of light in the visible range and the reciprocal of the radiance required to produce an equally bright visual sensation) courbe *f* de luminosité
~FACTOR - [light] (the ratio between the total luminous flux and the total energy emitted) efficacité *f* lumineuse
LUMINOUS - [gen] lumineux
~BEAM - [light] faisceau *m* lumineux
~DENSITY - [opt] (the luminous energy which is found in a volume unit of space) densité *f* lumineuse
~DIAL - [instr] (dial with figures, scale and pointer coated with self-luminous material) cadran *m* lumineux
~EDGE - [telev] (reflection-reducing light filter) filtre *m* optique encadrant
~EFFICIENCY - [el] (the quotient of the luminous flux in lumes by the total energy in watts) efficacité *f* lumineuse
[opt] (a measure of luminosity per unit of radiant power; the visibility factor) coefficient *m* de visibilité, efficacité *f* lumineuse
~EMITTANCE - [phys] luminance *f*, brillance *f* lumineuse
~FLAME - [th eng] (diffusion flame) flamme *f* lumineuse, flamme *f* de diffusion
~-FLAME BURNER - [th eng] (non-aerated burner) brûleur *m* à flamme de diffusion
~FLOW LINE DIAGRAM - [comput] tableau *m* synoptique lumineux
~FLUX - [light] (the amount of light emitted by a source) flux *m* lumineux
~HAND - [instr & horol] aiguille *f* phosphorescente
~INTENSITY - [light] (the degree of brilliance of a light source) intensité *f* lumineuse
~PAINT - [paint] (a paint containing sulphides of cal-

cium, barium and strontium, which glows after being exposed to light) peinture *f* luminescente
LUMINOUS PANEL - [photo] panneau *m* lumineux
~PAPER - [paper man] papier *m* phosphorescent
~REFLECTANCE - [opt] (the ratio between luminous emittance and illuminance of a reflecting surface) degré *m* de réflexion
~SENSITIVITY - [telev] (of a camera tube) sensibilité *f* lumineuse
[electron] (the quotient of the signal current developed by a camera tube, by the incident luminous flux emitted by an unfiltered incandescent source at a colour temperature of 2,854° k) photosensibilité *f*
~SIGNAL - [signal] signal *m* lumineux, phare *m* lumineux
~SOURCE - [opt] source *f* lumineuse
LUMP , to- [gen] mettre en bloc, mettre en tas
[el] (of electrical constants) concentrer
[geol] (of electrical constants) concentrer
[geol] former des mottes
LUMP - [gen] bloc *m*, motte *f*, motte *f* d'argile
[gen] (of pasty stuff) pâton *m*
[gen] (a projection) bosse *f*, excroissance *f*
[min] (of ore) gros *mpl*
[of fibres] motte *f* (de coton etc)
[metall] (a bloom of malleable iron) bloom *m*
[gen] (of sugar) morceau *m*
[nucl] (piece of uranium metal used a fuel element in a reactor) bloc *m*
~BREAKER - [metall] casse-gueuse *m*
~-FUEL - [mining] combustible *m* en morceaux
~ORE - [min] minerai *m* gros, minerai *m* en morceaux
~SCRAP - [rubber ind] (dried lumps) "lump scrap" *m*, grumeaux *mpl* desséchés
~-SORTING GRADING EQUIPMENT - [min] équipement *m* de triage caillou par caillou
~SUGAR - [sugar ind] sucre *m* en morceaux
~SUM - [comm etc] (agreed amount covering a number of items) forfait *m*
LUMPED CONSTANT - [el] (or concentrated constant; so called when its dimensions are small if compared to the wavelength propagation of currents in it) constante *f* localisée
~PARAMETER SYSTEM - [math] système *m* à paramètres localisés
~VOLTAGE - [electron] (a fictitious voltage) tension *f* localisée
LUMPER - [gen] (labourer employed in loading and unloading cargo, especially timber) déchargeur *m*, débardeur *m*
LUMPS - [paper man] (a defect caused by lumps in the pulp) pâton *m*
[rubber ind] (coagulated latex clods) grumeaux *mpl* de latex coagulés spontanément
LUMPY - [gen] grumeleux
[paper man] chantonné
LUNAR - [astr] (relating to the moon) lunaire
(of silver or containing silver) d'argent
~ATMOSPHERIC TIDE - [astr] marée *f* atmosphérique lunaire
~CRATER - [astr] cratère *m* lunaire
~DAY - [astr] jour *m* lunaire
~DISTANCES - [astr] (an ancient method of determining the ship's longitude by comparing the angular distance of the moon from a star at the given local time and the angular distance at a specified Greenwhich time) distance *f* angulaire entre la lune et un corps céleste
LUNAR MONTH - s. lunation
~ORBIT - [astronaut] orbite *f* circumlunaire
~PARALLAX - [astr] parallaxe *f* de la lune
~PROBE - [astr] sonde *f* lunaire
LUNATE - [gen] luné, en forme de croissant
LUNATIC - [med] d'aliéné
LUNATION - [astr] (synodyc period) lunaison *f*
LUNETTE - [arch] (small arched opening) lunette *f*
LUNG - [anat] poumon *m*
~MOTOR - [med] poumon *m* d'acier
~PLAGUE - [vet] péripneumonie *f* bovine
LUNGE NITROMETER - [instr] (an apparatus used to estimate oxides of nitrogen, and in other gas analyses) nitromètre *m* Lunge
LUNGWORM DISEASE - [vet] bronchite *f* vermineuse
LUNGWORT - [med plants] pulmonaire *f*
LUPANINE - [chem] (a poisonous alkaloid derived from varieties of Lupinus) lupanine *f*
LUPIN - [bot] lupin *m*
LUPINE - s. lupin
LUPOID - [med] lupoide
LUPOMA - [med] lupome *m*, tubercule *m* lupique
LUPULIN - [chem] (an aromatic bitter prepared from Humulus lupulus and used in medicine and brewing) lupuline *f*
~CONTENT - [chem] teneur *f* en lupuline
LURCH, to - [gen] embarder, marcher en titubant
[naut] (a sudden roll) faire une embardée
[auto] embarder
LURCH - [gen & naut] coup *m* de roulis, embardée *f*
LUSTERED LINEN YARN - [text] fil *m* de lin lustré
LUSTRACELLULOSE - [text] soie *f* artificielle au cuivre
LUSTRE - [gen, light etc] (term used to describe the quality and amount of light reflected by a body) brillant *m*, éclat *m*, lustre *m*
[min] (brilliance, gloss, of a surface) éclat
~WOOL YARN - [text] fil *m* de laine brillante
LUSTRELESS - [gen] (lacking in lustre, matt.) mat, terne, sans éclat
LUSTRING - [gen & text] (the operation of polishing threads or fibres) lustrage *m*, catissage *m*
~MACHINE - [text] machine *f* à catisser
LUSTROUS - [gen] (having lustre) lustré, brillant, éclatant
[text] satiné, brillant
~YARN - [text] fil *m* brillant
LUTE, to - [constr] (the levelling off of clay with a lute) luter
[th eng] (ovens) (the operation of sealing the tapping hole) boucher, luter
[metall] brasquer
LUTE - [impl] (a straightedge used to level off clay in brick mould) lut *m*
[ind chem] mastic
[metall] brasque *f*
[mus] (string instrument) luth *m*
[agric] mastic *m* à greffer
~A MOULD, to - [metall] luter un mou
LUTECIUM - [chem] (a rare earth element, symbol Lu, A.N. 71, A.W. 175,0, used in nuclear technology) lutécium *m*
LUTEIN - [chem] (a carotenoid present in animal fats) lutéine *f*
LUTEINIZATION - [med] lutéinisation *f*
LUTELESS JOINT - [mech] (metal-to-metal joint)

jonction *f* à sec
LUTEOMA - [med] lutéinome *m*
LUTHERN - [constr] (vertical window in a roof) lucarne *f*
LUTIDINE - [chem] (a derivative of pyridine with uses in the production of rubber chemicals, insecticides, drugs and dyestuffs) lutidine *f*
LUTING - [ceramics] (the operation of attaching preformed decoration to pottery by the use of liquid clay slip before firing) lutation *f*
[ind chem] (sealing of vessels or the like with clay, cement or other substances) applicage *m*
[mech] (of doors, a type of sealing) lutation *f*, lutage *m*
LUX - [light] (unit of illumination) lux *m*
LUXATION - [med] (dislocation) luxation *f*, déboîtement *m*
LUXMETER - [instr] (a type of portable photometer) luxmètre *m*
LUXURIANT - [gen] luxuriant, exhubérant
~GROWTH - [bot agric etc] végétation *f* luxuriante
LYCOMANIA - [med] lycanthropie *f*
LYCOPOD - [bot] lycopode *m*
LYCOPODIUM - [chem] (the spores of Lycopodium clavatum used as a dusting powder and diluent in medicine) poudre *f* de lycopode, soufre *m* végétal
LYDIAN STONE - [min] basanite *f*, jaspe *m* noir
LYE - [chem] (strong alkaline solution) lessive *f*
~-ASHES - [agric etc] charrée *f*
~BOILER - [ind chem] chaudière *f* à lessive
~-HOSE - [ind chem] tuyau *m* pour lessives alcalines
~-KEY - [mining] clef *f* de retenue
~-RESISTANT - [rubber ind] résistant aux lessives, résistant à la soude caustique
~-WATER - [ind chem] eau *f* seconde, lessive *f* faible
LYING - [gen] situé, placé
[ind chem] (to treat with lye) lessivage *m*
~-IN - [med] couches *f*pl
~PRESS - [mech mach] (or laying press; small portable press used for bookbinding) presse *f* pour relieurs
~SIDE - [mining] mur *m*, sol *m*
LYME-GRASS (UPRIGHT SEA) - [agric] élyme *m* des sables
LYMPH - [biol] (the fluid which circulates in the lymphatic vessels of vertebrates) lymphe *f*
[med] vaccin *m*
~GLAND - [anat] (aggregation of connecting tissue filled with lymphocytes) ganglion *m* lymphatique
~NODE - s. lymph gland
LYMPHADENECTOMY - [med] adénectomie *f*
LYMPHADENOCYST - [med] limphadénokyste *m*
LYMPHADENOGRAM - [med] adénogramme *m*
LYMPHADENOMA - s. lymphoma
LYMPHADENOPATHY - [med] lymphadénopathie *f*
LYMPHADENOSIS - [med] lymphadénose *f*
LYMPHANGIOMA - [med] lymphangiome *m*
LYMPHATIC CIRCULATION - [anat] circulation *f* lymphatique
~CONSTITUTION - s. lymphatism
~SARCOMA - s. lymphoma
~SYSTEM - [anat] (the system of lymphatic vessels pervading the body) système *m* lymphatique
LYMPHATIC VESSELS - [zool] vaisseaux *m*pl lymphatiques
LYMPHATISM - [med] lymphatisme *m*, constitution *f* limphatique
LYMPHEDEMA - [med] lymphoedème *m*
LYMPHENDOTHELIOMA - [med] limphangio-endothéliome *m*
LYMPHOBLAST - [biol] limphoblaste *m*, macrolymphocyte *m*
LYMPHOBLASTEMIA - [med limphoblastose *f*
LYMPHOBLASTOMA - [med] limphoblastome *m*
LYMPHOBLASTOSIS - s. lymphoblastemia
LYMPHOCOELE - [med] limphocèle *f*
LYMPHOCYST - s. lymphocoele
LYMPHOCYSTOBLAST - s. lymphoblast
LYMPHOCYTE - [biol] lymphocyte *m*
LYMPHOCYTHEMIA - [med] limphocytose *f*
LYMPHOCYTOMA - [med] lymphocytome *m*
LYMPHOCYTOSIS - [med] s. lymphocythemia
LYMPHODERMIA - [med] lymphodermie *f*
LYMPHOMA - [med] lymphome *m*
LYMPHOMYXOMA - [med] myxome *m* adénoïdien
LYMPHOPENIA - [med] (reduction of lymphocytes in the blood) lymphocytopénie *f*
LYMPHURIA - [med] lymphurie *f*
LYNDOCHITE - [min] (uranium which contains calcium-thorium-euxenite) lyndochite *f*
LYOLYSIS - [chem] (the process by which a salt interacts with a solvent to form an acid and a base) lyolise *f*
LYOPHIL - s. lyophilic
LYOPHILIC - [chem] (term used for substances which readily for colloïdal solutions) lyophile, aisément soluble
~SOL - [chem] (a colloïdal system) sol *m* lyophile
~SYSTEM - [chem] (colloïdal system in which there is a considerable attraction between the dispersed phase and the dispersion medium) système *m* lyophile
LYOPHILIZATION - [med] lyophilisation *f*
LYOPHOBIC - [chem] (term used for substances which form colloidal solutions with difficulty) lyophobe
~DISPERSION - [chem] dispersion *f* lyophobe
~SOL - [chem] (colloïdal system with little mutual affinity between dispersed phase and dispersion medium) sol *m* colloïdal lyophobe
~SYSTEM - [chem] (colloïdal system of the irreversible type) système *m* lyophobe
LYOTROPIC SERIES - [chem] (a series of anions or cations arranged in the order of their effect on reactions in colloïdal solutions) série *f* lyotropique
LYPEMANIA - [med] mélancolie *f*
LYRE - [mus] (a pluckes string instrument) lyre *f*
LYSERGIC ACID DIETHYLDIAMIDE (LSD) - [chem] amide *m* de l'acide lysergique, LSD
LYSINE - [chem] (a diaminomonocarboxylic acid with uses in food processing, the manufacture of pharmaceuticals and in biochemical research) lysine *f*
LYSIS - [biol] lyse *f*, dissolution *f* par une lysine (de une cellule)
LYSOCHINASE - [chem] lysokinase *f*
LYSOZYME - [chem] (a bacteria-dissolving enzyme) lysozime *m*
LYSSA - [med] rage *f*, hydrophobie *f*
LYTIC - [chem] lytique

M

M - [chem] (symbol for metal) métal *m*
[chem] (symbol for electropositive radical) (symbole de) radical *m* électropositif
[chem] (symbol for molar concentration) (symbole de) concentration *f* moléculaire
[chem] (in italics) (symbol for molecular weight) (symbole de) poids *m* moléculaire
Ma - [chem] (symbol for masurium) (symbole de) technécium *m*
MACADAM - [constr] (layer of broken stones of approximately uniform size) macadam *m*
MACADAMIZATION - [constr] macadamisation *f*
MACADAMIZE, to - [constr] (the operation of making a road according to macadam's system) macademiser, empierrer
[constr] (to level a road) macadamiser, niveler
[constr] (to make a metalled road) asphalter
MACADAMIZED - [constr] macadamisé
[gen] asphalté
~ROAD - [constr] macadam *m*
MACARONI - [food] (a paste made of wheat, of Italian origin) macaroni *m*
~PIPE - [oil ind] macaroni tube *m*
MACE - [gen] masse *f*
[bot] (the dried outer covering of the nutmeg) macis *m*, fleur *f* de muscade
~OIL - [chem] (an essential oil derived from Myristica fragrans and used in perfumery and the production of flavourants) huile *f* de muscade
MACERALS - [min] macérés *mpl*
MACERATE, to - [ind chem] (to soften by immersion in a liquid agent) faire macérer, lessiver, faire digérer, tremper
MACERATED - [ind chem] (softened through immersion in a liquid agent) macéré
~FABRIC - [ind chem] (filling material produced by cutting clean cotton cloth into small pieces) rognures *fpl* de tissu
~VENEER FILLED MOULDING COMPOUND - [plast ind] matière *f* à mouler avec copeaux de bois
MACERATION - [ind chem] (the process of softening a material by immersion in a liquid agent) macération *f*
[gen] (treatment of vegetable fibres) macération *f*, lessivage *m*
~OF FOETUS - [med] macération *f* du fœtus
MACERATOR - [mech] (used in rubber industry etc) première crêpeuse *f*
MACH - s. Mach number
~ANGLE - [aero] (the apex angle of a Mach cone) angle *m* de Mach
~CONE - [aero] (the wavefront of a Mach wave) cône *m* de Mach
MACH CRITERION - [phys] (principle stating that only those propositions should be admitted in physical theory from which statements about observable phenomena can be deduced) critère *m* de Mach
~CRITICAL PRESSURE NUMBER - [aero] (the value of the free stream Mach number at which the peak level of the velocity on the surface of a body reaches the local speed of sound) nombre *m* de Mach de pression critique
~FRONT - [phys] (the shock front which is formed by the fusion of the incident and reflected shock fronts from an explosion) front *m* de Mach
~INDICATOR - s. machmeter
~METER - s. machmeter
~NUMBER - [phys] (a number expressing the relation between the speed of a fluid and the local speed of sound in it, especially in regard to the speed of an aircraft) nombre *m* de Mach
~PRINCIPLE - [phys] (the inertia of a system is borne of the interaction of that system and the rest of the universe) principe *m* de Mach
~REGION - [phys] (the region on the surface at which the Mach front has formed following an explosion in the air) région *f* de Mach
~STEM - s. Mach front
~WAVE - [phys] (the shock wave arising from an object travelling at a Mach-plus speed) onde *f* de Mach
MACHE UNIT - [nucl] (unit of quantity of radioactive emanation) unité *f* de Mache
MACHICOLATIONS - [build] (openings between the corbels of a parapet, used in ancient castles) mâchicoulis *m*
MACHINABILITY - [mech] usinabilité *f*
MACHINABLE - [mech] usinable
~CAST IRON - [metall] fonte *f* douce
~CASTING - [metall] (casting made of machinable steel) pièce *f* de fonderie usinable
MACHINE, to - [mech] (to form by means of a machine tool) usiner, façonner, ajuster, travailler
MACHINE - [mech] (any meccanical device designed to act upon a resistance at one point by applying a measure of force at another point) machine *f*
~ACCOUNTING - [comput] comptabilité *f* mécanique
~AUGER - [mech] tarière *f* pour forage mécanique
~AVAILABLE TIME - [ind proc] (time during which a machine is available) temps *m* disponible d'une machine
[comput] (the time during which a computer has the power turned on) temps *m* machine
~BEAMER - [text] enrouleur *m* à la machine
~BED - [mech] fonture *f* de la machine
~BROKE - [paper man] (waste paper) déchets *mpl* de

fabrication
MACHINE CAST - [metall] coulé à la machine
~CASTING - [metall] moulage *m* pour pièces de machine
~CHEST - [paper man] cuve *f* d'alimentation
~COAL-MINING - [min] (a mechanized mining of coal) abattage *m* mécanique du charbon
~-COATED PAPER - [paper man] papier *m* glacé à la machine
~COMPOSITION - [print] composition *f* mécanique
~COMPOSITOR - [print] claviste *m*
~CONTROLS - [mech] commandes *fpl* de machine
~CORING - [metall] noyautage *m* mécanique
~-CUT TEETH - [mech] dents *fpl* taillées à la machine
~CUTTING - [metall] découpage *m* à la machine
~CYCLE - [comput] (the smallest complete process of action which repeats itself in order) cycle *m* de machine
~DIRECTION - [paper man] (of the paper fibres) sens *m* du papier
~-DIVIDED GEARS - [mech] engrenages *mpl* divisés à la machine
~-DOWN TIME - [gen] temps *m* de non-opération d'une machine
~-DRIED - [paper man] séché à la machine
~DRILL - [mining] marteau *m* perforateur
~DRILLER - [mining] ouvrier *m* sondeur
~DRILLING - [mech] perforation *f* mécanique, forage *m* mécanique
~DRIVE - [mech] commande *f* mécanique, entraînement *m* mécanique
~-DRIVEN - [mech] commandé mecaniquement
~DRYING - [text] séchage *m* mécanique
~EQUATION - [comput] (equation written out in a form acceptable to the computer) équation *f* de calculatrice, équation *f* de machine
~-FLACED FLANGE - [mech] bride *f* usinée
~-FINISHED - [paper man] (printing paper) (papier) apprêté dans la machine
[mech] finissage *m* à la machine
~FLUSH, to - [mech] (to finish to a uniformly smooth surface after assembly, by the use of suitable machine-tools, e.g. in the case of built-up external elements of high-speed aircraft) araser
~FOOT - [mech] (a stand) pied *m* de machine, socle *m* de machine
~FOR GRINDING BUSHES - [mach tool] rectifieuse *f* pour coussinets
~FOR MILLING GROOVES - [mach tool] fraiseuse *f* pour rainures
~FOR SPLITTING AND CUTTING TYRES - [rubber ind] machine *f* à détalonner et à découper les pneus
~FOR TAKING OFF THE BURR - [rubber ind] machine *f* à ébarber
~FOR TWISTING SILK - [text] machine *f* à confectionner la soie
~FOUNDATION - [mech] massif *m* de fondation d'une machine
~FOUNDATION PLATE - [mech] plaque *f* d'assise d'une machine
~FRAME - [mech] bâti *m* de machine
~-FUSE - [mining] amorce *f* à étincelle
~-GLAZED FINISH - [rubber ind] finissage *m* par lissage mécanique
~GREASER - [mech] graisseur *m*
~GRINDING - [mech] meulage *m* à la machine

MACHINE GUARD - [mech] (for protection purposes) volet *m*
~-GUN - [firearms] (an automatic weapon) mitrailleuse *f*
~HANDWHEEL - [mech] volant *m* de machine
~HEAD - [mech] tête du métier, têtière *f* de machine
~HOLDING - [mining] perforation *f* mécanique
~INSTRUCTIONS - [comput] instruction machine *f*
~INTERFERENCE - [radio] (the interference which is caused by electrical machines) perturbation *f* due aux machines électriques
~KEY RINGING - [telecomm] (in telephony) appel semi-automatique
~LANGUAGE - [comput] (information prepared in the form which the computer can handle) langage *m* machine
~LEG - s. machine foot
~LOAD - [mech] (the quantity of work carried out on a machine tool) charge *f* machine
[comput] charge *f* d'une machine
~LOADING SCHEDULE - [comput] plan *m* de charge d'une machine
~MADE - [chen] à la machine, à la mécanique
~-MADE BRICK - [build] brique *f* faite à la machine
~-MADE PAPER - s. machine paper
~MEMBER - [mech] organe *m* de machine
~MILKING - [agric] traite *f* mécanique
~MINDER - [gen] surveillant *m*
[print] conducteur *m* de presse
~MINING - [min] (mechanized mining) abattage *m* mécanique
~-MIXING - [ind proc] mélange *m* à la machine
~MOULDING - [metall] (the operation of making moulds by mechanical means) moulage *m* à la machine
~OIL - [gen] huile *f* à mécanisme
~OPERATION - [gen] opération *f* machine
[comput] (the operation required to obtain the result of setting the machine into action) fonction *f*, opération *f* machine
~PAPER - [paper man] (continuous web of paper made on a machine) papier *m* continu
~PLANE IRON - [mach tool] fer *m* de raboteuse
~POSITIONING PRECISION - [contr] précision *f* de positionnement d'une machine
~PROGRAM - [comput] programme-machine *m*
~PROGRAMMING - [comput] programmation *f*
~PUDDLING - [metall] puddlage *m* mécanique
~-PUNCH - [mech] poinçon *m* à la machine
~RATING - [mech] taux *m* de rendement d'une machine
~REAMER - [mach tool] alésoir *m* mécanique
~RINGING - [telecomm] (in telephony) appel *m* automatique, courant *m* d'appel cadencé
~RIVETING - [mech] rivetage *m* mécanique
~ROOM - [gen] atelier *m* des machines
[print] salle *f* des presses
[cin] (the room in which sound recording is made) cabine *f* d'enregistrement
~RUNNING UNDER LOAD - [mech] machine *f* marchant chargé
~SCREW - [mech] (parallel thread screw used in mechanical devices) vis *f* mécanique, vis *f* à métaux
~SET-UP TIME - [mech] temps *m* de positionnement
~SHEARING - [agric] tonte *f* mécanique

MACHINE SHED - [constr] hangar *m* des machines
~SHOP - [gen] atelier *m* de construction mécanique, atelier *m* d'usinage
~SIDE - [mech] (particularly of ovens, retorts etc; also called pusher side, ram side) côté *m* machine
~SIZER - [text] pareur *m* de chaînes, encolleur *m* à la machine
~SIZING - [paper man] surfaçage *m*
[text] encollage *m* à la mécanique
~SPUN YARN - [text] fil *m* mécanique
~STEEL - [metall] acier *m* à outils
~STITCHING - [shoe man etc] piquage *m* à la machine
~STRAIGHTENING - [metall] redressage *m* à la machine
~TAP - [mach tool] taraud *m* à la machine, taraud *m* pour machines
~TENDER - [gen] soigneur *m* de machines
~TOOL - [mech] (a tool designed to be used in a machine, e.g. a lathe) machine-outil *f*
~UNIT - [comput] unité *f* machine, unité *f* de la grandeur de calcul
~VICE - [mech] étau-tiroir *m*
~WARPER - [text] ourdisseur *m* à la machine, warpeur *m*
~WINDING - [text] bobinage *m* mécanique
~WIRE - [paper man] (the wire web of a paper machine) toile *f* métallique
~WITH CLOSED-CIRCUIT VENTILATION - [el mach] (machine in which the heat is transferred to the cooling fluid through an intermediate ventilation medium which circulates in a closed circuit) machine *f* à ventilation en circuit fermé
~WITH INHERENT SELF EXCITATION - [el mach] (machine with a commutator, in which the principal flux is produced by one of the rotor circuits) machine *f* à excitation interne
~WITH NATURAL COOLING - [el mach] (machine without device for producing air movements which are thus only due to the rotation or to differences in temperatures) machine *f* à refroidissement automatique
~WITH OPEN-CIRCUIT VENTILATION - [el mach] (machine in which the heat is given up directly to the cooling fluid which is replaced continuously) machine *f* à ventilation en circuit ouvert
~WORD - [comput] (a unit of information) mot *m* machine
~WORK - [mech] usinage *m*, travail *m* à la machine
~YARN - [text] fil *m* mécanique
MACHINED - [mech] usiné, façonné
~ALL OVER - [mech] complètement façonné
~FLANGE - s. machine-faced flange
~ROD - [mech] tige *f* usinée
~SURFACE - [metall] surface *f* usinée à la machine
MACHINERY - [mech] (machine or machine parts taken together) mécanisme *m*, machinerie *f*
[phys etc] (the dynamics of events) roulage *m*
~SEATING - [gen] installation *f* machines
MACHINING - [mech] (forming or finishing by manipulation in a machine) usinage *m*, ajustage *m* mécanique
[print] (the operation of printing by machine) impression *f* à la machine, tirage *m* à la machine
[mach] travail *m* à la machine-outil
~ACCURACY - [mech] exactitude *f* d'usinage
~ALLOWANCE - [mech] surépaisseur *f* pour usinage
~PRECISION - [mech] précision *f* d'usinage
MACHINING PROPERTIES - [mech] (the characteristic behaviour of a material during shaping by machine tools) propriétés *fpl* d'usinage
~REPRODUCIBILITY - [mech] reproducibilité *f* d'usinage
~TIME - [mech] temps *m* d'usinage
MACHINIST - [gen] machiniste *m*, mécanicien *m*
(of machine tools) ajusteur *m* de machines-outils
MACHMETER - [instr] (an instrument showing the Mach number corresponding to the speed of the aircraft) machmètre *m*
MACIES - [med] dépérissement *m*
MACKEREL SKY - [met] (sky which is largely-covered with cirro-cumulus or alto-cumulus clouds, usually patterned like the back of a mackerel) ciel *m* moutonné, ciel *m* pommelé
MACKEY TEST - [paint] (a method of measuring the tendency of an oxidizable oil to ignite spontaneously) épreuve *f* de Mackey
MACKINTOSH - [rubber ind] imperméable *m*, mackintosh *m*
MACKINTOSHITE - [min] (mineral which contains thorium, uranium, cesium, lanthanium and yttrium) mackintoshite *f*
MACKLE - [print] (defective printed sheet with a burred appearance) bavochure *f*, frison *m*
MACLE - [cryst] (a twin crystal) macle *f*
[min] (a dark spot in some minerals) macle *f*
MACLED - [min] maclé
MACRENCEPHALIA - [med] macrencéphalie *f*
MACRO - [gen] (or macr; prefix conveying the meaning and sense of large) macro
~AXIS - [cryst] (the longer axis in triclinic and orthorhombic crystals) macroaxe *m*
MACROBIOTIA - [med] longévité *f*
MACROBIOTIC - [biol] macrobiotique
MACROBLEPHARIA - [med] hypertrophie *f* des paupières
MACROCEPHALIC - [med] (of an abnormally large head) macrocéphale
MACROCEPHALY - [med] (abnormal largeness of the head) macrocéphalie *f*
MACROCHEMISTRY - [chem] (the study of the chemical properties of matter in bulk) macrochimie *f*
MACROCOSMOS - [astr] macrocosme *m*
MACROCYST - [med] macrocyste *m*, macrokyste *m*
MACROCYTES - [biol] (large uninuclear leucocytes having great power of mobility) macrocytes *mpl*
MACRODACTYLIA - [med] macrodactylie *f*
MACRODONTIA - [med] macrodontie *f*
MACRODONTISM - s. macrodontia
MACRODYSTROPHIA - [med] gigantisme *m* partiel
MACROETCH - [metall] (macrosopic etching test) macroattaque *f* à l'acide
MACROFOSSIL - [geol] macrofossile *m*
MACROGAMETE - [zool] (the larger of a pair of conjugating gametes) macrogamète *m*
MACROGAMETOCYTE - [zool] macrogametocyte *m*
MACROGASTRIA - [med] dilatation *f* de l'estomac
MACROGLOSSIA - [med] macroglossie *f*
MACROGNATHIA - [med] macrognathie *f*
MACROGRAPHY - [phys] macrographie *f*
MACROMASTIA - [med] hypertrophie *f* mammaire
MACROMAZIA - s. macromastia
MACROMELIA - [med] macromélie *f*
MACROMETER - [instr] (an instrument designed to measure siwe and distance of objects by means of

two reflectors on a common sextant) macromètre *m*
MACROMOLECULAR DISPERSION - [chem] (synonym for colloid) dispersion *f* macromoléculaire
MACROMOLECULE - [chem] (a molecule containing very large numbers of atoms, i.e. a polymer molecule) macromolécule *f*
MACROMYELOBLAST - [med] macromyéloblaste *m*
MACRONUCLEAR REGENERATION - [genet] régénération *f* macronucléaire
MACRONUCLEUS - [phys] (a nutrition nucleus) macronucléus *m*
MACROPHOTOGRAPHY - [photo] (the operation of making large prints from a negative) agrandissement *m* photographique
MACROPHYRIC - [geol] (term describing igneous rocks which contain phenocrysts of more than 2 mm in length) macrophyrique
MACROPIA - [med] macropie *f*, macropsie *f*
MACROPLASIA - s. macrodystrophia
MACROPLASTIA - s. macrodystrophia
MACROPODIA - [med] macropodie *f*
MACROPROGRAM - [automat] (automation plan) plan *m* d'automatisation, macroprogrammation *f*
MACROPSIA - s. macropia
MACRORHINIA - [med] hypertrophie *f* du nez
MACROSCOPIC - [opt] (visible to the naked eye) macroscopique
~CROSS-SECTION - [nucl] (the combined cross-section for all atoms per unit mass of a material) section *f* efficace volumique, section *f* efficace macroscopique
~REMOVAL CROSS-SECTION - [nucl] section *f* macroscopique de déplacement
~STRESS - [metall] effort *m* macroscopique
MACROSCOPICAL SCALE - [meas] échalle *f* macroscopique
MACROSEGREGATION - [metall] macroségrégation *f*
MACROSPORE - [metall] (the crystal structure of a metal which is visible to the naked eye or with low magnification) macrostructure *f*
MACROTUBE - [photo] tube-raccord *m* pour macrophotographie
MACULA - [zool & bot] (or macule; spot of colour; also a small tubercle) macule *f*, tache *f*
[mech] tache *f*, plaque *f*, stigma *m*
MACULATE - [bot & anat] maculé
MACULOSE STRUCTURE - [min] structure *f* tachetée
MADAPOLAM - [text] (or Madapolam, the Indian town from where it originates, a type of cotton cloth) madapolam *m*
MADAROSIS - [med] madarose *f*
MADDER - [dye] (the root of Rubia tinctorum, formerly used as a source of alizarin) garance *f*
~EXTRACT - [dye] extrait *m* de garance
~LAKE - [paint] (a fast red pigment made from the root of the madder plant) laque *f* de garance
~VARNISH - [dye] garancine, alizarine *f*
MADE - [gen] (part of to make) fait, produit
~CIRCUIT - [el] circuit *m* fermé
~GROUND - [constr] (a ground which is obtained by filling in natural or artificial pits) remblai *m*
~IN SECTIONS - [mech etc] démontable
~-TO-MEASURE - [gen] sur mesure
~UP RIBBONS - [text] rubans *mpl* confectionnés
~-UP WOOD - [carp] bois *m* compensé
MADEFACTION - [med] humidification *f*
MADESCENT - [med] humescent
MADHOUSE - [med] (house for the retention of the insane) maison *f* de fous
MADIA (SATIVA) - [bot] madi *m*, madie *f*
MADNESS - [med] (insanity; mental disease) folie *f*, démence *f*
MAE WEST - [naut] (colloq. for life-jacket) ceinture *f* de sauvetage
MAFIC MINERAL - [min] minerai *m* ferromagnésien
MAGAMP REGULATION - [electron] réglage *m* par amplificateurs magnétiques
MAGAZINE - [gen] dépôt *m* de marchandises
[naut] (in ships) soute *f* à munitions
[firearms] magasin *m*
[photo] (the light-tight enclosure for films) (chambre) à magasin
[print] revue *f*, périodique *m*
[print] (of a linotype) porte-matrices *m*
[comput] magasin *m* d'alimentation
~ARC LAMP - [light] (type of electric arc lamp fitted with a number of carbons which are automatically brought into operation as the others burn away) lampe à arc à électrodes de charbon de réserve
~ELECTRIC SOLDERING IRON - [metall] fer à souder universel
~FEED - [firearms] (of a machinegun) charge *f* automatique
~FOR BOBBINS - [text] récipient *m* de bobines, caisse *f* à bobines
~GRINDER - [text] broyeuse *f* à alimentation à magasin
~GUN - [firearms] fusil *m* à répétition
~LOADING-TYPE LATHE - [mach tool] tour *m* à chargeur
~MINING - [mining] exploitation *f* par chambres-magasins
~PISTOL - [firearms] pistolet *m* à répétition
~SIDE - [text] côté *m* du dispositif de changement
~STOVE - [th eng] poêle *m* à charge automatique
~TOOL HOLDER - [mach tool] porte-outils *m* multiple
~TRAIN DESCRIBER - [instr] appareil *m* enregistreur d'annonce des trains
~VALVE - [cin] (the film opening in the magazine of a projector) fente *f* de chargeur
MAGENTA - [dye] (fuchsine. A synthetic red dye) fuchsine *f* acide
[adj.; the colour] magenta *m*
~CONTACT SCREEN - [photo] écran *m* magenta
~WAX TEST - [ind chem] (for cables) épreuve *f* de la cire
MAGGOT - [zool] (worm or grub) ver *m*, asticot *m*, ver *m* de viande
[zool] (the larva of the cheese fly) larve *f* apode
MAGGOTY - [of plants] plein de vers
~FRUIT - [agric] fruits *mpl* véreux
MAGIC BALANCE - [electron] (instrument for measuring the strength of a magnetic field) balance *f* magnétique
~EYE - [electron] indicateur *m* d'accord cathodique, oeil *m* magique
~LANTERN - [cin] (obsolete, toy-like apparatus used to project lantern-slides) lanterne *f* de projection
~NUMBERS - [nucl] (2, 8, 20, 28, 50, 82 and 126; nuclides having these numbers of neutrons or pro-

tons have great stability) nombres *mpl* magiques
MAGIC TEE - [metall] épreuve *f* à baguette de Maddox
MAGMA - [geol] (the molten fluids and gaseous fractions generated within the earth) magma *m*
[chem] (crude mixture of organic matters in a thin paste) magma *m*
MAGMATIC - [geol] magmatique
~CYCLE - [geol] cycle *m* igné
~DIGESTION - [geol] assimilation *f* magmatique
~ORE DEPOSITS - [min] gîtes *mpl* magmatiques
~SEGREGATION DEPOSITS - [min] gisement *m* de ségrégation magmatique
~STOPING - [mining] effondrement *m* magmatique du toit
MAGNAFLUX CRACK DETECTOR - [instr] (detection of surface cracks by the magneflux method) détecteur *m* magnétique de fissures
~METHOD - [meas] (a non destructive method of testing ferromagnetic materials to detect surface or near-surface defects) essai *m* magnétique de particules ferromagnétiques
MAGNAL BASE - [electron] (a base for cathode-ray tubes with eleven prongs) culot *m* à II broches
MAGNALIUM - [metall] (aluminium-base alloy) magnalium *m*
MAGNASCOPE - [cin] (projection arrangement whereby the size of the picture can be altered without affecting its focus) système *m* de projection à loupes
MAGNESIA - s.magnesium oxide
~ALBA - [chem] (in commerce, basic magnesium carbonate) magnésie *f*, magnésie *f* blanche
~BRICK - [metall] brique *f* à magnésie
~GLASS - [metall] verre *m* à magnésie
~MIXTURE - [chem] (magnesium chloride, ammonium chloride and ammonia solution used in chemical analysis for the estimation of phosphorus) mélange *m* magnésien
~USTA - [chem] (in commerce, magnesien carbonate which is calcined at low temperatures for a long period) carbonate *m* de magnésie fritté
MAGNESIAN - [min & geol] magnésien
~LIMESTONE - [geol] (a dolomitic limestone found in the North-East of England) calcaire *m* magnésien, dolomie *f*
MAGNESIOCHROMITE - [min] (a refractory) magnésiochromite *f*
MAGNESITE - [min] (natural magnesium carbonate used as a starting material in the production of magnesium oxide and carbon dioxide) magnésite *f*
MAGNESIUM - [chem] (a metallic element, symbol Mg, A.N. I2, A.W. 24.32, with uses in alloys, as a catalyst and reducing agent in syntheses, in pyrotechnics, and in anti-corrosion systems) magnésium
~ACETATE - [chem] (a disinfectant and dyeing mordant) acétate *m* de magnésium
~AMMONIUM PHOSPHATE - [chem] (a fertilizer and a flameproofing agent for textiles) phosphate *m* ammoniaco-magnésien
~ARSENATE - [chem] (an insecticide) arséniate *m* de magnésium
~BORATE - [chem] (a fungicide) borate *m* de magnésium
~-BORON FLUORIDE - [chem] (a flux in metallurgy) fluorure *f* de magnésium et de bore
~BROMATE - [chem] (an analytical reagent) bromate *m* de magnésium
MAGNESIUM CARBONATE - [chem] (a component of cosmetics and rubber pigments; also used in medicine as an antacid) carbonate *m* de magnésium, carbonate *m* de magnésie
~CELL - [el chem] (primary cell of which the negative electrode consists of magnesium or an alloy in which magnesium predominates) pile *f* au magnésium
~CHLORATE - [chem] (a dessicant) chlorate *m* de magnésium
~CHLORIDE - [chem] (a textile processing agent, source of magnesium, a constituent of special cements and a filler for fire extinguishers) chlorure *m* de magnésium
~CITRATE - [chem] (a mild purgative used in medicine) citrate *m* de magnésium
~-COPPER SULFIDE RECTIFIER - [electron] redresseur *m* magnésium-sulfure de cuivre
~FLUORIDE - [chem] (a component of ceramic glazes) fluorure *m* de magnésium
~FLUOSILICATE - [chem] (a constituent of ceramic glazes) fluosilicate *m* de magnésium
~FORMATE - [chem] (an analytical reagent) formiate *m* de magnésium
~GLYCEROPHOSPHATE - [chem] (a source of phosphate in medicine and a component of plastics stabilizers) glycérophosphate *m* de magnésium
~HYDROXIDE - [chem] (has application in the extraction of sugar from molasses and in medicine as an antacid and laxative) hydroxyde *m* de magnésium, hydrate *m* de magnésie
~LIGHT - [photo] éclairage *m* magnésique
~METHYLATE - [chem] (a catalyst and crosslinking agent) méthylate *m* de magnésium
~MICA - [min] mica *m* magnésien
~NITRATE - [chem] (an oxidizing agent used in pyrotechnics) nitrate *m* de magnésium
~OLEATE - [chem] (a component of plasticizers and a drier for paints and varnishes) oléate *m* de magnésium
~OXIDE - [chem] (a filler for rubber, a refractory material, catalyst, electrical insulant, and an antacid used in medicine) magnésie *f*, oxyde *m* de magnésium
~PALMITATE - [chem] (a drier for paints and varnishes) palmitate *m* de magnésium
~PERBORATE - [chem] (a bleaching agent) perborate *m* de magnésium
~PERCHLORATE - [chem] (a dessicant for gases) perchlorate *m* de magnésium
~PERMANGANATE - [chem] (a disinfectant) permanganate *m* de magnésium
~PEROXIDE - [chem] (a bleaching agent and a deodorant and antiseptic with medicinal uses) peroxyde *m* de magnésium, superoxyde *m* de magnésium
~PHOSPHATE, DIBASIC - [chem] (dimagnesium orthophosphate; dimagnesium phosphate; magnesium phosphate, secondary; magnesium hydrogen phosphate; plastics stabilizer) phosphate *m* de magnésium dibasique
~PHOSPHATE, MONOBASIC - [chem] (a stabilizer for plastics) phosphate *m* monobasique de magnésium
~PHOSPHATE, TRIBASIC - [chem] (magnesium phosphate, neutral; trimagnesium phosphate; acid-soluble white powder with uses as a plastics stabilizer)

phosphate *m* tribasique de magnésium
MAGNESIUM RICINOLEATE - [chem] (a component of cosmetics) ricinoléate *m* de magnésie
~SALICYLATE - [chem] (an antipyretic used in medicine) salicylate *m* de magnésie
~SHOCK COOLING - [metall] refroidissement *m* par choc thermique
~SILICATE - [chem] (a filler for paints and rubber, catalyst, bleaching agent, and a medicinal antacid and absorbent) silicate *m* de magnésium
~STEARATE - [chem] (a drier for paints and varnishes, a dusting powder in medicine and cosmetics and a plastics stabilizer) stéarate *m* de magnésie
~SULPHATE - [chem] (a textile dyeing assistant and flameproofing agent, and an aperient used in medicine) sulfate *m* de magnésium, sel *m* anglais
~SULPHITE - [chem] (a pulping agent used in the production of paper from wood) sulfire *m* de magnésium
~TRISILICATE - [chem] (an antacid and absorbent used in medicine and an antioxidant, and deodorant) trisilicate *m* de magnésium
~TUNGSTATE - [chem] (a phosphor used in the preparation of cardiographic screens) tungstate *m* de magnésium
MAGNET - [magn] (any body producing a magnetic field external to itself) aimant *m*, magnète *m*
[electromagn] électro-aimant *m*
[impl] aimant *m*
~ALLOYS - [metall] alliages *mpl* magnétiques
~ARMATURE - [el] armature *f* d'aimant
~BRAKE - [mech] frein *m* magnétique
~CARRIER - [impl] support *m* de l'aimant
~COIL - [el] bobine *f* de champ
~CORE - [magn] (the iron core in the coil of an electromagnet) noyau *m* magnétique
~-CORE AERIAL - [radio] (an aerial which consists of a coil surrounding a core of ferrite material) antenne *f* à noyau magnétique
~-CORE ANTENNA - s. magnet-core aerial
~CRADLE - s. magnet carrier
~CRANE - [mech] pont *m* roulant à électro-aimant de levage
~FLUX TESTER - [instr] (for ferromagnetic materials) séparateur *m* magnétique (des ferrailles)
~FRAME - [mech] culasse *f* magnétique
~HOUSING - [acoust] boîtier *m* d'aimant
~-OPERATED BRAKE - [mech] frein *m* électromagnétique
~POLE - [metall] épanouissement *m* polaire
~STEEL - [metall] acier *m* pour aimants, acier *m* magnétique
~SUPPORT - s. magnet carrier
~SYSTEM - s. magnetic system
~YOKE - [telev] (a component part of television sets) culasse *f*
MAGNETIC - [magn] magnétique
~AIRBORNE DETECTOR - [instr] (magnetic detector designed to locate submarines from an aircraft) détecteur *m* magnétique de bord
~AMPLIFICATION - [radio] amplification *f* magnétique
~AMPLIFIER - [el] (electrical amplifier in which the amplification is obtained by transductors) amplificateur *m* magnétique
~ANISOTROPY - [cryst] (dependence of magnetic properties on direction) anisotropie *f* magnétique
MAGNETIC AREA - [comput] zone *f* magnétique, aire *f* magnétique
~ATTRACTION - [phys] (the force exerted by a magnetized body upon another which is capable of magnetization) force *f* d'attraction magnétique
~AXIS - [phys] (the line through the effective centres of the poles of a magnet) axe *m* magnétique
~AZIMUTH - [surv] azimut *m* magnétique
~BALANCE - [instr] (a magnetometer with a moving element rotating round an horizontal axis) balance *f* magnétique
~BALANCE TRACK - [el acoust] piste *f* magnétique de compensation
~BARRIER TRAP - [phys] (a configuration of magnetic fields confining a plasma) bouteille *f* magnétique
~BARRIERS - [phys] miroirs *mpl* magnétiques
~BAY - [met] baie *f* magnétique
~BEARING - [nav] (the angle in the horizontal plane between the direction of the magnetic north and the direction line of the ship or aircraft) relèvement *m* magnétique
[mining] direction *f* magnétique
~BIAS - s. magnetic biasing
~BIASING - [el acoust] (simultaneous conditioning of the magnetic recording medium during recording by superimposing an additional magnetic field on the one due to the signal) polarisation *f* magnétique
~BLOW-OUT - [el] (special magnet coil fitted to a circuit breaker etc to produce a magnetic field) soufflage *m* magnétique
~BRAKING - [el] (braking by means of an electromagnet) freinage *m* magnétique
~BRIDGE - [instr] pont *m* magnétique
~CARD - [comput] (a card coated with a magnetizing substance) carte *f* magnétique
~CELL - [comput] élément *m* magnétique de mémoire
~CHUCK - [el] (chuck with steel electromagnets) mandrin *m* magnétique
~CIRCUIT - [el] (the closed path which is taken by the magnetic flux) circuit *m* magnétique
~CLUTCH - [el] (a clutch in which the force required to hold the parts together is provided by an electromagnet) embrayage *m* magnétique
~COATING - [el acoust] enduit *m* magnétique
~COMPASS - [instr] (instrument indicating the direction of the horizontal component of a magnetic field) boussole *f*
~COMPUTER - [comput] calculateur *m* magnétique
~CONCENTRATION OF ORES - [min] traitement *m* magnétique des minerais
~CONSTANT - [phys] (the scalar dimensional factor which relates the mechanical force between two currents to their magnitudes and geometrical configurations) constante *f* magnétique
~CONTACTOR - [el] contacteur *m* magnétique
~CORE - s. magnet core
~-CORE MEMORY - [comput] (store using a number of small rings of ferromagnetic material having a rectangular hysteresis loop) mémoire *f* à noyau magnétique
~-CORE STORAGE - s. magnetic-core memory
~-CORE STORE - s. magnetic-core memory
~CRACK DETECTION - [metall] détection *f* magnétique de criques
~CREEP - s. magnetic viscosity
~CROTCHET - [met] variation *f* brusque de champ magnétique terrestre

MAGNETIC CURRENT - [el] courant *m* magnétique
~CUTTER - [el acoust] (a cutter in which the mechanical displacements of the recording stylus are caused by magnetic fields) graveur *m* magnétique
~DECLINATION - [el] déclinaison *f* magnétique
~DEFLECTION - [electron] (the deflecting of an electron beam by the action of a magnetic field) déviation *f* magnétique
~DELAY-LINE - [comput] (metallic medium on which the propagation velocity of magnetic energy is greatly smaller than the speed of light) ligne *f* à retard magnétique
~DETECTOR - [radio] (generic term for an early type of detectors depending on the demagnetizing effect of an alternating magnetic field on a magnetized iron core) détecteur *m* magnétique
~DIP - [phys] inclinaison *f* magnétique
~DIPOLE - [phys] (a fiction describing the first order properties of magnetic moment) dipôle *m* magnétique
~DIPOLE RADIATION - [phys] (the radiation occurring as the result of a variable magnetic dipole moment) rayonnement *m* d'un dipôle magnétique
~DISK - [comput] disque *m* magnétique
~-DISK MEMORY - [comput] (magnetic storage with the magnetizable surface distributed over a number of rotating plates) mémoire *f* à disque magnétique
~-DISK STORAGE - s. magnetic-disk memory
~-DISK STORE - s. magnetic-disk memory
~DISPLACEMENT - [el] déplacement *m* magnétique
~DISTURBANCE DAILY VARIATION - [met] perturbation *f* journalière magnétique
~DISTURBED-DAY DAILY RADIATION - [met] perturbation *f* magnétique journalière maximum
~DRESSING - [min] préparation *f* magnétique
~DRIVE - [cin] (a method of driving the film in the printing process) entraînement *m* magnétique
~DRUM - [comput] (cylinder coated with magnetic material on which instructions can be stored in the form of small polarized spots) tambour *m* magnétique
~DRUM FILE MEMORY - [comput] grande mémoire *f* à tambour magnétique
~DRUM MEMORY - [comput] mémoire *f* à tambour magnétique
~ELECTRON MICROSCOPE - [opt] microscope *m* électronique magnétique
~ELEMENT - [phys] élément *m* du champ magnétique terrestre
~ENERGY PRODUCT - [phys] produit *m* d'énergie magnétique
~EQUATOR - [phys] (a line passing through all points of zero magnetic dip) équateur *m* magnétique
~EXTRACTING DEVICE - s. magnetic separator
~EXTRACTION - s. magnetic separation
~FIELD - [phys] (the region of lines of force surrounding a magnet of electrical conductor) champ *m* magnétique
~FIELD STRENGTH - s. magnetizing force
~FILE - [comput] fichier *m* magnétique
~FILM - [telev] (in video recording) bande *f* magnétique perforée
~-FILM MEMORY - [comput] mémoire *f* à film magnétique
~FLIP-FLOP - [radio] (bistable circuit with one or more magnetic amplifiers having two discreet levels of output which are obtained through an adjustement of the control current) circuit *m* flip-flop à amplificateur magnétique
MAGNETIC FLUID CLUTCH - [mech] embrayage *m* à fluide magnétique
~FLUX - [phys] (the product of the area of a figure by the average component of the magnetic induction normal to that area) flux *m* d'induction magnétique
~FLUX DENSITY - s. magnetic induction
~FOCUSING - [electron] (focusing an electron beam by the action of a magnetic electron lens) focalisation *f* magnétique
~FREEZING - [mech] adhérence *f* magnétique
~GAP - [el] entrefer *m*
~GATE - [electron] gâchette *f* magnétique
~GRATE SEPARATOR - [min] (a grid of permanent magnets and steel bars, used to arrest material objects during conveyance of other material) séparateur *m* à grille magnétique
~HAMMER - [impl] marteau *m* magnétique
~HEAD - [comput] (electromagnetic device which reads, records or erases information from a magnetic drum or tape) tête *f* magnétique
~HUM - [electron] (due to the magnetic field of the heater current and tending to deflect electrons from their original path from cathode to anode) ronflement *m* magnétique
~HYSTERESIS - [el] (lagging of the magnetic flux behind the magnetixing force producing it) hystérésis *f* magnétique
~INCLINATION - [phys] (the acute angle between the horizontal plane and the direction of the earth's magnetic field) inclinaison *f* magnétique
~INDICATOR FOR LIGHTNING CURRENTS - s. magnetic link
~INDUCTION - [phys] (vector quantity which determines the electromotive force at any point in a magnetic field) induction *f* magnétique
~INK - [comput] (marking ink containing a magnetizable substance) encre *f* magnétique
~INTERFERENCE - [radio] (undesired magnetic effects cuased by electric currents or the presence of magnetic materials) interférence *f* magnétique
~IRON ORE - [min] fer *m* oxydulé, magnétite *f*
~IRON SHEET - [metall] feuille *f* magnétique
LAG - s. magnetic hysteresis
~LAMINATING TAPE - [telev] (in video recording) bande *f* pour piste marginale
~LAMINATION - [telev] (in video recording) coucher *m* de piste magnétique
~LATITUDE EFFECT - [phys] (the variation of cosmic-ray intensity with magnetic latitude) effet *m* d'inclinaison magnétique
~LEAKAGE - [el] (leakage flux) dispersion *f* magnétique
~LENS - [opt] (an arrangement of coils and electromagnets so arranged that the resulting magnetic fields produce a focusing force on a beam of charged particles) lentille *f* magnétique
~LINES OF FORCE - [phys] lignes *f*pl de force magnétique
~LINK - [instr] (a magnetic indicator for lightning currents, consisting of a bundle of fine wires or sheets of special steel and indicating the approximate strength of currents due to a nearby lightning stroke by changes in its magnetic characteristics) indicateur *m* magnétique de courant de foudre

MAGNETIC LOSS - [metall] perte *f* magnétique
~LUNAR DAILY VARIATION - [met] variation *f* magnétique diurne lunaire
~MASTER - [el acoust & telev] bande *f* magnétique originale
~MEMORY - [comput] (those portions of the store which make use of magnetic properties of materials) mémoire *f* magnétique
~MERIDIAN - [phys] (the direction in azimuth of the magnetic axis of a freely-supported magnetic body under the influence of the earth's field alonge) méridien *m* magnétique
~METHOD - [oil ind] (a method of exploration based on the measure of the intensity and direction of the earth's magnetic field and inferring the distribution of rocks with different magnetic properties from local variations in the field) méthode *f* d'exploration magnétique
~MICROPULSATIONS - [el acoust] micropulsation *f* en enregistrement magnétique
~MICROSCOPE - [instr] (electron microscope with magnetic lenses) microscope *m* électromagnétique
~MINE - [armament] (a submarine mine resting on the bottom in shallow water and actuated by a permanent magnet) mine *f* magnétique
~MODULATOR - [radio] modulateur *m* à variation de champ magnétique
~MOMENT - [phys] (of a current loop or a magnetized body; the measure of the magnetizing force produced by the current or by the magnetized body) moment *m* magnétique
~MOMENT DENSITY - [phys] (the volume density of magnetic moment) valeur *f* du moment magnétique
~MOMENT OF A CONSTANT CURRENT - [el] (an axial vector quantity associated with an electric circuit) moment *m* magnétique d'un courant
~MOMENT OF AN ORBITAL ELECTRON - [phys] moment *m* magnétique d'un électron satellite
~NEEDLE - [phys] (a thin strip of magnetized iron or steel so pivoted that it aligns its long axis North and South) aiguille *f* aimantée
~NORTH - [phys] (north as indicated by a magnetic compass) nord *m* magnétique
~PENDULUM - [phys] (bar magnet suspended by a thread in a magnetic field or poised on a pivot) pendule *f* magnétique
~PERMEABILITY - [phys] perméabilité *f* magnétique
~PERTURBATION - [phys] perturbation *f* magnétique
~PICK-UP - [instr] (measuring instrument used when direct contact between object and pick-up is not desirable) capteur *m* électromagnétique
[el acoust] (phonograph pick-up depending for its operation on the variation in the reluctance of a magnetic circuit) pick-up *m* électromagnétique
~PICK-UP COIL - [el acoust] bobine *f* magnétique enregistreuse
~PICTURE TRACING - [cin] (recording of pictures on magnetic tape) dessin *m* magnétique
[telev] enregistrement *m* vidéo, magnétoscopie *f*
~PLASMOID - [phys] plasmoïde *m* produit par décharge en champ magnétique
~PLATE MEMORY - s. magnetic-disk memory
~-PLATE STORAGE - s. magnetic-disk memory
~-PLATE STORE - s. magnetic-disk memory
~PLATED WIRE - [el acoust] (magnetic wire with a core of non-magnetic material) fil *m* à couche magnétique

MAGNETIC POLARIZATION - [chem] (obtained by placing an inactive substance in a magnetic field) polarisation *f* magnétique
[el & phys] (intrinsic induction) polarisation *f* magnétique
~POLE - [phys] (a term denoting certain magnetostatic phenomena) pôle *m* magnétique
[instr] (of a magnet) s. magnetic poles of a magnet
[geogr] (of the earth) pôle *m* magnétique
~POLE STRENGTH - [phys] intensité *f* du pôle magnétique
~POLE TIPS - [electron] pièces *f*pl polaires
~POLE OF A MAGNET - [phys] (the points near the ends of the magnet where it can be considered that two magnetic masses are situated) pôles *m*pl magnétiques (d'aimant)
~POTENTIAL - [phys] (pseudoscalar quantity existing only outside those spaces where the current density is not zero and with a gradient representing the magnetic field) potentiel *m* magnétique
~POTENTIAL DIFFERENCE - [phys] (the line integral of a magnetic field intensity between two points) différence *f* du potentiel magnétique
~POTENTIOMETER - [instr] (instrument designed to measure the magnetic potential between two points in a magnetic field) potentiomètre *m* magnétique
~-POWDER CLUTCH - [mech] embrayage *m* à poudre magnétique
~-POWDER COATED TAPE - [el acoust] ruban *m* à couche magnétique
~-POWDER IMPREGNATED TAPE - [el acoust] ruban *m* homogène
~POWER-CLUTCH - [el mech] embrayage *m* magnétique
~PRINT-THROUGH - [telev] (permanent transfer or recorded signals) echo *m* magnétique
~PRINTING - [comput] (the printing in magnetic ink of the cursive amount as this is keyed into the machine) impression *f* magnétique
[el acoust] (the permanent transfer of a recorded signal from a section of a magnetic recording medium to another of a different medium when these sections are brough near one another) effet *m* d'écho
~PROBE - [electron] (in waveguides) sonde *f* à raccordement magnétique
~PULLEY - [mech] (magnetic separation device in which the end pulley of a belt-conveyor is fitted with magnets, which carry tramp iron round to the point at which the belt no longer makes contact with the pulley face, while the rest of the material is discharged in the usual way) poulie *f* magnétique
~PUMPING - [phys] (method for heating a plasma) pompage *m* magnétique
~PYRITE - [min] pirrhotite *f*
~QUANTUM NUMBER - [phys] (quantum number determining the component of the angular momentum vector of an atomic electron along the externally applied magnetic field) nombre *m* quantique de magnétisme
~QUIET-DAY SOLAR DAILY VARIATION - [met] variation *f* magnétique diurne des jours calmes
~READING - [comput] magnétolecture *f*, lecture *f* de caractères magnétiques
~READING HEAD - [electron] tête *f* magnétique de lecture
~RECORDER - [el acoust] (machine recording and

reproducing speech etc on a magnetic tape) enregistreur *m* magnétique
MAGNETIC RECORDING - [el acoust] enregistrement *m* magnétique
~RECORDING HEAD - [el acoust] tête *f* d'enregistrement
~RECORDING MEDIUM - [el acoust] matériel *m* pour l'enregistrement magnétique
~RECORDING REPRODUCER - [el acoust] lecteur *m* de son magnétique
~RELUCTANCE - [phys] réluctance *f*
~REPULSION - [phys] répulsion *f* magnétique
~REPRODUCER - [electron] reproducteur *m* magnétique
~REPRODUCING HEAD - [el acoust] tête *f* de reproduction, tête *f* magnétique de reproduction
~RESIDUAL LOSS - [phys] (loss of energy, proportional to the frequency of the magnetic field, in a ferromagnetic material) perte *f* magnétique
~RESISTANCE - [phys] réluctance *f*
~RESONANCE - [el] résonance *f* magnétique
~RESONANCE SPECTRUM - [phys] spectre *m* de résonance magnétique
~REVOLVING FIELD - [phys] champ *m* magnétique tournant
~RIBBON RECORDING INSTRUMENT - [el acoust] enregistreur *m* à bande magnétique
~RIGIDITY - [phys] (of a particle) rigidité *f* magnétique
~RIPPLE - s. magnetic hum
~ROD - [mech] barre *f* magnétique
~SAND - [min] sable *m* magnétique
~SATURATION - [phys] (term denoting the condition of a magnetic material at high values of induction with small incremental permeability) saturation *f* magnétique
~SCREEN - [el] (screen or shield of ferromagnetic material protecting the operating part of a measuring instrument from the effects of external magnetic fields) écran *m* magnétique, cuirasse *f* magnétique
~SCREENING - [el] (the application of a magnetic screen) blindage *m* magnétique
~SEPARATION - [ind chem etc] (of ferromagnetic particles from a mixture) séparation *f* magnétique
~SEPARATOR - [el mach] (apparatus for separating ferrous particles from a mixture magnetically) séparateur *m* magnétique
~SHEET - [el acoust] (a magnetic plate-shaped record, mainly for dictating purposes) feuille *f* magnétique
~SHELL - [phys] (magnet in the form of an infinitely thin shell in which the magnetization is everywhere normal to the surface and in inverse ratio to the thickness) couche *f* double magnétique
~SHIELD - s. magnetic screen
~SHIELDING - s. magnetic screen
~SHUNT - [el] (piece of magnetic material in parallel with a portion of a magnetic circuit) dérivation *f* magnétique, shunt *m* magnétique
~SOLAR DAILY VARIATION - [met] variation *f* magnétique diurne solaire
~SOUND - [el acoust] son *m* magnétique
~SOUND RECORD COPYING MACHINE - [el acoust] tireuse *f* de copies d'enregistrements magnétiques
~SOUND TRACK - [el acoust] piste *f* magnétique de son

MAGNETIC SPECTROGRAPH - [instr] (electronic device based on the action of a constant magnetic field on the paths of electrons, and used to separate electrons with different velocities) spectrographe *m* magnétique
~SPECTROMETER - [instr] (instrument using a magnetic field) spectromètre *m* magnétique
~STANDARD - [instr] (a device used to calibrate galvanometers) aimant *m* étalon
~STORAGE - s. magnetic memory
~STORE - s. magnetic memory
~STORM - [phys] (disturbance in the earth's magnetic field) orage *m* magnétique
~STRAIN ENERGY - [phys] (component of potential energy in a magnetic sphere of action) énergie *f* magnétique de déformation
~STRIPE - [comput] filet *m* magnétique
~-STRIPE RECORDING - [comput] enregistrement *m* à filet magnétique
~-STRIPE STORAGE - [comput] mémoire *f* à filet magnétique
~SUSCEPTIBILITY - [phys] (of an isotopic material) susceptibilité *f* magnétique
~SYSTEM - [el acoust] (of an acoustic transductor) système *m* d'aimant
~TAPE - [el acoust & comput] (flexible tape coating with magnetic material on which instructions can be stored) ruban *m* magnétique
~-TAPE CONVERTER - [comput] convertisseur *m* de ruban magnétique
~TAPE FUNCTION GENERATOR - [contr] générateur *m* de fonctions à ruban magnétique
~TAPE LEADER - [el acoust & telev] amorce *f* de bande magnétique
~-TAPE READER - [comput] (electronic computer device) lecteur *m* de bande magnétique, lecteur *m* de ruban magnétique
~-TAPE RECORDER - s. magnetic ribbon recording instrument
~-TAPE-TO-CARD CONVERTER - [comput] convertisseur *m* ruban magnétique cartes perforées
~-TAPE-TO-PRINTER CONVERTER - [comput] convertisseur *m* bande magnétique impression
~TAPE WRITER - [comput] tête *f* d'écriture pour bande magnétique
~TEST COIL - [meas] bobine *f* d'essai magnétique
~THIN FILM - [comput] pellicule *f* mince magnétique
~TIME RELAY - [el] relais *m* magnétique temporisé
~TRANSFER - [telev] report *m* optique
[el acoust] s. magnetic printing
~TRANSITION TEMPERATURE - [phys] (Curie point) point *m* de Curie
~TWIST - [phys] torsion *f* magnétique
~VALVE - [el mech] soupape *f* magnétique
~VARIATION - [met] (the angle between the true meridian and the magnetic meridian at any given point) variation *f* magnétique, déclinaison *f* magnétique
~VARIOMETER - [instr] (instrument designed to measure differences in a magnetic field) variomètre *m* magnétique
~VIBRATOR - [el mech] (a conveying device in which the bottom plate of the trough containing the material is vibrated electro-magnetically) vibreur *m* magnétique
~VISCOSITY - [phys] (phenomena by which changes in magnetization of a ferromagnetic substance follow the changes in the field producing them) traîna-

ge *m* magnétique, viscosité *f* magnétique
MAGNETIC WAVES - [phys] (the spreading of magnetization from a small region of a specimen) ondes *fpl* magnétiques
~WELL - [phys] (used to contain hot plasma) puits *m* magnétique
~WIPING-DOWN - [aero] (process of removing magnetism produced by atmospheric electricity from the external parts of an aircraft) désaimantation *f*
~WIRE - [el acoust] (a magnetic recording medium in the form of wire) fil *m* magnétique
~WIRE RECORDER - [el acoust] enregistreur *m* à fil magnétique
MAGNETICALLY ATTACHED MONITORING THERMOMETER - [instr] thermomètre *m* de contrôle à champs magnétiques
~BIASED POLARIZED RELAY - [el] relais *m* à polarisation magnétique
~CONFINED ELECTRON BEAM - [electron] faisceau *m* électronique renfermé par champs magnétiques
~HARD MATERIAL - [magn] matériau *m* magnétique dur
~SCREENED TRANSFORMER - [el] transformateur *m* à blindage magnétique
~SOFT MATERIAL - [magn] matériau *m* magnétique doux
MAGNETISM - [phys] (the science dealing with the laws and conditions of magnetic force and its effects) magnétisme *m*
MAGNETITE - [min] (magnetic iron ore, ferrosferric oxide. An important ore of iron) magnétite *f*
~BLACK - [paint] (a black pigment made from magnatic iron oxide) noir *m* de magnétite
MAGNETIZATION - [phys] (the process whereby a ferrous material is rendered magnetic) aimantation *f*
[phys] (intensity of magnetization) (a vector associated with an element of a substance and equal in size and direction to the magnetic moment of this element divided by its volume) magnétisation *f*, aimantation *f*
~BY ELECTRICITY - [phys] aimantation *f* par l'électricité
~BY INFLUENCE - [phys] aimantation *f* par influence
~BY SEPARATE TOUCH - [phys] aimantation *f* par touche séparée
~BY SINGLE TOUCH - [phys] aimantation *f* par simple touche
~CURVE - [phys] courbe *f* de magnétisation, courbe *f* de magnétisme
~VECTOR - [phys] vecteur *m* d'aimantation
MAGNETIZE, to - [phys] (the operation of rendering magnetic) aimanter, magnétiser
MAGNETIZED SPOT - [comput] tache *f* magnétique, spot *m* magnétique, doublet *m* magnétique
MAGNETIZING - [el magn] aimantation *f*, magnétisation
[adj] magnétisant
~COIL - [el] bobine *f* de champ
~CURRENT - [el] (a current which is principally used to produce a magnetic field) courant *m* magnétisant
[el mach] (the current which is necessary to produce the magnetic flux in an electrical machine or apparatus) courant *m* magnétisant
~DEVICE - [el] dispositif *m* à aimant
~FIELD - [el] (magnetic field used to produce magnetization) champ *m* magnétisant
MAGNETIZING FORCE - [magn] (the ability of electric currents or magnetized bodies to produce magnetic induction) force *f* magnétique, intensité *f* de champ magnétique, intensité *f* magnétique
~ROASTING - [metall] grillage *m* magnétisant
MAGNETO - [el mech] (electro-mechanical device for generating high-tension ignition current of i.c. engines) magnéto *m*, machine *f* magnétoélectrique
~ALTERNATOR - [el mech] (for i.c. engines) aimant *m*
~-BELL - [telecomm] sonnerie *f* électromagnétique
~BRUSH - [el mech] (electrical brush used to pick up current in a magneto) balai *m* de magnéto
~CENTRAL OFFICE - [telecomm] (in telephony) bureau *m* central à appel magnétique
~-DYNAMO - [el] dynamo-magnéto *f*
~EXCHANGE - s. magneto central office
~IGNITION - [el mech] (ignition system based on a magneto) allumage *m* magnétique
~SPANNER - [mech] (a small spanner specially designed for use on magnetos) clef *f* de magnéto
MAGNETOCALORIC EFFECT - [phys] (the reversible heating and cooling of a substance by changes of magnetization) effet *m* magnétocalorique
MAGNETODYNAMIC - [phys] magnétodynamique
~RELAY - [el] relais *m* magnétodynamique
MAGNETODYNAMO - [el mech] (dynamo in which the exciting flux is obtained from permanent magnets) dynamo-magnéto *f*
MAGNETOELASTIC COUPLING CONSTANTS - [phys] constantes *fpl* de couplage magnétoélastique
MAGNETOELECTRIC - [el] magnéto-électrique
~INSTRUMENT - [instr] (a permanent-magnet moving coil instrument) instrument *m* magnéto-électrique
~RELAY - [el] relais *m* magnétoélectrique
MAGNETOFLUID DYNAMICS - [phys] magnétohydrodynamique *f*
MAGNETOGRAM - [instr] magnétogramme *m*
MAGNETOGRAPH - [instr] (instrument designed to record the variations of the earth's magnetic field) magnétographe *m*
MAGNETOGRAPHIC INSPECTION - [metall] examen *m* magnétographique
MAGNETOHYDRODYNAMIC CONVERSION - [phys] conversion *f* magnétodynamique
~INSTABILITY - [phys] instabilité *f* magnétodynamique
~WAVES - [phys] ondes *fpl* magnétodynamiques
MAGNETOHYDRODYNAMICS - s. magnetofluid dynamics
MAGNETOIONIC COMPONENT - [phys] composante *f* magnétoïonique
~DOUBLE REFRACTION - [el magn waves] biréfrigerence *f* magnétoionique
MAGNETOMECHANICAL DAMPING - [el mech] (component of energy loss associated with the elastic vibration of a magnetic material) amortissement *m* magnétomécanique
~FACTOR - [el mech] facteur *m* magnétomécanique
~RATIO - [el mech] rapport *m* magnétomécanique
MAGNETOMETER - [instr] (an apparatus for determining the strength of a magnetic field) magnétomètre *m*
MAGNETOMOTIVE FORCE - [phys] (along a closed curve; the line integral around the curve) force *f* magnétomotrice
[el] force *f* magnétomotrice (le long d'une ligne

fermée)
MAGNETOMOTIVE TENSION - [metall] tension *f* magnétomotrice
MAGNETON - [phys] (unit to measure the magnetic moment of atomic particles) magnéton *m*
MAGNETOOPTICAL ANALYSIS - [chem] (a method of chemical based on differences in the lag of the Faraday effect behind the magnetic intensity of different substances) analyse *f* magnétooptique
MAGNETOOPTICS - [phys] (the study of the action which is produced on light by magnetic fields) magnétooptique *f*
MAGNETOPHONE - [el acoust] (an apparatus for recording or reproducing sounds by means of a magnetic medium) magnétophone *m*
MAGNETORESISTANCE - [el] (changes in electrical resistivity due to changes of magnetization) magnétorésistance
MAGNETOSCOPE - [instr] (instrument designed to indicate the existance of magnetic fields and depending for its action on magnetic forces) magnétoscope *m*
MAGNETOSPHERE - [phys] magnétosphère *f*
MAGNETOSTATIC FIELD - [phys] champ *m* magnétostatique
MAGNETOSTATICS - [phys] (the study of the magnetic fields which do not vary with time) magnétostatique *f*
MAGNETOSTRICTION - [phys] (a magnetic field causing a change in physical dimensions) magnétostriction *f*
~DELAY LINE - [comput] ligne *f* à retard magnétostrictive
~LOUDSPEAKER - [el acoust] (loudspeaker depending for its operation on the magnetostrictive properties of a material) haut-parleur *m* à magnétostriction
~MICROPHONE - [el acoust] (microphone depending for its operation on the deformation of a material having magnetostrictive properties) microphone *m* à magnétostriction
~OSCILLATOR - [radio] (electronic tube in which the electrons are accelerated by a radial electric field and by an axial magnetic field) oscillateur *m* à magnétostriction
~STRAIN GAUGE - [instr] extensomètre *m* à magnétostriction
MAGNETOSTRICTIVE DELAY LINE - [comput] ligne *f* à retard magnétostrictive
MAGNETRON ARCING - [electron] décharge *f* disruptive de magnétron
~CITRICAL FIELD - [electron] intensité *f* de champ critique
~CRITICAL VOLTAGE - [electron] tension *f* critique d'un magnétron
~CUT-OFF - [electron] effet *m* de courant anodique nul, courant *m* de rupture
~FURNACE - [electron] (microwave furnace with a magnetron) four *m* à magnétron
~GAUGE - [electron] manomètre *m* magnétron
~GENERATOR - [radio] magnéto *m*
~PACKAGE - [electron] ensemble *m* magnétron, magnétron *m* en ordre de marche
~PERFORMANCE - [electron] diagramme *m* de fonctionnement d'un magnétron
~PULLING - [electron] glissement *m* aval d'un magnétron
MAGNETRON PUSHING - [electron] glissement *m* amont d'un magnétron
~STRAPPING - [electron] jumelage *m*
~VACUUM GAUGE - [instr] manomètre *m* de Redhead
MAGNIFICATION - [opt] (magnifying power) agrandissement *m*, grossissement *m*
[phys] (in oscillations) facteur *m* de résonance
[photo] grossissement *m*
~FACTOR - [radio] coefficient *m* de surtension
~INDICATOR - [photo] indicateur *m* du rapport d'amplification
MAGNIFIED - [opt] grossi, agrandi, amplifié
MAGNIFIER - [opt] (simple lens microscope, that is a converging lens of short focal length) loupe *f*
[radio] (any type of thermionic amplifier) amplificateur *m* à tubes électroniques
MAGNIFY, to - [opt] agrandir, grossir, amplifier, renforcer
MAGNIFYING GLASS - [opt] (a lens which increases the apparent size of an object viewed through it) loupe *f*, verre *m* grossissant
~LENS - [photo] lentille *f* dioptrique, oeilleton *m* de visée
~POWER - [opt] (the ratio of the apparent size of the image formed by a magnifying lens to that of the actual size of the object) agrandissement *m*, grossissement *m*
MAGNITUDE - [gen] grandeur *f*
[astr math etc] magnitude *f*
~OF CURRENT - [el] intensité *f* du courant
MAGNOLIA - [bot] magnolia *m*
~METAL - [metall] (an alloy of approximately 80% lead, the balance being mainly antimony) magnolia *m*, métal *m* magnolia
MAGPIE - [zool] pie *f*
MAGSLIP - [radar] (colloq. for automatic synchronizer) synchroniseaur *m* automatique
MAHOGANY - [bot] (an American tropical tree) acajou *m*
MAIDEN FIELD - [mining] gisement *m* vierge
~NUT - [mech] écrou *m* de serrage
MAIDENHOOD - [med] virginité *f*
MAIDISM - [med] pellagre *f*
MAIEUSIOMANIA - [med] folie *f* puerpérale
MAIEUSIOPHOBIA - [med] tocophobie *f*
MAIL, to - [gen etc] (to despatch letters, parcels etc) expédier, envoyer par la poste
MAIL - [gen] (an armour composed of interlaced rings) maille *f*
[gen] (the Post) courrier *m*, poste *f*, dépêches *mpl*
~BAG - [impl] sac *m* de poste
~BOAT - [naut] paquebot-poste *m*
~BOX - [gen] boîte *f* aux lettres
~CAR - [railw] wagon-poste *m*
~COACH - [gen] malle-poste *f*
~PACKET - s. mail boat
~STEAMER - s. mail boat
~TRAIN - [railw] train-poste *m*
~VAN - s. mail car
~WITH ONE OPENING - [text] maillon *m* à un seul œillet
MAILED EYE - [text] lisse *f* à maillon
~HEALD - [text] lisse *f* à maillon
MAILLE - [text] maillon *m*
MAILLECHORT - [metall] argent *m* allemand, maillechort *m*
MAIN - [gen] principal, essentiel, premier

MAIN - [plumb etc] (principal channel, duct etc for conveying water, gas, electricity sewage etc) conduite *f* principale, conduite *f* maîtresse, collecteur *m*
[el] conducteur *m* principal
[el] (in electric traction) conduit *m* de courant
~AIR CURRENT - [gen & mining] courant *m* d'air principal
~AIR RESERVOIR - [auto] réservoir *m* d'air principal
~AIRWAY - [gen & mining] galerie *f* principale d'aérage
~AND TAIL - [min] (a system of hauling by a main rope used to pull out the wagons and the tail rope used to draw back the empties) système *m* câble-tête et câble-queue
~ANGLE OF INCIDENCE - [opt] angle *m* d'incidence principal
~ANODE - [electron] (the anode carrying the load current) anode *f* principale
~BANG - [radar] (a pilot pulse; a large signal on a scope caused by the pulse) impulsion *f* pilote
~BATTERY - [el] (the principal electric battery in an aircraft) batterie *f* principale
~BEAM - [text] ensouple *f* de chaîne principale
[constr] (one of the beams in floor construction) maîtresse-poutre *f*, longeron *m*
~BEARING CAP - [mech] (in internal combustion engines) chapeau *m* de palier de vilebrequin
~BEARINGS - [mech] (crankshaft bearings) palier *m* de vilebrequin
~BLAST ENTRY - [mining] embouchure *f* principale du carneau
~BLOCK - [metall] bloc *m* principal
~BORE - [firearms] âme *f* rayée
~BOTTOM - [mining] roche de fond
~BRAKE CYLINDER - [auto] cylindre *m* principal de frein
~BRANCH - [gen & bot] branche *f* principale
~BRASSES - [mech] coussinets *mpl* de vilebrequin
~BREADTH - [naut] fort *m* (d'un navire)
~BRUSH - [el] balai *m* principal
~CABLE - [el & telecomm] câble *m* porteur principal, porteur *m* principal
~CAM - [text] excentrique *m* principal, came *f* principal
~CAPACITY - [el] (capacity of a distribution system) rendement *m* d'un réseau
~CASTING - [metall] crapaudine *f*
~CHAIN - [text] chaîne *f* principale
~CHUTE - [mining] maître couloir *m*
~CIRCUIT - [el] circuit *m* de courant, circuit *m* de série
~CONTACTS - [el] (the contacts of a current carrying switch) contacts *mpl* principals
~COCK - [hydr] robinet *m* général
~COLLECTOR - [hydr] (main sewer) grand collecteur *m*
~CONDUCTOR - [el] conducteur *m* principal
~CONNECTING ROD - [mech] bielle *f* motrice
~COUPLE - [constr] (the principal truss in a timber roof) ferme *f* principale
~CRANK-PIN - [mech] bouton *m* de manivelle de l'essieu moteur
~CROP - [agric] culture *f* principale
~CYLINDER BEARING - [mech] palier *m* du cylindre principal
~DECK - [shipbuild] (a full-length deck below the water deck) passerelle *f*
MAIN DISTRIBUTING FRAME - s. main distribution frame
~DISTRIBUTION FRAME - [telecomm] répartiteur *m* général, répartiteur *m* principal
[el] centre *m* de commutation
~DISTRICT EXCHANGE - [telecomm] bureau *m* suburbain
~DRAIN - [hydr] drain *m* collecteur, maître-drain *m*
~DRAW ROLL - [text] cylindre *m* d'appel principal
~DRIVE - [mech] transmission *f* principale
~DRIVING CONE - [mech] cône *m* principal de poulie
~DRIVING SHAFT - [mech] (the principal transmission shaft) arbre *m* de transmission principal
~DRUM STRIPPINGS - [text] débourrures *fpl* du grand tambour
~ENTRANCE SIGNAL - [radar] signal *m* de radiophare interne
~EXCHANGE - [telecomm] (telephone exchange with other exchanges depending on it) bureau *m* central principal, bureau *m* d'attache
~FIELD - [el] (in an electrical machine, the main exciting field) champ *m* principal
~FIELD COIL - [el] bobine *f* du champ principal
~FILE - [comput] (record with the basic data of a process to be programmed) fichier *m* principal
~FILM - [cin] (the essential part of a programme) long métrage *m*, grand film *m*
~FLOAT - [aero] (a buoyant structure designed to support a seaplane or amphibian when resting or moving on water) flotteur *m*
~FLUE - [constr] (of a chimney) carneau *m* collecteur
[metall] carneau *m* de la cheminée
~FORAGE AREA - [agric] surface *f* fourragère principale
~FRAME - [railw] (of a locomotive) longeron *m* principal
~FUEL TANK - [auto & aero] réservoir *m* principal du carburant
~GAP - [electron] (the conduction path between a cathode and a main anode) intervalle *m* principal de décharge
~GEAR WHEEL - [mech] roue *f* motrice
~GIRDER - s. main beam
~HARVEST - [agric] récolte *f* principale
~HATCH - [naut] panneau *m* principal
~HAULAGEWAY - [mining] galerie *f* principale de roulage
~HEAD - [mech] (the casting which forms the upper structure of a vertical hydraulic press) plateau *m* supérieur
~HONEY FLOW - [agric] grande miellée *f*
~HOOK - [text] crochet *m* principal
~INLET VALVE - [mech] (of a turbine) soupape *f* principale de turbine
~JET - [mech] (principal jet in a carburettor) gicleur *m* principal
[metall] coulée *f* principale
~JOURNAL - [mech] tourillon *m* principal
~LEAD SLEEVE FOR MULTIPLE JOINT - [telecomm] (of telephone installations) pièce *f* de division en plomb
~LEVEL - [mining] galerie *f* maîtresse
~LINE - [gen] ligne *f* principale
[railw] grande ligne *f*, ligne *f* d'artère
~-LINE COACH - [railw] voiture *f* de grandes lignes
~-LINE LOCOMOTIVE - [railw] locomotive *f* de route

MAIN LODE - [mining] filon *m* principal, filon *m* mère
~METER - [instr] (also called primary meter or master meter) appareil *m* de mesure globale
~NECTAR FLOW - s. main honey flow
~OSCILLATION - [radio] oscillation *f* fondamentale
~OUTFALL - [hydr] émissaire *m*
~-PIN - [mech] cheville *f* ouvrière
~PLANE - [aero] (the principal supporting surface of an aeroplane, consisting of port and starboard wings) plan *m* principal
~POINT - [railw] point *m* de croisement
~PROGRAMME - [comput] (the master routine) programme *m* principal
~PUSHER - [text] clavette *f* principale
~QUANTUM NUMBER - [phys] (or first quantum number; the number which gives the size of the electron orbit) nombre *m* quantique principal
~ROAD - [constr] (a highway) grande route *f*
[mining] galerie *f* maîtresse
[mech] (of a pump) maîtresse-tige *f*
~ROOT - [bot] racine *f* principale
~ROPE - s. main and tail
~ROTOR - [mech] (the complete assembly of the rotary parts of compressor and turbine or turbines) rotor *m* principal
[aero] (in helicopters) (a rotor which is primarily designed to provide lift) rotor *m* principal
~ROUTE - [gen] direction *f* principale
[railw] grande relation *f*
[telecomm] artère *f* principale
~RUNNER - [metall] (in foundry work) chenal *m* de coulée
[metall] fossé *m*, chenal *m*
~SADDLE - [mach tool] traînard *m*
~SEWER - [hydr] égout *m* collecteur
[constr] collecteur *m* des égouts
~SHAFTING - [mech] transmission *f*
~SIDEBAND - [radio] (the sideband which includes the components corresponding to the frequencies of the modulating signal) bande *f* latérale principale
~SLEEVE FOR MULTIPLE JOINT - [telecomm] pièce *f* de division, manchon *m* de distribution
~SPINDLE - [mach tool] mandrin *m* principal
~SPRING - [photo] (of a shutter) ressort *m* d'armement
~STAGE - [astronaut] étage *m* principal, période *f* de poussée maximale
~STATION PEG - [surv] jalon *m* principal (de station)
~SWITCH - [el] commutateur *m* principal
~SWITCH BOARD - [el] (large switchboard controlling the whole power of an installation) tableau *m* principal
~SYSTEM - [el] réseau *m* de distribution
~TANKS - [aero] réservoirs *mpl* principaux
~TIE - [constr] (the lower tensional member of a roof-truss) traverse *f* inférieure de ferme
~TRACK - [railw] ligne *f* principale
~TRANSDUCTOR - [radio] transducteur *m* principal
~TRANSFORMER - [el] (the transformer which is connected across two of the phases on the 3-phase side) transformateur *m* d'alimentation
~UNDERCARRIAGE - [aero] (a principal assembly of a landing gear) atterrisseur *m*
~VOLTAGE - [el] tension *f* du secteur, tension *f* du réseau

MAIN WHEEL - [horol] (the first wheel in the train) roue *f* motrice
~YARD - [naut] grande vergue *f*
MAILAND - [geogr] terre *f* ferme, continent *m*
MAINLAYING - [plumb gas installation el etc] pose *f* de canalisations
MAINMAST - [naut] grand mât *m*
MAINS - [el] (electrical conductors from which a subsidiary system takes its supply) réseau *m*
[hydr] conduite *f*, conduite *f* maîtresse, réseau *m*
~AERIAL - [radio] (built-in aerial connected with the mains) antenne *f* de secteur
~BAG - [gas ind] (gas network equipment) (a gas bag) ballon *m* obturateur
~CONNECTION - [radio] connexion *f* au réseau
~FILTER - [el] (an electrical circuit designed to reduce the leakage of radio frequency current into supply mains) filtre *m* de réseau
~FLUCTUATIONS - [el] variations *fpl* de la tension de réseau
~FREQUENCY - [el] fréquence *f* du réseau
~INTERRUPTIONS - [el] interruptions *fpl* de la tension du réseau
~JUNCTION - [of networks] (point of interconnexion) noeud *m* de canalisations
~LINE HUM - s. mains noise
~LOCKING - [telev] (the coupling of the field frequency with the mains frequency) réglage *m* de fréquence par un signal à fréquence constante
~NETWORK - [gen] réseau *m*
~NOISE - [el acoust] (the noise which is generated within the mains) bruit *m* de secteur, bourdonnement *m* du courant alternatif
~-OPERATED FLASH - [photo] lampe-éclair *f* à allumage sur secteur
~RECEIVER - [radio] récepteur *m* secteur, poste *m* secteur
~SUPPLY - [el] alimentation *f* secteur, alimentation *f* sur le secteur
~SUPPLY SWITCH - [el] interrupteur *m* de secteur
MAINSAIL - [naut] grand'voile *f*
MAINSPRING - [mech] grand ressort *m*
[horol] (the spring providing the motive power) ressort *m* moteur
~HOOK - [horol] (hook for attaching the mainspring to the barrel) crochet *m* du ressort moteur
~WINDER - [horol] (the tool used to coil the mainspring before inserting it into the barrel) bobinoir *m* du ressort moteur
MAINSTAY - [gen] soutien *m* principal
MAINTAIN, to - [gen] maintenir
[mech etc] entretenir
MAINTENANCE - [mech etc] (the operation of carrying out services and adjustements to keep a mechanism in working order) entretien *m*, maintien *m*
~CHARGE - [gen] coût *m* d'entretien, frais *mpl* d'entretien
~CIRCUIT - [el] circuit *m* de maintien
~DEPARTMENT - [gen] service *m* de surveillance
[railw] service *m* de l'entretien
~MANUAL - [gen] (book of instructions for maintenance) manuel *m* d'entretien
~TRUE BEARING - [radar] (keeping the indications of a true bearing of an object constant during a given time) azimut *m* constant
MAINWAY - [mining] galerie *f* maîtresse
MAITLANDITE - [min] (a mineral containing lead and

calcium) maitlandite *f*
MAIZE - [bot] mais *m*, turquet *m*
~COB - [agric] panouille *f*, épi *m* de mais
~FOR ENSILAGE - [agric] (animal feeding) mais *m* à ensiler
~HUSKER - [agric] dépouilleuse *f* de mais
~MEAL - [agric] farine *f* de mais
~OIL - [agric] (corn oil) huile *f* de mais
~PULP - [agric] (animal feeding) pulpe *f* de mais
~RUST - [bot] (plant disease) rouille *f* du mais
~SILAGE - s. maize for ensilage
~STRAW - [text] paille *f* de mais
~THRESHER - [agric] batteuse *f* à mais
MAJOLICA - [pottery] (decorated earthenware) majolique *f*
MAJOR - [gen] majeur
~AXIS - [opt] grand axe *m*
[geom] axe *m* transverse
~BED - [geogr] (of a river) lit *m* de crue, lit *m* majeur
~BEND - [electron] (of a waveguide) coude *m* de bout
~CYCLE - [comput] (a number of minor cycles) cycle *m* majeur
~DIAMETER - [mech] (of a thread) diamètre *m* externe
~ELEMENTS - [soil] macroéléments *mpl*
~FEEDBACK - [contr] (feedback signal monotoring the operation of a system) réaction *f* principale
~FOLD - [geol] pli *m* principal
~GRADUATIONS - [meas] (between the minor index graduations of a scale) graduations *fpl* principales
~INTERVAL - [mus] intervalle *m* majeur
~JOINT - [geol] diaclase *f* principale
~KEY - [mus] ton *m* majeur, mode *m* majeur
~LOBE - [radio] (of an aerial) lobe *m* principal
~LOOP - [contr] boucle *f* extérieure
~OVERHAUL - [ind proc] révision *f* totale, grandes réparations *fpl*
~PART - [gen] majeure partie *f*
[mech] partie *f* principale
~PRINCIPAL PLANE OF STRESS - [soil] plan *m* de la tension principale maximum
~SCALE OF EQUAL INTONATION - [mus] (equally tempered scale) gamme *f* tempérée
~SCALE OF EQUAL TEMPERAMENT - s. major scale of equal intonation
MAJORITY CARRIER - [electron] (the carrier which constituents more than half the total number of carriers) porteur *m* majoritaire
~CARRIER CONTACT - [electron] contact *m* de porteurs majoritaires
~CIRCUIT - [comput] circuit *m* majoritaire
~CONCENTRATION - [telecomm] concentration *f* majoritaire
~EMITTER - [electron] (electrode from which a flow of majority carriers enters the inter-electrode region) émetteur *m* de porteurs majoritaires
MAJUSCULE - [print] (a capital letter) majuscule *f*
MAKE, to - [gen] faire
[gen] (the action of constructing) construire, fabriquer
[gen] (to produce) produire
[el] (a contact) fermer
[naut] (of tide) se faire, monter, baisser
[constr] (the clay) rendre maigre, mélanger de sable
MAKE - [gen] (the manufacture's name or trademark of a device or mechanism) marque *f*, fabrication *f*
MAKE - [el] fermeture *f*
[telecomm] (the operation of a telephone relay) travail *m*
~A BEVELLED JOINT, to - [carp] bisauter
~-AND-BREAK CONTACT - [telecomm] contact *m* repos-travail
~-AND-BREAK DEVICE - [el] interrupteur *m*, rupteur *m*
~-AND-BREAK IGNITION - [auto] allumage *m* par ruptures
~-BEFORE-BREAK CONTACT - [el] (of a relay) contact *m* de séquence travail-repos
~-BREAK CONTACT - [el] contact *m* inverseur
~COMPOST, to - [agric] composter, fabriquer du compost
~CONTACT-OPERATING TIME - [el] temps *m* de fonctionnement du contact de travail
~-CONTACT UNIT - [telecomm] (of a relay) contact *m* de travail
~FAST, to - [gen etc] (to secure, especially by means of a rope or ropes) fermer, assurer
[naut] amarrer
~FOOTAGE, to - [oil ind] (in well drilling) avancer avec le forage au mètre
~FOR, to - [gen] se diriger
~HAY, to - [agric] faner
~HEAD, to - [gen] avancer
~HOLE, to - [oil ind] (in well drilling) faire un sondage, forer
~IMPULSE - [telecomm] (in automatic telephony) impulsion *f* de fermeture
~METER - [gas ind] (meter measuring the gas which is produced at a station) compteur *m* de fabrication
~PER RUN - [gas ind] volume *m* de gaz à l'eau fabriqué par cycle
~PULSE - s. make impulse
~-READY - [print] (the process of preparing a forme for printing) mise *f* en train
~-RUN - [gas ind] (or run) fabrication *f*, période *f* de fabrication
~TIGHT, to - [mech] rendre étanche, assurer l'étancheité
~UP, to - [print] mettre en page
[to prepare] préparer
[text] (of clothes etc) confectionner
~UP - [gen] maquillage *m*
[photo] maquillage *m*
[print] mise *f* en page
[comm] arrête *m*, confection *f* d'un bilan
[med] composition *f* d'un médicament
~-UP RAIL - [railw] rail *m* de compensation
~-UP TORQUE - [oil ind] couple *f* de blocage
~-UP WATER - [hydr] (water added to make up loss in a closed circuit system) eau *f* d'appoint
~WINE FROM, to - [agric] vinifier
MAKER - [gen] fabricant *m*, constructeur *m*
~-UP - [print] metteur *m* en page
MAKER'S BADGE - [comm] (also called name plate) plaque *f* signalétique
~INSTRUCTIONS - [gen] (printed or other directions issued by a manufacturer for the operation, maintenance, repair etc. of his products) manuel *m* de instructions
~NUMBER - [comm] (a number or numbers stamped or otherwise marked by a marker on his product

for reference when necessary) numéro *m* de série
MAKESHIFT - [gen] expédient *m*, dispositif *m* de fortune
~TOOLS - [tools] outils *mpl* de fortune
MAKING HOLE - [mining] profondeur *f* du sondage
~-UP - [text] finition *f*, exécution *f*
~WOOLLY - [text] lanification *f*
MALABSORPTION - [med] défaut *m* d'absorption
MALACHITE - [min] (naturally occurring hydrated copper carbonate used as an ore of copper, as a green pigment, and as an ornamental stone) malachite *f*
~GREEN - [dye] (a triphenylmethane dye) vert *m* malachite
MALACIA - [med] malacie *f*, ramollissement *m*
MALADJUSTMENT - [gen & med] défaut *m* d'adaptation
[mech] ajustement *m* défectueux, déréglage *m*
[contr] dereglage
MALAISE - [med] malaise *m*, indisposition *f*
MALARIA - [med] (febrile disease caused by infection) malaria *f*, paludisme *m*
MALARIOTHERAPY - [med] malariathérapie *f*, fièvre *f* artificielle
MALASSIMILATION - [med] assimilation *f* incomplète
MALATHION - [chem] (an insecticide) malathion *m*
MALAXATION - [med] malaxage *m*
MALCHITE - [geol] (a type of rocks) malchite *f*
MALDIGESTION - [med] digestion *f* difficile
MALE - [gen] mâle
~AND FEMALE - [mech] (trade term denoting inner and outer members of pipe fittings) mâle et femelle
~CONE - [mech] rodage *m* mâle, cône *m* mâle
~DIE - [metall & plast ind] (that section of the mould which enters the cavities in the negative mould) moule *m* mâle
~END - [mech] (of a pipe) bout *m* mâle
~FERN - [bot] fougère *f* mâle
~FLOWER - [bot] fleur *f* mâle
~GROUND CONE - [mech] rodage *m* mâle, cône *m* mâle
~GROUND CONE CONNECTION - [mech] raccord *m* à rodage mâle, ajutage *m* à cône mâle
~GROUND JOINT CONNECTION - [mech] raccord *m* à rodage mâle, ajutage *m* à cône mâle
~HORMONE - [biol] hormone *m* mâle
~MOULD - [plast ind] moule *m* mâle
~SCREW - [mech] filetage *m* extérieur
~TAPER JOINT - [mech] rodage *m* à mâle, cône *m* mâle
~THREAD - [mech] filetage *m* mâle
~TOOL - [tools] (tool which is forced into a female tool) mâle *m*
~VALVE - [mech] soupape *f* à mâle
MALEIC ACID - [chem] (a synthesis intermediate, starting material for synthetic resins and a textile processing agent) acide *m* maléique
~ANHYDRIDE - [chem] (a starting material in the production of synthetic resins, a textile processing agent, and a component of paper sizes) anhydride *m* maléique
~HYDRAZIDE - [chem] (a herbicide) hydrazide *f* maléique
~RESINS - [chem] (synthetic resins derived from maleic acid) résines *fpl* maléiques
MALFORMATION - [med] malformation *f*
MALFUNCTION - [gen] fonctionnement *m* défectueux
[mech etc] (a failure in the operation of a machine etc) défaut *m*, dérangement *m*, opération *f* défectueuse
~INDICATOR - [comput] indicateur *m* de dérangement, indicateur *m* d'incident
MALIC ACID - [chem] (a polybasic hydroxy acid used in food processing) acide *m* malique
MALIGNANCY - [gen] malignité *f*, virulence *f*
MALIGNANT - [gen & med] malin
~DISEASE - [med] (a synonym for cancer, sarcoma) cancer *m*, carcinome *m*
~ENDOCARDITIS - [med] (an infective endocarditis) endocardite *f* maligne
~OEDEMA - [vet] œdème *m* malin
MALLARD - [zool] malard *m*, canard *m* sauvage
MALLEABILITY - [metall] (possessing the property of being able to be extended etc., without fracturing) malléabilité *f*
MALLEABILIZATION - [metall] (the action of making malleable) malléabilisation *f*
MALLEABLE - [gen] (soft, easily formed) malléable, forgeable
[metall] (of iron) affiné
~CAST IRON - [metall] (a variety of cast iron exhibiting improved properties under tensile stress) fonte *f* malléable
~IRON - [metall] (wrought iron) fer *m* malléable
[foundry work] s. malleable cast iron
~IRON WIRE - [metall] fil *m* de fer recuit, fil *m* de fer malléable
~METAL - [metall] métal *m* malléable
~STEEL - [metall] acier *m* malléable, acier *m* doux
MALLEABLENESS - s. malleability
MALLEABILIZE, to - [metall] (annealing operation) malléabiliser
MALLEATE, to - [metall] marteler, forger
MALLEATION - [med] martellement *m*, chorée *f* malléatoire
MALLEIN - [chem] (used for inocula for the diagnosis of glanders in horses) malléine *f*
MALLEOLUS - [anat] (projections of the leg bone at the ankle) malléole *m*
MALLEOTOMY - [med] malléotomie *f*, section *f* du marteau
MALLET - [impl] (a wooden hammer) maillet *m*, mailloche *f*, batte *f*
MALLEUS - [anat] (one of the ear ossicles) marteau *m*
MALM - [geol] (an imitation of natural marl) malm *m*
[soil] marne *f*
MALMSTONE - [min] pierre *f* réfractaire
MALNUTRITION - [med] sous-alimentation *f*
MALOCCLUSION - [med] malocclusion *f*, anaraxie *f*
MALOJA - [met] (special type of mountain wind occurring on the Engadine-Bergell windshed, caused by overspill from an anabatic wind producing a katabatic flow on the other side of the pass) maloja *m*
MALONIC ACID - [chem] (an intermediate for the production of a number of hypnotic drugs) acide *m* malonique
MALPOSITION - [med] position *f* anormale
MALPRAXIS - [gen] faute *f* professionnelle
MALPRESENTATION - [med] présentation *f* anormale de l'enfant
MALT, to - [brew ind] (to promote artificial germination of cereals for brewing) malter

MALT - [bot & brew ind] (grain artificially germinated in the presence of moisture and afterwards kiln-dried) malt *m*
~ADJUNCT - [brew ind] succédané *m* de malt
~BIN - s. malt hopper
~CHARGE - [brew ind] versement *m*
~CLEANER - [brew ind] nettoyeuse *f*
~COFFEE - [agric] café *m* de malt, café *m* malté
~COMBS - [agric] (animal feeding) germes *mpl* de malt, touraillon *m*
~COUCH - s. malt piece
~CRUSHER - [brew ind] concasseur *m* pour malt
~CULMS - s. malt combs
~DEGERMINATING MACHINE - [brew ind] dégermeuse *f*
~EXTRACT - [brew ind] extrait *m* de malt
~EXTRACT BATH - [brew ind] (the process) bain-Marie *m*
[brew ind] (the apparatus) appareil *m* à tremper
~FLOOR - [brew ind] germoir *m*, aire *f*
~HOPPER - [brew ind] case *f* à malt, trémie *f* à mouture
~INFUSION - [brew ind] infusion *f* de malt
~KILN - [brew ind] touraille *f*
~MILL - [brew ind] concasseur *m* à malt
~PIECE - [brew ind] (term applied to a heap of malt, in brewing) tas *m* de malt
~PLOUGH - [agric] charrue *f* à malt
~POLISHER - [brew ind] polisseuse *f*
~PRODUCT - [brew ind] produit *m* à base de malt
~ROOTLETS - s. malt culms
~SILO - [agric] silo *m* à malt
~STARCH - [ind chem] fécule *f* de malt
~SUGAR - s. maltose
MALTASE - [biol] (an enzyme which converts maltose and sucrose into hexose sugar) maltase *f*
MALTED MILK - [food] lait *m* malté
MALTESE CROSS - [cin] (basic mechanism for feeding the film forward) croix *f* de Malte
[mech] croix *f* de Malte
~CROSS MOVEMENT - [cin] (of films) entraînement *m* à croix de Malte
MALTHA - [min] malthe *m*, bitume *m* glutineux, pissasphalte *m*
MALTHENES - [min] (bitumen constituents which are soluble in carbon disulphide) malthènes *mpl*
MALTHOUSE - [brew ind] (room or building used for malt production) malterie
MALTHUSIANISM - [gen] malthusianisme *m*
MALTING - [brew ind] maltage *m*, fabrication *f* du malt
~BARLEY - [agric] orge *f* de brasserie
MALTOBIOSE - s. maltose
MALTOSE - [chem] (a dextro-rotatory sugar used in food processing as a sweetening agent. It is formed during the germination of cereals) maltose *m*, sucre *m* de malt
MALTSTER - [brew ind] (person of firm engaged in producing malt) malteur *m*
MAM - s. mammatus
MAMELON - [med] mamelon *m*
MAMILLA - [anat] (nipple) bout *m* de sein, mamelon *m*
MAMMAL - [zool] mammifère *m*
MAMMALGIA - [med] mastodynie *f*
MAMMARY - [zool] mammaire
~ABSCESS - [med] abcès *m* du sein
MAMMARY ARTERY - [anat] artère *f* mammaire
~GLAND - [anat] mamelle *f*, glande *f* mammaire
~HORMONE - [med] hormone *m* mammaire
~VEIN - [anat] veine *f* mammaire
MAMMATOCUMULUS - [met] (type of cumulus cloud, in which more or less regular rounded masses are present on the underside of the cloud) mammatocumulus *m*
MAMMATUS - [met] (characteristic of a cloud, consisting in rounded forms on the underside of a cloud) mammatus *m*
MAMMECTOMY - [med] mastectomie *f*, ablation *f* du sein
MAMMILLARY BODIES - [med] tubercules *mpl* mamillaires
MAMMILOPLASTY - [med] mamilloplastie *f*
MAMMOGRAPHY - [med] (the radiological examination of the breast) mammographie *f*
MAMMOTH - [zool] mammouth *m*
[gen] géant, monstre
~AERIAL - [radio] (aerial of very large size) antenne *f* géante
~ANTENNA - s. mammoth aerial
MAN, to - [gen] garnir d'hommes
[naut] armer, équiper
MAN-ENGINE - [mining] échelle *f* mécanique
~-MADE FIBRE - [text] (a fibre made by artificial means, as distinct from natural fibre. The term includes fibres extruded from material regenerated from natural substances) fibre *f* artificielle
~-MADE ISOTOPE - [nucl] isotope *m* artificiel
~-MADE RUBBER - [rubber ind] caoutchouc *m* artificiel, élastomère *f* synthétique
~-MADE STATIC - [comput] radio-perturbations *f*
MANACLES - [impl] menottes *fpl*
MANAGE, to - [gen] diriger, gérer
MANAGEMENT - [gen & comm etc] direction *f*, gestion *f*, administration *f*
MANAGER - [gen] directeur *m*, gérant *m*
MANAGING - [gen etc] administration, gérance *f*
~DIRECTOR - [comm] (of an incorporated company) administrateur *m* délégué
MANCINISM - [med] gaucherie *f*, sinistralité *f*
MANDAMA - [med] phrynodermie *f*
MANDARIN - [bot] (a type of citrus fruit) mandarine *f*
~OIL - [chem] (an essential oil obtained from Citrus nobilis and used as a flavourant) essence *f* de mandarine
MANDATE - [gen & leg] mandat *m*, mandement *m*
MANDATORY INSTRUCTION - [autom] consigne *f*, ordre *m* impératif
MANDELIC ACID - [chem] (a synthesis intermediate and a bacteriostat used in medicine) acide *m* mandélique
MANDIBLE - [anat] (the lower jaw) mâchoire *f* inférieure
MANDOLINE - [mus] (plucked string instrument) mandoline *f*
MANDRAKE - [bot] mandragore *f*
MANDREL - [mech] (any rod-shaped tool upon which a workpiece is mounted for machining, assembly or other like purpose) mandrin *m*, arbre *m*
[mach tool] (the head stock spindle of a lathe or similar machine) mandrin *m* d'un tour
[electron] (the wire on which helical grids for electronic tubes are wound) mandrin *m*

MANDREL - [mining] pic *m* à deux pointes
[metall] mandrin *m*, noyau *m*
[oil ind] olive *f*
~CARRIER - [mech] croisillon *m*, lanterne *f*
~DRAWING - [metall] (of pipes) laminage *m* sur mandrin
~FORMING - [plast ind] (operation of shaping plastic sheet over a mandrel) formage *m* sur mandrin
~ROLL - [metall] cylindre *m* à mandrin pour tubes soudés
~TEST - [paint] (test for flexibility and adhesion of a coating, in which the composition is applied to metal strips, which are then bent round mandrels of various diameters) épreuve *f* de flexibilité
~WITH CUTTING TEETH - [rubber ind] (used for the retreading of tyres) cylindre *m* cardé et denté
~WRAPPED - [plast ind] (formed by coiling or wrapping round a rotating former of cylindrical or approximately cylindrical shape) bandelé sur mandrin
MANDRIL - s. mandrel
MANE - [anat] crinière *f*
~HAIR - [text] poil *m* de la crinière du cheval
MANEB - [chem] (manganese ethylenebisdithiocarbamate. A fungicide) maneb *m*
MANETON - [mech] (in a radial engine, the removable short stub-end of the crankshaft) maneton *m*
MANEUVER - s. manoeuvre
MANGAN-BLENDE - [min] (natural sulphide of manganese, occurring in massive form; a source of manganese. Also called Alabandite) alabandite *f*, sulfure *m* de manganèse naturel
MANGANATE - [chem] manganate *m*
MANGANESE - [chem] (a metallic element, symbol Mn, A.N. 25, A.W. 54.93, valency 2,3,4 or 7. Used in metallurgy and in the production of manganese compounds) manganèse *m*
~ACETATE - [chem] (a paint and varnish drier, catalyst and textile dyeing agent) acétate *m* de manganèse
~ALLOYS - [metall] alliages *mpl* de manganèse
~BLACK - [paint] (powdered manganese dioxide, used mainly as a drier) bioxyde *m* de manganèse
~BLENDE - [min] alabandite *f*
~BORATE - [chem] (a drier for paints and varnishes) borate *m* de manganèse
~BRONZE - [metall] laiton *m* à grande ténacité
~BROWN - [paint] (brown pigment produced as a by-product in chlorine manufacture) marron *m* de manganèse
~BUTYRATE - [chem] (a therapeutic agent in the treatment of some staphylococcal infections) butyrate *m* de manganèse
~CARBIDE - [metall] carbure *m* de manganèse
~CARBONATE - [chem] (a paint pigment, starting material for other manganese salts, and a dietary supplement) carbonate *m* de manganèse, blanc *m* de manganèse
~CAST IRON - [metall] fonte *f* manganésée
~CITRATE - [chem] (a source of physiological manganese) citrate *m* de manganèse
~COPPER - [metall] cuivre *m* manganésifère
~DIOXIDE - [chem] (an important oxidizing agent, a constituent of ceramic glazes and pyrotechnic mixtures, a feed additive, component of paint and varnish driers, and a starting material for other compounds of manganese) bioxyde *m* de manganèse
~DRIERS - [chem] (driers, including the dioxide, hydrated oxide, acetate, sulphate and borate of manganese as inorganic driers and linoleate, resinate, naphthenate and octoate as organic driers) siccatifs *mpl* au manganèse
MANGANESE GLUCONATE - [chem] (a dietary supplement) gluconate *m* de manganèse
~GLYCEROPHOSPHATE - [chem] (a source of glycerophosphate in medicine) glycérophosphate *m* de manganèse
~GREEN - [paint] (a green pigment made by roasting manganese dioxide with barium hydroxide in an oxidizing atmosphere) vert *m* de manganèse
~HYPOPHOSPHITE - [chem] (used in medicine as a source of phosphorous) hipophosphite *m* de manganèse
~-KILLED STEEL - [metall] acier *m* calmé au manganèse
~LINOLEATE - [chem] (a paint and varnish drier) linoléate *m* de manganèse
~NAPHTHENATE - [chem] (a paint and varnish drier) naphténate *m* de manganèse
~OLEATE - [chem] (a drier for paints and varnishes) oléate *m* de manganèse
~PIG - [metall] fonte *f* manganésée
~RESINATE - [chem] (a drier for paints and varnishes) résinate *m* de manganèse
~SPAR - [min] rhodochrosite *f*
~SPIEGEL - [metall] fonte *f* spéculaire au manganèse
~STEEL - [metall] (steel containing a proportion of manganese, giving improved physical characteristics) acier *m* au manganèse
~STEEL FROG - [metall] coeur *m* en acier au manganèse
MANGANIC FLUORIDE - [chem] (a source of fluorine in organic reactions) fluorure *m* de manganèse
~HYDROXIDE - [chem] (a textile pigment) hydroxyde *m* de manganèse
MANGANIFEROUS - [min] manganésifère
MANGANITE - [min] (a natural hydrated oxide of manganese, crystallizing in the orthorhombic system. An ore of manganese) manganite *f*
MANGANOUS - [min] manganeux
~CHLORIDE - [chem] (an intermediate for pharmaceuticals, a drier for paints and varnishes, a catalyst and a component of fertilizers) chlorure *m* manganeux
~NITRATE - [chem] (a catalyst and component of ceramic glazes) nitrate *m* manganeux
~OXIDE - [chem] (a catalyst, analytical reagent, textile processing agent and a component of ceramic glazes and primary cells) oxyde *m* de manganèse
~SULPHATE - [chem] (a catalyst in rayon manufacture; manganous sulphate also has applications in fertilizers, textile processing, ore extraction and in the preparation of other manganese compounds) sulfate *m* manganeux
MANGE - [vet] gale *f*, rogne *f*
MANGEL - [bot] betterave *f* champêtre
MANGER - [agric] mangeoire *f*, auge *f* (d'écurie)
~LATH - [constr] latte (de râtelier) *f*
MANGLE, to - [gen] calandrer, cylindrer, passer à la calandre
MANGLE - [impl] calandre *f*, calandreuse *f*
[mech] (machine for drying biscuit after application of glaze slip, consisting of racks carried through a drying chamber on a chain conveyor) ca-

landre *f*
MANGLE BOWL - s. mangle roller
~CLOTH - [text] toile *f* de mangle
~EFFECT - [text] effet *m* de mangle, taux *m* d'exprimage
~ROLLER - [text] rouleau *m* de mangle
~WHEEL - [mech] roue *f* à lanterne
MANGLING - [metall] (the operation of flattening plates by a mangle) dégauchissement *m*, redressage *m*
MANGO - [bot] mangue *f*
~-TREE - [bot] manguier *m*
MANGOLD AND BEET RUST - [bot] (plant disease) rouille *f* de la betterave
~FLEA BEETLE - [zool] altise *f* de la betterave
~FLY - [zool] pégomye *f*, mouche *f* de la betterave
MANGOSTEEN - [bot] mangouste *f*
~TREE - [bot] mangoustan *m*
MANGROVE - [bot] mangle *f*
[bot] (the tree) manglier *m*, palétuvier *m*
MANHOLE - [constr] (an orifice, usually fitted with a cover, permitting entry into a tank, etc) trou *m* d'homme, regard *m*, trou *m* de visite
[mining] passage *m* pour les hommes
[telecomm installations] chambre *f*
[railw] abri *m*, retraite *f*, niche *f*
~COVER - [gen] couvercle *m* de chambre
[ovens] plaque *f* de trou d'homme
~GUARD - [gen] garde-fou *m*
~HOOK - [mech] levier *m* pour soulever le couvercle
~JOINT - [mech] jointure *f* pour trou d'homme
~PLATE - [gen] plaque *f* de trou d'homme
~WALLS - [constr] piédroits *mpl* (de chambres ou de galeries)
MANIFEST - [naut] (a bill of lading) manifeste *m*, déclaration *f* d'expédition
MANIFOLD - [mech] collecteur *m*
[plumb, gas install etc] distributeur *m*, nourrice *f*, collecteur *m*, rampe *f* d'alimentation
[mech] (a branching passage to distribute material, especially in extrusion) distributeur *m*, branchement *m*
[paper man] (thin paper used for duplicating) papier *m* à copies multiples
[mech] (in internal combustion engines) tubulure *f*, collecteur *m*, culotte *f*
~ABSOLUTE PRESSURE - [mech] (the pressure in an engine manifold, measured in absolute terms) pression *f* absolu dans le collecteur
~OF ELECTRONIC STATES - [electron](the totality of the electronic terms of an atom or of a molecule) totalité *f* des états électroniques
~PRESSURE - s. manifold absolute pressure
~PRESSURE GAUGE - [instr] (an instrument designed to show the pressure prevailing in an engine manifold) manomètre *m* de pression d'alimentation
MANILLA - [bot] (a fibrous material used for ropes, paper, textile fabrics etc; also called Manila) manille *f*
[bot] (the tree) bananier *m* textile
~HEMP - [text] chanvre *m* de manille
~PAPER - [paper man] (strong paper made partly of manilla hemp) papier *m* de manille
~ROPE - [text] cordage *m* en manille
MANIOC - [bot] manioc *m*
MANIOPHOBIA - [med] peur *f* de la folie
MANIPHALANX - [anat] phalange *f* des doigts
MANIPULATE, to - [gen] manipuler
MANIPULATED VARIABLE - [contr] (regulating variable or correcting condition) grandeur *f* réglante
MANIPULATION - [gen] manipulation *f*
[med] exploration *f*
[nucl] (the handling of radioactive material by means of a manipulator) manipulation *f*
MANIPULATOR - [gen] manipulateur *m*
[nucl] (a mechanism used to handle radioactive material) télémanipulateur *m*
MANNA - [chem] (the dried juice of Fraxinusornus employed in medicine) manne *f*, manne *f* du frêne
~ASH - s. manne
MANNED - [gen] équipé, armé
~FLIGHT - [astronaut] (of a rocket with a man or men on board) vol *m* habité
MANNING - [railw] (of trains) accompagnement *m* des trains
MANNITE - s. mannitol
MANNITOL - [chem] (a synthesis intermediate and a diuretic used in medicine) mannite *f*, mannitol *m*
~HEXANITRATE - [chem] (a detonating explosive) hexanitrate *m* de mannitol
MANNOSE - [chem] (a carbohydrate with uses in biological)research) mannose *m*
MANOCRYOMETER - [instr] (instrument designed to indicate the melting point of a substance caused by a change of pressure) monocryomètre *m*
MANOEUVRABILITY - [gen] manoeuvrabilité *f*
[aero] (the quality in an aircraft determining the rate at which its direction of flight and attitude can be changed) maniabilité *f*
MANOEUVRE, to - [gen] manœuvrer
MANOEUVRE LOAD FACTOR - [aero] (the load factor corresponding to the total aerodynamic lift during a given manoeuvre) coefficient *m* de charge de manœuvre
MANOEUVRING AREA - [aero] (the part of the movement area of an airfield which is used for landing and take-off, and movements associated therewith) aire *f* de manœuvre
MANOGRAPH - [instr] (optical device used to make an indicator diagram for high speed) manographe *m*
MANOMETER - [instr] (pressure gauge) manomètre *m*
MANOMETRIC - [instr] manométrique
~BALANCE - [instr] balance *f* manométrique, manomètre *m* à poids mort
~FLAME - [phys] (small gas flame which oscillates by the variations of pressure caused by sound waves) flamme *f* manométrique
~HEAD - [hydr] hauteur *f* de refoulement manométrique, hauteur *f* manométrique
~SCALE - s. manometric balance
~SUCTION LIFT - [hydr] hauteur *m* manométrique de aspiration
MANROPE - [naut] garde-corps *m*, sauvegarde *f*
MANSARD - [constr] comble *m* en mansarde
~DORMER WINDOW - s. mansard window
~ROOF - [constr] (double-slope roof rising steeply from the eaves) toit *m* en mansarde
~ROOFED - [constr] mansardé
~WINDOW - [constr] lucarne *f*
MANTEL - [constr] (shelf to a fireplace, frequently ornamental) linteau *m*, manteau *m* de cheminée
~TREE - [constr] (the lintel of a fireplace) sous-pou-

tre *f* de cheminée
MANTELPIECE - s. mantel
MANTISSA - [math] (the decimal part of a logarithm) mantisse *f*
MANTLE, to - [mining] couvrir, revêtir
MANTLE - [gen] manteau *m*
[gas ind] (gas light) manchon *m*
[constr] (the outer covering of a wall, when it is of different material from the inner one) parement *m*
[zool] (of mollusca) manteau *m*
[metall] (of blast furnaces) chemise *f* extérieure, enveloppe *f* extérieure
[metall] (of a mould) surtout *m*
[metall] couronne *f*, marâtre *f* d'un haut fourneau
~RING - [metall] marâtre *f*
MANUAL - [print] (book of instructions) manuel *m*
[mus] (keyboard for the hands, generally in connexion with the organ) claviature *f* manuelle, clavier *m*, manuel *m*
[adj] manuel, fait à la main
~ACCESS - [comput] accès *m* manuel
~ADJUSTMENT - [contr] réglage *m* de main
~CENTRAL OFFICE - [telecomm] (in telephony) bureau *m* central manuel
~CLOSED LOOP CONTROL SYSTEM - [comput] système *m* de commande manuelle à asservissement
~CONTROL - [contr] commande *f* volontaire, régulateur *m* de main
[comput] (a component of a mechanism interpreting and carrying out manually initiated directions) commande *f* manuelle
~CONTROLLER - [contr] appareillage *m* de commande volontaire
~COUPLER - [mus] (coupling between the manuals of two organs) accouplement *m*
~EXCHANGE - s. manual central office
~INPUT - [comput] entrée *f* manuelle
~LABOUR - [gen] travail *m* manuel
~LIFT - [agric] élévation *f* à main
~OPERATION - [contr]commande *f* à main,commande *f* manuelle
[comput] fonctionnement *m* manuel, service *m* manuel
~OVERRIDE KEY - [comput] touche *f* d'effacement, touche *f* de déconnexion
~PATCHING - [comput] interconnexion *f* manuelle
~PROGRAMME - [contr] (a programme which is prepared by an operator by means of a device incorporated in the automatic equipment) programme *m* manuel
~RESET - [contr] retour *m* manuel
~RINGING - [telecomm] (in telephony) appel *m* manuel
~STARTER - [mech] démarreur *m* manuel
~SWITCH - [el] commutateur *m* à commande manuelle
~SWITCHBOARD - [telecomm] (in telephony] tableau *m* commutateur manuel
~SWITCHING SYSTEM - [telecomm] (in telegraphy) système *m* de commutation manuelle
~SWITCHROOM - [telecomm] salle *f* des téléphonistes
~TABULATING CARRIAGE - [mech] (of adding machine) chariot *m* tabulateur manuel
~TAPE RELAY - [telecomm] (a tape relay system in which the tape is transferred manually to a specified automatic transmitter position) transit *m* manuel par bande perforée
MANUAL TELEPHONE SYSTEM - [telecomm] système *m* de téléphonie manuelle, réseau *m* téléphonique manuel
~WORK - [gen] travail *m* manuel
MANUALLY - [gen] manuellement, à la main
~OPERATED - [gen & mech] à commande volontaire, à commande manuelle
MANUBRIOCLAVICULAR JOINT - [anat] articulation *f* sternoclaviculaire
MANUBRIUM - [anat] manche *m*, poignée *f*
MANUFACTURE, to - [gen & ind proc] fabriquer, manufacturer
[gen] (of industrial products) fabriquer
[text] confectionner
MANUFACTURE - [gen] fabrication *f*, confection *f*
[mech etc] (fabrication of objects or substances for use) fabriquer, manufacturer
[gen] (the objects or substances manufactured) produit *m* fabriqué
MANUFACTURED EDIBLE FAT - [food] graisse *f* comestible artificielle
~GAS - [gas ind] gaz *m* manufacturé
~GOODS - [gen] produits *mpl* fabriqués
~KNOP - [text] bouton *m* artificiel
MANUFACTURER - [gen] fabricant *m*
[mech etc] usinier *m*
MANUFACTURER'S CERTIFICATE - [comm] certificat *m* du fabricant
~PRICE - [comm] prix *m* de revient, prix *m* de fabrique
MANUFACTURING - [gen] fabrication, production *f*
~COSTS - [comm] frais *mpl* de fabrication
~MILLER - [mach tool] fraise *f* automatique
~PROCESS - [gen] (method of preparation) procédé *m* de fabrication
~SIZE - [photo] largeur *f* de fabrication
MANURE, to - [agric] fumer, engraisser
MANURE - [agric] (dung and/or decayed vegetable matter used to enrich the oil) engrais *m*
~BUCKET ON OVERHEAD MONORAIL - [agric] monorail *m* d'évacuation du fumier
~DISTRIBUTOR - [agric] épandeur *m* de fumier
~FORK - [agric] fourche *f* à fumier
~GAS - [chem] gaz *m* de fumier
~HANDLING - [agric] installation *f* d'évacuation du fumier
~HEAP - [agric] tas *m* de fumier
~LOADER - [agric] chargeur *m* de fumier
~MULCHER - [agric] émietteur *m* de fumier de ferme
~PILE - [agric] fumière *f*
~PIT - s. manure pile
~SPREADER - [agric] ébouseuse *f*
~SPREADING - [agric] épandage *m* du fumier, étalage *m* du fumier
~SPREADING FLOAT - [agric] ébouseuse *f*
~STACK - s. manure pile
MANURIAL REQUIREMENT - [agric] besoins *mpl* en engrais
~VALUE - [agric] valeur *f* fertilisante, valeur *f* engrais
MANURING - [agric] fumure *f*, apport *m* de fumier
MANWAY - [mining] (a passage) passage *m* pour les hommes
[gas ind] (in gasholders) couvercle *m*, plaque *f* pleine
~RAISE - [mining] montage *m* de circulation
~-UP - [oil ind] montage *m* de circulation

MANY-BODY FORCES - [nucl] (interaction between two particles which is altered by the presence of a third particle) forces *fpl* entre plusieurs corps
~-TO-FEW MATRIX - [comput] circuit *m* décodeur
~-VALUED FUNCTION - [math] fonction *f* polyvalente
M.A.P. - s. manifold absolute pressure
MAP, to - [geogr] (the operation of representing a part of the earth's surface on paper) dresser une carte
[surv] dresser un plan
[gas ind etc] (in mainlaying) reporter sur plan, tracer
MAP - [geogr etc] (the representation on paper of part of earth's surface) carte *f*
[town planning] (of a town, i.e. a very detailed representation) plan *m* d'une ville
~BOARD - [surv] planchette *f*
~GRID - [cartography] quadrillage *m*
~MAKER - [gen] cartographe *m*
~MAKING - [gen] cartographie *f*
~-MATCHING GUIDANCE - [radar] guidage *m* par référence cartographique
~MEASURER - [instr] (instrument designed to calculate the length of a route on a map) curvimètre *m*, molette *f* métrique
~OF NETWORK - [el] plan *m* du réseau, tracé *m* du réseau
~PAPER - [paper man] papier *m* pour cartes géographiques
~PLOTTING - [surv] cartographie *f*
~READING - [geogr & nav] lecture *f* de la carte
~SCALE - [geogr] échelle *f* d'une carte
~-SHEET - [geogr] coupure *f* de carte
MAPLE - [bot] érable *m*
~SUGAR - [chem] sucre *m* d'érable
~SYRUP - [chem] sirop *m* de sucre d'érable
MAPPER - s. map maker
MAPPING - [geogr] cartographie *f*
[surv] relèvement *m* topographique
~RADAR - [radar] (radar supplying data for the drawing of maps) radar *m* pour relèvements topographiques
M.A.R. - s. microanalytical reagent
MARABOU SILK - [text] marabout *m*
MARASMUS - [med] marasme *m*, maigreur *f* extrême
MARATHON-HOWARD PROCESS - [chem] (a recovery process for waste sulphite liquor) procédé *m* Marathon-Howard
MARBLE - [geol] (a crystalline limestone used as a structural and decorative material and as a source of carbon dioxide) marbre *m*
~BONES - [med] (osteopetrosis) os *m* de marbre, ostéosclérose *f* généralisée
~CUTTER - [gen] marbrier *m*
~CUTTING - [gen] marbrerie *f*
~DUST - [constr] sciure *f* de marbre
~FACING - [constr] revêtement *m* en marbre
~GRAVEL - [constr] gravier *m* de marbre
~PAPER - [paper man] papier *m* marbré
~POWDER - [constr] poussier *m* de marbre
~QUARRY - [geol] carrière *f* de marbre, marbrière *f*
~QUARRYING - [min] extraction *f* du marbre
MARBLED - [gen] marbré
~CLEAR THROUGH - [pottery] (effect resembling marble in appearance, not merely superficial but extending through the whole mass of the material in question) marbré sur toute l'épaisseur
MARBLED TILE - [constr] carreau *m* marbré
MARBLEIZATION - [paint & med] marbrure *f*
MARBLEIZE, to - [paint] (only USA; to colour so as to give an imitation of marble) marbrer
MARBLES - [glass man] (small spherical pellets of glass) billes *fpl*
MARBLING - [paint] marbrure *f*
[bookbind] racinage *m*, jaspage *m*
~BY CALENDERING - [rubber & plast ind] (of extruded ribbons) fabrication *f* de marbrés par calandrage
MARC - [bot] marc *m*
[agric] (for manure) tourte *f*
MARCASITE - [min] (white iron pyrites used as a source of iron, in jewellery and in the production of sulphuric acid. It is a natural disulphide or iron, crystallizing in the orthorhombic system) marcasite *f*, marcassite *f*
MARCELINE - [text] marceline *f*
MARCH - [met] (the change of a certain meteorological element during a given period of time, e.g. "diurnal march of temperature") marche *f*, changement *m*
MARCID - [gen] déchu, flétri
MARCONI AERIAL - [radio] (an aerial, in general not greater than three eighths of a wavelength, connected to earth by a reactance sufficient to make it resonate) antenne *f* Marconi
~ANTENNA - s. Marconi aerial
~BEAM AERIAL - s. Marconi aerial
~BEAM ANTENNA - s. Marconi aerial
~COHERER - [radio] (a type of detector) cohéreur *m* Marconi
~DETECTOR - [radio] (a form of magnetic detector) détecteur *m* magnétique de Marconi
MARCONIGRAM - [telecomm] radiogramme *m*, radiotélégramme *m*
MARE - [zool] jument *f*
MARE'S TAILS - [met] (cirrus; separate, usually white clouds, smooth and fibrous in appearance) cirrus *m*
[bot] pesse *f* d'eau
MAREKANITE - [min] (a type of perlite broken down in pebbles) marécanite *f*
MARGARIC ACID - [chem] (one of the three fatty acids) acide *m* margarique
MARGARINE - [food] (a substitute for butter manufactured from vegetable and other oils and having colouring and vitamins A and D added) margarine *f*
MARGARITE - [min] (natural hydrated calcium and aluminium silicate) margarite *f*
MARGIN - [gen] marge *f*, bord *m*, rive *f*
[constr] (the open strip of land alongside a road) marge *f*
[instr] (of instrument or apparatus) tolerance *f*, limite *f*, marge *f*
[telecomm] (of a telegraph apparatus, the maximum degree of distortion not affecting the translation of the signals) marge *f*
[agric] (of a wood) lisière *f*
[constr] (of a roofing slate) pureau *m*
~ADJUSTMENT OF THE NEGATIVE CARRIER - [photo] margeur *m* du porte-cliché
~OF SAFETY - [phys mech] (the difference between a given applied load and the ultimate load) marge *f*

de sécurité
MARGIN TILE - [constr] (exposed width of a tile) tuile *f* de bordure
MARGINAL - [gen] marginal
~ABCESS - [med] abcès *m* de la marge de l'anus
~CHECKING - [comput] (method of designing electronic circuits in a computer) contrôle *f* marginale
~CONTRAST - [photo] contraste *m* marginal
~CURRENT - [telecomm] (the adjusted current used for coding impulses in a coder) courant *m* marginal
~DEFINITION - [photo] netteté *f* marginale
~DISTORTION - [photo] distorsion *f* marginale
~FACIES - [min] faciès *m* marginal
~FARM - [agric] exploitation *f* marginale
~FOLD - [geol] pli *m* marginal
~GROOVE - [el acoust] (a blank groove) sillon *m* non-modulé
~LOSS - [comm] perte *f* marginale
~MAGNETIC TRACK - [el acoust] piste *f* couchée en marge
~PERFORATION - [comput] perforation *f* marginale
~PLANK - [constr] traversine *f*
~PUNCHED CARD - [comput] carte *f* à perforation marginale
~SHARPNESS - [opt] netteté *f* aux bords de l'image
~STABILITY - [phys] stabilité *f* marginale
~TESTING - s. marginal checking
~UNSHARPNESS - [photo] flou *m* marginal
MARIAHUANA - [bot] haschisch *m*
MARIAJUANA - s. mariahuana
MARIGOLD WINDOW - [build] (also called rose window) rosace *f*
MARIGRAPH - [instr] (a self-registering tide-gauge) marégraphe *m*, mérégraphe *m*
MARIGUANA - s. mariahuana
MARIHUANA - s. mariahuana
MARIJUANA - s. mariahuana
MARIMBA - [mus] (a xylophone like instrument) marimba *m*
MARINE - [gen & naut] marin, maritime
[naut] (shipping) marine *f*
~ACID - [chem] (hydrochloric acid) acide *m* chlorhydrique
~BAROMETER - [instr] baromètre *m* marin
~BOILER - [naut] (large and short cylindrical boiler with two or more furnaces) chaudière *f* marine
~BORER - [soil] taret *m*
~CHRONOMETER - [horol] (specially mounted chronometer used on board ships) chronomètre *m* marin
~CLAY - [geol] argile *f* (des polders)
~CLIMATE - [met] (type of climate prevailing over large areas of ocean) climat *m* maritime
~COMPASS - [instr] (a gyro compass) boussole *f*
~CONSTRUCTION - [naut] architecture *f* navale
~DENUDATION - [geol] (the erosive action of the sea) dénudation *f* marine
~DEPOSITS - [geol] (rock waste laid down under marine conditions) dépôts *mpl* marins
~ENGINE - [mech] (steam or oil engine used for ships) machine *f* de marine
~ENGINEERING - [mech] (the branch of mechanical engineering dealing with the design and production of machinery for use in ships) génie *m* maritime
~FORMATION - [geol] formation *f* marine
~GALVANOMETER - [instr] galvanomètre *m* marin
~GLUE - [chem] (a water resisting glue) glu *f* marine, colle *f* marine
MARINE INSURANCE - [leg & comm] assurance *f* maritime
~SCREW PROPELLER - [mech] (a boss fitted with blades and producing the thrust which drives a ship) hélice *f* marine
~SURVEYING - [surv] (in tidal waters) relèvement *m* hydrographique
~TRUMPET - [mus] (a one string instrument) trompette *f* marine
~-TYPE CONNECTING ROD - [med] bielle *f* à chapeau
~WATER-TUBE BOILER - [th eng] chaudière *f* aquatubulaire marine
MARINER'S COMPASS - s. marine compass
~NEEDLE - [instr] (the essential component of a compass) aiguille *f* du compas
MARISKA - [med] marisque *f*
MARITIME - [gen] maritime
MARJORAM OIL - [chem] (an essential oil obtained from origanum marjorana L. and used in perfumery and in the production of flavourants) essence *f* de marjolaine
MARK, to - [gen] marquer, estampiller, griffer
[metall] ajourer
[surv] repérer, indiquer
MARK - [gen] marque *f*, signe *m*, empreinte *f*, repère *m*, trace *f*
[comm] marque *f*
[gen] (aim) but *m*, cible *f*
[comm] poinçon *m* de garantie
~-BUOY - [naut] (buoyant structure used to indicate some feature on a sheet of water, e.g. the limit of a channel) bouée *f* d'avis, bouée *f* de marcation
~CHANNEL - [comput] (e.g. on a magnetic drum) piste *f* de marquage
~OF REFERENCE - [print] (any sign directing the reader to a footnote) signe *m* de renvoi
~OFF, to - [mech] (to mark on a workpiece the dimensions to which it is to be finished, or other points or lines needed in working it) tracer, séparer
~OUT, to - s. mark off
~SCANNING - [comput] méthode *f* "mark scanning"
~SCRAPER - [tool] pointe *f* à tracer
~-SENSED PUNCHING - [comput] perforation *f* correspondant à une marque graphitée sur une carte
~SENSING - [comput] lecture *f* de marques sensibles
~SENSING PUNCH CARD - [comput] carte *f* perforée à lecture graphique
~THE CHAIN, to - [text] marquer la pièce
MARKED - [gen] marqué, estampillé, signé
~CAPACITY - [railw] marque *f* de portée
~LEVELLING STAFF - [constr] mire *f* parlante
MARKER - [gen] marqueur *m*
[signal] marqueur *m*, signal *m*
[railw] signal *m*
[cin] dispositif *m* de marquage
[geol] (a formation which can be easily identified) guidon *m*
[comput] index *m*
~BEACON - [radio] (a low-power transmitter giving a characteristic audible signal to show course positions in respect to airfield) radiobalise *f*
~GENERATOR - [radar] générateur *m* marqueur
~HORIZON - [geol] horizon *m* repère
~LIGHTS - [aero] balises *fpl* de piste
[railw] lanternes *fpl* de queue

MARKER LIGHTS - [auto] (small amber and red lights indicating the overall clearance at night) feux *m*pl d'encombrement
~POST - [gen] (mainlaying work) poteau *m* de repérage, borne *f* de jalonnement
MARKERS - [gen] (indicators of officially-regulated colour and shape, intended to show certain areas or obstructions) balises *f*pl
MARKET, to - [gen & comm] vendre, lancer (des articles)
MARKET - [gen] marché *m*
[comm] marché
~ANALYSIS - [comm] analyse *f* du marché
~CONDITIONS - [comm] mouvement *m* des affaires, situation *f* du marché
~DAY - [gen] jour *m* de marché
~DEVELOPMENT - [comm] évolution *f* du marché
~ESTIMATE - [comm] évaluation *f* du marché
~FLUCTUATIONS - [fin] fluctuations *f*pl économiques, fluctuations *f*pl du marché
~FORECAST - [fin] prévisions *f*pl du marché
~GARDEN - [agric] jardin *m* maraîcher
~GARDENER - [agric] horticulteur
~GARDENING - [agric] culture *f* maraîchère
~GLUT - [comm] difficultés *f*pl d'écoulement
~-HALL - [agric] halle *f*
~INTELLIGENCE - [comm] rapports *m*pl sur la situation du marché
~PLACE - [gen] place *f* du marché
~POULTRY - [agric] poulets *m*pl de consommation
~PRICE - [comm] prix *m* courant
~RATE - [fin] taux *m* courant d'escompte
~REGULATIONS - [comm] réglementation *f* du marché
~RESEARCH - [comm] prospection *f*, démarchage *m*, étude *f* du marché
~SURVEY - [comm] sondage *m* d'un marché
~TREND - [comm] tendance *f* du marché, évolution *f* du marché
~VALUE - [comm] valeur *f* marchande
MARKETABLE - [gen & comm] négociable
~OUTPUT - [comm] production *f* susceptible d'écoulement
MARKETING - [comm] (the action of marketing) marketing *m*, service *m* commercial (d'un produit)
[comm] (the buying and selling of products) achat *m* au marché, vente *f* au marché
[comm] (the produce of products sold in a market) marchandise *f*
~RESEARCH - s. market research
MARKING - [gen] marquage *m*
(the marking of goods containers) marquage *m*
[mech] (the action of marking off) repérage *m*
[rail] (of coaches) marquage *m* (des voitures)
[telecomm] (in Morse telegraphy) travail *m*
[telecomm] (in printing telegraphy; the condition resulting in an active selecting operation in a receiving apparatus) travail *m*
[telecomm] (a signalling condition in isochronous systems) travail
~AWL - [tool] traceret *m*, traçoir *m*
~BAND - [text] cordon *m* de contrôle
~CHALK - [gen] craie *f* de marquage
~DEVICE - [mech] appareil *m* de marquage
~EQUIPMENT - [agric] marqueur *m*
~GAUGE - [carp] trusquin *m* à pointe, trousquin *m*
~INK - [gen] encre *f* à marquer
MARKING KNIFE - [tool] (steel tool with a chisel edge at one end and pointed at the other end) traçoir *m*
~MACHINE - [mech] machine *f* à marquer
~MOTION - [text] appareil *m* marqueur
~NEEDLE - [impl] aiguille *f* à marquer
~OFF - s. marking out
[metall] traçage *m*
~-OFF SLAB - [metall] (a massive cast-iron table on which a workpiece is placed for marking-off) plaque *f* de traçage
~OUT - [mech] (setting out centre lines and other marks to facilitate subsequent machining) traçage *m*
[surv] jalonnement *m*
~OUT OF TRACKS - [railw] jalonnement *m* de la voie
~PAINT - [paint] (special paint used for showing indications of position, etc. on a part) vernis *m* à marquer
~PERCENTAGE - [telecomm] taux *m* de travail
~POINTER - [instr] (stationary pointer which is adjustable for position on the scale) aiguille *f* fixe
~POST - [telecomm] borne *f* de repérage
~ROLLER - [agric] rouleau *m* marqueur
~TABLE - s. marking-off slab
~THE WARP - [text] marquage *m* de la chaîne
~TOOL - [tool] rouanne *f*, style *m* de repérage
~WAVE - [radio] (the electromagnetic wave which is radiated when marking signals are made in accordance with the code) onde *f* de manipulation
MARL, to - [agric] marner le sol
MARL - [geol] (general term denoting a fine-grained rock) marne *f*, caillasse *f*
[constr] (brick earth containing a large percentage of carbonate of lime) brique *f* à marne *f*
~-PIT - [mining] marnière *f*
~STONE - [constr] pierre *f* silico-marneuse
MARLIN'S SYNDROME - [med] anémie *f* phagocytaire
MARLINE - s. marling
~HITCH - [naut] transfilage *m* avec demi-clef
~HOLES - [naut] trous *m*pl à merliner
~-SPIKE - [naut] (or marlinspike; iron tool to separate the strands of rope in splicing) épissoir *m*, poinçon *m* à épisser
~TIE - [telecomm] fixation *f* (d'un câble porteur) au moyen d'une corde
MARLING - [naut] guirlande *f*
MARLPIT - [min] (pit from which the marl is dug out) marnière *f*
MARLY - [text] marli *m*
[geol] marneux
~CLAY - [geol] argile *f* marneuse
~LIME - [geol] chaux *f* marneuse
MARMALADE - [food] confiture *f* d'oranges
MARMATITE - [min] (ferruginous variety of blende) marmatite *f*
MARMON CLAMP - [astronaut] bride *f* de serrage "marmon"
MARMORATE - [bot] (coloured and marked like marble) marbré
MARMOREAL - [gen] de marbre
MARMOT - [zool] marmotte *f*
[text] (the dressed skin of the marmot) marmotte *f*
MAROON, to - [naut] abandonner dans une île déserte
MAROON - [dye] marron pourpré
[gen] (a paper shell) fusée *f* à pétard
MARQUEE - [gen] (projecting canopy) marquise *f*
[gen] (a large tent) tente-marquise *f*

MARQUENCHING - [metall] (a method of heat-treatment) refroidissement *m* rapide martensitique
MARQUETRY - [carp] (inlaid work used for decorating furniture) marqueterie *f*
MARQUISE - s. marquee
MARRIED JOINT - [el] jonction *f* à chevauchement
~PRINT - [cin] (film containing picture and sound tracks in their correct relationship) copie *f* finale
MARROW - [zool] (connective tissue in the central cavities of long bones) moelle *f*
[bot] courge *f* à la moelle, courgette *f*
~BONE - [anat] os *m* à moelle
MARROWSTEM KALE - [agric] chou *m* moellier
MARRY, to - [gen] marier
[constr] (the operation of lashing poles together) ancrage *m* à chevauchement
[mech] (to couple) raccorder, emmancher
[cin] (the operation of printing picture and sound track together) préparer la copie finale
MARS - [astr] (a planet) Mars *m*
MARSH - [geogr] marais *m*, marécage *m*
~DRAINAGE - [agric] dessèchement *m* des marais
~FEVER - [med] paludisme *m*
~FOXTAIL - [bot] vulpin *m* genouillé
~GAS - [chem] (synonym for methane) gaz *m* de marais, méthane *m*, formène *m*
~-MALLOW - [bot] guimauve *f*
~MARIGOLD - [bot] souci *m* d'eau, populage *m*
~ORE - [min] limonite *f*, minerai *m* des prairies
~PLANTS - [bot] plantes *fpl* palustres
~THISTLE - [bot] cirse *m* des marais
~TREFOIL - [bot] trèfle *m* d'eau
MARSH'S TEST - [chem] (for the estimation of arsenic) essai *m* de Marsh
MARSHAL, to - [gen] placer en ordre
MARSHAL A TRAIN, to - [railw] classer un train
MARSHALLING - [gen] disposition *f* en ordre
[railw] classement *m*, triage *m*
~BOX - [el] boîte *f* de dérivation
~TRACK - [railw] voie *f* de classement
~YARD - [railw] chantier *m* de triage, gare *f* de triage
MARSHY - [gen] marécageux
MARSUPIAL - [zool] marsupial *m*
MARSUPIALIZATION - [med] repliement, marsupialisation *f*
MARSUPIUM - [zool] (the pouch-like structure which is occupied by the young during the later development) marsupium *m*
MARTELINE - [impl] (a tool used for sculpting marble) martelet *m*
MARTEMPERING - [metall] (a form of heat-treatment) trempe *f* étagée martensitique
MARTENS DISTORTION TEST - s. Martens test
~TEST - [phys] (test for deformation at high temperature) essai *m* de Martens
~WEDGE - [opt] (quartzy wedge rotator for polarized light) coin *m* de Martens
MARTENSITE - [metall] (constituent formed when steel is cooled rapidily enough to prevent the austenite changing to pearlite and consisting of a solid solution of carbon in alpha-iron) martensite *f*
~CARBON - [chem] carbone *m* martensitique
~QUENCH - [metall] trempe *f* martensitique
MARTIN'S DISEASE - [med] périostéo-arthrite *f* du pied
MARVER - [glass man] (marble block on which glass is rolled during working by hand) plaque *f* de marbre
MARX EFFECT - [nucl] (the reduction in the energy of a photoelectric emission by the simultaneous incidence of radiation of a frequency lower than that of the emission) effet *m* Marx
MASHALADENITIS - [med] hidrosadénite *f* axillaire
MASHALEPHIDROSIS - [med] hyperhidrose *f* axilaire
MASCOT - [gen & auto] mascotte *f*, fétiche *m* (de bouchon de radiateur)
MASCULINE - [gen] masculin
MASCULINOVOBLASTOMA - [med] surrénalome *m* masculinisant
MASER - [nucl] (very stable low-noise oscillator operating by the interaction between radiation and atomic particles) maser *m*
MASH, to - [brew ind] (to crush, to knead) brasser, démêler
MASH - [gen] (material which has been kneaded or crushed) mélange *m*, mâche *f*, pâte *f*
[brew ind] (a mixture of coarse malt and hot water) trempe *f*, maische *m*, fardeau *m*
~COPPER - [brew ind] chaudière *f* à trempes
~FILTER - [brew ind] filtre *m* à moût
~FILTER PLATE - [brew ind] plateau *m* de filtre à moût
~GOODS - [brew ind] ensemble *m* de la trempe
~LIQUOR - [brew ind] eau *f* d'empâtage
~MACHINE - [brew ind] agitateur *m*
~SEAM WELDING - [metall] soudure *f* plastique continue
~SYSTEM - [brew ind] méthode *f* de brassage
~TUN - [brew ind] (a vessel in which mash is mixed and held at a temperature of about I50 deg. F. of about two hours, after which the sweet wort is drawn off) cuve-matière *f*
MASHER - [brew ind] (the vessel in which the malt is mixed with hot water before being passed to the mash tun) hydrateur *m*
MASHING - [brew ind] (the extraction of malt with water after grinding) brassage *m*
MASK, to - [gen & paint] (to apply masking material) masquer, cacher
(to produce a sound which causes loss of sensitivity by the ear for another specified sound) masquer
[mech] (to fit into a recess) masquer
MASK - [gen] masque *m*
[med] (device for supplying oxygen for respiration) masque *m*
[photo] (opaque print used to limit the printing of a negative) cache *m*, cache-cadre *m*, papillon *m*
[print] (opaque printing used to limit an area of colour) cache *m*
[comput] (auxiliary word deleting certain digits of another word) masque *m*
~MICROPHONE - [el acoust] (a microphone which is used inside a respiratory mask) microphone *m* de masque
~SHOT - [photo & cin] prise *f* de vue avec cache
MASKED SIGNAL - [gen] signal *m* masqué
~VALVE - [mech] (in internal combustion engines) soupape *f* incorporée
MASKING - [gen] pose *f* d'un masque, cache *m*
[photo] (in colour separation) cache *m*, obscuration *f*
[acoust] (the loss of sensitivity of the ear for certain sounds) masquage *m*
[chem] (the masking of those parts of a material

which are not to be subjected to a given treatment) masquage *m*
MASKING - [mech] masquage *m*
~AUDIOGRAM - [acoust] (graphical presentation of the masking which is due to a static noise) audiogramme *m* de l'effet de masque, audiogramme *m* de masquage
~FOR REPAIRS - [paint] couverture *f*
~FRAME - [photo] cadre-cache *m*
~LEVEL - [acoust] niveau *m* de masquage
~MATERIAL - [gen] (special material, usually adhesive, used to protect a surface from some form of treatment, e.g. paint spraying) matériel *m* pour masque
~PAPER - [paper man] (paper used to protect a surface from some treatment) papier-cache *m*
~PLATE - [telev] plaque *f* de masquage
~TAPE - [ind proc & ind chem] (adhesive tape used for convering parts of a workpiece which are to be protected from some form of treatment, e.g. paint spraying) ruban *m* pour masquer
MASLIN - [bot] méteil *m* mouture *f*
MASLINE FLOUR - [agric] farine *f* mélangée
MASON, to - [constr] (to build of stone, of bricks) maçonner, construire en maçonnerie
MASON - [gen] maçon *m*
~UP, to - [constr] maçonner, élever en briques
MASON'S BOLSTER - [constr] hachard *m*, ciseau *m* à froid (ou à chaud)
~CHISEL - [impl] ciseau *m* de maçon
~LEVEL - [impl] niveau *m* de maçon
MASONITE - [constr] (proprietary name used for a lining board because of its heat-insulation properties) masonite *f*
[min] masonite *f*
MASONRY - [constr] (the building in stone) maçonnerie *f*, maçonnage *m*
[metall] (of a furnace) muraillement *m*
~BOND - [constr] appareil *m* (de maçonnerie)
~DAM - [hydr] barrage *m* en maçonnerie, mur *m* de moellon
~SAND - [constr] sable *m* de mortier
~SHAFT - [mining] puits *m* maçonné
MASS - [gen] masse *f*, amas *m*
[phys] (the physical measure of the matter in a body) masse *f*
[el] masse *f*, terre *f*
~ABSORPTION - [chem] absorption *f* massique
~-ABSORPTION COEFFICIENT - [nucl] (the fractional decrease in intensity per unit surface density) coefficient *m* d'absorption massique
~ABUNDANCE - [nucl] concentration *f* en pourcentages de poids
~ACTION - [chem] (a chemical reaction the velocity of which is dependent on concentrations of the participating substances) action *f* de masse
~ASSIGNMENT - [nucl] (the determination of the mass of a radioactive species) détermination *f* du nombre de masse
~ATTENUATION - [phys] coefficient *m* d'atténuation massique
~BALANCE - [phys] équilibre *m* des masses
~-BALANCE WEIGHT - [aero] (a weight fixed to a control surface to modify the inertial coupling between the angular movement of the surface and some other motion of an aircraft) masse *f* d'équilibrage
MASS-BALANCING - [mech] équilibrage *m* des masses
~COEFFICIENT OF REACTIVITY - [nucl] coefficient *m* massique de réactivité
~-COLOURED GLASS FILTER - [photo] filtre *m* teinté dans la masse
~CONCRETE - [constr] béton *m* de ciment
~DECREMENT - [nucl] (the difference between the atomic mass and the mass number of a nuclide) défaut *m* de masse
~EFFECT - [metall] (the tendency of hardened steel to decrease in hardness from the surface to the centre) effet *m* de masse
[phys] effet *m* de tassement
~ENERGY ABSORPTION COEFFICIENT - [phys] coefficient *m* d'absorption d'énergie massique
~-ENERGY CONVERSION FORMULA - [phys] (the ratio between the atomic weight unit and the atomic mass unit determined experimentally) formule *f* de conversion énergie-masse
~-ENERGY EQUATION - s. mass-energy equivalence
~-ENERGY EQUIVALENCE - [phys] (the equivalence of a quantity of mass and a quantity of energy when both are related to the equation $e=mc^2$) équivalence *f* masse-énergie
~-ENERGY TOTAL - [phys] (the total of the mass and the energy of any particle) total *m* masse-énergie
~-ENERGY RELATION - [phys] équation *f* d'Einstein, relation *f* masse-énergie
~FILTER - [chem] filtre *m* de masse
~FLOW - [phys] débit-masse *m*
~FLUX - [phys] flux *m* de masses, flux *m* massique
~FORMULA - [phys] (equation giving the atomic mass of a nuclide as a function of its atomic number and its mass number) formule *f* massique
~HARDNESS - [metall] dureté *f* excessive
~-IMPREGNATED INSULATION - [el] isolation *f* au papier impregné
~-IMPREGNATED NON-DRAINED INSULATION - [el] isolation *f* à imprégnation stabilisée
~NUMBER - [phys] (the number nearest in value to the atomic mass when that quantity is expressed in atomic mass units) nombre *m* de masse, nombre *m* de nucléons
~NUMBER OF A NUCLEUS - [phys] (the total number of protons and neutrons contained by the atomic nucleus) nombre *m* massique d'un noyau
~OF THE ELECTRON - [electron] (quantity analogous to mass, characterizing the inertia effects of the electron in an electric or magnetic field) masse *f* de l'électron
~OF THE UNIVERSE - [phys] masse *f* de l'univers
~OPERATOR - [phys] (in the quantum theory) opérateur *m* de masse
~-PRODUCE, to - [gen] fabriquer en série
~-PRODUCED GOODS - [gen] articles *mpl* de série
~PRODUCTION - [gen] fabrication *f* en grande série
~-PRODUCTION WORK - [gen] travail *m* en série
~RADIATOR - [el] (source of electromagnetic radiation covering a large band) oscillateur *m* d'ondes amorties
~RANGE - [phys] (the ratio which is expressed in units of surface density) portée *f* massique
~REACTANCE - [acoust] (reactance due only to inertia) réactance *f* de masse
~SEPARATION - [nucl] séparation *f* en masses partielles sous-critiques
~SPECTRA - [phys] (the positive ray spectra which

are obtained by means of the mass spectrograph) spectres *m*pl massiques
MASS SPECTROGRAPH - [phys] (apparatus designed to sort streams of electrified particles according to their different masses by means of deflecting fields) spectrographe *m* de masse
~-SPECTROGRAPHIC METHOD OF ISOTOPE SEPARATION - [nucl] méthode *f* passe-spectrographie de séparation d'isotopes, méthode *f* électromagnétique
~SPECTROGRAPHY - [phys] spectrographie *f* de masse
~SPECTROMETER - [instr] (an apparatus for the analysis of positively charged particles) spectromètre *m* de masse
~SPECTRUM - [phys] spectre *m* de masse
~STOPPING POWER - [phys] pouvoir *m* d'arrêt massique
~TRANSFER - [phys] (the transfer of mass at the boundary face between two phases in a distillation columns)transfert *m* de masse
~-TRANSFER COEFFICIENT - [chem] coefficient *m* de transfert de masse
~UNIT - [phys] (atomic mass) masse *f* atomique
~VELOCITY - [phys] (the rate of flow of the mass per unit cross-sectional area) densité *f* de courant massique
~YIELD CURVE - [phys] courbe *f* de rendement massique
MASSAGE, to - [gen & med] masser, malaxer
MASSAGE - [gen & med] massage *m*
~APPARATUS - [mech] dispositif *m* à masser
MASSICOT - [chem] (a yellow pigment, identical with litharge the chemical standpoint) massicot *m*
MASSIF - [geol] (any considerable land-mass rising above the normal level of the country) massif *m*
MASSIVE - [gen] massif, aggloméré
~RISE - [constr] face *f* (en pierre) formant contremarche
MAST - [gen] mât *m*
[naut] (pole set upright on a ship's keel to support the sails) mât *m*, arbre *m*
[aero] (autogyro; the structural element which supports the rotor) pylône *m*
[oil ind] (the upright pole of a derrick) anche *f*
[railw] (of aerial ropeways) pylône *m* (pour téléphériques)
[el & telecomm] mât *m* en lattis
~AERIAL - [radio] (aerial in which the currents are carried by the metallic structure of the mast) antenne *f* à perche
~COAT - [naut] ton *m* de mât
~CRANE - [mech] mât-grue *m*
~FOR HOISTING - [med] mât *m* de levage
~HEEL - [naut] pied *m* de mât
MASTADENITIS - [med] mastite *f*, mammite *f*
MASTADENOMA - [med] adénome *m* mammaire
MASTATROPHIA - [med] atrophie *f* mammaire
MASTATROPHY - s. mastatrophia
MASTER, to - [gen] maîtriser, se rendre maître
MASTER - [gen] maître *m*, patron *m*
[naut] patron *m*
[mech] (term denoting special tools used for checking) calibre-mère
[el acoust] (the metal part obtained by electroforming, which is the negative of the recording) père *m*
~ALLOY - [metall] alliage-mère *m*
~AND ARTICULATED ASSEMBLY - [mech] (the whole connecting-rod assembly relating to those cylinders of an engine which lie in the same plane normal to the axis of the crankshaft) embiellage *m* à bielle maîtresse et biellette
MASTER BLACK CONTROL - [telev] dispositif *m* du réglage du noir
~BLADE - [el] (the blade of a wing-blade which cuts off the lights) lame *f* d'obturation
~BLOCK - [metall] porte-poinçon *m*
~BORER - [tool] maître-sondeur *m*
~BRAKE CYLINDER - s. master cylinder
~BRIGHTNESS CONTROL - [telev] dispositif *m* de réglage commun de la luminosité
~-BUILDER - [constr] constructeur *m*
~BUSHING - [oil ind] boisseau *m* principal
~CARD - [comput] (a card the perforations of which are automatically transferred on all the cards punched) carte *f* maîtresse
~CARD FILE - [comput] pile *f* de cartes maîtresses
~CARD GANG PUNCHING - [comput] perforation *f* en série
~CLOCK - [contr] (source of standard timing signals) horloge *f* mère, rythmeur *m*
[comput] timing *m* général
~CONNECTING-ROD - [mech] (in an engine having two or more cylinders in the same plane normal to the crankshaft axis, the principal connecting-rod which incorporates a big-end bearing, and to which the secondary connecting-rods are linked) bielle *f* maîtresse
~CONTROL - [cin telev etc] (a central switching point for amplifiers etc) régie *f*
~CONTROL VALVE - [oil ind] (at the well inlet) vanne *f* d'arrêt
~CONTROLLER - [contr] régulateur *m* direct , manipulateur *m*
~CYLINDER - [mech] (of the brake assembly of autos) cylindre *m* principal de frein, maître-cylindre *m*
~DIE PLATE - [metall] moule *m* maître, coquille-mère *f*
~DRAIN - [hydr] (of an irrigation field) conduit *m* de décharge principal
~FADER - [telev] atténuateur *m* principal
~FARMER - [agric] maître-agriculteur *m*
~FILE - [comput] données *f* directrices, données *f*pl pilotes
~FORM - [metall] (original form from which other forms can be prepared) forme *f* maîtresse
~FORMER - s. master die plate
~FREQUENCY METER - [instr] (instrument designed to integrate the number of cycles through with an alternative supply voltage has passed in a specified period of time) fréquencemètre *m* intégrateur
~GATE - [hydr etc] vanne *f* principale
~GAUGE - [instr] (a standard gauge from which others are made by comparison) calibre-mère *m*, comparateur *m*
[oil ind] rapporteur *m*
~HOB - [metall] (a printing tool) poinçon *m*
~KEY - [impl] (special key operating a number of locks having keys which are not interchangeable) passe-partout *m*
~LEAF - [mech] (of spring) lame *f* maîtresse
~LIFT CONTROL - [telev] régleur *m* principal du niveau du noir
~LOUDSPEAKER - [el acoust] haut-parleur *m* principal

MASTER MARK FEATURE - [comput] dispositif *m* de marquage de feuilles maîtresses
~MECHANIC - [gen] (chief mechanic or chief mechanical engineer) contremaître *m* mécanicien
~METER - [instr] compteur *m* général
~MODEL - [gen] (original or primary design from which others are produced or copied) modèle-mère *m*
~MONITOR - [telev] (the final monitor which controls the image broadcast) moniteur *m* d'émission
~MOULD - [metall & plast ind] (mould constituting a primary or original form, from which copies can be made) moule *m* maître
~OSCILLATOR - [radio] (oscillator generating a constant frequency from which the carrier frequency of a radio transmitter is derived) maître-oscillateur *m*, oscillateur *m* pilote
~PATTERN - s. master model
~PICTURE MONITOR - [telev] moniteur *m* principal d'image
~PRINT - [photo & cin] (the print from which a duplicate negative can be printed) copie *f* originale
~PROGRAMME - [comput] (routine controlling other routines) programme *m* principal
~ROD - [mech] bielle *f* maîtresse
~ROUTINE - [comput] programme *m* principal
~SCHEDULE-s. master programme
~SHOT - [cin] (a shot which reproduces the whole scene) plan *m* général
~-SLAVE MANIPULATOR - [nucl] manipulateur *m* asservi, robot *m*
~SOUND NEGATIVE - [cin] (the sound negative used to complete the picture negative) négatif *m* son primaire
~SOUND TRACK - [cin] (the complete sound track which is subsequently printed together with the picture film) piste *f* sonore primaire
~SWITCH - [el] disjoncteur *m* principal, interrupteur *m* général
~TAP - [mech] (screw tap used for accuracy) taraud *m* mère, taraud *m* matrice
~TAPE - [comput] (the magnetic tape with the main file) bande *f* matrice
~TAPER - [mech] taraud *m* conique mère
~TELEPHONE TRANSMISSION REFERENCE SYSTEM - [telecomm] (in telephony) système *m* fondamental americain de référence pour la transmission téléphonique
~TRANSMITTER - [radio & telev] (transmitter emitting the signal which is retransmitted by the slave transmitter) émetteur *m* pilote
~UNIT - [comput] calculateur *m* maître
MASTERBATCH - [ind chem] (a quantity of concentrated material from which batches for processing are made up by dilution) mélange *m* maître, mélange *m* mère
MASTHEAD - [naut] tête *f* de mât, mât *m* de flèche
~LIGHT - [naut] feu *m* de tête de mât
MASTIC - [chem] mastic *m*
[paint] (a pale yellow resin, used in protective varnishes for oil paintings) mastic *m*
[ind chem] (an adhesive, also used as a filler) mastic *m*
~ASPHALT - [min] (mixture of bitumen and fine mineral matter used for road surfacing, water-proofing constructions etc) mastic *m* d'asphalte
~BLOCKS - [constr] blocs *mpl* d'asphalte

MASTIC GUM - [chem] (a gum obtained from Pistachia lentiscus and used in medicine, in the production of varnishes and in a food manufacture) gomme *f* mastic
~TREE - [bot] lentisque *m*
MASTICATE, to - [gen] mâcher, mastiquer
[rubber ind etc] (the operation of converting rubber into a homogeneous mass) malaxer, plastifier
MASTICATION - [gen] mastication *f*, mâchement *m*
[rubber ind] mastication *f*, malaxage *m*
[mech] trituration *f*
MASTICATOR - [rubber ind] (a process machine with radial knives on a rotating shaft) masticateur *m*, plastificateur *m*, triturateur *m*
MASTICATORY - [zool] masticateur
MASTING - [naut] mâtage *m*
MASTITIS - [med] (inflammation of the mammary glands) mastite *f*, mammite *f*
MASTODYNIA - [med] (ache in the breast) mastodynie *f*
MASTOID - [anat] (nipple-shaped process of the otic capsule) mastoïde *f*
~ABSCESS - [med] mastoïdite *f* suppurée
~CAVITY - [anat] cavité *f* mastoïdienne
MASTOIDECTOMY - [med] trépanation *f* mastoïdienne
MASTOIDEOCENTESIS - [med] ponction *f* de la mastoïde
MASTOIDITIS - [med] mastoïdite *f*
MASTOPATHY - [med] mastopathie *f*
MASTOPEXY - [med] mastopexie *f*
MASTOSCIRRHUS - [med] squirre *m* du sein
MASURIUM - [chem] (an obsolete term for Element 43, now known as Technetium) technécium *m*
MAT, to - [brew ind] (the malt) feutrer
[gen] matir, mater
[hydr] clayonner
MAT - [gen] natte *f*, tapis *m*, carpette *f*
[plast ind] (for moulding) matte *f*
[naut] paillet *m*
[adj] mat
[text] (woven straw) carpette *f*, natte *f*
[print] empreinte *f* de clichage
[hydr] clayonnage *m*
~ETCHING - [metall] décapage *m* au mat
~FINISH - [paper man & text] papier *m* mat
~FOUNDATION - [soil] fondation *f* sur radier, radier *m* de fondation
~-GRASS - [bot] nard *m* raide
~MOULDING - [plast ind] (moulding from previously prepared sheets of glass-reinforced resin) moulage *m* de mats
~SINKING - [constr] (a shallow cavity housing a mat) about *m* de tapis
~WEAVE - [text] armure *f* natté toile
~WEED - [bot] stipe *m* plumeux
~WELL - s. mat sinking
MATCH, to - [gen] unir, égaler
[mech etc] adapter
[metall] (die casting) centrer
[carp] bouveter
MATCH - [gen] égal, pareil
[impl] allumette *f*
[expl] mèche *f*
[metall] (forging work) centrage *m*
~-LINING - s. matched boards
~-MERGE - [comput] interclassement *m* avec sélec-

tions
MATCH PLATE - [metall] plaque-modèle *f* double-face
~-PLATE JOB - [metall] travail *m* à plaque modèle
~ UP, to - [gen] assembler de manière *f* adéquate
~ WAGON - [railw] (a shock-absorbing wagon) wagon *m* de choc
MATCHBOARD - s. matched boards
MATCHBOARDING - s. matched boarding
MATCHED - [gen] couplé
[mech] bien adapté
~ BOARDING - [carp] planches *fpl* de recouvrement, jointes *fpl*
~ BOARDS - [carp] (boards cut at the edges so that close joints can be made) planches *fpl* bouvetées
~ DOUBLETS - [electron] doublets *mpl* assortis
~ ELECTRON TUBES - [electron] tubes *mpl* électroniques à caractéristiques identiques
~ JUNCTION - [radio] (loss-free junction) jonction *f* adapté
~ LENSES - [photo] (a pair of more or less equal lenses in a camera) objectifs *mpl* couplés
~ LOAD - [radio] (a load so adjusted as to accept the maximum power supplied to it) charge *f* adaptée, terminaison *f* adaptée
~ METAL DIES - [metall & plast ind] (a mould composed of two parts designed to close accurately) moule *m* et contre-moule rigides
~ REPEATING COILS FOR PHANTOM CIRCUIT - [telecomm] translateur *m* double pour circuit fantôme
~ REPEATING COILS FOR SIDE CIRCUIT - [telecomm] translateur *m* double pour circuit réel
~ TERMINATION - [telecomm] (termination to a line which does not reflect any energy) terminaison *f* adaptée
~ TRANSMISSION LINE - [electron] (waveguides) ligne *f* de transmission adaptée
~ WAVEGUIDE - [electron] guide *m* d'ondes adapté
MATCHING - [gen] adaptation, assortiment *m*, étalonnage *m*
[metall] centrage *m*
[carp] s. matched boarding
[text] triage *m* de la laine (suivant les qualités des toisons)
[el] (an adjustment of the electrical characteristics of the load circuit to ensure a maximum transfer of energy from the generator to the load) adaptation *f*
~ BLOCK - [electron] (solid body inserted in a waveguide for matching purposes) bloc *m* d'adaptation
~ DIAPHRAGM - [electron] (in waveguides) diaphragme *m* d'adaptation
~ IMPEDANCE - [el] impédance *f* d'adaptation
~ MACHINE - [mech] (machine for matched board) machine *f* à bouveter
~ OF A TELEVISION AERIAL CABLE - [telev] impédance *f* terminale d'un câble de télévision
~ OF BLOOD - [med] groupage *m* du sang
~ PIECE - [mech] raccord *m*
~ PILLAR - [electron] (a rod which projects from an interior face of a waveguide) pilier *m* d'adaptation
~ PLATE - [electron] (used for impedance matching) plaque *f* d'adaptation
~ POST - [electron] (of waveguides) tige *f* d'adaptation
~ REPEATING COILS - [telecomm] translateur *m* double
MATCHING STIMULI - [electron] stimuli *mpl* d'adaptation
~ STRIP - [electron] (waveguides) bande *f* d'adaptation
~ STUB - [radio] (of an aerial; a short loss-free conductor in a waveguide for adaptation purposes) adaptateur *m* d'impédance
~ TRANSFORMER - [el] (a coupling transformer so designed as to ensure maximum energy transfer even when the load impedance differs from that of the source) transformateur *m* d'adaptation
MATCHPLATE - s. match plate
MATE, to - [gen] unir, accoupler
[mech] accoupler
MATELASSE - [text] (compound cloth with a brocaded face) matelassé *m*
MATERIAL - [gen] matère *f*, matériaux *mpl*
[adj] matériel
[gen] (a substance) substance *f*
[text] étoffe *f*, tissu *m*
~ BALANCE - [phys] (the current relation between feed, product and waste) bilan *m* matière
~ BILL - [comm] inventaire *m* des matériaux
~ BUCKLING - [nucl] (property of the material on the multiplying medium) laplacien *m* matière
~ CHECKING - [comm] contrôle *m* des matériaux
~ CODE - [comput] code *m* matières
~ CONSTANT - [phys] constante *f* de la matière
~ ECONOMY - [nucl] (the efficiency with which fissionable material is used) économie *f* des matériaux
~ HANDLING - [gen] distribution *f* des matériaux
[ind prot] manipulation *f* de matériel
~ HANDLING EQUIPMENT - [mech] installation *f* pour le mouvement des matériaux
~ INSPECTION - [gen] essai *m* des matériaux
~ INVENTORY - [nucl] (the quantities of fissionable materials) inventaire *m* de matériel
~ MASS - [phys] (the momentum divided by the velocity) masse *f* matérielle
~ PARTICLE - [phys] (a concept relating to the principles) point *m* matériel
~ TESTING REACTOR - [nucl] (or MTR; nuclear reactor destined to test materials under high radiation fields) réacteur *m* d'essai de matériaux
~ WELL - [plastic ind] (the space in the mould allowing for the bulk factor) espace *m* de contraction
~ WITH LAID PILE - [text] tissu *m* à poil couché
~ WITH UPRIGHT PILE - [text] tissu *m* à poil droit
MATERIALIZATION - [nucl] (transformation of a photon into an electron-positron pair) matérialisation *f*
MATERIALS TESTING - [gen] essai *m* de matériaux
MATHEMATIC - [math] mathématique
MATHEMATICAL - [math] mathématique
~ ANALYSIS - [math] analyse *f* mathématique
~ CHECK - [comput] (check based on mathematical identities) contrôle *m* arithmétique
~ EXPRESSION - [math] expression *f* mathématique
~ FORMULA - [math] formule *f* mathématique
~ INSTRUMENT - [instr] instrument *m* de mathématiques
~ INTERPRETATION - [statist] interprétation *f* mathématique
~ LOGIC - [comput] (exact reasoning about non-numerical relations using efficient symbols) logique *f* mathématique
~ PENDULUM - [phys] pendule *m* simple
MATHEMATICS - [math] mathématiques *fpl*

MATING - [gen] union *f*, accouplement
[zool] accouplement *m*
[mech] ajustement *m*, accouplement *m*
~FLANGE - contre-bride *f*
~GEARS - [mech] engrenages *mpl* accouplés
MATRASS - [ind chem] (glass vessel used by chemists for distilling) ballon *m*, flacon *m*
MATRICULATION - [gen] immatriculation *f*
MATRIX - [anat] (the uterus, or womb) matrice *f*, utérus *m*
[zool] (the intercellular ground-substance of connective tissues) matrice *f*
[el] (array of circuit elements) matrice *f*
[metall] (a mould for casting) matrice *f*, moule *m*
[electron] (model used as a cathode in electroforming) matrice *f*
[min] gangue *f*, roche *f* mère, minière *f*, roche *f*
[math] matrice *f*
[print] (the mould from which a type is cast) matrice *f*
[metall] masse *f* microcristalline
[geol] matrice *f*, magma *m* de second temps
~COLUMN - [comput] colonne *f* de matrice
~ENCORDER - [comput] circuit *m* de codage
~INVERSION - [comput] inversion *f* de matrice
~MEMORY - [comput] (a number of magnetic cores arranged like the elements of a matrix) mémoire *f* à matrice
~NOTATION - [math] notation *f* matricielle
~PAPER - [paper man] papier *m* pour stéréotipie
~PRINCTING - [comput] impression *f* à bloc, impression *f* à fils
~STORAGE - s. matrix memory
~STORE - s. matrix memory
MATRIXING - [telev] (in colour television) matrixation *f*
MATT, to - [text] ternir
MATT - [gen] mat
~CALENDER - [text] calandre *f* de matage
~DIP - [metall] (solution designed to produce a dull or "matt" surface on a metal object) bain *m* de matage
~FINISH - [paint] (preparation of a surface to give a dull appearance, as opposed to a smooth or glossy one) finissage *m* mat
MATTE - [miner] (mass of fused sulphides produced in the smelting of sulphide ores) matte *f*
~SURFACE - [metall] surface *f* matte
MATTED - [text] feutré
MATTER, to - [gen] importer
[med] (of a wound) suppurer
MATTER - [gen] matière *f*, substance *f*
[phys] (an aggregate of material particles) matière *f*
[print] matière *f*, copie *f*, composition *f*
MATTING - [gen] nattes *fpl*, tapis *m*
[text] tissu *m* en paille
MATTNESS - [text] manque *m* de brillant
MATTOCK - [impl] pioche *f* à défricher, décintroir *m* à talus
MATTRESS - [gen] matelas *m*, sommier *m*
[of dried leaves] paillasse *f*
[constr] (sheet expanded metal for reinforcement of concrete roads) matelas *m* d'enrochement
[hydr] (of brush material for supporting a bank) clayonnage *m*
MATURATION - [zool bot etc] (final stage of development-ripening) maturation *f*
MATURATION OF LATEX - [rubber ind] maturation *f* du latex
MATURE - [gen] mûr, adulte
[paper man] échu
MATURING - [gen] maturation *f*, développement *m*
[paint] maturation *f*
[agric] (of cheese etc) affinage *m*
MATURITY - [gen] maturité *f*
[comm] (of an obligation) échéance *f*
~OF GERMINATION - [agric] maturité *f* du grain
~PERIOD - [agric] saison *f*
MAUL - [impl] (a beetle) mailloche *f*, maillet *m*
[impl] (a tool used in road making) batterand *m*
MAUVE - [gen] mauve
[chem] s. mauvein
MAUVEIN - [chem] (the first of the synthetic "coal-tar" dyestuffs to be obtained. It is a trimethyl-derivative of pseudomauveine) mauvéine *f*
MAVAR - [electron] (parametric amplifier) amplificateur *m* paramétrique
MAW - [zool] caillette *f*, quatrième poche *f* de l'estomac
~WORM - [vet] ascaride *m*
MAXILLA - [anat] (os) maxillaire *m*
MAXILLARY - [anat] maxillaire
~GLANDS - [anat] glandes *fpl* maxillaires
~SINUS - [anat] sinus *m* maxillaire
MAXILLOTURBINAL - [anat] cornet *m* inférieur
MAXIMUM - [gen etc] (the highest value of a given quantity occurring during a certain period) maximum
~ADMISSIBLE POWER - [mech] puissance *f* maximum admissible
~ALLOWABLE CONCENTRATION - [ind chem] concentration *f* maximum admissible
~AND MINIMUM POWER RELAY - [el] relais *m* à maximum et à minimum
~AND MINIMUM THERMOMETER - [instr] (instrument designed to record the maximum and the minimum temperatures of the air) thermomètre *m* à maxima et minima
~ANGLE OF DUMPING - [mech] angle *m* maximum de basculement
~ANGULAR VELOCITY - [phys] (for any specified cycle) vitesse *f* angulaire maximale
~AVERAGE OUTPUT POWER - [radio] (the maximum radio-frequency output power occurring under any combination of signals which are transmitted on average over the longest repetitive modulation cycle) puissance *f* de sortie moyenne maximum
~AVERAGE POWER OUTPUT - s. maximum average output power
~AXLE LOAD - [mech] charge *f* limite sur l'essieu
~BACKING PRESSURE - [mech] pression *f* primaire tolérable
~BLADE WIDTH - [aero] largeur *f* maximum de la pale
~BOILING POINT - [phys] point *m* maximal d'ébullition
~BRAKING GRADIENT - [railw] déclivité *f* limite de freinage
~CAPACITY - [el] puissance *f* maximale
~CAPACITY OF A LINE - [telecomm] pleine capacité *f* d'une ligne
~CHARGE ON LOOM - [text] charge *f* complète du métier

MAXIMUM CLEARANCE - [mech] jeu *m* limite
[plast ind] (with mould open) course *f* d'ouverture du moule
~COMPACTED DRY DENSITY - [soil] poids *m* spécifique sec maximum après compactage
~CONTINUOUS POWER - s. maximum continuous rating
~CONTINUOUS RATING - [aero] (the maximum allowable power output for continuous operation) régime *m* de puissance
~CURRENT RELAY - [el] relais *m* à maximum
~DAILY DEMAND - [el gas etc] (daily peak) pointe *f* journalière
~DEFLECTION - [instr] (the maximum value of the instantaneous deflection of a moving element) déviation *f* maximale, élongation *f*
~DEFLECTION ANGLE - [electron] angle *m* maximal de déviation
~DELIVERY PRESSURE - [mech] (in a pump, the pressure above which the delivery pressure of a variable-delivery hydraulic pump, taken at a given speed, begins to diminish) pression *f* maximum de débit
~DEMAND - [el] (maximum load taken by an electrical installation) puissance *f* absorbée maximale
~DEMAND INDICATOR - [instr] (meter with a pointer indicating the highest value of the average power of the current used during successive equal intervals of time) compteur *m* à indicateur de maximum
~DEMAND METER - [instr] (instrument designed to indicate the maximum demand) compteur *m* de maximum
~DEMAND RECORDER - [instr] compteur *m* à registration de maximum
~DENSITY - [photo] densité *f* maximum
~DEVIATION - s. maximum deflection
~DISPOSABLE LOAD - [aero] (the value obtained by substracting the empty weight of an aircraft from the maximum take-off weight for which it is licensed) charge *f* utile maximum
~DURATION - [aero] autonomie *f*
~EQUIVALENT CONDUCTANCE - [el] (of an electrolyte) conductance *f* équivalente maximale
~FLYING SPEED - [aero] (term defined as the maximum air speed of a given aircraft in a standard atmosphere, and under conditions in level flight) vitesse *f* maximum de vol
~FORCE - s. maximum mechanomotive force
~FOREPRESSURE - [mech] pression *f* primaire tolérable
~FREEZING POINT - [phys] point *m* de congélation maximal, point *m* de solidification maximal
~GAS IN STORAGE - [gas ind] (gasholders) stock *m* maximum
~GAS RATE - [th eng] débit *m* maximum de gaz
~GRADIENT - [railw & road constr] inclinaison *f* limite
~GRINDING DIAMETER - [mach tool] diamètre *m* maximum admissible de la pièce à rectifier
~GRINDING LENGTH - [mach tool] longueur *f* maximum admissible de la pièce à rectifier
~IMPULSE INDICATOR - [radio] indicateur *m* de crête, indicateur *m* d'impulsions maxima
~INFLATED DIAMETER - [aero] (in a parachute, the diameter of a cicle having an area equal to the maximum projected area of the canopy) diamètre *m* maximum de voilure gonflée

MAXIMUM INPUT - [el etc] entrée *f* maximale
[th eng] s. maximum gas rate
~INTERFERENCE - [mech] interférence *f* limite
~LICENSED TAKE-OFF WEIGHT - [aero] (the greatest permissible weight at take-off, according to the aircraft's airworthiness certificate) poids *m* maximum admissible au décollage
~LOAD - [mech etc] charge *f* limite, maximum *m* de charge
~MECHANOMOTIVE FORCE - [phys] (the maximum force for any specified cycle is the maximum absolute value of the instantaneous force during that cycle) force *f* mécanomotrice maximale
~MOVING DIMENSIONS - [railw] gabarit *m* de libre passage
~OUTPUT - [radio] (in receivers) puissance *f* maximum de sortie
~OVERSHOOT - [meas] surélongation *f*
~PERIOD OF IDLENESS - [el etc] durée *f* maximale de non-opération
~PERMEABILITY - [phys] (the largest value of normal, or incremental permeability) perméabilité *f* maximale
~PERMISSIBLE BODY BURDEN - [radiat] charge *f* corporelle maximale admissible
~PERMISSIBLE CONCENTRATION - [radiat] (the recommended maximum limit for the concentration of a radioactive substance in air or in water which may enter the human body) concentration *f* maximale admissible
~PERMISSIBLE CONSTANT DOSE RATE - [radiat] taux *m* de dose maximale admissible
~PERMISSIBLE DOSE - [radiat] dose *f* maximale admissible
~PERMISSIBLE EXPOSURE - s. maximum permissible dose
~PERMISSIBLE LOAD - [el] charge *f* maximale admissible
~POINTER - [instr] indicateur *m* de maxima
~PRICE - [comm] prix *m* maximum
~RADIOGRAPHIC RATINGS - [electron] (the maximum load an X-ray tube is subjected in radiography) taux *m* maximal
~RANGE - [nucl] (the maximum distance at which the ionization of a group of ionizing particles can be deteched) parcours *m* maximal
~READING - [instr] déviation *f* maxima
~READING ACCELEROMETER - [instr] accéléromètre *m* à maximum
~RECORDING ATTACHMENT - [instr] (a device fitted to a meter and drawing a curve, the ordinates of which are proportional to the average power in a circuit during equal intervals of time) enregistreur *m* de maximum
~RETENTION TIME - [electron] (the longest time between writing into and reading an acceptable output from a storage element) temps *m* maximal de rétention
~REVERSE VOLTAGE - [el] tension *f* inverse maximale
~REVOLUTIONS - [aero] nombre *m* maximum des tours (de l'helice)
~SAFE AIR-SPEED INDICATOR - [instr] (type of air-speed indicator provided with an additional needle showing the indicated air speed which corresponds to a pre-set. Mach number, and marked to show the maximum permissible value) anémomètre *m* à indi-

cation de vitesse critique
MAXIMUM SCALE VALUE - [instr] (the greatest value of the measured quantity which the scale can indicate) valeur *f* maximale d'échelle
~SENSITIVITY - [radio] sensibilité *f* maximum
~SHUTTLE SPEED - [text] vitesse *f* maximum de la navette
~SLOPE OF STRATUM - [geol] déclivité *f* maximum d'une couche
~SOUND PRESSURE - [acoust] pression *f* sonore maximum
~SPEED - [gen] vitesse *f* maximale, vitesse *f* maximum
~STRESS - [phys] tension *f* maximum
~STRESS IN BEND - [metall] module *m* de rupture
~SYSTEM DEVIATION - [radio] (in frequency modulation) déviation *f* maximum de fréquence tolerée
~TENSILE STRENGTH - [mech] charge *f* ultime de rupture
~TENSILE STRESS - [metall] (the ultimate tensile stress) effort *m* principal primaire
~TORQUE - [el] couple *m* maximale
~TRACTION TRUCK - [railw] (special form of bogie used on trams) voie *f* de traction maximale
~UNDISTORTED OUTPUT - [radio] puissance *f* maximum de sortie sans distortion
~USABLE FREQUENCY - [el magnetic waves] fréquence *f* limite supérieure
~USABLE FREQUENCY FACTOR - [radio] facteur *m* de fréquence maximum utilisable
~VALENCE - [phys] (the highest valence shown by an element in a compound) valence *f* maximale
~VALUE - [gen] valeur *f* maximum, crête *f*
~VALUE OF ELECTROMOTIVE FORCE - [el] valeur *f* maximum de la force électromotrice
~VOLTAGE RELAY - [el] relais *m* à maximum
~WAVELENGTH - [radio] longueur *f* d'onde maximum
~WEIGHT PER CYCLE - [plast ind] capacité *f*
~WIDTH - [gen] largeur *f* maximum
~WORK - [mech] (the maximum amount of work obtained from a process) travail *m* maximal
MAXWELL - [el] (the unit of magnetic flux) maxwell *m*
~-BOLTZMANN DISTRIBUTION LAW - [mech] (the velocity distribution of gas molecules in thermal equilibrium) loi *f* de distribution de Maxwell-Boltzmann
~-BOLTZMANN STATISTICS - [phys] (study of the probabilities of the macroscopic state of a system of non-quantized states) statistique *f* quantique de Maxwell-Boltzmann
~-BOLTZMANN VELOCITY-DISTRIBUTION LAW - [phys] loi *f* de la distribution de la vitesse de Maxwell-Boltzmann
~DEMON - [phys] (imaginary figure made by Maxwell to explain concept in gas kinetic) diable *m* de Maxwell
~EQUATIONS - [el] (set of the four classic formulae of the electromagnetic theory) équations *fpl* de Maxwell
~EXPERIMENT - [photo] (the demonstration of the three-colour additive synthesis, using three black and white negatives) expérience *f* de Maxwell
~M-L BRIDGE - [instr] pont *m* de Maxwell à induction mutuelle
~RELATIONS - [phys] (the four thermodynamic relations relating to the equilibrium state of a system) relations *fpl* de Maxwell
MAXWELLIAN DISTRIBUTION - s. Maxwell-Boltzmann law
MAXWELLIAN FLUID - [mech] (visco-elastic fluid in which the stress tensor is related to the rate of strain) fluide *m* maxwellien
~VIEW - [opt] (the method of observing an integrating photometric sphere) méthode *f* d'observation de Maxwell
MAYBUSH - [bot] aubépine *f*, épine *f* blanche
MAYDAY - [radio] (the radio-telephone international distress call) signal *m* de détresse
MAYER THEORY OF CONDENSATION - [phys] théorie *f* de condensation de Mayer
MAZUT - [min] mazout *m*
MC - s. millicurie
MC.P.S. - [el] Megacycle *m* par seconde
mcps - s. MC.P.S.
mc/s - s. MC.P.S.
M.C.S. - [telecomm] (abbrev. for mile of standard cable) mile *f* de câble étalon, MCE
MCW - s. modulated continuous waves
M-DERIVED L-SECTION FILTER - [radio] (a reactance network) filtre *m* en L dérivé en M
M DISPLAY - [radar] indicateur *m* type M
Me - [metall] (general symbol for a metal) métal *m*
[chem] (symbol for methyl radical) symbole de radical de méthyle
MEADOW - [agric] pré *m*, prairie *f*
~BLEACHING - [text] blanchiment *m* sur pré
~GRASS - [agric] pâturin *m*, herbe *f* des prés
~GRASS DERMATITIS - [med] dermatite *f* des prés
~HAY - [agric] foin *m* de prairie
~PLANT - [bot] plante *f* d'herbage
~ROLLER - [agric] rouleau *m* à prairie
~RUE - [bot] pigamon *m*
~-SWEET - [bot] épi *m* du vent
~VETCHLING - [bot] gesse *f* sauvage, vesceron *m*
MEAL - [gen] repas *m*
[agric] farine *f* (de mais, seigle etc)
MEALY - [gen] farineux
[bot] (of fruit) cotonneux, duveteux
~BUGS - [zool] cochenille *f* des serres
MEAN - [gen] moyen *m*, milieu *m*
[math] (the mean of a set of quantities) moyenne *f*
~ACTIVITY - [el] (of an electrolyte) activité *f* moyenne
~CHORD - [aero] (standard mean chord) profondeur *f* moyenne de l'aile
~CHORD OF A WING - s. mean chord
~DEVIATION - [statist] écart *m* moyen
~DURATION RATE - [statist] (accident severity rate) taux *m* de gravité des accidents
~EFFECTIVE PRESSURE - [mech] (M.E.P.) (the average pressure on the piston-head of a reciprocating engine during the whole of the working stroke) pression *f* moyenne effective
~FREE PATH - [phys] (the average distance travelled by a particle (e.g. a molecule) before a collision occurs with another similar particle) libre parcours *m* moyen
[acoust] (the average distance traversed by a sound wave between successive reflections) libre parcours *m* moyen
~FREE TIME - [nucl] (the average time between collisions) libre temps *m* moyen, libre moyen temps *m*
~HEMISPHERICAL CANDLE POWER - [light] intensité *f* moyenne hémisphérique
~HORIZONTAL CANDLE POWER - [light] intensité *f*

moyenne horizontale
MEAN IMPULSE INDICATOR - [telecomm] indicateur *m* d'impulsions moyennes
~JET PIPE TEMPERATURE - [aero] température *f* moyenne du jet
~KINETIC ENERGY - [phys] énergie *f* cinétique moyenne
~LATITUDE - [astr] moyenne latitude *f*
~LIFE - [nucl] (the average of the individual life of all the atoms of a radioactive substance) vie *f* moyenne, durée *f* moyenne
~LINE - [aero] ligne *f* moyenne
~MATURITY - [agric] maturité *f* normale
~MOLECULAR VELOCITY - [phys] vitesse *f* moléculaire moyenne
~MOTION - [phys] mouvement *m* angulaire moyen
~NOON - [astr] midi *m* moyen
~NUMBER - [math] moyenne *f*
~PICTURE LEVEL - [telev] niveau *m* moyen du signal vidéo
~POWER - [gen] puissance *f* moyenne
[radio] (of a radio transmitter, the power which is supplied during normal operations) puissance *f* moyenne
~PRESSURE ANGLE - [mech] (of a spiral bevel gear) angle *m* de pression moyen
~PULSE TIME - [radio] durée *f* moyenne d'impulsion
~RANGE - [nucl] parcours *m* moyen
~RISE AND FALL - [geogr] moyenne *f* du flot et jusant de la mer
~SEA LEVEL - [geogr] (M.S.L.) (the average level of the ocean surface, taking account of tides and other variations) niveau *m* moyen de la mer
~SOLAR DAY - [astr] (the time interval between two successive transits of the mean sun across the same meridian) jour *m* solaire moyen
~SOLAR TIME - [astr] temps *m* solaire moyen
~SPHERICAL CANDLE POWER - [light] intensité *f* moyenne sphérique
~SPRING RANGE - [astr] (of the tides) étendue *f* moyenne de la marée de syzygie
~SPRING RISE - [astr] hauteur *f* moyennne de la marée de syzygie
~SQUARE ERROR - [math] erreur *f* moyenne quadratique
~SQUARE LENGTH OF MODERATION - [nucl] (the mean distance which a thermal neutron travels between formation and capture) carré *m* moyen du parcours de modération
~SQUARE MOLECULAR VELOCITY - [phys] vitesse *f* moléculaire moyenne quadratique
~SQUARE OF VELOCITIES - [phys] vitesse *f* quadratique moyenne
~SQUARE VALUE - [math] valeur *f* quadratique moyenne
~SUN - [astr] (fictitious point describing the celestial equator) soleil *m* moyen
~TEMPERATURE DIFFERENCE - [phys] (for heat exchange) différence *f* moyenne de températures
~TENSILE STRAIN - [paper man] résistance *f* moyenne à la traction
~UPPER HEMISPHERICAL CANDLE POWER - [light] intensité *f* moyenne sphérique supérieure
~VALUE METER - [instr] (instrument designed to measure the mean value of radiation activity) appareil *m* de mesure de la valeur
~VALUE OF A PERIODIC QUANTITY - [el] (the mean value of a quantity during a period) valeur *f* moyenne d'une grandeur périodique
MEAN YEARLY HEIGHT - [phys] (of barometric pressure) hauteur *f* moyenne de l'année
~YIELD - [gen] rendement *m* moyen
MEANS - [gen] moyens *mpl*, voies *fpl*
[fin etc] moyens *mpl*, ressources *fpl*
~OF CONVEYANCE - s. means of transport
~OF TRANSPORT - [transp] moyens *mpl* de transport
MEANTONE - [mus] accord *m* pur
MEASLES - [med] rougeole *f*
[vet] (of swine) ladrerie *f*
MEASURABLE - [gen] mesurable
~VARIABLE - [meas] (physical quantity of condition which must be measured) grandeur *f* à mesurer
MEASURE, to - [gen] mesurer
MEASURE - [gen] mesure *f*
[instr] appareil *m* de mesure, mètre *m*, mesure *f*
[naut] cubage *m*, jaugeage *m*
[print] format *m*, largeur *f* de justification
[chem] dose
~IN THE CLEAR - [constr] cote *f* de passage
~OF CAPACITY - [mech] mesure *f* de la capacité
~OF CONTRACTION - [rubber ind] mesure *f* de retrait
MEASURED FEEDBACK - [contr] (feedback signal based on an output determined by a measuring instrument) réaction *f* mesurée
~VALUE - [contr] valeur *f* mesurée
~VARIABLE - [contr] grandeur *f* mesurée
MEASUREMENT - [gen] mesure *f*, dimension *f*
[gen] (the action of measuring) mesurage *m*, métrage *m*
~ACCURACY - [meas] exactitude *f* de mesure
~COMPONENT - [meas] (parts or assemblies used for the construction of measuring instruments or appliances) composante *f* d'un appareil de mesure
~ENERGY - [instr] (the energy which is required to operate a measuring instrument) dépense *f* d'énergie
~EQUIPMENT - [meas] appareillage *m* de mesure
~MECHANISM - [meas] organes *mpl* de mesure
~OF ELEVATION - [radar] (the determination of the height of an object by a mechanism actuated by the echo) mesure *f* de l'hauteur
~OF FLOW - [mech] (along a pipe or a channel) mesure *f* de l'écoulement
~OF TENSION - [el] mesure *f* de tension
~OF VOICE FREQUENCY RINGING CURRENT - [telecomm] mesure *f* de courant d'appel à fréquence vocale
~RANGE - s. measuring range
~SETUP - [electron] montage *m* de mesure, banc *m* de mesure
~SYSTEM - [meas] système *m* de mesure
MEASURING - [gen] mesurage *m*, mesure *f*
[surv] arpentage *m*
~AMPLIFIER - [meas] (used with oscilloscopes, voltmeters etc) amplificateur *m* de mesure
~AND KNOCKING OFF MOTION - [text] indicateur *m* de longueur
~APPLIANCES - [surv] appareillage *m* d'arpentage
~BLOCK - [mach tool] calibre *m* prismatique
~BRIDGE - [instr] (a Wheatstone bridge circuit used for electrical measurement) pont *m* de Wheatstone, pont *m* de mesure
~BY SIGHT - [gen] estimation *f* à vue d'oeil

MEASURING CABLE - [telecomm] câble *m* de mesure
~CHAIN - [surv] chaîne *f* d'arpentage
~CIRCUIT - [el] circuit *m* de mesure
[telecomm] montage *m* de mesure
~CLOCK - [instr] compteur *m*
~COIL - [el] bobine *f* de mesure
~CYLINDER - [ind chem] (graduated open vessel for approximate measurement of liquids in the laboratory) vase *m* gradué
~DEVICE - [meas] (assembly of components by which electrical or magnetic measurements can be made) appareillage *m* de mesure
~DIODE - [electron] (electronic tube used in measuring circuits) diode *f* de mesure
~DRUM - [paper man] (of the cutter) tambour *m* mesureur
~EYEPIECE - [opt] micromètre *m* oculaire
~FLASK - [ind chem] (flat bottomed flask marked so as to measure a given quantity of liquid) ballon *m* gradué
~FREQUENCY - [el] fréquence *f* de mesure
~GLASS - [ind chem] (a graduated glass or cylinder for dispensing specific quantities of fluid) verre *m* gradué
~HEAD - [meas] sonde *f*
~INSTRUMENT - [instr] (a device for the measurement of electrical conditions in a circuit) instrument *m* de mesure
~LOOP - [meas] boucle *f* de mesure
~MACHINE - [meas] (machine for the precision measuring of standard gauges) banc *m* micrométrique
~MACHINE WITH CLOTH INSPECTING TABLE - [text] métreuse *f* avec table de visitage
~MEANS - s. measuring unit
~MICROPHONE - [acoust] microphone *m* de mesure
~MOTION - [text ind] compteur *m*
~POINT - [electron] point *m* de mesure, prise *f* de mesure
~POINTER - [instr] aiguille *f* de mesure
~POTENTIOMETER - [instr] potentiomètre *m* de mesure
~PRECISION - [meas] précision *f* de mesure
~RANGE - [meas] étendue *f* de mesure
~RANGE SWITCH - [el] commutateur-sélecteur *m* des zones de mesure
~RELAY - [telecomm] relais *m* de mesure
~REPRODUCIBILITY - [contr] reproducibilité *f* de mesure
~RESISTOR - [el] résistance *f* de mesure
~ROD END GAUGE - [meas] jauge *f* à tête de bielle
~ROLLER - [text ind] rouleau *m* mesureur
~RULE - [meas] (a rule used to ascertain the proportions of an object) règle *f* graduée
~SCOOP - [photo] cuiller *f* de dosage, louche *f* de dosage
~SPARK-GAP - [meas] (a spark-gap used to measure peak voltages) éclateur *m* de mesure, spintermètre *m*
~SPRING TAPE - [meas] roulette *f* en ruban d'acier
~STAFF - [surv] mire *f* de mesure
~STICK - [meas] tige *f* de mesurage
~TANK - [meas] (a tank of specific capacity) jaugeur *m*
~TAPE - [meas] mesure *f* à ruban, roulette *f*, ruban *m*
~TOOL - [meas] instrument *m* de mesure

MEASURING TRANSDUCER - [el] transducteur *m* de mesure, convertisseur *m* de mesure
~TRANSFORMER - [el] transformateur *m* de mesure
~UNIT - [contr] (of an automatic controller) ensemble *m* de mesure
~VALUE - [instr] valeur *f* de mesure, valeur *f* mesurée
~VESSEL - [ind chem] vase *m* gradué
~WHEEL - [surv] roulette *f*
~WIRE - [instr] fil *m* de mesure
MEAT - [food] viande *f*
~BROTH - [food] bouillon *m*
[chem] bouillon *m* de culture
~CHOPPER - [impl] hacheviande *m*
~COOLING PLANT - [ind proc] installation *f* frigorifique
~EXTRACT - [food] extrait *m* de viande
~INSPECTION - [food] inspection *f* de la viande
~MARKET - [comm] marché *m* de la viande
~MINCER - [impl] hacheviande *m*
~-PACKING PLANT - [food] industrie *f* de la viande
~POULTRY - [food] poulets *mpl* de consommation
~PRESERVES - [food] conserves *fpl* de viande
~PROCESSING - [food] transformation *f* de la viande
~PROCESSING INDUSTRIES - [food] industries *fpl* de la viande
~PRODUCTION - [food] production *f* de viande
~REFRIGERATION - [food] réfrigération de la viande
~SAW - [impl] scie *f* de boucheur
~-SLICER - [impl] machine *f* à trancher la viande
~YIELD - [zool] rendement *m* en viande
MEATORRHAPHY - [med] méatorraphie *f*
MEATOTOMY - [med] méatotomie *f*
MEATUS - [zool] (a duct, or a channel) conduit *m*
[anat] meat *m*, conduit *m*
~OF THE NOSE - [anat] méat *m* des fosses nasales
MECHANIC - [gen] (workmann skilled in mechanical operations) mécanicien *m*, serrurier *m* mécanicien, monteur *m*
[adj] mécanique
MECHANIC'S TOOL BAG - [impl] boîte à outils
MECHANICAL - [mech] mécanique
~-ACOUSTICAL COUPLING - [acoust] (the interconnexion of mechanical and acoustical elements) accouplement *m* mécano-acoustique
~ADMITTANCE - [el acoust] (the reciprocal of the mechanical impedance) admittance *f* mécanique
~ADVANTAGE - [mech] (the ratio between load and applied force in a machine) bras *m* de levier
~AGITATOR - [mech] agitateur *m* mécanique
~ATOMIZER - [th eng] brûleur *m* automatique
~BALANCE - [phys] (the equilibrium of masses) équilibre *m* mécanique
~BOND - [constr] (in reinforced concrete structures) liaison *f* mécanique
~BOOSTER - [mech] dépresseur *m* radial
~BRAKE - [mech] frein *m* à commande mécanique
~BREAKDOWN - [mech] panne *f* mécanique
~BURRING - [text] échardonnage *m* mécanique
~CENTRING - [telev] (of the picture on the screen) centrage *m* mécanique
~CHARGER - [th eng] chargeur *m* mécanique
~CHARGING MACHINE - s. mechanical charger
~CLEANING - [text] nettoyage *m* à sec (de la laine)
~COAL PICK - [mining] marteau piqueur *m*
~COMPLIANCE - [mech] élasticité *f* mécanique
~CONDUCTOR LOADING - [el] charge *f* mécanique

d'un conducteur
MECHANICAL CONTROL SYSTEM - [contr] régulation *f* mécanique
~COUNTER - [comput] compteur *m* mécanique
~COUPLING - [mech] accouplement *m* mécanique
~CREEPER - [impl] sommier *m* roulant
~DAMPING RING - [constr] anneau *m* amortisseur mécanique
~DEPOLARIZATION - [el] (dissipation by mechanical means of the hydrogen bubbles causing polarization in an electrolytic cell) dépolarisation *f* mécanique
~DEPOSIT - [geol] (deposit of sediments cuased by some mechanical process) dépôts *mpl* dus à causes mécaniques
~DRESSING - [min] préparation *f* mécanique
~DRIVE - [mech] transmission *f* mécanique
~EFFICIENCY - [mech] (of an engine, a motor etc) rendement *m* mécanique
~EJECTION - [metall] (die casting) (removal of a moulding from the mould by a mechanical device) éjection *f* mécanique
~ELECTRO-PLATING - [el metall] (plating operation in which the cathodes are moved by mechanical means during the process) galvanoplastie *f* mécanique
~ENERGY - [mech] énergie *f* mécanique
~ENGINEER - [gen] mécanicien *m*
~ENGINEERING - [gen] art *m* de la mécanique
~EQUIVALENT OF HEAT - [phys] (ratio of the mechanical energy transformed into heat to the heat generated) équivalent *m* mécanique de la chaleur
~EQUIVALENT OF LIGHT - [opt] (the ratio between the radiant flux and the luminous flux) équivalent *m* mécanique de la lumière
~EXPANSION INSTRUMENT - [mech] instrument *m* à dilatation mécanique
~FIGURING - [text ind] brochage *m* mécanique
~FILTER - [cin] (in a drive to balance variations in the required constant speed) filtre *m* mécanique
~FIRING - [th eng] chauffage *m* mécanique
~FOAMING - [plast ind] (the production of cellular plastics by means of a mechanical action) production *f* mécanique de mousse
~FORCE - [phys] (any physical force which modifies the condition of movement or of rest of a body, or is capable of deforming it) force *f*, force *f* mécanique
~GRATE - [th eng] (a self-clinkering grate) grille *f* mécanique
~HOE - [agric] piocheuse *f*, houe *f* mécanique
~HYSTERESIS - [mech] hystérésis *f* mécanique
~IMPEDANCE - [phys] (the quotient of a force applied to a linear mechanical system divided by the resulting velocity in the direction of the force at its point of application) impédance *f* mécanique
~INDUSTRY - [gen] industrie *f* mécanique
~INTERLOCKING MECHANISM - [el] mécanisme *m* à verrouillage mécanique
~JACK - [mech] cric *m*
[mining] vérin *m* mécanique
~JOINT - [mech] (rubber joint) joint *m* à anneau de caoutchouc
~KNOCK-OUT BY VIBRATION - [metall] décochage *m* par vibrations
~LIFE - [instr] (the maximum number of complete cycles an instrument may complete without mechanical failure) durée *f* mécanique
MECHANICAL LINKAGE - [mech] liaison *f* mécanique, jonction *f* mécanique
~LOADER - [railw] chargeur *m* mécanique
~LOSS - [el] (loss by windage and by friction in bearings and brushes) pertes *fpl* mécaniques
~MASS - [phys] (that part of the particle mass which is supposed to be an intrinsic property of the particle) masse *f* mécanique
~MILKER - [agric] trayeuse *f* mécanique
~MIXING - [mech] mélange *m* à la machine
~MOMENT OF INERTIA - [mech] moment *m* mécanique d'inertie
~OPERATION OF A VALVE - [mech] commande *f* d'une soupape
~OPERATOR - [contr] clavier *m* à poussoirs
~OSCILLATOR - [mech] (system of components capable of oscillatory motion) oscillateur *m* mécanique
~PASSIVITY - [phys] (the abnormally slow ionization of a metallic anode due to the visible presence of an oxide film) passivité *f* mécanique
~PHONOGRAPH RECORDER - [acoust] graveur *m* mécanique
~PICKER - [text] machine *f* à cueillir
~PILOT - [aero] (automatic pilot) pilote *m* automatique
~POINT INDICATOR - [railw] indicateur *m* mécanique de position d'aiguille
~POLISHING - [metall] (the smoothing of a metal by means of abrasive particles) polissage *m* mécanique
~POWER - [phys] puissance *f* mécanique
~PRESS - [mech] (press operated by mechanical means) presse *f* mécanique
~PRODUCER - [gas ind] (generator with automatic grate) gazogène *m* à grille tournante
~PUDDLER - [metall] puddleur *m* mécanique
~PULP - s. mechanical wood-pulp
~RAMMER - [mech] pilette *f* mécanique
~REACTANCE - [mech] (imaginary component of the mechanical impedance) réactance *f* mécanique
~RECORDER - [el acoust] graveur *m* mécanique
~RECORDING HEAD - [el acoust] graveur *m*, burin *m*
~RECTIFIER - [el] (rectifier with a rotating commutator which operates synchronously with the a.c. supply) redresseur *m* mécanique
~RECTILINEAL RESISTANCE - [mech] (real part of the mechanical reactance) résistance *f* mécanique rectiligne
~REFRIGERATOR - [mech] (a complete refrigerating plant) installation *f* frigorifique
~REFRIGERATOR CONTAINER - [railw] container-isotherme *m* réfrigérant
~REGISTER - [nucl] (electromechanical device for the recording of a count) enregistreur *m* mécanique, numéroteur *m* électromécanique
[electron] compteur *m* mécanique
~RELEASE - [photo] déclencheur *m* mécanique
~RELEASE OF RELAY ARMATURE - [el] décollage *m* mécanique de l'armature d'un relais
~REPRODUCER - [el acoust] (a phonograph pick-up) phonocapteur *m*, pick-up *m*
~RESISTANCE - [mech] (a real component of the mechanical impedance) résistance *f* mécanique
~RESONANCE - [acoust] résonance *f* mécanique
~RESONATOR - [acoust] résonateur *m* mécanique
~ROTATIONAL COMPLIANCE - [mech] élasticité *f* de

torsion
MECHANICAL ROTATIONAL IMPEDANCE - [mech] impédance *f* de torsion
~ROTATIONAL REACTANCE - [mech] (imaginary part of the mechanical rotational impedance) réactance *f* de torsion
~ROTATIONAL RESISTANCE - [mech] (the real part of the mechanical rotational impedance) résistance *f* de torsion
~ROTATIONAL SYSTEM - [mech] (system adapted for the transmission of rotational vibrations) système *m* de torsion
~ROUTE INDICATOR - [railw] indicateur *m* mécanique d'itinéraire
~SCANNING - [telev] (system based on a beam of light controlled by a rotating mirror and a rotating scanning disk, to separate an image into a succession of narrow lines) analyse *f* mécanique
~SHORT-TIME CURRENT - [el] (of a current transformer; instantaneous short-circuit current) courant *m* limite dynamique
~SHOVEL - [mech] (a type of excavator) pelle *f* équipée en butte, pelle *f* à godet
~STABILITY - [mech] stabilité *f* mécanique
~STAGE - [instr] (stage provided with a mechanical device for adjusting the position of the object) platine *f* réglable, platine *f* à chariot
~STOKER - [th eng] (an automatic stoker; a device for stoking a boiler by automatic means) stoker *m*, foyer *m* mécanique
~STOKING - [th eng] chauffage *m* mécanique
~STRESS - [phys] (a mechanical force acting on a unit area in a solid) effort *m* mécanique
~STRIPPING - [metall] (removal of metal by mechanical means) dépouillage *m* mécanique
~SWITCH INDICATOR - s. mechanical point indicator
~TELEPHONE - [telecomm] téléphone *m* acoustique, téléphone *m* à ficelle
~TEST - [mech] essai *m* mécanique
~TRACK MAINTENANCE - [railw] entretien *m* mécanique de la voie
~TRANSLATION - [comput] traduction *f* automatique d'une langue
~TRANSMISSION SYSTEM - [mech] (assembly of elements for the transmission of mechanical power) système *m* de transmission mécanique
~TREATMENT - [rubber ind] traitement *m* mécanique
~TRIP - [contr] (automatic brake actuator) activateur *m* du frein automatique
~TWINS - [metall] macle *f* de déformation
~UNDERCUTTER - [mining] haveuse *f*
~UNIT - [mech] (of a machine etc) unité *f* mécanique
~VACUUM PUMP - [mech] pompe *f* mécanique
~VALVE OPERATOR - [mech] organe *m* mécanique d'actionnement de soupape
~VENTILATION - [mech] aérage *m* mécanique, ventilation *f* mécanique
~VOLUME CONTROL PUMP - [mech] pompe *f* à réglage mécanique de volume
~WEAVING - [text] tissage *m* mécanique en particulier
~WOOD-PULP - [paper man] (pulp made by mechanical disintegration, as distinct from methods using chemical reagents) cellulose *f* de bois
~ZERO - [instr] (the point on the scale at which the index stops when the moving element takes up its equilibrium position) zéro *m* mécanique
MECHANICALLY DRIVEN ORE GRINDER - [min] triturateur *m* de minerai commandé mécaniquement
~OPERATED SLAY - [text] battant *m* actionné mécaniquement
MECHANICS - [mech] (the study of the action of forces on bodies and of the motions produced) mécanique *f*
~OF LUIDS - [mech] mécanique *f* des fluides
~OF SOLIDS - [mech] mécanique *f* des (corps) solides
MECHANISM - [mech] mécanisme *m*, organes *mpl*
MECHANIZATION - [mech] mécanisation *f*
MECHANIZE, to - [mech] mécaniser
MECHANIZED FARMING - [agric] motoculture *f*, culture *f* motorisée
~WINNING - [mining] abattage *m* mécanique
MECHANO-ACOUSTIC COUPLING - s. mechanical-acoustical coupling
MECONIC ACID - [chem] (an aliphatic hydroxydicarboxylic acid present in the juice of Papaver somniferum) acide *m* mécanique
MECONIUM - [physiol] (the first faeces of a newborn infant) méconium *m*
MED - s. mediocris
MEDALLION - [gen] médaillon *m*
[arch] médaillon *m*
MEDIAL - [gen] médian, intermédiaire, médial
[math] médian
MEDIAN - [gen] médian
~LETHAL DOSE - [radiat] (the dose of radiation which is required to kill fifty percent of the individuals of a group) dose *f* létale 50 p.c.
~LETHAL TIME - [radiat] (the time which is required for the destruction of fifty percent of the individuals of a group) temps *m* létal 50 p.c.
~LINE - [aero] (a line drawn through an aerofoil, every point on which is equidistant from the upper and lower surfaces of such an aerofoil, measured at right angles to the line) ligne *f* médiane
~PLANE - [opt] plan *m* médian
MEDIANT - [mus] (the third note of any scale) médiante *f*
MEDIASTINITIS - [med] (inflammation of the tissues of the mediastinum) médiastinite *f*
MEDIASTINUM - [anat] (membrane separating the pleural cavities) médiastin *m*
MEDIATOR - [chem] médiateur *m*
MEDICAL - [med & chem] médical
~ELECTROLYSIS - [el biol] galvanisme *m*
~EXAMINATION - [med] visite *f* médicale
~IONIZATION - [el biol] ionothérapie *f*
~JURISPRUDENCE - [leg] médecine *f* légale
MEDICAMENT - [med] médicament *m*
MEDICATED SOAP - [ind chem] savon *m* médicinal
MEDICATION - [med] médication *f*
MEDICINAL - [med] médicinal, médicamenteux
~PLANT - [bot] plante *f* médicinale
~DROPPER - [med] compte-gouttes *m*
MEDICINEREA - [anat] substance *f* grise du noyau lenticulaire et du claustrum
MEDIOCARBAL - [anat] médiocarpien
MEDIOCRE - [gen] médiocre
MEDIODORSAL - [anat] médiodorsal
MEDIOFRONTAL - [anat] médiofrontal
MEDIONECROSIS - [med] aortite *f* scléro-atrophique
MEDIOPECTORAL - [anat] médiopectoral

MEDIUM - [adj] moyen, milieu *m*
[gen] moyen *m*
[phys] (substratum in which a specified system of physical entities exist) médium *m*, milieu *m*, moyen *m*
[chem] agent *m*
[paint] (liquid or semiliquid vehicle) véhicule *m*, milieu *m* de suspension
[paper man] papier *m* à imprimer (46 x 58 cm)
[print] (of a typeface) quarte-gras
~-ANGLE LENS - [photo] objectif *m* de demi-grand angle
~BURR WOOL - [text] laine *f* à teneur moyen de graterons
~CAPACITY - [ind proc] capacité *f* moyenne
~CAPACITY MIXER - [mech] mélangeur *m* de moyenne capacité
~CARBON STEEL - [metall] acier *m* demi-doux
~CHERRY RED - [metall] (at about 750 centigrades) rouge *m* cerise moyen
~CLOSE-UP - [cin & telev] (shot between a close-up and a medium shot) plan *m* rapproché
~CLOUDS - [met] (clouds of which the mean height is between 8000 and 20,000 feet.) nauges *mpl* moyens
~COARSE - [text] mi-gros
~COUNT - [text] numéro *m* moyen
~DRAWING - [metall] tréfilage *m* moyen
~EARLY - [agric] mi-hâtif
~FINE - [text] mi-fin
~FIT - [mech] montage *m* libre normal
~FORCED FIT - [mech] couplage *m* forcé
~FREQUENCY - MF - [radio] (term applied to radiowaves between 300 kc/s and 3 Mc/s (I000 m. to I00 m. wave-length) fréquence *f* moyenne
~GRADE - [metall] (USA only; having a tensile strength of 70,000lbs per square inch) jet *m* d'acier ayant une résistance à la rupture de 70.000 lbs
[gen] degré *m* de dureté moyen
~-GRADE ORE - [min] minerai *m* à teneur moyenne
~GRAINED - [geol] (a stage of the grain-classification) à grain moyen
~-GRAINED PIG IRON - [metall] fonte *f* à grain moyen
~-GRAINED WHEEL - [mech] meule *f* mi-douce
~GRINDING - [mech] broyage *m* moyen
~HACKLE - [text] gills *m* intermédiaire
~HARD STEEL - [metall] acier *m* demi-dur
~HEAVY LOADING - [el] charge *f* mi-forte
~HIGH VACUUM - [phys] vide *m* moyen
~IMPEDANCE TRIODE - [electron] (triode with an internal impedance of average value) triode *f* à impédance moyenne
~IRON - [metall] fer *m* métis
~LAMPHOLDER - [el] douille *f* à pas normal
~LATE - [agric] mi-tardif
~-LIFT - [hydr] à hauteur *f* moyenne d'élévation
~LONG SHOT - [cin & telev] (shot between a medium and a long shot) plan *m* demi-général
~-OIL ALKYDS - [chem] résines *fpl* alkydes à teneur moyen d'huile
~PITCH - [text] division *f* moyenne
~-PLATE MILL - [metall] laminoir *m* à tôles moyennes
~PRESSURE - [mech] moyenne pression *f*, moyenne tension *f*
~-PRESSURE CURRENT - [el] courant *m* à moyenne tension
~QUALITY - [gen & comm] qualité *f* moyenne
~SECTION - [metall] profil *m* moyen

MEDIUM SEED WOOL - [text] laine *f* à moyen teneur de graines
~SHOT - [cin & telev] (a shot between close-up and long shot) plan *m* moyen
~-SIZED FARM - [agric] exploitation *f* moyenne
~SOFT-STEEL - [metall] acier *m* (de)mi-doux
~STAPLE - [text] fibre *f* de moyenne longueur
~STEEL - [metall] acier *m* mi-dur
~-TERM - [fin] à moyen terme *m*
~TO PREVENT MILDEW - [chem] moyen *m* anti-rouille
~-TONE NEGATIVE - [photo] contre-type *m* correspondant aux tons moyens
~VACUUM - [phys] vide *m* moyen
~VOLTAGE - [el] (voltage between 250 and 650 volts) moyenne tension *f*
~WAVES - [radio] (electromagnetic waves of a wavelength between 200 ans I,000 m) ondes *fpl* moyennes
MEDLEY - [gen] mélange *m*, confusion *f*
[mus] (mixture of unrelated tunes) pot-pourri *m*
MEDULLA - [anat] (bone marrow) moelle *f*
[bot] médulle *f*
MEDULLARY - [anat] médullaire
~BUNDLE - [bot] (vascular bundle in the pith) faisceau *m* médullaire
~CANAL - [anat] (the cavity of the central nervous system) canal *m* médullaire
~FOLDS - [anat] (the lateral folds of the medullary plate) valvules *fpl* médullaires
~GROOVE - [anat] (on the surface of the medullary plate) gouttière *f* médullaire
~NAILING - [med] enclouage *m* médullaire
~PLATE - [anat] (during development, the dorsal plate-like surface of ectoderm later giving rise to the central nervous system) plaque *f* médullaire
~RAYS - [bot] (vascular rays) rayons *mpl* vasculaires
~SHEATH - [anat] (surrounding the axons of the central nervous system) gaine *f* médullaire
~SUBSTANCE - [anat] substance *f* médullaire
MEDULLECTOMY - [med] médullectomie *f*
MEDULLOBLAST - [biol] médulloblast *m*
MEDULLOCELL - [biol] myélocyte *m*
MEDULLOID - [anat] médullaire
MEDUSA - [zool] méduse *f*
MEERSCHAUM - [min] (a hydrated silicate of magnesium) écume *f* de mer, sépiolite *f*
MEETING POST - [constr] montant *m* de buse, pateau-battant *m*
~RAILS - [constr] (of a sliding sash) traverses *fpl* formant le joint horizontal d'une fenêtre à guillotine
~STILES JOINT - [constr] battée *f* à gueule-de-loup, gueule-de-loup *f*
~THE LOAD - [el & gas ind] couverture *f* des besoins
MEGACHROMOSOMES - [zool] mégachromosomes *mpl*
MEGACOLON - [med] (usually large colon) mégacôlon *m*
MEGACYCLE - [el] (unit of electrical frequency equal to one million cycles per second) mégacycle *m*, mégahertz *m*
MEGAELECTRONVOLT - [electron] (one million electron-volts) mégaélectron-volt *m*
MEGAFARAD - [el] mégafarad *m*
MEGAGAMETE - [biol] macrogamète *m*
MEGALINE - [el] (I,000,000 lines magnetic flux) flux *m* magnétique d'un million de maxwell
MEGALOBLAST - [biol] mégaloblaste *m* orthochromatique
MEGALOESOPHAGUS - [med] méga-œsophage *m*

MEGALOURETER - [med] méga-uretère *m*
MEGAMETER - [instr] (instrument designed to determine the longitude by the observation of the stars) mégamètre *m*
MEGAPHONE - [acoust] (large speaking-trumpet) mégaphone *m*
MEGAPHONING - [mech] (of motorbicycles) effet *m* de mégaphone
MEGATON - [nucl] (hydrogen bomb having an explosive force corresponding to 1,000,000 tons of dynamite) (bombe *f*) mégatonne
MEGAVOLT - [el] mégavolt *m*
MEGAVOLTAGE THERAPY - [el biol] thérapie *f* à rayons X à trés haute tension
MEGOHM - [el] (1,000,000 ohms) mégohm *m*
MEGOHMMETER - [instr] (instrument designed to measure very large resistance) mégohmmètre
MEIBOMIANTIS - [med] meibomiite *f*
MEIBOMITIS - s. meibomiantis
MEIOSIS - [biol] (nuclear division of the chromosomes) méiose *f*
MEKER BURNER - [th eng] (type of Bunsen burner in which a special head produces a large number of small flames) bec *m* Méker
MEL - [acoust](unit of pitch) mel *m*
MELACONITE - [min] (a curpic oxide) mélaconise *m*
MELAENA - [med] mélaena *m*
MELAMINE - [chem] (a synthesis intermediate and a component of melamine resins) mélamine *f*
~-FORMALDEHYDE RESINS - [chem] (thermosetting resins with good temperature and moisture resistance and dielectric strength) résines *f*pl à base de mélamine-formaldéhyde
~RESINS - [chem] (synthetic thermosetting resins produced from formaldehyde and melamine) résins *f*pl à la mélamine
MELANCHOLY - [med] (a mental disease) mélancolie *f*
[gen] mélancolie *f*, (vague) tristesse *f*
MELANEMESIS - [med] hématémèse *f* noire
MELANEURISIS - [med] mélanurie *f*
MELANGE YARN - [text] fil *m* Vigoureux
MELANIN - [chem] (a brown pigment occurring in animal tissue) mélanine *f*
MELANODERMA - [med] mélanisme *m*, mélanodermie *f*
MELANOMA - [med] chromatophorome *m*, mélanome *m*
MELANOSIS - [med] noirceur *f* de la peau
MELANOTYPE - [photo] (synonym of ferrotype) ferrotypie *f*
MELDOMETER - [instr] (Joly's apparatus for the determination of melting-points) meldomètre *m*
MELDWEED - [bot] ansérine *f* blanche
M-ELECTRON - [nucl] (an electron which is characterized by having a principal quantum number of value 3) électron *m* M
MELINITE - [chem] (an explosive) mélinite *f*
MELISMA - [mus] mélisme *m*
MELISSA OIL - [chem] essence *f* de mélisse
MELISSIC ACID - [chem] (a fatty acid with uses in research) acide *m* mélissique
MELLITE - [med] mellite *m*
MELLITOSE - [chem] (a naturally occurring trisaccharide) mélitose *m*
MELLOW - [gen] mûr, moelleux
[paper man] conditionné
[soil] meuble
MELLOWING - [agric] aoûtement *m*
MELLS BALM - [bot] mélisse *f*, citronelle *f*
MELODY - [mus] mélodie *f*, chant *m*
~SECTION - [mus] (in jazz bands) section *f* mélodique
MELON - [bot] melon *m*
MELOPLASTY - [med] méloplastie *f*, autoplastie *f* de la face
MELT, to - [gen] fondre, se fondre, dissoudre
[metall] fondre
[phys] (to pass from the solid to the liquid state) fondre, fusionner
MELT - [metall] (a cast) fonte *f*
[gen] fondu
[glass man] (quantity of glass manufactured at one time) fusion *f*
[metall] piquée *f*
~-BACK DIFFUSED TRANSISTOR - [electron] transistor *m* à diffusion à jonction récristallisée
~-BACK TRANSISTOR - [electron] transistor *m* à jonction récristallisée
~-EXTRACTOR SCREW - [mech] (extruder screw arrangement in which the melt is separated by passing over the lands from one channel to another) vis *f* d'extraction en fusion
~INDEX - [plast ind] (the amount, in grammes, of a thermoplastic material which can be forced through an orifice of .0825 in. in 10 minutes when subjected to a force of 2160 gms.) indice *m* de fluage, indice *m* de fusion
~OFF, to - [ind chem] séparer par fusion
~-QUENCH TRANSISTOR - [electron] transistor *m* à jonction fusée et récristallisation rapide
~ZONE - [plast ind] (the region of an extruder barrel in which the material assumes a plastic condition) zone *f* d'homogénéisation
[metall] (in blast furnaces) zone *f* de fusion
MELTED CEMENT - [ind chem] ciment *m* fondu
~WAX - [chem] cire *f* fondue
MELTER - [metall] creuset, bassin *m* de coulée
[glass man] pot *m* de fusion
MELTING - [gen] fonte *f*, fusion *f*
[adj] fondant
[plast ind] plastification *f*
~BATH - [metall] bain *m* de fusion
~CAPACITY - [plast ind] (the amount of material which an injection moulding machine can plastify) capacité *f* de plastification
~CRUCIBLE - s. melting pot
~DIAGRAM - [metall] diagramme *m* de fusion
~DOWN - [metall] fusion *f*
~DOWN PROCESS - [metall] procédé *m* par fusion de fonte et riblons
~FURNACE - [metall] four *m* de fusion
~HEAT - [phys] chaleur *f* de fusion
~KETTLE - [metall] creuset *m* de fonderie
~LOSS - [metall] pertes *f*pl de fusion
~OF SLAGS - [metall] fusion *f* du laitier
~POINT - [phys] (the temperature at which a solid begins to change to the liquid state) point *m* de fusion
~-POINT TEST - [constr] essai *m* du point de fusion (du bitume)
~POT - [metall] (the iron pot in which lead and solder are melted) creuset *m*, calebasse *f*
[glass man] pot *m*, creuset *m*

MELTING RANGE - [metall] intervalle *m* de fusion
~RATE - [metall] vitesse *f* de fusion
~SECTION - [plast ind] (the region in the barrel of an extruder in which plastification occurs) zone *f* de plastification
~STOCK - [metall] charge *f*
~TABLE - [metall] table *f* de coulée
~TEMPERATURE - [phys] température *f* de fusion
~TUBE - [metall] petit tube *m* à fusion
~UNIT - [metall] four *m* de fusion
~ZONE - [metall] (of a blast furnace, the region in the charge where fusion takes place) zone *f* de fusion
MEMBER - [gen] membre *m*
[mech etc] organe *m*, élément *m*
[carp] pièce *f* (de charpente)
[constr] barre *f* de membrure, longeron *m*, jambe *f* (de force)
MEMBRANE - [anat] (thin sheet-like structure) membrane *f*, tunique *f*
[gen] membrane *f*, pellicule *f*
~GAUGE - [instr] (pressure gauge in which the rate of damping of a very thin membrane of light silica is used as a measure of pression) manomètre *m* à amortissement
~POTENTIAL - [el biol] (the potential difference between the two sides of a membrane) potentiel *m* de membrane
MEMORIAL - [gen] mémorial *m*
[arch] (a monument) monument *m* commémoratif
MEMORIZE, to - [comput] retenir, mettre en mémoire
MEMORY - [gen] mémoire
[comput] (any device into which information can be introduced and stored) mémoire
~BLOCK - [comput] bloc *m* de mémorisation, bloc *m* mémoire
~CAPACITY - [comput] (the amount of information stored by a unit) capacité *f* de mémoire
~CELL - [comput] (the storage of one unit of information) mot *m*, cellule *f* de mémoire
~CONTENTS - [comput] (the information in any part of the computer store) contenu *m* de mémoire
~CONTROL - [comput] sélection *f* d'une mémoire
~DUMP - [comput] extraction *f* programmée des données
~FUNCTION - [comput] fonction *f* de mémoire
~LOCATION - [electron comput] (storage position) emplacement *m*
~OPERATION - [comput] (reading, writing, transferring or holding information) opération *f* de mémoire
~REGISTER - [comput] (register in the storage of the computer) registre *m*
~SELECTION CIRCUIT - [comput] sélecteur *m* de mémoire
~STACK - s. memory control
~STACK - [comput] (assembly of eight I6 x I6 arrays of ferrite cores) pile *f* refoulée
~TUBE - [electron] (electron tube in which information can be stored) tube *m* à mémoire
MENARCHE - [med] première menstruation *f*
MEND, to - [gen] raccommoder, corriger, réparer
[text] repriser
[constr] (a road) repiquer
[metall] surcharger
MENDEL'S LAWS - [biol] lois *fpl* de Mendel
~SECOND LAW: LAW OF SEGREGATION - [biol] loi *f* (de Mendel) de la ségrégation des caractères
MENDEL'S THIRD LAW: LAW OF INDEPENDENT ASSORTMENT - [biol] loi *f* (de Mendel) de l'indépendance des caractères
MENDELEEVITE - [min] (a mineral which contains calcium, uranium, titanium and niobium) mendéleyévite *f*
MENDELEEV'S LAW - [phys] (the periodic law; the elements may be arranged in a periodic table) loi *f* du système périodique
~TABLE - [phys] (periodic table) tabelle *f* du système périodique
MENDELEVIUM - [chem] (a transuranic radioactive element, symbol Mv. A.N. I0I) mendélévium *m*
MENDELISM - [biol] mendélisme *m*
MENDING CUP - [text] "coquetier" *m* raccoutrage
MENDIPITE - [min] (oxychloride of lead) mendipite *f*
MENILITE - [min] (liver-opal) ménilite *f*
MENINGES - [anat] (the connective covering the brains and spinal cord) méninges *fpl*
MENINGIOMA - [med] méningiome *m*, endothéliome *m* méningé
MENINGITIS - [med] méningite *f*
MENINGOBLASTOMA - [med] méningoblastoma *m*
MENINGOCELE - [med] méningocèle *f*
MENINGOCEPHALITIS - [med] méningo-encéphalite *f*
MENINGOCOCCEMIA - [med] méningococcémie *f*
MENINX - [anat] méninge *f*
MENISCHESIS - [med] rétention *f* des règles
MENISCUS - [opt] (the curved surface formed at the top of a column of liquid in a tube, especially of mercury in a mercurial barometer) ménisque *m*
~OF MERCURY - [chem] ménisque *m* de mercure
MENOPAUSE - [med] (change of life) ménopause *f*, retour *m* d'âge
MENORRHAGIA - [med] ménorragie *f*
MENSTRUAL - [med] menstruel
MENSTRUATION - [med] menstruation *f*
MENSURATION - [meas] mesurage *m*, mesure *f*
[math] (in geometry) mensuration *f*
MENTAL DISEASE - [med] aliénation *f* mentale
MENTALIA - [med] état *m* hallucinatoire
MENTHANE - [chem] menthane *m*
MENTHANEDIAMINE - [chem] (a synthesis intermediate and a crosslinking agent for epoxy resins) menthanediamine *f*
MENTHOL - [chem] (a naturally occurring alicyclic alcohol, also prepared synthetically and used in perfumery and medicine) menthol *m*
MENTHONE - [chem] (a constituent of peppermint oil) menthone *f*
MENTHYL ACETATE - [chem] (a perfumery agent) acétate *m* de menthyle
M.E.P. - s. mean effective pressure
MEPHITIC - [gen] méphitique
MERALGIA - [med] meralgie *f* paresthésique
MERCANTILE - [gen] mercantile, marchand
[comm] commercial
MERCAPTANS - [chem] (a synonym for thiols. Mercaptans are monosubstituted derivatives of hydrogen sulphide. They have highly offensive odours) mercaptans *mpl*, thioalcools *mpl*
MERCAPTOBENZOTHIAZOLE - [chem] (a major vulcanization accelerator) amercaptobenzothiazol *m*
MERCAPTOETHANOL - [chem] (an intermediate for pharmaceuticals, dyes, plasticizers, rubber chemicals, insecticides and textile chemicals) mer-

captoéthanol *m*
MERCAPTOTHIAZOLINE - [chem] (an intermediate for pharmaceuticals) mercaptothiazoline *f*
MERCATOR'S PROJECTION - [nav] (a cylindrical projection on a cylinder tangential to a Great Circle of the earth) projection *f* de Mercator
MERCERISED CLOTH - [text] tissu *m* mercerisé
~LINEN YARN - [text] fil *m* de lin mercerisé
MERCERISING AGENT - [text] agent *m* de mercerisage
~MACHINE - [text] machine *f* à merceriser
~PADDER - [text] foulard-merceriseur *m*
MERCERIZATION - [text] (the process of increasing appreciably the lustre of cotton yarns and fabrics) mercerisage *m*
MERCERIZE, to - [text] (the operation of increasing the lustre of cotton yarns) merceriser
MERCERIZED - [text] mercerisé
~COTTON - [text] coton *m* mercerisé
~YARN - [text ind] fil *m* mercerisé
MERCERIZING - s. mercerization
~MACHINE - [text] machine *f* à merceriser
MERCHANDISE - [gen & comm] marchandise *f*
MERCHANT BAR - [metall] fer *m* marchand, fer *m* commercial
~MARINE - [naut] marine *f* marchande
~MILL - [metall] laminoir *m* à fers marchands
~ROLL - [metall] cylindre *m* à fer marchand, cylindre *m* finisseur
~SERVICE - s. merchant marine
~SHIP - [naut] navire *m* marchand
~STEEL - [metall] acier *m* marchand
MERCHANTMAN - [naut] navire *m* marchand
MERCURIAL - [chem] mercuriel, hydrargyrique
[chem] (pharmacy) préparation *f* mercurielle
~PENDULUM - [horol] (compensation pendulum in which mercury is used as compensation medium) pendule *m* à mercure
MERCURIC AMMONIUM CHLORIDE - [chem] (also called ammoniated mercury. It has application in medicine) chlorure *m* double de mercure et d'ammonium
~ARSENATE - [chem] (a component of anti-fouling paints for ships) arséniate *m* mercurique
~BARIUM IODIDE - [chem] (an analytical reagent) iodure *m* de mercure et de barium
~CHLORIDE - [chem] (also called Corrosive Sublimate. A catalyst in organic reactions, starting material for the production of mercury compounds, a fungicide and wood preservative. Mercuric chloride is also used in medicine as an antiseptic and disinfectant) chlorure *m* mercurique
~CYANIDE - [chem] (an antiseptic and a photographic chemical) cyanure *m* mercurique
~FLUORIDE - [chem] (a synthesis intermediate) fluorure *m* mercurique
~IODIDE - [chem] (an analytical reagent and an antiseptic) iodure *m* mercurique
~NITRATE - [chem] (a caustic used in medicine and a nitrating agent in organic reactions) nitrate *m* mercurique
~OLEATE - [chem] (an antiseptic) oléate *m* mercurique
~OXIDE (RED) - [chem] (a pigment for protective coatings and ceramics, an oxidizing agent, and an antiseptic) oxyde *m* de mercure rouge
~OXIDE (YELLOW) - [chem] (an antiseptic used in pharmaceuticals) oxyde *m* de mercure jaune
MERCURIC OXYCYANIDE - [chem] (an antiseptic used in medicine) oxy-cyanure *m* de mercure
~POTASSIUM IODIDE - [chem] (a water-soluble compound with uses in medicine) iodure *m* de mercure et de potassium
~STEARATE - [chem] (a disinfectant) stéarate *m* de mercure
~SULPHATE - [chem] (a catalyst in organic synthesis) sulfate *m* mercurique
~SULPHIDE - [chem] (a red pigment for plastics, rubber and paints) sulfure *m* mercurique
~THIOCYANATE - [chem] (a chemical used in photography) thiocyanate *m* de mercure
MERCUROUS - [chem] mercureux
~CHLORIDE - [chem] (a fungicide and a purgative with medicinal uses) chlorure *m* mercureux
~CHROMATE - [chem] (a pigment for ceramics) chromate *m* mercureux
~NITRATE - [chem] (an analytical reagent) nitrate *m* mercureux
~SULPHATE - [chem] (a catalyst in organic synthesis) sulfate *m* mercureux
MERCURY - [chem] (quicksilver. A liquid, silvery, metallic element, symbol Hg, A.N. 80, A.W. 200.6I, used in electrical and physical instruments and apparatus amalgams and pesticides) mercure *m*
~ABSOLUTE PRESSURE BAROMETER - [instr] manobaromètre *m*
~ACCELEROMETER - [instr] accéléromètre *m* à mercure
~ARC - [el] arc *m* à mercure
~-ARC CONVERTER-[el](converter using the rectifying properties of the mercury arc) convertisseur *m* à vapeur de mercure
~-ARC RECTIFIER - [el] (gas-filled rectifier tube in which the gas is mercury vapour) mutateur *m*, soupape *f* à vapeur de mercure
~BAROMETER - [instr] baromètre *m* à mercure
~BOILER - [th eng] chaudière *f* à mercure
~-BREAK SWITCH - [el] interrupteur *m* à mercure
~CELL - [el chem] (used in hearing aids) cellule *f* au mercure
~CHLORIDE - s. mercuric chloride
~COLUMN - [phys] colonne *f* de mercure
~COLUMN GAUGE - [instr] (type of pressure gauge in which the pressure to be measured is made to balance a column of mercury, the result being given in millimetres) manomètre *m* à colonne de mercure
~CONTACT - [el] contact *m* à mercure
~-CONTACTS RELAY - [el] relais *m* avec interrupteur à mercure
~COULOMETER - [instr] voltmètre *m* à mercure
~CUT-OFF - [mech] robinet *m* à colonne de mercure
~CYANIDE - s. mercuric cyanide
~DELAY LINE - [comput] (sonic delay-line in which mercury is used as a medium for the propagation of acoustic vibrations) ligne *f* à retard à mercure
~DIFFUSION PUMP - [mech] pompe *f* à diffusion à mercure
~DISCHARGE LAMP - [light] (electric lamp in which the discharge occurs through mercury vapour) lampe *f* à vapeur de mercure
~ELECTROLYTIC CELL - [el chem] (type of electrolytic cell in which the cathodes consist of mercury, which absorbs the deposited metal as an amalgam) cellule *f* électrolytique au mercure

MERCURY FILLING - [ind chem] remplissage *m* de mercure, charge *f* de mercure
~FULMINATE - [chem] (a detonating explosive) fulminate *m* de mercure
~GAUGE - [instr] (manometer containing mercury) manomètre *m* à mercure
~JET MAGNETOMETER - [instr] (instrument for the measurement of magnetic field intensity without the introduction of a ferromagnetic body) magnétomètre *m* à jet de mercure
~MEMORY - [comput] (delay-lines using mercury as medium) mémoire *f* à mercure
~MENISCUS - [chem] ménisque *m* de mercure
~MOTOR METER - [instr] (motor meter with some of the parts of its moving system immersed in mercury) compteur *m* à mercure
~NAPHTHENATE - [chem] (a fungicide used in paints) naphténate *m* de mercure
~ORE - [min] minerai *m* de mercure, cinabre *m*
~OXIDE CELL - [el chem] (primary cell having a mercury oxide depolarizer, a sodium hydroxyde electrolyte and a zinc electrode) pile *f* à oxyde de mercure
~-POOL CATHODE - [electron] (pool cathode consisting of mercury) cathode *f* à bain de mercure
~PUMP - s. mercury vapour pump
~RELAY - [el] relais *m* à mercure
~SALTS - [chem] sels *mpl* de mercure
~SEAL - [mech] obturateur *m* à mercure
~-SEALED JOINT - [mech] joint *m* au mercure
~SEALED STOPCOCK - [mech] robinet *m* d'arrêt à mercure
~SENSITIZATION - [photo] sensibilisation *f* au mercure
~STEEL BOTTLE - [metall] récipient *m* à mercure
~STORAGE - s. mercury memory
~STORE - s. mercury memory
~SWITCH - [el] (switch in which the fixed contact consist of mercury cups) interrupteur *m* à mercure
~TANK - [comput] (holding delay-lines storing information) magasin *m* à mercure
~THERMOMETER - [instr] thermomètre *m* à mercure
~VAPOUR - [el] vapeur *f* de mercure
~VAPOUR CONCENTRATION - [el chem] concentration *f* de vapeurs de mercure
~VAPOUR LAMP - [light] (a lamp in which the light source is an electrical discharge through mercury vapour) lampe *f* à vapeur de mercure
~-VAPOUR PUMP - [ind chem] (a pump in which mercury is continuously vaporized and recondensed) pompe *f* à vapeur de mercure
~-VAPOUR RECTIFIER - [el] (a rectifier utilizing the rectifying properties of an arc in mercury vapour) redresseur *m* à vapeur de mercure
~-VAPOUR TUBE - [electron] (a tube in which the discharge occurs through mercury vapours) tube *m* à vapeur de mercure
~-VAPOUR TURBINE - [mech] turbine *f* à vapeurs de mercure
~WATTHOURMETER - [instr] wattmètre *m* à mercure
~-WETTED CONTACT - [electron] contact *m* mouillé au mercure
MERGE, to - [gen] fondre, fusionner
[comput] (to produce a single sequence of items) fusionner, intercaler
MERIDIAN - [geogr] (in the general use of the term, the half of a Great Circle passing through the poles, which lies on the side of the earth in question) méridien *m*
MERIDIAN ALTITUDE - [astr] (the altitude of a celestial body at the position of upper transit) hauteur *f* méridienne
~CIRCLE - [astr] (of a telescope moving in the meridian plane) cercle *m* méridien
~LINE - [geogr] méridienne *f*
~PASSAGE - [astr] passage *m* au méridien
~PLANE - [surv] (the vertical plane in the direction of true north-south) plan *m* méridien
MERIDIONAL - [geogr] (southern) méridional
[astr] méridien
~FOCAL LINE - [opt] (the focal line at right angles to the plane passing the principal axis of the lens and the axis of the beam of light) ligne *f* focale tangentielle
~PARTS - [nav] (the sum of the secants of latitude between the equator and the latitude in question, taken at intervals of one minute) polygonal *m* méridien
MERINO - [zool] (a breed of sheep) mérinos *m*
[text] (hosiery made of merino wool and cotton) mérinos *m*
~SHEEP - [zool] brebis *f* mérinos
MERISM - [bot] (development of more than one member of the same kind) reproduction *f* mérismatique
MERISPORE - [bot] (segment of a multiple spore) mérispore *f*
MERISTELE - [bot] (strand of vascular tissue) méristèle *f*
MERISTEM - [genet] (group of cells each of which is capable of division) méristème *m*
MERISTEMATIC LAYER - [bot] couche *f* méristematique
MERISTIC - [biol] (divided into parts) méristique
MERISTOGENETIC - [bot] (formed from a meristem) méristogénétique
MERIT FACTOR - [gen] facteur *m* de mérite, rendement *m* volumétrique
~RATING - [gen] qualification *f* du travail
MERLON - [arch] (on a battlement, a projecting part) merlon *m*
MEROCHROME - [chem] (mixed crystals consisting of two different coloured isomers) mérochrome *m*
MEROGENESIS - [genet] (formation of parts or segmentation) mérogénèse *f*
MEROGONY - [genet] mérogonie *f*
MEROPARESTHESIA - [med] troubles *mpl* de sensibilité du tact
MEROPIA - [med] cécité *f* partielle
MEROSMIA - [med] anosmie *f* élective
MERYCISM - [med] mérycisme *m*, rumination *f*
MESA - [geogr] mesa *f*
~TECHNIQUE - [electron] (for semiconductors) technique *f* mésa
MESENCHYMA - [med] mésenchyme *m*
MESENTERIC - [anat] mésentérique
MESENTERIOLUM - [med] méso-appendice *f*
MESENTERIOPEXY - [med] mésopexie *f*
MESENTERITIS - [med] mésentérite *f*
MESENTERY - [anat] (fold of tissue which supports part of the viscera) mésentère *m*
MESH, to - [mech] (term used of gears engaging with each other) engrener, endenter, s'engrener, être en prise
MESH - [gen] maille *f*
[text & mech] (the size of the openings in screening

cloth or wire gauge, usually taken in terms of the number of openings in unit length or area) maille *f*
MESH - [el] (combination of elements forming a complete circuit) (couplage *m*) polygonal
[constr] (expanded metal used for reinforcing concrete) réseau *m* d'armature
[mech] (of gears) prise *f*, engrènement *m*, engrenage *m*
[electron] maille *f*
~CEILING - [constr] plafond *m* à réseau
~CONNECTION - [el] (method of connecting the winding of an electrical machine) connexion *f* polygonale (à phases reliées entre elles)
~EFFECT - [telev] (type of blemish) effet *m* de maille
~IMPEDANCE - [el] (the ratio between voltage and current in a mesh) impédance *f* de maille
~IRON SHEET - [metall] tôle *f* striée
~NETWORK - [el] (network formed from impedances in series) réseau *m* maillé
~OF SCREEN - [telecomm] élément *m* d'un quadrillage
~SIEVE - [impl] crible *m* à mailles
~SIZE - [gen] ouverture *f* de maille
~-STAR - [el] (of a mesh connexion) étoile-triangle *f*
~STRUCTURE - [geol] (the alternation of olivine to serpentine) structure *f* maillée
~VOLTAGE OF A POLYPHASE SYSTEM - [el] tension *f* polygonale d'un système polyphasé
MESHED - [gen] maillée, réticulé
~ANODE - [electron] (anode made from wire gauze) anode *f* en mailles
MESHING - [mech] prise *f*, engrènage *m*
~GEAR - [mech] engrènement *m*
~OF TEETH - [mech] engrènement *m* des dents
MESIOCCLUSION - [med] mésiclusion *f*, mésiocclusion *f*
MESITITE - [min] (variety of magnesite) mésitite *f*
MESITYL OXIDE - [chem] (a solvent for resins, cellulose derivatives, and surface coatings) oxyde *m* de mésityle
MESITYLENE - [chem] (a synthesis intermediate) mésitylène *m*
MESLIN - [agric] méteil *m*
~FLOUR - [agric] farine *f* mélangée
MESOAORTITIS - [med] mésoaortite *f*
MESOCARDIUM - [anat] mésocarde *m*
MESOCARP - [bot] mésocarpe *m*
MESOCOLLOID - [phys] (colloid composed of medium-sized particles) mésocolloïde *m*
MESOCOLON - [anat] mésocôlon *m*
MESOCOLOPEXY - [med] mésocolopexie *f*
MESOCYST - [med] mésocyste *f*
MESODESMIC STRUCTURE - [cryst] structure *f* mésodesmique
MESODUODENUM - [anat] mésoduodénum *m*
MESOGASTER - [anat] mésogastre *m*
MESOMERISM - [chem] (phenomenon of optical isomers which do not posses rotation) mésomérie *f*
MESOMORPHIC - s. mesomorphous
MESOMORPHOUS - [phys] (existing in a state of aggregation between crystalline and amorphous state) mésomorphe
MESON - [phys] (high-energy particle of mass 8.99×10^{-26}) méson *m*
~FIELD - [nucl] champ *m* mésonique
MESONIC ATOM - [nucl] atome *m* mésonique
MESOPAUSE - [astr] (the upper limit of the mesosphere) mésopause *f*
MESOPEXY - s. mesenteriopexy
MESOPHYLL - [bot] mésophylle *m*
MESOPHRAGMA - [anat] bande *f* claire, mésophragme *m*
MESOPIC VISION - [opt] vision *f* mésopique
MESORECTUM - [anat] mésorectum *m*
MESOSALPINX - [anat] mésosalpinx *f*
MESOSPHERE - [astr] (space region between I,063 and I,I57 times the earth's radius) mésosphère *f*
MESOSYPHILIS - [med] syphilis *f* secondaire
MESOTHELIOMA - [med] mésothélioma *m*, pleurome *m*
MESOTHELIU - [med] mésothélium *m*
MESOTHORIUM - [phys] (a member of the thorium series) mésothorium *m*
MESOTRON - [nucl] (a heavy electron) mésotron *m*
MESOVARIUM - [med] mésovarium *m*
MESOZOIC - [geol] mésozoique, secondaire
~ERA - [geol] âge *m* mésozoïque
MESSAGE - [gen] message *m*, communication *f*
[telecomm] communication *f*, conversation *f*
~BEGINNING CHARACTER - [comput] signal *m* de début d'un message
~ENDING CHARACTER - [comput] signal *m* de fin
~HEADING - [comput] en-tête *m* d'un message
~REGISTER - [instr] (electromechanical device registering counts) compteur *m* d'abonné
[telecomm] compteur *m* de conversations
~REPORTER - [el acoust] (magnetic tape instruments giving information in shops etc) annonceur *m* magnétique
~SWITCHING OFFICE - [comput] central *m* de retransmission
~-WAITING INDICATOR - [comput] indicateur *m* de délai d'attente des messages
MESSENGER CABLE - [el] (strong suspension wire for holding aerial cables) câble *m* horizontal de suspension
~WIRE - s. messenger cable
MESTIZO - [gen] (the offspring of a Spaniard and an American Indian) métis *m*
[text] (a South American type of wool from mixed breeds of sheep) métis
~WOOL - [text] laine *f* métis
METABIOSIS - [med] métabiose *f*, symbiose *f*
METABISULPHITE - [chem] métabisulfire *m*
METABOLISM - [biol] (the whole of the chemical and physical changes continuously occurring in living organisms) métabolisme *m*
METABOLITE - [biol] (a substance produced and used in a living cell) métabolite *m*, produit *m* du métabolisme
METACENTRE - [hydr] (the intersection of the vertical through the centre of buoyancy after the displacement of the body and the line joining the centre of gravity and the equilibrium centre of buoyancy) métacentre *m*
METACENTRIC - [hydr] métacentrique
~HEIGHT - [hydr] (the distance between the centre of gravity of a floating body and its metacentre) hauteur *f* métacentrique
METACHEMICAL - [phys] (pertaining to the study of atomic and subatomic phenomena) métachimique
METACOELE - [anat] toit *m* du quatrième ventricule
METADYNE - [el] (type of machine with inherent self-excitation having more than one set of brushes per pole) métadyne *f*

METADYNE CONVERTER - [el] convertisseur *m* métadyne
~DRIVE - [radar] système *m* métadyne
~GENERATOR - [el] génératrice *f* métadyne
METAL, to - [metall] (to cover or furnish with metal) métalliser, doubler de métal
[constr] (a road) macadamiser, empierrer, ferrer
[railw] (to furnish a line under construction with ballast) empierrer
[metall] réguler
METAL - [phys & chem] (a member of the class of elements which easily forms positive ions) métal *m*
[constr] (broken stone forming the surface of a macadamized road) matériaux *mpl* d'empierrement
[railw] s. metals
[glass man] (the material in a molten state for making glass) verre *m* en fusion
[road & railw] (the broken stone used for the macadamizing of road or ballasting of railway lines) ballast *m*, pierraille *f*, chaille *f*
[print] caractères *mpl*
[min] pierre *f* de mine, roc *m*
~-ARC CUTTING - [metall] découpage *m* à l'arc à électrodes métalliques
~-ARC WELDING - [metall] (arc-welding process in which the arc is maintained between metal electrodes and the pieces to be worked) sousage *m* à l'arc métallique
~ARMOURING - [mech] armure *f*
~BACKING - s. metallized screen
~-BACKING SCREEN - [electron] écran *m* métallisé
~-BAND CONVEYOR - [transp] conveyeur *m* à courroie métallique
~BAND SAWING MACHINE - [mech] scie *f* à ruban pour métaux
~BEAD - [text] perle *f* métallique
~BEARING - [min] métalifère
~BOUND MALLET - [impl] marteau *m* à têtes rapportées
~BRUSH - [mech] balais *m* métallique
~BUMPING - [metall] (the treatment of plates, generally of a car body etc) battement *m* de métaux
~CASING - [mech] gaine *f* métallique
~CEILING - [constr] plafond *m* à réseau métallique
~CERAMIC - [metall] (a substance consisting of a mixture of a metal in a ceramic to give some ductility to the ceramic itself) céramique *f* métallique
~-CERAMIC SEAL - [metall] scellement *m* métal-céramique
~CHIPS - [metall] copeaux *mpl*
~-CLAD - [gen] blindé
~-CLAD BUILDING - [constr] édifice *m* à couverture métallique
~-CLAD SWITCHGEAR - [el] (switchgear in which each part is completely surrounded by earthed metal casing) appareillage *m* blindé
~-CORED CARBON - [light] (an arc-lamp carbon having a metal core which improves its conductivity) charbon *m* à noyau fusible
~COVERED EMBROIDERY YARN - [text] fil *m* de broderie recouvert de métal
~COVERED YARN - [text] fil *m* recouvert de métal
~CRATE - [gen] cageot *m* métallique
~-CUTTING SAW - [impl] scie *f* à métaux
~-CUTTING SEGMENTAL SAW - [impl] scie *f* à métaux à segments rapportés
~-CUTTING WITH OXYGEN - [metall] (also called oxy-gas cutting) oxycoupage *m*
METAL DARK SLIDE - [photo] châssis *m* métallique
~-DEGASSING - [ind chem] dégazage *m* de métaux
~DETECTOR - [instr] (a device which reveals the presence of metal in materials) détecteur *m* de métal
~DINGING - s. metal bumping
~DIP BRAZING - [mech] soudure *f* forte par immersion
~DISTRIBUTION RATIO - [metall] (the relative weight per unit area of metal deposited on different punts of a cathode) rapport *m* de distribution du métal
~DRIFT - [mining] galerie *f* en rocher, galerie *f* au stérile
~-EDGED - [metall] à bord *m* de métal
~EDGING - [aero] (to furnish wooden propeller with metal) blindage *m*
~ELECTRODE - [el] (the type of electron used in metal-arc welding) électrode *f* métallique
~-ENCLOSED SWITCHGEAR - [el] (a type of switchgear in which the whole equipment is enclosed in earthed metal casing) appareillage *m* à enveloppe métallique
~EYELET - [metall] oeillet *m* de métal
~FILAMENT - [el] (fine metal conductor which can be heated to incandescence) filament *m* métallique
~-FILAMENT LAMP - [el] lampe *f* à filament métallique
~-FILM RESISTOR - [el] résistance *f* à pellicule métallique
~FOG - [el metall] (metal finely dispersed through a fused electrolyte) brouillard *m* métallique
~FOIL - [metall] (thin, elongated piece of material) feuille *f* de métal
~-FOUNDING - [metall] moulage *m* des métaux
~GAUGE - [impl] jauge *f* pour tôles
~-GAUGE HEAD - [meas] cellule *f* de mesure métallique
~GAUZE - [metall] gaze *f* métallique, toile *f* métallique
~HARDENING - [metall] trempe *f* des métaux
~HEALD - [text] lisse *f* métallique
~-JOINTING - [mech] joint *m* métallique
~LACQUER - [paint] vernis-laque *m* pour métaux
~LAMINATES - [metall] laminés *mpl* de métal
~LATHING - [constr] (expanded metal used to cover surfaces before plastering) tôle *f* étirée
~LINE - [glass man] (of a tank furnace) ligne *f* du métal
~-LINED SHAFT - [mining] puits *m* blindé
~-MAN - [mining] mineur *m* au rocher
~MASTER - [el acoust] (the master which is produced by electro-forming from the face of a wax recording) original *m*, père *m*
~MINING - [mining] exploitation *f* minière de métaux
~MIST - s. metal fog
~MIXER - [metall] mélangeur *m* de fonte
~MOULD - [metall] (in diecasting) moule *m* métallique
~NEGATIVE - s. metal master
~NOTCH - [metall] trou *m* de coulée
~PACKING - [mech] (packing consisting of metal elements) garniture *f* métallique
~PAPER - [paper man] (specially treated paper produced in foils) papier *m* métallisé
~PATTERN - [metall] (in foundry work; a pattern in

cast iron or light alloy to obtain durability and permanence of form) modèle *m* métallique

METAL PENETRATION - [metall] (a defect) coquille *f* d'oeuf abreuvante

~PENETRATION FLASH - [metall] défoncement *m*

~PLATING - [el metall] (electro-plating) galvanoplastie *f*

~QUOIN - [print] (a metal wedge used to lock up forms) outillage *m* à serrer

~RECTIFIER - [el] (rectifier making use of the properties of a layer of oxide on a metallic disk) redresseur *m* sec

~-RECTIFIER TYPE ECHO SUPPRESSOR - [telecomm] suppresseur *m* d'echo à action continue

~REFINERY - [metall] usine *f* métallurgique d'affinage

~RIBS - [text] jumelles *fpl* métalliques

~RULE - [print] tiret *m*

~SAW - [impl] scie *f* à métaux

~SCREEN - [radiat] (intensifying screen of metal foil) écran *m* métallique

~SCREW - [metall] vis *f* à métaux

~SECTION - [metall] profilé *m* metallique

~SHEATHING - s. metal armouring

~SHUTTLE - [text] espolin *m* métallique

~SIEVE - [min] tamis *m* à mailles

~SMELTING - [metall] fusion *f* du métal

~SPINNING - [metall] (the operation of shaping thin metal-disks into cup-shaped forms by the lateral pressure of steel rollers) emboutissage *m* au tour

~SPRAY COATING - [ind proc] métallisation *f* par projection, métallisation *f* par aspersion

~SPRAYING - s. metal spray

~STRIP - [metall] feuillard *m*

~-TANK MERCURY ARC RECTIFIER - [el] soupape *f* à cuve d'acier

~-TO-GLASS SEAL - [metall] scellement *m* verre-métal, joint *m* verre-métal

~-TO-METAL CLUTCH - [auto] embrayage *m* par friction métal contre métal

~-TO-METAL JOINT - [mech] (a self-sealing joint) serrage *m* à sec

~-TO-METAL POSITION - [auto] (of shock absorbers; in compression) position *f* de butée
(of shock absorbers; in rebound) position *f* de butée

~TRESTLE - [mech] chevalet *m* métallique

~TRIM - [constr] (finishings made out of pressed metal sheeting) finissage *m* en métal

~TUBE SIZING - [metall] passage *m* par cylindres de finissage

~TURBIDITY - [ind chem & brew ind] trouble *m* de métal

~TURNING LATHE - [mach tool] tour *m* à métaux

~V-COLLAR - [el] (in a commutator construction) cône *m* de serrage

~V-RING - s. metal V-collar

~VALLEY - [plumb] (V-shaped gutter lined with a metal) noue *f* en métal

~WELDING - s. metal-arc welding

~WORKING - [metall] serrurerie *f*, travail *m* des métaux

METALDEHYDE - [chem] (a derivative of acetaldehyde with uses as an easily portable fuel) métaldéhyde *m*

METALIMNION - [limnology] thermocline

METALLIC - [gen] métallique

METALLIC BOND - [chem] (bond existing in metals, in which the valence electrons of the constituent atoms are free to move in the periodic lattice) liaison *f* métallique

~CIRCUIT - [el] (a line with a complete copper circuit without earth return) circuit *m* par conducteur séparé
[telecomm] circuit *m* bifilaire, ligne *f* bifilaire

~CLOTH - [text] tissu *m* en fil métallique

~CONDUCTION - [el] (the conduction of electricity through metals by the migration of electrons) conduction *f* métallique

~CRYSTALS - [metall] (the crystals of the metals and alloys) cristaux *mpl* métalliques

~DEOXIDIZER - [chem] désoxydant *m* métallique

~FIBRE - [text] (metal, metal-covered plastic or plastic-coated metal threads used in decorative yarns) fibre *f* métallique

~FUEL ELEMENT - [nucl] élément *m* combustible métallique

~LUSTRE - [metall] (the degree of lustre shown by some opaque minerals) éclat *m* métallique

~MIRROR - [photo] (mirror of highly-polished metal) glace *f* métallique

~OXIDE - [chem] oxyde *m* métallique

~PACKING - [mech] (packing formed by a number of rings of soft metal, or by a helix of metallic yarn, encircling the piston rods) garniture *f* métallique

~PAINT - [metall] couleur *f* métallifère

~PAPER - [paper man] papier *m* pour impression d'encres métallisées

~RECTIFIER - [electron] redresseur *m* métallique

~RECTIFIER STACK - [electron] (structure of metallic-rectifier cells) empilage *m* de redresseurs métalliques

~RETURN CIRCUIT - [el] circuit *m* de retour par conducteur séparé

~SALT - [chem] sel *m* métallique

~SIGNALLING SYSTEM - [telecomm] système *m* à boucle

~SURFACE LAYER - [metall] revêtement *m* métallique

~WOVEN FABRIC - [text] tissu *m* en fils métalliques

METALLIFEROUS - [metall] (containing metal, yielding metal) métallifère

~VEINS - [geol] (fissures in rocks containing ores of metals) veines *fpl* métallifères

METALLINE - [metall] (relating to, or resembling metal) métallique
[ind chem] (impregnated with metals, or metallic salts) métallin

METALLING - [constr] macadam *m*, empierrement *m* (the material used in road and railway work) ballast *m*
[metall] métallisation *f*, métallisage *m*

METALLIZATION - [metall] (process of depositing a thin layer of metal upon another substance) métallisation *f*
[chem] (the conversion of a substance into a metallic form) métallisation *f*
[mech] (the spraying of atomized metal on plates etc) métallisation *f* par projection

METALLIZE, to - [gen] (to render metallic, or to give a metallic form) métalliser
(to impregnate with metallic salts) métalliser
[rubber ind] vulcaniser

METALLIZED YARN - [text] fil *m* métallisé

METALLIZED BUSHING - [el] traversée *f* isolée métallisée
~COATING - [el] (of electrical cables) revêtement *m* métallique
~DYES - [dyes] (dyestuffs containing metal in chemical combination) teintures *fpl* métallisées
~FILAMENT - [el] (heat-treated carbon-lamp filament, which is thus partially converted into graphite) filament *m* métallisé
~GLASS - [el] verre *m* métallisé
~SCREEN - [electron] (of a cathode-ray tube) écran *m* luminescent métallisé
~TUBE - s. metallized valve
~VALVE - [electron] (tube having the exterior of the envelope coated with a conducting metallic film) tube *m* métallisé
METALLIZING - [mech] (coating with a thin film of metal) métallisation
~GUN - [impl] (a spray gun to metallize plates, more generally surfaces) atomiseur *m* pour la métallisation
~PROCESS - [metall] procédé *m* de métallisation
~TORCH - [impl] chalumeau *m* pour métallisation
METALLO-ORGANIC - [chem] (organic compound containing a metal) organométallique
~-ORGANIC PIGMENT - [chem] (a pigment which contains carbon and a metallic salt) pigment *m* semi-minéral
METALLOGRAPHIC - [metall] (of metallography) métallographique
~ANALYSIS - [metall] analyse *f* métallographique
~EQUIPMENT - [metall] (the equipment for the microscopic study of the structure of metals and alloys) installation *f* métallographique
~ETCHANT - [metall] réactif *m* métallographique
METALLOGRAPHY - [metall] (the study of the constitution and structure of metals) métallographie *f*
METALLOID - [chem] (substance having both metallic and non-metallic properties) métalloïde *m*
METALLOTHERAPY - [med] (a treatment by means of metal salts) métallothérapie *f*
METALLURGIC - s. metallurgical
METALLURGICAL - [chem] (relating to the science of extracting metals from ores) métallurgique
~CHEMISTRY - [ind chem] chimie *f* métallurgique
~COKE - [metall] (coke manufactured expressly for metallurgical use) coke *m* métallurgique
~MICROSCOPE - [instr] microscope *m* métallurgique
~NUCLEUS - [metall] (a small cluster of atoms) germe *m*
~PLANT - [gen] industrie *f* métallurgique
METALLURGIST - [gen] métallurgiste *m*
METALLURGY - [metall] (the science of metals) métallurgie *f*
METALS - [railw] (the rails) rails *mpl*
METALUETIC - [med] métasyphilitique
METALWORK TECHNOLOGY - [metall] technologie *f* métallurgique
METALWORKING - [metall] travail *m* sur métaux
~TOOLS - [tools] outils *mpl* pour le travail sur métaux
METAMERIC MATCH - [opt] (the visual equivalence of different colours) équivalence *f* métamère
METAMERISM - [chem] (a form of isomerism in which the compounds have the same percentage of composition and the same molecular weight) métamérie *f*
METAMERS - [opt] (radiations with different spectral compositions, yet producing the same visual effect) métamères *mpl*
METAMIC - [el] (cermet; a class of materials consisting of ceramic substances in a fine state of division bonded with a metallic medium) métal *m* céramique
METAMICT CRYSTALS - [cryst] cristaux *mpl* métamictes
METAMORPHISM - [geol] (transformation, e.g. of rocks, through natural causes) métamorphisme *m*
METAMORPHOSIA - [opt] (a vision defect making objects appear distorted) métamorphopsie *f*
METAMORPHOSE, to - [gen etc] (to change the form; to alter, to transmute) métamorphoser, transformer
METAMORPHOSIS - [gen & zool] (the passing from one form into another) métamorphose *f*
(a complete change in form or character) métamorphose *f*, altération *f*
METANEPHROS - [zool] (of the genital apparatus, the posterior one of three similar organs) métanéphros *m*
METANILIC ACID - [chem] (an intermediate for pharmaceuticals and dyestuffs) acide *m* métanilique
METAPHASE - [genet] (middle state of mitotic cell division) métaphase *f*
[nucl] (a stage in the nuclear division) métaphase *f*
METAPHOSPHATE - [chem] (a salt of metaphosphoric acid) métaphosphate *m*
METAPHOSPHORIC ACID - [chem] (commercial glacial phosphoric acid) acide *m* métaphosphorique
METAPHYSIS - [anat] métaphyse *f*
METAPLASIA - [anat] métaplasie *f*
METAPLEX - [med] plexus *m* choroïde
METASILICATE - [chem] (a detergent) métasilicate *m*
METASOMATOSIS - [geol] (form of metamorphism by which a rock or a mineral undergoes chemical change) métasomatose *f*
METASTABLE - [chem] (an apparently stable state) métastable
~ATOMIC STATE - s. metastable state
~ATOMS - [nucl] (cause of secondary ionization) atomes *mpl* métastables
~EQUILIBRIUM - s. metastable state
~LEVEL - s. metastable state
~NUCLEI - [nucl] (nuclei in an excited nuclear state which have measurable lifetimes) noyaux *mpl* métastables
~STATE - [phys] (the state from which an atom does not return by radiation to a lower state, in particular to the normal state) état *m* métastable, niveau *m* métastable
~SYSTEM - [phys] (system capable of undergoing a quantum transition to a state of lower energy) système *m* métastable
METASTASIS - [phys] (fundamental change in the position of a particle) métastase *f*
[radiat] (a growth in the body of a malignant cell at some distance from the original cancer) métastase *f*
METASTATIC ABSCESS - [med] abcés *m* métastatique
~ELECTRON - [phys] (electron moving from one atom to another) électron *m* métastatique
METASYPHILIS - [med] parasyphilis *f*
METATARSUS - [anat] metatarse *m*

METATELA - [anat] toile *f* chroïdienne du quatrième ventricule
METATHALAMUS - [anat] corps *mpl* genouillés externe et interne
METE, to - [gen] (to distribute by measure) distribuer, assigner
[gen] (to measure) mesurer
METE - [gen] (a boundary line) bornes *fpl*
METENCEPHALON - [anat] cerveau *m* postérieur, métencéphale *m*
METEOGRAPH - s. meteorograph
METEOR - [astr] (small body striking the earth's atmosphere with great speed; a shooting star) météore *m*
~CRATER - [geogr] cratère *m* météorique
~PATH - [astr] trajectoire *f* d'un météore
~SHOWER - [astr] (a number of meteorites arriving from the same radiant point) pluie *f* de météorites, averse *f* de météores
~-SWARM - [astr] (a meteor shower consisting of an exceptionally large number of individuals) essaim *m* de météorites
~TRAIN - [astr] trainée *f* météorique
~WAKE - [astr] sillage *m* du météore
METEORIC - [astr] (relating to meteors) météorique
~IRON - [min] sidérolithe *f*
~WATER - [met] (water precipitated through the atmosphere) eau *f* météorique
METEORITE - [astr] (a mass of solid matter arriving in the earth's atmosphere from space) météorite *f*
METEOROGRAPH - [instr] (a combined instrument recording several magnitudes simultaneously, e.g. temperature, pressure, humidity) météorographe *m*
METEOROLOGIC - s. meteorological
METEOROLOGICAL - [astr] (relating to the science dealing with atmospheric phenomenons) météorologique
~ROCKET - [astronaut] fusée *f* météorologique
~STABILITY - [met] (the conditions governing the establishment of vertical air currents due to convection) atabilité *f* météorologique
METEOROLOGY - [met] (the scientific study of weather) météorologie *f*
METEOROPATHOLOGY - [med] météoropathologie *f*
METER, to - [meas] mesurer
[ind chem] doser
METER - [meas] (or metre, in GB; a unit of measure) mètre *m*
[instr] (general term for an instrument or a unit by which a thing is measured) compteur *m*, compteur-jaugeur *m*
[el] (an electrical measuring instrument) compteur *m*
[photo] (regular grouping of notes with regard to their accent and duration) mètre *m*
~-ADJUSTING DEVICES - [el] (devices allowing correct adjustment of a meter under specified conditions) dispositif *m* de régrage d'un compteur
~AMPERES - [radio] mètre-ampères *mpl*
~BASE - [instr] (the back of the meter, by which it is fixed) socle *m* d'un compteur
~BOOK - [comm] carnet *m* de relevé
~BOTTOM BEARING - [instr] (the support on which the moving element of a meter rests) crapaudine *f*
~BRAKING ELEMENT - [instr] (that part of the meter which produces a braking torque by its action on the moving element) élément *m* de freinage d'un compteur
METER CASE - [instr] (base and cover of the meter) boîtier *m* d'un compteur
~CHANGE-OVER CLOCK - [instr] (clock changing the interconnexion between moving element and counting mechanism at given times) horloge *m* de commutation pour compteur
~COCK - [gas ind] (service cock or consumer's control) robinet *m* de compteur
~CONSTANT - [instr] (the energy corresponding to one revolution of the meter disk) constante *f* d'un compteur
~COVER - [instr] couvercle *m* d'un compteur
~CREEPING - [instr] (the rotation of the disk of a meter under no-load conditions) marche *f* à vide
~FRAME - [instr] (the part on which are affixed the moving element, the counting mechanism and the braking magnet) bâti *m* d'un compteur
~HAND - [instr] dispositif *m* indicateur d'un compteur
~INDEX - [instr] dispositif *m*, indicateur, minuterie
~INDEX CARD - s. meter book
~KEY - [instr] (in telephony) bouton *m* de comptage
~-KILOGRAM - [meas] (unit of measure for energy) kilogrammètre *m*
~LAMP - [instr] lampe *f* pilote de compteur
~PROVER - [gas ind] gazomètre *m* de contrôle, gazomètre *m* d'expérience
~READING - [instr] relevé *m*, lecture *f* d'un compteur
~RELAY - [instr] (relay comprising the elements of a meter) relais *m* compteur
~SHUNT - [instr] (shunt resistance increasing the measuring range of moving coil instruments) résistance *f* en dérivation d'un compteur
~SUPPORT - s. meter frame
~TERMINAL COVER - [instr] (the part of the cover covering the terminals of the wires) cache-bornes *mpl*, couvre-bornes *m*
~WITH MAXIMUM DEMAND INDICATOR - [instr] compteur *m* à indicateur de maximum
~WITH MAXIMUM DEMAND RECORDER - [instr] compteur *m* à enregistreur de maximum
METERING - [meas etc] mesure *f*, dosage *m*
~DEVICE - [meas] (any arrangement which continuously measures a flow, e.g. a metering screw) dispositif *m* de mesure
~JET - [mech] (a calibrated opening designed to regulate fluid flow through it to a determinate value, e.g. in a carburettor) injecteur *m* doseur
~NEEDLE - [mech] (a tapered stem inserted in a metering jet for adjustment) aiguille *f* de dosage
~ORIFICE - [mech] (of the carburettor) ouverture *f* de dosage
~PUMP - [mech] (type of pump which supplies a measured quantity of fluid) pompe *f* volumétrique
~RELAY - [instr] relais *m* compteur
~ROD - [auto] (special rods used in certain types of carburettors) aiguille *f* de dosage
~SCREW - [mech] (type of extrusion screw having a constant depth, usually shallow, and constant pitch section over the last 3 or 4 flights) vis *f* à section homogène
~TANK - [ind chem] réservoir *m* jauge
~ZONE - [plast ind] (the part of an extruder barrel in which the material is sheared, rendered uniform in composition and temperature and controlled to a constant flow) zone *f* de plastification, zone *f* de

homogénéisation
METHACROLEIN - [chem] (methacrylaldehyde. An intermediate for a number of synthetic resins) méthacroléine *f*
METHACRYLATE ESTERS - [chem] (starting materials for the thermoplastic acrylate resins) esters *mpl* de métacrylate
METHACRYLATOCHROMIC CHLORIDE - [chem] (a vinyl polymer insolubilizer also with uses as a water-repellent) chlorure *m* méthacrylatochromique
METHACRYLIC ACID - [chem] (an important monomer for a large number of polymers and resins) acide *m* méthacrylique
METHANE - [chem] (marsh gas. The first in the openchain (aliphatic) hydrocarbon series and an important starting material for a number of organic syntheses) méthane *m*, gaz *m* de marais
~FIRING SYSTEM - [mech] installation *f* de combustion à méthane
~GAS DEPOSIT - [oil ind] dépôt *m* méthanifère
~HYDROPEROXIDE - [chem] (a polymerization catalyst) hydroperoxyde *m* de méthane
~IONIZATION CHAMBER - [nucl] chambre *f* d'ionisation à méthane
~PIPELINE - [oil ind] pipeline *m* pour méthane, conduite *f* de méthane
~SERIES - [chem] (a group of saturated aliphatic hydrocarbons) série *f* du méthane
METHANEDIAMINE - [chem] (epoxy resin curing agent) méthanediamine *f*
METHANOL - [chem] (methyl alcohol. An intermediate for a large number of compounds and a fuel and general solvent) méthanol *m*, alcool *m* méthylique
~AND WATER INJECTION - s. methanol injection
~INJECTION - [aero] (the introduction of methyl alcohol and water into the first compressor stage of a gas turbine to compensate for loss of power at take-off due to ambient air temperature in excess of the designed value) injection *f* de méthanol
METHENAMINE - [chem] hexaméthylènetétramine *f*
METHILEPSIA - [med] méthomanie *f*
METHIONINE - [chem] (an essential, sulphur-containing amino-acid with uses in medicine and as a food additive) méthionine *f*
METHOD - [gen] méthode *f*
~ELECTRIC IMAGES - [meas] (used in capacitance measurements) méthode *f* des images électriques
~OF FERMENTATION - [brew ind] conduite *f* de la fermentation, procédé *m* de fermentation
~OF INEQUALITY THEOREMS - [cryst] (a method for the x-ray analysis of crystal structures) méthode *f* des théorèmes de l'inégalité
~OF ITERATION - [telecomm] méthode *f* itérative
~OF PEGGING OUT - s. method of stacking out
~OF RANDOM SAMPLING - [telecomm] méthode *f* de contrôle par sondage
~OF REED BINDING - [text] manière *f* de fabriquer le peigne
~OF REVERSALS - [meas] (a method for determining the magnetization curve) méthode *f* des inversions
~OF STACKING OUT - [railw] procédé *m* de jalonnement, procédé *m* de piquetage
~OF TRANSLATION OF IMPULSES - [telecomm] méthode *f* de traduction des impulsions
~OF WORKING - [gen] méthode *f* d'exploitation
METHODOLOGY - [gen] (the science of method, or of arranging in due order) méthodologie *f*
METHODS OF HARDNESS TESTING - [chem] (brinell method; vickers pyramid method; moh's scale of comparisons, rockwell method, etc.) méthodes *fpl* d'essai de la dureté
~MULTIPLEXING - [telecomm] (time division, frequency division or phase division) méthodes *fpl* de multiplexage
METHOXYACETIC ACID - [chem] (a synthesis intermediate) acide *m* méthoxyacétique
METHOXYACETOPHENONE - [chem] (an intermediate for plasticizers and a solvent for paints and pharmaceuticals) méthoxy-acéto-phénone *f*
METHOXYBUTANOL - [chem] (an intermediate for plasticizers and a solvent for paints and pharmaceuticals) méthoxybutanol *m*
METHOXYCHLOR - [chem] (an insecticide) méthoxychlore *m*
METHOXYETHYL ACETYL RICINOLEATE - [chem] (plasticizer for cellulose derivatives, polyvinyl acetate, polyvinyl butyral, vinyl chloride and vinyl chloride acetate) ricinoléate *m* acétométhoxyéthylique
~OLEATE - [chem] (solvent and plasticizer) oléate *m* méthoxyéthylique
~RICINOLEATE - [chem] (plasticizer for cellulose derivatives polyvinyl acetate, polyvinyl butyral, vinyl chloride and vinyl chloride acetate) ricinoléate *m* de méthoxyéthyle
~STEARATE - [chem] (a solvent and plastics plasticizer) stéarate *m* de méthoxyéthyle
METHOXYETHYLMERCURY ACETATE - [chem] (an agricultural disinfectant) acétate *m* méthoxyéthylmercurique
METHOXYETHYL OLEATE - [chem] (a solvent and plasticizer) oléate *m* de méthoxyéthyle
~STEARATE - [chem] (a solvent and plasticizer) stéarate *m* de méthoxyéthyle
METHOXYMETHYLPENTANOL - [chem] (a solvent for resins) méthoxyméthylpentanol *m*
METHOXYMETHYLPENTANONE - [chem] (a resin solvent) méthoxyméthylpentanone *f*
METHOXYPROPYLAMINE - [chem] (a synthesis intermediate) méthoxypropylamine *f*
METHOXYTRIGLYCOL ACETATE - [chem] (an agent used in the control of finely powdered substances) acétate *m* de méthoxytriglycol
METHYL - [chem] (univalent organic acid, forming esters with acids) méthyle *m*
~ABIETATE - [chem] (a plasticizer and solvent for surface coatings) abiétate *m* de méthyle
~ABIETATE, HYDROGENATED - [chem] (plasticizer for cellulose nitrate, ethyl cellulose, polymethyl methacrylate, polystyrene, polyvinyl butyral, vinyl chloride, and vinyl chloride acetate) abiétate *m* de méthyle hydrogéné
~ACETATE - [chem] (a solvent for paints, cellulose derivatives, and a number of plastics) acétate *m* de méthyle
~ACETOACETATE - [chem] (a solvent for cellulose derivatives) acéto-acétate *m* de méthyle
~ACETONE - [chem] (an extractant for perfumes and a solvent for rubber, resins, plastics and cellulose derivatives) méthyl-acétone *f*
~ACETYLRICINOLEATE - [chem] (a plasticizer for a number of plastics) acétylricinoléate *m* de méthyle
~ACETYLSALICYLATE - [chem] (a derivative of ace-

tylsalicylic acid with uses in medicine) méthyl-acétylsalicylate *m*

METHYL ACRYLATE - [chem] (an intermediate for a number of polymers) acrylate *m* de méthyle

~ALCOHOL - [chem] (methanol; an alcohol derived from methane, used in methanol injection) alcool *m* méthylique

~ANTHRANILATE - [chem] (a component of perfumes and flavourants) anthranilate *m* de méthyle

~ARACHIDATE - [chem] (a synthesis intermediate) arachidate *m* de méthyle

~BROMIDE - [chem] (a synthesis intermediate, agricultural fumigant, and filling for fire extinguishers) bromure *m* de méthyle

~BUTANOL - [chem] (amyl alcohol. A solvent and synthesis intermediate) méthylbutanol *m*

~BUTENE - [chem] (a synthesis intermediate) méthylbutène *m*

~BUTENOL - [chem] (intermediate for drugs and perfumery components) méthylbuténol *m*

~BUTYRATE - [chem] (a solvent for cellulose compounds) butyrate *m* de méthyle

~CAPRATE - [chem] (intermediate for resins, stabilizers, plasticizers and detergents) caprate *m* de méthyle

~CAPROATE - [chem] (organic intermediate used in the production of caproic acid, stabilizers, synthetic resins, and plasticizers) caproate *m* de méthyle

~CAPRYLATE - [chem] (methyl octanoate. An intermediate resins, plasticizers, stabilizers, lubricants and detergents) caprylate *m* de méthyle

~CARBONATE - [chem] (a synthesis intermediate) carbonate *m* de méthyle

~CEROTATE - [chem] (a synthesis intermediate) cérotate *m* de méthyle

~CHLORIDE - [chem] (an aerosol propellant, refrigerant (with uses in medicine to produce local anaesthesia), catalyst carrier in polymerization reactions and a methylating agent in synthesis) chlorure *m* de méthyle

~CHLOROACETATE - [chem] (a solvent) chloroacétate *m* de méthyle

~CHLOROFORMATE - [chem] (a synthesis intermediate) chloroformiate *m* de méthyle

~CHLOROPHENOXYPROPIONIC ACID - [chem] (a weedkiller) acide *m* méthyl-chlorophénoxy-propionique

~CHLOROSULPHONATE - [chem] (a synthesis intermediate) sulfonate *m* de méthyle chloré

~CINNAMATE - [chem] (a component of perfumes and flavouring agents) cinnamate *m* de méthyle

~CITRATE - [chem] (a solvent and plasticizer) citrate *m* de méthyle

~ETHYL KETONE - [chem] (a synthesis intermediate and a solvent) méthyléthylcétone *f*

~FORMATE - [chem] (a solvent for cellulose derivatives and a synthesis intermediate) formiate *m* de méthyle

~FUROATE - [chem] (a synthesis intermediate and solvent) furoate *m* de méthyle

~GLUCOSIDE - [chem] (a plasticizer for a number of synthetic resins) méthyl-glucoside *m*

~HENEICOSANOATE - [chem] (a synthesis intermediate) hénéicosanoate *m* de méthyle

~HEPTADECANOATE - [chem] (a synthesis intermediate) heptadécanoate *m* de méthyle

~HEXYL KETONE - [chem] (a solvent for synthetic resins) méthylhexylcétone *f*

METHYL IODIDE - [chem] (an analytical reagent and synthesis intermediate) iodure *m* de méthyle

~LACTATE - [chem] (a solvent for cellulose derivatives) lactate *m* de méthyle

~LAURATE - [chem] (an intermediate for resins, plasticizers, detergents and stabilizers) lactate *m* de méthyle

~LIGNOCERATE - [chem] (a synthesis intermediate) lignocérate *m* de méthyle

~LINOLEATE - [chem] (an intermediate for resins, plasticizers, surface active agents, and lubricants) linoléate *m* de méthyle

~MERCAPTAN - [chem] (a synthesis intermediate) méthylmercaptan

~METHACRYLATE - [chem] (a monomer of major importance in the production of a number of plastics and resins) méthacrylate *m* de méthyle

~MORPHOLINE - [chem] (a solvent, stabilizing agent and intermediate for pharmaceuticals) méthylmorpholine *f*

~MYRISTATE - [chem] (intermediate for synthetic resins and plastics, detergents and plasticizers) myristate *m* de méthyle

~MYRISTOLEATE - [chem] (a synthesis intermediate) myristoléate *m* de méthyle

~NAPHTHYL KETONE-[chem] (a perfumery agent) méthyl-naphtyl-cétone *f*

~NONANOATE - [chem] (a synthesis intermediate) nonanoate *m* de méthyle

~NONYL KETONE - [chem] (a perfumery agent) méthylnonylcétone *f*

~OLEATE - [chem] (intermediate for rubber and textile chemicals, stabilizers and detergents) oléate *m* de méthyle

~PALMITATE - [chem] (intermediate for plasticizers, synthetic resins and surface active agents) palmitate *m* de méthyle

~PALMITOLEATE - [chem] (a synthesis intermediate) palmitoléate *m* de méthyle

~PENTACHLOROSTEARATE - [chem] (plasticizer for cellulose derivatives, polymethyl methacrylate, polystyrene, polyvinyl acetate, vinyl chloride, and vinyl chloride acetate) pentachlorostéarate *m* de méthyle

~PENTADECANOATE - [chem] (a synthesis intermediate) pentadécanoate *m* de méthyle

~PHENYLACETATE - [chem] (a perfumery agent) acétate *m* de méthylphényle

~PHENYLDICHLOROSILANE - [chem] (an intermediate for silicones) méthyl-phényldichlorosilane *m*

~PHTHALYL ETHYL GLYCOLATE - [chem] (plasticizer for cellulose derivatives, polymethyl methacrylate, polystyrene, polyvinyl acetate, polyvinyl butyral, vinyl chloride, and vinyl chloride acetate) méthyl-phtalyl-éthyl-glicolate *m*

~PIPERAZINE - [chem] (an intermediate for pharmaceuticals) méthylpipérazine *f*

~PROPIONATE - [chem] (a solvent for cellulose derivatives) propionate *m* de méthyle

~PROPYL ETHER - [chem] (an inhalation anaesthetic) éther *m* méthyl-propilique

~PROPYL KETONE - [chem] (a solvent) méthylpropylcétone *f*

~RED - [chem] (an analytical indicator) rouge *m* de méthyle

~RICINOLEATE - [chem] (a wetting agent and plasticizer) ricinoléate *m* de méthyle

METHYL RUBBER - [rubber ind] méthylcaoutchouc *m*
~SALICYLATE - [chem] (a solvent for cellulose derivatives, flavourant, and a counterirritant used in medicine) salicylate *m* de méthyle
~STEARATE - [chem] (an intermediate for synthetic resins, stabilizers, plasticizers, detergents and lubricants) stéarate *m* de méthyle
~THIOURACIL - [med] (an anti throid compound used in the control of thyrotoxicosis) méthyl-thiouracile *m*
~TRIDECANOATE - [chem] (a synthesis intermediate) tridécanoate *m* de méthyle
~UNDECANOATE - [chem] (a synthesis intermediate) undécarboate *m* de méthyle
~VINYL ETHER - [chem] (modifier and plasticizer for copolymers and plastics) éther *m* méthylvinylique
~VINYLPYRIDINE - [chem] (a synthesis intermediate) méthyl-vinyl-pyridine *f*
~VIOLET - [chem] (a pigment, textile dye, analytical indicator and a microscopic stain. In medicine it has been replaced as an antiseptic by crystal violet) violet *m* de méthyle
METHYLACETANILIDE - [chem](an analgesic andantipyretic) méthyl-acétanilide *f*
METHYLACETOPHENONE - [chem] (a perfumery component) méthyl-acétophénone *f*
METHYLACETYLENE - [chem] (a synthesis intermediate) méthyl-acétylène *m*
METHYLAL - [chem] (formal. A synthesis intermediate and solvent) méthylal *m*
METHYLALLYL CHLORIDE - [chem] (a synthesis intermediate and insecticide) chlorure *m* de méthylallyle
METHYLALUMINIUM SESQUIBROMIDE - [chem] (a synthesis intermediate and a polymerization and hydrogenation catalyst) sesquibromure *m* de méthylaluminium
METHYLAMINE - [chem] (an intermediate for pharmaceuticals, rubber chemicals, dyestuffs and insecticides and a photographic chemical, solvent and polymerization inhibitor) méthylamine *f*
METHYLAMINOPHENOL - [chem] (a photographic chemical and synthesis intermediate) méthyl-aminophénol *m*
METHYLAMYL ACETATE - [chem] (a solvent for certain cellulose derivatives) acétate *m* de méthylamile
~ALCOHOL - [chem] (a synthesis intermediate and solvent) alcool *m* méthylamylique
~CARBINOL - [chem] (a solvent for synthetic resins) méthyl-amyl-carbinol *m*
~KETONE - [chem] (a solvent for cellulose derivatives) méthyl-amyl-cétone *f*
METHYLANILINE - [chem] (a synthesis intermediate) méthylaniline *f*
METHYLANTHRACENE - [chem] (a synthesis intermediate) méthyl-anthracène *m*
METHYLANTHRAQUINONE - [chem] (a synthesis intermediate) méthyl-anthraquinone *f*
METHYLATED SPIRITS - [chem] (common name for commercial ethyl alcohol which has been made unfit for drinking by the addition of methanol and other substances) alcool *m* dénaturé
METHYLBEHENATE - [chem] (a synthesis intermediate) béhénate *m* de méthyle
METHYLBENZOATE - [chem] (a solvent for rubber, cellulose derivatives and resins) benzoate *m* de méthyle
METHYL-ortho-BENZOYLBENZOATE - [chem] (a plasticizer) benzoate *m* de méthyl-ortho-benzoyle
METHYLBENZYLAMINE - [chem] (a synthesis intermediate) méthyl-benzylamine *f*
METHYLBENZYLDIETHANOLAMINE - [chem] (intermediate for textile chemiclas) méthylbenzyldiéthanolamine *f*
METHYLBENZYLDIMETHYLAMINE - [chem] (a polymerization catalyst) méthylbenzyldiméthylamine *f*
METHYLBENZYL ETHER - [chem] (a solvent) éther *m* méthylbenzylique
METHYLBROMOACETATE - [chem] (intermediate for drugs, dyestuffs and agricultural chemicals) bromoacétate *m* de méthyle
METHYLBUTYL BENZENE - [chem] (a synthesis intermediate and a solvent) méthylbutylbenzène *m*
METHYLBUTYL KETONE - [chem] (propylacetone. A solvent) méthylbutylcétone *f*
METHYLBUTYNOL - [chem] (a solvent for a number of polyamide resins and a synthesis intermediate for pharmaceuticals and perfumery components) méthylbutynol *m*
METHYLCELLULOSE - [chem] (a stabilizer and thickening agent for cosmetics and pharmaceuticals) méthylcellulose *f*
METHYLCLOTHIAZIDE - [chem] (a long-acting diuretic) méthyl-clothiazide *f*
METHYLCOUMARIN - [chem] (a component of synthetic perfumes and flavours) méthylcoumarine *f*
METHYLCYANOACETATE - [chem] (a synthesis intermediate for dyes and drugs) cyanoacétate *m* de méthyle
METHYLCYANOFORMATE - [chem] (a synthesis intermediate) cyanoformiate *m* de méthyle
METHYLCYCLOHEXANE - [chem] (a synthesis intermediate and a solvent for cellulose derivatives) méthylcyclohexane *m*
METHYLCYCLOHEXANOL - [chem] (a lubricant additive and a solvent for cellulose derivatives) méthylcyclohexanol *m*
METHYLCYCLOHEXANONE - [chem] (a solvent) méthylcyclohexanone *f*
~GLYCERYL ACETAL - [chem] (plasticizer) glycéryl acétal *m* de méthylcyclohexanone
METHYLCYCLOHEXENE CARBOXALDEHYDE - [chem] (a synthesis intermediate) méthyl-cyclohexène-carboxaldéhyde *m*
METHYLCYCLOHEXYL ISOBUTYL PHTHALATE - [chem] (plasticizer cellulose derivatives) phtalate *m* de méthylcyclohexylisobutyle
METHYLCYCLOHEXYLAMINE - [chem] (a solvent and synthesis intermediate) méthyl-cyclohexylamine *f*
METHYLCYCLOPENTADIENE DIMER - [chem] (a component of plasticizers and stabilizers) dimère *m* de méthyl-cyclopentadiène
METHYLCYCLOPENTANE - [chem] (a synthesis intermediate) méthylcyclopentane *m*
METHYLDICHLOROACETATE - [chem] (a synthesis intermediate) dichloroacétate *m* de méthyle
METHYLDICHLOROSTEARATE - [chem] (a plasticizer component and a synthesis intermediate) dichlorostéarate *m* de méthyle
METHYLDIETHANOLAMINE - [chem] (an intermediate for synthetic resins, dyes, textile and agricultural chemicals) méthyl-diétanolamine *f*
METHYLDIOXOLANE - [chem] (a solvent for cellulo-

se derivatives) méthyldioxolane *m*
METHYLENE - [chem] (bivalent organic radical) méthylène *m*
~BLUE - [chem] (a dyestuff, analytical indicator and microscopical stain. Methylene blue also has uses in medicine as a mild antiseptic and in the treatment of methaemoglobinaemia) bleu *m* de méthylène
~BROMIDE - [chem] (a solvent and synthesis intermediate) bromure *m* de méthylène
~CHLORIDE - [chem] (a solvent, local anaesthetic and extractant for a large number of substances) chlorure *m* de méthylène
~DITANNIN - [chem] (used in medicine for the treatment of dysentery and aczema) méthylène-ditannine *f*
~IODIDE - [chem] (a synthesis intermediate) iodure *m* de méthylène
METHYLENEDIANILINE - [chem] (a synthesis intermediate) méthylène dianiline *f*
METHYLERGONOVINE MALEATE - [chem] (a drug having application in obstetrics) maléate *m* de méthylergonovine
METHYLETHYLPYRIDINE - [chem] (an intermediate for textile chemicals, disinfectants and copolymers) méthyléthylpyridine *f*
METHYLFORMANILIDE - [chem] (a synthesis intermediate) méthyl-formanilide *f*
METHYLFURAN - [chem] (a synthesis intermediate) méthyl-furanne *m*
METHYLGLUCAMINE DIATRIZOATE - [chem] (an X-ray contrast medium) diatrizoate *m* de méthylglucamine
METHYLHEPTANE - [chem] (a synthesis intermediate) méthylheptane *m*
METHYLHEPTENONE - [chem] (a perfumery compound and synthesis intermediate) méthylhepténone *f*
METHYLHEXANE - [chem] (a synthesis intermediate) méthylhexane *m*
METHYLHEXANEAMINE - [chem] (a sympathomimetic amine with uses in medicine) méthyl-hexane-amine *f*
METHYLHYDRAZINE - [chem] (an intermediate, solvent and propellant for rockets) méthylhydrazine *f*
METHYLHYDROXYBUTANONE - [chem] (a solvent and intermediate for flavourants) méthyl-hydroxybutanone *f*
~METHYLIC - [chem] méthylique
METHYLIONONE - [chem] (a perfumery constituent) méthyl-ionone *f*
METHYLISOAMYL KETONE - [chem] (a solvent for acrylic and vinyl plastics and cellulose derivatives) méthylisoamylcétone *f*
METHYLISOBUTYL KETONE - [chem] (a solvent for natural and synthetic resins, cellulose derivatives, waxes and oils and a synthesis intermediate) méthylisobutylcétone *f*
METHYLISOEUGENOL - [chem] (a perfumery component) méthylisoeugénol *m*
METHYLISOPROPENYL KETONE - [chem] (a starting material for a number of plastics) méthylisopropénylcétone *f*
METHYLMAGNESIUM BROMIDE - [chem] (a Grignard reagent) iodure *m* de méthyle et de magnésium
~IODIDE - [chem] (a Grignard reagent) iodure *m* de méthyle
METHYLNAPHTHALENE - [chem] (a synthesis intermediate) méthylnaphtalène *m*
METHYLNONYLACETALDEHYDE - [chem] (a perfumery agent) méthyl-nonylacétaldéhyde *m*
METHYLOLDIMETHYL-HYDANTOIN - [chem] (a textile chemical) méthylol-diméthyl-hydantoïne *f*
~UREA - [chem] (a textile chemical and starting material for urea-formaldehyde resins) méthylol-urée *f*
METHYLPENTADIENE - [chem] (intermediate for polymers and synthetic resins) méthyl-pentadiène *m*
METHYLPENTALDEHYDE - [chem] (an intermediate for drugs and dyestuffs) méthyl-pentadéhyde *m*
METHYLPENTANE - [chem] (a synthesis intermediate) méthylpentane *m*
METHYLPENTANEDIOL - [chem] (a solvent and synthesis intermediate) méthyl-pentanediol *m*
METHYLPENTENE - [chem] (an intermediate for pharmaceuticals, dyestuffs, flavourants and synthetic resins) méthyl-pentène *m*
METHYLPENTYNOL - [chem] (a solvent for polyamides and an intermediate for pharmaceuticals and perfumery components. Methylpentynol also has uses in medicine as a tranquillizer) méthylpentynol *m*
METHYLPHLOROGLUCINOL - [chem] (a coupling agent for the production of dyestuffs) méthyl-phloroglucine *f*
METHYLPHOSPHORIC ACID - [chem] (a catalyst and polymerization agent for synthetic resins) acide *m* méthyl-phosphorique
METHYLPYRROLE - [chem] (a synthesis intermediate) méthyl-pyrrole *m*
METHYLPYRROLIDONE - [chem] (a solvent and synthesis intermediate) méthylpyrrolidone *f*
METHYLSTYRENE - [chem] (a monomer for styrene) méthyl-styrolène *m*
METHYLTAURINE - [chem] (an intermediate for drugs and dyestuffs) méthyl-taurine *f*
METHYLTETRAHYDROFURAN - [chem] (a synthesis intermediate) méthyl-tétrahydrofuranne *m*
METHYLTRICHLOROSILANE - [med] (an intermediate for silicones) méthyltrichlorosilane *m*
METHYLVINYL DICHLOROSILANE - [med] (an intermediate for silicones) méthyl-vinyl-dichlorosilane *m*
METOL - [chem] (derivative of cresol, used as a photographic developer) métol *m*, génol *m*
METOPE - [arch] (frontal surface) métope *f*
METOTIC - [anat] (posterior to the auditory vescicle) métotique
METRALGIA - [med] métralgie *f*, hystéralgie *f*
METRATONIA - [med] atonie *f* utérine
METRATROPHIA - [med] atrophie *f* de l'utérus
METRE - s. meter
~CANDLE - [el] (lux) bougie-mètre *f*
~-KILOGRAM-SECOND SYSTEM OF UNITS - [meas] unité *f* mètre-kilogramme-seconde
METRIC - [meas] métrique
~CARAT - [meas] (standard weight for precious stones) carat *m* métrique
~GEAR - [mech] engrenage *m* métrique
~HORSE-POWER - [mech] cheval-vapeur *m*
~PLUG - [auto] bougie *f* à filet métrique
~ROD - [surv] mire *f* métrique
~SYSTEM - [meas] système *m* métrique
~SCREW THREAD - s. metric thread
~THREAD - [mech] filetage *m* métrique
~THREAD SCREW - [mech] vis *f* à filetage métrique
~TON - [meas] (1,000kgs) tonne *f* métrique, tonne *f*

METRIC WAVES - [radio] (waves between I and I0 metres long) ondes *fpl* métriques
METRICAL - s. metric
METRITIS - [med] (womb inflammation) métrite *f*
METRO - s. metropolitan railway
METROCAMPSIS - [med] obliquité *f* de l'utérus, courbure *f* de l'utérus
METROCOLPOCELE - [med] métrocolpocèle *f*
METROLOGY - [meas] (the study of weights and measures) métrologie *f*
METROMORPHIC - [zool] (which resembles the mother) métromorphe
METRONOME - [instr] (instrument designed to mark exact times in music) métronome *m*
METROPHLEBITIS - [med] métrophlébite *f*, phlébite *f* utérine
METROPOLITAN RAILWAY - [railw] métro *m*, chemin de fer *m* urbain
METRORRHAGIA - [med] métrorragie *f*, hémorragie *f* utérine
MEV - [nucl] (mega-electron-volts; unit used in nuclear science) MeV *m*, méga-électron-volt *m*
MEWING - [acoust] (the sound made by cats) miaulement *m*
MEZZANINE - [constr] (intermediate floor built between two floors) entresol *m*, mezzanine *f*
MEZZO-SOPRANO - [mus] mezzo-soprano *m*
MEZZOTINT - [print] mezzo-tinto *m*, gravure *f* à la manière noire
MF - s. medium frequency
MFR - s. manufacture
mg - [abbrev. for milligram] milligramme *m*
MHO - [el] (the practical unit of electric conductance; the reciprocal of ohm) mho *m*
MIASMA - [gen] (a polluting exhalation) miasme *m*
MICA - [min] (generic name for a group of silicate minerals which can be easily split into thin sheets) mica *m*
~BOARD - [gen] (material in sheets obtained by bonding crap mica with shellac) mica *m* agglomeré
~CAPACITOR - [radio] (in radio receiver) condensateur *m* au mica
~CONE - [el] (mica V-ring) bague *f* de mica, cône *m* de mica
~CONTENT - [min] teneur *f* en mica
~FLAKES - [geol] feuilles *fpl* de mica, lamelles *fpl* de mica, paillettes *fpl* de mica
~FLAKING - [electron] (in electron tube supports of mica) écaillement *m* de mica
~INSULATION - [el] isolation *f* au mica
~PLATE - [electron] plaque *f* en mica
~PLUG - [auto] bougie *f* d'allumage au mica
~PUNCHING MACHINE - [electron] (machine used in the manufacture of electron tubes) poinçonneuse *f* de mica
~SCHIST - [geol] (shist composed essentially of micas and quartz) micashiste *m*, schiste *m* micacé
~SLATE - s. mica schist
~SPACER - [electron] lame *f* de mica, rondelle *f* de guide de mica
~TAPE - [el] ruban *m* isolant au mica
~WASHER - [mech] rondelle *f* de mica
MICACEOUS - [min] micacé
[soil] micacé
~CHALK - [geol] tuffeau *m*
~IRON ORE - [min] (variety of specular iron ore) minerai *m* de fer micacé
MICACEOUS IRON OXIDE - [min] (a mineral resembling mica in structure, but not chemically, used to produce a black pigment for anti-corrosive paints) oxyde *m* de fer micacé
~ROCK - [geol] roche *f* micacée
~SANDSTONE - [geol] (a sandstone which contains some mica flakes) grès *m* micacé
MICAFOLIUM - [el] (insulating material consisting of a paper backing covered with mica flakes) feuille *f* de mica
MICANITE - [el] (mica splittings bonded by shellac and used as insulating material) micanite *f*
MICELLA - s. micelle
MICELLAR BUNDLE - [chem] faisceau *m* de micelles
~FORCE - [chem] force *f* micellaire
MICELLE - [chem] (a particle of colloidal dimensions, e.g. a colloidal ion) micelle *f*
MICHELL BEARING - [mech] (thrust bearing in which pivoted pads support the trust collar and tilt under the wedging action of the lubricant, thus producing fluid lubrication conditions with a very low power loss in the bearing) palier *m* de butée type Michell
MICHELSON-MORLEY EXPERIMENT - [opt] (experiment made to detect the motion of the earth' through the ether) expérience *f* de Michelson-Morley
~ROTATING MIRROR - [opt] miroir *m* rotatif de Michelson
MICRO - [gen] (prefix used in compound terms; very small) micro
MICRO- - [chem] (prefix signifying $I0^{-6}$ units; symbol u) micro-
MICROAEROPHILE - [biol] (organism which grows well only when the oxygen concentration is low) microaérophile
MICROALLOY TECHNIQUE - [electron] technique *f* de microalliage
~TRANSISTOR - [electron] transistor *m* à microalliage
MICROAMMETER - [el] microampèremètre *m*
~-RATEMETER - [nucl] mesureur-ictomètre *m* de charges
MICROAMPERE - [el] microampère *m*
MICROANALYSIS - [chem] (the chemical analysis and/or identification of very small quantities) microanalyse
MICROANALYTICAL REAGENT - [chem] (abbrev. M.A.R. A standard of purity for microanalysis) réactif *m* microanalytique
MICROATTACHMENT - [photo] adapteur *m* photomicrographique
MICROBALANCE - [chem] microbalance *f*
MICROBALLONS - [ind chem] (very small vinyl spheres used as a protective layer on liquid surfaces to minimise evaporation) microballons *mpl*
MICROBAR - [acoust] (unit of pressure) microbar *m*
MICROBAROGRAPH - [instr] (a special instrument designed to measure and rapid changes in the pressure of the atmosphere) microbarographe *m*
MICROBE - [biol] (a living organism, animal or vegetable, of microscopic dimensions) microbe *m*
MICROBIAL - [biol] microbien, microbique
MICROBICIDE - [chem] microbicide *m*
MICROBIOLOGICAL - [biol] microbiologique
~CONTAMINATION - [nucl] contamination *f* microbiologique
MICROBIOLOGY - [biol] (the study of the structure, development etc. of bacteria, viruses etc) micro-

biologie *f*

MICROCANONICAL ASSEMBLY - [mech] (assembly in statistical mechanism in which the variations in energy lies within an infinitesimal range) ensemble *m* microcanonique

MICROCELLULAR RUBBER - [rubber ind] (rubber of a special structure with cavities of microscopic dimensions) caoutchouc *m* microcellulaire

MICROCEPHALIA - s. microcephaly

MICROCEPHALY - [med] (an abnormal development of the cranium resulting in unusual smallness) microcéphalie *f*

MICROCHEMISTRY - [chem] (the chemistry of substances existing in a system in imponderable quantities or in exceedingly low concentrations) microchimie *f*

MICROCHROMOSOME - [zool] microchromosome *m*

MICROCOCOCCUS - [bact] (any member of a genus of spherical bacteria) microcroque *m*, micrococcus *m*

MICROCOPY - [photo] microcopie *f*

MICROCOSM - [gen] (the world, or the universe on a very small scale) microcosme *m*

MICROCOSMIC SALT - [chem] (a name used for sodium ammonium hydrogen phosphate) sel *m* microcosmique, phosphate *m* sodio-ammonique

MICROCRYSTALLINE - [cryst] (crystals which can be seen as such only under the microscope) microcristallin

~WAXES - [oil ind] (waxes manufactured from residual slack waxes and having a very fine crystal structure. They are characterized by a high plasticity) cires *f*pl microcristallines

MICROCRYSTALLOGRAPHY - [cryst] (the study and description of microcrystalline textures) microcristallographie *f*

MICROCURIE - [nucl] (one millionth of a curie) microcurie *m*

MICROELECTRONICS - [electron] (design, production and application of devices of very small dimensions) microélectronique *f*

MICROFARAD - [el] (unit of capacitance equal to one millionth of a farad) microfarad *m*

MICROFILM - [photo] (highly reduced photograph of objects, documents, pages etc., for ease in transmission and storage) microfilm *m*
[photo] (the film used for microphotography) pellicule *f* pour microphotographie

~FOOTAGE COUNTER - [photo] compteur *m* de pieds de microfilm

MICROFLASH - [photo] micro-éclair *m*

MICROFOAM RUBBER - [rubber ind] (rubber of microporous structure) mousse *f* de latex microporeux

MICROFRACTOGRAPHY - [metall] (the use of an electron microscope for the examination of metal surface fractures) théorie *f* des microcassures

MICROGLIA - [anat] mésoglie *f*, microglie *f*

MICROGLIACYTE - [biol] microgliacyte *m*

MICROGNATHIA - [med] micrognathie *f*

MICROGAMIE - [zool] (syngamy between two of the smallest individuals produced by gemmation) microgamie *f*

MICROGRANITE - [geol] (a microcrystalline rock with the same mineral composition of granite) microgranit *m*

MICROGRAPH - [instr] (instrument designed to record and to magnify photographically very small movements, e.g. of a diaphragm) micrographe *m*

MICROGRAPH - [instr] (a pantograph instrument for minute drawing) micrographe *m*
[photo] (microscopic picture) microphotographie *f*

MICROGRAPHIC - [instr] (relating to an instrument for microscopic writing) micrographique
[instr] (relating to an instrument designed to magnify very small movements) micrographique
[geol] (of texture in which the crystallization of quartz and feldspar has made the former appear as isolated fragments) micrographique

MICROGRAPHY - [opt] (the description of microscopic objects) micrographie *f*

MICROGRINDING - [mech] micromeulage *m*

MICROGROOVE - [el acoust] (a groove of very small depth and width in mechanical recording) microsillon *m*

~RECORD - [el acoust] disque *m* de longue durée, disque *m* microsillon

MICROHENRY - [el] (the practical unit of inductance, equal to one millionth of one henry) microhenry *m*

MICROHM - [el] (the practical unit of resistance, equal to one millionth of one ohm) microhm *m*

MICROIRRADIATION - [nucl] microradioexposition *f*

MICROLITE - [min] (a mineral, essentially calcium pyrotantalate, often containing niobium, crystallizing in the cubic system. A source of tantalum) microlithe *m*, microlite *f*

MICROMANIA - [med] délire *m* de petitesse, micromanie *f*

MICROMANIPULATOR - [nucl] (device for making fine adjustments of nuclear apparatus from a remote station) micromanipulateur *m*

MICROMANOMETER - [instr] (an instrument designed to measure very small pressure differences) micromanomètre *m*

MICROMASTIA - [med] (the failure of the female breast to develop after puberty) micromastie *f*, petotesse *f* des seins

MICROMAZIA - s. micromastia

MICROMECHANICS - [mech] micromécanique *f*

MICROMETALLOGRAPHY - [metall] micrométallographie *f*

MICROMETEORITE - [astr] (meteorite of very small dimensions) micrométéorite *m*

MICROMETER - [instr] (instrument fitted with a telescope designed to measure very small distances) or angular separations) micromètre *m*

~ADJUSTMENT - [instr] ajustage *m* micromètrique

~ATTACHMENT - [instr] appareillage *m* pour l'enregistrement micromètrique

~CALIPER - [instr] (a caliper fitted with a micrometer screw used for precision measurements) micromètre *m*, calibre *m* à vis micrométrique, palmer *m*

~COMPARATOR - s. micrometric comparator

~DEPTH GAUGE - [instr] (instrument designed to measure the depth of minute cavities) micromètre *m* de profondeur

~DIAL - [instr] cadran *m* du micromètre

~DRIVE - [mech] commande *f* micrométrique

~EYE-PIECE - [opt] oculaire-micromètre *m*

~GAP - [el] éclateur *m* micrométrique

~GAUGE - s. micrometer

~MICROSCOPE - [instr] microscope *m* micrométrique

MICROMETER SCREW - [mech] (screw with a graduated head) vis *f* micrométrique
~SPARK-GAP - s. micrometer gap
~THEODOLITE - [surv] (theodolite fitted with micrometers for reading horizontal and vertical circles) théodolite *m* micrométrique
~VERTICAL ADJUSTMENT - [instr] ajustement *m* vertical micrométrique
MICROMICROCURIE - [nucl] (called sunshine unit in the USA; a millionth of a millionth of a curie) micrmicrocurie *m*
MICROMICROFARAD - [el] (equal to IO^{-I2} farad) micromicrofarad *m*
MICROMILLIMETER - [meas] micromillimètre *m*
MICROMINIATURE CIRCUIT - [electron] circuit *m* microminiature
MICROMINIATURIZATION - [comput] (the use of circuit of very small dimensions) microminiaturisation *f*
MICROMODULE - [electron] micromodule *m*
MICRON - [meas] micromillimètre *m*, millième *m* de millimètre
MICRONUCLEUS - [zool] micronoyau *m*
MICROORGANISM - [zool & bot] (any very small plant or animal organism which is not visible without the help of a microscope) micro-organisme *m*
MICROPHONE - [el acoust] (device for converting the energy of sound waves into electrical energy) microphone *m*
~AMPLIFIER - [radio] amplificateur *m* de microphone
~BLANKET - [acoust] (a cloth used for the projection of the microphone against drops of rain hitting the diaphragm) capuchon *m* de microphone
~BOOM - [impl] girafe *f*, perche *f*
~CURRENT SUPPLY - [el acoust] alimentation *f* de microphone
~DIAPHRAGM - [el acoust] membrane *f* du microphone
~DIRECTIVITY - [el acoust] directivité *f* de microphone
~HISS - [el acoust] souffle *m* microphonique
~INSET - [el acoust] (called transmitter unit in the USA) capsule *f* de microphone
~NOISE - [el acoust] bruit *m* de microphone
~OUTPUT - [el acoust] puissance *f* de sortie d'un microphone
~OUTPUT TERMINALS - [el acoust] bornes *f*pl de sortie de microphone
~PREAMPLIFIER PENTODE - [electron] pentode *f* préamplificatrice de microphone
~PUSH-TO-TALK BUTTON - [telecomm] (press-button switch which must be pressed to close the telephone circuit) bouton *m* d'alternat
~RELAY - [el acoust] relais *m* de microphone
~RESPONSE - [acoust] (the response over its operating frequency rangs) réponse *f* du microphone
~TRANSFORMER - [el] transformateur *m* microphonique
~TRANSMITTER - [el acoust] microphone *m*
MICROPHONIC - [acoust] microphonique
~EFFECT - [radio] (parasitic effect by which vibrations of certain component parts produce disturbing currents of the same frequency which modulate the useful currents) effet *m* microphonique
~MIXING - [el acoust] mixage *m* microphonique
~NOISE - s. microphonic effect
MICROPHONICITY - s. microphonic effect
MICROPHONICS - [el acoust] dérangement *m* microphonique
MICROPHONISM - s. microphonic effect
MICROPHONY - [radio] s. microphonic effect
[telev] (the varying position of a spot on the screen of a cathode-ray tube, causing the fluttering of the picture) effet *m* microphonique
MICROPHOTOGRAM - [spectrology] (enlarged photograph of a spectrum) image *f* agrandie d'un spectre
MICROPHOTOGRAPH - [photo] (a microscopic photograph) microphotographie *f*
MICROPHOTOGRAPHY - [photo] (the production of very small prints or transparencies from normal negatives) microphotographie *f*
[photo] (microscopic photography) photographie *f* microscopique
MICROPHYSICS - [phys] microphysique *f*
MICROPIA - s. micropsia
MICROPSIA - [med] (vision defect, whereby objects appear extremely small) micropie *f*, micropsie *f*
MICROPSYCHIA - [med] faiblesse *f* d'esprit
MICROPOROUS - [phys] (having pores of microscopic dimensions) microporeux
MICROPYROMETER - [instr] (optical instrument designed to measure the temperature of very small light- or heat- radiating bodies) micropyromètre *m*
MICRORADIOGRAPHY - [radiat] (the magnification of a radiogram) microradiographie *f*
MICRORADIOMETER - [instr] (thermosensitive detector of radiant power) microradiamètre *m*
MICROSCOPE - [instr] (optical instrument designed to produce magnified images of very small objects) microscope *m*
~LENS - [opt] lentille *f* du microscope
~OBJECTIVE - [opt] objectif *m* du microscope
~SLIDE - [opt] (slip of thin glass on which an objects is placed under the microscope) porte-objet *m*
MICROSCOPIC - [opt] (which can be seen only through a microscope) microscopique
~CROSS-SECTION - [nucl] (cross-section per atom or per molecule) section *f* efficace microscopique
~EXAMINATION - [opt] examen *m* microscopique
~MOBILITY - [phys] (the mobility of an untrapped particle in a semiconductor) mobilité *f* microscopique
~PREPARATION - [ind chem] préparation *f* microscopique
~STATE - [phys] (the state in which the electrons of a system are supposed to be individualized) état *m* microscopique
MICROSCOPICAL - s. microscopic
~ANATOMY - [anat] anatomie *f* microscopique
MICROSCOPY - [opt] (the art technique of examining objects with a microscope) microscopie *f*
MICROSECOND - [meas] (one millionth of a second) microseconde *f*
MICROSECTION - [metall] plaquette *f* polie micrographique
MICROSEGREGATION - [metall] ségrégation *f* mineure
MICROSEISM - [geol] (very minute vibration of the earth's surface) microsisme *m*
MICROSEISMOGRAPH - [instr] microsismographe *m*
MICROSHRINKAGE - [metall] (a foundry defect) microretassure *f*
MICROSOME - [genet] (very small granular inclusion in the cytoplasm) microsome *m*

MICROSPECTROSCOPE - [instr] (a combined microscope) microspectroscope *m*
MICROSPHERIC - [zool] microsphérique
MICROSPOROSIS - [med] microsporose *f*
MICROSTRAINER - [impl] microtamis *m*
MICROSTRIP - [telecomm] (transmission line in which the conductors are in the form of closely spaced strips) ligne *f* à bandes parallèles, ligne *f* microbande
MICROSTRUCTURE - [metall] (general term referring to the size, shape etc of the crystals of the constituents in metal or in an alloy) microstructure *f*
MICROSWITCH - [el] (special type of switch of small size, requiring only slight force to operate it) microinterrupteur *m*
MICROTASIMETER - [instr] (tasimeter used to measure very small extensions) microtasimètre *m*
MICROTELEPHONE - [telecomm] microtéléphone *m*
MICROTHELIA - [med] hypoplasie *f* des mamelons
MICROTHROWING POWER - [metall] capacité *f* de remplir les pores
MICROTOME - [instr] (instrument designed to cut very thin sections for microscopic observation) microtome *m*
MICROVOLT - [el] microvolt *m*
MICROVOLTMETER - [instr] (instrument for measuring voltage in microwolts) microvoltmètre *m*
MICROWAVE - [radio] (waves which are shorter than I metre or so in wavelength) micro-ondes *fpl*
~ABSORPTION SPECTRUM - [nucl] (that part of the spectrum of a molecule lying in the so-called microwave region of frequencies) spectre *m* micro-ondes
~AERIAL ASSEMBLY - [radio] réseau *m* d'antennes micro-ondes
~BEAM - [radio] faisceau *m* de micro-ondes
~EARLY WARNING - [radar] (powerful long-range I0 cm. early warning radar) radar *m* micro-ondes
~ELECTRON TUBES - [electron] tubes *mpl* micro-ondes
~HEATING - [electron] chauffage *m* à hyperfréquences
~LINK - [telev] faisceau *m* hertzien, liaison *f* micro-ondes
~REGION - [phys] hyperfréquences *fpl*
~RELAY SYSTEM - [radio] système *m* de retransmission à micro-ondes
~THERAPY - [el biol] (the therapeutic use of electromagnetic energy to generate heat in the body, the frequency being greater than I00 megacycles per second) thérapie *f* par ondes centimétriques
~TRANSMISSION - [radio] transmission *f* à micro-ondes
~TURBULENCE - [radio] turbulence *f* des hyperfréquences
MICTION - [med] miction *f*, micturition *f*
MICTURITION UROGRAPHY - [med] (radiological examination of the miction organs) urographie *f* de miction
MID-FEATHER - [paper man] (the partition in the beater inducing circulation of the wet pulp) paroi *f* divisoire, languette *f*
~-PART - [gen] partie *f* centrale
~-PART OF A CORE - [metall] (in foundry work) chape *f*
~-POINT CONNEXION - [electron] (center tap connexion made at midpoint of the primary or secondary of a transformer) prise *f* médiane
~-POSITION CONTACT - [el] contact *m* de position neutre
MID-SOLE - [shoe man] semelle *f* intermédiaire, patin *m*
~-WING MONOPLANE - [aero] (type of monoplane in which the main plane is fixed approximately half way between the lower and upper parts of the fuselage) monoplane *m* à aile médiane
MIDBRAIN - [anat] mésencéphale *m*
MIDCOURSE CORRECTION - [astronaut] correction *f* en course de route
~GUIDANCE - [astronaut] guidance *m* à mi-course
~MOTOR - [astronaut] moteur *m* de correction à mi-course
MIDDLE - [gen] moyen, intermédiaire, central
[gen] (the middle) centre *m*, milieu *m*
[metall] (a forging section between two larger ones) coupe *f* intermédiaire
[paper man] (a layer between two outer layers) couche *f* intermédiaire, noyau *m*
~BOX - [metall] châssis *m* intermédiaire
~CONDUCTOR - [el] (the neutral conductor) conducteur *m* médian
~-CUT FILE - [mech] lime *f* à taille moyenne
~-CUT RASP - [mech] râpe *f* à taille moyenne
~DECK - [naut] pont *m* intermédiaire
~EAR - [anat & acoust] (the part of the ear between the tympanic cavity and the inner ear) oreille *f* moyenne
~FRACTION - [chem] fraction *f* de recyclage
~-KEY PICTURE - [telev] image *f* normale
~KNIFE-EDGE - [mech] (of a balance) couteau *m* du fléau
~LATITUDE - [nav] (the parallel of latitude midway between the latitudes of the beginning and end of a journey) latitude *f* intermédiaire
~LIMB - [geol] flanc *m* central d'un pli
~MARKER BEACON - [aero] (instrument landing system marker, designed to show the second predetermined point in an approach) radiobalise *f* intermédiaire
~OILS - [chem] (carbolic oils obtained from coal-tar distillation) huiles *fpl* moyennes (de goudron)
~PASS - [metall] cannelure *f* intermédiaire
~-POST - [constr] (or king-post; a vertical timber tie connecting the ridge and the tie-beam of a roof) poinçon *m*, meneau *m*, poteau *m* intermédiaire
~PURLIN - [constr] panne *f* intermédiaire
~REGISTER - [acoust] (tenor and alto voices) voix *fpl* moyennes
~ROLL - [metall] cylindre *m* médian
~SOLE - s. mid-sole
~TAP - [mech] taraud *m* intermédiaire
~TAR OIL - [oil ind] (creosote oil) huile *f* moyenne de goudron
~THIRD - [constr] noyau *m* central
~TONE - [photo] ton *m* moyen
~VOICES - s. middle register
~WIRE - s. middle conductor
[geol] (a layer of rock between two coal seams) couche *f* intermédiaire
MIDDLING - [gen] médiocre
[metall] coupe *f* intermédiaire
[metall] forgeage *m* du goulot
[mining] coupe *f* intermédiaire
MIDDLINGS - [agric] (coarser part of the ground wheat) remoulage *mpl*
[min] (a partially concentrated product left after

the removal of clean concentrates) fragments *mpl* menus moyens
MIDGET - [gen] nain *m*
[electron] relais *m* miniature
~LIGHT - [cin] (a baby keg-light) petit projecteur *m* (500 W)
~MICROPHONE - [acoust] microphone *m* miniature
~SET - [radio] (a very small radio receiver) poste *m* miniature
MIDGETAPE - [el acoust] (a very small tape recorder) magnétophone *m* de poche
MIDNIGHT - [astr] minuit *m*
~SUN - [astr] soleil *m* de minuit
MIDPOINT CIRCLE - [math] cercle *m* à mi-pente
MIDSHIP - [naut] (in the middle of the vessel hull) milieu *m* du navire
~SECTION - [shipbuild] maître couple *m*
MIDSTREAM - [hydr] ligne *f* médiane, thalweg *m*
MIDWIFE - [med] sage-femme *f*
MIDWIFERY - [med] obstétrique *f*
MIGRAINE - [med] migraine *f*
MIGRATION - [gen] migration *f*
[chem] (the movement of the atom from one position of the molecule to another) migration *f*
[phys] (the movement of ions under the influence of electromotive force, towards an electrode) migration *f*
[plast ind] (of a plasticizer; loss of plasticizer from an elastomeric plastics compound with subsequent absorption by an adjacent medium of lower plasticizer concentration) migration *f*
[zool] (the removal from one region to another) migration *f*
~AREA - [nucl] (the area of a medium measured by one-sixth of the mean-square distance travelled by a neutron from its origin until its absorption) aire *f* de migration
~COEFFICIENT - [chem] coefficient *m* de migration
~FASTNESS - [paint] (property of resisting colour migration) résistance *f* à la migration de couleur
~LENGTH - [nucl] (the square root of the migration area) longueur *f* de migration
~OF COLOUR - [text] migration *f* du colorant
~OF FISSION PRODUCTS - [nucl] migration *f* de produits de fission
~OF IONS - [el chem] (ionic migration) transport *m* des ions, migration *f* des ions
~OF PLASTICIZER - s. migration (plast ind)
~SPEED OF AN ION - [phys] (the speed reached by an ion under the action of an electric field of one volt per cm) vitesse *f* de migrations d'un ion
~TUBE - [ind chem] (glass electrolytic apparatus used for the study of ionic migrations) tube *m* analyseur de migration
MIGRATORY - [gen & zool] migratoire
MIKE - [acoust] (colloq. for microphone) microphone *m*
~BLANKET - s. microphone blanket
~STEW - [acoust] (colloq. for microphone hiss) souffle *m* microphonique
MIL - [meas] (term sometimes used for onethousandth of an inch) millième *m* (de pouce)
[ind chem] millilitre *m*
MILD - [gen] doux, bénin
[met] (of the weather, or climate) doux, tempéré
[metall] (of steel) doux
[brew ind] (terms for beer which is not strongly flavoured with hops) (bière) légère, peu houblonnée
MILD CARBON STEEL - [metall] acier *m* demi-doux
~CAULIFLOWERING - [metall] rochage *m*
~STEEL - [metall] acier *m* doux
MILDEW, to - [agric] rouiller, moisir
MILDEW - [gen] (any type of mould on walls, food etc)rouille *f*, mildiou *m*
[bot] (a fungous disease of plants) rouille *f*
[agric] (of wine) oïdium *m*, mildiou *m*
~INHIBITOR - [chem] anticryptogamique *m*
~OF CEREALS AND GRASSES - [agric] blanc *m* de graminées
~ON OATS - [agric] charbon *m* nu de l'avoine
~-PROOF - [chem] résistant à la moisissure
~-PROOF AGENT - [chem] antirouille *m*
~RESISTANCE - [ind chem] résistance *f* à la rouille
MILDEWY - [gen] moisi, atteint de mildiou
[gen] (of paper) taché d'humidité, piqué
MILDNESS - [brew ind] douceur *f*
MILE - [meas] (1609) mille *m*
[naut] (1853) mille *m* marin
~OF STANDARD CABLE - [telecomm] (unit of attenuation) mille *m* de câble étalon
~-OHM - [el] (the weight of a 1-ohm cable which is one mile long) poids *m* d'un mille de câble à résistance de 1 ohm
MILEAGE - [meas] (the length of anything which is measured in miles) distance *f* en milles, kilométrage *m*
~ANALYZER - [instr] indicateur *m* instantané de consommation
~COUNTER - [instr] (in autos) compteur *m* kilométrique
~POINT - [railw] point *m* kilométrique
~RECORDER - [instr] (in autos etc) compteur *m* kilométrique
MILEPOST - [meas] (post, pillar or stone indicating a distance in miles at any point) borne *f* kilométrique
MILESTONE - [constr] borne *f* miliaire
MILFOIL - [agric] mille-feuille *f*
MILIARY TUBERCOLOSIS - [med] (tuberculous lesions in various parts of the body) tuberculose *f* miliaire
MILIUM - [med] acné *m* miliaire, concrétion *f* calcaire sous-cutanée
MILK, to - [gen] (to draw milk from) traire
[el chem] (the emission of gas bubbles during charging) devenir perlé
MILK - [gen] lait *m*
~BOTTLE - [impl] bouteille *f* à lait
~CAN - [agric] boîte *f* à lait, berthe *f*
~CHURN - [agric] bidon *m* à lait
~CONTROL - [agric] contrôle *m* laitier
~COOLER - [agric] refroidisseur *m* à lait
~COW - [zool] vache *f* laitière
~CRUST - [med] croûte *f* de lait, gourme *f*
~ENZYME - [chem] ferment *m* lactique
~FAT - [agric] matière *f* grasse du lait
~FEEDING - [agric] régime *m* lacté
~FEVER - [vet] fièvre *f* vitulaire
~FOOD - [agric] préparation *f* à base de lait
~GAUGE - [agric] lacto-densimètre *m*, pèse-lait *m*
~-GLASS - [glass man] opaline *f*
~GOAT - [zool] chèvre *f* laitière
~OF LIME - [chem] lait *m* de chaux
~OF MAGNESIA - [chem] (a suspension of magne-

sium hydroxide in water used in medicine) magnésie *f* blanche
MILK OF SULPHUR - [chem] lait *m* de soufre
~PASTEURIZER - [agric mach] pasteurisateur *m* de lait
~POWDER - [food ind] lait *m* en poudre
~PRESERVATION - [agric] conservation *f* du lait
~PRODUCTION - [agric] production *f* laitière
~PRODUCTS - [agric] produits *mpl* laitièrs
~PROTEIN - [chem] lactalbumine *f*, protéine *f* du lait
~PUMP - [agric] pompe *f* à lait
~RIPENESS - [agric] maturité *f* laiteuse, grains *mpl* "en lait"
~SECRETION - [zool] sécrétion *f* lactique
~SHEEP - [zool] brebis *f* à lait
~SUGAR - [chem] (the sugar contained in milk) lactose *f*, sucre *m* de lait
~TEETH - [zool] dents *fpl* de lait
~TESTER - [agric] lactomètre *m*
~TESTING - [agric] examen *m* du lait
~THISTLE - [bot] lait *m* d'âne, laiteron *m* maraîcher
~VEINS - [anat] veines *fpl* mammaires
~VETCH - [bot] tragacanthe *f*, astragale *m*
~YIELD - [agric] rendement *m* laitier
MILKINESS - [glass man] (strong couldiness) lactescence *f*
[paint] (a defect) aspect *m* laiteux
MILKING - [agric] traite *f*, mulsion *f*
~BAIL - [agric] abri *m* à traite
~BOOSTER - [el] (special motor-generator used to give an individual charge of a "backward" cell) survolteur *m* supplémentaire
~CELL - [el] élément *m* supplémentaire de batterie
~MACHINE - [agric] trayeuse *f* mécanique
~PAIL - [agric] seau *m* à traire
~SHED - [agric] parloir *m* de traite
~TRUCK - [agric] installation *f* mobile de traite
MILKY LIQUID - [gen] liquide *m* laiteux
~RIPENESS - [agric] maturité *f* laiteuse
~SECRETION - [rubber ind] sécrétion *f* laiteuse
~WAY - [astr] (a band of distant stars and nebulae which are individually invisible to the eye) Voie *f* lactée, Galaxie *f*
MILL, to - [mech] moudre, broyer
[rubber ind] mastiquer, plastifier
[mach tool] fraiser, tailler des engrenages
[agric] moudre
[mech] moleter, frotter, fraiser, godronner
[timber] (the sawing of timber in a mill) scier
[text] (preliminary process in the finishing of woolen fabrics) fouler
[min] broyer, bocarder
MILL - [mech] (a grinding or disintegrating machine) moulin *m*, broyeur *m*, concasseur *m*, malaxeur *m*
[agric] (a building fitted with the machinery required for grinding) moulin *m*
[mach tool] fraise
[text] (e.g. a cotton mill) usine *f*, manufacture *f*
[metall] (a rolling mill) laminoir *m*, équipage *m*, train *m*
[rubber & plast ind] broyeur *m*
[min] installation *m* de préparation mécanique
[mining] (for dressing) cheminée *f* à minerai
[text] dévidoir *m* vertical
[sugar ind] raffinerie *f*
~ACCESSORIES - [rubber ind etc] accessoires *mpl* de laminoir
MILL AWAY, to - [oil ind] détruire une section de casing
~BAND FEEDER - [mech] (in cement works) doseur *m* à courroie pour broyeur
~BAR - [metall] fer *m* en barre, fer *m* ébauché
~BASTARD FILE - [mech] lime *f* bâtarde pour scie
~BOARD - s. millboard
~COLLETS - [mach tool] collets *mpl* porte-fraises
~-COURSE - [mech] bief *m* de moulin
~CYLINDER WITH PINS - [text] tambour *m* avec chevilles
~-DAM - [hydr] (dam across the current of a stream to divert it into a mill-race) barrage *m* de moulin
~EDGE - [paper man] (the roughness of the paper edges) bord *m* inégal
~ENGINE - [mech] (large slow steam engine used in factories to drive machinery through ropes) locomotive *f* pour usine
~ENGRAVING MACHINE - [text] machine *f* à moleter, moletteuse *f*
~FINISH - [paper man] apprêt *m*
[metall] finissage *m* au laminoir
~-FINISHED - [paper man] apprêté
~FINISHING - [text] finissage *m* intégré
~-FITTING - [light] (factory-fitting) éclairage *m* pour usine
~-GEARING - [mech] rouage *m*
~LAYOUT - [ind] croquis *m*, esquisse *f*
~LIMIT - [metall] tolérance *f* de laminage
~-MIXTER - [mech] (a mixing machine in which some degree of reduction or disintegration is also carried out) broyeur-malaxeur *m*
~OFF, to - [metall] réduire à la fraise
~PACK - [metall] (a pack of hot-rolled sheets) produit *m* de lamination multiple
[metall] paquet *m* de tôles
~POND - [hydr] réservoir *m* de moulin, retenue *f*
~PROCESS - [metall] procédé *m* direct
~RACE - [hydr] (the channel made to lead water to a mill wheel) bief *m*, biez *m* de moulin
~RUGS - [text] (creases produced in woollen fabrics during milling) plis *mpl* de foulage
~RUN - [min] campagne *f* du bocard
[hydr] s. mill race
[mech] essai *m* au moulin
~SCALE - [metall] calamine *f* de recuit
~SPUN YARNS - [text] fil *m* mécanique
~STEEL WIRE - [metall] fil *m* d'acier fondu
~TABLE FEEDER - [mech] (in cement works) doseur *m* à plateau pour malaxeur
~-TAIL - [hydr] (the channel conveying water away from the mill) bief *m* d'aval, biez *m* de fuite
~TRAIN - [metall] train *m* de laminoir
~WARPING - [text] ourdissage *m* par nappe
~WASHING - [rubber & plast ind] lavage *m* mécanique
~WHEEL - [mech] (water-wheel driving the machinery of a mill) roue *f* de moulin
~WIDTH - [metall] largeur *f* de cage
~WRAPPING - [paper man] papier *m* d'emballage
MILLBOARD - [paper man] (a board manufactured from wood pulp, fibre refuse etc) carton *m* écru
MILLDAM - s. mill-dam
MILLED - [mech] (by a grinding machine etc) moleté, godronné
[min] (by a mixer) broyé
[mech] (by a cutter) fraisé, bocardé

MILLED - [mech] (with fluted or grooved edges) moleté
[metall] laminé
~CLOTH - [text] tissu *m* foulé
~EDGE - [mech] (fluted or grooved) crénelage *m*
~-EDGE THUMB SCREW - [photo] vis *f* micrométrique
~GLASS FIBRE - [glass ind] (small nodules of filamented glass made by treating glass fibre in a hammer mill) fibre *f* de verre broyée
~GROOVE - [mech] rainure *f* fraisée
~-HEAD SCREW - [mech] (adjusting screw with a milled edge) vis *f* à tête moletée, vis *f* moletée
~LEAD - [metall] plomb *m* laminé
~METAL - [metall] métal *m* usiné
~WHEEL - [mech] molette *f*
MILLER - [mech] (a milling machine) fraiseuse *f*
[mach tool] fraiseuse *f*
~BRIDGE - [instr] (bridge designed to measure the amplification factor of vacuum tubes) pont *m* de Miller
~CIRCUIT - [radio] (tube network in which the Miller effect is used) circuit *m* de Miller
~CRYSTAL INDICES - [crystal] (three indices used to represent any crystal face) indices *mpl* de Miller
~EFFECT - [electron] (alteration in the admittance of the control-grid circuit) effet *m* Miller
~INDICES - s. Miller crystal indices
~INTEGRATOR - [el] (capacity integrator utilizing the capacity which is produced by the Miller effect) intégrateur *m* de Miller
~TIME-BASE - [electron] (time base which is used to produced a time-base voltage more linear than any time-base circuit with the same voltage) base *f* de temps de Miller
MILLERITE - [min] (a sulphide of nickel) millérite *f*, trichopyrite *f*
MILLET - [bot] (a grass or its seeds, used as a cereal) millet *m*, mil *m*
MILLIAMETER - [instr] milliampèremètre *m*
MILLIAMPERE - [meas] milliampère *m*
~-SECOND - [radiat] (measure of X-ray exposure) milliampère-seconde *f*, mAs *f*
MILLIARD - [math] (one thousand millions; called a billion in the USA) milliard *m*, billion *m*
MILLIBAR - [met] (unit of atmospheric pressure, equal to one-thousandth of a bar) millibar *m*
MILLICURIE - [radiat] (one thousandth of a curie) millicurie *m*
~-DESTROYED - [radiat] (the amount of radiation by a radioactive nuclide during the time in which its activity falls by one millicurie) millicurie *m* détruit
~-HOUR - [radiat] (a measure of gamma-ray exposure) millicurie-heure *m*, mCih *m*
MILLIFARAD - [el] millifarad *m*
MILLIGAL - [meas] (unit of acceleration) milligal *m*
MILLIGRAM - [meas] milligramme *m*
MILLIHENRY - [el] millihenry *m*
MILLIKAN METHOD - [electron] (method of measuring the charge of an electron) méthode *f* de Millikan
MILLILAMBERT - [light] (unit of brightness) millilambert *m*
MILLILITER - [meas] millilitre *m*
MILLILUX - [photo] (unit of illuminating intensity) millilux *f*
MILLIMASS UNIT - [meas] (one thousandth of an atomic mass unit) unité *f* millimasse
MILLIMETER - [meas] millimètre *m*
MILLIMETER PITCH - [mech] (metric screw-thread) filetage *m* millimétrique
MILLIMICRON - [meas] millimicron *m*
MILLING - [mech] (the removal of material from a work-piece by a revolving cutter) fraisage *m*, dressage *m* à la fraise
[min] (dressing) bocardage *m*, broyage *m*
[text] (with soap, alkali, acid, pressure and friction; a preliminary process in the finishing of woolen fabrics) foulage *m*
[gen] (grinding) moulage *m*, mouture *f*
[metall] frottage *m*, moletage *m*, laminage *m*
~AGENT - [text] adjuvant *m* de foulage
~ANGLE - [mach tool] angle *m* de fraisage
~BY-PRODUCTS - [agric] sous-produits *mpl* de meunerie
~CUTTER - [mach tool] (a toothed cutter used in milling) fraise *f*
~-CUTTER SHARPENING MACHINE - [mach tool] machine *f* à affûter les fraises
~CUTTER WITH INSERTED TEETH - [mach tool] fraise *f* à dents rapportées
~EFFECT - [text] effet *m* de foulage
~FEED - [mach tool] amenage *m* à la fraise
~-FILE - [mech] lime *f* fraiseuse à main
~FLOCKS - [text] déchets *mpl* de foulage, bourre *f* de foulage
~GUIDE - [metall] cloche *f* à fraise
~-HEAD - [mach tool] tête *f* porte-fraise
~INDUSTRY - [agric] meunerie *f*, minoterie *f*
~MACHINE - [mach tool] (a miller) fraiseuse *f*
[text mach] (machine used for the preparation of woollen fabrics for subsequent finishing processes) machine *f* à fouler
~MACHINE STANDARD - [mach tool] bâti *m* de la fraiseuse
~MACHINE UPRIGHT - s. milling machine standard
~OF GROOVES - [mach tool] fraisage *m* des rainures
~ORE - [min] minerai *m* de broyage
~PIPE CUTTER - [mach tool] fraise *f* conique femelle
~PIT - [mining] cheminée *m* à minerai
~PRODUCTS - [agric] produits *mpl* de meunerie
~QUALITY WOOL - [text] laine *f* à propriété feutrante
~ROOM - [rubber ind] atelier *m* de mélange
~SYSTEM - [mining] exploitation *f* par entonnoirs
~TIME - [mech] (the period during which a batch of material is treated in a mill or like machine) durée *f* de malaxage
~TOOL - [tools] (for knurling) godronnoir *m*, molette *f*
~TOOLS - [mech] outils *mpl* de fraisage
~WASTE - [agric] sous-produits *mpl* de meunerie
~WORK BETWEEN CENTRES - [mach tool] fraisage *m* entre les pointes
MILLION ELECTRON-VOLT - s. megaelectronvolt
MILLIROENGTEN - [radiat] (one thousandth of a roentgen) millirőntgen
MILLISECOND - [meas] milliseconde *f*
MILLIVOLT - [meas] millivolt *m*
MILLIVOLTMETER - [instr] (instrument measuring voltage in millivolts) millivoltmètre *m*
MILLIWATT - [el] milliwatt *m*
MILLS-PACKARD CHAMBER - [ind chem] (conical type of lead chamber used in sulphuric acid manufacture) chambre *f* de plomb Mills-Packard
MILLSTONE - [mech] (one of the pair of heavy stone

disks used for grinding) meule *f* de moulin
MILLSTONE GRIT - [min] grès *m* meulier, silex *m* meulier
[geol] grès *m* houiller
MILPHAE - [med] atrichie *f* palpébrale
MIMEOGRAPH - [print] (apparatus in which a thin fibrous paraffin-coated paper is used as a stencil) autocopiste *m*, duplicateur
[paper man] (duplicating paper) papier *m* à polycopier
MIMESIS - [gen] mimétisme *m*
MIMETIC - [gen] (imitative) imitatif
[zool] (characterized by a resemblance to other animals or plant etc) mimétique
MIMETITE - [min] (natural arsenate of lead, with lead chloride, usually found in lead ores which have been subjected to secondary alternation) mimétèse *f*, mimétite *f*
MIMOSA - [bot] mimosa *m*
MINARET - [arch] minaret *m*
MINCE, to - [gen] hacher, mincer
MINCING MACHINE - [gen] hache-vinade, hachoir *m*
MINE, to - [min] exploiter, faire des travaux miniers
(the operation of digging out minerals) exploiter
[gen] fouiller la terre, fouiller
MINE - [min] (any excavation for digging out a useful product) mine *f*
[gen] (explosive charge) mine *f*
~ADITS - [mining] galeries *f*pl
~BUILDINGS - [min] bâtiments *m*pl de l'exploitation
~CONCESSION - [comm] concession *f* minière
~DAM - [mining] serrement *m*
~DETECTING SET - [electron] (electronic mine detector) détecteur *m* électronique de mines
~DETECTOR - [instr] (electromagnetic instrument used to locate the position of a mine) cherche-mines *m*
~DIAL - [instr] boussole *f* pour mines
~DRAGGER - [gen] dragueur *m* de mines
~DRAINWAY - [mining] albraque *m*
~DREDGER - s. mine dragger
~EARTH - [min] gîte *m* de minerai de fer
~FACE - [mining] front *m* de taille
~FIELD - [gen] (an area, on land or in water, planted with mines) champ *m* de mines
~GALLERY - [min] galerie *f* (de mine)
~GAS - [min] grisou *m*
~HEAD FRAME - [mining] chevalement *m*
~LAYER - [gen] (naval vessel used for the laying of mines) vaisseau *m* porte-mines
~LEVEL - [mining] galerie *f* de niveau de mine, galerie *f* de mine
~OUT, to - [mining] exploiter
~PROP - [mining] étancon *m* de mines
~PUMP - [impl] pompe *f* de mine, pompe *f* d'épuisement
~RUN - [min] (run-of-mine coal) charbon *m* tout-venant, tout-venant *m*
~SHAFT - [mining] puits *m* de mine
~SHIP - s. mine layer
~SLACK - [min] menus *m*pl non classés
~SURVEY - [mining] levé *m* de mines
~SWEEPER - [gen] (naval vessel fitted with the necessary equipment to detect and destroy, or remove, mines) relève-mines *m*
~VENTILATION - [min] aérage *m* de mines

MINER'S ELBOW - [med] boursite *f* olécrânienne
~NYSTAGMUS - [med] nystagmus *m* des mineurs
~RAMMER - [mining] refouloir *m*
MINERAL - [min] (any ore, rock etc) minerai
[gen] minéral
~ACID - [chem] acide *m* minéral
~ALKALI - [chem] (obsolete for sodium carbonate) carbonate *m* de sodium
~BLACK - [dye] (pigment made by fine grinding from shale or slate containing about 30 p.c. of carbon) noir *m* minéral
~CHARCOAL - [min] (a brittle form of coal resembling charcoal) charbon *m* fossile
~CLEAVAGE - [min] clivage *m* des minéraux
~CONCENTRATION - [min] (the amount of mineral in an ore) minérai *m* concentré
~CONSTITUENTS - [geol] composants *m*pl minéraux
[min] inertes *m*pl
~CONTENT - [chem] composants *m*pl minéraux
~FEEDS - [agric] aliments *m*pl additionnés de substances minérales
~FERTILIZER - [agric] engrais *m*pl minéraux
~FLAX - [min] amiante *f*
~LICK - [agric] bloc *m* à lécher
~MATTER - [min] stériles *m*pl
~NAPHTHA - [min] naphte *m* minéral, huile *f* de roche
~OIL - [min] (any oil derived from inorganic matter, in particular petroleum) huile *f* minérale
~OIL FRACTION - [oil ind] fraction *f* d'huile minérale
~ORANGE - [paint] (term sometimes applied to red lead oxide) orangé *m* minéral
~PIGMENT - [chem] (inorganic pigment) pigment *m* minéral
~PITCH - [min] (asphalt) asphalte *m*, bitume *m* minéral
~RUBBER - [min] (generic name for a number of asphaltites. The term is also applied to blown asphalts) caoutchouc *m* minéral, élatérite *f*
~SALTS - [chem] sels *m*pl minéraux
~SOIL - [geol] sol *m* minéral
~SPIRIT - [chem] alcool *m* dénaturé
~SPOT - [min] inclusion *f* minérale
~SPRING - [geol] source *f* minérale
~TAR - [min] (maltha) bitume *m* glutineux, malthe *f*, goudron *m* minéral
~TRASS - [min] trass *m* minéral
~TURPENTINE - [chem] (white spirit) essence *f* de térébentine artificielle
~VEIN - [min] (a crack in a rock filled with minerals) veine *f* minérale
~VIOLET - [paint] (pigment consisting essentially of manganese ammonium pyrophosphate) violet *m* minéral
~WATER - [min] (natural water containing some minerals in solution) eau *f* minérale
~WAX - [min] cire *f* minérale, cérésine *f*
~WHITE - [paint] (alternative name for gypsum, natural calcium sulphate) gypse *m*, pierre *f* à plâtre
~WOOL - [min] (fine filaments produced by blowing air through molten slag and used as an insulant, reinforcing material with synthetic resins, and as a filtrant) laine *f* de scorie, coton *m* minéral
MINERALIZATION - [bot] (the deposition of calcium salts, silica and other inorganic substances in a cell wall) minéralisation *f*

MINERALIZING - [chem etc] minéralisateur
~AGENTS - [min] agents *mpl* minéralisateurs
~FAULT - [geol] faille *f* nourricière
MINERALOGY - [gen] (the science of minerals) minéralogie *f*
MINETTE - [geol] minette *f*
MINIATURE - [gen] miniature *f*
[mech etc] maquette *f*
[cin] prise *f* de vue avec maquette
~BALL-BEARING - [mech] coussinet *m* miniature
~CAMERA - [photo] appareil *m* de petit format
~EDISON SCREW-CAP - [el] (Edison screw-cap with a screw-thread having a diameter of approx 3/8 of an inch) culot *m* Edison miniature
~FILM RADIOGRAPHY - [radiat] radiographie *f* à microfilm
~HOLDING - [agric] exploitation *f* minuscule
~LAMP HOLDER - [el] douille *f* miniature
~SOCKET - [el] douille *f* miniature
~THERMOCOUPLE WIRE - [meas] (used for the measurement of a wide range of temperature under extreme conditions) microconducteur *m* thermoélectrique
~TUBE - s. miniature valve
~UPRIGHT PIANO - [mus] piano *m* droit
~VALVE - [electron] (a valve in which all dimensions are greatly reduced) tube *m* miniature
MINIATURIZATION - [gen] (tendency to reduce the size of component parts) miniaturisation *f*, dimensionnement *m* miniature
MINIATURIZE, to - [electron etc] miniaturiser
MINIM - [mus] blanche *f*
MINIMUM - [gen] (the lowest value of a given quantity occurring during a certain period) minimum *m*
[adj] minimum, minimal
~ANGLE OF DEVIATION - [opt] angle *m* minimal de déviation
~AUDIBILE - [acoust] seuil *m* auditif
~BLOWING CURRENT - [el] (the minimum current causing the melting of a fuse) courant *m* limite de fusion
~BOILING POINT - [phys] point *m* d'ébullition minimal
~BURNER PRESSURE VALVE - [th eng] (a device to limit the minimum fuel pressure in a burner) soupape *f* de pression minimale d'opération
~CARRIER LEVEL - [electron] niveau *m* minimal de porteuse
~CHARGE - [comm] taxe *f* de consommation minimum
~CLEARANCE - [mech] jeu *m* minimum
~CLEARANCE BETWEEN POLES - [el] espace *m* interpolaire minimal
~CONTACT SURFACE - [el] contact *m* de minima
~CURRENT RELAY - [el] relais *m* à minimum
~DELAY CODING - [comput] codification *f* à retard minimal
~DEVIATION - [light] (angle of minimum deviation) déviation *f* minimale
[mech] déviation *f* minimale
~DISTANCE CODING - [comput] code *m* à distance minimale de signal
~FLIGHT ALTITUDE - [aero] (the lowest heigh above M.S.L. at which it is safe for an aircraft to operate on instruments) altitude *f* minimum de sécurité
~FLYING SPEED - [aero] (the lowest airspeed at which an aircraft can maintain level flight) vitesse *f* minimum de vol
MINIMUM GLIDING ANGLE - [aero] angle *m* minimal de plané
~INTERFERENCE - [mech] interférence *f* minimum
~IONIZATION - [nucl] (the smallest possible value) ionisation minimale
~LATENCY - [comput] (the smallest obtainable period of waiting) temps *m* d'attente minimal
~NUMBER OF THEORETICAL PLATES - [nucl] (the minimum number of theoretical stages) nombre *m* minimal de plaques théoriques
~PHASE/FREQUENCY CHARACTERISTIC - [radio] caractéristique *f* phase minimum/fréquence
~PHASE NETWORK - [el] réseau *m* à phase minimum
~PHASE SHIFT SYSTEM - [el] système *m* à déphasage minimal
~PRICE - [comm] prix *m* minimum
~SCALE VALUE - [meas] (the smallest value of the measured quantity which the scale is graduated to indicate) valeur *f* minima de l'échelle
~SELECTIVITY - [phys] sélectivité *f* minimale
~SENSIBLE - [acoust] seuil *m* de perception
~SPEED - [mech] vitesse *f* minimale
~SPEED OF AN INTERNAL COMBUSTION ENGINE - [mech] vitesse *f* de ralenti minimale
~THERMOMETER - [instr] (a thermometer provided with an indicator which remains at the lowest temperature registered until re-set) thermomètre *m* à minima
~VOLTAGE RELAY - [el] relais *m* à minimum
~WAVELENGTH-[phys] (the shortest wavelength in a spectrum) limite *f* quantique, longueur *f* d'onde limite
~WORKING CURRENT - [el] courant *m* minimal d'un relais
MINING - [min] minier
[mining] (the action of excavating a mine) exploitation *f* minière, exploitation *f*
[mining] (the action of extracting the mineral) abattage *m*, abatage *m*, travaux *mpl* miniers
~AREA - [mining] domaine *m* minier
~BUCKET - [mining] benne *f* d'extraction
~BY BLASTING - [mining] abattage *m* par explosifs
~CAMP - [min] (a temporary settlement of miners or prospective miners) camp *m* minier
~CIRCLES - [mining] milieux *mpl* miniers
~CLAIM - [mining] concession *f* minière
~COMPANY - [comm] societé *f* de mines
~DRILL STEEL - [mining] acier *m* pour perforation de roches
~ENGINE - [mining] engin *m* de mine, échelle *f* mécanique
~ENGINEERING - [mining] technique *f* minière
~EQUIPMENT - [mining] équipement *m* minier
~FLOOR - [mining] niveau *m* d'abattage
~GEOLOGY - [geol] géologie *f* minière
~INDUSTRY - [mining] industrie *f* minière
~LOCOMOTIVE - [mining] locomotive *f* minière, locomotive *f* du fond
~MACHINE - [mining] houilleuse *f*
~METHOD - [mining] méthode *f* d'exploitation des mines
~POINT - [mining] point *m* d'abattage
~RETREATING - [mining] exploitation *f* en rabattant
~SIDING - [railw] gare *f* de mine
~TOOLS - [mining] outils *mpl* de mines
~TRANSIT - [mining] théodolite *m* à boussole pour

mines
MINION - [print] (a size of type) mignonne *f*, corps 7 *m*
MINITRACK - [radar] réseau *m* minitrack
MINIUM - [paint] (term for red lead) minium *m*, plomb *m* rouge
MINOR - [gen] mineur, petit
[mech] secondaire
~BED - [geogr] (of a river) lit *m* mineur
~BEND - [electron] (in a waveguide) coude *m* à plat
~CONTROL - [comput] contrôle *m* mineur, contrôle *m* à l'étage inférieur
~CYCLE - [comput] (the word time of a serial computer) cycle *m* mineur
~DIAMETER - [mech] (of a screw thread) diamètre *m* interne
~EXCHANGE - [telecomm] bureau *m* secondaire
~GRADUATION - [meas] (the shortest division of a scale) graduation *f* secondaire
~INTERVAL - [mus] intervalle *m* mineur
~LOBE - [radio] (of aerial) lobe *m* secondaire
~MINERAL - [min] minéral *m* accessoire
~SCALE - [mus] gamme *f* en mineur
MINORITY - [gen] minorité *f*
~CARRIERS - [electron] (carriers constituting less than half of the total number of carriers) porteurs *mpl* minoritaires
~CIRCUIT - [comput] circuit *m* minoritaire
~EMITTER - [electron] (electrode from which a flow of minority carriers enters the inter-electrode region) émetteur *m* de porteurs minoritaires
MINSTER - [arch] (a monastery church) église *f* de monastère
[arch] (in GB) cathédrale *f*
MINT, to - [gen] (the making and stamping of money) frapper (de la monnaie), battre
MINT - [gen] (the place where money is coined) Hôtel *m* de la Monnaie
[adj] (unused, in the original condition) à l'état neuf
[gen] source *f*, origine *f*
[bot] (aromatic herb) menthe *f*
MINTING - [gen] frappe *f* de la monnaie, monnayage *m*
~DIE - [metall] matrice *f* à battre monnaie
MINUEND - [math] (the number from which a number is subtracted) nombre *m* à soustraire
MINUS - [math] moins
[gen] moins
~ACCELERATION - [mech] accélération *f* négative
~AXIS - [opt] axe *m* négatif
~BLUE - [dye] jaune *m*, moin bleu *m*
~CHARGE - [el] charge *f* négative
~COEFFICIENT - [math] coefficient *m* négatif
~COLOUR - [photo] (a colour which, when added to another colour, produces white light) couleur *f* moins
[dye] couleur *f* soustractive
~CYLINDER - [opt] cylindre *m* négatif
~GREEN - [dye] pourpre *m*, magenta *m*
~PLATE - [el chem] plaque *f* négative
~RED - [dye] bleu-vert *m*
~SIGHT - [surv] coup *m* avant
~SPHERE - [opt] sphère *f* négative
MINUSCULE - [gen] minuscule
[print] (lower case or small letters) minuscule *f*
MINUTE - [meas] (of time) minute *f*
MINUTE - [geom] (unit of angular measure) minute *f*
~FOLDING - [geol] plissement *m*
~HAND - [horol] (the hand which makes one complete turn per hour) grande aiguille *f*
~OF ARC - [meas] (unit of circular measurement, equal to one-sixtieth of a degree) minute *f* d'arc
~OF TIME - [meas] (unit of duration equal to one-sixtieth of an hour) minute *f* de temps
~PINION - [horol] (the pinion in the motion wheel driving the hour wheels) pignon *m* de roue des minutes
~WHEEL - [horol] (the wheel driven by the cannon pinion) roue *f* des minutes
~WHEEL PIN - [horol] (the vertical pinion which the minute wheel turns) axe *m* de la roue des minutes
~WHEEL STUD - s. minute wheel pin
MINVERITE - [geol] (basic intrusive rock, essentially a dolerite containing a hornblende rich in soda) minvérite *f*
MIOCENE PERIOD - [geol] miocène *m*
MIOPRAGIA - [med] miopragie *f*
MIOSIS - [med] (contraction of the eye pupil) miosis *f*
MIOTIC - [med] miotique
MIRABILITE - [min] (naturally occurring hydrated sodium sulphate) mirabilite *f*, sel *m* de Glauber
MIRAGE - [opt] (optical illusion caused by a marked difference in the temperatures of higher and lower strata of air) mirage *m*
MIRBANE OIL - [chem] (nitrobenzene) essence *f* de mirbane
MIRE - [gen] boue *f*, bourbier *m*, fange *m*
[hydr] (river deposit) vase *f*
MIRROR, to - [gen] refléter
MIRROR - [gen] (a looking glass) miroir *m*, glace *f*
[radar] (reflector which is obtained by eliminating parts of the paraboloid, so that the shape is approximately evensided) réflecteur *m* parabolique
[cin] (the reflecting means behind the arc lamp) miroir *m*
[electron] (a conducting surface reflecting the energy from, and modifying, the radiation pattern of a primary radiator-) réflecteur *m* (d'antenne)
~ALTITUDE - [radar] altitude *f* du miroir magnétique
~ARC - [cin] (a projecting arc) lampe *f* à arc pour projection
~COATING - [metall] métallisation *f* inférieure hémisphérique
~DRUM - s. mirrror wheel
~ELEMENTS - [phys] éléments *mpl* miroirs, éléments *mpl* spéculaires
~FINISH - [gen & mech] finissage *m* très brillant
~GALVANOMETER - [instr] (galvanometer with a mirror attached to the moving part so that deflections can be observed by directing a beam of light on to the mirror) galvanomètre *m* à miroir
~GLASS - [glass man] verre *m* à glaces
~IMAGE - [opt] image *f* spéculaire
~IMAGE NUCLEI - s. mirror nuclei
~INSTRUMENT - [instr] (instrument in which the reading is obtained by the movement of a beam of light reflected from a mirror which is attached to the moving element) instrument *m* à miroir
~IRON - [metall] fonte *f* spéculaire
~LENS - [opt] objectif *m* à lentilles spéculaires
~MOVEMENTS - [phys] mouvements *mpl* spéculaires
~NUCLEI - [nucl] (pairs of isobaric nuclei having a

number of protons which is one greather than the number of neutrons) noyaux mpl miroirs
MIRROR NUCLIDES - [nucl] (pairs of nuclides, each of which would be transformed into the other by exchanging all neutrons for protons and vice versa) nucléides mpl miroirs
~RATIO - [phys] rapport m de miroir magnétique
[phys] (in plasma technique) rapport m de miroir
~REFLECTION - [opt] réflexion f spéculaire
~REFLECTOR - [photo] réflecteur m à miroir
~REFLEX CAMERA - [photo] chambre f reflex à miroir
~SCALE - [instr] (scale with a mirror so places that the reflection of a knife edge is in line with the pointer when the eye is in the correct position for the reading of the instrument) échelle f à miroir
~SCREW - [telev] (arrangement of mirror on a rotating shaft, used for mechanical scanning) hélice f à miroir
~SHUTTER - [photo] obturateur m à miroir
~-STONE - [min] muscovite f
~SYSTEM WITH CORRECTION PLATE - [instr] système m à miroir avec plaque de correction
~TELESCOPE - [opt] télescope m à miroir
~VAULT - [constr] voûte f à miroir
~VIEWFINDER - [photo] viseur m à miroir
~WHEEL - [telev] (wheel on which mirrors are attached in mechanical television systems) roue f à miroirs, tambour m à miroirs
MIRRORING - [geol] dépôt m de couches réfléchissantes (ou spéculaires)
MIRY - [gen] fangeux, boueux, vaseux
MISACTION - [gen & med] action f manquée
MISALIGNED - [gen] non centré
~SHAFT - [mech] mauvais alignement m de l'arbre
~WHEELS - [auto] roues fpl désalignées
MISALIGNMENT - [gen] défaut m d'alignement
[comput] (of magnetic tape) mauvais ajustage m, alignement m incorrect
MISCARRIAGE - [gen] égarement m, insuccès m
[med] fausse couche f, avortement m
[comm] égarement m, perte f
MISCEGENATION - [genet] croisement m, métissage m
MISCELLANEOUS - [gen] (consisting of more than one kind) varié, divers, mélangé
~HYDRAULIC PRESS - [mech] presse f hydraulique pour travaux divers
~SECTIONS - [metall] fers mpl divers
MISCIBILITY - [chem] (the property of two or more liquids to mix and form one phase) miscibilité f
~GAP - [chem] (the range of values under which miscible liquids mix only partially) intervalle m de miscibilité
MISCIBLE - [chem] miscible
~SOLVENT - [chem] solvant m miscible
MISFIRE, to - [firearms] manquer, rater
[mech] (int comb engine) avoir des ratés d'allumage
MISFIRE - [firearms] (the failure to explode at the desired time) coup m raté
[electron] (failure to strike an arc from the anode during the normal conduction period) raté m d'amorçage, raté m d'allumage
[mech] (in int comb engines) raté m d'allumage
MISFIRING - [mech] (failure of the compressed charge to fire normally) raté m d'allumage
MISFRAMING - [photo] (incorrectly framing of picture) défaut m de cadrage
MISHAP - [gen] (accident) accident m
MISLIFT - [text] erreur f de lève
MISMATCH - [radio] désaccord m
[metall] variation f, déplacement m
[electron] (relating to the impedance of a connected load) inadaptation f
~FACTOR - [radio] (the reflection factor between two impedances) coefficient m des pertes dues aux réflexions, coefficient m de réflexion
~IN MOULD - [metall] variation f de moule
MISMATCHED - [gen] inégal
[radio] (when the impedance of the load does not match the impedance of the source to which it is connected) en désaccord
[mech] décentré
MISMATCHING - [gen] accouplement m défectueux
[mech] (failure to adjust correctly) déplacement m, variation f
MISPICK - [text] fausse fuite f
MISPICKEL - [min] (natural sulpharsenide of iron, crystallizing in the rhombic system; an important source of arsenic) mispickel m, arsénopyrite f
MISPLUG, to - [telecomm] enficher (dans le faux jack)
MISPRINT, to - [print] (to print erroneously) imprimer incorrectement
MISPRINT - [print] (error in printing) erreur f typographique, coquille f
MISREGISTRATION - [telev] fausse superposition f
MISROUTE, to - [telecomm] mal diriger
MISROUTING - [telecomm] fausse direction f
MISRUN - [metall] (fusion) mal venue
MISS, to - [gen] manquer, rater
[mech] (int comb eng) avoir des ratés
MISS - [gen] coup m manqué, manque m, absence f
~FIRE, to - [mech] (an engine misses fire when any one charge is not ignited) avoir des ratés d'allumage
~OF WEFT - [text] feinte f, clair m
MISSED-APPROACH ALTITUDE - [aero] (the least height at which a final approach should be broken off it completion cannot be effected) hauteur f d'approche manquée
~-APPROACH PROCEDURE - [aero] (the course of action followed by a pilot who has missed normal approach) procédure f d'approche manquée
~THREAD - [text] fil m sauté
MISSILE - [gen] (any object which is thrown or discharged) projectile m, missile m
~LAUNCHING SITE - [astronaut] aire f de lancement d'un missile
~RANGING - [radar] trajectographie f de missiles
~TRACKING STATION - [radar] station f d'observation
MISSILRY - [astronaut] (the study and science of missiles) techniques fpl des missiles
MISSING - [gen] absent, égaré
~TOOTH - [mech] interruption f des dents, brèche f, creux m d'un engrenage
MIST, to - [gen] couvrir de buée
MIST - [met] (slight decrease in visibility due to condensation of moisture in the atmosphere) brume f, buée f
[chem] (colloidal suspension of a liquid in a gas) suspension f colloïdale dans un gaz

MIST-COAT - [paint] voile *m*
~DRILLING - [oil ind] forage à l'air humide
~FILTER - [oil ind] filtre *m* d'échappement
~LADEN GAS - [chem] gaz *m* chargé de vapeur
~OF OIL - [chem] brume *f* huileuse
~PROJECTOR - [mech] pulvérisateur *f* d'eau
~SPRAYER - [agric] nébuliseur *m*, atomiseur *m*
~SPRAYING - [bot] nébulisation *f*, atomisation *f*
~TRAP - [mech] refroidisseur *m* pour le vide primaire
MISTAKE - [comput] erreur *f* humaine
MISTING - [phys] (condensation of moisture on windows, windscreens, etc) embuage *m*
MISTLETOE - [bot] gui *m*
MISTRAL - [met] (a storm cold Northerly wind occurring along the Northern shores of the Mediterranean) mistral *m*
MISTRIMMED FORGING - [metall] pièce *f* forgée à ébavurage excessif
MISTUNING - [mus] jeu *m* à cordes ravalées
MISTY - [gen] brumeux, embrumé
~WEATHER - [met] temps *m* brumeux
MISUSE, to - [gen] faire mauvais usage
MISUSE - [gen] mauvais usage *m*, abus *m*
MITE - [zool] acarien *m*, ciron *m*
~-KILLER - [bot] antimites *m*, miticide *m*
MITER - s. mitre
MITERING - [carp] assemblage *m* à onglet
MITHRIDATISM - [med] mithridatisme *m*
MITICIDE - s. mite-killer
MITIS CASTING - [metall] (castings of wrought iron at a lowered melting point) malléabilisation *f*
MITOSIS - [genet] mitose *f*
MITRAL - [anat] (relating to the mitral valve of the heart) mitral
~INSUFFICIENCY - [med] insuffisance *f* mitrale
~REGURGITATION - [med] regurgitation *f* mitrale
~STENOSIS - [med] rétrécissement *m* mitral
~VALVE - [anat] valvule *f* mitrale
MITRE, to - [carp] (to meter in the USA; to make a joint between two pieces at an angle to one another) assembler à onglet
[carp] (to saw at an angle) tailler à onglet
MITRE - [carp] (joint between two pieces at an angle) onglet *m*
~BEVEL GEARS - s. mitre gears
~BLOCK - [carp] (block of wood used to cut the mouldings for a mitre joint) boîte *f* à onglets
~CUT - [carp] coupe *f* à 45°
~-CUTTING MACHINE - [mech] machine *f* à couper d'onglet
~GEARS - [mech] engrenages *mpl* à onglet, engrenages *mpl* d'équerre
~JOINT - [carp] s. mitre
~PLANE - [impl] guillaume *m* à onglets
~POST - [hydr] poteau *m* battant, montant *m* de busc
~RULE - [impl] biveau *m*
~SHOOT - [carp] (or mitre board; block of wood used to hold the mitre face of a moulding at the right angle to the plane) boîte *f* à onglets
~-SILL - [hydr] (the raised part of a canal-lock bed where the lower sections of the gates abut) busc *m*, seuil *m* d'écluse
~SQUARE - [carp] (bevel-like tool with a blade fixed at 45 degrees to the stock) équerre *f* à onglet, angle *m* oblique
~VALVE - [mech] (valve with its seat at 45 degrees) soupape *f* conique
MITRE WHEEL - [mech] couronne *f* de couple conique
~WHEEL GEARING - [auto] couple *m* conique
~WHEELS - s. mitre gears
MITRED JOINT - [constr] joint *m* de l'épaulement à onglet
MITRIFORM - [zool] (shaped like a mitre, as the mitral valve) mitriforme
MITT - [gen] (glove which does not extend over the fingers) moufle *f*
MITTEN - [gen] (glove encasing the four fingers together and the thumb separately) mitaine *f*
MIX, to - [gen] mêler, mélanger
[gen] (to blend) mélanger, mixtionner
[text] (the operation of blending cotton of various types) mélanger
[metall] allier
MIX - [gen] (the act of mixing) mélange *m*
[ind chem] composition *f*
[constr] (mixing of concrete) malaxage *m*
[rubber ind] mélange *m*, composition *f*
[radio] (the correct blending of the sound input of microphones) mélange *m*
[cin] (colloq. for fading) fondu *m* enchaîné, enchaînement *m*
MIXED - [gen] mêlé, mélangé, mixte
~ADHESIVE - [ind chem] (an adhesive composed of several adhesive compounds) colle *f* de mélange
~ANILINE POINT - [oil ind] point *m* d'aniline
~-BASE NOTATION - [comput] numération *f* à base mixte
~-BASE OIL - [oil ind] pétrole *m* de base mixte
~BLANKING SIGNALS - [telev] signaux *mpl* de suppression mélangés
~COLOUR EFFECT - [text] effet *m* de couleur mélangé
~COLOUR STRIPE - [text] rayure *f* mêlée
~CONDUCTOR - [el chem] (conductor with a conduction which is both electrolytic and electronic) conducteur *m* mixte
~COUPLING - [radio] (inductive and capacitive coupling between two resonant circuits) couplage *m* mixte
~CRYSTAL - [cryst] (crystal consisting of two or more chemical compounds) cristal *m* mixte
~FELT - [text] feutre *m* mélangé
~-FLOW WATER TURBINE - [hydr] (an inward flow reaction turbine with runner vanes so curved as to be operated by the water as it enters radially and leaves axially) turbine *f* mixte, turbine *f* hélico-centripète, turbine *f* américaine
~GAS - [gas ind] gaz *m* mixte
~GLUE - [chem] (a synthetic resin adhesive to which a hardening agent has been added) colle *f* préparée
~-GRAINED - [soil] granulométrie *f* variée (ou complexe)
~HIGHS - [telev] (a process making it possible for coloured television to be obtained in a frequency band hardly wider than the normal one) hautes fréquences *fpl* mixtes
~HOPSACK WEAVE - [text] armure *f* natté melangé
~LIGHT - [light] lumière *f* mélangée
~MATRIX OIL COOLER - [mech] (a combined cooler and radiator, in which alternate tubes are used for oil and coolant) échangeur *m*
~MATRIX RADIATOR - [mech] (a combined radiator and oil-cooler, in which cooland and oil pass through alternate tubes) radiateur *m* à faisceaux mixtes

MIXED MOHAIR YARN - [text] fil *m* mohair mixte
~PASS - [text] remettage *m* amalgamé
~POLYELECTRODE POTENTIAL - [el chem] (potential of an electrode at which several different electrode reactions are occurring at the same time) tension *f* mixte d'une polyélectrode
~-PRESSURE TURBINE - [mech] (steam turbine operated from two or more sources of steam at different pressures) turbine *f* mixte
~RADIX - [math] base *f* mixte
~RIB - [text] cannelé *m* irrégulier
~SYNCS - [telev] sugnaux *mpl* de synchronisation mixtes
~TIE UP - [text] empoutage *m* mixte
~TRAIN - [railw] train *m* mixte
~YARN - [text] fil *m* mixte
~YEAST - [brew ind] mélange *m* de levures
MIXER - [mech] (a machine used to mix or incorporate materials) mélangeur *m*, malaxeur *m*
[el acoust radio etc] (a device with two or more adjustable inputs and a common output) mélangeur *m*
[telev] mélangeur *m* image/son, pupitre *m* de mélange
[el] (a frequency charger) variateur *m* de fréquence
[cin] (an expert in combining sounds from two or more sources into a single output) mélangeur *m* de sons
[electron] (in waveguides) mélangeur *m* à cristal
~FACE - [th eng] entrée *f* du mélangeur
~FOR LATEX PASTE - [rubber ind] mélangeur *m* à pâtes de latex
~HEAD - [th eng] tête *f* du mélangeur, convergent *m*
~METAL - [metall] fonte *f* brute de mélangeur
~THROAT - [th eng] col *m* du canon, col *m* du mélangeur
~TUBE - [th eng] divergent *m*, tube *m* mélangeur
~VALVE - [electron] (a valve in which two different frequency currents are combined for the purpose of modulation) tube *m* mélangeur
~WITH BLADES - [mech] (a paddle mixer) mélangeur *m* à palettes, agitateur *m* à ailettes
~WITH PROPELLER - [mech] mélangeur *m* à hélice
~WITH STIRRER - [mech] malaxeur *m* à agitateur
MIXING - [gen etc] mélange *m*, malaxage *m*
[rubber ind] mélange *m*, composition *f*
[text] (blending cotton of different types, but of similar staple and colour, to obtain the desired yarns) mélange *m* (du coton)
[mech] (in engines) mélange *m*
[nucl] (in separation by gaseous diffusion) homogénéisation *f* du résidu
~ACTION - [el] effet *m* convertisseur
~BASIN - [hydr] bassin *m* d'homogénéisation
~BED - [text] mélangeur *m*, claie *f* à mélange
~BOOTH - s. mixing room
~BOX - [paper man] (in a paper machine) boîte *f* de malaxage
~CHAMBER - [mech] (of a gas welding torch) chambre *f* de mélange
[auto] chambre *f* de carburation, chambre *f* de mélange
[mech] carburateur *m*
~CHANNEL - [hydr] rigole *f* de mélange
~CONDUIT - s. mixing channel
~CONE - [metall] trémie *f* de mélange
MIXING EFFICIENCY - [nucl] rendement *m* de homogénéisation
~FORMULA - [rubber ind] formule *f* de mélange
~FURNACE - [metall] (four) mélangeur *m*
~HEAD - s. mixer head
~HOPPER FEEDER - [text] chargeur mélangeur *m*
~IN PLACE - [rubber ind & constr] malaxage *m* sur place
~INTERSECTING GILL BOX - [text] banc *m* d'étirage mélangeur
~LADLE - [metall] poche *f* mélangeuse
~LENGTH THEORY - [mech] théorie *f* de la distance de mélange
~MACHINE - [ind chem] malaxeur *m*
~MILL - [rubber ind] (an open mill) malaxeur *m*, mélangeur *m* à cylindres
~PANEL - [el acoust] (for the microphones) panneau *m* de mélange
~PLATFORM - [constr] aire *f* de gâchage
~PROCESS - [hydr] processus *m* de mélange
~RATIO - [chem] rapport *m* de mélange
~ROLLS - [plast ind] mélangeur *m* à cylindres, malaxeur *m* à cylindres
~ROOM - [el acoust] (the room in which the mixing of sound records is carried out) salle *f* de mixage
[text] (the room in which the mixing of cotton takes place) salle *f* de mélange
~SCREW - [mech] vis *f* mélangeuse
~SHED - [metall] halle *f* de mélange
~STOPS - [mus] fourniture *f*, plein jeu *m*
~TABLE - s. mixing panel
~TAP - [th eng] (of a water heater) robinet *m* mélangeur, batterie *f* mélangeuse
~TUBE - [electron] (frequency changer) tube *m* mélangeur
[auto] (of the carburettor) tube *m* de mélange
~UNIT - [radio] montage *m* changeur de fréquence
~VALVE - s. mixing tube
MIXTURE - [gen] mélange *m*
[chem] mélange *m*, mixture *f*, amalgame *m*
[chem] dosage *m*
[metall] (the mixture of ore and flux) lit *m* de fusion
[text] (a mixture of yarns, or colours or of different materials) étoffe *f* mélangée
[mech] (int comb eng] (the combined inflammable gas and air) mélange *m* explosif
[phys] (matter containing two or more substances not in chemical combination and thus capable of separation by physical methods) mélange *m*
~CHARGE - [mech] (int comb eng) charge *f* de mélange
~CHAMBER - [mech] (int comb engines) chambre *f* de mélange
~CONTROL - [aero] (automatic device in a fuel-maltering system for regulating the fuel-air ratio) correcteur *m* de mélange
~FABRIC - [text] tissu *m* mélangé
~METHOD LUBRICATION - [auto] graissage *m* par mélange
~OF ISOTOPES - [nucl] (a chemical substance in which a number of isotopes are present) mélange *m* isotopique
~RATIO - s. mixing ratio
[mech] (int comb eng) rapport *m* de mélange
~RECIPE - s. mixing formula
~STOPS - [mus] fourniture *f*
~STRENGTH - [mech] (int comb eng) dosage *m* du mé-

lange
MIXTURE YARN - [text] filé *m* de mélange
MIZZEN - [naut] (a mizzenmast) artimon *m*
[naut] (triangular sail occasionally on the mizzenmast) perruche *f*
~COURSE - s. mizzen sail
~ROYAL MAST - [naut] cacatois *m* de perruche
~-TOPGALLANT MAST - [naut] mât *m* de perruche
~-TOPMAST - [naut] mât *m* de fougue
~YARD - [naut] vergue *f* de perroquet
MIZZENMAST - [naut] mât *m* d'artimon
MIZZONITE - [min] (one of the series of minerals forming the scapolite group) mizzonite *f*
MLD - (initials of Median Lethal Dose) s. median lethal dose
MLT - [nucl] (initials of median lethal time) s. median lethal time
mm - [meas] mm, millimètre
mmf - [meas] micromicrofarad *m*
M.M.F. - [el] (initials of magneto-motive force) force *f* magnéto-motrice
mm HG - [meas] (abbrev. for "millimetres of mercury"; a measure of pressure equal to that exerted by a column of mercury of the height specified) millimètres *mpl* de mercure
MNEMONIC - [gen & comput] (relating to memory) mnémonique
~CODE - [comput] code *m* mnémonique
~INSTRUCTION CODE - [comput] code *m* d'instructions mnémoniques
MNEMONICS - [gen] (the technique of enhancing the power of the memory) mnémotechnie *f*
MNEMOTAXIS - [zool] (the movements of an animals dictated by memory) mnémotaxie *f*
MOAN - s. moaning
MOANING - [acoust] (a sound expressing pain or suffering) gémissement *m*, geignement *m*
MOAT - [gen] fossé *m*, douve *f*
MOBILE - [gen & phys] mobile
~AERIAL - [radio] (a portable aerial) antenne *f* mobile
~BELT ELEVATOR - [mining] sauterelle *f*
~CLINIC - [med] camion *m* hôpital
~CRANE - [mech] grue *f* mobile, grue *f* roulante
~DARKROOM - [photo] camionnette *f* laboratoire
~FILM COLUMN - [ind chem] colonne *f* de distillation pelliculaire
~FILM DISTILLATION - [ind chem] distillation *f* pelliculaire de surfaces
~FILM STILL - [ind chem] installation *f* de distillation pelliculaire à surfaces mixtes
~RADIO SERVICE - [radio] liaison *f* mobile radio
~RIG - [mining] jig *m* mobile, tarière *f* mobile
~STATION - [radio] (transmitter set on a truck etc) station *f* mobile
~SURFACE STATION - [radio] (radio communication station in a vehicle or ship, designed to be used at any convenient point) station *f* émettrice au sol
~TRANSMITTER - [radio] émetteur *m* mobile
MOBILITY - [gen & phys] mobilité
[nucl] (the average drift velocity given to a charged particle by a unit potential gradient) mobilité *f*
~ANALOGY - [acoust] (an acoustical-mechanical dynamical analogy in which velocity corresponds to a voltage and force to a current) analogie *f* de mobilité
~OF A CHARGED PARTICLE - [electron] (in an electric field) mobilité *f* d'un porteur électrisé
MOBILITY OF AN ION - [el chem] (characteristic constant of an ion in an infinitely dilute electrolyte) mobilité *f* d'un ion
~OF IONS IN SOLIDS - [cryst] mobilité *f* d'un ion dans les solides
~RATIO - [geol] coéfficient *f* de mobilité
MOBILIZATION - [gen] mobilisation *f*
MOBILIZE, to - [gen] mobiliser
MOBILOMETER - [instr] (instrument for determining viscosity, in which a weighted plunger is made to pass down a cylinder containing the sample, the time taken to move through a given distance giving a measure of the viscosity) viscosimètre *m*
MOCASSIN - [shoe man] (shoe made of very soft leather) mocassin *m*
MOCEZUELO - [med] tétanos *m* des nouveau-nés
MOCK - [gen] feint, contrefait
~FOG - [met] brouillard *m* apparent
~LEAD - [min] (sphalerite) fausse galère *f*, sphalérite *f*
~-LENO WEAVE - [text] armure *f* fausse gaze
~MOON - [met] (paraselene; phenomenon analogous to a parhelion, but produced by the moon) parasélène *f*
~ORANGE - [bot] seringa *m* des jardins
~RIB EFFECT - [text] aiguille *f* tirée
~-ROSE - [bot] ciste *m* de Crète
~SEAM - [rubber ind & shoe man] fausse couture *f*
~SUN - [met] (parhelion; image of the sun formed at about 22 degrees in azimuth from the sun itself under certain atmospheric conditions) parhélie *m*, faux soleil *m*
~TEST - [astronaut] essai *m* de fonctionnement sans mise à feu
~-UP - [gen] (a trial or experimental structure of an incomplete character) maquette *f*
[mech] maquette *f*, modèle *m*
~-UP REACTOR - [nucl] maquette *f* de réacteur
~WATER - [text] fil *m* mi-chaîne
~WORSTED - [text] cardé-peigné *m*
MODACRYLIC FIBRE - [ind chem] (generic name for synthetic fibres containing a specific proportion of acrylonitrile units) fibre *f* modacrylique
MODE - [gen] (manner of operating, of doing; method) mode *m*, méthode *f*, manière *f*
[geol] (the actual mineral composition of a rock expressed quantitatively) composition *f*
[electron] (any possible field configuration in a guided way) mode *m*
[mus] (characteristic succession of intervals forming a scale) mode *m*
[statist] moyenne *f* prédominante
~CHANGER - [electron] (device for transforming an electromagnetic wave from one mode of propagation to another) transformateur *m* de mode
~CONVERSION - [comput] conversion *f* de mode
~FILTER - [electron] filtre *m* de mode
~FILTER SLOT - [electron] (mode filter in the form of a slot) filtre *m* de mode en fente
~JUMP - [electron] (in a magnetron) saut *m* du mode
~NUMBER - [electron] (in magnetrons and in reflex klystrons) nombre de mode
~OF ACTION OF THE BRAKE - [mech] manière *f* d'action du frein
~OF OPERATION - [gen] mode *m* opérationnel, mode *m* d'opération

MODE OF PROPAGATION - [electron] mode *m* de propagation
~OF RESONANCE - [electron] mode *m* resonnant
~OF VIBRATION - [radio] (pattern of motion of the individual particles of a vibratory body) mode *m* de vibration
~SEPARATION - [electron] (in an oscillator) différence *f* de fréquence de mode
~SHIFT - [electron] (frequency slip, frequency leap of a magnetron during the existence of a pulse) glissement *m* du mode
~SKIP - [electron] manque *m* d'amorçage
MODEL - [gen] (an object accurately representing something to be made) modèle *m*
[tool] (a moulding tool) larget *m*, platine *f*
~AERONAUTICS - [aero] technique *f* des avions miniature
~AIRCRAFT - [aero] modèle *m* d'avion
~AIRCRAFT STRIP - [rubber ind] élastique *m* pour moteurs de modèle d'avion
~FARM - [agric] ferme *f* modèle
~FLAPPER - [aero] (a model aircraft with flapping wings) modèle *m* d'avion à ailes battantes
~GLIDER - [aero] modéle *m* de planeur
~ROOM - [gen & comm] atelier *m* des modèles
~SAILPLANE - [aero] modèle *m* de planeur
~TESTING - [gen] (tests carried out with models) essai *m* sur modèle
~ZONE - [agric] zone-témoin *f*, zone-pilote *f*
MODELLING - [photo] relief *m*, effet *m* de profondeur
MODERATE ACCELERATOR - [chem] accélérateur *m* d'action moderée
~BREEZE - [met] jolie brise *f*
~GALE - [met] vent *m* frais
~PRICE - [comm] prix *m* modéré, prix *m* modique
MODERATED HYDRAULIC LIME - [ind chem] chaux *f* modérément hydraulique
~REACTOR - [nucl] (reactor in which a moderator is used) réacteur *m* à modération
MODERATING RATIO - [nucl] (the ratio between the slowing-down power and the macroscopic cross-section) rapport *m* de modération
MODERATION - [nucl] (reduction in the speed of a particle by collisions with nuclei) modération *f*
MODERATOR - [nucl] (any substance used to reduce the speed of neutrons by producing collisions, e.g. graphite) modérateur *m*, ralentisseur *m*
[mech] (regulating mechanism) régulateur *m*
[gen] modérateur *m*
~CONTROL - [nucl] (control of a nuclear reactor by the adjustement of the position or quantity of the moderator) commande *f* par le modérateur
~-COOLANT - [nucl] (a substance which acts as a moderator and also serves to abstract heat) modérateur *m* réfrigérant
~COOLING SYSTEM - [nucl] système *m* de refroidissement du modérateur
~DUMPING SAFETY MECHANISM - [nucl] mécanisme *m* de sécurité par vidange du modérateur
~LATTICE - [nucl] (a reactor lattice consisting only of moderator elements) réseau *m* moderateur
MODES OF OSCILLATION - [phys] modes *mpl* d'oscillation
MODIFICATION - [gen] modification *f*
[zool & bot] (variation due to external influences) modification *f*
~PLATE - [gen] (an indication plate attached to a unit or part to show that it has been subject to modification) plaque *f* de modification
MODIFIED CONSTANT VOLTAGE CHARGE - [el chem] (charge in which the bus voltage of the charging circuit is held substantially constant, but a fixed resistance is present in the battery circuit) charge *f* à tension constante modifiée
~GLYCEROPHTALIC RESINS - [chem] résines *fpl* glycérophtaliques modifiées
~INDEX OF REFRACTION - [opt] (in the troposphere) indice *m* modifié de réfraction
~POLYESTER - [chem] polyester *m* modifié
~REFRACTION INDEX - s. modified index of refraction
~RESINS - [ind chem & plast ind] résines *fpl* modifiées
MODIFIER - [ind chem] (influencing the terminal product of a chemical reaction) modificateur *m*
[comput] (quantity used to alter the address of an operand) modificateur *m*
MODIFY, to - [gen] modifier
[comput] modifier
MODIFYING AGENT - s. modifier
~FEEDBACK - [comput] réaction *f* réglable
~FEEDFORWARD - [comput] action *f* directe réglable
MODILLION - [arch] (or mode, ornate horizontal bracket used in series under a cornice) modillon *m*
MODULAR - [math] modulaire
~ANALOG COMPUTER - [comput] calculateur *m* analogique modulaire
~CONSTRUCTION - [electron] construction *f* modulaire
~DESIGN - [comput] construction *f* modulaire
~RATIO - [mech] rapport *m* des modules
MODULATE, to - [phys] (the operation of altering certain characteristics of a wave in accordance with a signal to be transmitted) moduler
MODULATED - [phys] modulé
~AERIAL - [radio] (periodic antenna in the US) antenne *f* modulée
~AMPLIFIER - [radio] (modulation stage) étage *m* modulateur
~AMPLIFIER ELECTRON TUBE - [electron] tube *m* modulé
~ANTENNA - s. modulated aerial
~CONTINUOUS WAVE - [radio] onde *f* entretenue modulée
~CURRENT - [el] courant *m* modulé
~DEPTH - [electron] profondeur *f* de modulation, facteur *m* de modulation
~HIGH FREQUENCY ENERGY - [el] énergie *f* de haute fréquence modulée
~OUTPUT POWER - [telecomm] puissance *f* modulée
~PHOTOELECTRIC SYSTEM - [electron] système *m* photoélectrique à interruptions préfixées
~STAGE - [electron] étage *m* modulé
~TELEGRAPHY - [telecomm] télégraphie *f* modulée
~VOICE FREQUENCY SIGNALLING CURRENT - [telecomm] courant *m* d'appel modulé à fréquence vocale
~WAVE - [phys] (wave with some of its characteristics varying in accordance with the value of a modulating wave) onde *f* modulée
MODULATING - [radio] modulateur, modulatrice, de modulation
~ALTERNATING FIELD - [telecomm] champ *m* alternatif modulateur
~-ANODE KLYSTRON - [electron] klystron *m* à anode

de commande
MODULATING ELECTRODE - [electron] (electrode to which a potential is applied to control the magnitude of the beam current) électrode *f* de modulation
~FREQUENCY - [radio] (of a modulating wave) fréquence *f* de modulation
~LOW-FREQUENCY OSCILLATION - [telecomm] oscillation *f* modulatrice à basse fréquence
~WAVE - [radio] (wave causing the variation of some characteristics of a carrier) onde *f* modulatrice
MODULATION - [radio] (the process of modifying a carrier wave in accordance with the signal to be transmitted) modulation *f*
[telecomm] (in telegraphy; the variation in time of one or more characteristics of an electromagnetic wave or of a direct current) modulation *f*
~AMPLIFIER - [radio] (amplifier in the modulation circuit) amplificateur *m* de modulation
~BAND - [radio] (the band of frequencies over which modulation is carried out) bande *f* de modulation
~CAPABILITY - [radio] (the maximum possible percentage modulation without disturbing distortion) taux *m* maximum de modulation
~CHARACTERISTIC - [radio] caractéristique *f* de modulation
~COMPONENT - [phys] composante *f* de modulation
~CONDITION - [telecomm] condition *f* de modulation
~CONTROL - [radio] réglage *m* de la modulation
~DEFOCUSING - [telecomm] déconcentration *f* par modulation
~DEGREE/RESPONSE CHARACTERISTIC - [radio] rapport *m* fréquence/taux de modulation
~DIRECTION - [telecomm] sens *m* de modulation
~DISTORTION - [telecomm] distorsion *f* de modulation
~ELEMENT - [radio] (condition assumed by the appropriate device in the transmitting apparatus, associated with the interval of time corresponding to its duration) élément *m* de modulation
~ENVELOPE - [radio] (of an amplitude-modulated) wave) enveloppe *f* de la modulation
~FACTOR - [radio] coefficient *m* de modulation
~-FACTOR METER - [radio] modulomètre *m*
~FREQUENCY - s. modulating frequency
~FREQUENCY HARMONIC DISTORTION - [radio] distorsion *f* de fréquences harmoniques
~FREQUENCY HARMONIC DISTORTION CHARACTERISTIC - [radio] caractéristique *f* de distorsion de fréquences harmoniques
~FREQUENCY HARMONIC DISTORTION FACTOR - [radio] facteur *m* de distorsion de fréquences harmoniques
~FREQUENCY-RESPONSE CHARACTERISTIC - [radio] caractéristique *f* de modulation
~INDEX - [radio] (in frequency modulation) indice *m* de modulation
~KEYING - [telecomm] s. modulation
~LEVEL - [electron] niveau *m* de modulation
~METER - [instr] (measuring instrument designed to control the contrast of a broadcast or recording) modulomètre *m*
~METHOD - [telecomm] méthode *f* de modulation
~MONITOR - [radio] (device enabling the modulation to be supervised) moniteur *m* de modulation, contrôleur *m* de modulation
~NOISE - [radio] souffle *m* dû au signal
MODULATION OF A CARRIER CURRENT - [telecomm] modulation *f* d'un courant porteur
~PERCENTAGE - [radio] (the modulation factor expressed as a percentage) pourcentage *m* de modulation
~POWER - [radio] (the power which is applied by the source of modulation voltage) puissance *f* de modulation
~PRODUCTS - [radio] (waves resulting from modulation) modulats *mpl*, produits *mpl* de modulation
~RATE - [telecomm] (telegraph speed) rapidité *f* de modulation
[comput] vitesse *f* de modulation
~SIGNAL - [radio & telecomm] signal *m* de modulation, signal *m* modulant
~STAGE - [radio & telecomm] étage *m* modulateur
~STANDARD - [electron] norme *f* de modulation
~SUPPRESSION - [radio] (the demodulation effect) suppression *f* de la modulation
~SUPPRESSION CHARACTERISTIC - [radio] caractéristique *f* de suppression de modulation
~SUPPRESSION RATIO - [radio] (the ratio between the magnitude of the modulation frequency response and a single excitation to its magnitude when there is a simultaneous excitation at a different frequency) taux *m* de suppression de modulation
~TRANSFORMER - [radio] transformateur *m* de modulation
~VELOCITY - [telecomm] (telegraph speed) rapidité *f* de modulation
MODULATOR - [radio] (device to carry out the process of modulation) modulateur *m*, étage *m* modulateur
~CELL - [telecomm] cellule *f* de modulateur
~ELECTRODE - [electron] (control electrode in the case of an electron gun) électrode *f* de modulation
~TUBE - [electron] (an electronic tube employed for the amplification of the signal applied to the microphone) tube *m* modulateur
~VALVE - s. modulator tube
MODULE - [arch] (the radius near the base of a column) module *m*
[mech] (in a gear wheel, the pitch circle diameter per tooth) module *m*
[hydr] (the volume of water which is required for an irrigation) module *m*
[meas] (unit of measure for a water flow per second) module *m*
MODULO-N CHECK - [comput] contrôle *m* modulo N
MODULUS - [meas] (coefficient or quantity for the measure of a force, a function or an effect) module *m*, coefficient *m*
[math] module *m* d'un nombre complexe
~OF A SECTION - [constr] coefficient *m* de résistance
~OF COMPRESSIBILITY - [phys] module *m* de déformation volumétrique sous pression hydrostatique, module *m* de compression
~OF A CUBIC COMPRESSIBILITY - [phys] module *m* d'élasticité cubique
~OF CUBIC ELASTICITY - s. modulus of cubic compressibility
~OF DECAY - [acoust] (of a vibrating system) module *m* d'extinction
~OF DEFORMATION - [phys] module *m* de déformation
~OF EFFICIENCY - [phys & mech] coefficient *m* de

rendement
MODULUS OF ELASTIC COMPRESSION - [phys] module *m* de compression élastique
~OF ELASTICITY - [phys] (the ratio of stress to strain in a material over the range for which this value is constant) module *m* d'élasticité
~OF ELONGATION - [mech] (stress at a specified elongation) module *m* d'allongement
~OF GEAR - [mech] module *m* d'engrenage
~OF PLASTICITY - [soil] module *m* de plasticité
~OF RIGIDITY - [mech] (the coefficient of elasticity in shear; the elastic modulus corresponding to a shear stress a pair of orthogonal planes) module *m* de cisaillement, module *m* de glissement
~OF RUPTURE - [mech] (a measure of the breaking load per unit area of a specimen, as determined from a bending test) module *m* de rupture
~OF SHEAR DEFORMATION - [soil] module *m* de déformation par glissement
~OF SHEARING - s. modulus of transverse elasticity
~OF STRAIN HARDENING - [metall] module *m* d'écrouissage par déformation
~OF TORSION - [soil] module *m* de torsion
~OF TRANSVERSE ELASTICITY - [mech] module *m* de glissement
~OF VOLUME ELASTICITY - [mech] (the measure of pressure to a material medium to change its volume) module *m* d'élasticité cubique
MOGUL - [railw] (US for a type of freight locomotive) locomotive *f* pour train de marchandises
MOH HARDNESS SCALE - s. Moh's scale
MOH'S SCALE - [meas] (a scale of hardness from 1 to I0. Talc is 1 and diamond I0) échelle *f* de dureté de Moh
MOHAIR - [text] (hair obtained from crossbred animals) mohair *m*, angora *m*
[text] (in its original meaning, the fleece-like hair obtained from the Angora goat) poil *m* de chèvre de Angora
~COVERING - [text] couverture *f* mohair
~PLUSH - [text] peluche *f* mohair
~YARN - [text] fil *m* mohair
MOHR'S CLIP - [mech] (spring clip for closing rubber tubes by compression) pince *f* à ressort
MOIL - [gen] (a spot, a soiling) souillure *f*, salissure *f*
[mining] pince *f*
[glass man] (waste in the manufacture of glassware) surépaisseur *f* (de verre)
[mining] coin *m*, pic *m*
MOILING - [mining] travail *m* à la massette
MOIRE - [text] moire *f*
[text] (fabric having a wavy, or watered appearance) moiré *f*
[telev] (the appearance of curved lines in an area of the picture) moirage *m*, moirure *f*
~CLOTH - [text] tissu *m* moiré
~CREPE - [text] crêpe *m* ondé
~LINING - [text ind] toile *f* doublure moirée
MOIST - [gen] humide, mouillé
[med] purulent
~ADIABATIC LAPSE - [met] gradient *m* vertical de température adiabatique saturée
~WEFT - [text] trame *f* mouillée
MOISTEN, to - [gen] humecter, mouiller, humidifier
MOISTENING AGENT - [text] produit *m* à humidifier
~CHAMBER - [text] chambre *f* de mouillage, chambre *f* d'humidification
MOISTENING DEVICE - [text] dispositif *m* humidificateur
~THE WEFT - [text] mouillage *m* de la trame, vaporisage *m* de la trame
MOISTMETER - s. moisture meter
MOISTURE - [gen & phys] humidité *f*
~AND ASH FREE COAL - [min] (pure coal substance) charbon *m* pur
~BLOW - [rubber ind] (in tyres) bulle *f* d'humidité, galerie *f* d'humidité
~CONTENT - [phys] (the degree of humidity of a substance) teneur *f* en humidité, état *m* hygrométrique
~DETERMINATION - [ind chem] détermination *f* de l'humidité
~EXPANSION - [phys] (the increase of volume due to moisture) dilatation *f* par l'humidité
~EXTRACTION - [chem] élimination *f* de l'humidité
~FILM - [chem] film *m* de liquide
~LOSS - [phys] (the evaporated quantity of moisture in the drying of solids) perte *f* d'humidité
~METER - [instr] (instrument designed to measure moisture content) hygromètre *m*
~-PROOF - [gen] à l'épreuve de l'humidité
~-RESISTANCE - [gen] résistance *f* à l'humidité
~TELLER - [instr] (instrument designed to measure the humidity content in a moulding sand) indicateur *m* d'humidité
~TESTER - s. moisture teller
~VAPOUR TRANSMISSION - [chem] (the rate at which a material will transmit water vapour under specified conditions) perméabilité *f* à l'humidité
MOITS - [text] (burrs, seeds, straw etc found in wool) débris *mpl* végétaux
MOITY WOOL - [text] laine *f* chardonneuse, laine *f* pailleuse
MOLAL - [phys & chem] (pertaining to moles) molal
~CONCENTRATION - [chem] (the number of moles per unit mass of a phase) concentration *f* molale
~SOLUTION - [chem] solution *f* molale
~VOLUME - [phys] (the volume of one mole of a substance in a given state under given conditions) volume *m* molal
MOLAR - [chem] (containing a grammomolecular weight; in particular, a solution with a concentration of one mole of the solute to I,000cc) molaire
~CONCENTRATION - [chem] (the concentration of a solution expressed as the number of gram-molecules of dissolved substance per I,000 cc of solution) concentration *f* molaire
~CONDUCTANCE - [chem] (the conductance of a solution containing one gramme mole of the solute, measured in the same way as for equivalent conductance) conductance *f* molaire
~DILUTION - [chem] dilution *f* molaire
~DISPERSIVITY - [opt] (the difference in molare refraction at two wavelenghts) dispersivité *f* molaire
~HEAT - s. molecular heat
~QUANTITIES - [chem] (quantities proportionate to the molecular weights of a substance) quantités *fpl* molaires
~REFRACTION - [phys] (the product of the specific refraction by the molecular weight) réfraction *f* moléculaire
~ROTATION - [phys] (the product of the specific rotation by the molecular weight, divided by I00) puissance *f* moléculaire rotative

MOLAR ROTATORY POWER - s. molar rotation
~SOLUTION - [chem] (a solution containing I mole of a solute in I litre) solution *f* molaire
~SURFACE - [phys] (the surface of a sphere, the mass of which is one mole) surface *f* molaire
~SURFACE-ENERGY - [phys] énergie *f* superficielle molaire
~VOLUME - s. molecular volume
MOLARC LIGHT - [cin] lampe *f* à arc I50 A
MOLASSE - [geol] (soft sandstone and grey and red sandy marls found in deposits in Switzerland) molasse *f*, mollasse *f*
MOLASSED FEED - [agric] fourrage *m* mélassé
MOLASSES - [ind chem] (the final mother-liquor in sugar crystallization, from which it is no longer possible to obtain crystalline sugar by any simple process. An important starting material for alcohols, rum being prepared from it) mélasse *f*
MOLD - [for mold and all derivatives, moulding etc] s. "mould"
MOLE - [chem] (the number of grammes of a substance which equal its molecular weight) mole *m*
[constr] môle *m*, digue *f*, jetée *f*
[constr] (a breakwater) brise-lames *m*
[zool] (insectiverous mammal having a velvety fur) taupe *f*
[med] (morbid mass in the womb giving rise to a false pregnancy) mole *f*
~-CRICKET - [zool] taupe-grillon *f*
~DRAINAGE - [agric] drainage *m* à taupinières, drainage *m* en galeries
~FRACTION - [nucl] (used in isotope separation) fraction *f* molaire
~PLOUGH - [agric] charrue *f* taupe
MOLECULAR - [phys & chem] (relating to, or consisting of, molecules) moléculaire
~ABUNDANCE - [phys] abondance *f* moléculaire
~AERODYNAMICS - [phys] aérodynamique *f* moléculaire
~ASSOCIATION - [phys] (the relatively loose coherence of groups of molecules in a liquid or gas) association *f* moléculaire
~ATTRACTION - [phys] force *f* d'attraction moléculaire
~BEAM - [phys] (unidirectional stream of neutral molecules passing through a vacuum) faisceau *m* moléculaire
~BEAM MASER - [electron] (type of gas maser) maser *m* à faisceau moléculaire
~BOND - [chem] (valence linkage between two atoms) liaison *f* moléculaire
~COLLISION - [phys] (collision between molecules, resulting in chemical reaction) choc *m* moléculaire
~COMPOUND - [chem] (compound formed by the union of already saturated molecules) composé *m* moléculaire
~CONDUCTIVITY - [el] conductibilité *f* moléculaire
~CRYSTAL - [cryst] cristal *m* moléculaire
~CURRENT - [el] courant *m* moléculaire
~DEPRESSION OF FREEZING POINT - [phys] (the depression in the freezing point of a liquid produced by the dissolution of one grammolecule of a substance in I00 grammes of solvent, under the laws governing dilute solutions) dépression *f* moléculaire du point de congélation
~DIAGRAM - [phys] représentation *f* graphique de une molécule
MOLECULAR DIAMETER - [phys] diamètre *m* moléculaire
~DIFFUSION - [phys] (process in which a substance gradually penetrates into another as a result of the continuous thermal motion of the individual molecules) diffusion *f* moléculaire
~DIPOLE MOMENT - [el] (the permanent dipole moment of certain molecules) moment *m* dipôle de molécules
~DISTILLATION - [nucl] (a process for the separation of isotopes, in which a substance is evaporated under extremely low pressure and the vapour condensed before collisions occur) distillation *f* moléculaire
~DRAG PUMP - [mech] pompe *f* moléculaire
~EFFUSION - [phys] effusion *f* moléculaire
~ELECTRONICS - [electron] électronique *f* moléculaire
~EXCITATION - [phys] (the process of putting an atom or a molecule into a condition in which the total energy of its inner mechanism is greater in the normal state) excitation *f* moléculaire
~FIELD APPROXIMATION - [phys] (a method for treating problems of ferromagnetism) approximation *f* du champ moléculaire
~FLOW - [phys] écoulement *m* moléculaire
~FLUX - [phys] flux *m* de molécules
~FORMULA - [chem] (indicates the type and number of atoms in each molecule of a compound) formule *f* moléculaire
~FREE PATH - [phys] (the average distance travelled by a molecule between collisions) libre moyen parcours *m* moléculaire
~GAGE - s. molecular gauge
~GAUGE - [instr] (special form of vacuum meter) videmètre *m* moléculaire
~HEAT - [phys] (value obtained by multiplying the molecular weight of a substance by its specific heat) chaleur *f* moléculaire
~ION - [phys] (a charged molecule) ion *m* moléculaire
~ION BREAKUP - [nucl] décomposition *f* d'ion moléculaire
~MASS - [phys] masse *f* moléculaire
~NUMBER - [phys] (the sum of the atomic numbers in a molecule) nombre *m* moléculaire
~ORBITAL - [phys] (the wave function of an electron moving in the field of the other electrons and nuclei forming a molecule) orbite *f* moléculaire
~ORIENTATION IN SURFACE - [phys] orientation *f* moléculaire dans la couche superficielle
~POLARIZABILITY - [el] polarisabilité *f* moléculaire
~PUMP - [phys] pompe *f* moléculaire
~ROTATION - [phys] (value obtained by multiplying the specific rotation by the molecular weight of an optically-active compound and dividing the product by I00) rotation *f* moléculaire
~SIEVES - [chem] (synonym for natural and synthetic zeolites. They have uses as desiccants and carriers for reactive compounds) zéolithes *fpl* moléculaire
~SIEVE SORPTION PUMP - [mech] pompe *f* à absorption
~SOLUTION - [chem] (one in which the molecules of the solute are separated from each other by those of the solvent. This is a true solution) solution *f* moléculaire

MOLECULAR SOLUTION VOLUME - [phys] volume *m* de solution moléculaire
~SORBENT BAFFLE - [mech] piège *m* à sorption
~SORBENT TRAP - [mech] piège *m* à absorption, piège à sorption
~SPECTRUM - [phys] spectre *m* de molécule
~STOPPING POWER - [nucl] pouvoir *m* d'arrêt moléculaire
~STRUCTURE - [phys] (the linkage arrangement of the atoms forming a molecule) structure *f* de la molécule
~VACUUM PUMP - [mech] pompe *f* moléculaire
~VELOCITY - [phys] (the velocity of a molecule caused by random thermal motion) vitesse *f* moléculaire
~VIBRATION - [phys] vibration *f* moléculaire
~VOLUME - [phys] (the volume occupied by one gram-molecule of a substance at its boiling-point under a pressure of 760 mm Hg.) volume *m* moléculaire
~WEIGHT - [phys] (the weight of a molecule in terms of that of an atom of oxygen taken as I6) poids *m* moléculaire
MOLECULE - [phys] (generally defined as the smallest particle of a substance which can exist alone and retain its chemical properties.) molécule *f*
MOLESKIN - [text] velour *m* de coton, molesquine *f*
MOLIMEN - [med] trouble *m* subjectif
MOLLESCENCE - [med] ramollissement *m*
MOLLETON - [text] (a type of heavy cotton fabric) molleton *m*
[print] (molleton used for dampening the litho rollers) feutre *m* à mouiller, molleton *m*
MOLTEN - [metall] fondu
~BISMUTH COOLING - [nucl] (the use of bismuth as a coolant in a nuclear reactor) refroidissement *m* par bismut fondu
~SALT REACTOR - [nucl] réacteur *m* à sels fondus
~TEST SAMPLE - [metall] éprouvette *f* de métal liquide prélevée
MOLYBDATE ORANGE - [chem] (a pigment for plastics and paints) orangé *m* de molybdate
MOLYBDENITE - [min] (natural molybdenum disulphide, crystallizing in the hexagonal system. A major ore of molybdenum) molubdénite *f*
MOLYBDENUM - [chem] (a metallic element, symbol Mo, A.N. 42, A.W. 96, used in metallurgy and in the production of catalysts, lubricants, dyestuffs, ceramic glazes and analytical reagents) molybdène *m*
~DISULPHIDE - [chem] (a hydrogenation catalyst and a component of extreme pressure lubricants) bisulfure *m* de molybdène
~ORANGE - [chem] (a pigment for plastics) orangé *m* de molybdène
~PENTACHLORIDE - [chem] (a catalyst and intermediate) pentachlorure *m* de molybdène
~SESQUIOXIDE - [chem] (a catalyst) sesquioxyde *m* de molybdène
~SULPHIDE - [chem] (compound of sulphur and molybdenum, used as a lubricant) sulfure *m* de molybdène
~TRIOXIDE - [chem] (an analytical reagent, component of ceramic glazes and pigments and a catalyst) trioxyde *m* de molybdène
~WASHER - [mech] (in a reactor) rondelle *f* en molybdène
MOLYBDIC - [chem] (relating to, or containing, molybdenum) molybdique
MOLYBDITE - [min] (natural hydrous ferric molybdate, crystallizing in the orthorhombic system) molybdine *f*, molybdite *f*
MOLYSMOPHOBIA - [med] mysophobie *f*, rupophobie *f*
MOMENT - [gen] moment *m*
[mech] (the product of a quantity and a distance to a significant point connected with that quantity) moment *m*, couple *m* moteur
~ARM - [mech] bras *m* du couple
~AT FIXED END - [mech] moment *m* au point fixe
~COEFFICIENT - [phys] (aerodynamical factor) coefficient *m* de moment
~CURVE - [mech] diagramme *m* des moments
~OF A DIPOLE - [el] (the product of one of the quantities of electricity or magnetism and the vector passing from the second to the first quantity) moment *m* de dipôle
~OF COUPLE - [mech] (the product of the force and the arm of a couple) moment *m* de couple
~OF FLEXURE - [mech] moment *m* de flexion
~OF FORCE - [mech] moment *m* de force
~OF INERTIA - [mech] (a measure of the resistance offered by a body to angular acceleration) moment *m* d'inertie
~OF MOMENTUM - [mech] (ambular momentum) moment *m* cinétique
~OF RESISTANCE - [mech] moment *m* résistant
~OF ROTATION - [phys] moment *m* de rotation
~OF SPARKING - [mech] (internal combustion engines) point *m* d'allumage
~PLANIMETER - [instr] (integrometer; instrument designed to assess the moment of inertia about a specified line of the area enclosed by a curve) intègremètre *m*
MOMENTARY - [gen] momentané, passager
~CONTACT - [electron] contact *m* de passage
~VALUE - [el] valeur *f* instantanée
MOMENTUM - [mech] (the impetus of a mobing body) moment *m*, force *f* d'impulsion
[phys] (the quantity of motion in a body resulting by the product of its mass by its velocity) quantité *f* de mouvement, force *f* vive
~PUMP - [mech] pompe *f* intermittente
MONADIC OPERATION - [math] (operation on one operand) opération *f* monadique
MONARTHRITIS - [med] monarthrite *f*
MONASTER - [anat] plaque *f* équatoriale, monastère *m*
MONATHETOSIS - [med] athétose *f* unilatérale
MONATOMIC - [phys] (possessing only one atom) monatomique
[chem] (univalent) monovalent, univalent
~GAS - [chem] (a gas in which the molecule contains only one atom) gaz *m* monatomique
MONAURAL RECEPTION - [telecomm] (reception by listening with only one ear) captation *f* monoauticulaire
MONAZITE - [min] (an ore of thorium and other rare earth metals) monazite *f*
~SAND - [min] (phosphate of cerite earth) sable *m* monazitique
MOND GAS - [chem] (gas produced by passing a measured flow of air through red-hot coal, forming a mixture of carbon monoxide, nitrogen and some carbon dioxide) gaz *m* de Mond
~PROCESS - [chem] (method of extracting nickel, ba-

sed on the formation of nickel carbonyl) procédé *m* Mond
MONEL METAL - [metall] (an alloy of nickel and copper) métal *m* monel
MONERGOL - [astronaut] (monopropellant) monergol *m*
MONEY-WORT - [bot] (trailing herb of the primrose family) monnayère *f*, herbe *f* aux écus
MONGOLIAN SPOT - [med] tache *f* mongolienne
MONGOLISM - [med] (form of arrested physical and mental development) mongolisme *m*
MONGOLOID - [gen] (resembling a mongol) mongoloïde *m*
[med] (suffering from mongolism) mongoloïde *m*
MONGOOSE - [zool] (small ferret-like animal which attacks venomous serpents) mangouste *f*
MONGREL - [zool] (the progeny of crossed breeding) métis *m*, bâtard *m*
MONILIASIS - [med] moniliase *f*, candidiase *f*
MONILIOSIS - s. moniliasis
MONISM - s. monogenesis
MONISTIC - s. monogenetic
MONITOR, to - [gen] contrôler
[radio] (to listen in on a radio broadcast with a view to controlling the quality etc) contrôler
[telecomm] (to listen in) se porter en écoute
[nucl] (the operation of determining the amount of ionizing radiation or radioactive contamination) contrôler
MONITOR - [el] (a checking device) moniteur *m*
[telecomm] (arrangement used to check a transmission without interfering with the transmission itself) appareil *m* de surveillance, appareil *m* de contrôle
[electron] (a cathode-ray tube with associated circuit, used to view the picture as scanned by the camera tube) moniteur *m*
[nucl] (instrument designed to measure a condition which must be kept within given limited) moniteur *m*
[el acoust] (a high fidelity loudspeaker in the control room of a studio, to ensure adequate transmission) moniteur *m*
[hydr] (arrangement, consisting of a nozzle and a holder, to change easily the direction of a stream) lance *f* hydraulique, moniteur *m*
[mech] (a nozzle which can be swung horizontally) lance *f*
[metall] (an ironclad furnace for mercury ore) four *m* pour minerai de mercure
[mach tool] revolver *m*, tourelle *f*
~A CIRCUIT, to - [telecomm] se mettre en surveillance sur un circuit
~A LINE, to - s. monitor a circuit, to
~CASING - [sugar ind] (of the turbine) cuve *f*, carter *m*
~CHECKING ROUTINE - [comput] programme *m* d'analyse
~EARPHONE - [cin] (the ear-phone which is used for monitoring) écouteur *m* de contrôle
~HEAD - [telev] (in video recording) tête *f* de reproduction de contrôle
~JONIZATION CHAMBER - [nucl] chambre *f* d'ionisation de contrôle
~MAN - [radio & cin] (a sound engineer) ingénieur *m* du son
~PRINTER - [comput] imprimante *f* de contrôle
~PROGRAMME - [comput] moniteur *m*
MONITOR RECEIVER - [radio] récepteur *m* de contrôle, récepteur *m* moniteur
~ROOF - [arch] lanterneau *m*
~ROOM - [el acoust] (the room where the recorded sound is tested for quality) cabine *f* d'écoute, cabine *f* d'enregistrement
~SPEAKER - s. monitoring loudspeaker
~TOP - s. monitor roof
~TUBE - [electron] tube *m* controleur d'images
MONITORED CONTROL SYSTEM - [contr] système *m* de réglage à boucle fermée
~MANUAL CONTROL SYSTEM - [contr] système *m* de réglage volontaire surveillé
MONITORING - [contr] (supervision of a process or operation by means of instruments) contrôle *m*
[nucl] (the periodic determination of the radio-active contamination in a region or a person) contrôle *m* de l'activité
[radio & cin] (the technique of controlling the transmission of picture or sound) commande *f*, réglage *m*, contrôle *m*
~AERIAL - [radio] (aerial picking up radio frequency output for checking the output of a radar system) antenne *f* du moniteur de reception
~AMPLIFIER - [radio] amplificateur *m* de contrôle
~ANTENNA - s. monitoring aerial
~DEVICE - [telecomm] dispositif *m* d'écoute
~ELEMENT - [contr] (resetter) élément *m* de contrôle
~FEEDBACK - [contr] réaction *f* de contrôle
~FILMING - [telev] filmage *m* d'images du monitor
~KEY - [telecomm] clé *f* de surveillance, clé *f* de écoute silencieuse
~LOUDSPEAKER - [radio & cin] (control loudspeaker in the recording room) haut-parleur *m* de cabine, haut-parleur *m* témoin
~PICTURE - [telev] image *f* de contrôle
~POINT - [contr] point *fm* de mesure
~SERVICE - [radio] (the listening in on radio broadcasts for the purpose of preparing a summary or precis of relevant news items therein) service *m* d'écoute
~TUBE - [telev] (a cathode-ray tube on the picture control desk, representing the vision signal as a function of time) tube *m* contrôleur (d'images), tube *m* moniteur
MONKEY - [zool] singe *m*
[constr] (the falling weight of a pile driver) mouton *m*
[metall] trou *m* à laitier, trou *m* de la scorie
[min] petite galerie *f* d'aérage
[glass man] creuset *m*
~-BLOCK - [mech] retour *m* de palan
~BOLT - [mech] verrou *m* à onglet, verrou *m* à bascule
~CARRIAGE - [constr] chariot *m* de roulement
~-CHATTER - s. monkey-talk
~COOLER - [metall] refroidisseur *m* du trou de laitier
~DRIFT - [mining] galerie *f* de prospection
~ENGINE - [mech] (the engine operating the pile driver) moteur *m* pour le mouton de battage
~GRASS - [text] (a brazilian fibre) piazzava *f* brésilienne
~HAMMER - [mech] (drop hammer) mouton *m* de battage
~HAND - [med] main *f* de singe
~LINK - [auto] (of the wheel chain) anneau *m* de fixation de la chaîne antidérapante

MONKEY PAW - s. monkey hand
~POT - [glass man] (small pot used on top of another pot) alcarazas *m*
[bot] marmite *f* de singe
~PRESS - s. monkey hammer
~PUZZLE - [bot] araucaria *m*
~SPANNER - s. monkey wrench
~TAIL BOLT - [mech] loqueteau *m* à pompe, loqueteau *m* à paillette
~-TALK - [radio] (peculiar sound in radio reception resulting from cross-modulation between from cross-modulation between sidebands of a strong adjacent channel and the carrier of the incoming signal) interférence *f* des bandes latérales
~WAY - [min] (small ventilation shaft in a coal mine) petite galerie *f* d'aérage
~WRENCH - [impl] (term sometimes used for adjustable spanners in general, but properly applicable to an obsolete sliding-jaw type) clef *f* à molette, clef *f* à mâchoires mobiles, clé *f* anglaise
MONOACCELERATION CATHODE-RAY TUBE - [electron] tube *m* cathodique à préaccélération
MONOACETINE - [chem] (a solvent for basic dyes) monoacétine *f*
MONOANODIC - [el] monoanodique
MONOAXAL - [phys] (of crystals) monoaxifère
MONOBASAL - [cryst] monobasal
MONOBASIC - [chem] (an acid with one displaceable hydrogen atom) monobasique
~ACID - [chem] (an acid having one displaceable hydrogen atom in the molecule) acide *m* monobasique
~AMMONIUM PHOSPHATE - [chem] (fertilizer and fire-proofing agent) phosphate *m* monobasique de ammonium
~CALCIUM PHOSPHATE - [chem] (dietary supplement, plastics stabilizer and component of agricultural chemicals) phosphate *m* calcique
~MAGNESIUM PHOSPHATE - [chem] (a plastics stabilizer) phosphate *m* monobasique de magnésium
~POTASSIUM PHOSPHATE - [chem] (used in medicine as a mild aperient) phosphate *m* monobasique de potassium
MONOBLOC - [mech] (the integral casting of all the cylinders of an engine) monobloc *m*
~PUMP - [mech] pompe *f* monobloc
MONOCARP - [bot] plante *f* monocarpienne
MONOCARPIC - [bot] (which bears fruit only once in its existence) monocarpien, monocarpique
MONOCELLED - [biol] unicellulaire
MONOCHLORIDE - [chem] (chloride containing one chlorine atom in each molecule) monochlorure *m*
MONOCHORD - [instr] (acoustical instrument with one string and a movable bridge used to measure intervals) monocorde *m*, sonomètre *m*
MONOCHROMATIC - [gen & phys] (of one colour) monochromatique
~FILTER - [photo] filtre *m* monochromatique
~ILLUMINATOR - [instr] (instrument which supplies a beam of light for a definite purpose) source *f* lumineuse monochromatique
~LIGHT - [opt] (light consisting of only one wavelength) lumière *f* monochromatique
~RADIATION - [phys] (electromagnetic radiation of a single wavelength in which all photons have the same energy) rayonnement *m* monochromatique
~X-RAYS - [radiat] (homogeneous X-rays) rayons *mpl* X homogènes
MONOCHROMATOR - s. monochromatic illuminator
MONOCHROME - [photo] (photographic print in one colour or varying brightness) monochrome *m*
~BANDWIDTH - [telev] largeur *f* de bande du signal monochrome
~FILTER - [photo] écran *m* monochromatique
~LIGHT - s. monochromatic light
~TRANSMISSION - [telev] (transmission of signal wave representing brightness values, but not the chromaticity values) transmission *f* en noir et blanc
MONOCLINAL - [geol] monoclinal
~VALLEY - [geol] vallée *f* monoclinale, combe *f*
MONOCLINIC - [phys] (a crystal system in which the crystals axes are of different lengths) monoclinique
MONOCOQUE - s. monocoque structure
~STRUCTURE - [aero] (a structure in which all loads are taken by the outer shell) structure *f* monocoque
MONOCRYSTALS - [phys] monoscristaux *mpl*
MONOCULAR - [opt] (one-eyed) monoculaire
~LENS - [photo] lentille *f* monoculaire
~VISION - [photo] (the viewing of the scene to be photographed by one eye only) vue *f* monoculaire
MONOCYCLIC - [bot] (of annual plant) monocyclique
MONOCYTE - [biol] (large, white blood corpuscle) monocyte *m*
MONOCYTOPENIA - [med] monocytopénie *f*
MONOCYTOSIS - [med] monocytose *f*
MONODIPLOPIA - [med] diplopie *f* monoculaire
MONODISPERSE SYSTEM - [phys] (of colloidal particles) système *m* monodisperse
MONODROMIC - [math] (théorème) monodrome
~FUNCTION - [math] fonction *f* monodrome
MONODY - [mus] monodie *f*
MONOECISM - [genet] (with male and femal organs in the same individual) monoécie *f*
MONOENERGETIC - [nucl] (said of particles having the same energy) monoénergétique
MONOFACTORIAL - [genet] monofactoriel
MONOFIL - [plast ind] (contraction of monofilament) monofilament *m*
MONOFILAMENT - [plast ind] (thread consisting of a single filament) monofilament *m*
~NETTING - [plast ind] (netting made by locally welding filaments at the time of extrusion, so as to form various types of net) toile *f* en monofilament
MONOGENETIC - [genet] (deriving from a single cell) monogénique
[geol] (resulting from one genetic process) monogénique
[zool] (multiplying by asexual reproduction) à reproduction asexuée
[chem] (of a dyestuff; producing only one colour on textile fabrics) monogénique
MONOGERM SEED - [agric] semences *fpl* monogerminales
MONOGONY - [genet] monogenèse *f*, reproduction *f* asexuée
MONOGRAPH - [gen] (a dissertation, a treatise on a given subject) monographie *f*
MONOHYDRATE - [chem] (the union of a single molecule of water with an element, or compound) monohydrate *m*
MONOHYDRIC ALCOHOL - [chem] (alcohol containing one hydroxyl group only) alcool *m* monohydrique
MONOLAYER - [phys] (a molecular film) couche *f* monomoléculaire

MONOLITH - [arch & constr] (a single block of stone) monolithe *m*
MONOLITHIC - [arch etc] monolithe
~CIRCUIT - [comput] (circuit consisting of a single block) unité *f* de couplage monolithique
~LINING - [metall] granissage *m* monolithe
~LINING MATERIAL - [metall] pisé *m*
MONOLOCULAR - [med] uniloculaire
MONOMANIA - [med] monomanie *f*, délire *m* partiel
MONOMER - [chem] (a simple organic molecule, usually containing carbon and capable of addition polymerization) monomère
MONOMETALLIC - [chem] (consisting of a single metal) monométallique
MONOMETRIC - [cryst] (isometric) isométrique
MONOMOLECULAR LAYER - [chem] (a layer of a substance which is only one molecule thick) couche *f* monomoléculaire
~REACTION - [chem] (reaction with a speed which is proportional to the concentration of one reactant only) réaction *f* monomoléculaire
MONOMORPH - [cryst] substance *f* monomorphe
MONOMORPHOUS - [phys] (term used for a substance which exists in only one crystalline form) monomorphe
MONOMYOSITIS - [med] myosite *f* localisée
MONOPENIA - s. monocytopenia
MONOPHOBIA - [med] monophobie *f*
MONOPHONY - [mus] (with one line of notes) monophonie *f*
MONOPIA - [med] monophthalmie *f*, cyclopie *f*
MONOPLANE - [aero] (an aeroplane having a single main supporting surface, i.e. one pair of wings) monoplan *m*
MONOPLEGIA - [med] monoplégie *f*
MONOPOD - [agric] siccateur *m* monopied
MONOPOLE - [nucl] (a particle with an isolated magnetic charge) monopôle *m*
~MASS FILTER - [opt] spectromètre *m* monopole
~MASS SPECTROMETER - [instr] spectromètre *m* monopole
MONOPOLIZE, to - [comm etc] (to assure a monopoly or to take up complete control) monopoliser
MONOPOLY - [gen & comm] (the exclusive right of dealing with a particular trade) monopole *m*
MONOPROPELLANT - s. monergol
[chem] (used for rocket motors) monergol *m*
~ROCKET - [astronaut] fusée *f* à monergol
MONORAIL - [railw] (railway system in which carriages are suspended from, and travel along, a single suspended rail) monorail *m*
(the actual track for carriages) monorail *m*
MONORAILWAY - [railw] chemin *m* de fer monorail
MONOSACCHARIDES - [chem] (simple sugars which cannot be decomposed by hydrolysis) monosaccharides *mpl*
MONOSACCHAROSES - [chem] (the simplest of the carbohydrates hydrates, containing from three to six carbon atoms: represented by aldehyde alcohols or ketone alcohols) monosaccharoses *mpl*
MONOSCOPE - [telev] (cathode-ray tube containing a stationary pattern which supplied an electrical signal for the testing and adjusting of the equipment) monoscope *m*
MONOSES - [chem] (synonym for the monosaccharoses) monoses *mpl*
MONOSILICATE - [chem] monosilicate *m*
MONOSTABLE CIRCUIT - [electron] circuit *m* monostable
~FLIP FLOP - [electron] basculeur *m* monostable
~MULTIVIBRATOR - [telev] (flip-flop generator) basculateur *m* monostable
~TRIGGER ELEMENT - [contr] basculeur *m* monostable
MONOSTATIC REFLECTIVITY - [phys] réflectance *f* monostatique
MONOTONIC - [math] (of a variable function which does not increase or decrease during a given interval) monotone
~FUNCTION - [math] fonction *f* monotone
MONOTRON - s. monoscope
MONOTROPIC - [phys] (term used of a substance which exhibits only one crystal form and is unstable under all conditions in other forms) monotropique
~ALLOTROPY - [chem] (allotropic phenomena in which the transition is irreversible) allotropie *f* monotropique
MONOTROPY - [phys] (of a substance having only one crystal form) monotropie *f*
MONOTYPE METAL - [metall] métal *m* monotype
MONOVALENT - [chem] (capable of combination with one atom of hydrogen or the equivalent) monovalent, univalent
MONOVARIANT - [chem] (or univariant; having one degree of freedom) monovariant
MONOXIDE - [chem] (compound with a single atom of oxygen) protoxyde *m*
MONSOON - [met] (seasonal winds, especially in and around the Indian ocean) mousson *m*
MONTABRASITE - [min] (a rare mineral; a fluophosphate of aluminium and lithium with an increase of hydroxyl) montebrasite *f*
MONTAGE - [cin & telev] (the rapid revolving of various images round a central image, to convey more than the actual material could, e.g. the passage of time etc) montage *m*
[radio] (the bridging of speeches by a unifying sound effect) liaison *f* sonore
[photo] s. montage photograph
~PHOTOGRAPH - [photo] (composite photograph, or the super-imposition of several pictures, so as the blend them into one another) photomontage *m*
MONTAN WAX - [chem] (a hard wax obtained from lignite and used as a substitute for beeswax) cire *f* de lignite
MONTANT - [carp] (vertical member in a panel or frame) montant *m*
MONTEBRASITE - s. montabrasite
MONTE-CARLO METHOD - [phys] (a method of solving physical problems by a series of statistical experiments performed by applying mathematical operations to random numbers) méthode *f* Monte Carlo
MONTEJUS - [sugar ind] (juice lifter) monte-jus *m*
MONTGOLFIER - [aero] (primitive type of balloon raised by heated air) montgolfière *f*
MONTH - [astr] (unit of time) mois *m*
~OF THE PHASES - [astr] mois *m* synodique
MONTHLY - [gen] mensuel
[print] (a monthly publication) revue *f* mensuelle
MONTICELLITE - [min] (a silicate of calcium and magnesium; a rare mineral) monticellite *f*
MONTMORILLONITE - [min] (an aluminium silicate present in bentonite and in Fuller's earth and used as an extender for powerds and as a carrier) montmorillonite *f*

MONUMENT - [arch] (statue or important structure) monument *m*
MONZONITE - [geol] (coarse-grained igneous rock containing some coloured silicates) monzonite *f*
MOOD - [med] disposition *f*
[mus] mode *f*
MOOING - [acoust] (the sound made by cows) mugissement *m*
MOON - [astr] (the earth's satellite) lune *f*
[astr] (in general, the satellite of any planet) lune *f*
~BLINDNESS - [opt] (nyctalopia) amblyopie *f* lunaire
[vet] oeil *m* lunatique (de cheval)
~PROBE - [astr] (unmanned capsule containing instruments designed to send information on the moon back to earth) sonde *f* lunaire
~ROOM - [astr] (experimental astronautical contrivance) chambre *f* lunaire
~-SHAPED FACE - [med] (or mooning) faciès *f* lunaire
MOONBLINK - s. moon blindness
MOONEY VISCOSITY - [chem] (viscosity measured by the Mooney system) viscosité *f* de Mooney
MOONRAKER - [naut] (sail above the sky-sail) papillon *m*
MOONSAIL - s. moonraker
MOONSTONE - [min] (variety of adularia with a pearly opalescence) pierre *f* de lune, adulaire *f*
MOOR, to - [naut] (to make fast a floating object, especially a seaplane or boat) amarrer, affourcher, mouiller
MOOR - [geogr] (tract of wasteground, a heath) bruyère *f*, lande *f*, terrain *m* marécageux
~BATH - [med] bain *m* de boue
MOORAGE - [naut] (mooring place) amarrage *m*, affourchage *m*
[leg] (payment for mooring) droits *mpl* d'attache
MOORING - [naut & aero] amarrage *m*, affourchage *m*
~ANCHOR - [naut] ancre *f* d'amarrage
~BITT - [naut] (post on a ship's deck for the mooring rope) bitte *f* d'amarrage
~-BUOY - [naut] (buoyant structure designed to provide a fixed object to which floating bodies can be moored) bouée *f* d'amarrage
~CONE - [aero] (on an airship) cône *m* d'amarrage
~GUY - s. mooring rope
~MAST - [aero] (small tower to which an airship is secured) mât *m* d'amarre
~PILE - [naut] pieu *m* d'amarrage
~PIPE - [naut] (a hawsepipe) écubier *m*
~POINT - [aero] (the section of the airship which is strengthened for mooring purposes) point *m* d'alarre
~RING - [gen] anneau *m* d'amarrage
~ROPE - [naut] amarre *f*, câbleau *m*
~SWIVEL - [naut] émerillon *m* d'affourche
~TOWER - [aero] (permanent tower for the mooring of airships) mât *m* d'amarre
MOOSE - [zool] élan *m*, orignac *m*
MOP, to - [gen] éponger, essuyer
MOP - [impl] balai *m* à laver
[impl] disque *m* en lisière de tissu, meule *f* flexible en chiffon
~-BOARD - [carp] (a skirting board) plinthe *f*
~-END BRUSH - [impl] brosse *f* dite poignon
~UP, to - [gen] éponger, rafler
~-UP WAVEFORM CORRECTOR - [electron] postcorrection *f* de la forme d'onde
MOPED - [mech] (a motor-assisted bicycle) cyclomoteur *m*
MOQUETTE - [text] (woollen fabric with a velvety pile, mainly used for upholstery and carpets) moquette *f*
~WITH ROUGH PILE - [text] moquette *f* frisée
MORAINE - [geol] (rock waste at the margins or base of a glacier) moraine *f*
~BELT - [geol] ceinture *f* morainique
~LODGE - [geol] loge *f* de moraine
MORAINIC - [geol] morainique
MORASS - [gen] (tract of soft, wet ground; a marsh) marais *m*, morasse *f*, fondrière *f*
~ORE - [min] minerai *m* de fer des marais, fer *m* des marais
MORBID - [med] (of a diseased, or abnormal state) morbide
[med] (relative to a disease, pathological) pathologique
~ANATOMY - [med] (the anatomy of diseased organs) anatomie *f* pathologique
MORBIDITY - [med] morbidité *f*, état *m* maladif
MORBIFIC - [med] (which caused disease) morbifique
MORCELLATION - [med] morcellement *m*
MORCELLEMENT - s. morcellation
MORDANT, to - [ind chem] (to treat with a mordant) mordancer
MORDANT - [chem] (a substance which unites with a dyestuffs and which the textile fibre to form an insoluble coloured compound) mordant *m*, (acide) caustique
~PRINT - [text] impression *f* au mordant
~TONING - [photo] virage *m* par mordançage
MORDANTING - [photo] (the addition to a silver image of a substance having the required affinity for a given dye) mordançage *m*
~BATH - [text] bain *m* de mordançage
~BY STEEPING - [text] mordant *m* adoucissant
~DYES - [dye] colorants *mpl* sur mordant
MOREEN - [text] moreen *f*
MORELLO CHERRY - [bot] (a variety of cherry) griotte *f*
MORGANITE - [min] (a pink variety of beryl; used as a gemstone) morganite *f*
MORGEN - [meas] 2,II acres (mesure hollandaise dans l'Afrique du Sud
MORIN - [chem] (an analytical reagent and dyeing mordant) morin *m*
MORION - [min] (an almost black variety of smoky quartz) morion *m*
MORLING - [text] laine *f* morte, moraine *f*
MORNING GLORY SHAFT SPILLWAY - [hydr] évacuateur *m* de crues en forme de puits
~STAR - [astr] (popular name for Venus or Mercury, seen on the eastern sky at about sunrise) étoile *f* du matin, Vesper *f*
MOROCCO - [leather ind] (fine leather made of goatskin) maroquin *m*
~GOAT-SKIN - s. morocco
~LEATHER - s. morocco
MORON - [med] faible d'esprit
MORONISM - [med] débilité *f* mentale
MORONITY - s. moronism
MOROSIS - s. moronism
MORPHEA - [med] (or morphoea; a localized hardness and rigidity of the skin due to an overgrowth of

fibrous tissues) morphée *f*
MORPHINE - [chem] (an alkaloid derived from opium and important as an analgesic in medicine) morphine *f*
MORPHINISM - [med] (the condition of the system which is caused by excess or habitual use of morphine) morphinisme *m*
MORPHINOMANIA - [med] morphinomanie *f*
MORPHOEA - s. morphea
MORPHOGENESIS - [biol] (the evolution of organs or organisms) morphogénèse
MORPHOLINE - [chem] (a solvent for resins and dyestuffs and a synthesis intermediate) morpholine *f*
MORPHOLOGY - [biol] (the study of the form and structure of organisms) morphologie *f*
[geol] (the study of the shape of objects, in particular of the earth's surface) morphologie *f*
MORPHOSIS - [biol] (the development of structural characteristics) morphose *f*
MORPHOTROPY - [cryst] (the change in the ratio in the length of the axes due to changes in molecular structure) morphotropie *f*
MORSE ALPHABET - s. morse code
~CODE - [telecomm] (system of telegraphic signals consisting of dots and dashes, or long and short flashes) code *m* Morse
~DASH - [telecomm] (signal element of the Morse code) trait *m* Morse
~DOT - [telecomm] (signal element of the Morse code) point *m* Morse
~KEY - [telecomm] manipulateur *m* Morse
~PRINTER - [telecomm] (apparatus scanning the recorded signals on perforated tape and prints corresponding characters) imprimeur *m* Morse
~RECEIVER - [telecomm] récepteur *m* Morse
~SPACE - [telecomm] espace *m*
~TELEGRAPHY - [telecomm] télégraphie *f* Morse
MORTALITY - [gen] mortalité *f*
(the rate of deaths in a given number of human beings) mortalité *f*
MORTAR - [firearms] (a type of gun) mortier *m*
[constr] (pasty substance gradually hardening on exposure) mortier *m*, enduit *m*
[ind chem] (bowl-like vessel for crushing small samples of material) mortier *m* (pour piler)
~BED - [constr]fondation *f* du mortier
~BOARD - [constr] (small square board with handles underneath used to carry mortar) taloche *f*, planche *f* à mortier
~BOND - [constr] appareil *m* avec mortier
~BOX - [impl] (the box in which the mortar is mixed) mortier *m* (de bocard)
~MILL - [constr] malaxeur *m* de mortier *m*, broyeur *m* à auge tournante
~MIXER - [constr] malaxeur *m* à mortier, tonneau *m* à mortier
~MIXING - [constr] gâchage *m* du mortier, malaxage *m* du mortier
~MIXING MACHINE - s. mortar mixer
~OF CEMENT - [constr] mortier *m* de ciment
~TUB - [constr] cuve *f* à mortier
MORTGAGE, to - [leg] (to make over a property to a creditor on condition that once the debt is paid the property is returned) hypothéquer
MORTGAGE - [leg] (the conveyance of a property from a debtor to a creditor) hypothèque *f*
~BOND - [leg] titre *m* hypothécaire
MORTGAGE LOAN - [leg] prêt *m* hypothécaire
MORTGAGEE - [leg] créancier *m* hypothécaire
MORTICE - s. mortise
MORTISE, to - [carp] (the operation of joining by a tenon and mortise) mortaiser, emmortaiser
MORTISE - [carp] (space, generally rectangular, hollowed out in a framework to receive the projection of a mating member) mortaise *f*
~AND TENON - [carp] (assemblage) à tenon et mortaise
~BLOCK - [carp] poulie *f* à mortaise
~CHISEL - [impl] (strong type of chisel used for cutting mortises) bédane *f*, ciseau *m* bédane, ciseau *m* à larder
~GAUGE - [carp] (marking gauge with an additional adjustable marking pin) trusquin *m* à double traçoir
~JOINT - [carp] (framing joint between a mortise and a tenon) assemblage à tenon et à mortaise
~LOCK - [mech] serrure *f* à larder, serrure *f* à mortaiser
~WHEEL - [carp] (gear with mortised wooden teeth) rouet *m*
MORTISED SLAY SWORD - [text] épée *f* de battant à mortaise
MORTISER - [mech] (machine designed to cut square or rectangular holes in wood) mortaiseuse *f*
MORTISING - [carp] joint *m* à tenon et mortaise
(the operation of cutting a mortise in wood) mortisage *m*
~MACHINE - s. mortiser
~STROKE - [mach tool] (the travel of the mortise cutting tool) course *f* de mortisage
MORTON WAVE CURRENT - [el biol] (obsolete treatment by interrupted current) courant *m* d'onde de Morton
MORULA - [med] morula *f*
MOSAIC - [gen] (inlaid work on plaster or stone) mosaïque *f*
[bot] s. mosaic disease
[telev] s. mosaic electrode
[photo] (combination of photographs taken from the air) mosaïque *f*
~COAT - [electron] couche *f* en mosaïque
~DISEASE - [bot] (virus disease causing mottling of the leaves) mosaïque *f*
~ELECTRODE - [telev] (electrode coated with photoelectric granules which are insulated from each other) mosaïque *f*
~FILTER - [photo] filtre *m* mosaïque
~GOLD - [chem] (a complex stannic sulphide also used as a pigment) or *m* mussif
~LAYER - [photo] couche *f* mosaïque
~PLATE - s. mosaic electrode
~SCREEN - [photo] (screen used for separating the component colours in colour photography) réseau *m* mosaïque, écran *m* en mosaïque
~STRUCTURE - [cryst] (the mosaic of blocks forming a single crystal, as shown by X-ray analysis) structure *f* en mosaïque
~TELEGRAPHY - [telecomm] (system of alphabetic telegraphy in which the characters are formed as mosaics formed by units transmitted as individual signal elements) télégraphie *f* par décomposition des signes, télégraphie *f* à mosaïque
~VISION - [opt] (the mode of vision of a compound eye) vision *f* à mosaïque
MOSELEY'S LAW - [nucl] (law of the relationbetween

the atomic number of an element and the frequency of its characteristic radiation) loi *f* de Moseley
MOSQUITO - [zool] (an insect) moustique *m*
MOSS - [bot] (delicate hydrophytic plant growing on the ground, trees, rocks etc) mousse *f*
[soil] marécage *m*, tourbière *f*
~AGATE - [min] agate *f* mousseuse
~RUBBER - [rubber ind] (micro-cellular rubber) caoutchouc *m* alvéolaire
MOST ECONOMICAL RANGE - [aero] (the range of an aircraft when flown so as to give maximum economy under the given conditions) distance *f* franchissable économique
~FAVOURABLE ROUTE - [railw & transp] (the cheapest route) itinéraire *m* économique
~PROBABLE MOLECULAR VELOCITY - [phys] vitesse *f* moléculaire la plus probable
MOSSY ZINC - [min] grenaille *f* de zinc
MOTE - [phys] (a minute particle) atome *m* de poussière
[text] picot *m*
[text] s. motes
~KNIFE - [text] couteau *m* détacheur
MOTES - [text] (or moits q.v.) matières *fpl* végétales, noeuds *mpl* de laine
MOTET - [mus] (a short vocal composition) motet *m*
MOTH - [zool] mite *f*, lépidoptère *m*
~BALL - [ind chem] boule *f* de naphtaline
MOTHER - [gen] mère *f*
[el acoust] (in record production, the copper electroplate positive made from the master) mère *f*
~BLANK - [el chem] (starting sheet blank, a sheet of metal on which the metal to be refined is deposited and from which is subsequently stripped)plaque-support *f* de feuille de départ
~CELL - [biol] (a cell which divides to produce daughter-cells) cellule *f* mère
~CRYSTAL - [cryst] (the raw material from which piezo-electric devices are made) cristal *m* naturel
~-GATE - [min] (in a coal mine) mère-galerie *f*, galerie *f* principale
~HUBBARD - [oil ind] trépan *m* type normal
~LIQUOR - [chem] (the solution from which a substance has been crystallized out) eau *f* mère, liqueur-mère *f*
[sugar ind] sirop-mère *m*
~LODE - [min] (the principal lode) filon *m* mère
~LYE - [chem] eau *f* mère
~METAL - [metall] métal *m* mère, métal-mère *m*
~OF COAL - [min] charbon *m* fossile, fusain *m*
~-OF-PEARL - [zool] (the iridescent calcareous layer of certain shells) nacre *f*
~-OF-PEARL PAPER - [paper man] papier *m* nacré
~OIL - [oil ind] pétrole *m* brut
~PLANT - [bot] plante-mère *f*
~ROCK - [geol] roche *f* mère
~TREE - [bot] (the parent tree) arbre *m* porte-graine
~WATER - [geol] eaux *fpl* mères
MOTILE - [phys & bot] (able to move) doué de mouvement
MOTILITY - [gen] (capacity for movement in a micro-organism) motilité *f*, mobilité *f*
MOTION - [phys] (a change of position in reference to an assumed point of a material particle) mouvement *m*, déplacement *m*
[gen] mouvement *m*
[mech] (the action or interaction of parts in any mechanism) mécanisme *m*
MOTION - [cin] (the action of the camera) mouvement *m*
[leg] (application to a Court of Justice to obtain an order) motion *f*
[mus] (a melodic progression) mouvement *m*
[med] évacuation *f*
~BAR - [mech] (of an engine crosshead) guide *m*, règle *f*
~DIRECTION OF THE ELECTRONS - [phys] direction *f* de déplacement des électrons
~OF REVOLUTION - [astr] mouvement *m* de révolution
~OF ROTATION - [phys] mouvement *m* de rotation
~OF TRANSLATION - [phys] mouvement *m* de translation
~PICTURE - [cin] (the sequence of pictures on a single film) projection *f* animée
~PICTURE CAMERA - [cin] (camera for moving pictures) caméra *f*, ciné-caméra *f*
~PICTURE EQUIPMENT - [cin] (the complete set of sound and image reproducing apparatus) installation *f* de cinéma
~PICTURE MECHANICS - [mech] (the mechanics of motion picture taking) mécanique *f* cinématographique
~PICTURE PROJECTOR - [cin] (projector used for the reproduction of films in a theatre etc) projecteur *m* de cinéma
~PICTURE STUDIO - [cin] studio *m* cinématographique
~PICTURE TECHNOLOGY - [cin] cinématographie *f*
~SICKNESS - [aero & astronaut] mal *m* des transport
~STUDY - [mech] (the study of the necessary movements of workers in performing a certain task) analyse *f* du mouvement
~UNSHARPNESS - [cin] (a fault which is limited to moving objects) flou *m* de mouvement
MOTIONAL ELECTROMOTIVE FORCE - [el] (the electromotive force which is induced in a circuit by the motion of the conductor across a magnetic field) force *f* électromotrice induite par un mouvement
~IMPEDANCE - [el acoust] (complex difference between the loaded impedance and the blocked impedance of an electromechanical transducer) impédance *f* cinétique
MOTIVE COLUMN - [mining] dépression *f* de la mine
~POWER - [phys & mech] (the power causing motion) force *f* motrice
~POWER UNIT - [railw] (tractive unit) véhicule *m* moteur, engin *m* de traction
[el] unité *f* motrice
MOTIVITY - [mech] (motive energy) énergie *f* cinétique
[biol] motricité *f*
MOTOR - [gen] (something imparting motion) moteur *m*
[adj] moteur, motrice
[mech] (machine imparting motion) moteur *m*
[el] (any device converting electrical into mechanical energy) moteur *m*
~AMBULANCE - [auto] ambulance *f* automobile
~AREAS - [zool] (the nerve centres of the brain connected with the correlation of movement) centres *mpl* moteurs
~ARMATURE - [el] induit *m*
~BASE - [auto] carter *m* moteur
~BICYCLE - [mech] cyclomoteur *m*, motocyclette *f*

MOTOR BLOCK - [auto] bloc-moteur *m*
~-BOATING - [telecomm] (self oscillation, generally of a pulse type, in an amplifier at low audio-frequency) chocs *mpl* de basse fréquence
~-BOGIE - [el] (the bogey on an electric locomotive or a motor-coach, carrying the electric motor or motors) bogie *m* moteur
~BRANCH CIRCUIT - [el] distribution *f* force
~CELL - [bot] (one of the expanding or contracting cells which cause movement in a plant element) cellule *f* motrice
~CIRCUIT - [el] marche *f* de moteur
~-CIRCUIT SWITCH - [el] commutateur *m* de marche de moteur
~-COACH - [transp] autocar *m*
[railw] (a passenger coach for use on electrified railways; fitted with own motors) rame *f* automotrice
~-COACH TRAIN - [railw] automoteur *m*, train *m* automoteur
~COMBINATION - [el] couplage *m* (des moteurs)
~CONNECTION - [el] couplage *m* du moteur électrique
~-CONTROLLED TAP - s. motorized valve
~-CONTROLLER - [contr] (a device for the adjustment of the speed of an electric motor) combinateur *m*, contrôleur *m*
~CONVERTER - [el] (form of convertor in which an induction motor is combined with a synchronous convertor connected with the d.c. circuit) convertisseur *m* en cascade
~CRANE TRUCK - [mech] autogrue *f*
~CUE - [cin] (marks with transparent outlines printed on a film) marque *f* de démarrage
~DRIVE - [el] commande *f* électrique
~DRIVE PUNCH - [comput] perforatrice *f* à moteur
~DRIVE VERIFIER - [comput] vérificatrice *f* à moteur
~DRIVEN - [el] actionné par moteur électrique
~-DRIVEN CONTROLLING ELEMENT - [contr] élément *m* final motorisé
~-DRIVEN PUMP - [mech] (pump operated by an electric motor) électropompe *f*
~-DRIVEN RELAY - [el] relais *m* type moteur
~-DRIVEN SWITCHGROUP - [el] combinateur *m* à moteur
~-DUSTER - [agric] poudreuse *f* à moteur
~-FIELD CONTROL - [el] réglage *m* du champ de moteur
~FOR RAPID SCANNING - [radar] (slewing motor in the USA; a motor at high speed used for scanning) moteur *m* pour analyse rapide
~FUEL - [gen] carburant *m*, essence *f*
~GENERATOR - [el] (convertor which consists of a motor connected to a supply and a generator providing output power to a system of a voltage, frequency, or number of phases different from those of the supply) groupe *m* moto-générateur
~-GENERATOR FOR WELDING - [el] (coupled rotating machines producing electric energy suitable for arc welding) groupe *m* convertisseur de soudage à l'arc
~-GENERATOR LOCOMOTIVE - [railw] (type of electric locomotive fitted with a motor generator) locomotive *f* à moto-générateur
~-GENERATOR SET - s. motor generator
~GRADER - [constr] (a power driven grader) niveleuse *f* à moteur
~GROUPING SWITCH GROUP - [el] combinateur *m* de couplage des moteurs
MOTOR HIGHWAY - [transp] autoroute *f*
~HOE - [agric] motobineuse *f*
~HOIST - [mech] benne *f* pour lever le moteur
~LAWN MOWER - [agric] tondeuse *f* à gazon à moteur
~METER - [instr] (integrating meter incorporating some form of motor) compteur-moteur *m*
~MOUNT - [mech] carter-moteur *m*
~MOWER - [agric] motofaucheuse *f*
~NERVE - [anat] voie *f* motrice, nerf *m* moteur
~OIL - [auto] huile *f* pour moteurs
~-OPERATED SWITCH - [el] (large circuit-breaker closed by means of an electric motor) interrupteur *m* à moteur
~-OPERATED VALVE - [mech] s. motorized valve
~OPERATOR - [contr] (part of the controlling means, applying power for operating the final control element) élément *m* final motorisé
~PLOUGH - [agric] charrue *f* à moteur
~POINTS - [railw] (power operated points) aiguille *f* à moteur
~PUMP - [mech] (pump operated by a motor) motopompe *f*
~RATING - [mech] puissance *f* nominale d'un moteur
~REDUCTION UNIT - [el] démultiplicateur *m* de vitesse
~ROLLER - [constr] rouleau *m* compresseur à moteur
~ROTATION - [mech] (term denoting the sense of rotation of a hydraulic hoist pump) sens *m* de rotation du moteur
~SHIP - [naut] (vessel propelled by power derived from oil burning) vaisseau *m* à moteurs
~SHOW - [auto] (exhibition of new models) salon *m* de l'automobile
~SPIRIT - [chem] (any fuel used for spark-ignition) essence *f* pour automobiles
~-SPRAYER - [agric] pulvérisateur *m* à moteur
~SPRINKLER- [mech] (a street flusher truck) arroseuse *f* automobile
~SPRINKLING MACHINE - s. motor sprinkler
~STARTER - [el] (device operating the necessary circuits for starting a motor and accelare it to full speed) démarreur *m* de moteur
~STROKE - [mech] temps *m* de moteur
~-THRESING MACHINE - [agric] batteuse *f* à moteur
~TORQUE - [mech] couple *m* moteur
~TRAIN SET - s. motor coach train
~TRAIN UNIT - [railw] élément *m* automoteur
~TRANSPORT - [transp] traction *f* automobile
~TRUCK - [auto] camion *m*
~-TYPE RELAY - [el] relais *m* type moteur
~UNDER LOAD - [mech] moteur *m* sous charge
~VEHICLE - [mech] (any vehicle driven by a motor or engine) engin *m* moteur, véhicule *m* à moteur
~VEHICLE WITH COUPLED AXLES - [mech] engin *m* moteur à essieux accouplés
~VEHICLES WITH INDEPENDENT AXLES - [mech] engin *m* moteur à essieux indépendants
~WASHER - [auto] appareil *m* de nettoyage du moteur
~WINCH - [mech] mototreuil *m*
~WITH COMMUTATING POLES - [el] moteur *m* à pôles de commutation
~WITH COMPOUND CHARACTERISTIC - [el] (a motor the speed of which decreases when the torque increases and the speed of which, at no load, is limited) moteur *m* à caractéristique compound
~WITH RECIPROCATING MOVEMENT - [el] moteur *m* à

mouvement alternatif

MOTOR WITH SERIES CHARACTERISTIC - [el] (a motor the speed of which decreases when the torque increases and the speed of which, at no load, is unlimited) moteur *m* à caractéristique série

~WITH SHUNT CHARACTERISTIC - [el] (a motor having a speed which is not affected by the torque) moteur *m* à caracteristique shunt

MOTORBOAT - [naut] (small boat propelled by an engine or a motor) canot *m* automobile

MOTORBUS - [auto] (autobus) autobus *m*

MOTORCARD - [auto] (self-propelled vehicle, almost universally driven by a multi-cylinder petrol engine) automobile *m*, voiture *f* automotrice
[railw] (a car driven by motors in the car itself) automotrice *f*

~ENGINE - [auto] moteur *m* d'automobile

MOTORCYCLE - [mech] motocyclette *f*, vélomoteur *m*

~WITH SIDECAR - [mech] motocyclette *f* à sidecar

MOTORDROME - [auto] (a racing track) autodrome *m*

MOTORED - s. motorized

MOTORING - [auto] automobilisme *m*
[gen] (travel) tourisme *m* en automobile

MOTORIST - [gen] automobiliste *m*

MOTORIZE, to - [mech] (to supply with motor or motors) motoriser

MOTORIZED - [mech] (supplied with, or driven by, motor) motorisé

~ADJUSTING SCREW - [mech] (adjusting screw in a roller machine, which is driven by an electric motor) vis *f* à réglage électrique

~KEYBOARD - [telecomm] (a keyboard in which the energy to move the combination bars derives from the motor driving the instrument of which the keyboard is a part) clavier *m* à moteur

~PAN AND TILT HEAD - [telev] (for reproducing technique) tourelle *f* universelle motorisée

~TAP - s. motorized valve

~VALVE - [mech] (valve operated by motor) robinet *m* motorisé

MOTOSCOOTER - [mech] scooter *m*

MOTT SCATTERING FORMULA - [nucl] (the formula which determines the differential cross-section of identical particles caused by a Coulomb interaction) formule *f* de diffusion de Mott

MOTTLE, to - [text] marbrer

MOTTLE - [gen] (discoloration or marking of an uneven nature) tache *f*, tacheture *f*

MOTTLED - [gen] truité, diapré, marbré

~ENAMEL - [dent] fluorose *f* dentaire

~IRON - [metall] fonte *f* truité

~MAHOGANY - [timber] acajou *m* madré

~PIG IRON - [metall] fonte *f* brute, demi-blanche

~STRUCTURE - [min] structure *f* mouchetée

~TAPESTRY CARPET - [text] moquette *f* bouclée à chaîne jaspée

~TEETH - [dent] fluorose *f* dentaire

~WOOD - [constr] bois *m* flammé

~YARN - [text] fil *m* marbré, fil *m* zébré

MOTTLING - [paint] (defect in a sprayed coating, consisting in a pattern of circular marks) marbrure *f*, diaprure *f*, tiqueture *f*
[metall] tiqueture *f*
[dent] (of teeth) réticulation *f*
[text] chinage *m*, chiné *m*

MOULD, to - [gen] (to shape, to model) mouler, façonner

MOULD, to - [chem] (to be affected by a fungous growth) se moisir, moisir
[arch] (to shape an ornamental band projecting from a wall) mouler
[rubber & plast ind] (to form by means of a heated die) mouler
[metall] (in foundry work) mouler
[shipbuild] gabarier
[print] prendre l'empreinte (d'une page)
[food] (the dough) métrir

MOULD - [bot] (a fungous growth) moisi *m*, moisissure *f*
[geol] (loose, friable earth) moule *f* externe, terreau *m*
[constr] (the provisional construction designed to support the concrete while setting) coffrage *m*
[constr] (a template) calibre *m*, profil *m*
[metall] (in foundry work) moule *m*
[plast ind] (the recessed structure which receives molten or plastic material in casting or injection moulding and gives it the desired shape) moule *m*
[metall] (of a press die) estampe *f*, matrice *f*
[paper man] (a frame covered with woven wire) châssis *m*
[el acoust] (metal part derived from a master by electro-forming, which is a positive of a recording) positif *m*, épreuve *f*
[print] (papier-maché impression of types from which stereotypes for printing can be made) empreinte *f*, matrice *f*
[arch] (a moulding; ornamental band projecting from a wall etc) moulure *f*
[gen] (auto, machines, ships etc; the outer form) profil *m*, gabarit *m*
[glass man] (a metal form for shaping glass) estampe *f*
[shipbuild] (form of wood) gabarit *m*
[geol] (impression in the earth of the convex side of a fossil shell) moule *m* externe (de coquillage fossile)

~ALIGNMENT - [plast ind] guidage *m* du moule

~-BOARD - [metall & plast ind] batte *f*
[agric mach] s. mouldboard

~BODY FOR INTERCHANGEABLE MATRICES - [rubber ind] (used in tyre retreading) étuve *f* chauffante à introduire des matrices interchangeables

~BREAKER - [metall & plast ind] vérin *m* d'ouverture des moules, dispositif *m* pour ouvrir les moules

~BREAKING JACK - s. mould breaker

~CASTING - [metall] moulage *m* en châssis

~CAVITY - [metall & plast ind] (the female section of a mould) empreinte *f*

~CHARGE - [plast ind] charge *f*

~CLAMP - [metall & plast ind] (a screw or lever device designed to force and hold the parts of a mould together) crampe *f* de moule

~CLAMPING DEVICE - [plast ind & rubber] dispositif *m* de serrage du moule

~CLEANING GUN - [rubber & plast ind] sableuse *f* pour nettoyage d'un moule

~CLEARING JACK - s. mould breaker

~DESIGNER - [plast ind] mouliste *m*

~DEVELOPMENT - [bot] (the development of fungous growth on plants) développement *m* de moisissure *f*

~DILATATION - [metall] dilatation *f* du moule

~DRYER - [metall & plast ind] four *m* de séchage des moules

MOULD DRYING - [foundry] séchage *m* des moules
~ENGRAVING MACHINE - [mach tool] machine *f* à graver les moules
~FOR BASE OF PILLAR - [constr] coffrage *m* de plinthe
~FRAME - [metall] moule *m*
~IMPRESSION - [plast ind] (that part of a mould which imparts shape to the moulding) empreinte *f* d'un moule
~IN A PIT, to - [metall] mouler en fosse
~IN A SNAP FLASK, to - [metall] mouler en motte
~INSERT - [plast ind] (any separate part of a mould which is fitted into the mould impression to facilitate manufacture of the mould) élément *m* interchangeable
~INSERT FOR MARKING - [plast ind] insertion *f* de marquage
~JACKET - [metall] jaquette *f*
~LOCKING FORCE - [plast ind & metall] (the pressure, usually in tons, applied to the moulds by the locking mechanism) pression *f* de serrage, force *f* de fermeture
~LOFT - [shipbuild] (extensive floor on which the lines of a ship are drawn in full scale) salle *f* de gabarits
~LUBRICANT - [plast ind] (a substance used to facilitate movement of material as it fills a mould, or to make removal of the moulded part easier) lubrifiant *m* pour les moules
~MISALIGNMENT - [rubber & plast ind] guidage *m* défectueux des moules
~OIL - s. mould lubricant
~ON END, to - [metall] mouler debout
~ON THE FLAT, to - [metall] mouler à plat
~PARTING AGENT - [plast ind] agent *m* de séparation (des moules)
~PARTING LINE - [plast ind] plan *m* de joint du moule
~PEAT - [geol] tourbe *f* moulée
~REACTION - [metall] formation *f* de soufflures bleutées
~RELEASE - [mech] démoulage *m*, extraction *f* du moule
~RELEASE AGENT - [ind chem] (any preparation applied to a mould to facilitate the removal of the moulded object) agent *m* de démoulage
~RETAINING FLANGE - [plast ind] bride *f* porte-moule, plateau *m* porte-moule
~SEAM - [metall & plast ind] ligne *f* de moulage
~SHIFT - [metall] variation *f* de moule
~SLIDE - [plast ind] coulisse *f*
~SOLUTION - s. mould lubricant
~STOOL - [metall] base *f* de lingotière, fromage *m*
~STRIPPING - s. mould release
~WEIGHTS - [metall] poids *m* de charge
~WITH CONICAL SPLITS - [plast ind] moule *m* à coins, moule *m* à coquilles
MOULDBOARD - [agric] (the metal plate in a plough which turns over the furrow slice) versoir *m*
[metall & plast ind] s. mould-board
~PLOUGH - [agric] (plough implement which detaches a slice of soil vertically from the undersoil by means of a coulter and a share) charrue *f* à soc
MOULDED - [gen] moulé, façonné
[plast ind etc] moulé
~ARTICLE - [plast ind etc] (any object formed from plastic material by moulding)pièce *f* moulée
~BRAKE LINING - [rubber ind] fourrure *f* de frein moulée
MOULDED DEPTH - [shipbuild] (the depth of a ship from the top of keel to the top of beam at side) creux *m* sur quille
LAMINATE - [plast ind] moulage *m* stratifié
~LAMINATED PLASTICS TUBE - [plast ind] tuyau *m* stratifié moulé
~LAMINATED TUBE - s. moulded laminated plastic tube
~MATERIAL - [plast ind etc] moulages *mpl*
~-ON TYRE - [rubber ind] bandage *m* plein fixé à la jante par vulcanisation
~PIECE - s. moulded article
~SECTION - [metall] profilage *m* de feuillards
~SHOE - [shoe man] chaussure *f* à semelle vulcanisée directement
~THREAD - [foundry & plast ind] (a screw thread formed in plastic material by a moulding process) filet *m* moulé
~WORK - [constr] moulure *f*
MOULDER, to - [gen] (to turn to dust by natural decay) s'éffriter, tomber en poussière
MOULDER - [mech] (machine making moulds for casting) machine *f* à moulurer
[gen] (in foundry or plastic ind) mouleur *m*, façonneur *m*
[metall] dimension *f* primitive
~BENCH - [impl] établi *m* de mouleur
MOULDER'S ADJUSTABLE DEPTH GAUGE - [metall] sonde *f* de mouleur
~HAMMER - [metall] marteau *m* de mouleur
~PEEL - [metall] pelle *f* de mouleur
MOULDERING - [metall] cassant
[gen] (the operation of moulding) moulage *m*
[arch] (ornamental band on a building) moulure *f*
[constr] (ornamental strip projecting from a wall) socle *m*
[plast ind] (the process of introducing the plastic material into the die) estampage *m*, emboutissage *m*
[plast ind] (the complete operation of moulding) moulage *m*
[plast ind & die casting] (object made by a moulding operation) object *m* moulé, matière *f* moulée
[constr] (any object or element cast in a mould) moulure *f*, profil *m* mouluré
[auto] (ornamental strip on a car body) enjoliveur *m*
[shipbuild] gabariage *m*
~ABRASION - [plast ind] (abrasion of moulds by the plastic material passing over internal surfaces) abrasion *f* de moulage
~BAY - [metall] atelier *m* de molage
~BENCH - [mech] banc *m* de moulage
~BOARD - [metall] plaque *f* pour moulage
~BOX - [metall] (used for holding the sand mould in which casting is made) châssis *m* (de fonderie)
~BOX PIN - [metall] goupille *f* de châssis
~CLAY - [metall] argile *f* à mouler
~COMPOUNDS - [plast ind] matière *f* à mouler avec copeaux de bois
~CUTTER - [carp] (specially shaped cutting tool for cutting a moulding profile) outil *m* à profiler
~CYCLE - [plast ind] (the complete series of operations when making a moulding) cycle *m* de moulage
~DEFECT - [plast ind etc] défaut *m* de moulage
~IN DRY SAND - [metall] moulage *m* sec
~IN FLASK - [metall] moulage *m* en châssis
~IN GREEN SAND - [foundry] moulage *m* en sable vert
~KNIFE GRINDER - [tool] affûteuse *f* multiple pour

fers à profiler
MOULDING LINE - [rubber ind] ligne *f* de moulage
~LOAM - s. moulding sand
~LOOP - [metall] carrousel *m*
~MACHINE - [metall & plast ind] (machine making moulds for casting) machine *f* à mouler
[carp] machine *f* à moulurer
~MACHINE WITH ROTARY TABLE - [foundry] machine *f* à mouler à table rotative
~MATERIALS - [plast ind] matières *fpl* à mouler
~METHOD - [plast ind] technique *f* de moulage
~OF GEAR WHEELS - [mech] formage *m* d'engrenages
~OPERATION - [plast ind etc] opération *f* de moulage
~PIT - [metall] fosse *f* de moulage
~PLANE - [carp] (tool for planing mouldings) rabot *m* à moulures, mouchette *f*
~PLANT - [plast ind] atelier *m* de moulage
~PLATE - [plast ind & foundry] plaque *f* pour moulage
[metall] plaque-support *f* de moule
~POWDER - [plast & rubber ind] (injection or compression moulding material in finely-divided form) poudre *f* à mouler
~PRESS - [mech mach] (a press designed for compression moulding) presse *f* à compression
~PRESSURE - [mech] (the designed pressure under which a compression moulding is made) pression *f* de moulage
~SAND - [metall & plast ind] (siliceous sand possessing plasticity, adhesiveness permeability and strength) sable *m* de moulage, sable *m* de fonderie
~SEAM - s. mould seam
~SHOP - [metall] atelier *m* de moulage
s. moulding bay
~SHRINKAGE - [metall & plast ind] (the difference in dimensions (inches per inch) between a moulding and the cavity in which it was moulded, measured at the same temperature) retrait *m* au moulage
~TECHNIQUE - [plast ind] (the special processes used in injection, compression and other types of moulding) technique *f* de moulage
~TIME - [plast ind] temps *m* de cuisson
~TOOL - [tool] (a compression moulding device) moule *m* à compression
MOULDINGS - [carp] (strips of wood used for decoration purposes) profilés *mpl*
[plast ind etc] pièces *fpl* façonnées, moulages *mpl*
MOULDMADE PAPER - [paper man] (imitation of handmade paper) papier *m* fait à la machine, cuvier *m* de machine
MOULDY - [gen] (covered with mould) moisi
~ROT - [bot] (fungous growth affecting the bark of trees) moisissure *f*
MOULT, to - [zool] muer, perdre (la peau, la carapace etc)
MOULT(ING) - [zool] mue *f*
~GALLERY - [mining] montage *m*
MOUND - [gen] (natural or artificial heap of earth) tertre *f*, monticule *m*
[geogr] (a hillock) petite colline *f*, monticule *m*, tertre *m*
[constr] (in excavation work, the undisturbed pile of earth left near the excavated site to indicate the depth of the work) remblai *m*
[gen] (orb of gold, representing the earth, often seen on a royal crown) globe *m*, monde *m*
[anat] mont *m*
MOUNT, to - [gen] monter, s'élever, gravir
MOUNT, to - [mech] enchâsser, emmancher
[gen] (a jewel) monter
[an instrument] monter, installer
MOUNT - [gen] montage *m*, support *m*
[anat] (of a hand-palm) mont *m*
[comput] (the introduction of information into a band reader) introduction *f*
[opt] (of lenses) monture *f*
[instr] (of a telescope) affût *m*, pied *m*
[mech] (of a machine) armement *m*
[firearms] (of a gun) affût *m*
[mech etc] (a framework) bâti *m*, ossature *f*
[gen] (of a painting) carton *m* de montage
[print] (a base for a plate or block) pied *m*, empattement *m*, socle *m*
[electron] raccord *m*
~A TYRE, to - [auto] monter un pneumatique
MOUNTAIN - [gen & geogr] montagne *f*
~ASH - [bot] sorbier *m* commun, arbre *m* à grives
~BAROMETER - [instr] (or orometer; an aneroid barometer giving the approximate elevation above sea level) baromètre *m* altimétrique
~BREEZE - [met] (a katabatic wind blowing down a mountain slope at night or in winter) brise *f* de montagne
~CHAIN - [geogr] chaîne *f* de montagnes
~CORK - [min] (type of asbestos consisting of thick interlaced fibres) liège *m* de montagne, liège *m* fossile
~CRYSTAL - [min] (rock crystal) cristal *m* de roche, cristal *m* de montagne
~FARM - [agric] exploitation *f* agricole de montagne
~FLAX - [min] amiante *f*, cuir *m* fossile
~FLESH - [min] chair *f* fossile, chair *f* minérale
~FLOUR - [min] farine *f* fossile
~GREEN - [min] vert *m* minéral
~LEATHER - [min] (type of asbestos consisting of flexible sheets of interlaced fibres) cuir *m* de montagne
~MEADOW - [agric] alpage *m*
~MEAL - [min] (a type of fossil meal) farine *f* fossile
~MILK - [min] lait *m* de montagne, lait *m* de lune
~PAPER - s. mountain cork
~RAILWAY - [railw] chemin *m* de fer de montagne
~RANGE - [geogr] chaîne *f* de montagnes
~RIDGE - [geogr] croupe *f*
~ROAD - [gen] chemin *m* de montagne
~TOBACCO - [bot] arnica *f*
~WOOD - s. mountain leather
MOUNTANT - [photo] colle *f*
MOUNTING - [mech] (preparing for use) montage *m*
[mech] (a framework, a support) bâti *m*
[firearms] (the gun carriage) affût *m*
[text] (of a loom) équipage *m*, lisses *fpl*
[radio] s. montage
[mech] (of a machine) armement *m*, entoilage *m*, encollage *m*
~-BASE TEMPERATURE - [electron] température *f* de la plaquette de montage
~BOARD - [photo] (flexible cardboard strips used to mount photographic prints) carton *m* de montage
~BRACKETS - [mech] (mounting angles, or bars) pattes *fpl* de fixation du mécanisme de benne
~CAR - [telecomm] voiture *f* d'équipe (à bras)
~CARDBOARD - [photo] carton *m* de montage d'épreuves

MOUNTING CLAMP - [mech] bras *m* de serrage
~CLIP - [gen] griffe *f* de flexion
~DRAWING - [gen & mech] dessin *m* de montage
~FAULT - [mech & telecomm] défaut *m* de montage
~FOIL - [photo] support *m* de montage
~FOR MAGNIFYING GLASS - [opt] monture *f* d'une loupe
~GUIDE RIBS - [mech] (centering ribs) nervures *fpl* de centrage
~HEIGHT - [auto] (the vertical distance between the top of the chassis rails and the bottom of the body floor sheet) hauteur *f* de benne au dessus du châssis
~MACHINE - [mech] machine *f* à monter
~OF THE WHEEL FLANGE - [railw] (on rail) montée *f* du boudin sur le rail
~PAD - [mech] coussinet *m* de montage
~PILLAR - [constr] colonne *f* montante
~PLATE - [plast ind] (a metal plate on which moulds are fixed in the moulding press) plateau *m* de fixation
[gen] (a support) platine *f*, support *m*
~PLATEN - [mech] (a press platen or table in which moulds are fixed) table *f* de la presse
~PLIERS - [mech] pince *f* de montage
~POSITION - [mech] (the position of a part in a machine etc) position *f* de montage
[electron] (the position, either horizontal, vertical etc. of an electronic tube in the apparatus) position *f* de montage
~TEST - [paper man] (test carried out to determine the absorbing power of blotting paper) essai *m* à adsorption
~WIRE - [el] fil *m* de montage
MOUSE - [zool] souris *f*
[constr] (colloq. for a weight attached to a sash-window cord) contre-poids *m*
[naut] (swelling on a rope to prevent slipping) guirlande *f* de cordage
~BARLEY - [bot] (a weed]) orge *f* queue de souris, orge *f* des murs
~HOLE - [oil ind] trou *m* de souris, avant-trou *m*
~ROLLER - [print] toucheur *m* auxiliaire
~TRAP - [oil ind] souricière *f* à clapet
MOUTH - [anat] bouche *f*
[geogr] (of a river) embouchure *f*
[gen] ouverture *f*, entrée *f*, orifice *m*, guichet *m*, pavillon *m*
(of a tool) mâchoire *f*
[min etc] (the entrance into a mine or a cavity) halde *f*, amorce *f*
[gen] (of a bag etc) gueule *f*, entrée *f*
[gen] (of a bottle) goulot *m*
[metall] (of a furnace) gueulard *m*, trou *m* de coulée, oeil *m*
[mus] (the opening between the lips of a flue organ pipe) bouche *f*
[acoust] embouchure *f*, bec *m*
[impl] (the slot into which the cutters of a plane are fitted) lumière *f*, mortaise *f* (de rabot)
[zool] gueule *f*
~BLOWPIPE - [ind chem] (metal tube fitted with a jet at right angles, designed to produce a hot atmospheric flame by means of a spirit lamp or the like, used in mineral analysis) chalumeau *m* à bouche
~DIAMETER - [aero] (in a parachute, the diameter assumed by the hem when the canopy is fully distended) diamètre *m* de bord d'attaque
MOUTH-GAG - [med impl] ouvre-bouche *m*
[vet] pas *m* d'âne
~LOCK - [aero] (of a parachute) accrochage *m* de la coupole
~OF A TUNNEL - [constr] entrée *f* de tunnel
~OF THE TONGS - [impl] (or jaws of the tongs) mâchoire *f* des pinces
~OPENER - [med impl] ouvre-bouche *m*
~ORGAN - [mus] harmonica *m*
~ROT - [vet] pyorrhée *f* des chiens
MOUTHPIECE - [gen] (part of any instrument or tool which is applied to the mouth) embouchure *f*
[acoust] (dictating machine etc) embout *m*
[mus] bec *m*, embouchure *f*
~OF RETORT - [ind chem] tête *f* de cornue
MOVABLE - [gen] (capable of being moved) mobile (which can be moved, or removed) amovible
~ANTENNA - [radio] (a portable aerial) antenne *f* mobile
~BEARING - [mech] support *m* mobile
~BOILER - [th eng] chaudière *f* mobile
~BRIDGE - [constr] (bridge which can be displaced to permit the passage of vessels) pont *m* mobile
~CARBON ROD - [nucl] (movable control rod of carbon) barre *f* de commande mobile en carbone
~-COMB HIVE - [agric] ruche *f* à cadres mobiles
~CONTACT - [el] (of a switch) contact *m* de commande
~DAM - [hydr] (a dam which can be removed) barrage *m* mobile
~HALF - [metall & plast ind] (the part of a compression or other mould which can be removed) partie *f* mobile de moule
~HALF OF THE MOULD - s. movable half
~JET - [astronaut] jet *m* orientable de stabilisation
~PLATE - [th eng] (of a cooker) plafond *m* mobile
~PLATEN - [plast ind] (of a mould) plateau *m* mobile
~SCENE - [cin & telev] (movable scenery decorations, called 'getaway' in the USA) coulisse *f* de studio
~SHEAVE - [mech] poulie *f* mobile
~TYPE - [print] (single types, as distinguished from blocks) caractère *m* mobile
~WEIR - [hydr] (a temporary weir, it is removed during period of flood) barrage *m* mobile
MOVE, to - [gen] déplacer, se mouvoir
[gen] (to move from one place to another) se déplacer
[mech] (to move a part etc) mettre en marche, déplacer
[gen] (the action of advancing) avancer, se mouvoir
[gen] (to take action) agir
[leg] (to submit a motion) proposer
MOVE - [gen] mouvement *m*, démarche *f*, déplacement *m*
~BACK THE BOUNDARIES - [gen] reculer les bornes
~TOGETHER, to - [foundry & plast ind] (the halves of the moulds) appliquer
MOVEMENT - [gen & phys] mouvement *m*, déplacement *m*
[mech] (a mechanism) mécanisme *m*
[horol] (the actual mechanism of watch or clock) mouvement *m*
[cin] (the main part of the mechanism of a camera or projector) mécanisme *m*, mouvement *m*
~AREA - [aero] (the part of an aerodrome specifically designed and used for landing, take-off and associa-

ted movements) aire *f* de mouvement
MOVEMENT BLUR - [radiat] (lack of sharpness caused by the relative movement of the source of radiation, or of the irradiated object, and the film or screen) flou *m* cinétique, flou *m* de mouvement
~HOLDER - [horol] porte-mécanisme (d'horlogerie)
~IN DEPTH - [cin] (the movement of an object to be photographed from or towards the lens) mouvement *m* en profondeur
~OF RATCHED WHEEL - [text] déplacement *m* de la roue à rochet
~OF THE ROLLER - [text] mouvement *m* du rouleau
~OF THE SHUTTLE - [text] mouvement *m* de la navette
~OF THE SLAY - [text] mouvement *m* du battant
~OF UNDERGROUND WATERS - [geol] écoulement *m* de l'eau souterraine
~OF WARP AND CLOTH - [text] marche *f* de la chaîne et du tissu
MOVER - [gen & mech] moteur *m*
MOVIE - [cin] (US colloquialism) film *m*
~CAMERA - [cin] (US only; a cinema camera) camera *f* cinématographique
~THEATRE - [cin] (US only) cinéma *m*
MOVING - [gen] (causing to move) mouvant, mobile
[mech etc] moteur, motrice
[gen] (the act of moving) mouvement *m*, déplacement *m*
~BOUNDARY METHOD - [el] (a suitable sequence of electrolytes in a tube will exhibit interfacial boundaries) méthode *f* des surfaces limites mobiles
~-COIL GALVANOMETER - [instr] (galvanometer in which the moving element, which comprise a coil, moves in a constant magnetic field, when a current flows through it) galvanomètre *m* à cadre mobile
~-COIL LOUDSPEAKER - [el acoust] (a moving-conductor loudspeaker in which the conductor has the form of a coil) haut-parleur *m* électrodynamique
~-COIL MICROPHONE - [el acoust] (moving-conductor microphone in which the conductor has the form of a coil) microphone *m* électrodynamique
~-COIL PICK-UP - [el acoust] (dynamic reproducer) phonocapteur *m* électrodynamique
~-COIL RELAY - [el] relais *m* à cadre mobile
~-CONDUCTOR LOUDSPEAKER - [el acoust] (an loudspeaker depending for its operation on the motion of a conductor, joined to a diaphragm and carrying a varying current, in a steady magnetic field) haut-parleur *m* électrodynamique
~-CONDUCTOR MICROPHONE - [el acoust] (microphone depending for its operation on the generation of an electromotive force in a conductor moving) microphone *m* électrodynamique
~CONTACT ROD - [el] tige *f* de contacteur
~CONTACT SPRING - [el] ressort *m* de contact de commande
~CORE - [metall] (die casting) noyau *m* mobile
~DIE PLATE - [plast ind] (of a mould) plaque *f* mobile de moule
~EFFECT - [radio] (the shadow effect) affaiblissement *m* de propagation
~ELEMENT - [instr] (of an instrument; the part moving as a direct result of any variation in the quantity measured by the instrument) équipage *m* mobile
~GRID - [radiat] (an anti- diffusion grid kept in motion during X-ray exposure, to eliminate its own shadows) grille *f* mobile, grille *f* oscillante
MOVING-IRON INSTRUMENT - [instr] (a ferromagnetic instrument) appareil *m* à fer mobile, appareil *m* ferromagnétique
~-IRON LOUDSPEAKER - [el acoust] (a loudspeaker depending for its operation on variations of the reluctance of a magnetic circuit) haut-parleur *m* électromagnétique
~-IRON MICROPHONE - [el acoust] (preferably called electromagnetic microphone, i.e. a microphone depending for its operation on the variation of the reluctance of a magnetic circuit) microphone *m* électromagnétique
~-IRON VOLTMETER - [instr] (cheap rough instrument used at power frequencies) voltmètre *m* à fer mobile
~-MAGNET GALVANOMETER - [instr] (galvanometer in which the fixed part comprises one or more coils carrying current, and the moving element one or more magnets) galvanomètre *m* à aimant mobile
~-MAGNET INSTRUMENT - [instr] (instrument in which one or more fixed coils carrying current actuate a moving magnet or a system of magnets) appareil *m* à aimant mobile
~MATTE - [telev] procédé *m* de transparence électronique
~PERIOD - [cin] (the period during which the film is in motion) période *f* de mouvement
~PICTURE - s. motion picture
~PICTURE CAMERA - [cin] (the shooting camera) camera *f* de prise de vue
~PICTURE LAMP - [cin] (cinema lamp in GB) lampe *f* de projection
~SHAFT - [text] lame *f* de formation du pas
~STAIRCASE - [constr] (synonym of escalator) escalier *m* à marches mobiles, escalier *m* roulant
~TAPE TRANSMITTER - [radio] (equipment for the transmission of words written or typed on a moving tape) émetteur *m* sur bande coulissante
~-TARGET INDICATION - [radar] (device limiting the display of radar information to moving targets) cibles radar *m* en mouvement, indicateur *m* des objectifs mobiles, suppresseur *m* d'échos fixes
~THE WARP FORWARD - [text] avancement *m* de la chaîne
~VAN - [auto] fourgon *m* de déménagement
~VANES OF A QUADRANT ELECTROMETER - [instr] secteurs *mpl* mobiles d'un électromètre à quadrants
MOW, to - [agric] (to cut down grass, grain etc) faucher, moissonner, tondre
~-BINDER - [agric] moissonneuse-liseuse *f*
~CURING - [agric] séchage *m* en grange
MOWER - [agric] faucheuse *f* à herbe
~CRANK - [agric] manivelle *f* de coupe
~DRIVE - [agric] entraînement *m* de barre de coupe
MOWING - [agric] coupe *f*, fauche *f*
~ATTACHMENT - [agric] faucheuse *f*, commande *f* de fauchage
~MACHINE - s. mower
~SPEED - [agric] vitesse *f* de fauchage
MOXA - [chem] (material used as a cautery) moxa *m*
m.p. - [chem] (abbrev. for melting point) point *m* de fusion
m.p.g. - [meas] (abbrev. for miles per gallon) milles *fpl* par gallon (litres aux cent kilomètres)
m.p.h. - [meas] (abbrev. for miles per hour) milles *fpl* à l'heure
M.S. - [mech] (abbrev. for maximum stress) effort

m principal primaire
M.S. - [metall] (abbrev. for milt steel) acier *m* doux
M/S. - [naut] (abbrev. for motor ship) vaisseau *m* à moteur
M.S.C.P. - [light] (abbrev. for mean spherical candle power) intensité *f* moyenne sphérique
M-SHELL - [nucl] (the third electron shell from the nucleus) couche *f* M
M.S.L. - s. mean sea level
MsTh - [chem] (the symbol of mesothorium) symbole de mésothorium
M.T. - (abbrev. for metric ton) tonne *f* métrique
MTD - [phys] (abbrev. for mean temperature difference; the average temperature difference causing heat exchange) différence *f* moyenne de température
M.T.R. - [nucl] (abbrev. for material testing reactor) réacteur *m* d'essai de matériaux
MU - [gen] (conventional transcription of the Greek letter) mu *m*
[meas] (abbrev. for micron) micro-
[telecomm] coefficient *m* d'amplification
~FACTOR - [electron & radio] (an amplification factor of an electronic tube; the voltage factor of the anode and the control electrode, the anode current remaining unchanged) coefficient *m* d'amplification
~METAL - [metall] (alloy containing iron, nickel, copper and manganese, having high magnetic permeability and low hysteresis loss) métal *m* mu (alliage *m* de fer, nickel et manganèse)
~OIL - [paint] (term sometimes used for tung oil derived from Aleurites Montana) huile *f* d'abrasin, huile *f* de Tung
MUCIC - [chem] mucique
~ACID - [chem] (acid obtained by the oxidation of galactose) acide *m* mucique
MUCILAGE - [paint] (a slimy deposit separated from unrefined vegetable oils on heating) mucilage *m*
MUCILAGES - [chem] (group of organic compounds) mucilages *mpl*
MUCILAGINOUS - [chem] mucilagineux
MUCINES - [chem] (group of glucoproteins found in mucus and saliva) mucines *fpl*
MUCK, to - [gen] salir, souiller
[agric] (to fertilize with muck) cultiver en sols tourbeux
[mining] charger une berline
MUCK - [gen] ourdures *fpl*, fumier *m*, fange *f*
[min] (broken rocks after blasting) morts-terrains *mpl* de recouvrement
[agric] (vegetable mould combined with earth) sol *m* tourbeux
[mining] déblai *m*, tas *m* à charger
~BAR - [metall] fer *m* brut, ébauché *m* de puddlage
~FARMING - [agric] culture *f* en sols tourbeux
~-ROLL - [mech] cylindre *m* dégrossisseur
~ROLLS - [metall] cylindres *m* ébaucheurs
~SCRAP - [metall] bocage *m* de poterie
~SOIL - [geol] terre *f* noire
~TRAIN - [metall] équipage *m* dégrossisseur, train *m* ébaucheur
MUCKING - [min] (lashing; layer of broken stones after blasting) enlèvement *m* du recouvrement
[mining] chargement *m*
MUCOCELE - [med] (localized accumulation of mucous secretion) mucocèle *f*
MUCOENTERITIS - [med] entérite *f* mucomembraneuse
MUCOIDS - [chem] (group of glucoproteins) mucoïdes *mpl*
MUCONIC ACID - [chem] (acid obtained by the cleavage of aromatic amino acids which occur naturally) acide *m* muconique
MUCOPROTEINS - [chem] (conjugated proteins containing a carbohydrate group) mucoprotéines *fpl*
MUCORMYCOSIS - [med] mucormycose *f*
MUCOSA - [anat] (the mucous membrane lining the air passages and the gastrointestinal tract) muqueuse *f*, tunique *f* muqueuse
MUCOUS - [gen] (secreting or producing mucus) muqueux
~GLANDS - [zool] (glands which secrete mucus) glandes *fpl* muqueuses
~MEMBRANE - [zool] (a tissue layer lining various cavities of the body, e.g. the uterus, the gut etc) membrane *f* muqueuse
MUCUS - [physiol] (the viscous fluid secreted by mucous glands) mucus *m*, mucosité *f*
MUD, to - [gen] (to cover with mud, to soil) encrotter, couvrir de boue
[oil ind] (the introduction of muds into an oil well to stop gas coming out during drilling) colmater, enduire
MUD - [gen] (liquid or semiliquid mixture of earth and water) boue *f*, bourbe *f* fange *f*
[geol] (fine-grained rock consisting of several minerals and with a high percentage of water) limon *m*, vase *f*, boue *f*
[mech] tartres *mpl* boueux, boue *f*
~AUGER - [oil ind] tarière *f* à glaise
~-BANK - [geol] banc *m* de sable, javeau *m*
~BLANKET - [hydr] couverture *f* de boues
~BOX - [mining] puisard *m*
~CRACKS - [geol] fentes *fpl* de dessèchement
~DRUM - [th eng] (vessel placed at the lowest part of a steam boiler, to retain insoluble matter or sludge) collecteur *m* de fanges
~ERUPTION - [geol] éruption *f* boueuse
~FLAP - [auto etc] (flexible attachment to the mudguard) bavolet *m*, pare-boue *m*
~-FLOW - [geol] (the flow of mud from a mud-volcano) coulée *f* boueuse
~GEYSER - [geol] source *f* boueuse
~GUN - [mech] boucheuse *f*
~-HARDENING - [ind chem] durcissement *m* de la boue
~-HARDENING AGENT - [ind chem] durcisseur *m* de boues
~-HARDENING PROCESS - [ind chem] (used in military operations etc) procédé *m* de durcissement des boues
~HOG - s. mud pump
~HOLE - [mech] (in a mud drum) orifice *m* du collecteur de boues
~-LADEN FLUID - [oil ind] boue *f* (de forage) de circulation
~LINE - [oil ind] conduite *f* à boues
~LINING - [oil ind] enduit *m* de boue, pâte *f* de boue
~LOGGING - [oil ind] diagraphies *fpl* de boues
~LOSS - [oil ind] perte *f* de circulation
~MIXING PLANT - [oil ind] centrale *m* de malaxage de boue
~PAN - s. mud pit
~PIT - [oil ind] bassin *m* de décantation, bassin *m* à

boue
MUD PLUG - [th eng] (in a boiler) porte *f* de vidange
~PRESSURE - [oil ind] (the pressure of the mud in drilling operations) pression *f* de la boue
~PUMP - [ind] (or slush pump) pompe *f* à boue, pompe *f* de circulation
~ROCK FLOW - [soil] torrent *m* de boue
~SCREEN - [oil ind] (vibrating screen over which the drilling fluid is conducted. The drilling cuttings are retained on the screen) tamis *m* vibrant
~SOCKET - [petr ind] (tool designed to clean mud in a well) douille *f* de nettoyage
~SPRING - [geol] fontaine *f* de boue, source *f* boueuse
~TANK - [oil ind] bac *m* à boue
~TRAP - [road constr] collecteur *m* de boue
~UP, to - [th eng] (the operation of sealing a joint or a stopper in a furnace by the application of wet clay) obturer les fissures
~VOLCANO - [geol] (conical hill formed the piling up of fine mud erupting from an orifice in the ground) soufflard *m*, soffione *m*
MUDDING - [oil ind] envasement *m*
MUDDLER - [impl] (stick used for stirring or churning a liquid) agitateur *m*
MUDDY - [gen] boueux, fangeux, bourbeux
[gen] (of liquid) trouble
[hydr] vaseux
MUDGUARD - [gen] (of vehicles) garde-boue *m*
~FLAP - s. mud flap
MUDSTONE - [geol] schiste *m* argileux
MUF - [radio] (abbrev. for Maximum Usable Frequency) fréquence *f* limite supérieure
MUFF - [gen] manchon *m*
[mech] (a cylindrical covering) manchon *m* d'accouplement
~COUPLING - [mech] (or box coupling; shaft coupling formed by a longitudinally split sleeve) accouplement *m* à manchon
~JOINT - [mech] slip joint *m*, joint *m* S
MUFFLE - [metall] (the heating chambers in a muffle furnace) moufle *m*
~CARBURIZING FURNACE - [metall] four *m* de carburisation à moufle
~FURNACE - [metall] (type of furnace used in assaying and similar operations, having fireclay chambers of semicircular section, closed at the inner end and fixed in the body of the furnace, to contain cupels) four *m* à moufle
~KILN - [ceram] (a furnace to which saggers are placed for firing pottery) four *m* de grillage à moufle
MUFFLER - [gen] (anything which is used for wrapping up) cache-col *m*, moufle *m*
[acoust] (for sound) silencieux *m*, pot *m* d'échappement
[mech] (of an exhaust pipe) silencieux *m* d'échappement
[mus] (of a piano) étouffoir *m*
MUG SHOT - [cin & telev] (the shot of a detail of a scene in image size) gros plan *m*, plan *m* très rapproché
MUGGY - [met] (colloquial term for warm moist atmospheric conditions) mou, chaud et humide
~WEATHER - [met] temps *m* chaud et humide
MUGWORT - [bot] armoise *f* commune, barbotine *f*
MULATTO - [anthropology] (of white and negro parentage) mulâtre
MULBERRY - [bot] mûre *f*
~MARK - [med] angiome *m* tubéreux
~TREE - [bot] mûrier *m*
MULCH - [agric] (loose material, straw etc. used to cover stalks of plants to protect their roots) maillis *m*, litière en décomposition
MULE - [zool] (the hybrid between an ass and a mare) mulet *m*, mule *f*
[text] (a cotton spinning machine which draws, stretches and twists in one operation) mule-jenny *f*, renvideur *m*
[mech] (a small electric engine or tractor) tracteur *m* électrique
[shoe man] (blackless slipper) mule *f*
~BAND - [text] corde *f* de renvideur
~COP - [text] cannette *f* trame
~-JENNY - s. mule [text]
~PULLEY - [mech] (adjustable belt pulley) poulie-guide *f* réglable
~SPINDLE - [text] broche *f* de renvideur
~SPINNING - [text] filage *m* au renvideur
~SPUN YARN - [text] fil *m* de renvideur
~TRACK - [gen] piste *f* muletière
~TWIST - [text] chaîne *f* du renvideur
~WARP - [text] fil *m* de chaîne de renvideur
~WARP COP - [text] canette-chaîne *f* de renvideur
MULEY - [zool] (only USA) (vache) sans cornes
~-AXLE - [railw] essieu *m* sans champignon
MULL, to - [photo] (in engraving) grainer
[gen] (of wine) chauffer (avec des épices)
[gen] (to grind, to reduce to powder) boyer
MULL - [text] (light, soft cotton fabric) mousseline *f*
[text] (for medical use) gaze *f*
~CURTAIN - [text] rideau *m* de mousseline fine
MULLER - [mech] (mill of edge runner or pan type) molette *f* (de broyeur), patin *m* (d'une cuve)
[metall] frotteur *m*
~MIXER - [mech] mélangeur *m* à meule
MULLER BRIDGE - [instr] (bridge used for measuring four terminal resistors, in particular resistance thermometers) pont *m* de Muller
~CIRCLE - [opt] (horopter) cercle *m* de Muller, horoptère *m*
MULLERIANOMA - [med] tumeur *mf* mullerienne
MULLING - [metall] frottage *m*
MULLION - [constr] (the vertical piece between window panels) meneau *m* vertical
MULLITE - [min] (an aluminium silicate with uses as a refractory material) mullite *f*
MULLOCK - [min] (accumulation of waste rock and earth) stériles *mpl*
MULLOCKING - [mining] travail *m* au rocher
MULLOCKY - [min] (of waste rock and earth about a mine) stérile
MULTI-ADDRESS INSTRUCTION - [comput] instruction *f* multiadresses
~-ANODE RECTIFIER - [el] soupape *f* polyanodique
~-COURSE ROTATION - [agric] assolement *m* multiple
MULTIANODE TUBE - [electron] tube *m* à plusieurs anodes
MULTIBAND AERIAL - [radio] (aerial which can work at several wavelenghts) antenne *f* à plusieurs bandes
~ANTENNA - s. multiband aerial

MULTIBOTTOM PLOUGH - [agric] charrue *f* polysoc
MULTIBREAK - [el] à coupure multiple
MULTIBULB FLOOD - [light] flood *m* composé de plusieurs lampes
MULTIBURST - [telev] (in colour television) salve *f* multiple
MULTICARRIER RADIO TRANSMITTER - [radio] radio-émetteur *m* à plusieurs porteuses
MULTICAVITY MOULD - [plast ind] moule *m* à empreints multiples
MULTICELL HORN - [acoust] pavillon *m* multicellulaire
MULTICELLULAR - [phys] multicellulaire
~HORN - s. multicell horn
~LOUDSPEAKER - [el acoust] (a horn loudspeaker in which the radiating element is coupled to the medium by two or more superimposed horns) haut-parleur *m* multicellulaire
~VOLTMETER - s. multiple electrometer
MULTICHANNEL LOUDSPEAKER - [el acoust] (system of two or more loudspeakers with dividing networks, designed to radiate simultaneously in particular frequency bands) haut-parleur *m* à plusieurs canaux
~RADIO TRANSMITTER - [radio] émetteur *m* radio à plusieurs canaux
~TELEGRAPH SYSTEM - [telecomm] télégraphie *f* à plusieurs voies
~TELEVISION - [telev] télévision *f* à plusieurs canaux
MULTICOMPONENT - [chem phys etc] à plusieurs éléments
~SIGNAL - [radio & telecomm] (signal consisting of more than one component) signal *m* à plusieurs éléments
MULTICONDUCTOR PLUG - [el] prise *f* de conducteur en faisceau
MULTICONE FRICTION CLUTCH - [mech] embrayage *m* à plusieurs cônes
MULTICORE - s. multipolar
MULTICOUPLER - [mech] coupleur *m* multiple
MULTICOURSE SYSTEM - [agric] polyculture *f*
MULTICUT - [mach tool] à outils multiples
~SEMIAUTOMATIC LATHE - [mach tool] tour *m* semiautomatique à outils multiples
MULTICYCLE FEEDING - [comput] alimentation *f* de cartes en plusieurs cycles
MULTICYLINDER - [mech] polycylindrique, à plusieurs cylindres
MULTIDAYLIGHT PRESS - [mech] (a press with floating platens between table and main head) presse *f* à plateaux multiples
MULTIDECK SCREEN - [mech] (type of sizing machine in which several screens of different mesh operate one above another) crible *m* à étages multiples
MULTIDISK BRAKE - [mech] (in landing gear, a disk-type brake comprising more than one friction disk) frein *m* à plusieurs disques
MULTIDROP - [comput] circuit *m* à dérivation multiple
MULTIELECTRODE - [electron] (of an electronic tube of more than three electrodes associated with a single electron stream) à plusieurs électrodes
~TUBE - [electron] (electronic tube containing more than three electrodes associated with a single electron stream) tube *m* à plusieurs électrodes
~VALVE - s. multielectrode tube
~VOLTAGE STABILIZER - [electron] tube *m* stabilisateur à plusieurs électrodes
MULTIELEMENT OIL COOLER - [mech] (type of cooler containing a number of units) radiateur *m* d'huile complexe
MULTIENGINED - [mech] à plusieurs moteurs
MULTIEXCHANGE AREA - [telecomm] réseau *m* à plusieurs bureaux centraux
MULTIFEEDER MACHINE - [mech] (machine fitted with several feeding devices) machine *f* à plusieurs dispositifs d'alimentation
MULTIFILAMENT YARN - [text] (yarn made by extrusion through spinnerets) fil *m* en multifilament
MULTIFILE REEL - [comput] bobine *f* à plusieurs fichiers
MULTIFLEX SCRATCHERS - [oil ind] gratteur *m* de tubage horizontal
MULTIFLOW COOLING SYSTEM - [nucl] système *m* de refroidissement à plusieurs voies
MULTIFOUNT READER - [comput] lecteur *m* de fontes multiples
MULTIFREQUENCY - [radio] (of a transmitter capable of operating on any of the present carrier frequencies) à plusieurs fréquences
~HETERODYNE GENERATOR - [radio] oscillateur *m* hétérodyne à plusieurs fréquences
~SYSTEM - [el] (system in which currents of different frequencies are superimposed) système *m* polycyclique
MULTIFUEL ENGINE - [mech] moteur *m* polycarburant
MULTIFUNCTION PUNCHED CARD MACHINE - [comput] machine *f* à cartes perforées à plusieurs fonctions
MULTIGRADE OIL - [mech] huile *f* multigrade
MULTIGRAPH - [print] (duplicating machine) polycopiste *m*
MULTIGRID MIXING VALVE - [radio] tube *m* convertisseur à plusieurs grilles
MULTIGRIP FLOOR - [metall] tôle bossée
MULTIGROUP MODEL - [nucl] (of a nuclear reactor in which the neutron flux is divided into parts which correspond to several discrete energy ranges) modèle *m* à plusieurs groupes
~THEORY - [nucl] (a theory which considers the neutron energy spectrum as consisting of a number of energy group) théorie *f* à plusieurs groupes
MULTIHOLE NOZZLE - [mech] (in internal combustion engines) injecteur *m* à plusieurs trous
MULTIMPRESSION MOULD - [metall & plast ind] (a mould with two or more mould impressions, i.e. a mould which produces more than one moulding per moulding cycle) moule *m* à plusieurs empreintes
MULTILAYER - [gen] à plusieurs couches
~COIL - [el] (coil consisting of several layers of wire) bobine *f* à plusieurs couches
~DIODE - [electron] diode *f* multijonctions
~WINDING - [el] (cylindrical winding in which several layers of wire are wound one over the other) enroulement *m* à plusieurs couches
MULTILAYERED CATHODE - [electron] (cathode, on the heater of which more than one layer of emitting material is applied) cathode *f* à plusieurs couches
MULTILIFT GASHOLDER - [gas ind] (telescopic gasholder, or multiple section gasholder) gazomètre *m* télescopique
MULTIMETER - [instr] (a test instrument) multimètre *m*
MULTINOMIAL - [math] (synonym of polynomial) polynôme *m*

MULTIPACTOR GAP LOADING - [electron] (electronic gap admittance) charge *f* de l'espace d'interaction

MULTIPASS DRYER - [mech] (type of dryer in which the material passes through the machine several times) séchoir *m* à plusieurs passages

~HEAT EXCHANGER - [th eng] échangeur *m* thermique à parcours multiple

MULTIPATH DISTORTION - [comput] distorsion *f* de signaux transmis par un système multivoies

~EFFECT - [telev] (double image, ghost) image *m* fantôme, double image *f*

~TRANSMISSION - [radio] transmission *f* sous plusieurs angles

MULTIPHASE - [el] multiphasé, polyphasé

MULTIPLANE - [aero] (an aeroplane having more than one main surface, i.e. more than one pair of wings) multiplan *m*

MULTIPLE - [gen & math] multiple
[el] (with two or more conductors connected in parallel) multiple, en faisceau
[el] (adj) en parallèle

~ACTION - [contr] (or compound action; of a controller which operates with more than one type of action) action *f* composée

~-ADDRESS CODE - [comput] instructions *fpl* à adresses multiples

~AERIAL - [radio] (aerial system with two or more separate aerials) antenne *f* multiple

~ANTENNA - s. multiple aerial

~-APERTURE CORE - [nucl] tore *m* à plusieurs trous

~ARCH DAM - [hydr] barrage *m* à voûtes multiples

~AVERAGING - [telev] (in colour television) détermination *f* multiple de valeurs moyennes

~BELTING - [mech] courroie *f* multiple

~-BLADE SAWING MACHINE - [mech] scie *f* à châssis à plusieurs lames

~BLOCK SIGNALLING - [railw] système *m* de signalisation à cantons multiples

~BOILER - [metall] chaudière *f* multitubulaire, chaudière *f* inexplosible

~BOX LOOM - [text] métier *m* à boîtes multiples

~BOX MOTION - [text] mouvement *m* à plusieurs boîtes (pour changement de navette)

~BRANCHING DELAY - [nucl] désintégration *f* à embranchement multiple

~BREAK CONTACTS - [el] contacts *mpl* de repos multiples

~-CAVITY DIE - [metall & plast ind] moule *m* à empreintes multiples identiques

~CHANNEL - [telecomm] (telecommunication system so arranged that more than one message can be passed over the same link as the same time) à plusieurs canaux

~-CIRCUIT WINDING - [el] enroulement *m* imbriqué

~COINCIDENCE - [comput] (coincidence of more than two signals or pulses) coincidence *f* multiple

~COLOURED YARN - [text] fil *m* à plusieurs couleurs

~-COLUMN CONTROL FEATURE - [comput] dispositif *m* de sélection sur positions multiples

~COMPLETION - [oil ind] (the completion of a single well exploiting more than one oil-bearing layer) complétion *f* multiple

~CONDUCTOR - [el] conducteur *m* en faisceau

~CONNECTION - [el] connexion *f* multiple

~CONTACT - [el] contact *m* multiple

~CONTAINER - [mech & el] châssis *m*

MULTIPLE CONVERTER - [el] convertisseur *m* multiple

~CRIT - [nucl] masse *f* critique exédante

~CROSSTALK - [acoust] (disturbing sounds in a telecommunication system due to crosstalk) diaphonie *f* multiple

~CRUCIBLE FURNACE - [metall] four *m* à plusieurs creusets

~-CURRENT GENERATOR - [el] (electrical machine capable of producing simultaneous currents of different voltages) génératrice *f* polymorphique

~DECAY - [nucl] (the occurrence of two or more modes by which a radionuclide can undergo radioactive decay) embranchement *m*, désintégration *f* multiple

~DECK OVEN - [gas ind] four *m* à plusieurs étages

~-DEGREE OF FREEDOM SYSTEM - [phys] système *m* à plusieurs degrés de liberté

~DIE - s. multi-cavity die

~DISINTEGRATION - s. multiple decay

~DISK CLUTCH - [mech] (clutch with multiple laminated disk) embrayage *m* multidisque

~-DISC SHUTTER - [photo] obturateur *m* à plusieurs disques

~DOME DAM - [hydr] barrage *m* à dômes multiples

~DRILLING MACHINE - [mach tool] perceuse *f* à broches multiples

~-DUCT CONDUIT - [telecomm] canalisation *f* multitubulaire, conduite *f* multiple de câbles

~ECHO - [acoust] (called flutter echo in the USA; a succession of separate echoes from one source) échos *mpl* multiples

~-EFFECT EVAPORATOR - [heat] (type of evaporator in which the heated vapour is passed to another unit and part of its heat transferred to the liquid therein) évaporateur *m* à multiple effet

~ELECTRODE - [el chem] (electrode at which several electrode reactions occur simultaneously) électrode *f* multiple, polyélectrode *f*

~ELECTROMETER - [instr] (electrometer in which several pairs of quadrants actuate several needles fixed to the same axis) électromètre *m* multicellulaire

~EXCITATION - [el] (system of excitation with several windings carrying separate currents) excitation *f* multiple

~-EXPANSION ENGINE - [mech] machine *f* à multiple expansion

~EXPOSURE - [photo] expositions *fpl* multiples

~FEED-RACK - [mech] rampe *f* de graissage à départs multiples

~FLUE - [th eng] (common, or shared flue; called common vent in the USA) conduit *m* unitaire

~GRATING - [acoust] (ultrasonic cross grating) réseau *m* de diffraction

~-GUN CATHODE RAY TUBE - [electron] tube *m* à rayons cathodiques à plusieurs canons

~INTENSITY RULES - [phys] règles *fpl* d'intensité pour un multiplet

~INTRUSIONS - [geol] (minor intrusions formed by several successive injections of approximately the same magma) intrusions *fpl* multiples

~IONIZATION - [electron] (the loss of an electron by an already positively ionized atom) ionisation *f* multiple

~JACK - [telecomm] jack *m* général, jack *m* de commutateur multiple

MULTIPLE JET NOZZLE - [auto] (of carburettor) injecteur *m* multijet
~-LENGTH ARITHMETIC - [comput] opération *f* à longueur de mot multiple
~LIFT GASHOLDER - s. multilift gasholder
~-LINE GARD - [comput] carte *f* multilignes
~-LINE PRINTING - [comput] impression *f* multilignes
~-LINE READ SELECTION - [comput] sélection *f* pour lecture multilignes
~LINEN - [text] toile *f* multiple
~LOOM SHUTTLE - [text] navette *f* pour métier à grille
~MOULD - [metall & plast ind] moule *m* multiple
~OBJECT PHASE TRACKING AND RANGING - [radar] système *m* de repérage et de yélémétrie d'objectifs multiples
~-OPERATOR WELDING UNIT - [el] (electric-arc welding generator which supplies current to two or more welding arcs operating in parallel) groupe *m* convertisseur de soudage pour deux ou plusieurs arcs
~PARALLEL WINDING - [el] (a winding of a drum armature in which the commutator pitch is greater than unity and less than the pole pitch) enroulement *m* parallèle multiplex
~PASS WELD - [metall] soudure *f* à passes multiples
~PILE-UP - [contr] pile *f* multiple
~PRINTING MACHINE - [comput] (machine printing the same information on several forms simultaneously) imprimate *f* multiple
~PRODUCTION - [nucl] production *f* multiple en collision unique
~PUNCHING - [comput] perforation *f* multiple
~RADIAL FEEDER - [el] ligne *f* en antenne multiple
~RECEPTION - [radio](simultaneous reception of two or more separate signals) réception *f* multiple
~RECTIFIER CIRCUIT - [el] circuit *m* redresseur multiple
~REFLECTIONS - [opt] (while some of the radiation of a light ray striking a transparent plane-parallel plate is reflected, some will enter the plate and be reflected back and forth between the two sides of the plate) réflexions *fpl* multiples
~RESISTANCE WELDING - [metall] soudage *m* par résistance multiples
~RHOMBIC AERIAL - [radio] (aerial system consisting of a number of stationary rhombic aerials, the composite major lobe of which is electrically steerable) antenne *f* en losange multiple
~ROPING - [el] entraînage *m* par câble à enroulement multiple
~SAND EXPLOITATION - [oil ind] exploitation *f* de faisceaux de sables
~SCANNING - [telev] analyse *f* multiple
~SCATTERING - [nucl] (the scattering of a particle in which the final displacement is the vector sum of several small displacements) diffusion *f* multiple
~-SCREW EXTRUDER - [mech] (type of extruder having more than one feed screw) extrudeuse *f* à vis multiples
~SECTION GASHOLDER - s. multilift gasholder
~SERIES CONNECTION - [el] connexion *f* en séries parallèles
~SOUND TRACK - [el acoust] (sound track divided into two or more separate tracks) piste *f* multiple
~SPARK-GAP - [el] éclateur *m* multiple
MULTIPLE-SPARK SYSTEM - [radio] (form of quenched spark system in which the spark discharge takes place across a series of gaps between metallic plates) système *m* à étincelles multiples
~-SPEED FLOATING CONTROL SYSTEM - [contr] système *m* de réglage flottant à plusieurs vitesses
~-SPEED MOTOR - [el] (a motor which can be operated at different given speeds by changing its electrical connexions) moteur *m* à vitesses multiples
~SPINDLE DRILLING MACHINE - [mach tool] perceuse *f* à broches multiples, machine *f* à percer multiple
~SPOT WELDING - [metall] soudure *f* continue par points
~-STAGE COMPRESSOR - [mech] compresseur *m* à plusieurs étages
~-STAGE PUMP - [mech] pompe *f* à plusieurs étages
~-STAGE TURBINE - [mech] turbine *f* multi-étages, turbine *f* à plusieurs étages
~-SWITCH STARTER - [el] (a starter in which separate hand-operated switches are provided for each step) démarreur *m* à interrupteurs multiples
~SYSTEM - [comput] système *m* de multitraitement [el chem] (electrolytic cell in which all the anodes are connected to one bus bar and all the cathodes to the negative) système *m* multiple
~THREADED SCREW - [mech] (a screw in which two or more threads are used to reduce the size of thread and maintain and adequate core strength) vis *f* multiple, vis *f* à plusieurs filets
~TOOL - [mech] outillage *m* multiple
~-TOOL LATHE - [mech] tour *m* à plusieurs outils
~TUBE COUNTS - [radiat] (spurous counts in radiation counter tubes induced by previous tube counts) comptages *mpl* parasites
~TUNED AERIAL - [radio] (antenna with a number of leads to earth connected at intervals, each lead having in series a tuning arrangement) antenne *f* à accords multiples
~TUNED ANTENNA - s. multiple tuned aerial
~TWIN-CABLE - [telecomm] (cable containing a number of multiple twin quads) câble *m* à paires combinables, quarte *m* DM
~-TWIN QUAD CABLE - s. multiple-twin-cable
~-UNIT CONTROL - [el] réglage *m* central de la commande
~UNIT RUNNING - [railw] marche *f* en unités multiples
~-UNIT SEMICONDUCTOR DEVICE - [electron] (semiconductor device with two or more sets of electrodes associated with independent carrier streams) système *m* semiconducteur multiple
~-UNIT STEERABLE ANTENNA - s. multiple rhombic aerial
~-UNIT TRAIN - [railw] (electric train set) rame *f* automotrice
[railw] (set of coaches) rame *f* à éléments multiples
~-UNIT TUBE - [electron] (electronic tube containing two or more groups of electrodes associated with independent electron streams within one envelope) tube *m* multiple
~-UNIT VALVE - s. multiple-unit tube
~VANE PUMP - [mech] pompe *f* multiple, pompe *f* à palettes multiples
~WEFT TRIMMING - [text] galon *m* à plusieurs trames
~WIRE AERIAL - [radio] (aerial consisting of a number of parallel wires) antenne *f* à brins en parallèle

MULTIPLE WIRE ANTENNA - s. multiple wire aerial
~-WIRE GLASSED HEADER - [electron] ensemble *m* base
MULTIPLET - [phys] (a group of several very closely packed spectrum lines, appearing as a single line) multiplet *m*
MULTIPLEX, to - [telecomm] (a transmit telegrams by multiplex) transmettre par multiplex
[radio] (to transmit by multiplex radio transmission) transmettre par multiplex
MULTIPLEX - [telecomm] (telegraph system providing a number of channels with an allocation of the line to the latter by means of distributors) (telegraphie) multiplex
[radio] s. multiplex radio transmission
~CEMENTING COLLAR - [oil ind] manchon *m* de cimentation étagée
~CHANNEL - [telecomm] voie *f* multiplex
~LAP WINDING - [el] enroulement *m* parallèle multiple
~RECEPTION - [radio] (the simultaneous reception of two or more separate signals using a specified common feature, e.g. a single aerial or a single carrier-frequency) réception *f* multiplex
~WAVE - [el] (series parallel winding; winding of a drum armature with more than two paths, in which the winding pitch is approximately double the pole pitch) enroulement *m* série-parallèle
MULTIPLEXED INFORMATION - [comput] informations *f* multiplexées
MULTIPLEXER - [comput] (or traffic pilot; device allowing concurrent operations) multiplexeur *m*
[opt] (optical device used to combine pictures from various sources into a common generating channel) multiplexeur *m*
~CHANNEL - [comput] canal *m* multiple
MULTIPLEXING - [contr] multiplexage *m*
~SOUND ON VISION - [telev] multiplesage *m* du son dans le canal vidéo
MULTIPLICAND - [math] multiplicande *m*
~REGISTER - [comput] (the register which contains the multiplicand and/or the divisor) registre *m* du multiplicande
MULTIPLICATION - [math] multiplication *f*
~BY CLONAL SEEDS - [genet] reproduction *f* asexuée
~BY GRAFTING - [agric] reproduction *f* par greffage
~POINT - [contr] (in feedback control systems) point *m* de multiplication
~TABLE - [math] table *f* de multiplication
MULTIPLICATIVE MIXING - [telev] mixage *m* à multiplication
MULTIPLICITY - [gen] multiplicité *f*
[nucl] (the numbers 2S + I, in which S represents the spin angular moment of an atom) multiplicité *f*
MULTIPLIER - [math] multiplicateur *m*
[electron] (or electron multiplier; section in which an initial electron current is amplified by successive dynodes) multiplicateur *m*
[comput] (device with two or more inputs and whose output is a representation of the product of the quantities represented by the input signals) multiplicateur *m* analogique
[el] résistance *f* additionnelle en série
~PHOTOTUBE - [electron] tube *m* photomultiplicateur
~QUOTIENT REGISTER - [comput] registre *m* multiplicateur-quotient
~REGISTER - [comput] registre *m* du multiplicateur

MULTIPLY, to - [math & gen] multiplier
MULTIPLY - (or multi-ply) [timber] (plywood with more than three layers of wood) contre-plaqué en plusieurs épaisseurs
~INSTRUCTION - [comput] instruction *f* de multiplication
MULTIPLYING - [math & gen] multipliant, multiplicatif
~CAMERA - [photo] (camera which makes it possible to take a number of small exposures on the same negative) camera *f* à objectif multipliant
~CONSTANT - [surv] (factor of distance computation by tacheometric method) constante *f* de multiplication
~FACTOR - [gen] facteur *m* de multiplication
[photo] facteur *m* de pose
~GAGE - s. multiplying gauge
~GAUGE - [instr] (a type of liquid manometer) manomètre *m* de précision
~LINKAGE - [mech] (a kinematic chain in which the original amplitude of movement is augmented) engrenage *m* multiplicateur
~MEDIUM - [phys] (a medium in which there occurs neutron multiplication) milieu *m* multiplicateur
~POWER OF GALVANOMETER SHUNT - [instr] (a factor by which the current passing through a galvanometer, connected to a shunt, must be multiplied, in order to obtain the total current passing through the shunt and the galvanometer) pouvoir *m* multiplicateur d'un shunt galvanométrique
~PULLEY - [mech] poulie *f* à plusieurs rainures
MULTIPOINT CONNECTOR - [electron] connecteur *m* multibroches
~IGNITION - [mech] (in internal combustion engines) allumage *m* en plusieurs points
MULTIPOLAR - [el] (of cables) multipolaire
[el] (relating to a machine to denote that the field magnet has more than two poles) multipolaire
~DYNAMO - [el] (dynamo with a field magnet of more than two poles) dynamo *f* multipolaire
MULTIPOLE - [el] (of a switch when it is suitable for breaking or making an electrical circuit on two or more poles simultaneously) multipolaire
~COMMON-FRAME CIRCUIT BREAKER - [el] commutateur *m* multipolaire à bâti commun
~MOMENT - [nucl] (the electric and magnetic multipole moment of a system in a specified state) moment *m* multipolaire
~RADIATION - [el] (the radiation field in free space expands into electric and magnetic multipole fields, as well as into plane waves) rayonnement *m* multipolaire
MULTIPOSITION ACTION - [contr] (control action in which predetermined control actions occur at two or more given values of the controlled conditions) action *f* à plusieurs positions
~CONTROL - [contr] (control system using at least three steps) réglage *m* à plusieurs positions
MULTIPROBE RADIATION METER - [nucl] polyradiamètre *m*
MULTIPROCESSING - [comput] multitraitement *m*
~MULTIPROCESS - [comput] multi-ordinateur *m*
MULTIPROPELLANT - [astronaut] polyergol *m*
MULTIPURPOSE EQUIPMENT - [agric] appareil *m* polyvalent
~MIXTURE - [bot] produit *m* polyvalent
~REACTOR - [nucl] réacteur *m* à plusieurs desseins

MULTIPURPOSE VEHICLE - [transp] camion *m* tous usages
MULTIRANGE INSTRUMENT - [instr] instrument *m* à gammes multiples
MULTIRATE METER - [instr] (integrating meter registering on one or another set of dials according to which the varying prices of electricity are operative) compteur *m* à tarif multiple
MULTIRATIO - [gen mech etc] à rapport multiple
MULTIREAD FEEDING - [comput] alimentation *f* à lecture multiple
MULTIREADING FEATURE - [comput] système *m* multilecture
MULTIREGION REACTOR - [nucl] réacteur *m* à plusieurs régions
MULTIROTATION - [chem] (or mutarotation, the change with time of the optical activity of a freshly prepared solution of a certain sugars) mutarotation *f*
MULTISEGMENT MAGNETRON - [electron] (a magnetron with a cylindrical anode divided into more than two segments) magnétron *m* à anode segmentée
MULTISEQUENTIAL SYSTEM - [comput] calculateur *m* à plusieurs sequences
MULTI-SHUTTLE - [text] à plusieurs navettes
MULTISPEED FLOTATING ACTION - [contr] action *f* flottante à plusieurs vitesses
MULTISPINDLE - [mach tool] à broches multiples
~DRILL - [mach tool] perceuse *f* à broches multiples
MULTISPIRAL - [telev] hélice *f* multiple
~SCANNING DISK - [telev] disque *m* d'analyse à hélice multiple
MULTISTABLE STORAGE - [comput] mémoire *f* multistable
MULTISTAGE - [mech etc] à plusieurs étages
~AMPLIFIER - [radio] (a cascade amplifier) amplificateur *m* en cascade
~CENTRIFUGAL PUMP - [mech] dispositif *m* d'épuisement en répétitions
~COMPRESSION - [phys] compression *f* à plusieurs étages
~DIFFUSION UNIT - [nucl] ensemble *m* de diffusion à plusieurs étages
~PROCESS - [nucl] (process of separating isotopes in a number of stages) procédé *m* à plusieurs étages
MULTISTART THREAD - s. multiple-thread screw
~WORM - [mech] (a worm in which two or more helical threads are used to increase the velocity ratio of the drive) vis *f* sans fin à filetage helicoïdal multiple
MULTISTEP ACTION - [contr] action *f* par plusieurs échelons
~CONTROL - [contr] action *f* par échelons multiples
~CONTROL SERVOMECHANISM - [comput] servomécanisme *m* de commande à commutateur
MULTITONE - [acoust] (tone having a plurality of simple harmonic tones) multiton *m*
~HORN - [auto] avertisseur *m* à sons multiples
MULTITRACK TAPE - [el acoust] bande *f* multipiste
~TAPE RECORDER - [el acoust] enregistreur *m* à plusieurs pistes
MULTITUBULAR - [th eng] multitubulaire
~BOILER - [th eng] chaudière *f* multitubulaire
MULTIVALENCE - [chem] polyvalence *f*, plurivalence *f*
MULTIVALENT - [chem] (polyvalent, i.e. having a valency greater than I) polyvalent, plurivalent
MULTIVALVE - [mech] multivalve *f*
MULTIVIBRATOR - [electron] (arrangement of thermionic valves which sustains a relaxation oscillator) multivibrateur *m*
MULTIVOLTAGE CONTROL - [el] régulation *f* par variation de courant
MULTIWHEEL LAYOUT - [aero] (landing gear in which more than four wheels are used) train *m* d'atterrissage à plusieurs roues
MULTIWIRE-TRIATIC AERIAL - [radio] (aerial array in which the aerials form the sides of a triangle) antenne *f* multifilaire en triangle
~-TRIATIC ANTENNA - s. multiwire-triatic aerial
MUMA - [med] pyomyosite *f* tropicale
MUMBLER - [glass man] (a blower) souffleur *m*
MUMPS - [med] oreillons *mpl*, parotidite *f* émidémique
MUNDIC - [min] (iron pyrite) mundick *m*, pyrite *f* de fer
MUNGO - [text] (the waste obtained from felted cloth) mungo *m*, laine *f* renaissance
~BEAN - [agric] haricot *m* velu
~YARN - [text] fil *m* mungo
MUNITION - [gen] (ammunition and more generally the necessary war material) munitions *fpl*
MUNSELL CHROMA - [opt] (the dimension of the Munsell system of colour which corresponds most closely to saturation) saturation *f*
MUON - [phys] (a mu meson) muon *m*, méson *m* mu
~NUMBER - [phys] nombre *m* muonique
MUONIUM - [phys] (bound system consisting of a positive mu meson and an electron) muonium *m*
MURAL ABSCESS - [med] abcés *m* de la paroi abdominale
~TYPE RADIATION BEACON - [nucl] balise *f* de rayonnement murale
MURIATIC - [chem] (obsolete for hydrochloric) chlorhydrique
~ACID - [chem] (old name for hydrochloric acid) acide *m* chlorhydrique, esprit *m* de sel
MURMUR - [gen & med] murmure *m*, bruit *m* (cardiaque), souffle *m*
MURRAIN - [vet] épizootie *f*
MURRAY LOOP - [instr] (bridge circuit used to find faults on communication circuits) boucle *f* de Murray
MUSA ANTENNA - s. multiple rhombic aerial
MUSCARINISM - [med] empoisennement *m* par les champignons
MUSCLE - [anat] muscle *m*
~PLATE - [anat] lame *f* musculaire, myotome *m*
~POISON - [med] poison *m* musculaire
MUSCOVITE - [min] (a native hydrous potassium aluminium silicate with uses as a filler for rubber and paints, as a dielectric and in metallurgical equipment) muscovite *f*, mica *m* blanc
MUSCULAR SYSTEM - s. musculature
MUSCULATURE - [anat] système *m* musculaire, musculature *f*
MUSH - [gen] bouillie *f*
[radio] (noise due to irregularities in the arc discharge) bruit *m* de friture, brouillage *m*
[electron] (soft confused background noise in electronic sound equipment) distorsion *f*, brouillage *m*
~AREA - [radar] (interference area, nuisance area)

zone *f* de brouillage
MUSH-ICE - [met] glace *f* sans consistance
~WINDING - [el] (single multilayer coil in which the layers are not definite, nor spaced from one another) enroulement *m* brouillé
MUSHROOM - [bot] champignon *m*
[gen] (having the shape of a mushroom) champignon, à champignon
~ANCHOR - [naut] crapaud *m* de mouillage
~-BOLT - [mech] boulon *m* à tête en goutte-de-suif
~BURNER - [th eng] brûleur à champignon
~CAM - [mech] came *f* à champignon
~CONSTRUCTION - [constr] (concrete construction consisting of columns and floor slabs) construction *f* à champignon
~FLOOR - [constr] plancher *m* champignon
~FOLLOWER - [mech] galet *m* à champignon
~HEAD - [mech] (of a valve) tête *f* en goutte-de-suif
~-HEAD VALVE - [mech] (type of valve in which the obturating body is of a round flattente shape, resembling a mushroom) soupape *f* à tête en goutte-de-suif
~INSULATOR - [el] isolateur *m* à cloche
~IRONING PRESS - [text] presse *f* bombée
~MIXER - [mech] (mixer having a mushroom-shaped closed bowl rotating about an axis inclined at about 60 deg. to the horizontal, used with or without balls for various types of mixing and grinding etc) mélangeur *m* à tonneau désaxé
~VALVE - [mech] soupape *f* en champignon
[oil ind] tiroir *m* à coquille
~VENT - [oil ind] soupirail *m* à vent
MUSHY STATE - [metall] état *m* pâteux
MUSIC - [mus] musique *f*
[chem] s. mucic
~ACID - [or mucic acid] (a synthesis intermediate) acide *m* mucique
~LOADING - [telecomm] charge *f* pour radiodiffusion musicale
~NEGATIVE - [acoust] musique *f* conservée
~PAPER - [paper man] papier *m* à musique
MUSICOLOGY - [mus] (the study of the art and history of music) musicologie *f*
MUSK - [bot] musc *m*
~-MALLOW - [bot] mauve *f* musquée
~MELON - [bot] melon *m* brodé
~-OX - [zool] boeuf *m* musqué, ovibos *m*
~-PEAR - [bot] muscat *m*, poire *f* musquée
MUSKET - [firearm] (hand-gun for infantry) mousquet *m*
MUSKMALLOW - [agric] ambrette *f*, ketmie *f* musquée
MUSKRAT - [zool] rat *m* musqué, ondatra *m*
MUSKY GOURD - [agric] courge *f* musquée
MUSLIN - [text] (very light cotton cloth) mousseline *f*
~BINDING - [text] (used for insulation of pipes) couche *f* de liage en mousseline
~GAUZE - [text] gaze *f* de mousseline, tissu *m* de coton écru
MUSQUASH - [zool] rat *m* musqué, castor *m* du Canada
MUSSEL - [zool] moule *f*
~POISON - [chem] toxine *f* des moules
~POISONING - [med] intoxication *f* par les moules
~SCALE - [zool] cochenille *f* virgule
MUSSITATION - [med] (a quiet delirium) mussitation *f*
MUST - [agric] (grape juice in course of fermentation) moût *m*
[gen] moisi *m*, moisissure *f*
~OPERATE VALUE - [contr] valeur *f* de mise au travail de consigne
~RELEASE VALUE - [contr] valeur *f* de décroissement de consigne
MUSTARD - [bot] moutarde *f*
~GAS - [chem] (a poison gas) ypérite *f*, gaz *m* moutarde
~-LEAF - [med] sinapisme *m*, papier *m* rigollot
~OIL - [chem] (an essential oil derived from varieties of Sinapis and used in medicine) sénévol *m*, essence *f* de moutarde
MUSTY - [gen] (smelling of mould) (odeur etc) de moisi
MUTABILITY - [genet] mutabilité *f*
MUTAFACIENT - [genet] mutateur
MUTAGENIC - [genet] mutagène
MUTANT - [genet] (the individual showing of a mutation) mutant
~GENE - [genet] (a gene which undergoes mutation) gène *m* mutant
MUTAROTATION - s. multirotation
MUTATION - [phys & genet] (the gradual variation towards a definite) mutation *f*
[genetics] mutation *f*
~BREEDING - [genet] sélection *f* par mutation
~RATE - [genet] taux *m* de mutation
~STOP - [mus] (register sounding a note of different name from that of they key pressed) jeu *m* de mutation
MUTATOR - [el] (a frequency converter) mutateur, soupape *f* à vapeur de mercure
~MOTOR - [el] moteur *m* à mutateur
~TRANSFORMER - [el] transformateur *m* de mutateur
MUTE - [gen] muet
[mus] sourdine *f*
~ANTENNA - [radio] (an artificial aerial, i.e. a load of resistive impedance connected to the output of a transmitter in place of the aerial) antenne *f* artificielle
MUTED DRUM - [mus] tambour *m* drapé
MUTILATED SELECTION - [telecomm] sélection *f* inachevée
MUTILATION - [med] mutilation *f*
MUTING - [radio etc] (the operation of preventing feedback from loudspeaker to microphone in an aircraft, or similar, internal telephone system) blocage *m* de réponse
~CIRCUIT - [telecomm] (circuit designed to obviate feed-back from loudspeaker to microphone in intercommunication telephone system) circuit *m* de réglage silencieux
~RELAY - [telecomm] (a relay used in a muting circuit to bring in a resistor when a pressbutton circuit is closed) relais *m* de blocage de réponse
~SWITCH - [radio] (silencing switch) commutateur *m* de réglage silencieux
~THRESHOLD - [radio] seuil *m* de réglage silencieux
MUTISM - [med] mutisme *m*
MUTOSCOPE - [cin] (early instrument for showing recorded continual motion) mutoscope *m*
MUTTON - [food] (the flesh of sheep) mouton *m*
[print] (em quad) cadratin *m*
~QUAD - s. mutton (print)

MUTTON RULE - [print] (em rule; dash with the width of one 'm') tiret *m*
MUTUAL - [gen] mutuel, réciproque
[el] mutuel
~ATTRACTION - [phys] (between particles in a system) attraction *f* mutuelle
~BRANCH - [el] (the common branch of a network) dérivation *f* commune
~CAPACITANCE - [el] capacité *f* mutuelle
~CHARACTERISTIC - [electron] (the relation between the current of one electrode and the voltage of another electrode) caractéristique *f* interélectrode
[radio] (relationship between output and input of any electronic amplifier or transducer) caractéristique *f* grille-anode
~CONDUCTANCE - [electron] (the control-grid-to-anode transductance) pente *f*
~CONDUCTANCE METER - [instr] appareil *m* de mesure de la pente
~INDUCTANCE - [el] (the magnetic flux which the current flowing in one circuit induces in another circuit, divided by the current in the first circuit) coefficient *m* d'induction mutuelle
~INDUCTION - [el] (the induction of an electromotive force in one circuit by the change in the current flowing through another circuit) induction *f* mutuelle
~INDUCTION BRIDGE - [instr] (for the measure of mutual inductance in terms of resistance and capacitance) pont *m* à induction mutuelle
~INTERACTION - [phys] (the complex of forces between particles in a system) interaction *f* mutuelle
~REPULSION - [phys] (the repulsive force between particles in a system) répulsion *f* mutuelle
~SURGE IMPEDANCE - [el] impédance *f* d'onde mutuelle
MUTULE - [arch] (rectangular block of a Doric column) mutule *f*
MUZZLE, to - [naut] (the operation of gathering the sails) haler bas
MUZZLE - [firearms] bouche *f*, gueule *f*
[zool] (the projecting snout of an animal) museau *m*
~LOADER - [firearm] pièce *f* se chargeant par la bouche
~-LOADING - [firearms] se chargeant par la bouche
~PROTECTOR - [firearm] couvre-bouche *m*
~SWELL - s. muzzle bell
~VELOCITY - [ballistics] (the velocity of the bullet, or projectile, when it leaves the muzzle of a gun) vitesse *f* initiale, vitesse *f* à la bouche
mw - [el] (abbrev. for megawatt) megawatt *m*
MYCELIUM - [bot] (the filamentous structure of a fungus) mycélium *m*
MYCETISM - [med] s. muscarinism
MYCOPUS - [med] muco-pus *m*
MYCOSIS - [bot] (plant disease) mycosis *f* fungoïde
MYDRIASIS - [med] midriase *f*, dilatation *f* pupillaire
MYELAPOPLEXY - [med] hématomyélie *f*, hémorragie *f* de la moelle épinière
MYELENCEPHALITIS - [med] myélencephalite *f*
MYELIN - [anat] myéline
~GLOBULES - [anat] globules *mpl* myéliniques
MYELINIZATION - [physiol] myélinisation
MYELINOLYSIS - [med] myélinolyse *f* centrale
MYELITIS - [med] myélite *f*
MYELOGRAM - [med] myélogramme *m*
MYELOGRAPHY - [med] (radiological examination of the space between the theca and the spinal cord following an injection of air) myélographie
MYELOPARALYSIS - [med] paralysie *f* spinale
MYELOPLAX - [biol] ostéoclaste *m*
MYELOSIS - [med] myélose *f*
MYELOSYPHILIS - [med] syphilis *f* médullaire
MYESTHESIA - [med] sensibilité *f* musculaire
MYIASIS - [vet] myase *f*
MYOBLAST - [biol] myoblaste *m*
MYOCARDITIS - [med] myocardite *f*
MYOCELIALGIA - [med] douleur *f* dans les muscles abdominaux
MYOCEROSIS - [med] dégénérescence *f* cireuse des muscles
MYODEMIA - [med] dégénérescence *f* graisseuse musculaire
MYOEDEMA - [med] oedème *m* musculaire
MYOFIBROSIS - [med] fibrose *f* musculaire, sclérose *f* musculaire
MYOMA - [med] (muscular tumour) myome *m*
MYOMETRIUM - [anat] (muscular layer of the uterus) couche *f* musculaire de l'utérus, myomètre *m*
MYONEUROSIS - [med] myoneurose *f*
MYOPATHIA - [med] myopathie *f*
MYOPIA - [med] myopie *f*
MYOPLASTY - [med] myoplastie *f*, greffe *f* musculaire
MIORRHAPHY - [med] myorraphie *f*
MIORRHEXIS - [med] rupture *f* d'un muscle, myorrexis *f*
MYRINGITIS - [med] myringite *f*
MYRINGODECTOMY - [med] myringectomie *f*
MYRISTIC ACID - [chem] (an intermediate for flavours and perfumes) acide *m* myristique
MYRISTOYL PEROXIDE - [chem] (a catalyst for vinyl monomers) peroxyde *m* de myristoyle
MYRISTYL ALCOHOL - [chem] (a fixative for perfumes, plasticizer and synthesis intermediate) alcool *m* myristylique
MYRRH OIL - [chem] (a perfumery agent) essence *f* de myrrhe
MYRTLE - [bot] myrte *m*
MYTILOTOXIN - s. mussel poison
MYZESIS - [physiol] (sucking)succion *f*, aspiration *f*

N

n - [phys] (symbol for mols) nombre de molécules (abbrev. for normal; containing an unbranched carbon chain in the molecule) normal.
[light] (symbol for refractive index) symbole de l'indice de réfraction

N - [chem] (symbol for nitrogen) symbole de l'azote
(abbrev. for Normal Concentration) concentration normale
[mech draw] normalisé
[mech] (symbol of Modulus of Rigidity) symbole du module de cisaillement
[print] demi-cadratin

N.A. - [constr] (abbrev. for Neutral Axis) axe neutre
[opt] (initials of Numerical Aperture, q.v.) ouverture numérique

NA - [chem] (symbol of sodium) symbole du sodium

N.A.C.A. COWLING - [aero] (recommended for a radial engine) capotage *m*

N-ADDRESS CODE - [comput] (form of coding in which each instruction is related to addresses) code *m* à n adresses

N-ADDRESS ELECTRONIC COMPUTER - [comput] calculateur *m* électronique à n adresses

N-ADDRESS INSTRUCTION FORMAT - [comput] format *m* d'instruction à n adresses

n-BODY PROBLEM - [mech] (three-body problem) probleme *m* à trois corps

n-CHANNEL TAPE - [comput] bande *f* à n canaux

N CORES PER BIT STORAGE - [comput] mémoire *f* à n tores par bit

N-CUBE - [math] cube *m* de n dimensionnel

N-DIGIT NUMBER - [math] nombre *m* à "n" chiffres

N DISPLAY - [radar] (a combination of K and M display) indicateur *m* type N

N-LEVEL ADDRESS - [comput] adresse *f* à niveaux

N-LEVEL LOGIC - [math] logique *f* à n niveaux

n-LEVEL TAPE - s. n-channel tape

N-LINE - [spectro] (one of the lines in the N-series of x-rays) ligne *f* N

N-PLUS-ONE ADDRESS INSTRUCTION - [comput] instruction *f* à (n+I) adresses

N-RADIATION - [nucl] (one of a series of x-rays due to the excitation of electrons of the N-shell) rayonnement *m* N

N SCOPE - s. display

NACELLE - [aero] (any enclosed structure forming part of an aircraft but not of the fuselage itself, and designed to accomodate crew, passengers, engines etc) nacelle *f*, carlingue *f*, fuseau *m*
[of an airship] nacelle *f*
[radar] (or radome, or blister; an insulated cover which is transparent to radio frequency energy for protection of radar aerials) dôme *m* radar, radome *m*

NACRE - [zool] (shellfish from which mother-of-pearl is obtained) pinne *f* marine, nacre *f*

NACREOUS - [gen & min] (lustrous, having a pearl-like sheen) nacré, perlaire
~CLOUD - [met] (cloud having a pearly appearance) nuage *m* nacré

NACRITE - [min] (variety of clay mineral) nacrite *f*

NADIR - [astr] (the pole of the observer's horizon which is vertically below his feet) nadir *m*
[gen meas] (the lowest possible point) nadir *m*

NADIRAL - [astr] (of nadir) nadirale

NAEGITE - [min] (zirconim yttrium, niobium, tantalum, thorium and uranium containing mineral) naégite *f*

NAG - [zool] petit cheval *m*

NAGANA - [vet] nagana *m*, trypanosomiase *f* (par la mouche tsétsé)

NAGELFLUH - [geol] (group of massive conglomerates forming the Rifi and Rossberg in Switzerland) nagelfluh *m*

NAGYAGITE - [min] (black Tellurium. Lead-gold sulphotelluride, occurring naturally and containing 6 to I2 p.c. of gold) nagyagite *f*, tellure *m* noir

NAIL, to - [gen] clouer, clouter

NAIL - [zool] (the horny plate at the end of a flinger or toe) ongle *m*
[impl] (a piece of metal with a point and a head) clou *m*
[meas] (two and a quarter inch) 2,25 pouces
~BED - [anat] lit *m* de l'ongle
~-CLAW - [impl] arrache-clou *m*, arrache-pointe *m*
~DRAWER - [impl] s. nail-claw
~EXTENSION - [med] extension *f* par broche
~FIDDLE - [mus] (also called iron fiddle, nail harmonica, or nail violin) violon *m* de fer
~FOR REED - [constr] clou *m* galvanisé fixant les lattes du plafond
~FOR ROOFING SLATES - [impl] clou *m* à ardoise
~FOR TARRED FELT - [constr] clou *m* à carton-pierre
~HARMONICA - s. nail fiddle
~HEAD - [gen] tête *f* de clou
~HOLE - [gen] clouure *f*, étampure *f*, onglet *m*
~INJURY - [auto] (tyres) crevaison *f* par clou
~PASS - [metall] cannelures *fpl* pour clous
~PLATE - [anat] corps *m* de l'ongle
~PULLER - [impl] s. nail-claw
~PUNCH - [impl] (steel rod tapering at one end almost to a point) pointeau *m*
~PUNCTURE - [rubber ind] (of tyres) perforation *f* par clou

NAIL ROD - [metall] fer *m* à clous
~SET - s. nail punch
~VIOLIN - s. nail fiddle
NAILHEAD SPAR - [min] variété *f* de calcite
NAILING - [gen] couage *m*, clouement *m*, cloutage *m*
[leather ind] (the stretching and nailing damped skins to a pattern) remplissage *m* du chargeur de clous
[metall] chauffage *m* progressif de creusets vers le rouge
~BLOCK - [carp] bloc *m* pour clouer
~STRIP - [carp] latte *f*, traverse *f*
NAILPROOF LINER - [rubber ind] (of tyres) pare-clous *m*, bouclier *m*
NAKED - [gen] nu
~BARLEY - [agric & brew ind] orge *f* nue
~FLOORING - [carp] (timbers without the boards) appareil *m* porteur du plancher
~LIGHT - [light] (unscreened, open light) feu *m* nu, flamme *f* nue
NAKER - [mus] (tuned basin-shaped drums) timbale *f*
NAME, to - [gen] nommer, donner un nom, désigner
NAME - [gen] nom *m*
~ADDRESS AND RESIDENCE CARD - [comput] carte *f* de nom, adress et domicile, carte *f* NAD
~PLATE - [gen] (plate bearing the name of the manufacturer) plaque *f*
[auto] emblème *m*, écusson *m* de marque
NAND-OPERATION - [comput] opération *f* NON-ET, opération *f* ON
NANISM - [med] (the condition of being abnormally small) nanisme *m*
NANKEEN - [text] nankin *m*
[dye] chamois, jaune pâle
NANNY GOAT - [zool] chèvre *f*, bique *f*
NAP, to - [text] (the operation of raising a soft downy surface on a fabric) lainer, garnir, molletonner
NAP - [text] (soft woolly surface on fabrics) poil *m*, duvet *m*, lainer *m*
~CLOTH - [text] tissu *m* floconneux
~CLOTH WEAVE - [text] armure *f* d'étoffe flocon
~OF A HAT - [text] peluche *f* long-poil
~WARP - [text] chaîne *f* de poil, chaîne *f* supérieure
NAPALM - [chem] napalm *m*
NAPEX - [anat] occiput *m*
NAPHTA - [oil ind] (petroleum fraction covering the end of the petrol and the beginning of the kerosene range. Used in reforming processes) naphte *m*
~SOLUTION OF RUBBER - [rubber ind] dissolution *f* benzolique de caoutchouc
NAPHTHALENE - [chem] (a fungicide and an intermediate for dyestuffs, synthetic resins and a number of other organic compounds) naphtalène *m*, naphtaline *f*
~BLACK AB - [chem] (a dyestuff derived from naphthalene) noir *m* acide 4B
~CRYSTALLIZER - [ind chem] (for the tar preparation) cristallisoir *m* à naphtaline
~DERIVATIVES - [chem] (substitution products of naphtalene) dérivés *mpl* du naphtalène
~DISULPHONIC ACID - [chem] (an intermediate for dyestuffs) acide *m* bisulfonique de naphtalène
~OIL - [chem] (creosote oil) huile *f* à naphtaline, huile *f* de créosote
NAPHTHALENESULPHONIC ACID - [chem] (a synthesis intermediate) acide *m* sulfonique de naphtalène
NAPHTHALINE - [chem] (tar camphor) naphtaline *f*
NAPHTHAZARIN - [chem] naphtazarine *f*
NAPHTHENES - [chem] (group of saturated ring hydrocarbons, obtained from petroleum) naphtènes *mpl*
NAPHTHENIC ACIDS - [chem] (the organic acids characterized by the presence of a naphtene ring and one or more carboxyl groups) acides *mpl* naphténiques
NAPHTHIONIC ACID - [chem] (an intermediate for dyestuffs) acide *m* naphtionique
NAPHTHOL - [chem] (an intermediate for perfumery compounds, dyestuffs, drugs and rubber chemicals) naphtol *m*
NAPHTHOL DYE - [dye] colorant *m* à la glace, colorant *m* naphtol
NAPHTHOLDISULPHONIC ACID - [chem] (an intermediate for dyestuffs) acide *m* bisulfonique de naphtol
NAPHTHOQUINONE - [chem] (an intermediate for drugs and dyes and a polymerization modifier for synthetic resins and rubbers) naphtoquinone *m*
NAPHTHYLAMINE - [chem] (an intermediate for azo-dyestuffs) naphtylamine *f*
~DISULPHONIC ACID - [chem] (a dyestuffs intermediate) acide *m* naphtylamin-disulfonique
~SULPHONIC ACID - [chem] (a dyestuffs intermediate) acide *m* naphtylamin-sulfonique
~TRISULPHONIC ACID - [chem] (an intermediate for dyestuffs) acide *m* naphtylamin-trisulfonique
NAPHTHYLENEDIAMINE - [chem] (a synthesis intermediate) naphtylène-diamine *f*
NAPHTHYL ETHYL ETHER - [chem] (a perfumery compound) éther *m* naphtyléthylique
~METHYLCARBAMATE - [chem] (an insecticide) naphtyl-méthyl-carbamate *m*
NAPHTHYLTHIOUREA - [chem] (a rodenticide) naphtyl-thio-urée *f*
NAPIER - [meas] (a unit expressing the scalar ratio of two voltages; preferably "neper") néper *m*
~GRAS - [bot] herbe *f* à éléphant
NAPIERIAN LOGARITHMS - s. natural logarithms
NAPIFORM - [anat] napiforme, en forme de navet
NAPLES YELLOW - [paint] (load antimoniate, a yellow pigment used in artists' colours) jaune *m* de Naples
[chem] antimoniate de plomb
NAPLESS FINISH - [text] apprêt *m* rasé
NAPOLEONITE - [min] (structures consisting of shells of hornblende and feldspar) napoléonite *f*
NAPPE - [geol] (a major structure of mountains) chaîne *f* de montagnes
[hydr] (the water above a weir crest) nappe *f*
[math] nappe *f*
~INLIER - [geol] fenêtre *f*
~OUTLIER - [geol] témoin *m* de chevauchement
NAPPED - [text] (of a fabric) pelucheux, poilu
NAPPER - s. napping machine
NAPPING - [text] (the operation of raising a downy surface on a fabric) garnissage *m*, lainage *m*, tirage *m* à poil
~EFFECT - [text] effet *m* de lainage, effet *m* de grattage
~MACHINE - [text] (the machine used for the napping of woollen fabrics) laineuse *f*
~ROLLER - [text] cylindre *m* à lainer
NAPS - [text] tissus *mpl* floconneux
[text] (loops) boucles *fpl*
NARCEINE - [chem] (an alkaloid occurring in opium) narcéine *f*

NARCISSISM - [med] narcissisme *m*, auto-érotisme *m*
NARCOANALYSIS - [med] narco-analyse *f*
NARCODIAGNOSIS - s. narcoanalysis
NARCOSIS - [med] narcose *f*
~INHALER - [med impl] inhalateur *m* pour narcose
NARCOTIC - [chem] (a substance which induces unconsciousness) narcotique *m*
NARCOTINE - [chem] (an alkaloid occurring in opium) narcotine *f*
NARRATION - [cin] (the reading of a story by the narrator) narration *f*
NARRATOR - [radio cin etc] (the announcer reading the story or the commentary) commentateur *m*
NARROW, to - [gen] resserrer, se resserrer
[gen] (to reduce the size, the extent) se rétrécir
NARROW - [gen] étroit, serré
[gen](limited in extent) restreint, étroit
[min] (of a gallery) à l'étroit
~-ANGLE LENS - [photo] objectif *m* à angle étroit
~BAND - [radio] (less than 300 KHz) bande *f* étroite
~-BAND AXIS - [telev] (in colour television) axe *m* de la primaire transmise à bande étroite
~-BANDPASS FILTER - [electron] filtre *m* à bande étroite
~-BASE TOWER - [el] (tower for overhead transmission lines with a base supported on a single foundation) pylône *m* simple
~BEAM - [radio] (of an aerial) faisceau *m* filiforme
~-BEAM ABSORPTION - [radiat] (absorption measured in such condition, that scattered radiation is excluded from the measuring medium) absorption *f* à champ étroit
~-CUT FILTER - [telev] filtre *m* chromatique à bande étroite
~DIMENSION - [electron] (non-critical dimension) côté *m* étroit, dimension *f* non-critique
~EARED - [agric] (of grain, barley etc) à épis étroits
~FLAME - [metall] flamme *f* pointante
~-GAUGE - [railw] (railways gauge which is less than the standard width) à voie étroite
~-GAUGE FILM - [photo] film *m* substandard
~GAUGELINE - [railw] ligne *f* à voie étroite
~-GAUGE RAILWAY - [railw] chemin de fer *m* à voie étroite
~-GAUGED FILM - [cin] (film smaller than 32 mm) film *m* réduit, format *m* réduit
~INDENTED CHISEL - [impl] équarrissoir *m* étroit
~MESHED - [mech etc] à mailles étroites
~PULSE - [electron] impulsion *f* fine
~SHOWER - [nucl] (shower of cosmic-ray particles over a limited area) gerbe *f* étroite
~WORK - [mining] travail *m* à l'étroit, travaux *mpl* étriqués
NARROWING - [text] diminution *f*
[gen] rétrécissement *m*
~CAM - [text] came *f* pour diminutions
~CHAIN - [text] chaine *f* à diminution
~FINGER - [text] poinçon *m* de diminution
~HEAD GUIDE BAR - [text] barre-guide *f* de coulisseau du chariot de diminution
~LIFT LEVER - [text] bielle *f* de la mécanique à diminution
~LINK - [text] grain *m* à diminution
~LOOP - [text] point *m* de diminution, maille *f* reportée
~MACHINE - [text] machine *f* à diminution, mécanique *f* de diminution, diminueuse *f*
NARROWING OF THE VISUAL FIELD - [opt] rétrécissement *m* du champ de vision
~RATCHET - [text] rochet *m* du chariot de diminution
~ROD - [text] barre *f* à poinçons, porte-poinçons *m*
~SPINDLE - [text] vis *f* du chariot de diminution
NASAL - [zool] (pertaining the nose) nasal
[anat] épine *f* nasale
~CAVITY - [anat] fosse *f* nasale
~FIELD - [physiol] champ *m* olfactif
~SYRINGE - [med] irrigateur *m* nasal
NASCENT - [chem] (just formed by a chemical reaction) naissant
~GAS - [chem] (probably in the atomic state) gaz *m* naissant
~RED - [metall] rouge *m* naissant
~STATE - [chem] (just formed and thus very reactive) état *m* naissant
NASOPHARYNGITIS - [med] rhino-pharyngite *f*
NASTURAN - [min] (mineral mainly consisting of uranium oxides, an important ore of uranium and radium) pechblende *f*, uranite *f*
NATAL - [med] natal
NATALITY - [med & gen] natalité *f*
NATATORIAL - [zool] (adapted for swimming) natatoire
~ORGANS - [zool] organes *mpl* natatoires
NATIMORTALITY - [med] (stillbirth rate) mortinatalité *f*
NATIONAL - [gen] national
~PHYSICAL LABORATORY - (N.P.L.) (British National establishment which, among other things, tests and issues performance certificates for scientific and operational instruments) Laboratoire *m* Britannique de Physique
~PLATE - [auto] plaque *f* de nationalité
~TAG - s. national plate
NATIONALIZATION - [leg] (the State control or owership of previously privately owned concerns) nationalisation *f*, étatisation *f*
NATIONALIZE, to - [leg] nationaliser, étatiser
NATIVE - [min] (term used of a metal occurring naturally in the free state, e.g. native copper) natif, naturel
[gen] indigène, natal, inhérent
~ALBUMIN - [biochem] albumine *f* naturelle
~ASPHALT - [min] roche *f* asphaltique
~COPPER - [min] cuivre *m* (à l'état) natif
~CUSHION GAS - [gas ind] (gasholders) gaz-coussin *m* natif, gaz-coussin *m* d'origine
~GOLD - [min] or *m* natif
~ROCK - [geol] (rock of no value forming the walls of a reef) gangue *f*
[geol] roche *f* mère
~RUBBER - [rubber ind] caoutchouc *m* indigène
NATRIUM - [chem] (synonym of sodium) sodium *m*
NATROLITE - [min] (hydrated silicate of sodium and aluminium) natrolite *f*, pierre *f* de soude
NATRON - [min] (a vitreous, alkaline, hydrous sodium carbonate) natron *m*
NATRURESIS - [med] natriurèse *f*
NATTA CATALYST - [chem] (a stereospecific catalyst) catalyseur *m* Natta
NATURAL - [gen] naturel
[mus] (the sign restoring a note to its natural pitch) bécarré *m*
~ABUNDANCE - [nucl] (the relative abundance of an

isotope in a natural isotopic mixture) teneur *f* isotopique naturelle
NATURAL ACTIVITY - [nucl] radioactivité *f* naturelle
~AGEING - [metall] (the process of change in material with the passage of time, as distinct from similar changes induced artificially) vieillissement *m* naturel
~ASPHALT ROCK - [geol] roche *f* asphaltique naturelle
~BACKGROUND RADIATION - [nucl] fond *m* naturel de rayonnement, rayonnement *m* ionisant naturel
~BED - [constr] lit *m*, lit *m* de carrière
~BONDED SAND - [metall] sable *m* à agglutinant naturel
~CEMENT - [min] (structural cement formerly much used, and made by calcining argillaceous limestone) ciment *m* naturel
~CIRCULATION REACTOR - [nucl] réacteur *m* à circulation naturelle
~CONVECTION - [heat] (the transport of the heat through the motion of a fluid which is locally heated) convection *f* propre
~COVER - [bot] (of plants) protection *f* naturelle
~-DRAUGHT - [metall] (the airflow through a furnace induced by a chimney and depending on the height of the latter) courant *m* d'air naturel aspiré
[gen] à tirage naturel
~-DRAUGHT BURNING - [th eng] combustion *f* à courant d'air naturel
~-DRAUGHT GAS-PRODUCER - [gas ind] gazogène *m* à tirage naturel
~ELEMENT - [phys] (as found in nature; that is, without its accompanying isotopes) élément *m* naturel
~EXCITATION - [el] fonctionnement *m* à courants harmoniques indépendants
~FIBRES - [text etc] (fibres of animal, vegetable or mineral origin) fibres *fpl* naturelles
~FREQUENCY - [radio] (the lowest possible frequency of a free oscillation in a circuit) fréquence *f* propre
~FREQUENCY OF CIRCUIT - [el] (a simple tuned-circuit responds to an impulse by ringing) fréquence *f* propre de circuit
~GAS - [min] (misture of hydrocarbon gas obtained from geological deposits, usually in oilfields) gaz *m* naturel
~GAS LINE - [gas ind] conduite *f* de gaz naturel
~GAS MAIN - s. natural gas line
~GAS PIPELINE - s. natural gas line
~GLASS - [geol] (magma of any composition in glassy condition owing to rapid cooling) verre *m* naturel
~HARBOUR - [geogr] port *m* naturel
~IRON ORE - [min] oxydes *mpl* naturels
~LATEX - [rubber ind] latex *m* naturel
~LEAK - [radiat] (the rate of loss of charge of a measuring instrument due to causes other than ionization by the radiation to be measured) fuite *f* propre
~LOGARITHM - [math] logarithme *m* naturel, logarithme *m* népérien
~MAGNET - [min] (lodestone) aimant *m* naturel
~MAGNETIZATION - [radio] (in a magnetic amplifier) magnétisation *f* libre
~NEUTRON - [nucl] (neutron generated from a natural source) neutron *m* naturel
~NOTE - [mus] (note which neither raised nor lowered) ton *m* naturel, note *f* naturelle
NATURAL NUMBER - [math] nombre *m* naturel, nombre *m* entier
~OSCILLATION - [phys] (a free oscillation) oscillation *f* libre
~OXIDES - [min] oxydes *mpl* naturels
~PERIOD - [el] (the free period of circuit, i.e. the reciprocal of the natural frequency of the circuit) période *f* propre
~RADIOACTIVITY - [nucl] (exhibited by naturally occurring substances) radioactivité *f* naturelle
~RADIONUCLIDES - s. naturally radioactive nuclides
~RAFT - [soil] radier *m* naturel
~RESINS - [chem] (semi-solid viscous substances of vegetable origin; used in paints, plastics adhesives etc. They consist in general of highly polymerized acids and neutral compounds, with varying of terpene derivatives) résines *fpl* naturelles
~RESOURCES - [gen] ressources *fpl* naturelles
~REVIVIFICATION - [ind chem] (revivification by exposure to air) revivification *f* par exposition à l'air, revivification *f* naturelle
~RUBBER - [rubber ind] caoutchouc *m* naturel
~SCALE - [surv] (said of a section which is drawn with equal vertical and horizontal scales) échelle *f* naturelle
~SCIENCES - [gen] (the science dealing with the physical universe) sciences *fpl* naturelles
~SEASONING - [timber] (seasoning by exposing cut timbers laid in stacks) assaisonnement *m* naturel, séchage *m* naturel
~SELECTION - [biol] (the theory of the mechanism of evolution which postulates the survival of the fittest) sélection *f* naturelle
~SERIES - [math] nombre *mpl* entiers
~SIZE - [draw] grandeur *f* naturelle
~SLOPE - [constr] (the maximum angle at which soil will solid will stand without slipping) inclinaison *f* naturelle
~SOIL STRATUM - [soil] couche *f* de sol naturel
~STABILITY LIMIT OF A TRANSMISSION SYSTEM - [el] (the maximum power the system can transmit at the rated frequency) limite *f* de stabilité naturelle d'un système de transmission
~TRANSIENT STABILITY LIMIT OF A TRANSMISSION SYSTEM - [el] (the maximum power a system can transmit at the time of a sudden disturbance) limite *f* de stabilité dynamique naturelle d'un système de transmission
~-URANIUM REACTOR - [nucl] (nuclear reactor in which unenriched uranium is the principal fuel) réacteur *m* à uranium naturel
~VOID RATIO - [soil] indice *m* des vides naturels
~WAVELENGTH OF THE AERIAL - [radio] (about I I/2 times the electric length) longueur *f* d'onde naturelle de l'antenne
~WAVELENGHT OF THE ANTENNA - s. natural wavelength of the aerial
NATURALIZATION - [bot] retour *m* à l'état sauvage
NATURALIZED - [bot] (introduced from a different region) acclimaté
[leg] (of a person who has adopted a different nationality) naturalisé
NATURALLY RADIOACTIVE NUCLIDES - [nucl] (nuclides which occur naturally and exhibit radioactivity) éléments *mpl* radioactifs naturels
NATURE - [gen] nature *f*
NAUGHT - [gen] néant *m*, rien *m*

NAUSEA - [med & gen] nausée *f*
NAUSEOUS - [gen] (of disgusting smell or tast) nauséeux, dégoûtant
NAUTICAL - [naut] nautique, marin
~ALMANAC - [astr etc] (astronomical ephemeris which is published annually by the Admiralty) almanach *m* nautique
~CHART - [naut] carte *f* nautique
~LOG - [instr] (a log; as used by ships) loch *m*, sillomètre *m*
~MILE - [meas] (the mean length of an arc of the meridian which subtends an angle of I minute at the centre of the earth) mille *m* marin, mille *m* nautique
NAVAL - [gen] naval
~ARCHITECTURE - [naut] architecture *f* navale
NAVE - [arch] (the main body of a cruciform church) nef *f*, vaisseau *m*
[mech] (the hub of a wheel) moyeu *m*
NAVEL - [anat] nombril *m*, ombilic *m*
~ILLNESS - [med] phlébite *f* ombilicale
NAVICULAR DISEASE - [vet] maladie *f* naviculaire
NAVIGABILITY - [naut] navigabilité *f*
NAVIGABLE - [naut] navigable
[aero] (of a balloon) dirigeable
~WATERWAY - [naut] voie *f* navigable
NAVIGATE, to - [naut] (to journey by ship) naviguer
[naut] (to steer and conduct a ship) gouverner, diriger
NAVIGATION - [naut & aero] (the science of determining the position, and of maintaining the required direction, of a ship or aircraft) navigation *f*
~ACT - [leg] loi *f* maritime
~AID - [radio] aides-radio *fpl* à la navigation
~BRIDGE - [naut] pont *m* de commande
~CHANNEL - [naut] passe *f* navigable
~FLAME FLOAT - [aero] (pyrotechnic device dropped from an aircraft and burning while floating) signal *m* lumineux flottant
~LIGHTS - [naut & aero] (coloured lights shown by aircraft during darkness) feux *mpl* de position
~OFFICER - s. navigator
NAVIGATIONAL MICROFILM PROJECTOR - [radar] (microfilm projector similar to a virtual Plan-Position-Indicator reflecto-scope) projecteur *m* de microfilm pour la navigation
~PLANETS - [astronaut] planètes *fpl* de navigation
~TRIANGLE - [astronaut] triangle *m* de position
~SATELLITE - [astronaut] satellite *m* de navigation
NAVIGATOR - [aero] (a member of the crew of an aircraft whose primary duty is navigation) navigateur *m*
~COMPARTMENT - [aero] cabine *f* de navigation
NAVIGATOR'S STAR - [astr] (one of 55 stars selected as being of use in navigation) étoile *f* de référence
~BLUE - [dye] bleu *m* marine, bleu *m* foncé
~YARD - [naut] arsenal *m*
N.D.B. - s. non directional beacon
NEAP, to - [naut] décroître
NEAP - [gen] (the lowest) le plus bas
~TIDES - [astr] (high tides below the maximum) marée *fpl* de morte eau, marée *fpl* de quadrature, marée *fpl* bâtardes
NEAPED - [naut] (of a vessel) retenu par manque de eau
~NEAR - [gen] proche, près
~ECHO - [radar] (echo so near the scanner, as to be hardly perceptible on the display) écho *m* de proximité
NEAR-END CROSS-TALK - [telecomm] (in telephony; cross-talk between two parallel circuits when speaker and listener are at the same end of the parallelism) diaphonie *f*, paradiaphonie *f*
~-END CROSS-TALK ATTENUATION - [telecomm] affaiblissement *m* diaphonique
FIELD - [acoust] (the acoustic radiation field near the source) champ *m* proche
~-FOCUSING DEVICE - [photo] mise *f* au point sur objects rapprochés
~POINT OF THE EYE - [opt] (the nearest point at which an object can be clearly seen) point *m* proche
~SIDE - [auto etc] côté *m* gauche
~SINGING - [telecomm] tendance *f* au sifflement
~ZONE - [radio] (of a transmitting aerial) zone *f* d'induction
NEAREST APPROACH - [radar] (the minimum distance to which two vessels will close if neither changes speed or course) distance *f* minimum d'induction
~NEIGHBOUR - [cryst] (the nearest atom to a given atom in a crystal lattice) atome *m* voisin le plus proche
~-POINT DISTANCE - [photo] distance *f* du point le plus rapproché
NEAT - [gen] pur, sans eau
[gen] (without additives) pur
[constr] (of cement) sans sable
[text] neat
~CHARRING - [constr] ciselure *f* nette
~FLAME BURNER - [th eng] (diffusion-flame burner) brûleur *m* à flamme de diffusion, brûleur *m* à flamme blanche
~-GAS BURNER - s. neat flame burner
NEATSFOOT OIL - [chem] (a fixed oil obtained from the leg bones of cattle and used in leather processing) huile *f* de pied de boeuf
NEB PEG - [text] broche *f* à crochet
NEBULA - [astr] (faint luminous patch among the stars) nébuleuse *f*
[med] (on the eye) taie *f*
NEBULIZE, to - [gen] (to spray) atomiser, pulvériser
NEBULIZER - [impl] (a sprayer) atomiseur *m*, pulvérisateur *m*
NECK, to - [mech] (to form a peripheral groove in a shaft or similar part) forer circulairement
NECK - [anat] cou *m*
[mech] (peripheral groove cut in a shaft or similar part) col *m*, collet *m*
[metall] (of a rolling mill) collet *m*, tourillon *m*
[mus] (the projecting section of a stringed instrument, carrying the fingerboard) manche *m*
[aero] (the tail of a balloon) appendice *m*
[of a bottle] col *m*, goulot *m*
[text] (of a dress) encoulure *f*
[ind chem] (of a retort) col *m* (d'une cornue)
[geogr] (a projection of the coastline) bras *m*
[electron] (the small tubular section of the envelope near the base) col *m*
[arch] (the moulding between the capital of a column and the shaft) gorge *f*, gorgerin *m*, colarin *m*
[mech] (of an axle) fusée *f*
[geol] (vertical body of igneous rock, shaped like a plug and representing the feeding channel of a volcano) cheminée *f* d'ascension

NECK - [metall] (in forging work) reduction *f* d'aire dans la section de striction
~-BAND - [text] tour-du-cou *m*
~BEARING BUSHING - [mech] coussinet *m* supérieur
~DISSECTION - [med] évidement *m* ganglionnaire du cou
~-IN - [plast ind] (reduction of width of extruded sheet caused by surface tension) striction *f*
~-MOULD - [arch] (the narrow neck round the top of the shaft of a column) colarin *m*
~MOULDING PLANE - [impl] congé *m*
~OF THE PANCREAS - [anat] col *m* pancréas
~REGION - [electron] zone *f* du col
~RING - [glass man] (used for the finish of a hollow glass article) anneau *m* de serrage
~SHADOW - [electron] ombre *f* du col
~STRAP - [gen] bretelle *f*
~TWINE - [text] corde *f*, cordon *m*, collet *m*
NECKING - [arch] s. neck-mould
[metall] (the cross-section reduction preceding a fracture in a ductile material under stress) réduction *f* d'aire dans la section de striction
[railw] (due to wear) réduction *f* de section
[mech] (undercutting cut into the corner of a piece) noulet *m*
[oil ind] réduction *f* de section
~DOWN - [metall] entaillage *m*
NECROBIOSIS - [med] nécrobiose *f*
NECROBIOTIC - [med] nécrobiotique
NECROPNEUMONIA - [med] gangrène *f* pulmonaire
NECROPSY - [med] autopsie *f*, nécropsie *f*
NECROSIS - [med] nécrose *f*
NECROTIZING FACTOR - [biol] nécrotoxine *f*
NECROTOXIN - [med] s. necrotizing factor
NECTAR - [agric] nectar *m*
~FLOW - [zool] miellée *f*
NECTARINE - [bot] nectarine *f*, brugnon *m*
NEEDLE - [gen] aiguille *f*
[text] (knitting needle) aiguille *f* (à tricoter)
[mech] aiguille *f*, pointeau *m*, broche *f*, axe *m*
[med] (for surgical operations) aiguille *f*
[arch] obélisque *m*
[min] particule *f* aciculaire
[comput] (probe used to sorting or selecting cards) aiguille *f* de triage
[el acoust] aiguille *f*
[mech] (of the injector nozzle) pointeau *m*
[instr] (the moving element) aiguille *f*
[glass man] (a plunger) poinçon *m*
[build] (beam used as temporary support during underpinnin) cale *f* d'étayage
[geogr] (a sharp pointed rock) aiguille *f* rocheuse
[meas] (of a balance) langue *f*, languette *f*
~ALIGNMENT - [photo] alignement *m* d'aiguilles
~BAR - [text] fonture *f*, tête *f*, section *f*
~BAR CAM LEVER - [text] levier *m* de barre à aiguilles
~BAR SHAFT - [text] arbre *m* de la barre à aiguilles
~BAR WRENCH - [text] clef *f* à tube pour barre à aiguilles
~BEAM - [constr] (transverse floor-beam) aiguille *f* (de pont)
[railw] (the crossbearer in a steel truck) élément *m* transversal
~BEARING - [mech] (a type of roller bearing in which the rollers are long in comparison with their diameter and are not fitted in a cage) roulement *m* à aiguilles
NEEDLE BED - [text] barre *f* à aiguilles, fonture *f*
~BED FRAME - [text] support *m* des fontures
~BOARD - [text] (of a loom) planchette *f* aux aiguilles
~BOTTOM - [metall] fond *m* en baguettes
~BUG - [zool] anthonome *m* du fraisier
~CAGE - [auto] cage *f* à aiguilles
~CAM - [text] came *f* succédante
~CARRIER PLATE - [text] guide *m* d'auiguilles
~CASE - [impl] porte-aiguilles *m*
~CHAMP - [text] plaque *f* de recouvrement
~CLEARING CAM - [text] came *f* d'ascension de l'aiguille
~COUNTER - [instr] (radiation counter fitted with a long needle of stainless steel) compteur *m* à aiguille
~COVERED ROLLER - [text] cylindre *m* recouvert d'aiguilles
~CURVE - [mech] (of injection carburettor) courbe *f* du pointeau
~DAM - [hydr] barrage *m* à fermettes
~DEVIATION - [radar] (the relative motion deviation of a direction indicator needle caused by a vehicle leaving the desired flight-path) déviation *f* de l'aiguille
~DIAL - [instr] cadran *m* à aiguille
~DRAG - [el acoust] (the force resulting from the friction between the stylus and the surface of the record) frottement *m* de l'aiguille
~DRESSING - [text] dressage *m* des aiguilles
~EYE - [text] oeil *m* de l'aiguille, trou *m* de l'aiguille
~FILE - [impl] lime *f* à aiguille
~FISH - [zool] aiguille *f* de mer
~FLAME BURNER - [th eng] (or pinhole burner) brûleur *m* à flamme filiforme
~FLUCTUATION - [instr] fluctuation *f* de l'aiguille, vibration *f* de l'aiguille
~FLUTTER - s. needle fluctuation
~FORCE - [el acoust] (the stylus pressure) effort *m* du style, pression *f* du style
~FRAME - [text] barre *f* à aiguilles
~FURZE - [bot] genêt *m* épineux
~GRID - [text] grillette *f*
~GUIDE - [text] guide-aiguille *m*
~HACKLE - [text] sérançoir *m* à peigne
~HOLDER - [el acoust] (the mechanism holding the grammophone needle) porte-aiguille *m*
~HOOK - [text] crochet *m* de l'aiguille
~IRONSTONE - [min] goethite *f*
~JACK - [text] platines *f* pl de métier à mailles retournées
~JET - [mech] (internal combustion engines) gicleur *m* à aiguille
[metall] jet *m* reglé à aiguille conique
~LIFTER - [text] came *f* d'ascension de l'aiguille
~LINK - [text] élément *m* d'aiguille
~LOCK - [text] serrure *f* d'aiguilles
~LOCKING - [photo] blocage *m* de l'aiguille
~LUBRICATOR - [mech] (primitive form of lubricator consisting of an inverted flask attached to a bearing and with a wire loosely fitted in a hole in the stopper) graisseur *m* à pointeau
~MOTION - [text] mouvement *m* de l'aiguille
~NOISE - [el acoust] (the noise caused by the friction of the needle against the record) bruit *m* de aiguille, grattement *m* d'aiguille
~NOZZLE - [mech] buse *f* à pointeau

NEEDLE PITCH - [text] division *f* des aiguilles, écartement *m* des aiguilles
~PLIERING - [text] dressage *m* des aiguilles
~PLIERS - [text] pince *f* à dresser les aiguilles
~POINT - [text] pointe *f* d'aiguille
~-POINT GAP - [el] (spark-gap in which the electrodes are in the form of needle points) éclateur *m* à aiguille
~PUSHER - [text] poussoir *m*, pousseur *m*
~RING - [text] couronne *f* d'aiguille
~RIVET - [mech] rivet *m* d'aiguille
~SCAFFOLD - [constr] échafaudage *m* en bascule
~SCRATCH - s. needle noise
~SHAFT - [text] tige *f* d'aiguille
~SHANK - [text] palette *f*, tige *f* d'aiguille
~-SHAPED - [gen] (acicular; having the form of a needle) aiguillé, aciforme, aciculiforme, aciculé
~SORTING - [comput] (method of sorting marginal punched cards by the insertion of needles) triage *m* à aiguille
~SPACING - [text] écartement *m* des aiguilles
~SPRING - [text] ressort *m* de l'aiguille
~STONE - [min] flèches *fpl* d'amour, quartz *m* rutilé
~STRIP - [text] barre *f* à aiguilles
~TALK - [el acoust] (the noise which is produced by the needle itself during reproduction) bruit *m* de surface, grattement *m* d'aiguille
~TEAR RESISTANCE - [rubber ind] (the stitch tear resistance) résistance *f* à la déchirure initiée par une aiguille
~THREAD - [text] fil *m* d'aiguille
~THROW - [text] point *m* zigzag
~TRAY - [text] boîte *f* à aiguilles
~VALVE - [mech] (a valve consisting of a spindle having a conical extremity which seats in a corresponding conical recess, usually designed for fine adjustment of liquid flow) pointeau *m*
~-VALVE LUBRICATOR - [impl] graisseur *m* à pointeau
~WEAR - [el acoust] usure *f* de l'aiguille
NEEDLELIKE - [gen] aiguillé, aciculaire
NEEDLES - [constr] blochet *m*, faux-entrait *m*
NEEDLING - [text] garnissage *m* en aiguilles
~MACHINE - [text] aiguilletteuse *f*
NEGATION - [comput] négation *f*
NEGATIVE - [gen] negatif
[math] (denoted by the minus sign) (signe) moins
[math] (the actual quantity) quantité *f* négative
[el] (term applied to one of two points between which a difference of potential exists, to distinguish the one which is at a lower electric potential) négatif
[photo] (the black and white reverse image) négatif *m*
[cin] négatif *m*
[opt] (synonym of laevorotary) négatif, (système optique) divergent
[el metall] (negative matrix; a matrix the surface of which is the reverse of the surface to be produced by electro-forming) matrice *f* négative
~ADSORPTION - [phys] (when the concentration of solute is less in the surface than throughout the solution) adsorption *f* négative
~AFTER-POTENTIAL - [el] (relatively prolungued negativity following a short pulse in a homogeneous fibre group) queue *f* de potentiel négatif

NEGATIVE ALTITUDE - [astronaut] altitude *f* négative
~AREA - [el] (of the stray currents) zone *f* d'entrée des courants vagabonds
~BAG - [photo] pochette *f* pour négatifs
~BIAS - [electron] (negative voltage applied to the control grid of an electronic tube) polarisation *f* négative
~BOOSTER - [el] (booster arranged to reduce the voltage supplied by another electrical source) dévolteur *m*
~CATALYSIS - [el chem] (the retardation of a chemical process by a substance which is unchanged by the reaction which it affects) catalyse *f* négative
~CATALYST - [el chem] (a catalyst which reduces the speed of a reaction) catalyseur *m* négatif
~CONDUCTOR - [el] conducteur *m* négatif
~CRYSTAL - [cryst] (birefrongent crystal in which the velocity of the extraordinary ray is greater than that of the ordinary ray) cristal *m* négatif
[metall] cavité *f* en forme de cristal
~DIE - [metall] (recessed section of a mould) moule *m* négatif
~DIFFERENTIAL CONDUCTANCE REGION - [electron] région *f* de conductance négative différentielle
~DISTORTION - [telev] (called 'pincushion distortion' in GB; distortion of the picture in which the sides bulge like a pincushion) distorsion *f* en coussinet
[photo] distorsion *f* en croissant
~DOBBY - [text] ratière *f* négative
~ECHO - [telev] image *f* fantôme négative
~ELECTRICITY - [el] (a state arising from an excess of electrons above normal) électricité *f* négative
~ELECTRODE - [el chem] (that electrode which forms the cathode when the cell is discharging) électrode *f* négative
~ELECTRON - s. negatron
~EYEPIECE - [opt] (eyepiece placed inside the principal focus of the objective) oculaire *m* négatif
~FEEDBACK - [radio] (feedback resulting in a decrease of the amplification) contreréaction *f*
~FEEDBACK AMPLIFIER - [radio] amplificateur *m* à contre-réaction
~FEEDER - [el] (in a system of electric traction, the feeder connecting the negative conductor rail to the negative busbars at a generating station) artère *f* de retour
~g - [phys] g *f* négative, décélération *f*
~GHOST - s. negative image
~GLOW - [electron] (luminous region following the cathode dark space in which the electrons acquire a kinetic energy which is sufficient to excite the gas) lueur *f* cathodique
~GLOWLAMP - [el] lampe *f* à effluves
~GRADER - [cin] (a laboratory operator entrused with the task of obtaining the best results from the negative) étalonneur *m*
~GRID CURRENT - [electron] courant *m* de grille inverse
~GRID OSCILLATOR - [electron] oscillateur *m* à réaction et tension négative de grille
~HARDENING - [metall] trempe *f* négative
~IMAGE - [telev] (reverse image on the screen caused by a reversed connexion in the receiver) image *f* négative
~INSIDE LAP - [mech] (of a slide-valve) découvert *m* intérieur

NEGATIVE ION - [phys] anion *m*, ion *m* négatif
~ION BLEMISH - [electron] tache *f* ionique
~-ION VACANCY - [nucl] (in the lattice of an ionic crystal from which a negative ion is absent) lacune *f* d'ions négatif
~LENS - [opt] (diverging lens, so called because it causes the light which is parallel to its axis to diverge, as it were coming from a point which is the virtual focus of the lens) lentille *f* divergente
~LEFT OFF MOTION - [text] régulateur *m* de déroulement négatif
~MAGNETOSTRICTION - [el] (the dilatation of a material by the application of a magnetic field) magnétostriction *f* négative
~MATRIX - [el metall] s. negative
~MODULATION - [telev] modulation *f* négative
~NODAL POINTS - [opt] points *mpl* nodaux négatifs
~OUTSIDE LAP - [mech] (of a slide-valve) découvert *m* extérieur
~PICTURE PHASE - [telev] polarité *f* négative du signal image
~PLATE - [el] (the electrode which acts as anode during discharge) plaque *f* négative
~PLATFORM - [photo] couloir *m* du passe-vues
~POLE - [el] pôle *m* négatif
~-POSITIVE COMBINATION - [photo] tirage *m* simultané d'un négatif et de son image positive superposée
~PRESSURE - [mech] (normal stress tending to increase the volume of the substance) pression *f* négative, dépression *f*
~-PRESSURE VENT VALVE - [mech] (a fuel-tank vent-valve designed to operate when the pressure in the tank falls below the prevailing atmospheric pressure) robinet *m* de purge à dépression
~PRINCIPAL POINTS - [opt] points *mpl* principaux négatifs
~PROTON - [phys] (a particle resembling a proton in all respects, except that it possesses a negative charge) antiproton *m*, proton *m* négatif
~RESISTANCE - [el] (a negative resistance occurs in a circuit when the derivative of voltage across the circuit has a negative value) résistance *f* négative
~RESISTANCE OSCILLATOR - [radio] (oscillator produced by connecting a parallel-tuned resonant circuit to a two-terminal negative-resistance device) oscillateur *m* à résistance négative
~-RESISTANCE REPEATER - [radio] (repeater in which the gain is supplied by a series) répéteur *m* à résistance d'entrée négative
~RETOUCHING - [photo] retouche *f* du négatif
~SCANNING - [telev] analyse *f* d'images négatives
~SEGREGATION - [metall] ségrégation *f* inverse
~SEQUENCE ACTIVE POWER - [el] puissance *f* inverse d'un système triphasé
~-SEQUENCE FIELD IMPEDANCE - [el] impédance *f* de champ inverse
~-SEQUENCE POLYPHASE SYSTEM - [el] système *m* polyphasé inverse
~SEQUENCE RESISTANCE - [el] résistance *f* de séquence négative
~SHAFT MOTION - [text] mouvement *m* négatif des lames
~SHEDDING MOTION - [text] mécanisme *m* de formation du pas négatif
~SHUTTLE BOX DRIVE - [text] mouvement *m* négatif des boîtes à navettes

NEGATIVE STAGE - s. negative platform
~TERMINAL - [el] (the terminal of a machine which is at the lowest potential) borne *f* négative, pôle *m* négatif
~THERMALS - [met] (term sometimes used for vertical air currents flowing downward) courants mpl thermiques descendantes
~-TRANSCONDUCTANCE OSCILLATOR - [radio] (a positive grid oscillator) oscillateur *m* à champs de freinage
~TRANSMISSION - [telev] (television transmission in which the amplitude of the modulated radio signal decreases with the increasing brightness of the picture) transmission *f* à modulation négative
~TRANSMISSION POLARITY - [telev] polarité *f* négative de la transmission
~VALENCE - [electron] (electron valence possessed by an atom) valence *f* négative
~VARNISH OF RESIN - [photo] vernis *m* de négatifs à base de résine
~WORK - [mech] travail *m* négatif
~WORK FACTOR - [mech] (in turbines, a parameter of the power consumed in operating compressing mechanism) facteur *m* de travail négatif
NEGATOR - [mech] (a return spring) ressort *m* de rappel
NEGATOSCOPE - [radiol] (a film illuminator) négatoscope *m*
NEGATRON - [electron] (an electron carrying a negative charge) négatron *m*
NEGISTOR - s. negative-resistance repeater
NEGLIGIBLE - [gen] négligeable
NEGOTIATE, to - [gen & comm] négocier, traiter [fin] placer, négocier
NEGOTIATION - [gen & comm] négociation *f*, pourparler *m*
NEIGHING - [acoust] (the sound made by horses) hennissement *m*
NEISSEROSIS - [med] gonococcie *f*
NELSON CELL - [el chem] (a type of electrolytic cell used for the commercial production of sodium hydroxide and chlorine from brine) élément *m* de Nelson
NEMATIC PHASE - [cryst] (a form of the mesomorphic state) phase *f* nématique
NEMATODE - [med] nématode *m*
~DISEASE - [bot] (plant disease) anguillulose *f*
NEMO - [radio telev etc] (transmission of events outside the studio) extérieur *m*, prise *f* de vue à l'extérieur
NEOARSPHENAMINE - [chem] (an antisyphilitic drug) néoarsphénamine *f*
NEOCINCHOPHEN - [chem] (an analgesic and antipyretic) néocinchophéne *fm*
NEODIATHERMY - [med] diathermie *f*, ondes *fpl* courtes
NEODYMIUM - [chem] (a metallic rare earth element, A.N. 60; A.W. I44.3 symbol Nd, with uses in metallurgy and in the manufacture of coloured glasses) néodymium *m*
NEOFORMATION - [med] néoplasma *m*, néoformation *f*
NEOHEXANE - [chem] (a fuel additive) néohexane *m*
NEOMYCIN - [chem] (an antibiotic with therapeutic uses) néomycine *f*
NEON - [chem] (inert gaseous element, symbol Ne, A.N. I0, A.W. 20.I83 used to fill gas discharge

tubes. It is obtained by fractionating liquid air) néon *m*

NEON INDICATOR - [phys] (a neon gas filled lamp which glows in the presence of relatively intense radio frequency radiation) signaleur *m* au néon, tube *m* indicateur à néon

~LAMP - [el] (electric discharge lamp in which the light is emitted from, or excited by, the discharge through neon) lampe *f* au néon

~STABILIZER - [telev] (discharge tube used for switching the discharge circuit of a capacitor) tube *m* stabilisateur au néon

~TUBE - s. neon lamp

NEOPLASM - [phys] (new and unrestrained growth of cells) néoplasme *m*

NEOPLASTIC CELL - [phys] (cell in a new tissue) néoformation *f*

NEOPRENE - [chem] (a synthetic rubber produced by the polymerization of chloroprene) néoprène *m*

~LATEX - [rubber ind] latex *m* néoprène

NEOSTIGMINE - [chem] (an anticholinesterase used in medicine) néostigmine *f*

NEP - [text] (in cotton fibres, small knots produced by an uneven growth of the plant) noeud *m*

NEPER - [meas] (the unit of attenuation) néper *m*
[math] néper *m*

NEPHELINE - s. nephelite

NEPHELINITE - [geol] (fine-grained igneous rock occurring as lava flow) néphélinite *f*

NEPHELITE - [min] (a native silicate of sodium used in ceramic glazes) néphéline *f*

~BASALT - [geol] (a basic lava carrying nepheline as an essential constituent) basalte *m* à néphéline

NEPHELOMETER - [instr] (instrument for determining the degree of turbidity of a fluid) néphélomètre *m*

NEPHELOMETRIC ANALYSIS - [chem] (method of quantitative analysis in which the concentration of suspended matter in a liquid is determined by optical means) analyse *f* néphélométrique

NEPHOSCOPE - [instr] (instrument located on the ground and designed to determine the direction of motion of a cloud and its velocity-height ratio) néphoscope *m*

NEPHRADENOMA - [med] adénome *m* rénal

NEPHRALGIA - [med] néphralgie *f*

NEPHRATONIA - [med] atonie *f* rénale

NEPHRECTOMY - [med] néphrectomie *f*

NEPHRITIS - [med] (inflammation of the substance of the kidneys) néphrite *f*

NEPHROCAPSECTOMY - [med] décapsulation *f* du rein

NEPHROCELE - [med] hernie *f* du rein

NEPHROLITH - [med] calcul *m* rénal

NEPHROPATHY - [med] (renopathy) néphropatie *f*

NEPHROPHTHISIS - [med] tubercolose *f* rénale

NEPHROPTOSIS - [med] rein *m* flottant

NEPHROPYELITIS - [med] pyélo-nephrite *f*

NEPHROTOMY - [med] néphrotomie *f*

NEPPING - [text] noppage *m*

NEPPY - [text] riche en neps

NEPTUNIUM - [chem] (a synthetic, metallic radioactive element, symbol Np, A.N. 93, possessing properties similar to those of uranium) neptunium *m*

~SERIES - [chem] (the series of nuclides resulting from the decay of the synthetic nuclide Np^{237}) famille *f* du neptunium

NEROL - [chem] (a perfumery compound) nérol *m*

NEROL OIL - [chem] (orange flower oil. An essential oil obtained from the flowers of Citrus Aurantium and used as a flavourant in perfumery) essence *f* de nérol

NEROLIDOL - [chem] (a constituent of a number of essential oils having uses in perfumery) nérolidol *m*

NERVE - [anat] (branch of the nervous central system passing to an organ or part of the body) nerf *m*
[bot] nervure *f*
[rubber ind] (elasticity of unvulcanized rubber) nerf *m* (du caoutchouc)
[text] (of fibres) élasticité *f*, résistance *f*
[arch] (a projecting rib on a vault) nervure *f*

~BLOCK - [med] blocage *m* du nerf, anesthésie *f* tronculaire

~CANAL - [dent] (aperture in the root of a tooth through which the nerve passes to the pulp) canal *m* du nerf

~CAVITY - [med] chambre *f* pulpaire, cavité *f* de la dent

~CELL - [zool] (neurocyte) cellule *f* nerveuse, cellule *f* ganglionnaire

~ENDING - [zool] (the free distal end of a nerve or nerve fibre) terminaison *f* nerveuse

~FIBRIL - [zool] (axon; neurite) axone *m*, prolongement *m* cylindraxile

~HILLOCK - s. neuromast

~IMPULSE - [zool] (the disturbance passing through a nerve when it is stimulated) impulsion *f* nerveuse, influx *m* nerveux

~NET - [zool] (the primitive type of nervous system found in Coelenterata) réseau *m* nerveux

~PLEXUS - [zool] (network of nerve fibres) plexus *m* nerveux

~ROOT - [zool] (the origin of the nerves in the central nervous system) racine *f* du nerf

~STRETCHING - s. neurotonia

~TRUNK - [zool] (bundle of nerve fibres) faisceau *m* de nerfs

NERVELESS - [bot] énerve, sans nervures

NERVOSISM - [med] nervosisme *m*, nervosité *f*

NERVOSITY - s. servosism

NERVOUS BREAKDOWN - [med] épuisement *m* nerveux, prostration *f* nerveuse

NERVOUSNESS - s. nervosism

NESIDIOBLASTS - [anat] cellules *fpl* insulaires (du pancréas)

NESISTOR - [electron] (a negative-resistance semiconductor device) fieldistor *m* bipolaire

NESSLER'S REAGENT - [chem] (a solution of mercuric iodide in potassium iodide used in the detection of ammonia) réactif *m* de Nessler

NEST, to - [gen mech etc] emboîter

NEST - [gen] nid *m*, nichée *f*
[naut] hune *f*
[mech] (a group of gears etc) série *f*, jeu *m*
[th eng] (a group or assembly of tubes) faisceau *m* (tubulaire)
[geol] (concentration of mineral within a rock of different nature) nid *m* (de minerai)
[railw] épi *m* de voies

~PLATE - [plast ind] (retainer plate with recesses for cavity blocks) plaque *f* creuse

~SPRING - [mech] (helical spring containing one or more spring coils) ressort *m* à hélice cylindrique multiple

NET, to - [gen] faire au filet, faire du filet

NET - [gen] filet *m*
[adj] net
[comm] (of prices or costs) net *m*
[aero] (of balloons) (rope system enclosing the envelope and designed to distribute the load uniformly over it) filet *m*
[fishing impl] filet *m*, carrelet *m*, puche *f*, bichette *f*
[el chem] (a group of electrolytic cells placed closed together and electrically connected in series) batterie *f*
~BALANCE COUNTER - [comput] compteur *m* à balance direct
~BOBBIN - [text] bobine *f* plate, bobine *f* à tulle-bobin
~CALORIFIC VALUE - [th eng] pouvoir *m* calorifique inférieur
~COOLING - [plast ind] (the extraction of heat from the barrel of an extruder to keep material temperature down) refroidissement *m* en circuit
~CURTAIN - [text] rideau *m* de tulle
~CUTTER - [naut] coupe-filets *m*
~EFFICIENCY - [mech] (the relation between the torque h.p. and the net thrust h.p.) rendement *m* net
~GAIN - [telecomm] (the total gain between the ends of a line or apparatus) équivalent *m* d'un circuit, gain *m* total
~GAUZE - [text] gaze-filet *f*, filet-gaze *m*
~HEATING - [plast ind] (system of temperature control in an extruder barrel by supplying heat continuously to the extrusion material) chauffage *m* en circuit
~HEATING VALUE - s. net calorific value
~LACE - [text] dentelle *f* à réseaux
~LIFT - [aero] (in balloons, the value obtained by subtracting the disposable and fixed weights from the gross lift) force *f* ascensionnelle nette
~LOSS - [telecomm] (the total loss between the ends of a line or apparatus) atténuation *f* générale
~LOSS FACTOR - [telecomm] facteur *m* d'équivalent
~LOSS MEASUREMENT - [telecomm] mesure *f* d'équivalent
~SAND - [oil ind] sable *m* net
~SILKS - [text] grandes soies *f*pl
~SLIP - [oil ind] rejet *m* net
~SPEED - [mech] débit *m* utile, débit *m* effectif
~THRUST - [aero] (the resultant, taken parallel to the propeller axis, of the thrust of the whole ensemble of a propeller and the associated fuselage or nacelle) traction *f* nette
~TRANSMISSION EQUIVALENT - [telev] (the total attenuation due to leakage, absorption or radiation) équivalent *m* de transmission
~TRANSPORT - [nucl] (the difference between the total rate at which the desired isotopes are transmitted, and the rate at which they would be carried in the same flow by material of natural abundance) transport *m* net
~TRIMMING - [text] galon *m* à gaze
~WEIGHT - [meas] (actual weight) poids *m* net
~WING AREA - [aero] (value obtained by deducting from the gross wing area the area of that part which is covered by the fuselage) surface *f* alaire nette
~WING LOADING - [aero] (the gross weight divided by the net wing area) charge *f* alaire nette
NETTED STRUCTURE - [geol] structure *f* maillée
NETTING HOOK - [text] crochet *m* à filet
NETTING MACHINE - [text] noueuse *f* pour filets, métier *m* à nouer les filets
NETTLE - [bot] ortie *f*
[text] (tissu d')ortie *f*
[text] (small rope made by twisting two or three yarns tightly) corde *f* d'ortie
~CLOTH - [text] tissu *m* d'ortie
~RASH - [med] urticaire *f*, fièvre *f* ortiée
~-TREE - [bot] micocoulier *m*, perpignan *m*
~YARN - [text] fil *m* d'ortie
NETWORK - [el] (aggregation of conductors intended for the distribution of electrical energy) réseau
[radio] (plurality of interrelated circuits) réseau *m*
[radio] (a system of transmitters) réseau
[gas ind] (mains network) réseau *m*, réseau *m* maillée
[gen] ouvrage *m* en filet
[railw] réseau *m* (de voies ferrées)
[road constr] réseau *m* routier
~ANALOG - [math] modèle *m* de réseau
~ANALYSIS - [el] (the derivation of the electrical properties of a network having specified electrical properties) analyse *f* de réseau
~ANALYZER - [comput] (analog computer which simulates and solves problems of the electrical behaviour of a network of power lines etc) analyseur *m* de réseaux
~CONTROL ROOM - [gen] salle *f* de contrôle du réseau
~ELEMENT - [telecomm] élément *m* de réseau
~FEEDER - [el] feeder *m* alimentant un réseau
~-MASTER-RELAY - [el] relais *m* protecteur de réseau
~OF PIPES - [gas; water etc supply] réseau *m*
~OF RAILS - [railw] réseau *m* ferré
~PARAMETER - [el] (any resistance, inductance or capacitance in a branch of a network) élément *m* de réseau, paramètre *m* de réseau
~PRIMARY DISTRIBUTION SYSTEM - [el] système *m* de distribution par primaires
~RELAY - [el] relais *m* disjoncteur de réseau
~STRUCTURE - [metall] (structure formed in alloys when one constituent exists in the form of a continuous round the boundaries of the grains of the other) structure *f* cellulaire
~SWITCHING CENTRE - [el telev etc] centre *m* de commutation du réseau
~SYNTHESIS - [comput] synthèse *f* des réseaux
NEUMANN PRINCIPLE - [phys] (the physical properties of a crystal cannot be of lower symmetry than the symmetry of the external form of the crystal) principe *m* de Neumann
NEURAL - [zool] (of nerve) neural
[chem] (of a medicine) nerval
NEURALGIA - [med] névralgie *f*
NEURASTHENIA - [med] neurasthénie *f*
NEURAXITIS - [med] névraxite *f*, encéphalite *f*
NEURECTASIS - s. neurotonia
NEURINE - [chem] (trimethylvinylammonium hydroxide, formed in the putrefaction of meat and occurring in brain tissues. A ptomaine base, related to choline) neurine *f*
NEURITIS - [med] névrite *f*
NEUROANATOMY - [med] anatomie *f* du système nerveux
NEUROCANAL - [med] canal *m* épendymaire
NEURODEATROPHIA - [med] atrophie *f* de la rétine
NEURODERMATOSIS - [med] neurodermatose *f*

NEURODOCITIS - [med] névrodocite *f*, funiculite *f* vertébrale (de Sicart)
NEUROELECTRICITY - [el biol] (any electric potential in the nervous system) neuroélectricité *f*
NEUROGANGLION - [med] ganglion *m* nerveux
NEUROGLIA - [med] névroglie *f*
NEUROKERATIN - [biochem] (a protein occurring in brain and nerve substance) neurokératine *f*
NEUROLUES - [med] neuro-syphilis *f*
NEUROMAST - [anat] protubérance *f* nerveuse
NEUROPATHY - [med] neuropathie *f*
NEUROPLEXUS - [med] plexus *m* nerveux
NEURORELAPSE - [med] neurorécidive *m*
NEUROSCLEROSIS - [med] sclérose *f* du tissu nerveux
NEUROSYPHILIS - s. neurolues
NEUROTABES - [med] neurotobès *f*, polynévrite *f* alcoolique
NEUROTENSION - s. neurotonia
NEUROTHELE - [med] papille *f* nerveuse
NEUROTOMIA - [med] neurotonie *f*, élongation *f* nerveuse
NEUROTONY - s. neurotonia
NEUROTROPISM - [med] neurotropisme *m*
NEUROTROPY - s. neurotropism
NEUROVACCINE - [med] nerovaccine *m*
NEUROVARIOLA - s. neurovaccine
NEUTRAL - [chem] (exhibiting neither acidity nor alkalinity) neutre
[el] (neutral point; of a system, the point with the same potential of the point of junction of a group of equal resistances, connected at their free ends to the appropriate main terminals or lines of the system) neutre
[photo] (which possesses no colour; grey) neutre
[mech] (of a gear) point *m* mort
[zool] (having no sex) neutre, asexué
[gen] neutre, intermédiaire, indeterminé
~ADJUSTED RELAY - [telecomm] relais *m* réglé à l'indifference
~ANODE MAGNETRON - [electron] (magnetron with an anode consisting of three cylindrical sectors, to of which are subject to high-frequency potential variations and the third one connected to the mid-point of the resonant circuit) magnétron *m* à anode neutre, neutrode *f*
~ATOM - [nucl] (an atom in which the positive charge of the nucleus is equal to the total negative charge of the electronics surrounding the nucleus) atome *m* neutre
~AXIS - [mech] (the line of zero stress in a beam subjected to bending) axe *m* neutre
~COLLOID - [chem] (the colloidal system in soap solutions at high concentrations of soap) colloïde *m* neutre
~COMBUSTION - [th eng] (combustion under theoretical conditions) combustion *f* neutre, combustion *f* stoechiométrique
~CONDUCTOR - [el] (the middle wire of a direct-current three-wires system which is connected to the neutral point of the supply transformer) conducteur *m* neutre
~DENSITY FILTER - [electron] (optical filter reducing the intensity of light without an appreciable change of its colour) filtre *m* gris
[photo] (grey filter) filtre *m* gris (neutre)
~EQUILIBRIUM - [mech] équilibre *m* indifférent
~FILTER - [telecomm] filtre *m* neutre
NEUTRAL GREASE - [mech] (special grease which is free from any corrosive properties, used for coating parts for storage etc) graisse *f* neutre
~GREY FILTER - [photo] (without colour or hue) filtre *m* gris
~INSTABILITY - [phys] (plasma instability) instabilité *f* neutre
~LINE - [phys] ligne *f* neutre
~MESON - [nucl] (meson without an electrical charge) méson *m* neutre
~MOLECULE - [nucl] (molecule without an electrical charge) molécule *f* neutre
~PLANE - [el] (on a polyphase commutator machine) lignes *fpl* neutres (d'une machine à collecteur polyphasé
[el] (on a direct current machine) lignes *fpl* neutres (d'une machine à collecteur à courant continu)
[el] (on a single phase commutator machine) lignes *fpl* neutres (d'une machine à collecteur monophasé)
~POINT - [met] (any point at which the axis of a wedge of high pressure cuts axis of a through of low pressure) point *m* neutre
[el] s. neutral
[aero] s. neutral point with stick fixed
~POINT WITH STICK FIXED - [aero] (the position of the C.G. of an aircraft which is such that the position of the control column to give correct trim at any given speed is not affected by the speed) point *m* neutre manche bloqué
~POINT WITH STICK FREE - [aero] (the position of the C.G. of an aircraft which is such that the control surface hinge moment to give correct trim at any given speed is not affected by the speed) point *m* neutre manche libre
~POSITION - [auto] (of the gears) point *m* mort
[el] (of a commutator machine; the position of the brushes which gives equal speeds at the same load in either direction or rotation) position *f* neutre
~RECLAIMING PROCESS - [rubber ind] procédé *m* de régénération neutre
~RED - [chem] (an analytical indicator) rouge *m* neutre
~RELAY - [el] (polarized relay so arranged that it operates in any direction from a normal neutral position, according to the direction of the current in the controlling circuit) relais *m* à trois positions
~ROCKS - [geol] roches *fpl* intermédiaires, roches *fpl* neutres
~ROPE - [rope man] (rope so made that it has no tendency to twist) corde *f* sans tendance au retordage
~SALT - [chem] sel *m* neutre
~SOIL - [geol] terrain *m* neutre
~SOLUTION - [chem] (one which contains equal amounts of hydrogen and hydroxyl ions and is thus neither acid nor alkaline, having a pH value of 7) solution *f* neutre
~STATE - [el] (the state of a ferromagnetic substance not yet magnetized, or artificially brought back to a virgin state) état *m* neutre
~TEMPERATURE - [phys] (the temperature at which the electromotive force produced by heating one junction of a thermocouple, when the other junction is constant at 0 centigrades, reaches a maximum) température *f* neutre
~TERMINAL - [el] (of a polyphase machine or apparatus) borne *f* neutre

NEUTRAL TRACK - [railw] (dead section) voie *f* neutre, section *f* neutre
~WEDGE - [photo] coin *m* neutre
~WIRE - s. neutral conductor
~ZONE - [el] (of a commutator machine; the zone in which when the machine is running at no-load, the voltage between two consecutive bars is sensibly zero) zone *f* neutre
[mech] (allowance) jeu *m*
[contr] (the largest range of values of the measured variable to which the instrument will nor respond; also called dead zone) zone *f* neutre
NEUTRALITY - [gen etc] neutralité *f*
NEUTRALIZATION - [chem] (the reaction of an acid with a base to form a salt) neutralisation *f*
[radio] (the use of a negative feedback in an amplifier) neutralisation *f*
[gen] neutralisation *f*
OF EFFLUENT LIQUOR - [ind chem] (by oxidation) neutralisation *f* des eaux usées
NEUTRALIZE, to - [gen & chem] (to render neutral) neutraliser
NEUTRALIZING - [gen] neutralisation *f*
~CAPACITOR - [el] condensateur *m* de neutralisation
~INDICATOR - [radio] indicateur *m* de neutralisation
~VOLTAGE - [radio] tension *f* de neutralisation
NEUTRETTO - [nucl] (a particle identical with an electron by without a charge) neutretto *m*
NEUTRINO - [nucl] (a neutral particle of very small rest mass, having a spin quantum number of I/2) neutrino *m*
NEUTRODYNE - [radio] (registered trade name for a type of neutralized high-frequency amplifier) neutrodyne *m*
NEUTRON - [phys] (elementary particle whose mass is practically equal to that of the proton but without electric charge) neutron *m*
~ABSORBER - [nucl] (a substance possessing a good neutron-absorptivity) absorbeur *m* de neutrons, absorbant *m* de neutrons
~-ABSORBING GLASS PLATE - [nucl] plaque *f* de verre à absorption de neutrons
~ABSORPTION - [nucl] (absorption of neutrons during a fission chain reaction) absorption *f* de neutrons
~ABSORPTION CROSS SECTION - [nucl] section *f* efficace d'absorption des neutrons
~AGE - [nucl] (Fermi age, q.v.) âge *m* de Fermi
~ATOMIC MASS - [nucl] masse *f* du neutron
~ATTENUATION - [nucl] (the reduction which occurs in the intensity of a beam of neutrons on passing through matter) atténuation *f* du faisceau neutronique
~BALANCE - [nucl] bilan *m* neutronique, principe *m* de conservation de l'équilibre neutronique
BEAM - [nucl] (stream of usually fast neutrons) faisceau *m* électronique
~BINDING-ENERGY - [nucl] (the energy which is required to remove a single neutron from a nucleus) énergie *f* de liaison neutronique
~BOMBARDMENT - [nucl] (the process of projecting neutrons against a fissile material) bombardement *m* de neutrons
~BURST - [nucl] bouffée *f* de neutrons
~CAPTURE - [nucl] capture *f* de neutrons
~CHAIN REACTION - [nucl] (process in which the fission of an atom by a neutron produces other neutrons, which in their turn give rise to further fissions in a continuous sequence) réaction *f* neutronique en chaîne
NEUTRON COLLISION RADIUS - [nucl] (as determined by fast-neutron transmission experiments) rayon *m* efficace de neutron pour collision
~CONSERVATION OF BALANCE - [nucl] (the main principle for any theory of nuclear reactors) conservation *f* de l'équilibre neutronique
~COUNTER - [instr] (apparatus for counting neutrons) compteur *m* de neutrons
~CROSS-SECTION - [nucl] (cross-section for reactions with neutrons) section *f* efficace de neutrons
~CRYSTALLOGRAPHY - [phys] analyse *f* cristallographique par diffraction neutronique
~CURRENT DENSITY - [nucl] densité *f* de courant neutronique
~CYCLE - [nucl] (the history of the neutrons in a reactor) cycle *m* neutronique
~DECAY - [nucl] (spontaneous beta decay in a neutron) désintégration *f* de neutron
~DEGRADATION - [nucl] (decrease in energy caused by scattering collisions) retard *m* neutronique
DENSITY - [nucl] (the number of neutrons in unit volume of a substance) nombre *m* volumique de neutrons
~DETECTION - [nucl] détection *f* de neutrons
~DIFFRACTION - [phys] diffraction *f* à l'aide de neutrons
~DIFFRACTOMETER - [instr] (instrument used in diffraction analysis for measuring with a neutron beam the intensity of the diffracted beams at different angles) diffractomètre *m* neutronique
~DIFFUSION - [nucl] diffusion *f* des neutrons
~-ELECTRON INTERACTION - [phys] interaction *f* neutron-électron
~ENERGY - [nucl] énergie *f* neutronique
~ENERGY DISTRIBUTION - [nucl] (distribution of the energy of generated neutrons) distribution *f* de l'énergie neutronique
~ENERGY GROUPS - [nucl] (a classification of neutron energies based of strongly neutronabsorbing materials) groupes *mpl* d'énergie de neutrons
~ENERGY RANGE - [nucl] (the distribution field) intervalle *m* d'énergie de neutrons
~ESCAPE - [nucl] (the unwanted escape of neutrons in a nuclear reactor) fuite *f* de neutrons
~EXCESS - [nucl] (the number by which the neutrons in a nucleus exceed the protons) excès *m* de neutrons
EXCESS NUMBER - [nucl] s. neutron excess
~-FISSION-SCINTILLATION DETECTOR - [instr] (used for the detection of slow neutrons) compteur *m* à scintillation de neutrons lents
~FLUX - [nucl] (the product of neutron density with speed) flux *m* neutronique
~FLUX DENSITY - [nucl] débit *m* de fluence de neutrons
~FLUX DISTRIBUTION - [nucl] distribution *f* de flux de neutrons
~FORMATION BY STRIPPING - [nucl] formation *f* de neutron par cassure
~GENERATOR - [nucl] (nuclear reactor which is used for the production of neutrons for isotopes) générateur *m* de neutrons
~HARDENING - [nucl] (the effect of the diffusion of thermal neutrons through a medium which has an absorption cross-section that decreases with the

increase of energy) durcissement *m* du spectre des neutrons
NEUTRON HOUWITZER - [instr] (collimating apparatus used to produce a stream of neutrons) collimateur *m* de neutrons
~-INDUCED PROCESS - [nucl] (a process which is induced by neutrons) processus *m* induit par neutrons
~-INDUCED REACTION - [nucl] (nuclear reaction caused by neutrons) réaction *f* induite par neutrons
~IRRADIATION - [nucl] irradiation *f* par neutrons
~LEAKAGE - s. neutron escape, fuite *f* de neutrons
~LETHARGY - [nucl] (the negative of the neutral logarithm of the energy of the neutron) léthargie *f*
~LIFETIME - [nucl] vie *f* d'une génération neutronique
~MAGNETIC MOMENT - [nucl] moment *m* magnétique du neutron
~MONITORING - [nucl] (monitoring of the neutron flux) contrôle *m* du flux neutronique
~MOISTURE METER - [instr] humidimètre *m* de sol par rayonnement ionisant
~MULTIPLICATION - [nucl] (production of new neutrons from fission neutrons) multiplication *f* de neutrons
~MULTIPLICATION FACTOR - [nucl] facteur *m* de multiplication de neutrons
~NUMBER - [nucl] (in a nucleus) nombre *m* de neutrons
~OPTICS - [nucl] (the study of optical phenomena in neutron behaviour) optique *f* neutronique
~PERIOD - [nucl] (the time constant of the neutron flux) constante *f* de temps de flux neutronique, période *f* neutronique
~PHYSICS - [nucl] physique *f* neutronique
~POPULATION - [nucl] densité *f* de neutrons libres
~PRODUCER - s. neutron generator
~-PROTON EXCHANGE FORCES - [nucl] forces *fpl* d'échange neutron-proton
~RADIATION - [nucl] (neutron emission) rayonnement *m* neutronique
~RADIATIVE CAPTURE - [nucl] (capture of slow neutrons by an atomic nucleus) capture *f* de neutrons radiative
~RADIOGRAPHY - [radiol] (radiography by neutrons) radiographie *f* neutronique
~RATE - s. neutron period
~REAÇTION - s. neutron-indiced reaction
~REFLECTION - [nucl] réflexion *f* de neutrons
~RESONANCE ESCAPE PROBABILITY - [nucl] probabilité *f* de fuite de neutrons par résonance
~REST MASS - [nucl] masse *f* de repos du neutron
~SCATTERING - [nucl] (the change of direction of a neutron caused by a collision) diffusion *f* de neutrons
~SLOWING DOWN LENGTH - [nucl] (the square root of the slowing-down area) longueur *f* du retard des neutrons
~SOURCE - [nucl] (a neutron emitting material) source *f* de neutrons
~SPECTROMETER - [instr] (instrument designed to measure a neutron spectrum) spectromètre *m* neutronique
~SPECTRUM - [nucl] (the energy distribution of the neutron) spectre *m* neutronique
~SPEED - [nucl] (in a nuclear reactor) vitesse *f* de neutrons
NEUTRON TEMPERATURE - [nucl] température *f* neutronique
~THERAPY - [radiat] (neutron irradiation for therapeutic purposes) neutronthérapie *f*
~THERMOPILE - [nucl] (simple unit designed to measure neutron flux) thermopile *f* à neutrons
~-TIGHT - [nucl] (term used of a structure or substance which does not absorb or allow the escape of neutrons) étanche aux neutrons
~VELOCITY - s. neutron speed
~VELOCITY SELECTOR - [instr] (instrument designed to single out neutrons of a given range of velocities) sélecteur *m* de vitesse des neutrons
~WAVELENGTH - [nucl] longueur *f* d'onde du neutron
~YIELD - [nucl] (the useful number of neutrons which are generated in a nuclear reaction) nombre *m* de neutrons
~YIELD PER ABSORPTION - [nucl] (the total emitted per neutron which is absorbed in fuel) facteur *m* êta
~YIELD PER FISSION - [nucl] (the total emitted per fission) facteur *m* nu
NEUTROPHIL GRANULE - [med] (white blood corpuscles having granules which only stain with neutral dyes) granule *m* neutrophile
NEVE-GLACIER - [geol] glacier *m* de névé
NEVER EXCEED MACH NUMBER (MNE) - [aero] (the Mach number for given aircraft which must in no circumstances be exceeded) nombre *m* critique de Mach
~EXCEED SPEED (VNE) - [aero] (the airspeed for a given aircraft which must in no cirumstances be exceeded) vitesse *f* maximum d'utilisation
NEVILLE AND WINTHER'S ACID - [chem] (I-naphthol-4-sulphonic acid. A dyestuff intermediate) acide *m* de Neville et Winther
NEVUS - [med] naevus *m*, tumeur *f* érectile
NEW - [gen] nouveau
[bot] (of fruit etc) frais
[agric] (of wine) jeune
~BEER - [brew ind] bière *f* jeune
~BUSINESS EXPENSE - [comm] (US for development cost) frais *mpl* de développement
~CANDLE - [light] (a proposed new candle unit, i.e. a luminous intensity of solidifying platinum equal to 60 candles per sq. cm.) nouvelle bougie *f*
~COMPOUND - [rubber ind] mélange *m* non vulcanisé
~LUMINOUS SENSITIVITY - [electron] photosensibilité *f* d'un tube de prise de vues
~MOON - [astr] nouvelle lune *f*
~STAR - [astr] (star making a sudden appearance in the sky) nova *f*
WINE - [wine ind] vin *m* jeune
NEWEL - [constr] (post at the end of a stair or handrail) noyau *m*, pilastre *m* de rampe d'escalier
[constr] balustre *m*
[arch] (pillar at the side of a wing wall) pilastre *m*
[mech] (of an Archimedean screw) noyau *m* (d'une vis d'Archimède)
~CAP - [constr] chapiteau *m* du montant d'escalier
~DROP - [constr] base *f* du montant d'escalier
~POST - [constr] montant *m* d'escalier, pilastre *m* de rampe d'escalier
~STAIRS - [constr] escalier *m* hélicoïdal
NEWS - [gen] nouvelles *fpl*, actualité *f*
~CAMERA - [telev] caméra *f* d'actualités
~CASE - [print] (the types used for the compsition of

newspapers) casse *f* en deux pièces
NEWS EDITING - [telev] montage *m* du journal filmé
~LETTER - [gen & comm] feuille *f*
~NETWORK - [telecomm] réseau *m* de télécommunication d'informations
~PIC - s. newsreel
~SHOT - s. newsreel
NEWSCAST, to - [radio] (the broadcasting of news) diffusion *f* du journal parlé
NEWSCAST - [radio] journal *m* parlé
NEWSIE - [colloq] s. newsreel
NEWSPAPER - [gen] journal *m*
NEWSPRINT - [paper man] (cheap printing paper used for newspaper or weeklies) papier *m* de journal, papier-journal *m*
NEWSREEL - [cin] (motion picture showing current events) actualités *f*pl
NEWTON - [meas] (unit of force in the Meter-Kilogram-Second system, i.e. the force which induces in one Kilogram an acceleration of one meter per second/second) newton *m*
~EMISSION THEORY - [opt] (the theory that light is die to an emission of luminous corpuscles from a source) théorie *f* d'émission de Newton
~LAW FOR COOLING - s. Newton law for heat loss
~LAW FOR HEAT LOSS - [phys] (the heat lost from one body to another is proportional to the temperature difference between the two bodies) loi *f* de refroidissement de Newton
~LAW OF HYDRODYNAMIC RESISTANCE - [mech] loi *f* de la résistance hydrodynamique de Newton
~LAW OF UNIVERSAL GRAVITATION - [phys] loi *f* de gravitation universelle de Newton
~LAWS OF MOTION - [phys] (a particle does not move unless acted on by an external force) lois *f*pl de mouvement de Newton
~RINGS - [opt] (interference phenomenon which can be observed by laying a convex lens on a flat glass plate) halo *m* de Newton, cercles *m*pl de Newton
NEWTON'S DISK - [opt] disc *m* de Newton
NEWTONIAN - [gen] (relating to Newton's laws etc) newtonien
~FLUID - [mech] (a fluid whose viscous stresses are a multiple of the rate of distortion) fluide *m* newtonien
~MECHANICS - [mech] (based on Newton's laws of motion) mécanique *f* de Newton
~TELESCOPE - [opt] (reflecting-type telescope with a 45 degree mirror) télescope *m* newtonien
NEXT-CHANNEL MIXER - [telev] mélangeur *m* de canaux adjacents
NIB - [gen] (projecting part) talon *m*
[gen] (of a pen) taillon *m*
[zool] (beak of a board) bec *m*
[constr] (a small projection on the underside at the top of tiles) talon *m*, crochet *m*
[tools] (the point of a crowbar) pointe *f*
[el] (a plate lug) queue *f* conductrice
[metall] compact *m* terminé
~OF A TILE - [constr] talon *m* d'une tuile
NIBBLE, to - [gen] grignoter, mordiller
NIBBLE - [gen] (a little bite) mordillure *f*, grignotement *m*
NIBBLER - s. nibbling machine
NIBBLING - [metall] grignotage *m*
~MACHINE - [mech] (machine fitted with a continuously-reciprocating punch used to cut irregular shapes from sheet) découpeuse *f*, machine *f* à grignoter
NIBBLING MACHINE - [metall] poinçonneuse *f*
NIBS - [plast ind] (short, thin pieces, cut from rods of thermoplastic material, used in the production of sheet materials) fractions *f*pl
NICCOLITE - [min] (natural nickel arsenide, usually containing some iron, cobalt and sulphur and crystallizing in the hexagonal system. One of the most important ores of nickel) nickéline *f*, nickel *m* arsenical
NICHE - [constr] (any recess in a wall surface) niche *f*
~VAULTING - [constr] voûte *f* en niche *f*
NICHOL'S VANE RADIOMETER - [instr] (a sensitive type of radiometer) radiomètre *m* à ailettes de Nichol
NICHROME - [metall] (alloy of nickel and chrome) nickel-chrome *m*
NICK, to - [mech & carp] entailler, encocher
[gen] (to make a small indentation) cocher, entailler, ébrécher
[print] créner
NICK - [gen] (a small indentation of incision) entaille *f*, encoche *f*
[mech & carp] entaille *f*, hoche *f*, cran *m*, onglet *m*
[print] (the groove in the shank of a type letter) cran *m* habituel, cran *m*
[metall] (defect on a metal or wooden workpiece) brèche *f*
[mech] (of lubricating oil) saignée *f*
NICKED FRACTURE TEST - [metall] essai *m* de fracture par choc sur barreaux entaillés
NICKEL, to - [el metall] (to coat thinly with nickel) nickeler
NICKEL - [chem] (a metallic element, symbol Ni, A.N. 28 A.W. 58.69 in the production of a number of alloys and also as a catalyst in organic reactions) nickel *m*
~ACETATE - [chem] (a deying mordant) acétate *m* de nickel
~-ALUMINIUM BRONZE - [metall] bronze *m* de nickel et aluminium
~AMMONIUM CHLORIDE - [chem] (a mordant in dyeing) chlorure *m* de nickel ammoniacal
~ANTIMONY GLANCE - [min] ullmannite *f*
~ARSENATE - [chem] (a catalyst in the hydrogenation of fats) arséniate *m* de nickel
~BLOOM - [min] (a natural hydrous nickel arsenate, crystallizing in the monoclinic system, and apple-green in colour. Also called Annabergite) annabergite *f*, nickelocre *f*
~BRASS - [metall] laiton *m* au nickel
~-CADMIUM ACCUMULATOR - [el] accumulateur *m* au nickel-cadmium
~CARBONATE - [chem] (a component of ceramics glazes and an intermediate in the manufacture of nickel catalyst) carbonate *m* de nickel
~CARBONYL - [chem] (a colourless liquid used for the purification of nickel by the Mond process) carbonyle *m* de nickel
~CAST IRON - [metall] fonte *f* au nickel
~CHLORIDE - [chem] (green crystalline compound) chlorure *m* de nickel
~-CHROME STEEL - [metall] (a steel containing specific amounts of nickel and chromium and having high tensile strength) acier *m* au nickel-chrome

NICKEL-COPPER ALLOY - [metall] bronze *m* blanc
~DELAY LINE - [comput] (delay-line with a bunch of nickel wires) ligne *f* à retard à nickel
~DIBUTYLDITHIOCARBAMATE - [chem] (a rubber additive) dibutyldithiocarbamate *m* de nickel
~HARDENING METHOD - [metall] procédé *m* de trempe au nickel
~-IRON BATTERY - [el chem] (type of electrical storage battery in which the plates are of iron and nickel in an alkaline electrolyte) batterie *f* au ferro-nickel
~LAYER - [electron] (intermediate layer in dry rectifiers) couche *f* de nickel
~MATTE - [min] matte *f* de nickel
~NITRATE - [chem] (a component of ceramics glazes) nitrate *m* de nickel
~ORES - [min] minerais *mpl* de nickel
~PELLETS - [metall] grains *mpl* de nickel
~-PLATE, to - [el metall] (the operation of coating with metallic nickel by electro-deposition) nickeler
~-PLATED - [el metall] (coated with metallic nickel by electro-deposition) nickelé
~PLATING - [el metall] (the electrolytic deposition on a conducting surface of a thin coating of nickel) nickelage *m*, placage *m* au nickel
~-SILVER - [metall] (alloy of Cu.N. and Zn, in various proportions) maillechort *m*, métal *m* blanc
~STEEL - [metall] (steel containing up to 30 p.c. Ni) acier *m* au nickel
~SULPHATE - [chem] (a mordant in dyeing, starting material in the production of nickel catalysts, component of plating baths and an ingredient of ceramics glazes) sulfate *m* de nickel
NICKING - [metall] entaillage *m*
NICKINGS - [mining] charbon *m* menu
NICOL PRISM - [opt] (crystal of Iceland spar so cut and joined the ordinary ray is totally reflected out through the side of the crystal, while the plane-polarized ray passes without interference) nicol *m*, prisme *m* de Nicol
NICOLAYITE - [min] (mineral containing lead and calcium) nicolayite *f*, thorogummite *f*
NICOPYRITE - [min] pentlandite *f*
NICOTINAMIDE - [chem] (niacinamide. The antipellagra component of vitamin B with uses in the treatment of dietary deficiens) nicotinamide *f*
NICOTINE - [chem] (a poisonous alkaloid obtained from tobacco and used as an insecticide) nicotine *f*
NICOTINIC ACID - [chem] (niacin. An antipellagra factor used in medicine) acide *m* nicotinique
NICTATION - [med] cillement *m*, nictitation *f*
NICTITATION - s. nictation
NIDATION - [med] nidation *f*
NIDGING - [constr] taille *f* de la pierre
NIELLO - [metall] (the art of decorating metal plates by the incision of a design which is the filled in with black alloy) nielle *m*
[metall] (black alloy containing sulphur, silver, copper etc) nielle *m*
(the actual work produced by niello) niellure *f*
NIGELLA - [bot] nigelle *f*
NIGGER - [cin] (adjustable panel to eliminate stray lights) écran *m* noir
NIGHT - [gen &astr] (the period between sunset and sunrise) nuit *f*
~-ALARM KEY - [telecomm] commutateur *m* de nuit
~BLINDNESS - [opt] (any degree of inability to see under conditions of low illumination) héméralopie *f*, nyctalopie *f*
NIGHT COOLING - [met] (loss of heat from the ground by radiation during the night) refroidissement *m* nocturne
~EFFECT - [radio & radar] (errors in radio direction, finding, due to interference by reflected sky-waves or other causes peculiar to nocturnal conditions) effet *m* de nuit
~POSITION - [telecomm] position *f* de nuit
~SHIFT - [gen] équipe *f* de nuit
~SOIL - [agric] déjections *fpl* humaines, engrais *m* humain
~SWEAT - [med] transpiration *f* nocturne
~VISION - [opt] vision *f* nocturne
~WATCH - [gen] garde *f* de nuit, veille *f*
~WATCHER - [gen] gardien *m* de nuit
NIGHTINGALE - [zool] rossignol *m*
NIGHTSHADE - [bot] morelle *f* noire, raisin *m* de loup
NIGRE - [ind chem] (in soap manufacturing, an impure solution settling at the last stage) crasses *fpl* insolubles
NIGRINE - [min] rutile *m*, schorl *m* rouge
NIGROSINE - [chem] (any of a group of black or dark-blue dyes obtained from aniline) nigrosine *f*
NIKETHAMIDE - [chem] (aminocordine. An analeptic with specific uses as a respiratory stimulant) nikéthamide *f*
NIMBOSTRATUS - [met] (a dark-grey rain-cloud layer, low-lying, shapeless and generally uniform) nimbo stratus *m*
NIMBUS - [met] (heavy dark cloud, characteristic of thunderstorms) nimbus *m*
[astr] aréole *f*
NINES COMPLEMENT - [comput] complément *m* à 9
NIOBITE - [min] (or columbite; mineral containing iron, manganese, niobium and tantalum) niobite *f*
NIOBIUM - [chem] (columbium. A metallic element, symbol Nb, A.N. 4I, used in the production of cermets in metallurgy and in nuclear technology) niobium *m*, columbium *m*
~CARBIDE - [chem] (a constituent of certain alloy steels) carbure *m* de niobium
NIP, to - [text] exprimer
NIP - [gen] goutte *f*, pince *f*
[mech] (interference) jeu *m*, écartement *m*, fente *f*
[mech] (pair of rolls in a calender) cylindres *mpl* presseurs
[mech] (a clamping) serrement *m*
[mech] (of rollers; the line along which the material enters between the rollers) intervalle *m* entre les cylindres
[naut] (in a rope) portage *m*, étrive *f*
[min] (of a coal seam) étreinte *f*
[geogr] échranchure *f*
[text] jeux *m* de peigne, gousset *m*, grisotte *f*
[geol] amincissement *m*
[mining] éboulement *m*
~ROLL - [mech] (in a film slitting machine) cylindre *m* presseur
[plast ind] s. nipper rolls
NIPPER - [tool] pince *f*, tenaille *f*, cisailles *fpl*
[zool] (a cutting tooth, e.g. of a horse) incisive *f* (p.e. d'un cheval)
[naut] garcette *f* (de tournevire)
[text] pince *f* à dresser les aiguilles

NIPPER CAM - [text] excentrique *m* des pinces
~HEAD - [mech] tête *f* de griffe
~PEG - [text] broche *f* fendue
~ROLLS - [plast ind] rouleaux *mpl* pinceurs, rouleaux *mpl* presseurs
~SHAFT - [text] arbre *m* de la pince
~SHUTTLE - [text] navette *f* à griffe
~TEMPLE - [text] templet *m* à griffes
NIPPERS - [tool] pinces *fpl* de serrage, tenailles *fpl*
NIPPING EFFECT - [text] effet *m* de pinçage
~FORK - [impl] clef *f* de retenue
NIPPLE - [mech] (short length of pipe threaded at both ends) mamelon *m*, raccord *m* fileté
[mech] raccord *m*, jonction *f*, douille *f*
[mech] (of bycicles) écrou *m*, douille *f*
[zool] (the protuberance of the mammary gland) bout *m* de mamelle, mamelon *m*
NISUS - [med] effort *m*, impulsion *f*
NITER, to - [chem] (to treat with nitric acid) nitrifier
NITER - s. nitre
NITON - [chem] (the original name for radium emanation; radon) radon *m*
NITRALIZING - [metall] immersion *f* en nitrate de sodium fondu
NITRATE, to - [chem] (the introduction of a nitro group NO_2 into a molecule) nitrer
NITRATE - [chem] (a salt formed by the action of nitric acid on metallic oxides, carbonates and hydroxides) nitrate *m*
~FERTILIZERS - [chem] engrais *mpl* nitriques
~OF AMMONIUM - [chem] nitrate *m* d'ammoniaque
~OF LIME - [chem] nitrate *m* de chaux
~OF POTASH - [chem] nitrate *m* de potassium, salpêtre *m*
~OF SILVER - [chem] nitrate *m* d'argent, azotate *m* d'argent
NITRATING APPARATUS - [ind chem] (apparatus in which nitration in carried out) nitreur *m*
NITRATION - [chem] (the addition of a nitro group to a component) nitration *f*
NITRATOR - [ind chem] (closed vessel fitted with cooling tubes, in which nitration is carried out) nitreur *m*
NITRE - [chem] (native potassium nitrate, saltpetre) nitrate *m* de potassium, salpêtre *m*
~BATH - [ind chem] bain *m* de salpêtre
NITRIC ACID - [chem] (corrosive, colourless of yellowish liquid with important uses in organic synthesis) acide *m* nitrique
~ACID TEST - [metall] essai *m* à l'acide nitrique
~ESTER - [chem] (an ester produced by the action of concentrated nitric acid on an alcohol) ester *m* nitrique
~OXIDE - [chem] (an intermediate for ammonia and nitric acid) oxyde *m* azotique
NITRIDES - [chem] (compounds of metals with nitrogen only) nitrures *mpl*
NITRIDING - [metall] (a special process for case-hardening steel by heating in gaseous ammonia, nitrides being formed in the surface layer) nitruration *f*
~ATMOSPHERE - [metall] atmosphère *f* de nitruration
NITRIFICATION - [chem] (the introduction of a nitro group into a molecule) nitrification *f*
NITRIFY, to - [chem] (to treat with nitric acid) nitrifier
NITRILE - [chem] (generic name for esters of hydrogen cyanide) nitrile *m*
NITRITE - [chem] (a salt of nitrous acid) nitrite *m*
NITROACETANILIDE - [chem] (a synthesis intermediate) nitroacétanilide *f*
NITROANILINE - [chem] (an intermediate for dyestuffs) nitroaniline *f*
NITROANISOLE - [chem] (an intermediate for drugs and dyestuffs) nitroanisole *m*
NITROBACTERIA - [bacteriol] (or nitrate bacteria; soil-inhabiting bacteria forming nitrites from compounds of ammonia) nitrobactérie *f*
NITROBENZALDEHYDE - [chem] (an intermediate for drugs and dyestuffs) nitrobenzaldéhyde *m*
NITROBENZENE - [chem] (mirbane oil. A solvent for cellulose derivatives and an intermediate for quinoline, aniline and other organic compounds) nitrobenzène *m*
NITROBENZENEAZORESORCINOL - [chem] (an analytical reagent) nitrobenzèneazorésorcinol *m*
NITROBENZOIC ACID - [chem] (an intermediate for pharmaceuticals and dyestuffs) acide *m* nitrobenzoïque
NITROBENZOYL CHLORIDE - [chem] (an intermediate for pharmaceuticals and dyestuffs) chlorure *m* de nitrobenzoyle
~CYANIDE - [chem] (an intermediate for drugs and dyes) cyanure *m* de nitrobenzoïle
NITROBIPHENYL - [chem] (a dyestuffs intermediate and a plasticizer) nitrobiphényle *m*
NITROCELLULOSE - [chem] (guncottone, collodion cotton. A nitric acid ester of cellulose with uses in the manufacture of explosives, plastics and lacquers) nitrocellulose
~SILK - [text] soie *f* artificielle de Chardonnet
NITROCHALK - [chem] nitrate *m* de clacium ammoniacal
NITROCOTTON - [chem] (a high explosive obtained by treating cotton linters with nitric acid) fulmicoton *m*, coton *m* nitré
NITRODIPHENYLAMINE - [chem] (a synthesis intermediate) nitrodiphénylamine *f*
NITROETHANE - [chem] (a solvent for resins, dyestuffs and cellulose derivatives) nitroéthane *m*
NITROFURANTOIN - [chem] (an antiseptic active against Gram-positive and Gram-negative bacteria) nitrofurantoïne *f*
NITROFURAZONE - [chem] (a bacteriostat used in medicine) nitrofurazone *m*
NITROGEN - [chem] (a gaseous element, symbol N, A.N. 7, A.W. I4.008, used as a raw material for a large number of organic and inorganic compounds) azote *m*
~CYCLE - [chem] cycle *m* de l'azote
~DESATURATION - [chem] dénitrogénation *f*
~FIXATION - [chem] (the process of obtaining nitrogen in combined form by the treatment of atmospheric air) azotation *f*
~PEROXIDE - [chem] (nitrogen dioxide. A brownish-red gas with uses as a catalyst, oxidizing and nitrating agent, and as an intermediate in the production of nitric acid) peroxyde *m* d'azote
~PURCING - [chem] (the introduction of nitrogen (or other inert gas) into the vacant space in a nearly empty fuel tank to avoid danger of formation of an explosive mixture) injection *f* d'azote

NITROGEN TRICHLORIDE - [chem] (explosive, poisonous yellow oil) trichlorure *m* d'azote
~TRIFLUORIDE - [chem] (an oxidizing agent) trifluorure *m* d'azote
~TRIHYDRIDE - [chem] (a synonym for ammonia) ammonique *m*
NITROGENASE - [chem] (a nitrogen-fixing enzyme present in the soil) nitrogénase *f*
NITROGENOUS - [chem] (containing nitrogen) azoté
~BODY - [ind chem] substance *f* azotée
~FERTILIZERS - [agric] engrais *mpl* azotés
~MANURE - [agric] engrais *m* azoté
NITROGLYCERINE - [chem] (glyceryl trinitrate. An important industrial explosive used in the production of dynamite and gelignite) nitroglycérine *f*
NITROGUANIDINE - [chem] (an explosive) nitroguanidine *f*
NITROMERSOL - [chem] (an antiseptic used in medicine) nitromersol *m*
NITROMETER - [instr] (instrument designed to determinate the contents of nitrogen in chemical substances) azotomètre *m*
NITROMETHANE - [chem] (a solvent for cellulose derivatives, a number of dyestuffs and natural and synthetic resins) nitrométhane *m*
NITROMURIATIC ACID - [chem] (mixture of concentrated nitric and hydrochloric acid) eau *f* régale
NITRON - [chem] (1,4-Diphenyl-3,5- endanilodihydrotriazole. It forms an insoluble salt with nitric acid and is used in the estimation of the latter) nitron *m*
NITRONAPHTHALENE - [chem] (an intermediate for dyestuffs) nitronaphtaline *f*
NITROPARACRESOL - [chem] (a synthesis intermediate) nitro-para-crésol *m*
NITROPARAFFINS - [chem] (compounds derived from paraffins by substitution of an $-NO_2$ group for an atom of hydrogen) nitroparaffines *fpl*
NITROPHENETOLE - [chem] (an intermediate for dyestuffs) nitrophénétol *m*
NITROPHENIDE - [chem] (an intermediate for pharmaceuticals) nitrophénide *f*
NITROPHENOL - [chem] (an indicator and a synthesis intermediate) nitrophénol *m*
NITROPHENYLACETIC ACID - [chem] (an intermediate for pharmaceuticals) acide *m* nitrophénylacétique
NITROPROPANE - [chem] (a synthesis intermediate and a solvent for pigments, dyestuffs, waxes and resins) nitropropane *m*
NITROSALICYLIC ACID - [chem] (an intermediate for dyes) acide *m* nitrosalicylique
NITROSO COMPOUNDS - [chem] (those which contain the monovalent radical -NO) composés *mpl* nitrosés
~RUBBER - [chem] (a copolymer of tetrafluoroethylene and trifluoronitrosomethane, having good chemical resistance and low-temperature properties) caoutchouc *m* nitrosé
NITROSODIMETHYLANILINE - [chem] (a vulcanisation accelerator) nitrosodiméthylaniline *f*
NITROSODIPHENYLAMINE - [chem] (a vulcanization modifier) nitrosodiphénylamine *f*
NITROSOGUANIDINE - [chem] (an explosive used in the manufacture of detonators) nitrosoguanidine *f*
NITROSONAPHTHOL - [chem] (a synthesis intermediate) nitrosonaphtol *m*
NITROSOPHENOL - [chem] (an intermediate for dyes) nitrosophénol *m*
NITROSTARCH - [chem] (a high explosive) nitroamidon *m*, nitrate *m* d'amidon
NITROSTYRENE - [chem] (a polymerization modifier) nitrostyrène *m*
NITROSULPHATHIAZOLE - [chem] (a sulphonamide with therapeutic properties) nitrosulfathiazol *m*
NITROSYL CHLORIDE - [chem] (a catalyst and synthesis intermediate) chlorure *m* de nitrosyle
NITROTOLUENE - [chem] (a synthesis intermediate for a number of compounds including dyestuffs) nitrotoluène *m*
NITROTOLUIDINE - [chem] (a dyestuffs intermediate) nitrotoluidine *f*
NITROTRIFLUOROMETHYLBENZONITRILE - [chem] (an intermediate for dyestuffs) nitro-trifluorométhylbenzonitrile *m*
NITROUREA - [chem] (a high explosive) nitrourée *f*
NITROUS OXIDE - [chem] (laughing gas. A short-acting inhalation anaesthetic) oxyde *m* nitreux
NITROXYLENE - [chem] (a synthesis intermediate) nitroxylène *m*
NIVENITE - [min] (mineral containing small quantities of uranium) nivénite *f*
NO-AIRFIELD AIRCRAFT - [aero] (aircraft capable of alighting where a prepared landing group does not exist) avion *m* capable d'atterrir sans avoir besoin d'un champ d'atterrissage
~-BOND RESISTANCE - [phys] (hyperconjugation; the properties of a molecule in terms of resonance structures) hyperconjugaison *f*
~-BOTTOM SOUND - [acoust] son *m* sans basses fréquences
~CONNECTION - broche *f* à circuit ouvert
~-DELAY SERVICE - [telecomm] (in telephony) service *m* à la demande
~-DRAFT - [th eng] (US for zero flue draught) absence *f* de tirage
~-DRAFT VENTILATION - [auto] (US expression) aération *f* sans courant d'air
~-DRIFT POSITION - [aero] position *f* en absence de vent
~-FEATHERING AXIS - [aero] (the axis through the centre of rotation of a helicopter rotor in respect to which the blade pitch angle does not vary with the azimuth angle) axe *m* de pas constant
~-FINES CONCRETE - [constr] béton *m* sans fines
~-HOME RECORD - [comput] enregistrement *m* indirect
~LIFT DIRECTION - [aero] (the airflow direction, for a parallel untwisted aerofoil section, at which there would be no lift on such a section) direction *f* de portance nulle
~-LOAD - s. no-load operation
~-LOAD CHARACTERISTIC - [el] (a curve representing the no-load voltage at the terminals, for a given speed, as a function of the excitation current) caractéristique *f* à vide
~-LOAD CURRENT - [el] courant *m* à vide
~-LOAD OPERATION - [el] (the operation of a machine, transformer etc., when it absorbs power without providing any power) fonctionnement *m* à vide
~-LOAD TAP - [el] prise *f* à vide
~-LOAD POWER - [el] puissance *f* à vide
~-PRESSURE LAMINATE - [plast ind] stratifié *m* moulé sans pression

NO-PRESSURE RESIN - [plast ind] résine *f* durcissant sans pression
~REPLY - [telecomm] (in telephony) non résponse *f*
~-VOLTAGE RELAY - [el] relais *m* à tension nulle
NOBELIUM - [chem] (a synthetic radioactive element, symbol No. A.N. I02) nobélium *m*
NOBLE - [gen] noble
[gen] (of metals, gases etc) noble, précieux
[el] (electropositive) électropositif, cathodique
~GASES - [chem] (the elements argon, helium, neon, xenon, krypton and radon, occurring in the atmosphere, characterised by closed electron shells or subshells and hence chemically inert) gaz *mpl* nobles, gaz *mpl* inertes
~METALS - [metall] (those with a relatively positive electrode potential, e.g. gold, silver, platinum, which do not readily combine with non-metals and have high corrosion resistance) métaux *mpl* nobles
~POTENTIAL - [el] tension *f* standard d'une électrode
NOCTILUCENT CLOUDS - [met] (luminous clouds of cirrus type sometimes seen in summer and supposed to consist of fine dust in suspension at great altitudes, illuminated by reflected sunlight) nauges *mpl* noctilucents
NOCTOVISION - [telev] (special infrared ray television) télévision *f* nocturne par rayons infrarouges
NOCTOVISOR - [telev] appareil *m* de télévision nocturne
NOCUITY - [med & gen] nocivité *f*, nocuité *f*
NODAL - [gen] (having intersecting points) nodal
~DIAGRAMME - [electron] (of waves in a waveguide) diagramme *m* de nodes
~DISTANCE - [photo]intervalle *m* nodal
~EQUATION - [math] équation *f* nodale
~LINE - [radio] (a specific line in a standing-wave system) ligne *f* nodale
~LINES - [acoust] (the lines along which no vibration takes place on a vibrating diaphragm) lignes *fpl* nodales
~PLANES - [opt] (the transverse planes passing through the nodal points) plans *mpl* nodaux
~POINT - s. node [radio & telecomm]
~POINTS - [opt] (two points on the principal axis of a lens, such that the incident ray directed towards one of them emerges from the lens as if from the other point) points *mpl* nodaux
~SLIDE - [opt] (small testing optical bench) analyseur *m*
~SURFACE - [radio] (a specific surface in a standing-wave system) surface *f* nodale
NODE - [gen] noeud *m*
[phys] (a point, surface or line in a standing-wave system) noeud *m*
[radio & telecomm] (any point, line or surface in a distributed field, in which a specified value-voltage, current etc - has a null- or near null-value) noeud *m*
[el] (a branch point) point *m* d'écart
[astr] (the point which is diametrically opposed to another point, through which the orbit of a heavenly body cuts a great circle) noeud *m*, point *m* nodal
[acoust] (point, surface or line, of an interference pattern, at which the amplitude of the sound pressure is zero) noeud *m*
[bot] (the place where the leaf is attached to the stem) noeud *m*, bracelet *m*, articulation *f*
NODE OF ADMISSION - [opt] point *m* nodal d'incidence
~OF EMISSION - [photo] point *m* nodal d'émergence
NODICAL MONTH - [astr] mois *m* draconitique
NODOSITY - [gen & med] (knot-like swellings) nodosité *f*, protubérance *f*
NODULAR - [gen, geol, metall etc] (with local thickenings) nodulaire
~CAST IRON - [metall] fonte *f* à graphite nodulaire
~FIRE CLAY - [metall] chamotte *f* nodulaire
~GRAPHITE - [metall] graphite *f* en crabes, graphite *f* nodulaire
~ORE - [min] minerai *m* en rognons
NODULE - [gen, metall etc] (a small round swelling) nodule *m*, rognon *m*
[ceramics] druse *f*
~BACTERIA - [bot] bactéries *fpl* des nodosités, rhizobium *m*
NODULES - [el chem] (rounded excrescences formed on the cathode during deposition) nodules *mpl*
NOG - [gen] (a wooden peg, or small block or wood) cheville *f* de bois
[bot] (on a tree) chicot *m*
NOGGIN - [gen] (a small mug) (petit) pot *m*
[meas] (liquid measure; about a gill) quart *m* de pinte
NOGGING - [constr] (brick filling of spaces between timbers in partitions) hourdage *m*, remplissage *m* en briques
NOGS - [mining] soliveaux *mpl* de soutènement
NOIL - [text] (short-staple fibres obtained during the combing process) blousse *f*
NOILS DOFFER - [text] cylindre *m* à blousse, doffer *m* à blousse
~ROLLER - [text] cylindre *m* à blousse, doffer *m* à blousse
NOISE, to - [gen] (to be noisy, or to make a noise) faire du bruit
NOISE - [acoust] (loud, unwanted sound) bruit *m*
[gen] bruit, tapage *m*, fracas *m*
~AUDIOGRAM - [acoust] (curve relating aural masking and frequency) audiogramme *m* de bruit
~AUDIOMETER - [instr] (instrument designed to asses the intensity of a sound) psophomètre *m* subjectif
~BEHIND THE SIGNAL - [radio] (modulation noise) souffle *m* dû au signal
[electron] (modulation noise) bruit *m* de modulation
~BLANKING - [telecomm] suppression *f* des parasites
~-CANCELLING MICROPHONE - [el acoust] (antinoise microphone; microphone having characteristic which discriminate against ambient noise) microphone *m* anti-bruits
~DIODE - [electron] (the principal electronic tube of a noise generating circuit, operating in the saturation area) diode *f* de bruit
~FACTOR - [radio] facteur *m* de bruit
~FACTOR SPOT - [radio] (term stressing that the noise factor is a point function of inout frequency) facteur *m* de bruit pour une fréquence
~FIGURE - s. noise factor
~FIGURE SPOT - s. noise factor spot
~FILTER - [el acoust] filtre *m* anti-parasites
~-FREE - [gen & radio] sans bruits
~GENERATOR - [instr] (instrument designed the mea-

sure the noise figure in microwave circuits) générateur *m* de bruit
NOISE INSULATION - [acoust] (any system to reduce or eliminate noise) insonorisation *f*
~ INTENSITY - [acoust] intensité *f* du bruit
~ INVERSER - [telecomm] inverseur *m* des parasites, limiteur *m* des parasites
~ INVERSION - [telecomm] inversion *f* des parasites
~ KILLER - [telecomm] (in a telegraph circuit) réseau *m* anti-parasites
~ LEVEL - [telecomm] (the strength of disturbing signals in a circuit) niveau *m* de bruit
~ LIMITER - [telecomm] (electronic tube in a circuit cutting off the peaks of noise signals which are stronger than the wanted signal itself) écrêteur *m* de bruit
~ MEASUREMENT - [acoust] mesure *f* du bruit
~ METER - [instr] (instrument to measure some parameter of the strength of a noise, e.g. loudness or level of sound pressure) sonomètre *m*
~ MODULATION - [electron] modulation *f* par signal de bruit
~ OF GEARS - [mech] bruit *m* des engrenages
~ OUTPUT - [radio] (mathematical definition used in the calculations for the effect of aerial noise in circuits) puissance *f* de bruit
~ POWER - s. noise output
~ -PROOF - [gen & acoust] à l'abri du bruit
~ QUIETING - [radio] (of a receiver) réduction *f* du bruit de fond
~ RATIO - [electron] rapport *m* signal/bruit
~ REDUCTION - [aero] (any action taken to reduce the noise produced by aircraft) atténuation *f* du bruit
[el acoust] (in photographic recording and reproduction, a process in which the transmission of the sound track of the print is decreased for low levels and increased for high-level signals) réduction *f* des bruits
~ RESISTANCE - [el acoust] résistance *f* de soufle
~ SELECTION - [acoust] élimination *f* des bruits superflus
~ SOURCE - [el acoust] source *f* de bruit
~ SUPPRESSION - [radio] (the limitation of the response of a receiver to impulsive radio noise) amortissement *m* du bruit
~ SUPPRESSOR - [radio] (a section of the receiver circuit reducing noise automatically when no carrier is being received) dispositif *m* éliminateur de bruits
~ TEMPERATURE - [radio] (mathematical definition used in the calculations for the effect of aerial noise) température *f* de souffle
~ TRACK - [el acoust] piste *f* bruitée
~ TRANSMISSION - [radio] réduction *f* de la qualité de transmission due aux bruits de circuit
~ TRESHOLD - [phys] seuil *m* du son
NOISELESS - [gen] (making no noise, silent) silencieux, insonorisé
~ CHAIN - [mech] (of transmission) chaîne *f* silencieuse
~ RECORDING - [el acoust] (the practice of keeping the photographic noise level as far below the recorded level as possible) enregistrement *m* silencieux
~ RUNNING - [mech] fonctionnement *m* silencieux
[auto] vitesse *f* silencieuse
NOISELESSLY - [gen] sans bruit, en silence, silencieusement
NOISELESSNESS - [gen & telecomm] absence *f* de bruit
NOISY - [gen] bruyant
[opt] (of colours) criard, voyant
~ BLACKS - [radio] (lack of uniformity in the black area of a picture due to level changes caused by noise) noir *m* perturbé
~ RAILS - [railw] (rail corrugation) déformation *f* de la voie
NOMA - [med] noma *m*, stomatite *f* gangréneuse
NOMINAL - [gen] nominal
[comm] (virtual; existing in name only) nominal, fictif
~ BLACK SIGNAL - [telecomm] (signal corresponding to the normal black signal in amplitude or frequency) signal *m* de noir normal
~ BORE - [metall] diamètre *m* intérieur nominal
~ CIRCUIT VOLTAGE - [el] (the highest circuit voltage on which an instrument will be used and which mat determine its insulation tests) tension *f* nominale d'isolement
~ COLLECTOR RING VOLTAGE - [el] tension *f* nominale de barre collectrice
~ CONTROL AMPERE-TURNS - [radio] (the difference between the control ampere-turns for minimum output and for rated output) tolérance *f* d'ampère-tours
~ -EXCITER RESPONSE - [el] réponse *f* nominale de l'excitatrice
~ FEEDBACK RATIO - [radio] taux *m* nominal de réaction
~ HORSEPOWER - [mech] (obsolete method of rating steam engines) cheval-vapeur *m* nominal
~ LINE WIDTH - [telev] (the reciprocal of the total number of lines per unit length in the direction of line progression) largeur *f* nominale de ligne
~ LOAD - [el] charge *f* nominale, pleine charge *f*
~ MAXIMUM CIRCUIT - [radio] (hypothetical reference circuit) circuit *m* fictif de référence
~ PULL-IN TORQUE - [el] couple *m* nominal d'accrochage d'un moteur synchrone
~ RANGE OF USE - [instr] (the range of values which each of the limiting quantities can assume when the instrument complies with given conditions) domaine *m* nominal d'utilisation
~ SIZE - [mech] (a dimension given as an indicator, which may be subject to variation due to special conditions of working, or of tolerances imposed) dimension *f* nominale
~ SPEED - [auto] vitesse *f* nominale
~ STRESS - [metall] effort *m* nominal
~ SWITCHING FREQUENCY - [el] fréquence *f* nominale de commutation
~ TRANSFORMATION RATIO - [instr] (ratio between the rated primary voltage and the rated secondary voltage) rapport *m* de transformation nominal
~ UPSET FORCE - [metall] pression *f* de refoulement nominale
~ VALUE - [gen & fin] valeur *f* nominale
~ VALUE OF AN ENERGIZING QUANTITY - [el] valeur *f* caractéristique d'une grandeur d'alimentation
~ VOLTAGE - [electron] tension *f* nominale
~ WHITE SIGNAL - [radio] (signal which corresponds to the normal white signal in amplitude or frequency) signal de blanc normal
NOMINATION - [gen] nomination *f*
NOMOGRAM - [instr] (a simple computing device consisting of two or more scales so graduated and

arranged that a relation between the quantities represented can be quickly found, e.g. by the use of a streight edge) nomogramme *m*, abaque *m*

NOMOGRAPHY - [math] (a methode of graphic representation) nomographie *f*

NON-ABSORBING MEDIUM - [radiat] (substances which does not absorb radiation) matériel *m* non perméable

NON-ACTINIC - [radiat] (electromagnetic rays which do not affect a sensitive photographic emulsion) inactinique

~-ACTINIC LIGHT - [photo] lumière *f* inactinique

~-ADECANE - [chem] (a synthesis intermediate) nonadécane *m*

~-ADIABATIC PROCESS - [phys] processus *m* non adiabatique

~-AERATED BURNER - [th eng] (a diffusion flame burner) brûleur *m* à flamme de diffusion, brûleur *m* sans mélange préalable

~-AGEING - [gen & phys] non-vieillissement *m*

~-ALDECANOIC ACID - [chem] (a synthesis intermediate) acide *m* nonaldecanoïque

~-AMBIGUOUS - [math] univoque

~-ARTICULATED ARCH - [constr] arc *m* sans articulation

~-ASSOCIATED LIQUOR - [phys] (liquid in which the molecules form no coordinate bonds) liquide *m* nonpolaire

~-AXIAL TROLLEY - [el] trolley *m* désaxé

~-BACKING COAL - [mining] houille *m* maigre

~-BATTERY LOOP - [el] circuit *m* de batterie non mise à terre

~-BEARING - [mech] (carrying no load) non-portant

~-BEARING PARTITION - [constr] (partition carrying no load apart from its own weight) paroi *f* de séparation nontant

~-BEARING WALL - [constr] (wall carrying no load apart from its own weight) mur *m* de refend non-portant

~-BLEEDING CABLE - [el] câble *m* sans perte de compound

~-BREAKABLE - [gen] infrangible

~-BREEDING MATERIAL - [nucl] (fuel material which does not produce fertile material) matériel *m* sans surrégénération

~-BURNING STEEL - [metall] acier *m* résistant à haute température

~-CAKING COAL - [min] charbon *m* non-collant, houille *f* maigre

~-CASTERING - [aero] (a wheel or wheel assembly which is not free to swivel as in a castering gear) train *m* d'atterrissage non-orientable

~-CENTRAL FORCE - [nucl] (nuclear force with a direction depending in part on the spin orientation of the nucleons) force *f* non-centrale

~-CENTRIFUGAL SUGAR - [ind chem] (sugar not separated as a crystalline entity) sucre *m* non-cristallisé

~-COHERENT - [phys] incohérent

~-COHERENT ECHO - [radar] écho *m* incohérent

~-COHESIVE - [gen] sans cohésion

~-COMBUSTIBLE - [gen] incombustible

~-COMPOSITE COLOUR-PICTURE SIGNAL - [telev] signal *m* incomplet de l'image couleur

~-COMPOSITE SIGNAL - [telev] signal *m* incomplet de l'image

~-COMPOSITE VIDEO MIXING - [telev] mélange *m* signal vidéo

NON-CONDENSABLE GAS - [chem] gaz *m* non condensable

~-CONDUCTING - [gen] (incapable of transmitting, e.g. electricity) non-conducteur, mauvais conducteur
[of a material] isolant

~-CONDUCTIVE MATERIAL - [gen] matériel *m* nonconducteur

~-CONDUCTOR - [gen el etc] (not capable of transmitting, e.g. electricity) non-conducteur *m*

~-CONGEALABLE - [phys] (incapable of solidification) non congelable

~-CONSTANT LEAD - [mech] pas *m* variable

~-CONTACT PISTON - [electron] (piston without metallic contact with the walls of the waveguide) piston *m* à piège

~-CRAWLING - [mech] (resisting undesirable flow) résistant au fluage

~-CRIMPED - [text] non frisé

~-CRITICAL DIMENSION - [electron] (nazzow dimension) côté *m* étroit, dimension *f* non-critique

~-CROSSING RULE - [nucl] (for an infinitely slow change of internuclear distances, two electronic states of the same species cannot cross each other) loi *f* de non-entrecroisement

~-DATA SET CLOCKING - [comput] régulateur *m* de vitesse de transmission de buits

~-DAZZLING LIGHTING - [auto] éclairage *m* non éblouissant

~-DEGENERATE GAS - [electron] (gas formed by a system of particles, the concentration of which is so great, that the Maxwell-Boltzmann law does not apply; e.g. electrons emitted by a hot cathode) gaz *m* non-dégénéré

~-DESTRUCTIVE READING - [comput] (the system of keeping the information in a store by copying it into a second store) lecture *f* non-destructive

~-DESTRUCTIVE TEST - [gen chem etc] (a test which does non involve the destruction of the specimen) essai *m* non-destructif

~-DIMENSIONAL - [gen] sans dimensions

~-DIRECTIONAL AERIAL - [radio] (aerial with substantially uniform response in azimuth and a directional pattern in elevation) antenne *f* non-directive

~-DIRECTIONAL BEACON - [radio] (a beacon for which a direction finding device is necessary to obtain a bearing) radiophare *m* non-directionnel

~-DIRECTIVE AERIAL - s. non-directional aerial

~-DIRECTIVE ANTENNA - s. non-directional aerial

~-DISJUNCTION - [comput] opération *f* NI, opération *f* NON-OU

~-DISSIPATIVE STUB - [waveguides] (coaxial stub; in a waveguide circuit) bras *m* de réactance

~-DRYING OIL - [chem] (an oil which does not form a film on exposure to air) huile *f* non siccative

~-ELASTIC CROSS SECTION - [nucl] section *f* efficace non-élastique d'interaction

~-EQUILIBRIUM DIFFUSION EQUATION - [nucl] (relating to nuclear reactor theory) équation *f* de diffusion non-équilibrée

~-EQUILIBRIUM THERMODYNAMICS - [phys] (the study of the thermodynamic relations in open systems in the steady state) thermodynamique *f* du déséquilibre

~-ERASABLE STORAGE - [comput] mémoire *f* non-effaçable

NON-EXPLOSIVE - [chem] (not capable of detonation) inexplosible
~-FANDING - [photo] (of a filter) résistant à la lumière, bon teint *m*
~-FERMENTABLE - [chem] non fermentescible
~-FERROUS - [min] (not containing iron) non ferreux
~-FERROUS COPPER WIRE - [el] fil *m* de cuivre exempt de fer
~-FERROUS MATERIAL - [chem] (material whichdoes not contain iron in any form) matériel *m* non ferreux
~-FERROUS METALS - [metall] (a metal other the iron and its alloys) métaux *mpl* non ferreux
~-FERROUS SCRAP - [metall] mitraille *f* non ferreuse
~-FLAM FILM - [cin] (colloq. fon non-flammable film; safety film) film *m* ininflammable
~-FLAMMABLE - [gen] (ignitible only with difficulty) ininflammable
~-FLOCCULATING YEAST - [brew ind] (a powdery form of yeast) levure *f* poussiéreuse
~-FLUFFY CLOTH - [text] tissu *m* non duveteux
~-FOAMING - [chem] (which does not produce foam) sans mousse
[of glass or lenses] anti-voile
~-FREEZING - [gen & phys] incongelable
~-FREEZING SOLUTION - [auto] (antifreeze) solution *f* anti-gel
~-FUSING ARC WELDING ELECTRODE - [el] (arc welding electrode which does not constitute the filler metal and may be considered non-fusible) électrode *f* réfractaire pour soudage à l'arc
~-GASSING COAL - [mining] charbon *m* pauvre en gaz
~-GLARE HEADLAMP - [auto] phare *m* non-éblouissant
~-GROWING - [mech] indilatable
~-HALATION - [photo] antihalo *m*
~-IDEAL SUPERCONDUCTOR - [el] (hard conductor) supraconducteur *m* non-idéal
~-IMPINGING INJECTOR - [mech] injecteur *m* à jets non convergents
~-INDUCTIVE - [el] (term applied to a circuit or winding to indicate that its self-inductance is negligible compared with its resistance, for a specified purpose) non inductif
~-INDUCTIVE CIRCUIT - [el] (electric circuit, the inductance of which is negligible in the specified condition under consideration) circuit *m* non inductif
~-INDUCTIVE RESISTOR - [el] (resistor with negligible inductance) résistance *f* non inductive, résistance *f* non-selfique
~-IONIC - [chem] non-ionique
~-IONIC DETERGENTS - [ind chem] (detergents which do not ionize or form electrically charged particles when dissolved in water; their use -as they do not produce a great deal of foam- is recommended for mechanical dishwashers and jet washing machines) détergents *mpl* non-ioniques
~-IONIZED ATOM - [nucl] atome *m* neutre
~-ISOTOPIC CARRIER - [nucl] (carrier in which the added substance is a different element from the tracer) porteur *m* non-isotopique
~-LEADED PETROL - [chem] (called straight-run gasoline in the USA) essence *f* H
~-LINEAR AMPLIFIER - [radio] (amplifier in which the output is not related to the input by a simple constant) amplificateur *m* non-linéaire
NON-LINEAR DISTORTION - [telecomm] (that form of amplitude distortion which occurs when the transmission properties of a system are dependent on the magnitude of the signal) distorsion *f* non-linéaire
~-LINEAR DISTORTION FACTOR - [telecomm] facteur *m* de distorsion non-linéaire
~-LINEAR HARMONIC DISTORTION COEFFICIENT - [electron] coefficient *m* de distorsion non harmonique
~-LINEAR OPTIMIZATION - [comput] optimisation *m* non-linéaire
~-LINEAR POTENTIOMETER - [instr] (graded, or tapered potentiometer) potentiomètre *m* non-linéaire
~-LINEAR RESISTANCE - [telecomm] (any device in which there is no linear relation between the applied voltage and the consequent current) résistance *f* non-linéaire
~-LINEAR-RESISTANCE ARRESTER - [el] parafoudre *m* à résistance variable
~-LINEAR RESONANCE - [radio] (of a system) résonance *f* non-linéaire
~-LINEAR SCALE - [instr] (scales in which divisions having the same value are of unequal widths) échelle *f* non-linéaire
~-LINEAR TIME BASE - [radar] (time-base causing unequal spacing between the range rings of the PPI) base *f* de temps non-linéaire
~-LINEARITY - [telev] (the result of picture elements when they are either widely separated or too near one another) non-linéarité *f*
~-LINEARITY OF THE EAR - [acoust] (if a pure tone is impressed on the ear, a series of overtones of the original frequency is heard) non-linéarité *f* de l'oreille
~-LINED CONSTRUCTION - [el chem] (method of making dry cells in which there is only a layer of paste between the negative electrode and the depolarizing mix) montage *m* sans habillage
~-LISTING CYCLE - [comput] cycle *m* de tubulation
~-LOCALIZED MOLECULAR ORBITALS - [phys] (not localized between two nuclei but spread over a larger part of the molecule) orbital *m* moléculaire non-localisé
~-LOCKING - [comput] à modification limitée de l'interprétation
~-LOCKING SHIFT CHARACTER - [comput] caractére *m* d'échange d'action singulière
~-LUMINOUS FLAME - [phys] (flame dark in colour, or the oxidizing part of the Bunsen flame) flamme *f* non-lumineuse
~-MAGNETIC - [el] (incapable of magnetization) amagnétique, non magnétique
~-MAGNETIC CAST IRON - [metall] fonte *f* amagnétique
~-MAGNETIC STEEL - [metall] (steel containing approximately I2°/。 of manganese, which exhibits no magnetic properties) acier *m* non magnétique
~-MAGNETIC WATCH - [horol] (a watch having a performance which is not affected by magnetic fields) montre *f* antimagnétique
~-METAL - [chem] (element readily forming negative ions) metalloïde *m*
~-METALLIC INCLUSIONS - [metall] (particles of non-metallic materials which are retained in a solid metal) inclusions *fpl* non metalliques
~-METALLIC SHEATHED CABLE - [el] câble *m* à gaine

non- métallique
NON-MINERALISED PRODUCT - [min] produit *m* stérile
~-MISCIBLE - [phys & chem] (non capable of being mixed) non miscible, immiscible
~-MOVING CONTACT SPRING - [el] ressort *m* de contact non activé
~-NUMERIC CHARACTER - [comput] caractére *m* non-numérique
~-OPENING SCUTTLE - [naut] hublot *m* fixe
~-PERIODIC - [gen & phys] apériodique
~-PHANTOMED CIRCUIT - [radio] (circuit which is not arranged to form part of a phantom circuit) circuit *m* non combinable (non prévu pour un circuit fantôme)
~-PHOSPHORIZED - [metall] non phosphorisé
~-PITTING - [mech] (in internal combustion engines) (valve) ne se piquant pas
~-PLANAR NETWORK - [el] (network which cannot be drawn on a plane without crossing of branches) réseau *m* non-planaire
~-PLASTIC - [th eng] (used in the construction of ovens) non plastique
~-POLAR - [el & plast ind] (material and plastics incapable of significant dielectric loss; polythene and polystyrene are examples) non polaire
~-POLAR COMPOUND - [chem] (compound in which the centres of the positive and negative charge almost coincide) composé *m* non polaire
~-POLAR LIQUID - s. non-associated liquid
~-POLAR SOLVENT - [chem] (solvent the constituent molecules of which have no permanent dipole moments and form no ionized solutions) dissolvent *m* non polaire
~-POLARIZED ELECTROLYTIC CAPACITOR - [el chem] (electrolytic capacitor in which the dielectric film is formed adjacent to both metal electrodes and in which the impedance to the flow of current is about the same in both directions) condensateur *m* électrolitique non polarisé
~-POLARIZED RELAY - [el] (relay operating in accordance with the magnitude of the current flowing in the controlling circuit, irrespective of the direction of the current) relais *m* apolaire
~-POLARIZED RETURN-TO-ZERO RECORDING-[comput] enregistrement *m* avec retour à zèro (non-polarisé)
~-POROUS - [phys etc] (impermeable) non poreux
~-POSITIVE DOBBY - [text] ratière *f* négative
~-PREGNANT - [zool] vide
~-PREMIXED BURNER - s. non-aerated burner
~-PRE-SETTING CONTROL - [el] commande *f* perdue
~-PRESSURE COWLING - [aero] (a cowling designed to exclude the surrounding air from a nacelle) capotage *m* sans pression, capotage *m* étanche
~-PRESSURE WELDING - [el] soudure *m* sans pression
~-PRINT IMPULSE - [comput] impulsion *f* de non-impression
~-PRODUCTIVE OPERATIONS - [comput] opérations *f*pl d'organisation
~-QUADDED CABLE - [telecomm] câble *m* à paires
~-QUANTIZED SYSTEM - [phys] (system of particles whose energies are assumed to be capable of varying in a continuous manner) système *m* classique, système *m* non quantifié
~-RAMMING INTAKE - [aero] (an air intake so designed as to neutralise the effect of forward speed on the intake pressure) prise *f* d'air sans effet dynamique
NON-REACTIVE - [el] (having a reactance which is negligible compared with the resistance) non réactif
~-REACTIVE LOAD - [el] (a load in which the current is in phase with the voltage at the terminals) charge *f* non réactive
~-REACTIVE RESISTOR - [el] (resistor having negligible reactance) résistance *f* non réactive
~-REFLECTING TERMINATION - [electron] (termination to a waveguide transmission system which reflects no energy) charge *f* adaptée, terminaison *f* adaptée
~-REFRACTORY ALLOY - [metall] alliage *m* non réfractaire
~REPETITIVE FORWARD CURRENT - [electron] courant *m* en sens direct non répétitif de surcharge accidentelle
~-RESONANT - [el radio acoust] non-resonnant
~-RESONANT LINE - [el radio] (transmission line termined at its receiving end by its characteristic impedance) ligne *f* non-resonnante
~-RESONANT TRANSFORMER - [el] transformateur *m* antiresonnant
~-RESONATING AERIAL - [radio] (an aerial which does not resonate; called 'dumb antenna' in the USA) antenne *f* non-resonnante
~-RETRACTABLE TAIL-WHEEL - [aero] roulette *f* de queue fixe
~-RETURN FLAP - [photo] clapet *f* de retenue
~-RETURN FLOW WIND TUNNEL - [aero] (a type of wind tunnel in which the air stream is not returned after passing through the experimental chamber) soufflerie *f* sans retour
~-RETURN VALVE - [mech] (a valve which allows flow in one direction only) soupape *f* de retenue, clapet *m* de retenue
~-REVERSIBLE - [gen] irréversible
~-REVERSIBLE PLUG - [el] prise *f* de courant irréversible
~-RIGID - [gen] non rigide, souple
~-RIGID AIRSHIP - [aero] (type of airship in which the rigidity of the envelope dependes on the gas pressure within it and not upon any special structure) dirigeable *m* souple
~-RIGID SHEETING - [plast ind] feuille *f* plastifiée
~-RISING SPINDLE VALVE - [mech] (internal screw valve) robinet *m* à vis intérieure
~-RISING STEM VALVE - s. non-rising spindle valve
~-ROAD TYRE - [rubber ind] (of tyres) pneu *m* tous-terrains
~-ROTATING ENGINE - [mech] moteur *m* à mouvement alternatif
~-ROTATING SCREW - [mech] vis-mère *f* fixe
~-RUSTING - [metall] (stainless, inoxidizable)inoxydable
~-SCALING - [gen] non s'ecaillant
~-SCREEN FILM - [photo] film *m* sans écran
~-SELECTIVE RADIATOR - [phys] (radiator with a constant spectral emissivity) radiateur *m* non-sélectif
~-SELF-LUMINOUS COLOUR - [telev] (colour perceived to belong to a non-self-luminous object)couleur *f* superficielle
~-SELF MAINTAINED DISCHARGE - [phys] (a discharge in which the charged particles are produced only

by the action of an external ionizing agent) décharge *f* de Townsend, décharge *f* non autonome

NON-SHARP PICTURE - [cin & telev] (picture lacking a good definition) image *f* floue

~-SHATTERABLE GLASS - [glass ind] verre *m* de sécurité

~-SHRINK BASE - [photo] support *m* irrétrécissable

~-SHRINKING - [text] irrétrécissable

~-SIZING - [metall] (in wire drawing, the progressive increase in size, due to the wear of the die) augmentation *f* excessive du diamètre

~-SKID - [rubber ind] (of tyres) antidérapant

~-SKID CHAINS - [auto] chaîne *f* antidérapant

~-SKID DECK COATING - [naut] revêtement *m* antidérapant pour pont de navire

~-SKID TYRE - [rubber ind] (of tyres) pneu *m* antidérapant

~-SLIP - s. non-skid

~-SLIP TREAD - [metall] plaque *f* striée

~-SPECTRAL COLOUR - [opt] (colour which is not present in the spectrum of white light) couleur *f* non-spectrale

~-SPECULAR REFLECTION - [acoust] (reflection from rough surfaces, producing no diffraction and scattering of sound waves) réflexion *f* diffuse

~-SPILLABLE ACCUMULATOR - [el] accumulateur *m* étanche

~-STICK ROUTE - [el] intinéraire *m* à tracé permanent

~-STOP - [gen] sans arrêt
[railw] direct

~-STOP FLIGHT - [aero] vol *m* sans escale

~-STOP SWITCH - [el] commutateur *m* pour marche sans interruption

~-STORAGE CAMERA TUBE - [telev] tube *m* de prise de vues sans emmagasinement

~SYMMETRICAL - [gen & geom] asymétrique

~-SYNC SIGNALS - [telev] signaux *mpl* sans information de synchronisation

~-SYNCHRONOUS - [gen etc] asynchrone

~-SYNCHRONOUS MOTOR - [el] (a.c. motor which does not run at synchronous speed) moteur *m* asynchrone

~-TENSION JOINT - [el] jonction *f* sans tension

~-TOXIC - [chem] (not liable to cause poisoning) non toxique

~-TRACKING - [el] (term used of an insulator in which a flash-over does not produce a permanently conducting path) sans cheminement

~-TRACKING QUALITY - [el] résistance *f* à l'effluation

~-TRANSPARENT - [gen] opaque

~-TRAVERSE NET - [text] réseau *m* indivisible

~-UNIFORM - [gen] non-uniforme

~-UNIFORM FLOW - [hydr] (the flow in a channel when the water surface is not parallel to the invert) écoulement *m* non-uniforme

~-UNIFORM PLANE CIRCULAR SURFACE SOURCE - [phys] source *f* non-uniforme de surface circulaire plane

~-UNIFORM STRAIN - [mech] (strain varying from point to point under the same stress) déformation *f* non-uniforme

~-UNIFORM WALL THICKNESS - [rubber ind] (of tyres) paroi *f* d'épaisseur non-uniforme

~UPSET - [oil ind] sans refoulement

~-VALENT - [chem] à valence nulle

NON-VIABLE - [zool] non-viable

~-VOLATILE - [phys chem] (not easily evaporable) non-volatil

~-VOLATILE MEMORY - [comput] (permanent memory) mémoire *f* permanente

~-WETTABLE SURFACE - [metall] surface *f* non mouillante

~-WOVEN MAT - [text & plast ind] (glass fibre material in a flat sheet with fibres in random arrangement, as distinct from woven material) mat *m* non tissé

NONANAL - [chem] (a perfumery component) nonanal *m*

NONANE - [chem] (a paraffin containing nine atoms of carbon in the molecule. A synthesis intermediate) nonane *m*

NONET - [mus] (composition for nine voices, or nine instruments) nonette *m*

NONOIC ACID - [chem] (synonym for pelargonic acid) acide *m* nonylique

NONOSES - [chem] (monosaccharoses containing nine oxygen atoms in the molecule) nonoses *fpl*

NONSENSE TOTAL CHECK - [comput] vérification *f* par total de non-sens

NONYL ACETATE - [chem] (a perfumery component) acétate *m* de nonyle

~ALCOHOL - [chem] (a component of perfumes and synthetic flavourants) alcool *m* nonylique

~LACTONE - [chem] (a component of synthetic flavourants) nonyllactone *m*

~PHENOL - [chem] (an intermediate for drugs, dyes, pesticides, detergents, rubber chemicals, and plasticizers) nonylphénol *m*

NONYLAMINE - [chem] (an intermediate for drugs, pesticides, dyes and rubber chemicals) nonylamine *f*

NONYLBENZENE - [chem] (an intermediate for surface-active agents) nonylbenzène *m*

NONYLENE - [chem] (a synthesis intermediate) nonylène *m*

NOON - [astr] (the moment of sun's upper culmination) midi *m*

NOOSE - [gen] (a loop with a running knot) noeud *m* coulant

NOR - [chem] (prefix denoting "normal", a term used in chemical technology to mean a compound which contains an unbranched chain of carbon atoms) nor-

~-CIRCUIT - [comput] circuit *m* NOR

NORIA - [hydr] (a water raising wheel with buckets on its rim) noria *f*, pompe *f* à chapelet, roue *f* à sabots

NORITE - [geol] (coarse-grained igneous rock containing coloured minerals) norite *m*

NORLEUCINE - [chem] (a naturally occurring non-essential amino acid) norleucine *f*

NORM - [gen] (a rule taken as standard; also a value of a quantity or a condition which is statistically very frequent) norme *f*

NORMAL - [gen] normal
[chem] (which contains an unbranched chain of carbon atoms) normal
[math] (normal to a surface, or a line, is a line perpendicular to it) normal, perpendiculaire
[met] (the mean value over a period of years of a given meteorological element at a specified date or over a specified period) conditions *fpl* météorolo-

giques normales
[chem] (of solutions) normal, titré
NORMAL ACCELERATION - [phys] accélération *f* normale
~ATOM - [nucl] (an atom without overall electric charge, in which all the electrodes are at their lowest energy levels) atome *m* normal
~AXIS - [aero] (a line drawn through the centre of gravity of an aircraft and normal to the longitudinal axis in the plane of symmetry) axe *m* de lacet, axe *m* normal
~BAND - [phys] (the lowest band of the energy diagram for a substance, corresponding to the normal state in the absence of any energy provided by an external source) bande *f* normale
[comput] (band on a section of a magnetic drum, the access time of which has not been reduced by applying extra reading heads) bande *f* normale, zone *f* normale
~BEAM - [light] faisceau *m* de rayons normal
~BED - [geol] (of a river) lit *m* mineur
~BEND - [el] (bend with a longer radius than an elbow, or sharp bend, serving to connect two lengths of conduit which are at an angle of 90 degrees) coude *m* à angle de 90°
~BUD-SLIP INCISION - [agric] greffe *f* normale par oeil détaché
~CALOMEL ELECTRODE - [chem] (calomel electrode containing a normal potassium chloride solution) électrode *f* de chlorure normal
~CARD LISTING - [comput] travail *m* en liste
~CLEAR - [telecomm] (line clear) voie *f* libre
~COLOUR SIGHT - [opt] perceptivité *f* normale des couleurs
~CONTACT - [el] contact *m* normal
~CURRENT - [el] courant *m* de régime
~CURVE - [math] courbe *f* normale
~DENSITY - [phys] (the mass of unit volume of a gas under standard conditions) densité *f* normale, densité *f* standard
~DIRECTION FLOW - [autom] (on a flow chart) fluence *f* en direction normale
~DISTRIBUTION - [math] distribution *f* normale
~DISTRIBUTION CURVE - [math] coùrbe *f* de répartition normale
~-DROP-OUT HUB - [comput] jack *m* de sortie normal
~ELECTRODE POTENTIAL - [el] (of a metal, i.e. the electrode potential of a metal immersed in a solution containing ions of that metal at a unit activity) potentiel *m* d'électrode normale
~ELEMENT - [chem] (a standard cell) élément *m* standard
~ENERGY LEVEL - [electron] (the state of an atom when all its electrons are at the lowest energy level) état *m* normal, niveau *m* normal
~FAULT - [geol] (a fracture in rocks) faille *f* normale
~FLIGHT - [aero] (all manoeuvres required for ordinary flying) vol *m* normal
~FOLD - [geol] pli *m* normal
~FORCE - [phys] (that component which acts along the normal axis of the resultant force, including the resolved gravity component) force *f* normale
~GAUGE - [railw] écartement *m* normal
~GLOW DISCHARGE - [electron] (glow discharge characterized by the decrease of the working voltage as the current increases) décharge *f* luminescente normale
NORMAL HABITUS - [min] faciès *m* normal
HARDENING CEMENT - [constr] ciment *m* à prise normale
~INDUCTION - [el] (the limiting induction in a material in symmetrically cyclically magnetized condition) induction *f* normale
~LANDING - [aero] atterrissage *m* normal
~LAW CURVE - [gen] courbe *f* normale de probabilité
~LEVEL - [gen] (ground level) niveau *m* du sol
~LIQUID - [phys & chem] (non-associated, or non-polar liquid) liquide *m* non-polaire
~LOAD - [aero] charge *f* normale
[radio] (the apparatus which is normally connected to the output of an amplifier) circuit *m* de charge réel
~LOOP - [aero] (a manoeuvre comprising the sequence: normal flight - climb inverted flight - dive - return to normal flight) looping *m* normal
~MAGNETIZATION CURVE - [el] (the lines which joins the tips of different hysteresis loops obtained by varying the limits of the magnetizing fields) courbe *f* d'aimantation normale
~MAGNIFICATION - [opt] agrandissement *m* normal
~MIXTURE JET - [mech] (of a carburettor) jet *m* de mélange normal
~MODE OF VIBRATION - [acoust] mode *m* normal de oscillation
~PERMEABILITY - [el] (the ratio of the normal induction to the corresponding magnetizing force) perméabilité *f* absolue, perméabilité *f* normale
~PITCH - [mech] (normal circular pitch) pas *m* normal
~POSITION - [telecomm] (of a contact) position *f* de repos (d'un sélecteur)
~POSITION OF POINTS - [railw] position *f* normale des aiguilles
~PRESSURE - [phys & mech] (the standard pressure to which measurements of volume are usually referred) pression *f* normale
~-PRESSURE DRAG - [phys] (the drag due to the resolved components of pressures which are normal to the surface in question) traînée *f* de pression normale
~PROPELLER STATE - [aero] (the state of a rotor when its thrust acts in the direction opposite to that of the axial flow through and round the area of the rotor disk) fonctionnement *m* propulsif normal
~RATED OUTPUT - [el] puissance *f* de régime
~REACTION - [mech] (the reactive thrust of a constraining surface on a contacting object which is in its turn subject to a force having a component perpendicular to the surface) force *f* normale
~-REVERSE SWITCH - [telev] commutateur *m* inverseur d'image
~RUNNING FIT - [mech] montage *m* à glissement
~SALTS - [chem] (those formed by the replacement of all the replaceable hydrogen of an acid by a metal) sels *mpl* neutres
~SCALE OF TEMPERATURES - [phys] échelle *f* normale des températures
~SEGREGATION - [metall] (the segregation in which the content of impurities tends to increase from the surface to the centre of cast metals) ségrégation *f* normale
~SENSITIVITY - [el] (obsolete term for factor of merit of a galvanometer) coefficient *m* de qualité

NORMAL SLIDE-VALVE - [mech] (a lapless valve) tiroir *m* normal
~SOLUTION - [chem] (one which contains the gram-equivalent weight of solute in one litre of solution) solution *f* normale, solution *f* titrée
~SOLVENT - [chem] (solvent which does not undergo chemical association) dissolvant *m* normal
~SPECTRUM - [phys] (spectrum produced by a diffraction) spectre *m* de diffraction
~SPEED - [gen & auto] vitesse *f* normale
~SPIN - [aero] (a spin voluntarily continued, from which recovery can be effectued within two turns by the adjustment of control surfaces) vrille *f* normale
~STATE - [nucl] (the lowest energy state of a quantized system) état *m* fondamental, niveau *m* normal
[electron] s. normal energy level
~TERMINAL-STOPPING DEVICE - [el] interrupteur *m* de sécurité automatique
~THROW - [geol] rejet *m*
~TO IODINE - [ind chem] saccharifié normalement, saccharifié sans réaction avec solution iodique
~TRESHOLD OF AUDIBILITY - [acoust] seuil *m* normal d'audibilité
~TRESHOLD OF FEELING - [acoust] (the modal value of the treshold of feeling of the majority of observers of normal hearing) seuil *m* normal de douleur
~TRESHOLD OF HEARING - s. normal treshold of audibility
~VALENCE - [chem] (the valence exhibited by an element in the majority of its compounds) valence *f* normale
~VELOCITY - [phys] (the resolved velocity element in relation to the undisturbed air flow in the direction of the normal axis) vitesse *f* normale
~VOLUME - [phys] (of gases) volume *m* normal
~WEIGHT - [gen] (standard weight) poids *m* normal
~WORKING SPEED - [mech etc] vitesse *f* de régime
NORMALITY - [gen & chem] normalité *f*
NORMALISED STEEL - [metall] acier *m* normalisé
NORMALISATION - [gen] (the action of reducing to a normal state) normalisation *f*
NORMALIZE, to - [gen] normaliser, rendre normal
[metall] (the action of relieving materials from stresses by heating above the transformation range followed by cooling) normaliser
[math] normer
NORMALIZED ADMITTANCE - [electron] (the reciprocal of normalized impedance) facteur *m* d'admittance d'onde
~FORM - [comput] notation *f* normalisée
~IMPEDANCE - [electron] (the ratio of the impedance to the characteristic impedance of the waveguide at a cross-section) facteur *m* d'impédance d'onde
~PLATEAU SLOPE - [nucl] (figure of merit for a counter tube) pente *f* normalisée de plateau
NORMALIZING - [metall] (the process of relieving material of internal stresses set up in working it) normalisation *f*, recuit *m* de normalisation
~FURNACE - [metall] four *m* de normalisation
NORMALLY CLOSED AUXILIARY CONTACT - [el] contact *m* auxiliaire fermé-ouvert
~CLOSED INTERLOCK - [el] contact *m* de repos
~OPEN AUXILIARY CONTACT - [el] contact *m* auxiliaire ouvert-ouvert
~-OPEN CONTACT - [electron] contact *m* de travail
~OPEN INTERLOCK - s. normally open auxiliary contact
NORMATRON - [comput] (normalized computer, purely imaginary, used for the teaching of programming) normatron *m*
NORTH - [astr] (cardinal point of the compass) nord *m*
~BY EAST - [gen] (one point east of north on the mariner's compass) nord-quart-nord-est
~BY WEST - [nav] (one point west of north on the mariner's compass) nord-quart-nord-ouest
~-EAST - [geogr] nord-est *m*
~LIGHT TRUSS - [constr] ferme *f* pour comble
~POLE - [geogr] (the northern extremity of the earth's axis) pôle *m* nord
[phys] (the part of a magnet from which the lines of magnetic flux apparently diverge) môle *m* nord, pôle *m* austral
~STAR - [astr] s. polaris
~-WEST - [geogr] nord-ouest *m*
NORTHER - [met] (cold northerly wind, often rising suddenly, occurring in the colder season over Texas and the adjacent areas of the Caribbean Sea) vent *m* du nord, norther *m*
NORTHERN - [gen] nord, septentrional
NORTHING - [surv] (a north latitude) latitude *f* nord
[nav] (difference of latitude measured toward the north between a position and the last one determined) chemin *m* nord
[astr] (north declination) déclinaison *f* positive
NOSE - [zool] nez *m*
[gen] (frontal end, usually tapered) nez *m*, avant *m*, bec *m*, tête *f*
[aero] (the prow of an aircraft) proue *f*
[firearms] (the tapering front end of a torpedo) cône *m* de choc
[firearms] (of a bullet) pointe *f*
[mach tool] pilote *m*, taillant *m*
[plumb] (of a pipe) ajutage *m*
[metall] buse *f*, calotte *f*
[railw] pointe *f* de l'aiguille
~CAP - [aero] (a spinner which does not extend aft of the propeller blades) carénage *m* du moyeu
[aero] (of the envelope of an airship) chapeau *m* de proue
~CONE - [astronaut] pointe *f*, bouclier *m*
~COWL - [aero] cône *m* avant
~-DIVE, to - [aero] piquer de nez
~-DIVE - [aero] vol *m* piqué
~DOOR - [aero] (of an aircraft-carrying cargo) porte *f* antérieure
~-DOWN, to - [aero] (to depress the nose of an aircraft in flight) piquer de nez, piquer à mort
~DRAG - [aero] traînée *f* frontale
HEAVINESS - [aero] (tendency for the nose of aircraft to bear downwards during flight) tendance *f* à piquer
~-HEAVY - [aero] (the condition of trim in which the nose of an aircraft tends to sink unless correction is applied by means of the controls) tendant à piquer
~IN SHUTTLE - [text] nez *m* de la navette
~KEY - [carp] (a small wedge) contre-clavette *f*
~OF KNIFE BOX - [text] nez *m* de la boîte à couteaux
~OF THE JACK - [text] doigt *m* en forme de crochet
~OF THE LAST STEP - [constr] nez *m*, dessus *m* de la dernière marche
~-OVER, to - [aero] (to turn over nose downward) (of an aircraft landing accident) capoter

NOSE-PIECE - [opt] (the mechanical system carrying the objective in a microscope)porte-objectif
[mech] (a rotating mechanical system) revolver *m*
[hydr etc] ajutage *m*, bec *m* (de tuyau d'arrosage)
[th eng] tuyère *f* de soufflet, buse *f*
~PIPE - [metall] (nozzle) bec *m*, buse *f*, tuyère *f*
~-PLANE - [aero] (a fixed or movable surface located forward of the mainplane, to give longitudinal control and/or stability) plan *m* avant
~RADIATOR - [aero] (a radiator located in the nose of the fuselage or of a nacelle) radiateur *m* frontal
~RIB - [aero] (any rib in aerofoil occupying a position between the front spar and the leading edge) nervure *f* de bord d'attaque
~STIFFENERS - [aero] (of an airship envelope) raidisseurs *mpl* de proue
~UNDERCARRIAGE - [aero] (a main undercarriage assembly located below the nose of an aircraft) atterrisseur *m*
~-UP, to - [aero] (to elevate the nose of an aircraft in flight) cabrer
~-UP ATTITUDE - [aero] (the attitude of an aircraft in which the nose is sharply elevated) assiette *f* à cabrer
~-WHEEL - [aero] (a single landing wheel located under the nose of an aircraft) roue *f* d'atterrissement avant, train *m* avant
~-WHEEL LANDING GEAR - [aero] (type of landing gear in which a wheel is located under the nose of an aircraft) train *m* tricycle, train *m* d'atterrissage, à roue avant
~-WHEEL STEERING - [aero] (the control of the direction of an aircraft on the ground by swivelling the nose-wheel) orientation *f* train avant
NOSING - [constr] (the exposed edge of the tread of a step) nez *m*, profil *m*, astragale *m*
[carp] (a bid on the edge of a board) profil *m*, gâche *f*
[constr] (of a bridge pier end) arrière-bec *m* de la pile
[railw] (the lateral movement of a steam locomotive due to the alternating thrusts in the cylinders) lacet *m*
~MOTION - [railw] (pitching) mouvement *m* de tangage
[text] serre-pointes *m*, doigt *m* pour serrer les pointes
~MOTION FINGER - [text] doigt *m* du secteur, serre-pointe *m*
NOSTRIL - [zool] narine *f*, naseau *m*
NOT-CIRCUIT - [comput] (a logical circuit in digital computers) circuit *m* NON, circuit *m* inverseur
~-GO GAUGE - [mech] minimum *m*
~TO SCALE - [draw] non (rapporté) à l'échelle
NOTAL - [anat] dorsal
NOTAM - [aero] ("Notice to Airmen"; information on air-traffic rules and procedures etc. published regularly by the Ministry of Aviation) avis *m* aux navigateurs aériens
NOTARY PUBLIC - [leg] (officer of the law empowered to administer oaths etc) notaire *m*
NOTATION - [gen] (designation by figures) notation *f*, figuration *f* (par un symbole)
[comput] (method for stating quantities in which numbers are represented by a sum of coefficients times multiplies of the successive powers of a chosen base number) notation *f*
NOTATION - [mus] (the tones used in music) notation *f* musicale
~PATTERN - [text] papier *m* à dessin, papier *m* à patron, papier *m* de mise en carte
NOTCH, to - [gen] (to make a small incision or indentation) entailler, encocher, denteler
[carp] (to cut a groove in the side of a timber, so as to fit in the side of another timber) hocher, assembler à entailles
NOTCH - [gen] entaille *f*, encoche, cran *m*
[carp] (a groove cut in the side of a timber) hoche *f*, enfourchement *m*, ruinure *f*
[mech] entaille *f*, encoche *f*, barbe *f*
[el] (one of the position of a controller) position *f* du commutateur
[hydr] (the flow of water over a weir) échancrure *f*
[mech] (in a lock bolt) barbe *f* de pêne
~BAR TEST - [metall] essai *m* de choc sur éprouvette entaillée
~BRITTLENESS - [metall] (susceptibility to fracture) fragilité *f* d'entaille
~FATIGUE FACTOR - [metall] facteur *m* de fatigue d'entaille
~FILTER - [telev] filtre *m* de réjection à flancs raides
~PLATE - [constr] (vertical barrier across a water stream, having a notch cut in its upper edge and used for measurements) traverse *f* à entaillage à angle droit
~SENSITIVITY - [metall] (the measure to which the endurance of metals is reduced by surface discontinuities) sensibilité *f* à l'entaille
~TOUGHNESS - [metall] résilience *f* d'entaille
NOTCHED - [gen] entaillé, à coches, à crans, à entailles, à dents
~BAR - [metall] (a bar of cast aluminium) barre *f* entaillée
~-BAR TEST - [metall] (a test in which a notched metal sample is given blow by a falling weight; the energy absorbed in breaking the specimen is measured) essai *m* au choc sur l'entaille
~BOWL - [text] disque *m* encoché
~FRAME - [text] cadre *m* à rainures
~INGOT - [metall] lingot *m* entaillé
~PIN - [text] goupille *f* cannelée, goupille *f* à encoches
~SET PIN - [text] goupille *f* cannelée
~STECIMEN - [metall] preuvette *f* entaillée
~WEFT BOARD - [text] planchette *f* en bois à entailles
~WHEEL - [text] disque *m* à entailles
NOTCHING - [gen mech & carp] (the operation of making a notch) entaillage *m*, encochement *m*
[metall] (sheet metal working) entaillage *m*
[carp] (the operation of joining timbers by fitting into a notch) assemblage *m* à entailles, ruinure *f*
[el] (of a controller, to indicate that a predetermined number of separate steps in required to complete the operation) commutation *f* échelonnée
[timber] griffage *m*
[constr] (of stone) bretture *f*
~CONTROLLER - [el] commutateur *m* à échelons
~FILTER - [radio] (band-rejection filter producing a sharp notch in the transfer characteristic of a system) filtre *m* bouchon
~MACHINE - [mech] machine *f* à encocher
~OF A SLEEPER - [railw] entaillage *m* de la traverse
~RATIO - [el] finesse *f* de réglage
~RELAY - [el] relais *m* intégrateur d'impulsions

NOTE, to - [gen] noter, prendre note
(to observe) remarquer, constater
[gen] (to mention separately) annoter
NOTE - [gen] note *f*
[gen] (written note) note *f*, mémorandum *m*
[comm] (list of goods to be despatched or received) bordereau *m*
[acoust (musical tone) note *f*
~-BOOK - [gen] carnet *m*, agenda *m*
~PAPER - [paper man] papier *m* à lettres
NOTICE, to - [gen] observer, remarquer
NOTICE - [gen] avis *m*, notification *f*, avertissement *m*
[comm] avis *m*
[leg] intimation *f*, notification *f*
~BOARD - [gen] écriteau *m*
NOTIFIABLE ACCIDENT - [leg] (any accident which regulations require to be notified officially) accident *m* à déclarer
~DISEASE - [med] maladie *f* à déclaration obligatoire
NOTIFICATION - [gen] avis *m*, notification *f*
NOTIFY, to - [gen & leg] notifier, annoncer
NOTOCHORD - [anat] corde *f* dorsale, notocorde *f*
NOUGHT - [gen & math] rien *m*, néant *m*
NOVA - [astr] (new star) nova *f*
NOVACULITE - [geol] (fine-grained rock consisting of quartz or other forms of silica) novaculite *f*, coticule *f*
NOVOLAK - [plast ind] (soluble fusible resins of the phenolformaldehyde type) novolak *m*
NOWEL - [foundry work] (of a moulding box) dessous *m* de châssis, corps *m*
NOXA - [med] principe *m* nuisible, micro-organisme *m* pathogène
NOXIOUS - [gen] (harmful) nocif
NOXIOUSNESS OF A PRODUCT - [bot] nocivité *m* d'un produit
NOZZLE, to - [hydr] hydrauliquer
NOZZLE - [hydr] (outlet tube through which a discharge of fluid finally passes) jet *m*, ajutage *m*, bec *m*
[plast ind] (of injector; hollow metal nose forming a seal between the heating cylinder or the transfer chamber and the mould) buse *f*
[mech] (of jet engine) diffuseur *m*
[mech] (of oil engines) (orifice controlled by the injection valve) tuyère *f* d'injecteur
[hydr] (the rigid tube at the end of a hose) lance *f*, lance *f* d'eau
[text] (of extruders or spinning machines) tube *m*, buse *f*, tuyère *f*
[electron] (the opening in a waveguide through which the radiation takes place) buse *f*
[radio] (elementary aerial consisting of a waveguide) buse *f*
[metall] (at the end of a blowpipe) busillon *m*
~ADAPTER - [mech] (a device fitted to an extrusion machine or injection moulding machine to enable different nozzles to be used) tête *f* du pot
~BLADES - [mech] (of a turbine) aubes *fpl* directrices
~BLOCK - [mech] porte-buse *m*
~BRICK - [th eng] brûleur *m* (pour four à coke)
~CLEARANCE - [mech] distance *f* ou écartement, interstice *m*
~CONTROL VALVE - [mech] soupape *f* de contrôle de tuyère
NOZZLE CUP - [metall] capot *m* de tuyère, capot *m* de diffuseur
~EXIT - [th eng] embouchure *f* de diffuseur
~FOR ACID CONTAINER - [ind chem] bouchon-verseur *m* pour flacons d'acide
~GAP - [th eng] aire *f* de diffusion
~GUIDE BLADES - s. nozzle blades
~GUIDE VANES - [aero etc] (annularly disposed fixed vanes designed to accelerate the gas flow in a turbine and guide it to the moving blades) aubages *mpl* distributeurs
~HOLDER - [mech] porte-diffuseur *m*
~-MIXING BURNER - [th eng] brûleur *m* à flamme de diffusion, brûleur *m* sans mélange préalable, brûleur *m* à filets parallèles
~MOUTH - [th eng] embouchure *f* de diffuseur
~NEEDLE - [mech] pointeau *m* de buse
~REGISTER - [plast ind] douille *f* de carotte
~RING - [mech] (of a gas turbine etc) distributeur *m*
~SPINNING - [metall] capot *m* de tuyère
~STAND - [railw] colonne *f* d'échappement
~THROAT - [metall] col *m* de diffuseur, col *m* de tuyère
~TIP - [metall] bec *m* de tuyère
NOZZLING - [metall] frettage *m*
N.P.L. - [phys] (initials of national physical laboratory) Laboratoire *m* National de Physique
NSA - (initials of National Standard Association) (USA; corresponding to the British Standard Institution) Association *f* Nationale pour la Standardisation
N.T.P. - [chem] (initials of Normal Temperature and Pressure) Température *f* à Pression Normale
nt. wt. - [gen] (abbrev. of net weight) poids *m* net
N-TYPE CONDUCTIVITY - [electron] (the conductivity which is associated with holes in a semiconductor) conductivité *f* type n
~SEMICONDUCTOR - [electron] (extrinsic semiconductor in which the majority carriers are electrons) sémiconducteur *m* n
NUB YARN - [text] filé *m* boutonné, filé *m* irrégulier
NUBBY - [text] boutonneux
NUBECOLA - [med] opacité *f* de la cornée
NUBILITY - [gen] nubilité *f*
NUCHAL - [zool] (pertaining to the back of the neck) de la nuque
~BONE - [anat] os *m* de la nuque
~CARTILAGE - [anat] cartilage *m* de la nuque
~PLATE - [anat] plaque *f* de la nuque
NUCIVOROUS - [zool] (mut-eating animals) nucivore
NUCLEAR - [phys] (pertaining to a nucleus) nucléaire
~ASH - [nucl] (reactor fuel material no longer fissile-) combustible *m* épuisé
~ASSOCIATION - [genet] association *f* nucléaire
~ATOM - [nucl] (a nucleus which has lost its surrounding electrons) atome *m* dépouillé, atome *m* nucléaire
~ATTRACTION - [nucl] (the force which is active in the atomic nucleus and overcomes the neutral repulsion of the positive charge of the protons) attraction *f* intranucléaire
~BARRIER - [nucl] barrière *f* de potentiel
~BINDING ENERGY - [nucl] (the energy which would be necessary to separate an atom into hydrogen atoms and neutrons) énergie *f* de liaison nucléaire

NUCLEAR BOMBARDMENT - [nucl] (any process in which sub-atomic particles are directed against atomic nuclei at high velocity) bombardement *m* nucléaire
~BREEDER - [nucl] réacteur *m* surrégénérateur
~CAP - [biol] capsule *f* nucléaire
~CASCADE - [nucl] cascade *f* nucléaire
~CHAIN REACTION - [nucl] (caused by neutrons emitted during uranium fission by inducing fission in additional uranium nuclei) réaction *f* nucléaire en chaîne
~CHARGE - [nucl] (the sum of the proton charges in a nucleus) charge *f* nucléaire
~CHEMISTRY - [nucl] (the chemical aspect of the study of atomic nuclei) chimie *f* nucléaire
~COLLISION - [nucl] collision *f* nucléaire
~CONSTANTS - [nucl] (the constants which are related to neutral science)constantes *fpl* nucléaires
~CROSS-SECTION - [nucl] (the cross-section of the atomic nucleus) section *f* efficace nucléaire
~DELAY - [nucl] temps *m* de mise en marche de la barre de commande
~DENSITY - [nucl] (nucleon density in a nucleus) densité *f* nucléaire
~DEPTH CHARGE - [nucl] grenade *f* sous-marine nucléaire
~DIAMETER - [nucl] (nucleon diameter) diamètre *m* nucléaire
~DISINTEGRATION - [nucl] (transformation involving nuclei) désintégration *f* nucléaire
~DISINTEGRATION ENERGY - [nucl] (or Q value) énergie *f* désintégration nucléaire
~DISRUPTION - [genet] dislocation *f* nucléaire
~ELECTRON - [phys] électron *m* nucléaire
~EMULSION - [nucl] (photographic emulsion which is especially prepared for the observation of the tracks of ionizing particles) émulsion *f* nucléaire
~ENERGY - [mech] (energy released in a nuclear reaction) énergie *f* nucléaire
~ENERGY LEVELS - [nucl] niveaux *mpl* énergétiques du noyau
~ENGINEERING - [gen] technique *f* nucléaire, genie *f* nucléaire
~EQUATION - s. nuclear reaction equation
~EVAPORATION - [nucl] (displacement of neutrons due to thermal agitation) évaporation *f* du noyau
~EXPLOSION - [nucl] (the ultimate result of an explosive nuclear reaction) explosion *f* nucléaire
~FIELD - [nucl] (effect of the sum total of intranuclear forces) champ *m* nucléaire
~FISSION - [nucl] (the division of a heavy nucleus into approximately equal parts) fission *f* nucléaire
~FLUID - [nucl] (term to denote the hypothetical binding mass of the nucleus) fluide *f* liaison nucléaire
~FORCES - [nucl] (forces pertaining to nucleons, which are non-electromagnetic) forces *fpl* nucléaires
~FUEL - [nucl] (the fissile material supporting a chain reaction in fission chain reactors) combustible *m* nucléaire
~FUSION - [nucl] (nuclear reaction in which light nuclei combine and form a nucleus of a higher mass number) fusion *f* nucléaire
~GYROMAGNETIC RATIO - [nucl] (the ratio between the magnetic moment of the nucleus and the nuclear angular momentum quantum number) relation *f* gyromagnétique
NUCLEAR HEAT - [nucl] chaleur *f* de réaction nucléaire
~IMPORTANCE FUNCTION - [nucl] fonction *f* importance
~INDUCTION - [nucl] (the magnetic induction borne of the magnetic moments of nuclei) induction *f* nucléaire
~INFLAMMABILITY - [nucl] (during a spontaneous fission) inflammabilité *f* nucléaire
~INTERACTION - [nucl] (the system of exchange forces in the nucleus which keep it together) action *f* mutuelle nucléaire
~ISOBAR - [nucl] (nuclide having the same number of nucleons in its nuclei) nucléides *mpl* isobares
~ISOMER - [nucl] (atomic nucleus having the same mass and charge but different radioactive properties) nucléides *mpl* isomères
~ISOMERISM - [nucl] (the occurrence of nuclear isomers) isomérie *f* nucléaire
~LIMITATIONS-[nucl] limitations *fpl* dues à la physique nucléaire
~MAGNETIC ALIGNMENT - [nucl] alignement *m* magnétique nucléaire
~MAGNETIC MOMENT - [nucl] (electrically charged particle which possesses angular momentum acts like a small magnet) moment *m* magnétique nucléaire
~MAGNETIC RESONANCE - [nucl] résonance *f* magnétique ,nucléaire
~MAGNETON - [nucl] magnéton *m* nucléaire
~MASS - [nucl] (the sum of the masses of the protons and neutrons composing the nucleus) masse *f* nucléaire
~MEMBRANE - [cytol] (the delicate binding membrane of the nucleus) membrane *f* nucléaire
~MODEL - [nucl] modèle *m* nucléaire
~MOMENTS - [nucl] (the various moments inherent to the nucleus) moments *mpl* nucléaires
~NEUTRON - [nucl] (forming the nucleon) neutron *m* nucléaire
~PACKING - [nucl] concentration *f* des particules
~PARAMAGNETIC RESONANCE - [phys] résonance *f* paramagnétique nucléaire
~PARAMAGNETISM - [nucl] (the paramagnetism which is associated with nuclear magnetic moments) paramagnétisme *m* nucléaire
~PARTICLE - [nucl] (particle assumed to exist in the nucleus of certain atoms) particule *f* nucléaire
~PERIODICITY - [nucl] périodicité *f* nucléaire
~PHOTODISINTEGRATION - [nucl] (nuclear reaction induced by a photon) réaction *f* photonucléaire
~PHOTOELECTRIC EFFECT - s. nuclear photodisintegration
~PHYSICS - [nucl] (that branch of physics dealing with nuclear structure) physique *f* nucléaire
~PILE - s. nuclear reactor
~PLATE - [nucl] (supporting the nuclear emulsion) plaque *f* nucléaire
~POISON - [nucl] (substance reducing reactivity) poison *m* nucléaire
~POLARIZATION - [nucl] polarisation *f* nucléaire
~POTENTIAL - [nucl] (the potential energy of a nuclear particle as a function of its position in the field of a nucleus) potentiel *m* nucléaire
~POTENTIAL ENERGY - [nucl] énergie *f* potentielle nucléaire

NUCLEAR POWER - [nucl] (the power which is released in nuclear reactions) énergie *f* nucléaire
~POWER PLANT - [nucl] centrale *f* à énergie nucléaire
~POWERED DESALINATION PLANT - [hydr] installation *f* de dessalage alimentée par énergie nucléaire
~PROCESS - [nucl] (any process involving the nucleus of an atom) processus *m* nucléaire
~PROJECTILE - [nucl] (any projectile having awarhead containing a nuclear reactor) projectile *m* nucléaire
~PROPERTIES - [nucl] propriétés *fpl* nucléaires
~PROPULSION - [astronaut] propulsion *f* nucléaire
~PROTON - [nucl] (the particle which, together with the neutron, form the nucleon) proton *m* nucléaire
~RADIATIONS - [nucl] (emitted by naturally radioactive substances) rayonnements *mpl* nucléaires
~RADIUS - [nucl] (the radius of a spherical volume within which the density of nucleons in a nucleus is effectively large) rayon *m* nucléaire
~REACTION - [nucl] (induced nuclear disintegration) réaction *f* nucléaire
~REACTION ENERGY - [nucl] énergie *f* de réaction, valeur *f* Q
~REACTION EQUATION - [nucl] (the equation which shows the changes in composition of an atomic nucleus in the course of a nuclear reaction) formule *f* de la réaction nucléaire
~REACTOR - [nucl] (complex plant in which controlled chain reactions can be sustained for specific purposes) réacteur *m*, réacteur *m* nucléaire
~REARRANGEMENT - [nucl] (the transition of a nucleus into one having a different configuration after a nuclear reaction) transformation *f* nucléaire
~REPULSION - [nucl] (the mutual repulsion shown by particles in the nucleus) répulsion *f* intranucléaire
~RESEARCH - [nucl] recherches *fpl* nucléaires
~RESONANCE LEVEL - [nucl] (the excited level of the compound system formed in a collision between two systems) niveau *m* de résonance nucléaire
~RETICULUM - [biochem] (mesh of threads of chromatin seen in stained preparations of metabolic nuclei) réticule *m* nucléaire
~SAP - [genet] suc *m* nucléaire
~SELECTION RULES - [nucl] (used to classify specified transitions) lois *fpl* de sélection nucléaire
~SITING POLICY - [gen] étude *f* d'élection de domicile d'un réacteur
~SPECIES - [nucl] (atom characterized by charge, mass number and quantum state of its nucleus) espèce *f* nucléaire
[nucl] (nucleus of specified charge, mass number and quantum state) espèce *f* nucléaire
~SPIN - [nucl] (the rotation of the nucleus of the atom) spin *m* nucléaire
~STABILITY - [nucl] (occurring when the internuclear forces are in equilibrium) stabilité *f* nucléaire
~STAR - [nucl] (a type of nuclear reaction) étoile *f* nucléaire, étoile *f* sigma
~STRUCTURE - [nucl] (the internal structure of the nucleus) structure *f* nucléaire
~SUPERHEATING - [nucl] (in a reactor) surchauffe *f* nucléaire
~SURFACE TENSION - [nucl] énergie *f* de surface, tension *f* superficielle nucléaire
NUCLEAR TARGET - [nucl] (layer of material for bombardment by fast particles) cible *f*
~TEMPERATURE - [nucl] température *f* de la réactivité
~TEMPERATURE COEFFICIENT - [nucl] (in a nuclear reactor) coefficient de température de la réactivité
~TEST - [nucl] (test of nuclear weapons) essai *m* nucléaire
~TRANSFORMATION - [nucl] transformation *f* nucléaire
~TRANSITION - [nucl] (taking place in the interior of the nucleus) transition *f* nucléaire
~WEAPON - [nucl] (any weapon provided with a nuclear warhead) arme *f* nucléaire
NUCLEATION - [phys] (the formation of germ cells in crystallization) formation *f* des germes de cristaux
NUCLEI OF CRYSTALLIZATION - [cryst] (small particles placed in solutions, on which crystals may form) noyaux *mpl* de cristallisation
NUCLEIC ACID STARVATION - [genet] inhibition *f* de la synthèse d'acides nucléiques
~ACIDS - [chem] (the non-protein constituents of nucleoproteins) acides *mpl* nucléiques
NUCLEOALBUMIN - [med] nucléoalbumine *f*
NUCLEOGENESIC - [phys] (large-scale formation of nuclei in nature) nucléogenèse *f*
NUCLEOLAR CONSTRICTION - [genet] constriction *f* nucléaire
~FRAGMENTATION - [genet] fragmentation *f* nucléolaire
~TRACK - [genet] tractus *m* nucléolaire
NUCLEOLUS - [genet] nucléole *m*
~SICKLE-STAGE - [genet] paranucléole *m*
NUCLEON - [nucl] (constituent particle of the atomic nucleus; i.e. a proton or a neutron) nucléon *m*
~NUMBER - [nucl] (mass number) nombre *m* de masse
NUCLEONICS - [nucl] (nuclear technology) technique *f* nucléaire
NUCLEOPLASMATIC RATIO - [genet] rapport *m* nucléo-plasmatique
NUCLEOPROTEINS - [chem] (compounds in which a protein molecule is combined with nucleic acid. They probably form almost the whole of the substance of a virus) nucléoprotéines *fpl*
NUCLEOR - [phys] (the core of a nucleus) coeur *m* du noyau
NUCLEUS - [phys] (the positively-charged central core of the atom, representing almost all the mass of the atom, but only a very small part of its volume) noyau *m*, germe *m*
[metall] (small cluster of atoms of a more stable phase formed within a less stable phase) noyau *m*
[cytol] (of a cell) nucléus *m* de cellule
[astr] (the luminous part of a comet) noyau *m*
[met] (the particle on which condensation in free air occurs) noyau *m* (de condensation)
~CRYSTAL - [cryst] (initial crystallization element) germe *m* cristallin
~SCREENING - [phys] blindage *m* du noyau
~SPLITTING - [cytol] division *f* du noyau
~WALL - [cytol] paroi *f* du noyau
NUCLIDE - [nucl] (atom characterized by the constitution of its nucleus) nucléide *m*
NUDATION - [bot] (the formation of an area bare of plants) érosion *f*, ablation *f*

NUDE GAUGE - [mech] jauge *f* nue
~OF A FRAME - [constr] nu *m* d'une huisserie
NUGGET - [mech] (the obturator element of a small valve or stopcock) obturateur *m* (de robinet)
[geol] (e.g. of gold) pépite *f*
[gen] (e.g. of metal) nugget *m*, pépite *f*
NUISANCE - [gen] dommage *m*, atteinte *f* aux droits
~AREA - [radar] (interference area) zone *f* d'interférence
NULL - [leg] (of no legal force) nul
[gen] de nul effet
[acoust] (a cone of silence) zone *f* de silence
[radio] (point where there is no reception on a receiver) zone *f* morte, zone *f* de silence
~ASTATIC MAGNETOMETER - [instr] magnétomètre *m* équilibré
~BALANCE DEVICE - [contr] (mechanism which restores a signal converter to its position of zero output) dispositif *m* à tarage sur zéro
~COIL MAGNETOMETER - [instr] magnétomètre *m* à bobine équibratrice
~DETECTOR - s. null indicator
~DRIFT - [instr] dérive *f* du zéro
~ELECTRODE - [el] (electrode with a thermodynamic potential of zero) électrode *f* zéro
~GEODESIC - [opt] (a possible space-time path of a light ray) ligne *f* géodésique
~INDICATOR - [instr] (instrument showing whether a signal is present) indicateur *m* à tarage sur zéro
~METHOD - [instr] (a method of measurement such that the perfect adjustment of a measuring circuit is indicated by the absence of any reading in the instrument used) méthode *f* du zéro
~OFF-SET - [contr] (control method in which the off-set characteristic is zero at all points of the input/output curve) réglage *m* astatique
~VALENCE - [phys] (a condition in which an element has no valence, as it has a complete outer electronic shell, as inert gases of the atmosphere) valence *f* zéro
~VOLTAGE - [el] tension *f* nulle
NULLIFY, to - [gen] annuler, infirmer
NULLIPARA - [med] (said of a woman who has never given birth to a child) nullipare
NULLODE - [electron] (electrodeless tube) tube *m* à décharge lumineuse sans électrodes internes
NUMBER, to - [gen] numéroter, compter
[to amount to] être au nombre, s'élever
[text] numéroter
NUMBER - [gen & math] nombre *m*, numéro *m*
[text] écheveau *m*
~CHECKING ARRANGEMENT - [telecomm] dispositif *m* de blocage d'une communication (pour déterminer le numéro du demandeur)
~DETECTOR - [comput] détecteur *m* de comptes
~NAIL - [telecomm] marque *f* distinctive
~NOTATION - [math] notation *f* des nombres
~OF ACTIVE LINES - [telev] nombre *m* de lignes utiles
~OF FIELDS - [telev] ordre *m* d'entrelacement
~OF GEARS - [auto] nombre *m* de vitesses
~OF LOOPS - [electron] nombre *m* d'antinodes
~OF NEUTRONS PER FISSION - [nucl] nombre *m* de neutrons par fission
~OF PARTICLES - [phys] (the total number of particles generated in a process) nombre *m* de particules
NUMBER OF PICKS AND DENTS IN THE PATTERN - [text] grandeur *f* du rapport
~OF POLES - [el] nombre *m* de pôles
~OF REVOLUTIONS - [mech] nombre *m* de tours
~OF SCANNING LINES - [telev] rapport *m* fréquence lignes, fréquence trame
~OF SCREW PITCHES - [mech] nombre *m* des pas d'une vis
~OF SHUTTLES - [text] nombre *m* de navettes
~OF SPINDLE SPEEDS - [mach tool] nombre *m* des vitesses du mandrin
~OF SPINDLES - [text] nombre *m* de broches
~OF TEETH - [mech] (in a gear wheel) nombre *m* des dents
~OF THREADS PER INCH - [text] nombre *m* de fils (dans la chaîne) par unité de longueur
~OF TRANSFER UNITS - [chem] (the total number of units in a diffusion process) nombre *m* des unités de transfert
~OF TURNS - [mech] nombre *m* de spires
~OF YARN - [text] titre *m* du fil, numéro *m* du fil
~ONE BAR - [metall] fer *m* ébauché, barre *f* de fer brut
~PERIOD - [comput] (the time required by a pulse train to pass a given point) période *f* de nombres
~PLATE - [auto] plaque *f* d'immatriculation
~SYSTEM - [comput] (numerical notation) notation *f* de nombres
~TRANSFER BUS - [comput] ligne *f* principale entre la mémoire et le calculateur
~-UNOBTAINABLE TONE - [telecomm] (in telephony) signal *m* de dérangement
NUMBERED CARDS - [comput] (those punched cards which are serially numbered) cartes *f*pl numérotées
NUMBERING - [gen] numérotage *m*, comptage *m*, dénombrement *m*
[text] numérotage *m*, détermination *f* du numéro
[text] (for silk yarns) titrage *m*
~MACHINE - [print] numéroteur *m*
[comm] (in a ledger) folioteuse *f*
[comput] numéroteur *m* automatique
~MALLET - [impl] (used for numbering timbers etc) marteau *m* compteur
~TRANSMITTER - [comput] numéroteur *m* automatique
NUMERAL - [gen] (expressing a number) nombre *m*
[adj] numérique
~POSITIONING CONTROL - [contr] (for automatic machine tools) commande *f* numérique
NUMERATION - [gen] numération *f*
NUMERATOR - [math] (of a fraction) numérateur *m*
NUMERICAL - [math] numérique
~ANALYSIS - [math] analyse *f* numérique
~APERTURE - [opt] (of a microscope) ouverture *f* numérique
~CODING - [comput] codage *m* numérique
~COMPUTER - [comput] calculateur *m* numérique
~CONSTANT - [math] (constant represented by numbers only) constante *f* numérique
~CONTROL - [autom] (a system of controlling machines and working processes by numerical data through the use of tapes or punched cards) commande *f* numérique
~CONTROL MACHINE TOOL - [mach tool] machine-outil *f* à commande numérique
~DATA CODE - [comput] code *m* de données numériques

NUMERICAL DIGITS - [math] digits *mpl* numériques
~ENTRY - [comput] entrée *f* numérique
~GAIN - [electron] gain *m* numérique
~INDEX OF VECTOR GROUP - [el] (conventional number indicating the phase displacement between low and high voltage) indice *m* numérique de couplage
~INTEGRATION - [math] intégration *f* numérique
~POSITIONING CONTROL - [comput] commande *f* numérique de positionnement
~PUNCHING - [comput] perforation *f* numérique
~SECTION - [comput] partie *f* numérique
~SELECTION - [telecomm] (in telephony) sélection *f* numérique
~WORD - [comput] mot *m* numérique
NUMERICALLY CODED INSTRUCTION - [comput] instruction *f* codée numériquement
NUMISMATICS - [gen] (the science of coins and medals) numismatique *f*
N-UNIT - [phys] unité *f* d'indice de réfraction
NUN MOTH - [zool] nonne *f*, bombyx *m* nonne
NUN'S MURMUR - [med] (the sound which is heard when the stethoscope is applied to the jugular vein) bruit *m* de nonnes, bruit *m* de rouet
NUPAC MONITORING SET - [nucl] (warning system for undesirable radiation) ensemble *m* moniteur
NUPTIAL FLIGHT - [zool] (the flight of the virgin queen bee during which fertilization takes place) vol *m* nuptial
NURAGH - [arch] (prehistoric stone structure; frequent in Sardinia) nuragh *m*
NURSE - [med] infirmière *f*, garde-malade *f*
~BALLOON - [aero] (balloon connected with an airship or another balloon on the ground and serving as reservoir) ballon *m* nourrice
~-BEE - [zool] nourrice *f*
~CROP - [agric] plante *f* protectrice, plante-abri *f*
NURSERY - [agric] pépinière *f*
NURSERYMAN - [agric] pépiniériste *m*
NUSSELT NUMBER - [phys] (a number expressing the increase of heat transfer caused by fluid movement) nombre *m* de Nusselt
NUT - [bot] noix *f*
[mech] écrou *m*
[print] demi-cadratin *m*
[mus] (of a violin) sillet *m*, hausse *f* (d'archet)
[horol] roue *f* à denture droite
~-AND-LEVER STEERING - [auto] direction *f* à vis et écrou
~ARBOR - s. nut mandrel
~BOX - [mech] manchon *m* du tour
~-COAL - [mining] petite gailleterie *f*
~GAUGE - [mech] calibre *m* pour écrous
~INSULATOR - [el] (in electric traction lines) isolateur *m* à noix
~LATHE - [mach tool] tour *m* pour écrous
~LOCK - [mech] frein *m* d'écrou
~-MAKING MACHINE - [mech] machine *f* à forger les écrous
~MANDREL - [mach tool] mandrin *m* pour écrous
~OIL - [chem] huile *f* de noix
~RING - [mech] anneau *m* fileté
~SCREW - [mech] mère-vis *f*
~TAP - [constr] taraud *m* d'écrous
~TAPPING MACHINE - [mach tool] machine *f* à tarauder les écrous
~UPSETTING MACHINE - [mech tool] presse *f* à refouler les écrous

NUT WASHER - [mech] rondelle *f* d'écrou
NUTATING FEED - [mech] alimentation *f* oscillante
NUTATION - [astr] (a cyclic variation in the precession of the earth, due to the attraction of the sun acting on the equational bulge) nutation *f*
[mech] (as in a spinning top, the inclination of the axis of the top to the vertical varies periodically between certain limiting angles) nutation *f*
[med] nutation *f*
NUT RACKER - [impl] casse-noix *m*
[zool] (a crow-like bird) casse-noix *m*
NUTGALL - [bot] noix *f* de galle
NUTMEG - [bot] (the aromatic kernel of the fruit of the genus Myristica) muscade *f*, noix *f* de Banda
~TREE - [agric] muscadier *m*
NUTRIENT - [gen & med] (providing nourishment) nutritif, nourissant
~BROTH - [chem] (used in bacteriology) bouillon *m* de culture
~SALT - [chem] sel *m* nutritif
NUTRITION - [biol] (nutriment; also the processes by which food is assimilated) nourriture *f*, alimentation *f*
NUTRITIONAL LEVEL - [agric] niveau *m* nutritionnel
NUTRITIVE VALUE - [agric] (animal feeding) valeur *f* nutritive
NUTSHELL - [bot] coquille *f* de noix
NUX VOMICA - [chem] (the dried seeds of Strychnos nux vomica and a source of strychnine and related alkaloids) noix *f* vomique
NYCTALOPIA - [med] (abnormal difficulty in seing objects in the dark; night blindness) héméralopie *f*, cécité *f* nocturne
NYCTOPHOBIA - [med] (a morbid fear of the night and, generally, of darkness) nyctophobie *f*, scotophobie *f*
NYLON - [ind chem] (generic term for long-chain polyamides containing recurrent -CONH-groups in the main chain. The nylons are an important group of plastics with application in the manufacture of fibres, bristles, monofilaments, moulding powders, resins, staple and yarns) nylon *m*
~4 - [ind chem] (a nylon based on pyrrolidone) nylon 4
~6 - [ind chem] (polycondensation product of caprolactam) nylon *m* 6
~7 - [ind chem] (a polymer of ethyl aminoheptanoate, used in the production of tyre cords) nylon *m* 7
~8 - [ind chem] (nylon manufactured from caprolactam) nylon *m* 8
~9 - [ind chem] (nylon based on 9-amino-nonanoic acid) nylon *m* 9
~II - [chem] (nylon produced from II-amino-undecanoic acid. Used for fibre-forming and moulding compounds) nylon *m* II
~66 - (nylon 6,6; nylon 6/6) - [ind chem] (nylon produced by condensation of hexamethylenediamine with adipic acid) nylon *m* 66
~6I0 (nylon 6,I0; nylon 6/I0) - [ind chem] (nylon produced by condensation of hexamethelenediamine and sebatic acid) nylon *m* 6I0
~BRISTLES - [text] (relatively thick nylon monofilaments) poils *mpl* de nylon
~FIBRE - [plast ind] (synthetic fibre produced from nylon, usually by melt-spinning process) fibre *f* de nylon
~MONOFILAMENTS - [text] (monofilaments availa-

ble in a number of diameters and used as bristles, sutures, fishing lines etc) monofilaments *m*pl de nylon

NYLON MOULDING POWDERS - [plast ind] (moulding powders giving products with hugh strength, good mechanical and electrical properties, and good dimensional stability at relatively high temperatures) poudres *f*pl à mouler de nylon

~PLASTIC - [plast ind] (thermoplastic material chemically similar to other nylons) plastique *m* en nylon

~STAPLE AND TOW - [plast ind] (varieties of crimped nylon fibres) nylon *m* en fibre et en brin

~YARN - [text & plast ind] (single and multifilament products for the textile industry) fil *m* de nylon

NYLONIZE, to - [text] nyloniser

NYMPHOMANIA - [med] (morbid, frequently ungovernable, sexual desire in women) nymphomanie *f*

NYQUIST DEMODULATOR - [electron] démodulateur *m* à talon

~INTERVAL BANDWIDTH - [electron] talon *m*

NYSTAGMUS - [med] nystagme *m*, nystagmus *m*

NYTRIL - [plast ind] (generic name for synthetic fibres containing not less than 85 p.c. of a vinylidene dinitrile polymer) nytril *m*

O

O - [chem] (symbol for Oxygen) symbole de l'oxygène
[met] (Beaufort letter indicating overcast) symbole de couvert
~GUIDE - [electron] (a type of surface wave transmission line) guide *m* d'ondes cylindrique diélectrique
OAK - [bot] (strong, tough hardwood) chêne *m*
~APPLE - [bot] cinelle *f*, pomme *f* de chêne, noix *f* de galle
~BARK - [bot] (bark from various species of Quercus) écorce *f* de chêne
~-BARK TANNED - [leather ind] tanné à l'écorce de chêne
~EXTRACT - [ind chem] extrait *m* d'écorce de chêne
~FERN - [bot] polypode *m* du chêne
~STAVE - [agric] (of a barrel) douve *f* en bois de chêne
~TANNING - [leather ind] tannage *m* à l'écorce de chêne
OAKUM - [text] (hemp fibre which is obtained by untwisting and picking old rope) filasse *f*, étoupe *f*
[naut] (as above, used for caulking joints) étoupe *f*
OAR, to - [naut & sport] ramer
OAR - [naut] rame *f*, aviron *m*
OAST - [brew ind] (a kiln used for drying hops) séchoir *m* à houblon, four *m* à houblon
~-HOUSE - s. oast
OAT - [agric] (edible grains of a cereal) avoine *f* commune
~BRAN - [agric] (animal feeding) son *m* d'avoine
~-CAKE - [gen] galette *f* d'avoine
~HULLER - [agric] écosseuse *f* pour avoine
~STRAW - s. oat bran
OATMEAL - [food] (a meal made from oats) farine *f* d'avoine
OBCECATION - [med] cécité *f* partielle
OBELISK - [arch] (stone shaft, generally square and with a pyramidal top) obélisque *m*
OBESITY - [med] obésité *f*
OBJECT - [gen] objet *m*
~CODE - [comput] code *m* résultant
~COMPUTER - [comput] machine *f* exécutrice
~DECK - [comput] programme *m* de travail sur cartes perforées
~DISTANCE - [photo] distance *f* ultranodale antérieure
~FINDER - [photo] viseur *m* de mise en image
~-GLASS - [opt] (the lens nearest to be object in an optical system) objectif *m*
~LANGUAGE PROGRAMME - [comput] routine *f* résultante
~METAMERISM - [opt] métamérisme *m* d'un objet
OBJECT MODULE - [comput] module *m* résultant
~PHASE - [comput] phase *f* d'exécution
~POINT - [opt] (the point of intersection of a small beam of rays incident on an optical system) point *m* d'objet
~SPACE - [photo] espace-sujet *m*
~STAGE - [photo] platine *f* porte-objet
OBJECTIVE - [opt] (the lens combination which forms the image to be examined with the eyepiece) objectif *m*
[gen (aim) objectif
~LENS - [opt] (the lens of a system which is towards the object) objectif *m*
~NOISE METER - [instr] (instrument designed to assess the equivalent loudness of noise by an objective method) psophomètre *m* objectif
~PHOTOMETER - [instr] photomètre *m* objectif
~PRISM - [opt] (of a telescope) prisme *m* d'objectif
~VARIABLE - [contr] (quantity controlled by its relation to the controlled variable and not directly measured for control) grandeur *f* de commande auxiliaire
OBLATE - [gen] (flattened at the poles) aplati
OBLIGATION - [gen] obligation *f*
[fin] (binding agreement) obligation *f*
OBLIQUE - [gen] oblique
[print] (a type) italique *m*
~AERIAL PHOTOGRAPH - [surv] (photograph taken from the air for survey work, with the optical axis of the camera inclined from the vertical) photographie *f* aérienne oblique
~ANGLE - [gen] (any angle which is not a right angle) angle *m* oblique
~ANODE - [electron] (X-ray tube anode having a surface forming an oblique angle with its support) anode *f* inclinée, anode *f* oblique
~ARCH - [constr] (an arch having an axis which is not normal to its face) voûte *f* biaise
~BEDDING - [geol] stratification *f* oblique
~CONE - [geom] cône *m* oblique
~COORDINATES - [math] coordonnées *fpl* obliques
~ELECTRODE STRUCTURE - [electron] (electrode structure of an electronic tube forming an oblique angle with the horizontal axis of the tube) montage *m* oblique
~FAULT - [geol] faille *f* diagonale, faille *f* oblique
~HEAD - [mech] (an extruder head fitted at an angle to the centre line of the barrel) tête *f* de boudineuse oblique
~INCIDENCE TRANSMISSION - [radio] transmission *f* sous incidence oblique
~JOINT - [carp] (angle joint between pieces which are not right angle to one another) assemblage *m*

oblique
OBLIQUE LAMINATION - [geol] stratification *f* oblique, stratification *f* entrecroisée
~MAGNETIZATION - [phys] magnétisation *f* oblique
~MOTION - [acoust] (motion of a second part against a stationary first part) mouvement *m* oblique
~PIANOFORTE - [mus] (upright piano with its strings arranged diagonally) piano *m* à cordes obliques
~PLANE - [math] plan *m* oblique
~SECTION - [mech] coupe *f* oblique
~STROKE - [print] barre *f* de fraction, barre *f* transversale
~SYSTEM - [cryst] système *m* binaire, système *m* monoclinique
~VIEW - [photo] vue *f* oblique
~WINDSCREEN - s. oblique windshield
~WINDSHIELD - [auto] pare-brise *m* incliné
OBLIQUITY - [gen] obliquité *f*, biais *m*
~FACTOR - [opt] facteur *m* cosinus
~OF FELVIS - [med] inclinaison *f* du bassin
~OF THE ECLIPTIC - [astr] (the angle at which the earth's orbital plane is inclined to the earth's equatorial plane) obliquité *f* de l'éliptique
~OF THE WHEELS - [auto] carrossage *m*
OBLITERATION - [gen med etc] oblitération *f*
OBLITERATIVE COLORATION - [zool] coloration *f* oblitérative
OBLONG - [gen] oblong
OBNUBILATION - [med] (clouded state of consciousness) obnubilation *f*
OBOE - [mus] (reed woodwind instrument) hautbois *m*
~SYSTEM - [radar] (system using two signal transmitting stations to give direction to an aircraft both as regards course and the dropping of the bombs) système *m* OBOE
OBSCURATION - [gen] obscurcissement *m*
[paint] (the area a quantity of paint will cover without becoming too thin) pouvoir *m* couvrant, capacité *f* de couverture
[astr] obscuration *f*, éclipse *f*
OBSCURE, to - [gen] obscurcir, assombrir
OBSCURE - [gen] obscur, sombre
OBSCURED GLASS - [glass man] verre *m* opaque
OBSCURITY - [gen] obscurité
OBSERVABLE - [gen & meas] (physically measurable) observable, visible
OBSERVATION - [gen] observation *f*
[astr] (finding the altitude of a celestial body to obtain the position) observation *f* (astronomique)
[surv] coup *m* de lunette, visée *f*, nivelée *f*
~BALLOON - [aero] (a balloon designed to carry observers) ballon-observatoire *m*
~CAR - [railw] (car fitted with glass sides or ends) voiture *f* panoramique
~CARRIAGE - s. observation car
~COACH - s. observation car
~DESK - [gen] table *f* de contrôle
~DEVICE - [telecomm] dispositif *m* d'écoute
~ERROR - [instr] (reading error) erreur *f* d'observation
~GLASS - [impl] lanterne *f*
~HOLE - [cin] (small window in the wall of the projection room) fenêtre *f* de visée
~PIPE - [hydr] tuyau *m* d'observation
~PORT - s. observation hole
POST - [astr] (an observatory) observatoire *m*
~SLIT - [opt] regard *m*
OBSERVATION STATION - s. observation post
~WELL - [hydr] puits *m* d'observation
OBSERVATORY - [astr etc] (a building constructed and equipped fot the observation of natural phenomena) observatoire *m*
OBSERVER - [gen] observateur *m*
[aero] (a member of the crew in a reconnaissance flight) observateur *m*
~BOMBER OVER ENEMY SYSTEM - s. oboe system
OBSESSION - [med] obsession *f*, idée *f* de contrainte
OBSESSIONAL NEUROSIS - [psycho] (a psycho-pathological condition) névrose *f* due à une obsession
OBSESSIVE-COMPULSIVE - [med] obsédant
OBSIDIAN - [geol] (volcanic glass. An igneous rock of high silica content and non-crystalline structure) obsidienne *f*, verre *m* volcanique
OBSIDIONAL DELIRIUM - [med] délire *m* obsidional
OBSOLESCENCE - [gen] (the going out of use) vieillissement *m*, tendance à tomber en désuétude
[comm] (material depreciation due to new developments) avilissement *m*
[biol] atrophie *f*, contabescence *f*
OBSOLESCENT - [gen] tombant en désuétude, vieillissant
OBSOLETE - [gen] (gone out of use) désué, obsolète, suranné, abrogé
[biol] obsolète
OBSTACLE - [gen] (a hindrance) obstacle *m*
[radar] (or target; the obstacle which reflects the emitted high-frequency wave) cible *f*, obstacle *m*
OBSTETRICIAN - [med] (medical specialist in childbirth) accoucheur *m*
OBSTIPATION - [med] constipation *f* chronique
OBSTRETICS - [med] (the science relating to childbirth) obstétrique *f*
OBSTRUCT, to - [gen] obstruer, encombrer entraver
[gen] (to block or stop, as of pipes) engorger, boucher
OBSTRUCTION - [gen] (an obstacle) empêchement *m*
[radar] (occasionally used instead of obstacle, or target) obstacle *m*
[hydr etc] engorgement *m*, colmatage *m*
[mech] (of a filter etc) colmater
~GAUGE LIMIT - [railw] gabarit *m* des obstacles
~LIGHT - [naut] (a luminous indication of an obstacle to navigation) feu *m* d'entrave
~MARKERS - [aero] (visual non-luminous indications of the presence of obstacles) balises *f*pl d'obstacles
~TO RUNNING - [railw] (traffic interruption) entrave *f* à la circulation
OBTAIN, to - [gen] obtenir, se procurer, recueillir
OBTUND, to - [med] émousser
OBTUNDENT - [med] émollient, calmant
OBTURATION - [gen] (prevention of an escape of gas) obturation *f*
OBTURATOR - [gen] obturateur *m*
~MEMBRANE - [anat] membrane *f* obturatrice
~PLATE - [gen] plaque *f* de fermeture
~-RING - [mech] (a piston ring designed to retain pressure and L-shaped in cross section) segment *m* d'étanchéité
[firearms] coupelle *f*
OBTUSE - [geom] (an angle exceeding 90 degrees) obtus

OBTUSE ANGLE - [geom] angle *m* obtus
~BUTT JOINT - [mech] joint *m* d'about obtus
OCARINA - [mus] (small wind instrument) ocarina *m*
OCCASIONAL - [gen] de circonstance, de situation, occasionnel
~SEAT - [auto] siège *m* de secours
OCCIDENTAL QUARTZ - [min] quartz *m* de Ceylon
~TURQUOISE - [min] (odontolite) turquoise *f* occidentale, turquoise *f* osseuse
OCCIPITAL - [anat] (a bone of the skull) occipital *m*
~ARTERY - [anat] artère *f* occipitale
~BONE - [anat] os *m* occipital
~CREST - [anat] crête *f* occipitale
~LOBE - [anat] lobe *m* occipital
OCCIPUT - [anat] occiput *m*
OCCLUDE, to - [chem] (of a metal, to retain a gas on its surface) absorber, condenser, occlure
[phys] (to encompass, to shut in) occlure
[gen] boucher, fermer, obstruer
OCCLUDED FRONT - [met] occlusion *f*
~GAS - [metall] gaz *m* inclus
[chem] (gas contained within the structure of, or condensed upon the surface of, a substance having occlusive properties) gaz *m* occlus
OCCLUSION - [gen] occlusion *f*, bouchage *m*, fermeture *f*, recouvrement *m*
[chem] (condition of uniform molecular adhesion between a precipitate and a soluble substance, or between a gas and a metal) occlusion *f*
[zool] (the closure of a duct) occlusion *f*
[met] occlusion *f*
[med] (in dentistry) emboîtement *m*, articulé *m*
~OF TEETH - [med] articulé *m* dentaire
OCCULTATION - [astr] (the concealment of one celestial body by another) occultation *f*, obscuration *f*
OCCULTING BEACON - [telecomm] balise *f* à occultations
~LIGHT - [signal] (a cyclically-flashing light in which the periods of light are equal to or longer than those of darkness) feu *m* à occultations
OCCUPANCY FACTOR - [phys] (in X-rays) surveillance *f* du temps d'exposition
OCCUPATION - [gen] (a person's regular business or work) occupation *f*, profession *f*, métier *m*, emploi *m*
[railw] (in signalling) occupation *f*
[leg] prise *f* de possession
~CODE - [comput] code *m* occupation
~NUMBER - [nucl] (the number of electrons in the individual shells of an atom) nombre *m* d'occupation
~ROAD - [gen] (road network) chemin *m* vicinal
OCCUPATIONAL DISEASE - [med] névrose *f* professionnelle
~EXPOSURE - [nucl] exposition *f* professionnelle
OCCUPIED AREA - [nucl] (space in which radiation hazard may exist) espace *m* occupé, zone *f* occupée
~SPACE - s. occupied area
OCCUPY, to - [gen] occuper
[surv] (to set an instrument over a station) faire la mise
OCCUR, to - [gen] avoir lieu, survenir, se trouver
OCCURRENCE - [gen] événement *m*
[geol] venue *f*, rencontre *f*
~OF GAS - [oil ind] venue *f* de gaz
OCEAN - [geogr] océan *m*
~BASIN - [geogr] bassin *m* océanique
~BOTTOM - [geol] fond *m* sous-marin
OCEAN CURRENT - [geogr] courant *m* océanique
~DEPTH - [geogr] (the greatest depths in the various oceans) fonds *mpl* sous-marins
~-GOING TUGBOAT - [naut] remorqueur *m* au long cours
~GOING VESSEL - [naut] navire *m* long courrier
~-RANGE VESSEL - [astronaut] navire *m* de mesure des performances de fusées
~TEMPERATURES - [geogr] (the mean temperatures of the surface of various oceans) températures *fpl* marines
OCEANIC - [gen] (of the ocean) océanique
OCEANITE - [geol] (basaltic igneous rock occurring in oceanic islands) océanite *f*
OCEANOGRAPHIC - [geogr] océanographique
~SHIP - [naut] (a vessel used for oceanographic study and observations) navire *m* océanographique
OCEANOGRAPHY - [geogr] (the part of physical geography dealing with oceanic life and phenomena) océanographie *f*
OCELLAR STRUCTURE - [geol] structure *f* kélyphitique
OCELOT - [zool] ocelot *m*
OCHER - s. ochre
OCHRE - [min] (earth pigment, yellow or brown in colour, consisting chiefly of silica and iron oxides) ocre *f*
OCHRODERMATOSIS - [med] ochrodermie *f*
OCHRODERMIA - s. ochrodermatosis
OCTADECANE - [chem] (a synthetis intermediate and solvent) octadécane *m*
OCTADECENE - [chem] (an intermediate for pharmaceuticals, plastics, perfumery agents, flavourants, and dyestuffs) octadécène *m*
OCTADECENYL ALDEHYDE - [chem] (an intermediate for insecticides and rubber chemicals) aldéhyde *m* octadécénylique
OCTAGON - [geom] (figure with eight sides and eight angles) octagone *m*
OCTAGONAL - [geom] octagonal
~COIL - [electron] bobine *f* octagonale
OCTAHEDRAL - [geom] (having eight sides) octaédrique
~BORAX - [min] borax *m* octaédrique
~IRON ORE - [min] magnétite *f*
~SULPHUR - [chem] (crystalline form of sulphur having an octahedral shape) soufre *m* octaédrique
OCTAHEDRITE - [min] anatase *m*, octaédrite *f*
OCTAHEDRON - [cryst] (a modification of the cubic system, having eight similar faces each of which is an equilateral triangle) octaèdre *m*
OCTAL BASE - [electron] (type of base in an electronic tube, in which there are eight contacts) base *f* octal
~DIGIT - [math] (one of the symbols from 0 to 7 when used as a digit in numbering in scale of 8) digit *m* octal
~NOTATION - [math] notation *f* octale
~TUBE - [electron] (electron tube with an octal base) tube *m* octal
OCTALUX BASE - [electron] (tube with a groove centre post locking firmly in a corresponding eight-pin loctal socket) culot *m* loctal
OCTAMETHYL PYROPHOSPHORAMIDE - [chem] (an insecticide) octaméthyl-pyrophosphoramide *f*
OCTANAL - [chem] (a flavourant and perfumery agent) octanal *m*

OCTANE - [chem] (a paraffin having eight carbon atoms in the molecule. A synthetis intermediate and solvent) octane *m*

~NUMBER - [chem] (the volumetric percentage of isooctane in a mixture of isooctane and normal heptane having the same "knocking" characteristics, when used in an i.c. engine, as the fuel to which the octane number refers; this gives a "knock-rating" for the fuel in question) indice *m* d'octane

~RATING - s. octane value

~VALUE - [chem] s. octane number

OCTANOL - [chem] (a synthesis intermediate and perfumery component) octanol *m*

OCTANOYL CHLORIDE - [chem] (a synthesis intermediate) chlorure *m* d'octanoyle

OCTANT - [geom & math] (the eight part of a circle) octant *m*

OCTASTYLE - [arch] (building with a colonnade of eight columns) octastyle *m*

OCTAVALENT - [chem] (capable of combining with eight atoms of hydrogen or the equivalent) octavalent

OCTAVE - [acoust] (tone of which the frequency is twice that of the given tone) octave *f*
[el] (in electric communications, the interval between two frequencies having a ratio of two to one) octave *m*

~ANALYZER - [meas] (a type of filter in which the upper cut-off frequency is twice the lower cut-off frequency) filtre *m* octave

~-BAND PRESSURE LEVEL - [acoust] niveau *m* de pression sonore d'une octave

~FLUTE - [mus] (the highest wood-wind instrument) petite flûte *f*

OCTAVO - [print] (book with I6 pages to a sheet) octavo *m*
[adj] in-octavo

OCTENE - [chem] (a synthesis intermediate) octéne *m*

OCTET - [mus] (composition for eight voices or instruments) octuor *m*

OCTODE - [electron] (electronic tube consisting of a cathode, six grids and one anode) octode *m*

OCTOIC ACID - [chem] acide *m* caprylique

OCTOSES - [chem] (monosaccharoses having eight oxygen atoms in the molecule) octoses *mpl*

OCTROI - [leg] (trade monopoly granted by a government) octroi *m*
[leg] (the tax which is levied at the grates of a city) octroi *m*

OCTUPLE-PHANTOM CIRCUIT - [telecomm] circuit *m* fantôme octuple

OCTYL ACETATE - [chem] (a flavourant and perfumery compound) acétate *m* d'octyle

~BROMIDE - [chem] (a synthesis intermediate) bromure *m* d'octyle

~DECYL ADIPATE - [chem] (plasticizer for cellulose acetate butyrate, cellulose nitrate, ethyl cellulose, polystyrene, polyvinyl acetate, vinyl chloride, and vinyl chloride acetate) adipate *m* d'octyldécyle

~IODIDE - [chem] (a synthesis intermediate) iodure *m* octyle

~MERCAPTAN - [chem] (a polymerization modifier) octylmercaptan *m*

~METHACRYLATE - [chem] (a polymerization monomer) méthacrylate *m* d'octyle

OCTYL PHENOL - [chem] (an intermediate for dyes, rubber chemicals, synthetic resins and disinfectants) octylphénol *m*

OCTYLAMINE - [chem] (an intermediate for pharmaceuticals, rubber chemicals, dyes, and insecticides) octylamine *f*

OCTYLBICYCLOHEPTENE DICARBOXIMIDE - [chem] (a component of insecticide mixtures) dicarboximide *f* d'octylbicycloheptène

OCTYLENE GLYCOL TITANATE - [chem] (a polymerization agent) titanate *m* d'octylène-glycol

~OXIDE - [chem] (a synthesis intermediate) oxyde *m* d'octylène

OCTYLPHENOXY POLYETHOXYETHANOL - [chem] (a wetting agent with uses in medicine) octylphénoxy-polyéthoxyéthanol *m*

OCULAR - [opt] (the eye-piece of an optical instrument) oculaire *m*

~ACCOMODATION - [opt] (the changed in the tension of the ciliar muscles) accommodation *f* de l'oeil

OCULENTUM - [med] pommade *f* ophtalmique

OCULIST - [med] oculiste *m*, ophtalmologiste *m*

OCULOGYRAL ILLUSION - [astronaut] illusion *f* oculogyre

OCULOGYRATION - [med] oculogyration *f*

OCULOGYRIC - [opt] oculogyre

OCULOMOTOR - [zool] (nerve causing movements of the eye) moteur *m* oculaire

~NERVE - [zool] nerf *m* moteur oculaire commun

OCULOMYCOSIS - [med] oculomycosis *f*

O/D - [mech etc] (initials of Outer Diameter) diamètre *m* externe

ODD - [math] impair

~-EVEN BIT - [comput] bit *m* de parité

~-EVEN CHECK - [comput] contrôle *m* de parité

~-EVEN NUCLEI - [nucl] (a nucleus which contains an odd number of protons and an even number of neutrons) noyaux *mpl* impair-pair

~-EVEN RULE OF NUCLEAR STABILITY - [nucl] loi *f* impair-pair de stabilité nucléaire

~FUNCTION - [math] fonction *f* impaire

~-LINE INTERLACE - [telev] analyse *f* entrelacée à nombre impair de lignes

~MOLECULES - [phys] (some rare molecules have an odd number of valence electrons) molécules *fpl* impair

~NUMBER - [math] nombre *m* impair

~-ODD NUCLEI - [nucl] (a nucleus which contains odd numbers of both neutrons and protons) noyaux *mpl* impair-impair

~PARITY CHECK - [comput] contrôle *m* impair

~PICKS - [text] duites *fpl* impaires

~TERM OF ATOM - [nucl] terme *m* atomique impair

~TWILLING BARS - [text] tringles *fpl* impaires

ODDMENT - [gen] article *m* en solde

ODDMENTS - [text] (of wool) articles *mpl* dépareillés
[gen] (of books) défets *mpl*

ODDSIDE - [metall] couche *f*, fausse couche *f*

ODEUM - [arch] (a hall; a roofed building similar to a theatre) odéon *m*

ODIOMETER - [instr] (type of olfactometer) olfactomètre *m*

ODOGRAPH - [electron] (automatic electronic map tracer) odographe *m*

ODOMETER - [instr] (mechanism recording the revolutions of a wheel and thus the distance covered) odomètre *m*

ODONTALGIA - [med] odontalgie *f*, mal *m* de dents
ODONTALYSIS - [med] examen *m* des dents
ODONTAPRISIS - [med] broyage *m* par les dents, bruxomanie *f*
ODONTIATRIA - [med] odontologie *f*
ODONTOCLASIS - [med] fracture *f* dentaire
ODONTOCOELE - [med] kyste *m* alvéolo-dentaire
ODONTODYNIA - s. odontalgia
ODONTOGENESIS - [med] odontogénie *f*
ODONTOGRAPH - [instr] (instrument designed to lay out correctly gear teeth) odontographe *m*
[med] (a device used by dentists to assess irregularities in the surface of the teeth) odontographe *m*
ODONTOLITHIASIS - [med] tartre *m* dentaire
ODONTOLOXIA - [med] irregularité *f* des dents, obliquité *f* des dents
ODONTOMA - [med] kyste *m* corono-dentaire, odontome *m*
ODONTONEURALGIA - [med] névralgie *f* dentaire
ODONTOPATHY - [med] maladie *f* dentaire, odontopathie *f*
ODONTORRHAGIA - [med] odontorragie *f*
ODONTOSCOPE - [med] miroir *m* dentaire
ODONTOSEISIS - [med] dents *fpl* branlantes
ODONTOTHECA - [med] follicule *m* dentaire
ODONTOTRYPSY - [med] trépanation *f* d'une dent
ODOR - s. odour
ODORIFEROUS - [chem] (having odour) odoriférant
ODORIMETRY - [ind chem] (measurement of the density and enduring qualities of odours) odorimétrie *f*
ODORIZATION - [ind chem] (of gases) odorisation *f*
ODORIZE, to - [ind chem] (the operation of adding a smell to natural gases, mostly for safety reasons) odoriser
ODORIZER - [ind chem] odoriseur *m*, substance *f* odorante
ODOROMETER - [instr] (instrument designed to assess the intensity of odours in the air) odorimètre *m*
ODOUR - [gen] odeur *f*
~CONTROL - s. odour removal
~REMOVAL - [ind chem] désodorisation *f*, élimination *f* de l'odeur
ODOURLESS - [gen] (lacking in odour) inodore
~SIZE - [ind chem] apprêt *m* inodore
OEDEMA - [med] (pathological accumulation of fluid in the tissue spaces of the body) oedème *m*
O-ELECTRON - [nucl] (an electron which forms part of the fifth shell surrounding the atomic nucleus) électron *m* 0
OEDOMETER TEST - [soil] essai *m* oedométrique
OEDOMETRIC CURVE - [soil] (consolidation curve) courbe *f* de consolidation
OENOCYANINE - [chem] (a colouring agent for wines) oenocyanine *f*
OENOLOGY - [agric] (or enology; the science relating to wines) oenologie *f*
OENOMETER - [ind chem] (an instrument for measuring the percentage of alcohol in wine) oenomètre *m*
OERSTED - [el] (the electromagnetic unit of magnetizing or magnetic force) oersted *m*
OESE - [anat] anse *f* de platine
OESOPHAGOSCOPE - [med] oesophagoscope *m*
OESTRADIOL - [chem] (a female sex hormone with medicinal uses) oestradiol *m*
OESTRIOL - [chem] (a sex hormone used in medicine) oestriol *m*
OESTROGENS - [biol] (generic term for female sex hormones) oestrogènes *mpl*
OESTRONE - [chem] (a sex hormone used in medicine) oestrone *f*
OESTRUM - [physiol] (the period of sexual desire of and acceptance of the male by female mammals) oestre *m*, chaleur *f*, rut *m*
~CYCLE - [physiol] cycle *m* oestral
~PERIOD - [zool] période *f* du rut, période *f* des chaleurs
OFF - [gen] loin, de distance, à distance, écarté de
[mech] (of a mechanical device) désassemblé
[el] (of a switch) hors circuit
[hydr] (of a water tap) fermé
[comm] (following the name of the article in question; "....off"; term defining the number of items to be made) à faire, à produire
[oil ind] (of a well) presque épuisé
[location] à ....de distance, plus loin
[gen] (in work) (to be off) libre, en dehors du service
[print] (the end of a printing stage) imprimé
[naut] au large
~AND ON SIGNAL PROVING - [el] contrôle *m* de signal à l'ouverture et à la fermeture
~-AXIS PARABOLIC MIRROR - [opt] miroir *m* parabolique partiel
~-BOARD - [bookbind] verse *m*, plat *m* inférieur
~-CAMERA FLASH - [photo] éclair *m* télédéclenché
~CENTRE - [mech] désaxé, décalé, décentré
[constr] en porte-à-faux
[mech] (of an i.c. engine) décalé
~-CENTRE BORE - [mech] forage *m* excentrique
~-CENTRE DIPOLE - [radio] (of the aerial) dipôle *m* excentrique
~CENTRE LOAD - [mech] charge *f* excentrique
~-CENTRE PLAN DISPLAY - [radar] indicateur *m* panoramique déporté
~-CIRCUIT - [el] (disconnected from the circuit) hors circuit
~-CIRCUIT RATIO ADJUSTER - [el] commutateur *m* du rapport de transformation hors circuit
~-COLOUR - [photo] (said of a colour photograph when the colours do not correspond to natural colours) à couleurs pâles, à couleurs peu naturelles
~-CUT - [text] (a piece cut off from stock in manufacture) coupon *m*
~GAUGE - [meas] mesure *f* ratée, écart *m* effectif
~GRADES - [rubber ind] (rubber sheet which does not conform to commercial standards) sheets *mpl* non conformes aux qualités commerciales
~-HEAT - [metall] piquée *f* défectueuse
~-HIGHWAY TYRES - [rubber ind] pneus *mpl* agraires
~-IRON - [metall] fer *m* brut de qualité inférieure
~ITS FEET - [print] incliné
~-LINE DATA PROCESSING - [comput] traitement *m* séquentiel des informations
~-LINE PRINTER - [comput] imprimeur *m* pour traitement séquentiel des informations
~LOAD - [el] à circuit ouvert
~-LOAD ISOLATOR - [el] (a disconnecting switch) sectionneur *m*
~-LOAD REFUELLING - [nucl] recharge *f* hors opération
~-LOAD VOLTAGE - [el] (potential difference between cell or battery terminals when the circuit is open) force *f* électromotrice, tension *f* à circuit ouvert

OFF-LOADING STATION - [transp] (of an aerial ropeway) station *f* de déchargement
~-NORMAL CONTACTS - [telecomm] contacts *mpl* de arbre dans le mouvement d'ascension
~-NORMAL SPRINGS OF A DIAL - [telecomm] ressort *mpl* de shunt d'un cadran d'appel
~-PEAK - [gen] (relating to electricity, gas etc) de faible charge
~-PEAK LOAD - [el] charge *f* normale
~PERIOD - [electron] (the period of time during an operating cycle, in which the electronic tube is non-conducting) temps *m* de non-conduction
[el & gaz ind] période *f* de faible charge
~PLUMB - [gen] hors d'aplomb
~POSITION - [el] position *f* de repas, position *f* de rupture de circuit
[mech] (of brakes) position *f* de desserrage
~SHADE - [text] décoloré
[text] déviation *f* de nuance
~-SHADE DYEING - [text] teinture *f* défectueuse
~-SHORE - [naut] au large, vers le large
~SIZE - [mech etc] hors dimensions
~STATE - [electron] état *m* bloqué
~-STATE CURRENT - [electron] courant *m* dans l'état bloqué
~-STATE VOLTAGE - [electron] tension *f* dans l'état bloqué
~-SWITCH - [telev] (in video recording) commande *f* de retraite de bande
OFFALS - [agric] entrailles *fpl*
OFFCUT - [ind proc] (a piece cut off and discarded when cutting material to size) découpure *f*
[bookbind] bande *f*, carton *m*
OFFENCE AGAINST THE SIDE CONDITION - [opt] violation *f* de la condition des sinus
OFFERING SIGNAL - [telecomm] signal *m* d'immixtion
OFFICE TEST OF A METER - [el etc] vérification *f* en usine d'un compteur
OFFICIAL - [gen leg etc] officiel
~DWELLING - [constr] habitation *f* de fonction
~MATERIALS TESTING BUREAU - [constr] office *m* de contrôle des matériaux
OFFING - [naut] (beyond anchorage but still on a visible part of the sea; offshore) (le) large *m*, pleine mer *f*
OFFLET - [road constr] (small channel) petit *m* canal *m* d'écoulement
OFFSET, to - [mech] (to displace intentionally from a given line of reference, e.g. the case of a reciprocating engine in which the C.L. of the cylinder does not intersect that of the crankshaft) déporter, désaxer, décaler
[gen] (to set off against another; to balance) contrebalancer
[gen] (to cancel by balancing) compenser
[constr] (to alternate) alterner les joints des matériaux
[surv] (to measure land by the offset method) relever par le système des diagonales orthogonales
[constr] (to make a ledge where part of a wall is set back from the face) former un ressaut, former une saillie, faire déborder
[geol] (of faulting) se déplacer horizontalement
[print] (to print by offset printing) offset
[agric] pousser des rejetons
[print] maculer

OFFSET - [gen] (anything corresponding to a counterbalance) compensation *f*
[constr] (a ledge built at a place where part of a wall is set back from the face) ressaut *m*, retrait *m*, portée *f*
[print] (the impression made by the offset printing) maculage *m*
[print] (a system of printing) offset *m*, impression *f* planographique
[mech] (bend in a pipe, so that one part is brought out, but kept parallel, with the line of another part) double coude *m*
[mech] (of non-intersecting axes) désaxage *m*, décalement *m*, déportage *m*, décentrement *m*
[geol] distance *f* horizontale des affleurements
[adj] désaxé, décalé, en porte-à-faux
[el] ligne *f* de dérivation
[mining] (a short drift) recoupe *f* viaille *f*, volée *f*
[surv] (the horizontal distance measured to a point from a main line, in a direction at right angle to the latter) perpendiculaire *f*, ordonnée *f*
[plumb] coude *m* de renvoi, baionnette
[phys] (a sustained deviation due to an inherent characteristic of positioning-controlled action) déviation *f*
[mining] forage *m* de limite
[constr] gradin *m* d'un empattement
[constr] (of a massive panel) plate-bande *f*
[constr] (in brickwork) ante *f*
[contr] (the steady-state of the deviation from the ordered value) écart *m* résiduel permanent, composante *f* constante d'erreur
[mech] (designedly displaced from some given line of reference) en porte-à-faux, inclinaison *f*
[agric] rejeton *m*, stolon *m*
~AGREEMENT - [oil ind] accord *m* limite
~ANGLE - [el acoust] angle *m* de frottement
~BAR - [mech] barre *f* déportée
~BEHAVIOUR - [contr] statisme *m*
~CARRIER SYSTEM - [telev] (a system in which the frequency of the image carrier waves of some transmitters has such a value that the mutual interference is very greatly reduced) système *m* à ondes porteuses décalées
~CARRIERS - [electron] porteuses *fpl* décalées
~-CENTRE P.P.I. - s. off-centre plan display
~CHARACTERISTIC - [contr] (the property of a control system, where by a horizontal portion of the input/output curve is avoided) caractéristique *f* de statisme, statisme
~COEFFICIENT - [contr] taux *m* de statisme
~-COURSE COMPUTER - [radar] (automatic computer used to translate reference navigational coordinates into coordinates which are required for a specified course) intégrateur *m* de route
~DEEP PRINTING - [print] procédé *m* offset grand creux, offset *m* en creux
~DEVIATION - [contr] écart *m* de statisme
~DIE - [plast ind] (type of die in which the final extrusion takes place in the same direction as the feed, but along a line offset from it) filière *f* déplacée
~DISC HARROW - [agric] herse *f* zigzag
~FLAT BED PRESS - [print] presse *f* offset plane
~GANG PUNCHING - [comput] perforation *f* série intercalée
~HARDWARE - [mech] (type of hinge having a hook

effect) charnière *f* en porte-à-faux
OFFSET HEADS - [el acoust] têtes *f*pl magnétiques décalées
~LITHOGRAPHY - s. offset printing
~NOZZLE - [mech] tuyère *f* oblique
~OF THE BED - [mining] ramification *f* d'une couche
~OIL - [mech] huile *f* pour presses offset
~PAPER - [paper man] papier *m* offset
~-PEAK PERIOD - [mech] (of motors and machines) période *f* de marche à puissance réduite
~PLANE - [constr] (a skew bull nose) rabot *m* à plate-bandes
~PRESS - [print] presse *f* offset
~PRINTING - [print] (method of printing from a lithographic plate to a rubber cylinder from which it is transferred to paper) impression *f* offset, impression *f* planographique
~PRINTING PRESS - s. offset press
~PRODUCTION - [oil ind] production *f* limite
~RATIO - [contr] taux *m* de statisme
~ROD - [auto] bielle *f* déportée
[surv] (wooden pole painted in bands of different colours) piquet *m* pour chaîne d'arpenteur, fiche *f*
~ROTARY PRESS - [print] presse *f* offset rotative
~SHAFT - [mech] (off its axis) arbre *m* désaxé
~SIGNAL - [telev] signal *m* décalé
~STACKER - [comput] récepteur *m* à décalage de cartes, empilage *m* décalé
~SUBCARRIER SYSTEM - [telev] (in colour television) système *m* à porteuse de chrominance décalée
~VALVE POSITION - [mech] position *f* de soupape excentrique
~VIEWFINDER - [photo] viseur *m* à monter sur griffe
~WELL - [oil ind] forage *m* de limite, puits *m* de limite
OFFSHORE - [naut] (moving away from the shore towards the open sea) vers le large
[naut] (occurring at a distance from the shore) au large
[gen] de terre
~DRILLING - [oil ind] sondage *m* submarin
~DRILLING PLATFORM - [min] (platforms which are siotable for drilling operations far from the shore; they are either fixed platforms, mobile platforms or floating drilling vessels) plate-forme *f* de forage en mer
~RING - s. offshore drilling platforms
[oil ind] sonde *f* submarine
~SEA BED MOVEMENT - [phys] (studied by using isotope traces) mouvement *m* au fond de la mer au large
~WELL - [min] puits *m* au large des côtes, puits *m* en mer
~-WIND - [met] (a wind blowing from the shore over the sea) vent *m* de terre
OFFSIDE - [auto] côté *m* droit
[adj] à droite
OFFSPRING - [genet] descendance *f*, fruit *m* produit
OFFSTREAM - [oil ind] hors de service
OFFTAKE - [hydr] (of a channel) prise *f* d'eau
[metall] (in a blast furnace) ouverture *f* du trou de coulée
[mining] galerie *f* d'écoulement, voie *f* d'écoulement
[metall] (air vent) traînée *f* d'air
[oil ind] débit *m*
~MAIN - [mech] (suction main) tuyauterie *f* d'aspiration
OGEE - [arch] (moulding showing a reverse curve in profile) doucine *f*, cimaise *f*
~ARCH - [arch] (arch with two S curves meeting at the apex) arc *m* en doucine, arc *m* en dos d'âne
~PLANE - [impl] bouvement *m*
[constr] doucine *f*, mouchette *f*
OGIVAL - [gen] ogival
OGIVE - [arch] (a pointed arch)ogive *f*
[firearms] (the tapering front part of a projectile) ogive *f*
[astronaut] (of a rocket) ogive *f*
O.H. - [metall] (initials of oil hardening) trempe *f* en huile
OH MARIE - [nucl] (slang term for an Organic Moderated Reactor, i.e. one which is moderated and cooled by hydrocarbon fluid) réacteur *m* à modérateur organique
OHM - [el] (the practical unit of resistance) ohm *m*
OHM LAW - [el] (the current in an electric circuit is directly proportional to the electromotive force in the circuit) loi *f* d'Ohm
OHMIC - [el] (adjective of ohm) ohmique
~CONTACT - [el] (contact between two material having the properties that the potential difference across the contact is proportional to the current passing through it) contact *m* ohmique
~DROP - [el] (voltage drop due to the current passing through the ohmic resistance of a circuit) chute *f* ohmique
~LOSS - [el] (the loss in an electric circuit due to the current passing through its resistance) pertes *f*pl ohmiques
~OVERVOLTAGE - [el chem] (the portion of the overvoltage which takes the form of an ohmic drop at the electrode-electrolyte interface) surtension *f* ohmique
~RESISTANCE - [el] (the resistance a circuit offers to the flow of a direct current; also called true resistance) résistance *f* ohmique
OHMMETER - [instr] (instrument designed to measure the electrical resistance of conductors and insulating materials)ohmmètre *m*
OIL, to - [mech] (to lubricate with oil) huiler, graisser, lubrifier
OIL - [gen] (of vegetable origin) huile *f*
[min] (of mineral origin; petroleum) pétrole *m*, huile *f* minérale
[chem] (volatile products, principally hydrocarbons with characteristic odours) huile *f*, essence *f*
[mech] (a lubricant) lubrifiant *m*, graisse *f*, huile *f*
~ABSORPTION - [paints] (the specific quantity of oil absorbed a pigment) absorption *f* d'huile
~ACCUMULATION - [oil ind] accumulation *f* de pétrole
~ADDITIVES - [chem] adjuvants *m*pl pour huiles lubrifiants
~-ADMISSION PERIOD - [mech] (in heat treatment) période *f* de carburation
~ARRESTING - [mech etc] étanche *f* à l'huile
~ASPHALT - [constr] (asphalt obtained from the distillation of petroleum- asphalte *m* artificiel
~BAFFLE - [mech] déflecteur *m* d'huile
~BAIZE - [text] frise *f* huilée
~BANK - [oil ind] front *f* d'huile
~BARREL - [impl] tonneau *m* à huile

OIL BASE MUD - [oil ind] boue *f* à l'huile
~BASIN - [geol] bassin *m* pétrolifère
[auto] bac *m* à huile
~BATH - [mech] (heated vessel containing oil and used to obtain a constant temperature) bain *m* d'huile
~BATH AIR CLEANER - [auto] filtre *m* à air à bain de huile
~-BEARING - [geol] pétrolifère
~BELT - [geol] zone *f* pétrolifère
~BINDING PROPERTY - [chem] pouvoir *m* de combinaison avec l'huile
~BLAST CIRCUIT BREAKER - [el] (type of oil circuit-breaker in which the pressure of the gases produced the arc causes a blast of oil across the contact space, thus ensuring a quick extinction of the arc) disjoncteur *m* dans l'huile
~BLOOM - [oil ind] fluorescence *f* de pétrole
~BLOOMER - [paint] (a paint defect) iridescence *f*
~-BONDED SAND - [metall] sable *m* agglutiné à l'huile
~BONDING - [metall] agglutinant *m* en huile
~BOX - [mech] boîte *f* à graisse, godet *m* à huile, boîte *f* d'essieu
~BRAKE - [mech] frein *m* à huile
~-BREAK - [el] (said of a switch, circuit breaker etc to denote that the circuit is opened in oil) interruption *f* dans l'huile
~BRUSH - [mech] balai *m* graisseur
~BUNKER - [oil ind] réservoir *m* à pétrole
~BURNER - [th eng] brûleur *m* à gas-oil, brûleur *m* à gazole
~CAKE - [agric] tourteau *m* oléagineux
~CAKE BREAKER - [agric] brise-tourteaux *m*
~CAN - [impl] burette *f* à graisser, burette *f*, burette *f* de graissage
[gen] bidon *m* à huile
~CATARACT - s. oil dashot
~CATCHER - [auto] récupérateur *m* d'huile
~CELLAR - [impl] godet *m* à huile
~-CHALK TEST - [metall] contrôle *m* par ressuage
~CHAMBER - [text] chambre *f* à huile
~CHANGE - [auto] vidange *m* d'huile
~CHANNEL - [mech] chenal *m* à huile
~CHANNELLING MACHINE - [mech] machine *f* à faire les rainures de graissage
~CHART - [auto etc] guide *m* de lubrification
~CIRCUIT BREAKER - [el] (device for breaking a circuit opened in oil) interrupteur *m* dans l'huile
~CLEANER - [mech] (filter or similar apparatus for the mechanical purification of oil) épurateur *m* de huile
~-COCK - [impl] robinet *m* graisseur
~COKE - [gas ind] (a liquid fuel) coke *m* de pétrole
~COLOUR - [dyes] couleur *f* à l'huile
~-COMBED - [text] peigné *f* l'huile
~CONCESSION - [leg] concession *f* pétrolifère
~CONDENSER - [el] condensateur *m* à huile
~CONNECTOR - [auto] raccord *m* d'huile
~CONSERVATOR - [el] (vessel connected to the tank of an oil-filled transformer to permit free expansion and contraction of the oil, so as to minimize the deleterious effect of contact between the oil in the main tank and the air) conservateur *m* d'huile
~CONSUMPTION - [mech] consommation *f* d'huile
~CONTROL RING - [auto] (of a piston) segment *m* râcleur d'huile

OIL CONTROL VALVE - [mech] (automatic regulating valve controlling flow to an oil cooler) soupape *f* à huile de réglage à commande directe
~-COOLED - [mech] (cooled by means of circulating oil) refroidi par huile
~-COOLED TRANSFORMER - [el] (transformer immersed in oil to ensure the cooling of the winding, core or other working parts) transformateur *m* dans l'huile
~-COOLED TUBE - [electron] (X-ray tube in which the heat produced on the target is dissipated by means of): tube *m* à refroidissement par l'huile
~COOLER - [aero] (device for extracting heat from oil, especially the lubricating oil used in an engine) radiateur *m* d'huile
[gen] réfrigérant *m* d'huile
~-COP SPINDLE - [text] broche *f* pour cannette à huile
~CUP - [mech] (or oilcup) godet *m* à huile, godet *m* graisseur
~DAMPER - [mech] (an energy-absorbing device or buffer, consisting of a piston working in a cylinder containing oil) amortisseur *m* à huile, clapet *m* d'étranglement à huile
~DASH-POT - [mech] frein *m* à huile, amortisseur *m* à huile
~DEFLECTOR - [auto] déflecteur *m*, chicane *f* d'huile
~DEPOSIT - [oil ind] gisement *m* de pétrole
~DERRICK - [oil ind] derrick *m*
~DILUTION - [aero] (dilution of lubricating oil in an engine by the undesired addition of fuel) dilution *f* de l'huile
~DILUTION SYSTEM - [aero] (a system for diluting the oil in an engine to facilitate starting) circuit *m* de dilution d'huile
~DIPPER - s. oil level gauge
~DISH - [auto] coupelle *f* pour recueillir les gouttes d'huile
~DISTRIBUTOR - [auto] distributeur *m* d'huile
~DRAIN - [oil ind] boue *f* à l'huile
~DRAIN PLUG - [mech] (a threaded plug which can be removed to allow oil to drain off) bouchon *m* de vidange d'huile
~DREG - [gen] dépôt *m* de pétrole
~DRILL - [mech] mèche *f* à lubrification
~DRIP - [auto] graisseur *m* à gouttes
~DROP - [biol](droplet of oil substance in the cytoplasm) goutte *f* d'huile dans le cytoplasme
~DRUM - [impl] fût *m* à huile
~DRUMMER - [comm] (colloq. USA; a salesman for oil products) marchand *m* de produits pétrolifères
~DUCT - [mech] tube *m* de graissage
~-ELECTRIC DRIVE - [el] propulsion *f* diesel-électrique
~ENGINE - [mech] (compression-ignition engine) moteur *m* à pétrole, moteur *m* à huile lourde
~-ENGINED - [mech] (said of heavy motors) à huile lourde
~-ENGINED VESSEL - [naut] vaisseau *m* à huile lourde
~-ENRICHED RUBBER - s. oil-extended rubber
~ESSENCE - [chem] essence *f*
~EXPANSION CHAMBER - s. oil conservator
~EXPANSION TANK - s. conservator
~-EXTENDED RUBBER - [rubber ind] (rubber which has been treated with reclaiming oils and catalysts) caoutchouc *m* étendu à l'huile, caoutchouc *m* enrichi à l'huile

OIL-FED DIESEL LOCOMOTIVE - [railw] locomotive *f* Diesel
~-FED MOTOR - [mech] moteur *m* à huile lourde
~FEEDER - [mech] burette *f* à huile, alimentateur *m* d'huile
~FEEDING RESERVOIR - [mech & el] reservoir *m* alimentateur d'huile
~FIELD - [min] champ *m* de pétrole, district *m* pétrolifère
~-FIELD BRINE - [geol] saumure *f* des champs de pétrole
~-FILLED CABLE - [el] (impregnated paper-insulated cable so designed that the impregnating medium is free to flow at all working temperatures) câble *m* à huile fluide
~-FILLED TYPE BUSHING - [el] traversée *f* dans l'huile
~FILLER - [mech] bouchon *m* d'alimentation d'huile
~-FILLER CAP - [auto] bouchon *m* de remplissage de huile
~FILM - [mech] (a thin layer of oil of molecular proportions) film *m* d'huile, couche *f* d'huile
~FILTER - [auto] filtre *m* à huile
~-FILTER TUBE - [auto] tube *m* de filtre à huile
~-FIRED - [gen] chauffé au pétrole, chauffé au mazout
~-FIRED BOILER - [metall] chaudière *f* à naphte
~-FIRED FURNACE - [metall] four *m* chauffé au mazout
~-FIRED WARM AIR CONDITIONER - [th eng] appareil *m* de conditionnement d'air au pétrole
~FIRING - [th eng] (the use of oil as a combustible in e.g. a boiler) au pétrole, au mazout
~FOGGING - [ind chem] huilage *m*, brumisage *m*
~FOUNTAIN - [oil ind] puits *m* jaillissant
~FUEL STOWAGE - [naut] dépôt *m* d'huile lourde
~GAS - [chem] (gas of high calorific value obtained from the destructive distillation of high-boiling mineral oils) gaz *m* d'huile
~GAS BURNER - [metall] brûleur *m* à huile lourde
~GAS TAR - [min] goudron *m* de pétrole
~-GAS LIGHTING - [gen] éclairage *m* par le gaz de pétrole
~GAUGE - [mech] jauge *f* de niveau d'huile
~GRINDER - [mech] machine *f* à meuler à huile
~-GROOVE - [mech] (a groove cut in the bearing surface of a plain bearing to allow more completer distribution of oil) rainure *f* de graissage, saignées *fpl* de graissage
~-GROOVES - [mech] pattes *fpl* d'araignée
~-GROOVING MACHINE - [mech] machine *f* à faire les rainures de graissage
~GUN - [mech] pistolet *m* de graissage
~-HARDEN, to - [metall] (the operation of hardening cutting tools by heating and quenching them in oil) tremper à huile, tremper en huile
~-HARDENED STEEL - [metall] acier *m* trempant à l'huile
~HARDENING - [metall] (the hardening of cutting tools by heating and quenching in oil) trempe *f* en huile, huilage *m*
~-HARDENING STEEL - [metall] acier *m* pour trempe en huile
~HEATING - [th eng] chauffage *m* au pétrole
~HOLE - [mech] (a hole drilled in a part for the introduction of oil) trou *m* de graissage, lumière *f*
~-HOLE DRILL - s. oil drill

OIL HORIZON - [geol] nappe *f* pétrolifère, horizon *m* pétrolifère
~HOUSE - [railw] (the lamp room) lampisterie *f*
~HUMIDIFYING - [ind chem] humidification *f* à huile
~-IMMERSED - [el] (said of any apparatus to denote that the principal conducting parts are immersed in an insulating oil) à bain d'huile
~IMMERSED APPARATUS - [el] appareil *m* à bain de huile
~-IMMERSED FORCED AIR-COOLED TRANSFORMER - [el] transformateur *m* à bain d'huile à refroidissement forcé par air
~-IMMERSED FORCED OIL-COOLED TRANSFORMER - [el] transformateur *m* à bain d'huile à refroidissement forcé par circulation d'huile
~-IMMERSED NATURAL COOLING TRANSFORMER - [el] transformateur *m* à bain d'huile à refroidissement naturel
~-IMMERSED TUBE - [electron] (X-ray tube for operation in oil) tube *m* à rayons X à bain d'huile
~-IMMERSED SWITCH - [el] (an electrical make-and-break mechanism immersed in oil to prevent arcing) interrupteur *m* à bain d'huile, interrupteur *m* dans l'huile
~-IMMERSION OBJECTIVE - [opt] (immersion objective q.v.) objectif *m* à immersion
~-IMMERSION TEST - [mech] (a test to assess the resistance of a specimen to lubricating or other oil) essai *m* d'immersion dans l'huile
~-IMPREGNATED LIMESTONE - [geol] pierre *f* calcaire impregnée de pétrole
~-IMPREGNATED PAPER - [el] (used for low-and-voltage cables) papier *m* huilé
~IN PLACE - [oil ind] huile *f* en place
~IN WATER EMULSION - [oil ind] émulsion *f* d'huile dans l'eau
~INDICATIONS - [geol] indications *fpl* de pétrole
~INDICATOR - [auto] indicateur *m* de pression d'huile lampe-témoin *f* de pression d'huile
~INDUSTRY - [min] industrie *f* pétrolière
~-INSULATED - s. oil-immersed
~INSULATOR - [el] isolateur *m* à huile
~-JET CIRCUIT BREAKER - [el] disjoncteur *m* à jet d'huile
~LAMP - [gen] lampe *f* à pétrole, lampe *f* à huile
~LAYER - [oil ind] nappe *f* pétrolifère
~LEAD - [mech] (a pipe conveying lubricating oil) tuyau *m* de lubrification
~-LENGTH - [paint] (the proportions of oil and resin in a composition) proportion *f* d'huile, rapport *m* huile/résine
~LENS- [oil ind] lentille *f* de sable pétrolifère
~LEVEL - [gen] niveau *m* de l'huile
~-LEVEL GAUGE - [auto] jauge *f* d'huile
~-LEVEL INDICATOR - [instr] indicateur *m* de pression d'huile
~LEVER - s. oil-level gauge
~LINE - [auto] canalisation *f* d'huile
~LIVER - [chem] (coagulations of oil) épaississement *m*
~LOG - [oil ind] journal *m* de forage
~MEAL - s. oil cake
~METER - s. oleometer
~MILL - [ind proc] huilerie *f*, tardoir *m*, fabrique *f* d'huile
~-MILL MACHINERY - [mech] appareillage *m* d'huilerie

OIL MINE - [oil ind] mine *f* de pétrole
~MINING - [oil ind] exploitation *f* du pétrole
~-MIST LUBRICATION - [mech] (lubrication system in which the parts to be lubricated are surrounded by a mist of oil) dispersion *f* d'huile par brouillard
~-MOISTENED AIR FILTER - [mech] (a type of air filter in which the element is impregnated with oil) filtre *m* à air impregné d'huile
~NIOBE - [chem] benzoate *m* de méthyle
~NOZZLE - [auto] (in diesel engines) gicleur *m* d'huile
~NUT - [bot] noix *f* oléagineuse
~OCCURRENCE - [geol] indication *f* de pétrole
~OF ALMONDS - [chem] huile *f* d'amandes
~OF APPLES - [chem] valérianate *m* d'amyle
~OF BANANAS - [chem] huile *f* de bananes
~OF BITTER ALMONDS - [chem] essence *f* d'amandes amères
~OF CLOVES - [chem] essence *f* de clou de girofle
~OF GARLIC - [chem] (allyl sulphide) huile *f* d'ail
~OF GRAPES - [chem] huile *f* de pépins
~OF MIRBANE - [chem] (nitrobenzene q.v.) nitrobenzène *m*
~OF MUSTARD - [chem] huile *f* de moutarde
~OF PEANUTS - [agric] (peanut oil) huile *f* d'arachide
~OF PEARS - [chem] (methyl acetate; a solvent for paints, cellulose derivatives and a number of plastics) acétate *m* de méthyle
~OF PINETAR - [chem] (pine-tar oil) huile *f* de goudron
~OF ROSES - [chem] essence *f* de roses
~OF TARTAR - [chem] huile *f* de tartre
~OF TURPENTINE- [chem] (colourless aromatic essential oil obtained by the steam distillation of rosin) essence *f* de térébenthine
~OF VITRIOL - [chem] (archaic name for sulphuric acid) huile *f* de vitriol, acide *m* sulfurique
~OF WINTERGREEN - [chem] (methyl salicylate) salicylate *m* méthylique
~-ON PERIOD - s. oil-admission period
~PAINT - [paint] (pigment ground in oil) peinture *f* à l'huile
~PALM - [agric] palmier *m* à l'huile, éléis *m* de Guinée
~PAN - [mech] poche *f* d'huile, carter *m* à huile, auge *f* de graissage
[auto] carter *m* inférieur
~-PAN BAFFLE PLATE - [mech] (in autos) diaphragme *m* du carter inférieur
~-PAN WITH PUMP AND PIPINGS - [auto] carter *m* inférieur avec pompe et tuyaux
~PASTE - [paint] (high concentrated mixture of pigment and oil, used for tinting or for making paint by adding oil, thinners and driers) concentré *m*
~PIPE - [auto] tuyau *m* d'huile
~PIPELINE - [min] canalisation *f* à pétrole, tuyau *m* pour pétrole
[mech] (for greasing) tuyau *m* de graissage
~PISTOL - [impl] pistolet *m* de graissage
~PIT - [oil ind] puits *m* pétrolifère
~POOL - [geol] gisement *m* pétrolifère
~POOL - [oil ind] accumulation *f* de pétrole
~PRESS - [agric] pressoir *m* à huile, tordoir *m*
~PRESS CLOTH - [text ind] drap *m* pour filtre à huile
~PRESSURE - [mech] pression *f* d'huile
~PRESSURE GAUGE - [instr] (a gauge indicating the pressure in a forced lubrication system) manomètre *m* de pression d'huile
~OIL PRESSURE RELIEF VALVE - [mech] (spring-loaded valve in the delivery side of a forced lubrication system) soupape *f* de sûreté de la pression d'huile
~PRESSURE WARNING DEVICE - [mech] (device to give warning of excessively high or low pressure of lubricating oil) dispositif *m* d'avertissement de pression d'huile
~PROCESS - [photo] procédé *m* à l'huile
~PROOF - [chem] étanche à l'huile
~PROOFING - [constr] (cement treatment) traitement *m* antiacide du ciment
~PROSPECTING - [min] recherches *fpl* pétrolifères
~PUMP - [mech] (a pump used to supply lubricating oil under pressure) pompe *f* à huile
~PUMP AND PIPE CONNECTIONS - [mach tool] pompe *f* de lubrification avec sa tuyauterie
~-QUENCHED FUSE - [el] (liquid-quenched fuse in which the liquid is oil) fusible *m* à extinction dans l'huile
~QUENCHING BATH - [metall] bain *m* de refroidissement rapide en huile
~RADIATOR - s. oil cooler
~RADISH - [bot] radis *m* oléifère
~-REACTIVE RESIN - [chem] (phenol-formaldehyde resins which react with drying oils on heating to give coatings with specific properties) résine *f* oléo-active
~REFINERY - [oil ind] raffinerie *f* de pétrole
~REFININF - [oil ind] raffinage *m* de pétrole, épuration *f* des pétroles
~REGENERATION - [ind chem] (in the debenzolization of oil) régénération *f* du pétrole
~REGION - [geol] district *m* pétrolifère
~RESISTANT - [mech] (resistant to penetration by oil) résistant à l'huile
~RETAINER - [auto] chicane *f*, déflecteur *m* d'huile
~RETURN BAFFLE - [mech] chicane *f*
~RIG - [oil ind] (the equipment required for drilling operations) appareil *m* de sondage
~RIGHTS - [leg] droit *m* d'exploiter le pétrole
~RING - [mech] (a scraper ring) anneau *m* graisseur, bague *f* de graissage
~RING SLOTS - [mech] rainures *fpl* de l'anneau graisseur
~ROCK - [geol] roche *f* pétrolifère
~-SAMPLING VALVE - [mech] échantillonneuse *f* d'huile
~SAND - [geol] grès *m* pétrolifère, sable *m* à huile
~-SATURATED SAND - [geol] sable *m* pétrolifère
~SAVER - [oil ind] racleur *m* d'huile
~SCOOP - s. oil catcher
~SCRAPER RING - s. oil ring
~SEAL - [mech] joint *m* tournant de retenue d'huile
[auto] joint *m* d'huile, joint *m* annulaire d'étanchéité
~SEAL RING - [mech] anneau *m* du joint de retenue d'huile
~SEEPAGE - [geol] dispersion *f* du pétrole dans le gisement
[oil ind] infiltration *f* pétrolifère
~SEPARATOR - [mech & auto] déshuileur *m*, séparateur *m* d'huile, dégraisseur *m* de vapeur
~-SHALE - [geol] (a sedimentary argillaceous rock from which hydrocarbons can be obtained by distillation) schiste *m* bitumineux

OIL-SHALE MINE - [min] minière *f* de schiste bitumineux
~SHEDDER - [mech] graisseur *m*, huileur *m*
~SHEET - [geol] nappe *f* pétrolifère
~SHIELD - [mech] joint *m* annulaire d'étanchéité
~SHOCK-ABSORBER - [mech] (in autos) amortisseur *m* à huile
~SHOWS - [oil ind] traces *fpl* de pétrole
~SKIN - [text] toile *f* vernie
~SLINGER - s. oil deflector
~SLIP - [impl] (a small shaped piece of oilstone) pierre *f* à aiguiser, affiloire *f*
~SOAP - [ind chem] savon *m* à l'huile
~-SOFTENED RUBBER - s. oil-extended rubber
~-SOLUBLE RESIN - [chem] (phenol-formaldehyde resins soluble in drying oils and used in the production of surface coatings) résine *f* oléo-soluble
~SPLASH GUARD - [mech] tôle *f* garde-huile
~SPRAY - [impl] pulvérisateur *m* d'huile
~SPRING - [geol] source *f* de pétrole
~SQUIRT - [impl] burette *f* à pompe
~STEAMER - [naut] navire *m* pétrolier
~STONE - [impl] pierre *f* à aiguiser, pierre *f* à huile, affiloire *f*
~STORAGE TANK - [oil ind] réservoir *m* à pétrole
~STRIKE - [oil ind] découverte *f* du pétrole
~STRING - [oil ind] (casing string cemented at the bottom of a completed bore-hole. The connection with the layer is obtained by perforating the casing and cement) colonne *f* d'exploitation
~STRIPPING - [ind chem] (debenzolization of oil) désessenciement *m* de l'huile, dégazolinage *m* de l'huile
~SUMP - [mech] (oil cup; carter) carter *m* inférieur
[aero] carter *m* de l'huile
~SUMP BREATHER - [mech] (in aeroengines) évent *m* du carter de l'huile
~-SUMP GASKET - [mech] garniture *f* du carter inférieur
~SUPPLY - [auto] alimentation *f* en huile
~SWITCH - [el] (circuit-breaker with its contacts immersed in oil for insulating purposes) interrupteur *m* dans l'huile
~SYRINGE - [mech] (auto) seringue *f*
~TANK - [mech] (reservoir for lubricating oil) réservoir *m* à pétrole
~TANK VENT - [mech] évent *m* de réservoir à pétrole
~TANKER - [naut] pétrolier *m*
~TAR - [oil ind] goudron *m*
~TEMPERATURE - [gen & mech] température *f* de l'huile
~TEMPERATURE INDICATOR - [instr] (auto etc) thermomètre *m* de l'huile
~TEMPERED - [metall] trempé à l'huile
~-TEMPERED WIRE - [metall] fil *m* trempé à l'huile
~TEMPERING - [metall] trempe *f* à l'huile, huilage *m*
~THIEF - [oil ind] (colloq. USA; a device to extract small quantities of oil from a tank sampling) jauge *f* d'échantillonnage
~THROWER - [mech] (a disk fixed on a shaft, so as to prevent oil from creeping along it, the oil being thrown off centrifugally) déflecteur *m* d'huile
~-TIGHT BEARING - [mech] coussinet *m* étanche à l'huile

OIL-TO-AIR HEAT-EXCHANGER - [heat] échangeur *m* thermique huile/air
~TRANSFORMER - s. oil-cooled transformer
~TRAP - [mech] (in machines) siphon *m* de pétrol, séparateur *m* de pétrole
~-TRACKS - [mech] araignée *f* de palier
~TREE - [bot] ricin *m* commun
~TROUGH - [mech] fût *m* d'huile
~TURP - s. oil of turpentine
~VAPOUR - [mech] vapeur *f* d'huile
~VARNISH - [paint] (varnishes containing a drying oil) vernis *m* à l'huile, vernis *m* gras
~VESSEL - [gen] (of a lamp) culot *m*
~WELL - [min] puits *m* pétrolifère, puits *m* à pétrole
~WELL PLUNGER PUMP - [oil ind] pompe *f* de fond pour puits de pétrole
~WELL PUMP - [oil ind] pompe *f* de fond
~-WELL RIG - s. oil rig
~WIPER - [mech] râcleur *m* d'huile
~WIPER RING - [mech] (a piston ring designed to remove excess oil from the cylinder wall) anneau *m* râcleur d'huile
~WITHDRAWAL - [oil ind] enlèvement *m* de pétrole
~YARD - [oil ind] chantier *m* d'exploitation
OILCAN, to - [astronaut] bomber
OILCAN - [impl] broc *m* à huile, burette *f* de graissage
OILCLOTH - [text] (a cotton fabric made waterproof by impregnating it with oxidized oils) tissu *m* huilé
OILCUP - s. oil cup
OILED COTTON - s. oilcloth
~PAPER - [paper man] (paper impregnated with non-drying oil) papier *m* huilé
~PICKER - [text] taquet *m* huilé
~-SILK TAPE - [el] (oil-varnished silk tape used in cable joints to keep spreaders in position) ruban *m* de soie huilée
OILER - [impl] burette *f* à huile, godet *m* graisseur
[auto] pistolet *m* à huile
[oil ind] (colloq. for oil well) puits *m* pétrolifère, source *f* de pétrole
OILINESS - [chem] (the property of an oil to reduce the coefficient of friction under boundary conditions) état *m* graisseux
OILING - [mech] lubrification *f*, graissage *m*, huilage *m*
[text] (of the wool) huilage *m*
[text] ensimage
~PERIOD - [metall] phase *f* de carburation
~PLUG - [mech] trou *m* de graissage
~RING - [mech] (simple device to feed oil to a journal bearing) anneau *m* graisseur
~WILLOW - [text] ensimeuse *f*
OILLESS - [mech] sans huile
~CIRCUIT BREAKER - [el] (circuit breaker using a much smaller quantity of oil than an ordinary oil circuit-breaker) disjoncteur *m* sans huile
OILOMETER - s. oleometer
OILPAPER - [paper man] (paper impregnated with oils) papier *m* imprégné d'huile
OILSEED(S) - [agric] graines *fpl* oléagineuses
OILSILK - s. oilcloth
OILSKIN - [text] (cotton fabric made waterproof by impregnation with oxidized oils) toile *f* cirée
[naut] ciré *m*, caban *m*
OILSLIP - s. oil slip
OILSTONE - s. oil stone

OILSTONE DUST - [mech] poudre *f* d'émeri
OILTIGHT - [mech] étanche à l'huile
OILWAY - [mech] (a passage formed in a part to conduct lubricating oil) rainure *f* de graissage, pattes *f*pl d'araignée
OILY - [gen] huileux, graisseux
~SCUM - [text] écume *f* d'huile
OINTMENT - [gen & chem] onguent *m*, pommade *f*
~BASE - [chem] excipient *m* pour pommades
~MILL - [ind chem] broyeur *m* à pommade
OLD - [gen] vieux
~ENGLISH - [print] gothique
~FACE - [print] caractère *m* médiéval, Elzévir *m*
~FUSTIC - [dyes] (yellow pigment obtained from Morus tinctoria, consisting of a mixture of moric and morintannic acids) maclure *f*
~STYLE - [print] s. old face
~WOMAN'S TOOTH - [constr] guillaume *m*, rabot *m* incliné
OLDHAM COUPLING - [mech] (coupling which allows misalignment of the shafts connected) joint *m* d'Oldham, joint *m* tournevis
OLEAGINOUS - [gen] (oily) oléagineux, huileux
OLEAMIDE - [chem] (a lubricant used in polythene extrusion) oléamide *f*
OLEANDER - [bot] oléandre *m*
OLEANDOMYCIN - [chem] (an antibiotic produced by certain strains of Streptomyces antibioticus and active against a wide range of Gram-positive bacteria) oléandomycine *f*
OLEATE - [chem] (a salt containing oleic acid) oléate *m*
OLEFIANT GAS - [chem] (a synonym for ethylene) gaz *m* oléifiant
OLEFIN FIBRES - [ind chem] (generic term for synthetic fibres having a specific content of olefin units) fibres *f*pl oléifines
~RESINS - [ind chem] résines *f*pl oléfines
OLEFINS - [chem] (homologous series of hydrocarbons with the general formula C_nH_{2n}) oléfines *f*pl, alcènes *f*pl
OLEIC - [chem] (derived from oil) oléique
~ACID - [chem] (an unsaturated fatty acid of wide distribution in nature. Oleic acid is used as a synthesis intermediate and cosmetics, pharmaceuticals and polishes) acide *m* oléique
OLEIFEROUS - [bot] oléifère, oléagineux
OLEIN - [chem] (a glycerine ester of oleic acid) oléine *f*
~GREASING EMULSION - [chem] ensimage *m* à l'oléine
OLEO - [aero] (telescopic oil-cataract element to absorb impact in a landing gear) amortisseur *m* de choc à huile
~LEG - s. oleo
~STRUT - s. oleo
OLEODYNAMIC - [mech] oléodynamique
OLEOGRANULOMA - [med] paraffinoma *m*
OLEOGRAPH - [print] (imitation of an oil painting by chromolithography) oléographie *f*
OLEOMARGARINE - [chem] oléomargarine *f*
OLEOMETER - [instr] (an instrument for measuring the specific gravity of oil) oléomètre *m*
OLEOPNEUMATIC BRAKE - [mech] frein *m* oléopneumatique
OLEOREFRACTOMETER - [instr] (refractometer for use with oils) oléorefractomètre
OLEORESIN - [paint] (general term for naturally-occurring mixtures of resins and essential oil) oléorésine *f*
OLEOSTRUT - s. oleo
OLEOYL CHLORIDE - [chem] (a synthesis intermediate) chlorure *m* d'oléolyle
OLEOTHORAX - [med] oléothorax *m*
OLEUM - [chem] (fuming sulphuric acid) oléum *m*
~LIVER - [oil ind] film *m* d'oléum
~SPIRIT - [oil ind] esprit *m* d'huile
OLEYL ALCOHOL - [chem] (an intermediate for detergents and synthetic resins) alcool *m* oleylique
OLFACTION - [zool] (the sense of smell) olfaction *f*, odorat *m*
OLFACTORY - [zool] (pertaining to the sense of smell) olfactif
~AREA - [anat] région *f* olfactive
~BULB - [anat] bulbe *m* olfactif
~CELL - [anat] cellule *f* olfactive
~LOBES - [anat] (part of the fore brain concerned with the sense of smell) lobes *m*pl olfactifs
~NERVE - [anat] nerf *m* olfactif
~NEUROEPITHELIOMA - [med] tumeur *f* de la placode olfactive
~ORGAN - [anat] organe *m* olfactif
~SYSTEM - [anat] système *m* olfactif
~TRACT - [anat] bandelette *f* olfactive
~TUBERCLE - s. olfactory bulb
OLIGISTE IRON - [min] fer *m* oligiste
OLIGO - [gen] (a prefix meaning small in number) oligo
OLIGOCENE - [geol] (the period following the Eocene period in the Tertiary Era) oligocène *m*
OLIGOCLASE - [min] (a type of feldspar found in more acid igneous rocks) oligoclase *f*
OLIGOPHRENIA - [med] oligophrénie *f*
OLIVE - [bot] (evergreen tree bearing oily fruit) olivier *m*
[bot] (the fruit of the olive tree) olive *f*
~CRUSHING - [agric] broyage *m* des olives
~FLY - [zool] mouche *f* de l'olive
~GREEN - [dye] vert olive
~GROVE - [agric] olivette *f*, olivaie *f*
~GROWING - [agric] oliviculture *f*, oléiculture
~OIL - [ind chem] (a fixed oil obtained from Olea europea and used as a foodstuff, in pharmaceuticals, and in soap manufacture) huile *f* d'olive
~-OIL SOAP - [ind chem] savon *m* de Marseille
~PRESS - [agric] moulin *m* à huile
~-SHAPED JEWEL - [horol] rubis *m* olivaire
OLIVENITE - [min] (natural basic copper arsenate, green in colour and occurring occasionally as a secondary mineral in copper deposits) olivénite *f*
OLIVER - [metall] marteau *m* à pédale
OLIVINE - [min] (chrysolite. Natural iron-magnesium silicate with uses as a refractory) olivine *f*, chrysolithe *f*
~ROCK - [min] dunite *f*
~SAND - [min] sable *m* à l'olivine
OLONA - [naut] (water resistant fibre used for making ropes) fibre *f* olona
OMARTHRITIS - [med] arthrite *f* de l'épaule
OMARTHROSE - [med] omalgie *f*, scapulalgie *f*
OMASITIS - [vet] impaction *f* du feuillet
OMBRE' DYEING - [text] teinture *f* ombrée
~PRINTING - [text] impression *f* flammée
~YARN - [text] fil *m* ombré
OMBROGRAPH - [instr] (self-registering rain gauge)

pluviographe *m*
OMBROMETER - [instr] (instrument designed to measure the amount of rain falling over a given period; also called pluviometer, rain-gauge or udometer) ombromètre *m*, pluviomètre *m*
OMBROPHILE - [bot] (said of plants which thrive where rainfall is copious) ombrophile
OMEGA LOOP - [telev] (in video recording) guidage *m* alpha
OMNI-AERIAL - [radio] (aerial with uniform response in azimuth and directional pattern in elevation) antenne *f* non-directive
OMNIBEARING DISTANCE FACILITY - [radar] (radio facility with omnidirectional range in combination with distance measuring equipment) radiobalise *f* non-directive avec mesure de distance
~DISTANCE NAVIGATION SYSTEM - [radar] système *m* de navigation à coordonnées polaires
~INDICATOR - [radar] (instrument designed to indicate azimuth by a phase-shifting arrangement driven by a motor) indicateur *m* azimutal automatique
OMNIBUS - [transp] (a passenger vehicle) omnibus *m*
[glass man] (a protective cover) tôle *f* de sécurité
~BAR - [el] (the original term from which the more common "bus-bar" derives) barre *f* collectrice, barre *f* omnibus
~CUE CIRCUIT - [telev] circuits *mpl* d'interphone en anneau
~LINE - [comput] ligne *f* omnibus
~TRAIN - [railw] (train stopping at all stations) train *m* omnibus
~WIRE - [el] fil *m* omnibus
OMNIDIRECTIONAL AERIAL - s. omni-aerial
~BEACON - [radio] (a radio beacon radiating in all directions, used for D.F. purposes) radiophare *m* omnidirectionnel
~MICROPHONE - [el acoust] (microphone having a response which is practically independent of sound incidence) microphone *m* non directionnel, microphone *m* omnidirectionnel
~RANGE - [radar] (a facility for navigators) radiobalise *f* omnidirectionnelle
~TRANSMITTER - s. omnidirectional microphone
OMNIDIRECTIVE ANTENNA - s. omni-aerial
OMNIGRAPH - [telecomm] (an apparatus reproducing audibly Morse-code signals from a perforated tape) appareil *m* de lecture de son
OMNIRANGE - s. omnidirectional range
OMNIVORE - s. omnivorous
OMNIVOROUS - [zool] (of animals including both animal and vegetable tissue in the diet) omnivore
[bot] (pertaining to a parasitic fungus which thrives on many species of host plants) omnivore
OMOSTERNUM - [anat] cartilage *f* sterno-claviculaire
OMOTOCIA - [med] accouchement *m* prématuré
OMPHACITE - [min] (an aluminous pyroxene) omphazite *f*
OMPHALELCOSIS - [med] ulcération *f* de l'ombilic
OMPHALOMA - [med] tumeur *f* de l'ombilic
OMPHALOTAXIS - [med] remise *f* en place du cordon ombilical
OMRE - [nucl] s. organic-moderated reactor
ON - [gen] (prep) sur, à
[el] fermé
[mech] (a tap, valve etc) ouvert, en marche
~-AND-OFF ATTACHMENT - [oil ind] attaque *f* à accrochage et décrochage debout
ON BOARD - [naut & aero] à bord
~BOARD STATION - [naut & aero] station *f* radio de bord
~DRAUGHT - [brew ind] du tonneau
~END - [gen mech etc] debout, sur bout
~-LINE DATA PROCESSING - [comput] traitement *m* continu des données
~-LINE DATA REDUCTION - [comput] réduction *f* des données en exploitation continue
~-LINE OPERATION - [comput] fonctionnement *m* en régime continu
~-LOAD TAP CHANGER - [el] (tap changer designed to operate while the transformer is connected to the circuit) combinateur *m* de prise, combinateur *m* de réglage en charge
~-LOAD VOLTAGE - [el] (closed circuit voltage (the voltage which exists when circuit is closed) tension *f* en circuit fermé
~-OFF ACTION - [control] action *f* à deux échelons
~-OFF CODE - [comput] code *m* par tout ou rien
~-OFF CONTROL - [control] action *f* par tout ou rien
~-OFF CONTROLLER - [contr] régulateur *m* par tout ou rien
~-OFF SERVO-MECHANISM - [mech] dispositif *m* de asservissement par tout ou rien
~PERIOD - [electron] (the time during an operating cycle, when the electronic tube is conducting) temps *m* de conduction
~POSITION - [gen mech etc] position *f* de travail
~-SIGNAL PROVING - [contr] contrôle *m* de signal à la fermeture
~-SITE CONCRETE - [constr] béton *m* coulé
~SITE OXYGEN PLANT - [ind chem] (installation for producing gaseous oxygen at the plant in which it is to be used, to avoid cost of transport) générateur *m* d'oxygène de chantier
~STREAM - [ind chem] (term denoting that a unit of a chemical plant is operating) en marche
[oil ind] en marche productive
~STRUCTURE - [oil ind] placé sur anticlinal productif
~-SWITCH - [el acoust] commande *f* d'insertion de bande
~THE AIR - [radio] en émission
~THE BIAS - [text] (diagonally, e.g. "cut on the bias" in the case of textiles, cut in a direction which is parallel neither to the warp nor the weft) (coupé) en biais
~-THE-FLY PRINTER - [comput] imprimante *f* avec rouleau porte-types tournant
~-THE-FLY READOUT - [comput] lecture *f* au cours du mouvement
~THE PUMP - [oil ind] en phase de pompage
~-TOP ALTITUDE CLEARANCE - [aero] (clearance given for VFR flying above cloud or the like) autorisation *f* de vol au-dessus des nuages
ONAGER - [zool] (wild ass, found in Central Asia) onagre *m*
ONANISM - [med] onanisme *m*, masturbation *f*
ONCE - [gen] une fois
~-RUN - [ind] à procédé direct
[ind chem] à passage unique
~-RUN BENZOLE - [ind chem] benzol *m* rectifié
~-THROUGH - s. once-run
~-THROUGH COOLING - [nucl] (system in which the cooling fluid is passed only once through the appa-

ratus) refroidissement *m* à passage unique
ONCOGENESIS - [med] oncogenèse
ONCOGENOUS - [med] tumorigène
ONCOSIMETER - [instr](instrument designed to measure the specific gravity of molten metals) oncosimètre *m*
ONDEE - [text ind] ondé *m*, soie *f* ondée
ONDOGRAM - [instr] (the curve of the wave shape of voltage or current as drawn by the ondograph) ondogramme *m*
ONDOGRAPH - [instr] (instrument designed to record the wave shape of voltage or current by the periodic charge and discharge of a capacitor) ondographe *m*
ONDOMETER - [instr] (instrument designed to measure the frequency of electric waves) fréquencemètre *m*, ondomètre *m*
ONDOSCOPE - [instr] (glow-discharge tube used as a wave detector) ondoscope *m*
ONDULATED - [gen] ondulé
ONDULE' - [text] ondulé *m*
ONE - [number] un, une
~-ADDRESS - [comput] (single-address) à une adresse
~-ADDRESS COMPUTER - [comput] calculateur *m* à adresse unique
~-ADDRESS INSTRUCTION - [comput] (instruction which consists of an operation and one address) instruction *f* à une adresse
~-AXLE TRAILER - [mech] semi-remorque
~BOTTOM PLOUGH - [agric] charrue *f* monosoc
~-BRICK WALL - [constr] mur *m* (de l'épaisseur) de une brique
~-COAT WORK - [constr] (a single coat of coarse stuff) enduit *m* à une seule couche
~CONDITION - [comput] position *f* un
~-CORE-PER-BIT STORAGE - [comput] mémoire *f* à un tore par bit
~-COURSE FARMING - [agric] monoculture *f*
~-CROP SYSTEM - s. one-course farming
~-DAYLIGHT PRESS - [mech] presse *f* à un étage
~-DEGREE-OF-FREEDOM SYSTEM - [contr] système *m* à un degré de liberté
~-DIGIT-ADDER - [comput] additionneur *m* monodigit
~-FLOOR KILN - [brew ind] touraille *f* à un plateau
~-FLUID CELL - [el chem] (primary cell in which both electrodes are in contact with the same electrolyte) pile *f* à un liquide
~-FURROW BOTTOM TRACTOR - [agric] tracteur *m* à un sillon
~FURROW PLOUGH - [agric mach] charrue *f* à un sillon
~-GRAINED WHEAT - [agric] engrain *m*
~-GRID VALVE - [radio] tube *m* monogrille
~-GROUP MODEL - [nucl] (model for the study of neutron behaviour) modèle *m* à un groupe d'énergies
~-GROUP THEORY OF REACTOR - [nucl] (theory assuming that the production, diffusion and absorption of neutrons occur at a single energy) théorie *f* de réacteur à groupe unique
~GROUP TREATMENT REACTOR - s. one group theory of reactor
~-HOLE MOVE - [mech] avancement *m* d'un trou
~-HOLE NOZZLE - [mech] (for i.c. engines) gicleur *m* à un seul trou
~-HOUR DUTY - [el] (of electrical machines; a short-time duty with a constant load during one hour) service *m* unihoraire
~-HOUR RATING - [el] (for of rating used for electrical machines supplying an intermittent load, as a traction engine) régime *m* unihoraire
~-HOUR SPEED - [el] vitesse *f* au régime unihoraire
~-KICK MULTIVIBRATOR - [electron] multivibrateur *m* monostable
~-MAN - [gen mach etc] (tache) pour un seul homme
~OFF - [gen] (term used for making a single example of a given type or pattern, as distinct from a series of identical pieces) unique
~-OFF CASTING - [metall] coulée *f* unique
~OUTPUT - [comput] (of a static magnetic storage) signal *m* de sortie d'un un
~-PHASE - [el] monophasé
~-PIPE - [gen & mech] à un seul tuyau, à raccord unique
~-PIPE HOT WATER SYSTEM - [th eng] système *m* de chauffage avec distribution à un tuyau
~-PIPE STEAM SYSTEM - [th eng] système *m* avec distribution à un tuyau
~-PIPE SYSTEM - [th eng] drainage *m* à raccord unique
~-PLUS-ONE ADDRESS - [comput] à (I + I) adresses
~-PLUS-ONE INSTRUCTION - [comput] (a two-address instruction) instruction *f* à (I + I) adresses
~-POLE PLUG - [el] fiche *f* unipolaire
~POSITION - [el] (of a winding; a winding is said to be "one position" or "two position winding", according to whether it contains one or two coil-sides superimposed in one slot) à une couche
~-PROCESS PICKER - [text] batteur *m* "one-process"
~-QUADRANT MULTIPLIER - [contr] multiplicateur *m* monoquadrant
~-SHOT LUBRICATION - [auto] graissage *m* au pistolet à pression
~-SHOT MULTIVIBRATOR - [radio] (monostable multivibrator) multivibrateur *m* à coup unique, multivibrateur *m* monostable, bascule *f*
~-SIDE PERFORATED FILM - [cin] pellicule *f* à perforation unilatérale
~-SIDED - [gen] unilatéral
~STAGE AMPLIFIER - [electron] amplificateur *m* à un étage
~-STAGE RESIN - [chem] (a thermosetting resin based on the reaction of an aldehyde with a phenol in the presence of an alkaline catalyst) résol *m*
~STATE - [comput] (of a static magnetic storage; a state of a magnetic cell in which the magnetic flux has a positive value, when determined from an arbitrarily given direction of positive normal to that area) position *f* un
~-STRAND ROPE - [telecomm] câble *m* à un seul brin
~-TO-ONE ASSEMBLER - [comput] traducteur *m* un par un
~-TO-ONE TRANSFORMER - [el] (transformer with the same number of turns on the primary and on the secondary) transformateur *m* un par un
~-TO-PARTIAL-SELECT RATIO - [comput] (the ratio between a one output and a partial-select output) rapport *m* signal un/signal de sélection partielle
~-TO-ZERO RATIO - [comput] (of a static magnetic storage; the ratio between a one output and a zero output) rapport *m* un-zéro
~-TRACK - [railw] à une voie
~-VALUED FUNCTION - [contr] fonction *f* définie de

une façon univoque,
[math] fonction *f* univalente
ONE-VALVE RECEIVING SET - [radio] appareil *m* récepteur à une lampe
~-WAY - [auto] à sens unique
[el] (of a switch etc to denote that it provides a single path for current) unidirectionnel
~-WAY PLOUGH - [agric] charrue *f* pour labour à plast
~-WAY PLOUGHING - [agric] labour *m* à plat
~-WAY RESTRICTOR - [mech] soupape *f* à étranglement unidirectionnel
~-WAY STREET - [road traffic] rue *f* à sens unique
~-WAY SWITCH - [el] interrupteur unidirectionnel
~-WAY VALVE - [mech] (a valve permitting flow in one direction only) soupape *f* unidirectionnelle
ONES COMPLEMENT - [math] complément *m* à un
ONESIDED LOOP - [text] boucle *f* à une seule aigrette
ONION - [bot] oignon *m*, ognon *m*
ONIONSKIN - [paper man] (very thin translucent paper) papier *m* pelure d'oignon
ONIUM COMPOUND - [text] composé *m* d'onium
ONKINOCELE - [med] cedème *m* de la gaine tendineuse
ONOMATOMANIA - [med] (morbid state of mind in which the forget-fulness of certain words is associated with a very strong impulse to utter other, often obscene, words) onamatomie *f*
ONSET - [gen] assaut *m*, attaque *f*
[el] (the application of full load) application *f*
[chem] départ *m* (d'une réaction)
ONSETTING - [mining] encagement *m*
ONTOGENESIS - [genet] (history of the development of an individual) ontogénèse *f*
ONTOGENY - s. ontogenesis
ONYCHATROPIA - [med] onychatropie *f*
ONYCHAUXIS - [med] hypertrophie *f* des ongles
ONYCHIA LATERALIS - [med] tourniole *f*, panaris *m* périunguéal
ONYCHITIS - [med] onychite *f*
ONYCHOLYSIS - [med] décollement *m* des ongles
ONYCHOPHAGY - [med] onychophagie *f*
ONYCHOPHYMA - [med] dégénérescence *f* de l'ongle
ONYXITIS - s. onychitis
ONYX - [min] (variety of silica consisting of layers of different colours) onyx *m*
~MABBLE - [min] marbre *m* onyx, onyx *m* calcaire
OÖGENESIS - [biol] (the development of ova) ovogénèse *f*
OÖLITE - [geol] (a sedimentary rock) oolithe *m*
~IRONSTONE - [min] fer *m* oolitique
OOZE, to - [gen] (to emit moisture slowly) suinter
OOZE - [geol] (fine-grained, soft deep-sea deposit) vase *m*, limon *m*
[gen] (the emission of moisture) suintement *m*, jus *m*, jusée *f*
[text] poil *m*, duvet *m*
~HOSE - [leather ind] tuyau *m* pour tannerie
~LEATHER - [leather ind] cuirs *mpl* velours, veaux *mpl* velours
OOZING - [phys] (the slow emission of moisture) suintement *m*
~WELL - [hydr] puits *m* drainant, puits *m* absorbant
OPACIFICATION - [med] opacification *f*
OPACIFIER - [ind chem & metall] substance *f* opacifiante
OPACITY - [paint] (the hiding or obliterating power of a coating composition) opacité *f*, intransparence *f*
[gen & phys] opacité *f*
OPAL - [min] (variously coloured hydrous silica) opale *f*
~AGATE - [min] (agata-like variety of opal) agate *f* opale
~DIFFUSER - [photo] disque *m* opalin
~GLASS - [glass man] (opalescent or white glass, obtained through the addition of fluorides to the mixture) verre *m* opalin
~JASPER - [min] (a type of opal containing yellow iron oxides and other impurities) jaspe *m* opalin
~LAMP - [el] (electric lamp with a bulb made of opalescent glass) lampe *f* opale
OPALESCE, to - [ind chem] opaliser
OPALESCENCE - [phys] (an iridescent milky appearance in a solution, caused by light reflection from minutes suspended particles) opalescence *f*
OPALINE - [glass man] (white, translucent glass) opalin
OPALIZED FILM - [photo] pellicule *f* mate
OPAQUE - [phys] (impervious to the passage of light) opaque
[photo] (paint used in photography for retouching negatives) épiscopique
~PROJECTION - [opt] projection *f* épiscopique
~PROJECTOR - [photo] épiscope *m*
OPAQUENESS - [gen & phys] opacité *f*
OPEN, to - [gen] ouvrir
OPEN - [gen] ouvert
[el] ouvert, coupé, rupture *f*
[mech] (of a belt) ouvert, libre, non serré
[ind chem] (of mixtures) granulaire
[chem] (of a chain) à chaîne ouverte
[telecomm] défaut *m* de continuité
~A CIRCUIT, to - [el] ouvrir un circuit
~A HILL, to - [brew ind] débutter
~AERIAL - [radio] (exterior, or outdoor aerial) antenne *f* ouverte
~AIR - [gen] plein air *m*
[adj] (open-air) au grand air, en plein air
~-AIR IONIZATION CHAMBER - [nucl] (air-filled ionization chamber) chambre *f* d'ionisation en air libre
~-AIR TRANSFORMER - [el] transformateur *m* à l'air libre
~-AISLE COACH - [railw] (coach with centre gangway) voiture *f* à couloir central
~AMORTISSEUR WINDINGS - [el] enroulements *mpl* en court-circuit indépendants les uns des autres
~AN ACCOUNT, to - [fin] (at a bank etc) ouvrir un compte
~ANNEALED - [metall] recuit à l'air
~ANNEALING - [metall] recuit *m* à l'air, recuit *m* en noir
~ANTENNA - s. open aerial
~ARC - [el] (carbon arc with free access to external atmosphere) arc *m* libre à l'air, arc *m* à feu libre
~ASSEMBLY - [mech] (method of assembly in which pressure is applied as soon as parts have been brought into contact) montage *m* ouvert
~ASSEMBLY TIME - [plast ind] (the time during which glue coated surfaces are exposed to the air before being brought into contact) temps *m* d'exposition

avant assemblage
OPEN BELT - [mech] courroie *f* ouverte
~BUBBLE - [rubber & plast ind] (a cavity in an expanded material which is in communication with the air) bulle *f* ouverte
~BUILDING - [constr] constructions *fpl* dégagées
~BUS - [el] barre *f* omnibus non protégée
~CAST - [mining] (exploitation) à ciel ouvert [metall] coulée *f* à découvert
~CAST MINING - [mining] exploitation *f* à ciel ouvert
~CENTRE CONTROL - [radar] (expanded centre plan display) opération *f* de dilatation du P.P.I.
~CHAIN - [chem] chaîne *f* ouverte
~CHANNEL - [hydr] (channel conveying liquid, the surface of which is at atmospheric pressure within the channel) canal *m* découvert, conduite *f* libre
~CIRCUIT - [el] (condition of an electric circuit in which the current path is interrupted) circuit *m* de courant ouvert
~-CIRCUIT ADMITTANCE - [el] admittance d'un circuit ouvert
~-CIRCUIT FAULT - [telecomm] (a cable break) rupture *f* d'un câble
~-CIRCUIT IMPEDANCE - [el] impédance *f* d'un circuit ouvert
~-CIRCUIT OPERATION - [telecomm] (single-current signalling system in which no current flows in the circuit in the intervals between the transmission of characters) exploitation *f* par envoi de courant, système *m* de circuit ouvert
~-CIRCUIT PARAMETER - [electron] (in semiconductors) paramètre *m* de circuit ouvert
~-CIRCUIT REVERSE VOLTAGE TRANSFER RATIO - [electron] réaction *f* de tension à vide
~-CIRCUIT SIGNALLING - [telecomm] (system of signalling in which no current flows when the circuit in an idle condition) système *m* à fermeture de circuit
~-CIRCUIT TRANSFER ADMITTANCE - [el] admittance *f* de transfert pour circuit ouvert
~-CIRCUIT TRANSFER IMPEDANCE - [el] impédance *f* de transfer pour circuit ouvert
~-CIRCUIT TRANSITION - [el] (used in the series-parallel controls of traction motors) système *m* de commutation de série à parallèle
~-CIRCUIT VOLTAGE - [el] (off load voltage. The voltage which exists when the circuit is interrupted) force *f* électromotrice, tension *f* à circuit ouvert
~-CIRCUITED - [el] à circuit ouvert
~-CIRCUITED ELECTRODE - [electron] électrode *f* sans connexion
~CIRCUITED LINE - [electron] ligne *f* ouverte
~COIL - [el etc] bobine *f* ouverte
~-COIL ANNEALING FURNACE - [metall] four *m* à recuire les bobines ouvertes
~-COIL WINDING - [el] bobinage *m* ouvert
~CONTACT - [el] contact *m* de repos
~-CORE TRANSFORMER - [el] transformateur *m* à circuit magnétique ouvert
~CURE - [rubber ind] vulcanisation *f* à nu, vulcanisation *f* libre
~-CUT MINING - [min] exploitation *f* à ciel ouvert
~CYCLE - [mech] (a cycle of operation of a heat engine in which the fluid is used only once) cycle *m* ouvert
~CYCLE OUTPUT - [metall] production *f* à cycle ouvert
OPEN DIAPASON - [mus] (organ stop, one octave lower than the principal) montre *f*
~-DIAPHRAGM LOUDSPEAKER - [el acoust] (loudspeaker having a radiating diaphragm not fitted with a horn; it operates directly into the low impedance of the air) haut-parleur *m* à diaphragme ouvert
~DIE - [metall] (in die casting) estampe *f* ouverte
~DIE FORGING - [metall] forgeage *m* libre
~DIGGING - s. open-cut mining
~DITCH - [constr] canal *m* à ciel ouvert, tranchée *f* découverte
~ENDED - [comput] (of a programme) extensible
~EAVE - [constr] gouttière *f* saillante
~END BAROMETER - [instr] (a baroscope) baroscope *m* à extremié ouverte
~-END LOCKING RING - [mech] bague *f* de fermeture ouverte avec tendeur à vis
~-END SPANNER - [impl] (a type of spanner having separate jaws) clef *f* à extremité ouverte
~-END WRENCH - s. open end spanner
~EXHAUST - [mech] (exhaust system in which the gases are discharged direct to the atmosphere without the intervention of a silencer) échappement *m* libre
~FAULT - [geol] faille *f* ouverte
~FEEDER - [text] bec-fil *m* ouvert
~FIELDS - [agric] terrain *m* dénudé
~FILAMENT - [electron] rupture *f* de filament
~-FIRE KILN - [brew ind] touraille *f* à feu direct
~FLOOR - [constr] (a floor covered by a ceiling, so that the joists are visible) (plafond *m*) en gites apparentes
~-FLOW DELIVERY - [gas ind] potentiel *m* d'un puits
~-FLOW OF WELL - [gas ind] débit *m* à ouverture totale
~-FLOW POTENTIAL - s. open-flow delivery
~FLOW TEST - [oil ind] essai *m* à plein
~FOLD - [geol] pli *m* ouvert
~-FRAME GIRDER - [constr] (girder which is not framed by any diagonal members) poutre *f* en treillis, poutre *f* cloisonnée
~FREIGHT-CAR WITH HIGH SIDES - s. open goods wagon with high sides
~-FRONT ECCENTRIC PRESS - [mech] presse *f* à excentrique à col de cygne
~FUSE - [el] coupe-circuit *m* à l'air libre
~-FUSE OUTPUT - [el] coupe-circuit à fusion libre
~-GAP PRESS - [mech] presse *f* à col de cygne (ouverte sur un côté)
~GOODS WAGON WITH HIGH SIDES - [railw] wagon *m* tombereau
~GRID - [electron] (electron-tube grid unconnected to a circuit) grille *f* flottante
~-HEAD AUGER - [ind chem] (used in the treatment of gas) sonde *f* tire-bouchon, mèche *f* à déboucher
~-HEARTH - [metall] (of a reverberator furnace) sole *f*
~-HEARTH FURNACE - [metall] (a type of reverberator furnace used in the manufacture of steel) four *m* Martin
~-HEARTH OPERATION - [med] opération *f* à coeur ouvert
~-HEARTH PROCESS - [metall] (process for making steel from varying proportions of pig-iron and scrap; also called Siemens-Martin process) procédé *m* Martin

OPEN-HEARTH SLAG - [metall] laitier *m* Martin
~-HEARTH STEEL - [metall] acier *m* Martin
~HOLE - [gas ind] (of a natural gas well) trou *m* non tubé, trou *m* nu, puits *m* non cuvelé
~-HOLE INSERT - [plast ind] moule *m* à prisonnier de par en par
~HOOK - [text] bec-fil *m* ouvert
~-JET WIND TUNNEL - [aero] (a type of wind-tunnel having an open experimental section) soufflerie *f* à veine libre
~JOINT - [metall] joint *m* de moulage ouvert
~LADDER CIRCUIT - [contr] circuit *m* à échelle ouvert
~-LAND TYRE - [rubber ind] (cross-country tyre) pneu *m* tous terrains
~LINE - [railw] pleine voie *f*
~LOOP - [electron] boucle *f* ouverte
~-LOOP CONTROL - [contr] (a system of control in which signals are transmitted without feedback) commande *f* en chaîne ouverte
~-LOOP CONTROL SYSTEM - [contr] système *m* de réglage en chaîne ouverte, système *m* de commande en boucle ouverte
~-LOOP GAIN - [electron] gain *m* sans contre-réaction
~-LOOP PROCESS CONTROL - [contr] commande *f* directe d'un processus
~-LOOP REMOTE CONTROL - [contr] télécommande *f*
~MARKET - [comm] marché *m* libre
~MATTER - [print] belle composition *f*
~MILL - [mech] mélangeur *m* à cylindres, mélangeur *m* ouvert
~MORTISE - [carp] (a slot mortise) mortaise *f* à l'extremité d'une pièce
~MOTOR - [el] (said of an air-cooled machine to denote that no mechanical protection is embodied and that there is no restriction to ventilation) moteur *m* ouvert
~MOULD - [metall] moule *m* à découvert
~NEWEL STAIR - [constr] (stair with successive flights rising in opposite directions over a rectangular well hole) escalier *m* à deux volées
~NOTES - [mus] (on brass wind instruments) tons *mpl* ouverts
~OSCILLATING CIRCUIT - [el] circuit *m* oscillant ouvert
~OVERHAND STOPE - [mining] gradin *m* renversé sans remblayage
~PEDAL - [mus] pédale *f* forte, pédale *f* grande
~PIPE - [acoust] (pipe open at the upper end so that the wavelength of the resonance is about twice the length of the air column) tuyau *m* ouvert
~PLATFORM - [railw] (an uncovered platform) quai *m* découvert
~POOL - [gen] (outdoor swimming pool) piscine *f* en plein air
~QUARRY - [mining] carrière *f* à ciel ouvert
~SAND - [geol] (sand of good porosity and permeability) sable *m* poreux
~SAND FILTER - [hydr] filtre *m* à sable ouvert
~SAND MOULD - [metall] moule *m* à découvert
~SHED - [text] (a passage between the upper and lower lines of warp threads) pas *m* stationnaire
~SHEDDING - [text] battage *m* à pas ouvert
~SHOP - [labour] (any factory to which workers are employed whether or not they belong to a trade union) chantier *m* qui admet les ouvriers non-syndiqués

OPEN SHOP - [comput] (mode of computing machine support where the applied programmes are written by the same group which originated the problem) service *m* parties
~-SIDE PLANER - [mach tool] machine *f* raboter ouverte sur le côté
~-SIDE PRESS - s. open-gap press
~SPIRAL SPRING - [mech] ressort *m* de compression
~-STEAM CURE - [rubber ind] vulcanisation *f* en vapeur libre
~-STEAM OVEN - [rubber ind] étuve *f* de vulcanisation en vapeur libre, autoclave *f* en vapeur libre
~-STEAM RECLAIMING PROCESS - [rubber ind] procédé *m* de régénération en vapeur libre
~STEEL - [metall] acier *m* à désoxydation incomplète
~STRING - [mus] (not stopped by the fingers) corde *f* à jour, corde *f* vide
[constr] limon *m* à gradins
~SUBPROGRAMME - [comput] (a directly inserted subrouting) sous-programme *m* ouvert
~SYSTEM - [phys] (system allowing interchange of heat and matter with its surroundings) système *m* ouvert
~TEXTURE FABRIC - [text] tissu *m* à jour
~THE THROTTLE, to - [atuo] ouvrir les gaz
~TO TRAFFIC, to - [railw] (to open a line to traffic) ouvrir à l'exploitation
~TOP - s. open-type hotplate
~TOP FEEDER - [metall] masselotte *f* directe
~TOP SIDE FEEDER - [metall] masselotte *f* ouverte à talon
~TRACK - s. open line
~TRAVERSE - [surv] (traverse in which the final line does not link up with the first line) traverse *f* ouverte, cheminement *m* ouvert
~-TYPE BOILING-PLATE - [el] foyer *m* de cuisson électrique ouvert
~-TYPE HOTPLATE - [gas ind & el] table *f* de travail à brûleurs découverts
~-TYPE MACHINE - s. open motor
~TYPE OF BENCHBOARD - [el] pupitre *m* de commande à panneau séparé
~UNDERHAND STOPE - [mining] gradin *m* droit non remblayé
~WARP - [text] nappe *f*
~WEAVE - [text & rubber ind] tissu *m* peu serré
~WELL - [constr] jour *m* de l'escalier
[hydr] puits *m* à poulie
~-WIDTH BLEACHING - [text] blanchiment *m* au large
~WIRE - [el] (conductor which is individually supported by insulators above the ground surface) fil *m* aérien, conducteur *m* aérien
~WIRE CIRCUIT - s. open circuit
~WIRE LINE - [comput] ligne *f* aérienne
~WORK - [mining] exploitation *f* à ciel ouvert
[text] bonneterie *f* à effets à jour
~-WORK REFLECTOR - [radio] (open-work metal structure providing a good reflecting surface) réflecteur *m* à grille
~YARN - [text] fil *m* à torsion douce
OPENBAND TWINE - [text] (right-hand twine) ficelle *f* à torsion droite
~TWISTING - [text] (right-hand twisting) torsion droite, torsion *f* inverse
OPENCAST - [min] (open pit from which ore can be extracted) carrière *f* exploitée à ciel ouvert
~COAL - [min] charbon *m* exploité à ciel ouvert

OPENER - [impl] (any tool used for opening) outil *m* à ouvrir
[text] ouvreur *m*
~ROLLER - [text] cylindre *m* effilocheur
OPENING - [gen] ouverture *f*, trou, jour *m*
[mech] (of a valve etc) orifice *m*, ouverture *f*, découvrement *m*
[gen] (width) largeur *f*
[gen] (a gap) percée *f*
[geogr] (in a forest) clairière *f*
[text] (the first operation in the cotton spinning process) ouvraison *f*, nettoyage
[hydr] gueule *f* bée
[el] décollement *m*
[constr] (in a wall) jour *m*
~AND SCUTCHING MACHINE - [text] machine *f* de nettoyage
~BIT - [impl] alésoir *m*, équarissoir *m*
~FORCE - [metall] (in die casting) force *f* d'ouverture
~IN TABLE - [text] ouverture *f* dans la table
~IN THICK WALL - [hydr] ajutage *m* cylindrique, gueule-bée *f*
~LIGHT - [constr] jour *m*, percée *f*
~MACHINE - [text] ouvreuse *f* mécanique
~OF A QUARRY - [mining] cloche *f* de carrière
~OF SWING BRIDGE - [mech] ouverture *f* d'un pont tounant
~OF TELEPHONE COMMUNICATIONS - [telecomm] admission *f* des relations téléphoniques
~OF THE SHED - [text] ouverture *f* du pas
~OF THE VALVE - [mech] ouverture *f* de la soupape
~OF WRENCH - [mech] ouverture *f* de la clef
~PIPE - [el] (in telecommunication work) isolateur *m* de raccordement, pipe *f*, isolateur *m* de transition
~ROLLER - [text] rouleau *m* élargisseur
~STROKE - [metall] (in die casting) course *f* du plateau mobile
~SURFACE - [metall & plast ind] surface *f* d'ouverture
~TIME - [aero] (of parachutes, the time between the beginning of the opening and its completion) temps *m* d'ouverture
[el] (applied to a circuit breaker) durée *f* d'ouverture
OPENNESS - [gen] situation *f* exposée
[mech] (e.g. of a radiator) pénétrabilité *f*
~OF GRAINS - [metall] structure *f* ouverte
OPENWORK - [text] à jour *m*
~MINING - [min] exploitation *f* à ciel ouvert
OPERA GLASS - [opt] (a small size binocular) lorgnette *f*
OPERAND - [math] (any quantity related to an operation) facteur *m*
OPERATE - [contr] prêt à fonctionner
OPERATE, to - [gen] agir, opérer
[mech] fonctionner
[med] (to perform a surgical operation) opérer
[min] (to work) exploiter
[comput] actionner
OPERATED - [gen] opéré, commandé, actionné
[min] exploité
~ADDRESS - [comput] adresse-facteur *f*
~BY ECCENTRICS (ARMS) - [text] (bras) mus par des excentriques
~CHANNEL - [comput] canal *m* données
~REGISTER - [comput] registre *m* d'opérandes
OPERATED STRAIGHT-WAY VALVE - [mech] soupape *f* commandée directement
OPERATING - [gen] fonctionnement *m*, manoeuvre *f*, commande *f*, exploitation *f*
[adj] de manoeuvre, de commande
~ACCIDENT - [gen] accident *m* d'exploitation
~ALTERNATING CURRENT VOLTAGE - [el] tension *f* alternative de régime
~AMPERE TURNS - [el] ampère-tours *mpl* actifs
~ANGLE - [el] angle *m* de flux
~ARM FOR STAR WHEEL - [text] roue *f* à goupille
~BREAK TIME - [el] temps *m* de réponse du contact de repos
~CAB - [gen el etc] cabine *f* de manoeuvre
~CHARACTERISTIC - [el] caractéristique *f* dynamique
~COIL - [el] bobine *f* excitatrice
~CONDITIONS - [el] (all the electrical and mechanical quantities characterizing the working or a machine) régime *m*
~CONTACT - [contr] (US; normally open interlock) contact *m* de travail
~COSTS - [comm etc] frais *mpl* d'exploitation
~CURRENT - [el] courant *m* de régime, courant *m* de fonctionnement
~DATA - [gen mech mach etc] données *fpl* de fonctionnement
~DELAYS - [comput] perte *f* de temps utile
~DEVICE - [el] (any device, a push-button, a rope, a wheel, a lever etc., employed to actuate the control equipment) dispositif *m* de manoeuvre
~DUTY - [el] (a defined sequence of making or breaking operations executed without any alteration of the circuit) charge *f* normale
~EXPENSES - [comm ind proc etc] frais *mpl* d'exploitation
~FEATURES - [gen mech mach etc] caractéristique *f* d'exploitation, régime *m*
~FLOOR - s. operating platform
~FLUID - [mech] liquide *m* moteur
~FREQUENCY - [el] fréquence *f* de fonctionnement
~GRIP - [photo] poignée *f* de manoeuvre
~HANDLES - [mech] bras *mpl* d'une pompe
~HOOK - [text] clanche *f*, loquet *m* du cylindre
~INFORMATION - [comput] information *f* de contrôle
~LEVER - [mech] levier *m* de manoeuvre
~LINE - [phys] (mathematical or graphical relation) courbe *f* de travail
~LOSS - [comm] perte *f* de travail
~MAKE-TIME - [el] temps *m* de fonctionnement du contact de travail
~MECHANISM - [el] mécanisme *m* actif
~MECHANISM FOR SPOTTING SHUTTLES - [text] mécanisme *m* de mouvement des espolins
~OVERLOAD - [el] surcharge *f* en fonctionnement
~PIPES - [el] tiges *fpl* d'actionnement
~PLATFORM - [metall] banc *m* de manoeuvre, plateforme *f* de manoeuvre
~POINT - [contr] point *m* de fonctionnement
[electron] (on a grid-voltage-anode current characteristic curve of a vacuum tube) point *m* de fonctionnement
~POLE - [el] perche *f* isolante
~POSITION - [photo] position *f* pour la prise de vue
~POTENTIAL - [el] potentiel *m* de fonctionnement
~POWER - [el] puissance *f* de régime
~PRATICES - [gen] règles *fpl* d'exploitation
~PRESSURE - [mech] pression *f* de fonctionnement

OPERATING PRESSURE RANGE - [mech] domaine *m* de pression de fonctionnement
~RACK - [text] crémaillère *f* motrice
~RANGE - [gen] étendue *f* d'action
~RATE - [el] régime *m*
~RESERVE - [contr] réserve *f* de fonctionnement
~RESULT - [comm] bénéfices *mpl* d'exploitation
~ROOM - [telecomm] (the room in which the automatic exchanges are located) local *m* de service, local *m* de travail
~SEQUENCE - [comput] déroulement *m* des opérations
~SPEED - [aero] vitesse *f* de croisière
~STATE - [comput] état *m* d'exécution
~STRESS - [metall] effort *m* de travail
~TEMPERATURE RANGE - [instr] limites *fpl* de température de fonctionnement
~TEST - [gen] essai *m* de fonctionnement
~THEATRE - [med] salle *f* d'opération
~THRESHOLD - [electron] seuil *m* de réponse
~TIME - [el] (of a relay) temps *m* de fonctionnement
~TIME RATIO - [el] facteur *m* de service
~VALUE - [contr] valeur *f* de réglage
[el] valeur *f* de fonctionnement
OPERATION - [gen] (the act of operating) opération *f*, fonctionnement *m*
[mech] fonctionnement *m*, action *f*
[metall] (of a furnace) conduite *f*
[med] (a surgical operation) opération
[telecomm] (of a telegraph printer) préparation *f* télégraphique
~CARD - s. operation line-up sheet
~CODE - s. operational code
~DECODER - [comput] décodeur *m* du code d'opération
~DRUM - [el] (of a master switch) tambour *m* démarreur
~LIFE - [mech etc] durée *f* de vie
~LINE-UP - [autom] (in ordinary organization and in automation system) disposition *f* en cycle des opérations
~LINE-UP SHEET - [ind proc] plan *m* de travail
~NUMBER - [comput] (number indicating the position of an operation) nombre *m* d'opération, numéro *m* d'opération
~RATE - [comput] vitesse *f* d'exécution des opérations
~RATIO - [comput] rendement *m*
~RECORD - [comput] feuille *f* d'opérations
~REGISTER - [comput] (register containing the operative part of an instruction) registre *m* d'opération
~TIME - [electron] time *m* d'établissement, temps *m* d'opération
OPERATIONAL - [gen] (relative to an operation) opérationnel
[aero] (relating to the performance of a specified task) opératif
~AMPLIFIER - [comput] (amplifier used in electronic computers) amplificateur *m* de calcul, amplificateur *m* opérationnel
~CALCULUS - [math] calcul *m* opérationnel
~CHART - [comput] diagramme *m* opérationnel
~CODE - [comput] (function code) code *m* fonctionnel
~COMMAND - [comput] instruction *f* d'opération
~ENGINEER - [aero] (a member of an aircraft crew employed on engineering duties) mécanicien *m* navigant
~KNOB - [mech] poignée *f* de commande
OPERATIONAL SERVO - [comput] servo-mécanisme *m* opérationnel
~UNIT - [comput] (part of the computer in which a given operation takes place) unité *f* opérationnelle
OPERATIVE - [gen] (which exerts a force, or power) opératif, actif
[gen] (engaged in a given work) en fonction
[gen] (workman or mechanic etc engaged in practical work) ouvrier *m*, artisan *m*
~FIELD - [med] champ *m* opératoire
~POSITION - [mech] position *f* de mouvement
~RANGE - [mech el] étendue *f* d'action
OPERATOR - [gen] opérateur *m*
[fin etc] (a broker) courtier *m*
[telecomm] (working on a telephone switchboard) téléphoniste *m*
[telecomm] (the person receiving or transmitting telegrams) télégraphiste *m*
[math] (symbol indicating an operation) opérateur *m*
[cin] (cief-operator assistant) opérateur *m*
OPERATOR'S TEAM - [telecomm] brigade *f*
OPERCULAR APPARATUS - [zool] (in Fish) appareil *m* operculaire
OPERCULUM - [zool] (bony or membranous flap) opercule *m*
OPERETTA - [mus] opérette *f*
OPHICALCITE - [geol] (a combination of dolomitic limestone and silica) ophicalce *f*
OPHITE - [geol] (variety of altered diabase found in the Pyrenees) ophite *f*
OPHRIITIS - s. ophritis
OPHRITIS - [med] dermatite *f* des sourcils
OPHRYOSIS - [med] spasme *m* de sourcils
OPHTHALMEN - [med] appareil *m* de perception optique
OPHTHALMIA - [med] (inflammation of parts of the eye) ophtalmie *f*
OPHTHALMIC - [med] (pertaining to ophtalmia) ophtalmique
~APPLICATOR - [nucl] (applicator with beta emitting silver foil elements embedded in silver or plastic cups) porte-émetteur-beta *m* pour buts ophtalmiques
OPHTHALMODYNAMOMETER - [instr] (instrument for measuring the eye) ophtaldynamomètre *m*
OPHTHALMOLITH - [med] calcul *m* lacrymal
OPHTHALMOLOGIST - [med] oculiste *m*, ophtalmologiste *m*
OPHTHALMOLOGY - [med] (the part of medicine dealing with eye diseases) ophtalmologie *f*
OPHTHALMOMETER - [instr] (instrument for measuring the eye) ophtalmomètre *m*
OPHTHALMOMYASIS - [med] myiase *f* oculaire
OPHTHALMONEURITIS - [med] névrite *f* optique
OPHTHALMOPLEGIA - [med] ophtalmoplégie *f*
OPHTHALMOSCOPE - [instr] (an optical instrument for the examination of the eyes) ophtalmoscope *m*
OPIATES - [chem] (substances which contain or are derived from opium and which induce narcosis) opiacés *mpl*, opiats *mpl*
OPISOMETER - [instr] (instrument designed to measure curves) curvimètre *m*
OPISTHENAR - [med] dos *m* de la main
OPIUM - [chem] (the dried lates obtained from varieties of Papaver somniferum and used in medicine as a narcotic. Opium is the source of a number of therapeutically important alkaloids) opium *m*
~HABIT - [med] opiumisme *m*, opiomanie *f*

OPIUM POPPY - [agric] pavot *m* somnifère, oeillette *f*
OPOSSUM HAIR - [text] poil *m* d'opossum
OPOTHERAPY - [med] organothérapie *f*, opothérapie *f*
OPPOSE, to - [gen] opposer
OPPOSED - [gen] opposé
~CYLINDER COMPRESSOR - [mech] (compressor with opposed cylinders) compresseur *m* à cylindres opposés
~CYLINDER ENGINE - [mech] (a reciprocating engine in which the cylinders lie in the same plane on opposite sides of a common crankshaft) moteur *m* à cylindres opposés
~PISTON ENGINE - [mech] (a type of reciprocating engine in which each cylinder contains pair of pistons acting in opposite directions) moteur *m* à pistons opposés
OPPOSING SPRING - [mech] ressort *m* de rappel
~TORQUE - [mech] couple *m* antagoniste
~TRAIN - [railw] (a train running in the opposite direction) train *m* de sens contraire
OPPOSITE - [gen] opposé, vis-à-vis, en face
[math] (in geometry) opposé
~FIELD - [telev] trame *f* jointive
~FORCES - [phys] forces *fpl* contraires
~PHASE - [el] opposition *f* de phase
OPPOSITION - [gen] opposition *f*, résistance *f*
[el] (phase difference of one half-cycle) opposition *f*, quadrature *f* de phase
~DUPLEX - [telecomm] duplex *m* par opposition
~METHOD - [meas] (a method of measurement) méthode *f* d'opposition
~PHASE QUADRATURE - [radio & telev] (a phase shift of 90 degrees of two voltages or currents) quadrature *f* de phase
OPPOSITIONS - [astr] (the instant of time in which the elongation of a celestial body is 180 deg.) opposition *f*
OPSIURIA - [med] opsiurie *f*
OPSONIC - [biol] opsonique
OPSONIN - [biol] (substance in the blood increasing the phagocytic properties of leucocytes) opsonine *f*
OPTESTHESIA - [med] sensibilité *f* visuelle
OPTIC - [opt] (relating to the eye or the vision) optique
~ATROPHY - [med] (resulting from the degeneration of the optic nerve) atrophie *f* optique
~AXIAL ANGLE - [min] (the angle between the two optic axes in a biaxial mineral) angle *m* de l'axe optique
~AXIS - [opt] (direction of a crystal, along which a light of ray suffers no double refraction) axe *m* optique
~DISK - [anat] (the expanded part of the optic nerve as it enters the retina) disque *m* optique
~LOBE - [anat] (part of the mid-brain which is concerned with the sense of sight) lobe *m* optique
~NERVE - [anat] (the nerve connecting the retina with the cerebral centres) nerf *m* optique
~NEURITIS - [med] névrite *f* optique
OPTICAL - s. optic
~ABERRATION - [opt] (the failure of an optical system to form the exact image of an image) aberration *f* optique
~ABSORPTIVE POWER - [opt] (the transmissivity, that is the ratio between the transmitted radiation and the normally-incident radiation, subtracted from unity) pouvoir *m* absorbant
OPTICAL ABSORPTIVITY - s. optical absorptive power
~ACTIVITY - [opt] (the property in certain substances of rotating the plane of polarization of plane-polarized light passing through them. See dextrorotatory, laevo-rotatory, optical isomers) activité *f* optique
~AIR MASS - [phys] masse *f* d'épaisseur optique d'air
~ALTIMETER - [instr] (altimeter using optical system of measurement) altimètre *m* optique
~AMMETER - [instr] (electrothermic instrument measuring the current in a filament lamp by comparing the illumination with that produced when a given current is used in the same filament) ampèremètre *m* optique
~ANOMALY - [opt] (relating to the behaviour of certain organic compounds) anomalie *f* optique
~ANTIPODES - [chem] (two compounds composed of the same atoms and atomic linkages, differing in their structural formula in so far as one is the mirror image of the other) antipodes *mpl* optiques
~ATROPHY - s. optic atrophy
~AXIAL ANGLE - s. optic axial angle
~AXIS - [opt] (imaginary line passing through the centre of curvature and the optical centre of a spherical lens) axe *m* optique
~BLEACHES - [ind chem] (colourless dyestuffs capable of absorbing UV radiation and reemitting this in the visible range and used to enhance or minimise shades) décolorants *mpl* optiques
~BIREFRINGENCE - [opt] (when a crystal of a substance like, e.g. calcite is held between the eye and a pinhole in a card, two bright dots are seen, resulting from two refracted rays) biréfringence *f* à caractère optique
~BRIGHTENER - s. optical bleaches
~CENTRE - [opt] (of a lens; a point on the axis of a lens so placed that any ray passing through this point has its incident and emergent parts parallel) centre *m* optique
~CENTRE OF A LENS - s. optic centre
~COMPARATOR - [instr] (instrument used in mechanical work) comparateur *m* optique
~CHARACTER FEADER - [comput] lecteur *m* optique de caractères
~CHARACTER RECOGNITION - [comput] lecture *f* optique de caractères
~COMPENSATION - [cin] (a means to compensate the picture displacement in continuous system) compensation *f* optique
~CONSTANT - [cryst] constante *f* optique
~CONTACT - [opt] contact *m* optique
~CONVERTER - [comput] convertisseur *m* photosensible
~DENSITY - [opt] (the common logarithm of opacity) densité *f* optique, intensité *f* de noircissement
~DISTORTION - [opt] distorsion *f* optique
~DOUBLE - [astr] (a pair of stars which appear very close to one another but have no physical connexion) étoile *f* double optique
~DOUBLE REFRACTION - s. optical birefrangence
~EFFECT - [opt] effet *m* optique
~ENCODER - [comput] (code signal produced by photocell) traducteur *m* de code photosensible
~EXALTATION - [chem] (when a compound possesses a refraction different from the value calculated)

exaltation *f* optique
OPTICAL EXPOSURE - [photo] exposition *f* optique
~FILTER - [opt] filtre *m* optique
~FLAT - [photo] (generally a glass surface lapped by rubbing on an optically flat surface) plan *m* optique
~FOCUS - [opt] foyer *m* optique
~GLASS - [opt] (glass expressly manufactured for its optical quality) verre *m* optique, verre *m* d'optique
~GRID - [photo] grille *f* optique
~HORIZON - [astr] horizon *m* optique
~IMAGE - [opt] (real or virtual image) image *f* optique
~INDICATOR - [el acoust] (component of a modulation meter) indicateur *m* optique
~INSTRUMENTS - [instr] instruments *mpl* optiques
~INTERMITTENT - [telev] appareil *m* d'intermittence optique
~ISOMERISM - [chem] (the existence of isomeric compounds which vary only in their optical activity) isomérie *f* optique
~ISOMERS - [chem] (compounds which differ only in optical activity) isomères *mpl* optiques
~LENGTH - [opt] (the product of the geometrical distance and the refractive index in a medium of a given refractive index) trajet *m* optique
~LEVER - [opt] (a device for amplifying and measuring small rotations) levier *m* optique
~LIGHT FILTER - [telev] filtre *m* optique
~MAGNETIC POLARIZATION - [phys] (the optical activity shown in a magnetic field by an optically inactive substance) polarisation *f* magnéto-optique
~MASER - [electron] (maser in which the pumping frequency is in the visible light) maser *m* optique
~-MECHANICAL SYSTEM - [telev] (any system of television which used mechanical scanning) analyse *f* mécanique optique
~MODE - [cryst] (thermal vibration of a crystal lattice having a frequency which is independent of wave number) mode *m* d'oscillation optique
~PATH - s. optical length
~PATTERN - [el acoust] (in mechanical recording) image *f* optique
~PLASTICS - [plast ind] (plastics designed to have special optical qualities, e.g. refractive index, transparency etc) plastiques *mpl* optiques
~POLISHING - [opt] (the polishing of a lens or mirror after grinding) polissage *m* optique
~PRINTING EQUIPMENT - [cin] (the equipment which is required to print positive frames by optical arrangements) tireuse *f* par réduction optique
~PYROMETER - [instr] (an instrument for measuring temperature by comparison of the brightness or colour of the object with that of a standard radiant source) pyromètre *m* optique
~RANGE - [opt] (the range of vision) portée *f* de la vue, portée *f* optique
[instr & radar] étendue *f* visuelle
~REDUCTION PRINTING EQUIPMENT - [cin] (the equipment required for the reduction printing process) tireuse *f* par réduction optique
~RESONANCE - [electron] (luminescence in which the frequencies of the exciting and emitted radiation are the same) résonance *f* optique
~ROTATION - [opt] rotation *f* optique
~ROTATORY POWER - [opt] (the ability of a substance to rotate the polarization plane of polarized light) pouvoir *m* rotatoire optique
OPTICAL SCANNING SYSTEM - [comput] exploration *f* optique
~SCRATCH - [cin] (a shadow on the screen due to an imperfection in the slit of the projector) parasite *m* dû à l'état de la fente
~SENSITIZATION - [photo] optique, sensibilisation *f* optique
~SLANT RANGE - [opt] portée *f* optique calculée en atmosphère homogène
~SOUND - [acoust] son *m* optique
~SOUND RECORDER - [el acoust] (equipment for producing a modulated light beam and moving a light-sensitive medium relative to the beam, thus recording signal of acoustical derivation) enregistreur *m* optique du son
~SOUND REPRODUCER - [el acoust] (the combination of a light source, an optical system and a mechanical system for moving a film relative to a light beam having the purpose of converting signals in a sound track into electrical signals) lecteur *m* électro-optique
~SQUARE - [surv] équerre *f* à réflexion, équerre *f* d'arpenteur à réflexion
~SYSTEM - [opt] (of a projector, the various lenses regarded as a whole) système *m* optique
~TRACKING - [opt] poursuite *f* optique
~TRAIN - s. optical system
~VIEW-FINDER - [photo] (optically working view-finder) viseur *m* direct, viseur *m* optique
OPTICALLY ACTIVE - [opt] (capable of rotating the plane of polarized light) optiquement actif
~FLAT - [opt] (deviating from a true plane by distances wavelength of light concerned) optiquement plan
~VOID LIQUID - [opt & chem] (liquid containing no suspended solids) liquides *mpl* optiquement purs
OPTICIAN - [gen] (anyone dealing in, or making, optical goods) opticien *m*, lunettier *m*
OPTICS - [opt] (the study of light) optique *f*
OPTIMAL - s. optimum
~DAMPING - [el] amortissement *m* optimal
~MAGNIFICATION - [opt] (the maximum useful magnification, which is 800 times for a dry objective and 1200 times for oil-immersed objectives) agrandissement *m* optimal
OPTIMALLY CODED PROGRAMME - [comput] (minimum access routine) programme *m* optimale
OPTIMIZATION - [telev] (in vider recording) optimisation *f*
OPTIMIZING - [comput] (the operation of spacing the information so as to make it accessible when needed) optimalisation *f*, optimisation *f*
OPTIMUM - [gen] (most favourable or best) optimum, optimal
~BUNCHING - [electron] (the bunching condition producing maximum power) groupement *m* optimal
~COUPLING - [el] (critical coupling) couplage *m* critique
~CODING - [comput] (programming with the minimum possible waiting before obtaining the information) codage *m* optimal
~CONTROL - [contr] commande *f* optimale
~CURE - [rubber ind] cuisson *f* optimum
~MOISTURE CONTENT - [soil] teneur *f* en eau optimum
~PERFORMANCE - [telev] représentation *f* parfaite

OPTIMUM PROGRAMMING - s. optimum coding
~TRAFFIC FREQUENCY - [telecomm] fréquence *f* optimum de trafic
~WORKING FREQUENCY - [radio] fréquence *f* optimum d'utilisation
OPTION - [gen] (the liberty, or right of choosing) option *f*, choix *f*
~SWITCH - [comput] (at the control desk, to select alternative routines during programme execution) commutateur *m* conditionnel
OPTIONAL - [gen] (which depends on choice) facultatif, conditionnel
OPTOGRAM - [zool] (retinal image) optogramme *m*
OPTOMETER - [instr] (instrument designed to measure the distance of clear vision) optomètre *m*
OPTOMETRY - [med] optométrie *f*
[opt] (the study of the optical performance of the individuam eye) optométrie *f*
OPTOPHONE - [electron] (photoelectric device which enables the blind to read) optophone *m*
OR-CIRCUIT - [comput] circuit *m* OU
~-GATE - s. or-circuit
ORA - [med] marge *f*, bord *m*
ORAL - [zool] (pertaining to or situated near the mouth) oral, buccal
[gen] (which consists of spoken words) oral
ORANGE - [bot] orange *f*
[dye] (reddish-yellow colour) orangé, orange
~BLOSSOM - [bot] fleurs *fpl* d'oranger
~CHROMES - [paint] (pigments consisting of basic lead chromate, or mixtures of it with normal chromate) orangé *m* de chrome
~-CYAN AXIS TRANSFORMATION - [telev] (in colour television) transformation *f* de l'axe orange-cyan
~-CYAN LINE - [telev] (in colour television) ligne *f* orange-cyan
~JUICE - [agric] jus *m* d'oranges
~LEAD - [paint] (a lead oxide pigment, resembling red lead except that it contains less litharge and more lead peroxide) rouge *m* au plomb
~OIL - [chem] huile *f* d'orange
~PEEL - [bot] zeste *m* d'orange
~PEEL OIL - s. orange oil
~PEEL SURFACE - [plast ind & paint] (appearance of a defective moulding the surface of which is pitted) surface *f* en peau d'orange
~PEELING - [paint] (defect in coatings appearing as circular craters, the result of defective flow; most common in sprayed coatings) formation *f* de cratères
~SHELLAC - [chem] (a general term for several types of shellac prepared from seedlac) laque *f* en écailles orangé
~TONERS - [chem] (yellow dyes for ink) colorants *mpl* jaunes pour encres
~TREE - [bot] oranger *m*
~WRAPPERS - s. orange wrapping paper
~WRAPPING PAPER - [paper man] papier *m* pour agrumes
ORATORIO - [mus] oratorio *m*
ORB - [radio] s. omni-directional radio beacon
[gen] (a sphere or a globe) orbe *m*, globe *m*, sphère *f*
[astr] (a circle or orbit) orbite *f* (d'une planète)
[astr] (the plane of the orbit) plan *m* de l'orbite
ORBIT, to - [astr] (to put into orbit) mettre (un satellite) en orbite
ORBIT, to - [astr] (the movement of satellite along an orbit) être en orbite
[aero] (the circular slow flight of an aircraft while awaiting clearance to land) voler sur trajet orbital
ORBIT - [nucl] (the path of a particle under the influence of a field of force) orbite *f*
[astronaut] (the path in source along which a heavenly body, or a man-made satellite moves) orbite *f*
~GEAR - [mech] (the outer, internally-toothed ring gear of an epicyclic train with which the planet gears mesh) engrenage *m* orbital
~SHIFT COILS - [nucl] (set of coils through which a pulse of current is passed to alter the guiding field so as to cause the orbit radius to increase or decrease, thus deviating the accelerated particles) bobines *fpl* déviation
ORBITAL - [astr etc] orbital
~ELECTRON - [phys] électron *m* satellite, électron *m* orbital
~ELEMENTS - [astronaut] paramètres *mpl* d'orbite
~FLIGHT - [astr] (of a man-made satellite or vehicle) vol *m* orbital
~GLIDER - [astronaut] planeur *m* hypersonique
~MOTION - [phys] (the movement of a body in its orbit) mouvement *m* orbital
~PERIOD - [astronaut] période *f* de révolution
~QUANTUM NUMBER - [electron] (the number giving the angular momentum of the electron in its orbital motion) nombre *m* quantique azimutal
[phys] nombre *m* quantique secondaire
~VELOCITY - [phys] (the velocity of a body in its orbit) vitesse *f* orbitale, vitesse *f* circulaire
ORBITING SATELLITE - [astronaut] mise *f* en orbite de satellite
ORCEIN - [chem] (reddish-brown pigment) orcéine *f*
ORCHARD - [agric] verger *m*
~-GRASS - [agric] dactyle *m* pelotonné
ORCHECTOMY - [med] castration *f*, orchidectomie *f*
ORCHESTRA - [mus] (generally, a combination of musical instruments) orchestre *m*
[gen] (of a theatre) fauteils *mpl* d'orchestre
~BOX - [theatre] (the space in a theatre for the orchestra) orchestre *m*
~PIT - s. orchestra box
ORCHESTRATION - [mus] (the scoring for an orchestra) orchestration *f*, instrumentation *f*
ORCHID - [bot] orchidée *f*, orchis *m*
ORCHIDECTOMY - s. orchectomy
ORCHIDONCUS - [med] tuméfaction *f* du testicule
ORCHIECTOMY - s. orchectomy
ORCHIOCELE - [med] hernie *f* scrotale, tumeur *f* du testicule
ORDER, to - [gen] ordonner, commander, arranger
[leg] statuer, ordonner
[comm] (to place an order) commissionner, commander
[med] prescrire
ORDER - [gen] ordre *m*, commande *f*
[leg] décret *m*, arrêté *m*, injonction *f*, ordonnance *f*
[comm] commande *f*, demande *f*
[fin] (order for payment) ordre *m*, mandat *m*
[mech] (or firing in an i.c. engines) rythme *m*
[arch] (e.g. the style of a column) ordre *m*
[math] (number expressing the degree of complexity of an algebraic expression) ordre *m*
[comput] (instruction; instruction word) ordre *m*, commande *f*

ORDER - [comput] (the sequence of the instructions in a programme) séquence *f*
~ANALYSIS - [comput] analyse *f* des commandes
~BOOK - [comm] carnet *m* de commandes
~CODE - [comput] (instruction code) code *m* d'instructions
~-DISORDER TRANSFORMATION - [cryst] transformation *f* ordre-désordre
~ELEMENT - [comput] (instruction element) élément *m* d'instruction
~LETTER - [comm] lettre *f* de commission
~NUMBER - [comm & ind proc] numéro *m* de la commande
~OF COLOURS - [text] ordre *m* des couleurs
~OF EQUILIBRIUM - [phys & chem] (system of classification of chemical equilibria) ordre *m* d'équilibre
~OF INTERFERENCE - [opt] ordre *m* d'interférence
~OF PHASE TRANSITION - [phys & chem] ordre *m* des transformations de phase
~OF PICKING - [text] ordre *m* de chasses
~OF REACTION - [chem] (classification of chemical reactions) ordre *m* de réaction
~OF TREADLING - [text] matchage *m*, embreuvage *m*
~OF WEFT - [text] ordre *m* des duites
~REGISTER - [comput] (instruction register) registre *m* d'instructions
~WIRE - [telecomm] ligne *f* de renvoi, ligne *f* d'ordre, circuit *m* de renvoi
~-WIRE CIRCUIT - [telecomm] (circuit for the transmission of instructions from one operator to another) circuit *m* de renvoi, ligne *f* de conversation
~WIRE DISTRIBUTOR - [telecomm] distributeur *m* automatique de lignes d'ordre
~-WIRE SPEAKING KEY - [telecomm] bouton *m* de conversation
~-WIRE WORKING - [telecomm] exploitation *f* avec lignes d'ordres
ORDERED SCATTERING - [nucl] (the Bragg scattering; elastic scattering from a crystal, in which the individual waves strengthen each other and form a constructive interference) diffusion *f* de Bragg
ORDERLY - [gen] (adj) ordonné, méthodique
ORDINAL - [math] (form of numeral showing the order in a series) ordinal
ORDINARY - [gen] ordinaire, normal, courant
~RAY - [phys] (of the two plane-polarized components into which a ray of light is split, the ray which is deviated at an index of refraction independent of the angle of incidence) rayon *m* ordinaire
~WAVE - [radio] (electro-magnetic wave) onde *f* ordinaire
~WOOL - [text] laine *f* ordinaire
ORDINATE - [math] (the number indicating the distance of a point from the axis of abscissas; also applied to the distance between points and this) ordonnée *f*
ORE - [min] (mineral from which a metal may be extracted) minerai *m*, mine *f*
~ADDING - [metall] addition *f* de minerai au bain
~ASSAY - [min] analyse *f* quantitative des minerais
~ASSAYING - [min] essai *m* des minerais
~AT GRASS - [min] minerai *m* à la surface
~-BEARING - [min] métallifère
~-BEARING SHALE - [geol] schiste *m* métallifère
~BED - [min] gisement *m* de minerais
~BENEFICIATION - [min] préparation *f* du minerai
~BIN - [metall] accumulateur *m* à minerais
ORE-BIN GATE - [min] porte *f* de trémie à minerai
~BLOCKED OUT - [min] minerai *m* découpé
~BODY - [min] massif *m* de minerai
~BREAK - [min] rupture *f* du filon
~BREAKING - [mining] abattage *m* de minerai
~BRIQUETTING - [min] agglomération *f* de minerais
~-BUNKER - [min] réservoir *m* à minerai, coffre *m* à minerai
~BURNER - [min] grilleur de minerai
~-CARRYING - [min] minéralisateur
~CHIMNEY - s. ore chute
~CHUTE - [min] couloir *m* à minerai, coulée *f* de minerai
~CLASSIFICATION - [min] triage *m* des minerais, classification *f* des minerais
~CONCENTRATION - [min] (the amount of minerai in rock) enrichissement *m* des minerais
~CONSTANT IN GRADE - [min] minerai *m* à teneur constante
~CONTENT METER - [min] teneurmètre *m* de minerais
~CRUSHER - [min] concasseur *m* de minerais
~DEPOSIT - [min] (natural accumulation of ore) gisement *m* de minerai
~DISTRIBUTOR - [metall] (device for spreading the charge evenly) distributeur *m* du minerai
~DRESSING - [min] (the crushing and concentration of minerals to obtain metal or other values) préparation *f* mécanique du minerai
~DUMP - [min] halde *f* de minerai
~DUST - [min] poussière *f* de minerai
~ENRICHMENT - [min] enrichissement *m*
~EXTRACTION - [min] exploitation *f* des minerais
~FORMATION - [min] formation *f* métallifère
~GRIZZLEY - [min] harpe *f*, grille *f* à barreaux
~HOPPER - [min] trémie *f* à minerai
~LEVEL - [min] voie *f* d'extraction du minerai, galerie *f* d'exploitation
~MASS - [min] amas *m* de minerai
~MINING - [mining] exploitation *f* des minerais
~PASS - [mining] cheminée *f* à minerai
~PILLAR - [geol] pilier *m* de minerai
~POCKET - [min] poche *f* de minerai, nid *m*, sac *m*
~PROCESS - [metall] (the Siemens process) méthode *f* d'oxydation par le minerai, procédé *m* Siemens
~PULP - [min] fines *mpl* de minerai
~REFINING - [min] (the operation of freing ore from impurities) raffinage *m* des minerais
~ROAD - [mining] voie *f* d'extraction du minerai
~ROASTING - [metall] grillage *m* du minerai
~SAMPLING - [min] échantillonnage *m* du minerai
~SEPARATOR - [min] séparateur *m* à minerai
~SHOOT - [min] cheminée *f* de minerais
~SINTERING - [min] agglomération *f* des minerais
~SIZING - [min] triage *m* des minerais
~SORTING - [min] triage *m* de minerai
~STREAK - [min] bande *f* de minerai
~STAMP - [min] bocard *m* à minerai
~TREATMENT - [min] traitement *m* du minerai
~WAGON - [railw] wagon *m* à minerai
~WASHERY - [min] laverie *f* du minerai
~YARD - [gen] parc *m* à minerai
ORFORD PROCESS - [chem] (separation of copper from nickel by fusing the matter with sodium sulphide) procédé *m* à couches inférieures et supérieures, procédé *m* d'Orford
ORGAN - [gen zool bot etc] (any part of an organism) organe *m*

ORGAN - [mus] (musical wind instrument) orgue *m*
[comput] organe *m*
~PIPE SCALE - [mus] (the size of an organ pipe) facture *f*
~REGISTER - [mus] registre *m*
~TOSH - [rubber ind & mus] tissu *m* pour orgue, caoutchouc *m* pour soufflet d'orgue
ORGANDIE - s. organdy
~FINISH - [text] finissage *m* transparent
ORGANDY - [text] (crisp, transparent cotton muslin) organdi *m*
ORGANIC - [gen] (of organism; of the nature of organism) organique
[chem] (containing carbon as an essential ingredient) organique
[gen] (depending on structure) organisé, systématisé
~CARBON CYCLE - [chem] (the cycle of processes by which living organisms utilize the carbon of the carbon dioxide of the atmosphere in their metabolism) cycle *m* du carbon
~CHEMISTRY - [chem] (the chemistry of the compounds usually taken to exclude those in which that element is of minor importance, e.g. carbonates of the metals) chimie *f* organique
~COOLED REACTOR - [nucl] réacteur *m* à réfrigérant organique
~DISEASE - [med] (a functional disease) maladie *f* organique
~MANURE - [agric] engrais *mpl* organiques
~MATTER - [soil] matière *f* organique (du sol)
~MODERATED REACTOR - [nucl] (a reactor which is moderated and cooled by hydrocarbon fluid) réacteur *m* à moderateur organique
~MODERATOR - [nucl] modérateur *m* organique
~NON-SUGAR - [chem] non-sucre *m* organique
~PIGMENTS - [dyes] (pigments based on natural or synthetic compounds and, usually, inferior in lightfastness to inorganic pigments) pigments *mpl* organiques
~QUENCHED COUNTER TUBE - [nucl] tube *m* compteur à vapeur organique
~SOFTENER - [rubber ind] plastifiant *m* organique
~SOIL - [soil] sol *m* marécageux, sol *m* tourbeux
ORGANISM - [gen] organisme *m*
ORGANIZATION - [gen] organisation *f*
[gen] régime *m* (du travail)
~CHART - [gen] organigramme *m*
ORGANIZE, to - [gen] organiser
ORGANIZED LABOUR - [gen] organisations *fpl* ouvrières
ORGANOGENESIS - [med] organogénèse *f*
ORGANOMETALLIC COMPOUNDS - [chem] (compounds in which a carbon atom and a metallic atom are linked directly) composés *mpl* organométalliques
ORGANOSOL - [chem] (colloidal system in an organic dispersion medium) organosol *m*
ORGANZINE - [text] (fabric made of silk threads twisted together) organsin *m*
ORGASM - [med] orgasme *m*
ORIEL - [constr] (a bay window) fenêtre *f* en saillie
ORIENT, to - [gen] (to find one's position, with reference to the east) orienter, s'orienter
[surv] (the rotating of a map in an horizontal plane so that its lines corresponds to ground elements) orienter
ORIENT - [gen] orient *m*
ORIENT - [gen] (the luster of a pearl) étincelant, brillant
ORIENTAL - [gen & geogr] oriental
~CAT'S EYE - [min] cymophane *m*
~RUG - [text] (rug or carpet hand-woven in one piece) tapis *m* d'Orient
ORIENTATE, to - [gen etc] (to arrange in a specific direction) orienter, s'orienter
ORIENTATION - [gen] orientation *f*
[biol] (the position of an organ with relation to the whole) orientation *f*
[surv] (the operation of rotating a map or a plan in an horizontal plane until the corresponding lines on the ground) orientation *f*
[nucl] (the spatial arrangement of atoms in a molecule) orientation *f*
[text] (of fibres) orientation *f*
[chem] (the preference of certain covalent bonds to lie in particular directions in relation to the bonded atoms) orientation *f*, direction *f* de la valence
[comput] fixation *f* du temps
~BEACON - [aero] balise *f* d'orientation
~EFFECT - [nucl] (method of calculating the attractive forces between molecules) effet *m* d'orientation
[photo] effet *m* d'orientation
~OF FIBRES - [text] disposition *f* des fibres
ORIENTED ABSORPTION - [chem] absorption *f* orientée
ORIENTING - [gen] orientation *f*
[surv] s. orientation
[comput] (the allocation of absolute addresses to symbolic addresses) attribution *f* d'adresses
~CURVATURE - [biol] (reflex response to an external stimulus involving movements of the whole body) tropisme *m*
ORIFICE - [gen] orifice *m*, trou *m*
[hydr] (the mouth of a pipe) bouche *f*, embouchure *f*
~CAP - [th eng] (injector nozzle) tête *f* d'injecteur
~DISTORTION - [packing] distorsion *f* du gicleur
~HOOD - s. orifice cap
~IN THICK WALL - [hydr] gueule-bée *f*, ajutage *m* cylindrique
~LAND - [plast ind] (term used in the USA for die land, i.e. the terminal part of an extrusion die) partie *f* de filière à section constante
~METER - [instr] (instrument designed to measure the flow) débitmètre *m* à diaphragme
~METER PLATE - [instr] diaphragme *m* à mince paroi
~PLATE - [mech] (of a meter) diaphragme *m*
~PLATE FLOWMETER - s. orifice meter
~SPUD - [th eng] (injection nozzle) s. orifice cap
ORIFICING THE COOLANT - [nucl] réglage *m* par variation d'orifices
ORIGIN - [gen] origine *f*
~DISTORTION - [electron] (apparent loss of deflection sensitivity near the underflected position of the spot in gas-focused tubes) distorsion *f* d'origine, anomalie *f* du point zéro
~OF A FORCE - [phys] point *m* d'application d'une force
ORIGINAL - [gen] original
~ADDRESS - [comput] adresse *f* origine
~FISSION - [nucl] (the first reaction in a nuclear reactor) fission *f* primordiale
~MASTER - [el acoust] original *m*, père *m*
ORIGINATING EXCHANGE - [telecomm] (telephony) bureau *m* d'origine

O-RING - [aero] (the circular sealing ring between the wheel halves in an aircraft landing gear wheel) bague *f* à cachet du train d'atterrissage
ORIOLE - [zool] (a bird) loriot *m*
ORION - [astr] (a constellation) Orion *m*
ORLOP - [naut] (the lowest deck of a ship) faux-pont *m*
ORNAMENT - [gen] ornement *m*
[auto] enjoliveur *m*
ORNAMENTAL - [gen] ornamental
~CONCRETE - [constr] béton *m* ornamental
~HUB CAP - [auto] enjoliveur *m* de roue
ORNAMENTED FABRICS - [text] tissus *mpl* façonnés
ORNAMENTING BY SPECIAL THREADS - [text] ornamentation *f* par fils spéciaux
~BY STRIPES - [text] ornamentation *f* par rayures
ORNATE - [gen] orné, chamarré
ORNITHINE - [chem] (a non-essential amino acid) ornithine *f*
ORNITHOLOGY - [zool] (the part of zoology dealing with birds) ornithologie *f*
ORNITHOPTER - [aero] (a heavier-than-aircraft in which the greater part of the lifts is derived from flapping wings) ornithoptère *m*
ORNITHOSIS - [med] ornithose *f*
OROGENESIS - [geol] (a phase in the building of mountains in which the sediments are compressed) orogénèse *f*
OROGRAPHIC CLOUD - [met] (cloud due to condensation of moisture in an airstream deflected upward by mountains) nuage *m* orographique
~RAIN - [met] (rain precipitated from orographic clouds) pluie *f* orographique
OROGRAPHY - [geogr] (the study of the native and development of mountain ranges) orographie *f*
OROMETER - [instr] (mountain barometer) baromètre *m* orométrique, baromètre *m* altimétrique
ORPHENADRINE HYDROCHLORIDE - [chem] (mephenamine hydrochloride. A synthetic spasmolytic used in medicine) chlorhydrate *m* d'orphénadrine
ORPIMENT - [paint] (arsenic trisulphide, used as a pigment, and for masking the natural colour of shellac) orpiment *m*, orpin *m*
ORSAT ANALYZER - s. orsat apparatus
~APPARATUS - [ind chem] (portable apparatus employed in the analysis of flue and exhaust gases) appareil *m* d'Orsat
~GAS ANALYSIS APPARATUS - s. orsat apparatus
ORTHICON - [telev] (camera-tube in which the image is projected on to a mosaic) orthiconoscope *m*
ORTHO - s. orthochromatic
ORTHO- - [chem] (prefix denoting one of three isomeric forms of a di-substitution product derived from benzene) ortho-
ORTHOBARIC DENSITY - [phys] (the density of a liquid and of the satured vapour in equilibrium with it at any temperature) densité *f* orthobarique
ORTHOCENTRE - [geom] orthocentre *m*
ORTHOCHROMATIC - [photo] (photographic emulsions which are sensitive to all colours with the exception of red) orthochromatique
~REPRODUCTION - [photo] (the right reproduction of colour in monochrome) reproduction *f* orthochromatique
ORTHOCLASE - [min] (silicate of potassium and aluminium) orthose *m*, orthoclase *f*
ORTHODIAGRAPH - [instr] (the instrument used in orthodiagraphy) orthodiagraphe *m*
ORTHODIAGRAPHY - [radiat] (the exact recording of the size and form of organs inside the body) orthodiagraphie *f*, orthoradioscopie *f*
ORTHODROMIC - [surv] orthodromique
~MAP - [surv] carte *f* orthodromique
ORTHOFORMING - [ind chem] (reforming process) procédé *m* d'orthoforming
ORTHOGONAL - [geom] (having right angles) orthogonal
ORTHOGRAPH - [arch] (view showing an elevation of a building or part of it) orthographe *f*
ORTHOGRAPHIC - [arch] (of orthography) orthographique
~PROJECTION - [geom] orthogonal
[surv] (of a map) projection *f* orthogonale, coupe *f* perpendiculaire
ORTHOHYDROGEN - [chem] (hydrogen molecule in which the two nuclear spins are parallel) orthohydrogène *m*
ORTHOPAEDIC PAD - [med] coussin *m* orthopédique
ORTHOPAEDICS - [med] (the correction of deformities) orthopédie *f*
ORTHOPEDICS - s. orthopaedics
ORTHOPHORIA - [opt] (normal, position of the eyes) orthophorie *f*
ORTHOPHOSPHOROUS ACID - [chem] (an analytical reagent and reducing agent) acide *m* orthophosphorique
ORTHOPHOTIC EMULSION - [photo] orthophotique, emulsion *f* orthophotique
ORTHOPIA - [med] (correction of strabismus) orthopie *f*, correction *f* du strabisme
ORTHOPTER - s. ornithopter
ORTHORADIOSCOPY - [radiat] orthoradioscopie *f*
ORTHORHOMBIC - [cryst] (crystal system in which three crystal axes of different length are arranged at right angles to each other) orthorhombique
~SYSTEM - [cryst] système *m* orthorhombique
ORTHOSCOPIC SYSTEM - [opt] (corrected for distortion and spherical aberration) système *m* orthoscopique
~VIEW - [opt] (three-dimensional view in normal perspective) projection *f* orthoscopique
ORTHOSE - s. orthoclase
ORTHOTOMIC SYSTEM - [opt] (system containing only rays which can be cut at right angle by a correctly constructed surface) système *m* orthotomique
OSCHEOHYDROCELE - [med] hydrocèle *f* scrotale
OSCHEOMA - [med] tumeur *f* scrotale
OSCIDUCER - [electron] transducteur *m* à modulation de fréquence
OSCILLATE, to - [gen] (to swing to and from) osciller
[phys] (to complete a period of vibratory or periodic motion) osciller
OSCILLATING - [gen phys etc] oscillant
~ABSORBER - [nucl] absorbeur *m* oscillant
~AMPLIFIER - [radio] amplificateur *m* oscillant
~BEARING - [mech] support *m* oscillant
~CHUTE - [mech] (an oscillating conveyor channel) cheminée *f* oscillante, cheminée *f* à vibration
~CIRCUIT - s. oscillation circuit
~CONVEYING CHANNEL - [mech] gouttière *f* de transport oscillante
~CONVEYOR - [mech] transporteur *m* à secousses
~CRYSTAL METHOD - [cryst] (technique for the x-ray analysis of crystal) méthode *f* à cristal oscillant

OSCILLATING CYLINDER METHOD - [chem] (technique for measuring the viscosity of a gas or a liquid) méthode *f* à cylindre oscillant
~CYLINDRICAL VALVE - [auto] boisseau *m* rotatif
~DEBREEZING SCREEN - [impl] (used in coke screening) tamis *m* à secousses
~EXPANDER - [text] templet *m* mobile
~HOOK - [text] crochet *m* oscillant
~KLYSTRON - [electron] (a klystron in which the oscillation of the resonators is kept on by the coupling of the resonators by a loop) klystron *m* oscillateur
~KNIFE - [text] couteau *m* oscillant
~LEVER - [mech] levier *m* à bascule
~METER - [instr] (meter which registers the oscillations of a moving coil subjected to the action of a fixed coil) compteur *m* oscillant
~MILL - [mech] (used in rubber industry) broyeur *m* oscillant
~MOTION - [mech] mouvement *m* pendulaire
~PIVOT - [mech] pivot *m* va-et-vient
~QUANTITY - [phys] grandeur *f* oscillante
~RAIL - [text] vibrateur *m*, tringle *f* oscillante
~RECEPTION - [telecomm] réception *f* oscillante
~ROLLER - [text] cylindre *m* oscillant
~SCREEN - [impl] crible *m* à secousses, table *f* à secousses
~SHUTTER - [cin] (type of shutter sometimes used in cine-cameras) obturateur *m* oscillant
~SPRINKLER - [agric] arroseur *m* oscillant
~STRAINER - [impl] (used in paper manufacture) filtre *m* à secousses
~TABLE - [min] (implement of ore dressing plant) table *f* à secousses
~TEMPLE - [text] templet *m* mobile
~THREAD GUIDE ROD - [text] tringle *f* guide-fil oscillante
OSCILLATION - [gen] oscillation *f*, vibration *f*
[phys] (the single swing of an oscillating body between two extremes) oscillation *f*
[el] (an electric oscillation may occurr when a circuit is distributed from its condition of electrical equilibrium and a current flows alternately in the opposite direction) oscillation *f*
[radio] (in a resonant circuit, the generation of alternating currents) oscillation *f*
[mech] oscillation *f*, va-et-vient *m*
~ABSORBER - s. oscillation damper
~CIRCUIT - [el] (a circuit in which electrical oscillations occur freely) circuit *m* oscillant
~CONSTANT - [radio] (of a resonant circuit, the square root of the product of the inductance and the capacitance) constante *f* oscillatoire
~CURRENT - [el] (electric current which alternately reverses its direction in a circuit in a periodic manner) courant *m* oscillant
~DAMPER - [mech] amortisseur *m* d'oscillations
OSCILLATOR - [el] (a conductor with self-inductance, capacitance and resistance of such values that electric oscillations can be set up) oscillateur *m*
[radio] (non-rotating device for the production of alternating currents; its output frequency is determined by the characteristics of the device) oscillateur *m*, génératrice *f* de radiofréquences
~CALIBRATION - [el] tarage *m* de l'oscillateur
~CIRCUIT - s. oscillation circuit
~COIL - [el] bobine *f* d'oscillateur
OSCILLATOR DRIFT - [el] (the variation on oscillator frequency which is produced by physical changes) dérive *f* de fréquence
~HARMONIC INTERFERENCE - [radio] interférence *f* d'harmoniques de l'oscillateur local
~HARMONIC RESPONSE - [radio] réponse *f* d'harmoniques de l'oscillateur local
~STAGE - [el] étage *m* d'entrée
~STATE - [el] état *m* oscillatoire
~TRIODE - [electron] (in a superheterodyne circuit) oscillateur *m* local
OSCILLATORY - [gen phys etc] oscillatoire
~CIRCUIT - s. oscillation circuit
~CURRENT - s. oscillating current
~MOTION - [phys] (continuous motion confined to a given region of space) mouvement *m* oscillatoire
~SCANNING - [telev] analyse *f* oscillante
~SPIN - [aero] (a spin in which sustained oscillation occurs) vrille *f* oscillatoire
OSCILLOGRAM - [instr] (the graph produced by an oscillograph) oscillogramme *m*
OSCILLOGRAPH - [instr] (instrument designed to produce a graph representing a rapidly varying electrical quantity as a function of time) oscillographe *m*
~TEST WAGON - [railw] wagon *m* d'essai piezo-électrique
~TUBE - [electron] (cathode-ray tube in which the visible image is the graphical representation of a rapidly carrying electrical quantity) oscilloscope *m* cathodique.
~WITH BIFILAR SUSPENSION - [instr] (or bifilar oscillograph; oscillograph with a moving element which is constructed on the principle of a moving-coil galvanometer, and consists of two parallel strips) oscillographe *m* bifilaire
OSCILLOMETER - [instr] instrument designed to measure the angle of a vessel's roll or pitch) oscillomètre *m*
[med] (instrument designed to measure the oscillations of blood pressure) oscillomètre *m*
OSCILLOPSIA - [med] oscillopsia *f*
OSCILLOSCOPE - [instr] (a low-voltage cathode-ray oscillograph) oscilloscope *m*
~TUBE - s. oscillograph tube
OSCULATING - [geom] (of curves) osculation *f*
~CURVE - [geom] (touching of two curves) courbe *f* osculatrice
O-SHELL - [nucl] (groups of electrons which are characterized by the quantum number 5) couche *f* 0
OSIER STRAP - [constr] hart *m*, harre *m*
~-WILLOW - [bot] osier *m*, saule *m* des vanniers
OSMESTHESIA - [med] hypersensibilité *f* de l'odorat
OSMIC ACID - [chem] (a catalyst and a microscope stain) acide *m* osmique
OSMIRIDIUM - [min] (also called iridosmine. A native alloy of osmium and iridium which also contains platinum, ruthenium and rhodium) iridosmine *f*
OSMIUM - [chem] (hard, heavy metallic element, symbol Os, A.N. 76, A.W. I9I.0 used in the preparation of catalysts and in the production of alloys with platinum) osmium *m*
OSMOMETER - [instr] (an instrument for the measurement of osmotiç pressure) osmomètre *m*
OSMONDITE - [metall] (a steel structure which takes place after quenching) osmondite *f*
OSMOSE, to - [chem] soumettre à osmose
OSMOSE - s. osmosis

OSMOSIS - [chem] (the passage of liquids and solids through a semi-permeable membrane) osmose *f*
OSMOTIC - [chem] (of osmosis) osmotique
~CELL - [el chem] (a cell in which osmotic pressure can be observed by means of a semi-permeable partition) pile *f* osmotique
~COEFFICIENT - [phys] coefficient *m* osmotique
~PRESSURE - [chem] (the pressure exerted by the movement of a component through a semipermeable membrane) pression *f* osmotique
OSPHIALGIA - [med] douleur *f* lombaire
OSSEIN - [chem] (collagen forming the chief organic constituent of the bone) ostéine *f*, osséine *f*
OSSEOUS - [anat] (bony, of the bones) osseux
OSSICLE - [anat] (a small bone) osselet *m*
[anat] (of the ear) (one of the three small bones which are joined together by an elastic cartilage and connect the outer and inner membrane) osselet *m*
OSSIFICATION - [zool] (formation of bones) ossification *f*
OSTEANABROSIS - [med] atrophie *f* osseuse, dissolution *f* de l'os
OSTEITIS - [med] (bone inflammation) ostéite *f*
OSTEOACUSIS - [med] conduction *f* cranienne, conduction *f* osseuse
OSTEO-ARTHRITIS - [med] (form of chronic arthritis in which the bones adjacent to the cartilages of the joint wear away) ostéo-arthrite *f*
~-ARTHROPATHY - [med] (disease affecting bones and joints) ostéoarthropathie *f*
OSTEOBLAST - [biol] ostéoblaste *m*, ostéoplaste *m*
OSTEOCAMPSIA - [med] incurvation *f* d'un os
OSTEOCHONDRODYSTROPHIA - [med] ostéochonodrodystrophie *f*
OSTEOCHONDRODYSTROPHY - s. osteochondrodystrophia
OSTEOCHONDROLYSIS - [med] ostéochondrite *f* disséquante
OSTEOCOPE - [med] douleurs *fpl* ostéoscopes
OSTEODESMOSIS - [med] ossification *f* d'un tendon
OSTEODYNIA - [med] ostéalgie *f*
OSTEOFIBROMA - [med] fibrome *m* ossifiant
OSTEOGENESIS - s. ossification
OSTEOGENIC - [biol] (composed of tissue connected with the growth of bones) ostéogénique
OSTEOLIPOMA - [med] lipome *m* ossifiant
OSTEOLOGY - [zool] (the study of bones) ostéologie *f*
OSTEOMYELITIS - [med] (inflammation of the bone-marrow and of the bone) ostéomyélite *f*
OSTEONECROSIS - [med] nécrose *f* osseuse
OSTEOPATHY - [med] (a method of healing) ostéopathie *f*
OSTEOPERIOSTITIS - [med] ostéopériostite *f*
OSTEOPHONE - [instr] (hearing apparatus used in medicine making use of bone conduction) ostéophone *m*
OSTEOPHONY - [anat] s. osteoacusis
OSTEOPHYTOSIS - [med] formation *f* d'ostéophytes
OSTEOPOROSIS - [med] ostéoporose *f*
OSTEORRHAPHY - [med] suture *f* d'un os
OSTEOSPONGIOMA - [med] ostéome *m* spongieux
OSTEOTOMY - [med] ostéotomie *f*
OSTRICH - [zool] autruche *f*
OSTRICH YARN - [text] fil *m* à poil frisé, fil *m* autruche
OSTWALD DILUTION LAW - [chem] loi *f* de dilution d'Ostwald
OSTWALD U-TUBE - [instr] (instrument for determining viscosity, depending upon the flow of the liquid specimen from one leg of a U-tube to the other through a restricted passage) tube *m* en U d'Ostwald
OTHMER PROCESS - [ind chem] (commercial method of manufacturing acetic acid employing azeotropic distillation of pyroligneous acid) procédé *m* Othmer
OTITIS - [med] (inflammation of the ear) otite *f*
OTOLITH - [biol] otolithe *m*
OTOLOGY - [med] (the part of medicine and surgery dealing with the organs of hearing) otologie *f*
OTOPLASTY - [med] otoplastie *f*, plastique *f* auriculaire
OTOPOLYPUS - [med] polype *m* de l'oreille
OTORHINOLARYNGOLOGY - [med] oto-rhino-laryngologie *f*
OTOSCLERECTOMY - [med] ablation *f* des osselets
OTTER - [zool] (weasel-like fish eating animal) loutre *f*
~BOARDS - [naut] boards used to keep a trawl spread) paravane *m* otter
~PLUSH - [text] peluche *f* imitation loutre
~TRAWL - [naut] (in fishing boats) chalut *m* à paravanes otter
OTTO CYCLE - [mech] (four stroke cycle; in reciprocating internal combustion engines) cycle *m* à quatre temps
OUABAIN - [chem] (a poisonous glucoside obtained from Strophanthus gratus and from varieties of Acokanthera shimperi and used in the treatment of heart disease) strophantine *f*, ouabaïne *f*
OUDIN CURRENT - [el] (brush discharge which is dense enough to evaporate tissue-water without charring) courant *m* d'Oudin
~RESONATOR - [el] (a coil of wire designed to be connected to a source of high-frequency current, for the purpose of applying an effluve to a patient) résonateur *m* d'Oudin
OULONITIS - [med] pulpite *f*
OULORRHAGIA - [med] hémorragie *f* des gencives
OUNCE - [meas] (one sixteenth of one pound) once *f* (28,35 g.)
~METAL - [metall] (alloy of copper with one sixteenth of tin, zinc and lead) alliage *m* de cuivre avec étain, zinc et plomb
OUT, to - [gen] dehors, exposé
[gen] (of a fire etc) éteint
[gen] (of a book, a newspaper etc) paru
[el] (of a connection) hors circuit
[print] (words or letter omitted in setting) bourdon *m*
[as a prefix] hors, sur-
[naut] (of the tide) (marée) basse
OUT-EDIT - [telev] (in videon recording) point *m* de sortie
~OF ALIGNMENT - [aero] (term used of a propeller of which the sweep of one blade is different from that of the others) différence *f* angulaire du déport
~OF BALANCE - [gen] balourd *m*
[mech] asymétrique
[of wheels] décentré
~OF BALANCE CURRENTS - [el] courants *mpl* homopolaires
~OF BALANCE FORCE - [mech] force *f* asymétrique
~OF BREAKDOWN - [gen] dépannage
~OF CENTRE - [gen] décentré
~OF COMMISSION - [gen] hors service

OUT OF DATE - [gen] (of a ticket etc) suranné
~OF FOCUS - [photo & opt] flou
~OF FOCUS PICTURE - [photo cin etc] image *f* floue
~OF FRAME - [photo cin] (of a picture which is not correctly framed) décadré
~OF FRAME CONDITION - [cin] (incorrectly framed picture) décadrage *m*
~OF GAUGE - [railw] (of a railway truck) engagement *m* du gabarit
~OF GEAR - [mech] désengrené
~OF ORDER - [gen] dérégle, dérangé
~OF PHASE - [el] (said of periodic quantities having the same frequency and the same wave-form when the do not reach corresponding values at the same time) déphasé
~OF PITCH - [aero] (term used of a propeller of which the blade angle of one blade differs from that of the others at the same distance from the axis) décalage
~OF PLUMB - [gen & mech] hors d'aplomb, dévers
~-OF-RANGE NUMBER - [comput] (capacity exceeding number; called 'infinity' in the USA) nombre *m* dépassant la capacité
~OF RANGE VALUE - [math] valeur *f* hors de marge
~OF REPAIR - [mech etc] en mauvais état
~OF ROUND - [mech] (not truly circular or cylindrical) faux-rond *m*, à faux-rond
~-OF-ROUND DETECTOR - [tool] (used to control the roundness of a tyre) dispositif *m* de contrôle de la rondeur
~OF SQUARE - [gen] hors d'équerre
~OF STEP PROTECTION - [el] dispositif *m* de protection contre les ruptures de synchronisme
~OF TRACK - [aero] (term used of a propeller in which the blade tilt of one blade is not the same as that of the others) voile *f*
~-OF-TRUE - [mech] (distorted from the designed form) flambé, voilé, décentré
~-OF-TRUTH RIMSEAT - [rubber ind] (in tyres) dérivation *f* du siège du talon
~OF USE - [gen] hors d'usage, hors de service
~OF WORK - [gen] chômeur *m*, sans-travail *m*
~TAKE - [cin] (said of parts of a film which cannot be used) prise *f* de vue défectueuse
OUTAGE - [gen & of a machine] période *f* de repos
[el] durée *f* d'interruption
[gen] (wastage or loss) perte *f*
[mech] reniflard *m*
[nucl] (of a reactor) période *f* hors service
OUTBALANCE, to - [gen] emporter (sur)
OUTBAND - [build] (jamb stone laid as a stretcher) en parement
OUTBID, to - [gen & comm] enchérir, surenchérir
OUBID - [gen & comm] surenchère *f*
OUTBOARD - [mech] (adj) (located outside an aircraft) hors-bord *m*
~BEARING - [mech] palier *m* en porte-à-faux, palier *m* à chevalet isolé
~MOTOR - [naut] moteur *m* hors-bord
~STABILIZING FLOAT - [aero] flotteur *m* stabilisateur hors de bord
~VALVE - [mech] soupape *f* hors de bord
OUTBREAK - [gen] éruption *f*, sortie *f* violente
[min] affleurement *m* (d'un filon)
OUTBREEDING - [genet] outbreeding *m*
OUTBUILDING - [constr] (any structure forming part of a building, though generally separated from it) dépendences *f*pl
OUTBURST - [gen] éruption *f*, explosion *f*
[min] s. outbreak
~BANK - [hydr] (the middle part of an embankment) berge *f*
OUTCOME - [gen] issue *f*, résultat *m*, conséquence
OUTCROP, to - [geol & min] affleurer
OUTCROP - [min & geol] (exposure of rock or mineral on the surface) affleurement *m*, tranche *f* de couche
~MAP - [geol] carte *f* des affleurements
OUTDISTANCE, to - [gen] distancer, dépasser
OUTDOOR - [gen] extérieur, en plein air
~AERIAL - [radio] (any aerial placed outdoors) antenne *f* extérieure
~ANTENNA - s. outdoor aerial
~APPARATUS - [telecomm] (in telephony) appareil *m* d'extérieur
~CROP(PING) - [agric] culture *f* de pleine terre
~INSULATOR - [el] isolateur *m* à l'extérieur
~LIFE - [plast ind] (the length of time a plastic product will endure weather conditions, especially sunlight) durée *f* en plein air
~PHOTOGRAPHY - [photo] photographie *f* en plein air
~SUB-STATION - [telecomm] (telephony) poste *m* extérieur
OUTER - [gen] extérieur, externe
~ANGLE - [constr] angle *m* extérieur
~BARK - [bot] écorce *f* extérieure
~BARREL - [mining] tube *m* carottier extérieur
~BOARD - [constr] planche *f* avant la dosse *f*
~BREMSSTRAHLUNG - [radiat] (bremsstrahlung by which the energy loss by radiation appreciably exceeds loss by ionization) rayonnement *m* de freinage
~CAPSULE - [nucl] capsule *f* extérieure
~COATING - [electron] (conducting coating on the outside of the bulb of an electronic tube) revêtement *m* extérieur
~CONDUCTOR - [el] (in a two-wire system, external conductor; the conductor which is effectively earthed) conducteur *m* extérieur
~COVER - [aero] (of an airship, the external covering of hull structure) revêtement *m* extérieure
[auto] (of tyres) surface *f* de contact
~DEAD CENTRE - [mech] point *m* mort externe
~DIE - [plast ind] (the external part of an extrusion die) filière *f* externe
~DISTANCE SIGNAL - [railw] signal *m* de préavertissement
~DOOR CASE - [constr] huisserie *f* ou chassis *m* de porte *f* extérieure
~EDGE - [mech] (of a slide valve or cylinder-port) rebord *m* extérieure, arête *f* extérieure
~ELECTRODE - [electron] (electronic tube electron placed outside the bulb) électrode *f* extérieure
[phys] électron *m* périphérique
~ENVELOPE - [el] (of cables) revêtement *m* extérieur
~FLANK - [mech] aile *f* marchante
~FORME - [print] (the forme from which the outside of a sheet is printed) première forme *f*
~GRID - [electron] (the grid which is nearest to the wall of the bulb) grille *f* extérieure
~-GRID INJECTION - [electron] conversion *f* multiplicative
~HARBOUR - [naut] avant-port *m*
~HOME SIGNAL - [railw] carré *m* éloigné
~JIB - [naut] grand foc *m*

OUTER MARKER - s. outer marker beacon
~ MARKER BEACON - [aero] (an approach system beacon marking the first predetermined point on a beam approach) radiobalise *f* extérieure
~ ORBIT - [nucl] (the orbit which is farthest away from the nucleus) orbite *f* externe
~ PACK - [aero] (a bag used to receive the static line attached to the pack of a parachute) sac *m* extérieur
~ PLATING - [shipbuild] bordé *m* (en fer)
~ PRODUCT - [el] produit *m* vectoriel
~ SECTION - [aero] (the outer part of a main supporting surface) plan *m* principal, aile *f* voilure
~ SHELL ELECTRON - [electron] (electron belonging to the outer shell and concerned with light and conduction phenomena, and also in the chemical properties of the atom) électron *m* de conduction, électron *m* de valence, électron optique, électron *m* périphérique
~ STILL - [constr] appui *m* extérieur
~ SPACE - [astr] espace *m* interplanétaire
~ STACK - [metall] chemise *f* extérieure
~ STOP - [railw] (deferred stop) arrêt *m* différé
~ TUBE OF A TELESCOPE - [opt] tube *m* porte-objectif (de lunette)
~ TURBINE - [mech] turbine *f* d'arrière
~ WORK FUNCTION - [electron] (the energy gap between the crest of the potential barrier at the surface and the potential plateau) travail *m* externe
OUTERMOST ELECTRONIC SHELL - [phys] couche *f* électronique extérieure
OUTERS - [el] conducteurs *mpl* extérieurs
OUTERWEAR - [text] vêtement *m* de dessus
OUTFALL - [hydr] (in sewage, the discharge point of a sewer) issue *f*, dégorgeoir
[nucl] (pipeline or other construction conveying radioactive waste to the sea) égout *m* d'évacuation
~ DITCH - [hydr] émissaire *m*
~ DRAIN - [agric] fossé *m* collecteur
OUTFIT, to - [gen] équiper, armer
OUTFIT - [gen] équipement, appareillage *m*
[gen] (the various items of a particular equipment) équipement *m*
[naut] armement *m*
OUTFLOW - [phys] (the process of flowing out) écoulement *m*
[gen] (an outlet) écoulement *m*, dépense *f*, décharge *f*
OUTFLOWING - [oil ind] écoulement *m*
OUTGASSING - [ind chem] dégazage *m*
OUTGATE - [metall] (riser) masselotte *f*
OUTGOING - [gen] sortie *f*
[telecomm] (in telephony) de départ
~ CALL - [telecomm] communication *f* de départ
~ CONNECTION - [telecomm] liaison *f* aller
~ DELAY POSITION - [telecomm] (in telephony) position *f* de départ pour trafic différé
~ FEEDER - [el] (feeder along which power is supplied from a generating station) conducteur *m* de sortie
~ JUNCTION - [telecomm] (in telephony) ligne *f* auxiliaire sortante
~ LINE CIRCUIT - [comput] circuit *m* de ligne de départ
~ SIGNAL - [telecomm] (in radio transmission) signal *m* de départ
[telev etc] signal *m* de sortie
~ TRUNK - s. outgoing junction
OUTGOING TRUNK MULTIPLE - [telecomm] (in telephony) sectionnement *m* général
OUTGOINGS - [comm] dépenses *fpl*, débours *mpl*
OUTGROWTH - [agric] germination *f* sur les pieds
OUTHAUL - [naut] (rope for extending a sail along a spar) drisse *f*
OUTHOUSE - [gen] (outbuilding) dépendance *f*
[gen] (in the USA, a privy) cabinet *m*
OUTLAST, to - [gen] durer plus longuement, survivre
OUTLAY - [gen & comm] débours *mpl*
OUTLET - [gen & mech] orifice *m* d'émission, sortie *f*
[hydr] orifice *f* de décharge, égout *m*
[comm] débouché *m*
[hydr] dégorgeoir *m*, bonde *f*
[plumb] tubulure *f*
~ ANGLE - [mech] (of a fluid) angle *m* de sortie
~ BOX - [el] (box containing terminals) boîte *f* de prise de courant
~ CHANNEL - [hydr] rigole *f* de décharge
~ CONTROL WATER HEATER - [th eng] chauffe-bains *m* à pression
~ DITCH - [hydr] rigole *f* de colature, émissaire
~ EDGE - [mech] (of a turbine) bord *m* de sortie
~ END - [gen & mech] point *m* de sortie
~ HOPPER - [hydr] entonnoir *m* à trop-plein
~ HOSE - [mech] manche *f* de décharge
~ MALE THREAD - [mech] filetage *m* mâle, filet *m* extérieur
~ OPENING - [hydr] bonde *f*, orifice *m* de décharge
~ PLUG - [el] (portion of an outlet plug-and-socket intended for attachment to a cable and supplied with metallic contact-pins) tampon *m* de prise de courant
[mech] bouchon *m* de vidange
~ PLUG AND SOCKET - [el] (plug and socket intended for use at a supply point) tampon *m* et prise de courant
~ PRESSURE - [mech] (initial pressure in gas network) pression *f* de sortie
[rubber & plast ind] (of an extruder) pression *f* de sortie (de boudineuse)
~ RILL - [agric] rigole *f* de colature
~ STATION PRESSURE - s. outlet pressure
~ SYNDROME - [med] compression *f* du plexus brachial
~ VALVE - [hydr] robinet *m* d'écoulement, robinet *m* de vidange
~-VENT - [hydr] tuyau *m* d'évent, tube *m* aérateur
~ WEIR - [hydr] déversoir *m*
OUTLIER - [geol] témoin *m*
OUTLIERS - [statist] observations *f* extrêmes aberrantes
OUTLINE, to - [gen & draw] tracer, esquisser, contourner
OUTLINE - [gen] contour *m*, profil *m*
[draw] tracé *m*, dessin *m* au trait
[gen] (of a plan) gabarit *m*, aperçu *m* d'un projet
[constr] modénature *f*
~ SKETCH - [text] contour *m* de l'esquisse
OUTPATIENT - [med] malade *m* ambulatoire
OUTPORT - [constr] lumière *f* de sortie
OUTPOURING OF LATEX - [rubber ind] écoulement *m* du latex
OUTPUT - [gen] (the amount of a substance produced by a process) production *f*, rendement *m*, débit *m*
[mech] (of machines) rendement *m*
[mech] (the energy or power developed by a machine) rendement *m*, puissance *f* développée

OUTPUT - [el] (the power given out by a plant) puissance *f* fournie
[mech] (of a pump) refoulement *m*
[comput] (the information transferred from the internal to an external storage) sortie *f*
[med] excrétion *f*, débit *m*
[contr] grandeur *f* de sortie
~AMPLIFIER - [electron] amplificateur *m* de sortie, amplificateur *m* final
~AND ACCOUNTING - [comm] calcul *m* des frais de production
~AREA - [comput] enregistrement *m* groupé, zone *f* de sortie
~AT THE DRAWBAR - [mech] puissance *f* au crochet
~AT THE WHEEL RIM - [mech] puissance *f* à la jante
~AXIS - [el] axe *m* de sortie
~BLOCK - [comput] (section of internal storage designed for receiving data to be transferred to an external storage) zone *f* de sortie
~BRUSHES - [el] balais *mpl* principaux
~BUFFER STORAGE - [comput] mémoire *f* tampon de sortie
~CAPACITANCE - [electron] (of an electronic tube) capacité *f* de sortie
~CAPACITY - [mech etc] (yield capacity) rendement *m*
~CIRCUIT - [el] circuit *m* de sortie
[electron] (of an electronic tube; the external circuit connected to the output electrode to provide the load impedance) circuit *m* de charge
~CODE TRANSLATOR - [comput] traducteur *m* de code de sortie
~COEFFICIENT - [el] (a specific torque coefficient) coefficient *m* de couple
~CURVE - [gas ind etc] (in the calculation of mains; a consumption or load curve) courbe *f* de charge
~DAMPING - [telecomm] amortissement *m* de sortie
~DATA - [comput] données *f* de sortie
~ELECTRODE - [electron] (the electrode from which the amplified, modulated etc voltage is received) électrode *f* de sortie
~ELEMENT - [comput] unité *f* de sortie
~EQUIPMENT - [comput] (the equipment which is employed to transfer information) équipement *m* de sortie
~FILTER - [radio] (filter connected to a wired radio channel amplifier to eliminate interference between parallel channels) filtre *m* de sortie
~GAP - [electron] (interaction gap by which power can be abstracted from an electron system) espace *m* de capture
~IMPEDANCE - [el] (the impedance presented to the load) impédance *f* de sortie
~INSTRUCTION - [comput] instruction *f* de sortie
~METER - [instr] (instrument designed to measure the output of audio amplifiers) outputmètre *m*
~MONITOR - [radio & telev] (monitor controlling the transmission from the studio to the transmitter) moniteur *m* de sortie
~NOISE RATIO - [electron] rapport *m* de température de bruit
~NOISE VOLTAGE - [radio] tension *f* de température de bruit
~PENTODE - [electron] (used as an output electronic tube) pentode *f* finale
~PER DAY - [ind proc] rendement par jour

OUTPUT PER HOUR - [ind proc] rendement *m* à l'heure
[el] puissance *f* fournie par heure, débit *m* par heure
~PER SHIFT - [ind proc] rendement *m* par équipe
~POWER - [electron] (of an electronic tube) puissance *f* de sortie
[contr] enroulement *m* de sortie
~POWER/FREQUENCY CHARACTERISTIC - [radio] (bandwidth characteristic) caractéristique *f* de l'onde passante, caractéristique *f* puissance de sortie/fréquence
~POWER/TUNING CHARACTERISTIC - [radio] caractéristique *f* puissance de sortie/syntonisation
~PULSE - [instr] (in the instrumentation for nuclear reactor, a pulse in a scaler when a given number of input pulses has been received) impulsion *f* de sortie
~QUANTITY - [contr] (the physical quantity, the value of which is determined by the operation of each element of a system) grandeur *f* de sortie
~RESONATOR - [electron] (a catcher; resonant cavity supplying energy to an external circuit) résonateur *m* de sortie
~ROUTINE - [comput] (a library programme used to prepare ready-use tables) programme *m* de sortie
~SHAFT - [auto] (in the gear box) arbre *m* secondaire
~SIGNAL - [contr] (the signal which is transmitted from an element to the next one in the loop) signal *m* de sortie
~STAGE - [radio] étage *m* final
~-TRANSFER FUNCTION - [contr] fonction *f* de transfert de sortie
~TRANSFORMER - [el] (transformer coupling the last stage in a valve amplifier with a loudspeaker or a line) transformateur *m* de sortie, transformateur *m* de modulation
~TERMINALS - [electron] bornes *fpl* de sortie
~TRIODE - [electron] (triode used as an output electronic tube) triode *f* de sortie
~UNIT - [comput] (the unit delivering information) unité *f* de sortie
~VALVE - [radio] (valve designed to deliver a comparatively large amount of alternating-current power, instead of being merely a voltage amplifier) tube *m* de sortie
~VARIABLE - [contr] grandeur *f* de sortie
~WINDINGS - [radio] (of a saturable reactor) enroulements *mpl* de sortie
~WORK QUEUE - [contr] file *f* de travaux de sortie
OUTRIGGER - [gen & mech] (part projecting beyond a given line) arc-boutant *m*, espar *m* en saillie
[mech] palier *m* extérieur
[naut] (the bracket projecting from the side of a narrow rowboat) outrigger *m*
[naut] (a projecting boat-like float braced to the side of a canoe to prevent capsizing) balancier *m*
[aero] (projecting contrivance to support the lifting planes) consoles *fpl*
~WHEEL - [aero] (wheel in a landing gear located under the wing and separate from the main gear) roue *f* extérieure
OUTRIGGERS - [auto] pattes *fpl* transversales, supports *mpl* en équerre
OUTSIDE - [gen] (adj) extérieur, externe, à l'extérieur
~AERIAL - s. outdoor aerial
~BROADCAST - [radio] (programme which does not

originate in a studio; generally, the broadcast report of an event) radiodiffusion *f* en extérieurs
OUTSIDE CALIPER - [instr] compas *m* d'épaisseur
~CIRCLE - [mech] (of a gear) cercle *m* extérieur, cercle *m* de couronne
~CLEARANCE - [mech] (of a slide valve) découvert *m* extérieur, recouvrement *m* extérieur négatif
~CYLINDERS - [mech] (of a locomotive; the steam cylinders carried outside the frame of a locomotive) cylindres *mpl* extérieurs
~DIAMETER - [mech] diamètre *m* extérieur
~GEAR - [mech] engrenage *m* extérieur
~GOUGE - [carp] (gouge with the bevel ground on the convex side of the cutting edge) gouge *f*
~HELIX ANGLE - [mech] (of a gear) angle *m* de l'hélice sur le cercle extérieur
~LAP - [mech] (of a slide valve) recouvrement *m* à l'admission
~PLATING - [shipbuil] bordé *m* en fer
~PRODUCER - [gas ind] gazogène *m* indépendant
~ROLL - [aero] (a roll in which the aircraft is both initially and finally in an inversed position) tonneau *m* renversé
~SCREW - [mech] vis *f* pleine, vis *f* mâle
~STRING - [constr] (of stairs) limon *m*
~TAPPET - [text] excentrique *m* extérieur
~TEMPERATURE - [met] température *f* externe
~THREAD - [mech] filet *m* extérieur
~TREADING MOTION - [text mach] mouvement *m* de marches extérieures
~WINDOW - [constr] contre-fenêtre *f*
OUTSIDES - [paper man] (the top and bottom quires of the I8 quires of a ream of paper) feuilles *fpl* de garde extérieures
OUTSKIRTS - [gen] limites *fpl*, lisière *f*, périphérie *f*
[gen] (of a town) banlieue *f*, faubourgs *mpl*
OUTSOLE - [shoe man] semelle *f* extérieure
~STOCK - [rubber ind] mélange *m* pour semelles
OUTSTANDING - [gen] en suspense, en retard
[comm] impayé, dû, arrière
~FEATURE - [gen] caractéristique *f* principale
~QUALITY - [gen] qualité *f* supérieure
OUTSTEP WELL- [oil ind] (a well drilled beyond the proved limits of a producing field, to ascertain whether the oil accumulation is extending) puits *m* d'extension
OUTSTROKE - [mech] (outward stroke; as the thrust of a piston toward the crankshaft) course *f* aller, course *f* avant
OUTTURN - [gen] (quantity and quality of goods produced) production *f*
OUTWARD CARGO - [naut] cargaison *f* d'aller
~FLOW TURBINE - [mech] turbine *f* centrifuge
~JOURNEY - [railw] voyage *m* d'aller
OUTWASH PLAIN - [geol] plaine *f* de lavage superficiel
OUTWEAR, to - s. outlast, to
OUTWEIGH, to - [gen] (to weigh more than, to be worth more than) dépasser en poinds, avoir plus d'influence
OVAL - [gen & geom] ovale
[metall] (short for oval bar) barre *f* ovale
[bot] (of leaves; flat at each end and with curved sides) ovalaire
~CASING SWAGE - [oil ind] redresseur *m* olive
~CATHODE - [electron] cathode *f* ovale
OVAL COMPASS - [instr] compas *m* d'ellipse, compas *m* elliptique
GRID - [electron] grille *f* ovale
~-HEAD SCREW - [mech] vis *f* à tête ronde
~IRON - [metall] ovales *mpl*, fers *mpl* à olive
~PISTON - [mech] (piston, which was originally round, worn to an oval shape through friction) piston *m* ovalisé
~-SHAPED TRACK - [telev] (in video recording) ovalisation *f* de la piste
~SLOTTED HEAD - [mech] tête *f* ronde fendue
OVALITY - [gen & mech] (the quality of being oval rather than cylindrical, as in worn cylinder bore) ovalité *f*
OVALIZATION - [mech etc] (becoming oval through wear etc) ovalisation *f*
OVALIZE, to - [gen & mech] ovaliser
OVALIZED - [mech] (worn oval through friction at the thrust faces) ovalisé
OVARIAN - [anat] (relating to the ovary) ovarien
OVARIOCELE - [biol] ovariocèle *f*
OVARIOCYESIS - [med] grossesse *f* ovarienne
OVARIORRHEXIS - [med] rupture *f* de l'ovaire
OVARIOTOMY - [med] ovariotomie *f*
OVARITIS - [med] (inflammation of the ovary) ovarite *f*, oophorite *f*
OVARY - [anat] (reproductive gland producing ova) ovaire *f*
OVEN - [gen] (insulated chamber fitted with a door and supplied with means of heating) four *m*
[electron] thermostat *m*
~BOTTOM - [th eng] (in the USA; base plate in G.B.) tôle *f* de fond
[metall] (the lining base) sole *f*
~-CURING - [ind chem] (oven polymerization etc) polymerisation *f* à chaud
-DRY, to - [metall etc] sécher au four
~-DRY - [ind proc] séché au four
~-DRY WEIGHT - [meas] (constant weight value of a fibre or yarn obtained by drying at a temperature of I05-II0 deg. C) poids *m* sec absolu
~DRYING - [ind proc] séchage *m* au four, séchage *m* à l'étuve
~FURNACE - [metall] (of the hearth type) four *m* à sole
OVER - [gen] sur, dessus
[gen] (a length of time) au cours de, pendant
(as a prefix or attached to another word) au-dessus de, plus de
~-AND-UNDER CURRENT DELAY - [el] relais *m* à maximum et à minimum
~-CAR AERIAL - [radio] (aerial on the proof of a car) antenne *f* sur le toit de la voiture
~-CAR ANTENNA - s. over-car aerial
~-COMPOUND EXCITATION - [el] (of a generator; a cumulative compound excitation with series winding so arranged that the potential difference at the terminals of the machine increases with the load) excitation *f* hypercompound
~-COMPOUNDED - [el] (said of a compound-wound generator, to denote that the series winding is so proportioned that the voltage increases with the load) hypercompoundé
~-COMPOUNDED GENERATOR - [el] générateur *m* à excitation hypercompoundée
~-THE-HORIZON LINK - [comput] liaison *f* transhorizon

OVER-THE-TOP FLYING - [aero] (flying above a cloud formation) vol *m* sur les nuages
OVERADVANCE TIMING - [auto] avance *f* à l'allumage excessive
OVERAGEING - [metall] (to prolong excessively the time of ageing temperature) survieillissement *m*
OVERALL - [gen] hors tout
[adj] complète, total
(of dimensions) encombrement *m*
[naut] hors tout
~AMPLIFICATION - s. overall gain
~ATTENUATION - [radio] atténuation *f* générale
~ATTENUATION LOSS - [radio] affaiblissement *m* composite
~BRIGHTNESS-TRANSFER CHARACTERISTIC - [telev] (the ratio of the brightness of a scene to the brightness of the image) contraste *m* des luminosités
~BRILLIANCE-TRANSFER CHARACTERISTIC - s.overall brightness-transfer characteristic
~COEFFICIENT OF HEAT TRANSFER - [phys] coefficient *m* total de transmission de chaleur
~CONTRAST RATIO - [telev] (ratio between the brightness of the light and the dark areas of a television screen) rapport *m* global de contraste
~DIAMETER - [mech] diamètre *m* externe
~DIMENSION - [meas] (a dimension taken over the extreme limits of a machine, especially to determine the space it will occupy) dimensions *f*pl d'encombrement
~EFFICIENCY - [el] (the ratio of the output from the final item of a plant to the input to the first) débit *m* total
[gen] efficacité *f* totale
~ENRICHMENT PER STAGE - [nucl] (the percentage of isotope enrichment in one stage of the separation process) facteur *m* d'enrichissement par étage
~ERROR - [meas] erreur *f* totale
~GAIN - [telecomm] (the total gain expressed in decibels between the ends of a line) gain *m* total, équivalent *m* d'un circuit
~GAMMA - [telev] gamma *m*
~HEIGHT - [meas] (height taken between the extreme limits of a part or unit) hauteur *f* d'encombrement
~LENGTH - [meas] (length taken between the extreme limits of a part or unit) longueur *f* d'encombrement
~LOSS - [telecomm] (the total loss expressed in decibels between the ends of a line) atténuation *f* générale
~NOISE FACTOR - [radio] facteur *m* moyen de bruit
~NOISE FIGURE - [electron] facteur *m* de bruit total
~THRUST - [astronaut] poussée *f* totale
~TRANSADMITTANCE - [electron] transadmittance *f* totale
~TREATMENT TIME - [radiat] durée *f* de traitement totale
~TYRE WIDTH - [auto] grosseur *f* totale du pneu, largeur *f* du pneu
~WIDTH - [meas] (width taken between the extreme limits of a part or unit) largeur *f* hors tout
OVERALLS - [clothing] (coarse trousers worn over a pair of trousers) vêtement *m* couvre-tout, combinaison *f*
OVERARM - [mach tool] chariot *m* porte-outil
OVERBENDING - [metall] dépassement *m* de l'angle de flexion, tolérance *f* du retrait élastique
OVERBLOW, to - [gen] s'épanouir
[metall] suroxyder
OVERBLOWING - [metall] suroxydation *f*
[mus] (production of overtones by forcing the wind in a pipe) quintage *m*
OVERBOARD - [naut] hors du bord
[aero] (out of an aircraft, e.g. "fuel dumped overboard") hors de bord
OVERBOILING THE THREADS - [text] cuisson *f* exagérée des fils
OVERBOOT - s. overshoe
OVERBRIDGE - [road & railw works] passage *m* en dessus
OVERBUNCHING - [electron] (the bunching condition resulting from the continuation of the bunching process over the optimum condition) groupement *m* excessif
OVERBURDEN, to - [gen] surcharger
OVERBURDEN - [geol] (verlying stratum of soil in brick-fields) terrain *m* de couverture
[soil] (top soil) sol *m* superficiel
OVERCARBONIZING - [fuel] surcuisson *f* du coke
OVERCAST, to - [met] (to become covered with clouds) se couvrir
[sewing & bookbind] (to sew the edge with long wrapping stitches) surfiler, surjeter
OVERCAST - [met] (weather condition when the sky is obscured by haze or cloud) couvert
~STITCH - [text] (a stitch used in embroidery) point *m* de surjet
OVERCASTING - [sewing & bookbind] surjet *m*
OVERCHARGE, to - [gen] surcharger
OVERCHARGE - [gen] surcharge *f*
[comm] prix *m* excessif
OVERCHARGING THE LOOM - [text] surcharge *f* du métier
OVERCHOKE, to - [auto] étrangler excessivement
OVERCOAT - [text] pardessus *m*, paletot *m*
OVERCOATING - [text] (heavy cloth) étoffe *f* pour paletot
OVERCOLOUR, to - [photo] surcolorier
OVERCOME, to - [gen] surmonter
OVERCOMPACTION - [soil] surcompactage *m*
OVERCOOLED - [mech] (of an engine) refroidi excessivement
OVERCORRECTION - [med] surcorrection *f*
OVERCOUPLING - [comput] couplage *m* serré
OVERCURE, to - (to vulcanize rubber to an excessive degree) survulcaniser
OVERCURING - [rubber & plast ind] (excessive vulcanization of rubber, or curing of a thermosetting resin, with resultant degradation of physical properties) survulcanisation *f*, surcuisson *f*
OVERCURRENT - [el] (abnormal current greater than the full load) surintensité *f* de courant
~CLASS OF A CURRENT TRANSFORMER - [el] (the group in which the ratio between the rated short-circuit and the rated primary current has the same value) classe *f* de surintensité d'un transformateur de courant
~FACTOR - [el] indice *m* de surcharge
~RELAY - [el] relais *m* à intensité maximum
OVERCUT - [el acoust] déraillage *m*
OVERCUTTING - [el acoust] (in disk recording) gravure *f* excessive
OVERDAMPING - [instr] (periodic damping having a degree greater than that required for critical damping) amortissement *m* excessif

OVERDEVELOPED - [photo] (said of a picture which, owing to faulty development, is short of white) surdéveloppé
OVERDOOR - [build] (ornamental door-head) dessus *m* de porte
OVERDOSAGE - [radiat] surdosage *m*
OVERDRIVE - [auto] (or overgear, i.e. a gear for a speed which is higher than that of the engine) vitesse *f* surmultipliée, overdrive *m*, surmultiplication *f*
AXLE - [auto] pont *m* comportant une surmultiplication
OVEREXCITATION - [el] surexcitation *f*
OVEREXPOSE, to - [photo] (to expose for too long, thus losing gradation in the developed picture) surexposer
OVEREXPOSED - [photo] surexposé
OVEREXPOSURE - [photo] (prolongued exposure resulting in loss of gradation in the developed picture) surexposition *f*, excès *m* de pose
OVERFALL - [gen] raz *m* de courant causé par les hauts fonds
[hydr] (catch basin) déversoir *m*, dégorgeoir *m*
~DISCHARGE - [hydr] ouverture *f* de dégorgeoir
~ORIFICE - [hydr] bouche *f* à déversoir
~WEIR - [hydr] barrage *m* déversoir
OVERFEEDING - [agric etc] suralimentation *f*
OVERFILL - [metall] (a defect in rolling) bavure *f* de lamination
OVERFLOW, to - [gen] déborder (de)
[hydr] inonder, dégorger
[gen] (of river) sortir du lit
OVERFLOW - [hydr] (excess liquid escaping from a vessel) débordement *m*, épanchement *m*
[gen] inondation *f*
[gen] (outlet) dégorgeoir *m*
[hydr] (pipe outlet) trop-plein *m*
[comput] (the production of a number which is beyond the capacity of the counter) dépassement *m* de capacité
[min] produit *m* d'affleurement
[radio] (in voice-frequency signalling) débordement *m*
~CAVITY - [mech] chambre *f* à déversage
~CHANNEL - [rubber ind] (channel for excess rubber) couloir *m* de trop-plein
~COCK - [hydr] soupape *f* de trop-plein
~CONDUIT - [hydr] conduite *f* de trop-plein
~CONE - [rubber ind] embout *m*
~CONTROL INDICATOR - [comput] indicateur *m* de changement de formulaire
~DAM - [hydr] (or drowned dam) barrage-déversoir *m*
~LIQUID - [hydr & gen] liquide *m* du trop-plein
~METER - [telecomm] (in telephony, a meter for the number of calls which fail to go through because of lack of outlets) indicateur *m* de surcharge, compteur *m* à dépassement
~OF WHITES - [telev] surintensité *f* des blancs
~OPENING - [mech] trou *m* de trop-plein
~PIPE - [hydr] (pipe used to carry off liquid overflowing from a vessel) tuyau *m* de trop-plein
~PLUG - [mech] bouchon *m* du trop-plein
~PROGRAMME TRANSFER - [comput] transfert *m* du programme en cas de débordement
~REPEAT INTERLOCK - [comput] reprise *f* du programme après l'impression d'un total de report
~REPERFORATOR - [comput] reperforatrice *f* de débordement
OVERFLOW SPRING - [hydr] source *f* d'émergence, source *f* de trop-plein
~SWITCH - [comput] commutateur *m* de débordement
~TRAP - [mech] bouchon *m* de trop-plein
~TRIM KNIFE - [tool] (a cutting device to remove excess material) couteau *m* à ébavurer
~VALVE - [mech] (a valve used to allow excess liquid to escape from a vessel) soupape *f* de trop-plein, clapet *m* de trop-plein
~WATER - [hydr] eau *f* d'écoulement
~WATER-BOX - [instr] (a water-box in meters) réservoir *m* de trop-plein
~WEIR - s. overflow dam
~WELL - [metall] (in die casting) talon *m* de lavage, poche *f*, masselotte *f*
OVERFLOWING - [gen & hydr] débordant
OVERFLUX RELAY - [el] (a sensitive relay for releasing the safety-rid latches and shutting down the reactor) relais *m* de sécurité
[nucl] relais *m* déverrouillage
OVERFOLDING - [geol] recouvrement *m*
OVERFOOTAGE - [cin] longueur *f* excessive
OVERFULL - [gen] trop plein
OVERGEAR - s. overdrive
OVERGLAZE COLOURS - [ceramics] colorants *mpl* pour émail
~DECORATION - [ceramics] (ornamentation applied to biscuit after glazing) décoration *f* sur émail
OVERGRAZING - [agric] surpâturage *m*, surgrattage *m*
OVERGROUND CONDUCTOR - [el] conducteur *m* aérien de contact
OVERGROWN - [gen] grandi trop vite
[agric] envahi, recouvert
~MALT - [agric] hussards *mpl*
~WITH HERBS - [agric] envahi par les mauvaises herbes
OVERGROWTH - [gen] surcroissance *f*
[cryst] (the growth of a crystal of one substance on the surface of a crystal of another substance) supercroissance *f*
OVERHAND - s. overcast (sewing)
~KNOT - [naut] demi-noeud *m*
~STOPES - [min] gradins *mpl* renversés, maintenages *mpl*
~STOPING - [min] abattage *m* montant, abattage *m* en gradins renversés
OVERHANG, to - [gen] (to hang over something) surplomber, porter à faux pencher sur
[arch] (to build a projection on a structure) faire saillie (au dessus de)
OVERHANG - [aero] (the distance between a wing tip and its outermost point of support) longueur *f* en porte-à-faux
[aero] (half the difference in span between the upper and lower planes of a biplane) décalage *m* latéral
OVERHANGING - [gen] en surplomb, en porte-à-faux, surplombement *m*
[arch] (of a projecting structure) déversé
~ARM - [mach tool] support *m* porte-fraise
~BEAM - [constr] poutre *f* en porte-à-faux, poutre *f* encastrée à une extremité et libre à l'autre
~CRANK - [mech] manivelle *f* en porte-à-faux
~FACE - [mining] front *m* de taille en surplomb
~FIRE-BOX - [railw] (of a locomotive) foyer *m* en porte-à-faux
~GIRDER - s. overhanging beam

OVERHANGING JOINT - [railw] (suspended joint of rail) joint *m* de rail en porte-à-faux
~PULLEY - [mech] poulie *f* en porte-à-faux
~PUMP - [mech] pompe *f* en porte-à-faux
OVERHAUL, to - [mech] (to dismantle and examine a machine in detail) examiner en detail, reviser, remettre en état
[gen] (to overtake) dépasser, gagner de vitesse
[naut] (to slacken a rope by pulling in the opposite direction) affaler
OVERHAUL - [mach] (the operation of examining a machine in detail) révision *f*, visite *f*
~LIFE - [mech] (of motors, machines etc) durée *f* de temps entre une révision et la suivante
~MANUAL - [gen] (a book of instructions for carrying out overhauls) manuel *m* de remise
OVERHAULING - s. overhaul
~POINT - [railw] point *m* de dépassement
OVERHEAD - [gen] (placed above) aérien, surélevé, en surplomb
[gen & comm] (frais) généraux, forfataire
[mech] (of valves expenses) dépenses *fpl* générales
[mech] (working above or aloft, as a crane etc) aérien, roulant
[ind chem] (the total vapours leaving the top of a column) vapeur *f* de tête
[comput] (colloq. USA; computer operations which do not directly contribute to solving the problem) organisation *f*
[oil ind] flux *m* de tête
[mining] en gradins renversés
~BEARERS - [constr] (steel members supporting the lift equipment on top of the lift well) supports *mpl* de ligne aérienne
~CABLE - [el] (of an overhead line) câble *m* aérien
~CAMSHAFT - [mech] (type of camshaft carried in bearings on the cylinder heads) arbre *m* à cames en tête
~CLEARANCE - [transp & railw] hauteur *f* libre
~COMPASS - [instr] boussole *f* renversée
~CONE PULLEY - [mech] cône *m* de renvoi, contre-cône *m*
~CONDUCTOR RAIL - [el] rail *m* aérien de contact
~CONTACT SYSTEM - [el] (system of supplying electric power to a vehicle by one or more overhead conductors, the contact being maintained by current collectors mounted on top of the vehicle) système *m* à ligne de contact aérienne
~CONVEYOR - [transp] transporteur *m* aérien
~COOLING SYSTEM - [oil ind] système *m* de refroidissement des produits de tête
~CRANE - [mech] pont *m* roulant, grue *f* roulante aérienne
~CROSSING - [el] (device at the crossing of two contact-wires for the passage of current-collectors along either wire) croisement *m* aérien, conduite *f* aérienne
[rail] aiguillage *m* croisé
~DRIVING MOTION - [mech] transmission *f* secondaire à placer en l'air
~ELECTRIC SYSTEM - [el] réseau *m* électrique aérien
~ENGINE - [mech] machine-pilon *f*
~EXPENSES - [comm] dépenses *fpl* générales
~FEEDER - s. overhead line
~GROUND WIRE - [el] fil *m* aérien de terre
~INLET VALVE - [mech] (an inlet valve placed in the cylinder head and opening downwards) soupape *f* d'admission en tête
OVERHEAD IRRIGATION - [agric] arrosage *m* en pluie, aspersion *f*
~JUNCTION - [el] aiguillage *m* tangentiel
~LIGHT - [light] éclairage *m* vertical
~LINE - [el] (electric line above ground, generally with the conductors supported on separate insulators) ligne *f* aérienne
~MONORAIL CONVEYOR - [transp] (used in workshops, stores etc) transporteur *m* aérien à monorail
~MOTION - s. overhead driving motion -
~POSITION - [metall](in welding) position *f* pour soudage au plafond
~PRICE - [comm] prix *m* forfataire
~PRODUCT - [ind chem] (in distillation) produit *m* de tête
~RAILWAY - [railw] chemin *m* de fer suspendu
~RUNWAY - [transp] monorail *m* aérien
~SHAFTING - [mech] transmission *f* à arbres suspendu
~STOPING - [mining] abattage *m* montant
~TANK - [mech] réservoir *m* surélevé
~TELEPHONE SYSTEM - [telecomm] réseau *m* téléphonique aérien
~TRACK - [transp] voie *f* aérienne
~TRANSMISSION LINE - s. overhead line
~TRANSMISSION SYSTEM - [el] canalisation *f* aérienne
~TRANSPORT - [transp] transport *m* par trolley, telphérage
~TRAVELLING CRANE - [mech mach] (workshop crane consisting of a girder along which a wheeled crab can be traversed) pont *m* roulant, grue *f* roulante aérienne
~UNWINDING - [text] déroulage *m* à la défilée
~VALVE - [mech] (a valve in an i.c. engine which is located in the cylinder head and opens downward) soupape *f* en tête
~WIRE - [el] fil *m* aérien
OVERHEAT, to - [gen & heat] surchauffer
OVERHEATED - [heat] surchauffé
[metall] (denoting a metal which has been heated in preparation for hot-working, to a temperature at which large grains are produced) surchauffé
OVERHEATING - [gen heat mech etc] surchauffage *m*, surchauffe *f*
[chem] (raising the temperature of a liquid above its boiling point) surchauffage *m*
OVERHOISTING - [mining] mise *f* aux molettes
OVERHUNG CRANK - [mech] manivelle *f* en porte-à-faux
~PULLEY - s. overhang pulley
OVERINFLATION OF TYRES - [auto] surpression *f*
OVERKNEE BOOT - [shoe man] (hip boot) botte *f* cuissarde
OVERLAND - [gen]de terre, par voie de terre
OVERLAP, to - [gen] chevaucher, recouvrir
[gen] (to cause to fold over on something) chevaucher
OVERLAP - [gen] recouvrement *m*, chevauchement *m*
[constr] (the part which overlaps) chevauchure *f*, imbrication *f*
[geol] (extension of younger rock strata beyond the limits of underlying strata) recouvrement
[mech] (the overlapping of gears) chevauchement *m*
[metall] (in welding) repliure *f* de laminage

OVERLAP - [mech] (a defect due to a metal portion being folded on itself) pli *m*
[radio] (in a magnetic amplifier) recouvrement *m*
[telev] (defect in reproduction which may occur when the width of a scanning line is greater than the scanning pitch) recouvrement *m*
[photo] (in aerial photographs) recouvrement *m*
[auto] (in the timing cycle, when exhaust and inlet valves are open at the same time) croisement *m*
[contr] (neutral zone) zone *f* neutre
~ACTION - [contr] (two-step control using two different step values) action *f* de recouvrement
~ANGLE - [electron] (the time interval, in angular measure, during which two consecutive arc paths carry current simultaneously) angle *m* de recouvrement, angle *m* d'empliètement
[el] angle *m* d'empiètement
~AREA - [gen mech rubber ind etc] surface *f* de recouvrage
~FAULT - [geol] faile *f* inverse
~JOINT - [metall] joint à recouvrement
~OF PLIES - [rubber ind] recouvrement *m* en dégradé des tissus cordes
~SEAM - [sewing] surjet *m*
~SPAN - [el] (for section-gap or air-gap; arrangement of line-work for dividing a contact wire electrically and mechanically into sections) sectionnement *m* à intervalle d'air
~TEST - [el] (test used for locating a fault in a cable) méthode *f* de recouvrement dans la recherche de dérangements
OVERLAPPED - [gen etc] recouvert, dépassé
~JOINT - [mech] assemblage *m* à mi-bois, entaille *f* à mi-bois
OVERLAPPING - [gen] chevauchement *m*, recouvrement, chevauchure *f*
[phys] (the coincidence of the long-wave end of a diffraction with the short-wave end of the spectrum of the next higher order) recouvrement *m*
[telecomm] (in telephony) recouvrement *m*, chevauchement *m*
~BUTT JOINS - [metall] joint *m* d'about à recouvrement
~CHANNELS - [radio & telev] canaux *mpl* partiellement superposés
~LAYERS - [rubber ind] (for endless vulcanization) couches *fpl* de recouvrement (pour la vulcanisation sans fin)
~MULTIPLE - [telecomm] (in telephony) multiplage *m* partiel
~PARTS - [shoe man] parties *fpl* de recouvrement
~SEAM - [gen] ligne *f* de recouvrement
~SET LAYERS - [el] couche *f* en paquet à recouvrement
OVERLAY, to - [text] superposer
[print] mettre des hausses sur (le tympan)
OVERLAY - [print] (piece of paper placed on the tympan of a printing machine to make the impression heavier, or to balance a depression in the form) hausse *f*, béquet *m*
[electron] procédé *m* de transparence électronique
~KNIFE - [print impl] couteau *m* de découpage, couteau *m* de mise en train
OVERLEAF - [gen] au verso, au dos
OVERLIGHTING - [cin] (the lighting of a scene by means of light sources over it) éclairage *m* de la scène

OVERLOAD, to - [gen] surcharger
OVERLOAD - [gen] surcharge *f*
[mech] (load on a machine which is greater than that designed) surcharge *f*
[el] (of a transformer, a machine etc; a load which is in excess of the rated load) surcharge *f*
[mech] (in internal combustion engines) cognement (par une charge excessive)
~CAPACITY - [el] (the measure of overload an electrical machine or motor can withstand) courant *m* maximal temporaire d'un contact
[instr] (the level beyond which an instrument or mechanism is permanently damaged) capacité *f* de surcharge
~CIRCUIT BREAKER - [el] interrupteur *m* à maximum
~CLUTCH - [auto] limiteur *m* de couple à friction
~CUTOUT - [el] (safety device for electric motors) coupe-circuit *m* de sécurité
~FACTOR - [contr] facteur *m* de surcharge
~FORWARD CURRENT - [electron] courant *m* en sens direct de surcharge prévisible
~INTERLOCK X-RAY UNIT - [radiat] (a unit which prevents the X-ray tube being energized when the permissible loading is exceeded) dispositif *m* automatique contre la surcharge
~LEVEL - [el etc] (the level at which the operation fails to be satisfactory as a result of signal distortion, overheating, damage etc) puissance *f* limite admissible
~PROTECTIVE SYSTEM - [el] (by means of over-current relays) dispositif *m* de protection contre les surcharges
~RELAY - [el] relais *m* de surcharge
~RELEASE - [el] (or overcurrent release; tripping device which operates when the current exceeds a predetermined value) déclenchement *m* à surintensité
~SPRING - [auto] (auxiliary spring for overloads) ressort *m* additionnel, ressort *m* de surcharge
~SWITCH - s. overload release
~VALVE - [mech] (an automatic pressure-release valve to relieve overload) soupape *f* de surcharge
OVERLOADED - [gen mech el etc] surchargé
OVERLOCK, to - [text] remailler
OVERLOCK - [text] en surjet
~SEAM - [text] couture *f* en surjet
OVERLYING - [gen] surjacent
~BOBBIN - [text] bobine *f* superposée
~STRATUM - [geol] couche *f* supérieure, ciel *m*, toit *m*
OVERMAKE - [paper man] (the amount exceeding the specified quantity to be produced) excédent *m* de production
OVERMASTICATED COMPOUND - [rubber] (a mixture which has been milled for too long a time) mélange *m* trop mastiqué
OVERMASTICATION - [rubber ind] (or overmilling) mastification *f* excessive
OVERMILLED COMPOUND - s. overmasticated compound
OVERMILLING - s. overmastication
OVERMODULATE, to - [el acoust] (to supply a modulation current which is greater than that giving a onehundred percent modulation) surmoduler
OVERMODULATED - [el acoust] (said of a recording) surmodulé
OVERMODULATION - [el acoust] (modulation to a

depth exceeding onehundred percent) surmodulation *f*
[radio] (the overload of an amplitude-modulated transmitter) surmodulation *f*
OVERNIGHT - [gen] (la) nuit *f*, pendant la nuit
OVERPASS - [constr] passage *m* supérieur
OVERPICK - [text] (occurring when the picking arm of a loom is above the shuttle box) chasse *f* à fouet
~LOOM - [text] métier *m* à chasse à fouet
~MOTION - [text] mécanisme *m* de commande de la chasse à fouet
OVERPICKLED - [metall] (prolongued process of removing a coating of oxide or tarnish from a metal object) surdécapé
OVERPICKLING - [metall] surdécapage *m*
OVERPITCHED - [build] (of a roof) à inclinaison excessive
OVERPLUS - [gen] excédent *m*
OVERPOLED COPPER - [metall] cuivre *m* superperché
OVERPOTENTIAL - [el] surpotentiel *m*
OVERPRESS - [glass man] (the projecting glass fragments due to faulty closing of the mould joints) bavure *f*
OVERPRESSURE - [phys] (excessive pressure) surtension *f*, excès *m* de pression
OVERPRINT, to - [print] (to print material on already printed sheets) surimprimer
OVERPRINT - [print] excédent *m* de feuilles imprimés
[gen] (any words or figure printed on a stamp to alter its value) surimpression *f*
OVERPRINTING - [photo] postlumination *f*
OVERPRIZE, to - [gen & comm] (to value more than its worth) surestimer
OVERPRODUCE, to - [ind proc] (to produce in excess) produire trop
OVERPRODUCTION - [ind proc] superproduction *f*
OVERPUNCH, to - [comput] perforer tous les moments d'un code à titre de correction
OVERPURCHASE, to - [comm] (to purchase at a price which is beyond the real worth of the goods) acheter trop cher
OVERRADIATION ALARM - [nucl] installation *f* d'alarme de radioactivité dangereuse
OVERRANGE - [contr] dépassement *m* de l'étendue de mesure
OVERREACH, to - [gen] (to reach beyond) dépasser
OVERREACH - [vet] (in horses, an error of gait) forgeage *m*
OVERREDUCED STEEL - [metall] acier *m* à désoxydation excessive
OVERRICH MIXTURE - [auto] mélange *m* trop riche
OVERRIDE, to - [gen] outrepasser, surmener
OVERRIDING - [gen etc] s. override, to
[med] chevauchement *m* (des fragments osseux)
~OF THE RAIL BY THE WHEEL FLANGE - [railw] ascension *f* du rail par le boudin
OVERRIGID - [mech] (said of any structural framework having more members than necessary to be perfect) trop rigide
OVERRIPE - [agric] trop mûr, blet
OVERRIPENESS - [agric] maturité *f* excessive
OVERROAD STAY - [constr] hauban *m* ancré à l'autre côté de la route
OVERRUN, to - [gen] (to spread over, to ravage) se répandre
[gen] (to infest) envahir
[gen] (to run beyond) dépasser
OVERRUN, to - [print] (to carry words from one of type to another until the matter fits) reporter à la ligne
[mech] (of a motor; to run too fast) surmener, fatiquer
[mech] (to exceed the trated speed) dépasser la vitesse limite
[el] survolter, pousser, surmener
OVERRUN - [print] excédent *m* de copies imprimées
[mech] (of electric lifts; the top or bottom clearance) tolérance *f*
~LAMP - [photo] (photoflood lamp) lampe *f* survoltée
OVERRUNNING - [gen] (for the various meanings, see overrun, to) incursions *f*pl, envahissement *m*
[auto] doubler *m*
[el] (of bulbs) survoltage *m*
~CLUTCH - [el] (of an electrical machine) roue *f* libre
~OF A SIGNAL - [railw] franchissement *m* d'un signal
~OF THE CYLINDER - [mech] rotation *f* excessive du cylinder
~OF THE REEL - [text] décancement *m* du dévidoir
~OF THE SPOOL - [text] avance *f* de la bobine
~TORQUE - [el] (of an electrical machine) couple *m* de décollage
~TYPE OF TRAILER BRAKE - [auto] frein *m* de remorque à inertie
OVERS - [print] (the excess number of printed sheets or forms) chaperon *m*, passe *f*, main *f* de passe
OVERSAIL, to - [gen] (to project beyond) déborder
OVERSAILING COURSE - [build] (a stone or brick course projecting from a wall) débordant *m*
OVERSCANNING - [electron] surbalayage *m*
OVERSEA - [geogr] d'outre mer
OVERSEAS - [gen & geogr] par delà les mers
~MARKET - [comm] marché *m* d'outre-mer
OVERSEE, to - [gen] (to supervise etc) surveiller
[gen] (to inspect) inspecter
OVERSEEING - [gen & ind proc] surveillance *f*
OVERSEER - [gen] surveillant *m*, inspecteur *m*
[gen] (in a factory etc) contremaître *m*, chef *m* de atelier
[print] (the foreman in a printing work) prote *m*
OVERSHOE - [shoe man] couvre-chassure *m*, galoche *f*
OVERSHOES - [aero] (colloq; mechanical deicers) dispositif *m* anti-givre
OVERSHOOT, to - [gen] (to shoot beyond the mark) outrepasser, dépasser
[aero] (to alight beyond the intended area, or to follow a path which would result in so alighting) atterrir trop long
[phys] (to flow swiftly shead of another moving body) outrepasser
OVERSHOOT - [gen & phys] dépassement *m* rapide
[telev] (the continuation of the signal beyond the final value of the output signal) dépassement, suroscillation *f*
[instr] (in an instrument, the measure of the overtravel of the indicator) suroscillation *f*
[radio] (temporary exaggeration of the magnitude of the edge of a steep-sided signal) survibration *f*
[aero] (a landing; or attempted landing which brings, or would have brought, the aircraft to the

ground beyond the intended area) atterrissage *m* trop long
OVERSHOOT - [contr] dépassement *m* de réglage
~CYLINDER TEDDER - [agric] épandeur-faneur *m* combiné
~RATIO - [telev] rapport *m* de suroscillation
~WHEEL - [mech] (a type of water-wheel) roue *f* en dessus
OVERSHOOTING - s. overshoot
OVERSIGHT - [gen] (error caused by lack of attention) omission *f*, oubli *m*
OVERSINTERED - [metall] surfritté
OVERSITE CONCRETE - [constr] (concrete layer covering a building site within the walls) radier *m* en béton
OVERSIZE - [gen] dimensions *fpl* au dessus de la moyenne
[min] (the material which is too coarse to pass a given screen) matières *fpl* refusées
[mech] cote *f* de réparation
[metall] écart *m* supérieur
~GRAIN - [metall] gros grain *m*
~PISTON - [mech] (a piston specially made in excess of the standard diameter, for fitting to a cylinder bore which has been bored out to correct wear) piston *m* de diamètre excessif
~TYRE - [rubber ind] pneu *m* surprofilé, pneu *m* surdimensionné
OVERSIZED - [gen] (larger than normal) surdimensionné
[mech] (in excess of the standard measure) au dessus des dimensions moyennes
OVERSLEEVE - [text] manchette *f*, fausse manche *f*
OVERSLUNG - [auto] (of a frame fitted above the axles) monté sur les essieux
~SPRING - [mech] ressort *m* monté sur les essieux
OVERSMOOKED SHEET - [rubber ind] feuilles *fpl* surfumées, "oversmoked sheets" *mpl*
OVERSPEED, to - [mech] travailler à allure excessive
OVERSPEED - [mech] (the excessive speed of a machine) excès *m* de vitesse, allure *f* excessive
[el] (of a turbine or an electrical machine) survitesse *f*, vitesse *f* d'emballement
~GEAR - [mech] modérateur *m* de vitesse
~GOVERNOR - [el] (mechanism operating the safety-gear in the event of the speed of a lift in a descending direction exceeding a given limit) dispositif *m* de coupure par augmentation de la vitesse
~LIMITER - [contr] limiter *m* de vitesse d'emballement
~OF AN INTERNAL COMBUSTION ENGINE - [mech] vitesse *f* de pointe d'un moteur thermique
~PROTECTION - [el] (generally a centrifugally operated mechanism to avoid excessive speed) protection *f* contre l'augmentation de la vitesse
OVERSPEEDER AND OVERWINDER - [mining] évite-molettes *mpl* (modérateurs des vitesse)
OVERSPILL TYPE DAM - [hydr] barrage *m* déversoir
OVERSPUN STRINGS - [acoust] (strings wound with fine wire to increase their weight) cordes *fpl* filées
~WIRE - s. overspun strings
OVERSTEER, to - [auto] survirer
OVERSTEERING - [auto] (faulty distribution of the masses causing the car the oversteer on bends) survirage *m*
OVERSTOCK, to - [agric] surcharger
OVERSTOPING - [min] (overhand stoping) abattage *m* en gradins renversés
OVERSTRAIN, to - [mech] (the operation of stressing an elastic material beyond its yield point) surtendre
OVERSTRAIN - [mech] tension *f* excessive, surmenage *m*
[med] surmenage *m*
OVERSTRENGTH - [chem] concentration *f* excessive
OVERSTRESS, to - [mech] surcharger
OVERSTRESS - [mech] surcharge *f*
OVERSTRETCHED - [gen & mech] trop tendu
OVERSTRUNG - [gen] (strung too tensely) surexcité
[mus] (of a piano) (piano)oblique, (piano) à cordes croisées
OVERSWING - [telev] s. overshoot
OVERSWUNG SLAY - [text] battant *m* libre
OVERTAKE, to - [gen] rattraper, atteindre
[auto] doubler, dépasser
OVERTAKING - [auto] doublure *f*
~OF A TRAIN - [railw] dépassement *m* d'un train
~POINT - [railw] point *m* de dépassement
~STATION - [railw] gare *f* de dépassement
~WHILE IN MOTION - [railw] (of a train) dépassement *m* en marche
OVERTAPPING - [rubber ind] saignée *f* excessive
OVERTHRUST - s. overthrust fault
~FAULT - [geol] (caused by earlier and lower strata which are pushed over later strata by faulting) charriage *m*, chevauchement *m*
~PLANE - [geol] surface *f* de carriage
~SHEET - [geol] charriage *m* de nappes
OVERTIME, to - [photo] (to expose too long) surexposer
OVERTIME - [gen] heures *fpl* supplémentaires
OVERTONE - [acoust] harmonique *m*
OVERTRAVEL SWITCH - [el] contacteur *m* de surcourse rupteur *m* de surcourse
OVERTURE - [mus] (instrumental prelude) ouverture *f*
OVERTURN, to - [gen] renverser, se renverser
[naut] chavirer
OVERTURNED - [gen] renversé, chaviré
~ANTICLINE - [geol] anticlinal *m* renversé
~FOLD - [geol] pli *m* déversé, flexure *f* renversée
OVERTURNING MOMENT - [phys] moment *m* de basculement
OVERTWISTING OF BINDING YARN HANKS - [text] surtorsion *f* des écheveaux des fils frisés
OVERVELOCITY - [phys] survitesse *f*
OVERVOLTAGE - [el chem] (the difference, for a given electrochemical reaction, between the dynamic electrode potential and the reversible electrode potential) surtension *f*
[met] (caused by atmospheric electricity) surtension *f* atmosphérique
~DUE TO RESONANCE - [el] surtension *f* de résonance
~FACTOR - [el] coefficient *m* de surtension, facteur *m* de tension maximale
~PROTECTIVE DEVICE - [el] (device protecting electrical machines against the possibility of damage due to excess voltage) dispositif *m* de protection contre la surtension
~RELAY - [el] relais *m* à tension maximale
~RELEASE - [el] (device tripping an electrical circuit when its voltage exceeds a specified value) déclenchement *m* de surtension

OVERVOLTAGE TRIPPING - [el] fonctionnement *m* à surtension
OVERWEIGHT - [gen] surpoids *m*, excédent *m*
[mech etc] surcharge *f*
OVERWIND, to - [gen] (to wind too tightly) remonter, trop tendre
[el] surbobiner
[mining] envoyer (la cage) aux molettes
OVERWRITING ERROR - [comput] (the storing of data in the same location when erasing programme steps) erreur *f* de chevauchement d'instructions
OVICELL - [zool] ovicellule *f*
OVICIDE - [bot] ovicide
OVIDUCT - [zool] (tube leading out of the ovary) oviducte *m*
OVIFORM - [gen] oviforme, ovoïde
OVINIA - [vet] variole *f* ovine
OVIPAROUS - [zool] (laying eggs) ovipare
OVISAC - [zool] (receptacle for the eggs) ovisac *m*
OVOID - [bot] (shaped like an egg and solid) ovoïde
OVOLO - [arch] (quarter-round convex moulding) ove *m*, ovicule *m*
OVOVITELLIN - [med] vitellus *m*, jaune *m* d'oeuf
OVOVIVIPAROUS - [zool] (producing eggs which hatch out in the mother's uterus) ovovipare
OVULATION - [zool] (the formation of the ova) ovulation *f*, ponte *f* ovarique
OVULE - [bot] (young seed in the process of development) ovule *m*
OVUM - [zool] (female gamete) ovule *m*, oeuf *m*
[arch] ove *m*
OWE, to - [gen] devoir
[comm] être redevable (de)
OWEN BRIDGE - [instr] (bridge used for the determination of inductance) pont *m* de Owen
OWL - [zool] hibou *m*, chouette *f*
OWN WEIGHT - [text] poids *m* propre
OWNER - [gen] propriétaire *m*
~DRIVER - [auto] conducteur *m* propriétaire de la voiture
~-OCCUPIER - [agric] propriétaire *m* exploitant
OWNER'S RISK - [insur] aux risques de l'expéditeur
OWNERSHIP- [gen & leg] propriété *f*, possession *f*
OX - [zool] boeuf *m*
~GAD-FLY - [vet] hypoderme *m* du boeuf
~GALL - [vet] fiel *m* de boeuf
~-STALL - [agric] bouverie *f*, étable *f* à boeufs
~-TONGUE - [agric] buglosse *f* officinale
~WARBLE FLY - [vet] s. ox gad-fly
OXALATES - [chem] (salts and esters of oxalic acid) oxalates *mpl*
OXALEMIA - [med] oxalémie *f*
OXALIC ACID - [chem] (a bleaching agent, catalyst, and a component of metal cleaning compounds) acide *m* oxalique
OXALURIA - [med] oxalurie *f*
OXAMIDE - [chem] (the diamide of oxalic acid. It is used as a stabilizer for certain cellulose derivatives) oxamide *f*
OXEN SHOES - [zool] fers *mpl* à boeuf
OXIDABLE - [chem] (subject to oxidation) oxydable
OXIDANT - [chem] (a substance which caused oxidation) oxydant *m*
[astronaut] (element in a rocket fuel designed to provide oxygen) (oxygène) oxydant *m*
~-FUEL RATIO - [chem] rapport *m* oxygène-combustible
OXIDATION RATIO - [metall] (degree of oxidation) degré *m* d'oxydation
OXIDASE - [biochem] (an enzyme which promotes oxidation) oxydase *f*
OXIDATE, to - [chem] s. oxidize, to
OXIDATION - [chem] (the addition of an electronegative atom to, or the removal of an electropositive atom from a molecule) oxydation *f*
[ceramics] (phase of firing during which the ferrous iron is converted to ferric iron and carbon and sulphur is oxidized) oxydation *f*
~CATALIST - [chem] (pro-oxygenic agent) catalyseur *m* d'oxidation
~FERMENT - s. oxidase
~POTENTIAL - [chem] potentiel *m* d'oxydation
~RECLAIMING PROCESS - [rubber ind] procédé *m* de régénération à l'oxygène
~-REDUCTION CYCLE - [nucl] (in an ionizing event) cycle *m* d'oxydoreduction
~-REDUCTION INDICATORS - [chem] (compound whose colour varies dependent on whether they are in the oxidized or reduced form) indicateurs *mpl* d'oxydoreduction
~-REDUCTION POTENTIAL - [chem] (the potential which is established at an inert electrode dipping into a solution containing equimolecular amounts of an ion or molecule in two states of oxidation) potentiel *m* d'oxydoreduction
~STABILITY - [ind chem] (in oil refining) stabilité *f* à l'oxydation
~STATE - [nucl] (of an atom when it has gained or lost electrons from its normal state) état *m* d'oxydation
~TEST - [ind chem] essai *m* d'oxydation
OXIDATIVE SLAGGING - [nucl] (method of removing fission products from molten metals) scorification *f*
OXIDE - s. oxides
~BED - [chem] (layer of material, layer of oxide in chemical purification) couche *f* d'oxyde
~BLUE - [dye] cobalt *m* d'outremer, bleu *m* de cobalt
~BOX - [chem] (in chemical purification) boîte *f* d'épuration
~CATHODE - s. oxide coated cathode
~-COATED CATHODE - [electron] (cathode with the active surface consisting of a coating of oxides of alkaline earth on a metal) cathode *f* à oxyde
~-COATED FILAMENT - [electron] (oxide-coated incandescent filament acting as an oxide-coated cathode) filament *m* à oxydes
~COATING - [ind chem] (the coating of metals with oxides of the alkali and alkaline-earth metals in thermionic applications) couche *f* d'oxyde, revêtement *m* d'oxyde
~OF IRON - [dye] oxyde *m* de fer
~OF MERCURY CELL - [el] pile *f* à oxyde de mercure-zinc
~RECTIFIER VOLTMETER - [instr] voltmètre *m* à redresseur à oxyde
~SHEDDING - [chem] (in video recording) poudrage *m*, perte *f* d'oxyde
~YELLOW - [dye] jaune *m* d'ocre
OXIDES - [chem] (compounds of oxygen with another element or organic radical) oxydes *mpl*
OXIDIZE, to - [chem] (the operation of adding oxygen to a compound) oxyder, oxygéner
[metall] calciner
~BACK, to - [metall] se reoxyder

OXIDIZED BITUMEN - [chem] (alternative term for Blown Bitumen. It is produced by forcing air through it) bitume *m* oxydé
~RUBBER - [rubber ind] (rubber which has been attacked by atmospheric oxygen) caoutchouc *m* oxydé
~TURPENTINE - [chem] (alternative name for Fat Turpentine, i.e. turpentine blown with air at atmospheric temperature for considerable periods to increase viscosity) térébenthine *f* oxydée
OXIDIZER - s. oxidizing agent
~UNIT - [nucl] (apparatus in which the oxidation of material used in a reactor occurs) dispositif *m* de oxydation
OXIDIZING - [chem] oxydant
[print] (treatment of a graphited wax surface with copper sulphate and iron filings, to produce a conducting copper coating) cuivrage *m*
~AGENT - [chem] (a substance which will produce oxidation) agent *m* oxydant
~FLAME - [chem] (the outer part of the Bunsen flame, in which oxidation occurs) flamme *f* oxydante, feu *m* d'oxydation
OXIDOREDUCTION - [chem] (oxidation-reduction) oxydoreduction *f*
OXIMATION - [chem] oximation *f*
OXIMES - [chem] (compounds containing the divalent oximino group: N.OH attached to the carbon) oximes *fpl*
OXIMETRY - [chem] oxymétrie *f*
OXIMIDE - [chem] oxymide *f*
OXINDOLE - [chem] oxindole *m*
OXONIUM - [chem] oxonium *m*
OXTRIPHYLLINE - [chem] (choline theophyllinate. A diuretic) oxtriphylline *f*
OXYACID - [chem] oxyacide *m*
OXY-ARC CUTTING - [metall] (a method of cutting metals combining the effects of heat from an arc and the physical and chemical effect of an oxygen jet) oxycoupage *m*
~-ARC CUTTING ELECTRODE - [metall] électrode *f* pour oxycoupage
OXYACETYLENE - [chem] (of a mixture of oxygen and acetylene) oxyacétylène *m*
~BLOWPIPE - s. oxyacetylene torch
~CUTTING - [mech] oxycoupage *m*
~CUTTING MACHINE - [metall] machine *f* d'oxycoupage oxyacétylénique
~TORCH - [impl] (apparatus for the combustion of oxygen and acetylene to produce a flame of extremely high temperature with uses in welding and the cutting of metals) chalumeau *m* oxyacétylénique
~WELDING - [metall] (welding with a flame which results from the combustion of oxyacetylene) soudage *m* oxyacétylénique
OXYBUTURIA - [med] oxybutyrémie *f*
OXYCALORIMETER - [instr] (calorimeter in which the energy of a substance is measured on the basis of the oxygen consumed) oxycalorimètre *m*
OXYCELLULOSES - [chem] (compounds formed by the oxidation of cellulose. They have strong reducing properties) oxycelluloses *fpl*
OXYCHLORIDE - [chem] oxychlorure *m*
OXYCHROMATIN - [biol] (form of chromatin with a light staining action) oxychromatique *f*
OXYCRESYL CAMPHOR - [chem] camphre *m* d'oxycréosole
OXYDIETHYLENEBENZOTHIAZOLE SULPHENAMIDE - [chem] (a vulcanization accelerator) sulphénamide *f* d'oxydiéthylène-benzothiazol
OXYGEN - [chem] (gaseous element, symbol O, A.N. 8, A.W. I6. It has application in metallurgy, as a starting material in a number of synthesis, and in medicine. It constitutes approximately one fifth of the atmosphere and is essential to all higher forms of life) oxygène *m*
~ACIDE - [chem] oxacide *m*
~BLEACHING - [text] blanchiment *m* à l'oxygène
~BLOWN CONVERTER STEEL - [metall] acier *m* à l'oxygène
~BOMB - [ind chem] bouteille *f* d'oxygène, bombe *f* à oxygène
~BOMB AGEING - [ind chem] vieillissement *m* en bombe à oxygène
~BOMB CALORIMETER - s. oxycalorimeter
~BOMB TEST - [ind chem] essai *m* en bouteille d'oxygène
~BOTTLE - [impl] bouteille *f* d'oxygène
~-BREATHING SET - s. oxygen mask
~CUTTING - [mech] oxycoupage *m*
~CYLINDER - s. oxygen bottle
~DEFICIENCY - [astronaut] déficience *f* en oxygène
~DESEAMING - [metall] décriquage *m*
~EFFECT - [phys] (the increased sensitivity to radiation of a biological material when in the presence of oxygen) effet *m* d'oxygène
~ENRICHED BLAST - [metall] vent *m* oxygéné
~-FREE HIGH CONDUCTIVITY COPPER - [metall] (copper obtained through a special treatment after electrolytic refining) cuivre *m* à haute conductivité
~GENERATOR - [ind chem] générateur *m* d'oxygène
~HOSE - [rubber ind] tuyau *m* à oxygène
~-HYDROGEN WELDING - [metall] soudage *m* au chalumeau oxhydrique
~LANCE - [metall] lance *f* d'oxygène
~-LANCE CUTTING - [mech] coupage *m* à la lance
~LANCING - s. oxygen-lance cutting
~MASK - [aero, med etc] (a device for the inhalation of oxygen) masque *m* à oxygène
~METER - [instr] oxygènemètre *m*
~PRESSURE-GAUGE - [instr] manomètre *m* de l'oxygène
~RESPIRATOR - s. oxygen mask
~SET - s. oxygen mask
~STEEL PLANT - [metall] aciérie *f* d'acier soufflé à l'oxygène
~SYSTEM - [mech] (arrangement of compressed oxygen cylinders, piping and inhalation equipment, to supply oxygen to aircraft crews) installation *f* pour l'oxygène
OXYGENATION - [chem] (treating, combining etc with oxygen) oxygénation *f*
OXYHEMOGLOBIN - [biol] oxyhémoglobine *f*
OXYHYDROGEN-[ind chem] oxhydrique
~BLOWPIPE - s. oxyhydrogen torch
~TORCH - [impl] chalumeau *m* oxydrique
~WELDING - [metall] soudage *m* au chalumeau oxydrique
OXYKRININ - [biochem] sécrétine
OXYMETHANDROLONE - [chem] (an anabolic agent with application in medicine) oxyméthandrolone *f*
OXYMORPHONE HYDROCHLORIDE - [chem] (a potent analgesic) chlorhydrate *m* d'oxymorphone
OXYOSPHRESIA - [med] hyperacuité *f* de l'odorat
OXYPHENONIUM BROMIDE - [chem] (a post-ganglio-

nic chlinergic blocking agent used in medicine) bromure *m* d'oxyphénonium
OXYPHILE GRANULE - [chem] (granule which can be stained by acid dyestuffs) granule *m* oxyphile
OXYPROLINE - [chem] (a derivative of proline) oxyproline *f*
OXYRYGMIA - [med] éructation *f* acide
OXYSALT - [chem] sel *m* oxygéné
OXYTETRACYCLINE - [chem] (an antibiotic with uses in medicine and food processing) oxytétracycline *f*
OXYTOGIA - [med] accouchement *m* rapide
OXYTOCIN - [biol] (alpha-hypophamine. The oxytocic principle of the posterior lobe of the pituitary, with application in obstetrics) oxytocine *f*
OYSTER - [zool] huître
~FITTING - [light] (bulkhead-fitting designed to emit light simultaneously on both sides of the partition to which it is attached) installation *f* (d'éclairage) à huître
~-PLANT - [agric] salsifis *m* blanc
~SHELL - [zool] écaille *f* d'huître
~SHELL SCALE - [zool] cochenille *f* ostréiforme
OZ - [abbrev for ounce] once *f*
OZALID - [print] héliographique
~PAPER - [print] papier *m* héliographique, papier *m* au photocalque
~PROCESS - [print] procédé *m* ozalid
~PROOF - [print] ozalid *m*
OZENA - [med] rhinite *f* atrophique, punaisie *f*
OZOCERITE - [min] (mineral wax. A naturally occurring waxy hydrocarbon with uses in the production of electrical insulants, polishes, and as a substitute for beeswax) ozocérite *f*, cire *f* minérale, ozokérite *f*
OZOKERITE - s. ozocerite
OZONATOR - [ind chem] (apparatus for converting ordinary oxygen into ozone) ozonizeur, ozonateur *m*
OZONE - [chem] (an allotrope of oxygen containing three atoms in the molecule. It has application as a bleaching agent, oxidant, and in the purification of water supplies) ozone *m*
~BATTERY - [ind chem] batterie *f* d'ozoniseur
~DEGRADATION PRODUCT - [chem] produit *m* dégradé par l'ozone
~LAMP - [impl] lampe *f* ozonisatrice
~LAYER - [astr] ozonosphère *f*, couche *f* d'ozone
~PAPER - [ind chem] (filter paper coated with starch and potassium iodide, turning blue when exposed to the action of ozone) papier *m* pour la détermination de l'ozone
~RESISTANCE - [chem] résistance *f* à l'ozone
OZONIZATION - [ind proc] ozonations *f*
OZONIZE, to - [chem] (to convert oxygen into ozone) ozoner
OZONIZED AIR - [chem] air *m* ozonisé
OZONIZER - [ind chem] (apparatus for the conversion of oxygen into ozone through an electric brush discharge) ozoniseur *m*
OZONIZING - s. ozonization
OZOMETER - [instr] (instrument designed to determine the quantity of ozone in the air) ozomètre *m*
OZONOSPHERE - s. ozone layer
OZONYSIS - [chem] ozonolyse *f*

P

P - [chem] (the symbol of phosphorus) symbole du phosphore

P.A. - [abbrev. for power amplifier] amplificateur *m* de puissance
[telecomm] (abbrev. for Public-Address) sonorisation *f* extérieure

P/A - [leg] (abbrev. of Power of Attorney) procuration *f*, mandat *m*

P.A.B.X. - [telecomm] (abbrev. for Private Automatic Branch Exchange) bureau *m* automatique privé

P.A.X. - [telecomm] (abbrev. for Private Automatic Exchange) installation *f* privée automatique

P.A.Y.E. - [fin] (abbrev. for Pay As Your Earn; the current system of direct taxation in G.B.) retenue *f* de l'impôt à la source

PACE, to - [gen] aller au pas, marcher

PACE - [gen] pas *m*, vitesse *f*, train *m* d'allure
[constr] (the area of a floor raised above the general level) emmarchement *m*
[zool] (of a horse) amble *m*

~OF THE WARP - [text] marche *f* de la chaîne

~VOLTAGE - [el] tension *f* de pas

PACHIMETER - [instr] (instrument designed to measure the elastic shear of a solid) pachomètre *m*

PACHULOSIS - [med] peau *f* sèche, peau *f* écailleuse

PACHYCARPOUS - [bot] (with a thick pericarp) pachycarpe

PACHYCEPHALY - [med] machycéphalie *f*

PACHYDERM - [zool] pachyderme *m*

PACHYDERMATOSIS - [med] pachydermie *f* chronique

PACHYDERMIA - [med] (or pachyderma; abnormal thickness of the skin) pachydermie *f*, hypertrophie *f* de la peau

PACHYLOSIS - s. pachulosis

PACHYMENINGITIS - [med] (inflammation of the outer membrane covering the brain) pachyméningite *f*

PACHYMETER - [instr] (instrument designed to measure the thickness of panes, paper sheets etc) pachomètre *m*

PACHYMUCOSA - [med] épaississement *m* des muqueuses

PACHYPLEURITIS - [med] pachypleurite *f*

PACHYSALPINGITIS - [med] salpingite *m* chronique hypertrophique

PACHYVAGINITIS - [med] pachyvaginite *f*

PACING MOTION - [text] mécanisme *m* de la chaîne et du tissu

~OF THE WARP - [text] guidage *m* des fils de chaîne

PACK, to - [gen] emballer; empaqueter
[gen] (to make a pack or a bundle) faire un paquet
[mech] (to apply a packing) garnir, étouper

PACK, to - [constr] (to fill in e.g. a crack etc) remplir, bourrer
[constr] tasser (dans un trou)
[zool] (to form a pack) ameuter
[comput] (the combining of more than one field of information into the same word) comprimer, condenser

PACK - [gen] paquet *m*, balle *f*, ballot *m*
[aero] (the bag in which a parachute is packed) sac *m* à parachute
[mining] muretin *m* de remblai, muretin *m* en pierres sèches
[zool] bande *f*, meute *f*
[metall] (a specified amount of plates) paquet *m*
[text] (weight measure, i.e. 240 lbs) balle *f* de 240 livres
[med] drap *m*, enveloppement *m*
[geogr] (large area of floating ice) banquise *f*

~A SLEEPER, to - [railw] bourrer une traverse

~ANIMAL - [zool] bête *f* de charge, sommier *m*

~ANNEALING - [metall] recuit *m* en paquet

~-CARBURIZING - [metall] cémentation *f* en caisse

~-CARRIER TELEVISION STATION - [telev] émetteur *m* de télévision portatif

~COVER - [aero] s. pack

~DUCK - [text] (a strong cloth used for packing) toile *f* d'emballage

~HARDEN, to - [metall] cementer, tremper en paquet

~-HARDENING - [metall] (hardening treatment preceded by case-carburizing) cémentation *f*, trempe *f* en parquet

~HORSE - [zool] cheval *m* de bât

~-ICE - [geogr] glace *f* de banquise

~OF CARDS - [comput] paquet *m* de cartes, pile *f* de cartes

~-OFF - [oil ind] porte-garnitures *m* complet

~-OFF ASSEMBLY - [oil ind] dispositif *m* de contrôle du débit d'un puits

~-OUT UNIT - [oil ind] partie en caoutchouc du "bag preventer"

~PROP - [min] (a prop supporting the pack in a mine) taquet *m* de mur de remblai

~ROLLING - [metall] (the rolling of two or more metal sheets) lamination *f* en paquet, colaminage *m*

~SADDLE - [impl] bât *m*

~SHEET - [text] toile *f* d'emballage

~THE HOPS, to - [brew ind] emballer le houblon

~-TRAIL - [agric] piste *f* mulettière

~UP, to - [constr] étanconner, étayer

~-UP BLOCK - [constr] billot *m* de support

~-WALL - [mining] mur *m* de ramblai, meurtiat *m* en pierres sèches

PACKAGE, to - [gen] emballer, empaqueter
PACKAGE - [gen] paquet *m*
[packing] empaquetage *m*, emballage *m*
[comm] (a set of some products) articles *mpl* prêts
[text] moche *f* de soie filée
[electron] sous-ensemble *m*, bloc *m* fonctionnel
[min] compact *m*
~BOILER - [th eng] (type of boiler which is completely autonomous, producing the required steam output at preset pressure and temperature by automatic built-in control) chaudière *f* automatique
~DYEING - [text] teinturerie *f* des bobines croisées
~FILLING MONITOR - [nucl] moniteur *m* isotopique de produits emballés
~IRRADIATION PLANT - [nucl] installation *f* d'irradiation de produite emballés
~MONITOR - [nucl] moniteur *m* de produits emballés
~REACTOR - [nucl] (reactor which can be transported) réacteur *m* préfabriqué, réacteur *m* compact transportable
~TYPE CONSTRUCTION - [hydr] construction *f* en ouvrage unique
PACKAGED CIRCUIT - [electron] circuit *m* enrobé
~COMPONENT - [nucl] (of a reactor) élément *m* de construction prêt à fonctionner
~MAGNETRON - [electron] (a structure including a magnetron, its magnetic circuit and output-matching device) ensemble *m* magnétron, magnétron *m* en ordre de marche
PACKAGING - [gen] (the process of making up articles into packages for shipment, sale or the like) emballage *m*
~MACHINE - [mech] machine *f* à emballer
~PLANT - [ind] installation *f* d'emballage
PACKCLOTH - s. pack duck
PACKED BED - [ind chem] lit *m* fixe
~COLUMN - [ind chem] (a distillation or other column containing permeable packing material to increase the area of contact) colonne *f* garnie
~DECIMAL - [math] décimal *m* condensé
~TOWER - [ind chem] (a vertical hollow structure filled with loose material through which counter currents of the substances to be treated are passed) tour *f* de percolation
~TOWER SCRUBBER - [ind chem] (a static washer) laveur *m* statique, tour *f* de lavage à surfaces mouillées
~WASHING TOWER - s. packed tower scrubber
PACKER - [gen] emballeur *m*
[mech] (a machine or device for packing) emballeuse *f*
[mining] remblayeur *m*
[food ind] fabricant *m* de conserves en boîtes
~BODY - [auto] benne *f* étanche à ordures
PACKET - [gen] paquet *m*, colis *m*
[naut] paquebot *m*
PACKHOUSE - [comm] magasin *m*
PACKING - [gen] emballage *m*, arrimage *m*, empaquetage *m*
[mech] (compressible material used to fill a stuffing box; also to make a joint) garniture *f*, garnissage *m*, bourrage *m*
[constr] (material used for filling) agglomération *f*, tassement *m*
[constr] (the action of filling or stuffing) tassement *m*, remblayage *m*
[naut] arrimage *m*
[constr] (the filling for a double wall) remplissage *m*, remblayage *m*
[aero] (of a parachute) pliage *m*
[med] enveloppement *m*
[comput] (of information on a storage medium) augmentation *f* de la densité
[mech] encastrement *m*
[soil] compacité *f*, densité *f*
[print] (of an offset press) habillage *m*
[constr] remblay *m*
[mech] (of a piston) garnissage *m*
[telev] (picture compression) compression *f* de l'image
[metall] serrage *m* (du sable d'un moule)
[nucl] (the compression of material particles into a confined space) compression *f*
PACKING BOLT - [mech] boulon *m* de serrage
~BOX - [mech] presse-étoupe *m*
~CANVAS - [text] toile *f* de jute commune
~CASE - [gen & transp] caisse *f* d'emballage, layette *f*
[mining] dame *f* de remblai
~COSTS - [comm] frais *mpl* d'emballage
~COUNTERBORE - [mach tool] fraise *f* pour garnitures
~CRATE - [gen] cadre *m* d'emballage
~CUTTER - [mach tool] porte-outil *m*
~DENSITY - [comput] (the relative number of units of required information within given dimensions) densité *f* d'enregistrement
[comput] (in a micro-miniature module) densité *f* de compacité
~DISK - [mech] rondelle *f* pour garnitures
~DRUM - [mech] cylindre *m* étanche
~EFFECT - [nucl] (or mass effect; the difference between the observed mass of a nucleus and the mass calculated with the addition of the masses of constituent elementary particles) effet *m* de masse, effet *m* de tassement
~EXPENSES - [comm] dépenses *fpl* d'emballage
~FACTOR - s. packing fraction
~FELT - [text] feutre *m* d'étoupage
~FRACTION - [phys] (the ratio between mass defect and mass number of a nuclide) fraction *f* de tassement
~FRACTION CURVE - [nucl] courbe *f* de la fraction de tassement
~GLAND - [mech] presse-étoupe *m*, chapeau *m* de presse-étoupe
~GLAND CROSSOVER - [oil ind] allonge *f* du presse-étoupe
~HOUSE - [food ind] (establishment organized for packing provisions) frabrique *f* de conserves
~INDUSTRY - [food ind] conserverie *f*
~JOINT - [mech] joint *m*
~LIST - [comm] bordereau *m* d'emballage
~LOSS - [nucl] défaut *m* de masse
~MACHINE - [mech mach] machine *f* à emballer
~NEEDLE - [impl] carrelet *m*, aiguille *f* d'emballage
~NUT - [mech] (a stuffing box follower which screws into the body, as distinct from a non-rotating follower forced down by nut) écrou *m* du presse-étoupe
~OF CARBON GRANULES - [el acoust] (a defect occurring in carbon microphone) tassement *m* de granules de charbon
~OF SLEEPERS - [railw] bourrage *m* des traverses

PACKING OF THE PISTON [mech] garnissage *m* du piston
~OF THE TRACK - [railw] (with ballast) garnissage *m* de la voie
~OFF ATTACHMENT - [oil ind] attaque *f* pour faire circuler le tube de repêchage à travers le poisson
~PAPER - [paper man] papier *m* d'emballage, papier *m* gris
~PIECE - [carp] cale *f*, cale *f* d'appui, semelle *f*
~PLANT - [agric] installation *f* d'emballage
~RING - [mech] bague *f* de garniture
~ROD - [impl] (in railway and construction works) barre *f* à bourrage
~ROOM - [gen] salle *f* d'emballage
~SET - [mech] ensemble *m* de garnitures
~SPECIFICATIONS - [comm] instructions *f*pl pour l'emballage
~STATION - s. packing plant
~STICK - [naut] (used to press the packings) cheville *f* à tourniquet
~STRIP - [mech] bande *f* antifriction, (pour rattraper l'usure), lardon *m* d'ajustage
~WARP - [text] chaîne *f* de fourrure
~WASHER - [mech] rondelle *f*
PACKS - [mining] remblai *m*
PACKSAND - [geol] grès *m* à grain fin
PACKWAY - [road] muletière *f*
PACO - [text] (synonym for alpaca) alpaca *m*
PACQUET LIVER - [med] foie *m* ficelé
PAD, to - [gen] bourrer, rembourrer
[mech] (to fit with pads) garnir de tampons
[leather ind] larder
[paint] imprégner de mordant
PAD - [gen] coussinet *m*, bourrelet *m*, tampon *m*
[mech] tampon *m*, support *m* amortisseur
[plumb] (branch connexion) anneau *m* de renforcement
(of a hand drill etc) mandrin *m* de vilebrequin
[el] (attenuator pad) dispositif *m* d'affaiblissement fixe
[gen] (for rubber stamps etc) tampon *m*
[paper man] bloc *m*
[astronaut] plate-forme *f* de lancement
[metall] (die casting) bloc *m* rapporté (en acier spécial dans lequel est taillée l'empreinte)
[zool] (the soft skin enlargment under the toe surface of some animals) pelote *f* digitale, pulpe *f*
[welding] saillie *f*
[cin] patin *m*, glissière *f*
[metall] nodule *m* de soudure
[oil ind] (rubber pad) tampon *m*
~CHARACTER - [comput] caractère *m* de remplissage
~CRIMP - [leather ind] presse *f* à cambrer
~DELUGE - [astronaut] aspersion *f* de la plate-forme de lancement
~-JIG PROCESS - [text] procédé *m* Pad-Jig
~SAW - [impl] (hand-saw with a narrow tapering blade) scie *f* à chantourner, égoïne *f*
~-STEAM PROCESS - [text] procédé *m* de foulardage et de vaporisage
~STONE - [constr] (template made of stone) gabarit *m* en pierre
PADDED - [gen] rembourré, matelassé
[mech] garni de bourre
~INSTRUMENT PANEL - [auto] planche *f* de bord rembourrée
PADDER - [el] (small adjustable condenser, generally for filters) condensateur *m* de filtrage
[text] foulard *m*
~FINISH - [text] apprêt *m* au foulard
PADDING - [gen] rembourrage *m*, remplissage *m*
[gen] (the material used for padding) ouate *f*, bourre *f*
[paint] (impregnation with mordant) imprégnation *f* de mordant
~CAPACITOR - [radio] (series capacitor in the oscillator circuit of a superheterodyne receiver) condensateur *m* padding, padding *m* d'équilibrage
~CONDENSER - s. packing capacitor
~STOCK - [rubber ind] (a cushion compound) mélange *m* gomme sur dernier pli, gomme *f* matelas
PADDLE, to -[naut] (to row slowly with a paddle) pagayer
[mech] (to beat with a paddle) battre
[mech] (to move by paddle wheels) actionner par des aubes
[leather ind] (the operation of tanning light skins, by which skins and liquor are moved by a paddle) mouliner
PADDLE - [naut] pagaie *f*
[impl] aube *f*, pale *f*, palette *f*
[impl] (implement used for mixing) palette *f*
[hydr] (of a water wheel) jantille *f*, volet *m*
[hydr] (the water wheel) roue *f* à aubes, vannelle *f*, alichon *m*
[mech] (the moving blade of a mixing machine) aube *f*, palette *f*
[mech] (the moving blade of an agitator) ailette *f*
[min] (straight iron tool used to mix ores in a furnace) crochet *m*
~BOARD OF A WATER WHEEL - [hydr] aube *f* d'une roue hydraulique, palette *f* d'une roue hydraulique
~BOAT - s. paddle steamer
~BOX - [naut] (of a paddle boat) tambour *m* de la roue à aubes, garde-roue *m*
~HOLE - [hydr] (opening in a lock-gate) vannelle *f*
~MECHANISM - [hydr] agitateur *m*
~MIXER - [mech] (mixer containing rotating flat blades of various form) mélangeur *m* à ailettes
~-PLANE - [aero] (rotorcraft in which the rotor loks like a marine paddle-wheel) hélicoplane *m*
~STEAMER - [naut] vapeur *m* à aubes, vapeur *m* à roues
~STIRRER - [text] malaxeur *m* à aubes
~TUMBLER - [leather ind] (the drum used for padding) moulinet *m*
~WHEEL - [naut] (propelling a vessel) roue *f* à aubes
~-WHEEL AERATION - [gen] aération *f* par roues à palettes
~WHEEL FAN - [el] ventilateur *m* à palettes
~-WHEEL ROTOR - [aero] (a rotor in a cyclogyro) rotor *m* d'helicoplane
PADDLEBLADES - [mech] aubes *f*pl, palettes *f*pl
PADDLING - [gen] s. paddle, to
[leather ind] (process in tanning light skins, by which skins and liquor are kept in movement by a revolving paddle) moulinage *m*
[glass man] (the rough shaping in the furnace) formage *m* préliminaire
PADDOCK - [gen] enclos *m*, paddock *m*
[agric] (a pasture lot) parc *m*
[min] (enclosed piece of land used as a temporary

deposit) halde *f* de minérai
PADDOCK - [mining] dépôt provisoire
PADDY - [agric] (rice in husk) paddy *m*, riz *m* non décortiqué
~FIELD - [agric] champ *m* de riz
~RICE - s. paddy
PADLOCK, to - [gen] cadenasser
PADLOCK - [impl] (detachable lock) cadenas *m*
PAEDOGAMY - [genet] (autogamy in which nucleus and cytoplasm divide and reunite) paedogamie *f*
PAEDOGENESIS - [genet] (reproduction by immature forms) paedogénèse *f*
PAGE, to - [print] (to make up in pages) paginer, folioter
[print] (to number the pages) numéroter
[gen] (to notify a person of a call etc, especially in hotels) envoyer chercher (par un chasseur)
PAGE - [gen] page *f*
[print] (the type prepared for one page) page *f*
~PAPER - [print] (strong sheet of paper on which the type is placed before printing) porte-page *m*
~PRINTING APPARATUS - [telecomm] appareil *m* imprimeur en page
~PROOF - [print] correction *f* en bon à clincher
~SIZE - [print] format *m* de la page
~TELEPRINTER - [comput] (printing machine placed at a distance from the computer) imprimante *f* par page
PAGEPRINTER - [comput] téléimprimeur *m* en page
PAGINA - [bot] (synonym of lamina) limbe *m*
PAGINATION - [print] (the numering of the pages) pagination *f*, foliotage *m*, numérotage *m*
PAGING SYSTEM - [print & gen] système *m* de numérotage
PAGODA - [arch] pagode *f*
PAGODITE - [min] (a type of ordinary massive pinite, i.e. a hydrous silicate of aluminium and potassium, containing more silica) pagodite *f*
PAID TIME RATIO - [telecomm] (telephony and telegraphy) coefficient *m* d'occupation d'un circuit, rendement horaire d'un circuit
PAIL - [impl] seau *m*
PAIN - [gen & med] douleur *f*, souffrance *f*
PAINT, to - [paint] peindre, enduire, couvrir
PAINT - [paint] (liquid which can be applied to a surface to form a protective or decorative film) peinture *f*, couche *f* de peinture
(the operation of painting) peinture *f*
~A COAT OF, to - [paint] donner une couche de peinture
~BASE - [chem] base *f* du couleur
~BRUSH - [impl] pinceau *m*, brosse *f* à peinture
~DRIER - [paint] siccatif *m*
~EXTENDER - [paint] (additive to a coating composition, to give some special quality) produit *m* d'addition pour couleurs
~FACTORY - [gen] fabrique *f* de peintures
~GRINDING - [mech] broyeuse *f* pour teintures
~MILL - [text] broyeur *m* de colorants
~OVER, to - [constr] cimenter à la brosse
~POT - [oil ind] source *f* de boue bouillonnante
~RECLAIMING - [paint] régénération *f* des peintures
~REMOVER - [chem] décapant *m* pour peintures
~ROCK - [geol] ocre *f* de fer
~SPRAY MASK - [impl] masque *m* de protection contre couleurs pulverisées
~SPRAYER - [paint] (implement for spraying paint) pistolet *m* à peindre, pistolet *m* vaporisateur
PAINT SPRAYER-BULB - [impl] poire *f* de vaporisateur
~STRIP - [constr] mordant *m*
~STRIPPER - [paint] décapant *m* pour peintures
~THINNER - [paint] diluant *m* pour peintures
PAINTABLE FINISH - [constr] (a surface capable or receiving paint) surface *f* prête pour l'application
PAINTER - [gen] peintre *m*, peintre *m* en bâtiments
[naut] (rope used to fasten a boat by its bow) bosse *f*, câbleau *m*, amarre
PAINTER'S PUTTY - [paint] (mixture of whiting and linseed oil) lut *m*, mastic *m*, enduit *m*
~ROLL - [impl] rouleau *m* pour peintres
PAINTING - [gen] peinture *f*
[med] badigeonnage *m*
PAIR, to - [gen] apparier, accoupler
PAIR - [gen] paire *f*
[el] (the two conductors which are associated to form a part of a communication channel) paire *f*
[mech] (combination of two elements forming a unit in the mutual production of motion, e.g. a piston and a cylinder) couple *m* (cinématique)
~COLOURS, to - [dye] apparier des couleurs
~CONVERSION - [nucl] (the conversion of a photon into an electron and a positron) conversion *f* d'une paire
~EMISSION - [nucl] (the production of an electron and a positron by collision of a gamma quantum) production *f* de paires
~FORMATION - [nucl] (the formation of an electron and a positron when high-energy photons strike a solid or a gas) production *f* de paires
~OF COMPASSES - [impl] compas *m*
~OF ROLLERS - [text] paire *f* de cylindres
~OF SCALES - [impl] balance *f*
~PRODUCTION - s. pair emission
~-PRODUCTION ABSORPTION - [nucl] (the absorption of gamma rays in the process of pair production) absorption *f* avec production de paires
~SPECTROMETER - [instr] spectromètre *m* aux paires
PAIRED CABLE - [el] (cable containing a number of twisted pairs) câble *m* à paires
~ECHO - [radio] (method of analysing the effects of phase distortion) écho *m* à paires
~ELECTRON - [phys] électron *m* commun
~ENGINES - [mech] (two engines mounted close together, by wholly independent) moteurs *mpl* jumelés
~RUNNING - [railw] jumelage *m* de véhicules (thermoélectriques)
PAIRING - [gen] (division into groups of two) appariement *m*, conjugaison *f*
[gen] (to bring two units or groups together) accouplement *m*
[telev] (the effect of defective interlacing; also called twinning) pairage *m*
~BLOCK - [genet] bloc *m* d'appariement
~ENERGY - [phys] énergie *f* de création d'une paire
~MATING - [genet] croisement *m* de paires
~OF GENES - [genet] couple *m* de gènes
~OFF - [telev] disparition *f* des trames
PAKTONG - [metall] (chinese alloy of zinc, nickel and copper) packfond *m*, packfong *m*
PAL COLOUR SYSTEM - [telev] (phase-alternation line system) système *m* de télévision couleurs PAL
PALATABLE - [gen] agréable au palais
PALATAL - [anat] (the roof of the mouth) palais *m*

PALATE - [anat] (the roof of the mouth) palais *m*
PALATINE - [zool] (of the palate) palatin
PALATOPLASTY - [med] uranoplastie *f*, palatoplastie *f*
PALATOPLEGIA - [med] paralyse *f* du voile du palais
PALATORRHAPHY - [med] staphylorraphie *f*
PALE - [constr] (fence enclosing an area) pieu *m* de clôture
[constr] (the stake used for the fence) pieu *m*
[adj] pâle, blême
[dye] sans couleur, clair, pâle
~ALE - [brew ind] (a quality of light ale) bière *f* pâle
~LEAF GOLD - [metall] (an alloy of gold and silver beaten out in very thin sheets) feuille *f* d'or et de argent
PALEA -[bot] (or palet; the inner bracteole enclosing a grass flower) bractée *f*
PALENCEPHALON - [med] palencéphale *m*
PALENESS - s. pallescence
PALEONTOLOGY - [geol] (or palaeontology; the study of animal life in geological periods) paléontologie *f*
PALETTE - [impl] palette *f*
~KNIFE - [impl] spatule *f*
PALILALIA - s. palinphrasia
PALINDROME - [gen] (any word which is read the same whether forward or backward, e.g. radar) palindrome *m*
[math] (a number which is read the same whether forward or backward) palindrome *m*
PALINDROMIA - [med] palindromie *f*, récidive *f*
PALINDROMIC - [math etc] palindrome
[med] récidivant, récurrent
PALINESTHESIA - [med] retour *m* à la sensibilité
PALING - [gen] pâlissement *m*, palis *m*, claire-voie *f*
PALINGENESIS - [genet] (the reproduction of ancestral characters in the development of an individual) palingénésie *f*
[geol] (the rebirth of granitic or similar magma by pure melting) palingénésie *f*
PALINPHRASIA - [med] palilalie *f*, palinphrasie *f*
PALIPHRASIA - s. palinphrasia
PALISADE - [gen] (a fence of strong timbers) palissade *f*
~CELL - [bot] (cell of the palisade layer of a leaf) cellule *f* à palissade
~LAYER - [bot] (layer of elongated cells under the upper epidermis) couche *f* à palissade
~TISSUE - [bot] (one or more layers of palisade cells) palissade *f*
PALISANDER - [bot] palissandre *m*
PALLADIAN - [arch] palladien
PALLADINIZED ASBESTOS - [constr] (asbestos permeated with palladium in a finely-divided condition) amiante *f* palladiée
PALLADIOUS IODIDE - [chem] (obtained as a precipitate from the reaction of potassium iodide with palladious chloride solution. A distinguishing test for iodine, since the other halides of palladium are relatively insoluble) iodure *m* palladeux
PALLADIUM - [chem] (a metallic element, symbol Pd. A.N. 46, A.W. 106,7 with uses as a catalyst, in the manufacture of electrical equipment, and in jewellery) palladium *m*
~ASBESTOS - [chem] amiante *f* au palladium
~CHLORIDE - [chem] (an analytical reagent, intermediate for photographic chemicals, and dyeing mordant) chlorure *m* de palladium
PALLADIUM NITRATE - [chem] (an analytical reagent) nitrate *m* de palladium
~OXIDE - [chem] (a catalyst in organic synthesis) oxyde *m* de palladium
~PLATING - [metall] placage *m* au palladium
PALLESCENCE - [gen & med] pâleur *f*
PALLET - [gen] (a light platform provided with ribs or bars on the underside to allow of convenient handling and stacking of boxes or the like, by fork trucks or other devices) plateau *m*, palette *f*, chargeur *m*
[impl] (used by potters etc) palette *f*
[mech] (of a ratchet wheel) cliquet *m*
[horol] (the surface on which the teeth of the escape wheel etc) palette *f* de l'arbre
[mus] (of organ) soupape *f* du sommier
[mech] glissière *f*
~ARBOUR - [horol] arbre *m* de la palette
~BRICK - [constr] (brick with a groove in one edge to house a fixing strip) brique *f* chanfreinée
~BRIDGE - [horol] pont *m* de l'ancre
~COCK - s. pallet bridge
~FOR ANVIL - [impl] tas *m*, tas *m* inférieur
~FOR TUP - [impl] (of a power-hammer) frappe *f*
~FORK - [horol] ancre *f*
~JEWEL - [horol] rubis *m* de contact
~KNIFE - [constr] spatule *f* du peintre
~TRUCK - [transp] gerbeur *m* à fourche
PALLETIZATION - [gen] (a form of internal transport) palletisation *f*
PALLETIZE, to - [transp] (to move by pallets) palletiser
PALLIATIVE - [med] (which affords a temporary relief) palliatif
~MEDICINE - [med] palliatif *m*
PALLID - [gen & med] pâle, décoloré
PALLIDOIDOSIS - [vet] syphilis *f* du lapin
PALLIUM - [zool] (part of the wall of the cerebral hemispheres) manteau *m*
PALLOR - [gen & med] pâleur *f*
PALM - [anat] (the inner surface of the hand) paume *f*
[bot] palmier *m*
[naut] (a kind of shield used by sailmakers to protect their hand while pushing the needle) paumelle *f*
[naut] (the expanding end of the arm of an anchor) patte *f*, oreille *f*
~FIBER - [text] fibre *f* de coco, coir *m*
~-KERNEL CAKE - [agric] tourteau *m* de palmiste
~NUT OIL - [chem] (an oil obtained from Elaesis guineensis and used in the production of cosmetics, pharmaceuticals, confectionery, soap and margarine) huile *f* de palmiste
~TREE - [bot] palmier *m*
~-WINE - [agric] vin *m* de palme
PALMAR - [anat] palmaire
~ARCH - [anat] arcade *f* palmaire cubitale
PALMAROSA OIL - [chem] (an essential oil obtained from varieties of Cymbopogon and used in perfumery) essence *f* de géranium des Indes
PALMATINE - [chem] (an alkaloid of the isoquinoline group) palmatine *f*
PALMER GUN RECLAIMING PROCESS - [rubber ind] (high-pressure steam reclaiming process) procédé *m* de régénération à la vapeur haute pression
PALMETTE - [arch] (a carved ornament) palmette *f*
~TRAINING - [agric] taille *f* en palmette
PALMIPED - [zool] palmipède *m*

PALMITATE - [chem] (salt or ester of palmitic acid) palmitate *m*
~SOAP - [chem] savon *m* de l'acide palmitique
PALMITIC ACID - [chem] (a widely distributed fatty acid with application in the production of soaps and palmitates) acide *m* palmitique
PALMITIN - [chem] (glycerine ester of palmitic acid) palmitine *f*
PALMITOLEIC ACID - [chem] (a synthesis intermediate) acide *m* palmitoléique
PALPABLE - [gen] palpable
PALPATE, to - [med] palper
PALPEBRA - [anat] (eyelid) paupière *f*
PALPEBRAL - [anat] palpébral
~FISSURE - [anat] (the space between the upper and lower eyelid) fente *f* palpébrale
PALPEBRATION - [gen] clignotement *m*
PALPITATION - [med] (the awareness of the heart's beat against the chest wall) palpitation *f*
PALSY - [med] (paralysis) paralysie *f*
PALUDIFICATION - [geol] transformation *f* en marais
PALUDISM - [med] (malaria) paludisme *m*
PALYNOLOGY - [palaeontology] (study of the fossil pollen and spores of plants) palynologie *f*
PAM - [electron] (Pulse Amplitude Modulation) modulation *f* d'impulsion en amplitude
PAMPAS GRASS - [bot] (type of tall ornamental grass growing in the Argentine pampas) gynérion *m* argenté, herbe *f* des pampas
PAMPERO - [met] (strong dry wind occurring in the Argentine pampas) pampero *m*
PAMPHLET - [print] (any brief printed work which is not bound) brochure *f*
PAN, to - [cin] (the operation of moving a camera across a scene to obtain a panoramic effect) panoramiquer
[min] (rough method of separating gold from sand) laver à la batée
PAN - [gen] cuvette *f*
[instr] (the dished plates on which weights and objects to be weighed are placed) plateau *m*, plat *m*, bassin *m*
[impl] (a shallow cooking vessel) casserole *f*, poêlon *m*
[ind chem] (a vessel used for boiling and evaporating) cuve *f*, cuvette *f*, bassine *f*, cuve *f* d'amalgamation
[telev] (or panning) panoramique *m*
[auto] (carter) carter *m*, tôle *f* inférieure
[instr] (of a meter; diaphragm ring) cadre *m* de fixation de la membrane
[geol] trémie *f*
[mining] batée *f*, pan *m*, sébile *f*
~AMALGAMATOR - [min] amalgamateur *m* à cuve
~AND TILT HEAD - [telev] tête *f* panoramique, tourelle *f* universelle
~BREEZE - [metall] (mixture of small coke and clinker from the ash-pan of the furnace) scorie *f* de coke
~COKE - [fuel] coke *m* récupéré
~DOWN - [cin] (tilting of the camera on its horizontal axis in a vertical direction) décentrement *m* vertical
~FILM - [photo] (abbrev. for panchromatic film) pellicule *f* panchromatique
~-HEAD - [mech] (of a screw or bolt) à tête tronconique
PAN-HEAD RIVET - [mech] rivet *m* à tête tronconique
~HEAD WITH TAPERED NECK - [mech] rivet *m* à tête cône tronqué avec collet
~MILL - [mech] (machine of the same type as the edge runner mill, but having stationary roller unit and revolving pan) broyeur *m* à meules verticales
~OF THE MILL - [min] sole *f* de frotteur
~OF THE SAND MILL - [min] cuve *f* de frotteur
~PIPE - [mus] flûte *f* de Pan, syrinx *m*
~SUPPORT - [th eng] (a burner grate) grille *f*
~UP - s. pan down
PANACHE - [gen] (bunch of feathers; a plume) panache *m*
PANANGITIS - [med] panartérite *f*
PANARIS - [vet] (inflammation of the horny covering of the claw of cattle) panaris *m*
PANARTHRITIS - [med] arthrite *f* généralisée
PANCAKE, to - [aero] (to make a pancake landing) atterrir à plat, descendre à plat
PANCAKE - [gen] crêpe *f*
[aero] (abrupt landing) atterrisage *m* à plat
~COIL - [radio] (pancake-shaped coil, i.e. arranged rate or low forward speed) atterrissage *m* à plat
~LANDING - [aero] (a landing in which the aircraft is levelled-off a few feet from the ground, so that alighting is completed along a steep path in a normal attitude) atterrissage *m* à plat
PANCAKING - [aero] (alighting at abnormally high descent rate or low forward speed) atterrissage *m* à plat
PANCARDITIS - [med] (inflammation affecting the three main structures of the heart) pancardite *f*
PANCHROMATIC - [photo] (uniformly to all colours of the visual spectrum) panchromatique
~EMULSION - [photo] émulsion *f* panchromatique
~FILM - [photo] pellicule *f* panchromatique
PANCREAS - [anat] (compound digestive gland) pancréas
PANCREATECTOMY - [med] pancréatectomie *f*
PANCREATIC - [med] (pertaining to the pancreas) pancréatique
~ARTERY - [anat] artère *f* pancréatique
~ENZYMES - [biochem] (the enzyme of the pancreatic juice) enzymes *f*pl pancréatiques
~JUICE - [biochem] suc *m* pancréatique
PANCREATIN - [chem] (a digestive enzyme with uses in medicine) pancréatine
PANCREATITIS - [med] pancréatite *f*
PANCYTOLYSIS - [med] pancytolyse *f*
PANDEMIC - [med] (of an epidemy; occurring over a vast area) pandémique
PANE - [gen] (flat rectangular piece) vitre *f*
[glass man] plat *m* de verre, carreau *m* de verre
[mech] (the lateral, generally rectangular face of a nut) pan *m*
[impl] (preferably pene, or peen, of a hammer) panne *f*
[carp] (a rectangular panel) panneau *m*
PANEL, to - [gen] diviser en panneaux, recouvrir de panneaux
PANEL - [gen & build] panneau *m*
[carp] (a thin flat wooden piece) panneau *m*
[el] (sheet of slate or other material on which instruments, switches etc are mounted) panneau *m*
[plast ind] planche *f*, panneau *m*
[min] (of a mine) chambres *f*pl isolées

PANEL - [railw] (of a track) travée *f* (de voie)
[aero] (a portion of an aircraft skin, e.g. in a wing, including the stiffening elements associated with it) panneau *m*
[aero] (parachute or balloon; a portion of a gore) fuseau *m*
~BEATER - [auto etc] tôlier *m*
~BOARD - [el] (on a machine etc) panneau *m*
[paper man] carton *m* pour panneaux
[print] encadrement *m*, contour *m*
~BODY TRUCK - [auto] fourgon *m*
~BRACES - [el etc] équerres *fpl* de fixation
~DESIGN - [text] modèle *m* pour tricot à différentes zones
~ENTRY - [mining] galerie *f* d'exploitation
~FILLING - [constr] masque *m*
~FIN - [mech] pointe *f* à tête très large
~FIRE - [el] radiateur *m* mural
~FRAME MOUNTING - [mech] montage *m* sur panneau
~GIRDER - [constr] (girder divided into frame units separated by vertical members) poutre *f* cloisonnée
~HEATING - [heat] (heating system in which the heating elements are concealed in special panels) chauffage *m* par panneaux
~LIGHTING - [el] (arrangements for illuminating an instrument panel) éclairage *m* du panneau
~MOUNTING - [mech] (term used of an instrument, switch or other component designed to be mounted with its bezel above the level of the panel, as distinct from "flushmounting") montage *m* sur panneau
~ROD - [constr] barre *f* de treille
~SELECTOR - [telecomm] (in telephony) sélecteur *m* plan
~-SWITCH - [el] (a flush switch) interrupteur *m* encastré
~SYSTEM - [mining] système *m* d'exploitation par panneaux
PANELLED - [gen] à panneaux, en panneaux
~DOOR - [constr] porte *f* à panneaux
~FRAMING - [carp] lambrissage *m*, panneautage *m*
~WOOD CEILING - [constr] plafond *m* panneaux
PANELLING - [gen] panneautage *m*, division *f* en panneaux
[mining] exploitation *f* par chambres isolées
~MATERIAL - [gen] matériau *m* pour panneaux
PANGAMIC - [zool] (mating indiscriminately) pangamique
PANGLOSSIA - [med] loquacité *f*
PANIC - [gen] panique *f*
~BOLT - [mech] (special safety bolt) boulon *m* de sécurité
PANICLE - [bot] (branched inflorescence) panicule *m*
~OATS - [agric] avoine *f* paniculée
PANIFICATION - [gen] panification *f*
PANMIXIA - [genet] panmixie *f*, croisement *m* au hazard
PANNICULITIS - [med] (inflammation of the subcutaneous fat) fibrosite *f*, panniculite *f*
PANNICULUS - [anat] (dermal muscles covering the trunk and part of the limbs) couche *f* sous-cutanée adipeuse
PANNIER - [gen] panier *m*, verveux *m*
[agric] hotte *f*, benne *f*
[med] pharmacie *f* portative
PANNING - [cin & telev] (the operation of rotating a camera in a vertical or horizontal plane, to keep a moving object within the picture range) panoramique *f*, survol *m*
PANNING SHOT - [cin & telev] prise *f* de vue panoramique
PANOPHTHALMIA - s. panophthalmitis
PANOPHTHALMITIS - [med] (inflammation of the entire structure of the eye) phlegmon *m* de l'oeil
PANORAMA HEAD - [photo] plate-forme *f* panoramique
~VIEW - [photo] vue *f* panoramique
PANORAMIC - [gen] panoramique
~CAMERA - [photo] (camera designed to take very wide-angle views) chambre *f* panoramique
~HEAD - [photo] (of a camera) tête *f* panoramique
~MONITOR - [electron] (a frequency spectrograph) spectrographe *m* de fréquence
~RECEIVER - [radio] (receiver periodically tuned through a specified band of frequencies) récepteur *m* panoramique
~TELESCOPE - [opt] télescope *m* panoramique
~VIEW - [gen & photo] vue *f* panoramique
PANREST - s. pan support
PANSY - [bot] pensée *f*
PANT, to - [gen & acoust] (to breathe with difficulty) panteler, haleter
PANT - [mech] (the sound made by the defective beat of an engine) battement *m* anormal
PANTECHNICON - [transp] fourgon *m* de déménagement
PANTHER - [zool] panthère *f*, couguar *m*
PANTING - [acoust] (the sound produced by heavy difficult breathing) dyspnée *f*, anhélation *f*
~BEAM - [naut] barrot *m* de coqueron
PANTILE - [build] (roofing tile shaped with a double curve) tuile *f* flamande, panne *f*
PANTOPHOBIA - [med] pantophobie *f*
PANTOGRAPH, to - [draw] dessiner au pantographe
PANTOGRAPH - [instr] (mechanism used to copy at any required scale the path traced by another point) pantographe *m*
[el] (sliding type of current collector) pantographe *m*
~FOOT PUMP - [el] (in electric traction) pompe *f* à pédale pour le soulèvement du pantographe
~PAN - [el] semelle *f* de pantographe
~TROLEY - [el] (in trams or railways) trolley *m* de pantographe
PANTOMETER - [instr] (instrument designed to measure all angles in measuring distances etc) pantomètre *m*
PANTOSCOPE - [photo] (a wide-angle lens) pantoscope *m*
PANTOSCOPIC - [photo] (of pantoscope) pantoscopique
PANTOTHENIC ACID - [chem] (a member of the Vitamin B complex with uses as a nutritional supplement) acide *m* panthénique
PANTRY - [gen] (room where provisions, table linen etc are kept) dépense *f*, garde-manger *m*
[naut] office *m* du maître d'hôtel
PAPAIN - [chem] (a proteolytic enzyme derived from unripe fruit of Carica papaya. It has uses in medicine and in food processing) papaïne *f*
PAPAVER - [bot] (poppy capsule. The dried fruits of Papaver somniferum with uses in medicine as a mild sedative) pavot *m*
PAPAVERINE - [chem] (an alkaloid derived from opium; has applications in medicine) papavérine *f*

PAPAYA - [bot] papaye *f*
~TREE - [bot] papayer *m*
PAPER - [gen] papier *m*
[gen] (a printed or written document) écrit *m*, document *m*
[gen] (a newspaper) journal *m*
[gen] (a sheet) feuille *f* de papier
~-BACK - [print] livre *m* broché, livre *m* à bon marché
~BACKED MICA - [el] ruban *m* au mica-papier
~BAG - [gen] cornet *m* en papier, sachet *m* en papier
[gen] (for industrial uses) sac *m* de papier
~BAND - [text] bande *f* de papier
~-BASE TINGE - [photo] teinte *f* du support
~BOARD - [print] plateau *m* à papier
~BOWL - [text] rouleau *m* de papier
~BREAK INDICATOR - [mech] indicateur *m* de rupture
~CABLE - [el] (a dry-core cable) fil *m* sous papier
~CAPACITOR - [el] (capacitor having a thin paperas the dielectric separating aluminium foil electrodes) condensateur *m* à papier
~CARD - [text] carton *m*, carte *f* en papier
~CARRIER - [paper man] sac *m* à anse, porte-sac *m*
~CASE - [gen] serre-papier *m*
~CHROMATOGRAPHY - [ind chem] chromatographie *f* sur papier
~CLIP - [impl] pince-feuilles *m*
~CONDENSER - s. paper capacitor
~CONDITIONING PLANT - [paper man] machine *f* de conditionnement
~CONVERTING - [paper man] façonnage *m* du papier
~CORD - [gen] corde *f* de papier
~COVERED CABLE - [el] câble *m* sous papier
~CUP - [paper man] gobelet *m* de papier
~CURRENCY - s. paper money
~CUTTER - [mech] coupe-papier *m*, tranche-papier *m*
~CYLINDER OF EMBOSSING MACHINE - [text] contre-cylindre *m* de gaufreuse
~DEFORMATION - [comput] (of punch cards) déformation *f* du papier
~DRILL - [mech] (used in printing works) machine *f* à forer le papier
~DRILLING MACHINE - s. paper drill
~FASTENER - s. paper clip
~FELT - [paper man] feutre *m* du papier
[text] feutre *m* pour papeterie
~FIBRE - [paper man] fibre *f* du papier
~FIBRE PLANTS - [bot] plantes *fpl* fibreuses pour papier
~FINGER - [mech] (of a typewriter) presse-feuille *m*
~FLEECE - [paper man] peluche *f*
~FOR COPPERPLATE PRINTING - [print] papier *m* pour taille douce
~GAUGE - [paper man] calibre *m* d'épaisseur du papier
~GRADE - [photo] dégré *m* de contraste d'un papier
~GUIDE - [mech] (of a typewriter) guide-papier *m*
~HANGER - [gen] tapissier *m*
~-INSULATED CABLE - [el] câble *m* sous papier
~-INSULATED WIRE - [el] fil *m* sous papier
~INTERLAYER - [gen] feuille *f* de protection
~INTERLEAVING - [el] (in a winding) isolement *m* des spires (d'un enroulement) sous papier
~LAPPING MACHINE - [mech mach] guipeuse *f* à papier
~LEADERS - [photo] tirants *mpl*
~LEADING ROLL - [print] rouleau *m* guide-papier, rouleau *m* pour guider la bande de papier
PAPER-LINED CONSTRUCTION - [el chem] (method of making dry cells in which a paper liner soaked in electrolyte lies between the negative electrode and the depolarizing mix) montage *m* au papier
~MACHINE - [paper man] (machine converting pulp into paper) machine *f* à papier
~MASK - [photo] cache-papier *m*
~MILDEW - [paper man] moisissure *f* du papier
~MILL - [ind] papeterie *f*, moulin *m* à papier
~MONEY - [fin] billets *mpl* de banque, papier-monnaie *m*
~MOULD - [paper man] flan *m*, moule *m* en papier
~NEGATIVE - [photo] négatif-papier *m*
~PACKING WARP - [text] chaîne *f* de renfort en papier
~PILE - [paper man] (ready for pressing) tympan *m*
~PULP - [paper man] (macerated wood fibre used as a starting material in the manufacture of newsprint) pâte *f* à papier
~REEL - [paper man] bobine *f* de papier, rouleau *m* de papier
~RELEASE LEVER - [mech] (of a typewriter) levier *m* libre-papier
~RESISTANCE TESTER - [paper man] appareil *m* à contrôler la résistance du papier
~REWINDING SPOOL - [print] bobine *f* d'enroulement du papier
~RIBBON - [paper man] ruban *m* de papier
~ROLL - [text] rouleau *m* de papier
~SAC - [paper man] sac *m* en papier
~SCALES - [paper man] balance *f* à papier
~SHAVINGS - [paper man] rognure *f* de papier
~SIZE - [paper man] format *m* du papier
~SIZING - [paper man] encollage *m* du papier
~SKIP - [print] saut *m* de papier
~SLEEVE - [el] manchon *m* de papier, cigarette *f* imprimeur *m* de papiers peints
~SLEEVE VARNISHING MACHINE - [paper man] machine *f* à enduite les tubes en papier
~SPOOL - [paper man] bobine *f* de papier
~STAINER - [paper man] imprimeur *m* de papiers peints
~STEREO PROCESS - [print] prise *f* d'empreinte sur papier
~STREAMER - [paper man] serpentin *m*
~STRETCH - [paper man] allongement *m* du papier
~STRIP - [telecomm] (in telegraphy) zone *f*
~TAPE - [comput] (strip of paper used in storage devices) bande *f* perforée
~TAPE PUNCH - [comput] perforateur *m* de bande de papier
~TAPE READER - [comput] (device restoring information punched on a paper tape to a sequence of electrical pulses) lecteur *m* de bande perforée
~-TAPE-TO-CARD CONVERTER - [comput] convertisseur *m* bande-carte
~TESTING - [paper man] analyse *f* du papier
~TEXTURE - [paper man] texture *f* du papier
~THICKNESS - [paper man] épaisseur *f* du papier
~THROW CHARACTER - [comput] caractère *m* d'alimentation du papier
~TOWER LEVER - [print] levier *m* de mise en marche de la tour
~TREE - [paper man] (any tree whose timber is used for paper making) bois *m* à papier
~TUBE GLUEING MACHINE - [paper man] colleuse *f*

de tubes en papier
PAPER YARN - [text] fil *m* de papier
~ YARN WEAVING - [text] tissage *m* du fil de papier
~ WEB - [paper man] (continuous sheet or band of paper, as distinct from cut sheets) bande *f* de papier
~ WEIGHT - [impl] presse-papiers *m*
~ WINDING REEL - [print] bobineuse *f*
~ WOOL - [paper man] déchets *mpl* de papier
~ -WRAPPED SECURING MACHINE - [paper man] machine *f* à râper les dos
PAPERBOARD - [paper man] carton *m*
PAPERHANGER - [gen] colleur *m* de papiers peints
PAPERHANGER'S PASTE - [chem] colle *f* pour papiers peints
PAPERMAKER - [paper man] fabricant *m* de papier
PAPERMAKER'S SOAP - [chem] savon *m* des papetiers
PAPIER-MACHE - [paper man] (a mixture of paper pulp and glue or some other hardening material) papier *m* mâché
PAPILLA - [zool] (small conical projection of soft tissue) papille *f*
PAPILLAR - [zool] papillaire
~ LAYER - [anat] papilles *fpl* dermiques
~ MUSCLES - [anat] muscles mpl papillaires, piliers mpl du coeur
PAPILLITIS - [med] (an optic neuritis) papillite *f*
PAPILLOMA - [med] papilloma *m*
PAPILLOMATOSIS - [med] papillomatose *f*
PAPIN'S DIGESTER - [ind chem] (vessel used for treatment of substances under moderate heat and pressure, obtained by boiling water under sealed conditions) marmite *f* de Papin, digesteur *m*
PAPRIKA - [agric] paprika *m*
PAPULOPUSTULE - [med] papule *f* pustuleuse
PAPYRINE - [paper man] papyrine *f*
PAPYRUS - [bot] papyrus *m*
P.A.R. - s. Precision-Approach Radar
PAR - [gen comm etc] pair *m*, égalité *f*
~ WITH - [gen] au niveau de, de pair avec
PARA - (prefix denoting an isomeric form of di-substitution product derived from benzene) para
~ -ANALGESIA - [med] para-analgésie *f*
~ POSITION - [chem] position *f* para
~ RUBBER - [rubber ind] caoutchouc *m* para, caoutchouc *m* de Para
PARAAMINO-BENZOIC ACID - [chem] acide *m* para-amino benzoique
PARABIOSIS - [biol] (congenital fusion of two individuals) parabiose *f*
PARABIOTIC TWINS - [biol] frères mpl siamois
PARABLASTIC - [zool] (of the yolk of a meroblastic egg) parablastique
PARABLE - [gen] parabole *f*
PARABLEPSIA - [med] vision *f* erronnée, perversion *f* de la vision
PARABOLA - [geom] (the conic section resulting from cutting a cone by a plane parallel to a generator of the cone) parabole *f*
PARABOLIC - [gen] parabolique
~ AERIAL - [radio & telev] antenne *f* parabolique
~ ARCH - [arch] (with the outline of a parabola) arc *m* parabolique
~ BRANCH - [math] branche *f* de la parabole
~ CATENARY - [math] caténaire *f* parabolique
~ COORDINATES - [math] coordonnées *fpl* paraboliques
~ CYLINDER - [mech] cylindre *m* parabolique
PARABOLIC GIRDER - [constr] poutre *f* parabolique
~ GOVERNOR - [contr] régulateur *m* parabolique
~ MIRROR - s. parabolic reflector
~ NOZZLE - [mech] (nozzle of parabolic section with high discharge ratio) gicleur *m* à section parabolique
~ ORBIT - [astronaut] orbite *f* parabolique
~ REFLECTOR - [radio] (of an aerial, a reflector which is a portion of a paraboloid of revolution) réflecteur *m* parabolique
~ -REFLECTOR MICROPHONE - [el acoust] (microphone fitted with a parabolic reflector to enhance its sensitivity) microphone *m* à réflecteur parabolique
~ SHADING - [telev] irrégularité *f* parabolique dans le noir
~ VELOCITY - [phys] vitesse *f* parabolique
[astronaut] vitesse *f* parabolique
~ WAVEFORM SWITCHING SIGNAL - [telev] signal *m* commutateur à forme d'onde parabolique
PARABOLOID - [geom] (surface generated by the rotation of a parabola about its axis) paraboloïde *m*
~ HEAD-LAMP - [auto] phare *m* à parabole
~ OF REVOLUTION - s. parabolid
PARABULIA - [med] paraboulie *f*
PARACANTHOMA - [med] tumeur *f* de la couche de Malpighi
PARACANTHOSIS - s. paracanthoma
PARACENTESIS - [med] (the puncture of hollow cavities by means of a hollow needle) ponction *f* abdominale
PARA-CHLOROTHIOPHENOL - [chem] (a component of plasticizers and rubber chemicals) parachlorothio-phénol *m*
PARACHOLIA - [med] paracholie *f*
PARACHROMA - [med] changement *m* de couleur
PARACHUTE, to - [aero] descendre en parachute
PARACHUTE - [aero] (umbrella-like structure designed to retard the fall of a body attached to it) parachute *f*
[brew ind] (automatic skinning device) écumeur *m* automatique
[aero] (device used for decelerating an aircraft, a minecage etc) parachute *f*
~ ASSEMBLY - [aero] (a parachute complete with all harness and deployment devices) ensemble *m* de parachute
~ BOAT - [naut & aero] bateau *m* pneumatique de sauvetage
~ BRAKE EXTRACTOR - [aero] (a small parachute designed to be released from the tail of an alighting aircraft to pull out and deploy the parachute brake) extracteur *m* du parachute de freinage
~ BRAKE OPENING PANEL - [aero] (the door which covers the parachute brake recess and opens to allow it to deploy) couvercle *m* de la cavité du parachute de freinage
~ BUCKET - [aero] (the container in which the parachute pack a supply-parachute assembly is packed) logement *m* à parachute
~ CANOPY - [aero] coupole *f* du parachute
~ DESCENT - [aero] descente *f* en parachute
~ FLARE - [aero] (a flare attached to a parachute so as to fall slowly and give a longer period of illumination) fusée *f* à parachute
~ HARNESS - [aero] (the system of straps, webs and buckles by which a parachute is attached to its load) ceinture *f* à parachute
~ PACK - s. pack

PARACHUTE PACKER - [aero] (a person specially skilled in stowing parachutes in their packs) ensacheur *m* de parachute
~RIGGING LINES - [aero] suspentes *f*pl du parachute
~RIP-CORD - [aero] câble *m* de déclenchement du parachute
~RIP-LEVER - [aero] levier *m* de déclenchement du parachute
~SHROUD - [aero] (one of the lines by which a parachute is attached to the harness) suspente *f* de parachute
~SUSPENSION LINES - s. parachute rigging lines
~TOWER - [aero] (a tower used for training personnel in parachute jumping) tour *f* d'entraînement de parachutistes
~TRAY - [aero] (special type of pack, consisting of a metal tray with a fabric cover) baquet *m* de parachute
~VENT - [aero] (an opening at the apex a parachute to allow air to escape and give stability) cheminée *f* du parachute
PARACHUTIST - [aero] (the person who descends by means of a parachute) parachutiste *m*
PARACINESIA - [med] parakinésie *f*
PARACINESIS - [med] s. paracinesia
PARACME - [med] période *f* de déclin d'une maladie, rémission
PARACOLPIUM - [anat] espace *m* paravaginal
PARACUSIA - [med] paracousie *f*
PARACUSIS - [med] s. paracusia
PARACYSTITIS - [med] paracystite *f*, phlegmon *m* prévésical
PARADENTITIS - [med] paradentite *f*
PARADONTITIS - s. paradentitis
PARAEPILEPSY - [med] para-épilepsie *f*
PARAESTHESIA - [med] (abnormal sensation) paresthésie *f*
PARAFFIN - [chem] s. paraffins
[chem](paraffin oil) pétrole *m*
[ind chem] (solid paraffin) paraffine *f*
[ind chem] (only in the USA) huile *f* pétrole
~BASE - [chem] base *f* paraffinique
~-BASE OIL - [oil ind] pétrole *m* paraffinique
~GO-DEVIL - [oil ind] passediable *m* à paraffine
~HYDROCARBONS - s. paraffins
~OIL - [chem] (in G.B.) huile *f* de pétrole, huile *f* lampante
~PAPER - [paper man] papier *m* paraffiné
~RING - [text] anneau *m* de paraffine
~SCRAPER - [oil ind] racleur *m* à paraffine
~TROUBLE - [oil ind] embarras *m* de paraffine
~TEST - [metall] contrôle *m* par ressuage
~WAX - [chem] (a white wax consisting of a mixture of the higher hydrocarbons and used in the manufacture of candles, polishes, pharmaceuticals, electrical insulants, lubricants, and waxed paper) cire *f* de paraffine
PARAFFINIC DISTILLATES - [chem] (a liquid fuel) produits *m*pl de distillation paraffiniques
PARAFFINS - [chem] (alakanes. A large family of hydrocarbons of the methane series) paraffines *f*pl
PARAFORM - s. paraformaldehyde
PARAFORMALDEHYDE - [chem] (a component of synthetic resins used for plywood adhesives and a fungicide and disinfectant) paraformaldéhyde *f*
PARAGENESIS - [geol] (the formation of minerals in contact in such way that it affects the development of the individual crystals) paragénèse *f*
PARAGLOSSIA - [med] paraglossite *f*
PARAGNEISS - [geol] (gneissose rocks derived from detrital sedimentary rocks) paragneiss *m*
PARAGON - [gen] paragon *m*, modèle *m*
[print] paragon *m*, texte *m* (d'une annonce)
PARAGONITE - [min] (a silicate of sodium, minium and hydrogen; a sodium mica) paragonite *f*
PARAGRAPHIA - [med] (a misplacement of letters or words due to a lesion in the brain) paragraphie *f*
PARAHEPATITIS - [med] périhépatite
PARAHIDROSIS - [med] transpiration *f* anormale
PARAHYDROGEN - [chem] (hydrogen molecule in which the two nuclear spins are anti-parallel) parahydrogène *m*
PARALALIA - [med] paralalie *f*
PARALDEHYDE - [chem] (a condensation product of acetaldehyde. It is an intermediate for dyestuffs and rubber chemicals and a hypnotic used in medicine) paraldéhyde *f*
PARALGESIA - [med] paresthésie *f* douloureuse
PARALLACTIC - [opt etc] (of parallax) parallactique
~ANGLE - [opt etc] (the angle of parallax) angle *m* de parallaxe
~DISPLACEMENT - [opt] déplacement *m* parallactique
~ELLIPSE - [astr] (the small ellipse of the celestial sphere seemingly described by every star in a year about its mean position) parallaxe *f*
~INEQUALITY - [astr] (periodic term in the mathematical expression of the moon's motion) inégalité *f* parallactique
PARALLAX - [opt, astr etc] (the difference between the apparent and the real positions of a celestial body) parallaxe *f*
~ADJUSTMENT - [photo] correction *f* de la parallaxe
~COMPENSATION - s. parallax adjustment
~-COMPENSATION MARK - [photo] repère *m* de compensation de la parallaxe
~CORRECTOR PRISM - [photo] prisme *m* pour la correction de la parallaxe
~ERROR - [opt] (an observer's error when measuring) erreur *f* de parallaxe
~-FREE MIRROR - [instr] (strip of mirror in the dial of an indicating instrument; parallax errors are avoided by aligning the pointer with its image in the mirror) miroir *m* sans parallaxe
~IN ALTITUDE - [astr] (the angle between a straight line from the centre of the earth to a celestial body and one from the body to an observer) parallaxe *f* d'altitude
~STEREOGRAM - [photo] (the use of a line screen in front of a positive transparency of alternate strips of two views of an object) stéréogramme *m* à parallaxe
PARALLEL, to - [gen] placer parallèlement
[el] synchroniser
PARALLEL - [geom] parallèle
[geogr] (a degree of latitude) parallèle *m*
[mech] (with a constant cross section) cylindrique
[impl] calibre *m* d'épaisseur
[el] (conductors are in parallel with one another when the current flowing in the circuit is divided between the conductors) en dérivation, en parallèle
~ACCESS - [comput] accès *m* parallèle
~ADDER - [comput] additionneur *m* parallèle
~AND LONGITUDINAL NERVES - [rubber ind] (of tyres) nervures *f*pl parallèles et longitudinales

PARALLEL ANTENNA TUNING - [radio] accord *m* d'antenne en parallèle
~ARITHMETIC MODE - [math] méthode *f* arithmétique parallèle
~ARITHMETIC UNIT - [comput] unité *f* arithmétique parallèle
~ARRANGEMENT - [el] montage *m* en parallèle
~BATTERY - [el] pile *f* en batterie
~BLOCK - [tool] parallèle *m*
~BROACH - [tool] équarissoir *m* cylindrique
~CASCADE ACTION - [contr] (the regulation of the set points of two or more automatic controllers) action *f* parallèle en cascade
~CHEESE - [text] bobine *f* croisée cylindrique
~CIRCUITS - [el] (or shunt circuit; circuits are in parallel when the current or flux is divided between them) circuits *mpl* en dérivation, circuits *mpl* en parallèle
~CONNECTED TRANSFORMER - s. parallel transformer
~CONNECTION - [el] (method of connecting two or more cells, so that all the positive terminals are joined, to give two terminals only) couplage *m* en parallèle
~CORDS - [text] cordes *fpl* parallèles
~DAMPING - [radio] amortissement *m* parallèle
~DIGITAL COMPUTER - [comput] (computers in which the digits are handles in parallel) calculateur *m* digital parallèle
~EQUALIZER - [el] connexion *f* équipotentielle en dérivation
~FEED - [electron] (method of connecting the anode of a thermionic valve to the high-tension supply through a high resistance or inductance, while the a.c. circuits are connected through a condenser) alimentation *f* parallèle
~FEEDER - [el] (connected in parallel with an existing feeder) feeder *m* à brins en parallèle
~FEEDWHEEL - [text] roue *f* de fournisseur cylindrique
~FIBRE FEED - [text] alimentation *f* en disposant les fibres dans le sens de la longueur
~FLOW - [electron] (the flow which occurs in the same direction of two streams within a system) flux *m* à courant parallèle
~-FLOW APPARATUS - [ind chem] appareil *m* à courants parallèles
~FLOW GASIFICATION - [ind chem] gazéification *f* en courants parallèles
~FLYING - [aero] vol *m* en parallèle
~FORCES - [mech] (system of forces so arranged that the lines of action of all forces are parallel to each other)forces *fpl* parallèles
~FULL ADDER - [math] additionneur *m* pour traitement parallèle
~FULL SUBTRACTER - [math] unité *f* de soustraction pour traitement parallèle
~GROOVED PROFILE - [rubber ind] (of tyres) sculpture *f* à sillons longitudinaux
~GROOVED TREAD - s. parallel grooved profile
~GROWTH - [min] surcroissance *f* parallèle
~HOP - [auto] (a wheel hop, in which a pair of wheels hop in phase) oscillation *f* parallèle
~IMPEDANCE - [el] (antiresonance) antirésonance *f*
~INDUCTION - [el] induction *f* propre en parallèle
~-JAW PLIERS - [impl] (special pliers incorporating a parallel motion in the jaws) pinces *fpl* parallèles

PARALLEL-JAW VICE - [tool] étau *m* parallèle
~LINES - [geom] lignes *fpl* parallèles
~LINK MOTION - [text] parallélogramme
~LINKS - [text] levier *m* en parallélogramme
~LYING THREADS - [text] fils *mpl* disposés parallèlement
~MAGAZINE APPARATUS - [cin] (camera supplied with two parallel magazines for the unexposed and the exposed films) appareil *m* à chargeurs parallèles
~MEMORY - [comput] (a storage in which all bits etc are equally available in space without time being one of the coordinates) mémoire *f* parallèle
~MOTION - [mech] (system of links by which the reciprocating motion of one point is copied to a larger scale by another) mouvement *m* parallèle
~OF ALTITUDE - [astr] parallèle *m* de hauteur, cercle *m* de hauteur
~OF DECLINATION - [astr] (a small circle on the celestial sphere the plane of which is parallel to that of the equinoctial)parallèle *m* de déclinaison
~OPERATION - [comput] (flow of information through the computer using more than one line simultaneously) opération *f* parallèle
~ORDER OF THREADS - [text] disposition *f* parallèle des fils
~PATHS - [el] dérivations *fpl* parallèles, voies *fpl* d'enroulement parallèles
~PERSPECTIVE - [draw] perspective *f* en vue de face
~PIN - [mech] goupille *f* parallèle
~PLANER - [tool] raboteuse *f* pour épaisseurs
~PLATE CHAMBER - [nucl] (ionization chamber with plane parallel electrodes) chambre *f* d'ionisation à plaques parallèles
~-PLATE COUNTER TUBE - [electron] tube *m* compteur à plaques parallèles
~-PLATE LENS - [electron] (of a waveguide) lentille *f* à lames parallèles
~PLATE OSCILLATOR - [el] oscillateur *m* à plaques parallèles
~PLATEGUIDE - [electron] (region limited by two parallel plates in which electromagnetic energy can be propagated) guide *m* d'ondes à plaques parallèles
~PRINTER - [comput] imprimante *f* parallèle
~RAY - [opt] (a ray from a very distant point reaches an optical system as a parallel ray) rayon *m* parallèle
~READOUT - [comput] lecture *f* parallèle
~RECTIFIER CIRCUIT - [el] couplage *m* en parallèle de redresseur
~REPRESENTATION - [comput] représentation *f* parallèle
~RESONANCE - [radio] (the steady-state condition in a circuit comprising inductance and capacitance connected in parallel, when the current entering the circuit in phase with the voltage across the circuit) résonance *f* parallèle
~ROUND FILE - [impl] lime *f* ronde parallèle
~RULERS - [instr] (a pair of rulers connected together in such a way as to remain parallel when moved a part) règle *f* à tracer des parallèles
~SEAM WELDING - [metall] soudure *f* continue parallèle
~SERIAL CONVERTER - [comput] convertisseur *m* parallèle-série
~SHANK TWIST DRILL - [tool] foret *m* hélicoïdal à queue cylindrique

PARALLEL SLOTS - [el] (the recesses in a core, in which the windings are laid, on parallel lines) encoches *f*pl parallèles
~SPOT WELDING - [metall] soudure *f* parallèle à points
~STORAGE - s. parallel memory
~STORE - s. parallel memory
~SYSTEM OF DISTRIBUTION - [el] (or shunt; a system of distribution in which the consuming devices are connected in such way as to have the same nominal voltage applied to them) réseau *m* à tension constante
~T-NETWORK - [radio] réseau *m* en T en dérivation
~TRANSDUCTOR - [el] transducteur *m* à couplage parallèle
~TRANSFER - [comput] (system of transfer in which the characters of an information are transferred simultaneously over a set of paths) transfert *m* parallèle
~TRANSFORMER - [el] (transformer having the terminals of the same polarity electrically connected together) transformateur *m* à couplage parallèle
~TRANSMISSION - [radio] (for telecommunication) transmission *f* parallèle
~V BLOCK - [tool] parallèle *m* à V
~VICE - [mech] étau *m* parallèle
~WINDING - [text] enroulement *m* parallèle
~-WIRE LINE - [telecomm] (transmission line composed of two parallel wires) ligne *f* à deux fils parallèles
PARALLELEPIPED - [geom] (a prism, the six faces of which are parallelograms) parallélépipède *m*
PARALLELEPIPEDON - [geom] parallélépipède
PARALLELING - [el] (the connecting together of the conductors of the same polarity) mise *f* en parallèle [text] s. parallelization of the threads
~OF ALTERNATION - [el] accrochage *m* d'une machine synchrone
~OF SYNCHRONOUS MACHINES - [el] mise *f* en parallèle de deux machines synchrones
PARALLELISATION OF THE THREADS - [text] parallélisation *f* des fils
PARALLELISM - [gen etc] (the condition of being parallel) parallélisme *m*
PARALLELOGRAM - [geom] (four-sided plane figure whose sides are parallel) parallélogramme *m*
~OF FORCES - [mech] (a geometrical representation by which the sim of difference of two concurrent forces can be considered the diagonals of a parallelogram) parallélogramme *m* de force
~OF VELOCITY - [phys] (graphical representation in which the lines of the velocities form the diagonals) parallélogramme *m* des vitesses
PARALLELS - [metall & plast ind] (parallel-sided spacing bars used in moulding) cales *f*pl d'épaisseur
PARALLERGIA - [med] parallergie *f*
PARALOGIA - [med] paralogie *f*
PARALYSIS - [med] paralysie *f*
[radar] (blocking which occurs when strong signals enter the receiver from the transmitter) blocage *m*
PARALYTIC - [med] paralytique
PARAMAGNETIC - [el] (capable of being attracted magnetically) paramagnétique
~ABSORPTION - [nucl] (the absorption of particles by paramagnetic substances) absorption *f* paramagnétique
PARAMAGNETIC ELEMENT - s. paramagnetic substance
~SUBSTANCE - [el] (any substance which has a permeability greater than that of a vacuum) substance *f* paramagnétique
PARAMAGNETISM - [el] paramagnétisme *m*
PARAMATTA - [text] (twill fabric made into waterproof garments) paramatta *m*
PARAMETER - [math] (a quantity constant under specific conditions) paramètre *m*
~SETTING INSTRUCTIONS - [comput] instruction *f* de substitution "paramètre"
PARAMETHADIONE - [chem] (an anticonvulsant used in the treatment of epilepsy) paraméthadione *f*
PARAMETRAL PLANE - [metall] plan *m* paramétrique
~RATIO - [metall] rapport *m* paramétrique
PARAMETRIC ABSCESS - [med] phlegmon *m* du ligament large
~EQUATIONS - [math] (of a curve or surface) équations *f*pl paramétriques
~VARIATION - [contr] variation *f* paramétrique
PARAMETRITIS - [med] (inflammation of the pelvic cellular tissue in the area of the uterus) pelvicellulite *f*
PARAMETRIUM - [anat] (subperitoneal connective tissue surrounding the uterus) paramétrium *m*, base *f* du ligament large
PARAMETRON - [comput] (digital computing element utilizing parametric oscillations) paramétron *m*
PARAMIMIA - [med] paramimie *f*
PARAMITOSIS - [zool] (a form of mitosis) paramitose *f*
PARAMNESIA - [med] paramnésie *f*
PARAMORPH - [min] (term applied to minerals which can change their molecular constitution without changes of their chemical substance) paramorphe
PARAMORPHISM - [min] paramorphisme *m*
PARANALGESIA - [med] analgésie *f* des membres inférieurs
PARANEPHRIC ABSCESS - [med] abcès *m* périnéphritique
PARANOIA - [med] (psychosis based on the development of a permanent delusional system and accompanied by the preservation of clear thought and action) paranoïa *f*, délire *m* systématisé
~SCHIZOPHRENIA - [med] (type of schizophrenia showing symptoms of paranoia) schizophrénie *f* paranoïde
~STATE - [med] (a condition in which a person shows some characteristics of paranoia, yet without disintegration of the personality) condition *f* paranoïde
PARANUCLEAR - [genet] (of any structure lying beside the nucleus) paranucléaire
PARANUCLEUS - [genet] (term denoting any structure lying beside the nucleus) paranucléus *m*
PARAPATHY - [med] psychonévrose *f*
PARAPET - [constr] (a low wall along the edge of a roof, bridge etc) parapet *m*, garde-feu *m*
PARAPHASE AMPLIFIER - [electron] (an amplifier a single output is divided into two inputs, coupled in opposite phase relationship) amplificateur *m* déphaseur
PARAPHASIA - [med] (defect of speech due to a lesion in the brain) paraphasie *f*, hétérophasie *f*
PARAPHIA - [med] anomalie *f* du sens de toucher
PARAPHIMOSIS - [med] paraphimosis *f*

PARAPHOBIA - [med] phobie *f* atténuée
PARAPHONIA - [med] paraphonie *f*
PARAPHRASE AMPLIFIER - [electron] amplificateur *m* en push-pull
PARAPHRENIA - [med] (a type of schizophrenia) état *m* paranoïde
PARAPLASM - [biol] (the inactive part of the cytoplasm) paraplasme *m*
PARAPLEGIA - [med] paraplégie *f*
PARAPOSITRONIUM - [nucl] parapositonium *m*
PARAPRAXIA - [med] parapraxie *f*
PARAPRAXIS - s. parapraxia
PARAPROCTIUM - [anat] espace *m* pararectal
PARAPSIS - s. paraphia
PARAROSANILINE - [chem] (a triamino-triphenylcarbinol) pararosaniline *f*
PARAROSOLIC ACID - [chem] (aurine) acide *m* pararosolique
PARARTHRIA - [med] pararthrie *f*
PARASELENE - [met] (also called moon-dog, mock moon; phenomenon analogous to a parhelion, but produced by the moon) parasélène *f*
PARASHEET - [aero] (a parachute made from pieces of material with the warp parallel, the rigging lines being attached to the angles of the polygon so formed) parasheet *m*, parachute *f* "Mouchoir"
PARASITE - [gen] (animal of plant living in or on another organism) parasite *m*
PARASITIC - [gen] parasite, parasitaire
[el] (said of losses in an electrical machine due to eddy currents) parasite
[electron] (said of undesired self-sustaining oscillations in a tube network) parasite
~ AERIAL - [radio] (that part of a directional aerial which is not directly connected to the transmitter or receiver) antenne *f* à alimentation indirecte
~ CAPTURE - [nucl] (absorption of a neutron which does not result in a fission or in the production of an element) capture *f* parasite
~ DISEASES - [bot] (plant disease) maladies *f*pl parasitaires
~ DRAG - [aero] (any form of drag other than the drag associated with the production of lift) traînée *f* passive
~ ELEMENT - [radio] (of aerials) élément *m* parasitaire
~ FREQUENCY - [electron] fréquence *f* parasite
~ INFESTATION - [vet] infestation *f* parasitaire
~ LOSS - [el] (term denoting loss in electric machines due to eddy currents in any part of the machine) pertes *f*pl parasitaires
~ NOISE - [telecomm] parasite *m*, brouillage *m*, interférence
~ OSCILLATIONS - [electron] (undesired oscillations in a tube network) oscillations *f*pl parasites
~ PLANT - [agric] plante *f* parasite
~ RESISTANCE - s. parasitic drag
~ STOPPER - [electron] éliminateur *m* d'oscillations parasites
~ SUPPRESSOR - [electron] résistance *f* antiparasite
~ WEEDS - [bot] herbes *f*pl gourmandes
PARASITICALLY EXCITED ANTENNA - s. parasitic aerial
PARASITICIDE - [chem] parasiticide *m*
PARASITISM - [biol] (partnership between two organisms which is harmful to one of them) parasitisme *m*

PARASOL - [impl] parasol *m*, ombrelle *f*
[aero] (said of an aircraft in which the main plane is mounted above the fuselage) parasol *m*
~ MONOPLANE - [aero] (type of monoplane in which the main plane is mounted above the fuselage on a special supporting structure) monoplan *m* parasol
~ WING - [aero] aile *f* parasol
PARASSOMNIA - [med] parasomnie *f*
PARATHYROID - [anat] parathyroïde *f*
[biol] (a ducteless gland controlling the level of calcium in the blood plasma) glande *f* parathyroïde
~ GLAND - s. parathyroid
PARATOMY - [genet] (form of reproduction by fission in Annelida) paratomie *f*
PARATRIPSIS - [med] écorchure *f*, attrition
PARATROOPER FRACTURE - [med] fracture *f* de parachutistes
PARATROOPS - [milit] (soldiers expressly trained to be dropped by parachute, e.g. for invasion) parachutistes *m*pl
PARAURETHRITIS - [med] péri-urétrite *f*
PARAVANE - [naut] (torpedo-shaped device having sharp projecting teeth for cutting the moorings of sunken mines) paravane *m*, pare-mines *m*
PARAWING - [aero] aile *f* parachute
PARAXIAL - [opt] (the path of a ray parallel to the axis of an optical system) paraxial
~ RAY - [opt] rayon *m* paraxial
PARBOIL, to - [gen] (to boil partially) échauder, faire bouillir à demi
PARBUCKLE - [impl] (sling made by passing both ends of a rope through its bight) trévire *f*
PARC NOISE - [telev] (noise produced by high-frequency luminance signals) bruit *m* de chrominance dû à la diaphonie de bruit de luminance
PARCEL, to - [gen] parceller, partager
[naut] (a rope) limander
PARCEL - [gen] paquet *m*, colis *m*
[comm] pièce *f*, morceau *m*, parcelle *f*
[metall] rouleau *m* de fil
~ GRID - [transp] (attached to a vehicle) porte-colis *m* à grille
~ OF ELECTRONS - [electron] paquet *m* d'électrons
~ OUT, to - [comm] parceller, partager
~ PLATING - [metall] dêpot *m* limite
~ POST - [gen] poste *f* aux paquets, service *m* de messageries
~ RACK - [railw aero] filet *m* à bagages
PARCELLING - [naut] limande *f*
PARCHMENT - [gen] (the skin of sheep, goats etc prepared with pumice stone for writing) parchemin *m*
[paper man] (limitation parchment made by treating paper with sulphuric acid and water) papier *m* parchemin
~ BINDING - [bookbind] reliure *f* en parchemin
~ DRESSING - [leather ind] tannage *m* en parchemin
~ GLUE - [ind chem] colle *f* de peau
~-INSULATED WIRE - [el] fil *m* isolé au parchemin
~ LEATHER - [leather ind] cuir *m* parchemin
~ PAPER - [paper man] s. parchment
PARCHMENTIZATION - [paper man] (the treatment of paper with sulphuric acid and water) parcheminerie *f*
PARCHMENTIZING EFFECT - [text] effet *m* parcheminé
PARENCEPHALITIS - [med] inflammation *f* du cerve-

let
PARENCEPHALOCELE - [anat] hernie *f* du cervelet
PARENCHYMA - [anat & bot] (a soft spongy tissue) parenchyme *m*
PARENT - [genet] parent *m*
[chem] substance *f* père
[comm] (of a firm) (compagnie) mère *f*
[gen] origine *f*, source *f*
~ATOM - [nucl] (the atom which contains the original nucleus) atome *m* père, père *m* atomique
~CELL - [biol & bot] cellule *f* mère
~MAGMA - [geol] magma *m* primaire
~MASS PEAK - [spectrol] raie *f* mère, crête *f* principale
~MATERIAL - [geol] matière *f* originale
~METAL - [metall] matrice *f*, métal *m* base
~NUCLIDE - [nucl] (radio nuclide which yields a specified nuclide on disintegration) nucléide *m* père
~PLANT - [agric] plante-mère *f*
~ROCK - [geol] roche mère *f*
PARENTAL GENERATION - [genet] génération *f* des parents
PARENTERAL - [med] (of a mode of assimilation other than through the alimentary canals) parentéral
~ADMINISTRATION - [med] introduction *f* par voie parentérale
PAREPICOELE - [med] recessus *m* latéral du quatrième ventricule
PARESIS - [med] (incomplete paralysis) parésie *f*, paralysie *f* incomplète
PARFOCAL - [opt] (having the lower focal points all in the same plane) parfocal
PARGASITE - [min] (monoclinic amphibole similar to horn-blende) pargasite *f*
PARGE-WORK - s. pargeting
PARGET - [constr] (lime and hair mortar mixed with cow dung) plâtre *m*, crépi *m*
PARGETING - [constr] (to plaster the interior of a flue with a lime and hair mortar mixed with cow dung) plâtrage *m*, crépissure *f*
PARHELIC CIRCLE - [astr] (the circular halo surrounding the sun when a parhelion occurs) cercle *m* parhélique
PARIDIGITATE - [zool] paridigité
PARIES - [anat] paroi *f*
PARIETAL - [gen & anat] (pertaining to the wall of a structure) pariétal
[anat] s. parietal bone
~BONE - [anat] (paired dorsal membrane bone of the skull) pariétal *m*
~CELLS - [biol] (oxyntic cells of a gastric gland) cellules *fpl* pariétales
~LOBE - [anat] lobe *m* pariétal
PARING - [gen] (the operation of cutting off a surface or edge etc) rognage *m*, parage *m*, ébarbage *m*
[gen] (the part thus cut off) rognure *f*
[bot] (a peel, skin etc) épluchement *m*
[leather ind] dolage *m* des peaux
~KNIFE - [impl] tranchet *m* de relieur
~MACHINE - [mech] (machine for cutting a continuous sheet from a log of cellular plastic) trancheuse *f* pour le bois de bout
PARIS BLACK - [paint] (alternative name for Lampblack, a fine black pigment obtained by substances such as tar, turpentine and the like in a restricted air supply) noir *m* de fumée
PARIS GREEN - [paint] (alternative name for Emerald Green, a complex acetate-arsenite of copper, occasionally used as a pigment because of its peculiar blue-green colour) vert *m* émeraude
~PLASTER - [soil] sulfate *m* de calcium (ou de chaux)
~VIOLET - [paint] (alternative name for methyl violet) violet *m* méthylique
~WHITE - [paint] (a filler consisting of purified chalk) blanc *m* de Paris, blanc *m* de Meudon
~YELLOW - [paint] (a synonym for neutral lead chromate) jaune *m* de Paris
PARISH ROAD - [gen] chemin *m* vicinal
PARISIAN BLUE - [paint] (a form of Prussian Blue, ferric ferrocyanide, used as a blue pigment) bleu *m* de Paris, bleu *m* de Prusse
PARISON - [plast ind] (an intermediate form in making hollow ware) paraison *f*
~SWELL - [glass & plast ind] (the ratio of the cross-sectional area of the parison to that of the die cavity) gonflement *m* de la paraison
PARITY - [gen] parité *f*, égalité *f*
~BIT - [comput] bit *m* de parité
PARK, to - [auto etc] parquer, garer
PARK - [gen] parc *m*
[gen] (of vehicles in general) parc *m*, stationnement *m*
[auto] garage *m* pour autos
PARK'S PROCESS - [ind chem] (a commercial method for the separation of silver from lead, based on the fact that zinc forms compounds with gold and silver which are not readily soluble in lead) procéde *m* Parke
PARKER'S CEMENT - [constr] ciment *m* romain
PARKERISE, to - [metall] (registered process to protect steel and iron with a protective coating) parkériser
PARKERIZED STEEL - [metall] acier *m* parkérisé
PARKERIZING - [metall] parkérisation
PARKING - [gen] (of vehicles) parcage *m*, stationnement *m*
~AREA - [aero] (space in an airfield set apart for aircraft or vehicles temporarily out of use) endroit *m* de stationnement
~BRAKE - [auto] (hand brake) frein *m* à main
~LIGHTS - [auto] feu *m* de stationnement
~LOT - [auto] parc *m* de stationnement
~METER - [auto] taximètre *m* pour stationnement
~ORBIT - [astronaut] orbite *f* d'attente
~PLACE - s. parking area
PARKINSON'S DISEASE - [med] maladie *f* de Parkinson, paralysie *f* agitante
PARKWAY - [gen] (wide throughfare lined with trees) allée *f*
PARLOUR - [constr] (room where callers are received) petit salon *m*, parloir *m*
PAROMPHALOCELE - [med] paromphalocèle *f*
PARONIRIA - [med] rêve *m* morbide
PARONYCHIA - [med] (a witlow) panaris *m* périunguéal
PAROSMIA - [med] hallucination *f* olfactive
PAROTID GLAND - [anat] (a salivary gland) glande *f* parotide
PAROTIDIS - s. parotitis
PAROTITIS - [med] (inflammation of the parotid gland) parotite *f*
PAROXYSM - [med] paroxysme *m*, crise *f*
PARQUET - [carp] (floor covering consisting of hard-

wood blocks) parquet *m*
PARQUETRY - [constr] parquetage *m*, parqueterie *f*
PARREL - [naut] (a sliding rope by which a yard is attached to a mast) racage *m*, matagot *m*
~TRUCK - [naut] pomme *f* de racage
PARROT - [zool] perroquet *m*
~COAL - [min] charbon *m* à longues flammes
~DISEASE - [med] (psittacosis) psittacose *f*
~-NOSE PIPE-WRENCH - [impl] clef *f* à fer creux
PARSLEY - [bot] persil *m*
~FERN - [bot] fougère *f* femelle
PARSNIP - [bot] panais *m*
PARSONSITE - [min] (lead, uranium and phosphor containing mineral) parsonsite *f*
PART, to - [gen] séparer (en deux), diviser
[gen] (to share out) partager, répartir
[mech] (a bar, rod) rompre, céder
[naut] (of a chain, or rope) rompre
[chem] départir
PART - [gen] partie *f*
[gen & mech] pièce *f*
~-LOAD EFFICIENCY - [mech] (efficiency of an engine at less than full load) rendement *m* à charge partielle
~-SPENT OXIDE - [chem] (in chemical purification) matière *f* usagée
~-TIME - [gen] en chômage partiel
~-TIME LEASED CIRCUIT - [telecomm] circuit *m* loué pour une partie de la journée
PARTHENOGENESIS - [genet] parthénogénèse *f*
PARTHENOGENETIC - [genet] parthénogénétique
PARTIAL - [gen] partial, partiel
~AUTOMATIC SUBSTATION - [el] poste *m* électrique semi-automatique
~BOILING BATH - [text] bain *m* d'assouplissage
~CARRY - [comput] (partial transfer of tens) report *m* partiel
~COMBUSTION - [chem] (incomplete combustion) combustion *f* incomplète
~CONDENSATION - [chem] (the partial condensation of a mixed vapour, thus increasing the concentration of the more volatile constituents in the remaining vapour) condensation *f* partielle
~DIFFERENTIAL COEFFICIENT - [math] coefficient *m* différentiel partiel
~DIFFERENTIAL EQUATION - [math] équation *f* différentielle partielle
~DISTURBED ONE-OUTPUT SIGNAL - [comput] signal *m* de sortie d'un préalablement perturbé à excitation partielle
~DISTURBED RESPONSE SIGNAL - [comput] signal *m* de sortie d'un élément de mémoire préalablement perturbé à excitation partielle
~DRAFT - [text] étirage *m* partiel
~DRIVE PULSE - [electron] impulsion *f* partielle de excitation
~EARTH - [el] défaut *m* à la terre
~ECLIPSE - [astr] éclipse *f* partielle
~FUSING IN ANNEALING - [metall] fusion *f* partielle au recuit
~HIP - [constr] (of a roof) (toit en) croupe raccourcie
~MISCIBILITY - [chem] (in binary liquid systems) miscibilité *f* partielle
~MULTIPLE - [telecomm] (in telephony) multiplage *m* échelonné, multiplage *m* partiel
~NODES - [acoust] (in a standing wave system) noeuds *mpl* partiels
PARTIAL PITCH AT THE COMMUTATOR - [el] partiel *m* d'un enroulement
~PLATING - [metall] (deposition upon part only of a cathode) dépôt *m* partiel
~PRESSURE - [mech] (in a mixture of gases, the pressure exerted by each component) pression *f* partielle
~READ CURRENT - [comput] courant *m* de lecture partiel
~-READ PULSE - [comput] (in static magnetic storage, any-one of the currents used causing the selection of a cell for reading) impulsion *f* partielle de lecture
~RESTORING TIME - [radio & telecomm] (of an echo suppressor) temps *m* de blocage d'un suppresseur d'écho à action discontinue
~ROASTING - [metall] (roasting for the partial elimination of sulphur in an ore) grillage *m* pour l'élimination partielle du soufre
~-SELECT OUTPUT - [comput] (in static magnetic storage) sortie *f* de noyaux partialement sélectionnés
~STORED FIELD SYSTEM - [telev] (in video recording) système *m* de mis en mémoire partielle de l'image
~TONE REVERSAL - [radio] (defect in which a reproduction tone moves from white to black and towards black again instead of proceedingly steadily from white to black) inversion *f* partielle des nuances
~VACUUM - [phys] vide *m* imparfait
~VALENCIES - [chem] (the residual affinity still prevailing in double bonds) valences *fpl* partielles
~WAVE - [opt] (the component of the wave function in a scattering process corresponding to a given angular moment) onde *f* partielle
~WIDTH - [nucl] (applied to a system having several alternate modes of disintegration) largeur *f* de niveau partielle
~-WRITE PULSE - [comput] (in static magnetic storage) impulsion *f* partielle d'écriture
PARTIALLY-AERATED FLAME - [th eng] flamme *f* aérée
~OCCUPIED BAND - [electron] (energy band in which not all the levels correspond to the energy of each of two electrons with opposite spins in a specified substance in a specified state) bande *f* partiellement occupée
~SELECTED CELL - [comput] élément *m* sélectionné partiellement
~SWITCHED CELL - [comput] élément *m* de mémoire perturbé
PARTICLE - [phys] (smallest non-atomic amount of matter obtainable) particule *f*
~ACCELERATOR - [electron] (electronic device giving charged particle a very high velocity) accélérateur *m* de particules
~CHARGE - [nucl] (the electrical charge of a particle) charge *f* d'une particule
~COUNTER - [nucl] compteur *m* de particules
~DISPLACEMENT - [phys] (in an elastic medium) déplacement *m* d'une particule
~FLUENCE - [phys] flience *f* des particules
~FLUX - [phys] flux *m* de particules
~FLUX DENSITY INDICATOR - [nucl] signaleur *m* de la densité de flux de particules
~FLUX DENSITY METER - [nucl] fluxmètre *m* de particules

PARTICLE FLUX DENSITY MONITOR - [nucl] moniteur *m* de la densité de flux de particules
~PATH - [phys] (the track followed by an infinitesimal element of a fluid) parcours *m* de la particule
~RADIATION - [nucl] (a stream of atomic sub-atomic particles, which may be uncharged or have a positive or negative charge) rayonnement *m* corpusculaire, rayonnement *m* de particules
~SHAPE - [phys] (the physical form of the average type of particle in a specimen of powdered or granulated material) forme *f* de particule
~SIZE - [metall] dimension *f* de grains de poudre
~SIZE ANALYZER - [metall] analysateur *m* des dimensions de particules
~-SIZE DISTRIBUTION - [min] granulométrie *f*
[metall] distribution *f* granulométrique
~-SIZE FRACTION - [metall] fraction *f* granulométrique
~-SIZE RANGE - [metall] intervalle *m* de classement granulométrique
~TRACK COMPUTER - [nucl] calculateur *m* analogique du parcours du faisceau particulier
PARTICULAR - [gen] particulier
PARTICULATE ACTIVITY - [phys] activité *f* de particules
~MATTER - [phys] matière *f* subdivisée
PARTICULATES - [nucl] (large-sized particles) macroparticules *fpl*
PARTING - [gen] séparation *f*, division *f*
[metall & plast ind] plan *m* de séparation
[metall] (the sand applied on patterns to eliminate adhesion) sable *m* sec
[mech] rupture *f*
[ind chem] (method of separating silver from gold, especially in assaying, the bead of silver and gold remaining after cupellation being treated with nitric acid, which dissolves the silver leaving a skeleton of fine gold) séparation *f*
[mining] intercalation *f* stérile
[geol] division *f*, limet *m*
~AGENT - [chem] (also release agent. A lubricant for coating the mould cavity to assist removal of the completed article) agent *m* de séparation
~BEAD - [carp] (bead fitted to the cased frame of a double window to separate the inner and outer sashes) baguette *f*
~LINE - [metall & plast ind] plan *m* de séparation, plan *m* de joint
~LINE GUARD - [metall] (a splash guard) protecteur *m* contre les éclaboussures
~-OFF - [mech] (by means of a cropper) rupture *f*, rompément *m*
~PLANE - [gen] plan *m* de séparation
~POWDER - [metall] (powder sprinkled on the parting face of a mould to prevent adhesion of the two surfaces) poussier *m* isolant, poncif *m*, isolant *m*
~SAND - [constr] (the layer of dry sand which separates two layers of damp sand) sable *m* isolant
[metall] (the sand sprinkled on the face of a mould to prevent adhesion) sable *m* sec, sable *m* à saupoudrer
~SLIDE WITH SCREW FEED - [impl] porte-outil *m* à tronçonner
~SLIP - [carp] (the thin lath of wood or zinc which keeps the sash-weights separated) baguette *f*
~STRIP - [constr] (the strip of wood or metal separating contiguous parts) baguette *f* (de panneaux)

PARTING TOOL - [tool] outil *m* à tronçonner
[metall] outil *m* à détalonner
PARTITION - [gen] partage *m*, répartition *f*
[constr] cloison *f*, cloisonnage *m*, cloison *f* verticale
[metall & plast ind] (of a mould) séparation *f*
[agric] éclatage *m* (des racines)
~CHROMATOGRAPHY - [ind chem] (in gas analysis) chromatographie *f* de partage
~COEFFICIENT - [chem] (the ratio of the equilibrium concentrations of a substance dissolved in two immiscible solvents) coefficient *m* de répartition
~FUNCTION - [phys] (relating to the distribution of molecules) fonction *f* de répartition
~LAW - [phys] loi *f* de répartition
~LINE - [metall & plast ind] (the line along which the parts of a mould separate) plan *m* de joint
~NOISE - [electron] (distribution noise) bruit *m* de répartition
~WINDOW - [auto] glace *f* de séparation
PARTLY - [gen] partiellement
~DOUBLE PRESELECTION - [telecomm] (in telephony) présélection *f* double partielle
PARTNER - [naut] (one of the framing pieces surrounding a mastro strengthen the deck) étambrai *m*
~OF THE CAPSTAN - [naut] etambrai *m* du cabestan
PARTRIDGE - [zool] perdrix *f*
PARTS SUBSTITUTION LIST - [gen] liste *f* de substitution
PARTURIENT - [med] en travail, en parturition
PARTURITION - [zool] travail *m*, parturition *f*, mise *f* bas
PARTY FENCE - [constr] (fence separating adjoining properties) clôture *f* mitoyenne
~LINE - [telecomm] (in telephony) poste *m* groupé
~WALL - [constr] (wall dividing two adjoining properties) mur *m* mitoyen
PARVIS - [constr] (the enclosed area in front of a church) parvis *m*
PARVULE - [med] boulette *f* médicinale, granule *m*
PASCAL LAW - [phys] (the law of the transmissibility of pressure) loi *m* de l'accroissement de la pression, loi *f* de Pascal
~RULES - [el] (the law relating to the diamagnetic susceptibility of complex molecules) règles *fpl* de Pascal
PASCHEN'S LAW - [electron] (at a constant temperature, the breakdown voltage is a function only of the gas pressure by the distance between parallel plane electrodes) loi *f* de Paschen
PASS, to - [gen] passer, dépasser
[leg] (a law) approuver
PASS - [geogr] col *m*, passage *m*, défilé *m*
[gen] (a permit) permis *m*
[metall] (in welding; a run) passe *f*, cordon *m* de soudure
[metall] (a groove) rainure *f*
[text] (of the warp threads) passage *m*
[metall] (in rolling) phase *f* d' usinage, passe *f*
[metall] (in lamination) laminage *m*
[min] (channel or shaft for the passage of ore from one level to a lower one) cheminée *f*
~A CALL, to - [telecomm](in telephony) transmettre une demande de communication
~-BAND - [radio] (band of frequencies in which gain or attenuation is greater or smaller than a given value) bande *f* passante

PASS BAND - [opt] (of optical glasses) bande *f* de transmission
~BOX - [photo] passe-cassettes *m*
~GROOVE - [metall] rainure *f* de cylindre
~-KEY - [mech] (key operating on several locks in a building) passe-partout *m*
~OFF, to - [chem] s'éliminer par distillation
~-OUT TURBINE - [el] turbine *f* à soutirage
~-OVER MILL - [metall] laminoir *m* du irréversible
~RANGE - [telecomm] zone *f* de filtrage
~SCHEDULE - [metall] séquence *f* de laminage
PASSAGE - [gen] passage *m*
[constr] couloir *m*, corridor *m*
[gen] (a voyage) traversée *f*
~CHANNEL - [text] canal *m* de passage
~OF THE SHUTTLE - [text] passage *m* de la navette
~SELECTOR - [text] sélecteur *m* de passage
~VALVE - [mech] valve *f* de passage
~WIDTH - [text] largeur *f* de la portée d'ourdissage
PASSAGEWAY - [gen etc] passage *m*, ruelle *f*, couloir *m*
[railw] corridor *m*
PASSBOOK - [fin] (the book given by a bank to a client and showing all deposits and withdrawals) livret *m* bancaire
PASSEMENTERIE - [text] (trimmings etc) passementerie *f*
PASSENGER - [gen] (a person carried in trains, aircraft etc) passager *m*, voyageur *m*
~CABLE RAILWAY - [railw] chemin *m* de fer funiculaire
~CAR - [auto] voiture *f* privée
[railw] voiture *f* à voyageurs
~COACH - [railw] wagon *m* à voyageurs
~ELEVATOR - s. passenger lift
~LIFT - [mech] ascenseur *m* à personnel
~PLANE - [aero] avion *m* passagers
~PLATFORM - [railw] trottoir *m*, quai *m*
~SEAT - [aero] fauteil *m* pour passagers
~TRAIN - [railw] train *m* de voyageurs
PASSING CONTACT - [el] contact *m* de passage d'un commutateur
PASSION FLOWER - [bot] passiflore *f*
~FRUIT - [bot] barbadine *f*, passiflore *f* comestible
PASSIVATE, to - [metall] (the operation of rendering the surface of a metal part chemically inactive) passiver
PASSIVATING DIP - [metall] solution *f* de passivation
PASSIVATION - [metall] (the process of rendering the surface of a metal part chemically inactive, to prevent corrosion) passivation *f*
PASSIVE - [gen] passif
[metall] (condition of a metal or alloy on the surface of which a thin film (e.g. of oxide) has been formed, making it relatively inert) passif
[chem] (without reaction) passif
~AERIAL - s. parasitic aerial
~ANTENNA - s. parasitic aerial
~BALANCE RETURN LOSS - [radio] affaiblissement *m* passif d'équilibrage
~CIRCUIT - [electron] circuit *m* passif
~DIKE - [constr] digue *f* dormante
~ELECTRIC NETWORK - s. passive network
~ELECTRODE - [el] (the earthed electrode of a precipitation apparatus) électrode *f* passive
~HOMING - [radio] radioralliement *m* passif, homing *m* passif
PASSIVE HOMING GUIDANCE - [radio] radio-guidage *m* passif
~METAL - [metall] (metal on which an oxide film preventing corrosion is readily formed) métal *m* passif
~NETWORK - [el] (network in which there is no secure of electromotive force) réseau *m* électrique passif
~RETRANSMISSION - [telev] réémission *f* passive
~SATELLITE REPEATER - [telev] répéteur *m* de satellite passif
~SINGING POINT - [telecomm] (the singing point of a circuit without any amplifiers and terminated in specified conditions) point *m* d'amorçage passif
~SOIL PRESSURE - [soil] pression *f* passive du sol
~TRANSDUCER - [electron] (transducer in which the output energy is derived only from the input) transducteur *m* passif
~TUNING UNIT - [telev] unité *f* de syntonie passive
PASSIVITY - [metall] (the condition in a metal in which electrochemical corrosion is inhibited by special conditions of the surface) passivité
PAST-POINTING - [med] déviation *f* de l'index
PASTE, to - [gen] coller
[gen] (to work into a paste) empâter
PASTE - [ceramics] (the material forming the body of porcelain, made of china stone and china clay. Paste for artificial porcelain is made from glass frit and white clay) pâte *f*
[el chem] (electrolyte-containing gelatinized layer lying against the negative electrode) pâte *f*
[glue] colle *f*
[food ind] pâte *f* de pâtisserie
[plast ind] (or plastisol) (fluid dispersion of a special vinyl chloride polymer in a plasticizer, used for making leather cloth, moulded and dipped articles) plastisol *m*
[geol] magma *m* de consolidation
~CATHODE - [electron] (cathode with emitter-carrier covered with a paste of emissive material) cathode *f* empâtée
~MIXER - [mech] malaxeur *m* pour pâtes
~PAINT - [paint] pigment *m* en pâte
~PRODUCTS - [agric] pâtes *fpl* alimentaires
~RESIN - [plast ind] résine *f* pour pâte
~SOLDER - [metall] pâte *f* à souder
PASTEBOARD - [paper man] carton *m*, carton-pâte *m*
~CARD - [paper man] carton *m*
[text] carte *f* de carton
PASTED - [gen] (with glue) collé
[el] (of a plate coated with oxides or salts of lead) empâté
[paper man] à la colle, empâté
~CORE - [metall] noyau *m* collé
~FILAMENT - [metall] (a filament obtained through squirting a paste of powdered metal through a die, together with a binding material) filament *m* empâté
~FRAME-PLATE - [el] plaque *f* à cadre, plaque *f* à masse
~PLATE - [el] (plate of lead-antimony alloy coated with oxides or salts of lead, which are subsequently converted to give the required active material) plaque *f* à oxyde rapporté, plaque *f* Faure
~SEAM - [rubber ind] (connecting two webs of material) ourlet *m* collé, joint *m* collé
PASTEL - [gen] (a picture made with pastel crayons

or pastel dyes) pastel *m*
PASTEL CRAYON - [draw] crayon *m* pastel
PASTELESS BOARD - [paper man] (used for water-colour paintings) carton *m* pour aquarelles
PASTEUR FLASK - [ind chem] ballon *m* de Pasteur
PASTEURELLOSIS - [vet] (a contagious infection) choléra *m* des poules
PASTEURIZATION - [med] (reduction in the micro-organism population of a substance by maintaining it at a temperature of 62.8 to 65.5 deg. C. for about thirty minutes) pasteurisation *f*
~FLAVOUR - [brew ind] goût *m* de pain cuit, goût *m* de pasteurisation
PASTEURIZE, to - [med] pasteuriser, stériliser
PASTEURIZED MILK - [food] lait *m* pasteurisé
PASTEURIZER - [ind chem] pasteurisateur *m*
PASTILLE - [med] (a troche or lozenge) pastille *f*
[el chem] (of an accumulator plate) pastille *f*
PASTING - [gen] collage *m*
[paper man] papier *m* peint
[ind chem] (saponifying) saponification *f*
~AND CUTTING MACHINE - [text] machine *f* à coller et à couper
~MACHINE - [text] machine *f* à coller, machine *f* à encoller
~PRESS - [mech mach] presse *f* à coller
PASTORAL - [gen] pastoral
[mus] (an instrumental composition) pastorale *f*
~PEST - [vet] (infectious disease affecting sheep) maladie *f* des pâturages
PASTRY - [gen] pâtisserie *f*
~OVEN - [th eng] four-coffre *m*
PASTURE, to - [agric] pâturer, paître
PASTURE - [agric] pâturage *m*, pré *m*, herbage *m*
~GROUND - [agric] lieu *m* de pâturage, pacage *m*
~HUSBANDRY - [agric] entretien *m* du pâturage
~LAND - [agric] pâturages *mpl*, pays *m* d'herbages
~MEADOW - [agric] pré *m* pâturé, prairie *f* pâturée
~PLANT - [agric] plante *f* d'herbage
~YIELD - [agric] rendement *m* des pâturages
PASTURING - [agric] pacage *m*
PASTY - [gen] pâteux, empâté
PATCH, to - [gen] rapiécer, raccommoder
[rubber ind] (the inner tube) poser une pastille
[mech] colmater
PATCH - [gen] pièce *f*, tache *f*, morceau *m*
[rubber ind] (of a rubber tube) pastille *f*, emplâtre *m*, corset *m*
[gen] (small piece of cloth used for repairs) pièce *f* à raccommoder
[gen] (a small piece of ground) parcelle *f* de terre
[aero] (fitted to a balloon envelope) renfort *m*
[med] tache *f*
[med] (a piece of material covering an injury) emplâtre *m*
[naut] placard *m* de voile
[el] connexion *f* provisoire
[comput] (superficial correction of a routine) raccord *m*
~AERA - [electron] (due to lack of perfect homogeneity of electron emitters) zone *f* de taches
~BAY - [comput] panneau *m* d'interconnexions
~EFFECT - [electron] (effect caused by the exhibition of different crystals facets making up the polycrystalline surfaces) effet *m* de taches
~OF ORE - [min] poche de minerai
~OUTFIT - [auto etc] nécessaire *m* de réparation pour chambre à air
PATCH PANEL - [telecomm] (a jack field) panneau *m* de commutation, panneau *m* de jacks, répartiteur *m*
~TEST - [med] test *m* épicutané
PATCHBOARD - [comput] (movable board with hundreds of electric terminals into which short connecting wire cords may be plugged in various patterns for different programmes) panneau *m* de raccordement
[el] (short connecting wire cord used in patchboards) raccord *m*
PATCHCORD - [el] cordon *m* de renvoi, cordon *m* de raccordement
PATCHINESS - [gen] moutonnement *m*
PATCHING - [gen] réparation *f*, rapiècement *m*, raccommodage *m*
[th eng] colmatage *m*, obturation *f* de fissures
[metall] (burning on) retorchage *m*
[text] raccommodage *m*
~CEMENT - [gen] (putty) mastic *m*
[rubber ind] (for tyres) dissolution *f* à coller les plaques de réparation
~COMPOUND - [plastic ind] (repair material for moulds) matière *f* à jointoyer
~CORD - [telecomm] cordon *m* de renvoi
~JACKFIELD - [comput] panneau *m* de jacks
~MATERIAL - [gen] (coating material used in ovens) enduit *m*, badigeon *m*
~PADDLE - [impl] crochet *m* à jointoyer
~RUBBER - [rubber ind] (for tyres) caoutchouc *m* réunissant
~SLURRY - s. patching material
~WORK - [text] travail *m* de stoppage
PATCHSTRIP - [rubber ind] (for tyres) emplâtre *m* (pour réparation de pneus)
PATCHULI - [bot] patchouli *m*
PATCHWORK - [gen] ouvrage *m* de pièces de rapport, rapiéçage *m*
PATELLAPEXY - [med] rotulopexie *f*
PATENCY - [med] état *m* ouvert, non-fermeture *f*
PATENT, to - [leg] faire breveter, proteger par un brevet
PATENT - [leg] (exclusive licence to manufacture and market an invention) brevet *m*, brevet *m* d'invention
~APPLICATION - [leg] (the request made to a government patent for the grant of a patent) demande de brevet
~BLOWPIPE - [welding] chalumeau *m* pour soudage à gaz à basse pression
~BOARD - [constr] (term denoting any proprietary building board) planche *f* brevetée
~CLAIM - [leg] revendication *f* de brevet
~CORE NAIL - [metall] pointe *f* à noyau brevetée
~FLATTENING - [metall] dressage *m* par étirage, planage *m* par étirage
~GLAZING - [build] (term applied to various devices for fixing glass sheets) vitrage *m* breveté
~GRANT - [leg] concession *f* d'un brevet
~INFRINGEMENT - [leg] (the unauthorized exploitation of a patent) contrefaçon *f*, violation *f* de brevet
~LEATHER - [leather ind] (enamelled leather obtained by spraying with a cellulose lacquer) cuir *m* vernis
~LOG - [instr] (log mounted on the taffrail and

consisting of a rotator, a log line and a recording apparatus) loch *m* enregistreur, sillomètre *m*
PATENT MEDICINE - [med] spécialité *f* pharmaceutique
~OFFICE - [leg] (the office of department where patents are registered) office *m* des brevets
~RIGHT - [leg] (the exclusive privilege to the uses and control of an invention) droits *mpl* de brevet; proprieté *f* industrielle
~SLIP - [naut] slip *m*
~SPECIFICATION - [leg] (a description of a product or process for which a patent is required) description *f* de brevet
~YELLOW - [chem] oxychlorure *m* de plomb
PATENTABILITY - [leg] (having the necessary premisses for being patented) brevetabilité *f*
PATENTABLE - [leg] (capable of being patented) brevetable
PATENTED - [leg] (protected by a patent) breveté
[metall] (treated) trempé en bain de plomb
~STEEL WIRE - [metall] fil *m* d'acier trempé en bain de plomb
PATENTEE - [leg] (the person or party in whom a patent is vested) détenteur *m* du brevet
PATENTING - [metall] trempe *f* au plomb
PATERA - [arch] (circular ornament on relief) patère *f*
PATH - [gen] sentier *m*, chemin *m*
[astr] (of a planet) orbite *f*
[mech] course *f*
[el] (of an armature winding) dérivation *f*, voie *f*, parcours *m*
[mach tool] course *f*
[astronaut] trajectoire *f* apparente
[draw] trace *f*
[gen] (a track) piste *f*
[anat] voie *f*, passage *m*
~LENGTH - [comput] (in static magnetic storage; the length of a magnetic flux line in a core) longueur *f* du canal
[phys] (the distance measured along the path of the particle) longueur *f* du parcours
~OF AN ARMATURE WINDING - [el] voie *f* d'enroulement
~OF CURRENT - [el] parcours *m* du courant
~OF INTEGRATION-[telecomm] trajet *m* d'intégration
~OF LIFT - [mech] course *f* de soulèvement
~OF PERCOLATION - [soil] ligne *f* de cheminement
~OF PROPAGATION - [el] parcours *m* de propagation
~OF RAY - [opt] parcours *m* du rayon
~OF THE BEATERS - [text] parcours *m* circulaire des batteurs
~TRIMMER - [agric] désherbeur *m*
PATHERGASIA - [med] trouble *m* du comportement
PATHERGY - [med] pathergie *f*
PATHETIC - [anat] pathétique
~MUSCLE - [zool] (the superior oblique muscle of the eye) muscle *m* pathétique (de l'oeil), grand oblique *m* de l'oeil
PATHOGENESIS - [med] (development of a disease process) pathogénie *f*
PATHOGNOMY - [med] pathognomie *f*
PATHOLOGICAL - [med] pathologique
~CHEMISTRY - [chem] chimie *f* pathologique
PATHOLOGY - [med] pathologie *f*
PATHOPHOBIA - [med] pathophobie *f*
PATHOPHORESIS - [med] transmission *f* des maladies
PATHWAY - [gen] sentier *m*
[gen] (a footway) trottoir *m*, accotement *m*
PATIENCE DOCK - [agric] patience *f*, oseille-épinard *f*
PATIENT - [med] patient *m*, malade *m*
PATINA - [gen] (the green rust covering metals) patine *f*
~FINISH - [metall] patine *f* artificielle
PATIO PROCESS - [metall] procédé *m* d'amalgamation
PATROL, to - [gen & milit] fouiller
PATROL GRADER - [mech] (used in road construction) machine *f* à aplanissement
PATRONITE - [min] (vanadium containing mineral) patronite *f*
PATTEN - [arch] (the base of a pillar) socle *m*, patin *m*
PATTERN - [mech] (a representation of a part used to impress its form in the moulding sand, in making a casting) modèle *m*
[gen] (anything shaped to serve as model) modèle *m*, type *m*, dessin *m*
[radio] (graphic illustration of the behaviour of electromagnetic radiations) diagramme *m* de rayonnement
[telev] (the stationary pattern of a cathode-ray tube) mire *f*
[text] dessin *m*
[chem] spectre *m*
[rubber ind] (of tyres) sculpture *f*, profil *m*, gravure *f*
[rubber ind] (for clickers) patron *m* de coupe
[gen] (a templet) gabarit *m*, calibre *m*, patron *m*
~BOARD - [metall] plaque-modèle *f*
~CARD - [text] carton *m* de changement, carte *f* de échantillons
~CARD CYLINDER - [text] cylindre *m* à cartons
~CARD RAILS - [text] glissière *f* des cartons, guide-cartons *m*
~CUTTING - [text] coupage *m* des échantillons
~CYLINDER - [text] prisme *m*
~DESIGNING - [text] exécution *f* du dessin
~DETECTION - [comput] détection *f* de caractères représentés par une mosaique
~DISC - [text] disque *m*
~DRAFT - [metall] (the taper given to the sides of a pattern, so as to withdraw it easily from the mould) dépouille *f*
~-DRAW MOULDING MACHINE - [metall] démouleur *m* par extraction du modèle
~DRAWER - [text] dessinateur *m*
~-DRAWING MACHINE - [metall] machine *f* à démouler
~DRUM - [text] tambour *m* à dessin
~DRUM LAY-OUT - [text] schéma *m* pour enfoncer les goupilles dans le tambour-mémoire
~EFFECT - [text] effet *m* de façonnage
~FOR CARD CUTTING - [text] armure *f* du perçage de cartons
~FOR CARD PUNCHING - [text] mise *f* en carte de percage
~GENERATOR - [electron] générateur *m* de mire
~HEATING - [metall] chauffage *m* du modèle
~HOLDER - [mech] porte-modèle *m*
~JACK - [text] platine à dessin
~LOOM - [text] métier *m* à échantillonner
~MAKER - [gen & foundry work] modeleur *m*
~METAL - [metall] alliage *m* de modèle

PATTERN MILLER - [metall] fraiseuse *f* à modèles
~NOISE - [electron] tensions *fpl* de bruit périodiques
~PAPER - [text] papier *m* pour ratiéres
~PLATE - [metall] plaque *f* modèle
~PROCESSING - [comput] traitement *m* d'informations obtenues par la lecture d'une mosaïque
~RECOGNITION - [comput] (in data processing) reconnaissance *f* de configuration
~ROLLER - [text] galet *m*, roulette *f* de carte
~SCREW - [metall] tire-fond *m* à modèle
~-SENSITIVE FAULT - [comput] défaillance *f* due aux données
~SHEET - [metall & plast ind] (the outside sheet of a laminated material on which the design appears) strate *f* décorée
~SHRINKAGE ALLOWANCE - [metall] double retrait *m*
~SINKER - [text] platine *f* à dessin
~SIZE GRADING PROGRAMME - [autom] programme *m* pour copie des patrons en mesures différentes
~STRIPE - [text] bande *f* de dessin
~THREAD - [text] fil *m* de dessin
~TIE BAR - [metall] barre *f* à reboucher
~UNIT - [rubber ind] élément *m* du profil
~WEAVE - [text] liage *m* du dessin
~WEAVING - [text] tissage *m* des étoffes à dessin
~WHEEL - [text] reproducteur *m*
~WITH LONG STRIPES - [text] dessin *m* en rayures le long
PATTERNED CLOTH - [text] tissu *m* faconné, étoffe *f* à dessin
~FABRICS - [text] tissus *mpl* façonnés
PATTERNING - [text] dessin *m*
[telev] surimposition *f* de fond faux
~MECHANISM - [text] dispositif *m* à dessin
PATTERNMAKER - s. pattern maker
PATTERNMAKER'S GLUE - [chem] colle *f* pour modeleurs
~ALLOWANCE - [metall] double retrait *m*
~RULE - [metall] mètre *m* à retrait
~SAW - [impl] scie *f* des modeleurs
PATTERNMAKING - [metall] modelage *m*
PATTINSON PROCESS - [metall] (a commercial method for the separation of silver from lead, by selective crystallization from the molten metal) procédé *m* Pattinson
PAULI-FERMI PRINCIPLES - [phys] (each level of a quantized system can include only none, one, or two electrons; when it includes two electrons, these have spins of opposite directions) principe *m* d'exclusion de Pauli et Fermi
PAULING CONCENTRATOR - [ind chem] (a packed tower fed with nitric and sulphuric acid at the top and live steam at the bottom) concentrateur *m* Pauling
PAULOCARDIA - [med] diastole *f* prolongée
PAUNCH - [zool] ventre *m*, panse *f*
[naut] paillet *m*
PAUSE - [gen] pause *f*, arrêt *m*
PAVE, to - [constr] (to cover a road with asphalt etc) paver, carreler
PAVED - [constr] pavé
~ROAD - [constr] rue *f* pavée
~WITH COBBLESTONES - [constr] rue *f* pavée en cailloutis
PAVEMENT - [constr] pavage *m*, carrelage *m*, dallage *m*
[gen] (footway) trottoir *m*
[mining] sole *f* de galerie
PAVEMENT COURSE - [constr] empierrement *m*
~LIGHT - [constr] (panel formed of glass blocks build into a pavement surface and opening to the basement of a building) verdal *m*, dallage *m* en verre
~OF A BRIDGE - [constr] tablier *m* de pont
PAVER - [mech] (used in road making) bétonnière *f*
[gen] paveur *m*, dalleur *m*
PAVILION - [constr] (ornamental structure) pavillon *m*
~ROOF - [constr] (roof forming a figure of more than four straight-sides) comble *m* en pavillon
PAVING - [constr] (of roads) pavage *m*, carrelage *m*, dallage *m*
[constr] (road covering made of hard bricks or cobblestones) pavé *m* en cailloutis
~BEETLE - [impl] hie *f*, dame *f*, demoiselle *f*
~BLOCK - [constr] pierre *f* à paver
~BREAKER - [mech] (pneumatic hammer) pic *m* à air comprimé
~CLINKER - [constr] brique *f* trés dure
~FLAG - [constr] (flat stone used for surfacing pavements) dalle *f*, cadette *f*, pierre *f* à paver
~SLAB - s. paving flag
~STONE - [constr] grés *m* à pavés
~TILE - [constr] (hard type of tile) carreau *m* de pavage
PAVIOUR - [constr] (hard brick used for pavings) dalle *f* à paver
[gen] paveur *m*, dalleur *m*, carreleur *m*
~BRICK - [constr] pavé *m* recuit
~SCOTCH - [constr] marteline *f* pour briques de pavage
PAWL, to - [mech] (to secure a capstan by means of a pawl) mettre les linguets
PAWL - [mech] (a mobile tooth or dog engaging a ratchet wheel to prevent movement of the latter in an undesired direction) cliquet *m* d'arrêt, linguet *m*, doigt *m* d'encliquetage
[naut] (of a capstan) saucier *m*
~-AND RATCHET MOTION - [mech] encliquetage *m* à rochet
~BEARING WHEEL - [horol] roue *f* porte-cliquet
~BEARING YOKE - [horol] bascule *f* porte-cliquet
~COUPLING - [radio] (device used in tuning means of receivers) accouplement *m* à cliquet
~MECHANISM - [text] système *m* de cliquets
~-RIM - [naut] (notched cast-iron ring for pawls) couronne *f* des linguets
~-SPRING - [mech] (spring used to press a pawl into engagement) ressort *m* de cliquet
~WHEEL - [radio] (of receivers) zone *f* à cliquet, zone *f* d'encliquetage
~WINDING WHEEL - [horol] roue *f* à cliquet
PAWPAW - [bot] papaye *f*
~-TREE - [bot] papayer *m*
PAY, to - [gen] payer
[comm] payer, solder
[naut] (to run out a rope or chain by slackening) laisser filer
PAY - [gen & comm] paye *f*, paie *f*, salaire *m*
~BED - [mining] couche *f* minéralisée
~-DAY - [gen] jour *m* de paie
~LOAD - [transp] (the part of the useful load which produces revenue) charge *f* utile
[mech] charge *f* utile
~ORE - [mining] minerai *m* payant
~ORE-BODY - [mining] gîte *m* justifiant l'exploitation

PAY OUT, to - [naut] laisser filer
~ROCK - [oil ind] roche *f* payante
[mining] formation *f* géologique payante
~ROLL - [comm] feuille *f* de paie, feuille *f* d'émargement
~SAND - [oil ind] sable *m* payant
~SHEET - s. pay roll
~STREAK - [oil ind] couche *f* de sable payant
~ZONE - [oil ind] zone *f* pétrolifère payante
PAYABLE - [gen & fin] payable
PAYDIRT - [min] (any soil containing sufficient metal, particularly gold, to warrant mining) alluvion *f* exploitable
PAYEE - [fin] bénéficiaire *m*
PAYER - [gen & fin] payeur *m*
PAYING LOAD - [comm] charge *f* utile
PAYLOAD - s. pay load
~MASS RATIO - [phys] rapport *m* de masses
~WEIGHT - [gen] poids *m* de la charge utile
PAYMENT - [gen & fin] payement *m*
~IN ADVANCE - [fin] payement *m* par avance
~IN KIND - [fin] payement *m* en nature
~ON ACCOUNT - [fin] versement *m* à compte
PAYROLL - [comm] feuille *f* des salaires, feuille *f* de paie
~PROCESSING - [comput] décompte *m* des salaires
Pb - [chem] (symbol for lead) symbole du plomb
P.B. - [metall] (initials of Phosphor Bronze) initiales de Bronze Phosphoré
P.B.X. - [telecomm] (initials of Private Branch Exchange) initiales de Installation d'abonné avec postes supplémentaires
P.C. - [draw] (initials of Pitch Circle) initiales de Cercle Primitif (ou ligne d'engrènement d'une roue)
P.D. - [el] (potential difference) initiales de Différence de Potentiel (ou tension)
Pd - [chem] (the symbol of Palladium) symbole du Palladium
PEA - [bot] pois *m*
[min] (a grain) grain *m*
[naut] (of the anchor) bec *m* d'ancre
~-GRIT - [min] pisolithe *f*
~IRON ORE - [min] fer *m* pisiforme
~MOTH - [zool] tordeuse *f* du pois, pyrale *f* du pois
~RUST - [bot] (plant disease) rouille *f* des pois
~WEEVIL - [zool] sitone *m* des pois, sitone *m* rayé
PEACH - [bot] pêche *f*
~COKE - [mining] grésillon *m*
~STONE - [min] chloritoschiste *m*
~TREE - [bot] pêcher *m*
PEACOCK - [zool] paon *m*
~COPPER - s. peacock ore
~ORE - [min] (bornite, so called because it becomes iridescent from tarnish) cuivre *m* panaché, bornine *f*, érubescite *f*
~-STONE - [min] (type of agate) plume *f* de paon
PEAHEN - [zool] paonne *f*
PEAK, to - [mech] (to reach a peak value) pousser au maximum de charge
PEAK - [mech] (the maximum value reached by a machine, e.g. peak load) charge *f* maximum, pointe *f*, apogée *m*
[gen] maximum
[geogr] (of a mountain, hill etc) pic *m*, piton *m*, cime *f*, sommet *m*
[gen] (of value) valeur *f* maximum
PEAK - [naut] (of a sail) coqueron *m* (de la cale)
[shipbuild] (the narrowed part of the hold of a vessel) bec *m* (d'un vaisseau)
[statist etc] apogée *m* d'un diagramme, crête *f*
~ALTERNATING GAP VOLTAGE - [electron] tension *f* de crête de l'espace d'interaction
~ANGULAR VELOCITY - [phys] (the maximum value of the instantaneous angular velocity in a specified time interval) vitesse *f* angulaire de crête
~ARCH - [arch] arc *m* gothique
~BLACK - [telev] (the maximum excursion of the signal in the black direction) crête *f* du noir
~BRIGHTNESS - [telev] hyperluminosité *f* d'aire réduite
~CATHODE CURRENT - [electron] (surge) courant *m* anodique de crête
[electron] (steady state) courant *m* de crête cathodique en régime périodique
~CATHODE FAULT CURRENT - [electron] courant *m* de crête anormal
~CONSUMPTION - [el gas etc] pointe *f*, pointe *f* de consommation
~CURRENT - [el] courant *m* de crête
~DELAY - [light] délai *m* d'attente de l'intensité maximum
~DEMAND - [el gas ind etc] pointe *f* d'émission
~DISTORTION - [telev] distorsion *f* de la crête de l'amplitude
~EFFICIENCY - [mech etc] rendement *m* maximum
~-ENVELOPE POWER - [radio] (the power in one radiofrequency cycle at the crest of the modulation envelope under given conditions) puissance *f* de pointe
~EXCURSION - [telev] excursion *f* maximale
~FACTOR - [el] (the ratio between the peak value of an alternating wave and its root-mean-square value) facteur *m* de crête
~FLUX DENSITY - [electron] (of static magnetic storage) débit *m* maximal de fluence
~FORCE - [mech] (peak mechano-motive force) force *f* mécanomotrice maximale
~FORWARD ANODE VOLTAGE - [electron] tension *f* anodique directe de crête
~FORWARD GATE CURRENT - [electron] courant *m* de crête en sens direct
~HOUR - [el gas etc] (maximum hourly demand) pointe *f* horaire
~INDICATOR - [el] indicateur *m* de crête, voltmètre *m* de crête
~INVERSE ANODE VOLTAGE - [electron] tension *f* de crête anodique inverse
~INVERSE VOLTAGE RATING - [electron] tension *f* inverse de crête nominale
~LIMITER - [radio] éxcréteur *m* de sortie, limiteur *m* de puissance
~LIMITING - [el acoust] limitation *f* de la dynamique
~LOAD - [mech etc] (the greatest load to which a system is subjected under normal conditions) charge *f* maximum
[gas el etc] s. peak consumption
~-LOAD GAS - [gas ind] gaz *m* de pointe
~MAKING-CURRENT - [el] courant *m* établi
~OF TRAFFIC - [telecomm] pointe *f* de trafic
~POINT - [electron] (in semi-conductors) point *m* de crête
~POWER - [mech] (of an engine, in particular a diesel engine) puissance *f* maximum au frein

PEAK POWER - [radar] (the very great transmitter power which is necessary to produce an adequate pulse in radar) puissance f de crête
~POWER OUTPUT - [electron] puissance f de sortie de crête
~PRESSURE - [mech hydr] pression f maximum
~PROGRAMME LEVEL - [telev] niveau m de la dynamique de crête
~PROGRAMME METER - [instr] (instrument designed to measure the volume of programme in a sound channel) modulomètre m de crête
~PULSE AMPLITUDE - [radio] valeur f de crête d'amplitude
~PULSE POWER - [radio] puissance f de l'impulsion de crête
~r.p.m. - [mech] (maximum number of revolutions per minute) régime m maximal
~RESISTANCE - s. peaking resistor
~RESPONSE - [radio] réponse f maximale
~REVERSE VOLTAGE RATING - s. peak inverse voltage rating
~SEND OUT - s. peak demand
~SERVICE - [el] émission f de pointe
~-SHAVING GAS - s. peak-load gas
~SIDEBAND POWER - [radio] puissance f de crête de bande latérale
~SOUND PRESSURE - [acoust] (the maximum absolute value of the instantaneous sound pressure for any given time interval) pression f sonore maximum
~SPEECH POWER - [acoust] (the maximum value of the instaneous speech power during a given time interval) amplitude f de la radiation sonore
~SPEED - [mech] vitesse f maximum
~TANK - [shipbuild] bec m
~-THROUGH RATIO - [electron] rapport m crête-creux
~-TO-PEAK AMPLITUDE - s. peak-to-peak value
~-TO-PEAK VALUE - [telev] (the difference between the maximum and the minimum values of a signal) amplitude f crête-à-crête
~-TO-VALLEY RATIO - [telev] (in the pass curve of a composite network) distance f crête-creux
~-TO-ZERO - [radio] (specified voltage range used to determine on which level a measurement has been effected) gamme f de tensions de mesure
~TORQUE - [mech] couple f maximum
~TRAFFIC - [telecomm] trafic m de pointe
~TRANSIENT REGULATION - [contr] régulation f à crête transitoire
~TRANSIENT REVERSE VOLTAGE - [electron] (in semiconductors) tension f inverse de pointe non-répétitive
~VALUE - [el] (of a wave; the maximum positive or negative value attained) valeur f de crête
[math] valeur f de crête
~VELOCITY - [phys] (the maximum absolute value of the instantaneous velocity for any specified time interval) vitesse f de crête
~VOLTAGE - [el] crête f de tension
~VOLTMETER - [instr] (instrument designed to measure potential difference by electrostatic means) voltmètre m de crête
~WHITE - [telev] (the maximum excursion of the signal in the white direction) crête f du blanc
~WHITE RASTER - [telev] canevas m à niveau du blanc
~WORKING INVERSE VOLTAGE - [electron] tension f de crête de fonctionnement inverse
PEAK WORKING OFF-STATE FORWARD VOLTAGE - [electron] tension f de crête de fonctionnement dans l'état bloqué en sens direct
PEAKER - [radar] (differentiating circuit) circuit m d'augmentation de la pente
~STRIP - [el] (saturable core and coil in a magnetic field to provide a short voltage pulse at the time of the flux reversal in the core) bobine f de différentiation de crêtes
PEAKING - [electron] accentuation f
~CIRCUIT - [telev] circuit m d'augmentation de la pente
~COIL - [telev] (in an amplifier to increase the gain in the higher frequencies) bobine f de crête
~NETWORK - [telev] (interstage coupling network in an amplifier) circuit m de différentiation
~RESISTOR - [telev] résistance f de différentiation
PEAKS - [radio] (of high volume levels) crêtes f pl du niveau
PEAKY CURVE - [draw] courbe f à pointe
PEAL - [acoust] (set of bells) carillon m
PEAN - s. peen
PEANING - [mech] martelage m, battage m
PEANUT - [bot] arachide f, pistache f de terre
~BULB - [light] (miniature bulb) ampoule f miniature
~OIL - [ind chem] huile f d'arachide
PEAR - [bot] poire f
~OIL - [chem] (a term for amyl acetate) essence f de poires, acétate m d'amyle
~-PUSH - [el] (pushbutton switch attached to a flexible) interrupteur m de cordon
~PUSHBUTTON - s. pear-push
~-SHAPED - [gen] à forme de poire
~SWITCH - s. pear-push
~TREE - [bot] poirier m
PEARL - [gen] perle f
[print] (obsolete name for a type size) corps 5, parisienne f
[min] nacre m de perle
~ASH - [chem] carbonate m de potasse brut
[min] perlasse f
~BARLEY - [bot] orge m perlé
~BEARING - [anat] (huitre) perlière
~BULB - [light] ampoule f doucie
~COTTON YARN - [text] fil m ondé
~DISEASE - [vet] tuberculose f bovine
~-GREY - [paint] gris perle, gris m de perle
~LAMP - [light] lampe f dépolie intérieurement
~MILLET - [agric] millet m à chandelles
~ONION - [agric] petit oignon m blanc
~SPAR - [min] (dolomite) dolomite f
~WHITE - [chem] blanc m de perle
PEARLASH - [chem] carbonate m de potasse brut
PEARLESCENT - [paint] perlé, nacré
~PIGMENT - [paint] (a pigment the particles of which are reflective, thus imparting a nacreous appearance to an article to which it is applied) pigment m nacré
PEARLITE - [metall] (constituent of steel and cast-iron, produced at the eutectoid point as ferrite and cementite are formed simultaneously) perlite f
PEARLITIC - [metall] perlitique
~CAST IRON - [metall] fonte f perlitique
~MALLEABLE IRON - [metall] fer m malléable perlitique
~STRUCTURE - [metall] structure f perlitique

PEARLY - [metall] grenu
PEASANT - [gen] paysan *m*
PEASANTRY - [gen] compagnards *mpl*
PEAT - [geol] (layer of dead vegetation, widely used as a fuel) tourbe *f*
~BED - s. peat bog
~BOG - [geol] tourbière *f*, marais *m* tourbeux
~COAL - [mining] charbon *m* de tourbe
~CUTTER - [mining] trieur *m* de tourbe
~CUTTING - [agric] tourbage *m*
~DUST - [geol] farine *f* de tourbe, poussière *f* de tourbe
~FIBRE - [text] fibre *f* de tourbe
~FORMATION - [geol] transformation *f* en tourbière
~MOOR - s. peat bog
~SOIL - [geol] terrain *m* tourbeux, sol *m* tourbeux
~WADDING - [text] ouate *f* de tourbe
~YARN - [text] fil *m* de tourbe
PEATY - [geol] tourbeux
PEBBLE, to - [leather ind] crépir, maroquiner
PEBBLE - [gen] caillou *m*, galet *m*
[gen] (semi-precious stone) cristal *m* de roche
[constr] galet *m*
[opt] lentille *f* en cristal de roche
[leather ind] maroquinage *m*
~BED - [geol] couche *f* de galets
~DASH - s. pebble-dashing
~-DASHING - [constr] (rough finish of a wall by coating it with plaster and throwing on this, while still wet, pebbles and liquid lime) crépissage *m*
~GRAVEL - [gen] gravier *m*
~GUARD - [mech] (on vehicles) garde-cailloux *m*
~MILL - [mech] (grinding machine similar to a ball mill, but charged with flint pebbles) tube *m* finisseur à galets, broyeur à boulets, moulin *m* à galets
~STONE - [geol] galet *m*
PEBBLEWORK - [constr] cailloutage *m*, cailloutils *m*
PEBBLING - [paint] crépissage *m*
PEBRINE - [vet] (disease affecting the silk worm) muscardine *f*
P.E.C. - [cin] (or pec; short for photoelectric cell) cellule *f* photoélectrique
PECAN - [bot] pacane *f*
PECK, to - [zool] becqueter, picoter
[agric] piocher
PECK - [zool] coup *m* de bec
[meas] (about 9 litres) boisseau *m*, picotin *m*
PECKER - [el] (a motor operating in steps) moteur *m* pas-à-pas
[comput] (sensing member performing the sensing mechanically) goupille *f* d'exploration
PECKLING MOTOR - s. pecker
PECLET NUMBER - [phys] (in the mechanics of fluids; number obtained by multiplying the Prandtl number by the Reynolds number) nombre *m* de Peclet
PECTASE - [chem] pectase *f*
PECTATE - [chem] pectate *m*
PECTEN - [zool] (comb-like structure) peigne *m*, coquille *f*
PECTINEAL - [zool] (adj. of pecten) pectinéal
PECTINIC ACID - [chem] acide *m* pectinique
PECTINS - [chem] (carbohydrates occurring in the tissues of fruits and some vegetables. They are non-crystalline and of high molecular weight, and are of importance in jam-making because of their power to form jellies with fruit juices under suitable conditions) pectined *fpl*
PECTIZATION - [chem] (the formation of jelly) pectisation *f*
PECTOCELLULOSE - [chem] pectocellulose *f*
PECTOLITE - [min] (silicate of calcium and sodium) pectolite *f*
PECTORAL - [zool] pectoral
[anat] (muscle) pectoral
~ARCH - [anat] (the skeletal framework with which the locomotor appendages articulate) arcade *f* pectorale
~FIN - [zool] (in fish, the anterior fins) nageoire *f* pectorale
PEDAL, to - [gen & mech] pédaler
[mus] jouer sur le pédalier
PEDAL - [gen & mech] pédale *f*
[mus] pédale *f*
~BOARD - [mus] pédalier *m*
~CIRCUIT - [el] circuit *m* de pédale
~DRIVE - [text] commande *f* à pédale
~LEVER - [text] levier *m* à pédale
~RUBBER - [mech] patin *m* de pédale
PEDERASTY - [med] pédérastie *f*
PEDESTAL - [gen etc] piédestal *m*
[mech] (a support) appui *m*, support *m*, socle *m*, palier *m*, chaise *f*
[aero] (of a landing gear on skis) support *m* pour skis
[constr] (base for the support of a statue etc) piédestal *m*, socle *m*
[el] base *f*
[railw] plaque *f* de garde
[constr] (used to support the barrel of a boiler) chandelier *m*
[telev] décollement *m* du niveau du noir
~ADJUSTMENT - [telev] ajustage *m* du décollement du noir
~BEARING - [mech] (a bearing carried on a support fixed to a floor, wall or like base) palier *m*
~COVER - [mech] chapeau *m* de palier
~FRAMING - [text] montant *m*
~-HORN - [mech] glissière *f* de la plaque de garde
~INSULATOR - [el] isolateur *m* à piédestal
~MOULDING - [build] moulure *f* d'embase
~PAD - [mech] (a mounting pad) palier *m*
~SET SCREW - [text] vis *f* de réglage du support du tambour
~TABLE - [carp] table *f* à piédestal
PEDESTRIAN - [gen] piéton *m*, pédestre
~CROSSING - [gen] passage *m* réservé aux piétons
PEDIATRIC - [med] pédiatrique
PEDIATRICS - s. pediatry
PEDIATRY - [med] (or paediatry; the medical study of childhood and the disease of children) pédiatrie *f*
PEDICATION - s. pederasty
PEDICELLATE - [bot] pédicellé
PEDICLE - [bot] pédicelle *f*, pédoncule *m*
[anat] pédicule *m*
PEDICULOSIS - [med] (infestation of the body with lice) phtiriase *f*
PEDIGREE - [gen & zool] arbre *m* généalogique, pedigree *m*
~BREEDING - [genet] élevage *m* généalogique, sélection *f* généalogique
~HERD - [zool] troupeau *m* reproducteur
~SELECTION - [genet] culture *f* de sippes
PEDIMENT - [arch] (segmental part over a portico) fronton *m*, fronteau *m*

PEDION - [cryst] (crystal form consisting of a single plane) pédion *m*
PEDIPHALANX - [anat] phalange *f* du pied
PEDOGAMY - [med] endogamie *f*
PEDOLOGY - [agric] (the study of the nature and properties of soils) pédologie *f*, étude *f* du sol
PEDOMETER - [instr] (instrument designed to record the number of steps in walking and thus measure the distance covered) podomètre *m*, compte-pas *m*
PEDUNCLE - [anat] (the portion joining thorax and abdomen) pédoncule *m*, scape *m*
PEDUNCULAR - [zool] pédonculaire
PEEL, to - [gen] peler, éplucher
[metall] (die casting) écailler
PEEL - [gen] écorce *f*, pelure *f*
[metall] (charge shovel) pelle *f* de chargement
~OFF, to - [gen] s'écailler
~RECLAIM - [rubber ind] (of tyres; the tread reclaim) régénére *m* de chapes
PEELABLE COATING - [paint etc] (a coating which can be peeled off) revêtement *m* pelable
~PROTECTIVE COATING - [plast ind] revêtement *m* de protection à peler
PEELED - [gen] pelé, épluché, décortiqué
[timber] écorcé
~-CALIBRATED BAR - [metall] barre *f* dégrossie-calibrée
~FIBRE - [text] fibre *f* écorcée
PEELER - [mach tool] épluchoir *m*
PEELING - [metall] (undesired detaching of a plated coating from the base) écaillage *m*
[gen] éçorcage *m*, épluchage *m*
[paint] (a defect) déroulage *m*
PEELING AND STRAIGHTENING MACHINE - [metall] machine *f* à écrouter et dresser
~OFF - [metall] délaminage *m*
~STRENGTH - [rubber ind] (of tyres) adhésion *f* entre plis et bande du roulement
PEEN, to - [mech] marteler, rabattre
PEEN - [impl] (of a hammer) panne *f* (du marteau)
~DOWN, to - [mech] (to reduce the thickness of a piece of metal by hammering) rabattre, mater
~HAMMER - [impl] marteau *m* à panne
PEENED - [gen] martelé, maté
PEENING - [mech] martelage *m*, matage *m*
[metall] matage *m*, écrouissage *m*
~MARK - [gen] marquage *m*
~SHOT - [metall] grenaille *f* d'acier
PEEP-HOLE - [gen] (a small closeable orifice in e.g. a furnace door) regard *m*, orifice *m* de visite
~-SIGHT - [firearms] oeilleton *m*
PEG, to - [gen] (to keep in place by means of pegs) cheviller
[leather ind] brocher
PEG - [gen] (a wooden pin used to fasten articles together) cheville *f*, fiche *f*, jalon *m*, goujon *m*
[gen] (projecting wooden pin used to mark a boundary etc) borne *f*, piquet *m*, jalonnette *f*
[mus] (wooden pin holding the end of a string) cheville *f*
[mech] goujon *m*, dent *f*
[el] fiche *f*
[impl] (a block used with forging tools) cale *f* de épaisseur
[metall] (block for forging tools) pièce *f* d'écartement
[photo] clef *f* de manoeuvre
PEG-AND-HOLE ADJUSTMENT - [mech] réglage *m* à cheville et trous
~AWL - [shoe man] passe-corde *m*
~BAR - [cin & telev] barre *f* à chevilles
[text] barre *f* de selecteurs à goupilles
~COUNT - [telecomm] comptage *m* par épreuves
~-COUNT METER - [instr] compteur *m* de statistique
~-DRUM THRESHER - [agric] batteuse *f* à pointes
~HOLE - [text] trou *m* de pédonne
[constr] enlassure *f*
~OUT, to - [surv] (to outline a piece of land, generally as claimed by a miner, with peds) piqueter, jalonner
~OUT A LINE, to - [surv] jalonner une ligne
~PLAN - [text] armure *f* du perçage de cartons
~RAMMER - [metall] pilette *f*
~STAND FOR HANKS - [text] banc *m* à chevilles pour écheveaux
~-TOOTH HARROW - [agric] herse *f* articulée, herse *f* à chaînons
~TRACK - [cin] voie *f* de chevilles
PEGBOARD - [comput] panneau *m* de jacks
PEGGED CYLINDER - [text] cylindre *m* à chevilles
~PATTERN CARD - [text] carton *m* à chevilles en bois
~PRICE - [comm] prix *m* de soutien, prix *m* subventionné
PEGGING PLAN - [text] fiche *f* de tissage, dessin *m* de rentrage
PEGMATITE - [geol] (a coarse igneous rock) pegmatite *f*
PELARGONIC ACID - [chem] (a synthesis intermediate for plastics, pharmaceuticals, perfumes and flavours) acide *m* pélargonique
PELARGONYL CHLORIDE - [chem] (a synthesis intermediate) chlorure *m* de pélargonyle
P-ELECTRON - [electron] (electron having a principal quantum number of 6) électron *m* P
PELICAN - [zool] pélican *m*
PELIDNOMA - [med] péliome *m*, tache *f* cutanée livide
PELIOM - [min] péliom *m*
PELIOMA - s. pelidnoma
PELIOSIS - [med] péliose *f*
PELITE - s. pelitic gneiss
PELITIC GNEISS - [geol] roche *f* gneissique
~SCHIST - [geol] (of sedimentary origin) schiste *m* pélitique
PELLAGRA - [med] (chronic disease affecting persons who eath maize) pellagre *f*
PELLET, to - [gen] (to make into pellets) agglomérer
PELLET - [gen] boulette *f*
[plast ind etc] (a compressed mass of moulding material of given form and weight) oastille *f*
[firearms] (of a cartridge) grain *m* de plomb
[firearms] plomb *m*
[chem] pastille *f*, pilule *f*
[electron] (pellet of geranium used in semiconductor devices) pastille *f*
[metall] boulette *f*
[ind chem] bâtonnet *m*, tube *m*, anneau *m*
~RUBBER - [rubber ind] caoutchouc *m* en forme de granules
PELLETED PURIFYING MATERIAL - [ind chem] substance *f* d'épuration granulée
PELLETIERINE TANNATE - [chem] (a mixture of the tannates of the alkaloids obtained from Puncia granatum having application as a taenifuge) tannate *m*

de pelletiérine
PELLETING - [plast ind] (forming plastic material into small pieces for transfer from one machine or process to another) pastillage *m*
~MACHINE - [mech] pastilleuse *f*, machine *f* à comprimés
~PRESS - [mech] (a special press for compressing extrusion stock into pellets) pastilleuse *f*
PELLETIZED RUBBER - s. pellet rubber
PELLETIZER - s. pelleting machine
PELLICLE - [bot] pellicule *f*, membrane *f*
PELLITORY - [bot] pyrèthre *m*
PELLUCID - [gen] (clear, transparent) transparent pellucide
PELMA - [anat] plante *f* du pied
PELORIA - [bot] (abnormal condition of plants) pélorie *f*
PELORUS - [instr] (a circular plate, graduated in degrees, and used to take the bearing of some objects) alidade *f* à réflexion
PELT, to - [gen] lancer
[gen] (of rain) tomber à verse
PELT - [leather ind] peau *f*, fourrure *f*, peau *f* verte
~OF SHEEP - [text] peau *f* de mouton
PELTIER COEFFICIENT - [el] (coefficient of the electromotive force generated at a junction between two metals at a given temperature) coefficient *m* de Peltier
~EFFECT - [el] (production or absorption of heat due to the passage of current across the junction of two different metals) effet *m* Peltier
PELTON WHEEL - [mech] (impulse turbine) turbine *f* à action, turbine *f* Pelton
PELTRY - [leather ind] pelleterie *f*, peausserie *f*
PELVEOPERITONITIS - [med] pelvi-péritonite, pelvimétrosalpingite *f*
PELVIC - [anat] (pertaining to the pelvis) pelvien
~ARCH - [anat] (the skeletal frame with which the posterior locomotor appendages articulate) ceinture *f* pelvienne
~CAVITY - [anat] cavité *f* pelvienne
~FINS - [zool] (the posterior pair of fins) nageoires *fpl* postérieures
~GIRDLE - s. pelvic arch
~ERITONITIS - s. pelveoperitonitis
~PLEXUS - [anat] plexus *m* pelvie, plexus *m* hypogastrique
PELVIMETER - [instr] (instrument designed to measure the dimensions of the pelvis) pelvimètre *m*
PELVIMETRY - [radiat] (radiological measurement of the pelvic outlet) pelvimétrie *f*
PELVIOLITHOTOMY - [med] pyélolithotomie *f*
PELVIOPERITONITIS - s. pelveoperitonitis
PELVIPERITONITIS - s. pelveoperitonitis
PELVIS - [anat] (in man, the hip bones and the sacrum) bassin *m*
PEMMICAN - [food] pemmican *m*
PEN - [gen] plume *f*
[agric] enclos *m*, parc *m*
[shipbuild] cage *f*
[zool] cygne *m* femelle
~STOCK - [metall] coude *m* du porte-vent
PENAL - [leg] pénal
~SERVITUDE - [leg] travaux *mpl* forcés
PENALTY - [gen] peine *f*, pénalité *f*
[leg] sanction *f*, amende *f*
[comm] amende *f*
PENCATITE - [geol] (crystalline limestone) marbre *m* à brucite
PENCIL, to - [gen] marquer au crayon, dessiner au crayon
PENCIL - [impl] crayon *m*
[opt] (homocentric bundle of light rays) faisceau *m*
[el] (of carbon) crayon *m* (de charbon)
~-BEAM AERIAL - [radio] (a type of unidirectional aerial) antenne *f* à faisceau filiforme
~-BEAM ANTENNA - s. pencil-beam aerial
~COMPASSES - [impl] compas *m* à crayon
~CORE - [metall] crayon *m* de masselotte
~GATE - [metall] attaque *f* par collier-douche, attaque *f* en pluie
~MARKING - [gen] marquage *m* au crayon
[radar] (reference mark) marques *fpl* de répère
~OF RAYS - s. pencil [opt]
~RING GATE - [metall] attaque *f* par collier-douche
~SHARPENER - [impl] taille-crayon *m*
~-STONE - [geol] pyrophyllite *f*
~TUBE - [electron] (a pencil-shaped disk-seal tube) tube *m* en forme de crayon
PENCILLING - [build] crayonnage *m*
PENDANT - [gen] pendentif *m*
[el] (a suspended electrical fitting) lustre *m* électrique, lampe *f* suspendue
[horol] pendant *m*, anneau *m* de montre
[arch] clef *f* pendante
[naut] (short rope hanging from the head of a mast) martinet *m*, pantoire *f*
[gen] pendant
~AIR HOIST - [mech] treuil *m* suspendu
~PIN BARS - [text] barres *fpl* oscillantes à aiguilles
~PUSH - [el] (device for closing a circuit by a push-button attached to a flexible cord) interrupteur *m* de cordon
~SOCKET - [el] (socket for attachment to a flexible cord) douille *f* de lampe suspendu
~SWITCH - s. pendant push
PENDENTIVE - [arch] (spherical triangle formed by a dome springing from a square basis) pendentif *m*, trompe *f*
~DOME - [constr] dôme *m* suspendu
PENDING - [gen] pendant
PENDULAR - [gen & phys] pendulaire
PENDULATION - [gen] pendulation *f*
PENDULOUS - [gen & bot] pendant, oscillant
~UDDER - [zool] pis *m* distendu
PENDULUM - [phys mech] (body suspended from a fixed point and free to swing) pendule *m*, balancier *m*
[oil ind] poteau *m* oscillant
~ARM - [mech] bras *m* du pendule
~BALL - [mech] lentille *f* de pendule
~BARKER - [paper man] écorceuse *f* à pendule
~BOB - [horol] (the weight at the end of the pendulum) lentille *f* de balancier
~CIRCULAR SAW - [impl] scie *f* circulaire à balancier
~CONVEYOR - [mech] balancelle *f*
~DAMPER - [aero] (short heavy pendulum attached to the crank of a radial aero-engine to neutralize the fundamental torque impulses) amortisseur *m* à balancier
~GOVERNOR - [mech] (engine governor having forms which involve the principle of the conical pendulum) régulateur *m* à balancier
~GRIP GEAR - [mech] (in lifts) dispositif *m* d'arrêt

à balancier

PENDULUM MAGNETOMETER - [instr] (magnetometer in which the force on specimen is balanced by current through solenoid carrying magnetic plunger) magnétomètre *m* à pendule

~METER - [instr] (oscillating meter in which a coil is attached to a pendulum) compteur *m* à balancier

~ROD - [horol] tige *f* du pendule

~SHUTTER - [photo] obturateur *m* oscillatoire

~SPRING - [horol] (the ribbon of steel used to suspend the pendulum) ressort *m* du balancier

~TEST - [rubber ind] (impact resilience test) essai *m* de rebondissement à pendule, essai *m* au mouton pendule

~THERAPY - [radiat] (type of therapy in which the source of radiation swings to and from in an arc) radiothérapie *f* pendulaire

~WEIGHT - [mech] poids *m* du pendule

PENE - [mech] (the extremity of a hammer-head opposite to the face, and having various forms for different kinds of work) panne *f*

PENEPLAIN - [geol] (gently rolling land) pénéplaine *f*

PENETRABILITY - [gen] pénétrabilité *f*
[radiat] (the susceptibility of an object to the penetrated by radiation) pénétrabilité *f*

PENETRANT - [gen] pénétrant

~DYE METHOD - [metall] méthode *f* des liquides pénétrants

PENETRATE, to - [gen] pénétrer
[soil] tasser

PENETRATING - [gen] pénétrant

~COMPONENT - [nucl] (component penetrating through the potential barrier) composante *f* pénétrante

~OIL - [ind chem] (special type of oil designed to release seized or corroded parts) huile *f* pénétrante

~POWER - [opt] (of a microscope) puissance *f* de pénétration
[radiat] dureté *f* de rayonnement

~RADIATION - [radiat] (hard radiation) rayonnement *m* dur

~SHOWER - [nucl] (cosmic shower in which constituent particles are capable of greater penetration than is possible for electromagnetic radiation) gerbe *f* pénétrante

PENETRATION - [gen] pénétration
[chem] (term used in the testing of bituminous materials) pénétration *f*
[firearms] (of a shell) profondeur *f* de pénétration
[radio] (depth of penetration) profondeur *f* de pénétration

~COEFFICIENT - [radio] transparence *f* de grille

~DEPTH - [metall] pénétration *f* de la fusion

~DRAG - [aero] traînée *f* de pénétration

~FACTOR - [radio] (the reciprocal of the amplification factor) coefficient *m* de grille
[nucl] facteur *m* de pénétration

~INDEX - [rubber ind] indice *m* de pénétration

~-LOAD CURVE - [soil] courbe *f* enfoncement-chargement

~POTENTIAL - [nucl] (the potential required by a particle to penetrate the potential barrier) potentiel *m* de pénétration

~PROBABILITY - [nucl] (the transmission coefficient) effet *m* tunnel

~RATE - [mining] vitesse *f* d'avancement

PENETRATION-RESISTANCE CURVE - [soil] diagramme *m* de pénétration

~RESISTANCE OF A PILE - [soil] résistance *f* au battage

~TEST - [rubber ind] essai *m* de pénétration

~VOLTAGE - [electron] (in semiconductors) tension *f* de pénétration

PENETRATIVE ROCK - [geol] roche *f* intrusive

PENETRATOR - [tool] (tool used to test hardness) pénétrateur *m*

PENETROMETER - [instr] (an instrument for measuring the hardness of materials) pénétromètre *m*
[med] (in radiology, a device used to judge the overall quality of a radiograph) pénétromètre
[ind chem] (instrument designed to measure the consistency of greases, bituminous materials etc) pénétromètre *m*

~SENSITIVITY - [radiat] (radiographic sensitivity) sensibilité *f* de pénétromètre *m*

~TEST - [soil] essai *m* de sondage

PENGUIN - [zool] manchot *m*
[aero] (aircraft fitted with a low-power engine and incapable of flying; used in early training) appareil *m* d'école

PENICILLIN - [chem] (generic term for a number of antibiotics of real therapeutic importance) pénicilline *f*

~-FAST - [med] résistant à la pénicilline

~-RESISTANT - s. penicillin-fast

PENICILLINASE - [chem] (an enzyme, obtained commercially from Bacillus cereus, which destroys penicillin and is used in medicine in the treatment of allergic conditions due to medication with penicillin) pénicillinase *f*

PENINSULA - [geogr] péninsule *f*, presqu'île *f*

PENIS - [anat] pénis *m*, verge *f*

PENKNIFE - [impl] canif *m*

PENNANT - [naut] (long narrow flag) flamme *f*, banderole *f*
[naut] (a pendant) pantoire *f*
[naut] (small flag used for signalling) guidon *m*

PENNINITE - [min] (a silicate of magnesium) pennine *f*, penninite *f*

PENNON - s. pennant

~CRESS - [bot] tabouret *m* des champs, herbe *f* aux écus

PENNYROYAL OIL - [chem] (pulegium oil. An oil distilled from Mentha pulegium and used as an emmenagogue in medicine) essence *f* de menthe Pouliot

PENNYWEIGHT - [meas] (one twentieth of an ounce) (approx) un gramme et demi

PENSTOCK - [hydr] (a conduit to a waterwheel gate) canal *m* d'amenée, bief *m* d'amont
[hydr] (a sluice controlling the discharge of water) vanne *f* de tête d'eau

PENT - [gen] (penned up) renfermé, resserré

~ROOF - [constr] (a single-slope roof) toit *m* en appentis, toit *m* à un seul égout

PENTABORANE - [chem] (boron hydride. A fuel for rockets) pentaborane *m*

PENTACHLOROETHANE - [chem] (a solvent) pentachloréthane *m*

PENTACHLORONITROBENZENE - [chem] (an agricultural disinfectant and a synthesis intermediate) pentachloronitrobenzène *m*

PENTACHLOROPHENOL - [chem] (a wood preservative

and fungicide) pentachlorophénol *m*
PENTAD - [math] groupe *m* de cinq, pentade *f*
PENTADACTYL - [zool] (with five digits) pentadactyle
PENTADECANE - [chem] (a synthesis intermediate) pentadécane *m*
PENTADECANOIC ACID - [chem] (a synthesis intermediate) acide *m* pentadécanoïque
PENTAERYTHRITOL - [chem] (an intermediate for synthetic resins pesticides, drugs and plasticizers) pentaérythritol *m*
~TETRANITRATE - [chem] (a vasodilator used in the treatment of angina pectoris) tétranitrate *m* de pentaérythritol
PENTAGON - [geom] (figure with five angles and five sides) pentagone *m*
PENTAGONAL - [geom] pentagonal
PENTAGRAM - [geom] (figure having five lobes) pentagramme *m*
PENTAGRID - [electron] (heptode) heptode *f*
~CONVERTER - [electron] heptode *f* oscillatrice-mélangeuse
~MIXER - [electron] heptode *f* mélangeuse
~TUBE - [electron] heptode *f*
PENTAHEDRON - [geom] (solid bounded by five faces) pentaèdre *m*
PENTAHYDRIC ALCOHOLS - [chem] (alcohols containing five hydroxyl groups) alcools *mpl* pentavalents
PENTAMETHYLENE - [chem] (saturated cyclic hydrocarbon) pentaméthylène *m*
PENTANE - [chem] (a low-boiling paraffin used as a refrigerant and anaesthetic. B.P. of n-pentane is 36° C.) pentane *m*
PENTANEDICOL - [chem] (an anti-freeze agent and an intermediate for synthetic resins) pentandiol *m*
PENTANOL - [chem] (a solvent and intermediate) pentanol *m*
PENTAPLOIDY - [genet] (the condition of having five times the haploid number of chromosomes) pentaploïdie *f*
PENTASTYLE - [arch] (row of five columns) pentastyle *m*
PENTATONIC SCALE - [mus] gamme *f* pentatonique
PENTATRIACONTANE - [chem] (a synthesis intermediate) pentatriacontane *m*
PENTAVALENT - [chem] (capable of combining with five atoms of hydrogen or the equivalent) pentavalent
PENTELIC - [arch] (a type of white marble) pentélique
PENTHOUSE - [constr] (a building erected on the roof of another building) maison *f* construite sur le toit
[constr] (shed or roof with a single slope attached to the wall of a building) appentis *m*, abrivent *m*
[mining] plancher *m* de sûreté
PENTLANDITE - [min] (a natural iron-nickel sulphide, often found intergrown with pyrrhotite. It crystallizes in the cubic system and can be distinguished from pyrrhotite by its octahedral cleavage) pentlandite *f*
PENTOBARBITAL - [chem] (a hypnotic used in medicine) pentobarbital *m*
PENTODE - [electron] (electronic tube consisting of a cathode, three grids and an anode) pentode *f*
~ELECTRON GUN - [electron] canon *m* électronique à trois anodes
PENTODE TRANSISTOR - [electron] transistor *m* pentode
~VALVE - s. pentode
PENTOLITE - [chem] (a high explosive) pentolite *f*
PENTOSE - [chem] (generic term for a sugar each molecule of which contains five carbon atoms) pentose *m*
PENTOSURIA - [med] pentosurie *f*
PENTOXIDE - [chem] pentoxyde *m*
PENTROUGH - s. pentstock
PENTYLENETETRAZOL - [chem] (leptazol. An analeptic used in the treatment of coma due to barbiturate overdosage and similar conditions) pentylènetétrazol *m*
PENUMBRA - [gen] pénombre *f*
[opt] pénombre *f*
PEONY - [bot] pivoine *f*, rode *f* de Notre-Dame
PEPPER - [bot] (the immature berries of the pepper plant) poivre *m*
~-AND-SALTS - [paint] (a mixed white and black, resulting in a speckled grey) poivre-et-sel
[text] (of a cloth) marengo
~BLISTERS - [metall] piqûres *mpl* de surface
~CORN - [bot] grain *m* de poivre
PEPPERGRASS - [bot] (cress) passerage *f*, cresson *m* alénoir
PEPPERMINT - [bot] (pungent aromatic herb; mentha piperita) menthe *f* poivrée
~OIL - [chem] (an essential oil obtained from varieties of Mentha and used as a flavourant) essence *f* de menthe
~TREE - [bot] eucalyptus *m* poivré
PEPSIN - [chem] (a proteolytic enzyme with uses in medicine) pepsine *f*
PEPTIC - [biol] peptique, gastrique
~ULCER - [med] (ulcer of the stomach or duodenum) ulcère *m* gastrique, ulcère *m* superficiel de l'estomac
PEPTISIZE, to - s. peptize, to
PEPTISIZED RUBBER - s. peptized rubber
PEPTIZATION - [chem] (the production of a colloidal solution) peptisation *f*
PEPTIZE, to - [chem] (to produce a colloidal solution) peptiser
PEPTIZED RUBBER - [rubber ind] caoutchouc *m* peptisé, caoutchouc *m* ramolli
PEPTIZER - [chem] (compound used to break down natural and synthetic rubbers) agent *m* peptisant
~RECLAIMING AGENT - [rubber ind] agent *m* régénérateur peptisant
PEPTIZING AGENT - s. peptizer
~AGENT RECLAIMING PROCESS - [rubber ind] procédé *m* de régénération catalytique aux agents peptisants
PEPTONE - [chem] (a mixture of amino-acids and peptones with uses in medicine in the non-specific desensitization of allergic states) peptone *f*
PEPTONIZED MILK - [food] lait *m* peptonisé
PEPTONIZING REST - [brew ind] relai *m* d'albumine
PERACETIC ACID - [chem] (a polymerization catalyst, oxidant, bactericide, and bleaching agent for paper and textiles) acide *m* peracétique
PERACID - [chem] (an acid formed by the reaction of hydrogen peroxide with an acid) peracide *m*
PERACUTE - [med] suraigu
PERAMBULATOR - [transp] voiture *f* d'enfant

PERAMBULATOR - [instr] (a surveyor's instrument designed to measure distances; a pedometer) podomètre *m*
PERBORATE - [chem] (a salt of perboric acid) perborate *m*
PERBORAX - [chem] (sodium perborate) perborate *m* de sodium
PERCALE - [text] (a cotton fabric, generally coloured or printed) percale *f*
~TAPE - [text] ruban *m* de percale
PERCALINE - [text] percaline *f*
PERCEIVED COLOUR - [opt] couleur *f* perçue
PERCENT - [math etc] (or per cent) pour cent
~CONSONANT ARTICULATION - [acoust] articulation *f* des consonnes, intelligibilité *f* des consonnes
~DRIFT - [radio] (drift percentage; of a balanced magnetic amplifier) poucentage *m* de dérive
~ELONGATION - [metall] allongement *m* pour-cent après rupture
~EXCESS CHARGE - [el] (coefficient by which the quantity of electricity delivered during the discharge must be multiplied to obtain the quantity of electricity required for the charge) coefficient *m* de charge
~HEARING - [acoust] (percent hearing equals I00 less the percent hearing loss) pourcentage *m* d'audition
~RIPPLE - [radio] taux *m* d'ondulation
~RIPPLE VOLTAGE - [electron] taux *m* d'ondulation résiduelle
~VOWEL ARTICULATION - [acoust] articulation *f* des voyelles, intelligibilité *f* des voyelles
PERCENTAGE - [math etc] (rate per hyndred) pourcentage *m*
~BEAM MODULATION - [electron] taux *m* de modulation du faisceau
~BY VOLUME - [math etc] pourcentage *m* en volume
~DEPTH DOSE - [radiat] (the percentage of radiation delivered at a given depth in tissue) pourcentage *m* de dose en profondeur
~ERROR - [instr] (the difference between the percentage registration of a meter and I00 percent) pourcentage *m* des erreurs
~LOSS - [nucl] (the number of neutrons lost by leakage etc) pourcentage *m* des pertes
~MODULATION - [radio] (modulation percentage; the modulation factor expressed as a percentage) pourcentage *m* de modulation
~OF ASH - [text] teneur *f* en cendres (de fibres de jute)
~OF FIBRE - [text] pourcentage *m* de fibres, rendement *m* de fibres
~OF REJECTS - [gen] (the proportion, expressed in units per hundred, of manufactured objects which are found on inspection or test to be below a given standard of dimensions, quality etc.) pourcentage *m* de déchets
~SYNCHRONIZATION - [telev] pourcentage *m* de cynchronisation
~TILT - [telev] taux *m* d'inclinaison, pourcentage *m* de déclivité
PERCEPTION - [gen] (the recognition of external objects through the senses) perception *f*
~TIME - [med] temps *m* de perception
PERCEPTIVITY - [gen] perceptivité *f*
PERCEPTUAL CONSTANCIES - [opt] constances *fpl* de perception
PERCH, to - [gen] percher, se percher, jucher, gîter
PERCH, to - [leather ind] étirer
PERCH - [gen] perchoir *m*
[agric] bâton *m*
[zool] perche *f*
[meas] perche *f* (approx. 5m)
[leather ind] chevalet *m* d'étirage
PERCHING - [gen] s. perch, to
PERCHLOR-ETHER - [chem] (a solid with an odour of camphor by chlorine) perchloréther *m*
PERCHLORATES - [chem] (the salts of perchloric acid) perchlorates *mpl*
PERCHLORIC ACID - [chem] (a powerful oxidizing agent. It is used as a catalyst and analytical reagent) acide *m* perchlorique
PERCHLOROETHYLENE - [chem] (a synthesis intermediate, solvent, and heat transfer medium) perchloroéthylène *m*
PERCHLOROMETHYL MERCAPTAN - [chem] (an intermediate for dyestuffs) perchlorométhyl-mercaptan *m*
PERCHLORYL FLUORIDE - [chem] (an oxidant and synthesis intermediate) fluorure *m* de perchloryle
PERCOLATE, to - [gen, constr work etc] (of a fluid, to pass slowly through a substance) filtrer (à travers)
PERCOLATION - [gen, chem etc] (the slow passage of a fluid through a porous solid) filtration *f*, filtrage *m*
~RATE - [chem] vitesse *f* de percolation
PERCOLATOR - [impl] filtre *m*
PERCRYSTALLIZATION - [chem] (crystallization of a substance from a solution which is being dualysed) percristallisation *f*
PERCUSSION - [gen & med] percussion *f*
~CAP - [expl] capsule *f* de fulminate, amorce *f*
~CAP COPPER - [metall] alliage *m* en cuivre pour capsules fulminantes
~COMPOSITION - [expl] (of the percussion primer) fulminate *m*
~DRILL - [impl] perforatrice *f* percutante, perforatrice *f* à percussion
[soil] (bailer) soupape *f*, curette *f*
~FUSE - [firearms] (mechanism fitted to high-explosive shells to produce detonation on impact) fusée *f* percutante
~INSTRUMENTS - [mus] instruments *mpl* à percussion
~JUMPER - s. percussion drill
~LOCK - [firearms] (gunlock exploding the charge by percussion) platine *f* à percussion
~PIN - [firearms] rugueux *m* de fusée
~PRIMER - s. percussion cap
~WELDING - [metall] (method of stored energy welding, used for wires, in which pressure is applied immediately after the passage of the current) soudage *m* par percussion
PERCUSSIVE BORING - [mech] sondage *m* percutant, sondage *m* de battage
~WELDING - s. percussion welding
PERDISTILLATION - [chem] (distillation through a dialysing membrane) perdistillation *f*
PERENNIAL - [gen] éternel, perpétuel
[bot] vivace, persistant
PERENNIALS - [bot] plantes *fpl* vivaces
~PLANT - [bot] plante *f* vivace
PERFECT, to - [gen] perfectionner, rendre parfait, achever
PERFECT - [gen] parfait

PERFECT COMBUSTION - [phys] (combustion under theoretical conditions) combustion *f* neutre, combustion *f* stoechiométrique
~FLUID - [hydr] (an ideal fluid, i.e. incompressible, with uniform density and offering no resistance to distorting forces) fluides *mpl* parfaits
~GAS - [phys] (a gas which would follow Boyle's law and Dalton's law in all respects) gaz *m* parfait
~INTERVAL - [mus] (the 4th, 5th and 8th of the scale) intervalle *m* juste
~MOSAIC CRYSTAL - [phys] cristal *m* mosaique parfait
~NUMBER - [math] nombre *m* parfait
~SHOT - [cin] prise *f* de vue excellente
~UP, to - [print] (to print the second side of a sheet of paper) imprimer recto-verso
PERFECTING - [gen] perfectionnement *m*, achévement *m*
[print] impression *f* recto-verso
~ENGINE - [paper man] (machine which reduces large particles and knots in the pulp, generally called 'refiner') raffineur *m* cônique
~PRESS - [print] machine *f* à retiration
PERFECTION - [gen] perfection *f*
PERFECTLY INELASTIC COLLISION - [mech] (a collision in which the coefficient of restitution is zero) collision *f* parfaitement inélastique
PERFORATE, to - [gen] perforer, percer
[mech] perforer, poinçonner, grillager
PERFORATED - [gen & mech] perforé, percé, grillagé, ajouré
[phys] (porous) poreux
~BOTTOM - [metall] fond *m* en baguettes
~BRICK - [constr] brique *f* creuse
~CAGE - [text] tambour *m* perforé
~CARD - [text] carton *m* piqué, carton *m* troué, carte *f* perforé
~CATHODE - [electron] (indirectly heated cathode with an active material surrounded by a perforated metal cylinder) cathode *f* perforée
~CONICAL TUBE FOR CHEESES - [text] tube *m* conique perforé pour bobine croisée
~FLANGE - [mech] plateau *m* à claire-voie
~FLEXIBLE BEND - [text] cintre *m* flexible troué
~HEAD - [text] ailette *f* à tête perforée
~PLATE - [metall] tôle *f* perforée
~PLATE OSCILLATING SCREEN - s. perforated plate reciprocating screen
~PLATE RECIPROCATING SCREEN - [min] (for coke) table *f* à secousses
~PUNCHING PLATE - [text] plaque *f* perforée de piquage
~RAIL - [text] rail *m* perforé
~SCREEN - [cin] (a screen with a large number of small holes to allow the sound to pass from the loudspeaker to the auditorium) écran *m* perforé
~STRIP - [text] bande *f* perforée
~TAPE TRANSMISSION - [telecomm] (automatic routing) émission *f* télégraphique à bande perforée, reperforateur-retransmetteur *m*
~TROUGH - [text] caisse *f* à trous
PERFORATING - [mech] (in sheet metals) perforation *f*, perforage *m*
~DIE - [impl] poinçon *m*
~JOB - [mining] forage *m* à canon perforateur
~MACHINE - [cin] (machine making the perforations in a coated film) perforatrice *f*
PERFORATING PINCERS - [impl] pince *f* à poinçonner
~PLIERS - [text] pince *f* à trous, poinçon *m* à emporte-pièce
PERFORATION - [gen & mech] perforation *f*, percement *m*, perforage *m*
[cin] (the holes punched in the film to engage the teeth of the sprocket) perforation *f*
~GAUGE - [photo] pas *m* de la perforation
~ON THE CARD - [text] trou *m* de carton
~PITCH - [gen] pas *m* de perforation
PERFORATOR - [telecomm] (instrument for the manual preparation of a perforated tape, used in telegraphy) perforatrice *f*
PERFORM, to - [gen] exécuter, accomplir, effectuer
[gen] (theatre) représenter
[mus] jouer
PERFORMANCE - [gen] exécution *f*, acte *m*, exploit *m*
[mech] (of a machine, a motor etc) fonctionnement, rendement *m*
[ind etc] (execution carried out with skill) exécution *f*
[auto etc] (the qualities of an aircraft, auto etc., which can be expressed as quantities e.g. speed, manoeuvrability, rate of climb of an aircraft ect) performance *f*
~AND PROGENY TESTING - [zool] jugement *m* animal
~CHARACTERISTICS - [mech] caractéristiques *fpl* de fonctionnement
~CHART - [mech electron etc] diagramme *m* de fonctionnement
~FACTOR - [el] qualité *f* de fonctionnement
~MONITOR - [radar] (an echo box used for checking purposes) moniteur *m* du rendement
~TEST - [el mech etc] (a test carried out to determine the efficiency of a machine, material, etc. under conditions of use) essai *m* de fonctionnement
PERFORMETER - [radar] (echo box actuator) circuit *m* resonnant pour échos artificiels
PERFORMIC ACID - [chem] (an oxidizing agent) acide *m* performique
PERFUME - [chem] (generally a volatile liquid emitting a fragrant smell) parfum *m*
~PLANT - [bot] plante *f* à parfum
PERFUSION - [med] perfusion *f*
PERGAMYN - [paper man] (vegetable parchment) pergamine *f*
PERGETING - s. pargeting
PERGOLA - [arch] (arbor of a structural nature) pergola *f*, treille *f* à l'italienne
PERHAPSATRON - [nucl] (apparatus for investigating controlled fusion of hydrogen) perhapsatron *m*
PERI POSITION - [chem] (in naphtalene derivatives, the I,8 position) position *f* péri
PERIANGITIS - [med] périvascularite *f*
PERIANTH - [bot] périanthe *m*
PERIAORTITIS - [med] périaortite *f*
PERIARTERITIS - [med] (a type of inflammation of the arteries) périartérite *f*
PERIARTHTRITIS - [med] périarthrite *f*
PERIARTICULAR - [anat] (of the tissues immediately around a joint) périarticulaire *f*
PERIASTRON - [astr] (in an orbit around a star, the point at which the body describing the orbit is nearest to the star) périastre *m*
PERIBLAST - [med] cytoplasma *m*
PERIBLEPSIS - [med] air *m* hagard des aliénés
PERICARDIECTOMY - [med] péricardectomie *f*

PERICARDIORRHAPHY - [med] suture *f* du péricarde
PERICARDIOSTOMY - [med] péricardiostomie *f*
PERICARDIOTOMY - [med] (the incision of the pericardium) péricardiotomie *f*
PERICARDITIS - [med] (inflammation of the pericardium) péricardite *f*
PERICARDIUM - [anat] (membrane surrounding and protecting the heart) péricarde *m*
PERICARP - [bot] (husk or fruit surrounding a seed, derived from the wall of the ovary) péricarpe *m*
PERICHOLECYSTITIS - [med] péricholécystite *f*
PERICHONDRITIS - [med] (inflammation of the perichondrium) périchondrite *f*
PERICHONDRIUM - [anat] (the vascular membrane surrounding cartilage) périchondre *m*
PERICLASE - [min] (natural magnesium oxide, used as a refractory) périclase *f*
PERICLINAL - [geol] (adj of pericline) périclinal
PERICLINE - [min] (variety of albite occurring in the Swiss Alps) péricline *f*
PERICOLITIS - [med] (inflammation of the peritoneum covering the colon) péricolite *f*
PERICORONITIS - [med] péricoronite *f*
PERIDERM - [bot] (protective layer) périderme *m*
PERIDIDYMIS - [anat] périddidyme *m*, tunique *f* vaginale du testicule
PERIDIDYMITIS - [med] vaginalite *f*
PERIDOT - [min] péridot *m*
PERIDURAL - [anat] (surrounding the dura mater) péridural
PERIFOCUS - [phys] péricentre *m*
PERIGASTRITIS - [med] (inflammation of the external surface of the stomach) périgastrite *f*
PERIGEE - [astr] (that point in the moon's orbit at which it is nearest to the heart) périgée *m*
PERIGONIUM - [bot] (the perianth) périgone *m*
PERIGYNOUS - [bot] (situated around the ovary) périgyne
PERIHELION - [astr] (the point in the orbit of a planet or comet at which it is nearest to the sun) périhélie *m*
PERIHEPATITIS - [med] (inflammation of the peritoneum covering the liver) périhépatite *f*
PERIKINETIC - [chem] (relating to the Brownian movement) périkinétique
PERILYMPH - [zool] (the fluid which fills the space between the membranous labyrinth and the bony labyrinth of the internal ear) périlymphe *f*
PERILYMPHATIC - [zool] périlymphatique
PERIMETER - [geom] périmètre *m*
[med instr] (instrument shaped like an arc designed to measure the field of vision) périmètre *m*, campimètre *m*
~ SHEAR - [soil] cisaillement *m* au contour
~ TRACK - [aero] (a taxying track round the perimeter of an aerodrome) voie *f* de circulation périphérique
PERIMETRITIS - [med] (inflammation of the uterus covering peritoneum) périmétrite *f*
PERIMETRIUM - [anat] (peritoneum covering the uterus) revêtement *m* péritonéal de l'utérus
PERIMETRY - [med] campimètrie *f*, périmétrie *f*
PERIMYELITIS - [med] pie-mérite *f* rachidienne
PERINEOPLASTY - [med périnéoplastique *f*
PERINEOTOMY - [med] périnéotomie *f*
PERINEPHRIUM - [anat] capsule *f* adipeuse du rein
PERIOD - [phys etc] (the duration of one complete oscillation) période *f*, temps *m*
PERIOD - [gen] (a definite portion of time) période *f*
[el] (the minimum interval of the independent variable after which the same characteristics of a periodic phenomenon recur) période *f*
[acoust] (repeating its value regularly at equal time intervals) période *f*
[chem] (the elements between an alkali metal and the rare gas of next higher atomic number) période *f*
[nucl] (the time required for one cycle of a repeated series of events) période *f*
~ DEMAND SIGNAL - [contr] signal *m* de la constante de temps de consigne
~ LUMINOSITY LAW - [astr] (a relationship between the period and the absolute magnitude for all Cepheid variables) loi *f* de période-luminosité
~ METER - [nucl] (instrument allowing an operator to read the reactor period in seconds) périodemètre *m*
~ OF AGEING - [phys] période *f* de vieillissement
~ OF DECAY - [nucl](half life) période *f* radioactive
~ OF DEFORMATION - [mech] (the time from the initial impact until bodies cease to approach) temps *m* de déformation
~ OF HARMONIC MOTION - [mech] période *f* de mouvement harmonique
~ OF ISOMETRIC RELAXATION - [med] période *f* de relâchement postsystolique
~ OF MOON'S NODE - [astr] période *f* de nutation
~ OF RESTITUTION - [mech] (the time from maximum compression until bodies are separated and moving with final velocities) temps *m* de restitution
~ RANGE - [nucl] (in a nuclear reactor, an intermediate range of reaction rate) domaine *m* de divergence
~ SCRAM - [nucl] dispositif *m* d'arrêt d'urgence d'un réacteur
~ SHUT-DOWN - [nucl] (of a reactor) arrêt *m* d'urgence par constante de temps insuffisante
PERIODATES - [chem] (formed by the oxidation of iodates) périodates *mpl*
PERIODIC - [gen etc] périodique
[chem] (of period) périodique
~ ACID - [chem] (an oxidizing agent) acide *m* périodique
~ AERIAL - [radio] (a tuned aerial) antenne *f* accordée
~ ANNEALING FURNACE - [metall] four *m* à recuire périodique
~ ANTENNA - s. periodic aerial
~ ARRANGEMENT - s. periodic system
~ CLASSIFICATION - s. periodic system
~ DUTY - [el] (duty at variable load changing periodically) service *m* périodique
~ FOCUSING FIELDS - [electron] (a method of electron beam focusing) champs *mpl* de focalisation à variations périodiques
~ FUNCTION - [math] fonction *f* périodique
~ LAW - s. periodic system
[phys] loi *f* du système périodique
~ LINE - [telecomm] (line of identical sections, similarly oriented) ligne *f* à sections électriques équivalentes
~ POLARITY REVERSAL - [telev] (in colour television) inversion *f* périodique de polarité
~ PULSE-TRAIN - [telecomm] (pulse train consisting

of identical groups of pulses) série *f* d'impulsions périodique
PERIODIC QUANTITY - [chem] grandeur *f* périodique [phys el etc] (a quantity identically reproduced at equal intervals of the independent variable) grandeur *f* périodique
~SPRING - [hydr] source *f* intermittente
~SYSTEM - [chem] (a classification of the elements showing the periodic variation of the physical and chemical properties of an element and its compound with the atomic number. It comprises nine groups of elements) classification *f* périodique
~TABLE - [chem] (a display of the Periodic System in tabular form) table *f* périodique
~TIME - [phys] (the period of time during which the independent variable is time) période *f* de temps
~WAVE - [radio] (wave in which the disturbance repeats self periodically) onde *f* périodique
PERIODICAL - [gen] périodique
~MOVEMENT - [phys] mouvement *m* périodique
PERIODICITY - [phys & gen] (the recurrence if an event at regular intervals of time or distance) périodicité *f*
PERIODIDE - [chem] (iodide with a larger percentage of iodine than any other iodide of the same series) périiodure *m*
PERIODONTITIS - [med] (inflammation of the tooth membrane in the jaw) périodontite *f*, péricementite *f*
PERIONYCHIUM - [anat] repli *m* péri-unguéal
PERIOPHTALMITIS - [med] ténonite *f*
PERIOST - s. periosteum
PERIOSTEUM - [anat] (vascular membrane covering the bone) périoste *m*
PERIOSTITIS - [med] (inflammation of the periosteum) périostite *f*
PERIPHERAL - [gen] (situated round the edges) périphérique
~ELECTRON - [electron] (outer-shell electron) électron *m* de conduction, électron *m* de valence, électron *m* optique, électron *m* périphérique
~EQUIPMENT - [comput] (additional equipment) équipement *m* périphérique
~HEM - [aero] (the hem round the periphery of a parachute) bord *m* d'attaque
~MEMORY - [comput] mémoire *f* périphérique
~REGION - s. perisphere
~SPEED - [mech] (the speed of the periphery of e.g. a wheel) vitesse *f* périphérique
~TRANSFER - [comput] transfert *m* périphérique
~VISION - [opt] vision *f* périphérique
~VISUAL FIELD - [opt] champ *m* de vue périphérique
PERIPHERICAL - [gen] périphérique
PERIPHLEBITIS - [med] (inflammation of the outer coat of a vein) périphlébite *f*
PERIPOSITION - s. peri position
PERIPROCTITIS - [med] (inflammation of the tissue round the rectum) périrectite *f*
PERIPROSTATITIS - [med] périprostatite *f*
PERIPTER - [arch] (building having a single row of pillars surrounding it) périptère *m* [phys] (the aerial region about a moving object, which is vortically disturbed) périptère *m*
PERIRHIZOCLASIA - [med] périradiculite *f* dentaire
PERISALPINGITIS - [med] (inflammation of the peritoneum of the Fallopian tube) périsalpingite *f*
PERISCOPE - [instr] (an optical instrument consisting of revolving prism reflecting light rays down a vertical tube) périscope *m*
PERISCOPIC - [opt] périscopique
~LENS - [photo] périscopique
~SEXTANT - [instr] sextant *m* périscopique
PERISH, to - [gen] périr, détériorer, altérer [rubber ind] décomposer, dégrader
PERISHABLE - [gen] périssable [comm] (of goods) périssable, sujet à s'altérer
~GOODS - [comm] (e.g. fruit during transport etc) marchandises *fpl* périssables
PERISHED - [gen] altéré, détérioré [rubber ind] dégradé [text] détérioré
~STAPLE - [text] soie *f* de coton détériorée
~STEEL - [metall] acier *m* dénaturé
PERISHING OF RUBBER - [rubber ind] (the degradation of rubber due to sunlight, ozone, UV light, etc. under natural conditions) dégradation *f* du caoutchouc , décomposition *f* du caoutchouc
PERISPERM - [bot] (a nutritive tissue) périsperme *m*
PERISPHERE - [phys] (the volume round an object, in which the electric, magnetic etc. fields of the object produce effects which can be observed) périsphère *f*
PERISPLENITIS - [med] (inflammation of the peritoneum of the spleen) périsplénite *f*
PERISTALTIC - [zool] (contracting in successive circles) péristaltique
PERISTRUMITIS - [med] péristrumite *f*
PERISTYLE - [arch] (a system of columns round a building of an internal court) péristyle *m*
PERITECTIC - [phys] (point at which incongruent melting takes place) point *m* péritectique
~CHANGE - [metall] (a type of phase change in which incongruent melting occurs) transformation *f* péritectique, réaction *f* péritectique
PERITECTOID - [metall] péritectoide
PERITECTOMY - [med] ablation *f* du pannus
PERITHELIOMA - [med] (tumor having its cells arranged around thin-valled blood-vessels) périthéliome *m*
PERITHELIUM - [anat] périthélium *m*
PERITONEUM - [anat] (serous membrane lining the abdominal cavity and, in higher vertebrates, forming a completely closed sac in males, in females, the fallopian tubes open into the cavity) péritoine *m*
PERITONITIS - [med] (inflammation of the peritoneum) péritonite *f*
PERIVESICULITIS - [med] périvésiculite *f*
PERIVISCERITE - [med] périviscérite *f*
PERLITE - [min] (a siliceous rock similar to obsidian and used as an insulant and in special plasters and cements) perlite *f*
~IRON - [metall] (or pearlite iron) perlite *f*
PERLITIC STRUCTURE - [geol] (a structure which consists of systems of spheroidal concentric cracks which were produced during cooling) structure *f* perlitique
PERMANENCE - [gen] permanence *f*, stabilité *f*
PERMANENCY - [paint] stabilité *f* des peintures
~OF VISION - [opt] (retention of image) permanence *f* des images
PERMANENT - [gen etc] permanent, inamovible, fixe
~ACTION - [contr] (method by which a control is held permanently) action *f* permanente
~ADJUSTMENT - [surv] (to a surveying instrument) régulation *f* permanente
~ALIGNMENT - [surv] alignement *m* permanent

PERMANENT COIL - [el] bobine *f* magnétique durable
~COMPRESSION SET - [mech] rémanence *f* de compression
~CONTRACTION - [metall] retrait *m* permanent, post-retrait
~DATA - [comput] données *f* constantes
~DEFORMATION - [phys] (the deformation remaining when the deforming stress has been removed) déformation *f* permanente
~DENTITION - [zool] (the set of teeth replacing the milk dentition) dentition *f* permanent
~DISPOSAL - [nucl] (permanently stored radio-active waste) élimination *f* permanente
~DYE - [text] colorant *m* solide, colorant *m* durable
~ECHO - [radar] (a fixed echo) écho *m* fixe
~ELONGATION - [rubber ind] allongement *m* résiduel, allongement *m* rémanent
~EXPANSION - [th eng] dilatation *f* permanente, post-dilatation *f*
~FRAME - [mech] châssis *m* à demeure
~GAS - [phys] (gases requiring low temperatures for their liquefaction; e.g. oxygen, nitrogen, hydrogen) gaz *m* permanent
~HARDNESS - [chem] (hardness due to dissolved mineral salts which are not precipitated by boiling) crudité *f* permanente
~IMPLANT - [radiat] (implant of radioactive material of short half-life; the prescribed dose is delivered by the time the material has decayed) implantation *f* permanente
~LOAD - [constr] (the dead loading on a structure, i.e. the weight of the structure and the fixed loading carried by it) charge *f* fixe
~LOOP - [telecomm] (in telephony) faux appel *m*
~MAGNET - [el] (hard-steel magnet which retains the greater portion of its magnetization) aimant *m* permanent
~MAGNET CENTRING - [telev] centrage *m* de l'image par aimants permanents
~-MAGNET LOUDSPEAKER - [el acoust] haut-parleur *m* à aimant permanent
~-MAGNET MOTOR - [el] (machine in which the field magnet is a permanent magnet) magnéto-moteur *m*, moteur *m* magnétoélectrique
~-MAGNET MOVING-COIL INSTRUMENT - [instr] appareil *m* à cadre mobile
~-MAGNET MOVING-IRON INSTRUMENT - [instr] appareil *m* à fer mobile et aimant
~-MAGNET STEEL - [metall] (the material which is used for the manufacture of loudspeakers etc) acier *m* d'aimant permanent
~MEMORY - [comput] (media retaining information in the absence of power, e.g. magnetic tapes) mémoire *f* non-effacable
~MOTOR - s. permanent-magnet motor
~MOULD - [metall] (a metal mould used for the production of castings) coquille *f*, moule *m* métallique, forme *f* fixe
~MOULD CASTING - [metall] coulée *f* en coquille
~PILOT - [gas ind] (constant burning pilot) veilleuse *f* permanente
~PUNCHEON - [constr] rallonge *f* de pilotis *m*
~SET - [phys] (any deformation of a material which remains after removal of the deforming stress) déformation *f* permanente
~STORAGE - s. permanent memory
~STORE - s. permanent memory

PERMANENT TWIST - [text] torsion *f* effective
~WAY - [railw] (the finished track for a railway) superstructure *f* de la voie
PERMANGANATES - [chem] (compounds of manganese with a strong oxidizing action, potassium permanganate being the most important) permanganates *mpl*
PERMATRON - [electron] (vacuum tube in which the control of plate current is not effected by a grid byt by a magnetic field) permatron *m*
PERMEABILITY - [phys] (the rate at which a gas or liquid under pressure diffuses through a porous substance) perméabilité *f*
[el] (the ratio between flux density and magnetizing force producing it) perméabilité *f*
[metall] (the rate of diffusion of gas through the sand mould) perméabilité *f*, porosité *f*
[gen] perméabilité *f*
[geol] (a rock's faculty to allow a liquid to flow through it) perméabilité *f*
~ALLOYS - [metall] alliages *mpl* à haute perméabilité
~BRIDGE - [instr] (device for measuring the magnetic properties of a sample of magnetic material) perméamètre *m* à pont magnétique
~CURVE - [phys] courbe *f* de perméabilité
~OF THE SOIL - [soil] perméabilité *f* du sol ou du terrain
~TUNING - [radio] (inductive tuning by alteration of the reluctance of the magnetic circuit of the coil) accord *m* par variation de perméabilité
PERMEABLE - [gen & phys] perméable, pénétrable, poreux
~COCOON - [text] cocon *m* perméable à l'eau
~MEMBRANE - [biol] membrane *f* poreuse
PERMEAMETER - [instr] (instrument designed to determine permeability in samples of bar form) perméamètre *m*
PERMEANCE - [el] (the reciprocal of reluctance of a magnetic circuit) perméance *f*
PERMEATE, to - [gen & phys] filtre à travers, pénétrer, saturer
PERMEATION - [phys] (separation process depending on preferential absorption of certain liquids by certain solids) pénétration *f*, saturation *f*, infiltration *f*
PERMIAN SYSTEM - [geol] système *m* permien
PERMISSIBLE - [gen] admissible, permissible, permis, autorisé
~CLEARANCE - [mech] jeu *m* tolérable
~DOSE - [radiat] (the amount of radiation which can be received by an individual without harmful results) dose *f* maximale admissible
~PEAK INVERSE VOLTAGE - [electron] tension *f* inverse maximale
~PRESSURE (ON THE WHEELS) - [auto] pression *f* autorisée sur les roues
~SIGNAL DISTORTION - [radio] distorsion *f* de signal admissible
~STRESS - [phys] effort *m* admissible
~TOLERANCE - [mech] jeu *m* admissible
~WEAR - [mech] usure *f* admissible
PERMISSION - [gen] permis, permission *f*, autorisation *f*
~CONTACT - [el] contact *m* de consentement
PERMISSIVE BLOCK - [railw] cantonnement *m* permissif
~DOWN SIGNAL - [telecomm] signal *m* d'arrêt permissif
~LIGHT - [railw] (stop and proceed light) feu *m* fran-

chissable
PERMISSIVE SIGNAL - [railw] (stop and proceed signal) signal *m* permissif
PERMIT, to - [gen] permettre
PERMIT - [gen] permis *m*, permission *f*, autorisation *f*
PERMITTIVITY - [el] (dielectric constant; a constant giving the influence of an isotropic medium on the forces of attraction or repulsion between electrified bodies) permittivité *f*, constante *f* diélectrique
PERMUTATE, to - s. permute, to
PERMUTATION - [math] (the difference arrangements which can be made of a specified number of items) permutation *f*
[chem] (change in the order of sequence) permutation *f*
~ CODE SWITCHING SYSTEM - [telecomm] (keyboard selection) numérotation *f* au clavier
PERMUTE, to - [gen etc] (to subject to permutation) permuter
PERMUTOID - [chem] (which involves a double decomposition between a soluble and an insoluble substance) permutoïdes *mpl*
PERNICIOUS - [gen] pernicieux
~ ANAEMIA - [med] (disease characterized by abnormalities in the size and shape of red blood corpuscles) anémie *f* pernicieuse
~ MALARIA - [med] paludisme *m* pernicieux
PERNITRATES - [chem] pernitrates *mpl*
PERONEAL - [anat] (pertaining to the fibula) péroné *m*
~ ARTERY - [anat] astère *f* péronière
~ NERVE - [anat] nerf *m* muscule-cutané
PEROVSKITE - [min] (titanate of calcium) pérovskite *f*
PEROXIDASES - [chem] (a group of enzymes which have the property of activating hydrogen peroxide and thus promoting reactions which would not occur otherwise) peroxydases *fpl*
PEROXIDES - [chem] (oxides in which the molecule contains two linked oxygen atoms. They react with acids form hydrogen peroxide) peroxydes *mpl*
PERPENDICULAR - [gen geom etc] perpendiculaire
[impl] (appliance used to indicate a vertical line from a point) niveau *m* à plat, fil *m* à plomb
~ AXIAL THERAPY - [med] irradiation *f* perpendiculaire axiale
~ MAGNETIZATION - [el acoust] (the magnetization of a recording medium in a direction which is perpendicular to its line of travel and plane) magnétisation *f* perpendiculaire
PERPENDICULARLY - [gen] perpendiculairement, verticalement, d'aplomb
PERPETUAL - [gen] perpétuel
~ CALENDER - [astr] calendrier *m* perpétuel
~ INVENTORY - [comput] inventaire *m* permanent
~ INVENTORY FILE - [comput] fichier *m* d'inventaire permanent
~ MOTION - [phys] mouvement *m* perpétuel
PERPHENAZINE - [chem] (a phenothiazine derivative with uses as a sedative) perphénazine *f*
PERQUISITE - [gen & leg] (incidental profit from service in addition to salary or wages) bénéfice *m*, profit *m* éventuel
PERRON - [constr] perron *m*
PERRY - [agric] (alcoholic beverage made by fermenting crushed pears) cidre *m* de poire
~ PEAR - [bot] poire *f* à cidre, poiré *m*
PERSALTS - [chem] (salts corresponding to peracids) persels *mpl*
PERSEVERATION - [med] (the repetition of an action without motive or meaning) persistance *f*, hantise *f*
PERSIAN COTTON - [text] coton *m* de Perse
~ GULF RED - [paint] (alternative name for Gulf Red, an iron oxide pigment) rouge *m* persan
~ RAW SILK - [text ind] soie *f* grège de la Perse
PERSISTENCE - [gen] persistance *f*
[phys] (the continuation of an effect) persistance *f*
[electron] (the afterglow of a luminous screen) persistance
~ CHARACTERISTICS - [electron] (of a luminous screen) caractéristique *f* d'illumination
~ OF RETINAL IMPRESSIONS - [opt] persistance *f* rétinienne
~ OF VISION - [opt] persistance *f* rétinienne
~ SCREEN - [telev] écran *m* à persistance
PERSISTENT - [gen etc] persistant
~ CHARACTERISTIC OF CAMERA TUBES - [telev] caractéristique *f* d'illumination, durée *f* du point lumineus
~ PHOSPHOR - [telev] substance *f* luminescente à persistance
~ SPECTRUM - [phys] (ultimate lines) lignes *fpl* ultimes
~ TRAIN - [astr] traînée *f* météorique persistante
PERSISTRON - [electron] (electroluminescent panel) panneau *m* électroluminescent
PERSONAL AIR SAMPLER - [nucl] appareil *m* de prélèvement d'echantillons d'air individuel
~ MONITORING - [nucl] contrôle *m* individuel
PERSONALTY - [leg] objet *m* mobilier
PERSONNEL - [gen] personnel *m*
PERSORPTION - [chem] (the very effective absorption of a gas by a solid) persorption *f*
PERSPECTIVE - [draw] (the effect of distance on the appearance of objects) perspective *f*
[adj] perspectif, en perspective
~ AEROPHOTOGRAPHY - [photo] perspectivité *f* aérienne
~ DISTORTION - [photo] distortion *f* de perspective
~ LINES - [photo] fuyants *mpl*
~ REPRESENTATION - [radar] représentation *f* en perspective
PERSPIRATION - [gen & med] transpiration *f*, perspiration *f*
~ TEST - [physiol] essai *m* de sudation
PERSPIRE, to - [gen & med] transpirer, suer
PERSTRICTION - [med] hémostase *f* par ligature
PERSUADER - [impl] bâton *m*
[electron] déflecteur *m* d'électrons
PERTHITE - [min] (megascopic intergrowths of potash- and soda- feldspars) perthite *f*
PERTHOSITE - [min] (type of soda-syenite) perthosite *f*
PERTURBATION - [gen] perturbation *f*, désordre *m*
[astr] (departure from regularity in the orbital motion of a planet) perturbation *f*
[met] perturbation *f*
[phys] (the effect of small changes on the behaviour of a system) affolement *m*
~ THEORY - [nucl] (the study of the effect of small changes on the behaviour of a system) théorie *f* des perturbations
PERTUSSIS - [med] coqueluche *f*
PERUVIAN BALSAM - [chem] (a resinous substance

derived from Myroxylon pereirae and used in medicine as a parasiticide and mild antiseptic) baume *m* du Pérou

PERUVIAN BALSAM TREE - [bot] baumier *m* du Pérou, myroxyle *m* du Pérou

~BARK - [chem] (used for the production of Peruvian balsam) écorce *f* de quinquina

~ROUGH (COTTON) - [text ind] coton *m* dur du Pérou

PERVEANCE - [electron] (of a diode; the quotient of the space-charge-limited cathode current by the three-halves power of the anode voltage in a diode) pervéance *f*

PERVERTED APPETITE - [vet] pica *m*, allotriophagie *f*

PERVIOUS SHELL - [soil] couche *f* drainante, couche *f* filtrante

PERVIOUSNESS - [gen & phys] (permeability) perméabilité *f*

PESSARY - [med] ovule *m* vaginal

PEST - [med] (a virulent epidemic) peste *f*

~CONTROL - [bot] lutte *f* contre les parasites

PESTHOUSE - [med] (hospital in which patients are kept isolated) lazaret *m*

PESTICIDE - [chem] pesticide *m*

PESTIFEROUS - [med] (carrying diseases) pestifère, pernicieux

PESTLE - [impl] (tool of hard material used to crush substances in a mortar) pilon *m*

PET COCK - [mech] (small plug-cock used to drain condensed steam from steam-engine cylinders) soupape *f* d'évent

[mech] (plug-cock used to test the water level in boilers) robinet *m* de purge

PETAL - [bot] (a leaf composing the corolla) pétale *m*

~CAP - [aero] (in a parachute, a cover formed from triangular flaps temporarily connected to the periphery of the bucket) ensemble *m* de pattes de fermeture du sac

PETALOID - [bot] (petal-shaped) pétaloïde

PETALS - [oil ind] segments *fpl*

PETECHIA - [med] (small red spot caused by a minute haemorrhage in the skin) pétéchie *f*

PETER, to - [gen] (to lose power) s'affaiblir

[min] (to become exhausted) mourir, s'épuiser

PETER OUT, to - [gen] disparaître, s'arreter

[min] s'épuiser

[auto] flancher

PETERSON CONCENTRATOR - [ind chem] (used in nitric acid manufacture; short packed tower fitted with falling film boiling-tubes of high-silicon iron) concentrateur *m* de Peterson

PETINET - [text] petinet *m*

~MACHINE - [text] métier *m* pour tricot à jour

PETRI DISH - [ind chem] (shallow saucer used for biological cultures and the like) boîte *f* de Pétri

PETRIFICATION - [geol] (term denoting organic remains changed in composition by molecular replacement) pétrification *f*

PETRIFIED WOOD - [min] (wood whose structure has been replaced by calcium carbonate etc) bois *m* pétrifié

PETRIFY, to - [geol] (to convert organic material into a substance of stony nature) pétrifier

PETROGRAPHIC CHARACTER - [soil] nature *f* pétrographique

~CONSTITUENTS - [min] constituants *mpl* pétrographiques, lithotypes *mpl*

PETROGRAPHY - [geol] (description of rocks) pétrographie *f*

PETROGRAPHY - [min] (the systematic classification of rocks) classification *f* des roches

PETROIL - [fuel] (a mixture of petrol and oil for small two-stroke engines) essence *f* mélangée d'huile de graissage

~LUBRICATION - [mech] lubrification *f* à mélange

PETROL - [fuel] (light petroleum fraction used for fuel in electro-ignition i.c. reciprocating engines) (called gasoline in the USA) essence *f* de pétrole, essence *f*

~CAN - [fuel] bidon *m* à essence

~-DRIVER - [mech] actionné à essence

~DRUM - [mech] fût *m* d'essence

~ENGINE - [mech] (internal combustion engine) moteur *m* à essence, moteur *m* à combustion interne

~FEED-PIPE - [auto] tuyau *m* d'alimentation en essence

~FILTER - [auto] filtre *m* de l'essence

~GAUGE - [auto] indicateur *m* du niveau de l'essence

~PRESSURE GAUGE - [auto] indicateur *m* de pression de l'essence

~PUMP - [auto] pompe *f* à essence

~TANK - [auto etc] réservoir *m* de l'essence

~TIN - s. petrol can

PETROLATED GAUZE - [chem] gaze *f* vaselinée

PETROLATUM - [chem] (mixture containing about one third microcrystalline petroleum, one third resins and one third oil) pétrolatum *m*, vaseline *f* jaune

PETROLEUM - [chem] (generic term for complex mixtures of hydrocarbons and paraffins of mineral origin. They are the source of many fuels and a large of organic compounds) pétrole *m*, huile *f* minérale

~ADDITIVE - [chem] additif *m* pour pétrole

~ASPHALT - [ind chem] asphalte *m* artificiel

~BEARING - [min] pétrolifère

~BLACK - [paint] noir *m* de pétrole

~COKE - [ind chem] (a by-product of the thermal cracking process; one of the purest forms of industrial carbon) coke *m* de pétrole

~DISTILLATE - [chem] produit *m* de la distillation du pétrole

~ENGINE - [mech] moteur *m* à essence, machine *f* à pétrole

~ETHER - [chem] (gasolene) éther *m* de pétrole, gazoline *f*, rhigolène *f*

~FIELD - [min] champ *m* de pétrole, champ *m* pétrolifère

~FURNACE - [th eng] four *m* chauffé au pétrole

~GAS - [min] (gas oil) gaz *m* de pétrole

~GREASES - [chem] (semisolids consisting of dispersion of 3 to 30 p.c. of solid thickening agents in petroleum oils; used for lubrication) graisses *fpl* de pétrole

~INDUSTRY - [petr ind] industrie *f* pétrolifère, industrie *f* pétrolière

~INJECTOR - [oil ind] injecteur *m* à pétrole

~JELLY - [chem] (petrolatum. A mixture of microcrystalline waxes and oil) vaseline *f* jaune

~MIGRATION - [geol] migration *f* du pétrole

~NAPHTA - [min] naphte *m* de pétrole

~PIPE-LINE - [oil ind] canalisation *f* à pétrole

~PITCH - [min] poix *f* de pétrole

~REFINERY - [ind chem] raffinerie *f* de pétrole, petrolerie

~REFINING - [ind chem] raffination *f* du pétrole

~RESIDUES - [oil ind] résidus *mpl* de pétrole

~RESINS - [chem] résines *fpl* de pétrole

PETROLEUM SPIRIT - [chem] essence *f* de pétrole
~TAILINGS - [oil ind] résidus *mpl* de la distillation du pétrole
~TAR - [fuel] (bitumen) bitumen *m* de pétrole
~TESTER - [oil ind] appareil *m* pour l'essai du pétrole
~WAXES - [chem] (hard white translucent solids with M.P. between I05 degrees F. and I80 degrees F.) cires *fpl* de pétrole
~WELL - [min] puits *m* à pétrole, puits *m* de pétrole
~-WORKINGS - [oil ind] exploitation *f* de pétrole
PETROLIFEROUS - [geol] pétrolifère
PETROLOGICAL - [geol] (pertaining to petrology) pétrologique
PETROLOGY - [geol] (the study of the mineral and chemical composition of rocks) pétrologie *f*
PETROMASTOID - [anat] os *m* péri-optique
PETROSAL BONE - [anat] rocher *m* de l'os temporal
PETROSITIS - [med] pétrosite *f*
PETROUS BONE - s. petrosal bone
~PYRAMID - s.petrosal bone
PETTICOAD - [text] jupe *f*, cotte *f*
[el] (electric insulator shaped like an inverted cup) cloche *f* d'isolateur
PEW - [gen] (in churches) banc *m* d'église
PEWTER - [metall] (an alloy of tin, containing lead, antimony and copper in carying proportions) potin *m* (un alliage d'étain et d'antimoine)
~GRASS - [bot] prêle *f* des champs
PEXIA - [med] pexie *f*, fixation *f*
PEXIS - s. pexia
pF - s. pico-farad
pH - [chem] (a method of notation used to express the degree of hydrogenion concentration of a solution) pH
pH METER - [instr] (instrument designed to measure the pH value of a solution) appareil *m* de mesure du pH
pH RANGE - [chem] intervalle *m* de pH
pH VALUE - [chem] (term denoting hydrogen-ion concentration. It represents the logarithm, to the base I0, of the reciprocal of the concentration of hydrogen ions in an aqueous solution)pH
PHACELLITE - [min] (of kaliophilite; a silicate of potassium and aluminium) caliophyllite *f*, phacellite *f*
PHACOCYST - [med] capsule *f* du cristallin
PHACOCYSTECTOMY - [med] capsulectomie *f* du cristallin
PHACOIDAL STRUCTURE - [geol] (a rock structure including lens-like minerals) structure *f* phacoïdale
PHACOLITE - [geol] (minor intrusion of igneous rock) phacolite *f*
PHACOLYSIS - [med] phacolise *f*
PHACOMALACIA - [med] remollissement *m* du cristallin
PHACOMATOSIS - [med] phacomatose *f*, neuroectodermose *f*
PHACOMETACHORESIS - [med] luxation *f* du cristallin
PHACOMETECESIS - s. phacometachoresis
PHACOPLANESIS - [med] hypermobilité *f* du cristallin
PHAGEDENIC ULCER - [med] ulcère *m* phagédénique
PHAGOCARYOSIS - [med] phagocaryosis *f*
PHAGOKARYOSIS - s. phagocaryosis
PHAGOMANIA - [med] phagomanie *f*
PHAGOPYRISMUS - [vet] fagopyrisme *m*, phagopyrisme *m*
PHAKOMATOSIS - s. phacomatosis
PHALLIN - [chem] (the hemolytic poison of the death-cup fungus) phalline *f*
PHALLUS - [zool] phallus *m*
PHANERIC - [min] (relating to igneous rocks in which the crystals of the essential minerals can be detected by the naked eye) fanérocristallin
PHANEROCRYSTALLINE - s. phaneric
PHANEROUS - [med] visible
PHANIC - s. phanerous
PHANOTRON - [electron] (hot-cathode gas discharge tube) phanotron *m*
PHANTASMATOMORIA - [med] démence *f* délirante
PHANTASTRON - [radar] (a circuit which either develope short sharp pulse at a timed interval, or pulses with a repetition rate in decrasing ration to the synchronization pulses) circuit *m* phantastron
PHANTOM - [gen] fantôme *m*
[print] voile *m*
~AERIAL - [radio] (a load connected to the output of a transmitter in place of the aerial) antenne *f* fictive
~CIRCUIT - [el] (superposed circuit from two pairs of wires) circuit *m* fantôme, circuit *m* combiné
~CIRCUIT REPEAT COIL - [el] (repeating coil used at a terminal of a phantom circuit) bobine *f* translatrice de circuit fantôme
~DRAWING - [draw] dessin *m* au trait
~GROUP - [telecomm] (group of four open-wire conductors suitable for the derivation of a phantom circuit) groupe *m* combinable
~LIMB PAIN - [med] douleur *f* des membres fantômes
~LOAD - [instr] (in testing meters, to avoid waste of power) charge *f* fictive
~LOADING - [telecomm] charge *f* du circuit fantôme
~MATERIAL - [radiat] matière *f* phantôme
~TRANSPOSITION - [telecomm] croisement *m* des lacets
PHANTOMING - [el & telecomm] combinaison *f* des circuits
PHARCIDOUS - [med] ridé, rugueux
PHARMACEUTICAL - [chem] pharmaceutique
~BALANCE - [impl] balance *f* de pharmacien
~CHEMISTRY - [chem] chimie *f* pharmaceutique
PHARMACEUTICS - [chem] pharmacie *f*, pharmaceutique *f*
PHARMACODYNAMICS - [chem] pharmacodynamie *f*
PHARMACOLITE - [min] (hydrous arsenate of calcium) pharmacolithe *f*
PHARMACOLOGY - [chem] (the study of drugs) pharmacologie *f*
PHARMACOPOEIA - [chem] (collection of standard formulas and methods for the preparation of medicines) pharmacopée *f*, codex *m* pharmaceutique
PHARMACOTHERAPY - [med] pharmacothérapie *f*
PHARMACY - [chem] pharmacie *f*
[comm] (chemist's shop, a drug store) pharmacie *f* de détail, boutique *f* de pharmacien
PHAROS - [light] (luminous flux) flux *m* lumineux
PHAROSAGE - [light] (luminous-flux density) éclairement *m*
PHARYNGECTASIA - [med] hernie *f* pharingienne
PHARYNGEMPHRAXIS - [med] obstruction *f* du pharynx
PHARYNGISM - [med] pharyngisme *m*
PHARYNGISMUS - s. pharyngism
PHARYNGITIS - [med] pharyngite *f*

PHARYNGOAMYGDALITIS - [med] pharyngotonsillite *f*
PHARYNGOLARYNGITIS - [med] pharyngolaryngite *f*
PHARYNGOLYSIS - [med] paralysie *f* des muscles du pharynx
PHARYNGORRHAGIA - [med] hémorragie *f* du pharynx
PHARYNGOSCOPE - [instr] (instrument for viewing the pharynx) pharyngoscope *m*
PHARYNGOSTENOSIS - [med] sténose *f* du pharynx
PHARYNGOTONSILLITIS - s. pharyngoamygdalitis
PHARYNX - [anat] pharynx *m*
PHASE, to - [phys etc] mettre en phase, caler en phase
PHASE - [phys chem el etc] (the whole of that part of a material system which is of the same chemical composition and in the same physical state) phase *f*
~ADJUSTMENT - [el & telev] mise *f* en phase
~ADVANCER - [el] décaleur *m* de phase, déphaseur *m*, régulateur *m* de phase
~AMPLITUDE DISTORTION - [radio] distorsion *f* phase/amplitude
~ANGLE - [el] (the angle between the vectors representing two variables in an alternating current) angle
~ANGLE ERROR - [instr] (the percentage error introduced by the transformer into a power measurement and due to the phase displacement of the transformer) erreur *f* de déphasage
~BALACE RELAY - [el] relais *m* à interruption de phase
~BANDWITH - [radio] (of an amplifier) plage *m* de linéarité de phase
~BELT - s. phase spread
~BOUNDARY - [phys] (the boundary layer between two substances in different phases) couche *f* limite de phases
~BREAK - [el] (gap section) section *f* de séparation
~-CHANGE COEFFICIENT - s. phase constant [radio]
~CHANGER - [el] convertisseur *m* de phase, déphaseur *m*
~COINCIDENCE - [el] concordance *f* de phases
~COMPARISON PROTECTION - [el] protection *f* par comparaison de phase
~COMPENSATING NETWORK - [el] circuit *m* de compensation de phase
~COMPENSATOR - [telecomm] filtre *m* compensateur de phases
~CONDUCTOR - [el] fil *m* de phase
~CONSTANT - [el] (of a travelling plane wave at a specified frequency) constante *f* de phase
[radio] (the imaginary part of the propagation coefficient) constante *f* de phase
~CONTOUR - [el] courbe *f* de déphasage
~-CONTRAST MICROSCOPE - [instr] microscope *m* à contraste de phase
~CONTROL - [contr] réglage *m* de phase
~-CONTROL COEFFICIENT - [contr] coefficient *m* de réglage
~CONVERTOR - [el] (a machine for converting alternating current which a specified number of phases into alternating current having a different number of phases, through the same frequency) convertisseur *m* de phase
~-CORRECTED HORN - [electron] (horn making the emergent electro-magnetic waves plane at the mouth) cornet *m* à correction de phase
~CORRECTION - [telecomm] (in telegraphy) compensation *f* des phases, correction *f* de phase
PHASE CROSSOVER - [contr] (in automatic control) coupure *f* de phase
~DELAY - [el] (in a circuit element, the time equal to the phase shift between its two ends divided by the frequency) retard *m* de phase, temps *m* de propagation de phase
~DEVIATION - [radio] déviation *f* de phase
~DIAGRAM - [metall] (of an alloy) diagramme *m* de phases
~DIFFERENCE - s. phase displacement
~DISPLACEMENT - [el] (between two sinusoidal quantities; the difference between the phases of these quantities at a given instant of time) déphasage *m*, décalage *m* de phase
~DISPLACEMENT ERROR - s. phase angle error
~DISTORTION - [radio] (distortion occurring when the phase shift does not vary in relation to the frequency) distorsion *f* de phase, distorsion *f* phase/fréquence
~DISTORTION INDEX - [radio] taux *m* de distorsion de phase
~EQUALIZER - [el] (network compensating for phase/frequency distortion) compensateur *m* de phase
[telev] compensateur *m* du temps de propagation de phase
~ERROR - [el] erreur *f* de phase
~EXCURSION - [electron] excursion *f* de phase
~FAILURE - [el] manque *m* de phase
~FOCUSING - [electron] focalisation *f* de phase à modulation de vitesse
~FREQUENCY - [el] fréquence *f* de phase
~/FREQUENCY DISTORTION - [radio] (phase distortion) distorsion *f* phase/fréquence
~FRONT - [phys] front *m* d'onde, front *m* de phase
~INDICATOR - [instr] (mechanism used as a rotor position marker in balancing operations) indicateur *m* de phase
~INTEGRAL - [mech] (action variable) intégrale *f* de phase
~INVERTER CIRCUIT - [electron] circuit *m* inverseur de phase
~-INVERTER VALVE - [el] tube *m* de déphaseur
~JITTER - [el acoust] bruit *m* de phase
~LAG - s. phase delay
~LAMP - [el] lampe *f* de phase
~LENGTH - [el] longueur *f* de phase
~LOCALIZER - [radar] (range measurements by a phase-shifting device) localisateur *m* de phase
~LOCK - [el] asservissement *m* de phase
~-LOCK LOOP - [el] boucle *f* d'asservissement de phase
~-LOCKED OSCILLATOR - [electron]oscillateur *m* à phase rigide
[comput] (parametron acting as storage cell) oscillateur *m* à blocage de phase
~LOCKING - [el] accrochage *m* de phase
~LOCUS - [math] lieu *m* des phases
~MARGIN - [contr] (in an open control circuit) marge *m* de phase
~METER - [instr] (power factor meter) phasemètre *m*, indicateur *m* de coefficient de déphasage
~-MODULATED TRANSMITTER - [radio] émetteur *m* à modulation de phase
~MODULATION - [radio] (angle modulation in which the angle of a sine-wave carrier departs from the carrier angle by an amount proportional to the ins-

tantaneous value of the modulating wave) modulation *f* de phase
PHASE MODULATION RECORDING - [comput] enregistrement *m* par modulation de phase
~MONITOR - [instr] (instrument designed for the adjustment of directional aerials) moniteur *m* de phase
~-NUMBER TRANSFORMER - [el] transformateur *m* à nombres de phase différents
~OF A PERIODIC QUANTITY - [contr] phase *f* d'une grandeur périodique
~OF A SINUSOIDAL QUANTITY - [el] (the variable angle in the sinusoidal representation of the quantity) phase *f* d'une grandeur sinusoïdale
~OF SINE WAVE - [el] (angle of sine wave) angle *m* de phase
~OF SINK - [electron] phase *f* de région d'instabilité électronique
~OF THE MOON - [astr] phase *f* de la Lune
~PLANE - [math] (a two-dimensional phase space) plan *m* de phase
~-PROPAGATION RATIO - [radio] (the propagation ratio divided by its magnitude) constante *f* de propagation de phase
~QUADRATURE - [cin] (a phase shift of 90 degrees) quadrature *f*
~RECOVERY TIME - [electron] temps *m* de rétablissement de phase
~RELATIONSHIP - [radio] (phase rule) relation *f* de phase
[chem] règle *f* des phases
~RESERVE - s. phase margin
~RESONANCE - [mech] (the velocity resonance) résonance *f* de phase
~REVERSAL - [phys] (change of phase equal to one-half cycle) inversion *f* de phase
~REVERSAL OF EMULSION - [chem] (the separation of an emulsion into its component parts) émulsion *f* réversible
~-REVERSAL PROTECTION - [el] protection *f* contre les inversions de phase
~-REVERSING ANODE - [electron] (in interdigital magnetrons) anode *f* à inversion de phase
~-ROTATION RELAY - [el] relais *m* à séquence de phases
~RULE - [chem] (a rule in the study of equilibria between phases) règle *f* des phases
~-SENSITIVE DETECTOR - [instr] (detector responding to input signals having the same frequency of the control signal and a given phase relative to that signal) détecteur *m* à sensibilité de phase
~-SEQUENCE INDICATOR - [instr] (instrument designed to determine the phase frequency at any point in a polyphase circuit) indicateur *m* de séquence de phase
~-SHAPED ANTENNA - [radio] antenne *f* à phase variable
~-SHIFT - [el & telecomm] déphasage *m*
~-SHIFT CONTROL - [contr] réglage *m* de déphasage
~-SHIFT MICROPHONE - [el acoust] (microphone employing phase-shift networks to produce directional properties) microphone *m* à variation de phase
~-SHIFT OSCILLATOR - [radio] oscillateur *m* à déphasage
~SHIFT RESPONSE - [el] réponse *f* de déphasage
~SHIFTER - [el] (transformer in which the phase angles of the secondary voltages can be made to vary in relation to the primary voltages) déphaseur *m*, décaleur *m* de phase
PHASE SHIFTING CIRCUIT - [el] circuit *m* déphaseur
~-SHIFTING TRANSFORMER - [contr] (a synchro which is energized by a polyphase system) synchrodéphaseur
~-SHIFTING UNIT - [comput] unité *f* de déphasage
~SLOWNESS - s. phase delay
~SPACE - [electron] (space of 6N dimensions, corresponding to the 3N co-ordinates of position and to the 3N kinetic moments of the N particles considered) espace *m* de phase, extension *f* en phase
~-SPACE CELL - [electron] (an elementary hypervolume in the phase space) cellule *f* de l'espace de phase
~SPLITTER - [el] (a device which from a single input wave produces two or more output waves differing in phase from one another) séparateur *m* de phase
~SPREAD - [el] largeur *f* de phase
~STABILITY - [el] stabilité *f* de phase
~SWINGING - [el] (periodic variations in the speed of a synchronous machine) oscillations *fpl* pendulaires
~-TUNED TUBE - [electron] tube *m* accordé à réglage de phase
~VELOCITY - [el] (the velocity of an observer moving along the normal to the plane of the wave at such speed that the wave characteristics appear to him to remain constant in phase) vitesse *f* de phase (d'une onde)
~VOLTAGE OF A MACHINE - [el] (the potential difference between the extremities of a phase winding of the machine) tension *f* de phase d'un moteur
~VOLTAGE OF A WINDING - [el] (the potential difference across one phase of an electrical machine) tension *f* par phase d'enroulement
~-WHITE - [radio] (a phasing signal corresponding to picture white) mise *f* en phase sur blanc
PHASELESS BOOST - [telev] récupération *f* d'énergie sans distorsion
PHASEMETER - s. phase meter
PHASER - [telev] cadreur *m*
PHASING - [el etc] mise *f* en phase, synchronisation *f*
[electron] phase *f*
~SIGNAL - [radio] signal *m* de mise en phase
PHASITRON - [electron] phasitron *m*
PHASMAJECTOR - [telev] (device designed to provide a standard video signal for testing) monoscope *m*, tube *m* de mire
PHASOR - [el] vecteur *m* de phase
PHATNORRHAGIA - [med] hémorragie *f* alvéolaire
PHEASANT - [zool] faisan *m*
PHEASANT'S EYE - [bot] adonis *m*, goutte-de-sang *f*
PHELLANDRENE - [chem] (a terpene) phellandrène *m*
PHENACEMIDE - [chem] (an anticonvulsant used in the treatment of epilepsy) phénacemide *f*
PHENACETIN - [chem] (an antipyretic) phénacétine *f*
PHENAKITE - [min] (natural beryllium orthosilicate, crystallizing in the hexagonal system, it is easily confused with quartz and is sometimes cut as a gemstone) phénacite *f*
PHENANTHRENE - [chem] (an intermediate for pharmaceuticals) phénanthrène *m*
PHENANTHRENEQUINONE - [chem] (a synthesis intermediate) phénanthrène-quinone *f*
PHENARSAZINE CHLORIDE - [chem] (a lachrymatory gas) chlorure *m* de phénarsazine

PHENATES - [chem] (the salts formed by phenols) phénates *mpl*

PHENAZINE - [chem] (a synthesis intermediate) phénazine *f*

PHENETIDINE - [chem] (an intermediate for pharmaceuticals and dyestuffs) phénétidine *f*

PHENINDIAMINE TARTRATE - [chem] (an antihistamine drug used in the treatment of allergic conditions) tartrate *m* de phénindamine

PHENOBARBITAL - [chem] (a hypnotic used in medicine) phénobarbital *m*

PHENOL - [chem] (carbolic acid. Generic name for hydroxyl derivatives of aromatic hydrocarbons in which the OH group is attached directly to the benzene ring) phénol *m*

~-FORMALDEHYDE RESINS - [chem] (an important class of synthetic resins produced from phenol acid formaldehyde. They have wide application as casting and moulding compounds, in laminates, as adhesives, and in the manufacture of surface coatings) résines *fpl* de phénol-formaldéhyde

~RED - [chem] (synonym for phenolsulphonphthalein, an acid-base indicator) rouge *m* de phénol

~RESINS - [chem] (an important class of thermosetting synthetic resins) résines *fpl* phénoliques

PHENOLDISULPHONIC ACID - [chem] (a synthesis intermediate for pharmaceuticals) acide *m* phénoldisulfonique

PHENOLIC - [chem] phénolique

~CEMENT - [ind chem] (cement based on a phenolic/formaldehyde synthetic resin) ciment *m* phénolique

~PLASTICS - [chem] (an important body of plastics based on phenol/formaldehyde, resorcinol/formaldehyde, or phenol/furfural synthetic resins) phénolplastes *mpl*

~RESINS - [chem] (thermosetting resins produced by condensation of phenols with aldehydes) résines *fpl* phénoliques

~VARNISH - [chem] (a solution of a heat-hardening phenolic resin; also a solution of a phenolic resin for use with an acid catalyst and hardening at room temperatures) vernis *m* phénolique

PHENOLOGY - [biol] (the study of the organism in relation to climate) phénologie *f*

PHENOLPHTHALEIN - [chem] (an analytical indicator, pharmaceutical and intermediate for dyestuffs) phénolphtaléine *f*

PHENOLSULPHONIC ACID - [chem] (an intermediate for pharmaceutical and dyestuffs) acide *m* phénolsulfonique

PHENOLSULPHONPHTHALEIN - [chem] (an analytical acid-base indicator) phénolsulfophtaléine *f*

PHENOMENON - [phys] (something which is perceived, any fact considered as the subject of observation) phénomène *m*

PHENOPLAST - [chem] (a synthetic phenolic resin) phénoplaste *m*

PHENOTHIAZINE - [chem] (an insecticide and an intermediate for dyes) phénothiazine *f*

PHENOTYPE - [genet] phénotype *m*

PHENOXYACETIC ACID - [chem] (a synthesis intermediate for pharmaceuticals, insecticides, and dyestuffs) acide *m* phénoxyacétique

PHENOXYBENZAMINE - [chem] (dibenzyline. An adrenolytic and sympatholytic agent used in the treatment of peripheral vascular disease) phénoxybenzamine *f*

PHENOXYPROPANEDIOL - [chem] (an intermediate for pharmaceuticals and a plasticizer for natural and synthetic resins. Phenoxypropanediol also has application in the manufacture of surface coatings) phénoxypropandiol *m*

PHENTOLAMINE HYDROCHLORIDE - [chem] (a peripheral vasodilator used in the treatment of hypertension) chlorhydrate *m* de phentolamine

PHENYLACETALDEHYDE - [chem] (a perfumery component) phénylacétaldéhyde *m*

PHENYLACETAMIDE - [chem] (an intermediate for pharmaceuticals) phénylacétamide *f*

PHENYL ACETATE - [chem] (a synthesis intermediate and solvent) acétate *m* de phényle

PHENYLACETIC ACID - [chem] (an intermediate for perfumery components) acide *m* phénylacétique

PHENYLAMINONAPHTHOL SULPHONIC ACID - [chem] (a dyestuffs intermediate) acide *m* phénylamino-naphtolsulfonique

PHENYLARSONIC ACID - [chem] (an analytical reagent) acide *m* phénylarsonique

PHENYLBUTAZONE - [chem] (an analgesic and antipyretic) phénylbutazone *m*

PHENYLBUTYNOL - [chem] (a synthesis intermediate) phénylbutynol *m*

PHENYLCARBETHOXYPYRAZOLONE - [chem (an intermediate for dyes) phénylcarbéthoxypyrazolone *f*

PHENYLCARBYLAMINE CHLORIDE - [chem] (a synthesis intermediate) chlorure *m* de phénylcarbylamine

PHENYL CHLORIDE - [chem] (a synonym for chlorobenzene) chlorure *m* de phényle

PHENYLCYCLOHEXANE - [chem] (an intermediate and solvent) phénylcyclohexane *m*

PHENYLCYCLOHEXANOL - [chem] (an intermediate) phényldiéthanolamine *f*

~DIMETHYLUREA - [chem] (a herbicide) phényldiméthylurée *f*

PHENYLENEDIAMINE - [chem] (a vulcanization accelerator, photographic chemical, and intermediate for dyes) phénylènediamine *f*

PHENYLETHANOLAMINE - [chem] (a synthesis intermediate for dyestuffs) phényléthanolamine *f*

PHENYLETHYL ACETATE - [chem] (a perfumery component) acétate *m* de phényléthyle

PHENYLETHYLACETIC ACID - [chem] (a synthesis intermediate) acide *m* phényléthylacétique

PHENYLETHYL ALCOHOL - [chem] (a synthesis intermediate, flavourant and perfumery agent) alcool *m* phényléthylique

~ANTHRANILATE - [chem] (a perfumery and flavouring agent) anthranilate *m* de phényléthyle

PHENYLETHYLETHANOLAMINE - [chem] (a synthesis intermediate and solvent) phényléthyléthanolamine *f*

PHENYLETHYL ISOBUTYRATE - [chem] (a constituent of perfumes) isobutyrate *m* de phényléthyle

~PROPRIONATE - [chem] (a flavourant and perfumery agent) propionate *m* de phényléthyle

PHENYLHYDRAZINE - [chem] (a synthesis intermediate and analytical reagent) phénylhydrazine *f*

PHENYL ISOCYANATE - [chem] (an analytical reagent) isocyanate *m* de phényle

PHENYLMERCURIC ACETATE - [chem] (a herbicide) acétate *m* phénylmercurique

~BORATE - [chem] (a fungicide) borate *m* phénylmercurique

~CHLORIDE - [chem] (a germicide and fungicide) chlorure *m* phénylmercurique

PHENYLMERCURIC HYDROXIDE - [chem] (a fungicide) hydrate *m* phénylmercurique
~NAPHTHENATE - [chem] (a wood preservative) naphténate *m* phénylmercurique
~PROPRIONATE - [chem] (a fungicide for protective coatings) propionate *m* phénylmercurique
PHENYLMERCURIETHANOLAMMONIUM ACETATE - [chem] (an agricultural fungicide) acétate *m* de phénylmercuriéthanolammonium
PHENYLMETHYLETHANOLAMINE - [chem] (a synthesis) phénylméthyléthanolamine *f*
PHENYLMORPHOLINE - [chem] (an intermediate for photographic chemicals, vulcanization accelerators and dyestuffs) phénylmorpholine
PHENYLNAPHTHYLAMINE - [chem] (an intermediate for dyes) phénylnaphtylamine *f*
PHENYLPHENOL - [chem] (an intermediate for rubber chemicals and dyestuffs) phénylphénol *m*
PHENYLPIPERAZINE - [chem] (an intermediate for pharmaceuticals) phénylpipérazine *f*
PHENYLPROPANOLAMINE HYDROCHLORIDE - [chem] (norephedrine. A sympathomimetic amine used in the treatment of allergic conditions) chlorhydrate *m* de phénylpropanolamine
PHENYLPROPYL ACETATE - [chem] (a perfumery component) acétate *m* de phénylpropyle
~ALDEHYDE - [chem] (a component of flavours and perfumes) aldéhyde *m* phénylpropylique
PHENYLPROPYLMETHYLAMINE - [chem] (phenpromethamine. A sympathomimetic amine used in medicine) phénylpropylméthylamine *f*
PHENYLTRICHLOROSILANE - [chem] (an intermediate for silicones) phényltrichlorosilane *m*
PHIAL - [gen & med] fiole *f*, ampoule *f*
PHILLIP'S SCREW - [mech] (type of screw having a cruciform slot) vis *f* à fente cruciforme
PHLEBECTASIA - [med] phlébectasie *f*, varice *f*
PHLEBECTASIS - s. phlebectasia
PHLEBECTOMY - [med] flébectomie *f*
PHLEBITIS - [med] phlébite *f*
PHLEBOGRAPHY - [radiat] (radiological examination of veins made by means of an injection of a contrast medium) phlébographie *f*, veinographie *f*
PHLEBOPEXY - [med] phlébopexie *f*
PHLEBOSCLEROZATION - s. phlebozation
PHLEBOTHROMBOSIS - [med] thrombophlébite *f*
PHLEBOZATION - [med] phlébosclérose *f*
PHLEGMON - [med] abcès *m* chaud, phlegmon *m*
PHLEGMONOUS - s. phlegmon
PHLOGISTIC - [med] inflammatoire
PHLOGISTON - [phys] (the principle once assumed to be an essential constituent of all combustible substances and be given up in burnings) phlogistique *m*
PHLOGOPITE - [min] (a silicate of potassium, magnesium, aluminium and hydrogen) phlogopite *f*
PHLOROGLUCINOL - [chem] (an intermediate for synthetic resins, drugs and dyes, and an analytical reagent) phloroglucine *f*
PHOBIA - [gen] phobie *f*
PHOBOPHOBIA - [med] phobophobie *f*
PHOBOTAXIS - [biol] (reaction of an organism to harmful) phobotaxie *f*
PHON - [acoust] (unit of objective loudness) phon *m*
~METER - s. phonometer
~SCALE - [acoust] (scale of the intensity of the 1000 cycles per second reference tone) gamme *f* des phons
PHONAUTOGRAPH - [acoust] (a type of earlier recording machine) phonoautohraphe *m*
PHONASTHENIA - [med] phonasthénie *f*
PHONENDOSCOPE - [med] phonendoscope *m*
PHONETIC - [acoust] (pertaining to phonetics) phonétique
PHONETICS - [acoust] (the study of the production of the sounds of speech and their representation) phonétique *f*
PHONIC - [acoust] (pertaining to sound) phonique
~DRUM - s. phonic wheel
~MOTOR - [el] (synchronous motor operating by means of teeth on rotor and stator) moteur *m* phonique
~WHEEL - [telecomm] (elementary synchronous device used to keep constant the speed of rotation of a motor) roue *f* phonique
PHONOCARDIOGRAM - [med] phonocardiogramme *m*
PHONOCARDIOGRAPHY - [med] phonocardiographie *f*
PHONOCHEMISTRY - [chem] (the science of the effect of sonic and ultrasonic waves on chemical reactions) phonochimie *f*
PHONOGRAM - [telecomm] (a telegram dictated over the telephone) sténogramme *m*, télégramme *m* téléphoné
~CIRCUIT - [telecomm] circuit *m* de télégramme téléphoné
PHONOGRAPH - [acoust] (any apparatus for the recording and the reproducing of sounds) phonographe *m*
~AMPLIFIER - [el acoust] (a low-frequency amplifier used for disk-reproducing circuits) amplificateur *m* phonographique
~PICKUP - [el acoust] (a mechanical reproducer) phonocapteur *m*, pick-up *m*
PHONOLITE - [geol] (fine-grained igneous rock) phonolite *f*
PHONOMANIA - [med] phonomanie *f*, manie *f* homicide
PHONOMETER - [instr] (instrument designed to measure the pressure and intensity of sound) phonomètre *m*
PHONOMETRY - [acoust] (the measurement of the pressure and intensity of sound) phonométrie *f*
PHONOPNEUMOMASSAGE - [med] massage *m* pneumatique de l'oreille
PHONOTELEMETER - [telecomm] (a military device) phonotélémètre *m*
PHONOVISION - [telev] (trade mark; a subscription television system registered by the Zenith Radio Corporation) phonovision *f*
PHORIA - [med] déviation *f* des yeux
PHOROMETER - [instr] (instrument designed to measure the strength and direction of the extrinsic muscles of the eye) phoromètre *m*
PHORONE - [chem] (a synthesis intermediate and a solvent) phorone *f*
PHOROPLAST - [med] tissu *m* conjonctif
PHOSGENE - [chem] (carbonyl chloride. A synthesis intermediate, military poison gas, and chlorinating agent) phosgène *m*
PHOSPHATE - s. phosphates
~ADDITIVES - [chem] additifs *mpl* phosphatiques
~COAT - [mech] (rust resisting agent) phosphatation
~CHALK - [geol] craie *f* phosphatée
~FERTILIZERS - [chem] engrais *mpl* phosphatés

PHOSPHATE GUANO - [agric] guano *m* phosphaté, phospho-guano *m*
~OF CALCIUM - [chem] phosphate *m* de chaux
~OF IRON - [chem] phosphate *m* de fer
~OF SODIUM - [chem] phosphate *m* de soude
~ROCK - [min] (natural calcium phosphate used directly as a fertilizer and in the production of "superphosphates" for agricultural purposes) roche *f* phosphatée
PHOSPHATES - [chem] (the salts of phosphoric acid) phosphates *mpl*
PHOSPHATIC - [chem] phosphatique
~CHALK - [min] craie *f* phosphatée
~DEPOSITS - [geol] (calcium phosphate containing beds) dépôts *mpl* phosphatiques
~IRON ORE - [min] minerai *m* de fer phosphoreux
~NODULES - [geol] (rounded masses containing calcium phosphate) nodules *mpl* phosphatés
~SLAG - [metall] scories *fpl* de déphosphoration
PHOSPHATIDE - [chem] (synonym for phospholipid) phosphatide *m*
PHOSPHATING - [ind chem] (treatment against corrosion) phosphatation *f*, phosphatisation *f*
PHOSPHATIZE, to - [chem] phosphoriser
PHOSPHATIZING - s. phosphating
PHOSPHIDE - [chem] (a compound of phosphorous united directly to a metal) phosphure *m*
~BANDING - [metall] bande *f* de phosphure
~SWEAT - [metall] diamant *m*
PHOSPHINE - [chem] (hydrogen phosphide. A poisonous, evil-smelling gas) phosphine *f*
PHOSPHITES - [chem] (the salts of phosphorous acid) phosphites *mpl*
PHOSPHOLIPID - [chem] (generic name for substituted fats containing phosphoric acid and a nitrogenous base) phospholipide *m*
PHOSPHOMOLYBDIC ACID - [chem] (an analytical reagent) acide *m* phosphomolybdique
PHOSPHONIUM BASES - [chem] (compounds produced by the combination of an alkyl halide with a tertiary phosphine) bases *fpl* de phosphonium
~IODIDE - [chem] (a synthesis intermediate) iodure *m* de phosphonium
PHOSPHOPENIA - [med] phosphoropénie *f*
PHOSPHOR - [chem] s. phosphorus
[chem] (any phosphorescent substance) substance phosphorescente
[opt] (luminescence exhibiting substance) phosphore *m*
~-BRONZES - [metall] (term used for bronzes containing ten to fourteen per cent of tin and from 0.I to 0.3 p.c. of phosphrous with minor additions, e.g. lead and nickel. They are specially suitable for parts subject to simultaneous wear and corrosion) bronzes *mpl* phosphorés
~CLUSTER - [electron] triade *f* à substance luminescente
~COLOUR RESPONSE - [electron] réponse *f* chromatique d'une substance luminescente
~DOT - [electron] (spot of phosphor material on the screen of a cathode-ray tube) point *m* à substance luminescente
~GRAIN - [electron] grain *m* de substance luminescente
~SATURATION - [electron] émission *f* maximale de une substance luminescente
~SCREEN - [electron] écran *m* luminescent
PHOSPHOR SCREEN BRIGHTNESS - [electron] luminosité *f* d'un écran luminescent
~STRIP - [electron] (component part of colour display tubes) ruban *m* luminescent
PHOSPHORESCENCE - [phys & opt] (luminescence delayed after excitation) phosphorescence *f*
PHOSPHORESCENT PIGMENTS - [chem] (pigments which emit light without activation) pigments *mpl* phosphorescents
PHOSPHORIC - [chem] (pertaining to phosphorous) phosphorique
~ACID - [chem] acide *m* phosphorique
PHOSPHORISM - [med] phosphorisme *m*
PHOSPHORITE - [min] (rock-phosphate; a variety of apatite) phosphorite *f*
PHOSPHORIZED COPPER - [metall] (copper which has been deoxidized with phosphorus) cuivre *m* phosphoreux
PHOSPHOROSCOPE - [instr] (instrument designed to assess the duration of the phosphorescent glow) phosphoroscope *m*
PHOSPHOROUS - [chem] (derived from phosphorous, particularly in its lower valence) phosphoreux
~ACID - [chem] (a reducing agent, formed by the action of cold water on phosphorous oxide) acide *m* phosphoreux
~OXIDE - [chem] (a poisonous substance) oxyde *m* phosphoreux
PHOSPHORS - [chem] corps *mpl* phosphorescents
PHOSPHORUS - [chem] (soft, non-metallic element, symbol P; it combines easily with oxygen) phosphore *m*
~CHLORIDE - [chem] chlorure *m* de phosphore
~CHLORONITRIDES - [chem] (result of the interaction of phosphorus pentachloride and ammonium chloride) chlornitrures *mpl* de phosphore
~OXYCHLORIDE - [chem] oxychlorure *m* de phosphore
~PENTACHLORIDE - [chem] pentachlorure *m* de phosphore
~PENTAHALIDES - [chem] (formed by the action of the dry halogen on the trihalide) pentahalogénures *mpl* de phosphore
~PENTOXIDE - [chem] (a drying agent) pentoxyde *m* de phosphore
~TRIHALIDES - [chem] (phosphorus trichloride, tribromide, triiodile and trifluoride) trihalogénures *mpl* de phosphore
PHOSPHORYL - [chem] phosphoryle *m*
~BROMIDE - [chem] bromure *m* de phosphoryle
~CHLORIDE - [chem] chlorure *m* de phosphoryle
~FLUORIDE - [chem] fluorure *m* de phosphoryle
PHOSPHORYLATION - [chem] (esterification of alcoholic hydrogen atom) phosphorylation *f*
PHOSPHURANYLITE - [min] (uranylorthophosphate) phosphuranylite *f*
PHOT - [light] (unit of light flux; the illumination on a square centimetre surface from one standard candle) phot *m*
PHOTESTHESIS - [med] sensibilité *f* à la lumière
PHOTISTOR - [electron] (or phototransistor; light-sensitive transistor) transistor *m* à effet photoélectrique
PHOTO - [photo] (abbrev. for photograph, of photographic) s. photograph
~FINISH - [sport] (the finish of a race in which the two or three leads are so close, that only the accuracy of a photograph can determine the winner) dé-

cision *f* par photo
PHOTOACTIVE - [photo] (photosensitive) sensible à la lumière
PHOTOBARRIER CELL - [electron] (photoelectric cell in which light causes the passage of electrons across a rectifying contact-gap) cellule *f* photoélectrique à couche de barrage
PHOTOCATALYSIS - [chem] (the acceleration or retardation of the rate of a chemical reaction by light) photocatalyse *f*
PHOTOCATHODE - [electron] (or photo-electric cathode; a cathode which functions primarily through the action of incident light) cathode *f* photoélectrique, photocathode *f*
PHOTOCELL - s. photoelectric cell
~PICK-OFF - [electron] capteur *m* photoélectrique
PHOTOCHEMICAL - [chem] (pertaining to photochemistry) photochimique
~CELL - [electron] (a photoelectric cell with two electrodes of the same metal immersed in an electrolyte) cellule *f* photochimique
~EQUILIBRIUM - [chem] (a position of equilibrium reached in a reversible chemical change in which the reactions are sensitive to light) équilibre *m* photochimique
~EQUIVALENCE - [chem] (the Einstein law of photochemical equivalence) équivalence *f* photochimique
~INDUCTION - [chem] (the delay between the absorption of light by a system and the ensuing chemical reaction) induction *f* photochimique
~REACTION - [chem] (a reaction whose rate is determined by the action of light) réaction *f* photochimique
PHOTOCHEMISTRY - [chem] (the study of the chemical effects of radiation) photochimie *f*
PHOTOCHROMY - [photo] (colour photography) photochromie *f*
PHOTOCHRONOGRAPH - [instr] (instrument designed to record time photographically) photochronographe *m*
PHOTOCHRONOGRAPHY - [astr] (the recording of time by photography) photochronographie *f*
PHOTOCOMPOSE, to - [print] (the operation of composing letters for printing by the projection of light images of the letters on to a photosensitive material) photocomposer
PHOTOCOMPOSING MACHINE - [print] photocomposeuse *f*, machine *f* à composer photographique
PHOTOCOMPOSITION - [print] composition *f* photographique
PHOTOCOMPOSITOR - s. photocomposing machine
PHOTOCONDUCTION - [phys] conductibilité *f* photoélectrique
PHOTOCONDUCTIVE CAMERA TUBE - [electron] tube *m* de prise à photoconduction
~CELL - [electron] (photocell in which the electrical resistance changes under the action of light) cellule *f* photoconductive
~DIODE - [electron] photodiode *f*
~EFFECT - [electron] (photoelectric effect revealed as a change in the electric conductivity of a solid or a liquid) effet *m* photoélectrique interne
PHOTOCONDUCTIVITY - [electron] (the property possessed by certain materials of varying their electrical conductivity under the action of light) photoconductibilité *f*
PHOTOCONDUCTOR - [electron] (semiconductor in which conductivity varies with illumination) photoconducteur *m*
PHOTOCURRENT - s. photoelectric current [electron] courant *m* photoélectrique
PHOTODERMATISM - [med] sensibilité *f* actinique
PHOTODERMATOSIS - [med] dermatose *f* actinique, lucite *f*
PHOTODISINTEGRATION - [nucl] (of the atomic nucleus ; breaking down caused by the action of radiation) photodésintégration *f*, réaction *f* photonucléaire
PHOTODISSOCIATION - [chem] (dissociation produced by the absorption of radiant energy) photodissociation *f*
PHOTOELASTIC STRESS ANALYSIS - s. photoelasticity
PHOTOELASTICITY - [mech] (the determination of stress distribution by passing polarized light through a nitrocellulose model) photoélasticité *f*
PHOTOELECTRIC - [el] (relative to those electric effects which are caused by the action of light) photoélectrique
~ABSORPTION - [phys] (the absorption of photons in the photoelectric effect) absorption *f* photoélectrique
~ALARM - [electron] (electronic alarm circuit with a photoelectric relay) circuit *m* photoélectrique d'alarme
~ATTENUATION COEFFICIENT - [electron] coefficient *m* d'atténuation photoélectrique
~CATHODE - [electron] (a cathode which functions mainly by the process of photoelectric emission) cathode *f* photoélectrique, photocathode *f*
~CELL - [el] (generally, a device the electrical state of which is altered by the incidence of light) cellule *f* photoélectrique
~CELL AMPLIFIER - [electron] (thermionic amplifier placed close to the photo-electric cell receiving the light-beam modulated by the sound-track) amplificateur *m* à cellule photoélectrique
~CHARACTER READER - [electron] (a light sensitive device) lecteur *m* de caractères
~COLOUR COMPARATOR - [electron] comparateur *m* photoélectrique de couleurs
~CONTROL - [contr] (control functioning through a change in incident light) commande *f* photoélectrique
~COUNTER - [electron] compteur *m* photoélectrique
~CURRENT - [electron] (the current which is due to a photoelectric effect) courant *m* photoélectrique
~DENSITOMETER - [instr] (instrument designed to measure the density of a material) densitomètre *m* photoélectrique
~DEVICE - [contr] dispositif *m* photoélectrique
~DOOR-OPENER - [contr] (device which automatically opens doors on the interruption of a light-beam, e. g. on the approach of a person or vehicle) ouvre-portes *m* photoélectrique
~EFFECT - [electron] (the interaction between radiation and matter which results in the absorption of photons and the liberation of electrons) effet *m* photoélectrique
~EMISSION - [electron] (emission of electrons due to the incidence of radiant energy) effet *m* photoélectrique externe, photoémission *f*
~EXPOSURE METER - [photo] (exposure meter incorporating a photoelectric cell) posemètre *m* photoélectrique

PHOTOELECTRIC FATIGUE - [electron] (decrease of sensitivity of photoelectric substances) fatigue *f* photoélectrique
~FLAME FAILURE DETECTOR - [electron] (automatic furnace- watching device, which gives a signal when the flame fails) détecteur *m* photoélectrique d'absence de flamme
~GLOSSMETER - [instr] (a device for the determination of glossiness in surfaces, using a photo-electric system) brillancemètre *m*
~GRAMMOPHONE PICK-UP - [electron] capteur *m* photoélectrique
~INSPECTION - [electron] (photoelectric device to check the quality of goods) examen *m* photoélectrique de la qualité
~INTRUSION DETECTOR - [electron] (electronic device used to detect illegal entry) avertisseur *m* photoélectrique
~LIGHTING CONTROLLER - [electron] régulateur *m* photoélectrique d'éclairage
~MATERIAL - [electron] (material emitting electrons when exposed to radiant energy in a vacuum) substance *f* photoélectrique
~OPACIMETER - [electron] turbimètre *m* photoélectrique
~PHOTOMETER - [photo] (a photometer incorporating a photoelectric cell) photomètre *m* photoélectrique
~PICK-UP - [electron] détecteur *m* photoélectrique
~PINHOLE DETECTOR - [electron] détecteur *m* de pores photoélectrique
~PLETHYSMOGRAPH - [instr] (instrument designed to examine the condition of blood vessels) électroartériographe *m* photoélectrique
~PYROMETER - [instr] (instrument designed to measure high temperature by means of photoelectric cell) pyromètre *m* photoélectrique
~READER - [comput] lecteur *m* photoélectrique
~RECORDER - [instr] enregistreur *m* photoélectrique
~REFLECTION METER - [instr] (instrument designed to measure the reflection of surfaces, powders etc) reflectomètre *m* photoélectrique
~REGISTER CONTROL - [contr] relais *m* de commande photoélectrique
~SCANNING - [electron] (the scanning of punched cards by photoelectric means) analyseur *m* photoélectrique
~SCLEROSCOPE - [instr] (a photoelectric instrument designed to determine the hardness of metals) scléroscope *m* photoélectrique
~SMOKE METER - [instr] indicateur *m* photoélectrique de la densité de fumée
~SORTER - [instr] (a photoelectric instrument designed to sort objects according to size, colour etc) assortisseur *m* photoélectrique
~SPECTROPHOTOMETER - [instr] (spectrophotometer having a photoelectric detector) spectrophotomètre *m* photoélectrique
~THRESHOLD - [electron] seuil *m* photoélectrique
~TUBE - s. photoelectric valve
~VALVE - [electron] (electronic valve in which one of the electrodes is a photoemissive cathode) tube *m* photoélectrique
~VISIBILITY METER - [electron] transmissomètre *m*
PHOTOELECTRICITY - [el] (electricity produced by the action of light) photoélectricité *f*
PHOTOELECTROLYTIC - s. photochemical
~CELL - s. photochemical cell
PHOTOELECTROMAGNETIC EFFECT - [electron] (in semiconductors) effet *m* photoélectromagnétique
PHOTOELECTROMOTIVE FORCE - [electron] force *f* photoélectromotrice
PHOTOELECTRON - [electron] (electron liberated by the photoemissive effect) photoélectron
PHOTOELECTRONICS - [electron] (the study of the interaction of electricity and light) photoélectronique *f*
PHOTOEMISSIVE - [electron] (pertaining to a body from the surface of which electrons are liberated through the action of light) photoémettrice
~CAMERA TUBE - [electron] tube *m* de prise de vues à photoémission
~CELL - [electron] cellule *f* photoémettrice
~DETECTOR - [instr] (instrument designed to measure radiant energy by the emission of electrons from a photo-cathode) détecteur *m* à photoémission
~EFFECT - s. photoelectric emission
[electron] (ejection of electrons solely due to the incidence of radiant energy) effet *m* photoélectrique externe, photoémission *f*
PHOTOENGRAVING - [print] (the production of relief blocks or plates for printing by means of photography) similigravure *f*, photogravure *f*, zincographie *f*
PHOTOERYTHEMA - [med] photo-érythème *m*
PHOTOFINISHING LABORATORY - [photo] laboratoire *m* de façonnage
PHOTOFISSION - [nucl] (nuclear fission induced by photons) photofission *f*
PHOTOFLASH LAMP - [photo] (electric bulb fitted with aluminium wire which ignites and supplies incandescent light with the passage of current) lampe-éclair *f* électronique
PHOTOFLOOD LAMP - [photo & cin] (or movieflood lamp; a tungsten-wire incandescent lamp run at excess voltage) lampe *f* à tungstène
[photo] lampe *f* survoltée
~LIGHT - [photo] lumière *f* douce
PHOTOFLUOROGRAPH - [radiat] (a photo-rongten unit) appareil *m* de radiographie
PHOTOGEN - [zool] (a light-generating organ) organe *m* phosphorescent
[ind chem] (oil distilled from bituminous shale and used for illumination) huile *f* de tourbe
PHOTOGENIC - [photo] (having the quality of registering a good image) photogénique
[zool bot etc] (which emits light) photogène, phosphorescent
PHOTOGLYPHY - [print] (photographic engraving) photoglyptie *f*
PHOTOGONIOMETER - [instr] (device used for the study of the phenomena of crystal X-ray diffraction and spectra) goniomètre *m* photoélectrique
[surv] (instrument designed to obtain the direction of a ray from the nodal point of the camera lens to the image of any point on the photograph) goniomètre *m* photoélectrique
PHOTOGRAM - [photo] (picture obtained by photography) photogramme *m*
PHOTOGRAM'S HORIZONTAL LINE - [photo] horizon *m* de l'image
PHOTOGRAMMETRIC SURVEYING - [surv] levé *m* photogrammétrique
PHOTOGRAMMETRY - [photo] (the technique of making surveys by means of photographs) photogram-

métrie *f*
PHOTOGRAPH, to - [photo] photographier
PHOTOGRAPH - [photo] photographie *f*
PHOTOGRAPHIC - [photo] photographique
~BAROMETER - [instr] (formalin-hardened bromide print used to indicate the humidity) baromètre *m* photographique
~DENSITY - [photo] (the opacity of exposed and processed films) densité *f* photographique
~DEVELOPERS - [photo] (reducing agents for silver salts) révélateurs *mpl* photographiques
~EMULSION - [photo] (the light-sensitive layer) émulsion *f* photographique
~EXPOSURE - [photo] exposition *f* photographique
~EYEPIECE - [photo] oculaire *m* photographique
~GRAIN - [photo] (particle of metallic silver in a photographic emulsion after development) grain *m*
~GRAININESS - [photo] granulation *f*
~PLOTTER - [photo] photocartographe *m*
~RECORDER - [cin] (film recorder) enregistreur *m* optique du son
~RECORDING - [el acoust] (the registration of a modulated track on photographic film) enregistrement *m* optique du son
~SOUND RECORDER - s. photographic recorder
~SOUND REPRODUCER - s. photographic recorder
~SURVEYING - [surv] levé *m* photographique
~TELESCOPE - [instr] (astronomical telescope in which a camera replaces the eye-piece) télescope *m* photographique
~TRANSMISSION DENSITY - s. photographic density
PHOTOGRAVURE - [print] (printing by means of etched copper plates; the etching is made through a gelatine relief print of the subject) impression *f* en héliogravure, rotogravure *f*
[print] (the picture produced by the method) héliogravure *f*
~ROTARY PRESS - [print] machine *f* à héliogravure rotative
PHOTOHALIDE - [chem] (light-sensitive halogen salt) photohalogénure *m*
PHOTOHELIOGRAPH - [instr] photohéliographe *m*
PHOTOIONIZATION - [nucl] (atomic photoelectric effect) photoionisation *f*
PHOTOLITHOGRAPH, to - [print] (printing by means of photolithography) photolithographier
PHOTOLITHOGRAPH - s. photolithography
PHOTOLITHOGRAPHY - [print] (the production on stone, mainly by photographic means, of a printing surface from which impressions are taken by lithographic process) photolithographie *f*
PHOTOLUMINESCENCE - [electron] (the luminescence which is caused by visible electromagnetic radiation) photoluminescence *f*
PHOTOLYSIS - [chem] (decomposition or dissociation of the molecule by the action of light) photolyse *f*
PHOTOLYTIC SILVER - [chem] argent *m* photolytique
PHOTOMA - [med] photome *m*
PHOTOMAP - [photo] (map composed of aerial photographs) photoplan *m*
PHOTOMECHANICAL - [print] photomécanique
~COPYING - [photo] phototirage *m*
~PROCESS - [print] (printing method in which photography is used in etching or engraving) procédé *m* photomécanique
PHOTOMESON - [nucl] (meson ejected by a nucleus by an impinging photon) photoméson *m*
PHOTOMETER - [instr] (instrument designed to measure the intensity of light) photomètre *m*
~BENCH - [light] (bench on which the apparatus carrying out photometric tests is mounted) banc *m* photométrique
PHOTOMETRIC - [light] (pertaining to photometry) photométrique
~FILTER - [photo] filtre *m* photométrique
~INTEGRATOR - [light] (the part of an integrating photometer which sums up the light flux) intégrateur *m* photométrique
~STANDARD - [opt] étalon *m* photométrique
~SURFACE - [light] (used for photometric comparisons) surface *f* photométrique
PHOTOMETRY - [chem] (a method of volumetric analysis) photométrie *f*
[light] (the measure of the luminous intensity of light sources) photométrie *f*
PHOTOMICROGRAPH - [photo] (photograph of a microscopic object taken by a camera attached to a microphone) microphotographie *f*
PHOTOMICROGRAPHIC CAMERA - [photo] chambre *f* microphotographique
PHOTOMONTAGE - [photo] (montage with photographs) photomontage *m*
PHOTOMOSAIC - [electron] mosaique *m* photoélectrique
[photo] mosaique *f* photographique
PHOTOMULTIPLIER - [electron] (sensitive detector of light) photomultiplicateur *m*
~COUNTER - [nucl] (scintillation counter) compteur *m* à scintillation
~TUBE - s. photoelectric cell
PHOTOMURAL - [photo] (greatly enlarged photographs used for wall decoration) photographie *f* murale
PHOTON - [nucl] (a quantum of electromagnetic radiation) photon *m*
~ENERGY - [phys] (the energy of a photon) énergie *f* de cellule photoélectrique
~ENGINE - [astronaut] moteur *m* à photons
~NOISE - [electron] bruit *m* de cellule photoélectrique
~ROCKET - [astronaut] fusée *f* photonique, fusée *f* à photons
PHOTONEGATIVE - [opt] (having negative conductivity) photorésistant
PHOTONEUTRON - [nucl] (a neutron released from a nucleus by a photonuclear reaction) photoneutron *m*
PHOTONUCLEAR REACTION - [nucl] (a nuclear reaction induced by a photon) réaction *f* photonucléaire
PHOTOPATHY - [biol] (negative phototaxis) photopathie *f*
PHOTO-PEAK - [electron] (of a pulse amplitude distribution) raie *f* photoélectrique
PHOTOPERCEPTOR - [bot] (part of eye spot sensitive to light) photopercepteur *m*
PHOTOPHOBIA - [med] (eye intolerance to light) héliophobie *f*, photophobie *f*
PHOTOPHORESIS - [phys] (the action of light in inducing the migration of suspended particles) photophorèse *f*
PHOTOPHTALMIA - [med] (a burning pain in the eyes caused by electric light) photophtamie *f*
PHOTOPIA - [med] adaptation *f* à la lumière
PHOTOPIC RESPONSE - [opt] réponse *f* photopique
~VISION - [opt] vision *f* photopique
PHOTOPLASM - [zool] (the cytoplasm of a luminous

cell) photoplasma *m*
PHOTOPLAY - [cin] (a play represented in motion picture) film *m* dramatique
PHOTOPOLYMER - [chem] (a polymer which is sensitive to light and which has application in photographic and similar processes) photopolymère *m*
PHOTOPOSITIVE - [electron] photoconducteur
PHOTOPRINTING PAPER - [photo] papier *m* héliographique
PHOTOPROTON - [nucl] (a proton released from a nucleus by a photonuclear reaction) photoproton *m*
PHOTORADIOGRAM - [telecomm] (photograph transmitted by radio) photoradiogramme *m*
PHOTORECEPTOR - [zool] (sensory nerve-ending which receives light-stimuli) photorécepteur
PHOTOSENSING MARK - [comput] marque *f* photosensible
PHOTOSENSITIVE - [phys] (the property of being sensitive to the action of visible or invisible light) photosensible
PHOTOSPHERE - [astr] (term denoting the visible surface of the sun) photosphère *f*
PHOTOSTAT - [print] (trade-name for a camera which rapidly reproduces drawings documents etc) photostat *m*
[print] (the actual print) photocopie *f*
~PAPER - s. photoprinting paper
PHOTOSYNTHESIS - [biol chem] (the natural process by which chlorophyll converts water and carbon dioxide into more complex molecules under the influence of sunlight) photosynthèse *f*
PHOTOSYNTHETIC CAPACITY - [bot] (the capacity of a plant or cell to carry out the process of photosynthesis) capacité *f* photosynthétique
~NUMBER - [bot] indice *m* photosynthétique
~QUOTIENT - s. photosynthetic ratio
~RATIO - [bot] (the ratio of the volume of carbon dioxide absorbed to the volume of oxygen set free) rapport *m* photosynthétique
PHOTOTECHNIC - [photo] phototechnique *f*
PHOTOTELEGRAPHY - [telecomm] (the transmission of still pictures over telegraph circuits) phototélégraphie *f*, radiotransmission *f* des images
PHOTOTELESCOPE - [instr] phototélescope
PHOTOTHEODOLITE - [surv] (camera having fixed known focal length and horizontal and vertical wires pressing against the sensitive plate) photothéodolite *m*
PHOTOTHERAPY - [med] (light treatment) photothérapie *f*
PHOTOTIMER - [radiat] (a timer for radiographic exposure control) photominuterie *f*
PHOTOTOPIC VISION - [opt] vision *f* photopique
PHOTOTOPOGRAPHY - s. photogrammetry
PHOTOTOXIS - [med] phototraumatisme *m*
PHOTOTRANSISTOR - s. photistor
PHOTOTRAY - [photo] cuvette *f* à développeur
PHOTOTROPY - [paint] (the loss of colour of a dyestuff under the influence of light) phototropisme *m*
PHOTOTUBE - [electron] (electronic tube in which one of the electrons is a photo-emissive cathode) tube *m* photoélectronique
PHOTOTYPE - [print] (a relief plate for printing by photo) phototype *m*
PHOTOTYPOGRAPHY - [print] (photochemical process of engraving which can be reproduced in connexion with a printing press) phototypie *f*
PHOTOVALVE - s. phototube
PHOTOVARISTOR - [electron] (varistor in which the current-voltage relation may be modified by illuminating cadmium sulphide etc) photodiode *f*
PHOTOVOLTAIC - [el] (capable of producing an electromotive force under the action of light) photovoltaïque
~CELL - [contr] (a device which produces an electromotive force under incident light) cellule *f* à couche d'arrêt, cellule *f* photovoltaïque
~EFFECT - [electron] (photo-electric effect resulting in the appearance of a voltage at the contact between an electrode and an electrolyte) effet *m* photovoltaïque
PHOTOXYLOGRAPHY - [print] photoxylographie *f*
PHOTOZINCOGRAPHY - [print] photogravure *f* sur zinc
PHRASE, to - [gen] exprimer
PHRASE - [gen] expression *f*, tour *m* de phrase
[mus] phrase *f*
~INTELLIGIBILITY - [telecomm] netteté *f* pour les phrases
PHREATIC - [geol] (pertaining to underground waters) phréatique
~ERUPTION - [geol] explosion *f* phréatique
~GASES - [geol] (those gases of atmospheric or oceanic origin which come into contact with ascending magma and possibly provide the motive force for volcanic eruptions) gaz *m*pl phréatiques
~SURFACE - [soil] niveau *m* de la nappe, surface *f* phréatique
~WATERS - [geol] eaux *f*pl phréatiques
PHRENALGIA - [med] phrénalgie *f*, diaphragmatalgie *f*
PHRENASTHENIA - [med] phrénasthenie *f*
PHRENECTOMY - [med] s. phrenicectomy
PHRENIC - [anat] (pertaining to the diaphragm) phrénique
~NERVE - [anat] nerf *m* phrénique
PHRENICECTOMY - [med] (excision of part of the phrenic nerve) exérèse *f* du phrénique
PHRENICLASIA - s. phrenemphraxis
PHRENICLASIS - s. phrenemphraxis
PHRENIOCARDIA - [med] phrénocardie *f*
PHRENODYNIA - s. phrenalgia
PHRENOPLEGIA - [med] paralysie *f* du diaphragme
PHRENOSIN - [biochem] (cerebroside obtained from the brain substance) phrénosine *f*
PHRENOSPASM - [med] spasme *m* du diaphragme
PHTHALAMIDE - [chem] (a synthesis intermediate) phtalamide *f*
PHTHALEINS - [chem] (derivatives obtained by the reaction of phthalic anhydride and phenol) phtaléines *f*pl
PHTHALIC ACID - [chem] (an intermediate for perfumes and dyestuffs) acide *m* phtalique
~ANHYDRIDE - [chem] (an intermediate for pharmaceuticals, dyestuffs, synthetic resins and plasticizers) anhydride *m* phtalique
~RESINS - [chem] (a class of synthetic resins) résines *f*pl phtaliques
PHTHALIMIDE - [chem] (an intermediate for dyestuffs) phtalimide *f*
PHTHALOCYANINE - [chem] phtalocyanine *f*
PHTHALONITRILE - [chem] (an intermediate for dyestuffs and pigments) phtalonitrile *m*
PHTHALYLSULPHACETAMIDE - [chem] (a sulphona-

mide used in medicine) phtalylsulfacétamide *f*
PHTHALYLSULPHATHIAZOLE - [chem] (a sulphonamide with medicinal uses) phtalylsulfathiazol *m*
PHTHISIS - [med] (consumption, mainly pulmonary; also a progressive emaciation) tuberculose *f* pulmonaire
PHUGOID OSCILLATION - [aero] (long-period cyclic variation in the longitudinal motion of an aircraft) oscillation *f* phygoïde
PHYCOCYANIN - [chem] (a blue pigment) phycocyanine *f*
PHYCOERYTHRIN - [chem] (a red pigment) phycoérythrine *f*
PHYCOLOGY - [bot] (the study of algae) phycologie *f*
PHYLETIC CLASSIFICATION - [bot] (plant classification based on the presumed evolutionary descent of organisms) classification *f* phylétique
PHYLLITE - [geol] (argillaceous rocks in a state of metamorphosis) phyllite *f*
PHYLLODE - [anat] foliacé
PHYLLOXERA - [bot] (a plant louse very harmful to grape vines) phylloxéra *m*
PHYLOGENESIS - [genet] (the evolution of a species) phylogenèse *f*
PHYLOGENY - [genet] (the history of the evolution of a species, a group etc) phylogénie *f*
PHYMA - [med] tumeur *f* cutanée
PHYSALIFEROUS CELL - s. physaliphore
PHYSALIPHORE - [biol] cellule *f* géante vacuolaire
PHYSIATRICS - [med] physiatrie *f*
PHYSICAL - [phys] (pertaining to the material universe or to the physical science) physique
[med] (of a symptom) somatique
~ASTRONOMY - [astr] astronomie *f* physique
~BARRIER - [biol] (any obstacle to the migration of animals or plants) barrière *f* physique
~CHANGE - [phys] transformation *f* physique
~CIRCUIT - [telecomm] (the normal loop circuit for telephony) circuit *m* réel, circuit *m* composant, circuit *m* combinant
~CONTACT - [el] contact *m* matériel
~FOAMING - [plast ind] (production of cellular plastics by incorporating volatile liquids, e.g. methyl chloride in them, foaming being set up during a rise of temperature in processing) production *f* physique de mousse
~LINE - s. physical circuit
~MAP - [geogr] carte *f* géographique physique
~MASS UNIT - [phys] (one sixteenth of the mass of the oxygen atom in the physical scale) unité *f* de masse atomique
~MEASUREMENTS - [phys] mesures *fpl* physiques
~METALLURGY - [metall] métallurgie *f* physique
~OPTICS - [opt] (that part of the study of light which deals with diffraction, interference etc) optique *f* physique
~PENDULUM - [mech] (pendulum consisting of a rigid body) pendule *m* physique, balancier *m* physique
~PHOTOMETER - [instr] (instrument designed to measure photometric quantities through the effect produced by a radiation on a physical receptor) photomètre *m* objectif
~PROPERTIES - [gen] caractéristiques *fpl* physique, propriétés *fpl* physiques
~RONGTEN EQUIVALENT - [radiat] rep *m*
~SCALE OF ATOMIC WEIGHTS - [nucl] table *f* physique de poids atomiques
PHYSICAL SYSTEM TIME - [comput] temps *m* de problème
~TRACER - [nucl] (tracer fitted by mechanical means to the object being traced) traceur *m* physique
PHYSICIAN - [med] médecin *m*
PHYSICIST - [gen] (a student of, or specialist in, physical science) physicien *m*
PHYSICO-ELECTRIC RELAY - [el] relais *m* physico-électrique
PHYSICS - [phys] (the science of physical phenomena) physique *f*
PHYSIOGNOMY - [zool & bot] (features) physionomie *f*
PHYSIOGRAPHY - [geogr] géographie *f* physique
[gen] (description of nature and natural phenomena) physiographie *f*
PHYSIOLOGICAL - [zool & bot] (pertaining to the functions of living organisms) physiologique
~ACCELERATION - [astronaut] accélération *f* physiologique
~ACOUSTICS - [acoust] (branch of acoustics dealing with the production and detection of sound by living organs) acoustique *f* physiologique
~ANATOMY - [biol] (study of the relationship between structure and functions) anatomie *f* physiologique
~DROUGHT - [bot] sécheresse *f* physiologique
~RACE - [zool] (group belonging to a species but differing from other groups in its habits) race *f* physiologique
~VARIETY - [biol] forme *f* biologique
~ZERO - [biol] (threshold temperature under which metabolism ceases) zéro *m* physiologique
PHYSIOLOGY - [biol] (the science of organic functions) physiologie *f*, zoonomie *f*
PHYSIOPATHOLOGY - [med] physiopathologie *f*
PHYSIOSIS - [med] météorisme *m* abdominal
PHYSIOTHERAPY - [med] physiothérapie *f*
PHYSOCELE - [med] physocèle *f*
PHYSOSTIGMINE - [chem] (an alkaloid obtained from Physostigma venenosum and used in medicine) physostigmine *f*, ésérine *f*
PHYTIC ACID - [chem] (a chelating agent) acide *m* phytique
PHYTIN - [chem] (a calcium-magnesium salt obtained from the reeds of various plants) phytine *f*
PHYTO-SANITARY CONTROL - [agric] contrôle *m* phytosanitaire
PHYTOCHEMICAL - [chem] (a chemical reaction or process carried out within a plant cell) phytochimique
PHYTOGENESIS - [bot] (the study of the generation and development of plants) phytogenèse *f*
PHYTOHORMONE - [med] phytohormone *m*
PHYTOL - [chem] (an intermediate in the synthesis of vitamins E and K) phytol *m*
PHYTONADIONE - [chem] (vitamin KI. It has application in medicine) phytonadione *f*
PHYTOPHARMACOLOGY - [chem] phytopharmacie *f*
PHYTOSTEROLS - [chem] (generic term for sterols which occur only in plants) photostérols *mpl*
PHYTOTOXIC - [bot] phytotoxique
pi - [geom] (the ratio of the circumference to its diameter; i.e. 3.I4I6) pi *m*
PIA MATER - [anat] (the inner most membrane surrounding the brain and spinal cord) pie-mère *f*
PIANETTE - [mus] pianino *m*

PIANO - [mus] (or pianoforte) piano *m*
~CARD CUTTING MACHINE - [text] piqueuse *f* à clavier
~WIRE - [mech] corde *f* à piano
~WIRING - [el] câblage *m* en fil nu
PIANOFORTE - s. piano
PIASSABA - [text] piassava *m*
PIAZZA - [constr] (a veranda, a porch) véranda *f*
PIC FACTORY - [cin] (colloq. for film studio) studio *m* cinématographique
PICA - [med] picacisme *m*, allotriophagie *f*
[print] (early name for a I2 point character) corps *m* I2, Cicéto *m*
PICCOLO - [mus] piccolo *m*, petite flûte *f*
PICK, to - [gen] choisir
[agric] (of fruit) cueillir
[mech] (a lock) crocheter
[gen] (to work with a pick) piocher
[text] (the wool) échardonner
[min] trier au marteau
[mining] localiser
PICK - [impl] pic *m*, pioche *f*
[text] chasse *f*
[text] (one of the threads across the width of a fabric) fil *m* de trame, duite *f*
[mining] rivelaine *f*, picot *m*
~-AND-CLAW CROWBAR - [impl] levier *m* à pied de biche et à pointe
~-AND-SHOVEL WORK - [mining] abattage *m* à la pioche et à la pelle, travail *m* au pic et à la pelle
~-AXE - [impl] (or pickaxe; a pick fitted with a cutting edge) pioche *f*, hoyau *m*
~COAL-CUTTIN MACHINE - [mining] haveuse *f* à pic
~COUNTER - [text] compteur *m* de duites
~FINDER - [text] dispositif *m* de recherche de la duite
~GLASS - [text] compte-fils *m*
~HAMMER - [impl] marteau *m* à pioche
~MACHINE - [mining] haveuse *f* à pic
~MATTOCK - [impl] (pick-axe with blades instead of points) décintroir *m* à talus
~OFF, to - [min] trier
~-OFF - [mech etc] (used and adective) démontable
[el acoust] capteur *m*
~-OFF BRUSH - [comput] balai *m* de prélèvement
~OUT, to - [gen] enlever, extirper
[min] abattre à la pioche
~THE BURRS, to - [text] échardonner
~UP, to - [gen] prendre, ramasser, cueillir, soulever
[mech] prendre de la vitesse, reprendre
[naut] prendre (son coffre)
[naut] (a submarine cable) relever
[radio] capter, accrocher
[aero] (an aircraft by searchlight) répérer
~-UP - [el acoust] (electromechanical transducer actuated by the modulation of the groove of the recording medium) phonocapteur *m*, pick-up *m*
[gen & mech] reprise *f*
[telev] (the generation of the electrical signal to correspond with the received optical image) prise *f* de vues
[photo] (of a camera) dispositif *m* de prise de vues
[metall] grippage *m*, rayage *m*
[transp] pick-up *m*
[nucl] (type of nuclear reaction in which the incident particle takes a nucleon from the target nucleus) enlèvement *m*, rapt *m*

PICK UP - [el] (of relays) mise *f* au travail
~-UP AMPLIFIER - [el acoust] amplificateur *m* phonographique
~-UP ARM - [el acoust] (a bar pivoted at one end to hold the pickup) bras *m* de pick-up
~-UP BALER - [agric] ramasseuse-presse *f*
~-UP BUSH - [el acoust] douille *f* de pick-up
~-UP CARTRIDGE - [el acoust] (removable portion of the pickup which comprises the electrochemical transducing elements and the stylus) tête *f* de pick-up
~-UP COIL - [el] bobine *f* exploratrice
~-UP ELEMENT - [electron] organe *m* de prélèvement
~-UP GRAB - [mining] pince *f* de repêchage
~-UP LOADER - [agric] ramasseuse-chargeuse *f*
~-UP PLATE - [electron] plaque *f* de signal
~-UP PLOUGH - [agric] charrue *f* alternative
~-UP PRESS - [agric] ramasseuse-presse *f*
~-UP ROLL - [mech] (a roller used to collect fluid from a tank or bath and apply it to a continuous band of material) rouleau *m* enducteur
~-UP ROLLER - [photo] rouleau *m* de léchage
~-UP SPECTRAL CHARACTERISTICS - [phys] caractéristique *f* spectrale d'un électrocapteur de rayonnement
~-UP TRUCK - [transp] camionnette *f* bâchée
~-UP TUBE - [telev] (the tube in which the electrical signal is generated, which must correspond to the received optical image) tube *m* de prise de vues, tube *m* analyseur
~-UP VELOCITY - [telev] (scanning speed) vitesse *f* d'analyse
~-UP VOLTAGE - [electron] tension *f* d'excitation
~-UP WINDING - [comput] (of a static magnetic storage) enroulement *m* détecteur
PICKED - [gen & ind] choisi
~ORE - [min] concentré *m* de triage
~SHUTTLE - [text] navette *f* lancée
PICKER - [text] (the implement which propels the shuttle across the loom) taquet *m*, chasse-navette *m*
[text] (in regenerating processes) éplucheuse *f*
[agric] grapilleuse *f*
[mining] trieur *m*, marteau *m* de triage, pic *m*
~ARM - [comput] bras *m* d'alimentation
~BELT - [electron] courroie *f* d'alimentation
~FORK - [text] fourchette *f* du taquet
~HEAD - [text] tête *f* du taquet
~KNIFE - [mech] (in electronic computer systems etc; a knife pushing a document from a pile into a machine) lame *f* d'alimentation
~PUSHER - [text] poussoir *m*
~SPINDLE - [text] tringle *f* du taquet
~TONGUE - [text] pied *m* du taquet
PICKET, to - [aero] (to secure an aircraft in the open against wind by anchoring it to stakes driven into the ground) attacher à piquet
PICKET - [surv] (short ranging rod) jalon *m*
[gen] piquet *m*
~BOAT - [naut] vedette *f*
~FENCE - [constr] palissade *f*, palis *m*
PICKING - [text] lancement *m*, chasse *f* de la navette
[text] tramage *m*, insertion *f* de la trame
[mining] triage *m* ă main
~ABOVE - [text] duitage *m* de dessus
~BAND - [text] fouet *m*, sonnette *f*, caribari *m*
~BAR - [text] barre *f* de chasse
~BELOW - [text] duitage *m* d'envers

PICKING BOWL - [text] galet *m* de chasse
~CATCH - [text] loquet *m* de chasse-navette
~CORD - [text] corde *f* du taquet, cordon *m* de tirage
~ECCENTRIC - [text] came *f* de chasse
~HEART - [text] excentrique *m* de chasse
~LEVER - [text] levier *m* de chasse
~MECHANISM - [text] mécanisme *m* de la chasse
~MOTION - [text] dispositif *m* d'insertion de la trame
~MOTION BREAKAGE - [text] dispositif *m* de sûreté contre les ruptures du mécanisme de chasse-navette
~NEB - [text] bec *m* d'excentrique, nez *m* de chasse
~NOSE - [text] bec *m* de chasse
~ORE - [min] minerai *m* trié au marteau (ou à la main)
~PERIOD - [agric] cueillette *f*
~REPEAT - [text] rapport *m* de duitage
~RESISTANCE - [paper man] (of coated paper) résistance *f* à l'arrachage
~SPINDLE - [text] tringle *f* du taquet
~STICK - [text] battant *m*, bras *m* de chasse
[text] sabre *m*
~STRAP - [text] fouet *m*, sonnette
~STRENGTH - s. picking resistance
~TAPPET - [text] excentrique *m* de chasse
~THE SHUTTLE - [text ind] lancement *m* de la navette
~-UP - [timber] (of wood leaving a fluffy surface when cut) cotonneux
PICKLE, to - [metall] (to cleanse a metallic object, e.g. a casting, by immersion in an acid bath) décaper
PICKLE - [metall] (solution designed to remove oxides or other compounds from metal) décapant *m*
[leather ind] (treatment of light skins with a weak solution of sulphuric acid and salt) bain *m* de picklage
~BRITTLENESS - [metall] fragilité *f* au décapage
~INHIBITOR - [metall] inhibiteur *m* de décapage
~STAIN - [metall] décoloration *f*, tache *f* de décapage
~TEST - [metall] (of steel surfaces) essai *m* d'immersion en acide
PICKLED - [metall] décapé
~SHEET - [metall] tôle *f* décapée
PICKLESS - [agric] légumes *mpl* au vinaigre
PICKLING - [leather ind] (treatment with sulphuric acid and salt after de-hairing) mise *f* en jusée
[metall] (immersion in an acid bath to remove scale, oxide etc) décapage
[el chem] (method of pickling in which a current is passed through the solution to the metal or from the metal) décalage *m* électrolytique
~BATH - [metall] bain *m* de décapage
~CONVEYOR FOR CASTINGS - [metall] transporteur *m* pour jets de fonderie
~CRACK - [metall] crique *f* de décapage
~INHIBITOR - [metall] (agent added to a pickling solution to reduce the rate of dissolution of a given metal) inhibiteur *m* de décapage
~LINE - [metall] installation *f* de décapage continu, ligne *f* de décapage
~POND - [ind chem] (pond in which calcium sulphure is crystallized out in a solar salt process) bac *m* de décapage
~VAT - s. pickling pond
PICKS PER CENTIMETER - [text ind] (nombre de) duites *fpl* au centimètre
PICOFARAD - [electron] (one-millionth of a microfarad) picofarad *m*
PICOLINE - [chem] (a solvent) picoline *f*
PICRAMIC ACID - [chem] (an analytical reagent and an intermediate for dyestuffs) acide *m* picramique
PICRATE - [chem] (salt or ester of picric acid) picrate *m*
PICRIC - [chem] picrique
~ACID - [chem] (trinitrophenol. An explosive, antiseptic, dyestuff and intermediate for dyestuffs) acide *m* picrique
PICRITE - [min] (a coarse grained igneous rock mainly consisting of olivine) picrite *f*
PICROCHROMITE - [min] (an igneous rock) picrochromite *f*
PICROGEUSIA - [med] saveur *f* amère
PICROLONIC ACID - [chem] (an analytical reagent) acide *m* picrolonique
PICROTOXIN - [chem] (cocculin. An analeptic agent and respiratory stimulant used in medicine) picrotoxine *f*
PICS - [comput] (Production Information and Control System) système *m* PICS
PICTORIAL PHOTOGRAPHY - [photo] photographie *f* artistique
~WIRING DIAGRAM - [electron] diagramme *m* schématique
PICTURE - [gen] tableau *m*, peinture *f*
[photo] photographie *f*
[telev] image *f*
[cin] film *m*
[comput] modèle *m*
AMPLITUDE - [telev] hauteur *f* d'image
~AMPLIFIER - [telev] amplificateur *m* d'image
~AREA - [cin] (the total surface covered by the image) cadrage *m*, champ *m* de l'image
~BLACK - [telev] (the level corresponding to the darkest part of the material to be transmitted) noir *m* d'une image, niveau *m* du noir le plus profond
~CARRIER - [telev] porteuse *f* vidéo
~CHECK PRINT - [cin] (a print made for checking purposes) copie *f* d'essai
~COMPRESSION - [telev] compression *f* de l'image
~CONTROL COIL - [telev] (device for centring the picture on the face of the picture tube) bobine *f* de cadrage
~COUNTER - [telev] (frame counter) compteur *m* d'images
~DUPE NEGATIVE - [cin] (picture negative made from a picture duping print) contre-type *m* négatif
~DUPING PRINT - [cin] (print on a special print for the production of a duplicate negative) copie *f* lavande
~EDGE - [telev] bord *m* de l'image
~ELEMENT - s. picture point
~FREQUENCY - [telev] (the number per second of complete picture) fréquence *f* d'image
~FREQUENCY BAND - [telev] bande *f* de fréquence d'image
~INFORMATION - [telev] information *f* d'image
~JOIN - [telev] égalisation *f* des vitesses d'images
~JUMP - [cin] oscillation *f* de l'image
~LOCK - [telev] (the mechanism by which the picture is kept in place) calage *m* de l'image
~MONITOR - [telev] (for the control of the technical quality of television pictures) moniteur *m* d'image
~OUTPUT - [telev] (video time basis) base *f* de temps vidéo
~PERIOD - [telev] période *f* d'image

PICTURE PICKUP SYSTEM - [telev] prise *f* de vues en télévision
~PLANE - [photo] plan *m* de l'image
~POINT - [telev] (one of the large number of minute areas which make up a television image) point *m* photographique
~RATIO - [telev] (the ratio between height and width of a television image) rapport *m* de format, rapport *m* largeur/hauteur de l'image
~-ROTATE CONTROL - [telev] (the manual control to place the picture correctly) réglage *m* de rotation de l'image
~SHIFTING - [photo] déplacement *m* de l'image
~SIGNAL - [telev] (the part of the video signal carrying the information of the picture) signal *m* d'image
~SIGNAL AMPLITUDE - [telev] amplitude *f* du signal d'image
~SIZE - [photo] format *m*
~STRIP - [photo] film *m* en bande
~SYNCHRONIZATION - [telev] (frame synchronization) synchronisation *f* d'image
~SYNTHESIS - [telev] (the method of building up the picture) synthèse *f* de l'image
~TELEGRAPH APPARATUS - [telecomm] appareil *m* phototélégraphique
~TELEGRAPHY - [telecomm] phototélégraphie *f*, transmission *f* des images, téléiconographie *f*
~TONE - [telecomm] (the frequency of the carrier which is used in amplitude modulated facsimile transmission) fréquence *f* porteuse de modulation d'image
~TRANSMISSION - [telecomm] (radiotelegraphic transmission of pictures) téléphotographie *f*, bélinographie *f*
~TRANSMITTER - [telev] (transmitter of television pictures) transmetteur *m* d'image
~TUBE - [telev] (cathode-ray tube for the reproduction of television pictures) tube *m* image
~WHITE - [telev] (the level corresponding to the whitest part of the transmitted picture) niveau *m* de blanc de la porteuse
~WITH A HIGH GAMMA - [cin] image *f* à grand contraste
~WITH A LOW GAMMA - [cin] image *f* à faible contraste, image *f* douce
PIE, to - [print] (to upset type-matter by accident) mettre en pâte
PIE - [print] (type matter which has been upset by accident) pâte *f*, pâté *m*
~WOOL - [text] laine *f* arrachée
PIEBALD (HORSE) - [zool] (cheval) pie
PIECE - [gen] pièce *f*, morceau *m*
[text] pièce *f*
~BEAM - [text] ensouple *f* de l'étoffe, ensouple *f* enrouleuse
~COMPOSITOR - [print] compositeur *m* à la tâche
~CUTTER - [mech] outil *m* à trancher
~-DYED - [text] teint en pièce
~-DYEING - [text] teinture *f* en pièce
~-GOODS - [comm] marchandises *fpl* à la pièce
~KNOCK-OUT - [mech] (US colloq. for an ejecting device) dispositif *m* d'éjection
~MARKER - [text] marqueur *m*
~RATE - [comm & ind] (the payment for a unit of production) payment *m* à la pièce
~-WAGES - [comm] salaire *m* à la tâche, salaire *m* à la pièce

PIECE WORK - [comm ind etc] travail *m* à la tâche
PIECER - [text] noueur *m*
PIECEWISE-LINEAR FUNCTION GENERATOR - [comput] générateur *m* de fonction à sous-tendante
PIECING - [text] rattachage *m*
~MACHINE - [text] lance-bouts *m*
~THE COCOON FILAMENT - [text] jet *m* du brin de coton
~THE FILAMENTS - [text] rattache *f* du filament
~THE YARN - [text] rattache *f* du fil
PIED - [zool] bariolé, bigarré
PIEDRA - [med] trichomycose *f* noueuse
PIEDROIT - [constr] (projecting pier without cap or base) pied-droit *m*, piédroit *m*
PIEPLANT - [bot] rhubarbe *f*
PIER - [gen] jetée *f*, môle *m*
[constr] (the wall parts between doors and windows) trumeau *m*
[mech] pilier *m*
[constr] piédroit *m*, pilier *m*, pilastre *m*
[constr] (of a bridge) palée *f*
[constr] (the support for an arch etc) pilastre *m*
[naut] (a breakwater adapted for landing) embarcadère *m*, appontement *m*
[naut] (a mole) digue *f*, môle *m*
~ARCH - [arch] (arch carried on piers) arc *m* sur piédroits
~BOND - [constr] appareil *m* de pile
~FOUNDATIONS - [constr] fondation *f* sur pilotis
~-GLASS - [gen] trumeau *m*
~-HEAD - [naut] musoir *m*, extrémité *f* de la jetée
~TEMPLATE - [constr] pile *f* d'échafaudage
~WALL - [ind chem] (in horizontal retorts) piédroit *m*, pilier *m*
PIERAGE - [comm] droits *mpl* de jetée
PIERCE, to - [gen] percer, pénétrer
[min] (the operation of excavating a tunnel) perforer
[mech] (to bore through) percer
[metall] (a mould) épingler
PIERCER - [mech] (a punch) porte-foret *m* à archet
[metall] aiguille *f* à trous d'air
PIERCING - [gen] aigu, percant
[acoust] aigu
[mech] perforation *f*
[metall] épinglage *m*
PIERFENDER - [naut] protection *f* de môle
PIERHEAD - s. pier-head
PIERRE PERDUE - [constr] (foundation work consisting of stone blocks deposited at random) empierrement *m*
PIERS - [gas ind] (US only; inlet or outlet well in GB) caisson *m* d'entrée, caisson *m* de sortie
PIESESTHESIA - [med] perception *f* de la pression
PIEZOCHEMISTRY - [chem] (the branch of chemistry concerned with reactions at extremely high pressures) piézochimie *f*
PIEZOELECTRIC - [el] (pertaining to piezoelectricity) piézoélectrique
~CRYSTAL - [phys] (crystal with strong piezoelectric properties) piézocristal
~CRYSTAL ELEMENT - s. piezoelectric crystal
~CRYSTAL PLATE - [el] (a piece of piezoelectric material cut to specified dimensions) lame *f* piézoélectrique
~CRYSTAL UNIT - [el] élément *m* piézoélectrique
~EFFECT - [el] (the property of certain crystals to

expand along one axis and contract along another when subjected to an electric field) effet *m* piézo-électrique
PIEZOELECTRIC LOUDSPEAKER - [el acoust] (loudspeaker in which the mechanical displacements are produced by piezoelectric action) haute-parleur *m* piézoélectrique
~MANOMETER - [instr] (a piezometric pressure gauge , i.e. a pressure gauge in which use is made of the piezoelectric effect) manomètre *m* piézoélectrique
~OSCILLATOR - [radio] (oscillator with a piezoelectric crystal in its tuned circuit) oscillateur piézo-électrique, oscillateur *m* piloté cristal
~PLATE - [el] lame *f* piézoélectrique
~RELAY - [el] relais *m* à électrostriction
~RESONATOR - [radio] (piezoelectric crystal used as a resonating standard of frequency) résonateur *m* piézoélectrique
~STRAIN GAUGE - [instr] extensomètre *m* piézoélectrique
PIEZOELECTRICITY - [el] (electricity resulting from pressure on certain bodies) piézoélectricité *f*
PIEZOID - [el] (finished piezoelectric crystal product after the completion of all processes) piézoïde *m*
PIEZOMETER - [instr] (an instrument for measuring pressures) piézomètre *m*
~TUBE - [mech] (static pressure tube) tube *m* piézométrique
PIEZOMETRIC - [phys] piézométrique
~HEAD - [soil] hauteur *f* piézométrique, charge *f* piézométrique
~LEVEL - [soil] niveau *m* piézométrique
~LINE - [soil] courbe *f* piézométrique
~PRESSURE - [soil] pression *f* piézométrique
PIEZOMICROPHONE - [el acoust] (microphone in which the generated electromotive force depends on the piezo effect) microphone *m* piézoélectrique
PIEZOTHERAPY - [med] pneumothorax *m* artificiel
PIG, to - [metall] couler les gueuses, couler la fonte en gueuses
PIG - [zool] (a hog) porc *m*, cochon *m*
[metall] (an oblong mass of metal, mainly iron or lead, just run from the smelter) gueuse *f*, saumon *m*
[metall] (short for pig iron) gueuse *f* de fonte
[pottery] (only in Scotland; an earthen article) cruchon *m*, bouteille *f* en grès
[metall] fosse *f* de coulée pour gueuset
~-BACK, to - [metall] (the operation of adding pig iron in steel making) faire une addition carburante, ajouter de la fonte brute
~BED - [metall] (a set of moulds for metal pigs formed in a bed of sand) lit *m* de coulée, moules *mpl* à gueusets
~BOILING - [metall] (puddling) puddlage *m* gras
~BREEDING - [zool] élevage *m* de porcs
~COPPER - [metall] cuivre *m* en saumons
~FITTED WITH WIRE BRUSH - [impl] (brush used to clean pipes) hérisson *m*
~HERD - [zool] troupeau *m* de porcs
~IRON - [metall] (cast iron an ingots as run from the blast furnace) fonte *f* brute
~IRON MIXER - [metall] mélangeur *m* de fonte
~-KEEPING - [zool] élevage *m*
~LEAD - [metall] saumons *mpl* de plomb
PIG MANURE - [agric] fumier *m* de porc
~SOW - [metall] chenal *m* de coulée à lingots
~-TAIL GUIDE - [text] guide-fil *m* en queue de cochon
~TIN - [metall] étain *m* en saumon
~WASHING - [metall] affinage *m* de fonte brute
PIG'S FOOT - [text] glissière-guide *f* pour la règle
PIGEON - [zool] pigeon *m*
~BREAST - [anat] poitrine *f* saillante
~-HOLE - [gen] (small compartment in a desk etc) case *f*, alvéole *m*
~-HOLED - [constr] (of a wall when build with regularly spaced gaps in it) creux
PIGGING - [metall] lingotage *m*
~UP - [metall] addition *f* de fonte brute
PIGGY-BACK - [aero] (large transport plane carrying a smaller one) avion *m* porte-avions
~-BACKING - [aero] transport *m* d'un avion
PIGLET - [metall] petit saumon *m*
[metall] (foundry) petite gueuse *f*
~SCOUR - [vet] diarrhé *f* des porcelets
PIGLING - [zool] porcelet *m*, goret *m*
PIGMENT, to - [dye] teindre, colorer
PIGMENT - [dyes] (generic term for natural and synthetic organic and inorganic substances capable of giving colour to materials, liquids and mixtures) pigment *m*
~-BINDER RATIO - [paint] (the ratio of the pigment to the binding agent) rapport *m* pigment/liant
~CELL - [zool] (a cell which contains pigment granules) cellule *f* pigmentaire
~PAPER - [paper man] (paper used in the pigment process) papier *m* pigmenté
~PASTE - [paint] (dry colouring material ground with a special oil) pigment *m* en pâte
~PROCESS - [photo] (process involving pigment suspended in a tissue, which is partially dissolved to form the required image) procédé *m* colorant
PIGMENTATION - [dyes] (coloration) pigmentation *f*
[med] (diseased condition of the pigment cells of the skin) pigmentation *f*
PIGMENTED LAYER - [ind chem] couche *f* pigmentée
PIGMENTING - [dyes] pigmentation *f*
~CAPACITY - [ind chem] (the rate of absorption of pigments) capacité *f* de pigmentation
~MACHINE - [mech] machine *f* à enduire
PIGS - [metall] gueuses *fpl* de fonte
PIGSKIN - [leather ind] (the skin of a wild boar) peau *f* de porc
~PICKER - [text] taquet *m* en peau de porc
PIGSTY - [agric] (a pen for pigs) porcherie *f*
[mining] (crib made of timber) pile *f* de bois
~TIMBERING - [mining] soutènement *m* par piles de bois
PIGTAIL - [gen] queue *f*, natte *f*
[el] câble *m* d'arrivée
PIGWEED - [bot] ansérine *f*
PIKE - [zool] brochet *m*
[impl] pic *m*, pioche *f*
[gen] (only USA) barrière *f* de péage
~POLE - [constr] pique *f*, haste *f*
PILASTER - [constr] (square pier engaged in a wall) pilastre *m*
PILBARITE - [min] (thorium, uranium, silicium and lead containing mineral) pilbarite *f*
PILE, to - [gen] entasser, mettre en pile, amasser
[constr] consolider avec pilotis, soutenir au moyen de pilots, empiler

PILE - [gen] tas *m*, amas *m*
[el] (a device for generating electric current) pile *f*
[constr] (column sunk into the ground to support a vertical loading) pieu *m*, pilot *m*
[text] poil *m*
[metall] (of wrought iron) paquet *m*
[print] (of sheets) pile *f* (de papier)
[med] (haemorrhoid) hémorroïde *f*
~BENT - [soil] groupe *m* de pieux
~BODY - [soil] fût *m* du pieu
~BRIDGE - [constr] pont *m* sur pilots
~CAISSON - [hydr] caisson *m*
~CAP - [constr] (horizontal beam over a row of piles) chapeau *m* de pieu, coiffe *f* de pieu
[soil] (driving cap) casque *m* de pieu
~CARPET - [text] tapis *m* velouté
~CUTTER - [text] dispositif *m* de coupage du poil
~DRAWER - [constr] (mechanism designed to extract driven piles) arrache-pieux *m*
~DRIVER - [constr] (heavy weight running in upright guides) sonnette *f*, bélier *m*, hie *f*
~-DRIVING - [constr] battage *m* de pilots, pilotage *m*
~DRIVING BY VIBRATION - [soil] fonçage *m* de pieux par vibration
~DRIVING RIG - [soil] sonnette *f* de battage
~ENGINE - s. pile driver
~EXTRACTOR - s. pile drawer
~FABRIC - [text] tissu *m* à poil
~FACTOR - [nucl] facteur *m* de pile
~FRAMING - [constr] patins *mpl* de pilotis
~GUN - [nucl] sonde *f* de réacteur
~HOOP - [constr] (the metal band fitted to the head of a pile) frette *f* de pilot
~LAYOUT - [soil] réseau *m* de pieux
~LOADING TEST - [soil] essai *m* de chargement *m* d'un pieu
~OF PLATES - [opt] (device for producing plane-polarized light) colonne *f* de plaques
~OSCILLATOR - [nucl] (or reactor oscillator; arrangement designed to maintain a neutro-absorbing body in periodic motion in a nuclear reactor) oscillateur *m* de pile
~PATTERN - s. pile layout
~PERIOD - [nucl] (the reciprocal of the time derivative of the natural logarithm of the reactor power) période *f*
~PICK - [text] trame *f* de poil
~PIER - [constr] (pier supported on piles) môle *m* sur pilots
~PLANK - [constr] palplanche *f*
~PLANKING - [constr] rideau *m* de palplanches
~POINT - [constr] pointe *f* de pilotis
~POISONING - [nucl] (poisoning of the reactor) empoisonnement *m* du réacteur
~PULLER - s. pile drawer
~SHAFT - s. pile body
~SHOE - [constr] (the steel point fitted to the foot of a pile) sabet *m* de pilot, lardoire *f*
~STEM - s. pile body
~THREAD - [text] fil *m* de poil
~TOE - [soil] pointe *f* d'un pieu
~-UP - [radiat] (in a radiation detector) empilement *m* d'impulsions
~WARP - [text] chaîne *f* de poil
~WEFT - [text] duite *f* de chenille, trame *f* de poil
~WIRE - [text] fer *m* à velours
PILED-UP GROUP - [cryst] encombrement *m* de dislocations
PILEUS - [met] (accessory cloud of cap- or hood-like shape rising from the upper part of a cumuliform cloud) pileus *m*
PILEWORT - [bot] ficaire *f*
PILGER MILL - [metall] laminoir *m* à pas de pélerin
~ROLL - [metall] cylindre *m* de laminoir à pas de pélerin
PILATION - [physiol] formation *f* des cheveux
PILIFEROUS - [bot] (ending in a hair-like point) pilifère
~LAYER - [bot] (the outermost cell layer of a young root) couche *f* pilifère
PILIMICTION- [med] pilimiction *f*
PILING - [constr] (a set of piles) pilotis *mpl*, pieux *mpl*
(the operation of sinking the piles) mise *f* en pile, pilotage *m*, empilage *m*
[paint] (a defect) entassement *m*
[constr etc] armature *f*
[metall] paquetage *m*, empilage *m*
[constr] battage *m* de pieux
[mining] poussage *m* par palplanches
~HAMMER - [impl] sonnette *f* à main
~-UP - [mech] (deposition of petrol in the induction manifold of a petrol-engine) entassement *m* d'essence
~WALL - [soil] batardeau *m* de palplanches
PILL - [med] (any medicinal substance put up in a pellet) pilule *f*
~PRESS - [mech] presse *f* à briquettes, presse *f* à comprimés
[metall] poussage *m* par palplanches
PILLAR, to - [gen constr etc] soutenir avec des piliers
PILLAR - [constr] (column or pier supporting a structure) pilier *m*, montant *m*, colonne *f*, pied-droit *m*
[el] (structure of pillar form for switches etc) montant *m*, colonne *f*
[el] (of a brush) pivot *m* de porte-balais
[horol] (a cylindrical piece of metal acting as distance piece) chandelle *f*
[mech] (rod part designed to fix flat parts at a certain separation, as between the plates of a clock movement) colonne *f*, montant *m*
[min] (any mass of ore supporting the roof of a mine) pilier *m*, massif *m*
[shipbuild] (upright bar of iron) épontille *f*, étançon *m*
[mech] (of a bicycle) tige *f* de selle
~-AND-ROOM - [mining] exploitation *f* par chambres et piliers
~-AND-STALL WORK - [min] méthode *f* des piliers et galeries, méthode *f* des massifs courts
~BLOCK - [mech] (or pillow block) colonne *f*
~BOX - [telecomm] (a letter box) boîte *f* au lettres, borne *f* postale
~COAL - [mining] charbon *m* en piliers
~CRANE - [mech] (crane rotating about a fixed pillar) grue *f* à colonne
~DIE SET - [metall] monture *f* d'estampe à guidage à colonnes
~DRAWING - [min] (the removal of ore pillars in a mine on completion of the work) enlèvement *m* des piliers, dépilage *m* (des piliers)
~DRILL - [mech] (drilling machine with spindle and table supported by brackets carried by a pillar) ma-

chine *f* à percer montée sur colonne
PILLAR FILE - [impl] lime *f* plate à côtés lisses
~MINING - [mining] exploitation *f* par piliers abandonnés
~OF THE HOLD - [naut] étance *f*
~PRESS - [mec] (column press) presse *f* à colonne
~ROBBING - s. pillar drawing
~STAND - [mech] colonne *f*
~WORKING - s. pillar drawing
PILLARING - [mining] exploitation *f* par piliers
[metall] colonne *f* froide
PILLARS OF THE ABDOMINAL RING - [anat] piliers *mpl* de l'anneau inguinal
PILLION - [gen] (the pad placed at the back of a horse, behind the saddle, so that a second person may ride) coussinet *m* de cheval
[transp] (on a motorbicycle etc) siège *m* arrière, selle *f* tandem
~SEAT - [transp] (of a motorcycle) selle *f* de passager
PILLOW - [gen] oreiller *m*
[mech] coussinet *m*, dé *m*, grain *m*
~BLOCK - [mech] palier-support *m*
~BUSH - [mech] coussinet *m*, dé *m*
PILOCARPINE - [chem] (an alkaloid derived from varieties of Pilocarpus and used in medicine) pilocarpine *f*
PILOMOTOR - [anat] (which causes movements of the hair) pilomoteur *m*
PILOSIS - [med] trichauxis *f*, hirsutisme *m*
PILOT, to - [gen] conduire
[aero & naut] piloter
[mech] guider
PILOT - [aero] (the person who controls the movements of an aircraft) pilote *m*, pilote *m* aviateur
[naut] pilote *m*
[el] (pilot wire, auxiliary line for measurements, signals etc in an electrical network) pilote *m*, fil *m* pilote
[radio] (a pilot signal) onde *f* pilote
[mach tool] partie *f* de guidage
[railw] chasse-bestiaux *m*
[th eng] (burners) s. pilot burner
[el] lampe *f* témoin
~-BALLOON - [met] (a small free balloon used to give information on wind currents by direct observation from the ground) ballon *m* pilote
~BAR - [mach tool] barre *f* de guidage
~BEARING -[mech] roulement *m* pilote
~-BIT - [tool] mèche *f* pilote
~BOAT - [naut] bateau-pilote *m*
~BOSS - [mech] bossage *m* de guide
~BRUSH - [el] (a brush used to measure the voltage between adjacent commutator bars) balais *m* d'essai
~BURNER - [th eng] (a small burner kept permanently alight, in order to initiate or ensure the maintenance of combustion in the main burner) veilleuse *f*
~BUSHING - [auto] (the journal bearing sleeve) bague *f* pilote
~CABLE - [telecomm] (auxiliary cable intended for telecommunication in an electric network) câble *m* pilote
~CARRIER - [radio] porteuse *f* pilote
~CELL - [el] (a cell in a battery selected as a type of the condition of the whole battery) élément *m* pilote
PILOT CHANNEL - [radio] (channel over which a pilot wave is transmitted) canal *m* d'onde pilote
~CHART - [naut] carte *f* de route
~CHUTE - s. pilote parachute
~CIRCUIT - [contr] (of a control system) circuit *m* pilote
~CONTROLLER - [el] (or master controller; multi-way switch controlling the operation of a set of contractors) combinateur *m* pilote
~ENGINE - [railw] (locomotive preceding a train) locomotive *f* pilote
~FARM - [agric] ferme *f* expérimentale, ferme-pilote *f*
~FATIGUE - [med] fatigue *f* psichosomatique des pilotes
~FLAME BURNER - [th eng] brûleur *m* à flamme pilote
~-HOUSE - [naut] (the structure, generally in the forward part of the ship, containing the steering wheel and compass) kiosque *m* de barre, kiosque *m* de navigation
~JET - [mech] (in internal combustion engines) gicleur *m* principal
~LAMP - s. pilot light
~LIGHT - [el] (a small lamp to show that a given circuit is alive) lampe *f* témoin, voyant *m*
~OPERATION - [comput] opération *f* de pilotage
~PARACHUTE - [aero] (small parachute assisting the opening of the main parachute) parachute *m* auxiliaire
~PIN - [cin] (a component part of printing apparatuses) cheville *f* de guidage, doigt *m* d'entraînement
~PLANT - [gen] (a small installation set up to a new process on a reduced, but still and industrial scale) installation-pilote *f*
~PROTECTION - [el] protection *f* par pilote
~PULSE - [radar] (large signal on A scope caused by the transmitted pulse) impulsion *f* pilote
~RELAY - [el] relais *m* pilote
~SIGNAL - [el] signal *m* de commande
~SPARK - [electron] (low-power spark creating ionization and thus preparing the way for a larger discharge) décharge *f* préliminaire
~SPREAKER - [el acoust] (control loudspeaker in a recording room) haut-parleur *m* témoin
~STICK - [aero] (the main control column of a helicopter) levier *m* de commande
~SWITCH - [el] interrupteur *m* pilote
~VALVE - [mech] (a relay valve, controlling a supply of oil under pressure to the piston of a servomotor) soupape *f* pilote
~WINE - [railw] (of a locomotive) roue *f* porteuse d'avant
~WIRE - s. pilot
~WIRE PROTECTION - [el] protection *f* par fils pilotes
~WIRE REGULATOR - [radio] (automatic device designed to control adjustable gain or loss) régulateur *m* à fil pilote
~WORKING - [railw] exploitation *f* par pilotage
PILOTAGE - [gen] pilotage *m*
[naut] droits *mpl* de pilotage
[aero] (contact flying) vol *m* par contact
PILOTLESS - [aero] sans pilote
PIMELIC ACID - [chem] (saturated dibasic acid of the oxalic acid series) acide *m* pimélique
PIMENTO - [bot] piment *m*

PIMPLE - [gen] bouton *m*, échauboulure *f*
[med] pustule *f*, babouin *m*
[el acoust] (a defect in a record) soufflure *f*
[metall] soufflure *f*
~METAL - [metall] matte *f* à soufflures
PIMPLES - [metall, plast ind etc] (surface flaws in a moulded article) soufflures *f*pl, granulation *f* superficielle
PIMPLING - s. pimples
[metall] granulation *f* superficielle
[nucl] (of a canned fuel element) gondolage *m*
PIN, to - [gen] épingler
[mech] fixer avec goupilles
PIN - [gen] épingle *f*
[mech] goupille *f*, cheville *f*, clavette *f*
[carp] broche *f*, cheville *f*, pivot *m*, tenon *m*
[mech] (of a lock) boulon *m*
[mech] (of a crank) clou *m*
[mech] (of a pulley) axe *m*
[draw] goupille *f*
[metall] (foundry work) (of a flask) broche *f*, goujon *m* (de châssis de fonderie)
[glass man] (of a mould) broche *f* du moule
[el] broche *f*, fiche *f*
[horol] goupille *f*
[firearms] percuteur *m*, aiguille *f* de fusil
[electron] (the metal pin connected with the electrodes of an electronic tube) broche *f*
~AND ARC INDICATOR - [radio] (a type of beam direction indicator) indicateur *m* géometrique
[radiat] rapporteur *m* d'angle
~BAR - [text] barre *f* à aiguilles
~BAR GRID - [text] grille *f* des barres à pointes
~BEARING - [auto] coussinet *m* à aiguilles
~BIT - [tool] foret *m* à téton cylindrique
~BOARD - [comput] tableau *m* de connexions à fiches
~BOX - [text] boîte *f* à aiguilles
~CHAIN - [mech] (short retaining chain attached to removable pins to avoid losing them when detached) chaînette *f* de fixation, chaîne *f* à fuseau
~CIRCLIP - [mech] (in internal combustion engines) bague *f* élastique
~COP - [text] (small package of yarn of a convenient size for a loom shuttle) cannette *f* trame
~CYLINDER - [text] tambour *m* garni d'aiguilles
~DISC - [text] disque *m* à goupilles
~DOWN, to - [gen] fixer
~DRILL - [tool] foret *m* à téton cylindrique
~ESCAPEMENT - [horol] échappement *m* à cheville
~EXPANSION TEST - [metall] essai *m* d'expansion à goupille conique
~EXTRACTOR - [impl] tire-goupille *m*
~-FEED DRUM - [comput] (in printing) cylindre *m* de entraînement par ergots
~FEED PLATEN DEVICE - [comput] dispositif *m* d'entraînement à griffes
~GEAR - [mech] roue *f* à goupilles
~HINGE - [constr] penture *f*
~-HOLE - s. pinhole
~-INSULATOR - [el] (insulator rigidly mounted on a pin) isolateur *m* rigide
~JOINT - [mech] assemblage *m* par goupilles
~-JOINTED LINKAGE - [mech] transmission *f* à articulation rotative
~KEY - [mech] clef *f* bénarde
~LEASE - [text] envergeure *f* du bas
~LIFT - [metall] démoulage *m* sur chandelles
PIN LIFTING THRESHER - [agric] batteuse *f* à pointes
~MOVEMENT - [cin] (movement caused by the claws engaging the perforation) système *m* à griffes
~PALLET ESCAPEMENT - [horol] échappement *m* à chevilles
~PICKER - [text] chasse-navettes *m* à chevilles
~PLUG - [el] tenon *m*
~-POINT GATE - [metall & plast ind] (an orifice through which material enters a mould cavity, less than 30/I000" dia.) orifice *m* en tête d'épingle
~PUNCH - [mech] repoussoir *m*, chasse-goupille *m*
~PUNCHING REGISTER - [comput] registre *m* à poinçon
~PUNCHING SYSTEM - [comput] système *m* à poinçon
~RAMMER - [metall] pilette *f*
~SPACING - [electron] distance *f* entre broches
~SPANNER - [impl] clef *f* à griffes
~STENTER - [text] rameuse *f* à picots
~STRIPE - [text] filet *m*
~TONGS -[horol] (hand-vices) étau *m* à main, étau *m* queue
~VALVE - [mech] pointeau *m*
~-VICE - s. pin tongs
~WEIR - [hydr] baragge *m* à aiguilles
~WHEEL - [horol] (of a striking clock) roue *f* chevilles
[mech] roue *f* à fuseaux, hérisson *m*
~-WHEEL GEAR - [mech] roue *f* à taquets, roue *f* à chevilles
PINACOID - [min] pinacoïde *m*
PINACOLINE - [chem] (methyl buthyl ketone) pinacoline *f*
PINACONE - [chem] (tetramethyl-ethylene glycol) pinacone *f*
PINAFORE - [text] tablier *m*
PINASTER - [bot] pinastre *m*, pin *m* maritime
PINBOARD - [electron] panneau *m* de commutation
~PROGRAMMING - [comput] (the operation of preparing a programme on a board, in which the instructions are given by means of plug-in pins) programmation *f* à tableau de connexions à fiches
PINCEMENT - [med] pincement *m*
PINCERS - [zool] pince *f*
[impl] tenaille *f*
[carp] tricoise *f*
PINCH, to - [gen] pincer
[min] (to become narrow) serrer, se rétrécir
PINCH - [gen] pincement *m*
[min] (narrow tapering section of a vein) étreinte *f*, étranglement *m*, serrée, serrement *m*
[impl] s. pinch bar
[metall] repliure *f* de laminage, onde *f*
[mech] encombrement *m*
[electron] (the part of the envelope of an electronic tube carrying the electrodes and through which pass the connexions to the electrodes) effet *m* de compression, pincement *m*
[geol] pli *m* longitudinal
~BAR - [impl] (crowbar used to move forward heavy objects) pince *f* à talon, levier *m* à griffe, arcanseur *m*
~BASE - [electron] (of an electronic tube) embase *f* en pincement
~-COCK - [impl] (manually-adjustable clamp used to compress a flexible pipe to control the flow of a fluid) pince *f* d'arrêt
~DOG - [mech] griffe *f*
~EFFECT - [el] (the constriction occurring when a

liquid conductor, e.g. mercury carries a heavy current, caused by magnetic attraction between the elements of the conducting material) effet *m* de compression
PINCH EFFECT - [nucl] (effect obtained by a doughnut in which a large current is built up in the tube) effet *m* de striction
[el acoust] (in disk recording) effet *m* de pince, irrégularité *f* de coupe
~OFF, to - [agric] évriller
~OFF - [electron] fréquence *f* de coupure
~PASS - [metall] dressage *m*, écrouissage *m*
~PASS MILL - [metall] laminoir *m* à dresser
~ROLL - [mech] (a roller used to draw sheet material away from a machine or process unit after treatment) cylindre *m* d'entraînement
~ROLLS - [metall] cylindres *mpl* de tension
PINCHCOCK - s. pinch-cock
PINCHING - [radio] (blocking of an electronic tube in a receiver) blocage *m*
PINCUSHION - [gen] pélote *f* à épingles
~DISTORTION - [telev] (picture distortion, in which the sides bulge out like a pincushion) distorsion *f* en coussinet, distorsion *f* en croissant
PINE - [bot] pin *m*
~CONE - [bot] pomme *f* de pin, cône *m* de pin, pigne *f*
~NEEDLE YARN - [text] fil *m* de laine de bois
~OIL - [chem] (a natural oil derived from Pinus palustris and having application as a disinfectant) essence *f* de pin
~SEED- [bot] noyau *m* de pin
~STRAWBERRY - [bot] fraise *f* ananas
~TAR - [chem] (viscous, dark brown liquid obtained by the destructive distillation of species of Pinus and having application in medicine, paint manufacture, metallurgy, plastics, and as a timber preservative) goudron *m* de pin, goudron *m* de bois
~WOOD - [bot] bois *m* conifère, bois *m* de pin
~WOOL - [text] laine *f* de bois
PINEAL BODY - s. pineal gland
~GLAND - [zool] glande *f* pinéale
PINEALOBLASTOMA - s. pinealoma
PINEALOMA - [med](tumor of the pineal gland) pinéalome *m*
PINEAPPLE - [bot] ananas *m*
~FIBRE - [text] fibre *f* de l'ananas
~LEAF FIBRE - [text] fibre *f* de l'ananas
~WINDING - [text] bobine *f* biconique
PINENE - [chem] (a solvent and a synthesis intermediate for resins and perfumery agents) pinène *m*
PING, to - [mech] (in internal combustion engines) (to knock) cogner, taper
PING - [mech] (in internal combustion engines) cognement *m*, détonation *f*
[radar] (pulse signal from echo-ranging sonar apparatus) impulsion *f* de sonar
PINGING - s. ping [mech]
PINGUID - [gen] (oily or greasy) gras, graisseux
PINHEAD - [gen] tête *f* d'épingle
PINHOLE - [gen] trou *m* de cheville, trou *m* de goujon
[metall] (die casting, a defect) piqûre *f*
[plast ind] (a flaw in a calendered thermoplastic sheet) trou *m* d'épingle, piqûre *f*
[paint] (a defect) picoture *f*
[photo] (lack of density) piqûre *f*
[constr] trou *m*
[opt] petite ouverture *f*
PINHOLE CAMERA - [photo] (camera used to determine the dimensions of the focal spot of an X-ray tube, by projecting it through a small hole on a film) sténoscope *m*
~DISK - [opt] ophtalmoscope *m*
~IMAGE - [opt] photographie *f* sans objectif
~POROSITY - (a defect) porosité *f* à piqûres
PINHOLES - [text] (in knitted fabrics) défaut *m* de maille
[paint] s. pinholing
[paper man] (a defect) marque *f* du fondeur
[metall] piqûres *fpl* de scorie
PINHOLING - [paint] (defect in a coating consisting of the presence of very small holes, due to a variety of causes) piquage *m*
PINIFORM - [med] conique
PINION - [mech] (a toothed gear, usually the smaller of a pair of gears) pignon *m*
~CALLIPERS - [horol] calibre *m* à compas
~HOUSING - [mech] (in a rolling mill) cage *f* à pignons
~LEAF - [horol] (a tooth of the pinion) dent *f* de pignon
~STAND - [mech] cage *f* à pignons
~STEEL - [metall] acier *m* à pignons
~TRUNNION - [mech] tourillon *m* à pignon
~WALL - [th eng] (a pier wall) culée *f*
~WIRE - [horol] fil *m* pour pignons
PINITE - [min] (hydrous silicate of aluminium and potassium) pinite *f*
PINK, to - [gen] percer, denteler, évider
[paint] teindre en rose
PINK - [gen] rose *m*
[naut] (sailing vessel with a narrow stern) pinque *f*
[bot] oeillet *m*
[zool] saumoneau *m*, véron *m*
~EYE - [vet] ophtalmie *f* périodique
~LEAF STALK - [bot] pétiole *f* rosée
~SALT - [chem] chlorure *m* d'étain ammoniacal
~SALT BATH - [text] bain *m* de chlorure d'étain ammoniacal
~SILK - [text] soie *f* rose
PINKED SILK - [text] soie *f* chlorurée
PINKIE - [naut] s. pink
PINKING - [mech] (high-pitched periodic sound in i.c. reciprocating engines caused by pre-ignition or detonation) cliquetis *m*
[mech] (knocking) cliquetis *m*
[text] dentelage *m*
~IRON - [leather ind] fer *m* à découper
~MACHINE - [text] machine *f* à denteler
~SHEARS - [impl] découpoir *m* à figures
PINKY - [gen] rosé, rosâtre
[naut] s. pink
PINNACE - [naut] (small, light sailing vessel) pinasse *m*
[naut] (an eight-oared boat carried on-board a man-of-war) grand canot *m*
PINNACLE - [arch] pinacle *m*, clocheton *m*
PINNED SWIVEL SHUTTLE - [text] navette *f* circulaire à chevilles
~VERTICAL FEEDING LATTICE - [text] tablier *m* alimentaire vertical à dents
PINNING - [text] embrochement *m*
~OF THE COMB STRIP - [text] peuplement *m* de la barrette à aiguilles

PINPOINT, to - [gen] indiquer exactement
PINPOINT - [nav] (a precise position marked on a map or chart) position *f* indentifiée
~FLAME - [th eng] (burners) flamme *f* en forme de dard
PINT - [meas] (a liquid measure; 0.568 l) pinte *f*
PINTA - [med] (tropical skin disease) pinta *m*, mal *m* de pinto
PINTLE - [mech] (the pin of a hinge) pivot *m* central
[auto] (hook used in towing) cheville *f* ouvrière
[naut] (metal brace on which the rudder swings) aiguillot *m*, vitonnière *f*
[mech] (the needle of an oil-engine injector valve) aiguille *f*
[railw] (the king-pin of a wagon) cheville *f* ouvrière
~-CHAIN - [mech] chaîne *f* à fuseau
~HOOK - [railw & auto] crochet *m* de cheville ouvrière
~INJECTOR - [mech] (the needle of an oil-engine injection valve, opened by the oil pressure on an annular face) injecteur *m* à téton
PIONEERING - [agric] défrichement *m*, essouchage *m*
PIOSCOPE - [instr] (instrument designed to determine the fat contents of milk by colorimetry) butyromètre *m*
PIP - [bot] pépin *m*
[mech] (a small projection frequently caused by operations on a lathe) pointe *f*
[radar] (or blip; pip-shaped range mark produced on the screen) top *m* d'écho
[electron] (or tip; a small projection on the envelope resulting from the sealing of it after evacuation) pointe *f*
PIPAGE - [gen] (system of pipes) tubulures *fpl*
[transp] (the transport of oil, water, gas etc through pipes) transport *m* par canalisation
[comm] frais *mpl* de transport par canalisation
PIPE, to - [mus] jouer de la flûte
[transp] transporter par canalisation
[radio] (to send a programme over wires) télétransmettre
[naut] (bringing about by piping) siffler
[metall] (to form a pip) retasser
[mining] hydrauliquer
[oil ind] refouler
PIPE - [gen] tuyau *m*, tube *m*, conduit *m*
[transp] (a long conducting passage for the transport of a fluid) conduite *f*, canal *m*
[metall] (depression formed in the top of an ingot due to unequal contraction on freezing) retassure *f*
[mus] (of an organ) tuyau *m* d'orgue
[horol] moyeu *m*
[glass ind] (a blowtube) tube *m* de souffleur
[min] colonne *f* (de richesse)
[metall] (foundry) cuiller *m*
[metall] (funnel) retassement *m*
~BED - [plumb] (in mainlaying work) lit *m* de pose
~BELL - [plumb] manchon *m*
~BEND - [plumb] (a short curved length of pipe, made as a fitting to join two straight lengths) coude *m*
~BENDER - [mech] (tool or machine for bending pipe without impairing its circular action) machine *f* à cintrer les tuyaux
~BENDING - [plumb] cintrage *m* des tuyaux
~BENDING MACHINE - s. pipe bender
~BOX - [plumb] châssis *m* à tuyaux
~BRACKET - [plumb] cavalier *m*
PIPE BRACKET - [plumb] (open bracket) gâche *f* à vis
~BREAKAGE - [plumb] rupture *f* de canalisation
~-BRUSH - [impl] (pig fitted with wire brush) hérisson *m*
~BURNER - [th eng] brûleur *m* rectiligne
~BURST - [plumb] rupture *f* d'un tuyau
~BUTT WELDING - [plumb] soudure *f* par encollage des tuyaux
~CHAPLET - [metall] support *m* de noyau
~CHOCKING - [oil ind] encrassement *m* d'un tuyau
~CLAMP - [mech] (a special type of curved spike for securing pipes to fixed objects) collier *m* de retenue, frein *m* de retenue
~-CLAY - [min] terre *f* de pipe, scoulérite *f*
~CLIP - [plumb] griffe *f* pour tuyaux
~COIL - [plumb] serpentin *m*
~CONNECTIONS - [plumb] tuyauterie *f*
~COUPLING - [plumb] (a short collar with female threads at both ends) raccord *m* de tuyau, manchon *m* pour tuyaux, lunette *f*
~CUTTER - [mech] coupe-tuyaux, coupe-tube *m*
~DIE - [mech] filière *f* pour tubes
~DOG - [impl] clef *m* à tubes
~DUCT - [constr] fourreau *m*
~EXTRUDER - [plast ind] presse *f* à extruder les tubes
~EXTRUSION DIE - [plast ind] (a special die used to form continuous extruded plastic piping) filière *f* pour tubes
~FITTER - [plumb] (a workman skilled in pipework, espec. iron, steel and copper pipes) tuyauteur *m*, installateur *m* de tuyauterie
~FITTING - [plumb] raccord *m*
~FITTINGS - [plumb] (collective term for sockets etc. used in pipe work) raccords *mpl*
~FORMER - [mech] (a hollow part used to limit the external size of extruded pipe) manchon *m* calibré
~FORMING - [plast ind] roulage *m* de tubes (à partir de plaques)
~FRACTURE - s. pipe breakage
~GRAB - [impl] accroche-tube *m*
~GRIP - [tool] (tool for turning a pipe about its axis) pince *f* à tube
~HANGER - s. pipe hook
~-HANGER BUFFER - [plumb] amortisseur *m* pour gâche à scellement
~HAULING - [transp] transport *m* de tuyauterie
~HEATER - [th eng] étuve *f* à tubes chauffants
~HOOK - [plumb] grappin *m*, crochet-étrier *m*
~INSTALLATION - [plumb] tuyauterie *f*
~INSULATION - [plumb] isolement *m* de tuyaux
~-JOINTS - [plumb] assemblages *mpl* de tuyauterie
~LAGGING - [plumb] enveloppe *f* de tuyaux
~LAYER - [mech] (sideboom tractor used for mainlaying) tracteur *m* à grue latérale
~-LAYING - [plumb] pose *f* de tuyaux
~LINE - s. pipeline
~LINER - [mech] tuyau *m* de revêtement
~LINING - [plumb] enveloppe *f*
~LOCATOR - [instr] (a metal detector) détecteur *m* de conduite
~MOULDING - [metall] (production of cast-iron pipes by moulding in green sand) roulage *m* de tubes
~NIPPLE - [plumb] mamelon *m*, raccord *m*
~OPENER - [plumb] appareil *m* à mandriner les tubes
[oil ind] machine *f* à élargir les tubes

PIPE PINCERS - [impl] pince *f* à tuyaux
~PLUG - [plumb] tournant *m* à vis
~PROVER - [instr] appareil *m* à vérifier l'étanchéité des tubes
~PUSHER - [mech] perforatrice *f*
~RACK - [oil ind] parc *m* à tiges et tuyaux
~RAMP - [oil ind] couloir *m* de la tour
~REAMER - [mach tool] fraise *f* conique mâle
~REDUCER - [plumb] réducteur *m* de tuyau
~RESONANCE - [acoust] (the acoustic resonance of a pipe) résonance *f* acoustique d'un tube
~RESONATOR - [acoust] (acoustic resonator in the form of a pipe) résonateur *m* à tube
~SADDLE - [plumb] collier *m* de branchement
~SECTION - [plumb] troncon *m* de tube
~SET BACK - [oil ind] gerbage *m*
~SLEEVE - [plumb] fourreau *m*
~SLICK - [metall] lissoir *m* à tuyaux
~SOCKET - [plumb] manchon *m* élargi, emboîtement
~SQUEEZER - [plumb] pince-tube *m* hydraulique, écrase-tube *m*
~STAND - [plumb] support *m*, collier *m*
~STILL - [oil ind] (type of crude oil still in which the crude oil is fed through externally heated pipe coils) alambic *m* tubulaire
~STRAIGHTENER - [mech] banc *m* de redressage de tubes
~STRAP - [impl] patte *f* à crochet
~SUPPORT - [plumb] support *m*
~SUPPORTING ROLLER - [plumb] support *m* à rouleau
~TAP, to - [mech] tarauder
~TAP - [plumb] taraud *m* pour tuyauteries
~THICKNESS - s. pipe wall thickness
~THREAD - [mech] (special screw thread standard used for pipes) filetage *m* de tuyaux
~THREADER - [mech] (a pipe die) machine *f* à tarauder les tuyaux, filière *f*
~THREADING MACHINE - s. pipe threader
~TRAIN - [plast ind] (the whole of a pipe-extrusion assembly, from extruder to cutter) tuyauterie *f*
~-TYPE CABLE - [el] câble *m* en tube
~UNION - [mech] (screw device for connecting pipes to other parts) raccord *m* de tuyaux
~VENTILATED - [el] à ventilation par canaux
~VEIN - [min] colonne *f* de richesse
~VICE - [tool] étau *m* à tubes
~VISE - s. pipe vice
~WALL - [plumb] paroi *f* du tuyau
~WALL THICKNESS - [plumb] épaisseur *f* de tube
~WELL - [constr] source *f* à tube *m*
~WRENCH - [tool] clef *m* pour tuyaux, clé *f* à tubes
~YARD - [oil ind] parc-tuyauterie *m*
PIPECLAY TRIANGLE - [ind chem] (support for flask or the like consisting of three pieces of wire twisted together at the ends to form a hollow triangle, on which are threaded thickwalled pipe-clay tubes) triangle *m* en fil de fer
PIPED KEY - [mech] clef *f* forée
~-LOCK - [mech] serrure *f* à broche
~PROGRAMME - [radio] (wire broadcasting) télédiffusion *f*
~TELEVISION - [telev] (community television) programme *m* diffusé par téléphone
~VEHICLE - [railw] véhicule *m* à conduite blanche
PIPELAYING - s. pipe-laying
PIPELINE - [transp] canalisation *f*, pipeline *f*
[transp] (for gas) conduite *f*, gazoduc *m*, pipeline *f*
PIPELINE - [transp] (for oil) pipeline *f*
[transp] (for natural gas) conduite *f* de gaz naturel
[transp] (a transmission main, or trunk main) conduite *f*
~CLEANER - [impl] appareillage *m* à curer les pipelines
~GAS - [gas ind] (grid, or long-distance gas) gaz *m* de réseau de transport
~MARKER - [constr] (a marker-post) borne *f* de jalonnement
~NETWORK - [plumb] réseau *m* de conduites
~RUN - [oil ind] (the quantity actually delivered to a pipeline) passe *f*
PIPELINING - [gen] (the laying of a pipeline) pose *f* des conduites
[transp] transport *m* par pipeline
PIPER - [min] (a fracture) soufflard *m* de grisou
PIPERAZINE - [chem] (diethylene-diamine. It has the useful medicinal property of forming soluble salts which uric) pipérazine *f*
~HYDRATE - [chem] (an antheimintic used in medicine) hydrate *m* de pipérazine
~OESTRONE SULPHATE - [chem] (a synthesis oestrogen conjugate with therapeutic properties) sulfate *m* de pipérazine-oestrone
PIPERIDINE - [chem] (a solvent and synthesis intermediate) pipéridine *f*
PIPERIDINOETHANOL - [chem] (a synthesis intermediate) pipéridinoéthanol *m*
PIPEROCAINE HYDROCHLORIDE - [chem] (a local anaesthetic) chlorhydrate *m* de pipérocaïne
PIPERONAL - [chem] (methylene-protocatechuic aldehyde. Used as a scent under the name of heliotropin) pipéronal *m*
PIPEROXANE HYDROCHLORIDE - [chem] (a diagnostic agent in medicine) chlorhydrate *m* de pipéroxane
PIPERYLENE - [chem] (an intermediate for polymers and synthetic resins) pipérylène *m*
PIPETTE - [ind chem] (a glass tube with a bulb in the middle and a jet at one end, for drawing up small samples of liquid by suction) pipette *f*
~STAND - [ind chem] (rack for holding a number of pipettes) étagère *f* à pipettes
~WITH GRADUATED SCALE - [ind chem] pipette *f* graduée
PIPING - [plumb] (any system of pipes) conduite *f*
[transp] réseau *m* de tuyauterie
[metall] (a pip) formation *f* de retassures
[mining] abattage *m* hydraulique, travail *m* à l'eau, abattage *m* à l'eau
~BY HEAVE - [soil] formation *f* de renards par soulèvement
~CORDAGE - [text] cordes *fpl* tressées pour tuyaux
~SYSTEM - [plumb] s. piping
PIQUE' - [text] (a type of good quality cotton fabric) piqué *m*
PIRATE VIEWER DETECTION - [telev] dépistage *m* de téléspectateurs clandestins
PIRIFORM - [med] piriforme
PIRN - [text] (small-size wooden bobbin on a shuttle) cannette *f*, sépoule *f*
~WINDER - s. pirn winding machine
[text] bobinoir *m*
~WINDING MACHINE - [text] cannetière *f*, coconneuse *f*, bobinoir *m*
~WITH CONICAL BASE - [text] bobine *f* à embase conique

PIRN WITH SLIT - [text] fuseau *m* fendu
PIROGUE - [naut] pirogue *f*
PIROPLASMOSIS - [vet] (an infection of the blood) piroplasmose *f* bovine, babésiose *f* bovine
PISCES - [astr] poissons *mpl*
PISCINA - [arch] (the stone basin, generally in a niche, in which the priest washes the chalice after communion) piscine *f*
[gen] vivier *m*
PISCIVOROUS - [zool] (fish-eating) piscivore
PISIKITE - [min] (essentially a niobatetantalatetitanate) pisikite *f*
PISIFORM - [zool] (shaped like a pea) pisiforme
~BONE - [anat] os *m* pisiforme
PISOLITE - [geol] (a type of limestone) pisolithe *f*
PISOLITIC - [geol] (said of the structure of certain sedimentary rocks containing pisolite) pisolitique
PISTACHIO - [bot] pistache *f*
~TREE - [agric] pistachier *m*
PISTACITE - [min] (iron-containing mineral occurring in igneous rocks) pistazite *f*
PISTIL - [bot] pistil *m*, dard *m*
PISTILLATE FLOWER - [agric] fleur *f* femelle
PISTILLODE - [bot] (non functional pistill) pistilloïde
PISTOL - [firearms] (a small arm, a revolver) pistolet *m*
[tool] (for spraying) pistolet *m*
~GRIP - [firearms] poignée *mf*
~LIGHT - [el] projecteur *m* à main
~PIPE - [metall] tuyau *m* à pistolet
PISTON - [mech] (the cylindrical part which reciprocates within the cylinder bore and transmits the impulse produced by the exploding charge to the connecting rod) piston *m*
[mech] (of a pump) piston *m*, sabot *m*
[electron] (of a waveguide) piston *m* de court-circuit
~ATTENUATOR - [electron] affaiblisseur *m* à piston
~BLOWER - [mech] soufflante *f*
~BODY - s. piston skirt
~BOSS - [mech] (the thickened part within the body of the piston) moyeu *m* du piston
~BRACING - [gas ind] (gasholders) chevalet *m* du piston
~CARRIAGES - [gas ind] (gasholders) rouleaux *mpl*
~CLEARANCE - [mech] espace *m* nuisible, espace *m* libre, espace *m* neutre
~COMPRESSOR - [mech] (type of positive displacement compressor in which a piston acts on the gas in a closed cylinder) compresseur *m* à piston
~CROWN - [mech] tête *f* du piston
~DECKING - [gas ind] (in gasholders; the top of the piston) toit *m* du piston
~DISPLACEMENT - [mech] (the swept volume of a working cylinder) cylindrée *f*
~DRILL - [mining] marteau *m* perforateur
~ELLIPSE - [mech] ovalisation *f* du piston
~ENGINE - [mech] (or reciprocating engine) moteur *m* alternatif
~FRAMING - s. piston bracing
~FRICTION - [mech] frottement *m* du piston
~GAGE - [instr] (only USA; manometric balance, or deadweight pressure gauge in GB) balance *f* manométrique, manomètre *m* à poids mort
~HEAD - [mech] (the face of a piston on which the pressure acts) tête *f* du piston
~LAND - [mech] plateau *m* du piston
PISTON MANOMETER - [instr] manomètre *m* à poids mort
~MEAN SPEED - [mech] vitesse *f* moyenne du piston
~PACKING - [mech] (any arrangement to prevent passing between a piston and the cylinder walls) garniture *f* de piston
~PIN - [mech] (the transverse pin in the piston which connects it to the connecting rod) axe *m* de piston
~-PIN BOSS - s. piston boss
~PIN BUSH - [mech] boîte *f* de l'axe de piston
~PLAY - [mech] jeu *m* du piston
~PUMP - [mech] (type of pump consisting essentially of a piston or pistons reciprocating in a cylinder or cylinders) pompe *f* à piston
~RING - [mech] (elastic ring fitted in a groove in a piston to act as a pressure seal) segment *m* de piston
[mech] (of a steam engine) garniture *f* de piston
~-RING CLEARANCE - [mech] (the space allowed between the ends of a piston-ring to provide for expansion etc) jeu *m* à la coupe d'un segment de piston
~-RING CUT - [mech] (the division between the ends of a piston ring) coupe *f* d'un segment de piston
~-RING FLUTTER - [mech] (an opening-and-closing motion of a piston-ring, caused by irregularity in cylinder bore diameter) tremblement *m* du segment de piston
~-RING GAP - [mech] coupe *f* d'un segment de piston
~-RING GROOVE - [mech] (the groove in a piston in which the piston ring is located) gorge *f* de segment dans le piston
~-RING SCRAPER - [mech] segment *m* râcleur d'huile
~-RING SIDE CLEARANCE - [mech] jeu *m* latéral du segment de piston
~-RING STICKING - [mech] blocage *m* du segment de piston
~ROD - [mech] (the rod connecting the piston of a hydraulic actuator to its linkage) bielle *f*
[mech](of a steam engine) verge *f* de piston
~ROD COLLAR - [mech] collier *m* de tige de piston
~ROD GLAND - [mech] bague *f* de la bielle
~ROD GUIDES - [mech] guides *mpl* de la tige
~ROD KNUCKLE - [mech] articulation *f* de la tige du piston, crosse *f* de piston
~ROD NUT - [mech] écrou *m* de la tige
~SCRAPER RING - [mech] segment *m* râcleur d'huile
~SEIZURE - [mech] grippage *m* du piston
~SKIRT - [mech] (the cylindrical part of a piston at the opposite extremity to the head and below the piston-pin bosses) corps *m* de piston, jupe *f* de piston
~SLAP - [mech] évasement *m* du piston
~SPEED - [mech] vitesse *f* du piston
~STROKE - [mech] (the distance travelled by a piston between successive points of reversal) course *f* du piston
~SUPERCHARGER - [mech] (type of supercharger in which compression is effected by pistons moving in cylinders) compresseur *m* à piston
~TRAVEL - [mech] s. piston stroke
~-TYPE COMPRESSOR - s. piston compressor
~-TYPE GASHOLDER - [gas ind] (a waterless gasholder) gazomètre *m* sec, gazomètre *m* à piston
~-VALVE - [mech] (a cylindrical part having the form of a piston and moving in a bore provided with ports, which it covers or uncovers as it moves) tiroir *m* à piston, tiroir *m* rond
~-VALVE MOTION - [mech] distribution *f* à piston
~VANE - [mech] ailette *f* formant piston

PISTON WITH SLIPPER - [mech] (of a steam engine) piston *m* à patin
PISTONPHONE - [el acoust] (an apparatus with a rigid piston which can be given a reciprocating motion of known frequency and amplitude, thus making it possible to establish a known sound pressure in a small closed chamber) chambre *f* de compression
PIT, to - [gen & metall] piquer, ronger, se piquer
[constr] (the formation of pit) former des fosses
[agric] (to introduce into a pit) ensiler
PIT - [gen] fosse *f*, trou *m*, puits *m*
[gen] (in a theatre) fosse *f*
[anat] creux *m* de l'estomac
[med] (the shaft of a mine) marque *f*, cicatrice *f*
[min] (the shaft of a mine) puits *m*
[min] (the place where minerals are dug) fosse *f*, mine *f*
[rock formation] fouille *f*
[gen] (a cavity) creux *m*
[agric] (artificial cavity where grain etc is stored) silo *m* souterrain
[auto] (in a garage) fosse *f*, emplacement *m*
[metall] trou *m*
[metall] (due to corrosion) piqûre *f*
[metall] (in foundry works) fosse *f* de coulée
[metall] (a soaking furnace) four *m* d'égalisation
[bot] (thin spot in the cell walls if some plants) favéole *f*
[mining] carrière *f*
~BANK - [mining] carreau *m* de la mine
~BAROMETER - [instr] baromètre *m* de mine
~BARRING - [mining] soutènement *m* du puits
~BOTTOM - [gen & mining] fond *m* du puits
~CAGE - [mining] cage *f* de mine
~CASTING - [metall] coulée *f* en fosse
~COAL - [min] houille *f*
~CRATER - [geol] cratère *m* d'effondrement
~-EYE - [min] fond *m* d'un puits
~FRAME - [mining] chevalement *m*
~FURNACE - [metall] (furnace used for heat treatment) four *m* d'égalisation
~GAS - [min] grisou *m*, brisou *m*
~-GEAR - [mining] appareillage *m* de puits
~HEAD - [min] orifice *m* de la fosse, bouche *f* du puits
~HOLDER - [gas ind] (a holder with an underground tank) gazomètre *m* à cuve enterrée
~KILN - [metall] four *m* à coke
~MOULDING - [metall] moulage *m* en fosse
~MOUTH - s. pit head
~PATTERN - [telev] (in video recording) configuration *f* de piqûres
~PLANER - [tool] raboteuse *f* à fosse
~PROP - [min] étai *m* de mine
~SAMPLE - [min] prise *f* d'essai de fonte
~SAND - [min] sable *m* de carrière
~SAW - [tool] scie *f* de long
~-SILO - [build] silo *m* souterrain
~SOAKING - [metall] égalisation *f* thermique
~TOP - [min] orifice *m* du puits
~WOOD - [mining] bois *m* de mines
PITA FIBRE - [text ind] fibre *f* d'agave
~HEMP - s. pita fibre
~PLANT - [bot] agave *m* d'Amérique
PITCH, to - [aero & naut] (the move in such a way as to change the angle which the longitudinal axis makes with the horizontal) tanguer, canarder
PITCH, to - [aero] (when the pitch occurs by the tail) tanguer
[aero] (when the pitch occurs by the nose) plonger du nez
[gen] (to pitch a tent) planter sa tente
[gen] jeter, lancer
[constr] empierrer, paver, établir la fondation
[naut etc] (to coat with pitch) enduire de poix, brayer, poisser
[mech] (to engage) engrener, s'engrener
[brew ind] (the yeast) mettre en levain
[mining] (of a coal mine) jouer dans une clef donnée
PITCH - [ind chem] (the residue of tar distillation) poix *f*
[mech] (of a screw; the distance between the threads) pas *m* d'une vis, filetage *m*
[mech] (of a single-start screw) (the distance between identical points of adjacent threads measured along the axis) pas *m* effectif
[mech] (of a multi-start screw) pas *m* apparent
[mech] (of a gear) (the distance between the centre-lines of adjacent gear teeth measured along the pitch circle) pas *m*
[mech] (of a rivetted joint) (the distance between the centres of adjacent holes in a series of such holes in a rivetted joint) pas *m* réel
[naut] (the pitching motion) tangage *m*
[aero] (the change of the angle made by the longitudinal axis with the horizontal) angle *m* d'inclinaison
[acoust] (the quality of sound which determined its position in the musical scale) hauteur *f* du son
[aero] (of a propeller) (the distance which any given element of a propeller would advance in one complete revolution if it were moving along the helix of angle equal to the blade angle) pas *m* géométrique
[el] (of a commutator) (the number of commutator segments per pole of the machine) pas *m* polaire, écartement *m* angulaire
[constr] (of a roof etc) degré *m* de pente
[railw] (of the sleepers) travelage *m*
[min] (the inclination of a vein of ore) pente *f*, inclinaison *f*
[mech] calibre *m* de cylindres, cannelure *f*
[metall] plan *m* de solidification
~ACCENT - [mus] accent *m* tonique
~ANGLE - [mech] (of a bevel gear) angle *m* du cône primitif
~APEX - [mech] (of a bevel gear) sommet *m* d'engrenage conique
~ARC - [mech] arc *m* d'engrenement
~ATTITUDE - [aero] (the angle which the longitudinal axis of an aircraft makes with the horizontal at any given instant) angle *m* d'inclinaison
~BALL - [text] gâteau *m* de résine
~CHAIN - [mech] chaîne *f* à articulations
[mech] (driving chain formed with internal teeth on the links, which engage teeth of gear wheels over which it runs) chaîne *f* calibrée
~CIRCLE - [mech] (a circle which passes through the centres of holes arranged equidistant from a given point, e.g. in a pipe-flange) cercle *m* primitif, trait *m* de division
~CONE - [mech] (of a bevel gear) cône *m* primitif
~CONTROL - [telev] réglage *m* de la distance inter-

ligne
PITCH CURVE - s. pitch line
~CYLINDER - [mech] (of a spur gear) cylindre *m* primitif
~DIAMETER - [mech] (of a gear) diamètre *m* primitif
~FACE - [constr] (rough stone surface) appareil *m* rustique
~FACTOR - [el] (component factor of the winding factor which takes into account the coil pitch) facteur *m* de raccourcissement
~FIR - [bot] pin *m* raide
~INDICATOR - [instr] (instrument to measure the angle of pitch of an aircraft) indicateur *m* de tangage
[naut] (of a propeller having a controllable pitch) indicateur *m* du pas de l'hélice
~INTERVAL - [acoust] (the pitch relation of two tones) relation *f* entre des hauteurs de son
~LINE - [mech] primitif *m*, cercle *m* de contact
~OF A SOUND VAVE - [acoust] hauteur *f* de son d'une sonde sonore
~OF AN ARCH - [constr] montée *f* d'un arc
~OF COIL - [text] pas *m* de l'enroulement
~OF JACQUARD - [text] nombre *m* de crochets
~OF RIVETS - [mech] pas *m* des rivets
~OF ROOF - [constr] pente *f* de comble
~OF TEETH - [mech] pas *m* des dents
~OF THE REED - [text] écartement *m* des dents
~OF WARP THREADS - [text] écartement *m* des fils de chaîne
~OF WEFT THREADS - [text] compte *m* en trame, duitage *m*
~-ON METAL - [metall] tôle *f* en acier goudronnée
~OPAL - [min] résinite *f*, silex *m* résinite
~ORE - s. pitch blende
~PAYING - [naut] enduit *m* de brai
~PINE - [bot] (strong heavy wood growing in the Southern States of America) pitchpin *m*
~POINT - [mech] (of gears) point *m* de contact des circles primitifs
~POLISHING - [plast ind & glass man] polissage *m* à résine
~RADIUS - [mech] rayon *m* primitif
~RATIO - [mech] (the relation between the pitch of a propeller and its diameter) pas *m* géometrique relatif
~ROOF - [constr] comble *m* à deux égouts, comble *m* à deux versants
~SETTING - [aero] (the blade angle of a propeller of which the pitch can be varied or adjusted measured at the standard radius) calage *m* de pas, réglage *m* de pas
~SURFACE - [mech] (of a gear) surface *f* primitive
~VARIATION - [acoust] (increase or decrease in the height of a tone) variation *f* de hauteur de son
~-VARIATION INDICATOR - [instr] indicateur *m* de variations de hauteur de sons
PITCHBLENDE - [min] (also called Uraninite. A mineral of variable composition, in general of oxides of uranium. A very important source of radium and uranium) pechblende *f*, uranine *f*, pechurane *m*
PITCHED - [constr] (of a road) empierré
~ROOF - [build] comble *m* à deux longs pans
PITCHBOARD - [build] (triangular board used for setting out stairs) gabarit *m* de marches d'escalier
PITCHFORK - [agric] fourche *f* à foin, bident *m*
[mus] diapason *m*
PITCHING - [aero & naut] (change of attitude about the lateral axis) tangage *m*
PITCHING - [brew ind] (the operation of pitching) mise *f* en levure *f*
[constr] empierrement *m*, pavage *m*
~AXIS - [aero] axe *m* de tangage
~FREQUENCY - [aero & naut] fréquence *f* de tangage
~MACHINE - [brew ind] (for barrels) goudronneur *m*
~MOMENT - [aero] (the component of the couple caused by the corresponding airflow, taken about the lateral axis) moment *m* de tangage
~RATE - [brew ind] (the amount of yeast added to a mash) dose *f* de levure
~VESSEL - [brew ind] cuve *f* guilloire
~YEAST - [brew ind] (yeast put into a mash to start fermentation) mise *f* en levain
PITCHOMETER - [naut] s. pitch indicator
PITCHOVER - [astronaut] basculement *m*
PITCHSTONE - [min] rétinite *f*
PITCHY - [gen] poisseux, gluant
~TASTE - [brew ind] goût *m* de poix
~WOOL - [text] laine *f* tenace
PITH - [anat] (cylinder of spongy tissue in the centre of the stem or branch in certain plants) médulle *f*
[timber] (the central core of the wood) cervelle *f*, moelle *f*
~OF ELDERWOOD - [bot] (used for cleaning watches and clocks) moelle *f* de sureau
PITCHING - [med] décérébration *f*
PITHODE - [anat] fuseau *m* nucléaire
PITMANS - [mech] (rod connecting a rotary with a reciprocating part) bielle *f*
PITOMETER - [instr] (instrument designed to record automatically any variation of the velocity of flowing) pitomètre *m*
PITOT COMB - [aero] (a set of Pitot tubes designed to provide simultaneous readings at different points) peigne *m* de Pitot
~HEAD - [instr] (an assembly designed to support a pitostatic device and attach it to an aircraft) sonde *f* de Pitot
~PRESSURE - [aero] (the pressure measured by a correctly aligned pitot tube) pression *f* dynamique
~-STATIC TUBE - [instr] (combination of a pitot and a static tube, designed to measure the difference between the impact and the static pressure) antenne *f* de Pitot
~TUBE - [instr] (a tube having an open end facing an oncoming airstream and designed to measure the pressure produced by the latter) tube *m* de Pitot
~-VENTURI TUBE - [instr] (a combination of a Pitot and a Venturi tube) antenne *f* Pitot-Venturi
PITS - [metall] (small depressions formed in an electrode during electro-chemical action) piqûres
[paper man] (under a paper machine) fosses *f*pl, alvéoles *f*pl
PITSAW - s. pit saw
PITTED - [gen & med] (covered with several small depressions) piqué, troué, rongé
[metall etc] (corroded in the form of many small depressions on the surface) alvéolé
[geol] (with a surface marked by small excavations) piqué
[biol & bot] favéolé
[firearms] chambré
~SURFACE - [metall] surface *f* grippée
PITTING - [med] formation *f* d'un godet
[metall] (corrosion in the form of numerous small

depressions in a surface) formation *f* d'alvéoles, piqûre *f*
PITTING - [mining] exécution *f* des fouilles de recherche
~CORROSION - [metall] corrosion *f* sélective
PITTINITE - [min] (impure variety of gummite) gummite *f*
PITUICYTE - [anat] cellule *f* hypophysaire
PITUITARISM - [med] trouble *m* hypophysaire
PITUITARY - [zool] (of a gland) pituitaire, hypophysaire
~BODY - s. pituitary gland
~GLAND - [anat] (an endocrine gland) glande *f* pituitaire, hypophyse *f*
PITUITECTOMY - [med] hypophysectomie *f*
PITUITRISM - s. pituitarism
PITYRIASIS - [med] (skin disease) pityriasis *f*
PIVOT, to - [gen & mech] (to turn on a pivot) pivoter, tourner
[mech] (to fit on a pivot) monter sur pivot
PIVOT - [mech] (pin or fulcrum about which a part can turn) pivot *m*, tourillon *m*
[mech] (of a door) aiguillon *m*, cheville *f* ouvrière
~ARM - [mech] bras *m* de pivot
~BRIDGE - [constr] (a form of swing bridge) pont *m* tournant
~BURNISHER - [mech] brunissoir *m* pour pivots
~CENTRE - [mech] centre *m* du pivot
~FRICTION - [instr] (friction of the moving system at a pivot of an instrument) friction *f* au pivot
~-HUNG SASH - [constr] (of a window) châssis *m* à pivot
~JAW - [el] (fixed jaw to which the blade of a switch is pivoted) contact *m* fixe d'un interrupteur à couteaux
~JOINT - [mech] (articulation for rotary movements) assemblage *m* à pivot
[anat] diarthrose *f* rotatoire
~OF A HORIZONTAL SHAFT - [mech] pivot *m* d'arbre horizontal
~OF A LYING SHAFT - [mech] pivot *m* à portée plane
~PIN - [mech] tourillon *m*
~POINT - [mech] point *m* d'articulation
~-SLEWING CRANE - [mech mach] grue *f* tournante
PIVOTED - [gen & mech] monté sur pivot
[mech] articulé
~ARM - [mech] bras *m* articulé
~VENTILATION WINDOW - [auto] volet *m* pivotant d'aération
PLACE, to - [gen] placer, mettre, poser
[gen] (to assign to a particular locality etc)placer
[gen] (to appoint) donner un rang, employer
PLACE - [gen] lieu *m*, endroit *m*, place *f*, localité *f*, emploi *m*
~IN OPERATION, to - [gen mech etc] mettre en marmarche
~IN POSITION, to - [gen] mettre en place
~OF ERECTION - [mech] endroit *m* de montage
~OF JUNCTION - [constr work] point *m* de jonction, point *m* d'assemblage
~THE CARD CHAIN, to - [text] placer la chaîne de cartons
~THE CARDS IN POSITION, to - [text] assujettir les cartons
~THE WORK-PIECE, to - [mech] fixer la pièce
PLACEBO -[med] placebo *m*, remède *m* factice
PLACENTA - [zool] (a structure formed by the union of the allantois and chorion with the uterine wall of the mother) placenta *m*, délivre *m*
PLACENTAL - [med] placentaire
PLACENTATION - [zool] (the method of union of the foetal and maternal tissues in the placenta) placentation *f*
PLACENTITIS - [med] placentite *f*
PLACENTOGRAPHY - [radiat] (radiological examination of the placenta) placentographie *f*
PLACER DEPOSITS - [geol] (superficial deposits rich in heavy ore minerals, including native gold,platinum etc) gisement *m* alluvionnaire
~GOLD - [min] or *m* alluvionnaire
PLACING THE CARD ON THE CYLINDER - [text] montage *m* du carton sur le cylindre
PLACUNTITIS - s. placentitis
PLAGIOCEPHALIC - [med] plagiocéphale
PLAGIOCEPHALISM - s. plagiocephaly
PLAGIOCEPHALY - [med] (asymmetrical condition of the head) plagiocéphalie *f*
PLAGIOCLASE - [min] (feldspar mainly consisting of the silicates of sodium, calcium and aluminium) plagioclase *m*
~FELDSPAR - [min] (series of triclinic minerals ranging from pure soda feldspar to pure lime feldspar) feldspath *m* plagioclase
PLAGIOTROPISM - [bot] (tendency of members to take up a position at right angle across the direction of a stimulus) plagiotropisme *m*
PLAGUE - [med] peste *f*
PLAID - [text] (woollen shawl or wrap) plaid *m*, couverture *f* en tartan, tartan *m*
PLAIN - [gen] plat, plan
[geogr] plaine *f*
[gen] (of wood) cru
[text] (of a cloth) uni, lisse
[glass man] (free from bubbles during melting) clair
~-AERIAL SENDER - [radio] (sender in which a close coupling exists between the main oscillatory circuit and the aerial) émetteur *m* à couplage direct d'antenne
~-AERIAL SYSTEM - [radio] (system of transmission by means of a plain-aerial sender) système *m* à couplage direct d'antenne
~-ANTENNA SYSTEM - s. plain-aerial system
~AUGER - [instr] sonde *f* lisse, sonde *f* droite
~BACK - [text] fond *m* uni
~BACK VELVETEEN - [text] velours *m* de coton à fond lisse
~BALANCE - [horol] balancier *m* plan
~BEAM FLANGE - [text] plateau *m* d'ensouple plein
~BEARING - [mech] (a bearing of which the running surfaces are smooth, as distinct from ball or roller bearings) palier *m* lisse, palier *m* à coussinets lisses
~BIT - [tool] trépan *m* simple
~CARBON STEEL - [metall] acier *m* non-allié
~CIRCULAR WEAVE - [text] armure *f* toile pour tissu creux
~COATING -[text] drap *m* d'habit uni
~COLUMN - [arch] colonne *f* lisse
~CONDUCTOR - [el] conducteur *m* homogène, fil *m* monométallique
~CONDUIT - s. plain steel conduit
~COTTON GOODS - [text] tissus *mpl* de coton unis
~COUPLER - [el] (coupler without threads connect-

ing the ends of two adjacent lengths of plain conduit) manchon *m*
PLEIN CRIMP - [text] crêpé *m* uni
~CYLINDRICAL BOILER - [th eng] chaudière *f* cylindrique simple
~CYLINDRICAL COCOON FILAMENT - [text] brin *m* de cocon cylindrique aplati
~EMERY LAYER - [gen] couche *f* d'émeri unie
~FABRIC - [text] tissu *m* lisse, tissu *m* uni
~FLAP - [aero] (a flap which forms a trailing portion of an aerofoil, and moves as a whole) volet *m* simple, aileron *m* de courbure
~FURNACE - [th eng] four *m* simple
~GAUZE - [text] gaze *f* unie
~GRINDER - [mach tool] machine *f* à rectifier les surfaces cylindriques extérieures
~GRINDING FILLET - [text] ruban *m* d'émeri uni
~INDEXING - [mach tool] division *f* simple
~JOURNAL BEARING - [mech] coussinet *m* lisse
~KNIFE - [impl] couteau *m* lisse
~KNOPS - [text] boutons *mpl* plats
~LATHE - [mach tool] tour *m* simple, bidet *m*
~LEAD-COVERED CABLE - [el] câble *m* sous plomb nu
~LINEN GOODS - [text] toiles *fpl* de lin
~LOOM - [text] métier *m* simple, métier *m* à une navette
~MILLER - s. plain milling machine
~MILLING MACHINE - [mach tool] fraiseuse *f* cylindrique
~MUSLIN - [text] mousseline *f* unie
~PIPETTE - [metall] pipette *f* jaugée
~RING GAUGE - [instr] calibre *m* à anneau
~-ROLL - [metall] cylindre *m* lisse
~SEGMENT - [text] segment *m* lisse
~SET - [text] boutage *m* plein
~SHEET LATEX FOAM - [rubber ind] mousse *f* de latex en nappes pleines, mousse *f* de latex
~SLIDE VALVE - [mech] tiroir *m* en coquille
~STEEL CONDUIT - [el] (conduit consisting of a light-gauge steel tubing, the ends of which are not screwed) tube *m* protecteur en acier
~-SURFACE - [el] (totally enclosed, self-cooled machine) à carcasse lisse
~THERMIT - [metall] thermite *f* normale
~TILE - [constr] (ordinary flat tile) tuile *f* plate
~TREADLE-LATHE - [mach tool] tour *m* à cylinder
~TUBING - [oil ind] tube *m* de pompage normal
~TURNING - [mach tool] tournage *m* droit
~TYRE - [rubber ind] bandage *m* dépourvu de mentonnet
~V SLIDE - [mech] guide *m* à un V
~WASHER - [mech] garniture *f* plate
~WEAVE - [text] (the simple interlacing of warp and weft threads) armure *f* toile
~WORSTED FABRICS - [text] tissus *mpl* de laine peignée unis
PLAINS - [text & rubber ind] étoffes *fpl* unies
PLAINTIFF - [leg] (the party beginning an action at law) poursuivant *m*
PLAINTILE - s. plain tile
PLAIT, to - [gen etc] tresser
[text] (the cloth) plier (le tissu)
PLAIT - [gen] (a braid, e.g. of hair) tresse *f*
[a fold] pli *m*
~POINT - [chem] (the point at which two conjugate solutions of partially liquids have the same composition, so that the two layers become one) point *m* de pliage
PLAITED - [gen] tressé, en nattes
~WORK - [text] tresse *f*
PLAITER - [text] ouvrier *m* plieur, plieur *m*
PLAITING - [nucl] (the total of undulations formed on a fuel can by thermal cycling) plissée *f*
~DOWN - [text] doublage *m*
~MACHINE - [text] machine *f* à tresser
~MACHINE FOR ROPES - [text] tresseuse *f* mécanique pour cordes
~OF THE CLOTH - [text] déchargement *m* du tissu
PLAN, to - [gen] faire le plan, dessiner le plan,projeter
PLAN - [gen] plan *m*, projet *m*
[draw] plan *m*
[surv] levé *m*
[text] armure *f*
~POSITION APPROACH - [radar] (special type of radar surveillance approach, given by traffic control to guide aircraft to a runway) approache *f* PPI
~POSITION INDICATOR - [radar] (equipment which continuously shows the position of the aircraft on a horizontal plane as calculated from instrument data) indicateur *m* de position panoramique PPI
~VIEW - [draw] projection *f* horizontale
PLANAR DIODE - [electron] (diode having a planar structure, used for experimental purposes) diode *f* à électrodes plan-parallèles
~IMPLANT - [radiat] (implant arranged in planes) implant *m* en surface
~MASK - [electron] (in a three-gun tricolour TV tubes) masque *m* d'ombre
~NETWORK - [el] (network on a plane without crossing of branches) réseau *m* planaire
~TRANSISTOR - [electron] transistor *m* planar
PLANCK CONSTANT - [phys] (quantum theory) constante *f* de Planck
~LAW - [phys] (quantum theory) loi *f* de Planck
~RADIATION FORMULA - [phys] formule *f* de rayonnement de Planck
PLANCHETTE - [constr] planchette *f*
PLANCKIAN COLOUR - [opt] chromaticité *f* de Planck
~LOCUS - [opt] lieu *m* des corps noirs
PLANE, to - [aero] descendre en vol plané
[aero] (of hydroplane) courir sur le redan
[mech] raboter, aplanir
[mech] (with a planing machine for large surfaces) surfacer, dresser
PLANE - [gen] (a surface) plan *m*, surface *f* plane
[aero] (an aerofoil which is primarily designed to contribute to the lift of an aircraft) surface *f* portante
[impl] rabot *m*
[mech] plan *m*
[bot] platane *m*
[aero] (short for airplane) aéroplane *m*
[mining] galerie *f* de roulage
~AERIAL - [radio] (flat-top aerial; called "sheet antenna" in the USA) antenne *f* à réflecteur plan
~ANGLE - [geom] angle *m* plan
~ASHLAR - [constr] (block of stone with the marks of the tool) moellonage *m*
~BIT - [carp etc] (the cutting part of a plane) fer *m* de rabot
~CIRCULAR SURFACE SOURCE - [phys] (plane circular surface having all its parts vibrating with the same strength and phase) source *f* à surface circu-

laire plane
PLANE COORDINATES - [math] coordonnées *f*pl planes
~EARTH FACTOR - [el] (the ration between the electric field strength that would result from propagation over an imperfectly-conducting plane earth and that from propagation over a perfectly-conducting plane) facteur *m* de terre plane
~FLOW - [soil] courant *m* à deux dimensions
~FLYING - [aero] vol *m* plan
~GEOMETRY - s. planar implant
~IRON - [mech] lame *f* de raboteuse, fer *m* à raboter
~IRON GRINDER - [mach tool] machine *f* affûter les lames de raboteuse
~-MILLING - [mech] fraisage *m* en plan
~MIRROR - [opt] miroir *m* plan
~OF CLEAVAGE - [min] plan *m* de clivage
~OF CONSTRAINTS - [phys] plan *m* de contraintes
~OF FRACTURE - [geol] plan *m* de fracture
~OF LEAST RESISTANCE - [soil] plan *m* de la moindre résistance
~OF OSCULATION - [math] plan *m* osculateur
~OF POLARIZATION - [opt] (the plane containing the incident ray, the reflected ray and the normal to the surface) plan *m* de polarisation
~OF PRINCIPAL SHEARING STRESS - [soil] plan *m* de cisaillement effectif
~OF PROJECTION - [draw] plan *m* de projection
~OF REFLECTION - [opt] plan *m* de réflexion
~OF REFRACTION - [opt] plan *m* de réfraction
~OF SEPARATION - [soil] plan *m* de séparation
~OF SIGHT - [firearms] plan *m* de mire
~OF SLIDING - [soil] plan *m* de glissement
~OF STRATIFICATION - [soil] plan *m* de stratification
~STRESS - [soil] efforts *m*pl bidirectionnels
~OF SYMMETRY - [phys] plan *m* de symétrie
~-TABLE SURVEY - [surv] levé *m* à la planchette
~OF THROAT - [metall] niveau *m* du gueulard
~OF VIBRATION - [opt] (plane containing the reflected ray and normal to the plane of polarization) plan *m* de vibration
~OFF, to - [mech] aplaner
~POLARIZATION - [radio] (of an electromagnetic wave) polarisation *f* plane
~POLARIZED LIGHT - [opt] lumière *f* polarisée dans un plan
~POLARIZED SOUND WAVE - [acoust] onde sonore *f* polarisée dans un plan
~POLARIZED WAVE - [el magnetic waves] onde *f* polarisée dans un plan, onde *f* à polarisation linéaire
~REFLECTOR AERIAL - [radio] antenne *f* à reflecteur plan
~REFLECTOR ANTENNA - s. plane reflector aerial
~SINUSOIDAL WAVE - [phys] (plane progressive wave such that the corresponding physical quantities vary sinusoidally with time) onde *f* sinusoïdale plane
~SOURCE - [nucl] (neutron source of plane form) source *f* plane
~SQUARE SURFACE SOURCE - [phys] (plane square having all its parts vibrating with the sale strength and phase) source *f* à forme de surface plane quadratique
~STABILIZER - [aero] plan *m* fixe horizontal
~TABLE - [surv] (drawing-board mounted on a tripod, so that it can be levelled and rotated about a vertical axis and clamped in position) planchette *f*
[min] table *f* inclinée à aire plane
PLANE TILE - [constr] tuile *f* plate
~TREE - [bot] platane *m*
~TRIANGLE - [geom] triangle *m* rectiligne
~TRIGONOMETRY - [math] trigonométrie *f* rectiligne
~WAVE - [phys] (wave such that the corresponding physical quantities are uniform in any plane perpendicular to a fixed direction) onde *f* plane
~WAVE FRONT - [phys] front *m* d'onde plan
PLANED FAULT - [geol] faille *f* nivelée
PLANER - [mach tool] raboteuse *f*
[print] (a flat piece of wood used to level the surface of a forme) taquoir *m*
~-TYPE HORIZONTAL BORING MACHINE - [mach tool] alésoir *m* horizontal à rabot
~-TYPE MULTIPLE-HEAD MILLING MACHINE - [mach tool] fraiseure *f* horizontale à plusieurs poupées
~-TYPE SINGLE-HEAD MILLING MACHINE - [mach tool] fraiseuse *f* horizontale à une poupée
PLANET - [astr] (the heavenly bodies recolving about the sun) planète *f*
~GEAR - [mech] (in an epicyclic train, one of the intermediate gears which mesh with both the orbit gear and the sun pinion) engrenage *m* planétaire, moiche *f*
~-SPINDLE - [mech] arbre *m* à mouvement planétaire
~WHEEL - [mech] (a gear in an epicyclic train which revolves round the sun wheel) roue *f* planétaire
~WHEEL CARRIER - [mech] (the part in epicyclic gear which carries the planet wheels) porte-planète *m*
~WHEEL MIXING HEAD - [plast ind] (also epicyclic mixing head; set of epicyclic gear carry out the mixing of the extrusion material before it reaches the die) malaxeur *m* à mécanisme épicyclique
~WHEEL PIN - [mech] (a pin fixed in the planet carrier in which the planet wheel revolves) moyeu *m* de la roue planétaire
PLANETARY - [astr etc] planétaire
~ABERRATION - [astr] aberration *f* planétaire
~ACTIVITY SPHERE - [astr] (the sphere in which a planet can hold a satellite) sphère *f* d'activité planétaire
~BOUNDARY LAYER - [astr] exosphère *f* planétaire, couche *f* limite atmosphérique
~CHANGE-CAN MIXER - [mech] (a type of mixer in which removable cans are stirred by paddles fixed on vertical shafts and driven by planetary gear train) agitateur *m* planétaire à cuve amovible
~CIRCULATION - [astr] circulation *f* planétaire
~CONFIGURATION - [astr] configuration *f* des planètes
~GEAR - s.planet gear
~GRAVISPHERE - [astr] sphère *f* de gravitation planétaire
~MIXER - [mech] (type of vertical-shaft mixer in which the paddle shafts rotate and move in a circular path at the same time) agitateur *m* planétaire
~NEBULA - [astr] (small nebula showing a disk which resembles a planet) nébuleuse *f* planétaire
~PRECESSION - [astr] précession *f* planétaire
~PROBE - [astr] sonde *f* planétaire
~STIRRING MACHINE - [mech] (type of agitator in which the paddles are rotated as a group and also revolved individually by means of an epicyclic gear train) agitateur *m* planétaire

PLANETARY STRANDER - [text] toronneuse *f* à cages roulantes
~TRANSMISSION - [mech] système *m* planétaire différentiel
PLANETOID - [astr] planétoïde *m*, astéroïde *m*
PLANIGRAPHY - [radiat] (body section radiography) planigraphie *f*
PLANIMETER - [instr] (instrument designed to measure the area of a plane) planimètre *m*
PLANIMETRIC - [geom] planimétrique
~AEROPHOTOGRAM - [photo] aérophotographie *f* planimétrique
~DISTORTION - [photo] déformation *f* planimétrique
~MAP - [surv] carte *f* planimétrique
~SURVEY - [photo] levé *m* planimétrique
PLANIMETRY - [geom] (the measurement of the area of a plane) planimétrie *f*
PLANING - [mech] rabotage *m*
~AND MOULDING MACHINE - [mech] machine *f* à raboter et à mouler
~AND THICKNESSING MACHINE - [mach tool] machine *f* à raboter tirant des bois d'épaisseur
~BENCH - [impl] banc *m* de menuisier
~BOTTOM - [naut] (the part of the under surface of a hull or boat which gives lift by hydrodynamic action) fond *m*
~GENERATOR - [gen] machine *f* à tailler les engrenages à rabot
~HEIGHT - [mech] épaisseur *f* de rabotage
~MACHINE - [mech mach] (machine designed to produce large flat surfaces) raboteuse *f*
~MACHINE FOR SURFACING - [mach tool] raboteuse *f* à surfacer
~MACHINE WITH CYLINDRICAL CUTTERS - [mach tool] raboteuse *f* à cylindres
~TOOL - [tool] lame *f* de raboteuse
~WIDTH - [mech] largeur *f* de rabotage
PLANISH, to - [mech] (to make wood smooth or plane) planer, aplaner
[mech] (to smooth out a sheet metal surface) dresser, égaliser, polir
[photo] glacer, satiner
PLANISHING - [metall] planage *m*
~HAMMER - [impl] (special hammer used for planishing sheet metal) marteau *m* à dresser
~MACHINE - [metall] planeuse *f*
~MILL - [metall] laminoir *m* polisseur
~ROLL - [metall] cylindre *m* espatard, polisseur *m*
~STAKE - [impl] tas *m* à planer
PLANISHED - [gen mech etc] dressé, égalisé
PLANISPHERE - [astr] (a polar projection of half the celestial sphere) planisphère *f*
PLANK, to - [gen] (to cover with planks) garbir de madriers
[shipbuild] vaigrer, border
[text] (the operation of joining the silver ends prior to drawing) jointoyer les rubans
PLANK - [gen] planche *f*, madrier *m*
[shipbuild] vaigre *f*, bordage *m*
~DAM - [mining] cloison *f* en planches
~FRAME - [constr] châssis *m* en madrier, huisserie *f* en planches
~LAGGING - [carp etc] revêtement *m* en madriers
~PARTITION - [constr] cloison *f* en madriers
~-SHEER - [naut] (the planks on deck) planche *f*
~TRUSS - [constr] (roof truss made of planks) ferme *f* en planches
PLANKING - [gen] planchéiage *m*, voligeage *m*
[shipbuil] vaigrage *m*, bordage *m*
[text] (in hat making) foulage *m* du feutre
[mining] coffrage *m*, planchéiage *m*
PLANKTON - [biol] (animals and plants floating in water) plancton *m*
PLANNING - [gen] tracé *m*, projet *m*
[comm etc] programmation *f*
[town planning] plan *m* d'aménagement d'une ville
~TABLE - [photo] table *f* de montage
PLANO-CONCAVE - [opt] (flat on one side and concave on the other) plan-concave
~-CONCAVE CYLINDER - [opt] cylindre *m* à plan concave
~-CONCAVE LENS - [opt] lentille *f* plan-concave
~-CONVEX - [opt] (flat on one side and convex on the other) plan-convexe
~-CONVEX LENS - [opt] lentille *f* plan-convexe
~-CYLINDRICAL - [opt] plan-cylindrique
PLANOGRAPHIC PRINTING - [photo] impression *f* à plat
PLANOMETER - s. planimeter
PLANOMILLER - [mach tool] fraiseuse *f* gerne raboteuse
PLANT, to - [gen etc] planter
[comput] placer
PLANT - [bot] pl ante *f*
[mech etc] installation *f*, appareillage *m*
[constr] usine *f*
[el] equipement *m* électrique
~ASHES - [agric] cendres *fpl* végétales
~BIOLOGY - [bot] biologie *f* végétale
~-BREEDING - [genet] phytogénétique *f*
~CHARACTERISTIC - [contr] (inherent characteristic) caractéristique *f* naturelle
~CONCRETE - [constr] béton *m* d'usine
~CONSTRUCTION - [constr] construction *f* d'usines
~DIBBLE - [agric] plantoir *m*
~DISEASE - [bot] maladie *f* des plantes
~-EATING - [zool] phytophage
~FACTOR - s. plant load-factor
~FOODS - [agric] éléments *mpl* nutritifs des plantes
~GELATINE - [chem] gelée *f* végétale
~GEOGRAPHY - [agric] phytogéographie *f*
~GROWTH - [bot] accroissement *m* des plantes
~HAIR - [text] duvet *m* végétal
~INJURIES - [bot] affections *fpl* des plantes
~LAYOUT - [ind] arrangement *m* des installations
~LOAD-FACTOR - [el] (the ratio between the total load supplied by a generating plant and the load which would be supplied if the generating plant were operated continuously at its maximum rating) coefficient *m* d'utilisation d'une centrale
~-LOUSE - [zool] puceron *m*, aphide *m*
~NUTRITION - [agric] nutrition *f* des plantes
~PATHOLOGY - [bot] pathologie *f* végétale
~PEST - [bot] parasite *m*
~POTATO - [agric] pomme *f* de terre de semence
~PROTECTION - [bot] protection *f* des plantes
~PROVING RUN - [gen] marche *f* d'essai d'une installation
~THERMAL EFFICIENCY - [th eng] (ratio of the actual heat delivered and the heat supplied by the fuel) rendement *m* thermique d'une installation
~WAX - [chem] cire *f* végétale
PLANTAR - [anat] (pertaining to the sole of the foot) plantaire

PLANTAR REFLEX - [med] reflexe m cutané plantaire
PLANTATION - [gen] fondation f, colonisation f
[rubber ind] plantation f
~LATEX - [rubber ind] latex m de plantation, latex m des champs
~RUBBER - [rubber ind] caoutchouc m de plantation
~WHITE - [ind] (centrifugal sugar made for local consumption in sugar-producing countries) sucre m colonial
PLANTE PLATE - [el chem] (very large plate on the surface of which the active material is formed from the lead of the plate itself) plaque f Planté, plaque f à grande surface,
PLANTER - [agric] plantoir m mécanique, planteuse f
PLANTER'S BALE - [text] balle f des planteurs
PLANTING - [agric] plantage m, plantation f
[constr] (the operation of laying the foundation courses) pose f des cours de fondation
[el] (of a battery) installation f
~DEPTH - [agric] profondeur f de plantation
~DISTANCE - [agric] écartement m de plantation
~HOLE - [agric] trou m de plantation
~IRON - s. plant dibble
~MACHINE - [agric] planteuse f
~PEG - [agric] plantoir m
~PIN - s. planting peg
~SEED AT STAKE - [rubber ind] plantation f directe en graines
~SPADE - [agric] bêche f à planter
~TIME - [agric] période f de plantation
~TOOL - [agric] outil m de plantation
PLAQUE - [gen] plaque f
PLASM - [biol] protoplasme m
PLASMA - [chem] (the clear, liquid part of the blood separable by centrifuging) plasma m
[min] (a bright-green translucent variety of chalcedony, used as a gem) plasma m, calcédoine f vert foncé
[phys] (the region in a gaseous medium in which electron and ion populations are approximately balanced, so that the space charge is virtually zero) plasma m
~BALANCE - [nucl] (the condition in which ionization balances diffusion) équilibre m de plasma
~CELL - [biol] plasmacellule f
~CLOUD - [phys] nuage m de plasma
~CONTAINMENT - [phys] (in a controlled thermonuclear programme) confinement m du plasma
~DIAMAGNETISM - [phys] diamagnétisme m de plasma
~EXTENDER - s. plasma volume
~FREQUENCY - [phys] fréquence f de plasma
~FREQUENCY PROBE - [phys] sonde f de fréquence de plasma
~GUN - [phys] canon m à plasma
~INSTABILITY - [phys] instabilité f de plasma
~MOTOR - [astronaut] moteur m à plasma
~PRESSURE - [phys] pression f du plasma
~PROPULSION - [astronaut] (propulsion by the use of a plasma which can reach velocities of I0^{7} CM. sec.) propulsion f par plasma
~PURITY - [phys] pureté f du plasma
~ROCKET - [astronaut] fusée f à plasma
~SUBSTITUTE - s. plasma extender
~TORCH - [phys] chalumeau m à plasma
~VOLUME - [med] succédané m volhémique du plasma
PLASMACYTOMA - [med] (bone tumor composed of cells which resemble plasma cells) plasmacytome m
PLASMAGENE - [biol] (postulated units of inheritance in the cytoplasm of the cell) plasmagène
PLASMALEMMA - [genet] absence f de membrane solide chez certaines bactéries, plasmalemma m
PLASMATIC - [biol] protoplasmique
~INHERITANCE - [genet] hérédité f plasmatique
PLASMO-DITROPHOBLAST - [med] plasmoditrophoblaste m
PLASMOCYTE - [zool] (leucocyte) leucocyte m
PLASMOCYTOSIS - [med] plasmocytose f
PLASMOGAMY -[biol] (fusion of cytoplasm) plasmogamie f
PLASMOID - [phys] (discrete piece of plasma) plasmoïde m
PLASMON - [phys] (hypothetical particle associated with the various waves which may exist in a plasma) plasmon m
PLASMOTROPHOBLAST - s. plasmoditrophoblast
PLASMOZYME - [biol] prothrombine f, thrombogène m
PLASTER, to - [constr] plâtrer, ravaler, enduire de plâtre
[med] mettre un emplâtre
PLASTER - [constr] (general term denoting a plastic substance used to coat walls etc) plâtre m enduit m
[constr] (the material itself) plâtre m
[med] emplâtre m
~BOARD - [constr] (board made of plaster with paper facings) panneau m en planches de plâtre
~CAST - [gen] enduit m
[med] plâtrage m
~CASTING - [metall] coulée f en plâtre
~CEMENT - [constr] plâtre-ciment m
~CUP - [impl] cuvette f à plâtre
~-DEPTH SWITCH - [constr] interrupteur m sous enduit
~-FINISH - [constr] revêtement m en plâtre
~KILN - [th eng] four m à plâtre, plâtrière
~LATHS - [th eng] lattes fpl pour enduit
~MOULD - [metall] moule m en plâtre
~MOULDING - [metall] moulage m en plâtre
~OF PARIS - [chem & constr] (dehydrated gypsum, finely-ground and used for making casts and the like, because of its property of taking up water and setting hard, with slight expansion) plâtre m de Paris, plâtre m de moulage, .sulfate m de calcium
~-OF-PARIS BANDAGE - [med] appareil m plâtré
~PRIMING - [build] apprêtage m
~REFUSE - [constr] débris mpl de maçonnage
~ROCK - s. plaster stone
~SPRAYER - [mech] pistolet m à enduite
~STONE - [constr] pierre f à plâstre, gypse m
~STUCCO - [constr] plâtrage m
~WITH LIME, to - [constr] enduire au lait de chaux
PLASTERED BICKWORK - [constr] maçonnerie f en briques crépie
PLASTERING - [constr] plâtrage m
[med] pose f d'un emplâtre
[agric] plâtrage m (d'un champ)
~SAND - [constr] sable m pour plâtre
~TROWEL - [impl] truelle f à plâtre
PLASTIC - [gen] (pliable) plastique
[phys] (capable of being moulded) plastique
[plast ind] (pertaining to plastics) plastique
~BRONZE - [metall] (bronze containing a high pro-

portion of lead) bronze *m* de coussinet, bronze *m* plastique

PLASTIC CLAY - [constr] (pure clay) argile *f* plastique

~COATED - [gen] à revêtement plastique

~COLLISION - [mech] (collision in which plastic deformation occurs) collision *f* plastique

~DEFORMATION - [phys] (that part of the strain in a stressed body which remains after the applied stress has been reduced to zero for a given time) déformation *f* plastique, déformation *f* permanente

~EFFECT - [telev] (effect of relief due to a distortion of the picture signal) plastique *f*

~FILAMENT FROM SODIUM CELLULOSE - [text] fil *m* plastique de cellulose esodique

~FLOW - [phys] (a measure of the degree of deformation in a material under load) écoulement *m* plastique

~-FOIL CAPACITOR - [electron] condensateur *m* à film plastique

~IMPACT - [chem] (collision between two bodies without appreciable coefficient of restitution) choc *m* inélastique

~LIMIT - [soil] (the water content of soil to a point at which it is no longer plastic) limite *f* de plasticité, limite *f* inférieure de plasticité

~MASS - [text] pâte *f* plastique

~MATERIAL - [gen & plast ind] masse *f* plastique

~MODEL - [gen & arch] modèle *m* plastique

~MOULDING FLAKES - [text] masse *f* à rognures comprimés

~MOULDING PRESS - [plast ind] presse *f* à compression pour masses plastiques

~PAINT - [paint] peinture *f* plastique

~RANGE - [phys] (stress range in which a material will suffer permanent deformation without failing) limite *f* de plasticité

~SHEETS - [plast ind] feuilles *f*pl plastiques

~SULPHUR - [chem] (formed when sulphur is distilled into water) soufre *m* mou

~SURGERY - [med] (the repair and restoration of damage of the body) chirurgie *f* plastique

~YIELD - [phys] (non-elastic deformation) déformation *f* non-élastique
[rubber ind] commencement *m* d'écoulement plastique

~YIELD TEST - [plast ind] essai *m* de seuil, essai *m* écoulement de contraint critique

~YIELD VALUE - [plast ind] seuil *m* de fluage

PLASTICATE, to - [gen & phys] (to soften by heating or kneading) plastifier

PLASTICINE - [ind chem] plasticine *f*

PLASTICITY -[phys] (term denoting the ability of a material to be easily formed by moulding or similar manipulation) plasticité *f*

~INDEX - [agric] (of a soil) index *m* de plasticité (de sol)

~OF SOIL - [geol] compacité *f* du sol, ténacité *f* du sol, plasticité *f* du sol

PLASTICIZE, to - [plast ind] (to increase the plasticity of a material by the introduction of a plasticizer. Also, the process of heating a thermoplastic material to make it plastic or fluid) plastifier

PLASTICIZED - [plast ind] plastifié

~MATERIALS - [plast ind] masses *f*pl souples

~PVC - [plast ind] PVS *m* plastifié

PLASTICIZER - [build] (substance added to Portland cement to render the concrete mix more plastic) plastifiant *m* de malaxage, agent *m* plastifiant

PLASTICIZERS - [plast ind] (high boiling point organic compounds which are incorporated in polymers to improve their physical properties) plastifiants *m*pl

PLASTICIZING AGENT - s. plasticizers

~CAPACITY - [ind chem] (the weight of material which a machine can bring to plasticizing temperature in a given time, usually per hour) capacité *f* de plastification

PLASTICS - [plast ind] (a general term for certain organic compounds, both natural and synthetic, which become plastic under heat and pressure and so can be formed by moulding, extrusion and other methods, into shapes which are permanent on cooling) matière *f* plastique, plastiques *m*pl

~PLUMBER AND FITTER - [plast ind] soudeur *m* et installateur de tuyauteries en plastiques

~WELDER AND FABRICATOR - [plast ind] soudeur *m* et transformateur de plastiques

PLASTID - [biol] (dense protoplasmic inclusion in a cell) plastide *m*

~MUTATION - [genet] mutation *f* plastidique

PLASTIMETER - [instr] (an instrument designed to measure the consistency of semi-fluid substances) consistomètre *m*, plastimètre *m*

PLASTISOL - [chem] (a liquid dispersion of a synthetic resin in a plasticizer) plastisol *m*

PLASTOGAMY - [biol] (the union of individual protozoa) plastogamie *f*

PLASTOMER - [chem] (a synthetic plastic material) plastomère *m*

PLASTOMETER - [instr] (an instrument for measuring the viscosity of a coating composition) plastomètre *m*

PLAT - [surv] (only USA; a map) plan *m* (d'un terrain etc)
[min] (short for platform) plate-forme *f*

PLATBAND - [arch] (flat projecting moulding) plate-bande *f*
[constr] voûte *f* plate

PLAT-BAND - [metall] moulure *f* de la face

PLATE, to -[metall] (to coat with metal) plaquer
(to cover or lag with plates) métalliser
[el chem] (the operation of electroplating) plaquer, faire déposer une couche de métal
[shipbuild] border
[print] clicher (les pages)

PLATE - [gen] plaque *f*, plateau *m*, lame *f*
[metall] tôle *f*, plaque *f*, taque *f* de fonte, tôle *f* grosse
[mech] (or disk; of a clutch) plaque *f*, plateau *m*
[photo] (glass used as support for the sensitive emulsion) plaque *f*
[el] (one of the conducting parts of a capacitor) plaque *f*
[el chem] (of a storage battery; the complete assembly of active material and support) plaque *f* d'accumulateur
[electron] (only USA; anode in GB; the electrode having the function of collecting electrons) anode *f*
[railw] (of a wheel) disque *m*
[mach tool] plateau *m*, table *m*
[print] (electrotype of stereotype) cliché *m*
[mech] (of a press) semelle *f* d'une presse
[radio] (a conducting surface) armature *f*
[ind chem] (of a distillation tower) plat *m*, plateau *m*

PLATE - [med] (a dental plate) dentier *m*, denture *f* artificielle
[metall] plaque-filière *f*
[comm] (metal plate thicker than 3 mm) tôle *f* forte, tôle *f* grosse
[tool] (used in forging) chemise *f*, séparateur *m*
[auto] (the registration plate) plaque *f* matricule
[constr] sablière *f*, poutre *f* sablière, semelle *f*
~ADAPTER - [photo] adapte-plaques *m*
~ANCHORING - [electron] ancrage *m* de l'anode
~ANEMOMETER - [instr] (anemometer which consists of a plate member against which the wind blows) anémomètre *m* à palette
~BATTERY - [electron] (power supply battery for the plate of an electronic tube) pile *f* d'anode
~BEARING - [mech] support *m* à plaque
~-BENDING ROLLER - [mech] machine *f* à cintrer les tôles
~BLOCK - [el] bloc *m* de plaques
~BOX - [photo] (for negatives) boîte-classeur *f*
~BRIDGE - [electron] cavalier *m* d'anode
~BYPASS CAPACITOR - [electron] condensateur *m* de découplage anodique
~CAM - [mech] came *f* à disque
~CAMERA - [photo] chambre *f* à plaques
~CAP - [electron] capuchon *m* d'anode
~CAPACITOR - [el] condensateur *m* à plaques
~-CARRIER - [photo] châssis *m*
~CHANGING BOX - [photo] châssis-magasin *m*
~CHARACTERISTIC - [electron] courbe *f* caractéristique de l'anode
~CHARGER - [firearms] chargeur *m* à plaque
~CIRCUIT - [electron] circuit *m* d'anode
~CLAMP - [mech] pince *f* d'anode
~CLOSER - [mech] presse-tôle *m*
~CLUTCH - [mech] (disk clutch) embrayage *m* à disques
~COLUMN - [ind chem] (a distillation column) colonne *f* à plateaux
~CONDENSER - s. plate capacitor
~COOLING CAPACITY - [electron] capacité *f* de refroidissement de l'anode
~CORE - [el] noyau *m* en tôle
~COUPLING - [mech] accouplement *m* à plateau, manchon *m* à plateaux
~CURRENT - [electron] (anode current, flowing from the cathode to the plate in an electronic tube) courant *m* anodique
~DECOUPLING RESISTOR - [radio] (anode decoupling resistor) résistance *f* de découplage dans le circuit anodique
~DISPERSION PLUG - [plast ind] (assembly of perforated plates fitted in the nozzle of an injection moulding machine to disperse colourant) jeu *m* de plateaux de dispersion
~DISSIPATION - [electron] dissipation *f* anodique
~-EDGE PLANER - [mech] chanfreineuse *f*
~EFFICIENCY - [electron] rendement *m* anodique
~FEED RESISTANCE - [electron] (anode feed resistance) résistance *f* de charge d'anode
~FEEDER - [mech] alimentateur *m* à palette
~FILTER PRESS - [mech] filtre-presse *f*
~FIN - [electron] ailette *f* d'anode
~FINISHING MILL - [metall] laminoir *m* finisseur à tôle forte
~FLATTENING AND BENDING MACHINE - [mech] machine *f* à rouler et à cintrer les tôles

PLATE FLOOR - [constr] plancher *m* en sablières
~-FOLDING AND BENDING MACHINE - [metall] machine *f* à plier et couder les tôles
~FOR ROVING - [mach] plaque *f* guide-mèche
~FRAME - [el] (of a nickel-iron-alkaline accumulator) cadre *m* de plaques
~GAUGE - [mech] (limit gauge formed by cutting slots of the required gauge width in a steel plate) calibre *m* d'épaisseur
~GIRDER - [constr] (built-up steel girder) poutre *f* en tôle
~GIRDER BRIDGE - [constr] pont *m* à poutres en tôle
~GLASS - [glass man] (high-quality glass more than I/4" in thickness, used chiefly for glazing large openings without intermediate support, e.g. shop windows) glace *f* de vitrage, verre *m* laminé
~-GLASS CUTTING SHAPE - [photo] calibre *m* à découper en glace forte
~GLAZER - [paper man] lamineur *m*
~GLAZING - [paper man] calandrage *m* à feuilles
~GRAFTING - [agric] greffe *f* en placage
~GRID - [mech] grille *f* à plaque
~GROUP - [el] (of an accumulator; a complete electrode consisting of positive or negative plates) faisceau *m* de plaques
[auto] (clutch) empilage *m* de disques d'embrayage
~-HEAD WOOD-SCREW - [mech] vis *f* à tête plate
~-HOLDER - [photo] châssis *m* à plaques, porte-plaques *m*
~-HOLDER SHUTTER - [photo] volet *m* de châssis
~HUM - [electron] renflement *m* d'anode
~INPUT POWER - [electron] puissance *f* d'entrée anodique
~IRON - [metall] tôle *f*, tôle *f* de fer
~-IRON GIRDER - [constr] poutre *f* en tôle
~KEYING - [radio] (or anode keying; keying of a transmitter by interrupting the plate-supply circuit) manipulation *f* anode, manipulation *f* dans la haute tension
~-LAYER - [railw] poseur *m* de voie
~-LAYING - [railw] pose *f* des voies
~LEAD - [electron] conducteur *m* d'alimentation de anode
~LIFE - [el] (of a storage battery) durée *f* des plaques
~-LINK - [mech] coulisse *f* à flasques
~LOAD - [electron] (the external circuit between the anode and the cathode of a tube) charge *f* d'anode
~LUG - [el] (projection on a plate for connecting it to a terminal bar) queue *f* conductrice
~MANGLE - [mech] dégauchisseuse *f* de tôles
~MILL - [metall] cylindre *m* à tôles
~MODULATION - [electron] (anode modulation) modulation *f* par l'anode
~MOULDING - [metall] (method of moulding the halves of a split pattern on opposite sides of metal or wood plates) moulage *m* sur plaque
~NEUTRALIZATION - [electron] (or anode neutralization, the method of neutralizing an amplifier) neutralisation *f* par anode
~OF A CAPACITOR - [el] (each of the two conductors separated by the insulating medium, the whole forming the capacitor) armature *f* d'un condensateur
~OF SPRING - [mech] (leaf of spring) lame *f* de ressort
~ORIFICE - [meas] (simple device to measure the amount of gas or liquid passing through a pipe) ori-

fice *m* mesureur
PLATE PENETROMETER - [instr] pénétromètre *m* de dureté
~PLANER - [mech] rabot *m* pour tôles
~POTENTIAL - [electron] tension *f* anodique
~POWER INPUT - [electron] (or anode power input; to a stage of a radio transmitter) énergie *f* d'anode, puissance *f* anodique
~PROOF - [print] (proof from a plate) épreuve *f* de cliché
~PULSE MODULATION - [electron] (or anode pulse modulation) modulation *f* d'anode par impulsions
~-RAIL - [metall] rail *m* plat
~RECTIFICATION - [electron] redressement *m* de plaque
~RECTIFIER - [electron] redresseur *m* de plaque
~RESISTANCE - [electron] résistance *f* interne
~REST CURRENT - [electron] courant *m* anodique permanent
~-ROLL - [metall] cylindre *m* à tôle
~ROLLING MILL - [metall] laminoir *m* à tôle
~SATURATION - [el] saturation *f* anodique
~SCREWS - [surv] (the screws connecting the head of a level etc with its base and are used for adjustments) vis *f*pl de réglage
~SECTION - s. plate group
~SHEARING MACHINE - [mech] cisailleuse *f* à tôles
~SHEARS - [metall] cisailles *f*pl à tôles
~SINGEING - [text] flambage *f* à plaque
~SPINDLE - [text] broche *f* à disque
~SPINNING MACHINE - [text] métier *m* à filer à assiette
~SPRING - [mech] ressort *m* à lames, ressort *m* à lames étagées, ressort *m* à disques
~STEEL - [metall] acier *m* à plaques
~SUPPLY - [electron] source *f* d'alimentation anodique
~SUPPORT - [el] (the support from which the plates of an accumulator are suspended or on which they rest) support *m* d'anode
~TENSION - [mech] frein *m* à ressort plat
~TERMINAL - [electron] (or anode terminal; the point at which connexion to the plate circuit of a set is made) borne *f* de plaque, borne *f* d'anode
~THICKNESS - [metall] épaisseur *f* de la tôle
~TOWER - s. plate column
~TURNER - [metall] appareil *m* de visite, retourneur *m* à tôles
~-TYPE - [gen] à plaques
~-TYPE CLUTCH - s. plate clutch
~-TYPE HEAT EXCHANGER - [heat] échangeur *m* thermique à plaques
~VOLTAGE - [electron] (anode voltage) tension *f* anodique
~WARMER - [impl] chauffe-assiettes
~WASHER - [mach] rondelle *f* plate
~-WEB GIRDER - [constr] poutre *f* à âme pleine
~WHEEL - [mech] roue *f* à toile, roue *f* à centre plein
~WITH A LARGE AREA - s. plante plate
PLATEAU - [geogr] plateau *m*
[med] plateau *m*
[electron] (portion of the plateau characteristics for which the counting rate is almost independent of voltage) palier *m*, plateau *m*
~BURNING - [astronaut] combustion *f* en plateau
~CHARACTERISTICS - [nucl] (or counting rate characteristics; the relation between counting rate and voltage applied to a counter tube for a specified constant source of radiation) courbe *f* caractéristique de palier
~LENGTH - [nucl] (the range of applied voltage over which the plateau of a radiation counter tube extends) longueur *f* du palier
~SLOPE - [electron] pente *f* du plateau
PLATED - [gen] (provided with plates, e.g. for defence) recouvert de plaques, garni de plaques, blindé
[metall] plaqué
~DECK - [naut] pont *m* cuirassé
PLATEHOLDER - s. plate-holder
PLATELET - [biol] (a thrombocyte) plaquette *f* sanguine, thrombocyte *m*
~COUNT - [biol] numération *f* thrombocytaire
PLATEN - [mach tool] (the work table) plateau *m*, table *f*
[mech] (a metal plate) plaque *f* métallique
[plast ind] (steel plate used to heat and to transmit pressure to the mould assembly in a press) plateau *m*
[print] (that part of a press, on which the paper is supported) platine *f*
[mech] (of a typewriter) rouleau *m* porte-papier
[cin] plaque *f* en verre
~MACHINE - [print] (printing machine in which the impression is taken with a flat surface) presse *f* à platine
~PRESS - s. platen machine
PLATER - [gen] plaqueur *m*
PLATFORM - [gen] plate-forme *f*, tablier *m*
[railw] quai *m*
[railw] (of a passenger coach) wagon *m* plate-forme
[carp] (a raised surface) terrasse *f*
[mech] (gasholders etc) passerelle *f*
[constr] plate-forme *f*, estrade *f*
[astronaut] (a lunching pad for missiles) table *f* de lancement
[geogr] plateau *m*, bande *f* continentale
[metall] (of a furnace) pont *m* de chargement
~BALANCE - [meas] (a type of scales for weighing large masses) bascule *f* à romaine
~BARRIER - [railw] barrière *f* d'accès aux quais
~BRACKETS - [metall] support *m* de la plateforme du gueulard
~CAR - [railw] wagon *m* sans parois
~LIFT - [mech] plate-forme *f* élévatrice
~REPEATING SIGNAL - [railw] signal *m* répétiteur de quai
~ROOF - [railw] marquise *f*
~SCALE - [meas] bascule *f* à romaine
~TRAILER - [transp] remorque *f* à plateau
~TRUCK - [transp] camion-plateau *m*
PLATFORMING - [ind chem] (reforming process which makes use of a catalyst containing platinum and excess of hydrogen) "platforming" *m*
PLATINATES - s. platinic hydroxide
PLATING - [metall] revêtement *m* en tôle, placage *m*, tôles *f*pl
[aero] (only USA) revêtement *m*
[shipbuild] bordé *m* en fer
[paper man] (the process of imparting a special surface to paper by rolling it between metal sheets, the design of which is thus transferred to the paper) clichage *m*

PLATING - [med] culture *f* sur plaques
~BATH - [metall] bain *m* de placage
[el chem] (plating through electro-deposition) bain *m* galvanique
~DESIGN - [text] dessin *m* par mailles vanisées
~DYNAMO - s. plating generator
~GENERATOR - [el] (special type of direct-current generator designed to give low voltages and large currents for electro-plating and the like operations) générateur *m* pour galvanoplastie
~-OUT TEST - [metall] épreuve *f* par rabattement
~RACK - [el chem] (frame for supporting electrodes in a plating bath and supplying current to them) support *m* d'accrochage
~SOLUTION - [chem] bain *m* électrolytique
PLATINIC ACID - [chem] acide *m* platinique
~HYDROXIDE - [chem] (forming platinic salts by dissolution in acids) hydroxyde *m* platinique
~OXIDE - [chem] (the dark-frey powder formed by heating platinic hydroxide) oxyde *m* platinique
PLATINIFEROUS - [min] (platinum bearing) platinifère
PLATINIRIDIUM - [min] (native alloy of platinum, iridium and other metals) platine *f* iridé
PLATINITE - [metall] (an alloy containing iron and nickel) platinite *f*
PLATINIZE, to - [el chem] (to coat with finely-divided platinum) platiner
PLATINIZED ASBESTOS - [ind chem] (asbestos containing platinum in a finely-divided state dispersed through its structure) amiante *m* platiné
PLATINOCHLORIDE - [chem] (used in photographic processes) chlorure *m* de platine
PLATINOCYANIC - [chem] (derived from compounds containing platinum and cyanogen) platinocyanhydrique
~ACID - [chem] acide *m* platinocyanhydrique
PLATINOCYANIDE - [chem] (a cyanide of platinum and other radicals) platinocyanure *m*
PLATINOID - [metall] (an alloy containing copper, zinc, nickel and tungsten) platinoïde *m*
PLATINOTYPE - [photo] (photographic process in which a positive is obtained by a deposit of finely - divided platinum in combination with iron) platinotypie *f*
PLATINOUS HYDROXIDE - [chem] hydroxyde *m* de platine
~OXIDE - [chem] (obtained through heating platinum hydroxide) oxyde *m* de platine
PLATINUM - [chem] (metallic element, symbol Pt, with A.N. 78, A.W. 195.23, applications as a catalyst, in jewellery, chemically-resistant laboratory ware, and in electrical equipment) platine *m*
~AMMINES - [chem] (compounds of platinum and ammonia) amines *f*pl de platine
~BARIUM CYANIDE - [chem] (a phosphor for radiographic screens) cyanure *m* de barium et platine
~BLACK - [chem] (a very finely divided form of platinum with uses as a catalyst and absorbent for gases) noir *m* de platine
~CHLORIDE - [chem] (an analytical reagent) chlorure *m* de platine
~CONTACT - [chem] contact *m* platiné
~DIOXIDE - s. platinic oxide
~FUSE - [el] coupe-circuit *m* à filament de platine
~GREY - [chem] gris *m* de platine
~LAMP - [light] lampe *f* à filament de platine
~PLATED - [el chem] platiné
PLATINUM POINT - [el] grain *m* platiné, grain *m* de platine
~REFORMING - s. platforming
~RESISTANCE THERMOMETER - s. platinum thermometer
~SILVER - [metall] alliage *m* de platine et argent
~SPONGE - [chem] (finely divided platinum having application as a catalyst and as an ignition agent for hydrogen) mousse *f* de platine
~TETRACHLORIDE - [chem] (formed by dissolving platinum in aqua regia) tetrachlorure *m* de platine
~THERMOMETER - [instr] (instrument used to measure changes of temperature in the body on the basis of the changes in the electrical resistance of a platinum coil immersed in that body) thermomètre *m* à résistance de platine
~TONING - [photo] virage *m* au platine
PLATT - [mining] recette *f*
PLATTER - [metall] masse *f* totale de fer à forger
PLATYBASIA - [med] platybasie *f*
PLATYCEPHALIC - [anat] (having a flattened head) platycéphale
PLATYDACTYL - [zool] (having the tips of the digits flattened) platydactyle
PLATYMETER - [instr] (instrument designed to measure the capacity of capacitors) platymètre *m*
PLATYPUS - [zool] ornithorynque *m*
PLATYSMA - [anat] (sheet of dermal musculature in the neck) peaucier *m*
PLAY - [mech] (the amount of free movement between two linked or engaging parts) jeu *m*, chasse *f*
~IN THE TENSION - [mech] jeu *m* dans la tension
~OF COLOURS - [opt] reflets *m*pl irisés, iris *m*
PLAYBACK, to - [el acoust] reproduire
PLAYBACK - [el acoust] (a general term denoting the reproduction of a recording) reproduction *f*
[comput] reproduction *f*
~BUTTON - [el acoust] touche *f* de reproduction
~CHARACTERISTICS - [el acoust] (reproducing characteristics) caractéristique *f* de reproduction
~DURATION - [el acoust] durée *f* d'audition
~HEAD - [el acoust] phonocapteur *m*, pick-up *m*
~LOSS - [el acoust] (a loss of recorded signal) erreur *f* de piste
~LOUDSPEAKER - [telev & cin] haut-parleur *m* de fond
~MACHINE - [el acoust] reproducteur *m* pour radiodiffusion
~ON MONO - [el acoust] reproduction *f* d'enregistrements monophoniques
~ON STEREO - [el acoust] reproduction *f* d'enregistrements stéréophoniques
~STRENGTH - [el acoust] intensité *f* de reproduction
PLEAT, to - [text] plisser
PLEAT - [text] pli *m*
PLEATED DIAPHRAGM LOUDSPEAKER - [el acoust] haut-parleur *m* à diaphragme ondulé
~FABRIC - [text] tissu *m* plissé
PLEDGET - [med] (strip of gauze or pad of lint for a wound) tampon *m*, bourdonnet *m*, tampon *m* de charpie
[naut] (oakum string used in caulking) étoupe *f*
PLEIADES - [astr] Pléiades *f*pl
PLEIOTROPISM - [genet] (a condition in which one factor has an effect on more than one character in the offspring) pléiotropie *f*
PLEISTOCENE PERIOD - [geol] (the period in which an ice-sheet covered North Europe and North America)

pléistocène *m*
PLENALVIA - [vet] (a filling up of the rumen of cattle) entassement *m*
PLENARY - [gen] complet, entier, plénier
~ CAPACITANCE - [el] capacité *f* entre deux fils de un système à plusieurs conducteurs
PLENILOQUENCE - [med] logorrhé *f*
PLENUM - [gen] plein *m*
~ CHAMBER - [th eng] (a vessel from which air is fed to an engine or other units, usually supplied from a ramming intake) chambre *f* de mélange d'air
~ CHAMBER METHOD - [plast ind] préforme *m* en chambre fermée
~ FAN - [mech] ventilateur *m* positif, ventilateur *m* soufflant
~ SYSTEM - [th eng] (a system of air conditioning in which the air in a building is maintained at a higher pressure than the atmosphere) système *m* de mélange d'air
PLEOCHROIC HALO - [min] halo *m* pléochroïque
PLEOCHROISM - [min] (the property of certain crystals to exhibit different colours in different crystallographic directions, due to the selective absorption of transmitted light) pléochroïsme *m*
PLEOCHROMOCYTOMA - [med] pléochromosytome *m*
PLEOCYTOSIS - [med] (increase in the number of white-blood cells) pléocytose *f*
PLEOMASTIA - [med] polymastie *f*
PLEOMAZIA - s. polimastia
PLEOMORPHISM - [cryst] (the property of crystallizing in more than one form) pléomorphisme *m*
PLEOMORPHOUS - [zool] (polymorphic) pléomorphe
PLEON - [zool] (the abdominal region in Crustacea) pléon *m*
PLEONASTE - [min] (oxide of magnesium, iron and aluminium) pléonaste *m*
PLEROMORPH - [min] pléromorphe
PLESSITE - [min] (an alloy of iron and nickel which is found in some iron meteorites) plessite *f*
PLETHORA - [med] pléthore *f*, polyémie *f*
PLETHYSMOGRAPH - [instr] (apparatus designed to measure variations in the size of bodily parts and in the flow of blood through them) pléthysmographe *m*
PLETHYSMOGRAPHY - [med] pléthysmographie *f*
PLEURA - [zool] (the serous membrane lining the pulmonary cavity) plèvre *f*
PLEURAL - [zool] (pertaining to the pleura) pleural
~ CAVITY - [anat] cavité *f* pleurale
~ EFFUSION - [med] épanchement pleural
~ MEMBRANE - [anat] membrane *f* pleurale
PLEURALGIA - [med] pleurodynie *f*, point *m* de côté
PLEURISY - [med] (inflammation of the pleura) pleurésie *f*
PLEUROCELE - [med] hernie *f* du poumon
PLEUROCLYSIS - [med] lavage *m* pleural
PLEURODYNIA - s. pleuralgia
PLEUROGRAPHY - [med] (radiological examination of the pleural cavity) pleurographie *f*
PLEUROLITH - [med] calcul *m* pleural
PLEUROLYSIS - [med] pneumolyse *f*, pleurolyse *f*
PLEUROPERICARDITIS - [med] pleuropéricardite *f*
PLEUROPERITONITIS - [med] pleuropéritonite *f*
PLEUROPNEUMONIA - [med] fluxion *f* de poitrine, pleuropneumonie *f*
PLEUROSCOPE - [med] pleuroscope *m*
PLEXALGIA - [med] plexalgie *f*
PLEXITIS - [med] (inflammation of a nerve plexus) plexite *f*
PLEXUS - [anat] (a mass of interwoven fibres) plexus *m*
PLIABILITY - [gen & phys] (the property of a substance which makes it easily bent or folded) flexibilité *f*, souplesse *f*
~ OF THE FIBRE - [text] flexibilité *f* de la fibre
~ OF THE ROPE - [text] flexibilité *f* d'une corde
PLIABLE - [gen] (easily bent, flexed or folded) pliable, flexible
~ FIBRE - [text] fibre *f* souple, fibre *f* flexible
PLIANT - s. pliable
PLICATION - [geol] plissement *m* de couches, plissotement *m*
PLIERS - [impl] pince *f*, pinces *f*pl
~-TYPE SPOT WELDER - [el metall] machine *f* à souder par points à pince
PLIMSOLL LINE - s. Plimsoll mark
~ MARK - [naut] (a mark on the hull showing how deeply a vessel may be loaded) ligne *f* de flottaison en charge
PLINTH - [constr] (projecting course at the base of a building) plinthe *m* d'un mur, socle *m*
[arch] (the cuboidal base of a column) plinthe *m*
~ TILE - [constr] carreau *m* de plinthe *f*
PLIOCENE PERIOD - [geol] Pliocène *m*
PLIODYNATRON - [electron] pliodynatron *m*
PLIOTRON - [electron] (high-vacuum tube used to control the space-current) pliotron *m*
PLODDER - [mech] (extrusion machine used in soap-making) emboutisseuse *f*
PLOMBAGE - [med] plombage *m*
PLOT, to - [gen] (to mark out the results of observation, e.g. course on a map or chart) lever, relever, tracers
[surv] (by a surveyor, the drawing on paper from field notes) dresser le plan
[town planning] fractionner en parcelles
PLOT - [gen] (a piece of ground) lot *m* de terrain
[surv] (a plan) plan *m*
[naut] tracé *m* en graphique
PLOTTER - [photogrammetry] (visual display in which a dependent variable is marked by a moving pen as a function of the independent variable) traceur *m*, restituteur *m*
[radar] (apparatus for transferring the data obtained on a plotting sheet) traceur *m* de route
[surv] appareil *m* pour le traçage des courbes
PLOTTING - [surv] levé *m*
[draw] tracé *m*, graphique *m*
[radar] (drawing on a chart the position of a moving object) traçage *m*
[photogrammetry] (visual display by a plotter) restitution *f*
[town planning etc] plan *m* parcellaire
~ DEVICE - [surv] appareil *m* de restitution
~ FROM PHOTOGRAPHS - [surv] restitution *f* des clichés
~ INSTRUMENT - s. plotter
~ INTERVAL - [radar] intervalle *m* de traçage
~ OF POINTS - [surv] topométrie *f*
~ PAPER - [draw & surv] papier *m* quadrillé
~ PLATE - [radar] (reflector tracker) plan *m* de traçage
PLOUGH, to - [agric] labourer, creuser

PLOUGH, to - [carp etc] bouveter
[bookbind] rogner le papier
[naut] fendre, sillonner
PLOUGH - [agric] charrue *f*
[carp] bouvet *m*
[el] frotteur *m* souterrain, sabot *m* de prise
[el] (underground collector) sabot *m* souterrain
[bookbind] couteau *m* à rogner
[astr] (Ursa Major) Grande Ourse
~BEAM - [agric] flèche *f* de charrue
~BIT - [tool] bec-d'âne *m* de rabot
~BODY - [agric] corps *m* de la charrue
~BOTTOM - s. plough body
~CULTIVATION - [agric] culture *f* à la charrue
~GRINDING - [text] aiguisage *m* latéral en ouvrant préalablement les canaux par un soc
~-HANDLE - [agric] mancheron *m*
~UNDER, to - [agric] enfouir à la charrue
~UP, to - [agric] retourner à la charrue
PLOUGHING - [agric] labour *m*, labourage *m*
[aero] flottaison
~EQUIPMENT - [agric] matériel *m* de labour
~IN MANURE - [agric] enfouissement *m* de fumier
~-SHARE - [agric] soc *m* de charrue
~SPEED - [agric] vitesse *f* de labour
~UNDER, to - [agric] enfouissement *m* à la charrue
~UP - [agric] retournement *m*
PLOW - s. plough
PLUCK, to - [gen] arracher, cueillir
[gen] (feathers) arracher (des plumes), plumer
[text] éplucher
PLUCK - [food] (the heart, liver and lungs of an animal) fressure *f*
[gen] (a sudden pull) petit coup *m* sec
~THE WOOL, to - [text] éplucher la laine
PLUCKED INSTRUMENT - [mus] (a string instrument which is played by plucking the strings) instrument *m* à cordes pincées
~WOOL - [text] laine *f* pelade, laine *f* avalie
PLUCKER - [text] (for long-staple wool) éplucheuse *f*
PLUCKING - [text] arrachage *m*
PLUG, to - [gen] tamponner, boucher, taper
[med] tamponner
[mech] (of pipes etc) taper, boucher
[el] (the operation of braking an electric motor by making it run in the reverse direction) freiner par contre-courant
[oil ind] boucher, tamponner
[constr] enfoncer des chevilles, sceller (un mur)
[mining] (in quarry work) enferrer, faire le chemin
PLUG - [gen] tampon *m*, bouchon *m*
[mech] (any part, usually cylindrical, used to close an opening) tampon *m*
[mech] (of a pipe) (externally threaded cylindrical part which is screwed into an opening) tampon *m* fileté, cheville *f*, fusible *m*
[impl] poinçon *m*
[el] (device consisting of two portions, a plug and a socket, having metallic contacts and arranged to engage each other) fiche *f* de connexion
[el] (a spark plug) bougie *f* d'allumage
[el] (a fusible plug) cheville *f* fusible, rondelle *f* fusible
[telecomm] (of a telephone exchange) fiche *f*, raccord *m*
[hydr] (discharge outlet) bouche *f* d'eau, hydrante *f*
PLUG - [hydr] (of a cock) canillon *m*, nois *f* d'un robinet
[hydr] (in a reservoir) bonde *f*, crapaudine *f*
[metall] (remains in a tapping hole of a furnace) carotte *f*
[tool] (used in plug drawing) coin *m*, cale *f*
[metall] (in foundry) bouchon *m* d'obturation de trou de coulée
~ADAPTER - [el] (lampholder plug) fiche *f* d'adaptation
~AND FEATHER, to - [mining] bosseyer, battre des aiguilles infernales
~AND FEATHER - [mining] aiguille-coin *f*, aiguille *f* infernale
~AND FEATHERING - [mining] bosseyage *m*
~-AND-FEATHERING MACHINE - [mech] bosseyeuse *f*
~AND JACK - [el] fiche *f* et mâchoire de contact
~-AND-SOCKET - [el] (device consisting of a plug and a socket, forming a ready means of connecting or disconnecting a current-using apparatus) prise *f* de courant
~ASSISTS - [plast ind] (plug-like parts used to introduce sheet into deep moulds) aides *mpl* de pressage
~BIB - s. plug cock
~-BOARD - [el] panneau *m* à accouplement à fiches
~BOX - [el] boîte *f* de contacts
~BOX-WRENCH - [impl] clef *f* à douille
~BUSH - [el] manchon *m* de contact
~CAPACITANCE BOX - [el] boîte *f* de capacités à fiches
~CENTRE-BIT - [tool] foret *m* à teton cylindrique
~COCK - [mech] (type of valve in which the obturating element is a cylinder or cone revolving in a corresponding recess in the body) robinet *m* à clef
~COMMUTATOR - [el] commutateur *m* à fiche
~CONNEXION - [el] connexion *f* à fiche
~CONTACT - [el] prise *f* de courant
~CORD - [el] cordon *m* de fiche
~DRAWING - [metall] étirage *m* de tubes à mandrin fixe
~-DRILL - [tool] perforatrice *f* percutante à main
~FUSE - [el] (form of fuse in which the fuse link is housed in a plug) coupe-circuit *m* à fiche
~GAUGE - [instr] (plug-shaped gauge used to test the diameter of an opening) calibre *m* à bouchon
~HOLE - [plumb] trou *m* de clef de robinet
~IN, to - [telecomm] enficher, établir une communication
~-IN CIRCUIT CARD - [comput] carte *f* imprimée enfichable
~-IN CIRCUITRY - [el] carte *f* enfichable
~-IN COIL - [radio] (inductance coil fitted with a set of contact pins on the base, thus making it easily interchangeable) bobine *f* interchangeable
~-IN MODULE - [comput] module *m* enfichable
~-IN RELAY - [el] relais *m* à fiches
~-IN UNIT - [electron] bloc *m* interchangeable, élément *m* à fiches
~-IN TRANSFORMER - [el] transformateur *m* interchangeable
~INDUCTANCE BOX - [el] boîte *f* à inductances à fiches
~KEY - [impl] canillon *m*
~LINE - [metall] traces *fpl* de mandrin
~-OPERATED RHEOSTAT - [el] rhéostat *m* à fiche
~RESISTANCE BOX - [el] boîte *f* de résistances à fiches

PLUG TAP - [mech] (the final tap required to finish a blind hole) taraud *m* demi-conique, taraud *m* finisseur
[tool] (only USA) taraud *m* intermédiaire
~-TENON JOINT - [carp] enture *f* à simple tenon
~THREAD GAUGE - s. plug gauge
~-TYPE DOOR - [aero] (type of aircraft door which fits into a recess for sealing) grande porte *f* à bouchon
~TYPE WAVEGUIDE TERMINATOR - [electron] bouchon *m* d'adaptation
~VALVE - s. plug cock
~WELD - [el metall] soudure *f* en bouchon
PLUGBOARD - s. plug-board
PLUGGED STEEL - [metall] acier *m* effervescent contrôlé
PLUGGER - [med] obturateur *m* dentaire, fouloir *m*
PLUGGING - [gen etc] tamponnement *m*, bouchage *m*
[el] freinage *m* par contre-courant
[constr] (sound insulating material) matériel *m* isolant
[metall] bouchage *m*
~CHART - [el] schéma *m* de connexions
~LOOP - [instr] (in reactor instrumentation) boucle *f* détectrice d'obstructions
~OF THE VOIDS - [soil] colmatage *m* des interstices
PLUM - [bot] prune *f*
~TREE - [bot] prunier *m*
PLUMAGE - [zool] plumage *m*
PLUMB, to - [gen] (to test a perpendicular with a plumb) plomber
[gen] (to make plumb) mettre à plomb
[gen] (to gauge the depth) sonder
[gen] (to seal) plomber
(only USA; to supply a building with pipes) poser les tuyauteries
PLUMB - [gen] (the load weight at the end of a line) plomb *m* (de fil à plomb)
[meas] aplomb *m*
[naut] (a sounding lead) sonde *f*, ligne *f* de sonde
~BOB - [impl] (the actual weight at the end of a line) plomb *m* de fil à plomb
~LEVEL - [meas] niveau *m* vertical
~LINE - [impl] fil *m* à plomb
[naut] ligne *f* de sonde
~RULE - [meas] (narrow board used to check verticals) niveau *m* vertical, niveau *m* de maçon
PLUMBAGO - [min] (a synonym for graphite, used in making crucibles and pencils, lubricants etc) plombagine *f*, mine *f* de plomb, graphite *f*
~CRUCIBLE - [metall] (crucible made of graphite and clay) creuset *m* en plombagine, creuset *m* en graphite
PLUMBATE - [chem] plombate *m*
PLUMBER - [gen] plombier *m*
PLUMBER'S CAULKING TOOL - [tool] matoir *m*
~FIRE - [constr] brasier *m* du plombier
~JOINTS - [plumb] assemblage *m* de plomberie
~WIPED SOLDERED JOINT - [plumb] (welding; soldering) soudure *m* de plombier
PLUMBIC - [chem] plombique
~OCHRE - [min] massicot *m*
PLUMBIFEROUS - [min] plombifère
PLUMBING - [hydr etc] (the installation of domestic water-supply systems, sanitary fittings etc) plomberie *f*, plombage *m*, tuyauterie *f*
[electron] (term denoting waveguides and accessory equipment for radio frequency transmissions) équipement *m* de guide d'ondes
PLUMBISM - [med] (lead poisoning, usually chronic, often exhibited by those who work with lead and its compounds) saturnisme *m*
PLUMBITE - [chem] (lead hydroxide) hydroxyde *m* de plomb
PLUMBONIOBITE - [min] (yttrium, ytterbium, iron, lead, uranium etc containing mineral) plumboniobite *f*
PLUMBOSOLVENCY - [chem] (the capacity of drinkable water to dissolve lead, espec. from pipes and the like) solubilité *f* du plomb
PLUME - [zool and bot] plume *f*
[gen] (of gas or smoke from a chimney) panache *m*
[radar] écho *m* artificiel en plume
[nucl] (or column) colonne *f* d'eau
PLUMIPED - [zool] (with feathered feet) plumipède
PLUMMER - [mech] palier *m*, empoise *f*
~BLOCK - [mech] (journal bearing for line shafting) palier *m*
[metall] coussinet *m*
PLUMMET - [gen] s. plumb bob
[naut] ligne *f* de sonde
[impl] niveau *m* de maçon
~DEVICE - [naut] sonde *f*
~-LAMP - [mining] lampe *f* à fil de plomb, fil *m* à plomb à lampe
~LEVEL - [impl] fil *m* à plomb
~-LINE - [impl] plomb *m*, aplomb *m*
PLUMOSE - [gen] (having feather-like characteristics) plumeux
PLUMP-GATE -[metall] chenal *m* de coulé en chute directe
PLUNGE, to - [gen] plonger, immerger
[metall] tremper
PLUNGE ACTION - [mech etc] (as a piston) action *f* plongeante
~CUT GRINDING - [mach tool] rectifieuse *f* à avancement à plongeon
PLUNGER - [mech] (a reciprocating part usually similar to and performing the functions of a piston) plongeur *m*, piston *m*
[el] (of an electric suction coil) noyau *m* mobile, piston *m* de douille
[el] plongeur *m*
[mech] piston *m*, plongeur *m*
[mech] (of a press) piston *m*, piston *m* plein
[metall] quenouille *f*
[auto etc] (of the valve fitted to the inner tube) pointeau *m*
[glass man] poinçon *m*
[oil ind] quenouille *f*
[electron] (in waveguides) piston *m* de court-circuit
~ARMATURE STUD - [mech] (of a spring) bouton *m* de poussée
~ATTENUATOR - [electron] (in waveguides) affaiblisseur *m* à piston
~ELEVATOR - [mech] ascenseur *m* à piston plongeur
~LEATHER - [auto] garniture *f* de cuir du plongeur
~LIFT - [mech] levée *f* du piston
~MOULDING - [plast ind] (transfer moulding) transfert *m*
~PLATE - [mech] plaque *f* positive
~PUMP - [mech] pompe *f* à piston
~ROD - [die casting] tige *f* de piston d'injection
~SET - [mech] jeu *m* foulant

PLUNGER SPRING - [mech] (in a fuel injection pump) ressort *m* de piston
~SPRINGING - [mech] (in motorbicycles) suspension *f* à ressorts
~-TYPE PUMP - s. plunger pump
~-TYPE RELAY - [electron] relais *m* à bobine plongeante
PLUNGING - [surv] plongée *f* (de la lunette d'un théodolite)
~MACHINE - [metall] (in die casting) machine *f* à chambre chaude à piston
PLURAL - [gen] plural, pluriel *m*
~PRODUCTION - [nucl] production *f* multiple en collisions successives
~SCATTERING - [nucl] diffusion *f* nombreuse
~SERVICE - [el] distribution *f* mixte
PLURICELLULAR - [biol] pluricellulaire
PLURIGLANDULAR - [med] (pertaining to several glands) pluriglandulaire
PLURIMOTOR - [aero] à plusieurs moteurs
PLURIVALENT - [biol] (formed of more than two chromosomes) plurivalent
PLUS - [gen & math] plus
~AXIS - [opt] axe *m* positif
~COUNT - [astronaut] compte *m* après lancement
~CYLINDER - [opt] cylindre *m* positif
~HUB - [comput] jack *m*
~PLATE - [el chem] (positive plate) plaque *f* positive
~SIGHT - [surv] coup *m* arrière
~SIGN - [math] plus *m*, signe *m* de l'addition
~SPHERE - [opt] sphère *f* positive
~-ZONING - [comput] sélection *f* de la zone "Plus"
PLUSH - [text] (fabric with a cut pile on one or both sides) peluche *f* panne *f*
~BOARD - [text] planche *f* à peluche
~COPPER ORE - [miner] (chalcotrichite) chalcotrichite *f*
~DRESS MATERIAL - [text] peluche *f* pour articles d'habillement
~NEEDLE BAR - [text] barre *f* à peluche
~THEARD CLEARER - [text] nettoyeur *m* en peluche
~WEAVING - [text] tissage *m* de la peluche
PLUTO - [astr] (the ninth major planet in the solar system) Pluton *m*
PLUTONIC INTRUSIONS - [geol] intrusions *f*pl plutoniques
~ROCKS - [geol] (a very large intrusions) roches *f*pl plutoniques
PLUTONITES - [geol] (rocks occurring in major intrusions) plutonites *f*pl
PLUTONIUM - [chem] (a synthetic element, symbol Pu, with chemical properties similar to those of uranium) plutonium *m*
~AEROSOL MONITOR - [nucl] moniteur *m* pour aérosols de plutonium
~BOMB - [nucl] bombe *f* au plutonium
~BREEDER - [nucl] réacteur *m* de production
~ENRICHED FUEL - [nucl] combustible *m* en plutonium enrichi
~FISSION - [nucl] fission *f* de plutonium
~PRODUCING REACTOR - [nucl] réacteur *m* de production de plutonium
~REACTOR - [nucl] (reactor in which plutonium is the main fissionable material) réacteur *m* au plutonium
~SEPARATION PLANT - [nucl] installation *f* de séparation du plutonium
PLUVIOGRAPH - [instr] (self-registering rain-gauge) pluviomètre *m* enregistreur
PLUVIOMETER - [instr] (a rain-gauge) pluviomètre *m*
PLY, to - [gen] manier avec force
[gen] (to fold) plier
[naut] louvoyer
PLY - [gen] pli *m*
[gen] (single sheet used in making laminated material) pli *m* de tissu
[of tyres] couche *f* de nappe de corde, plis *m*
~ADHESION - [rubber ind] (in tyres) adhérence *f* entre plis
~-GAUGE - [rubber ind] (in tyres) épaisseur *f* de couche
~LOOSENESS - [rubber ind] (in tyres) séparation *f* entre plis, décollage *m* entre plis
~RATING - [rubber ind] (the number of plies of a tyre) "ply rating" *m*, résistance *f* d'un pneu indiqué en nombre équivalent de plis en coton
~SEPARATION - s. ply looseness
~TURN-UPS - [rubber ind] (in tyres) retournement *m* des plis
~YARN - [text] (fancy type of yarn) fil *m* retors
PLYING - [rubber ind] (in tyres) application *f* des couches
PLYWOOD - [timber] (material composed of wooden lamellae bonded together, the grain of each lamella running at right angles to that of the preceding one) bois *m* contreplaqué
~ADHESIVE - [ind chem] (hot- and cold-setting urea/formaldehyde and pheno/formaldehyde resin based adhesives) colle *f* pour contreplacage
~GLUE - s. plywood adhesive
p-m ERASING HEAD - [el acoust] (erasing head the fields of permanent magnets) tête *f* d'effacement à aimants permanents
p-n BOUNDARY - [electron] (surface in the transition region between P-type and N-type material) zone *f* p-n
p-n HOOK TRANSISTOR - [electron] transistor *m* à dépression
p-n JUNCTION - [electron] (region of transistor between P- and N-type semiconductors) jonction *f* p-n
p-n JUNCTION PHOTOCELL - [electron] (photoelectric of the P-N-junction type) phototransistor *m* p-n
PNEU - [gen] (abbrev for pneumatic) pneu *m*
PNEUMATIC - [phys mech etc] (operated or operating by means of a gas under pressure) pneumatique
[rubber ind] (a tyre) pneumatique *m*
~BOAT - [naut] (inflatable boat) bateau *m* pneumatique
~BRAKE - [railw] (air brake) frein *m* à air comprimé
~BRAKE HOSE - [railw] manchon *m* pour freins à air comprimé
~CAISSON - [hydr] caisson *m* à air comprimé
~CARD STRIPPER - [text] débourreur *m* de carde pneumatique
~CARD STRIPPING - [text] débourrage *m* pneumatique de la carde
~CAULKING - [naut] calfatage *m* pneumatique
~CHIPPER - [tool] burineur *m* à air comprimé
~CHISEL - [tool] trépan *m* pneumatique
~CHUCK - [mach tool] mandrin *m* pneumatique

PNEUMATIC CIRCUIT - [aero etc] (a system of piping and equipment designed for a specified purpose, e.g. flap control) circuit *m* pneumatique
~CONTROLLER - [contr] régulateur *m* pneumatique
~CONVEYOR - [mech] (machine conveying loose materials through tubes by air) conveyor *m* pneumatique
~CUSHION - [mech] amortisseur *m* pneumatique
~CUSHIONING - [mech] amortissement *m* à air comprimé
~DE-ICING - [aero] (method of detaching ice from the leading edges of wings etc. by means of a boot which can be distended by air pressure, to break up and separate the skin of ice) dégivrage *m* pneumatique
~DIGGER - [mech] excavateur *m* pneumatique
~DRILL - [mech] (hard rock drill working by compressed air) perforatrice *f* pneumatique, perforateur *m* à air comprimé
~DRYER - [mech] (apparatus for drying filler ingredients and the like, by blowing hot gases through them separating in a cyclone) séchoir *m* à gaz
~ELEVATOR - [transp] ascenseur *m* à air comprimé
~EXCAVATOR - [mech] excavateur *m* à air comprimé
~-FEED COLUMN - [mining] colonne *f* à avancement pneumatique
~FLOAT - [aero] (for seaplanes) flotteur *m* pneumatique
~HAMMER - [tool] marteau *m* pneumatique, marteau *m* piqueur, marteau *m* brise-béton
~HAMMER-DRILL - [mech] marteau *m* perforateur pneumatique
~HOIST - [mech] treuil *m* à air comprimé, palan *m* pneumatique
~HOISTING - [mining] extraction *f* pneumatique
~JACK - [mech] vérin *m* pneumatique
~JIG - [mech] jig *m* à air
~KNOCK-OUT - [metall] décochage *m* pneumatique
~LIFT - [mech] monte-charge *m* pneumatique
~LOUDSPEAKER - [el acoust] (loudspeaker depending on controlled variations of an air stream) haut-parleur *m* pneumatique
~MACHINE - [mech] machine *f* pneumatique
~MICROMETER - [instr] (type of precision measuring instrument in which the variation of diameter of a bore from a preset value is determined by the rate of escape of compressed air from ports in a measuring head placed within the bore in question) micromètre *m* pneumatique
~MIXING - [text] mélange *m* pneumatique
~MOTOR - [mus] (of an organ) moteur *m* pneumatique
~MOTOR OPERATOR - [contr] élément *m* de réglage final à mouvement pneumatique
~OPERATION OF THE SHUTTLE - [text] mouvement *m* pneumatique de la navette
~PACKING - [mining] remblayage *m* pneumatique
~PICK - [mech] (tool used in road construction) fleuret *m* pneumatique
~POSITIONING RELAY - [contr] (relay device fitted to a diaphragm motor) relais *m* mis en position pneumatiquement
~POWER-HAMMER - [impl] marteau-pilon *m* atmosphérique
~-PROBE PYROMETER - [instr] canne *f* pyrométrique à aspiration
~PULLDOWN - [telev] transporteur *m* pneumatique
~PUMP - [phys] machine *f* pneumatique

PNEUMATIC RAILWAY - [railw] chemin *m* de fer atmosphérique
~RAMMER - [metall] fouloir *m* à air comprimé [mech] damoir *m* à air comprimé
~RAMMING - [metall] serrage *m* pneumatique
~RELAY - [contr] relais *m* pneumatique
~RELEASE - [photo] déclencheur *m* pneumatique
~RESERVOIR - [gen] récipient *m* pneumatique
~REVERSE - [mech] marche *f* arrière à air comprimé
~RIVETER - [mech] (high-speed riveting machine delivering over 1,000 blows per minute) rivoir *m* pneumatique
~RIVETING - [mech] rivure *f* pneumatique
~RIVETING HAMMER - [tool] marteau *m* pneumatique à river
~ROCK-DRILL - [mining] machine *f* à forer
~SANDER - [railw] sablière *f* à air comprimé
~SEAT - [gen] siège *m* pneumatique
~SEPARATION - [min] préparation *f* à air comprimé
~SEPARATOR - [min] séparateur *m* pneumatique
~SQUEEZE RIVETING MACHINE - [mech] rivoir à serrage pneumatique
~STAMP - [min] bocard *m* pneumatique
~STARTER - [aero] (starter for aircraft engines by compressed air) démarreur *m* pneumatique
~SUCTION ELEVATOR - [mech] ascenseur *m* à air comprimé
~SYSTEM - [mech] (a complete compressed-air system) système *m* pneumatique
~SYSTEM TO CONVEY COTTON - [text] installation *f* de transport pneumatique du coton
~TABLE - [min] table *f* d'épuration pneumatique
~TEST - [min] essai *m* pneumatique
~TRANSMISSION SYSTEM - [mech] système *m* pneumatique de transmission
~TROUGH - [min] cuve *f* pneumatique
~TUBE CONVEYOR - [mech] transporteur *m* pneumatique
~TYRE - [rubber ind] bandage *m* pneumatique, pneu *m*
~TYRE DRIVER - [rubber ind] batte *f* de pneu pneumatique
~VALVE - [mech] (used in gas installations a pressure controlled valve) soupape *f* pneumatique
~WRENCH - [tool] clef *f* pneumatique
PNEUMATICALLY-OPERATED CIRCUIT BREAKER - [el] (a circuit-breaker based on a piston operated by compressed air) interrupteur *m* pneumatique
~OPERATED SWITCH - s. pneumatically-operated circuit breaker
PNEUMATICS - [mech] (the branch of physics dealing with the mechanical properties of air and other gases) pneumatique *f*
PNEUMATINURIA - [med] pneumaturie *f*
PNEUMATOSIS - [med] pneumatose *f*
PNEUMECTOMY - [med] pneumectomie *f*
PNEUMOARTHROGRAPHY - [med] pneumoarthrographie *f*
PNEUMOCENTESIS - [med] ponction *f* du poumon
PNEUMOCHOLECYSTITIS - [med] cholécystite *f* gazeuse
PNEUMOCOCCEMIA - [med] pneumococcémie *f*
PNEUMOCOCCUS - [med] (the causative agent of pneumonia) pneumocoque *m*
PNEUMOCONIOSIS - [med] (potter's asthma or mason's lung) pneumoconiose *f*
~SIDEROTICA - [med] sidérose *f* pulmonaire
PNEUMODERMA - [med] emphysème *m* sous-cutané

PNEUMOGRAPHY - [med] pneumographie *f*
PNEUMOHYDROPERICARDIUM - [anat] pneumohydropéricarde *m*
PNEUMOHYDROTHORAX - [med] pneumohydrothorax *m*
PNEUMOHYPODERMA - s. pneumoderma
PNEUMOLITH - [med] pneumolithe *m*
PNEUMOMASSAGE - [med] massage *m* pneumatique de l'oreille moyenne
PNEUMOMETER - [instr] (air speed indicator) pneumomètre *m*
PNEUMOMYCOSIS - [med] mycose *f* pulmonaire
PNEUMONECTASIS - [med] emphysème *m* pulmonaire
PNEUMONEDEMA - [med] oedème *m* pulmonaire
PNEUMONEMIA - [med] congestion *f* pulmonaire
PNEUMONIA - [med] pneumonie *f*
PNEUMONIC PEST - [med] peste *f* pneumonique
PNEUMONICYTE - [biol] cellule *f* alvéolaire
PNEUMONOMONILIASIS - [med] moniliase *f* pulmonaire
PNEUMONOMYCOSIS - s. pneumomycosis
PNEUMOPERICARDIUM - [anat] pneumopéricarde *m*
PNEUMOPERITONEUM - [anat] pneumopéritoine *m*
PNEUMOPEXY - [med] pneumopexie *f*
PNEUMORACHIS - [med] pneumorachis *f*
PNEUMOTHORAX - [med] (injection of air or gas into the pleural cavity) pneumothorax *m*
~APPARATUS - [med] (an apparatus designed to inject air or gas into the pleural cavity for therapeutical purposes) appareil *m* pour pneumothorax
p-n-i-p TRANSISTOR - [electron] transistor *m* p-n-i-n
POACH, to - [gen] braconner
[agric] (of the ground) labourer, piétiner la terre
[soil] (of clay) (to reduce to uniform consistency by the addition of water) se détremper
[paper man] (to bleach) blanchir, décolorer
POACHER - [paper man] bac *m* de blanchissage
POCKET, to - [gen] empocher
[mech] monter en retrait
POCKET - [gen] poche *f*
[gen] (a bag) sac *m*
[adj] de poche, portatif
[metall] (die casting) couche *f* d'air
[mech] retrait *m*
[mech] (in internal combustion engines) chambre *f* de soupape
[geogr] (a glen) ravin *m*, gorge *f*
[el chem] (of an accumulator plate) pochette *f*
[min] nid *m* de minerai
[phys] (air pocket) cavité *f* remplie de gaz, cavité *f* remplie d'eau
[geol] (water containing cavity) cavité *f* remplie d'eau
[mining] fosse *f* de remplissage
[metall] (in a casting) cavité *f*
[comput] (a card stacker in a sorter) case *f*
~BATTERY MONITOR - [nucl] (a radiation counter of small size and operated by batteries) moniteur *m* portatif à pile
~CALCULATOR - [comput] cercle *m* à calcul
~CHAMBER - [nucl] (pocket-size ionization chamber used to measuring the radiation exposure of personnel) chambre *f* d'ionisation de poche
~COMPASS - [instr] boussole *f* de poche
~DRILLING - [agric] semis *m* en poquets
~FLASHLIGHT - [light] lampe *f* de poche
~GRID - [text] grille *f* en forme de poches
POCKET HAMMER - [geol] marteau *m* de géologue
~HANDKERCHIEF - [text] mouchoir *m*
~INSTRUMENT - [instr] (instrument which can be carried into a pocket) instrument *m* de poche
~ION CHAMBER - s. pocket chamber
~IONIZATION CHAMBER - s. pocket chamber
~MAKING DRUM - [rubber ind] (tyre pocket building machine) tambour *m* à confectionner les nappes de câbles, tambour *m* confection bracelets
~METER - s. pocket chamber
~OF FINE DAMP - [mining] sac *m* de grisou, poche *f* de grisou, ballon *m*
~OF INFECTION - [med & vet] foyer *m* d'infection
~OF INFESTATION - [bot] foyer *m* d'infestation
~OF ORE - [min] poche *f* de minerai
~OF WEB - [text ind] poche *f* du voile
~PRINT - [metall] portée *f* montante, portée *f* tirée à l'anglaise
~SPEED COUNTER - [instr] compte-tours *m* de poche
~STEEL TAPE - [impl] roulette *f* de poche à ruban de acier
~TAPE RECORDER - [el acoust] magnétophone *m* de poche
~TYPE PLATE - [el chem] (type of plate containing recesses filled with active material) plaque *f* à pochettes
~WHEEL - [mech] poulie *f* à empreintes
POCKETED VALVE - [mech] soupape *f* en retrait
POCKETING - [text] toile *f* pour poches
[mech] montage *m* en retrait
POCKETS - [rubber ind] (a pair of rubberized plies from which the carcass is built) plis *mpl* de nappes
POCKETY - [mining] poché
POCKMARK - [gen etc] marque *f*
[med] stigmate *m* (de la petite vérole)
[paint] (a defect) grain *m*
POCKMARKED - [gen paint etc] marqué, picoté
POCKWOOD - [timber] gaïac *m*
p.o.d. - [comm] (abbrev for Pay On Delivery) contre chèque
POD - [tool] mandrin *m* de vilebrequin
[aero] (an enclosed structure in an aircraft, separate from the fuselage to contain an engine or other unit) nacelle *f*
[mech & carp] (a groove) cannelure *f*
[bot] cosse *f*, gousse *f*
[astronaut] compartiment *m* détachable, nacelle *f*
[min] lentille allongée de minerai
~AUGER - [tool] tarière *f* à gouge
~BIT - [tool] foret *m* à gouge
~BOILER - [nucl] chaudière *f* encastrée
~DISEASE - [bot] maladie *f* des graines
PODAL - [zool] (pertaining to the foot) du pied
PODALGIA - [med] douleur *f* du pied
PODDED ENGINE - [aero] (engine closed in a pod) moteur *m* en nacelle
PODEX - [anat] (the anal region) fesses *fpl*, siège *m*
PODIUM - [arch] podium *m*
PODOBROMIDROSIS - [med] bromhidrose *f* des pieds
PODOCARP - [bot] podocarpe *m*
PODODYNIA - s. podalgia
PODOLITE - [min] podolite *f*
PODOMETER - [instr] (instrument to measure distances used by surveyors) podomètre *m*
PODZOL - [soil] podsol *m*
PODOZOLIZATION - [soil] podzolisation *f*
POGGENDORF CELL - [el chem] (a single-fluid type

of bichromate cell. Unit of absolute viscosity, such that a force of one gramme will maintain unit rate of sheat of a film of unit thickness between surfaces of unit aera, viz. one dyne-second per sq. cm.) élément *m* de Poggendorf

POGGENDORF COMPENSATION METHOD - [el] (method of determining the electromotive force a cell) méthode *f* à compensation de Poggendorf

POID - [acoust] (US for sinusoide) sinusoïde *f*

POIKILOCYTE - [biol] poïkilocyte *m*

POIKILODERMA - [med] poïkilodermie *f*

POIKILOTHERM - [biol] poïkilotherme, animal *m* à sang froid

POIKILOTHYMIA - [med] instabilité *f* d'humeur

POINT, to - [gen] tailler en pointe, faire une pointe
[constr] (of a wall) jointoyer, gobeter, bloquer un mur
[firearms] pointer, braquer
[surv] (an instrument) diriger, orienter
[gen] (to show, e.g. the way) désigner
[gen] (to mark with points) marquer de points
[med] (of an abscess) mûrir
[constr] rejointoyer

POINT - [gen] (a sharp end) pointe *f*, extremité *f*
[geom etc] point *m*
[print etc] (a dot, also a prick etc) point *m*
[el] contact *m*
[gen] (a characteristic) caractéristique *f*
[gas ind] point *m* d'alimentation, prise *f* de gaz
[railw] aiguille *f* de raccordement
[naut] queue *f* de rat
[math] (the dot marking the separation between the integral andthe fractional part of a number) virgule *f*
[tool] pointe *f*, poinçon *m*
[tool] (of a drill) mouche *f* d'un foret
[mus] (the end of the bow of a stringed instrument) pointe *f*
[zool] (of a horse) extremité *f*
[print] point *m* typographique
[paper man] unité *f* d'épaisseur
[instr] (e.g. of a thermometer) division *f*

~-BEARING PILE FOUNDATION - [soil] fondation *f* sur pieux chargés en pointe

~CATHODE - [electron] (an experimental cathode) cathode *f* à pointe, cathode punctiforme

~-CHUCK - [mach tool] plateau *m* à toc, plateau *m* pousse-toc

~CIRCLE - [mech] (of a gear) cercle *m* extérieur, cercle *m* de tête

~CONTACT - [electron] (pressure contact between a semiconductor and a metallic point) contact *m* de pointe

~-CONTACT DIODE - [electron] diode *f* à pointe

~CONTACT RECTIFIER - [electron] (contact rectifier in which the contact is between a metallic point and an extrinsic semiconductor) diode *f* à cristal, à contact de pointe, redresseur *m* à cristal à contact de pointe

~CONTACT TRANSISTOR - [electron] (a type of transistor which metal wires are in contact with regions in the bases) transistor *m* à contacts depointe

~CONTROL SWITCH - [railw] commutateur *m* de contrôle d'aiguilles

~COUNTER TUBE - [nucl] (counter tube in which the central electrode is a point or a small sphere) tube *m* compteur à pointe

POINT DETECTOR - [railw] dispositif *m* de contrôle de position d'aiguille

~DIAMOND - [impl] diamant *m* à pointes naïves

~DISCHARGE - [el] décharge *f* entre pointes

~DISCHARGE RECORDER - [instr] enregistreur *m* de décharges entre pointes

~ELECTRODE - [el] électrode *f* active, tâteur *m*

~FINGER - [text] came *f* à diminuer

~-FOCAL LENS - [opt] lentille *f* à image ponctuelle

~GROUP - [phys] (the symmetry classes) classes *fpl* de symétrie

~IMAGES - [opt] points-images *mpl*

~INDICATOR - [railw] (a diverging junction signal) indicateur *m* mécanique de position d'aiguille

~-JUNCTION TRANSISTOR - [electron] transistor *m* à contact de pointe et à jonction

~LATTICE - [nucl] (e.g. the sites of atoms in a crystal) réseau *m* de points

~LEVER - [railw] levier *m* d'aiguille

~LOAD - [soil] charge *f* ponctuelle

~LOCK - [railw] serrure *f* d'aiguille

~LOCKING - [railw] verrouillage *m* des aiguilles

~MEASURING - [photo] mesure *f* ponctuelle

~MECHANISM - [railw] mécanisme *m* de commande d'aiguille

~MOTOR - [railw] moteur *m* d'aiguille

~-OBSERVER - [phys] (idealized concept used in relativity theory) observateur *m* ponctuel

~OF ACTION - [mech] point *m* d'application

~OF APPLICATION OF FORCES - [phys] (the point at which the force is applied) point *m* d'application des forces

~OF BLADE - [railw] pointe *f* de l'aiguille

~OF CARD WIRE - [text] aiguilles *fpl* de garniture

~OF CONNECTION - [el] point *m* de raccordement

~OF CONTACT - [gen] point *m* de contact
[mech] (of gears) point *m* de contact

~OF CRITICATLITY- [phys] (critical point) point *m* critique

~OF DEPARTURE - [gen] point *m* de départ

~OF DISTANCE - [surv] point *m* de distance

~OF ENTRY - [aero] (the point at which an aircraft crossed the control zone boundary in approach) point *m* d'entrée

~OF FAILURE - [soil] point *m* de rupture

~OF FROG - [railw] pointe *f* réelle du coeur

~OF FUSION - [metall] point *m* de fusion

~OF IMPACT - [phys] point *m* du choc
[electron] point *m* d'impact

~OF INFLECTION - [geom] point *m* d'inflexion

~OF INTERCONNECTION - [plumb] (a mains junction) noeud *m* de canalisations

~OF NO FLOW - [plumb] (in gas installations) point *m* neutre

~OF NO RETURN - [aero] (the critical point) point *m* critique

~OF OSCILLATION - [electron] point *m* d'oscillation

~OF OSCULATION - [math] point *m* d'attouchement

~OF REFERENCE - [surv] repère *m*

~OF RUPTURE - [rubber ind] point *m* de rupture

~OF SIGHT - [surv] point *m* de vue

~OF STAPLE - [text] pointe *f* de la fibre

~OF SUPPORT - [gen & arch] point *m* d'appui

~OF TANGENCY - [geom] point *m* de tangence

~OF THE COMPASS - [instr] aire *f* de vent

~OF THE HORSE - [mining] point *m* de ramification d'un filon

POINT OF THE PIRN - [text] extrémité *f* en pointe du fuseau
~OF TONGUE - [railw] pointe *f* mathématique du coeur
~OF TOOTH - [mech] (of a gear) tête *f* de la dent d'engrenage
~OF VIEW - [gen] point *m* de vue
~OPERATING APPARATUS - [railw] appareil *m* de manoeuvre d'aiguille
~OPERATING STRETCHER - s. point rod
~PAPER - [text] carte *f*
~PAPER WITH PRINTED DRAFT - [text] papier *m* quadrillé avec pointé de liage imprimé
~PASS - [text] chevron *m* à point, remettage *m* à retour
~-PLOTTING - [comput] tracé *m* point par point
~POSITION - [comput] position *f* d'une virgule
~PRESSURE - [soil] pression *f* de pointe
~RAIL - [railw] pointe *f* de croisement, aiguille *f*
~RESISTANCE OF A PILE - [soil] résistance *f* à la pointe d'un pieu
~ROD - [railw] tringle *f* de manoeuvre
~SETTING - [comput] positionnement *m* d'une virgule
~SHAFT - [text] lame *f* de pointe
~SHIFTING - [comput] décalage *m* de la virgule
~SIZE - [print] corps *m*
~SOURCE - [opt] (a very small source of light) source *f* ponctuelle
~STRETCHER - [railw] tige *f* de manoeuvre
~-SWITCH - [railw] changement *m* à aiguilles
~SYSTEM - [print] système *m* de force de corps
~TO POINT CIRCUIT - [telecomm] (permanent circuit between two telegraph sets) circuit *m* poste à poste
~-TO-POINT CIRCUIT - [electron] liaison *f* poste à poste
~-TO-POINT POSITIONING SYSTEM - [comput] système *m* de positionnement
~-TO-POINT RADIO COMMUNICATION - [radio] (radio communication between two fixed stations) radiocommunication *f* entre deux points fixes
~-TOOL - [mach tool] outil *m* à pointe
~VORTEX - [phys] (the section of a straight-line vortex in two-dimensional motion) point-tourbillon *m*
POINTED - [gen] pointu, à pointe, aigu
~ARCH - [arch] ogive *f*
~-ARM SAFETY-CATCH - [mining] parachute *f* à bras pointus
~ASHLAR - [constr] pierre *f* taillée en angle aigu
~BOLT - [mech] boulon *m* à tige à pointe
~CENTRE - [tool] pointe *f* à outil
~CHISEL - [tool] burin *m* à pointe
~COCOON - [text] cocon *m* pointu
~COP NOSE - [text] pointe *f* de canette trop longue
~ENDS - [text] extrémités *fpl* pointues
~HAMMER - [constr] marteau *m* à pointe
~IRON END - [text] épée *f*
~LIGHTNING PROTECTOR - [el] paratonnerre *m* à pointes
~PROJECTION - [mech] cheville *f* pointue
~TIE UP - [text] empoutage *m* en pointe
POINTER - [gen and instr] aiguille *f*, index *m*
[gen] (signal) indicateur *m*
[instr] (of a balance) langue *f*, languette *f*
[railw] (point lever, or switch) levier *m* d'aiguille
[tool] (tool used for raking out old mortar from joints before pointing) fiche *f* de maçon
POINTER COUNTER - [instr] horlogerie *f* à aiguille indicatrice, indicateur *m* à cadran
~INSTRUMENT - [instr] (instrument in which readings are shown by a pointer or needle moving over a scale) appareil *m* de mesure)à aiguille
~THERMOMETER - [instr] (solid expansion thermometer) thermomètre *m* à index
POINTING - [constr] (the raking out brickwork jointing and the refilling with cement mortar) jointoiement *m*, gobetage *m*
[surv] (of an instrument) collimation *f*
[meas] (the operation of locating certain marks on the scale, so as to complete its division) calibrage *m*
[metall] taillage *m*, pointage *m*
~COLOUR - [text] couleur *f* de pointage
~IRON - [constr] bourroir *m*
~MORTAR - [constr] mortier *m* de rejointoyage
~NAIL - [metall] crochet *m* à jointoyer, polissoir *m*, fiche *f*
~TROWEL - [impl] fiche *f*
POINTS - [railw] aiguillage *m*
[print] (general term for punctuation marks) pointure *f*
[zool] (of a horse) extrémités *fpl*
~AND CROSSING - [railw] appareil *m* de voie
~CONTROL - [railw] contrôle *m* de la position des aiguilles
~IN REVERSE POSITION - [railw] aiguille *f* deviée
~OF THE COMPASS - [instr] (the thirty-two equal angular division of the Mariner's Compass card) quarts *mpl* de vent
POINTSMAN - [railw] aiguilleur *m*
POISE, to -[gen] équilibrer, tenir en équilibre
[chem] (to maintain the oxidation-reduction potential of a solution constant; a suitable compound is added to this purpose) maintenir constant
POISE - [gen] équilibre *m*
[mech] (a balance weight) contrepoids *m*
[horol] équilibre *m*
[phys] (the unit of absolute viscosity of a fluid) poise *f*, unité *f* de viscosité absolue
POISEUILLE EQUATION - [mech] (a relation between the volume flow along a cylindrical tube and the pressure difference between the ends) équation *f* de Poseuille
POISING - [gen etc] équilibrage *m*, équilibration *f*
POISON - [nucl] (any non-fissile element in a reactor which possesses an appreciable absorption cross-section) poison *m*
~AGENT - [electron] (in donor composition to balance the production) agent *m* d'empoisonnement
~BAIT - [bot] appât *m* empoisonné
~-BEARING - [bot] vénénifère
~BLACK CHERRY - [bot] belladone *f*
~CHANGES - [nucl] variations *fpl* d'empoisonnement
~FLOWER - [bot] célastre *m*
~GAS - [chem] gaz *m* toxique
~LIMIT - [nucl] (poison concentration which does not enable the reactor to become critic) limite *f* d'empoisonnement
~RANGE - [nucl] intervalle *m* d'empoisonnement
POISONING - [gen] empoisonnement *m*
~COMPUTER - [nucl] calculatrice *f* analogique de l'effet d'empoisonnement
~OF A REACTOR - [nucl] (the ratio of the number of

thermal neutrons absorbed by poison to those absorbed in fuel) empoisonnement *m* du réacteur
POISONING PREDICTOR - [comput] prédicteur *m* de empoisonnement xénon
POISONOUS PREPARATION - [chem] préparation *f* toxique
POISSON RATIO - [mech] module *m* de contraction latérale, rapport *m* de Poisson
POKE, to - [gen] tisonner, piquer
POKE - [gen] poussée *f*
~WELDING - [metall] (in welding) soudure *f* à extrusion
POKER - [impl] tisonnier, ringard *m*
[text] chandelle
[impl] pointe *f* métallique (pour pyrogravure)
~BUSH - [text] tibe *m* des chandelles
~DRAWING - [gen] (pirography) pyrographie *f*
~FOOT - [text] pied *m* de la tige
POKING - [gen] tisonnage *m*, attisage *m*
~HOLE - [gas ind] ouverture *f* de décrassage
POLAR - [gen el phys etc] (pertaining to the poles) courbe *f* polaire en coordonnées polaires
[phys] (of a molecule exerting local electrical force) polaire
~ADSORPTION - [phys chem] (adsorption of electrically unequal amounts of ions, so that the adsorbed film has an over-all electrical charge) adsorption *f* polaire
~AXIS - [astr] (the diameter passing through the poles) axe *m* polaire
[phys] (crystal axis to which no two- or four-fold axes are normal) axe *m* polaire
~BEAR - [zool] ours *m* blanc
~BODY - [biol] (small cell detached from the ovum during the maturation period) globule *m* polaire
~BOND - [phys] (a bond due to the transfer of an electron from one atom to another, the ions being held together electrostatically) liaison *f* polaire
~CAP - [biol] (plasmatic strands at the pole of a dividing nucleus) capsule *f* polaire
~CELL - s. polar body
~CIRCLE - [geogr] (the artic or antartic circle) cercle *m* polaire
~COMPOUND - [chem] (in general, a compound exhibiting polarity) composé *m* polaire
~COORDINATE - [math] coordonnée *f* polaire
~CURVE - [light] (curve draw in polar coordinates showing the distribution of light about a light source) courbe *f* photométrique
~DIAGRAM - [radio] (diagram showing the relative effectiveness of transmission or reception of an aerial system in different directions) diagramme *m* polaire
~DISTANCE - [astr] (the intercept on a celestial meridian between a celestial body and the celestial pole) distance *f* polaire
~FRONT - [met] (a front originating from a depression on the eastern seabord of N. America) front *m* polaire
~FUSION NUCLEUS - [genet] noyau *m* de fusion polaire
~GLOBULE - s. polar body
~LIGHTS - [astr] (aurora borealis or aurora australis) aurore *f* boréale, aurore *f* australe
~LIQUID - [phys] (a liquid exhibiting optical polarity) liquide *m* polaire
~LOW - [met] (a small depression moving southward in temperate regions) dépression *f* polaire
POLAR MOLECULE - [chem] (a molecule having an electrical dipole moment due the presence of polar valence bonds) molécule *f* polaire
~MOMENT OF INERTIA - [phys] moment *m* d'inertie polaire
~NUCLEI - [genet] (in the embryo sac) noyaux *mpl* polaires
~ORBIT - [astronaut] orbite *f* polaire
~PATTERN OVERBOOTS - [shoe man] galoches *fpl* de sécurité contre explosion
~PLANIMETER - [instr] (instrument designed to measure the area under a given curve) planimètre *m* polaire
~POTENTIOMETER - [instr] (a bias potentiometer) potentiomètre *m* à polarisation
~PROJECTION - [surv] projection *f* polaire
~PROPERTY - [phys chem] propriété *f* polaire
~RECIPROCATION - [math] transformation *f* polaire des réciproques
~RELAY - [electron] relais *m* polarisé
~RESPONSE - [acoust] (of a microphone or a loudspeaker, when measured at a single frequency for all directions round a circle) réponse *f* polaire
~RESPONSE CURVE - [acoust] (curve showing the distribution of the radiating energy from a sound reproducer) courbe *f* polaire de réponse
~SEQUENCE - [astr] (a scale for the determination of photographic stellar magnitude) séquence *f* polaire
~SIDEROSTAT - [instr] (instrument designed to reflect a portion of the sky in a fixed direction) sidérostat *m*
~SOLVENT - [chem] (solvent consisting of molecules which exert local electrical forces) solvant *m* polaire
~STAR - [astr] étoile *f* polaire
~VALENCE - [chem] valence *f* polaire
~VECTOR - [el] (vector representing a physical phenomenon having a direction) vecteur *m* polaire
POLARIMETER - [instr] (an instrument for the determination of the optical activity of a liquid by the use of Nicol prisms) polarimètre *m*
POLARIMETRY - [chem] (the measure of optical activity) polarimétrie *f*
POLARIS - [astr] (the Pole Star, the position of which is about I degree from the celestial Pole) étoile *f* polaire
POLARISCOPE - [instr] (optical instrument designed to measure the polarization of light) polariscope *m*
POLARISTROBOMETER - [instr] (polarimeter used in saccharimetry) strobomètre *m* polaire
POLARITY - [gen] (shown by a line segment when the two ends are distinguishable) polarité *f*
[phys] (a physical system has polarity when two points in the system have different characteristics) polarité *f*
~DIRECTIONAL RELAY - [el] relais *m* directionnel de tension
~INDICATOR - [instr] (instrument designed to determine the polarity of electric terminals) indicateur *m* de polarité
~OF PICTURE SIGNAL - [telev] polarité *f* du signal d'image
POLARIZABILITY - [el] (the dipole moment produced by unit electric field acting on a system) polarisabilité

POLARIZABILITY CATASTROPHE - [el] polarisation *f* spontanée
POLARIZATION - [phys] (a condition in which the positive and negative charges of a molecule are separated) polarisation *f*
[el] (condition in a primary cell, in which the coltage falls off after a certain period of working, because of an accumulation of reaction products at the electrodes) polarisation *f* électrolytique
[chem] (the concentration of sugar) degré *m* saccharimétrique
[radio] (the direction of inclination of the components of an electromagnetic wave) polarisation *f*
~CAPACITANCE - [el] capacité *f* de polarisation
~CHARGE - [el] (the net charge per unit volume) charge *f* de polarisation
~CURRENT - [el] (the current caused by the variation of the dielectric polarization) courant *m* de polarisation
~ELLIPSE - [el] (of a field vector) ellipse *m* de polarisation
~ERROR - [radio] (error in determining the direction of arrival of radio-waves by a direction-finder) erreur *f* de polarisation, effet *m* de polarisation
~FADING - [el] affaiblissement *m* de polarisation
~OF A MEDIUM - [el] (the change in the physical state of the medium by phenomena affecting this medium are given a vectorial character) polarisation *f* d'un milieu
~ OF FLUORESCENCE - [el] (of solutions) polarisation *f* de fluorescence
~OF LIGHT - [light] polarisation *f* de la lumière
~PHOTOMETER - [instr] (photometer fitted with Nicol prisms) photomètre *m* à polarisation
~POTENTIAL - [el biol] (boundary potential across any interface) potentiel *m* de polarisation
~REACTANCE - [el] réactance *f* de polarisation
~RECEIVING FACTOR - [radio] coefficient *m* de réception de polarisation
~RESISTANCE - [el biol] résistance *f* de polarisation
~UNIT VECTOR - [el] vecteur *m* unité de polarisation
POLARIZE, to - [phys el chem etc] polariser
POLARIZED - [gen phys el etc] polarisé
~BEAM - [radiat] (of an electromagnetic radiation) faisceau *m* polarisé
~BELL - [telecomm] sonnerie *f* magnétique
~ELECTROLYTIC CAPACITOR - [el chem] (electrolytic capacitor in which the dielectric film adheres to the only metal electrode) condensateur *m* électrolytique polarisé
~LIGHT - [opt] (obtained by reflecting ordinary light from a plane surface at the angle of polarization) lumière *f* polarisée
~RADIATION - [radiat] (radiation in which the condition shows certain asymmetries in respect of the axis of propagation) rayonnement *m* polarisé
~RELAY - [el] (relay in which there is a permanent flux) relais *m* polaire
~RINGER - s. polarized bell
~WAVES - [radio] (plane-polarized waves) ondes *f*pl électromagnétiques polarisées
POLARIZER - [el chem] (any substance which when to an electrolyte increases polarization) polarisant *m*
POLARIZING ANGLE - [opt] (the angle of incidence for which a wave polarized parallel to the plane of incidence is completely transmitted) angle *m* de polarisation, angle *m* de Brewster
POLARIZING FILTER - [photo] filtre *m* de polarisation
~MAGNETIZING FORCE - [phys] force *f* magnétisante moyenne
~MICROSCOPE - [opt] (microscope for the examination of objects under illumination by polarized light) microscope *m* polarisant
~POWER OF SILK - [text] pouvoir *m* de polarisation de la soie
~PRISM - [opt] (any light-polarizing prism) prisme *m* polarisateur
~SCREEN - [photo] écran *m* polariseur
POLAROGRAM - [chem] (the current-voltage curve obtained with a polarograph) polarogramme *m*
POLAROGRAPH - [instr] (apparatus for the automatic electroanalysis of a solution by means of the dropping mercury cathode) polarographe *m*
POLAROGRAPHIC - [instr] polarographique
~ANALYSIS - [chem] (the analysis of a solution by means of a dropping mercury cathode) analyse *f* polarographique
POLAROGRAPHY - [chem] (the method of measuring potential difference current relationships in solutions by means of a polarized microelectrode) polarographie *f*
POLAROID - [opt] (said of a glass which has the property of polarizing and therefore reducing the intensity of light passing through it) polaroïde
~LENS - [opt] lentille *f* polaroïde
POLDER - [geol] polder *m*
[constr] (diked country) polder *m*
POLE - [gen] poteau *m*, perche *f*
[geogr] pôle *m*
[el] pôle *m*
[phys] (one of the two points at which opposite physical qualities are concentrated) pôle *m*
[naut] flèche *f* de mât
[surv] mire *f*
[electron] (of a multiple-cavity magnetron, the part of the anode between two slots) dent *f*, pôle *m*
~ARC - [el] (the arc embraced by the pole shoe) arc *m* polaire
~ARM - [constr] traverse *f*
~ARMATURE - [el] induit *m* polaire
~AUGER - [mech] tarière *f*
~AXE - [impl] hache *f*, merlin *m*
~BEAN - [bot] haricot *m* à rames
~BEVEL - [el] (portion of the pole face of an electric machine, near the pole tip) biseau *m* polaire
~-BOX - [mining] boîte *f* de sonde
~BRACE - [constr] entretoise *f*
~-CHANGE BRIGHTENING - [text] avivage *m* par dépolarisation
~CHANGER - [el] (of an induction motor) inverseur *m* de polarité, vibrateur *m* de polarité
~-CHANGING CONTROL - [el] (method for obtaining two or more speeds from an induction motor) régulation *f* par changement du nombre de pôles
~-CHANGING STARTER - [el] démarreur *m* par changement du nombre de pôles
~-CHANGING SWITCH - [el] commutateur *m* de polarité
~CLIMBERS - [impl] griffes *f*pl pour l'ascension d'un poteau
~COMPASS - [surv] boussole *f* à poteau
~CONNECTION - [el] accouplement *m* entre pôles
~CORE - [el] (the part of the pole where the exciting

winding is placed) noyau *m* magnétique
POLE CRIBBING - [constr] semelle *f*
~DIAGRAM - [telecomm] loi *f* de croisement, loi *f* de rotation
~DYNAMO - [el] dynamo *f* à pôles
~EARTH WIRE - [el] fil *m* paratonnerre
~END-PLATE - [el] (thick plate at each end of the laminations of a laminated pole) joue *f* magnétique d'extrémité
~FACE - [el] (the surface of the pole piece of an electrical machine which faces the armature) face *f* polaire
~-FACE LOSS - [el] (iron losses occurring in the iron of a pole face) pertes *fpl* de face polaire
~FENDER - [constr] pieu *m* chasse-roue
~FINDER - [el] (a paper treated with a chemical solution and placed across the two poles of an electric circuit; a red mark is made when it touches the positive pole) cherche-pôles *m*
~FINDING PAPER - [el] (the paper used in the pole finder) papier *m* cherche-pôles
~GUY - [el] hauban *m* de poteau
~HOLE - [constr] fouille *f* pour l'implantation de poteaux
~HOOK - [el] repos *m* de perche
[mining] caracole *f*
~HORN - [el] (of an electrical machine) corne *f* polaire
~INDICATOR - s. pole finder
~JACK - [mining] vérin arrache-étai
~KEY - [el] clavette *f* polaire
~LATHE - [mach tool] tour *m* à pointes
~MAST - [naut] mât *m* à pible, mât *m* de flèche
~OF A PILE DRIVER - [constr] écoperche *f*
~PIECE - [el] (in an electric motor or generator, a part attached to the pole shank and forming part of the armature tunnel) épanouissement *m* polaire, armature *f* d'aimant
~PIECE FACE - [el] face *f* polaire
~PITCH - [el] (the peripheral distance between fixed points in corresponding positions on two consecutive poles) pas *m* polaire
~PITCH AT THE COMMUTATOR - [el] pas *m* polaire au collecteur
~RESERSER - s. pole changer
~RETRIEVER - [el] rattrapeur *m* de perche
~REVERSER - [el] inverseur *m* de pôles
~SHANK - s. pole core
~SHIM - [el] séparateur *m* de régulation
~SHOE - s. pole piece
~-SHOE COIL - [el] enroulement *m* de la pièce polaire
~SOCKET - [constr] socle *m* de poteau
~STEPS - s. pole climbers
~STRENGTH - [el] (the force excreted by a magnet pole) intensité *f* polaire
~TIPS - [el] cornes *fpl* polaires
~-TOOL BORING - [mining] sondage *m* à la tige
~TROLLEY - [el] trolley *m* à perche
~TURNER - [mining] manche *m* de manoeuvre, tourne-à-gauche *m*
~-TYPE TRANSFORMER - [el] transformateur *m* de poteau
~UNIT - [el] pôle *m* individuel
POLECAT - [zool] putois *m*
POLES OF A MAGNET - [el] pôles *mpl* d'un aimant
~OF ROTATION - [geogr] (the extremities of the earth's rotational axis) pôles *mpl* de rotation
POLES OF THE ELLIPSE - [geom] pôles *mpl* de l'ellipse
POLESTAR - [astr] étoile *f* polaire
POLHODE - [mech] (the line of intersection of the cone outlined by the angular velocity vector with the momental ellipsoid) polhodie *f*
POLICEMAN - [mining] chapeau *m* en caoutchouc
POLIENCEPHALITIS - [med] polioencéphalite *f*
POLING - [metall] (the fire-refining of copper by the introduction of gases which are produced when logs, or poles, are burnt in the molten metal) perchage *m*
[radio] (the interchange of the conductors at a point in a transmitting path) inversion *f* de conducteurs
[constr] étayage *m* avec des perches, blindage *m*
[mining] poussage *m*, enfilage *m*
~BOARD - [constr] (rough vertical planks) planche *f* de boisage, bois *m* de couchis
POLIO - s. poliomyelitis
POLIOMYELITIS - [med] poliomyélite *f*
~VIRUS - s. poliovirus
POLIOPLASM - [biol] (granular protoplasm) polioplasma *m*
POLIOSIS - [med] poliose *f*
POLIOVIRUS - [med] virus *m* de la poliomyélite
POLISH, to - [gen] polir, lisser
[mech] polir, brunir, lisser
[leather ind] astiquer
POLISH - [gen] poli *m*, brillant *m*
[ind chem] cire *f* à polir
~ATTACK - [metall] polissage *m* en bas-relief
~ETCH - [metall] attaque *f* chimique polisseuse
~POWDER - [metall] poudre *f* à polir
~WITH EMERY, to - [glass man] polir à l'émeri
POLISHABLE - [gen & metall] polissable
POLISHED BACK REST - [text] porte-fils *m* poli
~GLASS - [glass man] verre *m* poli
~GLASS PLATE - [glass man] plaque *f* de verre poli
~GUIDE ROLLER - [text] rouleau-guide *m* poli
~NEEDLE - [text] aiguille *f* polie
~ROD - [oil ind] tige *f* étanche
~STEEL - [metall] acier *m* bruni
~THREAD - [text] fil *m* poli
~YARN - [text] fil *m* glacé
POLISHER - [gen] polisseur *m*
[agric] (for rice) appareillage *m* à glacer le riz
POLISHING - [gen] polissage *m*
[mech] (the operation performed by a polishing machine) polissage *m*, brunissage
[carp] cirage *m*, encaustiquage *m*
[text] (of yarn) lissage *m*, frottage *m* au mouillé
~AND BUFFING MACHINE - [mech] polisseuse *f*
~BOB - [impl] (soft polishing material for mounting on a polishing spindle) disque *m* à polir, touffe *m* polisseur
~CYLINDER - [metall] cylindre *m* polisseur
~CLOTH - [text] chiffon *m* à cirer
~HEAD - [tool] touret *m* de polisseur
~LAP - [impl] disque *m* en lisière de tissu, meule *f* flexible en chiffon
~MACHINE - [mech] polisseuse *f*, polissoir *m*
~MOP - [impl] (group of disks of soft textile material for mounting on a polishing spindle) disque *m* à polir
~PLATES - [mech] (metal plates with a finely-polished surface, used in a press to produce a smooth surface) plaques *fpl* de brunissage
~ROLLS - [impl] (rollers having a finely-finished surface to originate a similar finish on the sheet

treated) rouleaux *mpl* à polir
POLISHING VARNISH - [paint] (a varnish which can be polished with fine abrasive and oil, to give a special type of finish) vernis *m* à polir
~WHEEL - [mech] disque *m* à polir, meule *f* polissoire
POLKA DOTS - [text] points *mpl* type Polka
POLL, to - [gen] voter, faire voter
[statist] faire voter
[agric] tondre, écorner
POLL - [gen] vote *m*, scrutin *m*
[tool] (the blunt edge of a hammer or an axe) tête *f*
[statist] résultat *m* du vote
~PICK - [impl] pic *m* ordinaire
POLLAKISURIA - [med] pollakiurie *f*
POLLAKIURIA - s. pollakisuria
POLLEN - [bot] (the material produced in anothers and consisting of grains which contain two male nuclei) pollen *m*
~CHAMBER - [bot] (the cavity on which the pollen grains lodge) sac *m* pollinique
~FLOWER - [bot] fleur *f* pollinifère
~-GRAIN - [bot] grain *m* de pollen
~LETHALS - [bot] léthaux *mpl* pour le pollen
~SAC - [bot] (the cavity on the another where the pollen is formed) sac *m* pollinique
~TUBE - [bot] (the tubular outgrowth from the pollen grain) tibe *m* pollinique
POLLEX - [anat] pouce *m*
POLLICIZATION - [med] pollicisation *f*
POLLINATE, to - [bot] (to transfer pollen) polliniser
POLLINATION - [bot] (the transfer of pollen from another to stigma) pollination *f*
POLLING - [comput] appel *m* sélectif
POLLINIFEROUS - [bot] pollinifère
POLLINIUM - [bot] (mass of pollen grains) grains *mpl* de pollen
POLLUCITE - [miner] (rare aluminium silicate of caesium) pollucite *f*, pollux *m*
POLLUTE, to - [gen] contaminer, polluer, souiller
POLLUTED BY FAT - [ind chem] (of acid ball) chargé de gras
POLLUTION - [gen] contamination *f*, pollution *f*
~ABATEMENT - [hydr] protection *f* des eaux naturelles, protection *f* contre la pollution des eaux
~CARPET - [ecology] couche *f* de polluants
~CONTROL - s. pollution abatement
~LOAD - [hydr] charge *f* de pollution
~OF WATER - [gen] pollution *f* de l'air
POLOCYTE - [biol] globule *m* polaire
POLONIUM - [chem] (a radioactive element, symbol Po, A.N. 84, A W. 210 which occurs as a decay of radium and has been synthesized) polonium *m*
POLY - [chem] (a prefix indicating "many") poly-
~DISC-PLOUGH - [agric] déchaumeuse *f* à disques
POLYACRYLAMIDE - [chem] (a thickening agent) polyacrylamide *f*
POLYACID - [chem] polyacide
POLYACRYLATE - [chem] (generic term for polymers formed from acrylic acid esters) polyacrylate *m*
POLYACRYLIC ACID - [chem] (a synthetic textile sizing agent) acide *m* polyacrylique
POLYADENIA - [med] pseudoleucémie *f*
POLYADENITIS - [med] polyadénite *f*
POLYAESTHESIA - s. polyeshesia
POLYALCOHOLS - [chem] polyalcools *mpl*
POLYALKANE - [chem] (a hydrocarbon consisting of long chain molecules containing only saturated carbon atoms in the main chain) polyalcane *m*
POLYALLELE CROSSING - [genet] croisement *m* polyallélique
POLYALLOMER - [chem] (generic term for a number of crystalline polymers obtained by the combination of two or more dissimilar monomers) polyallomère *m*
POLYAMIDE - [chem] (a polymer having the structural units linked by amide groups) polyamide *f*
~RESINS - [chem] (polymers produced by condensation dibasic acid with diamines and by polymerization of lactams and amino acids) résines *fpl* de polyamide
POLYAMINE-METHYLENE RESIN - [chem] (an ion-exchange resin) résine *f* de polyamine-méthylène
POLYAMINOTRIAZOLES - [chem] (generic name for polymers based on sebacic acid, acetamide and hydrazine. They have application in the production of synthetic fibres) polyaminotriazoles *mpl*
POLYANDROUS - [bot] polyandre
POLYANDRY - [bot] (the condition of having a large number of stamens) polyandrie *f*
[zool] (of the female consorting with more than one male) polyandrie *f*
POLYARCHY - [bot] polyarchie *f*
POLYATOMIC - [chem] polyatomique
POLYBASIC - [chem] (capable to combine with more than two univalent bases) polybasique
~ACIDS - [chem] (acids with two or more hydrogen atoms in the molecule) acides *mpl* polybasiques
POLYBASITE - [min] (natural silver-antimony sulphide, crystallizing in the monoclinic system) polybasite *f*
POLYBUTADIENE - [chem] (a synthetic rubber based on butadiene and having good heat dispersion characteristics and abrasion resistance) polybutadiène *m*
~-ACRYLIC ACID COPOLYMER - [chem] (has applications as a fuel binder for solid fuel rockets) copolymère *m* de l'acide polybutadiène-acrylique
POLYBUTENES - [chem] (polybutylenes) (polymers of isobutene) polybutènes *mpl*
POLYBUTYLENES - s. polybutenes
POLYCARDIA - [med] tachycardie *f*
POLYCARP - [biol] (sexgland on the inner surface of the mantle in some Urochorda) polycarpe
POLYCARPIC - [bot] (which yields fruit many times in succession) polycarpellé
POLYCARPOUS - [bot] (with an gynaeceum consisting of one or more carpels) polycarpe
POLYCENTRIC - [geom] polycentrique
POLYCHLORIDE - [chem] polychlorure *m*
POLYCHLOROPRENE - [chem] (intermediate in the production of GR-M synthetic rubber) polychloroprène *m*
POLYCHLOROTRIFLUOROETHYLENE - [chem] (see chlorotrifluoethylene resins) chlorotrifluoroéthylène *m*
POLYCHROMASIA - [med] polychromasie *f*
POLYCHROMATIC - [opt] (having more than one colour) polychrome
POLYCHROME - [gen] polychrome *m*
POLYCHROMIC - s. polychromatic
POLYCLINIC - [med] polyclinique *f*
POLYCONDENSATE - [chem] (the product of a condensation reaction) polycondensat *m*
POLYCONDENSATION - [chem] (formation of large molecules consisting essentially of recurring structural units from simpler molecules by their chemi-

cal combination with the elimination of a simple substance, such as water) polycondensation *f*

POLYCONIC PROJECTION - [surv] (method of map projection) projection *f* polyconique

POLYCRASE - s. polycrasite

POLYCRASITE - [min] (mineral containing a large number of metallic elements, including uranium) polycrase *f*

POLYCROSS - [genet] polycroisement

POLYCRYSTAL MINERAL SCINTILLATOR DETECTOR - [instr] détecteur *m* à scintillateur minéral polycristallin

POLYCRYSTALLINE - [phys] (said of a crystal body made up of several small crystals in a mass) polycristallin

POLYCYCLIC - [chem] (a term used of compound in which there are more than one ring structures in the molecule) polycyclique

POLYCYCLIC COMPOUNDS - [chem] composés *mpl* polycycliques

POLYCYTHEMIA - [med] (overproduction of erythrocytes) polycythémie *f*

POLYDACTYLIA - [med] polydactilisme *m*

POLYDACTYLISM - s. polydactylia

POLYDISPERSE SYSTEM - [chem] (colloidal system consisting of particles having different sizes) système *m* polydisperse

POLYDONTIA - [med] polyodontie *f*

POLYDYSPLASIA - [med] polydysplasie *f*

POLYELECTRODE - [el] (multiple electrode; electrode at which several electrode reaction take place at the same time) polyélectrode *f*, électrode *f* multiple

POLYELECTROLYTE - [el] polyélectrolyte *m*

POLYELECTRONS - [phys] polyélectrons *mpl*

POLYEMBRIONY - [genet] (the presence of more than one embryo in one fertilized ovum) polyembryonie *f*

POLYENERGETIC RADIATION - [radiat] (particle radiation in which particles have different energies) rayonnement *m* polyénergétique

POLYENERGID - [genet] (possessing several sets of chromosomes) polyénergide *m*

POLYESTER - [chem] (a polymer in which the structural units are linked by ester and amide (and or thio-amide) grouping) polyester-amide *m*

~FIBRE - [ind chem] (synthetic fibre containing not less than 85 p.c. of terephthalic acid and a dihydric alcohol) fibre *f* de polyester

~RESINS - [chem] (a very large class of synthetic resins formed from dihydric alcohols and dibasic acids. They have good resistance to corrosion and weathering and give products of considerable lightness and strength in combination with glass and asbestos reinforcement materials) résines *fpl* de polyesters

~RUBBER - s. polyurethane rubber

POLYESTERIFICATION - [chem] polyestérification *f*

POLYESTHESIA - [med] polyesthésie *f*

POLYETHER FOAMS - [chem] (polyether foams employing a polyether instead of a polyester resin) mousses *fpl* de polyéther

POLYETHYLENE - [chem] (a thermoplastic ethylene polymer, it is a valuable elastomer, with good electrical properties at H.F., and has many uses in the sheet form) polyéthylène *m*

~GLYCOL - [chem] (a synthesis intermediate) polyéthylène-glycol *m*

~TEREPHTHALATE - [chem] (a polyester resin used in the production of synthetic fibres) téréphtalate *m* de polyéthylène

POLYFORMALDEHYDES - [chem] (high-melting point, thermoplastic crystalline material produced by the polymerization of formaldehyde. Also known as acetal resins) polyformalaldéhydes *mpl*

POLYFORMING - [chem] procédé *m* de reforming et polymérisation

POLYGAMOUS - [biol] (mating with more than one of the opposite sex during the same breeding season) polygame *m*

POLYGAMY - [gen] polygamie *f*

POLYGASTRIC - [zool] (having more than one gastric cavity) polygastrique

POLYGENESIS - [biol] (the belief that organisms originate from cells of different kinds) polygenèse *f*

POLYGENETIC - [chem] (producing two or more shades with different mordants) polygénétique, polygénique

POLYGENY - [biol] (the origination of species from several independent pairs of ancestors) polygénie

POLYGLANDULAR - [biol] polyglandulaire

POLYGLYCOL DISTEARATE - [chem] (a plasticizer for a number of resins) distéarate *m* de polyglycol

POLYGON - [geom] polygone *m*

~OF FORCES - [mech] (geometrical representation by which the sum of two or more concurrent forces can be determined by the successive vector addition of the forces involved) polygone *m* de forces

~OF STRESSES - [mech] polygone *m* d'efforts

POLYGONAL - [geom] polygonal

~CATENARY - [el] (in electrical traction) caténaire *f* polygonale, caténaire *f* verticale

~NUMBERS - [math] nombres *mpl* polygonales

~ROOF - [constr] comble *m* polygonal

~RUBBLE - [constr] appareil *m* irrégulier

~SILK FILAMENT - [text] fil *m* de soie polygonal

~TRACE - [surv] réseau *m* polygonal

POLYGONIZATION - [phys] (occurring when plastically bent crystals are annealed) polygonisation *f*

POLYGRAPH - [print] (a machine used for copying) appareil *m* à polycopier

POLYHEDRON - [crystal] (solid having faces formed from plane polygons) polyèdre *m*

POLYHEXAMETHYLENEADIPAMIDE - [chem] (a variety of nylon) polyhexaméthylène-adipamide *f*

POLYHYDRIC - [chem] (consisting of a number of hydroxyl groups in the molecule) polyhydrique

POLYKARYOTIC - [genet] à noyaux multiples

POLYLINER - [plast ind] (perforated sleeve having lengthwise ribs, fitted in the cylinder of an injection - moulding machine instead of a torpedo) douille *f* perforée

POLYLOGIA - s. polyphrasia

POLYMASTIA - [med] polymastie

POLYMAZIA - s. polymastia

POLYMER - [chem] (generic term for an organic compound of high molecular weight and consisting of recurrent structural groups. The groups capable of undergoing polymerization, or self-addition, are known as monomers and are of relatively simple structure) polymère *m*

~STRUCTURE - [chem] (the position in three-dimensional space occupied by the atoms in a polymer molecule. The structure of a polymer determines a number of its properties, e.g. softening temperatures, flexibility, elasticity and second order tran-

sition temperature) structure *f* polymère
POLYMERIC - [chem] (pertaining to polymer) polymère
POLYMERIZATION - [chem] (self-addition, the ability of certain organic compounds to react together to form a single molecule of higher atomic weight. In theory, polymerization can continue indefinitely) polymérisation *f*
POLYMERIZE, to - [chem] polymériser
POLYMERIZED OILS - [chem] (blown oils. Oils which have been partially oxidised, deodorised and polymerized) huiles *fpl* polymérisées
POLYMERIZER - [chem] (polymerizing agent) agent *m* de polymérisation
POLYMERY - [genet] (condition occurring when a whole consists of several members) polymérie *f*
POLYMETER - [instr] (combination of various meteorological instruments) polymètre *m*
POLYMETHACRYLATES - [chem] (synthetic resins and plastics based on acrylic acid) polyméthacrylates *mpl*
POLYMETHOXY ACETAL - [chem] (has applications as a plasticizer; solvent; phenolic resin modifier; and mould lubricant) polyméthoxy-acétal *m*
POLYMETHYL METHACRYLATE - [chem](thermoplastics composed of polymers of methyl methacrylate) polyméthyle-méthacrylate *m*
POLYMETHYLENE - [chem] (methylene consisting of three or more groups in the ring) polyformaldéhyde *m*
~DERIVATIVES - [chem] (cyclic compounds with three or more methylene rings in the group) dérivés *mpl* polyméthyléniques
POLYMORPHIC - s. polymorphous
POLYMORPHISM - [phys] (the property of crystallizing in several genetically different forms) polymorphisme *m*
POLYMORPHNUCLEAR LEUCOCYTE - [biol] (leucocyte which is characterized by an irregular deep-stained nucleus) granulocyte *m*, leucocyte *m* granulé polynucléaire
POLYMORPHOUS - [phys] (pertaining to polymorphism) polymorphe
POLYMYXINS - [chem] (water-soluble basic antimicrobial polypeptides with application in medicine) polymyxines *fpl*
POLYNEURALGIA - [med] polynévralgie *f*
POLYNEURITIS - [med] névrite *f* multiple
POLYNOMIAL - [math] polynôme *m*
POLYNUCLEAR - [biol] polynucléaire
POLYOESTROUS - [zool] (having more than one oestrous cycle during the breeding season) polyoestral
POLYOLEFINS - [chem] (generic name for polymers obtained by the polymerization of olefins) polyoléfines *fpl*
POLYOPIA - [med] polyopsie *f*
POLYOPSIA - s. polyopia
POLYOSES - [chem] (term used collectively for polysaccharoses) polyoses *mpl*
POLYOXAMIDE - [chem] (generic name for nylon-like derivatives of diamines and oxalic acid) polyoxamide *f*
POLYOXYMETHYLENE - [chem] (synonym for paraformaldehyde and polyformaldehyde resins) polyoxyméthylène *m*
POLYOXYPROPYLENE GLYCOL ETHYLENE OXIDE - [chem] (plasticizer for cellulose derivatives, polymethyl methacrylate, polyvinyl acetate, polyvinl butyral, vinyl chloride and vinyl chloride acetate) oxyde *m* de polyoxypropylène et éthylèneglycol
POLYOXYPROPYLENE GLYCOLS - [chem] (polyether intermediates for elastomers and polyurethane foams) glycols *mpl* de polyoxypropylène
POLYP - [med] polype *m*
POLYPHASE - [el] (system or machine in which there are two or more alternating voltages) polyphasé
~CIRCUIT - [el] circuit *m* polyphasé
~COMPOUND COMMUTATOR MOTOR - [el] moteur *m* polyphasé compound à collecteur
~EQUILIBRIUM - [phys] (heterogeneous equilibrium) équilibre *m* hétérogène
~INDUCTION MOTOR - [el] moteur *m* à induction polyphasé
~NETWORK - [el] réseau *m* polyphasé
~SERIES COMMUTATOR MOTOR - [el] moteur *m* polyphasé série à collecteur
~SERIES SHUNT MOTOR - [el] moteur *m* polyphasé shunt à collecteur
~SYNCHRONOUS GENERATOR - [el] génératrice *f* synchrone polyphasée
~SYSTEM - [el] système *m* polyphasé
~TRANSFORMER - [el] (transformer for polyphase working in which the magnetic corresponding to the electrical phase circuits have parts in common) transformateur *m* polyphasé
~WATTHOURMETER - [instr] compteur *m* polyphasé de énergie active
POLYPHENOL - [chem] polyphénol *m*
POLYPHONY - [mus] (combination of two or more melodies) polyphonie *f*
POLYPHRASIA - [med] polyphrasie *f*
POLYPHYLETIC - [genet] (descending from different ancestors) polyphylétique
POLYPHYLY - [genet] (the theoretical origination from several distinct sources) polyphylie *f*
POLYPLASTOCYTOSIS - [med] thrombocythémie *f*
POLYPLOID - [genet] (pertaining to polyploidy) polyploïde
POLYPLOIDY - [genet] (the condition of having more than twice the normal haploid number of chromosomes) polyploïdie *f*
POLYPNEA - [med] polypnée *f*, tachypnée *f*
POLYPOSIS - [med] polypose *f*
POLYPROPYLENE - [chem] (a polymer derived from propylene and having good low temperature characteristics, abrasion and chemical resistance) polypropylène *m*
POLYPUS - s. polyp
POLYROD - s. polyrod aerial
POLYROD AERIAL - [radio] (a polystyrene dielectric wire aerial) antenne *f* diélectrique à tige
~ANTENNA - s. polyrod aerial
POLYSACCHARID - [chem] polysaccharide *f*
POLYSACCHAROSES - [chem] (group of complex carbohydrates, such as cellulose, starch etc) polysaccharoses *mpl*
POLYSTYLE - [arch] polystyle
POLYSTYRENE - [chem] (a synthetic resin of major importance having high mechanical strength and insulating power) polystyrène *m*
~CO-POLYMER - [chem] copolymère *m* de polystyrène
POLYSULPHIDE - [chem] polysulfure *m*
~RUBBERS - [ind chem] (synthetic polymers with

exceptional resistance to light degradation , oils and solvents. They are manufactured by the reaction of sodium polysulphide with organic dichlorides) caoutchoucs *mpl* de polysulfure
POLYSYNOVITIS - [med] polysynovite *f*
POLYTERPENE RESINS - [ind chem] (polymers of alpha-or beta-pinene with uses as rubber plasticizers) résines *fpl* de polyterpène
POLYTETRAFLUOROETHYLENE - [chem] (a non flammable polymer with an extremely low coefficient of friction and very high resistance to chemical attack. It retains its useful properties over a very large temperature range (450° to 550° F.) polytétrafluoroéthylène *m*
POLYTHELIA - [med] polythélie *f*
POLYTHELISM - s. polythelia
POLYTHENE - s. polyethylene
POLYTONALITY - [mus] polytonalité *f*
POLYTROPHIC - [biol] polythrophique
POLYTROPHY - [biol] (the obtaining of food from various sources) polythrophie *f*
POLYTROPIC - [phys] (said of the infinite number of ways in which gases expand from an initial pressure) polytropique
~CURVE - [phys] courbe *f* polytropique
~EFFICIENCY - [chem] (efficiency over an infinitesimal stage of compression or expansion) rendement *m* polytropique
~PROCESSES - [phys] procédés *mpl* polytropiques
POLYURETHANE - [plast ind] polyuréthane *m*
~FOAM - [plast ind] (flexible or rigid forms of polyurethane resins with application as insulants, upholstery materials, fabric backing, packaging materials, and as filters) mousse *f* de polyuréthane
~RESINS - [plast ind] (isocyanate resins)(thermosetting or thermoplastic polymers formed from a diamine and a compound containing a hydroxyl group) résines *fpl* de polyuréthane
~RUBBER - [rubber ind] (a solid, flexible polyurethane resin) caoutchouc *m* d' uréthane
POLYVALENCE - [chem] (a valence greater than unity) polyvalence *f*
POLYVALENT - [chem] (having a valence greater than unity) polyvalent
POLYVINYL - [chem] polyvinyle *m*
~ACETAL -[chem] (compression and injection moulding compounds also used as films and adhesives) acétals *mpl* polyvinyliques
~ACETAL RESINS - [plast ind] (generic term for resins obtained by reaction of polyvinyl alcohol with an aldehyde) résines *fpl* acétalpolyvinyliques
~ACETATE - [chem] (PVAc) (a transparent thermoplastic synthetic resin obtained by polymerization of vinyl acetate. Major application in latex paints; also as an intermediate for polyvinyl alcohol; in fabric finishing and in adhesives) acétate *m* de polyvinyle
~ALCOHOL - [chem] (an intermediate for a number of synthetic resins, a textile size and a component of emulsifying agents) alcool *m* polyvinylique
~BUTYRAL - [chem] (a polyvinyl resin produced by condensation of polyvinyl alcohol with butyraldehyde. Used for safety-glass interlayering and in other shatterproof laminates) butyral *m* de polyvinyle
~BUTYRAL RESIN SHEETING - [plast ind] (flexible, transparent sheeting with very good moisture and UV degradation resistance with wide applications in shatterproof glass and acrylics.) feuille *f* de résin butyral-polyvinylique
POLYVINYL CARBAZOLE - [plast ind](a thermoplastic resin used in the production of electrical components) polyvinylcarbazol *m*
~CHLORIDE - [plast ind](a thermoplastic resin of major importance and wide application) chlorure *m* de polyvinyle
~CHLORIDE-ACETATE - [plast ind] (copolymer of vinyl acetate and vinyl chloride. Properties are similar to polyvinyl chloride but flexibility is greater) chlorure-acétate *m* de polyvinyle
~DICHLORIDE - [plast ind] (vinyl resin derivative with improved temperature and chemical-resistance) dichlorure *m* de polyvinyle
~ESTERS - [chem] esters *mpl* polyvinyliques
~ETHYLETHER - [chem] (plasticizer for cellulose nitrate) polyéther *m* éthylvinylique
~FLUORIDE - [chem] (a vinyl fluoride polymer having good dielectric strength andresistance to chemical attack and weathering) fluorure *m* de polyvinyle
~FORMALS - [plast ind] (chiefly used in lacquers and impregnation but can also be compression or injection moulded. Water resistance is poor)formals *mpl* polyvinyliques
~ISOBUTYL ETHER - [chem] (PVI) (liquid or elastomer derived from vinyl isobutyl ether. Used in plasticizers, adhesives, laminating agents, and coatings) éther *m* polyvinyl-isobutylique
~KETONES - [chem] cétones *mpl* polyvinyliques
~METHYL ETHER - [chem] (PVM) (plasticizer, adhesive, and heat sensitizer for natural and synthetic rubbers, polythene, and paper, derived from vinyl methyl ether) polyvinylméthyléther *m*
~RESINS - [plast ind] (vinyl plastics,q.v.) résines *fpl* polyvinyliques
POLYVINYLIDENE CHLORIDE - [chem] (polymers of vinylidenechloride) (asym-dichlor-ethylene) chlorure *m* de polyvinylidène
~FLUORIDE - [chem](a thermoplastic synthetic resin) fluorure *m* de polyvinylidène
POLYVINYLPYRROLIDONE - [chem] (a thickening agent, textile chemical and a substitute for blood plasma) polyvinylpyrrolidone*f* *f*
POMACE - [gen] (the substance of fruit crushed by grinding) pulpe *f*, marc *m* de pommes
[gen] (fish scraps) tourteau *m* de poisson
POME - [bot] fruit *m* à pépins
POMEGRANATE - [bot] grenade *f*
POMICULTURE - [agric] pomiculture *f*
POMMEL - [gen] (ornament in the form of a ball) pommeau *m*
[plast ind] (the plunger of a transfer mould or an extrusion machine) piston *m* de transfert, poinçon *m*
~GIRTH - [text] courroie *f* de pommeau
POMOLOGY - [agric] (the study of fruits and fruit trees) pomologie *f*
POMPEIAN BLUE - [paint] (alternative name for Blue Frit, a mixture of copper and calcium silicates) bleu *m* pompéin
POMPET RED - [paint] (term used a type of ferric oxide pigment) rouge *m* pompéin
POMPHOID - [med] papuloïde
POND - [gen] étang *m*
[hydr] (artificial lake) bassin *m*
~RETTING - [text] rouissage *m* à l'eau croupissante
PONDERABLE AMOUNT - [nucl] (that amount obtained

in a nuclear reaction which can be weighted) quantité *f* pondérable
PONDERABLE QUANTITY - s. ponderable amount
PONDERATOR - [nucl] (a device used to produce high-energy particles when the speed of these approaches that of light and an increase of energy involves a considerable increase in mass) pondérateur *m*
PONDEROUS - [gen] massif, lourd
PONDING - [hydr] (of a trickling filter) encrassement *m*
PONDWEED - [bot] potamot *m* luisant
PONIARD - [gen] (a small dagger) poignard *m*
PONTIE - s. punty
PONTIL - [glass man] pontil *m*
PONTINE ANGLE - [anat] angle *m* pontin
PONTOON - [gen] (a floating vessel) ponton *m*
[aero] (of a seaplane) flotteur
[naut] (flat-bottomed barge) bateau *m*, ponton *m*
~BRIDGE - [constr] (temporary bridge on pontoons) pont *m* de bateaux
~PILE-DRIVER -[constr] sonnette *f* de battage sur ponton
~SWING BRIDGE - [constr] pont *m* de bateaux tournant
PONY - [zool] (a very small horse) poney *m*
~GIRDER - [constr] (a secondary girder) membrure *f* de poutre en porte-à-faux
~MIXER - [mech] (change can mixer) malaxeur *m* à cuve mobile
~MOTOR - [el] (an auxiliary motor) moteur *m* auxiliaire
~ROUGHER - [metall] cage *f* dégrossisseuse
~TRUCK - [mech] bissel *m*, truck *m* à un seul essieu
POOL, to - [gen] mettre en commun
[mining] haver, souchever
POOL - [gen] fontaine *f*, mare *f*
[hydr] trou *m* d'eau
[gen and comm] fonds *mpl* communs, syndicat *m*
[chem] mélange *m*
[constr] (swimming pool) piscine *f*
[oil ind] mare *f* de pétrole
[mining] travail *m* au coin
[oil ind] gisement *m* d'hydrocarbures
~CATHODE - [electron] (liquid arc-cathode) cathode *f* liquide
~OPENER - [mining] sondage *m* positif
~-RECTIFIER - [electron] (gas-filled rectifier with a pool cathode) soupape *f* à cathode liquide
~TUBE - s. pool rectifier
POOP - [naut] poupe *f*
[radar] (colloq. for pulse) impulsion *f*
~BREAK - [naut] fronteau *m* de dunette
~DECK - [naut] dunette *f*, pont *m* de dunette
~LANTERN - [naut] fanal *m* de poupe
~RAIL - [naut] rambarde *f*
POOR - [gen] pauvre, médiocre
[mech etc] inférieur, inefficace
[ind chem etc] (e.g. of soil, cement etc) pauvre, maigre
~AMMONIA WATER - [chem] (a weak ammoniacal liquor) eau *f* ammoniacale pauvre
~CONTACT - [el] contact *m* défectueux
~GAS - [gas ind] gaz *m* pauvre
~GEOMETRY - [nucl] (in a nuclear experiment) géométrie *f* fausse
~IN FAT - [agric] pauvre en matières grasses
~LANDING - [electron] (colloq. expression to describe a port hole, i.e. a defect in a camera tube) oblitération *f* des coins
POOR-MIXING LINES - [plast ind] (lines appearing in extruded sheets due to too-low pressure in the barrel) lignes *fpl* de mélange déficient
~MIXTURE - [mech] (in internal combustion engines) mélange *m* pauvre
~PACK - [mining] mauvais remblai *m*
POORNESS IN WOOL - [text ind] pauvreté *f* en laine
POP, to - [gen] éclater, péter
[mech] (in internal combustion engines) bafouiller
POP - [acoust] bruit *m* soudain
[mech] (a small-size boss) moyeu *m* à vis de serrage
[mining] petit coup *m* de mine
~CORN - [agric] mais *m* grillé et éclaté
~-GATE - [metall] (in goundry works) chenal *m* de coulée en chute directe
~JOINT - [oil ind] bout *m* de tuyau
~STRANDING - [telev] décalage *m* vertical de spires
~TEST - [paper man] (test designed to assess the bursting strength of paper) essai *m* de résistance à l'éclat
~VALVE -[mech] (a boiler safety-valve) soupape *f* de sécurité à action directe
POPLAR - [bot] peuplier *m*
~WOOD - [timber] bois *m* de peuplier
POPLES - [anat] creux *m* poplité
POPLIN - [text] popeline *f*
POPLITEAL - [anat] (of the back part of the leg,behind the knee) poplité
~ARTERY - [anat] artère *f* poplitée
~MUSCLE - [anat] muscle *m* poplité
~SPACE - s. poples
POPLITEUS - s. popliteal muscle
POPPET - [mech] (a vertical support) barre *f*
[naut] (bit of wood on the boat's gunwhale to support the rowlocks) colombier *m*
[mining] chevalement *m*
~BALL - [auto] bille *f* de désaccouplement
~HEAD - [min] (or poppethread, a pulley frame over a mineshaft) chevalement *m*, chevalet *m* d'extraction
[mech] (the lathe headstock) poupée *f*
[mech] (the lathe tailstock) contre-poupée *f*
~HOLES - [mech] (of a capstan) logements *mpl* des barres
~VALVE - [mech] (in internal combustion engines) soupape *f* en champignon, distributeur *m* à soupape
[mech] (USA only; a valve mounted on a poppet) soupape *f* soulevante
~VALVE ENGINE - [mech] moteur *m* avec soupape en champignon
POPPETHEAD - s. poppet head
POPPETS - [shipbuild) (vertical support used during launching) colombiers *mpl* de lancement
POPPY - [bot] pavot *m*
~OIL - [chem] huile *f* d'oeillette, huile *f* de pavot
~SEED CAKE - [agric] tourteau *m* d'oeillette
~-SEED OIL - s. poppy oil
POPULATION DENSITY - [statist] densité *f* de la population
~PRESSURE - [statist] pressure *f* de population
PORADENITIS - [med] podénite *f*
PORCELAIN - [gen] (hard, close-grained pottery ware, translucent or partially so. True or "hard paste" porcelain has a china stone and china clay body:

artificial or "soft paste" porcelain is made from and white clay) porcelaine *f*
PORCELAIN BARREL MILL - s. porcelain pot mill
~CLAY - [geol] (kaolin, or Chine; a hydrate silicate of aluminium) terre *f* à porcelaine, kaolin *m*, feldspath *m* argiliforme
~CRUCIBLE - [ind chem] (small open vessel of fine porcelain, used for ignition of substance etc) creuset *m* de porcelaine
~FOOTSTEP - [text] crapaudine *f* en porcelaine
~GLASS - [glass man] verre *m* porcelanique
~-INSULATOR - [el] isolateur *m* en porcelaine
~PAINT GRINDER - [mech] broyeur *m* de porcelaine pour vernis
~POT MILL - [mech] broyeur *m* à boulets de porcelaine
~SHELL - [gen] revêtement *m* de porcelaine
[zool] (of molluscs) cyprée *f*, porcelaine *f*
~TEETH - [text] dents *f*pl en porcelaine
~THREAD GUIDE - [text] guide-fil *m* en porcelaine
PORCH - [arch] porche *m*, portique *m*
[constr] (only USA) véranda *f*
[telev] palier *m*
PORCUPINE - [zool] porc-épic *m*
[text] porc-épic *m*, ouvreuse *f*
[impl] hérisson *m*
~BEATER - [text] tambour *m* porc-épic
~CREEL - [text] râtelier *m* oblique
~CYLINDER - [text] tambour *m* porc épic, tambour *m* égreneur
~DRAWING - [text] étirage *m* à hérissons
~DRAWING FRAME - [text] s. percupine drawing
~FEED ROLLER - [text] hérisson *m* alimentaire
~OPENER - [text] ouvreuse *f* préparatoire porc-épic
~ROLLER - [text] (roller drawing the lace fabric over the facing bar on to the work roller) hérisson *m* alimentaire
PORE - [gen] (a minute opening in a tissue) pore *m*
[bot] (the opening of a stoma) pore *m*
[metall] (a casting defect) pore *m*
~CONDUCTIVITY - [electron] (the conductivity in porous semiconductors) conductivité *f* des pores
~SPACE - [metall] volume *m* des pores
[soil] volume *m* des vides
~WATER - [min] (water contained in the pores of clay, equal to the difference between the shrinkage water and the total moisture content) eau *f* interstitielle
PORENCEPHALITIS - [med] porencéphalite *f*
PORENCEPHALY - [med] (cerebral porosis) porencéphalie *f*
PORES - [metall] (micro-discontinuities in a metal coating which extend through the underlaying coating) pores *m*pl
[geol] espaces *m*pl interstitiels
PORK - [food] porc *m*, viande *f* de porc
~PIG - [food] porker *m*
POROCELE - [med] hernie *f* scrotale calleuse
POROCYTE - [biol] porocyte *m*
POROGAMY - [bot] porogamie *f*
POROID - [bot] (having pores) poroïde
POROME - [med] porome *m*
POROMETER - [instr] (an instrument for measuring the relative dimensions of very small pores) poromètre *m*
POROSIMETER - [instr] (an instrument for the determination of porosity in relation to the passage of fluids through solids) porosimètre *m*
POROSIS - [med] callosité *f*
POROSITY - [phys] (permeability, possessing pores) porosité *f*
~METER - s. porosimeter
~RATIO - [soil] (of the soil) pourcentage *m* de pores
POROUS - [phys] (having pores) poreux, perméable
[metall] poreux, spongieux
~BARRIER - [nucl] (diffusion-type component of an isotope separating plant) barrière *f* poreuse
~BEARING - [mech] (a self-lubricating bearing) coussinet *m* autolubrifiant
~BED - [geol] assise *f* perméable, couche *f* poreuse
~BRICK - [constr] brique *f* poreuse
~CASTING - [metall] pièce *f* coulée poreuse
~CHROME HARDENING - [metall] chromage *m* dur poreux
~CLOTH - [text] étoffe *f* perméable
~DIFFUSION - [nucl] diffusion *f* gazeuse
~FUEL PARTICLE KERNEL - [nucl] noyau *m* poreux d'une particule de combustible
~LIMESTONE - [geol] pierre *f* calcaire poreuse
~MOULDS - [plast ind] (moulds of porous structure from which gas or liquid can escape) moules *m*pl poreux
~PIECE - [metall] morceau *m* poreux
~POT - [el] (an unglazed pottery vessel used in two-fluid primary cells as a diaphragm between the liquids) récipient *m* poreux
~REACTOR - [nucl] (nuclear reactor composed of a porous material) réacteur *m* poreux
~ROCK - [geol] roche *f* poreuse
~STONE - [soil] pierre *f* poreuse
~TEXTURE - [metall] structure *f* poreuse
PORPHYRINS - [chem] (important organic compounds occurring in plants and animals as the basis of the oxygen; carrying pigments) porphyrines *f*pl
PORPHYRITIC TEXTURE - [geol] (of the texture of igneous rocks containing large isolated crystals) texture *f* porphyritique
PORPHYROBLASTIC - [geol] (of metamorphic rocks containing large crystals) porphyroblastique
PORPHYRY - [geol] (igneous rock containing large isolated crystals) porphyre *m*
PORPOISING - [aero] (in a seaplane or amphibian, an undulating movement while travelling on the water, caused by lack of stability) tangage *m* à flot
PORT - [mech] (any opening into a cylinder or the like, through which fluid flow is conducted into or out of it) lumière *f*, orifice *m*
[metall] (die casting; filling port on a gooseneck machine) trou *m* de remplissage du cylindre (sur machines à chambre chaude à piston)
[naut aero] (the left-hand side, looking in the direction of forward movement) bâbord *m*
[gen] (a harbour) port *m*
[shipbuild] (opening on a side) sabord *m*
[electron] entrée *f*
[metall] carneau *m*
[electron] (of a waveguide) porte *f*
~BEAM - [naut] bau *m* de bâbord
~BRIDGE - [mech] barrette *f* du tiroir
~CHARGES - [comm] droits *m*pl de port
~-FACE - [mech] (seat of slife-valve) glace *f* du tiroir, table *f* des lumières
~FRONT SEAT - [aero] siège *m* antérieur de bâbord
~-HOLE - [mech] (of a machine) orifice *m*, lumière *f*

PORT-HOLE FAN - [impl] ventilateur *m* aspirateur mural
~ LIGHT - [naut] hublot *m*
~ OF VALVE - [mech] lumière *f* d'une soupape
~ OF CALL - [naut] port *m* d'escale
~ OF DESTINATION - [naut] port *m* de destination
~ -OPENING - [mech] (of a steam cylinder) ouverture *f* de la lumière
~ RADAR INSTALLATION - [radar] radar *m* du port
~ SIDE - [naut] côté *m* de bâbord
~ SPLIT FLAP - [aero] volet *m* d'intrados de bâbord
~ WIDTH - [mech] (of cylinder) largeur *f* de lumière
PORTABLE - [gen] portatif, mobile
~ ACCUMULATOR - [el] (light-weight accumulator suitable for easy handling and removal from place to place) accumulateur *m* portatif
~ AERIAL - [radio] (a movable aerial) antenne *f* mobile
~ APPLIANCE - [el] appareil *m* électrique portatif
~ BATTERY - [el] batterie *f* transportable
~ BELT CONVEYOR - [mech] (a belt conveyor, usually with its own moving power, mounted in a wheeled framework so that it can be moved from place to place as a complete unit) transporteur *m* à courroie mobile
~ BOILER - [th eng] chaudière *f* locomobile, chaudière *f* roulante
~ BORING MACHINE - [mach tool] aléseuse *f* transportable
~ BRIDGE - [constr] pont *m* amovible
~ CONVEYOR - [mech] sauterelle *f*, chargeuse *f* continue
~ CRANE - [mech] grue *f* volante
~ DERRICK-CRANE - [mech] grue *f* derrick volante
~ DOLLY - [impl] (for mainlaying works) diabolo *m*
~ DRILL - [tool] foret *m* portatif
~ DRILLING MACHINE - [mach tool] machine *f* à percer portative
~ DRYING OVEN - [text] étouffoir-séchoir *m* mobile
~ ELECTROMETER - [instr] (a portable electrometer of the absolute attracted-disk type) électromètre *m* portatif
~ ELEVATOR - [mech] élévateur *m* transportable
~ ENGINE - [mech] (internal combustion engine which is carried on road wheels but is not self-propelled) machine *f* locomobile
~ FORGE - [metall] forge *f* volante, forge *f* de campagne
~ GAMMA BACKSCATTER THICKNESS METER - [instr] (used for the measurement of pipe wall thickness) épaisseurmètre *m* portatif à rétrodiffusion gamma
~ GAS - [gas ind] gaz *m* portatif
~ GRAMMOPHONE - [acoust] phonographe *m* portatif
~ HAND-CRANE - [mech] grue *f* roulante manoeuvrée à la main
~ HAND-PUMP - [mech] pompe *f* volante à bras
~ HEATING CARPET - [el] (small carpet in which a heating conductor is worked in a regular pattern) tapis *m* chauffant portatif
~ HYDRAULIC RIVETTER - [mech] machine *f* riveter hydraulique
~ INSTRUMENT - [instr] (any measuring instrument which is carried by hand from one place to another) instrument *m* portatif
~ IONIZATION CHAMBER - [nucl] (a radiation-survey instrument) chambre *f* d'ionisation portative
~ KEYWAY CUTTER - [mech] machine *f* portative à rainer
PORTABLE LAMP - [el] portable fitting for inspection purposes) lampe *f* portative
~ LOGARITHMIC SCINTILLATOR EXPOSURE RATEMETER - [instr] débitmètre *m* d'exposition logarithmique portatif à scintillateur
~ MIXER - [mech] (mixer mounted on legs or a carriage, so as to be moved about the shop) agitateur *m* portatif
~ MOULD - [metall and plast ind] (a small mould designed to be removed by hand from the press for the purpose of tripping and loading) moule *m* portatif
~ PHOTOMETER - [instr] (a photometer easily carried about and used to measure illumination values) photomètre *m* portatif
~ PLANT - [gen] outillage *m* mobile
~ POWER STATION - [el] centrale *f* portative
~ PRESS BOX - [text] compartiment *m* d'alimentation sur rail
~ PROSPECTING RADIATION METER - [instr] radiamètre *m* de prospection portatif
~ RATCHET DRILL - [tool] cé *m*
~ SPOT-LIGHT - [el] spot *m* portatif
~ STANDARD - [el] lampadaire *m* portatif
~ STEAM-CRANE - [mech] grue *f* locomobile à vapeur
~ SUBSTATION - [el] (or portable power-station; a movable power-station used for feeding an electrical network in case of a breakdown or overloading) poste *m* électrique portatif
~ SUCTION FAN - [impl] (used for aerosols on built-in filters) aspirateur *m* étanche portatif
~ TELEVISION RECEIVER - [telev] téléviseur *m* portatif
~ TELEVISION STATION - [telev] appareil *m* de prise de vues portatif
~ TELEVISION TRANSMITTER - [telev] (portable television station principally used to pick up scenes away from the studio) émetteur *m* de télévision portatif
~ TOOL - [tool] (a tool which can be carried from place to place, and is driven by a built-in electric, air or other motor fed by cable or hose.) outil *m* portatif
~ VICE-BENCH - [mech] étau *m* roulant
PORTAL - [arch] portail *m*
[railw] (structural frame for lines) portique
[gen] (of a tunnel) entrée *f*
[anat] (portal vein) veine *f* porte
~ CRANE - [mech] grue *f* à portique
~ JIB CRANE - [mech] (jib crane mounted on a fixed or movable structure allowing the passage of trucks etc under the crane) grue *f* à portique
PORTATIVE ORGAN - [mus] (small organ used in processions) orgue *m* portatif
~ POWER - [mech] (bearing capacity) force *f* portante
PORTCULLIS - [arch] herse *f*, porte *f* coulante
PORTED SLEEVE - [mech] (a sleeve pierced with openings for the distribution of fluid in a machine, e.g. a sleeve-valve in an i.c. engine) manchon *m* à lumières
PORTER - [text] (a group of threads) faisceau *m* de fils, portée *f*
[metall] (iron bar with forged end) porte-lingot *m*
[brew ind] (a type of beer) porter *m*
[metall] (forging tool) cingleresse *f*
~ BAR - [metall] porte-lingot *m*
PORTERAGE - [comm] transport *m*, factage *m*

PORTFOLIO - [impl] serviette *f* (pour documents)
PORTHOLE - [naut] sabord *m*
[shipbuild] hublot *m*
PORTHOLE FAN - [el] ventilateur *m* aspirateur mural
~PACKING - [rubber ind] garniture *f* de hublot
PORTICO - [arch] (colonnade at one side of a building) portique *m*
PORTION - [gen] partie *f*, part *f*
[railw] rame *f*, tranche *f*
~OF A GIRDER - [build] pièce *f* de poutre
PORTIPLEX - [anat] toile *f* choroïdienne supérieure
PORTIPLEXUS - s. portiplex
PORTLAND BLAST FURNACE CEMENT - [metall] (cement made with blast furnace slag and limestone) ciment *m* métallurgique
~CEMENT - [constr] (structural cement made by calcinating a mixture of finely-ground clay and limestone followed by fine grinding: it sets to a hard mass on the addition of water) ciment *m* Portland
~CEMENT AND LIME MORTAR - [constr] mortier *m* au ciment et à la chaux
~CEMENT AND TRASS MORTAR - [constr] mortier *m* au ciment et au trass
PORTRAY, to - [gen] représenter graphiquement
POSIGRADE ROCKET - [astronaut] fusée *f* de séparation, fusée *f* auxiliaire
POSITION, to - [gen] (to place in position) placer en position
[gen] (to set in the correct position) mettre en position, mettre au point situer
[gen] (to find the place of, to locate) situer
POSITION - [gen] position *f*, place *f*, emplacement *m*
[geogr] situation *f*
[astr] point *m*, altitude *f*, position *f*
[naut] lieu *m*
[comput] rang *m*, emplacement *m* d'une perforation
~ANGLE - [astr] (measure of the orientation of one point of the celestial sphere with respect to another) angle *m* de position
~CIRCLE - [surv] (a circle on the surface of the earth, all points on which have the same zenith distance as the observer) cercle *m* de position
~COUPLING - [telecomm] (in telephony) groupement *m* de positions (d'opérateurs voisins)
~COUPLING KEY - [telecomm] (in telephony) clé *f* de liaison entre positions d'opérateur
~ENCODER - [comput] capteur *m* de position
~ERROR - [aero] (error in indicated airspeed due to the position of the pilot-static head on the aircraft) erreur *f* de position (d'antenne anémométrique
~FINDER - [instr] télémètre *m*
~FINDING - [radio] (the determination of the location of a transmitter) recherche *f* d'un poste de émission
~FIXED BY OBSERVATION - [aero] point *m* observé
~GROUPING KEY - s. position coupling key
~HEAD - [hydr] (elevation head) hauteur *f* de chute
~LIGHT - [naut] feu *m* de position
[aero] feu *m* de position d'avion
~LINE - [aero] (a line on which the observer is located, according to observation of terrestrial or celestial objects) ligne *f* de position
~MARK - [surv] repère *m*
~METER - [instr] compteur-totalisateur *m*
~OF EFFECTIVE SHORT - [electron] distance *f* du court-circuit
POSITION OF NIPPERS - [text] position *f* des mâchoires
~OF POINTS - [railw] position *f* de l'aiguille
~OF REST - [gen and mech] position *f* de repos
~OF THE HEALDS - [text] position *f* des lisses
~OF THE RING RAIL - [text] position *f* de la plate-bande
~OF THE SHAFT - [mining] emplacement *m* du puits
~OF THE SHUTTLE - [text] position *f* de la navette
~OF THE SLIDE VALVE - [mech] position *f* moyenne du tiroir
~OF THE TUFT - [text] position *f* de la barbe
~OF WIRE POINTS - [text] position *f* des aiguilles
~PILOT LAMP - [telecomm] lampe *f* pilote de groupe
~PULSE - [comput] impulsion *f* de position
~TARGET - [railw] signal *m* de position
~TELEMETER - [instr] (telemeter which transmits an indication of angular or linear position) appareil *m* de télémesure à comparaison
~TRACKER - [radar] (a plotting plate) radar *m* de poursuite
~WELDING - [metall] soudage *m* en position
POSITIONAL - [gen] de position
~ASTRONOMY - [astr] (the study of the position of heavenly bodies as points on the celestial sphere of the observer) astronomie *f* sphérique
~CONTROL SERVO - [opt & photo] dispositif *m* de téléréglage
~CROSSTALK - [electron] (in a multibeam cathode-ray tube) interpénétration *f* du parcours
~NOTATION - [comput] notation *f* de positions
POSITIONER - [tool] (used in welding) positionneur *m*
[contr] (part of a regulating unit) positionneur *m*
POSITIONING - [gen etc] mise *f* en position
~CONTROL - [constr] commande *f* desmodromique
[autom] commande *f* de positionnement
~CONTROL SYSTEM - [contr] système *m* de commande de position
~DOWEL - [gen] piquet-repère *m*
POSITIVE - [gen] positif
[el] (having a higher electric potential than another point) positif, matrice *f* positive
[phys math etc] positif
[mech] (of a movement which is due to mechanical means) positif, desmodromique, direct
[photo] (image in the same scale of contrast as the original) positif *m*, photogramme *m*
[cin] (the developed film as printed from the negative) épreuve *f*, positif *m*
[opt] (of a lens, eyepiece etc) positif, convergent
~ACCOMMODATION-[med]accommodation *f* pour la vision de près
~AFTER-IMAGE - [opt] (the image still perceived when the object is removed and the eye is closed, due to fatigue of the retina) traînage *m* positif
~AFTER-POTENTIAL - [el biol] (the prolonged positivity which follows the negative after-potential in a homogeneous fibre group) queue *f* de potentiel positif
~ALLOWANCE -[mech] jeu *m*
~AND NEGATIVE BOOSTER - [el] (booster arranged to increase and also to reduce the voltage supplied by another electrical source) survolteur-dévolteur *m*
~AREA - [el] zone *f* de sortie des courants vagabonds
~AREA OF STRAY CURRENTS - s. positive area

POSITIVE BIAS - [electron] (a positive voltage applied to the control grid of an electronic tube) polarisation *f* positive
~BLOWER - [mech] ventilateur *m* soufflant
~BOOSTER - [el] (booster arranged to increase the voltage supplied by another electrical source) survolteur *m*
~CHARGE - [el] (the condition of a body having a deficiency of electrons) charge *f* positive
~CLUTCH - [mech] embrayage *m* mécanique
~COLUMN - [electron] (the luminous region following the Faraday dark space. A plasma) colonne *f* positive
~COUNTER FALLER MOTION - [text] mouvement *m* positif de la contre-baguette
~CRYSTAL - [phys] cristal *m* positif
~DIHEDRAL ANGLE - [geom] angle *m* dièdre positif
~DISPERSION - [electron] (in a delay line) dispersion *f* positive
~-DISPLACEMENT BOOSTER - [gas ind] (in network equipment) surpresseur *m* volumogène
~-DISPLACEMENT METER - s. positive meter
~-DISPLACEMENT PUMP - [mech] pompe *f* volumétrique
~-DISPLACEMENT SUPERCHARGER - [aero] (the type of supercharger in which the air or gas is separated into closed chambers and positively compressed) compresseur *m* à déplacement positif
~DISTORTION - [telev] (a distortion in which the sides bulge out) distorsion *f* en barillet
~DOBBY - [text] ratière *f* positive, mécanique *f* d'armure positive
~DRIVE -[mech] (any type of drive in which slip cannot occur) transmission *f* positive
~DRIVEN SPINDLE - [text] broche *f* à commande positive
~ECHO - [telev] image *f* fantôme positive
~ELECTRICITY - [el] (electricity similar to that produced by rubbing a piece of glass with silk) électricité *f* positive
~ELECTRODE - [electron] (that electrode which forms the cathode when the cell is discharging) électrode *f* positive
~ELECTRON - s. positron
~EMULSION - [photo] (used for printing positives) émulsion *f* positive
~EYEPIECE -[opt] oculaire *m* positif
~FEED - [mech] entraînement *m* positif
~FEEDBACK - [radio] (or regeneration; feedback resulting in an increase of the amplification) réaction *f* positive
~FEEDER - [el] (the feeder which is connected to the positive terminal of a d.c.supply) feeder *m* positif, artère *f* d'alimentation positive
~FILM STOCK - s. positive stock
~G-[phys] (acceleration) g *f* positive
~GLOW - [el] (luminous phenomenon accompanying an electric discharge in a gas under certain conditions of pressure and distance between the electrodes) lumière *f* anodique, lumière *f* positive
~GOING SYNC PULSES - [telev] impulsions *fpl* de synchronisation à potentiel positif
~GRID OSCILLATOR - [radio] (the Barkhausen-Kurz oscillator) oscillateur *m* de Barkhausen et Kurz, oscillateur *m* à champs de freinage
~GROUP - [el] groupe *m* plaques positives
~GUIDING OF FALLER - [text] mouvement *m* forcé de la baguette
POSITIVE IMAGE - [telev] image *f* positive
~IMPULSE FLASH-OVER VOLTAGE - [el] tension *f* de contournement à impulsion positive
~INTEGER - [math] nombre *m* entier positif
~ION - [phys] (ion carrying a positive charge) cation *m*, ion *m* positif
~ION BEAM - [nucl] (beam of positive ions) faisceau *m* d'ions positifs
~-ION EMISSION - [electron] émission *f* d'ions positifs
~ION RAYS - [phys] rayons *mpl* canaux
~LET OFF MOTION - [text] régulateur *m* de déroulement positif
~LENS - [opt] (a converging lens) lentille *f* convergente
~MAGNETOSTRICTION - [phys] (expansion of a material by the application of a magnetic field) magnétostriction *f* positive
~MATRIX - [el metall] (matrix having a surface which is similar to that which must ultimately be produced by electroforming) matrice *f* positive
~METER - [instr] (volumetric meter) débitmètre *m* à déplacement
~MINERAL - [min] (mineral in which the ordinary ray velocity is greater than that of the extraordinary ray) cristal *m* positif
~MODULATION - [telev] modulation *f* positive
~MOMENT -[phys] moment *m* positif
~MOTION - [mech] mécanisme *m* commandé
~MOULD - [metall & plast ind] (the part of mould which enters the recesses in the negative mould) moule *m* mâle
~MOVEMENT - [mech] mouvement *m* commandé
~MOVEMENT OF SHAFTS - [text] mouvement *m* des lames positif
~-NEGATIVE THREE LEVEL ACTION - [comput] (bang-bang action in the USA) action *f* par plus ou moins
~NUCLEUS - [nucl] (the nucleus of an atom, which carries a positive charge) noyau *m* positif
~NUMBER - [math] nombre *m* positif
~PHASE-SEQUENCE - [el] (the order in which the three phases normally attain a maximum potential of a given sign) composante *f* positive de la phase
~PHASE-SEQUENCE RELAY - [el] relais *m* fonctionnant sur la composante positive de la phase
~PICTURE PHASE - [telev] polarité *f* positive du signal image
~PLATE - [el] (the electrode which acts as cathode during discharge) plaque *f* positive
~POSITION STOP - [mech] goupille *f* d'arrêt de position
~POTENTIAL - [el] potentiel *m* positif
~-PRESSURE VENT VALVE - [mech] (a vent-valve operated by a pressure above than of the ambient) robinet *m* de purge à pression positive
~RAKE - [mech] inclinaison *f* positive
~RATCHET MOTION - [text] mécanisme *m* à commande positive de la roue à cliquet
~RAY - [phys] (a positive particle in a vacuum tube escaping through bored holes in the cathode) rayon *m* canal
~RAY-CURRENT - [electron] courant *m* de rayons anodiques
~REACTION - [bot] réaction *f* positive
~RELEASE PRINTER - [cin] tireuse *f* de copie finale
~SAFETY - [gen] (fool-proof control) sécurité *f* to-

tale
POSITIVE SAFETY STOP - [mech] goupille *f* d'arrêt de sécurité
~SEQUENCE - [el] séquence *f* positive de phases
~SEQUENCE ACTIVE POWER - [el] puissance *f* directe d'un système triphasé
~SEQUENCE POLYPHASE SYSTEM - [el] système *f* polyphasé direct
~SHAFT MOTION - [text] mouvement *m* positif des lames
~SHEDDING MOTION - [text] mécanisme *m* pour la formation du pas positif
~SHUTTLE BOX DRIVE - [text] mouvement *m* positif des boîtes à navettes
~SIGN - [math] signe *m* positif
~STOCK - [cin] (positive film stock; unexposed film on which the prints are made) film *m* positif
~STOP - [mech] arrêt *m* de sécurité
~STRESS - [mech] effort *m* de compression, travail *m* à l'écrasement
~SUPPORT - [text] appui *m* positif
~TERMINAL - [el] (the terminal of a cell from which the current flows through the external circuit to the negative terminal) pôle *m* positif, borne *f* positive
~TRANSMISSION - [telev] s. positive video signal
~TUBE - [electron] (thyratron in which the discharge is triggered by a positive pulse on the grid) thyratron *m* à impulsion positive
~VALENCE - [chem] (valence state of an atom in which the valence number is positive) valence *f* positive
~VIDEO SIGNAL - [telev] (signal in which the increasing amplitude corresponds to increasing light-value in the image) signal *m* vidéo positif
POSITON - [electron] (a positive electron) positon *m*
POSITRON - [electron] (an electron carrying a positive charge) positron *m*
~DECAY - [nucl] (positron emission) désintégration *f* positogène
POSITRONIUM - [phys] (quasi-stable system which consists of a positron and an electron bound together) positonium *m*
POSOLOGY - [med] (the dosage of drugs) posologie *f*
POSSUOLANA - s. pozzuolana
POST, to - [gen] poster
[gen] (mail) mettre à la poste
[comm] (of accounts) passer (les écritures)
[comput] signaler, annoter
POST - [gen] poteau *m*, pieu *m*, pilier *m*, montant *m*
[gen] (the mail) poste *f*
[constr] (of a door) montant *m*
[constr] montant *m*, jambage *m*, dormant *m*
[mining] (a supporting pillar) butte *f* de boisage, chandelle *f*
[naut] sabord *m* d'arcasse
[el] borne *f* à vis
[metall] (in foundry work) charge *f* de minerai
[paper man] (standard size of printing paper) papier *m* affiche
[paper man] (sheets of wet pulp) feuille *f* de pâte engraissée
[gen] (a prefix meaning after) post-
[mech] (a spindle) arbre *m* de trousseau
[min] (charge of mineral) charge *f*
[min] (a pillar of ore) pilier *m*, massif *m*
[mech] (of a crane) arbre *m* (d'une grue), fût *m* (d'une grue)

POST-ACCELERATION ANODE - [electron] (additional electrode in a cathode-ray tube) anode *f* à post-accélération
~-ACCELERATION CATHODE RAY TUBE - [electron] tube *m* à rayons cathodiques à post-accélération
~-AND BAR - [mining] cadre *m* à un seul montant
~AND PANE WALL - [constr] mur *m* à colombages
~-AND-STALL - [mining] exploitation *f* par traçage et dépilage
~-BLOSSOM SPRAY - [agric] traitement *m* post-floral
~-DEFLECTION ACCELERATING ELECTRODE - [electron (intensifier electrode; electrode increasing the velocity of electrons at the end of their trajectory) électrode *f* post-accélératrice
~-DEFLECTION ACCELERATION - [electron] (the acceleration of the beam electrons after deflection) post-accélération *f*
~DRILL - [mining] perforatrice *f* à colonne
~-EDIT, to -[comput] préparer pour l'imprimante
~-EDITION - [comput] (in machine translation) post-édition *f*
~-FLIGHT INSPECTION - [aero] (inspection carried out after the conclusion of a flight) inspection *f* après le vol
~-FORMED MOULDING -[plast ind] moulage *m* post-formé
~-FORMING - [plast ind] (term used of laminated plastic products which can be shaped after manufacture) post-formage *m*, postformable *m*
~-FORMING SHEET - [plast ind] feuille *f* postformable
~-HANDLING - [plast ind] (operations on a plastic-coated wire, e.g. permanent curling, printing etc) traitement *m* ultérieur
~-HEAD - [el] (a terminal pillar) boulon *m* polaire
~-HOLE - [mining] potelle *f*
~INSULATOR -[el] (porcelain insulator having the shape of a post) isolateur *m* à poteau
~LUMINESCENCE - [phys] post-luminescence *f*
~-MORTEM - [med leg] après décès, autopsie [adj] d'autopsie
~-MORTEM DUMP - [comput] analyse-mémoire *f* d'autopsie
~-MORTEM EXAMINATION - s. post-mortem
~-MORTEM ROUTINE - [comput] (a diagnostic routine) programme *m* autopsie
~OFFICE - [gen] bureau *m* des postes
~OFFICE BOX - [gen] boîte *f* postale
[el] (assembly of calibrated resistors mounted in a box and forming three arms of a Whetstone bridge) boîte *f* à pont, pont *m* de résistances étalonnées
~OFFICE RED - [paint] (alternative name for Helio fast scarlet, made from diazotized 3 - nitro 4 - toluidine) hélio-rouge
~PULLER - [mining] arrache-étais *m*
~PUNCHER - [mining] haveuse *f* à pic sur affût-colonne
~-RECORD - s. post-scoring
~-SCORING - [cin] (the use of a separate sound film for a silent picture) post-synchronisation *f*
~STONE - [min] grès *m* à grain fin
~-STORAGE TEST - [gen] (test made on an item after withdraval from storage to decide if it has deteriorated) essai *m* après l'emmagasinage
~-SYNC FIELD-BLANKING INTERVAL - [telev] intervalle *m* de suppression de trame après synchronisation
~-SYNCHRONIZATION - s. post-scoring

POST-SYNCHRONIZATION - [telev] postsynchronisation *f*
~-TENSIONING - [metall] post-tension *f*
~-TERM BIRTH - [med] (retarded birth) accouchement *m* retardé
~-TYPE MIXER - [mech mach] (mixer designed to be mounted against a wall or structural column) agitateur *m* mural, agitateur *m* à console
POSTAGE - [mail] port *m*, affranchissement *m*
[comm] frais *mpl* de port
~STAMP - [mail] timbre *m*
POSTAL - [gen mail] postal
[paper man] (standard board size) carton *m* pour cartes postales
~AIRCRAFT - [aero] aéropostal *m*
~VAN - [railw] wagon *m* postal
POSTCARD - [paper man] (standard size of cut card) format *m* de carte postale
POSTCAVA - [anat] veine *f* cave inférieure
POSTCLAVICULAR - [anat] rétroclaviculaire
POSTDURAL - [anat] rétrodural
POSTEMPHASIS - [acoust] affaiblissement *m*, atténuation *f*
POSTER - [print] (placard or bill used for advertising purposes) placard *m*, affiche *f* murale
~-LIKE-EFFECT - [photo] effet *m* de gouache
POSTERIO-OCCLUSION - [med] distocclusion *f*
POSTERIOR - [gen] postérieur
~CORD SYNDROME - [med] syndrome *m* cordonal postérieur
~LOBE OF HYPOPHYSIS - [anat] s. posthypophysis
~PROJECTION - [radiation] (or dorsal projection; radiograph in which the ray beam crosses the body from front to back) projection *f* dorsale, vue *f* antéro-postérieure
POSTERN - [constr] (gate or door generally on the side of a building) poterne *f*, porte *f* de service
POSTEROCCLUSION- [med] s. posterio-occlusion
POSTEROLATERAL SYNDROME - [med] syndrome *m* cordonal postérolatérale
POSTETHMOID - [anat] rétro-ethmoïdal
POSTEQUALIZATION - s. postemphasis
POSTFRONTAL - [anat] rétrofrontal
POSTHEAD - [railw] (in electrical traction system) boulon *m* polaire
POSTHEATING - [metall] réchauffage *m* de la soudure
POSTHITIS - [med] posthite *f*
POSTHOLE - [constr] potelle *f*
~DIGGER - [impl] excavateur *m* pour pieux
POSTHORN - [mus] cor *m* de postillon
POSTHYPOPHYSIS - [anat] lobe *m* postérieur de l'hypophyse, neurohypophyse *f*
POSTICAL - [bot] postérieur
POSTICUS - [anat] postérieur
POSTIGNITION - [mech] (in internal combustion engines) postcombustion *f*
POSTING FLUID - [print] encre *f* à copier
POSTIRRADIATION SYNDROME - [med] syndrome *m* d'irradiation
POSTMATURITY - [med] postmaturité *f*
POSTOBLONGATA - [med] portion *f* bulbaire du plancher du quatrième ventricule
POSTORBITAL - [anat] postorbital
POSTPALATAL - [acoust] postpalatal, vélaire
POSTPARIETAL - [anat] postpariétal
POSTPONE, to - [gen] ajourner, remettre
[med] être en retard
POSTPONEMENT - [gen] remise *f* à plus tard
POSTSCRIPT - [gen] post-scriptum *m*, postface *f*
POSTULATE, to - [math] poser en postulat
[gen] postuler, demander
POSTULATE - [gen and geom] axiome *m*, postulat *m*
POSTURE - [gen] posture *f*, pose *f*
~SENSE - [med] sens *m* de position
POT, to - [gen] mettre en pot, conserver
[electron] (to enclose electronic components or the like in a solid mass of plastic or the like to prevent vibration effects) monter dans un creuset
POT - [gen] pot *m*
[gen] (cooking impl) marmite *f*
[metall] (a metal crucible) creuset *m*
[el] pot *m*, four *m* électrolytique, four *m*
[plast ind] (term used for the transfer chamber) pot *m* de transfert
[glass man] (container for molten glass) creuset *m*
[constr] (of a chimney) mitre *f* de cheminée, pot *m* de cheminée
[el] (of a battery) bac *m* de pile
[paper man] papier *m* pot
~ANNEALED - [metall] recuit en boîtes
~ANNEALING - [metall] (close annealing) recuit *m* en boîtes
~ANNEALING FURNACE - [metall] four *m* à pots pour recuire
~ARCH - s. pot furnace
~BRAZING -[metall] soudure *f* forte à immersion
~-CHEESE - [food] fromage *m* blanc, caillebotte *f*
~-CLAY - s. potter's clay
~CORE - [phys] noyau *m* en pot
~-DRESSING - [chem] vernis *m* pour creusets
~-EYE - [text] oeillet *m* en porcelaine
~-FURNACE - [metall & glass ind] (furnace in which are set a number of pots for preheating) four *m* à creuset
~HEAD JOINTING SLEEVE - [telecomm] manchon *m* tête de câble
~HOLE - [metall] marmite *f*
~LIFE - [ind chem] (the greatest time during which an adhesive prepared for application remains usable) vie *f* en pot
~PLUNGER - [plast ind] (the plunger in a transfer moulding machine) piston *m* de transfert
~SPINNING FRAME - [text] continu *m* à filer à pot
~STILL - [ind chem] (a batch-distillation still, consisting of a boiling vessel and condenser. A fractionating column may be added) alambic *f m* (chauffé directement par la flamme)
~WAGON - [glass man] (vehicle for the transport of pots from the pot furnace) diable *m*
POTABILITY - [hydr] potabilité *f*, esculence *f*
POTABLE - [gen] (drinkable) potable
~WATER - [gen] (water suitable for drinking) eau *f* potable
POTASH - [chem] (potassium hydroxide) potasse *f* caustique
[chem] (crude potassium carbonate obtained by leaching the ashes of plants) carbonate *m* de potasse, potasse *f*, alcali *m* vegetal
~ALUM - [chem] alun *m* de potasse, alun *m* potassique
~BULBS - [ind chem] (glass apparatus consisting of bulbs blown on tubes for passing gases through potassium hydroxide) tube *m* à potasse
~EXTRACTION PLANT - [ind chem] installation *f* pour

l'extraction de potasse
POTASH FELDSPAR - [min] (a silicate of aluminium and potassium) feldspath *m* potassique
~FERTILIZERS - [agric] engrais *mpl* potassiques
~GLASS - [glass man] verre *m* à base de potasse
~HARDENING - [metall] cémentation *f* au prussiate de potasse
~LYE - [chem] lessive *f* de potasse
~MANURE - [ind chem] engrais potassique
~MICA - [min] (muscovite) mica *m* potassique, muscovite *f*
~MINE - [min] mine *f* de sels de potasse
~RECOVERY - [chem] récupération *f* de la potasse
~SOAP - [chem] savon *m* à base de potasse
~SOLUTION - [chem] lessive *f* de carbonate de potasse
~-SYENITE - [geol] (syenitic rock with an excess of potash-feldspar over soda-feldspar) syénite *f*
~WATER - [chem] eau *f* gazeuse bicarbonatée
~WATER GLASS - [ind chem] silicate *m* de potasse
POTASSEMIA - [med] kaliémie *f* potassienne
POTASSIC - [chem] potassique
~CYANIDE - [chem] cyanure *m* de potassium
POTASSIUM - [chem] (a metallic element, symbol K, A N. I9, A.W. 39.096. It ignites spontaneously in moist air. It has many uses especially in the preparation of potassium salts) potassium *m*
~ABIETATE - [chem] (an insecticide) abiétate *m* de potassium
~ACETATO - [chem] (an analytical reagent, diuretic, diaphoretic, and dehydrating agent) acétate *m* de potassium
~ACID OXALATE - [chem] oxalate *m* acide de potassium
~ACID SACCHARATE - [chem] (an additive in rubber processing) saccharate *m* acide de potassium
~ACID SULPHATE - [chem] sulfate *m* acide de potassium
~ACID TARTRATE - [chem] tartrate *m* acide de potassium
~ALUM - [chem] (hydrous aluminium potassium sulphate) alun *m* de potasse
~ALUMINATE - [chem] (a mordant in dyeing) aluminate *m* de potassium
~ALUMINIUM FLUORIDE - [chem] (a pesticide) fluorure *m* de potassium et aluminium
~ANTIMONYL TARTRATE - [chem] tartrate *m* de potassium et antimonyle
~ARSENATE - [chem] (a component of insecticides) arséniate *m* de potassium
~ARSENITE - [chem] (an analytical reagent) arsénite *m* de potasse
~BICARBONATE - [chem] (a gastric antacid and a component of fire-extinguishing mixtures) bicarbonate *m* de potassium
~BINOXALATE - [chem] (salts of lemon. A bleaching agent) bioxalate *m* de potassium
~BISULPHATE - [chem] (an intermediate for the production of fertilizers) bisulfate *m* de potassium
~BISULPHITE - [chem] (a bleaching and reducing agent) bisulfite *m* de potassium
~BITARTRATE - [chem] (cream of tartar. A cathartic) bitartrate *m* de potassium, tartrate *m* acide de potassium
~BROMATE - [chem] (an oxidizing agent and analytical reagent) bromate *m* de potassium
~BROMIDE - [chem] (a sedative and photographic chemical) bromure *m* de potassium
POTASSIUM BROMOFLUORIDE - [chem] bromofluorure *m* de potassium
~CARBONATE - [chem] (an intermediate and textile processing agent) carbonate *m* neutre de potassium, potasse *f*
~CHLORATE - [chem] (a bleaching agent, constituent of explosives; and oxidizing agent) chlorate *m* de potassium, sel *m* de Berthollet
~CHLORIDE - [chem] (a fertilizer, and source of physiological potassium in medicine) chlorure *m* de potassium
~CHLOROCHROMATE - [chem] (an oxidizing agent) chlorochromate *m* de potassium
~CHLOROPLATINATE - [chem] (a photographic chemical) chloroplatinate *m* de potassium
~CHROMATE - [chem] (a textile dyeing mordant, analytical reagent, and intermediate for chromium compounds) chromate *m* de potassium
~CITRATE - [chem] (a diuretic) citrate *m* de potassium
~COBALTICYANIDE - [chem] cobaltocyanure *m* de potassium
~CYANATE - [chem] (a herbicide) cyanate *m* de potassium
~DISHROMATE - [chem] (an analytical reagent, tanning and oxidizing agent, and dyeing assistant) bichromate *m* de potassium
~FERRICYANIDE - [chem] (an intermediate for pigments and dyestuffs) ferricyanure *m* de potassium
~FERROCYANIDE - [chem] (a dyeing agent, analytical reagent, component of tempering baths in metallurgy, and an intermediate for pigments) ferrocyanure *m* de potassium
~FLUORIDE - [chem] (an etching agent for glass) fluorure *m* de potassium
~FLUOSILICATE - [chem] (a constituent of ceramic glazes) fluosilicate *m* de potassium
~GLUOCONATE - [chem] (a source of physiological potassium in medicine) gluconate *m* de potassium
~HALIDES - [chem] (compounds of potassium with the halogenes) halogénures *mpl* de potassium
~HYDROXIDE - [chem] (caustic potash. A caustic in medicine, analytical reagent, and a starting material in soap manufacture) potasse *f* caustique, hydroxyde *m* de potassium
~HYDROXIDE SOLUTION - [chem] (caustic potash solution) lessive *f* de potasse
~IODATE - [chem] (an analytical reagent) iodate *m* de potassium
~IODIDE - [chem] (an analytical reagent, photographic chemical, and a diuretic and expectorant) iodure *m* de potassium
~MAGNESIUM SULPHATE - [chem] (a fertilizer) sulfate *m* de potassium et magnésium
~MANGANATE - [chem] (a bleaching and oxidizing agent, photographic chemical, and disinfectant) manganate *m* de potassium
~METABISULPHITE - [chem] (a photographic chemical, food preservative, and antiseptic) métabisulfite *m* de potassium
~METAPHOSPHATE - [chem] (synonymous with monopotassium metaphosphate) métaphosphate *m* de potasse
~MICA - [min] mica *m* potassique
~MOLYBDATE - [chem] (an analytical reagent) molybdate *m* de potassium

POTASSIUM NAPHTHENATE - [chem] (a drier for paints and varnishes) naphténate *m* de potassium
~NITRATE - [chem] (saltpeter) salpêtre *m*, nitre *m*, nitrate *m* de potassium
~NITRITE - [chem] (an analytical reagent) nitrite *m* de potassium
~OLEATE - [chem] (an emulsificant) oléate *m* de potassium
~OSMATE - [chem] (an analytical reagent) osmiate *m* de potassium
~OXALATE - [chem] (a bleaching agent and analytical reagent) oxalate *m* de potassium
~PERCARBONATE - [chem] (an analytical reagent and textile processing agent) percarbonate *m* de potassium
~PERCHLORATE - [chem] (an analytical reagent oxidizing agent and component of explosives) perchlorate *m* de potassium
~PERIODATE - [chem] (an oxidizing agent) périodate *m* de potassium
~PERMANGANATE - [chem](an analytical reagent, disinfectant, oxidizing agent, and bactericide) permanganate *m* de potassium, permanganate *m* de potasse
~PEROXIDE - [chem] (a bleaching and oxidizing agent) peroxyde *m* de potassium
~PERSULPHATE - [chem] (a polymerization catalyst, photographic chemical, oxidizing and bleaching agent, and analytical reagent) persulfate *m* de potassium
~PHOSPHATE DIBASIC - [chem] (a fertilizer) phosphate *m* bipotassique
~PHOSPHATE (MONOBASIC) - [chem] (a mild aperient used in medicine) phosphate *m* mono-potassique
~PHOSPHATE (TRIBASIC) - [chem] phosphate *m* tripotassique
~POLYSULPHIDE - [chem] (a fungicide) polysulfure *m* de potassium
~PYROPHOSPHATE-[chem] (a textile processing agent and component of detergent liquids) pyrophosphate *m* de potasse
~QUADROXALATE - [chem] (a combination of the normal and acid oxalates: under the name of Salts of Sorrel, it is used as a stain-remover. Highly poisonous) quadroxalate *m* de potassium
~SELENATE - [chem](an analytical reagent) séléniate *m* de potassium
~SILICATE - [chem] (water-glass. A catalyst and component of adhesives) silicate *m* de potassium
~SODIUM TARTRATE - [chem] (a saline diuretic and cathartic) tartrate *m* de potassium et sodium
~STANNATE - [chem] (a textile dyeing mordant) stannate *m* de potassium
~STEARATE - [chem] (a starting material in the manufacture of textile softening agents) stéarate *m* de potassium
~SULPHATE - [chem] (a fertilizer and analytical reagent) sulfate de potassium
~SULPHIDE - [chem] (a depilatory and analytical reagent) sulfure *m* de potassium
~SULPHITE - [chem] (a photographic chemical) sulfite *m* de potassium
~SULPHOCARBONATE - [chem] (an analytical reagent) sulfocarbonate *m* de potassium
~TARTRATE - [chem] (a saline purgative) tartrate *m* de potassium
~TITANATE - [chem] (used in the form of fibres as a high-temperature insulant) titanate *m* de potassium
POTASSIUM TRIPOLYPHOSPHATE - [chem] (a component of fertilizers) tripolyphosphate *m* de potassium
~UNDECYLENATE - [chem] (a bacteriostat) undécylénate *m* de potassium
POTATO - [bot] pomme *f* de terre
~CLAMP - [agric] silo *m* à pommes de terre
~CLEANER - [agric] laveuse *f* de pommes de terre
~CROP - [agric] arrachage *m* des pommes de terre
~DIGGER - [agric]arracheuse *f* de pommes de terre
~DIGGER AND CLEANER - [agric] arracheuse-rangeuse *f* de pommes de terre
~FLUOR - [agric] farine *f* de pomme de terre
~GRADER - [agric] trieuse *f* de pommes de terre
~HARVESTER - [agric] arracheuse-chargeuse *f* de pommes de terre
~HAULM - [agric] fanes *f*pl de pommes de terre
~HAULM PLUCKER - [agric] arrache-fanes *m*, batteuse *f* de fanes de pommes de terre
~HOOK - [agric] croc *m*
~LEAF CURL - [plant disease) frisolée *f* de la pomme de terre
~LIFTER - [agric] arracheuse *f* de pommes de terre
~LIFTER PLOUGH - [agric] charrue-arracheuse *f* de pommes de terre
~LIFTING - [agric] arrachage *m* de pommes de terre
~MASHER - [agric] broyeur *m* de pommes de terre
~PEELS - [agric] (animal feeding) pelures *f*pl de pommes de terre
~PLANTER - [agric] planteuse *f* de pommes de terre
~PULP - [agric] (animal feeding) pulpe *f* de pomme de terre
~PULPER - s. potato masher
~RIDDING AND LIFTING PLOUGH - [agric] butteur-arracheur *m* combiné
~-STALK BEATER - [agric] arrache-fanes *m*
~STARCH - [sugar ind] fécule *f*
~STEAMER - [agric] cuiseur *m* de pommes de terre
~VINE - [agric] fanes *f*pl de pommes de terre
~WART - [bot] (plant disease) gale *f* noire de la pomme de terre, chancre *m* de la pomme de terre
~WASHING MACHINE - [agric] laveuse *f* de pommes de terre
~WILT - [bot] (plant disease) maladie *f* du jaune, verticilliose
~SPIRIT - [chem] (crude ethyl alcohol obtained by the fermentation of potatoes) eau-de-vie *f* de pommes de terre
POTCH, to - [paper man] (to bleach) blanchir
POTCHER - [paper man] (a pulp mixing and washing machine) cuvier *m* mélangeur
POTCHING ENGINE- s. potcher
POTELOT - [min] molybdénite *f*
POTENCY - [med] puissance *f*, efficacité *f*
POTENT - [gen] puissant, efficace
POTENTIAL - [gen] (possibility of power) potentiel *m*
[adj] potentiel
[el] (potential at a point is the potential difference between that point and earth) potentiel *m*
~ATTENUATOR - [telecomm] (attenuating potentiometer) atténuateur *m* de potentiel
~BARRIER - [phys] (potential energy curve delimiting two regions if the energy diagram)barrière *f* de potentiel
~CORRECTION - [contr] écart *m* temporaire
~CURVE - s. potential-energy curve
~DIAGRAM - [electron] (of an electron-optical sys-

tem; diagram showing the equipotential curves in a plane of symmetry of an electron-optical system) diagramme *m* d'énergie potentielle
POTENTIAL DIFFERENCE - (P.D.) - [el] (the difference of electric potential between two given points) différence *f* de potentiel, tension *f*
~DISTRIBUTION - [el] répartition *f* de potentiel
~DIVIDER - [el] (voltage divider) réducteur *m* de tension, potentiomètre *m*
~DROP - [el] chute *f* de potentiel, chute *f* de tension
~ENERGY - [phys] (the energy of a body by virtue of its position) énergie *f* potentielle, énergie *f* accumulée, énergie *f* intrinsèque
~-ENERGY CURVE - [phys] (in an energy diagram, a curve representing the variations of potential energy as a function of a parameter characterizing the degree of freedom of a particle) courbe *f* de potentiel, courbe *f* d'énergie potentielle
~EQUALIZER - [el] régulateur *m* de potentiel
~FALL - [electron] (anode fall; the fall of potential due to the space charge near the anode) chute *f* anodique
[electron] (when near the cathode) chute *f* cathodique
~FLOW - [mech] (flow in which the velocity is the gradient of a scalar function of position, the velocity potential) effluent *m* potentiel
~FUNCTION - [math] fonction *f* potentielle
[el] (in an irrotational field) potentiel *m* dans un champ irrotationnel
~FUSE - [el] (protecting the voltage circuit of a measuring instrument) fusible *m* de tension
~GRADIENT - [el] (the rate of charge of electric potential along a conductor) gradient *m* de potentiel
~HILL - s. potential barrier
~IN AN IRROTATIONAL FIELD - s. potential function [el]
~HOLE - s. potential well
~INDICATOR - [el] (voltage, or charge indicator) détecteur *m* de tension, indicateur *m* de tension
~JUMP - [el] (sudden change in the value of the potentiel) saut *m* de potentiel
~MEDIATOR - [chem] (substance added to an oxidation-reduction system to accelerate the establishment of a given potential) médiateur *m* de potentiel
~MINIMUM - [el] minimum *m* de potentiel
~-MINIMUM SURFACE - [electron] (a virtual cathode; the region in the space-charge where there is a potential minimum behaving as a source of electrons by reason of the space-charge density) cathode *f* virtuelle
~OF A BODY - [el] potentiel *m* d'un corps
~PEAK - [el] charge *f* forte
~PEAK PERIODS - [el] périodes *f*pl de charge forte
~PLATEAU - [phys] (the higher surface of the potential hills) plateau *m* de potentiel
~REGULATOR - [el] régulateur *m* de tension
~SCATTERING - [nucl] (that part of the nuclear scattering which has its origin in reflection from the nuclear surface) diffusion *f* potentielle, dispersion *f* potentielle
~TEMPERATURE - [phys] (the temperature which a given specimen of air or other gas would attain if brought adiabatically to a standard pressure, usually taken as I000 millibars) température *f* potentielle
~TRANSFORMER - [el] (instrument-transformer for the transformation of voltage) transformateur *m* de tension
POTENTIAL TROUGH - [phys] (the region of the energy diagram delimited by the sides of two neighbouring potential hills) puits *m* de potentiel
~VALUE - [contr] valeur *f* d'équilibre
~WELL - [nucl] (a region of minimum nuclear potential) puits *m* de potentiel
POTENTIOMETER - [instr] (an instrument for the measurement of potential differences) potentiomètre *m*
~-BRAKING - [el] (braking method used for series motors) freinage *m* potentiométrique
~-BRAKING CONTROLLER - [el] régulateur *m* pour freinage à blocage
~CIRCUIT - [el] circuit *m* de potentiomètre
~METHOD - [instr] (opposition method in which the numerical variable of the variable resistor is a simple multiple of the voltage to be measured) méthode *f* potentiométrique
~PICK-OFF - [contr] capteur *m* à potentiomètre
~SET - [el] ensemble *m* de potentiomètre
~STEP - [el] contact *m* du potentiomètre
~STUD - s. potentiometer step
~-TYPE RESISTOR - [el] résistance *f* potentiométrique
~-TYPE FIELD RHEOSTAT - [el] (field rheostat in which the resistor is suitable for connection across the source of supply) rhéostat *m* potentiométrique
POTENTIOMETRIC - [el etc] potentiométrique
~ANALYSIS - [chem] (electrometric titration) titrage *m* potentiométrique
POTHEAD - [el] (sealing end) bout *m* de câble, manchon *m* d'extrémité
POTION - [med] potion *f*
POTOMANIA - [med] potomanie *f*, dipsomanie *f*
POTSTONE - [min] pierre *f* ollaire, chloritochiste *m*
POTT - [paper man] (size of paper; I2 $^1/_2$ x I5) papier *m* pot
POTT'S ASTHMA - [med] asthme *f* thymique de Kopp
~DISEASE - [med] tuberculose *f* vertebrale, mal *m* de Pott
~GANGRENE - [med] gangrène *f* sénile
~PARALYSIS - [med] paraplégie *f* pottique
~PARAPLEGIA - [med] s. pott's paralysis
POTTER - [gen] potier *m*
POTTER'S CLAY - [geol] (a fine-textured, highly plastic clay) terre *f* de potier, argile *f* plastique
~EARTH - s. potter's clay
~LEAD - [min] alquifoux *m*
~WHEEL - [impl] (arrotating circular table on which hollow ware is made) tour *m* de potier, roue *f* de potier
POTTERY - [gen] (term used for a wide variety of ware made from clay alone or compounded with other materials) poterie *f*
[gen] (potter-making factory) fabrique *f* de céramiques
POTTING - [gen] fabrication *f* de céramiques
[el] (the system of impregnating and encapsulating an electrical assembly, generally with an epoxy resin) enrobage *m*
[food] mise *f* en pot
POUCH - [gen] bourse *f*, petit sac *m*
[zool] poche *f* ventrale
POUDRETTE - [agric] (fertilizer) poudrette *f*
POULARD - [zool] poularde *f*
POULTRY - [agric] race *f* avicole
~BREEDING - [agric] élevage *m* de volailles
~FARM - [agric] parc *m* avicole, ferme *f* avicole

POULTRY FARMING - [agric] s. poultry breeding
~FAT - [food] graisse *f* de volailles
~FEED - [agric] (animal feeding) aliments *mpl* pour volailles
~HUSBANDRY - s. poultry farming
~MANURE - [agric] poulaitte *f*, fumier *m* de poules
~MARKET - [comm] marché *m* de volailles
~-RUN - [agric] basse-cour *f*
POUND, to - [gen] concasser, broyer, piler
[min etc] (to pulverize) piler
POUND - [meas] (454 g) livre *f*
[fin] (pound sterling) livre *f* sterling
[agric] fourrière *f*
[hydr] bassin *m*, réservoir *m*, bief *m*
~-CALORIE - [phys] (heat unit, also called centrigrade heat unit; one hundredth of the heat require to raise the temperature one pound of water from 0 to I00 centigrades) calorie *f*
~-ROOM - [min] recette *f* à eau
POUNDAL - [mech] (the unit force in the footpound-second system of units) pied-livre *m*
POUNDER PESTLE - [impl] pilon *m*
POUNDING - [gen] broyage *m*, écrasage *m*, pilage *m*
~APPLIANCE - [mech] moulin *m* à pilons
~-IN - [mech] martellement *m*
~MACHINE - [mech] pilonneuse *f*,
[paper man] machine *f* pilonneuse
~TROUGH - [min] caisse du bocard
POUNDS PER SQUARE INCH - [meas] livres *fpl* par pouce carré
POUR, to - [gen] verser, se déverser
[met] (of rain) tomber à torrents
[metall] (in foundry work) couler
POUR - [gen] torrent *m*, deluge *m*
[metall] (poured quantity or material) quantité *f* de métal coulée, coulée *f*
[metall] (in foundry work) trou *m* de coulée
~IN, to - [gen] verser
~OFF, to - [ind chem] décanter
~ON END, to - [metall] couler débout
~POINT - [chem] (the lowest temperature at which a liquid, in particular a fuel oil, will flow under given conditions) point *m* de solidification
[metall] température *f* de coulée
~POINT DEPRESSANT - [oil ind] inhibiteur *m* de coagulation
~REVERSION - [metall] (pour instability) instabilité *f* à l'écoulement
POURABLE - [phys] (of such a consistency that pouring is possible) pouvant être déversé
~SEALING COMPOUND - [rubber ind] produit *m* de jointement coulant
POURED - [gen] versé
[metall] (in foundry work) coulé
~ASPHALT - [constr] asphalte *m* coulé
~SHORT - [metall] (in die casting) mal venu
POURING - [metall] coulée *f*
~AND CHARGING HOLE - [metall] (of a Bessemer converter) ouverture *f* de chargement et de coulée
~BASIN - [metall] bassin *m* de coulée, entonnoir *m* de coulée, cône *m* de coulée
~BED - [metall] lit *m* de coulée
~BUSH - [metall] godet *m* de coulée
~CHANNEL - [metall] chenal *m* de coulée
~CUP - [metall] poche *f* de coulée
~GATE - [metall] (in a mould) coulée *f*, trou *m* de coulée
POURING LADLE - [metall] poche *f* de coulée
~LEVEL - [metall] niveau *m* de coulée
~LIP - [metall] bec *m* de coulée
~PIT - [metall] fosse *f* de coulée
~PLATFORM - [metall] plate-forme *f* de coulée
~SLOT - [metall] (die casting) trou *m* de remplissage de la chambre froide
~TRUCK - [metall] chariot *m* de coulée
~VESSEL - [metall] (die casting) bassin *m* de coulée
POWDER, to - [gen] réduire en poudre, pulvériser (to reduce to powder) pulvériser, triturer
POWDER - [gen] (minute free particles of a dry substance) poudre *f*
[chem] (for explosions) poudre *f*
[med] poudre *f*
~ADHESIVE - [chem] (adhesive in the form of a powder) colle *f* en poudre
~BASE - [ind chem] excipient *m* en poudre
~BINDER (G.R.P.) - [chem] (dry binder in powder, used in preforming) liant *m* en poudre
~BURNING - [metall] oxycoupage *m* à la poudre de fer
~CAMERA - [phys] (camera used for powder method of analysis) chambre *f* Debye-Scherrer
~-CHAMBER - [firearms] chambre *f* à charge
~COMPACTING - [metall] moulage *m* par pression
~CORE - [el] noyau *m* de fer divisé
~DENSITY - [plast ind] (of moulding powders) masse *f* volumique des poudres à mouler, masse *f* volumique apparente
~DIFFRACTION - [metall] (a method of analysis) méthode *f* Debye-Scherrer
~FORM - [phys] état *m* pulvérisant
~FUMES - [chem] gaz *mpl* dégagés par la poudre
~MAGAZINE - [firearms & mining] magasin *m* à poudres
~METAL - [metall] métal *m* fritté
~METALLURGY - [metall] (the technique of producing components from powdered metals by the application of heat and pressure) métallurgie *f* des poudres
~METHOD - s. powder method of analysis
~METHOD OF ANALYSIS - [phys] (method for the x-ray analysis of crystals) méthode *f* Debye-Scherrer
~MILL - [mech] moulin *m* à poudre
~MINE - [mining] trou *m* de mine
[gen] poudrerie *f*
~MIXING - [metall] mélange *m* de poudres
~ORE - [min] minerai *m* disséminé
~PARTICLE - [metall] grain *m* de poudre
~PATTERN - [phys] cristallogramme *m* à poudre de cristal
~ROLLING - [metall] laminage *m* de poudre
POWDERED - [gen] pulvérisé
~BRICKS - [constr] briques *fpl* en poudre
~CHARCOAL - [min] poussier *m* de charbon de bois
~COAL - [min] charbon *m* pulvérisé
~COPAL - [text] (used in artificial silk spinning) copal *m* pulvérisé
~ELEMENT BLADES - [mech] (gas turbine blades made by sintering from metallic powder) aubes *fpl* agglomérées
~FUEL - [chem] combustible *m* pulvérisé
~GRAPHITE - [nucl] graphite *f* en poudre
~HORN - [text] poudre *f* de corne
~LIME - [chem] chaux *f* en poudre
~METAL PRESS - [metall] presse *f* à forger avec mé-

tal pulvérisé
POWDERED MILK - [food] lait *m* en poudre
~ORE - [min] farine *f* minérale, farine *f* de bocard
~PEAT - [geol] poudre *f* de tourbe
~PUMICE - [min] pierre *f* ponce en poudre
~STAINLESS STEEL - [metall] acier *m* inoxydable pulvérisé
~SUGAR - [food] sucre *m* en poudre
POWDERINESS - [gen & min] pulvérulence *f*
POWDERING - [gen] pulvérisation *f*, trituration *f*
POWDERY - [gen] (particulate consisting of small particles) pulvérulent
~YEAST - [brew ind] levure *f* poussiéreuse
POWER, to - [gen] (to fit with a motor) fournir de énergie
[math] (to multiply a quantity by itself) élever (un nombre)
POWER - [gen] puissance *f*, pouvoir *m*, énergie *f*
[leg] autorité *f*, pouvoir *m*
[mech] (the rate of doing work) force *f*, force *f* motrice
[phys] (any form of energy) force *f*
[el] (of electrical motor or machine) force *f*, pouvoir *m*, puissance *f*
[opt] (of a lens) puissance *f*
[math] puissance *f*
~AMPLIFICATION - [radio] (the ratio of power level at the output terminals) amplification *f* de puissance
~AMPLIFICATION RATIO - s. power amplification
~AMPLIFIER - [radio] (thermionic amplifier to give substantial power output rather than voltage gain) amplificateur *m* de puissance
~AMPLIFIER STAGE - [radio] étage *m* d'amplificateur de puissance
~ASSISTED CONTROL - [mech] (any control mechanism in which the physical force exerted by the pilot is aided by power-operated devices) régulation *f* indirecte
~ATTENUATION - s. power loss
~AUGMENTATION - [aero] (any method of temporarily increasing the power output of an aircraft engine special conditions, e.g. at take-off) augmentation *f* de puissance
~-BAND MERIT - [radio] produit *m* du facteur d'amplification et de la largeur de bande
~BRAKE - [mech] (a brake in which the physical force exerted by the user is aided by a power-operated device) servo-frein *m*
~BREAKING CAPACITY - [el] puissance *f* de coupure
~BREED REACTOR - [nucl] (a reactor which produces power and also breeds fuel) réacteur *m* surrégénérateur de puissance
~BREEDER - s. power breed reactor
~BRUSHING - [mech] nettoyage *m* par brosse rotative
~BURNER - [th eng] (high-pressure burner) brûleur *m* à mélange surpressé
~CABLE - [el] câble *m* de transmission
~CAPACITOR - [el] condensateur *m* de puissance
~CHUCKING - [mach tool] montage *m* sur plate-forme automatique
~CIRCUIT - [el] (the part of a wiring of an electrical installation which is used to supply various apparatus other than lighting) circuit *m* de puissance
~COEFFICIENT - [nucl] (the change of reactivity with increase in power) coefficient *m* de puissance
~COMPONENT - [el] (the component of an alternating voltage which is in phase with the current) composante *f* active du courant
POWER CONDUCTOR - [el] conducteur *m* d'énergie
~CONSUMPTION - [el] (the power required by a motor or similar apparatus over a given period) consommation *f* propre, consommation *f* d'énergie
~CONTROL - [nucl] réglage *m* de puissance
~CONTROL ROD - [nucl] (rod which controls the power level of a nuclear reactor) barre *f* de commande de la puissance
~CRANE - [mech] grue *f* à moteur
~CURRENT FUSE - [el] fusible *m* de courant de suite
~CURRENT SWITCH - [el] interrupteur *m* de courant
~CURVE - [mech] (of an engine or motor) courbe *f* de puissance
~CUT-OFF RELAY - [el] relais *m* de coupure de courant
~CUT-OUT - [el] coupe-circuit *m*
~CYLINDER - [mech] cylindre *m* générateur
~DAM - [hydr] barrage *m* pour centrale hydroélectrique
~DETECTOR - [radio] détecteur *m* de puissance
~DELIVERED - [el] puissance *f* fournie
~DIRECTIONAL RELAY - [el] relais *m* directionnel de puissance
~DISSIPATION - [electron] puissance *f* dissipée
~DISTRIBUTION - [el] distribution *f* de l'énergie
~DIVE - [aero] (a dive carried out under substantial engine power) piqué *m* avec moteur
~DIVIDER - [mech] (an auxiliary gear box or the like designed to distribute power from one driving shaft to several separate mechanism) repartiteur *m* de couple, prise *f* de mouvement multiple, boîte *f* de transfert
[electron] (in a waveguide system) répartiteur *m* de puissance
~DIVIDER ATTENUATOR - [electron] (in a waveguide) affaiblisseur *m* à répartiteur de puissance
~DOWN - [mech] (term used for power operated return in a hydraulic lift cylinder, as distinct from return by gravity) mécanisme *m* hydraulique à double effet, piston *m* à double effet
~DRAG LINE - [mech] excavateur *m* à benne automatique
~DRAIN - [electron] consommation *f* de puissance
~DRILL - [mech tool] perforatrice *f* mécanique
~-DRIVEN - [mech] mû par moteur, à entraînement mécanique
~ELECTRON TUBE - [electron] (used in the output stage of an amplifier) tube *m* de sortie
~END - [oil ind] arrière *m* d'une pompe à boue
~ENGINEERING - [gen] technique *f* du courant fort
~EQUALIZER - [telev] (only USA; echo trap in GB; device or circuit eliminating undesired echos) correcteur *m* d'écho, piège *m* d'écho
~EXPONENT - [math] exposant *m* de puissance
~FACTOR - [el] (of an insulating material; the power factor of an electrical capacitor of which the material in question forms the dielectric, when subjected to an alternating voltage) facteur *m* de puissance
~FACTOR ADJUSTMENT - [instr] (for an a.c. meter; a device for regulating the speed of the moving element for nominal values of voltage, current and frequency and for a phase difference between current and voltage other than zero) dispositif *m* de réglage en courant déphasé
~FACTOR CORRECTION - [el] correction *f* du facteur

de puissance
POWER FACTOR INDICATOR - s. power factor meter
~FACTOR METER - [instr] (an instrument giving a direct reading of the value of the power factor) indicateur *m* de coefficient de déphasage, phasemètre *m*
~FACTOR OF THE FUNDAMENTAL - [el] facteur *m* de déphasage
~FAILURE - [el] manque *m* de courant
~FEED - [mech] avance *f* automatique
~FEED MECHANISM - [mech] mécanisme *m* d'avance automatique
~FEEDER - [el] feeder *m* force
~FORGING - [metall] forgeage *m* mécanique, fer *m* forgé mécaniquement
~FORMULA - [math] formule *f* exponentielle
~-FREQUENCY WITHSTAND VOLTAGE - [el] tension *f* de tenue à fréquence industrielle
~FREQUENCY VOLTAGE TEST - [el] essai *m* de tension
~GAIN - [el] gain *m* de puissance
~GAS - [gas ind] (producer gas) gaz *m* carburant
~GATE - [mech] (endgate used in conjunction with hydraulic or mechanical hoisting mechanisms) hayon *m* élévateur
~GAUGE - [oil ind] manomètre *m* de pression
~GOVERNOR - [mech] régulateur *m* de la puissance
~GRID DETECTOR - [electron] détecteur *m* de puissance par la grille
~HACK-SAW - [impl] machine *f* à scier alternative au moteur
~HAMMER - [impl] marteau *m* à forger, pilon *m*
~HOP - [auto] (the hopping motion of one or a pair of wheels occuring when tractive force is applied) oscillation *f* de réaction de couple
~HOUSE - s. power station
~INDUCTION NOISE - [el] (circuit noise caused by electrostatic or electromagnetic induction from power or traction lines) bruit *m* induit
~INPUT - [el] puissance *f* absorbée, consommation *f* d'énergie
~INPUT TO A MACHINE - [el] puissance *f* absorbée par une machine
~JET - [mech] jet *m* de puissance
~LATHE - [tool] tour *m* marchant au moteur
~LEAD - [el] alimentation *f* de courant, amenée *f* de courant
~LEVEL - [el] (an expression of the power at any point in a system) puissance *f* effectivement transmise
~LEVEL DIAGRAM - [el] diagramme *m* de la puissance transmise
~LEVEL SAFETY SYSTEM - [nucl] appareil *m* de contrôle de sécurité à haut flux
~LIFT - [mech] monte-charge *m* au moteur
~LIMIT - [el] limite *f* de puissance
~LINE - [el] (in electrical traction, a train-line used for interconnecting collector shoes of like polarity throughout the train) ligne *f* omnibus
[el] ligne *f* à haute tension
~LINE HUM - [el] bruit *m* de secteur, bourdonnement *m* du courant alternatif
~LINE STRINGING DEVICE - [el] tendeur *m* des câbles
~LOADING - [aero] (value obtained by dividing the gross weight of an aircraft by the total engine h.p.) charge *f* au cheval
~LONGITUDINAL FEED - [mach tool] avance *f* automatique longitudinale

POWER LOOM - [text] métier *m* à tisser mécanique
~LOSS - [el] pertes *fpl* de puissance
~LOUDSPEAKER - [el acoust] (a loudspeaker having a great volume of sound) haut-parleur *m* de grande puissance
~MAINS - [el] réseau *m* de distribution (d'énergie électrique)
~MAKING-CAPACITY - [el] puissance *f* de fermeture
~MOWER - [agric] motofaucheuse *f*
~NAVVY - s. power shovel
~OF ATTORNEY - [leg] mandat *m*, procuration *f*
~OF COHESION - [phys] (cohesion energy) énergie *f* de cohésion
~OF IMPREGNATION - [chem] capacité *f* d'imprégnation
~OF LENS - [photo] (the relative focusing power of a lens) puissance *f* d'une lentille
~OF MAGNIFICATION - [opt] (magnifying power) pouvoir *m* de grandissement
~OFF INDICATOR - [el] indicateur *m* lumineux de mise hors circuit
~-OPERATED CHUCK - [mach tool] mandrin *m* à moteur
~-OPERATED CONTROL - [contr] (any control which is operated purely by power, the pilot making no contribution except to move the actuating device) régulation *f* indirecte
~OPERATED POINTS - [railw] aiguille *f* à moteur
~OPERATED SIGNAL BOX - [railw] poste *m* à pouvoir
~OPERATION - [contr] (the actuation of an apparatus by electrical or other power) fonctionnement *m* assisté
~OUTAGE - [electron] défaillance *f* du secteur
~OUTPUT - [el] puissance *f* fournie, puissance *f* de sortie
~OUTPUT SUPPLIED BY A MACHINE - [el] puissance *f* fournie par la machine, puissance *f* utile d'une machine
~OVERRIDE - [nucl] (of a reactor) outrepassage *m* de puissance
~PACK - [el electron etc] (an assembly for providing supply for electronic apparatus, comprising a battery or a transformer with associated items, such as rectifiers , filters etc) bloc *m* d'alimentation
~PACKAGE - [aero] groupe *m* moto-propulseur
[gen] groupe *m* moteur
~PER UNIT AREA - [nucl] (the power generated in a reactor core per unit area) puissance *f* par unité de surface
~PLANT -[mech] (a unit producing mechanical power) groupe *m* propulseur, moteur *m*
[el] (a power station) centrale *f* électrique
[el] (a generator set) groupe *m* électrogène, groupe *m* générateur
[aero] moteur *m*
~PRACTICAL CEILING - [aero] altitude *f* de plafond pratique
~PRESS - [mech] presse *f* mécanique
~PRODUCER - [mech & el] source *f* d'énergie
~PROTECTION - [el] dispositif *m* de protection de puissance
~RACK - [el] herse *f*, bâti *m* pour têtes de câble
~RANGE - [mech] gamme *f* de puissance
[nucl] (of a nuclear reactor) domaine *m* de puissance

POWER RATING - [mech] (the power allowable under regulations for a given purpose, e.g.take-off) régime *m* de puissance
[el & electron] puissance *f* de sortie
[electron] capacité *f* de charge
~RATIO - [el] gain *m*
~REACTOR - [nucl] (reactor providing useful mechanical power) réacteur *m* de puissance
~RECTIFIER - [el] redresseur *m*
~REEL - [text] dévidoir *m* mécanique
~RELAY - [el] relais *m* de puissance
~RESERVE - [auto] réserve *f* de puissance
~RIVETING - [mech] rivetage *m* mécanique
~SAW - [mech] scie *f* mécanique
~SEPARATION FILTER - [telev] filtres *mpl* séparateurs
~SERIES - [math] série *f* exponentielle
~SHOVEL - [mech] pelle *f* mécanique
~SOURCE - [el] source *f* d'énergie
~SPARK - [mech] (in internal combustion engines) étincelle *f* chaude
~STAGE - [electron] étage *m* de puissance
~STATION - [el] (or generating station; station where electrical energy is produced from some other form of energy) installation *f* de production
~STATION WITH RESERVOIR - [el] centrale *f* hydroélectrique à réservoir
~STEERING - [auto] servo-direction *f*
~STRENGTH TESTER - [text] dynamomètre *m*
~STROKE - [mech] (the piston stroke in a reciprocating engine during which expansion takes place) course *f* motrice, course *f* d'explosion
~SUPPLY - [el] alimentation *f*
~SUPPLY UNIT - [el] élément *m* d'alimentation, bloc *m* secteur
~SUPPLY VARIATION - [el] variation *f* d'alimentation
~SWITCHBOARD - [el] tableau *m*
~SWITCHGROUP - [el] combinateur *m* de puissance
~SYNCHRO - [contr] synchro-machine *f* de puissance
~SYSTEM - [el] système *m* électrique
~TAKE-OFF - [auto etc] prise *f* de force, prise *f* de mouvement
~TAKE-OFF LEVER - [mech] levier *m* de commande de la prise de force
~TAKE-OFF SHAFT - [mech] arbre *m* de prise de force
~TAMPER - [mech] (mechanical rammer) dame *f* mécanique, tasseur *m* à explosions, grenouille *f*
~TONG - [oil ind] clef *f* suspendue entrainée mécaniquement
~TOOL - [tool] outil *m* mécanique
~TRAIN - [mech] engrenage *m* de transmission
~TRANSFER FEED - [mach tool] avance *f* transversale automatique
~-TRANSFER RELAY - [el] relais *m* de transfert de puissance
~TRANSFORMER - [el] transformateur *m* de puissance
~TRANSISTOR - [electron] transistor *m* de puissance
~TRANSMISSION - [mech] transmission *f* de force, transmission *f* d'énergie
~TRUCK - [railw] (motor bogie) bogie *m* moteur
~TUBE - [electron] tube *m* de puissance
~TYRE PUMP - [auto] gonfleur *m* de pneumatiques à moteur
~UNIT - [mech] (the complete assembly of engine or engines and accessories forming a propulsion unit) groupe *m* moteur amovible
POWER UNIT - [el] (a motor generator set) groupe *m* électrogène
[radio] s. power amplifier
~VALVE - [electron] tube *m* de sortie
~VAN - [radio & telev] voiture *f* d'alimentation
~WASH MILL - [mech] cylindre *m* laveur à commande mécanique
~WATER - [hydr] eau *f* pour la production d'énergie
~-WEIGHT RATIO - [mech] puissance *f* massique
~WHEEL - [horol] roue *f* motrice
~WINDING CURRENT - [el] tension *f* absorbée
~WINDINGS - [el]enroulement *m* de puissance
POWERBOAT - [naut] bateau *m* à moteur
POWERED BICYCLE - [mech] bicyclette *f* à moteur, moto-vélo *m*
POWERFORMING - [ind chem] (a reforming process) reforming *m* continu
POWERFUL - [gen] puissant, énergique, efficace
[mech] (e.g. of engine) puissant
POWERHOUSE - [el] centrale *f* électrique
POYNTING VECTOR - [radio] (vector the flux which through a surface represents the instantaneous electromagnetic power transmitted through this surface) vecteur *m* de Poynting
POZZOLANA - [geol] (volcanic dust which forms a hydraulic cement when mixed with lime) pozzolane *f*, pouzzolane *f*
[constr] (a hydraulic cement) ciment *m* pouzzolanique
PPI - s. plan position indicator
P.P.I. PREDICTION - [radar] image *f* théorique
P.P.I. REPEATER - [radar] représentation *f* P.P.I.
p-p JUNCTION - [electron] (region of transition between two regions having different properties in P-type semiconducting material) jonction *f* p-p
PPL - s. visual flight rules
P.P.M. - [electron] modulation *f* à impulsions à variation de temps
PR - [met] (abbrev. for passing rain; in the Beaufort code) précipitation *f*
[rubber ind] (abbrev. for ply rating) "ply rating" *m*
~UNIT - [radiat] appareil *m* de radiographie
PRACTICABILITY - [gen] (the possibility of putting into practice) practicabilité *f*
PRACTICAL - [gen] pratique
~CONCENTRATION - [ind chem] charge *f* utile
~ELECTRICAL UNITS - [el] unités *fpl* électriques pratiques
~ENTROPY - [nucl] (or virtual entropy; the entropy of a system which neglets the contribution due to nuclear spin) entropie *f* virtuelle
~RANGE - [nucl] (the rectilinear distance of the path if an ionizing particle) portée *f* réelle
~SYSTEM - [el] (system in which the integral units are multiples or submultiples in powers of I0 of the corresponding units in the C.G.S. system) système *m* pratique
PRACTICE - [gen] pratique *f*, exercice *m*
~FACTOR - [telecomm] coefficient *m* de pratique expérimentale
~OF INTERLACING - [text] pratique *f* des armures
~POSITION - [telecomm] position *f* d'exercice
PRACTICED - [gen] expérimenté, versé
PRACTITIONER - [med] médecin *m*
PRAETARSUS - [zool] (tarsus outgrowth in some in-

sects) prétarse *m*
PRAIRIE - [gen] prairie *f*
PRAISE, to - [gen] louer
PRAISE - [gen] louange *m*
PRALINE - [comput] programme *m* praline
PRAMOXINE HYDROCHLORIDE - [chem] (a local anaesthetic) chlorhydrate *m* de paramoxine
PRANDTL-MEYER EXPANSION - [phys] (two-dimensional homentropic supersonic flow in which an expansion wave occurs at the junction of two plane boundaries) expansion *f* de Prandtl-Meyer
~NUMBER - [mech] (the kinematic viscosity of a fluid divided by its thermal conductivity) nombre *m* de Prandtl
PRASE - [miner] (translucent greenish variety of chalcedony) prase *m*
PRASEODYMIUM - [chem] (a metallic element, symbol Pr, A.N. 59, a A.W. I40.9 belonging to the rare earth group of elements.The salts are used in the manufacture of coloured glasses) praséodyme *m*
PRATIQUE - [naut] (the privilege granted a master to land passengers after complying with sanitary inspection) libre pratique *f*
PRAYING CARPET - [text] tapis *m* de prières
PREABDOMEN - [zool] (in scorpione) préabdomen *m*
PREADAPTATION - [zool] préadaptation *f*
[genet] préadaptation *f*
PREAMBLE - [gen] introduction *f*, préambule *m*
PREAMPLIFICATION - [radio] (amplification used ahead of the main amplification) préamplification *f*
~TRANSFORMER - [el acoust] transformateur *m* de préamplification
PREAMPLIFIER - [radio] (amplifier used ahead of the main amplifier) préamplification *m*
PRE-ARCING TIME - [el] durée *f* de préarc
PREARRANGE, to - [gen] arranger au préalable
PREAVIS CALL - [telecomm] (in telephony) conservation *f* avec pré-avis
PREBARIC CHART - [met] (an isobar chart giving a forecast for a specified period) carte *f* d'isobares prévues
PREBLEND - [rubber ind] mélanger au préalable, uniformiser
PRE-BLOSSOM SPRAY - [bot] traitement *m* préfloral
PREBONDING TREATMENT - [rubber ind] traitement *m* au préalable pour collage
PRE-CAMBRIAN - [geol] (geological era preceding the Cambrian age, i.e. the oldest so far defined) précambrien
PRECARIOUS - [gen] précaire
PRECARTILAGE - [anat] cartilage *f* embryonnaire
PRECAST - [constr] (concrete blocks cast separately before being fixed in position) coulé d'avance
~CEMENT - [constr] béton *m* prêt à être posé
~CONCRETE PILE - [constr] pieu *m* préfabriqué en béton
~PILES FOUNDATIONS - [build] pieux *mpl* de fondation prêts à être posées
PRECAUTION - [gen] précaution *f*
PRECAVAL VEIN - [zool] veine *f* cave antérieure
PRECEDE, to - [gen] précéder
PRECEDENCE - [gen] priorité *f*, préséance *f*
~RATING - [comput] degré *m* de priorité
PRECEDENT - [gen] précédent
PRECENTOR - [mus] maître *m* de chapelle
PRECESS, to - [mech] (to move according the phenomenon of procession) précesser
PRECESS, to - [comput] (to cause a procession) précéder
PRECESSION - [mech] (the movement exhibited by a rotating body, especially a gyroscope, when a torque is applied to it which is such as to tend to change the attitude of its axis of rotation) précession *f*
[astr] (of the equinoxes) (the conical motion of the earth's axis in space, caused by procession, and producing continuous change of the position of the points of intersection of the equator and the ecliptic) précession *f* des équinoxes
~IN DECLINATION - [astr] précession *f* en déclinaison
PRECESSIONAL MOVEMENT - [electron] mouvement *m* de précession
PRE-CHAMBER AIRTANK - [auto] anti-chambre *f* de reserve d'air
~-CHLORINATION - [hydr] préchloration *f*
PRECINCT - [gen] enceinte *f*, enclos *m*
PRECIOUS - [gen] précieux
~METAL - [metall] (a metal-gold, silver, and the platinum group of metals - having value as bullion) métaux *mpl* précieux
~STONE - [min] (a gemstone) pierre *f* précieuse
PRECIPICE - [gen] précipice *m*
PRECIPITABLE - [chem] (capable of being precipitadet) précipitable
PRECIPITANT - [chem] précipitant *m*
~FEED - [chem] dosage *m* des précipitants
~POWER - [chem] pouvoir *m* précipitant
~OF METHYLATED SPIRIT - [chem] pouvoir *m* précipitant de l'alcool méthylique
PRECIPITATE, to - [chem] (the throw suspended matter out of solution) précipiter
[gen] précipiter, accélérer
[met] se condenser
PRECIPITATE - [chem] (the relatively insoluble material produced by precipitation) précipité *m*
~OF ALUMINA HYDROXIDE - [chem] précipité *m* de hydroxyde d'aluminium
~OF CUPROUS OXIDE - [chem] précipité *m* d'oxyde de cuivre
~OF GYPSUM - [chem] précipité *m* de gypse
~OF LIME SOAP - [chem] précipité *m* de savon de chaux
~OF METALLIC SULPHIDE - [chem] précipité *m* de sulfure métallique
PRECIPITATED BARIUM CARBONATE - [chem] carbonate *m* de baryum précipité
~BARIUM SULPHATE - [chem] sulfate *m* de baryum précipité
~COPPER - [metall] cuivre *m* de précipitation
~DRIERS - [paint] (metallic soaps obtained by reacting alkali soaps of certain acids with metallic salts, in aqueous solution, as distinct from driers obtained by heat processes. See also Fused Driers) siccatifs *mpl* précipités
~SULPHUR - [chem] soufre *m* précipité
PRECIPITATING ACID BATH -[chem] bain *m* acide précipitant
~COLUMN - [ind chem] tour *f* de précipitation
PRECIPITATION - [met] (collective term for the deposition of water from the atmosphere in liquid or solid states, e.g. rain, snow, hail etc) précipitation *f*
~ANNEAL - [metall] recuit *m* de précipitation
~AREA - [met] (a region in which precipitation is

occurring or expected to occur) zone *f* de précipitation
PRECIPITATION DOWNDRAUGHT - [met] (a negative thermal causing sudden reversal of precipitation up draught and due to the presence of a saturated superadiabatic region in the lower part of the cloud concerned) précipitation *f* descendante
~HARDENING - [phys] (artificial ageing) durcissement *m* par précipitation, durcissement *m* par vieillissement
~HEAT TREATMENT - s. precipitation hardening
~STATIC INTERFERENCE - [radio] (noise in aircraft radio installations caused by static discharges from the aircraft to the atmosphere) interférences *fpl* statiques atmosphériques
~UNIT - [nucl] (in reactor instrumentation) capteur *m* de dépôt radioactif
~VALUE - [chem] produit *m* de solubilité
PRECIPITATOR - [ind chem] cuve *f* de précipitation
[electron] (apparatus used to remove smoke, dust etc. from the air) appareil *m* électrostatique de précipitation
PRECIPITINOGEN - [chem] (substance introduced into the blood plasma to call forth the precipitin) précipitinogène *m*
PRECISE - [gen] précis, exact
PRECISION - [gen] précision *f*, exactitude *f*
~APPROACH - [radar] (par approach; ground controlled approach using precision radar apparatus without surveillance radar) approche *f* du radar de précision, approche *f* PAR
~APPROACH RADAR-PAR - [radar] (primary radar equipment showing accurately the position of an aircraft during approach in relation to the required approach path) radar *m* d'approche de précision
~BALANCE - [instr] (weighing apparatus so constructed as to be of extreme accuracy) balance *f* de précision
~BRIDGE - [el] pont *m* de précision
~CASTING - [metall] moulage *m* de précision
~CASTING PROCESS - [metall] moulage *m* à la cire perdue
~CONTOUR GRINDING - [mach tool] meulage *m* de précision (profils)
~FORK - [instr] diapason *m* de précision
~GAUGE BLOCK - [mech] bloc *m* jaugeur de précision
~GRINDING - [mach tool] meulage *m* de précision
~GROUND BAR - [metall] barre *f* rectifiée à précision
~INSTRUMENT - [instr] (an instrument capable of accurate and delicate manipulations, e.g. weighing) instrument *m* de précision
~LATHE - [mach tool] tour *m* de précision
~MOULDING - [metall] (lost wax moulding) moulage *m* à la cire perdue
~REEL - [text] dévidoir *m* de précision
~TEST - [mech] essai *m* de précision
~YARN SCALE - [text] (precision scale used to weigh yarns) balance *f* de précision pour fil
PRECLEANER - [mech] préfiltre *m*
PRE-COAGULATION - [rubber ind] précoagulation *f*
PRECOAT - [oil ind] adjuvant *m* pour filtre
[sugar ind] précouche *f*
~FILTER - [sugar ind] filtre *m* à précouche
PRECOATED AGGREGATE - [constr] agrégat *m* préenrobé
~SAND - [metall] sable *m* synthétique
PRECOATING - [plast ind] (application of an adhesive to a wire before extruding a coating on to it) enduction *f* préliminaire
PRECOCIOUS REVERSION - [genet] réversion *f* précoce
PRECOMBUSTION CHAMBER - [mech] (in internal combustion engines) chambre *f* de précombustion
PRECOMMISSURE - [med] commissure *f* antérieure
PRECOMPRESSION - [soil] précompression *f*
~BY DESSICATION - [soil] précompression *f* par dessication
PRECONDITIONING - [hydr] traitement *m* préliminaire
PRECONDUCTION CURRENT - [electron] courant *m* de la décharge préliminaire
PRE-COOKED - [food] précuit
~-COOL, to - [th eng] préréfrigérer
PRECOOLER - [mech] (of a motor) préréfrigérateur *m*
PRECOOLING - [mech] (auxiliary cooling process) préréfrigération *f*
PRECORDIAL - [anat] (in front of the heart) précordial
PRECORDIALGIA - [med] précordialgie *f*
PRECORNU - [anat] corne *f* antérieure du ventricule latéral
PRE-CRUSHER - [min] broyeur *m* primaire
PRECUNEUS - [anat] précunéus *m*, lobule *m* quadrilatère
PRECURE - [ind chem] (the full or partial setting of a synthetic resin adhesive in a joint before the clamping operation is complete) précuisson *f*
PRECURING - [plast ind] (the result of allowing thermosetting moulding material to remain too long in the hot mould cavity. The material precures, loses some of its flow and can give rise to defective mouldings) précuisson *f*, prévulcanisation *f*
PRECURSOR - [nucl] (radioactive parent nuclide) précurseur *m*
[radio] (the initial transient response to a unidirectional change in input, which precedes the main transition and is opposite in sense) sousvibration *f*
PRECUT, to - [constr] (the operation of preparing the parts of a prefabricated building) précouper
~COLUMN - [ind chem] (in gas analysis) colonne *f* de dégrossissage
PREDAZZITE - [miner] (a mixture of calcite, brucite, periclase and hydromagnesite occurring near Predazzo in Italy) marbre *m* à brucite
PREDEFINED PROCESS - [comput] processus *m* à dénomination provisoire
PREDELLA - [arch] prédelle *f*
PRE-DETARRER - [ind chem] (tar batter in the USA, or decanter) prédégoudronneur *m*
PREDETERMINATION - [genet] (delayed inheritance) hérédité *f* retardée, prédétermination *f*
PREDETERMINED COUNT SCALER - [instr] compteur *m* (d'impulsions) prédéterminé
~COUNTER - [comput] compteur *m* à présélection
~SEQUENCE - [telev] séquence *f* prédéterminée
~TIME-SHARING - [comput] utilisation *f* collective prédéterminée
PREDETONATION - [chem] (of a bomb, before the sheduled time) prédétonation *f*
PREDICTOR - [firearms] (device used in anti-aircraft defence, which mechanically interprets a number of data, as height, speed etc of the moving target) dispositif *m* de commande
PREDIGESTION - [med] prédigestion *f*

PREDISPOSITION - [gen & phys] prédisposition *f*
PREDISSOCIATION - [chem] (the process by which a molecule which has absorbed energy dissociates before losing energy by radiation) prédissociation *f*
PREDISTILLATION - [chem] prédistillation *f*
PRE-DISTORTER - [radio] (a circuit used in frequency-modulation system to obtain a signal current which is inversely proportional to the signal frequency) prédéformateur *m*
PREDISTORTION - [telecomm] (the operation of altering the response of a circuit to compensate distortion) prédéformation *f*
PREDNISOLONE - [chem] (a synthetic steroid with uses in medicine) prédnisolone *f*
PREDNISONE - [chem] (deltacortisone. A synthetic steroid used in medicine) prednisone *f*
PREDORMITIUM - [med] stade *m* d'endormissement
PRE-DRAWING - [metall] préétirage *m*
PREDRIER - [mech] (an initial drying apparatus) présséchoir *m*
PRE-EDIT, to - [comput] préparer pour l'ordinateur
~-EDITION - [comput] (step in mechanical translation) pré-édition *f*, mise *f* au point
~-EMPHASIS - [el acoust] (the operation which consists in raising part of the response curve of a recording system so as to equalize the statistical position of the energy in the audiofrequency range before applying the signal to the recording medium) préaccentuation *f*, pré-emphase *f*
~EMPTION - [comput] prise *f* initiale
~-ENERGIZE, to - [el] préalimenter
~-EQUALIZATION - [el acoust] (the correction applied at a point in a recording system, preceding some particular element, the characteristics of which must be corrected) préégalisation *f*
~-ESTIMATE - [comm] prévisions *f*pl
~-EXCITATION - [el] excitation *f* préalable
PREFAB - [constr] (abbrev. for prefabricated) préfabriqué
PREFABRICATE - [constr] préfabriquer
PREFABRICATED - [gen] préfabriqué
~HOUSE - [constr] maison *f* préfabriquée
PREFABRICATION - [constr] préconstruction *f*
PREFACE - [gen] préface *f*
PREFADE LISTENING - [radio] (listening to a programma for control purpose before it is faded up for transmission) écoute *f* préparatoire, appareillage *m* de contrôle d'écoute
PREFERENTIAL - [gen] préférentiel
~MATING - [zool] (the selection of the male by the female according to Darwin's theory) accouplement *m* préférentiel
~RECOMBINATION - [chem] (recombination occurring immediately after the ion-pair is formed unless the components are separated) recombinaison *f* préférentielle
PREFERRED ORIENTATION - [crystal] (in the course of slip, when a certain amount of deformation has been performed, a partial rearrangement of the crystals occurs, whereby similar axes display a common preferred orientation) orientation *f* préférentielle
PREFILL VALVE - [mech] soupape *f* à préremplissage
PREFILTERING - [hydr] préfiltration *f*
PREFIRING - [radio] temps *m* de pré-démarrage
PREFIX - [gen] préfixe *m*
[comput] préfixe *m*
PREFIX - [radio] signal *m* de déclenchement
~NOTATION - [math] notation *f* par préfixes
PRE-FLIGHT INSPECTION - [aero] (inspection carried out on aircraft before a flight) inspection *f* avant le vol
PREFLORATION - [bot] (the disposition of flowers within the bud) préfloraison *f*, estivation *f*
PREFOCUSING - [opt] préfocalisation *f*
~LENS - [opt] lunette *f* de préfocalisation
PREFOGGING - [photo] (pre-exposure) prélumination *f*
PREFOLIATION - [bot] (the disposition of leaves within the bud) préfoliation *f*, vernation *f*
PREFORM - [plast ind] (piece of moulding composition, somewhat larger than the finished moulding and of the same shape) ébauche *f*, préforme *f*
[el acoust] (small slab of record stock material prepared for use in the record presses) tablette *f*
~BINDER - [plast ind] (bonding material, usually polyester resin, used in preforming to consolidate the chopped glass roving) liant *m* de préforme
~MACHINE - [plast ind] (machine in which articles are formed by deposition of chopped glass roving on a screen) machine *f* de préformage
~SCREEN - [plast ind] (perforated plate on which material is deposited in preforming) écran *m* de préforme
PREFORMATION - [biol] (the theory stating that an organism exists fully preformed in the germ) préformation *f*
PREFORMED PRECIPITATE - [chem] (precipitate used for coseparation of tracer) précipitate *m* préformé
~WINDING - [el] (winding made up of elements which are shaped before being placed in the slots) enroulement *m* sur gabarit
~WIRE ROPE - [oil ind] câble *m* préformé
PREFORMING - [plast ind] préformage *m*
PREFRACTIONNATOR - [oil ind] préfractionneur *m*
PREFRONTAL - [anat] préfrontal
PREGNANCY - [med] grossesse *f*, gestation *f*
PREGNANEDIOL -[chem] (a progesterone metabolite with uses as a diagnostic aid for pregnancy and in the synthesis of progesterone) prégnandiol *m*
PREGNENOLONE - [chem] (a naturally occurring steroid with therapeutic uses) prégnenolone *m*
PREHEAT, to - [gen] préchauffer
PREHEAT - [th eng] préchauffage *m*
~ROLL - [plast ind] (a hot roll used in extrusion coating to heat the substrate before application of the coating) cylindre *m* de préchauffage
PREHEATED CHILL - [metall] coquille *f* chauffée
PREHEATER - [th eng] (preliminary heating apparatus, often based on heat exchange principles) réchauffeur *m*
[plast ind] (used for the preheating of forms) préchauffeur *m*
[metall] rehausse *f*
~MILL - s. preheater [plast ind]
~BY INFRARED RADIATIONS - [phys] préchauffeur *m* infrarouge
PREHEATING - [th eng] (process of transferring heat to a substance before a process) préchauffage *m*
~BENCH - [electron] (an apparatus used to bring electronic tubes to the required temperature for checking purposes) banc *m* de préchauffage
~CYLINDER - [mech] cylindre *m* à réchauffage
~HOPPER - [plast ind] (type of extruder feed hopper in which the material is heated by a hot air blast

blown through the pellets before they reach the feed screw) trémie *f* de préchauffage
PREHEATING MILL - [rubber ind] laminoir *m* à réchauffage, réchauffeur *m*
~TABLE - [rubber ind] table *f* de réchauffage
~TIME - [el] (e.g. in a mercury arc valve; the time which is required for all parts of the valve to reach the operating temperature) temps *m* de préchauffage
~TORCH - [metall] (torch used to preheat component parts which must be welded) chalumeau *m* préchauffeur
PREHENSILE - [zool] (adapted for grasping) préhensile, préhenseur
~FOOT - [zool] pied *m* préhensile
PREHNITE - [miner] (pale-green fibrous acid orthosilicate of calcium and aluminium) prehnite *f*
PREHNITENE - [chem] (a synthesis intermediate) préhnitène *m*
PREIGNITION - [mech] (the ignition of a charge in an i.c. reciprocating engine at an earlier point in the cycle than that at which it should occur) allumage *m* prématuré
PREIMMUNIZATION - [med] préimmunisation *f*
PRE-IMPREGNATED CABLE - [el] câble *m* à préimprégnation
~-IMPREGNATED INSULATION - [el] isolation *f* au papier préimprégné
PREINSULA - [anat] région *f* céphalique de l'insula
PREIONIZATION - [phys] (auto-jonization) auto-ionisation *f*
PRELIMINARY - [gen] préliminaire, préalable
~ANESTHESIA - s. premedication
~CARDING OF JUTE - [text] cardage *m* en gros du jute
~CLEANING - [gen] nettoyage *m* préliminaire
~CLEANING OF COTTON - [text] nettoyage *m* préliminaire du coton
~DRAWING - [metall] tréfilage *m* grossier
~DRAWING FRAME - [text] banc *m* d'étirage de premier passage
~HEATING - [th eng] (preheating) préchauffage
~HEATING ZONE - [metall] (in a blast furnace, the region at the top of the charge, where it is first heated) zone *f* de préchauffage
~INTERLACING PLAN - [text] mise *f* en carte préliminaire
~MATTER - [print] feuille *f* de titre
~PROGRAM - [comput] programme *m* préliminaire
~READING - [comput] (the encoding of information before operation begin) lecture *f* indirecte
~REFINER - [rubber ind] mélangeur *m* déchiqueteur
~RETTING - [text] rouissage *m* préparatoire
~SCREEN - [impl] (used in the brewing industry) pré-tamis *m*
~TENSION DEVICE - [text] avant-tendeur *m*
~TEST - [gen] essai *m* préliminaire
~TREATMENT - [gen] traitement *m* préparatoire
PRELIMING - [sugar ind] préchaulage *m*
PRELOADING - [mech] pré-charge *f*
PRELUDE - [mus] prélude *m*
PREMASH, to - [brew ind] (to digest) empâter préalablement
PREMASHER - [brew ind] (converter) hydrateur *m*
PREMATURE - [gen] prématuré
[chem] (exploding before the scheduled time) avant le temps
~DELIVERY - [med] accouchement *m* prématuré, accouchement *m* avant terme
PREMATURE DISCONNECTION - [telecomm] rupture *f* prématurée
~IGNITION - s. preignition
~RELASE - s. premature disconnection
~SINTERING - [plast ind] (partial local fusion of pellets in the feed system of an extruder, causing them to adhere to one another and blocking the feed) égouttement *m* prématuré
PREMAXILLARY - [anat] prémaxillaire
PREMEDICATION - [med] médication *f* pré-opératoire, pré-anesthésie *f*
PREMISES - [gen] local *m*, immeuble *m*
[constr] beins *mpl* batis
PREMIUM - [insur] prime *m* d'assurance
[gen] (a reward) prix *m*, récompense *f*
~MOTOR FUEL - [auto] super-carburant *m*
~NOTICE WRITING - [comput] impression *f* des bordereaux de primes
~SYSTEM - [agric] système *m* de primes
PREMIX, to - [chem] (the preliminary blending of two or more substances) prémélanger
PREMIX - [chem] (of rockets) prémélange
~BURNER - s. premixed burner
~INJECTOR - [rocketry] (the mixing of the propellants before their injection into rocket motor) injecteur *m* de prémélange
~NOZZLE - s. premix injector
PREMIXED BURNER - [th eng] (aerated burner) brûleur *m* à mélange préalable
PREMIXING - [gen & ind chem] (preliminary blending of two or more substances) prémélange *m*
[el acoust] prémixage *m*
PREMODIFICATIONS - [comput] (modifications applied to programmes beforehand) modification *f* préalable
PREMOLAR - [zool] prémolaire *f*
PRE-MORDANTING - [text] mordant *m* préalable
PREMUNITION - [med] prémunisation *f*, immunité *f* relative
PREMUTATION - [genet] prémutation *f*
PRENATAL - [med] prénatal
PREOPERATIVE - [mach tool] (e.g. the headstock) présélecteur
~HEADSTOCK - [mach tool] poupée *f* sélectrice
PREORBITAL - [zool] (membrane bone in front of the orbit in some fish) préorbital
PREOSCILLATION CURRENT - [electron] (the value of the electron-stream current through an oscillator at which self-sustaining oscillations will begin) courant *m* au démarrage
PREPAID - [gen & comm] affranchi, payé d'avance
PREPARATION - [gen] préparation *f*
~FOR COMBING - [text] préparation *f* pour le peignage
~FOR FLAX SPINNING - [text] préparation *f* du lin pour le filage
~OF FEED - [agric] (animal feeding) préparation *f* du fourrage
~OF HEMP FOR SPINNING - [text] préparation *f* du chanvre pour le filage
~OF THE ROVING - [text] filage *m* en gros, préparation *f* de la mèche
~OF THE SLUBBING - s. preparation of the roving
~OF THE WARP - [text] préparation *f* de la chaîne
~OF THE WEFT - [text] tramage *m*
~OF WOOL FOR CARDING - [text] préparation *f* de

laine pour le cardage
PREPARATORY - [gen] préparatoire, préalable
~DRAWING - [text] étirage *m* préparatoire
~PROGRAM - [comput] programme *m* préparatoire
~SPINNING - s. preparatory of the slubbing
PREPARE, to - [gen] préparer, accomoder
[gen] (to condition) conditionner
[leather ind] apprêter
[gen] (a document) rédiger
PREPARED LUGHOLE - [metall] oreille *f* régulée
~SCRAP - [metall] bonne mitraille *f*
PREPAREDNESS - [gen] état *m* de préparation
PREPARER - [text] ouvreuse *f*
~GILL-BOX - [text] étirage *m* préparatoire
PREPARING - [gen etc] préparation *f*
~MACHINE - [text] machine *f* de préparation
PREPAY, to - [gen and comm] (to pay in advance) payer d'avance
PREPAYMENT - [gen and comm] payement *m* d'avance
[comm] (advance) préachat *m*
~METER - [instr] (a slot meter, fitted with a coin-feed mechanism) compteur *m* à payment préalable
~VALVE - [mech] (in meters) valve *f* de blocage
PREPONDERANCE - [gen] prépondérance *f*
PREPOTENCY - [biol] prépotence *f*
PRE-PRINTING - [text] impression *f* préalable
~ROLLER - [text] égoutteur *m*
PREPUBERTY - [med] prépuberté *f*
PREPUBESCENCE - s. prepuberty
PREPUCE - [anat] prépuce *m*
PREPUNCH, to - [comput] (pre-insertion of data in a punch-card) préperforer
PRE-RELEASE - [cin] avant-première *f*
PRE-RET, to - [text] rouir préalablement
PREROLL TIME - [telev] (in video recording) intervalle *m* de garde, temps *m* de mise en service
PREROTATION - [aero] (in landing gear, the practice of setting the wheels in rotation before the aircraft touches down, to avoid the effect on the tyres of the sudden acceleration when contact is made with the runway) prérotation *f*
PRESBYACUSIA - [med] surdité *f* sénile
PRESBYCUSIS - s. presbyacusia
PRESBYOPE - [med] (one who suffers from presbyopia) presbyte
PRESBYPHRENIA - [med] démence *f* sénile
PRESBYOPIA - [med] (impairment of vision causing long-sightedness) presbytie *f*
PRESBYOTIC - [med] (adj) presbytique
PRESBYTERY - [arch] (part of the church apart from the clergy) presbytère *m*
PRESCORE, to - [cin] (the operation of recording before a picture is made) présynchroniser
PRESCORING - [cin] présynchronisation *f*
PRESCRIBE, to - [gen] prescrire
[med] prescrire, ordonner
[leg] acquérir par prescription
PRESCRIPTION - [gen] prescription *f*
[med] ordonnance *f*
PRESEDIMENTATION - [hydr] décantation *f* primaire
PRESELECTING ROTARY LINE SWITCH - [telecomm] présélecteur *m* rotatif
PRESELECTION - [telecomm] (in telephony; the operation of selecting, by means of a preselector, before seizing an idle trunk) présélection *f*
~OF THE APERTURE - [photo] présélection *f* du diaphragme
PRESELECTIVE DRIVE - [auto] boîte *f* de vitesse présélective
PRESELECTOR - [radio] (preamplifier between aerial and receiver increasing the sensitivity and selectivity of the receiver) préamplificateur *m* acccordé
[telecomm] (in automatic switching the mechanism carrying out the selecting operation before seizing an idle trunk) présélecteur *m*
[auto] (of drive) présélecteur *m*
~CONTROL - [photo] commande *f* du diaphragme présélectionné
PRESENTATION - [gen] présentation *f*
[telev] (the impression the picture makes on the viewer) présentation *f*, régie *f* principale
[radar] (the form in which radar signals appear on cathode-ray tube) présentation *f* visuelle de l'écho
PRESERVATION - [gen] conservation *f*
~OF COLOURS - [text] conservation *f* des couleurs
~OF MILK - [food] traitement *m* du lait pour sa conservation)
~OF NATURE - [gen] protection *f* des sites, protection *f* de la nature
~RECTIFIER - [telecomm] rédresseur *m* conservateur
PRESERVATIVE - [gen] préservatif *m*
[chem] (substance added to foodstuffs to prevent decay) préservatif *m*, agent *m* de conservation
[rubber ind] (preserving agent) agent *m* préservant
[mech] (for stored component parts etc) matériel *m* de protection
~AGENT - [chem] agent *m* de conservation, antiputride *m*
PRESERVE, to - [gen] conserver, préserver
PRESERVE CAN - s. preserve tin
~THE STAPLE DURING RINSING, to - [ind chem] préserver la méche pendant le rinçage
~THE STAPLE IN WASHING, to - [text] préserver la méche au lavage
~TIN - [food] boîte *f* à conserves
PRESERVED FOOD - [food] conserves *f*pl
~IN SYRUP - [food] conservé au sirop
PRESERVES - [food] conserves *f*pl
PRESERVING JAR - [food] pot *m* à confiture
~RECTIFIER - s. preservation rectifier
PRESET, to - [gen and contr] préfixer, afficher au préalable
PRESET APPARATUS - [instr] appareil *m* affiché
~DIAPHRAGM - [photo] diaphragme *m* présélectionné
~FADERS - [telev] résistances *f*pl chutrices préréglées
~GOVERNOR - [contr] (a clock-controlled governor) régulateur *m* à programme
~PARAMETER - [comput] paramètre *m* affiché
~SHUTTER - [photo] obturateur *m* à armement préalable
~SPEED - [contr] vitesse *f* préfixée
PRESETTING - [gen] pré-régulation *f*
[radio] (resetting) réenclenchement *m*
~THE CHANNELS - [telev] présélection *f* des canaux
PRESHAVING - [mech] préparation *f* du rasage
~HOB - [mech] (a tool used for the preshaving of gears) fraise-mère *f* pour roues d'engrenage à raser
~SHAPER CUTTER - [mech] couteau *m* raseur pour roues d'engrenage
PRESIGNALLING DISTANCE - [railw] distance *f* d'avertissement

PRESINTERING - [metall] préfittage *m*
PRESOFTEN, to - [rubber ind] plastifier préalablement
PRESOFTENED RUBBER - [rubber ind] caoutchouc *m* pré-ramolli
PRESS, to - [gen] presser, appuyer, pressurer
[mech] presser, mettre sous presse, comprimer
(of a substance) presser, exprimer
[metall and glass man] matricer, emboutir
[metall] mouler
[ceramics] (the operation of making pottery ware by pressing into a mould) estamper
[leather ind] fouler
[text] catir, calandrer
[paper man] satiner, calandrer
PRESS - [mech] presse *f*
[print] (general term denoting the printing stage of the process) presse *f*, imprimerie *f*
[print] presse *f* d'imprimerie
[agric] (wine press) pressoir *m*
[gen] (general term denoting newspapers, journals etc) presse *f*, journaux *mpl*
[electron] (flat fused glass seal) pincement *m*
~BOX - [text] boîte *f* de presse
~BUTTON - [gen] bouton *m*, bouton *m* pressoir, poussoir *m*
[telecomm] touche *f*
~BUTTON BOARD - [el] tableau *m* de commande à boutons
~BUTTON KEY - [telecomm] bouton *m* à enclenchement
~CASTING - [rubber ind] coulage *m* sous pression
~CONTROL - [mech] réglage *m* de la pression
~-CURE - [rubber ind] (process of vulcanizing rubber in a press) vulcanisation *f* sous presse
~CURED ARTICLE - [rubber ind] article *m* vulcanisé sous presse
~CUTTING - [gen] coupure *f* de presse
~FILTER - [mech] filtre *m* à pression
~FINGER - [text] doigt *m* presseur
~FIT - [mech] montage *m* à pression
~FIT INSERT [mech] (inserts placed by forcing them into holes in which they have an interference fit) prisonniers *mpl* d'ajustage par pression
~FOR MOUNTING SOLID TYRES -[mech] presse *f* pour le montage de pneus pleins
~-FORGE, to - [metall] emboutir, forger à la presse
~-FORGED - [metall] forgé à la presse
~-FORGING - [metall] forgeage *m* à la presse
[metall] (the forged piece) pièce *f* forgée à la presse
~HEAD - [mech] sommier *m* supérieur, tête *f* d'une presse
~HOME, to - [gen mech etc] (to force a part which is inserted into another, into is final position)emmancher à fond
~INTO POSITION, to - [gen and mech] forcer en place
~KEY - [gen] touche *f*
~LEVER WEIGHT - [text] poids *m* du levier de pression
~OF A BENCH - [carp] étau *m* de banc de menuisier
~OF COLUMN TYPE DESIGN - [mech] (a column press) presse *f* en colonnes
~OF FRAME TYPE DESIGN - [mech] presse *f* à cadre, presse *f* à construction à cadre
~OF THE BORED PLATEN TYPE - [mech] presse *f* à plateaux forés, presse *f* à plateaux avec canaux forés
PRESS OF THE DRILL PLATEN TYPE - s. press of the bored platen type
~PART - [paper man] (the portion of a paper-making machine in which water is pressed out of the web of paper formed in the preceding "Fourdrinier part") section *f* de pression
~PISTON - [mech] piston *m* de la presse
~PLATE - [mech] plateau *m* de la presse, plaque *f* de pression
[metall] (die casting) plaque *f* porte-estampes
[auto] (clutch) disque *m* d'embrayage
~PLUNGER - [mech] piston *m* presseur
~POLISH - [plast ind] (fine finish produced on sheet by pressing between very smooth heated plate) vernis *m* étendu par pression
~PROOF - [print] épreuve *f* en bon à tirer, correction *f* à clicher, feuille *f* machine
~PUMP WITH DIRECT DRIVE - [mech] pompe *f* à pression à commande directe
~RAM - s. press piston
~ROLL FOR PAPER MACHINE - [paper man] rouleau *m* compresseur pour machine à papier
~-ROLLER - [text] cylindre *m* enfonceur, rouleau *m* calandreur
~ROLLS - [plast ind & paper man] rouleaux *mpl* compresseurs
~SEAM - [rubber ind] couture *f* à la presse
~SWITCH - [el] interrupteur *m* à poussoir
~TABLE - [text] cadre *m* à presses
~THE FLUTES INTO THE ROLLER, to - [text] presser les cannelures sur le cylindre
~THE NIPPERS TOGETHER, to - [text] serrer les pinces
~THE PATTERN HOME, to - [metall] asseoir le modèle
~TOOL - [plast ind] outillage *m*
~TROUGH - [text] caissette *f* de pression
~WHEEL FOR COILER - [text] roue *f* à canal oblique
~WELDING - [metall] soudage *m* à la presse
~WITH HEATED PLATENS - [mech] presse *f* à plateaux chauffants
~WITH SLIDING TABLE - [mech] presse *f* avec table à coulisse, presse *f* avec tablier de tirage
PRESSBOARD - [paper man] ais *m*
PRESSED - [gen] pressé, serré
[metall] forgé à la presse, embouti, estampé
~AMBER - [min] (synthetic amber formed by heating and compressing pieces of natural amber) ambre *m* comprimé
~BALE - [text] balle *f* pressée
~BAR - [metall] barre *f* en poudre comprimée
~BRICK - [constr] (good-quality brick moulded under pressure) brique *f* pressée
~CATHODE - [electron] cathode *f* compensée moulée
~COAL - [min] briquette *f* de houille
~COTTON - [text] coton *m* en balles
~GLASS - [glass man] (glass shaped in a mould by pressure) verre *m* moulé
~GLASS BASE - [electron] (base for electronic tube in which heated powered glass is pressed around the supports and leading in wires) base *f* en verre moulé
~KAPOK - [text] kapok *m* pressé
~NAP CLOTH - [text] tissu *m* floconneux pressé
~-ON TYRE - [rubber ind] (of tyres) bandage *m* plein emmanché à la presse

PRESSED PLUSH - [text] peluche *f* gaufrée
~PULP - [paper man] pâte *f* pressée
[sugar ind] pulpes *fpl* pressées
~SHEET- [plast ind] feuille *f* pressée
~TO SHAPE - [metall] façonné à la presse
~TYPE BOND - [el] connexion *f* coincée, connexion *f* goupillée
PRESSEL-SWITCH - [el] (switch attached to the end of a flexible cord) poire *f* de contact
PRESSER - [text] (of wool) presse-étoffe *m*, fourchette-presse *f*
[mech] (of a sewing machine) presseur *m* (de machine à coudre)
~ARM - [text] doigt *m* presseur
~BAR - [impl] pied-de-biche *m*
~DRUM - [text] cylindre *m* de pression
~FINGER - [text] doigt *m* comprimeur, presseur *m*
~FOOT - s. presser (of a sewing machine)
~PAD - [photo] (of a camera) presseur
~PLATE - [text] plaque *f* à guillocher
PRESSING - [gen] pressant, urgent
[ceram] (operation of making pottery ware by pressing into a mould) étampage *m*
[acoust] (disk record produced in a record-moulding press from a master or stamper) disque *m* moulé
[text] (the final finishing process for woollen and worsted fabrics) catissage *m*, calandrage *m*
[mech] étampage *m* sous presse
[metall] emboutissage *m*, pièce *f* matricée
[leather ind] foulage *m*
~BED - [text] cuvette *f* de presse
~CRACK - [metall] crique *f* de glissement
~DIE - [metall] matrice *f* pour presses
~MACHINE - [glass man] (machine performing the entire forming operation by pressing the plastic into a mould by means of a plunger) machine *f* à mouler le verre
~OF THE COTTON - [text] (com) pressage *m* du coton
~ON OF WHEELS - [railw] calage *m* à la presse des corps des roues (sur les essieux)
~OUT THE YARN, to - [text] essorage *m* des fils
~PUMP - [mech] pompe *f* foulante
~SCREW - [mech] vis *f* de pression
~SKIN - [metall] peau *f* du comprimé
~SPEED - [mech] vitesse *f* de fermeture d'une presse
~THE HEMP SEED - [text] pressage *m* des graines de chanvre
~WITH FLOATING DIE - [metall] compression *f* à matrice flottante
PRESSINGS - [metall] pièces *fpl* forgées à la presse
[print] feuilles *fpl* de couverture
PRESSMARK - [paper man] filigrane *m*, marque *f*
PRESSOR - [med] hypertenseur *m*, presseur *m*
~BASE - [med] substance *f* vasomotrice
~SUBSTANCE - s. pressor
PRESSORECEPTIVE - [med] pressoréception *f*, pressosensitif
PRESSORECEPTOR - [med] (pressure receptor) pressorécepteur *m*
PRESSOSENSITIVE - s. pressoreceptive
PRESSPAHN - [el] (fibrous insulating material made of woodpulp and used for the insulation of electrical equipment) presspahn *m*
PRESSROOM - [print] salle *f* de la presse
PRESSURE - [gen] pression *f*, poussée *f*
[mech] (type of stress having uniformity in all directions) pression *f*
PRESSURE - [hydr] (of water) pression *f*, poussée
[constr] compression *f*
[geol] (of the earth) poussée
[el] (electromotive force) force *f* électromotrice
~ACCUMULATION - [hydr] (a device in hydraulic mechanism designed to store energy by the compression of a spring or of elastic fluid) collecteur *m* à pression
~AIR BURNER - [th eng] brûleur *m* à air comprimé
~ALTIMETER - [instr] (barometer so graduated as to indicate altitude in respect to a standard atmosphere) altimètre *m* barométrique
~ALTITUDE - [aero] (the altitude at which a balloon becomes fully distended) altitude *f* de pression
~AMPLITUDE - [phys] pression *f* acoustique maximale
~ANGLE - [mech] angle *m* de pression
~-BAG MOULDING - [plast ind] (moulding in which pressure is applied by an inflated elastic bag) procédé *m* au sac sous pression
~BEAM - [text] rouleau *m* de pression
~BELL - [constr] aire *f* de pression en forme de cloche
~BETWEEN THE ROLLERS - [text] pression *f* de pinçage
~BLASTING - [metall] nettoyage *m* à jets d'air
~BOILER - [th eng] chaudière *f* à pression
~BOMB - [mining] mesureur *m* et enregistreur de la pression d'une couche
~BOOSTER - [mech] (an auxiliary device used to augment pressure in a system) piston-plongeur *m* additionnel, surpresseur
~BREATHING - [astronaut] respiration *f* sous pression
~BROADENING - s. pressure shift
[astronaut] élargissement *m* dû à la pression
~BULKHEAD - [shipbuild & aero] (bulkhead designed to resist fluid pressure without leakage) cloison *m* étanche
~BURNER - [th eng] brûleur *m* à mélange surpressé
~CABIN - [aero] (a cabin in which pressure is maintained at a level corresponding to normal respiration) cabine *f* pressurisée
~CABLE - [el] (cable in which the dielectric is maintained at a pressure which is in excess of the atmospheric pressure) câble *m* sous pression
~CALIBRATION - [mech] étalonnage *m* de pression
~CAPSULE - [mech] (this-walled cylindrical metal bellows) capsule *f* barométrique
~CAST PATTERN - [metall] modèle *m* coulé sous pression
~CASTING - [metall] coulée *f* sous pression
~CELL - [soil] cellule *f* manométrique
~CHAMBER - [hydr] chambre *f* de pression
~-CHARGED - [mech] (of motors) alimenté sous pression
~CIRCUIT - [el] (voltage circuit; the circuit of an instrument through which flows a current dependent on the voltage to be measured) circuit *m* en dérivation
~CIRCULATING SYSTEM OF LUBRICATION - s. pressure lubrication
~COMPOUNDING - [ind chem] réduction *f* graduelle de la pression de vapeur
~CONTOUR - [met] (a line on a chart which shows the altitude at which specified atmospheric pressures occur) isohypse *f*

PRESSURE-CONTROL VALVE - [mech] (type of valve which automatically maintains downstream pressure at a constant level irrespective of variations in upstream pressure) soupape *f* régulatrice de pression
[plast ind] (a device fitted in an extruder head to control the pressure dowstream from the breaker plate) vanne *f* de régulation de pression
~-CONTROLLED VALVE - [mech] (pneumatic valve) soupape *f* pneumatique
~COOKER - [impl] (closed vessel for cooking under steam pressure) marmite *f* à pression, marmite *f* express
~COWLING - [aero] (unsealed cowling) capotage *m* sous pression
~CURVE - [hydr] courbe *f* piézométrique
~CUT-OFF - [gas ind] coupe-gaz *m*
~-DEMAND OXYGEN SYSTEM - [astronaut] circuit *m* d'oxygène sous pression à la demande
~DIE CASTING - [metall] moulage *m* sous pression
[metall] (the product) pièce *f* coulée sous pression
~-DIFFERENCE TRANSDUCER - [instr] (an apparatus designed to measure the rate of flow through pipelines) détecteur *m* de différence de pression
~DIFFERENTIAL - [mech] (differential pressure) pression *f* différentielle
~DISTILLATE - [oil ind] condensat *m* de cracking
[oil ind] distillat *m* sous pression
~DISTILLATION CRACKING - [oil ind] cracking *m* sous pression
~DRAG - [aero] (that element of the total drag which is attributable to the resolved component of pressure normal to the surface))trainée *f* de pression
~DRAUGHT - [mech] aérage *m* positif
~DRILLING - [mech] (in mining etc) forage *m* sous pression
[mining] forage *m* sous pression
~DROP - [el] chute *f* de tension, perte *f* de charge, baisse *f* de potentiel
[hydr] dénivellation *f* piézométrique
~DWELL - [mech] (the period of continuance of pressure in a press, during which the platen remains motionless) période *f* de pression
~EQUALIZATION - [mech] équilibrage *m* des pressions
~EQUALIZING PLATE - [mech] (a plate used to distribute the force applied by a press) coussin *m* de pression
~FACE - [aero] (the side of a propeller which corresponds to the lower side of air aerofoils) intrados *m* de la pale d'hélice
~FEED - [mech] graissage *m* sous pression
~FEEDER - [metall] masselotte *f* à pression de gaz
~FIGURE - [crystal] figure *f* de pression
~FILTER - [mech] filtre *m* à pression
~FLUSH - [nucl] (in isotopes) coup *m* de pression
~FOOT - [text] (in a sewing machine) presseur *m* (d'une machine à coudre)
~FORGING - [metall] (drop forging) forgeage *m* sous pression
~FORMING - [metall & plast ind] (the production of a specific shape by means of pressure and, usually, heat) estampage *m*, gaufrage *m*
~FRAME - [photo] cadre *m* presseur
~FRONT - [nucl] (shock front of an explosion) front *m* de choc
~FUSION WELDING - [metall] soudage *m* par fusion avec pression
PRESSURE GAS - [gas ind] gaz *m* pauvre par pression
~-GAS ENGINE - [mech] moteur *m* à gaz pauvre par pression
~GAS PRODUCER - [gas ind] gazogène *m* à vent soufflé
~GATE - [cin] (part of the fil-guiding mechanism) contre-plaque *f*
~GAUGE - [instr] (instrument to measure fluid pressure) manomètre *m*
~GOVERNOR - [gas ind] (gas governor designed to supply gas at a constant pressure) régulateur *m* de pression
~GRADIENT - [plast ind] (curve showing the pressure at different points along an extruder feed screw) courbe *f* de pression
~GRADIENT MICROPHONE - [acoust] (microphone offering very little obstruction of the passage of sound-waves, depending for its operation on the resultant of sound pressures acting on both sides of the diaphragm) microphone *m* à gradient de pression
~GREASE CUP - [mech] graisseur *m* sous pression
~GUIDE - [cin] (of the film projector) galet *m* presseur
~HARNESS - [text] lames *f*pl à coulisses
~HEAD - [hydr] (the energy possessed by unit weight of a fluid, due to its pressure) hauteur *f* piézométrique, charge *f* d'eau
[aero] (combination of a static and a Pitot tube connected to the opposite sides of a differential pressure gauge, for a visual reading corresponding to the speed of an air current) antenne *f* de Pitot
~HEIGHT - [aero] (the altitude in a standard atmosphere at which the pressure is equal to the pressure in question) altitude-pression *f*
~HOLDER - [gas ind] (high-pressure gas-holder) réservoir *m* sous pression
~IGNITER - [impl] allumeur *m* à pression
~INCREASE - [gen] augmentation *f* de la pression
~INDICATOR - [instr] manomètre *m*
~LEACHING - [ind chem] lessivage *m* sous pression
~LINE - [mech] (of gears) ligne *f* d'action
[gen] conduite *f* à pression
~LOSS - s. pressure drop
~LUBRICATION - [mech] (any system of lubrication in which the lubricating medium is supplied under pressure from a pump) graissage *m* sous pression
~MICROPHONE - [acoust] (a microphone depending for its operation on the excess pressure in a sound-wave) microphone *m* à pression
~MULTIPLIER - [mech] (a device in which the pressure of a sample of gas in a high-vacuum system is amplified, so as to permit an easy determination of the pressure) amplificateur *m* de pression
~NOZZLE - [mech] (a jet supplied with fluid under pressure, to produce a spray) buse *f* de projection
[aero] (only USA) s. pressure head
~OF BLAST - [th eng] pression *f* du vent
~OF THE REED - [text] pression *f* du peigne
~OF WIND - s. pressure of blast
~ON THE SPINNING SOLUTION - [text] pression *f* de filage
~ORIFICE - [hydr] lumière *f* de refoulement
~PACKAGING - [packag] emballage *m* pressurisé
~PACKED - [packag] emballé sous pression
~PAD - [cin] (of projector) glissière *f*, patin-presseur *m*

PRESSURE PAD - [plast ind] (device designed to reduce the pressure on the land areas of a mould when the mould is closed) surface *f* d'appui, cales *f*pl de renfort
~-PATTERN FLYING - [aero] (planning and carrying out a long-distance flight in reference to the distribution of atmospheric pressures) navigation *f* isobarique
~PICK-OFF - [mech] capteur *m* de pression
[el] capteur *m* à signal proportionnel
~PIECE - [mech] pièce *f* de pression
~PIN - [mech] goujon *m* de pression
~PIPE - [hydr] conduite *f* en charge
~PIPING - s. pressure pipe
~PLATE - [auto] (the part of a friction clutch assembly which transmits the thrust of the actuating mechanism to the clutch plates) plateau *m* de pression
~PUMP - [mech] (a pump used to supply lubricating oil under pressure to an engine) pompe *f* de pression
~RECEIVER - [mech] réservoir *m* à air comprimé
~REDUCING VALVE - [mech] (an automatic valve which receives fluid at a given pressure and discharges it at a pre-set lower pressure) détendeur *m*, clapet *m* réducteur de pression
~REFUELLING SYSTEM - [aero] ravitaillement *m* en carburant sous pression
~REGISTER - [instr] mesureur *m* des pressions
~REGISTRATION ARM - [railw] antibalancant *m* poussant
~REGULATING VALVE - [mech] régulateur *m* de pression
~RELAY - [el] relais *m* pneumatique
~RELIEF VALVE - [mach] (an automatic valve designed to open and discharge the fluid concerned when the pressure reaches a preset level) soupape *f* abaisseuse de pression
~RESPONSE - [el acoust] (of a microphone, for a specified frequency) sensibilité *f* à la pression
~REVERSAL - [mech] inversion *f* de direction de mouvement
~RIGID AIRSHIP - [aero] dirigeable *m* semi-rigide
~RING - [mech] anneau *m* de pression
~RISE - [el] surtension *f*
~ROLLER - [el acoust] (a roller holding the magnetic tape against the capstan by pressure) rouleau *m* presseur, tambour *m* presseur
[mech] cylindre *m* presseur
~-SCAVENGED - [mech] (of motors) à lavage sous pression
~SCREW - [mech] vis *f* de pression
~-SENSITIVE ADHESIVE - [ind chem] adhésif *m* autocollant
~SENSITIVITY - s. pressure response
~SHELL - [electron] cosse *f* à pression
~SHIFT - [phys] (increasing by pressure and density of a gas, the lines of its spectrum lose sharpness without change of total intensity) élargissement *m* dû à la pression
~SIDE - [oil ind] côté *m* du refoulement
~SINTERING - [metall] frittage *m* sous charge
~SPECTRUM LEVEL - [acoust] (of a sound) niveau *m* de pression sonore spectrale
~SPRING - [mech] ressort *m* à pression
~SPRING ADJUSTING SCREW - [mech] vis *f* de réglage de pression
~STABILIZED - [phys] stabilisé par la pression

PRESSURE STAGE - [mech] étage *m* de pression
~-STAGE TURBINE - [mech] turbine *f* à étages de pression
~STAGING - [mech] étages *m*pl de pression
~SUIT - [astronaut] combinaison *f* pressurisée
~SUPPRESSION - [nucl] (a safety system) enlèvement *m* de pression
~SURGE - [met] (sudden development of local increase in atmospheric pressure) augmentation *f* rapide de la pression atmosphérique
~SWITCH - [el] (pressure-operated cut-off switch) automate *m* manométrique, manostat *m*
~TEST - [gen] essai *m* à pression
[el] essai *m* de rigidité
[plumb] essai *m* d'étanchéité, essai *m* hydraulique
~THERMIT WELDING - [metall] (alumino-thermic process) soudure *f* à thermite sous pression
~THROWN BY HOLDER - [gas ind] pression *f* gazométrique
~THRUST - [astronaut] poussée *f* relative
~-TIGHT - [gen] étanche
~TRANSDUCER - [instr] (transducer providing an electrical signal which is in proportion to the pressure) détecteur *m* de pression, transducteur *m* de pression
~TUBE - [nucl] (used in reactors) tube *m* de force
~TUBE ANEMOMETER - [instr] (wind-measuring instrument based on the principle of a pilot tube) anémomètre *m* hydrostatique
~TUBE REACTOR - [nucl] réacteur *m* à tubes de force
~TUNNEL - [aero] (type of wind tunnel in which the working circuit is closed and maintained above atmosphere pressure) soufflerie *f* aérodynamique à pression
~TURBINE - [mech] turbine *f* à pression, turbine *f* à réaction
~-TYPE GASHOLDER - s. pressure holder
~VARIATION - [mech] variation *f* de pression
~VENTILATION - [mech] ventilation *f* à air comprimé
~VESSEL - [mech] récipient *m* à pression
[nucl] caisson *m* de réacteur, cuve *f* de réacteur
~VESSEL REACTOR - [nucl] réacteur *m* à récipient sous pression
~WATER - [hydr] eau *f* sous pression
~WATER PIPE - s. pressure pipe
~WEIGHT - [gas ind] (in gasholders) (ballast blocks) lest *m*
~WELDING - [metall] (welding employing static or dynamic pressure to complete the union, but without fusion of the constituent parts) soudure *f* par pression
~WELL - [mining] puits *m* d'injection
PRESSURIZE, to - [gen] pressuriser
PRESSURIZED - [aero] (said of an aircraft cabin in which the atmospheric pressure is kept normal) pressurisé
~AIRCRAFT - [aero] avion *m* pressurisé
~CABIN - [aero] cabine *f* pressurisée
~CASING - [nucl] (capable of withstanding a high pressure) gaine *f* de force, gaine *f* résistante à la pression
~GAS FLOW IONIZATION CHAMBER - [nucl] chambre *f* d'ionisation à circulation de gaz pressurisé
~HEAVY WATER REACTOR - [nucl] réacteur *m* à eau lourde sous pression
~REACTOR - [nucl] réacteur *m* pressurisé

PRESSURIZED SUIT - s. pressure suit
~WATER REACTOR - [nucl] (power reactor using slightly enriched uranium as fuel and water under high pressure as coolant and moderator) réacteur *m* à eau sous pression
PRESSURIZING - [aero] (the operation of maintaining a pressure above that of the ambient atmosphere in a closed spaced, e.g. an aircraft cabin) pressurisation *f*
~VALVE - [mech] (valve used in pressurizing system) soupape *f* de manutention sous pression
PRESSWORK - [print] (work carried out on the hand press) passage *m*
[metall] étampage *m*
[metall] (the pressed product) pièce *f* étampée
~PROCEDURE - s. presswork system
~SYSTEM - [metall] système *m* d'emboutissage
PRESTAGE - [astronaut] amorçage *m*
PRESTERNUM - [anat] présternum *m*
PRESTORAGE TEST - [gen] (test made on parts or units before putting them in storage) essai *m* avant l'emmagasinage
PRESTORE, to - [comput] (the operation of setting an initial value for the address of a cycle index) préenregistrer
[gen] (the operation of storing a quantity in a suitable location before it is required) emmagasiner au préalable
PRESTRESS, to - [build] (of building material) précontraindre
PRESTRESSED - [constr] précontraint
~CONCRETE - [constr] béton *m* précontraint
PRESTRETCHED SHEET - [plast ind] (sheet material previously stretched) feuille *f* préétirée
PRESYSTOLE - [med] présystole *f*, systole *f* auriculaire
PRETARSUS - [anat] prétarse *m*
PRETREAT, to - [gen] (the operation of performing a preliminary treatment) traiter préalablement
PRETREATMENT - [gen] (a preliminary treatment, usually to impart desired qualities to a substance before further processing) traitement *m* préalable
~FOR BONDING - [ind chem] traitement *m* préalable pour le collage
PREVAIL, to - [gen] prévaloir, prédominer
PREVAILING WESTERLIES - [met] (belts of winds on the Northern and Southerm sides of the subtropical high pressure zones) vents *mpl* dominants de NE et SE
PREVALENT - [gen] dominant, général
[met] (said of winds) dominant
PREVENCEPTION - [med] contraception *f*
PREVENT, to - [gen] empêcher, prévenir
PREVENTER PLATE - [naut] fausse *f* écoute
PREVENTING BREAKAGE - [text] prévention *f* de la rupture
PREVENTION OF BREAKAGE - [text] sûreté *f* contre les ruptures
~OF MOULD - [ind chem] préservation *f* des moisissures
PREVENTIVE - [gen] préventif
~-CHOKE COIL - [el] (choking coil connected between the two halves of the moving contact used in varying the tapping of a transformer or a battery) bobine *f* d'arrêt
~MAINTENANCE - [mech etc] entretien *m* préventif
~MEDICINE - [med] médicament *m* préventif
PREVENTIVE RESISTANCE - [el] (resistance connected between the two halves of the moving contact used in varying the tapping of a transformer or a battery) résistance *f* protectrice
PREVENTORIUM - [med] préventorium *m*
PREVERMIS - [med] vermis *m* supérieur du cervelet
PREVIEW, to - [cin & telev] voir en première vision
PREVIEW - [cin & telev] (an advance showing) première vision *f*, preview *m*
~MONITOR - [telev] moniteur *m* de caméra
PREVIOUS - [gen] préalable, antérieur
PREVULCANIZATION - [rubber ind] prévulcanisation *f*, précuisson *f*
PREWIRED CIRCUIT - [electron] circuit *m* précâblé
PREZYGAPOPHYSIS - [med] apophyse *f* articulaire supérieure
PRIAPISM - [med] priapisme *m*
PRICE, to - [gen & comm] (to set a price) fixer un prix
[gen] (to value, to assess) estimer, évaluer
PRICE - [gen & comm] prix *m*
[comm] (cost price) prix *m* de revient, prix *m* de fabrique
~CONTROL - [fin] régularisation *f* des prix, contrôle *m* des prix
~CUT - [comm] abaissement *m* des prix
~DIFFERENTIAL SPREAD - [comm] éventail *m* des prix
~ESTIMATE - [comm] devis *m* de prévision des prix
~FLUCTUATION - [comm] fluctuations *fpl* des prix
~FORECAST - [comm] prévision *f* des prix
~FORMATION - [comm] formation *f* des prix
~FREE FRONTIER - [comm] prix *m* franco frontière
~FREEZE - [comm] blocage *m* des prix
~INDEX - [comm] indice *m* des prix
~LEVEL - [comm] niveau *m* des prix
~LIST - [comm] liste *f* de prix, tarif *m*, prix *m* courant
~MEMORY - [electron comp] (generally a memory, or storage, containing the prices of all articles manufactured or handled a firm) mémoire *f* des prix
~SCISSORS - [comm] éventail *m* des prix
~SLASHING - [comm] réduction *f* drastique des prix
~STORAGE - s. price memory
~STORE - s. price memory
PRICEITE - [min] (natural calcium borate used as a source of borax) pandermite *f*
PRICES SOARING - [comm] flambée *f* des prix
PRICING - [comm] fixation *f* du prix
[gen] évaluation *f*
PRICK, to - [gen] piquer
[zool] (to erect the ears, as a horse) dresser (les oreilles)
[naut] (to trace by puncturing, e.g. a ship's course) porter le point
[agric] (a horse) enclouer (un cheval)
PRICK-EAR - [zool] oreille *f* droite, oreille *f* pointue
~-OFF, to - s. prick, to [naut]
~PUNCH - [tool] poinçon *m*
PRICKED COCOON - [text] cocon *m* vermoulu, cocon *m* piqué
PRICKER - [metall] (sharp-pointed tool used for making holes in a mould) aiguille *f* à trous d'air, broche *f*
[th eng] (a furnace poker) ringard *m*, attisoir *m*
[tool] poinçon *m*, aiguille *f*
[leather ind] épée *f*, tire-point *m*

PRICKER - [impl] mouchette *f*
[mining] (a probe) lance *f* de sonde
~BAR - [impl] (used in the heating of retors) attisoir *m*, cure-feu *m*
PRICKERS - [impl] pinces *f*pl
PRICKING - [th eng] décrassage *m*
~AWL - [rubber ind] (a small pointed tool to piece air bubbles in rubber manufacturing) pique *f* à sonder
~MACHINE - [text] machine *f* à piquer les cartons
~OUT - [agric] repiquage *m*, transplantation *f*
~UP - [th eng] piquage *m*
PRICKLE - [bot] piquant *m*, épine *f*
~HEAT - s. prickly heat
PRICKLY HEAT - [med] lichen *m* vésiculaire, bourbouille *f*
~PEAR - [bot] figuier *m* de Barbarie
PRILL - [min] pépite *f*
PRIMAGE - [comm] (special allowance) primage *m*
(the premium paid to the owner of a vessel in addition to freightage) chapeau *m* de mérite
[hydr] (priming water) rentrée *f* d'eau, primage *m*
[oil ind] eau *f* d'amorçage
PRIMAQUINE PHOSPHATE - [chem] (an antimalarial drug) phosphate *m* de primaquine
PRIMARY - [gen] primaire
[geol] primatif
[chem] (a substance obtained directly from a raw material by extraction and purification; e.g. phenol, as a coal-tar primary) produit *m* primaire
[el] (contraction of "Primary Winding": the winding in a transformer into which the current to be transformed is fed) primaire *m*, enroulement *m* primaire
[astr] rayon *m* cosmique principal
~ACIDS - [chem] (acids in which the carboxyl group is attached to the end carbon atom of a chain) acides *m*pl primaires
~AIR - [heat] (in combustion) air *m* primaire
~AIR CLEANER - [mech] filtre *m* d'air primaire
~ALCOHOLS - [chem] alcools *m*pl primaires
~AMIDE - [chem] (an amide in which one hydrogen atom has been replaced by an acyl group) amide *f* primaire
~AMINES - [chem] (amines containing the amino group) amines *f*pl primaires
~BATTERY - [el] (two or more cells connected electrically) batterie *f* de piles, batterie *f* primaire
~BODY CAVITY - [zool] cavité *f* de segmentation
~BOW - [met] (rainbow) arc-en-ciel *m*
~CARRIER FLOW - [electron] (of a semiconductor) flux *m* primaire des porteurs
~CELL - [el chem] (an individual cell for the production of electrical energy by electromechanical means) élément *m* primaire, pile *f*
~CELL CARBON ROD - [el] charbon *m* de pile
~CIRCUIT - [el] circuit *m* primaire
~COIL - [el] (coil in which flows a current setting up the magnetic flux required by an electrical machine) bobine *f* primaire
~COIL SYSTEM - [telecomm] système *m* de bobines de selfinduction primaires
~COLOUR SIGNAL - [telev] (in colour television) signal *m* primaire de couleur
~COLOUR UNIT - [electron] point *m* de couleur primaire
[telev] élément *m* d'une couleur primaire
~COLOURS - [paint] (red, yellow and blue) couleurs *f*pl fondamentales
PRIMARY COLOURS - [photo] (set of three colours from which multicolour images are built up) primaires *f*pl, couleurs *f*pl primaires
~CONSTANT - [el] constante *f* primaire
~CONSTRICTION - [genet] (the place where a chromosome is attached to the spindle) constriction *f* primaire
~COOLANT - [nucl] fluide *m* de refroidissement primaire
~COSMIC RAYS - [nucl] (probably consisting of atomic nuclei) rayons *m*pl cosmiques primaires
~CREEP - [metall] (initial creep) fluage *m* primaire
~CRUST - [metall] croûte *f* primaire
~CRYSTALLIZATION - [metall] cristallisation *f* initiale
~CURRENT - [electron] (the current which is carried by primary electrons) courant *m* primaire
~CURRENT RATIO - [el] (the relation between current densities at two given points in an electrode in the absence of polarization) rapport *m* de courant primaire
~DARK SPACE - [electron] espace *m* sombre d'Aston
~DETECTOR - [instr] (the first system element performing the initial measuring operation) capteur *m* primaire
~DISCONNECTING DEVICES - [el] dispositifs *m*pl primaires de coupure
~DISTRIBUTION NETWORK - [el] réseau *m* primaire de distribution
~DYNAMO - [el] dynamo *f* primaire
~ELECTRON - [electron] (electron in the primary emission) électron *m* incident
~ELEMENT - [instr] (of a measuring instrument) capteur *m* primaire
~EMISSION - [electron] (electron emission due directly to the temperature of a surface, the irradiation, or the application of an electric field to a surface) émission *f* primaire
~EXTINCTION - [crystal] extinction *f* primaire
~FAULT - [el] percement *m* initial
~FEEDBACK - [contr] (major feedback) réaction *f* principale
~FERMENTATION - [brew ind] fermentation *f* principale
~FISSION YIELD - [nucl] rendement *m* de fission primaire
~FLEXURE - [zool] (the flexure of the mid-brain) courbure *f* primaire
~FLOW - [electron] courant *m* primaire
~HOLES - [th eng] (holes in the flame tube of a burner to admit air for the early stage of combustion) trous *m*pl primaires
~INCREASE - [bot] (the increase of a stem or root) croissance *f* primaire
~INDUCTANCE - [radio] inductance *f* primaire
~INSULATION - [el] (insulation extruded into wire for electrical insulation as distinct from mechanical protection) isolation *f* primaire
~ION - [nucl] (the ion having the greater energy after a collision between two ions) ion *m* primaire
~ION PAIR - [nucl] paire *f* d'ions primaire
~IONIZATION - [nucl] (in counter tubes) ionisation *f* primaire
~KEYING - [radio] manipulation *f* dans le primaire d'alimentation
~LIGHT SOURCE - [phys] source *f* primaire

PRIMARY LINE SWITCH - [telecomm] chercheur *m* primaire, chercheur *m* de lignes
~LUMINOUS STANDARD - [light] (standard of luminous intensity) étalon *m* d'intensité lumineuse
~MATERIALS - [gen] matières *f*pl premières
~MERISTEM - [bot] (meristem derived immediately from a promeristem and giving rise to the cells which build up the primary body of the plant) méristème *m* primaire
~METER - [instr] (main, or master meter) mesureur *m* principal
~NATURAL RADIONUCLIDES - [nucl] (those natural radionuclides having a lifetime of hundreds of million years) radionucléides *m*pl naturels primaires
~NETWORK - [el] réseau *m* primaire
~NITRO-COMPOUNDS - [chem] composés *m*pl azotés primaires
~PARTICLE - [soil] particule *f* élémentaire
[metall] grain *m* élémentaire
~PIPE - [metall] retassure *f* ouverte
~PROTECTIVE BARRIER - [nucl] écran *m* primaire de radioprotection
~QUANTUM NUMBER - [phys] nombre *m* quantique principal
~RADAR - [radar] (a system in which the pulses are reflected by the target) radar *m* primaire
~RADIATOR - [electron] (in waveguides) élément *m* d'antenne actif
~REACTOR - [nucl] réacteur *m* à double dessin
~ROUTE - [telecomm] voie *f* normale
~SEPARATION PROCESS - [chem] procédé *m* de séparation primaire
~SEQUENCE - [comput] séquence *f* mineure
~SERVICE AREA - [radio] zone *f* de réception primaire
~SIGNAL MONITOR - [telev] (in colour television) appareil *m* de contrôle des signaux primaires, oscilloscope *m* triple
~SKIP ZONE - [radio] zone *f* de silence primaire
~SOLID SOLUTION - [metall] (constituent of alloys formed when atoms of an element B is incorporated in the crystals of a metal A) solution *f* solide primaire
~SPECTRUM - [phys] (the first-order spectrum produced by diffracting grating) spectre *m* primaire
~STANDARD - [meas] prototype *m*
~STRUCTURE - [aero] (any part of an aircraft, such that its failure would cause serious danger in operation) oeuvres *f*pl vives
~TERMINAL - [el] borne *f* primaire
~TISSUE - [bot] (tissue formed from cells derived from primary meristems) tissu *m* primaire
~TRANSIT-ANGLE GAP LOADING - [electron] admittance *f* de l'espace d'interaction
~TRANSMISSION FEEDER - [el] feeder *m* d'un circuit de transmission principale
~VOLTAGE - [el] (the voltage supplied to the primary winding of an induction coil or transformer) voltage *m* primaire
~WINDING - [el] (the winding of an induction coil or transformer which receives the driving current) enroulement *m* primaire
PRIMASTIC COKE - [gas ind] coke *m* en morceaux allongés
PRIME, to - [mech] (to deliver water along with the steam output) abreuver, entraîner de l'eau
[hydr] (to fill a pump casting and suction line with liquid to ensure starting) amorcer
PRIME, to - [mech] (to inject fuel directly into the cylinders of an internal combustion engine to make starting easier) injecter du carburant dans les cylindres
[paint] (to apply the first coat of paint) imprimer, apprêter
[brew ind] donner du sucre
[firearms] amorcer
[gen] (to prepare for work) apprêter
PRIME - [math] premier
[gen comm etc] principal
~A MINE, to - [mining] étoupiller une mine
~A PUMP, to - [mech] amorcer une pompe
~BOND - [fin] obligation *f* de premier ordre
~COST - [comm] (the sums paid by the contractors to merchants for specific articles or materials) prix *m* de revient
~MATERIAL - [gen] matières *f*pl premières
~MERIDIAN - [geogr] méridien *m* origine
~MOTOR - s. prime mover
~MOVER - [mech] (engine or mechanism by which a natural source of energy is converted into mechanical power) machine *f* motrice, premier moteur *m*
~NUMBER - [math] nombre *m* premier
PRIMER - [mech] (device for easy starting in an i.c. engine, which sprays fuel directly into the induction system or the combustion chamber) dispositif *m* d'injection au démarrage
[paint] (special paint used for the first coating of a surface) apprêt *m*
[firearms] amorce *f*
[electron] (auxiliary electrode in a trigger tube to ensure safe triggering) électrode *f* activatrice
PRIMES - [metall] produits *m*pl de haute qualité
PRIMEVAL FOREST - [gen] forêt *f* vierge
PRIMIDONE - [chem] (primaclone. An anticonvulsant drug used in the treatment of epilepsy) primidone *m*
PRIMIGRAVIDA - [med] primigeste
PRIMING - [paint] (the work of priming) apprêt *m*
[paint] (the first coat) première couche *f*
[mech] primage *m*
[hydr] (of a pum) amorçage *m*
[th eng] (the delivery of water along with steam output) primage *m*, entraînement *m* de l'eau par la vapeur
[electron] activation *f*
[oil ind] injection *f*
[mining] préparation *f* de la cartouche
~COAT - [paint] couche *f* d'apprêt
~COCK - [hydr] robinet *m* d'amorçage
~NEEDLE - [impl] épinglette *f*
~PAINT - [paint] (special paint designed for a first coating, to give proper adhesion to previously uncoated materials and to allow for the high absorbing power of newly-wrought timber) peinture *f* d'impression
~PAPER - [constr] papier *m* de fond *m*
~PIPE - [mech] (of a pump) fourreau *m*
~PUMP - [mech] pompe *f* d'amorçage
~SPEED - [electron] vitesse *f* d'activation
PRIMIPARITY - [med] primiparité *f*
PRIMITIVE - [gen] primitif
~CIRCLE - [mech] (of gearing) cercle *m* primitif, cercle *m* de division
~FORM - [genet] forme *f* de souche
~RADIUS - [mech] (of a gear wheel) rayon *m* primi-

tif
PRIMITIVE ROCKS - [geol] roches *fpl* primitives
~STRATA - [geol] sol *m* primordial
~TRANSLATION - [crystal] (a space lattice repeats itself exactly if it is carried through one of the different translations in space) translation *f* primitive
PRIMORDIAL - [gen zool etc] primordial
~CELL - [bot] (cell prior to the formation of a cell wall) cellule *f* primordiale
~MERISTEM - [bot] (promeristem) proméristème *m*
PRIMORDIUM - [genet] primordium
PRIMROSE CHROMES - [paint] (light yellow pigments of complex composition,containing chromate and sulphate of lead with aluminium hydroxide) chromes *mpl* de primevère
~DERMATITIS - [med] dermatite *f* des primevères
PRIMULINE - [chem] (a thiazole dye, used for direct dyeing of cotton, primrose-yellow in colour) primuline *f*
~PROCESS - [photo] (diazotyne process) procédé *m* diazotypique
~RED - [chem] (a red dye obtained from primuline by diazotization and coupling with beta-naphthol) rouge *m* de primuline
PRINCIPAL - [gen] principal
[comm] directeur *m*, chef *m*, patron *m*
[gen] (schol) directeur *m*
[leg] mandant *m*, commettant *m*
[fin] (property or capital) donneur *m* d'ordre
~AXIS - [crystal] (the optical axis of a crystal) axe *m* principal
~AXIS FILTER - [radio] filtre *m* de bande principal
~AXIS OF A LENS - [opt] (the straight line passing through the centre of curvature of the faces of a lens) axe *m* principal d'une lentille
~BEAM - [constr] poutre *f* principale, longeron *m*
~BODY AXES - [phys] (those body axes which are coincident with the axes of inertia) axes *mpl* principals d'un corps
~BREED - [zool] race *f* principale
~E PLANE - [radio] (plane containing the direction of maximum radiation and in which the electric vector lies in the plane everywhere) plan *m* E
~FOCAL POINT - [photo] foyer *m* principal
~FOCUS - [opt] (the focus point) foyer *m*
~H PLANE - [radio] (plane containing the direction of maximum radiation and in which the electric vector is normal to the plane and the magnetic vector lies in the plane) plan *m* H
~MEMBER OF A FRAME - [mech] maîtresse-pièce *f* d'une charpente
~PATH - [telecomm] trajet *m* principal
~PLANE - [opt] tableau *m*
~PLANES - [opt] (two conjugate planes for which the lateral magnification is unity) plans *mpl* principaux
~POINTS - [opt] (the intersection of the principal planes with the optical axis) points *mpl* principaux
~QUANTUM NUMBER - [phys] (quantum number determining the energy of an electron in one of the allowed orbits around the nucleus) nombre *m* quantique principal
~RAFTER - [constr] (rafter forming part of the roof truss and supporting the purlins) chevron *m* principal, arbalétrier *m*
~RAY - [opt] rayon *m* visuel principal
~SECTION - [crystal] coupe *f* principale
PRINCIPLE - [gen] principe *m*
[phys etc] (a basic law) principe *m*, loi *f*
~OF ADDITIVITY - [mech] (the properties of the solution of a strong electrolyte are the sum of the individual properties of its ions) principe *m* de sommation
~OF CONTINUITY - [phys] (or continuity equation; the application of the principle of conservation of mass to fluid motion) loi *f* de continuïté
~OF LEAST ACTION - [mech] principe *m* d'action minimale
~OF LEAST TIME - [light] principe *m* de temps minimal
~OF MOBILE EQUILIBRIUM - [mech] (the tendency in a system of equilibrium to adjust to any change occurring in one of the conditions) principe *m* de l'équilibre mobile
~OF MOMENTS - [mech] principe *m* des moments
~OF OPTICAL SUPERPOSITION - [phys] principe *m* de superposition optique
~OF PHASE STABILITY - [electron] (in a particle accelerator) principe *m* d'autophasage
~OF REVERSEBILITY - [opt] principe *m* de réversibilité
~OF SEPARATE EQUILIBRIUM - [mech] (the equilibrium of several particles can be studied by examining the equilibrium of a single particle in the group) principe *m* de l'équilibre séparé
~OF THE EQUIPARTITION OF ENERGY - [chem] (the total energy of a molecule in the normal state is divided up equally between its different capacities for holding energy) principe *m* de l'équirépartition de l'énergie
~OF VIRTUAL WORK - [mech] principe *m* des travaux virtuels
PRINCIPLES OF HEREDITY - [zool] lois *fpl* de l'hérédité
~OF INTERLACING - [text] principes *mpl* d'entrecroisement des fils
PRINT, to - [print] imprimer
[photo] (to make a positive by exposing a negative image) tirer (une épreuve)
PRINT - [gen] impression *f*, caractères *mpl*
[print] (printed matter) matière *f* imprimée
[photo] épreuve *f*, copie *f*
[cin] (the developed film after being printed from the negative) positif *m*, épreuve *f*
[draw] reproduction *f*
[print] estampe *f*, gravure *f*
[gen] (an impression) marque *f*, trace *f*
[text] indienne *f*, cotonnade *f*
[metall] (a core print) portée *f* de moulage
~CONTRAST RATIO - [comput] rapport *m* de contraste d'impression
~CONTRAST SIGNAL - [comput] signal *m* de contraste d'impression
~COPY - [cin] positif *m*
~CUTTER - [photo] coupe-épreuves *m*
~GRADER - [cin] étalonneur *m*
~HAMMER - [paint] marteau *m* d'impression
~LIBRARY - [photo] photothèque *f*
~LINE - [comput] ligne *f* d'impression
~MEMBER - [comput] porteur *m* de types
~-OUT EFFECT - [opt] (the appearance of metallic silver in a silver halide emulsion after prolonged exposure to light) procédé *m* de copie
~PROOF - [print] épreuve *f* positive

PRINT ROLL - [comput] rouleau *m* d'impression
~SELECTION COMMON EXIT - [comput] plot *m* commun "imprimante"
~SELECTION ENTRY HUBS - [comput] plots *m* d'entrée "sélection imprimante"
~WHEEL - [comput] (of a tabulator) roue *f* d'impression
PRINTABILITY - [plast ind] (capacity of a plastic product to make and retain printing ink) qualité *f* permettant l'impression
PRINTED - [print] imprimé
[text] imprimé
~CALICO - [text] calicot *m* imprimé, indienne *f*
~CIRCUIT - [el] (electrical circuit formed photographically on copper sheat bonded to a dielectric, the unwanted copper being etched away) circuit *m* imprimé
~CIRCUIT CARD - [electron] carte *f* de circuit imprimé
~CIRCUIT GENERATOR - [el] traceur *m* de circuits imprimés
~CLOTH - s. printed fabric
~DRAFT - [text] pointé *m* de liage imprimé
~FABRIC - [text] tissu *m* imprimé
~FORM - [gen] imprimé *m*
~KNOP YARN - [text] fil *m* boutonneux imprimé
~MOQUETTE - [text] moquette *f* imprimée
~PANEL - [el] panneau *m* imprimé
~RECORD STORAGE - [comput] mémoire *f* tampon d'une imprimante
~RIBBON - [text] ruban *m* imprimé
~WINDING - [electron] bobinage *m* imprimé
~YARN - [text] fil *m* imprimé
PRINTER - [gen] imprimeur typographe *m*
[cin] (machine designed to print positive films) tireur *m* d'épreuves
[text] (a printed cotton fabric) imprimeur *m*
[rubber ind] laminoir-marqueur *m*
[photo] (printing apparatus) tireuse *f*
~FACTOR - [cin] (the difference in density on the positive copies) coefficient *m* de tirage
~KEYBOARD - [comput] imprimante *f* et clavier
~LIGHT - [cin] (the light source in a film printer) lumière *f* pour le tirage
~PERFORATOR - [comput] perforatrice *f* imprimante
~SPACING CHART - [print] table *f* d'espacement
~TAPE - [comput] bobineau *m*, bande *f* d'impression
PRINTER'S BLANKET - [print] blanchet *m* d'imprimerie, toile *f* d'encrage
~CLOTH - s. printer's blanket
~DEVIL - [print] demon *m* de la coquille
~ERROR - [print] faute *f* d'impression, coquille *f*
~METAL - [print] métal *m* à caractères
~PALSY - [med] paralysie *f* stibiée des imprimeurs
~PROOF - [print] épreuves *f*pl
~REAM - [paper man] rame *f*
~ROLLER - [print] rouleau *m* d'impression
PRINTERGRAM - [telecomm] printergramme *m*
PRINTING - [print] impression *f*, imprimerie *f*
[photo] tirage *m*
[gen] (presswork) étampage *m*
~BLOCK - [print] cliché *m*
~CLOTH - s. printer's blanket
~CYLINDER - s. printing roller
~DOWN ON STEEL - [print] copie *f* sur acier
~DOWN PROCESS - [print] copie *f* sur métal
~FILTER - [photo] filtre *m* de tirage
PRINTING FRAME - [photo] châssis-presse *m*
~HOUSE - [comm] imprimerie *f*
~INK - [print] (a mixture of colouring matter and linsead, or mineral, oil) encre *f* d'impression
~KEYBOARD PERFORATOR - [telecomm] (keyboard perforator in which the depression of a key will cause the code of the corresponding character to be punched in a tape with the simultaneous printing of the character) clavier *m* perforateur à impression
~LAMP - [photo] lampe *f* de tirage
~LOSS - [cin] (the loss, particulary in sound quality, caused by the copying process) pertes *f*pl de tirage
~MACHINE - [print] machine *f* à imprimer, presse *f* typographique
~MASK - [photo] cache *m*
~-OUT EMULSION - [photo] émulsion *f* à noircissement direct
~PAPER - [paper man] (various types and grades of paper used for printing) papier *m* à imprimer
~PATTERN - [text] dessin *m* d'impression
~PLANT - [ind] typographie *f*
~PRESS - [print] press *f* typographique
~PUNCH - [comput] perforatrice *f* imprimante
~-PUNCHING - [comput] perforation *f* imprimante
~READER - [comput] lecteur-imprimeur *m*
~REPERFORATOR - [telecomm] récepteur *m* perforateur imprimeur, réperforateur *m* imprimeur
~ROLL - s. printing roller
~ROLLER - [print] rouleau *m* d'impression
~SCREEN - [text] cadre *m* pour l'impression
~SHOP - [print] imprimerie *f*
~TELEGRAPHY - [telecomm] (telegraph system in which the signals are translated and give a readable message by operating a printing machine) impression *f* directe, télégraphie *f* par appareils imprimeurs
~TIMER - [photo] minuterie *f*
PRINTINGS - [paper man] (types and grades of paper used for printing books etc) papier *m* à imprimer
PRINTS - [metall] portée *f*pl
PRIONOTRON - [electron] (velocity-modulation tube) prionotron *m*
PRIORITY - [gen] priorité *f*
~ROUTINE - [comput] programme *m* de priorité
~TELEGRAM - [telecomm] télégramme *m* urgent
PRIORY - [arch] (monastic house presided over by a prior) priorat *m*
PRISM - [gen] (transparent structure usually glass, with two or more faces meeting at an edge) prisme *m*
~AERIAL - [radio] (electromagnetic horn with its sides forming a pyramid) antenne *f* en prisme
~ANTENNA - s. prism aerial
~BINOCULAR - [opt] jumelle *f* à prismes
~DRUM - [telev] (mechanical scanning device with a series of prism mounted on a rotating drum) tambour *m* à prismes
~POWER - [opt] puissance *f* du prisme
~SPECTROSCOPE - [instr] (spectroscope in which the resolving element is a prism) spectroscope *m* à prisme
~SQUARE - [surv] équerre *f* à prisme
~VIEWFINDER - [photo] viseur *m* à prismes
PRISMATIC - [gen] (shaped like or containing a prism) prismatique, prismé
~ASTROLABE - [instr] (instrument used for observing stars) astrolabe *m* prismatique

PRISMATIC CLEAVAGE - [geol] clivage *m* prismatique
~COMPASS - [instr] boussole *f* topographique à prisme
~EYE - [photo] viseur *m* à prisme
~GUIDE - [mech] guide *m* prismatique
~JOINTING - [geol] séparation *f* prismatique
~LAYER - [zool] couche *f* prismatique
~LENS - [opt] lentiprisme *m*
~REFLECTOR - [opt] prisme *m* à réflexion
~SIGHT - [cin] viseur *m* à prisme
~SPECTRUM - [opt] (the spectrum formed by refraction in a prism) spectre *m* prismatique
~SULPHUR - [chem] (the crystalline form of sulphur) soufre *m* prismatique
~SYSTEM - [crystal] (orthorhombic system) système *m* orthorhombique
~VIEWFINDER - s. prism viewfinder
PRISOMETER - [instr] (an apparatus for determining the setting time of Portland cement) prisomètre *m*
PRITCHEL - [tool] poinçon *m*
~HOLE - [mech] trou *m* rond de l'enclume
PRIVACY EQUIPMENT - s. privacy system
~SYSTEM - [telecomm] (system designed to make unauthorized reception impossibile or at least difficult) système *m* de transmission codé
PRIVATE ADDRESS SYSTEM - [telecomm] installation *f* privée automatique
~AUTOMATIC BRANCH EXCHANGE - [telecomm] bureau *m* privé automatique
~AUTOMATIC EXCHANGE - [telecomm] bureau *m* privé automatique
~BRANCH EXCHANGE - [telecomm] installation *f* de abonné avec postes supplémentaires
~EXCHANGE - [telecomm] installation *f* téléphonique privée
~EXPERIMENTAL STATION - [telecomm] station *f* expérimentale privée
~MANUAL BRANCH EXCHANGE - [telecomm] installation *f* privée manuelle
~PILOT'S LICENCE - [aero] (licence issued to person qualified to fly private aircraft only) licence *f* de pilote privé
~SIDING - [railw] embranchement *m* particulier, voie *f* industrielle
~UNDERTAKING - [gen & ind] entreprise *f* particulière
~WIPER - [el] frotteur *m* privé, frotteur *m* de test
~WIRE - [telecomm] fil *m* privé, fil *m* de test
~WIRE AGREEMENT - [telecomm] contrat *m* de location d'un circuit
~WIRE CIRCUIT - [telecomm] circuit *m* de location
PRIVATEER - [naut] corsaire *m*
PRIVILEGE - [gen] privilège *m*, prérogative *f*
PRIVILEGED DIRECTIONS - [opt] directions *fpl* priviligées
PRIZE, to - [mech] (to force with a lever) forcer avec un levier, ouvrir avec un levier
[comm] évaluer, estimer, priser
PRIZE - [gen] prix *m*
[mech] force *f* de levier, point *m* d'appui
PROACTINIDES - [chem] (name suggested for elements of the last row of the periodic system in oxidation state) proactinides *fpl*
PROBABILITY - [gen] probabilité *f*
[math] probabilité *f*
~CURVE - [statist] courbe *f* de la cloche
~DENSITY FUNCTION - [math] fonction *f* de densité de probabilité
PROBABILITY FACTOR - [phys] (correction factor applied to the values calculated for the reaction rates by collision theory) coefficient *m* de probabilité
~OF COLLISION - [nucl] (the probability that an electron collides with an atom or molecule when moving through a distance of I cm) probabilité *f* de choc
~OF DISINTEGRATION - [nucl] probabilité *f* de désintégration
~OF THE IONIZATION - [nucl] (the ration between the number of collisions and the total number of collisions during a given time) probabilité *f* d'ionisation
~SAMPLE - [statist] échantillon *m* probabiliste
PROBABLE - [gen] probable
~ERROR - [math] erreur *f* probable
PROBANG - [med] sonde *f* molle oesophagienne
PROBARBITAL SODIUM - [chem] (a hypnotic) probarbital *m* sodique
PROBE, to - [gen] sonder, explorer
PROBE - [med] sonde *f*, cathéter *m*
[radar] (in radar used for meteorology; instrument attached to a balloon and transmitting to a ground station a variety of information) sonde *f*, sonde *f* spatiale
[instr] (a sensing device located at a point at which measurement is to be made and connected to an instrument remote from it) sonde *f*
[mining] (tool for measuring the depth of holes) lance *f* de sonde
[electron] (small auxiliary electrode placed in a gas tube to determine the space potential) sonde *f*
[electron] (a straight rod used for coupling to an external circuit) sonde *f*
~COUPLING - [electron] (in a waveguide) accouplage *m* à sonde
~MICROPHONE - [el acoust] (a microphone used to explore a sound field without disturbing it) microphone *m* à sonde
~TRANSFORMER - [electron] (mode changer) transformateur *m* sonde
~UNIT - [radiat] (a form of head amplifier suitable for use with a Geiger-Müller counter) sonde *f* amplificateur
[radiat] sonde *f* de détection
PROBERTITE - [min] (hydrated oxide of sodium, calcium and boron) probertite *f*
PROBING - [mining] prise *f* d'échantillons du sol
~ELECTRODE - s. probe electron
PROBLEM DEFINITION - [math] énoncé *m* du problème
~PROGRAMME - [comput] programme-problème *m*
PROBOSCIS - [zool] proboscide *f* (d'insecte), trompe *f*
PROCAINAMIDE HYDROCHLORIDE - [chem] (a drug used in the treatment of cardiac diseases and also having application as an analgesic) chlorhydrate *m* de procaïnamide
PROCAINE - [chem] (a local anesthetic) procaïne *f*, novocaïne *f*
~HYDROCHLORIDE - [chem] (ethocaine hydrochloride. A local anaesthetic) chlorhydrate *m* de procaïne
PROCARTILAGE - [zool] (early stage in the formation of cartilage) procartilage *m*
PROCEED POSITION - [railw] position *f* de voie libre
~-TO-SEND SIGNAL - [radio & telecomm] signal *m* d'invitation à transmettre
PROCEEDS - [gen comm fin] (the material results) bénéfices *mpl*, produit *m*

PROCESS, to - [ind] (to treat in some special way in manufacture to subject to a process) traiter
[ind chem] traiter
[leg] (to proced against) poursuivre
[text] apprêter
PROCESS - [gen] procédé *m*, traitement *m*, développement *m*
[chem & ind chem] procédé *m*, processus *m*, réaction *f*, opération *f*
[photo] (the mechanical sequence up to the final positive) procédé *m*, développement *m*
[anat] excroissance *f*, procès *m*
[contr] (the collective functions in the equipment in which a variable is controlled) processus *m*
[leg] procès *m*, action *f* en justice
~ALLOYS - [metall] additions *fpl*
~ANNEALING - [metall] recuit *m* intermédiaire
~AUTOMATION - [autom] automation *f* de processus
~BACKGROUND - [cin] (projection of a scenary used as a background during shooting) projection *f* par transparence
~CAMERA - [photo] caméra *f* pour reproductions photomécanique
~CONTROL - [ind proc] (automatic control over industrial processes) contrôle *m* de procédé
[contr] conduite *f* des processus industrials
~-CONTROL COMPUTER - [comput] calculateur *m* pour la conduite des processus industrials
~DEVELOPER - [photo] développement *m* photomécanique
~ENGRAVER - [print] photograveur *m*
~ENGRAVING - [photo] phototypogravure *f*, similgravure *f*
~FILM - [photo] film *m* photomécanique
~HEAT REACTOR - [nucl] réacteur *m* de production de chaleur
~HOLD-UP TIME - [gen] (the time during which material is detained while undergoing a process of manufacture) période *f* improductive d'un procédé
~INFORMATION, to - [comput] traiter des données
~LAPSE RATE - [met] gradient *m* vertical de température d'un élément d'air
~LENS - [photo] objectif *m* pour reproductions photomécaniques, lentille *f* pour la reproduction photomécanique
~METALLURGY - [metall] métallurgie *f* des minerais
~OF FERMENTATION - [chem] procédé *m* de fermentation
~OF SPINNING BY STRETCHING - [text] procédé *m* de filage par étirage
~OILS - s. processing oils
~PLATE - [photo] plaque *f* photomécanique
~STEAM - [gen] (steam used expressly for an industrial operation, as distinct from power generation) vapeur *f* utilisée pour des usages industriels
~STOCK - [metall] charge *f*
~WATER - [ind chem] (water used in an industrial chemical process) eaux *fpl* résiduaires
PROCESSABILITY - [ind chem] (the extent to which a material is amanable to processing) aptitude *f* à subir un processus
PROCESSED - [ind chem] traité, développé, usiné
[rubber ind] artificiel
~RUBBER - [rubber ind] caoutchouc *m* artificiel
~SULPHUR - [ind chem] soufre *m* raffiné
PROCESSING - [ind proc] (the operation of treating a material) traitement *m*, usinage *m*, travail *m*, méthode *f* de fabrication
PROCESSING - [photo] (the chemical sequences) développement *m*
[agric] transformation *f*
[comput] traitement *m*
~INDUSTRY - [agric] industrie *f* de transformation
~OF HARVESTED CROPS - [agric] traitement *m* de la récolte
~OF MILK - [agric] transformation *f* du lait
~OILS - [ind chem] huiles *fpl* ramollissantes
~OVERLAP - [comput] chevauchement-calcul *m*
~STAGE - [agric] stade *m* de transformation
~STATION - [comput] centre *m* de traitement des données
~UNIT - [comput] unité *f* de traitement des données
PROCESSOR - [comput] appareil *m* de traitement des données
PROCHLORPERAZINE DIMALEATE - [chem] (a phenothiazine derivative used in the treatment of neuroses and labyrinthine disturbances) dimaléate *m* de prochlorpérazine
PROCREATION - [gen] procréation *f*
PROCTAGRA - [med] douleur *f* anale
PROCTALGIA - [med] proctodynie *f*, rectalgie *f*
PROCTATRESIA - [med] atrésie *f* de l'anus, imperforation *f* de l'anus
PROCTECTASIA - [med] dilatation *f* rectale
PROCTECTOMY - [med] amputation *f* du rectum
PROCTENCLEISIS - [med] rétrécissement *m* rectal
PROCTOCELE - [med] rectocèle *f*
PROCTOCOLPOPLASTY - [med] recto-colpoplastie *f*
PROCTOCYSTOTOMY - [med] cystotomie *f* par voie rectale
PROCTOELYTROPLASTY - [med] élytroproctoplastie *f*
PROCTOLOGY - [med] proctologie *f*
PROCTOPARALYSIS - [med] paralysie *f* du sphincter anal
PROCTOPEXY - [med] proctopexie *f*
PROCTOPLEGIA - s. proctoparalysis
PROCTOPTOMA - [med] prolapsus *m* incomplet du rectum
PROCTORRHAGIA - [med] proctorragie *f*
PROCTORRHAPHY - [med] suture *f* du rectum
PROCTORRHEA - [med] proctorrhée *f*
PROCTOSCOPE - [med] (instrument for examining the mucous membrane of the rectum) rectoscope *m*
PROCTOSCOPY - [med] rectoscopie *f*
PROCTOSIGMOIDOSCOPE - [med] (instrument used to examine the sigmoid flexure of the colon) proctoscope *m*
PROCTOSPASM - [med] spasme *m* du rectum
PROCTOSTASIS - [med] stase *f* rectale
PROCTOTOMY - [med] rectotomie *f*
PROCUMBENT - [med] couché, étendu sur le ventre
PROCURE, to - [gen] procurer, obtenir
PROCUREMENT - [gen] approvisionnement *m*
PROD, to - [gen] aiguillonner, pousser
PROD - [impl] (a pointed metal probe usually provided with an insulating sleeve, used to make contact with any desired point in an electric circuit in testing) aiguillon *m*
[metall] (of a loam-plate or mould) broche *f* (de une plaque de fond)
PRODROMAL SYMPTOMS - [med] signes *mpl* avant-coureurs
PRODROME - [med] prodrome *m*

PRODUCE, to - [gen] produire, présenter, montrer
[cin & theatre] mettre en scène, produire
[gen] (goods etc) fabriquer
[geom] prolonger
PRODUCE EXCHANGE - [comm] bourse *f* de commerce
~OF SOIL - [agric] rendement *m* du sol
PRODUCER - [gen] producteur *m*
[cin] (the person supplying the finaces for a film) producteur *m*
[cin & telev] (the man responsible for the production) metteur *m* en scène, directeur *m*, éditeur *m*
[gas ind] gazogène *m*
~BACK WALL - [th eng] (a thrust block) massif *m* de butée
~BOTTOM - [gas ind] cendrier *m*
~COAL - [gas ind] charbon *m* à gazogène, houille *f* de gazogène
~-FIRED SETTING - [gas ind] four *m* à gazogène
~GAS - [gas ind] (a fuel gas manufactured by the passage of air and steam through a bed of incandescent solid fuel, or by combustion of a solid fuel in a restricted air supply, to produce a mixture of hydrogen and carbon monoxide) gaz *m* pauvre, gaz *m* à l'air
~GAS EQUIPMENT - s. producer gas plant
~GAS MAIN - [gas ind] conduit *m* de gaz du gazogène
~GAS PLANT - [gas ind] atelier *m* des gazogènes
~WATER CHANNEL - [metall] (cinder quench) bac *m* à eau de cendrier, cuve *f* à mâchefer
PRODUCING A WELL - [oil ind] mise *f* en production d'un puits
~FORMATION - [oil ind] (a gas field) gisement *m* de gaz, formation *f* productive
PRODUCT - [gen] (the finished article or substance obtained by a manufacturing process) produit *m*
[chem] (the result of a chemical change) produit *m*
[math] produit *m*
~DEMODULATOR - [radio] démodulateur *m* de produit
~MODULATOR - [radio] modulateur *m* idéal
~OF COMBUSTION - [chem] (a substance formed by combustion, e.g. of coal) produit *m* de la combustion
~OF CROSSING - [genet] produit *m* du croisement
~RELAY - [el] relais *m* de produit
PRODUCTION - [gen] production *f*
[math] (in geometry) prolongement *m* (d'une ligne)
[cin] production *f*, mise *f* en scène
~AIRCRAFT - [ind] (an aircraft which being manufactured in a regular sequence is distinct from an experimental or special type) avion *m* de série
~COSTS - [ind] frais *mpl* de fabrication
~CURVE - [ind] diagramme *m* de production
~-CYCLE - [ind] cycle *m* de production
~DEFICIENCY - [ind] production *f* déficitaire
~DERRICK - [mining] chevalement *m* de production
~FACTOR - [nucl] (in nuclear reactors) facteur *m* de fission thermique
~IN SERIES - [ind] fabrication *f* en série
~INDEX - [ind] indice *m* de production
~INFORMATION AND CONTROL SYSTEM - [autom] système *m* PICS
~LINE - [ind] ligne *f* de production
~OF FILAMENT - [text] formation *f* du fil
~OF LOOM - [text] production *f* du métier
~OF PAPER YARNS - [text] fabrication *f* des fils de papier
~OF PULP ROVING - [text] préparation *f* pour fil de pâte
PRODUCTION OF THE COCOON FILAMENT - [text] production *f* du fil de cocon
~OF THE FLY FRAME - [text] production *f* du banc à broches
~PLANNING - [autom] étude *f* de la production
~REACTOR - [nucl] (a regenerative reactor) réacteur *m* de production
~SCHEDULE - [ind] programme *m* de production
~STRING - [oil ind] colonne *f* d'exploitation
PRODUCTIVE - [gen] productif, générateur
[soil] fertile, riche
PRODUCTIVITY - [gen] productivité *f*
PROFESSIONAL MICROPHONE - [acoust] (a microphone used for broadcasting) microphone *m* professionnel
PROFILATED RUBBER SOLE - [shoe man] semelle *f* caoutchouc façonnée
PROFILE, to - [gen] (to draw profile) profiler
[mech] profiler, contourner
[metall] fraiser en bout, profiler
[carp] moulurer
PROFILE - [gen] profil *m*
[draw] coupe *f* perpendiculaire, configuration *f*
[surv] (longitudinal section) profil *m* altimétrique
[arch] profil *m*
[ceramics] calibre *m* de tourneur
[rubber ind] (of tyres) sculpture *f*, profil *m*, gravure *f*
~BOARD - [mech] silhouette *f*
~CALENDER - [rubber ind] profileuse *f*, gaufreuse *f*
~CUTTER - [tool] fraise *f* à profiler
~CUTTING - [mach tool] profilage *m*
~DEPTH - [rubber ind] (of tyres; tread depth) profondeur *f* de sculpture
~DIE - [plast ind] (extrusion die for producing solid sections) filière *f* à profiler
~DRAG - [aero] (the sum of the drags due to surface friction and form) traînée *f* de profil
~-DRAG POWER LOSS - [aero] (the power consumed by the profile drag of the aerofoils forming the blades of a propeller) puissance *f* absorbée par la traînée de profil
~GRINDER - [mach tool] machine *f* à rectifier les surfaces profilées
~IRON - [metall] fer *m* profilé
~LENGTHWISE - [constr] profil *m* longitudinal
~MACHINE - s. profiling machine
~MAP - [surv] carte *f* des profils de la route
~MIDDLE LINE - [mech] axe *m* du profil d'une dent
~OF BOBBIN - [text] forme *f* de la bobine
~OF FLOW - [constr] profil *m* d'ecoulement
~OF THE CAM - [text] profil *m* de la came
~OF THE ROAD - [constr] profil *m* de route
~OF THE SUBSOIL - [soil] profil *m* du sol de fondation
~OF THE TAPPET - [text] courbe *f* de l'excentrique
~PAPER - [draw] papier *m* quadrillé à millimètres
~TESTING INSTRUMENT - [instr] appareil *m* pour les control des profils
~TURNING - [mach tool] tournage *m* à profiler
PROFILED JOIST - [constr] madrier *m* profilé
~PIN - [text] goupille *f* profilée
~RUBBER - [rubber ind] caoutchouc *m* profilé
PROFILER - s. profiling machine
PROFILING - [mech] fraisage *m* à profiler
[constr] profil *m*
~ATTACHMENT - [mach tool] dispositif *m* à copier

PROFILING LATHE - [mach tool] tour *m* à profiler
~MACHINE - [mach tool] (machine which cuts out shapes from sheet material from a master template) fraiseuse *f* à copier
~ROLLER - [mach tool] touche *f* d'une machine à fraiser
PROFILOMETER - [instr] (instrument designed to determine the profile of the land) profilomètre *m*
PROFLAVINE - [chem] (an acridine derivative with uses as an antiseptic) proflavine *f*
PROGAMY - [genet] progamie *f*
PROGENY - [genet] progéniture *f*, descendance *f*
~TEST - [genet] test *m* de descendance
PROGERIA - [med] progérie *f*, nanisme *m* sénile
PROGESTERONE - [biochem] (a natural hormone with therapeutic uses) progéstérone *f*
PROGESTIN - [chem] (substance promiting the gestational activity of the corpus lutem) progéstérone *f*, hormone *f* lutéinique
PROGLOSSIS - [anat] pointe *f* de langue
PROGLOTTID - [med] anneau *m* de ténia
PROGLOTTIS - s. proglottid
PROGNATHOUS - [zool] (with protruding jaws) prognathique
PROGNOSIS - [med] (the forecast of a probable course of an illness) pronostic *m*
PROGRAM, to - [comput] (the operation of making programme) programmer
PROGRAM - s. programme
PROGRAMME - [contr] (the information which anables a control system to perform a given series of operations) programme *m*
[comput] (the sequence of coded instructions for the computer) programme *m*
[gen] programme *m*
~-ADVANCE - [comput] progression *f* du programme
~AUTHORIZATION - [comm] (authorization granted to auditors to examine the programme arranged for bookkeeping) autorisation *f* du programme
~CARD - [comput] carte-programme *f*
~CHECKING - [comput] programme *m* de contrôle
~CIRCUIT - [radio] circuit *m* pour transmissions radiophoniques
~CIRCUIT LOADING - [radio] charge *f* pour radiodiffusion musicale
~CONTINUITY - [telev] continuité *f* du programme
~CONTROL - [comput] commande *f* par programme
~-CONTROLLED COMPUTER - [comput] calculatrice *f* à programme enregistré
~CONTROLLING ELEMENT - [comput] programmateur *m*
~COUNTER - [comput] (control instruction counter) registre *m* de commande
~DEBUGGING - [comput] élimination *f* des erreurs dans un programme
~DEVELOPMENT TIME - [comput] temps *m* d'essai global
~DISPLAY - [comput] (the operation of punching out the instructions as they are obeyed) représentation *f* du programme
~ELEMENTS - [comput] éléments *mpl* de programme
~EVALUATION AND REVIEW TECHNIQUE - [comput] méthode *f* PERT
~FILLER - [cin] (a fill-up in a cinema programme) film *m* complément de programme
~LEVEL - [comput] niveau *m* de programme, étage *m* de programme

PROGRAMME LEVEL - [telev] niveau *m* des signaux du programme
~LIBRARY - [comput] (or routine library; collection of programmes, routines and subroutines) bibliothèque *f* de programmes
~LISTING - [comput] liste *f* de programme
~LOOP - [radio] (transmission line carrying broadcast programme material) ligne *f* de radiotransmission entre deux stations d'émission
[comput] boucle *f* de programme
~MONITOR - [radio etc] moniteur *m* d'émission
~MUSIC - [mus] (the complementary of absolute music) musique *f* à programme, musique *f* descriptive
~PANEL - [comput] tableau *m* de programmation
~PARAMETER - [comput] paramètre *m* de programme
~REGISTER - [comput] registre *m* de commande
~ROUTING - [telev] acheminement *m* du programme
~-SENSITIVE ERROR - [comput] défaillance *f* due au programme
~-SENSITIVE FAULT - s. programme-sensitive error
~SEQUENCE - [comput] séquence *f* de programme
~SHEET - [comput] feuille *f* programme
~SKIP - [comput] saut *m* de programme
~STATE WORD - [comput] mot *m* d'état du programme
~STEP - [comput] (generally one instruction) pas *m* de programme
~STOP-SWITCH - [comput] clef *f* pour pas de programme
~STORAGE - [comput] mémoire *f* de programme
~SWITCHER - [telev] commutateur *m* du programme
~TAPE - [comput] (the tape containing the sequence of instructions to the computer) bande-programme *f*
~TESTING - [comput] contrôle *m* de programme
~TRANSMITTER - [contr] (apparatus designed to control different values according to a predetermined programme) transmetteur *m* de commande
PROGRAMMED CHECK - [comput] contrôle *m* programmé
~CHECKING - [comput] contrôle *m* programmé
~CONTROL - [contr] (control in which the desired value is automatically changed from time to time in accordance with a predetermined programme) réglage *m* programmé
~FLOATING-POINT OPERATION - [comput] calcul *m* à virgule flottante programmée
~OVERLAPPING CONTROL SAMPLE - [contr] échantillon *m* de commande programmé à recouvrement
PROGRAMMER - [contr] (the apparatus controlling the execution of a programme) programmateur *m*
[comput] (the persons who prepares the sequences of instructions for the computer) programmateur *m*
PROGRAMMING - [gen] (the preparation of a schedule of operations for a process or piece of work) programmation *f*
~LANGUAGE - [comput] langage *m* de programmation
PROGRESS, to - [gen] s'avancer, s'approcher
PROGRESS - [gen] avancement *m*, marche *f* en avant, progrès *m*
~CERTIFICATE - [constr] (certificate made out in favour of the contractor to enable the latter to obtain payment for the work done) certificat *m* d'avancement
~CHART - [gen] (chart showing the continuous record of the work done) graphique *m* de l'avancement
~REPORT - [gen] rapport *m* d'avancement

PROGRESS SCHEDULE - s. progress chart
PROGRESSION - [gen] progression *f*
[math] progression *f*
[surv] (traversing) levé *m* par intersection
[mus] marche *f*, progression *f*
~RATIO - [mech] rapport *m*
PROGRESSIVE - [gen] progressif
~ACTION - [contr] action *f* progressive
~AXIAL MOTION - [mech] course *f* axiale progressive
~BRAKING - [mech] serrage *m* progressif
~COLOUR PROOFS-s. progressive proofs
~DIES - [metall] (in die casting, follow dies) estampe *f* multiple
~INDUCTION SEAM WELDING - [metall] soudage *m* progressif à la molette par induction
~INTERLACE - [telev] analyse *f* entrelacée progressive
~METAMORPHISM - [geol] métamorphisme *m* progressif
~NUMBER - [text] découchement *m*
~PROOFS - [print] (in colour printing; set of proofs supplied to the printer as a guide to colour and registration) épreuve *f* en couleur
~PUNCH PRESS - [mech] presse *f* à découpage progressif
~SCANNING - [telev] (sequential scanning; a rectilinear scanning process) analyse *f* ligne par ligne non entrelacée
~RAISING - [text] élévation *f* progressive
~SPOT WELDING - [metall] soudage *m* progressif par points
~WAVE - [radio] (a wave which is propagated through an infinite homogeneous medium) onde *f* progressive
~-WAVE AERIAL - [radio] (travelling-wave aerial) antenne *f* à ondes progressives
~-WAVE ANTENNA - s. progressive-wave aerial
~WINDING - [el] enroulement *m* progressif
PROIOTIA - [med] puberté *f* précoce
PROJECT, to - [gen] projeter, lancer
[geom] projeter, tracer la projection
[ind chem] verser
[constr] faire saillie, déborder, porter à faux
PROJECT - [gen] projet *m*
PROJECTED AREA - [metall] (the area formed by the geometrical projection of a body on a given surface) surface *m* de moulage
~PEAK POINT - [electron] point *m* de crête projeté
~SCALE INSTRUMENT - [instr] (instrument the scale of which is projected on the a screen) instrument *m* à échelle projetée
PROJECTILE - [firearms] projectile *m*
[astronaut] engin *m* spatial
~LATHE - [mach tool] tour *m* à projectiles
PROJECTING - [gen & arch] saillant, en porte-à-faux, hors d'oeuvre
~EDGES - [mech] extrémités *f*pl saillantes
~LENS - [photo] objectif *m* de projection
~PART - [gen] partie *f* saillante
~STONE - [arch] heurt *m* saillant
~WIRE POINT - [text] cavalier *m*
~WOOL TUFT - [text] tête *f* de la touffe
PROJECTION - [surv] (the representation on a plane surface, e.g. a chart, of features on the surface of the terrestrial sphere) projection *f*
[cin] (of motion pictures) projection *f*
PROJECTION - [math] projection *f*
[arch] ressaut *m*, partie *f* qui fait saillie, forjet *m*, porte-à-faux *m*
[mech] (from a machine etc) mentonnet *m*, téton *m*
[gen] projecture *f*, prolongement *m*
~BOOTH - s. projection box
~BOX - [cin] (a room from which motion pictures are projected) cabine *f* de projection
~COMMUNICATION SYSTEM - [comput] (in data progressing) système *m* de communication à projection
~DISTANCE - [cin] (the distance between the projecting lens and the screen) distance *f* de projection
~FACTOR - [cin] rendement *m* de la tête sonore
~LAMP - [cin] lampe *f* de projection
~LANTERN - [cin] (a two-lens optical system designed to project on to a screen a magnified image of a transparency) lanterne *f* de projection
~LENS - [cin] (the objective lens in a camera projector) lentille *f* du projecteur
~OF CARD CYLINDER - [text] pédonne *f*, cheville *f*
~OF THE NEEDLE - [text] longueur *f* franche de l'aiguille
~ON FRAMING - [text ind] doigt *m* du bâti
~ON SPUR WHEEL - [text] nez *m* sur la roue droite
~OPTICS - [opt] (system of mirrors and lenses used in projection television) optique *f* de projection
~PERIOD - [cin] phase *f* de projection
~PORT - [cin] (opening in the projection box for the light beam) fenêtre *f* de projection
~RECEIVER - [telev] récepteur *m* de télévision à projection
~SCREEN - [cin] (the area receiving the picture from the projector) écran *m*, écran *m* de projection
~SIZE - [telev] dimensions *f*pl de l'image
~TELEVISION - [telev] (television on a screen) télévision *f* à projection
~TUBE - [telev] (cathode-ray producing a television picture suitable for enlarged projection) tube *m* de projection
~WELDING - [metall] (a type of resistance welding limited to predetermined points) soudage *m* par bossages
PROJECTIONIST - [cin] chef-opérateur *m* de cabine, projectionniste *m*
PROJECTIVE - [geom] projective
~FIELD THEORIES - [phys] (unified field theories of gravitation and electromagnetism) théories *f*pl projectives de champ
~GEOMETRY - [geom] géométrie *f* descriptive
~PROPERTY - [math] propriété *f* projective
~TRANSFORMATION - [math] transformation *f* projective
PROJECTIVITY - [math] projectivité *f*
PROJECTOR - [light] projecteur *m*
[cin] (the machine projecting the motion picture on the screen and reproduces the recorded sound) cinéprojecteur *m*
~CHARGING MACHINE - [th eng] (belt charging machine) chargeuse *f* à courroie
~EFFICIENCY - [acoust] (the transmitting of an electroacoustic transducer) rendement *m* de transducteur
~HEAD - [cin] bloc *m* optique
~POWER RESPONSE - [acoust] (transmitting power response) taux *m* de transmission
~RAKE - [cin] (the downward tilt to centre the pic-

ture on the screen) inclinaison *f* du projecteur
PROJECTOR-TYPE FILAMENT LAMP - [el] (filament lamp in which the filament is arranged in concentrated form for focusing purposes) lampe *f* de projection
PROLACTIN - [biol] prolactine *f*, lactostimuline *f*
PROLAMINE - [chem] (a simple vegetable protein) prolamine *f*
PROLAPSE - [med] (dislocation of an organ) prolapsus *m*, protrusion *f*
~ OF THE IRIS - [med] protrusion *f* de l'iris
~ OF THE RECTUM - [med] prolapsus *m* rectal
~ OF THE UTERUS - [med] prolapsus *m* utérin, descente *f* de la matrice
~ OF UMBILICAL CORD - [med] procidence *f* du cordon ombilical
PROLATE - [mech] (elongated in the direction of a line joining the poles) allongé, prolongé
~ CYCLOID - [math] cycloïde *f* allongée
PROLATION - [mus] (relation between semibreve and minium) prolation *f*
PROLEUCEMIA - [med] leucanémie *f*
PROLEUKEMIA - s. proleucemia
PROLIFERATE, to - [gen bot etc] proliférer
PROLIFERATION - [gen bot etc] prolifération *f*
PROLIFEROUS - [gen] prolifère
PROLIFIC - [gen] prolifique, fertile
PROLIFICATION - s. proliferation
PROLINE - [chem] (a biological culture medium) proline *f*
PROLONG, to - [gen] prolonger
PROLONGATION - [gen] prolongation *f*
PROMAZINE HYDROCHLORIDE - [chem] (a tranquillizer and anti-emetic) chlorhydrate *m* de promazine
PROMENADE DECK - [naut] (the upper deck on which passengers can walk) pont-promenade *m*
PROMERISTEM - [bot] (the meristem in an embryo) proméristème *m*
PROMETHAZINE HYDROCHLORIDE - [chem] (an antihistamine drug with local anaesthetic properties) chlorhydrate *m* de prométhazine
PROMETHIUM - [chem] (a rare earth element, symbol Pm. A.N. 6I. It has been prepared synthetically) prométhéum *m*
PROMINENCE - [gen & anat] proéminence *f*, saillie *f*
~ OF THE TWILL - [text] saillie *f* du sillon
PROMINENCES - [astr] (tongues of glowing gas out of the sun's disk) protubérances *fpl* solaires
PROMISCUITY - [gen] promiscuité *f*, mélange *m*
PROMONTORY - [geogr] (in high point of land extending into the sea) promontoire *m*
[anat] (rounded projection) promontoire *m*, protubérance *f*
PROMOTE, to - [gen] promouvoir
[gen] (to contributed to a development) favoriser
[chem] (to increase the action of a catalyst) amorcer, provoquer une réaction
PROMOTED MIXING - [nucl] homogénéisation *f*
PROMOTER - [gen] (in business) promoteur *m*, fondateur *m*
[chem] (a substance which augments the action of a catalyst) activeur *m*, substance *f* activante
PROMPT-CRITICAL - [nucl] (capable of sustaining a chain reaction without the aid of delayed neutrons) critique instantané
~ CRITICALITY - [nucl] criticité *f* instantanée
~-FISSION NEUTRONS - [nucl] (neutrons released with the fission process) neutrons *mpl* instantanés
PROMPT-GAMMA - s. prompt-gamma radiation
~-GAMMA RADIATION - [nucl] (gamma radiation emitted nuclear fission) rayonnement *m* gamma instantané
~ INDUSTRIAL SCRAP - [metall] ferrailles *fpl* de production propre
PROMPTER - [telev] autocue *m*, projecteur *m* du texte
PRONAOS - [arch] (portico of a temple) pronaos *m*
PRONATION - [physiol] (hand and forearm movement by which the palm of the hand is turned downwards and the radius and ulna brought into a crossed position) pronation *f*
PRONATOR - [anat] (the muscle which causes pronation by contracting) pronateur *m*
PRONAUS - [anat] vestibule *m* du vagin
PRONE - [gen] sujet à
PRONEPHROS - [anat] (the anterior portion of the kidney) pronéphros *m*, resin *m* primitif
PRONG - [mech] dent *f*, branche *f*, griffe *f*
[el] (of a plug) pince *f*, broche *f*
[electron] (a base pin) broche *f*
~ KEY - [tool] clef *f* à dents
~-TYPE INSTRUMENT TRANSFORMER - [el] transformateur *m* de courant à pince
PRONGED - [mech] à dents, à fourchons
~-CENTRE - [mach tool] pointe *f* à trois dents
~-CHUCK - [mach tool] mandrin *m* à tulipe, griffe *f*
~ WASHER - [mech] rondelle *f* à dents
PRONY BRAKE - [mech] (a type of absorption dynamometer) frein *m* dynamométrique
PROOF, to - [ind chem] (to render resistant or impervious to a substance or to an action) rendre étanche, imperméabiliser
PROOF - [gen] preuve *f*
[math] épreuve *f*
[print] épreuve *f*
[ind chem] éprouvette *f*
[chem] (spirit contents) preuve *f*
[adj](resistant or impervious against) à l'épreuve de, imperméable
[bookbind] témoins *mpl*
~ AGAINST DECAY - [ind chem] imputrescible
~ BAR - [metall] barre *f* de comparaison
~-BEND TEST - [metall] essai *m* de flexion à moment de flexion spécifié
~ CORRECTION - [print] correction *f* sur épreuves
~ FACTOR - [aero] (the factor of safety derived from the proof load) coefficient *m* d'épreuve
~ FIGURE - [comput] nombre *m* de test
~ HYDROSTATIC TEST - [phys] essai *m* de fonctionnement hydrostatique
~ IN SHEETS - [print] feuille *f* imprimée
~ LOAD - [aero] (a test load such that the structure remains fir for service after its amplification) charge *f* d'épreuve
~ MARK - [firearms] poinçon *m* d'essai
~ NEEDLE - [impl] aiguille *f* d'essai, toucheau *m*
~ PLANE - [el] (piece of conducting material used for receiving or removing charges in electrostatic experiments) plan *m* d'épreuve (pour essai électroscopique)
~ PRESS - [print] presse *f* à épreuves
~ PRESSURE TEST - [rubber ind] essai *m* d'imperméabilité sous pression
~ PULLING - [print] tirage *m* des épreuves
~ SAMPLE - [gen] éprouvette *f* d'essai

PROOF SHEET - [print] épreuve *f*
~SPIRIT - [chem] (alcoholic liquor of proof strength) trois-six *m*
~STRESS - [metall] (the stress required to produce a permanent set in metals lacking a sudden yield point) effort *m* d'essai, limite *f* conventionnelle de élasticité
~TEST - [metall] (test made on a unit under abnormal, but possible, load conditions) essai *m* de réception non-destructif
~TO BURSTING - [firearms] épreuve *f* d'outrance
PROOFCOCK - [sugar ind] robinet-sonde *m*
PROOFED FABRICS - [text] tissus *mpl* imperméabilisés
~SHEETING - [rubber ind] tissu *m* gommé
~TAPE - [el] (the wrapping of rubber-insulated cables) ruban *m* isolant imperméabilisé
PROOFING - [rubber ind] gommage *m*, caoutchoutage *m*
[gen] imperméabilisation *f*
~BY SPREADING - [rubber ind] (rubberizing process) imperméabilisation *f* par enduisage
~MIX - [rubber ind] mélange *m* à enduire, mélange *m* de gommage
PROOFREAD, to - [print] (to correct printer's proofs) corriger les épreuves
PROOFREADER - [print] correcteur *m* d'épreuves
PROOFREADING - [print] correction *f* sur épreuves
PROOFSTICK - [impl] sonde *f* à tiroir, sonde *f* de épreuve
PROP, to - [gen] (to support) soutenir, appuyer
[constr] (to sustain, to keep from declining) étayer, étançonner, chandeller, étrésillonner
[mining] boiser, buter
PROP - [gen] appui *m*, support *m*
[constr] chandelle *f*, étançon *m*, pied-droit *m*, pointal *m*
[mining] taquet *m*, clichage *m*
[agric] paisseau *m*, tuteur *m*, écuyer *m*, échalas *m*
[aero] (colloq. for propeller) hélice *f*
[shipbuild] épontille *f*, béquille *f*
~AND SILL - [mining] montant *m* et semelle
~DRAWER - [mining] déboiseur *m*
~DRAWING - [mining] déboisage *m*
~JET - [aero] turboréacteur *m*
~JET ENGINE - [aero] turboréacteur *m*
~PLOT - [cin telev etc] liste *f* des accessoires
~PULLER - s. drop drawer
~SLICING - [mining] exploitation *f* par tranches avec foudroyage
~STAY - [mining] étai *m*, butte *f*
~THE LOOM FRAMING, to - [text mach] étrésillonner le bâti du métier
~UP, to - [gen & mech] appuyer, soutenir
[constr] étayer
PROPAGATE, to - [gen] propager, disséminer, transmettre
[genet] faire reproduire
[bot] bouturer
PROPAGATE YEAST, to - [brew ind] reproduire la levure (second ensemencement)
PROPAGATED POTENTIAL - [el biol] (wave of change of potential involving depolarization progressing along excitable tissues) potentiel *m* propagé
PROPAGATION - [gen] propagation *f*, reproduction
[radio] (the travel of waves) propagation *f*
[bot] bouturage *m*
[phys] propagation *f* (de la lumière etc)
PROPAGATION BY CUTTINGS - [bot] bouturage *m*
~BY LAYERS - [bot] marcottage *m*
~BY ROOT STOCKS - [bot] propagation *f* par rhizomes
~BY ROOT SUCKERS - [bot] propagation *f* par drageons
~BY SEED - [bot] propagation *f* par graines
~COEFFICIENT - [radio] (the natural logarithm of the vector ratio of the steady-state amplitudes of a wave at a given frequency, at points in the direction of propagation separated by unit length) constante *f* de propagation
~CONSTANT - s. propagation coefficient
~FACTOR - s. propagation ratio
~OF LIGHT - [opt] (the propagation of the transverse electromagnetic space) propagation *f* de la lumière
~PATH - [phys] parcours *m* de propagation
~RATIO - [el] (for a wave from one point to another) taux *m* de propagation
~VELOCITY - [phys] (the speed at which a wave travels through a medium) vitesse *f* de propagation
PROPAGULE - [bot] propagule *m*
PROPAGULUM - s. propagulo
PROPANCREATITIS - [med] pancréatite *f* suppurée
PROPANE - [chem] (the third in the paraffin series. A fuel gas and synthesis intermediate) propane *m*
~-AIR MIXTURE - [gas ind] air *m* propané
PROPANOL - [chem] (normal propyl alcohol) propanol *m*
PROPANTHELINE BROMIDE - [chem] (an anticholinergic agent with therapeutic uses) bromure *m* de propanthéline
PROPARGYL ALCOHOL - [chem] (a synthesis intermediate) alcool *m* propargylique
PROPEL, to - [gen] propulser, actionner, donner une impulsion
[text] (the shuttle) chasser, lancer
PROPELLANT - [gen] (generally, the filling of cartridges) propulseur *m*, propulsif
[astronaut] (a combination of fuel and oxidant) propergol *m*
~GRAIN - [astonaut] grain *m* de propergol
PROPELLED - [gen] poussé en avant, mû, actionné
PROPELLER - [aero] (a power-driven system or radial aerofoils, designed to develop thrust when rotated in air) hélice *f*
[naut] (power-driven device, having blades mounted on a rotating shaft, propelling a craft through water) hélice *f*, propulseur *m* à hélice
[text] taquet *m*, chasse-navette *m*, rat *m*
~APERTURE - [naut] (the opening in which the propeller rotates) puits *m* de l'hélice
~BALANCING STAND - [mech] (apparatus used for the final balancing of propeller) appareil *m* pour l'équilibrage des hélices
~BLADE - [mech aero & naut] pale *f* d'hélice, aile *f* d'hélice
~BLADE ANGLE - [aero] (the angle between the chord of a propeller blade and the plane of rotation) angle *m* de calage de la pale
[aero] (in helicopters; the angle between the no-lift direction and the plane of rotation) déport *m* dans le plan de rotation
~BLADE AREA - [aero] superficie *f* de la pale de l'hélice
~BLADE ASPECT RATIO - [aero] (the relation of the tip radius to the maximum blade width of a propel-

ler blade) allongement *m* de la pale de l'hélice
PROPELLER BLADE LOADING - [aero] (in helicopters, the quantity obtained by dividing the rotor thrust by the total blade area) charge *f* de la pale
~BLADE SWEEP - [aero] (if one line is drawn through the centroids of the blade sections, and another tangentially to it from the propeller axis, the angle of rotation, is termed the blade swep) compensation *f* dans le plan de rotation
~BOSS - s. propeller hib
[mech etc] moyeu *m* d'hélice
~CAVITATION - [naut] (the formation of cavities in the water around the blades of the propeller, caused by excessive speed of rotation and resulting in loss of efficiency) cavitation *f* de l'hélice
~DE-ICER - [aero] (device for detaching ice from propeller blades) déguvreur *m* des pales de l'hélice
~DISK - [aero] (the circle described by the blade tips) disque *m* de l'hélice
~DISK-AREA - [aero] (the area swept by the baldes of a propeller) surface *f* du cercle balayé par les pales de l'hélice
~EFFICIENCY - [mech] (the relation between thrust horse-power and torque-power) rendement *m*
~FAN - [mech] (fan consisting of a rotor carrying blades of air-screw farm working in a cylindrical casing) ventilateur *m* hélicoïdal
~HUB - [mech] moyeu *m* d'hélice
~MIXER - [mech] (mixing machine in which the agitating elements is in the form of a screw propeller) agitateur *m* à hélices
~NOISE - [aero] bourdonnement *m* de l'hélice
~PITCH - [aero] (the distance which any given element of a propeller would advance in one complete revolution if it were moving along a helix of angle equal to the blade angle) pas *m* géométrique
~POST - [naut] étalingure *f* avant
~PUMP - [mech] pompe *f* à hélice
~RACE - [aero] écoulement *m* (aérodynamique) de l'hélice
~RAKE - [aero] déport *m* de la pale dans le plan de avancement de l'avion
~REDUCTION GEAR - [aero naut] engrenages *mpl* de reduction de l'hélice
~REMOVER - [aero] tire-hélice *m*
~ROOT - [aero naut] (that part of a propeller blade adjacent to the hub) racine *f* de la pale
~SHAFT - [auto] (the driving shaft, conveying the engine power) arbre *m* de transmission
[aero] arbre *m* porte-hélice
~SHAFT HOUSING - [auto] tunnel *m* de l'arbre de transmission
~SHAFT SLEEVE - [auto] manchon *m* de l'arbre de transmission
~SHAFT TAILPIECE - [aero] queue *f* de l'arbre porte-hélice
~SLIP - [aero] recul *m* de l'hélice
~SPEED CONTROL - [aero] réglage *m* de la vitesse de l'hélice
~THRUST -[aero naut] (the axial force developed by a propeller) traction *f*
~TIPPING - [aero] (protective covering on the tips of some types of propeller) recouvrement *m* de protection de l'hélice
~TORQUE - [aero naut] couple *m* de l'hélice
~TURBINE - [aero] turbine *f* pour propulseur à l'hélice
PROPELLER TURBINE ENGINE - [aero] (a gas turbine engine in which part of the power output is used to drive a propeller) turbopropulseur *m*
~TYPE FAN - s. propeller fan
~-TYPE WATER TURBINE - [mech] (water turbine fitted with a runner resembling a four-blade ship's propeller) turbine *f* hydraulique à hélice
~WASH - [aero] (the stream of air flowing aft from the propeller) remous *m* d'air de l'hélice
PROPELLING - [gen etc] propulsif
~CHARGE - [astronaut] charge *f* de propergol
~FORK - [text] fourche *f* de transport
~MOVEMENT - [railw] marche *f* en refoulement
~NOZZLE - [mech] (the nozzle at the after end of the exhaust cone or jet pipe) buse *f* propulsive
~POWER - [naut] puissance *f* propulsive
~RAKE - s. propelling fork
PROPEPTONURIA - [med] propeptonurie *f*
PROPER - [gen] propre, convenable, juste
~FRACTION - [math] fraction *f* propre
~MOTION - [astr] (the component of the motion in space of a star, which is at right angle to the line of sight) mouvement *m* propre
[radar] (the speed and course of one's own vessel or the other vessel, during a plotting interval as represented by the construction lines on the plotting diagram) mouvement *m* propre
~OPERATION - [el] fonctionnement *m* correct
~WOOL - [text] poil *m* (exclusivement) laineux
PROPERTIES - [cin & theatre] décors *mpl*, reserve *f* de décors
PROPERTY - [gen] propriété *f*
[phys chem etc] (a characteristic) propriété *f*, faculté *f*
[leg] biens *mpl*
~MAN - [cin & theatre] accessoiriste *m*, costumier *m*
~MANAGER - [cin telev theatre] ensemblier *m*
~SORT - [comput] (the selection of records from a file following a given criterion) sélection *f* suivant propriétés spécifiques
~TAX - [fin] impôt *m* foncier
PROPHASE - [genet] (the preliminary stages of mitosis or meiosis leading to the formation of the astroid) prophase *f*
PROPHYLACTIC - [med] (preventing, or protecting against, disease) prophylactique
PROPHYLAXIS - [med] (preventive treatment) prophylaxie *f*
PROPINE - [chem] (allylene) allylène *m*
PROPIOLACTONE - [chem] (a synthesis intermediate) propiolactone *f*
PROPIOLIC ACID - [chem] (or propine acid; it forms an explosive in silver salt) acide *m* propiolique
PROPIONALDEHYDE - [chem] (an intermediate for rubber chemicals and synthetic resins) propionaldéhyde *m*
PROPIONIC ACID - [chem] (an intermediate for pharmaceuticals, flavourants and perfumery agents) acide *m* propionique
~ANHYDRIDE - [chem] (an intermediate for synthetic resins, drugs, and dyes) anhydride *m* propionique
PROPIONYL GROUP - [chem] groupe *m* propionyle
PROPIOPHENONE - [chem] (a synthesis intermediate for pharmaceuticals and other compounds) propiophénone *f*

PROPLEX – [med] plexus choroïde latéral
PROPLEXUS – s. proplex
PROPORTION, to – [gen] proportionner
[gen] (to apportion) répartir, distribuer
[chem] doser
[mech etc] déterminer les dimensions
[mining] compasser (les feux)
PROPORTION – [gen] proportion *f*, partie *f*, portion *f*
[math] (identity between rations) proportion *f*
~OF ACID – [chem] (acid concentration) dosage *m* d'acide
~OF CELLULOSE – [text] teneur *f* en cellulose
~OF FAT – [text] teneur *f* en graisse
~OF FIBRE – [text] teneur *f* en filasse
~OF GREASE – [text] teneur *f* en suint (de la laine crue)
~OF POLARIZATION – [opt] proportion *f* de polarisation
~PUMP – [ind chem] pompe *f* de dosage
PROPORTIONAL – [gen] proportionnel
[math] proportionnelle *f*
~ACTION – [contr] (a method of operation in which the output signal variations are proportional to the corresponding variations in the input signals) action *f* proportionnelle
~ACTION COEFFICIENT – [contr] coefficient *m* de action proportionnelle
~BAND – [contr] (in a proportional-position controller) zone *f* proportionnelle
~CONTROL – [contr] régulation *f* P, régulation *f* proportionnelle
~CONTROL FACTOR – [contr] (the ratio between the potential correction and the deviation in proportional control) facteur *m* proportionnel de réglage
~CONTROLLER – [contr] régulateur *m* à action proportionnelle
~COUNTER – [instr] (instrument fitted with a proportional counter tube and its associated circuits) compteur *m* proportionnel
~COUNTER TUBE – [instr] (radiation counter tube operated in the proportional region) tube *m* compteur proportionnel
~DIVIDERS – [draw] compas *m* de proportion
~LIMIT – [mech] (the maximum unit stress which can be obtained in a structural material without causing a change in the ratio between the unit stress and the unit deformation) limite *f* proportionnelle
~MEAN – [math] moyenne *f* proportionnelle
~-PLUS-DERIVATIVE ACTION – [contr] action *f* PD, action *f* proportionnelle et par dérivation
~-PLUS-INTEGRAL CONTROLLER – [contr] (a two-term controller) régulateur *m* à action proportionnelle et par intégration
~-POSITION ACTION – [contr] réglage *m* à action proportionnelle
~REGION – [nucl] (the range of operating voltage for a counter tube or ionization chamber in which the gas amplification is greater than one and independent of the primary ionization) région *f* de proportionnalité
PROPORTIONATING PISTON – [plast ind] piston *m* doseur, sabot *m* de dosage
PROPORTIONALITY – [gen] (the property where by corresponding values a proportion) proportionnalité *f*
PROPORTIONATE, to – [gen] (to make proportional) proportionner
PROPORTIONING COAL MIXTURES – [gas ind] dosage *m* (d'un mélange de charbon)
~ELEMENTS – [contr] (the elements combining two or more signals of the same physical kind in preselected proportion) organes *mpl* de distribution
~PUMP – [hydr & ind chem] pompe *f* de dosage, pompe *f* doseuse
PROPOSITION – [gen] affaire *f*, entreprise *f*
[geom] proposition *f*, problème *m*
PROPPING – [gen] soutènement *m*
[constr] étayage *m*, étrésillonnement *m*
[shipbuild] épontillage *m*
[mining] consolidation *f*
PROPRIETARY – [gen] (pertaining to a proprietor) de propriétaire
[comm etc] (an article protected by a registered name) de propriété, breveté
~ARTICLE – [comm] spécialité *f*, article *m* breveté
~MEDICINE – [med] spécialité *f* pharmaceutique
~STIMULATION – [physiol] sensation *f* proprioceptive
PROPRIOCEPTOR – [physiol] (sensory receptor of stimuli produced within the organism) récepteur *m* proprioceptif
PROPS – [mining] bois *m* de mine
PROPTOSIS – [med] (displacement of part of the body, in particular of the eye) prolapsus *m*
PROPULSION – [gen] (the operation of propelling) propulsion *f*
[cin] (of a film) entraînement *m* (du film)
~ENGINE – [mech] moteur *m* principal
~-FAN – [mech] ventilateur *m* volumogène, ventilateur *m* déplaceur
PROPULSIVE – [gen & mech] propulsif, propulseur
~EFFICIENCY – [aero naut] (the relation between propulsive h.p. and torque h.p.) rendement *m* propulsif
~THRUST – [aero] (the thrust of propeller and nacells or fuselage taken as a combination, plus the force required to move the nacelle or fuselage at the same speed in the absence of the propeller) traction *f* propulsive
PROPYL – [chem] (a radical; derived from propane) propyle *m*
~ACETATE – [chem] (a synthesis intermediate, solvent for natural and synthetic resins and cellulose derivatives, and a perfumery agent and flavourant) acétate *m* de propyle
~ALCOHOL – [chem] (a solvent; for cellulose derivates, natural and synthetic resins, and a synthesis intermediate) alcool *m* propylique
~BUTYRATE – [chem] (a solvent for cellulose derivates) butyrate *m* de propyle
~CHLOROSULPHONATE – [chem] (a synthesis intermediate) chlorosulfonate *m* de propyle
~HYDROXYBENZOATE – [chem] (propylparaben. An antifugal and antibacterial agent) hydroxybenzoate *m* de propyle
~MERCAPTAN – [chem] (a synthesis intermediate) propyl-mercaptan *m*
~PROPIONATE – [chem] (a solvent for cellulose derivates) propionate *m* de propyle
PROPYLTHIOURACIL – [chem] (a drug used in the control of thyrotoxicosis) propylthiouracil *m*
PROPYLAEUM – [arch] (the structure forming a large entrance before an encient temple) propylée *m*
PROPYLAMINE – [chem] (a synthesis intermediate)

propylamine *f*
PROPYLENE - [chem] (a colourless, flammable gas with uses as a synthesis intermediate) propylène *m*
~CARBONATE - [chem] (a colourless odourless liquid with application as a synthesis intermediate, plasticizer, and solvent) carbonate *m* de propylène
~CHLOROHYDRIN - [chem] (a synthesis intermediate) chlorohydrine *f* propylénique
PROPYLENEDIAMINE - [chem] (a colourless, hydroscopic liquid with uses as an intermediate for rubber chemicals, and dyes) propylène-diamine *f*
PROPYLENE DICHLORIDE - [chem] (colourless flammable liquid with uses as a solvent) bichlorure *m* de propylène
~GLYCOL - [chem] (colourless, hydroscopic, viscous liquid with application as a solvent, antifreeze agent and synthesis intermediate) propylène-glycol *m*
~GLYCOL MONORICINOLEATE - [chem] (a solvent and plasticizer) monoricinoléate *m* de propylène-glycol
~GLYCOL PHENYL ETHER - [chem] (a perfumery fixative, solvent, bactericide, and intermediate) éther *m* phénylique de propylène-glycol
~OXIDE - [chem] (a colourless flammable liquid uses as an intermediate) oxyde *m* de propylène
PROPYLHEXEDRINE - [chem] (a sympathomimetic amine used in medicine) propylhexédrine *f*
PROPYLITE - [min] (variety of andesite altared by the action of hot water) propylite *f*
PRORATE, to - [gen] (only USA; to allocate, to apportion) répartir proportionnellement
PRORATED PRODUCTION - [oil ind] production *f* réglementaire
PRORATION - s. prorationing
~AGREEMENT - [oil ind] convention *f* de rationnement
PRORATIONING - [oil ind] (only USA; restriction of production by a Government commission, generally on the strength of market demand) rationnement *m*
[gen] (only USA; a proportional allocation) répartition *f* proportionnelle
PROSCENIUM - [theatre] (front part of the stage) proscénium *m*, avant-scène *f*
~LIGHTS - [theatre] lumières *f*pl d'avant-scène
PROSECTOR - [med] (the person who dissects dead bodies for the purpose of anatomical teaching) prosecteur *m*
PROSECUTE, to - [leg] poursuivre (en justice), traduire (en justice)
PROSECUTION - [leg] poursuites *f*pl judiciaires, accusation *f*
PROSECUTOR - [leg] poursuivant *m*, plaignant *m*
[leg] (public prosecutor) procureur *m* (du Roi, de la République)
PROSENCEPHALON - s. prosocoele
PROSOCOELE - [anat] cavité *f* du prosencéphale
PROSOPANTRITIS - [med] sinusite *f* frontale
PROSOPOSPASM - [med] spasme *m* facial, tic *m* convulsif
PROSPECT, to - [min] (to explore for minerals, oil etc) prospecter
PROSPECT HOLE - [mining] trou *m* de recherche
PROSPECTING - [min] recherches *f*pl, prospection *f*
~PIT - [min] trou *m* de prospection
~SHAFT - s. prospecting pit
~SITE - [constr] reconnaissance *f* du terrain
PROSPECTIVE CURRENT OF A CIRCUIT - [el] courant *m* propre d'un circuit
PROSPECTOR - [min] prospecteur *m*
PROSPERMIA - [med] éjaculation *f* précoce
PROSTATALGIA - [med] prostatalgie *f*
PROSTATAUXE - [med] hypertrophie *f* de la prostate
PROSTATE - [anat] s. prostate gland
~GLAND - [anat] glande *f* prostatique, prostate *f*
PROSTATECTOMY - [med] prostatectomie *f*
PROSTATERIA - [med] prostatisme *m*
PROSTATIC CALCULUS - s. prostatolith
PROSTATISM - s. prostateria
PROSTATODYNIA - s. prostatalgia
PROSTATOLITH - [med] calcul *m* prostatique
PROSTRATION - [med] prostration *f*, abattement *m*
PROSTYLE - [arch] (having a frontal row of columns) prostyle *m*
PROTACTINIDES - [phys] (term used for the elements of atomic number 89 to I03 when they are in the oxidation state +5) protactinides *f*pl
PROTACTINIUM - [chem] (a radioactive element, symbol Pa, A.N. 9I. It occurs in all ores of uranium) protactinium *m*
PROTALBUMOSE - [biol] protéose *f* primaire
PROTEAN - [anat] protéen, protéiforme
PROTEASE - [chem] (generic name for proteinsplitting enzymes) protéase *f*
PROTECT, to - [gen] protéger, défendre
PROTECTED - [gen] protégé
[mech etc] (of machines and apparatus) protégé, caché, cuirassé
~MOTOR - [el] moteur *m* cuirassé
~SCREW - [mech] vis *f* cachée
~SWITCH - [el] (protected from accidental contract) interrupteur *m* protégé
PROTECTING BAND - [rubber ind] bande *f* de fond de jante, flap *m*
~BOARD - [text] planchette *f* de protection
~CAP - [el] capot *m* de protection
~CASE - [el] enveloppe *f* protectrice
~CHOKE - [el] bobine *f* d'arrêt
~DISK - [photo] disque *m* de protection
~GLASS - [photo] verre *m* protecteur
~GRID - [photo] grille *f* protectrice
~NETWORK - [el] réseau *m* protection
~PLATE - [text] plaque *f* de fermeture
~ROLLER - [text] cylindre *m* entasseur
~SHIELD - [el] plaque *f* de protection
PROTECTION - [gen] protection *f*, défense *f*, abri *m*
~AGAINST DISEASES - [agric] protection *f* contre les épizooties
~AGAINST DROUGHT - [agric] protection *f* contre la sécheresse
~AGAINST EROSION - [soil] protection *f* contre l'érosion
~AGAINST HAIL - [agric] protection *f* contre la grêle
~BUSHING - [mech] manchon *m* de protection
~CAP - [el] (metal structure fitted to the frame of a machine so as to enclose the end and afford protection from accidental contact with the rotating parts) cage *f* protectrice
~CHANNEL - [comput] voie *f* de protection
~COVER - [text] couvercle *m* protecteur
~FOR INTERTURN SHORT-CIRCUITS - [el] dispositif *m* de protection contre les courts-circuits entre spires

PROTECTION HOLE - [mining] sondage *m* à l'avancement
~OF ANIMALS - [zool] protection *f* des animaux
~OF LABOUR - [gen] protection *f* du travail
~OF LEASEHOLDERS - [agric] contrôle *m* de baux à ferme
~OF NATURE - [gen] protection *f* des sites, protection *f* de la nature
~OF SOL - [chem] (protection of lyophobic systems against the coagulating effects of electrolytes) protection *f* du sol
~SURVEY - [radiat] contrôle *m* de protection
~WAGON - [railw] wagon *m* de protection
PROTECTIONISM - [gen] (the economic doctrine of protection) protectionnisme *m*
PROTECTIVE - [gen] protecteur
[gen] (measure insuring or intending to ensure protection) protecteur *m*, préservatif *m*
~APRON - [nucl] (a lead-rubber apron) tablier *m* en caoutchouc au plomb
~BARRIER - [nucl] écran *m* protecteur
~CAPACITOR - [el] condensateur *m* de protection
~CASING - [el] enveloppe *f* protectrice
~CIRCUIT - [el] circuit *m* de protection
~COATING - [chem] (layer of relatively inert substance to prevent chemical attack) couche *f* protectrice, enduit *m* protecteur
~COLLOID - [chem] (a liophilic colloid stabilizing a liophobic solution when present in small amounts) colloïde *m* protecteur
~DEVICE - [telecomm & el] (device designed to protect a piece of electrical equipment; a protector) dispositif *m* protecteur
~GAP - [el] (a surge diverter) éclateur *m* de protection
~GEAR - [mech] appareillage *m* de protection
~GLOVES - [radiat] (gloves incorporating lead rubber) gants *mpl* protecteurs
~GROUNDING - [el] installation *f* de mise à la terre de protection
~HEAD LEADER - s. protective leader
~HORN - [el] (an element of a spark-gap; conductor in the shape of a horn so arranged that it protects an insulator from damage by a power-arc) corne *m* de protection, électrode *f* de garde
~LEAD-GLASS - [radiat] (glass containing a high proportion of lead compounds) verre *m* au plomb protecteur
~LEADER - [cin] (a length of film preceding the live part) bande *f* amorce
~MATERIAL - [nucl] (material protecting against ionizing radiation) matière *f* protectrice
~MEMBRANE - [el] membrane *f* de sécurité
~REACTANCE COIL - [el] inductance *f* de protection
~REACTOR - [el] bobine *f* de protection
~RELAY - [el] relais *m* de protection
~RESISTANCE - [el] (resistance designed to keep large voltage, currents or power out of a given part of a circuit) résistance *f* de protection
~SCREEN - [radiat] (screen which contains elements for the absorption of radiation) écran *m* protecteur
~SLEEVE - [rubber ind] (of tyres) emplâtre *m*
[mech] manchon *m* de protection
~SYSTEM - [el] (system protecting an electrical installation from the effects which may result from over-voltage, current-surge etc by isolating the faulty section) système *m* protecteur
PROTECTIVE TAIL LEADER - s. protective trailer
~TRAILER - [cin] (a length of film at the end of the live part) amorce *f* en fin de bonine
~VALUE - [radiat] (lead equivalent) équivalent *m* de plomb
PROTECTOR - [gen] protecteur *m*
[el] (protective device) dispositif *m* protecteur
[constr & hydr] (any structure aiming at protecting against damage) dispositifs *mpl* de protection
~FRAME - s. protector rack
~FUSE - [el] fusible *m* protecteur
~GROUND - [el] mise *f* à la terre de protection
~RACK - [el] écran *m* des dispositifs protecteurs
~TUBE - [el] (expulsion gap) éclateur *m* à expulsion
PROTEIN CONTENT - [chem] teneur *f* en protéines
~FIBRE - [ind chem] fibre *f* de protéine
~GRAINS - [biol] aleurone *f*
~MATERIAL - [chem] matières *fpl* azotées
~TURBIDITY - [brew ind] trouble *m* de protéine
PROTEINS - [chem] (complex compounds containing carbon, hydrogen, oxygen, nitrogen and sulphur, and in specific cases other elements, and forming the chief constituent of protoplams. A number of proteins can now be synthesized) protéines *fpl*
PROTEINURIA - [med] protéinurie *f* albuminurie *f*
PROTEOLYTIC - [chem] (of enzymes causing the breakdown of proteins simpler substances) protéolytique
PROTEOPEPSIS - [med] digestion *f* des protéines
PROTEROGYNY - s. protogyny
PROTEURIA - s.proteinuria
PROTHALLIAL CELL - [bot] prothalle *m*
PROTHROMBIN - [chem] (protein-like substance in the blood plasma) prothrombine *f*
PROTIUM - [nucl] (the name which is sometimes applied to the hydrogen isotope of mass I) protium *m*
PROTOACTINIUM - s. protactinium
PROTOCHLORIDE - [chem] protochlorure *m*
PROTOCHLORIDE OF MERCURY - [chem] (mercurous chloride used in medicine) chlorure *m* mercureux, calomel *m*
PROTOCHONDRIUM - [med] cartilage *m* primitif
PROTOCOL - [gen] protocole *m*
PROTODERM - [bot] (dermatogen) dermatogène
PROTOGAMY - [biol] (the union of gametes without fusion of their nuclei) protogamie *f*
PROTOGENIC - [chem] (capable of supplying a hydrogen ion) protogène
PROTON - [phys] (a positively-charged particle, the charge of which is equal to that of an electron. It has about 1840 times the mass of the electron and so constitutes the greater part of the mass of an atom) proton *m*
~BINDING ENERGY - [phys] (the energy which is required to remove one proton from a nucleus) énergie *f* de liaison d'un proton
~MICROSCOPE - [instr] (a device which is similar to the electron microscope but in which the charged particles are protons) microscope *m* protonique
~NUMBER - [phys] numéro *m* atomique, nombre *m* de protons
~-PROTON CHAIN - [nucl] (a series of thermonuclear reactions initiated by a reaction between two protons) réaction *f* en chaîne proton-proton
~-SYNCHROTRON - [nucl] (cosmotron) cosmotron *m*, proton-synchrotron *m*
PROTOPATHIC - [med] (which responds only pronounced stimuli) protopathique

PROTOPATHIC SENSATION - [med] (sensation caused by the generally vague response to a pronounced stimulus) sensation *f* protopathique
PROTOPATHY - [med] (a primary disease, a lesion) protopathie *f*
PROTOPLASM - [biol] (the basis of all life; a transparent semi-fluid of complex chemical) protoplasme *m*
PROTOTYPE - [gen] (accepted standard) prototype *m*
~AIRCRAFT - [aero] (the original model of a series of aircraft) avion *m* prototype
~TESTS - [el] (the whole of the tests on a single unit to ascertain that the type complies with the specified operation requirements) essais *mpl* de qualification
PROTOVERATRINE - [chem] (a mixture of alkaloids derived from Veratrum album and used in the treatment of hypertension) protovératrine *f*
PROTOXIDE OF IRON - [chem] (ferrous oxide) protoxyde *m* de fer
PROTOZOA - [zool] protozoaires *mpl*
PROTRACT, to - [gen] prolonger, allonger
PROTRACTED TREATMENT - [radiat] étalement *m* de la dose
PROTRACTION - s. protracted treatment
PROTRACTOR - [instr] (instrument designed to measure and lay down angles) rapporteur *m*
PROTRUDE, to - [gen] pousser en avant, avancer, sortir
PROTRUDING-TYPE INSERT - [plast ind] broche *f* dépassante
PROTRUSION - [gen] sortie *f*, saillie *f*
[mech] saillie *f*, protubérance *f*
[anat] protrusion *f*
PROTUBERANCE - [gen] (anything that protrudes) protubérance *f*
~SHAPED CUTTER - [mech] outil *m* de limeuse à secteur denté
PROUSTITE - [miner] (an ore of silver, essentially silver-arsenic sulphide, commonly associated with pyrargyrite it crystallizes in the trigonal system) proustite *f*
PROVABLE - [gen] prouvable, démontrable
PROVE, to - [gen] éprouver
[math] faire la preuve, vérifier
[print] (to take a proof of or from) tirer une épreuve d'essai
[gen] (to put a test) prouver, démontrer
[gen] (to determine a test) prouver
[gen] (to determine capacity or power) vérifier
PROVENDER - [agric] fourrages *mpl* grossiers
PROVIDE, to - [gen] fournir, pourvoir (à)
PROVING - [gen] épreuve *f*, essayage *m*, démonstration *f*
[print] tirage *m* d'une épreuve d'essai
[mining] reconnaissance *f*
~CIRCUIT - [el] circuit *m* de contrôle
~HOLE - [mining] sondage *m* de recherche
~OF OPENING OF SWITCH BLADES - [railw] contrôle *m* d'entrebâillement d'aiguille
~PRESS - [print] presse *f* à tirer les épreuves de essai
~PROVISION - [gen] (a measure, or set of measure) provision *f*, prise *f* des dispositions nécessaires
[food] vivres *mpl*, comestibles *mpl*
[leg] article *m*, clause *f*, stipulation *f*
PROVISIONAL - [gen] provisoire
PROVISO - [gen] clause *f* conditionnelle
PROVISORY - [gen] (which depends on a proviso) conditionnel
PROVITAMIN - [chem] (any substance which may promote the formation of vitamins) provitamine *f*
PROZONE - [med] phénomène *m* de zone
PROW - [naut] proue *f*
[aero] (the nose of an airship) proue *f*
PROXIMAL - [biol] (nearest to the point of attachment) proximal
PROXIMITY - [gen] proximité *f*
~EFFECT - [el] (in a conductor, the change in current distribution caused by an adjacent conductor) effet *m* de proximité
PROXY - [leg] (any person which is given the authority to act behalf of another person) mandataire *m*, délégué *m*
(the actual instrument granting the authority) procuration *f*, mandat *m*
PRUDENT LIMIT OF ENDURANCE - [aero] (the time period during which an aircraft can remain in flight without reducing the fuel safety margin) autonomie *f*
PRUNE, to - [agric] écimer, élaguer, émonder, ébrancher
PRUNING - [agric] (to cut off superfluous branches) taille *f*, émondage *m*, ébranchage *m*
~HOOK - s. pruning knife
~KNIFE - [agric] serpette *f*
~SAW - [agric] ciseaux *mpl* de jardin
~SHEARS - [agric] sécateur *m*
PRURIGO - [med] dermatite *f* atopique
PRUSSIAN BLUE - [paint] (deep blue pigment, basically ferriferrocyanides of potassium sodium or ammonium) bleu *m* de Prusse
~BROWN - [paint] (brown iron oxide pigment obtained by heating Prussian Blue to decomposition) brun *m* de Prusse
PRUSSIATE - [chem] (a salt of prussic acid) prussiate *m*
PRUSSIC ACID - [chem] (solution of hydrogen cyanide in water) acide *m* prussique, acide *m* cyanhydrique
PRY THE BEAD OF THE RIM, to - [auto] décoller le talon
~THE BEAD OVER THE FLANGE, to - [auto] forcer le talon par-dessus le bord de jante
PRYING IRON - [auto] décroche-pneus *m*, burin *m* de décollage
PSALTERY - [mus] (an ancient string instrument) psaltérion *m*
PSAMMITE - [min] (fine-grained sandstone) psammite *m*
PSAMMITIC - [geol] psammitique
~GNEISS - [geol] (gneissose rock produced by the metamorphism of fine-grain sand sediments) gneiss *m* psammatique
~SANDSTONE - [geol] grès *m* psammite
~SCHIST - [geol] (a schist formed by arenaceous sedimentary rocks) schiste *m* psammite
PSAMMOTHERAPY - [med] arénation *f*, psammathérapie *f*
PSEUDOARTHROSIS - s. pseudoarthosis
PSEUDESTHESIA - [med] pseudesthésie *f*
PSEUDO-ACID - [chem] (a compound having two tautomeric forms; one of which acts as an acid) pseudo-acide *m*

PSEUDOADIABATIC EXPANSION - [met] détente *f* pseudoadiabatique
PSEUDOALUMS - [chem] (term denoting double sulphates of the alum type) pseudo-alumine *f*
PSEUDOARTHROSIS - [med] pseudoarthrose *f*, fausse articulation *f*
PSEUDOASIMMETRY - [chem] (of an atom) pseudo-asymétrie *f*
PSEUDO-BASE - [chem] (a compound having two tautomeric forms, one of which acts as an acid) pseudo-base *f*
PSEUDOBLEPSIA - s. pseudopsia
PSEUDOCARP - [bot] (a false fruit) pseudocarpe *m*
PSEUDOCHROMESTHESIA - [med] audition *f* colorée
PSEUDOCODE - [comput] (abstract code) pseudo-instruction *f*
PSEUDOCOELE - [zool] (the space between the inner walls of the opposed cerebral hemispheres) cavité *f* septale, ventricule *m* de la cloison
PSEUDOCROUP - [med] faux croup *m*, laryngite *f* striduleuse
PSEUDOCRYSTALLITES - s. pseudocrystals
PSEUDOCRYSTALS - [crystall] (in organic fibres) pseudo-cristaux *mpl*
PSEUDOCUMENE - [chem] (an intermediate for synthetic perfumes and dyestuffs) pseudo-cumène *m*
PSEUDO-EQUALIZING PULSES - [telev] pseudosignaux *mpl* d'égalisation
PSEUDOEQUIVALENT TEMPERATURE - [met] température *f* adiabatique équivalente
PSEUDOESTHESIA - s. pseudesthesia
PSEUDO-FIELD-SYNC PULSES - [telev] pseudoimpulsions *fpl* de synchronisation de trame
PSEUDOFRONT - [met] (term used for a temperature gradient or group of weather conditions when there is some doubt as to whether the term "front" is fully justified) pseudofront *m*
PSEUDOGESTATION - s. pseudopregnancy
PSEUDOHALOGEN - [chem] pseudo-halogène *m*
PSEUDOHYDRONEPHROSIS - [med] périnéphrose *f* traumatique
PSEUDO-IMAGE - [radar] pseudoimage *f*
PSEUDO-LOCK - [telev] (in colour television) pseudoenclenchement *m*
PSEUDOLOGIA - [med] fabulation *f*
PSEUDOMANIA - [med] manie *f* du mensonge
PSEUDOMONOTROPY - [phys chem] (the existence of a substance in a plurality of forms, all but one being unstable under all ordinary conditions) pseudomonotropie *f*
PSEUDOMORPH - [min] (a mineral formed by molecular replacement from another, and hence not exhibiting its normal conformation. Such a mineral is said to be "pseudomorphic after" the other, from which it was formed) pseudomorphe
PSEUDOMORPHIC - [min] (having the characteristic of a pseudomorph) pseudomorphe
PSEUDONUCLEOLUS - [biol] pseudonucléole *m*
PSEUDOPARALYSIS - s. pseudoplegia
PSEUDOPLEPSIA- [med] pseudoparalysie *f*
PSEUDOPREGNANCY - [med] gressesse *f* nerveuse, grossesse *f* imaginaire
PSEUDOPSIA - [med] pseudoblepsie *f*
PSEUDORACEMIC - [chem] (term used for a mixture of crystals of the d- and I- optically-active forms of a compound) pseudo-racémique
PSEUDORANDOM SEQUENCE - [comput] (said of numbers which can be produced by a definite calculation process, yet satisfy one or more of the standard test for statistical randsomness) séquence *f* pseudoaléatoire
PSEUDOSCALAR QUANTITY - [meas] (quantity represented by a single numerical value which associates it with a unit but the sign of which depends on the orientation of the axis) grandeur *f* pseudoscalaire
PSEUDOSCOPIC VIEW - [radar] (reversed stereoscopic view in which near objects seem to be distant and viceversa) vue *f* pseudoscopique
PSEUDOSOLUTION - [chem] (colloidal solution or suspension) pseudosolution *f*
PSEUDOSPHERE - [geom] pseudosphère *f*
PSEUDOSPHERICAL - [geom] pseudosphérique
PSEUDOSTEREOSCOPIC EFFECT - [cin] (an optical illusion which occurs when one looks at rotating wheels reproduced on the screen) effet *m* stéréoscopique
PSEUDOSYMMETRY - [crystall] (term relating to minerals having symmetry elements which place them on the borderline between two crystal systems) pseudosymétrie *f*
PSEUDOVECTOR - [math] vecteur *m* axial
~COUPLING - [phys] couplage *m* pseudovectoriel
PSEUDOVENTRICULE - [med] ventricule *m* de la cloison
PSEUDOYEAST - [brew ind] fausse levure *f*
P-SHELL - [nucl] (the collection of electrons which is characterized by the principal quantum number of 6) couche *f* P
PSI - [meas] (pounds per square inch) kilogrammes *mpl* par mètre carré
PSILOMELANE - [min] (a major ore of manganese, regarded as colloidal manganese dioxide) psilomélane *m*, manganèse *m* oxydé hydraté baritifère
PSITTACINITE - [min] (vanadate and hydroxide of lead and copper) psittacinite *f*
PSITTACOSIS - [med] (tha parrot disease) psittacose *f*
PSOPHOMETER -[instr](instrument designed to measure electrical noise) psofomètre *m*
PSOHOMETRIC - [instr] psofométrique
PSOPHOMETRIC ELECTROMOTIVE FORCE - [telecomm] (in a telephone circuit) force *f* électromotrice psofométrique
PSOPHOMETRIC FACTOR - [el] coefficient *m* psofométrique
~POWER - [el] (the power absorbed by a resistance of 600 ohms from a source of psophometric electromotive force) puissance *f* psofométrique
PSOPHOMETRIC VOLTAGE - [el] tension *f* psofométrique
PSORIASIS - [med] psoriasis *m*
PSP - [radio] (initials of Peak Sideband Power; the peak power which is applied by the sidebands) puissance *f* de crête de bande latérale
PSYCHANOPSIA - [med] cécité *f* psichique
PSYCHECLAMPSIA - [med] psichoéclampsie *f*, manie *f* aigue
PSYCHIATRY - [med] (the branch of medicine which deals with disorders of the mind) psychiatrie *f*
PSYCHISM - [biol] (the doctrine that living matter possesses attributes which are not recognized in non-living matter) psichisme *m*
PSYCHOANALYSIS - [med] (the method of treatment of mental disorders, originated by Freud) psycha-

nalyse *f*
PSYCHOBIOLOGY - [med] (the study of mental processes in relation to physiology and the nervous system) psychobiologie *f*
PSYCHOCOMA - [med] stupeur *f* mélanconique
PSYCHOGALVANOMETER - [instr] (galvanometer designed to detect psychogalvanic reflexes) psychogalvanomètre *m*
PSYCHOKINESIA - [med] impulsion *f* brusque
PSYCHOLEPSY - [med] psycholepsie *f*
PSYCHOMOTOR ABILITY - [physiol] aptitude *f* psychomotrice
PSYCHONEUROSIS - [med] (functional disorder of the mind in a subject who is legally sane and has some insight into his condition) psychonévrose *f*, parapathie *f*
PSYCHROMETER - [instr] (a hygrometer having both wet and dry bulbs) psychromètre *m*
PT - [gen] (abbrev. for point) point *m*
[meas] (abbrev. for pint) pinte *f*
PTERYGOID DEPRESSION - [anat] fossette *f* d'insertion du ptérygoïdien externe
PTM - s. pulse-time modulation
P.T.O. - [mech] (abbrev. for Power take off) prise *f* de force
PTOMAINES - [chem] (generic term for organic compounds, e.g. cadaverine, formed during the decay of animal proteins. They were formerly supposed to be the cause of types of food poisoning, now known to be due to specific bacteria) ptomaïnes *fpl*
PTYALIN - [chem] (digestive ferment in the saliva) ptyaline *f*
PTYALISM - [med] ptyalisme *m*, salivation *f*
PTYALOCELE - [med] ptyalocèle *f*
P-TYPE CONDUCTIVITY - [electron] (the conductivity associated with holes in a semiconductor) conductibilité *f* p
P-TYPE SEMICONDUCTOR - [electron] (extrinsic semiconductor in which the majority carriers are holes) semiconducteur *m*
PUBERTY - [physiol] (sexual maturity) puberté *f*
PUBESCENCE - [zool & bot] (covering of fine hairs) pubescence *f*
PUBESCENT - [zool & bot] pubescent, velu
PUBLIC ADDRESS AMPLIFIER - [telecomm] (amplifier used in a public address system) amplificateur *m* de sonorisation extérieure
~ADDRESS SYSTEM - [telecomm] (an installation comprising one or more microphones, amplifiers and loudspeakers, used for public announcements, e.g. in airport waiting-rooms or in aircraft) sonorisation *f* extérieure
~CRUSHING - [min] broyage *m* à façon
~CALL OFFICE - [telecomm] cabine *f* téléphonique publique
~DIGGINGS - [constr] placer *m* publique
~HEALTH SERVICE - [gen] service *m* sanitaire
~HOUSE - [gen] auberge *f*, taverne *f*
~WORKS - [gen] (works undertaken by a Government or local Government) travaux *mpl* publiques
PUCCOON - [bot] sanguinaire *f*
PUCHERITE - [min] (vanadate of bismuth) puchérite *f*
PUCK - [el acoust] (colloq. in the USA, for pressure roller) rouleau *m* presseur, tambour *m* presseur
PUCKER - [metall] pli *m*
[geol] pli *m*
PUCKERING - [paint] (a defect) gaufrage *m*
PUCKERING - [geol] froncement *m*, plissotement *m*
PUDDING - [food] pudding *m*, boudin *m*, pouding *m*
[naut] (plaited cordage round the mast and yards of a ship) emboudinure *f*
PUDDINGSTONE - [geol] (conglomerate) poudingue *f*
PUDDLE, to - [metall] (to purify molten iron by stirring in an open hearth) puddler
PUDDLE - [gen] flaque *f*, pétrin *m*, gâchis *m*
[hydr] glaise *f* corroi *m*
[metall] bain *m* de fusion
~BALL - [metall] (the mass of iron mixed with slag formed by puddling pig-iron) balle *f* en fer, loupe *m* de puddlage
~BAR - [metall] fer *m* ébauché, fer *m* brut
~ROLLS - [metall] laminoir *m* de puddlage
~SLAG - [metall] laitier *m* de puddlage
PUDDLED CLAY - [hydr] argile *f* corroyée, glaise *f*
~IRON - [metall] fer *m* puddlé
~PLATES - [metall] tôles *fpl* puddlées
~STEEL - [metall] acier *m* puddlé
PUDDLER'S CANDLES - [metall] flamme *f* de puddlage
PUDDLING - [metall] (the agitation of a bath or molten metal in an oxidizing atmosphere to produce wrought iron) puddlage *m*
~FURNACE - [metall] (the small reverberatory furnace in which the iron is puddled) four *m* de puddlage
~IRON - [metall] fer *m* puddlé
~MILL - [metall] train *m* de puddlage
PUERPERAL FEVER - [med] fièvre *f* puerpérale, septicémie *f* puerpérale
~SEPSIS - s. puerperal fever
PUERPERALISM - [med] pathologie *f* puerpérale
PUFF, to - [gen] souffler
[gen] (to inflate) gonfler, faire mousser
PUFF - [gen] souffle *m*, bouffée *f*
[mech] échappement *m* soudain (de vapeur)
~ADDER - [zool] vipère *f* clotho
~-BALL - [bot] chandelle *f*
PUFFER - [mech] (a small-size hoisting engine used in mining) moteur *m* d'extraction
PUFFING AGENT - [paint] épaississant *m*
PUG, to - [constr] (to work clay in a pugmill) glaiser
[constr] (a wall etc) hourder
[min] pétrir
PUG - [geol] salbande *f* argileuse, lisière *f*
[hydr] glaise *f*
~HOLE - [mining] cheminée *f*
~MILL - [mech] (horizontal semi-cylindrical tank with a longitudinal shaft fitted with blades; used in brickmaking) malaxeur *m*, pétrin *m*
PUGGING - [constr] (mixture between floor joints to insulate a room against sounds from below) hourdage *m* hourdis *m*
PULICOSIS - [med] piqûres *fpl* de puces
PULL, to - [gen] tirer, haler, traîner
[naut] (to row, to transport by rowing) ramer
[leather ind] (to plunk, as in tanning) abattre (dans le confit)
[print] (a proof) tirer (une épreuve)
[gen & mech] haler, extraire
[oil ind] détuber, extraire
[aero] cabrer
PULL - [gen] tirage *m*, traction *f*
[phys] (the force exerted to draw something) traction *f*, effort *m* de traction
[print] (a gallery proof) première épreuve *f*

PULL - [glass man] (from a furnace in a specified time) production *f*
[metall] (on an ingot) crique *f* superficielle transversale
[mech] crochet *m* de traction, crochet *m* d'attelage
[metall] (a pull crack) crique *f* par retrait contrarié
~A PROOF, to - [print] tirer une épreuve
~BACK, to - [gen] tirer en arrière
~-BACK - [mech] (a device for drawing something, back; specifically, a device to return the platen of a press after an operation) dispositif *m* de rappel
~-BACK RAM - [mech] (hydraulic ram used to return the platen of a press) bélier *m* de rappel
~-BACK SPRING - [mech] ressort *m* de rappel
~BOX - [el] boîte *f* pour tirage des conducteurs
[el] (transfer box) boîte *f* de dérivation
~BROACH - [mech] broche *f* de rappel
~DOWN, to - [gen] baisser, démolir, démonter
~FLAX, to - [text] arracher le lin
~IN, to - [gen] rentrer, entrer
~-IN-AND-SLIDE DOOR - [aero] (type of aircraft door which is closed by pulling it into the closed position and then sliding it laterally) grande porte *f* à coulisse
~-IN TORQUE - [el] (the maximum constant load-torque on the shaft under which the motor will reach synchronous speed at rated voltage and frequency, under given conditions of excitation and inertia of moving parts) couple *m* d'accrochage d'un moteur synchrone
~-IN VALUE - [electron] valeur *f* d'excitation
~LOCK - [constr] serrure *f* à tirage *m*
~OARS, to - [naut] ramer
~OF THE THREAD - [text] tirage *m* du fil
~OF THE WEIGHT - [mech] traction *f* par poids
~OF THE YARN - [text] entraînement *m* par le fil
~-OF SPRING - [mech] (a return spring) rappel *m* du ressort
~OF TRAVELLER - [text] tension *f* du curseur
~OUT, to - [gen] sortir, tirer
[aero] (of a dive) redresser après un piqué
[oil ind] relever
~-OUT - [aero] (a manoeuvre in which the aircraft is brought into horizontal flight after a dive) ressource *f*
~-OUT DISTANCE - [aero] (the distance travelled by an aircraft during the action of an arrester gear) distance *f* de freinage au crochet
~-OUT MECHANISM - [mech] dispositif *m* d'extraction
~-OUT TORQUE - [el] (the maximum torque on an induction motor) couple *m* de décrochage
~-OUT VALUE - [electron] valeur *f* de retombée
~-OVER MILL - [metall] (rolling-mill with a single pair of rolls) laminoir *m* duo irréversible
~ROD - [mech] (one of an assembly of bars used in a hanging bar ejection system, to connect the press head and ejector frame) tige *f* d'éjection, rouleau *m* tendeur
[oil ind] tige *f* d'entraînement
~ROD LINE - [oil ind] tringles *fpl* de transmission
~-ROLL - [mech] tringle *f* de manoeuvre, tringle *f* de connexion, tringle *f* de tirage
~ROLL STAND - [impl] dispositif *m* enrouleur
~-SHOVEL - [mech] pelle *f* rétrocaveuse
~-SWITCH - [el] (or ceiling switch, switch mounted on the ceiling of a room and operated by a cord) interrupteur *m* à cordon
PULL TEST - [metall] essai *m* de traction
~THE JUTE PLANT, to - [text] arracher le jute
~THE TOW, to - [text] étirer l'étoupe, démêler l'étoupe
~THE WELL, to - [oil ind] arracher le tubage d'isolement
~-THROUGH - [ind chem] (also called a pig) mouton *m*
~-THROUGH WINDING - [el] enroulement *m* à fils tirés
~UP, to - [gen] tirer en haut, remonter, hisser
~-UP - [aero] (manoeuvre in which an aircraft is brought suddenly from horizontal flight into a short steep flight) redressement *m*
~UP THE HEMP, to - [text] arracher le chanvre
~-UP TORQUE - [mech] couple *m* minimal au démarrage
PULLDOWN - [telev] entraînement *m* du film
~CLAW - [mech] (for a film) griffe *f* d'entraînement, griffe *f* d'escamotage
PULLED ELBOW - [med] subluxation *f* de la tête du radius
~SURFACE - [rubber ind] surface *f* rugueuse
PULLET - [zool] poulette *f*
PULLER - [impl] (a tool for drawing a part from a socket or shaft) outil *m* de démontage, extracteur *m*
[metall] (for cores) démouleur *m* de noyau
[text] (used to pull out the wool from the skin) arracheur *m*
~BELTING - [mech] courroie *f* de commande
PULLEY, to - [mech] envoyer aux molettes
PULLEY - [mech] (a wheel to carry a belt in a belt drive) poulie *f*
[mech] (a small grooved wheel; a sheave) molette *f*, molette *f* à gorge
~-BLOCK - [mech] moufle *f*, poulie *f* moufflée, palan *m*
~-BLOCK HOOK - [mech] crochet *m* de palan
~CASING - [mech] cage *f* de la poulie
~CORD - [radio] (simple mechanism used in tuning means of radio sets) corde *f* de poulie
~DRIVE - [mech] mécanisme *m* de manoeuvre de la poulie motrice
~FORK - [mech] fourche *f* de la poulie
~FRAME - [mining] châssis *m* à molettes, belle-fleur *f*
~LATHE - [mech] tour *m* à poulies
~RIM - [mech] jante *f* d'une poulie
~SHAFT - [mining] puits *m* d'extraction
~SHELL - [mech] chape *f* de poulie
~SUPPORT - [mech] chevalet *m* porte-poulie
~TACKLE - [mech] moufle *f*, palan *m*
~TRANSMISSION - [mech] transmission *f* à poulie
~WITH FRICTION DISH - [text ind] poulie *f* avec cloche de friction
~YOKE - s. pulley fork
PULLING - [gen] tirage *m*, traction *f*
[radio] (frequency pulling) entraînement *m* de fréquence
[telev] (the tendency of parts of the picture to pull or stretch) décalage *m* de ligne
~CORD - [mech] cordon *m* de commande
~EYE - [impl] (for cables) orin *m*
~FIGURE - [electron] (the total frequency change of an oscillator when the phase of the reflection coef-

ficient of the load impedance varies throug 360 degrees and the absolute value of this coefficient is constant and equal to 0.2) facteur *m* d'entraînement, indice *m* de glissement aval
PULLING-IN LINE - [constr] câble *m* de tirage
~OF FLAX - [text] arrachage *m* du lin
~ON WHITES - [telev] (a prolongation of the traling edge of the picture element) filage *m* horizontal
~PAWL - [mech] cliquet *m* de traction
~THE COTTON - [text] démêlage *m* du coton
PULLMAN VALVE - [mech] (an ascension pipe valve; used in the condensation of gas) vanne *f* de barillet
POLMONARY ARTERY - [anat] artère *f* pulmonaire
~CAVITY - [anat] cavité *f* pulmonaire
~CIRCULATION - [med] circulation *f* pulmonaire
~CIRRHOSIS - [med] cirrhose *f* pulmonaire
~PLEXUS - [anat] plexus *m* pulmonaire, plexus *m* bronchitique
~STENOSIS - [med] rétrécissement *m* pulmonaire
PULMOTOR - [med] (iron lung) poumon *m* d'acier
PULP - [gen] (a soft, slightly moist and weakly cohering mass or organic matter) pulpe *f*
[min] (powdered ore mixed with water) pulpe *f*
[ind chem] (fibrous material of high cellulose content, obtained by maceration and purification of wood. A raw material in viscose rayon, paper etc manufacture) pulpe *f*
[oil ind] eau boueuse *f* des bocards
~-BOARD - [paper man] carton *m* en jet
~BOX - [paper man] cuvier *m* mélangeur
~CATCHER - [hydr] puiseur *m* de pulpe
[sugar ind] épulpeur *m*
~COLOURING - [text & paper man] coloration *f* dans la masse
~DRIER - [sugar ind] (beet-pulp drier) séchoir *m* à pulpes
~ENGINE - [paper man] pile *f* raffineuse
~FILTER - [impl] filtre *m* à masse
~HYDRATION - [paper man] préparation *f* de la pâte
~IN REELS - [paper man] pâte *f* en rouleaux
~IN SHEETS - [paper man] pâte *f* en feuilles
~MOULDING - [plast ind] (a moulding formed from a combination of paper pulp and thermosetting resin) pièce *f* moulée en pulpe agglomérée
~OF SCALDING PROCESS - [sugar ind] pulpe *f* du procédé Stéffen
~-PIT - [sugar ind] fosse *f* à pulpes
~PRESS - [sugar ind] presse *f* à pulpes
~PRESS WATER - [sugar ind] eaux *fpl* de presse
~-SAVER - [paper man] (mechanism designed to prevent the loss of fibrous material) épurateur *m* grossier
~STRAINER PLATE - [paper man] filtre *m* plat pour pâte
~THICKENER - [oil ind] épaississeur *m* des boues
~WASHER - [brew ind] laveuse *f* de masse filtrante
~WATER - [paper man] eaux *fpl* de presse
PULPING - [gen] pulpage *m*, pulpation *f*, réduction *f* en pâte
~MACHINE - [paper man] presse-pâte *m*
PULPBOARD - s. pulp-board
PULPER - [paper man] (a pulp engine) pile *f* raffineuse
PULQUE - [brew ind] pulque *m*
PULSANTE, to - [gen] battre, palpiter
[phys] entrer en vibration
[min] cribler au berceau

PULSATING - [gen and el] pulsatoire
~CURRENT - [el] (electric current which undergoes regularly recurring variations in magnitude) courant *m* pulsatoire
~CURRENT FED TRACK CIRCUIT - [el] circuit *m* de voie à courant pulsé
~FORCE - [soil] force *f* de pulsation
~LOAD - [soil] charge *f* vibrante
~QUANTITY - [el] (periodic quantity the mean value of which is not zero) grandeur *f* ondulée, grandeur *f* pulsatoire
PULSATION - [med] pulsation *f*
[el] battement *m*
~WELDING - [el heat] (method of spot welding in which the current is repeatedly interrupted while the electrodes are kept stationary and pressing on the pieces) soudage *m* par pulsation
PULSATOR - [agric] (apparatus attached to a milking machine, causing alternations of suction and release of the cow's teets) pompe *f* à lait
[min] crible *m* à grille mobile
~JIG - [min] crible *m* à grille mobile
PULSE - [phys] (a transient change in the value of a quantity) pulsation *f*, impulsion *f*
[radio] (an electrical wave-form of relatively short duration) impulsion *f*
[med] pouls *m*
~AMPLIFIER - [nucl] (amplifier designed to amplify the signals of a nuclear detector) amplificateur *m* d'impulsions
~AMPLITUDE - [el] (the peak value of a pulse) amplitude *f* d'impulsion
~AMPLITUDE DISCRIMINATOR UNIT - [comput] élément *m* discriminateur d'amplitude
~AMPLITUDE MODULATION - [radio] (of a pulse carrier) modulation *f* d'amplitude d'impulsions
~BANDWIDTH - [radio] largeur *f* de bande d'une impulsion
~CARRIER - [radio] (pulse train used as carrier) porteuse *f* d'impulsions, train *m* porteur
~CHAMBER - [nucl] (an apparatus designed to measure the thermal neutron flux in the counter region) chambre *f* d'ionisation à impulsions
~CLIPPER - [telev] écrêteur *m* d'impulsion
~CLIPPING - [telev] écrêtage *m* d'impulsion
~CODE - [comput] (set of pulses having a particular meaning assigned to them) code *m* d'impulsions
[telecomm] (a pulse train so modulated that it conveys information) code *m* impulsionné, impulsions *fpl* codées
~CODE MODULATION - [radio] (modulation based on a pulse code) modulation *f* par impulsions codées
~CORRECTOR - [el] (a circuit designed to correct the edges of pulses) correcteur *m* d'impulsions
~COUNTING RATE ASSEMBLY - [nucl] ensemble *m* de mesure de taux de comptage
~CROPS - [agric] légumineuses *fpl*
~CROWDING - [comput] grande densité *f* d'impulsions
~DECAY TIME - [radio & radar] durée *f* d'affaiblissement d'impulsion
~DISTORSION - [telecomm] (in automatic telephony) distortion *f* d'impulsion
~DISTRIBUTION AMPLIFIER - [electron] distributeur *m* d'impulsions
~DISTRIBUTOR - [electron] distributeur *m* d'impulsions

PULSE DOPPLER SYSTEM - [radar] (anti-jamming system) radar *m* à impulsions utilisant l'effet Doppler
~DROOP - [telecomm] pente *f* négative du créneau
~DURATION - [el] durée *f* d'impulsion
~-DURATION MODULATION - [radio] modulation *f* pendant toute la durée d'une impulsion
~DURATION RATIO - [radio & radar] (relation between duration and repetition frequency) rapport *m* de durée des impulsions
~DUTY FACTOR - [el] (the ratio between the average pulse duration and the average pulse spacing) taux *m* d'impulsions
~ECHOMETER - [instr] (instrument designed to record echoes by pulses) échomètre *m* à impulsions
~EXCITATION - [el] excitation *f* par choc
~FORMING CIRCUIT - [electron] circuit *m* formeur d'impulsions
~FREQUENCY - [el] (the number of pulses per second in a train) fréquence *f* d'impulsions
~FREQUENCY MODULATION - [radio] (PFM; pulse-time modulation in which the pulse repetition rate is the characteristic varied) modulation *f* de la fréquence de l'impulsion
~GENERATOR - [el] (a device for producing very short pulses of high voltage) générateur *m* d'impulsions
~GLIDE PATH - [radar] guidage *m* d'atterrissage par impulsions
~GROUP - s. pulse train
~HEIGHT ANALYZER - [electron] (electronic circuit recording pulses according to height) analyseur *m* d'amplitude
~INTERLACING - [telev] transmission *f* simultanée d'impulsions différentes sur une seule voie
~INTERLEAVING - [radio] entrelacement *m* d'impulsions
~INTERROGATION - [radio] (the triggering of a transponder by a pulse) interrogation *f* par impulsions
~INTERVAL - [radio] intervalle *m* d'impulsions
~INTERVAL MODULATION - [radio] modulation *f* d'intervalle d'impulsions
~IONIZATION CHAMBER - [nucl] (ionization chamber used to detect single ionizing events) chambre *f* d'ionisation à impulsions, chambre *f* d'ionisation compteuse
~-JET - s. pulsojet
~JITTER - [el] vacillement *m* d'impulsion, instabilité *f* de l'impulsion
~-LENGHT MODULATION - s. pulse-duration modulation
~LIMITING RATE - [radio] (the rate of limiting the pulse by a limiting circuit) taux *m* de limitation des impulsions
~MACHINE - s. pulse generator
~MIXING - [telev] mixage *m* d'impulsions
~MODE - [radio] (sequence of pulses in a prearranged pattern used for selecting and isolating a communication channel) mode *m* d'impulsions
~-MODE MULTIPLEX - [telecomm] sélection *f* de voies par modes d'impulsions
~MODULATION RECORDING - [comput] enregistrement *m* par modulation d'impulsions
~-MODULATOR RADAR - [radar] radar *m* à modulation par impulsions
~-MODULATED WAVES - [radio] ondes *f*pl modulées par impulsions

PULSE MODULATION - [radio] (the modulation of a carrier by a pulse train) modulation *f* par impulsions
~MODULATOR - [radio] modulateur *m* à impulsions
~MULTIPLEX - [radar] multiplex *m* à impulsions
~NOISE - [el] parasite *m* de courte durée
~NUMBER - [el] indice *m* de pulsation
~OPERATION - [el] technique *f* des impulsions
~OSCILLATOR - [radio] oscillateur *m* à impulsions
~OVERLAPPING - [electron] superposition *f* d'impulsions
~PACKING - [comput] enregistrement *m* dense de impulsions
~PERIOD - [el] (the time between the corresponding points of two successive impulses in a train) période *f* d'impulsion
~PHASING - [telev] mise *f* en phase d'impulsions
~PIKE - [electron] fréquence *f* de récurrence des impulsions
~POSITION MODULATION - [radio] (form of pulse-time modulation in which the positions in time of pulses are varied) modulation *f* à impulsions à varied) modulation *f* à impulsions à variation de temps
~RADAR - [radar] (emission of waves by means of pulses) radar *m* à impulsions
~RATE - [med] fréquence *f* du pouls
~RATIO - [radio] (the ratio between the duration of an impulse and its impulse period) rapport *m* de impulsion
~RECURRENCE FREQUENCY - [radio] (the number of pulses per second) fréquence *f* d'impulsions
~REGENERATING CIRCUIT - [electron] régénerateur *m* d'impulsions
~REGENERATION - [radio] conformation *f* d'impulsions
[electron] régénération *f* des impulsions
~REGENERATION UNIT - [telev] dispositif *m* de rétablissement de la forme d'impulsion
~REPEATER - [radio] (device used for receiving pulses from one circuit and for transmitting corresponding pulses into another circuit) répéteur *m* de impulsions
~-REPETITION FREQUENCY - [electron] (the pulse repetition rate when this is independent of the interval of time over which it is measured) fréquence *f* de répétition des impulsions
~REPETITION RATE - [comput] (the number of electric pulses per unit of time experienced by a point in the computer) taux *m* de fréquence d'impulsions
~REPLY - [radio] écho *m* d'impulsion
~RESPONSE - [radio] caractéristique *f* de saut
~RISE TIME - [radio] durée *f* d'établissement d'impulsion, temps *m* de montée d'une impulsion
~SCALER - [electron] diviseur *m* d'impulsions
~SEPARATOR - [telev] (in TV receivers) séparateur *m* d'impulsions
~SHAPE - [electron] forme *f* d'une impulsion
~SHAPE DISCRIMINATOR - [electron] discriminateur *m* de forme
~SHAPER - [radio] (transducer used for changing the characteristics of a pulse) conformateur *m* de impulsions
~SHAPING - [radio] conformation *f* d'impulsions
~SHAPING CIRCUIT - [radio] circuit *m* configurateur d'impulsions
~SPACING - [radio] écart *m* d'impulsions

PULSE SPECTRUM - [radio] (the distribution, as a function of frequency, of the magnitudes of the Fourier components of a pulse) spectre *m* d'impulsions

~SPECTRUM BANDWIDTH - s. pulse bandwidth

~SPEED - [telecomm] (in automatic telephony) vitesse *f* d'impulsions

~SPIKE - [electron] (undesired pulse of short duration superimposed on the main pulse) impulsion *f* parasite

~SPRINGS - [telecomm] ressorts *mpl* d'impulsion

~STORAGE TIME - [electron] temps *m* d'emmagasinage des porteurs de charge

~STRETCHER - s. pulse corrector

~STRETCHING - [radar] (or tail; the tail of the trailing edge of a pulse) queue *f* de l'impulsion

~SWITCH - [radio & radar] commutateur *m* d'impulsions

~TILT - [radio] (distorsion characterized by a rise at the pulse top) pente *f* positive de créneau

~-TIME MODULATION - [radio] (PTM; modulation in which the time of the occurrence of some characteristics of a pulse carrier is varied from the unmodulated value) modulation *f* d'impulsions dans le temps

~TOP - [electron] palier *m*

~TRAILING EDGE - [electron] flar[illegible] arrière de l'impulsion

~TRAIN - [radio] série *f* d'impulsions

~-TRAIN FREQUENCY SPECTRUM - [radio] spectre *m* de fréquence de série d'impulsions

~-TRAIN SPECTRUM - s. pulse-train frequency spectrum

~TRANSMITTER - [radio] émetteur *m* à impulsions

~TRIGGERING - [electron] déclenchement *m* de circuit par impulsions

~VALLEY - [radio] vallée *f* d'impulsion

~WIDTH - [electron] largeur *f* d'impulsion

~-WIDTH MODULATION - s. pulse-duration

~-WIDTH RECORDING (NON-RETURN-TO-REFERENCE) [comput] enregistrement *m* par largeur d'impulsion sans retour à zéro

~-WIDTH RECORDING (RETURN-TO-BIAS) - [comput] enregistrement *m* par largeur d'impulsion avec retour à la magnétisation fondamentale

PULSED COLUMN - [chem] colonne *f* pulsée

~-DISCHARGE TUBE - [electron] tube *m* de décharge modulé par impulsions

~NEUTRON SOURCE - [phys] (apparatus used to produce short burste of neutrons) source *f* pulsée de neutrons

~NEUTRONS - [phys] neutrons *mpl* pulsés

~REACTOR - [nucl] (reactor used to produce intense bursts of neutrons for short intervals of time) réacteur *m* pulsé

PULSERS IN MICROWAVES - [radio] générateur *m* d'impulsions pour microondes

PULSES - s. pulse crops

PULSIFIER'S METHOD - [metall] préparation *f* de meulage, préparation *f* des surfaces polies

PULSIMETER - [instr] (instrument designed to assess the strength of the pulse beat) pulsomètre *m*

PULSING RELAY - [telecomm] relais *m* batteur, relais *m* d'impulsions

~SIGNAL - [telecomm] signal *m* de numérotage

~SYSTEM - [radar] radar *m* à impulsions

PULSION - [med] pulsion *f*

PULSOJET - [aero] (a propulsion unit similar to a ramjet but non-continuous in action, the jet consisting in a series of pulses) pulso-réacteur *m*

PULSOJET ENGINE - [aero] pulso-réacteur *m*

PULSOMETER - [mech] (pumping device operated by steam and without piston) pompe *f* à impulsion de vapeur

PULVERIZE, to - [gen] pulvériser, reduire en poudre, satomiser
[min] (e.g. coal) broyer
[chem] (for pharmaceuticals) porphyriser

PULVERIZE THE VEGETABLE MATTER, to - [text] pulvériser les matières végétales

PULVERIZED CHALK - [rubber & plast ind] (finely ground chalk used as a filler for rubber and plastics) craie *f* pulvérisée

~-COAL BURNER - [th eng] brûleur *m* à charbon broyé

~FUEL - [fuel] (any fuel first subjected to crushing and subsequently reduced to dust by pulverizers) combustible *m* pulvérisé

~VEGETABLE MATTER - [text] matières *fpl* végétales pulvérisées

PULVERIZER - [mech] pulvérisateur *m*, broyeur *m*

~HARROW - s. pulverizer plough

~PLOUGH - [agric] charrue *f* émotteuse

PULVERIZING - [ind chem] (operation of grinding sulphur and charcoal together) pulvérisation *f*, porphyrisation *f*

~MACHINE - [text] loup *m* broyeur

~THE CARBONISED BURS - [text] pulvérisation *f* des charbons carbonisés

PULVERULENCE - [gen] pulvérulence *f*

PULVERULENT - [gen] (having an appearance as though covered with fine dust) pulvérulent, poudreux

PULVINAR - [anat] pulvinar *m*, tubercule *m* postérieur du thalamus

PULVINATE - [bot] (cushion shaped) pulviné

PULVINATED - [carp] (having a bulging face) a face convexe

PUMA - [zool] puma *m*, couguar *m*

PUMICATE, to - [constr] poncer

PUMICE, to - [gen] (the operation of polishing with finely divided pumice) poncer, polir à la ponce

PUMICE - [min] (a light, porous igneous rock used as a filler for paints and plastics, in abrasives, as an insulant, and in lightweight concretes) ponce *f*, pierre *f* ponce, pumite *f*

~CONCRETE - [constr] béton *m* de ponce

~CONCRETE BLOCK - s. pumice stone

~SAND - [geol] sable *m* de ponce

~SLAB - [constr] dalle *f*

~STONE - [geol] pierre *f* ponce

PUMICING - [gen] (polishing or finishing by rubbing with finely divided pumice) ponçage *m*

PUMICITE - [min] cendre *f* volcanique

PUMP, to - [gen] pomper
[hydr] pomper, extraire de l'eau
[naut] pomper l'eau de la cale
[a tyre] gonfler

PUMP - [mech] (a machine for imparting energy to a fluid) pompe *f*
[auto] (in a filling station) pompe *f* à essence, distributeur *m* automatique

~BACK - [oil ind] reflux *m*

~BARREL - [mech] corps *m* de pompe, cylindre *m* de pompe

~BODY - [mech] corps *m* de pompe

~CASE - [mech] chapelle *f* de pompe, coquille *f*

PUMP CASING - [mech] carter *m* de pompe
~CHAMBER - [mech] corps *m* de la pompe
~CHECK VALVE - [mech] (regulating valve of a pump) soupape *f* régulatrice, clapet *m* de pompe
~CONNECTION - [mech] raccord *m* de pompe
~CONTROL - [mech] régulateur *m* de pompe
~CYLINDER - [mech] cylindre *m* de pompe
~DELIVERY - [mech] refoulement *m* de la pompe
~FEED - [mech] alimentation *f* à pompe
~GEAR - [mech] armature *f* de pompe
~HAND-LEVER - [mech] levier *m* d'amorçage d'une pompe
~HEAD - [mech] tête *f* de pompe
~HOSE - [rubber ind] tuyau *m* à pompe
~HOUSE - [gen] bâtiment *m* des pompes
~IMPELLER BLADE - [mech] aubage *m* de turbine de pompe
~INLET - [mech] entrée *f* de pompe
~ISLAND - [auto] (USA only) terre-plein *m* de station service
~LEATHER - [naut] maugère *f*
~-LINE - [el] (in electrical traction, a train-line used for the control of auxiliary apparatus) câble *m* de commande
~LOG - [hydr] tronçon *m* évidé
~MAIN - [hydr] conduite *f* de pompe
~NOZZLE - [mech] injecteur-pompe *m*
~OFF, to - [mech] pomper, épuiser, assécher
[oil ind] vider par une pompe
~OUT, to - [mining] assécher, dénoyer
~OUT THE OIL, to - [mech] épuiser l'huile par une pompe
~PIPING - s. pump main
~PLUNGER CYLINDER - [mech] cylindre *m* de plongeur de pompe
~POWER-END - [oil ind] corps *m* hydraulique de pompe
~ROD - [mech] tige *f* de pompe
~SHOT - [mech] coup *m* de pompe
~SPRING-BOX - [impl] compas *m* à pompe
~VALVE BALL - [mech] bille *f* de clapet de pompe
~WATER - [hydr] eau *f* de fontaine, eau *f* de puits
~WITH BALANCED HYDRAULIC THRUST - [hydr] pompe *f* à roue équilibrée
~WORKS - [hydr] station *f* de pompage
PUMPED ELECTRON TUBE - [electron] tube *m* à vide entretenu
PUMPING - [met] (unsteadiness in a mercurial barometer caused by wind gusts or movement of the instrument, as in a small ship) oscillations *fpl* de colonne barométrique
[electron] (in masers and lasers) pompage *m*
[electron] (of a tube etc) évacuation *f*
~BEAM - [oil ind] balancier *m* de pompage
~ELEMENT - [mech] (of the injection pump of a Diesel engine) élément *m* pompant
~ENGINE - [mech] machine *f* d'épuisement
[mining] épulse *f*
~HEAD - [hydr] hauteur *f* de refoulement
~JACK - [oil ind] chevalement *m* de pompage
~PLANT - [mech] matériel *m* d'exhaure, pompes *fpl* d'épuisement
~SHAFT - [mining] puits *m* d'exhaure, puits *m* de épuisement
~STATION - s. pump house
~STRING - [oil ind] colonne *f* de tubes de pompage
~TUBE - [phys] (in vacuum technique) queusot *m* de pompage
PUMPING WELL - [oil ind] puits *m* en pompage
PUMPKIN - [bot] courge *f*, potiron *m*
PUNCH, to - [mech] (to pierce material by forcing a special tool through it) poinçonner, percer, découper, perforer
[metall] (to impress a design; to stamp a die) découper, poinçonner
[gen] trouer, percer, perforer
[agric] conduire à l'aiguillon
[constr] serrer, pincer
PUNCH - [tool] (a tool for perforanting material by shearing action) poinçon *m*, pointeau *m*, découpoir *m*
[metall] (machine for stamping a die) étampe *m*, poinçonneuse *f*, dégorgeoir *m* conique
[print] (die used for the matrix for casting prints) matrice *f*
[impl] chasse-clou *m*, perce *f*, percoir *m*
[comput] perforation *f*, perforateur *m*
~AND PIN DRIFT - [mech] mandrin *m* à bec
~BRUSH - [comput] brosse *f* de lecture
~CAP - [impl] chapeau *m* du poinçon
~CARD - [comput] (or punched card, or card; card of a constant size and punched in a pattern which has a meaning) carte *f* perforée
~CARD UTILIT - [comput] générateur *m* de fonctions classiques
~CARRIAGE - [text] chariot *m* de la boîte aux poinçons
~CHECK - [comput] vérification *f* des perforations
~CLUTCH - [comput] accouplement *m* de perforatrice
~CUTTING MACHINE - [mech] machine *f* à graver des poinçons
~DRIVE - [comput] entraînement *m* de perforatrice
~DRUNK - [med] cranéo-encéphalite *f* des boxeurs
~FEED - [comput] piste *f* de perforation
~HOLDER - [mech] (of a punch press) porte-poinçon *m*
~KNIFE - [impl] goupille *f* perforatrice
~LEG - [cin] (of a film perforating machine) poinçon *m* de perforation
~POSITION - [comput] (the position of a punch on the card) position *f* de perforation
~PRESS - [mech] presse *f* à découper
~PROJECTION - [text] ressaut *m* du poinçon
~PROP - [mining] cale *f*, tasseau *m*
~RADIUS - [metall] rayon *m* de matrice
~-THROUGH - [electron] pénétration *f*
~-THROUGH VOLTAGE - [electron] tension *f* de pénétration
~WARE - [glass man] articles *mpl* de verre estampé
PUNCHED - [mech] poinconné, perforé, estampé
[comput] (of a card) perforé
~ARMATURE - [el] induit *m* denté
~CARD - s. punch card
~CARD EQUIPMENT - [autom] équipement *m* de cartes perforées
~-CARD INFORMATION - [comput] informations *fpl* mécanographiques
~CAVITY - [metall] découpage *m* à l'emporte-pièce
~CODE - [comput] perforation *f* codique
~HOLE - [gen] perforation *f*
[comput] perforation *f*
~NEEDLE - [text] aiguille *f* estampée
~PLATE - [metall] tôle *f* perforée
~TAPE - [electron] (paper tape so punched, that a pattern having a meaning is obtained) bande *f* per-

forée
PUNCHED-TAPE CODE - [comput] code-bande *m*
~-TAPE READER - [comput] lecteur *m* de bandes
PUNCHEON - [carp] (short post for intermediate support to a beam) poinçon *m*
[meas] tonneau *m* de 318 litres
[mining] cale *f*, tasseau *m*
[constr] tronçon *m* fendu
PUNCHER - [mech] (a punching apparatus) perforateur *m*, poinçonneuse *f*
PUNCHING - [gen] perçage *m*, poinçonnage *m*, découpage *m* à l'emporte-pièce
[el] (core plate, lamination, of an electrical machine or transformer. One of the thin sheets or iron or steel forming part of the core) tôle *f* de noyau
~BEAR - [mech] poinçonneuse *f* portative
~CARDBOARD - [print] carton *m* de découpage
~COMB - [text] peigne *m* fixe
~DIE - [mech] (a tool for making perforations or forming blanks by punching action) matrice *f* de perforation
[print] caractère *m* relieur
~HEAD - [mech] plateau *m* presseur
~MACHINE - [mech] poinçonneuse *f*
~MECHANISM - [comput] mécanisme *m* perforateur
~PIN - [comput] poinçon *m*
~PLATE - [mech] plaque *f* de découpage
~PLATEN - [mech] platine *f* de découpage, presse *f* à platine à découper
~PRESSURE - [mech] pression *f* du poinçon
~TONGS - [mech] poinçon *m* à emporte-pièce
~UNIT - [comput] unité *f* de perforation
PUNCTUATION - [gen & print] ponctuation *f*
~MARKS - [gen & print] signes *mpl* de ponctuation
PUNCTURE, to - [gen] ponctionner
[auto] (a tyre) perforer, crever
[el] (an insulator, through a gradually increasing voltage) percer
PUNCTURE - [gen] perforation *f*
[auto] (of tyre) perforation *f*, crevaison *f*
[el] (the passage of a disruptive discharge through an insulating material) percement *m*, perforation *f*
[med] ponction *f*
~PROOF - [rubber ind] (of tyres) increvable
~-PROOF TUBE - [rubber ind] (of tyres) chambre *f* à air auto-obturatrice, chambre *f* à air increvable
~SEAL - [rubber ind] (for tyres) emplâtre *m* de réparation
~SEALING COMPOUND - [rubber ind] composition *f* auto-obturatrice
~VOLTAGE - [el] (the voltage at which an insulator is electrically punctured when subject to a gradually increasing voltage) tension *f* de percement
PUNCTURED - [gen] crevé, perforé, composé de piqûres
[constr] (rough masonry work, characterized by irregularly punched holes in the faces of the stones) fait de piqûres
~CLOAT - [auto] (of the carburettor) flotteur *m* percé
PUNNER - [constr] hie *f*, dame *f*, pilon *m*
PUNNING - [constr] (ramming, compaction) damage *m*, pilonnage *m*
PUNT, to - [naut] conduire à la perche, yoler
PUNT - [naut] bateau *m* plat, bachot *m*
PUNTEE - s. punty
PUNTY - [glass man] (a short used in glass manipulation and blowing) pontil *m*
PUP - [zool] (diminutive of puppy) chiot *m*
[ovens] (a closer, a soap) mulet *m*, savon *m*, quart *m*
~JOINT - [plumb] (a short pipe) tube *m* court
PUPA - [zool] (the inactive stage in the life of an insect) chrysalide *f*, nymphe *f*
~BED - [zool] lit *m* de la chrysalide
~CASE - [zool] pupe *f*, enveloppe *f* de chrysalide
~STATE - [zool] état *m* de chrysalide
PUPAL STAGE - [zool] étage *m* nymphal
PUPATION OF THE SILK WORM - [zool] transformation *f* du ver en chrysalide
PUPIL - [anat] pupille *f*, prunelle *f*
PUPILLARY - [anat] pupillaire
~MEMBRANE - [anat] membrane *f* pupillaire
~REFLEX - [med] (contraction) réflexe *m* pupillaire, réflexe *m* lumineux
PUPILLATONIA - [med] atonie *f* pupillaire
PUPILLOMETER - [instr] pupillomètre *m*
PUPILLOMETRY - [med] pupillométrie *f*
PUPILLOPLEGIA - s. pupillatonia
PUPILLOSCOPY - [med] pupilloscopie *f*, sciascopie *f*
PUPIN CABLE - [el] (cable in which the conductors are coil-loaded at intervals, thus causing attenuation which is uniform up to a cut-off frequency and then rises rapidly) câble *m* de charge
~COIL - [el] (the loading coil devised by Pupin) bobine *f* de charge, bobine *f* Pupin
PUPINIZATION - [el] (coil loading) pupinisation *f*
PUPINIZE, to - [telecomm] (coil loading) pupiniser
PUPPET - [rubber ind] (term used for a roll uncured rubber) poupée *f*
~FILM - [cin] film *m* de marionnettes
PUPPY - [zool] jeune chien *m*, chiot *m*
PURBLIND - [gen] presque aveugle, myope
PURCHASE, to - [gen] acheter, acquérir
PURCHASE - [gen] achat *m*, acquisition *f*
[naut] cartahu *m*
[mech] (a mechanical hold; the amount of leverage) force *f* mécanique, prise *f*
[a tackle] appareil *m* de levage
~-BLOCK - [naut] moufle *f* à estrope double
~-FALL - [mech] courant *m* de palan
~INSPECTION GAUGE - [meas] (a device for checking the dimensions of purchased goods) jauge *f* de garantie
~PRICE - [comm] prix *m* d'achat
PURE - [gen] pur
[chem etc] (free from adulteration) pur
[opt] pur
[acoust] pur
[phys] (of torsion, bending etc) pur
~BINARY CODE - [comput] code *m* binaire naturel
~BINARY NOTATION - [comput] notation *f* binaire pure
~BREED - [zool] race *f* pure
~-BRED SHEEP- [zool] mouton *m* pur sang
~BREEDING - [zool] propagation *f* d'une même race
~CELLULOSE - [text] cellulose *f* pure
~CLAY - [constr] (a brick earth composed of silica and alumina and only a small percentage of lime, magnesia, soda etc) glaise *f*
~COAL SUBSTANCE - [min] (coal free from moisture and ash) charbon *m* pur (sans cendres et sec)
~COLOUR STRIPE - [text] rayure *f* de couleur pure
~CONTINUOUS WAVES - [radio] (waves which are modulated into trains) ondes *fpl* entretenues pures

PURE FIBRE - [text] fibre *f* pure
~GOLD - [metall] or *m* pur
~GUM MIX - [rubber ind] mélange *m* pure gomme
~GUM MIXTURE - s. pure gum mix
~GUM VULCANIZATE - [rubber ind] vulcanisat *m* pure gomme, vulcanisat *m* sans charge
~LINE - [genet] (homogeneous collection of individuals resulting from autogamous reproduction) lignée *f* pure
~MATHEMATICS - [math] mathématiques *fpl* pures
~MECHANICS - [mech] mécanique *f* rationnelle
~METAL - [metall] (any metal of high purity) métal *m* pur
~METAL CRYSTALS - [metall] (the crystals of which a solid pure metal is composed) cristaux *mpl* de métal pur
~MOHAIR YARN - [text] fil *m* mohair pur
~MUSIC - [mus] (or absolute music, music which depends on its structure for its apprehension) musique *f* absolue, musique *f* pure
~RESISTANCE LOAD - [electron] charge *f* ohmique pure
~RUBBER TAPE - [rubber ind] ruban *m* pure gomme
~SILK CLOTH - [text] étoffe *f* tout soie
~SPIRIT - [chem] alcool *m* rectifié
~SUBSTANCE - [phys] (substance having well defined properties which can be represented by numerical constants) substance *f* pure
~TONE - [acoust] (a tone in which the sound pressure varies sinusoidally with time) ton *m* pur
~WHITE VEGETABLE SILK - [text] soie *f* végétale d'un beau blanc
~YEAST - [brew ind] levure *f* pure
PUREBRED - [zool] de race
PURFLE, to - [text] orner d'une bordure brodée
[arch] orner, embellir
PURFLING - [mus] (strips of wood following the outline of stringed instrument) filet *m*
PURGA - [met] (cold north-easterly wind laden with drifting snow, occurring in Russia and Siberia) purga *m*
PURGATIVE - [med] purgatif *m*, cathartique *m*
PURGE, to - [gen] purger, nettoyer
[mech etc] épurer, clarifier
[med] purger
[sugar ind] claircer
PURGE AIR, to - [gas ind etc] (from a pipe etc) purger
~COCK - [plumb] (a vent, or bleed cock) robinet *m* de vidange, robinet *m* de purge
~THE SEDIMENTS, to - purger les boues
~WITH GAS - [plumb] purger au gaz
PURGING - [gen] purge *f*, purgation *f*
[plast ind] (the process of displacing the material in an injection moulding cylinder by another type) purge *f*
[min] lavage *m*
[nucl] (the scouring of nuclear reactor fuel elements) rodage *m*
~COMPOUND - [plast ind] (material used to purge an extruder) composé *m* pour purge
~CONTROL VALVE - [hydr] robinet *m* de purge
~FROM AIR TO GAS - [gas ind] mise *f* en gaz
~FROM GAS TO AIR - [gas ind] mise *f* à l'air
PURIFICATION - [gen & chem] purification *f*, épuration *f*
~STREAM - [gas ind] ensemble *m* des appareils d'épuration
PURIFIER BOX - [gas ind] (oxide box) cuve *f* d'épuration
~COVER - [ind chem] couvercle *m* de l'épurateur
~GRIDS - [ind chem] claies *fpl* d'épurateur
PURIFY, to - [gen] purifier
[chem] (to free from vitiating elements) épurer
PURIFYING COLUMN - [ind chem] colonne *f* d'épuration
~MATERIAL - [ind chem] (called 'sponge' in the USA) matière *f* épurante
~PLANT - [ind chem] installation *f* d'épuration
PURINE - [chem] (a synthesis intermediate) purine *f*
~BASE - [chem] base *f* purinique
~GROUP - [chem] (a group of cyclic diureides derived from one molecule of a dibasic hydroxy acid and two molecules of urea) group *m* purinique
PURITY - [gen] pureté *f*
[opt] (in colorimetry) pureté *f*
[aero] (the ratio of lifting gas to the volume of gas in a balloon) pureté *f*
~MAGNET - [telev] aimant *m* de pureté de couleur
~METER - [instr] (instrument designed to measure the percentage of lifting gas in a balloon) indicateur *m* de pureté
~OF THE BREED - [genet] pureté *f* de race
~OF THE FIBRE - [text] pureté *f* de la fibre
~OF THE STOCK - s. purity of the breed
PURL, to - [gen] murmurer, gazouiller
[text] engrêler, faire des mailles à l'envers
PURL - [gen] (a circling movement of water) gazouillement *m*
[text] engrêlure *f*, picot *m*, cannetille *f*
[brew ind] bière *f* chaude épicée avec genièvre
~FABRIC - [text] envers *m*
PURLIN - [constr] (member supporting the common rafters) panne *f*, filière *f*
~CLEAT - [constr] chantignolle *f*
~POST - [constr] jambette *f*
~ROOF - [constr] comble *m* à chevron
PURLING - [acoust] (a quiet sound) murmure *m*, clapotement *m*
PUROMYCIN - [chem] (an anti-protozoal antibiotic produced by Streptomyces Alboniger) puromycine *f*
PURPLE - [dye] pourpre *f*
~COPPER ORE - [min] (bornite) bornite *f*, cuivre *m* panaché
~ORE - s. purple copper ore
~-WOOD - [bot] palissandre *m*
PURPOSE - [gen] but *m*, fin *m*, dessein *m*
~-MADE BRICK - [constr] (specially moulded brick for use in particular positions) brique *f* spéciale
PURPURIN - [chem] (a dyestuff) purpurine *f*
PURR, to - [acoust] ronronner
PURR - [acoust] ronron *m*
[mech] (of engine) ronflement *m*
[aero] (of an aircraft) vrombissement *m*
PURRING SOUND - s. purr
PURSE SILK - [text] cordonnet *m*
~TWIST - s. purse silk
PURSER - [naut] commissaire *m*
PURSINESS - [vet] pousse *f*
PURSLAIN - [bot] pourpier *m*
PURSLANE - s. purslain
PURSY - [vet] poussif
PURULENCE - [med] purulence *f*, pus *m*
PURULENCY - s. purulence

PURULENT - [med] (forming pus) purulent
~ARTHRITIS - s. pyarthrosis
~PERITONITIS - s. pyoperitoneum
PURVEY, to - [gen] fournir
PURVEYANCE - [gen] approvisionnement *m*
PUS - [med] (the fluid formed by suppuration) pus *m*, boue *f*, sanie *f*
PUSH, to - [gen] pousser, presser
PUSH - [gen] poussée *f*, impulsion *f*
~ASIDE, to - [gen] écarter
~BACK, to - [gen] repousser, reculer
~-BACK - [mech] (of a press) piston *m* de retour
~BAR - [text] barre *f* de poussée
~BENCH - [metall] banc *m* d'étirage
~-BROACH, to - [mech] équarrir
~-BROACH - [mech] alésoir *m* à pression
~BUTTON - [el] (type of electric switch consisting of a spring loaded plunger) poussoir *m*, bouton *m* (de contact), bouton *m* poussoir
~-BUTTON CONTROL - [el] commande *f* par boutons poussoirs
~-BUTTON FAUCET - [mech] robinet *m* à repoussoir
~BUTTON PANEL - [el] panneau *m* des boutons poussoirs
~-BUTTON START - [mech] démarrage *m* à bouton poussoir
~-BUTTON SWITCH - [el] interrupteur *m* à bouton poussoir
~-BUTTON TUNING - [radio] accord *m* par bouton poussoir
~CAR - [railw] lory *m*
~-CART - [transp] charrette *f* à bras
~CONVEYOR - [mining] transporteur *m* à racloirs
~DOWN A STACK, to - [comput] refouler une pile
~DOWN BLASTING MACHINE - [mining] exploseur *m* à poignée
~-DOWN LIST - [comput] liste *f* refoulée
~-DOWN STORAGE - [comput] mémoire *f* à colonne
~FIT - [mech] montage *m* à frottement doux
~-JACK - [railw] pousseur *m*
~MORAINE - [geol] moraine *f* de poussée
~-ON END - [gas ind] (a hose nozzle) mamelon *m* porte tuyau souple
~-ON FILTER - [photo] filtre *m* à emboîtement
~ON SCREEN - [photo] écran *m* se montant en friction sur le barillet
~-OVER WIPE - [telev] fondu *m* effacé
~-PATTERN PEAR SWITCH - [el] allumeur-extincteur *m* à poussoir
~-PLATE CONVEYOR - [mech] (a draglink conveyor) transporteur *m* à palettes
~-PULL - [el] push-pull, en reversible
~-PULL AMPLIFICATION - [radio] amplification *f* push-pull, amplification *f* symétrique
~-PULL AMPLIFIER - [radio] (amplifier circuit consisting of two thermionic valves so connected that when the grid of one is positive, that of the other is negative) amplificateur *m* push-pull
[contr] (in automatic control) amplificateur *m* à sortie différentielle
~-PULL CARBON TRANSMITTER - [el acoust] microphone *m* à charbon à montage équilibré
~-PULL CIRCUIT - [radio] circuit *m* push-pull, circuit *m* symétrique
~-PULL CONNECTION - [el & telecomm] montage *m* push-pull
~-PULL DETECTION CIRCUIT - [telecomm] montage *m* détecteur push-pull
PUSH-PULL ENERGIZATION - [electron] excitation *f* symétrique
~-PULL MICROPHONE - [el acoust] (a microphone in which two carbon-granule cells are mounted on either side of a stretched diaphragm. The amplitude distorsion arising in one is well balanced by the opposite amplitude distorsion in the other) microphone *m* à montage équilibré
~-PULL MODULATOR - [radio] modulateur *m* push-pull
~-PULL OPERATION - [electron] fonctionnement *m* de dispositifs électroniques en disposition symétrique
~-PULL OSCILLATOR - [radio] (balanced oscillator using two similar tubes in phase opposition) oscillateur *m* push-pull
~-PULL RECORDING TRACK - [el acoust] système *m* symétrique d'enregistrement
~-PULL SOUND TRACK - [el acoust] enregistrement *m* symétrique en opposition
~-PULL TRANSFORMER - [el] transformateur *m* push-pull
~-PULL TUBE OPERATION - [radio] fonctionnement *m* push-pull
~-PUSH CIRCUIT - [radio] circuit *m* push-push
~-PUSH CURRENTS - [radio] courants *mpl* continus de polarité et phase égale
~-PUSH VOLTAGES - [radio] tensions *fpl* continues de polarité et phase égale
~-ROD - [mech] (in a reciprocating i.c. engine, a rod used to transmit motion from the cam-follower to the rocker of an overhead valve) poussoir *m* de soupape, tige-poussoir *f*
[mech] (control rod) tige *f* de commande de clapet
~-TO-TALK BUTTON - [telecomm] (press-button switch to close the speech circuit of an intercommunication telephone system) bouton-poussoir *m* de interphone
~-to-TEST BUTTON - [el] (press-button switch which pressed to test a circuit) bouton *m* test
~-to-TEST CIRCUIT - [el] (a circuit which, when closed by pressing a button, makes a test of a circuit and shows the result by means of a light or other indication) circuit *m* test
~-to-TEST LIGHT - [el] (indicator light in a push-to-test circuit) voyant *m* de circuit test
~-UP - [metall] affaissement *m* du moule
~UP A STACK, to - [comput] remonter une pile
~-UP LIST - [comput] liste *f* directe
~WAGONS ON THE CAGE, to - [mining] encager des wagons
~WELDING - [metall] soudure *f* à poussée
PUSHER - [railw] (onlyUSA; a banking engine) machine *f* de renfort, machine *f* pour la montée des côtes
[gas ind] poussoir *m* de défournement, déchargeuse *f*
~AEROPLANE - [aero] (an aeroplane in which the propeller or propellers are designed to produce compressive stress in the propeller shaft) avion *m* à hélice propulsive
~BEAM - [gas ind] (a ram bar) bras *m* de défournement, poutre *f* de défournement
~FURNACE - [gas ind] four *m* poussant
~HEAD - [gas ind] bouclier *m* de défournement
~LOCOMOTIVE - [railw] (a banking engine) machine *f* de renfort
~

PUSHER MACHINE - [gas ind]défourneuse *f*
~SIDE - [gas ind] (of the pusher machine) côté *m* machine
~-TYPE CONVEYOR FURNACE - s. pusher furnace
PUSHING - [gen] poussée *f*
[gas ind] défournement *m*
~AND LEVELLING MACHINE - [gas ind] défourneuse-repaleuse *f*
~CATCH - [mech] cliquet *m* de poussée
[text] cliquet *m* de poussée
~FIGURE - [electron] (the change of oscillator frequency with a specified change in anode current, excluding thermal effects) indice *m* de glissement amont
~OFF - [railw] lancement *m*
~STUD - [mech] cheville *f* du poussoir
~UP PRICES - [comm] hausse *f* illicite des prix
PUSHOVER - [aero] commencement *m* de descente en piqué
PUSTULA - [med] pustule *f*
PUSTULE - s. pustula
PUT, to - [gen] mettre, placer, poser
[mining] (the wagons) pousser (les wagonnets)
[leather ind] donner une main (de lustre)
PUT A PLUG IN A JACK, to - [telecomm] (in telephony) introduire une fiche dans un jack
~ABDUT, to - [naut] virer de bord
~BACK, to - [gen] replacer, remettre à sa place
[naut] rentrer au port
~DOWN A BORE-HOLE, to - [mining] pratiquer un trou de sonde
~DOWN A WELL, to - [oil ind] faire un sondage
~FORWARD, to - [gen] émettre, avancer, proposer
~IN A RISE, to - [mining] percer une remontée
~IN THE ROVING BOBBINS, to - [text] placer les bobines de préparation
~INTO GEAR, to - [mech] engrener, embrayer
~INTO OPERATION, to - [gen] mettre en opération
~OFF, to - [gen] renvoyer, différer, atermoyer
[naut] déporder du quai
~ON THE BRAKE, to - [mech] serrer le frein, freiner
~ON THE CURRENT, to - [el] mettre en circuit
~ON THE LAP, to - [text] placer le rouleau
~OUT A FIRE, to - [gen] éteindre un incendie
~OUT OF GEAR, to - [mech] désengrener, débrayer
~OUT OF SERVICE, to - [gen] mettre hors service, bloquer, immobiliser
~TO EARTH, to - [el] mettre à la terre
~TO PASTURE, to - [agric] mener au pâturage
~TENSION ON THE FILLET, to - [text] tendre le ruban de la garniture
~THE ENGINE IN STEAM, to - [railw] mettre la locomotive sous pression
~THE FLYER ON THE SPINDLE, to - [text] placer l'ailette sur la broche
~THE WARP INTO CHAINS, to - [text] mettre la chaîne en mailles
~THROUGH, to - [telecomm] mettre en communication
~UP, to - [gen] ériger, construire
[comm] hausser, majorer
~UP A RISE, to - [mining] remonter
PUTAMEN - [anat] (the lateral part of the lentiform nucleus of the cerebrum) putamen *m*, noyau *m* lentinulaire
PUTATIVE - [gen & leg] putatif
PUTLOG - [constr] (transverse bearer supporting the scaffold boards) boulin *m*
~HOLE - [constr] (the hole into the putlog is wedged) trou *m* du boulin
PUTREFACTION - [gen] (the decay of animal or vegetable matter) putréfaction *f*
PUTREFY, to - [gen] (to undergo putrefaction) putréfier, pourrir
PUTRESCIBILITY - [chem] putrescibilité *f*, digestibilité *f*
PUTRESCINE - [chem] (tetramethylene-diamine formed in the process of putrefaction of the flesh) putrescine *f*
PUTTIES - [text] bandes *f*pl molletières
PUTTING IN THE FULL BOBBIN - [text] mise *f* en place de la bobine pleine
~INTO OPERATION - s. putting into service
~ON THE BELT - [text] montage *m* de la courroie
~THE HANK ON THE REEL - [text] placement *m* de l'écheveau sur l'aspe
PUTTY, to - [gen] mastiquer, boucher au mastic
PUTTY - [gen] (a stiff paste of whiting and linseed oil, used to stop cracks and holes in woodwork before painting, in glazing windows etc) lut *m*, mastic *m*, enduit *m*
[glass man] mastic *m* de vitrier
[constr] mastic *m* à la chaux
~COAT - [constr] enduit *m*
~-JOINT - [constr] joint *m* au minium
~KNIFE - [tool] spatule *f* de vitrier, couteau *m* à mastiquer
~POWDER - [ind chem] potée *f* d'étain
~SEAM - [constr] bord *m* de mastic *m*
PUTTYING - [constr & glass ind] masticage *m*
PUY - [geol] pui *m*
PUZZLE LOCK - [mech] serrure *f* à combinaisons
PUZZOLANA - [geol] (vulcanic dust forming a hydraulic cement when mixed with lime) pouzzolane *f*, pozzolane *f*
~CEMENT - [constr] (cement made from Pozzolane and lime) ciment *m* pouzzolanique
P.V.C. - s. polyvinyl chloride
~ENVELOPE - [aero] (special envelopes of polyvinyl chloride used to pack aircraft parts in storage) enveloppe *f* de chlorure polyvinyle
P.WIRE - [telecomm] fil *m* de test, fil *m* privé
P.W.R. - s.Pressurized Water Reactor
P.X. - [telecomm] (private exchange) installation *f* téléphonique privée
PYARTHROSIS - [med] arthrite *f* purulente
PYCASTYLE - s. pycnostyle
PYCNOMETER - [instr] (special flask of known content used in determining specific gravities) picnomètre *m*, flacon *m* à densité
PYCNOSIS - [biol] (the shrinkage of the stainable material of a nucleus into a deeply stained knot) pycnose *f*
PYCNOSTYLE - [arch] (a type of colonnade) pycnostyle *m*
PYELOCYSTITIS - [med] pyélocystite *f*
PYELOGRAPHY - [med] (radiological examination of the upper urinary tract) pyélographie *f*, urographie *f*
~BY ELIMINATION - [med] pyélographie *f* descendante
PYELOMETER - [instr] pelvimètre *m*
PYELONEPHROSIS - [med] pyélonéphrose *f*

PYELOPLICATION - [med] plicature *f* du bassinet
PYEMIA - [med] (plebitic septicemia) pyémie *f*
PYGALGIA - [med] douleur *f* dans la région fessière
PYLEMPHRAXIS - [med] pyléthrombose *f*
PYLEPHLEBITIS - [med] pyléphlébite *f*
PYLETHROMBOSIS - s. pylemphraxis
PYLON - [constr] (self-supporting vertical structure) pylône *m*
[telecomm] (a type of mast) pylône *m* en treillis
[aero] (for airships and balloons) pylône *m*
~BEARING - [construct work] appui *m* de pylône
~LEGS - [telecomm] montants *mpl* d'un pylône
PYLORIC STENOSIS - s. pyloristenosis
PYLORISTENOSIS - [med] sténose *f* du pylore
PYLOROSTENOSIS - s. pyloristenosis
PYLORUS - [anat] (the point where the stomach is connected with the intestine) pylore *m*
PYOARTHROSIS - s. pyarthrosis
PYOCYTE - [med] cellule *f* du pus, pyocyte *m*
PYODERMATITIS - [med] pyodermite *f*
PYODERMATOSIS - s. pyodermatitis
PYODERMITIS - s. pyodermatitis
PYROGENESIS - [med] pyogénie *f*
PYOMETRITIS - [med] métrite *f* purulente
PYO-OVARIUM - [med] abcès *m* de l'ovaire
PYOPERICARDITIS - s. pyopericardium
PYOPERICARDIUM - [med] pyopéricardite *f*
PYOPERITONEUM - [med] péritonite *f* purulente
PYOPERITONITIS - s. pyoperitoneum
PYOPNEUMOTHORAX - [med] pyopneumothorax *m*
PYOSIS - [med] suppuration *f*
PYRAMID - [geom] pyramide *f*
~AERIAL - [radio] antenne *f* en pyramide
~ANTENNA - s. pyramid aerial
~BRACE - [auto] renfort *m* pyramidal
~OF LIGHT - [opt] cône *m* lumineux, triangle *m* lumineux
~PROFILE - [rubber ind] (of tyres) sculpture *f* à pyramides
~TREAD - s. pyramid profile
PYRAMIDAL - [gen] pyramidal
~HORN - [radio] (of a prism aerial) cornet *m* en pyramide
~ROOF - [constr] comble *m* en pavillon
~TEETH - [mech] dents *fpl* pyramidales
PYRARGYRITE - [min] (an ore of silver, essentially silver-antimony sulphide, crystallizing in the trigonal system) pyrargite *f*, argent *m* rouge antimonial
PYRATHIAZINE HYDROCHLORIDE - [chem] (an antihistamine drug) chlorhydrate *m* de pyrathiazine
PYRAZINAMIDE - [chem] (a tubercolostatic drug) pyrazinamide *f*
PYRAZINES - [chem] pyrazines *fpl*
PYRAZOLONE - [chem] (a synthesis intermediate) pyrazolone *f*
PYRENE - [chem] (a tetracyclic hydrocarbon obtained from the coal-tar fraction boiling above 360°) pyrène *m*
PYRETHRIN - [chem] (an important insecticide derived from Pyrethrum) pyréthrine *f*
PYRGEOMETER - [instr] (instrument designed to measure radiation from the surface of the earth) pyrgéomètre *m*
PYRHELIOMETER - [instr] (instrument to measure the effect of solar radiation) pyrhéliomètre *m*
PYRIDINE - [chem] (a heterocyclic compound with a ring of 5 carbon atoms and one nitrogen atom. An intermediate for pharmaceuticals and rubber chemicals) pyridine *f*
PYRIDINIUM BROMIDE PERBROMIDE - [chem] (a brominatin agent in organic synthesis) perbromure *m* de bromure de pyridinium
PYRIDOSTIGMINE BROMIDE - [chem] (an anticholinesterase used in medicine) bromure *m* de pyridostigmine
PYRIDOXINE HYDROCHLORIDE - [chem] (vitamin B.6. A member of the vitamin B complex and essential to health) chlorhydrate *m* de pyridoxine
PYRILAMENE MALEATE - [chem] (mapyramine maleate. An antihistamine drug) maléate *m* de pyrilamine
PYRIMETHAMINE - [chem] (an antimalarial drug) pyriméthamine *f*
PYRIMIDINES - [chem] (compounds consisting of six-membered heterocyclic rings with two nitrogen atoms in the meta position) pyrimidines *fpl*
PYRITE - [min] (iron pyrites, a natural sulphide of iron, crystallizing in the cubic system, an important source of iron and sulphur, used in sulphuric acid manufacture) pyrite *f*
~BURNER - [ind chem] (arrangement for burning pyrites, e.g. in sulphuric acid manufacture) four *m* à griller les pyrites
~DETECTOR - [electron] (crystal detector employing iron pyrites) détecteur *m* à pyrite
PYRITIC - [min] pyriteux
PYRITIFEROUS - [min] pyritifère
PYRITIZATION - [chem] pyritisation *f*
PYRITIZE, to - [chem & metall] pyritiser
PYRITOHEDRAL - [phys] (of crystals) pyritoèdre
PYRITOLOGY - [phys] pyritologie *f*
PYRITOUS - s. pyritic
PYRO - [chem] (prefix denoting a substance formed or obtained by heat) pyro-
~-ACID - [chem] acide *m* pyrogéné
PYROBORATES - [chem] (borates) pyroborates *mpl*
PYROBORACID ACID - [chem] acide *m* pyroborique
PYROBUTAMINE PHOSPHATE - [chem] (an antihistamine used in medicine) phosphate *m* de pyrobutamine
PYROCATECHIN - s. pyrocatechol
PYROCATECHOL - [chem] (an intermediate for drugs, dyes and plastics stabilizers) pyrocatéchine *f*
PYROCHLORE - [min] (a niobate of the cerium metals) pyrochlore *m*
PYROCLASTIC ROCKS - [geol] (fragmental deposits of volvanic origin) roches *fpl* pyroclastiques
PYROCONDENSATION - [chem] (molecular condensation caused by heating at high temperature) pyrocondensation *f*
PYROCONDUCTIVITY - [el] pyroconductivité *f*
PYROELECTRIC EFFECTS - [el] (effects caused by the generation of heat by the dielectric losses) effets *mpl* pyroélectriques
PYROELECTRICITY - [phys] (positive and negative charges of electricity which develop simultaneously on different parts of the same crystal when its temperature is charged) pyroélectricité *f*
PYROGALLIC ACID - [chem] (a synthesis intermediate and parasiticide) acide *m* pyrogallique, pyrogallol *m*
PYROGALLOL - s. pyrogallic acid
PYROGENESIS - [phys] (production of heat) pyrogénèse *f*

PYROGENETIC - [phys] pyrogénétique
PYROGENIC - [chem] (resulting from the application of high temperatures) pyrogène
PYROGENOUS - [geol] pyrogène
PYROGNOSTIC - [chem] pyrognostique
PYROGRAPHY - [print] (the process of producing a design, e.g. on leather by a red-hot point) pyrogravure *f*
PYROLIGNEOUS ACID - [chem] (a mixture of acetone, methanol, acetic acid, furfural, and tars, obtained by the destructive distillation of wood) acide *m* pyroligneux
~ALCOHOL - [chem] méthanol *m*
PYROLUSITE - [min] (natural manganese dioxide ; a major ore of manganese, crystallizing in the orthorhombic system, and often pseudomorphic after Manganite) pyrolusite *f*
PYROLYSIS - [chem] (the operation of breaking up large molecules into smaller ones by the application of heat) pyrolyse *f*
PYROLYTIC CARBON-SILICON CARBIDE MIXTURE - [nucl] mélange *m* de carbone pyrolytique et carbure de silicium
PYROMAGNETIC - [phys] (relating to the changes in magnetic intensity due to changes of temperature) pyromagnétique
PYROMANCY - [gen] (divination by fire) pyromancie *f*
PYROMANIA - [med] pyromanie *f*
PYROMELLITIC ACID - [chem] (an intermediate for plasticizers) acide *m* pyromellitique
PYROMETER - [instr] (instrument for measuring high temperatures, which does not depend on fluid expansion) pyromètre *m*
~IN PROTECTION TUBE - s. pyrometer rod
~PROBE - [instr] sonde *f* pyrométrique
~ROD - [instr] canne *f* pyrométrique
~SOCKET - [mech] (a recess in a part of a machine in which a pyrometer is fitted) prise *f* de pyromètre
~TUBE - [instr] (thermometer resistor fitted into a protective tube) canne *f* pyrométrique
PYROMETRIC CONE EQUIVALENT - [metall] point *m* de ramollissement
~CONES - [meas] (or fusion cones; used for determining temperatures in ceramic furnaces) cônes *mpl* pyrométriques
PYROMETRY - [meas] (the art of measuring high temperatures) pyrométrie *f*
PYRONES - [chem] (heterocyclic compounds with a ring of live carbon atoms and one oxygen atom, one of the carbon atoms forming part of a CO group) pyrones *fpl*
PYROPE - [min] (a silicate of magnesium and aluminium; perfectly transparent and often used as a gem) pyrope *m*, grenat *m* de Bohême
PYROPHORIC - [phys] (capable of spontaneous ignition or contact with air) pyrophorique
PYROPHORIC FUEL - [astronaut] (fuel igniting spontaneously in air) propergol *m* pyrophorique
~POWDERS - [chem] poudres *fpl* pyrophoriques
PYROPHOSPHORIC ACID - [chem] (a catalyst) acide *m* pyrophosphorique
PYROPHOTOMETER - [instr] (a pyrometer designed to measure high temperature by means of the luminosity of a substance) pyrophotomètre *m*
PYROPHYLLITE - [min] (a native hydrous aluminium silicate with application as a talc substitute and in ceramics) pyrophyllite *f*
PYROSCHIST - [geol] pyroschiste *m*
PYROSTAT - [instr] (thermostat for very high temperatures) pyrostat *m*
PYROSTATIC - [phys] pyrostatique
PYROSULPHURIC - [chem](disulphuric) pyrosulfurique
PYROSULPHURIC ACID - [chem] acide *m* pyrosulfurique
PYROTARTARIC ACID - [chem] (methylsuccinic acid. A synthesis intermediate) acide *m* pyrotartarique
PYROTECHNIC - [gen] (fireworks) pyrotechnique
PYROTECHNICS - [gen] (the making or using of fireworks) pyrotechnie *f*
PYROXENE - [min] (metasilicate of calcium, magnesium, iron with manganese etc) pyroxène *m*
PYROXILIN PLASTICS - [plast ind] plastiques pyroxyliniques
PYROXYLINS - [chem] (synonymous with gun cotton; nitrocellulose with two to four nitrate groups in the molecule) pyroxyles *mpl*, pyroxylines *fpl*
PYRRHOTITE - [min] (a feebly magnetic ore of iron) pyrrhotine *f*
PYRROLE - [chem] (an easily polymerized yellowish oil with applications as a starting material for a number of pharmaceuticals) pyrrole *m*
PYRROLIDONE - [chem] (an intermediate for rubber chemicals, pharmaceuticals, pesticides and curing agents for synthetic resins) pyrrolidone *f*
PYRUVIC ALDEHYDE - [chem] (a synthesis intermediate) aldéhyde *m* pyruvique
PYRVINIUM CHLORIDE - [chem] (an anthelmintic) chlorure *m* de pyrvinium
PYTHAGOREAN PROPOSITION - [math] théorème *m* de Pythagore
~SCALE - [acoust] (a musical scale) gamme *f* de Pythagore
~THEOREM - s. pythagorean proposition
PYTHIUM DISEASE - [bot] (plant disease) pythium *m* de la betterave, jambe *f* noire
PYURIA - [med] pyurie *f*
PYXIDIUM - [bot] (capsule which dehisces by means of a tranverse circular split; the upper part of the pericard thus forms a lid) pyxide *f*
PYXIS - [anat] cavité *f* cotyloïde

Q

Q - [phys] (symbol for dynamic pressure) symbole de la pression dynamique
~-AERIAL - [radio] (or stub-matched aerial; a dipole fitted with a matching stub) dipôle *m* antenne avec adapteur
~BAND - [radar] (band of wave-lengths between 36,000 and 46,000Mc/s used in radar) bande *f* Q
~-BRIDGE - [instr] (of Jaumann's differential bridge; a type of differential bridge) pont *m* de Jaumann
~-CORRECTION - [astr] (in latitude observations, the correction applied to correct for the lack of coincidence of Polaris with the celestial pole) correction *f* Q
~-ELECTRON - [electron] (electron with an orbit of such dimensions that the electron constitutes part of the seventh shell of electrons surrounding the atomic nucleus) électron Q
~EXTERNAL - [electron] (the quality of a loaded resonating system excluding the resonator loss resistance when considering the load resistance) facteur *m* de qualité pour circuit ouvert
~FACTOR - [el] (measure of the relationship between stored energy and rate of dissipation) coefficient *m* de qualité, facteur *m* de mérite
[el] (magnification factor) coefficient *m* de surtension
[instr] (factor indicating the mechanical strenght of an instrument) facteur *m* de qualité
~FEVER - [med] fièvre *f* Q, fièvre *f* de Mossman
~LOADED - [electron] (the quality of a loaded resonating system considering the load resistance and the resonator loss resistance) facteur *m* de qualité sous charge
~-METER - [instr] (apparatus for measuring Q factors) Q-mètre *m*
~MULTIPLIER - [telev] (type of filter) multiplicateur *m* Q
~-NUMBER THEORY - [phys] (quantized field theory in which wave functions are replaced by operators which in general do not commute) théorie *f* des nombres Q
~POINT - [electron] (quiescent point) point *m* de repos
~SHELL - [electron] (the seventh layer of electrons in motion round the nucleus of an atom) couche *f* Q
~UNLOADED - [electron] facteur *m* de qualité à vide
~VALUE - [nucl] (nuclear disintegration energy) énergie *f* de désintégration nucléaire
QUACK GRASS - [bot] chiendent *m* commun
QUACKING - [acoust] (the sound made by ducks etc) nasillement *m*, couin-couin *m*
QUAD - [paper man] cadrat *m*
[telecomm] (lead-covered cable having a unit group consisting of four paper-insulated conductors twisted together) étoile *f*, quarte *f*
QUAD - [geom] (short for quadrangle) figure *f* quadrangulaire
~CABLE - s. quad [telecomm]
~-PAIR CABLE - [telecomm] (cable containing a number of units, each consisting of four twisted pairs which in turn are twisted about a common axis) câble *m* à paires-étoile
QUADDED CABLE - [telecomm] câble *m* à quartes-étoile
QUADDING MACHINE - [mech] câbleuse *f*, tordeuse *f*
QUADRANGLE - [geom] quadrilatère *m*, tétragone *m*
QUADRANGULAR - [geom] quadrangulaire
QUADRANT - [geom] (a quarter of a circle) quadrant *m*, quart *m* de cercle
[mech] (slotted segmental guide) secteur *m* denté, secteur *m* crénelé
[instr] (an angle-measuring instrument of the sextant type) quadrant *m*, quart *m* de cercle
[mach tool] cavalier *m* (d'un tour à fileter), lyre *f*
~AERIAL - [radio] antenne *f* quadrant
~ARM - [mech] bras *m* du secteur
~CENTRE SHAFT - [mech] pivot *m* du secteur
~CHAIN - [mech] chaîne *f* du secteur
~DIVIDERS - [carp] (dividers in which one limb moves over an arc fixed rigidly to the second limb) compas *m* quart de cercle
~DRUM - [mech] tambour *m* de la chaîne du secteur, barillet *m*
~ELECTROMETER - [instr] (electrometer in which a flat metal needle is deflected between fixed elements shaped like quadrants) électromètre *m* à quadrants
~HANDLE - [mech] manivelle *f*
~IRON - [metall] fer *m* en quart de rond
~MECHANISM - [mech] mécanisme *m* du secteur
~PINION - [mech] pignon *m* du secteur
~PIVOT - [mech] pivot *m* du secteur
~PLATE - [metall] cavalier *m*
~ROD - [mech] arbre *m* des secteurs
~SCALES - [meas] secteur *m* pour échevettes, romaine *f*
~SLIDE - [mech] bras *m* du secteur
~SCREW - [text] vis *f* du secteur
~SPINDLE - [text] broche *f* du secteur
~TOOTHING - [mech] segment *m* denté du secteur
QUADRANTAL - [gen] (relative to a quadrant) quadrantal
~ALTITUDE - [aero] altitude *f* quadrantale
~BEARING - [surv] (the horizontal angle of less than 90 degrees, between a survey line and the magnetic

meridian) relèvement *m* quadrantal
QUADRANTAL COMPONENT OF ERROR - [radar] (the component of the error which varies sinusoidally with twice the bearing) composante *f* d'erreur quadrantale
~ CORRECTION - [aero] correction *f* quadrantale
~ CRUISING LEVEL - [aero] altitude *f* de croisière quadrantale
~ DEVIATION - [naut] (error in a ship's compass due to magnetism induced in the steel hull by the earth's field) déviation *f* quadrantale
~ ERRORS - [radio] (errors in radio D.F. due to the distorsion suffered by a wave on meeting a conductor) erreurs *fpl* quadrantales
~ HEIGHT SEPARATION RULE - [aero] (a rule laying down a specified height which an aircraft flying on a magnetic track within a given quadrant is to mantain) règle *f* des altitudes quadrantales
~ POINTS - [instr] points *mpl* interpolaires
~ SPHERES - [naut] (quadrantal correctors) sphères *fpl* quadrantales
QUADRANTS - [instr] (metal sectors, usually in the form of quarters of a circle, forming part of certain measuring instruments) quadrants *mpl*
QUADRATE - [geom] carré
[anat] (paired cartilage bone of the skull) muscle *m* carré
~ BONE - s. quadrate [anat]
QUADRATIC - [math] quadratique
[math] (an equation) équation *f* du second degré
~ PIEZOELECTRIC EFFECT - [phys] (strain in a crystal caused by a spontaneous polarization) effet *m* piézoélectrique quadratique
~ PRESSURE DROP - [el] chute *f* de tension quadratique
~ SYSTEM - [crystal] (the tetragonal system) système *m* quadratique, système *m* quaternaire
QUADRATRIX - [geom] (the curve by which the quadrature of other curves is obtainable) quadratrice *f*
QUADRATURE - [geom] quadrature *f*
[astr] (the instant of time at which the elongation of a given celestial body is 90 degrees) quadrature *f*
[el] (with the voltage or current) champ *m* transversal, quadrature *f* de phase
[el] (term applied to the two alternating phenomena of equal frequency when there is a phase difference of 90 degrees between them) en quadrature
~ AXIS - [el] axe *m* du champ transversal
~ AXIS COMPONENT OF THE ELECTROMOTIVE FORCE - [el] composante *f* transversale d'une force électromotrice
~ AXIS OF THE VOLTAGE - [el] composante *f* transversale d'une tension
~-AXIS SUBTRANSIENT ELECTROMOTIVE FORCE - [el] force *f* électromotrice subtransitoire transversale
~-AXIS SYNCHRONOUS IMPEDANCE - [el] (the quadrature voltage divided by the quadrature-axis current component in a steady-state condition) impédance *f* synchrone transversale
~-AXIS TRANSIENT ELECTROMOTIVE FORCE - [el] force *f* électromotrice transitoire transversale
~-AXIS VOLTAGE - [el] tension *f* du champ transversal
~ BRUSHES - [contr] balais *mpl* en quadrature
~ COMPONENT - [el] (the component of an alternating current which is in quadrature with the voltage) composante *f* réactive
QUADRATURE COMPONENT OF THE CURRENT - [el] composante *f* réactive du courant
~ COMPONENT OF THE VOLTAGE - [el] composante *f* active de la tension
~ INFORMATION CORRELATOR - [telev] (in colour television) quadricorrélateur *m*
~ OF THE CIRCLE - [geom] quadrature *f* du cercle
~ PHASE - [telev] phase *f* rectangulaire
~ REACTANCE - [el] réactance *f* en quadrature
~ TRANSFORMER - [el] (a transformer so designed that the second electromotive force is 90 degrees displaced from the primary one) transformateur *m* en quadrature
QUADRIC - [geom] (said of a surface, the equation of which is of the second degree) quadrique
[math] (applied when there are more than two variables) quadrique
QUADRICEPS - [anat] quadricéphale
~ MUSCLE - [anat] (muscle with four insertions) muscle *m* quadricéphale
QUADRIGEMINAL - [anat] (tubercules) quadrijumeaux
QUADRILATERAL - [geom] quadrilatère *m*
[geom] quadrilatéral, tétragone
QUADRIMOLECULAR - [chem] (associated with four molecules) quadrimoleculaire
QUADRINOMIAL - [math] quadrinôme *m*
QUADRIPOLE - [el] (network with only two pairs of terminals) quadripôle *m*
QUADRIVALENT - [chem] (tetravalent) quadrivalent, tétravalent
QUADRIVALENCY - [chem] (synonym for tetravalency) quadrivalence *f*
QUADROSONIC RECORDER - [el acoust] magnétophone *m* autonome tétraphonique
QUADRUPED - [zool] quadrupède *m*
QUADRUPLE - [math] quadruple
~-EXPANSION ENGINE - [mech] (steam-engine in which the steam is expanded successively in four cylinders) machine *f* à quadruple expansion
~ HAIR SIEVE BOTTOM - [text] fond *m* de tamis à quadruple crin
~ MULTIPLEX APPARATUS - [telecomm] appareil *m* quadruple
~ NOSEPIECE - [plumb] ajutage *m* quadruple
~-PHANTOM CIRCUIT - [telecomm] (in telegraphy) circuit *m* fantôme quadruple
~ POINT - [chem] (that point in a concentration-pressure temperature-diagram at which a two-component system can exist in four phases) point *m* quadruple
~ SCANNING - [telev] (television system requiring four frames to be scanned so as to give a complete picture) entrelacement *m* quadruple
~ THREAD - [mech] filetage *m* quadruple
QUADRUPLED SCANNING INTERLACE - s. quadruple scanning
QUADRUPLET - [mus] quartolet *m*
[mech] (a set of four elliptic springs) quadruplette *f*
[zool] quadrigémellaire
QUADRUPLEX - [telecomm] quadruplex *m*
~ SYSTEM - [telecomm] (system of telegraphy for the simultaneous transmission of two messages in each direction over one circuit) système *m* quadruplex

QUADRUPLEX VIDEOTAPE RECORDER - [telev] magnétoscope *m* à pistes transversales
QUADRUPLING OF A LINE - [railw] quadruplement *m* des voies
QUADRUPOLE - s. quadripole
~FILTER - [phys] spectromètre *m* quadripôle
~MASS FILTER - [phys] filtre *m* de masse quadripolaire
~MOMENT - [el] moment *m* quadripôle
~RADIATION - [el] (the radiation issued by a quadrupole) rayonnement *m* quadripolaire
QUAGGINESS - [timber] (a defective condition of the timber) fente *f* de coeur, cuadranure *f*
QUAIL - [zool] caille *f*
QUAKE, to - [gen] trembler, branler
QUALIFICATION - [gen] qualification *f*, réserve, aptitude *f*
~TEST - [gen] examen *m* d'aptitude
QUALIFIED - [gen] apte, propre, diplômé, compétent
~EXPERT - [gen] expert *m* diplômé
QUALIFY, to - [gen] qualifier
QUALIFYING - [gen] qualificatif
~PERIOD OF INSTRUCTION - [gen] apprentissage
~SHARES - [fin] actions *f*pl statutaires
~STATEMENT - [gen] déclaration *f* corrective
QUALIMETER - [instr] (a penetrometer; instrument designed to measure the hardness of X-rays) pénétramètre *m*
QUALITATIVE - [gen] (pertaining to quality) qualitatif
~ANALYSIS - [chem] (the operation of detecting and identifying the components of a substance, without regard to the quantities in which they are present) analyse *f* qualitative
QUALITY - [gen] qualité *f*
[acoust] (in sound reproduction) qualité *f*
[mus] (one of the attributes of a note) qualité *f*
~CHARACTERISTIC - [ind] caractère *m* de qualité
~CONTROL - [ind] (the operation of maintaining a standard of quality in a product) contrôle *m* de qualité
~FACTOR - [instr] (of a meter; the ratio of the torque-weight ratio to the angular velocity of the moving element for nominal power) facteur *m* de qualité
~GRADE - [agric] catégorie *f* de qualité
~MARK - [comm] marque *f* de qualité
~OF SLUBBING - [text] qualité *f* de la mèche
~OF TRANSMISSION - [telecomm & radio] (transmission performance) qualité *f* de transmission
~PREMIUM - [agric] prime *f* de qualité
~SEED - [agric] semences *f*pl de qualité
~STAMP - [ind etc] (mark of quality) marque *f* de qualité
~STANDARD - [text] qualité-type *f*
~TERM FOR YARNS - [text] désignation *f* de la qualité des fils
QUANTIFIED SYSTEM ANALYSIS - [electron] analyse *f* quantitative du système
QUANTIFY, to - [phys] quantifier
QUANTIMETER - [instr] (a dose meter) dosimètre *m*
QUANTITATIVE - [gen] (pertaining to quantity) quantitatif
~ANALYSIS - [chem] (the operation of measuring the quantities in which the components of a substance occur in it) analyse *f* quantitative
~EXAMINATION - [gen] analyse *f* quantitative

QUANTITY - [gen] quantité *f*
[math phys etc] quantité *f*
~GOVERNOR - [mech] (volumetric governor) régulateur *m* volumétrique
~METER - [instr] (ampere-hour meter) ampèreheuremètre *m*, compteur *m* de quantité
~OF ELECTRICITY - [el] (the excess of positive or negative electricity on a body or in a space) charge *f* électrique
[el] (the product of a current in a circuit and the time during which it flows) charge *f* électrique
~OF RADIATION - [radiat] (intensity of radiation) quantité *f* de rayonnement, intensité *f* de rayonnement
~PRODUCTION - [gen] fabrication *f* en série
~-SURVEYING - [surv] métrage *m*, toisé *m*
~SURVEYOR - [surv] métreur *m*
QUANTIZATION - [phys] (an observable quantity is quantized when its magnitude is restricted to a discrete set of values) quantisation *f*
[radio] (process in which the range of values of a wave is divided into an infinite number of sub-ranges, each of which is represented by a quantized value within the sub-range) quantification *f*
~DISTORTION - s. quantization noise
~LEVEL - [radio] niveau *m* de quantification
[comput] niveau *m* de quantification
~NOISE - [radio] (distorsion in the process of quantization) bruit *m* de fond de quantification
QUANTIZED - [phys] (pertaining to quantization) quantifié
~FIELD THEORY - [phys] (field theory in which the electro-magnetic and matter fields are represented by operators obeying certain commutation relations) théorie *f* de champ quantifié
~PULSE MODULATION - [radio] (delta modulation) modulation *f* delta, modulation *f* par impulsion quantifiée
~SYSTEM - [phys] (a system of particles, the energies of which can have discrete values only; they can only vary in a discontinuous manner) système *m* quantifié
QUANTIZER - [comput] (a device dealing with an analog quantity supplied by an observing instrument) quantificateur *m*, convertisseur *m* analogique-numérique
QUANTOMETER - [instr] (instrument designed to measure the magnetic flux) quantomètre *m*
QUANTUM - [phys] (a discrete indivisible amount of energy in the Quantum Theory) quantum *m*
~CONDITION - [nucl] (the mathematical condition which must be satisfied for any specified state of an atom) condition *f* quantique
~EFFICIENCY - [phys] (a measure of the efficiency of conversion of light or other energy) rendement *m* quantique
~ELECTRODYNAMICS - [phys] électrodynamique *f* quantique
~EMISSION - [nucl] (the emission of a quant, i.e. a photon) émission *f* quantique
~JUMP - [nucl](a sudden readjustment accompanied by the emission of a quantum of radiant energy) saut *m* quantique
~LEAKAGE - [nucl] (the tunnel effect) effet *m* tunnel
~LIMIT - [radiat] (the shortest wavelenght in a spectrum) limite *f* quantique, longueur *f* d'onde limite

QUANTUM MECHANICAL SYSTEM - s. quantum mechanics
~MECHANICS - [phys] (a mathematical physical theory dealing with the mechanics of atomic systems in terms of quantities which can be measured) mécanique *f* quantique, mécanique *f* des quanta
~NUMBER - [phys] (the number assigned to a value of a quantized quantity in its discrete range) nombre *m* quantique
~NUMBERS - [phys] (of an atom; numbers characterizing the degree of freedom of an atom) nombres *mpl* quantiques d'un atome
~STATE - [phys] niveau *m* d'énergie
~STATISTIC - [phys] statistique *f* quantique
~THEORY - [phys] (the concept of energy as not being continuosly variable in amount, but always existing in multiples of an indivisible discrete minimum termed a Quantum) théorie *f* de quanta
~THEORY OF LIGHT - [opt] théorie *f* des quanta de la lumière
~TRANSITION - s. quantum jump
~VOLTAGE - [electron] (the voltage through which the electron must be accelerated to acquire the enregy which corresponds to a specified quantum) tension *f* quantique
~YIELD - s. quantum efficiency
QUAQUAVERSAL - [geol] structure *f* périclinale
~DIP - [geol] pendage *m* rayonnant
QUARANTINE, to - [med] mettre en quarantaine
QUARANTINE - [med] quarantaine *f*
QUARREL - [glass man] (diamond-shaped pane of glass) carreau *m* de vitrail
[impl] (a glazier's diamond) diamant *m* de vitrier
[tool] (used by stonecutters) burin *m* de tailleur de pierres
[impl] (in engraving) burin *m* losange
QUARRY, to - [gen] extraire de la carrière
[min] exploiter une carrière
QUARRY - [gen] carrière *f*
[glass man] carreau *m* de verre
~BAR - [impl] barre *f* de carrière
~BODY - [mining] tombereau *m* de carrière
~DRAINAGE - [mining] drainage *m* ede carrière
~FACE - [mining] lit *m* de carrière
~-FACED - [constr] (building stone hammer-dressed in the quarry) (moellon) brut
~HEAD - [mining] carreau *m* de la carrière
~-PITCHED - [constr] (roughly squared stone leaving the quarry) brut
~REFUSE - [soil] déchets *mpl* de carrière, pierre *f* perdue
~RUN ROCKFILL - [geol] enrochement *m* tout-venant
~SPALL - [geol] résidu *m* de carrière
~-STONE - [constr] moellon *m*
~-STONE BOND - [constr] appareil *m* en moellons
~STRIPPING - [mining] enlèvement *m* des cosses
~WASTE - [mining] menus *mpl* de carrière
QUARRYING - [mining] exploitation *f* de carrières
QUART - [meas] (a quarter of a gallon) quart *m* de gallon (I,I36 l.)
QUARTAN (FEVER) - [med] fièvre *f* quarte
~AGUE - s. quartan (fever)
QUARTER, to - [gen] diviser en quarte
[mech] caler à 90°
QUARTER - [gen] quart *m*
[gen] (of a year) trimestre *m*
[fin] (currency) (in the USA and Canada) pièce *f* de 25 cents
QUARTER - [astr] (phase of the moon) quartier *m* (de la lune)
[mus] noire *f*
[gen] (a locality) localité *f*, quartier *m*
[naut] aire *f* de vent, côté *m*
[meas] (a quarter of a hundredweight) I2,7 kilogrammes
[mech] (perpendicularity) calage *m* à 90°
[mech] (rough vertical timber in a framework) montant *m*
[food] (of beef etc) quartier *m* (de boeuf etc)
~BEND - [plumb] (in a piece of pipe) coude *m* d'équerre
~BLOCK - [naut] poulie *f* de retour
~-BOUND - [book bind] à demi-reliure
~-CHORD LINE - [aero] (a line passing through the quarter-chord points) ligne *f* à 25 pourcent de la corde
~-CHORD POINT - [aero] (a point on the chord of an aerofoil so positioned as to be distant from the leading edge by an amount equal to one-quarter of the chord lenght) point *m* à 25 pourcent de la corde
~CLOTH BINDING - [book bind] demi-reliure *f*
~-DECK - [naut] gaillard *m* arrière
~GRADE - [text] quart degré *m*
~LEATHER BINDING - [book bind] demi-reliure *f*
~PANEL - [auto] panneau *m* latéral
~-PHASE - [el] diphasé
~-PHASE CIRCUIT - [el] circuit *m* diphasé
~-PHASE SYSTEM - [el] (system in which there are two or more alternating voltages) système *m* diphasé
~PLATE - [photo] plaque *f* (8,2 x I0,8 cm)
~ROUND - [carp] quart *m* de rond
~-ROUND MILLING-CUTTER - [mach tool] fraise *f* à quart de rond
~SAVER - [text] (in a knitting machine) dispositif *m* de blocage
~SAWING - [timber] coupe *f* en quatre
~SPACE LANDING - [constr] palier *m* d'angle
~-SQUARES MULPTIPLIER - [comput] multiplicateur *m* à quart de carrées
~SWING VALVE - [mech] clapet *m* d'étranglement
~TURN - [carp] tordu au quart, semi-croisé
~-TWIST BELT - [mech] courroie *f* tordue d'un quart
~-UNDULATION PLATE - [opt] (a quartz plate which retards one of the two refracted rays by one quarter of a wave length) lame *f* à quart d'onde
~-WAVE - [radio] (equal to one quarter of a wavelength) quart *m* d'onde
~-WAVE AERIAL - [radio] (an aerial having an electrical length corresponding to one quarter of the wavelength) antenne *f* quart d'onde
~-WAVE ANTENNA - s. quarter-wave aerial
~-WAVE LINE - [radio] (a transmission line the length of which is one quarter of a wave) ligne *f* quart d'onde
~-WAVE PLATE -[metall]lamelle *f* quart d'onde
~-WAVE SKIRT DIPOLE - [radio] antenne *f* pliée en quart d'onde
~-WAVE SLEEVE - [electron] (of a waveguide) symétriseur *m* à écran coaxial
~-WAVE TRANSFORMER - [el] transformateur *m* quart d'onde
QUARTERING - [gen] division *f* en quatre
[mech] calage *m* à 90°

QUARTERING THE SAMPLE - [metall] réduction *f* de la prise d'essai
~WAY - [metall] plan *m* de cassure
QUARTERLY - [gen] trimestriel
[print] (any publication which appears every three months) publication *f* trimestrielle
QUARTERMASTER - [naut] maître *m* de timonerie
QUARTERS - [astr] (terms used of the moon's appearance about seven days after the new and full moons, when about one quarter of its visible surface is illuminated) quartiers *mpl* (de lune)
QUARTET - [mus] quotuor *m*
QUARTETTE - [genet] (group of four related nuclei formed as a result of meiosis) quartette *m*
QUARTILE - [statist] (the portion of a frequency distribution curve which includes one quarter of the observed cases) quartile
QUARTZ - [min] (natural silicon dioxide, very widely distributed in many types of rock. It retains high resistivity at elevated temperatures and has many uses in electrical technology) quartz *m*
~BLOW- [min] affleurement *m* de quartz
~COOLED TRANSFORMER - [el] transformateur *m* en quartz
~CRYSTAL - [radio] (a disk of piezoelectric quartz accurately ground, so that its natural resonance occurs at a particular frequency) cristal *m* de quartz
~-CRYSTAL CLOCK - [horol] (synchronous electric clock of very high accuracy) horloge *m* à cristal de quartz
~DELAY-LINE - [electron] (acoustic delay-line in which quartz is used as the medium of sound transmission) ligne *f* à retard à quartz
~DIORITE - [min] (coarse-grained igneous rock consisting of quartz, hornblende, plagioclase feldspar and biotite) diorite *f* quartzifère
~-DOLERITE - [min] (variety of dolerite containing interstitial quartz) dolérite *f* quartzifère
~FIBER DOSE METER - [instr] (a dosimeter designed to record radiation doses; essentially, portable electroscope with a charged quartz-fibre as indicator) dosimètre *m* à fibre de quartz
~FIBRE GAUGE - [instr] manomètre *m* à fil de quartz
~GLASS - [glass man] verre *m* de quartz
~GRANULE - [min] grain *m* de quartz
~LAMP - [light] (a mercury-vapour lamp in a quartz tube, transmitting ultra-violet rays) lampe *f* à tube de quartz
~LENS - [opt] lentille *f* en quartz
~MEMBRANE GAUGE - [instr] manomètre *m* à lame de quartz
~OSCILLATOR - [radio] (a quartz crystal maintained in mechanical vibration at its natural frequency by a thermionic valve) oscillateur *m* de quartz
~PLATE - [el] plaque *f* de quartz piézoélectrique
~PORPHYRY - [min] porphyre *m* quartzifère
~REEF - [geol] filon *m* de quartz
~RESONATOR - [radio] (standard of frequency comparison using a piezoelectric quartz crystals) résonateur *m* à quartz
~ROCK - s. quartzite
~SAND - [min] sable *m* quartzeux
~SCHIST - [geol] schiste *m* siliceux
~SINTER - [geol] travertin *m* siliceux
~SPECTROGRAPH - [instr] (spectrograph detecting radiations in the ultraviolet region of the spectrum) spectrographe *m* à quartz
QUARTZ STRINGER - [min] cordon *m* de quartz
~SYENITE - [min] (potash- or soda syenite, including quartz as an accessory constituent) syénite *f* quartzifère
~TOPAZ - [min] (topaze *f* quartzifère, citrine *f*
~TRACHYTE - [min] (potash or soda trachyte carrying quartz as an accessory constituent) trachyte *f* quartzifère
QUARTZIFEROUS - [min] quartzifère
QUARTZITE - [geol] (a massive metamorphic rock formed by hardened siliceous sandstone or grit) quartzite *f*
QUARTZOSE - [geol] quartzeux
QUASI-ARC WELDING - [el] (arc-welding in which covered iron electrodes are used) soudage *m* au quasiarc
~-DIELECTRIC - [el] quasi-diélectrique
~-DUPLEX - [telecomm] (in telephony a circuit which apparently operates duplex, in fact, functions in one direction only) quasi-duplex *m*
~FERMI LEVELS - [el] (in a semiconductor) quasi-niveaux *mpl* de Fermi
~-OPTICAL WAVES - [radio] (electromagnetic wave which are so short that their propagation is similar to that of visible light) ondes *fpl* quasi-optiques
~-RANDOM ACCESS - [comput] accès *m* quasi instantané
~-STABLE STATE - [electron] (stability of a multivibrator for a certain time) état *m* métastable
~-STATIONARY ENERGY LEVEL - [nucl] niveau *m* quasi-stationnaire
QUASSATION - [med] broyage *m*
QUASSIA - [bot] (the dried stems of Aeschrion excelse with uses in the preparation of medicinal bitters) quassia *f*, bois *m* de quassia
QUATERNARY - [chem] (consisting of four components) quaternaire
[geol] (of the quaternary era) quaternaire
~ALLOY - [metall] alliage *m* quaternaire
~AMMONIUM COMPOUNDS - [chem] (alkylation products having the general formula $R_4N^+X^-$. They have application in textile dyeing, in fungicides and disinfectants, and as emulsion stabilizers) composés *mpl* d'ammonium quaternaires
~DEPOSITS - [geol] dépôts *mpl* quaternaires
~ERA - [geol] ère *f* quaternaire
~EUTECTIC ALLOY - [metall] alliage *m* quaternaire eutectique
~FISSION - [phys] (hypothetical break-up of a nucleus into four fragments) fission *f* quaternaire
~NOTATION - [math] notation *f* quaternaire
~PERIOD - s. quaternary era
~STEEL - [metall] (steel alloy consisting of iron, carbon and two more constituents) acier *m* quaternaire
QUATERNATE - [bot] (in groups of four) quaternifolié
QUATERNION - [math] (mathematical factor) quaternion *m*
[print & paper man] (paper folded in sets of four sheets) cahier *m* de quatre feuilles
QUATREFOIL - [arch] (ornament with four foil) quatre-feuilles *m*
QUAVER, to - [gen] chevroter
QUAVER - [mus] croche *f*
[acoust] chevrotement *m*
QUAY - [constr] quai *m*

QUAY - [naut] appontement *m*
~BERTH - [naut] place *f* à quay
~CRANE - [mech] grue *f* de port
~FRONTAGE - [constr] front *m* de quay
~RATES - [comm] quayage *m*, droits *mpl* de quai
QUAYAGE - [naut] quayage *m*, quais *mpl*
QUEBRACHO - [chem] (the dried bark of Aspidosperma quebrachiol having application in medicine as a febrifuge and bitter) québracho *m*
~EXTRACT - [chem] (a febrifuge and bitter) extrait *m* de québracho
QUEEN - [gen] reine *f*
~-BEE - [zool] abeille *f* mère
~CLOSER - [constr] (a brick cut in two lengthwise and laid headerwise to make up a course) clausoir *m*
~POST - [constr] (the vertical ties of the king-post) double poinçon
~-POST ROOF - [build] comble *m* à doubles poinçons
~-POST TRUSS - [constr] ferme *m* à armature double en arbalète
~TRUSS - [constr] ferme *f* à deux poinçons
QUENCH, to - [metall] (to cool rapidly by immersion in water) refroidir rapidement, tremper
[gen] (to extinguish) éteindre
[el] (to extinguish the spark in a spark-gap when the energy in the primary circuit first becomes zero) étouffer
[electron] amortir
QUENCH AGEING - [metall] vieillissement *m* par refroidissement rapide
~AND FRACTURE TEST - [metall] essai *m* de flexion sur éprouvette refroidie
~AND TEMPER, to - [metall] tremper
~CRACK - [metall] tapure *f* de trempe
~FREQUENCY - [radio] fréquence *f* de découpage
~HARDENING - [metall] durcissement *m* par refroidissement rapide
~IN WATER, to - [metall] éteindre dans l'eau
~TANK - [metall] cuve *f* de refroidissement rapide
~THE LIME, to - [constr] éteindre la chaux
QUENCHED GAP - s. quench spark-gap
~SPARK-GAP - [el] (a spark-gap having electrodes so arranged that they quench the spark) éclateur *m* à étincelles amorties
QUENCHER - [chem] (a substance affecting luminescence) inhibiteur *m*
QUENCHING - [gen] (sudden cooling of a body or substance, e.g. of hot ash, or the contents of a gas retort which would otherwise be too hot to handle) refroidissement *m* rapide
[metall] (sudden cooling used in a heat treatment, e.g. in tempering steel) refroidissement *m* rapide, trempe *f*
[of a fire] extinction *f*
[radio] (the suppression of oscillations) étouffement *m*
~BARROW - [gas ind] crouette *f* d'extinction
~BATH - [metall] bain *m* de refroidissement rapide
~CAR - [gas ind] (coke quenching car) chariot *m* à coke
~CHARGE - [metall] charge *f* refroidie
~CIRCUIT - [nucl] (circuit which diminishes, suppresses or reverses the voltage applied to a counter tube) circuit *m* coupeur
~CRACK - [metall] tapure *f* de trempe
~FURNACE - [metall] four *m* pour trempe
~GAS - [nucl] (in a Geiger-Muller counter tube) gaz *m* de coupage
QUENCHING MEDIA - [metall] bain *m* de trempe
~OF A FLAME - [gen] (extinction of a flame) extinction *f* d'une flamme
~OF ORBITAL ANGULAR MOVENTUM - [electron] extinction *f* du spin
~OIL - [metall] huile *f* de refroidissement rapide
~RAMP - [gas ind] rampe *f* d'extinction
~RESISTOR - [instr] (resistor capable of quenching the circuit of a counter) résistance *f* de coupure
~SALT - [metall] sels *mpl* de refroidissement rapide
~STRESS - [metall] effort *m* dû au refroidissement rapide
~TOWER - [gas ind] tour *f* d'extinction
~WATER - [metall] eau *f* de trempe
QUERCETIN - [chem] (tetrahydroxyflavonol) quercétine *f*
QUERCITOL - [chem] (a hexamethylene derivative, found in oak timber. It forms colourless dextrorotatory crystals) quercitol *m*
QUERCITRON - [chem] (a yellow colourant obtained from species of Quercus and used in the manufacture of yellow lakes) quercitron *m*
~OAK - [bot] quercitron *m*
QUERY, to - [comput] consulter
QUICK - [gen] rapide
~-ACCESS MEMORY - [comput] mémoire *f* rapide
~-ACCESS STORAGE - s. quick-access memory
~-ACCESS STORE - s. quick-access memory
~-ACTING - [gen] à action *f* rapide
~-ACTING FERTILIZER - [agric] engrais *m* à action rapide
~-ACTING REGULATOR - [contr] (for electrical motors) régulateur *m* rapide
~ACTING VALVE - [mech] vanne *f* à fermeture rapide
~-ACTION JOINT - [oil ind] joint *m* rapide
~-ACTION RELAY - [el] relais *m* à action rapide
~-ACTION TREADLE - [text] pas *m* dur
~-ACTION VICE - [mech] étau *m* à action rapide
~-AGEING - [metall] vieillissement *m* rapide
~-AGING - s. quick-ageing
~ASSEMBLY SYSTEM - [mech] (in vacuum technology) principe *m* des unités de montage pour l'agencement rapide des appareillages à vide
~BLOW - [metall] battage *m* rapide
~-BREAK - [el] à interruption rapide
~-BREAK SWITCH - [el] interrupteur *m* de coupure rapide
~CHANGE - [mach tool] changement *m* rapide
~CHANGE ADPTER - [mach tool] mandrin *m* à extraction rapide
~CHANGING MOUNT - [photo] monture *f* pour changement rapide
~CHARGE - [el] (or boost change; partial charge at a high rate for a short period) charge *f* rapide (de biberonnage)
~CLOSING VALVE - [mech] vanne *f* à fermeture rapide
~COMBUSTION - [phys] combustion *f* vive
~CONNECTION - [mech] (small flange) petite bride *f*
~CONNECTIONS - [mech] raccords *mpl* rapides
~COUPLING - [mech] accouplement *m* à débrayage rapide
~DEMOUNTABLE RIM - [auto] jante *f* à démontage rapide
~DISCONNECT COUPLERS - [auto] (hydraulic fittings)

raccords *mpl* rapides
QUICK DRYING PASTE - [ind chem] colle *f* à prise rapide
~FEED - [mach tool] avance *f* rapide
~-FIRE CAMERA - [photo] chambre *f* à tir rapide
~-FIRING GUN - [firearms] canon *m* à tir rapide
~-FREEZING - [ind chem etc] congélation *f* rapide
~GOUGE - [tool] (gouge with a curved cutting edge) gouge *f* à bec de corbin
~-GRIP REVOLVING BENCH-VICE - [mech] étau *m* à base tournante à serrage
~GRINDING - [mech] broyage *m* instantané rapide
~GROUP - [soil] terrain *m* coulant
~-LEVELLING HEAD - [surv] (special fitting supplied for certain levels to facilitate setting up the instrument) tête *f* à calotte sphérique
~-LEVELLING TRIPOD - [surv] pied *m* à calotte sphérique
~MAKE-AND-BREAK SWITCH - [el] (a snap switch: a switch which makes and breaks a circuit with a quick snap by means of a blade having a rate of motion which is independent of the operator's action) interrupteur *m* rapide
~-MATCH - [firearms] mèche *f* d'artilleur
~MOUNT - [photo] monture *f* rapide
~OPERATING RELAY - [telecomm] relais *m* rapide
~PAINT - [paint] (quick-drying paint) peinture *f* rapide
~-PITCH SCREW - [mech] vis *f* à pas allongé
~-RELEASE - [el] rupture *f* immédiate, déclenchement *m* rapide
~RELEASE ADAPTER - [mech] (a type of adapter designed to be detached quickly) adapteur *m* à démontage rapide
~-RELEASE VALVE - [mech] soupape *f* à ouverture
~RETTING - [text] rouissage *m* rapide
~RETURN - [mach tool] (a reciprocating motion for operating the tool) retour *m* rapide
[adj] à retour rapide
~-RETURN STROKE - [mech] course *f* de retour rapide
~-SET LEVEL - s. quick-setting level
~-SETTING - [chem] à prise rapide
~-SETTING CEMENT - [constr] ciment *m* à prise rapide
~-SETTING LEVEL - [surv] (level fitted with a quick-levelling head) niveau *m* à tête à calotte sphérique
~SOLDER - [metall] soudure *f* vive
~TENSIONING - [photo] armement *m* rapide
~-THREAD SCREW - [mech] vis *f* rapide
~-TRAVERSE - [mach tool] amenage *m* rapide
~TRAVERSE BOBBIN - [text] bobine *f* à fil croisé
~TRAVERSE WINDER FOR BOBBINS OF EMBROIDERING MACHINES - [text] bobineuse *f* croisée pour métiers à broder
~TRAVERSE WINDER WITH SLOTTED CAM DRUM - [text] bobinoir *m* croisé à came à rainure
~TRAVERSE WINDING FRAME - [text] bobinoir *m* à fil croisé
~TRIPPING RELAY - [el] relais *m* à action rapide
QUICKEN, to - [gen] vivifier, stimuler, s'accélérer
QUICKENING - [med] premiers mouvements *mpl* du foetus
QUICKENS - [bot] froment *m* rampant
QUICKING - [metall] couche *f* en mercure, placage *m* au mercure
QUICKLIME - [ind chem] (commercial calcium oxide, obtained by roasting limestone or chalk) chaux *f* vive, chaux *f* anhydre
QUICKSAND - [geol] (loose sand with a very low bearing pressure) sables *mpl* mouvants, lise *f*
QUICKSILVER, to - [glass man] étamer (une glace)
QUICKSILVER - [chem] (common name for mercury) mercure *m*, vif-argent *m*
QUIESCENCE - [gen] repos *m*, quiétude *f*
QUIESCENT - [gen] en repos
~AERIAL - [radio] (a dummy aerial) antenne *f* postiche
~ANTENNA - s. quiescent aerial
~CARRIER - [radar] (the suppressed carrier wave in radar during resting periods) suppression *f* de la porteuse
~-CARRIER MODULATION - [radio] (system of modulation in which the carrier is radiated only when modulation is taking place) modulation *f* à suppression de la porteuse dans les silences
~CURRENT - [electron] (the electrode current which corresponds to the electrode bias voltage) courant *m* de repos
~POINT - [electron] (the point on a characteristic curve which corresponds to the electrode bias voltage) point *m* de repos
~PUSH-PULL AMPLIFICATION - [radio] (a balanced operation in which both electronic tubes operate under class-B-condition) amplification *f* push-pull classe B
~PUSH-PULL AMPLIFIER - [telecomm] amplificateur *m* push-pull classe B
~STATE - [phys] état *m* de repos
~TANK - [hydr] (a type of sedimentation tank in which the sewage rests for a time) bassin *m* de sédimentation intermittente
QUIET - [gen] tranquille, calme
[mech] (of engine and motor) silencieux
~AUTOMATIC GAIN CONTROL - [radio] réglage *m* automatique silencieux de gain
~CIRCUIT - [telecomm] circuit *m* silencieux
~LIME - [ind chem] chaux *f* maigre
QUIET, to - [metall] laisser se calmer
QUIETING SENSITIVITY - [radio] (the minimum signal input required to give a specific output signal-to-noise ratio under certain conditions) seuil *m* de sensibilité
QUILL - [mech] (hollow shaft revolving on a solid spindle) arbre *m* creux (qui tourne autour d'un arbre plein), fourreau *m*
[mach tool] (of an internal grinder) quille *f*, fourreau *m*
[zool] (in birds) tuyau *m* de plume
[text] canette *f*, bobine *f* de trame
[text] (of a feather) cannette *f* de tisserand
[el] (form of drive used for electric locomotives in which the armature of the driving motor is mounted on a quill surrounding the driving axle) arbre *m* creux
[mus] (in the early harpsichord mechanism) plume *f*, pipeau *m*
~BIT - [tool] bit *m* à quille
~BOX - [text] boîte *f* à canettes, caisse *f* à canettes
~DRIVE - [railw] (in electrical locomotives; see quill) transmission *f* à arbre creux
~FEATHER - [zool] penne *f*
~GEAR - [mech] engrenage *m* à arbre creux
~-PEN - [impl] plume *f* d'oie

QUILLED - [bot] tubuliflore
QUILLER - [text] canetière *f*
QUILLING - [text] entoilage *m*, tuyautage *m*
QUILT, to - [gen & text] piquer, ouater
QUILT - [text] couverture *f* piquée, édredon *m* piqué
[plumb] (an insulating material for pipes) revêtement *m* isolant
[acoust] panneau *m* d'absorption acoustique
~WADDING - [text] ouate *f* pour couverture piquée
QUILTED COAT - [text] douillette *f*
~GOODS - [text] piqué *m*
QUILTING COTTON - [text] ouate *f*
~MACHINE - [text] machine *f* à piquer
QUINALDINE - [chem] (2-methyl-quinoline; occurs in quinoline derived from coaltar) quinaldine *f*
QUINARY NOTATION - [math] notation *f* quinaire
QUINCE - [bot] coing *m*
~TREE - [bot] cognassier *m*
QUINHYDRONE - [chem] (a compound obtained by the addition of one molecule of p-quinone to one molecule of hydroquinone) quinhydrone *f*
~ELECTRODE - [el] (a gold or platinum electrode immersed in a saturated solution of quinhydrone) (contained in a half-cell) électrode *f* à la quinhydrone
QUINHYDRONE HALF-CELL - [el chem] demi-cellule *f* à la quinhydrone
QUINIC ACID - [chem] (chinic acid; a synthesis intermediate) acide *m* quinique
QUINIDINE - [chem] (a stereo-isomer of quinine obtained from species of Cinchona and used in the treatment of cardiac conditions) quinidine *f*
~SULPHATE - [chem] (bitter, acicular crystals used in the treatment of cardiac disease) sulfate *m* de quinidine
QUININ - s. quinine
QUININE - [chem] (a natural alkaloid, of major importance as an antimalarial agent, obtained from species of Cinchona) quinine *f*
~CARBACRYLIC RESIN - [chem] (a polyacrylic carboxylic resin used as a diagnostic aid in gastric conditions) résine *f* carbacrylique de quinine
~ETHYLCARBONATE - [chem] (in an antimalarial) éthylcarbonate *m* de quinine
~HYDROCHLORIDE - [chem] (an àntimalarial) chlorhydrate *m* de quinine
~SULPHATE - [chem] (an antimalarial agent) sulfate *m* de quinine
QUINISM - [med] quinisme *m*
QUINOL - [chem] (reducing agent used as a developer) quinol *m*
QUINOLINE - [chem] (a heterocyclic compound, obtained by the condensation of a pyridine and a benzene ring. An intermediate for pharmaceuticals) quinoléine *f*
~YELLOW - [dyes] (a dyestuff formed from phthalic anhydride and quinaldine) jaune *m* de quinoléine
QUINONE - [chem] (an intermediate for dyestuffs) quinone *f*
QUINOXALINES - [chem] (heterocyclic compounds formed by the condensation of a benzene and a diazine ring. Synthesis intermediates) quinoxalines *fpl*
QUINQUEMOLECULAR - [chem] (associated with five molecules) pentamoleculaire
QUINQUIVALENT - [chem] (synonymous with pentavalent) pentavalent
QUINSY - [med] (acute inflammation of the tonsils) esquinancie *f*
QUINTAL - [meas] (100 kg) quintal *m*
QUINTESSENCE - [gen] (essential principles) quintessence *f*
QUINTET - [mus] (composition for five voices or instruments) quintette *m*
QUINTETTE - s. quintuplets
QUINTUPLE - [math] quintuple
~POINT - [chem] (the point at which a three-component system can exist in five phases) point *m* quintuple
[phys] (the temperature at which five phase are in equilibrium) point *m* quintuple
QUINTUPLETS - [genet] cinq *m* enfants nés d'une seule couche
QUIRE - [paper man] (25 sheets) main *f* de papier
QUIRED PAPER - [paper man] papier *m* en cahiers
QUIRK - [gen] (a short turn, a twist) tour *m*
[carp] (narrow groove along a bead) carré *m*
[impl] (a quirk float; plasterer's trowel) truelle *f* à plâtre
~-BEAD - [carp] gorgerin *m*
~FLOAT - [impl] truelle *f* pour gorges
~MOULDING - [carp] (moulding with a small groove in it) moulure *f* à gorge
~OGEE - [arch] doucine *f* à carré
~OVOLO - [arch] boudin *m* à carré
QUIT, to - [gen] quitter
[gen] (to stop, e.g. a job of work) abandonner
QUITTANCE - [gen & comm] quittance *f*
QUITTER - [metall] scorie *f* d'étain
QUITTOR - [vet] javart *m* cartilagineux
QUOIN - [constr] (salient angle of a building) pierre *f* d'angle, angle *m* de mur
[constr] (of an arch) (wedge; shaped stone of an arch) claveau *m*, voussoir *m*
[print] (wedge used to lock up type) coin *m*, cale *f*
[mech] coin *m* (pour caler)
~KEY - [print] clef *f*
~STONE - [constr] pierre *f* d'angle
QUOIT - [mech] (disk of iron or other material with a hole in the centre) palet *m*
QUORUM - [leg] (the number of members of a legal body which is required for the legal transaction of business) quorum *m*
QUOTA - [gen] (a share or specific part) quote-part *f*
QUOTATION - [comm] cotation *f*, prix *m*
[print] cadrat *m* creux
~MARKS - [print] guillemets *mpl*
QUOTE, to - [comm] établir un prix, coter
QUOTIDIAN - [gen] quotidien
QUOTIENT - [math] (the result obtained by dividing one number by another) quotient *m*
~METER - [instr] logomètre *m*
~RELAY - [el] relais *m* de quotient
QUOTIMETER - [electron] pénétromètre *m*

R

r. - [chem] (symbol for a specific refraction) symbole d'une réfraction spécifique
[met] (Beaufort letter for rain) symbole (Beaufort) de la pluie
[chem] (symbol for organic hydrocarbon radical) symbole du radical de l'hydrocarbure
[geom] (short for radius) radius *m*
[opt] (red primary) rouge *m* primaire
R.A. - [astr] (initials of Right Ascension) ascension *f* droite
R-ACID - [chem] (an acid used in the preparation of azo-dyes for wool) acide *m* R
~BLACK LEVEL - [telev] niveau *m* minimal pour le signal rouge
R.F. - [radio] (Radio Frequency) haute fréquence *f*
R PEAK LEVEL - [telev] niveau *m* maximal pour le signal rouge
R UNIT - [el] (the international unit of X-ray or gamma ray quantity) röntgen *m*
ra - [chem] (a prefix placed before the symbol for an element, to denote that a radio-active isotope of that element is meant, e.g. rana denotes radiosodium) ra-
Ra - [chem] (symbol of radium) symbole *m* de radium
RABBET, to - [carp] (to cut a groove) feuiller
[carp] (to unite parts in a close joint) assembler à feuillure
RABBET - [carp] (a groove, a recess) rainure *f*, feuillure *f*
[constr] (groove in the edge of a door) battant *m*
[impl] (of a forging hammer) rabat *m* de marteau pilon
~FOR GLAZING - [carp] battant *m* pour la pose de vitres
~IRON - [impl] fer *m* de guillaume
~JOINT - [carp] assemblage *m* à feuillure
~PLANE - [tool] (plane used to cut a groove) guillaume *m*
RABBETED LOCK - [carp] (lock fitted into a recess cut in the edge of a door) serrure *f* enfoncée
RABBETTING - [med] assemblage *m* à feuillure
RABBIT - [nucl] (a small container which can be passed through a tube in a nuclear reactor so as to expose its contents to irradiation in the active section) furet *m*
[nucl] (single stage recycle) recyclage *m* à étage unique
[zool] lapin *m*
[oil ind] jauge *f* pour tuyaux
~HAIR - [text] poil *m* de lapin
~HOLE - [nucl] tube *m* pneumatique
~HUTCH - [agric] cabane *f* à lapins
~-SKINS - [leather ind] peaux *f*pl de lapin
RABBIT STAND - [metall] poupée *f* de rabat
RABBLE, to - [metall] (to stir with a rabble) ringarder
RABBLE - [metall] (iron tool, usually bent at one corner, to skim melted iron) ringard *m*
[metall] (a tool used to agitate a metal bath) crochet *m*, râble *m*
RABBLING - [metall] brassage *m* d'un bain
~MECHANISM - [metall] mécanisme *m* à brasser
RABIES - [vet] (an acute disease of dogs, wolves etc. Hydrophobia) rage *f*, hydrophobie *f*
~VIRUS - [med] virus *m* rabique
RACCOON - [zool] raton *m* laveur
RACE, to - [mech] (to run an engine at high speed without load) emballer (un moteur), s'emballer
RACE - [gen] race *f*, descendance *f*
[gen] course *f*
[mech] (the ring of a ball bearing) bague *f* de roulement
[mech] (the raceway of a ball bearing) voie *f* de roulement
[mech] (of a pulley) cage *f*
[hydr] (channel conveying water, connected to a hydraulically operated machine) bief *m*, rigole *f*
[mech] (the groove along which a piece moves or slides) chemin *m* de roulement
[text] voie *f* de la navette
[aero] (the air stream produced by the propeller) courant *m* d'air (de l'hélice)
[bot] racine *f* de gingembre
[naut] (strong current) raz *m* de courant
~BOARD - [text] lit *m* de la navette
~-COURSE - [gen] champ *m* de courses
~-FINISH RECORDING - [photo] photographie *f* de l'arrivée des courses
~-HORSE - [zool] cheval *m* de course
~LEVEL - [text] plan *m* du battant
~OF SLAY - [text] semelle *f* du battant
~PLATE - [text] sommier *m* du battant
~ROTATION - [phys] rotation *f* de l'écoulement
~-TRACK - [phys] (an assembly of calutrons in the shape of a race track; they have a common magnetic field) tube *m* électronique de forme ovale
[nucl] piste *f*
~-WAY - [hydr] (channel conveying water) bief *m*
[mech] (of a ball bearing) piste *f*
[el] (in electrical traction system, a tube for the protection of wires) canalisation *f*
RACEMATION - s. synonymous with Racemization
RACEME - [bot] (an inflorescence without definite shape) racème *m*
RACEMIC - [opt] (optically inactive, but capable of resolutions into forms of opposite activity) racé-

mique
RACEMIC ACID - [chem] (a synthesis intermediate obtained as a by-product in the production of tartaric acid) acide *m* racémique
~COMPOUNDS - [chem] (those which contain equal quantities of the d- and l- forms of enantiomorphous stereoisomers, and are optically inactive) composés *mpl* racémiques
RACEMIZATION - [chem] (conversion of an optically active compound to its racemic form) racémisation *f*
RACEPHEDRINE HYDROCHLORIDE - [chem] (a sympathomimetic amine with therapeutic uses) chlorhydrate *m* de racéphédrine
RACER - [text] aspe *m*
RACEWAY - s. race-way
RACHITIC ROSARY - s.rachitic beads
~BEADS - [med] chapelet *m* costal
RACIAL - [gen] de race, d'une race
RACIALISM - [gen] racisme *m*
RACING - [mech] (the running of an engine at high speed without load) emballement *m*
RACK, to - [brew ind] (in general, to draw off clear liquor from lees etc, e.g. wine or beer) soutirer, entonner
(of ropes etc) (to stretch and tear apart) arracher
(of rubber) étirer
[mech] déplacer par crémaillère
[mech] (a machine) détraquer (une machine)
RACK - [gen] (a framework for the storage of materials) étagère *f*, casier *m*, râtelier *m*
[mech] (a bar or rod on which teeth like those of a gear wheel are cut, to allow it to be moved in a straight line by a rotating pinion) crémaillère *f*
[print] barre *f* dentée
[el] (vertical mounting frame used for the mounting of panels) bâti *m*
[railw] (for luggage) porte-bagages *m*
[photo] (a drying rack) chevalet *m*
[mech] (of a turbine) grille *f* (de turbine)
[th eng] barre *f* à crémaillère
[impl] (in an oven) plafond *m* mobile
[gen] (a shelf; only USA) clayette *f*
~AND PIN PROCESSING UNIT - [cin] (implement used for the processing of short lengths of films) système *m* à crémaillère, mécanisme *m* à crémaillère
~AND PINION - [mech] (a gear wheel engaging a toothed bar, to convert rotary to rectilinear motion) crémaillère *f* et pignon
~-AND-PINION DRIVE - [mech] transmission *f* à crémaillère
~-AND-PINION FOCUSING - [photo] mise *f* au point à crémaillère
~-AND-PINION GEAR - [mech] engrenage *m* à crémaillère
~-AND-PINION JACK - [mech] cric *m* à crémaillère
~AND PINION MECHANISM - [mech] (mechanical device consisting of a toothed bar engaging a pinion, used to convert rectilinear to rotary motion or vice-versa) mécanisme *m* à crémaillère
~-AND-PINION PLANING MACHINE - [mech] raboteuse *f* à commande par pignon et crémaillère
~-AND-PINION PRESS - [mech] (a press in which the platen is moved by a rack engaging a pinion on a shaft rotated by hand or power) presse *f* à crémaillère et pignon
RACK-AND-PINION STEERING-GEAR - [auto] mécanisme *m* de direction à crémaillère
~BRIGHT, to - [brew ind] entonner la bière clarifiée
~CABLING - [el] câblage *m* de panneau
~CUTTING MACHINE - [mech] fraise *f* pour tailler les crémaillères
~DRYING - [text] séchage *m* sur claies
~FEED - [mech] amenage *m* par crémaillère
~FOR DRAINING WET WOOL - [text] râtelier *m* pour l'égouttage de la laine mouillée
~FOR MOULDS - [metall] étagère *f* à moules
~FOR THE TENSION ROLLER - [text] tringle-guide *f*
~-GENERATING CUTTER - [mech] outil *m* à tailler par génération les crémaillères
~GREEN, to - [brew ind] traverser vert
~HOUSING - [text] châssis *m* à crémaillère
~LATH - [constr] latte *f* de râtelier
~LOADING FORK - [agric] fourche *f* à foin pour siccateurs
~LOCOMOTIVE - [railw] locomotive *f* à crémaillère
~MOUNTING - [el] (the use of standard racks for mounting panels etc) montage *m* sur panneau
~-PINION - [railw] pignon *m* engrenant sur la crémaillère
~POST - [text] colonne *f* support
~RAIL - [railw] (of a rack railway) crémaillère *f*
[mech] rail *m* denté
~RAILWAY - [railw] (mountain railway in which additional adhesion is obtained through a toothed rail placed between the regular rails) chemin *m* de fer à crémaillère
~RELEASE - [text] rappel *m* de la crémaillère
~ROD - [mech] crémaillère *f*
~SAW - [impl] (saw with wide teeth) scie *f* à bûches
~SECTION - [el] bâti *m* unitaire
~-SHAPED CUTTER - [tool] outil *m* à crémaillère
~-SHAPED TOOL - s.rack-shaped cutter
~SPANNER - [impl] clef *f* à crémaillère
~TOOTH - [mech] dent *f* de la crémaillère
~-TYPE DIFFERENTIAL - [contr] différentiateur *m* normalisé insérable
~VICE - [mech] étau *m* à crémaillère
~WHEEL - [mech] roue *f* dentée
RACKBOARD - [mus] (in an organ) faux-sommier *m*
RACKED BACK TOOTHING - [constr] gradine *m* d'attente de la maçonnerie
~RUBBER - [rubber ind] caoutchouc *m* étiré
RACKER - [brew ind] soutireuse *f*, bock *m* de soutirage
RACKET - [gen] raquette *f*
~POCKET - [impl] enveloppe *f* pour raquette
RACKING - [brew ind] (the operation of conveying the beer by a hose from the vat to the cask) soutirage *m*
[hydr] (the conveying of liquids from one container to another) transvasement *m*
[naut] bridure *f*
[min] (separation of ore by washing on an inclined plane) séparation *f* par plan incliné
~CAPACITY - [oil ind] capacité *f* de stockage du derrick
~CHAIN - [text] chaîne *f* pour chevalement
~COCK - [brew ind] (plug cock or valve used to draw off liquor in racking) canule *f*
~HOSE - [brew ind] (a hose used to empty a container of fluid) tuyau *m* de soutirage
~MACHINE - s. racker

RACKING MOTION - [text] dispositif *m* de chevalement
~ROOM - [brew ind] (room used for racking) local *m* de soutirage
~SQUARE - [brew ind] (tank used in racking beer etc) cuve *f* de soutirage
~TANK - s. racking square
~WHEEL - [text] roue *f* de chevalement
RACKWORK - s. rack mechanism
RACON - [radar] (a responder beacon, or transponder; a device enabling a receiver belonging to the transmitter to recognize the distance and the direction of the beacon) radar *m* de radionavigation, balise *f* répondeuse
RAD - [radiat] (the unit of absorbed dose; 100 ergs per gramme) rad *m*
RADAR - [radar] (system of locating distant objects by measuring at least two of the coordinates defining their position with respect to the receiver) radar *m*
~AERIAL - [radar] antenne *f* radar
~ALTIMETER - [instr] radioaltimètre *m*
~ALTITUDE - [radar] altitude *f* radioaltimétrique
~ANTENNA - s. radar aerial
~AUTOMATIC TRACKING - [radar] radar *m* poursuite automatique
~BALLOON - [radar] ballon *m* sonde avec radar
~BEACON - [radar] (racon) balise *f* répondeuse
~BEAM - [aero] faisceau *m* de radar
~BLIP - [radar] (colloq. for spike) top *m* d'écho
~CAMOUFLAGE - [radar] camouflage *m*
~CHARTS - [surv] (a method of producing charts by means of radar screen pictures) cartes *f*pl par radar
~COASTAL PICTURE - [radar] image *f* de ligne côtière
~CONTACT - [radar] (identification of a given object by radar echo) contact *m* radar
~CONTROL - [radar] (air-traffic control by radar) contrôle *m* radar
~CONTROLLER - [radar] contrôleur *m* radar
~DETECTION - [radar] localisation *f* radar
~DRAWING - [radar] (of the group) levé *m* du terrain radar
~DUCT - [radar] guide *m* radar
~ECHO - [radar] écho *m* radar
~EQUIPMENT - [radar] équipement *m* radar
~FREQUENCY - [radar] bande *f* de fréquence d'un radar
~HOMING SYSTEM - [aero] système *m* de homing radar
~HORIZON - [radar] horizon *m* radar
~INDICATOR - [radar] radaroscope *m*
~MONITORING - [aero] (air-traffic radar procedure to give warning of dangerous traffic situation to pilots) contrôle *m* radar
~MOSAICS - [radar] (a picture which is built up by combining a number of radar photographs on the same scale) mosaïque *f* du radar
~PHOTOGRAPH - [radar] image *f* de radar
~PILOTAGE - [radar] pilotage *m* électronique
~PLOTTING - [radar] diagramme *m* de l'information radar
~PRISM - [radar] (a prism which is used in radar system to balance the blind sectors of a picture) prisme *m* de radar
~RANGE - [radar] portée *f* d'un radar, distance *f* mesurée par un radar
RADAR RANGE CALCULATOR - [radar] calculateur *m* de la porté du radar
~REFLECTIVITY - [radar] coefficient *m* de réflexion d'un cible au radar
~REFLECTOR - [radar] réflecteur *m* de radar
~SCANNING - [radar] balayage *m* par radar
~SCREEN - [radar] rideau *m* de radars
~SCREEN PICTURE - [radar] image *f* radar sur écran
~SERVICE - [radar] service *m* radar
~SCOPE AFTERGLOW - [radar] (fluorescence on the screen of a radar cathode ray tube after the electron beam has ceased to excite it) persistance *f* sur l'écran de radar
~SET - [radar] équipement *m* radar
~SHADOW - [radar] angle *m* mort de radar
~SPEED METER - [instr] (used to check the speed of traffic) tachymètre *m* à radar
~STRIP MAP - [surv] levé *m* radar sur bande de papier
~SURVEILLANCE - [radar] (control of aircraft in the neighbourhood of an airfield or during approach and landing, by means of radar) surveillance *f* radar
~TARGET - [radar] cible *f* de radar
~TRACKING - [radar] poursuite *f* radar
~TRANSMITTER - [radio] (the transmitting portion of a radio detection and ranging system) émetteur *m* radar
~VOLUME - [radar] espace *m* balayé par un radar
~WAVE - [radar] onde *f* de radar
~WIND - s. radar balloon
RADARSCOPE - [radio] radarscope *m*
RADDLE - [text] (a half reed used to spread the threads) râteau *m*, peigne *m* répartiteur
RADECTOMY - [med] excision *f* de la racine d'une dent
RADESYGE - [med] gale *f* norvégienne
RADIAL - [gen] (pertaining to a radius) radial
[phys] (radiating out from a common centre) radial
~AND THRUST BEARING - [mech] butée *f* à billes radiale
~ARM - [mech] bras *m* radial
[mach tool] (as a screw-cutting lathe) lyre *f*, tête *f* de cheval
~ARTERY - [anat] artère *f* radiale
~AXLE - [mech] essieu *m* pivotant
~BALL BEARING - [mech] coussinet *m* à billes radial
~-BEAM TUBE - [electron] tube *m* à faisceau radial
~BEARING - [mech] coussinet *m* radial
~CAM - [mech] came *f* radiale
~-CHART RECORDER - [comput] enregistreur *m* de coordonnées polaires
~COMMUTATOR - [el] (direct-current machine commutator having bars arranged radially from the axis, so as to form a disk instead of a cylinder) commutateur *m* radial
~DEFLECTING ELECTRODE - [electron] électrode *f* radiale de déviation
~DEFLECTION - [electron] déviation *f* radiale
~DEPRESSION - [anat] fossette *f* condylienne
~DERIVATION OF RIMSEAT - [rubber ind] dérivation *f* radiale du siège du talon
~DISPLACEMENT - [mech] déplacement *m* radial
~DISTRIBUTION FUNCTION - [phys] (for a liquid) fonction *f* de répartition radiale
~DISTRIBUTION METHOD - [phys] (statistical method for the analysis of data obtained by measuring the intensity of x-ray diffraction at various angles) mé-

thode *f* de la répartition radiale

RADIAL DRILL - [mach tool] (large drilling machine, the head of which can be adjusted radially along a rigid arm carried by a pilar) machine *f* à percer radiale

~DRILLING MACHINE - s. radial drill

~DUCTS - [el] (ventilating ducts in an electrical machine, running radially from the shaft) conduite *f* de ventilation radiale

~ENGINE - [mech] (a type of reciprocating i.c. engine in which the cylinders are disposed radially round the crankshaft) moteur *m* en étoile

~FAULT - [geol] faille *f* radiale

~FEEDER - [el] (independent feeder) feeder *m* de sous-station

~-FLOW - [mech] écoulement *m* radial

~FLOW BASIN - [hydr] bassin *m* à écoulement radial

~-FLOW IMPELLER - [hydr] roue *f* à aube à écoulement radial

~FLOW PUMP - [med] pompe *f* à flux radial

~-FLOW TURBINE - [mech] (a gas turbine in which the general direction of gas flow is radial) turbine *f* radiale

~FORCE - [phys] force *f* centrifuge

~GATE - [hydr] vanne *f* à segment

~GRATING - [electron] filtre *m* radial des ondes parasites

~GRINDER - [mach tool] rectifieuse *f* radiale

~GUIDES - [gas ind] guidage *m* radial

~GUIDING - s. radial guides

~HOOK - [mech] (circular saw. The inclination, in respect to the radius, of the leading edge of the tooth) angle *m* d'inclinaison d'une scie circulaire

~INWARD-FLOW TURBINE - [mech] turbine *f* radiale centripète

~LOCATION - [gen] emplacement *m* radial

~MILLING CUTTER - [mach tool] fraise *f* à dents radiales

~NERVE - [anat] nerf *m* radial

~NETWORK - [el] réseau *m* radial

~PACKING - [mech] joint *m* radial

~PINS - [text] aiguilles *fpl* radiales

~PLAY - [mech] jeu *m* radial

~PRESSURE - [phys] pression *f* radiale

~PUMP - [mech] (pump with radially disposed cylinders) pompe *f* radiale

~QUANTUM NUMBER - [phys] nombre *m* quantique radial

~RAY - [opt] rayon *m* radial

~RECORDING - [el acoust] (lateral recording) enregistrement *m* latéral

~ROLLER BEARING - [mech] coussinet *m* à billes radiales

~ROTOR - [aero] (the rotor of a gyroplane or helicopter) rotor *m* radial

~SEAL - [mech] anneau-joint *m* radial

~SEAL RING - s. radial seal

~SHAKE - [timber] crevasse *f* radiale

~STRESS - [phys] effort *m* radial
[soil] contrainte *f* radiale

~SYMMETRY - [zool] symétrie *f* radiale

~SYSTEM - [el] (of cables, distribution system in which the cables radiate from a supply station) réseau *m* radial

~TAPPING SWITCH - s. radial commutator

~TRANSMISSION LINE - [radio] (of waveguides) ligne *f* de transmission radiale

RADIAL TRIANGULATION - [surv] triangulation *f* radiale

~TRUCK - [railw] bogie *m*

~VALVE GEAR - [mech] (steam-engine valve gear in which the side-valve is given independent component motions which are in proportion to the sine and cosine of the crank angle) distribution *f* radiale

~VELOCITY - [astr] (line of sight velocity) vitesse *f* radiale

~WALL - [bot] paroi *f* radiale

~WEIR - [hydr] barrage *m* à segment

RADIALLY OPERATED NETWORK - [el] réseau *m* à exploitation radiale

RADIAN - [math] (unit of circular measure, that is an arc equal in length to the radius of a circle of which it is a part) radian *m*

~FREQUENCY - [telecomm] (in telephony; angular frequency) pulsation *f*, vitesse *f* angulaire

RADIANCE - [gen] splendeur *f*
[phys] rayonnement *m*, radiation *f*, radiance *f*
[opt] (the radiant flux per unit solid angle per unit of projected area of the source) flux *m* énergétique spécifique

RADIANCY - s. radiance

RADIANT - [gen] radiant, rayonnant
[phys] radiant *m*, point *m* radiant
[astr] (the point of the heaven from which parallel tracks, like those of meteors in a shower, appear to originate) radiant *m*
[bot] rayonnant

~BURNER - [th eng] brûleur *m* radiant

~COOLING - [phys] refroidissement *m* par rayonnement

~DENSITY - [phys] densité *f* de rayonnement

~ELEMENT - [el] (an element in which the heat emission is radiant, in particular infra-red) élément *m* rayonnant

~ENERGY - [el] (the energy which is transferred by electromagnetic waves without a corresponding transfer of matter) énergie *f* de rayonnement

~ENERGY THERMOMETER - [instr] thermomètre *m* mesurant l'énergie rayonnée

~FLUX - [phys] (time rate of transfer of radiant energy) flux *m* de rayonnement

~FLUX DENSITY - [el] (irradiance; the measure of radiant power per unit area flowing across a surface) éclairement *m* énergétique

~HEAT - [phys] (the heat which is communicated to a body by radiation) chaleur *f* rayonnante

~HEATER - s. radiant panel

~HEATING - [el] chauffage *m* par rayonnement
[el] (panel heating) chauffage *m* par panneau radiant

~INTENSITY - [el] intensité *f* de rayonnement

~PANEL - [el] panneau *m* radiant

~POINT - [astr] (the general area of the heavens from which a given shower of meteors appears to come) zone *f* rayonnante

~REFLECTANCE - [opt] (the ratio between reflected radiant flux and incident radiant flux) facteur *m* de réflexion

~SENSITIVITY - [telev] sensibilité *f* du tube de prise de vues

~TUBE - [gas ind] tube *m* radiant

~-TYPE FURNACE - [metall] étuve *f* à tubes chauffants

RADIATE, to - [gen] (to send out in rays) rayonner, irradier
[phys] (of light, heat etc) irradier

RADIATED POWER - [radio] (the total power emitted by the transmission aerial) puissance *f* de rayonnement

RADIATING ATOM - [nucl] (radiation-emitting atom) atome *m* rayonnant

~BRICK - [constr] (compass brick) brique *f* de voûte

~BRIDGE - [constr] pont *m* en éventail

~CIRCUIT - [radio] (any circuit which sends out power in the form of electromagnetic waves into space) circuit *m* radiant

~ELEMENT - [th eng] (in steam heating) élément *m* rayonnant
[radio] (basic subdivision of an aerial) élément *m* rayonnant

~FLANGE - [auto] (cooling fin) ailette *f* de radiation

~GUIDE - [radio] guide *m* d'ondes rayonnante

~SURFACE - [th eng] (the effective area of a heater which transmits heat by radiation) surface *f* rayonnante

RADIATION - [phys] (emission of energy in the form of electro-magnetic rays) rayonnement *m*, irradiation *f*
[phys] (of radium) radiation *f*

~ABSORBER - [radio] (in an aerial) absorbant *m* de rayonnement

~ABSORPTION - [nucl] absorption *f* de rayonnement

~ANGLE - [light] angle *m* de radiation

~BARRIER - [nucl] barrière *f* de rayonnement
[th eng] panneau *m* cryostatique

~BEACON - [nucl] balise *f* d'alarme

~BELT - [astr] ceinture *f* de rayonnement

~BURN - [radiat] (a burn caused by-over-exposure) brûlure *f* par rayonnement

~CATALYSIS - [chem] (the use of radiation, e.g. atomic, to catalyse a process) catalyse *f* par rayonnement

~CHEMISTRY - [chem] (the branch of chemistry concerned with reactions induced by ionizing radiations) chimie *f* du rayonnement

~CONE - [radiat] (the cone which is formed by the emitted x-rays towards the irradiated subject) cône *m* de rayons

~CONTRAST - [radiat] (in a radiograph) contraste *m* de rayonnement

~COOLING - [th eng] (a system in which the heat is radiated into the ambient air) refroidissement *m* par rayonnement

~COUNTER - [radiat] ensemble *m* de comptage

~COUNTER TUBE - [instr] (a radiation detector) tube *m* compteur

~DAMAGE - [radiat] (effects of radiation on substances) dégâts *mpl* par rayonnement

~DAMPING - [radiat] amortissement *m* par rayonnement

~DANGER ZONE - [radiat] zone *f* de rayonnement dangereux

~DECOMPOSITION - [chem] (radilysis) décomposition *f* par rayonnement

~DENSITY CONSTANT - [opt] (the constant used in the Stefan-Boltzmann law) constante *f* de densité de rayonnement

~DETECTOR - [instr] détecteur *m* de rayonnement

~DOSE - [radiat] dose *f* de rayonnement

~DOSEMETER - [instr] (an instrument for recording or measuring the amount of radiation received by an object) dosimètre *m*

RADIATION DRYER - [text] séchoir *m* par rayonnement

~EFFECT - [radiat] (the effect of electromagnetic radiation on living organism) effet *m* de rayonnement

~EFFICIENCY - [radio] (the ratio between the power radiated from an aerial system and that delivered to it by the transmitter) rendement *m* du rayonnement

~ENERGY - [phys] énergie *f* rayonnante

~EXCITATION - [el] (of a gas; the excitation of a gas under the excitation of electromagnetic radiation) excitation *f* par rayonnement

~EXPOSURE - [radiat] exposition *f* aux rayonnements

~FIELD - [radiat] (the area over which energy is propagated) champ *m* de rayonnement

~FLUX - [radiat] (the radiation energy per unit time, passing through a surface) flux *m* de rayonnement

~FOG - [met] (condensation of atmospheric moisture caused by fall of temperature of the earth's surface due to radiation, giving rise to cloud and fog) brouillard *m* de rayonnement

~GROWTH - [radiat] grandissement *m* par rayonnement

~HARDNESS - [radiat] (the penetrating power radiation) dureté *f* de rayonnement

~HAZARD - [radiat] (the possible danger to health through exposure)risque *m* d'irradiation

~HEAT - [phys] chaleur *f* rayonnante

~HEATING - [th eng] chauffage *m* par rayonnement

~HEIGHT - [radio] (effective height of an aerial; the height of an ideal vertical radiator) hauteur *f* efficace

~HYGIENE - [radiat] (ways and means of avoiding radiation effects) hygiène *f* du rayonnement

~INDICATOR - [instr] (instrument designed to indicate the nature of a radiation) signaleur *m* de rayonnement

~INDUCED GENETIC EFFECT - [radiat] effet *m* génétique induit par rayonnement

~INJURY - [radiat] radiolésion *f*

~INTENSITY - [radiat] (the energy of a number of particles passing through a unit area perpendicular to the line of propagation) intensité *f* de rayonnement

~IONIZATION - [nucl] (of a gas or vapour; the ionization of atoms or molecules of a gas or vapour by the action of electromagnetic radiation) ionisation *f* par rayonnement

~LENGTH - [nucl] (the mean path length for the reduction of the energy of charged particles as they pass through matter) longueur *f* de rayonnement

~LETHALITY - [radiat] mortalité *f* par suite de rayonnement

~LEVEL - [radiat] niveau *m* de rayonnement

~LOBE - [radio] (portion of the radiation pattern limited by one or two cones of nulls) lobe *m* de rayonnement, pétale *f* de rayonnement

~LOSS - [radio] (that part of the transmission which is lost through radiation of radio-frequency power from a transmission system) perte *f* par rayonnement

~LOSSES - [phys] (heat lost through radiation) pertes *fpl* par rayonnement

~MAZE - s. radiation trap

~MEASUREMENT - [radiat] mesure *f* du rayonnement

~MONITOR - [instr] (device measuring radiation to determine the hazard) moniteur *m* de radioprotection

RADIATION OUTPUT - [radiat] puissance *f* de sortie d'une source
~PATTERN - [radio] (called directional pattern in the USA; a polar diagram) diagramme *m* de rayonnement
~POTENTIAL - [el] (the potential difference in volts which corresponds to the energy in electron volts required to excite an atom or molecule)potentiel *m* de rayonnement
~PRESSURE - [el] (the pressure on a surface exposed to electromagnetic radiation) pression *f* de rayonnement
~PROTECTION - [radiat] protection *f* contre le rayonnement
~PYROMETER - [instr] (pyrometer in which the radiat power from the object being measured is utilized in measuring its temperature) pyromètre *m* à rayonnement
~PYROMETRY - [meas] pyrométrie *f* à rayonnement
~RESISTANCE - [radiat] (property to with-stand the penetration of radiations) radiorésistance
~SHADOW - [nucl] (to ascertain the centre of a nuclear explosion) ombre *f* de rayonnement
~SHIELD - [radiat] écran *m* absorbant le rayonnement
~SICKNESS - [nucl] maladie *f* due à l'irradiation
~SOURCE - [radiat] source *f* de rayonnement
~STABILITY - [radiat] stabilité *f* sous rayonnement
~STERILIZATION - [radiat] (sterilization by electromagnetic radiation) stérilisation *f* sous rayonnement
~SURVEY - [radiat] contrôle *m* de protection
~TEMPERATURE - [phys] température *f* de rayonnement
~THERAPY - [med] (the treatment of disease by means of radiation) thérapie *f* par rayonnement
~TRAP - [nucl] (indirect means of access to any enclosed space containing a plant giving rise to radiation hazards) chicane *f* de rayonnement, piège *m* de rayonnement
~-TYPE BOILER - [th eng] chaudière *f* à rayonnement
~WINDOW - [radiat] fenêtre *f* transparente au rayonnement
RADIATIVE - [nucl] (pertaining to radiation) radiatif
~CAPTURE - [nucl] (nuclear capture process having the immediate result of the emission of electromagnetic radiation only) capture *f* radiative
~CAPTURE CROSS-SECTION - [nucl] section *f* efficace de capture radiative
~COLLISION - [nucl] collision *f* radiative
~CORRECTION - [phys] (in quantum theory) correction *f* radiative
~EQUILIBRIUM - [astr] (ideal state of a star) équilibre *m* radiatif
~INELASTIC SCATTERING CROSS SECTION - [radiat] section *f* efficace de diffusion inélastique radiative
~RECOMBINATION - [radiat] (recombination by radiation) recombinaison *f* par rayonnement
RADIATOR - [gen] (device for abstracting heat from a body or fluid and transferring it to the ambient atmosphere with the object either of cooling the former or heating the latter) élément *m* rayonnant
[auto etc] (device for cooling an engine by transferring heat from it to the ambient atmosphere) radiateur *m*
[phys] (a body emitting electromagnetic radiation) radiateur *m*
[radio] émetteur *m* de radiations
RADIATOR BOX - [auto] boîte *f* à eau
~BACKET - [auto] support *m* du radiateur
~CAP - [auto] bouchon *m* de radiateur
~CASE - s. radiator box
~CASING - s. radiator box
~CLEANER - [chem] (special fluid for washing out radiators) produit *m* de nettoyage pour radiateur
~CORE - [auto] corps *m* de radiateur
~CORE SECTION - [auto] élément *m* démontable du radiateur
~COVER - [auto] couvre-radiateur *m*
~COWL - [auto] déflecteur *m* d'air pour le radiateur
~COWLING - s. radiator cowl
~DAMPER - [auto] régulateur *m* de l'air de refroidissement
~FAN - [auto] ventilateur *m*
~FILLER CAP - s. radiator cap
~FLUSH - s. radiator cleaner
~FRAME - [auto] berceau *m* de radiateur
~GRILLE - [auto] grille *f* de radiateur
~HEADER TANK - [mech] (the tank at the upper part of a radiator into which the tubes are led) boîte *f* à eau
~HOSE - [auto] durite *f* de radiateur
~MASCOT - [auto] emblème *m* de radiateur
~RIM - [auto] entourage *m* de radiateur
~SCREEN - [auto] écran *m* de radiateur
~SHELL - [auto] faisceau *m* de radiateur
~SHELL UPPER CHAMBER - [auto] tête *f* du radiateur
~SHUTTER - [auto] dispositif *m* d'occultation du radiateur
~STAY - [auto] tirant *m* du radiateur
~TANK - s. radiator header tank
~TUBE - [electron] (air-cooled x-ray tube in which a fin radiator is fitted to the external end of the anode stem) tube *m* à rayons X avec ailettes de refroidissement
RADICAL - [chem] (a group of atoms which, though not normally able to exist in a separate state, can pass unchanged through a reaction or a series of reactions) radical *m*
[math] (pertaining to the root or roots of a number) radical *m*
~AXIS - [geom] axe *m* radical
~EXPRESSION - [math] (algebraic expression involving a surd) expression *f* radicale
RADICATION - [bot] (the root system of a plant) radication *f*
RADICIFEROUS - [bot] (bearing roots) radicifère
RADICIVOROUS - [zool] (root- eating) radicivore
RADICLE - [bot] (any very small root) radicule *f*
RADICOTOMY - [med] radicotomie *f*
RADICULALGIA - [med] radiculalgie *f*
RADICULAR - [bot] (pertaining to a radicle) radiculaire
~SYNDROME - [med] syndrome *m* radiculaire
RADICULECTOMY - s. radicotomy
RADICULITIS - [med] radiculite *f*
RADII - [geom] (the plural of radius) rayons *mpl*
RADIO - [radio] (prefix denoting radioactivity) radio -
[radio] (generic term denoting signalling through space by means of electromagnetic waves) radio *m*
[radio] (generic term for receiving set) radio *f*
~ALTIMETER - [instr] (altimeter using radio-wave reflection) radioaltimètre *m*
~-ASTRONOMY - [astr] radioastronomie *f*

RADIO-AUTOCONTROL - [radio] (the control of an object by radio reference from itself to other objects) radio-autocommande *f*
~AUTOPILOT COUPLER - [contr] groupe *m* de commande du pilote automatique
~BEACON - [radio] (a type of beacon which gives warning by means of automatic radio transmissions) radiophare *m*
~-BEACON RECEIVER - [radar] radiophare *m*
~-BEACON SYSTEM - [radar] réseau *m* de radiophares
~-BEACON WITH DOUBLE MODULATION - [radar] radiophare *m* à deux modulations
~BEAM - [radio] (radiowave, the energy of which is confined in at least one direction) faisceau *m*
~-BEARING - [surv] radiolevé *m*
~-BEARING INSTALLATION - [radar] radiogoniomètre *m*
~-BROADCAST, to - [radio] radiodiffuser
~BROADCASTING - [radio] radiodiffusion *f*
~-BROADCASTING STATION - [radio] station *f* de radiodiffusion
~CHANNEL - [radio] (a specific frequency band used in radio communications) canal *m* radio
~CHOKE-COIL - [radio] bobine *f* d'arrêt
~CIRCUIT - [radio] (for radio communication between two points) circuit *m* radio
RADIOCOMMUNICATION - [telecomm] radiocommunication *f*
~CIRCUIT - [radio] (radio system for a communication between two points) circuit *m* de radiocommunication
~COMPASS - [instr] (automatic direction finder) radio-compas *m*
~CONTROL - [contr] (remore control by electromagnetic waves) radiocommande *f*, télécommande *f* par radio
~-CONTROLLED AIRCRAFT - [aero] (an aircraft controlled from the ground by radio signals, e.g. for target practice) avion *m* radioguidé
~-CONTROLLED TARGET - [radio] cible *f* radiocommandée
~DETECTION - [radio] (the detection of the presence of an object by radio location) radiosignalisation *f*
~DIRECTION FINDER - [radio] (equipment for obtaining a bearing by means of radio transmission from a ground station) radiogoniomètre *m*
~-DIRECTION FINDING - [radio] radiolocalisation *f*
~DUCT - [radio] guide *m* radioélectrique
~ECHO - [radio] (the reception of a signal additional to the desired signal) radio-écho *m*
~-ENGINEERING - [gen] radiotechnique *f*
~EQUIPMENT - [radio] installation *f* radio
~FADE-OUT - [radio] (a phenomenon during which all radiowaves normally reflected by ionospheric layers suffer partial or complete absorption) effet *m* Dellinger
~-FIELD INTENSITY - [radio] intensité *f* de champ radioélectrique
~FLYING - [aero] vol *m* radioguidé
~-FREQUENCY - [radio] (a frequency within the range used for radio communication) haute fréquence *f*
~-FREQUENCY ALTERNATOR - [radio] (rotating-type generator for producing radio-frequency power) alternateur *m* haute fréquence
~-FREQUENCY AMPLIFIER - [radio] (amplifier designed to operate at the frequencies used for radio transmissions) amplificateur *m* HF

RADIO-FREQUENCY BRIDGE - [el] (a bridge designed to measure capacity, inductance and resistance over frequency range I00 K/cs - 20 M/cs) pont *m* pour hautes fréquences
~-FREQUENCY CHOKE - [radio] bobine *f* d'arrêt HF
~-FREQUENCY INTERMODULATION DISTORSION - [radio] distorsion *f* d'intermodulation HF, distorsion *f* par intermodulation dans les étages de haute fréquence
~-FREQUENCY LEAK DETECTOR - [electron] détecteur *m* de fuites à haute fréquence
~-FREQUENCY PULSE - [radio] (a radio-frequency carrier which is amplitude-modulated by a pulse) porteuse *f* modulée par impulsions
~-FREQUENCY TRANSFORMER - [radio] transformateur *m* HF
~-FREQUENCY TRANSPARENT - [radar] (denoting substances which let through waves without an adequate measure of attenuation) transparent à la radiofréquence
~-FREQUENCY WELDING - [metall] (a method of welding thermoplastic materials in which the necessary heat is generated by the application of a radio frequency field to the material) soudage *m* à haute fréquence
~GONIOMETER - [instr] (automatic airborne equipment which gives a continuous reading of the bearing of a C.W. ground beacon in relation to that of the aircraft) radiogoniomètre *m* automatique
~-GONIOMETRY - [radio] (the determination of the direction of distant objects by radio emission) radiogoniométrie *f*
~GRAMOPHONE - [radio] (combination of radio receiver and gramophone) radiophonographe *m*
~GUIDANCE SYSTEM - [radio] système *m* de guidage par radio
~HOLE - [phys] trou *m* radioélectrique
~HORIZON - [radio] (that locus of points where direct rays from the transmitter become tangential to the earth's) radiohorizon *m*
~INTERFERENCE - [radio] (disturbance of radio reception as a result of the radiation of radio frequency energy from miscellaneous sources) parasites *mpl*, perturbations *fpl* de radio
~-INTERFERENCE SUPPRESSOR - [radio] suppresseur *m* de parasites
~LICENCE - [gen] taxe *f* de radiodiffusion
~LINK - [radio] (radio communication circuit) liaison *f* radioélectrique
~-LOCATION - [radio] (determination of location by means of the propagation characteristics of radiowaves) radiorepérage *m*
~LOCATOR - [radar] radar *m*
~MAGNETIC INDICATOR - [el] indicateur *m* radiomagnétique
~MAST - [radio] (a mast used to support a radio aerial system) mât *m* d'antenne
~-METEOROGRAPH - s. radiometeorograph
~MONITOR - [radio] moniteur *m* d'émission radio
~NAVIGATION - [naut] (navigation by means of radio devices) radionavigation
~NAVIGATIONAL AIDS - [naut] (devices depending on radio transmission to aid in navigation) aides-radio *fpl* à la navigation
~NOISE - s. radio interference
~-NOISE FIELD INTENSITY - [radio] (measure of the field intensity of disturbing electromagnetic waves

at one given point) intensité *f* de champ perturbateur
RADIO OPERATOR - [aero naut etc] radiotélégraphiste *m*
~-OPTICAL LINE OF DISTANCE - [telev] portée *f* radiooptique
~-OPTICAL RANGE - [radar] portée *f* radiooptique
~ORIENTATION - [radio] radioorientation *f*
~PATROL CAR - [auto] (a police car fitted with radio) voiture *f* de patrouille équipée de radio
~POSITION-LINE DETERMINATION - [radio] (the determination of a position-line by radio location) radiorepérage *m* d'une route
~RACK - [radio] (a metal structure composed of partitioned shelves to contain radio equipment units) râtelier *m* à accessoires radio
~RANGE -[aero naut etc] (a radio beacon giving track guidance by transmissions) radiophare *m* d'alignement
~RANGE BEACON - [radio] (for course beacon) radiophare *m* directionnel
~-RANGE BEAM - [radar] faisceau *m* de radiophare directionnel
~RANGE-FINDING - [radio] radiotélémétrie *f*
~-RANGE LEG - s. radio-range beam
~RANGE ORIENTATION - [aero naut etc] radio-orientation *f*
~RECEIVER - [radio] (a set equipment for receiving radio transmissions) poste *m* récepteur, récepteur *m* radio
~RECORDING TAPE - [radio] bande *f* pour l'enregistrement de radio-transmissions
~RELAY - [telecomm] liaison *f* hertzienne
~-RELAY AERIAL - [radio] antenne *f* pour station-relais
~-RELAY SYSTEM - [radio] système *m* station-relais
~SET - [radio] poste *m* récepteur
[radio] (a transmitter) émetteur *m* radio
~SHIELDING - [radio] blindage *m*
~-SONDE - [instr] (a small free balloon carrying instruments for upper-air meteorological observations and apparatus for transmitting reading of these to ground stations) radiosonde *f*
~-SOUNDING - [meas] (measurements carried out by radio-sonde) radiosondage *m*
~SPECTROMETER - [instr] (apparatus designed to analyze the radio-frequency energy reaching an aerial) radiospectromètre *m*
~SPECTRUM - [radio] (the frequency which can be used for the transmission and reception of radio energy) spectre *m* de radiofréquences
~STATICS - [radio] parasites *mpl*
~STATION - [radio] station *f* radio
~-TELECONTROL - [contr] (the control of mechanism, machines etc by radiowaves) radiotélécommande *f*
~TELESCOPE - s. radiotelescope
~TRANSMISSION - [radio] (transmission of signals by electromagnetic waves) radiotransmission *f*
~TRANSMITTER - [radio] (a complete set of equipment for transmitting radio signals) émetteur *m* radio
~WARNING - s. radio detection
~-WAVE - [radio] onde *f* radioélectrique
~-WAVE PROPAGATION - [radio] (the transfer of energy by electromagnetic radiation) propagation *f* des ondes
~WELDING - s. radio-frequency welding

RADIO WIND FLIGHT - s. radio-sonde
RADIOACTINIUM - [chem] (a thorium isotope, radioactive, with a half-life of 18.9 days) radioactinium *m*
RADIOACTIVE - [nucl] (having the ability to emit alpha, beta, and gamma rays) radioactif
~BATTERY - [nucl] batterie *f* atomique
~BY-PRODUCT - [nucl] (any by-product created during a nuclear process) sous-produit *m* radioactif
~CELL - cellule *f* radioactive
~CHAIN - [nucl] (decay sequence) chaîne *f* de désintégrations
~CHAMBER - [nucl] chambre *f* radioactive
~CONCENTRATION - [nucl] concentration *f* radioactive
~CONTAMINATION - [nucl] (adhesion of radioactive material to a body) contamination *f* radioactive
~DECAY - [nucl] désintégration *f*
~DECAY LAW - [nucl] (the law governing the decrease with time of the number of atoms of a radioactive species) loi *f* de décroissance radioactive
~DECAY SERIES - [nucl] (decay chain) chaîne *f* de désintégrations
~DECONTAMINATION - [nucl] (the removal of radioactive contamination from a body) décontamination *f*
~DEPOSIT - [min] (any ore deposit containing radioactive material) gisement *m* radioactif
[nucl] (radioactive material deposited on a surface) dépôt *m* radioactif
~DISINTEGRATION - s. radioactive decay
~DISPLACEMENT LAW - [nucl] loi *f* de déplacement radioactif
~DRY FALL-OUT - [nucl] dépôt *m* radioactif sec
~DUST - [nucl] (dust which is deposited from the air on the earth etc) poussière *f* radioactive
~EFFLUENT - [nucl] (radioactive waste material discharged from a system) effluent *m* radioactif
~EFFLUENT DISPOSAL - [nucl] (the disposal of contaminated waste liquid in a nuclear power plant) élimination *f* des déchets radioactives liquides
~EFFLUENT DRAIN PIPE - [nucl] tuyau *m* d'écoulement d'eau usée radioactive
~ELECTRON TUBE - [electron] tube *m* électronique à substance radioactive
~EMANATION - [nucl] gaz *m* radioactif
~EQUILIBRIUM - [nucl] (or secular equilibrium, an expression relating to the half-life of a parent when it is counted in centuries) équilibre *m* radioactif
~FALL-OUT - [nucl] (the measure of contamination following the explosion of a nuclear bomb) retombées *fpl* radioactives
~FAMILY - [nucl] (succession of nuclides, each of which is transformed into the next one by radioactive disintegration, until a stable nuclide results) famille *f* radioactive
~FISSION PRODUCT - [nucl] produit *m* de fission radioactif
~GAS - [nucl] (gas created in a nuclear reactor) gaz *m* radioactif
~GO-DEVIL - [nucl] râcleur *m* radioactif
~GRAIN - [radiat] grain *m* radioactif
~HALF-LIFE - [nucl] (the time taken by an active isotope to reach half of its initial value) période *f* radioactive
~HEAT - [nucl] (the heat which is produced in the earth by the disintegration of active nuclides) chaleur *f* radiogénique

RADIOACTIVE INCINERATOR - [nucl] incinérateur *m* de déchets radioactifs
~IODINE - [nucl] isotope *m* de l'iode
~IONIZATION GAUGE - [nucl] jauge *f* à source radioactive
~ISOTOPE - [nucl] radio-isotope *m*
~MATERIAL - [nucl] (any material capable of emitting radiations) matière *f* radioactive
~NUCLEUS - [nucl] noyau *m* radioactif
~NUCLIDE-[nucl] (radioactive, minute, highly charged central portion of an atom) nucléide *m* radioactif
~PAINT - [paint] (paint containing mesothorium or radiothorium, which emits light by radioactivity, as distinct from luminous paint which only glows after exposure to light) peinture *f* radioactive
~PERIOD - [nucl] vie *f* moyenne
~POISON - [nucl] (parasitic absorber of thermal neutrons and acting as poison for the chain reaction) poison *m* radioactif
~PRODUCT - [nucl] (a product of the radioactive decay of a nuclide, which is itself radioactive) produit *m* de décroissance radioactif
~PURITY - [nucl] pureté *f* radioactive
~RELATIONSHIP - [nucl] (the relation between the parent substance, the daughter product and the end product) filiation *f* radioactive
~SERIES - s. radioactive family
~SAMPLING EQUIPMENT - [chem] installation *f* d'échantillonnage actif
~SOURCE - [nucl] (quantity of a substance capable of emitting radiations) source *f* radioactive
~STANDARD - [nucl] (a reference source) étalon *m* radioactif
~TRACER - [nucl] (any physical or chemical tracer with the property of being radioactive) traceur *m* radioactif
~TRANSFORMATION - [nucl] transition *f* radioactive
~VALVE - s. radioactive electron tube
~WASTE - [nucl] (waste from a nuclear reactor) déchets *mpl* radioactifs
~WATER - [nucl] (water activated by the waste of a nuclear reactor) eau *f* radioactive
RADIOACTIVITY - [nucl] (the spontaneous disintegration of a nucleus, accompanied by the emission of radiation) radioactivité *f*
~METER - [instr] activimètre *m*
~SIMULATOR - [instr] simulateur *m* de radioactivité
RADIOAUTOGRAPH - [nucl] (a record of radiation made by placing a radioactive object close to a photographic emulsion) autoradiogramme *m*
RADIOBALANCE - [instr] (instrument designed to measure absolutely the amount of incident radiation) balance *f* de rayonnement
RADIOBIOLOGIC ACTION - [radiat] effet *m* radiobiologique
RADIOBIOLOGY - [radiat] (the study of the effects of radiations on living matter) radiobiologie *f*
RADIOCARBON - [nucl] isotope *m* du carbone
~AGE - [nucl] (the age calculated from the specific activity of the carbon) datation *f* C-I4
RADIOCHEMICAL ANALYSIS - [chem] analyse *f* radiochimique
RADIOCHEMISTRY - [chem] (the chemistry of the radioactive elements) radiochimie *f*
RADIOCHROMATOGRAPH - [instr] (a measuring assembly) radiochromatographe *m*
RADIOCHRONOMETER - [instr] (used in radiology) radiochronomètre *m*
RADIOCOBALT - [nucl] isotope *m* du cobalt
RADIOCOLLOID - [chem] (group of radioactive atoms into colloidal aggregates) radiocolloïde *m*
RADIODE - [radiat] (Radium Capsule) capsule *f* à radium
RADIODERMATITIS - [med] (inflammation of the skin caused by over-exposure to radiation) actinodermatose *f*, radiodermite *f*
RADIODERMITIS - s. radiodermatitis
RADIODIAGNOSIS - [med] röntgendiagnostic *m*
RADIODONTIA - [med] radiographie *f* des dents
RADIOELEMENT - [nucl] (a radio-active atom resulting from an artificially-induced nuclear reaction) élément *m* radioactif
RADIOGENIC - [nucl] (produced by radioactive transformation) radiogénique
~HEAT - s. radioactive heat
RADIOGOLD - [nucl] or *m* radioactif
RADIOGONIOMETER - s. radio goniometer
RADIOGRAM - [med] (in radiology, a radiograph) radiogramme *m*
RADIOGRAPH - [med] (in radiology, the impression made on a sensitive plate by the passage of x-rays through an object) radiogramme *m*
RADIOGRAPHIC PUTTY - [radiat] substance *f* antidiffuseuse
RADIOGRAPHY - [phys chem etc] (the technique of making x-ray photographs) radiographie *f*
RADIOIRON - [nucl] fer *m* radioactif
RADIOISOTOPE - [nucl] (a radioactive isotope) radio-isotope *m*
~CONCENTRATION - [nucl] concentration *f* radioisotopique
RADIOLARIA - [zool] (an order of marine Sarcodina with a central capsule and a skeleton of siliceous spicules) radiolaires *mpl*
RADIOLARIAN CHERT - [geol] (cryptocrystalline siliceous rock, composed partly of the remains of radiolaria) radiolarite
RADIOLARITE - s. radiolarian chert
RADIOLOCATION - s. radio location
RADIOLOGICAL PHYSICS - [phys] (the physics pertaining to radiology) physique *f* radiologique
RADIOLOGIST - [med] radiologue *m*
RADIOLOGY - [med] (the science dealing with the examination of the human body by means of x-rays) radiologie *f*
RADIOLUCENT - [radiat] (permitting the passage of x-rays) radiotransparent
RADIOLUMINESCENCE - [radiat] (luminescence resulting from radiant energy bombardment) radioluminescence *f*
RADIOLYSIS - [radiat] (radiation decomposition) radiolyse *f*, décomposition par rayonnement
RADIOMETALLOGRAPHY - [metall] (the radiology of metals) radiométallographie *f*
RADIOMETEOROGRAPH- [instr] radiométéorographe *m*
RADIOMETER - [instr] (instrument designed to detect and measure radiant energy) radiomètre *m*
~GAGE - s. radiometer gauge
~GAUGE - [instr] manomètre *m* de Knudsen
RADIOMETRIC ANALYSIS - [chem] analyse *f* radiométrique
RADIOMETRY - [meas] (the measurement of radiant energy) radiométrie *f*

RADIOMICROMETER - [instr] (instrument designed to measure radiant energy) radiomicromètre *m*
RADIONUCLIDE - s. radioactive nuclide
RADIOPACITY - [radiat] opacité *f* aux rayonnements
RADIOPAQUE - [radiat] (preventing the passage of atomic radiation) opaque aux rayonnements
RADIOPHONE - [radio] (telephone system employing radio) radiotéléphone *m*
[acoust] (instrument for the production of sound by intermittent radiant energy, such as light, heat etc) radiophone *m*
RADIOPHOTOLUMINESCENCE - [radiat] (the luminescence which is shown by certain minerals when irradiated with beta and gamma rays, and subsequently exposed to light) radiophotoluminescence *f*
RADIORESISTANCE - [radiat] (the relative resistance of cells, tissues etc to radiation damage) radiorésistance *f*
RADIOSCOPY - [radiat] (the examination of objects by x-rays) radioscopie *f*
RADIOSENSIBILITY - [radiat] radiosensibilité *f*
RADIOSENSITIVE - [radiat] (sensitive to radiation) radiosensible
RADIOSENSITIVENESS - [radiat] s. radiosensibility
RADIOSENSITIVITY - [radiat] radiosensibilité *f*
RADIOSODIUM - [chem] isotope *f* radioactif du sodium
RADIOSONDE - s. radio sonde
RADIOSTRONTIUM - [nucl] (radioisotope of strontium) strontium-90 *m*
RADIOSULPHUR - [chem] soufre *m* radioactif
RADIOTELEGRAM - [telecomm] radiogramme *m*
RADIOTELEGRAPH - [telecomm] radiotélégraphe *m*
~RECEIVER - [radio] (radio receiver for the reception of radiotelegraph signals) récepteur *m* radiotélégraphique
~OPERATOR - [telecomm] radiotélégraphiste *m*
RADIOTELEGRAPHY - [telecomm] radiotélégraphie *f*
RADIOTELEPHONE - [telecomm] radiotéléphone *m*
~CIRCUIT - [telecomm] circuit *m* radiotéléphonique
~OPERATOR - [telecomm] radiotéléphoniste *m*
RADIOTELEPRINTER - [telecomm] radiotéléimprimeur *m*
RADIOTELESCOPE - [opt] (telescope consisting of a paraboloidal reflector for collecting and focusing radiowaves) radiotéléscope *m*
RADIOTELETYPEWRITER - s. radioteleprinter
RADIOTHALLIUM - [nucl] (a thallium isotope) isotope *m* du thallium
RADIOTHEODOLITE - [meas] (a goniometer designed to measure the angle of elevation) radiothéodolite *m*
RADIOTHERAPEUTICS - s. radiotherapy
RADIOTHERAPY - [med] (treatment of diseases by ionizing radiations) radiothérapie *f*
RADIOTHERMOLUMINESCENCE - [radiat] (luminescence shown by certain substances when irradiated by beta and gamma rays and subsequently heated) radiothermoluminescence *f*
RADIOTHORIUM - [nucl] (a thorium isotope with a half-life of 1.90 days) radiothorium *m*
RADIOVISION - [telev] (television system in which the link between the transmitter and receiver is via radio) télévision *f* à circuit fermé
RADIUM - [chem] (a radio-active metal, symbol Ra, A.N. 88, A.W. 226.05, important for its uncontrollable spontaneous disintegration . It resembles barium chemically. It is used in medicine, metallurgy, as an ionizing agent and in making radio-active paints) radium *m*
RADIUM AGE - [nucl] (the age calculated from the number of radium atoms originally present, now and when equilibrium is established with ionium or uranium) âge *m* de radium
~BROMIDE - [chem] (white, radioactive, crystalline with applications in medicine) bromure *m* de radium
~CAPSULE - [radiat] capsule *f* à radium
~CARBONATE - [chem] (a radioactive compound used in medicine) carbonate *m* de radium
~CELL - [radiat](sealed radium container) capsule *f* à radium
~CHLORIDE - [chem] (radioactive compound with uses in medicine) chlorure *m* de radium
~CONTAINER - [radiat] (container designed to hold radium, generally sealed to retain decay products) récipient *m* de radium
~CONTENT - [radiat] contenu *m* de radium
~EMANATION - [radiat] (a synonym for Radon) émanation *f* de radium
~MOULD - [radiat] (moulding containing radium) moulage *m* de radium
~NEEDLE - [radiat] (radio container shaped like a needle) aiguille *f* de radium
~PACK - [radiat] applicateur *m* de radium
~PLAQUE - [radiat] (radium container in which radium is distributed over the surface) plaque *f* radiofère
~SEED - [radiat] (permanent implant) semence *f* à radium
~SOURCE - [chem] preparation *f* de radium
~SULPHATE - [chem] (a radioactive compound with therapeutic applications) sulfate *m* de radium
~THERAPY - [med] (treatment by means of radiations from radium) radiumthérapie *f*
~TUBE - [radiat] (radium container shaped like a bluntended tube) tube *m* à radium
RADIUS, to - [mech] (to round off a sharp angle into an arc of a circle) raccorder une courbe
RADIUS - [geom] rayon *m*
[mech] (of a crane) portée *f* d'une grue
~A PLATE, to - [metall] recourber une tôle
~BAR - [mech] bielle *f* du tiroir
~OF ACTION - [aero] (one-half the total range of the aircraft in question in still air) autonomie *f*, rayon *m* d'action
~OF ATOM - [phys] diamètre *m* atomique
~OF CIRCLE - [geom] rayon *m* du cercle
~OF CURVATURE - [geom] rayon *m* de courbure
~OF GYRATION - [phys] rayon *m* de gyration
~OF INERTIA - s. radius of gyration
~OF INTRADOS - [constr] rayon *m* d'intrados
~OF PITCH CIRCLE - [mech] (of a gear) rayon *m* du cercle primitif
~OF SERVICE AREA - [radio] portée *f* d'un émetteur
~OF SOFFIT - s. radius of intrados
~OF THE WARP BEAM - [text] rayon *m* de l'ensouple à chaîne
~OF TURNING - [auto] rayon *m* de braquage
~PARAMETER - [phys] (the effective radius of a nucleus divided by the cube root of its mass number A) paramètre *m* du rayon
~PLANER - [tool] raboteuse *f* pour surfaces courbes
~ROD - [mech] (the rod attached to the die of a valve gear of the radial type) bielle *f* du tiroir

RADIUS TURNING TOOL - [tool] outil *m* pour le tournage de courbes
~VECTOR - [math & astr] (the line which joins the focus to the body moving about in an elliptic orbit) rayon *m* vecteur
RADIX - [zool] (in anatomy, the root, or point of origin of a structure) racine *f*
[math] (the radix in any scale of notation for numbers) base *f*
~NOTATION - [comput] notation *f* à base
~POINT - [math] virgule *f*, point *m* de base
[comput] rang *m* de la virgule
RADOME - [radio] (a cover over an aircraft radio aerial system designed to allow the passage of radio waves) radome *m*, dôme *m* radar
RADON - [chem] (the heaviest of the noble gases, symbol Rn, A.N. 86, A.W. 222, half-life 3.82 days. It is zero-valent and radio-active and is produced by the spontaneous disintegration of radium) radon *m*
~CONTAINER - [radiat] (radon holding sealed container) tube *m* à radon
~CONTENT - [radiat] contenu *m* de radon
~SEED - [radiat] (a short capillary tube containing radon and designed to interstitial therapeutic irradiation) semence *f* à radon
RAFFIA - [bot] (cultivated palm originating from Madagascar, the leaves of which supply a fibre used for making hats, baskets etc) raphia *m*
[text] fibre *f* de raphia
~FIBRE - [text] raphia *m*
RAFFINATE - [chem] (the refined material from a refining process) raffinat *m*
~LAYER - [chem] (the liquid layer in a solvent extraction system from which a solute has been extracted) couche *f* de raff
RAFFINOSE - [chem] (melitriose. A trisaccharose, without taste, dextro-rotatory, found in molasses, sugar beet etc)(It does not reduce Rehling's Solution. It is used in biology and medicine) raffinose *m*
RAFFLE - [mech] disque *m* de freinage
RAFT, to - [naut] transporter sur un radeau
RAFT - [naut] radeau *m*, ras *m*
[constr] radier *m*
~BRIDGE - [constr] (pontoon bridge, supported on raft instead of boats) pont *m* de radeaux
~FOUNDATION - [constr] (layer of reinforced concrete under the entire area of a building and projecting over the line of its walls) châssis *m* de fondation
~OF WAGONS - [railw] (a trains set) rame *f*
RAFTER, to - [constr] chevronner (un comble)
RAFTER - [constr] (beam supporting a roof) chevron *m* (d'un comble)
~CONNECTION - [constr] chevronnage *m*
~END - [constr] (the section of the rafter projecting beyond the face of the building) tête *f* de chevron
~ROOF - [constr] comble *m* à chevrons
~SET - [mining] cadre *m* complexe
~TIMERING - [min] boisage *m* à chevrons
RAG, to - [min] (the operation of breaking up ore with a hammer, before sorting) vorscheider
RAG - [gen] chiffon *m*, lambeau *m*
[paper ind] (the waste cloth used for paper making) drilles *f*pl, chiffons *m*pl
[min] impureté *f*
[constr] (coarse roof-slate, generally large) ardoise *f* grossière
RAG - [metall] pierre *f* à aiguiser
~BOILER - [paper man] bouilleur *m* pour chiffons
~-BOLT - [mech] (a bolt with barbs directed towards the head) boulon *m* de scellement à crans
~CHOPPER - [paper man] délisseuse *f*
~CUTTER - [paper man] dérompoir *m*
~-CUTTING MACHINE - s. rag cutter
~DEVIL - [text] effilocheuse *f* pour chiffons
~DUST - [paper man] poussière *f* de drilles
~ENGINE - [paper man] pile *f* défileuse
~ENGINE FOR COLLODION WOOL - [text] pile *f* à moudre le colloidion en laine
~GRINDER - [text] effilocheuse *f* pour chiffons
~OPENER - [paper man] ouvreuse *f* pour chiffons
~PICKER - [text] chiffonnier *m*
~PICKING - [paper man] triage *m* des chiffons
~PULLING - [text] effilochage *m* des chiffons
~PULLING MACHINE - [text] effilocheuse *f* pour chiffons
~PULPE - [paper man] (pulp for paper-making obtained by cooking cleaned rags with lime or a mixture of lime with soda ash under moderate steam pressure) pulpe *f* de chiffons
~SHAKER - [text] secoueur *m* à chiffons
~SHAKING - [text] dépoussiérage *m* des chiffons
~SORTER - [text] repasseuse *f* de chiffons
~SORTING TABLE - [text] table *f* de triage à chiffons
~STOCK - [rubber ind] déchets *m*pl de caoutchouc avec textile
~STONE - [geol] moellon *m* schisteux
~WASHING - [text] lavage *m* des chiffons
~WASHING MACHINE - [text] laveuse *f* mécanique à chiffons
~-WHEEL - [mech] pignon *m* de chaîne, poulie *f* à chaînes, hérisson *m*, bouc *m*
[metall] disque *m* en drap
RAGGED PICTURE - [telev] image *f* ondulante
~-ROBBIN - [bot] oeillet *m* des prés
~ROLL - [metall] cylindre *m* à surface rugueuse
RAGGING - [min] (rough washing, for a low ratio of concentration) vorscheidage *m*, scheidage *m* d'épuration
[metall] (in a rolling-mill roughing rolls) surface *f* rugueuse
[metall] lit *m* de grenailles
RAGLAN - [text] raglan *m*
RAGWORT - [bot] (a weed) jacobée *f*
RAIL - [mech] (one of an assembly of slotted, grooved or shouldered bars, fixed, or held by other means, in a press) rail *m*, barreau *m* de guide
[carp & mech] (the horizontal member in a framing) traverse *f*
[constr] (the upper member in a balustrade) balustrade *f*
[naut] (the bar of wood or iron capping the bulwarks of a ship) lisse *f*
[metall] (a bar) lierne *f*
[transp] (of a cart) ridelle *f*
~ANCHOR - [railw] dispositif *m* d'encrage des rails
[railw] (to prevent creeping) arrêt *m* de cheminement
~-ANCHOR CLIP - [railw] serre-rail *m*
~ATTACHMENT - [railw] attache *f* des rails
~BED - [railw] assiette *f* des rails
[railw] (rail seat) entaille *f* de la traverse
~BENDER - [mech] (a steel girder used by platelayers to bend rails to the wanted curvature) presse

f à cintrer et dresser les rails
RAIL BOND - [el] (electrical connection between two adjacent lengths of track) connexion *f* électrique de rail, railbond *m*
~-BOND TESTER - [instr] dispositif *m* à essayer les railbonds
~BONDING - [railw] connexion *f* de rail à rail
~BRAKE - [railw] (retarder) frein *m* de voie, rail-frein *m*
~-BRUSH - [railw] balai *m* frotte-rail
~BUCKLING - [railw] déjettement *m* de rail
~-CAR - [railw] autorail *m*
~CHAIR - [railw] (cast-iron support spiked to the sleepers for securing a bull-head in position) coussinet *m* de rail
~CHAIR WITH ONE JAW - [railw] coussinet *m* unilatéral de rail
~CLIP - [railw] attache-rail *m*
~CONTACT - [railw] contact *m* de rail
~CORRUGATION - [railw] usure *f* ondulatoire du rail
~CROSS-SECTION - [railw] section *f* transversale du rail
~-CURRENT TRANSFORMER - [el] transformateur *m* de courant à rail
~CUTTING MACHINE - [metall] coupeuse *f* de rails
~CYCLE - [railw] (an inspection trolley) lorry *m*
~DRILLING MACHINE - [mach tool] machine *f* à percer les rails
~EFFECT - [el] effet *m* inductif du rail
~FASTENING - [railw] attache *f* des rails
~FATIGUE - [railw] fatigue *f* du rail
~FILLER - [text] bâcleur *m*, varouleuse *f*
~FIXING - s. rail fastening
~-FOOT - [railw] patin *m* du rail
~FORK - [railw] fourchette *f* porte-rails
~GAUGE - [railw] gabarit *m* d'écartement de voie
~-GAUGE TEMPLATE - [railw] gabarit *m* d'écartement
~GUARD - [railw] (of a locomotive) chasse-pierres *m*
[railw] (a guard-rail) contre-rail *m*
~GUIDE - [text] guide-rail *m*
~GUIDE CAP - [text] guide-rail *m*
~HEAD - [railw] champignon *m* de rail
(a terminal station) gare *f* tête de ligne
~HOLDER - [text] support *m* des rails
~IN ADVANCE - [railw] rail *m* aval
~IN REAR - [railw] rail *m* amont
~INSULATION - [railw] isolement *m* des rails
~JACK - [railw] cric *m* relève-rails
~JOINT - [railw] joint *m* de rail
~-JOINT PLATE - [railw] éclisse *f*
~LAYING - [railw] pose *f* de rails
~LEVEL - [railw] assise *f* des plaques d'appui
~LEVER - [railw] levier *m* de rail
~LIFTER - [railw] soulève-rail *m*
~LIFTING JACK - [railw] cric *m* relève-rails
~MILL - [metall] laminoir *m* à rails
~MOTOR SET - [railw] (multiple unit train) rame *f* automotrice
~-PAD - [railw] patin *m* de rail
~PASS - [railw] tracé *m* pour rails
~PIECE - [text] morceau *m* de rail
~PINCH-BAR - s. rail lifter
~POST - [constr] (a newel post) potelet *m* de balustrade
~PROFILE - [railw] (the rail section) profil *m* du rail
RAIL-ROAD SERVICE - [transp] service *m* rail-route
~ROLLING MILL - [metall] laminoir *m* à rails
~SAW - [tool] scie *f* pour rails
~SAWING AND DRILLING MACHINE - [mech] machine *f* à scier et à percer les rails
~SEAT - [railw] entaille *f* de la traverse
~SECTION - [railw] coupon *m* de rail
[railw] (profile) profil *m* du rail
~SETTER - s. rail filler
~SHIPMENT - [transp] envoi *m* par chemin de fer
~SHOE - [railw] patin *m* de rail
~SLEWER - [railw] appareil *m* à tirer les rails
~SLEWING DEVICE - s. rail slewer
~SPIKE - [railw] crampon *m* (rail)
[mech] crampon *m*
~SQUARE - [railw] équerre *f* de pose
~STORES - [railw] dépôt *m* rails
~SUPPORT - [photo] rail *m* support
~TONGS - [tool] pinces *fpl* à rails
~WAGON - [railw] wagon *m* à rails
~WEB - [railw] âme *f* du rail
~WITH SCALE - [meas] rail *m* avec divisions
RAILS - [railw] chemin *m* de fer, voie *f* ferrée
RAILBUS - s. railcar
RAILCAR - [railw] autorail *m*
RAILING - [constr] balustrade *f*
[constr] (open form of fence consisting of iron rods and horizontal bars) grille *f*, clôture *f* en fer
[gen] guide-mains *m*
RAILINGS - [telev] brouillage *m* en palissade
RAILROAD - [railw] (US for railway) chemin *m* de fer
~CAR - [railw] wagon *m*
~TANK CAR - [railw] wagon-citerne *m*
~TRUCK - [railw] chariot *m* à bagages
RAILWAY - [railw] chemin *m* de fer
(a street railway, in the USA only) ligne *f* de tramways
~BRIDGE - [railw] pont *m* de chemin de fer
~CAR - [railw] wagon *m*
~CENTRE - s. railway junction
~CHAIR - [railw] coussinet *m* pour rails
~COACH - [railw] wagon *m*
~CONSTRUCTION - [railw] construction *f* de voie ferrée
~CROSSING - [railw] croisement *m* de voies
~CUTTING - [railw] déblai *m*, tranchée *f*
~ELECTRIFICATION - [railw] électrification *f* des chemins de fer
~EMBANKMENT - s. railway fill
~ENGINE - [railw] locomotive *f*
~FILL - [railw] remblai *m* pour voie ferrée
~GRADIENT - [railw] déclivité *f* de la voie
~HAUL - [railw] trajet *m* par voie ferrée
~JUNCTION - [railw] jonction *f*
~LINE - [railw] ligne *f* de chemin de fer
~MOUNT - [railw] affût *m* de voie ferrée
~NETWORK - [railw] réseau *m* de chemin de fer
~OVERBRIDGE - [railw] passage *m* en dessus
~PLATFORM - [railw] quai *m* de chemin de fer
~PRECINCTS - [railw] enceinte *f* du chemin de fer
~RATES - [railw] tarif *m* des chemins de fer
~REGULATIONS - [railw] règlement *m* des chemins de fer
~ROUTE - [railw] tracé *m* de la voie
~SECTION - [railw] district *m* de la voie
~SERVICE CAR - [railw] voiture *f* de service
~SIGNALLING REGULATIONS - [railw] (signal code)

code *m* des signaux
RAILWAY STATION - [railw] station *f* de chemin de fer, gare *f*
~SURROUNDINGS - s. railway territory
~SYSTEM - s. railway net
~TANKER - s. railroad tank car
~TARIFF - [railw] tarif *m* des chemins de fer
~TERRITORY - [railw] domaine *m* du chemin de fer
~TIPPING BRIDGE - s. railway weigh-bridge
~TRAFFIC - [railw] trafic *m* ferroviaire
~VEHICLE - [railw] véhicule *m* de chemin de fer
~WEIGH-BRIDGE - [railw] pont *m* à bascule ferroviaire
~WORKSHOP - [railw] atelier *m* de chemin de fer
~YARD - [railw] gare *f* de triage
RAIN, to - [met] pleuvoir
RAIN - [met] (precipitation in the form of drops of water) pluie *f*
[cin] s. rain effect
~AWNING - [gen] tente *f*
~-BAND - [met] (absorption band in the solar spectrum produced by water-vapour in the earth's atmosphere) bande *f* d'absorption de la pluie
~CHART - [met] carte *f* pluviométrique
~CLOUD - [met] (nimbus; a heavy dark cloud) nimbus *m*
~EFFECT - [cin] (a fault in the picture caused by a wrong relation between the maltese cross and the shutter blade) effet *m* de pluie, rayures *f*pl
~-GAUGE - [instr] (a device for measuring the amount of precipitation) pluviomètre *m*
~-GLASS - [instr] baromètre *m*
~GUN - [agric] canon *m* arroseur
~ICE - [met] (a layer of smooth ice formed by rain freezing on the surface of an aircraft) verglas *m*
~-OUT - [nucl] (fall-out deposited by rain) retombées *f*pl entraînées par la pluie
~OUTLET - [hydr] (a sewer overflow) déversoir *m* d'orage
~PILLAR - [geol] pyramide *f* coiffée
~SHADOW - [gen] zone *f* abritée de la pluie
~SQUALL - [met] (rain accompanying a strong wind, generally of short duration) grain *m*
~-TIGHT - [gen] protégé contre la pluie
~WATER - [gen] eaux *f*pl pluviales
~-WATER BASIN - [hydr] bassin *m* d'eau de pluie
~-WATER BYPASS - [hydr] bypass *m* pour l'eau de pluie
~-WATER PIPE - [constr] (a downpipe) descente *f*
RAINBOW - [met] arc-en-ciel *m*
~COLOURS - [metall] iris *m*
~DYEING - [text] teinture *f* ombrée
~GENERATOR - [telev] (type of signal generator) générateur *m* de mire en arc-en-ciel
~NEGATIVE - [cin] (the front negative in the Multicolour process) négatif *m* orthochromatique
~PRINTING - [text] impression *f* flammée
~QUARTZ - [min] pierre *f* d'iris
~SEASON - [met] saison *f* des pluies
~SYMPTOM - [med] halo *m* coloré (glaucome)
~TROUT - [zool] truite *f* arc-en-ciel
~VISION - s. rainbow symptom
~WHEEL - [light] disque *m* porte-filtres colorés
RAINCOAT - [text] imperméable *m*
RAINFALL - [met] (the recorded amount of precipitation for a given point or area) précipitation *f* atmosphérique
RAINPROOF, to - [text] imperméabiliser
RAINPROOF - [gen] étanche à la pluie, imperméable, imbrifuge
RAINPROOFER - [ind chem] agent *m* imperméabilisant
RAINSTORM - [met] (storm accompanied by rain) averse *f*
RAINY WEATHER - [met] temps *m* pluvieux
RAISE, to - [gen] dresser, mettre debour
[constr] faire bâtir
[zool] élever
[agric] cultiver
[gen] (to make higher) relever, hausser, remonter
[fin] se procurer de l'argent
[mech] (power etc) relever, monter
[mining] percer en montant
[text] lainer, frapper, carder, gratter
[brew ind] faire lever
RAISE - [gen] augmentation *f*
[min] (a vertical passage) montage *m*
~A BRIDGE, to - [constr] ériger un pont
~A WALL, to - [constr] relever un mur
~AN EMBANKMENT, to - [constr] élever une digue
~COAL, to - [mining] extraire le charbon (d'une mine)
~FUNDS, to - [fin] réunir des fonds
~PRODUCTIVITY, to - [gen] augmenter la productivité
~STEAM, to - [th eng] (to generate steam pressure in a boiler, espec. when starting with the boiler cold) produire de la vapeur
~STOPING - [mining] abattage *m* en remontage
~THE CONCENTRATION, to - [chem] augmenter la concentration
~THE FLAT, to - [text] enlever le chapeau
~THE PRICE, to - [comm] augmenter le prix
RAISED - [gen] levé, relevé
[text] cardé, gratté, lainé
[print] (for the blind) impression *f* anaglyptique (pour les aveugles)
[mech] (in relief) saillant, en relief
~EFFECT - [text] effet *m* de granité
~HEAD - [mech] (of a rivet) rivure *f* saillante
~-HEAD SCREW - [mech] vis *f* à tête saillante
~LOOP - [text] boucle *f* droite
~OIL - [leather ind] huile *f* récupérée
~PANEL - [carp] (panel standing out from the general surface) panneau *m* en relief
~PLAN - [surv] élévation *f*
~PRINTING - [print] impression *f* en relief
~SPOOL CARRIER - [text] support *m* de bobine relevé
~WARP THREADS - [text] fils *m*pl de chaîne levés
~WINDING FALLER - [text] baguette *f* levée
~WOOLLEN CLOTH - [text] drap *m* de laine gratté
~YARN GUIDE - [text] guide-fil *m* soulevé
RAISER - s. riser [constr]
RAISIN - [food] raisin *m* sec
~SEED OIL - [agric] huile *f* de pépin de raisin
~WINE - [agric] vin *m* de raisins secs
RAISING - [text] (the production of a pile surface on certain woollen and worsted fabrics and in cotton goods) garnissage *m*, grattage *m*
[leather ind] (process of oil recovery from chamois-leather wash-waters) récupération *f* de l'huile
[mining] remonte *f*, extraction *f*
[gen] élévation *f*, élèvement *m*, surhaussement *m*, montage *m*

RAISING BAND - [text] rayure *f* de lainage
~CAM - [mech] came *f* d'ascension
~JACK - [text] platine *f* auxiliaire
~MACHINE - [text] laineuse *f*
~OF THE BEAM - [text] élévation *f* de l'ensouple
~OF THE HOOKS - [text] levée *f* des crochets
~OF THE WATER LEVEL - [hydr] exhaussement *m* de la retenue
~ROLLER - [text] cylindre *m* à lainer
~STREAK - [text] rayure *f* de lainage
~TEASEL - [text] cardère *f*
~WASTE - [text] déchets *mpl* de lainage
RAKE, to - [agric] râteler
[gen] (to cause to stand out of the perpendicular) pencher, être incliné
RAKE - [impl] râteau *m*
[metall] (of a furnace) ringard
[naut] inclinaison *f*
[mech] (angle of inclination) angle *m* d'inclinaison
[mech] (of the propeller) pas *m* de l'hélice
[tool] dépouille *f*
[mach tool] dégagement *m* d'un outil
[shipbuild] élancement *m*
[brew] (forked tool for loading coal) râble *m*
[geol] inclinaison *f*
~CARRIAGE - [text] chariot *m* à râteau guide-fils
~FRAME - [text] bâti *m* du râteau guide-fils
~OF A MAST - [naut] inclinaison *f* d'un mât
~-OUT - [gas ind] (for the treatment of gas) tabatière *f*
~VEIN - [mining] filon *m* vertical
RAKED BOW - [naut] étrave *f* inclinée
RAKER - [tool] racloir *m*, grattoir *m*
[mining] curette *f* de mineur
~CLEANSER - [rubber ind] curette *f* de tasse à latex
RAKING - [mech etc] incliné, élancé
~SHORE - [constr] (inclined baulk of timber used to support a wall) contre-fiche *f*, étai *m* incliné
RALE, to - [acoust & med] râler
RALE - [med] râle *m*
RALEIGH DISK - [acoust] (a device for the measure of sound particle-velocity in a medium) disque *m* de Raleigh
RALLY, to - [gen] rallier
[gen] (to revive, to restore) se reprendre
RALLY - [gen] ralliement *m*
RAM, to - [gen] battre, tasser
[constr] (to consolidate by hammering and pressing down) damer, pilonner
[constr] (to solidify, to make compact) bourrer
[metall] fouler, battre
[metall] (the refractory lining) pousser
[naut] éperonner
[firearms] (to press down the bore of a gun) bourrer, refouler
[auto] tamponner
RAM - [zool] (a male sheep) bélier *m*
[impl] (instrument for driving, crushing, forcing) pilon *m*, mouton *m*
[hydr] (a hydraulic ram) piston *m* plongeur, bélier *m* hydraulique
[constr] (the monkey of a pile driver) mouton *m*, sonnette *f* de battage
[mach tool] (of a shaping machine) chariot *m* porte-outil, trompette *f*
[mech] (of a power hammer) pilon *m*, mouton *m*
[min] (of clay) bourre *f* d'argile

RAM - [shipbuild] longueur *f* de tête en tête
[astr] (the constellation of Aries) (le) Bélier
[aero] effet *m* de prise dynamique
[metall] mâchoire *f*, défourneuse *f* de coke
[mech] cric *m*
[text] support, tréteau *m*
~AREA - [hydr] (the area of the end of the ram in a hydraulic jack or press on which the fluid pressure acts) surface *f* du piston plongeur
~AIR - [astronaut] air *m* sous pression dynamique
~BAR - [metall] (or pusher ram) bras *m* de défournement
~BOLSTER - [mech] tête *f* du piston
~COMPRESSION - [aero] (compression of air for turbojet or ramjet engines obtained from the forward motion of the aircraft) compression *f* dynamique
~DRAG - [aero] (or sink drag, drag associated with the removal of fluid from the airflow in question) traînée *f* d'immersion
[astronaut] traînée *f* de prise d'air
~ENGINE - [mech] sonnette *f* de battage
~EXTRUDER - [plast ind] presse *f* à filer
~HEAD - [metall] bouclier *m* de défournement
~-JET - s. ramjet
~LAMB - [zool] agneau *m*
~PRESSURE - [hydr] (the total force applied by a hydraulic ram, equal to the line pressure multiplied by the ram area) pression *f* du piston plongeur
~SEAL - [med] scellement *m* à plongeur
[metall] scellement *m* métal-céramique
RAM'S HORN - [mech] (for a crane) tête *f* de bélier, crochet *m* double
~SIDE - [ovens] (machine side) côté *m* machine
~TRAVEL - [mech] (the maximum distance through which the ram of a hydraulic jack or press is designed to move) trajet *m* du piston plongeur
~WITH CLAY, to - [text] (a retting hole) damer d'argile (le routoir)
~WOOL - [text] laine *f* de bélier
RAMARK - [radar] (or raymark; a radio beacon which is identifiable by all radar receivers) radiophare *m* pour radar
RAMBLE - [mining] faux toit *m*
RAMBLER - [bot] rosier *m* grimpant
RAMICORN - [zool] (of insects which have branched antennae) ramicorne
RAMIE - [bot] (natural fibre with applications in papermaking) ramie *f*
~BAST CELL - [bot] cellule *f* élémentaire de la ramie
~CARDING MACHINE - [text] carde *f* à ramie
~CLOTH - [text] tissu *m* de ramie
~COVERED WIRE - [el] fil *m* sous ramie
~DECORTICATOR - [text] décortiqueuse *f* mécanique à ramie
~FIBRE - [text & paper man] fibre *f* de ramie
~NOIL - [text] blousse *f* de ramie
~SPINNING - [text] filature *f* de la ramie
~YARN - [text] fil *m* de ramie
RAMIFICATION - [bot & zool] ramification *f*
RAMIFIED HEMP STEM - [text] tige *f* de chanvre rameuse
RAMIFY, to - [gen] ramifier
RAMJET - [aero] (propulsion unit in which air compressed by the forward movement of the aircraft is mixed with fuel injected into a convergent-divergent duct to produce combustion and provide reac-

tion) stato-réacteur *m*
RAMJET ENGINE - [aero] (a jet engine in which the air supply is compressed by the forward motion of the aircraft, without the use of a compressor) statoréacteur *m*
~HELICOPTER - [aero] hélicoptère *m* à statoréacteur
RAMMED CONCRETE - [constr] béton *m* damé
RAMMER - [impl] pilon *m*, bourroir *m*
[constr] (a punner) mouton *m*
[plast ind] (hand tool used to pack the sand of a mould) barre *f* de répalage
~LOG - [impl] dame *f*
RAMMING - [constr] damage *m*
[metall] (the packing of sand in the mould) foulement *m*
[aero] (effect due to the position of the ramming intake) effet *m* dynamique
~BOARD - [metall] plaque *f* à fouler
~INTAKE - [aero] (an air intake opening in the direction of forward motion of the aircraft to obtain compression) prise *f* d'air dynamique
~PISTON - [metall] piston *m* plongeur
~PLATE - [metall] plaque *f* de serrage
RAMP - [gen] rampe *f*
[constr] (an inclined surface) rampe *f*, pente *f*
[aero] (a launching ramp) rampe *f*
[auto] (in a garage) ponton *m* élévateur
[el] (US) parachute *f* à coin
[el] (traction) plan *m* incliné
[contr] fonction *f* à accroissement linéaire
~-FORCED RESPONSE - [contr] réponse *f* forcée par pente
~-FORCED RESPONSE TIME - [contr] temps *m* de réponse forcée par pente
~FUNCTION - [telev] échelon *m* de vitesse
~GENERATOR - [electron] générateur *m* de dents de scie
~SHOE - [el] frotteur *m* de locomotive
~SIGNAL - [contr] signal *m* à pente, signal *m* à variation linéaire de valeur
RAMPANT - [arch] rampant
~ARCH - [arch] (an arch with abutment which are not in the same horizontal line) arc *m* rampant
~CENTRE - [arch] centre *m* d'arc rampant
RAMPART - [constr] rempart *m*
RAMPION - [bot] raiponce *f*
RAMUS - [anat] branche *f*, rameau *m*
~OF THE JAW - [anat] branche *f* montante du maxillaire inférieur
RANCH - [agric] (farm for rearing cattle) ranch *m*
[gen] (a large farm) ferme *f* d'élevage
RANCHING - [agric] pâture *f* maigre
RANCID - [gen] (offensive flavour or odour, as of stale fat or oil) rance
~OIL - [gen] huile *f* rance
RANCIDITY - [med] rancidité *f*
RAND - [shoe man] couche-point *m*
RANDING - [mining] recherche *f* par fouilles
RANDOM - [gen] (irregular, haphazard) au hasard
~ACCESS - [comput] (to the storage) accès *m* sélectif
~ACCESS MEMORY - [comput] mémoire *f* à libre accès, mémoire *f* à accés aléatoire
~ACCESS PROGRAMMING - [comput] (the programme of a problem without regard for the time of access to the information in the registers required in the programme) programmation *f* indépendante du temps d'accès
RANDOM ACCESS STORAGE - s. random access memory
~ACCESS STORE - s. random access memory
~ASHLAR WORK - [constr] maçonnerie *f* en moellons bruts
~COINCIDENCE - [phys] (accidental coincidence) coïncidence *f* accidentelle
~DRIFT - [comput] dérive *f* aléatoire
~DRIFT RATE - [contr] vitesse *f* de dérive aléatoire
~ERROR - [statist] erreur *f* aléatoire
~-INCIDENCE RESPONSE - [el acoust] (of a microphone, for a specified frequency) sensibilité *f* moyenne
~INTERLACE - [telev] analyse *f* entrelacée approximative
~NOISE - [el acoust] (background noise) bruit *m* de fond
[radio] effet *m* de grenaille
~NUMBER - [math] nombre *m* aléatoire
~OVERLAPPING CONTROL SAMPLE - [contr] échantillon *m* de commande à composition aléatoire
~PROCESSING - [comput] traitement *m* à choix libre
~RUBBLE - [soil] enrochements *mpl* tout-venant
~SENSITIVITY - s. random-incidence response
~VARIABLE - [math] variable *f* aléatoire
~VELOCITY - [electron] (the velocity at the peak of the distribution curve, equivalent to the mean energy of electron temperature) vitesse *f* complexe
~VIBRATION - [phys] vibration *f* aléatoire
~WALK - [electron] parcours *m* erratique
~WALK METHOD - [math] méthode *f* de la route aléatoire
~WINDING - [el] (single multilayer coil in which the layers are not definite, nor spaced from each other) enroulement *m* au hasard
RANDOMIZATION - [comput] (arrangement at random) ordination *f* au hasard
RANDOMIZE, to - [comput] ordination *f* à choix libre
RANDOMIZER - [comput] (a change machine, a machine capable of arranging figures at random) machine *f* à nombres aléatoires
RANDOMLY DIRECTED MAGNETIZATION - [comput] magnétisation *f* de direction quelconque
RANGE, to - [gen] ranger, disposer
[phys etc] (of temperatures etc) s'éntendre
[gen] (to arrange in a definite order) disposer en ordre
[print] aligner
[opt] (e.g. a telescope) braquer
[instr] (to place in position and adjust) braquer
[naut] ranger
RANGE - [gen] rangée *f*, direction *f*, portée *f*, étendue *f*
[aero] (the distance an aircraft can travel under conditions starting with fuel tanks full) distance *f* franchissable
[firearms] portée *f*
[radio] (the distance over which signals can be transmitted) portée *f*
[phys] (the distance a particle will penetrate a specified substance before its kinetic energy is reduced) portée *f*
[radar] (distance to the target) rayon *m* d'action
[math] (all the values of a function) marge *f*
[math] (a finite series of numbers having specifi-

cally determined sequence and value) série *f*
RANGE - [phys] (of frequencies, or colours etc) gamme *f*
[surv] champ *m*
[el] plan *m*
[naut] bitture *f* de câble
[acoust & mus] gamme *f*
[statist] étendue *f*
[geogr] (of mountains) chaîne *f* de montagnes
~AMPLITUDE DISPLAY - [radar] (a display; a scope) indicateur *m* type A
~AND BEARING DISCRIMINATION - [radar] (resolution) pouvoir *m* de résolution
~AT ECONOMIC SPEED - [aero] distance *f* franchissable économique
~AT FULL SPEED - [aero] autonomie *f* à vitesse maximum de vol
~AT MAXIMUM WEAK-MIXTURE POWER - [aero] (the range of an aircraft at a speed corresponding to the maximum power on weak mixture) distance *f* franchissable à puissance maximum avec mélange pauvre
~BEACON - [radio] (radio beacon used for the guidance of aircraft) radiophare *m*
~BEARING DISPLAY - [radar] (B display, B scope) indicateur *m* type B
~BOILER - [th eng] réservoir *m* à eau chaude
~CHANGE SWITCH - [radar] sélecteur *m* de distances
~-CLAMPING SWITCH - [electron] commutateur *m* de gammes
~CONTROL GEAR - [contr] (of torpedoes) régulateur *m* de distance
~CONTROL SWITCH - [el] commutateur *m* d'étendue
~DATA - [radar] informations *f*pl de distance
~DETERMINATION - [photo] profondeur *f* de champ
~DIAGRAM - [statist] diagramme *m* des champs de variation
~DIAL - [firearms] cadran *m* de pointage
~DISCRIMINATION - [radar] (resolution of a radar set in respect to range) sélection *f* des distances
~-ENERGY RELATION - [mech] relation *f* parcours-énergie
~FINDER - [instr] (or rangefinder; instrument designed to find the distance to the target) télémètre *m*
~-FINDER FIELD - [surv] champ *m* de mesure du télémètre
~-FINDER FOCUSING - [photo] mise *f* au point par télémètre
~-FINDER WINDOW - [photo] fenêtre *f* de mise au point
~GATING - [radar] sélection *f* de signaux pour utilisation du radar dans une gamme de distances
~HEIGHT INDICATOR - s. range bearing display
~HEIGHT MARKER - s. range amplitude display
~INDICATION - [radar] indication *f* de la distance
[radar] marqueur *m* d'étalonnage
~LINE - [surv] alignement *m*
~MARKER GENERATOR - [radar] générateur *m* de marqueurs d'étalonnage
~MARKS - [radar] (called distance marks in the USA; marks on a cathode-ray indicator indicating the distance from the set to the target) repères *m*pl de distance
~MEASUREMENT - [radar] mesure *f* de distance
~OF A TIDE - [astr] étendue *f* de la marée
~OF ACCOMMODATION - [med] amplitude *f* de l'accommodation

RANGE OF ADJUSTMENT - [photo] intervalle *m* de réglage
~OF APPLICATION - [gen] domaine *m* d'application
~AUDIBILITY - [acoust] (at a specified frequency) aire *f* d'audibilité
~OF BALANCE - [instr] portée *f* de mesure de la balance
~OF DISTRIBUTION - [comm] aire *f* de distribution
~OF ELECTROMAGNETIC OSCILLATIONS - [el] gamme *f* des oscillations électromagnétiques
~OF EXPOSURE - [photo] latitude *f* d'exposition
~OF HEADLAMP - [auto] portée *f* d'un phare
~OF LINEARITY - [contr] bande *f* proportionnelle
~OF NUCLEAR FORCES - [nucl] portée *f* des forces nucléaires
~OF PROPORTIONALITY - [contr] étendue *f* proportionnelle
~OF REVOLUTIONS - [mech] gamme *f* des régimes
~OF SHARPNESS - [photo] étendue *f* de la netteté
~OF SOFTNESS - [rubber ind] trajet *m* de ramollissement
~OF SPEEDS - [mech] gamme *f* de vitesses
~OF SPINDLE SPEEDS - [mach tool] gamme *f* des vitesses du mandrin
~OF STRESS - [metall] (the range between the upper and lower limit of a cycle of stress, as applied in a fatigue test) gamme *f* des efforts
~OF VARIATION - [contr] domaine *m* de variation
~OF VISIBILITY - [opt] portée *f* de la vue
~OF VOLTAGE - [el] champ *m* de la tension
~POLE - [surv] jalon *m*
~RINGS - s. range marks
~STRAGGLING - [nucl] (the variation in the range of particles with the same initial energy) dispersion *f* statistique du parcours
~STROBE - [surv] repère *m* de distance
~TRANSMISSION UNIT - [radar] indicateur *m* de distance
~SWITCH - [radio] commutateur *m* de gammes d'ondes
~WIND - [astronaut] composante *f* longitudinale du vent balistique
RANGER - [gen] garde *m* forestier
RANGING - [instr] télémétrie *f*
[meas] (calibration) réglage *m*
~PULSE - [phys] impulsion *f* de distance
~RADAR - [radar] radar *m* de réglage de tir
~ROD - [surv] (a wooden pole to mark a station) jalon *m*
~SYSTEM - [radar] système *m* de télémétrie radar
RANK WILDCAT - [oil ind] puits *m* de recherche en zone inconnue
RANULA - [med] grenouillette *f*
RAOULT'S LAW - [chem] (this states that in dilute solution the relative reduction of the vapour pressure of a liquid by the dissolution in it of a given substance in nearly equal to the mol fraction of such substance, irrespective of temperature or of the character of the solute and solvent) loi *f* de Raoult
RAP, to - [gen] frapper
[metall] (die casting) ébranler
~SHOT - [cin] (colloq. for a complicated shot) plan *m* difficile à réaliser
RAPE - [agric] (the refuse stalks and skins of grapes) râpe *f*, marc *m* de raisin
[brew ind] drêches *f*pl

RAPE - [bot] colza *m*
~CAKE - [agric] tourteau *m* de colza
~MEAL - [agric] farine *f* de colza
~OIL - [chem] huile *f* de navette
~-SEED OIL - [chem] (a starting material in the production of rubber substitutes) huile *f* de colza
RAPHE - [anat] (a broad junction) raphé *m*
RAPID - [gen] rapide
[photo] (of a lens) rapide
(of a river, a descent) rapide *m* (d'un fleuve)
~CONNECTIONS - [mech] raccords *mpl* rapides
~COOLING - [gen] refroidissement *m* rapide
~HARDENING PORTLAND CEMENT - [constr] (portland cement very finely ground to give high early strenght) ciment *m* Portland prompt
~RELEASE - [el] déclenchement *m* rapide
~RELEASE VALVE - [mech] soupape *f* à ouverture rapide
~RETURN - [mach tool] retour *m* rapide
~SETTING - [mech] ajustage *m* rapide
~TRAVERSE - [mach tool] chariotage *m* rapide
RAPIDITY - [gen] rapidité *f*
~OF LENSES - [photo] (the ratio between the working aperture diameter and focal length) ouverture *f* relative
~OF MODULATION - [telecomm] rapidité *f* de modulation
RAPIER - [text] crochet *m*, pince *f*
RAPLOT - [radar] (radar plotting method) raplot *m*
RAPPER - [metall] ébranloir *m*
RAPPING - [metall] (the process of loosening a pattern in a mould) ébranlage *m*
~BAR - [metall] barre *f* à ébranler
~DEVICE - [metall] dispositif *m* d'ébranlage
~HAMMER - [metall] maillet *m* conique
~HOLE - [metall] trou *m* d'ébranlage
~PIN - [metall] picot *m* démouleur
~PLATE - [metall] plaque *f* d'ébranlage
~SPIKE - [metall] pointe *f* à boucle
RAPTUS - [med] raptus *m*
RARE - [gen] rare
[chem etc] peu dense
~EARTHS - [chem] (the oxides of the rare earth elements) terres *fpl* rares
~EARTH ELEMENTS - [chem] (a group of elements having atomic numbers from 57 to 7I usually with the inclusion of scandium and yttrium, which, though outside this range, have similar properties. Their chemical similarity is such that they are considered to occupy the position of a single element in Group III) éléments *mpl* des terres rares
~GAS CARTRIDGE - [el] limiteur *m* de tension à gaz rare
~GASES - [chem] (the elements of the group 0 of the periodic system, helium, neon, argon, krypton, xenon and radon) gaz *mpl* rares
~MIXTURE - [auto] (of fuel) mélange *m* pauvre
RAREFACTION - [phys] (a decrease in pressure or density) raréfaction *f*
~WAVE - [phys] onde *f* de succion
RAREFY, to - [phys] (to diminish in density or pressure) raréfier
R.AS. - [aero] (rectified airspeed) vitesse *f* corrigée
RASHIG PROCESS - [ind chem] (a commercial method for the production of phenol by the catalytic reaction of benzene with oxygen and hydrogen chloride) procédé *m* Raschig
RASCHIG RINGS - [ind chem] (cylindrical rings used to pack columns) anneaux *mpl* Raschig
RASE, to - s. raze, to
RASH - [med] éruption *f*
[mining] charbon *m* terreux
RASP, to - [mech] râper, racler
RASP - [impl] râpe *f*
RASPATORY - [med] râpe *f*
RASPBERRY - [bot] framboise *f*
~BUSH - [bot] framboisier *m*
~TONGUE - [med] langue *f* framboisée
RASTER - [telev] (the pattern of the parallel lines on the fluorescent screen of the cathode-ray tube when the frame and line-scanning currents are applied simultaneously) canevas *m*, trame *f*
~BURN - [telev] endommagement *m* de canevas
~PITCH - [telev] espacement *m* des lignes
~SHADING - [telev] canevas *m* à luminosité inégale
RAT - [zool] rat *m*
~CONTROL - [gen] destruction *f* des rats
~HOLE - [oil ind] trou *m* pour la tige carrée
~POISON - [chem] mort *f* aux rats
~-RACE - [electron] (hybrid junction of a waveguide) anneau *m* hybride
~-TAIL BURNER - [th eng] (pinhole burner) brûleur *m* à flamme filiforme
~-TAILED FILE - [mech] lime *f* queue-de-rat
RATCH, to - [mech] encliqueter, denter
RATCH - [text] écartement *m* des cylindres
[mech] roue *f* à cliquet
~OF THE FALLERS - [text] distance *f* des barrettes
RATCHET - [mech] (a set of saw-like teeth on a bar or the rim of a wheel) cliquet *m*, chien *m*
[mech] (the mechanism) encliquetage *m*
~ADJUSTMENT - [mech] réglage *m* à rochet
~-AND-PAWL-MOTION - [mech] encliquetage *m* à rochet
~BIT BRACE - s. ratchet brace
~BRACE - [mech] (type of drilling brace) vilebrequin *m* à cliquet, cliquet *m* à canon, raccagnac *m*
~CATCH - [mech] cliquet *m* à rochet
[text] cliquet *m*
~DOG - [mech] levier *m* à cric
~DRILL - [tool] (a type of hand-drill) vilebrequin *m* cliquet, drille *f* à rochet
~FEED - [mech] mécanisme *m* de progression par cliquet de blocage
~GEAR - [mech] encliquetage *m*
~GEARING - [mech] mécanisme *m* de la roue à cliquet
~JACK - [mech] verin *m* à bouteille
~LEVER - [mech] levier *m* à cliquet
~MOTION - [mech] encliquetage *m*
~MOTION - [text] mécanisme *m* de la roue à cliquet
~PAWL - [mech] (pivoted tooth in a ratchet mechanism) cliquet *m*
[text] cliquet *m* d'accouplement
~SCREWING-STOCK - [mech] filière *f* à cliquet
~SPANNER - [mech] clef *f* à cliquet
~SPRING - [mech] ressort *m* du cliquet
~STOP - [instr] (of a micrometer) bouton *m* à friction
~TIME BASE - [telev] (time base giving a measure of delay to the light spot in a cathode-ray tube) base *f* des temps à encliquetage
~TOOTH ESCAPEMENT - [horol] échappement *m* à

cliquet
RATCHET WHEEL - [mech] (a wheel furnished with specially shaped teeth, used in a ratchet mechanism) roue *f* à rochet, roue *f* à cliquet
~WHEEL MECHANISM - [text] mécanisme *m* d'encliquetage
~WHEEL MOTION - [text] roue *f* à rochet
~WRENCH - [mech] clef *f* à rochet, cliquet *m* simple
RATE, to - [gen] estimer, évaluer
[comm] évaluer, fixer la valeur
[instr] régler
[meas] étalonner
[gen] (to classify) classer
RATE - [gen] taux *m*, raison *f*
[mech] (speed) taux *m* de vitesse
[comm] prix *m*
[comm & fin] taux *m*
[mech etc] régime *m*
[meas] (in relation to meters) tarif *m*
[el] (of a storage battery; the discharge in terms of current which a battery can give for a specified time under specified conditions) régime *m*
~ACTION - [contr] (derivate action) action *f* influencée par un réglage en dérivation
~CLASSIFICATION - [comm] (only USA, special-use tariff, a rate of payment applied for the supply of electricity, gas etc for special uses) tarif *m* par usages
~FEEDBACK - [comput] réaction *f* tachymétrique
~FIXER - [ind] chronométriste *m*
~-FREE - [leg] exempt d'impôt
~GYRO - [instr] fyromètre *m*
~MAKING - [comm] (establishing a tariff) tarification *f*
~-METER - [instr] (instrument designed to give a continuos indication of the mean rate of ionizing events) ictomètre *m*
~OF ABSORPTION - [soil] vitesse *f* d'absorption
~OF CAMBER CHANGE - [auto] valeur *f* de la variation de carrossage
~OF CAPILLARY RISE - [soil] vitesse *f* d'ascension capillaire
~OF CASTER CHANGE - [auto] valeur *f* de la variation de chasse
~OF CATCH - [aero] (the rate at which liquid precipitation strikes an aircraft usually measured in pounds per ft per hr) taux *m* de captation
~OF CHANGE - [radar] (the method of indicating the degree in which the registered baring data change in value) taux *m* de variation de l'orientation
~OF CLIMB - [aero] (the rate at which an aircraft rises in relation to the earth's surface) vitesse *f* ascensionnelle
[aero] (in aircraft testing; the vertical component of the air path of an ascending aircraft, corrected for the standard atmosphere) composante *f* verticale de la vitesse ascensionnelle
~-OF-CLIMB INDICATOR - [instr] (instrument used to show the rate at which the altitude of an aircraft is changing) variomètre *m*
~OF COALESCENCE - [rubber ind] vitesse *f* de coalescention
~OF COMBUSTION - [th eng] vitesse *f* de la combustion
~OF CRACK GROWTH - [rubber ind] vitesse *f* de croissance de craquelure

RATE OF CURE - [rubber ind] vitesse *f* de vulcanisation
~OF DECAY OF SOUND - [acoust] taux *m* d'extinction
~OF DEFORMATION - [soil] vitesse *f* de déformation
~OF DEPOSITION - [metall] vitesse *f* de dépôt
~OF DISAPPEARENCE - [med] vitesse *f* d'élimination
~OF DRAINAGE - [soil] vitesse *f* de drainage
~OF EXCHANGE - [fin] taux *m* du change
[nucl] (the rate of reaction at which atoms of two different molecular species exchange places) vitesse *f* d'échange
~OF FAILURES - [comput] taux *m* de défaillance
~OF FEED - [mech] vitesse *f* d'avance
~OF FLAME PROPAGATION - [phys] vitesse *f* de déflagration
~OF FLOW - [phys] débit *m*, vitesse *f* d'écoulement
~OF GROWTH - [gen] taux *m* d'accroissement
~OF HEATING - [th eng] loi *f* d'échauffement
~OF HEAVE - [soil] vitesse *f* de soulèvement
~OF ICING - [aero] (the rate of formation of ice on an aircraft surface normally expressed in inches of thickness per minute) vitesse *f* de givrage
~OF INFLOW - [phys] (the rate at which a quantity enters into a volume) vitesse *f* de débit à l'entrée, volume *m* d'apport
~OF INHERENT REGULATION - [comput] vitesse *f* d'autorégulation
~OF INJECTION - [plast ind] (the volume of material per second discharged through the nozzle during injection) vitesse *f* d'injection
~OF INTEREST - [fin] taux *m* d'intérêt
~OF KNITTING - [text] compte *m* en lisses
~OF PERFORATION - [comput] vitesse *f* de perforation
~OF PUNCHING - s. rate of perforation
~OF PULSE REPETITION - [comput] (the number of electric pulses per unit of time a point in the computer) taux *m* de répétition d'impulsions
~OF READING - [comput] vitesse *f* de lecture
~OF RECIPROCATION - [mech] taux *m* d'alternation
~OF SELF-REGULATION - [contr] vitesse *f* d'autorégulation
~OF SHEAR - [metall] vitesse *f* de cisaillement
~OF SIDE-SLIP - [aero] (velocity in the direction of the lateral axis, relative to the undisturbed air) vitesse *f* latérale
~OF STOCKING - [agric] charge *f* des parcelles en cours de pâturage
~OF TRAVEL - [th eng] (of the charge) vitesse *f* de passage
~OF TREAD CHANGE - [auto] valeur *f* de la variation de voie
~OF WORKING - [gen] allure *f* du travail
~SCHEDULE - [comm] (tariff scale) barême *m* de tarifs
~TEST - [contr] essai *m* de constantes de temps
~TIME - [contr] (derivative action time) temps *m* de doublage D-P
RATEBANE - [min] orpiment *m*
RATED - [mech el etc] nominal
~ALTITUDE - [aero] (the height at which the maximum power is delivered) altitude *f* nominale
~BREAKING CAPACITY - [el] (the maximum prospective current which may be associated with a fuse under given conditions) pouvoir *m* nominal de rupture
~BREAKING CURRENT - [el] courant *m* nominal de

rupture

RATED BURDEN - [el] (the apparent power which the transformer can supply to the secondary at its rated voltage without the errors which it introduces exceeding the guaranteed values) puissance *f* de précision

~CAPACITY - [el] puissance *f* utile absorbée nominale

~CURRENT - [el] (of a machine or an apparatus) courant *m* nominal

~CURRENT LOAD - [el] intensité *f* de courant nominal

~DUTY - [el] (the duty for which an electrical machine has been designed) service *m* nominal

~DYNAMIC CURRENT - [el] courant *m* de fonctionnement nominal

~ELECTRODE DISSIPATION - [electron] dissipation *f* normale d'électrode

~FREQUENCY - [el] (the frequency used in the specifications) fréquence *f* nominale

~FUSING CURRENT - [el] courant *m* nominal de fusion

~-GROWN JUNCTION - [electron] (grown junction produced by varying the rate of crystal growth) jonction *f* progressive

~HORSEPOWER - [mech] puissance *f* nominale

~IMPEDANCE - [el] (the impedance of the secondary circuit corresponding to the rated burden at the rated voltage) impédance *f* de précision

~INPUT - [el] (of an electrical or transformer) puissance *f* nominale

~INSULATING VOLTAGE - [el] tension *f* nominale de isolement

~LOAD - [el] (the output which is assigned as the maximum under given conditions) puissance *f* utile fournie nominale

~MAKING-CAPACITY - [el] (the making capacity, that is the capability of a device to make an electrical circuit under given conditions for which a circuit-breaker or similar apparatus is designed) pouvoir *m* nominal de fermeture

~MAKING CURRENT - [el] courant *m* nominal de fermeture

~OUTPUT - [mech] puissance *f* nominale
[el] s. rated load

~OUTPUT POWER - [el] puissance *f* de sortie

~PHASE ANGLE - [el] déphasage *m* nominal

~POWER - [mech] (the maximum power output or input of a machine as specified by the manufacturer) puissance *f* nominale

~POWER SUPPLY - [el] puissance *f* nominale d'alimentation

~PRESSURE - [mech] pression *f* nominale

~PRIMARY CURRENT - [el] (of a current transformer) courant *m* nominal primaire

~PRIMARY VOLTAGE - [el] (of a voltage transformer) tension *f* nominale primaire

~QUANTITY - [el](of an electrical machine; the rated current, voltage, frequency etc of an electrical machine of which the numerical value is included in the definition of rating) grandeur *f* nominale

~SHORT-CIRCUIT CURRENT - [el] (of a current transformer) courant *m* de court-circuit nominal

~SHORT-TIME CURRENT OF A SWITCH - [el] courant *m* de courte durée admissible d'un appareil

~SHORT-TIME THERMAL CURRENT - [el] courant *m* limite thermique

~SPEED - [el & mech] vitesse *f* nominale

RATED TEMPERATURE-RISE CURRENT - [instr] (of an instrument) courant *m* d'échauffement

~VOLTAGE - [el] (of a cable; the voltage for which the different parts of the dielectric of a cable are designed) tension *f* nominale

~~VALUE - [mech] valeur *f* de régime

~WATT CONSUMPTION - [el] puissance *f* absorbée normale

RATEEN - [text] étoffe *f* ratinée

RATINE - [text] (cotton or worsted fabric with a rough surface) ratine *f*

RATING - [gen] évaluation *f*, classement *m*
[mech] (the operating capacity of a piece of machinery) cheval *m* nominal
[el] (the period of time during which a motor or generator can produce electrical power at a given rate without excessive heating) régime *m* nominal
[el] (of a storage battery) (the number of amperehours a battery can supply from full charge under specified conditions) capacité *f* nominale (d'un accumulateur)
[el] (of a limiting quantity) calibre *m* d'un appareil
[naut] (the classification of a vessel) classement *m*, classe *f*
[the eng] (of a boiler) taux *m* de vaporisation
[mech] (of a motor) charge *f* prévue (d'un moteur)
[el mach] s. rating-plate

~CHART - [electron] (for x-ray tubes) courbes *fpl* de charge

~PLATE - [el] (a plate attached to an electrical machine by the makers, on which the characteristics and rating are shown) plaque *f* signalétique

RATIO - [gen & math] (the arithmetical relation between two quantities) raison *f*, rapport *m*
[el] (of an instrument transformer; the ratio between the primary terminal voltage and the secondary terminal voltage of a voltage transformer) rapport *m*

~ADJUSTER - [el] régulateur *m* de commande de proportion

~ARMS - [el] (two adjacent arms of a Wheatstone bridge, in which the resistance can be made to have one of several fixed ratios) rapport *m* des consoles

~BALANCE PROTECTIVE SYSTEM - [el] (special from of impedance protective system) système *m* de protection à rapport balancé

~COMPUTER - [comput] calculateur *m* de rapport

~CONTROL - [contr] régulation *f* de rapport

~CONTROL SYSTEM - [contr] système *m* de réglage à rapport

~CONTROLLER - [contr] (controller operating to maintain a predetermined ratio between two physical quantities) régulateur *m* de proportion

~DETECTOR - [telev] (a circuit for the demodulation of modulated signals) détecteur *m* de rapport

~DEVIATION - [radio] indice *m* de modulation

~DISTORSION - [telecomm] (in automatic telephony) distorsion *f* de rapport d'impulsion

~ERROR - [el] (a departure from the ratio of the primary to the secondary voltages of a transformer from the rated value) erreur *f* de rapport

~-METER - s. ratiometer

~OF ACTIVITY DENSITIES - [nucl] (in radioactive waste) rapport *m* d'activité

~OF CONVERSION - s. ratio of transformation

~OF DOUBLING - [text] rapport *m* de doublage

~OF DRAFTING - [text] rapport *m* d'étirage

RATIO OF GEARING - [mech] rapport *m* de démultiplication
~OF MIXTURE - [mech] (the proportion of fuel to air) proportion *f* du mélange
~OF TRANSFORMATION - [el] (of an instrument transformer) rapport *m* de transformation
~ROLLING - [metall] réduction *f* de laminage
~VARIATION - [mech] dispositif *m* de variation du rapport (de transmission)
RATIOMETER - [instr] (instrument designed to measure electrically the quotient of two quantities) logomètre *m*, quotientmètre *m*
RATION, to - [gen] rationner
RATION - [gen] ration *f*
RATIONAL - [gen] (pertaining to reason or possessing the faculty of reasoning) raisonnable
[math] (expressing the ratio of two whole quantities) rationnel
~ACTIVITY COEFFICIENT - [el] (in the theory of electrolytes) coefficient *m* d'activité rationnel
~FORMULA - [phys](or constitutional formula; diagram showing the relative positions of atoms and group in two dimensions) formule *f* de constitution, formule *f* rationnelle
~FRACTION - [comput] fraction *f* rationnelle
~HORIZON - [astr] (true horizon) horizon *m* rationnel
~INTEGRAL FUNCTION - [math] fonction *f* intégrale rationnelle
~MECHANICS - [mech] mécanique *f* rationnelle
~NUMBER - [math] nombre *m* rationnel
RATIONALIZATION - [math] (the process of clearing from irrational quantities) rationalisation *f*
[ind] (scientific organization of production to eliminate waste of labour) organisation *f* rationnelle
RATIONALIZE, to - [math] rationaliser, faire évanouir les quantités irrationnelles (d'une expression)
[gen] (to render conformable to reason) ratio - naliser
RATIONING - [gen] rationnement *m*
RATLING - s. ratline
RATLINE, to - [naut] (to fasten the ratlines on the shrouds of the vessel) fixer les enfléchures
RATLINE - [naut] (small line horizontally fastened to the shrouds, used as steps up and down the rigging) enfléchure *f*
~STUFF - [naut] quarantenier *m*
RATOON - [sugar uind] rejeton *m* de canne à sucre
RATRAN - [radar] (tricon radar system, a radar system in which the receiver records the coincidence of received pulses from a group of three ground stations pulses in variable time sequence) ratran *m*
RATTAN - [bot] jonc *m* d'Inde
RATTEENING - [text] ratinage *m*
RATTLE, to - [gen] (to produce a quick succession of short sounds) faire entendre un bruit sec, crépiter, ferrailler, claquer
[el acoust] (of records) cracher
[metall] dessabler au tonneau
RATTLE - [mus] (an instrument used to produce quick sharp noises) crécelle *f*
[mech] (irregular noise from an engine or motor) bruit *m*, ferraillement *m*
[acoust] crécelle *f*, fracas *m*
[paper] (a crackling noise heard when handling paper) crépitement *m*
RATTLE - [med] râle *m*
~JACK - [geol] schiste *m* houiller
~OF CHAIN - [gen & auto] ferraillement *m* de la châine
RATTLER - [mining] cannel-coal *m*
[min] tambour *m* dessableur
RATTLING - [acoust] (the sound made by a rattle) bruit *m* de crécelle
[metall] dessablage *m* au tonneau
~NOISE - [electron] (in an electronic tube) bruit *m* d de crécelle
RAUCHWACKE - [geol] cargneule *f*
RAUVITE - [min] (calcium, vanadium and uranium containing mineral) rauvite *f*
RAUWOLFIA - [chem] (the powdered root of Rauwolfia serpentina. It is the source of a number of alkaloids with uses as hypotensive agents. Rauwolfia is used as an extract in medicine) rauwolfia *m*
RAVEL, to - [text] s'embrouiller
[gen] embrouiller
RAVEL - [text] vautoir *m*
~COURSE - [text] série *f* protectrice
RAVELLING - [text] (the thread drawn out of a woven fabric) effilochure *f*
[constr] (a series of craks) rupture *f* due à érosion
RAVEN - [zool] corbeau *m*
RAVENOUS - [gen] rapace, vorace
RAVINE - [geol] (a hollow, a deep gorge) ravine *f*
[gen] (sudden and violent rush) ravin *m*, ravinée *f*
[draw] (of a diagram) creux *m* accusé
RAW - [met] (bleak cold weather conditions) temps *m* âpre (gris et froid)
[gen] cru
[timber] (of trees) décortiqué
[leather ind] vert, non apprêté
~AMMONIA WATER - [chem] (crude ammoniacal liquor) eau *f* ammoniacale brute
~BENZOL - [ind chem] (crude benzole) benzol *m* brut
~BLOCK - [metall] lingot *m* de départ
~COAL - [min] charbon *m* brut
~COCOON - [text] cocon *m* vert
~COKE - [fuel] (run-of-retort coke) coke *m* brut
~COMPOUND - [min] masse *f* brute
~COPPER - [min] cuivre *m* brut
~COTTON - [text] coton *m* brut, coton *m* en bourre
~CULLET CHARGE - [glass man] charge *f* de calcin
~DATA - [comput] données *f*pl brutes
~FILM - [cin] film *m* vierge
~FLAVOUR - [brew ind] bouquet *m* de jeune
~FLAX - [text] lin *m* brut
~FLESH - [gen & med] chair *f* à vif
~GAS - [gas ind] (crude gas; unpurified gas) gaz *m* brut, gaz *m* non épuré
~GAS OUTLET - [gas ind] sortie *f* de gaz brut
~GLASS - [glass man] verre *m* brut
~GRAIN - [brew ind] (unmalted grain) grain *m* cru
~GUM - [chem] gomme *f* brute
~-HIDE - [leather ind] (hide dressed without tanning) cuir *m* vert
~-HIDE PINION - [mech] pignon *m* en cuir vert
~JUTE - [text] jute *m* cru
~LINSEED OIL - [chem] (linseed oil which has not been polymerized, e.g. by blowing or boiling) huile *f* de lin brute
~MATERIAL - [gen] (basic material necessary to a specific manufacture) matière *f* première, matière

f brute
RAW MATTE - [min] matte *f* brute de cuivre
~MINE - [min] minerai *m* de fer
~OIL - [chem] (untreated linseed oil) huile *f* crue
~ORE - [min] minerai *m* brut
~PAPER - [paper man] papier *m* brut, papier *m* support
~PETROLEUM - [min] pétrole *m* brut
~PIG IRON - [metall] fonte *f* brute de moulage
~RUBBER - [rubber ind] caoutchouc *m* brut
~RUNNING STRIP ROLLING MACHINE - [rubber ind] machine *f* à rechaper
~SHALE - [geol] schiste *m* dur
~SIENNA - [paint] (a brownish-yellow earth pigment, chiefly oxide of iron, with silica, alumina, calcium carbonate and oxides of manganese) terre *f* de Sienne brute
~SILK - [text] soie *f* grège
~SILK BALE - [text] balle *f* de soie grège
~STEEL - [metall] acier *m* brut
~STOCK - s. raw film
[text] fibres *fpl* grèges
~STORAGE - [comput] informations *fpl* de base
~SUGAR - [sugar ind] sucre *m* brut
~TAPE - [el acoust] (unused tape) ruban *m* vierge
~VISCOSE - [text] viscose *f* brute
~WATER - [chem] eau *f* brute, eau *f* non épurée
~-WATER-WHITE - [oil ind] distillat *m* léger
~WOOL - [chem] laine *f* brute
~WOOL SWEEPINGS - [text] balayures *fpl* laine brute
~YARN - [text] fil *m* brut
~ZINK - [metall] zinc *m* d'oeuvre
RAWHIDE GEARS - [mech] engrenages *mpl* en cuir vert
~HAMMER - [impl] (used by fitters to avoid damaging a finished surface) marteau *m* à panne en cuir vert
RAWIN - [met] mesure *f* du vent par radiosonde
RAY - [gen & phys] rayon *m*, radiation *f*
~ACOUSTICS - [acoust] (the analysis of acoustical problems on the theory that sound travels along rays through homogeneous material) méthode *f* des rayons acoustiques
~BEAM - [gen & phys] faisceau *m* de rayons
~-CONTROL ELECTRODE - [electron] électrode *f* de commande du faisceau
~DEFLECTOR - [opt] déflecteur *m* de rayons
~DIVERGENCE - [phys] divergence *f* de rayonnement
~-FILTER - [photo] écran *m* orthochromatique
~GRASS - [bot] ivraie *f* vivace
~OF LIGHT - [phys] rayon *m* lumineux
~PATH - [photo] trajet *m* des rayons
~PROOFING - [radiat] (protection for people handling x-ray tubes against damage by scattered rays) protection *f* contre le rayonnement
~-SHADE - [opt] (of a telescope) couvre-soleil *m*
RAY'S MANIA - [med] folie *f* morale
~TRACING - [opt] (the calculation of the path of rays through a system) traçage *m* des rayons
RAYED HOPSACK WEAVE - [text] armure *f* natté rayé
RAYLEIGH BALANCE - [el] (current weigher in which one of the coils is suspended from one of the balance) balance *f* de Rayleigh
~DISK - [phys] (a disk on a torsion suspension designed to measure the sound particle velocity in a fluid) disque *m* de Rayleigh
RAYLEIGH DISTILLATION - [chem] (simple distillation, in which the composition of the residue changes continuously during the distillation) distillation *f* de Rayleigh
~LOOP - [el] boucle *f* de Rayleigh
~NUMBER - [phys] (quantity defined for the fluid-filled space between two parallel horizontal planes) nombre *m* de Rayleigh
RAYMARK - [radar] (ramark; continually emitting radio beacon) radiophare *m* pour radar
RAYMOND MILL - [mech] (a special type of ring-roll mill used in paint manufacture) broyeur *m* Raymond
RAYON - [text] (fibre made by the viscose and cuprammonium processes; the term is often used for cellulose acetate fibre) rayonne *f*
~CLOTH - [text] tissu *m* de rayonne
~STAPLE - [text] fibre *f* de rayonne
~STAPLE YARN - [text] fil *m* de rayonne
~TYRE CORD - [rubber] cablé *m* en rayonne
RAZE, to - [gen] raser, abattre à ras de terre
RAZOR - [impl] rasoir *m*
~STONE - [min] novaculite *f*
Rb - [chem] (the symbol of rubidium) rubidium (symbole de)
RBE - (the initials of Relative Biological Effectiveness of radiation) EBR, efficacité *f* biologique relative
R.C. COUPLING - [radio] couplage *m* par résistance-capacité
RdAc - [chem] (the symbol for radioactinium) radioactinium (symbole de)
RDF - [telecomm] (initials of Repeater Distribution Frame) répartiteur *m* de répéteur
Re - [chem] (symbol for Rhenium) rhénium (symbole de)
REABLEMENT - [med] réhabilitation *f*
REABSORPTION - [chem] réabsorption *f*
REACH, to - [gen] atteindre, arriver à, s'étendre
REACH - [gen] extension *f*, portée *f*
[mech] (of a machine) portée *f*, champ *m*
[hydr] (clear stretch of water) bief *m*, biez *m*
[mech] (bar attached to the rear axle) barre *f* de relevage
[mech] (the point or position reached) distance *f*
[mech] (of hoist) hauteur *f* de levage
[text] (of a drawing frame) écartement *m*
[naut] bordée *f*
~OF A CANAL - [constr] partie *f* d'un canal ou fleuve
~RODS - [auto] (the adjustable tension rods connecting the crank and handle of the body to the lower endgate at the rear) barres *fpl* réglables de manoeuvre
~-THROUGH - [electron] (in semi-conductors) pénétration *f*
~-THROUGH VOLTAGE - [electron] tension *f* de pénétration
REACHING-IN MACHINE - [text] machine *f* à donner les fils
REACT - [gen] réagir
[chem] (to change chemically) (to exert a mutual action) réagir
[phys] (to exert an equal and opposite force) réagir
[fin] (of prices) réactionner
REACTANCE - [el] (the component of the resistance in an alternating-current circuit, which does not oppose the current, but causes at time a difference

of phase between the current and the electromotive force) réactance *f*
REACTANCE - [acoust] (the imaginary component of the acoustic impedance) réactance *f* acoustique
~BOND - [el] connexion *f* à réactance
~COIL - [el] (inductor, piece of apparatus possessing the property of inductance) bobine *f* de réactance
~COUPLING - [radio] (coupling between two circuits by means of a reactance common to both circuits) accouplement *m* à réactance
~DROP - [el] (the decrease in the available voltage at the terminals of a circuit, caused by the reactance voltage within that circuit) chûte *f* de tension par réactance
~EARTHED - [el] mise *f* à la terre par enroulement
~ELECTRON TUBE - [electron] (a variable impedance tube) tube *m* de réactance
~FED ALTERNATING-CURRENT TRACK CIRCUIT-[el] circuit *m* de voie à courant alternatif à reactance
~MODULATION - [radio] (modulation produced by variable reactance) modulation *f* à réactance
~MODULATOR - [radio] modulateur *m* à réactance
~OSCILLATOR - [el] oscillateur *m* à réactance
~PROTECTION - [el] dispositif *m* de protection à réactance
~RELAY - [el](impedance relay operating as soon as the reactance of the circuit to which it is connected falls below a predetermined value) relais *m* de réactance
~RISE - [el] (the increase in the available voltage at the terminals of a circuit , caused by the reactance voltage within that circuit) montée *f* de réactance
~TRANSFORMER - [el] (an impedance-transforming device) transformateur *m* de réactance
~TUBE - s. reactance valve
~-TUBE FREQUENCY MODULATOR - [radio] (circuit containing a reactance valve used to produce frequency modulation) modulateur *m* à réactance
~TUBE MODULATOR - [electron] tube *m* de réactance modulateur
~-TUBE PHASE MODULATION - [radio] modulation *f* à réactance
~VALVE - [radio] circuit *m* à tube de réactance
~VOLTAGE - [el] (the voltage, being a result of the current flowing through the reactance of a circuit) tension *f* de réactance
REACTANTS - [chem] (the substances which take part in a chemical reaction) réactifs *mpl*
REACTED PRESSURE - [mech] (in a variable delivery hydraulic pump, the stabilized pressure at which delivery becomes zero) pression *f* de débit nul
REACTING WEIGHT - [phys] (the number of parts of any element, in terms of weight, which can enter into combination with one part, by weight, of hydrogen or eight parts of oxygen, or the atomic weight divided by the valence) poids *m* équivalent
REACTION - [chem] (a transformation about by the action of one substance upon another) réaction *f*
[mech etc] réaction *f*
~ALTERNATING-CURRENT GENERATOR - [el] génératrice *f* à réaction à courant alternatif
~-AND-IMPULSE TURBINE - [mech] turbine *f* d'action-réaction
~BALANCE - [astronaut] balance *f* de poussée
~BLOWHOLES - [metall] soufflures *fpl* bleutées
~CHAIN - [chem] (chain reaction) réaction *f* en chaîne
REACTION CHANNEL - [nucl] réaction *f* nucléaire en chaîne
~CIRCUIT - [electron] (part of anode circuit of a thermionic valve) circuit *m* de réaction
~COIL - [electron] (coil included in the anode circuit of a thermionic valve) bobine *f* de réaction
~CONDENSER - [radio] (variable condenser controlling the degree of reaction) condensateur *m* de réaction
~COUPLING - [radio] (coupling between the anode and grid circuits of a thermionic valve to obtain reaction) couplage *m* par réaction
~CROSS-SECTION - [nucl] section *f* efficace de réaction
~ENERGY - [nucl] (nuclear reaction energy) énergie *f* de réaction
~FORMULA - [nucl] (nuclear reaction equation) formule *f* de réaction nucléaire
~GENERATOR - [el] synchronous generator excited by alternating current) génératrice *f* à réaction
~INHIBITION - [chem] inhibition *f* de réaction
~JET - [astronaut] veine *f* de réaction
~LIMIT - [metall] limite *f* de corrosion
~MOTOR - [el] (a synchronous motor, the rotor of which consists of salient poles, without winding and without permanent magnet) moteur *m* synchrone à fer tournant
~NOZZLE - [astronaut] tuyère *f* de réaction
~ORDER - [chem] (order of reaction) ordre *m* de réaction
~PRODUCTS - [chem] (the substances which are formed in chemical reactions) produits *mpl* de réaction
~PROPULSION - [astronaut] propulsion *f* par réaction
~RATE - [nucl] (the rate at which fission takes place in a nuclear reactor) vitesse *f* de réaction
~SPRING - [mech] ressort *m* à réaction
~TEMPERATURE- [chem] température *f* de réaction
~TIME - [zool etc] (the time interval between a nervous stimulus and the consequent reaction) temps *m* de réaction
~TORQUE - [mech] couple *m* dû à une réaction
~TURBINE - [mech] (steam turbine in which the steam expands progressively through alternate rows of fixed and moving blades) turbine *f* à réaction, turbine *f* à pression
~VELOCITY - [gen & chem] (the rate at which a reaction proceeds) vitesse *f* de réaction
~VESSEL - [ind chem] chambre *f* de réaction
~WATER-WHEEL - [hydr] roue *f* à tuyaux, roue *f* à réaction
REACTIONAL BIAS - [radio] tension *f* de polarisation à réaction
REACTIVATE, to - [gen] réactiver
[chem] (to restore to the original strength) réactiver
REACTIVATION - [chem] (the restoration of a substance, e.g. charcoal, to an activated state) réactivation *f*
REACTIVE - [chem] (entering easily into combination with other substances) réactif
~ANODE - [el] (or sacrificial anode) anode *f* réactive, anode *f* soluble
~ATTENUATOR - [radio] (attenuator which does not absorb energy) affaiblisseur *m* réactif
~CIRCUIT - [el] circuit *m* réactif

REACTIVE COIL - [el] bobine *f* de réactance
~COMPONENT OF THE CURRENT - [el] (the component of an alternating current, considered as a vector quantity, which is in quadrature with the voltage) composante *f* réactive du courant
~COMPONENT OF THE VOLTAGE - [el] composante *f* active de la tension
~CURRENT - [el] courant *m* réactif
~-ENERGY METER - [instr] (Var-hour meter; integrating instrument measuring the reactive energy in var-hours) compteur *m* d'énergie réactive, varheuremètre
~FACTOR - [el] (the ratio between the reactive volt-amperes and the totalvolt-amperes) coefficient *m* de réactance
~IRON - [el] (in a transformer to increase leakage-resistance) fer *m* de réactance
~LOAD - [el] (a load in which the current is out of phase with the voltage at the terminal) charge *f* réactive
~POWER - [el] (the product of the voltage of a circuit and the reactive component of the current) puissance *f* réactive
~SPUTTERING - [electron] pulvérisation *f* cathodique réactive
~VOLT-AMPERES - [el] (the product of the active voltage and the current) volt-ampères *mpl* réactifs
~VOLTAGE - [el] (the component of an alternating voltage, considered a vector quantity, which is in quadrature with the current) tension *f* réactive
REACTIVITY - [nucl] (a measure of the departure of the reactor from critical) réactivité *f*
~DRIFT - [nucl] variation *f* de réactivité
~METER - [electron] réactimètre *m*
~POWER COEFFICIENT - [nucl] coefficient *m* de puissance de réactivité
~RATE - [nucl] taux *m* de variation de réactivité
~TEST - [nucl] essai *m* de réactivité
REACTOR - [chem] (any vessel in which a chemical reaction is carried out on an industrial scale) réacteur *m*
[el] (reactive coil) réactance *f*
[nucl] (an apparatus in which nuclear fission is sustained in a self-supporting chain reaction) réacteur *m* nucléaire
~CHEMISTRY - [nucl] (the chemical processes occurring in a nuclear reactor) chimie *f* du réacteur
~CONTAINMENT - [nucl] retenue *f* d'un réacteur
~CONTROL - [nucl] commande *f* d'un réacteur
~CONTROL-BOARD - [nucl] panneau *m* de commande de réacteur
~CORE - [nucl] (the area in the nuclear reactor where the nuclear reaction occurs) coeur *m* du réacteur
~CROSS-SECTION - [nucl] section *f* efficace du réacteur
~DESIGN - [nucl] projet *m* du réacteur
~DEVELOPMENT - [nucl] mise *f* au point du réacteur
~DOME - [nucl] dôme *m* du réacteur
~EXCURSION - [nucl] excursion *f* de puissance du réacteur
~FAMILY - [nucl] (set of characteristics) filière *f* de réacteurs
~LATTICE - [nucl] réseau *m* du réacteur, réseau *m* multiplicateur
~-LOADING - [nucl] charge *f* du réacteur
~LOOP - [nucl] boucle *f* de réacteur
REACTOR METALLURGY - [metall] métallurgie *f* des réacteurs
~OSCILLATOR - [nucl] (arrangement designed to maintain a neutron-absorbing body in periodic motion) oscillateur *m* de réacteur
~PERIOD - [nucl] (the time which is required for the neutrons flux to change by a factor of e) constante *f* de temps d'un réacteur
~POISONING - [nucl] empoisonnement *m* du réacteur
~POWER - [nucl] puissance *f* du réacteur
~SAFETY FUSE - [nucl] fusible *m* de sécurité d'un réacteur
~SHELL - [nucl] couverture *f* du réacteur
~SHIMMING - [nucl] compensation *f*
~SIMULATOR - [nucl] maquette *f* de réacteur
~SPECTRUM - [nucl] distribution *f* de l'énergie des neutrons dans le réacteur
~SPHERE - [nucl] (the spherical outer shell of a nuclear reactor) dôme *m* du réacteur
~-START MOTOR - [el] moteur *m* à démarrage par self
~THEORY - [nucl] (the physical laws set out for the construction of a nuclear reactor) théorie *f* du réacteur
~TRIP - [nucl] arrêt *m* rapide du réacteur
~VAULT - [nucl] (the concrete shield in a nuclear reactor containing the reactory and primary cooling circuit) voûte *f* du réacteur
~VESSEL - [nucl] (the stainless steel pot housing the core and breeder in a breeder reactor) caisson *m* de réacteur, récipient *m* à pression
READ, to - [gen] lire
[comput] (to copy from one form of storage to another) lire
READ AMPLIFIER - [comput] amplificateur *m* de lecture
~-AROUND NUMBER - s. read-around ratio
~-AROUND RATIO - [comput] (in a cathode-ray tube storage, the number of times the information can be read successively) nombre *m* de lectures
~BACK FROM STORAGE, to - [comput] demander hors d'une mémoire
~-BACK SIGNAL - [comput] signal *m* de lecture
~BUFFER STORAGE - [comput] mémoire *f* tampon de lecture
~COIL - [comput] enroulement *m* de lecture
~IN, to - [comput] stocker, mémoriser
~-IN DATA - [comput] données *fpl* introduites
~MARK - [comput] marque *f* de lecture
~NUMBER - [comput] (in charge-storage tubes the number of times a storage element is read without rewriting) nombre *m* de lectures
~-OFF THE NUMBER OF TWISTS, to - [text] lire le nombre de tours
~-ONLY MEMORY - [comput] mémoire *f* permanente
~OUT, to - [comput] extraire les retenus
~-OUT CIRCUIT - [comput] circuit *m* de lecture
~-OUT PULSE - [comput] impulsion *f* de lecture
~-OUT UNIT - [comput] unité *f* de lecture
~OUTPUT - [comput] signal *m* de lecture
~PRINTER'S PROOF - [comput] épreuve *f* lue
~PULSE - [comput] (in static magnetic storage) impulsion *f* de lecture, impulsion *f* lire
~PUNCH UNIT - [comput] unité *f* de lecture perforatrice
~WINDING - [comput] (in static magnetic storage) enroulement *m* lire

READ-WRITE DISTURBED ONE OUTPUT SIGNAL - [comput] signal *m* d'un perturbé par une impulsion de lire et une impulsion d'écrire
~-WRITE HEAD - [comput] tête *f* d'inscription-lecture
~-WRITE REGISTER - [comput] registre *m* lecture-écriture
READABLE - [gen] lisible
READABILITY - [instr] (of a balance; i.e. the smallest fraction of division to which the index scale can be read) lisibilité *f*
READER - [gen] lecteur *m*
[print] (proof reader) correcteur *m* d'épreuves
[comput] (a device reading information and transferring the result to other devices or machine parts) lecteur *m*
~INTERPRETER - [comput] partie *f* de programme contrôlant la lecture et l'interpretation
READIED - [gen] (made ready) prêt
READING - [gen] lecture *f*
[instr] (the indication of a graduated instrument) lecture *f*, relevé *m*, cote *f*
~BRUSH - [comput] (metal brush reading information from punched cards) balai *m* de lecture
~BY REFLECTION - [comput] lecture *f* à miroir
~DEVICE - [text] dispositif *m* de lisage
~ERROR - [instr] (instrument error) erreur *f* de cote
~GLASS - [opt] loupe *f* à lire
~HEAD - [comput] tête *f* de lecture, tête *f* lectrice
~-IN - [text] lisage *m*
~-IN BOARD - [text] bâti *m* de lisage
~-IN FRAME - [text] bâti *m* de lisage
~-IN MACHINE - [text] liseuse *f* mécanique
~-IN THREAD - [text] fil *m* de lisage
~LIGHT - [light] liseuse *f*
~LIGHT FLAP - [photo] volet *m* de lecture
~MICROSCOPE - [instr] loupe *f* à lire
~SCALE - [photo] échelle *f* de lecture
~STATION - [comput] position *f* de lecture
READJUST, to - [gen] (to repeat an adjustment) rajuster, rectifier
READJUSTMENT BY HAND - [gen] réajustement *m* à la main
~OF ZERO - [comput] remise *f* à zero
READJUSTING MECHANISM - [mech] mécanisme *m* de réajustage
~THE FRICTION SURFACE - [text] réglage *m* des plateaux de friction
READY, to - [gen] (to prepare, to make ready) apprêter
READY - [gen] prêt
~-MADE - [gen & text] (generally said of clothing; not made on order, but prepared for general demand) tout fait, confectionné
~-MADE CLOTHING - [gen] vêtements *mpl* tout faits
~-MIXED PAINT - [paint] peinture *f* prête à l'usage
~TO SPIN - [text] (of a silk worm) prêt *m* à filer
~FOR SPINNING - [chem] prête au filage (une solution)
~-TO-WEAR - [text] confectionné, prêt à porter
~-TO-WEAR CLOTHING - s. ready-made clothing
READZING - [railw] (of the sleepers) resabotage *m*
REAFFOREST, to - [agric] reboiser
REAFFORESTATION - [agric] reboisement *m*
REAGENT - [chem] (a compound employed in chemical analysis to detect or measure specific substance) réactif *m*
REAGENT BOTTLE - [ind chem] (special bottle, usually with glass stopper, to contain reagents for bench use) flacon *m* à réactifs
~PAPER - [chem] papier *m* réactif
REAL - [gen] réel, vrai
~ATTENUATION - [brew ind] (of the fermentation) atténuation *f* réelle
~CIRCUIT - [telecomm] circuit *m* réel, circuit *m* composant
~DEGREE OF FERMENTATION - [brew ind] s. real attenuation
~ESTATE - [leg] (a piece of land including whatever is part of it) propriété *f* immobilière
~IMAGE - [opt] image *f* réelle
~IRON CROSS-SECTION - [el] section *f* réelle du fer
~LINE - s. real circuit
~NUMBER - [math] (number relating to quantities) nombre *m* réel
~OBJECT - [opt] (in geometrical optics, an object from each point of which light diverges towards the optical system) objet *m* réel
~POWER - [electron] puissance *f* active
~PROPERTY - s. real estate
~RATIO OF EXPANSION - [mech] (of a steam engine) rapport *m* de détente effectif
~REVERSED IMAGE - s. real image
~SOLUBILITY - [chem] (the amount of non-ionized dissolved salt existing in unit volume of a solution) solubilité *f* réelle
~STRENGTH - [metall] charge *f* maximale
~TIME - [comput] (in solving a problem) temps *m* réel
~-TIME COMMUNICATION - [comput] trafic *m* direct
~-TIME COMPUTATION - [comput] opération *f* en temps réel
~-TIME COMPUTER - [comput] calculatrice *f* en temps réel
~TIME OPERATION - [constr] opération *f* en temps réel
~VALUE - [comm] valeur *f* effective
REALGAR - [min] (natural arsenic disulphide, also obtained synthetically, with uses as a red pigment in textile printing, in fireworks, and as a depilatory in leather manufacture) réalgar *m*
REALIGNING - [mining] réparation *f* du boisage
REALIZATION - [gen] conception *f* nette (d'un fait)
[fin] conversion *f* en espèces
[cin] réalisation *f*
REALIZE, to - [gen] réaliser, comprendre
[fin] convertir en espèces
REALTOR - [comm] agent *m* immobilier
REAM, to - [mech] (to bring the bore of a hole to the desired size and or finish by means of a special tool) aléser, équarrir
[mech] (to enlarge the mouth of a hole) fraiser, aléser
[naut] patarasser
REAM - [paper man] (twenty quires, or 480 sheets) rame *f*
~OUT, to - s. ream, to
REAMER - [tool] (tool used in reaming) alésoir *m*, équarissoir *m*
[metall] dégorgeoir *m*, épinglette *f*
[oil ind] trépan *m* aléseur
REAMING - [mech] alésage *m*, équarrissage *m*
[metall] brochage *m*, dégorgement *m*

REAMING BIT - [mech] trépan *m* aléseur
REAP, to [agric] (to cut down and gather) moissonner
REAPER - [agric] (machine for harvesting) moissonneuse *f*
~AND BINDER - [agric] (reaping machine with an attachment designed to bind the grain as it cuts it) moissonneuse-lieuse *f*
~YARN - [text] fil *m* pour moissonneuse-lieuse
REAPING - [agric] moisson *f*
~MACHINE - s. reaper
~SCYTHE - [agric] faux *f* à céréales
REAR - [gen] arrière *m*, derrière *m*
[adj] arrière (situé à l')
~AND PLATE - [mech] flasque *f* arrière du corps de pompe
~-ARCH - [constr] arc *m* intérieur, arc *m* en retrait
~AXLE - [auto] pont *m* arrière, essieu *m* arrière
~-AXLE DRIVING SHAFT - [auto] arbre *m* de transmission
~-AXLE HOUSING - [auto] carter *m* de pont arrière
~-AXLE PLANETARY GEAR - [auto] engrenage *m* du pont arrière
~-AXLE RADIUS ROD - [auto] jambe *f* de force essieu arrière
~-AXLE TUBE - [auto] tube *m* de pont arrière
~-BELLOW FRAME - [photo] cadre *m* à soufflet arrière
~BEVEL GEAR - [auto] train *m* planétaire arrière
~BOX - [railw] (block post in rear) poste *m* de block amont
~BRAKE - [auto] frein *m* arrière
~BRAKE-VAN - [railw] (called caboose in the USA) fourgon *m* de queue
~COVER - [aero] (of the engine) couvercle *m* arrière
~DRIVE - [mech] traction *f* arrière
~-END SUSPENSION - [auto] suspension *f* arrière
~-ENGINE - [auto] moteur *m* arrière
~-ENGINE CAR - [auto] voiture *f* à moteur arrière
~FOCAL LENGHT - [photo] longueur *f* focale postérieure
~FRAME - [railw] (of a locomotive) longeron *m* d'arrière (d'une locomotive)
~GUARD - [gen & milit] arrière-garde *f*
~HORSE - [agric] cheval *m* de main
~LIGHT - [auto etc] lanterne *f* arrière, feu *m* arrière
[cin & telev] contre-jour *m*
~LIGHTING - [photo] éclairage *m* en contre-jour
~-MOUNTED - [gen mech] monté arrière
~PANEL - [telev] panneau *m* arrière
~PROJECTION - [telev] (in this type of projection the picture is projected on a transparent screen, so that it can be viewed by watchers in front of that screen)projection *f* par transparence
~PULLEY - [mech] (of a tractor) poulie *f* arrière
~QUARTER PANEL - [auto] (of a car body) panneau *m* latéral arrière
~RADIUS ROD - [auto] (rear-axle radius rod) jambe *f* de force d'essieu arrière
~REFLECTING MIRROR - s. rear-view mirror
~SCANNING - [cin] (a sound-reproduction system whereby the film is placed between the exciting lamp and the scanning slit) balayage *m* postérieur
~SEAT - [auto etc] siége *m* arrière
~SIDE MEMBER - [auto] longeron *m* arrière
~SPRING - [auto] ressort *m* arrière
REAR STAND - [mech] (of a motorbicycle) béquille *f* arrière
~-TO-FRONT RATIO - [radio] (the ratio of the effectiveness of a directional aerial towards the front and towards the rear) rapport *m* avant-arrière
~TOW-HOOK - [auto] crochet *m* de remorquage
~-VIEW MIRROR - [auto] miroir *m* rétroviseur
~-VISION MIRROR - s. rear-view mirror
~WAGON - [railw] (or tail wagon) wagon *m* de queue
~WHEEL - [gen and auto] roue *f* arrière
~-WHEEL BRAKE - [auto] frein *m* arrière
~-WHEEL DRIVE - [auto] propulsion *f* par les roues arrière
~WHEELHOUSE PANEL - [auto] passage *m* de roue arrière
~WINDOW - [auto] lunette *f* arrière
REARER - [zool] éleveur
[min] dressant *m*
REARING - [zool] élevage *m*
~CATTLE - [zool] bovins *mpl* d'élevage
~CENTRE - [zool] centre *m* d'élevage
~STATION - s. rearing centre
REARRANGE, to - [gen] rarranger
REARRANGEMENT - [gen] rarrangement *m*
REARWARD - [gen] (coming last; towards the rear) arrière de (à l')
REASCEND, to - [gen] remonter
[astronaut] re-décollage *m*
REASON METER - [instr] (a type of electrolytic meter) compteur *m* de Wright
REASSEMBLE, to [mech] remonter
REASSEMBLY - [mech] remontage *m*
REAUMUR SCALE - [meas] (a temperature scale in which the boiling point of water is taken as 80° and its freezing point as 0°) échelle *f* Réaumur
~TEMPERATURE SCALE - s. Reaumur scale
~THERMOMETER - [instr] thermomètre *m* à échelle Réaumur
REBALE, to [gen] remettre en balles
REBALE - [text] balle *f* réemballée
REBALLASTING - [railw] (the track) rechargement *m*
REBATCH, to - [text] réenrouler
REBATE, to - [gen & comm] bonifier, faire un escompte
[carp] assembler en feuillure, feuiller
REBATE - [gen] rabais *m*, remise *f*, escompte *m*
[carp] (groove cut into the edge of a piece of timber) feuillure *f*
[comm] (repayment following an excess remittance) ristourne *f*
[constr] rainure *f*, battant *m*
~PLANE - [tool] guillaume *m*
REBATED FRAME - [text] cadre *m* à feuillure
~JOINT - [carp] assemblage *m* à feuillure
~WEATHER BOARD - [constr] planche *f* à feuillure
REBATING - [carp] assemblage *m* à feuillure
REBEAMING - [text] ensouplage *m*
REBECCA - [radar] (the airborne element of the Eureka responder system (Second World War), which actuates the responder beacon and displays the reply) Rébecca *f*
~-EUREKA SYSTEM - [radar] (a homing system in the Second World War, comprising an airborne interrogating element (Rebecca) and a ground responder beacon) compas *m* magnétique, compas *m* à répétiteurs
REBENZOLIZATION - [ind chem] rebenzolage *m*

REBOIL, to-[agric] (of wine) rentrer en fermentation
REBOIL - [glass man] récoction *f*
REBOILER - [ind chem] (a type of heat exchanger designed to supply heat to the bottom of a fractionating column) élément *m* chauffant
REBORE, to [mech] réaléser
REBORING - [mech] réalésage *m*
REBOUND , to - [gen] rebondir
REBOUND - [gen] rebond *m*, rebondissement *m*
[rubber ind] rebondissement *m*
~CABLE - [auto] (rebound stop) butée *f* de rebond de suspension
~DAMPERS - [auto] pare-choc *m*
~DEGREE - [rubber ind] degré *m* de rebondissement
~ELASTICITY - [rubber ind] résilience *f* dynamique, élasticité *f* de rebondissement
~RUBBER - [auto] butoir *m* de pare-choc
~STOP - [auto] (elastic member increasing the wheel rate toward the end of the rebound travel) butée *f* de rebond de suspension
~STRAP - [auto] sangle *m* de limitation de débattement
REBOUNDING OF SHUTTLE - [text] rebondissement *m* de la navette
REBREATHER - [astronaut] régénérateur *m* d'atmosphère
REBROADCAST, to - [radio] répéter une radioémission
REBROADCAST - [radio] (of a programme) répétition *f* d'une radioémission
REBUILD, to - [gen] rebâtir
REBUILDING - [constr] reconstruction *f*, relèvement *m*
[mech] réfection *f*
RECALESCENCE - [metall] (the evolution of heat occurring when iron or steel cools through the critical range) récalescence *f*
RECALIBRATION - [instr] (the second calibration after the calibration made in the factory) calibrage *m* successif
RECALL, to - [gen & telecomm] rappeler
[comput] appeler, demander
RECALL SIGNAL - [telecomm] signal *m* de rappel
RECALLING KEY - [telecomm] clé *f* de rappel
RECAP, to - [gen] (colloq. for recapitulate) récapituler
[rubber ind] (USA only) rechaper
RECAPITULATION - [gen] récapitulation *f*
[mus] réexposition *f*
RECAPPED TYRE - [rubber ind] pneu *m* regommé, pneu *m* rechapé
RECAPPING - [rubber ind] (of tyres) rechapage *m*
~MACHINE - [rubber ind] (for tyres) machine *f* à rechapage
RECARBURIZATION - [metall] recarbonisation *f*
RECARBURIZER - [metall] agent *m* récarburateur
RECASE A WELL, to - [mining] recuveler un puits
RECASING - [mining] recuvelage *m*
RECAST, to - [metall] refondre, remouler
RECEDE, to - [gen] reculer, se retirer
RECEDING OF CARRIAGE - [text] recul *m* du chariot
RECEIPT, to - [gen & comm] acquitter, quittancer
RECEIPT - [gen] recette *f*, réception *f*
[comm] acquit *m*, quittance *f*
~BOOK - [comm] quittancier *m*
RECEIVABLE - [gen] recevable
RECEIVABLE - [fin] (matured for payment) à recevoir
RECEIVE, to - [gen] recevoir
RECEIVED - [gen] recu, admis
~TRAFFIC - [railw] arrivages *mpl*
RECEIVER - [gen] récepteur *m*
[telecomm] récepteur *m*
[leg] (the person appointed by the court to deal with the property and funds of another while a judicial action is pending) administrateur *m* judiciaire
[ind chem] (a vessel used as a receptacle for gases or fluids) récipient *m*, cloche *f*
[ind chem] (a removable receptacle used to collect e.g. a distillate) matras *m*, ballon *m* à long col
[mech] (of an oil compressor) réservoir *m* (d'un compresseur d'air)
[mech] (expansion engine) réservoir *m* intermédiaire (d'une machine à détente)
[railw] (a receiver magnet) aimant *m* récepteur
[radio] (short for radio receiver) poste *m* récepteur
[metall] (the forehearth of a cupola) avant-creuset *m*
~BANDPASS - [telev] filtre *m* passe-bande de récepteur
~CAP - [telecomm] pavillon *m* de récepteur
~CASE - [telecomm] boîtier *m* du récepteur
~GATING - [radar] sélection *f* périodique de signal
[comput] attaque *f* d'un récepteur
~HOOK - [telecomm] crochet *m* de suspension du récepteur
~RADIATION - [telecomm] (the electromagnetic radiation from a receiver borne of a source of electric oscillation in it) rayonnement *m* d'un récepteur
~RECOVERY TIME - [radio] (the time which elapses after a specific excitation) durée *f* de rétablissement d'un récepteur
~RESPONSE - [el acoust] (the response of a telephone receiver operating into a real or artificial ear) réponse *f* du récepteur
~RESPONSE TIME - [radio] inertie *f* d'un récepteur
~SCREENING - [radio] blindage *m* de récepteur
~SHEEL - [telecomm] boîtier *m* du récepteur
~SPACE - [mech] (expansion engine) espace *m* intermédiaire
RECEIVERSHIP - [leg] (the office of a receiver) syndicat *m* de faillite
RECEIVING - [gen] réception *f*, récepteur *m*
~AERIAL - [radio] antenne *f* de réception
~ANTENNA - s. receiving aerial
~BAND PASS FILTER - [telecomm] filtre *m* passe-bande de réception
~CURRENT SENSITIVITY - [el acoust] (free-field current response) intensité *f* de courant d'émission
~HOPPER - [min] (unloading pit) fosse *f* de déchargement
~LOOP LOSS - [radio] (the part of the repetition equivalent which can be assigned to the station set etc. on the receivinq end) pertes *fpl* à la réception
~MECHANISM - [text] mécanisme *m* de réception
~PERFORATOR - [telecomm] (a telegraph instrument) perforateur *m* receveur
~ROLLERS - [text] cylindres *mpl* alimentaires
~SCREEN - [opt] écran *m* récepteur
~SET - [radio] poste *m* récepteur
~TERMINAL STATION - [telev] terminal *m* de réception
~TUBE - [text] tube *m* récepteur

RECEIVING VOLTAGE SENSITIVITY - [el acoust] (free field voltage response) intensité *f* de tension d'émission en champ nul
RECEPTACLE - [gen] réceptacle *m*
[gen] (a container, a vessel) récipient *m*
[el] (a socket for a plug) prise *f* de courant
[el] (board for terminals) tablette *f* à bornes
[bot] réceptacle *m*
~OUTLET - [el] prise *f* de courant multiple
RECEPTION - [gen] réception *f*
[radio] réception *f*
[telecomm] réception *f*
~BY BUZZER - [telecomm] lecture *f* au son
~BY SOUNDER - s. reception by buzzer
~BY TAPE - [telecomm] réception *f* sur bande
~COVERAGE - [comput] plage *f* de réception
~DIAGRAM - [radio] (of an aerial) diagramme *m* de réception
~LEVEL - [el acoust] niveau *m* de réception
RECEPTIVE - [gen] réceptif
~FIELD - [el biol] (the region the activity of which is observed by a pickup electrode) champ *m* réceptif
RECEPTIVENESS - [bot] (the condition of the stigma when effective pollination is possible) réceptivité *f*
RECEPTOR - [zool] (free nerve ending) nerf *m* récepteur
RECESS, to - [gen] évider, pratiquer un enfoncement, enfoncer
RECESS - [constr] (a depression, a niche) rentrant *m*, embrasure *f*, niche *f*
[gen] (a depression in a continuous line) évidement
[anat] recessus *m*, dilatation *f* ampullaire
[metall] (an undercut) gorge *f*
~IN THE CAM - [text] rainure *f* de la came
~IN THE FINGER - [text] encoche *f* dans la platine
~IN THE PUSHER - [text] encoche *f* dans le pousseur
~OF DRIVING SHAFT - [mech] creux *m* de l'arbre moteur
~OF PELVIC - [anat] recessus *m* intersigmoïde
RECESSED - [gen] enfoncé, renfoncé, en retrait
~ARCH - [arch] arc *m* renfoncé
~BALCONY - [constr] loggia *f*
~FLANGE-COUPLING - [mech] manchon *m* à plateaux à boulons noyés
~HEAD LAMP - [auto] phare *m* encastré
~OPENING - [photo] évidement *m* en retrait
~PINION GUIDE - [mech] rainure *f* pour guider le pignon
~SLEEVE - [mech] bague *f* d'arrêt
~SWITCH - [el] interrupteur *m* encastré
~STRIPE - [text] sillon *m*
~WALL HEATER - [th eng] (a panel fire) radiateur *m* mural
RECESSING MACHINE - [constr] machine *f* à defoncer, défonceuse *f*
~TOOL-[tool] outil *m* à chambrer
RECESSION - [gen] recul *m*, retraite *f*
[fin] (economic depression following a recovery from a depression) récession *f*
RECESSIVE CHARACTER - [biochem] (in genetics the one characteristic of a pair of contrasted characteristics, which does not appear in the hybrid if both are present) récessivité *f*
RECHAIR THE SLEEPERS , to [railw] resaboter les traverses
RECHARGE, to - [el] (to restore a secondary battery to the charged condition by passing a current through it) recharger
RECHARGE - [soil] apport *m* sur la teneur en eau
RECHUCK, to - [mach tool] remandriner
RECHUCKING - [mach tool] remandrinage *m*
RECIPROCAL - [gen] réciproque, inverse
[math] réciproque, inverse (fonction, raison)
~ACTION - [chem] action *f* réciproque
~COURSE - [aero & naut] (a course exactly opposite to that in question, i.s.differing from it by 180 degrees) route *f* réciproque
~DIAGRAM - [phys] diagramme *m* réciproque
~EQUATIONS - [math] équations *f*pl réciproques
~GEAR - [mech] engrenage *m* réciproque, engrenage *m* à retour
~INTERLOCKING - [el] enclenchement *m* réciproque
~LATTICE - [phys] (auxiliary construction for investigations in the field of space lattice theory and definition) réseau *m* réciproque
~LEG - [aero] (the phase of landing procedure during which the aircraft is flying on a course reciprocal to that of approach) parcours *m* d'éloignement
~LINEAR DISPERSION - [phys] dispersion *f* linéaire réciproque
~NETWORK - [el] (dual network) réseau *m* dual, réseau *m* réciproque
~OF AMPLIFICATION FACTOR - [electron] coefficient *m* de transparence de grille
~TRANSDUCER - [radio] (transducer satisfying the principle of reciprocity) transducteur *m* réciproque
~VELOCITY REGION - [nucl] zone *f* de vitesse réciproque
RECIPROCATE, to - [gen] échanger
[mech] (to cause to move to and fro) avoir un mouvement de va-et-vient
RECIPROCATING - [mech] (said of any engine employing a piston working in a cylinder, the piston is made to oscillate by the periodic pressure of the working fluid) alternatif, va-et-vient
~BALANCE - [phys] (obtained by opposing the shaking forces of a reciprocating mass by equal opposite forces obtained from another reciprocating mass) équilibre *m* de masses alternantes
~COMPRESSOR - [mech] compresseur *m* à piston
~CROSS-CUT SAW - [impl] scie *f* alternative à tronçonner
~DEVICE - [mech] va-et-vient *m*
~ENGINE - [mech mach] moteur *m* à mouvement alternatif
~GRID - [radiat] (moving grid) grille *f* mobile, grille *f* oscillante
~MASS BALANCE - s. reciprocating balance
~MILL - [metall] laminoir *m* à mouvement alternatif
~MOTION - [mech] mouvement *m* alternatif
~PUMP - [mech] pompe *f* à mouvement alternatif
~RIDER - [print] chargeur-balladeur *m*
~SAWING MACHINE - [mech mach] scie *f* alternative
~SCREEN - [min] (an oscillating screen, used in coke screening) crible *m* à secousses
~WHEEL - [mech] (of a reciprocating e.g. a sawing machine) volant *m* de commande de mouvement alternatif
~WHEEL SPINDLE MECHANISM - [mach tool] mécanisme *m* d'oscillation axiale du mandrin
RECIPROCATION - [mech] mouvement *m* alternatif
RECIPROCITY - [gen] réciprocité *f*
[math] réciprocité *f*

RECIPROCITY COEFFICIENT - [el] coefficient *m* de réciprocité
~PRINCIPLE - [el] (for an electracoustical transducer) principe *m* de réciprocité
~THEOREM - [phys] (the interchange of electromotive force at one point in a network and the current produced at any other point produces the same current for the same electromotive force) théorème *m* de réciprocité
RECIRCULATE, to [gen] recirculer
[hydr] (by pumping) pomper en retour
RECIRCULATED WATER - [hydr] eaux *f*pl de retour
RECIRCULATING STORE - [comput] mémoire *f* circulante
~TRACK - [comput] piste *f* d'enregistrement
RECIRCULATION - [ind chem hydr] (circulation of a fluid through a system more than once) recirculation *f*, recyclage *m*
~OF SEWAGE - [hydr] recirculation *f* des eaux d'égout
~RATIO - [hydr] taux *m* de recyclage
~SYSTEM - [hydr] recirculation *f* continue
RECITATIVE - [mus] récitatif *m*
RECKON, to - [gen] compter, calculer
RECKONER - [gen] calculateur *m*, compteur *m*
RECKONING - [gen] comptage *m*
[naut] (the determination of a ship's or aircraft's position) estime *f*
[naut] (the actual position after reckoning) point *m* estimé
RECLAIM, to - [gen] (to recover for subsequent reuse) gagner, rendre cultivable
[rubber ind] (the operation of recover rubber from waste) régénérer
[agric] (to restore to cultivation) défricher du terrain, mettre en valeur
RECLAIM COMPOUND - [rubber ind] mélange *m* à base de régénéré
~DISPERSION - [rubber ind] dispersion *f* de régénéré
~MACHINERY - [rubber ind] (machinery used for recovering rubber from scrap material) machines *f*pl pour régénérés
~MILL - [rubber ind] concasseur *m*
[rubber ind] (the factory) usine *f* de régénéré
RECLAIMABLE - [gen & ind] récupérable
RECLAIMED - [gen] récuperé
[rubber ind] régénéré
[agric] (of land) amendé (terrain)
~CELLULOSE - [ind chem] cellulose *f* régénéré
~GROUND - [agric] terrain *m* amendé
~OIL - [oil ind] huile *f* régénérée
~RUBBER - [rubber ind] (rubber obtained by the treatment of scrap material) régénéré *m*, caoutchouc *m* régénéré
~WOOL - [text] laine *f* régénérée
RECLAIMER - [paper man] ramasse-pâte *m*
RECLAIMING - [gen] récupération *f*
[rubber ind] régénération *f*
[agric] (of land) défrichement *m*, assèchement *m*
~AGENT - [rubber ind] agent *m* de régénération
~DIGESTER - [rubber ind] (a vessel used for digestion, in reclaiming scrap rubber) digéreur *m* à régénération
~OILS - [rubber ind] (oils used in the processing of reclaimed rubber) huiles *f*pl pour régénérer le caoutchouc
RECLAIMING OF LAND - [agric] mise *f* en valeur des terres incultes
~OF MARSHLAND - [agric] assèchement *m* de marécages
~OF WASTE LAND - s. reclaiming of land
~PLANT - [text] installation *f* de récupération
RECLINATION - [gen and med] réclinaison *f*
RECLINING BERTH - [railw] couchette *f*
~TWILL - [text] croisé *m* à angle obtus
RECLOSER - [el] interrupteur *m* à réenclenchement
RECLOSING RELAY - [el] relais *m* à réenclenchement
~TIME - [el] durée *f* de réenclenchement
RECOIL, to - [gen] reculer
[firearms] reculer, repousser
[mech] se détendre, repousser
RECOIL - [gen] rebondissement *m*
[firearms] recul *m*
[nucl] (the motion of an atom caused by the emission of an alpha particle) recul *m*
[mech] (of spring) détente *f*
~ABSORBER - [mech] amortisseur *m* de recul
~ATOM - [nucl] (atom undergoing a sudden change or reversal in its direction of motion as the result of its emission of a particle) atome *m* de recul
~CYLINDER - [firearms] frein *m* de tir
~ELECTRON - [nucl] (electron which is set in motion through interaction with photon in the Compton effect) électron *m* de recul
~ESCAPEMENT - [horol] échappement *m* à recul
~GEAR - [firearms] amortisseur *m* de recul
~NUCLEUS - [nucl] (nucleus recoiling following a collision with a nuclear particle) noyau *m* de recul
~PARTICLE - [nucl] (particle set into motion by a process which involves the ejection of another particle) particule *f* de recul
~PROTON - [nucl] proton *m* percuté
~PROTON COUNTER TUBE - [nucl] tube *m* compteur à protons de recul
~PULL - [mech & firearms] contre-coup *m*
~RADIATION - [nucl] rayonnement *m* de recul
~SLIDE - [mech] glissière *f*
~SPINDLE - [mech] tige *f* de butée
~SPRING - [mech] ressort *m* de recul
RECOILLES - [firearms] sans recul
RECOMB, to - [text] (the operation of combing worsted tops for a second time, so as to remove further poil) repeigner
RECOMBINATION - [chem] (the reentering into combination) recombinaison *f*
[electron] (the capture of an electron or of a negative ion by a positive ion with resulting neutralization of the charges) recombinaison *f*
~COEFFICIENT - [electron] (the quotient of the deionization rate by the square of the ion density of the recombining ions) coefficient *m* de recombinaison
~VELOCITY - [electron] (the quotient of the normal component of the electron current density at the surface by the excess electron charge density at the surface) vitesse *f* de recombinaison
RECOMBINE, to - [chem] (to enter into combination again) recombiner
RECOMBINER - [chem] catalysateur *m* recombinateur
RECOMMISSION, to - [naut] réarmer
[mech] réparer, réfectionner
RECOMPACTING - [metall] recompression *f*
RECOMPLEMENT, to - [math] recomplémenter

RECOMPRESSION STATION - [gas ind] poste *m* de surpression
RECONDITION, to - [gen & mech] rénover, remettre en état
[auto] reviser
[mech] (of valves) roder
RECONDITIONED CARRIER RECEPTION - [radio] (the method of reception in which the carrier is separated from the sidebands to eliminate amplitude variations and noise, and then added at increased level to the sideband to obtain a comparatively undistorted output) système *m* antifading à amplification constante de porteuse
RECONDITIONING - [gen] rénovation *f*, remise *f* à neuf, revision *f*
[oil ind] régénération *f*
RECONNAISSANCE - [gen] reconnaissance *f*
~AIRCRAFT - [aero] avion *m* de reconnaissance
~MAP - [surv] carte *f* de reconnaissance
~SATELLITE - [astronaut] satellite *m* d'observation
RECONSTITUENT - [med] reconstituant *m*
RECONSTRUCT, to - [gen] reconstruire, rebâtir
RECONSTRUCTION - [gen etc] reconstruction *f*
RECOOLER - [mech] (a secondary cooling device) machine *f* frigorifique à convection
RECOOLING - [th eng] réfrigération *f*
RECORD, to - [gen] enregistrer
[el acoust] (the operation of registering waveforms arising from sound sources) enregistrer
RECORD - [gen] enregistrement *m*
[el acoust] disque *m*
~CARD - [comput] carte *f* de fichier
~CHANGER - [el acoust] (a device which makes it possible to ply a number of records in succession) changeur *m* de disques
~CHART - [instr] (any recording medium, generally of paper, for use in a recorder) feuille *f* d'enregistreur
~CIRCUIT - [telecomm] ligne *f* d'enregistrement
~COUNT - [comput] nombre *m* d'enregistrements
~CURRENT - [telev] courant *m* d'enregistrement
~FORMAT - [comput] format *m* d'enregistrement
~GAP - [comput] intervalle *m*
~HEAD - [el acoust] tête *f* d'enregistrement
[comput] (magnetic head use to write) tête *f* d'écriture
~KEY - [el acoust] touche *f* enregistrement
~LAYOUT - [comput] structure *f* d'enregistrement
~LIBRARY - [gen] discothèque *f*
~MARK - [comput] marque *f* d'enregistrement
~PHOTOGRAPHY - [photo] photographie *f* documentaire
~PLAYER - [el acoust] tourne-disques *m*
~SEPARATOR - [comput] caractère *m* de séparation d'enregistrements
~STORAGE - [comput] (the placing of records in a storage cabinet or a storage room) emmagasinage *m* des enregistrements
RECORDED PROGRAM - [telev] programme *m* enregistré
RECORDER - [instr] (apparatus for providing a permanent record of e.g. a process) appareil *m* de mesure enregistreur
[telev] enregistreur *m*
RECORDERS - [mus] flûtes *f*pl à bec
RECORDING - [acoust] (the art and practice of registering wave forms arising from sources) enregistrement *m*
RECORDING - [cin] (camera shooting) prise *f* de vue
[telecomm] (of sounds) enregistrement *m*
~ACCELEROMETER - [instr] (type of accelerometer which gives a continuous record of readings) accéléromètre *m* enregistreur
~ALTIMETER - [instr] (an instrument which record altitude in respect to time) altimètre *m* enregistreur
~AMMETER - [instr] ampèremètre *m* enregistreur
~AMPLIFIER - [el acoust] amplificateur *m* d'enregistrement
~ANEMOMETER - [instr] (an instrument which makes a continuous record of wind velocity) anémographe *m*
~BAROMETER - [instr] barométrographe *m*
~BOARD - [telecomm] service *m* des annotatrices
~BRIDGE - [telecomm] (piece of equipment utilizing a bridge circuit of operating telegraph apparatus from a telegraph signal) pont *m* enregistreur
~CHAIN - [telev] chaîne *f* de lecture
~CHANNEL - [el acoust] (one of a number of independent recorders in a recording system) canal *m* d'enregistrement
~CHARACTERISTIC - [el acoust] caractéristique *f* d'enregistrement
~CHART - s. record chart'
~CIRCUIT - [el] (circuit producing an impulse capable of operating a mechanical recording device) ligne *f* d'annotatrice
~CUTTER - [el acoust] graveur *m*
~DENSITY - [comput] densité *f* d'enregistrement
~DEVICE - [el acoust] appareil *m* enregistreur
~ELEMENT - [instr] élément *m* enregistreur
~EQUALIZER - [telev] correcteur *m* de distorsion de l'enregistrement sonore
~FLOW-METER - [instr] débitmètre *m* enregistreur
~FREQUENCY METER - [instr] fréquencemètre *m* enregistreur
~GAP - [comput] entrefer *m*
~GAUGE - [instr] manomètre *m* enregistreur
~HAND - [instr] aiguille *f* du compteur
~HEAD - [el acoust] (the electro-mechanical device to which modulation currents are applied to operate the stylus in a wax record) tête *f* d'enregistrement
[el acoust] (the registering device containing magnetizing coils and pole-pieces, through which magnetic tape is drawn in magnetic recording) tête *f* d'enregistrement
~INSTRUMENT - [instr] (a device designed to make a record, e.g. on paper tape, of the quantity measured) appareil *m* enregistreur
~LAMP - [el acoust] lampe *f* d'enregistrement
~LEVEL - [telev] niveau *m* de modulation de la sortie de l'amplificateur
~LOSS - [el acoust] (in a recording system, loss of recorded signal as a function of frequency due to different causes) perte *f* à l'enregistrement
~MANOMETER - [instr] (instrument designed to make a continuous record of pressure) manomètre *m* enregistreur
~METER - [instr] compteur *m* enregistreur
~OPERATOR - [telecomm] (for telephone) annotatrice *f*
~OSCILLOGRAPH - [instr] oscillographe *m* enregistreur
~PEN - [impl] plume *f* à marquer

RECORDING PLAYBACK HEAD - [el acoust] (a magnetic head which is used for registration and reproduction) tête *f* d'enregistrement/lecture
~POSITION - [telecomm] position *f* d'annotatrice
~POTENTIOMETER - [instr] potentiomètre *m* enregistreur
~RAIN GAUGE - [met] pluviographe *m*
~ROOM - [cin] studio *m* d'enregistrement
[telecomm] local *m* de captation
~SLIT - [el acoust] fente *f* d'enregistrement
~SPEED - [el acoust] vitesse *m* de gravure
~SPEED INDICATOR - [instr] compte-tours *m* enregistreur
~SPOT - [el acoust] spot *m* d'enregistrement
~STYLUS - [el acoust] burin *m*, style *m* graveur
~TAPE - [telecomm] (in telegraphy) bande *f* d'enregistrement
~THERMOMETER - [instr] thermomètre *m* enregistreur
~TRACK - [comput] piste *f* d'enregistrement
~TRUNK - [telecomm] ligne *f* d'enregistrement
~UNIT - [telecomm] (for telegraph signals, which are converted into pulses capable of operating a teleprinter) dispositif *m* d'enregistrement
~VOLTMETER - [instr] voltmètre *m* enregistreur
~WATT AND VARMETER - [instr] (recording instrument registering on the same diagram active and reactive powers in a circuit) watt-varmètre *m* enregistreur
~WATTMETER - [instr] wattmètre *m* enregistreur
~WHEEL - [telecomm] (in telegraphy) roue *f* d'enregistrement
RECOVER, to - [gen] recouvrer
[med] guérir
[bookbind] renouveler la reliure
[ind chem] (to separate or otherwise collect a process medium after use) récupérer
RECOVERED ACID - [oil ind] acide *m* de récupération
RECOVERING PLANT - s. recovery plant
~THE WOOL FROM RAGS - [text] récupération *f* de la laine des chiffons
~UNIT - s. recovery plant
RECOVERY - [gen] recouvrement *m*, récupération *f*
[med] guérison *f*
[aero] (the return to normal flight after a manoeuvre) redressement *m*
[mech] (of a spring etc] course *f* de retour
[phys] (of physical properties) récupération *f*
[chem] (the operation in which a liquid or solid mixture is brought into contact with an immiscible or partially miscible liquid to achieve a redistribution of solute between the phases) extraction *f*
[metall] détente *f*
[min] captage *m*
~CAPSULE - [astronaut] capsule *f* récupérable
~CYCLE - [el biol] (the sequence of states of varying excitability following a conditioning stimulus) cycle *m* de rétablissement
~FLAP - [aero] (a flap designed to change the pitching characteristics of an aircraft to make recovery from a dive easier or even automatic) volet *m* de ressource, volet *m* antipiqué
~OF A LINE - [telecomm] démolition *f* d'une ligne
~OF SOLVENTS BY ABSORPTION - [chem] récupération *f* des solvants par absorption
~OF SOLVENTS BY ACTIVATED CARBON - [chem] récupération *f* des solvants par absorption par charbon actif
RECOVERY OF SOLVENTS BY CONDENSATION - [chem] récupération *f* des solvants par condensation
~PLANT - [ind chem] installation *f* de récupération
~PLASTOMETER - [rubber ind] plastomètre *m* de reprise élastique
~RATE - [med] (the rate at which recovery takes place after a radiation injury) vitesse *f* de restauration
~TIME - [electron] (the time required after interruption of anode current for the grid to regain control under given conditions) durée *f* de rétablissement, temps *m* de recouvrement
[nucl] (the time after the initiation of a count which must elapse before a counter can deliver a pulse of substantially full size at the next ionizing event) durée *f* de rétablissement
~VOLTAGE - [el] (the normal frequency appearing across the contacts of a circuit-breaker after it has interrupted the circuit) tension *f* de rétablissement
RECRUDESCENCE - [gen and med] recrudescence *f*
RECRUIT, to - [gen] recruter
RECRUITMENT - [gen] recrutement *m*
RECRUSHER - [min] broyeur *m* secondaire
RECRYSTALLIZATION - [chem] (the reforming of crystals by dissolution followed by concentration of the solution with subsequent formation of the crystals) récristallisation *f*
~ANNEALING - [metall] recuit *m* à récristallisation
~TEMPERATURE - [metall] (the temperature at which recrystallization takes place) température *f* de récristallisation
RECRYSTALLIZE, to - [phys chem] (to form crystals by recrystallization) recristalliser
RECTANGLE - [geom] rectangle *m*
~TRIANGLE - [geom] triangle *m* rectangle
RECTANGULAR - [geom] rectangulaire
~ABUTMENT - [constr] culée *f* rectangulaire
~AXES - [math] axes *mpl* rectangulaires
~BILLET - [metall] billette *f* méplate ou rectangulaire
~BROACH - [tool] alésoir *m* pour trous rectangulaires
~CATHODE - [electron] (cathode shaped like a rectangular bar) cathode *f* rectangulaire
~CO-ORDINATES - [surv] (cartesian co-ordinate system) coordonnées *fpl* cartésiennes
~INGOT - [metall] lingot *m* rectangulaire
~INTEGRATION - [math] intégration *f* rectangulaire
~MEASURING WEIR - [hydr] déversoir *m* rectangulaire de jaugeage
~NOTCH - [hydr] (a notch plate with a rectangular notch cut in it, used for the measurement of large discharges) plaque *f* à rainure rectangulaire de mesure
~PICTURE TUBE - [telev] tube *m* image à écran rectangulaire
~PIPE - [plumb] tube *m* à section rectangulaire
~PLATE SPRING - [mech] ressort *m* à lame rectangulaire
~PROTRACTOR - [impl] rapporteur *m* rectangulaire
~PULSE - [electron] impulsion *f* rectangulaire
~ROPE - [text] corde *f* de section rectangulaire
~SURFACE CONDENSER - [ind chem] condenseur *m* rectangulaire par surface
~WAVE - [radio] (a periodic electro-magnetic) onde *f* rectangulaire

RECTANGULAR WELL - [phys] puits *m* rectangulaire
~WIRE - [text] fil *m* plat
~WIRING - [telecomm] armement *m* en rectangle
RECTIFICATION - [gen] rectification *f*
[chem] (purification of a spirit by redistillation) rectification *f*
[el] (the conversion of an alternating current into a direct current) redressement *m*
[mech] redressement *m*
[radio] (the conversion of radio frequency into audio frequency) détection *f*, redressement *m*
[gen] correction *f*
[surv] (in aerial photography) redressement *m*
~EFFICIENCY - [electron] efficacité *f* de redressement
~FACTOR - [electron] rapport *m* de redressement
~OF A CURVE - [constr] rectification *f* d'une courbe
~OF AN ALTERNATING CURRENT - [el] redressement *m* d'un courant alternatif
RECTIFIED - [gen] rectifié, redressé
[el] redressé
[radio] démodulé
[mech] rectifié
[instr] corrigé
[chem] rectifié
~AIRSPEED (R.A.S.) - [aero] (indicated airspeed after correction for position and instrumental errors) vitesse *f* corrigée
~ALCOHOL - [chem] alcool *m* rectifié
~BENZOLE - [ind chem] benzol *m* rectifié et lavé
~CIRCUIT - [electron] montage *m* de redresseurs
~CURRENT - [el] (the low-frequency current output from a rectifier) courant *m* redressé
~VOLTAGE - [el] tension *f* redressée
RECTIFIER - [el] (device for changing alternating to unidirectional current) redresseur *m*
[ind chem] (apparatus in which redistillation is carried) separateur *m* d'eau
[surv] redresseur *m* aérophotographique
[electron] (a group of devices each of which has the property of conducting current in one direction only) redresseur *m*
[radio] (a device used for detection) détecteur *m*
~ANODE - [el] anode *f* d'une soupape
~BULB - [el] ampoule *f* redresseuse
~CATHODE - [el] cathode *f* d'une soupape
~CELL - [el] (in photo-electric cells) élément *m* redresseur
~DIODE - [radio] diode *f* à semiconducteur
~FILTER - [electron] (used to smooth out the voltage fluctuation on a thermionic rectifier) filtre *m* de redresseur
~FORM FACTOR - [electron] facteur *m* de forme de redresseur
~INSTRUMENT - [instr] (alternating-current instrument in which the current to be measured is rectified and thus measured on a direct-current instrument) appareil *m* à redresseur
~LEAKAGE CURRENT - [electron] (alternating current passing through a rectifier without being rectified) courant *m* de fuite de redresseur
~LOCOMOTIVE - [railw] locomotrice *f* à redresseurs
~MODULATOR - [radio] (modulator using a diode or diodes) diode *f* modulatrice
~RIPPLE FACTOR - [electron] taux *m* d'ondulation de redresseur
~STACK - [electron] (assembly of semiconductor rectifier disks or wafers) empilage *m* de redresseurs
RECTIFIER STACK ARM - [electron] section *f* d'empilage de redresseurs
~TANK - [el] cuve *f* de redresseur
~TRANSFORMER - [el] transformateur *m* de redresseur
~-TYPE ECHO SUPPRESSOR - [telecomm] suppresseur *m* d'écho à action continue
~UNIT - [el] appareil *m* redresseur
~VALVE - [el] soupape *f* électrique
~VOLTMETER - [instr] voltmètre *m* à redresseur
RECTIFY, to - [gen] rectifier, corriger
[chem] (to purify by redistillation) rectifier
[mech] rectifier, dresser
[el] redresser
[radio] détecter
RECTIFYING - [oil ind] (the separation of fractions of different boiling range from a vapour mixture by fractional condensation) rectification *f*
~COLUMN - [ind chem] (column in which rectification is carried out) colonne *f* à rectifier
~DETECTOR - s. rectifier
~ELEMENT - [el] (circuit element having the property of conducting current in one direction only) élément *m* redresseur
~SECTION - [oil ind] (the portion of a petroleum distillation column which is above the feed inlet) section *f* de rectification
~TUBE VOLTAGE DOUBLER - s. rectifying valve and voltage doubler
~VALVE - [radio] (thermionic valve in which direct use is made of the unilateral conductivity effect) tube *m* redresseur
RECTILINEAL - [gen] rectiligne
RECTILINEAR - [gen] rectiligne
~COMBER - [text] peigneuse *f* rectiligne
~DRESSING MACHINE - [text] peigneuse *f* mécanique plane
~FLOW ELECTRON GUN - [electron] canon *m* électronique à commande rectiligne du faisceau
~LENS - [photo] (lens giving images without distortion as far as parallel lines are concerned) objectif *m* rectiligne
~MOTION - [phys] (the motion of a particle in a straight line) mouvement *m* rectiligne
~PROPAGATION - [light] propagation *f* rectiligne (de la lumière)
~SCANNING - [telev] (the process of scanning an area in a predetermined sequence of straight parallel strips) analyse *f* par lignes
~SYSTEM - [opt] (orthoscopic system) ortoscope *m*, système *m* rectilinéaire
RECTITIS - [med] rectite *f*, proctite *f*
RECTO - [print] (the right-hand page of a book) recto *m*
RECTOCELE - [med] (a protrusion of the lower section of the posterior vaginal wall) rectocèle *f*
RECTOCOLITIS - [med] rectocolite *f*
RECTOSCOPY - [med] rectoscopie *f*
RECTOSTENOSIS - [med] sténose *f* rectale
RECTRICES - [zool] (in birds, the tail feathers used for steering) pennes *fpl* rectrices
RECTRIX - [zool] (singular of rectrices) rectrice *f*
RECTUM - [anat] (the portion of the alimentary canal leading to the anus) rectum *m*
~CATHETER - [med] sonde *f* rectale
~PROLAPS - [med] prolapsus *m* du rectum

RECULTURE, to - [chem] (to add further inoculum to a fermenting mix) inoculer
RECUMBENT ANTICLINE - [geol] anticlinal *m* incliné
~FOLD - [geol] pli *m* couché
RECUPERATED RUBBER - [rubber ind] (reclaimed rubber) régénéré *m*
RECUPERATING DEVICE - s. recuperator
RECUPERATION - [med] guérison *f*
[phys etc] régénération *f*, récupération *f*
RECUPERATIVE - [mech] de rétablissement
~AIR-HEATER - [th eng] (air-heater in which heat is transmitted from hot gases to the air through metallic walls; the flow of gas and the flow of air are continuous and unidirectional) réchauffeur *m* d'air à récupération
~FURNACE - [metall] four *m* à récupération de chaleur
RECUPERATOR - [firearms] (system of springs whereby the gun returns to its firing position) récupérateur *m*
[metall] (a system of flues enabling the hot gases leaving a furnace to be utilized in heating the incoming air) récupérateur, préchauffeur *m* à récupération
RECURRENCE - [med] récurrence *f*
~RATE - [comput] (of pulses) fréquence *f* de récurrence
RECURRENT - [gen & med] récurrent, périodique
~FEVER - [med] fièvre *f* périodique
~NETWORK - [telecomm] réseau *m* récurrent
~PULSES - [electron] impulsions *fpl* récurrentes
~RECIPROCAL SELECTION - [genet] sélection *f* récurrente réciproque
~SENSIBILITY - [biol] sensibilité *f* récurrente
~-SURGE OSCILLOGRAPH - [instr] (instrument used for research in connection with electrical surges) oscillographe *m* d'impulsions périodiques
RECUT, to - [gen] recouper
[mech] retailler
RECUTXING - [mech] retaillage *m*
RECYCLE, to - [ind chem] recirculer, recycler
RECYCLED FUEL - [nucl] (reprocessed fuel for a nuclear reactor) combustible *m* recyclé
RECYCLING - [chem] recyclage *m*
~DETECTOR - [radio] détecteur *m* commutateur piloté par la porteuse
~DEVICE - [comput] dispositif *m* régénérateur
~TIME - [photo] durée *f* de recharge
RED - [gen] rouge
~ADDER - [telev] circuit *m* mélangeur pour le rouge
~ALGAE - [bot] algues *fpl* rouges
~ANTIMONY - [min] (natural oxysulphide of antimony) antimoine *m* rouge, kermésite *f*
~ARSENIC - [min] (realgar) arsenic *m* sulfuré rouge, réalgar *m*
~BODY - s. red gland
~BRASS - [metall] laiton *m* rouge, tombac *m*
~-BRITTLE IRON - [metall] fragilité *f* du fer au rouge
~BRONZE - [metall] bronze *m* rouge
~CHALK TONE - [photo] ton *m* rouge sanguine
~CLAY - [geol] (a deep-sea deposit) argile *f* rouge
~CLOVER - [bot] trèfle *m* commun
~COBALT - [min] cobalt *m* arséniaté, érythrine *f*
~CONSCIOUS - [telev] hypersensible pour le rouge
~COPPER OXIDE - [min] (a pigment for glass, and ceramics and an insecticide) cuprite *f*
~CORPUSCLE - [biol] (erythrocyte) érythrocyte *m*
RED COUNT - [biol] (the number of red corpuscles per cubic millimetre of blood) numération *f* érythrocytaire
~-CROSS COACH - [railw] voiture *f* sanitaire
~-CROSS TRAIN - [railw] (ambulance train) train *m* sanitaire
~DEAL - [bot] pin *m* sylvestre
~EARTH - [geol] terre *f* rouge
~EBONY - [timber] bois *m* d'ébène rouge
~ELECTRON GUN - [telev] canon *m* du rouge
~FEVER - [vet] (swine erysipelas) rouget *m* du porc
~FIR - [bot] sapin *m* noble d'Amerique
~FOG - [photo] voile *m* dichroïte
~GLAND - [zool] (in fish) glande *f* rouge
~GRANITE - [geol] granit *m* rose
~-HARD STEEL - [metall] acier *m* rapide
~HARDNESS - [metall] dureté *f* au rouge
~HEAT - [metall] chaleur *f* rouge, chaude *f* rouge
~HIGHS - [telev] hautes fréquences *fpl* pour le rouge
~-HOT - [gen] chauffé au rouge
[metall] au rouge, porté au rouge
[metall] (noun) métal *m* chauffé au rouge
~IRON - [min] hématite *f* rouge, ferret *m*
~IRON OXIDE - [chem] (ferric oxide. Used as a pigment for paints and rubber) oxyde *m* rouge de fer
~LEAD - [paint] (bright red pigment, much used in priming paints for wood and steel) minium *m* de plomb
~LEAD ORE - [min] plomb *m* rouge, plomb *m* chromaté
~LITMUS PAPER - [chem] papier *m* de tournesol rouge
~LOWS - [telev] basses fréquences *fpl* pour le rouge
~MARBLE - [geol] marbre *m* rose
~MARLS - [geol] (red silts and calcareous clays) marnes *fpl* rouges
~MULBERRY TREE - [bot] mûrier *m* rouge
~MUSCLES - [zool] (muscles which are rich in sarcoplasm and haemoglobin) muscles *mpl* rouges
~NUCLEUS - [zool] (aggregation of red cells in the tegmentum) noyau *m* rouge
~OCHRE - [dyes] ocre *f* rouge
~OXIDE OF ZINC - [min] (zincite) zincite *f*
~PHOSPHORUS - [min] phosphore *m* amorphe
~POLE - [phys] pôle *m* nord
~PRIMARY SIGNAL - [telev] signal *m* du rouge primaire
~ROOT DISEASE - [bot] (wetrot) maladie *f* rouge des racines
~SANDSTONE - [min] grès *m* rouge
~SCREEN-GRID - [telev] grille-écran *m* rouge
~SENSITIZATION - [photo] sensibilisation *f* au rouge
~SHIFT - [phys] (the displacement towards the red end of the spectrum of absorption lines in spectra of light of stars etc) déplacement *m* vers le rouge
~-SHORT - [metall] cassant à chaud
~-SHORTNESS - [metall] fragilité *f* à chaud
~SILVER ORE - [min] (the dark ore) argent *m* rouge antimonial
[min] (the light ore) pyrargyrite *f*
~SPOT - [astr] (marking on the surface of the planet Jupiter) tache *f* rouge
~STAIN - [metall] tache *f* rougeâtre
~TAPE - [gen] paperasserie *f*
~-TAPE OPERATIONS - [comput] (non-productive operations) aménagement *m*
~VIDEO VOLTAGE - [telev] tension *f* vidéo pour le rouge
~-WATER - [vet] hématurie *f*

RED WATER TROUBLE - [hydr] eau *f* rouge
~WOOD - [timber] (the red deal) pin *m* sylvestre
~ZINC ORE - [min] zincite *f*
REDDISH - [gen] rougeâtre
REDDLE - [min] (read earthy variety of haematite with some admixture of clay) arcanne *f*, ocre *f* rouge
REDEVELOP, to - [photo] (to put through a second developing process) développer de nouveau
REDEVELOPMENT - [photo] redéveloppement *m*
REDISSOLVE, to - [chem] redissoudre
REDISTIL, to - [ind chem] redistiller
REDISTILLATION - [chem] redistillation *f*
REDISTILLED ZINC - [metall] (zinc from which impurities have been removed by selective distillation) zinc *m* redistillé
REDONDA PHOSPHATE - [miner] (a natural source of phosphorus, chiefly aluminium phosphate, occurring in the West Indies) phosphate *m* de Redonda
REDOUT - [astronaut] voile *m* rouge
REDOX - [chem] (abbrev for oxidation-reduction) oxydoréduction *f*
~ENZYME - [chem] enzyme *f* d'oxydoréduction
~POTENTIAL - [chem] potentiel *m* d'oxydoréduction
REDRAWING PRESS - [metall] presse *f* à réemboutissage
~TOOL - [tool] outil *m* à étirage
REDRESS, to - [gen] rétablir, corriger
[aero] redresser
REDRESS - [gen] redressement *m*, réparation *f*
REDRESSEMENT - [med] redressement *m*
REDRESSING DEVICE - [photo] dispositif *m* de redressement
~MIRROR - [photo] miroir *m* de redressement
REDRILL, to - [mach tool] reforer
REDRUTHITE - [min] (natural copper sulphide, crystallizing in the ortho-rhombic system and commonly associated with the more abundant ores of copper) chalcosite *f*
REDUCE, to - [gen] réduire
[chem] (to cause the reaction termed reduction) réduire, désoxyder
[phys] (to cause a change resulting in the addition of electron to an atom, e.g. the removal of oxygen from a molecule) réduire
[photo] affaiblir, atténuer
[metall] (the operation of extracting metals from ores) réduire l'épaisseur
[med] résoudre (une tumeur)
[el] abaisser (la tension)
REDUCED - [gen etc] reduit
~ADMITTANCE - [el] (normalized admittance) admittance *f* normalisée
~CRUDE - [oil ind] huile *f* réduite
~EQUATION OF STATE - [phys] (generalized equation of state containing as variables the reduced pressure, the reduced volume and reduced temperature) équation *f* d'état réduite
~FOCAL LENGTH - [opt] distance *f* focale réduite
~FREQUENCY - [aero] (frequency parameter) fréquence *f* réduite, paramètre *m* de fréquence
~IMPEDANCE - [el] impédance *f* normalisée
~LEVEL - [surv] niveau *m* réduit
~MASS - [phys] masse *f* réduite
~PASS - [text] remettage *m* réduit
~PRESSURE - [mech] dépression *f*
~RESISTANCE TO TAKING UP - [text] résistance *f* à l'enroulement diminuée (par la pression du peigne)
REDUCED RESIN - [plast ind] (a natural or synthetic resin which has been modified in a specific manner) résine *f* modifiée
~SCALE - [draw] échelle *f* réduite
~SHOT NOISE - [electron] bruit *m* de grenaille réduit
~SPEED - [mech auto etc] vitesse *f* réduite
~-VOLTAGE STARTER - [el] démarreur *m* statorique
REDUCER - [mech] réducteur *m*
[plumb] (pipe fitting of smaller diameter at the outlet that an the inlet) réducteur *m*
[chem] s. reducing agent
[photo] (solution acting on the silver image and dissolving it away, thus reducing the contrasts) affaiblisseur *m*
[paint] (a thinner) diluant *m*
REDUCIBLE EQUATION - [math] équation *f* réductible
REDUCING - [gen & mech] réduction *f*, diminution *f*
[chem] (causing reduction) réducteur
[photo] (of a bath) faiblisseur
[metall] réduction *f* d'épaisseur
~AGENT - [chem] (a substance which causes the reaction termed reduction) agent *m* réducteur
[photo] agent *m* d'affaiblissement
~AND SIZING MILL - [metall] réducteur *m* et calibreur
~ATMOSPHERE - [metall] atmosphère *f* réductrice
~BEND - [plumb] (a bend to connect to a different sized pipe at each end) coude *m* de réduction
~BLEACH - [text] blanchiment *m* par réduction
~BUSHING - [photo] écrou *m* intermédiaire
~CHAIN - [text] chaîne *f* de réduction
~COUPLING - [mech] manchonnage *m* réducteur
~CROSS - [plumb] (a pipe-fitting designed to connect four pipes at right angles, one or more of which is smaller in bore than the others) raccord *m* à croix
~DIE - [metall] matrice *f* à réductrice
~FITTING - [plumb] (a fitting to connect two pipes of different sizes) manchon *m* de réduction
~FLAME - [chem] (the inner part of a Bunsen flame, in which reducing action takes place) flamme *f* réductrice
~FLANGE - bride *f* de réduction
~FRAME - [photo] intermédiaire *m*
~FURNACE - [metall] four *m* de réduction
~GEAR - [mech] engrenage *m* démultiplicateur
~GLASS - [opt] loupe *f* de réduction
~JOINT - [mech] épissure *f*
~PIECE - [plumb] cône-réduction *m*, cône *m*
~PIPE - [plumb] (a special pipe to connect others of different diameters) tuyau *m* de réduction
~POWER - [paint] pouvoir *m* d'affaiblissement
~RETORT - [chem] cornue *f* de réduction
~SCREEN - [opt] (transparent screen used in photometry to absorb a given fraction of the luminous flux falling on it) écran *m* réducteur
~SOCKET - [plumb] (a pipe fitting consisting of a sleeve with internal threads at both ends, to provide connexion between two pipes of different diameters arranged in the same line) manchon *m* de réduction
~SUGARS - [chem] sucres *mpl* réducteurs
~SURFACE - [light] (a prepared surface, used in photometry, reflecting only a given fraction of the luminous flux falling on it) surface *f* à reflexion affaiblie
~TEE - [plumb] (fitting to connect pipes of different diameters, two of which run in the same direction

ar right angles to the third) joint *m* à té
REDUCING VALVE - [mech] (a valve which automatically reduces the pressure of the fluid passing through it) soupape *f* de réduction, détendeur *m*
REDUCTASES - [biochem] (enzymes which bring about the reduction of organic compounds) réductases *fpl*
REDUCTION - [phys chem] (any change which results in the addition of an electron to an atom or an iron, e.g. the removal of oxygen from a molecule) réduction *f*
[min] (the extraction of gold from ores) réduction *f*
[comm & gen] rabais *m*
[photo] affaiblissement *m*
[el chem] (of an accumulator plate) réduction *f*
[met] (to standard conditions) correction *f*
[gen] rapetissement *m* réduction *f*, diminuition *f*
~AREA - [metall] (the diminuition in the cross-sectional area of a test piece due to cold flow under tensile load) striction *f*, diminuition *f* de la section transversale
~CATALYST - [chem] catalyseur *m* réducteur
~COEFFICIENT - [radiat] (effective absorption coefficient) coefficient *m* d'absorption efficace
~DISCHARGE - [chem] rongeant *m* réducteur
~DIVISION - [genet] (meiosis) division *f* réductionnelle, division *f* hétérotypique
~FACTOR - [opt] coefficient *m* de réduction
~GEAR - [mech] (a train of one or more pairs of gears, arranged to reduce speed) engrenages *mpl* réducteurs
[mech] (the reduction unit) réducteur *m*
~GEAR BEVEL PINION - [mech] pignon *m* cônique du réducteur
~GEAR CASING - [mech] boîtier *m* du réducteur
~GEARS - s. reducing gear
~IN AREA - [metall] coefficient *m* de striction
~IN BULK - [nucl] (chemical process for reducing the volume of radioactive waste) réduction *f* massique
~OF CROSS SECTION - [metall] réduction *f* de section
~OF DETECTION EFFICIENCY - [telev] affaiblissement *m* du rendement détecteur
~OF GRADIENT - [surv] réduction *f* de pente
~OF LIFT - [text] diminuition *f* de la course de montée
~OF SECTION - [metall] striction *f* à la rupture
~OF STROKE - [text] diminuition *f* de la course
~OF THE CELLULOSE - [text] réduction *f* de la cellulose
~OF THE FLOOD - [hydr] atténuation *f* des crues
~OF THROW - s. reduction of lift
~POTENTIAL - [el] (the potential drop involved in the reduction of a cation to a neutral form) potentiel *m* de réduction
~PRINTING PROCESS - [cin] (a printing process for the conversion of standard film pictures to narrow-gauge film pictures) tirage *m* par réduction
~RATIO - [mech] (the relation between input and output speeds of a reducing gear train) degré *m* de réduction
~SPEED - [mech] taux *m* d'accélération négative
~TIME - [comput] temps *m* de réduction
~UNIT - [mech] réducteur *m*
~VALVE - [mech] (for pressure) valve *f* de réduction
~WHEEL - [mech] roue *f* réductrice
REDUCTION ZONE - [metall] zone *f* de réduction
REDUNDANCY - [gen] surabondance *f*, redondance *f*
~CHECK - [comput] (a check using extra digits in machine words for the detection of mistakes) contrôle *m* de redondance
~RATIO - [comput] taux *m* de redondance
REDUNDANT - [gen] redondant
~DIGIT - [comput] (a check digit) chiffre *m* de contrôle
REDUPLICATION - [genet] réduplication *f*
REDWOOD - [bot] sequioia *m* toujours vert
~VISCOSIMETER - [instr] (an instrument for determining viscosity) viscosimètre *m* de Redwood
REECHO, to - [acoust] renvoyer un son, résonner
REED - [bot] (slender stem of tall grass) roseau *m*
[mus] (a vibrating tongue of wood or metal) anche *f*
[mus] (part of reed pipe) rigole *f*
[text] (the component part of a loom which drives the filling against the woven fabric) peigne *m*
[arch] (semicylindrical ornamental moulding) rudentures *fpl*
[metall] (a defect due to an inclusion) cavité *f*, retassement *m*
[bot] couche *f* annuelle
[zool] caillette *f*
[mining] méche *f*
[electron] (of a dry-reed relay) languette *f*
~BINDING - [text] liage *m* des peignes
~BLADE - [text] lame *f* du peigne, passette *f* pour peigne
~CARRIER - [text] clinquette *f*, cadre *m* de peigne
~DENT - [text] dent *f* du peigne
~DEPARTMENT - [mus] (of an organ) tuyaux *mpl* à anche
~FOR BEAD WEAVING - [text] peigne *m* pour tissage des perles
~FOR TRIMMINGS - [text] peigne *m* pour passementerie
~FRAME - [text] cadre *m* du peigne
~-GRASS - [agric] laîche *f*
~GUIDING - [text] guidage *m* du peigne
~HOLDER - [text] support *m* du peigne
~HOCK - [text] passette *f* pour peigne
~HOOK LEVER - [text] levier *m* de la passette
~LOCKING DEVICE - [text] garde *f* du peigne
~LOUDSPEAKER - [el acoust] haut-parleur *m* à anche
~MAKING MACHINE - [text] machine *f* à fabriquer les peignes
~MATTING - [text] natte *f* de jonc
~MEADOW GRASS - [agric] glycérie *f* aquatique
~MOVEMENT - [text] mouvement *m* du peigne
~OF CANE - [text] peigne *m* en roseau
~OPENERS - [text] ouvreurs *mpl* de dents
~PIN - [text] boulon *m* de peigne
~PIPE - [mus] (of the organ) tuyau *m* à anche
~RELAY - [el] relais *m* à lame vibrante
~RELIEF MOTION - [text] mécanisme *m* de peigne mobile
~STAY - [text] jumelle *f*
~STOP - [mus] (of organ) jeux *mpl* d'anche
~SWEET GRASS - s. reed meadow grass
~WITH STEEL DENTS - [text] peigne *m* à dents d'acier
REEDINESS - [text] striage *m* en chaîne
REEDING - [text] empeignage *m*, piquage *m* en peigne
~MACHINE - [text] piqueuse *f*
~ORDER - [text] ordre *m* du piquage au peigne
REEDUCATION - [gen & med] rééducation *f*

REEDY FABRIC - [text] tissu *m* avec entrebats
REEF, to - [naut] (to reduce the extent of a sail by folding part of it and tying it round a yard or boom) prendre des ris
REEF - [geogr] (a ridge of sand or rock) récif *m*
[min] (a lode, or vein) reef *m*
[naut] (the part of sail which is folded to reduce its size) ris *m*
[geogr] (a shoal) banc *m*
~-BAND - [naut] bande *f* de ris
[mining] filon *m* tabulaire
~DRIVE - [mining] gallerie *f* de chassage
~KNOT - [naut] noeud *m* plat
~POINTS - [naut] (the short ropes fixed in a line along the reef-band) garcettes *f*pl de ris
~TACKLE - [naut] palanquin *m*
REEFER - [railw] (only USA) wagon *m* réfrigérant
[naut] (only USA) bateau *m* frigorifique
REEK, to - [gen] exhaler des vapeurs
REEK - [gen] exhalaison *f*, buée *f*
REEKING - [metall] noircissement *m* à suie
REEL, to - [text] (to wind on a reel or bobbin) dévider, bobiner
[paper man] bobiner
[gen] (of a building) s'ébranler
[zool] (of grasshoppers) grésiller
[med] (to have a sensation of giddiness) vaciller
REEL - [gen] (rotatory device for winding cord, rope etc) bobine *f*, tournette *f*
[impl] (wooden spool for wire etc) touret *m*
[text] dévidoir *m*, aspe *m*
[text] (sewing thread wound on a bobbin) bobine *f*
[photo] (spool on which the film is wound) bobine *f*
[cin] (standard length of a film supplied for exposure) rouleau *m* de film
[text] (ropemaking) caret *m* de corderie
[naut] (for lines) touret *m*
[el acoust] (of a magnetic tape) bobine *f*
[print] bobine *f* de papier
[comput] (of magnetic tape) bobine *f*
~AERIAL - [radio] (an aircraft aerial designed to be wound up on a drum when landing and run out when in flight) antenne *f* à rouet
~AND TRAY PROCESSING MACHINE - [cin] (used for films up to 200 ft long) tireuse *f* à cuves et rouleaux
~BRAKING - [text] freinage *m* de l'aspe
~CARRIER - [text] support *m* de dévidoir
~END WITH BRIDGE - [text] côté *m* du métier avec pont
~FOR COARSE YARNS - [text] dévidoir *m* pour fils gros
~HOLDDOWN - [comput] fixation *f* de bobine
~-LIKE BOBBIN - [text] bobine *f* à forme d'aspe
~OF YARN - [text] dévidoir *m* de fils
~OFF, to - [text] (the operation of winding yarn on to a reel) dévider, dérouler
[text] (the operation of unwinding the silk filament from cocoons and combing them to form a silk thread) tirer la soie du cocon
~-OFF BOBBIN - [text] bobine *f* à dérouler
~SPINDLE - [text] arbre *m* de l'aspe
~STAND - [text] support *m* d'aspe
~STICK - [text] (of the loom) porte-bobines *m*
~-TO-REEL RECORDER - [telev] (in video recording) magnétophone *m* à deux bobines
REELABLE COCOON - [text] cocon *m* dévidable
REELED SILK - [text] soie *f* en écheveaux
REELED YARN - [text] fil *m* dévidé
REELER - [mech] bobineuse *f*, dévideuse *f*
[metall] laminoir *m* lisseur pour tubes
REELER'S TRAY - [text] bassine *f*
~TROUGH - [text] bassine *f*
REELING - [text] dévidage *m*
~APPARATUS - [text] bassine *f*
~BASIN - [text] bassine *f*
~DEVICE - [text] dispositif *m* d'enroulement
~FRAME - [text] dévidoir *m*
~MACHINE - [text] (machine for winding sheet, strip, cord, thread, wire or the like on bobbins) tour *m*, machine *f* à tirer la soie
~METHOD - [text] méthode *f* de dévidage
~OF YARNS - [text] dévidage *m* des fils
~PROCESS - [text] procédé *m* de dévidage
~ROOM - [text] atelier *m* de dévidage
~SPEED - [text] vitesse *f* de renvidage
~THE COCOON BY WET PROCESS - [text] tirage *m* du cocon à l'eau
~THE COCOONS IN THE DRY WAY - [text] tirage *m* à sec des cocons
~WASTE - [text] déchets *m*pl de dévidage
~WATER - [text] eau *f* de dévidage
REENGAGE, to - [gen] rengager
[mech] rengrener
REENRICHMENT - [nucl] (the uraniumhexafluoride which is reclaimed from spent reactors fuel is converted into fuel elements and reenriched in U 235 by gaseous diffusion plant) réenrichissement *m*
REENTER, to - [gen] rentrer
[astronaut] (the reentry of a capsule into the atmosphere) rentrer
REENTER - [gen] rentrée *f*
[astronaut] (of a spaceship) rentrée *f*
REENTERABLE LOAD MODULE - [comput] module *m* de chargement toujours utilisable
~PROGRAMME - [comput] programme *m* invariant
REENTRANT - [gen] (extending inward) rentrant
~ANGLE - [geom] angle *m* rentrant
~GAS COOLING - [nucl] refroidissement *m* par gaz réentrant
~HORN - [acoust] (a horn for coupling the sound-reproducing diaphragm with the outer air) corne *f* rentrante
~OSCILLATOR - [radio] oscillateur *m* à cavités
~WINDING - [el] enroulement *m* à rentrée
REENTRY - [astronaut] s. reenter
~BODY - [astronaut] partie *f* rentrant dans l'atmosphère
~CORRIDOR - [astronaut] corridor *m* de rentrée
~NOSE CONE - [astronaut] pointe *f* de rentrée
~POINT - [comput] point *m* de retour
REEVE, to - [gen] (to pass a rope, or a rod through a hole) passer (un cordage)
REEXPORT - [comm] réexportation *f*
REFACE, to - [mech] réparer
[mech] (in internal combustion engines) rectifier (le siège des soupapes)
REFACE A BEARING, to - [mech] rectifier un coussinet
~THE CLUTCH, to - s. reface the clutch disk, to
~THE CLUTCH DISK - [mech] regarnir l'embrayage
REFERENCE ADDRESS - [comput] adresse *f* de référence
~AUDIO LEVEL - [telev] niveau *m* de référence audio
~AXIS - [electron] (in semiconductors) axe *m* prin-

cipal
REFERENCE BEAM - [telev] faisceau *m* de référence
~BLACK LEVEL - [telev] signal *m* du noir de référence
~BLOCK - [mech] (for control of equipment) bloc *m* de référence
~EDGE OF TAPE - [telev] (in video recording) bord *m* de référence de la bande
~ELECTRODE - [el] (a constant-potential half-cell used to measure other electrode potentials) électrode *f* de référence
~EQUIVALENT - [telecomm] (as indicated by the Master Telephone Transmission Reference System) équivalent *m* de transmission effective
~FILTER - [electron] filtre *m* de référence
~FREQUENCY - [el] fréquence *f* étalon
~GAGE - s. reference gauge
~GAUGE - [mech] (used for testing) calibre-étalon *m*
[meas] rapporteur *m*
~-GENERATOR PERFORMANCE- [telev] conduite *f* du générateur de la porteuse de chrominance
~INPUT ELEMENTS - [contr] organes *mpl* comparateurs
~INPUT SIGNAL - [contr] signal *m* d'entrée de référence
~INPUT VARIABLE - [nucl] (desired value, or set point; the independently set reference in a control system) grandeur *f* de référence
[telev] signal *m* d'entrée de référence
~JUNCTION - [meas] point *m* de référence
~LEVEL - [gen] niveau *m* de référence
~LINE - [radar] (base line) ligne *f* de base
~MARK - [surv] (a mark used to measure angular distances) repère *m*
[print] (sign directing a reader to a footnote) signe *m* de renvoi
~NOISE - [telecomm] bruit *m* de référence
~NOTE - [acoust] ton *m* de référence
~PERFORMANCE - [meas] rendement *m* de référence
~PILOT - [radio] (wave which is different from those transmitting telecommunication signals) onde *f* pilote
~PLANE - [surv] plan *m* de repère
~POINT - [gen] point *m* de référence
~POWER SUPPLY - [contr] alimentation *f* étalon
~PRESSURE - [mech] (value obtained by multiplying the square of the velocity of a fluid by its density and dividing the result by two) pression *f* de référence
~QUANTITY - [contr] (the value of the input quantity preceding the controlled condition) grandeur *f* de référence
~RANGE - [contr] domaine *m* de référence
~RECORD - [comput] registre *m* de référence
~RECORDING - [telev] enregistrement *m* pour consultation ultérieure
~ROLL - [mech] rouleau *m* d'étalonnage
~SIGNAL - [radio] signal *m* de référence
~SOURCE - [nucl] (a radioactive standard) source *f* de référence, étalon *m* radioactif
~SUPPLY - [el] tension *f* de référence
~TELEPHONIC POWER - [telecomm] niveau *m* téléphonique de référence
~TEMPERATURE - [gen] température *f* de référence
~TEST OF A METER - [instr] essai *m* de référence d'un compteur
~TIME - [comput] (in static magnetic storage; the moment near the beginning of switching used as an origin for time measurements) instant *m* de référence
REFERENCE TONE - s. reference note
~VALUE - [contr] grandeur *f* de référence
~VARIABLE - s. reference value
~VOLUME - [acoust] niveau *m* de référence
~WHITE LEVEL - [telev] niveau *m* du blanc de référence
~WINDING - [el] enroulement *m* de référence
REFERRIZATION - [hydr] réenrichissement *m* en fer
REFILE, to - [comput] retransmettre
[gen] (to fill again) remplir
REFILLING UNIT - [mech] (for liquid air or nitrogen) chargeur *m* automatique d'air liquide ou d'azote liquide
REFINE, to - [gen] raffiner
[metall] (to obtain a higher purity) affiner
[paper man] (the operation of reducing the knots and large particles in the pulp) raffiner
[metall] (e.g. cast iron) mazer
[oil ind] épurer
REFINED ASPHALT - [constr] asphalte *m* pur
~BAR IRON - [metall] fer *m* en barres affiné
~GREASE - [ind chem] graisse *f* raffinée
~IRON - [metall] fer *m* au bois, fer *m* affiné
~PETROLEUM - [oil ind] pétrole *m* raffiné
~PIG IRON - [metall] fonte *f* brute affinée
~STEEL - [metall] acier *m* affiné
~TAR - [ind chem] (dehydrated tar) goudron *m* préparé
REFINEMENT - [gen] raffinement *m*
[metall] affinage *m*
REFINER - [paper man] (perfecting machine; a machine designed to reduce knots and large particles in the pulp) pile *f*, raffineur *m* cônique
[glass man] (of a tank furnace) zone *f* d'affinage
[rubber ind] (a refining mill) malaxeur *m* de raffinage
~WASTE - [rubber ind] déchets *mpl* de raffinage
REFINERY - [oil ind] (plant designed to produce purified products from crude petroleum) raffinerie *f*
[sugar ind] (a plant for the purification of raw sugar) raffinerie *f*, sucrerie *f*
~GAS - [chem] (mixture of hydrocarbon gases produced as a by-product of petroleum refining. It is a raw material for a number of organic syntheses) gaz *m* de raffinerie
REFINING - [gen] raffinage *m*, affinage *m*
[metall] affinage *m*, épuration, procédé *m* d'affinage
[metall] (of cast iron) mazéage *m*, blanchiment *m*
~ANNEALING - [metall] recuit *m* d'affinage
~BOILER - [metall] chaudière *f* de raffinage
~FLUX - [metall] épurant *m*
~FOAM - [metall] épurant *m*
~HEAT - [metall] température *f* d'affinage
~MILL - [rubber ind] laminoir-raffineur *m*
~PUDDLING - [metall] puddlage *m*, affinage *m*
~SLAG - [metall] scories *fpl* de raffinerie
~STEP - [ind chem] étage *m* de raffinage
REFIT, to - [gen] remettre en état de service
[naut] radouber
[mech] rajuster (une machine), regarnir
REFIT - [gen] réparation *f*
[naut] radoub *m*
[naut] (of a ship) réarmement *m*
[mech] rajustement *m*, regarnissement *m*

REFIT - [firearms] remontage *m*
REFITTING WORKS - [gen] usine *f* de réparation
REFLECT, to - [phys] (to throw back, as rays of light etc) réflechir, refléter, renvoyer
REFLECTANCE - [opt] (the ratio between the reflected radiant flux and the incident radiant flux) facteur *m* de réflexion
~-BEAM KINESCOPE - [telev] tube *m* image à faisceau réfléchi
REFLECTED BINARY CODE - [comput] code *m* binaire réfléchi
(cyclical binary code) code *m* binaire-cyclique
~CODE - [comput] code *m* réfléchi
~CONTROL CIRCUIT RESISTANCE - [el] (the resistance of a control circuit referred to an arbitrary turns basis) résistance *f* transformée d'un circuit pilote
~FIELD - [el] (the difference between the impressed field and the actual field) champ *m* réfléchi
~LIGHT - [opt] lumière *f* réfléchie
~PRESSURE - [phys] (the totale pressure resulting instantaneously at the surface when a shock wave travelling in one medium strikes another medium) pression *f* réfléchie
~RAY - [light] rayon *m* réfléchi
~REACTOR - [nucl] réacteur *m* à réflecteur
~SHOCK WAVE - [phys] onde *f* de choc réfléchie
~WAVE - [phys] onde *f* réfléchie
REFLECTING - [phys] réflecteur, réfléchissant
~CIRCLE - [astr] cercle *m* à réflexion
~GALVANOMETER - [instr] (galvanometer in which the deflection is observed by the reflection of a beam of light projected on to a mirror mounted on the moving element) galvanomètre *m* à miroir
~GONIOMETER - [instr] goniomètre *m* à réflexion
~GRATING - [electron] (waveguides) filtre *m* radial des ondes parasites
~LAYER - [chem] (of a mirror) couche *f* réfléchissante
~LEVEL - [surv] niveau *m* à réflexion
~LUSTRE - [text] lustre *m* miroitant
~MAGNETOMETER - [instr] magnétomètre *m* à miroir
~MIRROR - [opt] glace *f* de réflexion
~POWER - [phys & opt] pouvoir *m* réflecteur
~POWER OF SILK - [text] pouvoir *m* réflecteur de la soie
~PRISM - [opt] (of a compass) prisme *m* de réflexion
~SCREEN - [cin] (a screen which is used to balance shadow parts with diffused light) écran *m* réflecteur
~TELESCOPE - [instr] (form of telescope invented by Newton and designed to overcome the difficulties of chromatic aberration) télescope *m* à miroir
~VIEW-FINDER - [photo] viseur *m* à chambre noir
~WATTMETER - [instr] (a dynamometer wattmeter for laboratory use) wattmètre *m* réflecteur
REFLECTION - [phys opt] (the return of light waves striking a surface) réflexion *f*
~ATTENTUATION - [telecomm] affaiblissement *m* d'adaptation
~COEFFICIENT - [electron] coefficient *m* de courants réfléchis
~COLOUR TUBE - [telev] tube *m* image en couleurs à réflexion
~EFFECT - [electron] (possible cause of deficiency of low-energy electrons) effet *m* de réflexion
~FACTOR - [opt] (for plane waves or transmission lines) coefficient *m* de réflexion
REFLECTION GAIN - [radio] (of aerial) gain *m* de réflexion
~GRATING - [electron] filtre *m* à sélection sélectif
[opt] réseau *m* à réflexion
~LOSS - [opt] (the transition loss which is due to the reflection of power at a discontinuity) perte *f* par réflexion
~MEASURING SET - s. reflectometer
~METER - s. reflectometer
~MODE FILTER - [electron] (waveguides) filtre *m* de mode à réflexion
~OPTICS - [opt] (system of mirrors and lenses used in projection television) optique *f* de projection
~PLANE - [crystall] (symmetry element possessed by certain crystals, whereby one half of the crystal is the reflection of the other half in a plane drawn through the centre of the crystal) plan *m* de réflexion
~PLOTTER - [radar] (reflective optics are used to eliminate parallax when plotting) réflectoscope *m*
~PREVENTING - [opt] (of lenses) antiréflechissant
~SEISMOGRAPH METHOD - [oil ind] (oil underground prospectingby means of reflected sound waves) méthode *f* sismographique à réflexion
~SHOOTING - [oil ind] (a seismic method of exploration based on the principle that vibrations caused by an explosion near the earth's surface are partly reflected at the boundary between two strata of different density) méthode *f* sismique
~TARGET - [electron](target so set that the useful X-ray beam emerges from the surface on which the electron beam is incident) cible *f* réfléchissante
REFLECTIVE SPOT - [comput] (to indicate end-of-tape) spot *m* réfléchissant
REFLECTIVITY - [opt] (the fraction of the incident radiant energy reflected by a surface exposed to uniform radiation from a source which fills its field of view) degré *m* de réflexion
REFLECTOGAGE - [instr] (ultrasonic flaw detector) détecteur *m* ultrasonore de fissures internes
REFLECTOMETER - [instr] (instrument designed to measure reflection) réflectomètre *m*
REFLECTOR - [light etc] (any device to intercept and return radiation) réflecteur *m*
[auto] (headlight) phare *m*
[telev] (a parasistic element applied to television aerials) réflecteur *m*
[opt] (reflecting surface covered with a thick layer of coloured glass) catadioptre *m*
[nucl] (part of the reactor, between the shield and the core, introduced to return neutrons to the core) réflecteur *m*
[electron] (reflector electrode) électrode *f* de réflexion
~ELEMENT - [radio & telev] (of an aerial) élément *m* réflecteur
~LAMP - [telev] (in a studio) lampe *f* à réflecteur
~OF AN AERIAL - [radio] réflecteur *m* d'antenne
~SAVINGS - [nucl] (decrease in critical core size obtained through the introduction of the reflector, or tamper) économie *f* due au réflecteur
~SPACE - [electron] (in a reflex klystron, the part of the tube following the buncher space, and terminated by the reflector) espace *m* de réflexion
~TANK - [nucl] récipient *m* de réflecteur liquide
~TRACKER - [radar] (plotting plate, or position traker) plan *m* de traçage
REFLECTOSCOPE - [metall] (used to test the sound-

ness of metal lingots etc) réflectoscope
REFLEX - [physiol] (not under the control of the will) réflexe *m*
~ACTION - [med] effet *m* réflexe
~AMPLIFICATION - [radio] amplification *f* réflex
~ARC - [med] arc *m* réflexe
~BAFFLE - [acoust] (a loudspeaker baffle) écran *m* réflecteur
~BUNCHING - [electron] (bunching occurring in an electron stream which has not been made to reverse its direction in the drift space) groupement *m* réfléchi
~CAMERA - [photo] chambre *f* reflex
~CIRCUIT - [telecomm] (circuit in which one tube simultaneously amplifies signals in two widely separate frequency bands) circuit *m* reflex
~KLYSTRON - [electron] (klistron incorporating a single resonator which, by the use of a reflector, is caused to act both as an input resonator and an output resonator) klystron *m* reflex
REFLEXOTHERAPY - [med] réflexothérapie *f*
REFLOAT, to - [gen & naut] renflouer, remettre à flot
[naut] (of ships) relever
REFLUX - [chem] (in distillation, condensed liquid allowed to flow down a fractionating column against the ascending vapour) reflux *m*
[hydr] reflux *m*, refluement *m*
[nucl] (the countercurrent recycle of a portion of an affluent) reflux *m*
[naut] (of the tide) jusant *m* de la marée
~BOILING - [ind chem] (a method of boiling a liquid in a vessel connected to a condenser so that the condensate flows back and the liquid is maintained at boiling point without loss by evaporation) ébullition *f* au reflux
~COOLER - [chem] réfrigérant *m* à reflux
~RATIO - [chem] (in a counter-current system) taux *m* de reflux
~TOWER - [oil ind] colonne *f* de distillation à reflux
~VALVE - [mech] (non-return type of valve used in pipe-lines) soupape *f* de reflux
REFORMATE - [chem] produit *m* de reforming
REFORMING - [oil ind] (pyrolysis of the lighter petroleum fractions with conversion of paraffins to olefins) reforming *m*, réformation *f*
REFRACT, to - [phys & opt] (to bend a ray of light or other radiation when it passes from one translucent substance to another of different density) réfracter
[acoust] (to vary the direction of sound transmission due to spatial variation of the wave velocity in the medium) réfracter
REFRACTED - [opt & acoust] réfracté, dévié
~LIGHT - [opt] lumière *f* réfractée, lumière *f* transmise
~RAYS - [phys] rayons *mpl* réfractés
~WAVE - [phys] (the part of an electromagnetic wave which travels from one medium into a second medium) onde *f* réfractée
REFRACTING ANGLE - [opt] (of a prism) angle *m* réfringent (d'un prisme)
~EDGE OF A PRISM - [opt] arête *f* réfringente d'un prism
~TELESCOPE - [opt] lunette *f*
REFRACTION - [opt & phys] (the bending of a light ray when it passes from one translucent substance to another of different density) réfraction
REFRACTION - [acoust & radio] réfraction
~ANGLE - [opt] angle *m* de réfraction
~CORRECTION - [astr] (the small amount which must be subtracted from the observed altitude of a heavenly body to allow for the refraction of light by the earth's atmosphere) correction *f* de réfraction
~LOSS - [opt] (the transmission loss which is due to refraction resulting from non-uniformity of the medium) perte *f* de réfraction
~OF LIGHT - [opt] réfraction *f* de la lumière
~SHOOTING - [geol] sismique *f* à réfraction
REFRACTIONATE, to - [chem] refractionner
REFRACTIONOMETER - [instr] (instrument designed to measure refractive defects of the eye) réfractionemètre *m*
REFRACTIVE - [opt etc] réfractif, réfringent
~CUTICLE - [text] cuticule *f* réfringente
~DISPERSITY - [opt] dispersivité *f* réfractive
~INDEX - [opt] (the ratio of the sine of the angle of incidence to the sine of the angle of refraction) indice *m* de réfraction
~MEDIUM - [opt] milieu *m* réfractif
~MODULUS - [opt] (in the troposphere) module *m* de réfraction
~POWER - [opt] pouvoir *m* réfringent
REFRACTIVITY - [phys chem opt] (the property of refraction) réfringence *f*
REFRACTOMETER - [instr] (an instrument for measuring refraction) réfractomètre *m*
~DISK - [opt] disque *m* réfractométrique
REFRACTOMETRY - [opt etc] (the art of measuring indices of refraction) réfractométrie *f*
REFRACTOR - [opt] (a device by which the direction of a beam of light is changed) milieu *m* réfringent, loupe *f*
REFRACTORIES - [metall] (bricks and other ceramic products made from fire-clay and designed to resist high temperatures without fusion, in some cases up to 1700° C) réfractaires *mpl*
REFRACTORINESS - [min] (the property of resisting high temperature without fusing) nature *f* réfractaire
REFRACTORY - [phys] (high temperatures without change of properties capable of resisting) réfractaire *m*
~ASBESTOS - [min] amiante *m* réfractaire
~BRICK - [th eng] brique *f* réfractaire
~CEMENT - [th eng] ciment *m* réfractaire
~CLAY - [min] argile *f* réfractaire
~CONCRETE - [constr] (concrete capable of withstanding very high temperatures) béton *m* réfractaire
~DRESSING - [metall] poteyage *m*
~LINING - [th eng] garniture *f* réfractaire
~MATERIALS - [gen] (non-metallic substances capable of with standing high temperatures) matériaux *mpl* réfractaires
~ORE - [min] minerai *m* réfractaire
~PERIOD - [biol] (the time interval in which an excitable tissue is incapable of response to a second stimulus applied after a previous one) période *f* réfractaire
~SLAG - [metall] laitier *m* réfractaire
~SPECIAL - [th eng] brique *f* de forme, pièce *f* spéciale
~WASH - [metall] poteyage *m*
REFRANGIBLE - [opt] (capable of being refracted) ré-

frangible
REFRIGERANT - [th eng] (substance used as a heat transfer vehicle in freezing processes) réfrigérant *m*
~INJECTION - [mech] (the injection of a cooling medium into the airstream to improve performance) injection *f* de liquide réfrigérant
REFRIGERATE, to - [gen] (the operation of producing cold artificially) réfrigérer, frigorifier
REFRIGERATED TRAILER - [transp] remorque *f* frigorifique
~TRAP - [chem] (trap consisting of a section of the vacuum line refrigerated to very low temperature, so that condensation of mercury takes place) piège *m* refroidi
REFRIGERATING - [gen] réfrigérant, frigorifique
~CAPACITY - [phys] pouvoir *m* frigorifique
~LIQUID - [chem] liquide *m* réfrigérant
~MACHINE - [th eng] machine *f* frigorifique
~PLANT - [th eng] appareils *mpl* frigorifiques
~UNIT - [meas] frigorie *f*
~VEHICLE - s. refrigerator vehicle
REFRIGERATION - [gen] (the artificial production of cold) réfrigération *f*
[food] (of meat) frigorification *f*
~COMPRESSOR - [mech] compresseur *m* pour une machine frigorifique
~CYCLE - [phys] (cycle taking heat at lower temperature and rejecting it at a higher one) cycle *m* de réfrigération
~INDUSTRY - [gen] industrie *f* du froid
~PLANT - [gen] appareils *mpl* frigorifiques
REFRIGERATIVE - [gen] réfrigérant, réfrigératif
REFRIGERATOR - [th eng] (a machine or plant producing and maintaining low temperatures by utilizing heat energy) machine *f* frigorifique, congélateur *m*
[gen] glacière *f*
[gen] (a freezer) congélateur *m*
~BODY - [transp] carrosserie *f* frigorifique
~CAR - [railw] wagon *m* frigorifique
~CONTAINER + [transp] container-isotherme *m* frigorifique
~SHIP - [naut] bateau *m* frigorifique
~TREATMENT - [med] réfrigération
~TRUCK - [transp] camion *f* frigorifique
~VEHICLE - [transp] véhicule *m* réfrigéré
~WAGON - [railw] wagon *m* réfrigérant
REFRINGENCE - [phys] (the power to refrect) réfringence *f*
REFRINGENT - [phys] réfringent
REFUEL, to - [fuel] (to replenish with fuel) faire du carburant
REFUELLING - [gen] ravitaillement *m* en combustible
REFUELLING PROBE - [aero] (tubular structure designed to be introduced into the receiving element of another aircraft for the transfer of fuel in flight) sonde *f* de ravitaillement en carburant
~IN FLIGHT - [aero] ravitaillement *m* en vol
~TENDER - [aero] ravitailleur *m* d'aviation
REFUGE - [gen] (shelter) abri *m*
~HOLE - [railw etc] (in a tunnel) niche *f*, refuge *m*
REFUND, to - [gen & fin] rembourser
REFUND - [gen & fin] remboursement *m*
REFUSAL - [constr] (the depth a pile is sunk to after repeated blows) refus *m*
~OF A PILE - [constr] refus *m* d'un pieu
~OF ACCEPTANCE - [comm] refus *m* de l'acceptation
REFUSE, to - [gen] refuser
REFUSE - [gen] refus *m*, rebut *m*
[comm] (of goods, objects etc) refus *m*, non-acceptation *f*
[text] coron *m* de la laine
[min] déblai *m*
~BODY - [transp] benne *f* à ordures
~COLLECTOR - [transp] camion *m* d'enlèvement des ordures
~CORK - [bot] liège *m* de rebut
~DISPOSAL - [gen] destruction *f* des ordures ménagères
~DUMP - [transp] benne *f* basculante)à ordures
~TIPPER - [transp] benne *f* basculante à ordures
REFUSION - [metall] refondre
~OF METALS - [metall] refonte *f* des métaux
REGAIN, to - [gen] récupérer, regagner
REGAIN - [text & plast ind] (the difference between the weight of a fibre and its ovendry weight) teneur *f* en humidité
~OF HUMIDITY - [text] degré *m* hygrométrique
~STANDARD - [text] humidité *f* normale
REGAP, to - [el] (in motors, to adjust the sparking plug electrode gap) régler l'écartement des electrodes
REGELATION - [phys] (the process by which ice melts when subjected to pressure and freezes again when pressure is removed) regélation *f*, regel *m*
REGENERATE, to - [chem] (to treat a chemical material so as to restore its original properties, when some of these have been lost in the course of a process) régénérer
[hydr] (the sludge) réactiver, réaérer
REGENERATED - [chem] régénéré
~CELLULOSE - [ind chem] (viscose. Cellulose by extruding cellulose xanthate through a spinneret into an acid solution) cellulose *f* régénérée
~FIBRE - [text & plast ind] (a man-made fibre produced by extrusion of material obtained from a natural substance, e.g. cellulose derived from vegetable sources) fibre *f* régénérée
~FUEL - [nucl] (reprocessed nuclear reactor material) combustible *m* régénéré
~GLACIER - [geol] glacier *m* remanié
~LEACH LIQUOR - [ind chem] (leach liquor which has been made fit for re-use by extraction of its dissolved metalliferous components) liqueur *f* de lessive régénérée
REGENERATING FURNACE-[metall]four *m* à récupérateur
REGENERATION - [gen] régénération *f*, épuration *f*
[radio] (positive feedback) réaction *f* positive
[biol] (renewal of a lost or damaged organ) régénération *f*
[electron] (in charge-storage tubes) régénération *f* de la charge
[hydr] (of the sludge) réactivation *f* réaération *f*
~AUTOCLAVE - s. regeneration boiler
~BOILER - [rubber ind] chaudière *f* à régénération, digéreur *m* à régénération
~CONTROL - [comput] commande *f* de régénération
~COUNTER - [comput] compteur *m* de cycles de régénération
~CUTTING - [agric] taille *f* de restauration
~CYCLE - [comput] cycle *m* de régénération
~OF ELECTROLYTE - [el chem] (the treatment of depleted electrolyte to make it again fit for use in an electrolytic cell) régénération *f* d'électrolyte

REGENERATION OF RUBBER - [rubber ind] (reclaiming of rubber) régénération *f* du caoutchouc
~PERIOD - [electron] durée *f* de régénération de la charge
~SWITCHGROUP - [el] combinateur *m* de récupération
REGENERATIVE- [gen] régénératif, régénérateur
~AIR HEATER - [mech] (air heater in which the heat transmitting surfaces are exposed alternatively to the heat-surrendering gases and to the air) aérotherme *m* à régénération
~AMPLIFIER - [radio] (a feedback amplifier) amplificateur *m* à réaction
~BRAKE - [el] frein *m* à récupération
~BRAKING - [el] (a system of electric braking in which energy is returned to the supply system) freinage *m* par récupération
~CONTROL - [el] (method of controlling electric motors in which regenerative braking is a feature) réglage *m* par freinage par récupération
~COOLING - [phys] refroidissement *m* par récupération
~DETECTOR - [radio] détecteur *m* à réaction
~DIVIDER - [radio] diviseur *m* de fréquence à réaction
~ENGINE - [astronaut] moteur *m* à refroidissement par récupération
~FEEDBACK - [electron] réaction *f* positive
~FURNACE - [th eng] (type of furnace in which heat is extracted from the burnt gases and returned to the combustion air) four *m* à récupération
~MEMORY - [comput] mémoire *f* régénératrice
~OVEN - [th eng] four *m* à régénération
~PROCESS - [biol] (the process by which damaged cells are replaced by new ones of the same type) processus *m* de régénération
~QUENCHING - [metall] refroidissement *m* rapide réitéré
~REACTOR - [nucl] (nuclear reactor for large scale production of transmutation products) réacteur *m* régénérateur
~REPEATER - [radio] (repeater performing pulse regeneration) répéteur *m* télégraphique automatique
[comput] trégénérateur *m*
~REPEATERING - [telecomm] (process in which each code element in a message is replaced by a new code element as given timing, waveform and magnitude) répétition *f* régénératrice
REGENERATOR - [heat] (apparatus for transferring heat from the product of a process to the material about to be submitted to such process) régénérateur *m*, récupérateur *m*
REGENT'S BUGLE - [mus] bugle *m* à clés
REGIMEN - [gen & phys] régime *m*
REGION - [gen & geogr] région *f*
[anat] région *f*
[nucl] (energy range, or energy region) portée *f*
~OF LIMITED PROPORTIONALITY - [nucl] (for a counter tube) proportionalité *f* limitée
~OF PARTIAL SHADOW - [radiat] (penumbra) pénombre *f*
REGIONAL ANATOMY - [anat] anatomie *f* des régions
~METAMORPHISM - [geol] (the changes in the mineral composition and texture of rocks due to stresses caused by intense earth movements) métamorphisme *m* général
REGISTER, to - [gen] enregistrer, immatriculer
REGISTER, to - [mech] engoujonner
REGISTER - [gen] registre *m*
[meas] (registering apparatus) compteur *m*
[acoust] (the range of a voice) registre *m*, étendue *f*
[print] (correct relation of colours) registre *m*
[mech] (valve for admitting or excluding air) registre *m*, rideau *m*, trappe *f*
[comput] (the device for storing one machine word) registre *m*
~LENGHT - [comput] capacité *f* de registre
~MARKS - [print] croix *f* de repère
~OF A METER - [instr] (the counting mechanism of the meter) élément *m* indicateur d'un compteur
~OF MATING - [agric] registre *m* des accouplements
~OF OPERATION - [el] domaine *m* de fonctionnement
~OF SHIPPING - [naut] tonnage *m* enregistré
~PLATE - [text] guide-fils *m*, matrice *f* perçoir
~ROTATION - [comput] décalage *m* cyclique
~SCREW - [mech] vis *f* d'enroujonnage
~TON - [meas] (I00 cubit feet) tonne *f* (de I00 pieds cubes)
REGISTERED - [gen] (recorded) enregistré, immatriculé
[fin] (of bonds) nominal
[gen] (of letters) (lettre) recommandée
~PATTERN - [leg] modèle *m* déposé
~STOCK - [fin] effets *mpl* nominatifs
~TRADE MARK - [comm] marque *f* déposée
REGISTERING BALLOON - [met] (a small free balloon carrying recording instruments, for observation of upper-air conditions) ballon *m* enregistreur
~CLOCK - [instr] horloge *f* enregistreuse
~DEVICE - [instr] mécanisme *m* enregistreur
~INSTRUMENT - [instr] appareil *m* de mesure enregistreur
~MACHINE - [text] toronneuse *f*
~STEAM-GAUGE - [instr] manomètre *m* enregistreur
~THERMOMETER - [instr] enregistreur *m* des variations de température
REGISTRATION - [gen] enregistrement *m*
[leg] immatriculation *f*
[print & photo] repérage *m*
[cin] (of the film) stabilisation *f* (du film)
[telev] (exact superimposition of the three colour images on the screen) superposition *f* des couleurs
~ARM - [railw] antibalançant *m*
~NUMBER - [gen auto etc) numéro *m* matricule, numéro *m* de police
~OF SEEDS - [agric] contrôle *m* officiel des semences
REGISTROGRAM - [el acoust] (the recording of a sound on a suitable medium) registrogramme *m*
REGISTRY - [gen] bureau *m* d'enregistrement
[naut] (certificat d') inscription *f*
REGLET - [arch] (flat narrow rectangular moulding) réglet *m*
[print] (thin strip of wood used for spacing) réglette *f*
REGRADING - [railw] (of a line) modification *f* du profil (d'une ligne)
REGRATING - [constr] (the operation of redressing the faces of old hewn stonework) regrattage *m*
REGRESS, to - [gen] retourner en arrière
[astr] rétrograder
REGRESS - [gen] rétrogression *f*, régression *f*
REGRESSION - [gen] (the act of moving back) retour *m* en arrière
[biol] régression *f*

REGRESSION - [math] rebroussement *m* (d'une courbe)
~OF A DISEASE - [med] décours *m* (d'une maladie)
REGRIND, to - [mech] rebroyer, remoudre
[mech] (of a valves) roder à nouveau (les soupapes)
REGRIND - [plast ind] (scrap plastic material ground for reuse) rognures *f*pl réutilisables
REGRINDING - [mech] rebroyage *m*
REGROOVE, to - [mech] (a cylinder) rainer à nouveau
REGROUND MATERIAL - s. regrind
REGROWTH - [agric] repousse *f*
[agric] (in forestry) repeuplement *m*
REGULAR - [gen] régulier
~AERODROME - [aero] (an aerodrome on a regular route, used as a scheduled stop) aérodrome *m* régulier
~-COURSED - [constr] (of a rubble wall built up in courses of the same height) à assises régulières
~-COURSED ASHLAR WORK - [constr] ouvrage *m* en pierres de taille
~FEED - [mech] alimentation *f* régulière
~GROWTH OF HAIR - [zool] accroissement *m* uniforme des poils
~LAY WIRE-ROPE - [metall] câble *m* métallique tordu spiroïdal
~REFLECTION - [opt] (a reflection in which a beam of light appears, after reflection, to emerge from an image of the source in the reflecting surface, and is reflected at an angle equal to that at which the beam falls on the surface) réflexion *f* spéculaire
~REFLECTION FACTOR - [opt] coefficient *m* de réflexion régulière
~REFLECTOR - [opt] réflecteur *m* spéculaire
~RIB - [text] cannelé *m* régulier
~SATEEN - [text] satin *m* simple
~SYSTEM - [crystal] (the cubic system) système *m* cubique, système *m* régulier
~TRANSMISSION - [opt] (transmission of light through a surface, so that the beam of light appears to emerge from the light source) transmission *f* (de la lumière) spéculaire
~TRANSMISSION FACTOR - [light] facteur *m* de transmission régulière
~YARN - [text] fil *m* régulier
REGULARITY - [gen] régularité *f*
~OF THE LAP - [text] régularité *f* de la nappe
~OF YIELD - [agric] régularité *f* de la productivité
~RETURN CURRENT COEFFICIENT - [el] coefficient *m* de régularité
~RETURN-LOSS - [el] affaiblissement *m* de régularité
REGULARIZE, to - [gen] régulariser
REGULARY SPUN COCOON - [text] cocon *m* confectionné régulièrement
REGULATE, to - [mech] (to adjust) régler, ajuster
REGULATE THE SPINNING PRESSURE, to - [text] régler la pression de filage
~THE TENSION OF A SPRING - [mech] régler la tension d'un ressort
~THE TENSION OF WEB, to - [text] régler la tension du voile
~THE THICKNESS OF THE THREAD, to - [text] régler le diamètre du fil
~VALVES, to - [mech] (to adjust the timing and or opening of valves) régler les soupapes

REGULATED FREQUENCY - [el] fréquence *f* asservie
~POWER SUPPLY - [el] alimentation *f* régulée
REGULATING - [gen & mech] (de) réglage
~ACTION - [comput] processus *m* de régulation
~BOWL - [text] galet *m* de réglage
~CHOKE COIL - [el] bobine *f* d'impédance de réglage
~CIRCUIT - [el] circuit *m* de régulation
~COCK - [plumb] robinet *m* de réglage
~DEVICE - [mech] dispositif *m* de réglage
~INDUCTOR - [el] (a device to regulate the current or voltage drop in the circuit) inductance *f* de réglage
~LAP MACHINE - [text] machine *f* à enrouler automatique
~LEVER - [mech] (a lever for effecting and adjustment) levier *m* de réglage
~PILOT - [radio] onde *f* pilote de réglage
~RELAY - [el] relais *m* de réglage
~RESERVOIR - [hydr] réservoir *m* compensateur, réservoir *m* régulateur du débit
~ROD - [nucl] (a control rod) barre *f* de pilotage
~ROLLER - [text] cylindre *m* de réglage
~SCREW - [mech] vis *f* de réglage
~SPINDLE - [auto] axe *m* de régulation
~SWITCH - [el] (of a battery; a switch whereby a number of cells in a battery connected with the external circuit is increased or decreased) interrupteur *m* de réglage
~TAP - [mech] robinet *m* de réglage
~THREAD GUIDE - [text] guide-fil *m* réglable
~TRANSFORMER - [el] enroulement *m* de réglage
~VALVE - [mech] (a valve for controlling fluid flow) soupape *f* régulatrice, soupape *f* de commande
~VARIABLE - [contr] grandeur *f* de commande
REGULATION - [mech] (adjustment) réglage *m*
[el & radio] variation *f* absolue de tension
[gen] (a rule) règlement *m*, arrêté *m*
[electron] (difference between the maximum and minimum anode voltage drop over a range of anode current) plage *f* de régulation
[contr] (USA only; automatic control) réglage *m* automatique
[mech] réglage *m* de charge
[mus] (the last step in voicing organs) égalisation *f*
~BY STOP COCK - [oil ind] distribution *f* par robinet
~CHAIN - [text] chaîne du régulateur
~DOWN - [el] (the change in voltage occurring when full load is put on to previously unloaded electric generator) variation *f* par réglage
~LIGHTS - [naut] feux *m*pl réglementaires
~MEASUREMENT - [naut] tonnage *m* réglementaire
~OF A RIVER - [hydr] correction *f* d'une rivière
~OF OUTPUT - [telev] réglage *m* de sortie
~TABLE - [telecomm] (in telephony) table *f* de tension des fils
REGULATOR - [mech] (an adjusting device, automatic or otherwise) régulateur *m*
[el] (a device for varying at will the voltage of a circuit ; also or automatically maintaining that voltage at or near a prescribed value) régulateur *m*
[mech] (of a steam engine) régulateur *m* de pression
[gas ind] (gas meters) régulateur *m* de niveau
[gas ind] (USA only; pressure regulator) détendeur *m*
~BOX - [paper man] (tank in which the flow of pulp to paper-making machine is controlled) cuve *f* ré-

gulatrice
REGULATOR-CELL - [el] (in a battery of accumulators, a cell which can be cut into or out of circuit to maintain the voltage of the supply constant during charge and discharge) élément *m* de réduction
~CONES - [text] cônes *mpl*
~CUT-OUT - [el] disjoncteur *m* de régulateur
~GATE - [hydr] vanne *f* lançoire, vanne *f* de travail
~MIXTURE - [chem] (buffer solution) solution *f* tampon
~POINTER - [horol] aiguille *f* du régulateur
~SPRING - [horol] ressort *m* du régulateur
~STATION - [gas ind] (USA only; governor house) salle *f* d'émission
REGULINE - [el metall] (term denoting electrodeposit which are firm and coherent) régulin
~DEPOSIT - [el metall] (a good electrodeposited metal) dépôt *m* régulin
REGULUS - [math] régule *m*
[metall] (intermediate product obtained in smelting ores of copper, lead, silver and nickel) culot *m*, matte *f* blanche de cuivre
~FURNACE - [metall] four *m* à régule
~OF ANTIMONY - [metall] (metallic antimony)(commercially pure) régule *m* d'antimoine
REGURGITATION - [med] régurgitation *f*
REHABILITATE, to - [gen] (to restore to a former state) réhabiliter
REHARDEN THE WIRES, to - [text] retremper les dents
REHEARSAL - [gen] répétition *f*
REHEAT, to - [gen] réchauffer
[metall] recuire
REHEAT JET PIPE - [aero] (afterburner; part of a turbojet engine where extra fuel is burnt between the gas turbine and the nozzle) dispositif *m* de post-combustion
~TREATING - [metall] traitement *m* thermique répété
REHEATER - [aero] (device by means of which further quantities of fuel are burnt in the excess air in the exhaust of the first stage of a turbine, in order to supply the second stage) réchauffeur *m*
REHEATING - [gen] réchauffage *m*, réchauffement *m*
[mech] (the process of passing steam which has been partially expanded in a steam turbine, back to a superheater before subjecting it to a further expansion) chaudière *f* pulmonaire
[mech] (in a gas turbine) chauffage *m* intermédiaire
[metall] recuisson *f*
~CHAMBER - [mech] (of a gas turbine) chambre *f* de réchauffement
[metall] chambre *f* de recombustion
~FURNACE - [metall] (the furnace in which metal ingots etc are heated to bring them to the temperature that is required for hot-working) four *m* de réchauffage
REICHMANN'S DISEASE - [med] gastrosuccorrhée *f*
REIGNITION - [el] nouvel allumage *m*
[electron] amorçage *m* parasitaire
~VOLTAGE - [electron] (in a gas-discharge device) tension *f* de réamorçage
REIMBURSE, to - [gen & comm] rembourser
REIMBURSEMENT - [gen & comm] remboursement *m*
REIMPOSE, to - [gen] réimposer
[print] réimposer, remanier
REIN, to - [gen] retenir
REIN - [gen] rêne *f*, guide *f*, bride *f*
REINDEER - [zool] renne *m*
REINDEER HAIR - [text] poil *m* de renne
REINFESTATION - [bot] réinfestation *f*
REINFORCE, to - [mech] (to strengthen mechanically by the inclusion of other material) renforcer
[plast ind] (the operation of strengthening resin with glass fibre) renforcer
[constr] (to strengthen concrete by the inclusion of iron rods) renforcer, consolider, armer
[metall] (the bridge of a furnace) armaturer
REINFORCE THE BEARINGS, to [mech] renforcer les supports
REINFORCED - [gen mech etc] renforcé
[constr] (of concrete) armé
~BEAM - [constr] poutre *f* armée, poutre *f* renforcée
~CEMENT - [constr] ciment *m* armé
~CONCRETE - [constr] (concrete strengthened by the inclusion of steel elements) béton *m* armé
~-CONCRETE GIRDER - [constr] poutre *f* en béton armé
~-CONCRETE GLASS - [constr] béton *m* vitreux
~-CONCRETE STRUCTURE - [constr] structure *f* en béton armé
~HOSE - [rubber ind] (flexible pipe strengthened with metal wire or ribbon) tuyau *m* renforcé
~LATEX - [rubber ind] latex *m* renforcé
~PLASTICS - [plast ind] (combinations of synthetic resins with usually, glass or asbestos fibres, giving products having good mechanical and physical properties) plastiques *mpl* renforcés
~REAR AXLE - [auto] pont *m* arrière renforcé
~STRUTTING - [constr] renforcement *m* de pilier
~TIMBERING - [mining] (supporting of roof and walls of mines with spacially strengthened structures) boisage *m* armé
REINFORCEMENT - [gen] (strengthening) renfort *m*, renforcement *m*
[constr] (for cement) armature *f*
[constr] renforçage *m*
~AGAINST INTERNAL PRESSURE - [el] frettage *m* d'un câble à pression interne
~BARS - [constr] barres *fpl* d'armature
~CLAMP - [mech] bride *f* de renforcement
~RING - [mech] (applied to flexible pipes) collier *m* de renfort
~ROD - s. reinforcing rod
REINFORCING AGENT - [ind chem] agent *m* renforçant
~CARBON BLACK COMPOUND - [rubber ind] mélange *m* chargé au carbon black
~FILLER - [plast ind] (a filler which also gives additional strenght to the compound of which it forms part) charge *f* active
~PAD - [metall] plaque *f* de renfort
~ROD - [constr] barre *f* d'armature
~STEEL - [metall] barres *fpl* d'acier pour armature
REINS - [impl] branches *fpl* de tenaille
[arch] (of arch) reins *mpl* d'une voûte
REINSTATE, to - [gen] réintégrer, rétablir
REINSURE, to - [gen & insur] réassurer
REINTRODUCE, to - [gen] réintroduire
REISSUE, to - [gen] rééditer
REISSUE - [gen] réédition *f*, nouvelle édition *f*
REITERATION - [gen] réitération *f*
[surv] (method of checking angular measurements) répétition *f*
[print] (the second side of a sheet to be printed) retiration *f*
REJECT, to - [gen] rejeter

REJECT, to - [comm] refuser (marchandises)
REJECT - [gen] (a part or unit found unacceptable on inspection) pièce *f* de rebut
~POCKET - [comput] (of a sorting machine, for unused punched cards) casier *m* d'éjection
REJECTION - [gen] réjection *f*, rejet *m*
~BAND - [electron] (in uniconductors waveguides) bande *f* de fréquence souscritique
~FILTER - [radio] (band eliminator filter) filtre *m* éliminateur de bande
~NUMBER - [comput] nombre *m* de rejet
~OF THE ACCOMPANYING SOUND - [telev] (sound rejection) réjection *f* de la porteuse son
REJECTOR - [radio] (combination of inductance and capacitance in parallel) réjecteur *m*
~CIRCUIT - [radio] (the circuit of a rejector) circuit *m* de l'éjecteur
~IMPEDANCE - [radio] (dynamic impedance) impédance *f* dynamique
REJECTS - [cin] déchet *m* de film
REJUVENATOR - [agric] herse *f* à prairie
REKINDLE, to - [gen] rallumer, renflammer
REL - [el] (unit of reluctance equal to one ampere-turn per maxwell) unité *f* de réluctance
RELAPSE - [gen & med] rechute *f*
RELAPSING FEVER - [med] (recurrent fever) fièvre *f* récurrente
RELATE, to - [gen] raconter, rapporter
RELATED - [gen] ayant rapport
[genet] (a relation) apparenté
~KEYS - [mus] clés *f*pl connexes
~TERMINALS OF A TRANSFORMER - [el] (the high and low voltage terminals related by convention so as to establish the phase displacement between the corresponding voltage) bornes *f*pl homologues d'un transformateur
RELATION - [gen] relation *f*, rapport *m*
[gen] (of family) parent *m*
RELATIONSHIP - [gen] rapport *m*
[genet] (the state of being related) parenté *f*
RELATIVE - [gen] relatif
~ABUNDANCE - [nucl] (the relative amounts in which isotopes are present in a given mixture) rapport *m* des teneurs isotopiques
~ADDRESS - [comput] (label used to identify the position of a storage location) adresse *f* relative
~ADDRESSING - [comput] codage *m* relatif
~APERTURE - [opt] (of a lens) (the ratio of the focal length to the diameter of the entrance pupil) ouverture *f* relative
~ATOMIC MASS - [surv] masse *f* atomique relative
~BEARING - [surv] orientation *f* polaire
~CHALK RATING - [ind chem] (figure indicating resistance to the chalking) coefficient *m* de résistance au farinage
~CODING - [comput] (a coding in which all addresses are represented symbolically) codage *m* relatif
~CONTROL RANGE - [contr] (ratio between the actual control range and an arbitrarily chosen value) étendue *f* relative de réglage
~DAMPING - [contr] facteur *m* d'amortissement
~DEAD ZONE - [contr] (relative insensitivity) zone *f* d'insensibilité relative
~DEAFNESS - [acoust] (of an ear at a specified frequency) surdité *f* relative
~DELAY - [radio] (delay of some signal components in filter circuits) retard *m* relatif
RELATIVE DENSITY - [phys] (the ratio between the mass of a specified volume of a substance and the same volume of another substance used as a standard) densité *f* relative
~EFFICIENCY - [mech] (of an internal combustion engine) rendement *m* relatif
~ERROR - [instr] (the ratio between the absolute error and the exact value of the quantity which is measured) erreur *f* relative
~EXIT VELOCITY - [mech] (in turbines) vitesse *f* relative de sortie
~HARMONIC CONTENT - [el] (the distortion factor) facteur *m* de distorsion, résidu *m* relatif
~HEARING LOSS - s. relative deafness
~HUMIDITY - [met] (the ratio of the humidity of a sample to that of saturated air at the same temperature) humidité *f* relative
~INLET VELOCITY - [mech] (in turbines) vitesse *f* relative d'entrée
~INTELLIGIBILITY - [telecomm] (of a telephone apparatus compared to another) intelligibilité *f* relative
~INTERFERING EFFECT - [radio] (of a single-frequency electric wave) interférence *f* relative
~ISOTOPIC ABUNDANCE - [nucl] abondance *f* isotopique relative
~KEYS - [acoust & mus] (a major and a minor key, both having the same key signature) clés *f*pl relatives
~LEVEL - [acoust] niveau *m* relatif
~LUMINANCE THRESHOLD - [opt] seuil *m* relatif de luminance
~LUMINOSITY - [opt] (the ratio between the value of the luminosity at a particular wavelength and the value at the wavelength of maximum luminosity) brillance *f* relative
~MOMENTUM - [phys] quantité *f* du mouvement relatif
~NUMBER - [math] nombre *m* relatif
~PERMEABILITY - [phys] (the ratio between the absolute permeability of a substance and that of a vacuum) perméabilité *f* spécifique
~PERMITTIVITY - [phys] (the ratio between the permittivity of a dielectric and that of a vacuum) constante *f* diélectrique relative
~PLATEAU SLOPE - [nucl] (the percentage change in counting rate per I00 volts increase of applied potential along the plateau) ente *f* relative de palier
~PLOT - [radar] (plot based on a line between successive positions of the echos as shown on a P.P.I.) pointage *m*
~PROGRAMMING - [comput] (programme making use of relative addresses) programmation *f* relative
~REFRACTIVE INDEX - [opt] (of two media) indice *m* de réfraction relative
~REGULATION - [el] (the ratio of the regulation between two working conditions and the voltage during the first condition) variation *f* relative de tension
~RESPONSE - [el acoust] (response under some particular conditions in relation to the response under reference conditions) efficacité *f* relative, réponse *f* relative
~SCATTERING FUNCTION - [phys] intensité *f* relative de diffusion
~SENSITIVITY - [instr] (the ratio between the change in the deflection and the corresponding relative change in the quantity which must be measured) sen-

sibilité *f* relative
RELATIVE SENSITIVITY - [el acoust] s. relative response
~SPEED - [gen] (with reference to the speed of the air) vitesse *f* relative
~SPEED DROP - [el] (ratio between the absolute speed drop between two working conditions and the speed during the first condition) chute *f* relative de vitesse
~SPEED RISE - [el] (the ratio of the absolute speed rise between two working conditions to the speed during the first condition) élévation *f* relative de vitesse
~SPEED VARIATION - [el] (the ratio of the absolute speed variation between two working conditions to the speed during the first condition) variation *f* relative de vitesse
~STOPPING POWER - [nucl] (of a specified substance in relation to that of a standard substance) pouvoir *m* d'arrêt relatif
~VELOCITY - [phys] (of a point with respect to a reference frame) vitesse *f* relative
~VISCOSITY - [chem] (the ratio of the absolute viscosity of a polymer in solution (of given concentration) and of the pure solvent at the same temperature) viscosité *f* relative
~VISIBILITY FACTOR - [opt] (the ratio between the apparent brightness of a monochromatic source and that of a source of wavelenght 5500 A.U. with the same energy) facteur *m* de visibilité relative
~VOLATILITY - [phys] volatilité *f* relative
~VOLTAGE DROP - [el] chute *f* relative de la tension
~VOLTAGE RISE - [el] augmentation *f* relative de la tension
RELATIVELY REFRACTORY STATE - [el biol] (the portion of the electrical recovery cycle during which the excitability is less than normal) état *m* relativement réfractaire
RELATIVISTIC - [phys math] (relating to relativity) relativiste
~MASS - [phys] masse *f* relativiste
~MASS EQUATION - [phys] équation *f* de masse relativiste
~PARTICLE - [phys] (particle having so great a velocity that its relativistic mass exceeds its rest mass by an amount which is significant for the computation) particule *f* relativiste
~RED SHIFT - [phys] déplacement *m* relativiste des raies vers le rouge, décalage *m* relativiste vers le rouge
RELATIVITY - [phys] (the principle postulating the equivalence of the description of the universe, in terms of physical laws, by various observers, alternatively in terms of different frames of reference) relativité
~THEORY - s. relativity
RELAX, to - [gen] relâcher, détendre
RELAXATION - [gen] relâchement *m*, détente *f*
[med] relaxation *f*
~BEHAVIOUR - [phys] conduite *f* de relaxation
~DISTANCE - [phys] (the distance in which the intensity of a beam of neutrons is reduced to a fraction of its initial value, the cause being the absorption of neutrons in the absence of scattering) longueur *f* de relaxation
~FREQUENCY - [phys] (the inverse of relaxation time) fréquence *f* de relaxation
RELAXATION INVERTER - [radio] (relaxation oscillator converting d.c. to a.c. power) oscillateur *m* de relaxation à conversion continu-alternatif
~OF RESISTANCE - [mech] perte *f* de résistance
~OF STRESS - [rubber ind] relaxation *f* de contrainte
~OSCILLATION - [radio] (oscillation characterized by a slow variation of current or voltage during part of the cycle, followed by a much faster return to the starting point) oscillation *f* relaxée
~OSCILLATION GENERATOR - [radio] générateur *m* d'oscillations relaxées
~OSCILLATOR - [telev] (oscillator generating relaxation oscillations) oscillateur *m* à relaxation
~TIME - [phys] (in many material phenomena, the response to a sudden change is often a time- measurable approach to equilibrium) temps *m* de relaxation
RELAY, to - [gen] (to forward by relay) relayer, transmettre
[contr] (to operate by a relay) commander par relais
[gen] (to lay again) reposer
RELAY - [contr] (any device through which a small amount of energy is made to control a larger amount by electrical, hydraulic, pneumatic or other means) relais *m*
[el] (a device which, when operated by current in one circuit, causes contacts to control the current in another circuit) relais *m*, contacteur *m*
[telecomm] (in telegraphy) répétiteur *m*
~BOBBIN - [el] bobine *f* de relais
~BOX - [el] armoire *f* à relais
~BRIDGE - [el] pont *m* à relais
~BROADCAST - [radio] émission *f* relais
~CAM - [el] came *f* de relais
~CHAIN CIRCUIT - [telecomm] circuit *m* de chaîne
~CIRCUIT - [el] circuit *m* relais
~COIL - [el] (the actuating coil of an electromagnetic relay) bobine *f* relais
~COMPUTER - [comput] (computer operating mainly by electromagnetic relays) calculateur *m* à relais
~CONTACT - [el] contact *m* du relais
~COUNTING CHAIN - [comput] chaîne *f* de comptage à relais
~CORE - [el] noyau *m* du relais
~GOVERNOR - [mech] régulateur *m* à servomoteur
~GROUP - [el] groupe *m* de relais
~LADDER - [el] échelle *f* de relais
~MACHINE - [mus] (of an organ) réservoir *m*
~NETWORK - [electron] circuit *m* de relais
~-OPERATED CONTROLLER - [contr] régulateur *m* à relais
~RADAR - [radar] (transmission of radar display from air to ground) retransmission *f* d'image radar
~RECEPTION - [radio] réception *f* relais
~-SET - [el] relais *m* amovible interchangeable, ensemble *m* de relais
~SHUTTER - [el] volet *m* de relais
~SPRING - [el] (the flexible part of a relay keeping it in an unoperated condition) ressort *m* du relais
~STATION - [radio] (radio station acting as an intermediate station for the transmission of programmes from the main station to other transmitters) émetteur *m* relais
~STORE - [comput] mémoire *f* à relais
~STUD - [el] goupille *f* de relais
~SWITCH - [el] contacteur-disjoncteur *m*

RELAY SYSTEM - [telecomm] (system using a central radio receiver for the acceptance of radio programmes which are then distributed to subscribers) système *m* tout à relais
~TELEVISION - [telev] (television system in which the transmitted programme is relayed to another transmitter for further broadcast) télévision *f* à relais
~TESTER - [instr] (apparatus designed to test switching times and distorsions of telegraph relays) mesureur *m* de relais
~TRANSMITTER - [radio & telev] (an automatic receiving and transmitting system in a link transmitter) émetteur *m* relais
~-TYPE ECHO SUPPRESSOR - [telecomm] suppresseur *m* d'écho à action discontinue
~-TYPE GRAPHIC INSTRUMENT - [el] (graphic instrument in which the writing device is operated by an auxiliary source of supply controlled by the current to be measured) appareil *m* enregistreur à relais
~UNIT - [telecomm] sélecteur *m* à relais
~WINDING - [el] bobinage *m* d'un relais
~WITH HOLDING WINDING - [el] relais *m* à collage
~WITH SEQUENCE ACTION - [el] relais *m* à action échelonnée
RELAYING FUNCTION - [telecomm] fonction *f* de relais
RELEASABLE - [mech] débrayable
[aero] largable
RELEASE, to - [gen] décharger, libérer
[mech] déclencher, débrayer
[mach tool] (e.g. the drilling spindle) débrayer
[aero] (a parachute) lancer
[mech] (the pressure) décharger
[mech] (the brake) dégager
[a spring] faire jouer, détendre
[photo] (to operate the trigger arrangement) déclencher (l'obturateur)
[chem] laisser échapper
[el & telecomm] déconnecter, retomber
RELEASE - [gen] libération *f*, décharge *f*
[mech] débrayage *m*, déclenchement *m*
[el] (a tripping device) déclencheur *m*
[photo] (the trigger arrangement for releasing the shutter and obtaining the exposure) déclencheur *m*
[chem] (of steam) dégagement *m*
[cin] (the handing of a film to cinema owners) distribution *f*
[telecomm] (in automatic telephony) rupture *f*
[mus] (of a piano) échappement *m*
~A CALL, to - [telecomm] déconnecter
~AGENT - [plast ind] (a substance used to facilitate the detachment of a moulding from the mould) agent *m* de démoulage
~BOLT - [mech] boulon *m* de débrayage
~BY CAM - [mech] débrayage *m* par excentrique
~CAM - [mech] (a cam used to disengage or uncouple a mechanism) came *f* à débrayage
~GUARD SIGNAL - [telecomm] signal *m* de libération de garde
~KEY - [comput] touche *f* à fonction
[telecomm] clé *f* de rupture, bouton *m* d'annulation
~MAGNET - [el] électro-aimant *m* de libération
~MOTION - [mech] dispositif *m* d'arrêt
~OF A BRAKE - [mech] déblocage *m* d'un frein
~OF THE POINTS - [railw] déverrouillage *m* de l'aiguille
RELEASE OF THE ROUTE LOCKING - [railw] libération *f* de l'itinéraire
~PEG - [text] loquet *m* de déclenchement
~PERIOD - [mech] (of exhaust of steam) période *f* d'émission
~POINT - [mech] (of steam) commencement *m* de l'émission
~PRINT - [cin] (the print sent out to cinema owners) copie *f* de distribution
~SPRING - [auto] (in the clutch assembly) ressort *m* de rappel, ressort *m* de relâchement
~THE BRAKE, to - [mech] (to remove the pressure exerted by friction surfaces of a brake) relâcher les freins, desserrer les freins
~THE CATCH, to - [text] déclencher le loquet
~THE COUPLING, to - [railw] desserrer l'attelage
~THE LATCH, to - [text mach] dégager le loquet
~THE NIPPER ACTION, to - [text] suspendre l'action de piçage
~THE SPRING, to - [mech] (to remove the pressure exerted on a spring) relâcher le ressort
~THE STOP MOTION, to - [text mach] manoeuvrer l'arrêt mécanique
~TRIGGER - s. release [photo]
~VALUE - [electron] valeur *f* de décollage
RELEASER - [mech] déclancheur *m*, démarreur
RELEASING CAM - s. release cam
~LEVER - [mech] came *f* de désaccouplement
~SPEAR - [oil ind] taraud *m* rattrappe-tube intérieur à coin dégageable
~SPRING - [mech] ressort *m* de rappel
RELEVELLING - [gen] renivellement *m*
RELIABILITY - [gen] sûreté *f*, sûreté *f* de fonctionnement
~OF OPERATION - [gen] sécurité *f* de fonctionnement
~OF SERVICE - [gen] sûreté *f* de service
~TEST - [instr] (a test to determine the degree of accuracy of instruments) épreuve *f* de régularité
RELIABLE - [gen] sûr, d'un fonctionnement sûr
RELIEF - [gen] soulagement *m*
[print] relief *m*
[mech] décompression *f*, détente *f*
[tool] (the angle of relief) angle *m* de dépouille
[geogr] relief *m* terrestre
[geogr] (representation of uneven ground) hypsométrie *f*
[arch] relief *m*
[mach tool] dépouille *f*, dégagement *m*
~ANGLE - [mach tool] angle *m* de dépouille
~ANNEALING - [metall] recuit *m* de relaxation
~BLOCK - [print] (printing block which can be used with normal printing type) cliché *m*
~CAM - [auto] (half compression cam) came *f* de décompression
~CHARACTERISTIC - [el] montagne *f*
~COCK - [mech] (compression cock) robinet *m* décompresseur
~HOLDER - [gas ind] gazomètre-tampon *m*
~HOLE - [mining] forage *m* de secours
~LINE - [el] ligne *f* auxiliaire
~MAP - [geogr] (showing mountains, uneven ground in relief) carte *f* en relief, plan-relief *m*
~PATTERN - [text] dessin *m* en relief
~PIPE - [gas ind] évent *m*
~PLATE - s. relief block
~POLISHING - [metall] polissage *m* en relief
~PRINTING - [print] impression *f* en relief

RELIEF ROLLER - [text] cylindre *m* à dessin en relief
~SEWER - [hydr] émissaire *m* de décharge
~-VALVE - [mech] (automatic valve designed to release excess pressure) soupape *f* de sûreté, clapet *m* de décharge, reniflard, ventouse *f* à air
~VALVE FOR STEAM - [mech] détenteur *m* de vapeur
~VENT - [mech] évent *m* de décharge
[el] lumière *f* anti-explosion
~WELL - [oil ind] (a directional well, drilled to intersect a well which is flowing "wildly") puits *m* de secours
[mining] puits *m* auxiliaire
RELIEVE, to - [gen] soulager, délivrer
[mech] dégager, dépouiller
[plast ind] (to reduce the area between the sealing faces of a mould or to provide a channel between those faces to facilitate the escape of the surplus material) dépouiller
[el metall] (the operation of the local removal of colouring from metal surfaces by mechanical methods) éliminer
[metall] (the mould) dédoubler (le moule)
RELIEVE A VALVE, to - [mech] soulager une soupape
RELIEVED MILLING CUTTER-[mach tool] fraise *f* à denture à dépouille
~TEETH - [mech] denture *f* à dépouille, dents *f*pl dégagées
RELIEVER - [mining] trou *m* auxiliaire
RELIEVING - [el metall] (the process of focal removal of colouring from metal surfaces by mechanical methods) élimination *f*
[mech] dégagement *m*
[metall] dédoublage *m* du moule
~ANODE - [electron] (an anode which provides an alternative conducting path to reduce the current to another electrode) anode *f* auxiliaire de soulagement
~ARCH - [arch] (arch built on the spandrel of a main arch in order to distribute the load) arc *m* de décharge
~ATTACHMENT - s. relieving device
~DEVICE - [mach tool] outil *m* à dépouiller
~DISCHARGE PATH - [el] trajet *m* de décharge de shuntage
~GEAR - [railw] (of a turnable) dispositif *m* de shuntage d'une plaque tournante
~HANDLE - [text] poignée *f* d'enlèvement
~LATHE - [mach tool] tour *m* à dépouiller
~RECTIFIER - [el] soupape *f* de shuntage
~ROLLER - [text] galet *m* racleur
~TEMPERATURE - [metall] (temperature eliminating internal stresses) température *f* de détente
~THE WEIGHTS FROM THE ROLLERS - [text] relevage *m* des poids
RELIGHT, to - [aero] (a jet engine) réallumer
RELIGHTING - [gen] réallumage *m*, rallumage *m*
RELINE, to - [gen] redoubler
RELINE - [metall] brasquage *m*
~THE BRAKES, to - [mech] (to renew the friction material in brakes) regarnir les freins
~THE CLUTCH, to - [auto] regarnir l'embrayage
RELINER - [rubber ind] (of tyres) pièce *f* de calage
RELINING - [gen] regarnissage *m*, rechemisage *m* (des cylindres)
RELINQUISH, to - [gen] abandonner, renoncer
RELINQUISHMENT - [gen] abandon *m*, renonciation *f*
[leg] répudiation *f*

RELOAD, to - [gen] recharger
RELOADING DIE - [metall] matrice *f* à recalibrer
RELOCATION - [railw] modification *f* du tracé
RELUCTANCE - [gen] répugnance *f*
[el] (the capacity of opposing magnetic induction) reluctance *f*, résistance *f* magnétique
~GENERATOR - [el] (a.c. generator including a field magnet with salient poles not provided with an exciting winding, the field of which is supplied only by the camature currents) alternateur *m* à réaction
~MOTOR - [el] moteur *m* à reluctance
RELUCTANCY - s. reluctance
RELUCTIVITY - [el] (the measure of the ability of magnetic material to conduct magnetic flux) magnétorésistance *f* spécifique
RELUXATION - [med] reluxation *f*
RELY, to - [gen] compter sur, avoir confiance
REMAGNETIZE, to - [phys] réaimanter
REMAIN, to - [gen] rester
REMAINDER - [gen] reste *m*
[math] reste *m*
[leg] réversion *f*
~THEOREM - [math] règle *f* du reste
REMAINING SPEED - [phys] vitesse *f* résiduelle
REMAINS - [gen] restes *m*pl
REMAKE, to - [gen] refaire
REMAKING - [gen & mech] réfection *f*
~JOINTS - [mech] réfection *f* de joints
REMAND, to - [gen] renvoyer
REMAND - [gen & leg] renvoi *m*
REMANENCE - [gen] remanence *f*
[phys] (the part of magnetic induction remaining in an electrical circuit after the removal of the applied magnetomotive force) magnétisme *m* résiduel, rémanence *f*
REMANENT - [gen & el] rémanent, résiduel
~FLUX DENSITY - [phys] densité *f* de flux rémanente
~MAGNETISM - [phys] magnétisme *m* résiduel
~RELAY - [el] (relay which is maintained in the operated position by remanence after the control current has been removed) relais *m* rémanent
REMANUFACTURED WOOL - [text] laine *f* renaissance, shoddy *m*
REMELTING - [metall] refonte *f*
REMETAL, to - [mech] (to renew the anti-friction metal in a bearing) regarnir de métal antifriction
REMEMBER, to - [gen] se rappeler
REMIGES - [zool] (the large contour feathers of the wing in birds) rémiges *f*pl
REMILL, to - [rubber ind] remalaxer
[metall] fraiser à nouveau
REMILLED CREPE - [rubber ind] crêpe *m* remalaxé, "blankets" *f*pl
REMILLING AND WASHING MACHINE - [rubber ind] déchiqueteur-laveur *m*
REMIPED - [zool] (with feet adapted for paddling) palmipède
REMITTENT FEVER - [med] (a fever characterized by remissions in which the temperature falls but not back to normal) fièvre *f* rémittente
REMNANT - [gen] reste *m*
[text] coupon *m* d'étoffe
[min] (an ore pillar) pilier *m* de minerai
REMODULATION - [el] rémodulation *f*
REMOTE - [gen] écarté, éloigné
[contr etc] télé-, à distance
[cin] (field pick-up) prise *f* de vue à l'extérieur

REMOTE ACCESS COMPUTING SYSTEM - [comput] système *m* ordinateur à accès à distance
~AIMING - [contr] télépointage *m*
~ATTITUDE DIRECTOR - [aero] télécommande *f* attitude de vol
~CONTROL - [contr] (control of a machine or the like from a more or less distant point) télécommande *f*, commande *f* à distance
~CONTROL OVERLOAD CUT-OUT - [contr] mise *f* hors circuit à distance
~CONTROL REVERSING SWITCH - [el] inverseur *m* de commande à distance
~CONTROL RIGS - [min] sonde *f* télécommandée
~CONTROL STATION - [el] téléterminal *m*
~CONTROL STEERING GEAR - [contr] direction *f* électrique
~CONTROL SWITCH - [el] interrupteur *m* de commande à distance
~CONTROLLED - [contr] télécommande, commandé à distance
~-CONTROLLED OPERATION - [contr] actionnement *m* télécommandé
~-CONTROLLED SUBSTATION - [el] poste *f* à télécommande
~-CUT-OFF TUBE - s. remote-cut-off valve
~-CUT-OFF VALVE - [electron] (vacuum valve in which the amplification factor varies in a predetermined way with control-grid voltage) tube *m* à pente variable
~GAIN CONTROL - [telev] téléréglage *m* de l'amplification
~GEAR-BOX CONTROL - [auto] commande *f* à distance de la boîte de vitesses
~HANDLING EQUIPMENT - [nucl] ensemble *m* de télémanipulation
~INDICATION - [instr] (system for the transmission of information concerning measured values, to a distance) téléindication *f*
~INDICATOR - [instr] téléindicateur
~LINE - [telev] ligne *f* de connexion entre le studio et l'extérieur
~LIQUID SAMPLING - [chem] échantillonnage *m* de liquides à distance
~MAINTENANCE - [comput] télémaintenance *f*
~MANIPULATION - [nucl] (the handling of the remote control elements) télémanipulation *f*
~MANIPULATOR - [nucl] télémanipulateur *m*
~MANUAL CONTROL - [contr] commande *f* volontaire à distance
~MASS-BALANCE WEIGHT - [mech] (a mass-balance weight coupled to a control surface by a mechanical system) masse *f* de compensation à distance
~-POSITION INDICATOR - [contr] (device for transmitting to a remote receiver the position of a mechanism) téléindicateur *m* de position
~r.p.m. INDICATOR - [instr] télétachymètre *m*
~-READING INSTRUMENT - [instr] (any instrument which shows the quantity measured on an indicator situated at some distance from the sensing device) instrument *m* à télélecture
~SIDE - [contr] plan *m* opposé
~STARTING CONTROL - [contr] commande *f* de démarrage à distance
REMOTELY CONTROLLED CAMERA - [photo] caméra *f* à téléoperation
~CONTROLLED CIRCUIT-BREAKER - [el] télérupteur *m*
REMOTENESS FROM MARKETS - [ind] éloignement *m* des marchés
REMOULD, to - [rubber ind] (of tyres) remouler de tringle à tringle
REMOULD - [rubber ind] rechapage *m* de talon à talon, remoulage *m* de tringle à tringle
REMOVABLE - [gen] (capable of being taken off) mobile, amovible, démontable
~CIRCUIT BREAKER - [el] rupteur *m* amovible
~CORE - [metall] noyau *m* détachable
~COUPLING - [mech] accouplement *m* démontable
~ELEMENT - [el] élément *m* amovible
~GRAPHITE BLOCKS - [nucl] blocs *mpl* en graphite transportables
~HATCH - [aero] panneau *m* amovible
~HOOD - [metall] chapeau *m*
~IRON SHAFT STAVE - [text] liseron *m* métallique enlevable
~MANWAY - [gas ind] (in gasholders) plaque *f* pleine
~PROGRAMME PANEL - [comput] panneau *m* de programmation amovible
~REED COVER - [text] chapeau *m* déplaçable du peigne
~TOP - s. removable hood
REMOVAL - [gen] déménagement *m*
[gen] (elimination) suppression *f*
[plast ind] démoulage *m*
[mech] démontage *m*
~OF A MINERAL DEPOSIT - [min] enlèvement *m* d'un gîte métallifère
~OF ACID - [chem] élimination *f* de l'acide
~OF AIR - [rubber ind] désaération *f*
~OF CLINKER - [metall] nettoyage *m* des grilles
~OF CUSTOMS BARRIER - [comm] suppression *f* des barrières douanières
~OF DESTROYED BURS - [text] élimination *f* des chardons détruits (de la laine)
~OF FLAWS - [metall] décriquage *m*
~OF STRESS - [mech] détente *f*
~OF THE BEAD - [rubber ind] (of tyres) suppression *f* du bourrelet
~OF THE VEGETABLE IMPURITIES - [text] élimination *f* des débris végétaux
~OF TREE STUMPS - [agric] essouchement *m*
~OF WATER - [text] (from the wool) essorage *m* (de la laine)
REMOVE, to - [gen] enlever, ôter, écarter, lever
[plast ind & rubber ind] démouler
REMOVE A MASH, to - [brew ind] prélever une maische
~A TYRE, to - [auto] démonter un pneu
~DUST, to - [agric] dépoussiérer
~FLAWS, to - [metall] décriquer
~FORMS, to - [constr] enlever le coffrage
~IRON, to - [hydr] déferriser
~MOULDS, to - [metall & plast ind] démouler
~NAILS, to - [carp] dériver, défaire la rivure
~PILLARS, to - [mining] dépiler
~RUST, to - [metall etc] dérouiller
~SIDE-SHOOTS, to - [agric] épamprer
~SLUDGE, to - [hydr] débourber, enlever les boues
~THE BAND FROM THE LEASE - [text] déverger
~THE BOBBINS, to - [text] enlever les bobines
~THE BURR - [metall] (from a casting) ébarber (une pièce moulée)
~THE COPS, to - [text] lever les canettes
~THE GRINDING DUST BY SUCTION, to - [text] aspirer la poussière d'aiguisage

REMOVE THE MACHINERY - [gen] déplacer les machines
~THE OVERBURDEN - [mining] décapeler le gîte
~THE SCUM, to - [hydr] enlever le chapeau
~THE SEAM, to - [metall] enlever les bavures de joint
~THE SHOVE FROM THE ROPE - [text] enlever les brindilles du câble
~THE STRIPS FROM THE CLOTHING - [text] enlever les débourrages des garnitures
~THE WOOL FROM THE BURS - [text] enlever la laine des chardons
REMOVER - [ind chem] décapant *m* pour vernis, dissolvant *m*
REMOVING OF KINKS - [telecomm] rectification *f* du fil
~THE WATER FROM THE WOOL - [text] essorage *m* de la laine
REMTRON - [electron] (gas tube often employed in computers and counters) remtron *m*
RENAL - [anat] rénal
~ARTERIES - [anat] artères *f*pl rénales
~COLIC - [med] colique *f* néphrétique
~PLEXUS - [anat] plexus *m* rénal
RENARDITE - [min] (a mineral which contains lead, uranium and phosphorus) rénardite *f*
RENCULUS - [anat] lobe *m* rénal
REND, to - [gen] déchirer, fendre
RENDER, to - [gen] rendre
[constr] (the operation of covering bricks or stonework with a coat of coarse stuff) enduire, gobeter
RENDER AND SET - [constr] crépi *m* et enduit
~ASTATIC, to - [instr] (to give astatic properties to measuring instruments) rendre astatique
~FLOAT AND SET - [build] enduit *m*, crépi et aplanissement
RENDERING - [constr] (covering bricks or stonework with a coat of coarse stuff) enduit *m*
RENDITION - [gen] (artistic, musical dramatic interpretation) interprétation *f*
RENDZINA - [soil] rendzine *f*
RENEW, to - [gen] renouveler
RENEWABLE FUSE - [el] fusible *m* interchangeable
RENEWAL - [gen] renouvellement *m*
[mech etc] remplacement *m*
RENICULUS - s. renculus
RENNET - [zool] (dried stomach of some young hoofed animals, which is capable of curdling milk) caille-lait *m*
~CASEIN - [chem] (paracasein. The major protein of milk, coagulated by the addition of rennet to skim milk. Casein is used in food processing and in the manufacture of adhesives and casein plastics) caséine *f* caillée
RENNETING - [food] emprésurage *m*
RENNIN - [chem] (chymosin; an enzyme occurring in gastric juice and responsible for the curdling of milk. Commercially it is used in the manufacture of cheese and the preparation of casein for plastics) rénine *f*
RENORMALIZATION - [gen] rénormalisation *f*
~OF MASS - [phys] (quantum theory) rénormalisation *f* de la masse
RENT - [gen] loyer *m*
[leg] rente *f*
[gen] (the result of rending) déchirure *f*
RENTERING SEAM - [text] rentrainure *f*
RENUNCULUS - s. renculus
RENVERSEMENT - [aero] (an aerial manoeuvre consisting of a half-roll followed by a half-loop) renversement *m*
REORDER - [gen & comm] commander à nouveau
REORGANIZE, to - [gen] réorganiser
[fin] assainir
REP - [text] (woven fabric with a corded surface) reps *m*
[radiat] (physical roetgen equivalent) rep *m*
~CARPET - [text] tapis *m* cordelé, tapis *m* en reps
REPACK, to - [gen] remballer
[mech] regarnir, remplacer la garniture
REPACK A PISTON, to - [mech] regarnir un piston
REPACKED BALE - [text] balle *f* réemballée
REPAINT, to - [paint] repeindre
REPAINTING - [paint] nouvelle peinture *f*
REPAIR, to - [gen & mech] (to carry out renewal or rectification) réparer, rajuster, refaire
REPAIR - [gen] (renewal or rectification of a worn or damaged part) réparation *f*
~A BEARING, to - [mech] refaire un coussinet
~A BREAKDOWN, to - [mech] (to remedy a fault) dépanner
~BASIN - [naut] bassin *m* de radoub
~COSTS - [gen] frais *m*pl de réparation
~KIT - [mech] trousse *f* de réparation
~LINK - [mech] fausse maille *f*
~PART - [gen] (a spare part used in repairs) pièce *f* de rechange, pièce *f* de réparation
~PATCH - [rubber ind] (for tyres) rustine *f*, plaque *f* de réparation
~PIT - [auto] fosse *f* de réparation
~SERVICE - [gen] service *m* réparations
~SHIP - [naut] navire-atelier *m*
~SHOP - [gen] (workshop devoted to repairs) atelier *m* de réparations
~TEST OF A METER - [instr] essai *m* de compteur lors de la révision
~TRUCK - [auto] camion-atelier de réparations
REPAIRING OUTFIT - [mech] nécessaire *m* à réparations
[auto] boîte *f* à réparations
REPARATION - [gen & leg] réparation *f*
REPAVE, to - [constr] repaver, recarreler
REPAY, to - [gen & comm] rembourser, rendre
REPAYMENT - [gen & comm] remboursement *m*
REPEAL, to - [gen & leg] abroger, révoquer
REPEAL - [gen & leg] abrogation *f*
REPEAT, to - [gen] répéter
[comm] renouveler
[opt] (e.g. the circle of a theodolite) réitérer
REPEAT - [gen] (repetition) répétition *f*
[mus] (the sign used to indicate that the music is to be played again) reprise *f*
~OF PATTERN - [text] rapport *m* de dessin
~OF THE SHUTTLES - [text] répétition *f* des navettes
~OF WARP THREADS - [text] rapport *m* en chaîne
~OF WEFT THREADS - [text] rapport *m* en trame
~POINT - [telecomm] (the location on the tuning dial of a superheterodyne receiver receiving an image-frequency transmission) point *m* d'accord image
REPEATABILITY - [instr] (a measure of deviation of test results from their mean value) aptitude *f* de répétition
REPEATED BEATING - [text] battage *m* répété
~BEND TEST - [metall] essai *m* de flexion multiple

REPEATED BLOW TEST - s. repeated impact test
~IMPACT TEST - [metall] essai *m* à chocs répétés
~PASS - [text] remettage *m* par plusieurs remises
~PASSAGE OF THE COTTON - [text] passage *m* répété du coton
~SOLIDIFICATION - [electron] (a term denoting germanium solidification when a batch of it is carried a number of times through a fusing channel) solidification *f* multiple
~-UNTIL-ACKNOLEDGED SIGNAL - [telecomm] signal *m* répété jusqu'à l'accusé de réception
REPEATER - [horol] (a type of watch) montre *f* à répétition
[instr] (instrument designed to retransmit electromagnetic signals) répéteur *m*, instrument *m* répétiteur
[telecomm](an amplifier for telephone circuits) amplificateur *m* téléphonique
[firearms] fusil *m* à répétition
[math] fraction *f* périodique
~BAY - [telecomm] travée *f* de répéteurs
~DISTRIBUTION FRAME - [telecomm] répartiteur *m* de répéteur
~SPACING - [comput] section *f* de répéteur
~STATION - [telecomm] station *f* de répéteurs
~TEST RACK - [telecomm] baie *f* de contrôle de répéteur
~TRANSMITTER - [radar] (slave transmitter) émétteur *m* répétiteur
REPEATING - [firearms] à répétition
~COIL - [telecomm] (transformer used for the interconnection of telephone circuits) translateur *m*
~DECIMAL - [math] (decimal fraction in which one figure is repeated indefinitely) décimal *m* périodique
~FIREARM - [firearms] revolver *m*
~INSTALLATION - [telecomm] installation *f* de translation
~INSTRUMENT - [instr] instrument *m* répétiteur
~MECHANISM - [mech] mécanisme *m* de répétition, dispositif *m* de répétition
[mus] (in pianoforte) double échappement *m*
~OF DESIGN - [text] rapport *m* de dessin
~POST - [railw] poste *m* répétiteur
~SELECTOR - [telecomm] (in automatic telephony; a selector which is operated by the first train of impulses received, and also repeats the received impulses to operate further selectors) sélecteur *m* répétiteur
~THE CARDS - [text] repiquage *m* des cartons
~TIMEBASE - [electron] base *f* de temps répétitrice
REPEL, to - [gen] repousser
[el] repousser
REPELLENCE FACTOR - [electron] facteur *m* de réflexion
REPELLENT - [gen] répulsif
REPELLER - [electron] (an electrode held at a negative potential with respect to a resonator for the purpose of returning to this resonator the electron beam which issued from its) réflecteur *m* de klystron
~SPACE - [electron] espace *m* de réflexion
~VOLTAGE - [electron] tension *f* de réflecteur
REPELLING OF ELECTRONS - [phys] répulsion *f* de électrons
REPERCUSSION - [gen] répercussion *f*
[med] dispersion *f* d'une tumeur
REPERFORATOR - [telecomm] (in telegraphy, a receiving perforator) réperforateur
REPERFORATOR SWITCHING - [telecomm] commutation *f* avec retransmission par bande perforée
REPETITION - [gen] répétition *f*
[a copy] réplique *f*, double *m*
~EQUIVALENT - [telecomm] (of a complete telephone connection; the quality of the transmission experienced by subscribers using the connection) équivalent *m* de répétition
~FREQUENCY - [telecomm] fréquence *f* itérative
~RATE - [comput] (the fastest rate of electronic pulses in the circuits of a computer) fréquence *f* itérative
[telecomm] (count of the number of repetitions in a conversation under service conditions over a telephone circuit) taux *m* de répétition
~WORK - [metall] moulage *m* en série
[gen] fabrication *f* en série
REPETITIVE ACCURACY - [electron] exactitude *f* de répétition
~ERROR - [contr] (or deviation; the maximum deviation of the controlled variable from the average value on successive return to specified operating conditions following a specified deviation) erreur *f* répétitive
~INSTRUCTION - [comput] (instruction repeated one or more times) instruction *f* de répétition
~INVERSE VOLTAGE - [electron] tension *f* inverse de pointe répétitive
~PEAK FORWARD CURRENT - [electron] courant *m* de pointe répétitif en sens direct
~PEAK OFF-STATE VOLTAGE - [electron] tension *f* de pointe répétitive à l'état bloqué
REPLACE, to - [gen] (to substitute one part for another) replacer
~THE FLYER, to - [text] replacer l'ailette
REPLACEABLE ELEMENTS - [mech] éléments *mpl* remplaçables
~HYDROGEN - [chem] (the hydrogen atoms in the molecule of an acid which can be replaced by atoms of a metal on neutralization with a base) hydrogène *m* remplaçable
REPLACEMENT - [gen] remplacement *m*
[mech etc] remontage *m*, remise *f* en place
[gen] (in industry) pièces *fpl* de rechange
~DIAGRAM - [telecomm] schéma *m* équivalent
~ENGINE - [gen] (a complete engine substituted for another as a unit) moteur *m* de remplacement
~RATE - [electron] vitesse *f* de remplacement
~SCHEME - s. replacement diagram
~TUBE - [electron] tube *m* identique
REPLACER - [railw] (inclined plane used for bringing back to a track a derailed truck) dispositif *m* de relevage
[railw] s. replacing switch
REPLACING - s. replacement
~SWITCH - [railw] rampe *f*
~THE FLYER - [text] replacement *m* de l'ailette
REPLANT, to - [agric] replanter
REPLATE, to - [metall] replaquer
REPLATING - [metall] replacage *m*
REPLENISH, to - [gen] (to fill again) remplir
[gen] (to supply in abundance) se ravitailler
[mech] regarnir, recharger
REPLENISH, to - [hydr] enrichir
REPLENISHER - [photo] cuve *f* de régeneration
[el] rechargeur *m*
REPLENISHING BASIN - [hydr] bassin *m* de réalimen-

tation, bassin *m* de reveinement
REPLENISHMENT - [gen] remplissage *m*
~OF STOCKS - [comm] renouvellement *m* des stocks
REPLETE - [gen] rempli, gorgé
REPLETION - [med] réplétion *f*, plénitude *f* d'estomac
REPLICA - [gen] reproduction *f*, copie *f*
[metall] (thin film of carbon or metal on a metal surface used for electron microscope analysis) copie *f* pelliculaire d'une surface
~GRATING - [phys] (in spectrology) réseau *m* de réplique
REPLY, to - [gen] répondre
REPLY - [gen] réponse *f*
~MESSAGE - [comput] message *m* d'accusé de réception
REPORT, to - [gen] rapporter, relater
REPORT - [acoust] (the sound of an explosion) détonations *fpl*, pétarde *f*
[gen] rapport *m*, exposé *m*
[leg] (a document) procès-verbal *m*
REPORTING POINT - [aero] (a geographic point used as reference in reporting the position of an aircraft) point *m* de compte rendu
REPOSE, to - [gen] reposer
REPOSITION - [med] remise *f* en position normale
REPOSITORY - [gen] (a place in which goods are stored) dépôt *m*
REPOUSSE - [metall] travail *m* de repoussé
~WORK - s. repousse
REPP - s. rep
REPPE PROCESSES - [chem] (high-temperature catalyzing processes for obtaining a variety of products from acetylene) procédés *mpl* de Reppe
REPRESENT, to - [gen] représenter
REPRESENTATIVE - [gen & comm] représentant *m*, représentatif
~ARRAY - [phys] disposition *f* type
~-CIRCULATING TIME - [comput] (method for assessing the speed performance of a computer) évaluation *f* de la vitesse d'un calculateur
~OBSERVATIONS - [met] (observations to obtain the true or typical meteorological conditions in a given air mass, unaffected by local conditions) observations *fpl* caractéristique
~SAMPLE - [gen] (average sample) échantillon *m* type
REPRESS, to - [gen] réprimer
REPRESSING - [metall] recompression *f*
REPRESSION - [gen] répression *f*
[med] (the unconscious mental mechanism by which complexes are kept out of consciousness) répression *f*, refoulement *m*
REPRESSURE LINE - [gas ind] (of natural gas) conduite *f* de réinjection de gaz
REPRINT, to - [print] réimprimer
REPRINT - [print] réimpression *f*, nouveau tirage *m*
REPROCESS, to - [gen] transformer en produit industriel
[nucl] régénérer
REPROCESSED COTTON - [text] coton *m* effiloché
~MATERIAL - [text] matière *f* effilochée
REPROCESSING - [nucl] (processing of the material recovered after use in a reactor, so that it may be used again) régénération *f*
~LOSS - [nucl] pertes *fpl* de régénération
REPRODUCE, to - [gen] reproduire
REPRODUCE, to - [draw] copier
REPRODUCER - [el acoust] (instrument designed to translate electric signals into sound waves) pick-up *m*
[comput] (reproducing punch) reproductrice *f*
REPRODUCIBILITY - [instr] (the closeness with which a measuring instrument repeats indications when measuring identical values of the measured variable under the same conditions) fidélité *f*, reproductibilité *f*
REPRODUCING - [gen] reproducteur, de reproduction
~CHARACTERISTICS - [el acoust] caractéristique *f* de reproduction
~PUNCH - [comput] reproductrice *f* de cartes
~SLIT - [el acoust] encoche *f* de reproduction
~STYLUS - [el acoust] aiguille *f*, saphir *m*
REPRODUCTION - [gen] reproduction *f*
[biol] (the process of generation whereby a species is perpetuated) reproduction *f*
[acoust] reproduction *f*
[print] reproduction *f*
~CUTTING - [agric] taille *f* de restauration
~FACTOR - [nucl] (the ratio between the number of neutrons present in a reactor at a specified time and the number present one life time earlier) facteur *m* de multiplication
~SET - [el acoust] appareil *m* de reproduction
REPRODUCTIVE CELL - [biol] cellule *f* de reproduction
~ORGANS - [zool] organes *mpl* reproducteurs
REPTILE - [zool] reptile *m*
~CALF - [leather ind] peau *f* factice de reptile
~SKINS - [leather ind] (the skins of snakes, alligators, crocodiles, lizards etc) peaux *fpl* de reptile
REPULSION - [gen] répulsion *f*
[phys] (the tendency to move apart) répulsion *f*
[biol] (the tendency shown by dominant characters to separate) répulsion *f*
~-INDUCTION MOTOR - [el] (a single-phase motor with a commutator which is started as a repulsion motor and is ultimately run as an induction motor with short-circuited commutator segments) moteur *m* à induction-répulsion
~MOTOR - [el] (a single-phase motor with a commutator in which one or several pairs of brushes are short-circuited) moteur *m* à répulsion
~MOTOR WITH DOUBLE SET OF BRUSHES - [el] (a repulsion motor with two sets of brushes, one of which is fixed and the other movable) moteur *m* à double jeu de balais, moteur *m* Deri
~-START INDUCTION MOTOR - [el] (repulsion motor with centrifugal device wich short-circuits all the commutators bars when the motor reaches a certain speed) moteur *m* à induction avec démarrage à répulsion
~-TYPE INSTRUMENT - [instr] (type of moving-iron instrument in which the moving is so shaped, that a highly uniform scale is obtained) instrument *m* à répulsion
REPULSIVE FORCES - [phys] (forces between bodies which tend to move them apart) force *fpl* de répulsion
~POTENTIAL - [nucl] (the force between two atoms keeping them apart at short distances is borne of the overlapping of the electron clouds) potentiel *m* de répulsion
REQUEST, to - [gen] demander

REQUEST - [gen] demande *f*, requête *f*
~BUTTON - [comput] touche *f* d'interrogation
~SIGNAL - [comput] signal *m* d'interrogation
~STOP - [comput] (stop enabling the operator to stop the programme on any specified instruction) arrêt *m* arbitraire
~TEST OF A METER - [instr] essai *m* à la demande d'un compteur
REQUIRE, to - [gen] demander, réclamer
REQUIRED - [gen etc] exigé, requis
~POWER - [mech etc] puissance *f* exigée
REQUIREMENT - [gen] demande *f*, réclamation *f*, exigence *f*
(that which is required) besoin *m*, nécéssité *f*
REQUISITES - [gen] fournitures *fpl*
REQUISITION, to - [gen] réquisitionner
REQUISITION - [gen] demande *f*, réquisition *f*
RERADIATION - [phys] (the scattering of incident radiation) rerayonnement *m*
RERAILING - [railw] remise *f* sur rails, relevage *m*
~RAMP - [railw] rampe *f* d'enraillement
RERECORD, to - [el acoust] (the operation of making a recording by reproducing a recorded sound source and recording this reproduction) réenregistrer
RERECORDING - [el acoust] réenregistrement *m*
~MIXING CONSOLE - [cin] table *f* de mixage de réengistrement
~ROOM - [el acoust] salle *f* de mixage
~SYSTEM - [el acoust] système *m* de réenregistement
REREEL, to - [text] redévider
REREELED RAW SILK - [text] soie *f* grège redévidée
REREELING - [gen] (the operation of rewinding) rebobinage *m*
[text] (of raw silk) redévidage *m* (de la soie grège)
~MACHINE - [mech] (mechanism for winding continuous material on to a reel or bobbin after it has been unwound for processing) rebobineuse *f*
RERING, to - [mech] (e.g. a piston) renouveler les segments
REROUTE, to - [telecomm] détourner, dévier
REROUTE THE TRAFFIC, to - [telecomm] détourner le trafic
REROUTING - [hydr] dérivation *f*
RERUN, to - [comput] (to run a programme over again on the computer) repasser
[mining] redescendre dans le puits
RERUN - [ind chem] redistillation *f*
~OIL - [oil ind] huile *f* redistillée
~POINT - [comput] point *m* de reprise, point *m* de répétition
~ROUTINE - [comput] (routine designed to be used following a mistake, in order to reconstitute a routine from the last previous rerun point) programme *m* de reprise
~UNIT - [oil ind] (in oil refineries) installation *f* de redistillation
~TOWER - [oil ind] colonne *f* de redistillation
RERUNNING - [oil ind] redistillation *f*
RESCINNAMINE - [chem] (an antihypertensive agent used in medicine) rescinnamine *f*
RESCREEN, to - [min] repasser au crible
RESCUE, to - [gen] sauver, porter secours
RESCUE - [gen] sauvetage *m*
~AIRCRAFT - [aero] avion *m* de secours
~CO-ORDINATION CENTRE - [gen] (a centre set up for search and rescue operations) centre *m* de coordination de sauvetage
RESCUE CRAFT - s. rescue aircraft
~SEAPLANE - [aero] hydravion *m* de secours
~SHIP - [naut] navire *m* de secours
~STATION - [gen] poste *m* de sauvetage
RESCUTCHED TOW - [text] étoupe *f* teillée
RESEARCH - [gen] recherche *f*
~CENTRE - [gen] centre *m* de recherche
~REACTOR - [nucl] (reactor designed for the study of scientific subjects) réacteur *m* de recherche
~STATION - s. research centre
RESEAT A VALVE, to - [mech] (to cut a new seat for a valve, as distinct from grinding-in or from renewing a removable seat) roder une soupape, fraiser le siège d'une soupape
RESEAU - [astr] (used for stellar photography) réseau *m* (de carrés de repère)
RESECTION - [med] (the cutting off of part an organ) résection *f*
RESEEDING - [agric] ressemis *m*
RESELL, to - [gen & comm] revendre
RESEMBLE, to - [gen] ressembler
RESEMBLANCE - [gen] (relative identity) ressemblance *f*
RESERPINE - [chem] (a natural alkaloid obtained from species of Rauwolfia and used in medicine as an antihypertensive and as a sedative in psychological medicine) réserpine *f*
RESERVATION - [gen] réserve *f*
[leg] réservation *f*
[gen] (a tract of government land) terrain *m* réservé
RESERVE, to - [gen] réserver
RESERVE - [gen] réserve *f*, restriction *f*
[fin] (of liquid assets etc) réserve *f* (d'argent), stocks *mpl*
[comm] réserve *f* de caisse
~BARS - [el] barres *fpl* de réserve
~BUOYANCY - [aero] (the amount by which the buoyancy of the floats of a seaplane exceeds its total weight, when the floats are wholly immersed) réserve *f* de flottabilité
~CELLULOSE - [bot] (cellulose present in endosperm) réserve *f* cellulosique
~CIRCUIT - [telecomm] circuit *m* de secours
~CYLINDER - [text] tambour *m* de réserve
~FACTOR - [aero] (the relation between the actual strength of a structure and the minimum strength called for to satisfy given conditions) marge *m* de sécurité
~FUND - [fin] fonds *mpl* de réserve
~PARACHUTE - [aero] (a second parachute for use if the normal parachute fails) parachute *m* de secours
~PILLARS - [mining] massif *m* en réserve
~PROTECTION - [el] protection *f* de réserve
~PROTEIN - [chem] protéine *f* de réserve
~SIDING - [railw] (a storage track) voie *f* de remisage
~TANK - [aero] (a tank used to contain extra fuel etc. for use if required) réservoir *m* de secours
RESERVED WORD - [comput] (in programming) mot *m* réservé
RESERVOIR - [gen] (the tank which contains the fluid on which a hydraulic system operates) réservoir *m*, bassin *m*
[el] (of a transformer; an oil conservator) réser-

voir *m* de niveau
RESERVOIR - [mech] (attachment to an air brake mechanism) réservoir *m*
[hydr] bassin *m* de retenue, décharge *f*
~PRESSURE - [gas ind & oil ind] pression *f* de gisement
~ROCK - [oil ind] (formations containing interconnected pores or fissures which may serve as reservoirs for oil or gas) roche *f* de transgression, roche-magasin *f* primaire
RESET, to - [gen] replacer, remonter, retourner
[print] recomposer
[instr] remettre à zéro
[comput] (to return a register to zero or to a given initial condition) effacer, remettre
RESET - [contr] (initial condition state) condition *f* initiale
~A SPRING, to - [mech] rebander un ressort
~A CLUTCH, to - [auto] refaire le réglage de l'embrayage
~ACTION - [contr] action *f* proportionnelle et par intégration
~BUTTON - [comput] touche *f* de rappel
~CONTROL CIRCUIT - [telecomm] (of a magnetic amplifier) circuit *m* de réglage de la reposition
~COUNTER - [instr] (a counting mechanism which can be returned to zero reading, usually by a manual control) compteur *m* avec remise au zéro
~DWELL TIME - [contr] temps *m* de remise à zéro
~FLUX LEVEL - [electron] niveau *m* de flux à la reposition
~LINE - [comput] ligne *f* de rappel
~MAGNET - [el] aimant *m* de réenclenchement
~PULSE - [comput] (in static magnetic storage; drive pulse tending to reset a magnetic cell) impulsion *f* de remise
~RACK DEVICE - [text] dispositif *m* de réajustage
~THE FLATS, to - [text] retoucher le réglage des chapeaux
~TIME - [electron] (the time which is required in a decade scaler tube to move the electron beam back to the starting position) durée *f* de retour
[contr] (integral action time) temps *m* d'action par intégration
~WINDING - [electron] (drive winding of a magnetic core) conducteur *m* d'excitation du core magnétique
[comput] enroulement *m* d'effacement
RESETTER - [contr] (monitoring element) convertisseur *m* de signal de sortie
RESETTING - [gen] remontage *m*, remise *f* en place
[print] recomposition *f*
[instr] remise *f* à zéro, réenclenchement *m*
~DEVICE - [contr] dispositif *m* de réarmement
~HALF-CYCLE - [el] (of a magnetic amplifier) demi-cycle *m* de reposition
~INTERVAL - [electron] (the interval of the cycle in which the saturable reactor core flux level is changing from the saturation level to the reset flux level) intervalle *m* de reposition
~PUSH BUTTON - [instr] bouton *m* poussoir de réarmement
~RESPONDING RATION - [contr] pourcentage *m* de retour
~THE SPRINGS IN TENSION - [mech] remise *f* sous tension du ressort
~VALUE - [contr] valeur *f* de retour
RESETTLE, to - [agric] réinstaller, transplanter
RESETTLEMENT - [agric] transplantation *f*
~ALLOWANCE - [agric] indemnité *f* de réinstallation
RESETTLING - s. resettlement
RESIDENCE TIME - [ind chem] (the period during which a material is allowed to remain in a certain device, e.g. a cooling bath) durée *f* du traitement
RESIDENTAIL AREA - [gen] quartier *m* résidentiel
~DENSITY - [gen] densité *f* d'habitation
RESIDUAL - [gen] résiduel, résiduaire
[gen] résidu *m*
[math] résidu *m*
~ACTIVITY - [nucl] (the radioactivity which remains in a substance after a period of decay) activité *f* résiduelle
~AFFINITY - [chem] (after the saturation of the normal valencies of the atoms forming a molecule, some chemical attractions remain, which give rise to molecular compounds. Such forces constitute Residual Affinity) affinité *f* résiduelle
~AIR - [physiol] air *m* résiduel
~BLUE - [opt] (phenomenon which is observed with white light scattered by small particles in suspension) lumière *f* bleue de Tyndall
~CHARGE - [el] charge *f* résiduelle
~CONCENTRATION - [ind chem] (in debenzolizing) charge *f* résiduaire, retentivité *f*
~CURRENT - [el] (the current of which it is possible to pass through an electrolyte without the production of apparent change in the electrodes) courant *m* résiduel
~CURRENT STATE - [electron] régime *m* de courant résiduel
~DEPOSITS - [geol] (accumulations of rock waste) dépôts *mpl* résiduaires
~ELONGATION - [rubber ind] (permanent elongation) allongement *m* rémanent
~ERRORS - [math] erreurs *fpl* résiduelles
~FIELD - [el] (the magnetic field remaining in a magnetic circuit after the magnetizing force has been removed) champ *m* de magnétisme rémanent
~FLUX DENSITY - [el] (the value of flux density which persists after the magnetizing force has been removed) induction *f* rémanente
~FUEL OIL - s. residual oil
~GAS - [electron] (the minute amount of gas inevitably present in a vacuum tube) gaz *m* résiduel
~INDUCTION - [el] (induction in a magnetic sample after a saturating magnetizing force has been removed) induction *f* rémanente
~IONIZATION - [nucl] (of air other gas) ionisation *f* résiduelle
~LOSS - [el] effet *m* magnétique secondaire
~MAGNETISM - [el] (the magnetism remaining in a substance after the magnetising force has been removed) rémanence *f*, magnétisme *m* rémanent
~MAGNETIZATION - [el] (remanence) aimantation *f* rémanente
~MODULATION - [radio] (carrier noise level) niveau *m* de bruit de porteuse
~NUCLEUS - [nucl] (the heavy nucleus which is the end product of transformation) noyau *m* résiduel
~OIL - [oil ind] résidu *m* de distillation, pétrole *m* restant
~ON EVAPORATION - [chem] résidu *m* sec
~PLATE - [el] pastille *f* antirémanente
~RADIATION - [opt] rayonnement *m* résiduel
~RANGE - [nucl] (the distance over which the parti-

cle can still produce ionization after the loss of some of its energy in passing through matter) portée *f* résiduelle

RESIDUAL RELAY - [el] relais *m* fonctionnant aupoint nul de la phase

~RESISTANCE - [el] (the part of the electrical resistance of a metal which is independent of the temperature) résistance *f* résiduelle

~STRESS - [metall] effort *m* résiduel

~STRESS CRACK - [metall] crique *f* par effort résiduel

~TACK - [chem] (a stage of drying in coating compositions, in which the coating can be handled without damage, but still retains a measure of stickiness) adhésivité *f* résiduelle

~VOLTAGE - [el] (the balance of interfering voltage in a communication circuit, caused by an adjacent power line, after both lines have been effectively transported) tension *f* résiduelle

RESIDUE - [gen] résidu *m*, reste *m*

~FUEL OIL - s. residual oil

~OF COMBED SLIVER - [text] restant *m* d'un ruban de peigné

RESIDUES - [chem] (e.g. used photographic solutions etc) residu *m*

RESIDUUM - [ind chem] (of distillation) résidu *m*

~AFTER COMBUSTION - [chem] résidu *m* de combustion

~OIL - s. residual oil

RESILIENCE - [phys] (the degree to which a body can resume its original shape after removal of a deforming stress) résilience *f*, résistance *f* vive

~METER - [instr] rebondimètre *m*

~TEST - [mech] essai *m* de résilience

RESILIENT - [phys] résilient, rebondissant, élastique

~RUBBER WEBBING BASE - [rubber ind] (for upholstery) support *m* résilient en sangles de caoutchouc entrelacées

~SLEEPER-PAD - [railw] semelle *f* de rail élastique

~TIE-PAD - s. resilient sleeper pad

~WASHER - [mech] rondelle *f* flexible

RESILIOMETER - s. resilience meter

RESIN - [chem] (natural; solid or semi-solid viscous materials derived from the secretions of trees and shrubs. Synthetic; semi-solid or solid amorphous substances obtained by the polymerization or condensation of one or more simple compounds) résine *f*

~BASED ADHESIVE - [ind chem] colle *f* à base de résine

~BINDER - [ind chem] charge *f*

~-BONDED - [ind chem] lié par la résine

~BONDED PLYWOOD - [constr] (plywood in which the veneers are bonded with a synthetic resin adhesive) contre-plaqué *m* à la résine

~CANAL - [bot] (intercellular space) canal *m* résinifère

~ESTERS - [chem] (ester gums) esters *mpl* résineux

~FLUX - [bot] (abnormal escape of resin) écoulement *m* de résine

~-IMPREGNATED - [ind chem] imprégné de résine

~-LIKE YOLK - [text] suint *m* résineux

~MILK - [ind chem] (an adhesive used in paper manufacturing) lait *m* de résine

~OIL - [ind chem] huile *f* de résine

~PLANT - [bot] plante *f* résineuse

~POCKET - [plast ind] (locally excess resin occurring in mouldings) poche *f* de résine

RESIN - RICH AREA - [plast ind] (region in a moulding in which there is an excessive proportion of resin) zone *f* riche en résine

~-SCUM - [brew ind] amers *mpl*

~SOAP - [chem] résinate *m* de soude, savon *m* de résine

~-STARVED AREA - [plast ind] (region in a preform moulding in which there is insufficient binding material) zone *f* pauvre en résine

~WOOL - [ind chem] (material which is used in exit air filtration) laine *f* de résine

RESINATES - [chem] (salts of rosin obtained by fusing it with the oxides of certain metals) résinates *mpl*

RESINIFICATION - [chem] résinification *f*

RESINOGENETIC - [bot] (giving rise to resin) résinogène

RESINOGENIC - s. resinogenetic

RESINOID - [chem] (term sometimes used for a synthetic thermosetting resin) résinoïde *m*

RESINOLIC ACID - [chem] acide *m* résinolique

RESINOLS - [chem] (a coal tar fraction containing phenols and soluble in benzene) résinols *mpl*

RESINOUS - [chem] (of the nature of resins) résineux

~COLOURING MATTER - [chem] colorant *m* résineux

~EXCHANGER - [ind chem] échangeur *m* à résine synthétique

~SOAP - [chem] savon *m* résineux

RESINS - [chem] (these may be Natural Resins or Synthetic Resins. They are semi-solid viscous substances consisting largely of high polymers) résines *fpl*

RESINTERING - [metall] refrittage *m*

RESIST, to - [gen] résister

RESIST - [el] (any substance used to render any part of a cathode or plating rack non-conducting) matière *f* de protection
[gen] (a protective coating for application to any region of a surface which is not to be affected by some process applied to the whole, e.g. in etching or selective dyeing) matière *f* de protection

RESISTANCE - [gen] résistance *f*
[el] (the property of a substance by which it resists the flow of electric current) résistance *f*
[el] (piece of apparatus possessing resistance) résistance *f*, rhéostat *m*
[mech] résistance *f*
[med & biol] (to attacks of a disease or of a parasite) résistance *f*
[med] (to pressure, of a tumour) rénitence *f* (d'une tumeur)

~AND INSULATION MEASUREMENT - [el] mesure *f* de la résistance et de l'isolement

~AT EXHAUST - [mech] résistance *f* à la décharge

~BOX - [el] (the box containing the resistors) boîte *f* de résistances

~BOX WITH PLUGS - [el] boîte *f* de résistance à fiches

~BRAZING - [metall] soudure *f* forte par résistance

~BRIDGE - [el] pont *m* de Wheatstone

~BUTT-WELDING - [metall] (resistance welding process in which the components are butted together and so maintained under pressure until the weld is complete) soudage *m* en bout par résistance

~CAPACITANCE COMPARISON BRIDGE - [instr] (instrument designed to compare resistors, capacitors

or inductors against specified standards) pont *m* de référence de résistance et capacité
RESISTANCE CAPACITANCE COUPLING - [el] couplage *m* par résistance-capacité
~-CAPACITANCE NETWORK - [electron] réseau *m* RC
~COIL - [el] bobine *f* de résistance
~COMPONENT - [el] composante *f* ohmique
~CONTACT - [el] contact *m* de résistance
~-COUPLED AMPLIFIER - [radio] (amplifier with resistance coupling between successive stages) amplificateur *m* à résistance
~COUPLING - [radio] (the association of two or more circuits by a resistance which is mutual to the circuits) couplage *m* par résistance
~-CUT OUT SWITCHGROUP - [el] combinateur *m* d'élimination des résistances
~DIAGRAM - [el] diagramme *m* des résistances
~DROP - [el] (with alternating current, the component of the voltage drop which is in phase with the current and equals the current in amperes multiplied by the resistance in ohms between the two points) chute *f* de tension par résistance ohmique
~EARTHED - [el] mise *f* à la terre par une résistance
~ENERGY - [mech] travail *m* résistant
~ETCHING - [metall] gravure *f* par résistance électrique
~FLASH-WELDING - [el metall] soudage *m* en bout par résistance
~FLASH-BUTT WELDING - [metall] (resistance welding in which the components are brought together after the voltage has been applied, so that sparking and local arcing take place heating the contact area progressively and burning off portions of them) soudage *m* en bout par étincelage
~FRAME - [el] (frame containing a number of resistors) résistance *f* à cadre
~FURNACE - [th eng] (electric furnace heated by resistance effect) four *m* à résistance
~GRADUATION - [el] graduation *f* d'une résistance
~GRID - [el] (resistance unit used for heavy currents) grille *f* de résistance
~HEATING - [el] (heating by means of the heat produced in a conductor by the Joule effect) chauffage *m* par résistance
~IN DIRECT CURRENT - [el] résistance *f* en courant continu
~IN PARALLEL - [el] résistance *f* en parallèle
~IN SERIES - [el] résistance *f* en série
~INSTRUMENT - [instr] instrument *m* de mesure à résistance
~LAP WELDING - [metall] soudage *m* par résistance par recouvrement
~MAGNETOMETER - [instr] (magnetometer depending for its operation on the variation of electrical resistance of material immersed in the magnetic field to be measured) magnétomètre *m* à résistance
~MANOMETER - [instr] manomètre *m* à résistance
~OF AN EARTHED CONDUCTOR - [el] résistance *f* de terre
~OF MATERIALS - [phys] résistance *f* des matériaux
~OF THE MINE - [mining] (to ventilation) résistance *f* de la mine
~OF THE THREAD - [text] résistance *f* du fil
~OVEN - s. resistance furnace
~PER UNIT LENGTH - [el] isolement *m* linéique
~PERCUSSIVE WELDING - [metall] soudage *m* par percussion par résistance
RESISTANCE PROJECTION WELDING - [metall] soudage *m* par bossages
~PYROMETER - [instr] (pyrometer functioning on the variation of the resistance of a wire with temperature) pyromètre *m* à résistance
~RELAY - [el] relais *m* de résistance
~SEAM WELDING - [metall] soudage *m* à la molette, soudage *m* en ligne continue, soudage *m* au galet
~SPOT WELDING - [metall] soudage *m* par résistance par points
~STANDARD - [el] résistance *f* étalon
~-START MOTOR - [el] moteur *m* à démarrage par résistance
~STRAIN GAGE - s. resistance strain gauge
~STRAIN GAUGE - [instr] (strain gauge in which use is made of the resistance variation) extensomètre *m* à résistance
~SWITCHGROUP - [el] éliminateur *m*
~-TEMPERATURE DETECTOR - [instr] indicateur *m* de résistance-température
~-TEMPERATURE INDICATOR - s. resistance-temperature detector
~TEMPERATURE METER - s. resistance pyrometer
~THERMOMETER - [instr] (electric thermometer using a resistance to which the heat is transmitted directly by conduction) thermomètre *m* à résistance
~TO ALTERNATING CURRENT - [el] résistance *f* effective, résistance *f* en courant alternatif
~TO ATMOSPHERIC CORROSION - [metall] résistance *f* à la corrosion atmosphérique
~TO BENDING - [metall] résistance *f* à la flexion
~TO BENDING STRESS - [mech] résistance *f* à l'effort de flexion
~TO BRAKING - [mech] résistance *f* au freinage
~TO BUCKLING - [metall] résistance *f* au flambage
~TO COMPRESSIVE STRESS - [phys] résistance *f* à la compression
~TO CRACKING - [rubber ind] résistance *f* aux craquelures
~TO CRUSHING - [mech] résistance *f* à l'écrasement
~TO FATIGUE - [mech] résistance *f* à la fatigue
~TO FLEXURE - [phys] résistance *f* à la flexion
~TO FLOW - [chem] résistance *f* à l'écoulement des liquides
~TO FRICTION - [mech] résistance *f* à la friction
~TO IMPACT - [metall] résistance *f* au choc
~TO LOAD - [phys] résistance *f* à la charge
~TO MOTION - [mech] résistance *f* passive
~TO OVERHEATING - [th eng] résistance *f* à la surchauffe
~TO OXYGEN - [rubber ind] résistance *f* à l'oxygène
~TO ROLLING - [rubber ind] résistance *f* au roulement
~TO RUBBING - [text] résistance *f* au frottement
~TO RUPTURE - [metall] résistance *f* à la rupture
~TO SCRAPING - [rubber ind] résistance *f* au grattage
~TO SEPARATION - [rubber ind] (in tyres) résistance *f* à la séparation (entre plis)
~TO SHEARING - [phys] résistance *f* au cisaillement
~TO SHEARING STRESS - [phys] résistance *f* à l'effort de cisaillement
~TO SLIDING - [mech] résistance *f* au glissement
~TO SOLVENTS - [ind chem] résistance *f* aux solvants

RESISTANCE TO SUCTION - [phys] résistance *f* à l'aspiration
~TO SWELLING - [rubber ind] résistance *f* au gonflement
~TO TAKING UP - [text] résistance *f* à l'enroulement
~TO TEAR - [rubber ind] résistance *f* à la déchirure
~TO TEARING - [mech] résistance *f* à l'arrachement
~TO TENSILE STRESS - [metall] résistance *f* à l'effort de traction
~TO TENSION - [mech] résistance *f* à la traction
~TO TENSIVE STRESS - s. resistance to tensile
~TO TORSIONAL STRESS - [metall] résistance *f* à la torsion
~TO TWISTING - [metall] résistance *f* à la torsion
~TO UNWINDING - [text] résistance *f* au dévidage
~TO WEAR - [gen & mech] résistance *f* à l'usure
~TO WEATHER - [gen] résistance *f* aux intempéries
~TRANSFORMER - [el] transformateur *m* accordé
~UNIT - [el] groupe *m* de résistances
~VOLTAGE - [el] (the voltage which is responsible for the resistance drop) tension *f* de résistance
~WELDER - [metall] machine *f* à souder par résistance
~WELDING - [metall] (pressure welding which the heat causing the fusion of the metals is produced by the welding current flowing through the contact resistance between the two surfaces to be welded) soudage *m* par résistance
~WELDING ELECTRODE - [el] (for resistance welding, an electrode which conveys the current to one of the pieces to be welded) électrode *f* pour soudage par résistance
~WIRE - [el] (wire of special composition used to provide electrical resistance in a circuit) fil *m* résistant
RESISTANT - [gen] résistant (à)
RESISTER - [gen & chem] corps *m* résistant, force *f* résistante
RESISTING - [gen] résistant
~COUPLE - [mech] couple *m* résistant
~MOMENT - [phys] moment *m* de résistance
~POWER - [phys] puissance *f* de résistance
RESISTIVE D.C. VOLTAGE DROP - [el] chute *f* ohmique de tension continue
~FEEDBACK - [electron] réaction *f* résistive
~LOAD - [el] charge *f* résistante
~WALL AMPLIFIER - [electron] (electron-beam amplifier in which the beam flows near a resistive wall) tube *m* à ondes progressives à paroi dissipative
RESISTIVITY - [el] (volume resistivity, or specific resistance; the resistance between opposite faces of a unit cube of a given material at a given temperature) résistivité *f*
~CURVE - [el] courbe *f* de résistivité
RESISTOR - [el] (a component in an electric circuit designed to contribute a known amount of resistance to such circuit) résistance *f*
~ALLOY - [metall] alliage *m* pour résistances
~CORE - [el] mandrin *m* de résistance
~ELEMENT - [el] élément *m* résistant
~FURNACE - [el] four *m* à chauffage indirect par résistance
~HOUSING - [el] carter *m* de résistance
RESNATRON - [electron] (electronic tube used for jamming a radar transmission) resnatron *m*
RESOJET ENGINE - [aero] (a pulsojet operated on resonance) pulsoréacteur *m* à résonance
[aero] (ramjet engine operated on resonance) stato-réacteur *m* à résonance
RESOL - [chem] (a synthetic resin) résol *m*
~RESINS - [chem] (synthetic resins obtained by reaction of a phenol and an aldehyde) résines *f*pl de résol
RESOLUTION - [chem] (the return into solution of metal already deposited) redissolution *f*, résolution *f*
[opt] (resolving power) résolution *f*
[phys] (of a force) résolution *f*, définition *f*
[phys chem] (the process of separating closely related forms or entities) résolution *f*
[gen & leg] délibération *f*
[telev] (the capacity of a television system to reproduce details) définition *f*
[radar] pouvoir *m* résolvant
[med] résolution *f* (d'une tumeur etc)
~OF THE FORCES - [phys] décomposition *f* des forces
~OF WIND PRESSURE - [met] décomposition *f* de la pression du vent
~PATTERN - [telev] mire *f* de définition
~TIME - [radiat] temps *m* de résolution
~WEDGE - [telev] coin *m* de définition
RESOLVE, to - [gen] résoudre
[chem] (to separate an optically inactive mixture into its optically active components) séparer, résoudre
[phys etc] décomposer
RESOLVER - [comput] (device for resolving a vector into two mutually perpendicular components) séparateur *m*
[contr] (asynchro principally designed as a calculating element) résolver *m* d'équations
RESOLVING - [opt etc] résolvant
~A DISCORD - [acoust] résolution *f* d'une dissonance
~POWER - [opt] (the ability of a lens to separate the images of close objects) pouvoir *m* séparateur
~POWER OF A LENS - [opt] (the ability of a lens to register very fine details) pouvoir *m* résolvant d'un objectif
~POWER OF THE EYE - [opt] (the angle subtended by a small object which can just determined visually) pouvoir *m* résolvant de l'oeil
~POWER TEST TARGET - [opt] mire *f* de résolution pour lentilles
RESONANCE - [phys] (the property of certain molecular structures to remain in an essentially fixed spatial configuration with their electrons satisfying one or more structural formulae) résonance *f*
[acoust] résonance *f*
[mech] (in a vibrating system, the synchronism of some harmonic of the forcing impulses with the natural frequency of vibration of the system) résonance *f*
[el] (state of balance between positive and negative reactance of a circuit) résonance *f*
~ABSORPTION - [nucl] (the absorption of neutrons having energies which correspond to a nuclear resonance level of the absorber) absorption *f* par résonance
~ABSORPTION SPECTRUM - [phys] spectre *m* de résonance
~AMPLIFIER - [radio] amplificateur *m* à résonance

RESONANCE BAND - [radio] bande *f* de résonance
~BRIDGE - [el] pont *m* à résonance
~CAPACITOR TRANSFORMER - [el] transformateur-condensateur *m* à résonance
~CAPTURE - [nucl] (capture of an incident particle into a resonant level of the resultant compound nucleus) capture *f* de résonance
~CONCEPT OF NUCLEAR REACTION - [nucl] concept *m* de résonance de la réaction nucléaire
~CROSS SECTION - [nucl] (the effective cross-section for the capture of an incident into a resonance level of the resultant compound nucleus) section *f* efficace de résonance
~CURVE - [radio] (curve showing the variations of current in a resonant circuit) courbe *f* de résonance
~ENERGY - [nucl] (the kinetic energy of a particle which will be captured preferentially owing to the presence of a suitable resonance level in the compound nucleus) énergie *f* de résonance
~ESCAPE PROBABILITY - [nucl] probabilité *f* d'échappement de résonance, facteur *m* antitrappe
~FACTOR - [radio] coefficient *m* de résonance
~FISSION - [nucl] fission *f* de résonance
~FLUORESCENCE - [phys] (the emission by an excited atom of a radiation having the same frequency as the exciting radiation) fluorescence *f* de résonance, rayonnement *m* de résonance
~FREQUENCY - s. resonant frequency
~FREQUENCY METER - [el] fréquencemètre *m* à résonance
~IMPEDANCE BOND - [el] connexion *f* inductive à résonance
~INTEGRAL - [nucl] (the negative of the natural logarithm of the resonance escape probability multiplied by the slowing-down power per atom of absorber) intégrale *f* de résonance
~LEVEL - [nucl] (nuclear resonance level, an excited level of the compound system capable of being formed in a collision between two systems) niveau *m* de résonance nucléaire
~NEUTRONS - [nucl] (intermediate neutrons) neutrons *mpl* de résonance
~OF THE ATOM - [phys] résonance *f* de l'atome
~PEAK - [nucl] (in the cross-section curve occurring at certain resonance energies) pic *m* de résonance
~PENETRATION - [nucl] pénétration *f* de résonance
~RADIATION - [phys] (fluorescence in which the exciting radiation is of the same frequency as that of fluorescence) rayonnement *m* de résonance
~REGION - [nucl] région *f* des énergies de résonance
~SCATTERING - [nucl] diffusion *f* résonante
~SPECTRAL LINE - [phys] raie *f* spectrale de résonance
~SPECTRUM - [spectrology] (spectrum excited by the interaction with a substance of radiation of a definite frequency) spectre *m* de résonance
~STATE - [phys] (a state from which the atom can return directly to the normal energy level by radiation) niveau *m* de résonance
~STEP-UP - [radio] (the ratio between the voltage appearing across the condenser of a parallel tuned circuit and the electromagnetic force acting around the circuit when this is resonant at the applied frequency) surtension *f* de résonance
~TEST - [aero] (test made to determine the mechanical natural frequency and type of oscillation of a structure) essai *m* de résonance
RESONANT - [gen] résonnant
[electron] résonnant, de résonance
~AERIAL - [radio] antenne *f* résonnante
~CAVITY - [electron] cavité *f* résonnante
~-CAVITY DIELECTROMETER - [instr] diélectromètre *m* à cavité résonnante
~-CAVITY MASER - [electron] (maser in which the paramagnetic active material is placed in a cavity resonator) maser *m* à cavité résonnante
~-CHAMBER SWITCH - [electron] (a waveguide switch) commutateur *m* à cavité résonnante
~CIRCUIT - [el] (circuit comprising an inductance coil and a condenser in series or parallel) circuit *m* résonnant
~-CIRCUIT DRIVE - [electron] (type of master oscillator) maître-oscillateur *m* à commande électronique
~DIAPHRAGM - [electron] diaphragme *m* résonnant
~FILTER - [electron] filtre *m* résonnant
~FREQUENCY - [el] fréquence *f* de résonance
~GAP - [electron] intervalle *m* résonnant
~IRIS - [electron] (resonant window in a circular waveguide) diaphragme *m* iris résonnant
~-IRIS SWITCH - [electron] commutateur *m* à diaphragme iris résonnant
~LINE - [radio] (used for stabilizing the frequency of short-wave oscillators) ligne *f* résonnante
[electron] (of a waveguide) ligne *f* accordée
~-LINE OSCILLATOR - [radio] oscillateur *m* à lignes résonnantes
~MODE - [electron] mode *m* résonnant
~RING DIPLEXER - [electron] (diplexer using a waveguide hibrid junction) mélangeur *m* d'antenne à anneau résonnant
~RISE OF A SIGNAL - [radio] (step-up ratio of a signal) accroissement *m* d'oscillation du signal
~SHUNT - [el] (tuned circuit placed in parallel with parts of the apparatus) shunt *m* résonnant
~WINDOW - [electron] fenêtre *f* résonnante
~-WINDOW SWITCH - [electron] commutateur *m* à fenêtre résonnante
RESONATE, to - [phys] résonner
RESONATING PIEZOID - [el] (piezoid used as a resonator) piézoïde *m* résonnant
RESONATOR - [acoust] (pipe or cavity etc exhibiting acoustic resonance) résonateur *m*
[radio] (any device which exhibits a sharply defined resonance) circuit *m* résonnant
[phys] (any device utilizing the effect of resonance) résonateur *m*
~CURRENT - [el] (a brush discharge having a 2 to 10 K-V drop in the air, from a monopolar electrode, generated by a special arrangement of transformers sparkgaps and capacitors, dense enough to evaporate tissuewater without charring) courant *m* d'Oudin
~GRID - [electron] (electrode, connected to a resonator, which is traversed by an electron beam and provides the coupling between the beam and the resonator) grille *f* de résonateur
~MODE - [electron] mode *m* de résonateur
RESONOSCOPE - [acoust] (instrument designed to show in visual terms any desired note) résonoscope *m*
RESORB, to - [chem] résorber
RESORCIN - s. resorcinol

RESORCINOL - [chem] (a very reactive phenol having two hydroxy groups, which when reacted with formaldehyde, gives resins suitable as cold setting adhesives. Resorcinol also has applications in the manufacture of rubber chemicals and camphor substitutes) résorcine *f*
~DIGLYCIDYL ETHER - [chem] (a starting material for epoxy resins) éther *m* diglycidylique de résorcine
~MONOACETATE - [chem] (an antipyretic with uses in medicine) monoacétate *m* de résorcine
~MONOBENZOATE - [chem] (an UV-light stabilizer for a number of plastics) monobenzoate *m* de résorcine
~TEST - [chem] (a method for detecting the presence of phthalic acid in alkyd resins and phthalate plasticizers) essai *m* à la résorcine
RESORCYLIC ACID - [chem] (an intermediate for pharmaceuticals, dyestuff, synthetic resins, and plasticizers) acide *m* récorcylique
RESORPTION - [phys chem] (the absorption by a body of material previously released from absorption by that same body) résorption *f*
RESORTING THE FLAX - [text] retriage *m* du lin
RESOUND, to - [acoust] resonner, retentir
RESOURCE - [gen] ressource *f*
RESOURCES - [fin] ressources *fpl*
RESOWING - [agric] ressemis *m*
RESPIRATION - [zool] (the interchange of oxygen and carbon dioxide associated with katabolic process in an aerobic organism) respiration *f*
~RATE - [physiol] vitesse *f* de respiration
RESPIRATOR - [med] masque *m* respiratoire
RESPIRATORY - [zool & bot] respiratoire
~CAVITY - [bot] (intercellular space beneath a stoma) cavité *f* respiratoire
~CENTRE - [zool] (nerve-centre of the hind-brain regulating the respiratory movements) centre *m* respiratoire
~CHAMBER - s. respiratory cavity
~CHROMOGEN - [bot] chromogène *m* respiratoire
~COMPLIANCE - [physiol] adaptation *f* respiratoire
~INDEX - [bot] (the number of milligrams of carbon dioxide freed from one gram of plant material when the temperature is I0 centigrades) indice *m* respiratoire
~MINUTE VOLUME - [astronaut] débit *m* respiratoire par minute
~MOVEMENTS - [zool] (the muscular movement which are associated with the supply of air to the respiratory organs) mouvements *mpl* respiratoires
~ORGANS - [zool] organes *mpl* respiratoires
~PIGMENT - [zool] (substance in the blood enabling an adeguate amount of oxygen to be conveyed to the tissues) pigment *m* respiratoire
~QUOTIENT - [zool & bot] (the ratio between the volume of carbon dioxide given off and that of the oxygen taken in during a specified period of time) quotient *m* respiratoire
~RATIO - s. respiratory quotient
~SYSTEM - s. respiratory organs
~TUBE - [zool] (median ventral tube by which water passes from the gullet to the gills) tube *m* respiratoire
~VALVE - [zool] (membranous folds preventing the water escaping through the mouth during respiration) valve *f* respiratoire
RESPOND, to - [gen] répondre
[leg] être responsable
RESPONDER - [el] (electric wave detector) répondeur *m*
~BEACON - [radar] (a radio beacon which emits informative signals when actuated by special equipment in an aircraft) radiophare *m* répondeur
RESPONSE - [el & telecomm] (the output of a device produced from it by a given signal) réponse *f*
[gen] réponse *f*
~CURVE - [radio] (a graph showing the voltage developed across an electrical network in relation to the signal voltage producing it) courbe *f* de réponse
~MARK - [print] (music) signe *m* de réponse
~PULSE SHAPE - [radio] facteur *m* de conformation d'impulsions
~TIME - [instr] (of a measuring instrument) inertie *f*, temps *m* d'établissement, temps *m* d'arrêt
[electron] (of a relay) temps *m* de réponse
~TO A CHEMICAL STIMULUS - [chem] réaction *f* à un stimulant chimique
~TO CURRENT - [el acoust] (of an electroacoustic transducer used for sound emission and for a given frequency) réponse *f* de courant
~TO POWER - [el acoust] (of an electroacoustic transducer) réponse *f* de puissance
~TO VOLTAGE - [el acoust] (of an electroacoustic transducer) réponse *f* de tension
RESPONSIBILITY - [gen] responsabilité *f*
RESPONSIBLE - [gen] responsable
RESPONSIVE - [gen etc] sensible
RESPONSIVENESS - [gen & med] sensibilité *f*
[mech] flexibilité *f*
[mech] (of motors) nervosité *f*
[instr] sensibilité *f*
[photo] (of the emulsion) réponse *f* (de l'émulsion)
RESPONSOR - [radio] récepteur *m*
REST, to - [gen] reposer, se reposer
[gen] (to be supported, or fixed) s'appuyer, être posé
[leg] conclure
[mining] clichage *m*
REST - [gen] repos *m*
[gen] (a shelter) abri *m*
[gen] (that which remains) reste *m*, restant *m*
[gen] (a support, a base) appui *m*, support *m*
[comm] arrêté *m* (de compte)
[hydr] (of a contact bed) chômage *m*
~BLOCKS - [gas ind] (in gasholders) dés *mpl* de repos
~BUSHING - [mech] manchonnage *m* d'appui
~DENSITY - [mech] (of a fluid) densité *f* au repos
~FRAME - [mech] (Lorentz frame in which the total momentum of a system vanishes) système *m* de référence baricentrique
~-HARROW - [bot] bugrane *f*
~MASS - [phys] (the mass of a particle at rest) masse *f* au repos
~-MASS ENERGY - [phys] énergie *f* au repos
~MASS OF THE ELECTRON - [electron] masse *f* au repos de l'électron
~-PERIOD - [agric] temps *m* de repos
~POSITION - [mech] position *f* de repos
RESTART, to - [gen] recommencer
[mech] remettre en marche
RESTART - [mech] remise *f* en marche
~AN INJECTOR, to - [mech] réamorcer un injecteur
~POINT - [comput] point *m* de reprise

RESTART THE LOOM, to - [text] remettre en marche le métier
RESTARTING INJECTOR - [mech] injecteur *m* de mise en marche automatique
~INTERLOCK - [mech] dispositif *m* de remise en marche
RESTIFORM - [zool] (rope-like) restiforme
~BODY - [anat] pédoncule *m* cérébelleux inférieur
RESTING CONTACT - [el] contact *m* de repos
~FREQUENCY - [electron] (frequency of the carrier wave in the absence of modulation) fréquence *f* nominale
~POTENTIAL - [el biol] (the voltage existing between the two sides of a living membrane or interface in the absence of excitation) potentiel *m* de repos
RESTITUTION - [gen] restitution *f*
~ELEMENT - [radio] (a certain condition assumed by the appropriate device in the receiver, associated with the interval of time corresponding to the duration) élément *m* de restitution
RESTLESS LEGS - [med] acroparesthésie *f* nocturne
RESTOCK, to - [gen] remonter, regarnir (un magasin)
[agric] (in forestry) reboiser
[hydr] rempoissonner
RESTOCKING - [comm] remontage *m*, réassortiment *m*
RESTORATION - [gen mech etc] restauration *f*
~CONSTANT - [instr] (one of the intrinsic constants of a galvanometer) constante *f* d'équilibre
RESTORATIVE - [med] fortifiant *m*
RESTORE, to - [gen] restaurer
[gen] (to repair) réparer
[gen] (to return) restituer
[mech etc] remettre sur pied
[metall] régénérer
[med] rétablir (la santé)
[aero] (to its position) redresser
[comput] (to return a computer word to its initial value) remettre en état initial
~A CIRCUIT TO SERVICE - [telecomm] remettre un circuit en service
~THE RECEIVER, to - [telecomm] (of a telephone) raccrocher le récepteur
~THE SHAFTS TO THE CLOSED POSITION, to - [text] ramener les lames
~THE VENTILATION, to - [gen & mining] rétablir l'aérage
RESTORED WASTE - [rubber ind] régénéré *m*, caoutchouc *m* régénéré
RESTORER - [gen] restaurateur *m*, réparateur *m*
[el] (a device designed to reset to the original state) mécanisme *m* pour remettre en état initial
RESTORING - [gen] réparateur
~FORCE - [mech] (the elastic force acting on a particle of a mechanical system when displaced from equilibrium; its direction is such as to return the system to equilibrium) force *f* de rétablissement; force *f* de rappel
~FORCE GRADIENT - [phys] gradient *m* d'équilibre des forces
~MECHANISM - [mech] mécanisme *m* de rétablissement
~MOMENT - [aero] (a moment dependent on any rotational displacement, which tends to return the aircraft to its original attitude) moment *m* redresseur, couple *m* de rappel
~SPRING - [mech] ressort *m* de rappel
RESTORING TORQUE - [instr] (the torque which tends to bring the moving element back to the mechanical zero of the instrument) couple *m* antagoniste
~TORQUE GRADIENT - [instr] gradient *m* de couple antagoniste
RESTRAIN, to - [gen] retenir
RESTRAINED - [constr] encastré
~CONTRACTION - [metall] retrait *m* contrarié
RESTRAINER - [photo] (chemical reducing the rate of action of a developer) retardateur *m*
RESTRAINT - [gen] contrainte *f*
[gen] (restriction) restriction *f*
~TANK - [nucl] récipient *m* enveloppant
RESTRICT, to - [gen] restreindre
RESTRICTED - [gen etc] restreint, limité, borné
~AREA - [nucl] zone *f* contrôlée
~GATING - [metall & plast ind] (means of reducing the gate left on the finished product) injection *f* capillaire
~HOUR MAXIMUM DEMAND INDICATOR - [instr] (indicator so arranged that it operates during certain hours of the day only) indicateur *m* de maximum intermittent
~HOURS' TARIFF - [el etc] rtarif *m* à horaire restreinte
~INTERNAL ROTATION - [phys] (restriction of the free rotation of the molecules in some substances) rotation *f* interne limitée
~PROPELLANT - [astronaut] propergol *m* partiellement inhibé
RESTRICTION - [gen] restriction *f*
~CRACK - [metall] crique *f* de retrait
RESTRICTIVE - [gen] restrictif
RESTRICTOR - [astronaut] limiteur *m* de combustion
RESTRIKE, to - [mech] (the operation to correct distortions) redresser
RESTRIKE - [mech] redressement *m*
RESTRIKING - [mech] redressement
[el] de rétablissement, de reamorçage
[metall] frappe *f* de finition
~VOLTAGE - [electron] (the anode voltage at which the discharge recommences when the supply voltage is increasing before substantial deionization has occured) tension *f* de réamorçage
[el] tension *f* transitoire de rétablissement
RESUING - [mining] attaque *f* par le mur
RESULT, to - [gen] résulter, découler
RESULT - [gen] résultat *m*, aboutissement *m*
RESULTANT - [phys] (the single force producing the same effect of two or more forces) résultante *f*
~INTERLOCKING - [el] enclenchement *m* indirect, enclenchement *m* résultant
~TONE - [acoust] résultant *m*
~WIND - [met] vent *m* résultant
RESUME, to - [gen] reprendre, regagner
RESUMPTION - [gen] reprise *f*
RESURFACING - [constr] refaire le revêtement (d'une route)
RESURGENT GASES - [geol] (superheating steam playing an active part in volcanic action) gaz *mpl* resurgeants
RESURVEY, to - [surv] réarpenter
RESURVEY - [surv] réarpentage *m*
RESUSCITATE, to - [gen & med] ressusciter
RESUSCITATION - [med] ressuscitation *f*
RET, to - [ind chem] (or to rot; the operation of steeping, or soaking flax straw in bacteria contain-

ing water, so as to loosen the flax fibres from the woody tissue) rouir
RET THE FLAX, to - [text] rouir le lin
~THE HEMP, to - [text] rouir le chanvre (en vert)
~THE RAMIE FIBRE, to - [text] rouir la fibre de ramie
RETAIL, to - [comm] vendre au détail
RETAIL - [comm] détail *m*, vente *f* au détail
~DEALER - [comm] marchand *m* au détail
~PRICE - [comm] prix *m* de détail
RETAILER - [comm] commerçant *m* détaillant
RETAIN, to - [gen] retenir, maintenir
[leg] conserver, garder
RETAINED IMAGE - [electron] (in a cathode-ray tube) image *f* retenue
RETAINER - [gen] arrhes *f*pl, honoraires *m*pl
[gen] (a device for keeping a part in its place) dispositif *m* de retenue
[mech] bague *f* d'arrêt, arrêtoir *m*
[mech] (bearing ball retainer) boîtier *m*
[auto] (US only; of a valve spring) cuvette *f*
[electron] (device designed to restrain the movement of an electron tube in its holder) collier *m* de tube
~LOCK - [mech] arrêtoir *m*
~PLATE - [mach tool] plaque *f* de blocage
~SPRING - [mech] (a spring used to retain a part in position) ressort-arrêtoir *m*
~WITH BALLS - [mech] arrêtoir *m* à billes
RETAINING BASIN - [hydr] bassin *m* de retenue de la crue
~DAM - [hydr] barrage *m* de retenue
~DEVICE - [mech] pièce *f* de fixation
~HOOK - [mech] crochet *m* de retenue
~MOTION - [text] dispositif *m* de retenue
~NOSE - [text] bec *m* de retenue
~PAWL-[mech] cliquet *m* de retenue
~PIN - [text] goupille *f*
~PLATE - [print] plaque *f* de maintien
~RING - [mech] anneau *m* de fixation
~RING OF RIM - [auto] anneau *m* de fixation du rebord de la jante
~WALL - [contr] (wall supporting earth at a higher level on one side than on the other) mur *m* de soutènement, mur *m* de terrasse
~ZONE - [telev] zone *f* de l'enclenchement
RETAKE, to - [cin] tourner à nouveau
RETAKE - [cin] réplique *f* d'une prise de vues
RETAP, to - [mech] retarauder
RETARD, to - [auto] (to adjust ignition timing to occur later in the cycle) retarder
~THE RETTING PROCESS, to - [text] retarder le procédé de rouissage
RETARDATION - [phys] accélération *f* négative
[med] retardement *m*
[mus] ralentissement *m*
[mech] retardation *f*, vitesse *f* retardée
~BASIN - s. retaining basin
~COIL - [el] (coil of high inductance used to separate alternating from direct current) bobine *f* d'arrêt
~METHOD - [el] (the method for measuring certain types of losses by plotting the slowing up of a machine as a function of time) méthode *f* de ralentissement
~OF THE REEL - [text] retard *m* du dévidoir
RETARDED CARRIAGE SPEED - [text] vitesse *f* retardée du chariot
RETARDED COMBUSTION - [mech] (in internal combustion engines) retard *m* à l'allumage
~DEVELOPMENT - [med] retardement *m* du développement
~FIELDS - [el] champs *m*pl retardés
~IGNITION - [mech] (in internal combustion engine) retard *m* à l'allumage
~MOTION - [mech] mouvement *m* retardé
~POTENTIAL - [phys] (a value occurring in electromagnetic wave theory) potentiel *m* retardé
~TIMING - [mech] (in internal combustion engines) calage *m* retardé
RETARDER - [chem] (an additive which slows down the rate of reaction) agent *m* retardant, retardeur *m* de prise
[railw] (rail brake) rail-frein *m*
~PARACHUTE - [aero] (a subsidiary parachute used to ensure the deployment of the rigging lines before that of the canopy) parachute *m* retardateur
RETARDING APPLIANCE - [mech] dispositif *m* de retenue
~EFFECT - [chem] effet *m* retardant
~ELECTRODE - [el] (in television systems, to retard the photoelectrons on their way towards the screen) électrode *f* de freinage
~FIELD - [electron] (electric field as it exists between a positively charged grid and a negatively charged outer electrode of a three-electrode thermionic vacuum tube) champ *m* de freinage
~FIELD DETECTOR - [electron] (type of detector employing a retarding field tube) détecteur *m* à champ de freinage
~FIELD OSCILLATOR - [electron] tube *m* oscillateur à champ de freinage
~-FIELD TUBE - [electron] tube *m* à champ de freinage
~FORCE - [phys] force *f* de freinage
~-FIELD VALVE - s. retarding-field tube
RETCHING - [med] vomiturition *f*
RETEMPER, to - [metall] retremper
RETEMPERING - [metall] retrempe *f*
RETENE - [chem] (occurring in coal-tar fraction boiling above 300 centigrades) rétène *m*
RETENTION - [gen & med] rétention *f*
[nucl] (the percentage of radioactive atoms which are not separable from the target compounds following the production of such atoms by nuclear reaction) rétention *f*
~BASIN - [nucl] récipient *m* de dépôt
~COEFFICIENT - [nucl] coefficient *m* de rétention
~OF COLOUR - [dyes] (colour fastness) solidité *f* de la couleur
~TIME - [hydr] durée *f* de séjour
RETENTIVE POWER - [soil] capacité *f* de rétention d'eau
RETENTIVITY - [phys] (the property of a magnetic material measured by the residual induction when a magnetizing force is removed) persistance *f*
RETETHELIOMA - [med] réticulo-sarcome *m*
RETICULE - [opt] (cross-hair lines) réticule *m*
RETICULAR - [gen zool med] (resembling a net) réticulaire
~CELL - s. reticulocyte
~DENSITY - [phys] (the number of points per unit area in a network; e.g. that of a plane in a crystal lattice) densité *f* de réseau
~FORMATION - [phys] formation *f* réticulaire
~STRUCTURE - [nucl] (crystal structure containing

a network) structure *f* réticulaire
RETICULAR TISSUE - [zool] (form of connective tissue in which the intercellular matrix is replaced by lymph) tissu *m* réticulaire
RETICULATE - [zool etc] (forming a network) réticulé, rétiforme
RETICULATED - [gen] réticulé, rétiforme
~STRUCTURE - [geol] structure *f* maillée
~VAULTING - [constr] voûte *f* réticulée
~WORK - [constr] appareil *m* réticulé
RETICULATION - [photo and paint] (fine markings on a film or a painted surface) réticulation *f*
RETICULE - [opt] (cross-hair cell for a surveying telescope) réticule *m*
RETICULIN -[chem] (a phosphorus containing collagen occurring in reticular fibrous tissues) réticuline *f*
RETICULOCYTE - [med] hématie *f* granuleuse
RETICULOENDOTHELIOSIS - [med] réticulose *f* leucémique
RETICULOMA - [med] réticulome *m*
RETICULUM CELL - s. retethelioma
RETIFORM - [bot & zool] (having the appearance of being netted) rétiforme
~TISSUE - s. reticular tissue
RETIGHTEN, to - [gen] retendre
[mech] resserrer
RETIMBER, to - [mining] reboiser
RETIME, to - [mech] (to readjust the timing of an engine) régler (l'allumage) à nouveau
RETIN, to - [metall] retamer
RETINACULUM - [med] rétinacle *m*
RETINAL FATIGUE - [med] fatigue *f* de la rétine
~ILLUMINANCE - [opt] (a psychophysiological quantity) illumination *f* de la rétine
RETINASPHALT - [min] rétinasphalte *m*, rétinellite *f*
RETINITIS - [med] rétinite *f*
~STELLATA - [med] rétinite *f* stellaire, rétinite *f* exsudative toxique
RETINNING - [metall] retamage *m*
RETINOID - [gen] résinoïde
RETINOL - [chem] (a member of the vitamin A complex) rétinol *m*
RETINOMALACIA - [med] rétinose *f*
RETINOSCOPE - [instr] rétinoscope *m*
RETINOSCOPY - [med] (a method of measuring the refractive state of the eye by reflecting light on to it from a mirror) rétinoscopie *f*
RETINOSIS - s. retinomalacia
RETINOPATHY - [med] rétinopathie *f*
RETIRE, to - [gen] se retirer
RETONATION WAVE - [phys] onde *f* rétrograde
RETOOTH, to - [mech] redenter
RETOOTHING - [mech] redentage *m*
RETORT, to - [ind chem] distiller en vas clos
RETORT - [chem] (a chamber of refractory material used for destructive distillation of coal. In modern practice, they are usually designed for continuous operation) cornue *f* à gaz
[ind chem] (a glass distilling apparatus, now replaced by special types of flask) cornue *f*
[metall] (a metal or refractory vessel used in metal distillation) cornue *f*
~CARBON - [min] (graphite) charbon *m* de cornue, graphite *f*
~CAULKER - [gas ind] calfeutreur *m*
~FURNACE - [metall] four *m* à cornue
~GRAPHITE - [min] graphite *f* de cornue
RETORT HOUSE - [gas ind] halle *f* de fours
~PATCHER - s. retort caulker
~STAND - [ind chem] (device consisting of a base with a standard provided with a movable clamp for holding apparatus) support *m* de cornue
~WITH TWO MOUTHPIECES - [gas ind] (a through retort) cornue *f* sans fond
RETORTING - [ind chem] distillation *f* à la cornue
~COAL - [ind chem] distillation *f* de la houille en vas clos
RETOTHELIAL SARCOMA - s. retethelioma
RETOTHELIOMA - s. retethelioma
RETOUCH , to - [photo] retoucher
RETOUCH - [photo] retouche *f*
RETOUCHING - [photo] retouche *f*
~DESK - [photo] chevalet *m* (pour retouche)
~DYE - [photo] teint *m* pour retouches
~KNIFE - [photo] grattoir *m*
~STAND - [photo] pupitre *m* à retouche
RETRACE - [telev] (the return of the scanning beam to the starting point after completion of a line or frame) retour *m* du spot
RETRACT, to - [gen] rétracter
RETRACTABLE - [aero] (capable of being withdrawn into a recess in the aircraft when not in use, especially of landing gear) escamotable, rentrant, relevable
[gen] rentrant, relevable
~UNDERCARRIAGE - [aero] (an undercarriage which can be withdrawn into a recess in the aircraft) atterrisseur *m* escamotable
RETRACTILE - [zool] (capable of being withdrawn) rétractile
RETRACTING SPRING - s. return spring
RETRACTION - [gen & phys] (e.g. in elasticity tests) rétraction *f*, retrait *m*, recul *m*
[med] rétraction *f*
~LOCK - [aero] (a device designed to prevent the undercarriage from being retracted by inadvertence) cliquet *m* d'arrêt du train d'atterrissage rétracté
RETRACTOR - [anat] (a muscle) muscle *m* rétracteur
[med] rétracteur *m*, écarteur *m*
RETRANSMISSION - [telecomm] retransmission *f*
RETREAD, to - [rubber ind] rechaper
RETREAD VULCANIZING COMPOUND - [rubber ind] croissant *m* de rechapage
RETREADED TYRE - [rubber ind] pneu *m* regommé, pneu *m* rechapé
RETREADING - [rubber ind] (of tyres) rajeunissement *m* de bande de roulement
~MOULD - [rubber ind] moule *m* pour rechapage
RETREAT, to - [gen] se retirer, s'éloigner
[gen & ind chem] retraiter
[mining] battre en retraite
RETREAT - [gen] retraite *f*
RETREATING METHOD - [mining] méthode *f* rétrograde
~SYSTEM - [mining] exploitation *f* en rabattant
RETREE - [paper man] (slightly damaged paper from reams) papier *m* de rebut
RETRIEVE, to - [gen] relever, rétablir
RETRIEVE - [gen] recouvrement *m*, réparation *f*
[mech] (caster action) réversibilité *f*
RETRIMMING - [metall] (in forming) réébarbage *m*
RETROACTION - [gen] (the operation of acting reciprocally) réaction *f*, contre-coup *m*

RETROACTION - [radio] (of an amplification circuit) réaction *f*
RETROACTIVE - [gen] (having a retrospective action) rétroactif
~AMPLIFICATION - [telecomm] amplification *f* par réaction
~AMPLIFIER - [radio] amplificateur *m* à réaction
~CIRCUIT - [telecomm] circuit *m* rétroactif
~ROCKET - [astronaut] fusée *f* de freinage
~TENACITY - [metall] résistance *f* à la compression
~TUBE - [el] (in a coupling) tube *m* réacteur
RETROACTIVITY - [gen] (retrospective effect; reversed action) rétroactivité *f*
RETROACTOR - s. retroactive tube
RETROCECAL - [anat] (placed behind the caecum) rétrocaecal
RETROCOLLIS - [med] torticolis *m* postérieur
RETRODIRECTIVE MIRROR - [opt] (mirror reflecting a beam of light parallel to its original path, independently of its original direction) miroir *m* central
RETROFLECTION - s. retroflexion
RETROFLEXED - [gen] (bent backward; turned backward) rétrofléchi
~UTERUS - [med] utérus *m* rétrofléchi
RETROFLEXION - [med] rétroflexion *f*
RETROGRADE - [gen] (moving backward) rétrograde
~MOTION - [astr] (the apparent motion of a planet from east to west among the stars) mouvement *m* rétrograde
~SOLUBILITY - [chem] (solubility which decreases with the rise in temperature) solubilité *f* rétrograde
RETROGRAPHY - [med] écriture *f* spéculaire
RETROGRESSION - [med] rétrogression *f*
[gen] rétrogradation *f*
[zool] dégénérescence *f*
[math] (of a curve) rebroussement *m* (d'une courbe)
RETROGRESSIVE METAMORPHISM - [geol] (the changes involved in the conversion of a rock of high metamorphic grade to one of lower grade) métamorphisme *m* régressif
~WINDING - [el] enroulement *m* rétrogressif
RETROMORPHOSIS - [biol] (during development; the tendency to degeneration) rétromorphose *f*
RETROPACK - [astronaut] unité *f* de freinage
RETROPERITONEUM - [anat] espace *m* rétropéritonéal
RETROPHARYNGEAL ABSCESS - [med] abcès *m* rétropharyngé
RETROPHARYNX - [med] espace *m* rétropharyngien
RETROPLASIA - [med] involution *f* cellulaire
RETROPULSION - [med] rétropulsion *f*
RETROROCKET - [astronaut] (the braking last stage of a rocket) rétrofusée *f*, fusée *f* de freinage
RETRORSE - [zool] (pointing backwards) renversé
RETROSEQUENCE - [astronaut] séquence *f* de freinage par rétrofusée
RETROTHRUST - [astronaut] poussée *f* rétroactive
RETROVERSE - s. retrorse
RETROSINUS - [med] cellules *f*pl mastoïdiennes postérieures
RETROVERSION - [med] rétroversion *f*
~OF UTERUS - [med] rétroversion *f* de l'utérus
RETRUSION - [med] rétropulsion *f* dentaire
RETTED FLAX - [text] lin *m* de rouissage
~HEMP - [text] chanvre *m* roui
RETTERY - [text] rouissoir *m*
RETTING - [text] rouissage *m*
~BOX - [text] ballon *m* à rouir, caisse *f* à jour
RETTING CRATE - s. retting box
~DAM - [text] routoir *m*
~HOLE - s. retting dam
~POND - s. retting dam
~PROCESS - [text] procédé *m* de rouissage
~RIPENESS - [text] rouissage *m* à point
~WATER - [text] eau *f* de rouissage
RETUBE, to - [min] rétablir le tube de pompage
[mech] (e.g. an oven) retuber
RETUBING - [oil ind] retubage *m*
RETURN, to - [gen] revenir, retourner
[gen] (to give back) rendre
RETURN - [gen] retour *m*, renvoi *m*
[fin] revenu *m*, recette *f*
[comm] profit *m*
[mech] (of a piston) course *f* de retour
[radar] (echo) écho *m*
~ADDRESS - [comput] (address in an instruction of a superroutine carried out after the completion of a subroutine) adresse *f* à rétroaction
~AIR - [th eng] air *m* de reprise
~AIR COOLING - [th eng] refroidissement *m* à air de reprise, refroidissement *m* par circulation
~AIR GRILLE - s. return air register
~AIR REGISTER - [th eng] (in space heating installations) bouche *f* de reprise
~ARC - [el] arc *m* de retour
~BAR - [horol] bascule *f*
~BEND - [plumb] (gooseneck) coude *m* double
~CABLE - [el] artère *f* de retour
~CIRCUIT - [el] circuit *m* de retour
~CODE - [comput] code *m* à rétroaction
~CONTROL TRANSFER - [comput] retour *m* au programme principal
~CRANK - [mech] (short crank on outside cylinder locomotives, fixed to the outer end of the main crank pin) manivelle *f* de renvoi
~CURRENT - [el] (current due to an impedance discontinuity in a transmission system) courant *m* de retour
~ELECTRONS - [electron] (in cathode-ray tubes) électrons *m*pl de retour
~FEEDER - [el] (negative feeder) conducteur *m* parallèle sur le circuit de retour
~FLAME BOILER - [metall] chaudière *f* à retour de flamme
~FLANGE - [mech] rebord *m*
~FLOW COOLER - [th eng] sous-refroidisseur *m*
~-FLUE - [th eng] tube *m* de retour de fumée
[metall] carneau *m* de retour
~-FLUE BOILER - [th eng] chaudière *f* à retour de flamme
~INSTRUCTIONS - [comput] instructions *f* de retour
~INTERVAL - [telev] durée *f* du retour du spot
~LIGHT - [telecomm] signal *m* réponse
~LINE -[hydr] (a pipeline carrying condensate from a steam trap, or a pipe carrying exhaust hydraulic fluid from a press) conduite *f* de retour
~LOSS - [electron] perte *f* par réflexion
[el acoust] affaiblissement *m* d'adaptation
~LOSS BALANCE - [telecomm] (the value of the return loss at a hybrid-coil terminating set) affaiblissement *m* d'équilibrage
~LOSS MEASURING SET - [instr] réflectomètre *m*
~PASS - [text] remettage *m* à retour
~PIN - [metall & plast ind] (a pin which effects the return motion of the mould ejector) butée *f* de ren-

voi de l'éjecteur, rappel *m* d'injection
RETURN PULLEY - [mech] poulie *f* de renvoi
~PUSHER - [mech] clapet *m* de rappel
~ROLLER - [mech] rouleau *m* de renvoi
~ROPE - [mech] câble *m* de renvoi
~SHAFT - [mining] puits *m* de sortie d'air
~SHOCK - [mech] contre-coup *m*, choc *m* en retour
~-SLUDGE - [hydr] boue *f* de retour
~SPEED - [mech] vitesse *f* de renvoi
~SPRING - [mech] (any spring used to restore a part to its place after movement. In a press, a spring used to return the platen of a press) ressort *m* de rappel
~STROKE - [mech] course *f* de retour, course *f* arrière
~TICKET - [transp] billet *m* d'aller et retour
~-TO-BIAS RECORDING - [comput] méthode *f* d'enregistrement par corrélation
~TRACE - s. retrace
~TRANSFER FUNCTION - [contr] (feedback transfer function) fonction *f* de transfert de retour
~TRAVERSE - [metall & plast ind] traverse *f* de retour
~VALVE POSITION - [mech] position *f* initiale de soupape
~VOLTAGE - [el] (due to an impedance discontinuity in a transmission system) tension *f* réfléchie
~WALL - [constr] (short lenght of wall built out from one end of a longer wall) mur *m* en retour
~WATER - [hydr] eaux *fpl* de retour
~WATER PUMP - [mech] pompe *f* pour eaux de retour
~WAVE - [hydr] ressaut *m* hydraulique
~WHEEL - [mech] poulie *f* de renvoi
~WING - s. return wall
RETURNING CHARGES - [metall] (the cost of smelting) frais *mpl* de transformation
RETURNS - [metall] (foundry scraps) jets *mpl* et débris de la coulée
REUSE, to - [gen] remployer
REUSE - [gen] remploi *m*
REUTILIZE, to - s. reuse, to
REV, to - [mech] faire s'emballer
REV - [mech] (short for revolution) tour *m*
REVACCINATE, to - [med] revacciner
REVALORIZATION - [comm] revalorisation *f*
REVALORIZE, to - [comm] revaloriser
REVALUATION - [comm etc] réévaluation *f*
REVALUE, to - [comm] réévaluer
REVAMP, to - [shoe man] remplacer l'empeigne (d'un soulier)
REVAMPING OF MATRICES - [rubber ind] (only USA) redressement *m* des matrices
REVARNISH, to - [paint] revernir
REVEAL, to - [gen] révéler
REVEAL - [arch] (the depth of the wall revealed in the sides of window, or door opening) jouée *f*
[auto] listel *m* d'encadrement
REVELLENT - [med] révulsi *f*
REVENUE - [fin] revenu *m*, rentes *fpl*
[fin] (the total current income of a government) trésor *m* publique
REVERBERATE, to - [phys] (to be reflected, as heat etc) renvoyer, réfléchir
[acoust] (of a reflected, repeated sound) réfléchir
[gen] réverbérer
REVERBERATION - [phys] réverbération *f*
[acoust] répercussion *f*, réverbération *f*
~ABSORPTION COEFFICIENT - [acoust] (of a surface at a given frequency) coefficient *m* d'absorption de la réverbération
REVERBERATION ABSORPTION FACTOR - [acoust] (of a surface or material at a given frequency and under specified conditions) coefficient *m* spécifique d'absorption de la réverbération
~BRIDGE - [acoust] (the measurement of reverberation time in an enclosed room) pont *m* de mesure de la réverbération
~CHAMBER - [acoust] chambre *f* réverbérante
~CONTROLLED GAIN CIRCUIT - [telev] montage *m* de expansion du contraste
~KEY - [telev] touche *f* de réverbération
~PERIOD - [acoust] (of an enclosure) temps *m* de réverbération
~REFLECTION COEFFICIENT- [acoust] (of a surface or material at a given frequency) coefficient *m* de réflexion de la réverbération
~REFLECTION FACTOR - [acoust] (of a surface or material at a given frequency and under specified conditions) coefficient *m* spécifique de réflexion de la réverbération
~RESPONSE - [acoust] (the response of a microphone for reverberant sound) sensibilité *f* du microphone pour la réverbération
~RESPONSE CURVE - [acoust] courbe *f* de sensibilité du microphone pour la réverbération
~ROOM - s. reverberation chamber
~TIME - s. reverberation period
~TIME METER - [instr] réverbéromètre *m*
~TRANSMISSION FACTOR - [acoust] (the acoustical transmission factor when the distribution of the incident sound is completely random) coefficient *m* de transmission de la réverbération
~UNIT - [el acoust] bloc *m* de réverbération
REVERBERATOR - [gen] (reverberating device, surface) réflecteur *m*
REVERBERATORY - [gen & phys] de réverbère, à réverbère
~CHAMBER - [metall] (of furnace) laboratoire *m*
~FLAME - [phys] feu *m* de réverbère
~FURNACE - [metall] (a type of shallow-hearth furnace in which the flame of the fuel impinges on a low roof, which heats the charge below chiefly by radiation) four *m* (à) réverbère
~MELTER - s. reverberatory furnace
REVERSAL - [gen] inversion *f*
[leg] annulation *f*
[mech] renversement *m*
[el chem] (a change in normal polarity of the cell or battery) inversion *f*
[photo] (of a negative) inversion *f* (de l'image)
[phys] s. reversal of spectrum lines
~FEED LEVER - [mech] levier *m* d'inversion de l'avance
~FILM - [photo] (in colour photography) film *m* inversible
~GRADIENT - [railw] contre-pente *f*
~LIGHT - [auto] feu *m* de marche arrière
~MATERIAL - [photo] (for colour photography) matériel *m* inversible
~MOTION - [mech] renversement *m* de marche, mouvement *m* réciproque
~OF CONTROL - [aero] inversion *f* de commande
~OF CURRENT - [el] inversion *f* du courant
~OF DAMPING - [el] désamortissement *m*
~OF DIRECTION - [mech] inversion *f* de direction

REVERSAL OF SIGN - [el] (change in potential) inversion *f* de charge
- OF SPECTRUM LINES - [phys] (lights which are darker than others, or than a continuous spectrum background, in a spectrum) lignes *fpl* de spectre d'inversion
- OF STRESS - [phys] inversion *f* d'effort
- OF THE REGENERATOR - [th eng] inversion *f* du régénérateur
- OF TRAVERSE MOTION - [mech] renversement *m* de la course
- PINION - [mech] pignon *m* inverseur
- POWER RELAY - [el] relais *m* à retour de courant
- PREVENTION - [instr] (a device for preventing the reversal movement of the rotating disk of an electricity meter) blocage *m* du mouvement de retour
- PROCESS - [photo] (a method whereby the original film is reversed in black and white chemically and becomes a positive transparency) procédé *m* d'inversion
- SPECTRUM - [phys] (spectrum containing some lines which are darker than others) spectre *m* d'inversion
- SPEED - [aero] (the speed at which the reversal of control occurs) vitesse *f* d'inversion de commande

REVERSALS - [telecomm] (in reception and transmission, a continuous series of alternate marking and spacing signal elements, all of equal length) roulement *m*

REVERSE, to - [gen] renverser
[mech] renverser (la marche)
[railw] marcher en arrière
[auto] faire marche arrière
[leg] révoquer

REVERSE - [gen] inverse *m*, contraire *m*, revers *m*
[mech] renversement *m*, marche *f* arrière
- ACTION - [cin] (the projection of a film from the end to the beginning) action *f* renversée
- BEND TEST - [metall] essai *m* de flexion alternée
- BIAS - [electron] polarisation *f* inverse
- BLOCKING CURRENT - [electron] courant *m* inverse de blocage
- BLOCKING STATE - [electron] état *m* bloqué inverse
- COMBUSTION - [metall] combustion *f* inverse
- CONDUCTING DIODE THYRISTOR - [electron] diode *f* thyristor à passage inverse
- CONTACT - [el] contact *m* d'inversion
- COUPLING - [radio] (negative feedback) réaction *f* inverse
- CURRENT - [el] courant *m* inverse
--CURRENT CLEANING - [chem] (method of electrolytic cleaning, in which the metal to be treated is made the anode) dégraissage *m* anodique
- CURRENT CUT-OUT - s. reverse-power release
--CURRENT RELAY - [el] relais *m* à retour de courant
- CURRENT RELEASE - s. reverse-power release
- CURVE - [surv] (curve composed of two arcs with their centres on opposite sides of the curve) contre-courbe *f*
- CYLINDER - [mech] cylindre *m* d'inversion
- DIRECT CURRENT RESISTANCE - [el] résistance *f* pour courant continu inverse
- DIRECTION - [electron] (the direction of greater resistance to current flow through the cell) direction *f* inverse
- DRUM FOR CONVEYOR BELTS - [mech] tambour *m* de rappel pour courroie transporteuse

REVERSE-ECCENTRIC - [mech] (of a locomotive) excentrique *m* de marche en arrière
- ELECTRODE CURRENT - [electron] courant *m* inverse d'électrode
- FAULT - [geol] faille *f* inverse
- FEEDBACK - [radio] contre-réaction *f*
--FLIGHTED SCREW - [plast ind] (type of extruder screw in which the material is fed at both ends of the barrel and extruded in the middle, half the screw being L.H. and half R.H.) vis *f* à pas inversé
--FLOW COMBUSTION SYSTEM - [mech] (type of combustion system in gas turbines, in which the flow in the combustion chambers is opposite in direction to that in the turbine itself) chambre *f* de combustion à écoulement inverse
- GATE CURRENT - [electron] courant *m* inverse de gâchette
- GATE VOLTAGE - [electron] tension *f* inverse de gâchette
- GRID CURRENT - [electron] (control grid circuit current due to electron emissions from the grid) courant *m* inverse de grille
- IDLER SHAFT - [auto] arbre *m* intermédiaire de marche arrière
- LAID ROPE - [rope man] câble *m* métallique à torons alternés
- LEAKAGE CURRENT - [electron] (current flowing through a cell in the reverse direction) courant *m* inverse de fuite
- LOOP - [text] maille *f* à l'envers
--MAKE VALVE - [mech] (three-way valve) vanne *f* à trois voies
- MOTION DEVICE - [text] dispositif *m* à rotation inverse
- MOULDING - [metall] moulage *m* à moule renversé
- OF DIP - [geol] changement *m* de pendage
- PHOTOELECTRIC EFFECT - [electron] (in light-sensitive devices) effet *m* photoélectrique inverse
- PITCH - [aero] (negative propeller setting) pas *m* négatif, pas *m* de reversion
- PLATING - [text] vanisage *m* renversé
--POWER PROTECTION - [el] dispositif *m* de protection à retour de puissance
--POWER RELEASE - [el] (a tripping device which operates when the power through the circuit has reversed its direction and has attained a predetermined value) déclenchement *m* à retour de courant
--POWER TRIPPING - [el] disjonction *f* par inversion de courant
- REACTION - [radio] (reaction opposing the production of self-oscillation) réaction *f* inverse
--REDUCTION GEAR - [mech] inverseur-réducteur *m*
- ROLL - [mech] (a roller running in the opposite direction to that of a band of material passing over it) cylindre *m* à marche inverse
--ROLL COATER - [paint] (machine for applying a coating, in which the application rolls run in the reverse direction to the material) machine *f* à enduire à rouleaux inversés
--ROLL COATING - [paint] (method of coating in which the material is metered between casting and doctor-rolls and transferred to the web by a wiping action) enduction *f* par rouleaux inversés
- ROTATION - [mech] antirotation *f*
- RUN - [text] marche *f* en arrière
- RUNNING - [railw] circulation *f* dans les deux sens
[railw] (the propelling movement) marche *f* en

refoulement
REVERSE STEAM, to - [mech] renverser la vapeur
~THE MOTION - [mech] renverser la marche
~THE POINTS, to - [railw] renverser l'aiguille
~TWIST - [text] torsion *f* inverse
~VOLTAGE - [el] (voltage applied to a metallic rectifier with the opposite of normal polarity) tension *f* en sens inverse
REVERSED - [gen] renversé, inverse
[mech] renversé
~ARCH -[hydr] radier *m*
~BLOCK - [print] cliché *m* négatif
~FAULT - s. reverse fold
~FOLD - [geol] pli *m* déversé
~IMAGE - [telev] (image appearing white where it should be black and vice-versa) image *f* négative
~NEGATIVE - [photo] (the negative is reversed in some mechanical photographic process, to obtain the correct orientation for the final print) négatif *m* dessus-dessous
~OGEE - [arch] doucine *f* renversée
~POLARITY - [el] polarité *f* inverse
~STEAM - [mech] contre-vapeur *f*
~VENTILATION - [mining] ventilation *f* rétrograde
REVERSER - [el] (a combination switch for changing the connections of traction-motors so as to reverse the direction of rotation) inverseur *m* de marche
[photo] inverseur *m*
REVERSIBLE - [mech] (capable of operating in either direction, especially in the sense of being capable of transmitting energy in either direction) réversible
[chem] (said of a reaction which can take place in both directions) réversible
~ABSORPTION CURRENT - [el] (current decreasing with time less rapidly than the geometrical charging current and is returned on short-circuiting the electrodes) courant *m* d'absorption réversible
~BOOK - [text] pince *f* réversible
~BOOSTER - [el] (a machine so arranged that its e.m.f. can be added or subtracted from the voltage supplied by another electrical source) survolteur-dévolteur *m*
~BURNER - [gas ind] brûleur *m* réversible
~CELL - [el chem] (an electro-chemical cell in which the conversion of electrical into chemical energy and vice versa is a reversible process) élément *m* réversible
~CHEMICAL ACTION - [chem] action *f* chimique réversible
~COLLOID - [chem] (lyophilic colloid) colloïde *m* réversible
~COUNTER - [comput] compteur *m* réversible
~ELECTRODE - [el] (electrode used in a reversible electrochemical reaction) électrode *f* réversible
~ELECTROLYTIC PROCESS - [el chem] réaction *f* réversible
~GEL - [chem] (a gel which may be-re-dispersed after coagulation) gel *m* réversible
~MAGNETOSTRICTION - [el] (the reversible change in dimensions which results from a small cyclic change in applied magnetic field superposed on a larger and steady field) magnétostriction *f* réversible
~MOTOR-COACH TRAIN - [railw] rame *f* automotrice réversible
~PATH - [phys] (in thermodynamics) parcours *m* réversible
REVERSIBLE PATTERN PLATE - [metall] plaque-modèle *f* réversible
~PENDULUM - [mech] (pendulum which is used for accurate determinations of the acceleration of gravity) pendule *m* réversible
~PERMEABILITY - [el] (the limit of incremental permeability as the incremental change in magnetizing force approaches zero) perméabilité *f* inversible
~PISTONS - [mus] (of an organ) pistons *mpl* réversibles
~POLARITY - [el] polarité *f* réversible
~POTENTIOMETER-TYPE FIELD RHEOSTAT - [contr] (potentiometer-type field rheostat arranged for reversing the polarity of a field winding) rhéostat *m* de champ à potentiomètre réversible
~POWDER - [rubber ind] poudre *f* réversible
~PROCESS - [el chem] (an electrochemical process which, at the equilibrium electrode potential, can take place in either direction) réaction *f* réversible
[phys] (a cycle of operations in which the different operations can be performed reversely with a reversal of their effects) procédé *m* réversible
~PUMP TURBINE - [mech] turbo-pompe *f* réversible
~RAKE - [text] râteau *m* réversible
~REACTION - [chem] (one which can go on in either direction, e.g. the formation of an ester and water by the reaction of an alcohol and an acid) réaction *f* réversible
~RING - [text] anneau *m* à deux rebords
~TRACK - [railw] voie *f* banalisée
~TRANSDUCER - [el acoust] (a transducer which is equally capable of transforming electrical energy into mechanical energy, or vice-versa) transducteur *m* réciproque
~VOLTAGE TRANSFORMER - [el] transformateur *m* de tension commutable
REVERSING - [gen] inverseur, inversion *f*
[mech] changement *m* de marche
~BATH - [photo] bain *m* d'inversion
~BELT MACHINE - [text] courroie *f* de renversement de la marche
~BEVELS - [mech] roues *fpl* coniques de renversement de marche
~BRACKET - [mech] pièce *f* à bascule
[text] pièce *f* à bascule
~CLICK MOTION - [text] virgule *f*
~COGGING MILL - [metall] laminoir *m* réversible à lingots
~COLD ROLLING MILL - [metall] laminoir *m* réversible pour laminage à froid
~COMMUTATOR - [el] (a reversing switch in which the contacts form two halves of a cylinder on which bear two brushes) commutateur *m* inverseur
~DEVICE - [mech] (any mechanical arrangement designed to cause a part or mechanism to move in the opposite sense to the more usual one) dispositif *m* d'inversion, dispositif *m* de renversement
~FIELD - [el] (in a commutator machine) champ *m* d'inversion
~FRAME - [photo] (of a camera) cadre *m* arrière réversible
~GEAR - [mech] (mechanism, especially gearing, designed to cause a machine or like device to move in the direction opposite to its more usual one) mécanisme *m* de renversement de marche
~HANDLE - [mech] levier *m* de renversement

REVERSING KEY - s. reversing switch
~LOOP - [text] glissoir *m* du dispositif pour la voltée
~MACHINE - [th eng] treuil *m* à inversion
~MILL - [metall] (type of rolling-mill in which the stock passes forwards and backwards between the same pair of tools which are reversed between each pass) laminoir *m* réversible
~MOTION - [text] dispositif *m* de renversement
~MOTION FOR FIBRE TUFTS - [text] dispositif *m* de renversement des barbes
~OF THE RIM SHAFT - [mech] rotation *f* inverse de l'arbre principal
~OF THE SPINDLES - [text] rotation *f* inverse des broches
~PRISM - [opt] (prism having the property of inverting a beam of light) prisme *m* à inversion
~ROD - [mech] (in a railway locomotive) barre *f* de relevage
~ROLLING MILL - s. reversing-mill
~ROUGHING MILL - [metall] laminoir *m* débaucheur réversible
~SCREW - [mech] vis *f* de changement de marche
~SHAFT - [text] axe *m* d'inversion
[mech] arbre *m* de relevage
~SINKER - [text] platine *f* tournante
~STARTER - [el] (for remote control) téléinverseur *m*
~STATION - [railw] gare *f* de rebroussement
~SWTCH GROUP - [el] combinateur *m* d'inversion
~SWITCH WITHOUT OFF POSITION - [el] permutateur *m*
~THE PIECES - [text] retournage *m* des poignées
~THE STRICK - [text] retournage *m* de la poignée
~THE TUFTS - [text] renversement *m* des barbes
~TRIANGLE - [railw] triangel *m* de raccordement
~-TYPE EXHAUST NOZZLE - [aero] (type of exhaust nozzle in which provision is made for directing the stream of gas in a direction opposite to the normal one, to obtain reversed thrust during a landing run) tuyère *f* d'inversion de l'échappement
~VALVE - [mech] valve *f* de renversement
~WINCH - s. reversing machine
REVERSION - [gen] réversion *f*, retour *m*
[fin] réversion *f*
[genet] (atavism) réversion *f*
[chem] retour *m*
[photo] inversion *f*
REVERT, to - [gen] retourner, revenir
[chem] revenir à l'état primitif
[mech] retourner
REVERTIVE CONTROL - [telecomm] commande *f* par impulsions inverses
~PULSE SYSTEM - s. revertive control
~PULSING CIRCUIT - [el] circuit *m* fondamental d'inversion
REVERTOSE - [chem] (a disaccharose) révertose *m*
REVETMENT - [constr] (a retaining wall) revêtement *m*
REVIVIFICATION - [chem] (the process of re-activating charcoal) revivification *f*
~BY EXPOSURE TO AIR - [ind chem] revivification *f* naturelle
REVOCATION - [gen] (the act of revoking) révocation *f*
[leg] (the cancellation of a decision) abrogation *f*
REVOKE, to - [gen & leg] révoquer, annuler
REVOLUTION - [gen] révolution *f*
[mech etc] (the act or state of revolving) révolution *f*, tour *m*
REVOLUTION - [astr] (term generally denoting orbital motion) révolution *f*
~COUNTER - [instr] (a device for counting the number of revolutions made by a part but not to integrate this quantity with respect to time: this is the function of a tachometer which gives direct readings of revolutions per unit of time) compte-tours *m*
~INDICATOR - [instr] indicateur *m* de vitesse, indicateur *m* de rotation
~OF MEASURING ROLLER - [text] révolution *f* du rouleau de mesurage
~PRESS - [mech] presse *f* à rotation
REVOLUTIONS PER MINUTE - [mech] tours *mpl* à la minute
REVOLVABLE - [mech] tournant
REVOLVE, to - [gen & mech] tourner
REVOLVER - [firearms] (pistol in which the ammunition is carried in a rotating magazine) revolver *m*
REVOLVING - [gen] tournant, pivotant
[mach tool] revolver
~ARMATURE - [el] induit *m* mobile
~BASKET - [text] bassine *f* rotative
~BEATER RODS - [text] ailettes *fpl* battantes rotatives
~BLADE - [mech] ailette *f* rotative
~BOBBINS - [text] bobines *fpl* à mouvement tournant
~BOILERS - [paper man] (large-size vessels in which rags, pulp etc are digested) chaudière *f* rotative
~BRUSH - [mech] brosse *f* circulaire
~CAGE - [text] panier *m* tournant
~CAN - [text] boîte *f* tournante
~CHAIR - [impl] fauteil *m* pivotant
~CRANE - [mech] grue *f* à pivot
~CREEL - [text] râtelier *m* tournant
~CUTTER - [mech] molette *f*
~-CUTTER MACHINE - [mech] machine *f* à cisailler à molettes
~CYLINDER - [text] tambour *m* tournant
~CYLINDER ENGINE - [auto] moteur *m* rotatif
~DIEHEAD - [mech] porte-lunette *m* revolver
~DOOR - [build] porte-revolver *f*
~DRUM - [mech] tambour *m* tournant
~DRYER - [mech] séchoir *m* rotatif
~FIELD - [el] champ *m* tournant
~FIELD GENERATOR - [el] génératrice *f* à champ tournant
~FLAT - [text] chapeau *m* marchant
~-FLAT CARD - [text] carde *f* à chapelets
~FURNACE - [metall] four *m* tournant
~GRATE - [th eng] grille *f* tournante
~-HEAD PUNCH - [tool] pinces *fpl* à emporte-pièce à revolver
~HEARTH FURNACE - [metall] (a rotary-hearth furnace) four *m* à sole tournante
~HOPPER - [mech] trémie *f* de chargement à pivot
~LETTER OF CREDIT - [fin] lettre *f* de crédit automatiquement renouvelable
~MASS - [mech] masse *f* tournante
~MOTION - [mech] mouvement *m* rotatif
~NOSE-PIECE - [opt] (of a microscope) revolver *m* porte-objectif
~NOZZLE - [text] filière *f* tournante
~PRESS - s. revolution press
~PUDDLING - [metall] puddlage *m* rotatif
~REED HOOK - [text] crochet *m* tournant
~RETORT FURNACE - [th eng] four *m* à cornue rotative

REVOLVING RING COMB - [text] peigne *m* annulaire rotatif
~SCREEN - [min] trommel *m*
~SHAFT - [mech] arbre *m* tournant
~SKIP BOX - [text] boîte *f* rotative sautante
~SPINNING CAN - [text] pot *m* tournant de filature
~SPOOL FRAME - [text] armature *f* tournante des bobines
~SPRINKLER - [agric] arroseuse *f* rotative
~STAGE - [opt] (of a microscope) platine *f* tournante (de microscope)
~TABLE - [mech] table *f* tournante
~TABLE PRESS - [mech] presse *f* à tableau tournant
~THE SPINDLE - [text] rotation *f* de la broche
~THREAD GUIDE - [text] guide-fil *m* à mouvement rotatif
~TIPPLER - [mech] culbuteur *m* rotatif
~TONGS - [impl] pincettes *fpl* tournantes
~TOOL-HOLDER - [mech] porte-outil *m* revolver
~WASHING DRUM - [min] tambour *m* laveur rotatif
REVULSION - [med] révulsion *f*
REVULSIVE - [med] révulsif
REWARD CLAIM - [mining] concession *f* de découverte
REWASH, to - [gen] relaver
REWELD, to - [metall] ressouder
REWELDING - [metall] ressoudage *m*
REWIND, to - [gen] rebobiner
[cin] (of a film) rembobiner
[mech] remonter
REWIND STAND - [mech] (a framework carrying reels and mechanism for rotating them, used to reel up material after a process) réenrouleuse *f*
~THE THREAD WASTE, to - [text] rebobiner des restes de bobines
~WARP BALLS INTO COPS, to - [text] remettre les pelotes de chaîne en bobines
REWINDER - [cin] (an apparatus which rewinds a film to make ready for reprojection) bobineuse *f*, réenroulesue *f*
[paper man] (a rewinding machine) rebobineuse *f*
REWINDING - [cin] (the operation of winding back a film after it has been projected) bobinage *m*
[gen] réenroulement *m*
[text] rebobinage *m*
~AND EDGE TRIMMING MACHINE - [mech mach] (a machine which winds sheet material on a reel after a treatment, and at the same time cuts the edge or edges of it to a straight line) enrouleuse-découpeuse *f*
~DEVICE - [text] dispositif *m* de réenroulement
~MACHINE - [text] rebobineuse *f*
~OF THE CLOTH - [text] déchargement *m* du tissu
~THE WARP - [text] rebobinage *m* de la chaîne
REWORK, to - [gen] retraiter
[mining] (an abandoned mine) reprendre (une ancienne mine)
REWORK - [plast ind] (plastic material cut off or otherwise discarded in a process, and later used again after treatment) produits *mpl* retraités
REWRITE, to - [electron] (in a storage device) régénérer
REXITE - [chem] (an explosive) rexite *f*
REYNOLDS NUMBER - [phys] (dimensionless ratio for assessing similarity of motion in viscous fluids, found by multiplying any typical length of a body by its velocity and dividing by the kinematic coefficient of viscosity of the fluid) nombre *m* de Reynolds

R.F. - s. radio frequency
~AMPLIFIER - [radio] (a radio-frequency amplifier for frequencies above 20 KHz) amplificateur *m* HF
~CHOKE - [radio] (radio-frequency choke) bobine *f* d'arrêt HF
~LEAK DETECTOR - [electron] détecteur *m* de fuites à haute fréquence
~SERVICE OSCILLATOR - [radio] (oscillator used for trimming and testing receivers) oscillateur *m* HF
~STANDARD SIGNAL GENERATOR - [radio] (an apparatus used for measurements in radio and carrier wave telephone technique) générateur *m* de signaux de base HF
~TRANSFORMER - [el] transformateur *m* haute fréquence
RGB SIGNALS - [telev] (in colour television; set of three signals representing one of the red, green or blue transmission primaries) signaux *mpl* RVB
Rh - [chem] (symbol for Rhodium) symbole du rhodium
~VALUE - [chem] (a value for the oxidizing power of a system, obtained by taking the common logarithm of that hydrogen pressure which would give rise to the same electrode potential as that of a given oxidation-reduction system in a solution having the same Ph value) rH *m*
RHABDITIFORM - [med] rhabdoïde
RHABDOID - s. rhabditiform
RHABDOMANCY - [gen] (divination; the discovery of springs or precious metals by means of a dividing rod) rebdomancie *f*, divination *f* à la baguette
RHABDOMYOBLASTOMA - [med] rhabdomyoblastome *m*
RHABDOSARCOMA - [med] rhabdosarcome *m*
RHAMNAZIN - [chem] (dye occurring in nature in the form of glucosides) rhamnazine *f*
RHAMNETIN - [chem] (dye occurring in nature in the form of glucosides) rhamnetine *f*
RHAMNOSE - [chem] (a methyl-pentose obtained from several glucosides) rhamnose *m*
RHAPSODY - [mus] rapsodie *f*
RHENIUM - [chem] (a metallic element, symbol Re, A.N. 75, A.W. I86. 3I, with applications in electrical equipment metallurgy, and as a dehydrogenation catalyst) rhénium *m*
RHEOBASE - [el biol] (the intensity of the steady cathodal current of adequate duration suddenly applied which is just sufficient to excite a tissue) rhéobase *f*, seuil *m* fondamental
RHEOGRAPH - [instr] rhéographe *m*
RHEOLOGICAL INSTRUMENTATION - [instr] (system of instruments for the measurement and control of flow) instruments *mpl* rhéologiques
RHEOLOGY - [phys] (the science of flow of matter) rhéologie *f*
RHEOMETER - [instr] (an instrument designed to measure the consistence of semi-fluid substances) consistomètre *m*
RHEOMORPHISM - [geol] (the process by which a preexisting rock is converted into magma by the introduction of migrating volatiles or liquids) rhéomorphisme *m*
RHEOPECTIC - [chem] (displaying the property of rheopexy) rhéopectique
RHEOPEXY - [chem] (the rapid solidification of a thixotropic fluid caused by slow continuous stirring) rhéopexie *f*

RHEOPHILY - [ecology] (the tendency shown by some aquatic animals to live in a current) rhéophilie *f*
RHEOPHORE - [el] rhéophore *m*
RHEOSCOPE - [instr] rhéographe *m*
RHEOSTAT - [el] (a resistor provided with means for readily varying the amount of resistance in a circuit) rhéostat *m*, résistance *f* à curseur
RHEOSTATIC BRAKING - [el] (a system of electric braking in which the motor is connected as a generator, the energy being dissipated in a rheostat) freinage *m* rhéostatique
~BRAKING CONTROLLER - [el] combinateur *m* de résistance de freinage
~CONTROL - [el] (a method of controlling electric motors consisting in the use of variable resistances in series with the motor armatures) régulation *f* rhéostatique
~CONTROLLER - [el] (a controller by means of which more or less resistance can be introduced into a circuit) combinateur *m* rhéostatique
~STARTER - [el] (starter comprising a resistor as means for readily reducing the resistance in a circuit) démarreur *m* régulateur
~STARTING - [el] démarrage *m* rhéostatique
RHEOSTRICTION - [el] (pinch effect; constriction occurring when a liquid conductor is made to carry a heavy current) effet *m* de pincement, rhéostriction *f*
RHEOTAXIS - [biol] (the reaction of an organism to the stimulus of a current) rhéotaxie *f*
RHEOTRON - [electron] (modified type of betatron) rhéotron *m*
RHEOTRONE - [el] (a commutator which reverses the direction of the current) commutateur *m* d'inversion
RHEOTROPIC BRITTLENESS - [metall] fragilité *f* rhéotropique
RHEOTROPISM - s. rheotaxis
RHEUMARTHRITIS- [med] rhumatisme *m* articulaire
RHEUMATALGIA - [med] douleur *f* rhumatismale
RHEUMATISM - [med] rhumatisme *m*
~OF THE HEART - [med] cardiopathie *f* rhumatismale
RHEXIS - [med] rupture *f*
RHIGOSIS - [med] frisson *m*
RHINALGIA - [med] rhinodynie *f*
RHINEDEMA - [med] oedème *m* nasal
RHINESTHESIA - [med] odorat *m*
RHINESTONE - [geol] caillou *m* du Rhin, strass *m*
RHINISM - [med]rhinophonie *f*
RHINITIS - [med] rhinite *f* hypertrophique
RHINOCANTHECTOMY - [med] rhinorraphie *f*
RHINOCEROS - [zool] rhinocéros *m*
RHINOCOELE - [med] ventricule *m* olfactif
RHINODYNIA - s. rhinalgia
RHINOLITH - [med] (nasal calculus) calcul *m* des fosses nasales
RHINOMMECTOMY - s. rhinocanthectomy
RHINOPATHY - [med] affection *f* nasale
RHINOPHARYNX - [anat] cavité *f* nasopharyngienne
RHINOPHYMA - [med] acné *m* hypertrophique
RHINOPLASTY - [med] rhinoplastie *f*
RHINORRHEA - [med] rhinorrhée *f*
RHINOSALPINGITIS - [med] rhinosalpingite *f*
RHINOSCLEROMA - [med] rhinosclérome *m*
RHINOSCOPE - [med] (speculum for viewing the interior of the nose) speculum *m* nasi
RHINOSCOPIC MIRROR - s. rhinoscope
RHIZANESTHESIA - [med] anesthésie *f* radiculaire
RHIZOID - [med] rhizoïde
RHIZOME - [bot] rhizome *m*
Rhm - [radiat] (roentgen-per-hour-at-one meter) rőntgen *m* par heure à I mètre
RHODAMINES - [chem] (generic name for a group of organic pigments which exhibit a red or an orange fluorescence under UV light) rhodamines *f*pl
RHODANINE - [chem] (a synthesis intermediate) rhodanine *f*
RHODANIZING - [el chem] (the process of electroplating with rhodium) galvanoplastie *f* au rhodium
RHODINOL - [chem] (a perfumery agent) rhodinol *m*
RHODINYL ACETATE - [chem] (a perfumery agent) acétate *m* de rhodinyle
RHODIUM - [chem] (a metallic element, symbol Rh, A.N. 45, A.W. I02.9I. A member of the platinum group of metals. It has high chemical inertness and has applications in electroplating and as a catalyst) rhodium *m*
RHODONITE - [min] (natural metasilicate of manganese; rose coloured; triclinic) rhodonite *f*
RHODCHROSITE - [min] (natural carbonate of manganese, crystallizing in the trigonal system in rose-pink rhombohedra) rhodochrosite *f*
RHODIUM GOLD - [min] rhodite *f*
RHODOLITE - [min] grenat *m* magnésien, pyrope *m*
RHODOPSIN - [biol] (the red-light-sensitive pigment of the eye) rhodopsine *f*
RHOMBIC AERIAL - [radio] (a directional aerial consisting of long wire radiators comprising the sides of a rhombus) antenne *f* en losange
~ANTENNA - s. rhombic aerial
RHOMBOHEDRON - [phys] (a form in the triagonal system, having six similar faces each of which is a parallelogram or a rhombus) rhomboèdre *m*
RHOMBOID - [geom] rhomboïdal
RHOMBUS - [geom] rhombe *m*, losange *m*
RHOTACISM - [med] rhotacisme *m*
RHO-THETA SYSTEM - [astronaut] système *m* de navigation en coordonnées polaires
RHUBARB - [chem] (the dried root and rhizome of species of Rheum having application in medicine) rhubarbe *f*
RHUMB - [naut] rumb *m*
~LINE - [surv] (a curved line on the surface of the terrestrial sphere, such that it intersects all meridians at the same angle) loxodromie *f*, ligne *f* de rumb
RHUMBATRON - [electron] (a resonator, generally in the form of a tours) rhumbatron *m*
RHYACOLITE - [min] (glassy type of orthoclase found in the lavas on Vesusius) rhyacolite *f*
RHYMER - [tool] alésoir *m*, équarissoir *m*
RHYOLITE - [min] (fine-grained igneous rock having a chemical composition similar to granite) rhyolite *f*
RHYPARIA - [med] fuliginosités *f*pl
RHYTHM - [gen]rythme *m*
~SECTION - [mus] (in a jazz band) section *f* rythmique
RHYTHMIC CRYSTALLIZATION - [crystal] (phenomenon shown by rocks of widely different composition and characterized by the development of orbicular structure) cristallisation *f* rythmique
~SEDIMENTATION - [geol] (regular interbanding of two or more of sedimentary rocks caused by a seasonal change in the conditions of sediment) sédi-

mentation *f* rythmique
RHYTHMIC TIME-SIGNALS-[radio] (special time-signals regulary broadcast by powerful stations for the determination of navigational data) signal *m* rythmiquement répété
RHYTHMOMETER - [instr] (instrument designed to mark musical rhythm) rythmomètre *m*
RHYTIDECTOMY - [med] excision *f* des rides
RHYTIDOSIS - [med] rhytidose *f*
RIB, to - [gen] nervurer
[mech] (of a pipe etc; to fit with fins) munir d'ailettes
[metall] rider
[agric] labourer à demi
RIB - [anat] côte *f*
[aero] (a structural element running longitudinally in a control surface or wing) nervure *f*, travée *f*
[rubber ind] (of tyres) nervure *f*
[naut] bigots *mpl*, taquets *mpl*
[mining] planche *f* (de charbon)
[print] coulisse *f* (de presse)
[mus] (of violin) éclisse *f* (de violon)
[aero] (of the fuselage) membrure *f*
[shipbuild] membrure *f*
[bot] projecture *f*, nervure *f*
[constr] (curved member of a ribbed arch) étançon *m*, entretoise *f*
[constr] (a frame) membrure *f*
[text] côte *f*
[bookbind] nervure *f*
[geogr] (of mountains) arête *f* (d'une chaîne de montagnes)
[agric] billon *m*, ados *m*
~-AND-PILLAR METHOD - [mining] (in coal mining) dépilage *m* en long avec planche protectrice
~FABRIC - [text] bords-côtes *mpl*
~FRAME - [text] métier *m* à double fonture
~HOSE MACHINE - [text] métier *m* à bas double cylindre
~OF A DOME - [arch] entretoise *f* de coupole
~LIKE INTERLACING - [text] armure *f* en forme d'arête
~OF COLUMN - [arch] nervure *f* de colonne
~PROFILE-[rubber ind] (of tyres) gravure *f* à nervures
~RADIATOR - [auto] radiateur *m* à ailettes
~STITCH - [text] boutage *m* en lignes
~TREAD - s. rib profile
~-WORK - [shipbuild] membrure *f*
~-WORT - [bot] plantain *m* lancéolé
RIBBED - [gen] à nervures, à côtes, cannelé
[mech] (provided with fins) à ailettes
[text] à côtes
[mech] cannelé, garni de nervures
[rubber ind] (of a tyre) strié, nervuré
~ARCH - [arch] voûte *f* à nervures
~BAR - [metall] fer *m* à nervures centrales
~CASTING - [metall] jet *m* nervuré
~CONVEYOR-BELT - [mech] (type of conveyor belt formed with ribs or ridges) courroie-transporteuse *f* cannelée
~COVERING - [text] couverture *f* en reps
~FABRIC - [text] tissu *m* à côtes
~FRAME - [constr] bâti *m* à nervures
~FUNNEL - [impl] entonnoir *m* cannelé
~GRANIT - [text] petit granité *m* en reps
~HEATING PIPE - [th eng] radiateur *m* à nervures
RIBBED PIPE - [mech] tuyau *m* à ailettes
~PLATE - [metall] fer *m* à nervures
~PLUSH - [text] peluche *f* côtelée
~PROFILE - s. rib profile
~-SURFACE - [el] (totally enclosed machine without ventilating ducts, in which the frame cooled by natural convection is provided with ribs intended to increase its surface of contact with the external cooling air) à nervures non ventilées
~TRANSFORMER TANK - [el] bac *m* à ailettes
~TREAD - s. rib profile
~TYRE - [rubber ind] pneu *m* nervuré
~VELVETEEN - [text] velours *m* cordelé
RIBBING OF PISTON - [mech] nervurage *m* de pistons
RIBBON - [text] (narrow strip of fabric) ruban *m*
[metall] (narrow flat strip of metal) ruban *m*
[min] rubané
[meas] roulette *f*
~BLENDER - [ind chem] mélangeur *m* à rubans
~BOX - [text] boîte *f* à ruban
~BRAIDING MACHINE - [text] tresseuse *f* pour rubans
~BRAKE - [mech] frein *m* à ruban, frein *m* à sangle
~BURNER - [th eng] brûleur-ruban *m*
~CONVEYOR - [mech] transporteuse *f* à ruban
~ELEMENT - [el] (heating element consisting of a heating resistor of rectangular section) élément *m* en ruban
~FEED - [text] alimentation *f* à rubans
~FEEDER - [text] transporteur *m* du ruban
[el] ligne *f* en ruban symétrique
[radio] (of an aerial) ruban *m* plat
~LAP MACHINE - [text] étirage *m* nappeur
~LIGHTNING - [met] éclair *m* en sillons
~LIKE FILAMENT - [text] filament *m* (de la soie) en forme de ruban
~LOOM - [text] métier *m* à ruban
~LOOM FRAME - [text] bâti *m* du métier de rubanerie
~LOOM LACE - [text] dentelle *f* au métier à ruban
~LOOM SLAY - [text] battant *m* pour métier de rubanerie
~LOUDSPEAKER - [el acoust] haut-parleur *m* à ruban
~MICROPHONE - [el acoust] (microphone in which the electromotive force is generated by the motion of thin metallic ribbon in the gap of a magnet) microphone *m* à ruban
~-MIXER - [mech] (type of mixer in which the blades consist of long narrow strips curved to a helical form) mélangeur *m* simplex
~OF CROSSED FIBRES - [text] ruban *m* de fibres en travers
~PARACHUTE - [aero] (a type of parachute in which the canopy is formed from ribbons instead of fabric) parachute *m* à rubans
~SAW - [impl] scie *f* à ruban, scie *f* à lame sans fin, scie *f* alternative
~SPRING - [mech] ressort *m* de torsion
~STRIP - [carp] (support for joists) membrure *f*
~WASP - [zool] ichneumon *m*
~WEAVING - [text] rubanerie *f*
~WITH FIGURING WARP - [text] ruban *m* façonné par chaîne
~WITH FIGURING WEFT - [text] ruban *m* façonné par trame
~WITH PLAIN EDGES - [text] ruban *m* à lisières unies
RIBOFLAVINE - [chem] (vitamin B 2. Riboflavine is a dietary essential and also has therapeutic applica-

tions) riboflavine *f*
RIBONUCLEIC ACID - [chem] (essential constituent of cytoplasm, of great significace in biology) acide *m* ribonucléique
RIBOSE - [chem] (a natural sugar with application in biological research) ribose *m*
RIBS - [mus] (the sides of string instruments of the violin type) éclisses *f*pl
RIBWORK - [mech] structure *f* à nervures
RICE - [bot] riz *m*
~CHAFF - [bot] paillettes *f*pl de riz
~FIELD FEVER - [med] fièvre *f* des rizières
~FLOUR - [agric] farine *f* de riz
~GROWING - [agric] riziculture *f*
~MEAL - [agric] (animal feeding) farine *f* riz pour bétail
~MILL - [agric] rizerie *f*
~PADDY - [agric] rizière *f*
~PAPER - [paper man] papier *m* de riz, papier *m* de Chine
~PAPER PLANT - [bot] aralie *f* à papier
~SEEDLING - [bot] mâs *m*
~STARCH - [chem] amidon *m* de riz
~STRAW - [bot] paille *f* de riz
~WATER STOOL - [med] selle *f* riziforme
~WEEVIL - [bot] charancon *m* du riz
RICER - [impl] (a kitchen utensil in which potatoes and other vegetables are pressed) presse-purée *m*
RICH - [gen] riche
[soil] fertile
~AMMONIA WATER - [chem] (strong gas liquor) eau *f* ammoniacale forte
~CLAY - [geol] argile *f* grasse
~COAL - [min] charbon *m* gras
~CONCRETE - [constr] béton *m* gras
~GAS - [gas ind] (coal gas of high calorific value produced from the first stages of carbonization of coal) gaz *m* riche
~IN CARBON - [chem] riche en carbone
~IN SILK - [text] riche *m* en soie
~LIME - [constr] chaux *f* grasse
~LIMESTONE - [min] calcaire *m* gras
~MIXTURE - [mech] mélange *m* riche, carburation *f* riche
~-MIXTURE KNOCK RATING - [mech] (a measure of the anti-knock value of fuel at high power) indice *m* d'octane en mélange riche
~SLAG - [metall] laitier *m* chaud
~SOLVENT - [oil ind](solvent- extract solution leaving a solvent extraction unit) solvant *m* riche
RICHARDSON-DUSHMAN EQUATION - [electron] (equation representing the saturation current of a metallic thermionic cathode in the saturation-current state) équation *f* de Richardson et Dushman
~PLOT - [electron] (by plotting the logarithm of the thermionic current per square Kelvin degree against the reciprocal of the absolute temperature, a straight line is obtained, the slope of which is a measure of the activation energy involved) caractéristique *f* de Richardson
RICHETITE - [min] (ore containing lead and uranium) richetite *f*
RICHNESS OF A GAS-AIR MIXTURE - [gas ind] (a measure of combustion) richesse *f* d'un mélange gazeux
RICIN - [chem] (poisonous principle obtained from the castor oil bean, with uses as an analytical reagent) ricine *f*
RICINOLEIC ACID - [chem] (an unsaturated fatty acid derived from castor oil and having application in soap manufacture and as a source material for other organic compounds) acide *m* ricinoléique
RICINOLEYL ALCOHOL - [chem] (an intermediate for plasticizers and pharmaceuticals and a component of surface coatings) alcool *m* ricinoléique
RICINUS COMMUNIS - [bot] (a castor oil plant) ricin *m*
RICK, to - [agric] ameulonner (le foin etc)
RICK - [agric] meule *f*
~-YARD - [agric] cour *f* de ferme
RICKETS - [med] (defective ossification and softening of the bones caused in children by faulty nutrition) rachitisme *m*
RICKETTSIA - [biochem] (bacteria-like bodies found in the blood and tissue of patients suffering from typhus) rickettsia *f*
RICKETTSIOSIS - [med] rickettsiose *f*
RIDDLE, to - [gen] cribler
RIDDLE - [impl] crible *m*, tamis *m*
[min] (a screen for coal) crible *m*
[metall] (implement for foundry work) tamis *m* de sable
RIDDLER - [metall] (a type of sand sifter) cribleuse *f*
RIDDLING - [gen] criblage *m*
RIDDLINGS - [min] refus *m*pl du crible
RIDE, to - [gen] chevaucher, monter
[naut] être au mouillage
[naut] flotter
~CLEARANCE - [auto] (the maximum displacement in compression of the sprung mass relative to the wheel centre permitted by the suspension system, from the normal load position) débattement *m* maximum de la roue vers le haut
RIDER - [impl] (of a balance; small device, such as a piece of wire or metal strip, placed on the beam of a balance for adjustement purposes) cavalier *m*
[leg] (in documents etc) papillon *m*, ajouté *m*
[math] exercice *m* d'application
[shipbuil] porques *f*pl
[print] (in an offset press) chargeur-balladeur *m*
[geol] (thin seam of coil above a thick one) nerf *m* de roche
[mining] nerf *m*
[min] crible *m*
~CASK - [brew ind]foudre *f* à la rangée supérieure
~SHEET - [metall] tôle *f* de tapis roulant
~SLIDE - [mech] tige *f* de manoeuvre du cavalier
~STRIP - [metall] ruban *m* métallique de protection
RIDER'S BONE - [med] (the ossification of either end of an adductor muscle of the thigh) exostose *f*
~SPRAIN - [med] étirement *m* des muscles adducteurs
~TENDON - [med] rupture *f* des tendons des adducteurs
RIDGE - [met] (a band of relatively high pressure between two regions of anticyclonic character) dorsale *f* de haute pression
[constr] (the line on which the rafters meet) faîte *m*, crête *f* de comble
[gen] crête *f*, arête *f*, ride *f*, strie *f*
[agric] billon *m*, butte *f*, couche *f* de fumier
[geogr] croupe *f*, crête *f*
~A ROOF, to - [constr] couronner le faîte d'un com-

ble
RIDGE-AND-FURROW AERATION - [hydr] aération *f* dans un bassin à radier en dents de scie
~-AND-FURROW IRRIGATION - [hydr] épandage *m* par sillons
~-AND-FURROW TANK - [hydr] bassin *m* à radier en dents de scie
~BEAM - [constr] poutre *f* faitière
~BOARD - [constr] (horizontal timber at the upper ends of the rafters) longrine *f* de faîtage
~CAPPING - [constr] (the covering which is applied over the ridge) couverture *f* du faîte
~CIRCUIT - [el] (a lightning conductor) circuit *m* des faîtes
~COVERING - s. ridge capping
~OF HILLS - [geogr] chaîne *f* de collines
~-PIECE - [constr] faîtage *m*, poutre *f* de faîte
~PLOUGH - [agric] buttoir *m*
~-POLE - s. ridge-beam
~PURLIN - [constr] panne *f* faitière
~RIB - [constr] nervure *f* de sommet
~ROLLER - [agric] presse *f* à paille
~ROOF - [constr] comble *m* à deux égouts
~-ROPE - [naut] filière *f* de beaupré
~-TILE - [constr] (tile specially shaped to cover the ridge of a roof) tuile *f* faitière
~WAVEGUIDE - [electron] guide *m* d'ondes à moulures
RIDGER - [agric] (a tool used to form ridges in the surface soil) billonneuse *f*, sillonneuse *f*
RIDGING - [agric] labour *m* en billons
[constr] enfaîtement *m*
~PLOUGH - s. ridger
RIDGLESS SEAM - [rubber ind & shoe man] joint *m* sans surépaisseur
~VARIABLE SPEED DRIVE - [mech] commande *f* à vitesse variable sans échelons
RIDGY - [gen] aréteux
RIDING - [mining] rupture *f* de boisage
~BOOT - [shoe man] botte *f*
~LIGHTS - [aero] feu *m* de position
RIEBECKITE - [min] (metasilicate of sodium and iron) riebeckite *f*
RIEKE DIAGRAM - [electron] (graph in polar coordinates showing the behaviour of electronic tubes as a function of the load impedance) diagramme *m* de Rieke
RIFFLE - [min] (a strip of wood across a sluice) rifle *f*
[paper man] (a strip of wood placed across the stream of pulp) traverse *f* du séparateur
[mining] (a groove or indentation in the bottom of an inclined sluice, to separate the gold contained in sand or gravels) riffle *f*, rifle *f*
RIFFLED PLATE - [metall] tôle *f* rainurée, tôle *f* striée
RIFFLER - [paper man] (shallow tank at the wet end of a paper-making machine, in which the pulp is made to flow round a partition, while water is added and sand allowed to settle out) sablier *m*
[tool] rifloir *m*
RIFFLES - [metall] rides *f*pl de bords
RIFLE, to - [firearms] (to make a spiral groove in the bore of a gun) rayer (un fusil)
RIFLE - [firearms] (firearm with spiral grooves in the surface of the bore) fusil *m*, carabine *f*
[firearms] (in the bore) rayure *f*
~BAR - [metall] barre *f* de rotation
RIFLE-BORE - [firearms] âme *f* rayée
~-BUTT - [firearms] crosse *f* de fusil
~DRILL - [mech] foret *m* à canon
RIFLED - [firearms] rayé
RIFLING - [firearms] (the spiral grooves in the surface of a bore) rayage *m*
RIFT - [gen] fente *f*, fissure *f*
[mining] (in quarrying) délit *m*
[metall] plan *m* de cassure
~-VALLEY - [geol] fossé *m*
RIG, to - [aero] (to adjust and align the parts of an aircraft) monter (avion)
[gen] garnir, armer, installer
[naut] gréer, équiper
[mech] (a derrick with a yard) pousser une vergue en bataille
[mining] appareiller
RIG - [gen] mécanisme *m* de manoeuvre
[ind] équipement *m*, installation *f*, équipage *m*
[oil ind] (the derrick and surface equipment of a drilling unit) appareil *m* de sondage, derrick *m* de puits à pétrole
[naut] gréement *m*
~FLOOR - [oil ind] plan *m* de la tour de sondage
~HOIST - [petr ind] élévateur *m* pour sondes
~IN, to - [naut] rentrer
~OUT, to - [naut] pousser dehors
[gen] équiper
~SHIFT - [oil ind] déplacement *m* de la tour
~TEST - [gen] épreuve *f* du banc
~THE CLOTH, to - [text] (to fold, to plait the cloth) plier le tissu
RIGGER - [aero] (workman specially skilled in the rigging of aircraft) monteur *m*
[mech] (a band or belt pulley) poulie *f* à courroie
[gen] mécanicien *m*
RIGGING - [aero] (adjustement or alignment of the parts of an aircraft) réglage *m*
[mining] mécanisme *m* de sondage
[telev] (in the studio) montage *m* préliminaire des lampes
(of a balloon, the system of wires and cords, by which the weight is distributed over the envelope) câblage *m*, haubanage *m*
[naut] manoeuvre *f*, gréage *m*
[mech] mécanisme *m* de manoeuvre, montage *m*
[shipbuild] garniture *f*, capelage *m*
~ANGLE OF INCIDENCE - [aero] (the angle between the cord line of the main plane and the horizontal, the aircraft being in the rigging position) angle *m* de calage des plans
~AT THE MASTHEAD - [naut] capelage *m* du mât
~BAND - [aero] (in balloons, a reinforced band fitted to the envelope for the rigging) courroie *f* d'attache du câblage
~BASIN - [shipbuil] bassin *m* d'équipement
~LINE - [aero] (in parachutes, a cord attached to the canopy to transmit the load to it) suspente *f*
~LOFT - [shipbuild] atelier *m* de garniture
~POSITION - [aero] (in attitude such that when the lateral axis of the aircraft is horizontal, an arbitrary longitudinal datum line is also horizontal) assiette *f* de réglage
~TENDER - [telev] voiture *f* de montage
~TIME - [oil ind] temps *m* de sondage
RIGHT, to - [gen] corriger, rectifier
[gen] (a boat, a car etc) redresser

RIGHT, to - [naut] relever
RIGHT - [gen] correct, exact
[gen] (direction) droit
[geom] droit
[mech] (normal to a base) normal
[gen & leg] droit *m*, privilège *m*
~-AND-LEFT COUPLING - [mech] lanterne *f* de serrage
~AND LEFT-HAND TURN-OFF - [railw] aiguillage *m* à deux voies symétriques
~-AND-LEFT SCREW - [mech] vis *f* à pas contraires
~ANGLE - [geom] angle *m* droit
~-ANGLE CROSSING - [railw] traversée *f* de voie à angle droit
~-ANGLED - [geom] rectangle, rectangulaire
~-ANGLED BEND - [plumb] coude *m* d'équerre
~-ANGLED PRISM - [opt] prisme *m* à réflexion totale
~-ANGLED TRIANGLE - [geom] triangle *m* rectangle
~ASCENSION - [astr] (the angular distance of a celestial body from the vernal equinox measured eastwards along the celestial equator in units of time) ascension *f* droite (AR)
~CIRCULAR CONE - [geom] cône *m* droit circulaire
~CONE - [geom] cône *m* droit
~GAUZE - [text] gaze *f* à croisement à droite
~-HAND - [gen] à droite
[mech] (of a gear) à droite, dans le sens des aiguilles d'une montre
~-HAND ACCESSORY - [mech] (an accessory unit which rotates clockwise as seen from the driven end) accessoire *m* tournant à droite (dans le sens des aiguilles d'une montre)
~-HAND COMBER - [text] peigneuse *f* à commande droite
~-HAND DRIVE - [mech] (a drive which rotates clockwise as seen from the driving end) entraînement *m* tournant à droite
~-HAND ENGINE - [aero] (an engine of which the propeller shaft revolves clockwise as seen by an observer facing towards the propeller at the engine end of the shaft) moteur *m* de rotation à droite
~-HAND HELICAL GEAR - [mech] engrenage *m* hélicoïdal à pas à droite
~-HAND LOCK - [mech] serrure *f* à droite
~-HAND MILLING CUTTER - [mach tool] fraise *f* avec denture à droite
~-HAND PATTERN - [mech] modèle *m* droit
~-HAND RULE - [el] (the Fleming's rule; a simple rule for relating the directions of the flux, motion and e.m.f. in an electrical machine) règle *f* de Fleming, règle *f* des trois doigts de la main droite
~-HAND STEERING - [auto] conduite *f* à droite
~-HAND SWITCH - [railw] changement *m* à droite, aiguillage *m* à droite
~-HAND TAP - [mech] taraud *m* à droite
~-HAND THREAD - [mech] filetage *m* à droite
~-HAND THREAD STRAND - [text] toron *m* à torsion droite
~-HAND TOOTH-FACE - [mech] profil *m* droit de la dent
~-HAND TWINE - [text] ficelle *f* à torsion droite
~-HAND TWIST - [text] torsion *f* droite
~-HAND TWISTED YARN - [text] fil *m* tordu à droite
~-HAND WORM - [mech] vis *f* sans fin à droite
~-HANDED - [gen mech] à droite, dextrorsum
[tool] (outil) pour la main droite
~-HANDED CRYSTAL - [crystall] cristal *m* droit
RIGHT-HANDED MOMENT - [mech] moment *m* dextrorsum
~-HANDED POLARIZED WAVE - [radio] (clockwise polarized wave; elliptically polarized transverse electromagnetic wave in which the rotation of the electric intensity vector is clockwise for an observer looking in the direction of the propagation) onde *f* à polarisation elliptique à droite
~-HANDED PROPELLER - [aero] (a propeller of which the rotation is clockwise as seen from behind the aircraft) hélice *f* à droite
~-HANDED SCREW LINE - [mech] spire *f* à droite
~-HANDED SPIRAL - [mech] hélice *f* à droite
~-HANDED STRAND - [text] toron *m* tordu à droite
~-HANDED THREAD - [text] filet *m* à droite
~LINE - [geom] ligne *f* droite
~OF FOREST PASTURE - [agric] droit *m* de pâture en forêt
~OF LIEN - [leg] droit *m* de gage
~OF PLEDGE - [leg] droit *m* d'hypothèque
~OF USE - [leg] droits *mpl* d'usage
~OF SEARCH - [leg] (in nautical law) droit *m* de visite
~OF WAY - [leg] (the right of a person to pass over the land of another) droit *m* de passage
[gen naut etc] droit *m* de priorité
[leg] (the actual land over which a road is build) servitude *f* de passage
~SIDE - [text] (of a cloth) beau côté *m* (du tissu)
[auto] (offside) côte *m* droit
~TO TIMBER - [leg] affouage *m*
~TRIANGLE - [geom] triangle *m* rectangle
RIGHTING MOMENT - [aero & naut] (restoring moment) moment *m* redresseur, couple *m* de rappel
RIGID - [gen] rigide, raide
~AIRSHIP - [aero] (a mechanically-propelled lighter-than-aircraft, of which the envelope is maintained in shape by a framework and not solely by gas pressure) dirigeable *m* rigide
~ARCH - [arch] (continuous arch without hinges or joints) arc *m* sans articulation
~AXLE - [auto] essieu *m* rigide
~BEARING PLUMMER-BLOCK - [mech] palier *m* à coussinets rigides
~BEND - [mech] segment *m* conique
~BODY - [phys] (aggregate of material particles in which the interaction forces of the particles are such that the distance between any two particles remains constant with time) corps *m* rigide
~BOX TYRE FRAME - [mech] bâti *m* en forme de caisson rigide
~CHECK OF SWITCH BLADES - [el] contrôleur *m* de lame d'aiguille conduit
~CONDUIT - [gen] tuyau *m* rigide
~CONNECTION - [plumb] reccord *m* rigide
~CONTACT - [el] contact *m* rigide
~DIE - [metall & plast ind] moule *m* rigide
~FASTENING - [el] fixation *f* rigide
~-FRAME BRIDGE - [constr] pont *m* en treillis rigide
~FROG - [railw] croisement *m* rigide
~LEG - [mech] (of a tripod) branche *f* droite
~MOUNTING - [mech] (e.g. an aeroengine) montage *m* rigide
~PIN-JOINTED PURLIN - [constr] lierne *f* sans articulation
~PLASTICS - [plast ind] (plastics having poor flexural qualities) plastique *m* rigide

RIGID PVC - [plast ind] (polyvinyl chloride or polyvinyl chloride acetate copolymer having a relatively high degree of hardness) PVC *m* rigide
~STAY - [constr] hauban *m* rigide
~SUPPORT - [el] (support designed to withstand without appreciable flexure the transverse and vertical loads of the line) potelet *m*
~SUSPENSION - [constr] suspension *f* rigide
~TRIPOD - [mech] pied *m* à trois branches droites
RIGIDITY - [gen] rigidité *f*, raideur *f*
[phys] (shear modulus; in a homogeneous isotropic elastic medium, the ratio between the shear stress and the shear strain) module *m* de cisaillement
~OF THE SPRING - [mech] rigidité *f* de ressort
~OF THE WIRE - [text] rigidité *f* de la dent
RIGIDLY-MOUNTED BLADE - [mech] (a blade which is not in any way pivoted to the shaft) pale *f* rigide
RILL - [geogr] (small stream of water) ruisselet *m*
[astr] (narrow straight valley on the face of the moon) rainure *f* (sur la face de la lune)
[geol] trace *f*
[mining] taille *f* à mi-pente
~EROSION - [geol] érosion *f* par ruissellement
~STOPE - [mining] gradin *m* incliné
RILLE - s. rill
RIM, to - [gen] border
[mech] janter
RIM - [gen] jante *f*, bord *m*, rebord *m*
[mech] (of a wheel) jante *f*, pourtour *m*
[mech] (of a flywheel) couronne *f* (de poulie)
~BAND - [rubber ind] (of tyres) bande *f* de fond de jante, flap *m*
~BASE - [rubber ind] (of tyres; the well of a rim) base *f* de la jante
~BLIGHT - [bot] "rim blight" *m*, maladie *f* des bords des feuilles
~BOLT - [auto] boulon *m* de jante
~BRAKE - [mech] (of a bicycle) frein *m* à la jante
~BRUISING - [rubber ind] (of tyres) endommagement *m* de la jante
~CHAFING - [rubber ind] (of tyres) frottement *m* de la jante
~CHANNEL - [auto] creux *m* de la jante
~CLAMP - [auto] crampon *m* de jante
~CUT - [rubber ind] coupure *f* de l'enveloppe par la jante
~DIAMETER - [rubber ind] (of tyres) calibre *m* de la jante
~EDGE - [rubber ind] (of tyres) collerette *f* de jante, voile *f* de jante
~FERMENTATION - [brew ind] fermentation *f* aux bords de la cuve
~-GRAVELS - [mining] bordures *f*pl du placer
~GROOVE - [mech] gorge *f* du volant
~GUTTER - [rubber ind] (of tyres) crochet *m* de jante, creux *m* de jante
~LOCK - [mech] (a lock screwed to the face of the door) serrure *f* encloisonnée
~MAGNET - [telev] aimant *m* à neutralisation du champ magnétique terrestre
~NUT - [mech] écrou *m* de jante
~OF FLYWHEEL - [mech] couronne *f* dentée du volant
~OF HEAD-LAMP - [auto] visière *f* de phare
~OF LANDING GEAR WHEEL - [aero] (a flange which retains the tire in position laterally) encadrement *m*, bordure *f*
~OF THE BOBBIN - [text] rebord *m* de la bobine
RIM PIVOT - [text] pivot *m* de la bande
~PROFILE - [rubber ind] profil *m* de jante
~PULLEY - [text] poulie *f* à bride
~SHAFT - [mech] arbre *m* moteur
~SHAFT PINION - [mech] pignon *m* de marche
~SHAFT SPUR WHEEL - [mech] pignon *m* de l'arbre moteur
~SHOULDER - s. rim edge
~SPEED - [mech] (the speed of translation of the periphery of a circular saw) débit *m* de rouage
~STAPLE - [mech] gâche *f*
~TOOL - [auto] outillage *m* de démontage des jantes
~WIDTH - [rubber ind] largeur *f* de jante
RIMA - [anat] scissure *f*
RIME, to - [gen] couvrir de givre
RIME - [met] (a deposit of feathery ice occurring during simultaneous frost and fog, usually on the windward side of an object) givre *m*, gelée *f* blanche
~-BREAK - [met] (on trees) bris *m* de givre
~FROST - [met] (hoar-frost) givre *m*
~ICE - [met] (thin white opaque ice-layer formed by small water droplets freezing on impact) givre *m* blanc
RIMER - s. reamer
RIMMED STEEL - s. rimming steel
~STRIP - s. rim band
RIMMING PRESS - [rubber ind] presse *f* pour mettre les jantes
~STEEL - [metall] (steel not completely deoxidized before casting) acier *m* effervescent
RIND - [bot] écorce *f*, peau *f*, pelure *f*
[agric] (of cheese) croûte *f*
[agric] (of bacon) couenne *f*
~GRAFTING - [bot] greffe *f* en couronne
RINDERPEST - [vet] (acute contagious disease affecting cattle) peste *f* bovine
RING, to - [acoust] (to emit a resonant sound) sonner, tinter
[gen] (to surround, to encircle) encercler
[telecomm] (to telephone) appeler
[mech] baguer, fretter
RING - [gen] anneau *m*
[gen] (enclosure) enceinte *f*
[zool] (of pigeons) collier *m*
[mech] (of a piston) segment *m*
[bot] (of trees) cercle *m*
[impl] rond *m*, cercle *m*, rondelle *f*, bague *f*
[geom] cercle *m*
[mech] couronne *f*, frette *f*
[mech] (of a ball bearing) cercle *m*, couronne *f*
[chem] chaîne *f* fermée
[text] (for a spinning frame) anneau *m* (d'un métier à filer)
[acoust] son *m* métallique, tintement *m*
[chem] (e.g. carbon) bague *f* (en charbon)
[naut] organeau *m* d'ancre
[telecomm] (telephone call) appel *m*
[radar] (indicating the range on the screen) cercle *m*
~ABSCESS - [med] infiltration *f* annulaire de la cornée
~AND WEDGES - [mining] anneau *m* de manoeuvre
~ANGEL - [radar] (spurious signal on radar screens probably due to birds) signal *m* parasite annulaire
~ARMATURE - [el] (armature with a ring winding) induit *m* en anneau
~AUGER - [impl] couronne *f*

RING BACK KEY - [telecomm] (telephony) clé *f* de rappel du demandeur
~-BACK SIGNAL - [telecomm] (in manual hold) signal *m* de rappel
~-BACK TELEPHONE APPARATUS - [telecomm] appareil *m* téléphonique mixte
~BALANCE - [instr] (for measuring pressures) tore *m* pendulaire
~BALANCE MANOMETER - [instr] (manometer in which the container for the liquid is circular) balance *f* manométrique
~BARK - [bot] écorce *f* annulaire
~BEVEL GEAR - [mech] couronne *f* conique
~-BOLT - [mech] boucle *f* d'amarrage
~BURNER - [th eng] brûleur *m* à anneau
~CABLE SYSTEM - s. ring main system
~CANAL - [zool] (of a medusa) canal *m* annulaire
~CHAIN - [chem] chaîne *f* fermée
~CIRCUIT - [electron] circuit *m* annulaire
~CLAMP - [plumb] contre-bride *f* de tête
~CLOTHING - [text] garniture *f* annulaire
~COLLECTOR - [el] (of induction motor) collecteur *m* en anneaux
~COMPLEX - s. ring dyke
~COMPOUND - [chem] composé *m* cyclique
~CONNECTION - [el] montage *m* en boucle
~CORE - [metall] noyau *m* annulaire
~COUNTER - [el] (a loop of interconnected bistable elements) compteur *m* annulaire
~COWLING - [aero] (annular fairing round a radial engine) bague *f* de carénage du capot
~CUTTER - [mech] fraise *f* à bagues
~DISTRIBUTION SYSTEM - [hydr] réseau *m* de ceinture
~DOFFER - [text] peigneur *m* à garniture annulaire
~DOUBLER - [text] métier *m* à retordre à anneau
~DOUBLING FRAME - [text] métier *m* à retordre à anneau
~DYKE - [geol] filon *m* éruptif annulaire
~DYNAMO - [mech] dynamo *f* à anneau
~FACE - [mech] surface *f* latérale du segment du piston
~FILAMENT - [el] (of an electric filament lamp) filament *m* en anneau
~FILLET - [text] peigneur *m* à garniture placée en anneau
~FINGER - [anat] annulaire *m*
~FOCUSING - [phys] (in particle spectrometers) focalisation *f* annulaire
~FOLLOWER - [mech] bague *f* filetée
~-FORWARD SIGNAL - [telecomm] (telephony) signal *m* d'appel
~FRAME - [text] métier *m* à anneaux
~FRAME TWIST - [text] fil *m* de chaîne continu à anneau
~GASKET - [mech] garniture *f* annulaire
~GATE - [metall] attaque *f* annulaire
~GAUGE - [mech] (steel ring with an internal diameter of a specified size used to check the diameter of finished cylindrical work) triboulet *m*
~GEAR - [mech] (an annular gear without a central structure) couronne *f* dentée
~GEOMETRY - [nucl] symétrie *f* annulaire
~GRINDER - [mach tool] machine *f* à rectifier les bagues
~GROOVE - [mech] gorge *f* du segment de piston
~HEAD - [el acoust] (magnetic head in which the magnetic material forms an enclosure with one or more air gaps) tête *f* annualire
RING JOINT FLANGE - [oil ind] bride *f* à joint circulaire
~LUBRICATING SYSTEM - [mech] coussinet *m* à bague (de graissage)
~LUBRICATION - [mech] graissage *m* par bague
~MAIN - [telecomm] câble *m* annulaire
[el] conducteur *m* de bouclage
~MAIN - [gas ind] ceinture *f*
[hydr] canalisation *f* bouclée
~MAIN SYSTEM - [el] (distribution system in which the cables form a closed ring) réseau *m* maillé
~MICROMETER - [instr] (annular micrometer) micromètre *m* en anneau
~MODE FILTER - [electron] (mode filter in the form of a resonant metalling ring) filtre *m* de mode annulaire
~MODULATOR - [electron] modulateur *m* annulaire
~MOTTLE - s. ring spot
~-NET - [fishing] épuisette *f*
~NOZZLE - [mech] buse *f* en anneau
~NUT - [mech] virole *f*
~"O" - [mech] joint *m* en tore
~OF GLAZE - [el] anneau *m* d'émail
~OF PLUG - [el] nuque *f* de fiche
~OF STRETCHED WOOL FIBRE - [text] sonorité *f* de la laine
~-OILED BEARINGS - [mech] palier *m* à graissage automatique par bagues
~OILER - [auto] anneau *m* graisseur
~ON THE BOBBIN BUTT - [text] anneau *m* sur l'embase de la bobine
~-OPERATED NETWORK - [el] réseau *m* à exploitation en boucles
~ORE - [min] minerai *m* en cocarde
~OSCILLATOR - [radio] oscillateur *m* annulaire
~PIRN - [text] canette *f* à anneaux
~PISTON - [mech] piston *m* à segments
~RADIATOR - [mech] (circular radiator for radial flow of cooling air) radiateur *m* annulaire
~RAIL - [text] plate-bande *f* porte-anneaux *m*
~ROLL MILL - [min] moulin *m* à cylindres annulaires
~SCALER - [instr] (a scaling circuit in which the asymmetrical condition is passed on to the next tube in line, with the last tube feeding back to the first) échelle *f* en anneau
~SCALER - [electron] circuit *m* démultiplicateur à réaction
~SEAL - [electron] (fused junction between the two halves of a bulb) scellement *m* annulaire
[mech] joint *m* en anneau
~SHAKE - [timber] roulure *f*
~-SHAPED - [gen] annulaire
~-SHAPED THERMOCOUPLE - [instr] (an apparatus designed to measure the temperature of the skin under the clothing) thermoélément *m* annulaire
~SLEEVE - [el] (of an induction motor, the collector ring hub) manchon *m* porte-anneaux
~SPANNER - [impl] clef *f* de calibre
~SPINDLE - [text] broche *f* pour continus à anneau
~SPINDLE OIL PUMP - [text] pompe *f* à huile pour broches à anneaux
~SPINDLE OILER - [text] huileur *m* pour broches à anneaux
~SPINNER - [text] fileuse *f* au continu à anneau
~SPINNING - [text] (method of spinning which pro-

duces a smooth worsted yarn) filage *m* à anneau
RING SPINNING MACHINE - s. ring spinner
~ SPINNING SYSTEM - [text] système *m* de filage à anneau
~ SPOT - [bot] (a plant disease) taches *f*pl annulaires
~ SPRING - [mech] ressort *m* annulaire, bague *f* élastique
~ SPRING LANDING GEAR - [aero] (type in which the impact is absorbed by a string of elastic steel rings engaging each other by conical surfaces formed on their inner and outer circumferences) train *m* de atterrissage à bagues élastiques
~ SPUN YARN - [text] fil *m* de chaîne du continu
~ STAND - [impl] baguier *m*
~ SWITCH - [electron] (a waveguide switch embodying a resonant metallic ring) commutateur *m* à anneau résonnant
~ TEMPLE - [mech] templet *m* à disques verticaux
~ THREAD GAUGE - [mech] calibre *m* à anneau pour filetage
~ THROSTLE - [text] métier *m* continu à filer à anneau
~ TRAVELLER - [text] (in cotton spinning) curseur *m*
~ TWISTER - [text] métier *m* à retordre à anneaux
~ -TYPE TRANSFORMER - [el] transformateur *m* annulaire
~ WARP - [text] fil *m* de chaîne du continu
~ WARP COP - [text] canette-chaîne *f* de continu à anneau
~ WINDING - [el] (winding formed of coils wound round a magnetic core of anular form, so that one side of each coil is looped through the ring) enroulement *m* à anneau
~ WIRE - [telecomm] fil *m* de nuque
~ WITH NEEDLE - [text] anneau *m* à aiguille
~ WITH STUD - [text] anneau *m* avec cheville
~ YARN - [text] fil *m* de continu à anneau
RINGBONE - [vet] (osteoarthritis of the coronary joint of horses) forme *f* sur le paturon
RINGER - [el acoust] sonnette *f*, sonnerie *f*
[telecomm] (telephone) machine *f* d'appel
RINGING - [el acoust] sonnerie *f*
[telecomm] (telephone) appel *m*
[radio] (damped oscillation in a network as a result of a sudden change in the input excitation) oscillation *f* parasite
[telev] (a damped oscillatory response to a picture pulse) dédoublement *m* d'image
~ CHANGE-OVER SWITCH - [telecomm] commutateur *m* de sonnerie
~ CURRENT - [telecomm] courant *m* d'appel
~ CYCLE - [telecomm] phase *f* d'appel
~ ENGINE - [constr] (a form of pile driven for small piles) sonnette *f* à tiraude
~ GUARD SIGNAL - [telecomm] signalisation *f* de connexion établie
~ KEY - [telecomm] bouton *m* d'appel
~ OSCILLATOR - [electron] auto-oscillateur *m*
~ PILOT LAMP - [telecomm] lampe *f* de contrôle d'appel
~ POSITION - [telecomm] position *f* d'appel
~ REPEATER - [telecomm] (device allowing a low-frequency ringing current to by-pass the two-wire repeater by regenerating the ringing current on the other side) signaleur *m* fréquence basse
~ TONE - [telecomm] (telephone) tonalité *f* de retour d'appel
~ -TRIP RELAY - [telecomm] relais *m* de coupure, relais *m* d'arrêt d'appel
RINGWORM - [med] (tines) teigne *f* annulaire
RINMANN'S GREEN - [paint] (alternative name for Cobalt green-usually a complex of zinc and cobalt oxides, but sometimes with phospates of these metals or with aluminium oxide, calcined at a high temperature. It has good light and heat fastness and alkali resistance. Also called Zinc Green and Cobalt Green) vert *m* de Rinmann
RINSE, to - [gen] rincer
RINSE - [gen] rinçage *m*
~ AWAY THE DIRT, to - [text] enlever les saletés par rinçage
~ THE FILAMENT, to - [text] rincer le fil
~ THE FLAX, to - [text] rincer le lin
~ THE RETTED HEMP - [text] laver le chanvre roui
~ THE WOOL, to - [text] rincer la laine
RINSER - [ind chem] (device for washing appliances by a spray of water) rinceuse *f*
RINSING - [gen] rinçage *m*
[gen] (the rinsing water) eau *f* de rinçage
~ BASIN - [hydr] rinçoir *m*, cuve *f* de rinçage
~ MACHINE - [mech] (used in textile industry etc) rinceuse
~ TANK - [impl] cuve *f* de rinçage
~ VAT - [min] tambour *m* laveur
RIOMETER - [instr] (instrument measuring the arrival of ionizing material near the earth) riomètre *m*
RIP, to - [gen] déchirer, fendre
[text etc] découdre
[carp] (to saw timber across the direction of the grain) refendre, scier le bois en long
[constr] (a roof) découvrir (un toit)
[rubber ind] (a tyre) éventrer (un pneu)
RIP - [gen] déchirure *f*, fente *f*
[tool] arpon *m*
~ CORD - [aero] (a cord which opens the pack and allows deployment, on being pulled) câble *m* de déclenchement
~ FENCE - [mach tool] (a saw rip) guide-pièce *m*
~ LINK - [aero] (of a balloon) anneau *m* du câble de déclenchement
~ PANEL - [aero] (of balloon) panneau *m* de déchirure
~ PIN - [aero] (of parachute; the pin which is withdrawn by the rip cord to release the flaps and allow deployment) broche *f* du câble de déclenchement
~ TOOL - s. rip bar
~ TRACK - [railw] (USA only; a repair track) voie *f* de réparations
RIPARIAN - [gen] riverain
~ LANDS - [surv] terrain *m* riverain
RIPCORD - s. rip cord
RIPE - [gen] mûr
~ CONDITION - [agric] maturité *f*
~ FOR CUTTING - [agric] bon à couper
~ FOR SOWING - [agric] mûr pour les semailles
~ HEMP SEED - [agric] chènevis *m* mûr
~ SLUDGE - [hydr] boue *f* mûre
~ VISCOSE - [text] viscose *f* mûre
RIPEN, to - [agric] mûrir
[agric] (wine, cheese etc) affiner
RIPENED CONDITION OF THE VISCOSE - [text] maturité *f* de la viscose
~ FILTER - [hydr] filtre *m* mûr
RIPENESS - [gen] maturité *f*
~ OF THE POD - [agric] maturité *f* de la capsule
~ OF THE SOLUTION - [chem] maturité *f* de la solu-

tion
RIPENESS OF THE STEM - [agric] maturité *f* des tiges
~OF THE VISCOSE - [text] maturité *f* de la viscose
RIPENING - [gen & agric] maturation *f*
[agric] (of wheat etc) jaunissement *m*
[agric] (of grapes) véraison *f* (du raisin)
~OF THE BAST - [text]maturation *f* du liber
~OF THE FRUIT - [agric] aoûtement *m* des fruits
~ROOM - [agric] (for bananas) mûrisserie *f* (de bananes)
RIPPER - [agric] scarificateur *m*
[tool] arpon *m*, fendoir *m*
[impl] (a rip-saw) scie *f* à refendre, arpon *m*
RIPPING BAR - [tool] bec-de-corbin
~CHISEL - [impl] ciseau *m* à planches
~KNIFE - [text] couteau *m* à découdre
~PANEL - [aero] (special strip in a balloon envelope which can be torn open for rapid deflation on landing) fuseau *m* de déchirure totale
~SLIT - s. ripping slot
~SLOT - [aero] ouverture *f* de déchirure
~THE ROOF - [mining] recoupage *m* du plafond
RIPPLE, to - [gen] onduler, se rider
[text] (e.g. the hemp) dréger (le chanvre), égrener (le chanvre)
RIPPLE - [gen] ride *f*, ondulation *f*
[acoust] murmure *m*
[text] (of cloth) ondulé *m*
[el] (term denoting the higher harmonics in an alternating current wave) ondulation *f*
[text] (used for cleaning) peigne *m* d'égrugeoir
~COMPONENT - [el] composante *f* ondulée
~CONTROL - [el] (method of controlling street lighting from a central point by means of a high-frequency ripple superimposed on the current-carrying conductors of an electrical power system) commande *f* à ondulation résiduelle
~FILTER - [radio] filtre *m* uniformisateur
~FINISH - [paint] (a special type of finish in which the whole surface is covered with wrinkles of small and uniform dimensions, usually obtained by stoving) finissage *m* ondulé
~FREQUENCY - [el] fréquence *f* d'ondulation
~MARK - [geol] (on the bedding planes of some sedimentary rocks) ripple-mark *f*
~NOISE - [telecomm] bruit *m* d'alimentation
~RATIO - [el] taux *m* d'ondulation résiduelle
~THE HEMP, to - [text] égrener le chanvre
~TRAY - [oil ind] (type of distillation column tray with perforated sinusoidal corrugations) plateau *m* ondulé
~VOLTAGE - [el] (alternating component of a pulsating voltage) tension *f* d'ondulation
~WELD - [metall] soudure *f* ondulée
RIPPLED - [gen] ondulé, ridé
~FINISH - [text] finissage *m* similisé
~FLAX - [text] drège *f* du lin
~SURFACE - [metall] (a surface defect) surface *f* ondulée
~WALL AMPLIFIER - [radio] amplificateur *m* à paroi ondulée
RIPPLES - [phys] (surface waves on a liquid) ondulations *f*pl, rides *f*pl
RIPPLING BENCH - [text] égrugeoir *m*
~COMB - [text] drège *f*, peigne *m* d'égrugeoir
~THE HEMP - [text] égrugeage *m* du chanvre
RIPRAP - [constr] (for hydraulic construction) enrochement *m*
RIPSAW - [impl] scie *f* à refendre
RISE, to - [gen] se lever, se soulever
[gen] (to ascend) monter, s'élever
[gen] (to increase) croître
[comm] (of prices) être à la hausse
[gen] (to originate) dériver
[geogr] (of streams) prendre sa source
[food] (of yeast) lever
[met] (of the wind) croître, forcer
RISE - [gen] (of the ground) montée *f*, côte *f*
[comm] (of prices) hausse *f*
[el] augmentation *f* (de tension etc)
[met] (of the barometer) hausse *f* (du baromètre)
[geol] inclinaison *f*
[aero] ascension *f*
[constr] flèche *f*, rampant *m* (d'une voûte)
[geogr] (of rivers) source *f*, naissance *f*
[hydr] crue *f*
[mining] montage *m*, remontage *m*
[constr] (of a step) hauteur *f* de marche
[constr] (the vertical distance between two points) élévation *f*
[naut] (of the tide) flot *m* (de la marée), flux *m* (de la marée)
[metall] (of a steam-hammer) volée *f* du marteau
~-AND-FALLFLOOR- [el] plancher *m* de cabine mobile
~-AND-FALL LEVELLING - [surv] nivellement *m* longitudinal
~-AND-FALL MILLER - [mach] fraise *f* pour profils
~AND FALL OF THE SEA - [met] flot *m* et jusant de la mer
~-AND-FALL PENDANT - [el] (light fitting the height of which can be adjusted by a pulley and counterweight) lampe *f* à contre-poids
~-AND-FALL SURFACING TABLE - [mach tool] plateau *m* à réglage vertical
~-AND-FALL SYSTEM - [surv] (system of reduction of levels) système *m* de nivellement longitudinal
~OF A CAMBER OF ARCH - s. rise of an arch
~OF AN ARCH - [arch] hauteur *f* sous clef
~OF CROWN - [gas ind] (of gasholders) flèche *f* de calotte
~OF FLOOR LINE - [shipbuild] (line joining the end of the flats of keel and the ship's bilge curve tangent) acculement *m* d'une varangue
~OF THE WING RAIL - [railw] surélévation *f* du contre-coeur
~TIME - [telev] (duration of the output signal of a network when a step signal is applied to the input) durée *f* d'établissement, temps *m* de montée
[radio] (of a receiver or amplifier) durée *f* d'établissement d'une implusion
~TIME CORRECTION - [telev] correction *f* du temps de montée
~TIME DISTORSION - [telev] (distorsion of the time during which an electron moves from one point to another in an electron tube) distorsion *f* du temps de montée
~WORKING - [mining] taille *f* montante
RISER - [constr] (the vertical part of a step) contremarche *f*
[plumb] tuyau *m* de montée, colonne *f* montante
[mech] broche *f* d'évent
[mining] remontée *f*, remontage *m*

RISER - [metall] (in a mould, a passage for the metal to flow after filling the mould cavity) évent *m*, trou *m* d'évent, masselotte *f*
[el] (a commutator lug) talon *m* du collecteur
[shoe man] gorge *f* du talon
[geol] faille *f* inverse
[oil ind] tuyau *m* de refoulement
~HEAD - [metall] main-courante *f*
~MAIN - [oil ind] tuyau *m* de refoulement
~PIN - [metall] (in foundry work) mandrin *m* d'évent, broche *f* d'évent
~PIPE - [plumb] (a common inlet and outlet pipe) pipe *f* (d'entrée ou sortie)
[plumb] (only USA) conduite *f* montante
RISING - [astr] (of the sun) levée *f*
[met] (of the tide) crue *f*
[gen] levant, qui monte
[soil] éminence *f* (du terrain)
[mining] (a raise) montée *f* (du pétrole dans le puits)
[instr] (of the barometer) hausse *f*
~AIR MASSES - [met] masses *fpl* d'air ascendantes
~AND FALLING SAW - [impl] (a type of circular saw the spindle of which can be moved in relation to the position of the working table) scie *f* à mouvement ascensionnel et de descente
~AND FALLING MOTION - [mech] mouvement *m* de monte et baisse
~ARCH - [arch] (arch having a springing line which is not horizontal) voûte *f* rampante
~BOX - [text] boîte *f* à navette montante
~BOX CHANGE MOTION - [text] changement *m* de navette par boîtes montantes
~GRADIENT - [railw] rampe *f*
~GROUND - [soil] élévation *f* de terrain
~HEAD - [metall] (of a mould) masselotte *f*
~MAIN - [el] (a main circuit running from one floor of a building to another) colonne *f* montante
[plumb] conduite *f* montante
[mining] colonne *f* d'exhaure
~NUMBER - [text] décochement *m*
~PLATFORM - [text] plateau *m* à hauteur variable
~ROLL BATCHER - [text] enrouler *m* à rouleau ascendant
~SHAFT - [constr] (shaft excavated from below) puits *m* foncé vers le haut
[text] lame *f* montante
~SPINDLE VALVE - s. rising stem valve
~STEM VALVE - [mech] (an external screw valve) robinet *m* à vis extérieure
~-SUN ANODE - [electron] (used for magnetrons) anode *f* de magnétron à double fréquence
~-SUN-TYPE MAGNETRON - [electron] magnétron *m* à double fréquence
~THREAD SPIRALS - [text] spires *fpl* de fil montantes
~TIDE - [astr] marée *f* montante
~VOLTAGE - [el] tension *f* en augmentation
~WOOD - [naut] (of the keel) contre-quille *f*
RISK, to - [gen] risquer
RISK - [gen] risque *m*
~AREA - [radar] (the region in which the receiver field strenght is greatly reduced by some obstruction) zone *f* d'ombre
~PROFIT - [comm] fonds *mpl* pour déficits de caisse
RISTOCETIN - [chem] (an antibiotic produced by Nocardia lurida and having therapeutic uses) ristocétine *f*
RIVE, to - [gen] fendre
RIVEL - [paint] (a wrinkle or fold in the surface of a coating, formed during drying) ride *f*
RIVELLING - [paint] (formation of wrinkless in a coating during drying) formation *f* de rides
RIVER - [geogr] fleuve *m*, cours *m* d'eau
~-BANK - [gen] bord *m* du fleuve
~-BAR - [hydr] barre *f* de rivière
~BASIN - [naut] bassin *m* fluvial
~BED - [gen] lit *m* de rivière
~BOAT - [naut] bateau *m* fluvial
~-BORNE - [transp] transporté par voie d'eau
~-BOTTOM DISEASE - [vet] (anaemia of horses) anémie *f* infectieuse du cheval
~CAPTURE - [geol] (the cutting off of a stream by another stream having greater power of erosion) capture *f* d'un fleuve
~CHANNEL - [hydr] chenal *m* de rivière, lit *m* d'un fleuve
~CLARIFYING BASIN - [hydr] bassin *m* d'épanouissement en rivière
~CLAY - [min] argile *f* fluviale
~CONSTRUCTION - [hydr] ouvrage *m* fluvial
~CROSSING - [plumb] (in mainlaying) conduite *f* sous-fluviale
~-CROSSING DITCH - [constr] (in mainlaying) souille *f*, cunette *f*
~DIGGINGS - [hydr] placer *m* fluvial
~DREDGE - [mech] drague *f* fluviale
~DREDGING - [hydr] dragage *m* de rivières
~DRIFT - [hydr] diluvium *m*, apport *m* des fleuves
~FALL - [hydr] cascade *f*
~GRAVEL - [constr] galets *mpl* roulés
~-HEAD - [geogr] source *f*
~MINING - [mining] exploitation *f* des alluvions immergées
~MOUTH - [geogr] bouche *f* de fleuve
~PLAIN - [geol] bas-fonds *mpl* d'une rivière
~POLLUTION - [ecology] pollution *f* des rivières
~-PORT - [naut] port *m* fluvial
~RETTING - [text] rouissage *m* à l'eau courante
~SAND - [geol] sable *m* fluviatile
~SOURCE - s. river-head
~STONE - [gen] galet *m*
~TERRACE - [geol] terrasse *f* de cours d'eau
~WALL - [hydr] bajoyer *m*
~WASHING - [text] lavage *m* en rivière
RIVERSIDE - [gen] rive *f*, bord *m* de l'eau
[gen] riverain
RIVERINE - [gen & bot] riverain
RIVET, to - [gen] river, clouer
[mech] clouer, river, riveter
RIVET - [mech] (a pin having a pre-formed head at one end used for permanently uniting other parts by passing it through them and forming a second head) rivet *m*, clou *m* à river
[ceramics] attache *f*
~-BAR - [mech] barre *f* à rivets
~CATCHER - [impl] corbeille *f* de retenue de rivets
~DOLLY - [impl] tas *m* à bouteroller
~DRIFT - [mech] broche *f* d'assemblage
~FORGE - [metall] forge *f* à chauffer les rivets
~FURNACE - [metall] fourneau *m* à chauffer les rivets
~GUN - [impl] marteau *m* à rivets
[mach tool] riveuse *f*

RIVET HEAD - [met] (the enlarged part or head at the end of a rivet) tête *f* de rivet
~-HEATING FURNACE - s. rivet furnace
~HOLE - [mech] (a hole punched or drilled to receive a rivet) trou *m* pour rivet
~IRON - [metall] fer *m* à rivets
~JOINT - [metall] assemblage *m* par rivets
~PIN - [text] rivet-goupille *m*
~-ROD - s. rivet-bar
~SET - s. rivet-snap
~SHANK - [mech] tige *f* d'un rivet
~-SNAP - [tool] (a tool for forming a partially spherical head on a rivet when the latter is closed) bouterolle *f*, chasse-rivet *m*
~STEEL - [metall] acier *m* à rivets
~TEST - [mech] essai *m* des rivets
~TONGS - [impl] tenaille *f* à rivets
RIVETED ANGLE IRON RIM - [metall] couronne *f* rivetée en fers cornières
~BOWL - [impl] cuve *f* rivée
~FRAME - [auto] châssis *m* rivé
~JOINT - [mech] (a butt, or lap joint) assemblage *m* rivé, rivure *f*
~-ON PIECE - [horol] bride *f* rivée
~STEEL PIPE - [metall] conduite *f* en tôle d'acier rivée
RIVETER - s. riveting machine
RIVETING - [mech] rivure *f*, rivetage *m*, clouure *f*
~-DIE - [tool] bouterolle *f*
~HAMMER - [impl] rivoir *m*, marteau *m* à river
~JOINT - s. riveted joint
~MACHINE - [mach] (a machine for closing rivets by power) riveuse *f*, riveteuse *f*
~PRESS - [mech] presse *f* à river
~SET - [tool] chasse-rivet *m*, bouterolle *f*
~SLIDE - [mech] coulisse *f* à river
~TONGS - [mech] pince *f* à rivets
RIVNUT - [mech] (a rivet which is drilled and internally threaded) rivet *m* foré et fileté
RIVULET - [geogr] ruisseau *m*
RMS INVERSE VOLTAGE RATING - [el] valeur *f* efficace d'une grandeur périodique
~REVERSE VOLTAGE RATING - s. rms inverse voltage rating
~PULSE AMPLITUDE - [radio] (the effective pulse amplitude) amplitude *f* d'impulsion efficace
~VALUE - [el] (the root mean square value of an alternating current or voltage) valeur *f* efficace
R-METER - [instr] (any ionization meter calibrated to read in roentgen) röntgenomètre *m*
ROACH, to - [gen] (the mane horse etc) couper en brosse (la crinière)
ROACH - [zool] gardon *m*
[hydr] (the brushing upwards of water by the float of a hydroplane) sillon *m*
[naut] échancrure *f*
ROAD - [gen] route *f*, voie *f*, chemin *m*
[naut] rade *f*
[mining] galerie *f*, voie *f*
~ADHERENCE - [auto] (of the tyres) adhérence *f* au sol
~APPROACH - [constr] voie *f* d'accès
~BALLAST - [constr] ballast *m* de route
~BED - [constr] plate-forme *f* des terrassements
~BLOCK - [constr] bloc *m* à paver
~BREAKER - [impl] (pneumatic hammer) marteau *m* pneumatique
ROAD BREAKING - [constr] défoncement *m* des routes
~BRIDGE - [constr] pont-route *f*
~BUILDER - [gen] constructeur *m* de routes
~BUILDING - [constr] construction *f* de routes
~BUMP - [gen] bosse *f*
~CENTRE - [road constr] axe *m* de la route
~CLEARANCE - [auto] garde *f* au sol
~CONDITION - [gen] viabilité *f*
~CREW - [railw] (USA only; train crew) personnel *m* roulant
~CROSSING - [constr] carrefour *m*
[railw] passage *m* à niveau
~CROWN - [constr] heurt *m*
~CURVE - [constr] virage *m* (dans la route)
~DRAIN WELL - [constr] ponceau *m*
~EMBANKMENT - [constr] banquette *f*
~ENGINE - [mech] machine *f* routière, locomotive *f* routière
~FINISHING MACHINE - [constr] finisseuse *f*
~FORK - [constr] fourche *f* de routes
~FOUNDATION - [constr] ballast *m*
~FOUNDATION MATERIAL - [constr] matériaux *mpl* d'assiette d'une chaussée
~GRADE - [constr] pente *f* de la route
~HAULAGE - [transp] transport *m* routier
~HEAD - [railw] tête *f* de voie
~HOLDING - [auto] tenue *f* de route
~-HOUSE - [gen] auberge *f*
~IMPROVEMENT WORKS - [constr] aménagement *m* des routes
~MAINTENANCE - [gen] entretien *m* routier
~MAKER - s. road builder
~MAKING - s. road building
~MAP - [surv] carte *f* routière
~MARKING - s. road sign
~MENDER - [gen] cantonnier *m*
~METAL - [constr] (the broken stone which form the surface of a macadamized road) matériaux *mpl* de empierrement pour routes
~METALLING - [constr] empierrement *m* des routes, cailloutage *m*
~MIXING - [road constr] malaxage *m* sur place
~NETWORK - [constr] réseau *m* routier
~PAD - [mech] (of a track) tampon *m*
~PATROL - [gen] police *f* de la route
~POST - [constr] poteau *m* indicateur
~-RAIL VEHICLE - [transp] véhicule *m* rail-route
~ROLLER - [mech] rouleau *m* compresseur
~ROLLING - [road constr] cylindrage *m* de la chaussée
~SCRAPER - [impl] rabot *m* d'ébouage
~SERVICE - [gen] patrouille *f* routière
~-SIDE - [gen] côté *m* de la route
~SIGN - [gen] signal *m* routier
~STABILITY - [auto] stabilité *f* routière
~STRAIN - [auto] effort *m* dû à un mauvais revêtement routier
~STUD - [constr] clou *m* pour routes
~SURFACE - [constr] revêtement *m* d'une route
~SURFACING - [constr] revêtement *m* routier
~SYSTEM - s. road network
~TAR - [constr] goudron *m* pour routes
~TAR MACHINE - [constr] répandeuse *f*, machine *f* à goudronner la couche de surface
~TEST - [auto] essai *m* routier
~TRACTION - [transp] traction *f* sur voies de terre
~TRACTOR - [mech] tracteur *m* routier

ROAD TRAFFIC CENTRAL STRIP - [constr] bande *f* médiane
~TRAIN - [tranps] train *m* routier
~TRANSPORT - [transp] transport *m* routier
~TRIAL - s. road test
ROADBED - [constr] assiette *f* de la route
[railw] superstructure *f* de la voie
~AND STATIONS - [railw] matériel *m* fixe
ROADBLOCK - s. road block
ROADING - [gen] (USA only) entretien *m* routier
[constr] (USA only) construction *f* de routes
ROADS - [naut] rade *f*
ROADSTEAD - s. roads
ROADSTER - [auto] roadster *m*
[naut] navire *m* en rade
ROADTRACK - s. roads
ROADWAY - [constr] chaussée *f*
[constr] (of a bridge) aire *f* de pont
[gen] (a road) passage *m* carrossable
[mining] galerie *f*
ROAK - [metall] (a seam due to slag pressed into the surface) soufflure *f* éclatée
ROAN - [zool] rouan
[bookbind] basane *f*
ROAR - [acoust] hurlement *m*
[zool] (of wild beasts) rugissement *m*
ROARING - [acoust] rugissant, rugissement *m*
~FORTIES - [met] (strong westerly winds occurring between latitude 40° and 50°S) bonne brise *f* de l'ouest, parages *mpl* océaniques entre les 40e et 50e degrés de latitude nord
ROAST, to - [gen] rôtir
[gen] (of coffee) torréfier
[metall] (to heat sulphide ores in air to convert to oxide) griller
~SINTERING - [metall] frittage *m*
ROASTED COPPER PYRITE - [min] pyrite *f* de cuivre grillée
~MALT - [brew ind] malt *m* noir, malt *m* torréfié
~ORE - [min] minerai *m* grillé
ROASTER - s. roasting furnace
ROASTING - [metall] (the operation of heating sulphide ores in air to convert to oxide) grillage *m*
[gen] (of coffee) torréfaction *f*
[gen] rôtissage *m*
[min] fritte *f*, calcination *f*
~BED - [min] lit *m* de grillage
~CORE - [min] noyau *m* de grillage
~FURNACE - [metall] (furnace in which fine ores and concentrates are roasted to eliminate sulphur) four *m* à griller
~JACK - [impl] tournebroche *m*
ROB, to - [mining] dépiler, déboiser
ROB A MINE, to - [mining] écrémer une mine
~THE PILLARS, to - [mining] déhouiller les piliers
ROBAND - [naut] (a piece of spun yarn used to fasten the head of a sail to a spar) raban *m* d'envergure
[text] (rope band) raban *m*
ROBBIN - s. roband
ROBBINS - [text] (wool fibres, longer than noils, removed during combing) fibres *fpl* éliminées pendant le peignage
ROBE - [gen] robe *f*, vêtement *m*
~RAIL - [auto] barre *f* porte-manteau
ROBIN - [zool] rouge-gorge *m*
ROBOMB - s. robot bomb
ROBOT PILOT - [aero] (automatic control device to keep aircraft on set course and in level flight) pilote *m* automatique
ROBUST - [gen & mech] robuste, rustique
ROCAMBOLE - [bot] rocambole *f*
ROCHELLE SALT - [chem] (sodium potassium tartrate used for making piezo-elastic crystals) sel *m* de Seignette
ROCK, to - [gen] (to swing about an axis) balancer, bercer
ROCK - [gen] roc *m*, rocher *m*
[geol] roche *f*
[naut] écueil *m*, roche *f*
~ALUM - [min] alun *m* de roche
~ASPHALT - [geol] roche *f* asphaltique
~-BASIN - [geol] bassin *m* géologique, bassin *m* de surcreusement
~BED - [geol] fond *m* de roche
~BENCH - [geol] plateforme *f* d'abrasion
~BEND - [geol] pli *m*
~BIT - [tool] (drilling bit used for hard formations) trépan *m*
~BODY - [transp] tombereau *m* basculant
~-BORING MACHINE - [mech] perforatrice *f* au rocher
~BREAKER - [constr] concasseur *m*, casse-pierres *m*
~BUCKET - [mech] benne *f* basculante
~BURST - [mining] (caused by a heavy pressure) coup *m* de toit
~BUTTER - [min] beurre *m* de pierre
~CAPPING - [mining] chapeau *m* de filon
~CHANNELER - [mech] trancheuse *f*
~CHERRY - [bot] cerisier *m* de Sainte-Lucie
~CHUTE - [mining] cheminée *f* à remblai
~CORK - [min] liège *m* de montagne
~CRYSTAL - [min] (colourless quartz, crystalline or otherwise) cristal *m* de roche
~-CRYSTAL MINE - [min] cristallière *f*
~CUTTING - [constr] deblai *m* de roches
~-DRIFT - [mining] galerie *f* au rocher
~DRILL - [mech] perforatrice *f* au rocher, bosseyeuse *f*
~DRILL HOSE - [mech] manchon *m* pour perforatrice
~DUST - [min] pulvérin *m* rocheux
~-DUST BARRIER - [mining] arrêt-barrage *m* des poussières incombustibles
~-DUSTING - [min] (of coal) schistification *f*
[geol] schistification *f*
~EXCAVATION - [constr] dérochage *m*
~FALL - [geol] éboulement *m* de rocher
~-FALL BASIN - [geol] bassin *m* d'éboulement
~FEVER - [med] (or undulant fever) fièvre *f* de Malte
~FILLING - [constr] remblai *m*
~FLINT - [min] silex *m* noir, phthanite *f*
~FLOUR - [geol] cilice *f*
~-FORMING MINERALS - [geol] (minerals occurring as the dominant constituent of igneous rocks) minérais *mpl* principales la roches
~GARDEN - [agric] jardin *m* de rocaille
~GAS - [geol] gaz *m* naturel
~GLACIER - [geol] coulée *f* d'éboulis mouvants
~INTRUSION - [geol] intrusion *f* rocheuse
~LEATHER - [min] cuir *m* de montagne
~LEVER - [mech] balancier *m* de suspension
~MARROW - [min] moelle *f* de roche
~MASS - [geol] masse *f* rocheuse
~MATRIX - [geol] gangue *f* rocheuse

ROCK MEAL - [min] (soft, white variety of calcium carbonate; it resembles cotton) farine *f* fossile
~MILK - [min] (soft white, easilybreaking variety of calcium carbonate) lait *m* de roche, agarice *f*
~OIL - [min] huile *f* de roche, naphte *m* minéral
~OUTCROP - [geol] affleurement *m* rocheux
~PHOSPHATE - [min] phosphorite *f*
~RUBBLE - [geol] brèche *f* de dislocation
~RUBY - [min] grenat *m* oriental
~SALT - [min] (halite) sel *m* gemme, halite *f*
~SHAFT - [mining] cheminée *f* à remblais
~SHEET - [geol] nappe *f*
~SOAP - [min] savon *m* de montagne, oropion *m*
~STEADINESS - [cin] (absence of movement of the image) stabilité *f*
~STEP - [geol] seuil *m* de cirque
~TERRACE - [geogr] terrasse *f* dans le roc
~WOOL - [min] (mineral wool) laine *f* minérale
ROCKER - [mech] (oscillating part transmitting the movement of a push-rod to the valve stem in an overhead valve) culbuteur *m*
[impl] branloire *f*
[photo] balance-cuvette *m*
~ARM - [mech] (mechanical oscillating about a pivot) basculeur *m*
[mech] (of a motor) balancier *m* de repartition, balancier *m* de renvoi
~BAR - [mech] bielle *f* de pont métallique
~BRACKET - [mech] (of a radial engine) boîte *f* de support des balanciers
~CONVEYOR - [mech] transporteur *m* à secousses
~DUMP - [transp] camion *m* à basculeur
~LEVER - [mech] (of motor) balancier *m*
~RING - [el] anneau *m* porte-balais
~SHAFT - [mech] balancier *m* de renvoi
ROCKERY - [agric] jardin *m* de rocaille
~ROCKET - [gen] fusée *f*
[mech] (a jet engine containing its own fuel and oxidant) moteur-fusée *m*
~AIRPLANE - [aero] avion-fusée *m*
~-ASSISTED AIRCRAFT - [aero] (type of aircraft in which additional thrust is obtained for take-off or for reaching the operating airspeed for ramjets) avion *m* à entraînement par fusée
~ASSISTED TAKE-OFF - [aero] décollage *m* assisté par fusée
~BOMB - [aero] bombe *f* à fusée
~CHARGE - [astronaut] charge *f* de propergol d'une fusée
~CLUSTER - [astronaut] groupe *m* de fusées
~ENGINE - [mech] (jet engine containing its own fuel and oxidant) moteur-fusée *m*
~GLIDER - [astronaut] planeur *m* à fusée
~LAUNCHER - [astronaut] dispositif *m* de lancement de fusées
~MISSILE - [astronaut] missile *m* à fusée
~MOTOR - [mech] moteur-fusée *m*
~NOZZLE - [astronaut] tuyère *f* de fusée
~PILE - [astronaut] réacteur *m* nucléaire de fusée
~PLANE - [aero] avion *m* fusée
~PROPELLANT - [chem] propergol *m*, fluide *m* propulsif
~-PROPELLED - [aero] à fusée
~PROPULSION - [aero] (reaction propulsion using internally sotred, instead of atmospheric, oxygen for combustion) propulsion *f* par fusée
~RAMJET - [aero] stato-réacteur *m* à fusée de démarrage
ROCKET RANGE - [astronaut] portée *f* d'une fusée
~RASP - [impl] (blunt nose rasp) râpe *f* conique
~SHIP - [astronaut] véhicule *m* à fusée
~SIGNAL - [signal] signal *m* à fusée
~SLED - [astronaut] traîneau *m* à fusée
~THRUST - [mech] poussée *f* d'une fusée
~VEHICLE - s. rocket ship
ROCKETRY - [gen] technologie *f* des fusées
ROCKING - [gen] oscillant, à bascule, oscillation *f*, tremblement, secousses *f*pl
[mech] dodinage *m*
[railw] mouvement *m* de lacet
~ARC FURNACE - [metall] (indirect arc furnace in which the body can be oscillated about the axis of the electrodes) four *m* à arc oscillant
~ARM - [mech] bras *m* oscillant
~BEAM - [text] porte-fil *m* mobile, porte-fils *m* oscillant
~FEEDER - [mining] alimentateur *m* oscillant
~FRAME - [text] cadre *m* oscillant
~FURNACE - [metall] four *m* oscillant
~GEAR - [mech] engrenage *m* à bascule
~GUIDE LATH - [text] latte *f* de guidage à mouvement oscillatoire
~LEVER - [mech] s. rocker
[mech] (of motor) s. rocker (motor)
~MILL - [metall] bocard *m* à pilons
~MOTION - [mech] mouvement *m* oscillatoire
~MOTION - [railw] mouvement *m* de lacet
~MOVEMENT - [text] mouvement *m* pendulaire
~MOVEMENT OF THE CLOTH - [text] mouvement *m* oscillant du tissu
~PIER - [constr] pile *f* oscillante
~RESISTOR FURNACE - [metall] four *m* à résistance oscillant
~TEMPLE CUTTER - [text] templet *m* coupe-fils oscillant
~TREE - [text] arbre *m* de l'épée de chasse
~YARN GUIDE - [text] guide-fil *m* oscillant
ROCKOON - [astronaut] fusée-sonde *f* emportée par un ballon
ROCKSHAFT - [mech] (of a rocking lever) arbre *m* de renversement de marche
ROCKWELL HARDNESS - [phys] (a measure of relative hardness of the surface of a substance) dureté *f* Rockwell
~HARDNESS TEST - [metall] (a hardness test based on the degree of indentation caused by a steel ball or point under specific conditions of load) essai *m* de dureté Rockwell
ROCKY DESERT - [geol] désert *m* rocheux
~GROUND - [geol] fond *m* rocheux
~HEADS - [brew ind] hautes mousses *f*pl
~POINT EFFECT - [electron] (a flash arc; sudden increase in space current caused by irregularities in the electrode surface) effet *m* Rocky-Point
ROD - [gen] canne *f*, baguette *f*
[mech] tige *f*, tringle *f*
[metall] (as used for reinforcing concrete) verge *f*, rondin *m*
[metall] (round iron) fer *m* rond
[anat] (of the retina) bâtonnet *m*
[fishing] canne *f* à pêche, gaule *f*
[surv] mire *f*
[mech] (of a valve) tige *f*, queue *f*, bielle *f* (du tiroir)

ROD - [mining] rallonge *f*
[oil ind] tige *f* de sonde
~AERIAL - [radio] (an aerial consisting of a bar of metal) antenne *f* à tige
~-ANODE TUBE - [electron] (X-ray tube in which the target is placed towards the end of a long tubular anode) tube *m* à rayons X à anode tubulaire
~ASSEMBLY - [auto] groupe *m* piston et bielle
[nucl] (the complete set of uranium rods in a nuclear reactor) ensemble *m* de barres
~BANK - [nucl] groupe *m* de barres
~BEARING - [mech] manchon *m* de bielle
~BENDING MACHINE - [mech] cintreuse *f*
~BODY - [mech] corps *m* de bielle
~BUSH - [mech] (in a locomotive) bague *f* de bielle d'accouplement
~CELL - [zool] (photosensitive cell of the retina) cellule *f* à bâtonnet
~CONTROL - [mech] commande *f* à tige
~COUPLING - [mining] emmanchement *m* de tiges
~CRACK - [metall] crique *f* longitudinale
~DROP EXPERIMENT - [nucl] expérience *f* de barre descendante
~ELEVATOR - [mech] (used in boring) pied-de-boeuf *m*
[oil ind] élévateur *m* de tige
~END - [mech] (in int.comb.engines) tête *f* de bielle
~END BEARING - [mech] (the bearing in the eyes of a clevis) roulement *m* à tige filetée
~-END STRAP - [mech] chape *f* de bielle
~EPITHELIUM - [zool] bâtonnets *mpl*
~FIBRE - [zool] (the fibre with which a retinal rod is connected internally) fibre *f* du bâtonnet
~GAP - [el] éclateur *m* à barreaux
~GRANULES - [zool] noyaux *mpl* des bâtonnets de la rétine
~GUIDE - [oil ind] lanterne *f* de guidage
~HEAD - [auto] (head of the steering rod) embout *m* de la bielle de direction
~HOISTER - s. rod elevator
~INSULATOR - [el] isolateur *m* à tige
~LATTICE - [nucl] (lattice composed of uranium rods in a nuclear reactor) réseau *m* de barres
~MILL - [metall] broyeur *m* à barres, laminoir *m* à barres
[mech] (a mill in which the triturating elements are of rod-form) train *m* de serpentage
~MILLING - [metall] laminage *m* d'un fil
~MIRROR - [electron] miroir *m* grille
~PASS - [metall] cylindre *m* de laminoir à fils
~POSITION INDICATOR - [nucl] indicateur *m* de la position d'une barre
~PRESS - [mech] presse *f* à colonnes
~PUMP - [metall] pompe *f* auxiliaire
~READING - [surv] côte *f* lue sur la mire
~REFLECTOR - s. rod mirror
~ROLLING - [metall] laminage *m* d'un fil
~SUPPORT - [mining] tourne-à-gauche *m* de support
~-TURNING TOOL - [mining] manche *f* de manoeuvre
~WIRE - [metall] fil *m* laminé, fil *m* machine
~WRENCH - [mining] tourne-à-gauche *m*
RODDING - [plumb] (the clearing of a stoppage in a pipe by inserting a rod to break down and remove the obstruction) déobturation *f* par verges
[oil ind] transmission *f* rigide
[mining] réparation *f* d'un guidage
RODDING EYE - [plumb] (an access eye, a removable plug fitted to an elbow for the elimination of obstructions) trou *m* pour verge de déobturation
RODENTICIDE - [chem] rodenticide *m*, raticide *m*
RODENT ULCER - [med] ulcère *m* rongeant
RODENTS - [zool] rongeurs *mpl*
RODMAN - [surv] (the man holding the staff in surveying work) porte-mire *m*
ROENTGEN - [radiat] (the quantity of x-or gamma radiation taken as an international unit of intensity) röntgen *m*
~RAYS - [radiat] (electromagnetic waves of very short wavelength which are set up when the velocity of electrons is altered suddenly) rayons *mpl* X
~THERAPY - [radiat] (radiotherapy with X-rays) röntgenthérapie *f*
ROENTGENOGRAPHY - [radiat] (radiography) röntgenographie *f*
ROENTGENOLOGY - [radiat] (the section of radiology dealing with X-rays) röntgenologie *f*
ROENTGENOSCOPE - [radiat] (fluoroscope) appareil *m* de radioscopie
ROGERSITE - [min] (decomposition product of samarskite) rogersite *f*
ROIL, to - [gen] (to make muddy by stirring the sediment) troubler
ROILED - [gen] trouble
ROKE - s. roak
ROLL, to - [gen] rouler
[metall] laminer
[naut & aero] (to change attidute about the longitudinal axis) rouler
[aero] (to make a complete revolution about the longitudinal axis) faire un tonneau
[mech] (to press with rollers) cylindrer
[mech] (to level with rollers) cylindrer
[mech] (to thread by means of rollers) fileter au laminoir
[paper man] (to calender) calandrer
[constr] cylindrer, passer au rouleau
[mech] (to spread under a roller) se laminer
[leather ind] calandrer
[print] (the operation of inking) encrer, charger
[gen] (to undulate) onduler
[mech] (to sway) osciller
[gen] (to envelope with) enrouler
ROLL - [gen] (of paper) rouleau *m*
[gen] (a list) liste *f*
[gen] (of drums) roulement
[food] (of bread) petit pain *m*
[gen] (a record) contrôle *m*, rôle *m*
[naut] (of the sea) coup *m* de roulis
[mech etc] (a cylinder) cylindre *m*
[metall] (of a rolling mill) cylindre *m*
[naut] roulis *m*
[anat] corde *f*
[aero] (of an aircraft; a complete revolution about the longitudinal axis) tonneau *m*
[photo] (of a film) bobine *f*
[acoust] roulement *m*
[arch] (of a column) (moulure) à rouleau
[text] (of a loom) ensouple *f* (d'un métier)
[gen] (a coil) galette *f*, bobine *f*
~ADJUSTMENT - [mech] (of a press) ajustage *m* du cylindre
~ANGLE - [aero] angle *m* de roulis
~ARRANGEMENT - [mech] arrangement *m* des cylin-

dres
ROLL ATTITUDE - [aero] (the attitude of an aircraft in in regard the longitudinal axis) position *f* de roulis
~AXIS - [auto] (the line joining the front and rear roll centres) axe *m* de roulis
~BACK, to - [comput] (to rerun) repasser
~CALIBER - [mech] calibre *m* de cylindre
~-CALL - [gen] appel *m* nominal
~CAMBER - [metall] bombage *m* du cylindre
~CASTING MACHINE - [metall] machine *f* pour la production des cylindres par centrifugation
~CENTRE - [auto] (the point in the transverse vertical plane through any pair of wheel centres and equidistant from them, at which lateral forces may be applied to the sprung mass without producing an angular roll displacement of it) centre *m* de roulis
~CHANGING - [metall] remplacement *m* des cylindres
~-COAT, to - [metall] revêtir le cylindre
[paint] enduire avec cylindres
~COATER - [impl] (a machine for applying a coating to sheet material by means of rollers) machine *f* à enduire sur rouleaux
~COATING - [metall] revêtement *m* du cylindre
~COMPACTING - [metall] laminage *m* de poudre
~CROSSING ARRANGEMENT - [mech] dispositif *m* de croisement des cylindres
~CRUSHER - [mech] concasseur *m* à rouleaux
~DEFLECTION - [mech] déflection *f* de cylindre
~DOCTOR - [impl] racle *f* rotative
~FACE WIDTH - [metall] (of a rolling mill) largeur *f* du cylindre
~FEED - [mech] dosage *m* à rouleau
~FEEDER - [metall] rouleau *m* de dosage
~FILM - [photo] (a film wound on a spool and wrapped in protective paper) bobine *f* de film, pellicule *f* en bobine
~-FILM DEVELOPING TANK - [photo] cuve *f* pour développer le film en rouleau
~FLUTING MACHINE - [mech] machine *f* à rainurer les cylindres
~FLY - [text] duvet *m* de hérisson
~FORGING - [metall] forgeage *m* par lamination
~FORMING MACHINE - [mach tool] machine *f* à profiler les feuillards
~GRINDER - s. roll grinding machine
~GRINDING-[metall] meulage *m* de cylindres
~GRINDING MACHINE - [mach tool] (machine for grinding the surfaces of rollers to a finish and or to given diameters) machine *f* meuler les cylindres
~GRIZZLY - s. roll screen
~HOLDER - [photo] châssis *m* à rouleaux
~IN THE SEED, to - [agric] passer le rouleau sur la graine
~INDICATOR - [instr] (in aircraft) indicateur *m* de roulis
~LAMINATED - [metall] laminé
~LATHE - [mach tool] tour *m* à cylindres
~MARKS - [metall] traces *f*pl de cylindre
~NECK - [mech] (the part of a roller which is reduced in diameter to form a journal carried in a bearing) tourillon *m* de cylindre
~-NIP ADJUSTMENT - [mech] (the operation of setting the distance between rollers) réglage *m* d'écartement des cylindres
~OF FIBROUS RIBBON - [text] rouleau *m* de ruban de matière fibreuse
~OF FILM - [photo] galette *f* de film
ROLL OF TWINE - [text] rouleau *m* de ficelle
~OF UNCURED RUBBER - [rubber ind] (unvulcanized sheet rubber in the form of a roll) manchonnage *m* de gomme crue, poupée *f*
~-OFF - [electron] augmentation *f* d'amortissement
~OUT, to - [comput] (to read out of a counter) extraire les retenus
~-OVER MACHINE - [mech] retourneur *m* de moule
~-OVER TABLE - [metall] table *f* de renversement
~PASS - [metall] phase *f* d'usinage
~SCALE - [metall] écailles *f*pl de laminage
~SHELL - [metall] (of crushing rolls) anneau *m* de cylindre (de broyeur)
~SCREEN - [impl] (for coke; a rotating grid) grille *f* à rouleaux
~SLITTING AND WINDING MACHINE - [mech] coupeuse-bobineuse *f*
~SPINDLE - [metall] allonge *f*, axe *m* du cylindre
~SPOT WELDING - [metall] soudure *f* par points à cylindres
~STRAIGHTENER - [mech] machine *f* à dresser les rouleaux
~SULPHUR - [chem] (commercial form of sulphur obtained in tapering moulds) soufre *m* en canon
~THE RUBBER, to - [rubber ind] laminer le caoutchouc
~-TOOTH CRUSHER - [mech] concasseur *m* à cylindres dentés
~-TOP DESK - [gen] bureau *m* à cylindre
~UP THE FLEECE, to - [text] rouler la toison
~WELDING - [metall] soudage *m* à cylindres
ROLLABLE - [metall] laminable
ROLLBACK POINT - [comput] (the rerun point) point *m* de reprise
~ROUTINE - [comput] programme *m* de reprise
ROLLED - [gen] roulé, doublé, en rouleau
[metall] laminé
~AND MOULDED LAMINATED SECTION - [plast ind] profilé *m* stratifié, roulé et moulé
~ASPHALT - [constr] mortier *m* bitumineux
~DEAD - [rubber ind] travaillé à mort
~-END TUBE - [mech] tube *m* à bouts roulés
~GLASS - [glass man] (flat glass formed by rolling) verre *m* roulé
~GOLD - [metall] (composite sheet made by welding a sheet of gold to a thicker sheet of silver and rolling the whole down to the required thickness) doublé *m* d'or
~IN EYELETS - [text] oeillets *m*pl enroulés
~-IN SCALE - [metall] battitures *f*pl enfoncées
~IRON - [metall] fer *m* laminé
~-IRON GIRDER - [constr] poutre *f* en fer laminé
~LAMINATED TUBE - [plast ind] (tube formed from resin-impregnated sheet by winding it on to a former under suitable conditions of heat, pressure and tension) tube *m* roulé stratifié
~PIPE - s. rolled tube
~RUBBER THREAD - [rubber ind] fil *m* de caoutchouc laminé
~SECTION - [metall] fer *m* laminé
~SHAPE - [metall] forme *f* laminée
~SHEET - [plast ind] (sheet prepared in a calender) feuille *f* laminée
~TUBE - [metall & plast ind] (tube made by winding sheet on to a former) tube *m* enroulé
~WIRE HEALD - [text] lisse *f* en fil laminé
ROLLER - [gen] rouleau *m*, cylindre *m*

ROLLER - [constr] cylindre *m* compresseur
[transp] cylindre *m* rouleur, rouleau *m*, roule *m*
[mech] (a cylindrical part forming an element of a friction reducing device or the like) cylindre *m*
[mech] (of a press) cylindre *m*
[print] cylindre *m* encreur
[paper man & text] calandre *f*
[mech] galet *m*, roulette *f*, grain *m* de came
[tool] (a threading tool) outil *m* à fileter
[naut] (a big wave) lame *f* de houle
[metall] lamineur *m*, laminoir *m*
[mining] rouleur *m*
[horol] disque *m* de balancier
[mech] (used to save a rope from friction) tourniquet *m* de cabestan
[text] (in ropemaking) virolet *m* de corderie
~AND CLEARER CARD - [text] carde *f* à hérissons
~AND PEDAL FEED - [text] alimentation *f* à auges à pédales
~ARRANGEMENT - [text] dispositif *m* de suspension par rouleau
~BEARING - [mech] (a bearing in which the rotating element is supported by a circle of rollers, taking the place of the balls in a ball-bearing) coussinet *m* à galet, palier *m* à rouleaux
~BEARING SUPPORT - [mech] cadre *m* de support
~BLIND - [constr] volet *m* roulant
~-BLIND SHUTTER - [photo] obturateur *m* à rideau
~-BLIND SLIDE - [photo] châssis *m* à rideau
~BODY - [mech] (tha main cylindrical part of a roller) corps *m* du cylindre
~BREAKING MACHINE - [mech] broyeuse *f* à cylindres
~BRIDGE - [constr] pont *m* roulant
~BRUSH - [text] cylindre *m* transporteur
~CALENDER - [text] machine *f* à cylindrer
~CAM - [mech] excentrique *m* des cylindres
~CARD WITH FLEECE DELIVERY - [text] carde *f* à hérissons avec livraison de ruban
~CARRIER - [mech] support-rouleau *m*
~CEMENT - [text] pâte *f* pour drap
~CHAIN - [mech] (type of chain in which the pins joining the links are provided with rollers engaging the sprockets to reduce friction) chaîne *f* à rouleaux
~CHAMBER - [text] compartiment *m* du plongeur
~CHANGE CARD - [text] carte *f* de changement à roulettes
~CLEARER - [text] cylindre *m* nettoyeur
~CLEARING APPARATUS - [text] dispositif *m* de nettoyage des cylindres
~CLOTH - [text] drap *m* pour cylindre
~-COATING ENAMEL - [paint] (special enamel for application by a coating machine, which applies it by means of a system of rollers) émail *m* pour application par machine
~COMPOSITION - [print] pâte *f* à rouleaux
~COMPOUND - [rubber ind] mélange *m* pour rouleaux
~CONVEYOR - [transp] transporteur *m* à rouleaux
~COUPLING - [mech] manchon *m* du cylindre alimentaire
~COVERER - [text] couvreur *m* des cylindres
~CRADLE - [plumb] diabolo *m*
~CRUSHER - [mech] broyeur *m* à cylindres
~CUTTING AND WIND-UP MACHINE - [mech] (a machine for slitting rolls of material lengthwise and re-rolling them) machine *f* à découper et enrouler
ROLLER DELIVERY - [text] livraison *f* des cylindres
~DIES - [tool] (used for threading) rouleaux *mpl*)à fileter
~END - [mech] tenon *m* carré du cylindre
~FADING - [radar] (variability of the signal due to the rolling of a ship) affaiblissement *m* par roulis
~FEED - [mech] amenage *m* par cylindres
~FEED MOTION - [mech] mécanisme *m* d'alimentation à cylindres
~FILLET - [text] garniture *f* des hérissons
~GAUGE - [mech] calibre *m* pour régler les cylindres cannelés
~GEAR - [mech] engrenage *m* à rouleaux
~GEAR BED - [mech] boîte *f* des engrenages à rouleaux
~GIN - [text] égreneuse *f* à cylindres
~HEARTH FURNACE - [metall] four *m* à rouleaux
~HOT-PRESS - [metall] calandre *f* à cylindres réchauffés
~INDEXING MECHANISM - [mach tool] mécanisme *m* diviseur à rouleau
~LAP-UP - [text] formation *f* du rouleau
~LEATHER CUTTING MACHINE - [mech] machine *f* à couper les cuirs pour cylindres
~LEATHER GRINDING MACHINE - [mech] machine *f* à meuler les cuirs
~LEVELLER - s. roller levelling machine
~LEVELLING MACHINE - [mach tool] niveleuse *f* à rouleaux
[metall] dresseuse *f* à galets
~MANGLE - [text] mangle *f* à rouleaux
~MILL - [mech] (grinding machine in which rollers running on verticals shafts follow a circular path inside a drum or pan) broyeur *m* à meules horizontales
~MIXER - [mech] mélangeur *m* ouvert, mélangeur *m* à cylindres
~MIXING MILL - s. roller mixer
~NECK - [mech] collet *m* de cylindre
~PAINTER - [text] vernisseur *m* des cylindres
~PRINTING - [text] impression *f* au rouleau
~PROCESS - [ind chem] (method of obtaining carbon black by deposition from natural-gas flames on rollers) procédé *m* à rouleaux
~SECTOR - [auto] secteur *m* à rouleaux
~SHAKER - [mining] couloir *m* oscillant à galets
~SHAPER - [text] rouleau-guide *m*
~SHUTTLE - [text] navette *f* à roulettes
~SHUTTLE FOR CLOTH WEAVING - [text] navette *f* à rouleaux pour drap
~SKATE - [gen] patin *m* à roulettes
~SKATING - [gen] patinage *m* à roulettes, skating *m*
~SPRING - [mech] ressort *m*
~SEQUEGEE - [photo] rouleau *m* pour le collage des épreuves
~STAND - [auto] (used for stationary tests of motocars) banc *m* à rouleaux
[mech] support *m* des cylindres
~STEP BEARING - [mech] crapaudine *f* pour rouleaux
~STREAKS - [print] (defects caused by uneven contact of inking rollers) taches *fpl* de graissage des rouleaux
~STUD - [text] tourillon *m* du hérisson
~SUPPORT - [gas ind] (in gasholders) chevalet *m* du piston
~SWEDGE - [mining] redresseur *m* à galets

ROLLER TAPPET - [mech] (a valve tappet furnished with a roller to reduce lateral thrust) poussoir *m* à galet
~TEMPLE - [text] templet *m* à rouleau
~TENSION - [mech] frein *m* à roulettes
~TESTER - [text] vérificateur *m* pour cylindres
~TESTING MACHINE - s. roller tester
~TOOTH BEARING - [auto] (for the steering post) roulement *m* à billes pour galet de direction
~-TRACK - [mech] piste *f* de roulement
~VARNISH - [text] vernis *m* pour les manchons de cuir
~VAT - [text] cuve *f* à rouleaux
~WASTE - [text] barbes *fpl* de hérisson
~WAY [mining] voie *f* de roulage
~WEIGHT - [mech] poids *m* de pression
~WEIR - [hydr] vanne *f* à rouleau
~WITH LOOSE BUSHES - [text] cylindre *m* à douilles mobiles
~WITH PLASTER LAGGING - [text] rouleau *m* avec garniture en plâtre
~WITH SPIRAL FLUTES - [text] cylindre *m* avec cannelures en spirale
~WITH SPIRAL RIDGES - [text] cylindre *m* à rainures en spirales
ROLLEY - [transp] wagonnet *m*
[mining] berline *f*, wagon *m* de mine
ROLLING - [gen] roulement *m*, roulant
[mech] (as the revolving of a wheel on its axis) roulement *m*
[aero] (change of attitude about the longitudinal axis) roulis *m*
[acoust] roulement *m*
[constr] cylindrage *m* (d'une chaussée)
[naut] roulis *m*
[text] (pressing by rolls) cylindrage *m*
[metall] laminage *m*
[mech] (threading) filetage *m*
[cin & telev] (tilting the camera on its horizontal axis) décentrement *m* horizontal
~AXIS - [aero] axe *m* longitudinal
~BALANCE - [instr] (device in a wind tunnel to measure aerodynamic moments and forces occurring when the model is rotating about a longitudinal axis) balance *f* de roulis
~BARRIER - s. rolling gate
~BEARING - [mech] roulement *m* à rouleaux
~BOBBIN - [text] bobine *f* à dérouler
~BRIDGE - [constr] pont *m* roulant
~CHOCK - [naut] quille *f* de roulis
~CIRCLE - [math] (of a cycloidal curve) roulante *f*
[mech] (of a gear) cercle *m* primitif, cercle *m* de contact, primitif *m*
~CONES - [mech] (of a gear) cônes *mpl* primitifs
~CONTACT - [mech] contact *m* de roulement
~CONTACT BEARING - [constr] appui *m* cylindrique
~CRUSHER - [min] broyeur *m* à cylindres
~CURVE - [mech] (of a gear) courbe *f* de roulement
~CYLINDER - [mech] (of a gear) rouleau *m* de roulement
~DOOR - [constr] porte *f* roulante
~DOWN MACHINE - [mech] (a stitching machine) appareil *m* à roulettes, machine *f* pour laminer les protecteurs sur les pneus
~DRAG - [aero] (the drag produced by the rolling resistance of the landing gear moving on the gound) traînée *f* de roulement
ROLLING FRICTION - [mech] frottement *m* de roulement, frottement *m* de la seconde espèce
~FURNACE - [metall] four *m* oscillant
~GATE - [constr] barrière *f* roulante
~GENERATING PRINCIPLE - [mech] (of a gear) méthode *f* de génération par roulement
~GENERATING PROCESS - [mech] (of a gear) procédé *m* de génération par roulement
~GROUND - [surv] terrain *m* ondulé
~INSTABILITY - [aero] (tendency in an aircraft to develop an increasing oscillatory movement after a disturbance in the sense of roll) instabilité *f* de roulis
~LINE - [mech] (of a gear) ligne *f* de roulement
~LOAD - [phys] poids *m* roulant
~LOCKER - [text] chariot *m* denté
~LOCKER DRIVE - [text] commande *f* par rouleaux pour les chariots
~MACHINE - [mach tool] machine *f* à fileter à rouleaux
~MILL - [mech] laminoir *m*
[gen] (the factory) laminerie *f*
[metall] laminoir *m* à tôle forte
~-MILL ENGINE - [mech] machine *f* pour trains de laminoir
~-MILL PRODUCTS - [metall] produits *mpl* de laminage
~MILL TRAIN - [metall] train *m* de laminoir
~MOMENT - [aero] (the component of the couple due to the corresponding airflow, taken about the longitudinal axis) moment *m* de roulis
~MOTION - [naut & aero] roulis *m*
~OVER - [metall] renversement *m*
~PIN - [impl] rouleau *m*
~PITCH DIAMETER - [mech] (of a gear) diamètre *m* primitif de roulement
~PLAIN - [surv] plaine *f* ondulée
~POINT - [mech] (the instantaneous centre of motion of a gear) point *m* de contact (des cercles primitifs)
~PRESS - [mech] (used in the paper industry) presse *f* à cylindre
~PRESSURE - [metall] pression *f* de laminage
~PROCESS - [metall] procédé *m* de laminage
~RESISTANCE - s. rolling friction
~SCALE - [metall] écaille *f* de laminage
~SHUTTER - [gen] rideau *m*
~STOCK - [railw] (general collective terms for coaches, trucks etc) matériel *m* roulant
~STOCK CLEARANCE GAUGE - [railw] profil *m* de libre passage du matériel roulant
~SURFACE - [mech] (of a gear) surface *f* de roulement
~TANK - [naut] caisse *f* à roulis
~TEXTURE - [metall] texture *f* de laminage
~THE FLAX - [text] cylindrage *m* du lin
~THE FLAX STEMS - [text] écrasement *m* des tiges de lin
~TITLE - [cin] titre *m* à rideau, titre *m* roulant
~TOTAL TABULATOR - [comp] tabulatrice *f* du total
~-UP - [print] (the inking of the forms) encrage *m*
~-UP DEVICE - [text] dispositif *m* d'enroulage
~WIDTH - [metall] largeur *f* de laminage
ROLLOWER BOARD - [metall] plaque *f* à fouler
ROLLS - [metall] machine *f* agglomératrice
ROLLY OIL - [oil ind] émulsion *f* naturelle d'huile et eau
ROMAN - [gen] romain

ROMAN - [print]caractère *m* romain
~ARCH - [arch] arc *m* plein cintre, arc *m* romain
~BALANCE - [instr] romaine *f*, balance *f* romaine
~CEMENT - [constr] (natural hydraulic cement obtained by calcining nodules found in the London clay) ciment *m* romain
~MOSAIC - [constr] (tessellated pavement) pavage *m* en mosaïque
~NUMERALS - [math] chiffres *mpl* romains
ROMEITE - [min] (antimonite of calcium) roméine *f*
RONDLE - [metall] (in foundry work) scorie *f*
RÖNTGEN - s. roentgen
~EQUIVALENT MAN - [nucl] (the unit of dose equivalent) rem *m*
~EQUIVALENT PHYSICAL - [nucl] (unit of absorbed doseequal to the absorbed dose in water subjected to an exposure dose of one röntgen) rep *m*
~RAYS - s. roentgen rays
ROOD - [gen] croix *f*
[meas] (a square land measure) rood *m*, quart *m* d'arpent
ROOF - [constr] toit *m*, comble *m*
[auto] (the top covering) toit *m*, capotage *m*
[constr] (the highest part) voûte *f*
[anat] (of the mouth) dôme *m* du palais
[mining] (of a gallery) ciel *m*, plafond *m*, toit *m*
[metall] dôme *m* (de four à reverbère)
~AERIAL - [radio] (aerial on the roof of a car) antenne *f* d'auto
~-AND-PILLAR SYSTEM - [mining] exploitation *f* par grands massifs
~ANTENNA - s. roof aerial
~BOARDING - [constr] planchéiage *m* du toit
~COVERING - [constr] (the material applied to the framework of a roof to form the outer upper covering of a building) couverture *f* des toits
[auto] pavillon *m*
~CRANE - [mech] grue *f* de toit
~FALL - [mining] éboulement *m* du toit
~FRAME - [constr] charpente *f* de comble
~FREIGHT CONTAINER - [auto] galerie *f* de toit
~GARDEN - [constr] jardin *m* sur toit en terrasse
~GLAZING - [constr] couverture *f* en verre
~HOOD - [auto] pavillon *m*
~HOOK - [impl] crochet *m* de couvreur
~LATHING - [constr] lattis *m* de toit
~LIMB - [mining] flanc *m* supérieur
~-PITCH - [constr] chute *f* de comble, inclinaison *f* de comble
~PLATE - [constr] poutre *f* sablière, sablière *f* de comble
~PRISM - [opt] (a type of total-reflection prism) prisme *m* en toit
~-ROCK - [geol] (the covering of a gas-bearing stratum) couche *f* de couverture
~SHEET - [railw] (in locomotive) berceau *m*
~SHINGLE - [constr] bardeaux *mpl* pour toit
~STANDARD - [telecomm] tourelle *f*, appui *m* sur toiture
~TILING - [constr] tuiles *fpl*
~TIMBERING - [mining] boisage *m* de tailles
~TRUSS - [constr] ferme *f* de comble
~VENTILATOR - [gas ind] (in gasholders) lanterneau *m*
~WITH VALLEY - [constr] toit *m* en retour
ROOFING - [constr] (the materials required for a roof) toiture *f*, garniture *f* de comble
ROOFING - [constr] (the operation of covering with a roof) pose *f* de la toiture
~BOARDS - [constr] (waterproof boards for roofing) carton *m* bitumé
~BOND - [constr] appareil *m* de toit
~FELT - [constr] carton *m* bitumé
~GLASS - [constr] verre *m* pour couverture
~PAPER - [constr] carton *m* bitumé
~SLAB - [constr] plaque *f* de toiture
~SLATE - [geol] ardoise *f* à tuiles
~TILE - [constr] tuile *f*
ROOFLESS - [gen] sans toiture
ROOFTREE - [constr] faux-entrait *m*
ROOK - [zool] corneille *f*
ROOM - [gen] espace *m*, place *f*
[constr] chambre *f*, salle *f*
~-AND-PILLAR SYSTEM - [mining] méthode *f* par chambres et piliers
~HEATING - [th eng] chauffage *m* indépendant
~NOISE - [acoust] (the general ambient noise in an enclosure, or the noise therein filtering from outside)bruit *m* de salle
~TEMPERATURE - [gen] (the general prevailing ambient temperature of a room in which a process or the like is taking place) température *f* ambiante
~THERMOSTAT - [th eng] thermostat *m* d'ambiance
ROOSTER - [zool] coq *m*
ROOT - [gen & bot] racine *f*
[carp] (section of a tenon) attache *f* (de tenon)
[mech] (of a blade) racine *f*, talon *m*
[mech] (of a gear tooth) pied *m*
[mech] (of a weld) base *f* (de la soudure)
[aero] (of a tailplane) attache *f*
~ABSCESS - [med] granulome *m* apical
~AMPUTATION - [med] résection *f* d'une racine dentaire
~ANGLE - [mech] (of a straight bevel gear) base *f* de la dent
~BALL - [bot] motte *f* (adhérant aux racines)
~-BEAD - [metall] (first weld) première passe *f*
~-CAP - [bot] coiffe *f*
~CIRCLE - [mech] (of a gear) cercle *m* de racine, cercle *m* d'évidement
~CLEANER - [agric] décrotteur *m* de racines
~COLLAR - [bot] collier *m* de la racine
~CONCAVITY - [metall] (in welding) concavité *f* à la base de la soudure
~CROP CULTIVATION - [agric] culture *f* sarclée
~-CROP-TRACTOR - [agric] tracteur *m* pour binage
~CROPS - [agric] plantes *fpl* sarclées
~CUTTER - [agric] coupe-racines *m*
~CUTTING - [bot] bouture *f* de racine
~DIAMETER - [mech] (of a gear) diamètre *m* intérieur
[mech] (of a screw) diamètre *m* à fond de filet
~EELWORM - [bot] anguillule *f* des racines
~END OF FLAX - [bot] queue *f* du lin
~FORM - [mech] forme *f* du fond de filet
~FUNGUS - [bot] parasite *m* des racines
~GAP - [mech] distance *f* entre les racines (d'une dent d'engrenage)
~GRAFTING - [agric] greffage *m* sur racine
~GROOVE - [metall] (a defect) affaissement *m*
~HAIR - [bot] poil *m* radiculaire
~INJECTOR - [agric] pal *m* injecteur
~LINE - [mech] (in gearing) ligne *f* de racine, droite *f* d'évidement

ROOT LINE SPIRAL ANGLE - [mech] (of a spiral bevel gear) angle *m* d'inclinaison de la ligne de racine
~LOCUS - [math] lieu *m* des pôles, lieu *m* des racines
~MEAN SQUARE - [math] moyenne *f* quadratique
~-MEAN-SQUARE ELECTROMOTIVE FORCE - [el] force *f* électromotrice efficace
~-MEAN-SQUARE ERROR - [instr] erreur *f* quadratique moyenne
~-MEAN-SQUARE VALUE - [el] (of an alternating current or voltage) valeur *f* efficace
~-MEAN-SQUARE VELOCITY - [phys] (the square root of the average of the square of the speed of the particles composing the system) vitesse *f* efficace
~OF A FOLD - [geol] racine *f* de charriage
~OF A THREAD - [mech] diamètre *m* de noyau
~OF SOAPWORT - [bot] racine *f* de savonnière
~OF TAIL - [anat] base *f* de la queue
~PARASITE - s. root fungus
~PLATEFORM - [constr] planche *f* de chemin
~POSITION OF A COMMON CHORD - [mus] base *f* d'un accord parfait
~PULPER - [agric] coupe-racines *m*
~RAKE - [agric] déracineur *m*
~ROT - [bot] (plant disease) piétin-verse *m*, jambe *f* noire
~RUBBER - [bot] (medicine plants) caoutchouc *m* de racines
~SHEATH - [agric] gaine *f* de la racine
~SHREDDER - [agric] hache-racines *m*
~-SUCKER - [bot] drageon *m*
~TIP - [bot] extrémité *f* radiculaire
~TUBER - [bot] (swollen root containing reserve food material) tubercule *m* radiculaire
~TUBERCLE - s. root tuber
~UP, to - [agric] déraciner, extirper
~VEGETABLES - [agric] légumes-racines *mpl*
ROOTED CUTTING - [bot] bouture *f*
~LAYER - [agric] marcotte *f* avec racine
ROOTER - [agric] scarificateur *m*
ROOTLET - [bot] radicelle *f*, radicule *f*
ROOTS BLOWER - [mech] (air compressor for delivering large volumes at comparatively low-pressure ratios) souffleur *m* Roots
~EXHAUSTER - [mech] extracteur *m* à pistons rotatifs
~SUPERCHARGER - [mech] (a supercharger having two meshing impellers of figure eight form) compresseur *m* Roots
ROPE, to - [gen] lier
[gen] (to fasten with ropes) corder
[naut] (a sail; to sew ropes on the border of a sail) ralinguer
ROPE - [gen] corde *f*, cordage *m*
[text ind] fil *m* retors, cordon *m*
[naut] filin *m*
[brew ind] graisse *f*
~AND CORD LAYING MACHINE - [text] commetteuse *f* pour cordes et ficelles
~BAND - [text] raban *m*
~BELT - [mech] corde-courroie *f*
~BLOCK - [mech] moufle *f* à corde
~BORING - s. rope drilling
~BRACING - [gen] rigidité *f* de la corde
~BRAKE - [instr] (a type of absorption dynamometer) frein *m* à corde
~BRAIDING MACHINE - [text] machine *f* à tresses pour cordes, tresseuse *f* mécanique
ROPE BROWN - [paper man] (paper obtained from old ropes) papier *m* gris
~CLAMP - [railw] (in aerial ropeways) griffe *f* de accrochage
[mech] agrafe *f* de jonction pour câbles
~CLIP - [mining] tenaille *f* d'attelage
~-CONTROLLED LIFT - [mining] ascenseur *m* contrôlé par câble
~CORE - s. rope heart
~COUPLING - [text] attache *f* de câbles, agrafe *f* de jonction pour câbles
~DRAG - [text] traîneau *m* de cordier
~DRILLING - [oil ind] sondage *m* à la corde, sondage *m* américain
~DRIVE - [mech] transmission *f* par câbles
~DRIVEN - [mech] actionné par câble, à commande par câble
~EQUALIZER - [el] compensateur *m* de corde
~FALL - [mech] (of a tackle) courant *m* d'un palan
~GRAB - [mining] gueule *f* de brochet
~GRAP - [impl] grappin *m* à câble
~GRIP - [text] pince *f* pour corde
~GROOVE - [text] moyeu *m* avec gorge
~GROUND - [text] corderie *f*
~GUIDING - [text] guide-corde *m*
~HEART - [text] âme *f* du câble, noyau *m* du câble
~HEMP - [text] chanvre *m* de cordier
~HOOK - [mining] crochet *m* pour câble
~KNIFE - [impl] (used in mining) cisaille *f* à câbles
~LADDER - [impl] échelle *f* de corde
~LIFT - [transp] monte-charge *m* à câble de traction
~MAKING - [text] corderie *f*, fabrication de cordages
~OF ANNEALED WIRE - [metall] câble *m* métallique en fil recuit
~OF WARP YARN - [text] chaîne *f* en boyau
~OF UNANNEALED WIRE - [metall] câble *m* métallique en fil non recuit
~OPENER - [text] étendeuse *f* de boyaux
~PILER - [text] introducteur *m* pour tissus en boyau
~PLAITING MACHINE - [text] tresseuse *f* mécanique
~PULLEY - [mech] poulie *f* à corde, volan *m* à gorge
~PULLEY-BLOCK - [mech] palan *m* à corde
~RAILWAY - [transp] voie *f* à câble aérien
~ROLL - [mining] tambour *m* d'extraction
~SCUTCHER - [text] étendeuse *f* de boyaux
~SHEAVE - [mech] poulie *f* à corde
~SLEDGE - s. rope drag
~SLING - [mech] élingue *f*, braye *f*
~SOCKET - [impl] douille *f* du câble
[mining] cosse *f* de câble
~SPEAR - [mining] harpon *m* pour câbles
~SPLICER - [text] épisseur *m*
~SPLICING - [text] épissure *f* des cordes
~SPOOL - [text] bobine *f* de corde
~STEP PULLEY - [mech] poulie *f* à gradins pour la corde
~TACKLE BLOCK - [impl] moufle *f* à corde
~TESTING - [text] essai *m* des cordes
~TESTING MACHINE - [text] dynamomètre *m* pour cordages
~THIMBLE - [mech] cosse *f* de câble
~TRANSMISSION - [mech] transmission *f* par câbles
~TWISTER - [text] commetteur *m*
~WHEEL - [mech] poulie *f* à câble
~WINCH - [mech] treuil *m* à câble
~WINDER - [el] enrouleur *m* de corde

ROPE YARD - [text] corderie *f*
~YARN - [text] fil *m* de caret
~WALK - [text mach] corderie *f*
ROPEMAKER - [gen] cordier *m*
ROPEMAKING - s. rope making
ROPEWAY - [gen] câble *m* aérien
[transp] (for goods) téléphérique *f*
ROPINESS - [paint] (defect in a coating composition, characterized by stringiness which makes it impossible to produce a uniform surface) tendance *f* à être filandreux
[gen] viscosité *f*
[brew ind] (in beer etc) graisse *f*
ROPING - [gen] cordage *m*
~NEEDLE - [impl] aiguille *f* de voilier
ROPY - [rubber ind & plast ind] filant
[brew ind] (of beer) graisseux
~FERMENTATION - [brew ind] fermentation *f* visqueuse
~LAVA - [geol] lave *f* cordée
RORQUAL - [zool] balénoptère *m*
ROSACE - s. rose-window
ROSANILINE - [chem] rosaniline *f*
ROSCOELITE - [min] (an unimportant ore of vanadium essentially muscovite in which the aluminium has been partially replaced by vanadium) roscoélite *f*
ROSE - [bot] rose *f*
[gen] (the colour) rose
[constr] (decorative circular escutcheon) rosace *f*
[instr] (of the compass) rose *f* (des vents)
[mech] crépine *f* d'aspiration, lanterne *f*, grenouillère *f* (d'une pompe), reniflard *m*
[impl] (of watering can) pomme *f* d'arrosoir
~APPLE - [bot] jambose *f*
~BIT - [mach tool] fraise *f* champignon, fraise *f* angulaire
~BURNER - [gas ind] brûleur *m* à couronne
~COPPER - [min] cuivre *m* rosette
~COUNTERSINK - [tool] fraise *f* taillée
~COUNTERSINK BIT - [tool] fraise *f* champignon
~CRUCIBLE - [impl] (a special type of crucible used for heating substances in a current of gas; it is fitted with an inlet tube in the cover) creuset *m* de Rose
~-CUT DIAMOND - [min] diamant *m* taillé en rose
~CUTTER - [mach tool] fraise *f* conique
~MALLOW - [bot] rose *f* trémière
~NAIL - [impl] clou *m* à tête de diamant
~OIL - [chem] (attar of roses. A fragrant volatile oil obtained by steam distillation of the flowers of varieties of Rosa and used in perfumery) essence *f* de rose
~OPAL - [min] (a red variety of common opal) opale *f* rose
~PIPE - [hydr] douche *f* en arrosoir
~QUARTZ - [min] (type of quartz of a rose-pink colour) quartz *m* rose
~REAMER - [mach tool] alésoir *m* angulaire
~TOPAZ - [min] (yellow-brown variety of topaz) topaze *f* rose
~-TYPE COUNTERSINK - [mach tool] fraise *f* conique
~WATER - [chem] eau *f* de rose
~-WINDOW - [constr] (circular window with radial bars) rosace *f*
ROSEMARY OIL - [chem] (an essential oil obtained from Rosmarinus officinalis and used in perfumery) essence *f* de romarin
ROSEOLA - [med] (a rose-coloured rash) roséole *f*
ROSET - [biol] rosette *f*, spirème *m*
ROSETTE GRAPHITE - [min] graphite *f* en rosettes
~PLATE - [el] rosace *f* de plafond
ROSEWOOD - [bot] palissandre *m*
~OIL - [chem] essence *f* de palissandre
ROSIN - [chem] (gum rosin; colophony; pine resin, wood rosin, common rosin complex, amorphous substance derived from pine trees and with uses as emulsifier for rubber, in protective coatings, paper sizes, and insulating compounds) colophane *f*
~OIL - [ind chem] (a high-boiling distillate obtained from rosin by destructive distillation) huile *f* de résine
~PITCH - [chem] (a dark-coloured residue from the destructive distillation of rosin) poix *f* de colophane
~SIZE - [ind chem] colle *f* de colophane
~SPIRIT - [chem] (a volatile solvent boiling below 260° C. obtained from the destructive distillation of rosin) essence *f* de colophane
ROSINED SOAP - [chem] savon *m* résineux
ROSOLIC ACID - [chem] acide *m* rosolique
ROSS, to - [agric] (USA only; to divest a tree of bark) décortiquer, écorcer
ROSSI-PEAKE TEST - [ind chem] (a test to determine the temperature at which a given moulding powder flows through a given orifice in a given time) test *m* de coulée
ROSSING - [agric] écorcage *m*
ROSTER - [comput] liste *f*
ROSTRUM - [gen] (raised platform for speakers) tribune *f*
[zool] bec *m*, rostre *m*
ROT, to - [gen] pourrir, se pourrir, se putréfier
ROT - [gen] pourriture *f*, putréfaction *f*, carie *f*
[vet] (of sheep) distomatose *f*
~OF WOOD - [bot] rot *m* du bois
ROTAMETER - [instr] (instrument for measuring the flow of liquids or gases) débitmètre *m* à flotteur
[draw] (in drawing, an instrument for measuring curved lines) gyromètre *m*
ROTOPLANE - [aero] (rotocraft) giravion *m*
ROTARY - [gen] rotatif, rotatoire, tournant
[mech] (of a movement) rotatoire
[tool] s. rotary drill
~ANEMOMETER - [instr] anémomètre *m* de rotation
~ARC - [el] lampe *f* à arc rotatif
~BEAM AERIAL - [radio] (aerial array radiating a rotating beam) antenne *f* à faisceau rotatif
~BEAM ANTENNA - s. rotay beam aerial
~BLOWER - [mech] soufflerie *f* à pistons rotatifs
~BOOSTER - s. rotary blower
~CLOTH-PRESS - [text] presse *f* à cuvette
~COLD TAGGING MACHINE - [plumb] machine *f* rotative à pointage à froid
~CONTINUOUS-CORE DRILL - [tool] sondeuse *f* rotative à carottage continu
~CONVERTER - [el] (a converter with a single armature winding having a commutator and sliprings, used to convert alternating current into direct current and vice-versa) commutatrice *f*, convertisseur *m* tournant
~CORE BIT - [mech] trépan *m* rotary
~CRANE - [mech] grue *f* pivotante
~CROSS CUTTER - [impl] coupeuse *f* transversale rotative
~CRUSHER - [mech] (machine in which a rotating

conical corrugated element crushes material against the inner surface of a steel conical body) concasseur *m* giratoire

ROTARY CURRENT - [hydr] courant *m* giratoire

~ CUT - [mech] (the operation of removing a single layer of veneer from a rotating log of wood) coupe *f* circulaire

~ CUTTER - [tool] fraise *f* rotative
[impl] coupeuse *f* rotative

~ DERIVATIVES - [aero] (those stability derivatives which are connected with rotational movements of an aircraft) dérivées *fpl* de rotation

~ DISCHARGER - [el] éclateur *m* rotatif

~ DISPLACEMENT METER - [instr] (for gas; called Roots meter in the USA) compteur *m* à pistons rotatifs

~ DISTILLATION - [oil ind] (process carried out in apparatus consisting of a heated shell containing a cooler rotary element, these parts being only about Imm distant from each other . Vapour condenses on the rotor and the condensate is thrown off into the hot shell) distillation *f* rotative

~ DRAWING - [text] étirage *m* à hérissons

~ DRAWING FRAME - [text] étaleuse *f* avec étirage à hérissons

~ DRILL - [tool] (used in mining) perforatrice *f* rotative

~ DRILLING - [oil ind] (a drilling process based on rotating a bit and drill pipe carrying a mud circulation system) forage *m* à système rotary

~ DRILLING RIG - [oil ind] installation *f* de sondage rotary

~ DRYER - [mech] (rotating cylindrical vessel for removing moisture from a material) séchoir *m* rotatif

~ ENGINE - [mech] (a radial engine in which the cylinders rotate as a whole round a fixed crankshaft) machine *f* rotative

~ EXHAUSTER - [gas ind] extracteur *m* à piston circulaire

~ FAN - [mech] ventilateur *m* rotatif

~ FILE - [tool] (hardened steel tool formed with file-teeth, used on a flexible shaft in die-sinking) lime *f* rotative

~ FILTER - [ind chem] (continuous-type filter in the form of a cloth-or screen-covered drum divided into compartments which are connected in succession to vacuum and pressure lines while the revolving drum dips into a bath of the liquid to be tread) filtre *m* rotatif

~ FISHING JAR - [oil ind] coulisse *f* rotary

~ FURNACE - [metall] four *m* rotatif, four *m* rotatoire

~ GAP - [el] éclateur *m* tournant

~ GRATE - [th eng] grille *f* tournante

~ GRATE WITH CLINKER ARMS - [th eng] grille *f* mécanique à barreaux rotatifs

~ HEADING MACHINE - [mining] coupeuse *f* rotative

~ HEARTH FURNACE - [th eng] four *m* à sole tournante

~ HOE - [agric] herse *f* à disques

~ HOSE - [oil ind] manche de rotary

~ IMPELLER EXHAUSTER - [gas ind] extracteur *m* à pistons rotatifs

~ JOINT - [mech] joint *m* tournant

~ KILN - [ind chem] (long cylindrical kiln rotating on rollers, fired from one end and fed with mixed raw materials at the other, for calcining the mixture of argillaceous and calcareous elements) four *m* tournant

ROTARY LAPPING - [mech] rodage *m* par rotation

~ LINE SWITCH - [telecomm] (in telephony) présélecteur *m* rotatif

~ MACHINE - [print] (printing machine in which the printing surface is a revolving cylinder) rotative *f*

~ MAGNET - [el] électroaimant *m* de rotation

~ MECHANICAL GRATE - s. rotary grate with clinker arms

~ METER - [instr] (for gas) compteur *m* à turbine

~ MOTION - [mech & phys] mouvement *m* de rotation, mouvement *m* rotatif

~ PELLETING MACHINE - [mech] pastilleuse *f* à couteau tournant

~ PHASE CHANGER - [electron] (phase changer using waveguide sections in line with metal and dielectric inserts) déphaseur *m* rotatif

~ PHASE CONVERTER - [el] (an electrical machine designed to transfer power from a system with a number of phases to a system having another number of phase) convertisseur *m* de phase

~ PHASE SHIFTER - s. rotary phase changer

~ PIERCING MILL - [metall] laminoir *m* à cylindres obliques

~ PISTON - [mech] piston *m* rotatif

~ PLANER - [mech] raboteuse *f* rotative

~ PRESS - s. rotary machine

~ PRESS - [mech] (also called dial feed press; one which is fed by a revolving platform) presse *f* à tourelle révolver

~ PUDDLING FURNACE - [metall] four *m* à puddler rotatif

~ PUMP - [mech] (a pump in which the impelling element rotates) pompe *f* rotative

~ SCREEN - [gas ind] (for coke; a revolving screen) trommel *m*, tamis *m* rotatif

~ SCREWING CHUCK - [mach] filière *f* tournante

~ SCRUBBER - [ind chem] laveur *m* rotatif

~ SELECTOR BANK - [telecomm] (in telephony) champ *m* radial de sélection

~ SHEARS - [mech] cisailles *fpl* rotatives

~ SHELF - [mech] plateau *m* tournant

~ SHELF DRYER - [mech] séchoir *m* à plateaux

~ SHOE - [oil ind] sabot *m* denté de rotary

~ SHUTTER - [cin] (shutter consisting of a holed flat circular disk rotating about its centre) obturateur *m* rotatif

~ SIEVE FOR MOULDING SAND - [metall] tamis *m* de fonderie

~ SLIDE VALVE - [mech] tiroir *m* rotatif

~ SNOW-PLOUGH - [railw] chasse-neige *m* à turbine

~ SOLENOID - [el] solénoïde *m* tournant

~ SPARGER - [brew ind] croix *f* écossaise

~ SPARK-GAP - [radio] (spark-gap consisting of a toothed or studded disk revolving between fixed electrodes) éclateur *m* tournant

~ SQUEEZER - [min] cingleur *m* rotatif

~ STABILIZER - [el acoust] (of a sound film projector) stabilisateur *m*

~ STEAM JOINT - [mech] accouplement *m* tournant à vapeur

~ STEAM PRESS - [text] calandre *f* à vapeur

~ STRAINER - [paper man] (a machine designed to eliminate foreign material from the pulp before it travels on to the paper-making machine) filtre *m* tournant

~ SUBSTATION - [el] (substation converting or trans-

forming electrical energy by means of rotating machines) sous-station *f* à groupe rotatifs
ROTARY SURFACE GRINDER - [mach tool] rectifieuse *f* à tableau tournant
~SWAGER - [metall] machine *f* à retreindre
~SWAGING - [metall] réduction *f* de section
~SWITCH - [el] (switch operated by a handle pivoted on the face of the switch; it is turned backwards and forwards in a plane which is parallel with the base) commutateur *m* rotatif
[telecomm] chopper *m*
~SWIVEL - [oil ind] tête *f* d'injection de rotary
~TABLE - [mach tool] plateau *m* tournant
[oil ind] table *f* de rotary
~TEMPLET - [text] templet *m* rotatif
~TONG - [oil ind] clef *f* de serrage
~TOWER CRANE - [oil ind] grue *f* pivotante géante
~TRANSFORMER - [el] (a rotary machine taking A.C. at one voltage and generating A.C. at one more other voltages) transformateur *m* à champ tournant
~TRIMMING SHEAR - [impl] cisailles *fpl* circulaires de rives
~VALVE - [mech] (a valve which performs its functions by rotating) vanne *f* à opercule tournant
~VANE PUMP - [mech] pompe *f* rotative à palettes
~VULCANIZING DRUM AUTOCLAVE - [rubber ind] chaudière-autoclave à tambour rotatif
~WALL-CRANE - [mech] grue *f* pivotante murale
~WASH OVERSHOE - [oil ind] sabot *m* pour tube de lavage
~WASHER - s. rotary scrubber
~WING - [aero] aile *f* tournante
~-WING AIRCRAFT - [aero] avion *m* à aile tournante
~WORKTABLE - [mech] plateau *m* porte-piéce tournant
ROTATABLE FLAP VALVE - [mech] vanne *f* tournante
~LOOP AERIAL - [radio] (a loop aerial which can be revolved in azimuth, used for direction-finding) cadre *m* tournant
ROTATE, to - [gen] tourner, pivoter, faire basculer
[agric] alterner (les cultures)
ROTATE THE BOBBIN, to - [text] faire tourner la bobine
ROTATING - [gen] tournant, rotatif
[mech] (of a mechanical element) tournant
~-ANODE TUBE - [electron] (x-ray tube in which a rotating anode is used) tube *m* à anode rotative
~BEACON - [radio] (radio beacon in which the transmitted beam is slowly revolved by mechanical or electrical means) balise *f* tournante
~COLOUR DISK - [cin] (the rotary colour filtres turning in synchronism with the frame in a projector) disque *m* chromatique tournant
~CROPS - [agric] cultures *fpl* alternantes
~CRYSTAL METHOD - [phys] (technique for the x-ray analysis of crystal structures) méthode *f* à cristal rotatif
~CYLINDER ENGINE - [mech] moteur *m* à cylindres tournants
~CYLINDER GAUGE - [instr] manomêtre *m* à entraînement
~CYLINDER METHOD FOR VISCOSITY - [ind chem] méthode *f* à cylindre rotatif
~DEAD CENTRE - [mech] point *m* mort rotatif
~DIRECTION FINDER - [radar] (a direction finder rotating about a vertical axis) radiogoniomètre *m* tournant
~-DISK CONTACTOR - [oil ind] (type of extractor in which a vertical cylindrical vessel contains a central shaft carrying disks which revolve close to stationary annular plates fixed to the shell) contacteur *m* à disques rotatifs
ROTATING DISK FIELD STORE - [telev] (in colour television) mémoire *f* de trame à disque tournant
~-DISK PROCESS - [ind chem] (preparation of carbon black by deposition from a natural-gas flame on a revolving disk) procédé *m* à disque rotatif
~DISK SHUTTER - s. rotary shutter
~DRYING DRUM - [rubber ind] séchoir *m* rotatif
~ELEMENT - [mech] (of a turbine etc) élément *m* tournant
~FIBRE HAMMERS - [mech] (used in oil refining) vibrateur *m*
~FIELD - [el] (magnetic field produced by a polyphase current flowing in the polyphase winding of an electrical machine) champ *m* tournant
~-FIELD AERIAL - [radio] (aerial generating a rotary field) antenne à champ tournant
~-FIELD ANTENNA - s. rotating field aerial
~FIELD INSTRUMENT - [instr] (induction instrument in which the moving element is subjected to the action of several alternating fields which have phase differences between them, so that the resulting magnetic field is a rotating field) instrument *m* à champ tournant
~FIELD MAGNET - [el] (the rotating portion of an electrical machine, in which the field poles rotate and the armature is stationary) aimant *m* à champ tournant
~FIELD TRANSFORMER - [el] (a polyphase transformer in which the windings are arranged to produce rotating magnetic fields) transformateur *m* à champ tournant
~GRATE - s. rotary grate
~GRID - [impl] (for coke; a roll screen) grille *f* à rouleaux
~IMPELLER - [instr] (of meters) piston *m* rotatif, mobile *m*
~JOINT - [radio] (joint between two waveguide elements) joint *m* tournant
~LOOP - [radio] cadre *m* tournant
~-LOOP AERIAL - [radio] (frame aerial so built, that a rotary field is generated) antenne *f* à cadre tournant
~-LOOP ANTENNA - s. rotating-loop aerial
~MAGNETIC FIELD - [phys] champ *m* magnétique tournant
~MASS - [mech] masse *f* tournante
~MOVEMENT - [geol] mouvement *m* de rotation
~RACK - [gen] casier *m* tournant
~RING - [mech] bague *f* tournante
~SCANNER - [telev] (any type of mechanical scanner in which the moving parts revolve) analyseur *m* à disque tournant
~-SCREEN TUBE - [electron] tube *m* cathodique à écran tournant
~SHUTTER - s. rotary shutter
~SOLDERING MACHINE - [electron] (used in the manufacture of component parts of electronic tubes) machine *f* rotative à souder
~SWITCH - [el] interrupteur *m* rotatif
~-WING AIRCRAFT - s. rotary-wing aircraft
~WIRE BRUSH - [mech] balais *m* métallique totatif
ROTATION - [gen] rotation *f*
[agric] rotation *f* (des cultures), assolement *m*

ROTATION - [el] (of a vector; a vector the flux of which across any infinitely small surface is equal to the circulation of the given vector round the contour of the surface) rotationnel *m* (d'un vecteur)
[metall] (of the hearth) basculage *m* (d'un creuset)
~ANGLE - [mech] angle *m* de rotation
~AXIS - [crystal] (a symmetry element possessed by certain crystals) axe *m* de rotation
~-INVERSION AXIS - [crystal] (symmetry element possessed by certain crystals) axe *m* de rotation-inversion
~OF CROPS - [agric] rotation *f* des cultures
~OF STOCKS - [agric] rotation *f* des stocks
~OF THE EARTH - [astr] rotation *f* de la terre
~OF THE PLANE OF POLARIZATION - [light] (property possessed by optically active substances) rotation *f* du plan de polarisation
~OF THE PRESENTING PART - [med] (in obstetrics) rotation *f* de la présentation
~OF THE REVOLVING BOX - [text] évolution *f* de la boîte tournante
~OF THE VIBRATION PLANE - [crystal] rotation *f* du plan oscillant
~PASTURE - [agric] pâturage *m* en rotation
~PHOTOGRAPH - [crystal] photographie *f* de cristal rotatoire
~-REFLECTION AXIS - [crystal] (symmetry element possessed by certain crystals) axe *m* de rotation-réflexion
~STRETCHING - [phys] allongement *m* de torsion
~THERAPY - [radiat] (radiation therapy in which either the source is rotated around the patient, or the patiend around the source) cyclothérapie *f*
ROTATIONAL - [gen] rotatif, de rotation
~CONSTANT - [phys] constante *f* de rotation
~CONTROL ELECTROMECHANISM - [contr] électromécanisme *m* rotatif de commande
~ELECTROMOTIVE FORCE - [el] force *f* électromotrice dynamique
~FIELD - [el] (a field in which the circulation is not, in some parts, always zero) champ *m* électrique tournant
~FINE STRUCTURE - [nucl] (the fine structure of an atomic spectrum supposed to be due to the rotation of the nucleus according to quantum conditions) structure *f* fine due à la rotation
~FLOW - [phys] (flow in a region in which vorticity occurs) effluent *m* rationnel
~HYSTERESIS - [el] histérésis *f* de rotation
~INERTIA COEFFICIENT - [phys] coefficient *m* de majoration de la masse du train
~MOTION - [phys] (the motion of a fluid with non zero vorticity) mouvement *m* de tourbillonnement
~MOULDING - [plast ind] (the operation of moulding materials by centrifugal force produced by rotation) moulage *m* par rotation
~QUANTUM NUMBER - [phys] nombre *m* quantique relatif à une impulsion de rotation
~SPEED - [mech] vitesse *f* rotationnelle
~TRANSFORM - [phys] transformation *f* rotationnelle
~VECTOR - [math] vecteur *m* rotationnel
~WAVE - [acoust] (a wave in an elastic medium which causes an element of the medium to change its shape without a change of volume) onde *f* de torsion
ROTATIVE - [gen & mech] rotatif, rotateur
ROTATIVE - [agric] en assolement
~MOMENT - [phys] moment *m* de rotation
ROTATOR - [el] (small electrical high-speed motor) dispositif *m* rotateur
[anat] (muscle) rotateur
[naut] hélice *f* du loch
ROTATORY - [opt] (optically active) rotatoire, optiquement actif
~DISPERSION - [phys] (the variation of the rotation of the plane of polarized light with wavelength for an optically active substance) dispersion *f* de rotation
~PHASE CONVERTER - s. rotary phase converter
~POWER - [phys] (the power of an optically active substance) pouvoir *m* rotatoire
ROTE - [text] bouton *m*
ROTENONE - [chem] (tubatoxin. An insecticide) roténone *f*
ROTOGRAVURE - [print] (the process of engraving a picture a cylindrical printing surface and run it through a rotary press which prints both sides of the paper at the same time) rotogravure *f*, héliogravure *f* impression *f* en creux
~PRESS - [print] rotative *f* pour l'impression en creux, rotative *f* hélio
ROTOR - [mech] (the revolving element) rotor *m*
[mech] (the revolving assembly of some types of mixing machines) rotor *m*
[mech] (of a turbine) rotor *m*
[aero] (a system of revolving aerofoils) rotor *m*
[el] (in a motor; of ignition distributors) rotor *m*, induit *m*
[instr] (the rotating portion of a motor meter) rotor *m*
[el] (of an electric motor or machine) induit *m*
~BLADES - [aero] pales *fpl* du rotor
~BLOCKING - [el] blocage *m* des bobines du rotor
~CORE - [el] (the assembly of laminations of a rotor) noyau *m* de rotór
~HEAD - [aero] (the hub on which the rotor blades are mounted) tête *f* de rotor
~HUB - [aero] (the rotary element on which the rotor blade are mounted) moyeu *m* du rotor
~MOMENT OF INERTIA - [constr] moment *m* d'inertie du rotor
~PLANE - s. rotorcraft
~RHEOSTAT - [contr] (a rheostat connected to the rotor windings of a motor) rhéostat *m* rotorique
~SHIP - [naut] (a vessel propelled by rotor operated by wind-power) navire *m* à rotors
ROTORCRAFT - [aero] (heavier-than-aircraft in which the lift is obtained from one or more rotors) giravion *m*
[aero] (of a helicopter) sustentateur *m* rotatif (de hélicoptère)
ROTOVATOR - [agric] rotovator *m*, fraiseuse *f* agricole
ROTTEN - [gen] pourri, putréfié, carié
~COCOON - [text] cocon *m* gaté
~LODE - [mining] filon *m* pourri
ROTTER - [radar] (colloq.for an automatic selective radar-jammer) dispositif *m* pour le brouillage automatique sélectif de radar
ROTTING - [geol] érosion *f*
ROTULA - [anat] rotule *f*
ROTUND - [bot] (approximately circular) rond, arrondi

ROTUNDA - [arch] (building circular in plan) rotonde *f*
ROTZ - [med] morve *f*, farcin *m*
ROUGE - [chem] (red ferric oxide used for polishing metals) rouge *m* à polir, rouge *m* d'Angleterre
~LAPPING - [mech] rodage *m* au rouge
ROUGH, to - [gen] ébouriffer
[mech & carp] dégrossir
[constr] (a wall etc) piquer, bretter
[glass man & opt] dépolir, dégrossir (une lentille)
ROUGH - [gen] rugueux, raboteux
[gen] (crude) brut
[of the sea] (mer) agitée
[gen] (of a road) raboteux
[gen] (of a piece of work) approximatif
[metall] balafré
~ARCH - [arch] arc *m* en briques profilées
~ASHALR - [constr] pierre *f* taillée grossièrement
~BORING - [mech] alésage *m* d'ébauche
~-CAST - [constr] (rough finish given to a wall by coating it with plaster containing gravel) gobeté, fouetté, crépi
[metall] brut de fonte
~CAST GEARS - [mech] engrenages *mpl* bruts de fonte
~CASTING - [constr] ravalement *m*, crépissage *m*
[metall] pièce *f* brute de fonderie
~CLOTH - [text] drap *m* brut, loden *m*
~COAL - [mining] tout-venant *m*
~COAT - [constr] (the first coat of plaster applied to a wall surface) ravalement *m*
~CUT - [mech] (of files) taille *f* grosse
[telev] montage *m* bout-à-bout
~-CUT FILE - [impl] lime *f* à grosse taille
~CUTTING - [cin] (preliminary edition of a film for inspection purposes) montage *m* préalable
~DIAMOND - [min] diamant *m* brut
~DRESSING - [constr] (of stone) taille *f* brute
~FLAX - [text] lin *m* en paille, lin *m* en tige
~FILE - [impl] lime *f* grosse
~FORGED - [metall] brut *m* de forge
~FORGING - [metall] pièce *f* brute de forge
~GLASS - [glass man] (glass obtained by cutting the original sheet of rolled glass into usable sizes) verre *m* dépoli
~GRAINED - [min metall etc] à gros grain, à grain grossier
~-GRIND, to - [mach tool] émoudre, dégrossir à la meule
~GROUND - [gen] terrain *m* raboteux
~HEALD - [text] lisse *f* brute
~-IN, to - [constr] (the operation of applying the first coat of plaster to a wall surface) piquer, bretter
~LEATHER - [leather ind] cuir *m* à grain grossier, gros cuir *m*
~LINEN - [text] grosse toile *f*
~LUMBER - [timber] bois *m* raboteux
~-MACHINED - [mech] ébauché
~-MACHINING - [mech] dégrossissage *m* à la machine
~PAPER - [paper man] papier *m* rugueux
~PILE BORDER - [text] bordure *f* frisée
~-PLANE, to - [mach tool] ébaucher à la raboteuse
~-ROOL, to - [metall] ébaucher au laminoir
~-ROLLED - [metall] brut de laminage
~RUG - [text] étoffe *f* pour couvertures grossières
ROUGH SEA - [met] mer *f* agitée
~-SHAPE, to - [mech] ébaucher
~SHEET - [plast ind] feuille *f* homogénéisé
~SKETCH - [draw] ébauche *f*, esquisse *f*
~STAMPED - [metall] brut d'estampage
~TO THE FEEL - [text] rude au toucher
~THE HEMP, to - [text] émoucheter le chanvre
~-TURN, to - [mech] ébaucher au tour
~-TURNED - [mach tool] ébauché, dégrossi
~-TURNING - [mach tool] ébauchage *m* au tour
~VACUUM - [phys] vide *m* grossier
~YARN - [text] fil *m* rude, fil *m* gros
~WEB - [text] voile *m* duveteux
ROUGHAGE - [agric] (animal feeding) fourrages *mpl* grossiers
[biol] matière *f* cellulosique
~FORGING - [metall] pièce *f* de forge dégrossie
ROUGHED HEMP - [text] chanvre *m* émoucheté
ROUGHEN, to - [gen] rendre rude
[constr] (the stone) boucharder (la pierre)
[naut] (of the sea) grossir
ROUGHEN A HORSE, to - [agric] ferrer à glace (un cheval)
ROUGHENED DISC - [text] disque *m* rugueux
~SHEET METAL - [metall] tôle *f* à picots
~SHOE - [agric] (ofr horses) fer *m* à glace
~TUBE FOR CHEESES - [text] tube *m* rugueux pour bobine croisée
ROUGHENING - [mech] rendre rugueux (le), dégrossissage *m*, brossage *m*
~MACHINE - [mech] (a buffing machine) machine *f* à brosser
~TOOL - [tool] appareil *m* à râper
~WHEEL - [impl] disque *m* à brosser, meule *f* de brossage
ROUGHER - [tool] ébaucheur *m*, outil *m* à dégrosser
[text] émoucheteuse *f*
ROUGHER'S HACKLE - [text] ébauchoir *m*
~TOOL - [text] s. rougher's hackle
~TOW - [text] émouchures *fpl*
ROUGHING - [mech] dégrossissage *m*
[metall] (mill operation) ébauchage
[phys] (reduction of pressure in a vacuum vessel by means of a pump as a preliminary to more complete exhaustion) pompage *m* préliminaire, prévidage *m*
[metall] (of ingots) laminage *m* de lingots
~CUT - [mech tool] (a primary machining operation, not to exact size) ébauchage *m*
~HACKLE - [text] s. roughr's hackle
~HOB - [tool] fraise *f* à ébaucher
~-IN - [constr] (first coat of plaster) crépissage *m*, piquage *m*
(the first coat of tree-coat plaster work applied on bricks) enduit *m*
~MILL - [metall] laminoir *m* ébaucheur
~-OUT - [mech & carp] (the operation of reducing a workpiece to approximate size) dégrossissement *m*
~PASS - [metall] (the first passage) passe *f* de dégrossissage
[metall] (in the body of the roll) cannelure *f* ébaucheuse
~ROLLER - [impl] cylindre *m* dégrossisseur
~ROLLS - [metall] train *m* ébaucheur
~SLOT MILL - [mach tool] fraise *f* pour dégrossir les rainures

ROUGHING TOOL - [tool] outil *m* à dégrossir, outil *m* à charioter
~WHEEL - [mech] meule *f* à dégrossir
ROUGHNESS - [gen] rudesse *f*, asperité *f*
[gen] (of a metal or wooden surface) rugosité *f*
[gen] (of soil) inégalité *f* (du sol)
~EFFECT - [phys] (of fluids) effet *m* de rugosité
~METER - [instr] (an apparatus designed to measure the degree of smoothness of a substance) dispositif *m* à mesurer la rugosité
ROULEAU - [biol] pile *f* de globules rouges
ROULETTE - [geom] roulette *f*
ROUND, to - [gen] (to make or to become round) rendre rond, arrondir
[gen & mech] arrondir
[bookbind] endosser (un livre)
[naut] doubler
[naut] garnir, fourrer
ROUND - [gen] (circular) rond, circulaire
[gen] (a series of concerted actions performed in succession by a number of persons) série *f*, tournée *f*
[firearms] cartouche *f*
[impl] (a rung of a ladder) échelon *fm*
[mining] (blasting of a series of drill holes) volée *f*
[mus] fugue *f* perpétuelle, canon *m*
[acoust] (full-toned) (voix) pleine, sonore
[carp] (of chair) roulon *m*, barreau *m*
[arch] rond (de moulure)
~ARCH - [arch] arc *m* plein cintre
~-ABOUT - [constr] (road junction consisting of a central island round which the traffic moves) carrefour *m* giratoire
~-ABOUT TRAFFIC SYSTEM - [gen] circulation *f* giratoire
~BALE - [text] (approximately 250 lbs) balle *f* ronde
~BAR IRON - [metall] fers *mpl* ronde
~BAR STEEL - [metall] acier *m* rond
~-BAR TONGS - [impl] tenaille *f* pour fer rond
~BASTARD FILE - [tool] lime *f* bâtarde ronde
~BELT - [rubber ind] courroie *f* ronde
~-BODY PACKING RING - [mech] anneau *m* de garniture en 0
~BLOCKS FOR GEAR WHEEL MANUFACTURE - [mech] ébauche *f* cylindrique pour la fabrication des engrenages par usinage
~BOTTOMED FLASK - [ind chem] (thin glass flask with spherical body) ballon *m* rond
~BROACH - [mech] équarissoir *m* pour le finissage de fers ronds
~BUDDLE - [min] round-buddle *m*
~BUSH - [mech] bague *f* ronde
~CABLE - [telecomm] câble *m* rond
~CHART - [instr] (the paper disk on which the results of a recording meter are registered) disque *m* de registration
~-CHART INSTRUMENT - [instr] enregistreur *m* à disque
~CHISEL - [tool] burin *m* rond
~COAL - [min] charbon *m* gros
[mining] gros *m*
~-COIL MEASURING INSTRUMENT - [instr] (a moving coil instrument with a round coil) dispositif *m* de mesure à bobine ronde
~COMBER - [text] peigneuse *f* à action continue
~CUTTER - [tool] fraise *f* à champ rond
~DIE - [metall] filière *f* ronde
ROUND EDGE MILLING CUTTER - [mach tool] fraise *f* à champ rond
~-END KEY - [impl] clavette *f* à angles ronds
~-FACED PULLEY - [mech] poulie *f* en dos d'âne
~FIGURES - [gen & comm] chiffres *mpl* ronds
~FILE - [tool] lime *f* ronde
~GRAIN - [text] grain *m* rond
~-HEAD BOLT - [mech] boulon *m* à tête ronde
~-HEAD RIVET - [mech] rivet *m* à tête ronde
~-HEAD WOOD SCREW - [mech] vis *f* à bois à tête ronde
~-HEADED SCREW - [mech] vis *f* à tête ronde
~-HEADED STAKE - [impl] (of anvil) tasseau *m* rond
~-HEADED SIEVE - [mech] tamis *m* à perforations rondes
~-HOUSE - [railw] rotonde *f*
~INGOT - [metall] lingot *m* rond
~IRON WIRE - [metall] fil *m* de fer rond
~KEY - [mech] clavette *f* ronde
~LACE - [text] galon *m* à coudre
~LEATHER BAND - [mech] corde *f* en cuir, courroie *f* ronde en cuir
~MOULD INSERT FOR MARKING - [plast ind] gravure *f* rapportée, macaron *m*
~-MOUTH TONGS - [impl] tenaille *f* à bec rond
~NEEDLE - [text] aiguille *f* ronde
~-NOSE BORING TOOL - [tool] alésoir *m* à bec rond
~-NOSE CHISEL - [tool] burin *m* grain d'orge, gouge *f* pleine
~-NOSE PLIERS - [mech] pinces *fpl* à becs ronds
~-NOSED CHISEL - [tool] s.rond-nose chisel
~NUT - [mech] écrou *m* cylindrique
~OF BEAM - [shipbuild] (a camber) cambrure *f* de bau
~OFF, to - [gen] arrondir
[mach tool] (a grinding wheel) arrondir
~-OFF ERROR - [comput] (error resulting from neglecting less significant digits of a quantity and applying some adjustement to the more significant digits retained) erreur *f* d'arrondissement
~PIVOT BEARING - [mech] support *m* de pivot rond
~PLANE - [tool] rabot *m* rond
~PLIERS - [impl] pince *f* ronde
~-POINT CHISEL - [tool] grain-d'orge *m*
~PUNCH - [tool] poinçon *m* rond
~RASP - [impl] râpe *f* ronde
~ROD - [metall] tige *f* ronde
~ROPE - [text] câble *m* rond
~RUBBER THREAD - [rubber ind] fil *m* rond de caoutchouc
~SEAM - [rubber ind] couture *f* ronde
~SLEEKER - [metall] pièce *f* à lisser ronde
~SPIRIT-LEVEL - [surv] niveau *m* rond à alcool
~-STEEPLE RIVET - [mech] trivet *m* à tête ronde
~THREAD - [mech] filet *m* rond
~-TOP - [naut] hune *f*
~-TOPPED - [gen etc] à sommet arrondi
~-TOPPED TOOTH - [mech] dent *f* à sommet arrondi
~TROWEL - [impl] truelle *f* ronde
~UP, to - [math] arrondir
~WIRE - [metall] fil *m* (métallique) rond
~YARN - [text] fil *m* rond
ROUNDED-APPROACH ORIFICE - [hydr] orifice *m* évasé
~COUNTERSUNK HEAD - [mech] (of screws) à tête fraisée arrondie
~EDGES - [text] coins *mpl* arrondis

ROUNDED NOTCH - [text] fente *f* arrondie
~-OFF GUIDE PLATE - [text] plateau-guide *m* arrondi
~-OFF NOTCH - [mech] échancrure *f* arrondie
~PAPER BAND - [text] bande *f* de papier arrondie
ROUNDEL - [arch] (a small round window; a semi-circular recess) rondeau *m*
ROUNDING - [bookbind] (the operation of giving the back of a book a convex shape before casing) endossage *m*
[naut] (the rope round a cable to prevent chafing) fourrure *f* (d'espar)
[shipbuild] bouge *m*
~ADZE - [impl] herminette *f* à gouge
~APPARATUS - [mech] mécanisme *m* d'arrondissement
~ERROR - s. round-off error
~FUNNEL - [impl] entonnoir *m* à arrondir
~MACHINE - [mach tool] tour *m* pour bâtons ronds
~-OFF - [mach tool] (of a grinding wheel) arrondissement *m*
~-OUT - [aero] (the change from approach to horizontal flight before alighting) arrondi *m*
~PULLEY - [mech] poulie *f* bombée
~TOOL - [tool] étampe *f* pour fers ronds
~WOODEN ROLLERS - [mach tool] (in a lathe) cylindrage *m* des rouleaux en bois
ROUNDISH CLODS - [text] pelotes *fpl*
~FORM OF SILK FILAMENT - [text] brin *m* arrondi au point de croisement
ROUNDNESS - [gen] rondeur *f*, arrondissement *m*
ROUNDS OF A LADDER - [impl] échelons *mpl* d'une échelle
ROUP - [vet] (diphteritic disease of poultry) diphtérie *f* des poules
ROUSE, to - [chem] (to stimulate fermentation) activer la fermentation
[naut] (to haul stringly) haler
[brew ind] agiter (le moût), aérer
~-ABOUT - [text] manoeuvre *m* du tondeur
ROUSER - [brew ind] balai *m*
ROUSING APPARATUS - [brew ind] appareil *m* à mélanger la levure
ROUSSIN'S BLACK SALT - [ind chem] (formed from Roussin's Salts by reaction with dilute acids) sel *m* noir de Roussin
~RED SALT - [chem] (this is formed by treating Roussin's Salts with sodium sulphide) sel *m* rouge de Roussin
~SALTS - [chem] (compounds formed by reacting sodium trisulphide with a solution of ferrous chloride which has been saturated with nitric oxide at 2° C) sels *mpl* de Roussin
ROUSTABOUT - [gen] (USA only; a navy) débardeur *m*, manoeuvre *m*
ROUT OUT, to - [carp] évider, rainurer (une planche)
[print] échopper
ROUTE, to - [gen] router
[ind] suivre le plan de travail
[comput] diriger
ROUTE - [gen] route *f*, itinéraire *m*, voie *f*
~CARD - [contr] (in work organization) fiche *f* suiveuse
~CONTROL - [railw] commande *f* des itinéraires
~DIAGRAM - [railw] tableau *m* d'acheminement
~LEVER - [railw] levier *m* d'itinéraire
~LOCKING - [railw] (after the signal) enclenchement *m* de transit
ROUTE LOCKING - [railw] (before the signal) enclenchement *m* de tracé d'itinéraire
~RELAY - [el] relais *m* d'itinéraire
~SELECTION - [comput] sélection *f* d'acheminement
~SURVEY - [plumb] (in mainlaying) nivellement *m* de la piste
ROUTER - [mach tool] (rotary cutting tool used to shape curved workpieces) couteau *m* de mèche à trois pointes
[impl] mortaiseuse
~HEAD - [mech] tête *f* du trusquin
~PLANE - [carp] guimbarde *f*
ROUTINE - [gen] routine *f*, travail *m* courant
[comput] (precise sequence of coded instructions) programme *m*
~ANALYSIS - [gen] analyse *f* périodique
~LIBRARY - [comput] bibliothèque *f* des programmes
~MAINTENANCE - [gen mech etc] maintenance *f* périodique
~MESSAGE - [comput] message *m* de routine
~REPLACEMENT - [nucl] (the normal replacement of fuel elements in a nuclear power plant) substitution *f* courante des combustible
~TESTS - [gen el etc] (approval tests carried out on all units of the same consignment) essais *mpl* individuels
~TEST EQUIPMENT - [telecomm] installation *f* d'essais systématiques
ROUTINER - s. routine test equipment
ROUTING - [mech] évidage *m*
[print] échoppage *m*
[comput] acheminement *m*
~CARD - [comput] (for industrial work organization) carte *f* d'acheminement
~CHART - [telecomm] tableau *m* d'acheminement des communications
~MACHINE - [print] (machine having a revolving point for the removal of unwanted metal from plates) machine *f* à échopper
~SHEET - s. routing card
~PLANT - s. routing chart
ROVE, to - [gen] rôder, errer
[text] (to draw into a thread) boudiner
ROVE - [text] mèche *f*, boudin *m*
[mech] (metal ring used in clinching a nail) rondelle *f*, contre-rivure *f*
ROVER - [text] banc *m* à broches
[naut] forban *m*
ROVING - [text] (continuous thread or stand of fibre, wound into a cylindrical package) mèche *f* de preparation
[plast ind] stratifil *m*, roving *m*
~BAR - [text] barre *f* guide-mèche
~BOBBIN - [text] tube *m* pour banc à broches
~EXHAUSTER - [text] système *m* d'aspiration de mèches cassées
~FRAME - [text] banc *m* à broches
~GUIDE - [text] guide-mèche *m*
~MACHINE - s. roving frame
~PREVIEW - [cin & telev] preview *m* superficiel
~WATE - [text] bouts *mpl* de mèche de préparation
ROW, to - [naut] ramer, nager
ROW - [gen] (rank; file) rang *m*, ligne *f*, file *f*
[gen] (a street) ruelle *f*
[gen] (noisy disturbance) vacarme *m*
[mech] (in a redial engine, a row of cylinders) étoile *f*

ROW - [comput] (on a punch card) rangée *f*
[agric] ligne *f*, rayon *m*
~BOAT - [naut] bateau *m* à rames
~CROP PLANTING - [agric] plantation *f* en lignes
~-CROPPING - [agric] culture *f* en lignes
~LINES - [crystal] (lines which are formed by the diffraction spots on a rotation photograph of a crystal, obtained by rotation of a crystal in a beam of x-rays, and photographing the diffraction pattern) file *f* de lignes
~OF BALLS - [mech] rangée *f* de billes
~OF CARRIAGES - [text] rangée *f* de chariots
~OF DENTS - [text] rangée *f* de dents
~OF HEALDS - [text] rangée *f* de mailles
~OF HOLES - s. row of peg holes
~OF HOOKS - [text] route *f* de crochets
~OF LOOPS - [text] rangée *f* de boucles
~OF NEEDLES - [text] rangée *f* d'aiguilles
~OF PARALLEL CORDS - [text] rangée *f* de cordes parallèles
~OF PEG HOLES - [text] rangées *fpl* de liage
~OF PINS - [text] jeu *m* de peignes
~OF RIVETS - [mech] rangée *f* de rivets
~OF SPINDLES - [text] rang *m* de broches
~SEEDING - [agric] semis *m* en lignes
~SPACING - [agric] écartement *m* des lignes
ROWAN-TREE - [bot] sorbier *m* commun
ROWELLING SCISSORS - [vet] ciseaux *mpl* pour l'application d'une ortie
ROWLOCK - [naut] porte-rame *m*, dame *f*
ROYALTY - [leg] (payment made to the proprietors of a patent for permission to make use of it) redevance *f*
[leg] (on books) droits *mpl* d'auteur
[comm] (the percentage which is paid to the owner of a machine for its use) redevance *f*
[mining] redevance *f* tréfoncière
R.P.M. - [mech] (revolution per minute) tours *mpl* à la minute
R.P.S. - [mech] (revolutions per second) tours *mpl* à la seconde
RS - [met] (Beaufort letters; rain-snow) pluie *f* et neige mêlées
R-THETA SYSTEM - [aero & astronaut] système *m* de navigation en coordonnées polaires
R-T UNIT - [radio] appareil *m* récepteur-émetteur
RUB, to - [gen] frotter
RUB - [gen] frottement *m*, friction *f*
[phys] friction *f*
~DOWN, to - [paint] poncer, adoucir
~DOWN THE CYLINDER WITH EMERY CLOTH, to - [text] nettoyer le grand tambour avec du papier émeri
~DOWN THE CYLINDER WITH SAND PAPER - [text] nettoyer le grand tambour avec du papier de verre
~RAILS - [auto] (running boards) marche-pied *m*
~OFF THE RUST, to - [gen] dérouiller
~OUT, to - [gen] effacer, gommer
~THE CYLINDER WITH GRAPHITE, to - [text] frotter le grand tambour avec graphite
~THE IMPURITIES OUT OF THE YARN, to - [text] enlever les impurités du fil par frottement
~THE TOP PRESSURE ROLLER WITH WHITING, to - [text] frotter le cylindre de pression avec du blanc d'Espagne
~WITH EMERY, to - [mech] nettoyer avec émeri
RUBBED - [gen] frotté, râpé
RUBBED DOWN - [paint] poncé, regratté
~HEMP - [text] chanvre *m* frotté
~-IN-JEWEL - [horol] rubis *m* monté
RUBBER - [rubber ind] (a natural or synthetic elastic substance) caoutchouc *m*
[gen] (anything used for rubbing) gomme *f*
[mech] (brake-shoe) sabot *m* de frein
~ACCELERATORS - [chem] (compounds which increase the rate of vulcanization or enable vulcanization to be carried out at a lower temperature) accélérateurs *mpl* pour le caoutchouc
~AFLOAT - [rubber ind] caoutchouc *m* flottant
~AND CORK SHEETING - [rubber ind] feuille *f* en caoutchouc-liège
~AND FABRIC APPLICATION - [rubber ind] application *f* de tissu gommé
~ASPHALT - [rubber ind] asphalte *m* caoutchouté
~BAG MOULDING - [plast ind] (moulding by means of an inflatable rubber bag) moulage *m* au sac en caoutchouc
~BANDS - [rubber ind] bandes *fpl* en caoutchouc
~-BASE PROPELLANT - [astronaut] propergol *m* à base de caoutchouc
~-BASED PAINT - [paint] peinture *f* au caoutchouc
~BEARING - [rubber ind] coussinet *m* en caoutchouc
~BLADE FAN - [rubber ind] ventilateur *m* à pales en caoutchouc
~BLANKET - [print] blanchet *m* de caoutchouc
~BLANKET SPREADING - [text etc] enduction *f* à la râcle sur tablier
~BLOCK - [print] cliché *m* en caoutchouc
~BOAT - [naut] canot *m* pneumatique
~BOND - [ind chem] lien *m* élastique
~-BONDED ABRASIVE GRINDING WHEEL - [mech] meule *f* à abrasif au moyen de caoutchouc
~-BONDED -TO-METAL COMPONENTS - [rubber ind] ensembles *mpl* métal-caoutchouc
~BONDED WHEEL - [mech] meule *f* à abrasif
~BREAST - [rubber ind] sein *m* en caoutchouc
~BUFFER - [auto] pare-choc *m* en caoutchouc
~BUSH - [rubber ind] (in a bearing) coquille *f* en caoutchouc
~CANVAS - [rubber ind] caoutchouc *m* toilé
~CEMENT - [ind chem] colle *f* au caoutchouc
~CEMENT MIXER - [ind chem] malaxeur *m* à dissolution
~CLOG - [rubber ind] sabot *m* en caoutchouc
~COATED - [rubber ind] gommé
~-COATED FABRIC - [rubber ind] tissu *m* gommé
~COMPOUND - [rubber ind] mélange *m* caoutchouc
~CONDENSER FOR BOURRETTE - [text ind] frottoir *m* pour bourrette
~CONTENT - [rubber ind] teneur *f* en caoutchouc
~-CONTENT OF LATEX - [rubber ind] teneur *f* en caoutchouc du latex
~CORDS FOR SHOCK DAMPING - [aero] cordes *fpl* élastique pour amortissement
~-CORE YARN - [text] fil *m* avec âme de caoutchouc
~COVERED ARTICLES - [rubber ind] articles *mpl* recouverts de caoutchouc
~COVERING - [rubber ind] revêtement *m* en caoutchouc
~CRUMB - [rubber ind] poudrettes *fpl* en déchets de caoutchouc vulcanisé
~DELIVERY HOSE - [rubber ind] tuyau *m* de refoulement
~DINGHEY - [naut] canot *m* pneumatique

RUBBER DISK - [rubber ind] rondelle *f* en caoutchouc
~DRAWING - [text] étirage *m* à frottoirs
~ECCENTRIC - [text] excentrique *m* des cuirs
~ENGRAVING - [print] cliché *m* en caoutchouc
~ERASER - [rubber ind] gomme *f* à effacer
~FACE - [text] couche *f* extérieure en caoutchouc
~FACTORY - [ind] usine *f* de caoutchouc
~FENDERING - [auto] tampons *mpl* en caoutchouc
~FILM - [bot] (on the cut) lanière *f* de caoutchouc (sur l'incision)
~FLOORING - [rubber ind] planchéiage *m* en caoutchouc
~GEAR - [text] frottoir *m*
~GOODS - [rubber ind] articles *mpl* en caoutchouc
~GREASE - [mech] graisse *f* à base de caoutchouc
~-HEADED TACK - [mech] clou *m* à tête en caoutchouc
~HOSE - [rubber ind] tuyau *m* en caoutchouc
~HYDROCHLORIDE - [chem] (non flammable thermoplastic obtained by treatment of a rubber solution with anhydrous hydrochloride. It has uses as a protective covering) hydrocarbure *m* de caoutchouc
~IMPREGNATED FABRIC - [text] tissu *m* impregné de caoutchouc
~INDUSTRY - [rubber ind] industrie *f* de caoutchouc
~INSULATED CABLE - [rubber ind] câble *m* sous caoutchouc
~JOINT - [mech] (a sealing element made of rubber) joint *m* de caoutchouc
~LATEX - [rubber ind] latex *m* de caoutchouc
~LEATHER - [text] manchon *m*, frottoir *m*
~-LINED - [rubber ind] garni en caoutchouc, à revêtement en caoutchouc
~LINED MILL - [rubber ind] moulin *m* à revêtement intérieur en caoutchouc
~LINING - [rubber ind] garniture *f* en caoutchouc, revêtement *m* en caoutchouc
~LINING OF A BEARING - [rubber ind] garniture *f* de caoutchouc d'un coussinet
~MANUFACTURER - [rubber ind] caoutchoutier *m*
~MASTIC - [rubber ind] mastic *m* pour pneus
~MIXER - [rubber ind] mélangeur *m* à cylindres pour caoutchouc
~MOUNTING - [mech] (of machines) montage *m* en caoutchouc, suspension *f* en caoutchouc
~-NECK - [mining] (investigator) rôdeur *m*
~-OIL - [chem] huile *f* de caoutchouc
~PAD - [mech] (a block of rubber used as a cushion in a press) coussin *m* de caoutchouc
~PATCH - [rubber ind] (for tyres) emplâtre *m* de caoutchouc, rustine *f*
~PAVING BLOCK - [constr] pavé *m* de caoutchouc
~PLANT - [bot] caoutchoutier *m*
[rubber ind] usine *f* de caoutchouc
~PLANTATION - [bot] plantation *f* de caoutchouc
~(IMPRESSION) PLATE - [print] cliché *m* en caoutchouc, plaque *f* d'imprimerie en caoutchouc
~-PLATED - [rubber ind] à revêtement en caoutchouc
~-PLUS AIR SHOCK ABSORBER - [aero] (type of shock absorber in a landing-gear, in which both rubber elements and pneumatic cushioning are used) amortisseurs *mpl* en caoutchouc et à air comprimé
~POWDER - [rubber ind] poudre *f* de caoutchouc
~PROCESSING INDUSTRY - [rubber ind] industrie *f* du caoutchouc

RUBBER-RESIN BLEND - [rubber ind] mélange *m* caoutchouc-résine
~-RESIN SOLE - [shoe man] semelle *f* en mélange caoutchouc-résine
~RING FOR METAL BARRELS - [rubber ind] cerclage *m* en caoutchouc pour fûts métalliques
~ROLLER - [mech] cylindre *m* garni de caoutchouc
~SEAL - s. rubber joint
~SEEDLING NURSERY - [bot] pépinière *f* de plantes satives
~-SEED-OIL - [chem] huile *f* de graines d'hévéa
~SHEETING - [rubber ind] feuilles *fpl* en caoutchouc
~SHOE VALVE - [rubber ind] (for tyres) valve *f* à siège en caoutchouc
~SLAB FOR THE RISER - [contr] recouvrement *m* de caoutchouc pour ais
~-SOLED CANVAS SHOE - [shoe man] soulier *m* en étoffe avec semelle en caoutchouc
~SOLUTION - [ind chem] (a solution of rubber used as an adhesive) dissolution *f* de caoutchouc
~SOLUTION HOMOGENIZING MIXER - [rubber ind] mélangeur *m* à homoguer une dissolution de caoutchouc
~SOLUTION MILL - [rubber ind] mélangeur *m* à dissolution
~SOLVENT - [chem] (a petroleum distillate used in the manufacture of rubber solution) solvant *m* à base de caoutchouc
~SPONGE - [rubber ind] (a cellular rubber produced either by the incorporation of a blowing agent in the latex before curing or by beating air into the latex) caoutchouc *m* spongieux, éponge *f* en caoutchouc
~SPREADER - [rubber & plast ind] machine *f* à enduire avec racle sur rouleau
~SPRING BUSHING - [mech] boîte *f* de ressort en caoutchouc
~SPRING MOUNTING - [mech] support *m* élastique en ressort de caoutchouc
~SPRING SHOCK ABSORBER - [aero] (type of landing-gear shock-absorber in which rubber is used as the cushioning medium) amortisseurs *mpl* en caoutchouc
~SPRUNG RAILWAY WHEEL - [railw] roue *f* de chemin de fer à ressort en caoutchouc
~-STAMP, to - [gen] timbrer
~STAMP - [impl] timbre *m*
~SUBSTITUTE - [rubber ind] substitut *m* du caoutchouc
~SUCTION HOSE - [rubber ind] tuyau *m* d'aspiration en caoutchouc
~-SULPHUR STOCK - [ind chem] mélange *m* gomme-soufre
~SURFACED PAVING BLOCK - [constr] bloc *m* de pavage en couche superficielle de caoutchouc
~TAPE - [rubber ind] ruban *m* caoutchouté
~THREAD - [rubber ind] fil *m* de caoutchouc
~THREAD COVERING MACHINE - [mech] machine à guipage de fil de caoutchouc
~-TO-METAL BONDING - [metall] composition *f* de caoutchouc et métal
~TOSH - [text] tissu *m* pour orgue, tissu *m* pour soufflet d'orgue
~TRACK - [rubber ind] (tyres) chenille *f* souple de caoutchouc
~TREE - [bot] arbre *m* à caoutchouc
~VARNISH - [paint] vernis *m* à base de caoutchouc
~WINNING - [rubber ind] exploitation *f* du caoutchouc

RUBBER WRAPPED BALE - [rubber ind] emballage *m* en feuilles de caoutchouc
RUBBERIZE, to - [rubber ind] (the operation of coating a surface with rubber) gommer, caoutchouter [text] (of silk) enduire de caoutchouc, caoutchouter à la calandre
RUBBERIZE WITHDOCTORKNIFEONROLL, to - [rubber ind] gommer à la racle sur cylindre
RUBBERIZED ASPHALT ROAD - [constr] route *f* en asphalte caoutchouté
~BITUMEN - [constr] bitume *m* caoutchouté
~CANVAS BELT - [rubber ind] courroie *f* en toile caoutchoutée
~CLOTH - [text] tissu *m* gommé
~CORK MOULDING - [rubber ind] pièce *f* moulée en liège caoutchouté
~FABRIC - [text] tissu *m* caoutchouté
~HAIR - [rubber ind] crin *m* caoutchouté
~RAINCOAT - [rubber ind] imperméable *m* en tissu caoutchouté
~STEEL CONVEYOR BELT - [mech] bande *f* transporteuse en acier caoutchouté
~WOVEN BELTING - [rubber ind] courroie *f* en tissu caoutchouté
~WOVEN - [rubber ind] en tissu caoutchouté
RUBBERIZING - [rubber ind](the coating of a surface with rubber, espec. of textiles) caoutchoutage *m*, enduisage *m* de gomme
~BY SPREADING - [rubber ind] imperméabilisation *f* par gommage au métier
~MACHINE - [rubber ind] machine *f* à enduire
RUBBERLIKE - [rubber ind] ressemblant au caoutchouc
RUBBERSEED OIL - [chem] (an oil derived from the seeds of Hevea Brasiliensis) huile *f* de graines d'hévéa
RUBBERTIP FOR PENCIL - [rubber ind] gomme *f* pour bout de crayon
RUBBERY - [rubber ind] gommeux
RUBBING - [gen] frottement *m*, frottage *m*
[paint] ponçage *m*
[naut] ripage *m* (d'un câble)
~ABRASION - [mech] abrasion *f* par frottement
~BAR - [gas ind] (for gasholders) patin *m* d'étanchéité
~COMPOUND - [ind chem] pâte *f* à polir
~LEATHER - [leather ind] cuir *m* à frotter
~NOISE - [acoust] bruit *m* de frottement
~PAPER - [gen] papier *m* abrasif
~PLATES - [text] plateaux *mpl* de frottoir
~RAG - [text] chiffon *m* rugueux
~ROLLER - [text] cylindre *m* de frottoir
~SURFACE - [gen] surface *f* frottante, surface *f* de frottement
[gen] (of a match-box) gratin *m*
~TEST - [paper man] essai *m* de résistance froissement
~VARNISH - [paint] (polishing varnish) vernis *m* à polir
RUBBISH - [gen] immondices *fpl*
RUBBLE - [gen] moellons *mpl*, brocage *m*
[constr] (rough uncut stones) moellons *mpl* bruts
~FILLING - [constr] moellonage *m*
RUBEANIC ACID - [chem] (dithio-oxamide.A reagent for the detection of small quantities of copper, with which it forms a black precipitate) acide *m* rubéanique
RUBELLITE - [min] (red, transparent variety of tourmaline, a semiprecious gemstone) rubellite *f*
RUBEOLA - [vet] (swine fever) rubéole *f*
RUBEDO - [med] rougeur *f* de la peau
RUBEOSIS - s. rubedo
RUBELLA - [med] (german measles) rubéole *f*
RUBESCENT - [bot] (turning ered) rubescent
RUBIDIUM - [chem] (a metallic element, symbol Rb, A.N. 37, A.W. 85.48; with applications in electronics and as a catalyst) rubidium *m*
~CHLORIDE - [chem] (an anlytical reagent) chlorure *m* de rubidium
RUBIGINOUS - [bot] (rust-coloured) rubigineux
RUBOR - [med] (redness) rougeur *f*
RUBSEN OIL - [food] huile *f* de navette
~SEED OIL - s. rubsen oil
RUBY - [min] (a gemstone consisting of a crystalline form of corundum, blood-red in colour) rubis *m*
[horol] (bearing of a watch) rubis *m*
~COPPER - [min] cuivre *m* vitreux rouge
~LASER - [phys] laser *m* à rubis
~MASER - [phys] maser *m* à rubis
~OF ARSENIC - s. ruby sulphur
~OF ZINC - [min] sulfure *m* de zinc
~SILVER ORE - [min] argent *m* rouge
~SPINEL - [min] rubis *m* balais
~SULPHUR - [min] rubis *m* d'arsenic, réalgar *m*
RUCHE - [text] ruche *f*
~EDGING - [text] ruchet *m*
RUCK, to - [gen] froncer
RUCK - [gen] fronçure *f*
RUCKSACK - [gen] sac *m* touriste
RUDDER - [aero] (the control surface in an aircraft which controls the direction of flight, when the machine is in a normal attitude) gouvernail *m* de direction
[instr] (type of vane) queue *f* d'orientation
[naut] gouvernail *m*
[agric] (of a windmill) queue *f* de moulin à vent
[brew ind] agitateur *m* mécanique
~ANGLE - [aero] (the angle which the rudder makes with the longitudinal axis of the aircraft) angle *m* de braquage
~ARMS - [naut] traverses *fpl* du gouvernail
~BALANCING SURFACE - [aero] angle *m* de compensation du gouvernail de direction
~BANDS - [naut] pentures *fpl* du gouvernail
~BAR - [naut] barre *f* du gouvernail
[aero] palonnier *m* du gouvernail de direction
~BLADE - [naut] lame *f* du gouvernail
~BOW - [naut] arc *m* du châssis du gouvernail
~BRACE - [naut] ferrure *f* du gouvernail
~CONTROL CABLE - [aero] câble *m* de contrôle du gouvernail
~FRAME - [naut] manchon *m* du gouvernail
~GUDGEON - s. rudder brace
~HEAD - [naut] tête *f* du gouvernail
~HEEL - [naut] talon *m* du gouvernail
~HINGE PIN - [naut & aero] cheville *f* du gouvernail
~HOLE - [naut] jaumière *f*
~INDICATOR - [naut] axiomètre *m*
~MOMENT - [naut] moment *m* d'évolution
~PEDALS - [aero] (the foot-bar controlling the rudder of an aircraft) pédales *mpl* de direction
~PENDANTS - [naut] pantoire *f* du gouvernail
~PINTLE - s. rudder hinge pin
~POST - [aero] (the principal vertical member of a

rudder, on which the hinges are mounted) longeron *m* de gouverne de direction, axe *m* du gouvernail
RUDDER POST - [naut] étambot *m* arrière
~SHAFT BEARING - [naut] support *m* de l'axe du gouvernail
~STOPS - [naut] cales *f*pl d'arrêt du gouvernail
~TAB - [aero] surface *f* stabilisatrice
~TILLER - [naut] béquille *f* du gouvernail
~TORQUE - [aero] (on the fuselage) moment *m* de torsion du gouvernail de direction
~TRIM TAB - [aero] actionneur *m* de trim
~TRUNK - [naut] manchon *m* de gouvernail
~UNIT - [aero] empennage *m* vertical
~WHEEL - [naut] roue *f* du gouvernail
RUDDERHEAD - [naut] (the top end of the rudder stock to which the tiller is attached) tête *f* du gouvernail
RUDDERHOLE - [naut] trou *m* de jaumière
RUDDERPOST - [naut] (additional sternpost) étambot *m* arrière
RUDDERSTOCK - [naut] âme *f* du gouvernail
RUDDLE - [min] arcanne *f*, ocre *f* rouge
RUDDLED - [gen] marqué à l'ocre rouge
~WOOL - [text] laine *f* teinte en rouge
RUDIMENT - [gen biol etc] rudiment *m*
RUDIMENTAL - [gen] rudimentaire
RUDIMENTARY - s. rudimental
RUE - [bot] rue *f*
~OIL - [chem] (an essential oil obtained from species of Ruta and having application in perfumery and as a synthesis intermediate) essence *f* de rue
RUFF, to - [text] peigner en gros
RUFF - [gen] collerette *f*
[zool] (of birds) cravate *f*, collier *m*
RUFFING - [text] peignage *m* en gros
RUFFLE, to - [gen] ébouriffer, froisser, chiffonner
RUFFLE - [gen] ruche *f*, fraise *f*
[naut] (on water) agitation *f*
~WITH SLOT - [mech] collier *m* du frein avec encoche
RUFFLED NOSE OF COP - [text] pointe *f* de cannette éboulée
RUG - [text] couverture *f*, tapis *m*
[text] (type of blanket) plaid *m*
RUGA - [anat] ride *f*, pli *m*
RUGGED - [gen] raboteux, inégal, anfractueux
~ELECTRON TUBES - [electron] tubes *m*pl de construction renforcée
RUGGERIZED CONSTRUCTION - [electron] (technique applied in electronic tube manufacturing to increase the mechanical strenght of the electrode system) construction *f* renforcée
RUGGEDIZATION - [electron] renforcement *m* de la construction
RUGOSITY - [gen] rugosité *f*
RUHMKORFF COIL - [el] (obsolete for a self make and-break induction coil) bobine *f* d'induction
RULE, to - [gen] gouverner, régir
[gen] (to control) régler
[leg] décider
[gen] (to mark with lines) rayer, tracer à la règle
RULE - [gen] règle *f*
[gen] (authority) autorité *f*
[surv] (a sighting rule) alidade *f*
[leg] décision *f*
[math] règle *f*
[print] filet *m*
[impl] règle *f* graduée, mètre *m*
RULE CUTTER - [print] taille-filets *m*
~OF COMPOUND THREE - [math] règle *f* du trois
~OF CONDUCT - [gen] directives *f*pl
~OF MAXIMUM MULTIPLICITY - [phys] (the energy of interaction between the electrons in any one atom is at minimum when their initial spin is greatest) loi *f* de la multiplicité maximale
~OF MUTUAL EXCLUSION - [phys] règle *f* de l'exclusion réciproque
~OF THE ROAD - [naut] règles *f*pl de route
~OF THREE - [math] règle *f* du trois
~OF THUMB - [gen] méthode *f* empirique
~OF THUMB METHOD - [phys] procédé *m* approximatif
RULER - [gen] (e.f. head of state etc) souverain *m*
[impl] règle *f*
[print] (a machine designed to rule paper) régleur *m*, rayeur *m*
RULES - [gen] normes *f*pl
RULING - [gen] réglage *m*, réglure *f*
[print] réglure *f*
[leg] ordonnance *f*
[photo] (of a screen etc) linéature *f*
~ENGINE - [opt] (a mechanism operated by a long micrometer screw for ruling the equally spaced lines on optical gratings) tire-ligne *m*
~PEN - [impl] tire-ligne *m*
~SECTION - [metall] (the combination of dimensions having the greatest influence on the mechanical properties obtained by heat treatment) dimensions *f*pl récurrentes
RUMBLE, to - [gen & acoust] gronder, bruire
[metall] (to polish with a tumbling machine) dessabler
RUMBLE - [gen & acoust] grognement *m*, vibration *f* à basse fréquence
[transp] siège *m* de derrière
[metall] tonneau *m* à dessabler
~SEAT - [auto] siège *m* de derrière
RUMBLER - [mech] (a revolving drum) tambour *m* dessableur
RUMBLING - [mech] (tumbling. Treatment in a revolving barrel) dessablage *m* au tonneau
[paint] (for small articles) vernissage *m* à tambour
~BARREL - [impl] tambour *m* à secoueurs
~MILL - [metall] tonneau *m* à dessabler
~POT - [mus] pot *m* à vessie
RUMEN - [anat] rumen *m*
RUMINANT - [zool] ruminant *m*
~CLOVENHOOFED ANIMAL - [zool] fissipède *m* ruminant
RUMINANTE, to - [zool] ruminer
RUMINATION - [zool] rumination *f*
RUMP - [zool] croupe *f*, croupion *m*, postérieur *m*
~WOOL - [text] laine *f* de la croupe
RUN, to - [gen] courir
[gen] (of water) écouler, couler
[dyes] (to melt) s'étendre, déteindre
[gen] (in industry) diriger
[naut] courir, filer
[naut] (before the sea) fuir devant la lame
[fishing] (of salmon) remonter la rivière
[surv] tracer (une ligne)
[mech] (to cause to run) faire fonctionner
[mech] (of a motor) fonctionner, travailler
[mech] (to revolve) tourner

RUN, to - [agric] mettre au vert (le betail)
[mining] exploiter
[metall] (from a mould) couler
RUN - [gen] course *f*
[gen] (a course) parcours *m*
[mech] (a single period of operation of a machine) course *f*, marche *f*
[mech] (of a machine) marche *f*
[mech] (of motor) marche *f* à vide
[gen] (a brook; only USA) ruisseau *m*
[min] (in ore dressing) intervalle *m* de travail entre deux nettoyages
[mach tool] (a deviation) déviation *f*
[constr] (of a step) distance *f* entre les montées de la marche
[min] éboulis *m*
[mining] chute *f* de la cage
[gen] (a kind of ramp) rampe *f*
[naut] (below the waterline, a wedgelike part of a vessel) échappée *f*, formes *f*pl arrière
[print] tirage *m*
[gen] (schedule of operations) cycle *m* de travail
[metall] (welding) passe *f*
[mining] direction *f*, cours *m*
[oil ind] longueur *f* du trou de mine
[metall] campagne *f* (d'un haut fourneau)
[mining] (inclined plane) plan *m* incliné
[mining] (a diagonal heading) thierne *f*
~A FARM - [agric] diriger une exploitation
~A SWAB, to - [oil ind] descendre un piston dans le puits
~A TANK, to - [oil ind] pomper un réservoir pour le vider
~AGROUND, to - [naut] échouer à la côte
~ASHORE, to - [naut] faire côte
~AWAY, to - [nucl] (of a nuclear reactor, to pass into an uncontrollable condition) s'emballer
~-BACK - [gen] recul *m*
~BADLY, to - [mech] tourner mal
~CASING, to - [oil ind] revêtir
~DOWN, to - [oil ind] distiller complètement
[hydr] (of water) ruisseler
[th eng] (the boilers) vider (les chaudières)
~DOWN - [el chem] (of a battery) déchargement *m*
[horol] (of a clock) (horloge) au bas
[oil ind] procédé *m* complet de reffinerie
~-DOWN LINES - [oil ind] tubes *m*pl d'adduction
~-DOWN TANK - [oil ind] réservoir *m* intermédiaire
~DRY, to - [gen] (e.g. of a tank) s'assécher, tarir
~EASILY, to - [mech] tourner facilement
~FOUL OF, to - [naut] s'aborder
~FREE, to - [mech] (of an engine) tourner à vide
~FROM THE BALL, to - [text] venir du boyau
~GATES, to - [metall] (in foundry work) appliquer les attaques de coulée
~HARD, to - [mech] tourner lourd
~HOT, to - [mech] (to become excessively hot in operation) chauffer
~IN, to - [mech] (to operate a new machine at reduced and or load until the parts have become accocommodated to each other) roder (un moteur)
[oil ind] insérer
~IN OIL, to - [mech] tourner dans l'huile
~INTO, to - [gen] rencontrer, tamponner, se tamponner
~LEAD JOINT - [plumb] joint *m* coulé, joint *m* au plomb

RUN LIGHT, to - [mech] (of machine, or engine) tourner au ralenti, marcher à vide
~LOADED, to - [mech] (of machine, engines etc) fonctionner sous charge
~OF CASING - [oil ind] opération *f* de tubage
~OF GOLD - [mining] veine *f* du placer
~-OF-MILL COKE - [min] coke *m* tour-venant
~OF MINE - s. run of-mine coal
~-OF-MINE COAL - [min] houille *f* tout-venant
~-OF-OVEN COKE - s. run-of-retort coke
~-OF-RETORT COKE - [gas ind] (coke as removed from a gas-retort before crushing or sizing) coke *m* brut
~-OF-RISE STATION - [el] centrale *f* hydroélectrique au fil d'eau
~OF STAIRS - [constr] montée *f* géométrique de la marche
~OFF, to - [print] imprimer
[metall] couler (le métal)
~-OFF - [metall] dégorgement *m*
~OFF THE END OF THE COP, to - [text] se défiler suivant l'axe de la canette
~-ON - [mech] (after cut-off) marche *f* à blanc
~-OUT, to - [hydr] découler, s'écouler
[mech] (in a circular saw, the extent to which the periphery of a saw blade diverges from a plane normal to the spindle, during rotation) dévier de la line verticale
~-OUT - [metall] (escape of molten metal from moulds) fuite *f* du métal
[plast ind] (incomplete casting due to the run-out of metal from the moulds) moule *m* vidé
~-OUT BEARING - [mech] coussinet *m* usé
~OUT OF TRUE - [mech] dévier
~-OUT OF TRUTH ON THE WHARVE, to - [text] courir sur la gorge suivant un certain angle
~-OUT TABLE - [metall] table *f* à rouleaux de sortie
~-OUT TRAILER - [cin & telev] amorce *f* de fin
~-OUT WEFT BOBBIN - [text] cannette *f* épuisée
~OVER THE FACING POINT, to - [railw] aborder l'aiguille en pointe
~PAST, to - [gen] dépasser, franchir
[railw & road traffic] dépasser, brûler (la station)
~SHORT, to - [gen] manquer, faire défaut, être à court
~STEEL - [metall] (malleable pig iron) acier *m* coulé
~STIFFLY, to - [text] se dévider avec difficulté
~THE FULL NUMBER OF PICKS, to - [text] faire tout le nombre de tours
~THE YARN ON THE BEAM, to - [text] faire passer le fil sur l'ensouple
~THROUGH, to - [gen] passer à travers, traverser
~TO THE REED AT AN ANGLE, to - [text] passer obliquement dans le peigne
~TO WASTE, to - [gen] se perdre
~TRUE, to - [mech] tourner en centrage parfait
~UNDER LOAD, to - [mech] marcher en charge
~UNTRUE, to - [mech] dévier
~UP, to - [gen] monter (en courant)
[metall] recouler (le métal aux coussinets)
[gen] (to increase) laisser grossir
~-UP TABLE - [metall] table *f* à rouleaux d'arrivage
~WAY - [oil ind] couloir *m* du portail de la tour
~WILD, to - [railw] partir à la dérive
RUNABOUT - [auto] (a roadster) voiturette *f*
[transp] (an uncovered truck) runabout *m*

RUNABOUT - [naut] (petit) canot *m* à moteur
RUNAWAY ELECTRONS - [phys] électrons *mpl* de fuite
~REACTION - [nucl] réaction *f* d'emballement
~SPEED - [el] vitesse *f* d'emballement
~SWITCH - [railw] aiguille *f* de déraillement
RUNCH - [bot] (a weed) ravenelle *f*, raifort *m* sauvage
RUNG - [gen] (of a ladder) échelon *m*
[mech] fuseau *m* de lanterne
[naut] varangue *f*
RUNITE - [min] (graphic granite) runite *f*
RUNNER - [mech] curseur *m*, coulant *m*
[metall & plast ind] (passage or channel through which material enters a mould) canal *m* d'alimentation, canal *m* de coulée, barre *f* d'alimentation
[mech] (a gear wheel between two others) roue *f* intermédiaire
[metall] (the metal still in a sprue-hole after casting) jet *m* (de coulée)
[mech] (of a water turbine) roue *f* mobile, couronne *f* mobile
[mech] moufle *f* mobile, poulie *f* mobile
[bot] marcotte *f*, stolone *f*
[mech] (of ropeway etc) galet *m* de roulement
[mining] patin *m*
[mech] (a guide-pulley) poulie-guide *f*
~BOWL - [mech] galet-guide *m*
~BOX - [metall] diffuseur *m*
~BUSH - [metall] godet *m* de coulée
~CLOTH - [rubber ind] tissu *m* entraîneur
~FOR CARRYING BEAMS - [text] chariot *m* porte-ensouple
~-GATE - [metall & plast ind] chenal *m* de coulée
~PIPE - [metall] chenal *m* de coulée
~PLATE - [cin] (a film-guiding part of camera and projector) couloir-film *m*, guide-film *m*
~SCRAP - [metall] galette *f*
~STICK - [metall & plast ind] broche *f* de coulée, mandrin *m* de coulée
~THROUGH - [metall] canal *m* de coulée
~WAGON - [railw] wagon *m* intermédiaire
~WHEEL - [mech] roue *f* de renvoi
RUNNERLESS INJECTION MOULDING - [plast ind] injection *f* à canaux chauffants
RUNNING - [mech] fonctionnement *m*, marche *f*, roulement *m*
[gen] (flowing) coulage *m*
~A LINE - [surv] tracé *m* d'une ligne
~AND FEEDING SYSTEM - [metall] système *m* d'alimentation
~BALANCE - [mech] équilibre *m* dynamique
~BLOCK - [mech] moufle *f* mobile, poulie *f* mobile
~BOARD - [auto] marche-pied *m*
[railw] (of a locomotive) tablier *m*
~-BOARD AERIAL - [radio] (a car aerial fixed under the running board) antenne *f* sous le marche-pied
~-BOARD ANTENNA - s. running-board aerial
~-BOARD COVER - [auto] revêtement *m* de marche-pied
~BRIDGE - [constr] pont *m* roulant
~CHARGE - [ind] frais *mpl* d'exploitation, tarif *m* à prix de base
~CHART - [railw] tableau *m* des temps de parcours
~COST - [ind] frais *mpl* d'exploitation
~COUPLER - [el] pièce *f* intermédiaire
~DOWN BY CIRCUIT - [el] (of an accumulator battery) marche *f* par inertie

RUNNING EDGE - [railw] bord *m* du rail
~FIT - [mech] montage *m* à glissement
~FIX - [aero] (the intersection of two position lines which have not been obtained simultaneously, but have been adjusted to a common time value) lignes *fpl* de position transportées
~FORWARD- [mech] marche *f* avant
~FROM RIGHT TO LEFT - [mech] fonctionnant de droite à gauche
~GATE - [metall & plast ind] (runner) canal *m* de coulée
~GEAR - [mech] (of a machine) organes *mpl* de roulement
~GROUND - [geol] terrain *m* ébouleux
~HEAD - [print] (current head) titre *m* courant
~IDLE - [mech] (of machines and engines) marche *f* à vide
~-IN - [mech] (of machines autos etc) rodage *m*
~-IN OIL - [mech] baignant dans l'huile
~LIGHT - [mech] marche *f* à vide
[railw] marche *f* haut-le-pied
~LINE - [railw] voie *f* courante
~LOOP - [gen] boucle *f* à noeud coulant
~MOULDS - [metall] jet *m* des moules
~NIPPLE - [plumb] mamelon *m* simple
~NUMBER - [gen] numéro *m* d'ordre
~OF THE LOOM - [text] marche *f* du métier
~-OFF FRAME - [text] réunisseuse *f* pour chaîne ourdie
~-OFF OVER THE SURFACE - [geol] ruissellement *m*
~-OFF REEL - [text] aspe *m* à déroulage
~OFF THE MOLTEN METAL - [metall] (from a blast surface) coulée *f* du métal en fusion
~ON OVERLOAD - [mech] marche *f* en surcharge
~ORDER - [auto] ordre *m* d'allumage
~OUT - [metall] (of a wire; progressive excess increase in size during drawing, due to the wear of the die) augmentation *f* excessive du diamètre
~-OUT FIRE - [metall] four *m* de premier affinage
~OUT OF SPOOL - [text] épuisement *m* de la bobine
~POSITION - [mech] position *f* de marche
~PROGRAMME - [comput] (programme immediately available) programme *m* en cours
~RABBITS - [radar] échos *mpl* parasites
~RAIL - [railw] rail *m* de roulement
[mech] (for overhead travelling crane) rail *m* de translation
~RECORDER - [instr] enregistreur *m* de fonctionnement
~RIGGING - [naut] manoeuvre *f* courante
~ROUND IN A CIRCLE - [mech] courant circulairement
~SHED - [railw] remise *f*
~SPEED - [mech] vitesse *f* de marche
~SPRING - [hydr] (of water) source *f* vive
~SURFACE - [railw] plan *m* de roulement
~TEST - [gen] essai *m* de fonctionnement
~TIME - [transp] durée *f* du parcours
[comput] (of a program) temps *m* de déroulement
[telev] durée *f* de projection
~UNDER LOAD - [mech] marche *f* en charge
~UNLOADED - [mech] marche *f* à vide
~UNTRUE - [text] (of the beam) faux-rond *m* de l'ensouple
~UP - [aero] (the operation of warming up aircraft engines by running them at more than idling speed while the aircraft is on the ground) réchauffement *m*

des moteurs
RUNNING VOLTAGE - [el] tension *f* de service
~WATER - [hydr] eau *f* courante
RUNNINGS - [brew ind] écoulement *m* (du moût)
RUNS - [paint] (a defect) dégouttement *m*
RUNT - [zool] nabot *m*
RUNWAY - [aero] (a straight path suitably surfaced used for take-of an landing) piste *f* (en dur)
[aero] (for airships) coursive *f* (de dirigeable)
[mech] (of bridge crane) chemin *m* de roulement, voie *f* de roulement
[transp] monorail *m* aérien, chemin *m* de fer suspendu
~CONTROLLER - [aero] (an air-traffic control officer at the down-wind end of the runway) contrôleur *m* de piste
~FLOODLIGHT - [aero] (a light designed to give intense illumination over a runway) projecteur *m* de piste
~LIGHTS - [aero] (lights used to define the area for take-off and landing on a runway) feux *mpl* de piste
[aero] (on the aircraft) feu *m* de position d'avion
~LOCALIZING BEACON - [aero] radiophare *m* d'alignement de piste
~SPEED - [el] (in a motor the speed reached at no-load at rated voltage) vitesse *f* d'emballement
~TRAIN - [railw] train *m* à la dérive
RUPPE'S DISEASE - [med] ostéite *f* fibreuse des maxillaires
RUPTURE, to - [gen] rompre, se rompre
RUPTURE - [gen] rupture *f*
[med] (hernia) hernie *f*
[med] (a fracture) rupture *f*, éclatement *m*
~STRENGTH - [metall] résistance *f* à la rupture
~STRESS - [metall] charge *f* de rupture, effort *m* de rupture
[auto] (of tyres) fatigue *f* provoquant d'éclatement
RUPTURED - [med] hernié
RUPTURING CAPACITY - [el] (breaking capacity) capacité *f* de rupture
~CURRENT - [el] courant *m* de rupture
RURAL - [gen] (pertaining to the country; also to agriculture) rural
~AUTOMATIC EXCHANGE - [telecomm] bureau *m* automatique rural
~PARTY LINE - [telecomm] ligne *f* rurale partagée
RUSH, to - [gen] se lancer, se jeter, pousser
RUSH - [gen] mouvement *m* rapide, hâte *f*
[gen] (traffic) bousculade *f*
[bot] jonc *m*
[metall] (in blasting) fétu *m*
[cin] (the first print of a scene prepared for approval) copie *f* rapide
[oil ind] affluence *f*
~OF AIR - [gen & metall] chasse *f* d'air, coup *m* de air
~OF CURRENT - [el] coup *m* de courant
~OF WATER - [hydr] coup *m* d'eau
~PRINT - [cin] copie *f* rapide
RUSHES - [cin & telev] épreuves *fpl* de tournage
[cin] production *f* journalière
RUSK - [food] biscotte *f*
RUSSET - [gen] roussâtre
[text] drap *m* de bure brunâtre
RUSSIAN LEATHER - [leather ind] cuir *m* de Russie
~SHEETING - [text] (type of duck) toile *f* à voile
RUST, to - [gen] se rouiller, s'oxyder
RUST - [chem] (the reddish coating caused on metals oxidation) rouille *f*
[biol] (micro-organism of the basidio-mycetes group, parasitic on plants especially cereals) rouille *f*
~CEMENT - [ind chem] mastic *m* de fonte
~FORMATION - [bot] formation *f* de rouille
~FUNGUS - [bot] mycète *m* de la rouille
~INHIBITOR - [chem] (any substance designed to reduce rusting) anti-rouille *m*
~JOINT - [mech] joint *m* au mastic de fonte
~PREVENTER - s. rust inhibitor
~-PREVENTING AGENT - s. rust inhibitor
~-PREVENTING MEDIUM - s. rust inhibitor
~-PROOF, to - [ind chem] rendre inoxydable
~-PROOF - s. rust resisting
~PROOFING - [chem] protection *f* anti-rouille
~-PROTECTIVE PAINT - s. rust-resisting paint
~REMOVER - [chem] dissolvant *m* de rouille
~RESISTING - [chem] (not liable to corrosion) inoxydable
~-RESISTING PAINT - [paint] peinture *f* anti-rouille
~-STAINED COCOON - [text] cocon *m* atteint par la rouille
RUSTED - [gen] rouillé, oxydé
RUSTIC - [gen] rustique
RUSTICATED ASHLAR - [constr] (ashlar work in which the face stands out of the joints, at which the arrises are bevelled) appareil *m* rustique
RUSTICATION - [constr] (surface treatment of ashlars) ouvrage *m* rustique
RUSTING - [metall] rouillement *m*, oxydation *f*
[metall] oxydation *f* après décapage
RUSTLE - [acoust] bruissement *m*
RUSTLING - [acoust] (undefined noise, e.g. as made by tree leaves in the wind) susurrement *m*, bruissement *m*
RUSTLESS - [metall] inoxydable
RUSTY - [gen] rouillé
[gen] (colour) rougeâtre
RUT, to - [gen] sillonner
[zool] être en rut
[mech] gripper
RUT - [gen] ornière *f*
[zool] (the noise made by certain animals when in heat oestrus) rut *m*
[mech] grippure *f*
RUTHENIC - [chem] ruthénique
RUTHENIUM - [chem] (a non-ductile metallic element of the platinum group, A.N. 44, A.W. 101 E, symbol Ru. It has uses in metallurgy and in electrical equipment) ruthénium *m*
~CHLORIDE - [chem] (an analytical reagent) chlorure *m* de ruthénium
RUTHERFORD - [nucl] (unit of radioactivity equal to 106 disintegrations per second) rutherford *m*
~ATOM - [nucl] (the atom as conceived by Rutherford consisting of a central dense nucleus containing a positive charge surrounded by planetary electrons. The ratio of the nuclear charge to the charge of a positron is the "atomic" number; of the element to which the atom belongs) atome *m* de Rutherford
~DISPERSION FORMULA - [nucl] formule *f* de diffusion de Rutherford
RUTHERFORDINE - [min] (rare alteration product of uraninite) didérichite *f*, rutherfordine *f*
RUTILATED QUARTZ - [min] sagénite *f*, flèches *fpl*

d'amour
RUTILE - [min] (natural titanium dioxide. It is a source of titanium and its compounds and also has uses as a pigment) rutile *m*
RUTIN - [chem] (rutoside. A microcrystalline powder obtained from species of Eucalyptus, and Fagopyrum esculentum and used in medicine in the treatmen of capillary fragility) rutine *f*
RUTTING PERIOD - [zool] saison *f* du rut

RYE - [bot] seigle *m*
~BRAN - [agric] son *m* de sigle
~FLOUR - [agric] farine *f* de seigle
~GRASS - [bot] fromental *m*
~MEAL - [agric] farine *f* de sigle fourragère
~SMUT - [bot] (plant disease) maladie *f* des stries noires
~STRAW - [agric] paille *f* de seigle

S

s. - [chem] (symbol for solubility) symbole de la solubilité
s. - [chem] (the symbol for Sulphur) symbole du soufre
[chem] (in dyestuffs, the symbol for black) symbole du noir (colorants)
[met] (Beaufort letter for Snow) symbole de la neige (Beaufort)
s. ACID - [chem] (an intermediate for dyestuffs) acide *m* bisulphonique de naphtol
~DISTORSION - [electron] (in electron optical systems) distorsion *f* en S
~EFFECT - [electron] effet *m* de charge superficielle
~ELECTRON - [electron] (electron with an orbital angular momentum quantum number of zero) électron *m* S
~-HOOK - [mech] crochet *m* en S
~PISTON - [electron] (of a waveguide) piston *m* en S
~STATE - [phys] (state of zero orbital angular momentum) état *m* S
~TWIST - [text] direction *f* de la torsion à droite
SA - [chem] (a symbol for Samarium) symbole de samarium
SAALBAND - [geol] salbande *f* argileuse
SABADILLA - [chem] (a parasiticide used in medicine) sabadilline *f*
SABATIER EFFECT - [opt] (reversal phenomenon observer with photographic materials under certain conditions) effet *m* Sabatier
SABER - s. sabre
~SHIN - [anat] tibia *f* en lame de sabre
SABIN - [acoust] (the unit of acoustic absorption) sabin *m*
SABINE LAW - [acoust] (experimental formula for the reverberation time in a room) loi *f* de Sabine
SABINENE - [chem] (terpene derivative occurring in marjoram oil) sabinène *m*
SABLE - [zool] zibeline *f*
[impl] (a sable brush) pinceau *m* à poil de martre
SABOT - [shoe man] (a wooden shoe) sabot *m*
SABOTAGE, to - [leg] (to cause malicious damage) saboter
SABOTAGE - [leg] (an act of malicious damage) sabotage *m*
[railw] (chairing sleepers) sabotage *m*
SABRE - [gen] (heavy cavalry sword) sabre *m*
SABUGALITE - [min] (rare secondary mineral containing about 54°/$_{\circ}$ of uranium) sabugalite *f*
SABULOUS - [gen] (gritty like sand) sablonneux
SAC - [anat] (a cavity; a membranous pouch) sac *m*, poche *f*
SACCATE FRUIT - [bot] (fruit with a bag-like envelope containing it) fruit *m* sacciforme
SACCHARASE - [chem] (invertase, an enzyme with uses as an analytical reagent in the production of invert sugars) invertase *f*, saccharase *f*
SACCHARATE - [chem] (a salt of saccharic acid) saccharate *m*
SACCHARIC - [chem] (derived from sugar) saccharique
SACCHARIC ACID - [chem] (obtained by the oxidation of glucose and other sugars) acide *m* saccharique
SACCHARIDE - [chem] (a carbohydrate containing sugar) saccharide *m*
[chem] (a saccharate) saccharate *m*
[chem] (a saccharose) saccharose *m*
SACCHARIFEROUS - [chem] (saccharin bearing) saccharifère
SACCHARIFIABLE - [chem] saccharifiable
SACCHARIFICATION - [ind chem] (the conversion of starched into sugar) saccharification *f*
~REST - [brew ind] pause *f* de saccharification
SACCHARIFIER - [brew ind] (apparatus designed to convert into or impregnate with, sugar) cuve *f* de trempage
SACCHARIFY, to - [ind chem] (to convert starches into sugar) saccharifier
SACCHARIMETER - [instr] (instrument for measuring the sugar content of a solution utilising the rotatory power of the latter upon a ray of polarized light) saccharimètre *m*
SACCHARIMETRY - [ind chem] (the determination of the percentage of sugar in a solution) saccharimétrie *f*
SACCHARIN - [chem] (O-Sulpho-benzimide. A synthetic sweetening agent with a sweetening power four hundred times that of cane sugar) saccharine *f*
SACCHARINEISH - [chem] (containing saccharin) saccharineux
SACCHAROBIOSE - [chem] (cane- sugar, or sucrose) saccharose *m*
SACCHAROIDAL TEXTURES - [geol] (granular textures resembling loaf sugar) structures *f*pl saccharoïdes
SACCHAROMETER - [instr] (special type of hydrometer designed to measure the strength of a sugar solution) saccharimètre
SACCHAROSES - [chem] (a group of hydrocarbons, classified according to their complexity by the use of the prefixes mono- , di-, tri-, and poly) saccharoses *m*pl
SACCULE - [anat] saccule *m*
SACCULIFORM - [biol] (shaped like a small bag) sacculiforme
SACK, to - [gen] (to put into a sack) ensacher,

mettre en sac
SACK, to - [gen] (to plunder, to pillage) saccager
SACK - [gen] (kind of bag) sac *m*
[agric] (a strongish white wine from S. Europe) vin *m* des Canaries
~BORER - [mech] tarière *f* à sac
~CHUTE - [agric] (in a flour mill) cheminée *f* à ensacher
~CLEANER - [mech] brosseur *m* pour sacs
~CONVEYOR - [mech] transporteur *m* pour sacs
~DRILL - [text] treillis *m* à sac
~ELEVATOR - [mech] monte-sac *m*
~FOR WOOL - [text] sac *m* pour laine
~HOIST - s. sack elevator
~LIFT - s. sack elevator
~LIFTER - s. sack elevator
~TIE - [text] cordon *m* de sac
~TRUCK - [transp] bérot-diable *m*, haquet *m* à main
~UNLOADING FUNNEL - [mech] (in a flour mill) tuyau *m* de déchargeur
SACKBUT - [mus] (a trombone) trombone *m*, saquebute *f*
SACKCLOTH - [text] (coarse fabric used for making sacks) toile *f* à sacs
SACKED WOOL - [text] laine *f* ensachée
SACKHOLDER - [agric] (in flour mills etc) bouche *f* de décharge
SACKING - [gen] toile *f* à sacs
~MECHANISM - [mech] (machine used for filling saks) installation *f* d'ensachage
SACRAL - [anat] (near the sacrum, or pertaining to the sacrum) sacral
~RIB - [anat] côte *f* sacrale
SACRALIZATION - [med] sacralisation *f*
SACRECTOMY - [med] opération *f* de Kraske
SACRIFICE, to - [gen] sacrifier
SACRIFICIAL ANODE - [el] (a reactive anode) anode *f* de protection
SACROILIAC - [anat] sacro-iliaque
~DISEASE - [med] sacro-coxalgie *f*
SACROILIITIS- [med] arthrite *f* sacro-iliaque
SACRUM - [anat] (composite bone constituting the dorsal part of the pelvis) sacrum *m*
SADDENING-DOWN - [paint] ternissement *m*
SADDLE, to - [gen] seller
SADDLE - [gen] (of a horse harness) selle *f*
[gen] (of a bicycle) selle *f*
[geogr] (of a mountain) col *m* de montagne
[mach tool] (saddle-like support, e.g. a bearing for a car axle) selle *f*, reposoir *m*
[telecomm] chapeau *m* de poteau
[tool] (a forcing tool used for enlarging an internal diameter; also called becking stand) sellette *f*
[plumb] gâche *f* pour tubes
[metall] support *m* de dégorgeoir
[constr] sabot *m* (de poutre verticale)
[geol] anticlinal *m*
~BACKING - [mech] gâche *f*
[constr] surélévation *f*
~BAR - [constr] (metal bar fixed across a window to give support to glazing secured in lead cames) fer *m* à vitrage
~BEARING - [mech] coussinet *m* à selle
~BLOCK ANESTHESIA - [med] anesthésie *f* en selle
~BOILER - [th eng] (inverted U-shaped boiler fitted in a kitchener to supply hot water) chaudière *f* horizontale

SADDLE CLAMP - [mach tool] blocage *m* du chariot
~-CLAMP LEVER - [mech] levier *m* pour le blocage du chariot
~CLIP - [hydr] (tapping sleeve) collier *m* de prise
~COVER - [impl] housse *f* de selle
~CUSHION - [impl] torche *f*
~FEED HANDWHEEL - [mech tool] volant *m* pour la marche du chariot
~FEED SHAFT - [mach tool] arbre *m* d'alimentation du chariot
~FLANGE - [mech] bride *f* à selle
~GIRTH - [gen] (of a horse harness) sangle *f* de selle
~GRAFTING - [agric] greffe *f* anglaise épaulée
~HORSE - [zool] cheval *m* de selle
~KEY - [mech] (key having a concave face which bears on the surface of the shaft which it grips by friction only) clavette *f* évidée, clavette *f* à friction
~LEAD SCREW - [mach tool] vis-mère *f* de commande du chariot
~NOSE - [anat] nez *m* affaissée, nez *m* concave
~PAD - s. saddle cushion
~PILLAR - [plumb] gâche *f* à selle
~POINT - [math] col *m*
[math] point *m* de fléchissement
~REEF - [geol] filon *m* en selle
[mining] couche *f* ondulée
~ROOF - [constr] comble *m* en batière, toit *m* à deux pentes
~SCAFFOLD - [constr] (scaffold erected over a roof from standars on both sides of the building) échafaudage *m* pour combles
~TANK - [naut] contre-carène *f*
~TILE - [constr] tuile *f* en dos d'âne
~WEIGHTING - [mech] pression *f* à selle
SADDLEBACK PIG - [zool] porc *m* croisé
SADDLER - [gen] sellier *m*
SADDLER'S WAX - [chem] cire *f* de sellier
SADDLERY - [gen] sellerie *f*, bourrellerie *f*
SADDLING THE WEIGHTS - [text] placement *m* des poids
SADIRON - [impl] (a flat iron for smoothing clothes) fer *m* à repasser
SAFE - [gen] sauf, sûr
[gen] (strong receptable for protecting valuables) coffre-fort *m*
~AREA - [telev] cadre *m*, format *m* d'image vu sur un récepteur
~CONCENTRATION - [nucl] concentration *f* sûre
~DISTANCE - [nucl] distance *f* de sécurité
~EDGE - [tool] à côté lisse
~EJECTOR - [metall] (return pin) rappel *m* d'éjection
~-END PIPE-SMOOTHER - [tool] lissoir *m* à tuyaux à chapeau
~FLIGHT INDICATOR - [instr] (instrument designed to indicate the stalling speed of an aircraft) indicateur *m* aérodynamique de la perte de vitesse
~GEOMETRY - [nucl] (of a reactor) géométrie *f* sûre
~-GUARD - [mech] (any safety device) sauvegarde *f*
[railw] (the check rail) chasse-pierres *m*, garde *f* (de locomotive)
~-LIGHT - [photo] (special light-filter used to lighten dark-rooms) lumière *f* sûre
~-LIGHT LAMP - [photo] (for dark-rooms) lampe *f* inactinique

SAFE LOAD - [mech] (factor of safety) charge *f* de sécurité
[aero] facteur *m* de sécurité
~MASS - [nucl] (product of the critical mass multiplied by an appropriate safety coefficient) masse *f* sûre
SAFEGUARD - s. safe-guard
SAFEGUARDING - [gen] protection *f*
SAFETY - [gen] sûreté *f*, sécurité *f*
[mech] (e.g. for firearms) cran *m* de sûreté
[auto] serrure *f* à condamnation
~APPLIANCE - [gen & mech] dispositif *m* de sécurité
~ARCH - [constr] (a discharging arch; arch built in a wall to protect a space from a weight above) arc *m* de décharge
~BANK - [nucl] (in a reactor) groupe *m* de barres de sécurité
~BARRIER - [aero] (a net to receive an aircraft if the arrester gear is missed) barrière *f* de sécurité
~BASE - [cin] (the type of slow-burning film used in the motion picture industry) support *m* en acétate
~BELT - [aero & auto] (belt to hold person in seat, reducing danger of injury in case of accident) ceinture *f* de sécurité
[naut] ceinture *f* de sauvetage
[impl] essartement *m* de protection contre le feu
~BOILER - [th eng] chaudière *f* inexplosible, chaudière *f* multitubulaire
~BOLT - [firearms] arrêt *m* de sûreté
~BOOT - [shoe man] boote *f* de sécurité
~BRAKE - [el] (on lifts) parachute *m*
~CABLE - [railw] (of cable railways) câble *m* de sécurité
~CAGE - [min] (cage fitted with a safety catch to prevent it form falling in the event of a hoisting rope breaking) cage *f* à parachute (de mine)
~CATCH - [mech] (device to prevent unintentional operation of a mechanism) fermoir *m* de sûreté
[mech] (of a hoisting machine) dispositif *m* d'arrêt de sécurité
[mining] parachute *m*
[firearms] cran *m* de sûreté
~CENTRE - [aero] (a combined area control and rescue co-ordination centre) centre *m* de sécurité
~CHAIN - [railw] (a chain connecting a truck body and a truck) chaîne *f* de sûreté
~CHANNEL AMPLIFIER - [nucl] amplificateur statique de sécurité
~CIRCUIT - [contr] tuyauterie *f* de sécurité
~CLAMP - [mech] bride *f* de sûreté
~CLIP - [mech] (of a draw-bar) agrafage *m* de sûreté (de la tige de traction)
~COCK - [mech] robinet *m* de sûreté
~CONTROL - [nucl] (the measures taken to safeguard the personnel and buildings of a nuclear reactor) mesures *fpl* de sécurité
~CORD - [mech] (cord stretched across the bite of a pair of rolls and connected to a device which stops and opens the rolls if the operator comes in contact with the cord) câble *m* de sécurité
~COUPLING - [railw] attelage *m* de sûreté
[mech] accouplement *m* de sûreté
~CURVE - [aero] courbe *f* de sécurité
~CUT-OUT - [el] (overload protecting device) coupe-circuit *m* de sécurité
~CUTTER-BLOCK - [mach tool] porte-outil *m* à base pivotante
SAFETY DEVICE - [mech] dispositif *m* de sécurité
~DOG - [mining] (of a mine-car) reculoire *f* (d'un chariot de mine)
~DOOR - [gen] porte *f* de sûreté
~DOOR-LOCK - [mech] (applied to autos, railway-carriages etc) serrure *f* de sûreté
~FACTOR - [mech el etc] coefficient *m* de sécurité
~FACTOR FOR DROP-OUT - [el] facteur *m* de sécurité pour la mise au repos
~FACTOR FOR HOLDING - [el] facteur *m* de sécurité au maintien
~FACTOR FOR PICK-UP - [el] facteur *m* de sécurité pour la mise au travail
~FENCE - [gen] barrière *f* de sûreté
~FILM - [photo cin] film *m* ininflammable, film *m* en acétate
~FIRE TRAP - [cin] (means to localize the burning of a film in the projector to only one picture) étouffoir *m* pare-feu
~FUNNEL - [ind chem] (long glass tube with an open bulb at the top, used for pouring liquids into closed vessels) tube *m* de sûreté
~FUNNEL WITH TUBE - [ind chem] tube *m* à boule
~FUSE - [el] fusible *m*, plomb *m* fusible
[mining] mèche *f* de sûreté
~GAP - [el] parafoudre *m*, limiteur *m* de tension
~GEAR - [el] (in lifts, a mechanical device fitted to the car-frame to stop and hold the car in the event of a free fall, or in the event of overspeed in a descending direction) frein *m* de sécurité
~GLASS - [glass man] (glass so designed and made as to minimise danger in case of breakage) glace *f* de sécurité, verre *m* de sécurité
~GOGGLES - [gen] lunettes *fpl* protectrices
~HEIGHT - [aero] (in instrument flying the minimum safe height having regard to the existence of high ground) altitude *f* de sécurité
~HOOK - [railw] (a hook for coupling a steam locomotive to a tender) crochet *m* de sûreté, mousqueton *m*
~ISLAND - [gen] (safe passage for pedestrians) île *f* de sécurité, refuge *m*
~JOINT - [mech] joint *m* de sécurité
~KEPS - [mining] taquets *mpl* de sureté
~LAMP - [mining] (miner's lamp which does not immediately ignite firedamp or gas in a coal-mine) lampe *f* de sûreté
~LIGHTER - [impl] allumeur *m* de sûreté
~LINTEL - [constr] (lintel doing the work of a relieving, or safety arch) arrière-linteau *m*
~LOCK - s. safety catch
~LOOP - [constr] (on stairs) étrier *m* de sûreté
[gas ind] (a back cage) crinoline *f*
~MARGIN - [gen] marge *f* de sécurité
~MEASURES - [gen] mesures *fpl* de sécurité
~MOTION - [mech] accouplement *m* de sûreté
~NET - [gen] filet *m* protecteur
~NUT - [mech] écrou *m* de sûreté, contre-écrou
~OUTLET - [el] prise *f* de courant mise à la terre
~PAPER - [paper man] (a paper difficult to duplicate) papier *m* de sûreté
~PILOT - [gas ind] (a flame-failure device) veilleuse *f* de sécurité
~PIN - [gen] épingle *f* de sûreté
[mech] goupille *f*
~PLATFORM - [constr] (of a scaffold) plancher *m* de

secours de l'échafaud
SAFETY PLUG - [el] bouchon *m* fusible, rondelle *f* fusible
~PRECAUTIONS - [gen] mesures *f*pl de prévoyance, protection *f* contre les accidents
~PULLEY - [mech] poulie *f* de sécurité
~RAIL - [railw] contre-rail *m*
~RAM - s. safety tamping
~RANGE - [gen] distance *f* de sécurité
~RAZOR - [impl] (shaving implement) rasoir *m* de sûreté
~RELAY - [el] relais *m* de sécurité
~RING - [mech] bague *f* de sécurité
~ROD - [nucl] (a control rod controlling a large amount of reactivity and capable of bringing the reactor below critical in a very short time) barre *f* de sécurité
~ROLLER - [horol] disque *m* du balancier
~ROPE - [gen] câble *m* de sûreté
~RULES - [gen] normes *f*pl de sécurité
~SHUT-OFF - [mech] arrêt *m* de sûreté
~SHUTTER - [photo] obturateur *m* de sécurité
~SLIDING CLUTCH - [mech] joint *m* de sûreté à frottement
~SPARK - [el] éclateur *m* de sécurité
~SPECIFICATIONS - [gen] prescriptions *f*pl de sûreté
~SPEED - [aero] (the lowest speed of an aircraft (above stalling) which gives a margin of safety for control in flight) vitesse *f* de sécurité
~SPRING - [mech] ressort *m* de sûreté
~STAKE - [metall] pieu *m* de guidage
~STOCK - [cin] (uninflammable film) film *m* ininflammable
~SWITCH - [el] (emergency stop) commutateur *m* de sûreté
~TAMPING - [metall] bourre *f* de sûreté
~THREAD - [aero] (a thread connecting the rip cord of a parachute to some point on the pack to prevent accidental release) ficelle *f* de sécurité
~THREAD COUNTER - [text] compte-fils *m* de sûreté
~THROW-OUT BAR - [impl] barre *f* débrayeuse de sécurité
~TONGS - [impl] griffes *f*pl de sûreté
~TRACK - [railw] (trap siding) voie *f* de sûreté
~TREAD - [railw auto etc] (a step covered with rubber or the like to prevent slipping) marche *f* antidérapante
~TUBE - [rubber ind] (of tyres) chambre *f* à air de sécurité
~VALVE - [mech] (a sping-loaded valve designed to release fluid when the pressure exceeds a preset level) soupape *f* de sécurité, soupape *f* de sûreté
~WEDGE - [mech] cale *f* de sécurité
~YARN GUIDE - [text] guide-fils *m* de sécurité
SAFFIAN LEATHER - [text] cuir *m* maroquin
SAFFLOWER - [bot] (thistle-like herb with spiny heads of orange-red flowers) carthame *m*, safran *m* bâtard
[chem] (a red dyestuff obtained from the herb) safranum *m*
~OIL - [chem] (an oil with semi-drying properties obtained from Carthamus tinctonius and used in paint manufacture) huile *f* de carthame
SAFFRON - [bot] (crocus. The stigmas of Crocus sativus used as a flavourant and colourant) safran *m*
SAFRANINES - [chem] (diamino-azine dyestuff with uses in the textile industry and as microscopic stains) safranines *f*pl
SAFROLE - [chem] (a pale yellow oil derived from camphor or sassafras oil and used in perfumery and medicine) safrol *m*
SAG, to - [gen] fléchir, s'affaisser, plier
[constr] (to settle) s'affaisser, peiner
[gen] (of a cable) se relâcher
[shipbuild] contre-arquer
[naut] tomber sous le vent
[constr] (e.g. a bridge) s'incliner, gauchir
[gen] (of a rope, a line etc) faire ventre, faire flèche, faire des rideaux
SAG - [gen] affaissement *m*
[constr] flèche *f*
[constr] (depression formed by two gradients) contre-arc *m*
[el] (the vertical distance between the lowest point on an overhead line and a point of suspension) flèche *f*
[aero] (of an airship) fléchissement *m* central de l'axe longitudinal
[naut] mouvement *m* de dérive, dérive *f*
[geogr] (depression place in a flat land) affaissement *m* (du sol)
~CORRECTION - [surv] (the correction for the sag of the measuring tape) correction *f* du fléchissement du mètre à ruban
~MAGNIFICATION - [instr] (in commercial hotwire instrument) grandissement *m* de la flèche
~POINT - [paint] (in lacquer overspraying etc) point *m* d'écoulement
~TO LEEWARD, to - [naut] tomber sous le vent, être dépalé
~TO LEEWARD - [naut] (sidewise drift) dérive *f* latérale
SAGE - [bot] sauge *f*
~OIL - [chem] huile *f* de sauge
SAGGAR - s. sagger
SAGGED - [gen] affaissé, fléchi, plié, gauchi
[mech] (e.g. a coil) écrasé
[naut] contre-arqué
SAGGER - [ceram] (pottery-ware container in which biscuit is placed in the biscuit kiln, to protect it from direct flame action and the weight of ware stacked above) casette *f*
[metall] boîte *f* de cementation, boîte *f* refractaire
SAGGING - [gen] affaissement *m*, fléchissement *m*
[constr] inclinaison *f*, tombée *f*
[naut] contre-arc *m*
[aero] (of airships) fléchissement *m*
[fin] (of the market) (marché) creux, baisse *f*
~AT THE CROWN - [constr] affaissement *m* du sommet
~OF THE BELT - [mech] flèche *f* de la courroie
~OF THE VAULT - [constr] affaissement *m* de la voûte
~POINT OF THE CHAIN - [text] point *m* de travail de la chaîne
SAGITTA - [constr] (the central voussoir at the crown of an arch) flèche *f* de sommet d'arc
[constr] (a rise; e.g. of a bridge) flèche *f*
~METHOD - [photo] méthode *f* de la flèche
SAGITTAL - [anat] (elongate in the median vertical longitudinal plane of an animal) sagittal
~FOCAL LINE - [telev-cin] focale *f* sagittage
SAGO - [bot] sagou *m*

SAGO PALM - [bot] sagoutier *m*
SAIL, to - [naut] naviguer, faire voile
[gen] (to leave port, to begin a voyage) faire route (sur)
[aero] voler
SAIL - [naut] voile *f*
[naut] (the complete set) voilure *f*
[impl] (of a windmill) aile *f*, toile *f* (de moulin)
[zool] nageoire *f* dorsale
~OUTFIT - [naut] voilure *f*, toile *f*
~OVER, to - [constr] être saillant
~-PLANE - [aero] planeur *m*
~ROOM - [naut] voilerie *f*
SAILBOAT - [naut] canot *m* à voiles
SAILCLOTH- [text] (very strong cotton, canvas) toile *f* à voile, canevas *m*
SAILER - [naut] voilier *m*
[naut] (a sailer ship, having good sailing qualities) marcheur *m*
SAILFLYING - [aero] vol *m* à voile
SAILING - [naut] navigation *f*
[gen] (to leave in a boat) départ *m*
~COURSE - [constr] (brick or stone course projecting from the wall) assise *f* en épi
~VESSEL - [naut] voilier *m*
SAILMAKER - [naut] voilier *m*
SAILPLANE, to - [aero] voler à voile, planer
SAILPLANE - [aero](a glider designed to maintain protracted flight by the use of air currents) planeur *m* (de vol à voile)
SAINFOIN - [bot] sainfoin *m*, luminelle *f*
SAINT ELMO FIRE - [opt] (brush-like discharge from charged objects in the atmosphere) feu *m* Saint-Elme
SAL - [chem] (salt) sel *m*
~ACETOSELLA - [min] sel *m* d'acétoselle
~AMMONIAC - [chem](ammonium chloride) sel *m* ammoniac, chlorure *m* d'ammonium
~AMMONIAC CELL - [el chem] (a cell in which the electrolyte consists mainly of a solution of ammonium chloride) pile *f* au sel ammoniac
~GEMMA - [min] sel *m* gemme, halite *f*
~MIRABILE - [min] mirabilite *f*, sel *m* admirable de Glauber
~SODA - [min] (sodium carbonate) carbonate *m* de sodium
~VOLATILE - [chem] (ammonium carbonate) sels *mpl* volatils anglais
SALAMANDER - [zool] salamandre *f*
[metall] (mass of hardened metal remaining in the hearth of a furnace) loup *m*, cochon *m*
[constr] (portable stove used by builders to prevent the freezing of the plaster) tisonnier *m* ardent
~GRILL - [gas ind] (a cooking implement) salamandre *f*
SALARY - [gen] traitement *m*, appointements *mpl*
SALE - [gen & comm] vente *f*
[comm] (in a store, shop etc) liquidation *f*, vente *f* au rabais
~PRICE - [comm] prix *m* de solde
~VALUE - [comm] valeur *f* marchande
SALEABLE - [gen & comm] (which may be sold; marketable) vendable, marchand
SALEITE - [min] (mineral containing magnesiumuranylorthophosphate) saléite *f*
SALERATUS - [chem] (potassium or sodium bicarbonate for use in cooking) bicarbonate *m* de potassium
SALFATORY EVOLUTION - [genet] mutation *f*
SALICIN - [chem] (an antipyretic and analgesic, also having application as an analytical reagent) salicine *f*
SALICIONAL - [mus] (soft open metal organ stop) salicional *m*
SALICYLALCOHOL - [chem] (O-Hydroxybenzyl alcohol. A local anaesthetic) alcool *m* salicylique
SALICYLALDEHYDE - [chem] (O-hydroxybenzaldehyde. A perfumery agent and analytical reagent) salicylaldéhyde *f*
SALICYLAMIDE - [chem] (O-hydroxybenzamide. An antipyretic, analgesic and antirheumatic agent) salicylamide *f*
SALICYLANILIDE - [chem] (an antimycotic with uses in medicine) salicylanilide *f*
SALICYLATE - [chem] (a salt or ester of salicylic acid) salicylate *m*
SALICYLAZOSULPHAPYRIDINE - [chem] (a sulphonamide with therapeutic uses) salycylazosulfapyridine *f*
SALICYLEMIA - [med] salicylémie *f*
SALICYLIC - [chem] (derived from certain willows) salicylique
~ACID - [chem] (additive in rubber compounding to prevent precuring) acide *m* salicylique
SALICYLISM - [med] salicylisme *m*
SALIENT - [gen] (standing out, prominent) saillant
[zool] bondissant
[hydr] (of water) jaillissant
[surv] (jutting out piece of land) saillant, en projection
~ANGLE - [gen] angle *m* saillant
~INSTRUMENT - [instr] (instrument mounted above the mounting plate) instrument *m* à montage extérieur
~POLE - [el] (the part of a magnetic circuit of a machine between the yoke and the air gap) pôle *m* saillant
~-POLE GENERATOR - [el] génératrice *f* à pôles saillants
~-POLE ROTOR - [el] rotor *m* à pôles saillants
SALIFEROUS - [geol] (salt bearing) salifère, salicole
SALIFY, to - [chem] (to form a salt, or to combine with a salt) salifier
[ind chem] (to impregnate with a salt) salifier
SALIGENIN - [chem] (phenolic alcohol) saligénine *f*
SALINA - [min] (pool or marsh containing salt) saline *f*, marais *m* salant
SALINE - [chem] (containing dissolved salts, especially sodium chloride) salin, salé
[min] (a salina) saline *f*
~LAKE - [geol] lac *m* salé
~SOLUTION - [chem] solution *f* saline
SALINITY - [chem] (specific salt content, especially of water) salinité *f*
[chem] (of the sea water) salure *f*
SALINOMETER - [instr] (instrument for measuring the saline content of boiler water) salinomètre *m*
SALIVA - [zool] (the secretion of the salivary glands) salive *f*
SALIVARY GLANDS - [zool] (glands having a duct opening into or near the mouth) glandes *fpl* salivaires
SALIVIN - [chem] ptyaline *f*
SALLE - [paper man] (spoiled paper) bardot *m*
SALLY, to - [gen] sortir, faire une sortie

SALLY - [gen] sortie *f*
[carp] (reentrant angle cut into the end of a timber) entaillure *f* d'extrémité
SALMIAC - [chem] (ammonium chloride) chlorure *m* d'ammonium
SALMINE - [chem] (protamine obtained from fish testicles) salmine *f*
SALMON - [zool] saumon *m*
~OIL - [chem] huile *f* de saumon
SALMONELLA - [biochem] (group of Gram-negative non-sporing bacilli) salmonella *f*
~INFECTION - [vet] s. salmonellosis
SALMONELLOSIS - [vet] (a disease affecting animals and caused by an infection by Salmonella bacilli) salmonellose *f*
SALOL - [chem] (phenylsalicylate. A component of light stabilizers for plastics and a preservative) salol *m*
SALOON - [gen] salle *f*, salon *m*
[auto] voiture *f* à conduite intérieure
[naut] salon *m*, cabine *f*
[railw] wagon-salon *m*
[railw] (USA only; compartment of a "parlor car") cabine *f*
~CARRIAGE - [railw] (parlor car in the USA) wagon-salon *m*
~COACH - s. saloon carriage
SALPINGECTOMY - [med] salpingectomie *f*
SALPINGEMPHRAXIS - [med] obstruction *f* tubaire
SALPINGITIS - [med] salpingite *f*
SALPINGOGRAPHY - [med] salpingographie *f*
SALPINGOPEXY - [med] salpingopexie *f*
SALPINGORRHAPHY - [med] salpingorrhaphie *f*
SALPINGOSTOMY - [med] salpingostomie *f*
S.A.L.R. - s. superadiabatic laps rate
SALSE - [geol] (mud volcano) salse *f*, salinelle *f*, volcan *m* de boue
SALSODA - s. sal soda
SALT, to - [gen] saler
SALT - [chem] (a compound in which all or a proportion of the hydrogen has been replaced by an element, radical or metal) sel *m*
[min] (common salt, sodium chloride) sel *m*
~BATH - [metall] (bath of molten salts used for heating steel, for hardening or tempering) bain *m* de sel
~BATH FURNACE - [metall] (heat-treatment furnace in which the material to be treated is immersed in a tank of molten salt) four *m* à bain de sel
~BATH QUENCH - [metall] refroidissement *m* rapide en bain de sel
~BEARING - [min] salifère
~BED - [min] gisement *m* de sel
~BINDING AMMONIA - [chem] sel *m* ammoniac
~BLOCK - [min] saline *f*
~BOTTOM - [min] bassin *m* salifère, chott *m*
~BRIDGE - [chem] (a buffer solution) pont *m* de gélose, pont *m* de sel
~CAKE - [chem] (crude sodium sulphate, having application in paper making, ceramics, and in the manufacture of glass and soaps) sulfate *m* de sodium commercial
~-CAT - [agric] (for pigeons) salègre *m* (pour pigeons)
~CELLAR - [impl] salière *f*
~CONTENT - [chem] teneur *f* en sel
~DOME - [geol] dôme *m* salicole

SALT EFFECT - [chem] effet *m* sel
~GLAND - [bot] glande *f* saline
~GLAZE - [ceramic] (a simple type of glaze used for drain-pipes and the like earthenware, produced by throwing salt over the articles while being fired) vernissage *m* par salage
~GRAINER - [ind chem] (open-pan evaporator heated by steam coils, used in concentrating salt solutions) évaporateur *m* de sel
~LAKE - [geol] lac *m* salé
~LICK - [agric] terrain *m* salifère
~-MARSH - [geol] marais *m* salant
~MEADOW - [agric] pré *m* salé
~MINE - [min] mine *f* de sel
~-PAN - s. salt-marsh
~OF AMBER - [chem] acide *m* succinique
~OF ARSENIOUS ACID - [chem] sel *m* de l'acide arsénieux
~OF OPIUM - [chem] narcotique *f*
~OF PHOSPHORUS - [chem] sel *m* microcosmique, sel *m* de phosphore
~OF PHOSPHORUS-BEAD TEST - [ind chem] essai *m* au sel de phosphore
~OF TARTAR - [chem] sel *m* de tartre
~OF TIN - [chem] sel *m* d'étain
~OF VITRIOL - [chem] sulfate *m* de zinc
~RESIDUE - [chem] résidu *m* de sel
~SCREEN - [radiat] (an intensifying screen) écran *m* radioscopique
~SPRAY TEST - [paint] (test for corrosion resistance in coatings, in which a saline solution is repeatedly sprayed on the specimen) essai *m* à la vapeur saline
~VEIN - [min] veine *f* de sel
~WATER - [glass man] (molten sulphates floating in a glass melting unit) fiel *m* de verre
~[gen] eau *f* salée
~-WATER POOL - [geogr] lagon *m*
~-WATER TREATMENT - [med] thalassothérapie *f*
~-WATER PLATING - [el chem] placage *m* sans courant extérieur
~WELL - [min] puits *m* d'extraction du sel
SALTATION - [biol] (sudden, discontinuous variation) mutation *f*
SALTERN - [mining] (a complex of shallow pools wherein sea water is allowed to evaporate to produce salt) marais *m* salant, saline *f*
SALTIGRADE - [zool] (progressing by jumps) saltigrade
SALTINESS - [gen] salinité *f*, salure *f*
SALTING - [gen] salaison *f*, salage *m*
[ceram] s. salt glaze
[photo] salage *m* (des papiers)
~OUT - [el metall] (term used in electroplating for precipitation) précipitation *f* par un sel
[chem] (the removal of an organic compound from an aqueous solution by the addition of salt) salaison *f*
[chem] relargage *m*
~-OUT AGENT - [chem] agent *m* relargant
SALTPETER - s. saltpetre
SALTPETRE - [chem] (common name for crude sodium nitrate) salpêtre, nitrate *m* de potassium
~-BED - [min] salpêtrière *f*
~WORKS - [min] salpêtrerie *f*
SALTPETROUS - [chem] salpêtreux
SALTS OF SORREL - [chem] (an old pharmaceutical

name for potassium quadroxalate) sels *mpl* d'oseille
SALTWORKS - [ind] salin *m*, salanque *f*
SALTY - [gen] (tasting like salt) salé, saumâtre
[chem] (containing salt) salin
~DEPOSITS - [min] grumeaux *mpl* de sel
SALVAGE, to - [gen] (to recover for further use) récupérer
SALVAGE - [gen and naut] sauvetage *m*
[gen] (the saved cargo) matériel *m* récupéré
[leg] (the reward granted for the salvage operations) indemnité *f*, prime *f* de sauvetage
~LORRY - [transp] (a break-down lorry) camion *m* de dépannage
~MONEY - [leg] droit *m* de savetage
~OPERATION - [mech] opération *f* de sauvetage
~-TUG - [naut] remorqueur *m* de sauvetage
SALVATELLA - [anat] veine *f* salvatelle
SALVE - [gen & chem] onguent *m*, pommade *f*
~FOUNDATION - [chem] base *f* d'onguent
SALVER - [impl] plateau *m*
SAMARIUM - [chem] (one of the rare earth metals, A.N. 62, A.W. 150.4, symbol Sm. It occurs in samarskite, cerite and other minerals) samarium *m*
~-DOPED CALCIUM - [chem] (used in lasers) fluorure *m* de calcium à dopage de samarium
SAMARSKITE - [min] (an ore of the rare earth metals) samarskite *f*
SAMIRESITE - [min] (mineral containing niobium, tantalum and uranium) samirésite *f*
SAMPLE, to - [gen] échantillonner
[food etc] goûter, déguster (un vin)
SAMPLE - [gen] (a specimen of a substance taken to assess or demonstrate its properties) échantillon *m*, spécimen *m*
[metall] (a material sample from which subsequent pieces are obtained) échantillon *m*
[comm] échantillon *m*, contre-échantillon *m*
~BAG - [gen] sac *m* à échatillons
[mining] boîte *f* à carottes
~BOOK - [comm] échantillonnage *m*
~BORING - [soil] sondage *m*
~COCK - [brew ind] robinet *m* de prise
~GLASS - [brew ind] verre *m* d'épreuve
~GRINDING - [min] broyage *m* d'échantillons
~INTELLIGENCE - [telev] échantillon *m* de signal
~LOT INSPECTION - [ind] inspection *f* d'un lot d'échantillons
~POINT - [opt] (the point on a chromaticity diagram representing the chromaticity of the sample) point *m* étalon
~REDUCER - [mech] (device for automatically dividing a sample of material into equal fractions and delivering one of these for examination) découpeuse *f* d'échantillons
~ROOM - [text etc] salle *f* d'échantillonnage
~SHOVEL - [min] pelle *f* à échantillons
~SIZE - [gen] (in statistical work) dimension *f* d'échantillon
~SURVEY - [gen] recensement *m* d'essai, enquête *f* pilote
~TEST - [ind, ind chem, etc] essai *m* expérimental
~UNIT - [comm] unité *f* d'échantillonnage
~WARP - [text] chaîne *f* d'échantillonnage
~WARPING FRAME - [text] ourdissoir *m* droit
SAMPLED DATA - [contr] données *fpl* échantillonnées
[comput] - échantillons *m* mesurés
SAMPLED-DATA CONTROL SYSTEM - [contr] système *m* de réglage à données échantillonnées
~-DATA FEEDBACK - [contr] système *m* de réglage à réaction à données échantillonnées
~-DATA SYSTEM - [contr] système *m* à données échantillonnées
SAMPLER - [min etc] (device for automatically taking a representative specimen from a quantity of material) échantillonneur *m*
[telev] (electronic switch scanning the momentary amplitude of the video signal of each colour and passing these impulses to the modulator) discriminateur *m* chromatique
[el] circuit *m* discriminateur
SAMPLING - [ind, chem etc] (the operation of taking a specimen from a quantity of a product or substance, espec. of so taking it as to ensure that it is tryly representative of the whole amount) échantillonnage *m*, prélèvement *m* d'échantillons
[telev] (in colour television, a progressively selective process) discrimination *f* chromatique
[comput] (of an analog function) échantillonnage *m*
~ACTION - [nucl] (for a nuclear reactor) mise *f* au point périodique d'un processus
[contr] mise *f* au point périodique
~BORER - [ind chem] (a device for taking semisolid samples; e.g. soap, wax etc) sonde *f* à échantillons
~COCK - [ind chem] (small valve used to draw off samples from a pipe or vessel) robinet *m* de prise
~CONTROL SYSTEM - [contr] système *m* de commande à échantillonnage
~ERROR - [gen] erreur *f* d'échantillonnage
~FREQUENCY - [comput] fréquence *f* d'échantillonnage
~OSSERVATIONS - [telecomm] contrôle *m* par épreuves
~PAPER - [paper man] papier *m* pour échantillonnage
~PERIOD - [contr] intervalle *m* d'échantillonnage
~PULSE GENERATOR - [telev] générateur *m* discriminateur
~RATE - [comput] vitesse *f* d'exploration
~SPOON - [min] carottier *m*
[metall] louche *f*
~TAP - [hydr] robinet *m* de prise des échantillons
~TEST - [gen] (test on a few samples taken at random out of a consignment) essai *m* expérimental
~THIEF - s. sampling cock
~THERMOCOUPLE ASSEMBLY - [el] ensemble *m* de couple thermoélectrique influencé par un mélange de gaz
~TOOL - [min] outil *m* de prélèvement
~TUBE - [min] tube *m* carottier à paroi mince
SAMSON POST - [naut] (of the hold) étance *f* à coches
SANATORIUM - [med] sanatorium *m*, infirmerie *f*
SANCTION, to - [gen] sanctionner
SANCTION - [gen] sanction *f*, autorisation *f*
SAND, to - [carp etc] (the operation of cleaning up wood and other surfaces with sandpaper) sabler, nettoyer au sable
[constr] (to mix with sand) mélanger avec du sable
SAND - [geol] (a natural, loose fine-grained substance composed mainly of particles of quartz and used in cements, abrasives and foundry moulds) sable *m*
[metall] (for foundry) sable *m*
~ADDITIVE - [metall] noir *m* de fonderie

SAND AND GRAVEL WASHER - [mech] laveuse *f* à sable et gravier
~ASPHALT - [constr] mortier *m* asphaltique
~BACKFILL - [constr] remblai *m* en sable
~BALLAST - [gen] ballast *m* en sable
~BAR - s. sandbank
~BATH - [ind chem] (tray containing sand on which glass vessels can be placed to protect them from direct contact with a flame during heating) bain *m* de sable
~BEAM - [text] rouleau *m* garni de sable
~BED - [metall] couche *f* de coulée
~-BELT MACHINE - [mech] machine *f* à poncer
~-BLAST, to - [metall] sabler
~-BLAST - [metall] (a stream or jet of sand driven forcibly e.g. by compressed air, against an object to remove scale, improve the surface or the like) sablage *m*, jet *m* de sable
~-BLAST MACHINE - [mech] (a machine in which objects can be treated by means of a sand-blast) sableuse *f*
~-BLAST NOZZLE - [mech] (the jet from which the sand is projected in sand-blasting) buse *f* à jet de sable
~-BLAST ROOM - [metall] chambre *f* de dessablage au jet de sable
~-BLAST SHEETING - [metall] feuille *f* pour decapage au sable
~-BLASTED FINISH - [metall] finissage *m* au sablage
~-BLASTING - [metall] projection *f* de sable
~-BOBBING - [mech] sablage *m* par abrasifs
~BOTTOM - [metall] sole *f* en sable
~BOX - [gen] caisse *f* à sable, sablière *f*
~BOX-LEVER -[railw] levier *m* de la manoeuvre de la sablière
~-BOX PIPE - [railw] tuyau *m* de la sablière
~BUCKLE - [metall] (irregular projection on a casting surface, containing sand and due to uneven ramming) gale *f* en sable
[metall] échilles *fpl*
~BURNING - [metall] coquille *f* d'oeuf abreuvante
~CALCITES - [geol] calcite *f* sableuse
~CARPETING - s. sand asphalt
~-CAST, to - [metall] couler en sable
~-CAST - [metall] coulé en sable
~-CAST PIG IRON - [metall] fonte *f* coulée en sable
~-CASTING - [metall] (the operation) coulée *f* en sable
[metall] (the cast) pièce *f* fondue en sable
~CONTROL - [metall] examen *m* de sables
~-CRACK - [vet] fissure *f* du sabot, seime *f*
~CREPE - [text] sablé *m*
~CRUSH - [metall] broiement *m* du sable, division *f* du sable
~CULTURE - [bot] (experimental method) culture *f* dans le sable
~CUSHION - [gas ind] (in gasholders) galette *f* d'assise
~CUTTER - [metall] diviseur *f* du sable
~DEWATERING MACHINE - [mach] machine *f* à assécher le sable
~DISTRIBUTOR - [railw] diffuseur *m* des sablières
~DREDGE - [mech] drague *f* à sable
~-DUNES - [geol] (rounded mounds of loose sand) dunes *fpl* de sable
~DUST - [soil] poussière *f* de sable
~EQUIVALENT TEST - [soil] essai *m* d'équivalent de sable
SAND FILTER - [mech] (filter in which graded sand is used to arrest solids) filtre *m* à sable
~FINISH - [carp] rectification *f*
~FLOOR - [metall] chantier *m* de moulage en sable
~FLOWABILITY - [metall] fluidité *f* du sable
~FOUNDATION - [constr] remblai *m* de sable
~FRITTING - [metall] frittage *m* du sable
~-GALL - [geol] tubulure *f*
~GRAINS - [min] grains *mpl* de sable
~GUTTER - [metall] (in foundry works) rigole *f* du sable
~-HILL - [geol] dune *f* de sable
~HOLE - [metall] (in a casting) soufflure *f*
~HOLES - [metall] (a casting defect; irregular cavities containing sand) trous *mpl* de sable, nids *mpl* de sable
~INCLUSION - [soil] lentille *f* de sable
~JET - s. sand blast
~LEACHING - [min] lessivage *m* de sable aurifère
~LEVER - s. sand-box lever
~-LIME BRICKS - [constr] (bricks made from a mixture of damp sand and some slaked lime mouled under pressure) briques *fpl* de sable et chaux
~LINE - [text] câble *m* à six torons
[oil ind] corde *f* de curage
~LOAD - [electron] (in waveguides) affaiblisseur *m* au sable
~MEASURE BOX - [constr] sablier *m*
~MILL - [min] broyeur *m* à sable
[metall] frotteur *m*
~-MIXING MACHINE - [mech] mélangeur *m* de sable
~MOULD - [metall] moule *m* de sable
~MOULDING - [metall] moulage *m* au sable
~PACK - [mining] remblai *m* d'ensablage
~PACKING - [metall] obturation *f* au sable
[mining] remblayage *m* par ensablage
~PATTERN - [metall] faux-modèle *m*
~PILLAR - [met] tourbillon *m* de poussière
~PIPE - [railw] tuyau *m* à sable
~PIT - [min] sablière *f*, sablonnière *f*
~PUMP - [oil ind] (a sludger) cloche *f* de curage, tube-cuiller *m*
~QUARRY - s. sand pit
~RECLAMATION - [metall] (in foundry works) régénération *f* du sable
~RECLAMATION PLANT - [metall] installation *f* pour la régénération du sable
~REEL - [oil ind] treuil *m* de curage
~RIDDLER - s. sand sifter
~ROLLER - s. sand beam
~ROLLER WHEEL - [text] roue *f* de commande du cylindre sablé
~ROLLING - [metall] traitement *m* dans le tambour à sable
~SHEAVE - [oil ind] poulie *f* de curage
~SHELL MOULDING - [plast ind] moulage *m* en coquille, moulage *m* en carapace
~SHOP - [metall] sablerie *f*
~SIEVE GRADING - [metall] granulométrie *f*
~SIFTER - [impl] cribleuse *f*
~TAILINGS - [metall] tailings *mpl* sableux
~TEMPERING - [metall] humectation *f* du sable
~TRAP - [paper man] (cavity in the floor of a hollander trough) sablière *f*
~UP, to - [oil ind] s'ensabler (d'un puits)
~VENT - [metall] (of a mould) trou *m* d'air

SAND WASH - [metall] (in foundry works) gale *f* volante par érosion
SANDAL BRICKS - [constr] brique *f* demi-cuisson
SANDALWOOD - [bot] (fine-grained East-Indian tree) bois *m* de santal
SANDALWOOD OIL - [chem] (an essential oil obtained from the wood of Santalum album and used in perfumery and medicine) essence *f* de bois de santal
SANDARAC - [min] réalgar *m*
[ind chem] s. sandarac gum
~GUM - [ind chem] (a resin obtained from Callitris quadrivalvis and used in the manufacture of lacquers, varnishes and cements) sandaraque *f*, gomme *f* de genévrier
SANDBANK - [geol] dune *f* de sable
[naut] banc *m* de sable
SANDBLAST BARREL - [metall] tambour *m* à jet de sable
~ROOM - [metall] chambre *f* de dessablage
~SAND - [metall] sable *m* de nettoyage
SANDBLASTED PEBBLE - [geol] pierre *f* à facettes
SANDBLASTING GUN - [metall] pistolet *m* de sablage
SANDBOX - s. sand box
SANDER - [mech] sableuse *f*
[mech] (a sanding device, e.g. a sandpapering machine) machine *f* à poncer à ruban
[railw] (a sand-box, as on a locomotive, used to spread sand to avoid slipping) sablière *f*
SANDGLASS - [horol] horloge *f* de sable
SANDING - [gen carp etc] (the operation of cleaning up surfaces with sandpaper) sablage *m*, nettoyage *m* au sable
[mech] polissage *m* mécanique
[railw] sablage *m*
~BLOCK - [impl] tampon *m* de papier-verre
~DISK - [mech] disque *m* d'émeri
~GEAR - s. sander
~MACHINE - [carp] machine *f* à nettoyer au sable
~UP OF THE WELL - [oil ind] ensablage *m* du puits
~VEHICLE - [transp] véhicule *m* de distribution de sable
SANDLINE SPOOL - [oil ind] treuil *m* de curage
SANDLING - [metall] polissage *m* mécanique
SANDPAPER - [gen & carp] papier *m* verré, papier *m* sablé, papier *m* de verre
~EFFECT - [electron] (the effect of random noise on an intensity-modulated oscilloscope) bruit *m* de fond pour modulation dans l'axe Z
SANDPAPERING MACHINE - [mech] ponceuse *f*
SANDSLINGER - [metall] (in foundry work, a machine for reproducing the action of a moulder in filling a mould by hand) projecteur *m* de sable
SANDSLINGING - [metall] projection *f* de sable
SANDSHEET MIXTURE - s. sand asphalt
SANDSTONE - [geol] (a type of sedimentary rock composed chiefly of grains of quartz) grès *m*, roche *f* psammitique
~VAT - [impl] cuve *f* en grès
SANDSTORM - [met] (strong wind laden with sand and dust, of considerable extent) simoun *m*, pluie *f* de sable
SANDWICH, to - [plast ind] (to place a flat piece of material between two other flat pieces, usually of different substance of type) laminer en trois couches
[gen] intercaler, serrer
SANDWICH - [gen aero etc] (structural element consisting of a core of special from, e.g. cellular covered on each side by skins of other material) sandwich *m*
SANDWICH BEAM - [constr] (of flitch beam; a built-up beam consisting of an iron plate between two timber beams) poutre *f* à plaques rapportées
~CONSTRUCTION - [mech etc] (method of building up a structure or material by cementing successive layers of different materials together) construction *f* sandwich
~DIE - [metall] matrice *f* à éléments empilés
~HEATING - [plast ind] (method of heating both sides of a sheet before forming) chauffage *m* en sandwich
~IRRADIATION - [radiat] (irradiation of tissues from opposite sides) irradiation *f* en sandwich
~LAMINATE - [plast ind] stratifié *m* sandwich
~PHOTOCELL - [electron] (photovoltaic cell; photocell in which the photo-voltaic effect is utilized) cellule *f* à couche d'arrêt
~ROLLING - [metall] lamination *f* conjointe de métaux différents
~SEAL - [electron] scellement *m* à chemise de cuivre
~STRUCTURE - [mech plast etc] (the form produced by superimposing sheets or layers of similar or materials and attaching them in that position) structure *f* sandwich
~FILM - [photo] (film having two emulsions, one on each side of the base; used for release prints in the Multi-color system) film *m* à deux émulsions
SANDY - [gen] sableux, arénacé, sablonneux
~BOTTOM - [naut] fond *m* de sable
~CLAY - [geol] argile *f* maigre
~COTTON - [text] coton *m* sableux
~EARTH - [geol] terre *f* sableuse
~GRAVEL - [geol] gravier *m* sableux
~LIMESTONE - [geol] pierre *f* à chaux sableuse
[soil] calcaire *m* arénacé
~MARL - [geol] marne *f* sableuse
~ORE - [min] minerai *m* sablonneux, roussier *m*
~SOIL - [geol] sol *m* sableux, sol *m* sablonneux
SANE - [gen] sain
[leg] sain d'esprit
SANFORIZE, to - [text] (the operation of treating a fabric to eliminate shrinking) sanforiser
SANFORIZED FABRIC - [text] tissu *m* sanforisé
SANFORIZING MACHINE - [text] machine *f* à sanforiser
SANGUICOLOUS - [zool] (living in the blood) sanguicole
SANGUIFEROUS - [zool] (carrying blood) sanguifère
SANGUINARY - [gen & med] sanguinaire, sanglant
SANGUINE - [gen] sanguin
SANGUINEOUS - [med] sanguin, pléthorique
SANGUINOLENT - [med] sanguinolent
SANGUIVOROUS - [zool] (feeding on blood) sanguivore
SANIDINE - [min] (a form of potash feldspar) sanidine *f*, feldspath *m* vitreux
SANITARIUM - [med] (US for sanatorium) sanatorium *m*
SANITARY - [gen] (relating to the preservation of health) hygiénique, sanitaire
~CARE - [gen] précautions *fpl* hygiéniques
~COTTON - [med] ouate *f*
~ENGINEERING - [gen] génie *m* sanitaire, constructions *fpl* et matériel sanitaires
~FITTINGS - [hydr] installation *f* sanitaire
~GOODS - [gen] articles *mpl* sanitaires

SANITARY SEWER - [hydr] égout *m* pour les eaux vannes
~TOWEL - [med] bande *f* périodique, serviette *f* hygiénique
~WALLPAPER - [paper man] (wallpaper with designs printed in oil colours, which can be sponged) papier *m* lavable
~WARE - [ceram] (glazed earthenware used for sanitary fittings) articles *mpl* de terre sanitaires
SANITATION - [gen] hygiène *f*, système *m* sanitaire
SANSERIF - [print] (type without serifs) caractères *mpl* sans obit et sans empattement
SANSEVIERA - [bot] sansevière *f*
SANTALOL - [chem] (a perfumery agent) santalol *m*
SANTALYL ACETATE - [chem] (a perfumery agent) acétate *m* de santalyle
SANTONIN - [chem] (santalactone. A lactone obtained from species of Artemisia and used in medicine and as an anthelmintic) santonine *f*
SANTONINE - s. santonin
SAP, to - [gen] saper, miner
SAP - [bot] (aqueous solution of mineral salts, sugar and other organic substances) sève *f*
[bot] (the soapwood) aubier *m*
~-GREEN - [dyes] vert *m* de vessie
~WOOD - [bot] aubier *m*
SAPHEN - [anat] (one of the two large surface veins of the leg) saphène *f*
SAPHENECTOMY - [med] saphénectomie *f*
SAPID - [gen] sapide, savoureux
SAPLING - [bot] (a young tree) jeune arbre, baliveau *m*
SAPODILLA - [bot] sapotille *f*
~TREE - [bot] sapotillier *m*
SAPONACEOUS - [chem] saponacé, savonneux
SAPONIFIABLE - [ind chem] saponifiable
~FATS - [chem] graisses *fpl* saponifiables
SAPONIFICATION - [chem] (the reaction which fats and oils are hydrolyzed to soaps by reaction with an alkali) saponification *f*
~NUMBER - [chem] (the amount in miligrams of potassium hydroxide necessary for the complete saponification of I gram of fat or oil) nombre *m* de saponification
~OF THE WOOL GREASE - [text] saponification *f* de la graisse de laine
~PROCESS - [chem] procédé *m* de saponification
SAPONIFIED BEES'WAX - [text] graisse *f* de laine saponifiée
SAPONIFY, to - [ind chem] (to convert into soap) saponifier
SAPONIFY THE YOLK, to - [text] saponifier la graisse de laine
SAPONIFYING - [ind chem] saponification *f*
~AGENT - [chem] (a substance which promotes saponification) agent *m* de saponification
~MEDIUM - [ind chem] moyen *m* de saponification
SAPONIN - [chem] (generic term for certain plant glucosides with uses as synthesis intermediates, detergents and emulsifying agents) saponine *f*
SAPONITE - [min] (amorphous silicate of magnesium and aluminium) saponine *f*
SAPPHIRE - [min] (a transparent, blue corundum valued as a gemstone. Synthetic sapphires are also produced) saphir *m*
[gen] saphirin
~NEEDLE - [acoust] (record reproducing needle ground from natural sapphire) aiguille *f* de saphir
SAPPHIRE QUARTZ - [min] (a rare indigo-blue variety of silicified crocidolite; used as a semi-precious stone) quartz *m* saphirin
SAPPHIRINE - [min] (a rare aluminio-silicate of magnesium) saphirine *f*
SAPPINESS - [bot] teneur *f* en sève
SAPPING - [gen] sapement *m*
[constr] sape *f*, mine *f*
SAPPY - [gen & bot] plein de sève
[bot] (of timber) vert
~WOOL - [text] (wool containing an excessive amount of grease) laine *f* en suint riche en graisse
SAPRAEMIA - [med] (the presence of toxic products in the blood) saprémie *f*
SAPREMIA - s. sapraemia
SAPROBIOTIC - [biol] (feeding on dead or decaying animals or plants) saprobiotique
SAPROGENOUS - [biol] (which grows on decaying matter) saprogène
SAPROPELITE - [geol] (term denoting coals derived from algal materials) sapropélite *f*
SAPROPELITIC - [geol] sapropélitique
SAPROPHILE - [zool] (animal capable of breeding in stagnant and polluted waters) saprophyle
SAPROPHYTIC - s. saprobiotic
SAPROZOIC - s. saprobiotic
SAPWOOD - [bot] aubier *m*
[carp] aubour *m*
SARAH - [radar] (colloq. for emergency transmitter beacon; initials of Search And Radar And Homing) radiophare *m* de sécurité
SARAN - [plast ind] (generic name for vinylidene chloride based thermoplastics having good chemical resistance) saran *m*
SARCENET - [text] taffetas *m* léger
SARCINE - [chem] sarcine *f*
SARCOBLAST - [zool] (elongate multinucleate muscle-cell) sarcoblaste
SARCOCELE - [med] sarcocèle *m*
SARCODERM - [bot] sarcoderme
SARCODIC - [zool & med] (relating to flesh; resembling flesh) sarcodique
SARCODOUS - s. sarcodic
SARCODY - [bot] (the conversion into a substance of fleshy texture) sarcodie *f*
SARCOID - s. sarcodic
SARCOIDOSIS - [med] sarcoïdose *f*
SARCOLEMMA - [med] myolemme *m*
SARCOMA - [med] (malignant tumour of connective tissue origin) sarcome *m*
SARCOMATOSIS - [med] (the presence of several sarcomata in the body) sarcomatose *f*
SARCOPHAGUS - [arch] (stone coffin or tomb) sarcophage *m*
SARCOSINE - [chem] (methyl aminoacetic acid. A synthesis intermediate) sarcosine *f*
SARCOSIS - [physiol] sarcose *f*
SARCOSTOSIS - [med] ossification *f* d'un muscle
SARCOTRIPSY - [med] sarcotripsie *f*
SARCOUS - [zool] (pertaining the flesh or to muscle tissues) sarco
SARD - [geol] (a brownish red variety of quartz employed as a gemstone) sarde *m*
SARDINE - [zool] sardine *f*
~STONE - [min] sardoine *f*
SARDONYX - [min] (a natural crystalline silica em-

ployed as a gemstone) sardonyx *m*, sardoine *f*
SARGASSO WEED - [bot] (one of a large genus of brown algae found in tropical seas) sargasse *f*
SARKINE - [chem] sarcine *f*
SARKING FELT - [constr] (bituminous underlining placed beneath tiles) feutre *m* bitumineux
SARSAPARILLA - [bot] (the dried root of species of Smilax with uses in medicine and beverages) salsepareille *f*
SASH - [gen] écharpe *f*
[gen] (a belt) ceinture *f*
[constr] (framing for window panes) cadre *m* (de fenêtre à guillotine)
[constr] (the movable part of window) fenêtre *f* à coulisse
[constr] (a casement window) châssis *m* mobile
~-AND-FRAME - [constr] (cased frame in which the counterweighted sashes slide vertically) fenêtre *f* à guillotine
~BALANCING WEIGHT - [constr] contrepoids *m* de châssis à guillotine
~BAR - [carp] (a mullion) petit bois *m* (de fenêtre)
[metall] fer *m* à vitrage
~CENTRES - [carp] (points about which the pivoted sash is moved) centres *mpl* du châssis de fenêtre
~CHAIN - [constr] chaîne *f* du contrepoids
~CORD - [constr] corde *f* de châssis à guillotine
~DOOR - [constr] (door having its upper part glazed) porte *f* vitrée
~FASTENER - [impl] crampon *m* de fermeture
~FILLISTER - [tool] (special plane for cutting grooves for sash bars) feuilleret *m*
~FRAME - [carp] dormant *m* de fenêtre à guillotine
~GATE - [hydr] vanne *f* à coulisse
~LIFT - [gen & railw] poignée *f* de fenêtre
~LINE - [constr] corde *f*
~LOCK - [mech] (e.g. a fastener for railways window) crampon *m* de fermeture
~POINT - [constr] tourillon *m*, pivot *m*
~PULLEY - [mech] poulie *f* pour la corde de châssis à guillotine
~STUFF - [carp] (the timber prepared for making sashes) bois *m* pour dormants
~TOOL - [impl] brosse *f* inclinée
~WEIGHT - [constr] (a weight used as a counterpoise in balancing the sashes of window) contrepoids *m* de fenêtre à guillotine
~WINDOW - [constr] fenêtre *f* à guillotine, fenêtre *f* à coulisse
SASSAFRAS - [bot] (tree of the laurel family) sassafras *m*
~OIL - [bot] (an oil obtained from the root of Sassafras albidum and used in perfumery and medicine) essence *f* de sassafras
SASSOLINE - [min] (naturally-occurring boric acid, taking the form of tabular triclinic crystals, found in volcanic regions) sassoline *f*
SATEEN - [text] satin *m* de coton, satinette *f*
~BINDING POINT - [text] point *m* de liage de l'armure satin
~GROUND - [text] fond *m* en satin
~RIB - [text] sillon *m* du satin
~WEAVE - [text] armure *f* satin
SATELLITE - [astr] (heavenly body revolving round a larger one, generally a planet) satellite *m*
[astronaut] (a man made satellite) satellite *m*
~COMPUTER - [comput] calculateur *m* satellite
SATELLITE EXCHANGE - [telecomm] bureau *m* satellite
~ORBITS - [telev] orbites *fpl* de satellites
~PULSE - [radiat] (pulse following a scintillation pulse) impulsion *f* satellite
~STATION - [telev] station *f* relais
~TRANSMITTER - [radio] (or slave transmitter; an auxiliary transmitter) émetteur *m* répéteur
[telev] émetteur *m* satellite
SATIN - [text] (silk fabric having a lustrous smooth surface) satin *m*
~BRAID - [text] galon *m* de satin
~FINISH - [mech] (of metals, e.g. a honed cylinder bored) finissage *m* à brosse métallique
~PAPER FOR CARDS - [text] carton *m* glacé
~RIBBON - [text] ruban *m* de satin
~SPAR - [min] (the fibrous varieties of calcite and gypsum) stapth *m* satiné
~TOPS - [text] futaine *f* satinée
~TWILL - [text] sergé-satin *m*
~WEAVE - [text] (a weave in which the surface consist almost entirely of either warp or weft) armure *f* satin
~WHITE - [paint] (a filler obtained by co-precipitation of calcium sulphate and aluminate, used as a base for lakes) blanc *m* satin
SATINY - [gen] satiné
~COCOON - [text] cocon *m* satiné, soufflon *m*
SATURABLE MAGNETOMETER - [instr] magnétomètre *m* saturable
~REACTOR - [radio] (electromagnetic device used in A. C. circuits to secure amplification or control) réacteur *m* saturable
~-REACTOR-TYPE COMPASS - [instr] (compass operating on the principle of the magnetic modulator) boussole *f* à induction terrestre
~TRANSFORMER - [radio] (a saturable reactor in which the output and the power winding are used to secure voltage transformation) transformateur *m* saturable
SATURANT - [chem] (any substance with which another is impregnated) agent *m* d'imprégnation
SATURATE, to - [gen phys etc] (the action of bringing about a state of being satisfied) saturer, imprégner, tremper
SATURATED - [gen phys etc] saturé, imprégné
~ADIABATIC LAPSE RATE - [met] (the rate of decrease of temperature of saturated air with height under adiabatic conditions) gradient *m* vertical de température adiabatique de saturation
~ATMOSPHERE - [phys] atmosphère *f* saturée
~CALOMEL ELECTRODE - [chem] (calomel electrode which contains saturated potassium chloride solution) électrode *f* de chlorure mercureux saturé
~COLOUR - [gen] couleur *f* riche, couleur *f* intense
~COMPOUND - [chem] (a compound in which there are not free valencies) composé *m* saturé
~DIODE - [electron] (diode operating in certain conditions, so that the anode-current characteristic is flat, due to the saturation) diode *f* saturée
~HYDROCARBONS - [chem] (synonym for alkanes or paraffin hydrocarbons) hydrocarbures *mpl* saturés
~MIXTURE - [ind chem] mélange *m* saturé
~RADIOACTIVITY - [nucl] activité *f* à saturation
~SOIL - [soil] sol *m* saturé
~SOLUTION - [chem] (a solution capable of existence in equilibrium with excess of the solute) solution

f saturée
SATURATED STEAM - [phys] (steam at the same temperature as the corresponding liquid phase, i.e. unsuperheated) vapeur *f* saturée
~TRANSFORMER - [electron] transformateur *m* saturé
~VAPOUR - [phys] (a vapour in equilibrium with its liquid at a given temperature) vapeur *f* saturante
~VAPOUR PRESSURE - [phys] (the vapour pressure of a vapour in contact with a liquid phase. It increases with the temperature and is sometimes confused with Vapour Pressure) pression *f* de vapeur saturante
~WASH OIL - [ind chem] huile *f* de lavage saturée
SATURATION - [phys] (the state in which all the bonds of an atom are attached to other atoms) saturation *f*
[chem] (a solution containing the maximum possible amount of solute) saturation *f*
[electron] (of electrons) saturation *f*
~ACTIVITY - [nucl] (the maximum activity which can be obtained by activation in a definite flux) activité *f* à saturation
~ANGLE - [electron] (gate angle) angle *m* de saturation
~BANDING - [telev] saturation *f* striée
~CONCENTRATION - [chem] charge *f* de saturation
~CURRENT - [el] (the steady current in a winding of an ironcored transformer causing the inductance in the winding to be appreciably reduced) courant *m* de saturation
[electron] (the value of the current in the saturation state) courant *m* de saturation
~CURVE - [el] (the characteristic curve relating magnetic flux density to the strength of the magnetic field) caractéristique *f* de saturation
~DEFICIENCY - [met] (of air) insuffisance *f* de saturation
~FACTOR - [el] (the ratio between the increase of field excitation and the increase of the generated voltage it produces) coefficient *m* de saturation
~FLUX - [phys] flux *m* de saturation
~GAIN - [electron] gain *m* de saturation
~INDUCTION - [el] (the maximum possible induction in a material) induction *f* de saturation, induction *f* magnétique maximale
~INTENSITY - [telecomm] intensité *f* de saturation
~INTERVAL - [radio] intervalle *m* de saturation
~LIMIT - [el] (the maximum flux density which is economic and attainable) limite *f* de saturation
~MAGNETIZATION - [el] (the application of an increasing magnetizing force to a ferromagnetic substance yields a resulting intrinsic induction approaching a constant value, i.e. the saturation magnetization) magnétisation *f* de saturation
~MAGNETOSTRICTION - [el] (the limiting value of magnetostriction) magnétostriction *f* de saturation
~OF THE AIR - [met] (the limit at which air, at a given temperature, can contain water vapour) saturation *f* de l'air
~PERMEAMETER - [instr] (permeameter having a large, artificially cooled magnetizing coil) perméamètre *m* à saturation
~POINT - [phys & chem] point *m* de saturation
~POTENTIAL - s. saturation voltage
~PRESSURE - [phys] (of a vapour) pression *f* de saturation
~REACTANCE - [radio] (the reactance of the gate winding of a magnetic amplifier during the saturation interval) réactance *f* de saturation
SATURATION RESISTANCE - [electron] (in semiconductors) résistance *f* de saturation
~SCALE - [opt] (series of visual stimuli having equal differences in saturation) échelle *f* de saturation
~STATE - [electron] (the state of working of an electronic tube in which the current is limited by the emission from the cathode) état *m* de saturation
~TIME FACTOR - [electron] (in semiconductors) constante *f* de temps pour saturation
~VALUE - [el] valeur *f* de saturation
~VAPOUR PRESSURE - [met] (the partial pressure of water vapour in equilibrium with plane surface of water or ice) tension *f* de vapeur de saturation
~VAPOUR TENSION - s. saturated vapour pressure
~VOLTAGE - [el] (the minimum value of applied potential necessary to produce saturation current) tension *f* de saturation, tension *f* résiduelle
SATURATOR - [ind chem] saturateur *m*
~BELL - [ind chem] cloche *f* de saturateur
SATURN - [nucl] (a thermonuclear device) appareil *m* saturne
SATURNISM - [med] saturnisme *m*, intoxication *f* par le plomb
SATYRIASIS - [med] satyriasis *m*
SAUCEPAN - [impl] casserole *f*
SAUCER - [gen] soucoupe *f*
[naut] (a flat caisson used to refloat sunken vessels) saucier *m*
[mech] cavité *f* pour la garniture
SAUCERIZATION - [med] mise *f* à plat d'une plaie
SAUNA BATH - [med] bain *m* de vapeur
SAURIAN - [zool] (lizard-like) saurien *m*
SAURIASIS - [med] peau *f* de crocodile
SAUSAGE - [food] saucisse *f*, saucisson *m*
~ANTENNA - [radio] (antenna with wires connected in parallel and arranged in a parallel formation around circular spreaders) antenne *f* Zeppelin
~BALLOON - [aero] ballon *m* d'observation
~INSTABILITY - [phys] (plasma instability) instabilité *f* à coques
~POISONING - [med] (botulism) botulisme *m*
SAUSSURITE - [min] (an aggregate of calcium aluminium silicates and calcite) saussurite *f*
SAVANNA - [geol] savane *f*
SAVE, to - [gen] sauver
[comm] économiser
SAVE-ALL - [paper man] (tank for the collection of escaping pulp) épurateur *m* de pâte
[naut] petite voile *f* supplémentaire
[mech] appareil *m* économiseur
SAVIN JUNIPER - [bot] genévrier *m*, savine *f*
SAVING - [gen] économie *f*, épargne *f*
[ind chem etc] récupération *f*, captage *m* des sous-produits
SAVINGS BANK - [fin] caisse *f* d'épargne
SAVOY - [bot] chou *m* frisé de Savoie
~-CABBAGE - s. savoy
SAW, to - [gen carp etc] scier, débiter, débiter à la scie
[constr] (stone or marble) sciotter, débiter
SAW - [impl] scie *f*
~ARBOR - [mech] arbre *m* porte-scie
~BENCH - [impl] scierie *f*
~BLADE - [mech] lame *f* de scie
~BOW - [mech] cadre *m* de scie

SAW BUCK - s. saw-horse
~BUCKLE - [mech] boucle *f* de la scie
~BUCKLE WITH ECCENTRIC ADJUSTMENT - [mech] boucle *f* de scie à excentrique
~CARRIAGE - [mech] chariot *m* d'une scie mécanique
~CLAMP - [mech] entaille *f* à affûter les scies
~-CUT - [mech] trait *m* de scie
~DOCTOR - [mech] (a skilled workman expert in sharpening, setting and trueing saws) affûteur *m* de scies
~DOCTORING - [mech] affûtage *m* de la scie
~-DUST - s. sawdust
~FILE - [mech] lime *f* à scies, lime *f* pour l'affûtage des scies
~-FILER'S VICE - [mech] étau *m* d'affûtage et d'avoyage pour scies
~FRAME - [impl] châssis *m* porte-scie
~GATE - s. saw frame
~GRINDER - [mech] machine *f* à affûter les lames des scies
~GUARD - [mech] appareil *m* protecteur pour scies
~-GUIDE - [mech] guide-lame *m*
~GUMMER - [mach tool] affûteuse *f* pour scies
~-HORSE - [mech] chevalet *m* de sciage, chèvre *f*
~-JACK - s. saw-horse
~JUMPER - s. saw set
~-KERF - s. saw-cut
~-KNIFE - [rubber ind] (used in latex-foam cutting) scie *f* de tailleurs pour le découpage de la mousse de latex
~-LOG - [mech] bille *f*
~MANDREL - [mech] arbre *m* de scie
~PULLEY - [mech] volant *m* porte-lame
~RIP - [impl] (on a machine tool) guide-pièce *m*
~SET - [tool] (instrument to adjust the teeth of a saw) tourne-à-gauche pour donner la voie aux scies
~SETTING MACHINE - [mech mach] machine *f* à avoyer les scies
~-SHARPENING MACHINE - [mech] machine *f* à affûter les lames de scies
~SPINDLE - s. saw mandrel
~SWAGE - s. saw set
~-TEETH TREAD - [rubber ind] sculpture *f* à dents de scie
~-TIMBER - [timber] bois *m* de sciage
~TOOTH - [mech] dent *f* de scie
~-TOOTH AMPLIFIER - [el] amplificateur *m* à dents de scie
~-TOOTH BIT - [mech] couronne *f* dentée
~-TOOTH CARD CLOTHING - [text] garniture *f* en dents de scie
~-TOOTH COUPLING - [el] couplage *m* en dents de scie
~-TOOTH CURRENT - [el] courant *m* à dents de scie
~-TOOTH DIRT TRAP - [mech] piège *m* à crasses en dents de scie
~-TOOTH GENERATOR - [radio] (generator of oscillations each cycle of which consists of a variation of voltage or current which is approximately proportional to time, followed by a comparatively quick variation of opposite sense) générateur *m* de dents de scie
~-TOOTH KEYBOARD - [telecomm] clavier *m* à action directe
~-TOOTH LICKER-IN - [text] avant-train *m* à garniture en dents de scie
~-TOOTH OSCILLATOR - [radio] (signal generator in which the frequency is varied continuously over a given range) oscillateur *m* à dents de scie
SAW-TOOTH PROFILE - s. saw-tooth tread
~-TOOTH PULSE - [electron] impulsion *f* en dents de scie
~-TOOTH ROOF - [constr] (a roof presenting a serrated profile when viewed from the end) comble *m* en dent de scie, shed *m*, toiture *f* en shed
~-TOOTH TAKER-IN - [text] avant-train *m* à garniture en dents de scie
~-TOOTH TRUSS - [mech] (truss used for small-span roofs of a saw-tooth form) ferme *f* de toit en shed
~-TOOTH WAVE - [radio] (periodic wave, the amplitude of which varies between two values, the interval required for one direction of progress is longer than that for the other) onde *f* en dents de scie
~TOOTH WIRE FILLETING - [mech] garniture *f* Garnett
~-TOOTHED - [gen] (serrate; with teeth like a saw) en dents de scie
~WEB - [mech] lame *f* de scie
~-WORT - [bot] serrette *f*, serratule *f*
SAWBUCK - [impl] (a rack for sawing wood) chevalet *m* de sciage
SAWDUST - [gen] (of wood) sciure *f* de bois, bran *m* de scie
[min] (of marble) poudre *f* de marbre
SAWFLY - [zool] mouche *f* à scie, tenthrède *f*
SAWHORSE - [mech] (a frame consisting of a long wooden plank supported by four extended legs) chècre *f*, bidet *m*
SAWING MACHINE - [mech] scierie *f* mécanique, machine *f* à scier
SAWMILL - [gen] scierie *f*
[mech] s. sawing machine
SAWN TIMBER - [timber] bois *m* débité
SAWYER - [gen] scieur *m*
SAXHORN - [mus] (a brass wind instrument) saxhorn *m*, bugle *m* à pistons
SAXON MERINO BREED - [text] race *f* mérino-saxonne
~WOOL - [text] laine *f* de Saxe
SAXOPHONE - [mus] (a wind instrument) saxophone *m*
S.B. - [radio] (initials of Simultaneous Broadcasting) émission *f* relayée
S.B.A. - [radio] (standard Beam Approach System) système *m* d'approche standard à faisceau, SBA
S.B.C. - s. small bayonet cap
SCAB, to - [metall] dartrer
SCAB - [gen] croûte *f*
[med] (crust formed on the surface of a wound) eschare *f*, croûte *f*
[vet] (contagious disease affecting sheep) gale *f*
[bot] (plant disease) gale *f*
[metall] (foundry work, a surface defect caused by adhesion of scale) dartre *f*, balèvre *f*
[metall] (die casting; sand on the surface area of a steel casting) écharde *f*, éclat *m*
[metall] (on an iron casting) gale *f* volante
[gen] (a blackleg) renard *m*
SCABBARD - [gen] gaine *f*, fourreau *m*
SCABBARD - [oil ind] enveloppe *f* de la tige carrée d'entraînement
SCABBLE, to - [constr] smiller (la pierre de carrière)
SCABBLING - [constr] (the rough dressing of a stone face with an axe) smillage *m*

SCABBLING HAMMER - [impl] marteau *m* à smiller
SCABIES - [med] gale *f*
~OF THE LEGS - [vet] gale *f* des pattes
SCABROUS - [gen & bot] scabreux, rugueux
SCABS - [constr] bavures *fpl*
SCABWORT - [bot] inule *f*
SCAEVOLISM - [med] automutilation *f* par le feu
SCAFFOLD, to - [constr] échafauder
SCAFFOLD - [constr] échafaud *m*, échafaudage *m*
[metall] (in a blast furnace) engorgement *m* d'haut fourneau
~BRANCH - [bot] branche *f* charpentière
~BRIDGE - [constr] pont *m* monté sur échafaudage
~CLAMP - [constr] étrier *m* d'échafaud
~FLOOR - [constr] plancher *m* volant
~LIMB - s. scaffold branch
~POLE - [constr] perche *f* d'échafaudage, pointier *m*
SCAFFOLDING - [constr] (temporary erection of timber or steelwork used in construction work) échafaudage *m*
[metall] (formation of arch-like structure in a furnace charge, which prevents its downward movement during combustion) accrochage *m* des charges, accrochage *m* du cubilot
[constr] (the material for scaffolding) matériel *m* d'échafaudage
[constr] (of a roof) charpente *f* du toit
[mining] échafaudage *m*
~STANDARD - [constr] baliveau *m*, grosse *f* pièce d'échafaudage
SCAGLIOLA - [constr] (imitation marble obtained by adding colouring matter to a hard cement) scagliol *m*
~SLAB - [constr] carreau *m* scagliol
SCALAR - [math etc] (of scale) scalaire
~FIELD - [el] champ *m* scalaire
~FORCE - [nucl] (nuclear force depending only on distance) force *f* scalaire
~FUNCTION - [math] fonction *f* scalaire
~POINT FUNCTION - s. scalar function
~POTENTIAL FIELD - [el] champ *m* de potentiel scalaire
~PRODUCT - [el] (in multiplying vector quantities) produit *m* scalaire
~QUANTITY - [phys] (quantity completely specified by its magnitude) grandeur *f* scalaire, scalaire *f*
SCALD, to - [ind chem] (to clean with boiling water or steam) échauder
[gen](to warm in hot liquid) échauder, ébouillanter, blanchir
[text] (the flax) lessiver (le lin)
SCALD - [med] échaudure *f*, brûlure *f*
~OUT, to - [ind chem] (to clean the inside of a vessel with boiling water or steam) ébouillanter
SCALDING - [text] (of flax) cuisson *f* (du lin), lessivage *m* (du lin)
~VAT - [text] cuve *f* de débouillissage
SCALE, to - [gen] (to climb) escaler
[gen] (of quantities) avoir une échelle commune
[mech etc] (to grade) graduer
[th eng] (to form scales) incruster, entartrer
[th eng] (to remove scales) écailler, détartrer, désincruster
[agric] (to asses the measure of timber) évaluer la mesure du bois d'oeuvre
[comm] peser
[th eng] (to come off in scales) s'entartrer, s'incruster
SCALE, to - [surv] (a map) tracer (une carte) à l'échelle
[agric] (a tree) exfolier
[med] (in dentistry) ruginer (les dents)
SCALE - [zool] (thin, flat horny membrane) écaille *f*
[bot] (thin flat semi-transparent plant member) écaille *f*
[mus] (series of tones progressing from a given tone to its octave by prescribed pitch intervals) gamme *f*
[zool] (a pest) cochenille *f*
[th eng] (coating formed on heat transfer surfaces, e.g. rust, calcium sulphate, oil etc) incrustation *f*, tartre *m*
[surv] (of a map or drawing) (the ratio which the dimensions shown on a drawing bear to the real ones) échelle *f*
[text] (of wool) écaille *f*
[metall] barbure *f*
[metall] (a coating of iron oxide) batitures *fpl*, écailles *fpl* de fer
[instr] (e.g. of a thermometer) graduation *f* (d'un thermomètre etc), série *f*
[bot] (appearing on very old trees) teigne *f*
[phys] (of a measuring instrument) échelle *f*
[oil ind] paraffine *f* brute en écailles
[meas] balance *f* plateau *m* de balance
[impl] (e.g. of a razor) châsse *f* (d'un rasoir etc)
~A BOILER, to - [gen] désincruster une chaudière, piquer une chaudière
~A SPRING - [mech] (in dynamometry) tarer un ressort
~BASE - [meas] base *f* d'une échelle
~-BEAM - [mech] fléau *m* de balance, verge *f* de balance
[meas] romaine *f*
~-BOARD - [gen] feuille *f* mince
[carp] (a sheet of wood used for veneering) lame *f* mince de bois
~BREAKER - [metall] laminoir *m* casse-oxyde
~-COATED - [th eng] (of a boiler) entartré, encrusté
~COPPER - [metall] cuivre *m* en écailles
~-CRUST - [mech] croûte *f* de tartre
~DEPOSIT - [oil ind] calcination *f*
~DOOR - [metall] porte *f* d'aérage à guichet
~-DOWN, to - [meas] (to reproduce a process on a smaller scale) réduire à l'échelle
~DRAWING - [draw] dessin *m* à l'échelle
~EFFECT - [aero] (the effect of charges in the dimension of a body on the non dimensional coefficients associated with it) effet *m* d'échelle
~ERROR - [instr] erreur *f* d'échelle
~FACTOR - [instr & comput] facteur *m* d'échelle
~FERN - [bot] cétérac *m*
~FOCUSING - [photo] mise *f* au point sur échelle
~HAIR - [bot] poils *mpl* écailleux
~INSECT - [zool] coccide *f*
~INTERVAL - [instr] (the increment of the measured quantity corresponding to the scale spacing) intervalle *m* d'échelle
~LEAF - [bot] feuille *f* squameuse
~LENGTH - [instr] (the length of the arc passing through the centre of the shortest markings on the graduated scale) longueur *f* de graduation
~MARKS - [instr] (the marks which make it possible

to identify the position of the moving element of an instrument) graduation *f*
SCALE NUMBERING - [instr] (the series of numbers forming part of the scale) graduation *f* d'un système numérique
~OF FUSIBILITY - [min] échelle *f* de fusibilité
~OF HARDNESS - [min] échelle *f* de dureté
~OF SPRING - [mech] tare *f* du ressort
~OF TEMPERATURE - [meas] graduation *f* thermométrique
~OF THE BALANCE - [meas] plateau *m* des poids de la balance
~-OF-TWO COUNTER - [el] (binary divider) compteur *m* binaire, échelle *f* binaire
~-OF-TWO MULTIVIBRATOR - [electron] multivibrateur *m* bistable
~OF WIRE - [metall] écaillage *m* du fil d'acier
~PAN - [impl] plateau *m* de la balance
~PIT - [metall] impureté *f*
~PLATE - [instr] (of an instrument) cadran *m*
~PREVENTING - [ind chem] antitartre *m*, tartrifuge *m*
~RANGE - [instr] gamme *f* d'échelle
~RESISTING STEEL - [metall] acier *m* inoxydable
~SETTING - [electron] positionnement *m* d'une échelle
~SHAPE - [instr] (the mathematical form of the scale of an instrument) forme *f* d'échelle
~SOLVENT - [ind chem] désincrustant *m*
~SPAN - [instr] portée *f* de l'échelle
~-TIGHT FINISH - [metall] finissage *m* résistant à l'écaillage
~TWILL - [text] sergé *m* à échelle
~UNITS - [instr] unités *f*pl d'échelle
~-UP, to - [meas] (to reproduce a process on a larger scale) augmenter à l'échelle
~WAX - [ind chem] (the paraffin which is obtained from slack wax or waxing distillate by deoiling) paraffine *f* en écailles
SCALEBOARD - s. scale-board
SCALENE - [geol] scalène
[geom] (a scalene triangle) triangle *m* scalène
SCALENECTOMY - [med] scalénotomie *f*
SCALEPAN - s. scale pan
SCALER - [electron] (electronic demultiplier of pulses) démultiplicateur *m*
[mech] piqueur *m* de chaudières
[impl] écailleur *m*
SCALES - [meas] balance *f*
SCALING - [metall] (in boilers) écaillage *m*, écaillement *m*
[metall] (formation of scales on heated steel) formation *f* d'écaillage, entartrage *m*
[metall] (the operation of removing the scale) désincrustation *f*
[mech] décapement des feuilles de fer
[mining] purge *f* du toit
[bot] exfoliage *m*
[comm] (of prices etc) graduation *f*
[dent] détartrage *m*
[comput] modification *f* de l'échelle des nombres
~CIRCUIT - [electron] (or scaler; device producing an output pulse whenever a given number of input pulses has been received) démultiplication *f* d'impulsions
~FACTOR - [instr] (the number of input pulses per output pulse of a scaling circuit) facteur *m* d'échelle
~FURNACE - [metall] fourneau *m* à décaper

SCALING HAMMER - [impl] (a boilermarker's hammer) marteau *m* à ébarber
~OFF - [metall] (the formation of minute portions of metal on a chilled surface) formation *f* d'écaillage
~TEMPERATURE - [metall] température *f* critique de écaillage
SCALL - [med] dermatose *f* squameuse
SCALLOP, to - [gen] denteler, mouler
[text] festonner
SCALOPP - [zool] pétoncle *m*, peigne *m*
[gen] feston *m*, dentelure *f*
SCALLOPED - [arch] en écailles
~EDGE - [text] bord *m* à languettes
~TRIMMING - [text] ruban *m* festonné
SCALLOPING - [telev] feston *m*
~BY FIGURING WEFT - [text] festonnage *m* par trame de dessin
~PATTERN - [text] échantillon *m* à zig-zag
SCALP, to - [gen] scalper
[med] (a bone) ruginer (un os)
SCALP - [anat] épicrâne *m*, cuir *m* chevelu
[gen] (of a mountain) sommet *m* pelé (d'une montagne)
[min] (of coke) précriblage *m*
SCALPEL - [impl] scalpel *m*
[med] bistouri *m*
SCALPER - [impl] (in milling) crible *m*
SCALPING - [min] (in the milling ind) précriblage *m*
[rubber ind] élimination *f* de la chape
[metall] enlèvement *m* de la couche superficielle
~MACHINE - [rubber ind] (machine for stripping off the treads of old tyres) machine *f* à déchaper, refendeuse *f*
SCALY - [gen] écailleux, squameux
SCAN, to - [gen] examiner, scruter
[telev] (the operation of traversing the surface of a picture by a beam of light or electrons, to transmit or to reproduce the image) analyser, balayer, explorer
[radar] (to explore a region by the continuous variation of the direction of a beam) explorer
SCAN - [telev] (the trace on a cathode-ray tube) trace *f*
[telev] (the forward motion of the scanning spot, controlled by the increasing voltage or current) balayage *m*
[radar] (the exploration of a region by the continuous variation of the direction of a beam) exploration *f*
~BURNS - [telev] brûlures *f*pl d'analyse
~CONVERSION DEVICE - [telev] convertisseur *m* du nombre de lignes
~GENERATOR - [telev] générateur *m* de dents de scie pour l'analyse
~LINEARITY - [telev] linéarité *f* de balayage
~MATRIX - [comput] grille *f* codée
~PERIOD - [comput] (in cathode ray-tube storage, the generation period) temps *m* de régénération
~PROTECTION - [telev] interruption *f* de courant du faisceau
~RINGS - [telev] suroscillation *f* en début de ligne
~SCANDIUM - [chem] (the least basic of the rare earth metals, A.N.2I, A.W.45.IO, symbol Sc) scandium *m*
SCANNER - [telev] dispositif *m* de balayage
[radar] (a rotating aerial and reflector for microwave radars) projecteur *m* d'onde d'écho

SCANNING - [telev] (at the transmitter end; the dissection of the image into elements which are then converted into electrical signals. At the receiver end; a synthesis of the picture from the elements obtained from the electrical signals) analyse *f*, balayage *m*, exploration *f*
~AMPLIFIER - [telev] amplificateur *m* d'analyse
~APERTURE - [telev] (one of the holes in the scanning disk) diaphragme *m* d'analyse
~BEAM - [telev] (the beam of light or of electrons scanning an image) faisceau *m* analyseur
~COIL - [telev] bobine *f* de balayage, bobine *f* de déviation
~CYLINDER - s. scanning drum
~DISK - [telev] (rotaing disk with a series of apertures, lenses etc. used for mechanical scanning) disque *m* analyseur
~DRUM - [telev] cylindre *m* analyseur
~FIELD - [telev] (the pattern of closely spaced lines formed on the screen of a cathode-ray tube when the frame and line scanning current are applied simultaneously) champ *m* d'analyse
~GATE - [cin] (in the film reproduction apparatus) fenêtre *f* de balayage
~GENERATOR - [telev] (or sweep generator; generator which generates the deflection voltage) générateur *m* de balayage
~HEAD - [electron] (a light-sensitive device) tête *f* de balayage
~HOLE - s. scanning aperture
~IN RECEPTION - [telev] restitution *f* de l'image
~IN TRANSMISSION - [telev] analyse *f* de l'image
~INTERFERENCE - [telev] interférence *f* dans le canevas
~LENS - [opt] optique *f* de balayage
~LINE - [telev] (the trace of a single traverse of the picture by the scanning spot) ligne *f* d'image, ligne *f* d'analyse
~LINEARITY - [telev] (the uniformity of scanning speed during the trace interval) linéarité *f* de balayage
~-LINES NUMBER - [telev] nombre *m* de lignes de balayage
~MOTION - [telev] mouvement *m* d'exploration
~PATTERN - [telev] diagramme *m* de balayage
~PITCH - [telev] (the distance of the scanning between the successive lines) espacement *m* des lignes d'analyse
~RASTER - [telev] canevas *m* d'analyse
~SIGNAL - [telev] signal *m* d'analyse
~SLIT - [cin] (a slit in the scanning gate) fente *f* d'analyse
~SLOT - s. scanning slit
~SPEECH - [med] (adisturbance of speech) scansion *f*
~SPEED - [telev] (the speed at which the scanning is made) vitesse *f* d'analyse
~SPOT - [telev] (the spot of light formed by the scanning beam on the screen) spot *m* analyseur
~STAGE - [telev] (the stage in which the transformation of the electrical image intoelectrical signals occurs) étage *m* d'analyse
~TRAVERSE - [telev] (the longitudinal translation of the scanning device relative to the drum during the scanning process) translation *f* de l'analyse
SCANSION - [telev] (the operation scanning) analyse *f*, balayage *m*
SCANT, to - [naut] (to haul the wind) refuser
SCANT - [gen] insuffisant, rare
SCANTLING - [constr] volige *f*
[carp] menu bois *m* de sciage
[gen] (a sample) échantillon *m*, équarrissage *m*
[constr] échantillon *m* de construction
[shipbuild] (dimensions of frames etc) échantillons *mpl*
SCAPE - [arch] (of a column) escape *f*, fût *m* de colonne
~WHEEL - [horol] roue *f* de rencontre
SCAPHOCEPHALY - [anat] scaphocéphalie *f*
SCAPHOID - [zool] (shaped like a boat) scaphoïde
SCAPOLITE - [min] (group of minerals varying from silicate of aluminium and calcium with calcium carbonate, to silicate of aluminium and sodium with sodium chloride) scapolite *f*
SCAPULA - [anat] omoplate *f*
SCAPULAR - [anat] scapulaire
SCAPUS - [anat] tige *f*, manche *f*
SCAR - [gen & med] cicatrice *f*
SCARCE - [gen] rare, insuffisant
SCARCEMENT - [constr] (a ledge formed where part of a wall is set back from the main face of the wall) ressaut *m*
SCARF, to - [carp] (to unite with a scarf joint) assembler à mi-bois, enter
[metall] (to chamfer) chanfreiner
[mech] (to taper) effiler
[shipbuild] (to overlap adjacent timbers) écarver, enter
[metall] décalaminer au chalumeau
SCARF - [carp] (a notched and lapped joint between two timbers placed end to end and secured with bolts) joint *m* en bec, assemblage *m* à mi-bois
[text] écharpe *f*, cache-col *m*, cravate *f*
[metall] biseau *m*
~CLOUD - [met] (a thin cirrus-type cloud draping the summit of tall cumulonimbus clouds) cirrostratus *mpl* en forme de plumes
~JOINT - [carp etc] (a joint made by cutting the ends of the adherents at an angle to their principal surfaces and placing these angular cuts in contact) assemblage *m* à mi-bois
~WELDING - [metall] soudure *f* par amorces, soudure *f* à recouvrement
SCARFING - [carp] assemblage *m* à mi-bois
[metall] (the removal of surface defects from ingots by oxy-gas flame) amorçage *m*, décalaminage *m* au chalumeau
[metall] (chamfering) biseautage *m*
[mech] effilement *m*
[shipbuild] assemblage *m* à mi-bois
SCARFS - [carp & mech] organes *mpl* d'assemblage à entaille
SCARIFICATION - [med] (a slight incision) scarification *f*
[agric] (the breaking up of the soil by means of a scarifier) scarification *f*
SCARIFIER - [impl] (a spiked mechanical device for breaking up soil, road surfaces etc; also used in surgery) scarificateur *m*, extirpateur *m*
SCARIFY, to - [gen] scarifier
SCARIOSE - s. scarious
SCARIOUS - [bot] (dry and as if scorched in appearances) scarieux
SCARLATINA - s. scarlet fever

SCARLATINIFORM - [med] (resembling the rash of scarlet fever) scarlatiniforme
SCARLET - [dye] écarlate
~BEAN - s. scarlet-runner bean
~CHROME - [paint] (also called Chrome Scarlet: complex pigments consisting of co-precipitated chromate, molybdate and sulphate of lead, brilliantly red in colour) écarlate *m* de chrome
~FEVER - [med] (acute infectious fever caused by an infection of the throat with a heomolytic streptococcus) scarlatine *f*
~LAKE - [paint] (pigment obtained by precipitating Scarlet 2R on gloss white or aluminium hydroxide) laque *f* écarlate
~-RUNNER BEAN - [agric] haricot *m* d'Espagne
SCARP - [gen] escarpement *m*
~FACE - [geol] (an escarpement) escarpement *m*
SCARRED - [gen] couturé de cicatrices, cicatrisé
SCATACRATIA - [med] défecation *f* involontaire
SCATEMIA - [med] toxémie *f* intestinale
SCATOPHILIA - [med] coprophilie *f*
SCATS - [comput] (Sequential Controlled Automatic Transmitter Start) amorce *f* automatique du transmetteur à commande séquentielle
SCATTER, to - [gen] disperser, éparpiller
[phys] (causing the random distribution of a group entities) disperser
SCATTER - [gen] (term used in statistical work) dispersion *f*
~ABSORPTION COEFFICIENT - [phys] coefficient *m* d'absorption de diffusion
~COEFFICIENT OF THE POPULATION - [statist] distribution *f* démographique
~LOADING - [comput] chargement *m* avec éclatement
~READING - [comput] lecture *f* avec éclatement
~THE KNOPS, to - [text] semer les boutons
~UNSHARPNESS - [radiat] flou *m* dû à la dispersion
SCATTERED BEAM - [radiat] faisceau *m* diffusé
~CLOUDS - [met] (clouds which are few in number and widely separated over the visible sky) nuages *mpl* épars
~LIGHT - [telev] lumière *f* dispersée
~NEUTRONS - [nucl] (a nucleus the direction of which has been deviated during its passage through a substance) neutrons *mpl* diffusés
~PARTICLE - [nucl] (a particle the direction of which has been deviated during its passage through a substance) particule *f* diffusée
~RADIATION - [radiat] (a radiation the direction of which has been deviated during its passage through a substance) rayonnement *m* diffusé
~RADIATION PROXIMITY INDICATOR - [meas] signaleur *m* de proximité par rayonnement ionisant diffusé
~WAVE - [phys] (a wave the direction of which has been deviated during its passage through a substance) onde *f* diffusée
~X-RAYS - [radiat] rayons *mpl* X de dispersion
SCATTERING - [phys] (to cause a random distribution of a group of entities) diffusion *f*
[nucl] (the change in direction of a particle or photon owing to a collision with another particle or system) diffusion *f*, scattering *m*
~AMPLITUDE - [nucl] (quantity which is closely related to the intensity of scattering of a wave by a central force field, e.g. of a nucleus) amplitude *f* de diffusion
SCATTERING ANGLE - [nucl] (the angle between the initial and final lines of motion of a scattered particle) angle *m* de diffusion
~CENTRE - [nucl] centre *m* de diffusion
~CIRCLE - [el] cercle *m* de dispersion
~COEFFICIENT - [phys] coefficient *m* de diffusion
~COLLISION - [nucl] collision *f* de diffusion
~CONE - [el] cône *m* de diffusion
~CROSS-SECTION - [nucl] (the cross-section for scattering) section *f* efficace de diffusion
~CURVE - [phys] courbe *f* de dispersion
~FREQUENCY - [nucl] (the actual frequency when neither the forward nor the backward scattering is counted) fréquence *f* de diffusion
~LOSS - [acoust] perte *f* de diffusion
~MATRIX - [electron] (in waveguides) matrice *f* de diffusion
~MEAN FREE PATH - [nucl] (the average scattering length) libre parcours *m* moyen de diffusion
~OF NEUTRONS - [nucl] diffusion *f* de neutrons
~POLARIZATION - [el] polarisation *f* par dispersion
SCAVENGE, to - [gen] balayer
[mech] (to remove exhaust gases, as from cylinders of an internal combustion engine) refouler, balayer les gaz brûlés
[metall] (to remove impurities) éliminer les impurités
SCAVENGE OIL - [mech] huile *f* de récupération
~OIL FILTER - [mech] filtre *m* de l'huile de récupération
~OIL PIPES - [mech] tuyaux *mpl* de l'huile de récupération
~OIL PRESSURE GAUGE - [mech] mesureur de pression de l'huile dans le filtre
~OIL PUMP - [mech] pompe *f* de l'huile de récupération
~PIPES - [mech] (pipes to carry oil back from engine to oil tank) tuyauteries *fpl* de récupération
~PUMP - [mech] (a pump used to withdraw oil from the engine and return it to the oil tank) pompe *f* de vidange, pompe *f* de récupération
SCAVENGER - [ind chem] (chemical additive to remove undesirable radionuclides) coprécipitant, entraîneur *m*
~CELLS - [biol] phagocytes *mpl* migrateurs
~PUMP - s. scavenge oil pump
SCAVENGING - [mech] évacuation *f*, refoulement *m*
[nucl] coprécipitation *f*, entraînement *m*
~AGENT - [ind chem] épurateur *m*
~AIR PORT - [mech] (of a Diesel engine) ajutage *m* de l'air de balayage
~MACHINE - [mech] machine *f* à balayer (les roues)
~PERIOD - [mech] (in two-stroke engines) phase *f* de balayage
~PUMP - s. scavenge oil pump
~STROKE - [mech] (exhaust stroke) course *f* d'échappement
~TUBE - [mech] tube *m* d'évacuation, tube *m* de vindage
SCENARIO - [cin] (the written plot of a dramatic play) scénario *m*, canevas *m*
SCENE - [gen] scène *f*
~PAINTER - [cin & theatre] chef-décorateur *m*
~SLATING - [cin] marque *f* de synchronisme
~SHIFTING - [cin & theatre] changement *m* de décors
SCENARY - [gen etc] mise *f* en scène, décors *mpl*
SCENOGRAPHY - [art] scénographie *f*

SCENT, to - [gen] odorer, parfumer
SCENT - [gen & chem] parfum *m*, odeur *f*
~GLAND - [bot] glande *f* à sécrétion odoriférante
~SPRAY - [radar] (colloq. term for the H_2S system) radar *m* type X H_2S
SCENTLESS - [gen] inodore
SCHAFFER'S SALT - [chem] (the sodium salt of 2-naphthol 6-sulphonic acid, with application as a synthesis intermediate) sel *m* de Schaffer
SCHAPPE - [text] (silk thread, from which a proportion of the natural gum, sericin, has been removed) schappe *f*, fleuret *m*
~SILK TWIST - [text] assemblé *m* de schappe
~SILK YARN - [text] fil *m* de fleuret
~SILK YARN SPINNING - [text] filature *f* des déchets de soie
~SPINNING - s. schappe silk yarn spinning
SCHEDULE, to - [gen] dresser un plan, dresser un programme
SCHEDULE - [gen] liste *f* officielle, inventaire *m*, nomenclature *f*, bordereau *m*
[gen] (in industry) plan *m* d'exécution de travail
[gen] (for applications) questionnaire *m*
[transp] indicateur *m*
[comm] plan *m*
[railw] (of a tran) horaire *m*
[comput] programme *m*
[autom] plan *m* d'échelonnement
~CONTRACT - [comm] contrat *m* de travail de régime
~OF DILAPIDATION - [constr] (list of necessary repairs after a period of tenancy) liste *f* des réparations
~OF PERIODIC TESTS - [gen] programme *m* de mesures périodiques
~SPEED - [railw] (the average speed of a train between two terminal stations) vitesse *f* commerciale
SCHEDULED - [gen] indiqué, au programme
[gen] (included in a list) inscrit *m* sur l'inventaire
~ITEM - [gen] article *m* inscrit sur une liste
SCHEDULER - [contr] fonction *f* de programme de commande, programmateur *m*
SCHEDULING - [autom] programmation *f*, ordonnancement *m*
SCHEELITE - [min] (natural calcium tungstate, crystallizing in the tetragonal system; one of the most important sources of tungsten) scheelite *f*
SCHEMA - [gen] (a summary, a synopsis) schéma *m*, diagramme *m*
[autom] (a diagrammatic representation) plan *m*, projet *m*
SCHEMATIC DIAGRAM - [el] (diagram of the general schema of an electrical circuit) diagramme *m* schématique
SCHEME, to - [gen] projeter
SCHEME - [gen] (a plan or design) arrangement *m*, plan *m*, projet *m*
(a diagrammatic representation) diagramme *m*
~ARCH - [constr] arc *m* surbaissé, voûte *f* surbaissée
~DIAGRAM - [gen] diagramme *m* schématique
SCHERING BRIDGE - [instr] (bridge designed to measure losses in cables and insulators at high working voltages) pont *m* de Schering
SCHIFF BASES - [chem] (compounds obtained by a condensation reaction between acryl aldehydes and amines. Schiff bases are important in the synthesis of a number of organic compounds, including dyes) bases *f*pl de Schiff
SCHIROIDISM - [med] schizothymie *f*
SCHIST - [geol] (metamorphic rocks having a tendency to split) schiste *m*
SCHISTOGLOSSIA - [med] langue *f* bifide
SCHISTOSE - [geol] (of schist) schisteux
SCHISTOSITY - [geol] (tendency in some rocks to split easily) schistosité *f*
SCHISTOSOME - [med] schistosomiase *f* cutanée
SCHISTOUS - s. schistose
SCHIZOGENESIS - [genet] (reproduction by fission) scissiparité *f*
SCHIZOGONY - [genet] (vegetative reproduction by fission) scissiparité *f*
SCHIZOID - [med] (showing qualities of a schizophrenic personality) schizoïde
SCHIZOIDISM - [med] schizoïdisme *m*
SCHIZOPHASIA - [med] schizophasie *f*
SCHIZOPHRENIA - [med] (dementia praecox) schizophrénie *f*
SCHIZOPHRENIC - [med] schizophrène
SCHIZOTRICHIA - [med] schizotrichie *f*
SCHLIEREN - [phys] (regions of varying refraction due to varying pressure in a fluid) schlieren *f*pl
~PHOTOGRAPH - [aero] (photographic interferometer observation of shock waves in a wind tunnel) photographie *f* Schlieren
SCHMIDT NUMBER - [phys] (the ratio of the diffusivity of mementum at the diffusivity of matter through a fluid) nombre *m* de Schmidt
SCHNAPPS - [gen] genièvre *m*, schnaps *m*
SCHOEPITE - [min] (mineral containing uranium sesquioxide) schoepite *f*
SCHOOL - [schol] école *f*, académie *f*
[mus] livre *m* d'instruction
[zool] (of fish) banc *m* voyageur
~SHIP - [naut] vaisseau-école *m*
SCHOOLING - [gen] instruction *f*, éducation *f*
SCHOONER - [naut] schooner *m*, goelette *f*
SCHORL - [min] (aggregates of blanck tourmaline and quartz) schorl *m*, tourmaline *f*
~ROCK - [min] tourmaline *f* quartzeuse
SCHORLACEOUS - [min] schorlacé
SCHORLIFEROUS - [min] schorlifère
SCHOTTKY EFFECT - [electron] (variation in electron current in a thermionic valve, caused by the effect of changes in anode voltage on the work function of the cathode material) effet *m* Schottky
~LINE - [electron] caractéristique *f* de Schottky
SCHRAGE MOTOR - [el] (shunt-characteristic polyphase commutator motor with double set of brushes) moteur *m* polyphasé à collecteur à caractéristique shunt à double jeu de balais
SCHREINER FINISH - [text] (lustrous finish on cotton satins and sateens by rollers engraved with fine lines which are imprinted on the fabric by heat and pressure) finissage *m* similisé, finissage *m* Schreiner
SCHREINERIZE, to - [text] donner un effet Schreiner
SCHROEKINGERITE - [min] (an ore of uranium) schroekingérite *f*
SCHWEINFURT GREEN - [dye] (pigment consisting of cupric subacetate) vert *m* de Schweinfurt
SCIATIC - [anat] (situated in the ischial region) sciatique
~NERVE - [anat] nerf *m* sciatique

SCIATICA - [med] (inflammation of the fibrous elements of the sciatic nerve) sciatique *f*
SCIENCE - [gen] science *f*
SCIENTIFIC - [gen] scientifique
SCIENTIST - [gen] homme *m* de science, savant
SCIEROPIA - [med] vision *f* ombrée
SCINDE COTTON - [text] coton *m* de Scinde
SCINTILLA - [phys] (a spark) étincelle *f*
SCINTILLATE, to - [phys] (to emit a luminous burst of very short duration) scintiller, étinceler
[opt] scintiller
SCINTILLATION - [phys] (a luminous burst of very short duration produced by the impact of a high-energy particle) scintillation *f*, scintillement *m*
~COUNTER - [instr] (instrument designed to count oscillations) compteur *m* à scintillation
~-COUNTER TIME DISCRIMINATION - [electron] intervalle *m* minimal de temps
~CRYSTALS - [instr] (used in scintillation counters) cristaux *mpl* scintillateurs
~DECAY TIME - [phys] temps *m* de descente de scintillation
~DURATION - [phys] durée *f* de scintillation
~OF WILD SILK - [text] scintillement *m* de la soie sauvage
~PROBE - [min] (a probe used in prospecting uranium) sonde *f* à scintillation
~RISE TIME - [electron] temps *m* de montée de scintillation
~SPECTROMETER - [instr] (scintillation counter for the study of energy distribution) spectromètre *m* à scintillation
SCINTILLATOR - [chem] (a kind of phosphor which emits a brief flash of light when a fast charged particle passes through it) scintillateur *m*
~CONVERSION EFFICIENCY - [electron] rendement *m* de conversion d'un scintillateur
~EXPOSURE RATEMETER - [instr] débitmètre *m* d'exposition à scintillateur
~FAST NEUTRON FLUXMETER - [instr] fluxmètre *m* de neutrons rapides à scintillateur
~PHOTON DISTRIBUTION - [electron] courbe *f* répartition des photons d'un scintillateur
SCINTILLOMETER - [instr] scintillomètre *m*
SCINTILLOSCOPE - [instr] scintilloscope *m*
SCINTISCANNER - [radiat] scintigraphe *m*
SCINTISCANNING - [radiat] scintigraphie *f*
SCINTLING - [constr] (of bricks, placing half-dry bricks a little distance a part in order to admit air between them) aérage *m* pour la dessiccation des briques
SCIOGRAPH - [draw] (drawing showing a sectional view of a structure) vue *f* en coupe
SCION - [gen] descendant *m*
[bot] (portion of a plant inserted into a rooted stock in grafting) greffon *m*
SCIROCCO - s. sirocco
SCIRRHOPHTHALMIA - [med] squirrhe *m* du globe oculaire
SCISSILE - [bot] (which can be split) scissile, fissile
SCISSION - [gen] (division cutting with a sharp implement) cisaillement *m*, coupage *m* avec un instrument tranchant
~REACTION - [rubber ind] réaction *f* de séparation
~SHOOT - [bot] bourgeon *m* greffé
SCISSIPARITY - [med] fissiparité *f*
SCISSOR, to - [gen] couper aves des ciseaux, cisailler
SCISSORS - [impl] ciseaux *mpl*
~CROSSING - [railw] (a double cross-over) traversée-bretelle *f*
~LINKAGE - [photo] liaison *f* par ciseaux
~TRUSS - [carp] ferme *f* à double contre-fiche
SCLERA - [anat] (the fibrous outer coat of the vertebrate eye) cornée *f* opaque
SCLERADENITIS - [med] adénite *f* scléreuse
SCLERATITIS - [med] sclérite *f*
SCLERECTASIA - [med] sclérectasie *f*
SCLEREMA - [med] sclérème *m*
SCLERENCEPHALY - [med] sclérose *f* cérébrale
SCLEROCHOROIDITIS - [med] sclérite *f* postérieure
SCLEROCONJUNCTIVITIS - [med] scléroconjonctivite *f*
SCLEROCORNEA - [anat] sclérotique *f* et cornée
SCLERODESMIA - [med] sclérodesmie *f*
SCLEROIRITIS - [med] scléro-iritis *f*
SCLEROMA - [med] sclérome *m*
SCLEROMETER - [instr] (an instrument for measuring hardness of materials) scléromètre
SCLEROMETRIC - [phys] sclérométrique
~HARDNESS - [phys] dureté *f* sclérométrique
SCLERONYCHIA - [med] scléronychie *f*
SCLEROPLASTY - [med] scléroplastie *f*
SCLEROSARCOMA - [med] sclérosarcome *m*
SCLEROSCOPE - [instr] (apparatus for determining the hardness of a material by measuring the rebound of a standard ball dropped on it from a predetermined height) scléroscope *m*
SCLEROSIS - [med] sclérose *f*
SCLEROTIC - [anat] (pertaining to the sclera) sclérotique
SCLEROTOME - [med] sclérotome
SCLEROTOMY - [med] sclérotomie *f*
SCOLECITE - [min] (hydrated silicate of calcium and aluminium) scolécite *f*
SCOLIOSIS - [med] scoliose *f*
SCONCE - [gen] (a type of wall bracket) applique *f*, bougeoir *m*
SCONCHEON - s. scontion
SCONE - [food] pain *m* au lait (cuit en galette)
~BRICK - [constr] (used in the construction of retorts) briquette *f*, galandage *m*
SCONTION - [gen mech etc] évider
SCOOP - [impl] (open shovel-like tool for picking up crushed or powdered solids) cuiller*f*; cuillère *f*
[impl] (for liquids) pelle *f* à main
[naut] épuisette *f*, sasse *f*
[mech] (of a conveyor) cuiller *f*, godet *m*
[mech] (concave curved element designed to collect air by the forward motion of an aircraft) buse *f* d'admission d'air
[med] (used in surgery) curette *f*
~CHAIN - [mech] chaîne *f* des godets
~CHARGING MACHINE - [mech] chargeuse *f* à godets
~DREDGER - [mech] drague *f* à godets
~KNIFE - [impl] couteau *m* en forme d'aube
~-NET - [fishing] drague *f*
~WAGON - [railw] wagonnet *m* à bec
~WITH AUGER - [mining] cuillère *f* ouverte à mouche de tarière
~-WHEEL - [hydr] tympan *m*
SCOOPS - [cin] (incandescent flood-light used in mo-

tion picture studios) lampes *f*pl à réflecteur ouvert
SCOOTER - [mech] scooter *m*
[naut] bateau *m* à voile et à patins
SCOPARIUM - [chem] (broom. The dried tops of Sarothammus scoparius with uses in medicine) scoparium *m*
SCOPE - [gen] portée *f*, étendue *f*
[gen] (a length) rayon *m*
[opt] (range as of view) champ
[radar] (radarscope) indicateur *m*
SCOPOLAMIN - s. scopolamine
SCOPOLAMINE - [hyscine] (a liquid alkaloid used in medicine) scopolamine *f*
SCOPOLINE - [chem] (oscine. An alkaloid derived from scopolamine) scopoline *f*
SCOPOMETER - [instr] (instrument designed to measure range etc) scopomètre *m*
SCOPOMETRY - [phys] (a system of turbidimetry based on the use of a scopometer) scopométrie *f*
SCORBUTIC - [med] (affected by scorbutus, or scurvy) scorbutique
SCORBUTUS - [med] scorbut *m*
SCORCH, to - [gen & rubber ind] (to burn at the surface only) brûler superficiellement, griller, roussir
SCORCH - [gen & rubber ind] (a surface burn) grillage *m*, brûlure *f* superficielle
~RETARDER - [rubber ind] antigrilleur *m*
~TEMPERATURE - [rubber ind] température *f* de grillage
SCORCHED IN DYEING - [text] (of yarn) (fil) affaibli par la teinture
~RUBBER - [rubber ind] (rubber which has lost essential properties by excessive heating) caoutchouc *m* brûlé
SCORCHING - [rubber ind] grillage *m*
SCORCHY ACCELERATOR - [rubber ind] accélérateur *m* précoce
SCORE, to - [carp mech] (to scratch or incise a line on a surface) érafler, inciser, encocher, rayer
[gen] (to scratch) érafler
[geol] strier
[el acoust] (to record a sound on a silent portion of a film) sonoriser
SCORE - [gen] trait *m*, éraflure *f*
[mech] (of a pulley) gorge *f* (de poulie)
[leather ind] incision *f*
[meas] vingtaine *f*, vingt *m*
[mech] entaille *f*, encoche *f*
[naut] (a groove in a block for the strap) engoujure *f*
[mus] (individual part in a composition) partition *f*
[geol] (on a rock) strie *f*
~CUT, to - [ind proc] (to divide material by making a deep scratch with a scriber and bending it along the line of the scratch) entailler, couturer
SCORED - [gen] éraflé, couturé, rayé
[firearms] affouillé
~CYLINDER - [mech] cylindre *m* rayé
~DRUM - [auto] tambour *m* rayé
~PULLEY - [mech] poulie *f* à gorge
SCORIA - [geol] (meas of volcanic rock simulating a clinker) scorie *f* volcanique
[metall] (slag) crasse *f* de fonte
~CONE - [geol] cône *m* de scories
SCORIFICATION - [metall] (a method in the dry assay for silver and gold, in which the powdered ore is heated strongly with litharge and fluxes in a shallow fire-clay dish. The noble metals dissolve in the lead, which is separated from the luxed gangue after cooling) scorification *f*
SCORIFIER - [metall] (a shallow dish of fireclay used in Scorification) scorificateur *m*
SCORIFY, to - [metall] scorifier
SCORING - [mech] (scratches etc) éraflement *m*, rayage *m*, entaillage *m*
[metall] (surface spot brittleness) grippage *m*, rayage *m*
[geol] abrasion *f*
[acoust] (the recording of found on the silent portion of a film) sonorisation *f*
[metall] formation *f* de rainures
SCORODITE - [min] (a minor ore of arsenic) scorodite *f*
SCORPION - [zool] scorption *m*
SCOTCH, to - [mech] caler
SCOTCH BLOCK - [railw] (a stop block) taquet *m* d'arrêt
[gen] cale *f*
~BOND - [constr] appareil *m* d'une boutisse et de deux panneresses
~CLEANER - [metall] (in foundry work) crochet *m* de mouleur
~DRESSING MACHINE - [text] pareuse *f* écossaise
~GAUZE - [text] marli *m*
~HEARTH - [metall] (furnace for smelting high-grade lead ores without previous roasting) four *m* à minerai de plomb
~REEL - [text] dévidoir *m* écossais
~STAFF - [surv] (a type of self-reading staff) mire *f*
~TOPAZ - [min] (a yellow transparent quartz used for ornamental purposes) topaze *f* d'Ecosse
~TROWEL - [tool] truelle *f* à lisser
~WARPING MILL - [text] ourdissoir *m* à tambour horizontal
SCOTIA - [arch] scotie *f*, nacelle *f*
SCOTOMA - [med] (blind, or partially blind area in the visual field) scotome *m*
SCOTOMETER - [instr] (instrument designed to measure a scotoma) scotomètre *m*
SCOTOPHOR - [telev] substance *f* noircissante
SCOTOPIC VISION - [telev] vision *f* scotopique
SCOTT-CONNECTION - [el] (of transformers, a method of interconnecting two transformers or windings to convert three-phase voltages to two-phase voltages and viceversa) système *m* Scott
SCOUR, to - [gen] (to clean or polish usually with the aid of an abrasive) décaper, dégraisser, nettoyer
[gen] (to rub for polishing) frotter, fourbir
[text] (of wool) dégraisser, dessuinter
[geol] (to erode) êroder
[metall] (a surface) dérocher, décaper
[hydr] dégrader (l'eau); donner une chasse à (un égout)
SCOUR - [geol] érosion *f*
SCOORED - [gen] nettoyé, décapé
[text] (of wool) dégraissé
~SILK - [text] soie *f* décreusée
~WOOL - [text] laine *f* dégraissée
~YARN - [text] fil *m* dégraissé
SCOURER - [agric] (flour-milling machine in which the wheat is subjected to the action of revolving beaters in a ventilated housinq for cleaning purpo-

ses) égreneuse *f* avec ventilateur
SCOURING - [text] (of wool) dégraissage *m*, dessuintage *m*
[text] (of silk) décrusage *m*
[geol] érosion *f*
[gen] nettoyage *m*, récurage *m*
[hydr] curage *m*, nettoyage *m* à grande eau
[hydr] (of rivers etc) affouillement *m* (d'un fleuve)
[leather ind] balayage *m* (du cuir)
~ACTION - [soil] effet *m* d'attrition
~AGENT - [text] agent *m* débouillissage
~BARREL - [mech] (or tumbling barrel; revolving drum used to polish small components) tonneau *m* à dérocher
~BATH - [text] bain *m* de dessuintage
~BOWL - [text] bac *m* de lavage
~CINDER - [metall] laitier attaquant le revêtement d'un four
~CLOTH - [text] tissu *m* à nettoyer
[text] serpillière *f*
~LIQUOR - [chem] lessive *f*, liquide *m* laveur
~LOSS - [text] perte *f* en poids au lavage
~MEDIUM - [text] agent *m* de décreusage
~VESSEL - [text] cuve *f* à décreuser
SCOVEN - [impl] (a blacksmith's shovel) palette *f* de forge
SCOW - [naut] chaland *m*
~END - [auto] (the raised portion of floor in the rear of a body to retain the load instead of using an endgate) benne *f* à cuillère
SCRAG - [zool] (of mutton) collet *m* de mouton
[agric] bête *f* efflanquée
[gen] (of a rock) éperon *m* (de roche)
~END - [food] (of muttons) bout *m* saigneux
SCRAM, to - [gen] (colloq. for an away) décamper
SCRAM - [nucl] (sudden shutting down of a nuclear reactor) arrêt *m* d'urgence
~BUTTON - [nucl] commutateur *m* de sécurité
~ROD - [nucl] barre *f* de sécurité
SCRAMBLE, to - [telecomm] brouiller
SCRAMBLED SPEECH - [telecomm] (speech made unintelligible for reasons of secrecy) parole *f* démodulée
~SPEECH SYSTEM - [telecomm] système *m* à parole démodulée
SCRAMBLER - [radio] (device used to achieve secrecy in radiotelephone conversations) dispositif *m* à brouiller les conversations radiotéléphoniques
SCRAP, to - [gen] mettre au rebut
[gen machine etc] se débarasser
[mech] réformer (le matériel)
[naut] (a ship) mettre hors de service
SCRAP - [gen] fragment *m*, morceau *m*
[gen] (metal scraps) mitraille *f*, débris *m*, déchets *mpl*
[metall] bocage *m*, débris *mpl* de fonte
[rubber ind] scrap *m*, déchets *mpl* de caoutchouc brut
~BALER - [mech] machine *f* à paqueter les ferrailles
~BALING - [metall] paquetage *m* des ferrailles, paquetage *m* de chutes
~BALING PRESS - [metall] presse *f* à paqueter les ferrailles
~CHUTE - [metall] (of a die set) couloir *m* à chutes
~CUTTER - [impl] (used in metallurgical work) cisaille *f* à scraps
~FLATTENING ROLL - [metall] laminoir *m* à scraps
SCRAPE FLUTTER - [telev] effet *m* de grattement
~GRANULATOR - [mech mach] (machine for reducing scrap plastics to a granulated form for re-use) réducteur *m* de déchets
~IRON - [metall] ferraille *f*, vieux fer *m*, mitraille *f*
~LOSSES - [metall] perte *f* due aux déchets
~MATERIAL - [gen] déchets *mpl*, ferrailles *fpl*, matériel *m* de rebut
~METAL - s. scrap iron
~PAPER - [gen] rognure *f* de papier
~PROCESS - [metall] (steel making proces) système *m* à chutes
~RAIL - [railw] rail *m* de rebut
~VIEW - [draw] champ *m* visuel partiel
~WASHER - [rubber ind] laveuse *f* de scraps
~WASTE - [gen & nucl] résidus *mpl*
SCRAPE, to - [gen] gratter, érafler
[mech] racler, gratter, curer
[constr] (to remove plaster) ravaler
[metall] décaper
SCRAPE-FINISHED - [mech] fini au racloir
~OFF, to - [gen] enlever au racloir
~TO BARE, to - [mech] racler complètement
SCRAPED - [leather ind] (of hides) écorché
SCRAPER - [mech] (tool for scraping any bearing and other surfaces) grattoir *m*, rognoir *m*, curette *f*
[mech] (a type of tractor) grattoir *m*
[mech] (int comb. eng; of the ring) grattoir *m*
[mining] curette *f* pour trou de mine
[leather ind & carp] allumelle *f*
[bookbind] paroir *m*
[impl] (for cleaning mains) piston *m* racleur, raclette *f*
[rubber ind] couteau *m* racleur
[oil ind] furet *m*, passe-diable *m*
~BLADE - [impl] lame *f* de racleur
~CHAIN - [mech] (a hydraulic main agitator) chaîne *f* à raclettes
~CONVEYOR - [mech] (a drag-link conveyor) transporteur *m* à palettes, transporteur *m* à raclettes
~DISC - [mech] racloir *m*
~KNIFE - [mech] lame *f*, grattoir *m*
~LOADER - [mech] (earth moving machine) grattoir *m*
~MAT - [gen] décrottoir *m*
~PLANE - [tool] rabot-racloir *m*
~RING - [mech] (piston ring designed to remove surplus oil from the cylinder all) segment *m* racleur d'huile
~-RING GROOVE - [mech] gorge *f* du segment racleur d'huile
~ROLLER - [text] égaliseur *m*
~TRAP - [plumb] sas *m* pour piston-racleur, gare *f* de piston-racleur
SCRAPING - [gen] raclage *m*, décapage *m*, décrottage *m*
[mech] raclage *m*, curage *m*
[metall] grattage
[leather ind] dépilage *m*, drayage *m* (d'une peau)
~AWL - [tool] poinçon-grattoir *m*
~BELT - [mech] ruban *m* à racloirs
~BLOCK - [text] chevalet *m* à ébourrer
~BRUSH - [impl] brosse *f* métallique
~CUTTER - [tool] outil *m* à rainer
~KNIFE - [impl] couteau *m* racleur
~MACHINE - [mech] machine *f* à racler

SCRAPING OUT IRON - [constr] dégorgeoir *m*
~OUT OF HOLES - [mech] équarissage *m* des trous
~PLANE - [tool] rabot-racloir *m*
~TOOL - [mech] grattoir *m*
SCRAPINGS - [metall] grattures *f*pl
SCRAPPAGE - [gen] (scrapped material) matériel *m* de rebut
SCRAPPED - [gen] de rebut, hors service
SCRATCH, to - [gen] gratter, frotter
SCRATCH - [gen] rayure *f*, frottis
[geol] striation *f*
[cin] titre *m* provisoire
~AWL - [impl] aiguille *f* à tracer, traceret *m*
~BOARD - [print] papier *m* procédé
~-BRUSH - [impl] (a wire brush designed to clean iron castings) brosse *f* métallique, gratte-bosse *f*
~-BRUSH FINISH - [metall] finissage *m* à brosse métallique
~COAT - [constr] enduit *m* brut
~FILTER - [radio] (a low-pass filter, for the elimination of noise due to needle scratching when the radio receiver is used for the reproduction of gramophone records) filtre *m* pour l'élimination des bruits de fond
~GAUGE - [tool] trusquin *m* à main
~HARDNESS - [phys] (sclerometric hardness) dureté *f* sclérométrique
~TEST - [metall] essai *m* à rayure
~TOOL - [constr] fer *m* à orner
~WORK - [constr] (a type of plaster work) graffite *m*
SCRATCHBRUSHING - [mech] (method of preparing a surface by the application of a wire brush) gratte-bossage *m*
SCRATCHER - [impl] grattoir *m*, gratteau *m*
[oil ind] gratteur *m* de tubage
SCRATCHING - [gen] rayage *m*, frottement *m*
[photo] (of films) (damage to the film surface) raie *f*, éraflure *f*
[el acoust] (a loudspeaker defect) bruit *m* de surface, bruit *m* de fond
~NOISE - [acoust & el acoust] bruit *m* de fond
SCREAMING - [acoust] bruits *m*pl stridents, retentissant *m*
SCREE - [geol] éboulis *m*
SCREECH, to - [acoust] pousser des cris perçants
SCREECH - [acoust] cri *m* perçant
[aero] (strong high-frequency vibration occurring in ram-jet engines, followed by mechanical failure) instabilité *f* de combustion à bruit aigu
SCREED - [constr] (a strip of wood or mortar laid on wall at interval to gauge the thickness of the plastering) guide *m* pour plâtrage
SCREEN, to - [gen] munir d'écran
[mech etc] (to cover or shelter with a screen) blinder
[gen] (to sift) tamiser, cribler
[cin] (to project on a screen) mettre à l'écran
[min] passer au crible, passer à la claie
SCREEN - [gen] écran *m*, rideau *m*
[arch] rideau *m*
[cin] écran *m*
[photo] écran *m*
[impl] tamis *m*, crible *m*, sas *m*
[telev] écran *m*
[el] (an earthed enclosure of metal sheet or mesh completely surrounding the generator and associated conductors or electrodes and designed to prevent radiation and to protect personel) écran *m*
SCREEN - [print] (a meshwork of lines at right angles on glass used to translate the subject of a halftone illustration into dots) trame *f*, réseau *m*
[aero] (an imaginary obstacle of a specified height used for the taking-off and landing performance) obstacle *m*
[telev] blindage *m*
[electron] (the chemically coated inner surface of the large end of a cathode-ray tube which becomes luminous when struck by an electron beam) écran *m* luminescent
[electron] (a device for supporting a number of electronic tubes to obtain the stabilization of the emission) rampe *f* d'activation
[el] (material so disposed in relation to a field as to reduce its penetration into a given region) blindage *m*, matériel *m* de blindage
~ANALYSIS - [ind proc] analyse *f* granulométrique
~ANGLE - [photo] angle *m* de trame
~BAR - [metall] (rolled sections) fer *m* à barreaux de grille
~BOARD - [gen] écran *m*
~BRIGHTNESS - [cin & telev] luminance *f* de l'écran
~BURNING - [electron] (the fluorescent power is burned or poisoned if the electron beam is left on a spot of the screen for a certain period of time) brûlure *f* d'écran
~CLASSIFIER - [mech] (machine for classifying material by means of metal network) crible-classeur *m*
~CLOTH - [gen] toile *f* pour tamis
~CREDIT - [cin] générique *m*
~DECK - [gen] surface *f* criblante
~DISTANCE - [photo & print] (the distance from the sensitive surface) distance *f* de trame
~EFFICIENCY - [electron] rendement *m* d'écran
~ETCHING - [text] rougement *m* à réseau
~EXAMINATION - [radiat] fluoroscopie *f*
~FACTOR - [electron] (of a grid; the ratio between the actual area of the grid structure and the total area of the surface containing the grid) facteur *m* d'écran, rapport *m* d'ombre
~FRAME - [text] cadre *m* d'impression
[photo] cadre *m* de l'écran
~GRID - [electron] (a grid placed between the control grid and an anode, usually maintained at a fixed positive potential, for the purpose of reducing the electrostatic influence of the anode in the space between the screen grid and the cathode) grille-écran *f*
~-GRID BIAS - [electron] (the negative potential applied to the screen-grid) polarisation *f* de grille-écran
~-GRID CURRENT - [electron] (the current which is caused by the electrons deposited on the screen-grid) courant *m* de grille-écran
~-GRID MODULATION - [radio] (system of modulation in which the potential of the screen in a multi-electrode valve is varied in accordance with the impressed modulating currents) modulation *f* par la grille-écran
~-GRID VOLTAGE - [electron] (the voltage between the screen-grid and the cathode in an electronic tube) tension *f* de grille-écran
~HEAD - [mech] tête *f* filtrante
~HEAD EXTRUDER - [rubber & plast ind] boudineuse-filtreuse *f*, boundineuse *f* à tamiser
~HOLDER - [print] châssis *m* porte-trame

SCREEN HOLDER - [opt] porte-écran m
~PACK - [ind chem] (wire gauze used for straining liquids) tamis m
~PATTERN - [telev] image f de l'écran
~PHOTOGRAPHY - [photo] photographie f tramée
~PLANT - [mech] installation f de criblage
~PLATE - [photo] plaque f à réseau
~PLAY - [cin] (the manuscript containing all details of the shots to be taken) découpage m, scénario m
~-PLAY EDITOR - [cin] auteur m du découpage
~PRINTING - [text] impression f au cadre
~PROCESS - [print & photo] (process of colour photography in which colour analysis and synthesis are carried out additively by the use of a mosaic screen of minute primary colour filters) serigraphie f
~-PROTECTED - [gen] protégé, grillagé
~-PROTECTED MOTOR - [el] moteur m grillagé
~ROLLERS - [print] cylindres mpl pour sérigraphie
~ROOM - [paper man] (department for the screening of the wood pulp) local m d'épuration
~SATURATION - [telev] limitation f de la luminosité de l'écran
~SETTLING - [telev] précipitation f d'une substance luminescente
~TILT - [photo] inclinaison f de l'écran
~TOWER - [cin] porte-écran m
~-TYPE CENTRIFUGE - [mech] essoreuse f à panier perforé
~VENT - [metall] filtre m de soufflage
~WIPER - [auto & aero] (mechanism for wiping water etc from a wind screen) essuie-glace m
SCREENAGE - [radiat] (the filtration afforded by a container of radioactive material) autoblindage m
SCREENED - [gen] blindé, protégé
[el etc] grillagé, protégé
[print & photo] tramé
~AERIAL - [radio] (aerial screened to eliminate disturbances) antenne f compensée
~ANTENNA - s. screened aerial
~APPARATUS - [el] appareil m protégé contre les contacts accidentels
~BROKEN STONE - [constr] pierraille f criblée
~CABLE - [el] câble m blindé
~COKE - [min] (graded coke) coke m classé, coke m criblé
~CONDENSER - [auto] condensateur m blindé
~FOIL - [photo] feuille f tramée
~GRID - [el] grille-écran f
~HANDLE - [nucl] (used to manipulate isotope preparations) poignée f blindée
~HOUSING - [auto] boîtier m blindé
~IGNITION SYSTEM - [auto etc] (an ignition system surrounded by a metal screen to obviate interference with radio signals) système m d'allumage blindé
~LINER - [oil ind] tube m perdu filtrant
~MAGNETO - [mech] (for motors) magnéto f blindée
~PENTODE - [electron] (pentode valve with a finemeshed auxiliary grid for use at high frequencies) pentode f haute fréquence
~RESISTANCE - [el] résistance f blindée
~SPARK PLUG - [auto] bougie f d'allumage blindée
~WIRE - [el] câble m blindé
SCREENER - [impl] crible m, tamis m
SCREENHOLDER - s. screen holder
SCREENING - [gen] mise f à l'abri, protection f
[phys] blindage m
[el radio etc] blindage m
SCREENING - [statist] classification f
[radio] (of the aerial) compensation f de l'antenne
[min] criblage m, passage m à la claie, sassement m
[telev] manufacture f de l'écran
~BOX - [min] caisse f de criblage
~CAGE - [electron] (a perforated metal screen used in some electronic tubes in order to avoid undesired charges on the bulb wall) cage f de Faraday
~CAN - [radio] (a metal receptacle in radio receivers to earth electronic tubes for screening purposes) boîte f de blindage
[telev] chemise f de blindage
~CONSTANT - [phys] (a quantity occurring in the relationship between the frequency of a line in one x-ray series, and the atomic number of the element emitting the rays) constante f de blindage
~DRUM - [impl] tambour m de tamisage
~EFFECT - [el] (the property of a metal envelope to screen an envelopped magnetic field from its outside surroundings) effet m d'écran
~HEAD - [el] capot m de blindage
~MATERIALS - [gen] matières fpl de blindage
~NUMBER - s. screening constant
~OF NUCLEUS - [nucl] (the reduction of the electric field about a nucleus by the space charge of the surrounding electron) effet m d'écran du noyau
~PLATE - [text] tôle f de protection
~REACTOR - [el] bobine f d'arrêt, bobine f de choc
~TUBING - [el] (used for the electrical screening of conductors) tube m isolant
SCREENINGS - [gen] déchets mpl de criblage
[min] déchets mpl de criblage
SCREENPLAY - s. screen play
SCREEVE - [metall] cordon m d'étanchéité
SCREW, to - [mech] visser
[gen] fixer, serrer
[mech] (a bolt etc) fileter
[mech] (a pipe etc) tarauder
SCREW - [mech] vis f
[aero & naut] (propeller) hélice f
~ADJUSTING CALIPER - [mech] calibre m à vis de réglage
~AND BOLT HEADER - [mach tool] machine f à frapper les vis et boulons
~-AND-NUT STEERING-GEAR - [auto] direction f à vis et écrou
~-AND-SOCKET JOINT - [mech] assemblage m à enture à vis
~APERTURE - [naut] (the opening in the stern of a ship, in which the screw propeller revolves) puits m de l'hélice
~AUGER - [impl] tarière f torse, tarière f à tire-bouchon
~AXIS - [crystal] (type of symmetry element possessed by certain space groups, in which the lattice is unaltered after a rotation about the axis, and a simultaneous translation along it) axe m hélicoïdal
~BAR - [impl] taranche f
~BEAR - [tool] poinçonneuse f à vis simple
~BELT FASTENER - [mech] vis m pour agrafes de courroie
~BLADE - [naut] pale f d'hélice
~BLANK - [mech] vis f en blanc,
~BOLT - [mech] boulon m à vis, boulon m à écrou
~BOSS - [naut] moyeu m de l'hélice
~BOX - [mech] manchon m à vis
~BRAKE - [mech] frein m à vis

SCREW CAP - [mech] couvercle *m* à vis, gobelet *m* à vis
[el] (of an incandescent lamp) culot *m* à vis
~CASING-HEAD - [mech] tête *f* de tube à vis
~CHANNEL DEPTH - [mech] (the depth of the core of a feed screw below the outer edge of the flight) profondeur *f* de filet
~CHANNEL WIDTH - [mech] (the distance between adjacent flights of a feed screw) largeur *f* du filet
~CHASE - [print] châssis *m* à vis
~CHASING - [mech] (accurate finish of the screw thread by means of a chaser) peignage *m*
~CHUCK - [carp] mandrin *m* à queue de cochon
~CLAMP - [impl] (clamp in which the pressure is applied by a screw or screws) vis *f* de serrage, presse *f* à vis
~CLIP - [mech] tenaille *f* d'attelage à vis
~CONNECTION - [mech] connexion *f* à vis
~CONNECTOR - [el] domino *m*
~CONVEYOR - [mech] transporteur *m* à vis sans fin, vis *f* transporteuse
~CORE PIN - [mech] (a core pin fixed by being screwed into a hole) broche *f* à trou fileté
~COUPLING - [mech] manchon *m* à vis, union *f* à vis
[railw] attelage *m* à vis
~COUPLING BOX - [mech] manchon *m* à vis
~CURRENT METER - [instr] moulinet *m* hydrométrique
~CUTTING - [mech] filetage *m*, taraudage *m*
~CUTTING DIAL - [mach tool] indicateur *m* du filet
~-CUTTING LATHE - [mech tool] tour *m* à fileter
~CUTTING MACHINE - [mach tool] machine *f* à fileter, décolleteuse *f*
~-CUTTING REVERSE - [mech] changement *m* de marche pour fileter à droite et à gauche
~DIE - [tool] filière *f*
~DISLOCATION - [crystall] dislocation *f* hélicoïdale
~DOG - [impl] toc *m* à vis
[mach tool] poupée *f* à pompe
~DOWN, to - [mech] visser
~-DOWN - [metall] vis *f* de réglage des cylindres
~-DOWN BIBCOCK - [hydr] robinet *m* d'écoulement, robinet *m* de vindage
~-DOWN COCK - [mech] robinet *m* à vis de pression
~-DOWN STOP-VALVE - [mech] robinet *m* d'arrêt à soupape
~-DOWN VALVE - [mech] robinet *m* à soupape
~-DRIVER - [impl] tournevis *m*
~-DRIVER BIT - [impl] tournevis *m* pour vilebrequins, tournevis *m* au fût
~ELEVATOR - [mech] élévateur *m* à vis sans fin
~EXTRUDER - [mech] (a machine in which plastic material is extruded by the action of a screw) machine *f* à extrusion, extrudeuse *f*, machine à vis sans fin
~EYE - [carp] piton *m* à vis, laceret *m*
~FAN - [mining] vis *f* pneumatique
~FEED - [mech] pression *f* à vis
~-FEED CHARGER - [mech] chargeuse *f* à vis sans fin
~FEEDER - [mech] limace *f*
~FERRULE - [mech] manchon *m* taraude
~FLANGE COUPLING - [mech] boulonnage *m* des brides
~FLIGHT CONVEYOR - [mech] (type of materials conveyor in which a deep helical fin revolves in a trough, carrying the material forward by a screwing action) transporteur *m* à vis
SCREW FLOWMETER - [instr] (gas meter) compteur *m* à moulinet
~FOCUSSING EYEPIECE - [opt] oeilleton *m* à réglage à vis
~GAUGE - [tool] calibre *m* pour la vérification des vis
~GEAR - [mech] engrenage *m* hélicoïdal
~GEARING - s. screw gear
~GILL - [text] barrette *f* commandée par vis
~GILL DRAWING FRAME - [text] étirage *m* à barrettes commandées par vis
~GOVERNOR - [mech] régulateur *m* à vis
~GRIP - [transp] tenaille *f* d'attelage à vis
~HAMMER - [impl] clef *f* à marteau
~HEAD - [mech] tête *f* de vis
~-HEAD FILE - [impl] lime *f* à fendre, lime *f* dossière
~HOOK - [mech] crochet *m* à vis
~IMPELLED PUMP - [mech] pompe *f* à hélice
~-IN SCREEN - [photo] écran *m* se vissant à l'intérieur du barillet
~INJECTION - [plast ind] (injection moulding in which the material is first plasticised by a screw like that in an extruder) injection *f* avec préplastification
~JACK - [mech] (lever arm connected to a screw) cric *m* à vis, vérin *m*
~JOINT - [mech] assemblage *m* à vis
~-KEY - [impl] clef *f* à vis
~-LIKE CONVOLUTION - [text] torsion *f* en vis (du fil de soie sauvage)
~LOCKING DEVICE - [mech] pince *f* à vis
~MACHINE - [mach tool] (type of turrest lathe) machine *f* à tarauder
~MICROMETER - [tool] (micrometer gauge) micromètre *m*
~MIXER - [mech] (mixer comprising a vertical helical flight fed from the lower end) mélangeur *m* à vis
~NAIL - [carp] (a nail with some helical depressions on its surface, so as turn like a screw when it is driven in place by a hammer) vis *f* à bois
~OFF, to - [gen & mech] dévisser
~ON, to - [mech] visser
~-ON LENS - [photo] objectif *m* à monture filetée
~OUT, to - [mech] dévisser
PACKING-JACK - [mech] vérin *m* de calage
~PEG - [mech] broche *f* spirale
~PILE - [constr] (pile with a wide projecting screw at the foot, used in alluvial ground) pieu *m* à vis, pilotis *m* à vis
~-PINE - [bot] pandanus *m*
~PITCH - [aero & naut] pas *m* de l'hélice
[mech] pas *m* de vis
~-PITCH GAUGE - [tool] (tool for determining the pitch of a screw) jauge *f* de pas
~PLATE - [mech] (a hardened steel plate in which screwing dies of different sizes are formed) filière *f* simple, filière *f* à cage
[mech] contre-platine *f*
~PLUG - [mech] tampon *m* à vis
~-PLUG CARTRIDGE FUSE - [el] fusible *m* à vis
~PRESS - [mech] (a press actuated by a screw turned by hand or power) presse *f* à vis
~PROPELLER - [aero & naut] hélice *f*
~PROPELLER HUB - [mech] moyeu *m* de l'hélice

SCREW PULLEY - [mech] poulie *f* à vis, poulie *f* de renvoi avec tige filetée
~PUMP - [mech] pompe *f* à limance
~PUNCHING-GEAR - [mech] poinçonneuse *f* à main
~PUTTY - [mech] mastic *m* pour filet
~RAIL BENDER - [mech] presse *f* à vis pour cintrer les rails
~RING - [mech] anse *f* à vis
~SHACKLE - [railw] (a device for hooking railway cars) manille *f* d'attelage à vis
~SHAFT - [aero & naut] arbre *m* porte-hélice
~SHAFT-PIPE - [mech] tuyau *m* pour arbre d'hélice
~SHOE - [constr] sabot *m* à vis
~SLIP - [naut] recul *m* de l'hélice
~-SLOT CUTTER - [mech tool] fraise *f* à entailler la tête de vis
~SOCKET - [el] douille *f* à vis
~SPANNER - [mech] clef *f* anglaise, clef *f* à vis
~SPEED - [mech] (the rotary speed of the screw in an extruder) vitesse *f* de la vis
~-SPIKE - [railw] (a rail-spike) crampon *m*, tire-fond *m*
~SPINDLE - [mech] arbre *m* fileté
~STAIR - [constr] escalier *m* en vis, escalier *m* tournant
~STAKE - [constr] pieu *m* à vis
~STAY - [mech] tirant *m* à vis
~STEEL - [metall] acier *m* à boulonnerie
~STOCK - [tool] filière *f* à cage, porte-filière *m*
~STOPPER - s. screw plug
~STUD - [mech] prisonnier *m*
~TAP - [tool] taraud *m*, quille *f*
~TERMINAL - [el] borne *f* à vis
~THREAD - [mech] (the helical ridge formed on the cylindrical core) filet *m* de vis
~-THREAD DOWEL - [railw] tire-fond *m* à vis
~-THREAD MICROMETER CALIPER - [tool] calibre *m* micrométrique pour la vérification des vis
~THRUST - [aero & naut] poussée *f* de l'hélice
~TIGHT, to - [gen & mech] serrer
~TIGHTENER - [impl] tendeur *m* à vis
~TOOL - [tool] peigne *m* à fileter
~TUNNEL - [naut] tunnel *m* de l'hélice
~TYPE EXTRUSION MACHINE - [rubber & plast ind] boudineuse *f* à vis
~-TYPE SPREADER ROLL - [mech] cylindre *m* largeur à vis filetée
~UP, to - [mech] visser, serrer
~-VICE - [mech] étau *m* à vis
~WASHER - [min] lavoir *m* à vis
~WHEEL - [mech] roue *f* hélicoïdale
~WINE PRESS - [agric] pressoir *m* à vis
~WITH EQUAL PITCH - [mech] vis *f* à pas constant
~WORM - [mech] vis *f*
~WRENCH - [tool] clef *f* à vis, clef *f* anglaise, clef *f* à mâchoires mobiles
SCREWED - [gen] vissé
~AUGER - [mech] (an open-head auger) sonde *f* tire-bouchon
~BOOTS - [shoe man] (boots in which the sole is attached to the upper by screws) bottes *fpl* à semelle vissée
~CONDUIT - [el] assemblage *m* par tube fileté
~CORE PIN - [mech] broche *f* à trou fileté
~COUPLER - [el] manchon *m* taraudé
~FITTINGS - [mech] tubulures *fpl* filetées
~HOOK - [mech] crochet *m* à vis
SCREWED-IN PIN - [mech] dent *f* vissée
~LAMPHOLDER - [el] douille *f* à vis, Edison *m*
~NIPPLE - [mech] raccord *m* fileté
~-ON CAM - [mech] excentrique *m* vissé
~PIN - [mech] broche *f* filetée
~PIPE-JOINT - [plumb] assemblage *m* de tuyauterie à vis
~SLEEVE - [mach] manchon *f* filetée
~SOCKET - [mech] douille *f* filetée
~SPINDLE - [mech] arbre *m* fileté
~WIRE - s. screwed spindle
SCREWDRIVER - s. screw-driver
SCREWER - [mech] turaudeuse *f*
SCREWING - [mech] vissage *m*, serrage *m*, filetage *m*, boulonnage
[mech] (screw-cutting) creusage *m* des filets
~DIE - [mech] (internally threaded steel block, sometimes in two halves, in which cutting edges are formed by longitudinal slots; used for cutting external threads) filière *f*, coussinet *m*
~DOWN - [metall] changement *m* de profil
~HEAD - [mech] porte-coussinet *m*, porte-lunette *m*, filière *f*
~MACHINE - [mech tool] (type of lathe adapted for the continuous production of screws by means of dies) machine *f* à tarauder
SCREWLESS - [gen] sans vis
~ADAPTER SLEEVE - [mech] manchon *m* de secours sans vis
SCREWPINE - [bot] pandanus *m*
SCREWS ROTATING IN OPPOSITE DIRECTIONS - [mech] vis *fpl* tournant en sens contraires
~ROTATING IN THE SAME DIRECTION - [mech] vis *fpl* de même pas, vis *fpl* tournant dans le même sens
SCREWSHIP - [naut] navire *m* à hélice
SCREWSTOCK - s. screw stock
SCRIBBLE, to - [text] scribler, drousser, écharper
SCRIBBLER - [text] (a carding machine) machine *f* à carder
SCRIBBLING - [text] (term used in the woollen-trade for carding) écharpage *m*, scriblage *m*
SCRIBE, to - [mech] (the operation of marking out and cutting thin metal sheets) trusquiner
SCRIBE - s. scriber
~-AWL - s. scriber
SCRIBER - [tool] (a sharp-pointed tool used for marking out and cutting thin sheet) tracelet *m*, pointe *f* de traçage, traçoir *m*
SCRIBING - [mech] traçage *m*
~AWL - s. scriber
~BLOCK - [tool] (a tool for gauging the height of a point on a piece of work, above a surface plate or a machine table) trusquin *m*, trusquin *m* à marbre
~COMPASS - [tool] rouanne *f*
~TOOL - [tool] traçoir
SCRIEVE BOARD - [shipbuild] (formation of portable parts of flat wooden boards on which are scrieved, or scribed the ship's transverse frame sections and lines indicating the shell seams, stringers etc) plate-forme *f* pour le repérage des opérations de traçage
SCRIM - [text] (a low-quality linen tow fabric used as reinforcement in book-binding etc) canevas *m* léger
SCRIPT - [cin] (the manuscript dealing with all the details of the shots to be taken etc) découpage *m*,

scénario *m*
SCRIPT - [print] (a type imitating handwriting) cursive *f*
SCROFULA - [med] (tuberculous condition of the lymphatic glands) scrofule *f*
SCROFULIDE - s. scrofuloderma
SCROFULODERMA - [med] scrofulide *m*
SCROLL - [gen] (of paper) rouleau *m*
[paper man] (a roll of parchment) rouleau *m* de parchemin
[arch] volute *f*, spirale *f*
[print] cartouche *f*, arabesque *f*
[mech] (of a water turbine, a blower etc) spirale *f*
[mech] (a curved cut on a sheet material) chantournage *m*
[mach tool] (of a chuck) mandrin *m* à spirale
[text] scroll *m*
[mus] (the curved head of a string instrument) crosse *f*, volute *f*
[telecomm] (conductor used in continuous recorders in conjunction with the writing bar) hélice *f* d'impression
[mech] vis *f* sans fin
~CAM - [text] excentrique *m* pour lisière
~CHUCK - [mach tool] (type of self-centering chuck) mandrin *m* à spirale
~DRUM - [text] tambour *m* de la chaîne du secteur
~GEAR - [mech] engrenage *m* à dents à spirale
~SAW - [impl] (a narrow-bladed saw, hand or power operated, for cutting curved forms from sheet) scie *f* à chantourner, sauteuse *f*
~THREAD - [mech] filet *m* à spirale
SCROLLER - [text] ouvreur *m* de lisières
SCROOP, to - [text] aviver
SCROOP OF SILK - [text] frou-frou *m* aigre (de la soie)
SCROOPING - [text] avivage *m*
~FEEL - [text] toucher *m* craquant
SCROTAL SAC - [anat] bourse *f*
SCROTUM - [anat] scrotum *m*
SCRUB, to - [gen] (to rub strongly) frotter, nettoyer
[ind chem] laver
SCRUB - [gen] friction *f*, nettoyage *m*
[bot] (a stunded shrub etc) arbusté *m* rabougri, brousse *f*
~CLEARING - [agric] débroussaillement *m*
~-CLEARING MACHINE - [agric] débroussailleur *m*
SCRUBBER - [gas ind] (the apparatus in which the gas is freed from tar, ammonia and sulphuretted hydrogen) laveur *m*
[oil ind] fermeture *f* étanche
~TOWER - [ind chem] (vertical cylindrical structure through which gases are passed upward against a stream of liquid in order to clean them) tour *f* de lavage
SCRUBBING - [ind chem] (the process of removing a component or components from a gaseous mixture by passing it upwards and counter to a stream of liquid capable of selective absorption of the desired component) épuration *f*, lavage *m*
[gas ind] (cleaning of gas, e.g. by passage through a tower) barbotage *m*
SCRUFF - [gen] nuque *f*
[gen] (a thin coating) pellicule *f*
[metall] (of a thinning bath) laitier *m* de bain d'étamage, scorie *f* de bain d'étamage
[hydr] (the surface of water) surface *f* (d'eau)

SCRUNITIZE, to - [gen] scruter, examiner
[gen] (to examine closely) sonder
SCUD, to - [naut] (to flee) fuir devant le temps
SCUD - [met] (fractostratus; cloud formation consisting of isolated masses of stratus) diablotins *mpl*, fractostratus *m*
SCUDDING - [leather ind] (the removal of the hair roots, pigment cells and lime salts from the grain side of a hide before tanning) passage *m* à la pierre, décrassage *m*
SCUFF, to - [leather ind] (to wear rough on the surface) érafler, racler
[mech] (for lack of lubrification) prégripper
[gen] effleurer
SCUFFING - [mech] prégrippage *m*
~AWAY - [mech] usure *f* (de la bande de roulement)
~RESISTANCE - [paper man] résistance *f* à l'éraflement
SCUFFLER - [mach tool] (for saws) aiguiseur *m* (pour scies)
[agric] extirpateur-scarificateur *m*
[agric] (of a horse) bineuse *f* à cheval
SCULL, to - [naut] (to propel by using a scull) ramer à couple
SCULL - [naut] (an oar worked over the stern of a boat) aviron *m* de couple
[naut] (small boat for sculling) aviron *m*
[metall] (of molten metal) fond *m* de poche
SCULLER - [naut] (boat propelled by a pair of scull) rameur *m* de couple
~HOLE - [naut] trou *m* de la rame
SCULLERY - [constr] arrière-cuisine *f*
SCULLS - [metall] fond *m* de poche
SCULPTURE - [gen] sculpture *f*
SCUM, to - [gen] (to remove floating impurities from the surface of a liquid) écumer
SCUM - [gen & ind chem] (a layer of impurities rising to, and floating on, the surface of a liquid) écume *f*, mousse *f*
[metall & constr] (surface formation of lime crystal appearing on new cement work) crasse *f*, scories *fpl*
[glass man] (layer of unmelted material on the molten glass surface) scorie *f*
[agric] (on wine) chapeau *m* (du vin)
~-COCK - [mech] robinet *m* d'extraction à la surface, robinet *m* de purge
SCUMBLE, to - [paint] frotter, blaireauter
[draw] (a line) estomper (une ligne)
SCUMBLING - [paint] (the blending of colours by rubbing them with a brush charged with dry or opaque colour) blaireautage *m*
SCUMP - [rubber ind] (the coagulated scum of rubber latex) "scump" *m*, coagulum *m* de latex écumé
SCUPPER, to - [naut] (to sink a ship in a surprise attack) couler *f* à fond
SCUPPER - [naut] (gutter bordering a ship's deck) dalot *m* de pont
[constr] (an outlet for water overflow) dalot *m*
SCURF, to - [gen] enlever les pellicules
SCURF - [metall] (in ovens) charbon *m* de cornue, graphite *f*
[gen] (on the skin) incrustation *f*
[gen] (in the hair) pellicules *fpl* du cuir chevelu
~FORMATION - [metall] graphitage *m*, engraphitage *m*
SCURFING - [mech] (the operation of removing

roughnesses by means of a scraper or emery cloth) grattage *m*, raclage *m*
SCURFING - [gas ind] (of ovens and retorts) dégraphitage *m*
~BAR - [impl] pince *f* à dégraphiter
SCURFY - [gen] pelliculeux
SCURVIED - [gen & med] scorbutique
SCURVY - [med] (scorbutus) scorbut *m*
SCUTCH, to - [text] (to separate the woody parts from the fibres) écanguer, teiller, écoucher
SCUTCH - [text] (device for beating flax, hemp etc) écang *m*, écouche *f*, teilleuse *f*
~BLADE - [text] brisoir *m*, écang *m*, dague *f*
~THE FLAX, to - [text] écoucher le lin
~THE HEMP, to - [text] écoucher le chanvre
SCUTED FLAX - [text] lin *m* écouché
~HEMP - [text] chanvre *m* écouché
SCUTCHEON - [mech] entrée *f* de serrure
SCUTCHER - [text] batteur *m*, étaleur *m*
~AND LAP MACHINE - [text] batteur *m* et étaleur
~FLY - [text] duvets *mpl* de batteur
~RAKE - [text] râtelier *m* oscillant
~ROLLER - [text] cylindre *m* briseur
~WASTE - [text] déchets *mpl* de batteur
~WITH CARDING BEATER - [text] batteur *m* avec volant cardeur
SCUTCHING - [text] écantage *m*, teillage *m*
~BLADE - [text] écang *m*
~BOARD - [text] planche *f* à écangeur, chevalet *m*
[text] poisset *m*
~CYLINDER - [text] tambour-teilleur *m*
~MACHINE - [text] briseuse *f*, teilleuse *f*
~MILL - [text] atelier *m* de teillage
~STAND - [text] poste *m* de l'écangueur
~STOCK - s. scutching board
~TOW - [text] étoupe *f* de teillage, étoupe *f* d'espadage
SCUTE - [anat] écaille *f*
SCUTELLATE - [bot] (rounded; nearly flat) scutelliforme
SCUTELLATION - [bot] (arrangement of scales) scutellation *f*
SCUTELLUM - [bot] (flattened portion of the embryo of a grass) scutelle *f*
SCUTIFORM - [bot] (shaped like a shield) scutiforme
~SCALE - [zool] écaille *f* scutiforme
SCUTTLE, to - [naut] (to sink a ship by making holes in the bottom or opening the sea valves) saborder
SCUTTLE - [naut] (a small hatchway with movable cover) écoutillon *m*, hublot *m*
[gen] seau *m* à charbon
[constr] tabatière *f*
[constr] (only USA) trappe *f* (de toit etc)
~VENTILATION CONTROL - [auto] commande *f* de l'aération d'auvent
SCYBALUM - [anat] scybale *m*
SCYTHE, to - [agric] faucher
SCYTOBLASTEMA - [anat] ébauche *f* de la peau de l'embryon
SCYTHE - [agric] faux *f*
~BLADE - [agric] lame *f* de faux
~-HAMMERING BENCH - [agric] machine *f* à battre les faux
~-STONE - [agric] pierre *f* à aiguiser les faux
S-DISTORTION - [telev] (horizontal lines which tend to twist up or down as the camera moves across) distorsion *f* en S

SEA - [gen] mer *f*, océan *m*
[naut] (the swell of the ocean) lame *f*, houle *f*
[naut] (a wave striking a vessel) coup *m* de mer, paquet *m* de mer
~ABEAM - [naut] mer *f* en belle
~ANCHOR - [naut] (a float to which a ship may be attached by a hawser to ride out a storm) ancre *f* flottante
~-ARM - [geogr] bras *m* de mer
~-BOARD - [geogr] littoral *m*
~BOAT - [naut] (said of a boat suitable for the high seas) bateau *m* marin
~BOTTOM - [gen] fond *m* de la mer
~-BOUND - [geogr] borné par la mer
~BREAM - [zool] (a fish) pagel *m*
~BREEZE - [met] (wind blowing from the sea to the shore during the day, and due to heating of air over the warmer land-mass) brise *f* de mer
~BUCKTHORN - [bot] argousier *m*
~-CHART - [naut] carte *f* marine
~CLAM - [zool] peigne *m*, palourde *f*
~CLUTTER - [radar] (or wave clutter, pictures on a radar screen caused by reflection on sea waves) signaux *mpl* parasites
~COAL - [min] charbon *m* de terre
~COCK - [plumb] robinet *m* de prise d'eau à la mer
[zool] coq *m* de mer
~COW - [zool] vache *f* marine
~-DACE - [zool] loup *m* de mer
~DAMAGE - [comm] fortune *f* de mer
~-DAMAGED BALE - [transp] balle *f* avariée par l'eau de mer
~-DAMAGED COTTON - [text] coton *m* avarié par l'eau de mer
~DISTURBANCE - [met] (local condition of the sea surface caused by wind) état *m* de la mer
~-FAN - [bot] gorgone *f* éventail
~FOAM - [naut] écume *f* de la mer
[min] (meerschaum) écume *f* de mer, magnésite *f*
~FOG - [met] (fog occurring at sea, commonly caused by warm air meeting the colder surface of the water) brouillard *m* marin
~-FRONT - [gen] esplanade *f*
~-GOING TUG - [naut] (a tug suitable to sail the high seas) remorqueur *m* de long cours
~-HARBOUR - [geogr] port *m* de mer
~HORIZON - [gen] (the line at which sea and sky appear to meet in clear weather) horizon *m*
~-ISLAND COTTON - [text] coton *m* à longue soie
~KALE - [bot] crambe *m*, chou *m* marin
~LEVEL - [surv geogr etc] (the standard taken for the measurement of heights) niveau *m* de la mer
~MARK - [naut] balise *f*, amer *m*
~MATWEED - [agric] carex *m* des sables
~MULE - [naut] (colloq. for tug-boat) remorqueur *m*
~OTTER - [zool] loutre *f* marine
~-PACKING - [transp] emballage *m* d'outre-mer
~-PLANE - s. seaplane
~POWER - [gen] puissance *f* maritime
~RETURNS - [radar] échos *mpl* de la mer
~ROOM - [naut] (the space available for manoeuvring) évitage *m*
~SALT - [min] sel *m* marin
~SAND - [min] sable *m* de mer, sable *m* marin
~SILK - [text] soie *f* marine, byssus *m*
~STORM - [met] coup *m* de mer
~URCHIN - [zool] hérisson *m* de mer, oursin *m*

SEA WALL - [constr] (embankment for preventing encroachments) digue *f*, endiguement *m*
~WATER - [gen] eau *f* de mer
~WATER INTAKE - [naut] (sometimes used for cooling an engine) prise *f* d'eau de mer
~-WAY - [naut] sillage *m* d'un navire
[naut] levée *f* de la mer
~WEED - [bot] algue *f*
~WOOL - [text] fil *m* d'alginate
~WRACK - [bot] fucus *m*
[naut] balayures *fpl* de la mer
SEABOARD - [geogr] littoral *m*, rivage *m*
SEACOAST - [geogr] (the seashore) côte *f* de la mer
SEADROME - [aero] aéroport *m* flottant
SEAFARING - [naut] de mer, marin
[naut] (the travelling over the oceans) voyages *mpl* par mer
SEAGIRT - [gen & geogr] (surrounded by waters) entouré par la mer
SEAGOING - [naut] (adapted for use on the seas) de haute mer, de long cours
SEAL, to - [gen] sceller
[mech] (to render tight against fluid leakage) assurer l'étanchéité
[el] plomber
[constr] (to fix with cement etc) appliquer une couche de scellement
[gen] (to seal a hole) boucher un trou
[metall] (to seal with lead) plomber
[mining] obturer (un puits)
SEAL - [gen] sceau *m*, cachet *m*
[mech] (element or material used to close any opening to prevent leakage) joint *m* d'étanchéité, rondelle *f* étanche
[hydr etc] obturateur *m*
[plumb] (the water contained in a trap and preventing the flow of air or gases from one side to the other) siphon *m* à étanchement
[hydr] joint *m* hydraulique, fermeture *f*
[comm] sceau *m*
[zool] phoque *m*
[leather ind] phoque *m* pour gainerie
[mining] barrage *m*
[electron] (in a waveguide) fenêtre *f* étanche (d'un guide d'ondes)
~-FIN DEFORMITY - [anat] main *f* phocomélique
~-IN COIL - [el] enroulement *m* de blocage
~-IN CONTACT - [el] contact *m* scellé
~-IN RELAY - [el] relais *m* à contacts scellés
~OIL - [ind chem] huile *f* de phoque
~-RING - [mech] (an annular element designed to prevent the passage of fluid leakage through a joint) rondelle *f* d'étanchéité
~THE SAMPLE, to - [comm] plomber l'échantillon, cacheter l'échantillon
~WELD - [el] soudure *f* d'étanchéité
~WIRE - [el] traversée *f* de courant
~WITH LEAD - [gen] plomber
SEALABLE EQUIPMENT - [el] équipement *m* scellable
SEALED - [gen & comm] cacheté, clos
[mech hydr etc] étanche
~AGAINST DUST - [mech] étanche à la poussière
~BEAM LAMP - [auto] (bulb, reflector and lens sealed together) phare *m* sealed beam
~CHAMBER TERMINAL - [el] tête *f* de câble, boîte *f* de raccordement
~COMBUSTION CHAMBER - [gas ind] (gas-tight circuit) circuit *m* étanche
SEALED COMBUSTION CHAMBER SPACE HEATER - [th eng] radiateur *m* à circuit étanche, radiateur *m* blindé
~COVER - [constr] (air-tight cast-iron cover used to cover a manhole) tampon *m* étanche
~COWLING - [aero] (type of cowling intended to prevent air entering a nacelle) capotage *m* étanche
~END - [el] extrémité *f* isolée
~END OF A CABLE - [el] embout *m* protecteur, manchon *m* d'extrémité
~GLASS AMPOULE - [med] ampoule *f* en verre scellée
~-IN FUEL UNIT - [nucl] (fuel element hermetically enclosed in a jacket) élément *m* combustible scellé
~RECTIFIER - [el] soupape *f* scellée
~RETORT FURNACE - [metall] four *m* à résistance autoréglée
~SWITCH - [el] interrupteur *m* étanche
~WORKING - [mining] chantier *m* barré
SEALER - [paint] (a liquid composition for application over a coating which may bleed through a subsequent coating as when the first coat consist of a bituminous stain, creosote or copper naphthenate) peinture *f* isolante
SEALING - [gen] (the impression of a seal) scellage *m*, plombage *m*
[metall] soudure *f* dans le verre
[mech] ferleture *f* hermétique
[metall] étanchement *m* de pores
~-BOX - [el] (a closed box fitted to one end of a cable where connection is made with an external conductor, in such manner as to protect the insulation of the cable from air or moisture) boîte *f* à masse de remplissage
~CEMENT - [ind chem] (a setting material designed to make a joint tight) mastic *m* hermétique
~CHAMBER - s. sealing-box
~COAT - s. sealer
~COMPOUND - [mech] compound *m*, masse *f* de remplissage
~CURRENT - [el] courant *m* de retenue
~-END - s. sealing-box
~FILM - [plast ind] feuille *f* d'isolement, feuille *f* d'étanchement
~GLAND - [mech] presse-étoupe *f*
[el] chapeau *m* étanche
~GROOVES - [rubber ind] (of tubeless tyre; grooves designed to make an airtight joint between beading and the wheel bim) rainures *fpl* à étanchéité
~-IN - [el] (of electrical cables) scellement *m*
~-IN MACHINE - [electron] (machine used in the manufacture of electronic tubes) machine *f* de scellement
~LIQUID - [ind chem] liquide *m* de garde
~LUTE MEDIUM - [ind chem] (a jointing material) lut *m*
~MACHINE - s. sealing-in machine
~MATERIAL - [ind chem, mech] (any product used for sealing) produit *m* d'étanchéité
~MEDIUM - [mech etc] (e.g. used to seal up parts of a nuclear reactor) substance *f* d'étoupage, substance *f* de scellement
~-OFF - [electron] scellement *m*
~-OFF BURNER - [electron] (component used in the manufacture of electronic tubes) chalumeau *m* de coupage

SEALING PLUG - [el] tampon *m* de fermeture
~PRODUCT - s. sealing material
~PROFILE - [plast ind] profilé *m* d'étanchéité
~RIDGES - [rubber ind] (of tyres) stries *f*pl d'obturation
~RING - [mech] rondelle *f* d'étanchéité
~ROCK - [gas ind] (of a natural gas deposit) couverture *f*
~RUN - [metall] (a run welding to make a joint tight) cordon *m* d'étanchéité
~SOLUTION - [plast ind] (solution of a thermosetting resin to seal porous castings) solution *f* bouche-pores
~STRIP - [gen] bande *f* de fermeture
~TAPE - [gen] bande *f* de papier gommé
~VOLTAGE - [el] tension *f* de retenue
~WASHER - [mech] (a ring of soft material used to prevent leakage) rondelle *f* d'étanchéité
~WAX - [ind chem] cire *f* à cacheter
~WELD - s. seal weld
~WIRE - [metall] fil *m* pour soudure dans le verre
SEALOADING LINE - [oil ind] conduite *f* sous-marine
SEALSKIN - [text] peau *f* de phoque, loutre *f*
SEAM, to - [gen] faire une couture
[glass man] (grinding the sharp edges of glass very slightly) meuler
[metall] agrafer
SEAM - [gen] couture *f*
[gen] (junction line) ligne *f* de jonction, couture *f*
[metall] (in foundry work) ébarbure *f*
[metall] (surface defect in worked metal) repliure *f* de laminage
[mining] (stratum or bed; generally a flat deposit of coal or mineral) couche *f*, gisement *m*, filon *m*
[plast ind] (point made in sheet material) couture *f*, joint *m*
[gen] (a scar) cicatrice *f*, balafre *f*
[geol] ligne *f* de séparation
~FOLDING MACHINE - [mech] machine *f* à agrafer
~OF COAL - [min] couche *f* de houille
~OF HIGH DIP - [geol] dressant *m*
~OF MEDIUM THICKNESS - [geol] couche *f* de moyenne puissance
~RESISTANCE WELDING - [metall] soudage *m* continu par résistance
~-TEST - [gen] essai *m* de la soudure
~-WELD, to - [metall] souder à la molette
~WELDING - [metall] (the process of forming a welded seam in the thermoplastic materials) soudure *f* à la molette
~-WORK - [mining] travaux *m*pl en couche
SEAMAN - [naut etc] marin *m*, matelot *m*
SEAMANSHIP - [gen] manoeuvre *f* et matelotage
SEAMARK - [naut] (any mark serving as a guide for navigation) balise *f*
SEAMING - [text] couture *f*
[metall] agrafage *m*
~DIE - [metall] moule *m* pour coutures
~DIES - [mech] estampes *f*pl pour agrafage
SEAMLESS - [gen] (made in a single piece, without a joint or seam) sans couture
~COATING WITH AN EXTRUDER - [rubber & plast ind] recouvrement *m* sans soudure dans la boudineuse
~POT - [ind chem] creuset *m* sans soudure
~SACK - [text] sac *m* sans couture
~STEEL CYLINDER - [gas ind] bouteille *f* d'acier sans soudure
SEAMLESS STEEL PIPE - [gas ind] tube *m* d'acier sans soudure
~STEEL TUBE - s. seamless steel pipe
~TUBE - [plumb] (tube formed by a drawing in a single operation without a joint) tube *m* sans soudure
SEAMS OF THE PLANKS - [shipbuild] coutures *f*pl
SEAPLANE - [aero] (an aeroplane fitted with floats to allow it to operate from a water surface) hydravion *m*
~CARRIER - [naut] (special type of warship designed to carry seaplanes) navire *m* porte-avions
~DOLLY - [aero] (a wheeled carriage for removing seaplanes from the water and supporting and carrying them on land) chariot-remorque *m* d'hydravions
~HULL - [aero] coque *f* d'hydravion
~STATION - [aero] hydroaéroport *m*, hydrobase *f*
~TANK - [aero] (a testing tank for seaplane models) bassin *m* de carène
~TROLLEY - s. seaplane dolly
SEAPORT - [naut] port *m* de mer
SEAQUAKE - [geol] (agitation of the sea caused by a submarine earthquake) tremblement *m* sousmarin
SEAR, to - [gen] cautériser
[gen] (of heat or frost) flétrir, dessécher
[agric] marquer au fer rouge
SEAR - [gen] flétri
[mech] (of a gunlock) gâchette *f*
SEARCH, to - [gen] faire des recherches
[leg] faire une perquisition
[med] (a wound) sonder (une plaie)
[comput] (a memory) chercher (une mémoire), compulser
SEARCH - [gen] recherche *f*
[leg] perquisition *f*
[gen] (as of customs authorities) visite *f* (de douane)
[comput] compulsation *f*
[mining] fouille *f*
~AND RADAR HOMING - [radar] (miniature beacon-transmitter built-in life-belts to facilitate rescue operations) radiobalise *f* de sécurité
~AND RESCUE - [naut aero etc] recherches *f*pl et sauvetage
~-AND-RESCUE AREA - [aero etc] (the area allotted for control to a given search and rescue centre) région *f* de recherches et de sauvetage
~CARD - [comput] carte *f* chercheuse
~COIL - [el] (a coil designed to measure magnetic flux by the phenomenon of induction) bobine *f* exploratrice, chercheur *m*
~CYCLE - [comput] cycle *m* de compulsation, cycle *m* de recherche
~FOR LEAKS - [plumb etc] recherche *f* de fuites
~KEY - [comput] critère *m* de recherche
~RADAR - [radar] (searchtracking radar; installation used to explore the neighbourhood for any aircraft) radar *m* de détection
~TIME - [comput] temps *m* de recherche
SEARCHER - [med] sonde *f*
SEARCHLIGHT - [aero & naut] projecteur *m*
~PROJECTOR - [opt] (projector employing a parabolic reflector) projecteur *m*
SEARCHLIGHTING - [radar] (the tracking of a moving target) balayage *m* horizontal
SEARCHTRACKING RADAR - s. search radar
SEARING - [metall] lissage *m*

SEASHORE - [geogr] côte *f*, bord *m* de la mer
SEASICKNESS - [med] mal *m* de mer, naupathie *f*
SEASIDE - [geogr] bord *m* de la mer
~RESORT - [gen] station *f* balnéaire
SEASON, to - [timber] (the process in which the moisture content of timber is reduced to a suitable amount) dessécher, conditionner
[leather ind] (the process of coating leather after dyeing, with a liquid albumen, prior to glazing and polishing) apprêter, flancher
[food] assaisonner
SEASON - [gen] saison *f*
[gen] (of time) période *f*
~A CASK, to - [agric] abreuver un tonneau
~CRACKING - [metall] corrosion *f* intergranulaire spontanée
SEASONABLE - [gen] de saison
SEASONAL - [gen] saisonnier
~POLYMORPHISM - [genet] (of lake faunas; different forms of the same species at different seasons) polymorphisme *m* saisonnier
SEASONED - [timber] sec
[leather ind] apprêté
[agric] (of wine) mûr
~SEED - [bot] graine *f* bien conditionnée
~TIMBER - [timber] bois *m* sec
SEASONING - [timber] séchage *m*
[leather ind] apprêt *m*, lustre *m*
[text] (of fibres) séchage *m*
[agric] (of the wine) maturation *f* (du vin)
[ind chem] laisser reposer
[electron] (of a magnetron) temps *m* de rodage (d'un magnétron)
SEASONINGS - [food] assaisonnement *m*
SEAT, to - [gen] asseoir
[mech] caler, faire reposer, caler sur son siège
SEAT - [gen] siège *m*
[gen] (chair) chaise *f*
[carp] (of a chair) fond *m* (d'une chaise)
[mech] (of a valve) siège *m* (d'une soupape)
[mech] (of a bearing) surface *f* d'appui, chaise *f*
[mech] (of a machine) surface *f* de contact (d'une machine)
[mech] (in internal combustion engines, of a cylinder) assiette *f* (de cylindre)
[mech] (of a slide-valve) glace *f* de distribution (du tiroir)
[gen] (of a fire) foyer *m* (d'incendie)
[med] (of a disease) foyer *m*
[auto] siège *m*, assise *f*
[horol] appui *m*
~COVER - [gen & auto] housse *f* de siège
~CUSHION - [gen] coussin *m* de siège
~DIVISION - [railw transp etc] (metal or wooden bar dividing a seat into sections) assise *f*
~-MILE - [transp] (ratio between number of passenger and transportation costs) place-kilomètre *f*
~RAIL - [auto] glissière *f* de siège
~RUNNER - s. seat rail
~SPRING - [auto railw etc] ressort *m* pour sièges
~STOP - [hydr] (of a lavatory) butoir *m* de siège de W.C.
SEATING - [mech] (of a valve etc) siège *m*
[mech] (of a boiler) surface *f* d'appui, bâti *m*
[text] (cloth used for upholstering) étoffe *f* de crin
[mech] montage *m*, ajustage *m*
[metall] (a core-print) portée *f* de noyau
SEATING RING - [mech] rondelle *f* d'ajustage
SEATINGS - [constr] (of a boiler) supports *mpl* réfractaires d'un chaudron
SEAWARDS - [gen] (towards the sea) vers la mer, vers le large
SEAWATER - [gen] eau *f* de mer
SEAWAY - [naut] (a way over the sea) route *f*
SEAWEED - s. sea-weed
SEAWORTHINESS - [naut] (of a vessel, a condition fit for a voyage) bon état *m* de navigabilité
[naut] (the quality) valeur *f* nautique
~CERTIFICATE - [naut] certificat *m* de navigabilité
SEAWORTHY - [naut] navigable, en état de navigabilité
SEBACEOUS - [chem] (secreting fat) sébacé
~CYST - [med] (cyst produced by the blockage of the duct of a sebaceous gland) kyste *m* sébacé
~GLAND - [anat] glande *f* sébacée
SEBACIC ACID - [chem] (an intermediate in the production of certain types of nylon, plasticizers and synthetic resins) acide *m* sébacique
SEBACONITRILE - [chem] (intermediate for a number of polymers) sébaconitrile *m*
SEBORRHEA - [med] séborrhée *f*
~DERMATITIS - [med] séborrhée *f* sèche
SEBUM - [zool] (the fatty secretion for a sebaceous gland) sébum *m*
SEC - [math] (a prefix denoting secant) sec
SECAM COLOUR SYSTEM - [telev] système *m* SECAM
SECEDE, to - [gen] faire scission
SECESSION - [gen] sécession *f*, scission *f*
SECLUDE, to - [gen] tenir éloigné
SECLUSION OF THE PUPIL - [med] séclusion *f* pupillaire
SECOBARBITAL - [chem] (a hypnotic used in medicine) sécobarbital *m*
~SODIUM - [chem] (a hypnotic used in medicine) sécobarbital *m* sodium
SECOND, to - [gen] seconder, appuyer
SECOND - [gen] (next from the first) second, deuxième
[gen] (the sixtieth part of a minute) seconde *f*
[math] seconde *f* (de degré)
~ANODE - [telev] anode *f* accélératrice
~-BEATING HAND - [horol] aiguille *f* trotteuse, aiguille *f* des secondes
~BOBBIN - [text] seconde bobine *f*
~BOTTOM CAGE - [text] deuxième tambour *m* métallique inférieur
~BREAKER - [text] carde *f* à tambour enrouleur
~CHANNEL INTERFERENCE - [radio] (interference arising in supersonic heterodyne receiver because of insufficient attenuation of signals of image frequency before the frequency changing stage) interférence *f* des signaux des images
~CLASS LEVER - [mech] levier *m* du second genre, levier *m* inter-résistant
~CONTRACTION - [biol] (shortening and thickening of the threads in the diplonema stage of meiosis when diakinesis comes on) deuxième contraction
~CUT - [mech] (of a file) taille *f* demi-douce
~-CUT FILE - [impl] lime *f* à taille demi-douce
~-CUT RASP - [impl] râpe *f* à taille demi-douce
~DERIVATIVE - [math] seconde derivée *f*
~DERIVATIVE ACTION - [constr] action *f* par double dérivation
~DETECTOR - [radio] (the detector following the in-

termediate frequency amplifier in a supersonic heterodyne receiver) deuxième détecteur *m*
SECOND GROUP SELECTOR - [telecomm] sélecteur *m* de groupe secondaire
~-HAND - s. secondhand
[horol] aiguille *f* trotteuse
~HARMONIC BAND - [phys] (the spectral band produced when the vibrational energy of a molecule changes from an initial level in which the vibration quantum number is 0 to a level in which the vibrational quantum number is 2, or viceversa) bande *f* de la deuxième harmonique
~HARMONIC MAGNETIC MODULATOR - [contr] modulateur *m* magnétique à deuxième harmonique
~INVERSION OF A CHORD - [mus] deuxième inversion *f* d'un accord
~-IONIZATION-POTENTIAL - [nucl] (relative to the removal of the most loosely bound electron from an atom from which one electron has already been removed) deuxième potentiel *m* d'ionisation
~LAMB - [text] (term used in the wool trade to denote lamb wool of second quality) laine *f* de la seconde sorte
~LINE FINDER - [telecomm] chercheur *m* secondaire
~LOOK - [med] revision *f* chirurgicale
~MEAN CHORD - [aero] (aerodynamic mean chord) corde *f* aérodynamique moyenne
~MINING - [mining] dépilage *m*
~MOTION SHAFT - [text] arbre *m* de dépointage et de rentrée
~-MOTION SHAFT - [mech] arbre *m* secondaire
~OF ARC - [meas] (unit of angular measurement, equal to one sixtieth of a minute of arc (1/360 degree) seconde *f* d'arc
~OF TIME - [meas] (unit of duration, equal to one-sixtieth of a minute of time) seconde *f*
~-ORDER PRISM - [phys] (in crystallography) prisme *m* de seconde espèce
~-ORDER PYRAMID - [phys] (in crystallography) pyramide *f* de seconde espèce
~-ORDER SYSTEM - [math] système *m* du deuxième ordre
~OVERTONE BAND - [phys] (third harmonic band) bande *f* de la troisième harmonique
~PASSAGE OF DRAWING - [text] étirage *m* intermédiaire
~PICKINGS - [agric] récolte *f* intermédiaire
~PIECES - [text] pièces *fpl* de seconde sorte
~PILOT - [aero] (an additional, fully-qualified pilot carried in large aircraft to assist or relieve the captain, and usually occupying a seat beside him) second pilote *m*
~QUANTIZATION - [phys] (the process by which a classical field is analyzed as a group of particles) deuxième quantification *f*
~QUANTUM NUMBER - [phys] (orbital quantum number) nombre *m* quantique secondaire
~STATE CREEP - s. secondary creep
~-SURFACE DECORATION - [plast ind] (application of decoration to the back of a transparent sheet, so that it can be seen but is protected in use) décoration *f* en seconde surface
~TAP - [tool] (a tap used to finish a through hole) taraud *m* demi-conique, second taraud *m*
~TOP CAGE - [text] deuxième tambour *m* métallique supérieur
~-TRACE ECHOES - [radar] échos *mpl* de traces secondaires
SECOND VENTRICLE - [anat] (the cavity in the right lobe of the cerebrum in vertebrates) ventricule *m* droit
~YARN GUIDE - [text] second guide-fil *m*
SECONDARY - [gen] (of second rank etc) secondaire
[el] (the secondary winding of a transformer) enroulement *m* secondaire
[chem] (subsequent in origin, involving a chemical or physical change of the original mineral) secondaire
~AIR - [metall] (introduced over the firebed for combustion) air *m* secondaire
~AMIDE - [chem] (an amide in which two atoms of hydrogen have been replaced by acyl groups) amide *f* secondaire
~AND CONTROL WIRING - [el] câblage *m* auxiliaire
~BATTERY - [el] (two or more secondary cells connected electrically) batterie *f* d'accumulateurs
~BEAM - [constr] solive *f* secondaire, solive *f* reposant sur une maitresse-poutre
~BOW - [met] (a rainbow with a radius of 52° having the red inside and the blue outside) arc-en-ciel *m* secondaire
~CABLE - [auto] câble *m* secondaire (allumage)
~CELL - [el chem] (a galvanic cell which can be restored to its original condition after discharge by passing a current through it in the direction opposite to that of the discharging current) élément *m* secondaire, accumulateur *m*
~CHAIN - [geogr] chaînon *m*
~CIRCUIT - [el] circuit *m* secondaire
~CLOCK - [el] horloge *f* secondaire
~COIL - [el] bobine *f* secondaire
~CONSTITUENT - [metall] métal secondaire
~COSMIC RAYS - [nucl] (radiation produced as the result of the interaction of primary cosmic rays with atmospheric nuclei and electrons) rayons *mpl* cosmiques secondaires
~CREEP - [phys] (the creep which becomes constant some time after the initial or primary creep) fluage *m* secondaire
~CRUSHING - [min] (of coal) broyage *m*
~CULTIVATION - [bot] culture *f* secondaire
~DEPRESSION - [met] (a smaller depression developing in the neighbourhood of a larger, primary one) dépression *f* secondaire
~DISCONNECTING DEVICES - [el] dispositifs *mpl* secondaires de coupure
~DISTRIBUTION FEEDER - [el] feeder *m* secondaire
~DISTRIBUTION MAINS - [el] ligne *f* de consommateur
~ELECTRODE - [el] (electrode without metallic connection with the anode and cathode, through which the current may pass, so that separate parts function as anode and cathode) électrode *f* intermediaire
~ELECTROMAGNETIC CONSTANTS - [el] (in transmission line theory) constantes *fpl* électromagnétiques secondaires
~ELECTRON - [electron] (the electron emitted from a surface by electronic bombardment, as distinct from the primary bombarding electrons) électron *m* secondaire
~ELECTRON GAP LOADING - [electron] admittance *f* de l'espace d'interaction par émission secondaire
~EMISSION - [electron] (the emission of electrons from a surface by its bombardment by electrons of

another source) émission *f* secondaire
SECONDARY EMISSION CHARACTERISTIC - [electron] (of a surface, the relation between the secondary emission rate of a surface and the voltage between the source of the primary emission and the surface) caractéristique *f* d'émission secondaire
[electron] (of a luminescent screen, the relation between the rate of secondary emission of the screen and its voltage) caractéristique *f* d'émission secondaire
~EMISSION MULTIPLIER - [electron] multiplicateur *m* d'électrons secondaires
~-EMISSION RATE - [electron] (of a surface, the number of secondary electrons detached from a surface by an incidental electron) émission *f* secondaire
~ENRICHMENT - [min] enrichissement *m* secondaire
~EXCHANGE - [telecomm] centre *m* de secteur, bureau *m* nodal
~EXTINCTION - [phys] (occurring with imperfect crystals) extinction *f* secondaire
~FAULT - [el] percement *m* secondaire
~FERMENTATION - [brew ind] fermentation *f* secondaire
~FILTER - [radiat] (filter used to remove the secondary radiation generated in the primary filter) filtre *m* secondaire
~FLOW - [phys] (the flow in pipes and channels often possesses components at right angle to the axis) effluent *m* secondaire
~GEAR - [auto] engrenage *m* secondaire,
~GRID EMISSION - [electron] émission *f* secondaire de la grille
~HARDENING - [metall] durcissement *m* secondaire, trempe *f* secondaire
~HARDNESS - [metall] (a further increase in hardness produced in tempering high-speed steels after quenching) dureté *f* secondaire
~HEAT EXCHANGER - [th eng] (in a chemical process) échangeur *m* de chaleur secondaire
~HOLES - [mech] (holes located in a flame tube downstream of the primary holes, to stabilize the flame and complete combustion) orifices *mpl* secondaires
~IMAGE - [opt] (ghost image) image *f* secondaire
~INGOT - [metall] lingot *m* de deuxième fusion
~ION - [nucl] (the ion which has the smaller energy after a collision between two ions) ion *m* secondaire
~IONIZATION - [nucl] (ionization which includes that due to delta rays) ionisation *f* secondaire
~LEAKAGE - [el] (the magnetic leakage which is associated with the secondary winding of a transformer) dispersion *f* secondaire
~LINE SWITCH - s. second line finder
~LOAD - [el] (burden; of an instrument transformer) charge *f* (du transformateur d'instrument)
~MEMORY - [comput] (a storage which is not an integral part of a computer, but directly linked to, and thus controlled by the computer) mémoire *f* auxiliaire
[comput] mémoire *f* secondaire
~MERISTEM - [bot] (meristem formed from permanent tissue) méristème *m* secondaire
~METER - [instr] (a check meter) compteur *m* divisionnaire, compteur *m* en décompte
~NATURAL RADIONUCLIDES - [nucl] (decay products of primary natural radionuclides) radionucléides *mpl* naturels secondaires
SECONDARY NEUTRAL GRID - [el] neutre *m* commun
~NEUTRON - [nucl] (a nucleus produced by the interaction with matter of a radiation regarded as primary) neutron *m* secondaire
~NITRO-COMPOUNDS - [chem] nitrocomposées *mpl* secondaires
~NOISE - [acoust] (undesired noise inherent to sound reproducing system) bruit *m* secondaire
~PIPE - [metall] retassure *f* intérieure
~RADAR - [radar] (radar system in which the pulses cause an automatic emission of pulses by the target) radiodétection *f* secondaire, radar *m* secondaire
~RADIATION - [nucl] (particles or photos produced by the interaction with matter of a radiation considered primary) rayonnement *m* secondaire
~RADIATOR - [radio] (the portion of an aerial which is energized neither directly nor through a feeder) élément *m* passif
~REACTION - [el chem] (chemical or electrochemical reaction between electrode-reaction products in an electrolyte) réaction *f* secondaire
~REACTOR - [nucl] (a nuclear chain reactor the fissile material of which is operated from the products of primary separation plants) réacteur *m* secondaire
~RELAY - [el] relais *m* indirect
~RIPENING - [geol] (obsolete; rocks deposited between the triassic and the cretaceous era) roches *fpl* secondaires
~SERVICE AREA - [radio] (the area of a radio broadcast transmitter within which satisfactory reception can be obtained only under favourable conditions) zone *f* de réception secondaire
~SHAFT - [auto] arbre *m* intermédiaire
~SPRING - [mech] ressort *m* secondaire
~STANDARD - [el] (copy of primary standard for general use in a standardizing laboratory) étalon *m* de laboratoire
~STANDARD LAMP - [el] étalon *m* photométrique
~STORAGE - s. secondary memory
~STORE - s. secondary memory
~SURFACE RADIATOR - [th eng] (a radiator provided with fins to increase the radiating surface) radiateur *m* à transmission indirecte
~TONE - [acoust] (a tone produced simultaneously with a pure tone) son *m* accessoire
~TRIANGULATION - [surv] triangulation *f* secondaire
~TRIANGULATION BEACON - [surv] point *m* géodésique d'un réseau secondaire
~VOLTAGE - [el] (the voltage in the output winding of a transformer) voltage *m* secondaire
~WINDING - [electron] (the output winding of a transformer) enroulement *m* secondaire
~X-RAYS - [radiat] rayons *mpl* X secondaires
~YEAST - [brew ind] levure *f* pour la fermentation secondaire
SECONDHAND - [gen] de seconde main
SECONDS COUNTER - [horol] compte-secondes *m*
SECRECY - [gen] secret *m*
~RELAY - [telecomm] relais *m* de secret
SECRET - [gen] secret
~DOVETAIL - [carp] (angle joint between two members) queue-d'aronde *f* cachée
~SWITCH - [el] (a locked-cover switch) interrupteur *m* fermé à clef
SECRETE, to - [gen] soustraire, cacher
[biol] (to elaborate, collect and discharge, as of a

gland) sécréter
SECRETIN - [biochem] (a hormone produced by certain cells, forming part of the intestine when stimulated by hydrochloric acid from the stomach) sécrétine *f*
SECRETION - [physiol] (a substance which is elaborated, collected and discharged by a gland) sécrétion *f*
SECRETORY - [gen] (of secretion) sécréteur
~CELL - [bot] (a cell in which oils, resins etc are formed) cellule *f* sécrétrice
~DUCT - [bot] (elongated intercellular space for the accumulation of secretion) canal *m* sécréteur
~PASSAGE - s. secretory duct
~TISSUE - [bot] (group of secretory cells) tissu *m* sécréteur
SECTION, to - [gen] (to separate into sections) sectionner, diviser en sections
SECTION - [gen] section *f*, division *f*, coupage *m*
[draw] (a representation, drawing etc shown as if cut by an intersecting plane) section *f*, coupe *f*
[railw] (a portion of a railway track under the care of a set of men) tronçon *m* (de voie), secteur *m*
[railw] (of a sleeping car) compartiment *m*
[el] (the elementary part of a complete wave-filter) élémént *m* d'un filtre
[el] (of a commutator winding, the shortest part of a winding which is included between two successive front connections) section *f*
[el] (a group of electrolytic cells placed close together and electrically connected in series) batterie *f*, système *m* multiple
[mach tool] (group of machines employed in the same operation) ligne *f*
[metall] (section iron etc) profil *m*, profilé *m*
[metall] (the dimension of the cut) échantillon *m*
[comput] (part of a magnetic tape) section *f* de bande
[th eng] (of a sectional boiler) élément *m*
[gen] (part of a road etc) tronçon *m*
[print] (folded sheet of a book) cahier *m*
[print] (the reference mark "S") paragraphe *m*, alinéa *m*
[instr] (a very thin slice of a substance, e.g. for microscopic examination) plaque *f* mince, lame *f* mince
[bot] (of an orange) tranche *f*
[zool] (a division of an animal group) groupe *m*
[bot] (division of a genus consisting of a number of closely related species) groupe *m*
[metall] (of a tube) coupon *m*
[constr] profilé *m* en métal
~BAR - [metall] profilé *m*
~BLOCK - [text] bloc *m* sectionnel
~CIRCUIT BREAKER - [el] interrupteur *m* de section
~CUTTING - [opt] (in microscopy) coupe *f* en section
~DRAWING - [draw] dessin *m* de coupe
~FUSE-BOARD - [el] (a distribution-board having a fuse or fuses only for each of the branch circuits) panneau *m* de fusibles de section
~-GAP - [el] (arrangement of line-work for dividing a contact-wire electrically and mechanically into sections while maintaining a continuous path for the current collector) sectionnement *m* à intervalle d'air
~GAUGE - [oil ind] diamétreur *m*
SECTION HEAD STOCK - [text] dispositif *m* d'enroulement pour ourdissoirs à sections
~INSULATOR - [el] (a device for dividing a contact-wire into electrical sections while maintaining mechanical continuity and a continuous path for the current collectors) isolateur *m* de section
~IRON - [metall] profilé *m* en fer
~LINEMAN - [telecomm] surveillant *m* de lignes
~LINING - [draw] hachure *f*
~MILL - [metall] laminoir *m* à profilés
~MODULUS - [mech] (a ratio in flexure tests reading between the moment of inertia of cross section and the distance of the farthest stressed element from the neutral axis) module *m* résistant de la coupure, couple *m* résistant
~OF A COMMUTATOR WINDING - [el] section *f* d'un enroulement à collecteur
~OF COKE CAKE - [metall] tampon *m* de coke
~OF HOT COKE - s. section of coke cake
~OF MAIN - [gas ind] (a lenght of pipeline) tronçon *m* de canalisation
~OF PIPELINE - s. section of main
~OF RECURRENT STRUCTURE - [telecomm] (in telephony) quadripôle *m* élémentaire d'un réseau récurrent
~SENSITIVITY - [metall] sensibilité *f*
~STRIP - [metall] acier *m* laminé à chaud
~SWITCH - [el] (a switch for dividing circuits or conductors into sections) disjoncteur *m* de bouclage, disjoncteur *m* de couplage
~TUBE - [metall] tube *m* profilé
~WARPING - [text] (the making of a warp in sections) ourdissage *m* à sections
~WARPING MACHINE - [text] ourdissoir *m* à sections
SECTIONAL - [gen] sectionnel, fractionné
[gen] (divided into sections, parts etc) en sections
~AIRBAG - [rubber ind] (for tyres) sac *m* à air sectionnel, bag *m* de reparations locales
~DRAWING - [draw] (a drawing made to show a part or machine as if cut along a given line or plane) dessin *m* de coupe
~FLYWHEEL - [mech] volant *m* à sections
~GRATE - [th eng] grille *f* à plusieurs panneaux
~IRON - [metall] fer *m* profilé
~KIER - [text] chaudière *f* sectionnée
~ORGAN - [mus] (one of the component of the grand organ) clavier *m*
~PANEL - [el] panneau *m* fractionné
~PATTERN - [metall] modèle *m* en deux parties
~PONTOON DOCK - [naut] bassin *m* flottant à sections
~PRESS - [rubber ind] (for tyres) presse *f* à réparation
~RECTIFIER - [el] redresseur *m* à vapeur de mercure à sections
~RELEASE ROUTE LOCKING - [railw] enclenchement *m* de transit souple
~REPAIRBAG - s. sectional airbag
~REPAIR MOULD - [rubber ind] (for tyres) moule *m* pour réparations locales
~TYPE RUBBER BEARING - [rubber ind] coussinet *m* en caoutchouc de douilles longitudinales
~VIEW - [draw] vue *f* en coupe
~WARP - [text] ourdissage *m* à sections
~WEIGHTS - [mech] contrepoids *mpl* de scie à ruban
SECTIONALIZED SPHERICAL CAVITY - [electron] (spherical cavity resonator, at the top and at the bottom of which segments have been cut away) ca-

vité *f* sphérique aplatie
SECTIONALIZED VERTICAL AERIAL - [radio] antenne *f* verticale en sections
~VERTICAL ANTENNA - s. sectionalized vertical aerial
SECTIONING - [el] (of a distributing network) sectionnement *m*
SECTOR - [geom] (plane figure enclosed by two radii of a circle and the arc cuttoff by them) secteur *m*
[instr] (mathematical instrument consisting of two arms marked with various scales and hinged together at one end) compas *m* de proportion
[gen] secteur *m*
[railw] (of a steam locomotive) secteur *m*
[mech] secteur *m*, couronne *f*
[cin] secteur *m* (de l'obturateur)
~ALIGNMENT INDICATOR - [contr] indicateur *m* de valeur de seuil
~AREA - [comput] zone *f* d'enregistrement normale
~CABLE - [el] câble *m* à conducteurs en forme de secteur
~CONNECTING ROD - [railw] (of a steam locomotive) bielle *f* de secteur
~DISK - [opt] (device designed to secure an accurately known control of the intensity of a beam of light) disque *m* à secteurs
~EXCHANGE - [telecomm] bureau *m* nodal
~GEAR - [mech] secteur *m* denté
~ON FALLER SHAFT - [text] secteur *m* sur l'arbre des baguettes
~-PATTERN INSTRUMENT - [el] (switchboard instruments contained in sector-shaped cases; used to save space on the switchboard) tableau *m* à secteur
~PHOTOMETER - [instr] (photometer in which sector disks are used) photomètre *m* à secteurs
~REGULATOR - [hydr] (a form of drum-weir) barrage *m* à secteurs
~SCALE - [mech] cadran *m* gradué
~SCANNING - [radar] (scanning through a limited plane angle about any axis) balayage *m* de secteur
~SHAPED LIFTER - [text] secteur *m*
SECTORAL ALIGNMENT INDICATOR - [contr] indicateur *m* d'excursion du signal
~E HORN - [electron] (of a waveguide) cornet *m* multicellulaire en E (d'un guide d'ondes)
SECTORING - [radar] (a set fault) formation *f* de secteurs
SECTORMETER - [instr] (potentiometer for electrometric titrations in which the microammeter is replaced by a cathode-ray tube) potentiomètre *m* à tube à rayons cathodiques
SECULAR - [gen] séculaire
~ACCELERATION - [astr] (relating to the moon's motion) accélération *f* séculaire
~CHANGES - [geol] (extremely slow changes) variations *f*pl séculaires
~EQUILIBRIUM - [nucl] (radioactive equilibrium) équilibre *m* radioactif séculaire
~PARALLAX - [astr] (a very slow effect by which, owing to the motion of the solar system as a whole, the apparent places of the stars will entirely change in the course of time) parallaxe *f* séculaire
SECURE, to - [gen] mettre en sûreté, assurer
[gen] (to fasten or confine) fixer, retenir
[gen] (to get) obtenir, acquérir
[bot] (a bud) épincer (un bourgeon)
[leg] mantir
SECURE, to - [mech] fixer
SECURE - [gen] sûr, sauf, à l'abri, fixé
~A CONCESSION, to - [mining] obtenir une concession
~IN POSITION, to - [mech] ancrer dans une position
~THE ANCHOR, to - [naut] saisir l'ancre
SECURED BY MORTGAGE - [fin] nanti de gages
SECURITY - [gen] sécurité *f*, sûreté *f*
[leg] (a pledge etc) garantie *f*, caution *f*
[fin] nantissement *m*, valeurs *f*pl
~PAPER - [paper man] papier *m* pour l'impression de valeurs
SEDAN - [auto] voiture *f* à conduite intérieure
~CHAIR - [transp] chaise *f* à porteurs
~COUNT - [text] numéro *m* de Sedan
SEDATIVE - [chem] (a drug which produced a relation of emotional tension) sédatif *m*
SEDENTARY - [zool & gen] sédentaire
SEDGE - [bot] carex *m*, laîche *f*
~GRASS - [bot] joncs *m*pl
~HAY - [agric] foin *m* de marais
SEDIMENT - [chem hydr etc] (solid deposited from a suspension) sédiment *m*, résidu *m*
[gen] sédiment *m*, boue *f*
[th eng] (of a boiler) vidange *m* d'une chaudière
[agric] (of the wine) lie *f* (du vin)
~BOWL - s. sediment chamber
~CHAMBER - [auto] (a receptable to catch sediment, e.g.in a carburettor) chambre *f* de sédimentation
SEDIMENTARY - [chem min etc] sédimentaire
[geol] terrain *m* de sédiment
~CLAY - [geol] argile *f* sédimentaire
~DEPOSITS - [geol] dépôts *m*pl sédimentaires
~-ROCKS - [geol] (rocks resulting from the degradation of older formation or from the remains of marine organisms) roches *f*pl sédimentaires
SEDIMENTATION - [chem] (the settling out from a solution of insoluble material) sédimentation *f*
~ANALYSIS - [chem] analyse *f* granulométrique par sédimentation
~BASIN - s. sedimentation tank
~CONSTANT - [phys] constante *f* de sédimentation
~PLANT - [hydr] installation *f* de décantation
~POTENTIAL - [el chem] (that electrokinetic potential gradient which occurs when a suspension or colloid is moved by gravity or centrifugal action through a liquid electrolyte at unit velocity) potentiel *m* de sédimentation
~TANK - [hydr] (in sewage works, a tank into which sewage is passed, so that suspended matters may sink to the bottom) décanteur *m*, bassin *m* de sédimentation
~TEST - [med] (the measurement of the rate of sinking of red blood cells in drawn blood placed in a tube) essai *m* de sédimentation
SEDOHEPTOSE - [chem] (heptose obtained from the leaves and stem of Sedum spectabile) sédoheptose *m*
SEE, to - [gen opt etc] voir
SEEBECK EFFECT - [el] (when two different metals are joined and the two junctions are kept at different temperatures, an electromotive force is developed in the circuit) effet *m* Seebeck, effet *m* thermoélectrique
SEED, to - [bot & agric] porter semence, semencer, grener
SEED - [agric & bot] grain *m*, graine *f*, semence *f*
[bot] (a grain) grain *m*

SEED - [radiat] (a short glass or metal capillary tube containing radon and designed for interstitial therapeutic irradiation) tube *m* à radon
[nucl] (fuel assembly) semence *f*
[electron] (of a semiconductor) germe *m* de cristal
~BAG - [mining] couronne *f* en toile remplie de graine de lin
~BED - [agric] couche *f* de semis, germoir *m*
~BOX - [agric] caisette *f* à semis
~BREEDER - [agric] sélectionneur *m* de semences
~BROADCASTER - [agric] semoir *m* à la volée
~CHARGE - [ind chem] agent *m* de précipitation, précipitant *m*
~CLEANER - [agric] appareil *m* à nettoyer les graines
~CORE REACTOR - [nucl] réacteur *m* à coeur à germes
~COTTON - [text] (unginned cotton) coton *m* en graine, coton *m* brut
~COTTON OPENER - [text] ouvreuse *f* pour coton en graine
~CRYSTAL - [chem] (a crystal introduced into a supersaturated solution in order to accelerate crystallization) cristal *m* de précipitation
~DENSITY - [agric] densité *f* de semis
~DRESSER - [agric] poudreuse *f* pour semences
~DRILL - [agric] semoir *m*
~DUSTER - s. seed dresser
~ENCLOSURE - [agric] pépinière *f*
~FIBRE - [text] duvet *m* de la graine
~FLAX - [agric] lin *m* pour semailles
~FOR SOWING PURPOSES - [agric] graine *f* d'ensemencement
~FREED FROM HAIRS - [bot] graine *f* privée de son duvet
~GRID - [agric] grille *f* à graines
~GROWER - [agric] multiplicateur *m* de semences
~GROWING - [agric] sélection *f* de semences
~HAIR - [text] duvet *m*
~HARROW - [agric] herse *f* à semences
~HOLE - [agric] poquet *m*
~LEAF - [bot] cotylédon *m*
~LOSS - [agric] perte *f* par infiltration
~MIXTURE - [agric] mélange *m* de semences
~POD - [bot] péricarpe
~PODDED JUTE - [bot] jute *m* à fruit capsulaire
~PLOT - [agric] champ *m* ensemencé, semis *m*
~POTATO - [agric] pomme *f* de terre à semence
~RIPENESS - [agric] maturité *f* complète du grain
~-ROW - [agric] ligne *f* de semis
~SHOOT - [bot] pousse *f* du grain
~SORTER - [agric mach] calibreur *m* de semences
~TESTING - [agric] essai *m* de semences
~TESTING STATION - [agric] station *f* d'essai de semences
~WINNOWER - [agric] nettoyeur *m* de grains
SEEDER - [agric] semoir *m*
SEEDINESS - [paint] (a term used for the appearance in a coating of a rough texture, as if it contained very fine particles, a result of incompatibility of the ingredients) rugosité *f*
SEEDING - [agric] ensemencement *m*, grenaison *f*
[agric] (of grapes) égrugeage (des raisins)
[agric] (of melons) épépinage *m* (des melons)
SEEDLING - [bot] (young plant from a germinated seed) élève *f*, jeune brin *m*
~FOREST - [agric] futaie *f*
~NURSERY - [agric] centre *m* de sélection (de semences)
SEEDY - [text] (said of wool which contains grass seeds offering difficulties for their removal) plein de graines
[glass man] (of a glass containing small bubbles) plein *m* de soufflures
~GLASS - [glass man] verre *m* avec soufflures
~WOOL - [text] laine *f* pleine de graines
SEEING - [astr] qualité *f* de l'image
SEEK, to - [gen] chercher, rechercher
[comput] chercher l'emplacement indiqué
SEEM, to - [gen] sembler, paraître
SEEMING STEM - [bot] tige *f* apparente
SEEP, to - [phys] (to percolate slowly through; to ouze out slowly) suinter, s'infiltrer
SEEP - [oil ind] dégagement *m* de gaz
SEEPAGE - [gen] infiltration *f*, suintage *m*, fuite *f*
~WELL - [hydr] puits *m* drainant, puits *m* absorbant
SEESAW, to - [gen] basculer, se balancer
[mech] (to cause to move in the manner of a seesaw) faire basculer
SEESAW - [gen] bascule *f*
[gen] basculaire
[mech] (of a reciprocating motion) mouvement *m* de va-et-vient
~CIRCUIT - [telev] (coupling with an amplifying tube the cathode of which is connected to a constant voltage) amplificateur *m* à cathode à la masse
~MARMUR - [med] frottement *m* péricardique à deux temps
~ROTOR - [aero] (type of rotor in which the blades are rigidly fixed to a head which is mounted in gimbals on the shaft) rotor *m* de va-et-vient
S-EFFECT - [electron] (surface charge effect) effet *m* de charge superficielle, effet *m* S
SEGER CONE - [meas] (a mixture of clay and oxides which melt at specific temperature; formed into a cone and used in the measurement of temperatures of from II00° - 3700°F.) montre *f* de Seger
SEGMENT, to - [gen] couper en segments
SEGMENT - [geom] (part of a figure cut off by a line or a plane) segment *m*
[gen] (a section) segment *m*, tranche *f*
[el] (one of the elements, insulated from one another, forming a commutator) lame *f* du commutateur
[comput] demi-mot *m*
~BEND - [plumb] (also called lobster-back, a pipebend formed of several short straight lengths welded together at an angle to make the required change of direction) coude *m* soudé
~CORES - [metall] (in diecasting) noyaux *mpl* en segments
~DIE - [metall] estampe *f* démontable
~DRUM - [rubber ind] (a collapsible former for tyres) tambour *m* escamotable, tambour *m* repliable à secteurs séparés
~-FEED LOOP - [electron] (a coupling loop in a multicavity magnetron mounted at the end space near one end of one of the resonant cavities of the tube) boucle *f* d'accouplement de segments
~PITCH - [el] (the peripheral distance between the centre lines of two adjacent segments) pas *m* des lames
~SHAPED COMB - [text] peigne *m* échancré
~TOOTHED CYLINDER - [text] tambour *m* à segment denté

SEGMENTAL - [gen] segmentaire
~ARCH - [arch] (arch having the shape of a circular arc struck from a point below the springings) arc *m* surbaissé, voûte *f* en segment de cercle
[arch] (of a bridge) arche *f* surbaissée
~CONDUCTOR - [el] conducteur *m* disposé en segments
~FACE ARCH - [arch] arc *m* de front surbaissé
~HORN - [electron] (of a waveguide) cornet *m* en forme de segment
~WHEEL - [mech] meule *f* à segments
SEGMENTATION - [gen] segmentation *f*
[comput] (of a programme) fractionnement *m*
~CAVITY - [biol] nucléole *m* (de cellule)
~NUCLEUS - [zool] (the nucleus of a fertilized ovum formed by the union of the male pronucleum with the female pronucleum) noyau *m* d'oeuf fertilisé
SEGMENTED CAM RING - [text] plaque *f* à cames du plateau
~EXTRUDER BARREL - [plast ind] (a barrel made up of a series of annular portions, to enable different lengths of barrel to be used) fût *m* segmenté d'extrudeuse
~WORD FEATURE - [comput] dispositif *m* de mot fractionné
SEGREGATE, to - [gen] isoler, séparer (l'un de l'autre)
SEGREGATED OILS - [paint] (drying oils obtained by removing non-drying constituents from semi-drying types, e.g. by solvent extraction, vacuum distillation or selective crystallization) huiles *fpl* séparées
SEGREGATION - [gen] ségrégation *f*
[metall] (non uniform distribution of impurities, inclusions and alloying constituents in metals) liquation *f*, ségrégation *f*
SEINE, to - [fishing] (to fish with a seine) pêcher à la seine
SEINE - [fishing] (a long shallow net with floats at the top edge and a weighted bottom rope) seine *f*, senne *f*
~-NET - s. seine
SEINING - [fishing] pêche *f* à la seine
SEISMAL - s. seismic
SEISMIC - [geol] (pertaining to earthquake) séismique, sismique, sismal
~DETECTOR LOCATION - s. seismic observation point
~METHODS - [oil ind] (artificial earthquake produced by explosives, thus generating energy waves in the surface layers, which are recorded by seismometers placed at various distances from the explosion) méthode *f* sismique
~OBSERVATION POINT - [geol] point *m* d'observation séismique
~PROSPECTING - s. seismic methods
~REFRACTION - [geol] (a method for the determination of the depth of rocks) réfraction *f* séismique
~SITE - [nucl] (area where a reactor is tested to assess its resistance to vibratory motion) site *m* résistant aux tremblements de terre
~SOUNDING - s. seismic methods
~WAVES - [geol] ondes *fpl* sismiques
SEISMOGRAM - [instr] (graphic record of the vibrations recorded by seismometers) sismogramme *m*
SEISMOGRAPH - [instr] (an instrument designed to register earthquake shocks and concussions) sismographe *m*
SEISMOLOGY - [geol] (the study of earthquake phenomena) sismologie *f*
SEISMOMETER - [instr] (instrument used in seismic survey to record vibrations) sismomètre *m*
SEISMOTHERAPY - [med] massage *m* vibratoire
SEISMOTHESIA - [med] perception *f* d'une secousse
SEISMS - [geol] (oscillations of the earth's crust) séismes *mpl*
SEIZE, to - [mech] (mechanical parts which, from overheating or lack of lubrication, grip each other instead of moving freely, are said to seize) gripper, coincer
[mech] (of a journal in a bearing) se coincer
[naut] (to secure by binding together) amarrer
[gen] confisquer, saisir
SEIZING - [mech] (of a piston) grippage *m*
[mech] (of a journal in a bearing) grippure *f*, grippement *m*
[metall & plast ind] (the gripping of one part of a mould by another, preventing opening) conglomération *f*, écorchure *f*
[naut] amarrage *m*
~SIGNAL - [telecomm] signal *m* de prise
SEIZURE - [gen] saisie *f*, prise *f*
[leg] appréhension *f*
[mech] s. seizing
[med] (sudden violent attack) attaque *f* d'apoplexie
SELECT, to - [gen] choisir
[min] trier
~THE COCOONS FOR BREEDING, to - [text] trier les cocons pour l'élevage
~THE REED, to - [text] choisir le peigne
SELECTANCE - [radio] (term sometimes used to denote selectivity, especially numerically) sélectivité *f*
SELECTED - [gen] choisi
~RANGE INDICATOR - [radar] indicateur *m* type C
~THREAD - [text] fil *m* choisi
SELECTING CIRCUIT - [comput] circuit *m* de sélection
~MECHANISM - [telecomm] organe *m* de sélection
~NEEDLE - [text] aiguille *f* de sélection
~TRANSFER CAM - [text] came *f* Jaquard mailles retournées
SELECTION - [gen] sélection *f*, choix *m*
[biol] (the process by which certain organisms are favoured in the struggle for survival and perpetuation) sélection *f*
~RATIO - [comput] (in static magnetic storage) rapport *m* de sélection
~RULES - [nucl] (a set of statements used to classify transitions of a specified type in terms of quantum numbers of the initial and final states of the system involved in the transitions in a certain way) règles *fpl* d'exception nucléaire
~WIRE - [comput] (a component of magnetic-core storages) fil *m* de sélection
SELECTIVE - [gen] sélectif
[radio] (designating a method by which certain radio frequencies can be transmitted or received to the exclusion of others) sélectif
~ABSORBER - [radiat] (a concentrated substance in an organ tissue which absorbs radiation to a high degree) absorbeur *m* sélectif
~ABSORPTION - [radiat] (absorption varying in amount with wavelength) absorption *f* sélective

SELECTIVE AMPLIFIER - [el acoust] amplificateur *m* sélectif
~ANNEALING - [metall] recuit *m* sélectif
~CALLING - [comput] appel *m* sélectif
~CARBURIZING - [metall] cementation *f* sélective
~COLLECTIVE AUTOMATIC CONTROL - [contr] auto-liftier *m* sélectif automatique
~DUMP - [comput] image-mémoire *f* sélective
~EMISSION - [el] (the property of an incandescent body whereby it emits radiation, predominantly of one frequency) émission *f* sélective
~FADING - [radio] (in short-wave transmission over long distances) fading *m* sélectif
~FERTILIZATION - [bot] fertilisation *f* sélective
~FILTER - [telev] filtre *m* sélectif
~FLOTATION - [metall] flottation *f* sélective
~HARDENING - [metall] (producing different degrees of hardening in different areas) endurcissement *m* sélectif
~HEATING - [metall] chauffage *m* sélectif
~JAMMING - [radar] (a type of jamming whereby a single channel only is jammed) interférence *f* sélective
~LIST CONTROL - [comput] contrôle *m* de travail en liste facultatif
~LOCALIZATION - [nucl] (accumulation of a particular isotope to a greater degree in certain cells or tissue) affinité *f* différentielle, localisation *f* sélective
~MATING - [zool] (preferential mating) accouplement *m* sélectif
~PROTECTION - [el] (a term relating to methods of protecting power transmission networks in which an automatic disconnection of the faulty occurs without disturbance of the remainder of the network) protection *f* sélective
~QUENCHING - [metall] refroidissement *m* rapide sélectif
~RADIATOR - [opt] (emitter of radiation yielding radiation of different spectral energy distribution from that of a black body at the same temperature) émetteur *m* secondaire
~RESONANCE - [el] (in electricity, resonance with a harmonic, instead of the fundamental. In radio, resonance occurring in one or more discrete frequencies, instead of extending over a band of frequencies) résonance *f* sélective
~SCATTERING OF SOUND - [acoust] (scattering of sound depending on frequency) dispersion *f* sélective du son
~SUMMARIZING - [comput] (summarizing of information in comparatively few summary cards) résumé *m* sélectif
~TRANSMISSION - [auto] changement *m* de vitesse à présélecteur
SELECTIVELY ADDRESSABLE MEMORY - [comput] mémoire *f* à adressage facultatif
SELECTIVITY - [radio] (the quality in a radio receiving circuit of responding more readily to signals to which it is tuned to others) sélectivité *f*
~AUTOMATIC CONTROL - [radio] accord *m* automatique de sélectivité
~CHARACTERISTIC - [radio] (the relation of the selectivity ratio to some given variable, generally the frequency separation of the simultaneous excitations) caractéristique *f* de sélectivité
~DISCRIMINATION - [telecomm] sélectivité *f* d'un filtre
SELECTIVITY FACTOR - s. selectivity ratio
~RATIO - [radio] taux *m* de sélectivité
SELECTOR - [telecomm] (in automatic telephony) sélecteur *m*
[comput] sélecteur *m*
~BANK - [telecomm] champ *m* radial de sélection
~BUTTON - [telev] bouton-poussoir *m* sélecteur
~CARRYING CAPACITY - [telecomm] charge *f* d'un sélecteur
~FORK - [auto] fourchette *f* de commande de changement de vitesse, fourchette *f* de baladage
~HUNTING TIME - [telecomm] temps *m* de recherche libre
~PULSE - [telecomm] (pulse actuating a time selector) impulsion *f* sélectrice
~RELAY - [electron] relais *m* pas à pas
~REPEATER - [telecomm] commutateur *m* discriminateur
~ROD - [el] arbre *m* porte-balais
~SHAFT - [telecomm] arbre *m* actionnant des contacts
~SHAFT GUIDE - [telecomm] fente-guide *f*
~SWITCH - [el] (a switch allows of connecting one circuit to any one of a number of others) commutateur *m* pas à pas
~VALVE - [mech] (a valve by means of which any required hydraulic circuit can be chosen) soupape *f* sélectrice
SELECTRON - [comput] (electron tube used as a computer memory) sélectron *m*
SELENATE - [chem] (a slat of selenic acid) séléniate *m*
SELENIATE - s. selenate
SELENIC ACID - [chem] acide *m* sélénique
SELENIOUS ACID - [chem] (an analytical) acide *m* sélénieux
SELENITE - [geol] (colourless and transparent variety of gypsum) sélénite *f*
SELENITIC CEMENT - [constr] (a mixture of weakly hydraulic cement with 5% of plaster of Paris) ciment *m* séléniteux
~LIME - s. selenitic cement
SELENIUM - [chem] (a non metallic element, A. N 34, A.W. 78 96 symbol Se with several allotropic forms. It has photoelectric properties and is used in glass manufacture) sélénium *m*
~BARRIER-LAYER PHOTOCELL - [electron] cellule *f* photoélectrique à couche d'arrêt au sélénium
~CELL - [el] (photo-electric cell depending for its action on the influence of light on the conductivity of selenium) cellule *f* à sélénium
~DIETHYLDITHIOCARBAMATE - [chem] (a vulcanization agent and accelerator) diéthyldithiocarbamate *m* de sélénium
~DIOXIDE - [chem] (a catalyst, oxidizing agent and analytical reagent) bioxyde *m* de sélénium
~GLASS - [photo] (red-orange glass filter used in colour cinematography) filtre *m* au sélénium
~HALIDES - [chem] halogénures *mpl* de sélénium
~LAYER - [electron] (active layer in dry rectifiers) couche *f* de sélénium
~RECTIFIER - [el] (a dry-contact rectifier employing a metal-to-selenium surface) redresseur *m* à sélénium
~RED - [paint] (alternative name for Cadmium red, a complex of cadmium sulphide and selephide. A pigment with good light, heat and alkali resistance,

but sensitive to acids) rouge *m* de sélénium
SELENIUM SULPHIDE - [chem] (reddish-orange, poisonous powder used for the treatment of skin conditions) sulfure *m* de sélénium
SELENIZING - [metall] revêtement *m* au sélénium
SELENOGRAPHY - [astr] (the description of the moon's surface) sélénographie
SELENOLOGY - [astr] (the science which treats of the moon in the way geology treats of the earth) sénélogie *f*
SELENOPHONE - [acoust] (photographic record of a sound on paper) sélénophone *m*
SELENOSIS - [med] intoxication *f* sélénique
SELENOUS ACID - s. selenious acid
SELF - [el] self *m*, bobine *f* de choc, inductance *f* d'arrêt
~-ABSORPTION - [radiat] (absorption of radiation in the body of material where it originates) auto-absorption *f*
~-ABSORPTION COEFFICIENT - [phys] coefficient *m* d'auto-absorption
~-ABSORPTION EFFECT - [radiat] effet *m* d'auto-absorption
~-ACTING - [gen mech etc] (automatic) automatique
~-ACTING BAND - [text] corde *f* de renvideur
~-ACTING BORING BAR - [mech] barre *f* d'alésage automatique
~-ACTING BRAKE - [mech] frein *m* automatique
~-ACTING CIRCULAR TABLE - [mach tool] table *f* circulaire automatique
~-ACTING CLUTCH - [mech] préhension *f* automatique
~-ACTING CONTROL - [contr] réglage *m* direct
~-ACTING CONTROLLER - [contr] (controller deriving the power required for its operation solely from the controlled physical quantity) combinateur *m* direct
~-ACTING DISCONNECTION - [mech] débrayage *m* automatique
~-ACTING FEED WITH INSTANTANEOUS STOP-MOTION - [mach tool] pression *f* automatique à débrayage instantané
~-ACTING INCLINE - [mining] plan *m* incliné automoteur
~-ACTING INJECTOR - [mech] injecteur *m* en mise en marche automatique
~-ACTING LIFT OF THE TOOL - [mach tool] levée *f* automatique de l'outil
~-ACTING MULE - s. self-actor mule
~-ACTING REGULATOR - [mech] régulateur *m* automatique
~-ACTING SPINNER - [text] fileur *m* au renvideur
~-ACTING STRIPPER - [text] débourreuse *f* automatique
~-ACTING SWITCH - [railw] aiguille *f* automatique
~-ACTING TEMPLE - [text] templet *m* automatique
~-ACTING TWINER - [text] renvideur *m* à retordre
~-ACTOR - [text] renvideur *m*
[mech] (an automatic machine) machine *f* automatique
~-ACTOR MULE - [text] (automatic spinning machine for cotton or woolen yarns) renvideur *m* automatique
~-ACTUATED CONTROLLER - s. self-acting controller
~-ADHESIVE TAPE - [gen] bande *f* adhésive
~-ADJUSTING GUIDING CHEEK - [el] (in lift) patin *m* de guidage à réglage automatique
SELF-ADJUSTING SEAL - [mech] self-garniture *f* métallique
~-ADJUSTING STONE-DOGS - [constr] pinces *fpl* articulées pour monter les pierres de taille
~-ALIGNING - [mech] auto-alignement *m*, auto-centrant
~-ALIGNING BALL-BEARING - [mech] roulement *m* à rangée de billes et à rotule
~-ALIGNING BEARING - [mech] (a bearing designed to move automatically into alignment with the shaft it carries) palier *m* flexible
~-ALIGNING SYSTEM - [contr] système *m* à auto-alignement
~-ALIGNING TORQUE - [mech] couple *m* de redressement
~-ANNEALING - [metall] (term applied to lead, tin and zinc, which recrystallize at air temperature) recuit *m* naturel
~-BAKING ELECTRODE - [el] (electric arc furnace electrode consisting of plastic carbonaceous material fed in through a metal sleeve and baked by the heat of the furnace itself) électrode *f* à auto-cuisson
~-BALANCE PROTECTION - [el] (a method of protecting transformers and generators from internal faults) protection *f* à auto-équilibrage
~-BALANCING - [mech el etc] à équilibrage automatique
~-BIAS - [el] polarisation *f* automatique de grille
~BINDER - [agric] moissonneuse-lieuse *f*
~-BREAKING STRUT - [aero] (type of strut which carries its own actuator) montant *m* du train automatique
~-CAPACITY - s. self-capacitance
~-CAPACITANCE - [el] (the capacitance which is inherent in a resistor or conductor) capacité *f* propre
~-CATALYSIS - [chem] auto-catalyse *f*
~-CENTERING - [mech] à centrage automatique, à serrage concentrique
~-CENTRING - s. self-centering
~-CENTRING CHUCK - [mach tool] (or universal chuck; lathe-chuck for cylindrical work in which the jaws are always kept concentric by a scroll) mandrin *m* à serrage concentrique
~-DIES - [mech] coussinets *mpl* à rapprochement concentrique
~-CENTRING PUNCH - [mech] poinçon *m* à centrage automatique
~-CENTRING PUSHER - s. self-centring punch
~-CHARGE - [el] (additional contribution to the electric charge of a charged particle due to vacuum polarization arising from the field produced by the original charge) charge *f* propre
~-CHECKING - [comput] contrôle *m* automatique
~-CHECKING NUMBER - [comput] indicatifs *mpl* numériques à autocontrôle
~-CHECKING SHUTTER - [photo] (self-setting shutter) obturateur *m* toujours armé
~-CLEANING - [mech] (of an oil filter) à nettoyage automatique
~-CLEANING FLAT - [text] chapeau *m* à débourrage automatique
~-CLEANSING - [hydr] (in sanitary engineering) autolavage *m*
~-CLINKERING GENERATOR - s. self-clinkering producer

SELF-CLINKERING GRATE - [th eng] grille *f* mécanique
~-CLINKERING PRODUCER - [gas ind] gazogène *m* à décrassage automatique
~CLOSING - [mech & el] à fermeture automatique
~-CLOSING DOOR - [el] (door which is opened manually and closed automatically by means of an electrical device) porte *f* automatique
~-CLOSING GATE - [el] (a gate which is opened manually and closed automatically by means of an electrical device) porte *f* à fermeture automatique [mining] barrière *f* automatique
~-COCK - [hydr] robinet *m* à ressort
~-COCKING - [firearms] à armement automatique
~COLOUR EFFECT - [chem] effet *m* de couleur naturelle
~-COMBUSTION - [chem] autocombustion *f*
~-COMPENSATED MOTOR - [el] (with primary rotor) moteur *m* autocompensé à alimentation rotorique
~-CONSISTENT FIELD METHOD - [electron] méthode *f* de Hartree
~-CONTAINED - [mech etc] independant, autonome
~-CONTAINED ACCUMULATOR - [hydr] accumulateur *m* indépendant
~-CONTAINED CARDAN JOINT - [auto] joint *m* de cardan incorporé
~-CONTAINED COOLING UNIT - [th eng] (in air conditioning)appareil *m* de conditionnement d'air autonome
~-CONTAINED COUNTERSHAFT - [mech] renvoi *m* adhérent (au bâti)
~-CONTAINED DOUBLING SPINDLE - [text] broche *f* de retordage Rabbeth
~-CONTAINED FRICTION SCREW PRESS - [mech] presse *f* à vis à friction indépendante
~-CONTAINED INSTRUMENT - [instr] appareil *m* autonome
~-CONTAINED PEDAL OPERATED COUNTERSHAFT - [mech] renvoi *m* adhérent avec débrayage par pédale
~-CONTAINED PRESS - [mech] (a press which contains a built-in mechanism for generating the necessary hydraulic pressure for operating it) presse *f* indépendante, presse *f* à commande directe
~-CONTAINED UNIT - [mech] groupe *m* d'assemblage autonome
~-COOLED - [th eng] autorefroidi
~-CORING CHISEL - s. self-coring mortising chisel
~-CORING MORTISING CHISEL - [impl] bédane *m* à joues
~-CORRECTING CODE - [comput] (a coding whereby errors are checked and corrected) code *m* autocorrecteur
~-CURING - [rubber ind] autovulcanisant
~-DEMAGNETIZATION - [phys] (a disadvantage occurring when magnetically testing a specimen of material in rod or strip form) autodésaimantation *f*
~-DISCHARGE - [el chem] (loss of chemical energy in a cell by internal currents unrelated to any external circuit) décharge *f* spontanée
~-DISCHARGING WATER-BUCKET - [mech] cuffat *m* à vidange automatique
~-DOCKING DOCK - [constr] (a floating dock built in sections, so that each section can be unbolted for repair purposes) bassin *m* de desserte à sections démontables
~-DRIVE CAR - [auto] voiture *f* en location sans chauffeur
SELF-DUMPING BUCKET - [mech] benne *f* à culbutage automatique
~-DUMPING CAGE - [mining] cage *f* à déchargement automatique
~-DUMPING SKIP - [mining] skip *m* à déversement automatique
~-ELECTRODE - [el] (electrode composed of the material being analyzed) électrode *f* autoémission
~-EMPTYING BORER - [mech] sondeuse *f* à soupape [constr] sonde *f* à clapet
~-ENERGY - [phys] (the energy equivalent for the rest mass of a particle) énergie *f* propre
~-ERECTING SCREEN - [photo] écran *m* à érection automatique
~-EXCITATION - [el] (a form of machine excitation in which the supply to the field system is obtained from the machine itself) auto-excitation *f*
~-EXCITATION CURRENT - [el] courant *m* d'auto-excitation
~-EXCITED - [el] (relating to a machine to denote that the field magnets are wholly, or substantially excited from the machine itself) auto-excité
~-EXCITED OSCILLATOR - [radio] (normal oscillator in which the excitation of the grid circuit is derived from the alternating current flowing in the anode circuit) oscillateur *m* auto-excité
~-EXCITED SENDER - [radio] émetteur *m* à auto-excitation
~-EXCITING - [el] auto-excitatrice *f*
~-EXCITING DYNAMO - [el] dynamo *f* auto-excitatrice
~-EXTINGUISHING - [ind chem] (term used of a substance which ceases to burn itself after ignition) auto-extinctif
~-FACED - [constr] (of a stone, which splits along natural cleavage planes, leaving faces which do not recuire dressing) à face en pierre
~-FEEDER - [agric] engreneur *m* automatique
~-FEEDING - [agric] alimentation *f* automatique
~-FEEDING REAMER - [tool] alésoir *m* à bout fileté pour l'amorçage
~-FERTILIZATION - [genet] (in hermaphrodite animals) autofécondation *f*
~-FLUXING GANGUE - [min] gangue *f* fusible
~-FLUXING ORE - [min] minerai *m* à gangue fusible
~-FOCUSED PICTURE TUBE - [telev] tube *m* image à autofocalisation
~-FRACTIONATING OIL PUMP - [mech] pompe *f* à diffusion à huile à autofraction
~-GRIPPING SCREWDRIVER - [mech] tournevis *m* automatique
~-HARDENING - [metall] trempe *f* à l'air [metall] (adj) autotrempant
~-HARDENING STEEL - [metall] (steel hardening on cooling in air and does not need being quenched in oil or water) acier *m* autotrempant
~-HEAL - [bot] brunelle *f* vulgaire
~-HEALING CAPACITOR - [electron] condensateur *m* autocicatrisant
~-HETERODYNE - [radio] (autoheterodyne) autodyne *m*
~-IGNITION - [phys] auto-inflammation *f* [mech] (of a motor) auto-allumage *m* [astronaut] (of the fuel in rockets) autocombustion *f*
~-IGNITION TEMPERATURE - [phys] (the temperature at which a substance ignites spontaneously, without

the application of a flame) température *f* d'auto-inflammation

SELF-IMPEDANCE - [radio] (at any pair of terminals of a network) auto-impédance *f*

~-INDUCED CURRENT - [el] courant *m* d'induction propre

~-INDUCED VIBRATION - [phys] vibration *f* auto-excitée

~-INDUCTANCE - [el] (the property of a circuit by which self-induction occurs, also called coefficient of self-induction) inductance *f* propre; coefficient *m* d'induction propre

~-INDUCTION - [el] (the property of a conductor which gives rise to a reserve E.M. F when a change takes place in the current flowing in it) auto-induction *f*, self- induction *f*

~-INFECTION - [med] auto-infection *f*

~-INFLICTED INJURIES - [med] automutilation *f*

~-INJURY - s. self-inflicted injuries

~-INSTRUCTED CARRY - [comput] report *m* automatique

~-LIMITING CHAIN REACTION - [nucl] (a chain reaction in which the moderator automatically keeps the reaction within preset limits) réaction *f* nucléaire en chaîne à automodération

~-LOADING - [mech] à chargement automatique

~-LOCKING - [mech] serrure *f* automatique

~-LOCKING COLLAR - [mech] bague *f* à serrure automatique

~-LOCKING NUT - [mech] écrou *m* de fermeture automatique

~-LOCKING SETSCREW - [mech] contre-écrou *m* à blocage automatique

~-LOCKING TRIPOD - [photo] trépied *m* se dressant automatiquement

~-LUBRICATING - [mech] (term used of a part which contains its own lubricant, espec. in pores in the material) auto-lubricant, à graissage automatique

~-MAINTAINED DISCHARGE - [electron] (discharge characterized by the fact that it maintains itself after the external ionizing agent is removed) décharge *f* autonome

~-MAINTAINING - [gen]chem nucl] (of a reaction) autonome

~-MAINTAINING GAS DISCHARGE - s. self-maintained discharge

~-MAINTAINING NUCLEAR CHAIN REACTION - [nucl] réaction *f* en chaîne auto-entretenue

~-MULE FOR FINE YARNS - [text] renvideur *m* pour fils fins

~-MULTIPLYING CHAIN REACTION - [nucl] (chain reaction in which the number of neutrons generated is automatically multiplied without external interference) réaction *f* en chaîne automultiplicatrice

~-OILING - [mech] graissage *m* automatique

~-OILING PLUMMET BLOCK - [mech] palier *m* à graissage automatique

~-OPENING - [mech] à déclenchement automatique

~-OPENING DIE HEAD - [mech tool] filière *f* à déclenchement automatique

~-OPERATED CONTROLLER - [contr] régulateur *m* à auto-activation

~-OSCILLATING SENDER - s. self-excited sender

~-OSCILLATION - [radio] (generation of continuous oscillations by a regenerative receiver when the degree of reaction is increased beyond a certain limit) auto-oscillation *f*

SELF-POLLINATION - [bot] (the transfer of pollen from the anthers to the stigma of the same flower) autogamie *f*, autopollinisation *f*

~POTENTIAL - [oil ind] polarisation *f* spontanée

~-POWERED RECORDER - [el acoust] (a battery-fed recorder) magnétophone *m* à batterie

~-PRIMING - [hydr] auto-amorçage *m*

~-PRIMING PUMP - [mech] (a pump incorporating an airexpelling device to eliminate the need for priming) pompe *f* à auto-amorçage

~-PROPAGATING NUCLEAR CHAIN REACTION - s. self-maintaining nuclear chain reaction

~-PROPELLED - [mech] autopropulsé

~-PROPELLED ELECTRIC LOCOMOTIVE - [railw] locomotive *f* électrique autopulsée

~-PROPELLING - [mech] automoteur, automobile

~-PROTECTED TUBE - [electron] (X-ray tube so constructed that a protection against excessive emission of radiation is obtained) tube *m* à autoprotection

~-PULSE MODULATION - [radio] (modulation by an internally generated pulse) modulation *f* par impulsions locales

~-PUNISHMENT - [med] autopunition *f*

~-QUENCHED COUNTER TUBE - [radiat] tube *m* compteur autocoupeur

~-QUENCHING - [radiat] (internally, terminating a pulse of ionization current in a Geiger- Muller counter) autocoupure *f*, étouffement *m*

~-QUENCHING DETECTOR - [radio] détecteur *m* à auto-extinction

~-QUENCHING OSCILLATOR - [radio] oscillateur *m* à extinction

~-READING ROD - s. self-reading staff

~-READING STAFF - [surv] (a levelling staff the graduations of which are so arranged that the observer at the level may read the value at which his line of sight intersects the staff) mire *f* parlante

~-RECORDING - [el acoust] enregistreur *m* automatique

~-RECTIFYING X-RAY TUBE - [radiat] tube *m* autoredresseur

~-RESITERING - [instr] à enregistrement automatique

~-REGISTERING APPARATUS - [instr] appareil *m* enregistreur

~-REGISTERING RAIN-GAUGE - [instr] pluviomètre *m* enregistreur

~-REGULATING - [mech] (of a petrol feeding pump in autos or aircraft) autorégulateur

~-REGULATING ARC-WELDING TRANSFORMER - [el] (arc-welding transformer in which the voltage drop increases substantially with the load current) transformateur *m* autorégulateur de soudage à l'arc

~-REGULATING CATTLE BOWL - [agric] abreuvoir *m* automatique

~-REGULATING D.C. WELDING GENERATOR - [el] génératrice *f* autorégulatrice de soudage à l'arc

~-REGULATION - [nucl] (of a nuclear reactor) autorégulation *f*
[el] autorégulation *f*
[contr] automatisme *m*

~-REGULATION AT BOTH OF BEAM - [text] réglage *m* automatique des deux côtés

~-RESET - [el] à réenclenchement automatique

~-RESETTING - [electron] rappel *m* automatique

~-RESONANT FREQUENCY - [electron] fréquence *f* de résonance propre

~-RESTORING COHERER - [radio] (a coherer in which

the coherer contact reverts automatically to its original condition after a signal) cohéreur *m* auto-régénérateur

SELF-RESTORING DROP - [telecomm] (of a telephone exchange) volet *m* à relèvement automatique par enfoncement de la fiche

~-RESTORING INDICATOR - s. self-restoring drop

~-SATURATING RECTIFIER - [radio] (a half-wave rectifying circuit element connected in series with output windings of a saturable reactor in the self-saturating magnetic amplifier circuit) redresseur *m* à autosaturation

~-SATURATION - [el] (in magnetic amplifiers, the connection of half-wave rectifying circuit elements with the output windings of the saturable reactors) auto-excitation directe

~-SCATTERING - [nucl] (the scattering of radioactive radiations by the substance emitting the radiation) autodiffusion *f*

~-SCREEN PLATE - [photo] plaque *f* orthochromatique sans écran

~-SEALING - [mech] autoscellement *m*

~-SEALING DOOR - [metall] porte *f* sans lutage, porte *f* autolutante

~-SEALING INJECTION NOZZLE - [mech] tuyère *f* d'injection à encliquetage automatique

~-SEALING JOINT - [mech] joint *m* sec, joint *m* sans lut

~-SEALING REPAIR PATCH - [rubber ind] (for tyres) plaque *f* de réparation autovulcanisant

~-SEALING TRIMMING PAD - [rubber ind] (for tyres) emplâtre *m* autocollant à vérifier l'équilibrage

~-SELECTING FEED MECHANISM - [mech] mécanisme *m* d'avancement autosélecteur

~-SEQUENCING REGISTER - [comput] registre *m* à auto-séquence

~-SERVICE - [comm] libre-service *m*

~-SHIELDING - [radiat] (the shielding from neutrons arising outside the region of the inner portion of material by its outer portion) autoprotection *f*

~-SHIELDING FACTOR - [nucl] facteur *m* d'autoprotection

~-STARTER - [auto] (small electrical motor, fed from the battery and used to start the engine) auto-démarreur *m*, moteur *m* de lancement

~-STARTING - [auto] à mise en marche automatique

~-STARING ROTARY CONVERTER - [el] (synchronous converter requiring no separate starting motor) commutatrice *f* à démarrage automatique

~-STEAMING PRODUCER - [gas ind] gazogène *m* autoproducteur de vapeur

~-STERILITY - [genet] (in hermaphrodites, the condition in which self-fertilization is impossible) autostérilité *f*

~-STOPPING MOTION - [mech] mécanisme *m* d'arrêt automatique

~-STOWING - [naut] (e.g. an anchor) sans arrimage

~-SUFFICIENCY - [gen] indépendance *f*

~-SUPPORTING - [gen] suffisant à ses besoins [radiat] s. self-maintaining

~-SUPPORTING AERIAL CABLE - [el] câble *m* aérien auto-porteur

~-SUPPORTING AERIAL MAST - [radio] (aerial mast supported by its own structure) tour *f* d'antenne autoporteuse

~-SUPPORTING ANTENNA TOWER - s. self-supporting aerial mast

SELF-SUPPORTING PARTITION - [constr] cloison *f* en décharge

~-SURGE IMPEDANCE - [el] impédance *f* d'onde

~-SUSTAINING - s. self-maintaining

~-SUSTAINING FRICTION-BRAKE HOIST - [mech] monte-charge *m* à frein de sûreté à friction

~-SYNCHRONIZER - [el] autosynchroniseur *m*

~-SYNCHRONIZING - [el] (term denoting a synchronous machine which can be switched on to the a.c. supply without being in exact synchronism with it) à autosynchroniseur

~-TAPPING SCREW - [mech] (hardened steel screw designed to form its own thread, espec. in thin sheet material) vis *f* se frayant un filet

~-THREADING - [text] enfilage *m* automatique

~-THREADING SHUTTLE - [text] navette *f* à enfilage automatique

~-TIME LEVER - [photo] levier *m* du retardateur

~-TIMER - [photo] déclencheur *m* automatique

~-TRIPPING - [mech] (of railways-or-road-trucks) à culbutage automatique

~-TIPPING WAGON - [railw] wagon *m* à culbutage automatique

~-TONING PAPER - [photo] papier *m* autovireur

~-TRIMMING - [naut] à autoarrimage

~-VENTILATED - [gen] à autoventilation

~-VENTILATED MOTOR - [el] moteur *m* autoventilé

~-VENTILATION - [gen & railw] autoventilation *f*

~-VULCANIZING - s. self-curing

~-WEIGHTED BACK TOP ROLLER - [mech] cylindre *m* de pression postérieur

~-WELDING - [phys] (of frozen mercury) auto-soudage *m*

~-WHISTLES - [telev] (interference caused by a signal between the oscillator signal or its harmonics, and harmonics of the desired radio-frequency vision or sound signals) autosifflements *mpl*

~-WINDING - [gen] à enroulement automatique [horol] à remontage automatique

~-WINDING MOVEMENT - [horol] mouvement *m* à remontage automatique

~-WINDING WATCH - [horol] (watch winding itself while being worn) montre *f* à remontage automatique

SELL, to - [comm] vendre

~BY AUCTION, to - [comm] vendre à l'enchère

~OFF, to - [comm] solder, liquider

~WHOLESALE, to - [comm] vendre en gros

SELLER - [comm] vendeur *m*

SELLING - [gen & comm] vente *f*, liquidation *f*

~OFF - [comm] liquidation *f*

SELSYN - [el] (self synchronizer) auto-synchroniseur

SELVAGE - [text] (a fabric edge) lisière *f* [mech] (the edge plate of a bolt lock) rebord *m* de serrure [min] (a layer of detrital rocks along a vein or seam) salbande *f*, salbande *f* argileuse

SELVAGEE - [naut] (a skein of yarns) erse *f* en bitord

SELVEDGE - [text] (the edge of a fabric) lisière *f*

~BOBBIN - [text] bobine *f* de lisière

~CAM - [text mach] excentrique *m* pour lisière

~CREEL - [text] râtelier *m* pour fils de lisières

~FEELER - [text] tâteur *m* de lisières

~GUIDE - [text] guide *m* des bords

~HEALD FRAME - [text] lame *f* de lisière

~HEALDS - [text] lisses *fpl* de lisières

~KNITTING MACHINE - [text] machine *f* à tricoter

des lisières
SELVEDGE OF THE SLIVER - [text] bord *m* du ruban
~PRINTING - [text] impression *f* sur lisières
~SINKER - [text] platine *f* cueillante de lisière
~TAPPET - s. selvedge cam
~UNCURLER - [text] ouvreur *m* de lisières
~YARN - [text] fil *m* à lisière
SEMANTICS - [gen] (the branch of glottology which deals with meaning) sémasiologie *f*, sémantique *f*
SEMAPHORE, to - [gen] transmettre par sémaphore, transmettre par signaux à bras
SEMAPHORE - [railw] (traffic lights) sémaphore *m*
~ARM - [railw] bras *m* du sémaphore
SEMEIOLOGY - [med] (the branch of medical science dealing with the symptoms of disease) sémiologie *f*
SEMEIOTIC - [med] (relating to the symptoms of disease) sémiotique
SEMEN - [biol] sperme *m*, semence *f*
SEMENURIA - [med] spermaturie *f*
SEMESTER - [meas] semestre *m*
SEMIACTIVE HOMING GUIDANCE - [radar] radioguidage *m* semi-actif
~TRACKING SYSTEM - [radar] système *m* de poursuite semi-actif
SEMIANTHRACITE COAL - [min] houille *f* maigre anthraciteuse
SEMIAUTOMATIC - [mech] (automatic to a limited extent) semi-automatique
~ADVANCE - [auto] avance *f* semi-automatique
~CONTROLLER - [contr] (electric controller in which the influence directing the performance of some of its basic functions is automatic) combinateur *m* semi-automatique
~CYCLE - [contr] (a cyclic operation which is only to some extent automatic) cycle *m* semi-automatique
~ELECTROPLATING - [metall] (a method of plating in which the cathodes move automatically through a single plating bath) galvanoplastie *f* semi-automatique
~EXCHANGE - [telecomm] (in telephony) bureau *m* semi-automatique
~REPORFORATOR SWITCHING - [telecomm] commutation *f* semi-automatique avec retransmission par bande perforée
~SWITCHING SYSTEM - [telecomm] (in telegraphy) système *m* de commutation semi-automatique
SEMI-AXIS - [geom] demi-axe *m*
SEMIBEAM - [constr] poutre *f* en porte-à-faux
SEMIBITUMINOUS COAL - s.semicoke
SEMIBUTTERLY CIRCUIT - [electron] circuit *m* semi-papillon
SEMIBREVE - [mus] ronde *f*
SEMICALCAREOUS - [geol] demi-calcaire
SEMICANTILEVER WING - [aero] aile *f* en porte-à-faux
SEMICARBAZIDE - [chem] (a base forming salts, it can be prepared from potassium cyanate and hydrazine hydrate) semi-carbazide *f*
~HYDROCHLORIDE - [chem] (an analytical reagent) chlorhydrate *m* de semi-carbazide
SEMICARBAZONES - [chem](the reaction products of aldehydes or ketones with semicarbazide) semicarbasones *mpl*
SEMICELLULOSE - s. semichemical pulp
SEMICHEMICAL PULP - [paper man] (pulp made by a combination of mechanical and chemical processes) pâte *f* semi-chimique
SEMICHORD - [geom] semicrode *f*
SEMICIRCLE - [geom] demi-cercle *m*
SEMICIRCULAR - [gen] demi-circulaire
~ARCH - [arch] voûte *f* plein cintre
[arch] (of a bridge) arche *f* de plein cintre
~CANALS - [anat] (the canals of the internal ear) canaux *mpl* du muscle du marteau
~COMPONENT OF ERROR - [radar] (the component of error which varies sinusoidally with the bearing) composante *f* d'erreur semi-circulaire
~DEVIATION - [naut] (a deviation of iron ships caused by the permanent magnetization of the ship itself or to the effect of soft iron contained in it) déviation *f* semi-circulaire
~FOCUSSING MAGNETIC SPECTROMETER - [instr] spectromètre *m* de masse à focalisation semi-circulaire
~PATH - [nucl] (the path of a particle which is semicircular in shape) parcours *m* semi-circulaire, trajectoire *f* semi-circulaire
~SEGMENT - [instr] (a component part of the Hoffman electrometer) segment *m* chargé
SEMICOKE - [gas ind] (coke obtained by low-temperature carbonization, to obtain a smokeless fuel of higher calorific value than fully carbonized coke) semi-coke *m*, coke *m* de distillation à basse température
SEMICOLON - [print] point-virgule *m*
SEMI-COMBINED CARBON DIOXIDE - [chem] acide *m* carbonique semi-combiné
SEMICONDUCTIVE MATERIAL - [electron] (material acting as semiconductrice
SEMICONDUCTOR - [electron] (electronic conductor with resistivity in the range between metals and insulators, in which the electrical charge carrier concentration increases with increasing temperature over a temperature range) semi-conducteur *m*
~COUNTER - [instr] compteur *m* à semiconducteur
~CRYSTAL SLICER - [electron] tailleur *m* de plaquettes semiconductrices
~DETECTOR DOSEMETER - [nucl] dosimètre *m* à détecteur semiconducteur
~DEVICE - [electron] dispositif *m* à semi-conducteur
~DIODE - [electron] (a two-electrode semiconductor device with an symmetrical voltage-current characteristic) diode *f* à semi-conducteur
~JUNCTION - [electron] jonction *f*
~STRAIN GAUGE - [instr] jauge *f* de contrainte semiconductrice
SEMICYCLIC BONDS - [chem] (the double linkage between a carbon atom in a ring and a carbon of a side chain) liaisons *fpl* semicycliques
SEMICYLINDRICAL LEAD BASIN - [metall] auge *f* semi-cylindrique en plomb
SEMIDIAMETER - [astr] (the radius of an observed celestial body which is added to or subtracted from an observation made on the limb of the body) demi-diamètre *m*
SEMIDIRECT SULPHATE RECOVERY - [ind chem] sulfation *f* semidirecte
SEMI-DROP CENTRE RIM - [rubber ind] jante *f* à base excentrée, jante *f* demi-creuse
~-DUAL CONTROL - [contr] commande *f* semi-alternative
SEMIELLIPTIC SPRING - [mech] ressort *m* semi-elliptique, ressort *m* à demi-pincette
SEMIEMPIRICAL MASS FORMULA - [nucl] (mass for-

mula based on the liquid-drop model of the nucleus) formule *f* semi-empirique de masse
SEMIENCLOSED - [el] (of electrical machines) semi-enfermé
~FUSE - [el] coupe-circuit *m* à fusion semienfermée
SEMIFINISHED - [gen] semi-ouvré
~FLAT - [metall] larget *m*, platine *f*
~GOODS - [ind] demi-produits *mpl*
~PRODUCTS - s. semifinished goods
~STEEL - [metall] acier *m* en barres
SEMIFIXED GIRDER - [constr] poutre *f* semi-encastrée, poutre *f* encastrée à une extrémité et sur appui simple à l'autre
SEMIFLEXIBLE - [phys & gen] semi-rigide
SEMI-FLOATING - [mech] semi-flottant
~-FLOAT - [rubber ind] régénéré *m* dont les fibres textiles sont enlevées
SEMIFLOATING AXLE - [mech] (an axle with the shaft transmitting torque and also carrying load) essieu *m* semi-porteur
SEMIFLUID - [phys] semi-fluide
SEMI-FLUSH SWITCH - [el] interrupteur *m* encastré partiellement
SEMIGIRDER - [constr] (a cantilever) poutre *f* en porte-à-faux
SEMI-GLOSS - [gen] demi-mat
SEMIIMMERSED LIQUID-DRENCHED FUSE - [el] (a liquid quenched fuse in which the fuse is above the liquid before operating but drawn down into it during or after fusion) fusible *m* à extinction par liquide à immersion partielle
SEMIINDIRECT LIGHTING - [el] éclairage *m* semi-indirect
SEMIKILLED STEEL - [metall] acier *m* semi-calmé
SEMILIQUID - [phys] semi-liquide
~EXTRACT - [chem] extrait *m* mou
SEMILUNAR - [gen] (crescent-shaped) à croissant
SEMIMACHINED - [gen] mi-ouvré
SEMIMAGNETIC CONTROLLER - [contr] (electric controller with only part of its basic functions performed by electromagnets) combinateur *m* semi-magnétique
SEMIMANUFACTURED - [gen] mi-ouvré
~POWER LOOM - [text] métier *m* à tisser mi-mécanique
SEMIMILD STEEL - [metall] acier *m* demi-doux
SEMINARCOSIS - [med] sommeil *m* crépusculaire
SEMINIFEROUS - [genet] (semen producing) séminifère
SEMINURIA - [med] s. semenuria
SEMI-PERFORATED TAPE - [comput] bande *f* perforée "chadless"
SEMIPERMEABLE - [phys] semi-perméable
~DIAPHRAGM - s. semipermeable membrane
~MEMBRANE - [chem] (a membrane which will permit the passage of solvent molecules but not those of the solute) membrane *f* semi-perméable
SEMI-PERMEABLE PARTITION - [el] membrane *f* semi-perméable
SEMIPOLAR BOND - [chem] (a bond in which a pair of electrons supplied by one atom is shared between it and another) liaison *f* semi-polaire
SEMIPOSITIVE MOULD - [plast ind] (type of mould which allows a certain amount of material to escape) moule *m* semi-positif
SEMIPRESSED BALE - [text] balle *f* demi-pressée
SEMIPRONATION - [med] demi-pronation *f*
SEMIPROTECTED - [el] (of electrical machine and apparatus) semi-enfermé
SEMIQUAVER - [mus] double-croche *f*
SEMIREMOTE CONTROL - [radio] semi-télécommande *f*
SEMIRIGID - [gen] semi-rigide
~AIRSHIP - [aero] (an airship which has a rigid structural member to distribute the weight of the load) dirigeable *m* semi-rigide
~THEORY - [aero] (the theory of elastic structures using a finite number instead of the theoretic infinite number of degrees of freedom) théorie *f* de la semi-rigidité
SEMI-SELF-MAINTAINED DISCHARGE - [electron] décharge *f* semi-autonome
SEMISPAN - [aero] (one-half of the span of an aircraft) demi-envergure *f* de l'aile
SEMISTEEL - [metall] fonte *f* aciérée
~CASTING - [metall] coulée *f* de fonte aciérée
SEMI-STRAIN INSULATOR - [el] suspension *f* à division de la tension
SEMISUPINATION - [med] demi-supination *f*
SEMITONE - [acoust] (the pitch interval between successive toned in a chromatic scale) demi-ton *m*, blanche *f*
SEMITRAILER - [auto] semi-remorque *f*
SEMIVALENCE - [chem] (a singlet linkage) semi-valence *f*
SEMI-WATER GAS - [chem] (a mixture of carbon monoxide, carbon dioxide, hydrogen and nitrogen) gaz *m* mixte de gazogène
SEMIWILD SILK SPINNER - [zool] papillon *m* de ver à soie demi-sauvage
SEMIWORSTED SPINNING - [text] filature *f* de peigné-cardé
SENARMONTITE - [min] (native antimony trioxide with uses in the manufacture of protective coatings) sénarmontite *f*
SEND, to - [gen] envoyer
[gen] (to forward) expédier
[radio] transmettre
[naut] tanguer fortement
SEND - [naut] (the flow of the waves; the impetus of the waves) fort tangage *m*
~-OUT - [gas ind] (USA only; a gasholder distribution) émission *f*
~-OUT CHART - [gas ind] (district pressure chart) diagramme *m* d'émission
SENDER - [gen] envoyeur *m*, expéditeur *m*
[telecomm] (in telegraphy and radiotelegraphy) signaleur *m*
[radio] (a transmitter; the equipment for generating and sending radio signals) émetteur *m*
~SELECTION - [telecomm] présélection *f*
~SELECTOR - [telecomm] (in automatic telephony) chercheur *m* d'enregistreur
SENDING - [gen] expédition *f*
[telecomm] émission *f*, transmission *f*
~-END IMPEDANCE - [telecomm] (of a line) impédance *f* de sortie
~KEY - [telecomm] manipulateur *m*
SENEGA - [chem] (the dried root of Polygala senega used in medicine as an expectorant) polygala *m* de Viriginie
SENESCENCE - [biol] (the declining of powers prior to death) sénescence *f*
SENESCENT - [biol] sénescent
SENGIERITE - [min] (rare, secondary mineral con-

taining approximately 43% of uranium) sengiérite *f*
SENILE - [gen] (of old age and infirmity) sénile
~DECAY - [gen & biol] dégénérescence *f* sénile
~DELIRIUM - [med] délire *m* sénile
SENILISM - [med] sénilisme *m*
SENILITY - [biol] (the condition of exhaustion and degeneration due to old age) sénilité *f*, caducité *f*
SENNA - [chem] (dried leaflets of leguminous plants used in pharmaceutical products for their purgative properties) séné *m*
~TEA - [chem] infusion *f* de séné
SENNIT - [naut] tresse *f* de paille, tresse *f* de chanvre
SENONIAN - [geol] (the highest of the three stages into which the British Chalk is divided) sénonien
SENSATION - [gen] (the awareness of a physical experience) sensation *f*, impression *f*
~AREA - [acoust] (the auditory sensation area) aire *f* de sensation auditive
~LEVEL - [acoust] (the pressure level of a sound in decibels above its threshold of audibility) niveau *m* de sensation auditive
~UNIT - [acoust] (obsolete term for decibel) décibel *m*
SENSE, to - [gen] sentir intuitivement
[comput] (to determine the arrangement of an element of the mechanical, magnetic, electrical and electronic devices of a computer) explorer
[comput] (to read the holes punched in paper) lire
SENSE - [gen] (the faculty of sensation) sens *m*
[gen] (direction) direction *f*
~AERIAL - [radio] (auxiliary aerial employed to determine the real direction) antenne *f* de levée de ambiguité
~AMPLIFIER - [comput] amplificateur *m* de lecture
~ANTENNA - s. sense aerial
~FINDER - [radar] (instrument designed to determine the direction of the shorter great-circle path) appareil *m* pour la levée de doute
~FINDING - [radar] (the method for determining the direction of the shorter great-circle path and thus remove the 180° ambiguity associated with direction finding) levée *f* du doute
~OF ABSOLUTE PITCH - [acoust] (the faculty of being able to sing immediately any asked for) oreille *f* absolue, ouie *f* absolue
~OF CURRENT - [el] (direction of current) sens *m* du courant
~OF ROTATION - [mech] sens *m* de rotation
~ORGAN - [zool] (any structure which is adapted for the reception of stimuli) organe *m* du sens
~RESEARCH - s. sense finding
~WINDING - [comput] (a pick-up winding in static magnetic storage) enroulement *m* de lecture
~WIRE - [comput] fil *m* de lecture
SENSELESS - [gen & med] inanimé, sans connaissance
SENSIBILISIN - [biochem] (anaphylactic antibody) sensibilisine *f*
SENSIBILITY - [gen] sensibilité *f*
[instr] (of instrument) sensibilité *f*
[bot] (the condition of being liable to parasitic attack) susceptibilité *f*
~OF SILK TO ALKALIES - [text] sensibilité *f* de la soie aux alcalis
~RECIPROCAL - [meas] (of a balance; the change in load which is required to change the equilibrium position by one division at any load) réciproque *m* de sensibilité
SENSIBILIZATION - s. sensitization
~AGENT - [chem] agent *m* de sensibilisation
SENSIBLE - [gen] sensible, perceptible, coscient, sensé
~HEAT - [phys] (the amount of heat which, when added or subtracted, alters the temperature of the body) chaleur *f* sensible
~HORIZON - [surv] (the visible horizon) horizon *m* visible
SENSILLA - [anat] (small sensory structure) papille *f* tactile
SENSING - [radar] (the relative direction of motion of a deviation indicator needle, caused by a vehicle leaving the desired flight-path) déviation *f* de l'aiguille
~COIL - [comput] enroulement *m* de lecture
~CONTACT - [comput] contact *m* de lecture
~HEAD - [comput] tête *f* de lecture
SENSITIVE - [gen & zool] (capable of receiving stimuli) sensible, sensitif, perceptible
[anat] (of a nerve) sensitif
[instr] (of instruments, photographic material etc) sensible
~BALANCE - [meas] balance *f* sensible
~DRILL - [mech] mèche *f* sensitive
~DRILLING MACHINE - [mach tool] mèche *f* sensitive
~FLAME - [phys] (gas flame changing its height and shape when sound waves fall on it) flamme *f* sensible
~GANG DRILL - [mach tool] machine *f* à percer sensitive multiple
~LAYER - [photo] (of photographic material) couche *f* sensible
~MATERIAL - [photo] matériel *m* sensible
~PAPER - [photo] papier *m* sensible
~PLATE - [photo] plaque *f* sensible
~PRECISION DRILL - [tool] machine *f* à percer sensitive de précision
~REGION - [radiat] (part of a cell which is particularly sensitive to radiation) région *f* sensible, part *f* sensible
~STOP MOTION - [mech] mécanisme *m* d'arrêt sensible
~TIME - [nucl] (the duration of supersaturation sufficient for track formation following expansion of a cloud chamber) durée *f* de sensibilité
~VOLUME - [radiat] (that part of a counter tube or ionization chamber which responds to a given radiation) volume *m* utile, volume *m* sensible
SENSITIVENESS - s. sensitivity
~OF THE SCALES - [instr] sensibilité *f*, de la balance
SENSITIVITY - [gen] sensibilité *f*, sensivité *f*
[chem] (the degree to which an explosive responds to initiation by shock) sensibilité *f*
[electron] (the quotient of the signal current developed by a camera tube divided by the incident radiating energy evenly distributed over the photocathode) sensibilité *f*
~DECREASE - [acoust] (loss in quality during recording or reproducing sound) perte *f* de sensibilité
~OF DEFLECTION - [radar & telev] (the displacement of the spot on the screen caused by the appli-

cation of unit potential difference between a pair of deflector plates for a specified voltage on the final accelerator) sensibilité *f* de balayage
SENSITIVITY TIME CONTROL - [radar] (a system to reduce the gain by a decreasing amount for a short period after the pulse has been transmitted) commande *f* différentielle de gain
~TO LIGHT - [phys] (the property of undergoing some change under the influence of light) sensibilité *f* à la lumière
SENSITIZATION - [chem] (the process by which a sol of liophilic colloid becomes liophobic in character) sensibilisation *f* (de sol)
[photo] (the process by which a material is made sensitive to light) sensibilisation *f*
[med] (the process of making susceptible to the action of a drug) sensibilisation *f*
[electron] (activation; process in the manufacture of an electron tube for image pick-up in television) activation *f*
~OF SOL - [chem] (the coagulation of a liophobic sol by the addition of a liophilic sol) sensibilisation *f* de sol
SENSITIZE, to - [chem photo etc] sensibiliser
SENSITIZED DECOMPOSITION - [chem] (chemical decomposition caused by the presence of a second substance which absorbs an exciting radiation) décomposition *f* sensibilisée
SENSITIZER - [chem] (a substance not the catalyst facilitating the start of a catalytic reaction) sensibilisateur *m*
[photo] (generally a dye, used to increase the sensitivity of photographic emulsions) agent *m* sensibilisateur
[electron] (an impurity, or displaced atom, which produces a new spectral-region absorption and excitation in the luminescent material) activateur *m*
[telev] (the agent used for the sensitization of an electron tube for image pick-up) activateur *m*
SENSITIZING BATH - [photo] bain *m* de sensibilisation
~PULSE - [radar] (or indicator gate; a rectangular voltage to the grid of an indicator cathode-ray tube to sensitize it during the desired time of the operating cycle) impulsion *f* de sensibilisation de l'indicateur
~SOLUTION - [photo] solution *f* sensibilisatrice
SENSITOMETER - [instr] (instrument designed to test and measure the sensitivity to light of photographic material) sensitomètre *m*
SENSITOMETRIC FILTER - [photo] filtre *m* sensitométrique
SENSITOMETRY - [photo] (the measurement of the light-responsed characteristics of photographic film) sensitométrie *f*
SENSOR - [instr] organe *m* sensitive, détecteur *m*, capteur *m*
SENSORIAL - [physiol] sensoriel
~POWER - [physiol] énergie *f* sensorielle
SENSORIUM - [physiol] (the seat of sensations; corresponding to the nervous system) sensorium *m*
SENSORY - [physiol] (conveying or producing sense impulses) sensoriel
~DEAFNESS - [med] surdité *f* de perception
~DISORDER - [med] affection *f* sensorielle
~MOTOR - [anat] nerve *f* sensitivo-moteur
~NERVE - [anat] nerf *m* sensoriel
SENSUALISM - [med] sensualité *f*
SENTINEL - [gen] sentinelle *f*
[comput] (symbol marking the beginning or the end of a piece of information) caractère *m* "Fin de bloc"
~PILE - [med] (oedematous mass of anal mucose) prolapsus *m* incomplet du rectum
SEPARATE, to - [gen & phys] (to divide into constituent parts) séparer
SEPARATE APPLICATION - [plast ind] (process of bonding surfaces with a synthetic resin adhesive, in which the resin is applied to one surface and the hardener to the other) application *f* séparée
~COMB - [text] cloison *f* de rainure
~EXCITATION - [el] (form of machine excitation in which the supply to the field system is given by a separate d. c. current) excitation *f* indépendante
~RECOVERY OF SOLVENTS - [chem] séparateur *m* des dissolvants
~SELF-EXCITATION - [el] auto-excitation *f* indirecte
~INTO STRICKS, to - [text] séparer en poignées
~SLUDGE DIGESTION - [hydr] digestion *f* séparée de boues
~SYSTEM - [hydr] (of a sewage) système *m* séparatif (d'assainissement)
~THE STRANDS, to - [text] partager les torons
SEPARATELY COOLED - [el] (of motors) à refroidissement séparé
~INSTRUCTED CARRY - [comput] report *m* adressé
~LEADED CABLE - [el] câble *m* multiplomb
~VENTILATED MOTOR - [el] moteur *m* à ventilation separée
SEPARATING BLADE - [text] plaque *f* de séparation
~COMB - [text] peigne *m* répartisseur
~FILTER - [photo] filtre *m* séparateur
~FUNNEL - [ind chem] entonnoir *m* à séparation
~KNIFE - [text] couteau *m* séparateur
~NEEDLE - [text] aiguille *f* pour la décomposition
~PINS - [text] chevilles *fpl* de séparation
~PLANT - [min] installation *f* de triage
~PLATE - [text] plaque *f* de séparation
~POWER - [soil] pouvoir *m* séparateur
~THE BURS FROM THE FIBRE - [text] séparation *f* des chardons de la fibre
~THE CREAM - [agric] écrémage *m*
~UNIT - [nucl] (component part of an isotope separating plant) groupe *m* de séparation
~WALL - [mining] cloison *f*
SEPARATION - [gen] séparation *f*
[chem] séparation *f*, liaison *f*
[aero] (the detachment of a flow from a surface with which it has been in contact) séparation *f*
[aero] (in manoeuvring during landing; the actual spacing of several aircraft) espacement *m*
[photo] (colour separation) sélection *f*
[electron] séparation *f* d'impulsions de synchronisation
~BUBBLE - [phys] (the region comprised between a re-attaching flow and the solid surface intervening between positions of separations and re-attachment) bulle *f* de séparation
~BY IMPINGEMENT - [min] séparation *f* par choc
~BY SPECIFIC GRAVITY - [min] séparation *f* par densité
~CIRCUIT - [electron] (circuit which will separate signals having different properties, e. g. amplitude, frequency etc) circuit *m* de triage de signaux
[el] circuit *m* bouchon
~COLUMN - [ind chem] (for separation processes)

colonne *f* de séparation
SEPARATION EFFICIENCY - [nucl] (in isotopes) rendement *m* de séparation
~ENERGY - [phys] (the energy per unit charge necessary to remove an electron from a given to an infinite distance) énergie *f* de liaison, énergie *f* de séparation
~FACTOR - [nucl] (the ratio of isotopic ratios after enrichment to that before enrichment) facteur *m* d'enrichissement
[chem] facteur *m* de séparation
~FROM REFLECTION COPY - [print] sélection *f* de l'original opaque
~FROM TRANSPARENCY - [photo] sélection *f* du dispositif
~NEGATIVE - [photo] négatif *m* sélectif
~OF FLOW - [phys] (the flow of slightly viscous fluid past a solid body resembles that of a non-viscid fluid until the thin layer of retarded fluid brought near the wall is brought to rest by the pressure gradients of the flow) détachement *m* de l'effluent
~OF IRON LOSSES - [el] (during magnetic measurements; the separation of iron losses in hysteresis loss and eddy current loss) séparation *f* des pertes
~OF LOSSES - [el] (the determination of the individual losses from the combined losses revealed during the testing of a machine) individualisation *f* des pertes
~OF SIGNALS - [telev] triage *m* des signaux
~OF THE DEAD WOOL - [text] séparation *f* de la laine morte
~PLANT - [nucl] (the chemical plant in which isotopes are separated) installation *f* de séparation
~POINT - [phys] (the point at which streamline flow leaves the surface of a body) point *m* de détachement
~POSITIVE - [photo] positif *m* sélectif
~POTENTIAL - [nucl] (a measure of the difficulty of preparing a quantity of an isotope mixture) potentiel *m* de séparation
~PRINTING - [photo] compression *f* des valeurs moyennes au tirage
~PROCESSES - [ind proc] (manufacturing processes based on differences in the physical properties of the components of a mixture; distillation, absorption, crystallization, absorption and solvent extraction) procédés *mpl* de séparation
~TOWER - s. separation column
~TUBE - [nucl] (thermal diffusion tube) colonne *f* de diffusion thermique
SEPARATIVE ELEMENT - [nucl] élément *m* séparateur
~POWER - [nucl] (a measure of the useful amount of separation completed in unit time by a separative element) pouvoir *m* de séparation
SEPARATOR - [ind chem] séparateur *m*, entonnoir à séparation
[el chem] (a spacing-piece of inactive material used to keep plates from touching each other) séparateur *m*
[mech] (the cage of a ball-bearing) cage *f* à billes
[mech] (oil separator) déshuileur *m*, dégraisseur *m*
[min] (used to separate valuable minerals from one another) trieur *m*
[carp] (a distance piece) pièce *f* d'écartement
~[telecomm] (a buffer stage) étage *m* intermédiaire, étage *m* tampon
[gas ind] colonne *f* d'épuration

SEPARATOR BOX - [oil ind] bassin *m* à décantation
~DIAPHRAGM - [mech] diaphragme *m* à séparation
~FOR BATTERY CASE - [mech] séparateur *m* pour accumulateur
~HEAD - [metall & plast ind] (the conical element torpedo in an extrusion die) torpille *f*
~VANE - [mech] ailette *f* de séparation
SEPIA - [dye] (a dark-brown pigment used in artist's colours, obtained from the cuttle-fish) sépia *f*
~COLOUR - [dye] sépia *f*
~PAPER - [paper man] papier *m* bistre
~TONING - [photo] virage *m* sépia
SEPIOLITE - [min] (meerschaum) sépiolite *f*, écume *f* de mer
SEPSIS - [med] (bacterial invasion of the body tissues) putréfaction *f*, septicémia *f*
SEPT - [gen] enclos *m*
[constr] clôture *f*
SEPTAL DEFECT - [med] inocclusion *f* du septum
SEPIATE - [anat] (divided into two or more chambers) à septum
[bot] (spore) cloisonnée
~WAVEGUIDE - [electron] guide *m* d'ondes cloisonné
SEPTATION - [bot] (the division of a plant member into separate parts) cloisonnement *m*
SEPTECTOMY - [med] résection *f* de la cloison du nez
SEPTFOIL - [arch] en sept lobes
SEPTIC - [gen] (putrid; productive of putrefaction) septique
~TANK - [hydr] (a tank in which sewage is left until purified by the action of anaerobic bacteria) fosse *f* septique
SEPTICAEMIA - [med] (the invasion of the bloodstream by bacteria) septicémie *f*
~OF NEWBORN CALVES - [vet] septicémie *f* des veaux
SEPTICAEMIC - [med] (pertaining to septicaemia) septicémique
~PLAGUE - [med] peste *f* septicémique
SEPTICEMIA - s. septicaemia
SEPTICITY - [med] septicité *f*
SEPTICOPYEMIA - [med] septico-pyémie *f*
SEPTUM - [anat] (partition between two cavities) septum *m*
[bot] cloison *f* (d'une spore)
SEPTUPLE - [math] septuple
SEQUEL - [gen] suite *f*
SEQUENCE, to - [comput] mettre en séquence
SEQUENCE - [cin] (section of a film representing an episode) séquence *f*
[gen] (order of succession) ordre *m* naturel, succession *f*
[el] (the order in which the phases of a polyphase alternating-current supply undergo their cyclic variation of electromotive force) séquence *f*
[mech] (in a working process) ordre *m*, séquence *f*
~-ACTION RELAY - [electron] relais *m* à fonctionnement séquentiel
~ALTERNATOR - [comput] (device alternating the sequence of the instructions) alternateur *m* de fréquence
~CHART - [comput] organigramme *m* de circulation
~-CHECKING ROUTINE - [comput] programme de pas à pas
~CONTACTS - [telecomm] contacts *mpl* échelonnés
~-CONTROL REGISTER - [comput] registre *m* de contrôle de séquence

SEQUENCE-CONTROL TAPE - [comput] bande *f* de programme
~-CONTROLLED CALCULATOR - [comput] calculatrice *f* à commande séquentielle
~-CONTROLLED CONTACTS - [el] contacts *mpl* à séquence imposée
~COUNTER - [comput] compteur *m* de phases de programme
~NUMBER - [comput] numéro *m* d'ordre
~SIGNAL - [telecomm] (signal containing more than one signal component and in which no space components are present) signal *m* complexe
~SWITCH - [el] (a switch, actuated by the tap-changer mechanism, designed to ensure that, once the tap changer is set in motion, it continues to move until a change in tapping is completed) commutateur *m* à échelons
~TIMER - [instr] (of the welding cycles) interrupteur *m* automatique de réglage de la séquence
~VALVE - [mech] (a valve which initiates the movement of a part only when the associated movement of another has been completed) vanne *f* de séquence
SEQUENCER - [electron] (a mechanism designed to put items of information in sequence) mécanisme *m* de séquence
SEQUENCING - [comput] (pertaining to the chronological order of the execution of the instructions) mise *f* en séquence
SEQUENTIAL - [gen] (consequent, resultant) séquentiel, continu
~ACCESS - [comput] accès *m* séquentiel
~ANALYSIS - [contr] analyse *f* séquentielle
~COLOUR SYSTEM - [telev] système *m* de télévision couleur à séquence de trames
~COMPUTER - [comput] calculateur *m* séquentiel
~CONTROL - [contr] automatisme *m* de séquence
~INTERLACE - [telev] analyse *f* entrelacée séquentielle
~PROGRAMMING - [comput] programmation *f* pas à pas
~RATE - [comput] fréquence *f* de récurrence
~SAMPLING - [autom] échantillonnage *m* progressif
~SCANNING - [telev] (progressive scanning) analyse *f* ligne par ligne non entrelacée
~SCHEDULING SYSTEM - [contr] système *m* d'exécution pas-à-pas
~SELECTION - [comput] sélection *f* séquentielle
SEQUESTER - [anat] séquestre *m*
SEQUESTRATION - [phys] (modification of the properties of an ion due its inclusion in an added substance) séquestre *m*
SEQUESTRUM - [anat] séquestre *m*
SEQUIN - [gen] sequin *m*
SEQUOIA - [bot] (giant tree growing in the western United States) sequoia *m*
SEREIN - [met] (the phenomenon of rainfall out of a serene sky) pluie *f* à ciel clair
SERGE - [text] (a type of dress and suiting fabric) serge *f*, sergé *m*
SERIAL - [gen] en série, de série
[cin & telev] à épisodes
~ACCESS - [comput] accès *m* en série
~DIGITAL COMPUTER - [comput] (a computer in which the digits are handled serially) calculateur *m* série
~NUMBER - [gen] (maker's number on a unit or part, showing the individual manufacturing number of it within a batch or series) numéro *m* matricule
SERIAL OPERATION - [comput] (the flow of information through the computer using only one line or more channel at a time) convertisseur *m* série-parallèle
~-PARALLEL CONVERTER - [comput] convertisseur *m* série-parallèle
~RADIOGRAPHY - [radiol] radiographie *f* en séries
~RIGHTS - [leg] droits *mpl* de reproduction
~SELECTION - [comput] mémorisation *f* en série
~TAP - [tool] taraud *m* finisseur
~TRANSFER - [comput] transfert *m* successif, transfert *m* en série
SERIATE, to - [gen] sérier
SERICEOUS - [bot] (with a silky sheen) soyeux
SERICIN - [chem] séricine *f*, gomme *f* de la soie
~COATING - [text] enveloppe *f* de séricine
~SURFACE - [text] s. sericin coating
SERICITE - [min] (a white potash-mica) séricite *f*
SERICULTURE - [text] sériculture *f*
SERIES - [gen] série *f*, suite *f*
[el] (applied to a d.c. machine to denote that the field magnets are excited by a winding connected in series, that is traversed by the same current) série *f*
[photo] (of lenses) trousse *f* (d'objectives)
[chem] série *f*
~BATTERY - [el] pile *f* en série
~CAPACITOR - [el] (capacitor connected in series with a transmission line) capacité *f* additionnelle
~CAPACITOR CLAMP - [telev] circuit *m* de verrouillage à condensateur en série
~CASCADE ACTION - [contr] action *f* en cascade en série
~CHARACTERISTIC - [el] (the characteristic graph relating terminal voltage and load current) caractéristique *f* en série
~CHARACTERISTIC MOTOR - [el] moteur *m* série
~CIRCUIT - [instr] (that part of a meter or measuring instrument supplied by the voltage of the circuit to be measured, or by proportion voltage supplied by a transformer or a voltage divider) circuit *m* de courant, circuit *m* série
~CIRCUITS - [el] (circuits so connected that the same current flows through them) circuits *mpl* en série
~CONNECTION - [el] (connection of a number of circuit elements in such a way that a current flows through all of them in sequence) montage *m* en série, embrochage *m*
[electron] branchement *m* en série
~DECAY - [nucl] (the process of successive radioactive transformation in radioactive series) désintégration *f* en chaîne
~DISTRIBUTION SYSTEM - [el] système *m* de distribution en série
~-EFFICIENCY DIODE - [telev] (a booster diode) diode *f* de récupération
~EXCITATION - [el] excitation *f* en série
~EXPANSION - [math] développement *m* en série
~-FED VERTICAL ANTENNA - [radio] (an end-fed vertical aerial) antenne *f* verticale alimentée en série
~FIELD - [el] (the main field winding of a motor) champ *m* en série
~LOADING - [telecomm] (loading in which reactances are inserted in series with the conductors of a

transmission circuit) charge *f* au moyen de bobines en série
SERIES MODULATION - [radio] (constant-current modulation) modulation *f* à tension constante
~MULTIPOLAR DYNAMO - [el] dynamo *mf* multipolaire-série
~NUMBER - s. serial number
~OF FALLERS - [text] série *f* de barrettes
~OF NIPPERS - [text] rangée *f* de pinces
~OF REACTIONS - [chem] réactions *fpl* caténaires
~OF SPECTRUM LINES - [phys] série *f* spectrale
~OIL COOLER - [mech] (a ducted oil cooler fitted on the after side of a radiator) réfrigérant *m* d'huile en série
~-PARALLEL - [el] (a method of connection in which machines may be connected alternatively in series or in parallel) série-parallèle
~-PARALLEL CONNEXION - [el] (combination of series and parallel connexion) couplage *m* en série-parallèle
~-PARALLEL CONTROL - [contr] (a method of control which is usually employed in electric traction in which d.c. motors initially are connected in series and finally in parallel) régulation *f* par couplage des moteurs
~-PARALLEL CONTROLLER - [el] combinateur *m* en série-parallèle
~-PARALLEL STARTER - [el] démarreur *m* série-parallèle
~-PARALLEL SWITCH - [el] commutateur *m* série-parallèle
~-PARALLEL SWITCHING STARTER - [el] commutateur *m* série-parallèle d'amorçage
~-PARALLEL WINDING - [el] (a winding of a drum armature with more than two paths, in which the winding pitch is approximately double the pole pitch) enroulement *m* série-parallèle
~PRODUCTION - [gen] (manufacture of large numbers of identical objects, using the same design without modification, quantities being greater than in batch/and less than in mass-production) fabrication *f* en série
~RADIATOR - [mech] (a ducted radiator fitted in a position forward of an oil cooler) radiateur *m* en série
~REACTOR - [el] (a reactor connected in series with a voltmeter, or the voltage circuit of a measuring instrument, especially to alter its voltage range) réactance *f* additionnelle
~RECTIFIER CIRCUIT - [electron] couplage *m* en série de redresseurs
~REGULATOR - [el] régulateur-série *m*
~RESISTOR - [el] (the resistor connection in series with some other component) résistance *f* additionnelle
~RESONANCE - [radio] résonance *f* série
~RESONANT CIRCUIT - [electron] circuit *m* de résonance série
~-SHUNT NETWORK - [el] réseau *m* en échelles
~SPECTRA - s. series of spectrum lines
~SYSTEM - [chem] (arrangement of multi-electrode electrolytic cells, in which one pair of electrodes in each cell are connected in series, the unconnected electrodes acting as bipolar elements) système *m* série
~SYSTEM OF DISTRIBUTION - [el] réseau *m* à courant constant
SERIES TRANSDUCTOR - [el] transducteur *m* à couplage série
~TRANSMISSION - [telecomm] transmission *f* série
~TRIP - [el] déclencheur *m* par bobine en série
~TRIPPING - [el] déclenchement *m* par bobine en série
~-TUNED CIRCUIT - s. series resonant circuit
~TWO-TERMINAL PAIR NETWORK - [el] quadripôle *m* série
~VENTILATION - [el] ventilation *f* simple
~WELDING - [metall] soudage *m* série
~WINDING - [el] (a winding of a drum armature in which there are two paths, and of which the winding pitch is approximately double the pole pitch) enroulement *m* série
~-WOUND MOTOR - [el] moteur *m* série
SERIF - [print] (light line or stroke) obit *m*, empattement *m*
SERIMETER TEST - [meas] essai *m* sérimétrique
SERINE - [chem] (a non-essential amino acid with application in biochemical research) sérine
SERINETTE - [mus] (small barrel organ used to teach birds to sing popular melodies) serinette *f*
SERIOSCOPY - [med] sérioscopie *f*
SEROLEMMA - [anat] membrane *f* externe de l'amnios
SERORELAPSE - [med] rechute *f* sérologique
SEROSITIS - [med] sérosite *f*
SEROSITY - [med] sérosité *f*
SEROTHERAPY - [med] sérothérapie *f*
SEROTONIN - [chem] (a naturally-occurring vasoconstrictor which may also be synthesized) sérotonine *f*
SERPENT - [zool] serpent *m*
[mus] (a wind instrument) serpent *m*
SERPENTINE - [min] (native hydrated magnesium silicate with use as ornamental stone in building) serpentine *f*
[gen] sinueux, serpentant
~ASBESTOS - [min] asbeste *m* serpentin
~CUCUMBER - [bot] cocombre *m* serpent
SERPENTINIZATION - [geol] serpentinisation *f*
SERRATE - s. serrated
SERRATED - [gen] (having longitudinal grooves, and corresponding teeth) dentelé, en dents de scie
~EDGE - denture *f*
[text] dent *f* suraiguisée
~KNIFE - [impl] couteau *m* dentelé
~PULSE - [electron] impulsion *f* en dent de scie
[telev] impulsion *f* à crête fractionnée
~SHAFT - [mech] arbre *m* cannelé
~VIBRATOR - [mech] vibrateur *m* denté
SERRATION - [gen] dentelure *f*
[mech] denture *f* en scie
[anat] engrenure *f* (du crâne)
SERRATURE - s. serration
SERRE-NOEUD - [med] (in surgery) serre-noeud *m*
SERRIED - [gen] serré
SERRIFORM - [gen] serriforme
SERRULATED - [gen anat etc] dentelé finement
SERUM - [physiol] (a watery secretion, separating from blood in coagulation) sérum *m*
~ALBUMIN - [chem] (an albumin obtained from serum and nutritive fluids) séroalbumine *f*
~RASH - [med] exanthème *m* sérique
~SICKNESS - [med] (the reaction sometimes occurring eight or so days after an injection of serum)

réaction *f* sérique
SERUM THERAPY - s. serotherapie
SERVE A ROPE, to - [naut] fourrer un cordage
SERVICE, to - [mech] (to carry out routine work on a car, aircraft or other machine or unit, e.g. hydraulic systems and the like) entretenir et réparer
SERVICE - [gen] service *f*
[comm] service *m*
[mech] entretien *m* et réparations, soin *m*, service *m*
[naut] fourrage *m*
~AERIAL - [radio] (the region surrounding a broadcast station in which the signals can be satisfactorily received by an average receiver) zone *f* utile
~BAND - [radio] (a band of frequencies accommodating several channels allocated to a given class or radiocommunication service) bande *f* allouée
~BOARD - [el] tableau *m* d'arrivée de secteur
~BRAKE - [auto] frein *m* à pied
~CABLE - [el] (an electric cable connecting the consumer's installation to the general supply system) câble *m* d'utilisation
~CALL - [telecomm] (in telephony) conversation *f* de service
~CAPACITY - [el] (of an electrical motor or machine) capacité *f* utile
~CAR - [auto] voiture *f* de service
~CEILING - [aero] (the altitude at which the rate of climb falls to some specified low value) plafond *m* pratique
~-CHANNEL - [comput] canal *m* de service
~CHARGE - [el gas ind etc] (a charge per consumer, which is independent of the amount actually supplied, but is generally related to the capital cost of making available the supply) taxe *f* d'installation
~CIRCUIT BETWEEN EXCHANGES - [telecomm] ligne *f* de service entre bureaux centraux interurbaines
~CLIP - [plumb] collier *m* de prise
~COACH - [railw] voiture *f* de service
~COCK - [gas ind] (meter cock) robinet *m* de compteur
~CONDITIONS - [el] (the external factors, e. g. altitude, air temperature, voltage changes etc which may influence the operation of an electrical machine) conditions *fpl* de fonctionnement
~CONNECTION - [gas ind] (or service pipe) branchement *m*
~CRACKS - [metall] craquelures *fpl* de travail
~DRIP - [gas ind] siphon *m* de branchement
~CORROSION - [el] corrosion *f* en service
~DROP - [el] branchement *m* sur ligne aérienne
~EARTH - [el] prise *f* de terre d'abonné
~ELL - [plumb] (ell with a male thread at one end) coude *m* à extrémité filetée
~ENGINEER - [mech] chef *m* de service
~GOVERNOR - [gas ind] régulateur *m* d'abonné
~HATCH - [constr] trappe *f*
~HOIST - [mech] monte-plats *m*
~INSTRUCTION - [gen] règlement *m* relatif à l'exploitation
~INSTRUCTIONS - [telecomm] mentions *fpl* de service, consigne *f*
~LEAD - [el] branchement *m* d'abonné
~LIFE - [gen & med] (the time during which an object will last in the normal use for which it was designed) durée *f*
SERVICE LIFT - [gen] monte-charge *m*
~LINE - [el] (a line connecting a consumer's installation to a distributor) prise *f*
[telecomm] branchement *m*, dérivation *f*
~MAINS - [el] (cables of small conductor cross-section leading the current from a distributor to the consumer's installation) ligne *f* de distribution
~MESSAGE - [telecomm] message *m* de service
~METER - [meas] compteur *m*
~OBSERVING DESK - [telecomm] table *f* de contrôle
~OUTPUT - [el chem] (the useful service of a cell or battery under given conditions) capacité *f* utile
~PIPE - [gas ind etc] (branch pipe drawing supplies from a main) branchement *m*
[el] conduit *m* de jonction
~PLATFORM - [aero] (a mobile staging used to give access to the higher parts of an aircraft on the ground, for servicing etc) plate-forme *f* d'accès
~REGULATOR - s. service governor
~RESERVOIR - [hydr] (small reservoir supplying a district) réservoir *m* d'eau d'usage
~ROAD - [railw] chemin *m* de service
~ROUTINE - [comput] (programme set out to assist in the actual operation of the computer) programme *m* de contrôle
~SADDLE - s. service clip
~SHAFT - [mining] puits *m* auxiliaire
~STATION - [auto] station *f* service
~STICKER - [comput] fiche *f* d'entretien
~STRESS - [phys] effort *m* de travail
~SWITCH - [el] interrupteur *m* de service
~TANKS - [aero] (the fuel tanks) réservoirs *mpl* du combustible
~TEE - [plumb] (tee with a female thread on the branch and one end of the run and a male thread on the other end of the run) té *m* de nettoyage, té *m* de branchement
~TEST - [el chem] (test designed to measure the capacity of a cell or battery under given conditions) essai *m* de décharge
[gen] essai *m* pratique d'emploi
~VALUE - [el] (term denoting the overall efficiency of an electric fan) rendement *m* d'un ventilateur
~VALVE - [plumb] (a shut-off cock) robinet *m* de barrage, robinet *m* extérieur, robinet *m* sous trottoir
~VOLTAGE - [el] tension *f* d'utilisation
~WATER - [gen] (water used in a factory) eau *f* de usage industriel
~WIRES - [el] circuit *m* d'alimentation
SERVICEABILITY - [gen] durée *f*, utilité
[mech etc] état *m* satisfaisant
SERVICEABLE - [gen] utilisable, en état de fonctionner
~TIME - [comput] temps *m* utile
SERVICEABLENESS - s. serviceability
SERVICING - [gen & mech] entretien *m* et réparations, soin *m* (de voiture)
SERVING MALLET - [naut] mailloche *f*
SERVITUDE - [gen] servitude *f*
[leg] (the right of using the land of another for a special purpose) servitude *f*
SERVO-ACTUATED MECHANISM - [mech] mécanisme *m* servocommandé
~-AMPLIFIER - [mech] (device in a servo-system to contribute additional power) servo-amplificateur *m*
~-BRAKE - [mech] (power-assisted brake) servo-

frein *m*
SERVO-CONTROL - [aero] (a control device in which the pilot's effort is augmented by means of a relay) servo-commande *f*
~-CONTROL MECHANISM - s. servo-mechanism
~-DRIVEN POTENTIOMETER - [comput] potentiomètre *m* d'asservissement
~-MECHANISM - [contr] (a power-driven device designed to provide a substantial controlling force in direct relation to a relatively small signal transmitted to it) servomécanisme *m*
~-MOTOR - [mech] (the part of a servo-mechanism supplying the required mechanical force) servo-moteur
~-OPERATED RECORDER - [meas] (a recorder driven by an amplifier in a servo-system) enregistreur *m* à servomoteur
~-SPEED CONTROL - [contr] servo-commande *f* de vitesse
~-SYSTEM - [contr] (a monitored automatic control system including a power amplifier in the main forward path) servo-système *m*
~-TAB - [aero] (a secondary tab which serves to move the main tab when adjusted by the pilot) servo-tab *m*
SESAME - [bot] (East Indian herb, the seeds of which are used as food and as a source of sesame oil) sésame *m*
~CAKE - [agric] (animal feeding) tourteau *m* de sésame
~OIL - [chem] (a fixed oil obtained from Sesamum indicum and used in the manufacture of soap, margarine and cosmetics) huile *f* de sésame
~OIL SOAP - [chem] savon *m* à l'huile de sésame
SESAMOID - [anat] sésamoïde
~BONE - [anat] os *m* sésamoïde
SESAMOLIN - [chem] (a pyrethrum synergist derived from sesame oil) sésamoline *f*
SESQUIT - [gen] (prefix denoting one and a half) sesqui
[chem] (indicating the presence of three atoms of one element and two of another in a compound) sesqui
SESQUIPLANE - [aero] (type of biplane in which the area on one wing is less than half that of the other) sesquiplan *m*
SET, to - [gen] placer, asseoir, poser, loger
[mech] (to adjust) régler, caler
[tool] doucir, affiler
[mech] (to adjust to a specified number of r.p.m.) ajuster
[gen] (to mount) sertir, enchâsser
[mech] (to give set to saw teeth) avoyer, donner de la voie
[constr] (to lower and fix in place) fixer à demeure, fixer
[ind chem etc] (to pass from a fluid to a firm or solid condition) faire prise
[print] composer
[photo] (a shutter) armer (un obturateur)
[astr] (to go down below the horizon) se coucher
[bot & agric] planter
[comput] afficher, régler
[med] (a bone) remettre (un os)
SET - [gen] fixe, rigide
[gen] (prescribed, specified) assigné
[phys] rigide
SET - [mech] (built in) rapporté
(formed or put together) formé
[gen] (a group, e g. of people) ensemble *m*
[gen] (a series) équipage *m*, ensemble *m*
[mech] (permanent deformation persisting after the cessation of the stress causing it) déformation *f*
[mech] (of saws; a permanent deflection, in alternate teeth, to one side and the other of the central plan of the saw blade, to ensure that the cut is wide enough to clear the blade itself) tourne-à-gauche pour donner la voie aux scies
[cin] (the properties, structures etc of a motion picture scene) décor *m*
[cin] (studio) studio *m*
[radio & telev] appareil *m*
[min] (the supporting timber frame) cadre *m*, châssis *m* de mine
[mech] (of a spring, the distance by which a spring deflects under normal load) flèche *f*, affaissement *m*
[text] (the number of reeds) compte *m* au peigne
[text] (the number of threads in a reed) nombre *m* de fils en dent
[constr] (setting coat) enduit *m*, dernière couche *f*
[print] (the width of a type character) approche *f*
[hydr] (the direction of a current of water) direction *f* du courant
[bot] (young plant, a seedling, a cutting) plantule *f*
[metall] orifice *f* de coulée
~A POLE, to - [constr] enforcer un pieu
~A SAIL, to - [naut] établir une voile
~A WATCH, to - [horol] régler une montre
~ACOUSTICS - [telev] acoustique *f* de plateau
~A PART, to - [gen] séparer
~ASIDE, to - [gen] mettre au rebut, rejeter
[leg] infirmer
~AT WORK, to - [mech etc] mettre en fonction, mettre en oeuvre
~AT ZERO, to - [instr] remettre
~BAR - [mech] clavette *f*
~BY, to - [gen] mettre à part
~BY THE COMPASS, to - [surv] relever au compas
~COLLAR - [mech] bague *f* d'arrêt, bague *f* de butée
~COPPER - [metall] cuivre *m* à 6‰ d'oxyde de cuivre
~DRESSER - [cin] metteur *m* en scène
~FAIR - [met] beau *m* fixe
~FOR CUTTING - [mach tool] trousse *f* d'outils coupants
~GAUGE - [mech] (for a saw) jauge *f* de la voie
~GOING, to - [mech] mettre en mouvement
~GREASE - [mech] graisse *f* consistante
~HAMMER - [impl] chasse *f* carrée de forgeron
~HDG BUTTON - [aero] (set heading button; press-button switch used to select the heading (automatic pilot) bouton *m* de sélection de cap
~IN, to - [gen] commencer
[naut] (of the tide) commencer à monter
~IN AIR, to - [constr] faire prise à l'air
~IN MOTION, to - [gen] mettre en marche, mettre en train
~LAYER - [el] couche *f* en paquet
~LIGHTING - [telev] éclairage *m* du plateau
~LIGHTS - [cin] (studio lights) éclairage *m* de prise des vues
~LINING - [acoust] (or studio lining; the use of materials to improve the acoustic qualities of a set) revêtement *m* de studio

SET NOISE - [radio] bruit *m* d'amplification, [cin] (the noise in the projecting system) bruit *m* propre du système
~NUT - [mech] contre-écrou *m*
~OF CASTINGS - [metall] série *f* de pièces de fonte
~OF DRAWINGS - [text] assortiment *m* d'étirages
~OF HACKLES - [text] rangée *f* de peignes
~OF LENSES - [photo] trousse *f* d'objectifs
~OF THE REED - [text] compte *m* au peigne
~OF THE WEFT - [text] duitage *m*, nombre *m* des duites
~OF TIMBER - [min] cadre *m* en bois
~OF TOOLS - [mech] assortiment *m* d'outils, équipage *m* d'outils
~OF WEIGHTS - [meas] série *f* de poids
~OF WIRES - [text] boutage *m* des dents
~OF WRENCHES - [tool] série *f* de clefs
~OFF - [gen] contraste *m*
[arch] saillie *f*, ressaut *m*
[leg] reconvention *f*
[print] (the smudging of ink from one printed sheet to another) maculage *m*
[constr] berme *f*
~OFF BLANKET - [print] (component part of an offset printing machine) garnissage *m* du cylindre
~ON, to - [gen] pousser en avant, inciter
~ON A PIVOT - [mech] à pivot
~ON FIRE, to - [gen] mettre en feu
~OUT, to - [gen] arranger, disposer
[gen] (to depart) se mettre en route
[naut] (of the tide) commencer à descendre
~OUT PLANTS, to - [agric] accoler
~OVER - [el] (contact line) désaxement *m*
~PIN - [mech] goupille *f* de calage, goujon *m* prisonnier
~POINT - [contr] (index value) valeur *f* de consigne
~PULSE - [comput] (in static magnetic storage; drive pulse which tends to set a magnetic cell) impulsion *f* de réglage
~SAIL, to - [naut] mettre une voile au vent
~SCREW - [mech] (a screw used to secure a part by pressing against it) vis *f* de pression
~-SCREW KEY - [mech] clavette *f* de la vis de pression
~SQUARE - [carp] équerre *f*
~SPARK, to - [auto] régler l'avance à l'allumage
~THE BEAM FLANGES, to - [text] régler les plateaux de l'ensouple
~THE COP, to - [text] embrocher la cannette
~THE SLAY, to - [text] régler la position du battant
~THE TEMPLE, to - [text] placer le templet
~TO ZERO, to - [instr] ramener à zéro
~TYPE - [print] composition *f* mécanique
~UNDER WATER, to - [constr] (of concrete) faire prise dans l'eau
~-UP, to - [gen] placer, fixer
[mech] (to assemble and install) monter, établir
[comm] (in business) s'etablir (dans le commerce)
[constr] ériger
[railw] établir (un itinéraire)
~-UP - [gen mech etc] (made ready for use) prêt
[plast ind] (group of moulds and or presses used consecutively in cycles by one more operators) ensemble *m*, groupe *m*
[print] (of type) composé
[instr] mise *f* en station

SET-UP - [mining] (of a rock-drill) mise *f* en position
[gen] (organization) organisation *f*
[gen] (only USA) structure *f*, édifice *m*
~-UP BOUNDARY STONES, to - [leg] borner
~-UP CURE - [rubber ind] (an operation preliminary to vulcanization) prévulcanisation *f*
~-UP ERROR - [contr] écart *m* maximal
~-UP FOR CUTTING - [tool] ensemble *m* d'outils coupants
~-UP GAUGE - [mech] jauge *f* de mise à point
~-UP INSTRUCTIONS, to - [comput] composer des instructions *f*
~-UP SCALE INSTRUMENT - [instr] (instrument in which the zero position is below the limit travel of the indicating means) appareil *m* à équipage mobile buté, appareil *m* de mesure zéro supprimé
~-UP THE FORMS, to - [constr] armer le coffrage
~-VALUE CONTROL - [contr] (a method of control in which the controlled quantity is maintained constant) réglage *m* à valeur de consigne
~WINDING - [electron] (one of the drive windings of a magnetic core in static magnetic storage) enroulement *m* de déclenchement, enroulement *m* de lecture
SETA - [zool] (small; bristle-like structure) poil *m* raide
SETACEOUS - [gen] sétacé
SETBACK - [gen] recul *m*
[med] rechute *f*
[constr] (in large buildings, the stepping of sections so that while the first one is erected on the street-line the subsequent ones are erected in step formation) reculement *m* du mur extérieur
SETON - [vet] (strip of linen drawn through an incision in the skin, to promote the drainage of an abscess) séton *m* à mèche
SETOSE - [gen] (bristly) séteux
SETPOINT - s. set point
SETSCREW - s. set screw
SETT - [constr] (a small rectangular block of stone) pavé *m*
[text] (of set the number of threads in a reed) compte *m*
~OF THE CLOTH - [text] compte *m* du tissu
~OF THE REED - [text] compte *m* en chaîne
~OF THE WARP - [text] densité *f* de la chaîne
SETTEE - [gen] canapé *m*
SETTER - [mech] (a machine designed to set the saw blade) machine *f* à affûter
[metall] plieuse *f*
SETTEWORT - [med plants) hellébore *m* fétide
SETTING - [gen & mech] montage *m*, pose *f*
[constr] (of cement) prise *f*
[print] composition *f*
[print] (an insertion) insertion *f*
[tool] doucissage *m*, affilage *m*
[mech] calage *m*
[gen] (of a jewel) monture *f*
[cin] mise *f* en scène
[med] réduction *f* (d'une fracture), clissage *m*
[chem] coagulation *f*
~ANGLE - [mach tool] angle *m* de position
[aero] (the angle of a wing, or tail-plane) angle *m* de calage (de l'empennage)
~ARCH - [metall] (of an oven) chapelle *f*
~AT WORK - [gen & mech] mise *f* en oeuvre

SETTING BACK - [railw] (turning back) rebroussement *m*
~BRICKWORK - [th eng] (internal masonry) maçonnerie *f* interne
~BURNISHER - [horol] outil *m* pour l'enfoncement des rubis
~COAT - [constr] (the finishing coat of plaster) dernière couche *f*
~COIL - [comput] enroulement *m* de déclenchement
~DEPTH - [oil ind] profondeur *f* de pose en puits
~DEVICE - [mech] dispositif *m* de réglage
~DIAL FOR DEPTH OF CUT - [mach tool] disque *m* gradué pour le réglage de la profondeur de la coupe
~GAUGE - [mech] jauge *f*
~LANDMARKS - [surv] pose *f* de bornes
~LATH - [text] latte *f* pour niveler
~LEVER - [mech] levier *m* de réglage
[horol] levie *m* de mise à l'heure
~MACHINE - [leather ind] machine *f* à étirer
~OF FRONT AXLE - [auto] (camber) réglage *m* du train avant
~OF THE RESPONDING VALUE - [contr] sélection *f* de la grandeur de fonctionnement
~OUT - [leather ind] (mechanical process designed to remove creases and marks from leather) étirage *m*
[surv] (pegging out, marking out) bornage *m*
~POINT - [chem] (of a lubricant oil) point *m* de coagulation
[metall] (only USA melting point) point *m* de fusion
~RATE - [glass man] (the time required for a glass surface to cool) vitesse *f* de refroidissement
~RULE - s. setting stick
~SCREW - [mech] vis *f* de serrage
~STEELWORK - [metall] (bracing) armature *f* d'un four
~STICK - [print] (a stick used for composition) composteur *m*
~TEMPERATURE - [rubber ind] (the temperature at which curing takes place) température *f* de prise
~THE BUBBLE - [photo] calage *m* du niveau
~THE ROLLERS - [text] réglage *m* des hérissons
~THRESHOLD - [comput] seuil *m* de polarisation
~TIME - [plast ind] (the time required for a synthetic resin casting mixture to reach optimum strenght) temps *m* de prise
[ind chem] (the time required for the glue to solidify and form an adhesive bond) temps *m* de prise
~UP - [mech] mise *f* à point
[plumb] (caulking) matage *m*
[rubber ind] engorgement *m* de la tête de boudineuse
~-UP TIME - [agric] période *f* de préparations
~WITH BUILT-IN PRODUCER - [gas ind] four *m* à gazogène accolé
SETTLE, to - [gen] établir, arranger
[comm] régler
[gen] (to cause to become steady) fixer, mettre bien en place, rendre stable
[gen] (a liquid) clarifier (un liquide)
[chem] (to deposit sediment and thus free from turbidity) deposer, sédimenter, se déposer
[leg] arranger, arbitrer
[gen] (to sink gradually) se tasser, tasser
[naut] s'enfoncer
SETTLE DOWN, to - [gen] s'établir
~OUT, to - [soil] se décanter
SETTLEMENT - [gen] arrangement *m*, solution *f*
[comm] liquidation *f*
[leg] transaction *f*
[soil] (subsidence) affaissement *m*
[constr] (road constr) tassement *m*
[gen] (only USA) petit village *m*
~OF ACCOUNTS - [comm] arrêté *m* des comptes
~STRESSES - [soil] contraintes *f*pl dues au tassement
SETTLEMENTS - [constr] (breaks in a structure caused subsidence) tassement *m*
SETTLER - [oil ind] appareil *m* de décantation
SETTLING - [constr] affaissement *m*
[ind chem] (the process of separation of solid particles from a suspension by the action of gravity) sédimentation *f*, défécation *f*
[chem] déposition *f*
~BASIN - [rubber ind] bac *m* de décantation
~CRACK - [constr] lézarde *f* de tassement
~DAY - [comm] jour *m* de liquidation
~DITCH - [oil ind] rigole *f* de décantation
~PIT - [ind chem] fosse *f* de décantation, fosse *f* de repos
~PITS - [hydr] puits *m*pl filtrants
~TANK - [hydr] (in sewage works) décanteur *m*, bassin *m* de sédimentation
[paper man] (a tank designed to recover the pulp from backwaters) décanteur *m*
~[oil ind] bassin *m* décanteur
~TIME - [contr] (the response time) temps *m* de stabilisation de réglage, temps *m* de récupération
~TUB - s. settling vat
~UP OFFICE - [comm] bureau *m* de liquidation
~VAT - [ind chem] (open vessel in which liquids are held to allow deposition of sediment) cuve *f* à défécation
~WELL - [hydr] (a clarification tank) bac *m* de décantation, chambre *f* de clarification
SETWALL - [bot] (used for pharmaceutical products) nard *m* de montagne
SEVEN SHAFT SATIN - [text] merveilleux *m*
SEVER, to - [gen] disjoindre, désunir
[constr] (a beam) sectionner (une poutre)
SEVERABLE - [gen] séparable
SEVERAL - [gen] plusieurs
SEVERALLY - [gen] responsables individuellement
SEVERALITY - [leg] (the holding of land in one's own name) propriété *f* individuelle
SEVERANCE - [gen] séparation *f*, disjonction *f*
SEVERITY FACTOR - [oil ind] facteur *m* de sévérité
SEW, to - [gen] coudre
[naut] laisser au sec
SEWAGE - [gen] (liquid domestic refuse) eau *f* d'égout
[constr] (sewerage) égout *m*
~DISPOSAL - [hydr] évacuation *f* des eaux d'égout
~FARM - [agric] champs *m*pl d'épandage
~FIELD - [agric] champ *m* d'épandage des eaux d'égout
~FLOW - [hydr] débit *m* d'eau d'égout
~FUNGUS - [bot] champignon *m* d'eaux usées
~GAS - [chem] gaz *m* de gadoues, gaz *m* d'eaux résiduaires
~LAGOON - [hydr] étang *m* d'eaux usées, étang *m* de stabilisation
~PLANT - [hydr] station *f* d'épuration des eaux d'égout
~POWER - [agric] poudrette *f*

SEWAGE PROCESSING - [agric] utilisation *f* des eaux d'égout
~PUMP - [hydr] pompe *f* à eau d'égout
~PUMPING STATION - [hydr] station *f* de relèvement des eaux d'égout
~SLUDGE - [ind chem] (coagulated, filtered and dried sewage used as a fertilizer) boue *f* d'égout, boue *f* des eaux d'égout
SEWER - [constr] (usually an underground conduit to convey drainage) égout *m*, égout *m* collecteur
[gen] couseur *m*
[bookbind] brocheur *m*
~BOTTOM - [hydr] radier *m* de l'égout
~CONDUIT - [constr] système *m* d'égouts
~GAS - [chem] miasme *m* égoutier, gaz *m* d'égout
~GATE - [constr] vanne *f* (d'égout)
~RAT - [zool] rat *m* d'égout
~ROD - [hydr] écouvillon *m*
~SLIME - [hydr] pellicule *f* biologique
~SLUICE VALVE - [hydr] vanne *f* pour canalisation
~TRUNK LINE - [constr] égout *m* collecteur
~WATER - [gen] eau *f* d'égout
SEWERAGE - [constr] (a system of sewers) système *m* d'égouts, canalisation *f*
SEWING - [gen] couture *f*, ouvrage *m* à l'aiguille
[bookbind] brochage *m*
~AWL - [shoe man] carrelet *m* de cordonnier
~MACHINE - [mech] machine *f* à coudre
~MACHINE SHUTTLE - [mech] navette *f* de la machine à coudre
~NEEDLE - [text] aiguille *f* à coudre
~SILK - [text] soie *f* à coudre
~THREAD - [text] fil *m* à coudre
~TWIST - [text] fil *m* à coudre
SEWN BAND ROPE - [text] câble *m* plat cousu
~CHENILLE - [text] chenille *f* cousue
~CORDAGE - [text] cordage *m* cousu
~ROPE WARE - [text] corderie *f* cousue
SEX - [biol] sexe *m*
~CELL - [biol] (gametes) gamète *m*
~CHROMOSOME - [biol] (the chromosome responsible for the initial determination of sex) chromosome *m* sexuel
~GLAND - [zool] (gonad) gonade *f*
~LINKAGE - [genet] (the inheritance of certain characteristics which are determined by genes located in the sex chromosomes) hérédité *f* liée au sexe, transmission *f* des caractères sexuels
~-LINKED CHARACTERS - [genet] caractères *mpl* héréditaires liés au sexe
~REVERSAL - [genet] (the gradual change of the sexual characters) inversion *f* du sexe
SEXADECIMAL NOTATION - [electron] notation *f* hexadécimale
SEXIVALENT - [chem] (synonym for Hexavalent) hexavalent
SEXTANT - [instr] (reflecting instrument designed to measure angles up to I20°) sextant *m*
SEXTET - [mus] (composition for six voices or six instruments) sextuor *m*
SEXTEDECIMO - [print] I/I6 in-seize
SEXTUPLE - [math] sextuple
SEXUAL - [biol] sexuel
~CELL - [biol] (a male or female germ-cell) cellule *f* sexuelle
~DIMORPHISM - [biol] (the structural differences between the male and female of a species) dimorphisme *m* sexuel
SEXUAL ORGANS - [zool] organes *mpl* sexuels
~SELECTION - [zool] (phase of natural selection based on the struggle for mating) sélection *f* sexuelle
SFERICS - [met] (a storm detector) étude *f* météorologique des atmosphériques
SFLOCS - [met] (hourly reports on the position of thundery outbreaks given by the Meteorological Office) bulletins *mpl* météorologiques
SGRAFFITO - [pottery etc] (or graffito; a type of surface decoration) sgraffite *m*
SHACK - [agric] mettre en pâture dans le chaume
[bot] gland *m* tombé
[leg] (colloq) droit *m* de pâture
[gen] hutte *f*
SHACKLE, to - [gen] entraver, mettre les fers
[naut] maniller, mailler
SHACKLE - [gen] fers *mpl* (de prisonnier)
[mech] maillon *m* de liaison
[mech] (of a chain; the ring supporting the hook) manille *f* d'assemblage, boucle *f*
[mech] (U-shaped metal link closed by a pin) maillon *m* de jonction
[mech] (of a leaf spring etc) huit *m* de ressort
[telecomm] amarrage *m* de câble
[naut] (of an anchor chain) maillon *m* de chaîne
[mech] (of a padlock) anse *f*, branche *f*
[railw] manille *f* d'attelage
~-BAR - [railw] bielle *f* d'attelage
~-BOLT - [mech] (a bolt used to close a shackle) cheville *f* d'assemblage
~INSULATOR - [el] (porcelain insulator with ends secured to metal shackles) isolateur *m* d'angle
~OF CHAIN - [naut] maillon *m* de chaîne
~-PIN - [mech] (a pin used to close a shackle but secured by a cotter instead of a threaded end, as in a shackle-bolt) cheville *f* d'assemblage
SHACKLER - [mining] (in mining; worker employed to couple and uncouple trains) accrocheur *m*
SHADDOCK - [bot] pamplemousse *f*
SHADE, to - [gen] ombrager, couvrir d'ombre
[gen] (light, colours etc) ombrer, atténuer, voiler
[paint] ombrer, nuancer
[comm] (colloq. USA; to reduce slightly) établir des prix dégressifs
[photo] ombrer
SHADE - [gen] ombre *f*, teinte *f*
[gen] (a nuance) nuance *f*
[impl] abat-jour *m*
[opt] dégradation *f* de couleurs
[constr] (only USA) store *m* (de fenêtre)
[naut] tente *f*
~BEARER - [bot] essence *f* d'ombre
~CARD - [comm] carte *f* de coloris
~-CARRIER RING - [el] (a screwed ring by which a shade-carrier is secured to a lampholder) bague *f* de support pour abat-jour
~DECK - [naut] pont *m* tente
~EFFECT - [opt] effet *m* d'ombrage
~LINE - [draw] trait *m* de force
~LINES - [draw] (used in map-making) hachures *fpl*
~OF GREY - [telev] marche *f* de gris
~TINT - [chem] teinte *f* à nuancer
SHADED - [gen] ombragé
[draw] ombré
~-POLE - [el] (a pole surrounded by a heavy copper ring) à enroulement en court-circuit

SHADED POLE INDUCTION INSTRUMENT - [instr] (an induction instrument in which the poles of the electromagnet are split into halves with a copper band round one half of each pole) appareil *m* à induction à enroulement en court-circuit
~ RULE - [print] filet *m* gras-maigre
~ SATEEN - [text] satin *m* ombré
SHADING - [gen] nuancement *m*
[opt] (of colours) dégradation *f*
[draw] nuance *f*
[surv] (for cartographic representation of relief) hachure *f*
[el acoust] (a method for controlling the directivity pattern of a transducer through control of the distribution of phase and amplitude of the transducer action over the active face) correction *f* de la directivité
[telev] (part of the camera aligning process for the control of noise level) réglage *m* de l'ombrage
[telev] (dark area in the picture caused by a difference in the field between different parts of the camera tube) effet *m* d'ombrage
~ COIL - [mech] anneau *m* de blindage
~ COMPENSATION SIGNAL - [telev] signal *m* compensateur d'ombrage
~ CORRECTION - [telev] correction *f* d'ombrage
~ DYESTUFF - [text] colorant *m* de nuançage
~ GENERATOR - [telev] générateur *m* de signaux correcteurs d'ombrage
~ OF BAND - [phys] atténuation *f* de la bande
~ TISSUE - [photo] papier *m* translucide interposé
~ VALUE - [telev] valeur *f* de la luminosité
SHADOW, to - [opt] ombrager
SHADOW - [gen] ombre *f*
[opt] (on the screen of an optical comparator) profil *m*
~ ANGLE - [electron] (the angle formed by the edges of the non-fluorescent parts with the centre of the magic eye) angle *m* d'ombre
~ AREA - [radar] (very strong reduction of the receiver field strength due to obstruction) zone *f* d'ombre
~ BANDS - [astr] (sometimes occurring before totality in a solar eclipse) bandes *fpl* d'ombre
~ BAR - [impl] écran *m* ombrageant directionnel
~ COLUMN INSTRUMENT - [instr] (an instrument in which the readings are given by a shadow column on an illuminated scale forming part of the instrument) appareil *m* à colonne d'ombre
~ CONE - [astr] cône *m* d'ombre
~-CRETONNE - [text] chiné *m*
~ DYEING - [text] teinture *f* flammée
~ EFFECT - [radio] (mountain effect) affaiblissement *m* de propagation
~ FACTOR - [el] (the ratio between the electric field strenght which would result from propagation over a sphere and that which would result from propagation over a plane) coefficient *m* d'ombre
~ LIGHT - [cin] projecteur *m* grand angle
~ MASK - [telev] masque *m* d'ombre
~ MASK HOLE - [telev] trou *m* de masque d'ombre
~ MASK TUBE - [telev] tube *m* à masque
~ REGION - s. shadow area
~ REP - [text] reps *m* ombré
~ SCATTERING - [phys] (diffraction scattering) diffusion *f* diffractive
~ SCRATCH - [cin] (in the slit of the projector caused by dirt) parasite *m* dû à l'état de la fente
SHADOW STOP - [photo] (the minium aperture for a half-tone negative) diaphragme *m* pour demi-tons
~ WELT - [text] sous-revers *m*
~ ZONE - [acoust] (a region, usually in the atmosphere or under water, where ray acoustics predicts zero penetraction of sound rays) zone *f* d'ombre
SHADOWGRAPH - [med] radiographie *f*, radiogramme *m*
SHADOWING - [el metall] (the interference of any part of an anode, cathode, rack or tank with uniform current distribution on a cathode) effet *m* d'écran
SHADRACH - [metall] (or salamander, a mass of hardened metal or slag remaining in the hearth of the furnace) loup *m*, cochon *m*, carcas *m*
SHAFT - [gen] flèche *f*, trait *m*
[arch] (of a column) tige *f* (de colonne), vif *m* (de colonne)
[constr] souche *f*
[impl] (of an axe etc) manche *m*
[mech] (an axle, a mandrel, an arbor etc) arbre *m*, axe *m*
[mining] (usually a vertical passage) puits *m*
[constr] (for a lift etc) puits *m*
[text] (of a loom) lame *f*
[anat] corps *m*
[mech] (of a vehicle) limon *m*
[of a blast furnace] (the upper tapering part of the inside of the furnace, from the belly to the throat) cuve *f* de haut fourneau
[opt] (of light etc) rayon *m*, éclair *m*
~ BEARING - [mech] palier *m* de transmission
~ BOTTOM - [mining] fond *m* du puits
~ BRACE - [mining] plancher *m* de manoeuvre
~ BRICK - [metall] brique *f* de cuve
~ CABLE - [mining] (a specially armoured cable of great mechanical strength) câble *m* pour puits de mine
~ CASING - [auto] tube *m* d'arbre, trompette *f* entourant *m* un arbre
~ COLLAR - [mech] collier *m* d'arbre
[mining] cadre *m* de superficie
~ CORD - [text] cordeau *m*
~ COUPLING - [mech] (any device used to connect two shafts end-to-end, so that the rotation of one is communicated to the other) arbre *m* d'entraînement, manchon *m* d'accouplement d'arbres, accouplement *m* d'arbres
~ CRUCIBLE FURNACE - [metall] four *m* à cuve pour creusets
~ DRIVE - [mech] transmission *f* par arbre
~ EYE - [text] anneau *m* à lame
~ FRAME - [text] cadre *m* à lisses, support *m* des lamettes
[mining] cadre *m* de puits
~ FURNACE - [metall] four *m* à cuve
~ GOVERNOR - [mech] (compact, spring-loaded governor designed to control the speed of small engines) régulateur *m* axial
~ HAMMER - [impl] marteau *m* à levier
~ HAULING - [mining] exploitation *f* à puits
-HEAD FRAME - [mining] charpente *f* du chevalement
~ HOLE - [mech] trou *m* d'arbre
[shipbuild] lunette *f* d'étambot
~ HOOK - [text] crochet *m* à liserons, crochet *m* de

lame
SHAFT HORSEPOWER - [naut] (of nautical engines) puissance *f* au frein
~HOUSE - [mining] chevalement-abri *m*
~KEY - [mech] clavette *f* d'arbre
~-LANDING - [mining] accrochage *m*
~-LATHE - [mach tool] tour *m* à arbres
~LEVER - [text] levier *m* de commande des lames
~LIFT - [text] contrôle *m* du remettage
~LINING - [metall] (blast furnace the refractory lining from the belly to the throat) maçonnerie *f* de la cuve
[mining] revêtement *m* de puits
~MINE - [mining] mine *f* exploitée par puits
~MOTION - [text] changement *m* des lames
~OF A RIVET - [mech] tige *f* d'un rivet
~OF BOBBIN - [text] arbre *m* de commande des bobines
~OF HAMMER - [impl] manche *m* de marteau
~OF HEALDS - [text] lame *f*
~OF THE HAIR - [zool] corps *m* du poil
~OF YARN GUIDE - [text] arbre *m* du guide-fil
~PACKING - [text] joint *m* d'arbre
~PASSAGE - [naut] (for the propeller) tunnel *m* de l'arbre de l'hélice
~PILLAR - [mining] (the area of the ore left unworked round the bottom of a pit for support) massif *m* de protection de puits
~POCKET - [mining] fosse *f* de remplissage
~POSITION ENCORDER - [comput] capteur *m* de position angulaire à sortie numérique
~PUMP - [oil ind] pompe *f* de puits
~RAISING MOTION - [text] dispositif *m* de relevage des lames
~ROD - [text] liseron *m*, lamette *f*
~ROLLER - [text] rouleau *m* des lames
~SEAL - [mech] scellement *m* à tige
~SET - [mining] cadre *m* de puits
~SIDING - [mining] recette *f* inférieure
~SINKING - [mining] fonçage *m* de puits
~STATION - [mining] recette *f* de puits
~STAVE BRACKET - [text] support *m* du liseron
~STEP - [mech] palier *m* de pied, crapaudine *f*
~SUPPORT BEARING - [mech] (in nautical motors) support *m* de l'axe
~SUSPENSION HOOK - [text] crochet *m* à liserons
~TIMBERING - [mining] boisage *m* du puits
~-TIMBERING ASCENDING METHOD - [mining] boisage *m* du puit par la méthode montante
~-TIMBERING DESCENDING METHOD - [mining] boisage *m* du puits par la méthode descendante
~TOP - [mining] recette *f*
~TUNNEL - s. shaft passage
~-TURNING LATHE - s. shaft-lathe
~WORKING - [mining] fonçage *m*
SHAFTING - [mech] transmission *f*, arbres *mpl* de transmission
[naut] (the shaft running from the propeller to the engine) ligne *f* d'axe
[arch] fûts *mpl*
SHAFTWAY - [mining] puits *m* de l'ascenseur
SHAG - [zool] poil *m* touffu
[text] peluche *f*
[gen] (a type of tobacco) tabac *m* fort (coupé fin)
[zool] cormoran *m* huppé
SHAGGY - [gen] poilu, hirsute
[bot] (covered with long and weak hairs) velu, poilu
SHAGGY - [text] peluché
~WOOL - [text] laine *f* velue
SHAGREEN, to - [text] chagriner
SHAGREEN - [leather ind] (type of leather obtained from the belly part of a shark skin) chagrin *m*, cuir *m* chagriné
SHAKE, to - [gen] secouer, ébranler
[gen] (of liquids) agiter (un liquide)
[gen] (to shiver, to quiver) trembler
[geol] (of the earth) trembler
[naut] faire ralinguer
[mech] ralentir (à cause de vibrations)
[bot] (of wood) se gercer
[mus] triller
SHAKE - [gen] secousse *f*
[geol etc] tremblement *m*
[shipbuild] longeron *m*
[mech] (backlash) jeu *m*, secousse *f*
[bot] (of timber) gerçure *f*
[mus] trille *m*
[constr] bardeau *m*
[paper man] (transverse vibration imparted to the wire screen of a Fourdrinier paper-making machine) secoueur *m*
~GRATE - [th eng] grille *f* mobile
~OUT, to - [min] cribler à secousse
~-PROOF WASHER - [mech] rondelle *f* éventail
~THE COTTON, to - [text] secouer le coton
~THE FLEECE, to - [text] secouer la toison
~-THE HANK, to - [text] battre l'écheveau
~-UP FLASK - [ind chem] (a conical glass flask in which the contents can be readily shaken up without danger of spilling) flacon *m* d'agitation
SHAKEOUT - [metall] (in foundry work, the removal of castings from the moulds) décochage *m*
~EQUIPMENT - [metall] décocheuses *fpl*
~MACHINE - [metall] (in foundry work) décocheuse *f*
~SAND - [metall] (in foundry work) sable *m* à décocher
SHAKER - [ind] transporteur-trembleur *m*
[agric] secoueur *m* (de paille)
[text] loup-batteur *m*
~CONVEYOR - [mech] transporteur *m* à secousses
~PINS - [text] aiguilles *fpl* à secousses
SHAKES - [med] frisson *m*, accès *m* de fièvre
SHAKING - [gen] secouage *m*, secouement *m*
[min] (of a screen) remuage *m* (d'un tamis)
[text] (of wool, to eliminate the burrs) secouement
[acoust] (of the voice) tremblotement *m* (de la voix)
~FEEDER - [mech] (a device for delivering material to a machine by a reciprocating motion slower than that of a vibrating feeder) distributeur *m* à va et vient
~GRID - [gas ind] (used for coke screening) grille *f* à secousses
~KILN - [brew ind] touraille *f* à secousses
~MACHINE - [metall] tonneau *m* dessableur
~SCREEN - [impl] (a vibrating screen) crible *m* à secousses
~SHOOT - [mech] (a jigging conveyor) transporteur *m* à secousses
~SIEVE - s. shaking screen
~TABLE - [gas ind] (an oscillating screen for coke) table *f* à secousses

SHAKING TEST - [geol] essai *m* par secouage
~TRAY - [impl] plateau *m* à secousses
SHAKINGS - [naut] (scraps of cordage) rognures *f*pl de vieux filin
SHALE - [geol] (consolidated clay-rock possessing well defined laminations) schiste *m*, argile *f* schisteuse
~DUSTING - [geol] schistification *f*
~NAPHTHA - [min] naphte *f* de schistes
~OIL - [chem] (an oil obtained from oil-bearing shale by destructive distillation) huile *f* de schiste
SHALLOON - [text] (a light, combed woollen cloth) chalon *m*
SHALLOP - [naut] chaloupe *f*
SHALLOT - [bot] échalote *f*
[mus] (component of reed pipes operated on by the tongue) rigole *f*
SHALLOW, to - [constr] s'ensabler
SHALLOW - [naut] bas-fond *m*, haut-fond *m*
[gen] peu profond, bas de fond
~CARBON STEEL - [metall] acier *m* demi-mur
~DRAUGHT - [naut] faible tirant *m*
~-DRAUGHT SHIP - [naut] navire *m* à faible tirant
~FLIGHT WORM - [mech] (an extruder screw in which the projection of the flight above the surface of the core is less than usual) vis *f* à filet peu élevé
~FOOTING - [soil] semelle *f* superficielle
~NUT - [mech] écrou *m* bas
~-WATER DEPOSITS - [geol] dépôts *m*pl de hauts fonds
~WELL - [hydr] (a shaft sunk almost to the bottom of a superficial permeable stratum so as to tap the water in it) puits *m* ordinaire à faible profondeur
SHALY - [geol] schisteux
~FEEDING - [med] repas *m* fictif
SHAMMY - [leather ind] peau *f* de chamois
SHAMROCK - [bot] trèfle *m* d'Irlande
SHANK - [anat] tibia *m*
[mech] bras *m*
[naut] verge *f* d'ancre
[tool] tige *f*
[impl] (of a spoon) manche *m*
[gen] (of a key) branche *f*
[of a column] (the shaft) fût *m*
[shoe man] cambrillon *m*
[geol] flanc *m* d'un pli
~CUTTER - [tool] couteau *m* à queue
~END MILL - s. shank mill
~LADLE - [metall] poche *f* à fourche
~MILL - [tool] fraise *f* à queue
~OF A BOLT - [mech] (the stem or body of a bolt between the threaded part and the head) tige *f* d'un boulon
~OF DRILL - [mech] queue *f* d'un foret
~OF LOOP - [text] côté *m* de maille
~OF THE CONNECTING ROD - [mech] corps *m* de la bielle
~PAINTER - [naut] (of an anchor) serre-bosse *m*
SHANKED - [gen & mech] à tige, à branche
SHANKINGS - [text] (wool obtained from the legs of sheep) laine *f* grossière courte soie
SHANTUNG - [text] (plain silk cloth of light brown colour having a rough surface) shantoung *m*
SHAPE, to - [gen] façonner, modeler
[mech] emboutir, profiler, gabarier
[pottery] contourner
[naut] (a course) donner une route
SHAPE - [gen] forme *f*
[metall] profil *m*, profilé *m*
[th eng] s. shape block
~BLOCK - [th eng] (a special refractory brick shape) brique *f* de forme
~CONSTANCY - [telev] constance *f* de forme
~CUTTING - [metall] formation *f* par enlèvement de copeaux
~CUTTING MACHINE - [mach tool] étau-limeur *m*
[metall] (a stamping press) emboutisseuse *f*
~OF TOOTH - [mech] (a gear tooth outline) profil *m* de la dent
~PRUNING - [agric] taille *f* de formation
~ROLL - [metall] cylindre *m* profilé
~ROLLING MILL - [metall] laminoir *m* à fers profilés
SHAPED - [gen] façonné, profilé, taillé, embouti
~-BEAM TUBE - [electron] tube *m* cathodique de générateur de fonctions
~BLOOM - [metall] bloom *m*
~CASTING - [plast ind] pièce *f* conformée
~CHARGE - [chem] (an explosive charge so shaped that the main blast is directed in a specific direction) charge *f* creuse
~CONDUCTOR - [el] (a conductor having a cross-section other than circular) conducteur *m* profilé
~-CONDUCTOR CABLE - [el] (a three-phase cable the conducting cores of which are specially shaped) câble *m* à conducteurs profilés
~IRON - [metall] fer *m* profilé
~PARACHUTE - [aero] (a type of parachute formed of shaped gores with rigging lines which are continuous over the canopy) parachute *m* en forme
~WIRE - [metall] fil *m* profilé
SHAPER - [mach tool] étau-limeur *m*, limeuse *f*
~BOWL - [text] (of a ring spinning frame) galet *m*
~CUTTER - [mech] (of a gear shaping machine) fraise *f* de machine à tailler les engrenages
~MOTION - [text] (copping motion) mouvement *m* de formation de la bobine
~RAIL - [text] pièce *f* de mise en forme
~SCREW - [text] vis-guide
SHAPING - [gen] formation *f* développement *m*
[gen] (the operation of forming into a particular shape) façonnage *m*
[mach tool] limage *m*
[metall] emboutissage *m*, formage *m*, profilage *m*
[pottery] contournement *m*
~BAG - [rubber ind] (for tyres) "bag" *m* de galbage
~CIRCUIT - [electron] circuit *m* formeur d'impulsions
~GROOVE - [metall] (of a rolling-mill roll) ébaucheur *m*
~MACHINE - [mach tool] (a machine tool for small flat surfaces, slots etc) limeuse *f*
~MACHINE - [text] dresseuse *f* mécanique
~NETWORK - [telev] réseau *m* conformateur
~OF A TYRE - [rubber ind] (cambering of tyres) galbage *m* d'un pneu
~PASS - [metall] cannelures *f*pl profilées
~PLANER - s. shaping machine
~UNIT - [comput] élément *m* de mise en forme
SHARE, to - [gen] partager
[gen] (to participate in) participer à
SHARE - [gen] part *f*, portion *f*
[comm & fin] action *f*
[fin] (of expenses etc) quote-part *f*
[agric] (a pointed wedge-shaped implement used

for the horizontal cut separating the furrow slice from the undersoil) soc *m*
SHARE CAPITAL - [fin] capital *m* social
~CERTIFICATE - [fin] titre *m* d'actions
~-FARMER - s. sharecropper
~OUT, to - [gen] distribuer, répartir
~PUSHING - [fin] marronnage *m*
~REGISTER - [fin] (the stock ledger) grand-livre *m* des titres
SHARECROPPER - [agric] métayer *m*
SHARED-CHANNEL BROADCASTING - [radio] radio-transmission *f* à fréquence commune
~ELECTRONS - [phys] électrons *mpl* partagés
~FLUE - [th eng] (a common flue) gaine *f* commune
SHAREHOLDER - [comm] actionnaire *m*
SHARING - [gen] partage *m*, distribution *f*
SHARK - [zool] requin *m*
~LIVER OIL - [chem] (a brownish, red, oily liquid obtained by expression of shark livers. It is a source of vitamins A and D) huile *f* de foie de requin
~PAINT - [paint] (fast drying flat paint, usually highly pigmented and containing a minimum amount of volatile binder, used for sealing and priming) peinture *f* à séchage rapide
SHARP - [gen] tranchant, aiguisé
[mech] affilé, affûté
[gen] (of a knife) aigu, aiguisé
[gen] (of an edge) tranchant, vif
[gen] (of a curve) (tournant) brusque
[gen] (of features) anguleux
[gen] (of smells) pénétrant
[mus] (the sign which raises a note one semitone from the original pitch) dièse *m*
[opt photo etc] net
[phonetics] (of a consonant) (consonne *f*) forte
~BEND - [el] (or elbow; a bend of short radius serving to connect two lengths of conduit which are at an angle of 90°) coude *m* de raccordement de tubes
~BOTTOM - [shipbuild] carène *f* fine
~-CRESTED WEIR - [hydr] déversoir *m* à crête mince
~CURVE - [gen] courbe *f* vive, courbe *f* à petit rayon
~CUT-OFF PENTODE - [electron] (a pentode with a very sharp cut-off and maximum anode current) pentode *f* à pente constante
~CUT-OFF TUBE - [electron] tube *m* à pente constante
~DIP - [geol] inclinaison *f* raide
~DIRECTIONAL EFFECT - [radio] effet *m* directionnel critique
~-EDGE ORIFICE - [hydr] orifice *m* en mince paroi
~-EDGE ORIFICE METER PLATE - [instr] (a thin-walled orifice plate, a flow-measuring meter) diaphragme *m* à mince paroi
~-EDGED - [mech etc] à arête vive
~-EDGED MEASURING WELL - [hydr] déversoir *m* de jaugeage à mince paroi
~-EDGED ORIFICE - [mech] (circuit orifice cut in a thin plate and placed in a pipe to measure the flow of air or gas) orifice *m* percé en mince paroi
~FLANGE - [mech] (sharpened through wear) bride *f* à vive arête
~-FREEZING - [th eng] (in cold, storage, a process of quick freezing) congélation *f* rapide
~IMAGE - [opt] image *f* nette
~MELTING POINT - [metall] point *m* de fusion fort
~PICK - [text] (a heavy or strong pick) chasse *f* dure
SHARP PICTURE - [cin] (a high definition picture) image *f* nette
~RING - [mech] bague *f* à arête vive
~SAND - [metall] sable *m* cru, gros sable *m*
~TUNING - [radio] syntonie *f* aigue
~TURN - [gen] courbe *f* raide, courbe *f* vive
SHARPEN, to - [gen] aiguiser, affiler
[gen] (a pencil) tailler (un crayon)
[mech] affûter, aiguiser, repasser
[mus] diéser
[naut] orienter à bloc
SHARPEN A PICTURE, to - [telev etc] accentuer la définition
SHARPENER - [mach tool] affûteuse *f*
SHARPENING - [gen & mech] aiguisage *m*, affilage *m*, affûtage *m*
~MACHINE - [mach tool] machine *f* à affûter, affûteuse *f*
~WHEEL - [tool] meule *f* affûteuse
SHARPITE - [miner] (uranium containing mineral) sharpite *f*
SHARPNESS - [gen] acuité *f*, acutesse *f*
[opt] (of an image) netteté
[mech] (of a blade) acuité *f* (du tranchant)
[railw] (of curves) exagération *f* (des courbes)
[gen] (of an angle) aiguïté
~INDICATOR - [photo] indicateur *m* de netteté
~LIMIT - [cin] (the boundary of the region which is in good focus for the camera) limite *f* de netteté
~OF DIRECTIVITY - [radio] (the extent to which the radiating properties of an aerial are concentrated within certain angular limits) acuité *f* de directivité
~OF RESONANCE - [radio] (the rapidity with which resonance phenomena are exhibited when the frequency of excitation of a constant driving force is varied) netteté *f* de résonance
~OF TUNING - [radio] (a term denoting selectivity) sélectivité *f*
SHATTER, to - [gen] briser en éclats
[gen] (of glass) voler en éclats, se briser
SHATTER - [gen] fragment *m*
~CRACKS - [metall] craquelures *fpl*, gerçures *fpl*
~INDEX - [phys] (figure of merit obtained from tests to determine liability of material to breakage in handling) indice *m* du point de rupture
~-PROOF - [gen] (so designed or made as not to disrupt into small pieces under impact) imbrisable
~-PROOF GLASS - [glass man] (a safety glass) glace *f* de sécurité
~TEST - [min] (test to determine the resistance of coke to fracture in handling) essai *m* de chutages des cokes
SHATTERING - [gen] écrasant, brisement *m*
[agric] égrenage *m*
SHATTERPROOF - s. shatter-proof
SHAVE, to - [gen] raser
[mach tool] planer
[mech] (of a gear etc) ébarber
[bookbind] (to trim the edges) écorner
SHAVE - [tool] plane *f*, racloir *m*
[gen] (a very thin slice, as cut by a tool) copeau *m*
~HOOK - [tool] grattoir *m*, racloir *m* en forme de coeur
SHAVING - [mech] (the operation of removing a very thin layer of material from a workpiece) rognure *f*
[metall] (the actual thin slice) rognures *fpl*

SHAVING - [el acoust] (the process of removing material from the surfaces of a recording medium, so as to obtain a new recording surface) shaving *m*
[carp] (of wood) planage *m*, dolage *m*
~CUTTER - [mech] (a tool used in gear shaving) fraise *f* à ébarber
~FIXTURE - [mech] outil *m* à ébarder
~HORSE - [impl] banc *m* d'âne
~MACHINE - [mach tool] ébavureuse *f*
~PRESS - [mach tool] presse *f* à ébavurer
~SCRUBBER - [ind chem] (a tar-fogged filter, used in washing) tour *f* à copeaux
~STOCK - [mech] (on gear tooth thickness) surépaisseur *f*
SHAVINGS - [carp etc] copeaux *mpl* de bois, rognures *fpl*
~EXHAUST PLANT - [carp] dispositif *m* d'aspiration des copeaux
SHAWL - [text] fichu *m*, châle *m*
SHAWM - [mus] (antique double reed instrument, a predecessor of the oboe) chalumeau *m*
SHEAF, to - [agric] gerber, enjaveler
SHEAF - [gen & agric] gerbe *f*, faisceau *m*
[geom] (aggregate of straight lines etc) faisceau *m* (de lignes etc)
[astronaut] gerbe *f* (de trajectoires)
~BINDER - [agric] machine *f* à lier en gerbes
~BINDING MACHINE - [bookbind] machine *f* à piquer des blocs
~CARRIER - [agric] ramasseuse *f* de gerbes
~LOADER - [agric] gerbeuse *f*
SHEAR, to - [gen] couper, trancher
[mech] faire subir un effort de cisaillement
[metall] cisailler
[text] tondre
[mining] pratiquer des rouillures
SHEAR - [impl] cisaille *f*
[mach tool] cisailles *fpl*
[phys] (any stress which displaces a plane of a solid body parallel to itself, relative to other parallel planes within the body) cisaillement *m*
[mining] (in coal mining) rouillure *f*
~A GATE, to - [plast ind] (to cut off the gate from a moulding) séparer, couper
~BLADE - [tool] lame *f* de cisailles
~BOLT - [mech] (a bolt designed to shear at a known load, as a stress-limiting device) boulon *m* de cisaillement
~CENTRE - [aero] (in a structural member of uniform cross section the point in the plane of a section at which a shear force applied in any direction produces only bending, no torsion being present) centre *m* de cisaillement
~CRACK - [metall] crique *f* par cisaillement
~CUTTER - [tool] cisailles *fpl*
~DEFORMATION - [metall] déformation *f* de glissement
~DOWEL - [mech] s. shear pin
~EDGE - [plast ind] (the edge of the mould which acts as the cut-off) bord *m* de cisaillement
~GRINDER - [gen] rémouleur *m*
~HULK - [naut] (pontoon fitted with shears for the lifting of heavy loads) ponton *m* à mâture
~LAG - [metall] (a special type of stress diffusion applicable tin particular to box structures) retard *m* de glissement
~-LEGS - [impl] bigue *f*, chèvre *f*, anches *fpl*
SHEAR-LEGS - [oil ind] chèvre *f* à trois pieds
~LINE - [metall] ligne *f* de cisaillage
~LOADING - [mech] charge *f* de glissement
~MACHINE - [mech] (power-driven machine for cutting metal especially sheet by a vertically-moving blade) cisailles *fpl*
~MODE OF VIBRATION - [radio] (one of the three common modes of vibrations) mode *m* à onde transversale
~MODULUS - [phys] (or rigidity; the ratio of the shear stress in an homogeneous isotropic elastic medium) module *m* de cisaillment, rigidité *f*
~PIN - [mech] (a pin connecting two parts, designed to shear when a fixed limit load is exceeded, for use as a safety device) goupille *f* de cisaillement
~RATE - [meas] (in viscosimetry) vitesse *f* de cisaillement
~RESISTANCE - s. shear strength
~SEAL - [mech] joint *m* à cisaillement
~STEEL - [metall] (a type of steel which is obtained by heating and hammering blister steel cut in short lengths) acier *m* corroyé
~STRAIN - [phys] (the strain which results from the application of a shear stress, by which parallel planes in a body suffer relative displacement) déformation *f* due au cisaillement, effort *m* de cisaillement
~STRENGTH - [mech] (the ability of a material to withstand shear stress, also the stress at which fails in shear) résistance *f* au cisaillement
~STRESS - s. shearing stress
~TEST - [mech] essai *m* de cisaillement
~THRUST - [geol] charriage *m* de cisaillement
~WAVE - [acoust] (or rotational wave, wave in an elastic medium causing an element of the medium to change its shape without a change of volume) onde *f* de torsion
~WAVES - [phys] (e.g. of an ultrasonic inspection appliance) ondes *fpl* transversales
SHEARED CLOTH - [text] tissu *m* tondu
~EDGE - [metall] bord *m* rogné, bord *m* cisaillé
~WOOL - [text] tonte *f*
SHEARINESS - [paint] (appearance in a dry coating as if an opalescent oily layer had been spread over the surface) effet *m* gras
SHEARING - [phys & mech] (the fracture produced by a shear stress) cisaillage *m*, cisaillement *m*
[text] (of sheep) tonte *f*
[geol] structure *f* cisaillée
[text] (the shortening of hair by cutting the tops of hairs) tonture *f*, affinage *m*
[metall] (cropping of wires) cisaillage *m*, coupure *f*
[mining] rouillure *f*
~ANGLE - [mach tool] (of a cutting machine) angle *m* de coupe
~AREA - [mech] surface *f* de cisaillement
~IN THE GREASE - [text] tonte *f* en suint
~MACHINE - [mech tool] machine *f* à cisailler
[mining] rouilleuse *f*
~OF CORE - [mining] coupe *f* de la carotte
~STRENGTH - s. shear strength
~STRESS - [phys] (intensity of a force acting parallel with the section of a rigid body) effort *m* de cisaillement
~TEXT - [metall] essai *m* de cisaillement
~THE FLEECE - [text] tonte *f* de la toison
~THE SKIN - [text] s. shearing the fleece

SHEARLING - [text] mouton *m* d'un an
SHEARS - [tool] cisailles *f*pl
[mach tool] glissières *f*pl
[naut] chèvre *f*
[mech] (ways, of a machine tool) glissières *f*pl, coulisses *f*pl
~AND SLITTERS - [mach tool] machine *f* à cisailler et à fendre
SHEATH, to - [gen] gainer, envelopper
SHEATH - [gen] gaine *f*, enveloppe *f*
[gen] (of an umbrella) couverture *f*
[bot] (the leaf base when it forms a vertical coating surrounding the stem) gaine *f*, enveloppement *m*
[anat] (a protective structure) enveloppe *f*, gaine *f*
[zool] (the elytron of some insects) élytre *m*
[constr] (a dyke made of stones to stem the waters) remblai *m* de pierres sèches
[el] (on an insulated cable) écran *m*
[photo] châssis *m* (négatif)
[electron] (metal wall of a waveguide) paroi *f*
~CURRENTS - s. sheath eddies
~EDDIES - [el] (in cables, currents induced in the sheath of a single cable and flowing even when the sheaths are isolated from each other) courant *m* parasite de gaine d'un câble
~EFFECTS - [el] (those phenomena which are associated with the metallic sheaths of cables carrying alternating currents) effets *m*pl de gaines métalliques
~-RESHAPING CONVERTER - [electron] transformateur *m* de mode à paroi flexible
~STRIPPING - [el] décapage *m*
SHEATHE, to - [gen] rengainer, revêtir, recouvrir
SHEATHED PYROMETER - [instr] (a type of pyrometer in which the junction is enclosed in a protective sleeve) canne *f* pyrométrique
SHEATHING - [gen] revêtement *m*
[gen] (coating) revêtement *m*, armure *f*
[mech] garniture *f*, chemise *f* (d'un cylindre)
[shipbuild] (plating) doublage *m*
[el] (of cable) armure *f*, cuirasse *f*
[carp] (a close boarding nailed to the framework of a building) revêtement *m*
[mining] cuvelage *m* (d'un puits)
~COMPOUND - [rubber ind] (an elastomeric plastics compound having properties which make it especially suitable for the outer covering of electric cables) matière *f* de gainage pour câbles
~PAPER - [constr] (flexible waterproof lining material made from bitumen reinforced with fibre) papier *m* pour voligeage
SHEAVE - [impl] rouet *m*
[mech] (a grooved wheel in a pulley block) poulie *f* à gorge, roue *f* à gorge
[mech] (of an excentris) plateau *m* d'excentrique
[mech] (for a pitched chain) poulie *f* à noix
~-BLOCK - [mech] moufle *f*
~GROOVE - [mech] gorge *f* de la poulie
SHED, to - [gen] verser, épandre
[gen] (to throw off) perdre, jeter
[text] (the warp) former (la foule)
SHED - [gen] (an outhouse) auvent *m*, baraque *f*
[constr] (a building with a lean-to roof) appentis *m*
[gen] (a hangar) hangar *m*
[geogr] (US only; a watershed, a divide) ligne *f* de partage des eaux
[text] (the horizontal opening formed between the warp threads in a loom, so as to let the through the shuttle) pas *m*, foule *f*
SHED - [railw] remise *f* à machines
[naut] tente *f* à marchandises
~A LOCOMOTIVE, to - [railw] remiser une locomotive
~ADJUSTING LEVER - [text] levier *m* de réglage de la foule
~CLEARING ARRANGEMENT - [text] dispositif *m* d'égalisation
~FOR SHUTTLE - [text] pas *m* de la navette
~FOR THE FIGURE - [text] foule *f* pour le dessin
~FOR WIRE - [text] pas *m* pour le fer
~ROD - [text] tige *f* d'égalisation
~ROOF - [constr] toit *m* en shed, toit *m* en forme de scie
SHEDDING - [text] (dividing the warp threads in a loom horizontally) formation *f* de la foule
[gen] perte *f*, chute *f*
[zool] mue *f*
~BY DOBBIES - [text] formation *f* de la foule par mécaniques d'armure
~BY HARNESS - [text] formation *f* de la foule par harnais
~BY HEALDS - [text] formation *f* de la foule par lames
~MECHANISM - [text] mouvement *m* du pas
~MOTION - [text] changement *m* des lames
~STRAIN - [text] tension *f* due à la formation de la foule
~WARP - [text] chaîne *f* formant le pas
SHEEN - [gen] lustre *m*, luminosité *f*
[paint] (the gloss which appears on a matt surface when viewed at glancing angles) brillant *m*
SHEENY - [gen] luisant, brillant
SHEEP - [zool] mouton *m*, ovins *m*pl
~BOTFLY - [zool] oestre *m* du mouton
~BREED - [zool] race *f* de moutons
~BREEDING - [agric] élevage *m* de moutons
~COT - [agric] bergerie *f*
~DOG - [zool] chien *m* de berger
~FARM FOR WOOL - [agric] ferme *f* pour l'élevage des moutons à laine
~FLOCK - [zool] troupeau *m* de moutons
~FOLD - [agric] bergerie *f*
~HUSBANDRY - [agric] élevage *m* de moutons
~LEATHER - [leather ind] peaux *f*pl de mouton tannées
~NOSE FLY - s. sheep botfly
~PEN - [agric] bercail *m*, parc *m* à moutons
~POX - [vet] (an epidermic disease of sheep) variole *f* ovine
~REARING - [agric] élevage *m* de moutons
~SHEARING - [agric] tonte *f* des moutons
SHEEP'S FESCUE - [agric] fétuque *f* ovine
~SCAB - [vet] gale *f* des moutons
~-SHEARING - [text] tondaison *f*
~SHEARING MACHINE - [text] machine *f* à tondre les moutons
~SORREL - [bot] oseille *f* de brebis
~-STATION - [agric] (a sheep farm in Australia) ferme *f* pour l'élevage des moutons
~STOCK - [agric] cheptel *m* ovin
~STUD - [agric] bergerie *f* d'élevage
~TICK - [vet] pou *m* de mouton
~WITH CURLY WOOL - [zool] mouton *m* à laine crépue
~WOOL - [text] laine *f* de mouton
SHEEPCOTE - s. sheep cot

SHEEPSHANK - [naut] jambe *f* de chien
SHEEPSKIN - [leather ind] peau *f* de mouton
[paper man] (parchment paper) parchemin *m*
SHEEPSKIN'S WOOL CARD - [text] carde *f* à laine de mouton
SHEEPWALK - [agric] pâturage *m* pour moutons
SHEER, to - [gen] (of rocks) se dresser verticalement
[naut] embarder
SHEER - [gen] pur, vraie
[gen] (of a beverage) pur
[text] (of a fabric) léger, fin
[geol] (of arrock, a wall etc) vertical, perpendiculaire
[naut] (of a deck of a ship seen from the side) embardée *f*
[shipbuild] (the shear strake) carreau *m*, vibor *m*
~LEGS - s, sheers
~OFF, to - [naut] largeur les amarres
~PLAN - [shipbuild] (projection of the lines of a ship on the median longitudinal vertical plane) plan *m* longitudinal
~RAIL BOAT FENDER - [naut] (of a boat) liston *m*
SHEERS - [mech] (large hoisting device used in shipyards) chèvre *f* à aubans
SHEET, to - [gen] couvrir d'un drap
[text] (to line) doubler
[mining]blinder(une galerie) limander (une galerie)
[constr] blinder
SHEET - [gen] feuillet *m*, feuille *f*
[gen] (of paper) feuille *f* (de papier)
[metall] plaque *f*, tôle *f* fine, lame *f*
[met] (of ice, or snow) couche *f*
[gen] (of water) nappe *f*
[geol] nappe *f*
[text] drap *m* (de lit)
[naut] écoute *f*
~ANCHOR - [naut] ancre *f* de veille
~ANTENNA - [radio] (a plane aerial) antenne *f* dipôle
~BANDS - [print] longement *m* des feuilles
~BAR - [metall] (intermediate roll section in flat bars) larget *m*, platine *f*
~BAR MILL - [metall] laminoir *m* de largets
~BEND - [naut] noeud *m* d'écoute
~BILLET - [metall] larget *m*
~BLOWING METHOD - [rubber ind] méthode *f* à former des pellicules thermoplastiques par soufflage
~-BRASS - [metall] (brass in a form commonly used for working purposes) tôle *f* fine de laiton
~CALENDER - [paper man & rubber ind] calandre *f* à tirer les feuilles
~CALENDERED PAPER - [paper man] papier *m* tiré en feuilles
~CALENDERING - [rubber ind & paper man] tirage *m* de feuilles à la calandre
~-COPPER - [metall] cuivre *m* en feuilles
~EROSION - [geol] érosion *f* de la nappe
~EXTRUDER - [plast ind] (an extrusion machine for producing plastic sheeting) boudineuse *f* pour feuilles
~-FED GRAVURE ROTARY - [print] machine *f* hélio pour feuilles
~-FED OFFSET ROTARY - [print] machine *f* offset pour l'impression de feuilles
~FILM - [photo] film *m* plan

SHEET FORMING - [plast ind] formage *m* des feuilles
~GASKET - [mech] (flat jointing material) joint *m* plat, rondelle *f* d'étanchéité
~GAUGE - [tool] jauge *f* pour tôles, calibre *m* d'épaisseur
[plast ind] épaisseur *f* de feuille
~GLASS - [glass man] (glass used for common glazing purposes and produced by drawing a continuous film of glass from a molten bath and cutting up the product into sheets) verre *m* en tables
~GRATING - [electron] (of a waveguide) filtre *m* à lames métalliques axiales
~HOME, to - [naut] (a sail) border à bloc
~IRON - [metall] tôle *f* de fer
~IRON BLADE - [mech] lame *f* de tôle en fer
~IRON CASING - [mech] gaine *f* en tôle en fer
~IRON DAMPER - [text] segment *m* plein
~IRON SHELL - [metall] enveloppe *f* de tôle en fer
~-IRON TONGS - [impl] pinces *f*pl pour tôles
~IRON UNDERGUARD - [auto] tôle *f* inférieure de protection
~JOINTS - [geol] diaclases *f*pl horizontales
~LEAD - [metall] (for common use) plomb *m* laminé, feuille *f* de plomb
~LIGHTNING - [met] (effect produced by the reflection of a lightning flash on clouds) éclair *m* diffus
~METAL - [metall] tôle *f*
~METAL AND TUBE BAND SAWING MACHINE - [mach tool] scie *f* à ruban pour tôle et tubes
~METAL GAUGE - [metall] jauge *f* d'épaisseur
~METAL ROLLER STRAIN RELIEVING MACHINE - [mach tool] tour *m* à dépuiller à rouleaux
~-METAL SCREW - [mech] vis *f* à tôle
~METAL SIEVE - [impl] grille *f* en tôle
~METAL SQUARING SHEARS - [mach tool] cisailles *f*pl pour tôle
~METAL SPOOL - [text] tube *m* métallique
~METAL TROUGH - [text] auge *f* en tôle
~METAL WORKING - [metall] travail *m* de tôle
~MICA - [min] mica *m* en feuilles
~MILL - [metall] laminoir *m* à tôles
~OF LAP - [text] feuille *f* de nappe
~PACK - [metall] tôle *f* en forme de paquets
~PAVEMENT - [constr] (road surfacing consisting of continuous material e.g. concrete) pavé *m* sans joints
~PILE, to - [constr] garnir de palplanches
~PILE - [constr] palplanche *f*
~PILED CURTAIN - [soil] batardeau *m* de palplanches
~PILING - [constr] (timber or steel sheeting used to resist lateral pressure) palplanches *f*pl, palée *f*
[hydr] encrèchement *m*
~PILING OF REINFORCED CONCRETE - [constr] file *f* de palplanches en béton armé
~PILING OF SECTION IRON - [constr] file *f* de palplanches en fer profilé
~PILLAR - [mining] pilier-abri *m*
~ROLL - [metall] cylindre *m* pour tôles
~ROLLER SHAFT - [text] arbre *m* du rouleau inférieur du tablier
~ROLLING - [metall] laminage *m* de tôles
~RUBBER - [rubber ind] caoutchouc *m* en feuilles
~SAW - [impl] scie *f* pour tôles
~SEPARATION - [mech] (the gap between lap parts after welding) écartement *m* entre les joints
~-STEEL - [metall] tôle *f* d'acier
~-STEEL SWITCH BOX - [el] coffret *m* extérieur

SHEET TENSION - [el] tension *f* de membrane
~TIN - [metall] fer-blanc *m*
~TRAIN - [plast ind] (the whole assembly of a plastics sheet producing plant from extruder to stacker) circuit *m* des feuilles
~VEIN - [mining] filon *m*, couche *f*
~WATER - [geol] eaux *f*pl des nappes
~ZINC - [metall] feuille *f* de zinc
~ZINC HEALD - [text] lisse *f* en feuilles de zinc
SHEETER - [print] (a cutting machine for a press) rogneuse *f*
~BOX - [text] nappeuse *f*
~LINES - [plast ind] lignes *f*pl de tranchage
SHEETING - [gen] (sheet material in general) feuilles *f*pl
[plast & rubber ind] feuille *f* continue
[plast ind] (thin plastic material) pellicule *f*
[metall] (covering) blindage *m*
[metall] blindage *m*
[text] (fabric used for sheets) toile *f* pour draps
[constr] (the wearing course) surface *f*
[constr] (rough horizontal boards supporting the sides of narrow trenches) blindage *m*
[mining] limande *f* (de galerie)
[naut] braie *f*
[geol] stratification *f*
[oil ind] tubage *m*
~BATTERY - [rubber ind] train *m* de laminoirs
~CALENDER - [plast ind] laminoir-tireur *m* de feuilles
~MACHINE - [rubber ind] laminoir *m* à feuilles
~MILL - [rubber man] laminoir *m* à sheets
~-OUT - [rubber ind] laminage *m* de feuilles
~-OUT MILL - [rubber ind] laminoir-tireur *m* de feuilles
~PROCESS - [rubber ind] tirage *m* en feuille
~RUBBER - [rubber ind] caoutchouc *m* pour préparation des feuilles
SHEETS - [naut] (the ropes used to adjust the angle of sails) écoutes *f*pl
[mining] tôle *f* d'aérage
SHELF - [gen] tablette *f*, planche *f*, rayon *m*
[geol] (of a rock) saillie *f*, rebord *m*
[geol] plateau *m*, seuil *m*, roche *f* de fond
[shipbuild] bauquière *f*
[anat] crête *f*
~LIFE - [gen] (maximum storage time for which a material remains usable) durée *f* limite de stockage
~TEST - [gen] essai *m* de conservation
SHELFPIECE - [shipbuild] gouttière *f* renversée
SHELL, to - [gen] écaler, décortiquer, écosser
[firearms] bombarder
SHELL - [gen] coquille *f*, écaille *f*, écorce *f*
[zool] (hard structure encasing an animal) carapace *f*
[zool] (of a mollusk) coquille *f*
[firearms] obus *m*
[firearms] (US only) cartouche *f*
[tool] cuiller *m*
[mech] (any curved hollow structure of stiffened sheet metal) corps *m*, coque *f* (de chaudière)
[phys] (a group of electrons, forming part of outer structure of an atom and having a common energy level) couche *f*
[el chem] (the external container in which the electrolysis of fused electrolyte is conducted) cuve *f*

SHELL - [shipbuild] (the outer covering of a ship) bordage *m*
[shipbuild] carcasse *f* de navire
[naut] (a small boat) canot *m*
[naut] (the casing of a block) caisse *f*
[mech] (of a pulley) chape *f* (de poulie)
[metall] (the metal part of a cupola supporting the refractory lining) enveloppe *f*, corps *m* cylindrique
[aero] (tube forming metal structure) coque *f*
[mech] (of a bearing) bague *f*, anneau *m*
[metall] (thin film of metal on the surface of steel) peau *f*
[el typing] (a layer of metal, either copper or nickel, deposited on, and separated from a mold) coquille *f*
[metall] (a sliver) écharde *f*, éclat *m*
[tool] (a concave grinding wheel) meule *f* concave
[tool] (a shell bit used with a brace) torpille
[text] porte-peigne *m*
[th eng] (boilers) corps *m* de chaudière
[mach tool] (a hollow shank) à manchon
[geol] écorce *f* (de la terre)
[hydr] boisseau *m* (de robinet)
[auto] calandre *f* (du radiateur), jupe *f*
~-AND-TUBE EXCHANGER - [th eng] (apparatus in which a nest of tubes carrying one fluid is arranged in a closed vessel containing the other) échangeur *m* à faisceaux
~AUGER - [tool] tarière *f* à cuiller
~BAFFLE - [mech] baffle *m* à cuvette, baffle *m* à écuelle
~BEARING - [mech] coussinet *m* à manchon
~BOILER - [th eng] chaudière *f* cylindrique
~BUCKET - [hydr] (used in well-drilling) foret *m* à soupape
~-CAST - [metall] moulé en coquille
~CASTING - [metall] moulage *m* en coquille
~CHUCK - [mach tool] mandrin *m* à vis
~CIRCUIT - [radio] (a coaxial resonator) circuit *m* en coquille
~CORE - [metall] noyau-carapace *m*
~DRILL - [tool] (used for rough reaming) mèche-cuiller *f*
~EDGE - [text] bord *m* à coquille
~END CUTTER - [mach tool] fraise *f* pour fraisage en bout
~FEED PLATE - [text] auge *f* d'alimentation
~FEED ROLLER - [text] cylindre *m* alimentaire
~FEEDER - s. shell feed roller
~-FLOUR FILLER - [plast ind] (a filling material for e.g. plastics, consisting of finely ground coconut shells) coquille *f* de charge
~-FRUITS - [agric] fruits *m*pl à coques
~GIMLET - [tool] vrille *f* à gouge
~GOLD - [metall] or *m* de coquille
~GRAVEL - [min] gravier *m* coquillier
~-LAC - s. shellac
~LIGAMENT - [zool] (the dorsal ligament joining the valves of the shell) ligament *m* de coquille
~LIMESTONE - [min] calcaire *m* coquillier
~MARBLE - [min] marbre *m* coquillier
~MODEL OF NUCLEUS - [nucl] (a nuclear model in which structure is postulated) modèle *m* des couches
~MOULD - [metall & plast ind] (a mould produced by covering heated metal patterns with sand and synthetic resin and baking the shell) moule-carapace *m*

SHELL MOULDING - [metall & plast ind] (the process of moulding with a shell mould) moulage *m* en coquille
~MOULDING MACHINE - [mech mach] machine *f* à mouler en coquille
~OF BOSH - [metall] enveloppe *f* des étalages
~OF WELL - [constr] couronne *f* de puits
~OUTAGE - [oil ind] hauteur *f* du vide
~PLATING - [shipbuild] (the outer covering of a metal vessel) tôle *f* de bordée
~-PROOF - [constr] blindé, à l'épreuve des obus
~PUMP - [mining] (a sludger) cloche *f* de curage, pompe *f* à sable
~REAMER - [tool] (a hollow reamer fitted on a shank) alésoir-fraise *m* creux
~ROLLER - [text] cylindre *m* à aiguilles
~ROLLER PLATE - [text] auge *f* du cylindre à aiguilles
~SEAM - [text] ourlet *m* à point-coquille
~SILK - [text] byssus *m*
~STRUCTURE OF NUCLEUS - [nucl] (the arrangement of the quantum states of nucleons of a specified kind in a nucleus in groups of approximately the same energy) structure *f* quantique du noyau
~TRANSFORMER - s. shell-type transformer
~-TYPE PATTERN - [text] dessin *m* cloqué
~-TYPE TRANSFORMER - [el] (transformer in which the magnetic circuit surrounds the windings) transformateur *m* cuirassé
~WING - [aero] (a wing formed of sheet metal in the form of a hollow stiffened structure) aile *f* monocoque
SHELLAC - [chem] (a natural alcohol-soluble resin, secreted by an insect, and used for spirit varnishes and as a thermosetting adhesive) gomme-laque *f*
~VARNISH - [ind chem] vernis *m* à gomme-laque
SHELLED - [agric] écalé, écossé
[zool] à coquille, à écaille, à cosse
SHELLING - [agric] égrenage *m*, décorticage *m*
~PEA - [agric] pois *m* lisse
SHELTER, to - [gen] abriter
SHELTER - [gen] abri *m*
~BELT - [agric] protection *f* contre le vent
~DECK - [naut] pont-abri *m*
SHELTERED - [gen] abrité
[comm] (against competition) garanti (contre la concurrence)
~INDUSTRY - [comm] industrie *f* garantie (contre la concurrence étrangère)
SHELVE, to - [gen] garnir de rayons
[gen] (to slope) aller en pente
SHELVED WAGGON - [mining] wagonnet *m* étagé
SHELVING - [gen] rayons *mpl*
[gen] (slope) en pente, incliné
SHEPHERD, to - [gen] surveiller les moutons, soigner les moutons
SHEPHERD - [agric] berger *m*
SHEPHERD'S NEEDLE - [bot] peigne *m* de Vénus
~PURSE - [bot] bourse-à-pasteur *f*
SHERARDIZE, to - [metall] (the operation of coating steel or iron with a thin coating of zinc by heat treatment) shérardiser
SHERARDIZING - [metall] (anti-corrosion process for steel parts, in which a thin coating of zinc is formed by a heat-treatment) shérardisation *f*, galvanisation *f* au gris de zinc
SHIDE - [constr] (shingle; a thin flat rectangular piece of wood laid like a slate or tile) bardeau *m*, tavaillon *m*
SHIELD, to - [gen] protéger
[radio] blinder
[paint] masquer
SHIELD - [gen] bouclier *m*
[zool] carapace *f*, écu *m*
[mech] tôle *f* protectrice
[el] écran *m*
[constr] (a kind of curb for use at the working face in driving a tunnel through loose or water-bearing ground) bouclier *m*
[gen] (of a lamp) cuirasse *f*
[bot] écusson *m* de greffe
[el chem] (piece of inert material e.g. methyl methacrylate sheet, used to shield the anodic region to areas of high current-density) écran *m*
[nucl] (any structure used to limit the passage or radiation) blindage *m*, matériel *m* de blindage
[firearms] (of a field piece) pare-balles *m*, masque *m*
~BASE - [electron] (device to carry the screen in an electronic tube) support *m* d'écran
~FACTOR - [telecomm] (the ratio of noise in a telephone circuit when a source of shielding is present, to the corresponding quantity when the shielding is absent) facteur *m* de blindage
~GRID - [electron] (a grid shielding the control electrode from the anode and or the cathode with respect to the radiation of heat and the deposition of thermionic activating material) grille *f* de protection
~-GRID VALVE - [electron] tube *m* à grille de protection
~WIRE - [telecomm] (a wire used to reduce the effects of external electro-magnetic fields on communication circuits) fil *m* de blindage
SHIELDED - [gen] protégé
[el] blindé
~BEARING - [mech] coussinet *m* blindé
~BOX - [nucl] enceinte *f* blindée contre rayons gamma
~CABLE - [el] câble *m* blindé
~CARBON ARC WELDING - [metall] soudage *m* à arc protégé
~CONDUCTOR CABLE - [el] (multicore cable in which the insulation of each conductor is separately enclosed in a conducting film in order to ensure a radial electric field surrounding the conductor) câble *m* à conducteurs blindés
~GALVANOMETER - [instr] (galvanometer provided with a magnetic screen protecting it from the action of external magnetic fields) galvanomètre *m* cuirassé
~IGNITION WIRING - [el] fil *m* de bougie blindé
~JOINT - [el] épuissure *f* cuirassée
~LAMP - [mining] lampe *f* protégée
~MAGNETO - [el] magnéto *m* cuirassé
~NUCLIDE - [nucl] (a nuclide having a charge higher by one unit than a stable nuclide of the same mass number) nucléide *m* blindé
~PAIR - [telecomm] (a two-wire transmission line surrounded by a metallic sheath) paire *f* blindée
~-POLE INSTRUMENT - [el] (induction instrument having a shaded pole) instrument *m* à pôle à neutralisation
~TRANSFORMER - [el] transformateur *m* cuirassé
~TRANSMISSION LINE - [telecomm] ligne *f* de transmission blindée

SHIELDED WIRE - [el] fil *m* blindé
~X-RAY TUBE - [electron] tube *m* à rayons X à gaine métallique
SHIELDING - [radio] blindage *m*
[el chem] (or shadowing; the interference of any part of an anode, rack etc. with uniforme current distribution upon a cathode) effet *m* d'écran
~CAN - [electron] pot *m* de blindage
~CONTAINER - [el] capot *m* de blindage
~EFFECT - [el] effet *m* d'écran
~FACTOR - [el] coefficient *m* de pénétration
~HARNESS - [el] cuirasse *f* de blindage
~POND - [nucl] réservoir *m* à blindage d'eau
~RATIO - [nucl] taux *m* de blindage
~WINDOWS - [nucl] fenêtres *fpl* blindées, hublots *mpl*
SHIFT, to - [gen] remuer, changer de place
[auto] (the gears) changer de vitesse
[el] (the brushes) décaler (le balai)
SHIFT - [gen] changement *m*
[agric] divagation *f*
[gen] (a relay of workmen) équipe *f*, relais *m* d'ouvriers
[phys] (of spectral lines) décalage *m*
[gen] (change of place or direction) renversement *m*
[geol] faille *f*
~CLUTCH - [mech] embrayage *m* à dent
~DOWN, to - [auto] rétrograder
~FAULT - [geol] faille *f* de rejet horizontale
~FORK - [auto] fourchette *f* de commande des vitesses
~KEY - [mech] (in typerwriters) touche *f* de manoeuvre
~LEVER - [auto] levier *m* de changement de vitesse
~LOCK - [mech] (in typewriters) dispositif *m* de blocage
~OF BRUSHES - [el] décalage *m* des balais
~OF SPECTRAL LINE - [phys] (small displacement in the position of a spectral line caused by a corresponding change in frequency due to a number of causes) décalage *m* de ligne spectrale
~PULSE - [comput] (in static magnetic storage) impulsion *f* de décalage
~REACTION - [chem] (catalytic conversion of carbon monoxide) conversion *f* catalytique
~REGISTER - [comput] registre *m* de déplacements
~RELAY - [comput] relais *m* de décalage
~RING - [hydr] (of a gate)boucle *f* de bardage
~THE TRACK, to - [railw] riper la voie
~THE TYRES, to - [auto] permuter les pneus
~UNIT - [comput] (a device for shifting numbers to adjacent positions) unité *f* de décalage
~WINDING - [comput] (one of the drive windings of a magnetic core) enroulement *m* de décalage
SHIFTED - [gen] déplacé, changé
[metall] (of a defective casting) déplacé
~CASTING - [metall] (a defect due to a shifting of the mould) variation *f* de moule
~CORES - [metall] variation *f* de noyau
SHIFTER - [mech] levier *m* de déplacement
~BAR - [text] barre *f* de déplacement
SHIFTING - [gen] déplacement *m*, changement *m* de place
[mech] mouvement *m*
[metall] déplacement *m*
[auto] changement *m* de vitesse
SHIFTING - [geol] (of sands) sables *mpl* mouvants
~CAMSHAFT - [auto] arbre *m* à cames de changement de vitesse
~CULTIVATION - [agric] divagation *f* des cultures
~DOGS - [auto] levier *m* de commande changement de vitesse
~FORK - [auto] fourchette *f* de commande de vitesse
~GUIDE PIN - [auto] guide *m* de changement de vitesses
~INSTRUCTION - [comput] instruction *f* de décalage
~OF FAULTS - [geol] déplacement *m* de failles
~REGISTER - [comput] registre *m* à décalage
~SPANNER - [impl] (type of spanner in which one jaw is adjustable) clé *f* réglable
SHIFTINGS - [min] (fly-ash) cendre *f* volante
SHIM, to - [mech & carp] caler
SHIM - [gen] pièce *f* d'épaisseur, cale *f* d'appui
[carp] épaisseur *f*
[el] (a plate inserted between two surfaces so as to alter the distance between them) séparateur *m*
[mech] (a piece of metal inserted between two parts to give accurate spacing between them) cale *f* de réglage
~ADJUSTMENT - [mech] réglage *m* par cale
~ELEMENT - [contr] élément *m* de compensation
~MECHANISM - [contr] (device allowing the fine control rod to operate efficiently) mécanisme *m* de compensation
~ROD - [nucl] (control rod controlling comparatively large amounts of reactivity) barre *f* de compensation
SHIMMER - [astr] scintillation *f*
SHIMMING - [el] (the adjustment of a magnetic field to achieve desired characteristics by means of thin spacers) ajustage *m* précis de champ
[nucl] compensation *f*
SHIMMY - [aero] (rapid mechanical oscillation of a castoring wheel in a landing gear) shimmy *m*
[auto] (a steering defect) shimmy *m*
~DAMPER - [aero & auto] (a damping device to reduce shimmy) amortisseur *m* de shimmy
SHIN - [anat] devant *m* du tibia
[zool] (the lower foreleg) canon *m*
[railw] éclisse *f*
~-BONE - [anat] tibia *m*
SHINE, to - [gen] luire, briller
(US only; to polish) cirer, polir
SHINE - [gen] lumière *f*, éclat *m*
SHINGLE, to - [constr] couvrir de bardeaux
[metall] (to eliminate slag and impurities from a mass of iron by hammering) cingler, tringler
SHINGLE - [constr] bardeau *m*, aissante *f*
[geol] (loose detritus, usually coarser than gravel) galets *mpl*
[soil] gravillon *m* roulé
SHINGLER - [metall] machine *f* à cingler, cingleur *m*
SHINGLES - [med] (eruption of crops of firm vesicles along the course of a nerve) ceinture *f*, zona *m*
SHINGLING - [metall] cinglage *m*
~ROLLS - [metall] cylindre *mpl* à cingler
~SQUEEZER - [metall] moulin *m* à cingler
SHINING - [gen] luisant, brillant
SHIP, to - [gen] embarquer
[comm] expédier, mettre à bord
[naut] monter, mettre an place
SHIP - [naut] navire *m*, bâtiment *m*
[naut] (colloq) bateau *m*

SHIP - [naut] (the chip of a log) bateau *m* de loch
[aero] (USA only) avion *m*
~-BISCUIT - [naut] biscuit *m* de mer
~-BOTTOM PAINT - [paint] peinture *f* sousmarine
~-BROKER - [comm] courtier *m* maritime
~CABLES - [gen] câbles *mpl* sousmarins
~CAISSON - [hydraulic] (ship-shaped floating caisson, capable of being floated into position at the entrance to a lock and sunk into grooves in the sides and bottom of the entrance) ascenseur-écluse *m*
~CANAL - [naut] canal *m* de navigation
~CHANDLER - [comm] fournisseur *m* de navires
~COMPANY - [naut] équipage *m*
~DRAUGHT - [naut] tirant *m* d'eau
~ELEVATOR-CONVEYOR - [mech] transporteur-élévateur *m* pour navires
~-FEVER - [med] (colloq. naut. for typhus) typhus *m*
~HUSBAND - [naut] (a shipping agent responsible for provisioning, repairs etc) gérant *m* à bord
~LOAD - [naut] cargaison *f*, fret *m*
~LYING AT ANCHOR - [naut] navire *m* au mouillage
~-OWNER - [naut] (any person a ship or shares in it) armateur *m*
~PAPERS - [naut] papiers *mpl* du bord
~PLATE - [shipbuild] bordage *m*, tôle *f* navale
~RADAR - [radar] radar *m* de bord
~SURVEYOR - [naut] inspecteur *m* de navires
~-WAY - [naut] couettes *fpl* de lancement
[a canal] canal *m* de navigation
SHIPBOARD - [naut] (the side of the ship) bord *m* de navire
~AERIAL - [radio] (aerial installed aboard a ship) antenne *f* pour navires
~AIRCRAFT - [aero] (aircraft designed to operate from an aircraft-carrier or other type of ship) avion *f* de bord
~ANTENNA - s. shipboard aerial
SHIPBUILDER - [naut] constructeur *m* de navires
SHIPBUILDING - [shipbuild] (the designing and building of vessels) architecture *f* navale, construction *f* navale
~YARD - s. shipyard
SHIPMENT - [naut] mise *f* à bord, embarquement *m*
[comm] chargement *m*
SHIPOWNER - s. ship-owner
SHIPPER - [comm] expéditeur *m*, affréteur *m*
[text] levier *m* d'embrayage
~BAR - [text] barre *f* de commande
~LEVER - [text] levier *m* d'arrêt
SHIPPERS - [constr] (sound and hard-burned bricks with irregular or defective shapes) briques *fpl* difformes
SHIPPING - [naut] marine *f* marchande
[comm] embarquement *m*, mise *f* à bord
~AGENT - [naut] agent *m* maritime
[comm] expéditeur *m*
~BILL - [comm] déclaration *f* de sortie
~COMPANY - [comm] compagnie *f* de navigation
~CONTAINER - [aero] (special container to protect aircraft or parts during transport) emballage *m* de transport
~COSTS - [comm] frais *mpl* de transport
~CUBAGE - [naut] cubage *m*
~DIMENSIONS - [naut] cube *m* maritime
~DOCUMENTS - [naut] connaissement *m*
~IN BULK - [comm] chargement *m* arrimé
~-MASTER - [naut] enrôleur *m* d'équipages

SHIPPING NOTICE - [comm] avis *m* d'expédition
~OFFICE - [comm] agence *f* maritime, bureau *m* de réception des marchandises
~PORT - [naut] port *m* d'embanquement
~SEAL - [el] (sealed end of a cable) bout *m* de câble, manchon *m* d'extrémité
~TONNAGE - [comm] tonnage *m*, capacité *f* de chargement d'un navire
~TRADE - [comm] commerce *m* maritime
~VOLUME - [naut] cubage *m* du navire
~WEIGHT - [comm] poids *m* total des marchandises
SHIPPLANE - [aero] (an aircraft designed to operate from the deck of a ship) avion *m* embarqué
SHIPSHAPE - [gen] bien arrangé, fin prêt
SHIPWAY - s. ship way
SHIPWORM - [zool] taret *m*
SHIPWRECK, to - [naut] faire naufrager
SHIPWRECK - [naut] naufrage *m*
[gen] (the remnants of a wrecked ship) restes *mpl* d'un navire naufragé
SHIPWRIGHT - [naut] (a ship carpenter or builder) charpentier *m* de navires
SHIPYARD - [naut] chantier *m* de construction
~CONJUNCTIVITIS - [med] kératoconjonctivité *f* épidémique
~DISEASE - s. shipyard conjunctivitis
~EYE - s. shipyard conjunctivitis
SHIRK, to - [gen] manquer, se soustraire à
SHIRT - [text] chemise *f*
[th eng] (the inner lining of a furnace) chemise *f* (de fourneau)
SHIRTING - [text] (closely woven cotton, linen etc fabric used for making shirts, blouses etc) toile *f* pour chemises
SHIVE - [gen] (a thin flat cork) bouchon *m*, bonde *f* (de tonneau)
SHIVER, to - [gen] (to tremble with cold or fear) frissonner
[naut] (to flutter in the wind) faseyer, ralinguer
[acoust] (to give a vibrating sound) vibrer
SHIVER - [gen] (the act of shivering, trembling) frisson *m*
[mech] (pulley) poulie *f*
[gen] (a splinter, a sliver) tranche *f*
SHIVES - [text] (vegetable matter found in wool) chènevottes *fpl*
SHIVY WOOL - [text] laine *f* chardonneuse
SHOAD - [geol] guidon *m*
SHOAL, to - [naut] (to become shallow) diminuer de profondeur
[zool] (to throng in shoals) se réunir en bancs
[zool] (to school, of fish) aller par bancs
SHOAL - [naut] (a shallow place) bas-fond *m*, eau *f* profonde
[naut] (a sandbanck or bar) haut-fond *m*, banc *m*
[zool] (of fish; a multitude) banc *m* voyageur
~OF FISH - [gen] banc *m* de poissons
SHOALY - [naut] plein de bancs de sable, plein de hauts-fonds
SHOCK, to - [gen] choquer
[med] souffrir de choc
[el] donner une secousse électrique
[agric] mettre en moyettes
SHOCK - [gen] choc *m*, secousse *f*
[gen] (a collision) impact *m* d'une collision
[med] choc *m*, traumatisme *m*
[el] secousse *f*

SHOCK - [geol] sisme *m*
~ABSORBER - [mech] (any device to take up the effect of violent impact) amortisseur *m* de choc
~-ABSORBER OIL - [auto] huile *f* spéciale pour amortisseurs
~ABSORBER PAD - [mech] coussinet *m* amortisseur
~-ABSORBING MOUNTING - [el acoust] (the mounting of any sound-emitting or receiving system which is so arranged as to evade spurious mechanical vibrations) montage *m* antivibratoire
~-ABSORBING WAGON - [railw] wagon *m* de choc, wagon *m* tamponneur
~BENDING TEST - [mech] essai *m* de flexion au choc
~CONCRETE - [constr] béton *m* choqué
~COOLING - [metall] refroidissement *m* par choc thermique
~CORD - [aero] (a rubber cord employed as a landing shock absorber for small aircraft) câble *m* élastique amortisseur
~-CORD TOW-OFF - [aero] remorqueur *m* à câble élastique
~CRACKS - [metall] criques *fpl* par choc
~ELASTICITY - [mech] (impact resilience) élasticité *f* au choc, résilience *f*
~EXCITATION - [radio] (the excitation of natural oscillations in an oscillatory system due to a sudden acquisition of energy from an external source) excitation *f* par choc
~-FORMING - [metall] (the forming of sheet metal by means of a high explosive) moulage *m* par choc
~FRONT - [phys] (the boundary between the pressure disturbance caused by an explosion and the ambient atmosphere) front *m* de choc
~HEATING - [phys] (of a plasma) chauffage *m* par onde de choc
~LOAD - [mech] charge *f* de choc
~LOAD ABSORBER - [mech] amortisseur *m* de charge de choc
~MOTION - [mech] (in a mechanical system, a transient motion characterized by suddenness and by significant relative displacements) mouvement *m* de choc
~MOUNTING - [mech] (any elastic fixing arrangement designed to absorb shock or vibration) montage *m* élastique
~PENDULUM - [impl] appareil *m* à marteau-pendule, mouton-pendule *m*
~-PROOF - s. shock resistant
~-PROOF LAMP-HOLDER - [el] douille *f* isolée
~-PROTECTED - [mech] anti-choc
~-PROTEXTING DEVICE - [horol] dispositif *m* amortisseur
~RESISTANT - [mech] résistant au choc
~RESISTANTS - [plast ind] (moulding materials incorporating a filler which improves the shock resistance of the finished article) produits *mpl* résistant au choc
~STRUT - [aero] (a type of landing gear strut designed to absorb shock) montant *m* amortisseur (du train d'atterrissage)
~TEST - [mech] essai *m* au choc
~TESTER - [electron] (an apparatus designed to measure the mechanical strength of electronic tubes) appareil *m* d'essai par chocs
~TUBE - [aero] (a tube in which shock waves can be generated for research purposes) tube *m* générateur d'ondes de choc

SHOCK TUNNEL - [aero] (a type of wind tunnel for hypersonic flight research) soufflerie *f* aérodynamique à ondes de choc
~WAVE - [phys] (a narrow belt transverse to the stream lines of a flow, in which abrupt increase in temperature, density and pressure occur and in which velocity falls sharply, with associated entropy increase) onde *f* de choc
~-WAVE DRAG - [aero] (drag arising from the presence of shock waves) traînée *f* due aux ondes de choc
SHOCKLESS STATIC BAR - [el] (induction static bar operated at a low voltage to avoid danger of electric shock) barre *f* statique d'induction
SHOCKPROOF - [mech] (e.g. an instrument, a watch etc) résistant au choc
~CASE - [transp] caisse *f* résistante au chocs
~SWITCH - [el] (a switch having all external metallic parts covered or protected by insulating material) commutateur *m* protégé contre les contacts accidentels
~TUBE - [nucl] (X-ray tube surrounded by an earthed shield) tube *m* antichoc
~TUBE HOUSING - [nucl] gaine *f* de tube antichoc
~WATCH - [horol] montre *f* antichoc
SHODDY - [text] (waste woolen material) shoddy *m*, drap *m* de laine d'effilochage
~MILL - [text] fabrique *f* de laine artificielle
~SILK - [text] shoddy *m* de soie
~TEASER - [text] effilocheur *m* des chiffons
SHOE, to - [gen] chausser
[agric] (a horse) ferrer
[mech] saboter
SHOE - [gen] soulier *m*
[of a horse] fer *m* de cheval
[mech] (of a ferrule) sabot *m*, patin *m*, semelle *f*
[mech] (of a brake) sabot *m* (d'un frein)
[el] (in an electrical motor) frotteur *m*, sabot (de tramway)
[el] (the device by which an electric tractor collects current from a live rail) patin *m*, sabot *m*
[plumb] (bending shoe) dauphin *m*
[impl] (of a sledge) sabot *m*
[arch] (of a bridge) lardoire *f*
[auto] (a blow-out shoe) guêtre *f* d'éclatement
[naut] savate *f* (d'ancre)
~A WHEEL, to - [auto] chausser une roue, bander une roue
~BARS - [mech] fers *mpl* cavaliers
~BRAKE - [mech] (type a brake in which lined blocks are applied to a drum (as distinct from a disk brake) frein *m* à sabot
~BUTT - [leather ind] cuir *m* fort, croupon *m*
~CANVAS - [text] canevas *m* pour chaussure
~DUCK - [text] toile *f* pour chaussure
~GUIDE - [oil ind] tampon guide *m* du sabot
~INSOLE LINING - [shoe man] revêtement *m* de premières (semelles)
~LEATHER - [leather ind] cuir *m* pour souliers
~MECHANISM - [el] mécanisme *m* de patin
~MECHANISM BRAKE - [el] mécanisme *m* de patin de freinage
~MECHANISM CONTROLLER - [el] combinateur *m* de mécanisme de patin
~NAILS - [metall] clous *mpl* pour chaussures
~PAN - [shoe man] autoclave *m* pour chaussures en caoutchouc

SHOE PLATE - [auto] (of brake) sabot *m*
~PLUSH - [text] peluche *f* pour chaussures
~SCRAPER - [gen] (a door-scrapermat) décrottoir *m*, tapis *m* gratte-pieds
~SOLE - [shoe man] semelle *f*
~SOLING - [shoe man] semelle *f*
~-STRING SAND - [geol] couche *f* filiforme
~TREE - [shoe man] tendeur *m* pour chaussures
~VELVET - [text] velours *m* pour pantoufles
SHOEBUTTON TUBE - [electron] (or acorn valve; a very small electronic tube) tube *m* bouton, tube *m* gland
SHOEMAKER - [gen] cordonnier
SHOEMAKING - [gen] cordonnerie *f*
SHOOK, to - [agric] mettre en botte
S-HOOK - [auto] crochet *m* en S
SHOOT, to - [gen] se lancer, s'élancer
[bot] pousser des bourgeons
[firearms] décharger
[firearms] (to hit with a bullet etc) tirer, faire feu
[cin] tourner (un film)
[astr] (a star etc) déterminer la hauteur
[mining] (the ore) bouter, culbuter
[mech] actionner
[carp] (to plane true) dégauchir, dresser, équarrir
SHOOT - [bot] pousse *f*, sarment (de vigne)
[constr] (a chute used for coal etc) conduit *m* incliné, couloir *m*
[min] (a body of ore) colonne *f* de richesse
[hydr] déversoir *m*
[arch] (the push of an arch) poussée *f*
[mining] (a shaft through which ore is passed on from one level to another) cheminée *f* à minerai
[text] (a thread of weft; a pick) duite *f*
[mining] (a small vein) caprice *m*
~A MOTION PICTURE, to - [cin] tourner un film
~A PICTURE, to - [photo] prendre un instantané
~AN OIL WELL, to - [oil ind] torpiller un puits à pétrole
~DOWN, to - [gen & aero] abattre, descendre
~INTO EAR, to - [agric] monter en épi, grener
~UP, to - [gen] jaillir, grandir
[comm] (of prices) augmenter rapidement
[gen] (of flames) jaillir
[bot] pousser
SHOOTING - [gen] action *f* de tirer
[min] culbutage *m*, chavirement *m*
[med] élancement *m* (de blessure)
[cin] prise *f* de vues
[carp] dressement *m*
~ANGLE - [cin] (the angle of view taken by the motion picture camera) angle *m* de prise de vue
~BOARD - [carp] (board prepared to steady a piece of timber while shooting edges) planche *f* à dresser
~BRAKE - [cin] (the vehicle used to carry the film apparatus) voiture *f* de reportage
[auto] camionnette *f*
[telev] car *m* de télévision
~CAMERA - [cin] camera *f* de prise de vue
~INTO A HEAP - [gen] (of coal, coke etc) mise *f* en tas
~PLANE - [impl] verlope *f* à équarrir
~RANGE - [cin] (the area which is covered by the lens of the camera) champ *m* visuel
~SCHEDULE - [cin] tableau *m* de travail
SHOE SCRIPT - [cin] découpage *m*, scénario *m*
~STAR - [astr] étoile *f* filante
~TRICKS - [cin] trucs *mpl* de prise de vue
SHOP - [gen] magasin *m*
[gen] (a workshop) atelier *m*
[glass ind] (team of workers) équipe *f*
[glass man] (the location at the furnace of a producing unit) emplacement *m*
~CRANE - [mech] grue *f* d'atelier
~EQUIPMENT - [mech] équipement *m* d'atelier
~FOREMAN - [gen] chef *m* d'atelier
~FRONT - [gen] devanture *f* de magasin
~LIFT - [mech] pont *m* élévateur
[auto] (a single, twin or four pole lift for autos) élévateur *m* à colonne
~OVERHEAD TRAVELLING CRANE - [mech] pont *m* roulant d'atelier
~SCALES - [gen] balance *f* à plateaux
~TEST - [gen] (test carried out in a factory or workshop) essai *m* en atelier
~-WINDOW - [gen] vitrine *f*, étalage *m*
SHOPPING CENTRE - [comm] place *f* marchande
SHORAN - [radar] (a system giving a precision position by using a pulse transmitter and receiver and two transponder beacons at fixed points) shoran *m*, système *m* de radar-navigation
SHORE, to - [naut] débarquer
[constr] (to support temporarily by shores the sides of excavations and unsafe buildings) étayer, buter, chandeller
[mining] chandeller (le toit)
[shipbuild] accorer, épontiller
[naut] caboter
SHORE - [gen] (of the sea) littoral *m*, rivage *m*
[geogr] (of a lake) bord *m*
[geogr] (of a river) rive *f*
[geogr] plage *f*
[constr] (a prop of timber or other material) étai *m*, butte *f*, contre-boutant *m*
[shipbuild] accore *m*, épontille *f*
~-BASED RADAR - [radar] (radar navigational aid in harbours etc) radar *m* côtier
~DEPOSIT - [geol] accumulation *f* littorale
~FAST - [naut] amarre *f* du quai
~LINE - [geogr] littoral *m*, ligne *f* du rivage
SHOREWARDS - [naut] vers la terre
SHORING - [constr & mining] (supporting by shores) étayement *m*, enchevalement, chandelles *fpl*
[soil] étrésillonnement *m*
SHORN - [gen & text] tondu, rasé
~CLOTH - [text] tissu *m* tondu
~WOOL - [text] laine *f* tondue
SHORT, to - s. short-circuit, to
SHORT - [gen] court, bref
[gen] (insufficient) insuffisant
[phys] (breaking easily) aigre
[cin] bande *f* annonce
[metall] (brittle) cassant à chaud
[metall] déchet *m* de tôle
[plumb] s. short pipe
~ACCESS TIME - [comput] court temps *m* d'accès
~-BASE RANGE-FINDER - [meas] (range finder with a home-base and one observation point) télémètre *m* monostatique
~BOBBIN RAIL - [text] chariot *m* des bobines court
~-BRISTLE STIPPLE BRUSH - [impl] (for rubber tyres) pinceau *m* à pointer, pinceau *m* à tapoter

SHORT CHECK STRAP - [text] courroie *f* d'arrêt courte
~-CIRCUIT, to - [el] (to connect two points by a conductor of negligible resistance) court-circuiter
~-CIRCUIT - [el] (a conducting path of negligible resistance) court-circuit *m*
~-CIRCUIT ADMITTANCE - [el] (the reciprocal of the short-circuit impedance) admittance *f* de court-circuit
~-CIRCUIT BRAKE - [el] frein *m* à court-circuit
~-CIRCUIT CALCULATOR - [el] (assembly of variable impedances or resistances which can be connected and thus represent in miniature the circuits of a power system) calculateur *m* de court-circuit
~-CIRCUIT CHARACTERISTIC - [el] (the characteristic graph relating electromotive force to load current) caractéristique *f* en court-circuit
~-CIRCUIT CURRENT - [el] (or flash current; initial value of the current obtained in a circuit of negligible resistance) courant *m* de court-circuit
~-CIRCUIT CURRENT LIMITING REACTOR - [el] bobine *f* de choc contre les court-circuits
~-CIRCUIT CURRENT TO EARTH - [el] perte *f* par court-circuit
~-CIRCUIT CURRENT TO GROUND - s. short-circuit current to earth
~-CIRCUIT FORWARD ADMITTANCE - [electron] admittance *f* directe en court-circuit
~-CIRCUIT IMPEDANCE - [el acoust] (on an electromechanical transducer, the electrical impedance at the input when the mechanical impedance of its load is made zero) impédance *f* en court-circuit
~-CIRCUIT-PROTECTION - [el] dispositif *m* de protection contre les court-circuits
~-CIRCUIT RATIO - [el] (of a synchronous machine, the ratio of the field current for rated open-circuit armature voltage and rated frequency to the field current for rated armature on sustained symmetrical short circuit at rated frequency) rapport *m* de court-circuit
~-CIRCUIT SPARK - [el] étincelle *f* de court-circuit
~-CIRCUIT TEST - [el] (the test carried out on an electrical machine with its output terminals short-circuited and full-load current flowing) essai *m* de court-circuit
~-CIRCUIT THE MOTOR, to - [el] mettre le moteur en court-circuit
~-CIRCUIT TIME - [el] durée *f* de court-circuit
~-CIRCUIT TO EARTH - [el] contact *m* à la terre
~-CIRCUIT TO GROUND - s. short-circuit to earth
~-CIRCUIT TRANSFER ADMITTANCE - [el] (between two meshes of a network) admittance *f* de transfert en court-circuit
~-CIRCUIT TRANSFER CAPACITANCE - [electron] capacité *f* mutuelle en court-circuit
~-CIRCUIT TRANSITION - [contr] (or shunt transition; a method of transition from series to parallel connection in the series-parallel control of d.c. electric motors) transition *f* par dérivation, transition *f* court-circuit
~-CIRCUIT TURN - [el] spire *f* en court-circuit
~-CIRCUIT VOLTAGE - [el] (the electromotive force required to cause full-load current to flow under short-circuit conditions) tension *f* de court-circuit
~-CIRCUITED ROTOR - [el] (or squirrel-cage rotor; a rotor of an induction motor the winding of which consists of a number of bars having their extremities at each end of the rotor connected by rings or plates) rotor *m* à cage d'écureuil
~-CIRCUITED WINDING - [el] enroulement *m* en court-circuit
~-CIRCUITING BRIDGE - [el] pont *m* de shuntage
~-CIRCUITING DEVICE - [el] (a switching device on the rotor of a slip-ring induction motor) dispositif *m* de court-circuit
~-CIRCUITING SWITCH - [el] interrupteur *m* de court-circuit
~CLAY - [min] argile *f* maigre
~COAL - [min] houille *f* à courte flamme
~CROP - [agric] récolte *f* pauvre
~CUT - [gen] raccourci *m*
~-DISTANCE SCATTER - [radio] (the direct return of signals to points within the skip zone by scattering from localized irregularities of free-electron distribution in the ionosphere) diffusion *f* à courte distance
~FIBRE - [text] fibre *f* courte
~FIBRED - [text] à fibre courte
~FLAX - [text] lin *m* à courts brins
~FLUCTUATION - [statist] fluctuation *f* court terme
~-FOCUS LENS - [opt] objectif *m* à court foyer
~-FORM VACUUM GAUGE - [instr] vacumètre *m* raccourci
~GROUND, to - s. short to earth
~-GROWN MALT - [brew ind] malt *m* peu germé
~HAND-TAP - [mech] taraud *m* court à main
~HAUL - [transp & railw] trafic *m* à courte distance
~-HAUL TOLL CIRCUIT - [telecomm] (in automatic telephony) circuit *m* régional
~-HEAD RATCHET-BRACE - [mech] cliquet *m* à manchon court
~HOB - [tool] taraud *m* mère
~HOLE - [metall] (in a furnace) trou *m* non complètement fermé
~HOOK - [text] crochet *m* court
~HORN - [mining] front *m* d'attaque à 60° avec les limets
~IRON - [metall] fer *m* aigre, fer *m* cassant
~-LANDED - [transp] (not corresponding to the quantity declared) livré partiellement
~LENGTH - [timber] (a length of timber less than 8 ft) mesure *f* incomplète
~LENGTH OF RAIL - [railw] coupon *m* de rail
~-LINK CHAIN - [mech] chaîne *f* serrée
~-LIVED - [nucl] (term denoting a radioactive element or an isotope having a very short lifetime) à vie courte
~-LIVED ISOTOPE - [nucl] isotope *m* de vie courte
~-LIVED RADIOACTIVE SUBSTANCE - [nucl] substance *f* radioactive à courte vie
~LOOM - [text] métier *m* court
~MOULDING - [plast ind] (a faulty moulding of density below normal, caused by insufficient mould charge or lack of pressure during moulding) moulage *m* court
~NIPPLE - [plumb] (a short length of pipe threaded at both ends, but with a short piece left plain in the middle) raccord *m* fileté
~-NOSE PLIERS - [impl] pinces *fpl* à becs, courts, béguettes *fpl*
~-OIL ALKIDS - [chem] alkydes *fpl* à basse teneur d'huile
~OIL VARNISHES - [paint] (oleo-resin varnishes in which the proportion of oil resin is I I/2 (or less)

to I, used when durability is not of primary importance, e.g. indoors) vernis *mpl* légers
SHORT-PATH DISTILLATION - [ind chem] distillation *f* à court trajet
~-PATH PRINCIPLE - [el] (an application of the Passchen law, stating that discharge between electrodes in a gas at a given pressure will not always occur between the closest points of the electrodes if the distance between these points corresponds to a point to the left of the minimum of the ignition potential curve) principe *m* de Hiffort
~-PATH STILL - [ind chem] installation *f* de distillation à court trajet
~PERIOD OF RISE - [nucl] (of a reactor, when only a short time is required to bring it to normal operation) période *f* courte de montée en puissance
~-PERSISTANCE CATHODE-RAY TUBE - [electron] tube *m* à rayons cathodiques de courte durée de persistance
~-PERSISTANCE SCREEN - [electron] écran *m* à courte durée de persistance
~PIPE - [plumb] tube *m* court
~-PITCH WINDING - [el] (a fractional-pitch winding in which the average coil pitch is less than the pole pitch) enroulement *m* à pas raccourci
~POINT-RAIL - [railw] branche *f* de pointe
~POURING - [metall & plast ind] coulée *f* incomplète
~-RADIUS CURVE - [geom] courbe *f* à petit rayon
~RANGE NAVIGATION SYSTEM - s. shoran
~RUBBER - [rubber ind] (weak rubber) short-rubber *m*, caoutchouc *m* friable
~-RUN - [metall] (a defective casting) manque *m*
~-RUN CASTING - [metall] (a misrun) pièce *f* non venue
~SHAFT PENDULUM TOOL - [tool] fouloir *m* d'établi
~SHOT - [metall & plast ind] (moulding operation in which the mould is not completely filled) pièce *f* incomplète, moulage *m* court
~-SIGHTED - [opt] myope
~-SIGHTEDNESS - [opt] myopie *f*
~SKEINED RAW SILK - [text] soie *f* grège en petits écheveaux
~-SLOT HYBRID COUPLER - [electron] (of a waveguide) coupleur *m* à 3dB à courte fente
~SPLICE - [text] épissure *f* carrée
~SPREADER - [text] étaleuse *f* pour fibres courtes
~STALKED LEAF - [bot] à pétiole court
~-STAPLED COTTON - [text] coton *m* courtesoie
~-STAPLED WOOL - [text] laine *f* courte-soie
~STEMMED - [timber] à tronc court
[bot] brévicaule
~STREAKING - [telev] traînage *m* court
~-TAKE-OFF AND LANDING AIRCRAFT - [aero] (an aircraft capable of alighting or leaving the ground with a much shorter run than normal types) avion *m* capable de décoller et atterrir en prenant peu de terrain
~-TERM - [comm] à court terme
~-TIME BREAKDOWN VOLTAGE - [el] (the voltage required to break down a cable in a period of time reckoned in minutes) tension *f* de désamorçage de courte durée
~-TIME DUTY - [el] (duty at constant load during a given time less than that required to attain constant temperature in continuous service at the same load, followed by a rest of sufficient duration to reestablish equality of temperature with the cooling medium) service *m* temporaire
SHORT-TIME MEASURING APPARATUS - [meas] (an electrical time-measuring apparatus for ranges between I second and 0,005 msec) appareil *m* de mesure de temps courts
~-TIME RATING - [el] (the output delivered by an electrical machine in a short period of time without exceeding a specified temperature) puissance *f* de brève durée
~TO EARTH - [el] (a conducting path to earth which is of negligible resistance) court-circuit *m* à la terre
~TO FRAME - [el] (a conduction path of negligible resistance to the frame or structure of the apparatus concerned) court-circuit *m* à la masse
~TON - [meas] tonne *f* courte
~TREADLE - [mech] marchette *f*
~VARNISH - [paint] vernis *m* léger
~-WAVE AERIAL - [radio] (aerial constructed for radiating or receiving short-wavelength radiation) antenne *f* pour ondes courtes
~-WAVE ANTENNA - s. short wave aerial
~-WAVE CONVERTER - [radio] (a device consisting of a heterodyne detector or mixer and an appropriate local oscillator) convertisseur *m* pour ondes courtes
~WAVES - [radio] (electromagnetic waves not exceeding a length of 50 metres) ondes *fpl* courtes
SHORTAGE - [gen] manque *f*, insuffisance
~OF CAPITAL - [comm] pénurie *f* de capitaux
SHORTBREAD - [food] sablé *m*
SHORTCOMING - [gen] imperfections *fpl*
[gen] (a defect) défaut *m*
SHORTED - [el] (short circuited) court-circuité
SHORTEN, to - [gen] raccourcir
[comm] (a price) réduire
[metall] (to make brittle) rendre fragile
SHORTEN SAIL, to - [naut] diminuer la voilure
SHORTENED BY TWISTING - [text] raccourci par la torsion
SHORTENING CONDENSER - [radio] (condenser inserted in series with an aerial to reduce its natural wavelength) condensateur *m* d'antenne
~OF THE ROPE - [text] raccourcissement *m* de la corde
~OF THE WARP - [text] raccourcissement *m* de la chaîne
SHORTEST WAVELENGTH - [phys] (quantum limit; the shortest wavelength in a spectrum, corresponding to the applied potential) limite *f* quantique, longueur *f* d'onde limite
SHORTHAND, to - [gen] sténographier
SHORTHAND - [gen] sténographie *f*
SHORTING - [el] mise *f* en court circuit
SHORTNESS - [metall & plast ind] (lack of tensile strength and ductility, causing easy tearing) friabilité *f*, aigreur *f*, fragilité *f*
SHORTS - [text] morceaux *mpl*
[gen] (the trimmings, clippings etc in a manufacturing process) déchets *mpl*, morceaux *mpl*
[timber] courçons *mpl*
[text] (in rope man) déchets *mpl* de chanvre
[agric] sous-produits *mpl* de meunerie
SHORTSTOP - [photo] (said of a bath used to stop the developping process) bain *m* d'arrêt
~BATH - s. shortstop
SHORTSTOPPER - [chem] (generic term for a polyme-

rization inhibitor) agent *m* d'arrêt, inhibiteur *m* de polymérisation
SHORTSTROKE PRESS - [mech] presse *f* à course courte
SHORTWEIGHT - [comm] faux poid *m*, poids *m* insuffisant
SHOT, to - [firearms] charger (une arme à feu)
[fishing] plomber (une ligne de pêche)
[metall] grenailler, se grenailler
SHOT - [gen] coup *m* de feu
[firearms] (a projectile) boulet *m*, balle *f*
[firearms] portée *f*
[cin] prise *f* de vue
[cin] (a scene) plan *m*
[opt] (of colours) changeant
[metall] (a single moulding operation) moulée *f*
[metall & plast ind] (the weight of material injected at each moulding operation) charge *f* d'injection
[text] (pick) chasse *f* de mine, pétard *m*
[metall] grenaille *f*
~-BAG MOULDING - [plast ind] (method of moulding in which a bag of shot is used as a male die) moulage *m* au sac chargé de plomb
~BLASTING - [mech] (treatment of workpiece by bombardment with hardened steel pellets propelled by a compressed-air blast) grenaillage *m*
~-BLASTING UNIT - [mech] dispositif *m* pour grenaillage
~BORING - [ind] sondage *m* à la grenaille d'acier
~CAPACITY - [plast ind] (the volume of material which a machine can inject into a mould at a given pressure) capacité *f* d'injection
~CORE DRILLING - [mining] carottage *m* à la grenaille
~-DRILL - [tool] sondeuse *f* à grenaille d'acier
~DRILLING - s. shot boring
~EFFECT - [radio] (background noise in amplifiers due to the variation in the emission of electrons from the cathode) bruit *m* de grenaille
[opt] effet *m* changeant
[electron] (the variations in the output of an electronic tube due either to random variation in the emission of electrons from the cathode, or instantaneous variations in the distribution of the electrons among the electrodes) effet *m* de grenaille
[text] effet *m* changeant
~FIRING - [mining] tirage *m* de coups de mine, allumage *m* des coups
~GUN - [firearms] fusil *m* de chasse
~-HOLE - [mining] (the hole bored in rock for a blasting charge) fourneau *m* de mine, trou *m* de mine
[timber] (the small hole bored in timber by a wood-boring insect) trou *m* de scolyte
~-HOLE BORER - [zool] scolyte *m*
~MOULDING - [metall & plast ind] (moulding with loose shot as the male die) moulage *m* avec charge de plomb en vrac
~NOISE - [electron] (noise in a thermionic valve anode circuit due to individual pulses as each electron reaches the anode) bruit *m* de grenaille
~-PEENING - [mech] (special treatment designed to increase the resistance to fatigue and the elasticity) grenaillage *m* d'écrouissage
~-PEENING INTENSITY - [mech] intensité *f* du grenaillage d'écrouissage
SHOT-PEENING MACHINE - [mech] machine *f* à grenailler
~POINT - [chem] point *m* initial d'explosion
~SAMPLE - [mech] échantillon *m* de grenaille
~TAFFETA - [text] taffetas *m* changeant
~WEIGHT - [plast ind] (the weight of material used in a single moulding operation) poids *m* injectable
~-WELD - [mech] (a system of spot welding) soudure *f* par points
SHOTTING - [metall] formation *f* de métal granulaire
SHOULDER, to - [gen] pousser de l'épaule
SHOULDER - [anat] épaule *f*
[gen] ressaut *m* (de terrain)
[geogr] (of a hill) épaulement *m*
[mech] (a portion of cylindrical part formed with a larger diameter than the adjacent one) embase *f*, arrêtoir *m* de tenon, arasement *m* talus *m* (d'une lettre)
[impl] cintre *m*, porte-vêtements *m*
[firearms] tenon *m* de recul
[aero] épaulement *m*
[mech] (of a hinge) charnière *f*
[draw] (of a curve, in a diagram) saillie *f*
[el acoust] (of a record) sillon *m* imparfait
[naut] épaulette *f* (de mât)
~BELT - [impl] bandoulière *f*
~BLADE - [anat] omoplate *f*
~BUSHING - [mech] manchon *m* avec bride
~CROSSPIECE OF FORK - [auto] épaulement *m* transversal d'une fourchette
~-CUT - [carp] arasement *m* de tenon
~GEAR - [mech] engrenage *m* à épaulement
~-GIRDLE SYNDROME - [med] syndrome *m* de la ceinture scapulaire
~GRINDING - [mach tool] émeulage *m* épaulement
~HALTER - [med] corset *m* maintien
~-HAND SYNDROME - [med] syndrome *m* épaule-main
~HOLES - [mining] mines *f*pl de dégraissage
~JOINT - [anat] articulation *f* de l'épaule
~KNOT - [text] (of a uniform) contre-épaulette *f*
~NIPPLE - [plumb] raccord *m* à épaulement
~OF A TYRE - [rubber ind] (of a tyre) épaulement *m* d'un pneu
~PAD - [gen] épaulette *f*
~PATCH - s. shoulder knot
~PIECE - [firearms] crosse *f*
~SCREW - [mech] vis *f* avec embase
~STRAP - [gen] bretelle *f*
~STUD - [mech] prisonnier *m* avec embase
~TENON - [carp] tenon *m* arrêtoir
~THRUST - [text] support *m* axial
~WASHER - [mech] rondelle *f* d'épaulement
SHOULDERED FORMER - [rubber ind] (for tyres) tambour *m* à épaulements
SHOVE, to - [gen] pousser
SHOVE - [gen] poussée *f*, coup *m* d'épaule
~CROSSWISE TO THE LENGTH, to - [text] fouler en largeur
SHOVEL, to - [gen] peller; pelleter
SHOVEL - [impl] pelle *f*
[mech] (of a mechanical digger etc) pelle *f*, cuiller *m*
[mech] (of a dredger) cuillère *f*
[mech] (an excavator) drague *f* à cuiller
~DREDGE - [mech] (a mechanical shovel) pelle *f* automatique
~EXCAVATOR - [mech] pelle *f* mécanique

SHOVEL-SHAPED - [text] en forme de pelle
SHOVELFUL - [gen] pellée *f*
SHOVELLING - [gen] pelletage *m*
[naut] paléage *m*
~ STOKER - [mining] chargeur *m* à pelletage automatique
SHOW, to - [gen] montrer
[gen] (to indicate, to represent) réprésenter, marquer
[gen] (to prove) démontrer, témoigner
[gen] (to lead to) indiquer
SHOW - [gen] mise *f* en vue, exposition *f*
[gen] (indication) trace *f*, indication *f*
[mining] (a safety lamp) auréole *f*
[gen] (theatre etc) spectacle *m*
~-BOARD - [gen] planche *f* à affiche
~-BOARD TESTER - [mech] plaque *f* noire, vérificateur *m* d'égalité (des filés)
~-CARD - [gen] pancarte *f*, étiquette *f* de vitrine
~-CASE - [gen] (e.g. in a museum) vitrine *f*
~-DOWN - [gen] mise *f* au jour, révélation *f*
~ OF OIL - [oil ind] traces *f*pl de pétrole
~ RAFTER - [constr] (a short decorated rafter projecting from a wall) console *f*
~ RING - [agric] concours *m* de bétail
~ WINDOW - [comm] (of a shop etc) vitrine *f*, étalage *m*
SHOWER, to - [gen] verser, faire tomber
SHOWER - [met] (precipitation in the form of rain for short periods, the sky usually clearing in the intervals) averse *f*
[hydr] (a shower bath) douche *f*
[nucl] (a large amount of charged particles) gerbe *f*
~-BATH - [hydr] bain-douche *m*
~ CURTAIN - [rubber ind] rideau *m* pour bain-douche
~ DECKS - [oil ind] (perforated plates, covering part of the fractionating column cross-section, which are sometimes used in place of bubble-cap trays) plaques *f*pl perforées
~ HEAD - [impl] pomme *f* d'arrosoir
~ PARTICLE - [nucl] (component of a nuclear shower) particule *f* de gerbe
~-PROOFING - [text] imperméabilisation *f*
~ UNIT - [nucl] (the mean path length required for the reduction, by the factor of I/2, of the energy of relativistic charged particles as they pass through matter) parcours *m* de gerbe
SHRAPNEL - [firearms] obus *m* à mitraille
SHRED, to - [gen] couper par petits morceaux en long
[mech] (to tear mechanically into small pieces) couper par languettes, déchirer
[paper man] effilocher, délisser
[sugar ind] (the cane) défribrer (la canne à sucre)
SHRED - [gen] fragment *m*, brin *m*
[mech] filament *m*, languette *f*
SHREDDED METAL PACKING - [mech] bourrage *m* de tournures métalliques caoutchoutées
~-WHEAT - [cin] (colloq. for a jammed film in a camera) bourrage *m* du film
SHREDDER - [mech] (machine for disintegrating soda cellulose in the preparation of viscose) délisseuse *f*
SHREDDING - [gen] coupage *m*, déchiquetage *m*
[mech] (the operation of tearing mechanically into small pieces) réduction *f* en fragments
[paper man] délissage *m*
SHREDDING MACHINE - [paper man] machine *f* à éffilocher
~ PRESS - [photo] presse *f* à nouilles
SHRINE - [gen] reliquaire *m*, châsse *f*
SHRINK, to - [gen] rétrécir, se rétrécir, se contracter
[phys] se contracter, se retirer
[constr] se rider
SHRINK - [gen] rétrécissement *m*, mouvement *m* de recul
[text] rétrécissement *m*
[metall] retrait *m*
~ FIT - [mech] serrage *m*
~-FITTING - [mech] montage *m* à chaud
~-HEAD - [metall] masselotte *f*
~ HOLE - [metall] retassure *f* interne
~ MARK - [metall] (in diecasting) retassure *f*
~ OF THE FABRIC - [text] rétrécissement *m* du tissu
~ ON, to - [metall] emmancher à chaud
~-ON - [metall] emmanchement *m* à chaud
~-RESISTING - [metall] frette *f*
~-RING COMMUTATOR - [el] (high-speed type of commutator in which the segments are held together by a steel ring shrunk on over a layer of insulation) commutateur *m* à frette
~ YARN - [text] fil *m* qui retrécit
SHRINKAGE - [gen] retrait *m*, rétrécissement *m*
[text] rentrée *f*(d'une étoffe), rétrécissement
[cin] (of a film) contraction *f* (du film)
[metall & plast ind] (the difference between the dimensions of the mould cavity and the moulding at room temperature) retrait *m*
~ ALLOWANCE - [mech] (the difference in diameter allowed in preparing parts to be united by shrinking, taken cold) serrage *m*
~ BLOCK - [plast ind] (a metal or wood block against which mouldings are held under light pressure while cooling, to reduce warping or distortion) conformateur *m*, dispositif *m* de refroidissement
~ CAVITIES - [metall] retassures *f*pl
~ CRACK - [metall] (cracks due to a collapse of the surface after solidification) crique *f* de retassure
~ FACTOR - [phys] (in the photographic emulsion method) facteur *m* de contraction
~ FIT - s. shrink fit
~ FROM MOULD DIMENSIONS - [plast ind] retrait *m*
~ HEAD - [metall] masselotte *f*
~ IN LENGTH - [text] rétrécissement *m* en longueur
~ IN POPULATION - [statist] diminution *f* de la population
~ IN WIDTH - [text] rentrage *m*, rétrécissement *m* en largeur
~ LIMIT - [soil] (the limit of water content in the soil, below which any further variation does not correspond to a volumetric change) limite *f* de retrait
~ OF THE TIMBER - [timber] contraction *f* du bois
~ OF THE YARN - [text] raccourcissement *m* des fils
~ POROSITY - [metall] (a casting defect caused by liquid shrinkage) porosité *f* de retassure
~ SPOT - [rubber ind] tache *f* de retrait
~ STOPINT - [mining] exploitation *f* par chambres-magasins
~ TESTER - [text] appareil *m* de contrôle du rétrécissement
~ WATER - [chem] (water lost by clay during drying) eau *f* colloïdale

SHRINKER - [mech] machine *f* à emmancher à chaud
SHRINKING - [gen] rétrécissement, repetissement *m*, retrait *m*
[metall] retrait *m*, contraction *f*
~EFFECT - [text] effet *m* de rétrécissement
~MACHINE - [text] machine *f* de rétrécissement
~ON - [metall] emmanchement *m* à chaud
~POWER - [text] force *f* de rétrécissement
~PROCESS - [text] procédé *m* de rétrécissement
SHRINKPROOF - [text] irrétrécissable
SHRIVEL, to - [gen] rider, déssécher
[gen] (of flowers etc) se dessécher
SHROD, to - [gen] envelopper d'un linceul
[arch] envelopper, protégér
[el] blinder
[mech] emboiter
SHROUD - [gen] linceul *m*
[naut] hauban *m*
[mech etc] bandage *m*, bouclier *m*, joue *f* (de pignon)
[auto] (iron sheet diaphragm between bonnet and cowl) panneau *m* de séparation
[of gears] (a circular web formed on each side of the teeth of a gear to give stiffness to them) enveloppe *f*
[aero] (of gas turbines) (peripheral strip used to strengthen rotor blades) anneau de renforcement de turbine
~DISC - [text] disque *m* échancré
~LAID ROPE - [text] câble *m* commis en quatre
~LINE - [aero] (a cord attached to the canopy of a parachute to connect it to the load) suspente *f*
~RING - [mech] (ring-shaped element in a gas turbine, designed to give additional strength to the blades) anneau *m* de renforcement de turbine
SHROUDED - [gen] enveloppé, voilé
[mech etc] à emboîtement
~BALANCE - [aero] (balance with control area forward of the hinge, and working within shrouds forming part of aerofoil contour) compensation *f* aérodynamique interne
~BLADE - [mech] (gas turbine blade reinforced with a shround) aube *f* renforcée
~BLADE TIP - [mech] (gas turbine blade tip strengthened with a shroud) aube *f* à extrémités renforcées
SHROUDING - s. shroud
SHRUB - [bot] arbuste *m*, arbrisseau *m*
[gen] (USA only, a drink) grog *m* à l'orange
SHRUBBERY - [gen] bosquet *m*
SHRUBBY PLANT - [bot] plante *f* arbustive
~ON - [mech] emmanché à chaud
~-ON SLEEVE - [mech] (a cylindrical part fixed on another by machining, so as to be a little too small, heating it until it expands and then putting it in place) manchon *m* fretté
~RING - [mech] anneau *m* de serrage
~SILK THREAD - [text] fil *m* de soie rétréci
SHUNT, to - [gen] (to turn aside) ajourner, mettre au rancart
[el] (to establish an additional path) shunter, dériver, mettre en dérivation
[railw] (to switch a train from one track to another) garer
[radio] connecter en dérivation
SHUNT - [railw] (a railway switch) changement *m* de voie, manoeuvre *f*, évitement *m*
[el] (a conducting path wich is alternative to another in an electric circuit) shunt *m*, dérivation *f*
SHUNT - [instr] (a resistor connected in parallel with an instrument in order to reduce the current which passes through it) shunt *m*
~-BOX - [el] boîte *f* de résistance shunt
~BRUSH - [el] balai *m* de régulation
~CABLE - [el] câble *m* de distribution
~CAPACITOR - [el] condensateur *m* shunt
~-CHARACTERISTIC MOTOR - [el] (a motor the speed of which remains practically unaffected by the load) moteur *m* à caractéristique shunt
~CIRCUIT - [el] (that part of a meter or measuring instrument, supplied by the voltage of the circuit supplied by a transformer or voltage divider) circuit *m* en dérivation, circuit *m* dérivé, circuit *m* de tension
~COIL - [el] (a coil or winding which is connected in shunt to some part of the main circuit) bobine *f* en dérivation
~CONNECTION- [el] montage *m* en dérivation
~DYNAMO - [el] dynamo *f* shunt, génératrice *f* shunt
~EXCITATION - [el] excitation *f* en dérivation
~-FED VERTICAL AERIAL - [radio] (vertical aerial connected to earth at the base and energized at a point suitably positioned above the earthing point) antenne *f* verticale alimentée en parallèle
~-FED VERTICAL ANTENNA - s. shunt-fed vertical aerial
~-FIELD RELAY - [el] (electromagnetic relay in which the magnetic flux can be routed outside the armature, so that the armature is insufficiently attracted) relais *m* à champ de shunt
~FIELD RHEOSTAT - [contr] (a field rheostat in which the resistor is suitable for connection across the source of supply, and means are provided whereby the field winding can be connected between various points of the resistor) rhéostat *m* de champ
~LEAD - [el] fil *m* de shunt
~LINE - [railw] voie *f* de garage
~LOADING - [radio] (loading in which reactances are applied in shunt across the conductors of a transmission circuit) charge *f* au moyen de bobines en parallèle
~METER - [instr] (a flow meter in which the moving parts are operated by a proportion of the total flow) débitmètre *m* à dérivation
~NEUTRALIZATION - [el] (a method of neutralizing an amplifier) neutralisation *f* inductive
~RECTIFIER CIRCUIT - [el] (or parallel rectifier circuit, a circuit in which two or more rectifiers operate in such way that their commutations coincide and their direct (forward) currents add) montage *m* en parallèle de redresseurs
~RESISTOR - [electron] résistance *f* shunt
~SYSTEM OF DISTRIBUTION - [el] (or parallel system of distribution, a system of distribution in which the consuming devices are so connected as to have the same nominal voltage applied to them) réseau *m* à tension constante
~T - [electron] (a tee junction) té *m* parallèle, té *m* plan H
~TRANSITION - [el] short circuit transition) transition *f* court-circuit, transition *f* par shunt
~TRIP - [el] (a tripping device operated by means of a trip coil energized from a low-voltage circuit and controlled by a circuit-closing relay, which may

be independent of the circuit-breaker and arranged for remote operation) déclencheur *m* à bobine en dérivation
SHUNT TRIPPING - [el] déclenchement *m* par bobine en dérivation
~WINDING - [el] enroulement *m* en dérivation
~-WOUND - [el] (applied to a direct-current machine to denote that the field magnets are excited by a winding connected in series with, or carrying a current proportional to that in the armature winding) shunt *m* à enroulement en dérivation
~-WOUND MOTOR - [el] moteur *m* shunt
~-WOUND OUTPUT - [el] puissance *f* fournie par un moteur shunt
SHUNTED INSTRUMENT - [instr] (a combination of a measuring instrument and a separate shunt) appareil *m* à résistance en dérivation
SHUNTER - [railw] décrocheur *m*
[railw] (a locomotive) locomotive *f* de manoeuvre
[el] (for an arc lamp) dérivateur *m*
SHUNTING - [el] shuntage *m*, dérivation *f*
[railw] (switching from one track to another) manoeuvre *f*, évitement, changement *m* de voie, garage *m*
~BOX - [el] boîte *f* de résistance shunt
~BY GRAVITATION - [railw] manoueuvre *f* par gravité
~CABLE RAILWAY - [mining] voie *f* funiculaire de manoeuvre
~DRUM - [mech] cabestan *m* de manoeuvre
~GRADIENT - [railw] plan *m* automoteur
~IMPACT - [railw] choc *m* de manoeuvre
~LOCOMOTIVE - [railw] locomotive *f* de manoeuvre
~MOVEMENT - [railw] mouvement *m* de manoeuvre
~POINTS - [railw] aiguille *f* de manoeuvre
~ROUTE - [railw] itinéraire *m* de manoeuvre
~SPEED - [railw] vitesse *f* de débranchement
~SPRINGS OF A DIAL - [telecomm] ressorts *mpl* de shunt d'un cadran d'appel
~TURNTABLE - [railw] plaque *f* tournante de manoeuvre
~WORK - [railw] (of a locomotive) service *m* de manoeuvre
~YARD - [railw] gare *f* de manoeuvre
SHUT, to - [gen] fermer, clore
SHUT - [gen] fermé, clos
[metall] (a gate shutter) écluse *f*
[metall] (the union line of welded metal piece) cordon *m* de soudure
[metall] (a defect in rolling and forging) chevauchement *m*
[mining] clichage *m*
~-DOWN, to - [mech] (a motor) fermer, couper (la vapeur)
[gen] (in a factory etc) chômer
[nucl] (a reactor) arrêter
~-DOWN - [mech] (of a motor, engine etc) arrêt *m*, blocage *m*
[gen] (of a factory etc) chômage *m*
[nucl] (the procedure of stopping the chain reaction by bringing the reactor to a subcritical condition) arrêt *m*, fermeture *f*
~-DOWN AMPLIFIER - [contr] amplificateur *m* du signal d'arrêt
~-DOWN PROCEDURE - [nucl] mesures *fpl* d'arrêt
~-DOWN REACTIVITY - [nucl] réactivité *f* à l'arrêt
~-DOWN ROD - [nucl] barre *f* de sécurité
~
SHUT-FRAME - [hydr] (of a sluice-gate) tableau *m* de vanne
~HEIGHT - [mech] (of a press die) hauteur *f* estampe fermée
~-IN PRESSURE - [gas ind] (of a natural gas well) pression *f* statique d'un puits
~OFF, to - [gen] interrompre, couper, fermer
[gen] (water, steam etc) fermer (l'eau)
~-OFF - [hydr] (of sluice-gate) pale *f*, palle *f*, bonde *f*
~-OFF COCK - [mech] (a valve used to cut off fuel from an engine in emergency) vanne *f* d'interception
~OFF STEAM, to - [mech] couper la vapeur
~OFF THE MOLTEN METAL, to - [metall] écluser le métal en fusion
~-OFF VALVE - s. shut-off cock
SHUTTER, to - [constr] mettre les volets à, fermer les volets
SHUTTER - [gen] volet *m*
[constr] (of a window) persienne *f*, volet *m*, contrevent *m*
[th eng] (device for regulating the cooling effect) organe *m* de réglage
[mining] (of a box regulator) guichet *m*
[photo] (the device in a camera for exposing the sensitized surface to the image of the object) obturateur *m*
[nucl] (movable plate of absorbing material used to cover a beam hole when radiation is not wanted) obturateur *m*
[hydr] vanne *f*, haussette *f*
[metall] écluse *f*
~APERTURE - [photo] ouverture *f* de l'obturateur
~AXIS - [photo] (the axis of the rotating shutter) axe *m* d'obturateur
~BLADES - [photo] lamelles *fpl* d'obturateur
~COCKING - [photo] armement *m* de l'obturateur
~CUT-OFF FREQUENCY - [cin] (usually at 48 interruptions per second) fréquence *f* d'obturation
~FREQUENCY - s. shutter cut-off frequency
~PRESETTING - [photo] réglage *m* sur obturateur présélectionné
~RELEASE - [photo] déclencheur *m* d'obturateur
~SPEED - [photo] vitesse *f* d'obturation
~SPEED RATE - [photo] échelle *f* des temps d'obturation
~SPEED RING - [photo] bague *f* des vitesses
SHUTTERING - [constr] (general term denoting temporary works for the support of reinforced concrete while it is setting) coffrage *m*
~BOARD - [constr] planche *f* de coffrage
SHUTTING CLACK - [auto] (throttle valve) papillon *f* des gaz
~FLAP - s. shutting clack
SHUTTLE, to - [gen] aller et venir, faire la navette
SHUTTLE - [text] (a device used in weaving to carry the warp to and for between the warp threads) navette *f*
[mech] (of a sewing machine) mouvement *m* alternatif
[transp] circulation *f* en navette
[cin] escomateur *m*
[nucl] (a container which can be passed through a tube in a nuclear reactor, so as to expose its contents to irradiation in the active section) furet *m*
[hydr] vanne *f*

SHUTTLE ARMATURE - [el] (simple form of armature having two slots, so that the armature stampings assume an H-shape) induit *m* en H
~BOX - [text] (of a loom) boîte *f* à navette
~BOX SWELL - [text] languette *f* de la boîte
~BRAKING - [text] freinage *m* de la navette
~BRAKING SWELL - [text] languette *f* de freinage de la navette
~BUFFER - [text] frein *m* à navette
~CARRIER - [text] support *m* d'espolin
~CARRYING PIN - [text] aiguille *f* portant la navette
~CHANGE - [text] changement *m* des boîtes à navette
~COURSE - [text] course *f* de la navette
~EYE - [text] oeillet *m* de navette
~EYE CUTTER - [text] coupe-fil *m* extérieur
~FILLING DEVICE - [text] dispositif *m* pour forcer la bobine dans la navette
~FOR BEAD WEFT - [text] navette *f* pour trame de perle
~FOR WEAVING BELTING - [text] navette *f* pour la fabrication des courroies
~FRAME - [text] partie *f* latérale de la navette
~-GATE INTERMITTENT MOTION - [cin] mouvement *m* intermittent oscillatoire
~GUARD - [text] (box-like extension at each end of the shuttle race board) garde-navette *m*
~GUIDE - [text] guide-navette, guidage *m* de la navette
~GUIDE RAIL - [text] rail-guide *m* pour navettes
~HOLDER - [text] entraîneur *m* de la navette
~KISSING - [text] enfilage *m* par aspiration
~MAGAZINE - [text] magasin *m* à navettes
~MECHANISM - [cin] mécanisme *m* du mouvement intermittent
~MOTION - [text] changement *m* de navettes
~PATH - [text] chemin *m* de la navette
~PEG - [text] brochette *f* de navette, pointicelle *f*
~PICK - [text] chasse *f* de la navette
~PICKING - [text] chasse *f* de la navette
~PLATE - [hydr] tablier *m* de vanne
~RACE - [text] (race board; that part of the slay in a loom along which the shuttle travels) lit *m* de la navette
~RACE OPENING - [text] ouverture *f* de la course de la navette
~RECTIFYING MACHINE - [text] machine *f* à dresser les navettes
~SERVICE - [transp] service *m* de navette
~-SERVICE TRAIN - s. shuttle train
~SHOT - [text] chasse *f* de la navette
~SPRING - [text mach] ressort *m* de navette
~THREADING DEVICE - [text mach] mécanisme *m* d'enfilage de la trame
~TIP - [text] bout *m* de la navette
~TRAIN - [railw] train-navette *m*
~-TYPE FEED - [mech] alimentation *f* à secousses
~-TYPE SPREADER ROLL - [rubber ind] rouleau *m* élargisseur oscillant, cylindre *m* déplisseur oscillant
~VALVE - [mech] (a piston-type valve actuated by pressure, i.e. not by direct mechanical means) clapet *m* baladeur, clapet-pilote *m*
~WINDING - [el] enroulement *m* en double T
~WITH NARROW WELL - [text] navette *f* à boîte étroite
~WITH PIRN - [text] navette *f* à défiler
SIAL - [geol] (discontinuous earth shell of granitic composition forming the foundation of the continental masses) sial *m*
SIALADENITIS - [med] (inflammation of the salivary gland) sialoadénite *f*
SIALAGOGUE - [med] (stimulating the flow of saliva) sialalogue
SIALOADENITIS - s. sialadenitis
SIALOGOGUE - s. sialagogue
SIALOGRAPHY - [radiol] (radiological examination of the salivary ducts and alveoli following the injection of a contrast medium) sialographie *f*
SIALOID - [zool] (resembling saliva) sialoïde
SIALOLITH - [med] calcul *m* salivaire, sialolithe *m*
SIAMESE BLOW - [glass ind] (operation of blowing two or more items simultaneously) soufflage *m* siamois
~TWINS - [biol] frères *mpl* siamois
SIBERIAN ANTICYCLONE - [met] (the persistance of anticyclonic system during winter in Siberia) anticyclone *m* sibérien
SIBILUS - [med] (whistling sound in the bronchial passages) râle *m* sibilant
SIBLINGS - [genet] enfants *mpl* des mêmes parents
SICCATIVE - [chem etc] (synonym for drier) siccatif *m*
~OIL - [ind chem] huile *f* siccative
SICK - [gen] malade
[bot & agric] malade, non productif
~BAY - [naut] (part of the ship earmarked for the care of the sick) infirmerie *f*
~BENEFIT - [gen] prestations-maladie *fpl*
~FLAG - [naut] pavillon *m* de quarantaine
~-LEAVE - [gen] congé *m* de maladie
SICKEN, to - [gen] tomber malade
[bot & agric] s'étioler, dépérir
SICKLE - [impl] faucille *f*
SICKLE-CELL ANAEMIA - [med] (drepanocytosis) anémie *f* à hématies falciformes
~MEDIX - [bot] luzerne *f* faucille
~SHAPED ARM - [text] pièce *f* en colimaçon
~SHAPED KNIFE - [text] couteau *m* à bout recourbé
~-SHAPED TRUGS - [constr] (a truss for roofs of very large span in which the upper and lower chords enclose a sickle-shaped area which is triangulated by diagonal members) ferme *f* triangulaire arcboutée
SICKLEMIA - s. sickle-cell anaemia
SICKNESS - [gen] maladie *m*, mal *m*
[gen] (nausea) nausée *f*
~BENEFIT - s. sick benefit
SIDA FIBRE - [text] fibre *f* de sida
~PLANT - [bot] sida *m*
SIDE, to - [gen] se ranger du côté
[carp] (of wood) dégrossir (le bois), débiter (le bois brut)
SIDE - [gen] côté *m*
[mech] (of a thread) flasque *m*
[geom] côté *m*
[geogr] flanc *m*, versant *m*
[geogr] (of a valley) paroi *f*, flanc *m*
[geol] lèvre *f*
[constr] (of a roof) pan *m* (d'un comble)
[mach tool] (of a lathe) flasque *m* (d'un tour)
[mining] (of lode) paroi *f* (de filon)
~-AND-FACE MILLING-CUTTER - [mach tool] fraise *f* à trois tailles
~APPROACH - [el] guide *m* de sabot collecteur

SIDE-ARCH BRICK - [constr] (of an oven) couteau *m*
~ARM - [constr] console *f* à bride
~ARM WATER HEATER - [gas ind] (a hot-water circulator) réchauffeur *m*
~ARMATURE RELAY - [el] relais *m* à armature latérale
~ARMOUR - [auto] renfort *m* latéral
~BAND - s. sideband
~BAR - [auto] barre *f* latérale
~BARS - [mech] (of the links of a roller-chain) flasques *mpl* des maillons d'une chaîne à rouleaux
~BENCHING - [mining] exploitation *f* par recoupes transversales
~BINDING - [el] ligature *f* latérale
~-BLOWN CONVERTER - [metall] convertisseur *m* à soufflage latéral
~BOARD - [auto] (vertical extension of the body sides) ridelle *f*, haut *m* de ridelle
[constr] volante *f*
~BRACING - [auto] (body side reinforcement) ceinture *f* de renfort
~BREAK SWITCH - [el] interrupteur *m* à couteau horizontal
~-BY-SIDE ASSEMBLY - [mech] (an arrangement of connecting rods having narrow big-ends running side-by-side on a common crankpin) embiellage *m* côte-à-côte
~-BY-SIDE TWIN ROTOR CONFIGURATION - [el] (arrangement of two main rotors in the same machine, placed laterally in respect to each other) configuration *f* d'un birotor à rotors parallèles
~CAP - [electron] téton *m* latéral
~-CENTERED - [metall] à faces centrées
~CHAIN - [nucl] (a subsidiary group of atoms attached to the atoms of the nucleus of a molecule) chaîne *f* latérale
~-CHAIN THEORY - [med] (theory by Ehrlich to explain the phenomena of the poisoning and immunity of living cells) théorie *f* des chaînes latérales
~CHAINS - [chem] (Alkyl group replacing hydrogen in ring compounds) chaînes *fpl* latérales
~CIRCUIT - [telecomm] (in telephony, the telephone circuit loop used as one leg the phantom) circuit *m* combinant, circuit *m* réel
~CIRCUIT LOADING COIL - [telecomm] bobine *f* de charge pour circuit réel
~CIRCUIT REPEAT COIL - [telecomm] bobine *f* translatrice de circuit réel
~CLEARANCE - [mech] jeu *m* latéral
[mech] (of a gear) jeu *m* entre les dents
~CONDUCTOR RAIL - s. side contact rail
~CONNECTING ROD - s. side rod
~CONTACT - [electron] (a small metal shell on the envelope of an electronic tube used to connect one electrode to an external circuit) téton *m* latéral
~CONTACT-RAIL - [el] (insulated rail used on electrical railways for conducting current, contact with the collector shoes, being made sideways) rail *m* latéral de contact
~-CORRIDOR COACH - [railw] voiture *f* à couloir latéral
~CRAMP - [metall] pression *f* latérale sur un côté de l'estampe
~CUTTER - [tool] fraise *f* à trois passes
~CUTTING - [constr] (excavation taken from the side of a railway or canal and used for work on the spot) cavalier *m* (excédent *m* deblai)

SIDE CUTTING-EDGE ANGLE - [mach tool] (of a lathe) angle *m* du profil du tranchant
~-CUTTING NIPPERS - s. side-cutting pliers
~-CUTTING PLIERS - [impl] (type of pliers having wire cutting jaws at the side of the flat jaws) pinces *fpl* coupantes sur côté
~-DELIVERY RAKE - [agric] moissonneuse *f* à décharge latérale, moissonneuse-javeleuse *f*
~DITCH - [constr] rigole *f* latérale
~DRIFT - [mining] galerie *f* horizontale d'accès
~DRUM - [mus] caisse *f* claire, tambour *m*
~DUMP - [transp] (hoist and body combination discharging on one side) benne *f* à basculement latéral
~-DUMP BODY - s. side dump
~DUMP CAR - [mining] wagon *m* basculant de côté
~ECHO - [radar] écho *m* faux
~EFFECT - [med] effet *m* secondaire
~ELEVATION - [draw] (an elevation drawing as viewed from one side) vue *f* de côté
~EMISSION - [electron] (the emission of electrons which do not interact with the H F. field of the anode of a magnetron) émission *f* vagabonde
~ENTRY - [mining] galérie *f* latérale
~EXTENSION - [gas ind] (in cookers) allonge *f*
~-FACE - [gen & photo] profil *m*
~FIELD - [surv] champ *m* optique
~FLANGE OF CYLINDER - [text] paroi *f* latérale du grand tambour
~FLUE - [metall] carneau *m* latéral
~FORCE - [aero] (the component of the total aerodynamic force which acts along the lateral axis of an aircraft) force *f* latérale
~FRAME - [railw] (of a truck body) paroi *f* latérale du châssis
[mech] face *f* latérale
[text] cadre *m*
~FRAMING - [text] bâti *m* latéral
~FREQUENCY - [radio] (the frequency of a side wave) fréquence *f* latérale
~GATE - [metall] chenal *m* de coulée à talon
[mining] entrée *f* latérale (d'une cage)
[metall] attaque *f* dirigée directe
~GEAR - [mech] pignon *m* soleil, pignon *m* planétaire
~GRAFTING - [agric] greffe *f* latérale
~GRINDING - [mech] meulage *m* latéral
~GUIDE - [mech] guidage *m* latéral
~HEAD - [print] titre *m* en hauteur
~JOINT - [constr] joint *m* vertical
~KEELSON - [naut] carlingue *f* latérale
~LACING - [mining] coffrage *m* latéral
~LAMP - [auto] feu *m* latéral, feu *m* de position
~LASH - [mech] jeu *m* latéral
~LEVER BRACKET - [text] bout *m*
~LIGHT - [photo] lumière *f* latérale
~-LIGHTED - [photo] éclairé par lumière latérale
~LIGHTS - [naut & aero] (green and red navigation lights) feu *m* de position
~LINE - [railw] voie *f* secondaire
~LOADING PLATFORM - [railw] quai *m* latéral
~LOADING TRUCK - [transp] chariot *m* à chargement latéral
~LOBE - [radio] (a minor lobe in the aerial) lobe *m* latéral
~-LOBE ECHO - [radar] (echo caused by a side lobe of a radar beam) écho *m* latéral, écho *m* faux
~-LOCKING RIM - [auto] jante *f* avec anneau latéral (fixe et démontable)

SIDE-LOCKING RING - [auto] anneau *m* de fermeture latéral
~MEMBER - [mech] (of a frame) longeron *m*
~-MILLING - [mach tool] fraisage *m* de côté
~MOTION - [railw] (a rocking motion) mouvement *m* de lacet
~-MOUTH TONGS - [impl] badine *f*
~OF CREEL - [text] bras *m* du cantre
~OF NECK - [text ind] longée *f* du cou (laine)
~OF TAPPET - [mech] flanc *m* de l'excentrique
~OF THE HEART - [constr] côté *m* coeur d'une pièce de bois
~PANEL - [auto] panneau *m* latéral
~PARTS - [carp] piles *fpl* (de portail)
~-PIECE - [gen] brancard *m* (de voiture), montant *m* (d'échelle)
[mining] montant *m* (d'une galérie)
~PIN - [mech] cheville *f* latérale
~PLANE - [impl] rabot *m* à lumière de côté
~PLANER-TOOL - [tool] outil *m* à raboter de côté
~-PLANING MACHINE - [mech] machine *f* à raboter latérale
~PLATE - [mech] plaque *f* de serrage, plateau *m* de serrage
[metall] flasque *f*
~-PLATE PRESS - [mach] presse *f* à parois latérales
~-POINT - [railw] branche *f* de pointe
~POURING - [metall] coulée *f* à talon
~PULLEY - [mech] (e.g. of a tractor) poulie *f* de renvoi à plat
~RAIL - [railw] (a check rail) contrerail *m*
[railw] (of locomotive) main *f* courante de la chaudière
~RAKE - [mech] (of a tool) angle *m* de dépouille latéral
~REBATE PLANE - [impl] guillaume *m*
~ROD - [railw] (of a steam locomotive) bielle *f* d'accouplement
~RODS - [text] (of loom) barres *fpl* latérales
~ROLL - [railw] (of a locomotive) bielle *f* d'accouplement
~RUBBER - s. sidewall rubber
~RUNNER - [metall] chenal *m* de coulée à talon
~-RUNNING TROLLEY - [el] (trolley-wheel arrangement in which contact is made sideways to the trolley-wire) trolley *m* à contact latéral
~SHEETING - [gas ind] (gasholders) robe *f*, jupe *f*
~SHEET - [el] (end and bottom insulating or protecting sheets of the container of an accumulator) écran *m* latéral
~SHIELD - [auto] écran *m* de protection latéral
~SHOOT - [agric] rebiot *m*, entre-coeur *m*
~SHOW - [cin] films *mpl* de première partie
~SILL - [railw, auto etc] flanc *m*, longeron *m* brancard *m*
~SLIP - [auto & aero] glissement *m* sur le côté, dérapage *m*
~-SLIP OF EMBANKMENT - [constr] affaissement *m* latéral du remblai
~SLOPE - [constr] pente *f* latérale
[mech] angle *m* de dépouille
~SPACE - [railw] accotement *m*
~-STABLE RELAY - [radio] relais *m* polarisé à deux positions stables
~STEP GATING - [metall] coulée *f* en échelons
~-STICK - [print] (tapering piece of wood placed at the sides of a page when locking it up) blanc *m* de prise de pinces
SIDE STOPING - [mining] abattage *m* par gradins latéraux
~STREAM - [oil ind] (a refinery by-product) fraction *f* latérale
[chem] (stream withdrawn from some point along a tower) coupe *f* latérale
~STRINGER - [aero] lisse *f* latérale
~-STRIPPER - [ind chem] (a vessel in which the light ends of a liquid drawn off a tray in a fractionating column are stripped off before it is run to storage) extracteur *m* latéral
~THRUST - [acoust] (in disk recording) force *f* de déplacement latéral
~-THRUST EFFECT - [phys] (the result of the force developed on a charged particle moving in a magnetic field) effet *m* de déplacement latéral
~THRUST OF PISTON - [auto] poussée *f* latérale du piston
~TIMBER - [constr] filière *f*, panne *f*
~TIP - [text] ergot *m*, pointe *f* latérale
~TIPPER - s. side dump
~TO-PHANTOM CROSSTALK - [telecomm] diaphonie *f* entre réel et fantôme
~-TO-SIDE CROSSTALK - [telecomm] diaphonie *f* entre réel et réel
~-TONE - [acoust] (the hearing of one's own voice in a reverberant enclosure) signal *m* local, effet *m* local
~-TONE REFERENCE EQUIVALENT - [acoust] équivalent *m* de référence de l'effet local
~TOOL - [tool] outil *m* de côté
~TRACKING - [railw] (turning off into a siding) voie *f* de garage
~-TRACKING SKATE - [aero] (device for moving an aircraft laterally on the ground) patin *m* de manoeuvre
~TRIM UNIT - [tool] (device for cutting the edges of material produced in sheet or slab) dispositif *m* de découpe latérale
~-VALVE - [mech] soupape *f* latérale
~-VALVE ENGINE - [mech] (a reciprocating e.c. engine in which the valves are placed at the side of the cylinders) moteur *m* à soupapes latérales
~-VIEW - [draw] (elevation as seen from the side) vue *f* de côté
~WALL - [constr] pied-droit *m*
[th eng] (ovens) façade *f*
[rubber ind] (of a tyre) bande *f* de côté (d'un pneu)
[mining] paroi *f* latérale (de galerie)
~-WALL CORING - [oil ind] carottage *m* latéral
~WAVE - [radio] (wave component lying within a sideband) onde *f* latérale
~WIND - [naut] vent *m* du travers
~WINDOW - [auto] glace *f* latérale
~WOOL - [text] laine *f* des flancs
SIDEBAND - [radio] (a band of frequencies above and below the carrier frequency, and equal to twice the highest modulation frequency) bande *f* latérale
~ATTENUATION - [radio] atténuation *f* d'amplitude
~COMPONENT - [radio] composante *f* de la bande latérale
~INTERFERENCE - [radio] interférence *f* par la bande latérale
~OSCILLATION - [radio] oscillation *f* additionnelle
~SPLASH - [telev] (colloq. for adjacent channel in-

terference) interférence *f* adjacente
SIDEBOOM TRACTOR - [mech] tracteur *m* à grue latérale, grue *f* latérale chenillée
SIDECAR - [of motorcycle] sidecar *m*
SIDERATION - [med] sidération *f*
SIDEREAL - [astr] (relating to celestial bodies) sidéral
~DAY - [astr] (the period of one revolution of the earth on its axis) jour *m* sidéral
~HOUR ANGLE - [astr] (the hour angle measured from the First Point of Aries as celestial meridian) angle *m* horaire sidéral, angle *m* sidéral origine
~PERIOD - [astr] (the period of one revolution of a satellite round its primary) période *f* sidérale
~TIME - [astr] heure *f* astronomique, temps *m* vraie
SIDERIOLITE - [min] (a meteorite composed of both metallic and silicate materials) sidériolithe *f*
SIDERITE - [min] (spathic iron ore, chalybdite. An ore of iron) sidérite *f*, fer *m* spathique
SIDEROPENIA - [med] sidéropénie *f*, déficience *f* de l'organisme en fer
SIDEROSCOPE - [instr] (instrument designed to detect small quantities of iron by the magnetic needle) sidéroscope *m*
SIDEROSIS - [med] (excessive deposits of iron in the body tissues) sidérose *f*
SIDEROSTAT - [instr] (an instrument designed to reflect a portion of the sky in a fixed direction) sidérostat *m*
~AND RUNGS OF A LADDER - [impl] montants *mpl* et échelons d'une échelle
SIDES OF RING GROOVE - [mech] surfaces *fpl* latérales de la gorge d'une poulie
SIDESHAKE - [horol] jeu *m* latéral
SIDESLIP, to - [aero] (of an aircraft, to move so that the mean airflow has a component along the lateral axis) glisser sur l'aile
SIDESLIP - [aero] (the component of the motion of an aircraft taken in the plane of the lateral axis) dérapage *m*
[auto] glissement *m* sur le côté
~ANGLE - [aero] (the angle between the plane of symmetry of an aircraft and the direction of the undisturbed flow) angle *m* de dérapage
~INDICATOR - [instr] indicateur *m* de dérapage
SIDESLIPPING - [aero] (motion of an aircraft such that there is a component of the air-flow along the lateral axis) glissade *f*
[aero] (of a parachute) glissement *m* latéral
[auto] dérapage *m*
SIDETRACK - [railw] voie *f* de garage
SIDETRACKING - [oil ind] (a deflection of the bore-hole in difficult drilling operations) forage *m* dévié
SIDEWALK - [constr] trottoir *m*, contre-allée *f*
~FLAG - [constr] dalle *f* de trottoir
~MANHOLE - [telecomm] chambre *f* sous trottoir
SIDEWALL - [rubber ind] (of tyres; the lateral part of a pneumatic tyre) flanc *m*
[glass man] (the longitudinal wall of a tank furnace above the metal line resting on the flux blocks) paroi *f* latérale
~RUBBER - [rubber ind] (of tyre the rubber forming the lateral part of a tyre) gomme *f* de flanc, chape *f* de flanc
SIDEWAYS - [gen & mech] latéralement, de côté
~RUNNING - [railw] position *f* de marche en crabe
SIDEWISE - [gen] latéralement
SIDING - [railw] voie *f* de garage, voie *f* d'évitement
[shipbuild] (the thickness of a beam) échantillon *m* sur le droit
SIEGE - [gen] siège *m*
[constr] (a mason's) banc *m* de maçon
[glass man] (the floor of a pot furnace) banc *m* de fourneau de fusion
SIEMENS DYNAMOMETER - [el] (term denoting a dynamometer type of instrument when arranged for measuring power) dynamomètre *m* de Siemens
~FURNACE - [metall] (open heath furnace) four *m* Martin
~MARTIN PROCESS - [metall] (open-hearth process) procédé *m* Siemens
~OZONE TUBE - [chem] (an apparatus used in the preparation of ozone by the silent discharge of electricity) tube *m* à ozone Siemens
SIENNA - [paint] (a yellowish-brown earth pigment. When calcined it is known as Burnt Sienna and gives an orangebrown pigment) terre *f* de Sienne
SIEVE, to - [gen] (to separate by passing through a sieve) cribler, tamiser, passer au tamis
[constr] (chalk) bluter (de la chaux)
SIEVE - [gen] (utensil or apparatus for sifting) tamis *m*, crible *m*, sas *m*
[constr] (an open container fitted with a mesh or gauze bottom) crible *m*
~ANALYSIS - [min] (the determination of the number of particles of different sizes in a granular material by passing it through sieves of successively finer mesh) analyse *f* granulométrique
~BOTTOM - [impl] fond *m* à tamis
~DRUM - [impl] tambour *m* cribleur
~FRACTION - [metall] fraction *f* granulométrique
~FRAME - [impl] cerce *f* de tamis
~GRATE - [metall] grille-tamis *f*
~MESH - [min] maille *f* de tamis
~PACK - [plast ind] (an assembly of fine screens and their mountings placed upstream of an extrusion die) tamis *m*
~PLATE - [impl] plaque *f* à cribler, crible *m*
~-PLATE COLUMN - [ind chem] (a column fitted with perforated plates) colonne *f* à plateaux perforés
~RESIDUE - [min] (material left on a sieve of given mesh after the undersize has passed through) résidu *m* de tamis
~-SHAKER - [ind chem] (machine for vibrating a nest of sieves used in sizing samples) vibreur *m* de filtre
~SPOON - [impl] cuiller *m* à tamis
~TRAY - [oil ind] (horizontal plates perforated in a regular pattern, placed in a distillation column to bring vapour into contact with down-flowing liquid) plateau *m* perforé
~WITH RUBBER TAPES - [impl] tamis *m* à bandes de caoutchouc
SIEVED ASBESTOS - [min] amiante *m* criblé
~GRAVEL - [gen] gravier *m* criblé
SIEVING CLOTH - [text] drap *m* de filtre
SIFT, to - [gen] (the action of separating fine from coarse particles) tamiser, passer au tamis
SIFTER - [metall] cribleuse *f*
SIFTING - [gen] criblage *m*, tamisage *m*
[gen] (the material on the sieve) produits *mpl* du tamisage
~OF THE BALLAST - [railw] réfection *f* du ballast

SIGHT, to - [gen] apercevoir, viser
[opt] (to bring into the field of observation with an instrument) viser
[instr] (to adjust the sight of an instrument) mettre au point
[firearms] (to find the right aim or elevation) mirer
[naut] relever (la terre), reconnaître (la terre)
SIGHT - [gen] vue *f*
[gen] (that which is seen) vue *f*
[opt] (the limit of the eyesight) vue *f*
[firearms] (a device to assist aiming) hausse *f*
[surv] coup *m* de lunette, visée *f*
[instr] lumière *f* de sextant, pinnule *f* (d'une alidade)
[instr] (of an optical instrument) appareil *m* de visé
~BAR - [surv] alidade *f*
[photo] aiguille *f* de mire
~DISTANCE - [visibility] visibilité *f*
~DRAFT - [comm] effet *m* à vue
~-FEED - [mech] débit *m* visible
~-FEED GLASS - [mech] (a type of lubricator) tube *m* en verre de débit visible
~-FEED LUBRICATOR - [mech] (a small glass tube through which oildrops from a reservoir can be seen) graisseur *m* à débit visible
~-FEED NEEDLE VALVE - [mech] pointeau *m* de débit visible
~GLASS - [mech] regard *m*
~HOLE - [metall] (an opening in a furnace to give a view of the interior) regard *m* de visite
~HOLE WITH PLUG - [metall] regard *m* à bouchon
~RULE - [surv] (alidade) alidade *f* à pinnules
~THROUGH, to - [surv] viser
~VANE - [surv] pinnule *f*, viseur *m*
SIGHTING - [firearms] visée *f*, pointage *m*
~ANGLE - [surv] angle *m* de mire
~BOARD - [surv] voyant *m*
~MARK - [gen] repère *m* à vue
~NOTCH - [photo] cran *m* de mire
~SLIT - [instr] voyant *m* d'instrument
~STAKE - [surv] jalon *m*
SIGMA AMPLIFIER - [nucl] (the source of the signal for operating the safety circuit) amplificateur *m* sigma
~NOTATION - [comput] notation *f* sigma
~PILE - [nucl] assembly of moderating material containing a neutron source, used in the study of the neutron properties of the material) colonne *f* sigma
~-TYPE KNEADER - [mech] (a kneading-machine having blades of the Greek capital letter sigma) mélangeur *m* à caoutchouc
SIGMOID - [gen] (S-shaped) sigmoïde
~COLUMN - [anat] anse *f* sigmoïde du côlon
~CURVE - [statist] courbe *f* sigmoïde
~FLEXURE - [zool] (an S-bend) anse *f* sigmoïde
SIGMOIDECTOMY - [med] (the excision of part of the sigmoid flexure of the colon) sigmoïdectomie *f*
SIGMOIDOPEXY - [med] sigmoïdopexie *f*
SIGMOIDORECTOSTOMY - [med] sigmoïdorectostomie *f*
SIGMOIDOSCOPY - [med] sigmoïdoscopie *f*
SIGMOIDOSTOMY - [med] sigmoïdostomie *f*
SIGMOSCOPY - [med] s sigmoïdoscopy
SIGN, to - [gen] signer, marquer
[comm] (a contract) passer (un contract)
SIGN - [the gesture] signe *m*
SIGN - [gen] (a trace) trace *f*, indice *m*
[med] symptôme *m*
[comm] (of a shop etc) enseigne *f*
[math] signe *m*
[comput] (a one or zero to designate the algebraic sign of a quantity plus or minus) signe *m*
[mus bot etc] symbole *m*
~CONTROL - [comput] commande *f* des signes
~CONVENTION - [opt] (it refers to the direction of measurement of conjugate distances along the axis and the sign to be allocated for image altitude in lateral magnification) convention *f* des signes
~DIGIT - [comput] digit *m* de signe
~-OFF - [telev] fin *m* des émissions
~-ON - [telev] commencement *m* des émissions
~REGISTER - [comput] registre *m* de signes
SIGNAL, to - [gen] (to make signals) signaler, donner un signal
SIGNAL - [gen] (a sign, a means of communications) signal *m*
[telecomm] (radiowave or electric current transmitting intelligence) signal *m*
~ALARM - [gen] avertisseur *m*
~AMPLIFIER - [electron] amplificateur *m* de signaux
~AND BELL APPARATUS - [el] dispositif *m* d'alarme à sonnerie
~AND POINT WIRE - [railw] fil *m* de commande des signaux et des aiguilles
~APPROACH - [railw] abords *mpl* du signal
~AREA - [aero] (a part of an aerodrome used for signalling to aircraft) aire *f* à signaux
~ARM - [el] (traffic lights) bras *m* de signal (de sémaphore)
~ASPECT - [railw] indication *f* fournie par un signal
~BACKLIGHT - [railw] feu *m* arrière du signal
~BEACON - [naut] fanal *m*
~BELL - [el] avertisseur *m* à sonnerie
~BOARD - [railw] panneau *m* des signaux
~BOMB - [aero etc] (pyrotechnic device in the form of a smoke bomb which ignites on reaching a water surface used for estimation of drift) bombe *f* à fumée
~BOOK - [naut] code *m* de signaux
~BOX - [railw] cabine *f* à signaux, cabine *f* de aiguillage
~BRACKET - [railw] porte-signal *m*
~CABIN - s. signal box
~CIRCUIT - [telecomm] (in voice-frequency signalling, part responding to the signalling frequencies) circuit *m* du signal
~CODE - [telecomm] code *m* de signaux
~COMPARATOR - [radio] (automatic radio monitor) comparateur *m* de signaux, moniteur *m* automatique d'émetteur
~COMPLEX - [telev] ensemble *m* des signaux
~COMPONENT - [radio] (the part of a signal which continues uniform in character throughout its duration) composante *f* de signal
~CONTROL - [contr] commande *f* par signaux
~CONTROL RELAY - [contr] relais *m* de commande du signal
~CONVERTER - [contr] (a device in which the input and output signals represent, in different form, the same quantity) conformateur *m*, convertisseur *m* de signal
~CURBING - [radio] (a network inserted in a circuit to reduce the higher-frequency components in the

keying signal wave-shape, thus reducing sideband spread) préfiltre *m*
SIGNAL DISC - [railw] disque *m* de signal
~DISTORSION - [telecomm] (of a voice frequency signal receiver) distorsion *f* de signal
~ELECTRODE - [electron] (of a camera tube; an electrode from which the signal output is obtained) plaque *f* de signal, électrode *f* de signal
[telev] anode *f* collectrice
~ELEMENT - [radio & telecomm] (the part of a signal occupying the shortest interval of the signalling code) élément *m* de signal
~FIELD - [telev] champ *m* du signal
~FLAG - [railw] drapeau *m*
~FLARE - [signal] (pyrotechnic device showing a light of distinctive colour) bombe *f* éclairante
~FOR COLOUR CHANGING - [text] signal *m* pour changement de couleur
~FREQUENCY - [el] fréquence *f* de signal
~GENERATOR - [radio] (an apparatus for the production of electrical signals) générateur *m* de signaux, hétérodyne *f*
[acoust] générateur *m* de fréquences musicales
~HORN - [mus] bugle *m*
~IMITATION - [radio] (in voice-frequency signalling, an unwanted response of a voice-frequency receiver to any condition other than a true signalling condition) signalisation *f* intempestive par imitation de signaux
~INDICATION - [contr] contenu *m* du signal
[constr] contrôle *m* de signal
~INTERPOLATION - [telecomm] (interpolation in submarine cable telegraphy) interpolation *f* de signal
~INTERVAL - [radio] intervalle *m* de signal
~INVERSION - [telev] inversion *f* de signal
~LAMP - [el & telecomm] (indicating lamp on a switch or a control-board) lampe *f* de signalisation
~LANTERN - [signal] fanal *m* de signal
~LEVEL - [telecomm] niveau *m* de signal
~LEVER - [railw] levier *m* de signal
~LIGHT PROVING - [contr] contrôle *m* d'allumage de signal
~MACHINE - [el] moteur *m* de signal
~MAGNITUDE - [contr] amplitude *f* d'un signal
~METER - [radio] (a meter used in some receiver to measure the signal strength) S-mètre *m*
~MIXER UNIT - [radar] (a distribution box designed to allocate the signals in an identification radar system) dispositif *m* mélangeur de signaux
~MIXTURE - [telev] mélange *m* de signaux
~-NOISE RATIO - [radio] (the ratio of the strength of the signal required to the noise interference experienced) rapport *m* signal-bruit
~ON OPEN LINE - [railw] signal *m* de pleine voie
~OPERATION - [el] actionnement *m* des signaux
~OUTPUT CURRENT - [telev] valeur *f* absolue du courant de sortie du signal
~PEAK - [telecomm] crête *f* de signal
~PEDAL - [railw] pédale *f* de commande des signaux
~-PIPE HOSE - [mech] boyau *m* d'accouplement de l'intercommunication pneumatique
~PLATE - [telev] (in a storage camera tube, the metal back plate of the mosaic) anode *f* collectrice
~PLUG - [telecomm] (in telephony) bouchon *m* placé dans le jack d'abonné suspendu
~POINT INDICATOR - [railw] signal *m* d'aiguille
SIGNAL POST - [railw] mât *m* de signal
~PROJECTILE - s. signal rocket
~READING - [contr] (means by which an output is reproduced in its original form) interprétation *f* de signal
~RECORDING TELEGRAPHY - [telecomm] télégraphie *f* par enregistrement de signaux sans traduction automatique
~RELAY - [telecomm] relais *m* de signalisation
~REPEATER - [railw] répétiteur *m* de signal
~RESTORING - [comput] restauration *f* d'un signal
~ROCKET - [naut, aero etc] fusée *f* de signalisation
~SCANNER - [contr] (device receiving signals from different sources and retransmitting anyone of them into a single output circuit) analyseur *m* de signaux
~SEPARATOR - [telev] séparateur *m* de signaux
~SHAPER - [telecomm] (network inserted in a circuit to produced the required change in the wave-shape of signals) conformateur *m* de signal
~-SHAPER AMPLIFIER - [telecomm] amplificateur *m* correcteur de forme
~SHAPING - [electron] mise *f* en forme des signaux
~-SHAPING NETWORK - [telecomm] (wave-shap set) réseau *m* correcteur de forme
~TAIL - [telev] queue *f* de signal
~-TO-CROSSTALK RATIO - [telecomm] écart *m* diaphonique
~TO REDUCE THE SPEED - [railw] signal *m* de ralentissement
~TOWER - [railw] poste *m* de manoeuvre
~TRACER - [instr] (apparatus designed to trace defect in receivers) analyseur *m* électronique
~VALUE - [telev] valeur *f* de signal
~VELOCITY - [telecomm] vitesse *f* de signal
~WAVE - [radio] (a wave having characteristics which makes it possible for a message to be conveyed) onde *f* signal
~WINDINGS - [radio] (input windigs) enroulements *mpl* d'entrée
~WIRE - [railw] fil *m* de transmission (de signaux)
~WIRE PULLEY - [railw] poulie *f* de transmission funiculaire
SIGNALLING - [gen] signalisation *f*
~APPARATUS - [gen] appareil *m* pour signaux
~FUNCTION - [comput] fonction *f* de transmission
~REGULATIONS - [railw] règlement *m* des signaux
~TEST - [telecomm] essai *m* de signalisation
SIGNATORY - [gen & leg] signataire *m*
SIGNATURE - [gen] signature *f*
[print] (a distinguishing mark on each page, or sheet of a book, as a guide to the binder) signature *f* (d'un cahier)
[mus] (the key, or the time signature) armure *f* de la clef, armature *f*
[radio telev] (the sound effect, or musical number opening or closing a programme) indicatif *m* musical
[med] (for pharmaceutical products) mode *m* d'administration
~MARK - s. signature [print]
~TUNE - [mus] indicatif *m* musical
SIGNPOST - [gen] (road traffic) poteau *m* de signal
SIGNET - [gen] (a seal) sceau *m*, cachet *m*
~RING - [impl] anneau *m* à cachet
SIGNIFICANT DIGITS - [comput] (digits appearing in the coefficient of a number when the number is written as a coefficient between I.000........ and 9.999 .... times a power of ten) chiffres *mpl* signi-

ficatifs
SILAGE - [agric] ensilage *m*, silotage *m*
[agric] (the grain thus stored) fourrage *m* ensiloté
~FERMENTATION - [agric] (animal feeding) fermentation *f* au silo
~FOODER - [agric] fourrages *mpl* ensilés
SILANE - [chem] (generic term for compounds containing silicon and hydrogen) silane *m*
SILENCER - [gen] (device for reducing noise, especially of engine exhaust) silencieux *m*
[for firearms] dispositif *m* silencieux
[el] (device to reduce the noise or hum of telegraph wires) silencieux, amortisseur *m* de bruit
[mech] (attached to an exhaust pipe) boîte *f* d'échappement
~MOUNTING - [mech] (an antivibration rubber metal connection) suspension *f* élastique en caoutchouc métal, monture *f* de silencieux
SILENCING - [gen & mech] (the process of reducing noise, especially of an engine) amortissement *m* du son
~SWITCH - [radio] (a switch designed to obtain intercarriage noise suppression) commutateur *m* de réglage silencieux
SILENT - [gen] silencieux, insonore
~CHAIN - [mech] (type of chain furnished with teeth to engage gear wheels)chaîne *f* silencieuse
~DISCHARGE - [el] décharge *f* obscure
~ENGAGEMENT - [mech] enclenchement *m* silencieux
~FILM - [cin] film *m* muet
~PERIOD - [telecomm] intervalle *m* de silence
~RUNNING - [mech] fonctionnement *m* silencieux
~SPEED - [cin] (the speed at which silent picture films are fed through the camera) vitesse *f* d'un film muet
[auto] vitesse *f* silencieuse
~TUNING - [radio] réglage *m* silencieux
~ZONE - [radio] zone *f* morte, zone *f* de silence
SILEX - [min] silex *m*
SILHOUETTE CLOCK - [text] renforcement *m* des lisières du haut talon
SILICA - [chem] (native silicon dioxide, occurring as quartz, flint, sand, agate and as a large number of other modifications. Silica is used in the manufacture of glass, ceramics and enamels, and as a filler and heat insulant) silice *f*
~BRICK - [constr] brique *f* de silice
~ELECTRON TUBE - [electron] tube *m* à silice
~FLOUR - [min] farine *f* siliceuse
~GEL - [chem] (amorphous silica with uses as a dessicant and catalyst carrier) gel *m* de silice
~-GEL HEATER - [ind chem] séchoir *m* d'air aux sels hygroscopiques
~GLASS - [min] (fused quartz occurring in shapeless masses on the surface of the Lybian desert etc) quartz *m* fondu
~SAND - [min] sable *m* siliceux
SILICANE - [chem] (a gas formed by the action of concentrated hydrochloric acid on magnesium silicide) monosilane *m*
SILICATE - [chem] (generic name for the large class of compounds containing silicon, oxygen, possibly hydrogen and a metal or metals) silicate *m*
~-BONDED WHEEL - [mech] meule *f* à agglutinant silicieux
~OF ALUMINA - [chem] silicate *m* d'alumine
~OF POTASH AND LEAD - [chem] silicate *m* de potasse et de plomb
SILICATE OF SODA - [chem] silicate *m* de sodium
~PAINTS - [paint] (water-vehicle paints consisting essentially of sodium silicate, one of the most important features of which is that they are not inflammable) peintures *fpl* au silicate
SILICEOUS - [min] silicieux
~BRICK - [constr] brique *f* silicieuse
~CALAMINE - [min] hémimorphite *f*
~CEMENTING MATERIAL - [constr] ciment *m* silicieux
~CLAY - [geol] (clay rock with an admixture of silica) argile *f* silicieuse
~DEPOSITS - [geol] (deposits containing a large proportion of silica in one or more of its modes of occurrence) dêpots *mpl* silicieux
~LIMESTONE - [geol] pierre *f* calcaire silicieuse
~SANDSTONE - [geol] molasse *f* silicieuse
~SINTER - [geol] (found in encrustations of fibrous growths) aggloméré *m* silicieux, geysérite *f*
~STONE - [geol] calcaire *m* silicieux
SILICIC ACID - [chem] (hydrated silica, with uses as a regenerative adsorbent and as a catalyst) acide *m* silicique
SILICIDES - [chem] (compounds formed by the combination of silicon with other elements, especially metals) siliciures *mpl*
SILICIFY, to - [geol] (to convert into silica) silicifer
[ind chem] (timber) imprégner d'un silicate
SILICO-ARSENIDE - [chem] silico-arséniure *m*
~-BRONZE WIRE - [metall] fil *m* de bronze silicié
~-CALCEREOUS - [geol] silico-calcaire
~-FLUORIC ACID - [chem] acide *m* fluorosilicique
~-MANGANESE - [min] silico-manganèse *m*
~-METHANE - s. silicone
SILICON - [chem] (a non-metallic element, symbol Si, and, after oxygen, the most abundant element. Silica is used in the manufacture of a large number of compounds, both organic and inorganic, and in the production of refractory ceramics and special metal alloys) silicium *m*
~BRASS - [metall] laiton *m* silicieux
~BRONZE - [metall] bronze *m* silicié
~CAPACITOR - [electron] diode *f* à silicium capacitive
~CARBIDE - [chem] (a black crystalline substance of exceptional hardness, 9 on Moh's scale, used as an abrasive) carbure *m* de silicium
~-CARBIDE RECTIFIER - [electron] (a semiconductor rectifier) redresseur *m* à carbure de silicium
~CONTROLLED RECTIFIER - [electron] thyristor *m* au silicium
~COPPER - [metall] cuivre *m* silicieux
~DETECTOR - [electron] (a type of crystal detector) détecteur *m* à silicium
~DIOXIDE - [chem] (the dioxide of silicon which occurs in crystalline forms as quartz etc) dioxyde *m* de silicium, silice *f*
~HYDRIDE - [chem] hydrure *m* de silicium
~IRON - [metall] (iron or low-carbon steel to which some silicon has been added; it is used for sheets for transformer cores) fer *m* silicieux
~-KILLED STEEL - [metall] acier *m* calmé au silicium
~MANGANESE - [metall] alliage *m* de silicium et manganèse
~NITRIDE - [chem] (a refractory material of major

importance) nitrure *m* de silicium
SILICON RECTIFIER – [electron] redresseur *m* au silicium
~ROLLING – [metall] traitement *m* en tambour à carbure de silicium
~SOLAR CELL – [electron] (element of a polar battery) cellule *f* solaire au silicium
~STEEL – [metall] (a variety of steel containing up to I6 p. c of silicon and used in chemical plants where high corrosion resistance is desirable. Silicon steels also have applications in electrical apparatus) acier *m* au silicium
~TETRACHLORIDE – [chem] (an intermediate for a number of silicones and a component of chemical smoke generating compounds) tétrachlorure *m* de silicium
~TETRAFLUORIDE – [chem] (an analytical reagent and an intermediate for fluosilic acid) tétrafluorure *m* de silicium
SILICONE – s. silicones
~FLUIDS – [chem] (organosiloxane polymers have good heat stability, electrical resistance and water repellency) fluides *mpl* silicones
~RESINS – [chem] (organosiloxane polymers with extremely good heat stability and chemical resistance. They have application in the production of electrical insulants for high temperature use and in the manufacture glass fibre reinforced components) résines *fpl* silicones
~RUBBER – [ind chem] (siloxane or silicone compounds having characteristics similar to those of natural rubber. Silicone rubbers can be used over a very wide temperature range and have good resistance to weathering and chemical attack) caoutchouc *m* silicone
SILICONES – [chem] (organo-silicon compounds of open-chain or cyclic form containing the typical group SIR_2O – and forming a very important body of substances, waxes, insulators and others) silicones *fpl*
SILICONIZING – [metall] (steel case hardening in a medium from which silicon can be absorbed) silicisation *f*
SILICOSIS – [med] (pneumokoniosis due to the inhalation of particles of silica) chalicose *f*, phtisie *f* des tailleurs de pierre
SILICOTUNGSTIC ACID – [chem] (a dyeing mordant and analytical reagent) acide *m* silicotungstique
SILIOTUBERCULOSIS – [med] silicotuberculose *f*
SILIQUA – [bot] silique *f*
SILK – [text] (a fibre secreted by the larve of Bombyx mori and used for the manufacture of fabrics) soie *f*
~AND COTTON FEATHER SHAG – [text] peluche *f* long-poil soie et coton
~BEAM – [text] ensouple *f* de soie
~BRAID – [text] galon *m* de soie
~BREEDER – [text] éducateur *m* de vers à soie
~CAMLET – [text] camelot *m* de soie
~CLOTH – [text] tissu *m* de soie
~COMBING – [text] peignage *m* de la soie
~COMBING MACHINE – [text] peigneuse *f* pour soie, dresseuse *f* pour soie
~CORD – [text] faille *f*
~COTTON – [text] bombycine *f*
~COTTON TREE – [bot] bombax *m*
~COUNT – [text] titre *m* de la soie
~COVERED – [el] isolé à la soie
SILK CROP – [text] récolte *f* des cocons
~CULTURE – [text] sériciculture *f*
~DEGUMMING – [text] dégommage *m* de la soie
~FABRIC – [text] étoffe *f* de soie
~FIBRE – [text] fibre *f* de soie, fil *m* de soie
~FIBRE FLUFF – [text] duvet *m* du brin de soie
~FIBRE FREE FROM FLOSS – [text] fibre *f* de soie sans duvets
~FLUID – [text] fibroïne *f*, matière *f* soyeuse
~FLY FRAME – [text] banc *m* à broches à soie
~GAUZE – [text] gaze *f* de soie, gaze *f* à blutoir
~GRAIN – [zool] oeuf *m* du ver à soie
~GRASS – [text] fibre *f* du bromélia karatas
~GROUND THREAD – [text] fil *m* de fond en soie
~GUM – [text] (the sericin; a main constituent of raw silk) séricine *f*
~HANK – [text] écheveau *m* de soie
~HEALD – [text] lisse *f* en soie
~KNOP – [text] bouton *m* de soie
~LOOM – [text] métier *m* à tisser la soie
~MILL – [text] filature *f* de soie
~MOTH – [zool] papillon *m* séricigène
~NOIL – [text] bourrette *f*
~NOIL INDUSTRY – [text] industrie *f* de la bourrette
~NOIL SPINNING – [text] filature *f* de la bourrette
~NOIL YARN – [text] fil *m* de bourrette, bourrette *f*
~NUMBER – [text] numéro *m* de la soie
~OF DULL LUSTRE – [text] soie *f* terne
~OF GOOD FEEL – [text] soie *f* ayant du toucher
~PLUSH – [text] peluche *f* de soie
~REEL MILL – [text] moulin *m* à guindres
~REELING – [text] tirage *m* de la soie, filature *f* des cocons
~RIBBON – [text] ruban *m* de soie
~RIBBON LOOM – [text] métier *m* pour ruban de soie
~ROVING FRAME – [text] banc *m* à ailettes à soie
~SCOURING – s. silk degumming
~SCREEN PRINTING – [text] impression *f* au cadre
~SCREEN PRINTING PLANT – [text] installation *f* de impression au cadre
~SCREEN PRINTING STENCIL – [text] stencil *m* pour impression au cadre
~SEED – s. silk grain
~SEED TRADE – [agric] commerce *m* des graines de vers à soie
~SHODDY SPINNING – [text] filature *f* du shoddy de soie
~SHODDY YARN – [text] fil *m* de shoddy de soie
~SKEIN – [text] écheveau *m* de soie
~SPINNING INSECT – [zool] insecte *m* séricigène
~STOCKING DISEASE – [med] érythrocyanose *f* des jambes
~THROWING MILL – [text] filature *f* de soie
~TITRE – [text] titre *m* de la soie
~TOP – [text] (nastro oil) coeur *m* de soie, déchets *mpl* de soie peignés
~VELVET – [text] velours *m* de soie
~VOILE – [text] voile *m* de soie
~WARP BOBBIN – [text] bobine *f* de chaîne soie
~WARPING MACHINE – [text] ourdissoir *m* pour soie
~WASTE – [text] déchets *mpl* de soie
~WEAVING – [text] tissage *m* de la soie
~WINDING FRAME – [text] bobinoir *m* pour soie
~YARN – [text] fil *m* de soie
SILKING – [paint] (fault in a coating, taking the form of fine Lining) éclat *m* soyeux
SILKS – [cin] (silk screens which are used to diffuse

light) écrans mpl diffuseurs en soie
SILKWORM - [zool] ver m à soie
~BOX - [text] claie f des vers à soie
~-BREEDING - [text ind] éducation f des vers à soie, magnanerie f
~FARMING - s. silkworm breeding
~HOUSE - [text] magnanerie f
~NURSERY - [text] coconnière f
~READY TO SPIN - [text] ver m à soie prêt à filer
~ROT - [text] (a disease) muscardine f
SILKY - [gen] soyeux
~FEEL OF THE WOOL - [text] toucher m soyeux de la laine
~FRACTURE - [metall] cassure f à grain fin
~LUSTRE - [text] lustre m soyeux
~TURBIDITY - [brew ind] trouble m soyeux
SILL - [constr] (the lower boundary of a door) seuil m, traverse f (de la porte)
[constr] (the lower boundary of a window opening) tablette f de fenêtre
[mining] (the floor of a deposit) semelle f d'une galerie
[mining] (of a timber set) sole f de galerie
[naut] (of a dry dock) seuillet m
[geol] (intrusive tabular sheet of igneous rocks) filon-couche m, couche f intrusive
[constr] (a fondation, a basis) sablière f basse
[hydr] (the top level of a weir) seuil m (d'écluse)
[railw] (US only; longitudinal members of a truck underframe) longrine f, longeron m (de wagon); brancard m
[mining] (of a mine) mur m
~-FLOOR TIMBERING - [mining] boisage m du niveau de fond
~PLATE - [impl] patin m (de palier)
[th eng] pare-ringard m (de porte de foyer)
SILLIMANITE - [min] (a native aluminium silicate with uses as a refractory) sillimanite f
SILLING - [mining] préparation f du niveau de base
SILLOMETER - [instr] (a device designed to measure the speed of a ship) sillomètre m
SILO - [agric] (a tall structure, generally of reinforced concrete, used as container for the storage of grain, or other loose material) silo m
SILOXANE - [chem] (generic name for compounds containing silicon, oxygen and hydrogen) siloxane m
SILPHA - [zool] silphe m
SILT, to - [hydr] (to obstruct or to become obstructed with silt or mud) envaser, ensabler
SILT - [hydr] (earthy sediment in finely divided form, which is carried by water) boue f, limon m, vase f
[geol] (the loose earthy sediment) apports mpl de ruissellement
[geol] (the material still suspended in water) colmatage m
[metall] (scum) écume f
~BOX - [hydr] (removable iron box used to accumulate deposited silt which is taken away periodically) collecteur m de boues
~MOVEMENT - [soil] mouvement m des alluvions
SILTED TRACK - [railw] voie f ensablée
SILTING - [agric] colmatage m
[mining] remblayage m hydraulique par embouage
~UP - [hydr] envasement m
SILTY - [gen] limoneux, envasé, vaseux
SILUMIN - [metall] (aluminium silicon alloy) alliage m d'aluminium et silicium
SILUNDUM - [ind chem] carbure m de silice
SILURIAN - [geol] silurien
~SYSTEM - [geol] système m silurien
SILVALIN PULP YARN - [text] fil m de pâte silvaline
SILVER, to - [gen] argenter
[gen] (a mirror) étamer
SILVER - [metall] (a metallic element, symbol Ag, used as bullion and in the production of photographic chemicals, as a lining material for chemical plant, in electrical equipment and in the manufacture of tableware and jewellery) argent m
~ACETATE - [chem] acétate m d'argent
~AMALGAM - [min] (a solid solution of mercury and silver crystallizing in the system) amalgame m d'argent
~ARSPHENAMINE - [chem] (a brownish-black powder with therapeutic uses) arsphénamine f d'argent
~-BARKED - [bot] à écorce luisante
~BATH - [chem] bain m de sels d'argent
~BEAD - [text] perle f argentée
~-BEARING - [min] argentifère
~BEET - [agric] bette f, poirée f
~-BONDED DIODE - [electron] diode f à connexions en argent
~BRAID - [text] galon m d'argent
~BRAZING - [metall] alliage m d'argent pour soudure forte
~BROCADE - [text] brocart m en broché argent
~BROMIDE - [chem] (a light-sensitive compound used in the manufacture of photographic film) bromure m d'argent
~BROMIDE PAPER - [photo] papier m au bromure de argent
~BRONZES - [metall] (a term denoting copper-zinc nickel alloys) alliages mpl de cuivre, zinc et nickel
~BULLION - [metall] matières fpl d'argent
~CHLORIDE - [chem] (a photographic chemical and a component of plating baths in silver plating) chlorure m d'argent
~CHLORIDE CELL - [el chem] (primary cell with a silver chloride depolarizer) pile f au chlorure d'argent
~CHROMATE - [chem] (an analytical reagent) chromate m d'argent
~CITRATE - [chem] citrate m d'argent
~COINAGE - [metall] argent m de monnaie
~CYANIDE - [chem] (a constituent of plating baths) cyanure m d'argent
~-DISK PYRHELIOMETER - [instr] pyrhéliomètre m à disque d'argent
~FIR - [bot] sapin m blanc, sapin m argenté
~FISH - [zool] argentine f
~FLUORIDE - [chem] (one of the silver halides) fluorure m d'argent
~FOG - [photo] voile m d'argent
~-FOIL - [metall] feuille f d'argent
~-FORK - [med] déformation f en dos de fourchette
~FOX - [zool] renard m argenté
~FULMINATE - [chem] fulminate m d'argent
~-GILT - [metall] vermeil m, argent doré
~GLANCE - [min] (argentite; silver sulphide) argentite f, argyrite f
~GREY - [gen] gris m argenté
~-HALIDE EMULSION - [photo] émulsion f à l'halogénure d'argent
~HALIDES - [chem] (compounds of silver with the halogens; some of these are light-sensitive and

used in photography) halogénures *mpl* d'argent
SILVER IODIDE - [chem] (a photographic chemical and an antiseptic used in medicine) iodure *m* d'argent
~LEAD - [min] plomb *m* argentifère
~LEAD ORE - [min] (term denoting galena containing silver) minerai *m* de plomb argentifère, galène *f* argentifère
~LEAF DISEASE - [bot] (a plant disease) maladie *f* du plomb
~MATRIX - [nucl] (carrier of radioisotopes in alpha foils) matrice *f* en argent
~MILL - [mining] installation *f* pour l'exploitation de l'argent
~NITRATE - [chem] (a germicide and caustic used in medicine a photographic chemical and an analytical reagent) nitrate *m* d'argent
~NITRITE - [chem] (an analytical reagent and synthesis intermediate) nitrite *m* d'argent
~ORE - [min] minerai *m* d'argent
~OXIDE - [chem] (a brown, odourless powder with uses as an oxidizing agent) oxyde *m* d'argent
~OXIDE CELL - [el chem] (primary cell using silver oxide as a depolarizer, with a potassium hydroxide electrolyte and a zinc electrode) pile *f* à oxyde d'argent
~PAPER - [paper man] papier *m* de soie blanc
~PHOSPHATE - [chem] (a catalyst and photographic chemical) phosphate *m* d'argent
~PICRATE - [chem] (yellow, light-sensitive crystals with uses as a trichomonacide in medicine) picrate *m* d'argent
~-PLATE, to - [metall] argenter
~PLATES - [metall] argenté, plaqué argent
~PLATED WIRE - [text] fil *m* d'argent demi-fin
~PLATING - [metall] argentage *m*, argenture *f*
~POTASSIUM CYANIDE - [chem] (an antiseptic and a constituent for plating baths) cyanure *m* d'argent et potassium
~PROTEIN - [chem] (a colloïdal form of silver with uses as an antiseptic) protéine *f* argentée
~SAND - [min] sable *m* silicieux
~SCREEN - [bot] (a plant disease) gale *f* argentée
~-SOLDER, to - [metall] souder à l'argent
~-SOLDER - [metall] (a copper-zinc-silver alloy used in fine brazing) soudure *f* à l'argent
~STEEL - [metall] acier *m* à l'argent
~STORAGE BATTERY - [el chem] (alkaline storage battery having a silver-oxide positive electrode and a calcium or zinc negative electrode) accumulateur *m* à l'argent
~SURFACE - [plast ind] (bright metallic appearance on the surface of a moulding, due to presence of moisture in the moulding powder) surface *f* d'argent
~THREAD - [text] fil *m* d'argent
~VOLTAMETER - [el chem] (electrolytic cell used to determine the average value of a current from the quantity of silver deposited from the silver nitrate solution forming the electrolyte) voltamètre *m* à argent
~-WEED - [bot] ansérine *f*, herbe *f* aux oies
~WHITE - [paint] blanc *m* d'argent
SILVERING - [metall] argentage *m*
[glass man] (the operation of pouring an ammoniacal silver solution mixed with Rochelle salts on a perfectly clean surface of glass) étamage *m*
SILVERING - [el metall] (in electro-typing; the application of a thin conducting film of silver by chemical reduction upon a plastic or wax matrix) argenture *f*
SILVERSKIN ONION - [bot] petit oignon *m* blanc
SILVERSMITH - [gen] orfèvre *m*
SILVERWARE - [gen] argenterie *f*
SILVERY - [gen] argenté
~IRON - [metall] fonte *f* grise (riche en silicium)
SIMA - [geol] (the continuous basaltic shell which underlies the continental masses beneath the sial and the ocean floor) sima *m*
SIMIAN - [zool] (resembling an anthropoid ape) simien, simiesque
SIMILAR - [gen] semblable, pareil, analogue
[geom] semblable, similaire
~TRIANGLES - [geom] triangles *mpl* similaires
SIMILARITY - [gen] similarité *f*
[geom] similitude *f*
SIMMER, to - [gen] frémir, bouilloter, mijoter
SIMMER - [gen & phys] frémissement *m*
~BURNER - [gas ind] champignon *m* central
SIMMERING - [gen] frémissement *m*, bouillottemant *m*
[acoust] (the sound made by a boiling thick fluid) bouillonnement *m*
SIMONS PROCESS - [el chem] (in electrolytic process for the production of fluorocarbons) procédé *m* de Simons
SIMPLE - [gen] simple
[chem] (which cannot be or has not been decomposed) simple
[text] semple *m* (de métier)
~BEAM - [constr] poutre *f* simple
~BELT - [mech] courroie *f* simple
~-BOARDED FLOOR - [constr] plancher *m* simple
~CURVE - [surv] (curve composed of a single connecting two straights) courbe *f* simple
~DISTILLATION - [chem] (distillation in which the vapour issued from the boiling liquid is removed as fast as it is formed without the return of condensed vapour to the boiling liquid) distillation *f* globalement équilibrée
~ELECTRODE - [el chem] (electrode at which only one electrode reaction occurs) electrode *f* simple
~ELLIPTIC HARMONIC MOTION - [phys] (a compounded oscillatory motion consisting of harmonic motion in two fixed perpendicular directions with equal frequencies) mouvement *m* elliptique harmonique simple
~-EXPANSION LOCOMOTIVE - [railw] locomotive *f* à simple expansion
~FRACTION - [math] fraction *f* simple
~FRACTURE - [med] fracture *f* fermée, fracture *f* simple
~FRUIT - [bot] (a fruit which is formed from one pistil) fruit *m* à pistil simple
~GEAR EXTRUDER - [rubber ind] boundineuse *f* à simple harnais d'engrenage
~HARMONIC CURRENT - [el courant *m* sinusoïdal
~HARMONIC MOTION - [phys] (motion in which the particle is attracted towards an origin by a force directly proportional to the instantaneous distance of the particle from the origin) mouvement *m* harmonique simple
~HARMONIC QUANTITY - [phys] grandeur *f* sinusoïdale
~HEAT TREAT - [metall] cycle *m* unique de traitement

thermique
SIMPLE INDEXING - [mach tool] division *f* simple
~INTERLACING - [text] liage *m* simple
~LEAF - [bot] (a leaf in which the lamina consists of one piece) feuille *f* simple
~LENS - [opt] (lens consisting of a simple piece of glass) lentille *f* simple
~LOOM - [text] métier *m* à simples
~MACHINE - [mech] (any elementary mechanical contrivance , e g. a lever, a wedge etc) machine *f* simple
~PARALLEL WINDING - [el] (winding of a drum armature in which the winding pitch is unity) enroulement *m* parallèle simple
~PENDULUM - [mech] pendule *m* simple
~POINT PASS - [text] remettage *m* à retour simple
~POINT SOURCE - [mech] (a small source which alternately injects fluid into a medium and then withdraws it) source *f* ponctuelle à alternance
~PROCESS - [phys] procédé *m* unitaire
~PROCESS FACTOR - [nucl] (the enrichment factor obtained in a single process, e.g. simple distillation) facteur *m* d'enrichissement unitaire
~RATIO CHANNELS - [telecomm] voies *f*pl de communication à rapport simple de période
~SIGNAL - [telecomm] signal *m* simple
~SOUND SOURCE - [acoust] (a source which radiates sound uniformly in all directions under free-field conditions) source *f* sonore omnidirectionnelle
~SOURCE OF SOUND - s. simple sound source
~STEAM ENGINE - [mech] (engine with one or more cylinders in which the steam expands from the initial pressure to the exhaust pressure in a single stage) machine *f* à vapeur à effet simple
~TISSUE - [bot] (tissue consisting of cells all of the same kind) tissu *m* simple
~TONE - [acoust] (pure tone) ton *m* pur
~VENTILATION - [el] ventilation *f* série
SIMPLEX CHANNEL - [telecomm] (channel of communications which transmits signals in one direction only at a time) canal *m* simplex
~CIRCUIT - [telecomm] (a circuit permitting the transmission of signals in one direction only) circuit *m* simplex
[telecomm] (a two-wire metallic circuit from which a simplex circuit is derived; the metallic and simplex circuit being capable of simultaneous use) circuit *m* approprié simultanément à la téléphonie et à la télégraphie
~ÇOMMUNICATION - [telecomm] communication *f* simplex, liaison *f* simplex
~LAP WINDING - s. simple parallel winding
~OPERATION - [telecomm] communication *f* simplex, trafic *m* simplex
~SYSTEM - [telecomm] (telegraph system using signals transmitted in one direction only) symplex *m*
~TELEGRAPHY - [telecomm] télégraphie *f* simplex
~WAVE WINDING - [el] (series winding) enroulement *m* série
~WINDING - [el] (armature winding through which there is only electrical path per pole) enroulement *m* simplex
SIMULATE, to-- [gen, aero, mech etc] (to imitate by artificial means, especially movements of aircraft or the reactions of controls) simuler
SIMULATED ASCENT - [aero] (in calibrating altimeter, artificial change of pressure to produce readings corresponding to increase of altitude in an aircraft) ascension *f* simulée
SIMULATED DESCENT - [aero] (in calibrating altimeters, artificial change of pressure to produce readings corresponding to decrease of altitude in an aircraft) descente *f* simulée
~PROGRAMME - [comput] programme *m* simulé
SIMULATION - [gen] simulation *f*
[comput] (the representation of physical systems and phenomena by computers, models and other equipment) simulation *f*
~EQUIPMENT - [comput] simulateur *m*
~TESTING - [comput] contrôle *m* de simulation
SIMULATOR - [gen] (a device representing systems, operating conditions of a unit, etc, by computers, models etc) simulateur *m*
SIMULCAST - s. simultaneous brodcasting
SIMULTANEOUS - [gen] simultané
~BROADCASTING - [radio] émission *f* relayée
~DRILLING - [oil ind] forage *m* simultané
SINAPISM - [med] (a mustard plaster) sinapisme *m*
SINCIPUT - [anat] sinciput *m*
SINE - [math] sinus *m*
~BAR - [tool] (a hardened and group steel bar fitted with two cylindrical plugs of known diameter at a fixed centre distance; used for setting out to close limits) règle *f* inclinée
~CONDITION - [opt] (the condition which must be satisfied by a lens to form an image free from aberrations) condition *f* des sinus
~-COSINE POTENTIOMETER - [instr] potentiomètre *m* sinus-cosinus
~CURVE - [math] sinusoïde *f*
~FUNCTION - [math] fonction *f* sinusoïdale
~GALVANOMETER - [instr] (galvanometer in which the coil and the scale are rotated to keep the needle at zero) boussole *f* des sinus
~-SHAPED - [phys] (of a wave) sinusoïdal
~TABLE - [tool] (a modification of a sine bar in which the upper surface forme a flat accurately ground table on which a workpiece can be mounted) règle *f* inclinée avec table de montage
~WAVE - [phys] (a wave in which the particles execute transverse vibrations of a simple harmonic type) onde *f* électromagnétique sinusoïdale
~WAVE CONVERGENCE - [telev] convergence *f* par courant sinusoïdal
SINEW - [anat] tendon *m*
SINEWY - [gen] tendineux
SING, to - [acoust] chanter
SINGE, to - [gen] brûler légèrement
[text] griller (une étoffe), flamber, gazer
~THE WOOL, to - [text] griller la laine
SINGED THREAD - [text] fil *m* gazé
SINGEING MACHINE - [text] machine *f* à gazer
~PLATE - [text] plaque *f* de flambage
~THE THREADS - [text] flambage *m* des fils, gazage *m* des fils
SINGING - [acoust] chant *m*
[zool] (of a bird) chanteur *m*
[gen] (of the wind) sifflement *m*
[el] (the continuous oscillation of audio frequency which is set up in a telephone circuit be energy feed-back round the circuit of a repeater) amorçage *m*, sifflement *m*
~ARC - [acoust] (a carbon arc made to play tunes

by continuously variable or step by step changes in inductance) arc *m* chantant
SINGING MARGIN - [telecomm] (of a telephone circuit) marge *m* d'amorçage
~PATH - [el] voie *f* de réaction
~POINT - [radio] (the maximum gain of a two-wire amplifier can be given when one hybrid coil is terminated by a line and its associated balancing network, and the other hybrid coil is terminated by 600 ohms and a disconnection) point *m* d'amorçage
~SUPPRESSOR - [telecomm] suppresseur *m* de réaction
SINGLE, to - [gen] séparer, distinguer
[agric] éclaircir
SINGLE - [gen] seul, unique, simple
[leg] célibataire
[text] poil *m* (de soie)
~ACCESS SATELLITE - [telev] satellite *m* à simple accès
~-ACTING - [mech] (operating in one direction only, as of an hydraulic cylinder in which power is developed only during piston movement in one specific direction) à simple effet
~-ACTING POWER HAMMER - [mech] marteau-pilon *m* à simple effet
~ACTION - [mech] simple effet *m*
~-ACTING AXIAL BEARING - [mech] palier *m* axial à simple effet
~-ACTION COMPACTING - [mech] compression *f* à simple effet
~-ACTING CYLINDER - [mech] cylindre *m* à simple effet
~-ACTING ENGINE - [mech] machine *f* à simple effet
~-ACTING PISTON PUMP - [mech] pompe *f* à piston à simple effet
~-ACTING PUMP - [mech] pompe *f* à simple effet
~ACTION PRINTER - [comput] (a printer which prints one character at a time) imprimante *f* de lettre à lettre
~-ADDRESS - [comput] à une adresse
~-ADDRESS-CODE - [comput] code *m* à simple adresse
~-ADDRESS INSTRUCTION - [comput] (an instruction of an operation and exactly one address) instruction *f* à une adresse
~-APERTURE CORE - [comput] tore *m*
~-ARMED LEVER - [mech] levier *m* à un bras
~-AXLE FULL TRAILER - [auto] remorque *f* à un seul essieu
~-AXLE SEMITRAILER - [auto] semi-remorque *f* à un seul essieu
~BALLING MACHINE - [text] pelottonneuse *f* simple
~-BANKED - [naut] armé en pointe
~-BASE PROPELLANT - [astronaut] propergol *m* monobase
~-BATTERY SWITCH - [el] (switch arranged to control the number of cells in a battery which are charging or discharging) réducteur *m* de charge
~-BEARING LIVE SHAFT - [mech] essieu *m* fou tenu par un seul palier
~BEAT - [text] coup *m* de battant simple
~BEATER SCUTCHER - [text] batteur *m* simple
~-BELTING - [mech] (a belt having a single layer of leather) courroie *f* simple
~-BEND TEST - [metall] essai *m* de flexion simple
~-BEND VALVE - [rubber ind] (of tyres) valve *f* coudée
SINGLE BEVEL - [mech] biseau *m* simple
~-BEVEL NOSE - [tool] (of a cutter) bec *m* à biseau simple
~BEVEL OF A CUTTER - [mech] biseau *m* oblique
~-BLADE MIXER - [rubber ind] mélangeur *m* à palette simple
~BLADE SHUTTER - [cin] (a rotating shutter with only one blade to cut off the light) obturateur *m* à lame unique
~-BLAST CIRCULAR BELLOWS - [mech] soufflet *m* cylindrique à simple vent
~-BLAST FORGE - [metall] forge *f* à simple vent
~BLOCK - [mech] poulie *f*
~-BOLSTER - [railw] (of a bogie) à une sellette
~-BOSS ROLL - [text] cylindre *m* à un seul fil par bosse
~-BREAK SWITCH - [el] (a switch in which the circuit is made or broken at one point only on each pole or phase) interrupteur *m* à rupture unique
~-BUTT STRAP - [mech] couvre-joints *m* simple
~CABLE - s. single-core cable
~CAPSTAN - [mech] cabestan *m* simple
~CARD - [text] carton *m* simple
~-CARD TOTAL ELIMINATION - [comput] élimination *f* du total pour groupes monocartes
~CASCADE ACTION - [contr] (the regulation of the set on the automatic controller by the action of another automatic controller) régulation *f* en cascade simple
~-CATENARY SUSPENCTION - [el] suspension *f* caténaire simple
~-CAVITY MOULD - [plast ind] (a mould containing only one recess for material) moule *m* à empreinte unique
~CHANNEL - [telecomm] (denoting a transmission system when it carries one channel of communication only) à un seul canal
~-CHANNEL MODULATION - [radio] modulation *f* simple
~-CIRCUIT DRIFT TUBE - [electron] (a drift tube with a single circuit, the electron beam of it interacting at least twice with the circuit) tube *m* de propagation à ligne simple
~CIRCUIT-LINE - [el] ligne *f* simple
~CLIP WOOL - [text] laine *f* du mouton tondu une fois par an
~-COATED FILM - [photo] pellicule *f* à couche unique
~-COIL LAMP - [el] lampe *f* à filament spiralisé, lampe *f* à simple boudinage
~-COIL REGULATION - [el] régulation *f* à deux barres régulation *f* à une inductance
~COLOUR EFFECT - [chem] effet *m* de couleur pure
~COLOUR JACQUARD PATTER - [text] dessin *m* Jacquard à une seule couleur
~-COLOURED - [gen] monochrome
~-COLUMN PLANER - [tool] raboteuse *f* à colonne simple
~COLUMNED - [print] (in a newspaper) à une colonne
~COMMUNICATION DIRECT CURRENT SIGNALLING - [telecomm] signalisation *f* à courant continu à commutation simple
~-COMPONENT SIGNAL - [telecomm] signal *m* à composante unique
~COMPOUND SWITCH - [railw] traversée-jonction *f* simple

SINGLE CONTACT - [el] contact *m* simple
~-CONTACT LAMP - [el] lampe *f* à contact simple
~-CONTACT SYSTEM - [el] ligne *f* de contact simple
~-CONTROL - [aero etc] monocommande *f*, commande *f* unique
~-CORE CABLE - [el] (a cable containing one core only) câble *m* unipolaire
~-CREAM CHEESE - [food] fromage *m* demi-gras
~-CROP FARMING - [agric] monoculture *f*
~CROP SYSTEM - s. single-crop farming
~CRYSTAL - [phys] (a macroscopic specimen of a solid in which all parts have the same crystallographic orientation) monocristal *m*
~-CRYSTAL CAMERA - [radiat] (an X-ray camera in which a single crystal is used) chambre *f* à monocristal
~-CRYSTAL GROWTH - [electron] (method used for the manufacture of semiconductors) croissance *f* de monocristaux
~-CURRENT SIGNALLING - [telecomm] (in telegraphy) signalisation *f* à courant simple
~CUT - [mech] taille *f* simple
~-CUT FILE - [tool] lime *f* à simple taille
~-CYLINDER - [mech] monocylindrique
~-CYLINDER MACHINE - [mech] machine *f* à cylindre unique, moteur *m* monocylindrique
~-CYLINDER PLANER - [mach tool] rabot *m* à fil
~DAMASK - [text] damas *m* simple
~-DECADE COUNTING UNIT - [comput] (a circuit designed to count to ten) compteur *m* monodécade
~DIFFUSION STAGE - [nucl] (single stage in an isotope separating process on the diffusion principle) étage *m* unique de diffusion
~DIODE - [electron] (electronic tube with only set of two electrodes) diode *f* simple
~DISK BRAKE - [aero] (type of disk-brake for landing gear in which there is only one friction disk) frein *m* à un disque
~DRAWING - [text] étirage *m* simple
~-DRIVER - [mech] (of a locomotive) à un seul essieu moteur (locomotive)
~-EFFECT EVAPORATOR - [ind chem] (evaporator in which heat is not transferred from the vapour to the liquor in another evaporator) évaporateur *m* à simple effet
~ELECTRODE POTENTIAL - [el] potentiel *m* spécifique d'électrode
~-ELECTRODE SYSTEM - [el] demi-cellule *f*
~-ELEMENT RELAY - [el] relais *m* à un élément
~-END CONTROL - [contr] commande *f* à sens unique
~-END VENTILATED MOTOR - [mech] moteur *m* ventilé à sens unique
~-ENDED AMPLIFIER - [radio] amplificateur *m* symétrique série
~-ENDED BOILER - [th eng] chaudière *f* à foyers d'un seul côté
~-ENDED CABLE GRIP - [telecomm] manchon *m* à mailles, grip *m* double
~-ENDED HAND-WRENCH - [impl] tourne-à-gauche *m* simple
~-ENDED OUTPUT - [electron] sortie *f* asymétrique
~-ENDED RAIL - [metall] rail *m* à simple champignon
~-ENDED RIBBON SPANNER - [impl] clef *f* simple à nervures
~-ENDED SPANNER - [impl] clef *f* à écrous simple, clef *f* simple à fourche

SINGLE ENDED SPOOL - [text] tube *m* à un seul plateau
~-ENDED TUBE - [electron] tube *m* avec toutes les sorties au culot
~-ENGINED - [mech aero etc] monomoteur
~-ENGINED AIRCRAFT - [aero] avion *m* monomoteur
~-ENTRY - [mining] galerie *f* simple, galerie *f* unique
~-ENTRY COMPRESSOR - [mech] (type of compressor in which air is admitted at one side only of the impeller) compresseur *m* à entrée unique
~-EXPANSION ENGINE - [mech] machine *f* à simple expansion
~-FACED VALVE - [mech] (a gate valve) vanne *f* à plateau
~FAUCET WATER HEATER - [th eng] (bath water heater with free outlet) chauffe-bains *m* à écoulement libre
~FEEDER - [el] feeder *m* unique
~FIBRE - [text] fibre *f* isolée
~FLANGED BOBBIN - [text] bobine-bouteille *f*
~FLEECE - [text] nappe *f* seule
~FLIGHT UNIFORM PITCH SCREW - [mech] vis *f* à un filet à pas constant
~-FLIGHTED - [plast ind] (term used of an extrusion screw having a single-start thread) à un seul filet
~-FLOAT - [aero] coque *f* flottante
~FLOOR - [constr] plancher *m* ordinaire
~FLOWER - [bot] (a flower with one set of petals and no indication of doubling) fleur *f* simple
~-FLUE BOILER - [th eng] chaudière *f* à un tube-foyer
~-FLUID CELL - [el] pile *f* à un liquide
~-FREQUENCY SIGNALLING SYSTEM - [telecomm] signalisation *f* à fréquence commune
~-FURROW PLOUGH - [agric] araire *m*
~-FURROW REVERSIBLE PLOUGH - [agric] double-brabant *m*
~-GEAR - [mech] simple harnais *m* d'engrenages
~-GEARED LATHE - [mach tool] tour *m* à simple harnais d'engrenages
~-GRAIN CHARGE - [astronaut] charge *f* en grain unique
~GRAVITY-HOIST - [mech] balance *f* sèche à simple effet
~-GUN COLOUR TUBE - [electron] tube *m* image en couleurs à canon électronique unique
~-HEAD DRAW FRAME - [text] banc *m* d'étirage à une tête
~-HEAD WRENCH - [mech] clef *f* simple
~-HEADED RAIL - [railw] rail *m* à champignon unique
~-HUNG WINDOW - [constr] (window with top and bottom sashes, of which only one is balanced by sash and weights so as to be moved vertically) fenêtre *f* à un ventail
~IMPRESSION MOULD - [plast ind] (a mould with a single cavity i.e. a mould which produces one moulding per moulding cycle) moule *m* à empreinte unique
~INDUCTIVE SHUNT - [el] shunt *m* inductif simple
~-INLET FAN - [th eng] ventilateur *m* à une seule ouie
~INTERLOCKING - [constr] emboîtement *m* simple
~-JET NOZZLE - [auto] injecteur *m* à jet unique
~KEY - [mech] clavette *f* simple
~KEYWAY BROACH - [mech] alésoir *m* des rainures de clavette simple

SINGLE-LAYER WINDING - [el] (a winding consisting of a single coil of one layer, usually having an axial length of large dimensions as compared with the diameter) enroulement *m* à couche unique
~-LEAF - [mech] (of a spring) à lame unique (ressort)
~-LEAF SPRING - [auto] ressort *m* à lame unique
~-LENS REFLEX - [photo] reflex *m* monoculaire
~LIFT DOBBY - [text] ratière *f* à une seule griffe
~-LIGHT UNIT - [opt] unité *f* lumineuse simple
~-LINE - [railw] à une seule voie
~-LINE RAILWAY - [railw] chemin *m* de fer à une seule voie
~LINEN - [text] toile *f* simple
~-LOBE CAM - [mech] came *f* à un bossage
~NEEDLE SYSTEM - [telecomm] (telegraph system using indicating galvanometers with single needles, deflections to the left or right indicating the coded signal) système *m* à aiguille simple
~NOTCH JOINT - [carp] assemblage *m* à entaille, trave *f*
~OR MULTISPINDLE AUTOMATIC LATHE - [mach tool] tour *m* automatique à un ou plusieurs mandrins
~PARTICLE MODEL OF NUCLEUS - [nucl] modèle *m* du noyau à particules individuelles
~-PASS TUNNEL OVEN - [rubber ind] tunnel-séchoir *m* à circulation en sense unique
~-PETTICOAT INSULATOR - [el] isolateur *m* à cloche simple
~-PHASE - [el] (qualifying term, applied to a system or apparatus, to denote one in which there is a single alternating voltage) monophasé
~-PHASE ARMATURE - [el] induit *m* monophasé
~-PHASE CABLE - [el] câble *m* pour courant monophasé
~-PHASE CIRCUIT - [el] circuit *m* monophasé
~-PHASE COMMUTATOR LOCOMOTIVE - [railw] locomotive *f* à moteur à collecteur monophasé
~-PHASE COMMUTATOR MOTOR - [el] moteur *m* à collecteur monophasé
~-PHASE COMMUTATOR MOTOR WITH SELF-EXCITATION - [el] moteur *m* à connecteur monophasé à excitation interne
~-PHASE CONCENTRIC SPIRAL WINDING - [el] enroulement *m* monophasé à spires concentriques
~-PHASE FURNACE - [metall] four *m* à courant monophasé
~-PHASE IMPEDANCE - [el] impédance *f* monophasée
~-PHASE INDUCTION REGULATOR - [el] (induction regulator for use on a single-phase circuit) régulateur *m* à induction monophasée
~-PHASE MACHINE - [el] machine *f* monophasée
~-PHASE METER - [instr] (induction type of watt-hour meter) compteur *m* de courant monophasé
~-PHASE OPERATION - [comput] service *m* à courant monophasé
~-PHASE RECTIFIER - [electron] (rectifier operating from a single-phase alternating current supply) redresseur *m* monophasé
~-PHASE SYNCHRONOUS GENERATOR - [el] génératrice *f* synchrone monophasée
~-PHASE SYSTEM - [el] (alternating current system in which only two conductors are used) système *m* monophasé
~-PHASE TRANSFORMER - [el] (a transformer intended for use on a single-phase system or on one phase of a polyphase system) transformateur *m* monophasé
SINGLE PICTURE MOVEMENT - [cin] projection *f* image par image
~-PIECE RIM - [auto] jante *f* d'une seule pièce
~PILE BINDING - [text] liage *m* par un seul poil
~PIPE RING - [plumb] (collar) collier *m* de fixation
~-PITCH ROOF - [constr] comble *m* à une seule pente
~-PLATE - [mech] monodisque *m*
~-PLATE CLUTCH - [mech] embrayage *m* monodisque
~-PLY BELT - [mech] courroie *f* à une seule couche
~-PLY CONSTRUCTION - [rubber ind] (of tyres) confection *f* pli par pli
~-POLARITY PULSE - [el] (a pulse in which the sense of the departure from normal is one direction only) impulsion *f* unipolaire
~POLE-PIECE MAGNETIC HEAD - [el acoust] (a magnetic headed with a single pole piece on one side of the recording medium) tête *f* magnétique à pièce polaire unique
~-POLE SWITCH - [el] (a switch for opening or closing only one of the two leads to an electric circuit) interrupteur *m* unipolaire
~-POLE TUBE - [electron] tube *m* unipolaire
~POTENTIAL - [el] potentiel *m* entre électrode et solution, potentiel *m* d'électrode
~PRICKER - [impl] piqueur *m* simple
~PRISM - [acoust] (a prism used in optical sound reproducing system) prisme *m* simple
~PULLEY - [mech] monopoulie *f*
~-PULLEY DRIVE - [mech] commande *f* par monopoulie
~-PULLEY HEAD - [mach tool] poupée *f* à monopoulie
~-PURCHASE CRAB - [mech] treuil *m* à simple engrenage
~-PURCHASE RACK-AND-PINION JACK - [mech] cric *m* à crémaillère à simple engrenage
~-PURPOSE COMPUTER - [comput] (a computer for a special purpose only) calculateur *m* à usage particulier
~-RAIL CRANE - [mech] grue *f* vélocipède sur un rail de translation
~RANGE INSTRUMENT - [instr] instrument *m* à gamme unique
~RASPING MACHINE - [text] défibreuse *f* simple
~-RATE PREPAYMENT METER - [el] (a prepayment meter for a flat-rate tariff, so that the circuit is broken after a predetermined number of units have been consumed) compteur *m* à prepaiement à tarif simple
~-REDUCTION GEAR - [mech] engrenage *m* à simple réduction
~RING BELL - s. single stroke bell
~-RIVETED - [mech] à un rang de rivets
~-RIVETED BUTT JOINT - [mech] rivure *f* à un rang de rivets
~-RIVETED LAP JOINT - [mech] rivure *f* à un rang de rivets à clin
~RIVETING - [mech] rivure *f* simple
~-ROPE CABLE-TRAMWAY - [transp] monocâble *m*
~-ROPE HAULAGE - [mining] plan *m* incliné de voie unique
~-ROTOR HELICOPTER - [aero] hélicoptère *m* monorotor
~ROVING - [text] mèche *f* simple
~ROVING CREEL - [text] râtelier *m* simple
~ROW BALL BEARING - [mech] (a ball bearing having only one circle of balls) roulement *m* à une rangée

de billes

SINGLE-ROW RADIAL ENGINE - [mech] moteur *m* en étoile simple rangée

~-SASHED WINDOW - [constr] fenêtre *f* à un battant

~SCATTERING - [nucl] (the deflection of a particle from its original path to one encounter with a single scattering centre in the material traversed) diffusion *f* unique

~SCREW - [plast ind] vis *f* unique
[naut] à une hélice

~-SEATER - [auto & aero] monoplace *m*

~SEED FRUIT - [bot] fruit *m* monosperme

~SERVICE - [el] distribution *f* unique

~SHEAR - [metall] cisaillement *m* simple
[mech] à un seul tranchant

~-SHEAR STEEL - [metall] (shear steel) acier *m* affiné

~SHEATH - [phys] potentiel *m* de simple couche

~-SHOT BLOCKING OSCILLATOR - [radio] oscillateur *m* à blocage à coup unique

~-SHOT MULTIVIBRATOR - [radio] (multivibrator with one stable and one quasi-stable state) multivibrateur *m* à coup unique, multivibrateur *m* monostable

~-SHOT TRIGGER CIRCUIT - [el] circuit *m* de déclenchement monostable à coup unique

~-SHUTTLE LOOM - [text] métier *m* à une navette

~-SIDEBAND MODULATION - [radio] (modulation whereby the spectrum of the modulating wave is transmitted and the other sideband is partially or totally suppressed) modulation *f* sur bande latérale unique

~-SIDEBAND TRANSMISSION - [radio] (method of operation in which one sideband is transmitted and the other partially or totally suppressed) émission *f* sur bande latérale unique

~-SIDEBAND TRANSMITTER - [radio] émetteur *m* sur bande latérale unique

~-SIGNAL RECEPTION - [radio] recepteur *m* à haute sélectivité

~SILK - [text] fil *m* poil

~SLIDE - [photo] châssis *m* simple

~-SLIDE SASH - [photo] glissière *f* pour diapositives séparées

~SLIP-SWITCH - [railw] traversée-jonction *f* simple

~-SLOT SCREW - [mech] vis *f* à fente simple

~SOUND TRACK - [acoust] (unilateral recording of sound by mechanical means) piste *f* sonore simple

~SPAR - [aero] monolongeron *m*

~-SPEED - [auto & mech] à une seule vitesse, a simple vitesse

~-SPEED FLOATING ACTION - [contr] action *f* flottante à vitesse unique

~-SPEED FLOATING CONTROL - [contr] réglage *m* flottant à vitesse unique

~-SPEED WINDSCREEN WIPER - [auto] essuie-glace *m* à simple vitesse

~-SPEED WINDSHIELD WIPER - s. single-speed windscreen wiper

~-SPINDLE - [mach tool] à un mandrin

~-SPOOL ENGINE - [mech] (a gas turbine having one spool only) turbine *f* à un seul rotor

~-STAGE COMPRESSOR - [mech] (a compressor in which there is only stage of compression) compresseur *m* simple, compresseur *m* monocylindrique

~-STAGE PUMPING - [mech] épuisement *m* en un seul jet

SINGLE-STAGE ROCKET - [astronaut] fusée *f* à un étage

~-STAGE TURBINE - [mech] (a turbine having only one stage of expansion) turbine *f* simple, turbine *f* à un étage d'expansion

~-STAMP MILL - [min] bocard *m* à simple pilon

~STIRRUP - [constr] (in reinforced concrete) étrier *m* simple

~-STROKE BELL - [el] (an indicator or alarm bell which makes only one stroke when energised) sonnerie *f* monocoup, cloche *f* électrique

~-STUB MATCHING - [electron] adaptation *f* à bras de réactance unique

~STUB TRANSFORMER - [electron] (in waveguides) transformateur *m* de mode à bras de réactance unique

~-STUB TUNER - [electron] (in waveguides) dispositif *m* de synchronisation à bras de réactance unique

~STUD - [metall] support *m* double

~-SUCTION IMPELLER - [mech] roue *f* à aube à simple aspiration

~TACKLE - [mech] palan *m* simple

~-TANK CIRCUIT BREAKER - [el] disjoncteur *m* à bac unique

~-THREAD - [mech] filet *m* simple

~THREAD WORM - [text] vis *f* sans fin à pas simple

~-THREADED - [mech] (of a screw) à un filet

~-THREADED HOB - [tool] vis *f* fraise à un filet

~-THREADED SCREW - [mech] vis *f* à un filet

~THROAT BURNER - [gas ind] brûleur *m* simple

~-THROW - [el] à commutation unipolaire

~-THROW CRANK - [mech] (of a crankshaft) manivelle *f* simple

~-THROW SWITCH - [el] interrupteur *m* à couteau unique

~TICKET - [transp] billet *m* simple

~-TONE KEYING - [radio] manipulation *f* à note unique

~TONGUING - [mus] (the normal method of tonguing in playing wind instrument) coup *m* de langue simple

~TRACK - [acoust] (the magnetic record has only one track over its whole width) piste *f* simple
[acoust] (or standard track, a variable-area sound track in which both positive and negative halves of the signal are linearly recorded) piste *f* simple
[railw] à une voie

~-TRACK HAULAGE ROAD - [mining] galerie *f* de roulage à voie unique

~-TRACK LINE - [railw] ligne *f* à voie unique

~-TRACK RECORDER - [el acoust] enregistreur *m* magnétique monopiste

~-TRIP TRIGGER CIRCUIT - s.single-shot trigger

~-TUBE CORE BARREL - [mining] tube *m* carottier simple

~-TUNED AMPLIFIER - [radio] (amplifier characterized by resonance at a single frequency) amplificateur *m* à résonance unique

~-TURN COIL - [el] bobine *f* à filament spiralisé

~-TURN TRANSFORMER - [el] (a current transformer in which the primary winding takes the form of a single straight conductor of heavy cross-section, to which the cable or the bus-bar is connected) transformateur *m* de courant à spire unique

~TURNOUT - [railw] branchement *m* à deux voies

~UNIT SEMICONDUCTOR DEVICE - [electron] (a semiconductor device having one set of electrodes associated with a single carrier stream) disposi-

tif *m* semiconducteur simple
SINGLE-V-BUTT WELD - [mech] (a butt weld in which the edges of the parts are chamfered on one side only) soudure *f* bout à bout sur chanfrein en V
~-VOLTAGE RATING OF A TRANSFORMER - [el] tension *f* nominale commune
~WARP BAGGING - [text] toile *f* à sac en jute simple
~-WAVE RECTIFICATION - [radio] (rectification by a circuit which operates by permitting the passage of current during only alternate half-cycles) redressement *m* demi-onde
~-WAY CONNECTION - [el] couplage *m* à simple voie
~WEB - [text] voile *m* simple
~WEFT PLUSH - [text] peluche *f* à une seule duite
~WEFT TRIMMING - [text] galon *m* à une seule trame
~-WEIGHT - [photo] force *f* papier
~-WING DOOR - [constr] porte *f* à un battant
~WIRE AERIAL - [radio] (an aerial which consists of a single wire) antenne *f* unifilaire
~-WIRE ANTENNA - s. single wire aerial
~-WIRE CIRCUIT - [el] (a circuit consisting of a single wire, the return being through the earth or frame) circuit *m* unifilaire
~-WIRE FEEDER - [radio] (for an aerial) feeder *m* unifilaire
~-WIRE GLASSED HEADER - [electron] ensemble *m* de scellement monofilaire
~-WIRE SYSTEM - [el] (a method of direct-current distribution employing a single outgoing conductor and using the earth as the return conductor) alimentation *f* à courant continu unifilaire
~WIRE TRANSMISSION LINE - [electron] (a waveguide consisting of a single metallic wire usually coated with dielectric material) guide *m* d'ondes unifilaire
~-WRAP ROPING - [el] entraînage *m* par cable à enroulement simple
~YARN - [text] fil *m* simple
SINGLES - [constr] (small roofing slates) tables *f*pl d'ardoise (de 45 x 20 cm.)
SINGLET - [chem] (a bond formed by a single shared electron) singlet *m*
[text] gilet *m* de corps
~LINKAGE - [chem] (a valence bond between two atoms which consists of a single electron) liaison *f* de singlet
SINGLING - [agric] éclairissage *m*, démariage *m*
~HOE - [agric] razette *f* à betteraves
SINGULAR - [gen] singulier
[math] singulier
~SOLUTION - [phys] solution *f* azéotrope
SINGULARITY - [gen & math] singularité *f*
SINIGRIN - [chem] (white, crystalline glucoside) sinigrine *f*
SINISTER - [gen] (situated on the left side) gauche, sénestre
SINISTRORSE - [bot etc] (twining from right to left) sinistrorse
SINISTROSIS - [med] névrose *f* de revendication
SINK, to - [gen & naut] tomber au fond, couler au fond, sombrer
[gen] (to enter a softer body) pénétrer, s'enfoncer
[gen] (to excavate downward) foncer, creuser
[gen] (to descend) descendre
[mech] (to cut the recess in a mould or die) grainer, graver
[soil etc] (to subside) se tasser, s'affaisser
SINK, to - [gen] (power etc] baliser (en puissance, s'affaiblir
SINK - [gen] évier *m*
[mining] percement *m* de galeries, cône *m* d'avancement
[hydr] (a sewer drain) souillard *m*, récepteur *m*
[astronaut] (the metal work of the combustion chamber) puits *m* thermique
[plumb] (in drawing, the reduction of a tube diameter) réduction *f* de diamètre
[metall] poquette *f*, tassement *m*
[metall & plast ind] (a depression in the surface of a moulding due to imperfect filling of the mould, e.g. by premature freezing) dépression *f*, tassement *m*
[electron] (of an oscillator) plage *f* d'instabilité électronique
[el] réceptrice *f*
~DRAG - [aero] (drag which is associated with removal of fluid from the airflow in question) trainée *f* d'immersion
~GRID - [hydr] grille *f* de bonde d'un évier
~HOLE - [constr] puisard *m*
[mining] puisard *m*, bouniou *m*
~MARK - [metall & plast ind] (shallow depression on the surface of a moulding) dépression *f* en surface
~PILES, to - [constr] foncer des pieux
~STOPPER - [gen] bouchon *m* d'évier
~STRAINER - [gen] filtre *m* d'évier
~TRAP - [hydr] siphon *m* d'évier
~WELL - [constr] puits *m* enfoncé
SINKAGE - [text] (the losses in weight occurring in wool materials during the various processes) perte *f* de poids (de la laine)
[mech] (of the wheels) enlisement *m* (des roues)
SINKER - [fishing] plomb *m* (de ligne de pêche)
[naut] (of a mine) crapaud *m* d'amarrage (d'une mine)
[text] (of a knitting machine) platine *f*
[naut] (the lead of a ground log) plomb *m* (d'une ligne de loch)
[brew ind] (said of grains which fall to the bottom of the vat) grain *m* plongeur
[mining] fonceur *m* de puits
[metall] larget *m*, platine *f*
~BAR - [oil ind] barre *f* de surcharge
~BOAT -[naut] catamaran *m*
~BREAST - [text] dessous *m* du bec de platine
~BURR - [text] roue *f* à maille
~CAM RING - [text] came *f* de platine
~CATCH BAR - [text] barre *f* à platines
~HEAD - [text] tête *f* de platine
~JACK - [text] platine *f* d'abattage
~THROAT - [text] engorgement *m* de la platine
~WHEEL - [text] mailleuse *f*
SINKHEAD - [metall] (in foundry work) masselotte *f* chaude
SINKHOLE - [hydr] (a drainage cavity) souillard *m*, collecteur *m* d'eaux
[geol] doline *f*, effondrement *m*
SINKING - [gen & naut] action *f* de couler, engloutissement *m*
[mining] (of a well) fonçage *m*, foncement *m*
[geol] affaissement, abaissement *m*
[mech] (an impression on a forging die block) gravure *f*

SINKING - [soil] tassement *m*
[metall] (the drawing of tubes without an internal mandrel) filetage *m* sans mandrin
[carp] (a recess cut below the general surface) entaille *f*
[opt] mirage *m* inférieur, réfraction *f* due à l'effet de prisme
~BY PILING - [mining] fonçage *m* au poussage
~DRUM - [mining] trousse *f* coupante
~FRAME - [min] cadre *m* de boisage
~FUND - [fin] caisse *f* d'amortissement
~HAMMER - [mining] marteau *m* à puits
~HEAD - s. sinkhead
~HEAD-FRAME - [mining] cadre *m* de superficie
~IN - [paint] (the absorption of the finishing coat) absorption *f* de la dernière couche
~-LIFT - [mining] (in pumping) jeu *m* de fonçage
~MILL - [metall] laminoir *m* réducteur
~PLAN - [fin] plan *m* d'amortissement
~PUMP - [mining] pompe *f* de fonçage
~SPEED - [aero] (the vertical velocity component of an aircraft descending in still air under given conditions) vitesse *f* verticale de descente
~THE FLOOR - [mining] rebanchage *m* du mur
~TIPPLE - [mining] chevalement *m* de fonçage
SINOATRIAL - [anat] sino-auriculaire
SINOAURICULAR - s. sinoatrial
~-BLOCK - [med] blocage *m* sino-auriculaire
SINOIDAL - s. sinusoïdal
SINOGRAPHY - [radiology] radiographie *f* des sinus
SINTER, to - [metall] (to bind together, e.g. metal or ceramic powder, by the application of heat, and, usually pressure) fritter, agglomérer
SINTER - [min] (calcareous or siliceous material deposited by springs) travertin *m*, sinter *m*
[chem] (the product obtained through sintering) aggloméré *m*
[metall] (dross of iron) sorne *f*
~LAYER - s. sinter skin
~SKIN - [metall] peau *f* de frittage
SINTERED - [metall] fritté
~CARBIDE - [metall] (used for the cutting tips) carbure *m* fritté
~COMPACT - [metall] comprimé *m* fritté
~DENSITY - [metall] densité *f* après frittage
~GLASS - [ind chem] verre *m* filtrant fritté
~IRON - [metall] fer *m* fritté
~MAGNET - [el] aimant *m* fritté
~PLATE - [el chem] (in an alkaline battery, a plate in which the support is made of sintered metal powder impregnated with active material) plaque *f* frittée
~TANTALUM CAPACITOR - [electron] condensateur *m* fritté au tantale
SINTERING - [ind chem] (process of heating a crushed material until incipient fusion causes agglomeration) frittage *m*
[el] (of an electric lamp filament) concrétionnement *m*
[min] agglomération *f*, agglutination *f*
~ACTIVITY - [metall] activité *f* de frittage
~COAL - [min] (non-coking) charbon *m* agglomérant par frittage
~FURNACE - [metall] (a furnace in which sintering is carried out) four *m* à fritter
~LINE - [min] ligne *f* d'agglomération
~METAL - [metall] métal *m* de frittage
SINUOUS FLOW - [hydr] (eddy flow) remous *m*
SINUS - [anat] (a cavity or depression of an irregular shape) sinus *m*, antre *m*
~OF THE KIDNEY - [med] sinus *m* du rein
SINUSAL - [med] (of the sinus) sinusal
SINUSITIS - [med] (the inflammation of one or more of the air-containing cavities communicating with the nose) sinusite *f*
SINUSOID - [math] sinusoïde *f*
SINUSOIDAL - [math] sinusoïdal
[el] (said of an alternating quantity when its trace, plotted to a time base, is a sine wave) sinusoïdal
~CURRENT - [el] courant *m* sinusoïdal
~QUANTITY - [el] (a quantity varying according to a sinusoidal function of the independent variable) grandeur *f* sinusoïdale
~THERAPY - [med] thérapie *f* sinusoïdale
SIPHON, to - [gen] siphonner, de transvaser
SIPHON - [zool] (tubular structure in some aquatic animals, for drawing in and expelling liquid) siphon *m*
[hydr] (a bent pipe used to transfer liquids from a higher to a lower level over an elevation) siphon *m*
~BRICK - [metall] (of a cupola, for the separating of slag) brique *f* à siphon
~OFF, to - [hydr] (to draw off liquid from a tank or the like by means of a siphon) siphonner
~CONDUIT - s. siphon piping
~PIPING - [hydr] conduite *f* en siphon
~RECORDER - [instr] (a device for making a continuous record, e.g on a moving paper band, of the reading of a instrument, using a pen fed with ink by a siphon device) enregistreur *m* à siphon
~SPILLWAY - [constr] (siphon connecting the upstream and the downstream sides of a reservoir dam, thus enabling flood to pass by) siphon *m* de passe-déversoir
~TRAP - [hydr] (in sanitary engineering) siphon *m* de décantation
SIPHONAGE - [hydr] (flow of liquid from a higher a lower level through a close passage which rises above both of these, but not above the absolute hydraulic gradient) siphonage *m*, siphonnement *m*
SIRE, to - [zool] engendrer, procréer
SIREN - [acoust] (strong source of noise) sirène *f*
SIROCCO - [met] (hot dry wind blowing from the African coast over the Mediterranean) siroc *m*
~FAN - [mining] ventilateur *m* de mine
SIRUP - s. syrup
SIRUPY - s syrupy
SISAL - [text] (a commercially important fibre obtained from Agave sisalana and used for the manufacture and sacking) sisal *m*, agave *f* d'Amérique
~HEMP - [text] chanvre *m* de sisal
SISMOGRAPH - s. seismograph
SISTER - [gen] soeur *f*
[med] (a head nurse) infirmière *f* en chef
~BLOCK - [naut] baraquette *f*, poulie *f* vierge
~CELL - [biol] (one of the two cells formed by the division of a pre-existing cell) cellule *f* soeur
~CHROMATIDS - [genet] chromatides *f*pl source
~ELEMENT - [chem] élément *m* affin
~HOOK - [mech] croc *m* à ciseaux
~KEELSON - [shipbuild] contre-carlingue *f*
~REUNION - [genet] réunion *f* des chromatides soeurs
SIT, to - [gen] s'asseoir, être assis

SIT, to - [zool] couver
~DOWN - [gen] grêve *f* avec occupation d'usine
SIT - [mining] écrasement *m* des piliers
SITE, to - [gen] placer, situer
SITE - [gen] emplacement *m*
[constr] (area of ground which is set apart for an engineering or building work) terrain *m* à bâtir, chantier *m*
~DELIVERED - [gen] à pied d'oeuvre
~ERROR -[radio] (in direction finding, a stable error due to presence of reflecting obstacles near the site of the receiving aerials) erreur *f* d'emplacement
~TEST - [ind] contrôle *m* local
SITFAST - [vet] (a small hard lump of the skin of a horse's back) cor *m*, induration *f* (sur le dos d'un cheval)
SITOMANIA - [med] boulimie *f*
SITOPHOBIA - [med] sitophobie *f*
SITOSTEROL - [chem] (a white, waxy solid with uses in medicine and cosmetics) sitostérol *m*
SITZ BATH - [gen] (small bathtub for bathing in a sitting posture) bain *m* de siège
SIX - [math] (a cardinal number) six *m*
~COLOUR RECORDER - [instr] (recording apparatus for six measured values, which are recorded by six different dots) enregistreur *m* à six couleurs
~COMPONENT BALANCE - [instr] (balance used in a wind tunnel to measure the complete system of forces and moments, i.e about three axes) balance *f* à six composantes
~-COUPLED LOCOMOTIVE - [railw] locomotive *f* à trois essieux couplés
~-FOOT WAY - [railw] (the distance between running lines) entre-voie *f*
~-PHASE CIRCUIT - [el] circuit *m* hexaphasé
~-PLY BELTING - [mech] courroie *f* à six plis
~-ROLLER MILL - [agric] moulin *m* à six cylindres
~-ROW BARLEY - [agric] orge *f* à six rangs
~-WHEELER - [auto] voiture *f* à trois essieux
~-WHEELER TRUCK - [transp] camion *m* à six roues
SIXTEENMO - [print] in-seize *m*
SIZABLE - [gen] (of comparatively large, or convenient size) assez grand
SIZE, to - [gen] (to classify, to graduate) classer par dimension
[mech etc] calibrer, graduer
[mech] (to bring to a given size) mettre *f* à dimension
[text] (to apply a coating of size) apprêter, maroufler
SIZE - [gen] dimension *f* , grandeur *f*
[min] (of coal) grosseur *f*
[paint etc] (animal glue in powder or jelly form) colle *f*
[text] (a component used to improve the physical properties of textiles during processing) apprêt *m*, empois *m*
[paper man] (as for textiles) colle *f*
[of a book] format *m*
[med] couenne *f* inflammatoire
~BLOCK - [mech] calibre *m*
~BOX - [text] cuve *f* à la colle, bac *m* à colle
~BOX ROLLER - [text] cylindre *m* d'encollage
~CONSTANCY - [telev] constance *f* de dimensions
~DISTRIBUTION OF COKE - [min] granulométrie *f* des cokes
SIZE DOWN, to - [instr] graduer en ordre déscendant
~GRADING - [min] granulométrie *f*
~MARGIN - [mech] tolérance *f* sur les dimensions
~MILK - [ind chem] (glue used in bookbinding) lait *m* de colle
~MIXING - [text] mélange *m* de l'apprêt
~MIXING APPARATUS - [text] appareil *m* de préparation de colle
~OF A WELL - [oil ind] capacité *f* de production par jour
~OF AN OPENING - [constr] dimension *f* de la baie
~OF A COAL - [min] (or grade of a coal) calibre *m* d'un charbon
~OF JAW - [mech] ouverture *f* de la mâchoire
~OF TYRE - [auto] mesure *f* du pneu
~PRESS - [paper man] presse *f* de la colle
SIZED COAL - [mining] charbon *m* calibré
~HEALD - [text] lisse *f* encollée à l'empois
~RAGS - [text] chiffons *mpl* amidonnés
~WARP - [text] chaîne *f* parée, chaîne *f* encollée
SIZER - [gen] classeur *m*, classeur-trieur *m*
[text] colleuse *f* de chaînes
[min] (a device separating the mineral in different particle sizes) classeur-trieur *m*
SIZER'S WASTE - [text] déchets *mpl* d'encollage
SIZER TAP - [tool] taraud *m* finisseur
SIZING - [min] (the operation of separating material in different known particle sizes) calibrage *m*, classement *m* par grosseur, triage *m*
[paper man] (the operation of giving paper a degree of water resistance) encollage *m*
[text] parage *m*, encollage *m*
[mach tool] (preselecting setting of the work on a grinding machine) calibrage *m* micrométrique
[meas] vérification *f* des dimensions
[mech] mise *f* à dimensions
[metall] finissage *m* à mesures finales
[mech] pression *f* en forme désirée
~AGENT - [ind chem] (used in the paper man) matière *f* collante
~AGENTS - [text] matières *fpl* pour l'encollage
~BRUSH - [impl] brosse *f* à parer
~IN HANKS - [text] encollage *m* des fils en écheveaux
~MACHINE - [text] encolleuse *f*, pareuse *f*
~MILL - [metall] laminoir *m* calibré, laminoir *m* calibreur
~ON THE LOOM - [text] encollage *m* sur métier
~PLANT - [meas] installation *f* de calibrage
~PLATE - [plast ind] (a plate, usually one of a series pierced with a hole of accurate size, and used to control the o/d of extruder pipe) plaque *f* de calibrage
~ROLL - [mech] cylindre *m* calibreur
~SCALE - [meas] échelle *f* de calibrage
~SCREEN - [min] crible *m* classeur
~SLEEVE - [plast ind] (hollow cylindrical part dowstream of the die, used to control d/d of extruded pipe) manchon *m* de calibrage
~SOLUTION - [paper man] solution *f* collante
SIZZLE - [acoust] crépitements *mpl*
[telecomm] friture *f*, choc *m* acoustique
SIZZLING - [gen] (very hot) tout chaud
~HEAT - [metall] température *f* entre 200° et 230°
SKATE MACHINE - [el] (shoe mechanism in GB) mécanisme *m* de patin
~MACHINE BRAKE - [el] (shoe mechanism brake in GB) mécanisme *m* de patin de freinage

SKATE MACHINE CONTROLLER - [el] (shoe mechanism controller in GB) combinateur *m* de mécanisme de patin
SKATOLE - [chem] (evil-smelling crystalline substance with application in perfumery) skatol *m*
SKEET - [naut] (long-handled scoop) écope *f* à long manche
SKEG - [naut] (projection of the afterpart of a vessel's keel) talon *m* de la quille
SKEIN - [text] (a quantity of yarn wound to a given length and subsequently doubled and knotted) écheveau *m*
[zool] (of wild geese) vol *m* d'oies sauvages
[med] spirème *m*
~DYEING - [text] teinture *f* en écheveau
~OF SILK - [text] flotte *f*, échevette *f* de soie
SKEINING - [text] liage *m*, échevettage *m*
SKELALGIA - [med] douleurs *fpl* dans les jambes
SKELASTHENIA - [med] faiblesse *f* des jambes
SKELETAL - [gen] (of skeleton) squelettique
~MUSCLE - [anat] muscle *m* strié
SKELETON - [anat] squelette *f*, charpente *f* osseuse
[gen] (of a structure) charpente *f*, carcasse *f*, châssis *m* de montage
[bot] nervures *fpl*
[surv] (the network of survey lines providing a figure from which shape and features of the survey can be determined) canevas *m*
[shipbuild] ossature *f*
~BRAID - [rubber ind] tissu *m* squelettique
~CASE - [transp] caisse *f* à claire voie
~CONSTRUCTION - [constr] structure *f* en charpentes
~CONTAINER - [railw] container *m* à claire voie
~CRYSTAL - [phys] cristal *m* imparfait
~CYLINDER - [text] tambour *m* à claire voie
~DRUM - [photo] tambour *m* à claire-voie
~DRUMS - [paper man] (drums over which paper is dried after sizing) tambours *mpl* de séchage
~FORMING - [plast ind] (process in which plastic sheet is drawn partially into a vacuum chamber and then snapped back into a skeleton mould in which the pressure is controlled by air admission) thermoformage *m*
~GIRDER - [constr] poutre *f* à jour
~HAND - [med] main *f* en squelette
~KEY - [mech] (a master key) crochet *m* de serrurier, fausse clef *f*
~SOIL - [geol] sol *m* squelettique
~-TYPE SWITCHBOARD - [el] (a switchboard in which the apparatus is mounted on an open metallic framework) tableau *m* de distribution type cadre
~WHEELS - [agric] roues *fpl* grilles
SKELLERED BEAM - [text] ensouple *f* déjetée
SKELP - [mech] (a strip of iron or steel for tubes) acier *m* en bandes, bande *f* à tube
SKENE ARCH - [constr] (or scheme arch; an arch having the shape of a circular arch subtending less than 180 degrees) voûte *f* surbaissée
SKENITIS - [med] squénite *f*
SKEP - [gen] panier *m*
SKEPTOPHYLAXIS - [med] skeptophylaxie *f*
SKETCH, to - [gen] esquisser
[draw] esquisser, ébaucher, dessiner au trait
SKETCH - [gen & draw] esquisse *f*
~DIVIDING GAUGE - [text] calibre *m* diviseur d'esquisse
~OF INTERLACING - [text] représentation *f* graphique des armures
~PLATE - [metall] tôle *f* façonnée
SKEW - [gen] biais, oblique
[arch] biais
[radio] distorsion *f* oblique
[comput] (of magnetic tape) défilement *m* en biais
[mining] apophyse *f* d'un filon
~AILERON - [aero] (aileron of which the hinge axis is not parallel to the transverse axis of aircraft) aileron *m* en biais
~ARCH - [constr] voûte *f* biaise
[constr] (of a bridge) arche *f* biaise
~BEVEL GEAR - [mech] engrenage *m* hyperboloïde
~BEVEL PINION - [mech] pignon *m* hyperboloïde
~BEVEL WHEEL - [mech] roue *f* de commande hyperbolique, roue *f* hyperboloïde
~BRICK - [constr] sommier *m*
~BRIDGE - [constr] pont *m* biais
~BUTT - [constr] (the projecting brickwork supporting the foot of a gable coping) console *f* en chanfrein
~-CHISEL - [impl] biseau *m*, fermoir *m* de tour néron
~COIL - [el] bobine *f* asymétrique
~CORBEL - s. skew butt
~CURVE - [math] courbe *f* asymétrique
~DOG - [auto] dent-de-loup *m*
~FACTOR - [el] (the component factor of the winding factor which takes into account the slot skewing) facteur *m* d'inclinaison
~GEARS - [mech] (skew bevel gears) engrenages *mpl* hyperboloïdes
~LINES - [math] lignes *fpl* biaises
~NAILING - [carp] rivure *f* en biais
~PLANE IRON - [impl] fer *m* de rabot oblique
~RABBET PLANE - [impl] (rabbet plane with its cutting edge arranged obliquely across the sole) guillaume *m* oblique
~RAYS - [opt] (rays which are not confined to a meridian plane, do not intersect the optical axis, and are difficult to trace) rayons *mpl* obliques
~TABLE - [constr] (stone bonded in with a gable wall as support for the foot of the coping) niveau *m* des naissances
~WALL - [constr] (a wall which does not form a face of a parallelepiped) mur *m* biais
SKEWBACK - [constr] (the part of a pier which immediately supports a segmental arch) naissance *f* d'une voûte
SKEWBALD - [zool] cheval *m* blanc et roux
SKEWED - [gen] oblique, de travers, en biais
~POLE - [el] pôle *m* incliné
SKEWER - [gen] brochette *f*
[mech] brochette *f*
[text] chandelle *f* de râtelier, brochette *f*
SKEWFOOT - [med] pied *m* valgus
SKEWNESS - [gen] (the state of being asymmetrical) obliquité *f*, biais *m*
SKI TYPE LANDING GEAR - [aero] (type of landing gear in which skis are used instead of wheel for landing on surfaces of snow or ice) train *m* d'atterrissage *m* skis
SKIAGRAPH - [photo] (a photograph made by exposing a sensitive emulsion to X-rays) skiagramme *m*, radiographie *f*
SKIASCOPE - [med] pupilloscope *m*
SKIASCOPY - [med] (retinoscopy) pupilloscopie *f*
SKIATRON - [electron] (a dark-trace tube, a catho-

de-ray tube having a special screen which changes colour but does not necessarily luminesce under electron impact) tube *m* cathodique à écran absorbant
SKIATRON DISPLAY - [electron] image *f* par tube skiatron
SKID, to - [gen] ensaboter, glisser
[auto] déraper, riper
[aero] déraper
[railw] (of wheels on the track) enrayer
SKID - [gen] embardée *f*, glissement *m*
[mech] sabot *m*, patte *f*
[mech] (a slideway) coulisse *f*
[naut] (for a boat) semelle *f* de lancement
[aero] (a runner in the landing gear) patin *m*
[constr] poutrelle *f* de rampe
~CHAIN - [auto] chaîne *f* antidérapante
~-DEFIED TYRE - [rubber ind] (US only) pneu *m* antidérapant
~FIN - [aero] (a fixed vertical longitudinal surface designed to reduce skidding) surface *f* stabilisatrice, plan *m* fixe de stabilisation
~TURN - [aero] (turn made with insufficient banking, causing outward lateral movement) virage *m* avec glissade
SKIDDING - [gen] (outward movement in a turn, caused by insufficent banking) ensabotage *m*, dérapage *m*
[aero] (the lateràl sliding in a turn) dérapage *m*
[railw] enrayage *m*
[auto] embardée *f*
SKIFF - [naut] esquif *m*, yole *f*
SKILFUL - [gen] habile, adroit
SKILFULNESS - [gen] habilité *f*
SKILL - [gen] habilité *f*, dextérité *f*
SKILLED - [gen] habile
[gen] (of a worker) spécialiste
SKILLET - [metall] creuset *m*
~CAST-STEEL - [metall] acier *m* fondu au creuset
SKIM, to - [gen] (to remove floating material, e.g. foam, from the surface of liquid) écumer
[gen] (to pass, or brush slightly over) effleurer
[constr] (plastering) gobeter, plâtrer
[the milk] écrémer
SKIM - [gen] écume *f*
~-BASIN - [metall] bassin *m* épurateur
~-COAT, to - [rubber ind] gommer un tissu fractionné
COAT - [rubber ind] couche *f* de gomme sur toile fractionnée
~COULTER - [agric] rasette *f*
~GATE - [metall] (in foundry work) chambre *f* d'épuration
~LATEX - [rubber ind] skim-latex *m*, latex *m* pauvre en caoutchouc
~MILK - [agric] lait *m* écrémé
~MILK POWDER - [food] lait *m* écrémé en poudre
~OFF, to - [ind gen] écrémer
[agric] prélever
~PLOUGHING - [agric] quasi-labour *m*
~RUBBER - [rubber ind] caoutchouc *m* de sérum, skim-rubber *m*
SKIMMER - [impl] écumoire *f*
[agric] (a skim-coulter) rasette *f*
[metall] (in foundry work, a device for preventing slag entering a mould) écumoire *f*, cuiller *f* écumoire, pelle *f* de retenue
SKIMMER - [mech] (in earth-moving machinery) pelle *f* niveleuse
[metall] dame *f* de laitier
[glass man] casse *f*
[mining] (a side hole) mine *f* de maizières
[metall] écrémoir *m*
~BRICK - [metall] brique-barrage *f*
~SCOOP - s. skimmer shovel
~SHOVEL - [mech] pelle *f* niveleuse
SKIMMING - [gen] écumage *m*, écrémage *m*
[ind chem] (topping) écrémage *m*
[metall] décrassage *m*
[oil ind] élimination *f* d'une fraction gazeuse
[metall] écrémage *m*
~AMOUNT - [agric] prélèvement *m* à l'importation
~COAT - [constr] gobetage *m*, barbotine *f*
~GATE - s. skin gate
~PLANT - [oil ind] installation *f* de distillation non poussée
~TOOL - [metall] écumoire *f*
SKIMMINGS - [brew ind] (mixture of beer and yeast taken off the top fermenting material) mélange *m* de bière et levure
[rubber ind] scump *m*, skimmings *mpl*
[ind chem] produits *mpl* d'écumage
SKIN, to - [gen] écorcher, dépouiller
[metall] décroûter, écrémer
SKIN - [gen] (the protective tissue layer) peau *f*, épiderme *m*
[bot] tunique *f*, pellicule *f*
[leather ind] peau *f*, cuir *m*, dépouille *f*
[agric] (a container for liquids) outre *f*
[bot] (of a fruit etc) peau *f*, pelure *f*
[agric] (of the milk) pellicule *f*
[naut] enveloppe *f*, coque *f*
[aero] (the sheet covering the internal structure of part an aircraft) revêtement *m*
[shipbuild] bordé *m* extérieur, chemise *f* (de voile)
[metall] (of a casting) croûte *f* de la fonte
[food] (of a sausage) robe *f* (de saucisson)
~BLEMISH - [agric] defaut *m* d'épiderme
~BOB - [metall] (a slag pocket) cul *m* d'oeuf
~-DEEP - [gen] superficiel, à fleur de peau
~DEPTH - [el] (for a conductor carrying currents at a given frequency, the depth below the surface at which the current density has decreased one neper below the current density at the surface) profondeur *f* de pénétration
~DISEASE - [med] dermatose *f*, maladie *f* cutanée
~DIVER - [naut] plongeur *m* sousmarin
~DRESSING - [leather ind] peausserie *f*
~-DRIED MOULDING - [metall] (in foundry work) moule *m* flambé
~DRYING - [metall] flambage *m*
~EFFECT - [electron] (the concentration of electric current on the outer skin of a conductor) effet *m* pelliculaire, effet *m* kelvin
~END OF THE STAPLE - [text] base *f* de la mèche
~ERUPTION - [med] éruption *f* cutanée
~FRICTION - [aero] (the tangential forces on an aerofoil due to the airstream) frottement *m* superficiel
~FRICTION DRAG - [aero] (drag which is due to the resolved tangential forces on the surface of a body in an airstream) traînée *f* due au frottement superficiel
~FULL OF FRILLS - [leather ind] peau *f* ridée

SKIN GRAFTING - [med] greffe f épidermique
~HARDENING - [metall] trempe f superficielle
~HARDNESS - [metall] dureté f de la croûte, dureté f superficielle
~HOLE - [metall] (on an ingot) cavité f superficielle
~LAYER - [metall] couche f marginale
~-MILLING - [aero] (the milling of a fuselage skin panel) fraisage m des panneaux de revêtement
~-MILLING MACHINE - [mach tool] fraiseuse f pour panneaux de revêtement
~OF CASTING - [metall] croûte f de la fonte
~-PANEL - [aero] panneau m de revêtement
~PASS - [metall] finissage m
~PASS MILL - [metall] laminoir m à dresser, laminoir m d'écrouissage
~RESISTANCE - [naut] (to friction) résistance f due au frottement
[el] résistance f superficielle
~SHRINKAGE - [metall] contraction f de la croûte
~STRAKES - [naut] bordés mpl extérieurs
~-TEST - [med] cuti-réaction f
~WOOL - [text] laine f pelade, laine f avalie
SKINNED GRAIN - [gen & bot] pelé, décortiqué, épluché, dépouillé
SKINNER - [leather ind] peaussier m
[el] (the length of insulated wire between the cable from which it emerges and the point of connection to a solder tag) bout m de fil isolé
SKINNING - [paint] (formation of "skin" or dried surface film on air-drying coating compositions left exposed to the air as in half filled containers) formation f de peau
[rubber ind] usure f de la bande de roulement
[el] enlèvement m de l'isolation, dénudation f
~COAT - [constr] gobetage m
~KNIFE - [rubber ind] (for tyres) couteau m diviseur
~MACHINE - [rubber ind] (machine for stripping off the treads of old tyres) refendeuse f, machine f à déchaper
SKINS - [gen & leather ind] peaux fpl, cuirs mpl
SKINSTRESSED COVERING - [aero] revêtement m résistant
~STRUCTURE - [transp] (of a vehicle a tubular construction) structure f tubulaire
SKINTLED - [constr] (said of a brickwork in which the bricks are laid irregularly) irrégulier (maçonnerie en brique)
SKIP, to - [gen] sauter, sautiller
[gen] (to move forward in leaps) gambader
[mech] (of a gear tooth) sauter
SKIP - [gen] saut m
[impl] (a bucket) skip m, godet m
[constr] (a container used to unload cement and other material into a mixer) bourrique f
[mining] (a guided steel box for hoisting mineral up a shaft) benne f, caisse f guidée, tonne f
[comput] saut m
[text] sautage m
~AREA - [radio] (the area surrounding a transmitter swept out by a complete rotation of a radius vector equal in length to the skip distance) zone f sautée
~BAR - [comput] barre f de saut
~BOX - [transp] caisse f guidée
[text] boîte f sautante
~BOX REPEAT - [text] rapport m de sautage
~CHANGE - [text] changement m sauté
~DISTANCE - [radio] (the minimum distance between the point at which direct wave transmission is attenuated to inaudibility and that at which indirect waves (reflected from the Appleton Layer) become audible) distance f de saut, largeur f de la zone de silence
SKIP DRAFT - [text] rentrage m sauté
~FOR YARN DAMPING - [text] panier m pour le mouillage du fil
~HOIST - [mech] monte-caisses m
~KEYING - [radar] manipulation f de la fréquence d'impulsions
~MAGNET - [comput] aimant m de saut
~PASS - [text] remettage m sauté
~POCKET - [mining] trémie f de chargement de skips
~PRESSURE TOOTH CUTTER - [mech] (of a hobbing machine) fraise-mère f pour dents intercalées
~TOOTH SAW - [tool] scie f sauteuse
~WELDING - [metall] soudage m à pas de pèlerin
SKIPPED DISTANCE - s. skip distance
~-THREAD - [text] fil m sauté
~TIE UP - [text] empoutage m sauté
SKIPPER - [naut] patron m (de bateau)
[aero] (USA only) commandant m, chef m de bord
SKIPPING - [comput] (of printing position) espacement m
~PRINTER - [photo] tireuse f par contact
SKIPS - [paper man] (thin paper used for lining crates, chests etc) papier m pour doublage de caisses
SKIRRET - [bot] chervi m, girolle f
SKIRT, to - [gen] contourner
[gen] (to run along a boundary) côtoyer
[naut] côtoyer, élonger (la côte)
SKIRT - [gen] jupe f
[gen] (a border) bord m
[aero] (the lower part of a parachute canopy) jupe f (de parachute)
[mech] (of piston the cylindrical part of a piston) jupe f (de piston)
[electron] (of a waveguide) collerette f
~BRAID - [text] galon m d'attache
~PLATE - [th eng] jupe f (d'un four)
~SHEET - [oil ind] cuvelage m
~-TYPE PISTON - [mech] plongeur m à marteau
SKIRTING - [gen] bord m, bordure f
[carp] plinthe f, socle m
[constr] (on the plaster wall where it meets the floor) bas m de lambris, plinthe f
[text] (fabric) tissu m pour jupes
~BOARD - [constr] (the board covering the plaster wall where it meets the floor) socle m de lambris
SKIVE, to - [leather ind] (to shave a sheet from a thicker one) doler, drayer
[rubber ind] fendre en feuilles minces, débrider au couteau
[gen] (a precious stone) polir (un diamant etc)
SKIVE - [impl] outil m à polir les diamants
~JOINT - [shoe man] assemblage m en fausse coupe
SKIVED BOOT - [rubber ind] (for tyre repairing) corset m de pneu, manchon-guêtre m
SKIVER - [tool] doloir m, drayeur m, doleur
[bookbind] (a thin slice of skin) parchemin m mince, peau f fendue
SKULL - [anat] crâne m
[metall] (the crust of solid material forming on the bottom of a ladle) fond m de poche
SKUNK - [zool] mouffette f

SKUTTERUDITE - [min] (an unimportant ore of nickel and cobalt) skutterudite *f*
SKY - [gen] ciel *m*
~BACKING - [cin] (backing with sky details) fond *m* à nuages
~CONDITION - [met] (the extent to which the sky is overcast by cloud) nébulosité *f*
~-FILTER - [photo] écran *m* de ciel
~PILOT - [aero] pilote *m* d'avion
~SHADE - [photo] écran *m* parasoleil
~SHINE - [nucl] (scattered gamma radiation escaping from an open container for radio-isotopes) rayonnement *m* diffusée
~TRAIN - [aero] train *m* aérien
~TRUCK - [aero] avion *m* de transport
~WAVE - [radio] (atmospheric wave; radiowave reflected from one of the ionospheric layers in the outer atmosphere) onde *f* à champ électrique horizontal
(ionospheric wave; electromagnetic wave propagated by way of the ionosphere) onde *f* ionosphérique, onde *f* d'espace
~WAVE SIGNAL - [radio] (signal reflected from the ionized region of the upper atmosphere) signal *m* à onde réfléchie
~-WRITING - [aero] (the production of writing against the background of the sky by emitting smoke from an aircraft which manoeuvres so as to leave a trail of the required form) inscription *f* sur le ciel
SKYLARK - [zool] alouette *f*
SKYLIGHT - [constr] (a glazed opening in the roof) lucarne *f* faitière, aéra *m*, gaine *f* de jour
[constr] (a window in the roof) lanterneau *m*, fenêtre *f* à tabatiére
[glass man] (poor-quality plate glass) verre *m* de basse qualité
~MICA - [electron] (component part of the electrode structure of electronic tubes) mica *m* blanc
~PURLIN - [constr] panne *f* de lanterneau
~TURRET - [arch] (of a dome) lanterneau *m*
SKYROCKET - [fireworks] fusée *f* volante
SKYSAIL - [naut] contre-cacatois *m*
SKYSCRAPER - [constr] (very tall, multistoried building) gratte-ciel *m*
SKYWRITING - s. sky-writing
SLAB, to - [gen] trancher
[constr] daller, paver
[timber] couper les dosses du bois
SLAB - [gen] plaque *f*, tranche *f*, dalle *f*
[metall] (intermediate rolled section having a width at least twice its thickness) brame *f*
[carp] (outer piece of a log cut away in the process of slabbing) dosse *f*, dosseau *m*
[mining] écoin *m*
[geol] pan *m* de rocher
[print] marbre *m* (à broyer les couleurs)
[th eng] (sole brick) écoin *m*, dosse *f*
[constr] table *f* (d'ardoise)
[chem] (for pyroxylin) carreau *m* (de fulmicoton)
[el] tableau *m*
~AND GIRDER FLOOR - [constr] (concrete floor in which the secondary beams are embodied with a thin slab for the floor surface) plancher *m* à dalles à nervures
~BEARER - [constr] sommier *m*
~BLOOM - [metall] brame *f*
~-CHARGING MACHINE - [metall] chargeuse-défourneuse *f* de brames
SLAB COIL - [el] (coil shaped like a flat spiral) galette *f*
~CONDITIONING - [metall] traitement *m* superficiel de la brame
~COPPER - [metall] cuivre *m* en brames
~DOWN, to - [mining] abattage *m* du stérile décollé
~EDGING PRESS - [metall] presse *f* à rogner les rives de la brame
~FURNACE - [metall] four *m* de rechauffage pour brames
~HOLE - [mining] trou *m* de mine auxiliaire
~INGOT - [metall] brame *f* brute
~JOIST - s.slab bearer
~LINES - [naut] dégorgeoirs *mpl* de voile
~OF MARBLE - [constr] tranche *f* de marbre
~PILE - [metall] paquet *m* de brames
~REHEATING FURNACE - [metall] four *m* de rechauffage pour brames
~SIDE PRESS - [mech] (side-plate press) presse *f* à paroi latérale
~STOCK - [gen] (material in large flat masses for further processing) matériel *m* en plaques
~OF TIMBER - [constr] flache *f*, dosse *f*
~SHEARS - [impl] cisailles *fpl* à brames
~TURNING DEVICE - [metall] tourne-brame *m*
~YARD - [metall] parc *m* à lingots
SLABBER - [mech] (machine for cutting stock material into thick flat blocks) fraiseuse *f* genre raboteuse
SLABBING -[constr] dallage *m*, tranchage *m* (du marbre)
[metall] tranchage *m*
[mining] décollement *m* de la roche
~MILL - [metall] laminoir *m* à brames
~-MILL STAND - [metall] cage *f* de laminoir à brames
SLACK, to - [gen] lâcher, détendre, donner du mou à
[gen] (to slow down) ralentir
[gen] (of a rope) prendre du lâche
SLACK - [gen] détendu, mou
[gen] (of ropes; not under tension) mou, lâche, étalé
[mech] (backlash) jeu *m*
[min] (coal dirt) fines *fpl* de charbon
[mech] (of a bolt etc) desserré
[auto] (of a tyre) dégonflé
[comm] (of a market) faible
[metall] déchet *m*
~BLOCKS - [constr] (striking wedges) supports *mpl* à coin
~-CABLE SWITCH - [el] interrupteur *m* de sécurité
~CHAIN - [text] chaîne *f* lâche
~LIME, to - [constr] éteindre de la chaux
~MALT - [brew ind] malt *m* trop humide
~MATERIAL - [text] tissu *m* sans force
~MELT COPAL - [paint] (copal which has been run to a small extent only, and usually at a rather low temperature) copal *m* à basse température
~PULLER - [tool] tendeur *m*
~SILK - [text] soie *f* floche
~STRAND - [text] toron *m* lâche
~TIDE - [met] marée *f* étale
~UP, to - [railw] (of a train) ralentir
~WARP - [text] chaîne *f* lâche
~WATER - [naut] mer *f* étale
~-WATER NAVIGATION - [constr] (canal and river

navigation made possible by the construction of a number of dams across the stream) navigation *f* sur eaux réglées par barrages
SLACK WAX - [oil ind] (crude petroleum wax) paraffine *f* brute
~WEFT - [text] trame *f* lâche
~YARN - [text] fils *mpl* floches
SLACKEN, to - [gen] ralentir
[gen] (a rope) prendre du mou
[chem] (of lime) s'éteindre
[naut] (of the tide) mollir
SLACKENING - [gen & mech] ralentissement *m*
[mech] (to become loose) desserrage *m*, amortissement *m*, relâchement *m*
[metall] dédoublage *m* du moule
SLACKLIME - [chem] (hydrated calcium produced by adding water to quicklime) chaux *f* éteinte
SLAG, to - [metall] (forming impurities on the surface of a metal during smelting) scorifier
SLAG - [metall] (impurities rising to the surface of a metal during smelting) crasses *fpl*, laitier *m*, scorie *f*
~BED - [metall] (slag bottom) fond *m* en scorie
~BREAKER - [metall] moulin *m* à scories
~BRICK - [constr] (brick cast from blast furnace slag) brique *f* de laitier, brique *f* à laitier
~CEMENT - [constr] (alternative term for blast furnace cement, i.e. Portland cement made from blast furnace slag) ciment *m* de laitier
~CONCRETE - [constr] béton *m* de laitier
~-FREE IRON - [metall] fer *m* exempt de scorie
~HOLE - [metall] (aperture through which slag is drawn off from a blast furnace) bec *m* du passage du laitier, trou *m* à crasse
~INCLUSION - [metall] (on the surface of a casting) inclusion *f* de laitier
~IRON - [metall] fer *m* riche en scories
~LADLE - [metall] cuve *f* à laitier, poche *f* à laitier
~NOTCH - [metall] trou *m* à crasse, chenal *m* de laitier
~-OFF - [metall] décrassage *m*
~OUT, to - [metall] (to draw off slag) from a flast furnace) délouper
~PLATE - [metall] tôle *f* d'écoulement de la scorie
~POCKET - [metall] chambre *f* à laitier, cul *m* d'oeuf
~PUDDLING - [metall] puddlage *m* gras
~RUNNER - [metall] rigole *f* de laitier
~SAND - [metall] sable *m* de laitier
~SEPARATION - [metall] écartement *m* du laitier
~SPOUT - [metall] chenal *m* de laitier
~TAPPER - [metall] trou *m* à laitier
~-TAPPING - [metall] coulée *f* de laitier
~TRAP - [metall] (an inclusion) canal *m* de coulée à deux niveaux, nid *m* à crasses, piège *f* à crasses, canal *m* de détente
~TUYERE COOLER - [metall] refroidisseur *m* du trou de laitier
~WAGON - [metall] (special vehicle to receive molten slag from a furnace) chariot *m* à laitier
~WOOL - [metall] (mass of fine filaments of blast furnace slag formed by blowing air through the molten material) laine *f* de laitier
SLAGGING - [metall] (the formation of slag) scorification
~MEDIUM - [metall] fondant *m* pour laitier
SLAGGING PRODUCER - [gas ind] gazogène *m* à fusion de cendres
SLAGGY - [metall] scoriacé
SLAKE, to - [chem] (to add water or unslaked lime, calcium oxide, to produce calcium hydroxide) éteindre
[metall] fuser
SLAKE LIME, to - [chem] éteindre de la chaux
SLAKED LIME - s. slacklime
SLAM, to - [gen] claquement *m*
SLAMMING STILE - [carp] (the upright member of a door case against which the door closes) battant *m* (de porte)
SLANT, to - [gen] incliner, être en pente, s'incliner
SLANT - [gen] inclinaison *f*, pente *f*
[gen] (a slanting direction, or plane) pente *f*, dénivellement *m*
[mining] (gallery) galerie *f* inclinée
~DISTANCE - [radar] (the distance from an object to another not at its own elevation) distance *f* réelle
~RANGE - [nucl] (the distance from a specified location to the point at which a nuclear explosion occurs) distance *f* oblique, distance *f* réelle
SLANTED WINDSCREEN - [auto] pare-brise *m* incliné
~WINDSHIELD - s. slanted windscreen
SLANTING - [gen] incliné, en pente, oblique
~ARM - [text] bras *m* oblique
~BEARING - [mech] palier *m* oblique
~COLLISION - [railw] prise *f* en écharpe
~CREEL - [text] râtelier *m* en toit
~FRAME - [text] bati *m* en forme de pupitre
~GRID - [mech] grille *f* placée obliquement
~OF THE HOOKS - [text] inclination *f* des crochets
SLAP, to - [mech] (of piston; a defect) claquer (d'un piston)
SLAP - [gen] coup *m*, tape *f*
[mech] (a defect) claquement *m* (d'un piston)
SLASH, to - [gen] taillader, cingler
SLASHER - [text] (machine designed to size, dry and beam yarn intended for warp) machine *f* à encoller
[text] pareur *m* encolleur
[text] (the cotton waste from the slasher) déchets *mpl* de parage
~SIZING MACHINE - [text] encolleuse *f* à cylindres
SLASHING - [text] encollage *m* au large
SLAT - [aero] (the portion of a slotted aerofoil which is forward of a slot, where the latter is forwardly-located) bec *m* d'aile à fente
[gen] planchette *f*, lamelle *f*
[constr] (of window blind) planchette *f* de jalousie
~CONVEYOR - [mech] (an apron conveyor) transporteur *m* à écailles
~-IRON - [metall] fer *m* en lattes
SLATE - [geol] (a metamorphic rock lying in parallel cleavage planes and used in constructional work and, powdered, as a filler, pigment and abrasive) ardoise *f*
[constr] (a tile) feuille *f* d'ardoise
[cin] (a clap-board) claquette *f*
~AXE - [tool] hache *f* d'ouvrage
~-CLAY - [geol] schiste *m*
~FLOUR - [rubber & plast ind] (finely ground slate used as a filler for rubber and plastics) poudre *f* d'ardoise
~HANGING - [constr] pose *f* verticale des ardoises
~LATH - [constr] volige *f*, latte *f* à ardoise
~QUARRY - [min] carrière *f* d'ardoise

SLATE QUARRYING - [mining] extraction *f* d'ardoise
~SPLITTING - [constr] refente *f* d'ardoise
SLATER - [constr] couvreur *m* en ardoises
SLATERS' HAMMER - [impl] asseau *m*, marteau *m* de couvreur
SLATING - [constr] (the operation of applying slates) couverture *f* en ardoise
[constr] (the material) ardoises *f*pl
SLATTED BASE - [gen] (in upholstery) support *m* à claire-voie
~BOX - [metall] boîte *f* à noyaux en douelles
SLATY - [geol] ardoiseux, schisteux
~CLEAVAGE - [geol] clivage *m* des roches
SLAUGHTER, to - [gen] abattre
SLAUGHTER - [gen] abattage *m* de bêtes de boucherie
~LIVESTOCK MARKET - [comm] marché *m* de bétail de boucherie
~TAPPING - [rubber ind] saignée *f* à mort
SLAUGHTERHOUSE - [gen] abattoir
SLAUGHTERING - [gen & food] abattage *m*
SLAVE - [telev] appareil *m* asservi
~CLOCK - [el] horloge *f* secondaire
~RELAY - [electron] relais *m* auxiliaire
~TRANSMITTER - [radio] (transmitter supporting the radiation of the main transmitter in a limited range) émetteur *m* répéteur
~UNIT - [comput] (component part which does not belong to the central part of the machine) organe *m* auxiliaire
SLAVELOCK - [telev] synchronisation *f* d'émetteurs secondaires
SLAY, to - s. sley
SLED, to - [transp] (USA only) transporter en traîneau
SLED - [transp] (USA only; sleigh in GB) traîneau *m*
SLEDGE - [transp] traîneau
[impl] (a heavy hammer for blacksmith's use) marteau *m* de forgeron, marteau *m* à frapper devant
[impl] (for stone breaking) massette *f*, têtu *m*
~-HAMMER - [mech] (heavy hammer, weighing up to 100 lbs, used with both hands) marteau *m* à deux mains, marteau *m* à frapper devant, frappe-devant *m*
SLEEK, to - [gen] lisser
[metall] (a mould) planer (un moule)
[metall] lisser
SLEEK - [gen] lisse, luisant
[metall] poli
SLEEKER - [impl] outil *m* à lisser
[metall] (tool used in die casting for smoothing the mould) lissoir *m*
SLEEKING - [metall] lissage *m*
~STICK - [shoe man] étire *f*
SLEEKNESS - [gen] luisant *m*
SLEEKSTONE - [metall] lissoir *m*
[tool] (in handicraft work) brunissoir *m*
SLEEKY - s. sleek
SLEEP, to - [gen] (to be asleep) dormir
SLEEP - [gen] (the state of complete or partial unconsciousness) sommeil *m*
~DISORDER - [med] trouble *m* du sommeil
~-WALKER - [med] noctambule
SLEEPER - [carp] (horizontal timber supporting a vertical shore or post) lambourde *f*, gîte *m*
[railw] (the beam passing transversely beneath the rails) traverse *f*
[railw] (a sleeping car) wagon-lit *m*

SLEEPER - [constr] (the horizontal timber supporting the floor joists) poutre *f* horizontale, sole *f*
[constr] (of staircase) patin *m* d'escalier
~ADZING MACHINE - [mech] machine *f* à saboter les traverses
~BEARING GIRDER - [constr] poutre *f* sous traverses
~BED - [railw] assiette *f* des traverses
[railw] (in the ballast) moule *m* de la traverse
~CHAIR - [railw] coussinet *m* pour traverses
~CLIP - [impl] serre-rail *m*, crapaud *m*
~GROOVE - [railw] entaille *f* de la traverse
~OF THE STAIRS - [constr] patin *m* d'escalier
~PAD - [railw] semelle *f* de rail
~PLATE - [constr] (wall plate on a sleeper wall) plaque *f* d'appui
~RAIL - [railw] longeron *m*
~SCREW - [railw] (screw spike) tire-fond *m*
~SPACING - [railw] (distance between sleepers) travelage *m*
~WALL - [constr] (a low wall built under the ground floor of buildings without basement as a support for the floor joists) mur *m* d'appui
SLEEPINESS - [paint] (defect of a coating in reduction of gloss as drying takes place) manque *m* de lustre
[gen & med] somnolence *f*
SLEEPING-BERTH - [railw] couchette *f*
~CAR - [railw] wagon-lit *m*
~SICKNESS - [med] trypanose *f* humaine, trypanosomiase *f* humaine
~[electron] (undesirable effect sometimes occurring in junction transistors) augmentation *f* graduelle des courants inverses
~SICKNESS OF THE CATTLE - [vet] nagana *m*, maladie *f* de la tsé-tsé
~TABLE - [min] table *f* dormante
~TOP - [mech] (a top is said to sleep when it rotates with constant angular speed about its axis in a vertical position) toupie *f* tournante à axe vertical
SLEEPY SICKNESS - [med] (epidemic encephalitis) encéphalite *f* léthargique
SLEET, to - [med] tomber de la neige fondue
SLEET - [met] (precipitation consisting of a mixture of rain and snow, or of partially melted snow) neige *f* à moitié fondue, pluie *f* et neige mêlées
SLEEVE - [mech] (a tubular part, especially enclosing another cylindrical part) manchon *m*, bague *f*, douille *f*, chemise *f*, fourreau *m*
[el] (metal structure of cylindrical form used for supporting the commutator) manchon *m*, gaine *f*
[gen] manche *f*
[aero] manche *f* à vent
[auto] (cylinder liner) chemise *f*
[rubber ind] (for a tyre) guêtre *f*
[oil ind] manchon *m* pour tuyaux
[metall] cuve *f* pour moulage sous pression
~AERIAL - [radio] antenne *f* à tube coaxial
~COLLAR - [mech] bride *f*
~CONNECTION - [mech] accouplement *m* à manchon
~COUPLING - [mech] accouplement *m* à douille, accouplement *m* à manchon
~-DIPOLE AERIAL - [radio] antenne *f* à dipôle à manchon
~-DIPOLE ANTENNA - s. sleeve-dipole aerial
~JOINT - [el] (a conductor joint formed by a sleeve fitting over the conductor end) manchon *m*
[mech] assemblage *m* à manchon, emmanchement *m*

SLEEVE NUT - [mech] manchon *m* taraudé
~PROTECTOR - [rubber ind] protège-manche *m*
~SPLICE - [metall] épissure *f* à manchon d'un câble
~STUB - [electron] (in waveguides) dipôle *m* à tube coaxial à quart d'onde
~TWISTERS - [tool] pinces *fpl* à tordre
~-TYPE BEARING - [mech] palier *m* à glissement
~VALVE - [mech] (a type of valve sometimes used in reciprocating i.c. engines and other machines, consisting of a tubular element pierced with ports which are brought into correspondence with others as it moves) soupape *f* à manchon, soupape *f* à fourreau, chemise-tiroir *f*
SLEEVES - [aero] (USA only protecting covering for aircraft wings) manches *fpl* de protection pour les ailes
SLEEVING - [el] (plastic tubing used to cover bare wires in electrical assemblies) tuyau *m* de recouvrement
SLEIGH - [gen & transp] traîneau *m*
SLENDER - [gen] mince, fusiforme
SLENDERNESS RATIO - [constr] (the ratio between the length or height of a pillar and its least radius of gyration) élancement *m* d'un poteau
SLEW, to - [gen] pivoter, virer
[mech] (to twist) tordre
SLEW - [gen & mech] déviation *f*, tour *m*
SLEWING - [mech] (of a machine, a crane etc) tournant
[telev] panoramique *m* latéral
~MOTOR - [radar] (motor for rapid scanning) moteur *m* pour balayage rapide
SLEY - [text] (the reed guiding the warp threads in a loom) battant *m*, chasse *f*
~BEAM - [text] masse *f* du battant
~CAP - [text] chapeau *m* du battant
~CHEEK - [text] semelle *f* du battant
~COMB - [text] peigne *m* preliminaire
~ECCENTRIC - [text] excentrique *m* du battant
~GROOVE - [text] rainure *f* dans le battant
~MOTION - [text] mouvement *m* du battant
~PIN - [text] tourillon *m* du battant
~PIVOTED BELOW - [text] battant *m* inférieur
~STOPPING LEVER - [text] levier *m* d'arrêt du battant
~SWORD - [text] épée *f* du battant, bras *m* du battant
~SWORD PIN - [text] goupille *f* du battant
~SWUNG FROM TOP - [text] battant *m* libre
~WITH SHUTTLE - [text] battant *m* avec navette
SLEYING - [text] compte *m* de chaîne en peigne
SLICE, to - [gen] trancher, découper en tranches
[mech] (to cut sheet from a block) trancher
SLICE - [gen] tranche *f*
[metall] écaille *f*
[mining] envelure *f*
[paper man] règle *f* d'épaisseur
~A FURNACE GRATE, to - [th eng] décrasser la grille d'un foyer
~BAR - [tool] (used for boilers) ringard *m* de chaufferie
SLICED FILM - [plast ind] (plastic material in sheet form cut from the block and less than 0.25 mm. thick) feuille *f* mince tranchée
~SHEET - [plast ind] (plastic material in sheet form cut from the block and more than 0.25 mm thick) feuille *f* tranchée
~STRUCTURE - [geol] structure *f* finement lamellaire
SLICER - [mech] (used in the food industry etc) machine *f* à trancher
[radio] (amplitude gate) découper *m*, éminceur *m*
SLICING - [metall] (the scraping of slag from the grate) décrassage *m*
[cin] mouvement *m* rotatif du microphone
[gen] tranchage *m*
~AND CAVING - [mining] exploitation *f* avec foudroyage en tranches horizontales
~LATHE - [mach tool] tour *m* à trancher
~MACHINE - s. slicer
SLICK, to - s. sleek, to
SLICK - [metall] schlich *m*
SLICKER - [metall] s. sleeker
[shoe man] s. sleeker
[agric] herse-trainoir *f*
SLICKING - s. sleeking
~-IN ROLLERS - [text] cylindres *mpl* d'alimentation
SLIDE, to - [gen] glisser
[mech] faire glisser, couler, faire couler, coulisser
SLIDE - [gen] glissement *m*
[mech] coulisse *f*, glissière *f*
[mech] (the motion) glissement *m*, coulant *m*
[mining] guide *m*
[instr] curseur *m*
[mach tool] tiroir *m*, coulisseau *m*
[phys] (shear stress) cisaillement *m*
[opt] (framed glass holding the object to be examined under a lens) porte-objet *m*
[cin] (glass plate used for projection) diapositive *f*
[photo] châssis *m* porte-plaques
[mus] (component part of brass instrument) coulisse *f*
[mining] (a vertical crack in a vein) fente *f* verticale
[mining] surface *f* de charriage
[soil] (short for landslide) éboulement *m*
[geol] dépôts *mpl* meubles sur une pente
~ARM - [mech] bras *m* de curseur
~-BACK - [radio] (the depression of the mean potential of the grid of a thermionic valve which occurs when a comparatively large alternating voltage is applied to it and the external grid to cathode path has a high resistance to direct current) glissement *m* du potentiel de grille
~-BACK VOLTMETER - [instr] (an electronic voltmeter, in which the unknown voltage is compared to a calibrated, adjustable voltage-source, the latter being adjusted until equal to the unknown) voltmètre *m* de comparaison
~BAR - [mech] glissière *f* de crosse, guide *m* de la tête de piston
~BASE - [el] base *f* glissante
~BEARING - [mech] coussinet *m* de glissement
~BINDING - [photo] montage *m* de diapositive
~-BLOCK - [mech] (of link-valve motion) coulisseau *m*, tasseau *m* de crosse
~BRACKET - [text] coulant *m*
~BRIDGE - [el] pont *m* à corde, pont *m* de Wheatstone
~CALIPER RULE - s. slide calipers
~CALIPERS - [tool] (measuring tool having one fixed and one sliding jaw, the distance between these being read on a suitable scale, usually with a vernier) pied *m* à coulisse, équerre *f* à coulisse, compas *m* à coulisse
~CHANGER - [photo] châssis *m* va-et-vien

SLIDE CONTACT - [el] frotteur *m*, curseur *m*
~COUPLING - [mech] accouplement *m* à glissement
~DEPTH GAUGE - [instr] calibre *m* à curseur pour profondeur
~DRUM FOR GIANT TYRES - [rubber ind] tambour *m* à charnières coulissantes pour pneus poids lourds
~-FACE - [mech] voie *f* à glissière
~FASTENER - [impl] (for clothing, a zipper) fermeture *f* éclair
~FEED - [mech] avance *f* du patin
~FOLLOWER - [plast ind] coquille *f* à glissière
~FRAME - [photo] cadre *m* pour diapositive
~-IN CHASSIS - [comput] caisson-tiroir *m*
~LATHE - [mach tool] tour *m* à charioter, tour *m* parallèle
~-LIFT DOOR - [auto etc] porte *f* à guillotine
~MASK - [photo] cache *m* pour diapositive
~-PRESERVER CHAIN - [rubber ind] (for tyres; US only; non-skid chain) chaîne *f* antidérapante
~PROJECTOR - [photo] projecteur *m* diascopique
~RACK - [text] coulisse *f*, coulisse *f* de guidage du chariot
~RAIL - [railw] lame *f* d'aiguille
[mech] glissière *f*
~-RAILS - [railw] (two or more rails on which an electric motor can be mounted in such way that its position may be altered by means of screws, or otherwise) chariot *m* transbordeur
~RESISTANCE - [el] (a rheostat the ohmic value of which is adjusted by sliding a contact over the resistance wire) résistance *f* à curseur
~REST - [mach tool] (of a lathe) support *m* à chariot
~-REST TOOLS - [mach tool] outils *mpl* à charioter
~-RULE - [instr] (simple calculator consisting of fixed and sliding scales logarithmically graduated) règle *f* à calcul
~SCANNER - [telev] analyseur *m* de diapositives
~SPLINE - [mech] cannelure *f* de glissement
~TRACK - [mech] (of a bridge crane etc) plan *m* de glissement
~VALVE - [hydr] (in hydraulic equipment, a distributor-valve consisting of a recessed block sliding on a face in which ports are formed) tiroir *m* de distribution
[mech] tiroir *m*, soupape *f* à tiroir
~VALVE HYDRANT - [hydr] borne-fontaine *f* avec robinet-vanne
~WIRE - [el] (a wire of uniform resistance with which a sliding contact makes connection at any desired point) fil *m* à contact glissant
~-WIRE BRIDGE - [el] (a bridge in which all or part of two adjacent arms is formed by a wire along which a sliding contact can be moved, so as to obtain a continuous variation in the ratio of the arms) pont *m* à fil
~-WIRE POTENTIOMETER - [el] (a potentiometer comprising calibrated resistors and calibrated wire) potentiomètre *m* à fil
SLIDER - [mech] coulant *m*, curseur *m*
[mus] (a board with holes in it running under each row of pipes in a windchest) registre *m*
[el] archet *m* (de tramway)
[transp] (in vehicles) sassoire *f* (de l'avant-train)
SLIDEWAY - [mech] guide *m*, coulisse *f* de tiroir
SLIDING - [gen] glissant
[mech] roulant, à glissière
[mech] (the action of sliding) coulissement *m*

SLIDING - [el acoust] (of inductance) variable *f*
[mach tool] chariotage *m* longitudinal
~AXLE - [mech] axe *m* de renvoi
~BAR - [mech] (in a door) verrou *m* de porte
~BEARING - [mech] coussinet *m* lisse, palier *m* à glissement
~BENCH - [mach tool] banc *m* coulissant
~BLOCK - [mech] (of a steam engine) patin *m*
~BOLT - [mech] verrou *m* à coulisse
~BOTTOM - [mech] fond *m* à traîneau
~BOW - [railw] (of the pantograph) archet *m*
~BRACKET - s. sliding copping plate carrier
~BRIDGE - [constr] pont *m* à coulisse
~CAISSON - [hydr] (of a floating body used to open or about the entrance to a dock or basin) caisson *m* mobile
~CAM - [mech] curseur *m*
[text] excentrique *m* glissant
~CARRIAGE MOULD - [plast ind] (a mould mounted on guides to allow of a sliding movement) moule *m* sur glissière
~CLIP - [impl] pince *f* coulissante
~CLUTCH - [mech] embrayage *m* à patin
[mech] embrayage *m* mobil
~COLLAR - [mech] (of a burner) bague *f* coulissante
~CONE - [text] cône *m* glissant (du tambour)
~CONTACT - [el] (contact in which there is relative movement with another surface) contact *m* glissant
[text] contact *m* par frottement
~COPPING PLATE CARRIER - [text] support *m* de la règle se déplacant dans une coulisse
~COUPLING - [mech] accouplement *m* à glissement
~CUT-OFF - [mech] fermeture *f* à guillotine
~DIAPHRAGM - [photo] diaphragme *m* coulissant
~DIE UPSET - [metall] refoulage *m* à estampe glissante
~DOOR - [constr] (a door which moves laterally instead of on hinges) porte *f* à coulisse
(of a furnace) porte-glissière *f*
~ELECTRODE - [electron] (in electronic tubes) électrode *f* ajustable
~FALSE WORK - [metall] couffrages *mpl* glissants
~FIT - [mech] accouplement *m* de glissement
~FRICTION - [mech] frottement *m* de glissement, frottement *m* de la première espèce
~GATE - [metall] grille *f* coulissante, barrière *f* roulante
[hydr] registre *m*
~GAUGE - s. slide calipers
~GEARS - [mech] engrenages *mpl* baladeurs, pignons *mpl* baladeurs
~GLASS - [auto] (of car windows) glace *f* ouvrante
~HEADSTOCK - [mach tool] poupée *f* mobile, poupée *f* courante
~HEALD - [text] lisse *f* mobile
~JAW - [mech] mâchoire *f* mobile
~JOINT - [mech] joint *m* glissant
~KEY - [mech] clavette *f* glissante
~LEG - [mech] (of a tripod) branche *f* à coulisse
~MEMBER - [mech] organe *m* mobile
~-MESH GEAR-BOX - [auto] (a gear box in which the ratio is changed by sliding one pair of wheels out of engagement and sliding another pair in) changement *m* de vitesse par engrenages baladeurs
~MOTION - [phys] mouvement *m* de coulissement
~NUT - [mech] écrou *m* glissant
~OF THE TOP - [text] glissement *m* du toupin

SLIDING-PANEL WEIR - [hydr] (a type of frame weir in which the wooden barrier consists of wooden panels, sliding in grooves between each pair of frames) barrage *m* à panneaux mobiles
~PARALLEL VICE - [mech] étau-tiroir *m*
~PIECE - [transp] traîneau *m*
~PINION - [mech] pignon *m* baladeur
~PLATE - [mech] pièce *f* glissante
~PULLEY - [mech] rouleau *m* glisseur
~PUNCH - [mech] (a punch mounted in a press on rails to permit its withdrawal from the press between moulding operations) poinçon *m* sur glissière
~RING - [mech] bague *f* coulissante
~ROD - [mech] allonge *f*
~ROOF - [auto] toit *m* ouvrant
~RULE - s. slide rule
~SASH - [carp] (sash moving horizontally on runners) châssis *m* à coulisse, châssis *m* à guillotine
~SCALE - [fin] échelle *f* mobile
~SCALE TARIFF - [comm] tarif *m* dégressif
~SCREEN-GRID VOLTAGE - [electron] tension *f* de grille-écran
~SCREEN PENTODE - [electron] pentode *f* à grille-écran à tension glissante
~SCREW - [mech] dispositif d'adaptation
~SEAT - [auto] siège *m* amovible
~SHOE - [mech] (of lifts) frotteur *m*, sabot *m*
~SHUTTER - [mining] (in mine ventilation) guichet *m* mobile
~SHUTTLE - [text] navette *f* sans roulettes
~SLEEVE - [auto] manchon *m* coulissant
~SLEEVE ENGINE - [auto] moteur *m* à chemises coulissantes
~SLEY - [text] battant *m* à glissière
~SPUR WHEEL - [mech] roue *f* droite coulissante
~STAFF - [surv] mire *f* à coulisse
~SURFACE - [mech] plan *m* de glissement
~SURFACING AND SCREW-CUTTING LATHE - [mach tool] (a lathe fitted to form from cylindrical parts, to surface disk-like pieces or to cut threads) tour *m* parallèle à fileter, charioter et surfacer
~SWITCH - [el] curseur *m*
~WAYS - [naut] sablières *f*pl de lancement
~WEIGHT BALANCE - [meas] (type of weighting machine in which a weight slides on a graduated beam) balance *f* à poids mobile, balance *f* curseur
~WINDOW - [constr] fenêtre *f* à châssis à coulisse
SLIGHT - [gen] léger, faible
~CRIMPS - [text] ondulation *f* faible
~DRAFT - [text] étirage *m* léger
~TWIST - [text] faible torsion *f*
~WIND - [met] vent *m* léger
SLIGHTLY CURLED WOOL - [text] laine *f* peu ondulée
~LIGNIFIED FIBRE - [text] fibre *f* légèrement lignifiée
~WAVED WOOL - s. slightly curled wool
SLIM, to - [gen & med] amincir
SLIM HOLE - [mining] forage *m* à faible diamètre
SLIME - [gen] (soft, adhesive mud or earth) limon *m*, vase *f*, boue *f*
[min] (the drain from a washing plant) limon *m*, schlamm *m*
[min] (liquid bitumen) bitume *m* liquide
[el chem] (fine insoluble material formed in electrolysis or in electro-deposition) boue *f* de l'anode
~FLUX - [bot] (exudation of a watery solution of sugars and other substances from trees which have been attacked by parasites) exsudation *f*
~SLIME PIT - [ind chem] bassin *m* à schlammes
~PULP - [ind chem] boue *f* à schlammes
~SEPARATOR - [ind chem] séparateur *m* de boues
SLIMES - [min] (material of less than 200 mesh, which settles very slowly and must be leached by agitation) boues *f*pl
SLIMING - [min] slimage *m*
SLIMY - [gen] vaseux, limoneux
[chem] visqueux
~FLAX TOW - [text] étoupe *f* de lin agglutinée
~SPINNING SOLUTION - [text] solution *f* à filer visqueuse
SLING, to - [gen] (to place in a sling for lifting) élinguer, brider
[gen] jeter, lancer
SLING - [gen] (a length of rope spiced to form a closed loop, used for handling loads) élingue *f*, braye *f*
[ceram] (a wire used for cutting clay) fil *m* à couper
[med] écharpe *f*
[el] étrier *m*, arcade *f* de cabine
[naut] suspente *f*
[vet] travail *m* (pour chevaux)
~CASE - [gen] sac *m* à courroie, étui-bandoulière *m*
~CHAIN - [mech] chaîne *f* de suspension
~DOG - [gen] griffe *f* d'élingue, patte *f*
~HOOK - [mech] (in a hoisting device) crochet *m* de levage
~GROOVE - [mech] (for lubrication) rainure *f* pour lubrification
~RING - [aero] (a disk fixed on a propeller shaft so as to prevent deicing fluid etc from creeping along it) bague *f* de projection d'huile
SLINGER - [metall] projecteur *m*
SLINGING WIRE - [el] (wire used to support one or more cathodes and to supply current) fil *m* de suspension
SLIP, to - [gen] glisser
[mech] glisser, patiner
[soil] s'ébouler
[naut] (the cable of an anchor) choquer
[aero] (to sideslip) glisser sur l'aile
[el] se décaler
SLIP - [gen] (of a pillow etc) taie *f* d'oreiller
[text] combinaison *f*
[mech] (of a belt etc) patinage *m*, glissement *m*
[shipbuild] (a sloping surface for the support of a vessel being built or repaired) cale *f*, chantier *m*
[geol] (the result of rock movements) éboulement *m*, affaissement *m*
[mining] (in the roof) filière *f* (dans le toit)
[geol] (in a fault) rejet *m*
[metall] (the process involved in the plastic deformation of metal crystals) glissement *m*
[aero] (sideslip) recul *m*
[ceram] (clay mixed with water to the consistency of cream) pâte *f*
[paint] (condition of a coating in which surface tackiness is virtually absent and the film seems as if lubricated) poli *m*
[plast ind] (the resistance of a plastics film to sliding movement over another film or over a metal surface) résistance *f* au glissement
[naut & aero] (the ratio of the actual speed to that obtained by a propeller acting on a solid) recul *m*

SLIP - [aero] (bank, as for turning) dérapage *m*
[hydr] cale *f* de chargement
[el] (the fraction by which the rotor speed of an induction motor is less than the speed of rotation of the stator field) glissement *m*
[constr] affaissement *m*
[min] grain *m*, fil *m*
[telecomm] (the continuous narrow strip of paper which is perforated with holes representing telegraphic signals) bande *f* perforée
[print] épreuve *f* en placard, placard *m*
[theathre] coulisse *f*
[railw] rame *f*
[mech] (in a pump) déperdition *f*, perte *f*
~ADDITIVE - [ind chem] (agent which acts as internal lubricant by blooming to the surface during and immediately after processing) additif *m* de lubrication
~BANDS - [metall] (the step formed on the polished surface of metal crystals as a result of the parts moving during slip) bandes *fpl* de glissement
~BOX - [mining] boîte *f* de dilatation
~BUSHING - [mech] manchon *m* à glissement
~CASTING - [metall] moulage *m* en pâte
~CLAMP - [impl] pince *f* à glissement
~CLEAVAGE - [geol] faux clivage *m*
~CLUTCH - [mech] embrayage *m* par glissement
~COUPLING - [mech] embrayage *m* par friction
~COVER - [gen] housse *f*
~CRACK - [metall] crique *f* de clivage, crique *f* de glissement
~DOCK - [naut] cale *f* de lancement
~FAULT - [geol] faille *f* d'effondrement, faille *f* normale
~FLASK - [metall] moule *m* coulissant
~FLOW - [phys] écoulement *m* glissant
~FUEL TANK - [aero] réservoir *m* largable
~FUNCTION - [aero] (the relation between the speed of forward motion of a propeller through undisturbed air to the product of the propeller diameter and the number of revolutions made in unit time) coefficient *m* de recul
~GAUGE - [meas] (a flat piece of steel, hardened and accurately ground to a specified thickness, used in precision measurements) cale-étalon *f*
~HANDLE - [metall] manche *m* du çoin
~HOOK - [mech] croc *m* à echappement
~JACKET - [metall] jaquette *f*, corbeille *f* de coulée
~JOINT - [mech] joint *m* glissant, joint *m* de dilatation
~JOINT PLIERS - [tool] pinces *fpl* à joint glissant
~KNOT - [naut] noeud *m* coulant
~LINE - [metall] ligne *f* de mouvement
~METER - [el] (a device for measuring the slip of an induction motor) indicateur *m* de glissement d'un moteur à induction
~MOUNT - [photo] carton *m* passe-partout
~OFF, to - [gen & mech] se détacher, tomber
~ON, to - [photo] coiffer de
~-ON FILTER - [photo] filtre *m* à emboîtement
~[telev] filtre *m* interchangeable
~-ON FLANGE - [mech] bride *f* glissante
~-ON SCREEN - [photo] écran *m* se montant à friction sur le barillet
~OVER CURRENT TRANSFORMER - [el] (a magnetic circuit carrying a winding and which can be threaded on an insulated cable to form a current) transformateur *m* de câble
SLIP PAGE - [print] page *f* spécimen
~PLANE - [crystal] (an atomic plane of a crystal along which slip may be supposed to have taken place in order to create an edge dislocation; the latter moves freely along its slip plane) plan *m* de glissement
~PROOF - [print] (proof taken from a gallery of type matter before it is made up into pages) épreuve *f* en placard
~REGULATOR - [el] (regulating resistance connected in series with the rotor of a slip-ring type induction motor) rhéostat *m* de glissement
~RING - [el] (or collector ring; a conducting ring rotating with a winding and connected to it; serving to make connection with an external circuit by a brush or brushes) bague *f* collectrice
~-RING LOSSES - [el] pertes *f* des bagues
~-RING MOTOR - [el] moteur *m* à bagues
~-RING ROTOR - [el] (rotor of an induction motor fitted with slip-rings connected to the winding) rotor *m* à bagues
~-RING SPIDER - [el] (a structure provided with arms and used for supporting slip-rings on a shaft) lanterne *f* pour bagues
~ROLLER - [text] cylindre-tracteur *m*
~-ROPE - [naut] amarre *f* passée en double
~-SHEET, to - [print] intercaler les macules
~-SHEET - [print] (a sheet inserted between the printed pages to avoid setoffs in offset printing) papier *m* macule
~-SHEETING - [print] intercalage *m* de macules
~SILL - [constr] (sill having a length equal to the distance between the jambs of the opening) appui *m* de fenêtre encastré dans le poteau
~SOCKET - [oil ind] souricière *f*
~SPEED - [mech] (the speed at which a supercharger will maintain the required pressure difference between intake and discharge at zero delivery) vitesse *f* critique de manutention de pression
~STONE - [carp] (small shaped piece of oilstone for putting an edge to gouges) pierre *f* à aiguiser les gouges
~-STREAM - s. slipstream
~-SWITCH - [railw] traverse-jonction *f*
~TANK - [aero] (a tank designed to be jettisoned when empty) réservoir *m* largable
~THE CABLE, to - [naut] filer la chaîne
~-TONGUE - [carp] languette *f* rapportée
~-TYPE CORE CATCHER - [oil ind] arrache-carottes *m* à coins
~WASHER - [mech] anneau *m* glissant
~-WAY - [el] cale *f* d'un bac
[shipbuild] slipway *m*, couettes *fpl*, cale *f* de lancement
SLIPE WOOL - [text] (wool which is removed from the skin by lime) laine *f* morte
SLIPKNOT - s. slip knot
SLIPPAGE - [mech] (of brakes etc) glissement *m*, patinage *m*
[mech] (the loss in power transmission) perte *f* par glissement
[cin] (a film defect) décalage *m* (de l'image)
[el] (e.g. of frequency) décalage *m* (de fréquence etc)
SLIPPER - [gen] pantoufle *f*
[mech] patin *m*, savate *f*

SLIPPER - [railw] (electromechanical brake acting directly on the rails of a tramway) frein *m* de voie
[metall] main-courante *m*
~-BLOCK - [mech] coulisseau *m*, crosse *f*, tête *f* de piston
~BRAKE - s. slipper (railw)
~PISTON - [mech] (a light piston with its lower part or skirt cut away between the thrust faces, to save weight and reduce friction) piston *m* à patin
~-TYPE CONNECTING-ROD ASSEMBLY - [mech] (assembly in which the individual rods are provided with bearing's external to the big-end bearing and located by flanges or grooves) embiellage *m* à glissière
SLIPPERY - [gen] glissant
~ROAD - [gen] route *f* glissante, route *f* grasse
SLIPPING - [mech] glissement *m*, patinage *m*
[mech] éraillement *m*
[vet] avortement *m*
[railw] décrochement *m*
~MAINSPRING DEVICE - [horol] dispositif *m* glissant du ressort moteur
~OF A VEHICLE - [railw] décrochage *m* en marche
~OF THE BELT - [mech] glissement *m* de la courroie
~OF THE CLUTCH - [auto] patinage *m* de l'embrayage
~OF THE RAIL - [railw] glissement *m* du rail
~OF THE THREADS - [text] glissement *m* des fils
~OF THE WARP - [text] glissement *m* de la chaîne
~SPRING - [horol] ressort *m* glissant
~WHEEL - [mech] accouplement *m* élastique
SLIPS - [oil ind] coins *mpl* grippeurs
[mining] surface *f* de glissement
SLIPSOLE - [shoe man] demi-semelle *f*
SLIPSTREAM - [aero] (the stream of air flowing aft from a propeller) sillage *m*
SLIPWAY - [aero] voie *f* de départ, piste *f*
[shipbuild] (a slip) chantier *m* de construction
SLIT, to - [gen] (to cut lengthwise) fendre
[metall] découper en bandes
SLIT - [gen] fente *f*
[mech] (straight cut) rainure *f*, entaille *f*
[cin] (the section of the projector system which is projected on the screen) fente *f*
[opt] (the long narrow opening through which radiation enters or leaves optical instruments) fente *f*
[mining] recoupe *m*
~AND TONGUE - [constr] embrèvement *m*
~AND TONGUE JOINT - [constr] embrèvement *m*
~CUP - [text] godet *m* à fente
~EDGES - [metall] bords *mpl* coupés par molettes
~GATE - [metall] attaque *f* à fente
~IMAGE - [cin] (the projected image of the slit) image *f* de la fente
~LEVER - [text] levier *m* à rainure
~MORTISE - [mech] enfourchement *m*
~-NOSED BORING BAR - [tool] barre *f* d'alésage fendue
~ORIFICE - [plast ind & rubber ind] (of an extruder) filière *f* en forme de fente
~PLATE - [text] guide-fil *m* à fente
~SHEARING - [metall] refendage *m*, coupe *f* longitudinale
~SOURCE - [radiat] (radiation source, in which the emission is slit-shaped) source *f* à fente
~SPEED - [photo] vitesse *f* de la fente
~WASHER - [mech] rondelle *f* fendue
SLITLESS SPECTROSCOPE - [opt] (objective prism) spectroscope *m* prismatique
SLITTER - s. slitting machine
[tool] fendoir *m*
[impl] pic *m*
[metall] cisailles *fpl* à bandes
~STUD - [mech] goujon *m* à collier
SLITTING - [gen mech etc] fendage *m*
~CUTTER - [tool] (a cutter for sheet metal) couteau *m* à découper en feuillards
~FILE - [tool] lime *f* à couteau, lime *f* à fendre
~GANG CUTTERS - [metall] cisailles *fpl* circulaires multiples
~MACHINE - [metall] (for sheet metal) machine *f* à découper les bandes de tôle
[cin] (machine designed to cut films in narrow pieces) découpeuse *f*
[paper man] (rotary machine cutting the web of paper lengthwise) guillotine *f* longitudinale
~MILL - [metall] fenderie *f*, machine *f* à refendre
~SAW - [mech] scie *f* circulaire à métaux
~SHEARS - [mach tool] cisailles *fpl* à bandes
~WHEEL - [metall] disque *m* à couper, disque *m* de coupe
SLIVER, to - [gen] couper en tranches
[carp] (of wood) éclater
[text] établir les rubans
SLIVER - [gen] (slender piece of wood, metal etc cut or turn off lengthwise) tranche *f*, éclat *m*
[text] (the continuous strands of fibres formed after carding) ruban *m*, bobine *f* de préparation
[metall] (in forging, a thin film of metal roughly attached to the surface of steel) écharde *f*
~BOBBIN - [text] bobine *f* de ruban
~CALENDER - [text] rouleaux *mpl* d'appel
~CAN - [text] pot *m* à ruban
~COUNT - [text] numéro *m* de ruban
~DELIVERY - [text] livraison *f* de ruban
~FEED MOTION - [text] dispositif *m* d'alimentation du ruban
~FORMING MACHINE - [text] machine *f* pour former une mèche continue
~FUNNEL - [text] entonnoir *m*, couloir *m* conique
~GUIDE - [text] guide-ruban *m*
~LAP MACHINE - [text] réunisseuse *f*, doubleuse *f*
~RESERVE - [text] approvisionnement *m* de rubans
~SCREEN - [paper man] (perforated steel plate mounted in a stationary, rotating or oscillating frame, to remove coarse dirt from pulp) crible *m* en acier
~STOP MOTION - [text] casse-ruban *m*
~TABLE - [text] table *f* à ruban
~-TO-YARN SPINNING - [text] filature *f* directe du ruban
SLOAM - [geol] couche *f* d'argile
SLOCOMB DRILL - [tool] (a special type of drill producing a blind hole combined with a countersink, used for centering lathe work) foret *m* de Slocomb
SLOE - [bot] prunelle *f*
SLOOP - [naut] sloop *m*
SLOP OIL - [oil ind] résidu *m* huileux
~-PADDING - [text] traitement *m* au foulard
SLOPE, to - [gen] pencher, incliner, être en pente
[gen] (to give a slanting direction) incliner, couper en pente
SLOPE - [gen] pente *f*
[railw & road] (of an embankment) rampe *f*, côte *f*
[constr] (a brick bevelled off on one side) brique *f*

biaise
SLOPE - [constr] (the angle of inclination) talus *m*, berge *f*, déclivité
[surv] pente *f*
[el radio etc] (mutual conductance) pente *f*
[radar] (the projection of a flight path in the vertical plane) inclinaison *f*
[mining] talus *m*
[hydr] (the sloping side of a hydraulic construction) talus *m*
~ANGLE - [radar] (angle of dip) angle *m* de pente
~CARRIAGE - [mining] truc *m* porteur
~CONDUCTANCE - [radio] pente *f*
~DEVIATION - [radar] (the difference between the projection in the vertical plane of the path of a vehicle and the planned slope of the vehicle) dérivation *f* de pente
~FROM CENTRE TO SIDE - [constr] (of a road) profil *m* transversal bombé
~LEVEL - [surv] (a type of clinometer designed to determine the slope of embankments) niveau *m* de pente
~OF A ROOF - [constr] pan *m* de toit, pan *m* coupé
~OF THE BOTTOM - [hydr] pente *f* du fond
~OF THE GROUND WATER - [hydr] pente *f* de la nappe
~RESISTANCE - [electron] (the resistive component of the electrode impedance when measured at a sufficient low frequency) composante *f* résistive de l'impédance d'électrode
~SHAFT - [mining] fourneau *m* de mine
~STEEPENING - [electron] raidissement *m*
SLOPING - [gen] incliné, en pente, en talus
[gen] (of a ditch) talutage *m* (d'un fossé)
~APPROACH - [constr] rampe *f* d'accès
~BARREL VAULT - [constr] berceau *m* rampant
~BOW - [naut] avant *m* devoyé
~FUNNEL - [naut] cheminée *f* dévoyée
~-HEARTH FURNACE - [gas ind] four *m* à sole roulante
~ROOFS WITH STEEL TRUSS - [constr] toitures *f*pl inclinées avec fermes en acier
~WALL - [constr] mur *m* rampant, mur *m* taluté
SLOSHING - [gen] ballottement *m* d'un liquide dans un réservoir
SLOT, to - [gen] tailler une fente, tailler une rainure
[mech] entailler, encocher, mortaiser, fendre
[mining] sous-caver
SLOT - [gen] fente *f*, entaille *f*, encoche *f*
[mech] (on a machine tool table) rainure *f*, entaille *f*
[mech] (of a screw) fente *f* (d'une tête de vis)
[mech] (for a sunk key etc) trou *m*, encoche *f*
[metall] (cut-out in a sheet metal) encochement *m*
[aero] (an air-passage formed in an aerofoil to improve flow conditions) fente *f*
[el] (a groove in the core: of a motor or generator armature in which the windings are placed) rainure *f*
[text] (of a yarn beam regulator) coulisse *f*
~AERIAL - [radio] (a radiating element consisting of a slot formed in the skin of aircraft) antenne *f* à fente
~ANTENNA - s.slot aerial
~ARRAY - [radio] (part of an aerial formed of an array of slot aerials) antenne *f* à fentes rayonnantes
~-CAST FILM - [plast ind] (flat film extruded as such, as distinct from tubular extrusion film) film *m* extrudé à plat
SLOT CUTTER - [mech tool] rainureuse *f*
~DIE - [plast ind] filière *f* plate, filière *f* droite
~-DIE EXTRUSION - [plast ind] extrusion *f* à filière plate, extrusion *f* de feuilles
~DISCHARGE - [el] décharge *f* dans l'encoche
~FOR KNIFE - [text] coulisse *f* du couteau
~FURNACE - [metall] (a forging furnace) four *m* de forge à fente
~IN THE BRACKET - [text] rainure *f* dans le coulant
~IN THE CLAMP - [text] fente *f* dans la mâchoire
~IN THE HOOK - [text] fente *f* du crochet
~LEAKAGE - [el] pertes *f*pl aux encoches
~MACHINE - [gen] distributeur *m* automatique
~METER - [gen] (common term denoting the coin in-the slot type of prepayment meter for gas and electricity) compteur *m* à paiement prealable, compteur *m* à jetons
~-MILING - [mach tool] fraisage *m* des rainures
~PERMEANCE - [el] (the total permeance of the several parallel portions of the slot-leakage flux path) perméance *f* de l'encoche
~PITCH - [el] pas *m* des rainures
~RADIATOR - [radio] (slot in the wall of a waveguide to act as primary radiator) fente *f* rayonnante
~SEPARATOR - [el] séparateur *m* pour encoches
~SPRAYER - [impl] pulvérisateur *m* à fente
~SUCTION - [mech] (boundary layer suction produced by means of a slot) zone *f* d'aspiration produite par les fentes, aspiration *f* à fente
~SYSTEM - [el] rail *m* de contact en caniveau, système *m* de tubes isolants
~WEDGE - [el] (a wedge of wood or other material which holds the winding in the slot of a slotted core) cale *f* d'encoche
SLOTH BEAR - [zool] ours *m* jongleur
SLOTTED - [gen & mech] rainé, à fente, à encoche
~AEROFOIL - [aero] (an aerofoil provided with slots) surface *f* portante à fente
~AILERON - [aero] (aileron similar to Frise type, but having air passage between its leading edge and the trailing edge of the wing) aile *f* à fente
~ARM - [mech] bras *m* à coulisse
~ARMATURE - [el] (an armature with the winding placed in slots) induit *m* à encoches
~BAR - [mech] tringle *f* à coulisse
~BEARING - [mech] support *m* à rainures
~BOBBIN HOLDER - [text] support *m* de bobines à glissière
~BOBBIN RAIL - [text] rail *m* de bobines avec rainures
~BURNER - [th eng] brûleur *m* à fentes
~CORE - [el] (core of a machine having slots for the reception of the winding) noyau *m* à encoches, noyau *m* à rainures
~CRANK - [mech] manivelle *f* à glissière
~CYLINDER AERIAL - [radio] antenne *f* à cylindre à fentes
~CYLINDER ANTENNA - s. slotted cylinder aerial
~DRAW ROD - [mech] bielle *f* à coulisse
~DROPPER - [text] lamelle *f* à coulisse
~FLAP - [aero] (a flap so arranged that when it moves a slot is opened between it and the aerofoil to which it is attached) volet *m* à fente
~FLEXIBLE BEND - [text] cintre *m* flexible troué
~-HEAD SCREW - [mech] vis *f* à tête fendue

SLOTTED LEVER - [mech] levier *m* à glissière
~MEASURING SECTION - [radio] (a length of waveguide, in the wall of which there is a nonradiating slot for measuring purposes) ligne *f* fendue de mesure, banc *m* de mesure
~NUT - [mech] écrou *m* à entailles, écrou *m* à dents
~RIVET - [mech] rivet *m* bifurqué
~ROUND-HEAD BOLT - [mech] boulon *m* à tête ronde fendue
~SHUTTER - [cin] (shutter which consists of a curtain in which a slit-shaped opening is made) obturateur *m* à rideau
~TABLE - [mach tool] table *f* à rainures de montage
~TUBE - [metall] tube *m* fissuré
~WING - [aero] aile *f* à fente
SLOTTER - s. slotting machine
SLOTTING - [mech] (the operation of keyway cutting etc) mortaisage *m*, encochement *m*, entaillage *m*
[mech] (of a tool) mortaiseur
~ATTACHMENT - s. slotting tool
~CUTTER - [tool] fraise *f* d'entrée
~MACHINE - [mach tool] mortaiseuse *f*, machine *f* à mortaiser
~TOOL - [tool] (a cutting tool used for keyway cutting etc) outil *m* à mortaiser
SLOUGH, to - [metall] décroûter
SLOUGH - [gen] bourbier *m*, fange *f*
[zool] (the cast-off outer skin of a snake) dépouille *f*
[med] (mass of dead tissue in an infected area) eschare *f*
~OFF, to - [text] s'ébouler
SLOW, to - [gen] ralentir, diminuer de vitesse
[auto] prendre le ralenti
SLOW - [gen] lent
[gen] en retard
[naut] doucement
[photo] (of a film) lent
~-ACCESS STORAGE - [comput] mémoire *f* à long temps d'accés
~-ACTING - [gen] à action lente
[mech] (e.g. a spring) paresseux
~ACTING RELAY - [el] relais *m* à action différée, relais *m* temporisé
~-BREAK SWITCH - [el] (a switch in which the speed of breaking is dependent upon the speed of action of the operator) interrupteur *m* à rupture lente
~BURNING - [gen] à combustion lente
[photo] ininflammable (film)
~COOLING - [metall] refroidissement *m* lente
~CURING - [rubber ind] vulcanisation *f* lente
~DOWN, to - [gen & mech] (to reduce speed) ralentir
~FEED - [mech] alimentation *f* lente
~FREQUENCY DRIFT - [telev & radar] (frequency change taking place gradually) glissement *m* graduel de la fréquence
~-HARDENING CEMENT - s. slow-setting cement
~HEATING - [metall etc] chauffage *m* lent
~IGNITION - [mech] allumage *m* progressif
~MATCH - [explos] mèche *f* à lente combustion
~MEMORY - [comput] mémoire *f* différée
~-MOTION ADJUSTING SCREW - [mach tool] vis *f* de rapper pour le mouvement lent
~-MOTION DEVICE - [mech] dispositif *m* de démultiplication
~-MOTION EFFECT - [cin] (projection showing a slow motion produced by accelerating the camera mechanism) effet *m* d'accélération de la camera
SLOW-MOTION PICTURE - [cin] film *m* tourné au ralenti
~-MOTION PROJECTION - [cin] ralentissement *m* de la projection
~-MOTION SCREW - [mech] vis *m* à mouvement lent de rotation
~-MOTION SHOT - [cin] prise *f* de vues au ralenti
~NEUTRON - [nucl] (a neutron having a kinetic energy of less than about I02 ev) neutron *m* lent
~NEUTRON CAPTURE - [nucl] capture *f* de neutrons lents
~NEUTRON FISSION - [nucl] (thermal neutron fission; nuclear fission produced by thermal neutrons) fission *f* par neutrons lents
~OPERATING DELAY - s.slow acting relay
~REACTOR - [nucl] (or thermal reactor; a nuclear reactor in which fission is induced primarily by slow neutrons) réacteur *m* à neutrons thermiques
~REGISTRATION - [meas] (in meters) retard *m* (d'un compteur)
~RELEASE RELAY - [el] (relay which has an intentional delay between deenergizing and release) relais *m* à relâchement différé, relais *m* à rupture retardée, relais *m* à fermeture différée
~RUNNING - [mech] (of a motor) à faible vitesse
~RUNNING JET - [mech] (carburettor jet designed to operate when the engine is running slowly) gicleur *m* de ralenti
~-RUNNING LOOM - [text] métier *m* à petite vitesse
~-SETTING CEMENT - [constr] ciment *m* à prise lente
~SPEED LOOM - s. slow-running loom
~STORAGE - s. slow memory
~STORE - s. slow memory
~-WAVE CIRCUIT - [radio] circuit *m* à ondes lentes
SLOWING DOWN - [gen etc] ralentissement *m*
~-DOWN AREA - [nucl] aire *f* de ralentissement
~-DOWN DENSITY - [nucl] densité *f* de ralentissement
~-DOWN KERNEL - [nucl] noyau *m* de l'intégrale de ralentissement
~-DOWN LENGTH - [nucl] (the square root of the slowing-down area) longueur *f* de ralentissement
~-DOWN POWER - [nucl] (the average loss in natural logarithm of energy of a neutron per unit distance travelled by the neutron in the substance) pouvoir *m* de ralentissement
~-DOWN STRENGHT - s. slowing-down power
~-DOWN TIME - [nucl] temps *m* de ralentissement
~SWITCH - [el] interrupteur *m* retardateur
SLOWLY - [gen] lentement
SLUB, to - [text] (of the yarn) boudiner
SLUB - [text] (a fault in cotton yarn appearing as a thicker part, with little twist) boudin *m*, mèche *f*, bouton *m* floche
~CATCHER - [text] (of a loom) épurateur *m* des filés
~EFFECT - [text] effet *m* flammé
~FLY YARN - [text] retors *m* boutonné
~IN THE YARN - [text] grosseur *f* dans le fil
~YARN - [text] fil *m* flammé, filé *m* boutonné
SLUBBER - [text] (slubbing machine) banc *m* à broches en gros
SLUBBING - [text] (the operation of twisting lightly) boudinage *m*
[text] (sliver fed from the draw frame to the slubbing machine, and there attenuated and wound on a

bobbin,with very slight twist) mèche *f* (de banc à broches)
SLUBBING BILLY - s. slubber
~BOX - s. slubber
~FRAME - s. slubber
~HEAD - s. slubber
~MACHINE - s. slubber
~MOTION - [text] compteur *m* d'alimentation
~ROLLER - [text] cylindre *m* dérouleur
~WASTE - [text] déchets *mpl* de préparation
SLUDGE - [gen] (general term for thick, mudlike mixture of finely-divided solids with water) boue *f*, vase *f*
[mech] (the thick mudlike mixture of lubricating oils etc) cambouis *m*
[el metall] boue *f*
[naut] glaçons *mpl* à moitié pris
[hydr] (of draining) boues *fpl*, eaux *fpl* usées
[min] slimes *fpl*, schlamms *mpl*
[mining] débris *m*
~ACTIVATION - [chem] activation *f* des boues
~BANK - [hydr] banc *m* de boues
~BARGE - [hydr] bateau-citerne *m* à boues
~-BLANKET FLOCCULATION - [hydr] floculation *f* à voile de boues
~BULKING - [hydr] gonflement *m* des boues
~CAKE - [hydr] gâteau *m* de boues
~CHANNEL - [hydr] rigole *f* à boues
~COLLECTOR - [hydr] collecteur *m* de boues
~CONDITIONING - [hydr] conditionnement *m* des boues
~DEWATERING - [hydr] séchage *m* des boues, déshydratation *f* des boues
~DIGESTION - [hydr] digestion *f* des boues
~DRYING - [hydr] dessication *f* des boues
~DRYING BED - [hydr] lit *m* de séchage des boues
~-HOLDING TANK - [hydr] bassin *m* à boues
~LAGOON - [hydr] étang *m* à boues
~PIPE - [hydr] tuyau *m* d'évacuation des boues
~PRESSING - [hydr] séchage *m* par pression
~REMOVAL - [brew ind] débourbage *m*
~RETURN - [hydr] recyclage *m* des boues
~RIPENING - [hydr] digestion *f* finale des boues
~RISING - [hydr] remontée *f* des boues
~SCRAPER - [impl] racleur *m* de boues
~SPEEDING - [chem] inoculation *f* à l'aide de boues, ensemencement *m* à l'aide de boues
~STIRRER - [hydr] agitateur *m* de boues
~STRIPPING-MACHINE - [hydr] machine *f* à enlever les couches de boues
~SUMP - [hydr] puisard *m* à boues
~THICKENING - [hydr] épaississement *m* des boues
SLUDGER - [mining] cloche *f* de curage, pompe *f* à sable, pompe *f* à boue
SLUDGY - [gen] bourbeux, fangeux, vaseux
SLUG, to - [firearms] charger (un fusil)
[firearms] (of a projectile; to alter the shape to pass through the bore) épouser les rayures (du canon)
SLUG - [zool] limace *f*
[fireams] lignot *m*
[metall] (any small chunk of metal) pièce *f* de métal
[metall] (in sheet metal working) barbe *f*, ébarbure *f*
[min] pépite *f*
[gen] (the chunk of metal used in slot machines) piécette *f*
SLUG - [meas] (unit of mass; the mass of a body which, when acted upon by a force of one pound, acquires an acceleration of one foot per second; also called geepound) slug *m*
[phys] (steam void) bulle *f* de vapeur
[min] trommel *m*
[metall] masselotte *f*
[nucl] (a piece of fissile material designed for insertion into a channel in a reactor lattice) barreau *m* de combustible
[print] (solid line of type as cast by the linotype process) lingot *m*
[print] (the strip of metal used for spacing) ligne-bloc *f*
[shoe man] (a type of nail) clou *m*
[firearms] (the lead core of a bullet) plomb *m* cylindre-conique
[metall] (in diecasting) galette *f*, pastille *f*
[electron] (in waveguide, a metallic or dielectric hollow cylinder inserted in a waveguide and forming part of a transforming section) goupille *f* d'adaptation
~A HEEL, to - [shoe man] bonbouter
~CASTING MACHINE - [print] fondeuse *f* de lignes
~COMPOSITION - [print] composition *f* ligne-blocs
~CUTTER - [print] coupoir *m* de lignes
~JUSTIFICATION - [print] justification *f* des lignes
~PUSHER - [print] pousse-ligne *m*
~TUNING - [radio] (the introduction of a slug of material into the electric and/or magnetic field to vary the frequency of a resonant circuit) accord *m* par noyau plongeur
~UP-ENDING TEST - [metall] essai *m* d'aplatissement
SLUGGER - [min] bandage *m* de cylindre broyeur
SLUGGING - [metall] (a type of pressworking of sheet metal) poinçonnage *m* sans détachement de la pièce
SLUGGISH - [gen] paresseux, léthargique
[mech] (slow to respond to a treatment etc) lent, mou, gommeux
[mech] (viscous) visqueux
[of a river] lent, paresseux
[metall] réfractaire, difficilement fusible
~FERMENTATION - [brew ind] fermentation *f* lente
~RING - [mech] bague *f* gommeuse
SLUGGISHNESS - [gen] (having little power of movement) lenteur *f*, mollesse *f*
[med metall etc] (slow to respond to a treatment) lenteur *f* de traitement
SLUICE, to - [hydr] vanner
[min] laver au sluice, laver le minerai
[hydr] canal *m* du trop-plein
SLUICE - [gen] (any artificial channel for conducting water) canal *m*
[hydr] (body of water controlled by a floodgate) écluse *f*
[hydr] (of a reservoir) pale *f* de réservoir
[hydr] (of a pond) bonde *f* (d'étang)
[hydr] (only USA; a waterways) pertuis *m*
[mining] sluice *m*
~BOARD - [hydr] vanne *f*
~BOX - [min] boîte *f* de sluice
~COLUMN - [hydr] corps *m* cylindrique de robinet-vanne
~GATE - [hydr] vanne *f*, porte *f* d'écluse
~HEAD - [hydr] tête *f* d'écluse

SLUICE WEIR - [hydr] traverse *f* de la vanne
~WITH CYLINDRICAL BODY - [hydr] robinet-vanne *m* à corps cylindrique
SLUICING - [hydr] (the deepening of a channel by discharging impounded water into it) vannage *m*, écoulement *m* à flots
[hydr] (sewers) débourbage *m*
SLUMP, to - [gen] tomber (comme une masse)
[geol] glisser
SLUMP - [gen] baisse *f* soudaine
[comm] (of prices, shares etc) effondrement *m*
[constr] (the sinking of the upper surface of freshly laid concrete when the form is removed) glissement *m*, tassement *m*
~TEST - [constr] (test for the consistence of concrete made with a metal mould in the form of a frustrum of a cone) épreuve *f* de consistance
SLUNG PUMP - [mining] pompe *f* volante
SLUNK OIL - [oil ind] huile *f* odoriférante
SLUPGALLING - [text] retrait *m* des mailles
SLUR - [mus] (curved line over under a group of notes) liaison *f*
[acoust] mauvaise articulation *f*
[text] cueillage *m*
~CAM - [text] chevalet *m* de cueillage
SLURRY - [gen] (a mixture of finely-divided solids with water, usually taken as being of an almost liquid consistency) boue *f*, bouillie *f*
[constr] (wet mixed raw materials ready for burning) coulis *m*, lait *m*
[chem] (a mixture, e.g. oxidizers and plastic fuels) suspension *f*, schlamm *m*
~FUEL - [chem] combustible *m* à particules en suspension
~PUMP - [mech] (special pump for handling fluid mixtures of solids with liquids) pompe *f* à boue
~REACTOR - [nucl] (a nuclear reactor having the fissionable material in a semifluid form) réacteur *m* à combustible en suspension, réacteur *m* à combustible fluidisé
SLURRYING - [chem] sédimentation *f*
SLUSH, to - [gen] (to cover with slush) crotter, éclabousser
[mech] (to daub, or cover with lubricant) graisser
[min] (to wash by throwing water on) laver à grande eau
[paper man] (to remove excess water from the pulp) éliminer l'eau de la pâte
[mech] (to coat with antirust agents) enduire avec anti-rouilles
[constr] crépir (un mur)
[mining] embouer
SLUSH - [gen] (sof material, as melting snow, soft mud etc) bourbe *f*, gâchis *m*
[met] (watery snow) neige *f* à demi fondue
[mech] (a mixture of lime with white lead or tallow, to protect metal parts from oxidation) graisse *f* lubrifiante
[naut] graisse *f* de coq
~CASTING - [metall & plast ind] (method of casting metal or plastic objects, which liquid material is poured into a mould and emptied out again, a portion adhering to the inside of the mould to form the casting) coulée *f* au renversé
~MOULD - [plast ind] (a mould designed for the slush process) moule *m* à plastisols
~MOULDING - [plast ind] (process of moulding in which a thin paste material is poured into a mould and the surplus poured out again, leaving a thin coating on the mould) travail *m* des plastisols, moulage *m* des pâtes
SLUSH PIT - [gas ind] bac *m* à boue
[oil ind] bassin *m* à boue
~PUMP - [oil ind] (a pump used in rotary drilling for the circulation of drilling fluid) pompe *f* à boue
SLUSHED-UP - [constr] (said of brickwork where the joints are filled with mortar) (maçonnerie *f* en briques) crépie
SLUSHING - [mining] remblayage *m* par embouage
[transp] transport *m* par scrapper
~COMPOUND - [ind chem] agent *m* anticorrosif
~OIL - [ind chem] huile *f* anti-rouille
[oil ind] huile *f* liquide
SMACK - [naut] bateau *m* de pêche
SMALL - [gen] petit, faible, menu
[brew ind] (of beer) faible
[agric] (of wine) léger
~ARMS - [firearms] (all weapons which can be fired with one hand, or from the shoulder) armes *fpl* portatives
~BAR-MILL - [metall] petit laminoir *m*, petit train *m*
~BAYONET CAP - [mech] (type of bayonet cap of about 5/8" diameter) petit culot *m* à baionnette
~BELL - [metall] petit cône *m*
~CALORIE - [meas] petite calorie *f*
~CAPITAL - [print] petite cap *f*
~CAP - s. small capital
~CIRCLE - [surv] (circle formed on the surface of a sphere by the intersection with it of a plane not passing through the centre of the sphere) cercle *m* mineur
~COAL - [min] (small-sized coal) charbon *m* menu
~CONTROL WHEEL - [mach tool] meule *f* portante
~-END - [mech] (the end the connecting-rod of a reciprocating i.c.engine which is connected to the piston) petite tête de bielle
~END BUSH - [mech] coussinet *m* de tête de bielle
~FILM - [cin] (any film of a size smaller than 35 mm°) film *m* reduit
~GOBOS - [cin] (only US; flags in GB) petits écrans *mpl*
~HAND-SAW - [impl] égoïne *f*
~INTESTINE - [anat] intestin *m* grêle
~MILL - [metall] petit train *m*
~-OIL VOLUME CIRCUIT BREAKER - [el] disjoncteur *m* à faible volume d'huile
~PICA - [print] corps *m* II
~-SHOT EFFECT - [electron] (shot effect) effet *m* de grenaille
~SHUTTLE - [text] espolin *m*, petite navette *f*
~-SIGNAL COUPLING COEFFICIENT - [electron] (for an electron stream) coefficient *m* de couplage pour signal faible
~-SIGNAL BEAM TRANSADMITTANCE - [electron] transadmittance *f* directe pour signaux faibles
~-SIGNAL BREAKDOWN IMPEDANCE - [electron] impédance *f* pour signaux faibles dans la zone de claquage
~-SIGNAL CAPACITANCE - [electron] capacité *f* différentielle, capacité *f* pour signaux faibles
~-SIGNAL DEPTH OF VELOCITY MODULATION - [electron] profondeur *f* de modulation pour signaux faibles
~-SIGNAL DIODE CAPACITANCE - [electron] capaci-

té *f* de diode pour signaux faibles
SMALL-SIGNAL FORWARD TRANSADMITTANCE - s. small-signal beam
~-SIGNAL GAIN - [electron] coefficient *m* d'amplification pour signaux faibles
~-SIGNAL OPEN-CIRCUIT OUTPUT ADMITTANCE - [electron] admittance *f* de sortie à circuit d'entrée ouvert pour signaux faibles
~-SIGNAL POWER GAIN - [electron] amplification *f* de puissance pour signaux faibles
~-SIGNAL SHORT-CIRCUIT FORWARD CURRENT TRANSFER RATIO - [electron] coefficient *m* d'amplification de courant en sens direct à sortie en court-circuit pour signaux faibles
~-SIGNAL SHORT-CIRCUIT FORWARD TRANSFER ADMITTANCE - [electron] coefficient *m* d'amplification de courant à circuit de sortie en court-circuit pour signaux faibles
~-SIGNAL SHORT-CIRCUIT INPUT ADMITTANCE - [electron] admittance *f* d'entrée à circuit de sortie en court-circuit pour signaux faibles
~-SIGNAL SHORT-CIRCUIT REVERSE TRANSFER ADMITTANCE - [electron] transadmittance *f* inverse à circuit d'entrée en court-circuit pour signaux faibles
~-SIGNAL THEORY - [radio] (the principle of using small excursions of current and voltage from their quiescent operating points in order to minimize difficulties due to the non-linearities) théorie *f* du signal faible
~-SIGNAL TRANSCONDUCTANCE - [electron] transconductance *f* pour signaux faibles
~-STRAIN INSULATOR - [el] isolateur *m* à faible isolement
~TOOLS - [tool] petit outillage *m*
~TUBE - [gen] tubule *m*
~WARE - [text] mercerie *f*
~WARE LOOM - [text] métier *m* à ruban
~WIRING - [el] câblage *m* auxiliaire
SMALLAGE - [bot] ache *f*
SMALLHOLDER - [agric] petit paysan *m*, petit propriétaire *m*
SMALLHOLDING - [agric] petite propriété
SMALLPOX - [med] (variola; acute infections disease) petite vérole *f*, variole *f*
SMALLS - [min] menus *mpl* de minerai
SMALT - [paint] (a blue glass made from silica and cobalt oxide and which powdered, is used as a pigment) smalt *m*, émail *m* de cobalt
SMALTINE - s. smaltite
SMALTITE - [min] (natural cobalt diarsenide; an ore of cobalt) smaltine *f*, cobalt *m* arsenical
SMARAGD - [min] (emerald, a precious stone) émeraude *f*
SMARAGDITE - [min] (a fibrous green amphibole) smaragdite *f*, diallage *f* verte
SMASH, to - [gen] briser en morceaux, fracasser
[gen] (a door) enfoncer
SMASH - [gen] coup *m* écrasant
[text] (breakage of the warp yarn) rupture *f*
SMASHBOARD SIGNAL - [railw] signal *m* de barre
SMASHING - [bookbind] (the pressing of the book in a machine after sewing) endossement *m*
SMEAR, to - [gen] salir, maculer
[gen] (to cover with ointment, grease etc) enduire
SMEAR - [gen] tache *f*, macule *f*
[telev] maculage *m*
SMEAR GHOST - [telev] image *f* fantôme à maculage
~STATE - [nucl] essai *m* de frottement
SMEARER - [telev] circuit *m* neutraliseur du dépassement balistique
SMEARING - [telev] effet *m* de traînage horizontal
SMECTIC PHASE - [crystall] (one of the forms of the mesomorphic state, or the liquid crystals) phase *f* smectique
SMEDDUM - [min] (fine ore particles, which have passed through a wire sieve) minerai *m* pulvérisé
SMEGMOLITH - [med] concrétion *f* préputiale
SMELL - [physiol] odorat *m*
[gen] (an odour, as perceived by the olfactory nerves) odeur *f*, senteur *f*
SMELT, to - [min] fondre
[metall] (to extract metal from an ore) extraite par fusion
[metall] (to refine, to scorify) fondre
SMELTER - [gen] fondeur *m*, métallurgiste *m*
[metall] (the works) fonderie *f*, usine *f* métallurgique
~COKE - [metall] coke *m* de fonderie
SMELTING - [metall] (extraction of a metal from an ore by fusion with fluxes) fusion *f*, fonte *f*, smeltage *m*
~FURNACE - [paper man] (chamber with refractory lining in the sulphate recovery process) four *m* de fusion
[metall] fourneau *m* de fonte
~PIT - [min] (for sulphur) four *m* de fusion du soufre
~PLANT - [metall] fonderie *f*
S METER - [instr] (signal meter, a meter designed to measure relative signal strength) S-mètre *m*
SMITH, to - [gen] forger
SMITH - [gen] forgeron *m*
SMITHAN - s. smeddum
SMITHERY - [ind] forge *f*, forgerie *f*
SMITHING COAL - [metall] charbon *m* de forge
SMITHS'BELLOWS - [impl] soufflet *m* de forge
~CONE - [impl] cône *m* de forgeron
~HEARTH - [metall] bâti *m* de forge
~JOINTS - [constr] assemblages *mpl* de charpente en fer
~MANDREL - s. smiths' cone
~PLIERS - [impl] tenaille *f* à forger
~PUNCH - [tool] poinçon *m* de forge
SMITHSONITE - [min] (naturally occurring zinc carbonate, an ore of zinc) smithsonite *f*
SMITHY - s. smithery
~COAL - [fuel] (forge coal) charbon *m* de forge
SMOCK - [text] blouse *f*
SMOG - [met] (term used for dense fog containing sulphur dioxide occurring under certain conditions over cities having little or no effective smoke and fume control) brouillard *m* enfumé, smog *m*
SMOKE, to - [gen] fumer
[ind chem] (to expose, e.g. rubber, to the action of smoke for curing purposes) fumer
SMOKE - [chem] (a suspension of finely divided solids in a gas or mixture of gases) fumée *f*
[gen] fumée *f*
~AMMUNITION - [chem] munitions *fpl* fumigènes
~BELL - [gen] fumivore *m*
~BLACK - [paint] noir *m* de fumée
~BOMB - [chem] bombe *f* fumigène
~BOX - [mech] (cylindrical extension on the front of a locomotive boiler through which the flue gases

pass from the tubes to the funnel) boîte *f* à fumée
SMOKE-BOX DOOR - [mech] (in a steam locomotive) porte *f* de la boîte à fumée
~-BOX NETTING - [th eng] grille *f* à flammèches, pare-étincelles
~-CONSUMING - [gen] fumivore
~CURING - [food] fumage *m*
~DEFLECTOR - [impl] (of a lamp) plateau *m* fumivore
~DENSITY INDICATOR - [instr] indicateur *m* de la densité de la fumée
~-FLOAT - [naut] (device designed to float on water and emit dense smoke, for marking purposes) dispositif *m* de signalisation par fumée
~FLUE - [constr] carneau *m* de fumée
[metall] carneau *m* à fumées
~HAZE - [met] (condition of poor visibility primarily due to smoke) brouillard *m* de fumée
~HOLE - [geol] fumerolle *f*
~HOUSE - [rubber ind] (building in which the smoking operation is carried on) fumoir *m*, séchoir-fumoir *m*
~JACK - [constr] moulinet *m*
~LIMIT - [mech] (of a diesel engine) limite *f* de la fumée
~OUTLET - [constr] sortie *f* pour la fumée
~PIPE - [constr] conduit *m* de fumée
~POINT - [chem] (the maximum height of flame measured in millimetres at which a kerosene will burn without smoking tested in a standard lamp) point *m* de fumée
~-PRODUCER - [gen] fumogène
~-PROOF - [gen] à l'épreuve de la fumée
~SCREEN - [gen] rideau *m* de fumée
~SHELL - [chem] (shell emitting a thick could of smoke on explosion) obus *m* fumigène
~SHIELD - [railw] pare-fumée *m*
~STACK - [gen] cheminée *f*
~-STACK LIP - [railw] visière *f* de la cheminée
~TEST - [constr] (a test made for low-pressure airtightness, by injecting smoke under slight pressure into the system concerned, leaks being made visible by the appearance of smoke from them) essai *m* par la fumée
~TUBE BOILER - [th eng] chaudière *f* tubulaire
~-WOOD - [bot] fustet *m*
SMOKED - [gen] fumé, enfumé
[glass man] (said of a glass which is discoloured in a reducing flame) enfumé, décoloré
[glass man] (glass covered with a smoky film) à teinte fumée
~GLASS - [glass ind] verre *m* fumé, verre *m* noirci à la fumée, verre *m* à teinte fumée
~SHEETS - [rubber ind] (raw sheet rubber partially cured by exposure to the smoke of an open fire) feuilles *fpl* fumées
SMOKELESS - [gen] sans fumée
~FUEL - [min] combustible *m* sans fumée
~FURNACE - [metall] four *m* sans fumée
~POWDER - [chem] (a propellant explosive based on nitrocellulose) poudre *f* sans fumée
SMOKESTACK - s. smoke stack
SMOKING - [gen] fumage *m*, émission *f* de fumée, fumant
[ceram] (the preliminary stage of firing) cuisson *f* préliminaire
[rubber ind] (process of coagulation and partial curing of raw rubber by exposure to the smoke of an open fire) enfumage *m*
SMOKING CAR - [railw] compartiment *m* pour fumeurs
~CHAMBER - [agric & bot] fumoir *m*
~COMPARTMENT - s. smoking car
~TOBACCO - [agric] tabac *m* à fumer
SMOKY QUARTZ - [min] quartz *m* enfumé
SMOOTH, to - [gen] lisser, planer, adoucir
[metall] (a grind) meuler, polir
SMOOTH - [gen] lisse, uni, doux, égal
[mech] lisse
[ind chem] (of a paste etc) bien amalgamé
~BORE - [firearms] à âme lisse
~BREAK - [constr] (a breaking plane) surface *f* de rupture
~CORE - [el] (old form of armature core in which the windings were laid on the smooth cylindrical surface of the core and secured by binding wires or small pegs) noyau *m* lisse
~-CORE ARMATURE - [el] (armature core without slots for the reception of the windings) induit *m* à noyau lisse
~CUT - [mech] taille *f* douce
~-CUT FILE - [tool] lime *f* à taille douce
~-CUT RASP - [tool] râpe *f* à taille douce
~CYLINDRICAL COUPLING - [el] manchon *m* de raccordement lisse, jonction *f*
~DRIFPIN - [tool] mandrin *m* lisse
~EDGED SCALE - [text] écaille *f* à bord lisse
~FIBRE - [text] fibre *f* lisse
~FILE - [mech] (a file used for finishing, having a large number of small teeth) lime *f* douce
~FINISH - [gen] finissage *m* lisse
~FRACTURE - [metall] cassure *f* nette
~HOLE - [mech] trou *m* lisse
~MATERIAL - [text] tissu *m* souple
~OPERATION - [mech] fonctionnement *m* doux
~PICK - [text] chasse *f* douce, coup *m* faible
~RUBBING LEATHER - [leather ind] cuir *m* de frottage lisse
~RUNNING - [mech] (of a motor etc) marche *f* douce, régulier
[mech] of a mechanism having no frictional resistance) à marche douce, bien roulant, fonctionnement *m* doux
~SEA - [met] mer *f* calme
~SILVERY LUSTRE - [text] lustre *m* tranquille et argenté
~STONE - [constr] caillou *m*
~TAPER DRIFT - [mech] broche *f* conique lisse
~THREAD - [text] retors *m* lisse
[rubber ind] (of tyres) bande *f* de roulement lisse
~TYRE - [rubber ind] pneu *m* lisse
~WATER - [naut] eau *f* calme
~WEB - [text] laine *f* lisse
~WOOL FIBRE - [text] fibre *f* de laine lisse
~YARN - [text] fil *m* lisse
SMOOTHING - [gen] lissage *m*, planage *m*, aplanissement *m*, adoucissement
[soil] (of the soil) égalisation *f* (dû terrain)
~CAPACITOR - [el] condensateur *m* de filtrage
~CHOKE - [el] (inductor used in a circuit designed to decrease the ripple in a direct-current power source) bobine *f* de filtrage
~CIRCUIT - [el] (circuit designed to reduce the amplitude of a ripple) circuit *m* de filtrage
~COIL - s. smoothing chocke
~DOWN THE FIBRES - [text] rabattage *m* des duvets

SMOOTHING FILTER - [el] filtre *m* de courant redressé
~HARROW - [agric] herse *f* à semences
~IRON - [impl] fer *m* à repasser
~PLANE - [tool] rabot *m* à repasser, rabot à polir
[constr] varlope *f*
~PLANER - [tool] tabot *m* plat carré
~ROLLS - [plast ind] rouleaux *mpl* à lisser
~TROWEL - [impl] truelle *f* à lisser
SMOOTHNESS - [gen mech etc] lisse *m*, égalité *f* (d'une surface)
[mech] (the quality of running smoothly) douceur *f*, bon fonctionnement
~CHECK - [comput] contrôle *m* de continuité
~OF THE FLUTE - [text] lisse *m* de la cannelure
SMUDGE, to - [gen] salir, souiller
(USA only; to produce a smoky fire) allumer un feu produisant une fumée épaisse
SMUDGE - [gen] (a smear, a stain) salissure *f*, tache *f*
[paint] (the mixture of residues in a paint) sédiments *mpl*
[gen] (USA only) fumée *f* épaisse
~OIL - [oil ind] huile *f* fumigène
SMUDGING - [telev] (irregularities in the picture) traits *mpl*
SMUT - [biol] (micro-organism belonging to the basidiomycetes, parasitic on certain plants; especially cereals) carie *f*
[agric] nielle *f* (des céréales), suie *f*, brûlure *f*, charbon *m*
[mining] charbon *m* terreux
SMUTTED - [bot & agric] charbonneux
SMYRNA CARPET - [text] tapis *m* de Turquie, tapis *m* de Smyrne
~COTTON - [text] coton *m* de Smyrne
~RUG - s. Smyrna carpet
~YARN - [text ind] fil *m* de Smyrne
SN - [chem] (the symbol for tin) symbole *m* de l'étain
S/N CURVE - [metall] (stress-number curve) courbe *f* du nombre des cycles d'effort
SNAFFLE - [impl] mors *m* de bridon
SNAG, to - [naut] (to cause a ship to hit a hidden obstacle) se heurter contre un obstacle
[gen] (to cut irregularly) couper irrégulièrement
[metall] ébarber
[agric] essoucher (un terrain)
SNAG - [gen] saillie *f*, dent *f*
[agric] (from a tree trunk) souche *f*, chicot *m*
[bot] porte-greffe *m*, sujet *m*
[tool] outil *m* à ébarber
SNAGGING BELT - [metall] transporteur *m* à ébarber
~MACHINE - [mach tool] ébarbeuse *f*
~OPERATION - [mech] ébarbage *m*
SNAIL - [zool] limacon *m*
[mech] came *f* à profil à spirale
[horol] (snail-wheel) limaçon *m*
SNAP BEAN - [agric] haricot *m* vert
~BOLT - [mech] verrou *m* à ressort
SNAKE - [zool] serpent *m*
~BITE - [med] morsure *f* de serpent
~CUCUMBER - [agric] concombre *m* serpent
SNAKING - [aero] (uncontrolled yaw oscillation of nearly constant amplitude) oscillation *f* de lacet
~-IN - [plumb] (lowering in) mise *f* en fouille
SNAP, to - [gen] se casse (net), casser
SNAP, to - [mech] (to spring back) faire ressort
[acoust] (to make or cause to make a short sharp sound) claquer, émettre un bruit sec
[gen] (of a dog) saisir
[photo] prendre un instantané de
[firearms] revenir brusquement
[mech] bouteroller (un rivet)
SNAP - [gen] (sharp quick sound) coup *m* sec, claquement *m*
[gen] (sudden rupture) rupture *f* soudaine
[mech] (a catch, a fastener) fermeture *f* automatique, fermoir *m*
[metall] (in foundry work, a snap flask) châssis *m* à démotter
[tool] (a punch used to form rivet heads) bouterolle *f*
[glass man] (a tool used for gripping a piece of glass for fire polishing) quenouille *f*
~ACTION - [radio] (in magnetic amplifiers, the abrupt jump in output current as a function of control current) saut *m* de courant de sortie
~-ACTION CONTACTS - [contr] contacts *mpl* à déclic
~-ACTION CONTROL - [contr] réglage *m* pour tout ou rien
~-ACTION MECHANISM - [mech] mécanisme *m* à déclic
~-BACK CHARACTERISTICS - [plast ind] (properties which favour the return of a plastic material to its original form after bending, without the appearance of blush marks) caractéristiques *fpl* de retrait
~DIE - s. snaphead
~FASTENER - [mech] bouton *m* à pression
~FLASK - [metall] (used in foundry work) châssis *m* articulé pour moulage en mottes, châssis *m* à démotter
~GAUGE - [tool] (a tool with fixed jaws determining the correctness of the thickness of a part) jauge *f* d'épaisseur
~GROOVE - [horol] rainure *f* pour la lanterne
~HAMMER - [tool] marteau *m* à river
~-HEAD - [tool] bouterolle *f*, contre-rivoir *m*
[metall] rivure *f* bouterollée
~-HEAD BOLT - [mech] (a bolt the head of which is formed as part of a sphere) boulon *m* à tête hémisphérique
~-HEAD RIVET - [mech] rivet *m* à tête ronde
~HOOK - [mech] crochet *m* à ressort, mousqueton *m*
~LOCK - [mech] (type of lock with a spring-loaded latch, to fasten automatically when the door is closed) serrure *f* à ressort
~MOULD - [metall] motte *f*
~MOULDING - [metall] moulage *m* en mottes
~OUT - [constr] sailli *m* du pêne
~RING - [mech] bague *f* élastique
~ROLL - [aero] (a roll carried out by a quick movement of the controls) tonneau *m* rapide
~SWITCH - [el] (a switch which makes and breaks a circuit with a quick snap of a blade or blades whose rate of motion is independent of the action of the operator) commutateur *m* à ressort
[el] interrupteur *m* rapide
SNAPHEAD - [tool] bouterolle *f*
SNAPPED-HEAD RIVET - [mech] tête *f* de rivet bouterollée
~HEADER - [constr] (a half-length brick) demi-brique *f*
SNAPPY - [gen] vif

SNAPPY - [mech] (of an engine or motor) nerveux
[photo] (said of a picture rich in halftones) riche en contraste
SNAPSHOT, to - [photo] prendre un instantané
SNAPSHOT - [photo] instantané *m*
SNARE - [gen] piège *m*, lacet *m*
[med] (a wire loop for the removal of soft tumours) serre-noeud *m*, anse *f*
SNARL, to - [gen] vriller
[acoust] gronder, grogner
[metall] travailler au repoussoir
SNARL - [gen] entortillement *m*, enchevêtrement *m*
[text] vrillage *m*, boucle *f*
[acoust] grondement *m*, grognement *m*
~CATCHER - [text] purgeur *m* de noeuds, leveur *m* de noeuds
~FORMATION - [text] formation *f* de frisures
~PREVENTER - [text] dispositif *m* anti-vrilles
~STRETCHING MOTION - [text] dispositif *m* pour étirer les vrilles
~WARP - [text] chaîne *f* de boucles, chaîne *f* de poil
SNARLING-IRON - [tool] (a tool consisting of a light hammerhead mounted on a spring shaft which maintains a rapid vibratory movement of the head, used for removing dents in inaccessible parts or thin hollow structures) marteau *m* à ressort
SNATCH, to - [gen] saisir
~BLOCK - [mech] poulie *f* coupée, galoche *f*
SNEAKERS - [shoe man] espadrilles *f*pl
SNEEZE, to - [gen] éternuer
SNEEZE - [gen] éternuement *m*
SNICK, to - [gen] entailler, encocher
SNIP, to - [gen] couper d'un coup de ciseaux
SNIP - [gen] petit morceau *m*, petite encoche *f*
[tool] cisaille *f* pour tôles
~PLIERS - [tool] (type of pliers with long slender jaws) pinces *f* à longues mâchoires
SNIPPED JUTE HARDS - [text] étoupe *f* de jute coupée
SNIPPERS - [tool] ciseaux *f*pl
SNIPPING - [text] coupage *m*
~MACHINE - [text] machine *f* à arracher (les pieds des tiges)
SNIPS - [tool] pince *f* à couper
SNOOT - s. snout [cin]
SNORING - [acoust] ronflement *m*
SNORKEL - [naut] (in submarines) schnorchel *m*, schnorkel *m*
SNORTING - [acoust] (the sound made by angry animals) reniflement *m*, ébrouement *m*
SNOUT - [zool] museau *m*
[mech] bec *m*, buse *f*, ajutage *m*
[cin] (a conical opaque object used on spotlights for light concentration) capuchon *m* conique
[mech] (tapering portion at the burner end of the flame tube in some turboprop engines) bec *m* de turbine
SNOW - [met] (precipitation consisting of feathery crystals of ice) neige *f*
[telev] (in the picture tube) neige *f*
~BANK SPOTS - [med] taches *f*pl blanches, rétinite *f* albuminurique
~BLINDNESS - [med] ophtalmie *f* des neiges, cécité *f* des neiges
~CHAIN - [auto] chaîne *f* à neige
~FLANGER - [railw] (a plate attached to a locomotive and designed to scrap off snow and ice from the sides of the rail heads) chasse-neige *m*, taille-neige *m*
SNOW-PLOUGH - [mech] (machine for removing snow from runways or roads) charrue *f* à neige
~REMOVER - s. snow-plough
~SCRAPER - s. snow flanger
~SKI - [aero] (ski-type landing gear for alighting on snow) patin *m* pour l'atterrissage sur la neige
~-WHITE WOOL - [text] ind] laine *f* nivéenne du Cap
SNOWDRIFT - [met] amoncellement *m* de neige
SNOWFALL - [met] chute *f* de neige
SNOWFIELD - [gen] champ *m* de neige
SNOWFLAKE - [met] (an agglomerated mass of feathery ice crystals) flocon *m* de neige
[metall] (defect appearing as a bright area and caused by fracturing a hair line crack) craquelure *f*, gerçure *f*
SNOWPLOW - s. snow plough
SNOWSHED - [railw] galerie *f* para-avalanche
SNOWSHORE - [gen] raquette *f*
SNOWSTORM - [met] tempête *f* de neige
SNUB, to - [mining] sous-caver
SNUBBER - [mech] (term denoting mechanism which employ dry friction to produce damping of suspension systems) amortisseur *m* par friction
SNUBBING UNIT - [oil ind] équipement *m* de forage sous pression
SNUFF - [gen] reniflement *m*, poudre *f* à priser
SNUFFLES - [med] jetage *m*, coryza *m* du nouveau-né
SNUG, to - [gen] se mettre au chaud
[mech] (to adjust accurately) monter à frottement doux
[naut] (to prepare for emergencies) parer (un navire)
[naut] (to snuf down the sails) ferler (les voiles)
SNUG - [gen] bien abrité, confortable
[mech] ergot *m*, oreille *f*
[metall] oreille *f* de châssis de fonderie
[mech] (a projection) dent *f*, ergot *m*
[naut] (said of a ship ready to sail) (navire) paré
~BOLT - [mech] (a bolt provided with a small projection under the head to prevent it revolving when the nut is screwed up) boulon *m* à ergot
SOAK, to - [gen] imbiber, imprégner, tremper
[gen] (to macerate) macérer
[phys] (to become saturated) saturer
[agric] (a cask) combuger (une futaille)
[el] (said of battery which is charged slowly) charger lentement
[leather ind] (the skins) reverdir (les peaux), détremper
SOAK - [gen] trempe *f*, imbibition *f*
[leather ind] (a bath in the tanning process) bain *m*
~THE CORDAGE, to - [text] mouiller le cordage
~THE RAGS, to - [text] tremper les chiffons
SOAKED ROBBIN - [text] bobine *f* trempée
~WOOL - [text] laine *f* imbibée
SOAKING - [metall] (term used for maintaining metal at a given temperature until this temperature is uniform throughout the whole body of metal) égalisation *f*
[rubber ind] imprégnation, imbibition *f*
[leather ind] (in the tanning process) reverdissement *m*
~OF THE CHARGE - [th eng] (in coke ovens) surcuisson *f* du coke
~PERIOD - [ceram] (period during which vitrification is continued) temps *m* d'imbibition

SOAKING PIT - [metall] (for steel ingots) four *m* d'égalisation
[glass man] (a conditioning furnace) four *m* de conditionnement
[min] puits *m* ordinaire
~POT - [metall] pot *m* de trempage
~THE COCOONS - [text] trempage *m* des cocons
~VAT - [rubber ind] cuve *f* d'imprégnation
SOAP - [chem] (general term for the alkaline salts of the fatty acids) savon *m*
[constr] brique *f* demi-épaisseur
~BARK - [bot] (the bark of the quillai, a large Venezuelan tree, used as a substitute for soap) bois *m* de Panama
~-BASE GREASE - [ind chem] graisse *f* à base de savons
~-CONSUMING POWER - [chem] (the specific amount of a given soap which will be precipitated by the (hard) water in question) capacité *f* d'absorption du savon
~CRUTCHER - [ind chem] mélangeur *m* pour savon
~-FLAKES - [ind chem] savon *m* en paillettes
~FRAME - [ind chem] boîte *f* à solidifier le savon
~MAKING - [ind chem] savonnerie *f*
~RUNNING - [ind chem] procédé *m* de neutralisation de l'excès d'acide graisse
~STOCK - [ind chem] pâte *f* huileuse
SOAPING MACHINE - [text] machine *f* à savonner
SOAPSTONE - [min] (steatite) pierre *f* de savon, stéatite *f*
~DISPENSER - [rubber ind] machine *f* à talquer
~POWER - [min] poudre *f* de stéatite
SOAPSTONING - [rubber ind] (dusting) poudrage *m*, talquage *m*
SOAPSUDS - [gen] lessive *f*
SOAPWORT - [bot] saponnaire *f*
SOAPY - [gen] savonneux
[bot] saponacé
SOAR, to - [gen] s'élever, monter
[aero] faire du vol à voiles
SOARER - [aero] appareil *m* sans moteur
SOARING - [aero] (unpowered flight, using ascending air currents) vol *m* à voile
SOCIOLOGY - [gen] (the science dealing with the origin and evolution of human society) sociologie *f*
SOCIOMETRY - [gen] (the study of human society from its psychological aspect) sociométrie *f*
SOCK - [text] chaussette *f*
[shoe man] semelle *f* intérieure
[aero] manche *f* à vent
~-LAMB - [zool] agneau *m* de lait
SOCKET - [gen] (recess or hollow part to receive another) douille *f*, emboîture *f*, crapaudine *f* (de gond de porte)
[el] (for an electric plug) douille *f* de lampe, douille *f* de jack
[el] (a lampholder) porte-lampe *m*, prise *f* de courant
[plumb] (internally-threaded sleeve for joining ends of threaded pipes) manchon *m* de tuyau
[naut] (of a steel rope) saucier *m*
[anat] alvéole *m*
[mech] boîte *f* à forets, baril *m* (de vilebrequin), mortaise *f*, godet *m*
[mining] culot *m* de trou de mine
~AERIAL - [radio] (a built-in aerial connected to the mains) antenne *f* de secteur
SOCKET AND SPIGOT BEND - [plumb] coude *m* à emboîtement et cordon
~CHISEL - [carp] (a robust chisel used mortising) burin *m* pour mortaises
~COUPLING - [mech] couplage *m* à manchon
~END - [plumb] bout *m* femelle
~FOR JOINTS - [plumb] joint *m* à emboîtement
~FOR PICKING STICK - [text] sabot *m* du bâton de chasse
~-HEAD - [mech] hexagonal, à six pans
~-HEAD SCREW - [mech] vis *f* à six pans
~HORN - [oil ind] tige *f* de repêchage
~JOINT - [plumb] joint *m* à rotule
[mech] assemblage *m* à douille
~OUTLET - [el] socle *m*
~-OUTLET ADAPTOR - [el] (accessory for insertion into a socket-outlet and containing metal contacts to which may be fitted one or more plugs) fiche *f* de dérivation
~OUTLET AND PLUG - [el] prise *f* de courant
~-PIPE - [plumb] conduite *f* à emboîtement
~-PUNCH - [tool] emporte-pièce *m*, découpoir *m*
~PUTTY - [constr] mastic *m* pour manchons
~SCREW - [mech] vis *f* héxagonale
~SET SCREW - [mech] vis *f* de pression à six pans
~SLUICE VALVE - [hydr] robinet-vanne *m* à emboîtement
~SPANNER - [impl] clef *f* à douille, clef *f* tubulaire
~WRENCH - s. socket spanner
SOCKETED FITTING - [plumb] tubulure *f* à emboîtement
~PIPE - [plumb] (a bell and spigot pipe) tuyau *m* à emboîtement
~TUBULURE - [th eng] (in ovens) tubulure *f* à emboîtement
SOCKLE BAR - [agric] barre-faucheuse *f*, barre *f* de coupe
SOCKLINING - s. sock [shoe man]
SOCLE - [arch] socle *m*
SOD - [agric] gazon *m*
~BURNING - [agric] brûlage *m* du terrain
~OIL - [ind chem] (degras) dégras *m* de peaux
SODA - [chem] (any of a number of talkaline compounds) soude *f*
~ASH - [chem] (anhydrous sodium carbonate, an important raw material) alcali *m* minéral, cendre *f* de soude, carbonat *m* de soude
~-CELLULOSE - [chem] (intermediate in viscose manufacture, produced by steeping cellulose in caustic soda) cellulose *f* à soude
~DEPOSITS - [geol] gisement *m* de soude
~-FOUNTAIN - [gen] bar *m* pour rafraîchissements non-alcooliques
~LAKES - [geol] (salt lakes having a water containing a high content of sodium salts) lacs *mpl* salés
~LIME - [chem] (a mixture of sodium, potassium and calcium hydroxides with uses as a drying agent and absorbent) chaux *f* sodée
~-LIME FELDSPAR - [min] feldspath *m* calcosodique
~LIME GLASS - [chem] (glass of alkali-lime-silica type) verre *m* de chaux sodée, crown-glass *m*
~NITRE - [min] (nitrate of sodium) nitrate *m* de sodium
~PROCESS - [paper man] (alkaline treatment of wood for the production of chemical wood pulp) procédé *m* à la soude
~PULP - [paper man] pâte *f* à la soude
~RECOVERY - [paper man] (liquor resulting from

the digestion of raw materials with caustic soda is subjected to a process of concentration and the organic matter is burnt off in a furnace The soda is thus recovered as soda ash and causticized for further use) récupération *f* de la soude

SODA SOFTENING - [hydr] adoucissement *m* au carbonate de sodium

~SYENITE - [geol] (a syenitic igneous rock containing soda-feldspar) roche *f* ignée syénitique

~WATER - [gen] eau *f* gazeuse bicarbonatée

SODALITE - [min] (a feldspathoid mineral employed as a gemstone) sodalite *f*

SODAMIDE - [chem] (compound obtained by passing ammonia gas over hot sodium) sodamide *m*

SODDING - [agric] curage *m*, faucardage *m*

SODERBERG ELECTRODE - [electron] électrode *f* de Soderberg

SODION - [chem] (a sodium ion) ion *m* de sodium

SODIUM - [chem] (a metallic element, symbol NA, which violently decomposes water on contact, with the formation of sodium hydroxide and evolution of hydrogen. Sodium has application as a reducing agent, in metallurgy pharmaceuticals, petroleum refining and in nuclear technology) sodium *m*

~ABIETATE - [chem] (a rosin soap with uses in paper and soap manufacture) abiétate *m* de soude

~ACETATE - [chem] (an intermediate for a wide range of compounds including pharmaceuticals, dyestuff and pigments) acétate *m* de sodium

~ACETRIZOATE - [chem] (an X-ray contrast medium used in medicine) acétrizoate *m* de sodium

~ACETRYLARSANILATE - [chem] (a white crystalline powder with application in the treatment of trypanosomiasis and malaria) acétylarsanilate *m* de sodium

~ALGINATE- [chem] (an alginic acid derivative used in the production of water soluble fibres. These have application in the manufacture of fine woollen threads. Sodium alginate is also used as a stabiliser, emulsificant and thickening agent) alginate *m* de sodium

~ALPHA-SODIOACETATE - [chem] (a synthesis intermediate) alpha-sodioacétate *m* de sodium

~ALUMINATE - [chem] (a mordant in dyeing) aluminate *m* de sodium

~ALUMINIUM HYDRIDE - [chem] (a reducing agent) hydrure *m* de sodium et d'aluminium

~ALUMINIUM SILICOFLUORIDE - [chem] (a dyeing assistant and insecticide) silicofluorure *m* de sodium et d'aluminium

~AMALGAM - [chem] (an analytical reagent) amalgame *m* de sodium

~AMIDE - [chem] (a synthesis intermediate) amide *f* de sodium

~AMMONIUM PHOSPHATE - [chem] (microcosmic salt. An analytical reagent) phosphate *m* de sodium et d'ammonium

~ANTIMONATE - [chem] (a component of ceramic glazes) antimoniate *m* de sodium

~ARSANILATE - [chem] (sodium aminoarsonate White, crystalline powder used in the treatment of malaria and trypanosomiasis) arsanilate *m* de sodium

~ARSENATE - [chem] (poisonous, colourless crystals with uses in medicine, insecticides, and as a dyeing assistant) arseniate *m* de sodium

~ARSENITE - [chem] (a dyeing assistant, herbicide, and leather chemical) arsénite *m* de sodium

SODIUM ASCORBATE - [chem] (a source of therapeutic vitamin C) ascorbate *m* de sodium

~AZIDE - [chem] (an intermediate in the production of initiating explosives) azide *f* de sodium, azothydrure *m* de sodium

~BENZOATE - [chem] (a dyestuff intermediate, preservative for foodstuff, and antiseptic) benzoate *m* de sodium

~BICARBONATE - [chem] (an analytical reagent, component of baking powders and beverages, and an antacid in medicine) bicarbonate *m* de soude

~BIFLUORIDE - [chem] (a preservative for histological specimen) bifluorure *m* de sodium

~BIPHOSPHATE - [chem] (colourless crystals with uses in medicine and as an analytical reagent) biphosphate *m* de sodium

~BISULPHATE - [chem] (a disinfectant, dyeing assistant , wool treatment chemical and a starting material for other sodium compounds) bisulfate *m* de sodium

~BISULPHITE - [chem] (an analytical reagent, bleaching agent and starting material for a large number of compounds) bisulfite *m* de sodium

~BITARTRATE - [chem] (an analytical reagent) bitartrate *m* de sodium

~BORATE - [chem] (a component of ceramic glazes and special glasses) borax *m*, borate *m* de sodium

~BORATE PERHYDRATE - [chem] (a bleaching agent) borate-perhydrate *m* de sodium

~BOROHYDRIDE - [chem] (a blowing agent for the production of foamed plastics) borohydrure *m* de sodium

~BROMATE - [chem] (an analytical reagent) bromate *m* de sodium

~BROMIDE - [chem] (a synthesis intermediate and a sedative used in medicine) bromure *m* de sodium

~CACODYLATE - [chem] (white, deliquescent crystalline powder used in the treatment of skin conditions) cacodylate *m* de sodium

~CAPRYLATE - [chem] (sodium octoate. An antimycotic used in medicine) caprylate *m* de sodium

~CARBONATE - [chem] (an agent for the pH control of water, also having application in the treatment of skin conditions) carbonate *m* de soude

~CARBONATE PEROXIDE - [chem] (a bleaching agent) peroxyde *m* du carbonate de soude

~CARBOXYMETHYLCELLULOSE - [chem] (a stabilizing and thickening agent for emulsion with application in medicine, food processing, surface coatings, paper and textile manufacture) carboxyméthylcellulose *f* de sodium

~CHLORATE - [chem] (a bleaching and oxidizing agent, dyeing mordant and herbicide) chlorate *m* de sodium

~CHLORIDE - [chem] (common salt. An important raw material for a large number of industrial chemicals such as chlorine, sodium compounds, hydrochloric acid, sodium hydroxide, and metallic sodium. Sodium chloride is also used as a ceramics glaze, foodstuffs preservative, in dyeing textiles, metallurgy, soap manufacture, and in refrigeration) chlorure *m* de sodium

~CHLORITE - [chem] (a bleaching and oxidizing agent) chlorite *m* de sodium

~CHLOROACETATE - [chem] (an intermediate for drugs, dyes, and herbicides) chloroacétate *m* de

sodium
SODIUM CHLOROTOLUENE SULPHONATE - [chem] (an intermediate for pharmaceuticals and dyestuff) chlorotoluène-sulfonate *m* de sodium
~CHROMATE - [chem] (a timber preservative and starting material for other chromates, dyes, leather chemicals and pigments) chromate *m* de sodium
~CHROMATE TETRAHYDRATE - [chem] (yellow deliquescent crystal with application in the production of pigments and other chromium compounds) chromate *m* de sodium tétrahydraté
~CITRATE - [chem] (an anticoagulant for physiological fluids) citrate *m* de sodium
~COOLED REACTOR - [nucl] (a reactor having a coolant consisting of liquid sodium) réacteur *m* refroidi au sodium
~-COOLED VALVE - [mech] (a sodium-filled exhaust valve) soupape *f* au sodium
~COOLING - [mech] (method of cooling the valves of high-performance i.c. engines, in which the exhaust valve stems and heads are hollow and partly filled with sodium) refroidissement *m* au sodium
~CYANATE - [chem] (an intermediate for pharmaceuticals, also has application in metallurgy) cyanate *m* de sodium
~CYANIDE - [chem] (an intermediate for pigments, dyes and chelating agents, also widely used in the extraction of gold and silver ores) cyanure *m* de sodium
~CYCLAMATE - [chem] (a sweetening agent with uses as a substitute for sucrose in the control of diabetes) cyclamate *m* de sodium
~DEHYDROACETATE - [chem] (a plasticizer and mildewcide) déhydroacétate *m* de sodium
~DIACETATE - [chem] (an intermediate and mould inhibitor) diacétate *m* de sodium
~DIATRIZOATE - [chem] (an X-ray contrast medium) diatrizoate *m* de sodium
~DICHLOROISOCYANUTRATE - [chem] (a disinfectant and bleaching agent) dichloroïsocyanurate *m* de sodium
~DICHLOROPHENOXYACETATE - [chem] (a herbicide) dichlorophénoxyacétate *m* de sodium
~DICHROMATE - [chem] (a starting material for pigments and other chromium compounds, and an oxidizing agent and dyeing mordant) bichromate *m* de sodium
~DIMETHYLDITHIOCARBAMATE - [chem] (a vulcanization accelerator) diméthyldithiocarbamate *m* de sodium
~DINITRO-ORTHOCRESYLATE - [chem] (a fungicide and weedkiller) dinitro-ortho-crésylate *m* de sodium
~DIURANATE - [chem] (a component of ceramic glazes) diuranate *m* de sodium
~DODECYLBENZENE SULPHONATE - [chem] (a constituent of detergents) dodécylbenzène-sulfonate *m* de sodium
~ETHYLATE - [chem] (a synthesis intermediate) éthylate *m* de sodium
~FERRICYANIDE - [chem] (a dyeing agent and intermediate for pigments) ferricyanure *m* de sodium
~FERROCYANIDE - [chem] (yellow, water-soluble crystals with application in metallurgy, photography and the production of pigments and dyestuffs) ferrocyanure *m* de sodium
SODIUM FLUORIDE - [chem] (a water treatment agent, component of ceramic glazes, pesticide and fungicide) fluorure *m* de sodium
~FLUOROACETATE - [chem] (a pesticide) fluoroacétate *m* de sodium
~FLUOROSILICATE - [chem] (a fluoridation agent, component of ceramic glazes, pesticide and intermediate) fluosilicate *m* de sodium
~FORMALDEHYDE SULPHOXYLATE - [chem] (a bleaching agent for foodstuffs) formaldéhyde-sulfoxylate *m* de sodium
~FORMATE - [chem] (a leather chemical, dyeing mordant, reducing agent, starting material for intermediates, also has application in medicine as a diuretic) formiate *m* de sodium
~GENTISATE - [chem] (an anti-rheumatic agent) gentisate *m* de sodium
~GLUCOHEPTINATE - [chem] (a sequestering agent) glucoheptinate *m* de sodium
~GLUCONATE - [chem] (a sequestering agent and food additive) gluconate *m* de sodium
~GLUTAMATE - [chem] (monosodium glutamate. An additive for foodstuffs) glutamate *m* de sodium
~GLYCEROPHOSPHATE - [chem] (white, monoclinic crystals with application in medicine) glycérophosphate *m* de sodium
~-GRAPHITE REACTOR - [nucl] réacteur *m* sodium-graphite
~GYNOCARDATE - [chem] (sodium hydnocardate. An antileprotic used in medicine) gynocardate *m* de sodium
~HEXYLENE GLYCOL MONOBORATE - [chem] (a lubricant additive and corrosion inhibitor) hexylèneglycolmonoborate *m* de sodium
~HYDRIDE - [chem] (an alkylating agent in organic reactions) hydrure *m* de sodium
~HYDROSULPHIDE - [chem] (a bleaching agent and intermediate) hydrosulfure *m* de sodium
~HYDROSULPHITE - [chem] (a bleaching agent and reducing agent with application in the textile and food industries) hydrosulfite *m* de sodium
~HYDROXIDE - [chem] (caustic soda. A caustic of major commercial importance with application in soap manufacture, textile processing, petroleum refining, and in the manufacture of other chemicals) hydrate *m* de sodium
~HYPOCHLORITE - [chem] (a bleaching agent, disinfectant, fungicide and intermediate) hypochlorite *m* de sodium
~HYPOPHOSPHITE - [chem] (white, granular, deliquescent powder with uses in medicine) hypophosphite *m* de sodium
~IODATE - [chem] (a disinfectant) iodate *m* de sodium
~IODIDE - [chem] (an analytical reagent, photographic chemical and a source of iodine in the treatment of thyrotoxicosis) iodure *m* de sodium
~IRON PYROPHOSPHATE - [chem] (a source of physiological iron as a foodstuff additive) pyrophosphate *m* sodique de fer
~ISOPROPYL XANTHATE - [chem] (a weedkiller) xanthate *m* isopropylique de sodium
~LACTATE - [chem] (a plasticizer and drying agent, also having application in medicine) lactate *m* de sodium
~LAMP - s. sodium-vapour lamp
~LAURYL SULPHATE - [chem] (a detergent) lauryl-

sulfate *m* de sodium
SODIUM LIGNOSULPHATE - [chem] (a stabilizing agent for emulsion) lignosulfate *m* de sodium
~METABORATE - [chem] (a weedkiller) métaborate *m* de sodium
~METANILATE - [chem] (a synthesis intermediate for pharmaceuticals and dyes) métanilate *m* de sodium
~METAPHOSPHATE - [chem] (a sequestering agent) métaphosphate *m* de sodium
~METASILICATE - [chem] (a component of detergents) métasilicate *m* de sodium
~METAVANADATE - [chem] (a photographic chemical) métavanadate *m* de sodium
~METHYLATE - [chem] (a catalyst and an intermediate for pharmaceuticals) méthylate *m* de sodium
~METHYL OLEOYL TAURATE - [chem] (a detergent) methyl-pléyl-taurate *m* de sodium
~MOLYBDATE - [chem] (an intermediate for pigments, corrosion inhibitor, analytical reagent, and a catalyst) molybdate *m* de sodium
~MONOXIDE - [chem] (a polymerization and condensation initiator) monoxyde *m* de sodium
~NAPHTHALENESULPHONATE - [chem] (a synthesis intermediate) naphtalène-sulfonate *m* de sodium
~NAPHTHENATE - [chem] (an intermediate for paint and varnish driers and an emulsifying agent) naphténate *m* de sodium
~NAPHTHIONATE - [chem] (an analytical reagent and an intermediate for dyestuffs) naphtionate *m* de sodium
~NITRATE - [chem] (chile saltpetre, caliche. An oxidising agent, starting material in the production of nitric and sulphuric acids, fertilizer, food preservative, analytical reagent, a starting material for drugs and dyestuffs, and a component of explosives and pyrotechnics) nitrate *m* de sodium
~NITRITE - [chem] (diazotising salts. A photographic chemical, analytical reagent food preservative, bleaching agent, and synthesis intermediate for dyestuffs. Sodium nitrate is also used in medicine) nitrite *m* de sodium
~NITROFERRICYANIDE - [chem] (an analytical reagent) nitroferricyanure *m* de sodium
~OLEATE - [chem] (a waterproofing agent for textiles and an ore flotation agent) oléate *m* de sodium
~ORTHO-PHENYLPHENATE - [chem] (a fungicide) ortho-phénylphénate *m* de sodium
~OXALATE - [chem] (a textile chemical and analytical reagent) oxalate *m* de sodium
~PARA-AMINOBENZOATE - [chem] (a white crystalline powder used in the treatment of rickettsial infections) para-aminobenzoate *m* de sodium
~PARA-AMINOHIPPURATE - [chem] (an aid in the assessment of kidney function) para-aminohippurate *m* de sodium
~PARA-AMINOSALICYLATE - [chem] (a tuberculostatic agent used in the treatment of tuberculosis) para-aminosalicylate *m* de sodium
~PENTABORATE - [chem] (a textile fireproofing agent and a weedkiller) pentaborate *m* de sodium
~PENTACHLOROPHENATE - [chem] (a wood preservative, herbicide and fungicide) pentachloraphénate *m* de sodium
~PERBORATE - [chem] (an oxidizing and bleaching agent and a dyeing assistant) perborate *m* de sodium
SODIUM PERCHLORATE - [chem] (an analytical reagent and a constituent of explosives) perchlorate *m* de sodium
~PERIODATE - [chem] (an analytical reagent) periodate *m* de sodium
~PERMANGANATE - [chem] (a disinfectant and oxidizing agent) permanganate *m* de sodium
~PEROXIDE - [chem] (a bleaching, oxidizing and deodorizing agent, bactericide, analytical reagent, and a source of oxygen) peroxyde *m* de sodium
~PEROXIDE CALORIMETER - [instr] (a calorimeter in which a weighed sample of solid or liquid is burnt by oxygen derived from sodium peroxide) calorimètre *m* à peroxyde de sodium
~PERSULPHATE - [chem] (a bleaching agent and analytical reagent) persulfate *m* de sodium
~PHENATE - [chem] (a synthesis intermediate and antiseptic) phénate *m* de sodium
~PHENYLPHOSPHINATE - [chem] (a stabilizer and antioxidant) phénylphosphinate *m* de sodium
~PHOSPHATE DIBASIC - [chem] (water-soluble translucer crystals with application in ceramics, food processing, fertilizers, the production of dyestuffs surface coating, and textile processing) phosphate *m* dibasique de sodium
~PHOSPHATE MONOBASIC - [chem] (an animal foodstuffs additive, dyeing assistant, and water treatment agent) phosphate *m* monobasique de sodium
~PHOSPHATE, TRIBASIC - [chem] (a component of water softening treatments, detergents, photographic and textile chemicals) phosphate *m* tribasique de sodium
~PHOSPHOMOLYBDATE - [chem] (an analytical reagent) phosphomolybdate *m* de sodium
~PHOSPHOTUNGSTATE - [chem] (an analytical reagent) phosphotungstate *m* de sodium
~-PHOTON COUNTER - [nucl] (a photon counter in which the sensitive substance contains pure sodium) compteur *m* de photons à sodium
~PICRAMATE - [chem] (an intermediate for dyestuffs and other organic compounds) picramate *m* de sodium
~POLYSULPHIDE - [chem] (an intermediate for dyestuffs, fuel additive and pesticides) polysulfure *m* de sodium
~POTASSIUM CARBONATE - [chem] (an analytical reagent) carbonate *m* de sodium et de potassium
~PROPIONATE - [chem] (a food preservative and fungicide) propionate *m* de sodium
~PYROPHOSPHATE - [chem] (emulsifying agent, dyeing assistant, water treatment chemical, and wool scouring chemical) pyrophosphate *m* de sodium
~REACTOR - s. sodium-cooled reactor
~RICINOLEATE - [chem] (an emulsifying agent) ricinoléate *m* de sodium
~SALICYLATE - [chem] (an analgesic and antipyretic in medicine) salicylate *m* de sodium
~SARCOSINATE - [chem] (an synthesis intermediate) sarcosinate *m* de sodium
~SELENATE - [chem] (an analytical reagent and a insecticide) séléniate *m* de sodium
~SELENITE - [chem] (a ceramic glaze and an analytical reagent) sélénite *f* de sodium
~SESQUICARBONATE - [chem] (a water softening agent and detergent) sesquicarbonate *m* de sodium
~SESQUISILICATE - [chem] (a textile chemical and constituent of industrial detergents) sesquisilicate

m de sodium
SODIUM SILICATE - [chem] (water glass.A constituent of special cements and adhesives and a textile and papermaking chemical) silicate *m* de sodium
~ SILICOFLUORIDE - s. sodium fluorosilicate
~ STANNATE - [chem] (a fireproofing agent and mordant for textiles and a component or ceramic glazes) stannate *m* de sodium
~ STEARATE - [chem] (a plastics stabilizer) stéarate *m* de sodium
~ SUCCINATE - [chem] (a respiratory stimulant with uses in medicine) succinate *m* de sodium
~ SULPHANILATE - [chem] (a synthesis intermediate, also having application in medicine in the treatment of upper respiratory tract infections) sulfanilate *m* de sodium
~ SULPHATE - [chem] (a component of ceramic glazes, and a paper, textile and leather chemical) sulfate *m* de sodium
~ SULPHIDE - [chem] (an analytical reagent, photographic chemical, intermediate for dyestuffs and a de-hairing agent in tanning; sodium sulphide also has application in metallurgy) sulfure *m* de sodium
~ SULPHITE - [chem] (a photographic chemical, synthesis intermediate, disinfectant, bleaching agent for textiles, food preservative, and rubber chemical) sulfite *m* de sodium
~ SULPHORICINOLEATE - [chem] (an emulsifying agent) sulforicinoléate *m* de sodium
~ SURAMIN - [chem] (a trypanocide) suramine *f* de sodium
~ TARTRATE - [chem] (an analytical reagent, diuretic and purgative) tartrate *m* de sodium
~ TETRACHLOROPHENATE - [chem] (a bactericide and fungicide) tétrachlorophénate *m* de sodium
~ TETRADECYL SULPHATE - [chem] (an anionic surface active agent with uses in medicine as an antiseptic and sclerosing agent) tétradécyl-sulfate *m* de sodium
~ TETRASULPHIDE - [chem] (a dyestuffs intermediate and ore floation agent) tétrasulfure *m* de sodium
~ THIOCYANATE - [chem] (a solvent for polyacrylates; also has application in rubber production) thiocyanate *m* de sodium
~ THIOGLYCOLATE - [chem] (an analytical reagent and a component of hair-waving lotions) thioglycolate *m* de sodium
~ THIOSULPHATE - [chem] (an analytical reagent, photographic chemical, antichlor and dyeing mordant, bleaching agent and flotation agent in metallurgy) thiosulfate *m* de sodium
~ TOLUENESULPHONATE - [chem] (a dyestuffs intermediate) toluène-sulfonate *m* de sodium
~ TRICHLOROACETATE - [chem] (a herbicide) trichloroacétate *m* de sodium
~ TRIPOLYPHOSPHATE - [chem] (a textile chemical and a component of soaps and detergents) tripolyphosphate *m* de sodium
~ TUNGSTATE - [chem] (an analytical reagent, fireproofing agent for textiles, and intermediate for tungsten and its compounds) tungstate *m* de sodium
~ UNDECYLENATE - [chem] (a bacteriostat in cosmetics and pharmaceuticals) undécylénate *m* de sodium
~ URANATE - [chem] uranate *m* de sodium
~ VALERATE - [chem] (unctuous, hygroscopic masses, used in the treatment of hysteria) valérianate *m* de sodium
~ -VAPOUR LAMP - [el] lampe *f* à vapeur de sodium
~ -VOID COEFFICIENT - [nucl] (in a fast-neutron reactor) coefficient *m* de sodium
~ XYLENE SULPHONATE - [chem] (a component of detergents) xylène-sulfonate *m* de sodium
~ ZIRCONIUM LACTATE - [chem] (an antiperspirant) lactate *m* de sodium et de zirconium
SOFAR - [radar] (abbrev for sound fixing and ranging) localisation *f* acoustique dans la mer par propagation guidée
SOFFIT - [arch] (the undersurface of a stair or of the head of an opening) soffite *m*, intrados *m*
SOFT - [gen] mou, doux
[chem] doux
[metall] doux
[plast ind] mou, souple
[photo] flou
[radiat] (denoting a less penetrating quality of Xrays etc) mou
[text] (of cloth, yarn etc) doux, à faible torsion
[ceramics] tendre
~ ANNEALING - [metall] recuit *m* complet
~ BOBBIN - [text] bobine *f* molle
~ BRASS - [metall] laiton *m* ductile
~ BRAZING-SOLDER - [metall] brasure *f* tendre
~ BROACHING - [mach tool] (broaching before hardening) alésage *m* à l'état doux
~ BRUSH - [impl] pinceau *m*
~ CLAY - [geol] argile *f* plastique
~ COAL - [min] houille *f* grasse
~ COMPONENT OF COSMIC RAYS - [radiat] (the portion of cosmic radiation which is absorbed in a moderate thickness of an absorber) composante *f* molle
~ COPPER - [metall] cuivre *m* doux
~ -COPPER WIRE - [metall] fil *m* de cuivre doux
~ DOWNY FEEL - [text] toucher *m* doux
~ FEEL - [rubber ind] douceur *f*
~ FIBRED HEMP - [text] chanvre *m* à filasse douce
~ FLOW - [plast ind] (ability of a material to flow easily under normal moulding conditions) haute fluidité *f*
~ -FOCUS LENS - [photo] objectif *m* anachromatique
~ GOODS - [comm] étoffes *fpl*, tissus *mpl*
~ GRADE - [mech] degré *m* de dureté tendre
~ GRINDING WHEEL - [mech] meule *f* tendre
~ HAIL - [met] (hail in the form of small soft opaque white pellets) neige *f* roulée
~ HAIR - [text] poil *m* doux
~ IRON - [metall] fer *m* doux
~ -IRON ARMATURE - [el] (the attracted part of an electromagnet retaining little residual magnetism) induit *m* en fer doux
~ -IRON CORE - [el] noyau *m* en fer doux
~ -IRON INSTRUMENT - [instr] (preferably moving-iron instrument; an instrument the operation of which depends on the force exerted by a fixed permanent magnet on a movable coil carrying a current) instrument *m* à fer doux
~ IRON ORE - [min] mine *f* de fer douce
~ -IRON OSCILLOGRAPH - [instr] (an oscillograph making use of the action of a coil on a strip of soft iron controlled by a permanent field) oscillographe *m* à fer doux
~ -KNITTING YARN - [text] (four-ply worsted yarns used for knitting stockings) fil *m* peigné doux

SOFT LANDING - [astronaut] (controlled landing, e.g. on the moon's surface) atterrissage *m* doux
~LEAD - [metall] plomb *m* doux
~LIGHTING - [cin] (the lighting of objects in such way that there is on strong demarcation between highlights and shadows) éclairage *m* donnant des ombres à contours flous
~NOSE OF COP - [text] pointe *f* de cannette éboulée
~PALATE - [anat] voile *m* du palais
~-PASTE PORCELLAIN - [ceram] (porcelain composed of glass frit and white clay) porcelaine *f* à pâte tendre
~PEDAL - [mus] (extra pedal on grand pianos) pédale *f* douce
~PICTURE - [photo] (picture without sharp contrasts) image *f* à faible contraste
~PIG IRON - [metall] fonte *f* douce
~PORCELAIN - [ceramics] porcelaine *f* tendre
~SIZE - [ind chem] colle *f* douce
~-SIZED - [paper man] (paper in which a small quantity of size has been used) demi-collé
~-SIZED PAPER - [paper man] papier *m* demi-collé
~SKIN - [gen] peau *f* lisse, peau *f* tendre
~-SKINNED - [metall] décarburé
~SOAPS - [chem] (potassium salts of the fatty acids) savons *mpl* mous
~-SOLDER, to - [mech] souder à l'étain
~-SOLDER - [metall] soudure *f* tendre
~SOLDERED - [metall] soudé tendre
~SOLDERING - [metall] (operation in which metallic pieces are joined by means of a molten filler metal having a melting temperature lower than that of the pieces to be joined) soudure *f* à l'étain
~SPUN YARN - [text] fil *m* filé floche
~STEEL - [metall] acier *m* doux
~-STAFF - [paper man] pâte *f* molle
~THREAD - [text] fil *m* souple
~TUBE - s. soft valve
~VALVE - [radio] (old type of thermionic valve not exhaust to a high vacuum) tube *m* mou
~WARES - s. soft goods
~WASTE - [text] déchets *mpl* de tissage
~WATER - [geol] eau *f* douce
~WAX - [ind chem] cire *f* molle
~WINDING - [text] bobinage *m* mou
~WOOD - s. softwood
~WORSTED YARN - [text] fil *m* peigné doux
SOFTEN, to - [gen] amollir, ramollir
[chem] adoucir, épurer
[metall] adoucir
[opt] (a colour etc) atténuer
[leather ind] (a skin) détremper, reverdir
~THE GREASE, to - [text] ramollir la graisse (de la laine)
~THE SERICIN, to - [text] amollir la séricine
SOFTENED RUBBER - [rubber ind] caoutchouc *m* ramolli
SOFTENER - [gen] émollient *m*, plastifiant *m*, plastificateur *m*
[impl] blaireau *m*
[chem] (substance for softening water) adoucisseur *m*
[rubber ind] plastifiant *m*
[text] (machine used to soften textile fibres) moulin *m* à piler (le chanvre)
[ceram] (tool used to spread colour on a biscuit) pinceau *m*
SOFTENER - [photo] (diffusing disk) disque *m* diffuseur
SOFTENING - [gen chem etc] amollissement *m*, ramollissement *m*
[chem] (of water) adoucissement *m*
[radiat] amollissement *m*
[metall] (reduction of hardness by heat treatment) adoucissement *m*
[text] pilage *m*, maillochage *m*
[paint] (a defect) ramollissement *m*
[metall] (of lead-removal of arsenic, antimony and tin from lead by oxidation in the reverberatory furnace) adoucissement *m*, adoucissage *m*
[leather ind] assouplissement *m*
~AGENTS - [rubber ind] (substances used in rubber manufacture to aid compounding and pigment dispersion, improve surface finish and prevent premature vulcanization) plastifiants *mpl*
~BY BEATING - [text] pilage *m* (du chanvre)
~IN PITS - [leather ind] assouplissement *m* en fosses
~MACHINE - [text] assouplisseuse *f*
~OF THE BRAIN - [med] cérébro-maladie *f*
~POINT - [plast ind] (temperature at which a substance e.g.a.plastic, becomes soft and semi-fluid) point *m* de ramollissement
~RANGE - [plast ind] (range of temperatures over which softening occurs in plastics) plage *m* de ramollissement
~TEMPERATURE - [phys] (an approximately definite physical constant of a substance which does not have a definite melting point) température *f* de ramollissement
SOFTNESS - [gen] douceur *f*, mollesse *f*
[rubber ind] fluidité *f*
[photo] (lack of strong contrasts) flou *m*
[text] (of fibres) douceur *f*
~INDEX - [chem] (a German official standard of fluidity) indice *m* de fluidité
SOFTWARE - [comput] (programming aids) aides *f* à la programmation
SOFTWOOD - [bot] bois *m* tendre
SOGASOID - [phys] (dispersed system of a solid in a gas) dispersion *f* d'un solide dans un gaz
SOIL, to - [gen] salir, souiller
[agric] alimenter à fourrage vert, mettre au vert
SOIL - [gen] (ground in general) sol *m*, terrain *m*
[geol] (finely divided rock mixed with decayed vegetables) terrain *m*, terre *f*, sol *m*
[agric] engrais *m*
~AERATION - [agric] aération *f* du sol
~AGGREGATE - [soil] agrégat *m* de sol
~ANALYSIS - [agric] analyse *f* du terrain
~BIOLOGY - [agric] biologie *f* du sol
~CHEMISTRY - [chem] chimie *f* des sols
~COMPACTION - [agric] durcissement *m* du sol, tassement *m* du sol
~CONCRETE - [constr] béton *m* d'argile
~CONDITIONING - [agric] amendement *m* du sol, amélioration *f* du sol
~CONDITIONS - [agric] état *m* du sol
~CONSERVATION - [agric] conservation *f* des sols
~CONSOLIDATION - s. soil compaction
~CONSTITUENTS - [agric] constituants *mpl* du sol
~COVER - [agric] couverture *f* du sol avec des déchets organiques
~DENSITY - [geol] densité *f* du sol
~DEVELOPMENT - [geol] pédogénèse *f*
~EMBANKMENT - [constr] remblais *mpl* en terre

SOIL ENGINEERING - [constr] technique *f* des ouvrages en terre
~EROSION - [geol] érosion *f* du sol
~EXHAUSTION - [agric] fatigue *f* du sol
~FILLING-UP - [constr] remblai *m*
~FORMATION - s. soil development
~FUMIGANTS - [agric] (chemicals which can be injected into the soil where their toxic vapour destroy harmful organisms) fumigants *mpl* pour l'assainissement du sol
~GRADER - [agric] niveleuse *f*
~GRAIN PROPERTIES - [soil] propriétés *fpl* du grain élémentaire
~INJECTOR - [agric] pal *m* injecteur
~LAYER - [geol] horizon *m*, couche *f* du sol
~LOADING - [agric] pression *f* sur le sol
~LOOSENING - [agric] ameublissement *m* du sol
~MAPPING - [surv] cartographie *f* pédologique
~MOSTURE - [agric] humidité *f* du sol
~PERMEABILITY - [geol] perméabilité *f* du sol
~PIPE - [hydr] (vertical pipe conveying waste matter to the drains) tube *m* de descente, tuyau *m* de chute
~POISONING - [soil] empoisonnement *m* du sol
~POROSITY - [geol] porosité *f* du sol
~PROFILE - [surv] profil *m* du sol
~REACTION - [agric] réaction *f* du sol
~-RETAINING STRUCTURE - [constr] rideau *m* de soutènement, mur *m* de butée
~SOLUTION - [bot] (or soil water; the dilute acqueous solution of mineral salts) eau *f* du sol
~SPECIES - [constr] désignation *f* de terre, nature *f* du sol
~SPECIMENT - [geol] échantillon *m* du sol
~STABILITY - [constr] stabilité *f* du sol
~STABILIZATION - [soil] stabilisation *f* du sol
~STABILIZATION BY FREEZING - [soil] stabilisation *f* du sol par congélation
~STABILIZER - [soil] agent *m* stabilisateur des sols
~STRAIN - [constr] compression *f*, tension *f*
~STRUCTURE - [geol] structure *f* du sol
~SURVEY CHART - [surv] carte *f* agronomique
~SWELL - [geol] foisonnement *m* du terrain
~TESTING - [geol] expériences *fpl* sur carottes, essais *mpl* de sol
~TEXTURE - [geol] texture *f* du sol
~VIBRATOR - [mech] compacteur *m* de sol vibrant
~WATER - s. soil solution
~WATER BELT - [hydr] zone *f* radiculaire
SOILING - [agric] fourrage *m* vert
~CROPS - [agric] (animal feeding) ration *f* de base
SOL - [chem] (a liquid colloïdal suspension) sol *m*
SOLAR - [gen & astr] (of the sun) solaire
~ACTIVITY - [astr] activité *f* solaire
~APEX - [astr] (on the celestial sphere; the point toward which the solar system is moving) apex *m* du soleil
~ATMOSPHERIC TIDE - [astr] marée *f* atmosphérique solaire
~BATTERY - [el] pile *f* solaire
~CELL - [el] cellule *f* photovoltaïque, cellule *f* solaire
~COMPASS - [instr] boussole *f* solaire
~CONSTANT - [phys] (the intensity of solar radiation in free space at the earth's mean solar distance) constante *f* solaire
~CONSTANT OF RADIATION - [astr] (the intensity or radiation from the sun (= I.34 x I0^{6} ergs per sq. cm. per sec.) constante *f* solaire de rayonnement
SOLAR CORONA - [astr] (luminous region seen round the sun during a total eclipse of that body) couronne *f* solaire
~COSMIC RAYS - [astr] rayons *mpl* cosmiques solaires
~DAY - [astr] (the period between successive transits of the sun across the same meridian) jour *m* solaire
~ECLIPSE - [astr] éclipse *f* solaire
~ENERGY - [phys] (the energy originating in the sun) énergie *f* solaire
~FLARE - [astr] éruption *f* solaire
~OIL - [chem] (gas, or diesel oil) huile *f* lourde, mazout *m*, huile *f* solaire
~PANEL - [astronaut] panneau *m* de cellule solaire
~PLEXUS - [anat] plexus *m* solaire
~POWER STATION - [el] centrale *f* héliothermique
~-POWERED - [astronaut] à énergie solaire
~PROMINENCE - [astr] protubérance *f* solaire
~RADIATION - [phys] (the radiation from the sun confined between the long infra-red rays and the ultraviolet rays, with a maximum intensity in the visible green at about 5000 angstroms) rayonnement *m* solaire
~SALT - [chem] (salt obtained by evaporation of seawater or other natural brine by the sun's rays) sal *m* marin
~STORMS - [astr] tempêtes *fpl* solaires
~SYSTEM - [astr] (the sun and the attendant bodies moving about it under gravitational attraction) système *m* solaire
~WIND - [astr] vent *m* solaire
SOLARIMETER - [instr] (instrument designed to measure the intensity of solar radiation) solarimètre *m*, pyranomètre *m*
SOLARIUM - [constr] (room or enclosed porch exposed to the sun) solarium *m*
SOLARIZATION - [photo] (the reversal of an image due to excessive exposure to light) solarisation *f*
SOLAROMETER - [instr] (instrument designed to find a ship's position) solaromètre *m*
SOLATION - [chem] (the process of liquefaction of a gel) gélatinisation *f*
SOLDER, to - [mech] (to join by means of solder) souder
SOLDER - [mech] (an alloy of relatively low melting point used to join other metals together. Solder commonly refers to a lead-tin alloy) produit *m* à souder [gen] soudure *f*, brasure *f*
~LUG - [metall] broche *f* à souder
~STRAP - [mech] pont *m* à souder
SOLDERED - [mech] soudé
~-IN DENTS - [mech] dents *fpl* soudées
~-REED - [text] peigne *m* soudé, peigne *m* en étain
SOLDERING - [mech] soudure *f*, soudage *m*
~ACID - s. soldering fluid
~BIT - [tool] (tool used in soldering, consisting of a shaped copper block fitted with a handle) fer *m* à souder
~BOLT - s. soldering bit
~BY INDUCTION - [metall] brasage *m* tendre par induction
~COPPER - [tool] fer *m* à souder, soudoir *m*
~FLUID - [chem] (a liquid flux for use in soldering) fondant *m* de soudage

SOLDERED FLUX - s.soldering fluid
~FURNACE - [metall] four *m* à souder
~IRON - [tool] (tool for soldering consisting of a small block of copper attached to a handle) soudoir *m*, fer *m* à souder
~LAMP - [impl] lampe *f* à souder
~LUG - [el] extrémité *f* soudée d'une broche
~SEAM - [mech] cordon *m* de soudure
~SPOT - [mech] joint *m* des enveloppes de plomb des câbles
~TAG - s.soldering lug
~TIN - [metall] étain *m* à souder
~TORCH - s.soldering lamp
~WITH THE BLOW-PIPE - [metall] soudure *f* au chalumeau
~WITH THE SOLDERING BIT - [metall] soudure *f* au fer à souder
SOLDERLESS WRAPPED - [electron] connexion *f* enroulée
SOLDIER - [constr] (a course of bricks, all standing on end) bahut *m*, rouleau-brique *m*
[metall] (reinforcement for a mould) renforcement *m*
[text] (a defective sheepskin) peau *f* lesée
[zool] (in insects, a strong specimen adapted for defending the community) soldat *m*, téléphore *m*
~COURSE - s.soldier [constr]
SOLE - [gen] (alone) seul, unique
[anat] plante *f* du pied
[zool] sole *f*
[shoe man] semelle *f*
[th eng] (of a furnace) sole *f*
[shipbuild] talon *m*, talonnière *f*
[gen] (of any implement) semelle *f*, plan *m* racinal de grue
[geol] socle *m*, substratum *m*
[agric] (of a plough) sep *m*
[constr] plate-forme *f*, patin *m* de fondation
~AND HEEL CUTTING MACHINE - [shoe man] machine *f* à poinçonner les semelles et talons
~-BAR - [mech] (in vehicles) longeron *m*
~BRICK - [constr] (a slab, or paving brick) radier *m*
~EDGE TRIMMING - [shoe man] fraisage *m* des bords des semelles
~FELT - [shoe man] feutre *m* pour semelles
~FLUE - [th eng] carneau *m* de sole
~LAYING - [shoe man] pose *f* des semelles
~LEATHER - [shoe man] cuir *m* pour semelles
~OF HOOF - [zool] sole *f* du sabot
~ROUNDER - [shoe man] (a tool) fraiseur *m* de semelles
~-PLATE - [mech] (the bed-plate of a machine) sole *f*, plaque *f* d'assise
[mech] (the bed-plate of a marine engine) taque *f* d'assise, plaque *f* de fondation
[mech] (adjustable plate for a slight adjustment of a bearing) patin *m* d'un palier
[railw] selle *f* d'arrêt
SOLEIL WEAVE - [text] armure *f* soleil
SOLENOID - [el] (a coil, usually of tubular form, for producing a magnetic field) solénoïde *m*
~BRAKE - [el] (electro-mechanical brake in which the closing force is provided by a solenoid) frein *m* à solenoïde
~BRAKING - [el] freinage *m* électromagnétique par solénoïde
~CURRENT - [el biol] (or d'Arsonval current; a current of intermittent and isolated trains of heavily damped oscillations, of high frequency, high voltage and relatively low aperage) courant *m* de solénoïde
SOLENOID-FOCUSED TRAVELLING-WAVE TUBE - [electron] tube *m* à ondes progressives à focalisation par solénoïde
~MAGNETIZATION - [el] magnétisation par solénoïde
~-OPERATED CONTROL - [contr] (control in which the acting force is provided by a solenoid) commande *f* actionnée par solénoïde
~-OPERATED VALVE - s. solenoid valve
~PULL - [el] puissance *f*, puissance *f* du soléinoïde
~RELAY - [el] relais *m* à plongeur
~STARTER - [el] (automatic starter in which a single operating electromagnet of the solenoid and core type is used for cutting out a resistor in a series of steps) démarreur *m* à soléinoïde
~SWITCH - [el] (switch in which the closing force is provided by a solenoid) interrupteur *m* électromagnétique à solénoïde
~VALVE - [el] (a valve controlled by a solenoid) soupape *f* à selénoïde
~VALVE OPERATOR - [contr] organe *m* d'actionnement de soupape à selénoïde
SOLENOIDAL - [el] (adjective of solenoid) solénoïdal
~FIELD - [el] (a field in which the divergence is zero) champ *m* solénoïdal
SOLEPLATE - s.sole-plate
SOLFATARA - [geol] (volcanic orifice in a dormant stage, from which sulphur dioxide and other gases are emitted) solfatare *f*, soufrière *f*
SOLID - [gen] solide, consistant
[math] (a geometric solid) solide *m*, corps *m* solide
[gen] (not hollow) plein, massif
[meas] (of measures) de volume (mesures)
[print] (set solid) non-interligné, serré
[print] aplat *m* de couleur, composition *f* pleine
[mech] (integral) solidaire
~ADSORBENT - [ind chem] (in gas analysis) adsorbant *m* solide
~ANGLE - [geom] angle *m* solide
~ANODE - [electron] (or heavy anode; an anode which consists of a solid piece of metal) anode *f* massive
~AXLE - [mech] essieu *m* plein
~BALL-TYPE CENTRIFUGE - [mech] essoreuse *f* à panier plein
~BEARING - [constr] (the support underneath a beam when it is supported along its whole length) appui *m* solide (des poutres)
~BIT - [mining] bit *m* plein
~BLOCKAGE - [aero] (wind tunnel blockage due to obstruction of the stream by the model itself) blocage *m* dû au model
~BRICK - [constr] brique *f* pleine
~CARBONS - [el] (carbon electrode for electric arc lamps, a core of softer material being absent) charbons *mpl* homogènes
~CARBURIZING - [metall] cémentation *f* par solides
~CASTING - [metall] coulage *m* plein
~COCOON - [text] cocon *m* solide
~CONDUCTOR - [el] (conductor consisting of a single wire, which may be composite) conducteur *m* massif, conducteur *m* simple
~CONTENT - [paint] residu *m* sec
~CONTRACTION - [metall] (in foundry work, from

the melting point to the ambient temperature) retrait *m* dans l'état solide
SOLID COP - [text] cannette *f* à dévidage intérieur, cannette-cocon *f*
~CORE - [el] noyau *m* massif
~CYCLIZED RUBBER - [rubber ind] caoutchouc *m* cyclisé en partant du caoutchouc solide
~CYLINDER - [geol] corps *m* solide cylindrique
~DIE - [metall] (in diecasting) filière *m* fermée, lunette *f*, lunette *f* à tarauder
~-DIE STOCK - [metall] filière *f* à lunettes
~DIELECTRIC - [electron] diélectrique *m* solide
~DIFFUSION - [metall] (the movement of atoms through the crystals of a solid metal, as when carbon diffuses into, or out of, steel during carburizing or decarburizing) diffusion *f* dans l'état solide
~DRAWING - [metall] étirage *m* de la loupe
~-DRAW - [metall] (of a weedles tube) étiré de la loupe
~END - [el] tête *f* de câble
~-EXPANSION THERMOMETER - [instr] (a thermometer in which used is made of the expansion of solid bodies as an indicator of temperature changes) thermomètre *m* à index
~FLOOR - [constr] (floor made of wood blocks laid on concrete) pavage *m* en bois
~-FORGED - [metall] monobloc
~FORM - [phys] état *m* solide
~FRICTION - [mech] frottement *m* de glissement
~FUEL - [min] combustible *m* solide, carburant *m* solide
~GEOMETRY - [geom] geométrie *f* à trois dimensions
~-HEAD - [mech] (of internal combustion engine) monobloc
~-HUB PULLEY - [mech] poulie *f* en une seule pièce
~HYDROCARBONS - [chem] hydrocarbures *mpl* solides
~INJECTION - [mech] injection *f* directe
~JIB - [mech] bras *m* en tôle pleine
~LINE - [draw] trait *m* plein
~MATTER - [print] (matter without leads) composition *f* pleine
~MEASURE - [meas] mesure *f* de volume
~NEWEL - [constr] (the centre post of a winding stair) noyau *m* (d'escalier) plein
~-NEWEL STAIR - [constr] escalier *m* à noyau plein, vis *f* à noyau plein
~OF REVOLUTION - [phys] (a solid generated by the revolution of a plane area about a line, the axis of revolution) corps *m* de révolution, solide *m* de révolution
~NOSE - [aero] cône *m* avant sans ouvertures
~OF UNIFORM STRENGTH - [mech] solide *m* d'égale résistance
~PANEL - [carp] (a panel the surface of which is in line with the faces of the stiles) pan *m* plein
~PHASE - s. solid state
~PHASE WELDING - [metall] soudage *m* par pression
~PISTON - [mech] plongeur *m*, piston *m* plein
~PULLEY - s. solid-hub pulley
~PUNCH - [impl] poinçon *m*
~ROCK - [geol] roche *f* compacte
~RUBBER-CARPED BLOCKS - [rubber ind] blocs *mpl* de pavé surmontés d'une couche en caoutchouc
~-RUBBER TYRE - s. solid tyre
SOLID SHAFT - [mech] arbre *m* plein
~SOLUBILITY - [metall] solubilité *f* dans l'état solide
~SOLUTION - [chem] (a component of an alloy which is produced by the incorporation of atoms of one element in the crystals of another. Usually there is a substitution of some of the atoms concerned) solution *f* solide
~STATE - [phys] (that state of matter in which molecules or ions have no motion of translation and can only move within limits about fixed mean positions) état *m* solide
~-STATE CIRCUIT - [electron] circuit *m* solide
~STATE COMPONENTS - [telev] éléments *fpl* d'état solide
~STATE PHYSICS - [phys] (the branch of physics dealing with the relation of molecular structure to the properties of solids) physique *f* de l'état solide
~STEEL - [metall] acier *m* calmé
~STRAND - [el] noyau *m* du conducteur
~TYRE - [rubber ind] bandage *m* plein, pneu *m* tout caoutchouc
~WHEEL - [mech] roue *f* pleine
~WIRE - s. solid strand
~YOLK - [text] suint *m* semi-liquide, suint *m* suiffeux
SOLIDIFIABLE - [gen] solidifiable, congelable
SOLIDIFICATION - [gen] (becoming solid) solidification *f*
[constr] (of the ground) compression *f* (du sol)
[chem] (of the oil) congélation *f* (de l'huile)
~DIAGRAM - [metall] (a diagram showing the temperatures at which solidification occurs in alloys) diagramme *m* de solidification
~OF THE SERICIN - [text] solidification *f* de la séricine
~POINT - [phys] point *m* de solidification
~RANGE - [chem & metall] (the range of temperature in which solidification occurs in alloys and silicate melts) intervalle *m* de solidification, zone *f* de solidification
SOLIDIFIED GASOLINE - [ind chem] gélatine *f* de essence de pétrole
~GREASE - [ind chem] graisse *f* (de laine) solidifiée
~SILK FILAMENTS - [text] fils *mpl* de soie solidifiés
SOLIDIFY, to - [gen] solidifier
[phys] (to bring a liquid to the solid state) se solidifier, se congeler,
[constr] (the ground) se comprimer, se concréter
SOLIDITY - [gen] solidité *f*
[aero] (of a propeller; the relation between the disk area and the total blade area) coefficient *m* de plénitude
~RATIO - [aero] (of a propeller) rapport *m* de plénitude
SOLIDUS - [math] solidus *m*
~CURVE - [phys] (a curve representing the equilibrium between the solid phase and the liquid phase in a condensed system of two components) diagramme *m* de solidification
SOLIFLUCTION - [geol] (soil-creep on sloping ground, mostly occurring in regions subjected to periods of alternating freezing and thawing) solifluction *f*
SOLIFLUXION - s. solifluction
SOLING MATERIAL - [shoe man] feuilles *fpl* à semelles
~PLATE - [shoe man] croupon *m*
~SHEETING - s. soling material

SOLIQUID - [phys] (dispersed system of a solid in a liquid) dispersion *f* d'un solide dans un liquide
SOLO FLIGHT - [aero] (flight in which the pilot is alone in the machine) vol *m* solo
SOLSTICES - [astr] (the two moments during the year when the sun is at its apparent maximum distance from the celestial equator) solstices *mpl*
SOLSTITIAL POINT - [astr] point *m* solsticial
~COLURE - [astr] grand cercle *m* des solstices
SOLUBILITY - [phys chem] (property of a substance by which it forms mixtures with other substances which are chemically and physically homogeneous throughout) solubilité *f*
~BOOSTER AGENT - [chem] agent *m* pour l'accélération de la solubilité
~COEFFICIENT - [phys] (the volume of gas dissolved by unit volume of solvent) coefficient *m* de solubilité
~CURVE - [phys chem] (a curve drawn to show the variation of solubility of a given substance with the temperature) courbe *f* de solubilité
~OF THE CELLULOSE - [text] solubilité *f* de la cellulose
~OF THE WOOL - [text] solubilité *f* de la laine
~PRODUCT - [chem] (the value at saturation of the product of the activities of the ions into which a dissolved substance concentrates) produit *m* de solubilité
~STARCH - [chem] (a product of the hydrolysis of starch obtained by treating the latter with dilute acids) amidon *m* soluble
SOLUBLE - [phys] soluble
~BLUE - [paint] bleu *m* à l'eau
~COLORANT - [dyes] (colouring matter which will dissolve in a solvent (usually organic pigments) colorant *m* soluble
~COTTON - [chem] (colloidal cotton) coton *m* de collodion
~CUTTING OIL - [oil ind] huile *f* de coupe soluble
~GLASS - [chem] (solid sodium or potassium silicates) silicate *m* de sodium solide
~IN WATER - [chem] soluble dans l'eau
~MATTER LOSS - [ind chem] perte *f* de la matière soluble
~NYLON RESIN - [chem] (applications as a coating and binder for paper and textiles and as an adhesive) nylon *m* sous forme de résine soluble
SOLUPLASTIC - [chem] (said of a material becoming plastic when moistened with a suitable solvent) soluplastique
SOLUTE - [chem] (in a solution, the substance with is dissolved) produit *m* dissous, corps *m* dissous, dissous, en solution
SOLUTION - [chem] (a very intimate mixture of two or more substances, one of which is generally a liquid. The constituents of a solution may be separated by simple physical methods) solution *f*
[gen] (of a problem etc) solution *f*, résolution *f*
~ADHESIVE - [chem] (an adhesive applied in the form of a solution, which hardens by evaporation of the solvent, as distinct from any chemical change) solution *f* adhésive
~CHEMISTRY - [phys-chem] (the chemistry of substances in dissolved state) chimie *f* des solutions
~-FEED CHLORINATION - [hydr] application *f* de chlore en solution
~-FEED DOSAGE - [hydr] dosage *m* par solution, alimentation *f* par solution
SOLUTION GAS DRIVE - [oil ind] production *f* par expansion de gaz dissous
~HEAT-TREATMENT - [metall] traitement *m* thermique à dissolution améliorée des métaux d'addition
~MILL - [rubber ind] mélangeur *m* à dissolution, pétrisseur-malaxeur *m* pour dissolution
~MIXER - [oil ind] mélangeur *m* pour solution
~POTENTIAL - [el] potentiel *m* entre électrode et solution
~PRESSURE - [chem] (the tendency of a substance to pass into solution) pression *f* de solution
~READY FOR SPINNING - [text] solution *f* prête au filage
~RECLAIMING PROCESS - [rubber ind] procédé *m* de régénération par dissolution
~STRAINER - [impl] passoire *f* à solution
~STRIPPING - [metall] dépouillage *m* chimique
~TANK - [hydr] bac *m* de dissolution
~TREATMENT - [metall] (the operation of heating suitable alloys in order to take the hardening constituent into solution) tempre *f* de solution
SOLUTROPE - [chem] (a mixture having two liquid phases in which the solute may be selectively dissolved dependent upon concentration of the solvent) solutrope *m*
SOLVATION - [chem] (the combination of a solute with a solvent) solvatisation *f*
~OF IONS - [phys] (the attachment of molecules of solvent to ions) solvatisation *f*
SOLVAY PROCESS - [chem] (production of sodium carbonate from common salt by reaction with ammonium bicarbonate and heating) méthode *f* Solvay
SOLVE, to - [gen] résoudre
SOLVENCY - [chem] solvabilité *f*
SOLVENT - [chem] (a substance which effects solution of another) solvant *m*, dissolvant *m*
[comm & leg] solvable
~BLUSHING - [paint] (temporary milky appearance in a coating film, due to rapid evaporation of solvent) turbidité *f*
~CLEANING - [el metall] (removal of grease by organic solvents) dégraissage *m* au solvant
[mech] dégraissage *m* au solvant
~DECARBONIZING - [oil ind] décarbonisation *f* au solvant
~DEWAXING - [oil ind] déparaffinage *m* par solvant
~DEWAXING PROCESS - [oil ind] procédé *m* de déparaffinage par solvant
~EMULSION DEGREASING - [mech] (method of removing grease by the use of a cresylic acid type of compound in emulsion in a petroleum fraction, in which the parts are soaked at room temperature) dégraissage *m* par émulsion
~EXTRACTION - [chem] (the extraction of the component in a solution, which is in excess) extraction *f* par dissolvant
~EXTRACTOR - [chem] extracteur *m* par dissolvant
~MATERIAL - [ind chem] matière *f* dissolvante
~NAPHTHA - [chem] (a mixture of xylene, benzene, and other coal-tar derivatives with uses as a solvent) solvant *m* naphta
~RECOVERY - [ind chem] (the operation of condensing and collecting the solvent evaporated during the process) récupération *f* du solvant
~RECOVERY PLANT - [mech] installation *f* pour la récupération des solvants
~REGENERATION PROCESS - [rubber ind] régénéra-

tion *f* par dissolution
SOLVENT RESISTANCE - [chem] (the ability of a material to resist the action of a solvent or solvents) résistance *f* à un solvant
~SEALING - [chem] (sealing of plastic articles by the use of a solvent) soudure *f* par solvant
~TOLERANCE - [chem] (the degree to which a concentrated solution can be diluted without forming a precipitate) tolérance *f* aux solvants
~-TYPE ADHESIVE - [ind chem] (an adhesive the action of which depends on the evaporation of the vehicle than on polymerization or other actions) colle *f* au solvant
SOLVOLYSIS - [chem] (generalized conception of the relation between a solvent and a solute whereby the solvent molecule gives a proton to, or both, forming one or more different molecules) solvolyse *f*
SOMA - [zool] (the body of an animal as distinct from the germ-cells) soma *m*, corps *m*
SOMATIC - [zool] somatique
~CELL - [zool] (one of the non-reproductive cells of the parent body) cellule *f* somatique
~MITOSIS - [biol] (division of the metabolic nucleus) mitose *f* somatique
~MUTATION - [genet] (mutation arising in a sometic cell and not in a reproductive structure) mutation *f* somatique, mutation *f* de bourgeon
~NUMBER - [biol] (the number of somatic cells in a body) nombre *m* somatique
~SEGREGATION - [genet] (a change in nuclear or hereditary constitution during vegetative growth) ségrégation *f* somatique
SOMATOBLAST - [genet] (in development, a cell which will produce somatic cells) somatoblaste *m*
SOMATOIDS - [chem] (small particles of definite shape and possessing a definite arrangement of matter, yet not homogeneous) somatoïdes *mpl*
SOMATOLOGY - [biol] (the science of organic bodies, in particular human bodies) somatologie *f*
SOMATOPLEURA - [anat] membrane *f* pariétale, somatopleure *f*
SOMERSAULT, to - [aero] capoter, faire panache
[auto] capoter
SOMITE - [biol] (a mesoblastic segment in a developing embroy) somite *m*, métamère *m*, segment *m*
~CAVITY - [anat] cavité *f* du somite
SOMMERING LINES - [constr] (the radiating lines relating to the direction of the bed-joints of the voussoir of an arch) lignes *fpl* de rayonnement de la clé d'une voûte
SOMNAMBULISM - [med] (the habit of walking in sleep) somnambulisme *m*
SOMNIAFACIENT - [med] somnifère, soporifique
SOMNIFIC - s. somnifacient
SOMNILOQUISM - [med] somniloquie *m*
SOMNOLENCE - [med] somnolence *f* assoupissement *m*
SONAR - [radar] (sound navigation and ranging) sonar *m*, radar *m* ultrasonique
~CAPSULE - [radar] réflecteur *m* sonar
SONATE - [mus] (extended instrumental composition) sonate *f*
SONDE - [met] (or probe; instrument attached to a balloon and transmitting information to a ground station) sonde *f*
SONE - [acoust] (unit of loudness) sone *m*
SONIC - [acoust] (relating to sound, especially relating to the speed of sound) sonique
SONIC AGGLOMERATION - [acoust] agglomération *f* par ondes acoustiques
~ALTIMETER - [instr] (altimeter using sound-wave reflection) altimètre *m* acoustique
~AREA RULE - [acoust] (this states that, at the speed of sound, wave drag is the same for all bodies having the same longitudinal distribution of cross-sectional area) loi *f* de l'aire sonique
~BARRIER - s. sound barrier
~BOOM - [acoust] (the sound produced by an aircraft etc when passing through the sound barrier) band *m* sonique
~DELAY LINE - [comput] (a device for transforming electric or electromagnetic oscillations in mechanical acoustic ones) ligne *f* de retard acoustique
~DEPTH FINDER - [instr] sonde *f* acoustique
~DETECTOR - [instr] (acoustic apparatus for detecting the presence and course of an aircraft by the sound emitted by it) détecteur *m* sonique
~DRILLING - [mech] usinage *m* aux ultrasons
~FREQUENCY - [phys] audiofréquence *f*
~LOG - [oil ind] log *m* sonique
~SOLDERING - [metall] soudage *m* au ultrasons
~SPEED - [phys] vitesse *f* du son
~TESTER - [instr] vérificateur *m* sonique
~TESTING - [metall] essai *m* acoustique
~WAVE - s.sound wave
SONIUM - [metall] (solid non-metallic inclusions in metal) inclusion *f* non métallique
SONITUS - [med] tintement *m* d'oreille
SONORITY - [acoust] (sonorous quality or state, resonace) sonorité *f*
SOOT, to - [gen] enduire de suie
[mech] (in engines) calaminer
SOOT - [gen] (a product of combustion of coal or wood consisting basically of carbon together with a proportion of tars salts and ash) suie *f*
[mech] calamine *f*, encrassement *m*
~AND WHITE WASH - [telev] (colloq. for dark and light spots) image *f* trop contrastée
~BLOWER - [th eng] ventilateur *m* pour suie
SOOTED - [mech] (of sparking plug)(term used of a sparking plug fouled with carbon deposit so as to be faulty) calaminé, encrassé
SOOTY - [gen] noir de suie, couvert de suie
~COAL - [mining] houille *f* fuligineuse
SOP, to - [gen] tremper, faire tremper
[ind chem] plonger dans un liquide, immerger
SOPHISTICATE, to - [gen] sophistiquer, falsifier
[gen] (a written document etc) altérer
[agric] (e.g. wine) frelater (un vin)
SOPHISTICATION - [gen] sophistification *f*
[food] falsification *f*
SOPORIFIC - [gen & med] soporifique, somnifère
SORB - [bot] (the tree) sorbier *m*, alisier *m*
[bot] (the fruit) sorbe *f*, alise *f*
SORBIC - [chem] sorbique
~ACID - [chem] (an intermediate in the manufacture of plasticizers) acide *m* sorbique
SORBITE - [metall] (term denoting fine pearlite or the structure produced by tempering steel at temperatures above 550 centigrades) sorbite *f*
[chem] (hexavalent alcohol) sorbite *f*
SORBITIC - [metall] sorbitique
~CAST IRON - [metall] fonte *f* sorbitique
~FRACTURE - [metall] cassure *f* sorbitique

SORBITOL - [chem] (intermediate in the production of synthetic resins) sorbitol *m*
SORBOSE - [chem] (an intermediate for vitamin C) sorbose *m*
SORE -[med] plaie *f*, blessure *f*
SOREL'S CEMENT - [constr] (a mixture of calcined magnesite with a strong solution of magnesium chloride, which sets in a short time to considerable hardness, and is used for flooring) ciment *m* de magnésie
SORET EFFECT - [nucl] (thermal diffusion) effet *m* Soret, diffusion *f* thermique
SORGHO - s.sorghum
SORGHUM - [bot] (canelike tropical grass) grand millet *m*, sorgo *m*
SORPTION - [phys chem] (general term for absorption and adsorption) sorption *f*
SORREL - [bot] (a weed) oseille *f*
~HORSE - [zool] cheval *m* saure
SORT, to - [gen] assortir
[gen] (to distribute) trier, classer, lotir, allotir
[text] (waste etc) classer
[telecomm] (letters etc) router (lettres etc)
[comput] trier
SORT - [gen] sorte *f*, genre *m*
~OUT, to - [print] dépatisser
SORTED HEMP - [text] chanvre *m* assorti
SORTER - [paper man] (a type of strainer for mechanical wood pulp) classeur *m*
[text] (of wool or waste) trieur *m*, assortisseur *m*
[comput] (a machine sorting cards according to the punches in a given column) trieuse *f*
SORTER'S HURDLE - [text] table *f* de triage, claie *f*
~SCREEN - s. sorter's hurdle
SORTING - [gen] triage *m*, classification *f*
[min] (of ore) triage *m*
~BOARD - [agric] volet *m*
~HOUSE - [text] établissement *m* de triage
~MACHINE - [mech mach] trieuse *f*
~PLANT - [min] installation *f* de classification
~SIDING - [railw] (marshalling track) voie *f* de classement
~TABLE - [text] table *f* de triage
S.O.S - [telecomm] (distress signal) S O S
SOUDOBRASAGE - [el metall] (braze welding; a brazing method in which a joint of the open type is obtained step by step, by an operating technique similar to fusion welding, with a filler metal the melting temperature of which is above 450 centigrades) soude-brasage *m*
SOUFFLE - [med] souffle *m*
SOUGH - [mining] galerie *f* d'écoulement de l'eau
SOUND, to - [gen] sonner, résonner, retentir
[med] ausculter, percuter
[naut] sonder, prendre le fond
SOUND - [acoust] (the sensation of hearing) son *m*, bruit *m*
[phys] (the physical cause of the sensation of hearing) son *m*
[geogr] (long and narrow body of water) bras *m* de mer, détroit *m*
[naut] (the sounding of depth) sonde *f*
[zool] (of fish) vessie *f* natatoire
[med] (a solid rod used for the exploration of hollow viscera) sonde *f*
[adj] (healthy, undmaged) sain *m*, solide, en bon état

SOUND ABSORBENT MATERIALS - [acoust] (substance capable of absorbing sound waves, to reduce noise) matériaux *mpl* d'isolation acoustique
~-ABSORBING - [acoust] amortisseur (de son)
~ABSORPTION - [acoust] (the process by which energy is reduced when passing through a medium or striking a surface) absorption *f* du son
~ABSORPTION COEFFICIENT - [acoust] coefficient *m* d'absorption acoustique
~ABSORPTION FACTOR - s.sound absorption coefficient
~AMPLIFICATION - [electron] amplification *f* du son
~ANALYZER - [el acoust] (an apparatus for the determination of the spectrum of a sound) analyseur *m* du son
~AND PICTURE RECORD - [cin] photo-phonogramme *m*
~AND WAVE TUNING - [telecomm] syntonisation *f* de son et d'onde
~ARRANGEMENTS - [telev] arrangements *mpl* sonores
~ARTICULATION - [acoust] (of any system used for transmitting or reproducing speech) articulation *f*
~ATTACHMENT - [el acoust] (of a sound film projector) tête *f* sonore
~BACKGROUND - [telev] illustrations *fpl* sonores
~BAND - [el acoust] (a band or tapeshaped carrier for sound recordings) ruban *m* sonore
~BANDWIDTH - [el acoust] (the frequency range for sound required to obtain a good quality of reproduction) largeur *f* bande de son
~BAR - [acoust] (in a bowed instrument, the strip of wood glued under the belly along the line of the lowest string and supporting one foot of the bridge) barre *f* d'harmonie
~BARRIER - [aero] (popular unscientific term for the sudden large increase in drag which occurs when an aircraft reaches sonic speed) mur *m* de son
~BEAM - [phys] pinceau *m* sonore
~-BOARD - [mus] table *f* d'harmonie, tamis *m* de orgue, caisse *f* de resonnance
~BOARDING - [constr] sous-plancher *m*
~BOOTH - [cin] (a soundproof cabin for the sound camera) cabine *f* sonore
~BOX - [el acoust] (acoustic pick up; a device which transforms groove modulations directly into acoustic vibrations) pick-up *m* acoustique, phonocapteur *m*
[cin] s. sound booth
~CAMERA - [cin] camera-son *f*
~CARRIER - [radio] (the carrier wave modulated by the sound signal) fréquence *f* porteuse son
~CARRIER ATTENUATOR - [telev] palier *m* son
~CARRIER SWEEP - [telev] excursion *f* de porteuse son
~CARRIER TRAP - [telev] piège *m* son
~CARTOON - [cin] dessin *m* animé sonore
~CHANNEL - [telev] (the carrier frequency involved in the transmission of sound) canal *m* son
~-CHROMINANCE BEAT - [telev] (interference between the chrominance channel and the co-channel sound signal) interférence *f* entre porteuse son et porteuse chrominance
~COLUMN - [el acoust] (a structure including a number of loudspeakers in a vertical line) colonne *f* acoustique
~CORRECTOR - [acoust] correcteur *m* de tonalité
~CUTTER - [acoust] (the implement for the sound

variations on a medium) graveur *m* de son
SOUND CUTTING - [telev & cin] montage *m* du son
~DAMPING MATERIALS - [acoust] (material capable of absorbing impinging sound waves) matériaux *mpl* insonores
~DEADENER - [acoust] (a sound deadening material) amortisseur *m* du son
[auto] (a compound sprayed inside the body) enduit *m* anti-bruit, enduit *m* anti-drumming
~-DEADENING - [acoust] (soundproof) absorbant le son, amortissant le son
~DEFLECTOR - [acoust] abat-son *m*
~DETECTOR - [telecomm] détecteur *m* son
[instr] géophone *m*
~DIFFUSER - [acoust] diffuseur *m* acoustique
~DISCRIMINATOR - [el acoust] redresseur *m* de la fréquence sonore
~DISTORTION - [acoust] distorsion *f* du son
~EFFECTS - [cin] (artificially produced sound) effets *mpl* sonores
~-EFFECT TRACK - [el acoust] piste *f* d'essais de bruit
~ENERGY - [acoust] énergie *f* sonore
~ENERGY DENSITY - [acoust] densité *f* d'énergie acoustique
~ENERGY FLUX - [el acoust] (the instantaneous acoustic power across a surface element; product of the instantaneous acoustic pressure and the volume velocity across the surface element considered) flux *m* d'énergie sonore
~ENGINEER - [cin] ingénieur *m* du son
~EQUIPMENT - [acoust] (the complete set of apparatus for recording or reproducing sound) installation *f* sonore, son *m*
~FADING - [cin] fondu *m* sonore
~FADING DEVICE - [cin] potentiomètre *m* de réglage
~FIELD - [acoust] (region of space containing sound waves) champ *m* sonore
~FILM - [cin] film *m* sonore
~FILM LAMP - [cin] (lamp used to make reproduction of optically recorded sound possible) lampe *f* phonique, lampe *f* d'excitation
~-FILM PRINTING MACHINE - [cin] machine *f* pour le tirage de films sonores
~FILTER - [acoust] filtre *m* de son
~FIXING AND RANGING - [radar] (a location system) radar *m* acoustique sousmarin
~GATE - [cin] (gate through which the sound record on the film is pulled past the reproducing light beam) fente *f* de lecture
~HEAD - s. sound attachment
~-HEAD AXIS - [acoust] (the central axis of the sound head) axe *m* de son
~HOLES - [acoust] (in bell towers) baies *fpl*, ouïes *fpl*
~IMAGE - [cin] image *f* sonore
~IMPULSE - [acoust] impulsion *f* sonore
~INSULATING BOARD - [acoust] (stiff sheet material designed for sound insulating) panneau *m* phono-isolateur
~INSULATION - [acoust] (provision for reducing or preventing the passage for sound waves) insonorisation *f*, isolation *f* phonique
~INTENSITY - [acoust] (the mean value of instantaneous acoustic power per unit area) intensité *f* sonore
~INTENSITY DECAY - [acoust] diminution *f* de l'intensité sonore
SOUND INTERCARRIER DETECTOR - [telev] détecteur *m* à interporteuse
~LEVEL - [acoust] niveau *m* du son
~LEVEL INDICATOR - [instr] volumemètre *m*
~LEVEL METER - [instr] (an instrument including a microphone, an amplifier, an output meter and frequency weighing networks for the measurement of noise in terms of sound levels in a specified manner) sonomètre *m*
~LIMITER - [acoust] limiteur *m* du son
~LOCATOR - [instr] (an electro-acoustic apparatus designed to locate a sound source) localisateur *m* de son
~LOCK - [telev] (in a studio) sas *m* son (de studio)
~METER - [instr] (instrument designed to measure the intensity of sound) sonomètre *m*
MIXER - [cin] (a device in which the sound of various sound-emitting apparatus is selected and mixed) mélangeur *m* de son
~MIXING - [cin] mixage *m* son
~-MODULATED WAVES - [radio] (waves resulting from the modulation of a carrier wave by frequencies corresponding to the voice, music or other sounds) ondes *fpl* modulées à fréquence acoustique, ondes *fpl* A3
~MODULATION - [el acoust] (the operation of controlling the intensity of the sound) modulation *f* son, commande *f* de l'intensité du son
~MOTION PICTURE FILM - [cin] film *m* sonore
~NAVIGATION AND RANGING - s. sonar
~NEGATIVE - [cin] (of the sound track film) négatif *m* de la colonne sonore
~-ON-DISK - [cin] disque *m* synchronisé avec le film
~-ON-FILM - [cin] film *m* sonore
~-ON-FILM PRINTING - [cin] tirage *m* du film sonore
~-ON-VISION - [telev] (effect of cross-modulation) son *m* dans l'image
~OVERSHOOTING - [acoust] surmodulation *f*
~PARTICLE VELOCITY - [acoust] (derivative with respect to time of the particle displacement) vitesse *f* acoustique, vitesse *f* d'une particule
~PERCEPTION - [acoust] sensation *f* du son
~PICK-UP - [el acoust] (of a sound-head) pick-up *m*
~PICTURE - [cin] film *m* sonore
~-POROUS - [cin] (of a screen) (écran) perméable du son
~POSITIVE - [cin] (of a sound track film) positif *m* de la colonne sonore
~-POST - [mus] (of a violin) âme *f*
~POWER OF A SOURCE - [acoust] énergie *f* sonore totale d'une source
~PRESSURE - [acoust] (the fluctuating pressure in air, or other fluid, constituting the presence of a propagating or stationary sound-wave) pression *f* acoustique
~PRESSURE INCREASE - [acoust] intensification *f* de la pression sonore
~PRESSURE LEVEL - [acoust] (expressed in decibels, twenty times the logarithm to the base ten of the ratio between the sound pressure and the reference pressure, which should be explicitly stated) niveau *m* de la pression sonore
~PROBE - [acoust] (a device for exploring a sound field without disturbing the filed in the explored region) sonde *f* acoustique

SOUND-PRODUCING - [acoust] sonore
~PROJECTION - [acoust] (the emission of sound waves from a source) rayonnement *m* du son
~PROJECTOR HEAD - [cin] (sound attachment) tête *f* sonore
~-PROOFED CABIN - [cin] cabine *f* insonorisée
~QUALITY - [acoust] qualité *f* acoustique
~RADAR - [radar] (location system on the principle of collecting measuring data of targets by means of acoustics) radar *m* acoustique
~RADIATION - s. sound projection
~RANGING - s. sound radar
~RECORDER - [el acoust] appareil *m* d'enregistrement sur bande magnétique
~RECORDING - [el acoust] (a technique whereby sound signals are embodied in a material base, with an aim to preserving them and reproducing them at will) enregistrement *m* sonore
~RECORDING EQUIPMENT - [acoust] appareillage *m* d'enregistrement du son
~RECORDING SYSTEM - [acoust] système *m* d'enregistrement sonore
~RECORDIST - s. sound engineer
~REDUCTION FACTOR - [acoust] (the reciprocal of the acoustical transmission factor) coefficient *m* d'atténuation coefficient *m* de réduction acoustique
~REFLECTION COEFFICIENT - [acoust] coefficient *m* de réflexion acoustique
~REFLECTION FACTOR - [acoust] coefficient *m* spécifique de réflexion acoustique
~REJECTION - [telev] (rejection of the accompanying sound) réjection *f* de la porteuse son
~REPRODUCING SYSTEM - [acoust] système *m* de reproduction sonore
~REPRODUCTION - [acoust] reproduction *f* sonore
~REVERBERATION - [acoust] réverbération *f* du son
~SCREEN - [telev] écran *m* sonore
~SIGNAL - [radio & telecomm] signal *m* de son, signal *m* audio, signal *m* basse fréquence
~SPECTRUM - [acoust] (representation of the magnitudes of the components of a complex sound arranged as a function of frequency) spectre *m* de fréquences acoustiques
~SPEED - [cin] vitesse *f* de film sonore
~STREAMING - [acoust] (the production of unidirectional flow currents in a medium arising from the presence of sound waves) génération *f* acoustique d'effluents
~TAKE-OFF - [telev] point *m* de dérivation du signal son
~TALK-BACK - [telev] (in a studio) interphone *m* du studio
~TRACK - [el acoust] (sound record on a narrow band along the margin of a film) piste *f* sonore
~TRANSMISSION COEFFICIENT - [acoust] coefficient *m* de transmission acoustique
~TRAP - [telev] (filter in TV receivers) piège *m* du son
~TRUCK - [cin] (the sound equipment carried on a truck) camion *m* de son
~TUNING - [radio] syntonisation *f* du son
~VELOCITY - [acoust] vitesse *f* du son
~VOLUME - [el acoust] (of a sound film projector) volume *m* acoustique
~WAVES - [acoust] ondes *f*pl sonores
~WOOL - [text] laine *f* saine
SOUNDBOARD - [mus] (the belly of a piano) table *f* d'harmonie
SOUNDBOARD - [mus] (the upper part of the wind-chest with holes for the pipes) chape *f*, tamis *m* d'orgue
[mus] (the upper surface of stringed instruments) table *f* d'harmonie
SOUNDER - [telecomm] (telegraph receiver instrument consisting of an armature and an electromagnet) parleur *m*, récepteur *m* acoustique
SOUNDING - [acoust] sonore
[naut etc] sondage *m*
[med] auscultation *f*
~BALLOON - [met] (a balloon sonde) ballon *m* sonde
~BOARD - [acoust] (above the speaker, to reinforce the sound) abat-voix *m*
~ELECTRODE - [electron] (a probe used to make measurements in a gas discharge) électrode *f* de sondage
~LINE - [naut] ligne *f* de sonde
~MACHINE - [naut] sondeuse *f*
~PIPE - [naut] tuyau *m* de sonde
~PROCESS - [railw & constr] (the operation of testing bridges, tracks etc by sounding) procédé *m* d'auscultation
~ROCKET - [astronaut] (a missile designed to measure the higher strata of the atmosphere) fusée-sonde *f*
~ROD - [mech] sonde *f* (d'une pompe etc)
SOUNDINGS - [naut] sondes *f*pl
SOUNDLESS - [phys] silencieux
[naut] (of the sea) insonsable
SOUNDNESS - [gen] état *m* sain, bonne condition *f*
[text] (of the wool) santé *f* (de la fibre de laine)
[comm] solidité *f*
SOUNDPROOF, to - [constr etc] insonoriser, isoler acoustiquement
SOUNDPROOF - [acoust] isolé acoustiquement, insonorisé
SOUNDPROOFED BOOTH - [cin] cabine *f* insonorisée
~CABIN - s. soundproofed booth
SOUNDPROOFING - [acoust] (insulation against sound-waves) isolement *m* acoustique, insonorisation *f*
SOUP, to - [aero] (US colloq for boosting up) surélever la puissance
SOUPLE - [text] (silk yarns fabrics from which the sericin has not been removed) (soie) souple, (soie) mi-cuite
SOUPY - [gen] épais
SOUR, to - [gen] surir, aigrir
[ind chem] laver à l'eau acidulée
SOUR - [gen] aigre, sur, acide
[text] vitrioler (un tissu)
[leather ind] mettre en confit
[chem] acidifié
[chem] (a light acid solution) eau *f* acidulée
[agric] (of the ground) (sol) trop humide
~CRUDE - [oil ind] (crude oils containing an abnormally large amount of sulphur and sulphur compounds which break down on refining) pétrole *m* acide
~GAS - [min] (natural gas containing more than I.5. grains of hydrogen sulphide per hundred cubic feet) gaz *m* naturel acide
~GASOLINE - [min] (US only; motor vehicle fuel containing appreciable quantities of mercaptans) essence *f* acide
~WELL - [oil ind] puits *m* de gaz corrosif

SOURCE - [gen] source *f*
~COMPUTER - [comput] calculateur *m* à programme original
~DECK - [comput] jeu *m* de cartes original
~DOCUMENT - [comput] document *m* de base
~IMPEDANCE - [radio] (the impedance presented by a source of energy to the input terminals of the device) impédance *f* de source
~HOLDER - [nucl] récipient *m* de source
~INTERLOCK - [nucl] (safety device in a reactor designed to inactivate the reactor space) verrouillage *m* de la source de neutrons
~OF CLOCK PULSES - [comput] source *f* d'impulsions d'horloge
~ON ENERGY - [phys] source *f* d'énérgie
~OF INFORMATION - [comput] source *f* d'informations
~OF PROJECTION - [surv] (the point of origin of projection lines) source *f* de projection
~PROGRAMME - [comput] programme *m* original
~RECORDING - [comput] enregistrement *m* de données originales
~REGION - [met] (an extensive region of the earth's surface in which meteorological conditions are essentially uniform, and so placed that air masses remain in it long enough to acquire definite characteristics) région *f* d'origine
~ROCK - [geol] roche-mère *f*
~-TO-FILM DISTANCE - [radiat] (the distance between focus and film) distance *f* foyer-film
SOURED WORT - [brew ind] (wort made acid by the addition of a special mash) moût *m* suri
SOURING - [text] (treatment of yarns or fabrics with dilute acid) vitriolage *m*
[leather ind] mise *f* en confit
[gen] aigrissement *m*
SOURSOP - [bot] corossol *m* anone *f* muriquée
SOUTH - [geogr] sud *m*, austral
[gen] sud, midi
~POLE - [geogr] pôle *m* sud
SOUTHDOWN - [text] (English breed of sheep) southdown
SOUTHERN - [gen] sud, méridional, du midi
~CROSS - [astr] (striking constellation of the Southern hemisphere) Croix *f* du Sud
~HEMISPHERE - [geogr] hémisphère *m* sud, hémisphère *m* austral
~LIGHTS - [astr] aurore *f* australe
SOUTHING - [gen] (difference of latitude measured toward the south between any position and last one determined) différence *f* de latitude ver le sud
[astr] (the passage across the meridian of a celestial body that culminates south of the zenith) passage *m* au méridian
[surv] (a south latitude) latitude *f* sud
[naut] chemin *m* sud
SOUTHWARD - [geogr] sud *m*
SOUTHWESTER - [met] (south-west wind) vent *m* du sud-ouest
[impl] chapeau *m* imperméable, suroît
SOU'WESTER - s. south-wester
SOW, to - [agric] semer, ensemencer
SOW - [metall] (of a blast furnace) mère-gueuse *f*, nourrice *f* des gueuses
[metall] (solidified metal in the casting bed) loup *m*, cochon *m*
[zool] truie *f*, coche *f*
SOW BLOCK - [metall] enclume *f* porte-matrice
[mech] (in forging work) enclume *f* porte-matrice
~-BREAD - [bot] cyclamen *m*
~-BROADCAST, to - [agric] semer à la volée
~CHANNEL - [metall] rigole *f* de coulée
~THISTLE - [bot] (a wedd) lait *m* d'âne, laiteron *m* *m* maraîcher
SOWE BOX - [impl] (sizing box) bac *m* à colle
SOWING BY HAND - [agric] semis *m* à main
~BY ROWS - [agric] semis *m* en bandes
~DISTANCE - [agric] distance *f* du semis
~MACHINE -[agric] machine *f* à semer
~OF FERTILIZERS - [agric] épandage *m* d'engrais
~PEA - [agric] pois *m* à semence
~-SEED - [agric] pois *m* à semence
~TIME - [agric] semaison *f*, temps *m* de semailles
SOXHLET APPARATUS - [chem] (laboratory apparatus for the continuous extraction of a solid substance with a solvent) appareil *m* de Soxhlet
SOY - [bot] soui *m*
SOYA - [bot] soya *m*
~-BEAN - [bot] soya *m*, pois *m* chinois
~BEAN CAKE - [agric] torteau *m* de soya
~BEAN OIL - [ind chem] (a fixed oil obtained from varieties of Soya and having application in the production of resins, surface coatings, soaps and foodstuffs) huile *f* de soya
~FLOUR - [agric] farine *f* de soya
S.P.A. - [radio] (Sudden Phase Anomaly) anomalie *f* brusque de phase
SPACE, to - [gen] espacer, échelonner
[mech] espacer
[print] espacer (les lettres etc)
SPACE - [gen] (the continuous and boundless extension) espace *m*, intervalle *m*, surface *f*
[print] espace *m*, blanc *m*
[mech] creux *m*, vide *m*
[telecomm] (one of the intervals during the transmission of a telegraph message when the key is not contact) espace *m*
~-AIR VEHICLE - [astronaut] véhicule *m* aérospatial
~BAR - [print] barre *f* d'espacement
~BETWEEN BEAMS - [constr] intervalle *m* des solives
~BETWEEN PLANTS - [agric] écartement *m* de plantation
~BETWEEN RAILS - [railw] entre-voie *f*
~BETWEEN TRUSSES - [constr] intervalle *m* des fermes
~BLINDNESS - [med] cécité *f* spatiale
~BOOSTER - [astronaut] (the launching rocket for a spacecraft) fusée *f* de démarrage, accélerateur *m* spatial
~BOX - [print] casseau *m* pour espaces
~CAPSULE - [astronaut] capsule *f* spatiale
~CHARGE - [phys] (charge in a region of space due to the presence of electrons and/or ions) charge *f* spatiale
~-CHARGE BARRIER - [phys] barrière *f* de charge spatiale
~-CHARGE DEBUNCHING - [electron] (any process in which the mutual interaction between electrons in the stream disperse the electrons of a bunch) dégroupement *m* spontané
~-CHARGE DENSITY - [phys] (the space charge per unit volume) charge *f* spatiale volumique
~-CHARGE DISTORTION - [electron] (possible source of error when studying secondary effects) distorsion *f* par charge spatiale

SPACE-CHARGE GRID - [electron] (a grid which controls the position and area of a virtual cathode and the value of its potential) grille *f* de charge spatiale

~-CHARGE LIMITED CURRENT - [electron] courant *m* de charge spatiale limité

~-CHARGE-LIMITED-CURRENT STATE - [electron] (the state of working of an electronic tube in which the current is limited by the space charge) régime *m* de charge spatiale

~-CHARGE-LIMITED OPERATION - [nucl] fonctionnement *m* limité par la charge spatiale

~-CHARGE REGION - [electron] zone *f* de charge spatiale

~CHEMISTRY - [chem] stéréochimie *f*

~CLOTH - [electron] (a carbon impregnated cloth used as waveguide termination) revêtement *m* absorbant

~-CRAFT - [astronaut] (a vehicle designed to travel outside the atmosphere) véhicule *m* spatial

~CURRENT - [electron] (or cathode current; the total emission current composed of all currents flowing from the various electrodes) courant *m* électronique

~ENCODER - [comput] traducteur *m* de code mécanique

~EXPLORATION - [astronaut] exploration *f* spatiale

~GROUP - [crystall] (grouping of identical space lattices, either by rotation about a given axis or by interpenetration) groupe *m* spatial

~GROUP EXTINCTION - [crystall] extinction *f* des groupes spatiaux

~HEATER - [el] radiateur *m* électrique à convection

~HEATING APPLIANCE - [th eng] appareil *m* de chauffage

~IMPULSE - [comput] impulsion *f* d'espacement

~LATTICE - [crystall] (the arrangement of the structure of a crystal, such that a straight line drawn through any two points or a line parallel to such a line, will pass through a series of similar points at equal intervals) réseau *m* cristallin

~LAW - [leg] droit *m* de l'espace

~-LINE - [print] interligne *f*

~MODULATION - [radio] modulation *f* d'espace

~MOTION - [astronaut] mouvement *m* dans l'espace

~-OCCUPYING LESION - [med] lésion *f* intracrânienne expansive

~PASS - [text] remettage *m* à plusieurs corps

~PATTERN - [acoust] (directional diagram; a graphical representation of the radiation of the loudspeaker or the reception of the microphone) diagramme *m* directionnel de radiation
[electron] mire *f* de linéarité

~PERMEABILITY - [phys] perméabilité *f* relative

~PHASING - [radio] décalage *m* de phase spatiale

~PLATFORM - [astronaut] plate-forme *f*

~PROBE - [astronaut] sonde *f* spatiale

~REDDENING - [astr] rougissement *m* interstellaire

~-RULE - [print] filet *m* maigre

~-SHIP - s. space craft

~SIGNAL - [telecomm] signal *m* d'espacement

~SIMULATOR - [astronaut] simulateur *m* spatial

~STATION - [astronaut] station *f* spatiale, plate-forme *f* spatiale

~SUIT - [astronaut] scaphandre *m* spatial

~TELEGRAPHY - [radio] radiotélégraphie *f*

~TIME - [phys] (a four dimensional continuum within which any magnitude having both extension and duration may be precisely located) continuum *m* espace-temps, monde-espace-temps *m*

SPACE-TIME CONTINUUM - s. space time

~-TO-MARK TRANSITION - [comput] transition *f* espace-marque

~TRAVEL - [astronaut] voyage *m* spatial

~VEHICLE - [astronaut] véhicule *m* spatial

~VELOCITY - [phys] (the volume of gas per hour per unit volume of catalyst which passes through a bed of catalyst material) vitesse *f* spatiale

~WASHER - [mech] entretoise *f*

~WAVE - [radio] (a wave reaching the reception point which is made up of ground-reflected direct) waves) onde *f* spatiale

SPACECRAFT - s. space craft

~TIME - [astronaut] temps *m* propre du véhicule spatial

SPACED - [gen] espacé, écarté
[mech] espacé, d'espacement

~AERIALS - [radio] (two or more aerials spaced part) système *m* d'antennes espacées

~ANTENNAS - s. spaced aerials

~CROSS WINDING - [text] bobinage *m* croisé ouvert

~FRAME LOOP AERIAL - [radio] (loop aerial for direction, finding consisting of two parallel aerials) radiogoniomètre *m* à cadre

~FRAME LOOP ANTENNA - s. spaced frame loop aerial

~LATHING - [mech] lattage *m* espacé

~-LOOM DIRECTION FINDER - [radio] radiogoniomètre *m* à éléments spatiaux

~REED - [text] peigne *m* à espacements

~WINDING - [text] bobinage *m* ouvert

SPACER - [mech] (of a typewriter) barre *f* d'espacement
[mech] rondelle *f* de réglage, entretoise *f*
[mech] (a spacing plate) plaque *f* d'espacement, pièce *f* d'épaisseur

~BLOCK - [mech] (a part used to preserve a given distance between two other) entretoise *f*

~RING - [mech] (annular spacer) anneau *m* intermédiaire, anneau *m* d'écartement

~SLEEVE - [mech] (a cylindrical part used as a spacer) manchon *m* entretoise

SPACIAL PARTIAL WAVE - [electron] (a wave of determined phase constants occurring in a delay line with periodical structure during the special analysis of the modus) onde *f* spatiale partielle

SPACING - [gen] espacement *m*, écartement
[print] espacement *m* (des lettres etc)
[mech] (an interval) pas *m*, intervalle *m*, répartition *f*
[telecomm] (in telegraphy) repos *m*
[radio] (the shortest distance between two neighbouring aerials) distance *f* dans l'espace
[med] intervalle *m* d'apyrexie
[text] (density of threads) compte *m*

~BAR - [mech] (of a typewriter) barre *f* d'espacement

~BATTERY - [el] batterie *f* tampon

~BETWEEN TRUSSES - s. space between the trusses

~BUSH - [text] douille *f* d'écartement

~COLLAR - [mech] manchon *m* entretoise

~CURRENT - [telecomm] courant *m* de repos

~DRILL - [agric] semoir *m* en ligne

~GAPS - [mech] intervalles *mpl* espacés

SPACING JIG - [electron] (for electronic tubes) calibre *m* d'espacement, jauge *f* de distance
~KEY - s. spacing bar
~OF CONDUCTORS - [el] distance *f* entre les conducteurs
~OF DRAINS - [agric] écartement *m* des drains
~OF SLEEPERS - [railw] (distance between sleepers) espacement *m* des traverses
~OF THE LEVELS - [mining] distance *f* entre les étages
~REED - [text] ros *m* extensible
~SLEEVE - [mech] douille *f* d'écartement
~TUBE - [horol] pièce *f* d'épaisseur
~WASHER - [mech] rondelle *f* d'espacement
~WAVE - [telecomm] onde *f* de contremanipulation
SPACIOUS - [gen] spacieux, vaste
SPADE, to - [gen] bêcher
[fishing ind] dépecer
SPADE - [impl] bêche *f*
~-BONE - [anat] scapule *f*
~HAND - [med] main *f* en battoir
~HARROW - [agric] herse *f* à bêches
~HUSBANDRY - [agric] bêchage *m*
~-WORK - [gen] travaux *mpl* préliminaires, travaux *mpl* à la bêche
SPADING - [agric] bêchage *m*
SPADIX - [bot] spadice *m*
SPAGHETTI - [food] spaghetti *mpl*
[cin] (or buckling; jammed film inside the camera) bourrage *m* du film
~TAPE - [cin] (colloq for film) film *m*, pellicule *f*
~TUBING - [el] (small bore plastic sleeving used in electrical work) tubage *m* en spaghetti, tubage *m* isolant
SPALL, to - [constr] (the chipping off or splinters from a stone block) dégrossir (la pierre), smiller
[min] broyer (le minerai)
SPALL - [min] scheider *m*
~-OFF - [metall] (the crumbling of a mould in forging work) fonçage *m*, déballage *m*
SPALLATION - [nucl] (a type of nuclear reaction in which several small particles are ejected from the nucleus) spallation *f*
~FRAGMENT - [nucl] fragment *m* de spallation
SPALLING - [constr] (the operation of breaking off splinters of stone from a block by slanting blows with a chisel) smillage *m*
[metall] (of thin flakes from the surface of case-hardened steel) exfoliation *f*, effeuillage *m*
[mech] (chipping) écaillement *m*
[mech] (breaking caused by heat or mechanical stress) effrittement *m*
[nucl] écaillement *m*, affouillement *m*
[min] scheidage *m*
[geol] chute *f* de la paroi
~HAMMER - [min] marteau *m* de scheidage, massette *f* de scheidage
~RESISTANCE - [metall] résistance *f* aux variations de température
~TEST - [metall] essai *m* d'effritement
~WEDGE - [impl] quille *f* de scheidage
SPAN, to - [gen] mesurer à l'empan
[constr] (to extend over) franchir, chevaucher
[constr] (to cross, as a bridge) franchir, traverser
[agric] (of horses, oxen) atteler
[naut] (to make fast) brider
SPAN - [gen] empan *m*
[constr] (the distance between the supports of an arch, a bridge etc) ouverture *f*, largeur (d'un arc) portée *f*
[meas] (originally the space over which the hand can be extended) 229 mm
[aero] (of an aerofoil; the length measured on a given line normal to the mean direction of air flow) envergure *f* de l'aile
[aero] (of an aeroplane, the distance from wing-tip to wing-tip) envergure *f* de l'avion
[mech] (of a press) largeur *f* libre d'une presse
[mech] (of a crane bridge) portée *f* (de grue roulante)
[constr] (of a beam etc) volant *m* (d'une poutre etc)
~LENGTH - [el] longueur *f* de la portée
~LOADING - [aero] (value obtained by dividing the cross weight of an aeroplane by the square of the span) charge *f* sur l'envergure
~OF A VAULT - [constr] largeur *f* d'une arche
~OF AN OVERHEAD LINE - [el] portée *f* d'une ligne aérienne
~PIECE - [constr] entrait *m* retroussé
~POLE - [el] (the pole to which the span wires are attached) pôle *m* de transmission
~-ROOF - [constr] comble *m* à deux égouts, comble *m* à double pente
~SAW - [tool] scie *f* ordinaire
~WIRE - [el] (one of the wires by which the trolley-wire of a trolley bus system is suspended from the street poles) fil *m* aérien, fil *m* tendeur
SPANDREL - [arch] (of an arch) tympan *m*
[constr] reins *mpl* d'une voûte
~VAULT - [constr] voûte *f* d'élégissement
~WALL - [constr] (wall built upon the extrados of an arch) mur *m* remplissant le tympan
SPANGLE, to - [gen] pailleter
SPANGLE - [gen] (small piece of brilliant metal foil) paillette *f*, lamé *m*
~GOLD - [metall] or *m* en feuilles
SPANGLES - [metall] fleurs *fpl* de zinc
SPANIOMENORRHEA - [med] espacement *m* des règles, spanioménorrhée
SPANISH BROOM - [bot] genêt *m* d'Espagne
~LEATHER - [leather ind] cuir *m* de Cordoue
~MOSS - [bot] crin *m* végétal
~RED OXIDE - [paint] (bright red pigment consisting of ferric oxide, obtained from Spain) oxyde *m* rouge d'Espagne
SPANISHING - [print] (process of inking up embossed film with a wipe-roll or the like) encrage *m*
SPANKER - [naut] brigantine *f*
~BROOM - [naut] bôme *m*
SPANNER - [tool] (a tool taking various forms, but primarly designed to grip and turn a nut without damage to the latter) clef *f*, clé *f*, clé *f* à écrous
[constr] entretoise *f*
[mech] tendeur *m*
~BOLT - [mech] boulon *m* à clé
~NUT - [mech] collier
SPANROOF FRAME - [agric] châssis *m* double
SPANWISE - [gen] (in the direction of the line along which the span is measured) transversalement
SPAR, to - [gen] clôturer avec des perches
[naut] mâter
SPAR - [gen] perche *f*, poteau *m*

SPAR - [naut] (round timber for extending a sail) bout *m* de mât, espar *m*
[min] (any non-metallic mineral with a good cleavage) spath *m*
[aero] (principal spanwise structural element of an aerofoil) longeron *m*
~BUOY - [naut] (a buoyant structure surmounted by a vertical mast, used for marking purposes on a surface of water) bouée *f* à fuseau
~DECK - [naut] pont *m* volant
~FRAME - [aero] (a specially strong frame located in the same plane with a front or rear spar) cadre *m* de liaison voilure-fuselage
~VARNISH - [paint] (special type of varnish formulated to give high resistance to water, for used in ships and boats) vernis *m* imperméable
SPARABLE - [shoe man] clou *m* carré sans tête
SPARE, to - [gen] épargner
SPARE - [gen] (adj) accessoire, de réserve, de remplacement
[gen] (something unused) de réserve, de rechange
[mech] s. spare part
[metall] (metal in excess of the required quantity in a cast) assurance *f*
~BOBBIN - [text] bobine *f* de réserve
~CABLE - [telecomm] (a laid reserve cable) câble *m* en attente
~CHANNEL - [comput] canal *m* de service
~COACH - [railw] voiture *f* de réserve
~ENGINE - [mech] moteur *m* de réserve
~HOOK - [mech] crochet *m* de réserve
~LINE - [telecomm] ligne *f* de réserve
~PART - [mech etc] (any part supplied or held for use as replacement) pièce *f* de rechange, parties *fpl* de rechange
~-PARTS DEPARTMENT - [gen] service *m* parties de rechange
~-PARTS LIST - [mech] (a systematic catalogue of spare parts) nomenclature *f*
~PROPELLER - [naut] hélice *f* de réserve
~TOOLS - [tool] outils *mpl* de rechange
~WHEEL - [auto] roue *f* de secours
SPARER - [med] substance *f* d'épargne
SPARGANOSIS - [med] (infestation of bodily tissues with the larvae of various tape-worms) sparganose *f*
SPARGE ARMS - [brew ind] tuyaux *mpl* de rinçage
~LIQUOR - [brew ind] lavages *mpl*, dernières eaux *fpl*
~PIPE - [mech] (pipe pierced with many small holes to provide dispersion of escaping gas or liquid) tuyau *m* perforé
[nucl] tube *m* d'arrosage
~WATER - [ind chem] (water supplied to a sprinkling device) eau *f* d'aspersion
[brew ind] eau *f* de lavage
SPARGER - s. space pipe
SPARGING - [nucl] (introduction of a gas under the surface of a liquid for the purpose of agitation heat transfer or stripping) barbotage *m*
SPARK, to - [gen] (to throw out sparks) émettre des étincelles
[el] (to form a spark discharge) cracher, jaillir (entre les bornes)
SPARK - [gen] étincelle *f*
[el] (electric discharge in the air or other insulating material) étincelle *f*

SPARK ABSORBER - [el] getter *m*
~ADVANCE - [mech] (of motors and engines) avance *f* à l'allumage
~ARRESTER - [mech] pare-étincelles *m*
[railw] grille *f* à flammèches
~AT BREAKING - [auto] étincelle *f* de rupture
~BALL - [el] boule *f* d'éclateur
~CAP - [el] éclateur *m*
~CAPACITOR - [el] condensateur *m* antiparasite, condensateur *m* d'extinction
~CATCHER - s. spark arrester
~CHAMBER - [nucl] (track chamber where the paths of ionizing particles are shown by a succession of sparks) chambre *f* à étincelles
~COIL - [el] (transformer in which the ferromagnetic core is open, and the primary winding is traversed by a current interrupted periodically) bobine *f*
~CONTROL - [auto] commande *f* de l'avance à l'allumage
~DISCHARGE - [el] décharge *f* disruptive
~EROSION - [metall] (removal of metal from electrodes by the passage of electric sparks) érosion *f* par étincelles
~-EROSION MACHINING - s. spark-machining
~EXTINGUISHER - [el] extincteur *m* de l'arc
~FAILURE - [mech] allumage *m* manqué
~-FIRED - [mech] (of a gas engine) allumage *m* à étincelle
~FORMATION - [auto] formation *f* d'étincelles
~FREQUENCY - [radio] (the frequency of repetition of the spark discharge in a spark transmitter) fréquence *f* des étincelles
~FUSE - [mining] amorce *f* de tension
~GAP - [el] (any special arrangement of electrodes between which it is intended that a disruptive discharge of electricity shall take place at some prescribed potential difference between the electrodes) éclateur *m*
[auto] (of a spark plug) écartement *m* des électrodes de bougie
~-GAP ARRESTER - [el] (type of surge arrester in which the overvoltage drives a spark discharge across an air-gap connected between the circuit and earth) éclateur *m*
~-GAP MODULATION - [radio] modulation *f* par éclateur
~-GAP TUBE - [electron] (or electrodeless tube; gas-filled glass envelope which can be inserted into a coaxial cable, the conductors thus acting as external electrodes) tube *m* à décharge lumineuse sans électrodes internes
~GENERATOR - [radio] générateur *m* à étincelle
~IGNITION - [el] allumage *m* par bougies
~INSTANT - [el] (of engines, motors) moment *m* d'allumage
~KNOCK - [mech] détonation *f*, cognement *m*
~LEAD - [mech] avance *f* à l'allumage
~LEVER - [auto] levier *m* d'allumage
~MACHINING - [mech] (erosion process for obtaining holes of small diameter in hardened steel) électroérosion *f*
[metall] s.sparking-machining
~METAL WORKING PROCESS - [metall] (or electro-erosion metal working process; metal working process in which sparks periodically flash through a liquid dielectric in the very narrow gap between the model electrode and the piece to be worked)

usinage *m* par électroérosion
SPARK-METER - [instr] (instrument designed to measure the strength of sparks) spinthéromètre *m*
~-OVER - [el] contournement *m*
~-OVER TEST - [el] (flash-over test) essai *m* de contournement
~PING - s. spark knock
~-PLUG - s.sparking-plug
~PLUG ADAPTOR - [auto etc] raccord *m* intermédiaire pour bougies
~PLUG ELECTRODE - [auto etc] électrode *f* de bougie
~-PLUG IGNITION - [auto] allumage *m* par bougies
~-PLUG LEAD - [auto] fil *m* de bougie
~-PLUG NIPPLE - [mech] écrou *m* de bougie
~-PLUG OPENING - [mech] écartement *m* des électrodes d'une bougie d'allumage
~PLUG RESISTANCE - [radio] (the resistance between the electrodes after the discharge has begun) résistance *f* de l'étincelle
~POINT - [auto] électrode *f* de bougie
~POTENTIAL - [auto] tension *f* d'allumage
~QUENCH - [el] circuit *m* anti-étincelles, circuit *m* éliminateur d'étincelles
~SCREEN - [metall] (of a cupola) para-étincelles *m*
~SPECTRUM - [phys] (spectrum produced by the passage of an electrical discharge through a gas) spectre *m* d'étincelle
~TEST - [metall] (a test designed to determine the approximate chemical composition of a steel by the kind of sparks ejected when holding a sample against a grinding wheel) essai *m* d'identification d'aciers par étincelles
~TRANSMITTER - [radio] (transmitter in which the waves are caused by the recurring discharges of a capacitor across the spark-gap) émetteur *m* à éclateur
~WELDING - [metall] soudage *m* à l'arc
SPARKER BOX - [mining] tableau *m* de mise
SPARKING - [el] (the occurrence of a spark discharge between the brushes and the surface of a commutator) émission *f* d'étincelles, crachement *m*
~ALLOY - [metall] alliage *m* pyrophore
~AT THE BRUSHES - [el] crachement *m* des balais
~BALL - [el] boule *f* d'éclateur
~CONTACT - [el] (auxiliary contact used on circuit-breakers) contact *m* auxiliaire de rupture
~LIMIT - [el] (the limit output of a d.c. machine as determined by considerations sparking) limite *f* d'émission d'étincelles
~-MACHINING - [metall] (method of removing metal from a workpiece by a controlled system of electric sparks passing from an electrode to the part) usinage *m* par étincelles
~-PLUG - [auto etc] (device for igniting a combustible mixture by the passage of an electric spark between electrodes fixed in it) bougie *f* d'allumage
~PLUG GAP - s. sparking plug opening
~PLUG-OPENING - [el] (the gap between the electrodes of a sparking plug) écartement *m* des électrodes de bougie
~PLUG POINT - [auto] électrode *f* de bougie
~PLUG TEST BENCH - [auto] banc *m* d'essai pour bougies
~PLUG SPANNER - [tool] (special spanner for sparking plugs) clé *f* à bougies
~PLUG SUPPRESSOR - [radio] (device attached to a sparking plug to prevent it causing radio interference) dispositif *m* éliminateur de bruits pour bougies, anti-parasite *m* pour bougies
SPARKING PLUG THREAD - [mech] (special thread to connect a sparking to the unit which it is used) filetage *m* de la bougie d'allumage
~POTENTIAL - [el] (initial voltage) tension *f* d'amorçage
SPARKLE, to - [gen] étinceler, scintiller
[astr] scintiller
[brew ind] (to become effervescent) pétiller, mousser
[naut] (of the sea) brasiller
SPARKLE - [gen] étincelle *f*, éclat *m*, étincellement *m*
~METAL - [metall] matte *f* de cuivre
SPARKLESS COMMUTATION - [el] (a system in which the reactance voltage is neutralized before actual commutation takes) commutation *f* sans étincelles
SPARKLING - [gen] étincelant, scintillant, brilliant
[agric] (of a wine) mousseux
[el] (the emission of sparkles) étincellement *m*
[astr] scintillement *m*
~HIGHLIGHTS - [photo] hautes lumières *f*pl pétillantes
SPARROW - [zool] moineau *m*, passereau *m*
SPARRY - [min] spathique
~IRON ORE - [min] minerai *m* carbonate grillé
~LIMESTONE - [geol] marbre *m* à gros grain
SPARS - [oil ind] tiges *f*pl de pompe
SPARTALITE - [min] (obsolete name for zincite) spartalite *f*, zincite *f*
SPARTEINE - [chem] (an alkaloid of the quinoclidine group) spartéine *f*
~SULPHATE - [chem] (an alkaloid derived from Spartium scoparium and used in the treatment of cardic conditions) sulfate *m* de spartéine
SPASM - [physiol] (an involuntary contraction of muscle fibres) spasme *m*
SPASMOLYGMUS - [med] hoquet *m* spasmodique
SPASMOPHILIA - [med] (term denoting the hypothetical heightened irritability of the nervous system of those who have a tendency to spasms, convulsions etc) spasmophilie *f*
SPASTIC - [med] (of the nature of spasm) spastique, spasmodique
SPASTICITY - [med] spasticité *f*
SPATHIC - [min] (resembling spar) spathique
~IRON ORE - [min] (a synonym for Chalybite) minerai *m* de fer spathique
SPATHOSE - s.spathic
SPATIAL - [gen] (relating to space) spatial
~CHARGE DENSITY - [telecomm] densité *f* de charge spatiale
~EFFECT - [acoust] (auditory perspective) perspective *f* auditive
~SCATTERING - [telecomm] dispersion *f* spatiale
SPATTER, to - [gen] éclabousser de boue
[gen] (of a liquid) jaillir, gicler
SPATTER - [gen] éclaboussure *f*
[metall] (in welding, drops of molten metal splashed near the seam) perle *f* de soudure
~DASH - [constr] enduit *m* rocaillé
~LOSS - [metall] rendement *m* de baguette de soudure
~WORK - [mining] abattage *m* hydraulique, hydrauliquage *m*, abattage *m* à l'eau
SPATULA - [impl] (flat tool for applying paste or the like) spatule *f*
SPATULATE - [gen] spatulé

SPAVIN - [vet] (chronic arthritis of the hock-joint of a horse) éparvin *m*
[min] chamotte *f*, argile *f* réfractaire, apyre *f*
SPAVINED - [vet] (of a horse) boiteux, atteint d'éparvin
SPAWN - [zool] (collection of eggs as deposited by some fish) frai *m*
[bot] (the mycelium of a mushroom) mycélium *m*
SPAWNING - [zool] frai *m*, moment *m* du frai
~GROUND - [zool] frayère *f*
~TIME - [zool] saison *f* du frai, montaison *f* (des saumons)
SPAY, to - [vet] châtrer
SPEAK, to - [gen] parler
[naut] héler
SPEAKER - [radio] speaker *m*
[el acoust] (a loudspeaker, sound-emitting device) haut-parleur *m*
~CLOTH - [radio] tissu *m* de haut-parleur
SPEAKING CLOCK - [telecomm] horloge *f* parlante
~KEY - [telecomm] clé *f* de conversation, clé *f* combinée d'écoute et de conversation
~TUBE - [aero, auto, naut] (an acoustic device for intercommunication) tube *m* acoustique, porte-voix *m*
SPEAR, to - [gen] percer, transpercer
SPEAR - [gen] lance *f*
[zool] (of a porcupine) aiguillon *m*
[zool] (of a fish) fouine *f*, trident *m*
[oil ind] tige *f* de pompe
[mech] tige *f*, verge *f*
[mining] taraud *m* rattrape-tube
[bot] jeune arbre *m*, brin *m* d'herbe
SPEARMINT - [bot] menthe *f* verte
~OIL - [chem] (an oil obtained from varieties of Mentha and used as a flavourant) essence *f* de menthe
SPECIAL - [gen] spécial, particulier
~CAST IRON - [metall] fonte *f* fine
~CODE SELECTOR - [telecomm] commutateur *m* discriminateur
~CONTROL POSITION - [telecomm] table *f* d'observation
~LOCKING - [el] enclenchement *m* conditionnel, enclenchement *m* multiple
~OBSERVATION POST - [telecomm] poste *m* d'écoute et de coupure
~-PURPOSE MOTOR - [el] moteur *m* à emploi spécial
~RELATIVITY THEORY - [phys] (theory developed by Einstein and based on the hypothesis that the velocity of light is the same as measured by anyone of a set of observers moving with constant relative velocity) théorie *f* spéciale de relativité
~ROOF BRACKET - [telecomm] (for cables) tourelle *f*, appui *m* sur toiture
SPECIALIST - [gen] spécialiste *m*
[med] spécialiste *m*
SPECIALIZATION - [gen med etc] spécialisation *f*
[biol] adaptation *f* spéciale
SPECIALIZED - [gen] spécialisé
~AUTOMATIC DATA PROCESSING - [comput] traitement *m* des données spécialisées
SPECIALITY - [gen] spécialité *f*, caractéristique *f*
SPECIALTY - [leg] (a sealed contract; a deed) contrat *m* formel (sous seing privé)
SPECIE - [fin] (coined money) espèces *fpl* monnayées

SPECIES - [biol] espèce *f*
[gen] genre *m*, sorte *f*
SPECIFIC - [gen] spécifique
~ABSORPTION - [radiat] (selective absorption) affinité *f* differentielle, localisation *f* sélective
~ACOUSTIC IMPEDANCE - [el acoust] (unit-area acoustic impedance) impédance *f* acoustique par unité de surface, impédance *f* acoustique spécifique
~ACOUSTIC REACTANCE - [el acoust] (unit-area acoustic reactance) réactance *f* acoustique
~ACTIVITY - [nucl] (the activity of a radioisotope on an element per unit weight of element present in the sample) activité *f* massique
~ADDRESS - [comput] adresse *f* absolue
~ADSORPTION - [phys] (preferential adsorption of one substance over another, also quantity of adsorbate held per unit area adsorbent) adsorption *f* spécifique
~BINDING ENERGY - [nucl] énergie *f* de liaison spécifique
~BURN-UP - [nucl] combustion *f* massique, taux *m* de combustion
~CAPACITY - [el] (the amount of energy stored, in a battery in relation to a given dimension of it) capacités *fpl* spécifiques
~CHARACTERS - [biol] (the constant characteristics distinguishing a species) caractéristiques *fpl*
~CHARGE - [electron] (the quotient of the electric charge by the mass) charge *f* spécifique d'un porteur
~CODING - [comput] (coding in which alls addresses refer to specific registers and locations) codage *m* en absolu
~CONDUCTANCE - [el] (obsolescent term for conductivity) conductivité *f*
~CONDUCTIVITY OF WOOD - [bot] (the rate at which water flows through a piece of wood of a give length and area in a given time) absorption *f* spécifique du bois
~CONSUMPTION - [mech] (the quantity of oil or fuel consumed for a given power output over a given period) consommation *f* spécifique
~DAMPING - [el] (the attenuation constant per kilometre of cable) amortissement *m* spécifique
~DEPRESSION - [phys] (depression of freezing point) abaissement *m* spécifique
~DIELECTRIC STRENGTH - [el] (the dielectric strength of an insulating material exoressed in volts per mm) rigidité *f* diélectrique spécifique
~DISPERSIVITY - [opt] (the difference between the specific refractions of two wavelengths of radiant energy) indice *m* spécifique de réfractivité différentielle
~ELECTRIC LOADING - [el] (the electric loading of the armature of a machine per cM of circumference) charge *f* électrique spécifique
~ELECTRONIC CHARGE - [electron] charge *f* spécifique de l'électron
~ELONGATION - [phys] élongation *f* spécifique
~EMISSION - [electron] (the rate of emission, that is the liberation of electrons from a surface into the surrounding space, per unit area) émission *f* spécifique
~ENERGY - [phys] (internal energy per unit mass) énergie *f* spécifique
~FUEL CONSUMPTION - [mech] (in internal combustion engines; the weight of fuel used by an engine

per unit horsepower per unit time) consommation *f* spécifique de carburant

SPECIFIC GRAVITY - [phys] (the ratio between a given volume of a substance and a equal of water) poids *m* spécifique

~GRAVITY BALANCE - [meas] (balance *f* hydrostatique

~GRAVITY BOTTLE - s. specific gravity flask

~GRAVITY CONCENTRATION - [chem] (concentration of chemical deposits by making use of the difference in specific gravity) concentration *f* par gravité

~GRAVITY FLASK - [instr] (a small flask of accurately known contents, used in determinations of specific gravity) pycnomètre *m*, flacon *m* à densité

~GRAVITY OF THE FIBRE - [text] poids *m* spécifique de la fibre

~HEAT - [phys] (the units of heat required to raise unit mass of a substance through I°C.) chaleur *f* spécifique

~HEAT OF GASES - [phys] chaleur *f* spécifique des gaz

~HUMIDITY - [met] (the weight of water vapour per ground of dry air) humidité *f* spécifique

~INDUCTIVE CAPACITY - [meas] (a dielectric constant) constante *f* diélectrique, permittivité *f*

~INERTANCE - [acoust] inertance *f* acoustique spécifique

~IONIC MOBILITY - [phys] mobilité *f* spécifique des ions

~IONIZATION - [phys] (the number of ion pairs formed per unit distance along the track of an ion passing through matter) ionisation *f* linéique, ionisation *f* spécifique

~IONIZATION COEFFICIENT - [electron] (the average number of pairs of ions with opposite charges that electrons with a specified kinetic energy produce in a gas at a specified pressure and temperature over a unit distance) coefficient *m* spécifique d'ionisation

~LUMINOUS INTENSITY - [phys] intensité *f* lumineuse spécifique

~MAGNETIC LOADING - [el] (the average flux density in the armature of a machine) charge *f* magnétique spécifique

~MAGNETIC RESISTANCE - [el] (obsolete term for the reciprocal of permeability) résistance *f* magnétique spécifique

~OUTPUT - [el] (of an electrical machine) rendement *m* spécifique

~PERMEABILITY - [el] (the ratio between the absolute permeability of a substance and the permeability of free space) perméabilité *f* spécifique

~POWER - [nucl] (power produced per unit mass of fuel present) puissance *f* massique

~REACTION RATE - [chem] (the velocity constant) vitesse *f* spécifique de réaction

~REFRACTION - [phys] réfraction *f* spécifique

~RESISTANCE - [el] résistivité *f* volumique

~RETENTION - [hydr] rétention *f* spécifique d'eau capillaire

~ROTATION - [opt] (the angular rotation of the plane of polarization as it passes through an optically-active material, divided by the length of the path and the density of the material) rotation *f* spécifique.

~ROUTINE - [comput] programme *m* absolu

~SLIDING - [mech] (of a gear) glissement *m* spécifique

SPECIFIC SPEED - [el] (a comparative quantity for the direct comparison of angular velocity between different sizes of hydraulic turbines) vitesse *f* spécifique

~STAIN - [zool] colorant *m* indicateur

~STRENGTH - [radiat] (the gamma-ray strength per unit volume of a radioactive source) intensité *f* spécifique

~SURFACE - [mech] (the surface of a substance per unit volume) surface *f* spécifique

~THERMAL CAPACITY - [phys] (the heat capacity of unit mass of a substance, interchangeable with specific heat) capacité *f* thermique spécifique

~TORQUE COEFFICIENT - [el] (a coefficient used in design of electrical machines) module *m* de torsion spécifique

~TRAIN RESISTANCE - [railw] résistance *f* en courbes, résistance *f* spécifique due aux courbes

~TRANSMITTIVITY - [opt] (the internal transmittance for unit thickness of a non-diffusing substance) nombre *m* spécifique de transmission

~VISCOSITY - [chem] (the relative viscosity of a solution of known concentration of the polymer minus one) viscosité *f* spécifique

~VOLUME - [phys] (the volume of unit weight of a substance) volume *m* spécifique

~WEIGHT - [phys] s. specific gravity
[aero] (of an aircraft engine) poids *m* au cheval

~YIELD - [hydr] débit *m* spécifique

~YIELD OF PORE SPACE - [hydrology] porosité *f* utile

SPECIFICALLY HEAVY FIBRE - [text] fibre *f* de densité élevée

~LIGHT FIBRE - [text] fibre *f* de densité faible

SPECIFICATION - [gen] spécification *f*, devis *m* descriptif
[gen] (norms and regulations) cahier *m* de charges
[leg] (of a patent) description *f* de brevet

~TAG - [gen] plaque *f* signalétique

SPECIFICATIONS - [comm etc] (definite and complete statement) spécifications *f*pl
[metall] (the chemical composition and the mechanical properties) prescription *f*

SPECIMEN - [gen] spécimen *m*, exemplaire *m*
[ind chem etc] (for tests) éprouvette *f*
[comm] échantillon *m*

~BAR - [metall] barre *f* de contrôle

SPECK - [rubber ind] (in a rubbersheet) tache *f*, mouche *f*
[gen] grain *m*, particule *f*

~-DETECTOR - [comput] détecteur *m* de grains

SPECKLE, to - [gen] tacheter, moucheter

SPECKLE - [gen] petite tache *f*, mouche *f*

SPECKLED - [gen] tacheté, tiqueté, moucheté, travelé
[agric] (of a hen) bariolé

~BACKGROUND - [radar] fond *m* pointillé

~FABRIC - [text] tissu *m* pointillé

~WOOD - [timber] bois *m* tacheté

SPECKS - [paper man] (defects on a sheet) taches *f*pl
[rubber ind] particules *f*pl d'impuretés

SPECTACLE - [gen] spectacle

~CONFIGURATION - [aero] (side-by-side twin rotor configuration) configuration *f* d'un birotor à rotors parallèles

SPECTACLES - [opt] lunettes *f*pl
SPECTRAL - [phys] (pertaining to spectrum) spectral
~ANALYSIS - [phys] analyse *f* spectrale
~CENTROID - [opt] centre *m* de gravité spectral
~CHARACTERISTIC - [electron] (of a luminescent screen) caractéristique *f* spectrale
~COLOUR - [opt] (a colour represented by a point on the chromatic diagram which lies on a straight line between the spectrum locus and achromatic point) couleur *f* spectrale
~DENSITY - [phys] (the relative distribution of radiant energy throughout the spectrum) distribution *f* spectrale
~DISTRIBUTION CURVE - [phys] courbe *f* de distribution spectrale
~DISTRIBUTION GRAPH - [phys] courbe *f* de répartition spéciale
~ENERGY DISTRIBUTION - [phys] distribution *f* spectrale de l'énergie
~LINE - [phys] (one of the lines forming a line spectrum) ligne *f* spectrale
~POSITION - [phys] (the effective wavelength or frequency of an essentially monochromatic beam) longeur *f* d'onde efficace
~PURITY - [opt] (the property of having a single wavelength) monochromie *f*, pureté *f* spectrale
~QUANTUM YIELD - [electron] (in a photoelectric device) rendement *m* quantique spectral
~REFLECTANCE - [opt] (the radiant reflectance for a specified wavelength of the incident radiant flux) pouvoir *m* réflecteur spectral
~RESPONSE CHARACTERISTIC - [electron] (of a photo electric device) caractéristique *f* spectrale de sensibilité
~SELECTIVITY - [electron] (of a photo-electric device; the change of photo-electric current with the wavelength of the irradiation) sélectivité *f* spectrale
~SENSITIVITY - [radiol] (the sensitivity of a detector measured for narrow spectral bands throughout the spectrum) sensibilité *f* spectrale
~SENSITIVITY CHARACTERISTIC - [electron] (of a camera tube) caractéristique *f* spectrale de sensibilité
~SERIES - [phys] (system of lines in a spectrum in which there is an obvious regularity) série *f* spectrale
~SHIFT CONTROL - [nucl] commande *f* par dérive spectrale
~SHIFT REACTOR - [nucl] réacteur *m* à dérive spectrale
~TRANSMISSION - [phys] (the transmission of a plate of material measured for narrow spectral bands throughout the spectrum) transmission *f* spectrale
SPECTROCHEMICAL ANALYSIS - [chem] analyse *f* spectrochimique
SPECTROGRAM - [phys] (the record produced by a spectrograph) spectrogramme *m*
SPECTROGRAPH - [instr] (an apparatus for photographing a spectrum) spectrographe *m*
SPECTROGRAPH TUBE - [electron] (electron tube used in a spectrograph) tube *m* de spectrographe
SPECTROHELIOGRAM - [astr] (a photograph of the sun and its prominences) spectrohéliogramme *m*
SPECTROHELIOGRAPH - [instr] (an instrument for photographing the sun by means of monochromatic light) spectrohéliographe *m*
SPECTROMETER - [instr] (instrument designed to measure spectra; also to determine the wavelengths of the various radiations) spectromètre *m*
SPECTROMETRY - [meas] (the measurement of spectra and of the wavelengths of the various radiations) spectrométrie *f*
SPECTROMICROSCOPE - [instr] microscope *m* spectroscopique
SPECTROPHOTOMETER - [instr] (instrument designed to analyse the spectral energy distribution of radiant energy) spectrophotomètre *m*
SPECTROPHOTOMETRY - [meas] (the analysis of spectral energy distribution of radiant energy) spectrophotométrie *f*
SPECTRORADIOMETER - [instr] (instrument designed to measure the spectral distribution of radiant energy) spectroradiomètre *m*
SPECTROSCOPE - [instr] (instrument to show the spectrum of light radiation) spectroscope *m*
SPECTROSCOPIC - [phys] (pertaining to spectroscopy) spectroscopique
~BINARY - [astr] (a double star the components of which are too close to be resolved visually) étoile *f* double
~PARALLAX - [astr] (indirect method of deducing the distance of stars which are too far away to have detectable annual parallaxes) parallaxe *f* spectroscopique
SPECTROSCOPICAL - s. spectroscopic
~TEST - [metall] recherches *f*pl au spectroscope
SPECTROSCOPY - [phys] (the branch of physical science dealing with the theory and interpretation of spectra) spectroscopie *f*
SPECTRUM - [phys etc] (a wide band of radiation in which the radiations are arranged in order of their frequencies) spectre *m*
~ANALYSIS - [phys] analyse *f* spectroscopique
~ANALYZER - [acoust] (a device for determining the frequency-energy distribution of a signal or a group of signals) analyseur *m* de spectre
~BAND - [phys] bande *f* du spectre
~LAMP - [phys] (lamp giving a non-luminous flame used for the spectroscopic examination of radiation from solids, liquids, or solutions introduced into the flame) lampe *f* pour spectroscopie
~LINE - [phys] raie *f* noire du spectre
~LOCUS - [phys] (the locus of points representing the chromaticities of spectrally pure stimuli in a chromatic diagram) lieu *m* des couleurs spectrales
~PRESSURE LEVEL - [el acoust] (band pressure level for a band-width of one cycle per second, centred at a specified frequency) niveau *m* élémentaire de spectre
~STABILIZER - [electron] stabilisateur *m* de spectre
SPECULAR - [opt] (pertaining to a speculum or a mirror; reflecting) spéculaire
~CAST-IRON - [metall] fonte *f* spéculaire
~DENSITY - [opt] (the logarithm of the reciprocal of the specular transmittance) densité *f* par réflexion, densité régulière, densité *f* spéculaire
~IRON - [min] (variety of haematite possessing a shining metallic lustre often showing iridescence) fer *m* spéculaire
~LIGHT - [photo] lumière *f* spéculaire
~PIG IRON - [metall] fonte *f* miroitante
~REFLECTANCE - [opt] (the ratio between the radiance

measured by reflection and the radiance measured directly) degré *m* de réflexion régulière
SPECULAR REFLECTION - [opt] (regular reflection) réflexion *f* régulière
~STONE - [min] mica *f*
~SURFACE - [gen] (reflecting surface) surface *f* miroitante
~TRANSMISSION DENSITY - [opt] densité *f* de transmission régulière, densité *f* de transmission spéculaire
~TRANSMITTANCE - [opt] transmission *f* spéculaire
SPECULUM - [med] (hollow or curved instrument for viewing a cavity in the body) spéculum *m*
[opt] miroir *m* de télescope
[zool] (iridescent area on the wings of certain birds, e.g. ducks) miroir *m* (sur l'aile d'un oiseau)
SPEECH - [gen] parole *f*, langue *f*
[gen] (an address, a discourse) discours *m*
[mus] sonorité *f*
~AMPLIFIER - [el acoust] (low-frequency amplifier) amplificateur *m* B.F.
~CHANNEL - [telecomm] liaison *f* téléphonique
~CIRCUIT - [telecomm] circuit *m* téléphonique
~CLIPPER - [radio] écrêteur *m*
~COIL - [el acoust] (the mobile loudspeaker-coil carrying the speech current) bobine *f* mobile
~CURRENT - [el acoust] (the current which flows through the winding on a cone of a loudspeaker) courant *m* de bobine mobile
~FILTER - [acoust] (a filer used to suppress bass notes) filtre *m* pour la reproduction de la parole
~FREQUENCY - [acoust] fréquence *f* vocale
~INPUT EQUIPMENT -[telecomm] émetteur *m* microphonique
~LEVEL - [acoust] niveau *m* de courants vocaux
~LEVEL MFTER - [instr] volumètre *m* de parole
~-NOISE RATIO - [acoust] rapport *m* bruit-parole
~OSCILLATIONS - [acoust] vibrations *f*pl vocales
~SOUND - [acoust] (the least distinctive element in speech, such as vowels and consonants) logatome *m*
~TEST - [telecomm] essai *m* de conversation
~VOLTMETER - [instr] (a type of volume indicator) voltmètre *m* de parole
SPEED, to - [gen] se hâter, aller vite
[auto] faire de la vitesse
[mech] (a motor, an engine) régler la vitesse (d'une machine)
SPEED - [mech] (a scalar quantity expressed in units of length divided by time) vitesse *f*, rapidité *f*
[mech] (of a motor) vitesse *f* de régime
[auto] vitesse *f*
[photo] (the measure of the exposure required by an emulsion) rapidité *f* d'une émulsion
[el] (the angular velocity of an electrical machine) vitesse *f*
[opt] luminosité *f* (d'un object)
~-ADJUSTING MOTOR - [mech] moteur *m* de réglage de vitesse
~-ADJUSTING RHEOSTAT - [el] (a rheostat arranged for varying, at will, the speed of a motor and suitable for continuous operation in any position) rhéostat *m* de réglage de vitesse
~-AND-FEED-CHANGING MECHANISM - [mach tool] mécanisme *m* pour le changement de vitesse et d'alimentation
~AND TORQUE CHARACTERISTICS - [el] courbe *f* caractéristique d'un moteur
SPEED AT CONTINUOUS RATING - [el] vitesse *f* au régime continu, vitesse *f* continue
~AT END OF RHEOSTAT STARTING PERIOD - [el] vitesse *f* de fin de démarrage rhéostatique
~AT ONE-HOUR RATING - [el] vitesse *f* au régime unihoraire
~BOX - [mech] boîte *f* de vitesse
~BRAKE - [aero] (a device for increasing the drag of an aircraft; also called airbrake) aérofrein *m*
~CHANGE - [auto] changement *m* de vitesse
~CHANGE DRUM - [auto] tambour *m* changement de vitesses
~CHANGE LEVER - [auto] levier *m* de changement de vitesses
~CHANGER - s. speed variator
~CHECKER - [mech] ralentisseur *m* de vitesse
~-CHECKING APPLIANCE - [mech] appareil *m* ralentisseur de vitesse
~CONE - [mech] (stepped cone pulley) cône *m* de vitesse, cône-poulie *m*
~CONTROL - [el] (the method by which the speed of an electric motor may be changed) réglage *m* de la vitesse
~CONTROLLER - [el] régulateur *m* de vitesse
~COUNTER - [instr] (of an engine etc) compteur *m* de tours, compte-tours *m*
~COURSE - [aero] (a determined course of known length used in measuring the ground speed of an aircraft) base *f* de vitesse
~CURVE - [phys] courbe *f* de la vitesse
~-DISTANCE CURVE - [el] (the curve indicating the relation between the speed of an electric tractor and the distance is has travelled) courbe *f* distance-vitesse
~DROOP - [mech] (of an engine speed governor) écart *m* permanent de tours
~FACTOR - [radiat] (the ratio between the exposure time without intensifying screens and that when screens are used) facteur *m* d'intensification
~FRAME - [text] banc *m* à broches
~GATHERING DISTANCE - [aero] (the distance required by an aircraft to reach a given speed) distance *f* de mise en vitesse
~GOVERNING - [el] (the method of keeping the speed of a prime mover independent of the electrical load on the generator which it is driving) réglage *m* de vitesse
~GOVERNOR - [mech] régulateur *m* de vitesse
~INDICATOR - [instr] (tachometer) tachymètre *m*, indicateur *m* de vitesse
~IN PICKS PER MINUTE - [text] nombre *m* de duites par minute
~-LATHE - [mach tool] tour *m* à marche rapide, tour *m* avec poupée fixe à cone
~LIMIT - [gen] vitesse *f* maxima
~LIMIT INDICATOR - [instr] indicateur *m* de limite de vitesse
~-LIMITING DEVICE - [contr] dispositif *m* de limitation de vitesse
~OF ANSWER - [telecomm] délai *m* de réponse
~OF BEAMING - [text] vitesse *f* d'ensouplage
~OF COMBUSTION - [auto] vitesse *f* de la combustion
~OF CONVERGENCE - [contr] vitesse *f* de convergence
~OF DISCHARGE - [mech] vitesse *f* d'écoulement
[el] vitesse *f* de décharge

SPEED OF DRUM - [text] vitesse *f* du tambour
~OF GERMINATION - [bot] rapidité *f* de la germination
~OF LIFT - [mech] (lifting speed) vitesse *f* de levage
~OF OPERATION - [comput] vitesse *f* de traitement
~OF REEL - [text] vitesse *f* du dévidoir
~OF RESPONSE - [el] vitesse *f* de reponse
~OF REVOLUTIONS - [mech & auto] régime *m* de rotation
~OF ROTATION - [mech] (the number of rotations about the axis of rotation divided by the time) vitesse *f* de rotation
~OF SPINDLE - [text] vitesse *f* de la broche
~OF THE FILAMENT - [text] vitesse *f* du fil
~OF THE SHUTTLE - [text] vitesse *f* de la navette
~OF TRAVERSE - [mach] vitesse *f* de translation
~OF WARPING - [text] vitesse *f* d'ourdissage
~OF WATER ABSORPTION - [chem] vitesse *f* d'absorption d'eau
~OF WINDING - [text] vitesse *f* de bobinage
~PLATE - [mach tool] plaque *f* des vitesses
~PULLEY - s.speed-cone
~RANGE - [mech] gamme *f*
~RATIO - [el] taux *m* de variation de vitesse
~RATIO CONTROL - [contr] réglage *m* de rapport de vitesse
~RECORDER - [instr] (for a machine) enregistreur *m* de vitesse
~REDUCER - [mech] réducteur *m* de vitesse
~REDUCTION - [mech] démultiplication *f*, réduction *f* de vitesse
~-REGULATING RHEOSTAT - s.speed-adjusting rheostat
~REGULATOR - [el mach] (the device for regulating the speed of a motor) régulateur *m* de vitesse
(the speed of the film) régulateur *m* de vitesse (du film)
~-RESTRICTION SIGNAL - [railw] signal *m* de ralentissement
~-RESTRUCTION WARNING SIGNAL - [railw] rappel *m* de ralentissement
~SELECTOR LEVER - [mech] levier *m* de changement de vitesse
~SELECTOR SWITCH - [el] (in video recording) commutateur *m* de vitesse
~-TIME CURVE - [el] (curve of train speed plotted against running time) courbe *f* vitesse-temps
~-TORQUE CHARACTERISTIC - [el] (the curve indicating the relation between the speed of a motor and the torque developed) courbe *f* vitesse-couple
~TRAP - [auto] zone *f* du contrôle de vitesse
~TRIAL - [naut] essai *m* de vitesse
~TUNING CIRCUIT - [telev] montage *m* de syntonisation accélérée
~UP, to - [gen & mech] accélérer
[gen] (the production) augmenter (la production)
~VALUE - [photo] valeur *f* de sensibilité
~VARIATOR - [mech] (device for controlling screw speed in an extruder) régulateur *m* de vitesse
SPEEDBOAT - [naut] motoglisseur *m*, bateau-glisseur *m*
SPEEDER - [mech] (a device for regulating speed) contrôleur *m* de vitesse
[railw] chariot *m* de service
[text] banc *m* à broches
SPEEDOMETER - [instr] (instrument designed to indicate miles or kilometres per hour) tachymètre *m*, compteur *m* de vitesse
SPEEDOMETER - [instr] (instrument used to register the number of miles or kilometres covered) compteur-indicateur *m* de vitesse
~CABLE - [auto] conduite *f* flexible pour tachymètre
SPEEDWAT - [gen] (USA only) autostrade *f*, autoroute *f*
SPEEDWELL - [bot] (a weed) véronique *f*
SPEEDY - [gen] rapide
SPEISS - [metall] speiss *m*
~COBALT - [min] cobalt *m* arsenical, smaltine *f*
SPEEL, to - [gen] épeler, orthographier
SPELLING - [gen & telecomm] épellation *f*
SPELT - [agric] (species of wheat) épeautre *m*
SPELTER - [metall] (a term for zinc of commercial purity) zinc *m* de commerce
[metall] (zinc solder) s.spelter solder
~SOLDER - [metall] soudure *f* de laiton
SPENCER - [naut] voile *f* goélette
SPEND, to - [gen] dépenser
[gen] (to waste, to squander) consumer, épuiser
SPENT - [gen] épuisé, apaisé
~ACID - [chem] acide *m* épuisé
~AMMONIA LIQUOR - [chem] eaux *f*pl résiduaires ammoniacales
~CARTRIDGE - [firearms] cartouche *f* vide
~FUEL - [nucl] combustible *m* épuisé
~FUEL HANDLING SYSTEM - [nucl] traitement *m* de combustible épuisé
~GAS LIQUOR - s. spent ammonia liquor
~GRAIN DOUGH - [brew ind] pâte *f* de la drêche
~GRAINS - [brew ind] (refuse grain after malting) drêche *f*
~HOPS - [agric] drêche *f* de houblon
~LIQUOR - [leather ind] jus *m* épuisé
~LYE - [chem] lessive *f* épuisée
~MASH - [ind chem] (distiller's mash) vinasses *f*pl de distillerie
~OIL - [mech] huile *f* épuisée
~OXIDE - [chem] (an iron oxide which is a highly valued raw material for the manufacture of sulphuric acid) oxyde *m* de fer usagé
~PICKLING LIQUOR - [ind chem] liqueur *f* de décapage épuisée
~SHALE - [min] sciste *m* épuisé
~SHUTTLE - [text] navette *f* épuisée
~STEAM - [mech] vapeur *f* d'échappement
~TAN - [leather ind] tannée *f*
~WASH - [chem] (alcohol-free liquid from a spirit still) lessive *f* résiduaire
SPERM - [physiol] (the male fertilizing fluid) sperme *m*, semence *f* (du mâle)
~OIL - [oil ind] (an oil derived from the sperm whale and used as a high-grade lubricant) huile *f* de baleine
~SULPHITE LIQUOR - [ind chem] lessive *f* résiduaire sulfitique
~WHALE - [zool] cachalot *m*
SPERMACETI - [chem] (a white wax, the main constituent of which is cetyl palmitate, obtained from the head of the sperm whale and used in the manufacture of standard candles and in pharmaceutical preparations) blanc *m* de baleine, spermaceti *m*
SPERMADUCT - [anat] tube *m* spermatique
SPERMARY - [anat] (testis) glande *f* séminale
SPERMATOCELE - [med] (or gonocele; a cyst of the tubules of the testis) spermatocèle *f*
SPERMATOCYST - [anat] vésicule *f* séminale, sperma-

tocyste *f*
SPERMATOCYSTITIS - [med] vésiculite *f*
SPERMATOCYTE - [biol] spermatocyte *f*
SPERMATOGONIUM - [anat](sperm mother cell) spermatogonie *f*
SPERMATOSPLASM - [biol] (the protoplasm of sperms) spermatoplasme *m*
SPERMINE - [chem] spermine *f*
SPERRY PROCESS - [ind chem] (a commercial process for the electrolytic production of lead carbonate) procédé *m* Sperry
SPERRYLITE - [min] (diarsenide of platinum) sperrylite *f*
SPEW, to - [med] vomir
[firearms] s'égueuler
SPEW - [med] vomissement
[el acoust] (the superfluous irregular rim of wax which must be removed from a record after it has been pressed between two stampers) dépôt *m* de cire
[rubber ind] (overflow) tropplein *m*
~GROOVE - [plast ind] (a groove in a mould designed to allow surplus material to escape during the moulding operation) gorge *f*
~RELIEF - [plast ind] (clearance provided between the mating surfaces of a mould through which any excess of material can escape and thus facilitate the closing of the mould) décharge *m* du dépôt
SP. Gr. - s. Specific Gravity
SPHAGNOPHILOUS - [ecology] (living in peaty waters) sphaignophile
SPHACELATION - [med] mortification *f*
SPHACELODERMA - [med] gangrène *f* cutanée
SPHACELUS - [med] sphacèle *m*
SPHALERITE - [min] (another name for Zinc Blende, natural sulphide of zinc, the chief ore of that metal) sphalérite *f*, blende *f*
SPHENOID - [gen] sphénoïdal
[anat] (the sphenoid bone) sphénoïde *m*
[crystall] (wedge-shaped crystal form consisting of four triangular faces) sphénoïdal
SPHENOIDITIS - [med] sphénoïdite *f*
SPHERE - [gen] sphère *f*, milieu *m*, domaine *m*
[geom] sphère *f*
~GAP - [el] (a spark-gap in which the electrodes are in the form of spheres) éclateur *m* à sphères
~GAP VOLTMETER - [instr] (a voltmeter which utilizes a sphere gap for operation) voltmètre *m* à sphères
~ORE - [min] minerai *m* sphérique
~POWER - [opt] puissance *f* de la sphère
~OF ATTRACTION - [phys] (of molecular forces) domaine *m* d'attraction
~OF INFLUENCE - [gen] zone *f* d'influence
[astr] sphère *f* d'influence
SPHERES - [th eng] (spheres of fused alumina used for air heaters) sphères *f*pl
SPHERIC - s. spherical
SPHERICAL - [gen] (pertaining to sphere; shaped like a sphere) sphérique
~ABERATION - [opt] (accurate calculation shows that a perfectly spherical surface can never form a perfect image of a finite object) aberration *f* sphérique
~ANGLE - [math] angle *m* sphérique
~ASTRONOMY - [astr] (positional astronomy) astronomie *f* sphérique
SPHERICAL BASALT - [constr] basalte *m* sphérique
~BODY - [geom] corps *m* solide sphérique
~BOWL - [geom] calotte *f* sphérique
~CANDLE POWER - [phys] (the illimitation of a sphere of unit radius with the source of light at its centre) intensité *f* lumineuse sphérique
~CAP - s. spherical bowl
~CONICAL SECTION - [math] section *f* conique sphérique
~CO-ORDINATES - [astr] (polar co-ordinates in space) coordonnées *f*pl sphériques
~CURVATURE - [math] courbure *f* sphérique
~EARTH FACTOR - [el] facteur *m* de terre sphérique
~-END MEASURING ROD - [meas] (caliber gauge with spherical ends) velte *f* à rotules
~EXCESS - [surv] (the amount by which the sum of the three angles of a spherical triangle exceed I80°) excès *m* sphérique
~FACEPLATE - [electron] (a type of TV picture tube faceplate) fenêtre *f* courbée
~FUNCTION - [math] fonction *f* sphérique
~GLASS GRANULES - [glass man] boulettes *f*pl de ferre, perles *f*pl de verre
~HEAD ROD - [mech] (a coupling element consisting of a rod terminating in a ball to engage a corresponding socket) barre *f* à tête sphérique
~INVERSION - [geom] inversion *f* sphérique
~LENS - [opt] lentille *f* sphérique
~LUNE - [geom] trochoïde *m*
~MIRROR - [opt] miroir *m* sphérique
~MIXING CHAMBER - [th eng] (of a burner) mélangeur *m* sphérique (de brûleur)
~PERSPECTIVE - [geom] perspective *f* sphérique
~POLYGON - [geom] polygone *m* sphérique
~RADIATOR - [radiat] (a radiator producing the same radiation intensity in all directions) émetteur *m* sphérique
~REACTOR - [nucl] réacteur *m* sphérique
~REDUCTION FACTOR - [light] (the ratio between the mean spherical candle-power of a lamp and the mean horizontal candle-power) facteur *m* de réduction sphérique
~ROLLER-BEARING - [mech] palier *m* à rouleaux sphériques
~SEAT - [mech] coussinet *m* sphérique, cuvette-rotule *f*
~SEAT OF THE BEARING - [mech] portée *f* sphérique dans le corps du palier secteur *m* sphérique
~SHELL - [geol] couche *f* concentrique
~SOCKET - [plumb] emboîtement *m* sphérique
~SURFACE - [mech] surface *f* sphérique
~TRIANGLE - [geom] (a triangle forming part of the surface of a sphere) triangle *m* sphérique
~TRIGONOMETRY - [math] (the branch of trigonometry dealing with spherical triangles) trigonométrie *f* sphérique
~VALVE - [mech] (a ball valve) vanne *f* sphérique
~WAVE - [radio] (a wave the equiphase surfaces of which form a family of concentric spheres) onde *f* sphérique
~-WAVE FRONT - [radio] front *m* d'onde sphérique
~WEDGE - [geom] onglet *m* sphérique
~WELL - [nucl] (potential well assuming a potential both constant and negative inside a certain radius, zero outside) puits *m* rectangulaire
SPHERICALLY SEATED COUPLING - [mech] joint *m* à rotule

SPHERICS - [math] géometrie *f* sphérique
[radio] (colloq for atmospherics) perturbation *f*, brouillage *m*
SPHEROID - [geom] (a body having nearly the form of a sphere) sphéroïde *m*
[metall] (of ductile cast iron) sphérule *f*
[geom] (ellipsoid) ellipsoïde
SPHEROIDAL - [gen & geom] sphéroïdal
~GRAPHITE - [min] graphite *m* sphéroïdal
~GRAPHITE CAST IRON - [metall] fonte *f* ductile, fonte *f* à graphite sphéroïdal
~JOINTING - [geol] (spheroidal cracks found in igneous and sedimentary rocks) séparation *f* sphéroïdale
~STATE - [phys] (the spheroidal drops which roll about like mercury drops when is dropped on a clean horizontal and red-hot plate, but do not boil, this being prevented by a cushion of steam on which the drop rides) état *m* sphéroïdal
~STRUCTURE - [geol] (structure exhibited by certain igneous rocks) structure *f* sphéroïdale
SPHEROIDITE - [metall] perlite *f* globulaire
SPHEROIDIZATION - [metall] sphéroïdisation *f*
SPHEROIDIZE, to - [metall] (to assume a globular form because of a particular heat-treatment) sphéroïdiser
SPHEROIDIZING - [metall] (prolongued heating near the transformation range followed by slow cooling) sphéroïdisation *f*
SPHEROMETER - [instr] (instrument designed to measure the curvature of spherical and other surfaces) sphéromètre *m*
SPHERULE - [anat] sphérule *f*
SPHERULITE - [min] (a crystalline body having fibres which radiate outwards from a centre and terminate on the surface of the sphere) sphérolithe *f*
SPHERULITIC IRON - s. spheroidal graphite cast iron
~TEXTURE - [geol] (a type of rock fabric consisting of spherulites) texture *f* sphérolitique
SPHINCTER - [anat] (a muscle which narrows or closes an orifice by its contraction) sphincter *m*, orbiculaire *m*
SPHYGMOCARDIOGRAPH - [med] sphygmocardiographe *m*
SPHYGMOGRAM - [med] sphygmogramme *m*
SPHYGMOMANOMETER - [instr] (instrument for measuring blood pressure) sphygmomanomètre *m*
SPHYGMOMETER - [instr] (or pulsimeter; instrument designed to measure the strength of the pulse beat) sphygmomètre
SPHYGMUS - [zool] (the beat of the heart and the corresponding beat of the arteries) pouls *m*, pulsation *f*
SPHYRECTOMY - [med] excision *f* du marteau
SPICA - [bot] épi *m*
[med] (a bandage) spica *m*
SPICE - [bot etc] épice *f*, aromate *m*
~BUSH-[bot]benjoin *m* odoriférant, arbre *m* à benjoin
~EXTRACTS - [agric] extraits *mpl* d'épices
~PLANT - [bot] plante *f* condimentaire
SPICES - [bot] épices *fpl*
SPICULAR - [gen] (covered with small, needle, -like processes-spiny) spiculaire, apiciforme
SPIDER - [zool] araignée *f*
[mech] (of a wheel) croisillon *m* brassure *f* de roue, étoile *f*

SPIDER - [el mech] (of a rotor) lanterne *f*, croisillon *m*
[metall] (a device used to stiffen a mould or core) armature *f*
[mech] (the component of an ejector mechanism which operates the ejector pins in a moulding press) lanterne *f*, support *m*
[plast ind] (the structure which supports and locates the torpedo in an extrusion machine) support *m* tournant, support *m* rotatif
[glass man] (an assembly of radiating tie-rods on the top of a furnace) curseur *m*
[el] (the centre part of an armature core, on which the core stampings are built up) support *m* de noyau d'armature
[cin] (in motion-picture studios, a portable distributing box for lighting cables) douille *f* multiple
[impl] (USA only; a three-legged iron stool for the support of pots over the fire) trépied *m*
[mining] bougeoir *m* de fer
[oil ind] collier *m* à coins
~AND SLIPS - [mining] anneau *m* de manoeuvre
~ANGIOMA - [med] angiome *m* stellaire
~-ANT - [zool] mutille *f*
~BOX - [el] (junction box for studio) douille *f* multiple
~BURSTS- [med] varicosités *fpl* capillaires
~CANCER - s. spider angioma
~GEARS - [auto] pignons *mpl* planetaires
~-LIKE SPRUE - [plast ind] (sprue connecting a number of mouldings, like a bunch of grapes) grappe *f* de moulage
~LINE - [opt] (for optical instruments) fil *m* d'araignée, réticule *m*
~NEVUS - s. spider angioma
~SHIM - [railw] cale *f* à joints
~SILK - [text] soie *f* d'araignée
~-WEB - [gen] toile *f* d'araignée
~-WEB AERIAL - [radio] (a type of unidirectional aerial) antenne *f* en fond de panier
~-WEB COIL - [el] bobine *f* en fond de panier
~WEBS - [surv] (across the reticule of a diaphragm) réticule
~SPIEGEL - [metall] spiegel *m*, fonte *f* spiegel, fonte *f* à facettes
SPIEGELEISEN - [metall] (pig iron with a content of I5 to 30 p.c. of manganese and 4 to 5 p.c. of carbon, added to steel in manufacture to increase manganese and deoxidize the melt) fonte *f* spéculaire
SPIGOT - [mech] (a cylindrical projection on a part, designed to enter a corresponding recess in another associated part) ergot *m*
[plumb] fausset *m*, broche *f*, clé *f* de robinet
[metall] (of a die set) bout *m* de centrage
[agric] (of a cask) cannelle *f* de tonneau
[hydr] (a cock) (USA only) robinet *m*
~AND FAUCET JOINT - s. spigot and socket joint
~-AND-SOCKET JOINT - [plumb] joint *m* à emboîtement, assemblage *m* à manchon
~AND SOCKET PIPE - [plumb] tuyau *m* à emboîtement et cordon
~AND SOCKET TEE - [plumb] emboîtement *m* de tubulure à emboîtement
~END - [plumb] bout *m* mâle
~JOINT - [plumb] joint *m* tulipé soudé
~-MOUNTED - [mech] (attached to another part

or unit by a projecting cylindrical portion, e.g. an electric motor, in which one end of the carcase is shaped to fit into a recess in the unit which it drives) monté avec ergot
SPIKE, to - [gen] (to fasten with spikes) clouer, cheviller
[gen] (a wall, a gate etc) barbeler une grille de pointes
[firearms] enclouer (un canon)
SPIKE - [mech] (stout piece of metal, roughly like a large nail) pointe *f* de métal, piquant *m* de fil barbelé, clou *m* à large tête
[el] (short-duration transient, comprising part of a pulse) impulsion *f* à pointe, pic *m*
[rail] crampon *m* d'attache
[shoe man] (on a shoe) clou *m*
[shoe man] (a very high heel on a woman's shoe) pointe *f*, talon *m* à pointe
[bot] (an ear of corn, wheat and other grains) épi *m*
[bot] (a flower cluster) hampe *f* florale
~-DRAWER - [mech] pied-de-biche *m*, pied-de-chèvre *m*
~FLOWER - [bot] fleur *f* à épis
~-LAVENDER - [bot] lavande *f* commune, spic *m*
~LEAKAGE ENERGY - [electron] puissance *f* de fuite d'impulsions à pointe
~-NAIL - [railw] chevillette *f*
~OIL - [ind chem] (synonym for Lavender spike oil) huile *f* d'aspic
~POTENTIAL - [el biol] (action spike, the greatest in magnitude and briefest in duration of the characteristic negative waves seen in records of the action potential) pointe *f* d'action, potentiel *m* de pointe
~-TONGS - [tool] pinces *fpl* pour chevillettes
~-TOOTH ROLLER - [agric] rouleau *m* à hérissons, rouleau *m* à disques étoilés
~TRAIN - [el biol] (a regular succession of pulses of unspecified shape, frequency, duration and polarity) courant *m* itératif, train *m* d'impulsions
SPIKED - [gen] à pointes, barbelé
BOBBIN BOARD - [text] planche *f* à bobines
~BRASS RING - [mech] disque *m* en cuivre jaune garni de pointes
~CHAIN - [text] chaîne *f* sans fin à picots, chaîne *f* à picots
~CHAIN TEMPLE - [text] templet *m* à chaîne à picots
~CYLINDER - [text] tambour *m* à dents
~ROLLER - [text] rouleau *m* garni d'aiguilles
SPIKELET - [bot] spicule *m*, épillet *m*
SPIKY - [metall] à tiges
SPILE, to - [gen] (to pierce a cask) boucher avec un fausset
[gen] (to provide with a spigot) pratiquer un trou de fausset
[constr] (to drive spiles into) piloter
SPILE - [gen] cheville *f*
[gen] (small plug) broche *f*
[constr] (a pile driven into the ground) pilotis *m* pieu *m*
[agric] (a spout driven into a sugar-maple tree to convey the sap to a bucket) perçoir *m* creux
[mining] (a forepole used in mining) palplanche *f*
SPLINING - [constr] pieux *mpl*, pilotage *m*
[shipbuild] flèche *f* d'une concavite, épaule *f* de bordage
[mining] (advancing method in loose ground) poussage *m*
SPILITE - [geol] (a fine-grained igneous rock) spilite *f*
SPILL, to - [gen] renverser, répandre
[hydr etc] (to allow fluid to overflow or escape) se répandre
[naut] étouffer une voile
SPILL - [gen] chute *f*
[gen] (of wood) éclat *m* de bois
[impl] (for plugging a hole) tampon *m*
[mech] (a pin) cheville *f*
[mining] palplanche *f*
[electron] (in charge-storage tubes, the loss of information from a storage element by redistribution) perte *f* de mémoire
[comput] (in electrostatic memory) dispersion *f*
~BURNER - [aero] (in a jet engine type of burner in which a part of the fuel is re-circulated instead of being supplied to the combustion chamber) brûleur *m* à retour
~-OVER - [radio] (USA only; overflow in GB; the part of a signal which passes from one section to another before the connection between the sections is split) débordement *m*
~-OVER ECHO - [radar] (echo due to superrefraction effect) écho *m* bizarre
~PIPE - [mech] (in a diesel engine) tuyau *m* de retour.
[oil ind] tube *m* de retour
~VALVE - [mech] (of an injection carburettor) clapet *m* de purge
SPILLAGE TRAY - [ind chem] cuvette *f* de propreté
SPILLING - [gen] renversement *m*
[mining] poussage *m*, enfilage *m*
[aero] (in a parachute) échappement *m* d'air
~OVER - [radio] (of an amplifier) à la limite d'accrochage
~SURGE CHAMBER - [hydr] réservoir *m* compensateur *m* à déversoir
~SURGE TANK - [hydr] chambre *f* d'équilibre avec déversoir
SPILLS - [metall] (defect appearing in iron bars) defauts *mpl* internes
SPILLWAY - [hydr] passe-déversoir *m*
~CHANNEL - [hydr] coursier *m*
~DAM - [hydr] (reservoir dam over which water is allowed to flow to a downstream channel at the foot of the dam) barrage *m* à déversoir
SPIN, to - [text] (to twist individual fibres or filaments together to form a yarn) filer
[plast ind] (to extrude material through a small orifice to form a continuous thread or filament) filer
[mech] (to shape) tourner, tournoyer
[mech] (to turn) pivoter
[mech] (to cause to turn, so as to test freedom of motion) faire tournoyer
[aero] (to descend by a continuous spiral) descendre en vrille
[metall] emboutir au tour, couler par force centrifuge
[auto] (of the wheels) glisser, patiner
SPIN - [aero] (continuous spiral descent during which the means angle of incidence exceeds the stalling angle) vrille *f*
[phys] (the rotation about an axis) spin *m*
[nucl] (the angular momentum) spin *m*

SPIN - [gen] tournoiement *m*, rotation *f*
~A COCOON, to - [text] coconner
~A SINGLE COCOON, to - [text] filer un seul cocon
~A TWIN COCOON, to - [text] filer un doupion
~ANGULAR MOMENTUM - [phys] moment *m* angulaire du spin
~-DEPENDENT FORCE - [nucl] force *f* dépendant du spin
~DIMPLING - [mech] (spinning) emboutissage *m* au tour
~EFFECT - [nucl] (the influence of the spin on the normal chemical properties of the molecules) effet *m* de spin
~HARDENING - [metall] indurcissement *m* par rotation
~IN LARGE LOOPS, to - [text] filer en larges enroulements
~-LATTICE RELAXATION - [nucl] relaxation *f* spin-réseau
~LOOSELY, to - [text] filer légèrement
~-MAGNETIC MOMENT - [nucl] (the magnetic moment of a magneton connected with the spin impulse moment of an electron) moment *m* magnétique du spin
~-MAGNETIC RESONANCE - [nucl] (the magnetic moment associated with the intrinsic spin of a particle) résonance *f* magnétique du spin
~MOMENT - [nucl] moment *m* quantique du spin
~OFF, to - [ind chem] (in sugar manufacturing to extract mother liquor by centrifuging) centrifuger
~-ORBIT COUPLING - [nucl] couplage *m* spin-orbite
~ORIENTATION - [nucl] orientation *f* du spin
~QUANTUM NUMBER - [phys] (a number which gives the angular momentum of the electron considered as a small charged sphere revolving about an axis) nombre *m* quantique du spin
~STATE - [phys] état *m* de spin
~THE VISCOSE, to - [text] filer la viscose
~TUNNEL - [aero] soufflerie *f* à courant d'air vertical
~VELOCITY - [phys] (angular velocity about an axis fixed in space or in the spinning body) vitesse *f* de rotation
~WELDING - [plast ind] (method of joining plastic materials by forcing them together while one part is revolving rapidly, the frictional heat melting them at the interface) soudage *m* par friction
SPINA - [anat] (small sharp-pointed process) épine *f*, apophyse *f* épineuse
SPINACH - [bot] (a vegetable) épinard *m*
~BEET - [bot] bette *f*, poirée *f*
SPINAL - [anat] spinal, vertébral
~COLUMN - [anat] colonne *f* vertébrale
~CORD - [zool] moelle *f* épinière
~FLUID - [zool] liquide *m* cérébrospinal
~REFLEX - [zool] (reflex situated in the spinal cord) réflexe *m* spinal
SPINBACK - [auto] (of the steering wheel etc) réversibilité *f*
SPINDLE, to - [mech] toupiller
[gen & bot] pousser en hauteur
SPINDLE - [el] support *m*, console *f* verticale
[text] (the rod on a spinning wheel, by the rotation of which the thread is twisted and wound) fuseau *m*
[mech] (light shaft, especially in small mechanism) arbre *m*, broche *f*, mandrin *m*, verge *f*
[mech] (of an axle) axe *m*, pivot *m*

SPINDLE - [metall] (in foundry work) lanterne *f*
[naut] (of a capstan) mèche *f*
[mach tool] (of a milling cutter etc) mandrin *m*, broche *f* de tour
[el] (coupling) tenon *m*
[mech] (of an injector) aiguille *f* d'injecteur
[print] (of a printing press) vis *f* de presse
[biol] (of a nucleus) fuseau *m* achromatique (d'une cellule)
[carp] toupie *f*, toupilleuse *f*
~AND SWEEP - [metall] appareil *m* à trousser
~-ARM - [tool] (used for loam work) trusquin *m* porte-planche, bras *m* porte-profil
~HAND - [text] corde *f* de la broche
~BEARING - [text] douille *f* entraîneuse de la broche
~BELL - [text] cloche *f* de filature
~BEVEL GAUGE - [mech] jauge *f* à régler l'inclinaison du mandrin
~BLADE - [text] fût *m* de la broche
~-BOLT LOCK-PIN - [mech] goupille *f* d'arrêt du boulon de fusée
~BORE - [mech] alésage *m* du mandrin
~BOX - [text] boîte *f* à broches
[metall] (of a rolling mill) cage *f* à pignons
~BUSH - [text] douille *f* de la broche
~CARRIAGE - [mech] chariot *m* porte-broche
~CLAMP - [mach tool] blocage *m* du mandrin
~CLUTCH LEVER - [mech] levier *m* d'embrayage du mandrin
~COLLAR - [text] collet *m* de la broche
~CONNECTING ROD - [auto] barre *f* d'accouplement
~CORE - [metall] broche *f*
~DRIVE - [text] conduite *f* de la broche
~DRIVING GEAR - [text] mécanisme *m* de commande des broches
~DRIVING SHAFT - [text] arbre *m* de commande des broches
~DRIVING WHEEL - [text] roues *f*pl de commande des broches
~FEED - [mach tool] avance *f* du mandrin
~FRAME - [text] plate-bande *f* des bobines
~GAUGE - [text] écartement *m* des broches
~GROOVE - [text] rainure *f* de la broche
~HEAD - [mech] tête *f* de la broche
~HEADSTOCK - [mach tool] (of a lathe) poupée *f* porte-broche
~HEEL - [text] pied *m* de la broche
~HOUSING - [metall] (a housing for the pinions of a rolling mill) cage *f* à pignons
~LIGHTNING PROTECTOR - [el] parafoudre *m* à bobine
~MOULDING MACHINE - [mach tool] toupie *f*, machine *f* à moulurer
~NOSE - [tool] nez *m* du mandrin
~OILS - [text] (low-viscosity lubricating oils, produced as distillates, suitable for use in the textile industries) huiles *f*pl à broches
~PITCH - [mech] écartement *m* des broches
~PLATE - [text] assiette *f* à broche
~PRESS - [mech] (a screw press) presse *f* à vis
~RAIL - [text] rail *m* des broches
~-SHAPED - [gen] fusiforme, fuselé
~-SHAPED PLASMOID - [phys] (discrete piece of plasma shaped like a spindle) plasmoïde *m* fuselé
~SHUTTER - [text] navette *f* à broche
~SHUTTLE - [text] navette *f* à broche
~SLEEVE - [mach tool] douille *f* de la broche
~SPEED - [mach tool] vitesse *f* de la broche

SPINDLE SPRING - [text] ressort *m* de la broche
~STAIRS - [constr] escalier *m* à noyau plein
~TAPE - [text] ruban *m* de broche
~TEMPER - [metall] (a term used by steel manufactures and denoting steel containing I,I2% carbon) acier *m* à I,I25 de C.
~-TRAINING - [agric] taille *f* en quenouille
~TREE - [bot] fusain *m* · bonnet *m* de prêtre
~TROUGH - [text] auget *m* de la broche
~WASHER - [text] assiette *f* à broche
~WAVE - [el biol] (a sharp, rather large wave considered of diagnostic importance in the electro-encephalogram) onde *f* en fuseau
~WHARVE - [text] noix *f* de la broche
~WHIRL - [text] noix *f* de la broche
~WRENCH - [text] clef *f* à broches
SPINDLEHEAD - [mech tool] (of a horizontal miller) poupée *f*
SPINE - [gen] épine *f*
[anat] épine *f* dorsale
[bookbind] dos *m* (d'un livre)
[geogr] arête *f*
~OF THE SPHENOID - [anat] épine *f* du sphénoïde
~OF THE TIBIA - [anat] épine *f* du tibia
~WALL - [constr] mur *m* de refend
SPINEL - [min] (a native aluminium magnesium oxide with uses as a gemstone) spinelle *m*, candite *f*
SPINESCENT - [bot] (bearing spines) spinescent
[zool] (having a tendency to become spinous during racial decline) spinescent
SPINET - [mus] épinette *f*
SPINIFEROUS - [bot] (bearing spines) épineux, spinifère
SPINK - [zool] pinson *m*
SPINNAKER - [naut] (triangular sail used for racing craft) spinnaker *m*
SPINNER - [text] fileur *m*
[text] s. spinning machine
[aero] (a fairing round the boss of a propeller, coaxial and rotating with it) cône *m* de pénétration de l'hélice
[radar] (rotating aerial and reflector for microwave radars; it includes mounts, motor etc) aérien *m* rotatif
~-BASKET - [mech] panier *m* de turbine
~SURVEY - [metall] contrôle *m* d'essuage
~TYPE SPRINKLER - [agric] arroseuse *f* à jet rotatif
SPINNERET - [text & plast ind] (a special form of extrusion head for producing fibres) filière *f*, buse *f* à filer
SPINNERULE - [zool] (the duct by which the fluid is discharged by spiders) filière *f* des araignées
SPINNEY - [gen] bosquet *m*
SPINNING - [text] filage *m*, filature
[gen & phys] rotation *f* mouvement *m* de rotation
[mech] (the process of forming hollow bodies from sheet metal by revolution in a lathe against a smooth tool) emboutissage *m* au tour
[instr] (of the magnetic needle) affolement *m* (de l'aiguille magnétique)
~APPARATUS - [text] (a spinning machine used for artificial silk) métier *m* à filer
~BRAKE - [mech] repoussoir *m*
~CAN - [text] pot *m* de filature
~CAPACITY OF THE FIBRE - [text] aptitude *f* à la filature de la fibre
~CARD - [text] carde *f* fileuse
SPINNING CASTING - [metall] coulée *f* sous pression centrifuge
~COUNT - [text] titre *m* de filature
~DISK - [rubber ind] (for tyres) disque *m* de roulage
~DIVE - [aero] vrille *f* verticale
~ELECTRON - [phys] électron *m* tournant
~FLUID - [text] solution *f* à filer
~FLY - [text] duvet *m* de filature
~FRAME - [text] métier *m* à filer
FRAME BOBBIN - [text] bobine *f* de filature
~GLAND - [zool] filière *f* (du ver à soie etc)
~HUT - [text] (for the silkworm) cabane *f* à filer, boisement *m*
~-IN OF THE SILK-WORM - [zool] coconnage *m* du ver
~JENNY - [text] jenny *f*
~LATHE - [mach tool] tour *m* à emboutir
~LIMIT - [text] limite *f* d'aptitude à la filature
~MACHINE - [text] machine *f* à filer
[el] (a machine used for insulating wires) machine *f* pour enroulements
~MACHINE WITH TRAVELLING SPINDLES - [text] métier *m* à files à broches voyageantes
~MATERIAL - [text] matière *f* filable, matière *f* textile
~MILL - [text] filature *f*
~MUSCLE - [zool] muscle *m* filant
~NIPPLE - [zool] mamelon *m* filant
~NOZZLE - s. spinneret
~PAPER - [text] papier *m* filable, papier *m* à filer
~PASTE - [text] pâte *f* à filer
~PRESSURE - [text] pression *f* sur la solution à filer
~ROOM - [text] filature *f*
~ROOM FLY - [text] duvet *m* de filature
~SILK WORM - [text] ver *m* à soie fileur
~SPEED - [el] (of an electrical centrifuge) vitesse *f* de régime
~SPRAYER - [text] filière *f* multiple
~TOP - [radiat] (disk of dense material with an eccentric aperture which, when spun above a film, enables a record to be made of variations of the X-ray output during the exposure) toupie *f* en plomb
~WASTE - [text] déchets *mpl* de filature
~WHEEL - [text] rouet *m*
~WOOL - [text] laine *f* filable
SPINOSE - [gen] épineux
~AGAVE LEAF - [bot] feuille *f* d'agave épineuse
SPINTHARISCOPE - [instr] (instrument in which scintillations are visually) spinthariscope *m*
SPINTHERISM - [med] spinthéropie *f*
SPINY LEAF - [bot] feuille *f* épineuse
SPIRACLE - [zool] (of insects) soupirail *m* (d'insecte)
[zool] (of Cetaces) évent *m* (de cétacé)
[geol] (of a volcano) évent *m* (de volcan)
SPIRAL, to - [gen] former une spirale, tourner en spirale
[aero] descendre en spirale
SPIRAL - [gen] spirale *f*
[gen] spiral, spiralé, hélicoïdal
[geom] hélicoïde *m*
[aero] (the trajectory of an aircraft) montée *f* en spirale, descente *f* en spirale
[astronaut] (of rockets) vriller
[comm] (of prices etc) montée *f* en flèche (des prix etc)
~ANGLE - [mech] (of a toothed gear) inclinaison *f* de la spirale
~AUGER - [oil ind] tarière *f* à vis cylindrique

SPIRAL BALANCE - [meas] peson *m* à hélice
~BARREL VAULT - [constr] voûte *f* en limaçon
~BEVEL GEAR - [mech] couple *m* conique d'engrenages, couple *m* à taille spirale
~BEVEL GEAR GENERATOR - [mach tool] vis-fraise *f* pour denture spirale
~BINDER - [bookbind] machine *f* pour reliure spirale
~BINDING - [bookbind] reliure *f* spirale
~BRAIDING MACHINE - [text] métier *m* à tresser en spirale
~BREAKING MACHINE-[text]dérompeuse *f* à spirales
~CASING - [mech] (of a turbine) enveloppe *f* en spirale
~CHUTE - [ind] plan *m* incliné-hélicoïdal
~CLUTCH - [mech] embrayage *m* à spirale, embrayage *m* à enroulement
~COILED SPRING - [mech] ressort *m* en spirale
~CONVEYOR - [transp] porteur *m* à vis sans fin
~COUPLING - [mech] joint *m* en spirale
~CURVE - [math] courbe *f* spirale
~CUT - [text] coupe *f* en spirale
~CUTTER - [mach tool] fraise *f* hélicoïdale
~DESCENT - [aero] (a banked continuous descent) descente *f* en spirale
~DOWEL - [mech] goujon *m* en spirale
~DRAWING FRAME - [text] étirage *m* à spirale
~DRILL - [tool] foret *m* hélicoïdal
~DRUM - [mining] tambour *m* spiraloïde
~END MILL - [mech] fraise *f* à queue à cannelures hélicoïdales
~FACE THREAD - [mech] spirale *f* d'Archimède
~-FLOW AERATION - [hydr] aération *f* dans un bassin à mouvement rotatif
~-FLOW TANK - [hydr] (used in the activated sludge process) bassin *m* à circulation, bassin *m* à mouvement rotatif
~FLUTE - [mech] cannelure *f* en spirale
~-FLUTED REAMER - [mach tool] alésoir *m* avec cannelures hélicoïdales
~-FOUR CABLE - [el] câble à paires-étoile
~-FOUR-QUAD - [el] (a quad cable; a cable containing a number of star quads) paires *f*pl cablées en étoile
~GANGLION - [zool] (a continuous ganglionic cord at the base of the spiral lamina, connected with the cochlear branch of the auditory nerve) ganglion *m* spiral
~GEAR - [mech] (a toothed gear for connecting two shafts having the axes at an angle) engrenage *m* hélicoïdal
~GLIDE - [aero] (a banked continuous gliding turn) descente *f* en spirale
~GROOVE BEARING - [mech] coussinet *m* rainure en spirale
~GUIDES - [gas ind] (in gasholders) guidage *m* hélicoïdal
~GUIDING - s. spiral guides
~INSTABILITY - [aero] (instability through which an aircraft tends to turn from straight flight with a combination of side-slipping and excessive banking) instabilité *f* spirale
~MILLING CUTTER - [mach tool] fraise *f* à denture hélicoïdale
~MOULD COOLING - [plast ind] (system of injection mould cooling using a spiral passage in the mould body) réfrigération *f* spirale du moule
~NEBULA - [astr] (one of the largest class of nebulae, so called because of their appearance in the telescope) nébuleuse *f* en spirale
SPIRAL PEG - [text] broche *f* spirale
~PUMP - [mech] pompe *f* spirale
~RIBBON MIXER - [mech] mélangeur *m* à spirale
~RIDGE - [mech] rainure *f* en spirale
~RIDGED ACCELERATOR - [nucl] accélérateur *m* à pôles magnétiques à rainures en hélices
~SCANNING - [radar] (type of cathode ray tube scan in which the spot moves spirally over the screen) exploration *f* circulaire, balayage *m* en spirale
~SCRAPER - [ind chem] (a sludge collector) racleur *m* spiral
~SHAPED DENT - [text] dent *f* en forme de spirale
~SHAPED HORN - [zool] corne *f* en spirale
~SLOT IN SPOOL CARRIER - [text] fente *f* du support de bobine en forme de spirale
~SPRING - [mech] ressort *m* à boudin
~STAIR - [constr] escalier *m* en limaçon, escalier *m* à noyau
~STRIATIONS - [text] stries *f*pl en spirales
~WELD PIPE - [plumb] tube *m* soudé en spirale
~WINDING - [el] enroulement *m* en spirale
~WOOD DRILL - [constr] boulonnière *f*
SPIRALS OF TWIST - [text] spires *f*pl de torsion
SPIRALLY-GUIDED GASHOLDER - [gas ind] gazomètre *m* à guidage hélicoïdal
SPIRE, to - [gen] s'élever en flèche
SPIRE - [constr] (a slender tapering tower) flèche *f* d'église, aiguille *f*
[mech] (of a propeller) tour *m* (d'hélice) spire *f*
[metall] (in blasting) fétu *m*
~-ROOF - [constr] toit *m* pyramidal, toit *m* de tour
SPIRELET - [constr] (a small spire) tourette *f*
SPIRIT - [gen] esprit *m*
[chem] (aqueous solution of ethyl alcohol, in particular obtained by distillation) alcool *m*, essence *f*
[mech] (motor spirit) carburant *m*
~-BLUE - [dye] bleu *m* à sec d'étain
~COMPASS - [instr] boussole *f* à alcool
~ENGINE - [mech] moteur *m* à alcool
~LEVEL - [instr] niveau *m* à bulle d'air, niveau *m* à alcool
~OF WINE - s. spirits of wine
~PAINT - [ind chem] (used in joinery) vernis *m* à alcool
~STAIN - [ind chem] (stain for wood; colouring matter dissolved in methylated spirit) mordant *m* à alcool
~TUBE - [constr] nivelle *f*
SPIRITS - [gen] spiritueux *m*pl, liqueurs *f*pl spiritueuses
~OF HARTSHORN - [chem] (obsolete term for an aqueous solution of ammonia) carbonate *m* d'ammoniaque
~OF SALT - [chem] (oil name for hydrochloric acid in aqueous solution) esprit *m* de sel
~OF TURPENTINE - [ind chem] essence *f* de térébenthine
~OF WINE - [chem] (an old term for ethyl alcohol) esprit *m* de vin, alcool *m* éthylique
SPIRITOUS - [gen] spiritueux
SPIRKETTING - [shipbuild] (of a wooden ship) rivure *f* bretonne
SPIROCHAETES - [bacteriology] (filamentous bacteria showing spirals or ondulations; some of them are causative agents of syphilis) spirochètes *m*pl

SPIROGRAPH - [med] spirographe *m*
SPIROMETER - [med] (instrument designed to measure the air inhaled and exhaled during respiration) spiromètre *m*, pnéomètre *m*
SPIROMETRY - [med] (the measurement of air inhaled and exhaled during respiration) spirométrie *f*
SPISSATED - [med] épaissi
SPISSITUDE - [med] épaississement *m*, condensation *f*
SPIT, to - [gen & med] cracher
[mech] (in internal combustion engines) avoir des retours de flamme
[cooking] mettre à la broche
[metall] (of silver) rocher
[mining] tirer
SPIT - [gen] salive *f*, crachat *m*
[impl] broche *f*
[geogr] (a long narrow shoal) langue *f* de sable, ras *m* de terre, bec *m* (de fleuves)
[mech] (in flash welding) crachement *m*
SPITTING - [mining] tir *m*
~EFFECT - [acoust] (undesirable noise in ultra-short wave receivers) effet *m* de crachement
SPLANCHNIC - [anat] (visceral) splanchnique
SPLANCHNICECTOMY - [med] splanchnicectomie *f*
SPLANCHNOLITH - [med] calcul *m* intestinal
SPLANCHNOPTOSIS - [med] (general dropping of the abdominal viscera) splanchnoptose *f*
SPLASH, to - [gen] éclabousser, faire jaillir
SPLASH - [gen] éclaboussement *m*, clapotage *m*
[naut] (of waves) clapotis *m* (des vagues)
[gen] (a spot) tache *f* (de lumière etc)
[metall] (of an ingot, surface roughness due to splashes from the casting stream) bavure *f*
[metall] (shell) peau *f*
~ARMS - [mech] (arms attached to a central spindle in an autoclave or the like, to ensure complete wetting of the material under treatment) agitateurs *mpl*
~BAFFLE - [electron] (in mercury-pool tubes) écran *m* de déviation, plaque *f* de déviation
BOARD - [impl] tablier *m* para-boue
~CORE - [metall] plaque *f* anti-érosion
~DOWN, to - [astronaut] (the landing on the sea of a space capsule) amerissage *m* d'un véhicule spatial
~FEED - s.splash lubrication
~GUARD - [mech] para-boue *m*, para-gouttes *m*
~-LUBRICATE, to - [mech] graisser par barbotage
~LUBRICATION - [mech] (transfer of lubricant from a well or sump to a moving part by the dipping of the part itself into the lubricant) graissage *m* par immersion, graissage *m* par barbotage
~PLATE - [hydr] déflecteur *m* d'eau
~RING - [electron] (of a mercury-pool cathode tube) bague *f* anti-projection
~SYSTEM - s. splash lubrication
SPLASHBOARD - [auto] tablier *m* para-boue *f*
SPLASHER- [mech] (in internal combustion engines) plongeur *m* d'huile
SPLASHING - [acoust] (the sound made by water waves) clapotement *m*, clapotis *m*
~SOUND - [med] bruit *m* de fluctuation
SPASHINGS - [metall] éclats *mpl* de métal
SPLASHPLATE - [impl] (in domestic cookers) dosseret *m*
SPLASHPROOF - [el] (said of an electrical apparatus, to denote that the live parts are enclosed by a cover, to exclude rain, snow and external splashings) protégé contre les projections d'eau latérales
SPLASHPROOF - [instr] (term applied to measuring instruments to denote that it will withstand occasional splashing with water without detriment to its performance) protégé contre les projections d'eau
SPLASHER - [metall] (near the gate of a furnace) plaque *f* de déviation (pendant la fusion)
SPLATTER - [telev] (adjacent channel interference) interférence *f* adjacente
SPLAY, to - [constr] (to give a divergent form) évaser, ébraser
[carp] couper en sifflet
SPLAY - [constr] ébrasement *m*, embrasure *f*
[arch] coupe *f* oblique, chanfrein *m*
[gen] évasement *m*
SPLAYED - [gen] évasé, ébrasé
[carp] (bevelled) chanfreiné, en sifflet
~BRICK - [constr] brique *f* ébrasée
~BUTT - [mech] assemblage *m* en sifflet désabouté
~INDENT SCARF - [carp] enture *f* à trait de Jupiter
~JAMB - [constr] montant *m* ébrasé
~JOINT - [carp] assemblage *m* en sifflet, bechevet *m*, joint *m* de paume
~MITRE JOINT - [carp] assemblage *m* à onglet en sifflet
~SCARF - [carp] enture *f* en sifflet
~TUYERE - [metall] (of a furnace) tuyère *f* évasée
SPLAYING OF WALL - [constr] ébrasement *m* du mur, biseau *m* du mur
SPLEEN - [anat] rate *f*
~RATE - [med] index *m* splénique
SPLENATROPHY - [med] atrophie *f* splénique
SPLENCULUS - [med] rate *f* accessoire, rate *f* surnumeraire
SPLENEOLUS - s.splenculus
SPLENITIS - [med] splénite *f*
SPLENOPEXY - [med] splénopexie *f*
SPLENOPTOSIA - [med] ptôse *f* splénique
SPLENOPTOSIS - s.splenoptosia
SPLENORRHAPHY - [med] suture *f* de la rate
SPLICE, to - [gen & naut] (to join a rope by means of a splice) épisser (un cordage etc)
[cin] (to join sections of film) monter (un film), coller (un film)
[paper man] (to join the two ends of a web of paper from a reel which has been broken) jointer
[carp] (two pieces of wood) enter (deux pièces de bois)
[auto] manchonner (une chambre à ait)
SPLICE - [gen] (of ropes; a joint in a rope formed by laying the strands together in a special way) épissure *f*
[gen] (of wire rope ends) ligature
[mech] enture *f*
[rubber ind] soudure *f* (d'un pneu)
~A ROPE, to -[naut] épisser un cordage
~BAR - [mech] éclisse *f*
~-BOLT - [railw] boulon *m* d'assemblage de la pointe et de la branche de pointe
~BUMB - [cin] (low-frequency sounds due to poorly finished blooping patches) bruit *m* des collures
~GRAFTING - [agric] greffe *f* en fente
~PIECE - [railw] éclisse *f*
~-RAIL - [railw] branche *f* de pointe
~TUBES, to - [auto] manchonner les chambres à air

SPLICER - [cin] (used to connect the separate film strips) presse *f* à coller, colleuse *f*
[tool] pince *f* à épisser
[gen] colleur *m*
SPLICING - [cin] (joining together different pieces of processed films) collage *m*
[cin] (the editing of a film) montage *m*
[gen] épissage *m* (des câbles)
~CEMENT - [cin] (glueing substance used to join together pieces of film) colle *f* à film
~COMPOUND - [ind chem] (used for strips of film) colle *f* à film
~EAR - [el] (a form of metallic fitting attached to a contact wire for the purpose of suspending or retaining the wire in position) connecteur *m* annulaire
~FITTING - [el] (a clamp) flûte *f* de jonction, griffe *f*
~SELVEDGE SPINDLE - [text] vis *f* de chariot d'anglaisage
~SLEEVE - [telecomm] (work) dalle *f* de raccordement, manchon *m* de raccord
~TABLE - [cin] table *f* de montage (de film)
~TAPE - [acoust] (special non-magnetic tape used for splicing magnetic tape) ruban *m* de collage
~UNIT - [cin] (film splicer) presse *f* à coller
SPLINE, to - [mech] (to make a slot or a groove for a spline) claveter
[mech] (to inset a spline) monter une languette
[mech] (to cut splines in a shaft) rainurer
SPLINE - [mech] (a feather key) languette *f*
[mech] (an integral shaft-key formed by cutting a series of parallel grooves in a shaft, the part left between them being the spline) cannelure *f*, rainure *f*
[carp] (thin tongue of wood used in matching grooved planks etc) cannelure *f*
[mech] (keyway) logement *m* de clef, rainure *f* de clavette
~BROACHING MACHINE - s. spline grinding machine
~GRINDING MACHINE - [mach tool] (a machine for grinding splines to precise dimensions) machine *f* à meuler les cannelures
~-HOBBING MACHINE - [mach tool] (a machine for cutting splines by means of a hob cutter) machine *f* à raboter les rainures des arbres à clavettes
~MILLING MACHINE - [mach tool] machine *f* à fraiser les rainures
~PIN - [mech] goupille *f* cannelée
~SHAFT - s. splined shaft
SPLINED - [gen] claveté, cannelé, à rainures
~SHAFT - [mech] (a shaft formed with alternating longitudinal grooves and ribs, engaging corresponding shapes in the centre of a wheel or the like, to allow endwise movement of the latter while the drive is continuously transmitted) arbre *m* cannelé
~SHAFT GRINDING MACHINE - [mach tool] machine *f* à meuler les arbres
SPLINES - [mech] (longitudinal grooves or teeth on a shaft or in the boss of a wheel) cannelure *f*
SPLINT, to - [med] éclisser, mettre une attelle
SPLINT - [med] clisse *f*, éclisse *f*, attelle *f*
(thin flexible piece of wood) éclisse *f*
[vet] (of horses) suros *m*
~COAL - [min] (steam coal) houille *f* flambante, houille *f* à longue flamme
SPLINTER, to - [gen] briser en éclats éclater, craquer
SPLINTER - [gen] éclat *m*, picot *m*
SPLINTER - [med] esquille *f* (d'un os fracturé)
[naut] écli *m* de bois
~BAR - [mech] volée *f*
~-BONE - [anat] péroné *m*
~HEMORRHAGE - [med] suffusion *f* hémorragique sous-unquéale
~-PROOF - [gen] (not liable to form splinters on fracture; or capable of resisting penetration by flying splinters) pare-éclats *m*, se brisant sans éclats
SPLINTERED FRACTURE - [med] fracture *f* esquilleuse fracture *f* comminutive
SPLIT, to - [gen] fendre, refendre, éclater
[chem] (to separate into components) séparer
[opt] (of images) séparer
[constr] (stone) cliver, déliter
[geol] se cliver, se dédoubler
[paint] se gercer
SPLIT - [gen] fente *f*, fissure *f*, crevasse *f*
[paint] gerçure *f*
[gen] (of a basket) lame *f* de gaulis
[text] (of a reed) dent *f* de ros
[th eng] (a firebrick of half standard thickness) brique *f* demi-largeur
[leather ind] (a hide split into two or more layers) couche *f* de peau fendue
[telecomm] (in Morse telegraphy) espacement *m* faux
[metall] crique *f* longitudinale
[mining] (in a mine, divided air-current) courant *m* partiel
[mining] (in a mine, the workings aired by a split air-current) dérivation *f*
~AILERON - [aero] (an aileron inset into the lower surface of the wing) aileron *m* fendu
~-ANODE MAGNETRON - [electron] (a magnetron with a cylindrical anode divided into two segments by divisions parallel to the axis) magnétron *m* à anode fendue
~AXLE - [auto] pont *m* à trompettes
~-BEAM CATHODE-RAY TUBE - [electron] tube *m* à rayons cathodiques à double faisceau
~BEARING - [mech] palier *m* en plusieurs pièces
~BEND - [plumb] coude *m* en deux pièces
~CABLE GRIP - [telecomm] amarrage *m* de câble, manchon *m* à mailles
~CASING - [el] boîtier *m* sectionné
~CAVITY - [plast ind] (a mould formed in several parts) empreinte *f* en plusieurs pièces
~CHUCK - [metall] manchon *m* de serrage
~COLLAR - [mech] manchon *m* en deux pièces
~COLLET CHUCK - [tool] mandrin *m* à collier fendu
~COLUMN CONTROL - [comput] éliminateur *m* d'X
~CONDUCTOR CABLE - [el] (a cable in which each conductor is divided into two or more sections insulated from each other and normally connected in parallel) câble *m* à conducteurs subdivisés, câble *m* à plusieurs brins en parallèle
~CONE - [mech] collier *m* conique fendu
~CORE BOX - [metall] boîte *f* à noyaux en deux parties
~-CORE TYPE TRANSFORMER - [el] (transformer without a primary, the magnetic circuit of which can be opened and the closed round the conductor carrying the current to be measured) transformateur *m* pince
~COTTER - [mech] (an annular part, divided into

two along a diameter, inserted into a groove in an i.c engine valve-stem to retain the spring) clavette *f* fendue
SPLIT CRANKCASE - [mech] (a crankcase in an i.c. engine which is divided into two horizontally, usually near the plane of the crank shaft centre-line) carter *m* en deux parties
~DIE - [metall] matrice *f* assemblée, estampe *f* démontable
~[mech] coussinet *m* fendu
~DRUM - [text] tambour *m* fendu
~DUCT - [el] caniveau *m* de raccordement (entre une conduite unitaire et une autre conduite)
~FEED - [metall] attaque *f* de coulée frontale
~-FIELD MOTOR - [el] moteur *m* à excitations inverses
~FIELD PICTURE - [telev] trame *f* à deux demi-images différentes
~-FIELD RANGE-FINDER - [instr] télémètre *m* à déplacement d'images
~FITTING - [el] (a bend, elbow or tee, split longitudinally so that it can be placed in position after the wires have been drawn into the conduit, the two parts being held together by screws) domino *m*
~FLANGE - [mech] plateau *m* en deux parties
~FLAP - [aero] (type of flap which forms a part of the lower surface only of the aerofoil) volet *m* d'intrados
~-FLOW REACTOR - [nucl] (reactor in which the coolant enters at the centre section and flows outward at both ends, or vice-versa) réacteur *m* à réfrigérant divisé
~FOCUS - [opt] ajustage *m* du foyer à un point entre deux objects
~FOLLOWER - [plast ind] empreinte *f* mobile, coquille *f* de moule
~-FOLLOWER MOULD - [plast ind] moule *m* à coins
~FOLLOWERS - [plast ind] coquilles *fpl*, coins *mpl* d'un moule
~GATE - [metall] attaque *f* de coulée sur le plan de jonction
~GRID - [radio] demie grille *f*
~HUB - [mech] moyeu *m* fendu
~HYDROPHONE - [acoust] (directional hydrophone) hydrophone *m* multicellulaire
~IMAGE - [telev] dédoublement *m* d'image
[cin] plan *m* de détail
~JOINT - [carp] joint *m* articulé
~-IN - [mech] (a locking device) goupille *f* de clavette
~LEAF - [bot] geuille *f* bifide
~-LENS STEREOSCOPE - [photo] stéréoscope *m* à lentilles prismatiques
~LINE SOURCE - [radiat] source *f* à raie spectrale subdivisée
~MOULD - [plast ind] (a mould in which the cavity is formed of two or more components held together by an outer chase. The components are known as splits) moule *m* en plusieurs pièces d'empreinte, moule *m* à coins
~MUFF-COUPLING - [mech] manchon *m* cylindrique en deux pièces
~-NUT - [mech] écrou *m* fendu
~OPEN-ENDED CORE BOX - [metall] boîte *f* à noyau-cadre
~PATTERN - [metall] (in foundry work) modèle *m* en deux parties
SPLIT PEG - [text] broche *f* fendue
~PEN - [constr] goupille *f* de sûreté
~PHASE - [el] (circuit arrangement for changing a single phase to a two-phase a supply) circuit *m* de déphasage
~-PHASE CIRCUIT - s. split phase
~-PHASE MOTOR - [el] moteur *m* à enroulements à pas brouillé
~-PHASE STARTING - [el] démarrage *m* par phase auxiliaire
~-PHASE WINDING - [el] enroulement *m* séparé des phases
~PICTURE - [cin] (a detail picture) détail *m*, plan *m* de détail
~-PIN, to - [mech] goupiller
~PIN - [mech] goupille *f* fendue, clavette *f* fendue
~PIN EXTRACTOR - [tool] (special tool for withdrawing split pins) extracteur *m* de goupille fendue, tire-goupille *m*
~PROJECTOR - [el acoust] pavillon *m* multicellulaire
~PULLEY - [mech] (a pulley which consists of two detachable halves) poulie *f* démontable
~REED - [text] roseau *m* fendue
~REEL - [cin] bobine *f* fractionnée
~RIM - [auto] jante *f* démontable en deux pièces
~RING - [mech] bague *f* fendue
~-RING CLUTCH - [mech] accouplement *m* à bague extensible fendue
~-RUN - [gas ind] (three-phase water-gas cycle) fabrication *f* à courant de vapeur alterné
~"S" - [aero] (a manoeuvre consisting of a half roll followed by pulling out to normal flight, giving a change to a reciprocal course, with loss of altitude) retournement *m*
~-SAW - [impl] scie *f* à refendre
~SCREEEN EFFECT - [telev] découpage *m* électronique
~SECOND - [gen] moins *m* de seconde
~-SECOND HAND - [horol] trotteuse *f* double (de chronomètre), aiguille *f* des secondes
~SERVICE COLLAR - [plumb] collier *m* à lunette
~SHAFT - [auto] éclisse *f*
~SKIRT PISTON - [auto] piston *m* à jupe fendue
~STRAW - [text] paille *f* fendue
~STRIPS, to - [text] fendre en lanières
~SYSTEM - [th eng] (hot air and radiating heating) installation *f* de chauffage et ventilation
~TEE - [plumb] té *m* fendu, té *m* de raccordement
~THE ATOM, to - [nucl] désintégrer l'atome
~UP, to - [gen] fractionner, se séparer
[railw] (a train) débrancher
[phys] (of ions) se dédoubler
~WINDING - [el] (or stepped winding; a two-position diamond winding where those coil sections which occupy the top positions in one slot, occupy the bottom position in two other slots) enroulement *m* à pas brouillé
~WIRE - [comput] câble *m* en fourche
~-WIRE TYPE TRANSFORMER - s. split-core type transformer
SPLITS - [plast ind] (the components of the cavity of a split mould) coins *mpl*
~AND TILES - [constr] dalles *fpl* et briques de hourdis
SPLITTER - [plast & rubber ind] (machine for dividing or slicing cellular rubber or plastic) machine *f* à fendage, machine *f* à refendre (le caoutchouc mousse)

SPLITTER - [impl] (of a saw-bench) couteau *m* fendeur
[hydr] arête *f* médiane
SPLITTERS OF A V-CORRUGATED KNIFE - [sugar ind] (ribs of a ridge knife) cloisons *fpl* d'un couteau faîtière
SPLITTING - [gen] fendage *m*, refente *f*, séparation *f*
[constr] (of stone) délitement *m*
[nucl] désintégration *f*
~COMB - [text] peigne *m* répartisseur
~MACHINE - [photo] découpeuse *f*
~ROD - [text] baguette *f* d'envergure
~THE POINTS - [railw] bivoie *f*
~TIME - [telecomm] temps *m* de coupure
~UP OF TRAINS - [railw] débranchement *m* des trains
SPLUTTERING - [telecomm] crépitements *mpl*, friture *f*
[metall] soufflure *f*
S.P.M. - [mech] (initials of Strokes per Minute) courses *fpl* à la minute
SPODIUM - [chem] (synonym for animal black) noir *m* animal
SPODUMENE - [min] (natural lithium aluminium silicate,occurring in large crystals in pegnatites; an important source of lithium) spodumène *m*, triphane *m*
SPOKING - [radar] effet *m* de roue
SPOIL, to - [gen] endommager, abîmer
[food] altérer
[comm] (of goods) avarier
SPOIL - [constr] (the excess of cutting over filling) déblais *mpl*
[min] (refuse excavated material) décombres *mpl*
~BANK - [constr] (earthwork made by depositing spoil) cavalier *m*, champbord *m*
[mining] halde *f* de déblais
~CAR - [railw] (small tip wagon) wagonnet *m*
SPOILED BOBBIN - [text] mauvaises bobines *fpl*
~COCOON - [text] galette *f*
~HEMP - [text] chanvre *m* de rebut
SPOILER - [aero] (device to reduce the lift of an aerofoil by changing the airflow round it) spoiler *m*
SPOILING OF YARN - [text] gaspillage *m* de fil
SPOKE - [mech] (of a wheel; the radial element connecting the rim and the boss) rais *m*, rayon *m*
[gen] (of a ladder) échelon *m* d'échelle
[naut] (of the steering wheel) manette *f*, poignée *f*
[impl] (in block the wheel of a vehicle) bâton *m* à enrayer
~WHEEL - [mech] (of a heavy vehicle, e.g. a locomotive) roue *f* à rais, roue *f* à bras
[mach tool] croisillon *m* à poignée
SPOKESHAVE - [tool] (a type of double-handled plane used for shaping concave wood surfaces) vastringue *f*
SPONDYL - [anat] (a vertebra) vertèbre *f*
SPONDYLARTHRITIS - [med] spondylarthrite *f*
SPONDYLITIS - [med] spondylite *f*
SPONDYLOLISTHESIS - [med] spondyloptose *f*
SPONGE - [gen] éponge *f*
[firearms] écouvillon *m*
[med] bourdonnet *m*, tampon *m*
[el chem] (soft, fluffy and spongy material formed in electro-deposition) éponge *f*
[food] pâte *f* molle
[metall] éponge *f* métallique
~BATH - [gen] tub *m*
~BIOPSY - [med] bipsie *f* à l'éponge
~CLOTH - [text] (a coarse cloth made of cotton yarn) chiffon *m* pour essuyage

SPONGE CHROMIUM - [metall] éponge *f* de chrome
~ELECTRODE - [metall] électrode *f* à éponge
~-HEADED DRUMSTICK - [mus] (the stick which is used for producing damped tones from percussion instruments) baguettes *fpl* d'éponge, mailloches *fpl*
~IRON - [metall] éponge *f* de fer
~PLATING -[metall] revêtement *m* électrolytique avec électrode à éponge
~PLATINUM - [metall] éponge *f* de platine
~RUBBER - [rubber ind] (cellular rubber obtained either by mechanical agitation of the latex or incorporation of a blowing agent) caoutchouc *m* spongieux (à alvéoles communiquantes)
SPONGINESS - [gen] spongiosité *f*
[metall] (a defect caused by intercrystalline local shrinkage) spongiosité *f*
~OF THE FIBRE - [text] spongiosité *f* de la fibre
SPONGING - [gen & med] nettoyage *m* à l'éponge, lotionnement *m*
[text] décatissage *m* (d'un drap)
[fishing] pêche *f* des éponges
~AGENT - [chem] (a substance used to develop bubbles of gas within a substance e.g. rubber, in order to give it a spongy structure) agent *m* gonflant
~MACHINE - [text] machine *f* à rétrécir
SPONGIOSIS - [med] spongiose *f*
SPONGY - [gen] spongeux
[anat] caverneux
~COKE - [min] coke *m* spongieux
~FIBRE - [text] fibre *f* spongieuse
~PLATINUM - [chem] (synonym for platinum black) mousse *f* de platine
~SPOT - [med] (vascular zone) zone *f* criblée rétroméatique
~TOP - [metall] masselotte *f* boursoufflée, champignon *m*
~VULCANIZATE - [rubber ind] vulcanisat *m* poreux
SPONSON - [aero] (a projection from a seaplane float or the hull of an amphibian aircraft, designed to increase stability) nageoire *f*
[naut] (a projection from a hull, to obtain lateral stability in water) encorbellement *m*
SPONSORED FILM - [cin] (any film which is made for an industrial company etc by the studio people) film *m* professionnel
~TELEVISION - [telev] (commercial television) télévision *f* commanditée, télévision *f* publicitaire
SPONTANEITY - [gen] (the tendency to actions which are independent of external forces or conditions) spontanéité *f*
SPONTANEOUS - [gen] (arising from inherent qualities without external efficient cause) spontané
~COMBUSTION - [phys chem] (the inititation of combustion in a substance through physical and chemical action within the substance itself) inflammation *f* spontanée
~COMBUSTION OF THE SILK - [text] autoinflammation *f* de la soie
~CREAMING - [rubber ind] crémage *m* spontané
~DECAY - [nucl] (of radioactive elements) désintégration *f* spontanée
~FISSION - [nucl] (fission which occurs without particles or photons entering the nucleus from the outside) fission *f* spontanée
~GENERATION - [biol] (the production of living matter from non-living matter) génération *f* spontanée

SPONTANEOUS IGNITION - [phys & chem] (combustion occurring from causes within the material, and not from external effects) autoinflammation *f*
- IGNITION TEMPERATURE - [phys] température *f* d'autoinflammation
- MAGNETIZATION - [el] (the magnetic saturation of the domains of a ferromagnetic material, even in the absence of an applied magnetizing force) magnétisation *f* spontanée
- NUCLEAR REACTION - [nucl] (radioactive decay) réaction *f* nucléaire spontanée
- POLARIZATION - [el] (the polarization of a domain of a ferro-electric crystal, independently of an applied field) polarisation *f* spontanée
- TRANSFORMATION - [phys] (transformation taking place through inherent forces or energy, as distinguished from processes carried out by the application of external forces) transformation *f* spontanée
- TRANSMUTATION - [biol] transmutation *f* spontanée

SPOOL, to - [gen & text] (to wind on a reel) bobiner, dévider, enrouler, canneter

SPOOL - [gen] bobine
[text] (a pirn, or bobbin) bobine *f*, canette *f*, dévidoir *m*
[photo] bobine *f* (de pellicule)
[cin] bobine *f*, rouleau *m*
[el] (a flange structure specially intended for the support of a coil) bobine *f*
[mech] (of turbines, the complete rotating assembly including compressor, turbine rotor and shaft) rotor *m*
[el] (of an electromagnet) semelle *f* d'un électroaimant
- CARRIER - [text] porte-bobine *m*, support *m* de bobine
- FLANGE - [photo] joue *f* d'une bobine
- HEAD - [text] base *f* de fuseau, pied *m* de fuseau
- OF COTTON - [text] bobine *f* de coton
- OFF, to - [text] débobiner
- PIN - [text] tringle *f* de tirage
- REEL - [text] bobinoir *m* à main
- REMOVING MACHINE - [text] machine *f* à enlever les tubes
- SAMPLE - [metall] échantillon *m* du four
- SLIDE - [text] chariot *m* du métier-tulle
- SPINDLE - [text] broche *f* des bobines
- THREAD - [text] fil *m* de dessin
- VALVE - [mech] (a type of small piston valve used in hydraulic mechanism) clapet *m* à boisseau, distributeur *m*
- WINDER - [text] bobinoir *m*, canneteuse *f*
- WITH ONE FLANGE - [text] bobine *f* à disque unique
- WITHOUT FLANGES - [text] bobine *f* sans disques

SPOOLED SILK - [text] soie *f* bobinée

SPOOLER - [text] (machine for winding monofilaments on spools or reels) bobineuse *f*
- WASTE - [text] (the waste discarded from the spooler) déchets *mpl* de bobineuse

SPOOLING - [text] bobinage *m*, dévidage *m*, envidage *m*
- WHEEL - [text] bobinoir *m*

SPOON, to - [gen] ramasser avec une cuiller
[mech] (in earth-moving work) excaver

SPOON - [gen] cuiller *f*, cuillère *f*
[med] cuiller *f*
[naut] aviron *m* à lame incurvée
[fishing] (the metallic lure attached to the fishing line) cuillère *f*

SPOON AUGER - [tool] tarière *f* à cuiller, laceret *m* à cuiller
- BIT - [tool] foret *m* à cuiller
- DREDGE - [mech] drague *f* à cuiller

SPOONING - [ind chem] (only USA ; rake-out in GB) tabatière *f*

SPOONWORT - [bot] (a medical plant) herbe *f* aux cuillers

SPORADIC - [gen] (occasionally here and there, isolated) isolé
[med] sporadique
- REFLECTIONS - [radio] (or abnormal reflections, ionospheric reflections of radiowaves at frequency higher than the critical frequency of the layer) reflexions *fpl* sporadiques

SPORADONEURON - [anat] cellule *f* nerveuse isolée

SPORATION - [med] sporulation *f*

SPORE - [bot] (reproductive bidy characteristic of plants) spore *f*
- FORMATION - s. sporulation

SPORES OF THE MOULD - [bot] spores *fpl* des fleurs

SPOROGONY - [med] sporogonie *f*

SPOROTRICHOSIS - [med] sporotrichose *f*

SPORT - [gen] sport *m*
[biol] variété *f* anormale, variation *f* sportive

SPORULATION - [bot] sporulation *f*

SPOT, to - [gen] tacher, souiller
[gen] (to locate) répérer
[mech] (to make a mark on a work-piece) marquer
[mech] centrer (un trou)
[phys] repiquer

SPOT - [gen] lieu *m*, endroit *m*
[gen] (a blot, a stain) tache *f*, macule *f*
[gen] (a flaw) défaut *m*
[gen] (a drop) goutte *f*
[instr] (image projected on to a scale and acting as an index) pinceau *m* lumineux
[cin] s. spotlight
[el acoust] (a recording light beam) élément *m* de piste sonore
[radio] (of transmission from a local station) local
[comm] (argent) comptant, au comptant
[electron] (the small area of the screen surface instantaneously affected by the impact of the electron beam) spot *m*; point *m* lumineux
[telev] (the focal point of the electron beam on the television picture screen) spot *m*
- ACCELERATION - [electron] accélération *f* du spot
- ANALYSIS - [chem] analyse *f* à la goutte
- BOARD - [constr] (square wooden board on which the plasterer works) truelle *f*
- CASH - [comm] argent *m* comptant
- CHECKS - [gen] contrôles *mpl* intermittents
- CLEANING - [paint] (of metal plates, e.g. autos) sablage *m*
- COTTON - [comm] (cotton available for immediate delivery) coton *m* payé comptant
- DIAMETER - [telev] (the dimension of a scanning line in horizontal direction) largeur *f* de ligne
- DISPLACEMENT - [telev] déplacement *m* du spot
- DISTORTION - [electron] (fault of a spot, the shape of which is not circular) distorsion *f* du spot
--DRILL, to - [mech] (the operation of drilling a small hole for locating purposes) centrer en trou
- ELEVATION - s. spot height
--FACE, to - [mech] dégrossir

SPOT FACING - [mech] dégrossissage *m*
~FACING TOOL - [tool] fraise *f* à dégrossir
~FILM DEVICE - [photo] sélecteur *m*
~FINISHING - [paint] retouche *f*
~-GLAZE - [paint] glacis *m* pour taches
~-GRINDING - [mech] meulage *m* par points
~HEIGHT - [surv] (the height above datum at a given spot on a map) point *m* coté
~LAMP - [auto] projecteur *m* orientable
~LEVEL - [surv] (the reduced level of a point chosen at random) cote *f* de niveau
~LIMITER - [telev] limiteur *m* de la visibilité d'interférence impulsive
~MARKET - [comm] marché *m* du disponible, marché *m* du comptant
~-MILLING - [mech] fraisage *m* par points
~NOISE FACTOR - [radio] (of a transducer at a selected frequency) facteur *m* propre de bruit
~ORDER - [comm] commande *f* au comptant
~PRESS WITH SHOULDER COUNTER-BLOCK - [rubber ind] (for tyres) presse *f* pour réparations locales avec bloc pour réparations à l'épaule
~PRESS WITH SIDEWALL BLOCK - [rubber ind] (for tyres) presse *f* pour réparations locales avec bloc pour réparations au flanc
~PRESS WITH TREAD COUNTER BLOCK - [rubber ind] (for tyres) presse *f* pour réparations locales avec bloc pour réparations à la bande de roulement
~PRICE - [comm] cours *m* en disponibles
~REMOVER - [gen] (stain remover) enlève-taches *m*
~REPAIR - [gen & mech] réparation *f* locale et superficielle
~SANDING - [paint] (for metal plates, as in autos) sablage *m* local d'une tôle
~-SEAM WELDING - [metall] soudure *f* continue et par points
~SHAPE CORRECTION - [telev] correction *f* de la configuration du spot
~SIZE - [electron] diamètre *m* du spot
~SPEED - [telev] vitesse *f* de balayage
~SUPPRESSOR - [telev] suppresseur *m* de tache de brûlage
~TERMS - [comm] conditions *fpl* du marché
~TEST - [metall] essai *f* à la goutte
~TESTS - [metall] essais *mpl* à la touche
~-WELD, to - [metall] (to weld locally, at separate points) souder par points
~-WELDING - [metall] (welding at single points, as distinct from a butt or seam weld) soudage *m* par points
~-WELDING MACHINE - [el metall] (machine for spot-welding, usually with a pair of electrodes which can be pressed together with the workpiece between them, the current passing from one to the other to make the weld) machine *f* à souder par points
~WOBBLE - [telev] (the oscillating movement of the cathoderay tube) vobulation *f* du spot
~YARN - [text] retors *m* boutonné
SPOTLIGHT - [opt] (a special lamp designed to project a concentrated beam of light) réflecteur *m* lenticulaire
[auto] projecteur *m* auxiliaire orientable
[cin] (as employed in studios) spot *m*, projecteur *m* convergent, projecteur *m* intensif
SPOTTED - [gen] tacheté, moucheté
[zool] maculé, madré, truité
SPOTTED COCOON - [text] mauvaise chaquette *f*, cocon *m* taché
~DISEASE - [bot] pébrine *f*
~DOG - [zool] chien *m* de Dalmatie
~FABRIC - [text] tissu *m* moucheté
~FEVER - [med] méningite *f* cérébro-spinale
~NET - [text] tulle *m* moucheté
~NET MACHINE - [text] métier *m* à tulle moucheté
~WEAVE - [text] article *m* à points brodés
SPOTTER - [aero] avion *m* de réglage de tir
[railw] surveillant *m* de la voie
[gen] observateur *m*
~PLANE - [aero] avion *m* observateur
SPOTTINESS - [telev] (the effect on the television screen caused by the variation of the instantaneous light value of the reproduced image due to electrical disturbances) parasites *mpl* en forme de tâches
SPOTTING - [gen] repérage *m*
[paint] (the appearing of small areas different in colour from the general one) tachetures *fpl*
[metall] (the appearance of spots on finished plate work) apparition *f* de tachetures
[mech] centrage *m* d'un trou
~DRILL - [tool] (flat drill with a point so shaped as to centre and face the end of a bar with one operation) foret *m* à centrer
~-IN - [paint] retouche *f*
~OUT - [metall] souillures *fpl*
~SHUTTLE - [text] espolin *m*
~THREADS - [text] fils *mpl* de mouchetures
~TOOL - [tool] outil *m* à centrer, outil *m* à marquer
SPOUT, to - [gen] jaillir, lancer (de l'eau)
[hydr] gicler, faire jaillir
SPOUT - [gen] (a discharging pipe) tuyau *m* de décharge
[gen] (of a tea-pot etc) bec *m*
[metall] (of a tilting furnace, a discharging lip) canal *m* de coulée, chenal *m* de coulée
[hydr] dégorgeoir *m*
[hydr] (an orifice) goulotte *f*
[radio] (opening in a waveguide through which radiation takes place) buse *f*
[gas ind] (in gas meters) siphon *m*
~FEED - [metall] alimentation *f* à chenal de coulée libre
~SLAG - [metall] scorie *f* de canal
SPOUTER - [oil ind] puits *m* jaillissant
SPOUTING HORN - [geol] caverne *f* karstique
SPRAG - [gen] (a timber prop, used to prevent the wheels of a vehicle from revolving) cale *f*, bâton *m*, carotte *f*
[mining] tasseau *m*
[auto] béquille *f* de recul
SPRAIN - [med] entorse *f*, foulure *f*
~FRACTURE - [med] fracture *f* par arrachement au cours d'une entorse
SPRAY, to - [gen] (to project a liquid from a nozzle in a fine mist) pulvériser, atomiser, projeter
[paint] peindre à l'air comprimé
[agric] arroser, bassiner
SPRAY - [gen] poussière *f* d'eau, pulvérin *m*
[gen] (the sprayed liquid) eau *f* vaporisée, eau *f* pulvérisée
[gen] (the liquid sprayed) liquide *m* pour vaporisation
[metall] (in foundry work; auxiliary gate) chenal *m* de coulée auxiliaire

SPRAY - [metall & plast ind] (a complete set of mouldings with the gates and feets attached, as removed from a multi-impression injection mould) grappe *f*
[min] apophyse *f* d'un filon
~ANGLE - [mech] (the apical angle of the cone formed by the fuel spray) angle *m* de pulvérisation
~-APPLIED WRAPPING - [transp] ("cobwebbing" or cocoon packing) coconisation *f*, enveloppage *m* par "cobwebbing"
~ARRESTER - [el] (of an accumulator; a sheet of glass, ebonite or other suitable material serving to prevent the escape of acid spray) protection *f* contre les jets d'acide
(a lightning arrester consisting of a spray of water from an earthed pipe impinging on a plate connected to the live circuit) parafoudre *m* à jet d'eau
~-BAR - [aero] (of a jet engine; radially-arranged bars provided with spray holes through which the fuel is delivered) pulvérisateur *m* radial
~BOOTH - [auto] cabine *f* pour la peinture à pistolet
~CARBURETTOR - [mech] carburateur *m* à diffuseur
~CASTING - [metall] coulée *f* en grappe
~COMPRESSOR - [mech] compresseur *m* à injection
~CONE - [mech] diffuseur *m*
~COOLING - [metall] rafraîchissement *m* par pulvérisation
~CUTTER - [paper man] (a device for cutting the web of paper in the paper machine) coupe-bande *m*
~DAMPER - [paper man] (device for damping the web of paper when it leaves the paper machine) appareil *m* à humidifier la bande
~-DAMPING MACHINE - [text] humecteuse *f*
~DIFFUSER - [impl] arroseuse *f* à poussière d'eau
~-DRIED RUBBER - [rubber ind] sprayed rubber *m* (caoutchouc en poudre)
~DRYER - [ind chem] (apparatus in which a solution or suspension is sprayed in droplets into a stream of hot gas) sécheur-pulvérisateur *m*
~DRYING - [ind chem] séchage *m* par pulvérisation
~DRYEING - [paint] peinture *f* au pistolet
~EMUSLION - [agric] bouillie *f* pour la pulvérisation
~ENAMELLING - [metall] émaillage *m* au pistolet
~GATE - [metall] (in die casting; an in gate consisting of a number of small separate gates, fed from the runner) grappe *f*
~GUN - [paint] (a spraying pistol) pistolet *m* de vernissage
~IRRIGATION - [agric] arrosage *m* en pluie, irrigation *f* en pluie
~-LINE IRRIGATOR - [agric] batterie *f* d'arrosage à jet horizontal, arroseur *m* à longue portée
~-LINE SYSTEM - [agric] installation *f* d'arrosage en pluie
~MIXTURE - s. spray emulsion
~NOZZLE - [mech] (a jet designed to reduce liquid forced through it to a fine spray) tuyère *f* de pulvérisation
~PAINT - [paint] peinture *f* au vaporisateur, peinture *f* au pistolet
~PAINTER - [rubber ind] (cement sprayer for tyres) pulvériseur *m* à dissolution
~PAINTING - [paint] (the application of paint by means of a compressed air jet) vernissage *m* au pistolet
~PIPE PASSAGE - [mech] (in a diesel engine) orifice *m* de pulvérisation
SPRAY POINTS - [el] (a row of sharp points charged to high d.c. potential, the purpose of which is to charge and discharge the conveyor belt in a Van de graaff generator) points *mpl* de décharge
~PRINTING - [text] impression *f* au pistolet
~PRODUCER - s. spray diffuser
~QUENCHING - [metall] refroidissement *m* rapide par eau atomisée
~-RINSE CHAMBER - [photo] chambre *f* de rinçage à arrosage
~SCRUBBER - s. spray tower
~TOWER - [ind chem] (an extraction or washing tower in which fluid is sprayed from the top, counter to gas or vapour admitted at the bottom) tour *f* de lavage à pulvérisation
~-TYPE FOAM-RUBBER WASHER - [rubber ind] machine *f* à laver le caoutchouc mousse par arrosage
~-UP - [plast ind] (method of contact moulding in which the resin and glass are applied by spray processes) moulage *m* au pistolet
~VAPOURIZER - [paint] vapoirsateur *m*, pulvérisateur *m*
~WASHING TOWER - s. spray tower
~WEBBING - [plast ind] (also called cocooning; the operation of enclosing an article in a closed envelope by spraying suitable plastic material over it) coconisation *f*
~WELDING - [th eng] obturation *f* par projection de coulis fondu
~WITH COPPER, to - [agric] (hops etc) arroser pour le traitement cuprique
SPRAYED CATHODE - [electrode] (a cathode for an electronic tube, the emissive coating being sprayed on a filamentary core) cathode *f* à revêtement pulvérisé
~RUBBER - [rubber ind] latex *m* défloculé
SPRAYER - [mech] (of a carburettor) diffuseur *m*
[gen] atomiseur *m*, vaporisateur *m*
[agric] arroseuse *f* à poussière d'eau
[th eng] brûleur *m* de mazout
~FOR MOULD LUBRICANT - [rubber ind & plast ind] pistolet *m* à badigeonner les moules de lubricant
SPRAYING - [gen] pulvérisation *f*, vaporisation *f*, arrosage *m*
[paint] peinture *mf* au pistolet
[metall] (the coating of a surface by projecting on it a spray of molten metal) pulvérisation *f* métallique
~AND FUSING - [metall] projection *f* et fusion
~APPARATUS - [brew ind] pulvérisateur *m*
~CHAMBER - [rubber ind] chambre *f* de pulvérisation
~COLUMN - [ind chem] tour *f* à pulvérisation
~HOSE - [gen] tuyau *m* d'arrusage
~MACHINE - [agric] pulvérisateur *m*
~OF THE FUEL - [th eng] (in a boiler furnace) pulvérisation *f* du combustible
~PISTOL - [print] (a device for spraying paint and the like; designed to be held in the hand) pistolet *m*
~PLANT - [ind] installation *f* pour la peinture au pistolet
SPREAD, to - [gen] étendre, s'étaler
[gen] (fire, disease, news etc) se répendre, se propager, semer (des nouvelles)
[gen] (to distribute) épandre, distribuer
[rubber ind] (to rubberize by spreading) enduire

par brossage
SPREAD, to - [rubber ind] (a tyre) écarter (un pneu)
[mus] (a chord) arpéger (un accord)
SPREAD - [gen] étendue *f*
[gen] (the limit of expansion of an object) développement *m*, ouverture *f*
[text] (a cloth or covering) couverture *f* de parade
[mech] (between two centres etc) ouverture *f*, distance *f*
[aero] envergure *f*
[paint] (the area covered by a given amount of fluid material) étendue *f*
[comm] différence *f* entre le prix de fabrique et le prix de vente
[metall] (margin) marge *f*
[oil ind] (arrangement of seismometers in relation to the shot point in seismic survey methods) distribution *f* des sismomètres
[el] dispersion *f*, fuite *f*
[electron] (with semiconductors, the range within which some values may vary) domaine *m* de définition
~APART, to - [gen & mech] écarter
~FABRIC - [rubber ind] tissu *m* gommé par enduction
~FACTOR - [el] (distribution Factor in the USA ; that component factor of the winding factor which takes into account the phase spread) facteur *m* de distribution
~FLEECE - [text] toison *f* étendue
~MANURE, to - [agric] épandre le fumier
~OF PARASITES - [bot] propagation *f* des parasites
~OUT, to - [gen] déployer
~OUT THE HANK, to - [text] étaler l'écheveau
~REFLECTION - [opt] réflexion *f* dispersée
~RING - [mech] bague *f* de tension
~SAILS, to - [naut] établir les voiles
~THE FLEECE, to - [text] étendre la toison
SPREADER - [gen] étendeur *m*
[agric] arrosoir *m*, éventail *m* (d'une lance d'arrosage)
[text] (a device designed to spread a protective coating over a fabric) table *f* à étaler, étaleuse *f*
[text] tendeur *m*
[mining] étrésillon *m*
[carp] couteau *m* fendeur
[radio] (a spar for keeping the wires of a multiwire aerial spaced apart from each other) traversier *m*, nappe *f* d'antenne
~BAR - [mech] (or spreader; a bar or beam used to provide two points of suspension for a load from a single hoisting rope) barre *f* d'écartement
~CHAINS - [auto] (used to limit the opening of a gate to control the flow of material being spread from the truck) chaînes *f*pl de hayon
~-JET - [hydr] jet *m* à éventail
~STOKER - [mech] (for steam engines) chargeur *m* mécanique à grille
SPREADING - [gen] développement *m*, déploiement *m*, étandage *m*, répandage *m*, enduction *f*
[gen] (propagation) propagation *f*, colportage *m*
[text] (of the flax) étandage *m*
[transp] (the dumping of a load at a uniform rate while the vehicle is in motion) distribution *f*, uniforme en marche, épandage *m* uniforme en marche
~BOX - [constr] (a machine designed to spread road material) machine *f* à répander (le goudron)
~CALENDER - [rubber ind] calandre *f* de gommage, calandre *f* d'enduction
SPREADING CAPACITY - [paint] (over a surface) pouvoir *m* couvrant
~CHEST - [rubber ind] table *f* chaude du métier à gommer
~COEFFICIENT - [phys] (thermodynamic expression for the work done in the spreading of one liquid on another) coefficient *m* de diffusion
~DOCTOR - [impl] racle *f*
~FLAME BURNER - [gas ind] (in domestic gas cookers) brûleur-pipe *m*
~KNIFE - [impl] couteau *m* racleur, contre-racle *f*
~MACHINE - [rubber ind] (mechanism for distributing coating material in a uniform layer on sheet) épandeuse *f*
[text] machine *f* à étaler
~MACHINE CHEST - s. spreading chest
~METHOD - [rubber ind] procédé *m* d'enduisage
~MILLET GRASS - [agric] millet *m* sauvage, millet *m* étalé
~MIX - s. spreading mixture
~MIXTURE - [rubber ind] mélange *m* pour revêtement, mélange *m* d'enduction, mélange *m* à gommer
~PASTE - [rubber ind] pâte *f* d'enduction
~POWER - [paint] (alternative term for Covering power, usually expressed as the area which a gallon of composition will cover properly) pouvoir *m* couvrant
~RATE - [paint] (the area coated with paint per unit of paint applied) rendement *m*
~RESISTANCE - [electron] résistance *f* extrinsèque
~ROLLER - [text] rouleau *m* d'enduction
~TABLE - [rubber ind] table *f* à enduire
SPRIG - [gen] brin *m*, petite branche *f*, broutille *f*
[carp] (small pointed implement) pointe *f* de Paris, semence *f* de tapissier
[metall] (small nail reinforcing the edge of a mould during pouring) clou *m* de renfort, épingle *f*
~NET - [text] tulle *m* faconné
SPRIGGING - [metall] épinglage *m*
SPRIGGY NET - s.sprig net
SPRILL - [metall] particules *f*pl cylindriques
SPRING, to - [gen] sourdre, descendre
[gen] (of water) jaillir
[gen] (to act suddenly as by an elastic reaction) sauter, bondir
[mech] (to recoil, rebound under the action of a spring) faire jouer, faire sauter
[mech] (to be elastic, as auto etc) munir de ressorts
[phys etc] (to cause to open) fendre
[mining] (to cause to explode) faire sauter
[arch] (an arch, a vault) établir le point de naissance
[auto] (to fit with springs) suspendre
[wood] gauchir, se déjeter
SPRING - [mech] (a mechanical part designed to act elastically) ressort *m*
[gen] (the season) printemps *m*
[gen] (a source of water) source *f* (d'eau)
[phys] (a source of energy, of power) source *f*, origine *f*
[gen] (a leap) saut *m*, bond *m*
[mech] (elasticity) élasticité *f*
[naut] embossure *f*, croupiat *m*
[naut] (the rope between ship and wharf) traversier *m*

SPRING - [shoe man] courbure *f* du dessous de la forme
[gen] (a crack) craqûre *f*
[arch] naissance *f*, retombée *f*, apophyge *f* de colonne
[mus] (of reed pipes) rasette *f*
~A LEAK, to - [naut] contracter une voie d'eau
~AMMETER - [el] ampèremètre *m* à ressort antagoniste
~ASSEMBLY - [el] (the collection of contact springs in a relay) ressorts *mpl* de contact
~BACK, to - [mech] (of sheet metal etc) faire ressort
~-BACK - [mech] (of sheet metal etc) bond *m* en arrière
[mech] (of a coil) déformation *f* remanente
~BALANCE - [instr] (weighing device in which the object to be weighed is suspended by a spring) peson *m* à ressort, peson *m* à hélice
~BALL JOINT - [mech] (a ball and socket joint in which the parts are held in contact by a spring) rotule *f* à ressort
~BAND - [mech] (of an elliptic spring) bride *f* du ressort
~BAR - [horol] barrette *f* à ressort
~BARLEY - [agric] orge *f* de printemps
~BLOCK - [text] grille *f* à ressorts
~BOLT - [mech] boulon *m* étoquiau
~BOWS - [instr] (small compasses whose two limbs are not hinged together, but connected by a bow of spring steel, the distance apart being adjusted by a screw) compas-balustre *m* à pincettes
~BOX - [text] étui *m*
~BRACKET - [mech] main *f* de ressort
~BRAKE FOR BOBBIN - [text] ressort *m* de bobine formant frein
~BRASS - [metall] laiton *m* de ressort
~BUCKLE - s. spring clip
~BUFFER - [auto] amortisseur *m* de direction, tampon *m* de ressort
~CALIPERS - [tool] compas *m* d'épaisseur à ressort avec écrou rapide
~CAMBER - [mech] flèche *f* de ressort
~CARRIER - s. spring bracket
~CARRIER ARM - s. spring bracket
~CATARRH - s. spring conjunctivitis
~CATCH - [mech] mousqueton *m*
~CENTER BOLT - (US only; spring tie bolt) boulon *m* étoquiau de ressort
~CENTRE - [mach; tool] pointe *f* à ressort
[mech] (the vertical line along which a vertical load applied to the sprung mass will produce only uniform vertical displacement) axe *m* vertical de ressort
~CHAPLET - [metall] support *m* de noyau de forme
~CHUCK - s. spring collet
~CLAMP - [mech] bride *f* de ressort
~CLICK - [horol] cliquet *m* à ressort
~CLIP - [mech] (of vehicles) étrier *m* de ressort
[impl] pince *f* à ressort
~CLIP BOBBIN HOLDER - [text] porte-bobine *m* élastique
~CLUTCH - [mech] accouplement *m* à ressort
~COCOON - [text] cocon *m* de printemps
~COLLET - [mach tool] (of a lathe) manchon-pince *m*
~CONJUNCTIVITIS - [med] conjonctivite *f* estivale, catarrhe *m* printanier

SPRING CONTACT - [el] contact *m* élastique
~-CONTACT PLUG - [el] fiche *f* à contact élastique
~-CONTACT STRIPS - [el] bandes *fpl* élastiques de frottement
~CONTROL - [el] (a method of controlling the movement of an indicating instrument by means of a spring) commande *f* à ressort
~COP SPINDLE - [text] broche *f* pour cannette à ressort
~CORN - [agric] céréales *fpl* de printemps
~COTTER - [mech] clavette *f* fendue
~COVER - s. spring gaiter
~CULTIVATION - [agric] travaux *mpl* de printemps
~CURTAIN - [gen] rideau *m* à enroulement
~CUSHIONING - [mech] amortissement *m* par ressort
~DIE - [tool] filière *f* réglable
~DIVIDERS - [instr] compas *m* à ressort, compas *m* à pointes à ressort
~DOWEL PIN - [mech] boulon *m* étoquiau de ressort
~DRIVE - [mech] (of transmission shaft) transmission *f* élastique
~DRIVE GEAR - [mech] engrenage *m* élastique
~EQUALIZING ROCKER ARM - [auto] culbuteur *m* compensateur des ressorts
~EYE - [mech] (of a leaf spring) oeuil *m* de ressort
~EYE BUSHING - [mech] bague *f* d'oeil de ressort
~FATIGUE - [mech] fatigue *f* du ressort
~FAUCET - [mech] robinet *m* à ressort
~FORK - [mech] fourchette *f* à ressort
~FORK WITH PIN - [text] fourchette *f* élastique avec goupille
~FRAME - [mech] (of motorcycle) châssis *m* à suspension à ressorts
~GAITER - [auto] gaine *f* de ressort
~GALVANOMETER - [instr] (a galvanometer containing a coil in which an iron element, suspended by a spring, moves in the hollow space of the coil) galvanomètre *m* à ressort
~GAUGE - [tool] jauge *f* à ressort
~GEAR - [mech] (auto etc) suspension *f*
~GOVERNOR - [mech] régulateur *m* à ressort
~GRAIN - s. spring corn
~-GRASS - [bot] flouve *f* odorante
~HANGER - [mech] tige *f* de suspension du ressort
~-HANGER PIN - [mech] boulon *m* de suspension
~HANKS - [naut] cosses *fpl*
~-HEAD - [hydr] source *f*
~HEEL - [mech] (of a motorcycle) suspension *f* postérieure télescopique
~HINGE - [carp] charnière *f* à ressort
[cart] penture *f* à ressort *m*
~HOLDER - [mech] longement *m* du ressort
~HOOK - [mech] crochet *m* de sûreté, mosqueton *m*
~INDEX - [mech] indice *m* d'enroulement d'un ressort
~INSULATOR - [el] isolateur *m* à ressort
~JOINT OF A VAULT - [constr] joint *m* de naissance *f*
~KEY - [impl] clavette *f* fendue
~LATCH - [mech] (of a door) cliquet *m*
~LEAF - [mech] lame *f* de ressort
~-LEAF OPENER - [tool] outil *m* pour écarter les lames de ressort
~LEG - [mech] branche-ressort *m*
~LEVER - [mech] levier *m* à ressort
~-LID OIL-CUP - [mech] graisseur *m* à couvercle élastique

SPRING LOAD - [mech] charge *f* du ressort
~-LOADED - [mech] (term used of a part against which a spring bears, to constrain it to move in a determined direction) à ressort
~-LOADED CULTIVATOR - [agric] cultivateur-vibroculteur *m*
~-LOADED ROPE BRAKE - [text] frein *m* à corde maintenu par un ressort
~-LOADED TAP - [mech] (of domestic cookers) robinet *m* à rattrappage de jeu
~-LOADED VALVE - [mech] soupape *f* à ressort
~LOCK - [mech] serrure *f* à ressort, serrure *f* à fouillot, serrure *f* à pêne à demi-tour
~-LOCK WASHER - [mech] rondelle *f* élastique
~MACHINE - [mech] machine *f* à fabriquer les ressorts
~MAIN LEAF - [mech] (auto etc) lame *f* maîtresse de ressort
~MEASURE - [impl] roulette *f* à ressort
~MOTOR - [cin] (a type of motor used in cinematography and sound recording installation) moteur *m* à ressort
~NEEDLE - [text] (for hosiery) aiguille *f* à ressort
~OUT OF SHAPE, to - [mech] se déformer
~PAD - [auto] tampon *m* intermédiaire de ressort
~PAWL - [mech] cliquet *m* d'arrêt à ressort
~PERCH - [mech] patin *m* de ressort
~PICK LOOM - [text] métier *m* à chasse-navette par ressort
~PICKING LOOM - [text] métier *m* à chasse par ressort
~PICKING MOTION - [text] mécanisme *m* de chasse à ressort
~PINCERS - [impl] pinces *fpl* à ressort
~PISTON RING - [railw] (of a steam locomotive) segment *m* de piston
~PIVOT SEAT - [mech] siège *m* pivotant de ressort
~PLATE - [mech] (of a leaf spring) lame *f* de ressort
~POINT - s .spring switch blade
~POINTS - [railw] aiguille *f* à ressort
~POWER-HAMMER - [impl] marteau-pilon *m* à ressorts
~PRESSURE - [mech] tension *f* du ressort
~PRESSURE-GAUGE - [instr] manomètre *m* à ressort, manomètre *m* métallique
~-RAIL FROG - [railw] croisement *m* à ressort
~RATE - [mech] flèche *f* de ressort
~REED - [text] peigne *m* à ressort
~RIGGING - [mech] suspension *f*
~RING - [mech] segment *m* de piston
[auto] (a detachable tyre-locking ring) jonc *m* de verrouillage fendu pour jante, cercle *m* de fermeture
~-RING JOINT - [auto] (piston ring lock) ergot *m* de positionnement d'un segment de piston
~ROLLER BLIND - [gen] store *m* à enroulement automatique
~SADDLE - [mech] étrier *m* de ressort, selle *f* à ressorts
~SAFETY VALVE - [mech] soupape *f* de sêreté à ressort
~SEAT - [mech] cuvette *f* de ressort, patin *m* de ressort
~SET - [mech] ensemble *m* de ressorts
~SHACKLE - [auto] jumelle *f* de ressort
~SHAFT - [text] tire-lisse *m*, marchette *f*
~SHOCK ABSORBER - [mech] amortisseur *m* de ressort
SPRING-SPOKED STEERING WHEEL - [auto] volant *m* de direction à rayons élastiques
~SPINDLE - [text] broche *f* à ressort
~STEM INTERCEPTION - [text] sélecteur *m* élastique
~STEEL - [metall] acier *m* de ressort, acier *m* à ressort
~STOP - [mech] butée *f* à ressort
~STUD - [contr] transmetteur *m* de mouvement d'armature
~SUSPENSION - [mech] (a method of suspending the measuring device by elastic means) suspension *f* à ressorts
~SUPPORTED ROLLERS - [text] support *m* de rouleaux élastique
~SUSPENSION BODIES - [mech] corps *mpl* suspendus du régulateur
~SWELL - [text] ressort *m* de la languette
~SWELL - [mech] ressort *m* de la languette
~SWITCH BLADE - [railw] lame *f* d'aiguille flexible
~SYSTEM - [mech] (of vehicles) système *m* de suspension
~TAB - [aero] (a type of tab provided with a spring device to reduce the effort to be exerted by the pilot at high speeds) tab *m* à ressort
~TENSION FOR SAW-BLADES - [mech] tension *f* élastique pour lame de scie
~TENSION TESTER - [mech] appareil *m* de tarage des ressorts
~TESTING MACHINE - [mech] machine *f* à tarer les ressorts
~TIDES - [astr] (the high tides occurring when the moon is new or full) grande marée *f*, marée *f* de syzygie, maline *f*
~TIE BOLT - s. spring dowel pin
~-TINED HARROW - [agric] herse *f* à dents flexibles
~TONGUE - s. spring switch blade
~-TOOTH CULTIVATOR - s. spring loaded cultivator
~-TOOTH HARROW - s. spring-tined
~TRANSMISSION - [mech] transmission *f* à ressorts
~U-BOLT - [mech] bride *f* de ressort
~VALVE - [mech]
~WATER - [geol] eau *f* de source, eau *f* de fontaine
~WASHER - [mech] rondelle *f* élastique
~WEIGHTED LEVER - [mech] levier *m* avec charge à ressort
~TIGHTENING BLOCK - [mech] bloc *m* de point d'appui du ressort
~WELL - [geol] source *f*, fontaine *f* montante
~WHEEL - [mech] roue *f* élastique
~WOOD PIRN - [text] fuseau *m* en bois avec ressort
~WOOL - [text] laine *f* de printemps
SPRINGBOARD - [gen] tremplin *m*
SPRINGER - [arch] sommier *m*, imposte *f* (d'arcade)
~STONE - [constr] coussinet *m*
SPRINGINESS - [gen] élasticité *f*
SPRINGING - [auto] suspension *f*
[arch] naissance *f*
[gen] (a cracking) crequement *m*
[mech] (of a rod etc) gauchissement *m* (d'une tige etc)
[mining] agrandissement *m* par explosion du fond d'un trou de mine
~LINE - [arch] ligne *f* de naissance
~SYSTEM - [mech] (of vehicles) suspension *f*
SPRINGY - [gen] flexible, élastique, moelleux
~FEEL OF THE WOOL - [text] laine *f* à toucher

élastique
SPRINKLE, to - [gen] jeter, répandre, arroser
[constr] saupoudrer
SPRINKLE MASK - [photo] masque *m* obtenu par saupoudrage pendant l'agrandissement
SPRINKLER - [impl] arrosoir *m*
[mech] arroseuse *f* à poussière d'eau
~IRRIGATION - [agric] arrosage *m* en pluie, irrigation *f* en pluie
~NOZZLE - [hydr] buselure *f* de pulvérisation, bec *m* de pulvérisateur
~SYSTEM - [fire fighting] extinction *f* automatique d'incendie
[agric] installation *f* d'irrigation en pluie
SPRINKLING - [photo] arrosage *m*
[min] inclusion *f* de minerai dans la roche en faible quantité
~FILTER - [biol] lit *m* percolateur, lit *m* bactérien
~HYDRANT - [hydr] bouche *f* d'arrosage
SPRINT - [mech] (of a bicycle) jante *f* en bois
SPROCKET - [mech] (an individual tooth of a spocket wheel) dent *f* de pignon
[mech] (device for intermittent movement, as in film projector) dent *f* de pignon de chaîne
[mech] (a sprocket wheel) pignon *m* de chaîne, pignon *m* Galle
~BIT - [comput] (the bit on a band indicating the position reached during an operation) impulsion *f* de rythme dérivée d'une roue dentée
~CHAIN - [mech] chaîne *f* à barbotin
CHANNEL - [comput] (on a magnetic or perforated tape) perforation *f* de contrôle
~HOLE - [cin] (of a film) perforation *f*
~-HOLE MODULATION - [electron] (used in video recording) entraînement *m* irrégulier par galets
~HUM - s. sprocket noise
~NOISE - [cin] bruit *m* causé par les perforations
~TEETH - [photo] dents *f*pl de l'axe d'entraînement
~WHEEL - [mech] (wheel furnished with teeth to engage a driving chain) roue *f* dentée, galet *m*, tambour *m*, pignon *m* Galle, poulie *f* à chicanes, bouc *m*
SPROCKETED TAPE - [electron] (used in video recording) bande *f* à perforations
SPROUT, to - [bot] (to begin the development of the first bud or shoot from a seed) pousser, germer
SPROUT - [bot] pousse *f*, germe *m*, jet *m*, bourgeon *m*
SPROUTING - [bot] germination *f*, bourgeonnement *m*
[metall] rochage *m*
~IN THE EAR - [agric] épiage *m*
SPRUCE - [bot] sapin *m*, spruce *m*
~FIR - [bot] épicéa *m*
SPRUE - [metall & plast ind] (the orifice of passage through which the material enters the mould) trou *m* de coulée, jet *m*, descente *f* de coulée
[metall] (the metal which solidifies in a sprue) jet *m* de coulée
[med] (psilosis; disease affecting the gastrointestinal track) psilosis *m*
~BASIN - [metall & plast ind] bassin *m* de coulée
~BUTTON - [metall] (locating cone) repère *m* de rempulage
~BUSH - [metall & plast ind] (the part of an injection mould in which the sprue is formed) buse *f* de carotte, cheminée *f* du moule
~BUSHING REAMER - [tool] alésoir *m* de la buse de coulée

SPRUE CONE - [metall & ind] carotte *f*
~CUP - [metall] godet *m* de coulée
~CUTTER - [metall & plast ind] (a tool for cutting sprues) coupe-coulées *m*
~CUTTING - [metall] tranchement *m* de la coulée
~EJECTOR - [plast ind] (an ejector pin in an injection mould or in a transfer mould which ejects that part of the feed stalk formed in the retaining recess provided for positively withdrawing the sprue from the sprue bush) éjecteur *m* de carotte
~EJECTOR BAR - [plast ind] (a rod or like part which expels the sprue from an injection mould) barre *f* d'éjection
~EJECTOR PIN - [plast ind] (device for ejecting sprue from a mould) extracteur-éjecteur *m*
~HOLE - [metall & plast ind] goulot *m* de coulée
~LOCK PIN - [metall & plast ind] éjecteur *m* de carotte
~PULLER - [metall & plast ind] (device for ejecting sprue from a mould) accroche-carotte *m*, éjecteur *m* de carotte, extracteur *m* de carotte
~PULLER PIN - s. sprue puller
SPRUEING - [metall & plast ind] (the operation of removing the gates from castings when the material is solidified) enlèvement *m* des entonnoirs de coulée
SPRUNG BASE - [mech] (in upholstery) support *m* à ressorts
~MASS - [mech] (a rigid body having equal weight, the same centre of gravity and the same moments of inertia about identical axes as the total sprung weight) masse *f* suspendue
~MASS VIBRATION - [mech] vibration *f* de la masse suspendue
~WEIGHT - [mech] poids *m* suspendu
~WEIGHT ON AXLES - [mech] charge *f* transmise aux essieux par ressorts
SPUD, to - [agric] sarcler, béquiller
[min] (the operation of drilling wells) forer par battage, travailler en spudding
[mech] (the operation of removing burrs) ébarber
SPUD - [agric] (an implement) sarcloir *m*
[mech] (a short projection) saillie *f*
[min] pilotis *m*, gouge *f* à repêchage
[agric] (colloq for potato) pomme *f* de terre
[mech] (dredging) piquet *m*
~IN, to - [oil ind] débuter
SPUDDER - [mining] treuil *m* léger au câble
[mining] perforation *f* à câble des sondages peu profonds
SPUDDING PULLEY - [mining] poulie *f* de forage au câble
~SHOE - [mining] sabot *m* de spudding
SPUE - [gen] (or spew; overflow) trop plein
SPUN - [gen & text] filé, cablé
[metall] repoussé
~CAST IRON PIPE - [plumb] (a centrifugal cast iron pipe) tuyau *m* en fonte netrifugée
~DYED - [text] teint *m* dans la masse
~GLASS - [glass man] coton *m* de verre, fil *m* de verre
~GOLD - [metall] fil *m* d'or
~RAYON - [text] fibrane *f*
~SILK - [text] (yarn made from silk waste spun like woollen yarns) fils *m*pl de déchets de soie
~YARN - [naut] bitord *m*
[text] filé *m*

SPUR, to - [gen] éperonner
SPUR - [gen] éperon *m*
[zool] ergot *m*, éperon *n*
[telecomm] embase *f* de poteau, socle *m* de poteau
[constr] largeur *f* de voie
[carp] contre-fiche *f*, entretoise *f*
[geogr] (of a mountain) contrefort *m*, embranchement *m* (d'une chaîne de montagnes)
[shipbuild] arc-boutant *m* de soutien
[geol] rameau *m*
[railw] embranchement *m*
~BAND - [el] bande *f* de fréquence pour ligne dérivée
~CENTRE - [tool] pointe *f* d'un tour
~CHUCK - [mach tool] mandrin *m* à trois pointes, griffe *f*, mandrin *m* à tulipe
~CUT - [mech] denture *f* à flancs droits
~GEAR - [mech] (gear of ordinary toothed type) engrenage *m* droit
~GEAR HOB - [tool] fraise *f* à tailler les engrenages droits
~GEARING - [mech] transmission *f* par engrenages droits
~PINION - [mech] pignon *m* droit
~PRUNING - [agric] taille *f* frutière
~RING - [mech] couronne *f* dentée
~TRACK - [railw] embranchement *m*
~WHEEL - s. spur gear
~WHEEL PULLEY BLOCK - [mech] moufle *f* à roues droites
~WHEEL REVERSING GEAR - [mech] (in machine tools) changement *m* de marche par engrenages droits
SPURGE, to - [gen] purger
SPURGE - [bot] (a weed) euphorbe *f*, épurge *f*
~-LAUREL - [bot] daphné *m* morillon, bois *m* joli
SPURIOUS - [gen] faux, falsifié
~COUNTS - [instr] (counts caused by an imperfection of the counter) coups *mpl* parasites
~DISSEPIMENT - [bot] fausse cloison *f*
~PULSE MODE - [electron] mode *m* d'impulsion parasitaire
~RADIATION - [radio] (emission from a radio transmitter at frequencies outside of its communication band) émission *f* parasite
[phys] rayonnement *m* parasite
~RESPONSE - [el] (any response, other the desired one, of an electric transducer or device)réception *f* non sélective
~SIGNAL - [telecomm] signal *m* de taches
~TRANSMITTER OUTPUT - [radio] sortie *f* parasite d'un émetteur
SPURN - [mining] jambe *f*
SPURRED CORN - [bot] ergot *m* du seigle
SPURRITE - [min] (a carbonate and silicate of calcium) spurrite *f*
SPUTTER, to - [electron] (in a gas discharge) pulvériser
SPUTTERING - [electron] (unwanted effect in gas-filled discharge tubes due to the cathode bombardment by positive ions) pulvérisation *f*
SPUTUM - [med] expectoration *f*
SPYGLASS - [opt] lunette *f* d'approche, longue-vue *f*
SPYHOLE - [gen] (in machines etc) regard *m*
[carp] (of a door) judas *m* de porte
SPYNDLE - [text] (unit of length used in counting jute yarn; I4, 400 yards) spindle *m*
SQ - [gen] (short for square) carré
SQ IN - [meas] (short for square inch) pouce *m* carré
SQUAB, to - [gen] rembourrer
SQUAB - [auto railw etc] coussin *m* de siège
[carp] canapé *m* rembourré
[gen] courtaud, boulot
[zool] pigeonneau *m* sans plumes
SQUAD - [gen] escouade, brigade *f*, équipe *f*
SQUALANE - [chem] (a component of high grade lubricants) squalène *m*
SQUALL - [met] (a strong wind, rising and dying away quickly; normally of short duration) grain *m*, rafale *f*
[met] (if accompanied by rain) grain *m* de pluie
[gen] bourrasque *f*
SQUALLY WEATHER - [met] temps *m* à rafales, temps *m* à grains
SQUAMA - [anat] squame *f*
[bot] pellicule *f*
SQUAMIFORM - [zool] (scale-like) squameux
SQUAMULE - [zool] (small scale) squamelle *f*, squamule *f*
SQUARE, to - [gen] carrer, équarrir
[comm] (of accounts) balancer, régler
[math] élever au carré
[mech] (to set accurately in position) équarrir
[mech] (to adjust) régler
[constr] (a stone etc) dresser (un bloc de marbre) équerrer (le bois)
[surv] (a map) quadriller
SQUARE - [gen] carré
[geom] carré *m*, quadrangulaire
[math] carré *m* (d'une expression)
[draw] équerre *f*
[gen] (of a city etc) place *f* (d'une ville etc)
[constr] (of houses) bloc *m* de maisons entre quatre roues
[metall] fer *m* carré, fer *m* rectangulaire
[constr] (an area of I00 sq feet of floor or roof) surface *f* de chevron
[constr] (of a roof; the number of tiles for an area of I00 sq feet) nombre *m* de tuiles par pied carré
[surv] (an instrument for setting out perpendicular alignments) équerre *f*, carreau-module *m*
[th eng] (ovens) brique *f* normale
[gas ind] (USA only) (a cock-plug head) carré *m* de manoeuvre
~ABUTMENT - [constr] culée *f* rectangulaire
~ACCOUNTS, to - [gen] régler les comptes, balancer les comptes
~AND PARALLEL - [constr] (of wood) (bois) équarré à vives arêtes
~AND RABBET - [arch] collier *m*
~BAR IRON - [metall] fer *m* carré
~BASTARD FILE - [tool] lime *f* bâtarde carrée
~BOLT - [mech] boulon *m* à tête carrée
~BROACH - [mech] équarissoir *m* pour trous carrés
~BUILT - [constr] bâti *m* en carré
~BUTT WELD - [mech] (a butt weld in which the edges of the parts are not chamfered) soudure *f* bout à bout sur bords droits
~CASCADE - [chem] (a cascade in which the circulation is the same in each stage) cascade *f* carrée, cascade *f* constante
~CORNER-SMOOTHER - [tool] lissoir *m* équerre, vif *m*
~CUTTER - [paper man] (a cross cutter) coupeuse *f* transversale

SQUARE DRIFT - [tool] mandrin *m* carré
~-DRILL, to - [mech] percer carrément
~EDGE - [mech] (of a cutter) fil *m* carré
~ELBOW - [plumb] coude *m* d'équerre
~-END PISTON RING - [mech] (in internal combustion engines) segment *m* de piston à extrémité droite
~ENGINE - [mech] (a reciprocating engine in which the bore is equal to the stroke) moteur *m* carré
~FILAMENT - [text] fil *m* de section carrée
~FILE - [tool] lime *f* carrée, carrelet *m*
~FOOT - [meas] pied *m* carré
~FOOT UNIT OF ABSORPTION - [acoust] sabin *m*
~GAUGE - [mach] calibre *m* carré
~-GLASS - [glass man] plate *f* de verre
~-HEAD BOLT - [mech] (a bolt of which the head is square as seen along the centre-line of the shank) boulon *m* à tête carrée
~-HEADED SCREW - [mech] vis *f* à tête carrée
~INCH - [meas] pouce *m* carré
~JOINT - [carp] assemblage *m* à plat-joint
~LAW - [phys] (the law of inverse squares expressing the relation between the amount of light falling on unit area of a surface and the distance of the surface from the light source) loi *f* des carrés inverses
~LAW DETECTION - [electron] (rectification in which the application of a sinusoidal input gives rise to an output which is proportional to the square of the input) détection *f* parabolique, détection *f* quadratique
~-LAW DETECTOR - [radio] détecteur *m* parabolique, détecteur *m* quadratique
~-LAW MODULATOR - [radio] modulateur *m* quadratique
~-LAW RECTIFIER - [radio] (rectifier in which the rectified output current is proportional to the square of the applied alternating voltage) redresseur *m* quadratique
~-LOOP FERRITE - [magn] ferrite *f* à cycle d'hystérésis rectangulaire
~MEASURE - [meas] mesure *f* de surface
~MILE - [meas] mille *m* carré
~MILL SAW FILE - [tool] lime *f* carrée pour scies
~MOTOR - s square engine
~NECK BOLT - [mech] (a bolt formed with a square section immediately below the head) boulon *m* épaulement carré
~NOSE BORING TOOL - [tool] outil *m* de forage à bout carré
~NOSED TROWEL - [impl] truelle *f* à bout carré
~NUT - [mech] écrou *m* carré
~OF THE CIRCLE - [math] quadrature *f* du cercle
~OFFSETS - [surv] ordonnées *fpl*
~OUT, to - [surv] faire le tracé des fondations
~PARACHUTE - [aero] (a parachute of which the canopy is approximately square when laid out on a flat surface) parachute *m* carré
~PIER - [constr] pilier *m* carré
~PULSE - [electron] impulsion *f* carrée
~RABBET-PLANE - [impl] guillaume *m* de fil
~-RIGGED SHIP - [naut] bateau *m* à gréement carré
~ROOT - [math] (the quantity of which a given quantity is the square) racine *f* carrée
~ROOT AND EDGE ANGLES - [metall] cornières *fpl* à à angles vifs
~-ROOT CALCULATOR - [comput] calculateur *m* de racines carrées

SQUARE SAIL - [naut] voile *f* carrée
~SCREW THREAD - [mech] filet *m* carré de vis
~-SET STOPING - [mining] exploitation *f* avec boisage parallélépipédique
~SHAFT SCREW OR BOLT - [constr] boulon *m* à tige carrée
~-SHANK DRILL - [tool] foret *m* à queue carrée
~STEEL - [metall] acier *m* carré
~THE VALVE, to - [mech] (of a steam engine) régler la tige du tiroir
~-THREAD, to - [mech] fileter à filet carré
~THREAD - [mech] (a screw thread which is square in section, instead of the more common triangular form; used chiefly for actuating devices rather than fixing parts, e.g. lathe leadscrews) filet *m* carré, filet *m* rectangulaire
~TIMBER - [wood] bois *m* carré
~TIMBER SCAFFOLDING - [constr] sapine *f*
~TIMBERING - [mining] boisage *m* parallélépipédique
~-TO ROOF - [constr] comble *m* en dents de scie
~TROWEL - [impl] truelle *f* carrée
~UP, to - [comm & gen] (of accounts) régler
[mech & carp] équarrir
~VOLTAGE - [electron] tension *f* en créneaux
~WAVE - [el] (a wave which alternately assumes two fixed values for equal lengths of time) onde *f* rectangulaire
~-WAVE RADIATOR - [radio] (signal generator for producing square or rectangular waves) générateur *m* d'ondes rectangulaires
~-WAVE RESPONSE - [telev] réponse *f* des signaux carrés
~-WAVE VOLTAGE - [telev] (voltage of square waveform with uniform intervals of time) tension *f* de onde rectangulaire, tension *f* crénelée
~WELL POTENTIAL - [nucl] potentiel *m* à puits rectangulaire
~WIRE - [metall] fil *m* carré
~WORK - [mining] exploitation *f* par piliers
SQUARED - [gen] carré
[math] élevé au carré
[wood & constr] équarri
~PAPER - [paper man] papier *m* quadrillé
~STONE - [constr] pierre *f* carrée
SQUARENESS - [mech] équarissage *m*, cathogonalité *f*
[gen] forme *f* carrée
~ERROR - [cin] (error in film perforating machines) défaut *m* de perforation
~RATIO - [comput] (in static magnetic storage) facteur *m* d'orthogonalité
SQUARER - [electron] (squaring circuit) circuit *m* conformateur
SQUARING - [gen] équarissage *m*
[constr] dressage *m* (de la pierre), mise *f* en équerre
[draw] quadrillage *m*
[math] quadrature *f*
[telev] conformation *f* d'onde rectangulaire
~CIRCUIT - [radio] (used for transforming a sine wave into a square wave) circuit *m* de quadrature d'onde
~SHEARS - [metall] cisailles *fpl* à ébouter
SQUARROUS - [med] squarreux
SQUASH, to - [gen] écraser, aplatir
SQUASH - [gen] écrasement *m*, aplatissement *m*
[agric] (of fruit) pulpe *f*

SQUASH - [acoust] bruit *m* mou
SQUASHING - [gen] écrasement *m*
SQUAWKING - [acoust] (sound made by some birds, e.g. gulls) piaillement *m*, cris *mpl* percants
SQUEAK, to - [gen & acoust] pousser des cris aigus
[zool] guiorer, vagir
[mech] (of a machine) grincer, crisser
[mus] (of musical instruments) faire des couacs
SQUEAKER - [zool] pigeonneau
SQUEAKING - [acoust] (sound made by some animals e.g. mice) couics *mpl*
SQUEAKING VOICE - [acoust] voix *f* flutée
SQUEAL, to - [acoust] pousser des cris aigus
SQUEAL - [acoust] cri *m* aigu, cri *m* percant
SQUEEGEE, to - [gen] essorer par compression *f*
SQUEEGEE - [impl] (rubber roller or brush for squeezing out surplus water) raclette *f*, rouleau *m* en caoutchouc, racloir *m*
[hydr] (used for desludging operations) rabot *m* en bois
~ROLL - [impl] (a roller, usually of rubber, used to remove water) rouleau *m* essoreur, rouleau *m* à pressurer
SQUEEZABLE WAVEGUIDE - [electron] guide *m* d'ondes à section variable
SQUEEZE, to - [impl] (to compress e.g. expel moisture) presser, exercer une pression, pressurer
SQUEEZE - [gen] compression *f*, serrage *m*
[food] (of fruit eg. a lemon) expression *f*
[mining] tassement *m*, affaissement *m* du toit
~BOX - [electron] (in waveguides) tronçon *m* à section variable
~CEMENTATION - [metall] cémentation *f* par pression
~MOULDING - [metall] serrage *m* par pression
~OUT, to - [gen] (to expel a liquid from a material by pressure) presser, épreindre, exprimer
~PUDDLED IRON, to - [metall] macquer le fer puddlé
~THE WOOL - [text] laminer la laine, exprimer la laine
~THE YARN, to - [text] essorer les fils, exprimer les fils
~TIME - [metall] (in resistance welding, the interval of time between the application of pressure and the application of the welding current) phase *f* d'approche
~TRACK - [el acoust] piste *f* sonore à densité et largeur variables
SQUEEZED WOOL - [text] laine *f* exprimée
SQUEEZER - [gen] appareil *m* à compression
[agric] pressoir *m*
[metall & plast ind] (a moulding machine operated by hand, compressed air and usually hydraulic power, in which the sand is squeezed or compressed into the box and round the pattern by a ram) machine *f* à compression
[metall] (alligator squeezer) maque *f*, macque *f*, presse *f* à macquer
[metall] (or rotary squeezer) cingleur *m* rotateur
SQUEEZING - [gen] compression *f*, etreinte *f*
[metall] macquage *m*
[food] (of an orange, a lemon etc) expression *f*
~MACHINE - [text] exprimeuse *f*
~MANGLE - [text] foulard *m* d'exprimage
~OUT THE WOOL - [text] pressage *m* de la laine
SQUEEZING ROLLER - [text] cylindre *m* exprimeur, rouleau *m* exprimeur
~ROLLS - [paper man] rouleaux *mpl* presseurs
~THE YARN - [text] essorage *m* des fils
SQUEGGING - [radio] (a mode of oscillation in a thermionic valve, in which H.F oscillations build up to a considerable amplitude and then suddenly cease, the cycle being repeated periodically) mode *m* d'oscillation *f* à extinction, mode *m* d'oscillation *f* de relaxation
~OSCILLATOR - [electron] (a form of linear time base which includes an oscillator operating in the squegging condition) oscillateur *m* à extinction
SQUELCH CIRCUIT - [radio] (a circuit reducing background noise) circuit *m* de réglage silencieux
SQUELCHING - [radio] (the elimination of noise reception when no signal is transmitted) réglage *m* silencieux
SQUIB, to - [gen] lancer des pétards
SQUIB - [gen] (tube-shaped device containing powder and made to explode with a crack) pétard *m*
[mining] canette *f*, raquette *f*, mèche *f*
SQUIBBING - [mining] agrandissement *f* par explosion du fond d'un trou de mine
SQUID, to - [fishing] pêcher en employant le calmar
SQUID - [zool] calmar *m*, encornet *m*
[aero] (not fully distended canopy of a parachute) coupole *f* à distension partielle
SQUELGEE - s. squeegee
SQUILL - [zool] cigale *f* de mer, sauterelle *f* de mer
[bot] scille *f*, squille *f*
SQUINT, to - [med] loucher
SQUINT - [med] (strabismus) strabisme *m*
[constr] (a brick suitable for a squint quoin) brique *f* d'angle
~ANGLE - [med] angle *m* de strabisme
~DEVIATION - s. squint angle
SQUINTING EYE - [med] oeil *m* strabique
SQUIRREL - [zool] écureuil *m*
[comm] (the fur) petit-gris *m*
~CAGE - [el] (a circuit consisting of a number of conducting bars having their extremities connected by metal rings or plates at each end) cage *f* d'écureuil, enroulement *m* à cage d'écureuil
~-CAGE ARMATURE - [el] induit *m* à cage d'écureuil
~-CAGE GRID TUBE - [electron] tube *m* à grille en cage d'écureuil
~-CAGE MAGNETRON - [electron] magnétron *m* à cage d'écureuil
~-CAGE MOTOR - [el] (induction motor in which the winding of the rotor has the form of a squirrel cage) moteur *m* à cage, moteur *m* à cage d'écureuil
~-CAGE ROTOR - [el] (rotor of an induction motor the winding of which consist of a number of bars having their extremities at each end of the rotor connected by rings or plates) rotor *m* à cage d'écureuil
~-CAGE WINDING - [el] (cage winding) enroulement *m* à cage d'écureuil
SQUIRT, to - [gen] lancer en jet, faire jaillir
SQUIRT - [impl] seringue *f*
[gen] (of water etc) jaillissement *m*
[gen] (fire fighting) extincteur *m* d'incendie
~-CAN - [impl] (an oil can with a flexible bottom) seringue *f* à graisse, pompe *f* à graisse
~GUN - s. squirt-can

SQUIRT-HOSE - [impl] tuyau *m* d'arrosage
SQUIRTED FILAMENT - [el] filament *m* passé par pression à la filière
SQUITCH - [el] interrupteur *m* à amorce
SQUITTER - [electron] fonctionnement *m* intempestif d'un répondeur
Sr - [chem] (symbol of strontium) symbole du strontium
SR - [radar] (abbrev of secondary radar) radar *m* secondaire, radiodétection *f* secondaire
[telecomm] (abbrev. of selective ringing) appel *m* sélectif
[el] (abbrev of slip-ring motor) moteur *m* à bagues
[mech] (abbrev of slow running) à marche lente
[radio] (abbrev. of short range) à courte portée
[telecomm] (abbrev. of sound ringing) signal *m* de son
S.R E. - [radar] (surveillance radar element) élément *m* radar de surveillance
[nucl] (sodium Reactor Experiment) réacteur *m* expérimental à sodium
SS - [radio] (abbrev. of single signal) signal *m* simple
~LORAN - [radio] loran *m* synchronisation par l'onde ionosphérique
SSB - [radio] (abbrev of single sideband) bande *f* latérale unique
SSF - [el] (abbrev. of supersonic frequency) fréquence *f* ultra-sonore
SSR - [radio] (abbrev. of single-signal receiver) récepteur *m* à haute sélectivité
SST - [telecomm] (abbrev. of supersonic telegraphy) télégraphie *f* supersonique
St - [telecomm] (abbrev of sound telegraphy) télégraphie *f* à fréquences acoustiques, télégraphie *f* harmonique
STAB, to - [gen] poignarder
[gen] (to pierce through) percer d'un coup de couteau
[bookbind] (of the sheets) piquer
[constr] repiquer (une surface)
[oil ind] guider (une tige dans le raccord)
STABBING - [bookbind] piqûre *f* métallique
~-AWL - [tool] tire-point *m*
~BOARD - [oil ind] plate-forme *f* d'accrochage
STABILITY - [gen] (the tendency to remain in a given state or condition, without spontaneous change) stabilité *f*
[mech] (a general; property of a mechanical system, whereby the system returns to a state of equilibrium after a disturbance) stabilité *f*
[aero] (quality in virtue of which any disturbance in the steady motion of an aircraft tends to diminish) stabilité *f*
~ABOUT THE AXIS OF YAW - [aero] stabilité *f* de route
~AUGMENTATION SYSTEM - [astronaut] système *m* stabilisateur
~DERIVATIVES - [aero] (quantities expressing in forces and moments acting on an aircraft, due to disturbance of steady motion) dérivée *f* de stabilité
~DOMAIN - [contr] domaine *m* de stabilité
~IN STORAGE - [gen] (shelf-life) stabilité *f* au stockage
~IN THE CURVE - [auto] stabilité *f* dans le virage
~METER - [instr] stabilomètre *m*
STABILITY OF DYES - [chem] solidité *f* des colorants
~OF EMULSIONS - [chem] stabilité *f* des émulsions
~TEST - [gen] (of cables) essai *m* de stabilité
~TO LIGHT - [phys] stabilité *f* à la lumière
STABILIZATION - [gen] stabilisation *f*
~OF MARKETS - [comm] stabilisation *f* du marché
~OF PRICES - [comm] stabilisation *f* des prix
~PERIOD - [mech] (of the speed of an engine following a load variation) temps *m* de stabilisation
STABILIZE, to - [gen] stabiliser
[aero] (an aircraft, by means of fixed surfaces etc) stabiliser
[fin] stabiliser, valoriser
STABILIZED FEEDBACK - [radio] réaction *f* stabilisée
~-FEEDBACK AMPLIFIER - [radio] amplificateur *m* à réaction stabilisée
~GASOLINE - [oil ind] (gasoline after subjection to fractionation by which the vapour pressure has been reduced to a specified maximum) essence *f* à pression de vapeur specifiée
~GLASS - [radiat] verre *m* non-décolorant
STABILIZER - [gen & mech] (any device designed to increase stability) stabilisateur *m*, équilibreur *m*
[chem] (substance reducing the spontaneous combustion of an explosive) stabilisant *m*
[rubber ind] anti-coagulant *m*
[aero] (tail plane) stabilisateur *m* fixe, plan *m* fixe de queue, empennage *m*
~BAR - [auto] barre *f* de compensation
STABILIZING - [gen] stabilisation *f*
~ANNEAL - [metall] recuit *m* de détente
~CHOKE - [el] (reactive choke coil inserted in series with an electric discharge lamp to balance its negative resistance characteristic) bobine *f* de stabilisation
~ELEMENT - [el] filtre *m* de stabilisation
~FEEDBACK - [contr] (correcting feedback of a function of the output quantity with the object of minimizing the tendency to self-oscillation) contre-réaction stabilisatrice
~FEEDBACK NETWORK - [contr] réseau *m* à réaction stabilisateur
~PARACHUTE - [aero] (a parachute used to render a load stable which would not otherwise be so) parachute *m* stabilisateur
~RESISTOR - [el] résistance *f* de stabilisation
~TREATMENT - [metall] traitement *m* de stabilisation
~TUBE - [electron] (a tube used when it is necessary to keep the voltage in a receiver as constant as possible) tube *m* stabilisateur
~WINDING - [el] (of a metadyne) enroulement *m* stabilisateur
STABISTOR - [electron] (silicon diode providing a fixed voltage drop) stabisteur *m*
STABLE - [gen] stable, solide
[phys] constant, stable
[constr] (bulding for lodging and feeding horses, cattle etc) écurie *f*
[nucl] (said of atomic or nuclear system not capable of spontaneous changes) stable
~A TRAIN, to - [railw] garer un train
~AND YARD OPERATIONS - [agric] travaux *mpl* à la ferme
~CATTLE - [agric] bétail *m* en stabulation
~COMMUNITY - [bot] (plant community which remains unaltered for a long period of time) collectivité *f* végétale stable

STABLE DUNG - [agric] fumier *m* de cheval
~EQUILIBRIUM - [mech] (equilibrium) équilibre *m* stable
~FATTENING - [agric] engraissement *m* en stabulation, engraissement *m* à l'auge
~GROUND - [constr] terrain *m* résistant
~ISOTOPE - [nucl] (an isotope which is not radioactive) isotope *m* stable
~ORBIT - [nucl] (the circle of constant radius described by accelerated particles) orbite *f* d'équilibre, orbite *f* stable
~OSCILLATION - [phys] (an oscillation tending constantly to decrease) oscillation *f* amortie
~PLATFORM - [astronaut] plate-forme *f* stabilisée
~REACTOR PERIOD - [nucl] constante *f* de temps stable du réacteur
~STATE - [phys] (system incapable of spontaneous changes) état *m* stable
~TRACER ISOTOPE - [nucl] (stable isotope used as a tracer) isotope *m* traceur stable
~TRIGGER CIRCUIT - [el] circuit *m* de déclenchement bistable
STABLING - [railw] (turning off wagons into a siding) garage *m*, stationnement *m*
[zool] stabulation *f*, étables *f*pl
~LIMIT SIGNAL - [railw] signal *m* limite de garage
~SIDING - [railw] voie *f* de garage
STACCATO - [acoust] staccato *m*
~SPEECH - [acoust] voix *f* saccadée
STACHYOSE - [chem] (a tetrasaccharose) stachyose *m*
STACK, to - [gen] entasser, emmeuler, empiler
[aero] (s. stacking) échelonner verticalement
STACK - [gen] meule *f*, tas *m*, pile *f*
[agric] meule *f* permanente
[gen] (of weapons) faisceau *m* (d'armes)
[meas] (of wood etc) mesure *f* de I08 pied cubes
[constr] (chimney stack) souche *f*, corps *m* de cheminée
~BOXES, to - [metall] gerber le châssis
~CUTTING - [metall] coupage *m* des tôles en paquets
~EFFECT - [phys] (the tendency of a heated gas to rise in a vertical passage) effet *m* cheminée
~FUNNEL - [constr] cheminée *f*
~GASES - [gas ind] (waste gases) gaz *m*pl brûlés
~HEIGHT - [heat] hauteur *f* du tirage
[agric] hauteur *f* d'empilage
~LOSSES - [phys] (flue loss) pertes *f*pl à la cheminée
~MIXING - [text] (a method of mixing cottons of different types by piling it in horizontal layers) entassement *m* par couches
~MOULD - [metall] grappe *f*
~MOULDING - [metall] coulée *f* en grappe
~OF CARDS - [comput] jeu *m* de cartes
~-ON OVEN - [gas ind] (a separate oven) four *m* indépendant
~-PIPE - [constr] (a downpipe for rain-water) descente *f* d'eau
~THE FLAX, to - [text] faire des cahoutes de lin
~VALVE - [constr] clapet *m* de cheminée, clapet *m*
STACKED - [gen] entassé, empilé
~AERIAL - [radio] antenne *f* à éléments superposés
~ANTENNA - s. stacked aerial
~-CERAMIC ELECTRON TUBE - [electron] tube *m* à disques céramiques scellés
~DIPOLE ARRAY - s. stacked aerial
STACKED MOULD - [metall] moule *m* à éléments suporposés
STACKER - [mech] empileur *m*
[mech] (a portable conveyor) élévateur *m* d'empilage
[min] élévateur *m* de tailings
[cin] (USA only; a camera crane) chariot *m* élévateur
[comput] empileur *m* de cartes
~DRUM - [comput] tambour *m* d'empileur de cartes
STACKLING - [aero] (procedure in the case of a number of aircraft awaiting their turns to land, in which each is allotted a different altitude at which to fly until it can be accepted) échelonnage *m* vertical
[gen] empilage *m*, entassement *m*
[agric] emmeulage *m*, mise *f* en meulé
~A RING - [oil ind] démonter une sonde
~FAULT - [crystall] (a deviation from the correct order of stacking of the atomic planes in the construction of a face-centered cubic close-packed lattice) défaut *m* de disposition
~SPACE - [gen] espace *m* volumétrique d'empilage
STACTOMETER - [instr] (small-bore pipette used for counting the drops) compte-gouttes *m*, stalagmomètre *m*
STADIA - [surv] (sighting instrument for measuring distances used in connection with a graduated rod) stadia *m*
~CONSTANT - [surv] rapport *m* stadiométrique
~HAIRS - [instr] (the two additional hairs fitted to the diaphragm of a telescope for tacheometric purposes) fils *m*pl stadimétriques
~LINES - [surv] traits *m*pl de stadia
~ROD - [surv] mire *f* stadia, mire *f* pour lecture au stadia
~SURVEYING - [surv] (a tacheometric method) levé *m* des plans au stadia
STADIUM - [gen] stade *m*
[surv] s. stadia
[med] phase *f*, stade *m*, période *f*
STAFF, to - [gen] fournir de personnel
STAFF - [gen] (personnel) personnel *m*
[gen] (a rod, for support etc) bâton *m*, bourdon *m*
[metall] ringard *m*, crochet *m*
[railw] bâton *m* pilote
[surv] mire *f*, jalon *m*
[med] sonde *f* cannelée
[constr] (material consisting of plaster of Paris cast in moulds with hemp fibres) staff *m*
[mus] portée *f*
~-AND-BALL - [naut] balise *f*
~ANGLE - [carp] (angle-staff) filet *m* d'angle
~BEAD - [constr] (an angle-bead) boudin *m* d'arête
STAG - [zool] cerf *m*
[agric] boeuf *m* châtré après pleine croissance
STAGE, to - [gen] monter (une pièce), mettre sur la scène
[gen] (a demonstration etc) organiser (une démonstration)
STAGE - [gen] estrade *f*, échafaud *m*
[gen] (a shelf for holding material) plateau *m*
[gen] (of a travel, a voyage) étape *f*
[theatre] scène *f*
[biol] stade *m*, période *f* (de développement)
[mech] (of pressure, in a turbine) saut *m*
[constr] (a scaffold) échafaudage *m*

STAGE - [railw] (a coaling stage) quai *m*
[instr] (the platform arranged at right angles to the line of vision, on which slides are placed) platine *f*
[astronaut] (of a rocket) étage *m*
[radio] (a single electronic tube and its associated circuits, or a circuit of two or more tubes, having the same function as a single-tube unit) étage *m*
[med] s. stadium [med]
[naut] (a landing stage) embarcadère *m* flottant
[min] passe *f*, opération *f*
[mining] niveau *m*, palier *m* de repos, travée *f*
~-AND-A-HALF - [astronaut] étage *m* à dispositif de propulsion auxiliaire largable
~BOX - [theatre] loge *f* d'avant-scène
~-COACH - [transp] diligence *f*
~COMPRESSOR - [mech] (type of air or gas compressor in which the pressure of the fluid treated is raised by successive steps in several cylinders or casings) compresseur *m* étagé, compresseur *m* étagé, compresseur *m* compound
~CRUSHING - [min] broyage *m* étagé
~EFFICIENCY - [radio] (the ratio between the useful power delivered to the load and the anode power input) rendement *m* d'étage
~FLOOD - [theatre] lampes *f*pl pour éclairage par projection
~HEATING - [rubber ind] (heating carried out in successive operations steps) vulcanisation *f* échelonnée, chauffage *m* par gradins, cuisson *f* échelonnée
~OF INVASION - [med] période *f* d'invasion
~OF MATURITY - [agric] degré *m* de maturité
~PASS - [text] remettage *m* en escalier
~PUMPING - [mining] épuisement *m* en répétitions
[oil ind] épuisement *m* en répétitions
~WORKING - [mining] exploitation *f* à ciel ouvert
STAGEING - s. staging
STAGELESS - [gen] sans étage
~COMPRESSION - [mech] compression *f* sans étage
STAGGER, to - [gen] chanceler, vaciller
[gen] (to vibrate) vibrer, osciller
[gen] (to arrange in groups) échelonner
[aero] décaler (les ailes)
[mech] (to place in rows, the articles being in one row alternating with the spaces in the next) placer alternativement, disposer en quinconce, alterner
[mech] (in internal combustion engines) tiercer
[el] (the brushes) disposer (les balais) en gradins
[mach tool] étager (les lames)
STAGGER - [aero] (in a multiplane, the distance by which the leading edge of an upper plane oversails that of a lower one (measured as an angle in U S A) décalage *m* des ailes
[gen] chancellement *m*
~ANGLE - [mech] (the angle formed by the blade chord with the turbine axis of rotation) angle *m* de calage de pas
~-TUNED AMPLIFIER - [radio] amplificateur *m* décalé
~TUNING - [radio] (a method of securing a wide bandwidth in a multistage intermediate frequency amplifier by detuning pairs of the tuned circuits in opposite direction by a given amount) accord *m* décalé
STAGGERED - [gen] en quinconce, placé alterné, disposé alterné
~CIRCUITS - [radio] circuits *m*pl décalés
~PILING - [constr] distribution *f* des pieux en quinconce
~POLES - [el] poles *m*pl alternés
~RIVETED JOINT - [mech] disposition *f* des rivets en quinconce
~SPOTS - [text] mouchetures *f*pl contre-semplées
~TEETH - [mech] dents *f*pl placées en quinconce
~TRAVELLING LATTICES - [text] tabliers *m*pl superposés et désaxés (d'un séchoir à laine)
~WEFT PILE WEAVE - [text] armure *f* contre-semplée de poil
STAGGERING - [el] (term denoting the displacement of the brushes of a commutator motor from the neutral zone) échelonnage *m* (des balais)
[radio] (the offsetting of two channels of different carrier systems from exact sideband frequency coincidence, so as to avoid mutual interference) décalage *m* de canaux
[photo] (of a view) échelonnement *m* en profondeur
~ADVANTAGE - [radio] gain *m* de décalage
STAGGERS - [vet] avertin *m*, tournis *m*
STAGING - [gen] mise *f* à la scène
[constr] (a scaffolding) échafaudage *m*
[naut] appontement *m*
[print] (the operation designed to avoid further reetching by protecting certain areas of a plate with varnish) vernissage *m* des clichés
STAGNANT - [gen] stagnant
~POOL - [hydr] flaque *f* d'eau stagnante
~WATER - [gen] eau *f* morte
STAGNANTE, to - [gen] être stagnant
[hydr] (of water) croupir
STAGNATION - [gen] stagnation *f*
[med] (accumulation and retardation of circulating fluids in the body) stase *f*, stagnation *f*
~MASTITIS - [med] engorgement *m* mammaire
~OF BLOOD - [med] stase *f* du sang
~POINT - [phys] (in a viscous fluid, the point at which pressure on a body is a maximum) point *m* de stagnation, point d'arrêt
~PRESSURE - [mech] (the pressure at a stagnation point in the flow) pression *f* de stagnation
STAIN, to - [gen] tacher, souiller
[gen] (to impart a colour) peindre, mettre en couleur
[text] (a fabric) imprimer (une étoffe)
[biol] (in bacteriology) prendre le gram
STAIN - [chem] (transparent solution of a pigment or dyestuff in solvent used to impart a desired colour to a substance with obscuring its surface) mordant *m*
[gen] (blemish or discoloration) tache *f*
[dyes] colorant *m*, couleur *f*
~SPOTS - [metall] (spots caused by the exudation of cleaning agents etc from pores in the base) souillures *f*pl
STAINED GLASS - [glass man] verre *m* de couleur, verre *m* coloré
~WOOL - [text] crottins *m*pl, parties *f*pl tachées par l'urine
STAINING - [gen] souillure *f*
[chem] coloration *f*, teinture *f*

STAINING - [rubber ind] (on sheets) formation *f* de tâches (sur sheets)
~AGENT - [chem] révélateur *m*
~POWER - [paint] (the intensity of a colour) pouvoir *m* colorant
STAINLESS - [gen] sans tache, pur
[metall] inoxydable, inrouillable
~IRON - [metall] fer *m* résistant à la corrosion
~STEEL - [metall] (a chromium-nickel steel which is proof against corrosion by a large variety of agents) acier *m* inoxydable
~STEEL REACTION VESSEL - [ind chem] récipient *m* en acier inoxydable pour réaction chimiques
STAIR - [build] (a step) marche *f* (d'escalier)
(a series of steps constructed to give access to a part of a building at a different level) escalier *m*
~-BALUSTER - [constr] rampe *f* d'escalier
~-CARPET - [text] tapis *m* d'escalier
~CHEEK - [constr] limon *m*
~CLIP - s.stair rod
~-HEAD - [constr] palier *m*, haut *m* de l'escalier
~HORSE - [constr] montant *m* d'escalier
~NOSING - [constr] bord *m* de la marche d'escalier
~RAIL - s. stair-baluster
~RISE - [constr] montée *f* de la marche, hauteur *f* de la marche
~-ROD - [metall] tringle *f* d'escalier
~TREAD - [constr] emmarchement *m*, giron *m*, foulée *f*
~WELL - [constr] cage *f* d'escalier
STAIRCASE - [constr] (the space containing the stairs) cage *f* d'escalier
[constr] (the actual stair) escalier *m*
~GENERATOR - [el] (a generator the output of which, if displayed on a linear-time base oscilloscope presentation, has the appearance of a staircase) générateur *m* de tension en forme d'escalier
~PHENOMENON - [med] phénomène *m* de l'escalier
~SIGNAL - [telev] signal *m* dégradé
~WELL - [constr] cage *f* d'escalier
STAIRHEAD - [constr] (the top of a flight of stairs) palier *m*, haut *m* de l'escalier
STAIRS - [constr] escalier *m*
STAIRWAY - s.staircase
~ACTUATOR - [aero] (hydraulic device to raise or lower a passenger stairway in an aircraft) appareil *m* de levage hydraulique de l'escalier
STAKE, to - [surv] jalonner, bornoyer (une route etc)
[constr] garnir de pieux, soutenir avec des pieux
[agric] échalasser, ramer
[leather ind] palissonner (les peaux)
[mining] (a claim) liqueter, borner (une concession)
STAKE - [gen] pieu *m*, poteau *m*
[carp] (implement) piquet *m*, fiche *f*, bigorne *f* à queue
[impl] tas *m*, enclumette *f*, tasseau *m*
[surv] jalon *m*, jalonnette *f*, piquet *m*
[leather ind] palisson *m*
[railw] (bar fitted to the sides and ends of flat wagons for keeping the load in place) rancher *m*
[agric] tuteur *m*, échalas *m*
~NET - [fishing] étente *f*, gord *m*
~POCKET - [auto] (an aperture in the sides or floor of bodies for the reception of stakes) bride *f* de fixation des pieux, douille *f* pour piquet de fixation de ridelles
STAKES - [auto] (metal or wood posts by which sides are attached to platforms) montants *mpl* de ridelles, piquets *mpl* de fixation de ridelles
STAKING - [surv] jalonnement *m*, bornage *m*
[agric] (of vinyards) échalassage *m* (d'une vigne)
~MACHINE - [leather ind] machine *f* à palissonner
~OUT - [surv] jalonnement *m*
STALACTITE - [geol] (concretionary deposit of calcium carbonate, formed by percolating solutions and hanging icicle-like from the roofs of caverns) stalactite *f*
STALACTITED - [geol] scalactifère
STALACTITIC - [geol] scalactitique
STALAGMITE - [geol] (concretionary deposit of calcium carbonate precipitated from dripping solutions on the floors and walls of limestone caverns) stalagmite *f*
STALAGMOMETER - [instr] (a stactometer) stalagmomètre *m*
STALAGMOMETRY - [chem] (the analysis of solutions by measurements of the surface tension) stalagmométrie *f*
STALE - [gen] rassis, vicié, vieux
[agric] (of wine) (vin) éventé
[comm] (of a market) marché *m* lourd, marché *m* plat
[leg] périmé
[zool] urine *f* des bestiaux
~SEWAGE - [hydr] eaux *fpl* usées fades
STALK - [bot] tige *f*, chaume *m*, queue *f*
[glass man] (of a glass) pied *m* (de verre)
[constr] (a tall chimney) souche *f*, cheminée *f* d'usine
[metall] (in foundry work, of a core) boîte *f* à noyau
[metall & plast ind] (the part of the feed next to the nozzle, which connects the latter to the runner) carotte *f*
[constr] (the upright part of a reinforced concrete retaining wall springing from the horizontal base) partie *f* verticale d'un mur de talus en béton armé
~-EYED - [zool] (of some insects) podophthalme
~FIBRE - [text] fibre *f* du liber
~OF THE PLANT - [bot] tige *f* de la plante
STALL, to - [gen] établer, mettre à l'étable
[mech] (of an engine) bloquer, caler
[aero] ralentir au dessous de la vitesse critique
[aero] (to reach or pass the stalling angle) décrocher, se mettre en perte de vitesse
STALL - [gen] stalle *f*, box *m*, case *f* d'étable
[gen] (in an open market) étalage *m*
[gen] (a stand) stand *m*
[min] chambre *f* de grillage du minerai
[theatre] fauteuil *m* d'orchestra
[mining] (in coal mining) taille *f*
[auto] (a parking place) box *m*
[aero] décrochage *m*
[mech] (in aerodynamics, the stall of a fluid stream from the airfoil) détachement *m* de la veine
~WARNING INDICATOR - [aero] (an instrument to give warning of the approach of stalling conditions) avertisseur *m* de décrochage
~WITHOUT POWER - [aero] décrochage *m* sans puissance
STALLED PRESSURE - [mech] (the pressure at which the discharge of a variable-delivery hydraulic pump

falls automatically to zero) pression *f* de débit nul
STALLED TURN - [aero] virage *m* lent
STALLING - [mech] arrêt *m*, blocage *m*, calage *m*
[aero] décrochage *m*, perte *f* de vitesse
[agric] stabulation *f*
~ANGLE - [aero] angle *m* d'incidence critique
~FLUTTER - [aero] (flutter near the stalling angle; it may occur in more than one degree of freedom) flutter *m* de décrochage
~SPEED - [aero] (the equivalent airspeed of an aircraft which is associated with its maximum lift coefficient) vitesse *f* de décrochage
~TORQUE - [el] (the overload torque which is sufficient to slow down to zero the speed of an electric motor operating under load) couple *m* de calage
STALLION - [zool] étalon *m*
STAMINA - [gen] force *f* vitale
STAMINATE - [bot] staminé, mâle (fleur)
STAMMERING -[med] balbutiement *m*, bégaiement *m*
STAMP, to - [mech] (to form a part by compressing it in a matrix) étamper, estamper, travailler à la press
[gen] (to impress a mark) timbrer
[metall] (the operation of marking after inspection) poinçonner, frapper
[gen] (the mail) affranchir
[min] (to crush) bocarder
STAMP - [gen] (a mark made by stamping) empreinte *f*, marque *f*, timbre *m*
[comm] cachet *m*
[mech] (for sheet-metal working) étampeuse, emboutisseuse *f*
[comm] (a seal) estampille *f*
[min] (a weight for crushing) pilon *m* de bocard
[mech] (a die) estampe *f*
[metall] chasse-pointes *m*, perceuse *f*, poinçon *m*
[min] (a mill for stamping) bocard *m*
~BATTERY - [min] batterie *f* de bocards
~BOSS - s. stamp head
~-DUTY - [min] rendement *m* du bocard en tonnes par 24 heures
~GUIDE - [min] guide *m* de pilon
~GUIDE PLATE - [text] plaque-guide *f* des poinçons
~HEAD - [min] surcharge *f* de pilon, tête *f* de pilon
[constr] tête *f* de pilon
~-MILL - [min] moulin *m* à brocards
~MILLING - [min] bocardage *m*, broyage *m* au bocard
~MORTAR - [constr] mortier *m* de bocard
~SCREEN - [min] tamis *m* de bocard
~SHOE - [min] sabot *m* de pilon
~STEM - [min] tige *f* du pilon
~THE HEMP SEED, to - [text] piler les graines de chanvre
STAMPED BUCKET - [mech] godet *m* embouti
~CONCRETE - [constr] béton *m* damé
~PACKING -RING -[mech] anneau *m* de garniture poinçonné
STAMPER - [mech] (a crushing machine) étampeuse *f*, machine *f* à étamper
[el acoust] (in disk recording, a negative from which positives are moulded) fils *m*, matrice *f*
STAMPING - [metall] (the operation of making a part by compression between dies) estampage *m*, emboutissage *m*
[metall] (the pressed part) pièce *f* estampée, pièce *f* matricée
STAMPING - [gen] (the impressing of marks) timbrage *m*
[min] (crushing) bocardage *m*
[el] (lamination, of a machine or transformer, each of the thin sheets of iron or steel forming part of the core) tôle *f* de noyau
[th eng] (coke ovens) pilonnage *m*
[mech] poinçonnage *m*, estampillage *m*
[constr] damage *m*, pilonnage *m*
[metall] (the impressing of marks, letters or numbers after inspection) estampage *m*, étampage *m*, emboutissage
~CARRIAGE - [text] chariot *m* de la boîte aux poinçons
~DIE - [metall] matrice *f* à estamper
~FOIL - [metall] (thin sheet embossing) feuille *f* estampée
~FORM - [metall] moule *m* de pilonnage
~KNIFE - [impl] emporte-pièces *m*
~MACHINE - [mech] estampeuse *f*, poinçonneuse *f*
[constr] pilonneuse *f*
~MACHINE FOR VALVE DISKS - [rubber ind] (for tyres) poinçonneuse *f* de rondelles de valves, découpeuse *f* de rondelles de valves
~MILL - [min] moulin *m* à bocards
~PAD - [rubber ind] tampon *m* à timbrer
~PRESS - [mech] (a press designed for embossing thin sheet) presse *f* à découper, presse *f* emboutisseuse
~SHEET - [rubber ind] plaque *f* à éstamper
~TOOL - [metall] matrice *f*
STANCE FATIGUE - [med] fatigue *f* orthostatique
STANCHION, to - [gen] étayer, épontiller
[constr] garnir de montants
STANCHION - [mech] (vertical stress-carrying part of a structure) étai *m*, étançon *m*, appui *m*
[carp] montant *m*
[shipbuild] épontille *f*
[agric] (in a stable) dispositif *m* d'attaque (dans une étable)
[railw] ranchet *m*
~STABLE - [agric] étable *f* avec dispositif d'attache
~-STIFFENING PLATE - [gas ind] (in gasholders) plaque *f* de raidissement et de guidage (dans les gazomètres)
STANCHNESS - [gen] étanchéité *f*
STAND, to - [gen] être debout
[gen] (to support) supporter, résister à
STAND - [gen] (a structure on which one may stand) place *f*, position *f*
[mech] (of a dial gauge) support *m*
[impl] pied *m*, socle *m*
[gen] (a stall on which marchandise is displayed) étalage *m*, boutique *f* en plein air
[metall] (housing) cage *f*
[instr] (of a microscope) affût *m*, pied *m*, statif *m*
[photo] chevalet *m*, support *m*
[ind chem] support *m* (de laboratoire), valet *m* de laboratoire
[mech] (of a motorbicycle rear-wheel) support-béquille *m*
[gen] (in an exhibition) stand *m*
[agric] récolte *f* sur pied
[agric] (in forestry) peuplement *m*
~-BY, to - (to be ready for immediate action) être prêt, se tenir prêt

STAND-BY, to - [naut] se tenir paré
[metall] (of a furnace) être au repos
~-BY - [gen] (used for anything kept in reserve for emergency use in place of a similar thing which is no longer effective) auxiliaire, de réserve, de secours
[gen] (unit or equipment in reserve to replace a similar item in case of breakdown or the like) ressource *f*, soutien *m*
[telecomm] à l'écoute
~-BY BATTERY - [el] (emergency battery) accumulateur *m* de réserve
~-BY GAS - [gas ind] (peak-load gas) gaz *m* de pointe
~-BY LOSS - [th eng] (the heat energy which must be supplied to a furnace to maintain its operating temperature when the process for which it is used is not being carried on) pertes *f*pl à vide
~-BY UNATTENDED TIME - [comput] temps *m* d'attente
~-BY UNIT - [gen] installation *f* auxiliaire
~COCK OF A LAVATORY - [hydr] robiner *m* de lavabo
~-IN - [cin] doublure *m*
~-IN BUNCHES, to - [text] se former en faisceaux
~-IN MAN - s. stand-in
~-INSULATOR - [el] (insulator on which the stand or stillage rests) isolateur *m* de chantier
~MICROMETER - [instr] (a micrometer mounted on a base, so that both hands can be used for the work) micromètre *m* avec support
~OF ROLLS - [metall] cage *f* de laminoir
~OFF, to - [naut] courir au large
~OFF - [oil ind] distance *f* entre l'épaulement du joint et le calibre
~-OFF CAPACITOR - [electron] condensateur *m* à fixation verticale par vis
~OIL - [paint] (drying oil which can be polymerized by heating) huile *f* siccative
[oil ind] huile *f* épaissie
~ON, to - [naut] faire route
~OUT, to - [gen] résister, tenir ferme, faire saillie
[naut] gagner le large
~PIPE - [mech] (open vertical pipe connected to a pipeline, to ensure that the pressure head at that point cannot exceed the length of the stand pipe) colonne *f* montante
~SHEET - [constr] (a window with no frame) fenêtre *f* sans bâti
~UP, to - [gen] se lever, se mettre debout
STANDARD - [gen] (an agreed dimension, property or other characteristic, fixed upon to preserve uniformity) norme *f*
[gen] (a model taken as standard) modèle *m*, type *m*
[gen] (which can be use for measuring) étalon, type, normal
[gen] (standard, original sample) étalon *m*
[mech] (a support) pied *m*, affût *m*, support *m*
[gen] (a flag, ensign etc) bannière *f*
[min] titre *m*
[mech] (a base) bâti *m*
[mach tool] montant *m*, jumelle *f*
[transp] (of an aerial ropeway) pylône *m*
[metall] (of a rolling mill) cage *f*, colonne *f* (de laminoir)
[constr] (a scaffold pole) perche *f*, échasse *f* de échafaud
[mining] (floor hanger for shafting) chaise *f* de sol

STANDARD AMPERE - [el] ampère *m* international
~AND DOUP - [text] lame *f* gaze
~ATMOSPHERE - [met] (a hypothetical atmosphere having arbitrary qualities, within the range of normal mean conditions, assumed for some specific purpose) atmosphère-type *f*
~BAR - [metall] (bar used for testing the resistance of materials) éprouvette *f* étalon
~BATTERY - [el] pile *f* étalon
BEAM APPROACH SYSTEM - [radar] (a system of radionavigation for providing aircraft with lateral and beacon guidance during approach) système *m* d'approche standard à faisceau
~BOGIE - [railw] bogie *m* normal
~BOTTLE - [gas ind] (in gas meters) clepsydre *f*
~CABLE - [el] câble *m* étalon
~CANDLE - [el] bougie *f* anglaise
~CELL - [el] (primary cell designed to act as a standard electromotive force) pile *f* étalon
~COLLAR - [plumb] manchon *m* droit, bout *m* à deux emboîtements
~COMPASS - [instr] compas *m* de contrôle
~CONDITIONS - [phys] (for a gas, a temperature of 0 centigrades and a pressure of I standard atmosphere) conditions *f*pl normales
~COPPER - [metall] cuivre-type *m*
~DEVIATION - [mech] écart-type *m*
~ELECTRODE POTENTIAL - [el chem] (the value of the equilibrium electrode potential when standard conditions prevail) tension *f* standard d'une électrode
~ENTROPY - [phys] (the total entropy of a substance in a state defined as standard) entropie *f* standard
~EQUIPMENT - [gen] équipement *m* standard, équipement *m* de série
~EXCESS MODIFIED REFRACTIVE INDEX GRADIENT - [phys] gradient *m* étalon normal du module de réfraction
~FILM - [cin] (film of 35 mm width) film *m* format standard, film *m* normal
~FILM STOCK - s. standard film
~FILTER - [photo] (filter giving a standard light) filtre *m* normal
~FINENESS - [metall] (of gold and silver) titre *m* standard
~FORM - [metall & plast ind] (a mould with standardized and interchangeable parts) moule *m* normalisé
~FORMULA - [gen] formule *f* classique
~FRAME - [plast ind] (for unit moulds) moule *m* normalisé
~FREQUENCY - [el] (between 50 and 60 cycles per second) fréquence-type *f*
~FREQUENCY GENERATOR - [radio] générateur *m* de fréquence-type
~GAS - [gas ind] (calibrating gas) gaz *m* de référence
~GAUGE - [instr] calibre *m* de référence, jauge *f* étalon
[railw] voie *f* normale, écartement *m* normal
[mech] gabarit *m* type
~-GAUGE LINE - [railw] ligne *f* à voie normale
~-GAUGE TRACK - [railw] voie *f* à écartement normal
~GOLD - [min] or *m* au titre
~GRADE - [comm] type *m* standard
~HEAT OF FORMATION - [phys] (the heat required to form one mole of a compound from its elements

in their standard state) chaleur *f* étalon de formation

STANDARD HOLE - [mech] trou *m* normal

~HYDROGEN ELECTRODE - [el] (a reversible hydrogen electrode of standard hydrogen ion state, with hydrogen at one atmosphere pressure) électrode *f* standard à hydrogène

~ILLUMINANT - [opt] (said of special lamps and filters which are used as standard sources for colorimetry) lumière *f* des étalons calorimétriques

~INSTRUMENT - [instr] (said of measuring instruments of high precision which are used particularly for calibrating other measuring instruments) instrument *m* étalon

~INTERNATIONAL ATMOSPHERE - [met] (international standard under conditions of sea level temp 15° c. , press. 1.0132 millibars, lapse rate of 6.5°C per km. from sea level to 11 km and thereafter a constant temp of 56.5°C.) atmosphère *f* standard internationale

~IONIZATION CHAMBER - [nucl] (an air-filled ionization chamber) chambre *f* d'ionisation étalon

~LAMP - [el] lampe *f* étalon

~LIGHT SOURCES - [opt] (a number of light sources adopted by the International Commission on Illumination in 1931 and 1951) sources *f*pl normalisées de lumière

~LUMINOSITY FUNCTION - [opt] (function established by the international Commission on Illumination to meet the variable sensitivity of the eye to radiation of different wavelength) fonction *f* étalon de luminosité

~MAGNET - [meas] (device used in the calibration of galvanometers) aimant *m* étalon

~MEAN CHORD - [aero] (the value obtained by dividing the gross wing area by the span) corde *f* moyenne géométrique, profondeur *f* moyenne

~MEASURES - [meas] mesures *f*pl étalons

~METER - [instr] étalon *m* du mètre [meas] compteur *m* étalon

~MICROPHONE - [el acoust] (a microphone whose response in specified conditions is accurately known) microphone *m* étalon

~MIX - [constr] (concrete mixed in the following proportions; 1 of cement to 2 of sand to 4 of coarse material) mélange *m* standard

~MOISTURE - [paper man] (the moisture coefficient used for calculating the weight of cellulose in wood pulps in air-dried conditions) humidité *f* de base

~NEEDLE - [text] aiguille *f* du fil fixe

~N.P L WIND TUNNEL - [aero] (a wind tunnel in which the experimental section is provided with rigid walls) soufflerie *f* à veine guidée

~NUT - [mech] écrou *m* ordinaire

~OF ILLUMINATION - [opt] (for photometry) étalon *m* d'éclairement

~OF LIGHT - [phys] étalon *m* de lumière

~OF PURITY - [min] titre *m* (de l'or etc)

~OHM - [el] (international ohm) ohm *m* étalon

~ORIFICE - [hydr] orifice *m* en mince paroi

~PAPER - [chem] papier *m* à réactif

~PARALLELS - [geogr] (the parallels of latitude on which the cone of the Lambert conformal projection intersects the terrestrial sphere) parallèles *m*pl sécants

~PART - [gen] pièce *f* standard

~PITCH - [mech] (of gears) pas *m* normal

STANDARD PITCH - [aero] (of a propeller, measured at two thirds of the radius) pas *m* de référence [acoust] (based on the tone "A" of 440 cycles per second) étalon *m* a, hauteur *f* du ton-étalon

~PLANE - [crystall] plan *m* cristallographique à indices de Miller

~POTENTIAL - [el] tension *f* standard

~PRESSURE - [phys] (atmospheric pressure) pression *f* atmosphérique

~PRESSURE-GAUGE - [instr] manomètre *m* étalon

~PRICE - [comm] prix *m* indicatif, prix *m* d'orientation

~PROPAGATION - [radio] (the propagation of radiowaves over a smooth spherical earth of specified dielectric constant and conductivity, under conditions of standard refraction of the atmosphere) propagation *f* idéale

~RADIUS - [aero] (an arbitrary radius used for specifying propeller characteristics) rayon *m* de référence

~REFLECTOR - [opt] (reflector conforming with the B.S.I. specification for industrial reflectors) réflecteur *m* type standard

~REFRACTION - [phys] (the refraction which would take place in an idealized atmosphere) réfraction *f* idéale

~RESISTANCE - [el] (specially designed and constructed resistor having a resistance which is accurately known and stable against drift) résistance *f* étalon

~SAMPLE - [gen] échantillon-type *m*

~SECTION - [metall] profilé *m* normal

~SEED - [agric] semence *f* naturelle

~SHAFT - [mech] arbre *m* normal

~SILVER - [min] argent *m* au titre

~SIZE - [mech] dimension *f* type, grandeur *f* normale

~SOLENOID - [instr] (device used for the calibration of galvanometers) solénoïde *m* étalon

~SOLUTION - [chem] (a solution of known strength, e.g a normal solution) solution *f* standard, liqueur *f* titrée, liqueur *f* normale

~SPECIFICATION - [gen] (the specification to which a machine must conform) spécifications *f*pl unifiées, cahier *m* de charges unifié

~SQUARE - [constr] brique *f* normale

~STATE - [phys] (the stable form of a substance at unit activity) état *m* normal

~STEELWORK MOULD - [metall] moule *m* normal pour aciérie

~TEMPERATURE - [phys] (the temperature established by some unvarying process, e.g. a melting or boiling point, or pressure etc of a substance under fully-defined conditions) température *f* standard

~TEST SPECIMEN - [metall etc] (a piece of material prepared for test according to specified rules and sizes) éprouvette *f* étalon

~THERMOMETER - [instr] thermomètre *m* étalon

~TIME - [gen] (arbitrary time used over a defined area, the difference between which and G.M.T. is usually an integral number of hours) heure *f* légale

~-TIME ZONE - [telecomm etc] fuseau *m* horaire

~TONE - [acoust] (the tone generated by a standard sound source) ton-étalon *m*

~TONE GENERATOR - [acoust] (a sound source generating sound of known and constant intensity) gé-

nérateur *m* de ton-étalon
STANDARD TRACK - [el acoust] (single track) piste *f* simple
~TRANSMITTER - [el acoust] microphone *m* étalon
~TWISTED YARN - [text] fil *m* tordu normal
~VOLUME - [phys] (the volume of one mole of a substance, in the form of a gas at a temperature of 0 centigrades and a pressure of I standard atmosphere) volume *m* standard
~WARP - [text] chaîne *f* normale
~WEAVES - [text] armures *f*pl fondamentales, armures *f*pl classiques
WHEEL SET - [metall] train *m* normal
~WIRE - [metall] fil normal
~WIRE GAUGE - [mech] échelle *f* des diamètres des fils métalliques
~WOOL FIBRE - [text] brin *m* type, brin *m* standard
~YARN - [text] fil *m* type
STANDARDIZATION - [gen] (the manufacture of articles to a pre-arranged set of rules and dimensions, to ensure interchangeability and compatibility) standardisation *f*, unification *f*, uniformisation *f*
[instr] étalonnage *m*
[chem] titrage *m*
~OF PRODUCTS - [ind] uniformisation *f* des produits
~OF TARIFFS - [transp etc] péréquation *f* des tarifs
STANDARDIZE, to - [gen] standardiser, unifier, étalonner
[chem] titrer (une solution)
STANDARDIZED - [gen mech etc] standardisé, normalisé, unifié
STANDBY TRANSMITTER - [telev] émetteur *m* de secours
STANDING - [gen] position *f*
[transp] stationnement *m*, arrêt *m* d'un service
[gen] (adj) debout
~BAFFLE - [hydr] chicane *f* de retenue
~BALANCE - [mech] équilibre *m* statique
~CHARGE - [comm] tarif *m* unitaire
~CLOSET - [hydr] latrine *f*
~END OF A TACKLE - [mech] dormant *m* d'un palan
~ENGINE - [gen] machine *f* en chômage, machine *f* inactive
~GROUND - [gen] point *m* d'appui
~MATTER - [print] forme *f* à tirer, marbre *m*, composition *f* permanente
~MINE - [mining] mine *f* en chômage
~OFF DOSE - [radiat] dose *f* de changement de poste, dose *f* maximale admissible professionnelle
~-ON-NINES CARRY - [comput] (high-speed carry) report *m* bloque à neuf
~PANEL - [carp] panneau *m* vertical
~PLANT - [bot] plante *f* vive
~POOL - [hydr] flaque *f* d'eau stagnante
~PRESS - [print] presse *f* à levier
~RIGGING - [naut] manoeuvre *f* dormante
~ROOM - [transp] places *f*pl debout
~START - [auto] (for speed tests) départ *m* arrêté
~THREAD - [text] fil *m* fixe
~TIMBER - [wood] bois *m* en étant
~TORQUE - [mech] couple *m* à l'arrêt
~TREE - [bot] arbre *m* en état, arbre *m* sur pied
~VALVE - [oil ind] clapet *m* de pied
~VICE - [impl] étau *m* à table

STANDING WATER - [hydr] eau *f* dormante, eau *f* morte
~WAVE - [phys] (a state of vibration in which the oscillatory phenomena at all points are governed by the same time function, with the exception of a numerical factor, varying from one point to another) onde *f* stationnaire
~WAVE AERIAL - [radio] antenne *f* à onde stationnaire
~WAVE ANTENNA - s. standing aerial
~-WAVE DETECTOR - s. standing-wave indicator
~-WAVE GENERATOR - [electron] générateur *m* de ondes stationnaires
~WAVE INDICATOR - [instr] (an instrument designed to measure the standing wave ratio in waveguides) appareil *m* de mesure du taux d'ondes stationnaires
~-WAVE LOSS FACTOR - [electron] facteur *m* de perte d'ondes stationnaires
~-WAVE METER - s.standing-wave indicator
~-WAVE RATIO - [electron] taux *m* d'ondes stationnaires
~-WAVE RATIO BRIDGE - [instr] pont *m* de mesure sur taux d'ondes stationnaires
~WAVE SYSTEM - [el acoust] (interference pattern characterized by stationary nodes and antinodes) système *m* d'ondes stationnaires
~-WAVE VOLTAGE RATIO - [electron] taux *m* de tension d'ondes stationnaires
STANDPIPE - [hydr] tube *m* piézométrique
[nucl] (in a reactor) tube *m* ascendant ouvert
s.stand pipe
STANDPOINT - [gen] point *m* de vue
STANDSTILL - [gen] arrêt *m*, immobilisation *f*
STANK - [hydr] étang *m*
STANNAME - [chem] (hydride of tin) hydrure *m* de étain
STANNATES - [chem] (the salts of stannic acid) stannates *m*pl
STANNIC - [chem] (containing tin, especially in its higher valence) stannique
~ACIDS - [chem] (two forms exist viz. alphastannic and beta-or metastannic acid, the former by reacting alkalies with stannous chloride and the latter by the action of nitric acid on metallic tin) acides *m*pl stanniques
~CHLORIDE - [chem] (a plastics stabilizer, bleaching agent, dyeing assistant for textiles, intermediate for pigments, and a starting material in the manufacture of other tin compounds) chlorure *m* stannique
~CHROMATE - [chem] (a ceramic glaze) chromate *m* stannique
~OXIDE - [chem] (a textile chemical, constituent of ceramic glazes, catalyst, and starting material for other tin salts) oxyde *m* stannique
STANNITE - [min] (also called Bell Metal Ore and Tin Pyrites; a natural sulphide of copper, tin and iron) stannine *f*, étain *m* pyriteux
STANNIZING - [metall] étamage *m* en vapeur de chlorure stanneux
STANNOUS - [chem] (pertaining to tin) stanneux
~CHLORIDE - [chem] (a bleaching agent, analytical reagent, textile dyeing assistant, reducing agent and intermediate for pigments) chlorure *m* stanneux
~CHROMATE - [chem] (a component of ceramic glazes) chromate *m* stanneux

STANNOUS ETHYLHEXOATE - [chem] (a polymerization catalyst) éthylhexoate *m* stanneux
~OLEATE - [chem] (polymerization catalyst) oléate *m* stanneux
~OXALATE - [chem] (a textile dyeing assistant) oxalate *m* stanneux
~OXIDE - [chem] (an intermediate for stannous salts) oxyde *m* stanneux
~SULPHATE - [chem] (a dyeing assistant) sulfate *m* stanneux
~SULPHIDE - [chem] (catalyst for hydrocarbon polymerization) sulfure *m* stanneux
~TARTRATE - [chem] (a textile dyeing assistant) tartrate *m* stanneux
STANNUM - [chem] (tin) étain *m*
STANTON NUMBER - [chem] (the ratio of the Nusselt number to the product of the Reynolds and Prandtl numbers) nombre *m* de Stanton
STAPEDECTOMY - [med] (excision of the stapes) excision *f* de l'étrier
STAPES - [anat] (auditory ossicle) étrier *m* de l'oreille
STAPHYLECTOMY - [med] excision *f* de la luette, uvulectomie *f*
STAPHYLEDEMA - [med] oedème *m* de la luette
STAPHYLOCOCCEMIA - [med] staphylococcémie *f*
STAPHYLOCOCCUS - [bacteriology] (a gram-positive coccus of which the individuals tend to form irregular clusters) staphylocoque *m*
STAPHYLOMA - [med] (local bulging of the weakened sclera of the eye) staphylome *m*
STAPHYLONCUS - s. staphyledema
STAPHYLODERMA - s. staphylodermatitis
STAPHYLODERMATITIS - [med] dermatite *f* staphylococcique
STAPLE, to - [mech] (to fix by a staple) fixer avec une agrafe
[text] (to sort according to the length of the fibres) classer les fibres
[comm] (to sort goods in accordance with the rules and requirements of a storehouse, or market) sélectionner suivant des normes
[bookbind] brocher
STAPLE - [gen] (the principal production, or commodity, of a region) produits *mpl* principals
[adj] principal
[raw material] matières *fpl* premières
[text] (the carded or combed fibre of wool, cotton or flax) brin *m*, fibre *f*
[text] (the length and fineness of fibres) caractéristique *f* de la fibre
[mech] (U-shaped piece of metal with pointed ends) crampon, crampon *m* à deux pointes, cavalier
[carp] (loop of thin wire, used as fastening) crampe *f*
[mech] (a lock staple) gâche *f*
[metall] (a chaplet, for holding moulds or cores in place) support *m* simple
[mining] bure *f*, puits *m* intérieur
[bookbinding] broche *f* en fil métallique
[geol] (a layer of decayed vegetation on a rock) couche *f* de végétation décomposée
[mining] entrepôt *m*
~ANALYSER - [text] appareil *m* pour établir le diagramme des fibres
~DRAFT - [text] étirage *m* des fibres coupées
~FIBRE - [text] (rayon yarn) fibre *f* coupée
STAPLE LENGTH - [text] longueur *f* des fibres
~OF COTTON - [text] soie *f* de coton
~PIT - [mining] (internal shaft connecting two coal-seams) bure *f*
STAPLER - [impl] brocheuse *f* mécanique
STAPLING - [mech] fixage *m* à l'aide d'agrafes, fixage *m* à l'aide de crampons
[bookbind] (the stapling of sheets of paper) brochage *m* au fil de fer
~MACHINE - [bookbind] (device for binding papers together) machine *f* à brocher au fil de fer
[mech] (a machine designed to staple hard, thick materials) brocheuse *f*
STAR - [astr] (self-luminous celestial body) étoile *f*
[cin] star *f*, vedette *f* de cinéma
[nucl] (a group of tracks due to ionizing particles originating at a common point) étoile *f*
~-ANISE - [bot] badiane *f*, anis *m* étoilé
~ANTIMONY - [metall] antimoine *m* pur
~BIT - [mech] trépan *m* en couronne
[mining] trépan *m* en croix, trépan *m* à tranchant
~BURNER - [gen] brûleur-étoile *m*
~CHAIN - [radar] (group of location beacons in y-formation in such way that three slave-transmitters are around the master transmitter in the middle) chaîne *f* à étoile
~CLUSTER - [astr] amas *m* d'étoiles
~-CONNECTED - [el] monté en étoile
~CONNECTION - [el] (a method of connection, in three-phase, six-phase or other A. C. working, in which three or more conductors or windings meet at a common junction known as the star point) montage *m* à étoile, connexion *f* en étoile
~DAY - [astr] (sidereal day) jour *m* sidéral
~-DELTA STARTER - s. star-delta switching starter
~-DELTA STARTING - [el] (a method of starting applicable to three-phase motors connected in delta under normal working conditions; it consists in connecting the three phases temporarily in star connection while starting) démarrage *m* étoile-triangle
~-DELTA SWITCH - [el] commutateur *m* étoile-triangle
~-DELTA SWITCHING STARTER - [el] (for a three phase induction motor. A switching starter arranged in such way that with the switch in the starting position the stator windings are connected in star, and with the switch in the running position they are connected in delta) démarreur *m* étoile-triangle
~DISTRIBUTION SYSTEM - [gas ind] réseau *m* en étoile
~DRILL - [tool] (for rock drilling) foret *m* en couronne, ciseau *m* de maçon
~FOR TOUCHSTONE - [metall] toucheau *m* à plusieurs alliages
~FRACTURE - [rubber ind] (of tyres) rupture *f* en étoile, rupture *f* en croix
~FRAME - [text] abaisseur *m* de cuve
~GRASS - [bot] grand chiendent *m*
~-GROUP WINDING METHOD - [el] méthode *f* d'enroulement en groupe étoile
~LOCK - [mech] rondelle *f* à étoile
~LOT - [text] (wool offered for sale at London sales, in lots usually not exceeding three bales) lot *m* spécial
~NETWORK - [el] (a set of three or more branches with one terminal of each connected at a common node) réseau *m* à étoile
~OF SWIFT - [text] croisillon *m* de l'aspe
~PERFORATED GRAIN - [astronaut] grain *m* de proper-

gol perforé en étoile
STAR POINT - [el] (the point at which branches of a star-connected winding or system are connected together and sometimes to earth) point m neutre
~-QUAD CABLE - [el] (a quad cable) câble m en paires-étoile, câble m à quartes-étoile
~QUARTZ - [min] quartz m à astéries
~REEL - [print] (of a rotary press) étoile f porte-bobines
[text] tourniquet m en étoile
~SHAKE - [timber] (number of shakes radiating from the heart of a log) cadranure f, cadran m
~SHAPED CARD - [text] carte f en étoile de fil à coudre
~-SHAPED ENGINE - [mech] moteur m à étoile
~SPOOL - s. shaped card
~-STONE - [min] astérie f
~STREAMING - [astr] mouvement m des étoiles
~VOLTAGE - [el] (in a three-phase or six-phase system, the voltage between any line and the neutral point of the system) tension f étoilée
~WASHER - [mech] (star-shaped washer, the points of which are turned up against the flats of the nut to secure it) rondelle-frein f en étoile
~WHEEL - [mech] (a wheel with pointed triangular teeth) croisillon m à poignées, étoile f à arrêtoir
[cin] croix f de Malte
STARBOARD - [naut & aero] (the right-hand side of aircraft or ship, looking in the direction of forward motion) tribord m
STARCH, to - [gen] amidonner, empeser
STARCH - [chem] (a high molecular weight carbohydrate found in all green plants. It can be hydrolysed to dextrin and glucose and is converted by diastase to maltose) amidon m
~BOX - [text] bac m à colle
~CONTENT - [chem] (the proportion of starch in a substance) teneur f en amidon
~EQUIVALENT - [chem] valeur f amidon
~-FLOUR - [chem ind] fécule f
~GUM - [chem] (dextrin) dextrine f
~PASTE - [chem] pâte f d'amidon
~-SUGAR - [chem] glucose f
STARCHED RAGS - [text] chiffons mpl amidonnés
STARCHING MACHINE - [text] machine f à amidonner
STARE - [gen & med] regard m fixe
STARK EFFECT - [el] (the effect of a strong, transverse electric field on the spectrum lines of a gas subjected to its influence) effet m Stark
STARLIGHT - [astr] lumière f stellaire
STARLING - [hydr] éperon m, pilotis m, brise-glace m en pilotis
[zool] étourneau m
STARLINGS - [constr] enrayure f
STARSHAKE - s. star-shake
START, to - [gen] partir
[gen] (to begin) commencer
[gen] (to set going) mettre en marche
[med] déboîter
[naut] se délier, s'ouvrir
[mech] mettre en marche, mettre en mouvement
[chem] mettre en fonction
[mech] (a pump) amorcer, allumer
[carp] disjoindre (des tôles etc)
[mining] désancrer, déhourder
START - [gen] commencement m, début m, départ m
START - [mech] démarrage m, mise f en marche
[aero] envol m
~AN ARC, to - [el] faire démarrer un arc
~AND STOP LEVER - [mech] levier m de démarrage et d'arrêt
~AN INJECTOR - [mech] amorcer un injecteur
~BOOSTER - [auto] dispositif m pour faciliter le démarrage
~LAYING, to - [zool] entrer en ponte
~OF LOOM - [text] mise f en marche du métier
~SIGNAL - [telecomm] (in telegraphy) impulsion f d'ouverture, signal m de démarrage
~-STOP - s. start-stop system
~-STOP APPARATUS - [telecomm] (a telegraph apparatus operating according to the start-stop system) appareil m arhythmique
~-STOP DISTORSION - [telecomm] distorsion f de système arhythmique
~-STOP MULTIVIBRATOR - [telev] (a multivibrator operating as a single-cycle trigger circuit) multivibrateur m monostable
~-STOP SYSTEM - [telecomm] (in machine telegraphy, a system in which each group of code elements corresponding to an alphabetic signal is preceded by a start signal which prepares the receiving mechanism for the reception and registration of a character and is followed by a stop signal in order to bring the receiving mechanism to rest in preparation for the reception of the next character) système m arhythmique
~THE BOBBINS, to - [text] mettre en route les bobines
~THE SPINDLE, to - [text] embrayer la broche
~-UP - [nucl] (of a nuclear reactor) démarrage m
~-UP ACCIDENT - [nucl] (accident occurring in starting a reactor before the period meters have started to function) accident m au démarrage
~-UP PROCEDURE - [nucl] méthode f de démarrage
~-UP TIME - [nucl] temps m de démarrage
STARTER - [el] (apparatus for starting an electric motor and bringing it up to speed) démarreur m, starter m, rhéostat m démarreur
[mech] (in internal combustion engines) démarreur
[electron] électrode f d'amorçage
~BATTERY - [el] (storage battery designed for starting i.c. engines) batterie f de démarrage
~BREAKDOWN VOLTAGE - [electron] tension f de rupture de l'électrode d'amorçage
~BUTTON - [auto] bouton m de commande du démarreur
~CABLE - [auto] câble m de démarreur
~CLUTCH - [auto] (the clutch of the starting handle) embrayage m de la manivelle de démarrage
~CONTROLLER - [el] (a starting switch) combinateur m de démarrage
~CURRENT - [electron] tension f de l'électrode de amorçage
~GAP - [electron] (the conduction path between a starting electrode and the other electrode to which starting voltage is applied) intervalle m d'amorçage
~MOTOR - [el] (an electric motor used for starting i.c. engine) starter m, démarreur m
~PEDAL - [auto] pédale f de commande du démarreur
~WITH INTERNAL REDUCTION GEAR - [auto] starter m avec engrenages de reduction
STARTING - [gen] commencement m

STARTING - [el & mech] démarrage *m*, amorçage *m*, mise *f* en marche
~AIR - [mech] air *m* de démarrage
~AND REVERSING SWITCH - [el] démarreur-inverseur *m*
~ANODE - [el] électrode *f* d'allumage
~AUTOTRANSFORMER - [el] auto-transformateur *m* de démarreur
~BOX - [el] s. starting resistance
~BY COMPRESSED AIR - [mech] (of a motor) démarrage *m* à air comprimé
~CAPACITOR - [el] condensateur *m* de démarrage
~CIRCUIT - [el] circuit *m* de démarrage
~COILS - [text] spires *f*pl de départ
~CRANK - [mech] manivelle *f* de mise en marche
~CURRENT - [el] (the current taken by an electric motor at starting) courant *m* de démarrage
~CYCLE - [mech] (of a jet engine) cycle *m* de démarrage
~DOGS - [mech] crabots *m*pl d'accouplement
~ELEMENT - [el] relais *m* de démarrage
~ELECTRODE - [electron] (auxiliary electrode used to initiate conduction) électrode *f* d'amorçage
~FRICTION - [mech] frottement *m* au départ
~GEAR - [mech] appareil *m* de mise en marche
~HANDLE - s. starting crank
~HUM - [telev] ronflement *m* au démarrage
~IGNITION - [auto] allumage *m* de démarrage
~IMPULSE - [el] impulsion *f* de démarrage
~JET - [mech] (carburettor jet designed to furnish a special mixture for starting) gicleur *m* de starter
~KEY - [telecomm] bouton *m* de démarrage, bouton *m* d'appel
~LEVER - [mech] levier *m* de mise en marche, manette *f* de mise en marche
~LOAD - [el] (of an induction meter) charge *f* de démarrage
~MOTOR - [mech] démarreur *m* du moteur
~NEWEL - [constr] balustre *m* de départ
~OF A BOILER - [th eng] mise *f* en feu
~OF A MACHINE - [mech] démarrage *m* d'une machine
~OF OSCILLATIONS - [electron] amorçage *m*
~PEDAL - [auto] pédale *f* de démarrage
~PLATFORM - [gen] quai *m* de départ
~POINT - [gen] point *m* de départ
~POINT COUNTER - [comput] compteur *m* d'instant de démarrage
~POWER OF A METER - [el] (the lower limit of power which will cause the moving element to make a complete revolution) puissance *f* de démarrage d'un compteur
~PRESSURE - [mech] pression *f* au départ
~RATCHET - [auto] rochet *m* de démarrage
~REACTOR - [el] self *m* de démarrage
~RELAY - [el] relais *m* de démarrage
~RESISTANCE - [el] (fixed resistance connected in series with the main circuit of a motor during starting-up) rhéostat *m* de démarrage
[mech] résistance *f* au démarrage
~RESISTOR - [el] (in electric-traction machines) résistance *f* de démarrage
~RHEOSTAT - s. starting resistance
~SHEET - [el] (thin sheet of refined metal used as the initial cathode) feuille *f* de départ
~SHEET BLANK - [el] (flat conductor used as an initial cathode to produce a thin coating of refined metal, which is then stripped off and used as a starting sheet) plaque-support *f* de feuille de départ
STARTING SHUTTER - [auto] volet *m* de départ
~SPEED - [gen mech etc] vitesse *f* initiale
~STEP - [constr] marche *f* de départ
~SWITCH - [el] interrupteur *m* de démarrage
~THE PUMP - [mech] amorçage *m* de la pompe
~TIME - [nucl] (equilibrium time) période *f* de mise en train
~TORQUE - [el] (the torque which is developed by a motor at starting) couple *m* de démarrage, couple *m* au démarrage
~TRANSFORMER SET - [el] groupe *m* transformateur de démarrage
~UNDER LOAD - [el] démarrage *m* sous charge
~UP FROM REST - [mech] (of a motor or engine) lancement *m*
~VOLTAGE - [el] (of an electric lamp) tension *f* d'amorçage
~WINDING - [el] enroulement *m* de lancement
STARVATION - [gen] affamement *m*, inanition *f*
STARVE, to - [gen] manquer de nourriture
[mech] (USA only) manquer de combustible
STARVED JOINT - [mech] (a joint in which there is not enough adhesive to ensure full attachment) joint *m* avec une quantité de colle insuffisante
STARVING - [gen] inanition *f*
[mech] (particularly at high speeds) alimentation *f* insuffisante
STASIBASIPHOBIA - [med] stasobasophobie *f*
STASIS - [med] (a stoppage of the circulation of blood through the capillaries and smaller blood-vessels) stase *f*, stagnation *f*
[zool etc] interruption *f* du développement
STATE, to - [gen] déclarer, affirmer
[to establish] régler, fixer
STATE - [gen] état *m*, condition *f*
~AIRCRAFT - [aero] (aircraft owned by the state and used for military, customs and police service) aéronef *m* d'Etat
~LAND - [leg] propriété *f* agricole de l'Etat
~OF AGGREGATION - [phys] (physical condition expressed as solid, liquid or gaseous) état *m* d'agrégation
~OF CURE - [rubber ind] (degree of vulcanization) degré *m* de vulcanisation
~OF THE MARKET - [comm] situation *f* du marché
~OF VULCANIZATION - s. state of cure
~SUBSIDY - [gen] aide *f* accordée par l'Etat
~TRADING - [comm] commerce *m* d'Etat
STATEDLY - [gen] à intervalles réglés
STATELESS - [leg] sans-patrie *mf*
STATEMENT - [gen] exposé *m*, exposition *f*, relation *f*
[comm] (of accounts) état *m* de compte, relevé *m* de compte
STATEROOM - [naut] cabine *f* de luxe
[railw] (US only) compartiment *m* privé
[gen] salle *f* de réception, chambre *f* d'apparat
STATIC - [phys] (relating to bodies at rest or to forces in equilibrium) statique
[phys] (acting as a weight, but not moving) statique
[met] (atmospheric electrical disturbance causing interference in radio signals) parasites *m*pl
~AND DYNAMIC EQUILIBRATION - [auto] (of tyres) balourdage *m* statique et dynamique, équilibrage *m* statique et dynamique
~BACKGROUND - [cin] (in animation) fond *m* perma-

nent

STATIC BACKING PRESSURE - [mech] (in vacuum technology) pression *f* statique de vide primaire

~BALANCE - [mech] équilibre *m* statique
[aero] (state of a control surface when the centre of mass lies on the hinge axis) équilibre *m* statique
[aero] (propeller condition in which a propeller accurately mounted on a shaft which is supported on knife-edges, will remain at rest at any point in its rotation) équilibre *m* statique

~BALANCED SURFACE - [aero] (a control surface which is in static balance) gouvernail *m* statique

~BALANCER - [el] (alternating-current balancer: an autotransformer or reactor with windings so interconnected as to divide the total voltage equally between the wires of a multiple-wire system) bobine *f* égalisatrice
[el] (an alternating current balancer) équilibrateur *m* statique

~BALANCING MACHINE - [mech] machine *f* pour l'équilibrage statique

~BED - [nucl] milieu *m* statique

~BOTTOM HOLE PRESSURE - [oil ind] pression *f* statique de fond

~BREAKDOWN CONDITION - [electron] (breakdown condition of a gap when the current grows infinitely slowly) condition *f* statique

~BREAKDOWN FOREPRESSURE - [mech] (in vacuum technology) pression *f* primaire limite

~BREEZE - [el] (or electric wind, or convective discharge; the movement of a visible or invisible stream of particles carrying away charges from a body which has been charged to a sufficiently high voltage) brise *f* statique, vent *m* électrique, effluvation *f*

~CABLE - [aero] (a cord attached to a parachute and to the aircraft from which it is dropped with the object of deploying it automatically) câble *m* de parachutage

~CALCULATION - [phys] calcul *m* statique

~CEILING - [aero] plafond *m* statique

~CHAMBER - [aero] (enclosed space in a pilot static unit in which the pressure is that of the ambient static level) chambre *f* statique

~CHARACTERISTIC - [electron] (the characteristic with zero load impedance and at zero or low frequency) caractéristique *f* statique

~CHARACTERISTIC OF AN ARC-WELDING SET - [el] (the relation between the output voltage of the set and its output current to a practically non-inductive load) caractéristique *f* de réglage d'un appareil de soudure à l'arc

~CHARGE - [el] (an electric charge developing on material in some processes) charge *f* électrostatique

~CHARGE GAUGE - [instr] (instrument for showing and measuring static electric charges on plastic materials during processing) détecteur *m* de charge électrique

~CONDENSER - [el] (static piece of an electrical apparatus having the property of drawing a leading current from an a. c. supply) condensateur *m* statique

~CONSTRAINT - [rubber ind] contrainte *f* statique

~CONVERGENDE - [telev] (in colour television) convergence *f* statique

~CURRENT CHANGER - [el] convertisseur *m* statique

STATIC CURVE - [phys] courbe *f* statique

~DETECTOR - [instr] (device for showing the presence of a static charge on material being processed) détecteur *m* de charge électrostatique

~DISCHARGE-HEAD - [mech] (in pumping) hauteur *f* de refoulement

~ELECTRICITY - [el] (of an electric current in which the charges are normally stationary) électricité *f* statique

~ELECTRODE POTENTIAL - [el] (the potential difference between the electrolyte and the electrode when no current is flowing) tension *f* statique d'une électrode

~ELIMINATOR - [el] éliminateur *m* d'électrostatique s. static screen

~FIRING - [astronaut] (of a rocket) tir *m* au point fixe, tir *m* static

~FOCUS - [electron] foyer *m* statique

~FOREPRESSURE - s. static backing pressure

~FREQUENCY CHANGER - [el] convertisseur *m* statique de fréquence

~FRICTION - [mech] (the value of the limiting friction just before slipping occurs) frottement *m* au départ

~GROUND - [aero] (automatic connexion of an aircraft earth system to the earth itself) mise *f* à la terre automatique

~HEAD - [hydr] hauteur *f* piézométrique, pression *f* d'eau

~HEAD OF A RESERVOIR - [hydr] charge *f* d'un réservoir

~IMPEDANCE - [el] (the electrical impedance of a machine or a tranducer when it is stopped from moving) impédance *f* statique

~INDUCED CURRENT - [el] courant *m* induit statique

~INPUT VOLTAGE - [electron] tension *f* continue à l'entrée

~JET THRUST -[aero] (the net thrust of a jet engine at standard sea level when it is not moving forward) poussée *f* statique standard

~LIFT - [aero] (of an aerostat) portance *f* statique

~LINE - s static cable

~LOAD - [constr] charge *f* statique

~LOADED RADIUS - [rubber ind] (of tyres) rayon *m* sous charge statique

~LUMINOUS SENSITIVITY - [electron] (of a photoelectric device) sensibilité *f* lumineuse statique

~MACHINE - [el] (of an electrostatic generator) machine *f* électrostatique à influence

~MEMORY - [comput] (a type of storage in which information is fixed in space and available at any time provided the power is on) mémoire *f* statique

~MOMENT - [phys] moment *m* d'une force

~OPENING - [mech] (of a pump) orifice *m* d'aspiration, orifice *m* d'admission

~OVER-VOLTAGE - [el] surtension *f* statique

~PHASE ERROR - [electron] erreur *f* statique de phase

~PIN - [aero] (pin attached to the static line and passing through the petals of the petal cap of a parachute, so that when withdrawn, it allows the parachute to deploy) broche *f* de fermeture du câble automatique

~PRESSURE - [phys] (the pressure at a point on a body moving with the fluid surrounding it) pression *f* statique

~PRESSURE ERROR CORRECTION - [phys] (correction

applied to static tube observations) correction *f* d'erreur de pression statique

STATIC-PRESSURE TUBE - [aero] (a tube of which the open end is so arranged that a still-air pressure reading can be obtained when the aircraft is in motion) prise *f* de pression statique
[mech] tube *m* piézométrique

~PROGRAMME - [comput] programme *m* statique

~RADIAL ENGINE - [aero] (with stationary cylinders) moteur *m* en étoile à cylindres fixes

~RAIL - [aero] (a metal rail attached inside the aircraft and serving as a static cable) rail *m* de parachutage

~RATE - [mech] (static rate of an elastic member is the rate measured between successive stationary positions at which the member has settled to substantially equilibrium conditions) taux *m* de flexibilité statique, taux *m* de rigidité statique

~REACTION - [mech] (the static force exerted on a body by other bodies by which it is supported in equilibrium) réaction *f* statique

~RELAY - [el] relais *m* statique

~SCREEN - [radio] anti-parasitage *m*

~SENSITIVITY - [electron] (the static energy sensitivity of a photo-electric device) sensibilité *f* statique

~STABILITY - [aero] (quality in an aircraft existing when the static margin is positive) stabilité *f* statique

~STORAGE - s. static memory

~STORE - s. static memory

~STRAIN TEST - [metall] épreuve *f* statique à la déformation

~STRESS - [phys] effort *m* statique

~SUBROUTINE - [comput] (subroutine involving no parameters other than the addresses of the operands) sous-programme *m* statique

~-SUCTION LIFT - [hydr] hauteur *f* d'aspiration

~SWITCHING - [electron] commutation *f* statique

~TEST - [aero] (a test reproducing the stresses on an aircraft and its parts by the application of static loads) essai *m* statique
[astronaut] (of a rocket) essai statique

~THRUST - [aero] (the thrust developed by a propeller which is revolving but is not moving forward) traction *f* au point fixe

~TOE-IN - [auto] (of a pair of wheels) pincement *m* statique, parallelisme *m*

~TOE-OUT - [auto] (of a pair of wheels) baillement *m* statique, ouverture *f*

~TORQUE - [el] couple *m* initial de démarrage

~TRANSCONDUCTANCE TEST - [electron] essai *m* statique de conductance de transfert

~VACUUM SYSTEM - [mech] (a sealed pumpless vacuum system) installation *f* de vide statique, installation *f* de vide scellée

~VALUE - [phys] valeur *f* statique

~VALUE OF THE FORWARD CURRENT TRANSFER RATIO - [electron] coefficient *m* d'amplification de courant en sens direct

~VENT - [aero] (an opening in an aircraft fuselage for the measurement of ambient static pressure) reniflard *m* statique

~WASHER - [min] (a packed tower scrubber) laveur *m* statique

STATICAL - s. static

~TRANSVERSE FIELD - [electron] champ *m* transversal statique

STATICIZER - [comput] convertisseur *m* série-parallèle

STATICS - [phys] (the science of bodies at rest and of the relations required to produce equilibrium) statique *f*
[cin] (film defect) effluves *mpl*
[radio] perturbations *fpl* atmosphériques, parasites *mpl* atmosphériques

STATION, to - [gen] placer, se poster

STATION - [gen] endroit *m*, poste *m*, place *f*
[railw] gare *f*
[mach tool] station *f*
[mech] (a specific point or stage in a process at which a device or machine operates) étage *m*
[radio] poste *m* émetteur
[telecomm] poste *m* téléphonique
[min] (in a shaft) recette *f*, accrochage *m*, chambre *f* d'envoyage
[el] centrale *f* électrique
[surv] (the point at which the instrument is set) station *f*
[agric] (Australia only) élevage *m* de moutons
[med] position *f* debout
[zool & bot] habitat *m*

~ALTIMETER - [instr] (an altimeter at an aerodrome, set for the height above M. S. L. of the point at which it is fixed) altimètre *m* d'aerodrome

~BILL - [naut] rôle *m* de manoeuvre

~BREAK - [radio] interruption *f* de l'émission

~CAPTION - [telev] indicatif *m* de la station

~DIAL - [telev] diapositive *f* d'identification de la station
[telecomm] cadran *m* de stations

~GOVERNOR - [gas ind] régulateur *m* d'émission

~KEEPING - [astronaut] (the sequence of manoeuvres to keep a vehicle in a predetermined orbit) maintien *m* en station

~LINE - [telecomm] (subscriber's line) ligne *f* d'abonné

~LOG - [radio] journal *m* de station

~-MASTER - [railw] chef *m* de gare

~OF DESTINATION - [telecomm] station *f* de destination

~OF ORIGIN - [telecomm] station *f* d'origine

~ON BOARD - [telecomm] station *f* de bord

~PEG - [surv] jalon *m*

~POINT - [surv] point *m* de station, point *m* de vue

~POINTER - [surv] (instrument for obtaining a solution of the three-point problem) stigmographe *m*

~POLE - [surv] (a wooden rod used to mark a survey station) mire *f*

~PRESSURE - [el] tension *f* au départ

~REGULATOR - s. station governor

~RINGER - [telecomm] sonnerie *f* de poste téléphonique

~ROD - [surv] (a levelling rod) mire *f*

~ROOF - [constr] (used for roofing railway stations; roof cantilevered out to one side or to both sides of a single line of stanchions) toit *m* en encorbellement

~TIMING - [telev] minutage *m* des signaux d'information

~WAGON - [auto] familiale *f*, canadienne *f*

STATIONARY - [gen] stationnaire, immobile
[mech etc] fixe, à demeure

~-ANODE TUBE - [electron] (an X-ray tube with a

stationary anode) tube *m* à anode fixe
STATIONARY BATTERY - [el] (storage battery designed for service in a permanent location) batterie *f* stationnaire
~BOBBIN - [text] grosse bobine *f*
~BOILER - [th eng] chaudière *f* placée à demeure, chaudière *f* fixe
~CAN - [text] pot *m* fixe
~CARRIAGE - [text] chariot *m* fixe
~CONTACT MEMBER - [el] contact *m* fixe
~CRANE - [mech] grue *f* fixe
~DIESEL ENGINE - [mech] diesel *m* fixe
~DREDGER - [mech] (a bucket-ladder dredger discharging dredged materials into attendant vessels) drague *f* stationnaire
~ENGINE - [mech] (a steam engine) machine *f* à vapeur fixe
~EXHAUST SYSTEM - [mech] banc *m* de pompage (avec dispositifs de pompage stationnaire)
~FEEDING NIPPERS - [text] pince *f* alimentaire fixe
~FLAT - [text] chapeau *m* fixe
~FLAT CARD - [text] carde *f* à chapeaux fixes
~FURNACE - [metall] four *m* fixe
~GEARS - [mech] engrenages *mpl* fixes
~GRID - [radiology] (a grid in which the opaque strips are so thin and close together that it can remain stationary without the shadow of the strip interfering with the interpretation of the film) grille *f* fixe, grille *f* de Lysholm
~HAND CRANE - [mech] grue *f* à main fixe
~HYDRAULIC RIVETING MACHINE - [mech] machine *f* à river hydraulique fixe
~INSTRUMENT - [instr] (measuring instrument fixed to a support) instrument *m* immobile
~JAW - [mech] mâchoire *f* fixe
~LIQUID PHASE - [ind chem] (in gas analysis) phase *f* liquide stationnaire
~NUCLEUS - [phys] noyau *m* au repos
~ORBIT - [astronaut] orbite *f* stationnaire
~PLANT - [mech] outillage *m* fixe
~PLATEN - [metall & plast ind] plaque *f* fixe du moule
~SHAFT - [text] lame *f* stationnaire
~SPINDLE - [text] broche *f* fixe
~STATE - [phys] (discrete energy state in which a quantized particle or system may exist, according to the quantum theory) état *m* stationnaire
~STRUCTURE - [el] (a supporting framework) ossature *f* de support, monture *f*
~TRANSFORMER - [el] transformateur *m* statique, transformateur *m* à action instantanée
~TWISTER - [text] fileuse *f* en gros
~VORTEX - [phys] (a vortical motion which is steady in space and time) tourbillon *m* stationnaire
~WARP - [text] chaîne *f* fixe, chaîne *f* des fils droits
~WAVE - [acoust] (a standing wave in which the energy flux is zero at all points) onde *f* stationnaire idéale
~WAVE SYSTEM - [acoust] (interference pattern characterized by stationary nodes and antinodes) système *m* d'ondes stationnaires
STATIONERY - [gen] papeterie *f*
STATIONMASTER - s. station master
STATISTICAL - [gen] statistique
~ACCELEROMETER - [instr] (an instrument to record the number of occasions on which predetermined acceleration values have been exceeded) accélérocompteur *m*
STATISTICAL ANALYSIS - [statist] (analysis of combination of data obtained from individual events) analyse *f* statistique
~COUNTER TIME-LAG - [nucl] (the period of time between the primary ionizing event and the occurrence of the count in the counter) retard *m* du compteur
~COUNTING ERROR - [nucl] erreur *f* de comptage
~ERROR - [meas] (the random error occurring in a measurement) erreur *f* statistique
~FLUCTUATION - [nucl] (the variation in range, ionization, or direction, which is due to fluctuations in the distance between collisions in the stopping medium and in the energy loss and deflection angle per collision) dispersion *f* statistique, fluctuation *f* statistique
~METHOD - [nucl] (a method of separating isotopes depending on small differences in the average behaviour of the molecules of the different isotopes) méthode *f* statistique
~MODEL - [nucl] (nuclear model) modèle *m* statistique
~NOISE - [radio] (a noise due to the aggregate of a large number of elementary disturbances with random occurrence in time) effet *m* de grenaille
~STRAGGLING - s. statistical fluctuation
~UNCERTAINTY - [phys] (an estimated amount by which the calculated value of a quantity may differ from the true value) incertitude *f* statistique
~WEIGHT - [phys] (of a macroscopic state; the number of microscopic states contained in a macroscopic state. The macroscopic state of which the statistical weight is a maximum, is a state of equilibrium) poids *m* statistique
[nucl] (in a nuclear reactor) fonction *f* de pondération
STATISTICS - [statist] (the science which deals with the collection, tabulation and systematic classification of facts) statistique *f*
STATOMETER - [instr] (an apparatus designed to measure the degree of exophtalmos) statomètre *m*
STATOR - [el] (the portion of a machine which includes the stationary magnetic parts with their associated windings) stator *m*, induit *m* fixe
~BLADES - [mech] (the fixed blades of a turbine) aubes *fpl* fixes
~CORE - [el] (the assembly of laminations of a stator) couronne *f* de tôle
~CORE LAMINATIONS - s. stator core
~EARTH LEAKAGE RELAY - [el] relais *m* de mise à la terre pour un stator
~FRAME - [el] carcasse *f* de stator
~LAMINATIONS - s. stator pack
~PACK - [el] paquet *m* de tôles (du stator)
~PLATES - s. stator pack
~PUMP - [th eng] pompe *f* à induction magnétique
~RESISTANCE-STARTER - [el] (combined stator-circuit switch and rotor circuit regulating resistance for use with slip-ring induction motors) démarreur *m* statorique à résistances
~WINDING - [el] (the part of the winding of a machine housed in the stator) enroulement *m* statorique
STATOSCOPE - [instr] (an instrument indicating small changes in altitude) statoscope *m*
STATUARY MARBLE - [min] marbre *m* statuaire
STATUE - [gen] statue *f*
STATURE - [gen] stature *f*, taille *f*
[med] hauteur *m* de l'homme debout

STATUS - [gen & leg] statut *m*, condition *f*
[med] état *m*
~OF THE PELVIS - [med] statique *f* pelvienne
STATUTE - [gen & leg] loi *f*, ordonnance *f*, réglements *mpl*
~MILE - [meas] (unit of distance on land equal to 5280 feet) mille *m* anglais, mille *m* terrestre
STATUTORY - [leg] (dependent on a legislative enactement) établi par la loi, réglementaire
~INTESTATE SUCCESSION - [leg] succession *f* légale, succession *f* ab intestat
STAUNCH, to - [med] (to stop or check the flow of blood) étancher
STAUNCH TIGHT JOINT - [mech] rivetage *m* hermétique
STAUROLITE - [min] (silicate of aluminium and iron with chemically combined water) staurolite *f*
STAUROSCOPE - [instr] (polariscope) stauroscope *m*
STAVE, to - [gen] garnir de douves, assembler les douves
[naut] s'effondrer
STAVE - [carp] (one of the strips of wood from which a cask is built up) douve *f*, longaille *f*
[gen] (of a ladder) échelon *m*
[mus] (the five horizontal lines on which music is written) portée *f*
[text] (heddle) liseron *m*
~FLY - [text] guindre *m* à bâton
~IN - [gen] défoncer, effondrer, emboutir
~OF SWIFT - [text] batte *f* du dévidoir
~SWIFT - s. stave fly
~UP, to - [metall] refouler
STAVING - [mech] (the thickening of the ends of pipes) refoulement *m*, épaississement *m* de l'extremité d'un tube
STAY, to - [gen] s'arreter
[gen] (to stand still) se tenir chez, rester
[gen] (to have a temporary abode) séjourner, démeurer
[gen] (to stop) arrêter
[leg] (to postpone, to suspend) ajourner, remettre
[constr] (to support, to hold) étayer, entretoiser, ancrer
[constr] (to stiffen, e.g. with sties) haubanner
[naut] virer de bord vent devant
STAY - [gen] (sojourn, visit) séjour *m*
[gen] (a pause) pause *f*, arrêt *m*
[leg] suspension *f*
[constr] support *m*, montant *m*, étai *m*, tirant *m*
[constr] (for stiffening) hauban *m*
[mach tool] (of a lathe) lunette *f*
[naut] (guy rope) draille *f*
[telecomm] hauban *m* de mât d'antenne
~ANCHOR - [el & telecomm] ancre *f*
~BLOCK - [constr] semelle *f* d'ancrage
~BOLT, to - [mech] entretoiser par des boulons
~BOLT - [mech] tirant *m*, entretoise *f*
~-BOLT TAP - [mech] taraud *m* pour entretoises
~CLAMP - [el & telecomm] bride *f* de jonction de câbles porteurs, mâchoire *f* de jonction
~CRUTCH - [el & telecomm] console *f* d'ancrage pour haubans
~FRAMING - [text] bâti *m* du métier
~PLATE - [metall] plaque *f* d'appui
~-PUT SWITCH - [telecomm] appareil *m* téléphonique sans position de repos
~ROD - [mech] tirant *m*
STAY ROPE - [el & telecomm] câble *m* de haubanage
~THIMBLE - [el & telecomm] casse *f* pour haubans
~TIGHTENER - [el & telecomm] tendeur *m* de hauban
~TIME - [gen] (the flow velocity of propellants in a rocket combustion chamber) durée *f* de séjour dans la chambre de combustion
~-TUBE, to - [th eng] entretoiser par tubes-tirants
~TUBE - [th eng] (in boilers) tube-tirant *m*
~WIRE - [el] (a steel cable by which a transmission-line pole is secured to the ground) hauban *m* en fil métallique, hauban-fil *m*
STAYED POLE - [el & telecomm] appui *m* haubanné
~TERMINAL POLE - [el & telecomm] appui *m* tête de ligne, appui *m* terminal
STAYER - [oil ind] puits *m* à grande stabilité de production
STAYING OF THEGROWN - [mining] armature *f* du ciel
STAYSAIL - [naut] voile *f* d'étai
STD -(short for standard) standard
STEADY, to - [gen] assurer, affermir
[naut] retrouver l'équilibre
[mech] (a machine etc) stabiliser, régulariser
STEADY - [gen] solide, ferme, fixed
[mech] constant, stabilisé
[gen] (regular) régulier
[comm] (of the market) (marché) soutenu
[naut] (of a boat) qui tient bien la mer
[mach tool] lunette *f*
~ARM - s. steady brace
~BEARING - [mech] support *m* de guide
~BRACE - [el] (a fitting used in catenary construction to maintain the correct lateral position of the contact wire) bras *m* de rappel, bras *m* de retenue
~CONDITION - [el radio etc] régime *m* permanent
~CURRENT - [el] courant *m* constant
~FLIGHT - [aero] (of a helicopter ; hovering) vol *m* à point fixe
~FLOW - [mech] (or viscous flow; flow in which the flow velocity at a point fixed with respect to the coordinates is independent of time) effluent *m* stationnaire
~FLOW REACTOR - [nucl] réacteur *m* à effluent stationnaire
~GRADIENT - [surv] inclinaison *f* constante
~LIGHT - [opt] lumière *f* constante
~LOAD - [phys] charge *f* permanente
~PIN - [mech] (for mechanical parts which must be fixed together accurately with one fixing screw) clavette *f* de calage
[metall] (of a moulding box) goujon *m* (de châssis de moulage)
[metall] (core-print) portée *f* de noyau
[metall] portée *f*
~REST - [mach tool] (of a lathe or grinding machine) lunette *f* fixe
~REST BEARING - [auto] (intermediate bearing support for the drive shaft) palier *m* intermédiaire, palier-relais *m*
~SOURCE - [radiat] source *f* de rayonnement constant
~SPAN - [el] (a transverse wire rope used in bridge or cross-span construction to maintain the correct lateral position of the contact wire) fil *m* tendeur pour support latéral
~STATE - [phys] (a condition of dynamic balance as in an equilibrium reaction, where at equilibrium

the concentration of each of the reactants remains constant) état *m* permanent
STEADY STATE - [el] (of an electric wave) régime *m* stabile
~-STATE COSMOLOGY - [phys] (the hypothesis of continuous creation) hypothèse *f* de la création continue
~-STATE DEVIATION - [phys] écart *m* permanent
~-STATE ERROR - [meas] erreur *f* permanente
~-STATE OSCILLATION - [phys] (the oscillation of a system in which the motion at each point is a periodic quantity) oscillation *f* stationnaire
~-STATE OUTPUT - [contr] valeur *f* prescripte, valeur *f* de sortie dans l'état stationnaire
~-STATE POWER LIMIT - [el] limite *f* de stabilité statique
~-STATE RESPONSE - [contr] réponse *f* de régime permanent
~-STATE STABILITY - [el] stabilité *f* normale
~-STATE STABILITY FACTOR - [el] facteur *m* de stabilité statique
~-STATE VALUE - [contr] valeur *f* en régime établi
~-STATE VIBRATION - s. steady-state oscillation
STEADYING BRACKET FOR THE BEAM - [text] support *m* du levier de balance
~FACTOR - [gen] (in industrial production) volant *m*
~RESISTANCE - [el] (the ballast resistance placed in series with a direct-current arc lamp, to counteract the negative resistance of the arc) résistance *f* de stabilisation
STEAL, to - [gen] voler, soustraire
STEAM, to - [gen & ind] passer à la vapeur, vaporiser
[gen] (to issue steam) exhaler de la vapeur, jeter de la vapeur
[text] délustrer, vaporiser (un drap etc), fixer à la vapeur (un colorant)
[carp] (the wood) traiter à la vapeur, vaporiser
STEAM - [phys] (water vapour, at a temperature at or above the boiling point of water; by analogy the vapour of any liquid) vapeur *f*
~ACCUMULATOR - [mech] (large pressure vessel into which surplus high-pressure steam is blown and condensed) accumulateur *m* de vapeur
~ADMISSION - [mech] admission *f* de vapeur
~-AIRBAG - [rubber ind] (used for retreading) chambre *f* de vulcanisation pour air et vapeur d'eau, air-vapeur bag *m*
~ATOMIZER - [mech] pulvérisateur *m* à pression de vapeur
~AUTOCLAVE - [ind chem] (an autoclave in which the contents are heated by steam) autoclave *m* à vapeur
~BALANCE - [mech] (a type of safety valve) balance *f* à poids
~BATH - [ind chem] bain *m* de vapeur
~BENT - [carp] (of timber) cintré à la vapeur
~BLAST - [mech] jet *m* de vapeur
~BOILER - [th eng] chaudière *f* à vapeur, générateur *m* de vapeur
~BOILING - [brew ind] cuisson *f* à la vapeur
~BOX - s. steam chest
~BRAKE - [mech] frein *m* à vapeur
~CALORIMETER - [instr] (calorimeter designed by Joly to measure the specific heat of solids) calorimètre *m* à vapeur
~CAR - [auto] (an automobile which is propelled by steam) voiture *f* à vapeur
STEAM CATAPULT - [aero] (steam-actuated device used on aircraft-carries for launching aircraft at speed) catapulte *f* de lancement
~CAUTERY - [med] atmocausis *f*, cautérisation *f* par jet de vapeur
~CHAMBER - [mech] boîte *f* à vapeur, boîte *f* de distribution de vapeur
~CHANNEL - [plast ind] (a continuous channel in a mould through which steam can be circulated) voie *f* à vapeur
~CHEST - [mech] (the chamber in which the slide-valve of a steam engine works and to which the steam pipe is connected) chapelle *f* du tiroir, réservoir *m* de vapeur
~CLEANER - [mech] (a machine used for washing mechanical parts) nettoyeuse *f* à jet de vapeur
~COAL - [min] houille *f* de chaudière, charbon *m* de chaudière, charbon *m* à vapeur
~COCK - [min] prise *f* de vapeur, robinet *m* de prise de vapeur
~COIL - [mech] serpentin *m* à vapeur
~COLOUR - [text] teinture *f* fixée à la vapeur
~CONDENSER PIPE - [mech] tube *m* de condenseur de vapeur
~CONDUIT - [mech] tuyautage *m* de vapeur
~CONE - s. steam nozzle
~CONNECTION - [mech] prise *f* de la vapeur
~CONSUMPTION - [mech] consommation *f* de vapeur
~COPPER - [brew ind] chaudière *f* pour cuisson à vapeur
~CORED MOULD - [plast ind] (a mould, designed to be heated by steam, in which the steam passages are incorporated as an integral part of the mould) moule *m* à voies calorifères
~CRANE - [mech] grue *f* à vapeur
~CULTIVATOR - [agric] cultivateur *m* à vapeur
~CURE - [rubber ind] (steam vulcanization) vulcanisation *f* à la vapeur, cuisson *f* en vapeur
~CURTAIN - [gas ind] écran *m* de vapeur
~CUSHION - [mech] matelas *m*
~CYLINDER - [mech] cylindre *m* à vapeur
~DISTILLATION - [ind chem] (method of distillation in which the liquid is heated by passing steam through it) distillation *f* à la vapeur
~DOME - [mech] dôme *m* de prise de vapeur, réceptacle *m* de la vapeur
[railw] dôme *m*
~-DOME PRESS - [mech] presse-autoclave *f* avec dôme à vapeur
~DRIER - [mech] sécheur *m* de vapeur, déssécheur *m* de vapeur
~-DRIVEN - [mech] à vapeur, actionné par la vapeur
~-DRIVEN PLANT - [mech] installation *f* à vapeur
~-DRIVEN RIVETING MACHINE - [mech] riveuse *f* à vapeur
~DROP HAMMER - [mech] marteau-pilon *m* à vapeur
~DRUM - [gas ind] ballon *m* de vapeur
[nucl] collecteur *m* de vapeur
~DRYER - s. steam drier
~ECONOMIZER - [mech] économiseur *m*
~EDGE - [mech] (of a slide valve) rebord *m* extérieur, arête *f* extérieure
~EJECTOR - [mech] (apparatus in which air is sucked out of a vessel by entrainment in a steam jet) éjecteur *m* à la vapeur

STEAM-ELECTRIC GENERATING SET - [el] (generating set in which the prime mover is a steam engine) groupe *m* électrogène à vapeur
~EMULSION NUMBER - [chem] (deemulsification time) durée *f* de désémulsionnement
~ENGINE - [mech] machine *f* à vapeur
~-ENGINE INDICATOR - [meas] indicateur *m* de pression, indicateur *m* dynamométrique
~ESCAPE - [mech] échappement *m* de vapeur
~EXCAVATOR - [mech] excavateur *m* à vapeur
~EXHAUST - [mech] échappement *m* de la vapeur
~EXHAUST PORT - [mech] lumière *f* d'échappement de la vapeur
~EXHAUST SIDE - [mech] côté *m* d'échappement de la vapeur
~FINISH - [paper man] calandrage *m* à jet de vapeur
~FOG - [met] (sea fog) brouillard *m* marin
~FORMATION - [phys] (in boilers etc) formation *f* de vapeur
~FUNNEL - [ind chem] (funnel with tubular steam, jacket, for filtration at temperatures above the ambient level) entonnoir *m* à tuyau de vapeur
~-GENERATING HEAT - [th eng] chaleur *f* à produire de la vapeur
~GENERATING REACTOR - [nucl] réacteur *m* générateur de vapeur
~-GENERATING TUBE - [th eng] tuyau *m* de vaporisation
~GENERATION - [phys] (per unit measure of a heated surface) génération *f* de vapeur
~GAUGE - [meas] manomètre *m* de pression de vapeur
~GOVERNOR - [mech] régulateur *m* de vapeur
~GUN - [mech] pistolet *m* à vapeur
~HAMMER - [mech] marteau-pilon *m* à vapeur
~HEAD - [plumb] (a tee-piece for steam pipes) culotte *f* de bifurcation du tuyau de prise de vapeur
~HEATED CIRCLE - [text] plaque *f* annulaire chauffée par la vapeur
~HEATED DRYING RANGE - [text] armoire *f* à sécher chauffée à la vapeur
~HEATED WATER - [th eng] eau *f* chauffée à la vapeur
~HEATING - [th eng] (the transfer of heat to a material, body or other object by means of steam) chauffage *m* à vapeur, chauffage *m* à la vapeur
~HOOTER - [mech] sirène *f* à vapeur
~HOPPER - [metall] trémie *f* à vapeur
~HOSE - [mech] (hose formed of interlocking spirals of bronze or steel, to convey live steam) tuyau *m* de vapeur
~INJECTOR - [mech] (device in which the kinetic energy of a steam jet is converted into pressure and imparted to water, e.g. for feeding boilers) injecteur *m* de vapeur
~INLET - [mech] conduit *m* d'admission (de la vapeur)
~JACKET - [mech] (an enclosed space through which steam is circulated to transfer heat to a structure within it) chemise *f* de vapeur, chapelle *f*, cylindre-enveloppe *m*
~JACKETED VULCANIZING PAN - [rubber ind] (a vessel used for vulcanizing rubber, heated by passing steam through an enclosed space surrounding it) chaudière *f* à vulcaniser chemisée de vapeur
~-JET BLOWER - [th eng] réchauffeur *m* à jet de vapeur
STEAM-JET CLEANING MACHINE - [mech] nettoyeuse *f* à jet de vapeur
~JET SPRAYER - [mech] pulvérisateur *m* à vapeur
~KETTLE - [med] bouilloire *f* pour humidifier l'atmosphère
~KILN - [brew ind] touraille *f* à vapeur
~LAUNCH - [naut] canot *m* à vapeur
~LOCOMOTIVE - [railw] locomotive *f* à vapeur
~LORRY - [mech] camion *m* à vapeur
~MAIN - [mech] conduite *f* maîtresse de la vapeur
~METER - [meas] compteur *m* de vapeur
~MOTOR - [mech] moteur *m* à vapeur
~NAVIGATION - [naut] navigation *f* à vapeur
~NAVVY - s. steam shovel
~NOZZLE - [mech] injecteur *m* de vapeur, tuyère *f* à vapeur, ajutage *m* à vapeur d'un injecteur
~OUTLET - [mech] orifice *m* de sortie de vapeur
~PACKET - [naut] paquebot *m* à vapeur
~PACKING - [mech] garniture *f* de vapeur
~PAN - [rubber ind] (a vessel provided with an enclosed space through which steam can be passed for heating it) chaudière *f* pour vulcanisation en vapeur
~-PILE DRIVER - [mech] sonnette *f* à vapeur
~PIPE - [mech & th eng] conduite *f* de vapeur, tuyau *m* de prise de vapeur
~PIPE AGITATOR - [text] agitateur *m* mécanique
~-PIPE-LINE - [mech] canalisation *f* de vapeur
~-PIPE TEE PIECE - s. steam-head
~PIPING - [plumb] tuyautage *m* de vapeur
~PISTON - [mech] piston *m* à vapeur
~PLATE - [plast ind] (a force-plate provided with passages for steam heating) plateau *m* chauffant, plaque *f* chauffante
~PLATEN PRESS - [plast ind] (a press designed in such a way that the platen or platens can be heated by passing steam through ducts formed in them) presse *f* à plateaux à vapeur
~PLOUGH - [agric] charrue *f* à vapeur
~POINT - [mining] (used to thaw frozen ground) pointe *f*
~PORTS - [mech] orifices *mpl* de vapeur
~POT - [metall] autoclave *m*
~POWER - [mech] énergie *f* de la vapeur
~POWER PLANT - [el] installation *f* de production à vapeur
~PRESS CURING - [rubber ind] vulcanisation *f* dans presse à caisson vapeur
~PRESSURE - [phys] (the specific pressure exerted by steam in a given vessel or system) pression *f* vapeur, tension *f* de vapeur de l'eau
~PRESSURE GAUGE - [meas] manomètre *m* de pression de vapeur
~PRESSURE PUMP - [mech] pompe *f* foulante à vapeur
~PROCESS REGENERATION - [rubber ind] (thermal reclaiming process) régénération *f* à la vapeur surchauffée, procédé *m* de régénération thermique
~PUMP - [mech] pompe *f* à vapeur
~PURGE - [gas ind] (for water gas) purge *f* à la vapeur, purge *f* de fin de soufflage
~-RAILCAR - [railw] automotrice *f* à vapeur
~RAISING - [mech] production *f* de vapeur
~RAM HOSE - [mech] tuyau *m* de mouton à vapeur
~RECLAIMED - [rubber ind] régénéré *f* à la vapeur
~REGULATOR - [mech] régulateur *m* de vapeur

STEAM RELIEF-VALVE - [mech] détendeur *m* de vapeur
~RETTING - [text] rouissage *m* à la vapeur d'eau
~REVERSING GEAR - [mech] (power reversing gear used in steam locomotives) mécanisme *m* de changement de marche à vapeur
~ROCKET - [astronaut] fusée *f* à vapeur
~-ROLLER, to - [constr] (roads) cylindrer (une route)
~-ROLLER - [mech] (used in road construction) cylindre *m* compresseur à vapeur, rouleau *m* compresseur
~ROLLS - [paper man] (rollers in a paper calender heated by steam passed through them internally) cylindres *mpl* chauffés à la vapeur
~SATURATION - [th eng] saturation *f* de la vapeur
~SAWMILL - [mech] scierie *f* à vapeur
~SEAL - [mech] (a special type of stuffing box through which steam is passed to prevent the entry of undesired micro-organism along a shaft to the inside of a machine) joint *m* de vapeur
~SECTIONAL BAG - [rubber ind] steambag *m* section
~SEPARATOR - [ind chem] (a device for removing condensate from steam) purgeur *m* de vapeur, séparateur *m* d'eau et de vapeur
[mech] déshuileur *m* de vapeur, dégraisseur *m* de vapeur
~SHOVEL - [mech] excavateur *m*, pelle *f* à vapeur, terrassier *m* à vapeur, cuiller *f* à vapeur
~SINKING PUMP - [mech] pompe *f* d'avaleresse à vapeur
~SLIDE VALVE - [mech] tiroir *m* à vapeur
~SPACE - [th eng] (of boilers) chambre *f* de vapeur (d'une chaudière)
~STERILIZER - [chem] (apparatus for sterilizing objects with steam) étuve *f*
~STILL - [oil ind] distillateur *m* à vapeur
~STOP-VALVE - [mech] robinet *m* d'arrêt de vapeur
~STRIPPING - [chem] (used to remove the light ends of sidestream products) distillation *f* fractionnée à vapeur
~SURGE DRUMS - [nucl] (in boiling-water reactors) récipient *m* de compensation à vapeur
~SUPERHEATER - [mech] surchauffeur *m* de vapeur
~SUPPLY PIPE - [mech] tuyau *m* de prise de vapeur
~SUPPLY WAY - [mech] canal *m* d'admission de la vapeur
~TAP - [mech] soufflante *f* de suie
~THE COPS, to - [text] vaporiser les canettes
~THROTTLE - [mech] régulateur *m* de la vapeur
~-TIGHT - [mech] (sealed in such a way that steam cannot pass) étanche à la vapeur, étanche
~TRACTION - [railw] traction *f* à la vapeur
~TRAP - [mech] (device for automatically separating and ejecting condensate from a steam system) purgeur *m* automatique, purgeur *m* d'eau de condensation
~TRAWLER - [naut] chalutier *m* à vapeur
~TURBINE - [mech] (a machine in which steam is working by expanding so as to create kinetic energy) turbine *f* à vapeur
~-UP, to - [mech] mettre la vapeur, pousser les feux
~-VALVE - [mech] soupape *f* d'admission de vapeur, soupape *f* de prise de vapeur
~VULCANIZATION - [rubber ind] (method of vulcanization in which steam heating is used) vulcanisation *f* à la vapeur
STEAM VOID - [phys] (in heat transfer) bulle *f* de vapeur
~WAY - s. steam channel
~WINCH - [mech] treuil *m* à vapeur
STEAMBAG - [rubber ind] steambag *m*, vapeur bag *m*
STEAMBOAT - [naut] bateau *m* à vapeur
STEAMED MECHANICAL WOOD PULP - [ind chem] pâte *m* mécanique brune
STEAMER - [naut] navire *m* à vapeur
[mech] machine *f* à vapeur
[impl] cuiseur *m* à vapeur
[transp] (steam-driven vehicle) véhicule *m* à vapeur
STEAMINESS - [phys] exhalation *f* de la vapeur
STEAMING - [gen] passage *m* à la vapeur
[text] fixation *f* (du colorant) à la vapeur
[nucl] injection *f* de vapeur
~CABINET - [text] chambre *f* à vaporiser
~IN OPEN BOXES - [text] vaporisation *f* dans des chambres ouvertes
~OF THE COPS - [text] vaporisage *m* des canettes
~OF YARNS - [text] vaporisage *m* des fils
~OVEN - [impl] étuve *f*
~THE SILK SKEINS - [text] vaporisation *f* des écheveaux de soie
~UNDER PRESSURE - [text] vaporisage *m* sous pression
~WASTE - [text] déchets *mpl* de vaporisation
STEAMSHIP - [naut] navire *m* à vapeur
STEAMTIGHT - s. steam tight
STEARATE - [chem] (generic term for derivates of stearic acid) stéarate *m*
STEARIC - [chem] (derived from stearin) stéarique
~ACID - [chem] (the most widely distributed fatty acid, with application in the production of stearates, polishes, pharmaceuticals, cosmetics, lubricants, candles, and plasticizers) acide *m* stéarique
STEARIN - [chem] (glyceryl tristearate. Colourless, tasteless crystals with uses in the production of adhesives, polishes, textile finishing agents, soap and candles) stéarine *f*
STEARINE - s. stearin
STEARYL ALCOHOL - [chem] (intermediate in the production of a number of plastic and resins) alcool *m* stéarylique
~MERCAPTAN - [chem] (octadecyl mercaptan) (intermediate for rubber chemicals) stéaryl-mercaptan *m*
~METHACRYLATE - [chem] (monomers which can be polymerized to a number of plastics) méthacrylate *m* de stéaryle
STEATITE - [min] (massive form of talc, a natural acid metasilicate of magnesium) stéatite *f*
STEATODENOMA - [med] adénome *m* des glandes sébacées
STEATONECROSIS - [med] granulome *m* lipophagique
STEATOPATHY - [med] affection *f* des glandes sébacées
STEATOPYGIA - [med] (accumulation of fat in the buttocks) stéatopygie *f*
STEATORRHOEA - [med] (excess of fat in the stool) diarrhée *f* graisseuse, stéatorrhée
STEATOSIS - [med] stéatose *f*, dégénérescence *f* graisseuse
STEED - [zool] coursier *m*
STEEL, to - [metall] aciérer, armer
STEEL - [metall] (an alloy of iron and carbon, contain-

ing less than 2 p.c. of the latter, less than I p.c. of manganese (excluding special manganese steel which contains up to I4 p.c.) and small quantities of silicon, sulphur, oxygen and phosphorus . A wide range of alloy steels in which other metals play an essential part are now in general use) acier *m*
STEEL - [min] (a steel rod for boring) fleuret *m*
~ALLOYS - [metall] alliages *mpl* d'acier
~ANNEALING - [metall] recuit *m* de l'acier
~BALL - s. steel lump
~BALL PEENING - [metall] traitement *m* par jet de balles en acier
~BAND - [metall] ruban *m* d'acier
[rubber ind] (on tyres) bandage *m* de roulement d'acier
~BEARER - [constr] poutre *f* en acier
~BELT - [metall] courroie *f* d'acier
~-BELT LACING - [mech] agrafes *fpl* à griffes pour courroies
~BLOOM - [metall] loupe *f* d'acier
~BODY - [auto] carrosserie *f* tout acier
~BOILER - [metall] chaudière *f* d'acier
~BRIDGE - [constr] pont *m* en fer
~BULB RECTIFIER - [el] redresseur *m* métallique
~CABLE - [metall] câble *m* en acier
~CAGE WHEELS - [mech] barillets *mpl* de jumelage
~CASING - [metall] gaine *f* en acier, revêtement *m* en acier
~CAST - [metall] coulé *m* en acier
~CASTING - [metall] moulage *m* d'acier, pièce *f* en acier moulé
~CEMENTING FURNACE - [metall] four *m* à acier cementé
~CHIP CRUSHER - [mech] concasseur *m* de copeaux métalliques
~CHUTE - [mining] couloir *m* en tôle d'acier
~-CLAD - [gen] revêtu d'acier, couvert d'acier
~CLIP - [mech] bande *f* d'acier
~COMB - [text] peigne *m* en acier
~CONSTRUCTION - [constr] structure *f* en fer
~CONTAINER - [el] (the container for the element and electrolyte of an alkaline storage cell) bac *m* en acier
~-CORD CASING - [rubber ind] (for tyres) carcasse *f* à fils métalliques
~-CORED ALUMINIUM WIRE - [el] (electrical conductor consisting of layers of aluminium wire surrounding a core of galvanized steel strands) fil *m* d'aluminium avec noyau en acier
~-ENGRAVED - [print] gravé sur acier
~ENGRAVING - [print] gravure *f* sur acier
~-FACED - [metall] aciéré
~-FACED ANVIL - [metall] enclume *f* à surface aciérée
~FASTENER - [text] lunette *f* en acier
~FORGING - [metall] pièce *f* d'acier forgé
~-FOUNDING - [metall] fonte *f* d'acier
~-FOUNDRY - [metall] fonderie *f* d'acier
~GIRDER - [constr] poutre *f* en acier
~GREY COLOUR - [gen] gris *m* d'acier
~GRIT - [metall] grenaille *f* d'acier
~HARDENING - [metall] cémentation *f* de l'acier
~HINGE - [metall] charnière *f* d'acier
~HOOP - [metall] bande *f* d'acier
~INCLINE - [text] nez *m* à bec oblique
~INGOT - [metall] lingot *m* d'acier
~LADLE - [metall] poche *f* à acier
STEEL LINING - [text] pièce *f* d'acier insérée
[nucl] (of a reactor vessel) peau *f* d'étanchéité
~LUMP - [metall] bloom *m* d'acier
~MAGNET - [metall] aimant *m* d'acier
~MAIN - s. steel pipe
~-MAKING - [metall] (the process of making steel from solid or molten pig-iron) aciérie *f*
~MANTLE - [metall] enveloppe *f* de tôle en fer
~MELTING CRUCIBLE - [metall] creuset *m* pour la fusion de l'acier
~MILL - [ind] aciérie *f*
~MIX CAST IRON - [metall] fonte *f* aciérée
~NEDDLE - [text] aiguille *f* d'acier
~OF SMITHING QUALITY - [metall] acier *m* de forge
~PATTERN - [rubber ind] (for tyres) sculpture *f* reproduite dans le moule
~PIANO-WIRE - [metall] corde *f* métallique de piano
~PIG - [metall] lingot *m* d'acier
~PILING - [metall] palplanches *fpl*
~PINNING - [text] aiguillage *m* en acier
~PIPE - [metall] tube *m* d'acier
~PLANT - [ind] aciérie *f*
~PLATE - [metall] tôle *f* d'acier
~-PLATED - [metall] aciéré
~PLATING - [metall] aciérage *m*
~POWDER - [metall] poudre *f* d'acier
~-PUDDLING - [metall] puddlage *m* du fer
~PULLEY - [mech] poulie *f* en acier
~RAIL - [metall] rail *m* en acier
~RINGS - [text] anneaux *mpl* en acier
~ROD - [metall] tige *f* d'acier
~ROOF TRUSS - [metall] charpente *f* métallique de toiture
~ROLLING MILL - [metall] laminoir *m* d'acier
~-RULE DIE - [metall & plast ind] (die formed by bending a flexible steel ribbon to the desired shape and mounting it in a wooden block) emporte-pièce *m* en bande d'acier
~SCRAPS - [metall] ferraille *f* d'acier
~SECTIONS - [metall] profilés *mpl* en acier
~SHARPENING - [mech] affûtage *m* des fleurets
~SHEET - [metall] tôle *f* d'acier
~SHELL - [metall] (used for hydraulic installations etc) enveloppe *f* en tôle
~SHUTTERING - [mining] boisage *m* en acier
~SHUTTLE - [text] navette *f* en acier
~SOFTENING - [metall] adoucissement *m* d'acier, détrempe *f* d'acier
~SPIKE STUDDING - [text] aiguillage *m* en acier
~-SPOKE WHEEL - [auto] rayon *m* de roue en acier
~SPRING - [mech] ressort *m* en acier
~SQUARE - [carp] équerre *f* en acier
~STRIP - [metall] bande *f* d'acier
~STRUCTURAL WORK - [constr] charpente *f* métallique
~STUDDED TYRE - [rubber ind] pneumatique *m* à clous d'acier
~TANK - [brew ind] cuve *f* en acier, tank *m* en acier
~-TANK RECTIFIER - [el] soupape *f* à cuve d'acier
~TAP HOLE - [metall] trou *m* de coulée d'acier
~TAPE - [meas] ruban *m* métrique d'acier
~TAPPING SPOUT - [metall] chenal *m* de coulée de acier
~TIMBERING - s. steel shuttering
~TOWER - [el & telecomm] pylône *m* en treillis, tour *f* en treillis, mât *m* en lattis
~TRAVELLER - [text] curseur *m* en acier

STEEL TUB - [mining] berline *f* en acier
~TUBE - [metall] tube *m* d'acier
~VESSEL - s. steel tank
~WEDGE - [text] coin *m* en acier
~WIRE - [metall] fil *m* d'acier
~WIRE BRUSH - [impl] (brush with bristles of steel wire used in cleaning metal parts) brosse *f* métallique
~WIRE CARD CLOTHING - [text] garniture *f* de carde en fil d'acier
~-WIRE ROPE - [metall] câble *m* métallique
~WOOL - [metall] (used for polishing) laine *f* de acier
STEELING - [el] (the electroplating of copper plates with nickel steel) aciérage *m*, aciération *f*
[metall] (casehardening) aciérage *m*, cémentation *f*
STEELWORK - [gen] profilés *mpl* pour construction
[auto] tôleries *fpl*
STEELWORKS - [ind] aciérie *f*
STEELY - [gen] aciéreux, acérain
[brew ind] (said of malt) vitreux
~PIG-IRON - [metall] fonte *f* aciéreuse
STEELYARD - [meas] romaine *f*, balance *f* romaine, peson *m* à contre-poids
STEENING - s. steining
STEEP, to - [gen] (to immerse in a liquid) tremper, mettre en trempe, infuser, mettre à macérer, imbiber
[brew ind] (to immerse barley for about 50 hours at a temperature of about 50 to 55°F to induce germination for malting) tremper
[text] (the cocoons) baigner (les cocons)
[text] (the wool) tremper
[leather ind] chauler, pelaner, confire
STEEP - [gen] raide, escarpé
[gen] (of prices) exorbitant
[ind chem] (the immersion in a liquid) trempage *m*
[chem] (the liquid in a liquid) immersion *f*, trempage *m*
[chem] (the liquid used for steeping) bain *m* de macération
[text] rouissage *m* (du chanvre)
~BATH - [text] bain *m* de rouissage
~EDGE OF A PULSE - [electron] front *m* raide d'une impulsion
~GRADIENT - [gen] pente *f* raide
~IN ALUM, to - [leather ind] aluner
~ROAD - [gen] route *f* escarpée
~SEAM - [mining] couche *f* en dressant
~SPIRAL - [mech] spirale *f* à grand pas, spirale *f* allongée
~THE LEAVES IN SALT WATER - [text] plonger les feuilles dans l'eau salée
~WINDING - [text] spires *fpl* allongées
STEEPED BARLEY - [brew ind] orge *f* gonflée
STEEPING - [ind chem] (soaking of cellulose board in caustic soda solution) trempage *m*, macération *f*
[text] (of cocoons) baigner (les cocons)
[text] trempage (de la laine), roui *m* du chanvre
[agric] (of herbs) infusion *f* à froid
~BOWL - [text] dessuinteuse *f*
~IN ALUM - [leather ind] alunage *m*
~PAN - [text] bassine *f* de trempage
~TANK - [brew ind] cuve *f* à tremper
~TIME - [gen] durée *f* de trempage
~WATER - [brew ind & text] eau *f* de trempage
STEEPLE - [constr] (structure surmounted with a spire) clocher *m*
STEEPLE - [constr] (of a roof) toit *m* en pyramide
~ENGINE - [mech] (of a boiler) machine *f* à vapeur verticale auxiliaire
~ROOF - [constr] toit *m* de tour
STEEPNESS - [gen] raideur *f*, escarpement *m*, rapidité *f*
~OF A CURVE - [geom] degré *m* d'inclinaison d'une courbe
~OF A GRADIENT - [gen] rapidité *f* d'une pente
STEER, to - [gen] gouverner
[auto] conduire, mener
[naut] diriger, manoeuvrer
[radio] (the operation of altering the direction of maximum sensitivity of a directional aerial) orienter
STEER - [zool] jeune boeuf *m*, bouveau *m*
[zool] (USA only) toreau *m*
~CLEAR OF, to - [gen] s'écarter de
STEERABLE AERIAL - [radio] (directional aerial whose major lobe can be readily shifted in position) antenne *f* orientable
~ANTENNA - s. steerable aerial
~SHOCK ABSORBER STRUT - [aero] (a strut in a landing gear comprising a shock-absorber system and an arrangement for steering) amortisseur *m* orientable
STEERAGE - [gen] direction *f*
[naut] manoeuvre *f* de la barre
[naut] (section of a vessel providing for cheap accommodation) emménagements *mpl* (pour émigrants etc), entrepont *m*
~PASSENGER - [naut] passager *m* de troisième classe
~-WAY - [naut] (sufficient movement of a vessel to enable it to answer the helm) erre *f* pour gouverner
STEERING - [gen] direction *f*
[naut] timonerie *f*, manoeuvre *f* de la barre
[auto] conduite *f*
~AND THIRD ARM - [auto] levier *m* d'accouplement des roues
~ANGLE - [auto] angle *m* de braquage
~ARM - [auto] (attached to the stub-axle) bielle *f* pendante
~ARM AND SWIVEL - [auto] levier *m* d'attaque de la fusée
~BOOSTER - [auto] (power-assisted steering) servomécanisme *m* de direction
~BOX - [auto] (the housing enclosing the steering-gear and providing an oil bath for the working surfaces) boîtier *m* de direction
~CAM - [auto] vis *f* de direction
~CLUTCH - [mech] (of a tractor etc) embrayage *m* de direction
~COLUMN - [auto] colonne *f* de direction
~COLUMN TUBE - [auto] jupe *f* de colonne de direction
~COMPARTMENT - [naut] timonerie *f*
~COMPASS - [instr] (a nautical instrument) compas *m* de route
~CONNECTING ROD - [auto] barre *f* d'accouplement de direction
~CYLINDER - [aero] (hydraulic cylinder used to actuate a landing gear steering system) cylindre *m* de direction à commande hydraulique
~DAMPER - [auto] (spring buffer) amortisseur *m* de direction
~DROP ARM - [auto] bielle *f* pendante

STEERING ENGINE - [naut] servo-moteur *m* du gouvernail
~GEAR - [auto] (the two geared members attached to the steering-column and the drop-arm spindle) boîtier *m* de direction, timonerie *f*, mécanisme *m* de direction
[naut] appareil *m* à gouverner
~-GEAR ARM - [auto] levier *m* de direction, bielle *f* pendante
~-GEAR BOX - s. steering gear housing
~-GEAR HOUSING - [auto] boîtier *m* de direction
~HEAD - [mech] (of a motorbicycle) tête *f* de fourche
[auto] tête *f* d'essieu avant
~JACK - s. steering cylinder
~KNUCKLE - [auto] porte-fusée *m*
~KNUCKLE ARM - s. steering kunckle
~-KNUCKLE PIN - [auto] axe *m* de fusée, pivot *m* de fusée, rotule *f* de direction
~-KNUCKLE PIVOT - [auto] pivot *m* de l'essieu avant
~LEVER - [auto] bielle *f* pendante
[mech] levier *m* de commande
~LINKAGE - [auto] timonerie *f* de direction
~LOCK - [auto] braquage *m* de la direction, serrure *f* de blocage
~MAST - [auto] colonne *f* de direction
~NUT - [auto] écrou *m* de direction
~PILLAR - s. steering column
~PIVOT - [auto] pivot *m* de direction
~PIVOT PIN - [auto] pivot *m* de fusée
~PROGRAM - [comput] (a program designed to process and control other programs) programme *m* directeur
~RATIO - [auto] rapport *m* de démultiplication de la direction
~REPEATER - [instr] (a nautical instrument) répétiteur *m* de route
~ROD - [auto] (the drag link) barre *f* longitudinale de commande de direction
~ROUTINE - s. steering program
~SCREW - [auto] vis *f* de direction
~SECTOR - [auto] secteur *m* de direction à vis
~SHAFT - [auto] axe *m* de direction, colonne *f* de direction
~STOP - [auto] butée *f* de direction
~STUB AXLE - [auto] porte-fusée *m*
~SWIVEL PIN - s. steering pivot pin
~TIE ROD - [auto] barre *f* d'accouplement de direction
~TRACK ROD - s. steering tie rod
~WANDER - [auto] (a defect) tendance *f* de la direction à tirer latéralement
~WHEEL - [auto] (the spoked handwheel attached to the top of the inner steering-column) volant *m* de direction
[naut] roue *f* du gouvernail
~-WHEEL HORN-CONTROL RING - [auto] couronne *f* de commande de l'avertisseur
~WHEEL PLAY - [auto] jeu *m* du volant
~WHEEL SPIDER - [auto] moyeu *m* du volant, rayons *mpl* du volant
~WORM - [auto] vis *f* de direction
~WORM GEAR - [auto] direction *f* à vis
~WORM SECTOR - [auto] secteur *m* de direction à vis
~WORK SECTOR SHAFT - [auto] axe *m* de secteur de direction à vis
~-YOKE SECTOR - [auto] axe *m* de fusée

STEERSMAN - [naut] timonier *m*
STEEVING - [naut] estivage *m*
STEFAN BOLTZMANN LAW - [phys] (a law which states that the total radiation from a black body is proportional to the fourth power of the absolute temperature) loi *f* de Stefan-Boltzmann
STEFFENS PROCESS - [ind chem] (method of recovery of sugar from beet molasses, by dilution with water and addition of lime, carried out at low temperatures of the order of I0 deg. C.) procédé *m* Steffens
STEGNOSIS - s. stenosis
STEINERT'S DISEASE - [med] myotonie *f* atrophique, maladie *f* de Steinert
STEINING - [constr] (the operation of lining a well with bricks, stone, timber or metal, to prevent the sides from caving in) revêtement *m* intérieur
STEINMETZ COEFFICIENT - [el] (coefficient by which the I.6th power of the flux density must be multiplied in order to give the hysteresis loss in ergs per cycle, when a sample of iron is taken through successive cycles of magnetization) coefficient *m* de Steinmetz
STELE - [bot] (the central region of the stem or root) stèle *f* cylindre *m* central
[arch] stèle *f*
STELLAR - [astr] (of the stars) stellaire
~ENERGY - [phys] (the energy originated by the stars) énergie *f* stellaire
~GENERATOR - [phys] (used in plasma physics) stellarator *m*
~GUIDANCE - [astronaut] guidage *m* stellaire
~INERTIAL GUIDANCE - [astronaut] équipement *m* inertiel stellaire, guidage *m* inertiel stellaire
~INTERFEROMETER - [instr] (a device by which it is possible to measure the angular diameter of certain giant stars by observation of interference fringes at the focus of the telescope) interféromètre *m* stellaire
~MAGNITUDE - [astr] magnitude *f* stellaire
~RADIATION - [astr] rayonnement *m* stellaire
STELLARATOR - s. stellar generator
STELLATE - [zool & bot] (radiating from the centre) étoilé, radié
~BLOCK - [med] infiltration *f* stellaire
STELLECTOMY - [med] stellectomie *f*
STELLITE - [metall] (series of alloys containing cobalt, chromium, tungsten and molybdenum in various proportions) stellite *f*
STELLITED - [metall] garni de stellite
~VALVE - [mech] (in internal combustion engines; poppet valves of high-duty petrol engines having facings of stellite so as to resist wear and corrosion) soupape *f* garni de stellite
ST. ELMO'S FIRE - [met] (a glow, appearing at the end of masts etc. in stormy weather and caused by electric discharge) décharge *f* en aigrettes, effluve *f* électrique
STEM, to - [gen] arrêter, contenir
[hydr] (to provide with a dam) endiguer
[naut] lutter contre la marée
[naut] (to move against the current) remonter (le courant)
[min] bourrer (un trou de mine)
[mining] bourrer
STEM - [bot] tige *f*, queue *f*
[naut] étrave *f*, avant *m*
[mech] (of a valve) queue *f* (de soupape), arbre *m*
[el] (the element holding the filament) broche *f*

STEM - [mech] (cylindrical prolongation of a part, usually for guidance) tige *f*, broche *f*
[print] (of a type) plein *m*, jambage *m*
[instr] (the metal pin on which the compass rose and the needle rest) pivot *m*
[electron] (part of the envelope of an electronic tube) embase *f*, pied *m*
[hydr] (of a sluice-gate) épée *f*
~CHAPLET - [metall] support *m* de noyau à tige
~CORRECTION - [instr] correction *f* de la tige d'un thermomètre
~FIBRE - [text] fibre *f* d'écorce
~GEAR - [mech] engrenage *m* avec tige
~OF JACK - [text] tige *f* du sélecteur
~POST - [shipbuild] (the portion of material forming the extreme forward end of a ship) étrave *f*
~RADIATION - [radiat] (X-rays given off from parts of the anode other than the target) rayonnement *m* extrafocal
~ROT - [bot] maladie *f* sclérotique du trèfle
~RUST - [bot] (a plant disease) rouille *f* noire
~THE CURRENT, to - [naut] étaler le courant
~-TYPE THERMOMETER - [instr] thermomètre *m* à tige
STEMMED PLANT - [bot] plante *f* à tige, plante *f* caulescente
STEMMER - [tool] (used in mining) bourroir *m*
STEMMING - [min] (the stopping material used to tamp the explosive in a shot-hole) bourrage *m*
[mining] bourrage *m*
STEMPLE - [min] étai *m*, butte *f*
STEMPOST - s. stem post
STEMSON - [shipbuild] marsouin *m* avant
STEN-GUN - [firearm] fusil-mitrailleur *m*
STENCH - [gen] puanteur *f*, odeur *f* infecte
STENCH-TRAP - [hydr] (in sanitary engineering) siphon *m*
STENCIL, to - [print] (to reproduce by stencils) imprimer au poncif, poncer
[paint] marquer (un ballot etc)
STENCIL - [gen] (thin sheet or plate in which a pattern is cut by means of spaces or dots, through which paint or ink is applied) pochoir *m*, poncif *m* cliché *m*
[gen] (the pattern obtained) tracé *m*, peinture *f* au pochoir
~-BRUSH - [print] brosse *f* à caractères à jour
~LETTERS AND FIGURES - [print] lettres *f*pl et chiffres à jour
STENCILLER - [gen] peintre *m* au pochoir
STENCILLING - [gen] peinture *f* au pochoir
STENCILMAN - s. stenciller
STENOGRAPHER - [gen] sténographe *m*
STENOGRAPHY - [gen] sténographie *f*
STENOPAEIC - [photo] (term applied to pin-hole photography) sténopéique
STENOSIS - [med] (constriction of a duct, an orifice or tubular passage as a result of a disease) sténose *f*
STENOTHERMY - [ecology] (the tolerance of a narrow range of temperatures only) sténothermie *f*
STENOTYPIST - [gen] (short-hand typist) sténodactylo *m* & *f*
STENTER - [text] (a frame on which fabrics are stretched between hooks, or by rollers) rame *f*, rameuse *f*
STENTON - [mining] recoupe *f*
STEP, to - [gen] aller, marcher
STEP,to - [meas] mesurer au pas
(to arrange in a series of steps or degrees) disposer en échelons
[naut] (to insert the lower end of a mast in a socket) arborer (un mât)
STEP - [gen] (the progressive motion, a pace) pas *m*
[meas] (the distance passed over, as in kinematics) mesure *f*
[constr] marche, échelon *m*
[aero] (an abrupt break in the undersurface of a seaplane float or amphibial hull, to improve take-off performance) redan *m*, retrait *m* (de flotteur)
[naut] (the socket into which the mast is inserted) collet *m* de pied de mât
[astronaut] (a stage) étage *m*
[radar] (a bright spot or movable range marker generated by a strobe generator under the control of some variable delay pulse) marque *f* stroboscopique
[nucl] (an off-set in the side of a hole through a shield or in the corresponding plug or other closure, so that when the hole is closed the mating steps will form a zig-zag joint) marche *f*
[railw] (footboard tailboard) marchepied *m*
[mus] intervalle *m*
[med] phase *f*
[mech] étage *m* de cône-poulie, gradin *m*
[carp] (of a ladder) barreau *m* d'une échelle, échelon *m*
~ACTION - [contr] action *f* par échelons
~AHEAD, to - [gen & mech] avancer
~-AND-REPEAT CAMERA - [photo] caméra *f* à répétition
~-AND-REPEAT CONTACT PRINTING MACHINE - s. step-and-repeat machine
~-AND-REPEAT MACHINE - [print] (a photocomposing machine) machine *f* à copie en répétition
~-BAR - [metall] éprouvette *f* en gradins
~BEARING - [mech] (of a vertical shaft) palier *m* de pied, crapaudine *f*
[constr] crapaudine *f*
~BIT - [mech] trépan *m* à redans
~BOLT - [mech] boulon *m* à épaulement
~BOX - s. step bearing
~BRACKET - [railw] (footboard bracket) support *m* de marchepied
~-BY-STEP CONTACT PRINTER - [cin] (intermittent contact printer) tireuse *f* simple par contact
~-BY-STEP EXCITATION - [phys] (the successive transition of an atom or a molecule by stages to higher levels of excitation, the last stage is ionization) excitation *f* par degrés
~-BY-STEP METHOD - [el] (a method of determining the hysteresis curve of a magnetic material, in which the field strength is increased and reversed in steps) méthode *f* pas à pas
~-BY-STEP MOTION - [cin & telev] défilement *m* saccadé
~-BY-STEP MOTOR - [el] moteur *m* pas à pas
~-BY-STEP OPTICAL PRINTER - [cin] tireuse *f* optique simple
~-BY-STEP SWITCH - [telev] mécanisme *m* de transport saccadé
~CHANGE - [contr] saut *m* du signal
~-CONE - [mech] cône-poulie *m*, cône *m* de transmission, poulie *f* à gradins
~-CONE DRIVE - [mech] commande *f* par cône

STEP COVER - [constr] couvre-marche *f*
~COVER STRIP - [constr] renforcement *m* métallique du nez
~CUT - [min] (the dressing of precious stones with long flat facets, which are slightly inclined one above the other) taille *f* à gradins
~-CUT RING - [mech] (in internal combustion engines) segment *m* à gradin
~DOWN, to - [el] (to change a high-voltage supply into a low-voltage supply) reduire la tension
~-DOWN - [gen] réducteur *m*
~-DOWN AMPLIFIER - [radio] (an inverted amplifier) amplificateur *m* de seuil à anode negative
~-DOWN GEAR - [mech] engrenage *m* de réduction
~-DOWN TRANSFORMER - [el] (transformer for changing high-voltage supply into low-voltage supply) transformateur *m* dévolteur
[telecomm] (an audio-frequency transformer which couples a circuit of high-impedance level to one of low-impedance level) transformateur *m* abaisseur, transformateur *m* réducteur
~-DOWN WHEELS - [mech] engrenages *mpl* réducteurs
~-FACED OVERHAND STOPES - [mining] gradins *mpl* renversés
~FAULT - [geol] (a series of normal faults having a parallel arrangement, so that they progressively step down a particular bed) faille *f* en gradins, faille *f* en escalier
~FLAG - [constr] grande dalle *f* en pierre
~FOLD - [geol] pli *m* monoclinal
~-FORCED RESPONSE - [contr] réponse *f* de pas
~FUNCTION - [el] (a function which is zero for all the time preceding a certain instant and has constant finite value thereafter) fonction *f* de pas
~-FUNCTION RESPONSE - [contr] réponse *f* de fonction de pas
~-FUNCTION RESPONSE CHARACTERISTIC - [el] (the relation to time of the response to an excitation which is a step-function of time) caractéristique *f* de réponse de fonction de pas
~GATING - [metall] attaque *f* de réchauffage
~GRATE - [th eng] grille *f* à gradins, grille *f* à étages
~-GRATE PRODUCER - [gas ind] gazogène *m* à grille à gradins
~HANGER - [auto] marche-pied *m*
~-HARDENING - [metall] trempe *f* étagée
~IRON - [constr] grappin *m*
~JOINT - [mech] (of a packing ring) joint *m* à recouvrement, assemblage *m* à recouvrement
~-LADDER - [impl] échelle *f* double, marchepied *m*
~LIGHT - [auto] éclairage *m* de seuil de porte
~MELTING - [metall] (in foundry work) fusion *f* à gradins
~OF CONE PULLEY - [mech] gradin *m* de cône-poulie
~OF FIELD SHUNTING - [el] cran *m* de shuntage réduit
~PAD - [auto] garniture *f* de seuil
~PENETROMETER - [instr] (a penetrometer of similar material to the specimen under examination, having steps ranging usually from 1 to 5 percent of the specimen thickness) pénétramètre *m* comparatif
~PLATE - [text] levier *m* à crans, disque *m* à gradins
~PULLEY - [mech] poulie *f* à étages
~QUENCHING - [metall] trempe *f* étagée martensitique

STEP RACK - s. step plate
~-RAIL - [railw] rail *m* de marchepied
[text] rail *m* porte-broches
~-RATE - [comm] (for the sale of gas or electricity) tarif *m* à échelons
~-RATE TARIFF - s. step-rate
~RESISTANCE - [el] résistance *f* graduée
~RESPONSE - [radio] (the waveform response of a transducer to a step-function input at time zero) réponse *f* de pas
~-SALE PREPAYMENT METER - [comm] compteur *m* à prépaiement pour tarif dégressif
~SWITCH - [el] commutateur *m* pas à pas
~TAP CHANGER - [el] commutateur *m* à gradins
~-TIME - [electron] (the time needed in a decade scaler tube to move the electron beam from one position to the next one) durée *f* du pas
~TWIST - [electron] (in waveguides) torsade *f* binomiale
~-UP, to - [el] (to change a low-voltage supply into a high-voltage supply) élever la tension
~-UP - [gen] survolteur *m*, élévateur *m* de tension
[mech] multiplication *f* d'un engrenage)
~-UP CURE - [rubber ind] (vulcanization of rubber carried out in stages) vulcanisation *f* échelonnée
~-UP GEAR - [mech] engrenage *m* multiplicateur
~-UP RATIO OF A SIGNAL - [radio] accroissement *m* d'oscillation du signal
~-UP TRANSFORMER - [el] (transformer for changing low-voltage supply into high-voltage supply) transformateur *m* survolteur
[telecomm] (audio-frequency transformer coupling an impedance to a higher impedance) transformateur *m* élévateur
~-UP WHEELS - [mech] engrenages *mpl* multiplicateurs
~-VEIN - [mining] filon *m* en gradins
~VOLTAGE REGULATOR - [el] régulateur *m* de tension échelonné
~-WEDGE - [radiat] (block of material in the form of a series of steps used to compare the radiographic effect of X-rays under various conditions) coin *m* gradué, coin *m* à degrés
~WHEEL - [mech] poire *f*
STEPHANITE - [min] (natural silver antimony sulphide, it usually occurs in association with other silver ores) stéphanite *f*, psaturose *f*
STEPLADDER - [impl] échelle *f* double, escabeau *m*, marchepied *m*
STEPLESS - [gen & mech] (operating with a continuous variation or motion, not in stages) sans gradins, sans étages, progressif
STEPPE - [geogr] (vast plain devoid of forests) steppe *f*
~PLANT - [bot] plante *f* des steppes
STEPPED - [gen mech etc] à gradins, en gradins, échelonné, en échelons
~ABUTMENT - [constr] culée *f* en escalier, culée *f* en gradins
~ANNEAL - [metall] recuit *m* échelonné
~CURVE DISTANCE-TIME PROTECTION - [el] dispositif *m* de protection de distance à caractéristique discontinue
~DRIVING PULLEY - [mech] (a pulley consisting of a series of pulleys of different, gradually increasing diameters formed as a single unit, to give different speed ratios by transferring a belt from one step to

another) cône *m* de transmission, poulie *f* de commande à gradins étagés
STEPPED EXTRUSION - [metall] extrusion *f* à pas de mèlerin
~FISSURE - [metall] fente *f* en échelons
~FOUNDATION - [constr] fondation *f* en gradins
~GABLE - [constr] pignon *m* à redans
~GEAR - [mech] engrenage *m* échelonné, engrenage *m* à denture croisée
~-GEAR GENERATOR - s. step-gear generator
~IRON CORE - [el] noyau *m* de fer échelonné
~JOINT - [metall] joint *m* décroché
~JOINT CHECKED CUT - [mech] (of a piston ring) coupe *f* crantée de segment
~LABYRINTH SEAL - [nucl] garniture *f* en labyrinthe échelonnée
~LEADER STROKE - [el] prédécharge *f* échelonnée
~LODE - [mining] filon *m* en gradins
~PISTON RING - [mech] (in internal combustion engines) segment *m* à extrémités à recouvrement
~ROLL - [metall] cylindre *m* à gradins
~ROOF - [constr] toit *m* en escalier, toit *m* en gradins
~RUNNER - [metall] canal *m* d'alimentation en cascade
~SPRUE - s. stepped runner
~TRANSFORMER - [electron] transformateur *m* à prises
~TWILL - [text] sergé *m* en escalier
~WINDING - [el] (or split winding; a two-position diamond winding where the coil sections which occupy the top position in one slot, occupy the bottom position in two other slots) enroulement *m* à pas brouillé
STEPPING - [constr] (the laying of foundations in horizontal steps on a sloping ground) fondation *f* en gradins
[gen] marche *f*, échelonnement *m*
[mining] exploitation *f* en gradins renversés
~BLOCK - [constr] échantignol *m*
~CONTROL - [contr] (control method operating in discrete steps) commande *f* pas-à-pas
~COUNTER - [comput] compteur *m* à pas à pas
~MECHANISM - [comput] mécanisme *m* à pas à pas
~MOTOR - [el] moteur *m* pas-à-pas
~RELAY - [contr] relais *m* pas-à-pas
~SWITCH - [comput] commutateur *m* à pas à pas
STERADIAN - [math] (the unit of solid angular measure) stéradiane *f*
STERCOLITH - [med] (hard faecal concretion in the intestine) coprolithe *m*
STERCORACEOUS - [med] (pertaining to or consisting of faeces) stercoraire, stercoral
STERCOREMIA - [med] stercorémie *f*, coprémie *f*
STERCOROLITH - s. stercolith
STERE - [meas] stère *m* (I mètre cube)
STEREO - s. stereoscopic - s. stereotype
~ATTACHMENT - [photo] duplicateur *m* stéréographique
~CASTER - [print] moule *m* à clichér, fondeuse *f*
~EFFECT - [photo] effet *m* de relief
~MASK - [photo] cache *m* stéréoscopique
~-MIXER - [el acoust] pupitre *m* de mélange stéréophonique
~MOUNT - [print] lingots *mpl* pour planches stéréotypiques
~-POWER - [opt] (for stereo systems, e.g. prism binoculars, the ratio of the distance between the objective axes to the distance between eyepiece axes multiplied by the magnifying power) pouvoir *m* stéréoscopique
STEREO-RECORDED TAPE - [el acoust] bande *f* à enregistrement stéréophonique
~REFLEX CAMERA - [photo] chambre *f* reflex stéréophotographique
~RUBBER - [print] caoutchouc *m* à graver, caoutchouc *m* pour clichés
~-TAPE RECORDER - [el acoust] appareil *m* d'enregistrement stéréophonique
STEREOAUTOGRAPH - [surv] stéréoautographe *m*
STEREOBATE - [constr] (a substructure or solid platform without columns) mur *m* de fondation
STEREOBLOCK POLYMER - [chem] (a polymer in which stereospecific structural sections are separated by sections of different structure) polymère *m* en blocs isotactiques
STEREOCAMERA - [photo] chambre *f* stéréophotographique
STEREOCARDIOGRAPHY - [med] vectographie *f* spatiale
STEREOCHEMISTRY - [chem] (the study of the spatiale arrangements of the atoms in a molecule) stéréochimie *f*
STEREOCHROMY - [paint] (a mural painting process) stéréochromie *f*
STEREOCOMPARATOR - [instr] (instrument used for topographic measurements) stéréocomparateur *m*
STEREOFLUOROSCOPY - [radiat] (fluoroscopic technique making it possible for screen images to be viewed three-dimensionally) stéréoradioscopie *f*
STEREOGNOSIS - [med] (the ability to recognize similarities and differences of size, weight etc of objects brought into contact with the surface of the body) sens *m* stéréognostique, stéréognosie *f*
STEREOGNOSTIC SENSE - s. stereognosis
STEREOGRAM - [opt] (a diagram or picture giving the impression of a solid in relief) stéréogramme *m*
STEREOGRAPHIC PROJECTION - [surv] (a method of zenithal projection) projection *f* stéréographique
STEREOGRAPHY - [geom] (the drawing on a flat surface of lines representing solids) stéréographie *f*
STEREOISOMERS - [chem] (of identical atomic and molecular compounds structure but different spatial arrangement) stéréoisomères *mpl*
STEREOISOMERISM - [chem] (the existence of different substances the molecules of which possess an identical structure but different arrangements of their atoms in space) stéréoisomérisme
STEREOMETER - [instr] (a volumenometer) stéréomètre *m*
STEREOMETRY - [geom] (the art of measuring the volume and other spatial elements of solids) stéréométrie *f*
STEREOMICROSCOPE - [instr] microscope *m* stéréoscopique
STEREOPHONIC - [acoust] (said of a sound reproduced in such way that the illusion of auditory perspective is realized) stéréophonique
STEREOPHONIC RADIOTELEPHONY - [radio] (system employing two complete communication channels, so disposed as to give a three-dimensional effect to listeners) radiotéléphonie *f* stéréophonique
~SOUND SYSTEM - [el acoust] (sound system in which a plurality of microphones, transmission

channels and loudspeakers are so arranged as to provide listeners with a sensation of the spatial distribution of the sound sources) système *m* stéréophonique
STEREOPHONY - [acoust] stéréophonie *f*
STEREOPHOTOGRAMMETRY - [surv] stéréophotogrammétrie *f*
STEREOPHOTOGRAPHY - [photo] photographie *f* stéréoscopique
STEREOPLANIGRAPH - [instr] (an optical instrument) stéréoplanigraphe *m*
STEREOPLASM - [zool] (the viscous part of protoplasm) stéréoplasma *m*
STEREOPLOTTER - [photo] stéréorestituteur *m*
STEREOPSIS - [opt] vision *f* stéréoscopique
STEREOPTICON - [photo] (a double-projection magic lantern arranged for fading on image into the next) appareil *m* double, appareil *m* à projection double
STEREORADIOGRAPH - [radiol] stéréoradiographie *f*
STEREORANGEFINDER - [instr] stéréotélémètre *m*
STEREOREGULAR POLYMERS - [chem] (polymers in which substituting groups are arranged all above, all below or all on the same side as the backbone chain) polymères *mpl* stéréospécifiques
STEREOSCOPE - [instr] (an apparatus for realizing the reproduction of views in three dimensions) stéréoscope *m*
STEREOSCOPIC - [opt] stéréoscopique
~BINOCULAR LENS - [opt] jumelles *fpl* stéréoscopiques
~CAMERA - s. stereocamera
~FILM - [cin] (a film giving the impression of depth) film *m* en relief
~LENS - [opt] objectif *m* stéréoscopique
~PLOTTING - [photo] stéréorestitution *f*
~RADIUS - [opt] distance *f* stéréoscopique
~RANGE FINDER - [instr] (range finder which gives a stereoscopic image) télémètre *m* stéréoscopique
~RECORDER - [acoust] enregistreur *m* binauriculaire
~SLIDE - [photo] diapositive *f* pour la stéréoscopie
~TELEVISION - [telev] (television pictures in three dimensions) stéréotélévision *f*
~VISION - [opt] vision *f* stéréoscopique
STEREOSPECIFIC CATALYSTS - [chem] catalyseurs *mpl* stéréospécifiques
~POLYMER - [chem] (a polymer with a specific spatial arrangement of its molecular constituents) polymère *m* stéréospécifique
STEREOSPECTROGRAM - [phys] (in spectrology) stéréospectrogramme *m*
STEREOTAXIC BRAIN OPERATION - [med] opération *f* stéréotaxique du cerveau
STEREOTAXIS - [biol] (response of an organism to the stimulus of contact with a solid body) stéréotaxie *f*
STEREOTELEMETER - [instr] (binocular telescope used as a range-finder) stéréotélémètre *m*
STEREOTROPISM - s. stereotaxis
STEREOTYPE, to - [print] (to cast a plate in a type metal from a mould or a matrix) stéréotyper, clicher
STEREOTYPE - [print] (a plate taken in type metal from a matrix, e.g paper; reproducing the surface from which the matrix was made) cliché *m*
~PLATE - [print] cliché *m*, stéréotype *m*
~PRINTING - [print] stéréotypie *f*
STEREOTYPY - [print] (the art of making stereotypes) stéréotypie *f*
STERIC FACTOR - [phys] (or probability factor; correction factor applied in case of slow reactions, to correct the values calculated for the reaction rates by the collision theory) facteur *m* de probabilité
~HINDRANCE - [chem] (spatial arrangement of atoms in a molecule which retards or prevents reaction with another molecule) obstacle *m* stérique
STERILE - [chem] (free from micro-organisms or moulds) stérile, aseptique
[biol] (having no reproductive power) stérile
[agric] (of the soil; lacking fertility) stérile
[bot] (of a plant) (plante) scarpe
~AIR - [chem] (air which has been freed from micro-organisms, e.g. for use in submerged aerobic fermentation) air *m* stérile
~BREED OF SHEEP - [zool] race *f* ovine stérile
~CELL - [bot] cellule *f* stérile
~SEED - [bot] (hemp) chènevis *f* stérile
STERILITY - [biol] stérilité *f*
STERILIZATION - [gen] (the act of sterilizing) stérilisation *f*
~BY IRRADIATION - [radiat] stérilisation *f* par irradiation
~TOWER - [ind chem] tour *f* de stérilisation, tour *f* de contact
STERILIZE, to - [gen & med] (to deprive of productive or reproductive power) stériliser
[chem] (to destroy bacteria; to free from germs) stériliser
STERILIZER - [instr] stérilisateur *m*
[ind chem] étuve *f* à stérilisation, autoclave *m*
STERILIZING AGENT - [ind chem] agent *m* de stérilisation
~EFFECT - [chem] effet *m* bactéricide
~FILTER - [hydr] filtre *m* stérilisateur
~TOWER - s, sterilization tower
STERLING - [fin] (the official standard of fineness for British coins) d'aloi
[fin] (the legal currency) livre *f* sterling
STERLINGITE - [min] (zincite) zincite *f*
STERN - [naut] arrière *m*
[aero] arrière *m*
~-BUSH - [shipbuild] tube *m* d'étambot
~END - [naut] arrière *m*
~-FAST - [naut] amarre *f* arrière
~-FRAME - [naut] casse *f*
~-FRAMING - s, stern frame
~-GLAND - [naut] presse-étoupe *m* arrière
~-HEAVY - [naut] à queue lourde
~HOOD - [naut] capot *m* d'habitacle
~KNEE - s, sternson
~LIGHT - [naut] feu *m* d'arrière
~LINE - [naut] amarre *f* arrière, croupière *f*
~-POST - [aero] (a single vertical member at the after end of a fuselage (also applicable to a seaplane float or amphibian hull) étambot *m*
[naut] étambot *m*
~-RAIL - [naut] garde-corps *m* arrière
~TUBE - [naut] (tube supporting the after part of the propeller shaft) tube *m* d'étambot
~-TUBE BULKHEAD - [shipbuild] cloison *f* arrière
~-WAY - [naut] culée *f*, marche *f* arrière
~-WHEEL - [naut] roue *f* arrière
~-WHEELER - [naut] vapeur *m* à roue arrière
STERNPOST - [shipbuild] étambot *m* arrière

STERNSON - [shipbuild] (inner sternpost attached to the centre keelson, so as to strengthen the stern frame) marsouin *m* arrière
STERNUM - [anat] (the breast bone of the Vertebrates) sternum *m*
STERNUTATION - [med] (the act of sneezing) sternutation *f*, éternûment *m*
STERNWAY - [naut] acculée *f*, marche *f* arrière
STEROIDS -[chem] (a large group of compounds with a common polycyclic ring structure with important uses in medicine) dérivés *mpl* stéraniques
STEROLS - [chem] (naturally-occurring secondary alcohols, many of which have important metabolic roles) stérols *mpl*
STET, to - [print] (let it stand; direction used in proof reading) maintenir (un mot etc)
STET - [print] (in proofreading; a direction indicating that a correction must be disregarded) à maintenir, bon
STHETOSCOPE - [instr] (apparatus designed for listening to sounds produced in the body) stéthoscope *m*
STEVEDORE, to - [naut] arrimer
STEVEDORE - [naut] arrimeur *m*, entrepreneur *m* d'arrimage
STEVENSON SCREEN - [instr] (a box-like structure of which the sides consist of downwardly inclined vanes, to contain group meteorological instruments and protect them from direct action of atmospheric conditions) grille *f* de Stevenson
STEW, to - [gen] faire cuire à la casserole
[brew ind] faner
STEW - [telecomm] (colloq for an undesired sound in recording) friture *f*
[gen] ragoût *m*
STEWARD - [aero] (a man carried by an airliner to attend upon passengers) steward *m*
[gen] intendant *m*, économe *m*
[naut] steward *m*, garçon *m*
STEWARDESS - [aero] (a woman carried by an airliner to attend upon passengers) hôtesse *f* de l'air
[naut] femme *f* de chambre
STEWING - [min] mijotage *m*
ST-FR - [met] (abbreviation of fractostratus; cloud formation consisting of isolated masses of stratus) fractostratus *m*
STHENIC - [med] sthénique
STHENOSIZE ARTIFICIAL SILK, to - [text] (to increase the strength) sthénoser la soie artificielle
STHENOSIZED SILK - [text] soie *f* sthénosée
STIBIALISM - [med] (antimony poisoning) intoxication *f* stibiée, stibialisme
STIBIATION - [med] impregnation *f* stibiée
STIBINE - [chem] (synonym for antimony hydride, a poisonous gas, less stable than arsine) stibine *f*, antimoine *m* sulfuré
STIBIUM - [chem] (synonym of antimony) antimoine *m*
STIBNITE - [min] (natural antimony trisulphide crystallizing in the orthorhombic system, in grey prism of metallic lustre. It often carries gold and silver values and is a very important ore of antimony) stibnite *f*
STIBOPHEN - [chem] (a colourless, crystalline powder used in the treatment of schistosomiasis) stibophène *m*
STICK, to - [gen] (to cause to pierce or enter) piquer, enfoncer
STICK, to - [gen] (to fix in place by inserting) fixer, planter
[gen] (to cause to cleave by some adhesive) coller, adhérer
[carp] (e.g. a bead on a joint) pousser
[mech] (to jam) coincer, gommer
[mech] (said of piston rings, of valves etc) se coller, rester coller
STICK - [gen] bâton *m*, canne *f*
[gen] (of an umbrella) manche *m*, canne *f*
[carp] pièce *f* de bois
[el] baguette *f* de charbon
[metall] (of solder) bâton *m* de soudure
[aero] (control lever) levier *m* de commande
[print] (a composing tool) composteur *m*
[mech] (of a honing machine) patin *m*
[mus] (the wooden part of the bow) baguette *f*, bois *m*
[chem] bâton *m* (de soufre), canon *m* de soufre, crayon *m* de potasse
~ACNE - [med] acné *f* bouchée, comédon *m*
~BRUSH - [impl] pinceau *m* à tige
~CIRCUIT - [el] circuit *m* d'auto-collage, circuit *m* d'auto-maintien
~FAST, to - [mech] bloquer, se coller
~INSULATOR - [el] isolateur *m* à tige
~LAC - [paint] (raw shellac, as taken from the tree on which it is deposited by insects) laque *f* en bâton
~ON, to - [gen] coller
~PAD - [rubber ind] bout *m* de canne
~PLANE - [impl] rabot *m* à tige
~SULPHUR - [min] soufre *m* en canon
STICKER - [metall] (in foundry work, a surface defect due to the adhesion of scale) adhérence *f*
[gen] (US only; a sticking label) affiche *f*
[impl] couteau *m* de boucher
[fishing] gaffe *f*, harpon *m*
STICKFUL - [print] paquet *m*
STICKING - [gen] adhésif, collant
[telev] (image retention) rémanence *f* d'image
[gen] adhérence *f*, collement *m*, collage *m*
[mech] coincement *m*
~OF VALVES - [mech] gommage *m* des soupapes
~PLASTER - [med] taffetas *m* gommé, emplâtre *m* résineux
~POTENTIAL - [electron] (the limiting value of screen-to-cathode potential on a cathode-ray tube with a non conducting backing for the phosphor) tension *f* de blocage
~PROBABILITY - [nucl] (in a nuclear reaction) probabilité *f* d'adhérence
~TAPE - [gen] (adhesive tape) ruban *m* adhésif
~VOLTAGE - [electron] (of a luminescent screen; the voltage applied to the electron beam below which the rate of secondary emission from the screen is less than unity. The screen then has a negative charge which repels the primary electrons) tension *f* de blocage
STICKY - [gen] gluant, visqueux, collant
~HEALD - [text] lisse *f* visqueuse
~LIQUID - [gen] liquide *m* gluant
~WOOL - [text] laine *f* visqueuse
STIFF - [gen] rigide, raide
[mech] dur
[soil] (tightly packed) tenace

STIFF - [med] (of joint etc) ankylosé
[naut] (of a rope) (filin) engourdi
[met] (of a breeze, a wind etc) fort, carabiné
[phys] (of a flow) rigide, lent
[chem] (of oil etc) consistant, épais
~BRUSH - [impl] brosse *f* dure
~CLAY - [min] argile *f* tenace
~FIBRE - [text] fibre *f* raide
~FLOW - [phys] (flow under conditions of high viscosity) grande dureté *f*
~HAIR - [text] poil *m* raide
~-PAPER BINDING - [bookbind] reliure *f* en carton, (volume) à couverture rigide
~SLAG - [metall] scorie *f* pâteuse
~YARN - [text] fil *m* raide
STIFFEN, to - [gen & mech] (to render more rigid) raidir, renforcer
[constr] (to make tightly packed) rendre ferme, se figer
[naut] (of the wind) se carabiner
STIFFENED SUSPENSION BRIDGE - [constr] pont *m* suspendu renforcé
STIFFENER - [mech] (any element in a structure designed to give additional rigidity) pièce *f* de renfort
[aero] raidisseur *m*
[mech] raidisseur *m*
[metall] (tôle de) fourrure *f*
STIFFENING - [gen] raidissement, renforcement *m*
[constr] consolidation *f*
[mech] durcissement *m*
[text] apprêt *m*
[carp] liernes *f*pl, étrésillons *m*pl
[naut] lest *m* de stabilité
~ANGLE - [mech] (a rolled or extruded member of angular section, used to give stiffness to a structure) cornière *f* de raidissement
~ARCH - [constr] arc *m* de renforcement
~BEAD - [mech] (on sheet metal panels etc) nervure *f* de renfort
~GIRDER - [constr] traverse *f* raidisseuse, poutre *f* intermédiaire, poutre *f* de renfort
~PIECE - [mech] (piece added to a structure to increase rigidity)plaque *f* de renfort
~PLATE - [mech] tôle *f* de renfort, renfort *m*
~RIB - [mech] (a rib on a casting or moulding to give greater rigidity) nervure *f* de renfort
STIFFNESS - [mech] (of a mechanism undesired resistance to movement, due to excessive friction) rigidité *f*, résistance *f* au mouvement
[gen] (structural capacity of a body or structure to resist distortion) rigidité *f*, consistance *f*, ténacité *f*
~COEFFICIENT - [mech] (in a linear mechanical system, the ratio between the applied force and the displacement from equilibrium) coefficient *m* de rigidité, coefficient *m* de raideur
~CONSTANTS - [mech] (elastic moduli) modules *m*pl d'élasticité
~CONTROL - [aero] (in a mechanically vibrating system) régulation *f* de la sensibilité
~CRITERION - [aero] (a measure of the relation of the stiffness of a structure to its other properties) critère *m* de rigidité
~IN FLEXING - [rubber ind] (flexural rigidity) raideur *f* de flexion
~IN TORSION - [rubber ind] rigidité *f* en torsion
STIFFNESS OF A CABLE - [mech] raideur *f* d'une corde
~REACTANCE - [acoust] (reactance due solely to stiffness) réactance *f* de rigidité
~TESTER - [instr] (used in paper manufacture to test paper) appareil *m* de mesure de la rigidité
STIFLE, to - [gen] suffoquer, étouffer
[gen] (a fire, a flame) éteindre
[chem] (a reaction) arrêter
STIFLE THE PUPA, to - [text] étouffer la chrysalide
STIFLED COCOON - [text] cocon *m* séché, cocon *m* étouffé
STIFLING - [text] (of the silkworm pupa) étouffement *m* (de la chrysalide)
[gen] étouffant, suffocant
~BY STEAM - [chem] étouffement *m* par vapeur d'eau
STIGMA - [gen] stigmate *m*, tache *f*
[med] stigmate *m*
[bot] stigmate *m* (du pistil)
STIGMASTEROL - [chem] (an intermediate for sterodis) stigmastérol *m*, stigmastérine *f*
STIGMATA OF BENECKI - [med] ulcérations *f*pl multiples de la grande courbure
~VENTRICULI - s. stigmata of Benecki
STIGMATIC - [gen] stigmatisé
[opt] (optical system having equal focal power in all meridians) stigmatique
[med] (of the skin) stigmatisé
~FOCUSING - [opt] (in spectrology) focalisation *f* stigmatique
~LENS - [opt] objectif *m* stigmatique, objectif *m* anastigmate
STIGMATISM - [opt] stigmatisme *m*
STIGMATIZATION - [med] stigmatisation *f*
STILB - [illum] (a unit of illumination) stilb *m*
STILBAMIDINE ISETHIONATE - [chem] (a trypanocide used in medicine) iséthionate *m* de stilbamidine
STILBENE - [chem] (an intermediate for dyestuffs) stilbène *m*
~DYES - [chem] colorants *m*pl au stylbène
STILBITE - [min] (a natural zeolite) stilbite *f*
STILE - [impl] échalier *m*, échalis *m*
[carp] montant *m* vertical
[oil ind] poteau *m* de support du derrick
STILL - [ind chem] (apparatus in which destillation is carried out) appareil *m* de distillation, alambic *m*, cornue *f*
[impl] (a retort for distillation) appareil *m* à distiller
[gen] (adj) (without movement, motionless) immobile, tranquille
[gen] (silent; making no sound) silencieux
[photo] photo *f* de plateau
[cin] photo *f* de publicité
[cin] (studio scenery) décor *m*, vue *f* fixe
~AIR - [gen & met] air *m* calme, air *m* tranquille
~AIR RANGE - [aero] (the theoretical distance which can be flown by an aircraft under given programmed conditions of speed and altitude in an International Standard Atmosphere in still air, and certain other conditions) rayon *m* d'action en air tranquille
~-BIRTH - s. stillbirth
~BODY - [ind chem] (the vessel in a still which contains the liquid to be treated) corps *m* d'alambic
~BOTTOM - [chem] (the residue which remains after a batch distillation) residu *m*
~GAS - [gas ind] gaz *m* de raffinerie

STILL LIFE - [gen] nature *f* morte
~-PICTURE PROJECTION - [cin] (projection of a single image for checking purposes) projection *f* arrêtée
~POT - s. still body
~PROJECTION - [photo] projection *f* fixe
~-RESIDUE - [chem] residu *m* de distillation
~WATER - [hydr] eau *f* morte
STILLAGE - [el] (a stand for accommodating accumulator cells) chantier *m*
[impl] (a platform for the transport and storage of parts in a factory) plateforme *f*
[gen] banc *m* de peu de hauteur
STILLBIRTH - [med] mise *f* au monde d'un enfant mort-né
STILLBORN - [med] mort-né *m*
STILLGOUT - [med] stilligoutte *m*, compte-gouttes *m*
STILLING - s. stillion
~WELL - [mining] puits *m* de repos
STILLINGIA OIL - [paint] (a drying oil of considerable importance, obtained from the Chinese tree Stillingia sebifera; it has the property of polymerizing more rapidly than linseed oil, for which it is an effective substitute) huile *f* de stillingie
STILLION - [impl] (special device for supporting a cask) bock *m* de soutirage
STILLSON WRENCH - [tool] (a tool for rotating cylindrical parts, having serrated jaws, one of which is so attached as to jam against the work when force is applied) clef *f* à tube
STILLSTAND - [chem] état *m* d'équilibre
STILT, to - [gen] monter sur des échasses
[constr] surhausser
STILT - [impl] échasse *f*
[constr] pilotis *m*
[mech] mancheron *m* (de charrue)
[constr] surhaussement *m*
STILUS - [med] stilet *m*, sonde *f*
STIMULANTS - [chem] (any stimulating agent; a drug a beverage etc) stimulants *mpl*
STIMULATING FREQUENCY - [electron] (used in lasers and masers to stimulate the emission of radiation) fréquence *f* d'excitation
STIMULATION - [zool etc] (the application of stimuli) stimulation *f*
~FATIGUE - [med] fatigue *f* de stimulation
STIMULUS - [zool etc] (an agent which will provoke active response or reaction in a living organism) stimulus *m*
[phys] grandeur *f* à mesurer, excitation *f*
[electron] impulsion *f* de déclenchement
STING - [aero] (in a wind tunnel, an arm supporting the model and transmitting forces from it to the balance) dard *m*
[zool] (of a wasp etc) aiguillon *m*
[gen] piqûre *f*
STINGER - [mining] combinaison *f* de trépan pour roches et alésoir
STINK, to - [gen] puer, sentir mauvais
STINK - [gen] puanteur *m*, odeur *f* fétide
~DAMP - [mining] (sulphuretted hydrogen) gaz *m* d'explosion de grisou
~TRAP - [constr] garde *f* hydraulique
STINKING - [gen] puant, fétide, infect
~ROOT ROT - [bot] (a plant disease) rouille *f* de la racine
STINT, to - [gen] imposer des restrictions, réduire
STINT - [gen] restriction *f*, limite *f*
[zool] bécasseau *m*
STIPE - [bot] (general term for the stalk of the fruit body) stipe *m*
STIPEL - [bot] stipelle *f*
STIPEND - [gen] traitement *m*
STIPITATE - [bot] (having a stipe) stipité
STIPPLE, to - [paint] pointiller, figurer en pointillé
[print] graver à réseau
STIPPLE - [print] grenure *f*
[paint] pointillé *m*
STIPPLER - [constr] brosse *f* à tamponner
STIPULATE, to - [comm & gen] stipuler, convenir
STIPULATION - [gen & leg] stipulation *f*
STIR, to - [gen] mouvoir, remuer
[ind chem] (to agitate a fluid by moving a rod, blade or the like through it) agiter, activer, brasser
[brew ind] (the mash) vaguer (le fardeau)
STIRRER - s. stirring mill
STIRRING BY AIR INJECTION - [rubber ind] agitation *f* par injection de l'air
~MILL - [ind chem] (another term for an agitator, viz. a machine in which a substance is mixed or subjected to internal movement) pétrisseur *m*, agitteur *m*, cuve *f* à brasser
~VANE COLUMN - [chem] colonne *f* à ailettes mélangeuses.
STIRRUP - [impl] (of a saddle) étrier *m*
[mech & carp] (a metal loop or strap) bride *f*, lien *m* de fer en U
[constr] (for reinforced concrete) étrier *m*, armature *f* en étrier
[med] étrier *m* (de la table d'opération)
[anat] (stirrupbone) étrier *m* de l'oreille
[mech] (of a pulley) moufle *f*, chape *f*, flasque *f* de poulie
~BOLT - [mech] lien *m* en fer à U, armature *f*
~BONE - [anat] étrier *m*
~-JOINT - [mech] assemblage *m* à étrier
~LENS PANEL - [photo] porte-objectif *m* en U
~STRAP - [constr] étrier *m*
~TREADLE - [text] pédale *f*
STIRRUPED CONCRETE - [constr] béton *m* fretté
STIRRUPS - [mech] (of a balance; the links connecting the beam with the pans) étriers *mpl*
STITCH, to - [gen] coudre
[med] suturer
[rubber ind] (a tyre) presser à la roulette, rouleter
[bookbind] brocher
STITCH - [gen] point *m*
[med] point *m* de suture
[text] (in knitting) maille *f*
[bookbind] brochage *m*
~ADJUSTMENT - [mech] (of a sewing machine) réglage *m* de la serre
~TEAR STRENGTH - [rubber ind] résistance *f* à la déchirure initiée par une aiguille
~WELDING - [metall & plast ind] (the progressive welding of thermoplastic materials by successive applications of two small mechanically operated electrodes, connected to the output terminals of a radio frequency generator, using a mechanism similar to that of a normal sewing machine) soudage *m* discontinu, soudure *f* à la molette, soudure *f* à points continue

STITCHED HARNESS CORDS - [text] fils *mpl* d'arcades cousus
~PIQUE' - [text] étoffe *f* avec effet de piqûre en travers
~RIB - [text] cannelé *m* lié
~SATEEN - [text] satin *m* renforcé, satin *m* lié
~TWILL - [text] croisé *m* à cordons multiples
STITCHEL - [text] brin *m* court et raide
STITCHER - [rubber ind] (for the retreading of tyres) roulette *f* lisse, roulette *f* à appuyer
[leather ind] piqueur *m*
[bookbind] brocheur *m*
~DISK - [rubber ind] roulette *f* à accrocher, roulette *f* à confectionner
~STAND - [rubber ind] (for tyres) support *m* à molette, rouleuse *f* de gomme
STITCHING MACHINE - [bookbind] machine *f* à brocher
[rubber ind] machine *f* pour laminer les protecteurs sur les pneus, machine *f* à roulettes
~TWINE - [text] ficelle *f* à brocher
ST. MARTIN'S SUMMER - [met] (period of fair, summer like weather, sometimes occurring in early autumn in temperate zones) été *m* de Saint-Martin
STOA - [arch] portique *m*
STOCK, to - [gen] stocker, approvisionner
[comm] avoir en magasin, tenir en dépôt
[firearms] monter (un fusil)
[mech] (the operation of rough-machining) dégrossir
[naut] jaler (l'ancre)
STOCK - [bot] souche *f*, estoc *m*, porte-greffe *m*, sujet *m*
[timber] tronc *m*
[a breed] race *f*, famille *f*
[gen & comm] (available material) marchandises *fpl*, dotation *f*, stock *m*
[gen] (raw material) matières *fpl* premières
[gen] (support) appui *m*, support *m*
[firearms] fût *m* de fusil, monture *f* de fusil
[mech] (machining allowance) surépaisseur *f* d'usinage
[fin] (the stock capital of a company) fonds *mpl*, valeurs *fpl*
[fin] actions *fpl*
[naut] jas *m* d'ancre
[gen] (in industry) production *f* journalière
[paper man] (term applied to wet pulp during processing) pâte *f* à papier
[metall] (forging stock; part of a bar to be upset with a forging machine or drop-forged) bout *m* à forger
[rubber ind] pâte *f*
[cin] (film stock) film *m* non-exposé
[carp] (of the plane) bois *m* de rabot, fût *m* de rabot
[tool] (in die casting; die holder for threading) filière *f* à cage, porte-filière *m*
[auto etc] de série
[metall] (of a blast furnace) charge *f*
[mech] (a bit-stock) vilebrequin *m*
[railw] (the transport equipment) matériel *m*
~ALE - [brew ind] ale *m* vieux, ale *m* de réserve
~AND BOWL MOTION - [text] dispositif *m* de suspension par poulies
~AND DIE - [metall] filière *f* garnie de coussinets
STOCK CAR - [railw] (U.S.A only) wagon *m* à bestiaux
~COAL - [gas ind] charbon *m* de stock
~COLUMN - [metall] colonne *f* des charges
~CRACKING - [oil ind] matière *f* première du craquage
~CULTURE - [bacteriology] souche *f* pour culture
~DISTRIBUTOR - [metall] appareil *m* de changement distributeur
~EXCHANGE - [fin] bourse *f* (des valeurs)
~FARM - [agric] élevage *m* de bestiaux
~FEED - [agric] fourrage *m*
~FEEDER - [rubber & paper ind] dispositif *m* d'alimentation
~GUIDE - [mech] (of a mill) flasque *f* de guidage latérale (d'un malaxeur)
~HOOPS - [naut] anneaux *mpl* de jas d'ancre
~INDICATOR - [metall] bécasse *f*, indicateur *m* de charge
~KETTLE - [impl] (US only, a boiling pan in GB) marmite *f*
~LINE - [metall] (of a blast furnace; the level of the top of the charge) niveau *m* de la charge
~LINE GAUGE - [metall] (in a blast furnace device for showing the position of the stock line) indicateur *m* de la descente des charges
~OF SHEEP - [agric] nombre *m* de moutons
~OF THE SEEDLING - [bot] tronc *m* de plantule
~OF THE SILK WORM - [text] race *f* du ver à soie
~OF WOOL - [text] existences *fpl* en laine
~ON HAND - [comm] stock *m* en magasin, stock *m* en dépôt
~RAIL - [railw] rail *m* contre-aiguille
~RAISING - [agric] élevage *m* du bétail
~ROOM - [ind chem] (room in a laboratory building for stocking chemical materials) magasin *m*
~SOLUTION - [chem] solution *f* concentrée
~-STONE - [leather ind] pierre *f* à poncer
~STRAINER - [mech] (a straining machine) boudineuse *f* à tamiser, boudineuse *f* à tête filtrante
~TURNOVER - [comm] roulement *m* du stock
~TURRET - [rubber ind] (for tyres) touret *m* de matériau
STOCKADE, to - [gen] palissader
[hydr] garnir d'une estacade
STOCKADE - [gen] palissade *f*, palanque *f*
[carp] estacade *f*
STOCKAGE - [gen] (storage of material in considerable quantities) emmagasinage *m*, stockage *m*
STOCKBROKER - [fin] agent *m* de change, courtier *m* de bourse
STOCKCAR - [railw] (US only, a cattle wagon) wagon *m* à bestiaux
STOCKED ANCHOR - [naut] ancre *f* avec jas
STOCKFISH - [zool] merluche *f*, morue *f* séchée
STOCKHOLDER - [fin] (US only, shareholder in GB) actionnaire *m*
STOCKING - [text] bas *m*
[mech] (in machining) dégrossissage *m*
[zool] (of a horse) balzane *f*
~LOOM - [text] métier *m* à bas
~YARN - [text] fil *m* à tricoter
STOCKIST - [comm] stockiste *m*
STOCKPILE, to - [comm] stocker en grandes quantités
[min] (to heap up) entasser
STOCKPILE - [gen & comm] stock *m*, réserve *f*

STOCKPILING - [gen] (the operation of accumulating materials in large quantities as a reserve for technical or commercial reasons) stockage *m*
[min] (heaping up) entassement *m*
STOCKRAIL - [railw] (of a switch) contre-aiguille *f*
STOCKS - [constr] (sound and hard-burned bricks, somewhat uneven in colour) carreaux *mpl*, briques *fpl* recuites
[shipbuild] cale *f* de lancement, chantier *m*
STOCKTAKING - [comm] inventaire *m*, levée *f* de inventaire
STOCKWORK - [min] (an irregular mass of interlacing veins of ore) stockwerk *m*, masse *f* irrégulière de filons
STOCKYARD - [constr] parc *m* à matériau
[agric] parc *m* à bétail
~FEVER - [vet] septicémie *f* hémorragique du bétail
STOICHIOLOGY - [physiology] (the science of the constituent processes in the physiology of animal tissues) stoechiologie *m*
STOICHIOMETRIC - (pertaining to stoichiometry) stoechiométrique
~COMBUSTION - [phys] (combustion under theoretical conditions) combustion *f* stoechiométrique
~CRYSTAL IMPURITY - s. stoichiometric impurity
~IMPURITY - [phys] (crystalline imperfection arising from a deviation from stoichiometric composition) impureté *f* stoechiométrique
~MIXTURE - [chem] (a mixture in which the constituents are in chemical combining proportions) mélange *m* stoechiométrique
STOICHIOMETRY - [chem] (the chemistry of the quantities entering into and produced by chemical reactions) stoechiométrie *f*
STOKE, to - [gen] alimenter le feu
[th eng] (boilers) entretenir le feu (d'un four)
[th eng] (to stoke a furnace with fuel) charger (un foyer)
STOKE - [phys] (unit of kinematic viscosity) stoke *m*
~-HOLE - [ovens] ouverture *f* de foyer, trou *m* de chargement, tisard *m*
STOKEHOLD - [naut] (the boiler room) chambre *f* de chauffe, chaufferie *f*
STOKEHOLE - s, stoke-hole
STOKER - [gen] chauffeur *m*, chargeur *m* (de foyer)
[mech] chargeur *m* automatique
STOKES LAW - [opt & electron] (the law stating that the wavelength of luminescence excited by radiation is always greater than that of the exciting radiation) loi *f* de Stokes
STOKING - [gen] chauffage *m*, chargement *m*
[metall] (of a furnace) alimentation *f* (d'un foyer)
~DIAGRAM - [th eng] diagramme *m* du chargement du four
STOLON - [bot] stolon *m*, stolone *f*
STOLONIFEROUS - [bot] (producing stolons or runners) stolonifère
STOMA - [anat & bot] (aperture in the walls of a blood vessel or in a serous membrane) stomate *m*
STOMACH - [zool] estomac *m*
~-ACHE - [med] douleurs *fpl* d'estomac
~-COUGH - [med] toux *f* gastrique
~INSECTICIDE - [chem] insecticide *m* par ingestion
~POISON - s, stomach insecticide
~-PUMP - [med] pompe *f* stomacale
~-TOOTH - [zool] canine *m* (de lait)
~TUBE - [med] sonde *f* stomacale

STOMALGIA - [med] stomatodynie *f*
STOMATALGIA - s. stomalgia
STOMATITIS - [med] (inflammation of mucous membrane of the mouth) stomatite *f*
STOMATODYNIA - s. stomatalgia
STOMATODYSODIA - [med] haleine *f* fétide, mauvaise haleine *f*
STOMATOMY - [med] incision *f* de l'orifice externe du col de l'utérus
STOMATOPLASTY - [med] (plastic surgery on the mouth) stomatoplastie *f*
STOMATORRHAGIA - [med] stomatorragie *f*
STOMATOTOMY - s, stomatomy
STONE, to - [gen] lapider
[constr] revêtir de pierres, paver de pierres
[mech] (to grind) passer à la pierre, meuler
[leather ind] poncer (une peau)
STONE - [gen] pierre *f*
[min] (colloq. for ore) minerai *m*, pierre *f* de mine
[geol] pierre *f*, rocher *m*
[mech] (implement) meule *f* à repasser
[print] (a lithographic stone) marbre *m*
[bot] (the hard endocap of a drupe, e.g. a cherry stone) noyau *m* (de fruit), pépin *m*
[meas] stone *m* (6,348 kg)
[constr] moellon *m*, pierre *f* de taille
[med] calcul *m*
~AGE - [geol] âge *m* de la pierre
~AT GRASS - [min] minerai *m* à la surface
~BASIN - [impl] bassin *m* en pierre
~BEDDING - [constr] plate-forme *f* en pierres, perré *m*, enrochement *m*
~BLUE - [ind chem] bleu *m* pour laver
~BOAT - [agric] traîneau *m*
~BOLT - [mech] vis *f* à scellement
~BOND - [constr] appareil *m* de pierres
~-BREAK - [bot] saxifrage *f*, casse-pierre *m*
~BREAKER - [mech] casse-pierres *m*, concasseur *m*
~BURNISHER - [mech] (a flint glazing machine) brunissoir *m* à pierre
~BUTTER - [min] beurre *m* de roche, beurre *m* de montagne
~CELL - [bot] (thick-walled cell with lignified walls) cellule *f* pierreuse
~CHANNELER - [constr] trancheuse *f*
~CHISEL - [impl] burin *m* de maçon, grain *m*, smille *f*, ciseau *m* de taille
~-CONCRETE - [constr] béton *m* de pierre, béton *m* de cailloux
~-CORBEL - [constr] console *f*
~CRUSHER - [mech] casse-pierre *m*, concasseur *m*
~CUTTER - [constr] tailleur *m* de pierres
~-CUTTER'S TOOL - [tool] outils *mpl* de taille
~CUTTING - [constr] taille *f* de pierre
~DEFLECTOR - [auto etc] pare-pierre *m*
~DOWEL - [constr] goujon *m* en pierre, fenton *m* en pierre
~DRESSER - [constr] dresseur *m* de pierres, tailleur *m* de pierres
~DRESSING - [constr] taille *f* de pierres, dressage *m* des pierres
~-DRIFT - [mining] galerie *f* en rocher
~-DUST, to - [geol] schistifier
~-DUST - [mining] pulvérin *m* rocheux, poussières *fpl* incombustibles
~FILLING - [constr] empierrement *m*

STONE FLOOR - [constr] dalle *f* en pierre, hourdis *m* en pierre
~FRUIT - [bot] (drupe, e.g. a fruit such as a cherry a plum, a peach etc) drupe *f*, fruit *m* à noyau
~GRINDING - [mech] meulage *m* par pierre
~GUARD - [aero & auto] (a wire mesh screen fitted to an air intake to exclude pebbles) grille *f*, pare-pierres *m*
~HAMMER - [tool] casse-pierres *m*, marteau *m* à concasser la pierre
~HEAD - [mining] (the solid rock which is first met while sinking a shaft) tête *f* de faille
[mining] (a tunnel in stone) galerie *f* en rocher
~HINGE - [constr] articulation *f* en pierre
~HOLDER - [impl] (for a cylinder grinder etc) porte-meule *m*
~INSULATOR - [el] (type of low-voltage insulator made from stoneware) isolateur *m* en porcelaine
~LATTICE - [geol] réseau *m* d'érosion
~LEEK - [bot] ciboule *f* d'hiver, oignon *m* d'Espagne
~LINING - [constr] revêtement *m* en pierre
~LINING OF THE LANTERN OPENING - [constr] couronne *f* de l'oeil
~MASON - [constr] macon *m*, tailleur *m* de pierre
~-MASON HAMMER - [tool] marteau *m* de tailleurs de pierre
~MASONRY - [constr] ouvrage *m* en pierre, ouvrage *m* de tête
~MILL - [mech] (grinding machine in which the triturating elements are of stone) concasseur *m*, casse-pierre *m*
~OIL - [oil ind] pétrole *m*
~OUT, to - [shoe man] poncer
~PARSLEY - [bot] sison *m*
~-PAVED ROAD - [constr] route *f* pavée en pierre
~PAVEMENT - [constr] pavé *m* en pierre
~PICK - [impl] pic *m* au rocher
~PILLAR - [constr] pilier *m* en pierre
~-PINE - [bot] pin *m* pignon, pin *m* parasol
~PIT - s. stone quarry
~PITCHING OF SLOPE - [constr] pierrée *f*, revêtement *m* du talus en pierres
~PLUG - [mech] (in internal combustion engines) bougie *f* à isolant de porcelaine
~PUTTY - [ind chem] mastic *m* à pierre, lithocolle *f*
~QUARRY - [min] carrière *f*
~SAW - [impl] scie *f* à pierre, passe-partout *m*
~-SCREENING PLANT - [mech] cribleuse *f* des pierres
~SCROLL - [constr] volute *f* en pierre d'un pignon
~SLIDE - [gen] éboulement *m*, traînée *f* d'éboulis
~STRUTTING - [constr] étayement *m* en pierre
~-SWEEPING - [geol] (falling stones) chute *f* de pierres
~TANK - s.stone basin
~TONGS - [impl] (used for hoisting blocks of stone) pince *f* à moellons
[constr] pince *f* à moellons
~TUBING - [mining] cuvelage *m* en pierre
[constr] cuvelage *m* en maçonnerie
~WALL - [constr] mur *m* en pierre, moellon *m*
~WASHING MACHINE - [mech] laveur *m* de pierres
~WAX - [min] cire *f* de lignite
~WEDGE - [constr] coin *m* à pierre
~WINCH - [impl] vérin *m* pour pierres, treuil *m* pour soulever les pierres
~-WORK - s. stonework
~-YARD - [constr] chantier *m* de pierre

STONECUTTING - s. stone cutting
STONES - [constr] pierraille *f*
STONEWARE - [ceram] (fine earthenware, fired at a high temperature) grès *m*, poterie *f* de grès
~CLAY - [geol] argile *f* à grès
~CRUCIBLE - [impl] creuset *m* en grès
~PIPE - [hydr] tuyau *m* en grès
~VESSEL - [impl] récipient *m* en grès
STONEWORK - [mining] travail *m* au rocher
[constr] ouvrage *m* en pierre, maçonnage *m*
[print] correction *f* sur le marbre
STONINESS - [geol] nature *f* pierreuse
STONING - [agric] énucléation *f*, épépinage *m*, dénoyautage *m*
~MACHINE - [mech] machine *f* à dénoyauter
STONY - [gen] pierreux, rocailleux
~GROUND - [soil] terrain *m* pierreux
~METEORITES - [geol] (meteorites mainly consisting of rock-forming silicates) météorites *fpl* pierreuses
STOOK, to - [agric] mettre en moyettes
STOOK - [agric] moyette *f*, meulette *f*, tas *m* de blé
[mining] pilier *m* de charbon
~THE FLAX, to - [text] faire des cahoutes de lin
STOOL, to - [bot] rejetonner
STOOL - [gen] (backless and armless seat) tabouret *m*
[metall] (a cylindrical brick for supporting the crucible) base *f* de lingotière, fromage *m*
[shipbuild] (support for machinery) appui *m*
[med] fèces *fpl*
[agric] souche *f*, pied *m* mère, plante *f* mère, tallage *m*
[constr] rebord *m* de fenêtre
STOOP, to - [gen] se baisser, se pencher
STOOP - [gen] penchement *m* en avant
[arch] (USA and Canada only) véranda *f*, terrasse *f* surélevée
[mining] (a pillar supporting the roof) pilier *m*, lopin *m*, massif *m*
[mining] pilier *m* de protection
~SYSTEM - [mining] exploitation *f* par piliers
STOOPING - [mining] dépilage *m*
STOP, to - [gen] (to cease from motion) arrêter, s'arrêter, cesser
[gen] (to halt progress) faire cesser, mettre fin
[naut] stopper
[gen] (to keep back, to retain, to withhold) empêcher
[gen] (to obstruct) boucher, obturer, fermer
[mech] arrêter, suspendre, désamorcer
[gen] (to prevent regress from) boucher, étancher, tamponner
[metall] (in a blast furnace, to close the hole through which molten iron is draw off, using a plug of fireclay) boucher (le trou de coulée)
[photo] disphragmer
[mus] presser (une corde)
[mus] (to shorten the length of a vibrating string) boucher (un son)
[naut] (to secure with a cable) genoper
STOP - [gen] (a halt; the act of stopping) arrêt *m*, interruption *f*
[gen] (interruption, as of a train etc) halte *f*, pause *f*
[mech] (any device which limits motion) dispositif *m* de blocage, butée *f*, taquet *m*
[photo] (the aperture of a lens) diaphragme *m*

STOP - [gen] (road traffic) arrêt *m* (de circulation)
[plast ind] (a device not forming part of the mould, used to limit the amount of closure) cale *f*, cale *f* d'épaisseur
[mus] (a rank of organ pipes) jeu *m* d'orgue
[mus] (a register, handle, tab etc used on an organ for control) registre *m*
[mus] (on the harpsichord, an implement to bring the suitable rank of strings into play) registre *m*
[el] limiteur *m* d'arrêt
~AND PROCEED LIGHT - [railw] feu *m* franchissable
~AND PROCEED SIGNAL - [railw] signal *m* d'arrêt franchissable
~AND START LEVER - [mech] levier *m* d'arrêt et de démarrage
~BAND - [radio] affaiblissement *m* pour la bande de fréquences non transmises
[text] ficelle *f* d'arrêt
~BAND ATTENUATION - [electron] affaiblissement *m* de blocage
~BAR - [mech] levier *m* d'arrêt, barre *f* de blocage
~BATH - [photo] bain *m* (acide) d'arrêt (du développement)
~BLOCK - [mech] taquet *m* d'arrêt
[railw] blochet *m* d'appui, blochet *m* d'arrêt
~BOLT - [mech] goupille *f* d'arrêt, arrêtoir *m*
~BRACKET - [text] levier *m* du dispositif à butoir
~BUFFER - s. stop block [railw]
~BUTTON - [comput] bouton *m* d'arrêt
~CATCH - [mech] cliquet *m*
~CLICK - [horol] cliquet *m* d'arrêt
~COIL - [el] bobine *f* d'arrêt en câble coaxial
~COLLAR - [mech] (for a tool etc) collier *m* d'arrêt, bague *f* d'arrêt, bague *f* de butée
~-COCK - [mech] (a shut-off cock) robinet *m* d'arrêt, robinet *m* de fermeture
~-COCK PLUG - [mech] clef *f* du robinet, canillon *m* du robinet, noix *f* du robinet
~DISK - [mech] disque *m* d'arrêt
~DISTANCE - [auto] distance *f* de freinage
~-DOWEL - [mech] gujon *m* d'arrêt
~DOWN, to - [photo] fermer le diaphragme
~-DRILL - [mech] foret *m* à repos
~-ENDED RETORT - [ind chem] cornue *f* avec fond
~FILTER - [radio] (band-elimination filter) filtre *m* d'élimination de bande
~FINGER - [horol] doigt *m* d'arrêt
~FRAME - [telev] arrêt *m* d'image, tirage *m* en image fixe
~-GAP - [gen] bouche-trou *m*
~-GEAR - [mech] organes *mpl* d'arrêt
~-GROOVE - [photo] cran *m* d'arrêt
~-KEY - [mus] bouton *m* d'appel (de jeu d'orgue)
[el] touche *f* arrêt
~KNOB - [instr] (in measuring apparatus) bouton *m* d'arrêt
~LAMP - s. stop light
~LEVER - [mech] levier *m* d'arrêt
~LIGHT - [auto] signal *m* d'arrêt, stop *m*
[railw] feu *m* non franchissable
~LOG - [hydr] (of a dam) batardeau *m*
~MOTION - [mech] débrayage *m*, organes *mpl* d'arrêt
~MOTION FOR DOFFING - [text] mécanisme *m* d'arrêt pour faire la levée
~MOTION ROCKING LEVER - [text] levier *m* oscillant du mouvement d'arrêt
~MOTION ROD - [text] tige *f* d'arrêt

STOP MOTION ROLLER - [text] galet *m* d'arrêt
~MOTION SPRING - [mech] ressort *m* d'arrêt
~NOSE - [text] nez *m* d'arrêt
~NUT - [mech] (of an adjusting screw) écrou *m* de blocage
~-OFF LACQUER - [el metall] (varnish used to prevent deposition in areas where it is not required) vernis *m* isolant
~ON PICKING LEVER - [text] déclic *m* du levier de chasse
~PIN - [mech] tige *f* d'arrêt, ergot *m* d'arrêt
~PLANK - [hydr] (of a dam) hausse *f* (de vanne etc)
~PLATE - [mech] platine *f* d'arrêt
~POSITION - [railw] (of a signal) (disque) fermé, (disque) à l'arrêt
[mech] position *f* d'arrêt
~PRESS - [gen] informations *fpl* de dernière heure
~RING - [mech] anneau *m* de blocage
[photo] anneau *m* de réglage du diaphragme
~ROD - [text] tige *f* d'arrêt, tige *f* de déclenchement
~SCREW - [mech] vis-butoir *f*, vis *f* d'arrêt
~-SEND SIGNAL - [telecomm] signal *m* d'arrêt de envoi
~SHOULDER - [mech] saillie *f* d'arrêt
~SIGNAL - [telecomm] émission *f* d'arrêt, signal *m* d'arrêt
[railw] signal *m* d'arrêt, signal *m* carré
~-SPEED VALUES - [photo] valeurs *fpl* diaphragme-vitesse d'obturation
~SPRING - [mech] ressort *m* d'arrêt, ressort *m* à cran d'arrêt
~-TAIL LAMP - [auto] lanterne *f* de feu arrière et feu de stop
~UP, to - [gen] obturer, étancher, boucher, tamponner
~VALUE - [photo & opt] ouverture *f* relative
~VALVE - [mech] (the main steam valve fitted to a boiler to control the steam supply) robinet *m* d'arrêt, soupape *f* d'arrêt
~-WATCH - [horol] (an accurate type of watch, fitted with a hand which can be started and stopped at will, for precise measurement of short periods of time) chronomètre *m* à déclic, chronographe *m* à pointage, compte-seconds *m*
~WEIGHT - [mech] poids *m* pour arrêter
~-WORK - [horol] mécanisme *m* d'arrêt
STOPCOCK - s. stop-cock
STOPE, to - [mining] (to excavate ore from a reef, vein etc) abattre, exploiter en gradins
STOPE - [mining] (the space left after the excavation of mineral from a tabular deposit) chantier *m* d'abattage
[mining] gradin *m*, chantier *m* en gradins
~DRILL - [tool] perforatrice *f* pour creusement en montant
~FACE - [mining] taille *f* d'abattage
~FLOOR - [mining] sole *f* du chantier
~HOLE - [mining] trou *m* en semelle
~ORE PASS - [min] cheminée *f* à minerai
~-PLAN - [mining] plan *m* des chantiers d'abattage
~TRUCK - [mining] wagonnet *m* pour mine
STOPED UP - [mining] foudroyé
STOPER - s. stope drill
STOPING - [mining] exploitation *f* en gradins, abattage *m* du minerai
STOPPAGE - [gen] arrêt *m*, interruption *f*
[mech] (of a motion) arrêt *m*

STOPPAGE - [plumb] (of a pipe) obturation *f*, bouchage *m*, occlusion *f*
[hydr] engorgement *m*
~OF A VEHICLE - [railw] immobilisation *f* d'un véhicule
~OF DRAW FRAMES - [text] débrayage *m* des bancs d'étirage
~OF LOOM - [text] arrêt *m* du métier
~OF THE DOFFER BY HAND - [text] arrêt *m* du peigneur à la main
~OF THE ROLLERS - [text] arrêt *m* des cylindres
~OF THE SPINDLE - [text] débrayage *m* de la broche
~OF TRAFFIC - [railw] entrave *f* à la circulation
[road traffic] suspension *f* de la circulation
STOPPED PARTICLE - [phys] particule *f* arrêtée
~SECTION CIRCUIT - [telecomm] circuit *m* de section tampon
~SPINDLE - [text] broche *f* au repos
~SPINDLE SECTION - [text] groupe *m* de broches arrêtées
STOPPER, to - [gen] boucher
[paint] (to fill holes, or to apply a coat) mastiquer
STOPPER - [gen] (a plug to close the mouth of a bottle or the like) bouchon *m*
[constr] (material used to fill up cracks in a surface) mastic *m*
[ind chem] (used in rubber manufacture etc) agent *m* stabilisant
[metall] (a peg) pièce *f* d'écartement
[metall] (a plunger) quenouille *f*
[plast ind] (plastic material used to fill holes) mastic *m*
[hydr] obturateur *m*, pointeau *m* d'arrêt
[mech] taquet *m* d'arrêt (de mouvement)
[naut] bosse *f*
~BELL - [metall] petite cloche *f*, petit cône *m*
~CIRCUIT - [telecomm] (anti-resonant circuit) circuit *m* bouchon, piège *m* électrique, circuit *m* antirésonnant
~OF THROAT - [metall] fermeture *f* du gueulard
~RING - [mech] (rubber ring on a tap to obtain a tight fit) rondelle *f*
~ROD - [metall] quenouille *f*
STOPPING - [gen] arrêt *m*, suspension *f*
[paint] mastic *m*, bouchage *m*
[mining] interruption *f* de l'aérage, cloison *f* d'aérage
[metall] (in furnaces, the closing of the tapping hole of a cupola) bouchage *m*
[phys] (the decrease in kinetic energy of an ionizing particle as a result of energy losses along its path through matter) freinage *m*, arrêt *m*
[constr] mastiquage *m*
[mech] (of an injector) désamorçage *m*
~ALCOHOLIC FERMENTATION - [chem] arrêt *m* de la fermentation alcoolique
~BRAKE - [mech] frein *m* d'arrêt
~CONDENSER - [el] condensateur *m* d'arrêt
~CROSS-SECTION - [nucl] (atomic stopping power, the energy loss per atom, per unit area to the particle's motion) pouvoir *m* d'arrêt atomique
~DISTANCE - s. stop distance
~DOWN - [photo] (to reduce the aperture of the diaphragm) diaphragmation *f*
[metall] (of a furnace) arrêt *m* du haut fourneau
~EQUIVALENT - [nucl] (for a specified thickness of a substance, the thickness of a standard capable of producing the same energy loss) équivalent *m* de arrêt
STOPPING HANDLE - [mech] poignée *f* d'embrayage
~KNIFE - [impl] (a knife used for spreading putty) couteau *m* à mastiquer
~OFF - [el] (operation of coating a cathode or rack with a resistance to make it non-conducting) isolation *f*
~POTENTIAL - [el] (the potential required to bring an emitted electron to rest) potentiel *m* d'arrêt
~POWER - [phys] (a measure of the effect of a substance on the kinetic energy of a charged particle passing through it) pouvoir *m* d'arrêt
~TIME - [railw] durée *f* d'arrêt
~TRAIN - [railw] (slow train) train *m* omnibus
STOPWATER - [naut] tampon *m*, bouchon *m*
STORAGE - [gen] (the deposition of goods in a warehouse for safe keeping) emmagasinage *m*
[gen] (the space for storing goods) magasins *mpl*, entrepôts *mpl*
[el] (of an electric storage battery) accumulation *f*
[comput] (the retention of information for subsequent reference) mémorisation *f*
[comput] (any device into which information can be introduced and subsequently extracted) mémoire *f*
~ALLOCATION - [comput] attribution *f*
~BATTERY - [el chem] (a set of accumulators or secondary cells) accumulateur *m*, pile *f* secondaire
~BATTERY CELL - s. storage cell
~BIN - [constr] (a silo) coffre *m*, récipient *m*
~BUILDING - [constr] dépôt *m*
[agric] locaux *mpl* de stockage
~BUTTER - [agric] beurre *m* de conserve
~CAMERA TUBE - [telev] (image storing tube) tube *m* d'emmagasinage d'image, tube *m* à mémoire
~CAPACITY - [comput] (the amount of information a storage can store) capacité *f* de mémoire
[gas ind] capacité *f* gazométrique
~CAVERN - [nucl] (underground space for radioactive waste) caverne *f* d'emmagasinage
~CELL - [el chem] (galvanic cell which can be restored to its original condition after discharge by passing a current through it in the opposite direction to that of discharge) accumulateur *m*, élément *m* secondaire
[comput] (storage for one unit of information) mot *m*, cellule *f* de mémoire
~CONTAINER - [nucl] récipient *m* de stockage
~CONTENTS - [comput] (memory contents) contenu *m*
~CORE - [comput] tore *m* de mémoire
~DECAY - [gen] pourriture *f* d'emmagasinage
~DENSITY - [comput] densité *f* de mémorisation, densité *f* d'enregistrement
~DEPRECIATION - [el chem] (the depreciation in service output of a primary cell as measured by a storage test) usure *f* en magasin
~EFFECT - [electron] (in a semiconductor junction) effet *m* d'emmagasinage de porteurs
~ELEMENT - [electron] élément *m* de mémoire
~HOPPER - [mech] trémie *f* de chargement
~HORIZON - [gas ind] (of a gasholder) couche *f* de stockage souterrain, horizon *m* de stockage
~KEY - [comput] clé *f* de mémoire
~KEYBOARD - [telecomm] (a keyboard in which the combination set up by the depression of a key does not directly control the transmitter but is transferred to one or more sets of storage members

for subsequent control of the transmitter) clavier *m* à mémoire
STORAGE LEVEL - [comput] étage *m* de mémorisation
~LIFE - [el chem] (the duration of storage under specified conditions, at the end of which a cell retains its ability to give a specified performance) durée *f* de conservation
~LOCATION - [comput] (a storage position holding one machine word and usually having a specific address) emplacement *m*
[comput] adresse-lieu *f*
~LOSSES - [oil ind] pertes *f*pl dans les réservoirs de stockage
~OPERATION - [comput] (memory operation) opération *f* de mémorisation
~OSCILLOGRAPH - [instr] (oscillograph storing events and capable of being cleaned in milliseconds) oscillographe *m* à mémoire
~PERIOD - [comput] période *f* de mémorisation
~POND - [hydr] étang *m* d'emmagasinement
~POSITION - [comput] compartiment *m* de mémoire
~PUMP - [hydr] pompe *f* d'emmagasinage hydraulique
~PUNCH EXITS - [comput] jacks *m* de sortie "Mémoire"
~READ-OUT - [comput] lecture *f*, extraction *f*
~REGISTER - [comput] (memory register) régistre *m* de mémoire
~RESERVOIR - [gas ind] (of a gasholder) réservoir *m* de stockage
~RING - [nucl] (of a particle-acceleration installation) anneau *m* de stockage
~RING SYNCHROTRON - [nucl] synchrotron *m* à anneau de stockage
~SET - [th eng] groupe *m* accumulateur
~SIDING - [railw] voie *f* de garage
~SIDINGS - [railw] faisceau *m* d'attente, gare *f* de remisage
~SPACE - [gen] espace *m* disponible
[nucl] (the space in which nuclear fuel is stored before being used in a reactor) dépôt *m*, magasin *m*
~STACK - [comput] (memory stack) empilage *m*
~TANK - [hydr] bassin *m* d'emmagasinement, réservoir *m* d'approvisionnement
~TEMPERATURE - [electron] (the temperature at which transistors should be kept in store) temperature *f* d'emmagasinage permise
~TEST - [el chem] (or shelf test; a test designed to measure retention of service output under specified conditions of storage)essai *m* de conservation
~TIME - [electron] (for a storage element; maximum retention time) durée *f* de conservation
~TRACK - [railw] (a reserve siding) voie *f* de remisage
~TUBE - [electron] (electron tube in which information can be introduced and subsequently extracted) tube *m* à mémoire
~VAT - [brew ind] foudre *m* de garde
[brew ind] (for sedimentation) cuve *f* de repos
~WATER HEATER - [th eng] chauffe-eau *m* à accumulation
~YARD - [ind etc] dépôt *m*
STORE, to - [gen] pourvoir, emmagasiner
[gen] (into reserve) mettre en réserve, ammasser
[comput] (to transfer a piece of information to a device from which it can later be obtained) mémoriser, mettre en mémoire
STORE - [gen] (of goods) (an amount on hand) approvisionnement *m*
[gen] (a storehouse, i.e. a place where commodities can be stored) magasin *m*, entrepôt *m*
[gen] (a shop) magasin *m*, bazar *m*
[comput] (storage) mémoire *f*
~AND FORWARD, to - [comput] mémoriser et faire suivre plus tard
~CALF - [agric] veau *m* d'engrais, veau *m* blanc
~CATTLE - [agric] jeune bétail *m* maigre
~LAMB - [agric] agneau *m* d'engraissement
~PIG - [agric] porc *m* coureur
~RECEIPT CARD - [comput] carte *f* de réception en magasin
STORED BASE CHARGE - [electron] charge *f* emmagasinée en base
~ENERGY - [el] énergie *f* accumulée, énergie *f* intrinsèque
~-ENERGY WELDING - [el mech] (a method of resistance welding in which the heating energy is stored in an inductor, a capacitor, an electric accumulator or a flywheel during a period of time relatively long compared to the welding time) soudage *m* à accumulation d'énergie
~FIELD SYSTEM - [telev] système *m* à emmagasinage de trames alternantes
~PROGRAM - [comput] programme *m* enregistré
~-PROGRA COMPUTER - [comput] calculatrice *f* équipée d'un programme de commande
~PROGRAMME - [comput] (a characteristic of certain machines whereby instructions in the form of numbers or other symbols are held within the machine) programme *m* mémorisé
~-PROGRAMME COMPUTER - [comput] ordinateur *m*
STOREHOUSE - [gen] dépôt *m*, entrepôt *m*, magasin *m*
STOREKEEPER - [gen] garde-magasin *m*
[naut] cambusier *m*
STOREROOM - [gen] halle *f* de dépôt, dépense *f*
[naut] cambuse *f*, soute *f* aux vivres
STOREY - [constr] (the part of a building which is included between two adjacent floors) étage *m* d'une maison
STORIED - [gen] historié
[constr] à étages
~FOREST - [timber] forêt *f* à arbres d'âge differents
STORK - [zool] cicogne *f*
STORM, to - [gen] se déchainer
[met] faire rage
STORM - [met] (in general, a marked atmospheric disturbance) tempête *f*, dépression *f*
[naut] orage *m*
~-AREA - [met] étendue *f* d'une dépression
~-BELT - [met] zone *f* des tempêtes
~CENTRE - [met] centre *m* du cyclone
~CLOUD - [met] nuée *f* d'orage
~DOOR - [constr] (in the USA only) contre-porte *f*
~DRAIN - s. storm sewer
~-GUYED-POLE - [telecomm] appui *m* consolidé, poteau *m* d'arrêt
~-LANTERN - [gen] lanterne-tempête *f*
~-PROOF - [agric] resistant au vent
~SAIL - [naut] voile *f* de cape
~SASH - [constr] contre-fenêtre *f*
~SEWER - [hydr] conduite *f* des eaux pluviales
~-WARNING - [met] (any signal used to give notice of the approach of storm conditions) avertissement *m* de tempête

STORM-WATER FLOW - [hydr] débit *m* d'eau de pluie
~-WATER TANK - [hydr] décanteur *m* pour les eaux pluviales
~WIND - [met] (wind equal to or exceeding force II in the Beaufort Scale) vent *m* d'orage
~WINDOW - s, storm sash
STORMY - [met] tempétueux, orageux
STOUT - [gen] fort *m* vigoureux
[mech] (of a machine, a structure etc) solide
[brew ind] stout *m*, bière *f* noire forte
STOVE, to - [mech] (to heat in a closed oven) étuver
[bot] élever en serre chaude
[text] (operation of bleaching by bringing wet wool in contact with sulphur fumes) soufrer
[gen] (of clothing) désinfecter
STOVE - [gen] (for heating) poêle *m*
[gen] (cooking implement) fourneau *m*
[ind chem] étuve *f*, four *m*
[gen] (drying room or box) séchoir *m*
[gen] (a pottery kiln) four *m* à briques
[metall] (for foundry work) étuve *f*
[agric] (a heated greenhouse) serre *f* chaude, forcerie *f*
~BLACKING - [ind chem] noir *m* d'étuve
~BOLT - [mech] boulon *m* à tête refoulée
~BOLT NUT - [mech] écrou *m* de boulon à tête carrée
~CHEST - [anat] enfoncement *m* du thorax
~COAL - [min] anthracite *f*
~DRYING - [gen] séchage *m* à l'étuve
~ENAMEL - [gen] émail *m* au four
~-ENAMELLING - [metall etc] (process for producing a hard glossy protective or decorative coating, in which a special enamel is applied in the usual way and is afterwards fused in a stove) émaillage *m* au four
~FINISHING - [paint] (production of a finish by applying a liquid coating and subsequently baking the whole part) finissage *m* à émail, apprêtage *m* à émail
~-KILN - [ceram] (old type of brick kiln of intermittent, up-draught pattern) four *m* à briques
~WITH WATER JACKET - [ind chem] étuve *f* à chemise d'eau
STOVEPIPE - [th eng] tuyau *m* de poêle
STOVING - [metall ceramic etc] (the operation of drying in a stove) étuvage *m*
[metall] (baking) étuvage *m*, cuisson *f*
[ind] étuvage *m*
~FINISHES - [paint] (finishing coatings which are designed to be dried by heat) émail *m* à four
~VARNISH - [paint] (type of varnish which is finished by heating in an oven) vernis *m* à four
STOW, to - [gen] mettre en place, ranger
[naut] arrimer
[transp] charger
[mining] remblayer
STOWAGE - [gen] magasinage *m*, espace *m* util
[mech] (the manner of stowing goods, parts etc) mise *f* en place
[comm] (the charge for stowing goods) frais *mpl* d'arrimage
[naut] arrimage *m*
~LOOPS - [aero] (loops used in stowing the rigging lines in a parachute pack) boucles *fpl* de lovage
STOWAWAY, to - [naut aero etc] s'embarquer clandestinement
STOWAWAY - [naut aero etc] passager *m* clandestin
STOWAWAY - [gen] cache *f*
~ARM REST - [auto] accoudoir *m* escamotable
STRABISMUS - [med] (a condition in which the visual axes of the eyes assume an abnormal position) strabisme *m*
STRABISMIC - [med] strabique
STRABISMOMETER - [instr] (instrument designed to measure the amount of strabismus) strabismomètre *m*
STRABOMETER - s, strabismometer
STRABOTOMY - [med] (the operation of cutting the eyeball muscles to correct strabismus) trabotomie *f*
STRADDLE, to - [gen] écarter les jambes, marcher les jambes écartées
[firearms] tirer à la fourchette
[mech] (of a crane etc) chevaucher
STRADDLE - [gen] écartement *m* des jambes
[mech] (of a crane etc) chevauchement *m*
~AND SLABBING CUTTERS - [tool] fraise *f* trois tailles
~CARRIER - [transp] (used for mainlaying) chariot *m* cavalier
~PACKER - [oil ind] packer *m* composé de deux éléments
~TRUCK - s, straddle carrier
STRAFILATO - [text] organsin *m* à fort apprêt
STRAGGLE, to - [gen] marcher à la débandade
STRAGGLING - [gen] marche *f* à la débandade
[phys] (the random variation or fluctuation of a property associated with ions of a given kind in passing through matter) dispersion *f*, fluctuation *f*
[gen] disséminé, éparpillé
[nucl] dispersion *f* aléatoire
~PARAMETER - [nucl] (equal to about I or 2 percent of the range for radioactive particles) paramètre *m* de dispersion
STRAIGHT - [gen] droit, rectiligne
[gen] (free of foreign matter) pur
[gen] (road, railw etc) (straight length connecting curves) rectiligne *m*
~ANGLE - [geom] angle *m* droit
~ARCH - [constr] (a flat arch) arc *m* en platebande, arc *m* linteau
~AXIAL FLOW PUMP - [mech] pompe *f* à flux axial directe, dépresseur *m* à flux axial
~BED - [mach tool] (of a lathe) banc *m* droit (d'un tour)
~BEVEL GEAR - [mech] engrenage *m* conique à dents droites
~BEVEL GEAR GENERATOR - [mach tool] fraise-mère *f* pour engrenages coniques à dents droites
~BIT - [metall] fer *m* droit (de fer à souder)
~BITUMEN - [oil ind] bitume *m* sans solvants
~BLADES - [mech] palettes *fpl* droites
~BOBBIN - [text] bobine *f* cylindrique
~BOW CENTRALIZER - [oil ind] centralisateur *m* à éléments longitudinaux
~BOX TUYERE - [metall] tuyère *f* rectangulaire
~BRICK - [constr] brique *f* rectangulaire
~BRIDGE - [constr] pont *m* droit
~BUTT-JOINT - [metall] joint *m* d'about droit
~CARBON STEEL - [metall] acier *m* au carbone
~CHAIN - [chem] chaîne *f* droite
~CHROMIUM - [ind chem] chrome *m* pur
~-COAL GAS - [gas ind] (a rich gas) gaz *m* riche
~COMBER - [text] peigneuse *f* rectiligne
~-CUP GRINDING-WHEEL - [impl] meule *f* à disque

STRAIGHT-CUTTING CIRCULAR SHEARING MACHINE - [impl] cisailles *f*pl circulaires pour métaux
~DRAWING IN - [text] rentrage *m* à la course
~EDGE - [impl] règle *f*, règle *f* à araser
~-EDGED - [mech] à tranchant droit
~EIGHT - s. straight eight engine
~EIGHT ENGINE - [mech] (an engine having eight cylinders in line) moteur *m* à huit cylindres en ligne
~EXTRUDING NOZZLE - [plast & rubber ind] tête *f* de tirage droite
~-FACED PULLEY - [mech] poulie *f* plate
~FEED - [text] alimentation *f* sans croisement
~FENCE - [carp] guide *m* rectiligne
~FIBER - [text] brin *m* lisse
~-FLANK GEAR - [mech] engrenages *m*pl à flancs rectilignes
~FLIGHT - [constr] (of stairs) (escalier) à rampe droite
~-FLOW COMBUSTION CHAMBER - [aero] (in jet engines) chambre *f* de combustion à écoulement direct
~-FLOW SYSTEM - [mech] (in a gas turbine; a combustion system in which the inflow of air follows the same direction as that of emergent gas flow) système *m* à écoulement direct
~-FLUTED - [mech] à rainures droites
~FLUTED DRILL - [tool] (a conical pointed drill with backed-off cutting edges, formed by cutting straight longitudinal flutes in the shank) foret *m* à rainures droites, mèche *f* evidée
~-FLUTED REAMER - [tool] alésoir *m* à rainures droites
~-FORWARD JUNCTION WORKING - [telecomm] méthode *f* de l'appel direct sur les lignes auxiliaires
~-FORWARD TRUNKING METHOD - s. straight-forward junction working
~-FRACTIONATION - [oil ind] fractionnement *m* primaire
~GANG PUNCHING - [comput] perforation *f* directe
~GAS UTILITY - [gas ind] (USA only; a gas undertaking in GB) société *f* gazière
~GRAINED WOOD - [bot] bois *m* à fibre droite
~GROUP WARP - [text] chaîne *f* de fond droite
~HALVED JOINT - [constr] joint *m* saillant
~HALVING - [carp] assemblage *m* à mi-bois
~HALVING - [constr] mi-bois *m*
~HEAD - [plast ind] (an extrusion head set in line with the axis of the barrel) tête *f* droite
~HOLE - [mech] trou *m* cylindrique
[constr] perforation *f* rectiligne
~HOLE HOB - [tool] fraise-mère *f* à trou cylindrique
~IN APPROACH - [aero] (approach along the axis of the runway) approche *f* directe
~INSULATOR PIN - [el] console *f* verticale
~JOINT - [carp] assemblage *m* à plat, joint *m* à plat
~KNIFE - [impl] couteau *m* droit
~-LEG-RAISING TEST - [med] signe *m* de Lasègue
~-LINE - [mech] (of a motion) ligne *f* droite
[mech] rectiligne, en tandem
~COMPRESSOR - [mech] compresseur *m* monocylindrique simple
~-LINE CONDENSER - s. straight-line frequency condenser
~-LINE DETECTION - [telecomm] détection *f* linéaire
~-LINE DETECTOR - [telecomm] détecteur *m* linéaire
~-LINE FREQUENCY CONDENSER - [radio] (variable condenser in which the capacitance is inversely proportional to the square of the scale reading) condensateur *m* orthométrique
STRAIGHT-LINE INSULATOR - [el] isolateur *m* pour alignements droits
~-LINE MOTION - [mech] mouvement *m* rectiligne
~-LINE RECTIFICATION - [radio] (linear rectification) redressement *m* linéaire
~LINE SOURCE - [phys] (large number of simple point sources of equal strength and phase arranged on a straight line and separated by equal and very small distances) source *f* rectiligne
~-LINE SPOT WELDING - [metall] soudure *f* par points en ligne
~-LINE SUPPORT - [el] support *m* d'alignement
~-LINED - [gen & mech] rectiligne, en ligne
~LINK CHAIN - [mech] chaîne *f* droite
~MATTER - [print] composition *f* à pleine justification
~MOTOR - [mech] moteur *m* à cylindres en ligne
~MOVING SHUTTLE - [text] navette *f* mue en ligne droite
~MULTIPLE - [telecomm] multiplage *m* droit
~-ON ANGLE SHOT - [cin] (a shot taken under a certain angle in a horizontal direction) prise *f* de vue angulaire directe
~OUTLET DIE - [rubber & plast ind] tête *f* droite
~-PANE HAMMER - [tool] marteau *m* à panne en long
~PASS - [text] remettage *m* suivi, remettage *m* à la course
[metall] cannelure *f* d'étirage
~-PEEN - [tool] (type of hammer) panne *f* en long
~-PEEN SLEDGE HAMMER - [tool] marteau *m* à devant
~PIPING - [oil ind] tubage *m* à diamètre constant
~POLARITY - [el] polarité *f* normale
~POLE BRACE - [telecomm] entretoise *f* droite
~RADIAL FLOW PUMP - [mech] pompe *f* à flux radial direct
~RAZOR - [impl] rasoir *m* à manche
~-READING INDEX - [instr] (USA only; direct-reading index in GB) totalisateur *m* à lecture directe
~REAMER - [tool] alésoir *m* pour trous cylindriques
~RECEIVER - [radio] (a receiver in which the high-frequency is carried out at the same frequency as that of the original signal) récepteur *m* à amplification directe, récepteur *m* sans fréquence intermédiaire
~REFINING - [petr ind] raffination *f* non-fractionnée
~RIBBON FEED - [text] alimentation *f* en rubans à fibres en long
~ROLLER - [mech] (of a roller bearing) rouleau *m* cylindrique
~-ROLLER BEARING - [mech] palier *m* à rouleaux cylindriques
~ROPE CORE - [text] âme *f* du câble droite
~RUN - [ind chem] distillé directement
[chem] (of petrol) essence *f* de distillation
~-RUN FUEL - [ind chem] combustible *m* de distillation
~-RUN GASOLINE - [oil ind] (USA only) essence *f* de distillation directe
~-RUN PITCH - [rubber ind] brai *m* de première distillation
[oil ind] résidu *m* primaire
~RUN STOCK - [oil ind] huiles *f*pl primaires
~SET - [text] boutage *m* droit, boutage *m* en ligne
~SHANK - [mech] queue *f* cylindrique
~-SHANK MILLING CUTTER - [tool] fraise *f* à queue

cylindrique
STRAIGHT-SHANK TWIST DRILL - [tool] (a metal working tool) foret *m* hélicoïdal à queue cylindrique
~-SIDE RIM - [auto] jante *f* à bords droits
~-SIDE TYRE - [auto] pneu *m* straight, enveloppe *f* à tringles
~SLAY - [text] battant *m* droit
~SLEEVE - [plumb] bout *m* à deux emboîtements, manchon *m* droit
~SLEEVE WITH BELL BRANCH - [hydr] T *m* à deux emboîtements avec tubulure à emboîtement
~SLEEVE WITH FLANGED BRANCH - [hydr] T *m* à deux emboîtements avec tubulure à bride
~SPANNER - [impl] clef *f* droite
~SPINDLE - s. straight insulator pin
~SPUR GEARS - [mech] engrenage *m* droit
~-TAIL LATHE-DOG - [mach tool] toc *m* de tour à tige droite
~TAP - [mech] taraud *m* cylindrique, taraud *m* finisseur
~-THROUGH JOINT - [el] jonction *f*, manchon *m* de raccordement lisse, à issure *f* de fil
~-THROUGH MANIFOLD - [plast ind] (a type of extrusion manifold) tubulure *f* à angle droit
~TIE UP - [text] empoutage *m* suivi
~TIP - [text] pointe *f* droite
~-TIP TONGS - [impl] tenaille *f* droite de forgeron
~TONGS - [impl] pinces *fpl* droites
~TONGUE OF A POINT - [railw] aiguille *f* droite
~-TOOTH WHEEL - [mech] roue *f* à dents droites
~-TOOTHED - [mech] (of a gear) à denture droite
~TRACK - [railw] alignement *m* droit, voie *f* en ligne droite
~TURNING - [mach tool] cylindrage *m*
~TYPE CABLE - [telecomm] (or solid type cable; cable using oil-impregnated paper as dielectric) câble *m* compact
~VACUUM FORMING - [plast ind] (process of forming sheet by drawing into a female mould by vacuum) formage *f* sous vide simple
~WARP - [text] chaîne *f* droite, chaîne *f* tendue
~WILLOW PIN - [text] dent *f* droite du loup
~WIRING - [telecomm] armement *m* plan
~WOOL FIBRE - [text] brin *m* de laine lisse
STRAIGHTAWAY MEASUREMENT - [meas] mesure *f* de transmission, mesure *f* en direct
STRAIGHTEDGE - [impl] règle *f*
STRAIGHTEN, to - [gen] redresser, rendre droit
[mech] se redresser, se rectifier, se dégauchir
[text] (tops etc) lisser
STRAIGHTEN WHILE COLD, to - [mech] redresser à froid
STRAIGHTENER - [mech] (a row of stator blades) redresseur *m*
[aero] (honeycomb; a system of intersecting surfaces resembling the cell-wals in a comb of honey) filtre *m* en nid d'abeilles
[mach tool] banc *m* de redressage, rédresseur
~BLADE - [aero] (of a jet engine) palette *f* du redresseur
STRAIGHTENING - [gen & mech] redressement *m*, dégauchissement *m*
[metall] redressage *m*
~BAR - [railw] pince *f* à riper
~MACHINE - [mach tool] machine *f* à redresser
~OF WIRE - [telecomm] rectification *f* du fil
~PRESS - [metall] presse *f* à dresser
STRAIGHTENING ROLLS - [mech] (machine for straightening rods, pipes of the like by means of adjustable rolles) rouleaux *mpl* à dresser
STRAIGHTWAY VALVE - [mech] (USA only; a full-bore valve in GB) robinet *m* à passage intégral, vanne *f* à passage intégral
STRAIN, to - [gen & phys] tendre, fatiguer
[mech] déformer, se déformer
[med] se surmener, se fatiguer
[ind chem] (to percolate, to purify by means of a strainer) filtrer, couler (un liquide)passer
STRAIN - [gen] tension *f*
[mech] (the deformation produced in a solid as a result of stress) contrainte *f*, allongement *m*
[mech] (the change in shape and or size of a body which accompanies a stressed condition) déformation *f* sous charge
[med] surmenage *m*, effort *m*
[biol] (hereditary disposition) tendance *f* héréditaire
[med] (in bacteriology) souche *f* bactérienne
[med] (a sprain) entorse *f*, foulure *f*
[biol] (line of descent, race, stock) race *f*, lignée *f*
[phys] déformation *f* spécifique
~AGE EMBRITTLEMENT - [metall] croissance *f* de la fragilité par efforts de longue durée
~AGEING - [metall] vieillissement *m* par travail à froid
~GAGE - s. strain gauge
~GAUGE - [instr] (device for measuring displacement between selected points) extensomètre *m*, jauge *f* de contrainte
~-GAUGE MEASURING BRIDGE - [instr] pont *m* de mesure de déformation
~-GAUGE ROSETTE - [instr] extensomètre *m* radial
~HARDENING - [metall] (single crystals of pure metals show a rapid plastic deformation at first but this is often followed by a considerable increase in shear strength) écrouissage *m*
~INSULATOR - [el] (an insulator inserted in the span wire of an overhead trolley-wire system) isolateur *m* d'ancrage
~ON THE WARP THREADS - [text] effort *m* sur les fils de chaîne
~ON YARN - [text] effort *m* sur le fil
~PLATE PRESS - [mech] presse *f* à cadre
~RODS PRESS - [mech] presse *f* à colonne
~-VIEWER - [opt] (a viewer utilizing the passage of polarized light through glass in order to observe strained regions) polariscope *m*, détecteur *m* de tensions
STRAINER - [impl] (any device for removing relatively large solid bodies from a liquid passing through it) filtre *m*, tamis *m*
[rubber & plast ind] (a breaker plate; a perforated plate through which the material passes to the extruder die) filtreuse *f*, boudineuse *f* à tamiser, grille *f*
[mech] (a device for tightening, strengthening or stretching) tendeur *m*, raidisseur *m*
[mech] (of a pump) crépine *f*, grenouillère *f*, reniflard *m*
[gen] (for air) épurateur *m*
[paper man] (of a paper machine) épurateur *m*
[brew ind] (a false bottom) faux fond *m*
~CAP - [auto] bouchon-filtre *m*
~CARRIER - [mech] (in engines, motors etc) porte-

filtre *m*
STRAINER CORE - [metall] noyau-filtre *m*
~SCREEN - [auto] tamis *m*
STRAINGAUGE - s. strain gauge
STRAINING - [gen] tension *f*, effort *m*, fatigue *f*
[ind chem] filtrage *m*
[rubber & plast ind] boudinage *m*
~BEAM - [constr] (the horizontal beam secured between the heads of the queen-post in a timber roof with no king-post, so as to prevent them from being forced inwards) contre-fiche *f* d'arc-boutement, entrait *m* retroussé
~CLAMP - [el] borne *f* d'arrêt
~HEAD - [rubber & plast ind] tête *f* de boudineuse filtreuse
~MACHINE - s. strainer [rubber & plast ind]
~PIECE - s. straining beam
STRAIT - [gen] étroit
[geogr] détroit *m*
~JACKET - [med] camisole *f* de force
STRAITS - [geogr] détroit *m*
STRAITWORK - [mining] travail *m* à l'étroit
STRAKE - [shipbuild] (line of plating on a vessel's hull from stem to stem) virure *f*, lisse *f*
[min] (term used in Cornwall and denoting a shallow well used for washing minerals) trait *m*, bande *f*
[mech] (of a tractor) languette *f*
[th eng] (shell plate of a boiler) virole *f*
STRAMONIUM - [chem] (the dried leaves and flowering tops of varieties of Datura used in the treatment of respiratory conditions) stramoine *f*
[bot] stramoine *f*
STRAND, to - [naut] échouer, s'échouer
[text] toronner
STRAND - [el] (a group of wires of fibres forming one element of a cable) toron *m*
[text] brin *m*, toron *m* de cordage, cordon *m*
[geogr] estrand *m*
~BOBBINS - [text] bobines *f*pl de torons
~CABLE - [el] câble *m* métallique
~DRAG MOTION - [text] dispositif *m* de freinage des torons
~OF WEB - [text] bande *f* du voile
~SPOOL - [text] bobine *f* de toron
~WORKS - [constr] défenses *f*pl des côtes
STRANDED - [naut] échoué
[text] à torons, à brins
~AERIAL WIRE - [radio] fil *m* multiple d'antenne
~CABLE - s. strand cable
~CAISSON - [constr] (watertight box with a solid floor which is floated over the side where a bridge pier is to be built) caisson *m* mobile
~CONDUCTOR - [el] (conductor woven from individual wires or strands) conducteur *m* cablé
~TWO-WIRE LEAD - [el] câble *m* à deux conducteurs torsadés
STRANDER - s. stranding machine
STRANDING - [naut] (of a vessel) échouement *m*
[text] toronnage *m*, réunion *f* des fils de caret en torons
~MACHINE - [text] toronneuse *f*, machine *f* à fabriquer les torons
STRANGLE WEED - [bot] (a weed) cuscute *f*, barbe *f* de moine
STRANGLER - [mech] (valve designed to restrict carburettor air intake to give a rich mixture at starting) volet *m* d'air, étrangleur *m*
STRANGLES - [vet] gourme *f*
STRANGLING OF GAS - [auto] (choking) étranglement *m* des gaz
STRANGULATION - [med] strangulation *f*, étranglement *m*
~OF A HERNIA - [med] étranglement *m* herniaire
STRAP, to - [gen] mettre une courroie
[naut] estroper
[med] mettre de l'emplâtre adhésif
[metall] polir à la bande de toile émeri
STRAP - [gen] courroie *f*, sangle *f*, attache *f*, lien *m*
[gen] (in a tram, bus etc) poignée *f* d'appui, bricole *f* de voiture
[mech] (for a shock absorber) plate-bande *f*, bride *f*, bande *f*
[mech] (for fastening) lien *m*, armature *f*
[packaging] (the flat thin strip of metal used for fastening a crate etc) braye *f*, ruban *m*
[mech] (a belt) courroie *f*
[el] (in a transformer, for high-voltage connections) lame *f* de contact à ressort
[shoe man] barrette *f* de soulier
[naut] estrope *f*
[electron] pont *m*, strappage *m*
[constr] (for gutters) collier *m*
[med] emplâtre *m* adhésif
[plumb] couvre-joint *m*, bande *f* de recouvrement
[mech] chape *f* de bielle, branche *f*
~-BAR - [mech] barre *f* de débrayage
~BOLT - [mech] (bolt having a flat section, so that it can be bent into a U-shape) lien *m* en fer en U, étrier *m*
~-BRAIDED - [el] (of a conductor) (conducteur) méplat
~BRAKE - [mech] frein *m* à bande, frein *m* à ruban
~BUCKLE - [mech] boucle *f* de courroie
~CLUTCH - [mech] embrayage *m* à ruban
~COIL - [el] bobine *f* spirale
~-END CONNECTING-ROD - [mech] bielle *f* à chape
~FORK - [mech] fourche *f* de manoeuvre des courroies
~FORK CAM - [mech] excentrique *f* de la fourche guide-courroie
~FORK CATCH - [mech] cliquet *m* de la fourche
~FORK MECHANISM - [mech] mouvement *m* de la courroie
~FORK ROD - [mech] fourche *f* guide-courroie
~FORK SLIDE - [mech] glissière *f* de la fourche
~FRAME GRID - [electron] grille *f* à cadran
~FREAKS - [electron] (discontinuities in the strapping system of a magnetron) discontinuité *f* de jumelage
~GUIDE - [mech] guide-courroie *f* à fourche
~HINGE - [mech] couplet *m*
~-IRON - [metall] fer *m* feuillard, feuillard *m*
~JOINT - [mech] couvre-joint *m*
~KEY - [el] contacteur *m* à lame de contact à ressort
~OIL - [mech] huile *f* de cotret
~-ON BOOSTER - [astronaut] propulseur *m* auxiliaire accolé
~RAIL - [metall] rail *m* méplat
~STRETCHER - [tool] (for packaging) tire-courroie *m*
~WEAVING - [text] fabrication *f* de sangles
STRAPPED - [mech] lié avec une courroie
[text] (in tailoring) garni de bandes
STRAPPING - [gen] (adj) solide

STRAPPING - [mech] (loops of metal) rubans *m*pl de métal
[med] emplâtre *m*
[constr] (the battens fixed to the internal face of a wall to support laths and plaster) estropes *f*pl d'attache des tiraudes
[oil ind] (the measurement of the external diameter of a tank by stretching a steel tape around each course of the tank's plates and recording the measurement) mesurage *m* d'un réservoir
[el] (in telecommunication work) câblage *m* en fil nu
[electron] (of multi-cavity magnetron; a coupling between the cavities by conductors connecting poles of the same polarity) jumelage *m*
[metall] polissage *m* à la bande de toile émeri
~MOTION - [text] (in cotton spinning) régulateur *m* de fil
~MOTION LEVER - [text] levier *m* relié à la contre-baguette
STRAPS - s. strap-iron
STRASS - [glass man] (lead of glass of considerable brilliancy) strass *m*, stras *m*
STRATAMETER - [instr] (an apparatus for surveying a borehole to determine the dip and strike of the strata and the deviation from the vertical) stratomètre *m*
STRATIFICATION - [geol & bot] stratification *f*
~PLANE - [geol] plan *m* de stratification
STRATIFIED DISCHARGE - [el] décharge *f* stratifiée
~ROCKS - [geol] roches *f*pl stratifiées
~RUBBER - [rubber ind] caoutchouc *m* stratifié
STRATIFORM - [gen] stratifié
[anat] stratiforme
[met] (of the general shape of stratus) stratiforme
STRATIFY, to - [geol] stratifier
STRATIGRAM - [radiat] (of X-rays) stratigramme *m*
STRATIGRAPH - [radiat] dispositif *m* pour radiographie en coupes, tomographe *m*, stratigraphe *m*
STRATIGRAPHIC AGE - [geol] (the relative age based on the position of a specimen in the sequence of geologicstrata) âge *m* stratigraphique
STRATIGRAPHICAL LEVEL - [geol] (geological horizon) niveau *m* stratigraphique
STRATIGRAPHY - [geol] (the historical study of the rocks of the earth's crust) stratigraphie *f*
STRATOCIRRUS - [met] cirro-stratus *m*
STRATOCRUISER - [aero] avion *m* pour vol stratosphérique
STRATOCUMULUS - [met] (patches, sheets or layers of cloud composed of rounded masses of cumulus form) strato-cumulus *m*, cumulo-stratus *m*
STRATOLINER - s. stratocruiser
STRATORTO - [text] organism *m* à faible apprêt
STRATOSCOPE - [met] (an astronomical telescope mounted in a stratosphere balloon to obtain observations of celestial bodies from outside the troposphere) stratoscope *m*
STRATOSPHERE - [met] (the layer of the atmosphere extending approximately from 30.000 to 100.000 feet above M.S.L., in which temperature variation in respect to altitude is relatively small) stratosphère *f*
STRATOSTAT - [aero] (a lighter-than-aircraft designed for operation in the stratosphere for research purposes) stratostate *m*
STRATOVISION - [telev] (rebroadcasting from an airborne television station) stratovision *f*
STRATUM - [gen] strate *m*
[geol] (a single bed of rocks bound by divisional planes) strate *m*, couche *f*
[anat] (a layer of cells) couche *f*
STRATUS - [met] (a uniform cloud layer with a flat base, usually grey in colour) stratus *m*
STRAW - [gen] paille *f*
[impl] (a slender tube used to suck up a beverage) paille *f*
~BALER - [agric] presse *f* à paille
~BALING PRESS - s. straw baler
~BAND - [text] corde *f* en paille
[metall] s. straw rope
~BAST FIBRE - [text] fibre *f* de paille
~BINDER - [agric] lieuse *f* à paille, botteleuse *f* à paille
~BOARD - [paper man] carton *m* paille
~CELLULOSE - [paper man] cellulose *f* de paille
~COLOURED COCOON - [text] cocon *m* paille
~COLOURED TINT - [dyes] teinte *f* paille
~CUTTER - [agric] hache-paille *m*
~-CUTTING MACHINE - s. straw cutter
~ELEVATOR - [agric] élévateur *m* de paille
~ENVELOPE - [gen] (a bottle wrapper) paillon *m*, revêtement *m* de paille
~FIBRE - [text] fibre *f* de paille
~FLY - [zool] mouche *f* jaune des chaumes
~HIVE - [agric] ruche *f* en paille
~LOFT - [agric] pailler *m*
~MANURE - [agric] fumier *m* pailleux
~MAT - [text] paillasson *m*
~MATERIAL - [text] tissu *m* en paille
~MATTRESS - [gen] paillasse *f*
~OIL - [oil ind] huile *f* de paille
~PAPER - [paper man] papier *m* paille
~-PRESS - [agric] presse *f* à paille
~PULP - [paper man] (pulp for paper-making obtained by cooking wheat, rye or oat straw with milk of lime in rotary digesters) pulpe *f* de paille
~ROPE - [metall] torche *f* de paille, corde *f* en paille
~SHAKER - [agric] secoueur *m* de paille
~SHREDDER - [agric] défibreur *m* de paille, broyeur *m* de paille
~SILO - [agric] meule *f* de paille
~SPLITTER - [text] fendoir *m* à paille
~STUFF - [paper man] pâte *f* de paille
~THREAD - [text] fil *m* de paille
~TRUSSER - [agric] lieuse *f* à paille, botteleuse *f* à paille
~YARN - s. straw thread
~YIELD - [agric] rendement *m* en paille
STRAWBERRY - [bot] fraise *f*
~EELWORM - [zool] anguillule *f* du fraisier
~MITE - [bot] tarsonème *m* du fraisier
~PLANT - [bot] fraisier *m*
~TREE - [bot] arbousier *m*, arbre *m* aux fraises
~YELLOW EDGE VIRUS - [bot] (a plant disease) jaunisse *f* des bords des feuilles
STRAWBOARD - [paper man] (cheap boards made from crude straw pulp) carton *m* paille
STRAY, to - [gen] vaguer, errer
STRAY - [gen] égaré
[gen] (isolated) épave
[min] (in oil drilling) formation *f* géologique imprévue

STRAY - [el] dispersion *f*
[leg] succession *f* tombée en deshérence
~ANGLE - [opt] angle *m* de diffusion
~CAPACITANCE - [radio] capacité *f* parasite
~CURRENTS - [el] courants *mpl* vagabonds
~EMISSION - [electron] (emission of electrons which do not interact with the R.F. field of the anode of a magnetron) émission *f* vagabonde
~-END LOSS - [el] perte *f* par dispersion
~FIELD - [el] (magnetic field set up near electrical machines or current-carrying conductors) champ *m* de dispersion
[radio] atténuation *f*, fuite *f*
~-FIELD EFFECT - [meas] effet *m* du champ de dispersion
~FLUX - [el] (the leakage flux in an a.c. machine or transformer) flux *m* de dispersion
~IMPEDANCES - [el] impédances *f* pl parasitaires
~LOAD LOSS - [el] (the additional losses, wherever occurring, caused by the load current due to changes in flux distribution and to eddy currents) pertes *fpl* supplémentaires
~LOSSES - [el] (the stray load losses of an electrical machine) pertes *fpl* par dispersion
~RADIATION - [radio] rayonnement *m* vagabond
[radiat] (a radiation serving no useful purpose) rayonnement *m* parasite
STRAYLIGHT - [telev] lumière *f* parasite, lumière *f* diffuse
STRAYS - [radio] (atmospherics) parasites *mpl*, atmosphériques *mpl*
STREAK, to - [gen] rayer, strier
STREAK - [gen] trait *m*, strie *f*, bande *f*
[gen] (of marble) veine *f*
[min] filon *m*, bande *f*
[shipbuild] (strake) virure *f*
[metall] atrie *f* colorée
STREAKED - [gen] strié, rayé, bariolé
[gen] (marble) veiné
[timber] (bois) filandreux
~CASTING - [metall] fusion *f* striée
~PLATE - [min] plaque *f* de biscuit
STREAKINESS - [paint] embus *mpl*
STREAKING - [telev] (pulling on whites) traînage *m*, filage *m* horizontal
~EFFECT TEST CARD - [telev] mire *f* de traînage
STREAKY - [gen] rayé, strié, en bandes
[food ind] (e.g bacon) entrelardé
STREAM, to - [gen] couler, ruisseler
[phys] (in the mechanics of fluid) couler à flots
[min] (to wash) laver, débourder
[naut] (the buoy) mouiller
STREAM - [phys] (current of flow of water or other fluids) flux *m*, jet *m*, courant *m*
[geogr] fleuve *m*, cours *m* d'eau, ruisseau *m*
[oil ind] production *f*
[metall] coulée *f*, jet *m* de métal
[astr] courant *m* météorique
~ANCHOR - [naut] ancre *f* d'embossage
~DAY - [oil ind] jour *m* de travail
~FEEDER - [print] margeur *m* à nappe
~FLOW - [hydr] débit *m* d'un cours d'eau
~GOLD - [min] or *m* alluvionnaire, or *m* de lavage
~-LINE - [hydr] fil *m* de l'eau, courant *m* naturel
~-LINED - s. streamlined
~OF LAVA - [geol] coulée *f* lavique, coulée *f* de lave
STREAM OF TRAFFIC - [railw] courant *m* de trafic
~PIRACY - [geogr] capture *f* de cours d'eau
~POLLUTION - [hydr] pollution *f* des rivières
~ROBBER - [geogr] voleur *m* de rivières
~SURVEY - [surv] relevé *m* de cours d'eau
~TERRACE - [geogr] terrasse *f* de cours d'eau
~THE BUOY, to - [naut] mouiller la bouée
~TIN - [min] étain *m* d'alluvion
STREAMER - [gen] banderole *f*
[naut] flamme *f*
[print] (USA only; headline taking up the entire width of the newspaper) manchette *f*
[electron] décharge *f* incohérente
STREAMING - [acoust] (the production of unidirectional flow currents in a medium, arising from the presence of sound waves) flux *m*
~FACTOR - [nucl] facteur *m* d'inhomogénéité
~POTENTIAL - [el biol] (the electrokinetic potential gradient resulting from unit velocity of liquid forced to flow through a porous structure or past an interface) potentiel *m* d'écoulement
STREAMLINE, to - [gen] (to design a structure so that it conforms to the pattern of natural streamline flow) caréner, fuseler
STREAMLINE - [aero] (in aerodynamics, the path of an element of fluid in steady flow) ligne *f* de courant
~FLOW - [aero] (steady flow without large-scale turbulence) écoulement *m* aérodynamique
[phys] (or laminar flow; fluid flow such that the local velocity of the fluid medium is steady in both direction and magnitude) écoulement *m* à courant naturel, flux *m* à vitesse uniforme
[hydr] écoulement *m* laminaire
~LOOP AERIAL - [radio] (a loop aerial enclosed in a streamlined casing) cadre *m* profilé
~MOTION - [phys] (laminar flow of a fluid past a body, such that no element moves in a closed curve or follows a sharply-changing path) mouvement *m* laminaire
~SHAPE - [gen] forme *f* aérodynamique
~WIRE - [aero] (a wire draw to a special cross-section to reduce drag) hauban *m* fuselé
STREAMLINED - [gen] caréné, fuselé, aérodynamique
STREAMLINER - [auto etc] véhicule *m* aérodynamique
STREAMLINING OF THE ENGINE - [railw] carénage *m* de la locomotive
~OF THE LOWER PARTS - [railw] carénage *m* des parties basses
STREET - [gen] rue *f*
~BOX - s. street manhole
~-CAR - s. streetcar
~CLEANING - [gen] nettoyage *m* des rues
~CROSSING - [gen] passage *m* de route
~DOOR - [constr] porte *f* d'entrée
~DRAIN - [hydr] caniveau *m*
~ELL - [plumb] (elbow pipe with a male thread at one end and a female thread at the other end) coude *m* de réduction
~FITTING - [plumb] (U.S.A.only) (a pipe to connect two pipes of different sizes) réduction *f*
~FLUSHER TRUCK - [auto] arroseuse *f* automobile
~GULLEY - [hydr] bouche *f* d'égout,
~INLET - s. street gulley
~-LEVEL - [constr] rez-de-chaussée *m*
~LINE - [constr] alignement *m* de voirie
~MANHOLE - [telecomm] chambre *f* sous chaussée

STREET ORGAN - [mus] orgue *m* de Barbarie
~RAILWAY - [transp] tramway *m*
~REFUGE - [road constr] (small safety island in the middle of a street for pedestrians) refuge *m*
~SPRINKLER TRUCK - s. street flusher truck
~SPRINKLING - [gen] arrosage *m* des rues
~SWEEPER - [mech] balayeuse *f* mécanique
~SWEEPINGS - [gen] balayures *fpl* de rues
~VIRUS - [med] virus *m* des rues
~WATERING STRANDPOST - [hydr] bouche *f* à clé sur trottoir
STREETCAR - [transp] (USA only) tramway *m*
STRENGTH - [gen] force *f*, énergie *f*
[mech] résistance *f*
[gen] (solidity) solidité *f*
[phys] (of materials) résistance *f* (des matériaux)
[chem] (of a solution) titre *m*
[chem] (quality in an explosive depending on the gas pressure developed by unit weight or volume) puissance *f* (d'explosion)
[phys] (of light, sounds) intensité *f*
[glass man] épaisseur *f*
~BULK RATIO - [rubber ind] résistance *f* par unité de volume
~CHANGE - [ind chem] (in soap manufacturing, the operation by which saponification is complete) cuisson *f*
~-DURATION CURVE - [el biol] (graph of the intensity of applied electrical stimuli as a function of the duration just needed to elicit responses in an excitable tissue) courbe *f* force-durée
~FUNCTION - [phys] fonction *f* densité
~OF A SIMPLE SOURCE OF SOUND - [acoust] intensité *f* d'une source sonore omnidirectionnelle
~OF A SOUND SOURCE - s. strength of a simple source of sound
~OF ACID - [chem] titre *m* de l'acide
~OF ARTIFICIAL SILK - [text] résistance *f* de la soie artificielle
~OF CURRENT - [el] intensité *f* de courant
~OF INDUCED FIELD - [el] intensité *f* de l'inductance
~OF LAYER OF CHARGE - [nucl] intensité *f* de couche de charge
~OF MAGNETIC POLE - [el] intensité *f* d'un pôle magnétique
~OF MATERIALS - [constr] résistance *f* des matériaux
~OF SHELL - [el] (the product of the magnetization by the thickness of the shell) puissance *f* d'un feuillet
~OF THE FABRIC - [text] solidité *f* du tissu
~OF THE FIBRE - [text ind] force *f* de la fibre, résistance *f* de la fibre
~OF THE SEWAGE - [hydr] concentration *f* des eaux d'égout
~OF THE SPRING - [mech] force *f* du ressort
~OF THE STEM - [bot] grosseur *f* des tiges
~OF THE STROKE - [text] force *f* du coup de chasse
~OF THE WARP THREAD - [text] résistance *f* du fil de chaîne
~OF WOOL FIBRE - [text] résistance *f* de la laine (à la rupture)
~TEST - [el] essai *m* de résistance
~TESTER - [paper man] essayeur *m* de la résistance
[text] dynamomètre *m*
~TESTER FOR CLOTHS - [text] dynamomètre *m* pour tissu
STRENGTH TESTER FOR HANKS - [text] dynamomètre *m* pour écheveaux
~TO WEIGHT RATIO - [phys] rapport *m* résistance poids
STRENGTHEN, to - [gen] renforcer, consolider, assurer, fortifier
[chem] augmenter la concentration
STRENGTHEN THE SOLUTION, to - [chem] renforcer la teneur d'une solution
STRENGTHENED BATH - [ind chem] lessive *f* remontée, lessive *f* rafraichie
STRENGTHENING - [gen] renforcement *m*, consolidation *f*
[adj] fortifiant, remontant
[constr etc] consolidation *f*
~PIECE - [mech] renfort *m*
~RIB - [mech] nervure *f* de renfort
~RING - [mech] disque *m* de renfort, anneau *m* de renfort
STREPHENOPODIA - [med] (pigeon toe) pied *m* bot varus
STREPTICEMIA - [med] streptococcémie *f*
STREPTOCOCCEMIA - s. streptocemia
STREPTOCOCCOLYSIN - [med] streptolysine *f*
STREPTOCOCCUS - [bacteriol] (a gram positive coccus of which the individuals tend to be grouped in chains) streptocoque *m*
STREPTODERMA - s. streptodermatitis
STREPTODERMATITIS - [med] dermatite *f* à streptocoques
STREPTODORNASE - [chem] (an enzyme with uses in medicine) streptodornase *m*
STREPTODUOCIN - [chem] (a mixture of equal parts of streptomycin sulphate and dihydrostreptomycin sulphate with therapeutic uses) streptoduocine *f*
STREPTOKINASE - [chem] (an enzyme-activating plasminogen used in medicine) streptokinase *m*
STREPTOLIN - [chem] (an antibiotic) streptoline *f*
STREPTOLYSIN - s. streptococcolysin
STREPTOMYCIN - [chem] (an antibiotic of major therapeutic importance) streptomycine *f*
STREPTOMYCOSIS - [med] streptomycose *f*
STREPTOTHRICIN - [chem] (an antibiotic with application in medicine) streptothricine *f*
STREPTOTRICHOSIS - [med] (rare infection, especially of the lungs) streptotrichose *f*
STRESS, to - [phys] (to subject to stress) charger, fatiguer, faire travailler
STRESS - [phys] (force acting upon a body and tending to produce strain) effort *m*, travail *m*, tension *f*
[gen] force *f*, contrainte *f*
[acoust] accent *m* tonique
[mech] (of materials) fatigue *f*, sollicitation *f*
[naut] (of water) temps *m* forcé
[med] "stress" *m*, agent *m* agressif
~ACCELERATED CORROSION - [metall] corrosion *f* accélérée par effort
~AGEING - [metall] vieillissement *m* par déformation plastique
~ANALYSIS - [phys] (in construction theory) analyse *f* des efforts
~AT BREAK - [mech] charge *f* ultime de rupture
~AT FAILURE - s. stress at break
~-BEARING PARTS - [mech] organes *mpl* soumis à efforts
~CONCENTRATION FACTOR - [phys] facteur *m* de

concentration d' efforts
STRESS CONE - [mech] (in cables) cône *m* d'usure
~CORROSION - [metall] corrosion *f* sous tension
~CORROSION CRACKING - [mech] fissuration *f* par corrosion par l'état latent d'efforts
~CRACK CORROSION - [metall] corrosion *f* par fissures dues à des tensions
~CRACKS - [metall] criques *fpl* de tension
~CYCLE - [phys] alternance *f* des efforts
~DEFORMATION DIAGRAM - [metall] courbe *f* effort-déformation
~DIAGRAM - [phys] (force diagram) diagramme *f* des efforts, diagramme *m* des forces
~DISTRIBUTION - [phys] distribution *f* des efforts
~-EQUALIZING ANNEAL - [metall] recuit *m* pour égaliser les tensions internes
~-FREE ANNEALING - [metall] recuit *m* de détente
~IN BENDING - [rubber ind] contrainte *f* par flexion
~IN TORSION - [rubber ind] contrainte *f* de torsion
~LAMP - [acoust] (used in connection with hearing aids in teaching deaf children) lampe *f* indicatrice d'accent
~LIMIT - [mech] limite *f* de fatigue, limite *f* de travail
~NUMBER CURVE - [metall] courbe *f* du nombre des cycles d' effort
~OF FLEXURE - [phys] effort *m* de flexion
~RAISER - [metall] inclusion *f* amplifiante de la fatigue
~RELAXATION - [phys] relaxation *f* de contrainte
~RELEASE - [rubber ind] relâchement *m* de contrainte
~RELIEF - s. stress relieving
~RELIEF ANNEAL - [metall] recuit *m* de détente
~-RELIEVED - [metall] recuit
~RELIEVING - [metall] (heating steel at a specified temperature and keeping it at such level of temperature for a given time in order to reduce internal stress) stabilisation *f*, recuit *m* pour l'élimination des tensions
~RUPTURE TEST - [metall] essai *m* à charge de rupture
~-STRAIN COMPRESSION-CURVE - [mech] courbe *f* de déformation par compression
~-STRAIN CURVE - [phys] (graphical representation of the relation between unit stress and unit deformation in a stressed body as a gradually increasing load is applied) courbe *f* charge-allongement, courbe *f* tension-allongement, courbe *f* de déformation
~TENSOR - [mech] (the components of the stress tensor in a continuous material are the stresses exerted across the surfaces normal to the directions of variation of a single coordinate) tenseur *m* des efforts
~-TO-RUPTURE - [mech] essai *m* à charge de rupture
~UNDER COMPRESSION - [rubber ind] contrainte *f* de compression
~UNIT - [phys] unité *f* de charge
STRESSED BOX - [aero] (a box-form structure designed to resist stresses) caisson *m* résistant
~PLATE - [mech] (plate having a bearing function) plaque *f* portante
~SKIN - s. stressed skin construction
~-SKIN CONSTRUCTION - [aero] (system in which a load-bearing skin takes a proportion of the total stresses, the remainder being borne by the internal structure) construction *f* à revêtement porteur
STRESSED-SKIN STRUCTURE - [aero] (a structure in which part of the stresses imposed is taken by a specially designed skin) structure *f* à revêtement travaillant
STRETCH, to - [gen] tendre, tirer, s'étendre
[gen] (to extend) étendre, élargir
[metall] étirer
[mech] (due to pull) allonger, retendre, bander (une courroie)
STRETCH - [gen] allongement *m*
[gen] (the extent of whatever is stretched) extension *f*
[metall] étirage *m*
[phys] (the elongation due to pull) allongement *m*, élasticité *f*
[mech] (strain) déformation
[text] (of yarn) aiguillée *f*
[naut] bordée *f*
[cin] (slow motion effect) effet *m* d'accélération de la camera
[paper man] (the elongation of paper due to the application of a breaking strain to the moment of fracture) allongement *m* de rupture
~A SPRING, to - [mech] bander un ressort, tendre un ressort
~FACTOR - [telev] facteur *m* de proportionalité
~FORMING - [plast ind] (another term for drape forming, using sheet material and a positive mould) drapage *m*, formage *m* sur moule positif
~-FORMING MACHINE - [mach tool] machine *f* à étirer et moulurer
~HAMMER - [constr] aplatissoir *m*
~MODULUS - [phys] module *m* d'élasticité
~OF LINE - [railw] tronçon *m* de voie
~OF SHOOTING FLOW - [hydr] course *f* précipitée
~OF STAPLE - [text] coefficient *m* d'allongement de la mèche
~-OUT - [gen] (a system of industrial operation designed to increase productivity) accroissement *m* de la productivité
[draw] (of a drawing) développement *m*
~SILK, to - [text] faire bouillir la soie
~THE FIBRES, to - [text] redresser les fibres
~THE HANK, to - [text] tendre l'écheveau
~YARN, to - [text] tendre les fils
STRETCHED - [gen] allongé, détendu
[mech] raide
~BEAM - [metall] poutrelle *f* étirée
~DIAPHRAGM - [acoust] (in a microphone or loudspeaker which has its rigidity increased by radial stretching, usually by screwing on it a rim near its edges) diaphragme *m* à extension
~GIRDER - s. stretched beam
STRETCHER - [impl] brancard *m*, civière *f*
[mech] tendeur *m*, tenseur *m*
[mech] (a device for expanding) extenseur *m*
[mech] (tie of a framework) tirant *m*
[constr] (a brick laid with its longer axis parallel with the length of the wall) panneresse *f*
[naut] traversin *m*, barre *f* des pieds
[leather ind] étire *f*
~BAR - [rubber & plast ind] (a roller used to prevent sheet material folding or creasing) rouleau *m* antiplis
~-BEARER - [gen] brancardier *m*, ambulancier *m*
~LEVELLER - [metall] planeuse *f*, dresseuse *f* par

étirage
STRETCHER LEVELLING - [metall] (of sheet material) dressage *m* par étirage, planage *m* par étirage
~STRAINS - [metall] (a defect in steel making) lignes *f*pl d'Hartmann
STRETCHING - [gen] tension *f*, élargissement *m*
[gen] (extension) allongement *m*
[metall] étirage *m*
~CORD - [mech] corde *f* de tension
~COURSE - [constr] assise *f* de carreaux
~IRON - [leather ind] étire *f*
~MACHINE - [mach tool] machine *f* à étirer
~MOTION - [text] étirage *m* supplémentaire par le chariot
~OF CORD - [rubber ind] (in tyre manufacturing) étirage *m* de cord
~OF THE COUPLING - [railw] allongement *m* de l'attelage
~OF THE WARP - [text] tension *f* de la chaîne
~PROPERTY - [gen] extensibilité *f*, élasticité *f*
[text] qualité *f* d'allongement
~PULLEY - [mech] poulie *f* de tension
~ROLL - [paper man] rouleau *m* tendeur
~SCREW - [mech] tendeur *m* à vis, vis *f* de tension
STREW, to - [gen] répandre, jeter
STREW GRAVEL, to - [constr] répandre du gravier
STREWN FIELD - [geol] terrain *m* parsemé de tectites
STRIA - [geol] (faint ridge, or furrow) strie *f*, striure *f*
[arch] listeau *m* (de colonne)
[bot] cannelure *f*
STRIAE - [phys chem] (gradients or variations in gas density) stries *f*pl
STRIATAL SYNDROME - [med] syndrome *m* strié
STRIATE - [gen] strié
STRIATED - [gen & anat] strié
~MUSCLES - [anat] muscles *m*pl striés
STRIATION - [gen] striation
~TECHNIQUE - [acoust] (a method for rendering sound waves visible by using their individual ability to refract light waves) méthode *f* des stries
STRICK - [text] botte *f*, riste *f*
~OF FIBRES - [text] barbe *f*, poignée *f* de fibres
~OF FLAX - [text] poignée *f* de lin
~OF HEMP - [text] botte *f* de chanvre
~OF JUTE - [text] riste *f* de jute
STRICKLE, to - [metall] trousser
STRICKLE - [gen] (a template) trousse *f*, gabarit *m*
[impl] pierre *f* à aiguiser (les faux)
[metall] (for striking a mould) trousse *f*, râble *m*, trousseau *m*
[meas] racloire *f*
~BOARD - [metall] planche *f* à trousser, planche *f* principale
~MOULDING - [metall] moulage *m* au trousseau
STRICKLED CASTING - [metall] pièce *f* troussée
~CORE - [metall] noyau *m* troussé
STRICKLING - [metall] troussage *m*
STRICT - [gen] exact, strict
STRICTION - [gen] striction *f*
[metall] (in metal testing) striction *f*
STRICTNESS - [gen] exactitude *f*, précision *f*
STRICTURE - [med] rétrécissement *m*, étranglement *m*
[metall] (in metal testing) striction *f*
STRIDE, to - [gen] marcher à grands pas
STRIDE - [gen] pas *m*, enjambée *f*
STRIDENT - [acoust] strident
STRIDING LEVEL - [surv] (sensitive level-tube fitted at each end with a leg at right angle to the tube, so that it may be placed astride a theodolite) niveau *m* à cheval
STRIDOR - [med] (harsh noise caused by an obstruction in the respiratory tubes) strideur *f*
STRIDULATION - [acoust] (the sound produced by the friction of one part of the body against another, as in some insects) stridulation *f*
STRIKE, to - [gen] percer, frapper, buter, se cogner
[metall] (in fouding) trousser, araser
[mus] frapper, toucher
[metall] (to impress with a die) trousser
[naut] piquer (l'heure)
[naut] (e.g. a rock) toucher le fond, talonner
[naut] (to run aground) heurter
[gen] (to quit work) se mettre en grève
[mining] suivre la direction, se diriger
[el] (an arc) amorcer (un arc), produire (un arc)
[leather ind] rebrousser (les peaux)
STRIKE - [gen] coup *m*
[gen] (the refusal to work) grève *f*
[geol] (the horizontal direction which is at right angles to the dip of a rock) direction *f* (d'un filon etc)
[min] rencontre *f*
[metall] trousse *f*, gabarit *m*
[el metall] (an initial thin layer of metal formed as an initial process in an electrodeposition operation) dépôt *m* amorce
[oil ind] découverte *f* de pétrole
~A BALANCE, to - [fin] établir une balance, dresser le bilan, arrêter un compte
~AN ARC - [el] amorcer un arc
~BATH - [el metall] (electrolyte used to deposit a thin layer of metal as an initial part of an electrodeposition process) bain *m* d'amorçage
~BATTEN, to - [text] pousser le battant
~-BREAKER - [gen] renard *m*
~-FAULT - [geol] faille *f* en direction, faille *f* longitudinale
~FIRE, to - [gen] allumer du feu
~GEAR - [mech] débrayage, passe-courroie *m*
~GOLD, to - [min] rencontrer de l'or
~GROUND, to - [naut] toucher le fond
~JOINT - [geol] joint *m* parallèle à la direction
~LINE - [surv] ligne *f* de direction
~NOTE - [acoust] (the note, in many ways subjective, which is prominent when a bell is struck) note *f* d'impulsion
~OF THE BEDS - [geol] direction *f* des couches
~OFF, to - [gen] abattre, trancher
[leg] (from a register) radier
[print] tirer (un nombre d'exemplaires)
~OIL, to - [min] rencontrer le pétrole, atteindre une nappe pétrolifère
~ON, to - [gen] faire entrer en frappant, ficher
~-PLATE - [impl] gâche *f*
~ROOTS, to - [bot] prendre racine
~SLIP - [geol] rejet *m* horizontal
~-SLIP FAULT - [geol] faille *f* à rejet horizontal
~THE ARC, to - [el heat] (in welding) amorcer l'arc
~THE CENTRE, to - [constr] décintrer (une voûte)
~WHEELS - [mech] roues *f*pl coniques de renversement
STRIKER - [impl] frappeur *m*

STRIKER - [firearms] percuteur *m*
[leather ind] rebrousseur *m*
[auto] (a device holding a door in alignment) gache *f* de serrure
[horol] (the hammer of a striking clock) marteau *m*
[gen] (a smith's assistant) frappeur *m*
~FORK - [auto] (striker clutch fork) fourchette *f* de débrayage
~PIN - [text] doigt *m* de contact
~ROD - [auto] doigt *m* de commande de vitesses
STRIKING - [el metall] (the process of depositing a strike) amorçage *m*
[text] battage *m*
[el] (of arc, or spark) amorçage *m*
[gen] frappement *m*
[bot] reprise *f* (d'une bouture)
[horol] sonnerie *f*
~BAR - [metall] (in foundry work) arbre *m* de trousseau
~BOARD - [metall] planche *f* à trousser
~CLOCK - [horol] (clock striking the hours only) horloge *f* à sonnerie
~CURRENT - [el] (the starter-gap current required to initiate conduction across the main gap for a specified anode voltage) courant *m* d'amorçage
~EDGE OF THE SHELL - [text] couteau *m* de l'auge
~EFFECT - [text] effet *m* de battage
~GEAR - s. strike gear
~HEAT - [brew ind] (initial temperature) température *f* d'empâtage, degré *m* initial
~PIN - [text] pointe *f* de batteur
~PLATE - [carp] (metal plate screwed to the jamb of a door) gâche *f*
~POST - [impl] poteau *m* battant
~POTENTIAL - [el] (the potential required to initiate an arc; in electronics, the starter anode potential for a cold-cathode gas triode initiating anode current flow) potentiel *m* d'amorçage
~SURFACE - [metall] (of a die) plateau *m* de battage
~THE CENTRING OF AN ARCH - [constr] décintrage *m*
~-UP - [metall] (in die casting) réglage *m* du sable
~WORK - [horol] sonnerie *f*
STRING, to - [gen] ficeler, garnir de cordes
[mus] (an instrument) mettre les cordes
[gen] (beads etc) enfiler
[chem] (of a glue) filer
[el] (cables) poser
STRING - [gen] ficelle *f*, cordon *m*, corde *f*
[anat] filet *m* de la langue
[mus] (of a string instrument) corde *f*
[bot] (e.g. of beans) fibre *f*, filament *m*
[arch] (a stringcourse) (a projecting-course in a wall) bandeau *m*, cordon *m*
[constr] (of a stair) limon *m*
[el] (series of insulator units forming a suspension insulator) isolateur *m*
[glass man] (straight or curled line) fil *m*
[geol] filet *m* de houille, petite veine *f*
[oil ind] installation *f* de forage
[comput] groupe *m* de données, chaîne *f*
[mining] petit filon *m*
~BAND - [mus] orchestre *f* à cordes
~BEAD - [metall] cordon *m* à passe étroite
~BEAN - [agric] haricot *m* vert
~CHART - [el] (diagram from which the relation between the sag of an overhead line and the temperature may be obtained) diagramme *m* d'abaissement d'une caténaite
STRING EFFICIENCY - [el] (the ratio of the flash-over voltage of a suspension insulator string to the product of the flash-over voltage of each unit and the number of units forming the string) rendement *m* d'un isolateur
~ELECTROMETER - [instr] électromètre *m* à corde
~GALVANOMETER - [instr] (a vibrator used in variable-area sound-film recording, which deflects the image of the slit across the track) galvanomètre *m* à corde
~GAUGE - [instr] (an apparatus for measuring the diameter of strings) cordomètre *m*
~INSULATOR - [el] isolateur *m* à éléments
~MANIPULATION - [comput] manipulation *f* de chaînes
~MILLING - [mach tool] fraisage *m* en série
~OF CASING - [oil ind] colonne *f* de tubage
~OF PIPE - [oil & gas ind] (a drill column) train *m* de tiges
~ORCHESTRA - s. string band
~OVER, to - [oil ind] mesurer la profondeur d'un puits pétrolifère
~PLATE - [mus] (the iron plate supporting the further ends of the strings of a pianoforte) sommier *m*
~-SHADOW INSTRUMENT - [instr] (an instrument in which the indicating means is the shadow of a filamentary conductor) appareil *m* à ombre de filament
STRINGCOURSE - [arch] (projecting course in a wall) bandeau *m*, cordon *m*
STRINGER - [aero] (a longitudinal structural element designed to stiffen the skin and to carry part of the direct load on it) lisse *f*, raidisseur *m*
[carp] (long horizontal member in a structural framework) longeron *m*
[railw] longrine *f*
[carp] (the tie of a truss) entrait *m*, tirant *m*
[min] veinule *f*, crin *m*, filet *m*
[shipbuild] (a girder) gouttière *f* renversée
[metall] (a defect) inclusion *f* linéaire dans la direction du travail
[constr] (of starir) limon *m*
[nucl] (a long structure occupying a hole through a shield and sometimes into the active section of a nuclear reactor, the removal of which allows entrance to the core for the insertion of experimental materials) grappe *m* d'éléments combustibles, faisceau *m* d'éléments combustibles
~BEAD - [mech] (first weld, first pass) première passe *f*
~PLATE - [shipbuild] tôle *f* gouttière
STRINGING - [gen] montage *m*, cordage *m*, bandage *m*
~THE BEADS - [text] enfilage *m* des perles
~PIPE - [oil & gas ind] bardage *m* des tubes
STRINGINESS - [paint] (resistance to the motion of the brush) filage *m*
STRINGY WOOL - [text] laine *f* feutrée
STRIP, to - [gen] arracher, dépouiller
[mech] (a thread) arracher le filet
[mech] (an engine; motor etc) démonter
[metall] dépouiller, dégalvaniser
[plast ind] (to remove products from the mould) démouler, décocher
[paint] (to remove paint by chemical agents) enle-

ver la peinture avec des solvants
STRIP, to - [min] décapeler (un gite)
[photo] (a negative) pelliculer (un clichet)
[el] (a cable) dénuder
[mech] rectifier
[photo] (to remove a film from its base) dépouiller
[mining] découvrir, exploiter
STRIP - [gen] bande *f*, ruban *m*
[el metall] (a bath used to remove metal from a base) bain *m* de dépouillage
[aero] (a landing strip) bande *f* d'atterrissage
[aero] (a portable runway made of steel sheet) bande *f* d'atterrissage métallique
[metall] feuillard *m*, bande *m*, feuillard *m* d'acier
[carp] languette *f*, tasseau *m*
[carp] (a strap for securing crates etc) bande *f*, feuillard *m*
[gen & rubber ind] tranche *f*
~A THREAD, to - [mech] (to destroy a screw thread by excessive tightening, incorrect insertion of one threaded part in another, or by abnormal axial loading) arracher le filet, foirer des filets
~ATTENUATOR - [radio] (absorptive attenuator in which the dissipative material has the form of a movable sheet) affaiblisseur *m* à lame
~CHART - [instr] abaque *m* pour enregistrement de variables
~-CHART RECORDER - [instr] appareil *m* d'enregistrement à bande
~COAL - [min] (opencast coal) charbon *m* d'exploitation en découverte, charbon *m* exploité à ciel ouvert
~CROPPING - [agric] culture *f* en bandes
~CULTIVATION - [agric] billonnage *m*
~FILM - [photo] film *m* en bande
~FUSE - [el] fusible *m* à lame
~GRAZING - [agric] pâturage *m* rationné, pâturage *m* par bandes
~HEATER - [el] élément *m* à ruban
~HEATING - [carp] (method of heating glued wood joints by means of an electrically-heated bare metal strip) chauffage *m* à bandes
~IRON - [metall] feuillard *m*, fer *m* en barres
~LATEX, to - [rubber ind] (to remove the monomeric substance from rubber latex) écarter les monomères du latex
~LINE - [telecomm] (or microstrip; a transmission line in which the conductors are in the form of closely-spaced strips, opposed face to face) ligne *f* à bandes parallèles
~MILL - [metall] train *m* à bandages
~MILL TRAIN - [metall] train *m* à feuillards
~-MOUNTED SET - [el] ensemble *m* interchangeable d'appareils
~OF FELT SATURATED WITH OIL - [mech] bande *f* de feutre saturée d'huile
~OF FUSES - [el] conducteurs *mpl* fusibles
~OF LEAF - [bot] lanière *f* de la feuille
~OF NEEDLES - [text] rangée *f* d'aiguilles
~PIT - [mining] carrière *f*, exploitation *f* à ciel ouvert
~PLATE - [metall] large plat *m*
~PRINTER - [photo] tireuse *f* sur bande
~PROJECTION - [photo] projection *f* de film en bande
~ROLLING MILL - [metall] laminoir *m* à feuillards
~STEEL - [metall] acier *m* feuillard
STRIP SUSPENSION - [instr] (a method of supporting the mirror in a moving-coil vibration galvanometer etc) suspension *f* à ruban
~THE CARD, to - [text] débourrer la carde
~THE CHROMIUM PLATING, to - [metall] déchromer
~THE COPS, to - [text] lever les canettes
~THE FLAT, to - [text] débourrer le chapeau
~THICKNESS - [metall] épaisseur *m* du feuillard
~TRANSMISSION LINE - [radio] guide *m* d'ondes à rubans, ligne *f* microbande
~WIDTH - [telev] largeur *f* de ligne
~WINDING MACHINE - [rubber ind] (for tyres) enrouleuse *f* de pneus
STRIPE, to - [gen] rayer, batrer
STRIPE - [gen] raie *f*, rayure *f*, bande *f*
[paint] filet *m* (de peinture)
[bot] (on a flower) panache *m*
[zool] zebrure *f*
~DISEASE ON BARLEY - [bot] maladie *f* des stries
~FILTER - [opt] filtre *m* à filets
~LOOM - [text] métier *m* à une navette
~PATTERN - [text] dessin *m* à rayures
~SIGNAL - [telev] signal *m* de filet marginal
~SMUT - [bot] maladie *f* des stries noires
STRIPED - [gen] à raies, à rayures
[anat] strié
[zool] zébré, rubané
~CRIMP - [text] crêpé *m* rayé
~CROSS - [text] rayé dans le sens de la largeur
~NET - [text] tulle *m* rayé
STRIPING - [plast ind] (longitudinal coloured markings on plastic insulation, e.g. for identification) rayures *fpl*
[gen] rayage *m*
STRIPPABLE COATING COMPOUND - s. stripping compound
STRIPPED - [metall] (of a product from the mould) démoulé
~ATOM - [nucl] (a nucleus which has lost its surrounding electrons) atome *m* nucléaire, atome *m* dépouillé
~THREAD - [mech] (a screw of which the upstanding part has been sheared off by excessive tightening, incorrect entry or undue axial loading) filet *m* foiré
STRIPPER - [mech] (a device fitted to a punch or the like, to clear the work from the punch on the return stroke) poinçon *m* éjecteur
[mech] excavateur *m*
[nucl] section *f* d'épuisement, section *f* de séparation
[chem] (chemical substance used to remove unwanted paint) décapant *m*
[mech] grue-démouleuse *f*
[text] (card cylinder) racloir *m*, débourreur *m*
[metall] (a device for extracting ingots from the mould) dispositif *m* à démouler
[mach tool] tour *m* de rectification
~COMB - [text] peigne *m* basculant
~CUTS - [rubber ind] défaut *m* de dépouillage
~PLANT - [oil ind] tour *f* de rectification
~PLATE - [plast ind] (a part fixed in such a way as to detach a moulding from the mould on opening) plaque *f* de dévêtissage
~PLATE MOULD - [plast ind] (a mould fitted with a fixed plate to detach the moulding) moule *m* à plaque de dévêtissage, moule *m* à extracteur
~STRAP - [text] courroie *f* des nettoyeurs

STRIPPER TANK - [el] (electrolytic cell in which starting sheet blanks are used) cuve *f* de dépouillage
[oil ind] cuve *f* de dépouillage
~WELL - [oil ind] puits *m* presque épuisé
STRIPPING - [mech] (of engine, motor etc) démontage *m*
[metall] (of an ingot, a product from a mould) démoulage *m* du lingot, retrait *m* de la matrice
[mech] écartement *m* du mandrin
[el metall] (the removal of plating from the base metal) dépouillage *m*, dégalvanisation *f*
[el metall] (the removal of chromium plating) déchromage *m*
[photo] (the operation of removing the negative emulsion from its glass support) pelliculage *m*
[mining] attaque *f* par le mur, dépilage *m*
[text] (the cutting of furs into strips) raclage *m*
[oil ind] distillation *f* primaire du pétrole
[nucl] (an effect observed in bombardment with deutrons or heavier nuclei, whereby only part of the incident particle emerges with the target nucleus, and the remainder proceeds with most of the momentum in its original direction) cassure *f* en vol, stripage *m*
[chem] réextraction *f*
[ind chem] lavage *m*
[gas & oil ind] (the separation of fractions of different boiling range from a liquid mixture by fractional evaporation) dégazolinage *m*, désessenciement *m*
[med] tringlage *m* (des varices)
~BLADE - [text] racloir *m*
~CASCADE - [nucl] (the group of solvent extraction stages used to wash the desired component back into the aqueous phase) cascade *f* de réextraction
~COLUMN - [ind chem] colonne *f* à désessencier
~COMPOUND - [el chem] (substance used to cover the surface of a cathode, so that the deposited metal can be easily stripped from it) composé *m* de dépouillage
[rubber ind] mélange *m* pélable, mélange *m* anticorrosif de trempage
~CRANE - [mech] (used in foundries) grue-démouleuse *f*
~DEVICE - [metall & plast ind] dispositif *m* à démouler
~DRUM - [text] tambour *m* défibreur
~FILLET - [text] ruban *m* de débourrage
~FILM - [photo] émulsion *f* à détacher, émulsion *f* pelliculable
~FORCE - [metall] force *f* de retrait
~HEAT - [oil ind] chaleur *f* de rectification
~KNIFE - [rubber ind] couteau *m* racleur
~LATTICE - [text] tablier *m* détacheur
~OF LEAVES - [bot] effeuillage *m*
~OFF THE BUR RESIDUES - [text] enlèvement *m* des débris de graterons
~OFF THE LEAF PULP - [text] enlèvement *m* de la pulpe des feuilles
~PLATE - [metall] plaque *f* de garde, guide *m*
~-PLATE MOULDING MACHINE - [metall] machine *f* à mouler avec plateau-peigne (pour engrenages)
~PLIERS - [el] pince *f* à dénuder
~ROLLER - [text] cylindre *m* débourreur
~SHEET - [text] garniture *f* de débourrage
~STILL - [ind chem] distillateur-finisseur *m*
[oil ind] appareil *m* de rectification

STRIPPING THE CARD BY SUCTION - [text] débourrage *m* pneumatique de la carde
~THE COPS - [text] levée *f* des canettes
~TOWER - [oil ind] tour *f* de rectification
~VAT - [el metall] cuve *f* de dépouillage
STRIPPINGS - [geol] terrains *mpl* de couverture
STROBE, to - [phys] (to select a desired epoch of a recurrent phenomenon) sélectionner par la méthode stroboscopique
[el] sélectionner un signal
STROBE - [radar] (or step, or gate; a bright spot generated by a strob generator under the control of some variable delay pulse) spot *m* stroboscopique
~PULSE - [comput] impulsion *f* de rythme, impulsion *f* de porte
~WHEEL - [acoust] roue *f* phonique
STROBIC - [opt] strobosique
STROBILACEOUS - [bot] (resembling a cone) strobilacé
STROBILATE - [bot] (of the nature of a cone) strobilacé
STROBILE - [bot] strobile *m*
STROBING - [radar] sélection *f* de signal
~CIRCUIT - [electron] circuit *m* modulateur
~PULSE - [radar] (a pulse, of duration shorter than the period of a recurrent phenomenon, used for scrutinizing a particular epoch of that phenomenon) impulsion *f* de sélection
~PULSE GENERATOR - [radar] (or gate pulse generator) générateur *m* d'impulsions de sélection
STROBOSCOPE - [el] (speed-measuring device consisting of a slotted disk driven at synchronous speed; rotating objects appear to rotate at a speed equal to the actual speed difference) stroboscope *m*
STROBOSCOPIC - [phys] stroboscopique
~CALIBRATING OF A METER - [instr] (a method of calibration based on stroboscopic observation of the motion of the moving element) étalonnage *m* stroboscopique d'un compteur
~CHECKING OF A METER - s. stroboscopic calibrating of a meter
~DISK - [instr] disque *m* stroboscopique
~EFFECT - [opt] (in fluorescent lamps) effet *m* stroboscopique
~FLASH LAMP - [opt] lampe *f* électrique stroboscopique
~METER DISK - [instr] (a disk having marks at equal distances on its periphery edge so that its movement can be observed stroboscopically) disque *m* stroboscopique d'un compteur
~SYNCHRONIZER - [cin] synchroniseur *m* stroboscopique
~TUBE - [electron] (a gas tube designed for the periodic production of short light flashes) tube *m* stroboscopique
STROBOTRON - [electron] (a type of cold-cathode gas tube) strobotron *m*
STROKE, to - [naut] nager, donner la nage
[gen] lisser avec la main
STROKE - [gen] coup *m*
[naut] nage *f*
[mech] (of a piston) coup *m*, battement *m*, excursion *f*
[print] trait *m*
[mech] (of an internal combustion engine) temps *m*
[mach tool] (of the tool) course *f*
[draw] trait *m*

STROKE - [horol] coup *m* (d'horloge)
[med] coup *m*(de sang), apoplexie *f*
[mech] (the distance travelled by a reciprocating part in one cycle of operation in either direction) course *f*
~-BORE RATIO - [auto] rapport *m* course/alésage
~CAPACITY - [auto] (cylinder capacity, swept volume) cylindrée *f*
~EDGE - [comput] bord *m* d'un segment
~OF PISTON - [mech] (distance travelled by a piston between successive reversals) course *f* du piston
~OF THE PAWL - [text] levée *f* du cliquet
~OF THE SLAY - [text] coup *m* du battant
~VOLUME - s. stroke capacity
~WIDTH - [comput] largeur *f* d'un segment
STROKER - [text] roue *f* à soleil
STROKING - [mech] (the piston movement) mouvement *m* du piston
STROMA - [biol] stroma *m*
STROMEYERITE - [min] (a natural metallic sulphide of copper and silver) stromeyérite *f*
STRONG - [gen] fort, robuste, puissant
[gen] (resistant) résistant
~ACIDS - [chem] acides *mpl* forts
~ALKALINE LYE - [chem] lessive *f* alcaline forte
~AMMONIACAL LIQUOR - [chem] eau *f* ammoniacale riche
~BREEZE - [met] vent *m* frais
~CARCASS - [zool] squelette *m* osseux
~CLAY - [constr] argile *f* grasse
~COLOUR - [min] (of gold) couleur *f* forte
~-COUPLING MODEL - [nucl] modèle *m* à couplage fort
~DETONATOR - [mining] capsule *f* renforcée
~ELECTROLYTE - [el] (electrolyte which is largely dissociated at low dilutions) électrolyte *m* fort
~FABRIC - [text] tissu *m* solide
~FIRMER-CHISEL - [impl] ciseau *m* renforcé
~GALE - [met] vent *m* volant
~GAMMA RAYS - [nucl] rayons *mpl* gamma durs
~GAP-LATHE - [mach tool] tour *m* parallèle renforcé
~GAS LIQUOR - s. strong ammoniacal liquor
~GRIP - [mech] serrage *m* énergique
~LODE - [mining] filon *m* puissant
~MULLIONS - [constr] montants *mpl* majeurs
~PICK - [text] chasse *f* dure, coup *m* dur
~PILLAR - [constr] contrefort *m*
~POINT - [aero] (a strong fitting in an aircraft to which a parachute static line can be attached) point *m* d'amarrage
~POST - s. strong pillar
~PULP - [paper man] (hard pulp) pâte *f* dure, pâte *f* solide
~ROOM - [constr] chambre *f* blindée, cave *f* forte
~RUBBER - [rubber ind] caoutchouc *m* nerveux, caoutchouc *m* haut module
~SAND - [metall] sable *m* fort
~SEWAGE - [hydr] eaux *fpl* usées concentrées
~SOLUTION - [chem] solution *f* concentrée
~TWISTED YARN - [text] fil *m* à torsion forte
~VENTILATION - [mining] aérage *m* intensif
~WOOL - [text] laine *f* qui à du nerf
~YARN - [text] fil *m* résistant
STRONGBACK - [shipbuild] (a stiffening beam) poutre *f* de renfort
STRONGHOLD - [constr] forteresse *f*, redoute *f*
STRONGLY DAMPED - [instr] (or dead beat; term denoting a measuring instrument in which the damping is such that any oscillating motion of its moving parts is rapidly eliminated) complètement apériodique
STRONGYLIASIS - s. strongyloidosis
STRONGYLOIDOSIS - [med] (the infestation of humans with a worm living in the intestines and causing diarrhoea; a tropical disease) strongyloïdose *f*
STRONTIANITE - [min] (natural strontium carbonate; a minor source of strontium compounds) strontianite *f*
STRONTIUM - [chem] (one of the alkaline earth metals, A.N. 38, A.W. 87.63 symbol Sr) strontium *m*
~BROMIDE - [chem] (a sedative used in medicine) bromure *m* de strontium
~CARBONATE - [chem] (a component of special glasses) carbonate *m* de strontium
~CHLORATE - [chem] (a white, crystalline powder used to produce red fire in pyrotechny) chlorate *m* de strontium
~CHROMATE - [paint] (bright yellow pigment, also called Strontium yellow) chromate *m* de strontium
~HYDROXIDE - [chem] (a plastics stabilizer) hydrate *m* de strontium
~NITRATE - [chem] (a source of red fire in pyrotechny) nitrate *m* de strontium
~OXIDE - [chem] (a starting material for other strontium compounds) oxyde *m* de strontium
~PEROXIDE - [chem] (a bleaching and oxidizing agent) peroxyde *m* de strontium
~SULPHATE - [chem] (a component of ceramic glazes) sulfate *m* de strontium
~SULPHIDE - [chem] (a depilatory) sulfure *m* de strontium
~TITANATE - [chem] (a source of strontium 90) titanate *m* de strontium
~UNIT - [nucl] unité *f* de strontium
~WHITE - [paint] (a filler, and base for lakes, obtained from celestite) blanc *m* de strontium
~YELLOW - [paint](alternative name for strontium chromate) jaune *m* de strontium
STROP, to - [naut] estroper
[gen] (to sharpen a razor with a strop) affiler, repasser sur le cuir
STROP - [naut] (a ring, or band, or hide or rope, with its ends, spliced together, used as fastening on a mast, a yard etc) estrope *f*
[gen] (strip of leather used for sharpening a razor) cuir *m* à repasser, affiloir *m*
[aero] (a length of webbing used to lengthen a parachute static line to give sufficient tail clearance of the load) prolongateur *m* de S.O.A.
STROPHANTHIN - [chem] (a glycoside obtained from Strophanthus komé and used in the treatment of cardiac conditions) strophantine *f*
STROPHANTUS SILK - [bot] soie *f* du strophantus
STROPHOTRON - [electron] (transit-time tube) strophotron *m*
STRUCK - [gen] (p.p. of to strike) frappé
[ind] (USA; shut down by a strike) chômage *m*
[constr] (dismantled, taken away) démoli, démantelé
[constr] (of a scaffolding) démonté
~CORE - [metall] (a loam formed by revolving the built-up core, covered with loam, against a strickle board) noyau *m* d'argile
~PARTICLE - [nucl] (in a nuclear process, a parti-

cle which is hit by an incident particle) particule *f* bombardée
STRUCK-UP CORE - [metall] noyau *m* troussé
STRUCTURAL - [gen] structural, de construction
~ANNEALING - [metall] recuit *m* structural
~ARRANGEMENT - [mech] plan *m* de construction
~COLOURS - [zool] (colour effects produced by structural modifications of the surface of the integument) couleurs *f*pl structurales
~CREST - [geol] crête *f* de la structure
~DAMPING - [mech] (the total damping in a structure) amortissement *m* structural
~DEFINITION - [telev] (in colour television) définition *f* structurelle
~DUAL NETWORK - [telecomm] réseau *m* dual
~FITTINGS - [constr] garnitures *f*pl structurales
~FLEXIBILITY - [constr] flexibilité *f* de la structure
~FORMULA - [phys] (represents the manner in which the atoms in a molecule are bonded by valency) formule *f* de constitution
~HARDENING ALLOY - [metall] alliage *m* de durcissement structural
~HYBRIDISM - [genet] (an organism in which the two seats of chromosomes in the diploid complement are different in composition) hybridisme *m* de structure
~IRON - [metall] fer *m* de construction, charpentes *f*pl métalliques
~MILL - [metall] laminoir *m* à fers profilés
~RESISTANCE - [aero] résistance *f* de pénétration
~RETURN LOSS - [telecomm] (regularity return-loss) affaiblissement *m* de régularité
~SHAPE - [metall] profilé *m*
~STEEL - [metall] acier *m* de construction
~SURFACE - [geol] surface *f* structurale
~TIMBER - [constr] charpentes *f*pl en bois
~WEIGHT - [mech] poids *m* à vide
STRUCTURALLY SYMMETRIC NETWORK - [telecomm] réseau *m* à structure symétrique
STRUCTURE - [gen] (that which is constructed) structure *f*
[metall] (of steel etc) structure *f*
[constr] construction *f*, batisse *f*
[constr] (construction of a particular work, e.g. a bridge, a road, a railway line) ouvrage *m* d'art, travail *m* d'art
[aero] structure *f*
[railw] gabarit *m*
~AMPLITUDE - [crystall] facteur *m* de structure
~CONFLICT - [el] rapprochement *m* dangereux de deux lignes aériennes
~FACTOR - s. structure amplitude
~GAUGE - [railw] (maximum moving dimensions) gabarit *m* de libre passage
~OF THE CLOTH - [text] contexture *f* du tissu
~OF THE FIBRE - [text] structure *f* de la fibre
~OF THE FLEECE - [text] structure *f* de toison
STRUCTURELESS - [geol] amorphe
STRUM - [mech] (strainer for the suction pipe of a pump) crépine *f* de pompe
STRUMA - [med] strume *f*, scrofules *f*pl
STRUMIPRIVAL - [med] strumiprive
STRUMITIS - [med] thyroïdite *f* typhique
STRUMMING - [acoust] (sound made on musical percussion instruments) pianotage *m*
STRUT, to - [constr] entretoiser, étrésillonner
[constr] (to stiffen) renforcer, contre-ficher
STRUT, to - [aero] moiser
[mining] (in excavation work) étayer (une tranchée)
STRUT - [aero] (a structural element designed to resist compressive stresses) mât *m*, pilier *m*
[constr] (framework member) entretoise *f*, étrésillon *m*, montant *m*, cale *f*
[telecomm] contre-fiche *f*
[el] soutien *m*
[constr] (in a wooden roof-truss) jambe *f* de force
[constr] (of an iron roof-truss) bielle *f*
[constr] (between floor joists) lierne *f*
[mech](strength of materials) pièce *f* chargée debout
~-ACTION PAWL-MOTION - [mech] encliquetage *m* à frottement
~BRACING - [constr] poutre *f* en U
~CAMERA - [photo] chambre *f* à tendeurs
~FRAME - [constr] treillis *m* avec pièces inclinées
~GIRDER - [constr] poutre *f* en treillis
~JACK - [constr] vérin *m* de serrage, vérin *m* d'écartement
~OF A TRUSS - [constr] jambe *f* de force
~OF TRUSS - [constr] bielle *f*
STRUTTED - [mining] étayé
~POLE - [el & telecomm] poteau *m* d'arrêt, appui *m* consolidé
~TERMINAL POLE - [el & telecomm] appui *m* tête de ligne, appui *m* terminal
STRUTTING - [constr] (the process of using props to give temporary support between two surfaces) entretoisage *m*, entretoisement *m*, renforcement *m* par contre-fiches, étrésillonnement *m*
~BEAM - [constr] entretoise *f*, lierne *f*, étrésillon *m*
~OF BEAMS - [constr] entretoisement *m* des poutres
STRYCHNINE - [chem] (an intensely poisonous alkaloid obtained from Nux vomica and with uses in medicine and as a rodenticide) strychnine *f*
STUB, to - [gen] arracher
[agric] essoucher
[gen] (to strike; in particular a toe against an obstacle) se cogner, se heurter
STUB - [gen] souche *f*
[comm] (of a cheque, the inner end) talon *m* (de cheque)
[gen] (any short part or piece, remnant) bout *m*, tronçon *m*
[mech] mentonnet *m*, ergot *m*, goujon *m*
[aero] (a projecting part, of which the length is small in relation to its other dimensions; e.g. the structure on which the engine pod is mounted close to the fuselage) nageoire *f*
[radio] (of an aerial; small stretch of resonating feed line) adaptateur *m*, plongeur *m*
[mining] pilier *m* de galerie
~AERIAL - [radio] (an aerial to which a stub has been attached for matching purposes) antenne *f* à bras de réactance
~ANTENNA - s. stub aerial
~AXLE - [auto] (a dead axle supporting a steerable wheel) fusée *f*
~BOLT - [mech] goujon *m*
~CABLE - [telecomm] câble *m* de raccordement
~CARD - [comput] (a card with a perforated extension serving as printer counterfoil) carte *f* à volet
~-END - [mech] (of a connecting rod, the crank pin) tête *f* de bielle
~-END OF CONNECTING ROD - [mech] grosse tête *f*

de bielle
STUB-END STATION - [railw] (USA only; reversing station in GB) gare *f* de rebroussement
~GUY - [telecomm] hauban *m* ancré à l'autre côté de la route
~-MATCHED AERIAL - [radio] dipôle *m* avec adapteur
~-MATCHED ANTENNA - s. stub-matched aerial
~MATCHING - [telev] (of aerials; matching with a stub of a quarter-wave length) adaptation *f* à bras de réactance
~MORTISE - [mech] mortaise *f* aveugle
~MULTIPLE FEEDER - [radio] (of an automatic station) ligne *f* en antenne multiple
~OF A TOWER - [telecomm] embase *f* de pylône
~PIPE - [aero] (a short exhaust pipe from an i.c. engine cylinder, opening directly into the atmosphere without a manifold) pipe *f* d'échappement direct
~PLANE - [aero] (a short plane forming part of fuselage or hull, to which the wing is attached) moignons *mpl*
(of a flying boat) (a projection from the hull to provide stability) nageoire *f*
~-REINFORCED POLE - [telecomm] poteau *m* d'exhaussement
~REINFORCEMENT - [telecomm] tuteur *m* en bois pour poteau
~SUPPORT - [radio] (of waveguides) bras *m*
~SUPPORTED COAXIAL - [electron] (in waveguides) ligne *f* coaxiale à support en bras de réactance
~TENON - [carp] (very short tenon for fitting into a blind mortise) tenon *m* invisible
~TOOTH - [mech] (of a gear; a gear tooth of smaller height and more robust form than that usually employed) dent *f* abaissée, stub *m*
~TOOTH GEAR - [mech] engrenage *m* à stub
~TOOTH HOB - [tool] fraise-mère *f* pour stubs
~TUNER - [radio] (of waveguides) bras *m* de réactance syntonisateur
~WING - [aero] demi-aile *f*, moignon *m*
STUBBLE - [gen & agric] chaume *m*, éteule *f*
~CLEANER - [agric] charrue *f* déchaumeuse
~-CLEANING - [agric] déchaumage *m*
~CLOVER - [bot] trèfle *m* semé sur chaumage
~CROP - [agric] culture *f* de chaumage
~FIELD - [agric] chaumes *mpl*
~MANURE - [agric] fumier *m* à courte paille
~PASTURE - [agric] pâturage *m* de chaume
~PLOUGH - [agric] déchaumeuse *f*
~PLOUGHING - [agric] déchaumage *m*
~SEED - [agric] semis *m* sur chaume
STUBBORN - [gen] obstiné
[mech etc] (difficult to handle) réfractaire
~ORE - [miner] (difficult to treat) minerai *m* réfractaire, minerai *m* rebelle
STUBBY - [gen] trapu
[bot] (of a plant) tronqué
[agric] (of a stretch of land) couvert de chicots
STUCCO, to - [constr] (to apply a smooth-surfaced plaster) stuquer, enduire de stuc
STUCCO - [constr] (fine plaster for walls etc) stuc *m*
STUCCOWORK - [constr] stucage *m*
STUCK-IN - [mech] (e, g. piston ring) collé
STUD, to - [mech] goujonner, garnir de clous
[constr] soutenir au moyen de poteaux
STUD - [mech] (a metal rod having threads at both ends, or a single thread extending over its whole length designed to be screwed into one part and to secure another to it by means of a nut) goujon *m* prisonnier, boulon *m* prisonnier
[mech] (projecting at one end) goujon *m*, tourillon *m*
STUD - [mech] (a pin) cheville *f*, goujon *m*
[constr] montant *m*, poteau *m*, tournisse, quille *f*
[metall] (claplet with round head and bottom plate) support *m* d'âme
[el] plot *m* de contact, tige *f* de fixation
[metall] (in patternmaking for foundry work) tige *f*
[mech] (crosspiece in a link) traverse *f*, tenon *m*
[zool] (collection of horses and mares for breeding) écurie *f*
[zool] (a stallion) étalon *m*
[shoe man] caboche *f*, clou *m* à souliers
~BOLT - [mech] goujon *m* prisonnier
~-BOLT BOSS - [mech] bossage *m* de prisonnier
~-BOOK - [agric] (in animal breeding) livre *m* de origines, stud-book *m*
~-BOX - [mech] (a tool for inserting studs) porte-prisonniers *m*
~BULL - [zool] taureau *m* de souche
~BUSH - [mech] douille *f* du tourillon
~CHAIN - [mech] chaîne *f* à fuseaux
~CHAPLET - [metall] support *m* double
~COUPLING - [mech] nez *m* de raccord
~-EXTRACTOR - [tool] (a tool for removing broken studs) extracteur *m* de goujons
~FARM - [agric] haras *m*
~FOR CLICK - [mech] tourillon *m* du cliquet
~HORSE - [zool] étalon *m*
~JOINT - [mech] joint *m* à prisonniers
~LATHE - [mech tool] tour *m* pour boulons
~-LINK - [mech] (of a chain) maille *f* à étai, maille *f* étançonnée
~-LINK CHAIN - [mech] chaîne *f* à étais
~-LINK CHAIN CABLE - [el] chaîne-câble *f* à étais
~NUT - [mech] écrou *m* prisonnier
~PARTITION - [carp] (wooden partition based on rough timber framing) mur *m* de séparation latté et plâtré
~PLATE - [mech] disque *m* du mécanisme d'arrêt
~REAMER - [mach tool] alésoir *m* à queue
~RIVET - [mech] (a screw rivet) rivet *m* prisonnier
~SETTER - s. stud box
~SYSTEM - [el] système *m* à plots
~TEMPERATURE - [electron] (of a power transistor) température *f* du goujon de contact
~UNION - s. stud coupling
~WHEEL - [mech] roue *f* intermédiaire
STUDDED TYRE - [rubber ind] enveloppe *f* à tétons
STUDDING - [mech] (metal rod having thread cut on it from end to end, used for repair work etc) rondin *m* fileté
[metall] soudure *f* à prisonniers
~SAIL - [naut] bonnette *f*
STUDDLE - [mining] (timber post of a frame) poteau *m* dans les angles (pour les cadres)
STUDIO - [gen] (of an artist, a photographer etc) atelier *m*
[cin] théatre *m* de prise de vues
[radio] auditorium *m*, studio *m* d'émission
~BROADCAST - [radio] émission *f* du studio
~CAMERA - [photo] chambre *f* d'atelier
~DECORATION - [cin] décoration *f* imaginative d'un studio, décors *mpl* du studio
~EQUIPMENT - [cin & telev] équipement *m* du studio

STUDIO EXTERIORS - [cin & telev] simulation *f* d'extérieurs dans le studio
~FACILITY - s. studio equipment
~FLOOR - [cin & telev] plateau *m*, scène *f*
~LIGHT BOARDS - [cin] banc *m* de lampes fluorescentes dans le studio
~LIGHTS - [cin] éclairage *m* de prise de vues
~LINING - [acoust] (the use of materials to improve the acoustic qualities of a studio) revêtement *m* de studio
~PICK-UP - [radio] s. studio broadcast
~SCENERY - [cin] décor *m*, vue *f* fixe
STUDTITE - [min] (secondary ore containing uranium and lead) studtite *f*
STUFF, to - [gen] bourrer, rembourrer
[constr] (to fill with material) remblayer
[gen] (to fill in the skin of a beast or bird) empailler
[mech] (to plug, to obstruct) boucher, garnir, rembourrer
STUFF - [gen] substance *f*, matière *f*
[text] (woven material) étoffe *f*, tissu *m*
[carp] planches *f*pl
[paper ind] pâte *f* à papier
[gen] (collection or more or less worthless things) fatras *m*
~CHEST - [text] bassin *m* d'agitation
~FIBRE - [paper] fibre *f* de la pâte à papier
~RUN INTO SMALL KNOTS - [paper man] pâte *f* granuleuse
STUFFER - [mech] (U.S. term for a hydraulic-ram extrusion machine) presse *f* à filer, machine *f* à extrusion
STUFFING - [gen] bourrage *m*, rembourrage *m*
[gen] (of animals) empaillage *m*
[agric] (the fattening of geese) gavage *m*
[mech] étoupage *m*, garniture *f*, garniture *f* de presse-étoupes
~BOX - [mech] (also termed a gland. An annular device surrounding a shaft or rod and filled with compressible packing material on which a nut or ring-shaped part presses, to form a seal through which the shaft can work) presse-étoupe *m*, boîte *f* à bourrage
~-BOX FOLLOWER - [mech] (the annular element in a stuffing-box which compresses the packing) anneau *m* de presse-étoupe
~-BOX NUT - [mech] écrou *m* de presse-étoupe
~DRUM - [leather ind] tonneau *m* à mettre en suif
STUFFY - [gen] (of a building) privé d'air
[gen] (of the weather) étouffant, suffocant
STUKE - s. stucco
STULL - [mining] (timber prop between the walls of a stope) pont *m* de travail
~TIMBERING - [mining] boisage *m* étayé
STULLED STOPE -[mining] taille *f* étayée
STUM, to - [agric] soufrer (le vin)
[agric] (in the USA; the operation of adding wort to wine in order to renew fermentation) renforcer le vin avec du moût
STUM - [agric] moût *m* muet
STUMP - [gen] souche *f*, tronçon *m*
[anat] (of a tooth) racine *f*
[med] moignon *m*
[fin] (of a cheque) souche *f* de cheque
[draw] estompe *f*
~HALLUCINATION - [med] membre *m* fantôme
STUMP TAILED SHEEP - [zool] mouton *m* à queue rudimentaire
~TENON - [carp] tenon *m* court conique
~WOOD - [timber] bois *m* de souche
STUMPER - [agric] machine *f* à essoucher
STUNT, to - [aero] (USA only) faire des acrobaties en vol
[gen] empècher de croître
STUNT - [gen] coup *m* d'épate
[gen] (in journalism) article *m* à fracas
[aero] acrobatie *f*
[bot & biol] arrêt *m* dans la croissance, avorton *m*
~MAN - [cin] casse-cou *m*
STUNTED WOOL FIBRE - [text] brin *m* de laine rabrougri
STUNTING - [text] (of wool fibres) rabrougrissement *m*
~OF THE WOOL FIBRE - [text] rabrougrissement *m* du brin de laine
STUPE - [med] (piece of cloth soaked in hot water, wrung out dry, used for external applications) compresse *f* pour fomentation
STUPEMANIA - s. stupor
STUPOR - [med] (state of mental and physical inertia) stupeur *f*
STURDINESS - [gen] robustesse *f*, vigueur *f*
STURDY - [gen] fort, robuste, vigureux
[vet] tournis *m*
~BAR - [metall] lingot *m*
~FABRIC - [text] tissu *m* solide
STURGEON - [zool] esturgeon *m*
STURINE - [chem] (a protamine isolated from fish testicles) sturine *f*
STUTTERING - [med] bégaiement *m*, balbutie *f*
S-TWIST - [text] (twisting of a yarn in a left-handed sense) torsion *f* droite, torsion *f* "S"
STYLE, to - [gen] (to name) dénommer
[gen] (to give a title to) donner le titre
STYLE - [gen] (the mode of expression) style *m*
[med] (used in surgery) stylet *m* à bouton olivaire
[el acoust] (of a record player) style *m*, aiguille *f*
[gen] (title) titre *m*, nom *m*
[tool] burin *m*, style *m*
[bot] style *m* (de l'ovaire)
STYLET - [impl] stylet *m*
[med] stylet *m* à bouton olivaire
STYLING - [gen] ornementation *f* tracée au style
[text] bordure *f* (de tapis)
STYLISH - [gen] élégant
STYLIST - [gen] (adviser on style in clothes, industrial products etc) styliste *m*
STYLIZE, to - [gen] (to conform to a set of rules etc) styliser
STYLOBATE - [arch] (continuous base for a number of columns) stylobate *m*, soubassement *m* de colonnade
STYLOID - [anat] (pillar-shaped) styloïde
STYLOMETER - [instr] (instrument designed to measure columns) stylomètre *m*
STYLUS - [gen] (pointed writing instrument) style *m*
[acoust] (the cutter in a gramophone recording-head) style *m*, aiguille *f*
[acoust] (the needle of a sound recording instrument) style *m* enregistreur, graveur *m*
[mach tool] (the tracer point, e.g. of a copy milling machine) traçoir *m*
[horol] (the gnomon of a sun-dial) gnomon *m*, style

m de cadran solaire
STYLUS DRAG - [acoust] frottement *m* de l'aiguille
~FORCE - [acoust] pression *f* du style, effort *m* du style
~-GROOVE RESONANCE - [el acoust] résonance *f* sillon-aiguille
~JUMP - [acoust] déraillage *m* du graveur
~PRESSURE - s. stylus force
STYPAGE - [med] stypage *m*
STYPHNIC ACID - [chem] (a derivative or resorcinol with uses as a detonating agent) acide *m* styphnique
STYRALLYL ACETATE - [chem] (a perfumery agent) acétate *m* de styrallile
~ALCOHOL - [chem] (dyestuff intermediate and component of perfumes) alcool *m* styrallique
STYRAX - [bot] (gum obtained from varieties of Liquidamar and used in perfumery) styrax *m*, aliboufier
STYRENE - [chem] (intermediate and starting material for polystyrene plastics, synthetic rubbers, resins and protective coatings) styrol *m*
~-BUTADIENE RUBBER - [chem] (a synthetic rubber obtained by the copolymerization of styrene and butadiene. It has properties and uses similar to natural rubber) caoutchouc *m* de styrolène-butadiène
~NITROSITE - [chem] (a product of the reaction of nitrogen dioxide and styrene. Its production is the basis for a test for monomeric styrene in other hydrocarbons) nitrosite *f* de styrol
~OXIDE - [chem] (organic intermediate miscible in benzene, acetone, ether and methanol) oxyde *m* de styrol
~RESINS - [chem] (synthetic resins obtained by the polymerization of styrene, with very good mechanical properties and corrosion resistance) résines *f*pl de styrolène
STYROLENE BROMIDE - [chem] bromostyrol *m*
SUB - (prefix used in compound words) sub-
SUB - [telecomm] (short for subscriber) abonné *m*
[telecomm] (short for subscriber's apparatus) poste *m* d'abonné
[gen] (short for submarine) sous-marin *m*
SUBACETATE - [chem] sous-acétate *m*
SUBACID - [chem] acidule, aigrelet
SUBACRID - [chem] (not quite acrid) aigre-doux
SUBACUTE - [med] (between acute and chronic) subaigu
SUBAERIAL - [geol] subaérien
SUBALIMENTATION - [med] sous-alimentation *f*
SUBALTERN - [gen] subalterne
SUBAQUEOUS - [gen] subaquatique
~PUMP - [mech] pompe *f* fonctionnante noyée
SUBARCTIC - [geogr] presque arctique
SUBASSEMBLY - [mech] (a group of parts assembled as a unit of a larger assembly) sous-ensemble *m*
SUBASTRAL - [geogr] terrestre, sublunaire
~FIXTURES - [mech] équipement *m* pour le montage d'un sous-ensemble
SUBATOMIC - [nucl] (within the atom) subatomique
~PARTICLE - [nucl] (particle yielded by reactions in which atoms undergo disintegration) particule *f* subatomique
SUBATOMICS - [nucl] (the study of changes occuring within the atom) subatomique *f*
SUBAUDIO FREQUENCY - [telecomm] fréquence *f* infra-acoustique
SUBAUDIO TELEGRAPH SET - [telecomm] installation *f* de télégraphie infra-acoustique
~TELEGRAPHY - [telecomm] télégraphie *f* infra-acoustique
SUBBAND - [comput] bande *f* partielle
SUBBASE - [constr] fond *m*, sous-dalle *f*
[arch] (of a column, a pillar) soubassement
[mech] (a base supporting engine and generator in a Diesel engine) soubassement *m*
SUB-BOUNDARY FIGURES - [metall] figures *f*pl de corrosion
SUBBUTUMINOUS COAL - [min] lignite *f*
SUBCARBONATE - [chem] sous-carbonate *m*
SUBCARRIER - [telev] (carrier signal inserted within the video-frequency pass-band to provide a channel for the transmission of additional information) sous-porteuse *f*
~BALANCE - [telev] équilibre *m* de la sousporteuse de la chrominance
~BEAT - [telev] intérférence *f* dans la sousporteuse de la chrominance
~BUZZ - [telev] bourdonnement *m* dans le signal de luminance
~DOTS - [telev] moiré *m* par le signal de chrominance
~DRIVE - [telev] signal *m* sinusoidal à la fréquence de la sousporteuse de référence
~FREQUENCY MODULATION - [radio] modulation *f* de fréquence d'une sousporteuse
~OFFSET - [telev] décalage *m* de la sousporteuse de la chrominance
~REGENERATOR - [telev] régénérateur *m* de la sous-porteuse de chrominance
SUBCAUDAL - [zool] subcaudal
SUBCHLORIDE - [chem] sous-chlorure *m*
SUBCIRCUIT - [el] (lighting circuit supplied from a common branch distribution fuse-board) circuit *m* secondaire
SUBCLAVIAN - [anat] (situated under the clavicle) sous-clavier
SUBCLAVICULAR - [med] sous-claviculaire
SUBCLOUD LAYER - [met] (a stable layer of convection clouds below the general cloud base) couche *f* de convection au dessous des nuages
SUBCOMMUTATION - [el] sous-commutation *f*
SUBCONSCIOUS - [gen] (denoting those phenomena of mental life which are not attended by full consciousness) subconscient
SUBCONSCIOUSNESS - [gen] subconscience *f*
SUBCONTRACT, to - [comm] (to make a contact which is subordinate to a main contract) sous-traiter
SUBCONTRACT - [comm] sous-traité *m*
SUBCONTRACTOR - [comm] sous-entrepreneur *m*
SUBCONTROL OFFICE - s. subcontrol station
~STATION - [telecomm] station *f* sous-directrice
SUBCOOLING - [chem] refroidissement *m* au dessous de la condensation
SUBCORTEX - [anat] sous-cortex *m*
SUBCRITICAL - [nucl] (having effective multiplication-constant less than one, so that a self-supporting chain reaction cannot be maintained) sous-critique
~ANNEALING - [metall] recuit *m* sous-critique
~MASS - [nucl] masse *f* sous-critique
~REACTOR - [nucl] réacteur *m* sous-critique

SUBCRUST - [constr] (in road construction) couche *f* inférieure
SUBCULTURE - [bot] (a culture prepared from a preexisting culture) sous-culture *f*
[med] (in bacteriology) repiquage *m*
SUBCUTANEOUS - [anat] (just below the skin) sous-cutané
~BLOWHOLE - [metall] soufflure *f* sous la surface
SUBDELIRIUM - [med] subdélire *m*
SUBDIVIDE, to - [gen & math] subdiviser
SUBDIVIDED CAPACITOR - [el] boîte *f* de capacités
SUBDIVISION - [gen] sudivision *f*, fraction *f*, sectionnement *m*
~RULE - [print] couillard *m*
SUBDORSAL - [anat] (situated below the dorsal surface) sous-dorsal
SUBDRIFT - [mining] voie *f* intermédiaire
~CAVING - [mining] exploitation *f* en sous-étages avec foudroyage
SUBDUE, to - [light] atténuer (la lumière)
SUBEDITOR - [gen] secrétaire *m* de la rédaction
SUBERIC ACID - [chem] (an intermediate for polymers, pharmaceuticals and dyestuffs) acide *m* subérique
SUBERIFICATION - s. suberisation
SUBERIN - [bot] (mixture of fatty substances in the cell walls of corky tissue) subérine *f*
SUBERISATION - [bot] (the impregnation of cell walls with suberin, with subsequent formation of cork) subérification *f*
SUBEXCHANGE - [telecomm] tableau *m* secondaire
SUBFAMILY - [bot] sous-embranchement *m*
SUBFEEDER - [el] artère *f* secondaire, feeder *m* secondaire
SUBFLARE - [astr] sous-éruption *f*
SUBFRAME - [meas] trame *f* secondaire
[auto] soubassement *m*
SUBGRADE - [constr] (the earth surface on which the bottom pavement of a building is built) plate-forme *f* des terrassements
SUBGRAVITY - [phys] subgravité *f*
SUBGROUP - [gen] sous-genre *m*
SUBHARMONIC - s. subharmonic vibration
~VIBRATION - [acoustic] oscillation *f* subharmonique, vibration *f* subharmonique
SUB-HEAD - s. subheading
SUBHEADING - [print] sous-titre *m*
SUBICULUM - [anat] crochet *m* de la circonvolution de l'hippocampe
SUBINVOLUTION - [med] subinvolution *f*
SUBIRRIGATION - [agric] irrigation *f* profonde, irrigation *f* souterraine
SUBJACENT - [geol] subjacent, sous-jacent
SUBJECT, to - [gen] assujettir, soumettre
[chem med etc] (to treatment) soumettre
SUBJECT - [gen] sujet *m*
[photo] sujet *m*, objet *m*
~CONTRAST - [photo] intervalle *m* des brillances extrêmes
[radiat] (contrast arising from variation in radiation opacity within an irradiated object) contraste *m* de l'objet
~CONTRAST RANGE - [radiat] latitude *f* du contraste de l'objet
~TO DUTY - [fin] passible de droits de douane
~TO OFFICIAL APPROVAL - [leg] sujet à autorisation
~TO PRESCRIPTION - [leg] prescriptible
SUBJECTIVE - [gen] (relating to mental states) subjectif
~BRIGHTNESS - [opt] (luminosity) luminosité *f*, brillance *f* subjective
~CONTRAST - [radiat] (qualitative contrast in a radiograph or fluorescence screen reproduction as estimated visually) contraste *m* subjectif
~NOISE METER - [instr] (instrument for the measurement of loudness by aural comparison with a standard) psophomètre *m* subjectif
~PHOTOMETER - [instr] (visual photometer) photomètre *m* subjectif
~SHADOW SERIES - [opt] série *f* de teintes subjectives
~TONE - [acoust] (aural sensation of a tone of a particular frequency in the absence of stimulation by sound of that frequency) ton *m* subjectif
SUBLATIO - [med] ablation *f*
SUBLEASE - [leg] sous-location *f*
SUBLESSEE - [leg] sous-locataire *m*
[comm] sous-preneur *m*
SUBLET, to - [leg] sous-louer
SUBLETHAL LEVEL - [nucl] niveau *m* sous-létal
SUBLEVEL - [nucl] (the part of a nuclear reactor which is built below ground level) sous-niveau *m*
SUBLIMATE - [chem] (a substance obtained as the result of sublimation) sublimé *m*
SUBLIMATED SULPHUR - [chem] fleurs *fpl* de soufre
SUBLIMATION - [phys chem] (change from the solid to the vapour state without an intermediate liquid state) sublimation *f*
~CURVE - [phys chem] (the graphical representation of the variation with temperature of the vapour pressure of a solid) courbe *f* de sublimation
~PRESSURE - [phys chem] tension *f* de sublimation
~VEIN - [geol] filon *m* de sublimation
SUBLIMATORY - [gen] sublimatoire
SUBLIME, to - [phys chem] (to pass from the solid to the gaseous state without melting) sublimer
SUBLIMED WHITE LEAD - [chem] (commercial term for basic lead sulphate used in paint manufacture) plomb *m* blanc sublimé
SUBLIMER - [chem] sublimatoire *m*
SUBLIMINAL - [gen] (below the threshold of consciousness) subliminal
SUBLIMING ABLATOR - [astronaut] matériau *m* d'ablation par sublimation
SUBLINGUITIS - [med] inflammation *f* de la sublinguale
SUBLUNAR POINT - [astr] point *m* sublunaire
SUBLUXATION - [med] (partial dislocation of a joint) luxation *f* incomplète, subluxation *f*
SUBMACHINE GUN - [firearms] mitraillette *f*
SUBMARINE - [gen] sous-marin
[gen] (in the navy) sous-marin *m*
~BELL - [naut] cloche *f* sous-marine
~BLASTING GELATINE - [chem] (special type of gelatine designed for use under water) gélatine *f* explosive sous-marine
~CABLE - [el] câble *m* sous-marin, câble *m* sous-fluvial
~EARTHQUAKE - [geol] sisme *m* sous-marin
~GATE - [plast ind] (edge gate located below the parting line or the top of the mould) entrée *f* inférieure
~INTERMEDIATE REACTOR - [nucl] réacteur *m* intermediaire de sous-marin
~MINE - [chem] (container for high explosives floa-

ted either on the surface or at predetermined depth) mine *f* sous-marine
SUBMARINE REACTOR - [nucl] réacteur *m* de sous-marin
~TELEGRAPH CABLE - [telecomm] câble *m* sous-marin télégraphique
~TENDER - [naut] ravitailleur *m* de sous-marins
~THERMAL REACTOR - [nucl] réacteur *m* à neutrons thermiques de sous-marin
SUBMINIATURIZATION - [electron] subminiaturisation *f*
SUBMAXILLARY - [anat] sous-maxillaire
SUBMAXILLARITIS - [med] sous-maxillite *f*
SUBMAXILLITIS - s. submaxillaritis
SUBMERGE, to - [gen] (to plung under water) immerger, submerger
SUBMERGED - [gen hydr naut] submergé, noyé
[naut] (of a submarine) en plongée
~AERIAL - [radio] (an underwater aerial, or antenna) antenne *f* sous-marine
~ARC WELDING - [el heat] (fusion welding process in which one or more electrodes melt under a powered flux) soudage *m* sous flux électro-conducteur
~BOOSTER PUMP - [mech] (a booster pump, so placed as to be wholly or partly below the surface of the liquid which it is to deliver) pompe *f* de suralimentation noyée
~COMBUSTION BURNER - [th eng] brûler *m* immergé
~CONCRETE - [constr] béton *m* immergé
~DYKE - [hydr] (a ground sill) barrage *m* noyé, barrage *m* submergé
~FOUNDATIONS - [constr] fondations *fpl* submergées
~MELT WELDING - s. submerged arc welding
~PLANT - [bot] plante *f* qui pousse sous l'eau
~PLUNGER MACHINE - [metall] machine *f* à chambre chaude à piston
~PUMP - [mech] pompe *f* centrifuge verticale immergée
~SPEED - [naut] (of submarine) vitesse *f* en immersion
SUBMERSE, to - s. submerge, to
SUBMERSIBLE - [gen] submersible
(of an electrical machine etc) étanche à l'immersion, étanche à la submersion
~MACHINE - [el] (a machine so constructed that it can work when submerged under a specified head of water for an indefinitely long period) moteur *m* étanche à l'immersion
~TRANSFORMER - [el] transformateur *m* étanche à l'immersion
SUBMERSION - [gen] submersion *f*
[mech] (in internal combustion engines; of the jet) noyage *m* (du gicleur)
SUBMICRON - [meas] (a particle of which the diameter is between 50 and 2000 Angstrom units, and is only visible with the ultra-microscope) submicron *m*
SUBMICROSTRUCTURE - [metall] (the structure of an alloy which is beyond the range of the microscope and must be inferred by other means) submicrostructure *f*
SUBMUCOSA - [anat] sous-muqueuse *f*
SUBMULTIPLE - [math] sous-multiple
SUBNITRATE - [chem] sous-nitrate *m*
SUBOCTAVE - [acoust] sous-harmonique *f*
SUBOPERCULUM - [med] segment *m* orbitaire de la deuxième frontale
SUBORBITAL - [anat] (said of the membrane bones surrounding the eye) sous-orbitaire
~FLIGHT - [astronaut] vol *m* sous-orbital
SUBOSCILLATOR - [electron] oscillateur *m* secondaire
SUBOXIDE - [chem] (oxide with a minimum amount of oxygen) sous-oxyde *m*
SUBPERIOSTAL ABSCESS - [med] parulie *f*, abcés *m* sous-périosté
SUBPERMANENT MAGNETISM - [phys] remanence *f*
~SET - [rubber ind] remanence *f* non permanente, transformation *f* provisoire
SUBPOENA, to - [leg] citer, assigner comme témoin
SUBPOENA - [leg] citation *f*, assignation *f* de témoins
SUBREFRACTION - [radio] (of electromagnetic waves, refraction less than standard refraction) infraréfraction *f*
SUBROUTINE - [comput] (part of a programme) sous-programme *m*
~CALL - [comput] sous-programme *m* d'appel
SUBSATELLITE POINT - [astronaut] point *m* sous-satellite
SUBSCRIBE , to - [gen] souscrire, signer
[gen] (to give one's consent) consentir
[gen] (to pledge oneself to take a book, a periodical etc) s'abonner
SUBSCRIBER - [gen and comm] signataire *m*
[leg] abonné
SUBSCRIBER'S CABLE - [telecomm] (cable connecting the telephone cable and the receiver) artère *f* de distribution, câble *m* d'abonné
~LINE - [telecomm] ligne *f* d'abonné, ligne *f* principale
~METER - [telecomm] (call-counting meter)compteur *m* d'abonné, compteur *m* de conversations
~RELEASE - [telecomm] libération *f* par le premier abonné qui raccroche
SUBSCRIPT - [comput] indice *m* inférieur
SUBSCRIPTION - [gen] (to a newspaper, a magazine etc) abonnement *m*
~CALL - [telecomm] conversation *f* par abonnement
~TELEVISION - [telev] télévision *f* à prépaiement
SUBSET - [telecomm] (subscriber's set) poste *f* d'abonné
SUBSIDE, to - [gen] tomber au fond, s'affaisser
[chem] précipiter
[chem] (of sediments) tomber au fond
[gen] (to sink) s'effondrer
[met] se calmer, s'apaiser
SUBSIDENCE - [aero] (a disturbance which diminishes without oscillations) subsidence *f*
[gen] affaissement *m*
[constr] (of the ground) tassement *m*, dénivellement *m*
[el] affaiblissement *m*
[med] (of a tumor) delitescence *f*
[geol] fondis *m*, effondrement *m*
[mining] tombée *f*,affaissement *m*
~OF A RIVER - [geogr] décrue *f* d'une rivière
~OF THE TRACK - [railw] affaissement *m* de la voie, déformation *f* de la voie
~RATIO - [contr] (the ratio between the amplitude of two successive oscillations, of the same sign) rapport *m* d'abaissement
SUBSIDIARY - [gen] subsidiaire, auxiliaire
~DRAIN - [agric] petit drain *m*
~SIGNAL - [railw] signal *m* secondaire
SUBSIDIZE, to - [gen] subventionner

SUBSIDY - [gen and fin] subvention *f*
[comm] prime *f*
SUBSILL - [railw] longrine *f* de renfort
SUBSISTENCE - [gen] existence *f*, subsistance *f*
[gen] (means of support) entretien *m*
~DIET - [med] régime *m* minimum de subsistance
SUBSOIL, to - [agric] sous-soler (la terre)
SUBSOIL -[geol] (the layerof earth beneath the surface soil)sous-sol *m*
~DRAIN - [hydr] (drain just below the ground level to carry off water from saturated ground) drainage *m* en sous-sol
~IRRIGATION - [hydr] irrigation *f* en sous-sol
~PLOUGH - [agric] charrue *f* défonceuse
~WATER - [hydr] eaux *f*pl souterraines
SUBSOILER - [agric] (machine to turn up the subsoil) charrue *f* sous-soleuse
SUBSOILING - [agric] sous-solage *m*
SUBSOLAR POINT - [astr] point *m* subsolaire
SUBSONIC - [acoust] (said of sound waves beyond the lower limit of human audibility) subsonique
[aero] subsonique
~FLOW - [aero] (flow of subsonic speed) écoulement *m* subsonique
~SPEED - [aero] (speed less than the focal speed of sound) vitesse *f* subsonique
SUB-SOW BLOCK - [metall] plaque *f* d'assise de l'estampe
SUBSPECIES - [zool] (category of individuals within a species distinguished by some characteristics from the typical members of the species) sous-espèce *f*
SUBSTAGE - [geol] sous-étage *m*
SUBSTANCE - [gen] (the material of which anything is made) matière *f*
[chem] (term denoting a pure chemical compound) substance *f*
[fin] bien *m*, fortune *f*
[paper man] (the relative weight of paper) poids *m* au mètre carré
[fin] solidité *f*
SUBSTANDARD - [gen] de qualité inférieure
[comm] de type courant
~FILM - [cin] (or narrow-gauge film ; a film of less than 35 mm) film *m* réduit, format *m* réduit
~INSTRUMENT - [instr] (measuring instrument whose accuracy and conditions of use are intermediate between those of a standard instrument and those of a precision instrument) appareil *m* de laboratoire, appareil *m* de precision
~PIPE - [plumb] tube *m* réformé
~PROPAGATION - [phys] propagation *f* avec sous-réfraction
~STOCK - s. substandard film
SUBSTANTIAL - [gen] substantiel
[gen] (wealthy) riche, solide
~DERIVATIVE - [astronaut] dérivée *f* individuelle
SUBSTANTIATION - [gen] justification *f*
SUBSTANTIVE - [gen] (a noun) substantif *m*
[gen] (self-subsisting person or thing) indépendant, autonome
[ind chem] substantif
~DYE - [text] colorant *m* substantif
SUBSTATION - [el] (switching, transforming or converting intermediate station) poste *m* électrique
SUBSTELLAR POINT - [astr] (the point at which the earth's surface is cut by a line from the centre of the earth to a given star) point *m* substellaire
SUBSTITUTE, to - [gen] substituer
SUBSTITUTE - [gen] remplaçant
[comm] contrefaçon *f*
[gen] (representative or delegate) suppléant *m* , intérimaire *m*
[food ind] succédané *m*
~SIGNAL - [railw] (an emergency signal) signal *m* de remplacement
SUBSTITUTION - [gen] substitution *f*, remplacement *m*
[chem] (the replacement in a compound of one element by another) substitution *f*
[leg] novation *f*, subrogation *f*
~(BACK) CROSSING - [biol] recroisement *m*
~READING - [photo] mesure *f* par substitution
SUBSTITUTIONAL ALLOY - [chem] (a class of metallic alloys in which atoms of one element have been replaced by atoms of another, without changing the basic crystal structure) alliage *m* de remplacement
SUBSTRATE - [chem] (the substance acted upon by an enzyme) sous-couche *f*
[ind chem] (material on which adhesive is spread) couche *f* inférieure
[agric] sous-sol *m*
SUBSTRATOSPHERE - [met] (region immediately below the stratosphere) sous-stratosphère *f*
SUBSTRATUM - [geol] sous-couche *f*, couche *f* inférieure
[constr] sous-sol *m*
SUBSTRUCTURE - [constr] (understructure or foundation) substruction *f*, fondement *m*
[constr] (in road or railway construction work) infrastructure *f*
SUBSURFACE DRAINAGE - [agric] drainage *m* en sous-sol
~IRRIGATION - [agric] irrigation *f* en sous-sol
SUBTANGENT - [geom] (the part of the axis of a curve cut off between the tangent to a given point and the ordinate of that point) sous-tangente *f*
SUBTENANCY - [leg] sous-location *f*
SUBTENANT - [leg] sous-locataire *m*
SUBTEND, to - [geom] (to extend under, e.g. the cord of an arc or the side of a triangle opposite to an angle) sous-tendre
SUBTERRANEAN - [gen] souterrain
~RIVER - [geol] rivière *f* souterraine
SUBTILIN - [chem] (an antibiotic with therapeutic uses) subtiline *f*
SUBTITLE - [cin] sous-titre *m*
SUBTRACT, to - [math] soustraire
SUBTRACTION - [math] soustraction *f*
SUBTRACTIVE - [gen & math] (serving to diminish) soustractif
~COLOUR PROCESS - [opt] méthode *f* subtractive
~COLOUR SYSTEM - [telev] système *m* soustractif de couleurs
~COLOURED LIGHT - [opt] (the monochromatic illumination obtained from a polychromatic light source by means of a suitable absorption screen) lumière *f* subtractive
~COMPLEMENTARY COLOURS - [telev] couleurs *f*pl soustractives complémentaires
~PRIMARIES - [opt] couleurs *f*pl primaires de synthèse soustractive
~PROCESS - [photo] (the printing of images corresponding to the three primary colours, in their subtractive or complementary colours) procédé *m* sous-

tractif
SUBTRACTIVE REDUCER - [photo] faiblisseur *m* superficiel
SUBTRAHEND - [math] nombre *m* à soustraire
SUBTRANSIENT - [el] subtransitoire
~TIME-CONSTANT - [el] (the smallest or the time-constants which appear during the transient conditions of current) constante *f* de temps subtransitoire
SUBTRANSLUCENT - [opt] subtranslucide
SUBTRANSPARENT - [opt] semi-transparent
SUBTREAD - [rubber ind] (of tyres) sous-couche *f* de la bande de roulement
SUBTROPICAL PLANT - [bot] plante *f* subtropicale
SUBURB - [geogr] (of a city) faubourg *m*
SUBURBAN - [gen] suburbain
~COACH - [railw] train *m* de banlieue
SUBVENTION - [gen] subvention *f*
[comm] prime *f*
SUBVERTICAL - [gen] (almost vertical) presque vertical
SUBWAY - [constr] passage *m* souterrain, galerie *f*
(in the USA, an underground railway) métro *m*, chemin *m* de fer souterrain
SUBZERO - [meas] sous le zéro, sous-zéro
~TEMPERATURE - [met] température *f* sous-zéro
SUCCEDANEUM - [chem] (a substitute) succédané *m*
SUCCINIC ACID - [chem] (an intermediate for perfumery components, dyestuffs, pharmaceuticals, and protective coatings) acide *m* succinique
~ANHYDRIDE - [chem] (an intermediate for pharmaceuticals and other organic compounds) anhydride *m* succinique
SUCCINIMIDE - [chem] (a synthesis intermediate) succinimide *f*
SUCCINYLCHOLINE BROMIDE - [chem] (suxamethonium bromide. A muscle relaxant used in medicine) bromure *m* de succinylcholine
SUCCINYLSULPHATHIAZOLE - [chem] (a sulphonamide with application in the treatment of dysentry and similar conditions) succinylsulfathiazole *m*
SUCCORY - [bot] chicorée *f*
SUCCUSSION - [med] (shaking a patient to discover the presence of fluid in a pleural cavity already containing air) succussion *f*
SUCK, to - [mech] sucer
[gen] (to draw air) aspirer
[ind chem] (to abstract fluid by reducing pressure upon it below that of the atmosphere) aspirer
SUCK DOWN WIND TUNNEL - [aero] (vacuum tunnel) soufflerie *f* à vide
~OUT, to - [mech] (to extract fluid from a vessel by means of a suction pump or like means) aspirer
SUCKER - [mech] (piston of a valve) piston *m* de pompe aspirante
[mech] (valve of a pump bucket) clapet *m* de pompe
[bot] rejeton *m*
[plumb] (a suction pipe) tuyau *m* d'aspiration
[oil ind] tuyau *m* d'aspiration
[mech] (a feeding sucker, cup-shaped rubber device used in a folding or gathering machine) ventouse *f*
~APPARATUS - s.sucker foot
~ELEVATOR - [oil ind] élévateur *m* de tige
~FOOT - [med] tentacule *m* de l'astrocyte
~PROCESS - s.sucker foot
~ROD - [mech] (of a well pump) tige *f* de commande du piston
SUCKER ROD ELEVATOR - [oil ind] (of a well pump) élévateur *m* de tige
~ROD GUIDE - [oil ind] lanterne *f* de guidage
~ROD PUMPING - [oil ind] pompage *m* par barre de pompage
~SAND - [oil ind] sable *m* absorbant
~WRENCH - [mech] clef *f* pour tiges
SUCKING - [metall] réduction *f* locale du diamètre
~AND FORCING PUMP - [mech] pompe *f* aspirante et foulante
~BOOSTER - [el] (booster having the function of overcoming voltage drop in a feeder) dévolteur *m*
~-CALF - [zool] veau *m* de lait
~COIL - [el] bobine *f* suceuse
~CUSHION - s. sucking pad
~EFFECT - [mech] (of a pump) effet *m* d'aspiration
~PAD - [med] boule *f* graisseuse de Bichat
~PUMP - [mech] pompe *f* aspirante
~THE THREAD - [text] aspiration *f* du fil par la bouche
SUCROSE - [chem] (cane sugar) saccharose *m*, hexobiose *f*
~MONOSTEARATE - [chem] (a constituent of detergents) monostéarate *m* de saccharose
~OCTOACETATE - [chem] (a plasticizer for synthetic resins and cellulose derivatives) octa-acétate *m* de saccharose
SUCKLING - [zool] nourisson *m*
[gen] allaitement *m*
SUCTION - [gen] (the act of sucking]) succion *f*, aspiration *f*
[mech] (by a pump) aspiration *f* de l'eau par une pompe
~AIR CHAMBER - [metall] chambre *f* à air de la conduite d'aspiration
~AND PRESSURE PUMP - [mech] (a rotary pump) pompe *f* foulante et aspirante
~AT GATE - [metall] (in foundry work) retassure *f* dans le chenal de coulée
~BASKET - [mech] (of a pump) filtre *m* d'aspiration
~BOTTLE - [hydr] bouteille *f* d'aspiration
~BOX - [mech] chambre *f* d'aspiration
~BOXES - [paper man] (transverse inverted troughs connected to a depression system, which abstract water from the pulp on the wire screen) caisses *f*pl aspirantes
~BRUSH - [impl] (in electro-domestic appliances) brosse *f* aspirante
~CHAMBER - [mech] chambre *f* d'aspiration, chambre *f* de la pompe, chambre *f* de compression, chambre *f* de détente
~CLEANER - [mech] dispositif *m* de nettoyage par le vide
~CONDUIT - [mech] conduit *m* d'aspiration
~CONVEYOR - [transp] (a pneumatic conveyor) convoyeur *m* pneumatique
~CUP - [gen] ventouse *f*
~CUP FOR MILKING MACHINE - [agric] tétine *f* pour machine à traire
~CUTTER DREDGER - [mech] drague *f* rotative à succion
~DISK - s.suction plate for denture
~DRAIN - [agric] petit drain *m*
~DRAINAGE - [med] drainage *m* par aspiration d'une caverne
~DREDGE - [mech] drague *f* suceuse, drague *f* aspirante

SUCTION EXTRACTOR - [med] ventouse *f* eutocique, ventouse *f* obstétricale
~FACE - [aero] (the side of a propeller blade which corresponds to the upper face of an aerofoil) extrados *m*
~FAN - [el] ventilateur *m* aspirant, ventilateur *m* négatif
~FILTER - [ind chem] (used in sugar industry) filtre *m* à aspiration
~GAS - [gas ind] gaz *m* pauvre d'aspiration
~GAS-ENGINE - [mech] moteur *m* à gaz pauvre d'aspiration
~GAS PRODUCER - [mech] (equipment for producing combustible gas from air, steam and some form of carbon, as fuel for an i.c. engine, flow through the apparatus being induced by the suction of the engine on the induction stroke) gazogène *m* à aspiration
~HEAD - [hydr] (of a pump) hauteur *f* d'aspiration
~HOLE - [mech] orifice *m* d'aspiration
~HOSE - [hydr] (special hose for use under pressure below the ambient, e.g. on the suction side of a pump, and reinforced against collapse by an internal coil of wire) tuyau *m* d'aspiration
~LIFT - [hydr] (the height to which a pump or like device raises liquid by suction (below the pump) hauteur *f* d'aspiration
~LINE - [hydr] (a type of tubular connection) tube *m* d'aspiration
~MAIN - [gas ind] tuyauterie *f* d'aspiration
~MANIFOLD - [auto] (intake manifold) collecteur *m* d'admission
~NOZZLE - [mech] buse *f* d'aspiration
~OF INTAKE PRESSURE - [auto] dépression *f* de l'admission
~PAD - s,suction cup
~PIPE - [hydr] tuyau *m* d'aspiration
~PLATE FOR DENTURE - [dent] rondelle *f* en caoutchouc pour dentiers
~POLISHER - [impl] (in electro-domestic appliances) machine *f* à polir aspirante
~PORT - [mech] orifice *m* d'aspiration
~PRESSURE - [bot] (the avidity with which cells take in water) pression *f* d'aspiration
~PUMP - [mech] (pump which acts by sucking) pompe *f* d'aspiration
~PYROMETER - [instr] (a high-velocity pyrometer) pyromètre *m* à aspiration
~RAM - [mech] bélier *m* aspirateur
~REFLEX - [med] réflexe *m* d'aspiration
~ROLLER - [text] cylindre *m* d'aspiration
~ROSE - [mech] (of a pump) aspirant *m*
~SCREEN - [mech] (intake screen) crépine *f* d'aspiration
~SHAFT - [min] puits *m* d'appel d'air
~SIDE - [mech] (of a pump) côté *m* de l'admission
~STRAINER - [mech] crépine *f* d'aspiration
~STROKE - [mech] (in engines and motors) course *f* d'admission
~SWEEPER - [mech] (for road cleaning) balayeuse *f* aspirante
~SYSTEM - [comput] système *m* d'aspiration
~TANK - [hydr] bâche *f* d'aspiration
~-TYPE HYDROMETER - [instr] (small hydrometer contained in a glass tube fitted with a bulb and suction tube, for testing the specific gravity of battery electrolyte) hydromètre *m* à aspiration
~VALVE - [mech] (of a pump) soupape *f* d'aspiration, clapet *m* d'aspiration
SUCTION VALVE CONE - [mech] siége *m* conique de la soupape d'admission
~VANE - [mech] aspirateur *m* à ailettes
SUD - [chem] (soap solution) lessive *f*
~AND DIRT COLLECTOR - [text] collecteur *m* des boues
SUDAMINA - [med] (whitish vesicles on the skin) sudamina *mpl*
SUDATION - [bot] exsudation *f*
[med] transpiration *f*, sudation *f*
SUDATORIUM - [med] bain *m* de vapeur
SUDDEN - [gen] soudain, brusque
~-CHANGE RELAY - [el] relais *m* de variation brusque
~-COMMENCEMENT MAGNETIC STORM - [astronaut] orage *m* magnétique soudain
~DEATH - [electron] (sudden reduction of the current multiplication factor in point-contact transistors) rupture *f* spontanée
~IONOSPHERIC DISTURBANCE - [radio] perturbation *f* ionosphérique à début brusque
~PHASE ANOMALY - [radio] anomalie *f* brusque de phase
~RELEASE - [el] déclenchement *m* brusque
SUDOKERATOSIS - [med] kératose *f* des canaux sudoripares
SUDOR - [med] (perspiration, sweat) sueur *f*
SUDORIFEROUS - [biol] (sweat producing) sudorifère, sudoripare
SUDORIFIC - [med] (connected with the secretion of sweat) sudorifique, diaphorétique
SUDORIPAROUS ABSCESS - [med] abcès *m* tubéreux, hidrosadénite *f*
SUDS - [chem] (soapy water stirred into bubbles and froth) eau *f* de savon, lessive *f*
[mech] (workshop term for a machining coolant-lubricant consisting of an emulsion in water of a special oil) liquide *m* réfrigérant
~PUMP - [mach] (a pump mounted on a machine-tool for circulating coolant) pompe *f* pour lubrifiants réfrigérants
SUE, to - [leg] poursuivre en justice, intenter un procès
SUEDE - [leather ind] (skin used for gloves and shoes) peau *f* de suède, daim *m*, suède *m*
~CALFSKIN - [leather ind] veau *m* chamoisé
~FINISH - [paint] (finish resembling suede leather) fini *m* daim
[leather ind] chamoisé
SUET - [zool] graisse *f* de rognon, graisse *f* de boeuf
SUFFIX - [gen] (letter or letters added to the end of a word or to root) suffixe *m*
[telecomm] (the significant portion of a signal) signal *m* actif
SUFFOCATING GASES - [chem] gaz *mpl* asphyxiants
SUFFOCATION - [med] suffocation *f*, asphyxie *f*
SUFFUSION - [med] suffusion *f*
SUGAR - [chem] (the common term for sucrose, or cane-sugar) sucre *m*
[chem] (water-soluble crystalline disaccharide) sucre *m*
~BEET - [bot] (sugar-producing varieties of the common beet) betterave *f* à sucre
~BEET-CHIPS - [agric] cossettes *fpl*
~BEET MILL - [ind] fabrique *f* de sucre de betterave
~-BEET PICK-UP LOADER - [agric] arracheuse-décolleteuse-chargeuse *f* de betteraves
~-BEET PULP - [agric] cossettes *fpl* épuisées, pulpe *f*

de betteraves
SUGAR BEET SEED - [agric] graine *f* de betteraves sucrières
~BOILER - [sugar ind] sucrier *m*
~BUSH - [agric] plantation *f* d'érables à sucre
~CANE - [bot] (strong perennial tropical grass rich in sugar) canne *f* à sucre
~-CANE BORER - [zool] chenille *f* mineuse de la canne à sucre
~CANE PLANTATION - [agric] plantation *f* de cannes à sucre
~COATED PILL - [med] dragée *f*
~CONTENT - [chem] teneur *f* en sucre
~CONE - s. sugar loaf
~CRUSHER - [sugar ind] (set of steel rollers in which the cane is crushed to extract the juice) broyeur *m* de cannes à sucre
~DETERMINATION - [chem] détermination *f* du sucre
~DYE - [chem] (or caramel, the brown dye which is formed when cane-sugar is heated above its melting point) caramel *m*
~LOAF - [sugar ind] pain *m* de sucre
~MAIZE - [bot] maïs *m* sucré
~-MAPLE - [bot] érable *m* à sucre
~OF LEAD - [chem] (obsolete term for lead acetate) acétate *m* de plomb
~OF MILK - [chem] lactose *f*, sucre *m* de lait
~PALM - [bot] palmier *m* à sucre
~PEA - [agric] mange-tout *m*, pois *m* sans parchemin
~PRESS CLOTH SHODDY - [text] shoddy *m* de drap de presse à sucre
~REFINERY - [sugar ind] raffinerie *f* de sucre
~SCALE - [sugar ind] bascule *f* à sucre
~SIFTER - [impl] saupoudroir *m* à sucre
~SILO - [sugar ind] silo *f* à sucre
~SOAP - [paint] (cleansing preparation for paint surfaces) solution *f* pour le nettoyage du vernis
~SOLUTION - [chem] solution *f* de sucre
~TWINE - [text] ficelle *f* pour sucre
SUGARED SILK - [text] soie *f* chargée de sucre
SUGGEST, to - [gen] suggérer, proposer
SUGGESTION - [gen] suggestion *f*, proposition *f*
SUGGILLATION - [med] suggilation *f*
SUIDA PROCESS - [ind chem] (a commercial method for the separation of acetic acid from pyroligneous acid) procédé *m* Suida
SUINT - [text] (natural wool grease from wool-washings) suint *m*
~GAS - [chem] gaz *m* de suint, gaz *m* de suintine
~SALT - [chem] sel *m* de suint
~WATER BOWL - [text] cuve *f* d'eau de suint
SUINTER - [chem] précipité *m* de graisse de laine, suintine *f*
~GAS - s. suint gas
SUIT, to - [gen] adapter, accomoder
[gen] (to render suitable, appropriate) approprier
SUIT - [gen] (the act of suing, a petition) requête *f*, demande *f*
[leg] (action in law) poursuite *f* en justice
[text] (of clothes) complet *m*
~OF SAILS - [naut] jeu *m* de voiles
SUITABLE - [gen] convenable, adapté
~FOR CULTIVATION - [agric] labourable, arable
~FOR MILLING - [text] laine *f* foulable, laine *f* à foulon
SUITASE - [impl] valise *f*, mallette *f*
SUITING - [text] tissus *mpl* de confection
SULFA - s. sulpha
~DRUG - [chem] (organic compound consisting mainly of substituted sulphanilamide derivatives , sulfa drugs have a wide range of therapeutic effects in the treatment of bacterial infections) sulfamide *f*
SULFANEMIA - [med] anémie *f* sulfamidique
SULFANURIA - [med] sulfamidurie *f*
SULFATE - s.sulphate
SULFONAMIDEMIA - [med] sulfamidémie *f*
SULFOPYRETOTHERAPY - [med] sulfopyrétothérapie *f*
SULL - [metall] couche *f* d'oxyde ferreux
SULLA - [bot] sulla *f*, sainfoin *m* d'Espagne
~SWEET VETCH - s, sulla
SULLAGE - [hydr] (the mud and slit deposited by flowing waters) eaux *fpl* d'égoutvase d'alluvion
[metall] crasses *fpl*, scories *fpl*, laitier *m* de poche
~-PIECE - [metall] masselotte *f*
SULLING - [metall] oxydation *f* après décapage
SULPHACETAMIDE - [chem] (yellowish-white, odourless, crystalline powder with uses in medicine as a bacteriostat) sulfacétamide *f*
SULPHADIAZINE - [chem] (a sulphonamide with uses in medicine) sulfadiazine *f*
SULPHADIMETHOXINE - [chem] (a sulphonamide with uses in medicine) sulfadiméthoxine *f*
SULPHAEMOGLOBINAEMIA - [med] (the presence in the blood of sulphaemoglobin) sulfahémoglobinanémie *f*
SULPHAGUANIDINE - [chem] (white, crystalline powder, darkening on exposure to light and used in the treatment of bacillary dysentery and similar conditions) sulfaguanidine *f*
SULPHALDEHYDE - [chem] sulfaldéhyde *m*
SULPHAMERAZINE - [chem] (sulphamethyldiazine, a sulphonamide used in medicine) sulfamérazine *f*
SULPHAMETHAZINE - [chem] (sulphadimidine. Creamy white, bitter-tasting powder with the general therapeutic properties of the sulphonamides) sulfaméthazine *f*
SULPHAMETHIAZOLE - [chem] (a sulphonamide used in medicine) sulfaméthiazole *m*
SULPHAMETHOXYPYRIDAZINE - [chem] (a sulphonamide used in the treatment of urinary tract infections) sulfamétoxypyridazine *f*
SULPHAMIC ACID - [chem] (intermadiate in the manufacture of plasticizers for cellulose derivatives) acide *m* sulfamique
SULPHAMIDE - [chem] sulfamide *f*
SULPHAMIDIC DRUG - s. sulfa drug
SULPHAMMONIUM - [chem] sulfammonium *m*
SULPHANILAMIDE - [chem] (a sulphonamide with application in medicine) sulfanilamide *f*
SULPHANILIC ACID - [chem] (an intermediate for pharmaceuticals and dyestuffs) acide *m* sulfanilique
SULPHANTIMONIDE - [chem] sulfantimoniure *m*
SULPHAPYRIDINE - [chem] (a sulphonamide with uses in medicine) sulfapyridine *f*
SULPHAQUINOXALINE - [chem] (a veterinary sulphonamide) sulfaquinoxaline *f*
SULPHARSENATE - [chem] sulfarséniate *m*
SULPHARSPHENAMINE - [chem] (a spirochaeticide used in medicine) sulfarsphénamine *f*
SULPHATE, to - [chem] (to become coated with lead sulphate) sulfater, se détériorer par sulfatation
SULPHATE - s. sulphates
~DRIER - [ind chem] séchoir *m* du sulfate

SULPHATE OF AMMONIA - [chem] sulfate *m* d'ammonium
~OF ANILINE - [chem] sulfate *m* d'aniline
~OF BARIUM - [chem] sulfate *m* de baryum
~OF CALCIUM - [chem] sulfate *m* de calcium, gypse *m*
~OF COPPER - [chem] vitriol *m* bleu
~OF IRON - [chem] sulfate *m* de fer, vitriol *m* vert
~OF LIME - [chem] sulfate *m* de chaux, chaux *f* sulfatée
~OF MANGANESE - [chem] sulfate *m* de manganèse
~OF POTASSIUM - [chem] sulfate *m* de potassium
~OF SODA - [chem] sulfate *m* de sodium
~OF ZINC - [chem] vitriol *m* blanc, sulfate *m* de zinc
~PAPER - [paper man] papier *m* au sulfate
~PULP - [ind chem] pâte *f* à la soude, pâte *f* au sulfate
~PROCESS - [paper man] (chemical pulp process in which alkali and sulphur loss is made up by the addition of salt cake (crude sodium sulphate) procédé *m* au sulfate
~-REDUCING BACTERIA - [biochem] bactéries *fpl* sulfato-réductrices
SULPHATED - [chem] sulfaté
~BATTERY - [el chem] accumulateur *m* sulfaté
~FATTY ALCOHOLS - [chem] alcools *mpl* sulfatés
~OIL - [chem] huile *f* sulfatée
SULPHATES - [chem] (the salts of sulphuric acid) sulfates *mpl*
SULPHATHIAZOLE - [chem] (a sulphonamide used in medicine) sulfathiazole *m*
SULPHATING - [el chem] (the formation of insoluble white lead sulphate on the plates of a lead-acid cell, with the effect of reducing capacity and efficiency) sulfatage *m*
~ROASTING - [metall] (roasting so carried out as to ensure the retention of a substantial amount of sulphur in the form of sulphate) grillage *m* de sulfatation
SULPHATION - [el chem] sulfatation *f*
SULPHATIZE, to - [el chem] se sulfater
SULPHIDE CELLULOSE PAPER - [chem] papier *m* de cellulose au bisulfite
~DYESTUFFS - [chem] (dyestuffs of unknown constitution containing sulphur) colorants *mpl* au sulfure
~EMBRITTLEMENT - [metall] fatigue *f* par l'hydrogène sulfuré
~OF IRON - [chem] sulfure *m* de fer
~ROAST - [metall] grillage *m* sulfatant
~TONING - [photo] virage *m* au sulfure, virage *m* sépia
SULPHIDES - [chem] (compounds formed by the combination of sulphur with a metal or a group acting as a metal) sulfures *mpl*
SULPHINIC ACID - [chem] (acid containing the monovalent sulphinic acid group SO. OH) acide *m* sulfinique
SULPHINPYRAZONE - [chem] (sulphoxyphenylpyrazolidone . A phenylbutazone analogue used in the treatment of chronic gout) sulfinylpyrazone *m*
SULPHITE - s. sulphites
~BOND - [paper man] liant *m* au sulfite
~CELLULOSE - [chem] cellulose *f* au bisulfite
~DIGESTER - [paper man] lessiveur *m* au sulfite
~LIQUOR - [paper man] (a solution of calcium and magnesium bisulphites used in the preparation of wood pulps for papermaking) lessive *f* sulfitée
~PROCESS - [paper man] (method of making pulp by cooking chips in a solution of calcium bisulphite containing free sulphur dioxide) procédé *m* au sulfite
SULPHITE WOOD PULP - [paper man] pâte *f* au sulfite
SULPHOBENZOIC ACID - [chem] (an intermediate for dyestuffs) acide *m* sulfobenzoïque
SULPHOBROMOPHTHALEIN SODIUM - [chem] (a diagnostic agent in medicine, used in assessing liver function) sulfobromophtaléine-sodium *m*
SULPHOCYANIDES - [chem] (synonym for rhodantes or thiocyanates) sulfocyanures *mpl*
SULPHONAL - s. sulphonmethane
SULPHONATED - [chem] sulfoné
~FATTY ACIDS - [chem] acides *mpl* graisses sulfonés
~OIL - [chem] (animal or vegetable oils which have been treated with sulphuric acid to yield products of the $R.O.SO_3H$ type. They have application in detergents, lubricants and emulsifying and scouring agents) huile *f* sulfonée
SULPHONATION - [chem] (the substitution of the SO_2OH group for hydrogen in a molecule) sulfonation *f*
SULPHONATOR - [ind chem] (cast iron or steel vessel fitted with agitators, for sulphonation operations) sulfonateur *m*
SULPHONETHYLMETHANE - [chem] (methylsulphonal. A hypnotic used in medicine) sulfonéthylméthane *m*
SULPHONIC - [chem] sulfonique
~ACID - [chem] (compound consisting of an organic radical in combination with the sulphonic radical) acide *m* sulfonique
SULPHONMETHANE - [chem] (sulphonal. A hypnotic used in medicine) sulfométhane *m*
SULPHONYLDIANILINE - [chem] sulfonyldianiline *f*
SULPHOPHTHALIC ACID - [chem] (an intermediate for wetting agents) acide *m* sulfophtalique
SULPHOSALICYLIC ACID - [chem] (an analytical reagent) acide *m* sulfosalicylique
SULPHUR - [chem] (brimstone. A non-metallic element, symbol S.A.N. I6, A.W 32.06 a raw material for sulphuric and agricultural chemicals, vulcanizing agents, dyestuffs and papermaking chemicals) soufre *m*
~ACID - [chem] sulfacide *m*
~ALKALI - [chem] sulfalcali *m*
BACTERIA - [biochem] bactéries *fpl* du soufre, bactéries *fpl* sulfureuses
~BASE - [chem] sulfobase *f*
~BATH VULCANIZATION - [rubber ind] vulcanisation *f* en bain de soufre
~BEARING - [min] sulfurifère
~BED - [min] gisement *m* sulfurifère
~BLACK - [chem] noir *m* de soufre
~BLOOMING - [chem] (the appearance of free sulphur on the surface of vulcanized products) efflorescence *f* du soufre
~BURNER - [metall] four *m* à soufre
~CANDLE - [chem] cannel *m* de soufre
~CHLORIDE - [chem] (an intermediate for other sulphur compounds and also having application in rubber vulcanization, pharmaceuticals, textile dyeing, and finishing, insecticides, metallurgy, and in sugar manufacture) chlorure *m* de soufre, sulfochlorure *m*
~CHLORIDE VULCANIZATION - [rubber ind] vulcanisation *f* au sulfochlorure
~COMBINATION - [chem] combinaison *f* sulfurée
~CONTENTS OF THE ORE - [min] teneur *f* en soufre

du minerai
SULPHUR CONTENTS OF THE WOOL - [text] teneur *f* en soufre de la laine
~DICHLORIDE - [chem] (a synthesis intermediate, vulcanizing and chlorinating agent) bichlorure *m* de soufre
~DIOXIDE - [chem] (a bleaching agent, synthesis intermediate, fumigant, and preservative for foodstuffs) oxyde *m* sulfureux, acide sulfureux, anhydride sulfureux
~DIOXIDE COMPRESSOR - [ind chem] (used in refrigerating plants) compresseur *m* à anhydride sulfureux
~DUSTING - [bot] saupoudrage *m* de soufre
~DYE - [chem] colorant *m* au soufre
~FREE NATURAL METHANE - [min] méthane *m* naturel sans soufre
~FURNACE - [metall] four *m* à soufre
~HEXAFLUORIDE - [chem] (a colourless, odourless gas with uses as a dielectric in electrical equipment) hexafluorure *m* de soufre
~KILN - s. sulphur furnace
~LIME - [chem] (an insecticide) mélange *m* de chaux sulfurée
~MATCH - [gen] allumette *f* soufrée
~MINE - [min] soufrière *f*
~ORE - [min] pyrite *f*, pyrite *f* de fer
~OXIDES - [chem] oxydes *mpl* de soufre
~PRINTING TEST - [metall] essai *m* Baumann
~REFINING - [min] raffinage *m* du soufre
~SALT - [chem] sulfosel *m*
~SPRAYER - [agric] soufreuse *f*
~SQUIB - [metall] (used in blasting) mèche *f* soufrée
~-TOLERANT FUNGI - [chem] moisissures *fpl* résistantes au soufre
~TRIOXIDE - [chem] (an oxidizing and sulphonating agent. It exists in three forms : alpha - melting at 62.2° C, beta - melting at 32.5° C, gamma - melting at I6.8° C) anhydride *m* sulfurique
~WATER - [min] eau *f* sulfureuse
~WORT - [bot] (a med plant) peucédan *m* officinal
~YELLOW - [chem] jaune *m* de soufre
SULPHURATE, to - s. sulphurize, to
SULPHURATED LIME - [chem] (a mixture of calcium sulphate and sulphide used as a depilatory in tanning and having some application in the treatment of skin diseases) chaux *f* sulfurée
SULPHURATION - [chem] sulfuration *f*, sulfurisation *f*
SULPHUREOUS - [min] sulfureux
SULPHURETTED HYDROGEN - [chem] (hydrogen sulphide, a gas used as an analytical reagent etc) acide *m* sulfhydrique, hydrogène *m* sulfuré
SULPHURIC - [chem] sulfurique
~ACID - [chem] (highly corrosive fluid and probably the most important industrial chemical . Applications include cellulose and rayon manufacture) acide *m* sulfurique
~ACID BATH - [ind chem] bain *m* en acide sulfurique
~ETHER - [chem] éther *m* sulfurique
SULPHURING CHAMBER - [ind chem] chambre *f* de sulfuration
SULPHURIZATION - [chem](treating, bleaching, fumigating etc with sulphur) sulfuration *f*, sulfurisation *f*
SULPHURIZE, to - [chem] (to treat, to impregnate, to bleach, to fumigate etc with sulphur) sulfurer
SULPHURIZE, to - [text] (the wool) soufrer (la laine)
SULPHURIZED ASPHALT - [constr] (type of asphalt treated with sulphur at high temperatures) asphalte *m* sulfuré
SULPHUROUS - [chem] sulfureux
~ACID - [chem] (a solution of sulphur dioxide in water, with application as a foodstuffs preservative, analytical reagent) acide *m* sulfureux
~IMPURITY OF THE VISCOSE - [text] impureté *f* sulfureuse de la viscose
SULPHURYL - [chem] (the bivalent radical SO2) sulfuryle *m*
~CHLORIDE - [chem] (a colourless, pungent-smelling liquid with application as a catalyst in synthesis and in the manufacture of plastics, dyes and drugs) chlorure *m* de sulfuryle
SULPHYDRATE - [chem] sulfhydrate *m*
SULTRY - [gen] étouffant
~WEATHER - [met] temps *m* lourd
SUM, to - [math] additionner
SUM - [gen] somme *f*, montant *m*
[math] exercice *m* d'arithmétique
~OF INFINITE SERIES - [math] (sum to infinity) somme *f* des séries
SUMBUL - [chem] (musk root. The dried root or rhizome of species of Ferula, with application in medicine) sumbul *m*
SUMAC - [bot] (dried and powdered leaves of the sumac plant used for dyeing and tanning) sumac *m*
~BATH - [ind chem] bain *m* de sumac
~EXTRACT - [chem] extrait *m* de sumac
SUMMARIZED - [gen] en résumé
SUMMARY - [gen] sommaire *m*
~CARD - [comput] carte *f* récapitulative, carte *f* sommaire
~PUNCH - [comput] perforateur *m* récapitulatif, perforatrice *f* récapitulatrice
SUMMATION - [math] sommation *f*
[med] sommation *f*
~ACTION - [contr] action *f* de sommation
~AMPLIFIER - [comput] amplificateur *m* de sommation
~BAND - [phys] (a combination spectral band for which the lower state is the vibrational ground state of the molecule) bande *f* de sommation
~CHECK - [comput] contrôle *m* de sommation
~ELEMENT - [comput] additionneur *m*, sommateur *m*
~INSTRUMENT - [instr] (instrument which measures the sum of the values assumed by quantities of the same kind in several circuits) appareil *m* totalisateur
~LOUDNESS - [acoust] (total loudness resulting from the loudness of all bands of the noise spectrum) intensité *f* sonore de sommation
~METER - [instr] (meter used to register the total energy consumed in two or more separate circuits) compteur *m* totalisateur
~TONE - [acoust] son *m* de sommation
SUMMER - [gen] (the season) été *m*
[constr] (heavy horizontal girder) poutre *f* de plancher
[constr] (horizontal beam resting on the external frame of a building) sommier *m*, poitrail *m*
~CHAFER - [zool] petit hanneton *m* de St.Jean
~CLOTH - [text] étoffe *f* d'été
~DIKE - [constr] digue *f* submersible
~FEEDING - [agric] alimentation *f* d'été

SUMMER LEVEL - [constr] (of water in a polder) niveau *m* estival
~LIGHTNING - [met] (popular, unscientific term for lightning which is visible but too distant for the resulting thunder to be audible) éclaires *mpl* d'été
~-LOAD WATERLINE - [shipbuild] (the waterline to which a ship may be loaded in summer) ligne *f* d'eau d'été
~MONSOON - [met] (monsoon wind blowing during the summer) (from the sea to the land) mousson *m* d'été
~PRUNING - [agric] taille *f* en vert
~SOLSTICE - [astr] solstice *m* d'été
SUMMERHOUSE - [constr] pavillon *m*, gloriette *f*
SUMMERING - [agric] estivage
SUMMING METER - s. summation meter
SUMMIT - [gen] sommet *m*
[geogr] (of a hill) faîte *f*, cime *f*
[hydr] point *m* de partage (d'un canal)
SUMMON, to - [gen] appeler, convoquer
[leg] sommer de comparaître, citer
SUMMONS - [leg] citation *f*
SUMNER LINE - [astr] (a line of position obtained by the observation of the altitude of a celestial body) ligne *f* de Sumner
SUMP - [mech] (recess at the lower part of a closed vessel specially an engine crankcase, to collect liquid especially lubricant) carter *m*, cuvette *f* d'egouttage, carter *m* inferieur
[hydr] (cesspool) fosse *f* d'aisance
[mining] (depression below the lowest level in a mine shaft) puisard *m*, collecteur *m* d'eau
~BREATHER - [mech] réniflard *m*
~FILLER - [auto] (a pump for the oil) pompe *f* de remplissage du carter
~HOLE - [oil ind] puisard *m*, puits *m* perdu
~PUMP - [mining] pompe *f* de mines
~SHAFT - [mining] puits *m* d'exhaure
~SHOT - [min] coup *m* de bouchon
~TANK - [oil ind] réservoir *m* de dépôt
SUMPTER - [zool] bête *f* de some
SUN, to - [gen] exposer au soleil
SUN - [astr] soleil *m*
~-AND-PLANET MOTION - [mech] mouvement *m* planétaire, engrenage *m* planétaire, mouche *f*
~ARCH - [cin] (arc lamp of adequate capacity to imitate the sun in a studio) lampe *f* à arc 225 A
~-ARC LIGHT - s. sun arc
~CHECKING - s. sun cracking
~CHECKING AGENT - [ind chem] (only USA, light-stability in GB) stabilisant *m* à la lumière du soleil
~COMPASS - [instr] (a device for obtaining orientation from the sun's position) boussole *f* solaire
~CRACKING - [metall etc] (the development of small fissures under the influence of solar radiation) fissures *fpl* de retrait dues au soleil, craquelure *f* par la lumière solaire
~CRACKS - [geol] craquelures *fpl* dues au soleil
~CURING - [agric] séchage *m* en champ
~DIAL - [instr] cadran *m* solaire
~-DOG - [met] (parhelion) parhélie *m*
~DRIED BRICK - [constr] brique *f* crue
~-DRY, to - [agric] sécher en champ
~FOLLOWER - [electron] (a type of photoelectric pickup) dispositif *m* photoélectrique à orientation automatique vers le soleil)
~GEAR - [mech] (the central gear of an epicyclic train) engrenage *m* principal
SUN GLASS - [glass man] verre *m* ardent
~LAMP - [cin] soleil *m*, grand réflecteur *m*
[med] s. sunlamp
~PARLOR - [arch] (USA only, a veranda in GB, a room enclosed in glass and exposed to the sun) solarium *m*
~PINION - [mech] (in an epicyclic train, the central gear which meshes with the planet pinions) pignon-soleil *m*
~RADIATION - [phys] rayonnement *m* solaire
~SCREEN - [auto] pare-soleil *m*, visière *f* anti-éblouissante
~SENSOR - [instr] détecteur *m* solaire
~SPOT - [astr] tache *f* solaire
~TRACKER - [instr] pointeur *m* solaire
~VISOR - s. sun screen
~WHEEL - [mech] engrenage *m* central
SUN'S WAY -[astr] trajectoire *f* du système solaire
SUNBEAM - [gen] rayon *m* solaire
SUNBURN - [med] érythème *m* solaire
SUNDIAL - [meas] (time-measuring device, showing the time by means of the shadow of a style thrown on the dial) gnomon *m*
SUNDOWN - [gen] (sunset) le couchant *m*
SUNDRIES - [gen] divers *mpl*
[comm] frais *mpl* divers
SUNFISH OIL - [chem] huile *f* de môle
SUNFLOWER - [bot] tournesol *m*, girasol *m*
~CAKE - [agric] (animal feeding) tourteau *m* de tournesol
~OIL - [chem] (a pale yellow oil extracted from the seeds of Helianthus annus and with uses as an edible oil, in the production of soap and protective coatings) huile *f* de tournesol
~SEEDS - [food] graines *fpl* de tournesol
SUNK - [gen] (p.p. of to sink) submergé, coulé
[naut] sombré
[mech etc] encastré
[constr] (of a road) route *f* creuse, cavée *f*
~FILLET - [arch] listel *m* encastré
~HYDRANT - [hydr] bouche *f* à clé sous trottoir
~KEY - [mech] (a key sunk into keyways in the shaft and in the hub) clavette *f* à rainure, clavette *f* encastrée
~MOULDING - [arch] moulure *f* encaissée
~MOUNT - [photo] monture *f* rentrante
~PANEL - [constr] panneau *m* encaissé
~PISTON HEAD - [mech] (depression of the head of a piston due to excessive temperature and or detonation) tête *f* de piston affaissée
~SHAFT FOUNDATION - [constr] fondation *f* par cuve, fondation *f* par tubage
~SPOT - [metall] dépression *f*
~STONE - [constr] pierre *f* perdue
~SWITCH - [el] interrupteur *m* encastré
~TURNTABLE - [railw] plaque *f* tournante encaissée
~WELL FOUNDATIONS - [constr] fondation *f* sur des puits foncés
SUNKEN - [gen] noyé, submergé, affaissé, creux
~ROAD - [constr] route *f* enterrée
~WRECK - [naut] épave *f* sous-marine
SUNLIGHT - [gen] lumière *f* solaire
~STABILITY - [paint] (ability to resist sunlight without deterioration) stabilité *f* sous l'effet du soleil
~TREATMENT - [med] héliothérapie *f*
SUNN HEMP - [text] chanvre *m* du Bengale, sunn *m*

SUNRISE - [astr] lever *m* du soleil
SUNSET - [astr] coucher *m* du soleil
SUNSHADE - [photo] (of a camera) cache-lumière *m*
SUNSHINE - [gen] lumière *f* du soleil
~RECORDER - [instr] (instrument for recording the duration and, in some types, the intensity of sunshine) héliographe *m* enregistreur
~UNIT - [radiat] (US only, a micromicrocurie) unité *f* de strontium
SUNSPOT - [astr] macule *f*, tache *f* solaire
SUNSTONE - [min] pierre *f* de soleil
SUNSTROKE - [med] insolation *f*, coup *m* de soleil
SUNUP - s. sunrise
SUNWISE - [gen] (clockwise) dans le sens des aiguilles d'une montre
SUPER - [radio] (short for superheterodyne receiver) récepteur *m* à changement de fréquence, superhétérodyne *m*
SUPERACIDITY - [chem] (excessive acidity) acidité *f* excessive
SUPERACTIVITY - [biol] suractivité
SUPERADIABATIC LAPSE RATE - [met] (lapse rate obtaining when insolation supplies heat to the ground more rapidly than it can be removed by convection, turbulence and ground conduction) gradient *m* superadiabatique
~LAYER - [met] (unstable layer of air close to the ground in which the lapse rate is superadiabatic) couche *f* super-adiabatique
SUPERAERODYNAMICS - [aero] (the science of fluid flow conditions such that the continuum flow conditions no longer apply) super-aérodynamique *f*
SUPERALLOY - [metall] alliage *m* spécial
SUPERANNUATE, to - [gen] périmer, suranner
SUPERAUDIO FREQUENCY - [telecomm] (a frequency which is above those commonly transmitted through an audio-frequency reproducing system) fréquence *f* ultratéléphonique
~TELEGRAPHY - [telecomm] télégraphie *f* ultratéléphonique
SUPER-BALLOON TYRE - [rubber ind] pneu *m* superballon
SUPERCALENDER - [paper man] (a calender separated from the paper machine) calandre *f*
SUPERCALENDERED - [paper man] satiné
~PAPER - [paper man] papier *m* pour illustrations
SUPERCALENDERING - [paper man] (operation of giving paper a high finish by means of a calender in which the rolls are alternately of chilled iron and of paper or cotton) supercalandrage *m*
SUPERCARGO - [naut] subrécargue *m*
SUPERCENTRIFUGE - [mech] super-centrifugeuse *f*
SUPERCHARGE, to - [gen] surcharger
[mech] suralimenter; alimenter à compression
[el] suralimenter
[aero] (to pressurize) maintenir sous pression
SUPERCHARGED - [mech el etc] suralimenté, surcomprimé
[aero] (of a cabin etc) pressurisé
[gen] surchargé
~CABIN - [aero] cabine *f* pressurisée
~ENGINE - [mech] moteur *m* suralimenté, moteur *m* à compresseur
~IGNITION HARNESS - [aero] (a device used to avoid electrical leaks at high altitude) installation *f* d'amorçage dans une canalisation étanche pressurisée
SUPERCHARGER - [mech] (device fitted to an internal combustion engine to supply it with more air than would normally be induced at atmospheric pressure usually a fan or blower) surcompresseur *m*, surpresseur *m*
SUPERCHARGING - [mech etc] suralimentation *f*
SUPERCHLORINATION - [hydr] surchloration *f*
SUPERCHOPPER - [nucl] (a device used in time of flight method, chopping the neutron beam into pulses) hacheur *m* ultrarapide
SUPERCOMPRESSION ENGINE - [mech] (an unsupercharged engine having a high compression ratio and designed to be run at full throttle only at altitudes above some fixed level) moteur *m* surcomprimé
SUPERCONDUCTING COMPUTER DEVICE - [comput] (computer in which the principle of superconductivity is used) élément *m* supra-conducteur
~TRANSITION - [phys chem] transition *f* à la supraconductivité
SUPERCONDUCTIVITY - [el] (abnormally high electrical conductivity suddenly appearing in mercury, magnesium, lead etc when they are cooled to a very low temperature) supra-conductivité *f*
SUPERCONDUCTOR - [el] supra-conducteur *m*
SUPERCONTROL TUBE - [electron] tube *m* à pente variable
SUPERCOOL, to - [chem] (to cool a liquid below the normal freezing point without changing its state) sous-refroidir
SUPERCOOLED - [chem] surfondu
~GRAPHITE - [metall] graphite *f* de surfusion
SUPERCOOLING - [chem] (the cooling of a liquid below the normal freezing point without change of state) sous-refroidissement *m*, surfusion *f*
SUPERCRITICAL - [phys & chem] (having an effective multiplication constant greater than one so that the rate of the reaction rises) surcritique
SUPERCUSHION TYRE - s. superballoon tyre
SUPERCUT FILE - [mech] lime *f* à taille très fine
SUPERCUTTING STEEL - [metall] acier *m* à coupe rapide
SUPERDISTENTION - [med] surextension *f*
SUPERDOMINANT - [mus] (the sixth note of any scale in the modern key system) sus-dominante *f*
SUPERELEVATED - [constr] surélèvé
SUPERELEVATION - [constr] surhaussement *m*, surélévation *f*
[railw] dévers *m*
~GAUGE - [railw] règle *f* à devers
SUPEREXCITATION - [med] surexcitation *f*
SUPEREXTENSION - s. superdistension
SUPERFATTED - [ind chem] (of soap) surgras (savon)
SUPERFICIAL - [gen] superficiel
~EXPANSION - [metall] coefficient *m* de dilatation superficielle
~EXTENT - [gen] étendue *f* superficielle
~LAYER -[geol] placage *m*
~VELOCITY - [chem] (the flow rate through a tower) vitesse *f* superficielle
~X-RAY THERAPY - [radiat] röntgenthérapie *f* superficielle
SUPERFICIES - [gen] (surface) superficie *f*
SUPERFINE FILE - [mech] lime *f* superdouce
SUPERFINES - [metall] fraction *f* fine
SUPERFINISHING - [metall] superfinissage *m*
SUPERFLEXION - [med] hyperflexion *f*
SUPERFLUID - [phys] (fluid having a very high ther-

mal conductivity and capillarity, e.g. helium at I° and 2° absolute temperature) suprafluide *m*
SUPERFLUIDITY METHOD - [nucl] méthode *m* de suprafluidité
SUPERFUSE, to - s. supercool, to
SUPERGLACIAL DRIFT - [geol] moraine *f* superficielle
~TILL - s. superglacial drift
SUPERGROUP - [telecomm] (in carrier telephony) groupe *m* secondaire
~BAND FILTER - [telecomm] filtre *m* de bande de groupe secondaire
~DISTRIBUTION FRAME - [telecomm] répartiteur *m* de groupe secondaire
~LINK - [telecomm] liaison *f* en groupe secondaire
~REFERENCE PILOT - [telecomm] (a reference pilot applied where the supergroup is assembled and accompanying the supergroup over the system until it is broken down to its carrier group)onde *f* pilote de groupe secondaire
~SECTION - [telecomm] (part of a supergroup link between two adjacent supergroup distribution frames) section *f* de groupe secondaire
~TRANSFER POINT - [telecomm] point *m* de transfert de groupe secondaire
~TRANSLATING EQUIPMENT - [telecomm] installation *f* de modulation de groupe secondaire
SUPERHEAT, to - [gen & phys] surchauffer
SUPERHEAT - [heat] (the amount of heat by which steam has been superheated) surchauffe *f*
[aero] (the increase or decrease of temperature of the gas in an aerostat, compared with the temperature of the surrounding air) chaleur *f* différentielle, surchauffe *f*
~RESISTANT GLASS - [glass man] verre *m* résistant aux hautes températures
SUPERHEATED - [gen] surchauffé
~STEAM - [phys] (steam which has been raised to a higher temperature than that of the vapour phase in the steam generator) vapeur *f* surchauffée, vapeur *f* non-saturée
~STEAM BOILER - [th eng] chaudière *f* à vapeur surchauffée
~STEAM RECLAIMING PROCESS - [rubber ind] procédé *m* de régénération à la vapeur surchauffée
~WATER - [chem] eau *f* surchauffée
SUPERHEATER - [th eng] (apparatus for raising the temperature of steam after it leaves the generating elements of a boiler) surchauffeur *m*,surchauffeur *m* de vapeur
[ind chem] (conical tube coil for superheating steam by means of a bunsen burner) surchauffeur *m*
[rubber ind] purgeur *m* pour vapeur surchauffée
~COILS - [th eng] (helical or other arrangement of steel tubes used for heat transfer in a superheater) serpentin *m* surchauffeur
~HEAD - [th eng] collecteur *m* du surchauffeur
~UNIT - [el] élément *m* surchauffeur
SUPERHEATING - [gen] surchauffage *m*, surchauffe *f*
SUPERHEAVY NUCLEUS - [phys] noyau *m* contenant des hypérons
SUPERHET - s. superheterodyne receiver
SUPERHETERODYNE RECEIVER - [radio] (a type of receiver in which predetector amplification is carried out at frequency between that of the station and that of the modulation) récepteur *m* super-hétérodyne
~RECEPTION - [radio] (a method of reception in which the signal carrier-frequency is changed to another radio-frequency by a heterodyne process) réception *f* superhétérodyne
SUPERHIGH FREQUENCY - [radio] (any radio frequency between 3000 and 30.000 megacycles per second) hyperfréquence *f*
SUPERHIGHWAY - [constr] (USA only, a highway built with the greatest possible regard for safety and for scenic beauty) grande route *f*
SUPERIMPOSE, to - [gen] superposer, surimposer
SUPERIMPOSED-IMAGE RANGEFINDER - [photo] télémètre *m* à images superposées
~IMAGES - [telev] images *f*pl superposées
~INTERFERENCE - [telev] transmodulation *f*
~KNIVES - [text] couteaux *m*pl superposés
~RINGING - [telecomm] signalisation *f* multiple
SUPERIMPOSER - [telev] appareil *m* de surimpression
SUPERIMPOSING - [gen] superposition *f*
[telev] (the superimposing of two or more images) superposition *f*, surimpression *f*
~THE WEB - [text] (in layers) superposition *f* de voiles
SUPERIMPOSITION - [gen] superposition *f*, surimpression *f*
~APPROXIMATION - [phys] principe *m* de superposition
SUPERINCUMBENT BED - [mining] couche *f* du toit
SUPERINFECTION - [med] surinfection *f*
SUPERINTEND, to - [gen] diriger, surveiller
SUPERINTENDENT - [gen] directeur *m*, chef *m* des travaux
SUPERIOR - [gen] supérieur
~CONJUNCTION - [astr] conjonction *f* supérieure
~FIGURES - [print] (small figures printed above the level of the line and directing the reader to a footnote) chiffre *m* supérieur
~LETTERS - [print] (small letters printed above the level of the line and directing the reader to a footnote) exposant *m* supérieur
~WOOL - [text] laine *f* prime
SUPERLATIVE - [gen] superlatif
[gen] (of objects etc) suprême
SUPERLATTICE - [nucl] (a type of arrangement of atoms in a multi-component solid system) réseau *m* superposé
SUPERMARKET - [comm] super-marché *m*
SUPERNATANT DRAW-OFF PIPE - [hydr] tuyau *m* d'extraction du liquide surnageant
~LIQUID - [chem] (the clear liquid above a precipitate which has just settled out) liquide *m* surnageant
SUPERNATE - [chem] couche *f* surnageante
SUPERNUMERARY - [gen] surnuméraire
[theatre] figurant *m*
SUPERPHANTOM CIRCUIT - [telecomm] circuit *m* combiné double, circuit *m* surcombiné, superfantôme *m*
SUPERPHOSPHATE - [chem] (a phosphorus fertilizer of major importance) superphosphate *m*
SUPERPOSE, to - [gen] superposer
[gen] (to lay in layers) étager
[math] (to place one figure upon another) superposer
SUPERPOSE THE SILK THREADS ON THE REEL, to -[text] superposer les fils de soie sur le guindre
SUPERPOSED CIRCUIT - [telecomm] (additional circuit, as a phantom on a telephone circuit) circuit *m* virtuel, circuit *m* superposé
~CYLINDERS WITH RADIAL PINS - [text] cylindres *m*pl superposés avec aiguilles radiales

SUPERPOSED SHUTTLE BOXES - [text] boîtes *f*pl à navette superposées
SUPERPOSITION - [gen] superposition *f*
~OF WEB - [text] superposition *f* du voile
SUPERPOWER - [el] (the combined sources of electric power in a given area) puissance *f* maximale
SUPERPRESSURE - [aero] (the difference between the pressure of a gas in an aerostat and that of the surrounding air) pression *f* differentielle
~BALLOON - [aero] ballon *m* pressurisé
SUPERREACTION - [radio] (reaction in a receiver to a degree which would cause self-oscillation) superréaction *f*
SUPERREFRACTION - [opt] (the transmission of microwave frequencies beyond line of sight due to an atmospheric duct) super-réfraction *f*
SUPERREGENERATION - s. superreaction
SUPERREGENERATIVE RECEPTION - [radio] réception *f* à super-réaction
SUPERRETROACTION - s. superreaction
SUPERSATURATE, to - [chem] (to saturate a solution or vapour to a point at which the concentration is greater than that of saturation) sursaturation *f*
SUPERSATURATED - [chem] sursaturé
~SOLUTION - [chem] solution *f* sursaturée
SUPERSATURATION - [chem] (the condition of a solution or vapour when the concentration is greater than that of saturation, a metastable state) sursaturation *f*
SUPERSEDE, to - [gen] remplacer
SUPERSENSITIZATION - [med] hypersensibilisation *f*
SUPERSONANT - [acoust] ultrason *m*
SUPERSONIC - [acoust] (denoting the sound waves which are beyond the upper limits of human audibility) ultra-sonore
[aero] (of aircraft) (flying beyond the speed of sound) supersonique
~AIRCRAFT - [aero] avion *m* supersonique
~AMPLIFICATION - [radio] amplification *f* ultrasonore
~AREA RULE - [aero] (an extension of the sonic area rule to supersonic speed) loi *f* de l'aire supersonique
~BANG - [aero] (sudden loud sound caused by shockwaves propagated from an aircraft flying at supersonic speed) détonnement *m* supersonique
~DELAY-LINE - [comput] (delay-line operating in the supersonic range of frequency) ligne *f* à retard ultrasonore
~DETECTOR - [radio] (or ultrasonic detector device for the detection and measurement of supersonic waves) détecteur *m* supersonique
~EFFUSER - [aero] (a nozzle in a supersonic wind tunnel in which the speed of flow is accelerated to supersonic levels) diffuseur *m* supersonique
~FLOW - [phys] (type of flow in which speeds are supersonic at every point outside the boundary layer) écoulement *m* supersonique
~FLIGHT - [aero] vol *m* supersonique
~FREQUENCY - [radio] (a frequency above the audio range) fréquence *f* ultrasonore
~GENERATOR - [radio] (a generator for the production of waves of supersonic frequency) générateur *m* supersonique, générateur *m* ultrasonore
~HETERODYNE RECEIVER - [radio] (type of receiver in which predetector amplification is carried out at a frequency between that of the station and that of modulation) réception *f* superhétérodyne
SUPERSONIC INSPECTION - [metall] examen *m* ultrasonore
~PLANE - s. supersonic aircraft
~REFLECTOSCOPE - [instr] (reflectometer designed to test the soundness of a metal bar or ingot by comparing reflexion of high-frequency waves) reflectoscope *m* ultrasonore
~SHAPE - [aero] (of wings) ligne *f* supersonique
~SOUNDING - [meas] sondage *m* par ondes supersoniques
~SPEED - [aero] (speed exceeding the local velocity of sound) vitesse *f* supersonique
~STROBOSCOPE - [instr] (light interruptor having an action based on the modulation of a light beam by a supersonic field) stroboscope *m* ultrasonore
~WAVE - [radio] (elastic wave, having a frequency above the audible range) onde *f* supersonique, onde *f* ultrasonore
~WIND TUNNEL - [aero] (a wind tunnel designed to deal with supersonic speeds) soufflerie *f* supersonique
SUPERSONICS - [acoust] (a term denoting phenomena associated with speed higher than the speed of sound) ultrasons *m*pl
SUPERSTANDARD PROPAGATION - [phys] propagation *f* supranormale
~REFRACTION - [phys] supra-réfraction *f*, réfraction *f* supra-normale
SUPERSTRUCTURE - [gen] (the part of a structure above any principal supporting level) superstructure *f*
[railw] (rails, sleepers etc of a railway, as distinguished from the road bed) superstructure *f* de la voie
[shipbuild] (those parts of ships and warships above the main deck) accastillage *m*
SUPERTURNSTILE AERIAL - [radio] (term denoting a broadband VHF transmitting aerial used in television) antenne *f* croisée multiple
~ANTENNA - s. superturnstile aerial
SUPERVISE, to - [gen] surveiller, diriger
SUPERVISION - [gen] surveillance *f*
[comm] direction *f*
SUPERVISOR - [gen] surveillant *m*, directeur *m*
SUPERVISORY CONTROL - [el] (of a generator set) surveillance *f*
[contr] (USA only, remote or distant control in GB) télécommande *f* surveillée
~INDICATOR - [telecomm] (in telephony) annonciateur *m* de fin de conversation
~LAMP - [gen] lampe *f* de contrôle, lampe *f* pilote
[telecomm] (in telephony) lampe *f* de clôture
~RELAY - [telecomm] relais *m* de clôture, relais *m* de fin
[el] relais *m* de supervision
SUPERVOLTAGE - [radiol] (term denoting radiation generated by X-rays operating at voltages ranging from 500 to 2000 KV) tension *f* trés élevée
~THERAPY - [el biol] (X-ray therapy by very hard X-rays usually generated at one million volts or more) rœntgenthérapie *f* à trés haute tension
SUPINATION - [med] (of the hand) supination *f*
[med] (position) décubitus *m* dorsal
SUPPLE - [gen] souple, pliable
~FIBRE - [text] fibre *f* flexible
SUPPLEMENT, to - [gen] ajouter un supplément

SUPPLEMENT - [gen] supplément *m*
[math] (in trigonometry)(of an angle) supplément *m*
~OF AN ANGLE - [math] supplément *m* d'un angle
SUPPLEMENTAL - [math] (in geometry) supplementaire *f*
SUPPLEMENTARY - [gen] supplémentaire
~AERODROME - [aero] (an aerodrome officially designated for use when the scheduled aerodrome cannot be reached) aérodrome *m* supplémentaire
~ANGLE - [geom] angle *m* supplémentaire
~LENS - [popt] lentille *f* supplémentaire
~LOSS - [el] (the excess of the actual losses over the sum of the separately measurable losses) pertes *fpl* supplémentaires
~VALENCE - [phys] (residual valence connecting atoms, groups or molecules in which the ordinary valences are already saturated) valence *f* supplémentaire
SUPPLENESS - [gen] souplesse *f*, flexibilité *f*
~OF THE FIBRE - [text] flexibilité *f* de la fibre
SUPPLIER - [gen] fournisseur *m*
SUPPLIES - [gen] approvisionnements *mpl*, apports *mpl*
~OF LIVESTOCK - [comm] arrivées *fpl* de bestiaux
SUPPLY, to - [gen] fournir, pourvoir
[gen] (to replace) remplir, réparer
SUPPLY - [gen] fourniture *f*, approvisionnement *m*
[comm] fourniture *f*, provision *f*
~BY ACCUMULATORS - [el] alimentation *f* par accumulateurs
~CIRCUIT - [el] circuit *m* d'alimentation
~CONNECTION - [gas ind & el] raccord *m* d'alimentation
~CURRENT, to - [el] débiter du courant
~FREQUENCY - [el] fréquence *f* du courant d'alimentation
~LEADS - [el] câble *m* de livraison
~MAIN - [el] (conductors connecting a generator to the distribution busbars) conducteur *m* de réseau, câble *m* de distribution
~PIPE - [hydr] conduite *f* d'amenée
~POINT - [el] point *m* de distribution
~PRESSURE - [gas ind] pression *f* d' alimentation
~REEL - [text] bobine *f* d'alimentation
[cin] (the spool containing the film to be inserted in the feeding mechanism) bobine *f* débitrice
~SPOOL - s. supply reel
~-STATION - [el] (a power station) centrale *f* électrique
~TERMINALS - [el] point *m* de livraison
~VOLTAGE - [el] tension *f* au point de livraison
[electron] (of an electrode, the voltage applied by an external source to the circuit of an electrode) tension *f* d'alimentation
SUPPORT, to - [gen] supporter, soutenir, appuyer
SUPPORT - [gen] soutien *m*, appui *m*
[mech] support *m*, chaise *f*
[constr] (of a beam, a girder) appui *m*, soutien *m*, pied *m*, soutenement *m*, assiette *f*, potence *f*
[carp] (a prop) étai *m*, pied *m*
[photo] (of a film) support *m*
[agric] tuteur *m*
[th eng] (boiliers) sabot *m*
~ARM - [mech] bras *m* support
~ARM SHAFT - [auto] axe *m* de bras support
~BRACKET - [mech] console *f* de soutien
~EQUIPMENT - [astronaut] équipement *m* au sol
~HOOK - [el] crochet *m*
SUPPORT INSULATOR - [el] isolateur *m* portant
~OF RIDGE PURLIN - [constr] appui *m* de panne de faîte
~PILLAR - [gen] colonne *f*
~PLATE - [mech] plaque-support *f*
~PRICE - [comm] prix *m* de soutien, prix *m* subventionné
~STAYS - [aero] balancines *fpl*
~STRAND - [el] câble *m* de décharge, câble *m* porteur longitudinal
~TUBE - [mech] tube-support *m*
SUPPORTED - [gen] appuyé, soutenu
~BEAM - [constr] poutre *f* appuyée
~CATALYST - [chem] catalyseur *m* soutenu
~FLANGE - [mech] joint *m* conique à bride
~IN BEARINGS - [mech] porté par des paliers
~JOINT - [railw] (of the rails on the sleeper) joint *m* appuyé
~SCREWED JOINT - [mech] joint *m* conique à manchon
SUPPORTER OF COMBUSTION - [gas ind] (a combustion agent) comburant *m*
SUPPORTING - [gen] d'appui, de soutènement
~ARM - [mech] bras *m* de soutènement
~AXLE - [auto] axe *m* support
~BAR - [text] selle *f*, support *m*
~BASE - [mech] base *f* de sustentation
~BEAM - [constr] poutre *f* portante
[mech] (of a winch) soupente *f* (d'un treuil)
~BLOCK - [mech] bloc *m* d'appui
~COLUMN - [metall] colonne *f* de soutènement
~CURB - s. supporting frame
~ELECTRODE - [meas] (in emission spectroscopy, an electrode, which is not a self-electrode, on which the sample is supported) électrode *f* porteuse
~FRAME - [mining] cadre *m* porteur, cadre *m* à oreilles
~FRAMEWORK - [gen] monture *f*, ossature *f* de support
~GRAPHITE SLEEVE - [nucl] (in a nuclear reactor) manchon *m* de support en graphite
~JOURNAL - [mech] tourillon *m* de support
~LEG - [photo] (of a camera) support *m* de pied
~LEVER - [mech] levier *m* de support
~MATERIAL - [gen] support *m*
~PACK - [mining] arête *f* de remblai
~PILLAR - [mining] pilier *m* de soutènement
~PLATE - [electron] (component part of electronic tubes) plaque *f* de support
~POST - [constr] pieu *m* de soutènement
[mining] butte *f*
~PROGRAMME - [cin] films *mpl* de première partie
~REINFORCEMENT - [metall] (in concrete) armature *f* longitudinale
~RIB - [mech] nervure *f* de renfort
~RING - [metall] rondelle *f* de support de ballon
~ROD - [electron] (component part of electronic tubes) tige *f* de support
~ROLL - [mech] (a roller used to support a band of material) support *m* à rouleaux
~ROPE - [mech] câble *m* porteur
~SADDLE - [gas ind] support *m*, sabot *m*
~STANCHIONS - [metall] (on ovens) pilier *m* d'ancrage
~SURFACE - [aero] (those surfaces essentially designed to provide lift for an aircraft) surfaces *fpl* portantes
~TABLE - [mech] plateau *m*
~TOWER - [constr] pyramide *f* de soutien, pylône *m*

de support
SUPPORTING TRESTLE - [railw] (in aerial ropeways) pylône *m* pour téléphériques
~WALL - [constr] mur *m* de talus, mur *m* de soutènement, piedroit *m*
SUPPOSITORY - [med] suppositoire *m*
SUPPRESS, to - [gen] supprimer, réprimer
[med] étouffer
SUPPRESSED - [radio] (term commonly used of apparatus fitted with an interference suppressor) supprimé
~AERIAL - [radio] (a radio aerial so constructed that it does not project from the structure of an aircraft, and so causes no drag) antenne *f* noyée
~-CARRIER OPERATION - [radio] transmission *f* à porteuse supprimée
~-CARRIER SYSTEM - [radio] (a system of transmission in which the carrier wave is not radiated but is supplied by an oscillator at the receiving end) système *m* à porteuse supprimée
~CARRIER TRANSMITTER - [radio] émetteur *m* à suppression de la porteuse
~FIELD SYSTEM - [telev] système *m* de suppression de trames alternantes
~SIDEBAND - [telev] bande *f* latérale supprimée
~ZERO INSTRUMENT - [instr] (an instrument in which the indicating element does not show readings below a minimum value) appareil *m* à équipage mobile buté, appareil *m* de mesure à zéro supprimé
SUPPRESSION CONTROL - [radar] (anti-clutter gain control) contrôle *m* differentiel de gain
SUPPRESSOR - [radio] (a device designed to eliminate or reduce radio interference by an electrical unit) résistance *f* antiparasite, résistance *f* suppresseuse
~AREA - [med] aire *f* suppressive
~GRID - [electron] (a grid the primary function of which is to reduce the effect of secondary emission) grille *f* d'arrêt, grille *f* suppresseuse
~-GRID KEYING - [electron] (keying effected by changing the bias potential applied to the suppressor grid of a tube) manipulation *f* dans la grille suppresseuse
~-GRID MODULATION - [radio] modulation *f* par la grille suppresseuse
SUPPURATE, to - [med] suppurer
SUPPURATION - [med] (the softening and liquefaction on inflamed tissue) suppuration *f*
SUPRAELEVATION - [constr] surélévation *f*
SUPRARENAL - [anat] surrénal
SUPRARENALECTOMY - [med] surrénalectomie *f*
SUPRASTERNAL BONE - [anat] osselet *m* épisternal
SUPRAVERGENCE - [med] strabisme *m* sursumvergent
SURAH - [text] (a soft silk fabric) surah *m*
SURBASE - [arch] (a dado rail) corniche *f* de piédestal
SURBASED ARCH - [arch] arc *m* surbaissé
SURCHARGE, to - [gen] surcharger
[fin] surtaxer
SURCHARGE - [gen] surcharge *f*
[fin] surtaxe *f*
[comm] prix *m* excessif
[constr] (the earth which is supported by a retaining wall) surcharge *f*
SURCINGLE - [anat] queue *f* du noyau caudé
[agric] (of harness) sous-ventrière *f*
SURD - [math] (an irrational number) irrationnel, quantité *f* incommensurable
SURD - [phonetics] consonne *f* sourde
SURE - [gen] sûr, certain
SURETY - [leg] (a pledge of money or goods) sûreté *f*, garantie *f*
SURF - [gen] barre *f* de plage, brisants *mpl* sur la plage
[naut] (the sea breaking on the shore) ressac *m*
SURFACE, to - [mech] lisser, polir, apprêter la surface
[carp] dégauchir
[constr] (a road etc) revêtir
[paper man] calancrer, satiner
[railw] relever (la voie)
[mach tool] surfacer
SURFACE - [gen] surface *f*, extérieur *m*
[math] surface *f*
[constr] (of a road) revêtement *m* (d'une route)
[mining] jour *m*
~ACTION - [phys] action *f* superficielle
~ACTIVE AGENTS - [chem] (compounds which alter the surface tension when dissolved in water. Surface active agents find application in detergents and wetting and dispersing agents) agents *mpl* tensio-actifs, agents *mpl* mouillants
~ACTIVE COMPOUNDS - [phys chem] (substances which lower the surface tension of a liquid when exposed to gas, e.g. water in air) composés *mpl* à activité superficielle
~ACTIVITY - [phys chem] (in the surface tension of liquids) activité *f* superficielle, action *f* tensio-active
~ACTIVITY OF A MINERAL - [min] activité *f* superficielle d'un minéral
~ANALGESIA - [med] analgésie *f* de surface
~ANALYZER - [instr] (a very sensitive instrument used to measure surface irregularities) analyseur *m* de surface
~APPLICATOR - [radiat] capsule *f* superficielle
~AUGER - [mining] tarière *f* à glaise
~BALANCE - [instr] (apparatus designed to measure surface pressure and surface area of monomolecular films on water) balance *f* de Langmuir
~BARRIER DIODE - [instr] (of a junction particle detector) diode *f* à barrière de surface
~BARRIER SEMICONDUCTOR DETECTOR - [electron] détecteur *m* semiconducteur à barrière de surface
~-BARRIER TRANSISTOR - [electron] transistor *m* à barrière superficielle
~BEARING - [mech] appui *m* à surface de contact
~BELT CONVEYOR - [mech] transporteur *m* à courroie de matériaux en surface
~BLUCH - [paint] trouble *m* superficiel
~BOUNDARY LAYER - [phys] couche *f* limite atmosphérique
~BOX - [gas ind] bouche *f* à clé, carter *m* de robinet
~BRIGHTNESS - [opt] (intrinsic brightness) luminance *f*
~BROACHING - [mach tool] équarissage *m* extérieur
~BROACHING MACHINE - [mach tool] équarissoir *m* pour surface extérieures
~BURST - [nucl] (the explosion of a nuclear weapon at the surface of land or water) explosion *f* nucléaire superficielle
~CASING - [mining] tube-guide *m*
~CHARGE DENSITY - [el] (the quantity of electric charge per unit area of a charged surface) densité

f de la charge superficielle
SURFACE CHARGE EFFECT - [electron] effet *m* de charge superficielle, effet *m* S
~CHECKING - [metall] fissuration *f* superficielle
~CHILL - [metall] refroidisseur *m* de surface
~CLEARING - [soil] défrichement *m*
~-COATED - [paper man] couché (papier)
~COMBUSTION - [phys] (the combustion of a gas-air mixture in contact with a refractory without actual flame, the refractory being kept in an incandescent state) combustion *f* de surface
~COMBUSTION BURNER - [th eng] brûleur *m* à combustion superficielle intersticielle
~COMPACTION - [soil] compactage *m* en surface
~CONDENSATION - [mech] (of steam) condensation *f* par surface
~CONDENSER - [mech] (steam condenser maintaining a vacuum at the exhaust pipe of a steam engine or turbine) condenseur *m* par surface, condenseur *m* à sec
~CONTACT-RECTIFIER - [electron] (contact rectifier where the contact surface has a finite area; it is a barrier-layer cell) redresseur *m* à barrière superficielle
~CONTACT-SYSTEM - [el] système *m* à plots
~CONTAMINATION METER - [nucl] contaminamètre *m* surfacique
~CONTAMINATOR INDICATOR - [nucl] signaleur *m* de contamination surfacique
~CONVERTING - [metall] trempe *f* de surface, cémentation *f*
~CONVEYANCE - [mining] transport *m* à ciel ouvert
~COOLER - [th eng] réfrigérant *m* à ruissellement, réfroidisseur *m* à surface
~CRACK - [metall] gerçure *f*
~CRUSTING - [geol] encroûtement *m* du sol
~CULTIVATION - [agric] façons *f*pl superficielles
~CURRENT - [gen] (of the sea) courant *m* de surface
~CUT - [soil] fouille *m* à la surface
~DAMPING - [el acoust] (additional damping at the edge of a membrane of a loudspeaker)amortissement *m* marginal
~DENSITY - [phys] (the quantity per unit area of anything that is distributed over a surface) densité *f* superficielle
~DEPOSIT - [geol] gîte *m* de surface
~DOSE - [nucl] (the dose of radiation supplied to a given area of the surface of an object) dose *f* en surface
~DRAIN - [hydr] tranchée *f* à ciel ouvert
[agric] saignée *f* d'irrigation
~DRAINAGE - [agric] drainage *m* superficiel
~DRAINING - [agric] drainage *m* à ciel ouvert
~-DRESSED WATER-BOUND MACADAM - [constr] empierrement *m* avec enduit
~DRESSING - [constr] (of roads) traitement *m* superficiel, enduction *f* de surface, goudronnage *m*
~DRIVE - [mining] fendue *f*
~DUCT - [radio] (a tropospheric radio duct having the earth as its lower boundary, and in which the modified refractive index is everywhere greater than the value at the upper boundary) conduit *m* prés du sol, conduit *m* de surface
~EFFECTS - [metall] effets *m*pl superficiels
~ENERGY - [phys] (the energy per unit area required to increase the surface of a solid or liquid)énergie *f* superficielle
SURFACE ENGINE - [mech] (used in mining) moteur *m* à la surface, moteur *m* au jour
~EQUIPMENT - [mining] installation *f* de surface
~FEET PER MINUTE - [mach tool] vitesse *f* périphérique
~FILM - [electron] (film of insoluble substance on the surface of a liquid consisting of a monolayer of closely packed molecules) couche *f* de surface
~FILTER ELEMENT - [mech] élément *m* de filtre à la surface
~FOLDING - [metall] rides *f*pl, peau *f* de crapaud, friasses *f*pl
~FRICTION - s. surface friction gdrag
~-FRICTION DRAG - [aero] (that part of the drag which is due to the forces acting tangentially to a body in an airstream) traînée *f* de frottement
~GAUGE - [tool] (instrument for measuring heights from a surface plate in setting out machining) trusquin *m*
~GEOLOGY - [geol] géologie *f* de la surface
~GRINDER - [mach tool] (machine for producing accurate plane surfaces by abrasion) machine *f* à rectifier les surfaces planes extérieures, rectifieuse *f* plane
~GRINDING ATTACHMENT - [mach tool] dispositif *m* à rectifier les surfaces planes extérieures
~GRINDING MACHINE - s. surface grinder
~GUIDE - [electron] (surface wave transmission line) ligne *f* de transmission de Goubau
~HARDENING - [metall] trempe *f* superficielle
~HAULAGE - [mining] transport *m* à la surface
~HAZE - [metall & plast ind] (dullness in a polished surface, especially of plastic sheet) mat *m* de surface, trouble *m* superficiel
~HOLE - [oil ind] trou *m* pour colonne de tubage
~INTEGRAL - [math] intégrale *f* superficielle
~ION DENSITY - [phys] nombre *m* surfacique d'ions
~IRRADIATION - [radiat] irradiation *f* superficielle
~IRRIGATION - [agric] irrigation *f* en surface
~LATENT IMAGE - [photo] image *f* latente superficielle
~LATHE - [mach tool] tour *m* à surfacer
~LAYER - [electron] (thin layer of electron emitting material on a metallic carrier) couche *f* superficielle
~LEAKAGE - [phys] (an unwanted phenomenon occurring when measuring high resistance and due to resistance offered to the passage of current along the surface of the insulation) fuite *f* superficielle
~LEAKAGE CURRENT - [el] courant *m* de dispersion superficielle
~LOAD - [el] (relation between the power consumed and the emitting surface of the element under constant conditions) puissance *f* surfacique
~LOADING - [aereo] (the normal mean force per unit area borne by an aerofoil under given conditions) charge *f* alaire
~LOUDSPEAKER - [el acoust] (loudspeaker which is partly buried in the ground) haut-parleur *m* enterré
~MIGRATION - [electron] (the movement of the electrons at the surface of an emitter) migration *f* superficielle
~MILLING - [mach tool] fraisage *m* en plan
~MINE - [mining] mine *f* à ciel ouvert, fondrière *f*
~MINING - [min] exploitation *f* à ciel ouvert, exploitation *f* au jour
~NOISE - [acoust] (the sound accompanying the sound

reproduced from records) bruit *m* de surface
SURFACE OF CONTACT - [auto] (of the tyres) surface *f* de contact
~OF EVAPORATION - [soil] surface *f* d'évaporation
~OF FRACTURE - [mech] surface *f* de rupture
[constr] surface *f* de la cassure
~OF LEAST RESISTANCE TO SLIDING - [soil] plan *m* de la moindre résistance au glissement
~OF OUTCROPPINGS - [geol] affleurements *mpl*
~OF REVOLUTION - [math] surface *f* de révolution
~OF SHEAR - [soil] surface *f* de glissement
~OF SLIDING - [soil] plan *m* de glissement
~OF STRIPPER - [text] surface *f* de couteau
~OF THE CLOTH - [text] surface *f* du tissu
~OF THE FLEECE - [text] surface *f* de la toison
~OF THE ROPE - [text] surface *f* du câble
~OF THE WOOL FIBRE - [text] surface *f* du poil
~OF THE YARN - [text] surface *f* du fil
~OF WEB - [text] surface *f* de la nappe
~OIL COOLER - [aero] (a device in which a part of the external surface of an aircraft is used to dissipate heat from the oil) radiateur *m* d'huile de revétement
~ORIENTATION - [phy] (the occupation of such positions of certain molecules in the surfaces of a liquid that one part of the molecule is turned toward the liquid) orientation *f* superficielle
~PASSIVATION - [electron] (used to reduce surface leakage in high-frequency transistors) passivation *f* superficielle
~PHOTOELECTRIC EFFECT - [electron] effet *m* photoélectrique superficiel
~PIT - [mining] fosse *f* à ciel ouvert , carrière *f* à ciel ouvert
~PITTING - [metall] formation *f* d'alvéoles
~PLANING - [carp] dégauchissage *m*
~PLANING MACHINE - [mach tool] machine *f* à dégauchir, dégauchisseuse *f*
~PLANT - [mining] carreau *m*
~PLATE - [mech] (a heavy iron casting ground to an accurate plane surface, used for marking out and measuring machine-shop work) table *f* à croquis, marbre *m* à dresser, marbre *m*
~PLAY OF IRON - [metall] miroitement *m* de la fonte
~PRINTING - [print] impression *f* par planches gravées en relief
~PYROMETER - [instr] (pyrometer used for measuring the temperature of surfaces) pyromètre *m* à surface
~RADIATOR - [aero] (type of radiator in which a part of the external surface of an aircraft is used to dissipate heat) radiateur *m* de revêtement
~RECOMBINATION RATE - [electron] (the time rate at which free electrons and holes recombine at the surface of a semiconductor) vitesse *f* de recombinaison superficielle
~RESISTANCE - [el] (the electrical resistance between opposite edges of a unit square of insulating material, measured under specific conditions) résistance *f* de surface
~RESISTIVITY - [el] (the electrical resistance between opposite edges of a unit square of insulation material under prescribed conditions) résistivité *f* de surface
~ROUGHNESS - [metall & plast ind] rugosité *f* de surface
[metall] rudesse *f* de la surface
SURFACE SANDPAPERING MACHINE - [mach tool] machine *f* à poncer à ruban
~SHAFT - [mining] puits *m* de surface
~SHEET - [metall & plast ind] (the uppermost sheet of a lamination) feuille *f* de couverture, couche *f* superficielle
~SHRINKAGE - [metall] retassure *f* ouverte
~-SIZED - [paper man] collé en surface
~SIZING - [paper man] collage *m* en surface, apprêt *m* de surface
~SOIL - [geol] sol *m* superficiel, terre *f* sus-jacente
~SPEED INDICATOR - [instr] compteur *m* de vitesse linéaire
~SPRINKLING - [agric] irrigation *f* en surface par ruissellement
~SURVEY - [surv] levé *m* des planes de surface, levé *m* au jour
~SWITCH - [el] interrupteur *m* sur crépi
~TENSION - [phys] (capillarity; the tension of a surface resulting from all attractive forces) tension *f* superficielle
~TERMINATION - [geol] affleurement *m*
~TESTER - [instr] appareil *m* pour le contrôle des surfaces planes
~THERAPY TUBE - [radiat] tube *m* pur traitement en surface
~-TYPE INSTRUMENT - [instr] instrument *m* à montage extérieur
~VITRIFICATION - [metall] grédage *m*, frittage *m*
~WATER - [hydr] eau *f* superficielle, eaux *fpl* folles, eaux *fpl* sauvages
~WAVES - [phys] ondes *fpl* superficielles
~WAVINESS - [plast ind] (wavy form in plastic sheet) surface *f* ondulée
~WINNING - [mining] exploitation *f* à ciel ouvert
~WIRING - [el] (wiring installation in which the insulated conductors are attached to the surface of a building) montage *m* en saillie, montage *m* sur crépi
~WORKING - [mining] exploitation *f* au jour
~WRINKLING - [nucl] (effect in uranium crystals due to elongation in one of its principal directions) craquellement *m*
~ZERO - [nucl] (hypocentre; the area on the surface of land or water vertically below or above the centre of a nuclear explosion) hypocentre *m*
SURFACED - [mech & carp] dressé, dégauchi
[carp] bois *m* dégauchi
SURFACER - [mech] dégauchisseuse *f*
[metall] machine *f* à surfacer
SURFACTANT - [chem] (an abbreviation for surface active agent) agent *m* tensio -actif
SURFACING - [metall] surfaçage *m*, planage *m*
[constr] revêtement *m* (d'une route)
[carp] dégauchissage *m*
[paper man] calandrage *m*
[gen] apprêtage *m*, polissage *m*
~AND BORING LATHE - [mach tool] tour *m* en l'air à plateau vertical
~COAT - [paint] couche *f* d'apprêt
~MOTION - [mach tool] mouvement *m* de surfaçage
SURFICIAL - [geol] subaérien
SURFUSE, to - [phys] surfondre
SURFUSIBILITY - [phys] surfusibilité *f*
SURFUSION - [phys & chem] surfusion *f*
SURGE, to - [gen] se soulever, rebondir
[gen] (to rise) monter
[naut] (to slacken) dériver, filer (in câble)

SURGE, to [naut] (to let go) choquer
[mech] glisser
[mech] (of internal combustion engine) tourner irrégulièrement
SURGE - [gen] houle *f*, levée *f* de la lame
[el] (transient and abnormal rush of electricity along a conductor, e.g. as caused by lightning, switching operations, faults etc) surtension *f* transitoire, onde *f* de surtension, choc *m*
[el] (a travelling wave) onde *f* à front raide
[met] (a general change in atmospheric pressure appearing as superposed upon cyclonic and normal diurnal changes) saut *m* de pression
[phys] coïncidence *f* de vibrations, battement *m*
[naut] saut *m* de cordage, coup *m* de fouet (au cabestan)
~ABSORBER - s. surge modifier
~ARRESTER - s. surge diverter
~CHAMBER - [hydr] chambre *f* d'équilibre
[oil ind] chambre *f* d'une pompe
~CHARACTERISTIC - [radio] (transient response) réponse *f* au signal unité
~CREST AMMETER - [el] ampèremètre *m* des valeurs de crête
~CURRENT GENERATOR - [el] générateur *m* de courant d'une impulsion
~CURRENT INDICATOR - [el] (magnetic indicator for lightning currents) indicateur *m* magnetique de courant de foudre
~CURRENT RATING - [electron] (of a semiconductor) courant *m* maximal d'une impulsion
~DAMPING VALVE - [phys] soupape *f* amortisseuse d'onde à front raide
~DIVERTER - [el] (lightning arrester) parasurtension *f*
~DRUM - [oil ind] accumulateur *m* de compensation
~ELECTRODE CURRENT - [electron] (fault electrode current) courant *m* anormal d'électrode
~ELEVATION - [el] élévateur *m* piézométrique
~GAP - [el] éclateur *m* à surtension
~GENERATOR - [el] (impulse generator) générateur *m* de choc
~GUARD - [el] dispositif *m* anti-pompage
~IMPEDANCE - [el] (self-surge impedance; the ratio of the voltage to the current in a travelling wave) impédance *f* d'onde
~LEVEL - [hydr] niveau *m* piézométrique
~-LIMITING ELECTROLYTIC CAPACITOR - [el] condensateur *m* électrolytique limiteur de tension de choc
~MODIFIER - [el] (a device connected to any conductor of a system and having for its object the modification of the wave-shapes of surges) absorbeur *m* d'ondes
~NON-REPETITIVE FORWARD CURRENT - [electron] (of a semiconductors) courant *m* en sens direct non-répétitif de surcharge accidentelle
~OF CURRENT - [el] impulsion *f* de courant
~POINT - [mech] (in internal combustion engines) point *m* initial
~PRESSURE - [hydr] pression *f* piézométrique
~-PROOF ELECTROLYTIC CAPACITOR - [el] condensateur *m* électrique résistant les ondes de choc
~-PROTECTIVE APPARATUS - [el] (lightning protective apparatus) installation *f* de paratonnerre
~TANK - s. surge chamber
~[oil ind] réservoir *m* de compensation

SURGE VOLTAGE - [el] surtension *f* transitoire
SURGEON - [med] chirurgien *m*
SURGEON'S AGARIC - [med] amadou *m*
SURGERY - [med] chirurgie *f*
[med] (doctor or surgeon's office) salle *f* de consultation
SURGICAL - [med] chirurgical
~CASE - [med] trousse *f* de chirurgien
~NEEDLE - [med] aiguille *f* chirurgical
~OPERATION - [med] intervention *f* chirurgicale
SURGING - [mech] (a free vibration of valve springs caused by resonance) fouettement *m*, battement *m*
[plast ind] (variation in dimensions of extrudate) modification *f* de dimension
[naut] (slackening) relâche *f*
[aero] (of the delivery pressure of a supercharger when the airflow is decreased) oscillation *f*, battement *m*
[hydr] contre-foulement *m* (de l'eau)
~LAP - [metall] repliure *f* de solidification
~OF A WELL - [hydr] décolmatage *m* d'un puits
SURPASS, to - [gen] surpasser, excéder
SURPLUS - [gen & comm] surplus *m*, excédent *m*
~PRODUCTION - [gen] surproduction *f*
~SHEETS - [bookbind] défets *mpl*
~STOCK - [comm] excédent *m*
[metall] surépaisseur *f*
SURPRINT, to - [print] surimprimer
SURPRINT - [print] surimpression *f*
SURROUND, to - [gen] entourer
SURROUNDING AREA - [photo] zone *f* entourant le champ photographié
~NOISE - [acoust] (ambient noise) bruit *m* de l'ambiance, bruit *m* de salle
SURROUNDINGS - [gen] entourage *m*, environnement *m*, milieu *m*
SURSUMDUCTION - [med] rotation *f* oculaire vers le haut
SURTAX - [fin] surtaxe *f*
SURVEILLANCE - [gen] surveillance *f*, contrôle *m*
~RADAR ELEMENT - [radar] (ground-controlled approach unit using primary radar surveillance for directing aircraft to the approach path) élément *m* radar de surveillance
SURVEY, to - [gen] (to view its entirety) regarder, examiner
[surv] (to determine accurately the area, or boundaries of the area, by measuring lines and angles according to principles of geometry and trigonometry) relever, lever le plan
[gen] (to inspect) inspecter, visiter
[constr] toiser, métrer
SURVEY - [surv] levé *m* de plan, arpentage *m*
[comm] (inspection) inspection *f*, visite *f*
[gen] apercu *m*, regard *m*, étude *f*
[naut] (of a ship) expertise *f*
[constr] métrage *m*
~BOOK - [surv] carnet *m* de levé
~ELEVATIONS, to - [surv] coter
~INSTRUMENT - s. survey meter
~LAND, to - [surv] arpenter
~MAP - [constr] plan-masse *m*
~METER - [nucl] (a portable instrument measuring radio-activity) appareil *m* de surveillance, instrument *m* de contrôle
~OF AN AREA - [surv] (the operation of finding the contours, area, boundaries etc, of a given surface)

levé *m* d'un plan
SURVEY OF CABLE ROUTE - [telecomm] étude *f* du tracé d'un câble
~OF HEIGHTS - [surv] levé *m* altimétrique
~POLE - [surv] mire *f*
SURVEYING - [surv] levé *m* de plans
~CHAIN - s. surveyor's chain
~DATA - [surv] données *f* de triangulations
~INSTRUMENTS - [surv] instruments *mpl* topographiques
~POLYGON - [constr] canevas *m* polygonal
~SPOT OF THE UNDERGROUND WATER - [hydr] endroit *m* ou l'on mesure l'eau souterraine
~VESSEL - [surv] navire *m* hydrographique
SURVEYOR - [gen] surveillant *m*, inspecteur *m*
[surv] arpenteur *m*, métreur *m*, géomètre *m*
[naut] inspecteur *m* de navires
[town planning & agric] cadastreur *m*
SURVEYOR'S CHAIN - [meas] (a measuring device for survey, equal to 20,II7 m) chaîne *f* d'arpenteur
~STAFF - [surv] jalon *m*
~COMPASS - [instr] boussole *f* d'arpenteur
~LEVEL - [instr] (an instrument consisting of a telescope to which is attached a sensitive level tube) niveau *m* à lunette
~MEASURE - [surv] mesure *f* d'arpentage
~ROD - [surv] mire *f*
~SQUARE MEASURE - [surv] mesure *f* d'arpentage
~TABLE - [surv] planchette *f*
~TAPE - [surv] roulette *f* d'arpenteur
~TRANSIT - [surv] théodolite *m* à boussole
SURVIVAL CURVE - [radiat] (curve obtained by plotting the percentage of organism surviving at a given time or at different intervals against a dose of radiation) courbe *f* de survie
~EQUIPMENT - [naut & aero] (any apparatus or device designed to prolong life in persons cast away at sea or in inaccessible regions on land as the result of an aircraft accident) matériel *m* de sécurité
SUSCEPTANCE - [el] (the component of the current in quadrature with the applied voltage divided by the voltage) susceptance *f*
SUSCEPTIBILITY - [gen] susceptibilité *f*
[el] (the ratio of the intensity of magnetization to the magnetizing force) susceptibilité *f* (magnétique)
~METER - [instr] (a device for measuring low values of susceptibility) mesureur *m* de susceptibilité
SUSCEPTIBLE - [gen] susceptible
SUSPEND, to - [mech] (to cause to hang down from a support) suspendre
[gen] (to interrupt for a time) suspendre, interrompre
[phys] (to sustain, as fine dust in the air etc) suspendre
SUSPENDED - [gen] suspendu
~BAND BRAKE - [mech] frein *m* à bande suspendue
~BOBBIN CARRIAGE - [text] chariot *m* à bobines suspendu
~BUCKET - [mech] benne *f* suspendue
~-COIL AMMETER - [el] ampèremètre *m* à courant mobile
~-COIL GALVANOMETER - [el] galvanomètre *m* à courant mobile
~COMB - [text] peigne *m* suspendu
~CREEL - [text] cantre *m* suspendu au plafond
~-FRAME WEIR - [hydr & eng] vanne *f* suspendue
~HANK - [text] écheveau *m* suspendu
SUSPENDED HOOK - [text] crochet *m* de suspension
~INSULATOR - [el] isolateur *m* d'alignement
~JOINT - [railw] (of the rails on a sleeper) joint *m* en porte à faux
[carp] assemblage *m* suspendu
~LATTICE SKIP - [text] panier *m* à lattes suspendu
~LEVER - [mech] levier *m* à cliquet oscillant
~LIFT - [mech] (mine pumps) jeu *m* volant
~MOTOR - [el] (frame-suspended motor) moteur *m* suspendu
~SCAFFOLD - [constr] échafaudage *m* suspendu
~SET - s. suspended lift
~SLAY - [text] battant *m* libre
~-SOLIDS CONTACT REACTOR - [hydr] bassin *m* de floculation à lit de boues
~SPAN - [constr] (of a bascule bridge) travée *f* d'un pont à bascule
~TRANSFORMATION - [mech] (the failure of a system to readjust itself immediately when conditions are changed) action *f* suspendue
~WEIGHT - [mech] poids *m* de suspension
SUSPENDER CLIP - [el] pince *f* à ressort pour le bout d'un cordon souple
~FOR AERIAL CABLES - [telecomm] bague *f* de suspension pour câbles aériens
SUSPENDING MEDIUM - [chem] agent *m* de suspension
SUSPENSION - [gen] suspension *f*
[comm] suspension *f* (des payments)
[phys chem] (a solution containing uniformly dispersed solid particles) suspension *f*
[instr] (of a moving element in an electrical measuring instrument) suspension *f*
[el] (a pear or pendant push) interrupteur *m* de cordon
~ARM - [auto] bras *m* de la suspension
~BAND - [aero] (of an aerostat) sangle *f* de suspension
~BOLT - [railw] (a swinging link) bielle *f* de suspension
~BOW -[constr] anse *f* de suspension
~BRIDGE - [constr] pont *m* suspendu
~CABLE - [constr] câble *m* porteur
~CABLE ANCHOR - [constr] ancrage *m* des câbles
~CHAIN - [mech] chaîne *f* de suspension, croc *m* de suspension
~CLAMP - [el] griffe *f* de suspension
~-FORK - [auto] fourchette *f* à ressort
~GRID - [el chem] (electrode framework) cadre *m* de suspension
~HOOK - [mech] crochet *m* de suspension
~INSULATOR - [el] isolateur *m* suspendu
~INSULATOR WEIGHTS - [el] poids *mpl* d'isolateurs à suspension
~LINE - [aero] corde *f* de suspension
~LUG - [el] crochet *m* de suspension
~POINT OF SPRING - [mech] point *m* de montage du ressort
~PUSH - s. suspension (el)
~RAILWAY - [railw] (overhead railway) chemin de fer *m* suspendu
~RATE - [auto] (the change of wheel load, at the centre of tyre contact, per unit vertical displacement of the sprung mass relative to the wheel at a specified load) taux *m* de flexibilité rapporté à la roue
~RING - [aero] (a ring to which the net ropes and the car suspension ropes of a balloon are both attached) cercle *m* de suspension

SUSPENSION REACTOR - [nucl] réacteur *m* à combustible en suspension
~ROD - [photo] (for films) tige *f* de suspension
~SCALES - [meas] bascule *f* en l'air
~SPRING - [mech] ressort *m* de suspension
~STAY - [constr] contre-fiche *f* de suspension
~STRAND - [el] câble *m* porteur (longitudinal)
~STRUT - [constr] contre-fiche *f* inclinée suspendue
~STUD - [mech] tourillon *m* de suspension de la coulisse
~SWITCH - [el] (a pear switch) poire *f* de contact
~TUBE - [hydr] tuyau *m* porteur
~WIRE - [el] (in a contact line for electrical traction) pendule *f* (de ligne)
SUSPENSOID - [chem] (substance capable of remaining in suspension in a liquid) suspensoïde *m*
SUSSULTORY - [geol] sussultoire
SUSTAIN, to - [gen] soutenir, supporter
SUSTAINED ELECTRICAL STRESS - [el] effort *m* électrique entretenu
~GROUND - [el] mise *f* à la terre entretenue
~OSCILLATION - [el] (sustained vibration) oscillation *f* entretenue
~OVERLOAD - [el] surcharge *f* de tension
~OVERVOLTAGE - [el] surélévation *f* de tension
~REACTION - [chem] (chemical reaction maintained without interruptions) réaction *f* entretenue
~SHORT CIRCUIT CURRENT - [el] (the current in the armature when short-circuited on its phases under steady conditions) courant *m* de court-circuit
~TWO-PHASE SHORT-CIRCUIT CURRENT - [el] courant *m* de court-circuit biphasé
~WAVE - [radio] onde *f* continue, onde *f* non-amortie
SUSTAINER - [astronaut] moteur *m* de vol
SUTHERLAND EQUATION - [phys] (a relationship between the mean free path of a molecule and the molecular diameter) équation *f* de Sutherland
SUTLER - [gen] cantinier *m*, vivandier *m*
SUTTLE - [comm] (of a weight; indicating that allowance has been made for the container) net
~WEIGHT - [comm] poids *m* net
SUTURAL - [med] sutural
~JOINT - [geol] stylolithe *f*
SUTURATION - [med] souture *f*, couture *f*
SUTURE - [gen & med] (the junction of two contiguous surfaces or edges by sewing) suture *f*
[anat] (the interlocking of two bones at their edges-) suture *f* (osseuse)
~-BAND - [med] ligament *m* de suture
~LINE - [geol] ligne *f* de suture
S V - [draw] (indicating a stop valve in a drawing) soupape *f* d'arrêt
S.V. - [mech] (side valve) à soupapes latérales
SVEDBERG EQUATION - [phys] (a relationship between the amplitude of a particle which exhibits Brownian movement and its period of vibration) équation *f* de Svedberg
SWAB, to - [gen] nettoyer, essúyer
[naut] fauberder, écouvillonner
[med] tamponner
[oil ind] pistonner
SWAB - [gen] torchon *m*
[naut] faubert *m*, vadrouille *f*
[tool] (utensil consisting mainly of a soft absorbent substance on the end of a handle) blaireau *m*
[impl] (cylindrical brush for cleaning firearms) écouvillon *m*, queue *f* de vache
SWAB - [med] (bit of sponge or cloth used for applying a medicine to a sick person, or to cleanse the mouth of a patient) tampon *m* d'ouate
[min] (a piston used for lifting liquids from a well) piston *m*
~OUT, to - [firearms] écouvillonner
SWABBING - [gen] nettoyage *m*
[metall] enduit *m*
[oil ind] pistonnage *m*
~WELL - [oil ind] puits *m* en pistonnage
SWADDLE, to - [gen] emmailloter
SWADDLING - [gen] (the bands of cloth wound around a new-born baby) emmaillotement *m*
SWAG - [mining] affaissement *m* du toit
SWAGE, to - [metall] (to shape metal with a swage) étamper, emboutir, matricer
[mech] suager
[tool] (of the saw teeth; slight widening of the cutting edges to give the same effect as "set")donner la voie (aux scies) par tourne-à-gauche
SWAGE - [metall] (a form for shaping metal)étampe *f*
[tool] (a smithing tool) emboutissoir *m*, emboutisseuse *f*
[decor] cannelure *f*
[mech] (in sheet-metal working, a moulding for stiffening sheet metal) effilement *m*, moulure *f*
~BLOCK - [tool] (heavy iron block or anvil with grooves or holes for shaping metal, heading bolts etc) tas-étampe *m*
~DOWN, to - [mech] riveter
SWAGED - [metall] matricé
SWAGING -[metall] étampage *m*, emboutissage *m*
[mech] (of saw teeth) tourne-à-gauche *m*
[plumb] (a reduction of the diameter at the tube end) reduction *f* de section
~DIE - [metall] matrice *f*
SWALLOW, to - [gen] avaler
SWALLOW - [zool] hirondelle *f*
[hydr] (a natural well) puits *m* naturel
[geol] (an abyss, a deep groove) gouffre *m*
[mech] (the channel in a hoisting block) gorge *f* de poulie
~HOLE - [geol] doline *f*, creux *m*, bétoire *f*, abîme *m*
~-TAIL - [carp] queue *f* d'hironde
~-TAIL JOINT - [carp] assemblage *m* à queue d'ironde, assemblage *m* à queue
SWALLOWING CAPACITY - [mech] (the quantity of air which a compressor intake system can take in) capacité *f* absorption
SWAMP, to - [gen] inonder, submerger
SWAMP - [gen] marais *m*, marécage *m*
~-FEVER - [med] fièvre *f* paludéenne, paludisme *m*
~FORMATION - [geol] transformation *f* en marais
~MEADOW-GRASS - [agric] pâturin *m* des marais
~-ORE - [min] minerai *m* de fer des marais
SWAMPLAND - [agric] bas-fond *m*, terrain *m* uliginaire
SWAMPING RESISTANCE - [instr] (a series resistance in a moving-coil instrument with a negligible temperature coefficient) résistance *f* en série à coefficient de température négligeable
~RESISTOR - [electron] (resistor placed in the emitter head of a transistor circuit to minimize the effects of temperature on the emitter-base junction resistance) résistance *f* réductrice
SWAMPY - [gen] marécageux, palustre, uligineux
~SOIL - [soil] sol *m* marécageux, sol *m* tourbeux

SWAN - [zool] cygne *m*
~NECK - [plumb] col *m* de cygne, col *m* d'oie
[gas ind] (of a gas lamp) crosse *f*
[naut] aiguillot *m*
~-NECK BEND - [plumb] (an S bend) coudes *mpl* de renvoi, baïonnette *f*
~-NECK COCK - [hydr] robinet *m* à col de cygne
~-NECK FLY-PRESS - [mech] découpoir *m* à col de cygne
~-NECK INSULATOR - [el] (a pin-insulator in which the pin is so shaped that the point of attachment of the phase conductor to the insulator is in the same horizontal plane as the point of attachment of the pin to its support) isolateur *m* à ferrure
~-NECK PRESS - [mech] presse *f* à col de cygne, presse *f* ouverte sur un côté
~-NECK SCREW-PRESS - s.swan-neck fly-press
~-NECK SPINDLE - [mech] arbre *m* à baïonnette
~-NECKED MULTIDAYLIGHT PRESS - [mech] presse *f* à col de cygne à plateaux multiples
~SOCKET - [el] (bayonet socket) douille *f* à baïonnette
SWANSDOWN - [text] futaine *f* à poil, piqué-molleton *m*
SWANSKIN - [text] molleton *m*
SWARF - [mech] (chips and the like produced in removing material from a workpiece) limaille *f* de métal, ribbots *mpl*
[mech] (fine particles cut on a workpiece from a grinding wheel) riblons *mpl*
SWARM, to - [gen] essaimer
SWARM - [zool] (large number of insects or small living things) essaim *m*, jetée *f*
~CLUSTER - [agric] barbe *f*
~OF PARTICLES - [nucl] (comparatively small number of particles keeping together) essaim *m* de particules
~-SPORE - [zool] (zoospore) zoospore *f*
SWARMING - [zool] essaimage *m*, essaimement *m*
[cin] grouillement *m* de l'image
~SEASON - [sool] époque *f* d'essaimage
SWARTZITE - [min] (rare secondary ore containing 33% of uranium) swartzite *f*
SWASH, to - [gen] faire jaillir, faire gicler
[gen] (of water) clapoter
SWASH - [gen] clapotis *m*, clapotage *m*
[hydr] (USA only) canal *m* entre un banc de sable et une rive
[mech] incliné sur l'axe (du tour) .
~BANK - [hydr] (the upper part of the slope of a sea embankment) talus *m*
~LETTERS - [print] (ornamental italic letters) lettres *fpl* ornées
~PLATE - [mech] (a plate, usually a disk, fixed on a shaft at an angle to the centre line of the latter, so as to give rectilinear reciprocating motion to elements pressed against or attached to it) plateau *m* oscillant
~-PLATE ENGINE - [mech] machine *f* à plateau oscillant
~-PLATE PUMP - [mech] (a pump in which the reciprocating parts are actuated by a swash-plate) pompe *f* à disque flottant, pompe *f* à plateau oscillant
SWAT, to - [gen] frapper
SWATH - [agric] fauchée *f*, andain *m*, javelle *f*
~RAKE - [agric] râteau *m* andaineur
SWATH TURNER - [agric] râteau *m* vire-anfains
SWATHE - [agric] coupe *f*
[text] bandage *m*, bandelette *f*
SWAY, to - [gen] osciller, de balancer
[mech] osciller lentement, faire osciller
[gen] (to deflect) détourner
[naut] (to swing into place) guinder, hisser
SWAY - [gen] oscillation *f*, balancement *m*
[auto] roulis *m*
[mech] mouvement *m* de va-et-vient
[railw] mouvement *m* de lacet
SWAYING - [cin] (a slow regular side-to-side shifting of the picture) déplacement *m* latéral faible
SWEAT, to - [phys] (to exude moisture) suer, ressuer, suinter
[physiol] (to perspire) transpirer, suer
[mech] (to melt solder between the surfaces to be jointed) souder à l'étain
SWEAT - [gen] sueur *f*, transpiration *f*
[leather ind] échauffe *f*
~BAND - [text] (of a hat) cuir *m* intérieur
~CANAL - [anat] canal sudorifère
~CLOTH - [gen] (of a saddle) tapis *m* de selle
~DUCT - [anat] conduit *m* sudorifère
~GLANDS - [anat] glandes *fpl* sudoripares
~HOUSE - [leather ind] étuve *f*, échauffe *f*
~OUT - [plast ind] (separation of the liquid constituents of a plastic to the surface) exsudation *f*
SWEATED JOINT - [mech] joint *m* soudé au feu
SWEATER - [metall] chaudière *f* de ressuage
SWEATING - [paint] (the separation of liquid material from the surface of a coating after drying) suintement *m*
[chem] (exudation of liquid nitroglycerine from dynamite, usually due to excessive temperature or age) exsudation *f*
[text] suée *f*, resuage *m*
[metall] soudure *f* à l'étain
[gen] suant, suintant
[min] ressuage *m* du minerai
[leather ind] étuvage *m*
[constr] (of a wall etc) suintement *m*
~HEAT - [metall] température *f* de soudure au fer
~IRON - [impl] couteau *m* de chaleur
~OUT - [metall] apparition *f* de globules
~OVEN - [ind chem] échauffe *f*, étuve *f*
~PIT - [leather ind] echauffe *f*
~SOCKET - [el] broche *f* à souder
~STOVE - s.sweating oven
SWEDE-LIKE RAPE - [bot] colza *m*
SWEDISH BLACK - [chem] (pigment made by carbonizing birch bark) noir *m* de Suède
~FILTER - [paper man] papier *m* suédois pour filtres
SWEEP, to - [gen] balayer
[gen] (a chimney) ramoner
[draw] (e.g. a curve) tracer (une courbe etc)
[metall] trousser
SWEEP - [gen] coup *m* de balai
[gen] (a circular movement) mouvement *m* circulaire
[aero] (of the wings) envergure *f*
[metall] (template for moulding) trousseau *m*, calibre *m* de moulage
[metall] planche *f* à trousser
[mech] (of a paddle; the curve or curved plane traced out by the paddle of a mixing machine as it moves) courbure *f* de braquage, courbe *f* de braquage

SWEEP - [arch] courbure *f*
[auto] galbe *m* d'une auto
[phys] (irreversible physical process in which a substance settles into thermal equilibrium or tends to do so) stabilisation *f* thermique
[naut] (an apparatus used for sweeping) câble *m* balayeur
[shipbuild] façons *f*pl d'un navire
[draw] compas *m* à verge
[agric] (of a windmill) aile *f* (de moulin)
[hydr] balancier *m* (de porte d'écluse etc)
[astr] (a systematic search of the sky) exploration *f* circulaire
[telev] mouvement *m* du faisceau, analyse *f*, balayage *m*
[radar] (the radial line made by the moving spot on the P.P.I.) balayage *m*
[hydr] (to draw water) bascule *f*
~A CHIMNEY, to - [gen] ramoner une cheminée
~AMPLIFIER - [telev] amplificateur *m* de balayage
~ANGLE - [aero] s. sweep back and sweep forward
~AWAY, to - [gen] emporter
~-BACK - [aero] (the angle between a line at right angle to the plane of symmetry and the plan projection of a given spanwise line in the wing when the latter lies above the former) flèche *f* arrière
~-BOARD - [metall] planche *f* à trousser
~CIRCUIT - [telev] circuit *m* de balayage linéaire
~DEVICE - [telev] dispositif *m* de balayage
~-FORWARD - [aero] (the angle between a line at right angles to the plane of symmetry and the plan projection of a given spanwise line in the wing, when the latter lies ahead of the former) flèche *f* avant
~GENERATOR - [telev] (scanning generator) générateur *m* de balayage
~-MOULDING - [metall] moulage *m* au trousseau
~NET - [naut] (for fishing) seine *f*
~OSCILLATOR - [radio] (signal generator in which the frequency is changed continuously over a given range) oscillateur *m* à dents de scie
~-OUT OF TELEVISION SCREEN - [telev] dépassement *m* de l'écran
~-RAKE - [agric] ramasseur *m* de foin, râteau-ramasseur *m*
~SAW - [tool] scie *f* à chantourner
~UNIT - [telev] partie *f* de balayage
~UP, to - [metall] trousser
~VOLTAGE - [telev & radar] (deflection voltage) tension *f* de base de temps
SWEEP'S ROD - [impl] tige *f* de traction
SWEEPBACK - s. sweep-back
SWEEPER - [mech] balayeuse *f*
~AND SPRINKLER - [mech] balayeuse-arroseuse *f*
~-FLUSHER - [auto] arroseuse-balayeuse *f*
SWEEPFORWARD - s. sweep-forward
SWEEPING - [metall] troussage *m*
[text] (for wool) balayures *f*pl
[ind chem] (in gas analysis) balayage *m*
[electron] (of an electron beam) balayage *m*
~COIL - [telev & radar) (a deflection coil) bobine *f* de balayage
~SYSTEM - [telev & radar] (deflection system) bloc *m* de déviation, bloc *m* de balayage, culasse *f* de balayage
~THROUGH - s. sweeping [ind chem]
SWEEPS - [telev] (scanning voltage) tension *f* d'analyse

SWEET - [gen] doux
[chem] adouci, non corrosif
~-BITTER - [gen] doux-amer
~CRUDE - [oil ind] brut *m* adouci
~NATURALGAS - [min] gaz *m* naturel non corrosif
~OIL - [oil ind] (free from sulphur compounds) pétrole *m* à faible teneur de soufre, pétrole *m* adouci
[agric] huile *f* (d'olive) douce
~POTATO - [bot] patate *f*
~SOP - [bot] atte *f*, anone *f* écailleuse
~SPIRITS OF NITRE - [chem] (spirit of ethyl nitrite; clear, slightly yellow, volatile liquid with uses as a stimulant and diaphoretic in medicine) esprit *m* de nitre dulcifié
~WORT - [brew ind] (the first running from a malt tun in brewing) moût *m* adouci
SWEETEN, to - [gen] adoucir, sucrer
[food ind] sucrer
[ind chem] (to purify motor-vehicle fuel from mercaptan content) adoucir
[soil] s'assainir
[mech] (of gears) se faire
[chem] édulcorer
[hydr] (of water) épurer, assainir
SWEETENED GASOLINE - [ind chem] (motor vehicle fuel which has been purified from mercaptan content) essence *f* adoucie
SWEETENER - [chem] édulcorant *m*
SWEETENING OF RUBBER - [rubber ind] amollissement *m* du caoutchouc
SWEETGUM - [bot] copalme *m*, gommier *m* d'Amérique, noyer *m* satiné
SWELL, to - [gen] gonfler, enfler
[phys] renfler
[constr] (of plaster) bouffer
[med] se tuméfier
[naut] (of the sea) se soulever
SWELL - [met] (wave motion in the sea which continues after the wind which gave rise to it has ceased) levée *f* de la lame, houle *f*
[gen] bosse *f*, renflement *m*
[mus] crescendo *m* et diminuendo
[mus] (of an organ) soufflet *m* (d'orgue)
[mech] (of a belt pulley) bombement *m*
[metall] (a casting defect) forçage *m*
[text] (of the shuttle box) languette *f* (de la boîte)
[hydr] (of a river etc) crue *f* (d'un cours d'eau)
[acoust] augmentation *f* d'un son
~BOX - [mus] (of an organ) boîte *f* expressive, caisse *f* d'expression
~ORGAN - [mus] récit *m*, orgue *m* expressif
~OF THE BOBBIN - [text] bombement *m* de la bobine
~PEDAL - [mus] pédale *f* d'expression
SWELLED LODE - [mining] filon *m* renflé
SWELLING - [gen] (overall increase in size) gonflement *m*, bombement *m*
[plumb] (increase in the diameter of a pipe end) élargissement *m*
[ind chem] foisonnement *m* (de la chaux)
[med] tuméfaction *f*, engorgement *m*, tumeur *f*
[nucl] (of the fuel rods) dilatation *f* des barres du réacteur
~AGENT - [ind chem] agent *m* gonflant
~COAL - [min] houille *f* gonflante, charbon *m* gonflant
~FATTY TISSUE - [zool] (of the skin of the silkworm) tissu *m* adipeux gonflant

SWELLING IN THE WINDING - [text] bourrelet *m* dans l'enroulement
~INDEX - [min] (in the coal industry) indice *m* de gonflement
~NUMBER - s. swelling index
~OF THE FIBRE - [text] gonflement *m* de la fibre
~-OUT OF SILK - [text] gonflement *m* de la soie
~PRESSURE - [min] (of coal) pression *f* de gonflement
SWEPT - [aero] (of the wing) en flèche
~-BACK WING - [aero] (a wing in which a spanwise reference line lies to the rear of the normal to the plane of symmetry) aile *f* en flèche
~-FORWARD WING - [aero] aile *f* en flèche négative
~GAIN - [radar] (anti-clutter gain control) contrôle *m* différentiel du gain
~VOLUME - [auto] cylindrée *f* unitaire
~WING - [aero] aile *f* en flèche
SWERVE, to - [gen] dévier, faire un écart
[auto] embarder
SWERVE - [gen] déviation *f*, écart *m*
[auto] embardée *f*
S.W.G. - [mech] (initials of Standard Wire Gauge issued by the Board of Trade in G B) jauge *f* standard pour fils métalliques
SWIFT, to - [naut] raidir
SWIFT - [gen] (adject) rapide, prompt
[zool] scélopore *m*
[text] secteur *m* oscillant
[text ind] (for winding) dévidoir *m* vertical
[metall] (a rotating device designed to unwind wire coils) tambour *m* de déroulement
~ENGINE - [text] bobineuse *f*
SWIFTER - [mech] ceinture *f*
[naut] (of a boat) hauban *m* bâtard
SWILL, to - [gen] laver à grande eau
SWILLING TANK - [ind chem] cuve *f* de lavage à grande eau
SWILLINGS - [agric] eaux *f*pl grasses
SWIM, to - [gen] nager
SWIM - [gen] nage *f*, action *f* de nager
[fishing] partie *f* de rivière riche en poissons
SWIMMERS - [brew ind] (floating grains of barley) orge *f* flottante, orge *f* folle
SWIMMING - [gen] nageant
~BLADDER - [zool] vessie *f* natatoire
~POOL - [constr] piscine *f*
~-POOL REACTOR - [nucl] (a thermal heterogeneous reactor moderated with light water) réacteur *m* piscine
SWINE - [zool] cochon *m*, porc *m*
~ERYSIPELAS - [vet] rouget *m* du porc
~-FEVER - [vet] peste *f* porcine
~PLAGUE - [vet] septicémie *f* hémorragique des porcins
~POX - [vet] variole *f* porcine
~-STONE - [min] anthraconite *f*
SWING, to - [gen] osciller, se balancer
[gen] (to cause to turn) faire balancer, faire tourner
[gen] (to cause to oscillate) faire osciller
[gen] (to hang by hinges) suspendre
[mech] (to turn a piece a few degrees to make it possible for another piece to be removed) faire basculer
[radio] (to undergo a variation in the wave frequency) dévier (de la fréquence)
SWING, to - [naut] rappeler (sur son ancre)
[auto] braquer (les roues avant), couper un virage
SWING - [gen] oscillation *f*, balancement *m*
[aero] (undesired deviation from course in an aircraft when moving on the ground) embardée *f*
[mach tool] hauteur *f* de la pointe d'un tour
[mech] va-et-vient *m*
[constr] (of a swing bridge) partie *f* tournante
[mach tool] (maximum of work on a lathe) diamètre *m* d'un tour
[radar] (needle deviation) déviation *f* de l'aiguille
[naut] évitage *m*
~A PROPELLER, to - [aero] (to revolve a propeller by pulling on a blade by hand, in order to start the engine) faire tourner à main
~ACROSS THE FACE, to - [mech] papillonner
~-ARM RADIUS - [auto] (the horizontal distance from the swing centre to the centre of tyre contact) rayon *m* d'oscillation
~AXLE - [auto] essieu *m* oscillant
~BAR - [aero] palonnier *m* de direction
~BEAM - [railw] poutre *f* oscillante
~-BOB LEVER - [mech] levier *m* à contrepoids
~BOLSTER - s. swing beam
~BOX - [text] boîte *f* sautante
~BRIDGE - [constr] pont *m* tournante
~CARRIER - [text] roue *f* intermédiaire de la commande des bobines
~CENTRE - [auto] (the instantaneous centre in the transverse vertical plane through any pair of wheel centres about which the wheel moves relative to the sprung mass) centre *m* de roulis
~CHECK VALVE - [mech] soupape *f* à battant dulaire
~CRANE - [mech] grue *f* pivotante
~CYLINDER ARM - [text] battant *m* du cylindre, balancier *m*
~DIAMETER - [mach tool] diamêtre *m* d'un tour
~DIFFUSER - [hydr] aérateur *m* à mouvement pendulaire
~DOOR - [constr] porte *f* va-et-vient
~FORK - [text] fourche *f* oscillante
~FRAME - [agric] (of a tractor) cadre *m* oscillant
[mach tool] tête *f* de cheval, cavalier *m*, lyre *f*
~FRONT - [photo] bascule *f* avant (de la chambre)
~-GATE - [constr] barrière *f* à pivot
~HANGER - [mech] menotte *f* de suspension
[railw] tige *f* de suspension du ressort, tirant *m* de suspension
~HEAD - [photo] (of a tripod) tête *f* de pied à bascule horizontale
~HOIST - [mech] treuil *m* pivotant
~-IN FILTER - [photo] filtre *m* escamotable
~JAW - [tool] mâchoire *f* oscillante
~-JIB RADIAL DRILL - [mech] machine *f* à percer radiale à potence
~JOINT - [plumb] joint *m* tournant
~LEVER - [mech] levier *m* oscillant, balancier *m*
~LINK - s. swing hanger
~-MOTION TRUCK - [transp] bogie *f* à menottes
~NEEDLE BOX - [text] (a needle box on a Jacquard loom) boîte *f* à aiguilles oscillante
~OVER BED - [mach tool] (of a lathe) diamètre *m* se rapportant au banc
~PIPE - [ind chem] (a pipe provided with a pivoted joint to allow the free end to follow the level of a liquid in a vessel, e.g. in decantation) tuyau *m* articulé, tube *m* à genouillère

SWING PLOUGH - [agric] araire *m*, charrue *f* araire
~RAIL - [text] arbre *m* de l'épée de chasse, porte-battant *m*
~REACTOR - [oil ind] (in oil refineries) réacteur *m* de réserve
~TOW - [text] étoupe *f* de teillage, étoupe *f* d'espadage
~WHEEL - [horol] balancier *m*
SWINGING - [gen] oscillant
[radio] évanouissement *m*
[radio] (of the frequency) fluctuation *f* de fréquence
~ARM - [mech] potence *f*, grue *f* pivotante
~BACK RAIL - [text] porte-fils *m* oscillant
~BAR - [auto] bras *m* oscillant
~BRACKET - [mech] support *m* du bras latéral
~CHOKE - [radio] (variable inductance choke frequently used as the input choke for a smoothing filter of a power supply) bobine *f* à fer saturée
~CHUTE - [min] couloir *m* oscillant
~CONE - [mech] cône *m* oscillant
~CRUSHER - [mech] broyeur *m* à pendule
~DOOR - [constr] porte *f* va-et-vient, porte *f* battante
~FORK - [mech] fourche *f* oscillant
~-JAW CRUSHER - [mech] (a type of crusher in which the moving jaw swings on a pivot) broyeur *m* à mâchoire à pendule
~JOINT - [mech] joint *m* pivotant
~OF THE REED - [text] échappement *m* du peigne
~SHUTTLE GUARD - [text] garde-navette *m* oscillant
SWINGLE, to - [text] (the flax; to cleanse by beating with a swingle) écanguer, teiller
SWINGLE - [text] (large, knife-like wooden implement for beating flax) écangue *f*
[agric] battoir *m* (d'un fleau)
~BENCH - [text] écangueuse *f* à main
~WASTE - [text] étoupe *f* de teillage
SWINGLED FLAX - [text] lin *m* teillé, lin *m* espadé
SWINGLETREE - [mech] palonnier *m*
SWINGLING MACHINE - [text] teilleuse *f*, moulin *m* flamand
~TOOL - [text] écang *m*, dague *f*, espade *f*
SWIRL, to - [gen] tourbillonner, tournoyer
[mech] faire tournoyer
SWIRL - [gen] remous *m*, tournoiment *m*
[mech] (in a motor, a turbine) turbulence *f*
[carp] ronce *f*
[ind chem] brassage *m* (de mélange gazeux)
~CHAMBER - [mech] (of a gas turbine) (enclosed space between the swirl plate and the burner orifice) chambre *f* de combustion à turbulence
~INJECTOR - [mech] (of a rocket motor) injecteur *m* à tourbillon
~NOZZLE - s. swirl injector
~VANE - [mech] (spirally-curved vanes designed to give a swirling motion to air as it enters the combustion chamber or flame tube) aube *f* de turbulence
SWIRLING - [gen] tourbillonnant
SWIRLY - [gen] tourbillonnant
SWISH PAN - [telev & cin] (USA only; a very fast framing across a stationary scene) panoramique *f* rapide
SWISH - [gen] bruire, bruit *m* rêche
SWISHING - [acoust] (rustling noise) frou-frou *m*
SWISS LAPIZ - [min] imitation *f* de lapis
SWITCH, to - [el] (to operate a switch) commuter le courant
SWITCH, to - [gen] faire mouvoir brusquement
[railw] (a train) manoeuvrer (un train)
[railw] aiguiller
SWITCH - [el] (mechanical device for opening and closing an electric circuit) interrupteur *m*, commutateur *m*
[railw] (USA only) aiguille *f*, appareil *m* de voie
[el] prise *f* de courant
~AND FUSE - [el] interrupteur *m* à fusible
~AND LOCK MOVEMENT - [el] dispositif *m* de déverrouillage
~BAR - [railw] tringle *f* de manoeuvre
~BASE - [el] (the mounting of a switch breaker) socle *m* d'interrupteur
~BAY - [el] cellule *f* de coupure, travée *f*
~BLADE - [el] (a blade-like member which makes contact with a contact jaw in closing the circuit) lame *f* de contact, couteau *m* de contact
~BLOCK - [railw] aiguillage *m*, branchement *m*
~BOARD - [el] tableau *m* d'interrupteurs, tableau *m* de commutateurs
[auto] tableau *m* de bord
~BOLT - [railw] verrou *m* d'aiguille
~BOX - [railw] boîte *f* de manoeuvre des aiguilles
~CAPACITANCE BOX - [el] (a box containing capacitors connected or disconnected by means of a switch) boîte *f* de capacités à commutateur
~CHAIR - [railw] coussinet *m* de changement
~-COCK - [plumb] robinet *m* à trois voies
~COLUMN - s. switch pillar
~CONNECTING ROD - [railw] tringle *f* de connexion
~CONTROL RELAY - [el] relais *m* d'asservissement d'aiguillage
~CORRESPONDENCE RELAY - [el] relais *m* de contrôle d'aiguillage
~FUSE - [el] coupe-circuit *m* à interrupteur
~-GROUP - [el] (or control group; a group of contactors for performing a particular operation in the control of the motors) combinateur *m*, combinateur *m* de couplage des moteurs
~HANDLE - [el] manette *f* d'interrupteur
~HOOK - [el] (hook switch) crochet *m* commutateur
~HOUSE - [el] coffret *m* extérieur
~IN - [el] intercaler
~INDICATOR - [railw] indicateur *m* d'aiguillage
~INDUCTANCE BOX - [el] (a box containing inductors connected or disconnected by means of a switch) boîte *f* d'inductances à commutateur
~KEY - [auto] (ignition key) clé *f* de commutateur, clé *f* de contact
~LAMPHOLDER - [el] douille *f* à interrupteur
~LAMPSOCKET - s, switch lampholder
~LEVER - [railw] levier *m* d'aiguille
~LOCOMOTIVE - [railw] locomotive *f* de manoeuvre, locomotive *f* de gare
~MACHINE - [railw] (point mechanism) mécanisme *f* de manoeuvre des aiguilles
~MACHINE LEVER - [railw] levier *m* du mécanisme d'aiguille
~MACHINE LEVER LOCK - [railw] verrouillage *m* du levier
~MACHINE POINT DETECTOR - [railw] (point control switch) commutateur *m* de contrôle d'aiguille
~MACHINE POLE CHANGER - [el] changeur *m* de pôles
~MOTOR - [railw] (point motor) moteur *m* d'aiguille

SWITCH OFF, to - [el] (to open an electrical circuit) interrompre, mettre hors circuit, couper le courant
~-OFF POSITION - [el] position *f* de mise hors circuit
~-OFF THE LIGHT, to - [el] mettre une lampe hors circuit
~ON, to - [el] (to close an electric circuit) mettre en circuit
~-ON PEAK - [el] crête *f* de courant transitoire
~-ON POSITION - [el] position *f* de mise en circuit
~-ON THE LIGHT, to - [el] mettre une lampe en circuit
~OPERATING POINT - [contr] point *m* d'opération d'un commutateur
~OVER, to - [el] commuter
~-PANEL - [el] panneau *m* des interrupteurs
~-PILLAR - [el] (or pillar; an assemblage of switchgear arranged in the form of a pillar) colonne *f*
~PLANT - [bot] plante *f* ligneuse à tige grêle
~PLATE - [railw] plaque *f* de manoeuvre
[el] plaque *f* d'interrupteur encastré
~PLUG - [el] broche *f* à interrupteur
~-POINTS - [railw] aiguilles *fpl*
~RAIL - [railw] lame *f* d'aiguille
~RESISTANCE BOX - [el] (a box containing resistors connected or disconnected by means of a switch) boîte *f* de résistances à commutateur
~ROD - s. switch bar
~-ROD COVERING - [railw] boîte *f* de recouvrement de tringles
~SIGNAL - [railw] (point signal) signal *m* de position d'aiguille
~SIGNAL LEVER - [railw] levier *m* de signal de position d'aiguille
~STAND - [railw] appareil *m* de manoeuvre d'aiguilles
~STARTER - [railw] (hand-operated switching starter, suitable for direct, series-parallel, or star-delta switching) commutateur *m* de démarrage
~TENDER - [railw] aiguilleur *m*
~TOGGLE - [el] interrupteur *m* articulé
~TONGUE - [railw] lame *f* d'aiguille
~TRACK - [railw] voie *f* de branchement
~-TYPE VOLTAGE REGULATOR - [el] régulateur *m* de tension du type interrupteur
~VALUES - [contr] valeurs *fpl* de réglage
~VALVE - [mech] soupape *f* à trois voies
SWITCHBACK - [gen] route *f* qui monte et descend
[railw] (back shunt) rebroussement *m*
[gen] en montagnes russes
SWITCHBOARD - [el] (a panel on which switches are mounted) tableau *m*
[telecomm] tableau *m* commutateur, meuble *m*, section *f*
~PANEL - [el] panneau *m* de tableau de distribution
~PLUG - [el] broche *f* de tableau de distribution
SWITCHER - [railw] aiguilleur *m*
[railw] locomotive *f* de manoeuvre
SWITCHGEAR - [el] (a switch) appareillage *m*
~CUBICLE - [el] armoire *f*
~OIL - [chem] huile *f* de disjoncteurs
~PEDESTAL - [el] boîtier *m* de commutation
SWITCHING - [telecomm] commutation *f*
[railw] aiguillage *m*, garage *m*
~AMPLIFIER - [el] aplificateur *m* commutateur
~CENTRE - [comput] centre *m* de communication
~COEFFICIENT - [comput] (in static magnetic storage) coefficient *m* de commutation
SWITCHING CONTROL PILOT - [radio] onde *f* pilote de commutation
~DIODE - [electron] (crystal diode providing the same function as a switch) diode *f* de commutation
~ELEMENT - [el] élément *m* de contact
~ENGINE - [railw] locomotive *f* de manoeuvre
~EQUIPMENT - [el] équipement *m*
~FUNCTION - [electron] fonction *f* de commutation
~HYSTERESIS - [el] fourchette *f*, hystérésis *f* de commutation
~-OFF - [el] mise *f* hors circuit
~-ON - [el] mise *f* en circuit
~-ON COIL - [el] bobine *f* de fermeture
~POINT - [telecomm] bureau *m* de transit, centre *m* de transit
~SELECTOR REPEATER - [telecomm] commutateur *m* discriminateur
~-SEQUENCE DIAGRAM - [el] (a diagram showing the sequence of contact operations in an apparatus) diagramme *m* de séquence de commutations
~SIGNAL - [el] signal *m* de commutation
~STATION - [el] (a substation for controlling the distribution of electrical energy by means of a switchgear without transformation or conversion) poste *m* de distribution
~STRUCTURE - [el] bâti *m* d'appareillage
~SURGE - [el] surtension *f* de commutation
~THRESHOLD - [electron] seuil *f* de commutation
~TIME - [el] durée *f* de commutation
~TORQUE - [el] couple *m* minimal
~TRANSISTOR - [electron] (transistor providing essentially the same function as a switch) transistor *m* commutateur
~TUBE - [electron] (gas switching tube) tube *m* commutateur à gaz
~VALVE - [electron] tube *m* commutateur
~VALUE - [contr] (for a step-action control, the value of input signal for which the step value changes) valeur *f* de commutation
~YARD - s. switchyard [railw]
SWITCHYARD - [el] sous-station *f*
[railw] gare *f* de triage
SWIVEL, to - [mech] (to revolve, to turn a pivot) tourner, pivoter
SWIVEL - [mech] rotule *f*, émerillon *m*
[mech] (of a chain) maillon *m* tournant
[mach tool] (a revolving table) plateau *m* tournant
[text] bras *m* articulé
[metall] (of a foundry flask) tourillon *m* (de chassis de fonderie)
[mining] touret *m*, tête *f* de sonde
~-BASE VICE - [mech] (a type of vice so mounted that it can be revolved on its base and clamped in any desired position) étau *m* pivotant, étau *m* à base tournante
~BEARING - [mech] palier *m* à rotule
~BLOCK - [mech] poulie *f* à émerillon
~BOLT - [mech] boulon *m* à émerillon
~BRIDGE - [constr] pont *m* tournant, pont *m* pivotant
~-CAP LUBRICATOR - [mech] graisseur *m* tournant
~CASTOR - [photo] roulette *f* orientable
~CHAIR - [carp] chaise *f* pivotante
~CLAMP - [mech] serre-pivot *m*
~CONNECTION - [mech] raccord *m* orientable
~COUPLING - [mech] joint *m* double
~EMBROIDERY SLAY - [text] battant *m* brocheur

SWIVEL HOOK - [mech] croc *m* à émerillon, crochet *m* tournant
~INSULATOR - [el] isolateur *m* pivotant
~JOINT - [mech] joint *m* à rotule
~LEVER - [text] levier *m* oscillant
~LINK - [naut] maillon *m* tournant
~LOOM - [text] métier *m* à brocher
~LOOM SHUTTLE - [text] navette *f* pour métier à grille
~-PIN - [mech] (kingpin) tourillon *m* d'articulation
~PLATE - [mech] plateau *m* orientable, plaque *f* tournante
~PLOUGH - [agric] charrue *f* balance
~-ROD - [oil ind] tête *f* de sonde
~ROLLERS - [mech] rouleaux *mpl* orientables
~SEGMENT GEAR - [mech] secteur *m* dentée pour l'avance angulaire
~SLAY - [text] battant *m* à changement horizontal des boîtes
~SLIDE-REST - [mech] support *m* à chariot pivotant
~SLOTTING HEAD - [mach tool] tête *f* orientable pour rainures
~SPRING HOOK - [mech] crochet *m* tournant à ressort
~SPRINKER - [agric] arroseur *m* oscillant
~VICE - [mach tool] étau *m* tournant, étau *m* à base tournante
~WEAVING - [text] tissage *m* au broché
SWIVELLING - [gen] pivotant, tournant
[gen] pivotation *f*
~BOBBIN CARRIER - [text] support *m* à charnière
~REED - [text] ros *m* à échappement
~WORKTABLE - [mach tool] plateau *m* porte-pièce orientable
SWOLLEN - [gen] (p. p. of to swell) enflé, gonflé
[of a river] en crue
[metall] (a defective casting) forcé
SWOON - [med] évanouissement *m*, syncope *f*
SWORD - [impl] épée *f*, sabre *m*
~BEAN - [agric] haricot *m* sabre, haricot *m* de Madagascar
~-SHAPED LEAF - [bot] feuille *f* en forme d'épée
SWORDFISH - [zool] espadon *m*, sabre *m*
SYCAMORE - [bot] sycomore *m*, faux platane *m*
[timber] faux plane *m*
SYCOSIS - [med] (inflamed infection of the skin) sycosis *m*
SYENITE - [geol] (igneous granular rock, principally composed of feldspar) syénite *f*
SYLLABLE - [gen] syllabe *f*
SYLLABUS - [gen] (a concise statement of the main points of a subject) programme *m*, sommaire *m*
SYLLOGISM - [gen] syllogisme *m*
SYLPHON BELLOWS - [instr] (in a pressure governing system) capsule *f* barométrique
SYLVIAN FOSSA - [anat] partie *f* transversale de la scissure de Sylvius
SYLVIN - s. sylvite
SYLVINE - s. sylvite
SYLVITE - [min] (natural potassium chloride, an important source of potassium compounds) sylvine *f*, sylvite *f*
SYLVINITE - [min] (a mixture of sylvite and halite, chiefly the latter) sylvane *m*
SYM - (prefix denoting organic compounds having a symmetrical structure in relation to the carbon ring or a functional group) sym-
SYMBIOSIS - [biol] (mutually beneficial association of living organisms) simbiose *f*
SYMBLEPHARON - [med] symblépharon *m*, sinéchie *f* palpébrale
SYMBOL - [gen math etc] symbole *m*
~WEIGHTS - [phys] poids *m* équivalent
SYMBOLIC ADRESS - [comput] (label chosen to identify a particular word, function etc, in a routine, independent of the location of the information within the routine) adresse *f* symbolique, pseudo-adresse *f*
~CODING - [comput] (coding making use of symbolic adresses) codage *m* symbolique
~REPRESENTATION - [phys] (the use of symbols and notations in nuclear chemistry and physics) représentation *f* symbolique
SYMMETRIC - s. symmetrical
~TOP MOLECULE - [phys] molécule *f* à cellule symétrique
SYMMETRICAL - [math chem etc] (susceptible of division into two or more parts of identical size and shape and located similary in relation to the line or point of division) symétrique
~ADJUSTMENT OF A POLARIZED RELAY - [el] ajustage *m* symétrique d'un relais polaire
~ALTERNATING CURRENT - [el] courant *m* alternatif symétrique
~ALTERNATING FUNCTION - [math] grandeur *f* sinusoïdale
~ALTERNATING QUANTITY - [el] grandeur *f* alternative symétrique
~DEFLECTION - [electron] déviation *f* symétrique
~FLUTTER - [aero] (a type of flutter which affects similar components on both sides of an aircraft to the same degree, in the same sense and at the same instant) flutter *m* symétrique
~HETEROSTATIC CIRCUIT - [instr] (by Mascart; an arrangement of the quadrant electrometer) montage *m* hétérostatique
~PATTERN - [text] dessin *m* symétrique
~POLYPHASE SYSTEM OF QUANTITIES - [el] système *m* de grandeurs poliphasé équilibré
~TRANSDUCER - [el acoust] (a transducer having equal input and output image impedances) transducteur *m* symétrique
SYMMETRICALLY CYCLICALLY MAGNETIZED CONDITION - [comput] (in static magnetic storage) condition *f* de magnétisation cyclique symétrique
SYMMETRIZATION - [math] transformation *f* mathématique faisant apparaître les symétries
SYMMETRY - [phys] (of molecules; the spatial arrangement of the atoms in reference to a plane of symmetry) symétrie *f*
[gen & draw] symétrie *f*
~AXIS - [math] (in geometry) axe *m* de symétrie
~CLASSES - [cryst] (the 32 different classes of symmetry groups to one of which each crystal belongs) classes *fpl* de symétrie
~ELEMENT - [cryst] (an operation which brings a crystal into a position which is indistinguishable from its original position) élément *m* de symétrie
~PLANE - [math] plan *m* de symétrie
SYMPATHECTOMY - [med] sympathectomie *f*
SYMPATHETIC - [gen] prompt à comprendre
[anat] sympathique *m*
~INK - [chem] (ink which is colourless and invisible until it is brought out by light, heat or a chemical action) encre *f* sympathique

SYMPATHETIC STRING - [phys] corde *f* qui vibre par résonance
~TONE - [acoust] (a sound produced by resonance) vibration *f* sympathique
SYMPATHICECTOMY - s. sympathectomy
SYMPATHICOBLAST - [biol] cellule *f* sympathique primaire
SYMPATHOBLAST - s. sympathicoblast
SYMPATHOGONIA - s. sympathicoblast
SYMPHONY - [mus] symphonie *f*
SYMPHYSIS - [med] symphise *f*
SYMPHYSODACTYLIA - s. syndactylism
SYMPIEZOMETER - [instr] (sensitive barometer, in which atmospheric pressure acting on a liquid in the lower part compresses an elastic gas in the upper) sympiézomètre *m*
SYMPLASM - s. syncytium
SYMPLASTIC TISSUE - s. syncytium
SYMPOSIUM - [gen] (meeting for discussion) symposium *m*, conférence *f*
SYMPTOM - [gen] symptôme *m*, indice *m*
[med] (any functional or organic condition indicating the presence of a disease) symptôme *m*
SYNALGIA - [med] synalgie *f*
SYNAPSE - [el biol] (the junction between two neural elements which has the property of one-way propagation) synapse *f*
SYNAPTIC JUNCTION - s. synapse
SYNC AMPLITUDE - [telev] amplitude *f* du signal de synchronisation
~COMPARATOR - [telev] appareil *m* de contrôle du synchronisme
~-IN PULSE - [telev] impulsion *f* d'entrée de la synchronisation
~LIMITER - [telev] limiteur *m* de l'amplitude du signal de synchronisation
~-OUT PULSE - [telev] impulsion *f* de sortie de la synchronisation
~PIP - [el acoust] signe *m* de synchronisation
~PULSE - [electron] impulsion *f* de synchronisation
~STRETCHING - [telev] expansion *f* du signal de synchronisation
SYNCHONDROSIS - [med] synchondrosis *f*
SYNCHRO - [contr] synchro-machine *f*
~-ANGLE - [contr] angle *m* électrique
~CONTROL DIFFERENTIAL TRANSMITTER - [contr] synchro-transmetteur *m* différentiel à commande
~CONTROL RECEIVER - s. synchro differential receiver
[contr] synchro-comparateur *m* d'angles
~CONTROL TRANSFORMER - [contr] synchro-transformateur *m*, synchro-comparateur *m*
~DIFFERENTIAL RECEIVER - [radio] (a synchro which produces a torque proportional to the difference between the electrical angles of two transmitters to which it is connected) synchro-comparateur *m* d'angles
~PHASE SHIFTER - [contr] synchro-déphaseur *m*
~RECEIVER - [radio] récepteur *m* synchrone
~RESOLVER - [contr] synchro-trigonomètre-résolveur *m*
~TORQUE DIFFERENTIAL RECIVER - [contr] synchro-récepteur *m* différentiel
~TORQUE DIFFERENTIAL TRANSMITTER - [contr] synchro-transmetteur *m* différentiel
~TORQUE RECEIVER - [contr] synchro-récepteur *m*
~TORQUE TRANSMITTER - [contr] synchro-transmetteur *m*
SYNCHROCYCLOTRON - [electron] (a cyclotron in which the radio-frequency of the electric field applied between the dees is frequency-modulated, to permit the acceleration of particles to relativistic energies) cyclotron *m* modulé en fréquence, synchrocyclotron *m*
SYNCHROFLASH - [photo] synchroflash *m*
SYNCHROMESH GEAR - [auto] (a gear system whereby driving and driven members are brought to the same speed before engaging) vitesse *f* synchronisée
~-MESH TRANSMISSION - [auto] boîte *f* de vitesses synchronisée
SYNCHRONISM - [gen] (the state of being synchronous) synchronisme *m*
SYNCHRONIZATION - [phys mech el etc] (the adjustment of dissimilar elements so as to make coincide in time) synchronisation *f*
~BAY - [telev] baie *f* de synchronisation
~COMPRESSION - [telev] (the reduction in the percentage synchronization resulting from a reduction in the relative synchronization amplitude) compression *f* du signal de synchronisation
~CONTROL - [telev] (of the electro-beam deflection) contrôle *m* de synchronisation
~LOSS - [electron] perte *f* de synchronisation
~OF A SYNCHRONOUS MACHINE - [el] accrochage *m* d'une machine synchrone
~STRETCHER - [telev] dispositif *m* d'expansion du signal de synchronisation
SYNCHRONIZE, to - [phys, mech, el etc] synchroniser
[aero] (to adjust the revolutions and propeller settings of two or more engines to a uniform level) synchroniser
SYNCHRONIZE AND CLOSE - [el] opération *f* de couplage
SYNCHRONIZED - [gen] synchronisé
~MULTIVIBRATOR - [radio] multivibrateur *m* piloté
~SHIFTING - [auto] changement *m* de vitesse synchronisé
~SWEEP - [telev] analyse *f* synchronisée
SYNCHRONIZER - [mech, aero, cinema etc] (device designed to obtain synchronization) synchroniseur *m*
[el] (instrument for indicating the phase-relation of two alternating voltages, and employed particularly in paralleling alternators) synchroniseur *m*
SYNCHRONIZING - [gen] synchronisation *f*
[el] accrochage *m* en phase
~CIRCUIT - [contr] tuyauterie *f* de synchronisation
~CURRENT - [el] (the circulating current, between two alternating-current generators connected in parallel, which tends to maintain two machines in synchronism and in phase) courant *m* synchronisant
~GEAR - [mech] mécanisme *m* synchroniseur
~LEADER - [cin] amorce *m* de lancement
~LEVEL - [telev] niveau *m* du signal de synchronisation
~PILOT - [radio] onde *f* pilote de synchronisation
~POTENTIAL - [telev] (the signal employed for the synchonizing of transient visual images) signal *m* de synchronisation
~PULSE - [telev & radar] impulsion *f* de synchronisation
~REACTOR - [el] réactance *f* synchronisatrice
~RELAY - [el] relais *m* synchronisé

SYNCHRONIZING SHAFT - [aero] (of a twin-engines twin-rotor helicopter) arbre *m* synchroniseur
~SIGNAL - s. synchronizing potential
~VOLTMETER - [instr] synchronomètre *m*
SYNCHRONOSCOPE - [instr] (an instrument designed to indicate whether two periodic phenomena are synchronous) synchronoscope *m*
SYNCHRONOUS - [gen] synchrone
[phys etc] (having the same period or rate of vibration) synchrone
~ADMITTANCE - [el] admittance *f* synchrone
~ALTERNATOR - s. synchronous generator
~BOOSTER CONVERTER - [el] convertisseur *m* synchrone à autorégulation
~COMPUTER - [comput] (automatic digital computer where the performance of all ordinary operations starts with equally spaced signals from a master clock) calculateur *m* synchrone
~CONDENSER - [el] (a synchronous machine, running without active load, for providing leading or lagging current) compensateur *m* synchrone
~CONVERTER - [el] commutatrice *f*
~DIGITAL COMPUTER - [comput] calculateur *m* numérique synchrone
~DRIVE - [mech] mouvement *m* synchrone
~GENERATOR - [el] (a synchronous machine working as an alternating-current generator) alternateur *m* synchrone
~IMPEDANCE - [el] impédance *f* synchrone
~INDUCTION MOTOR - [el] (a synchronous motor which starts as an induction motor and ultimately in synchronism using direct current excitation) moteur *m* asynchrone synchronisé
~MACHINE - [el] (an alternating-current machine in which the frequency of the e.m.f. generated is proportional to the speed) machine *f* synchrone
~MOTION - [mech] mouvement *m* simultané
~MOTOR - [el] (a synchronous machine working as a motor) moteur *m* synchrone
~RECTIFIER - [electron] détecteur *m* à sensibilité de phase
~SATELLITE - [telev] (a communication satellite in a fixed position above the earth) satellite *m* synchrone
~SCANNING - [telev] balayage *m* synchrone
~SPEED - [el] (of an alternating-current machine; the speed of rotation which corresponds to the rotation of the magnetic flux) vitesse *f* de synchronisme
~TIMER - [instr] (a counting device driven by a synchronous motor) chronomètre *m* synchrone
~WORKING - [contr] opération *f* synchrone
SYNCHRONY - [phys] synchronisme *m*
~DISTANCE - [cin] distance *f* de synchronisation
~MARK - [cin] marque *f* de synchronisme, repère *m*
SYNCHROPHASOTRON - [nucl] (a proton accelerator) synchrophasotron *m*
SYNCHROSCOPE - [instr] (a device for indicating the synchrony condition of two or more engines in an aircraft) synchroscope *m*
SYNCHROTRANSMITTER - [contr] synchrotransmetteur *m*
SYNCHROTRON - [nucl] (apparatus for accelerating particles in a circular orbit by means of an electric field synchronized with the orbital motion) synchrotron *m*
SYNCINESIS - s. synkinesis
SYNCLINAL - [geol] (of syncline) synclinal
SYNCLINE - [geol] (a downward fold in sedimentary formations) synclinal *m*
SYNCLINORIUM - [geol] (a series of synclines and anticlines) éventail *m* composé inverse
SYNCOPATION - [mus] syncope *f*
SYNCOPE - [med] syncope *f*, évanouissement *m*
SYNCYTIUM - [biol] syncytium *m*
SYNDACTYLIA - s. syndactylism
SYNDACTYLISM - [med] syndactylie *f*
SYNDECTOMY - [med] péritomie *f*, péridectomie *f*
SYNDESMITIS - [med] pied *m* forcé
SYNDESMOPLASTY - [med] plastie *f* ligamentaire
SYNDET - [chem] (contraction for "synthetic detergent") détergent *m* synthétique
SYNDIOTACTIC - [chem] (a stereospecific polymer in which the atomic groups, not part of the backbone, are in a specific spatial arrangement) syndiotactique
SYNDROME - [med] syndrome *m*
SYNECHIA - [med] synéchie *f*, adhérence *f*
SYNERESIS - [chem] (the contraction of a gel on standing, with expulsion of liquid) synérèse *f*
SYNERGIC ADDITIVES - [chem] (substances such that when two or more are added to a mixture, their total effect is greater than the sum of their separate effects) additifs *mpl* synergiques
~ASCENT - [astronaut] ascension *f* synergique
SYNERGIST - [chem] (a substance which increases the specific action of a drug, insecticide, or similar compound) synergiste *m*
SYNESTHESIA - [med] synesthésie *f*
SYNESTHESIALGIA - [med] synesthésie *f* douloureuse
SYNIZESIS - [med] occlusion *f* pupillaire
SYNKINESIS - [med] mouvement *m* associé
SYNODIC PERIOD - [astr] (the time interval between two similar positions of the moon or a planet in relation to a line joining the centres of earth and sun) période *f* synodique
SYNOPHRYS - [med] coalescence *f* des sourcils
SYNOPTIC CHART - [met] (a complet weather chart combining plotted and processes observations into a comprehensive picture of the general situation) carte *f* synoptique du temps
~METEOROLOGY - [met] (the collection and charting of observations made over a wide area for the production of a synoptic chart) météorologie *f* synoptique
~REPORT - [met] (a weather report giving an overall view of conditions over a wide area at a given time) bulletin *m* synoptique
~WEATHER CHART - s. synoptic chart
SYNOSTEOSIS - [med] synostose *f*
SYNOVIA - [med] synovie *f*
SYNOVIAL FLUID - s. synovia
SYNOVITIS - [med] synovite *f*
SYNTACTIC FOAMS - [plast ind] (cellular plastics produced by incorporating microballoons in epoxy or polyester resins) mousses *fpl* syntactiques
SYNTHESIS - [chem] (the building up of a substance artificially from simpler ones) synthèse *f*
[gen & photo] synthèse *f*
~GAS - [chem] (mixture of hydrogen and carbon monoxide used in production of petrochemicals from methane) gaz *m* de synthèse
SYNTHETIC - [gen & chem] (artificial;not occurring naturally) synthétique
~FIBRES - [ind chem] (synthetically produced threads

the raw material of which can be a wide variety of substances) fibres *f*pl synthétiques
SYNTHETIC FINISH - [paint] émail *m* synthétique pour finissage
~RESIN AHESIVE - [chem] (phenol and urea-formaldehyde based adhesives, either hot or cold-setting) colle *f* synthétique
~RESIN CEMENT - [ind chem] (phenol-formaldehyde and urea-formaldehyde based adhesives used in the production of laminates) ciment *m* synthétique
~RESIN MOULDED MATERIAL - [plast ind] article *m* moulé à base de résine synthétique
~RESIN MOULDING COMPOUND - [plast ind] matière *f* à mouler
~RESINS - [chem] (compounds artificially formed by condensation or polymerization from simpler substances; they are numerous and of great industrial importance) résines *f*pl synthétiques
~RUBBER - [rubber ind] (synthetic high polymers of similar structure to natural rubber. Their properties vary and can be tailored to fit specific requirements) caoutchouc *m* artificiel, caoutchouc *m* synthétique
~SAND - [metall] sable *m* synthétique
SYNTHETICS - [chem] (substances made artificially, as opposed to those which occur naturally) produits *m*pl synthétiques
SYNTHON - [chem] (term formerly used for all synthetic fibres, now going out of use) fibres *f*pl synthétiques
SYNTONIZATION - [radio] (the operation of placing in resonance with each other, as radio frequencies) syntonisation *f*
[instr] (the operation of tuning or toning together electrical instruments etc) syntonisation *f*, accordage *m*
SYNTONIZE, to - [radio & el] syntoniser, accorder
SYNTONY - [radio] (obsolete; the possession of a common resonance frequency by two or more circuits) syntonie *f*
SYPHILIS - [med] syphilis *f*
SYPHILODERM - s. syphilid
SYPHON - [gen & ind chem] (or siphon. A bent tube employed in transferring fluid from one vessel to another at a lower level) siphon *m*
~BOX - [ind chem] s. syphon
~BRICK - [metall] (of siphon bick; of a cupola, for separating slag) brique-siphon *f*
~HEAD - [impl] tête *f* de siphon
~PIPE - [mech] tube *m* en trompette
~POT - [ind chem] s. syphon
~RUNNER - [metall] coulée *f* en siphon
SYREN - [acoust] (or siren; electrical apparatus emitting a powerful noise) sirène *f*
SYRIAN RAW SILK - [text] soie *f* grège de la Syrie
SYRINGE, to - [gen] seringuer, injecter
SYRINGE - [tool] seringue *f*
[med] seringue *f*
~HYDROMETER - [instr] (a device for determining the sp. gr. of battery acid in which a syringe draws up a specimen of acid into a bulb containing a small hydrometer) hydromètre *m* à seringue
SYRINGING - [gen] seringage *m*
SYRINGOCOELE - [anat] canal *m* épendymaire de la moelle épinière
SYRINGOCYSTOMA - [med] syringocystome *m*
SYRINGOMYELITIS - [med] syringomyélite *f*
SYRINGOSYSTROPHY - [med] torsion *f* de la trompe de Fallope
SYRINX - [anat] trompe *f* d'Eustache
[zool] organe *m* phonateur
[mus] (pan pipe) flûte *f* de Pan
SYROSINGOPINE - [chem] (an analogue of reserpine) syrosingopine *f*
SYRUP - [chem] (liquid synthetic resin or solution, particularly of the aminoplastic type) sirop *m*
~-LIKE SOLUTION - [chem] solution *f* sirupeuse
[gen] sirupeux
SYSTEM - [gen phys chem astr etc] système *m*
[gen] (in industry) installation *f*
[el] (assembly) installation *f*
[gen] (method) méthode *f*
[mus] distribution *f* de la partition
[leg] régime *m*
~ANALYSIS - [comput] (analysis of a book-keeping) analyse *f* du système
~DEVIATION - [gen] écart *m* de système
~DYNAMICS - [contr] dynamique *f* des systèmes réglés
~EARTH - [el] installation *f* de mise à la terre de service
~ELEMENT - [meas] (one or more basic elements with other components and necessary parts to form one of the general functional groups into which a measurement system can be classified) élément *m* de système
~IMPROVEMENT TIME - [comput] temps *m* de mise au point du calculateur
~OF A RIVER - [hydr] bassin *m* fluvial
~OF BEAMS - [telev] (in colour television) système *m* de faisceaux
~OF BREEDING - [agric] système *m* d'élevage
~OF CRYSTALS - [cryst] (the seven large divisions into which all crystallizing substances can be placed) systèmes *m*pl de cristaux
~OF DISTRIBUTION - [el] système *m* de distribution
~OF FITS - [mech] système *m* de couplages
~OF PARTICLES - [phys] système *m* de particules
~OF SCREW THREAD - [mech] système *m* de filetage
~OF STALLING - [agric] système *m* de logement des animaux
~OF TRENCHES - [agric] réseau *m* de fossés
~OF UNITS - [meas] (a coordinated group of units of measurement) système *m* d'unités
~OF WORKING - [mining] méthode *f* d'extraction
~OVERSHOOT - [contr] dépassement *m* de circuit
~PARAMETER - [el] (any resistance, inductance or capacitance in a branch of a network) élément *m* de réseau
~PROGRAMMING - [comput] programmation *f* générale
~-SENSITIVE DEVICE - [el] dispositif *m* de palpage
~-SENSITIVE RELAY - [el] relais *m* palpeur
~TRANSFER FUNCTION - [contr] fonction *f* de transfert du système
~WITH CATENARY SUSPENSION - [el] ligne *f* à suspension caténaire
~WITH SOLIDLY EARTHED NEUTRAL - [el] réseau *m* à neutre directement à la terre
SYSTEMATIC - [gen] systèmatique
~ERRORS - [gen] (errors having an orderly character which can be corrected by calibration) erreurs *f*pl systèmatiques
SYSTEMIC - [zool] (pertaining to the body as a whole)

du système, de l'organisme
SYSTEMIC CIRCULATION - [physiol] circulation *f* générale
~REACTION - [radiat] (of the whole body) réaction *f* générale
SYSTEMATICS - [gen] (the principles of classification and nomenclature) systématique
SYSTOLE - [physiol] systole *f*
SYSTREMMA - [med] crampe *m* des muscles du mollet
SUSTYLE - [arch] systyle *m*
SYZYGY - [astr] (term denoting the position of the moon when in conjunction or opposition) syzygie *f*

T

T - [met] (short for thunder) tonnerre *m*
[mech] T *m*, té *m*

T - [phys] (symbol of Transport Number) nombre *m* de transport des ions
[metall] (in mechanical drawing, short for tempered) revenu
[mech] (in mechanical drawing, short for turn) tourner

T AERIAL - [radio] (a horizontal aerial to which the lead is attached at the middle point) antenne *f* à T

T ANTENNA - s. T aerial

T.B. - [med] (short for tubercolosis) tubercolose *f*

T BAR - [metall] fer *m* à T

T-BEAM - [constr] (a beam forming part of the construction of a reinforced concrete floor) poutre *f* en simple T

T-BOLT - [mech] (a bolt having a head in the form of a short cross-piece) boulon *m* à T, boulon *m* à tête de marteau

T BRANCH - [plumb] tuyau *m* en T, tube *m* en T

T CIRCULATOR - [electron] (a circulator in which three identical rectangular waveguides are joined asymmetrically to form a T-shaped structure, with a ferrite post or wedge at its centre) circulateur *m* en T

T CONNECTION - [mech] té *m* de raccordement

T.D.C. - [mech] (initials of Top Dead Centre; of a reciprocating engine or pump) point *m* mort haut (PMH)

T EQUIVALENT - [electron] (a transistor equivalent circuit in which the resistances of the electrodes are connected together at a common point in the form of a T) circuit *m* équivalent en T

T HEAD - [railw] culotte *f* de bifurcation du tuyau de prise de vapeur

T-HEAD CYLINDER - [mech] (of internal combustion engine) moteur *m*, culasse *f* en T, culasse à soupapes latérales opposées

T HINGE - [mech] penture *f* à T

T-HINGE STRAP - [mech] couvre-joint *m* à T

T-IRON - [metall] (wrought-iron or rolled-steel structural member having a T-shaped cross-section) fer *m* à T, équerre *f*

T-JOINT - [mech] noeud *m* d'empattement
[el] couplage *m* en T

T JUNCTION - [electron] (in a waveguide) jonction *f* en T

T-PIECE - [mech] T *m*, té *m*

T-PIECE UNIT - [mech] raccord *m* à T

T-PLATE - [carp] plaque *f* à T

T-REST - [mech] (a T-shaped rest clamped to the bed of a wood-turning lathe to support the tool) support *m* à éventail

T.S. - [mech] (short for two-strokes) à deux temps

T-SECTION - [el] (electrical network consisting of a series arm, a shunt arm, and another series arm equal to the first series arm) réseau *m* en T

T-SLOT - [mach tool] (of a table) rainure *f* à T

T-SLOT CUTTER - [mach tool] fraise *f* pour rainures à T

T SQUARE - [draw] (a drawing implement) équerre *f* à T

T.T. - [mech] (short for time taken) temps *m* employé

T TUBE - [metall] tube *m* en T

T.U. - [leg] (short for Trade Union) syndicat *m* ouvrier

T-WELDING - [metall] soudure *f* en T

T-WRENCH - [tool] clef *f* à T

Ta - [chem] (symbol of Tantalum) tantale *m*

TAB, to - [comput] imprimer en liste

TAB - [gen] (thin metal piece projecting from a piece) patte *f*, étiquette *f* (pour bagages), onglet *m*
[aero] (a hinged portion of the trailing edge of a control surface or flap) tab *m*

TABAGISM - [med] nicotinisme *m*, tabagisme *m*

TABBY, to - [text] tabiser

TABBY - [text] soie *f* moirée, moire *f*

~BACK - [text] fond *m* uni

~CAT HEART - [anat] coeur *m* tigré

~WEAVE - [text] armure *f* toile

TABERNACLE - [arch] (fixed or portable shelter, tent etc; a portable sanctuary) tabernacle *m*
[naut] cornet *m* de mât

TABES - [med] tabes *m*, consomption *f*

TABESCENT - [bot] (shrivelling) tabescent
[med] tabescent

TABLATURE - [mus] (a system of musical notation) tablature *f*

TABLE, to - [carp] assembler, emboîter
[leg] déposer un projet de loi, (US only) ajourner un projet de loi

TABLE - [gen] table *f*
[gen] (set of items) table *f*, tableau *m*, abaque *m*
[geogr] plateau *m*
[min] (flat top of a table-cut stone) table *f* (d'un diamant)
[anat] table *f*
[mach tool] plateau *m*
[math] (of logarithms) table *f*
[instr] (for the support of specimens) tablette *f*
[constr] (of a girder) semelle *f* (d'une poutre)
[mech] (of a rail) table *f* de roulement

~ASSEMBLY - [mach tool] groupe *m* table

~CLAMP - [mach tool] étau *m* d'établi, griffe *f*

~CLAMPING - [mech] blocage *m* de la table

~CLAMPING LEVER - [mach tool] levier *m* blocage

table
TABLE CONTROL PRESELECTOR LEVER - [mach tool] levier *m* pour prédisposition commande table
~CONTROL SELECTOR - [mach tool] poignée *f* pour prédisposition commande table
~COOKER - [impl] (domestic gas cooker) réchaud-four *m*
~CREEL FOR SINGLE ROVING - [text] râtelier *m* à une seule rangée
~CROSS ADJUSTMENT - [mach tool] régulation *f* du coulisseau croisé de la table
~-CUT DIAMOND - [min] diamant *m* taillé en table
~DRIVE-MOTOR CONTROL BUTTONS - [mach tool] tableau *m* de commande du moteur déplacement table
~FEED ENGAGE LEVER - [mach tool] levier *m* du mécanisme de l'avance
~FLAP - [carp] abattant *m* de table
~INCLINATION ADJUSTING SCREW - [mach tool] vis *f* de réglage inclinaison table
~JOINT - [constr] entaillage *m* droit à demi-bois
~LAMP - [el] lampe *f* de table
~-LEAF - [carp] rallonge *f* de table
~LIFTING SCREW - [mach tool] vis *f* pour le levage de la table
~LINEN - [text] linge *m* de table
~LOOK-UP - [comput] consultation *f* de table
~LOOK-UP INSTRUCTION - [comput] instruction *f* de consultation de table
~LOWER LIMIT STOP - [mach tool] blocage *m* inférieur de la table
~MAGNETIZATION CONTROL SWITCH - [mach tool] interrupteur *m* pour commande magnétisation table
~MEASURING MACHINE - [text] métreuse *f* à table
~REVERSE CONTROL LEVER - [mach tool] levier *m* pour commande renversement déplacement de la table
~REVERSE DOGS - [mach tool] butées *fpl* pour commande renversement déplacement à fin de course
~ROLLS - [paper man] (series of small rollers in a papermaking machine, which support the wire screen) rouleaux *mpl* de support
~ROTATION CONTROL - [mach tool] commande *f* de la rotation de la table
~SCARF - [carp] trait *m* de Jupiter horizontal
~SET - [radio & telev] (a set supported by a separate table) récepteur *m* de table, téléviseur *m* de table
~SPEED HYDRAULIC CONTROL KNOB - [mach tool] poignée *f* pour réglage vitesse déplacement table avec commande hydraulique
~SUPPORT - [mach tool] support *m* de la table
~TELEPHONE - [telecomm] appareil *m* mobile, appareil *m* portatif
~THRUST BEARING - [oil ind] coussinet *m* portant de la table rotary
~TILTING ADJUSTMENT - [mach tool] réglage *m* inclinaison de la table
~TOP - [carp] dessus *m* de table
~TRAVERSE HANDWHEEL - [mach tool] volant *m* pour commande déplacement table
~TRIPOD - [photo & cin] (a very low camera stand, colloquially called high hat) pied *m* de table
~VERTICAL ADJUSTMENT - s. table height adjustment
~-VICE WITH CLAMP - [impl] étau *m* à agrafes d'établi
~WINE - [gen] vin *m* de table
~WITH CLAMPS - [mech] table *f* à presses
TABLECLOTH - [text] nappe *f*
TABLED JOINT - [carp] emboîtement *m*
~SCARF - [constr] entaillage *m* avec crochet
TABLELAND - [geogr] plateau *m*
TABLESPOON - [impl] cuiller *f* à soupe
TABLESPOONFUL - [med etc] cuiller *f* à bouche
TABLET - [chem] (small, compressed mass of specific form, e.g. soap tablet) comprimé *m*, tablette *f*
[arch] entablement *m*
[gen] tablette *f*
~PROTECTOR - [el] paratonnerre *m* à plaques, paratonnerre *m* à cuivre et mica
TABLETING MACHINE - [ind chem] machine *f* à comprimés, pastilleuse *f*
[rubber ind] (pelleting machine for rubber) machine *f* à réduire la gomme en boulettes
TABLETS PRESS - [mech mach] machine *f* à comprimer des tablettes
TABLEWARE - [impl] articles *mpl* de table
TABLING - [carp] assemblage *m*
[naut] (of a sail) gaine *f*, doublage *m*,
[gen] (USA only) ajournement *m* d'un projet de loi
[min] triage *m* du minerai par tables inclinées
TABLOID - [print] tabloid *m*, journal *m* de petit format
[chem] comprimé *m*
[gen] en raccourci
TABS OF FILMPACK LEADERS - [photo] languettes *fpl* des tirants
TABULAR - [min bot geogr] (horizontally flattened) tabulaire, aplati
[gen] (relating to a table or list) tabulaire
~INSERT - [comput] cavalier *m* de tabulateur
~INTERPRETIVE PROGRAMME - [comput] programme *m* interprétatif tabulaire
~MATTER - [print] tableaux *mpl*, ouvrage *m* à filets et à chiffres
TABULATE, to - [gen] (to arrange in a table, in a list) disposer en forme de table
[gen] imprimer en liste
[comput] tabuler
TABULATION SEQUENTIAL FORMAT - [comput] disposition *f* à tabulation
TABULATOR - [mech] (of a typewriter) tabulateur *m*
TABWASHER - [mech] rondelle *f* d'arrêt
TACAN - [aero] (tactical Air Navigation; a polar coordinate system type of radio aerial navigation system) Tacan *m* (système de navigation aérienne)
TACCIOMETER - [instr] (an instrument for measuring the tackiness of a dried coating) viscosimètre *m*
TACHEOMETER - s. tachymeter
TACHEOMETRIC - s. tachymetric
TACHEOMETRY - [surv] tachéométrie *f*
TACHO-ALTERNATOR - [el] (alternator generating a voltage of amplitude and frequency proportional to the speed of its rotor) dynamo *f* tachymétrique
TACHOGRAPH - [instr] (instrument designed to record a speed as a function of time) tachygraphe *m*, enregistreur *m* de vitesse
TACHOMETER - [instr] (an instrument for indicating revolutions per minute of e.g. a shaft) tachymètre *m*, indicateur *m* de vitesse
~DRIVE - [mech] mécanisme *m* du compte-tours
TACHOMETRIC RELAY - [el] automate *m* tachymétrique
TACHYLALIA - [med] tachyphémie *f*, tachyphrasie *f*
TACHYMETER - [instr] (an instrument designed to

measure distance from any given point by telescopic observation of a staff held at the point) tachymètre *m*
TACHYMETER - [instr] (speed indicator) indicateur *m* de vitesse
TACHYMETRIC - [surv] tachymétrique
TACHYMETRY - [mech] tachéométrie *f*
TACHYPHAGIA - [med] tachyphagie *f*
TACHYPHASIA - s, tachylalia
TACHYSTEROL - [chem] (a precursor of vitamin D 3) tachystérine *f*
TACHYSYSTOLE - [med] tachysystolie *f*, extrasystole *f*
TACITRON - [electron] (a type of thyratron in which the low of anode current can be interrupted by the voltage on the grid) tacitron *m*
TACK, to - [gen] (to fasten with tacks) clouer avec de la semence
[text] (to secure two pieces of cloth) faufiler, baguer
[naut] (to change the course of a vessel) virer de bord
[naut] (of sailboats) louvoyer
[gen] (to attach, e.g. two pieces together) attacher, joindre
[chem etc] (to stick together) coller
TACK - [gen] pointe *f*, clou *m* de bouche, petit clou *m*
[text] point *m* de bâti
[naut] amure *f*
[naut] (of sailboats) bordée *f*
[ind chem] (stickiness, e.g. of an adhesive) état *m* collant
[mech] (welding) point *m* de soudure
[shoe man] cheville *f*, pointe *f*
~COAT - [paint] couche *f* de liaison
~DETECTOR - [electron] (electronic apparatus designed to detect the tacks in a shoe which may hurt the wearer) détecteur *m* de semences
~DIE - [mech] (for welding) montage *m* de pointage
~-FREE - [paint] (not exhibiting tackiness) sec hors-poisse
~WELD, to - [mech] (welding operation for holding two parts together temporarily) souder par points
~WELDING - [mech] soudure *f* par points
TACKER - [tool] (for driving tacks) marteau *m* de tapissier
TACKIFIER - [chem] (a substance inducing tackiness) agent *m* poisseux
TACKINESS - [phys] (the property of being tacky) viscosité *f*
TACKING - [naut] virement *m* de bord
[gen] couage *m*, jonction *f*
~IRON - [photo] fer *m* à coller
~STRIP - [rubber ind] (for upholstery) jupe *f*, bande *f* de montage
TACKLE, to - [gen] saisir à bras le corps
[gen] (a horse) atteler, harnacher
[mech] (to operate a tackle) actionner un appareil de levage
TACKLE - [mech & naut] (lifting device using ropes or chains and pulleys) palan *m*, engins *mpl* de levage
[mech] (shifting gears) agrès *mpl*
[mining] treuil *m* d'extraction
~BLOCK - [mech] moufle *f*, poulie *f* de palan
~BOARD - [text] traîneau *m* de cordier
TACKLESS FOOTWEAR - [shoe man] chaussures *fpl* sans pointes de montage
TACKY - [paint] (tending to adhere) poisseux, collant
TACTIC - [chem] (term denoting symmetry of regularity in the molecular structure of a polymer) tactique
TACTICAL - [gen] tactique
TACTICS - [gen] tactique *f*
TACTILE - [zool] (pertaining to the sense of touch) tactile, tangible
~BRISTLE - [zool] (stiff hair transmitting a contact stimulus) poil *m* tactile
~END ORGAN - s. tactor
~PERCEPTION - [acoust] (the perception of vibrations by the sense of touch, developed particularly by the deaf) perception *f* tactile
TACTOMETER - [instr] (instrument designed to measure the quality of the sense of touch) tactomètre *m*
TACTOR - [anat] organe *m* sensoriel tactile
TACTUS - [zool] toucher *m*
TAENIA - [zool] (ribbon - shaped structure) taenia *m*, ténia *m*
[arch] ténie *f*
[med] bandage *m* en ruban
TAENIASIS - [med] (infestation of the human bodies with tape-worms) taeniase *f*
TAFFAREL - s. taffrail
TAFFEREL - s. taffrail
TAFFETA - [text] (fine, glossy silk fabric) taffetas *m*
TAFFRAIL - [naut] (ornamental top of a stern) lisse *f* de couronnement
~LOG - [instr] (log mounted on the taffrail) sillomètre *m* mécanique
~ROUND PLATFORM - [metall] garde-corps *m* du gueulard
TAG, to - [nucl] (to include a radio-active substance in a compound so that the behaviour of the compound in a reaction or subsequent process may be studied) marquer
[gen] attacher une fiche
[carp] ferrer, embouter
TAG - [el] (a flat metal part attached to the end of an electric wire or cable, for making connexion to a terminal) broche *f* de contact, tige *f* de contact
[gen] fiche *f*, étiquette *f*
[metall] (metal fragments adhering to a casting) fragment *m* métallique
[text] (of wool) bout *m* de la queue
[el] (used for soldering) cosse *f*, patte *f*
[comput] (in an instruction word) index *m*
~A PACKAGE, to - [comm] attacher une fiche à un paquet
~PAPER - [paper man] papier *m* à étiquettes
TAGGED ATOM - [nucl] (an atom which is made radioactive and used, with other like, but not radioactive atoms, as a tracer for biological or other processes) atome *m* marqué
~COMPOUND - [chem] composé *m* marqué
TAGGER - [metal] (very thin sheet metal) tôle *f* très mince
TAGGING - [plumb] effilement *m*
[metall] pointage *m*
TAGLOCK - s. tag [text]
TAIL, to - [gen] mettre un queue (à)
[agric] couper la queue (aux agneaux)
[bot] enlever les queues, égrapper
[el] (the tailing off of a current) s'affaiblir

TAIL, to - [constr] encastrer
TAIL - [gen] queue *f*
[aero] (the whole of the stabilizing and controlling surfaces at the after end of an aircraft) queue *f* de l'avion, empennage *m*
[print] (the bottom of a page) bas *m* de page, ligne *f* de pied
[gen] (of a coin) pile *f*, revers *m*
[bookbind] marge *m* inférieur
[astr] (of a comet) queue *f*, chevelure *f*
[oil ind] (front and tails) produits *mpl* légers
[oil ind] (heavy and tails) produits *mpl* lourds
[min] (talings, waste) résidus *mpl*, refus *m* de broyage
[radar] (small pulse following the main pulse and in the same direction) queue *f* de signal
[radar] (or pulse stretching; the terminal of the trailing edge of a pulse) comète *f*
[anat] (of men) coccyx *m*
~BAY - [constr] travée *f* contigue au mur
[hydr] biez *m* d'aval
~BEARING - [mech] coussinet *m* à queue
~BLOCK - [mach tool] (or tailstock, of a lathe) contre-pointe *f*, poupée *f* mobile
[mech] (a pulley block) poulie *f* à fouet
~BOOM - [aero] (structure carrying the tail unit of an aircraft in which such a unit is not mounted on the fuselage) poutre-fuselage *f*, longeron *m* de fuselage
~BOX - [hydr] boîte *f* de queue
~BUSHING - [mech] manchonnage *m* à queue
~CENTRE - [mach tool] pointe *f* de la poupé mobile
~CHAIN - [mining] chaîne *f* de contrepoids
~CONE - [aero] (the conical structure terminating the fuselage of an aircraft) cône *m* de queue
~CORDS - [text] cordes *fpl* de rame
~CURRENT - [electron] (the constant local value of the anode current at a specific value of the grid bias) courant *m* d'amortissement
~-DOWN ATTITUDE - [aero] (in landing, the attitude in which the tail wheel is in contact with the ground as well as the main undercarriage wheels) position *f* d'atterrissage de queue sur sol, cabrage *m*
~-DOWN LANDING - [aero] atterrissage *m* cabré
~ELEVATOR - [mech] élévateur *m* en queue
~-END TREATMENT - [nucl] opérations *fpl* terminales
~FIN - [aero] (of an aircraft) dérive *f*
~FLOAT - [aero] (a buoyant structure designed to support the after part of a seaplane when on the water) flotteur *m* de queue
~GATE - [auto] (of a truck body) hayon *m*
[auto] (of an estate car) porte *f* postérieure
~-GATES - [hydraulic] (at the low-level end of a lock) portes *fpl* d'aval (d'une écluse)
~GIRDER - [aero] poutre *f* de liaison
~HAIR - [text] crin *m* de la queue (du cheval)
~HAMMER - [impl] martinet *m* à queue
~HEAD - [anat] base *f* de la queue
~-HEAVINESS - [aero] (tendency in an aircraft for the tail to sink in flight) tendance *f* à cabrer
~HEAVY - [aero] (the condition of an aircraft in which the tail tends to sink in normal flight, unless correction is applied by the controls) avec la tendance à cabrer
~JOURNAL - [mech] (of a shaft etc) tourillon *m* à l'arrière

TAIL LAMP - [auto etc] feu *m* arrière, lanterne *f* arrière
~LEADER - [cin] amorce *f* de fin
~LIQUOR - [ind chem] (in the production of soda) liqueur *f* de chlorure de calcium, solution *f* finale
~LOCK - [hydr] écluse *f* de fuite
~MILL - [mach tool] (end mill) fraise *f* en bout
~PIECE - [mech] crépine *f* d'aspiration, reniflard *m*, grenouillère *f*
~PIPE - [aero] (an exhaust-pipe leading the gases away from the manifold) tubulure *f* d'évacuation
[instr] (a barometric pipe) tuyau *m* à siphon barométrique
[mech] (of a pump) aspirant *m*
[mech] (of a jet engine) tuyau *m* de sortie
~PIPE - [oil ind] tube *m* de queue
~PIPE-UNIT - [aero] (of a jet engine) groupe *m* d'échappement
~PISTON-ROD - [mech] contre-tige *f* de piston
~PLANE - [aero] (for stability along the longitudinal axis) plan *m* stabilisateur de l'empennage
~PRINT - [metall] portée *f* montante, portée *f* tirée à l'anglaise
~PRODUCT - [oil ind] produit *m* de queue
~PRODUCTION - [oil ind] production *f* de queue
~PULLEY - [mech] poulie *f* de queue
~PUMP - [oil ind] pompe *f* actionnée par le groupe moteur
~RACE - [hydr] (the discharging channel of a hydraulic turbine) bief *m* d'aval, canal *m* d'évacuation
~REACTOR - [petr ind] réacteur *m* de queue
~-ROD - [mech] guide *m* de piston, tige *f* de piston prolongée, contre-tige *f*
~ROPE - [mining] câble-queue *m*, câble *m* d'équilibre
~[mining] câble *m* de renvoi
~ROPE HAULAGE - [mining] traction *f* par câble tête et câble queue
~ROTOR - [aero] (a small rotor used to counteract the reaction torque of the main rotor) rotor *m* de queue
~RUDDER - [aero] gouvernail *m* de queue
~SCALE - [metall] battitures *fpl* infiltrées
~SCREW - [mach tool] vis *f* de la contre-pointe
~-SETTING ANGLE - [aero] (the angle between the chord lines of main and tail planes) angle *m* de calage de l'empennage
~SHAFT - [aero] (of a radial engine) arbre *m* de la manivelle
[naut] (of a ship) extrémité *f* de l'arbre
~SHEAVE - [mech] poulie *f* de retour, poulie *f* de renvoi
~SKID - [aero] (a skid placed under the tail of an aircraft) patin *m* de queue
~SLIDE - [aero] (motion of an aircraft in flight in a direction contrary to that of normal forward movement) glissade *f* sur l'empennage
~SLUICE - [hydr] sluice *m* de décharge
~SPIN - [aero] vrille *f*
~SPINDLE - [mech] arbre *m* de la poupée mobile
~STICK - [text] bâton *m* de rame
~-STOCK - s. tailstock
~SURFACE - [aero] (a stabilizing or control surface in the tail of an aircraft) empennage *m*, surface *f* de queue
~TACKLE - [naut] palan *m* à queue
~TRIM - [aero] (adjustment of tail control surfaces to give a normal attitude in flight without the use of

the elevator) réglage *m* du plan fixe
TAIL UNDERCARRIAGE - [aero] (a main landing gear assembly designed to support the tail) atterrisseur *m*
~UNIT - [aero] (the complex of control and stabilizing surfaces forming the tail of an aircraft) empennage *m*
~VOCE - [impl] étau *m* à main, étau *m* à queue
~WAGON - [railw] wagon *m* de queue
~WATER - [hydr] (downstream water) eau *f* d'aval
~-WHEEL LANDING GEAR - [aero] (a landing gear which comprises a tail-wheel) train *m* d'atterrissage à roulette de queue
~-WIND - [aero] (a wind blowing in the direction in which the aircraft is flying) vent *m* arrière
~-WIND LANDING - [aero] (a landing carried out downwind) atterrissage *m* vent arrière
TAILBAND - [bookbind] (strip of cloth at the bottom of a book back, between the cover and the back) coiffe *f*
TAILERON - [aero] (an aileron attached to the tail of an aircraft) empennage *m* de profondeur
TAILING - [constr] chef *m* de base
[radio] (a defect in reproduction) traînage *m*
~OUT - [mech] amincissement *m* en coin
TAILINGS - [min] (residue of low metal content, remaining after a mineral concentration or extraction process) résidus *mpl*, schlamms *mpl*, résidus *mpl* stériles
[brew ind] (refuse from barley) déchets *mpl* d'orge
[chem] (or bottoms) résidus *mpl* de distillation
~-DAM - [min] arrêt-barrage des tailings
~WHEEL - [min] roue *f* à tailings
TAILESS - [aero] (avion) sans queue
~AIRCRAFT - [aero] (an aircraft in which the longitudinal controlling and stabilizing surfaces are incorporated in the main plane) avion *f* sans queue
TAILLIGHT - [auto etc] lanterne *f* arrière
TAILOR, to - [metall] (to cut sheet material to shape as a separate skilled operation for each job) façonner
TAILOR - [gen] tailleur *m*
~MADE - [gen etc] (costume) tailleur
TAILORS' CLIPPINGS - [text] découpures *fpl* de tailleurs
~TWIST - [text] cordonnet *m*
TAILORED - [gen] (cut to shape expressly for a given piece of work) façonné
~VERSION - [gen] exécution *f* spéciale
TAILPIECE - [aero] (of the propeller shaft of a radial engine) empennage *m* d'avion
[print] (ornamental design on the lower blank portion of a page) cul-de-lampe *m*
[mus] (of a violin) cordier *m*
TAILRACE - s. tail race
~TUNNEL - [hydr] galerie *f* de décharge
TAILSHAFT - s. tail shaft
TAILSTOCK - [mach tool] contre-pointe *f*, poupée *f* mobile, contre-poupée *f*
~ASSEMBLY - [mach tool] groupe *m* contre-poupée
~CLAMPING LEVER - [mach tool] levier *m* blocage contre-poupée
~MANDREL - [mach tool] arbre *m* de la contre-pointe
TAILWARD ACCELERATION - [astronaut] accélération *f* dans le sens tête-siège
TAINT, to - [gen] infecter
TAKE, to - [gen] prendre, attraper
TAKE, to - [mech] supporter, soutenir
[photo cin] prendre (une vue)
TAKE - [cin] vue *f*
[print] paquet *m* de composition
~A BEARING, to - [surv] relever
~A BEAUTIFUL POLISH, to - [constr] (of stone) prendre en beau poli
~A CURVE, to - [auto etc] franchir une courbe
~A DIRECTION, to - [gen] se diriger
~A POINT, to - [railw] franchir une aiguille
~A SWITCH, to - s. take a point, to
~-ABOUT CHUCK - [mach tool] mandrin *m* à toc, plateau *m* poussetoc
~AIM, to - [firearms etc] viser
~-ALL OF CEREALS - [bot] (plant disease) piétin *m*
~APART, to - [mech] démonter
~AWAY, to - [gen] enlever, emporter
~AWAY BELT - [mech] (delivering belt conveyor) transporteur *m* à courroie
~DIMENSIONS, to - [gen] mesurer
~DOWN, to - [gen] décrocher, descendre
[constr] démolir
[mech] démonter
[print] crocheter
~-DOWN - [mech] démontable
~DOWN THE ENGINE, to - [mech] démonter le moteur
~DOWN THE FALSEWORK, to - [constr] enlever les échafaudage, déséchafauder
~FIRE, to - [gen & phys] (to ignite) s'enflammer, prendre feu
~-HOLD PRESSURE-[mech] (holding pressure) vide *m* d'entretien, pression *f* à maintenir
~IN SAIL, to - [naut] carguer, ramasser
~IN TOW, to - [gen] remorquer, donner la remorque
~OFF, to - [aero] décoller, s'envoler
[print] desserrer les garnitures
[gen] enlever, ôter
[math] (to subtract) soustraire
[el] brancher (un courant)
[comm] défalquer
~-OFF - [plast ind] (mechanism for drawing extruded or rolled material away from the machine producing it as it is delivered) dispositif *m* de réception
[telev] (only USA; sound rejection in GB, suppression of the sound - carrier wave in the vision receiver without impairing its pass characteristic on the side where the sound-carrier wave is situated) réjection *f* de la porteuse son
~OFF A TRIAL BALANCE, to - [comm] faire un bilan de contrôle
~OFF ACCIDENT - [aero] (an accident occurring during take-off) accident *m* au décollage
~-OFF BEVEL GEARS - [mech] engrenages *mpl* coniques de transmission
~-OFF BOARD - [print] plateau *m* à papier
~-OFF DISTANCE - [aero] (the distance travelled by an aircraft from a state of rest on the ground until it is airborne) distance *f* de décollage
~-OFF POINT - [aero] (the point on the ground at which the pilot opens the throttle to take off) point *m* de décollage
~-OFF POWER - [aero] puissance *f* au décollage
~-OFF REEL - [comput] bobine *f* débitrice
~-OFF ROCKET - [aero] (a booster rocket to add thrust) fusée *f* auxiliaire de décollage
~-OFF ROLL - [plast ind] dérouleur *m*, rouleau *m* dévideur

TAKE-OFF RUN - [aero] (the distance which an aircraft travels during take-off while still in contact with the ground) course *f* au décollage
~-OFF RUNWAY - [aero] (a runway used for taking off) trajectoire *f* de décollage
~-OFF SAFETY SPEED - [aero] (the lowest speed which will ensure a safe control margin after failure of the engine most affecting control when the aircraft is taking off) vitesse *f* de décollage de sécurité
~OFF SCALES, to - [th eng] (of boilers) détrarter
~-OFF SLUR - [print] papillotage *m*, regiflage *m*
~-OFF SPEED - [aero] (the airspeed at which an aircraft becomes entirely airborne) vitesse *f* de décollage indiquée
~-OFF SPEED - [comput] vitesse *f* de défilement
~-OFF SPROCKET WHEEL - [cin] tambour *m* débiteur
~OFF THE BOBBINS, to - [text] enlever les bobines
~OFF THE BURN - [mech] ébarber
~OFF THE BURR, to - [metall] ébarber
~-OFF TRACK - [comput] (on an analog-to-digital converter) piste *f* de prélèvement
~-OFF WEIGHT - [aero] poids *m* au décollage
~ON, to - [gen] entreprendre, se charger
~ON LEASE, to - [leg] prendre à bail, prendre à ferme
~OUT, to - [gen] sortir, enlever
[leg] (a patent) obtenir (un brevet etc)
[leg] (an insurance policy) contracter (une assurance)
[constr] faire le devis, relever le cubage
~-OUT - [mech] (mechanical device for removing a finished article from a forming unit) extracteur *m*
~OUT OF WINDING, to - [mech] dégauchir
~OUT THE PILLARS, to - [mining] déhouiller les piliers
~OVER, to - [gen] prendre la suite, prendre la succession
[comm] recevoir, accepter
~PART, to - [gen] participer, s'associer
~PLACE, to - [gen] avoir lieu, survenir
~READINGS, to - [instr] relever
~REFUGE, to - [gen] se réfugier
~ROOT, to - [bot] s'enraciner, prendre racine
~SOUNDINGS - [naut] sonder, prendre le fond
~STEPS, to - [gen] prendre des dispositions
~THE AIR, to - [aero] (USA only) s'envoler
~THE EDGE OFF A CHISEL, to - [mech] émousser un ciseau
~THE PERCENTAGE OF FIRE-DAMP, to - [mining] relever la teneur en grisou
~THE STRESS, to - [phys] résister à l'effort
~THE TEMPERATURE, to - [gen & med] mesurer la température
~TO PIECES, to - [gen] démonter, déconstruire
~UP, to - [gen] se mettre à, se mettre en
[med] (an artery) serrer
~-UP - [mech] (of a belt conveyor; device for maintaining the necessary tension in a conveyor belt) tendeur *m*
[cin] (the winding of a film on a reel) enroulement *m* du film
[cin] (the implement used to wind up a film on a reel) enrouleuse *f*
[mech] rattrapage *m*, serrage *m*
[mech] (the device for taking up slack) dispositif *m* à rattraper le jeu

TAKE-UP - [mech] embrayage *m*
~-UP BOBBIN - [text] bobine *f* réceptrice
~-UP CASSETTE - [cin] cassette *f* enrouleuse
~-UP DEVICE - [cin] système *m* enrouleur
~-UP ELEVATOR - [cin] (tank lifter; component of the processing apparatus at the removing end) élévateur *m* de cuve
~-UP REEL - s. take-up spool
[comput] (of a magnetic tape unit) bobine *f* réceptrice
~-UP ROLLER - [text] cylindre *m* d'enroulement
~-UP ROLLS - [mech] (rollers arranged to draw extruded material away from the machine) cylindres *mpl* récepteurs
~-UP SPEED - [comput] (of a magnetic tape) vitesse *f* d'enroulement
~-UP SPOOL - [cin] bobine *f* enrouleuse
~-UP SPROCKET - [cin] tambour *m* inférieur
~-UP UNIT - [plast ind] (mechanism which winds extruded cable on to drums as it comes from the extruder) unité *f* de réception
~UP WEAR, to - [mech] (to adjust to a closer fit to compensate for wear) compenser l'usure, rappeler l'usure
TAKER - [impl] preneur *m*
[text] tambour *m* briseur
~-IN - [text] (the cylinder of a carder taking the lap from the feed rollers) briseur *m*
~-IN COVER - [text] couvercle *m* du briseur
~IN GRID - [text] grille *f* sous le briseur
TAKING - [gen] prise *f*
[mech] relevé *m*
[cin] prise *f* de vues
[med] prélèvement *m*
~CHARACTERISTIC - [telev] (camera tube spectral characteristic) caractéristique *f* spectrale du tube de prise de vues
~OFF OF THE CLOTH - [text ind] déchargement *m* du tissu
~OFF THE TUFT - [text] enlèvement *m* de la barbe
~OUT THE PILLARS - [mining] dépilage *m* (des piliers)
~ROOT - [bot] enracinement *m*
~SCROLL - [text] scroll *m* de rentrée du chariot
~SLANTWISE - [railw] (a slanting collision) prise *f* en écharpe
~THE SWINGS - [el] stabilisation *f* de la fréquence
~UP - [text] (of the cloth) enroulement *m*
~UP FRAME - [text] bâti *m* d'enroulement
~-UP LEVER - [text] levier *m* d'enroulage
~UP MOTION - [text] mécanisme *m* d'enroulement du tissu
TAKINGS - [comm] recette *f*
TALBOT BANDS - [phys] (used in spectrology) bandes *fpl* de Talbot
~LAW - [opt] loi *f* de Talbot
TALBUTAL - [chem] (amylobarbitone, 5-ethyl-5-isopentylbarbituric acid. A hypnotic used in medicine) talbutal *m*
TALC - [chem] (talcum; mineral graphite; steatite; naturally-occurring hydrous magnesium silicate with applications as a filler for plastics and rubber) talc *m*
~APPLICATOR - [rubber ind] (a dusting device) dispositif *m* à talquer, poudreuse *f*
TALCUM - s. talc
~POWDER - [chem] poudre *f* de talc
TALK, to - [gen] parler

TALK-BACK CIRCUIT - [telev] (intercommunication) circuit *m* d'ordres
~-LISTEN SWITCH - [telecomm] (the switch on an intercommunication unit which allows the loudspeaker to be used as a microphone) commutateur *m* parle-écoute
TALKIE - [cin] (colloq. for sound film) film *m* sonore
TALKING - [gen] parlant
~BEACON - [radar] (radio range facility whose courses are normally followed by interpretation of the aural signal transmitted) radiophare *m* directionnel acoustique
"TALKING DOWN" - [aero] (landing procedure in which ground control gives verbal instructions to the pilot by radio-telephone) atterrissage *m* contrôlé par radiophone
TALL - [gen] grand, de haute taille
~OIL - [paint] (a by-product of paper manufacture, consisting of a mixture of fatty and rosin, acids, and sometimes used as a substitute for linseed oil. Used in the production of synthetic resins, and in pharmaceuticals, and lubricants) résine *f* liquide
TALLBOY - [gen] commode *f* à secrétaire
[constr] (of a chimney) mitre *f* en tôle
TALLOW - [chem] (fat obtained from animal fats and used in the manufacture of soap, candles, and lubricants) suif *m*
[naut] suage *m*, flore *m*
~-LIKE WOOL GREASE - [text] graisse *f* de laine suiffeuse
~-LIKE YOLK - [text] suint *m* suiffeux
~POT - [metall] chaudière *f* à graisse
~SEED OIL - [chem] (a synonym for stillingia oil) huile *f* de suif végétale
~SOAP - [chem] savon *m* de suif
TALLY, to - [comm] pointer, contrôler
[gen] correspondre, s'accorder
[gen] (to apply a tally) étiqueter
TALLY - [gen] taille *f*, marque *f*, baguette *f* à encoches, étiquette *f*, pointage *m*
[surv] (of a measuring chain) encoche *f*
[comm] jeton *m* de présence
[comput] liste *f* d'additions
~LIGHTS - [cin] (cue lights) lampes *fpl* de signalisation
TALLYSHOP - [comm] boutique *f* faisant les ventes à tempérament
TALON - [zool] serre *f*, griffe *f*
[mech] ergot *m* (de pêne)
[arch] doucine *f*, talon *m*
[comm] talon *m* de souche
TALUS - [surv] talus *m*
[geol] talus *m* d'éboulis, clapier *m*
[anat] astragale *m* du tarse
[med] pied *m* bot talus
~SLIDE - [soil] rupture *f* de talus
TAMARACK - [bot] mélèze *m* d'Amérique, épinette *f* rouge
TAMARIND - [bot] fruit *m* du tamarinier
[bot] (the tree) tamarinier *m*
TAMARISK - [bot] tamaris *m*
TAMBOUR - [mus] grosse caisse *f*
[text] (for embroidery) métier *m* à broder
[arch] tambour *m* (de colonne etc)
~LACE - [text] dentelle *f* sur tulle
~-WORK YARN - [text] fil *m* de broderie au tambour
TAMBOURINE - [mus] tambour *m* de basque
TAME, to - [gen] apprivoiser, dompter
TAMMY - [impl] tamis *m*, passoire *f*
[text] (a wool and cotton fabric) étamine *f*
TAMP, to - [constr] damer, pilonner
[gen] bourrer (un fourneau de mine)
[soil] (the ground) damer
TAMPER, to - [gen] altérer, trifouiller
TAMPER - [impl] bourroir *m*
[constr] (a road compactor) dame *f*, pilonneuse *f*
[nucl] (or reflector; portion of a reactor between the shield and the core, designed to return neutrons to the core) réflecteur *m* de bombe
~MATERIAL - [nucl] matière *f* réflectrice
~-PROOF CABLE - [telecomm] câble *m* inviolable
TAMPERING - [gen] altération *f*, falsification *f*, adultération *f*
TAMPICO FIBRE - [text] tampico *m*
~HEMP - [text] chanvre *m* de Tampico
TAMPING - [constr] damage *m*, pilonnage *m*
[soil] (of the ground) blinage *m*
[mining etc] bourre *f* (d'un fourneau de mine)
~BAR - [impl] bourroir *m*
~OF SLEEPERS - [railw] bourrage *m* des traverses
~ROD - [impl] batte *f* à bourrer
~ROLLERS - [mech] (for road construction) cylindres *mpl* vibrants
TAMPON - [gen] tampon *m*
[med] tampon *m*
[mus] mailloche *m* double
TAMPONADE - [med] (the surgical use of the tampon) tamponnement *m*
TANPONMENT - s. tamponade
TAN, to - [leather ind] (to convert raw hide into leather) tanner
TAN - [chem] (short for tannin) tanin *m*
[math] (short for tangent) tangente *f*
[leather ind] tan *m*, tanné
~LIQUOR - [chem] jus *m* tannant
TANBARK - [bot] écorce *f* à tan
TANDEM - [mech] (a tandem bicycle) tandem *m*
[gen] en tandem
~AXLE - [auto] ponts *mpl* montés en tandem
~BOOSTER - [astronaut] moteur-fusée *m* à grains de propergol brûlant en tandem
~CALENDER - [mech] (a type of calender in which two or more pairs of rollers are arranged to receive material in succession) calandre *f* en tandem
~COMPRESSOR - [mech] (type of steam-driven compressor in which the air cylinders are placed in line with the steam cylinders, with common piston-rods) compresseur *m* en tandem
~CONNECTION - [el] (cascade connection) couplage *m* en cascade
~DIALLING - [telecomm] sélection *f* en tandem
~DISC HARROW - [agric] pulvérisateur *m* tandem
~ENGINE - [mech] machine *f* tandem
~GENERATOR - [el] (double Van de Graaff machine) générateur *m* de Van de Graaff en série
~KNIFE SWITCH - [el] (a switch in which two or more blades are mechanically coupled in order to separate simultaneously as a multiple pole-switch) interrupteur *m* à deux couteaux, interrupteur *m* à deux leviers
~LAUNCH - [astronaut] lancement *m* multiple
~MILKING PARLOUR - [agric] salle *f* de traite tandem
~MILL - [metall] laminoir *m* tandem

TANDEM MOTOR - [el] moteur *m* tandem
~NETWORK - [telecomm] réseau *m* en étoile
~PISTONS - [mech] pistons *mpl* en tandem
~PLANE - [aero] (a plane having wings in tandem, i.e. in front of each other) avion *m* à ailes en tandem
~POTENTIOMETER - [instr] potentiomètre *m* avec résistances en tandem
~PROPELLERS - [aero] (propellers placed in front of each other on a common shaft) hélices *fpl* en tandem
~ROCKET - [astronaut] fusée *f* à étages
~SYSTEM - [el] (a method of controlling two inductions-motors for the purpose of obtaining a speed lower than synchronism, by mechanically coupling the motors and supplying one motor from the rotor or secondary circuit of the other motor) commande *f* en cascade
~TWIN-ROTOR CONFIGURATION - [aero] (the general layout of a helicopter having two main rotors located in tandem formation) configuration *f* d'un hélicoptère à deux rotors en tandem
~WIPER - [auto] essuie-glace *m* double
TANG, to - [gen & mech] façonner une soie
[acoust] retentir, rendre un son aigu, faire resonner
TANG - [gen] soie *f*
[mech] (part inserted in the handle of a knife a file etc) queue *f*, soie *f*
[mech] (a tongue) languette *f*
[acoust] son *m* aigu, tintement *m*
TANGENCY - [math] tangence *f*
TANGENT - [math] tangente *f*
~CHORD CATENARY SUSPENSION - [el] suspension *f* caténaire à tendeurs
~COMPASS - [instr] (or tangent galvanometer) boussole *f* des tangentes
~DISTANCE - [surv] (the distance between the intersection point and one of the tangent points of a railway or a highway) distance *f* de la tangente
~GALVANOMETER - [instr] (a galvanometer with a moving magnet, so arranged that the tangent of the angle of deflection is proportional to the current to be measured) boussole *f* des tangentes
~OGIVE - [astronaut] ogive *f* tangente
~PADDLES - [mech] aubes *fpl* en tangentes
~PLANE - [math] plan *m* tangent
~PLATE - [mach tool] lyre *f*, tête *f* de cheval, cavalier *m* (d'un tour)
~POINT - [constr] (between a straight road and a curve) point *m* de raccord
~SCALE - [el] (of an electrical instrument) échelle *f* des tangentes
~SCREW - [instr] (a screw by which a fine adjustment may be made to the setting of an instrument about its axis) vis *f* micrométrique
[mech] vis *f* sans fin, vis *f* tangente
~SIGN - [surv] borne *f* de la tangente
~TRACK - [railw] (a straight) alignement *m* droit, voie *f* en alignement
~WEDGE - [mech] clavette *f* tangentielle
~WHEEL - [mech] (a worm wheel) roue *f* hélicoïdale
TANGENTIAL - [math] tangentiel, tangent
~ACCELERATION - [phys] accélération *f* tangentielle
~BURNER - [th eng] brûleur *m* tangentiel
~CASTING - [metall] coulée *f* tangentielle
~DISTORTION - [opt] distorsion *f* tangentielle
TANGENTIAL FIELD - [opt] champ *m* tangentiel
~-FLOW TURBINE - [mech] turbine *f* tangentielle
~FOCUS - [opt] foyer *m* tangentiel
~FORCE - [phys] (a force associated with a wheel or disk always acting perpendicular to the radius, e.g. frictional force between rolling wheel and a surface, or frictional force between a belt and a pulley wheel) force *f* tangentielle, force *f* de rotation
~GUIDES - s. tangential guiding
~GUIDING - [mech] guidage *m* tangentiel
~INCIDENCE - [el acoust] (said of a microphone when the angle of incidence differ from 90 degrees) incidence *f* tangentielle
~KEY - [mech] clavette *f* tangentielle
~PLANE - [opt] plan *m* tangentiel
~PRESSURE DIAGRAM - [el] diagramme *m* des efforts tangentiels
~PROJECTION - [radiat] (radiograph for which the central ray is tangential to the surface) vue *f* tangentielle, projection *f* tangentielle
~SCREW - s. tangent screw
~SPIRAL AGITATOR - [rubber ind] agitateur *m* à spirales tangentielles
~SPOKES - [mech] rayons *mpl* tangents
~STRAIN - [phys] déformation *f* tangentielle
~STRAINING - [phys] contrainte *f* tangentielle
~STRESS - [phys] (shearing stress) effort *m* tangentiel
~SURFACE - [math] surface *f* tangente
~TEST - [metall] essai *m* en direction tangentielle
~THRUST - [mech] charriage *m* tangentiel
~WAVE-PATH - [radio] trajectoire *f* directe, parcours *m* tangentiel d'une onde
~YARN TENSION - [text] tension *f* tangentielle du fil
TANGENTIALLY - [gen] tangentiellement
TANGERINE - [bot] mandarine *f*
~OIL - [chem] huile *f* de mandarine
TANGIBLE - [gen] tangible, palpant
TANGLE, to - [gen] embrouiller, mêler
TANGLE - [gen] embrouillement *m*, emmêlement *m*, fouillis *m*
TANK, to - [gen] mettre en réservoir
TANK - [gen] (a closed vessel to contain fuel, oil or the like) réservoir, *m*, citerne *f*, bache *f*
[el] épaississeur *m*
[el] (of a transformer) (a vessel containing oil, or other cooling insulating medium, in which a transformer is immersed) cuve *f*
[gas ind] (of a gasholder) cuve *f*
[hydr] (US only) bassin *m*
[naut] cale *f* à eau, charnier *m* à eau douce
[naut] (for the fuel) soute *f*
[constr] (of a roller) caisson *m* de chargement
[el chem] (for electrolytic bath) cuve *f*
[glass man] (a melting unit) bassin *m* de fusion
[photo] (container for processing films) cuve *f*, cuvette *f*
[cin] cabine *f* insonorisée
~AND HOSE FITTINGS - [mech] (assembled kit consisting of a reservoir, oil lines and connecting couplers for the supply of hydraulic power between tractor and dump trailer combination) canalisations *fpl* d'huile
~BALCONY - [gas ind] (in gasholders) margelle *f*
~BARGE - [transp] chaland-citerne *m*
~BATTERY - [oil ind] groupe *m* de réservoirs
~BOTTOM - [hydr] radier *m* d'un réservoir

TANK BOTTOMS - [oil ind] résidus *mpl* de réservoir
~-BREATHER TUBE - [oil ind] tube *m* d'aérage du réservoir
~CAPACITY - [railw] capacité *f* de la soute à eau
~CAR - [transp] voiture *f* citerne
[railw] wagon-citerne *m*
~CIRCUIT - [radio] (the oscillatory circuit in a thermionic amplifier in a radio transmitter) circuit *m* bouchon
~CLINCHING MACHINE - [mech] machine *f* à river les réservoirs
~DOME - [oil ind] regard *m* de réservoir
~ENGINE - [railw] (an engine carrying fuel and water) machine-tender *f*
~FARM - [nucl] (a place where waste-storage tanks are placed) entrepôt *m* de tanques, dépôt *m* de cuves
~FARM - [oil ind] parc *m* de réservoirs
~FILLER CAP - [auto] bouchon *m* de réservoir de carburant
~FITTINGS - [el] accessoires *mpl* de cuve
~FOR CONTINUOUS COAGULATION - [rubber ind] bac *m* de coagulation continue
~FRAMING - [gas ind] (in gasholders, interior landing platform) charpente *f* intérieure fixe
~FURNACE - [glass man] four *m* à bassin
~GAUGE - [oil ind] jauge *f* de niveau
~IRON - [metall] (iron plate of a thickness between boiler plate and sheet iron) tôle *f* moyenne
~LIFTER - [cin] (of a processing apparatus) élévateur *m* de cuve
~LINING - [rubber ind] revêtement *m* de cuve
[el] couche *f* isolante intérieure de cuve
~LOCOMOTIVE - s. tank engine
~OIL COOLER - [aero] (a combination of oil tank and oil cooler) réservoir-radiateur *m*
~OUTAGE - [oil ind] perte *f* de pétrole du réservoir
~PARTITION - [gen] cloison *f* de réservoir
~SHIP - [naut] bateau-citerne *m*
~STEAMER - [naut] bateau-citerne *m* à vapeur
~STRAINER - [photo] tamis *m* de cuve
~STRAPPER - [oil ind] mesureur *m* de réservoirs
~TEST - [naut] (of ship's models) essai *m* en bassin
~TESTING - [aero] (for pressure fatigue etc) essai *m* à pression
~THIEF - [oil ind] sonde *f* de prise d'échantillons
~TOP - [mech] plafond *m* du réservoir
~TOP CURB - s. tank balcony
~TRAILER - [auto] remorque-citerne *f*
~TRUCK - [auto] camion-citerne *m*
~VENT PIPE - [aero] (a pipe leading from the air space in a tank to the atmosphere or to another vessel) mise à l'air libre
[plumb] tube *m* d'égout
~VOLTAGE - [el] (total fall of potential between anode and cathode bus bars during a plating operation) tension *f* de cuve
~WAGON - [auto] camion-citerne *m*
[railw] wagon-citerne *m*
~WITH COOLING TUBES - [el] bac *m* à tuyaux de refroidissement
TANKAGE - [gen] (the operation of putting in tanks) stockage *m* du pétrole, emmagasinage *m*
[oil ind] (the capacity of a tank) capacité *f* du réservoir
[comm] (the price of storage in tanks) frais *mpl* d'emmagasinage du pétrole
[ind chem] résidus *mpl* de graisse

TANKAGE - [oil ind] stockage *m* du pétrole
TANKARD - [gen] pot *m* en étain, pot *m* à bière
TANKER - [transp] (ship or vehicle designed expressly for the bulk transport of liquids, espec. petroleum products) bateau-citerne *m*, navire *m* citerne
[transp] camion-citerne *m*
TANNAGE - [leather ind] tannage *m*
TANNATE - [chem] (salt or ester of tannin) tannate *m*
TANNED - [leather ind] tanné
~LEATHER PICKER - [text] taquet *m* en cuir tanné
~SKIN - [leather ind] cuir *m* tanné
TANNER'S SUMAC - [bot] sumac *m* des corroyeurs
TANNERY - [leather ind] tannerie *f*
TANNIC ACID - [chem] (naturally occurring substance with applications in rubber substitutes) acide *m* tannique
~ACID BATH - [ind chem] bain *m* d'acide tannique
~COMBINATION - [chem] combinaison *f* tannique
TANNIN - [chem] (the raw material of tannic acid, consisting of a mixture of polyhydroxy-benzoic acids, found especially in oak-galls).Used for tanning leather for its property of forming an insoluble precipitate with gelatine) tanin *m*
TANNING - [leather ind] (the process of converting raw hides into leather) tannage *m*
~AGENTS - [ind chem] substances *fpl* tannantes
~BARK - [leather ind] écorce *f* tannante
~LIQUOR - [ind chem] tannée *f*
~MATERIALS - s. tanning agents
~PLANT - [bot] plante *f* tannante
TANNOFORM - [chem] (methyleneditammin. A condensation product of formaldehyde and tannin, with uses in medicine) tannoforme *m*
TANTALITE - [min] (an ore of tantalum, essentially a tantalate or iron and manganese) tantalite *f*
TANTALUM - [chem] (extremely hard, abrasive compound used in the manufacture of high speed cutting tools) tantale *m*
~CARBIDE - [chem] carbure *m* de tantale
~CHLORIDE - [chem] (a chlorinating agent in organic synthesis) chlorure *m* de tantale
~DETECTOR - [electron] (detector consisting of the tip of thin tantalum wire touching the surface of a mercury pool) détecteur *m* à tantale
~LAMP - [opt] lampe *m* au tantale
~OXIDE - [chem] (a constituent of special glasses) oxyde *m* de tantale
TAON - [med] béribéri *m* infantile
TAP, to - [mech] (to cut a thread in a hole by means of a tap) tarauder
[metall] (to open a furnace in order to draw off molten metal) couler, percer
[brew ind] (to insert a cock into a vessel, e.g. a cask, in order to draw off liquid from it) mettre en perce
[gen] (to strike lightly) taper, frapper légèrement
[el] (to connect a circuit) faire une prise
[glass man] (to drain a furnace) drainer
[med] (a growth, a tumor) ponctionner
[mining] recouper, percer à
[telecomm] (e.g. telephone wire) dériver, bifurquer, brancher
[bot] (the trees) tailler, inciser, saigner
TAP - [gen] robinet *m*, fausset
[tool] (a mechanical tool for threading) taraud *m*, filière *f*
[el] (a point of connection) prise *f*, conducteur *m*

auxiliaire
TAP - [impl] (a plug) bouchon *m*
[metall] (the quantity of metal which is tapped from a furnace) coulée *f*
[gen] (a light blow) tape *f*, coup *m* léger
[shoe man] demi-semelle *f*
[telecomm] branchement *m*, bifurcation *f*
~AND REAMER WRENCH - [mech] tourne-à-gauche *m* pour tarauds et alésoirs
~BAR - [metall] barre *f* de contrôle de cémentation
~BOLT - [mech] boulon *m* taraudé
~BORER - [tool] bondonnière *f*
~BOX - [mus] (chinese wood box, used in jazz bands) caisse *f* chinoise
~CHANGER - [el] (piece of apparatus which by selection of tappings changes the ratio of a transformer) commutateur *m* à prises de réglage
~CINDER - [metall] scorie *f* de puddlage
~CUTTER - [tool] fraise *f* à tarauds
~DENSITY - [metall] densité *f* de la poudre
~FAN - [mech] (a handle; a thumb-piece) manette *f*
~-FIELD CONTROL - [el] commande *f* de vitesse de moteurs en série
~FUNNEL - [ind chem] entonnoir *m* à robinet
~GRIP - [mech] poignée *f* du robinet
~HOLDER - [mach tool] porte-taraud *m*
~HOLE - [metall] (of a blast furnace) trou *m* de coulée, oeil *m*
~HOLE GUN - [metall] machine *f* à obturer le trou de coulée
~LINE - [railw] tronçon *m* de voie de raccordement
~-OUT BAR - [metall] barre *f* à piquer
~PLATE - [tool] filière *f* à truelle
~ROOT - [bot] racine *f* pivotante
~SCREW - [mech] vis *f* à tête
~SLAG, to - [metall] décrasser les scories
~SWIRL - [mech] coupe-jet *m*
~-SELECTOR SWITCH - [el] (a switch in which the contacts are connected to the transformer ratio adjusting tappings, and by means of which tappings are selected) commutateur *m* sélecteur de prise
~SYSTEM - [plumb] batterie *f* de robinets
~THE HEARTH, to - [metall] vider le creuset
~THE TREES, to - [bot] saigner les plantes, gemmer
~WATER - [ind chem] (undistilled water) eau *f* de robinet
~-WRENCH - [tool] (a special tool for holding a tap and revolving it) tourne-à-gauche *m*
TAPA FIBRE - [text] fibre *f* du mûrier à papier
TAPE, to - [gen] ficeler, attacher, border
[bookbind] coudre sur ruban
[el] (to cover with tape) rubaner, guiper
[el acoust] enregistrer sur bande
[aero] maroufler (une couture)
TAPE - [text] (narrow strip of woven cloth) ruban *m*, cordon *m*, ganse *f*
[metall] (narrow strip of metal) ruban *m*
[telecomm] (the strip of paper used in the receiving instrument of a recording telegraph system) bande *f*
[el] (insulating tape) ruban *m* isolant, feuillard *m*
[el acoust] (for recorder) ruban *m* (magnétique)
~ARMOURING - [el] armure *f* en feuillard
~BACKING - [gen] dorsale *f* de bande
~BACKSPACING - [comput] (the backward movement of the tape after reading) espacement *m* en arrière
TAPE BAND STITCHER - [text] couseuse *f* de courroies
~BASE - [gen] support *m* de bande
~CARTRIDGE - s. tape cassette
~CASSETTE - [el acoust] cassette *f* à bande
~CLATTER - [el acoust] fracas *m*
~COATING - [gen] couchage *m* de bande
~CODE - [comput] code *m* de bande
~-CONTROLLED CARRIAGE - [comput] chariot *m* commandé par une bande pilote
~CORE - [el] noyau *m* en ruban
~COUNTER - [gen] compteur *m* de longueur de bande
~CUPPING - [el acoust] courbure *f* transversale de la bande
~CURLING - [el acoust] curling *f* de bande
~CURVATURE - [el acoust] courbure *f* de la bande
~DECK - [el acoust] châssis *m*
~DISTORTION - [el acoust] distorsion *f* de bande
~DIVIDER - [text] lanière *f*
~DRIVE -[text] commande *f* par sangle
~DRIVEN SPINDLE - [text] broche *f* avec commande par ruban
~FEED - [comput] (a mechanism feeding the tape to be read or sensed by the computer) alimentation *f* de bande
~FEEDING DEVICE - [comput] entraînement *m* de ruban, dispositif *m* d'entraînement d'un ruban
~FILE - [comput] données *f*pl enregistrées sur un ruban magnétique
~GRASS - [bot] vallisnérie *f*
~GUIDANCE - [telev] guidage *m* de piste
~GUIDING DRUM - [telev etc] tambour *m* de guidage de bande
~HANDLING UNIT - [comput] unité *f* de ruban magnétique
~LEADER - [telev] bout *m* mort
~LIFTER - [gen] décolleur *m* de bande
~-LINE - [meas] décamètre *m* à ruban
~LOOM - [text] métier *m* à rubans
~LOOP - [el acoust] (endless tape) ruban *m* magnétique ininterrompu
~MEASURE - [meas] mesure *f* à ruban
~-OPERATED ELECTRIC TYPEWRITING MACHINE - [comput] machine *f* à écrire automatique commandée par bande perforée
~OPERATED PRINTER - [comput] machine *f* à écrire automatique commandée par bande perforée
~PLAYER - [el acoust] dérouleur *m* son
~PRINTER - [telecomm] imprimeur *m* à bande
[comput] imprimeur *m* à bande
~READER - [telecomm] tête *f* de lecture de bande
[comput] lecteur *m* de bande
~READING-HEAD - s. tape reader
~-RECORD, to - [el acoust] enregistrer sur ruban magnétique
~RECORDER - [el acoust] (magnetic tape recorder) enregistreur *m* à ruban magnétique
~RECORDING - [el acoust] enregistrement *m* à ruban magnétique
~REEL - [comput] bobine *f* de ruban
~RELAY - [telecomm] (a system of retransmitting traffic from one channel to another) transit *m* par bande perforée
[comput] relayage *m* à bande
~RELAY WORKING - [telecomm] régulation *f* du trafic
~RETRANSMISSION - [telecomm] (perforated tape

transmission) émission *f* télégraphique à bande perforée, reperforateur-transmetteur *m*
TAPE SCRATCH - [gen] rayure *f* de bande
~SIZING MACHINE - [text] encolleuse *f* mécanique pour chaînes en boyau
~SLIPPAGE - [el acoust] glissement *m* de la bande
~SPEED - [el acoust] (the speed at which the tape moves past the recording head) vitesse *f* de ruban
~SPLICER - [telev] colleuse *f* de bande
~START - [comput] démarrage *m* du ruban, démarrage *m* de la bande
~TELEPRINTER - [comput] téléimprimeur *m* à bande
~TENSION - [gen] tension *f* de la bande
~TENSION CONTROL - [gen] réglage *f* de la tension de la bande
~TENSION ROLLER - [text] rouleau *f* de tension des lanières
~TENSIONING ARM - [mech] bras *m* tendeur
~THICKNESS - [comput] (for a static magnetic storage) épaisseur *f* de bande
~THREADING - [comput] enfilage *m* du ruban magnétique
~-TO-CARD - [comput] bande-à-carte *f*
~TRANSPORT - [comput] transport *m* de la bande
~TRANSPORT LOCKING CATCH - [gen] verrouillage *m* du transport de la bande
~TRAVEL - [comput] (the direction in which the magnetic tape is read) direction *f* de lecture
~WEAVING - [text] fabrication *f* de sangles
~WIDTH - [comput] largeur *f* de bande
[gen] largeur *f* de bande
~-WOUND CORE - [comput] (for static magnetic storage) tore *m* bobiné, tore *m* coupé
TAPELINE - s. tape-line
TAPER, to - [gen] effiler, amincir
[arch] fuseler, contracturer, diminuer
[mech] côner, rendre conique
[metall] dépouiller
TAPER - [gen] (diminishing in cross-section with length) cône *m*, conicité *f*
[mech] (a form given to a part, such that its diameter, or one or both of its other linear dimensions, decreases with length) conique
[impl] pic *m*, pointerolle *f*
[plumb] raccord *m* conique
[metall] angle *m* de retrait
~ATTACHMENT - [mach tool] (of a lathe) tirette *f*
~BASE RIM - [rubber ind] jante *f* à repos de talon conique
~-BELLOWS CAMERA - [photo] chambre *f* à soufflet conique
~BIT - [tool] alésoir *m* conique
~BOLT - [mech] boulon *m* conique
~BROACH - [tool] équarissoir *m* conique
~CHECK - [mech] contrôle *m* de la conicité
~FILE - [tool] lime *f* pointue, lime *f* conique
~FIT - [mech] couplage *m* conique
~FORGING AND ROLLING MACHINE - [mech] machine *f* à forger et à laminer les pièces coniques
~GAUGE - [tool] (mechanical tool designed to check the taper) calibre *m* conique
~GIB - [mech] contre-clavette *f* conique
~HOLE - [mech] trou *m* conique
~PIN - [mech] (a pin, usually cylindrical, of which the diameter decreases with its length) goupille *f* conique
~-PIN DRILL - [tool] foret *m* à teton conique
TAPER-PIN REAMER - [tool] alésoir *m* pour trous de goupilles coniques
~PIPE - [plumb] raccord *m* conique
~PLUG COCK - [plumb] (a plug cock) robinet *m* à tournant, robinet *m* à boisseau
~PLUG VALVE - [plumb] (inverted pattern) robinet *m* à tournant inversé
~RATIO - [mech] conicité *f*
~REAMER - [tool] alésoir *m* conique
~ROLLER BEARING - [mech] (a roller bearing in which the rollers are tapered, to take some degree of axial thrust) roulement *m* à rouleaux coniques
~-SAW FILE - s. taper file
~-SCREW CHUCK - [mach tool] mandrin *m* à queue de cochon
~SHANK - [tool] queue *f* conique
~SHANK MILL - [mach tool] fraise *f* à queue conique
~-SHANK TWIST-DRILL - [tool] foret *m* hélicoïdal à queue conique
~SLEEVE - [mech] manchon *m* conique
~SQUARE SHANK-TWIST DRILL - [tool] foret *m* hélicoïdal à queue carrée
~TAP - [mech] (a tap of tapering form, used to make the first cut in forming an internal thread) taraud *m* conique
~THREAD - [mech] pas *m* conique de vis, filet *m* conique
~-TURNING - [mach tool] (the operation of forming a tapered part in a lathe) tournage *m* conique
~-TURNING ATTACHMENT - [tool] appareil *m* à charioter conique
~WASHER - [mech] (a washer which tapers in a direction at right angles to the axis of the hole in it) rondelle *f* conique
~WINDING - [text] renvidage *m* conique
TAPERED - [gen & mech] conique, côné, diminué, à section décroissante
~BARS - [metall] barres *fpl* coniques, barres *fpl* effilées
~BOBBIN - [text] bobine *f* bouteille
~BONNET - [auto] capot *m* en coupe-vent
~CASCADE - [nucl] cascade *f* à volume d'étage croissant et décroissant
~CHEESE - [text] bobine *f* croisée en pointe
~COMPRESSION SPRING - [mech] ressort *m* conique de compression
~DOUBLE-CUT FILE - [tool] lime *f* à taille croisée à angles arrondis
~FLASK - [metall] châssis *m* à démoulage intérieur
~JOURNAL -[mech] tourillon *m* conique
~OVERLAP - [mech] recouvrement *m* fuselé
~PAPER TUBE - [text] tube *m* conique en papier
~PLUG - [mech] (a plug of truncated conical form) bouchon *m* conique, tige *f* finissant en cône
~POTENTIOMETER - [instr] potentiomètre *m* non-linéaire
~RIBBON - [text] ruban *m* conique
~-RIDGE JUNCTION - [radio] (a tapered coaxial waveguide junction between two coaxial lines of different size which employs ridges to reduce the standing-wave ratio) jonction *f* à moulures progressives
~ROLLER - [mech] (of a roller bearing) rouleau *m* conique
~SECTION - [radio] (a waveguide element having a cross-section changing progressively in size and possibly in shape) section *f* à raccord progressif

TAPERED SHANK - [mech] queue *f* conique (d'outil)
~-SHANK DRILL - [tool] foret *m* à queue conique
~SPINDLE - [tool] broche *f* creuse
~SPLINE - [mech] clavette *f* conique
~STONE - [tool] (a rotating abrasive tool of conical form used for trimming a moulding) meule *f* conique
~STRAIGHT LINE SOURCE - [phys] (a non-uniform straight-line source in which the strength varies linearly from its value at the centre to zero at either end) source *f* rectiligne à intensité décroissante
~STRING - [oil ind] (of rods) train *m* de tiges
~SUNK KEY - [mech] clavette *f* encastrée conique
~TAP - [oil ind] taraud *m* aléseur de repêchage
~TENON - [carp] tenon *m* passant conique
~WAVEGUIDE - [radio] (a waveguide in which a physical or electrical characteristic changes continuously with distance along the axis of the guide) guide *m* d'ondes à raccord progressif
~WING - [aero] aile *f* trapezoïde
TAPERING - [gen mech etc] effilement *m* progressif, conique, fuselé, à section décroissante
~CURVE - [railw] courbe *f* de raccordement
~EQUIPMENT - s. taper attachment
~POD AUGER - [oil ind] tarière *f* en entonnoir
~ROPE - [mining] câble *m* diminué
TAPESTRY - [gen] tapisserie *f*
~BRAID - [text] galon *m* de tapissier
~CARPET - [text] tapis *m* bouclé
~HEALD - [text] lisse *f* pour tapisserie
~LOOM - [text] métier *m* pour tapisserie
~WEAVING - [text] tapisserie *f*
TAPEWORM - [zool] ténia *m*, ver *m* solitaire
~DISEASE - [vet] taeniase *f*
TAPHOLE - s. tap hole
TAPING - [el] revêtement *m* en ruban
TAPIOCA - [food] tapioca *m*
TAPIOLITE - [min] (niobate and tantalate of iron and manganese) tapiolite *f*
TAPIR MOUTH - [med] bouche *f* de tapir
TAPPED COIL - [el] enroulement *m* fractionné
~CONDENSER - [el] condensateur *m* à prises multiples
~FACEPLATE - [mech] plateau *m* à trous filetés
~POTENTIOMETER - [comput] potentiomètre *m* à prises
~VARIABLE INDUCTOR - [el] inductance *f* à prises variables
TAPPER - [telecomm] (of a telegraph key) manipulateur *m*
TAPPET - [mech] (a sliding element, interposed between a cam follower and a valve stem or push rod, or other reciprocating part) taquet *m*, mentonnet *m*, came *f* de distribution, poussoir *m* de tige de culbuteur
[text] (on a power loom, a cam for moving heddles) excentrique *m*, came *f*
[auto] poussoir *m*
~ADJUSTMENT - [mech] réglage *m* des poussoirs
~BOWL - [mech] galet *m* de l'excentrique
~BUILT UP IN SECTIONS - [text] excentrique *m* construit à segments
~CHAIN - [text] chaîne *f* des excentriques
~CLEARANCE - [mech] (the distance allowed between the valve tapped and the valve stem) jeu *m* des poussoirs
~DRIVING - [mech] mécanisme *m* de commande des excentriques
TAPPET FOR CLOSED SHEDDING - [text] excentrique *m* pour pas fermé
~GUIDE - [mech] (the bore in which the tappet reciprocates) guide *m* de soupape
~LEVER - [mech] levier *m* oscillant
~MOTION - [mech] mécanisme *m* des excentriques
~NOSE - [mech] bec *m* d'excentrique, nez *m* de chasse
~NOSE - [text] bec *m* de chasse
~ON CAM SHAFT - [text] came *f* sur l'arbre à excentriques
~ROLLER - [mech] galet *m* de poussoir
~SHAFT - [mech] arbre *m* inférieur, arbre *m* à cames
[text] arbre *m* à cames
~SHAPE - [mech] forme *f* de l'excentrique
~SHEAVE - [mech] élément *m* de l'excentrique
[text] élément *m* de course
~SLAB - [mech] plaque *f* pour poussoirs
~STEM - [mech] tige *f* de poussoir
~STEM GUIDE - [mech] guide *m* de la tige de poussoir
~SWITCH - [el] interrupteur *m* commandé par le moteur
~WHEEL - [mech] disque *m* à excentriques
~WRENCH - [tool] clef *f* pour le réglage des poussoirs
TAPPING - [el] (a lead brought out from a phase-winding at any point between the winding ends) prise *f*, soutirage *m* de courant
[bot] (of trees) incision *f*, gemmage *m*
[mech] taraudage *m*
[metall] coulée *f* du metal en fusion, débouchage *m*
[med] ponction *f*
[agric] (of wine) tirage *m* du vin
[hydr] perçage *m*, prise *f* d'eau, puisage *m* d'eau
~BAR - [metall] barre *f* à piquer, périer *m*, pince *f* de débouchage de trou de coulée
~CHUCK - [mach tool] mandrin *m* porte-taraud
~CONTACTOR - [el] contacteur *m* de prise
~FLOOR - [metall] (of a blast furnace) plateforme *f* de coulée
~HEAD - [mach tool] tête *f* à tarauder
~HOLE - [metall] s. tap hole
~-HOLE DRILLING - [metall] ouverture *f* du trou de coulée
~IRON - [metall] (steel bar used to open a furnace for drawing off slag or metal) ringard *m*
~KNIFE - [bot] couteau *m* à incision
~LOSS - [telecomm] perte *f* de branchement
~MACHINE - [mach] taraudeuse *f*, machine *f* à tarauder
[hydr] machine *f* à percer les conduites
~OF A SPRING - [hydr] captage *m* d'une source
~OF THE GROUND-WATER - [hydr] captage *m* des eaux souterraines
~OF THE SLAG - [metall] lâchage *m* de la scorie
~POINT - [el] prise *f* de dérivation, prise *f* de réglage
~SAMPLE - [metall] (specimen of molten metal taken as the furnace is tapped) prise *f* d'essai à la coulée
~SCREW - [mech] vis *f* auto-taraudeuse
~SHOOT - [metall] chenal *m* de coulée
~SHOVEL - [metall] pelle *f* de barrage
~SIDE - [metall] face *f* arrière
~SLAG - [metall] scorie *f* de coulée
~SLEEVE - [hydr] collier *m* de prise
~SPOUT - [metall] canal *m* de coulée, chenal *m* de coulée, rigole *f* de coulée
~STREAMS - [hydr] prises *f*pl d'eau sur cours d'eau
~TRANSFORMER - [el] transformateur *m* à prises

TAPPING SWITCH - [el] commutateur *m* de prise
~TEMPERATURE - [metall] température *f* de coulée
~UNIT - [oil ind] (of an oil refinery) installation *f* de distillation non-poussée
TAPPINGS - [metall] (the metal tapped from a blast surface) coulée *f*
TAR - [ind chem] (dark, bituminous substance obtained by the destructive distillation of wood, coal etc. It is a source material for a number of organic compounds) goudron *m*, bitume *m*, poix *f* liquide
~ACIDS - [chem] (generic term for compounds, such as phenols and cresols derived from coal-tar distillate) acides *mpl* du goudron
~BATTER - [ind chem] (pre-detarrer) prédégoudronneur *m*
~BITUMEN - [ind chem] bitume *m*
~BRUSH - [impl] brosse *f* à goudronner
~CAMPHOR - [chem] naphtaline *f*, naphtalène *m*
~CANCER - [med] cancer *m* du goudron
~CHISEL - [tool] raclette *f* à goudron
~COATING - [gen] enduit *m* de goudron
~COLLECTING MAIN - [ind chem] collecteur *m* de goudron
~DISTILLATE - [oil ind] distillat *m* de goudron
~EMULSION - [constr] (a road-paving material) émulsion *f* de goudron
~EXTRACTOR - [ind chem] séparateur *m* de goudron, dégoudronneur *m*
~-FOG FILTER - [ind chem] (for the washing of gas) tour *f* à copeaux
~FROM LIGNITE - [min] goudron *m* de lignite
~FROM SHALE - [min] goudron *m* de schiste
~IN THE HOT STATE, to - [text] goudronner à chaud
~ITCH - [med] dermatite *f* du goudron
~-MACADAM - [constr] macadam *m* de goudron
~MASTIC - [ind chem] mastic *m* de goudron
~OIL - [ind chem] huile *f* de goudron
~PAPER - [constr] papier *m* asphalté, carton *m* bitumé, papier *m* goudronné
~PASTEBOARD - [constr] carton *m* goudronné
~PLUG - [oil ind] vanne *f* pour résidus
~RABBLE - [tool] refouloir *m*
~RAKE - [tool] griffe *f* à goudron
~-RESISTING PAINT- [constr] couleur *f* résistante au goudron
~SCOOP - [impl] cuiller *f* à goudron
~SEAL - [mech] (in gasholders) garde *f* de goudron
~SEPARATION - [ind chem] dégoudronnage *m*
~SPRAY HOSE - [rubber ind] tuyau *m* pour le goudronnage
~-SPRAYING CAR - [auto] (for road construction work) véhicule *m* de goudronnage
~SPRAYING MACHINE - [constr] machine *f* arroseuse de goudron
~STILL - [oil ind] distillateur *m* de goudron
~THE HEMP YARN, to - [text] goudronner le fil de chanvre
~TOWER - [ind chem] (for the treatment of gas) garde *f* de goudron du barillet, garde *f* d'évacuation du goudron
TARANTISM - [med] tarentisme *m*, tarentulisme *m*
TARANTULA - [zool] tarentule *f*
TARAXACUM - [chem] (dandelion. The root of Taraxacum officinale, with applications in medicine) pissenlit *m*, taraxacum *m*
TARE, to - [meas] (to weight in order to determine the amount of tare) tarer, faire la tare
TARE - [meas] (the weight of any container, vehicle or the like when empty, subtracted from the weight of the loaded container or vehicle to obtain the contents) tare *f*
[meas] plomb *m* à tarer
~WEIGHT - [aero] (the standard weight of an aircraft, for flight, less crew, fuel, oil, pay load or officially removable equipment) poids *m* à vide de construction
TARED FILTER - [chem] filtre *m* taré
TARGET - [gen] cible *f*
[nucl] (substance or object exposed to bombardment or irradiation) cible *f*
[telev] (the electrode in the camera tube which is subject to bombardment by the electron beam) cible *f*
[radar] (the obstacle which reflects the emitted high-frequency wave) plaque *f* à accumulation
[railw] disque *m*
[surv] (of a target rod) voyant *m*
[paper man] (plate fitted in the blow-pit against which the softened chips are blown to disintegrate them) plaque *f* d'arrêt
[el] anticathode *f*
~ANGLE - [electron] (anode angle) angle *m* de l'anode, angle *m* du foyer
~AREA - [nucl] aire *f* de la cible
~BOARD - [photo] planchette *f* d'orientation de caméra
~CAPACITANCE - [electron] (the capacitance between the scanned area of the target and the backplate) capacité *f* de cible
~CHAMBER - [nucl] chambre *f* de bombardement
~COMPUTER - [comput] (computer for the target programme) machine *f* exécutrice
~CUTOFF VOLTAGE - [electron] (in camera tubes) tension *f* de seuil de la cible
~DEUTERON - [nucl] (deuteron emitted from a target) deutéron *m* de cible
~DISCRIMINATION - [radar] pouvoir *m* séparateur, discrimination *f* de cible
~GLINT - [radar] (scintillation; on a radar display, a rapid apparent displacement of the target from its mean position) déplacement *m* d'écho
~IMPACT BURNER - [th eng] brûleur *m* à surface d'impact
~LANGUAGE - [comput] language *m* d'exécution
~LAYER - [electron] couche *f* de la cible
~MATERIAL - [nucl] matière *f* de cible
~MESH - [electron] gaze *f* de cible
~NUCLEUS - [nucl] (the initially stationary atom or nucleus in a nuclear reaction) noyau-cible *m*
~OF A RELAY - [el] indicateur *m* d'un relais
~PARTICLE - [nucl] (a bombarded particle) particule *f* bombardée
~PHASE - [comput] phase *f* d'exécution
~PLANE - [gen] (vertical plane containing the target line) porte-cible *m*
[aero] avion-cible *m*
~PRICE - [comm] prix *m* indicatif, prix *m* d'orientation
~ROD - [surv] mire *f* de nivellement, jalon *m* d'arpentage
~SPOT - [electron] tache *f* sur la cible
~STRENGTH - [acoust] (in decibels) intensité *f* de la cible

TARGET THEORY - [radiat] (a theory explaining some biological effects of radiation on the basis of ionization occurring in a very small sensitive region within the cell) théorie *f* de la cible
~VANE - [surv] pinnule *f*, viseur *m*
~VOLTAGE - [electron] (the potential difference between the thermionic cathode and the backplate in a camera tube with low-velocity scanning) tension *f* de cible
TARIFF - [comm etc] tarif *m*
~BARRIER - [comm] barrières *f*pl douanières
TARLATAN - s. tarlton
TARLTON - [text] tarlatane *m*
TARMAC - [constr] (short for tarmacadam) tarmac *m*
[aero] piste *f* en dur
TARMACADAM - s. tar-macadam
TARNISH, to - [gen] ternir, salir
[metall] se ternir
TARNISH - [gen] ternissure *f*
[metall] (discolouration produced on an exposed metal surface due to the formation of oxides and sulphides) ternissage *m*
TARNISHING - [paint] dédorage
TARO - [bot] taro *m*
TARP HOOK - [auto] (body fitting to which ropes are secured when lashing down tarpaulin coverings) crochet *m* de bâche, boucleteau *m* de bâche
TARPAULIN - [text] (sheet or heavy waterproof canvas used as a temporary covering) bâche *f*, toile *f* goudronnée
[naut] prélart *m*
~COATED HOSE - [rubber ind] tuyau *m* à robe extérieure en prélart
TARRAGON - [bot] estragon *m*
TARRED - [gen] goudronné, enduit *m* de goudron
~BOARD - [constr] carton *m* bitumé
~FELT - [constr] carton *m* bitumé
~OAKUM - [text] étoupe *f* noire, étoupe *f* goudronnée
~ROPE - [text] cordage *m* noir, cordage *m* goudronné
~STEEL SHEET - [metall] tôle *f* en acier goudronnée
TARRING - [gen] goudronnage *m*
~COLD - [text] goudronnage *m* à froid
~IN THE ROPE - [text] goudronnage *m* du câble entier, goudronnage *m* du câble en paquet
~IN THE YARN - [text] goudronnage *m* par fils
TARRY - [gen] (covered with tar) goudronneux, bitumeux
[gen] (like tar) bitumineux
[gen] attente *f*, séjour *m*
[mach tool] (dwell control) commande *f* d'arrêt momentané
~RESIDUE - [ind chem] residu *m* bitumeux
TARSECTOMY - [med] tarsectomie *f*
TARSOMETATARSAL - [anat] tarso-métatarsien
TARSUS - [anat] tarse *m*
TARTAN - [text] (a Scottish woollen fabric having multicoloured lines or stripes at right angle) tartan *m*
[naut] tartane *f*
TARTAR - [chem] (potassium antimonyl tartrate, it has application as a mordant in dyeing and as a cathartic in medicine) tartre *m*
[dent] tartre *m*
~DEPOSIT - [agric] dépôt *m* de tartre brut
TARTAR EMETIC - [chem] (dihydroxysuccinic acid. Colourless acid-tasting crystals, with application in photography, medicine, and leather and textile processing) tartre *m* stibié, tartrate *m* de potasse et d'antimoine
TARTARIC - [chem] tartarique, tartrique
~ACID - [chem] (dihydroxy-succinic acid: occurs in many plants and fruits; exists in four modifications, d-tartaric acid, M.P. 170° C.; l-tartaric acid, M.P. 170 ° C.; racemic or dl-tartaric acid M.P. 206°C. and mesotartaric acid M.P. 143 ° C) acide *m* tartarique
TARTRATE - [chem] (salt or ester of tartaric acid) tartrate *m*
TASCHENGURT - [aero] (short length of tape or webbing, run outside and across rigging lines to join, adjacent lobes of the hem in a parachute) martingale *f*
TASIMETER - [instr] (modified microphone designed to detect or measure minute extensions of movements of solid bodies) tasimètre *m*, microtasimètre *m*
TASK, to - [gen] assigner une tâche
[naut] (to test the solidity of a vessel) mettre à l'épreuve
TASK - [gen] tâche *f*, devoir, travail *m*
~CONTROL BLOCK - [contr] bloc *m* de commande de tâche
~DISPATCHER - [comput] distributeur *m* de tâches
~MANAGEMENT - [comput] distribution *f* de tâches
~QUEUE - [comput] file *f* de tâches
~TIME - [gen] temps *m* d'une opération
~WAGE - [gen] salaire *m* de travail à la tâche
~WORK - [gen] travail *m* aux pièces, travail *m* à la tâche
TASKTREE - [bot] (the trees of a tapping task) arbre *m* d'une tâche de saignée
TASKWORK - s. task work
TASMANIAN WOOL - [text] laine *f* de la Tasmanie
TASMANITE - [min] (a type of practically pure spore coal) tasmanite *f*
TASSEL, to - [bot] former des aigrettes
TASSEL - [gen] gland *m*, houppe *f*, floc *m*
[bookbind] signet *m*
[bot] aigrette *f* du maïs, panicule *f* terminale
TASTE, to - [gen] percevoir la saveur
[gen] (to have the taste of) sentir
[food] goûter
TASTE - [gen] goût *m*, saveur *f*
~BUDS - [anat] papilles *f*pl linguales
~CORPUSCLES - [anat] cellules *f*pl gustatives
TASTELESS - [gen] insipide, sans goût
TASTER - [gen] (one who tests the quality of something for trade) tâteur *m*
[impl] tasse *f* à déguster, sonde *f* à vin
TAURINE - [chem] (an amino-acid with uses as a synthesis intermediate) taurine *f*
TAUT - [gen] roide, raide
[gen] (of a rope) tendu
[naut] (of a sail) étarque
~PICK - [text] trame *f* tachée
~TAPE ATTACHMENT - [comput] (mechanism attached to a tape reader to prevent the further transport of the tape when the part to be read is in a taut condition) dispositif *m* d'arrêt pour bande raraidie
TAUTEN, to - [gen] (e.g. of cables, ropes etc) rai-

dir, roidir, embraquer, tendre
TAUTEN, to - [naut] étarquer
TAUTOMERIC - [chem] (of tautomerism) tautomère
~TRANSFORMATION - [phys chem] transformation *f* tautomère
TAUTOMERISM - [chem] (the phenomenon of certain compounds which exist as equilibrium mixtures of two isomeric or tautomers) tautomérie *f*
TAW, to - [leather ind] mégir, chamoiser
TAWERY - [leather ind] mégisserie *f*
TAWING - [leather ind] (the process by which lamb, kid and deer skins are tanned) tannage *m* à l'alun, mégisserie *f*, chamoisage *m*
TAX, to - [gen & leg] taxer, mettre un impôt
TAX - [gen & leg] impôt *m*, taxe *f*
~COLLECTOR - [leg] percepteur *m*, receveur *m*
~FREE - [comm etc] exempt d'impôts
~-EXEMPT - s. tax free
~EXEMPTION - [leg] exonération *f* fiscale
~-PAYER - [gen] contribuable *m*
~REMISSION - [leg] réduction *f* des impôts
TAXABLE - [leg] imposable, à la charge de
TAXAMETER - s. taximeter
TAXATION - [leg] imposition *f*, charges *fpl* fiscales
TAXI, to - [aero] (of an aircraft, to travel on the ground under its own power(excluding take-off and landing runs) rouler au sol
[aero] (of a seaplane) hydroplaner
[auto] aller en taxi
TAXI - [auto] taxi *m*
~CHANNEL - [aero] (a channel in a water aerodrome reserved for aircraft taxying) chenal *m* de circulation
~-CHANNEL MARKERS - [aero] (markers demarcating a taxi channel) balises *fpl* de chenal de circulation
~HOLDING POSITION - [aero] (a position at which a taxying aircraft is required to await permission for further movement) point *m* d'attente de circulation
~LIGHT - s. taxi track lights
~TRACK - s. taxiway
~TRACK LIGHTS - [aero] (lights showing the position of a taxi track) feux *mpl* de voie de circulation
TAXICAB - s. taxi
TAXIDERMY - [gen] (the art of stuffing and mounting the skins of dead animals) taxidermie *f*
TAXIMETER - [instr] (instrument used in taxicabs to indicate the fare and the distance) taximètre *m*
TAXIPLANE - [aero] avion-taxi *m*
TAXIS - [zool] (movement of an organism towards or away from a stimulus) taxis *m*
[med] (of hernia) taxis *m*
[biol] (tropism) tropisme *m*
TAXIWAY - [aero] (a marked path for taxing aircraft on a land aerodrome) voie *f* de circulation
TAXYING - [aero] (movement of an aircraft under its own power on the ground, excluding take-off and landing runs) roulement *m* au sol
TAXONOMIC SERIES - [biol] (the range of extant living organisms) série *f* taxonomique
TAXONOMY - [biol] (the science of classification as applied to living organisms) taxonomie *f*, taxologie *f*
TAYLORISM - [ind] (a method of work organization) système *m* de Taylor
T.C.TUBBER - [rubber ind] (Technically Classified Rubber) caoutchoucs *mpl* spécifiés

TE MODE - [electron] (Transverse Electrode mode, in a waveguide) mode *m* électrique transversal
TE WAVE - [electron] (Transverse Electrode Wave, in a homogeneous isotropic medium) onde *f* H, onde *f* électrique transversale, onde *f* T
Tb - [chem] (symbol for Terbium) terbium *m*
Te - [chem] (symbol for tellurium) tellure *m*
TEA - [bot] (the tea plant) thé *m*
TEACH, to - [gen] instruire, enseigner
TEAK - [bot] (East-Indian tree yielding a very hard and durable timber) teck *m*, chêne *f* des Indes
TEAM, to - [agric] atteler
TEAM - [gen] attelage *m*
[gen] (set of workers or players) équipe *f*
~OF OXEN - [zool] attelage *m* de boeufs
~PLOUGH - [agric] charrue *f* à traction animale, charrue *f* à attelage
TEAMWORK - [gen] travail *m* d'équipe
[gen] (the system of production, whereby the suggestions of the workers are taken into account) collaboration *f*
[telecomm] entr'aide *f*
TEAMSTER - [agric] charretier *m*
TEAPOY - [carp] petite table *f* trois pieds
TEAR, to - [gen] déchirer, arracher
TEAR - [gen] déchirement *m*, déchirure *f*
[gen] (a drop) goutte *f*
[physiol] (the drop of saline liquid secreted by the lacrimal gland) larme *f*
[metall] rompre par traction
[glass] inclusion *f*
[metall] inclusion *f*
~APART, to - [gen] séparer, arracher
~CHIPS - [carp] éclats *mpl*
~DUCT - [anat] conduit *m* lacrymal
~FIBRES, to - [text] déchirer les fibres
~GAS - [chem] gaz *m* lacrymogène
~GLAND - [anat] glande *f* lacrymale
~INITIATION - [rubber ind] (the beginning of a tear) initiation *f* du déchirement
~-OFF CAP - [aero] (a piece of fabric closing a parachute pack, designed to be torn away by the static line to allow deployment) patte *f* de déchirure
~OUT, to - [mining] démonter
~PROPAGATION - [rubber ind & text] (the development of a tear in a material) propagation *f* de la déchirure
~PROPAGATION TEST - [rubber ind] (a test made to determine the manner or rate of increase of a tear) essai *m* de propagation de la déchirure
~RESISTANCE - [gen] (the strength of a material against tearing action) résistance *f* à la déchirure
~STRENGTH - s. tear resistance
~TEST - [mech] essai *m* de déchirure
~TEST PIECE CUTTER - [rubber ind] machine *f* à découper les éprouvettes pour essais de résistance au déchirement
~THE WOOL FIBRE, to - [text] déchirer la fibre de laine
TEARABILITY - [phys] (the property of yielding to tearing action) propriété *f* du déchirement
TEARDROP BALLOON - [aero] ballon *m* en forme de larme
~MANIFOLD - [plast ind] (manifold in a sheet extruder die of approximately pear-shaped cross section) tubulure *f* en forme de poire
TEARING - [gen] (removal by force) déchirure *f*

TEARING - [telev] (break up of a section of a television picture by intermittent failure of the line synchronizing signal) déchirage *m* de l'image, déchiquetage *m* de l'image
~CYLINDER - [text] tambour *m* à pointes
~RESISTANCE - s. tear resistance
~TEST - [mech] (a trial of the extent to which a material can be torn) essai *m* de déchirure
~UP ROLLER - [text] briseur *m*
TEASE, to - [text] carder
TEASEL, to - [text] lainer, chardonner, garnir
TEASEL RAISING MACHINE - [text] laineuse *f* à chardons roulants
TEASELING MACHINE - [text] laineuse *f* à chardons roulants, laineuse *f* à hérissons, chardonneuse *f*
TEASER - [text] batteuse *f*, machine *f* de battage
[glass man] (worker regulating charging of batches and adjusting fires when operating a furnace in glass manufacturing) enfourneur *m*
~TRANSFORMER - [el] (main or Scott transformer) transformateur *m* principal, transformateur *m* Scott
TEASING MACHINE - s. teaser [text]
TEASPOONFUL - [med] cuillerée *f* à thé
TEAT - [zool] mamelon *m*, téton *m*, bout *m* de sein
[rubber ind] tétine *f*
[zool] (of cows) trayon *m*
~ACCESSORY - [rubber ind] tétine *f* supplémentaire
~END - [gas ind] (of an appliance) embout *m* cannelé
TECHNETIUM - [chem] (a radio-active, metallic, element, symbol Tc, A.N.43 occuring as a fission product of plutonium and uranium) technétium *m*
TECHNIC - s. technical
TECHNICAL - [gen] technique, technologique
~AERONAUTICS - [aero] aérotechnique *f*
~CHEMISTRY - [ind chem] chimie *f* industrielle
~DATA - [gen] données *f*pl techniques
~DIFFICULTY - [leg] question *f* de procédure
~DIRECTOR - [cin & telev] directeur *m* de production
~FEATURES - [mech] caractéristiques *f*pl de construction
~FIBRE - [text] fibre *f* technique
~INSTITUTE - [gen] école *f* des arts et métiers
~JOURNAL - [print] journal *m* technique
~MANAGER - [gen] directeur *m* technique
~OPERATIONS MANAGER - [cin] directeur *m* technique, directeur *m* de production
TECHNICIAN - [gen] technicien *m*
TECHNICOLOUR - [cin] (a system of colour cinematography) technicolor *m*
TECHNIQUE - [gen] technique *f*
[mech] mécanisme *m*
TECHNOLOGICAL - [ind] technologique
TECHNOLOGIST - [gen] technologue *m*
TECHNOLOGY - [ind] (theoretical knowledge of industry and industrial art) technologie *f*
TECNETRON - [electron] (semiconductor device similar to a triode in that it has cathode and anode connections at opposite end of a small rod of germanium) tecnétron *m*
TECTONIC - [constr] (pertaining to building of construction) conception *f* architectonique
[geol] (said of rock structures which are directly attributable to earth movements)tectonique (formé par accumulation)
TECTONICS - [constr] architectonique *f*
[geol] (the study of rock structures which are attributable to earth movements) tectonique *f*
TECTORIUM - [anat] membrane *f* recouvrante de l'organe de Corti
TECTUM - [anat] toit *m*
TED, to - [agric] sauter (le foin), faner
TEDDER - [agric] faneuse *f*
TEDGE - [metall] entonnoit *m* d'alimentation
TEE - [plumb] (a pipe fitting provided with two internal threads in line and one at right angles to these, for connecting two pipes in line with one at right angles to them) té *m*
[el] (a coupler with an additional opening serving to connect three adjacent lengths of conduit, two being in line with one another and the third at an angle of 90° to the other two) T *m*
~BOLT - [mech] boulon *m* à T
~FOR CLEARING SERVICES - [gas ind] (for service pipes and risers) té *m* de nettoyage, té *m* de branchement
~JOINT - [el] té *m* de dérivation
~-JOINT OF A CABLE - [el] boîte *f* de dérivation en T, embranchement *m* d'un câble
~-PIECE - [mech] raccord *m* à té
~SLOT - [mech] encoche *f* à té
~SQUARE - [impl] (a drawing implement) équerre *f* à té
~-WELDING - [mech] soudure *f* à té
TEED FEEDER - [el] feeder *m* à sorties multiples
TEEM, to - [gen] foisonner, abonder
[gen] (to pour out) verser
[metall] (the steel in a mould) couler, verser
TEEMING - [metall] (in foundry work) coulée *f*
~ARREST - s. teeming lap
~LAP - [metall] pli *m* de coulée
~NOZZLE - [metall] orifice *m* de coulée
~PUNCH - [impl] chasse-boulon *m*
~TROUGH - [mining] agoge *m*, arrugie *f*, cunette *f* d'écoulement
~TEMPERATURE - [metall] température *f* de coulée
TEETH - [gen] (the plural of tooth) dents *f*pl
[mech] denture *f*
~-CUTTING MACHINE - [mach tool] machine *f* à tailler les engrenages
~-OF LIFTER - [text] denture *f* de la platine
~OF THE CROWN - [text] denture *f* de la couronne
~OF THE WIND - [naut] lit *m* du vent
~PER INCH - [meas] (the number of teeth in one inch of blade length of saw, used as a measure of fineness) dents *f*pl par pouce
TEETHING - [med] dentition *f*
TEG - [zool] agneau *m* antenais
TEGMEN - [bot] tegmen *m*
TEGMENTAL - [med] tégumentaire
~SYNDROME - [med] syndrome *m* de la calotte
TEGUMENT - [bot] (the layer of cells covering the surface of a plant) tégument *m*
TEGULA - [zool] tégule *f* (de l'aile antérieure)
TEHUANTPECER - [met] (very strong northerly winter wind occurring chiefly over the Gulf of Tehuantepec) tehuantpecer *m*
TEL - [chem] (short for tetraethyl lead) plomb *m* tétraéthylique
TELAMON - [arch] télamon *m*, atlante *m*
TELANGITIS - [med] inflammation *f* des capillaires
TELANGIOMA - [med] tumeur *f* capillaire
TELAUTOGRAPH - [telecomm] télautographe *m*, télé-imprimeur
TELAUTOGRAPHY - [telecomm] (the transmission of

half-tone images by scanning electrically a gum print on metal foil: the reception is either by photographic scanning with a modulated light beam, or by electrolysis in prepared paper) télautographie *f*
TELEAMMETER - [instr] (instrument with which it is possible to measure current in a circuit from a distance) téléampèremètre *m*
TELEBREAKING - [el] télérupture *f*
TELECAST, to - [telev] (to transmit a programme by television) diffuser par la télévision
TELECAST - [telev] (a programme broadcast by television) télétransmission *f*
TELECASTING - s. telecast
TELECENTRIC LENS - [photo] objectif *m* télécentrique
~SYSTEM - [opt] (a telescopic system with the aperture stop placed at one of the foci of the objective lens) système *m* télécentrique
TELECINE PROJECTOR - [telev] (an apparatus which projects an enlarged image of a television receiver on a screen) analyseur *m* de film, télécinema *m*
TELECOMMUNICATION - [telecomm] (any communication of information by electrical means, either by wire or by radio) télécommunication *f*
~NETWORK - [telecomm] réseau *m* de télécommunications
~SYSTEM - s. telecommunication network
~TECHNICS - [telecomm] technique *f* des télécommunications
~TECHNIQUE - s. telecommunication technics
~SERVICE - [telecomm] service *m* de télécommunications
TELECONTROL - [contr] (control by mechanical devices remotely, either by radio, sound-waves or beams of light) télécommande *f*, télémécanique *f*
~OF STEERING GEAR - [contr] appareil *m* à gouverner électrique
TELECONTROLLED SUB-STATION - [telecomm] sous-station *f* commandée à distance
TELECURIETHERAPY UNIT - [radiat] appareil *m* de télécuriethérapie
TELEDIFFUSION - [telecomm] télédiffusion *f*
TELEDYNAMIC - [telecomm] télédynamique
TELEFILM - [telev] (a film projector and television camera used to pick up scenes from motion picture films) téléfilm *m*
TELEGENIC - [telev] (of an actor, or television personality) télégénique, photogénique
TELEGONY - [genet] (the alleged influence of a previous sire on the progeny of the same mother from subsequent mating with other males) télégonie *f*, hérédité *f* d'influence
TELEGRAM - [telecomm] télégramme *m*
TELEGRAPH, to - [telecomm] télégraphier
TELEGRAPH - [telecomm] télégraphe *m*
~ALPHABET - [telecomm] alphabet *m* télégraphique
~CABLE - [telecomm] câble *m* télégraphique
~CHANNEL - [telecomm] voie *f* télégraphique
~CHARACTER - [comput] signal *m* télégraphique
~CIRCUIT - [telecomm] circuit *m* télégraphique, voie *f* de télécommunications
~CODE - [telecomm] code *m* télégraphique
~FORM - [telecomm] formule *f* de télégrammes
~LANGUAGE - [comput] code *m* télégraphique
~LINE - [telecomm] ligne *f* télégraphique
~MAGNIFIER - [telecomm] préamplificateur *m* télégraphique
TELEGRAPH MODULATED WAVES - [telecomm] ondes *fpl* à modulation en télégraphie
~MODULATOR - [telecomm] modulateur *m* télégraphique
~NETWORK - [telecomm] réseau *m* télégraphique
~NOISE - [telecomm] bruit *m* de télégraphie
~OFFICE - [telecomm] bureau *m* télégraphique
~OPERATOR - [telecomm] télégraphiste *m*
~POLE - [telecomm] poteau *m* télégraphique
~POST - [telecomm] poteau *m* télégraphique
~PRINTER - [telecomm] téléimprimeur *m*
~RATE - [telecomm] tarif *m* de télégrammes
~REGENERATIVE REPEATER - [telecomm] répéteur *m* régénérateur télégraphique
~REPEATER - [telecomm] translateur *m*
~TRANSMISSION POTENTIAL - [telecomm] tension *f* émettrice télégraphique
TELEGRAPHIST - [telecomm] télégraphiste *m*
TELEGRAPHY - [telecomm] (electrical communication system whereby messages are transmitted by coded signals) télégraphie *f*
TELEICONOGRAPHY - [telecomm] phototélégraphie *f*, radiotransmission *f* des images, téléiconographie *f*
TELEINDICATOR OF LEVEL - [telecomm] indicateur *m* à niveau à distance
TELEMECHANICS - [contr] (the theory and practice of operating mechanisms from a distance) télécommande *f*, télémécanique *f*
TELEMETER, to - [meas] (to transmit measurements over a distance) télémesurer
[comput] télétransmettre de données de mesure
TELEMETER - [meas] (instrument designed to measure the distance of an object from an observer) télémètre *m*
[instr] (an instrument for the remote indication of electrical quantities) appareil *m* à télémesurer
TELEMETERING DEVICE - [meas] appareil *m* à télémesurer
TELEMETRY - [meas] télémétrie *f*
~ELSSE - [astronaut] indicateur *m* d'erreur de trajectoire déterminé par télémesure
TELEMOTOR - [mech] (hydraulic or electrical device by which power is applied at a distance) télémoteur *m*
[naut] transmission *f* de barre hydraulique
TELEOBJECTIVE - [photo] téléobjectif *m*
TELEOLOGY - [biol] (the study of animal and plant structures in terms of purpose and utility) téléologie *f*
TELEPHONE, to - [telecomm] (to converse by telephone or to send a message by telephone) téléphoner
TELEPHONE - [telecomm] (a combination of apparatus for conveying speech over a distance) téléphone *m*
[telecomm] poste *m* d'abonné, poste téléphonique
~AMPLIFIER - [telecomm] (combination of one or moreamplifiers, together with their associated equipment, for use in a telephone circuit) amplificateur *m* téléphonique
~APPARATUS WITH HOMING POSITION - [telecomm] appareil *m* téléphonique à position de repos
~BLOCK SYSTEM - [telecomm] cantonnement *m* téléphonique
~BOOTH - [telecomm] cabine *f* téléphonique
~BOX - s. telephone booth
~CALL - [telecomm] communication *f* téléphonique

TELEPHONE CHANNEL - [telecomm] voie *f* de communication téléphonique
- ~CIRCUIT - [telecomm] (a permanent electrical connection permitting the establishment of a telephone communication in both directions between two telephone exchanges) circuit *m* téléphonique
- ~CIRCUIT WITH TELEGRAPH CIRCUIT SUPERPOSED - [telecomm] circuit *m* approprié simultanément à la télégraphie et à la téléphonie
- ~CONDENSER - [radio] condensateur *m* téléphonique
- ~CONVERSATION - [telecomm] communication *f* téléphonique, conversation *f* téléphonique
- ~CONVERSATION RECORDING - [telecomm] enregistrement *m* des communications téléphoniques
- ~CORD - [telecomm] cordon *m* de téléphone
- ~CURRENT FORM FACTOR - [telecomm] facteur *m* téléphonique de forme du courant
- ~DIRECTORY - [print] annuaire *m* des abonnés au téléphone
- ~DROP - [telecomm] branchement *m* d'abonné
- ~EARPHONE - s. telephone receiver
- ~EXCHANGE - [telecomm] bureau *m* central téléphonique, bureau *m*
- ~FREQUENCY - [telecomm] (voice frequency) fréquence *f* vocale, fréquence *f* téléphonique
- ~HEADGEAR RECEIVER - [telecomm] casque *m* téléphonique, récepteur *m* serre-tête
- ~INFLUENCE FACTOR - [telecomm] (of a voltage or current wave in an electric supply circuit, the ratio of the square root of the sum of the squares of the weighted rms values of all the sine wave components to the rms value of the entire wave) facteur *m* TIF, facteur *m* téléphonique de forme de la tension
- ~INSTALLATION - [telecomm] installation *f* téléphonique
- ~KIOSK - [telecomm] cabine *f* téléphonique
- ~LIGHTNING PROTECTOR - [telecomm] parafoudre *m* téléphonique
- ~OPERATION - [telecomm] exploitation *f* téléphonique
- ~OPERATOR - [telecomm] téléphoniste *m*
- ~PICK-UP - [el acoust] (a type of induction coil device which slips over a telephone receiver and picks up both voices during a telephone conversation for recording on tape) bobine *f* d'écoute pour téléphone
- ~PLANT - [telecomm] installation *f* téléphonique
- ~POLE - [telecomm] poteau *m* téléphonique
- ~POST - s. telephone pole
- ~RECEIVER - [telecomm] (component part of a telephone in which a diaphragm is caused to vibrate by electric impulses, converting the varying current into sound) écouteur *m*
- ~REPEATER - [telecomm] répéteur *m* téléphonique, amplificateur *m* téléphonique
- ~SET - [telecomm] poste *m* téléphonique, appareil *m* téléphonique
- ~SET WITH DIAL - [telecomm] poste *m* téléphonique à cadran d'appel
- ~SIGNAL - [telecomm] signal *m* téléphonique
- ~SWITCHHOOK - [telecomm] crochet *m* de commutation
- ~-TELEGRAM - [telecomm] télégramme *m* transmis par téléphone
- ~-TELEGRAM CIRCUIT - [telecomm] (a circuit between two telegraph offices, over which telegrams are passed by telephone) circuit *m* de transmission de télégrammes par téléphone

TELEPHONE TRAFFIC - [telecomm] trafic *m* téléphonique
- ~TRANSMITTER - [telecomm] (a microphone for use in a telephone system) émetteur *m* microphonique, microphone *m*
- ~VOICE RECORDER - [telecomm] dispositif *m* d'enregistrement des conversations téléphoniques
- ~VOLTAGE FORM FACTOR - [telecomm] facteur *m* téléphonique de forme de la tension
- ~WIRE - [telecomm] fil *m* téléphonique
- ~WORKING - [telecomm] exploitation *f* téléphonique

TELEPHONIC - [telecomm] téléphonique
- ~CONNECTION - [telecomm] communication *f* téléphonique
- ~ECHO - [telecomm] écho *m* téléphonique

TELEPHONOGRAPH - [el acoust] (apparatus designed to record telephone messages) téléphonographe *m*

TELEPHONY - [telecomm] (the process of communication by telephone) téléphonie *f*
- ~AMPLIFICATION - [telecomm] (implying that the transmitting value is excited by an A.F. modulated excitation voltage) amplification *f* téléphonique
- ~TRANSMITTER - [telecomm] émetteur *m* radiophonique

TELEPHOTO - s. telephotography
- ~LENS - [opt] (a lens of long focal-length and narrow angle, for obtaining images of distant or very distant objects) téléobjectif *m*

TELEPHOTOGRAPH - s. telephotography

TELEPHOTOGRAPHY - [photo] (branch of photography which involved the use of a camera having a lens analogous to a telescope) téléphotographie *f*, photographie *m* à grandes distances

TELEPOSITIVE LENS - [opt] lentille *f* télépositive

TELEPRINTER - [telecomm] (form of telegraph transmitter having a typewriter keyboard and a type-printing telegraph receiver) téléimprimeur *m*, télétype *m*
- ~CODE - [telecomm] (internationnaly agreed method of writing decimal digits) code *m* télex

TELEPRINTING SYSTEM - [telecomm] système *m* téléimprimeur

TELERADIOGRAPHY - [med] (radiography with the X-ray tube at a distance from the body) téléradiographie *f*

TELERADIUM UNIT - [radiat] (a telecurietherapy containing radium) appareil *m* pour téléradiumthérapie

TELERAN - [radar] (television radar navigation) téléran *m*

TELERECORDING EQUIPMENT - [telev] vidigraphe *m*

TELESCOPE, to - [gen] télescoper
[mech] (to slide into) s'emboiter
[gen] (to shorten) raccourcir
[transp] (in a collision, to enter lengthwise into one another) têlescoper, se télescoper

TELESCOPE - [instr] (an optical instrument for making distant objects appear nearer) lunette *f*, lunette *f* d'approche
[surv] (a topographical instrument) lunette *f* viseur
- ~JACK - [mech] vérin-télescope *m*
- ~JOINT - [mech] joint *m* télescopique
- ~TUBE - [electron] télescope *m* électronique à rayons infrarouges

TELESCOPIC - [gen] télescopique
- ~ALIDADE - [surv] alidade *f* à lunette

TELESCOPIC CASING - [mining] tubage *m* télescopique
~FINDER - [photo] viseur *m* à longue vue
~FUNNEL - [naut] cheminée *f* à télescope
~GASHOLDER - [gas ind] (a multiple lift gasholder) gazomètre *m* télescopique
~JACK - [mech] (a type of jack having a series of extensible concentric sleeves) vérin *m* télescopique
~LEG - [mech] (of a tripod) branche *f* coulissante
~LEG LANDING GEAR - [aero] (type of landing gear leg in which the shock absorber system acts by telescopic movement) train *m* d'atterrissage à système amortisseur télescopique
~LEVELLING STAFF - [surv] mire *f* à emboîtement
~MAST - [radio] (a type of aerial mast; an extension mast) mât *m* d'antenne télescopique
~SYSTEM - [opt] (a combination of objective and ocular with which distant objects may be observed visually, photographically or by other detecting means) système *m* télescopique
~TOOLHOLDER - [mach tool] porte-outils *m* à télescope
~TRIPOD - [photo] trépied *m* à emboîtement
~VIEW-FINDER - [photo] viseur *m* à lunette
TELESCOPING - [gen] (as a consequence of a collision etc) télescopage *m*
[adj] télescopique
~FENDER - [gasholder] virole *f* télescopique
TELESPECTROSCOPE - [opt] téléspectroscope *m*
TELESCREEN - [telev] écran *m* de téléviseur
TELETHERMOMETER - [instr] téléthermomètre *m*
TELETICKETTING - [comput] téléréservation *f*
TELETYPE, to - [telecomm] téléimprimer
TELETYPEWRITER - [telecomm] téléimprimeur *m*
TELETYPING - [telecomm] télétypie *f*
TELEVIEWER - [telev] téléspectateur *m*
TELEVISE, to - [telev] (to transmit an image by television) téléviser
TELEVISION - [telev] (a system whereby an optical image is translated into electrical signals for transmission by radio or other wires, after which the signals are translated back to light rays to form a reproduction of the original image on the receiver screen) télévision *f*
~BAND - [telev] (television broadcast band) bande *f* de télévision
~BROADCAST BAND - s. television band
~BROADCAST STATION - [telev] station *f* émettrice de télévision
~CAMERA - [telev] (apparatus with camera tube and the electrical and mechanical equipment) caméra *f* de télévision
~CHANNEL - [telev] (the frequency range accommodating a particular television broadcast) canal *m* de télévision
~CHART - [telev] mire *f* de télévision
~CINEMA - [telev] (cinema in which televised pictures are projected into a screen) film *m* pour la télévision, téléfilm *m*
~COMMERCIAL - [telev] émission *f* de propaganda
~DISTRIBUTION SATELLITE - [telev] satellite *m* de répartition par télévision
~EYE - [telev] oeuil *m* de télévision
~FOLDED PICTURE - [telev] recouvrement *m* d'image
~GAMMA - [electron] (the exponent of the power law which is used to approximate the curve of output magnitude versus input magnitude over the region of interest) gamma *m* de tube de prise de vues
TELEVISION INTERCOMMUNICATION NETWORK - [telev] réseau *m* d'intercommunication en télévision
~LEVEL - [telev] niveau *m* du signal de télévision
~LINK - [telev] liaison *f* par lignes de télévision
~NETWORK - [telev] réseau *m* de télévision
~PATTERN GENERATOR - [telev] (an apparatus suitable for use on all frequencies of the present television transmitters) générateur *m* de mire
~PICTURE TUBE - [electron] (a cathode-ray tube for transforming electrical television signals into pictures on a luminescent screen) tube *m* image, cinescope *m*
~RADAR NAVIGATION - [radar] (a system of air navigation combining radar and television) téléran *m*
~RECEIVER - [telev] téléviseur *m*
~RECONNAISSANCE - [telev] reconnaissance *f* par télévision
~SET - [telev] (the receiver) s. television receiver
[telev] (the studio) studio *m*
~SHOOTING SCRIPT - [telev] scénario *m* pour télévision
~SIGNAL - [telev] (radio frequency signal produced by modulation of a carrier wave by the video signal) signal *m* de télévision
~SOUND - [telev] son *m* de télévision
~-SOUND TRANSMITTER - [radio] émetteur *m* son
~STATION LINK - [telev] chaîne *f* de stations émettrices de télévision
~STUDIO - [telev] (room for taking television pictures for transmission) studio *m* pour télévision
~SYSTEM CONVERTER - [telev] convertisseur *m* de norme
~TRANSMITTER - [telev] émetteur *m* de télévision, transmetteur *m* de télévision
~TUBE - s. television picture tube
~-VISION TRANSMITTER - [radio] (a radio transmitter providing the visio-modulated wave of a complete television service) émetteur *m* vidéo
~WAVEFORM - [telev] (the composite wave consisting of the picture information modulated carrier and the synchronization pulses) forme *f* d'onde de télévision
~WAVES - [radio] (waves resulting from the modulation of a carrier wave by frequencies produced at the time of scanning of fixed or moving objects) ondes *f*pl pour télévision
TELEVISOR - [telev] (television apparatus) téléviseur *m*
TELEVOLTMETER - [instr] (an instrument with which it is possible to measure voltage from a distance) télévoltmètre *m*
TELEWATTMETER - [instr] (an instrument with which it is possible to measure power from a distance) téléwattmètre *m*
TELEX - [telecomm] (a telgraph service enabling its subscribers to communicate directly and temporarily among themselves by means of start-stop apparatus and of circuits of the public telegraph network) télex *m*
TELLER - [gen] (in a bank) caissier *m*
[leg] (the person appointed to count ballots) scrutateur *m*
[paper man] compteuse *f* des feuilles
TELLTALE - [instr] (indicator of the postion of the

rudder of a boat) axiomètre *m* de gouvernail
TELLTALE - [instr] contrôleur *m*, aiguille *f* indicatrice
[auto] temoin *m* d'allumage
[mus] (an organ implement indicating the amount of wind in the bellows) indicateur *m* du vent
[el] lampe *f* temoin, lampe *f* avertisseuse
[oil ind] regard *m*
~CLOCK - [horol] horloge *f* de contrôle
~HOLE - [oil ind] regard *m*
~LIGHT - [el] lampe *f* signalisatrice
TELLURETTED HYDROGEN - [chem] acide *m* tellurhydrique
TELLURHYDRIC - [chem] tellurhydrique
TELLURIC ACID - [chem] (an analytical reagent) acide *m* tellurique
~BISMUTH - [min] (bismuth occurring in the trigonal system and containing a trace of tellurium) tétradymite *f*
~OCHRE - [min] tellurine *f*, tellurite *f*
~SILVER - [min] argent *m* telluré
TELLURIDE - [chem] tellurure *m*
~ORE - [min] minerai *m* telluré
TELLURIFEROUS - [min] tellurifère
TELLURIUM - [chem] (rare, non-metallic element, used in electrolytic refinement of zinc to eliminate cobalt, also as a vulcanized agent) tellure *m*
~DIOXIDE - [chem] (white, crystalline powder with uses as an antiseptic) bioxyde *m* de tellure
~LEAD - [metall] (lead resistant to corrosion and capable of being work-hardened) plomb *m* telluré
TELOBLAST - [biol] (a large cell from which many smaller cells are produced by budding) téloblaste *m*
TELOCOELE - [anat] cavité *f* du télencéphale
TELPHER - [transp] (light car suspended from cables, usually propelled by electricity) (ligne) téléphérique
~RAILWAY - [railw] ligne *f* de telphérage
TELEPHERAGE - [transp] telphérage *m*
TELSON - [med] piqûre *f* de scorpion
TELSTAR - [telev] (the first of the experimental satellites)telstar *m*
TEM MODE - [electron] (short for Transverse Electric and Magnetic Mode) mode *m* transversal électrique et magnétique, mode *m* TEM
~WAVE - [electron] (short for transverse Electro-Magnetic Mode) onde *f* transversale électrique, onde *f* TEM
TEMPER, to - [metall] (to give metal special heat-treatment for specific physical qualities) tremper, faire revenir, recuire, durcir
TEMPER - [metall] (reheating of hardened steel below the critical range) revenu *m*, trempe *f*
[glass man] refroidissement *m* rapide
[metall] (the mixture of metals which are added to obtain an alloy) mélange *m* de poudres
[metall] (due to cold rolling) allongement *m*
[metall] (the percentage of carbon content in steel) coefficient *m* de dureté
[ind chem] (the substance used to modify properties) agent *m* d'addition
~BRITTLENESS - [metall] (tendency to break easily, due to tempering) fragilité *f* de revenu
~CARBON - [metall] (of malleable iron) carbone *m* de revenu, pourcentage *m* de charbon
~COLOUR - [metall] couleur *f* de trempe, couleur *f* de revenu
TEMPER-HARDENING - [metall] (term applied to alloys which increase in hardness after rapid cooling) durcissement *m* par revenu
~NUMBER - [metall] (USA only) taux *m* de dureté
~PASS MILL - [metall] laminoir *m* à dresser, laminoir *m* d'écrouissage
~ROLLING - [metall] (USA only, a method of cold rolling) laminage *m* d'endurcissement
~THE LIME, to - [ind chem] gâcher la chaux
TEMPERA - [paint] (pigments ground up in liquid glue or yolk of egg) peinture *f* à la colle, détrempe *f*
TEMPERATE - [met] tempéré
TEMPERATURE - [phys] (the thermal condition of a body which determines whether heat will flow from it or into it) température *f*
~AT SPOUT - [metall] température *f* au moment de la coulée
~COEFFICIENT OF CAPACITY - [el] (the change in delivered capacity per degree centigrade relative to the capacity of the cell or battery at a specified temperature) coefficient *m* de température de la capacité
~COEFFICIENT OF E.M.F.- [el] (the variation per unit temperature in the open-circuit voltage in terms of the cell E.M.F. at a given temperature) coefficient *m* de température de la force électromotrice
~COEFFICIENT OF VOLTAGE DROP - [electron] (in glow-distance tubes) coefficient *m* de température de la chute de tension
~COLOUR SCALE - [opt] (the relationship between the temperature of an incandescent substance and the colour of light emitted) échelle *f* des températures de couleur
~-CONTROLLED WAGON - [railw] (an insulated wagon) wagon *m* isotherme
~CYCLE STRESSING - [metall] sollicitation *f* aux chocs thermiques
~DETECTOR - [instr] thermomètre *m*
~DROP CONTROL - [metall] (of an electric furnace during continuous operation) régulation *f* de l'abaissement de la température
~ERROR - [meas] (error in measuring caused by a rise or fall of the ambient temperature) erreur *f* de température
~FACTOR - [nucl] (the factor which expresses the reduction of intensity of a reflection due to the thermal vibrations of the atom in a crystal) facteur *m* de température
~GRADIENT - [met] gradient *m* de température
~GRADIENT STRESS - [phys] (stress arising from the existence of temperature gradients in a structure) sollicitation *f*
~IMPACT CURVE - [metall] courbe *f* température/résilience
~-INDICATING CRAYON - [meas] (thermometric crayon) crayon *m* thermométrique
~INVERSION - [phys] (in the troposphere, an increase in temperature with height) inversion *f* de température
~-LIMITED EMISSION - [electron] émission *f* limitée par la température
~OF COKING - [metall] température *f* de carbonisation
~OF POLYMERIZATION - [ind chem] température *f* de polymérisation
~RECORDER - [instr] thermographe *m*
~REGULATOR - [mech] régulateur *m* thermique

TEMPERATURE RELAY - [el] relais *m* thermique
~-RISE - [phys] échauffement *m*
[el] (the excess of temperature of any particular part over the temperature of the surrounding atmosphere) échauffement *m*
~-RISE VOLTAGE - [el] (the voltage at which the resistor complies with the temperature-rise conditions) tension *f* d'échauffement
~SATURATION - [electron] (in a thermionic vacuum tube) saturation *f* cathodique
~SCALE - [meas] (a sequence of values assigned to temperature) échelle *f* thermométrique
~SCANNER - [instr] sonde *f* thermométrique
~SELECTOR - [instr] (a thermostat) sélecteur *m* de température
~-SENSING SYSTEM - [el] circuit *m* de détection de la température
~SENSIBILITY - [phys] sensibilité *f* à la température
~SIMULATOR - [el] simulateur *m* de la température
TEMPERED - [metall] (term used of steel which has been given a designed degree of hardness by heat treatment) trempé, durci, attrempé, revenu, recuit
~SCALE - [acoust] (the musical scale of keyboard instruments) gamme *f* tempérée
TEMPERING - [metall] (the process of adjusting the hardness of steel, after rapid cooling, by controlled reheating) trempe *f*, recuit *m*
[glass man] trempe *f*
[ceramics] (process of mixing clay and water to give a homogeneous mass) revenu *m*
[ind chem] broyage *m*, delayage *m*, gachage *m*
~BATH - [metall] bain *m* de trempe
~CHARGE - [metall] charge *f* de revenu
~COLOUR - [metall] couleur *f* de revient, couleur *f* de revenu
~FURNACE - [metall] four *m* à tremper
~HARDNESS - [metall] trempe *f* de recuit
~HEAT - [metall] chaleur *f* de revenu
~POWDER - [metall] poudre *f* de durcissement
~SAND - [metall] sable *m* pour coulée
~TEMPERATURE - [metall] température *f* de trempe
TEMPEST - [met] tempête *f*
TEMPESTUOUS - [met] tempétieux
TEMPLATE - [constr] (long flat stone supporting the end of a beam) sablière *f*
[mech] (a pattern or gauge used as a guide in shaping something) gabarit *m*, calibre *m*, patron *m*
[arch] centre *m* d'une arche
[shipbuild] (a wedge for a building block under a ship's keel) coin *m* d'arrimage
[text] s. temple
[metall] trousse *f*, trousseau *m*
~MOULDING - [metall] moulage *m* au trousseau
TEMPLE - [text] (of a loom) temple *f*, rame *f*, templet *m*
[arch] temple *m*
[anat] tempe *f*
~CUTTER - [text] coupe-fil *m* de templet
~HOLDER - [text] support *m* de templet
~RING - [text] molette *f*, plateau *m* d'aiguille
TEMPLESTICKS - s. temperature-indicating crayon
TEMPLET - s. template
TEMPO - [mus] tempo *m*
TEMPORAL - [anat] (cartilage bone of the skull) temporal
TEMPORAL AND SPATIAL FACTORS - [opt] facteurs *mpl* de temps et d'espace
~BONE - [anat] os *m* temporal
~LOBE - [anat] lobe *m* temporal, lobe *m* sphénoïdal
~MUSCLE - [anat] muscle *m* temporal
TEMPORARY - [gen] temporaire, provisoire
~ADJUSTMENT - [surv] (adjustment to a survey instrument) régulation *f* provisoire
~ARRANGEMENT - [gen] installation *f* provisoire, montage *m* provisoire
~BRIDGE - [telecomm] jonction *f* nodale pour transfert
~DUTY - [gen & el] (short-time duty) service *m* temporaire
~EARTH - [el] installation *f* de mise à la terre pour travaux, terre *f* pour travaux
~HARDNESS - [chem] (of water hardness due to dissolved calcium carbonate, which can be precipitated by boiling) crudité *f* temporaire
~SET - [mech] déformation *f* momentanée, déformation *f* élastique
~STORAGE - [comput] (internal storage locations reserved for intermediate partial results) mémoire *f* temporaire
~TRACK LAYING - [railw] pose *f* volante
~WAY - [constr] voie *f* provisoire
TEN - [math] dix *m*
~-CHANNEL STRAIN-GAUGE EQUIPMENT - [meas] (equipment used to display in one cathode-ray tube the strain at ten different points in a test piece) jauge *f* de contrainte à dix canals
~-STATE SYSTEM - [comput] système *m* avec I0 états stables
TENACITY - [text] (of fibres tensile strength, usually expressed in grammes per denier) ténacité *f*
[metall] (ultimate tensile strength) cohésion *f*
TENACULUM - [med] érigne *f*, érine *f*
TENALGIA - [med] ténalgie *f*
TENANCY - [leg] location *f*, bail *m*
TENANT - [leg] locataire *m*, tenancier *m*
~FARM - [agric] ferme *f* cédée à bail
~FARMING - [agric] baux *m* à ferme
TEND, to - [gen] tendre, tourner, se diriger, soigner
~ONE LOOM, to - [text] conduire un seul métier
~SEVERAL LOMMS, to - [text] conduire plusieurs métiers
TENDED SPRING - [mech] ressort *m* tendu
TENDENCY - [gen] tendance *f*, inclination *f*
~TO RISE - [fin] poussée *f* vers la hausse
TENDER, to - [comm] soumissionner
[fin] faire une offre réelle
[paper man] affaiblir (le papier)
TENDER - [comm] offre *f*, soumission *f*
[railw] tender *m*, allège *f*
[naut] navire *m* ravitailleur
[fin] (the legal currency) cours *m* légal
~FIBRE - [text] fibre *f* tendre, fibre *f* peu résistante
~SHIP - [naut] navire *m* ravitailleur
~SPECIFICATIONS - [soil] cahier *m* des charges
TENDERER - [gen] offreur *m*, commissionnaire *m*
TENDERING FORM - [leg] billet *m* de soumission
TENDERNESS - [med] douleur *f* à la pression
TENDING ANIMALS - [agric] soins *mpl* aux animaux
TENDINITIS - [med] ténosite *f*, tendinite *f*
TENDON - [anat] (a band of fibrous connective tissue) tendon *m*

TENDON ADVANCEMENT - [med] avancement *m* tendineux
TENDOPLASTY - [med] ténoplastie *f*
TENDOSYNOVITIS - [med] synovite *f* crépitante, ténosynovite *f*
TENDOVAGINITIS - [med] tendovaginite *f*, ténosynovite *f*
TENDRIL - [bot] (of a climbing plant) vrille *f*, cirre *m*, nille *f*
[bot] (of vine) griffe *f* (de vigne)
TENEMENT - [constr] (one or more rooms for one family) appartement *m* dans une maison de rapport
[leg] fonds *mpl* de terre
~HOUSE - [constr] (building or house, generally of an inferior type, where several families can live independently from one another) maison *f* de rapport, logements *mpl* ouvriers
TENIA - s. taenia
TENIAFUGE - [med] ténifuge *m*
TENIASIS - s. taeniasis
TENNANTITE - [min] (natural copper arsenic sulphide, often containing antimony) tennantite *f*, cuivre *m* gris arsenical
TENNIS ARM - [med] crampe *f* du tennis
~CLOTH - [text] tissu *m* tennis
~ELBOW - [med] épicondylite *f* des joueurs de tennis
TENODESIS - [med] ténodèse *f*
TENODYNIA - s, tenalgia
TENON, to - [carp] (to form a tenon) tenonner
[carp] (to join by mortise and tenon) assembler à tenon (et mortise)
TENON - [carp] (projecting lug or tongue on a part of a structure, designed to enter a mating recess in another part, for position and fixing) tenon *m*
[metall] ailette *f*
~-AND-SLOT MORTISE - [carp] assemblage *m* à tenon
~-AND-TUSK JOINT - [carp] assemblage *m* à tenon avec renfort
~CUTTER - [mach tool] tenonneuse *f*
~JOINT - [carp] assemblage *m* à tenon
~SAW - [tool] scie *f* à tenon, scie *f* à dos
[carp] scie *f* à raccourcir
TENONER - [mech] machine *f* à tenons
TENONING - [carp] assemblage *m* à tenon, empattement *m*
~ATTACHMENT - [mach tool] (on a spindle a moulding machine) dispositif *m* à faire les tenons
TENONTITIS - s. tendinitis
TENOR - [mus] ténor *m*
[mus] (the instrument) alto *m*
[leg] copie *f* conforme
[comm] échéance *f*
~DRUM - [mus] caisse *f* roulante
TENORITE - [min] (oxide of copper) ténorite *f*
TENOSITIS - s, tenontitis
TENOTOMY - [med] ténotomie *f*
TENS TRANSFER - [comput] transfert *m* des dizaines
TENSE - [gen] tendu, rigide
TENSIBILITY - [rubber ind] extensibilité *f*
TENSILE - [gen] extensible, élastique
[metall] ductile
[mech] extensif, de traction
~LOADING - [mech] effort *m* de traction
~STRAIN - [mech] déformation *f* due à la traction
~STRENGTH - [phys] (the ability of a material to withstand tensile stress) résistance *f* à la traction
TENSILE STRENGTH - [text] résistance *f* à la rupture
[metall] (the stress at which a material fails in tension) charge *f* de rupture
~STRENGTH AT BREAK- [mech] résistance *f* à la rupture (par traction)
~STRESS - [phys] (the force per unit area directed perpendicular to one surface only in a material medium) effort *m* de traction
~TEST - [mech] essai *m* de traction
~[metall] essai *m* de traction
~-TEST PIECE - [metall] éprouvette *f* de traction
~TESTING MACHINE - [mech] machine *f* pour essais de traction
TENSIMETER - [instr] (instrument designed to measure gaseous tension) tensimètre *m*
TENSIOACTIVE - [med] tensio-actif
TENSIOMETER - [instr] (a device for determining tension, as the surface tension of liquids, or tautness, as in the wires of an aircraft) tensiomètre *m*, compteur *m* de tension
TENSION - [phys] (a force, usually in a wire, string or rod, supporting a weight or otherwise stretched between two points) tension *f*
[mech] (of a spring) tension *f*, bande *f*
[metall] (of materials) traction *f*
[el] (electric potential) tension *f*, potentiel *m* électrique
~APPARATUS - [mech] amortisseur *m*
~BAR - [auto] tendeur *m*
~BLOCK - [mech] renvoi *m* tendeur
~BRIDGE - [constr] pont *m* bow-string
~CARRIAGE - [mech] chariot *m* tendeur
~CONTROL - [plast ind] (system e.g. a capstan, for controlling the tension of a wire being passed through an extruder for coating) contrôle *m* de tension
~CRACK - [mech] fissure *f* due à la traction
~-CRACKING - [rubber ind] craquelures *fpl* par tension
~DIAGONAL - [mech] diagonale *f* de traction
~EQUIPMENT - [railw] (of a catenary) équipement *m* tendeur (caténaire)
~FLANGE - [bot] (mechanical tissue developed on the concave side of a coiled tendril) bride *f* tendue
~FUSE - [metall] (in blasting) amorce *f* de tension, amorce *f* à étincelle
~IMPACT - [phys] énergie *f* du choc
~IN THE ROVING - [text] tension *f* de la mèche
~INSULATOR - [el] (an insulator which transmits to a support the entire line-conductor tension) isolateur *m* d'ancrage
~OF A BELT - [mech] tension *f* de la courroie
~OF THE WEB - [text] tension *f* du voile
~OPTICS - [photo] optique *f* des tensions
~PIECE - [constr] tirant *m*
~POINT - [el] jonction *f* d'arrêt
~PULLEY - [mech] galet *m* de tension
~RATCHET - [mech] tendeur *m* à cliquet
~RING - [mech] anneau *m* tendeur
~ROD - [constr] (a structural member subject to tensile stress only) tirant *m*
[metall] barre *f* de chariotage, bielle *f* de traction
~ROLL - [mech] (a roller designed to maintain a given tension in a continuous strip of material) rouleau *m* tendeur, cylindre *m* de serrage
~ROLLER - [mech] (of a belt) galet *m* de tension
~SCREW - [mech] vis *f* de tension

TENSION SHACKLE - [mech] étrier *m* de tension
~SHOE - [constr] patin *m* presseur
~SLEEVE - [mech] lanterne *f* de serrage
~SPRING - [mech] (a spring designed to be extended in action) ressort *m* de traction
~SPRING BASE - [mech] (in upholstery) support *m* à ressorts à tension initiale
~TRUCK - [railw] chariot *m* tendeur
~-UNIT ELONGATION CURVE - [rubber ind] courbe *f* tension-allongement
[metall] courbe *f* charge-allongement
~WASHER - [mech] (type of washer consisting of an interrupted ring which can be sprung into a circular groove) jonc *m* d'arrêt élastique
~WEIGHT - [mech] contre-poids *m* de tension, poids *m* tendeur
TENSIONAL BAR - [mech] (tension rod) tirant *m*
TENSIONING ARM - [text] bras *m* de tension
~BAR - [text] barre *f* de tension
~CABLE - [mech] fil *m* tendeur
~CHAIN - [text] chaîne *f* de tension
~DEVICE - [mech] tendeur *m*
~DEVICE FOR LIGHT WIRES - [el] tendeur *fm* à levier coudé
~EQUIPMENT - [el] équipement *m* tendeur
~LEVER - [text] levier *m* de tension
~SCREW - [text] écrou *m* de tension
TENSIVE - [med] tensif
~STRESS - [phys] effort *m* de tension
TENSOR - [anat] tenseur *m*
[math] (vector quantity) tenseur *m*
~FIELD - [phys] champ *m* tensoriel
~FORCE - [nucl] (or non-central force; a nuclear force whose direction partly depends on the spin orientation of the nucleons) force *f* nucléaire tensorielle
~TENT, to - [med] tamponner, introduire une mèche
TENT - [gen] tente *f*
[med] mèche *f*, tampon *m*
~DUCKS - [text] grosse toile *f* à tente
TENTACLE - [gen] tentacule *m*
[bot] tentacule *m*
TENTAGE - s. tent ducks
TENTATIVE - [gen] expérimental
~EXPERIMENT - [gen] essai *m* expérimental
~OFFER - [comm] offre *f* pour entamer le négociations
~OUTLINE - [draw] dessin *m* d'essai
TENTCLOTH - s. tent ducks
TENTER, to - [gen] tendre, ramer, élargir
TENTER - [impl] (a device for stretching materials during some operation, e.g. drying) élargisseuse *f*
~DRYER - [mech] séchoir *m* à tension
TENTERHOOK - [text] (temple) clou *m* à crochet
TENTERING - [text] élargissage *m*
~MACHINE - [text] élargisseuse *f*
TENTH-VALUE THICKNESS - [radiat] (the thickness of a specified substance which, when introduced in the path of a given beam of radiation, reduces its effect to one tenth) couche *f* d'atténuation au dixième, épaisseur *f* d'atténuation au dixième
TEPHIGRAM - [met] (graph showing air temperature (T) in relation to entropy (phi) téphigramme *m*
TEPHRITE - [geol] (a rock) téphrite *f*
TEPHROITE - [miner] (orthosilicate of manganese) téphroïte *f*
TEPID - [gen] tiède
TEPOR - [gen] tiédeur *f*
TERATOLOGY - [med] tératologie *f*
TERATOMA - [med] tératome *m*, tumeur *f* tératoïde
TERBIUM - [chem] (a rare earth element, symbol Tb, A.N. 65 A, W.159.2) terbium *m*
TEREBENE - [chem] (a mixture of dipentene and other hydrocarbons and having application in medicine) térébène *m*
TEREBINTH - [bot] (a med plant) térébinthe *m*
TEREBRANT - [med] térébrant
TEREBRATING - s. terebrant
TEREPHTHALIC ACID - [chem] (an intermediate for synthetic resins) acide *m* téréphtalique
TEREPHTHALOYL CHLORIDE - [chem] (a crosslinking agent and an intermediate for synthetic resins, dyestuff, and pharmaceuticals) chlorure *m* de téréphtaloyle
TERM, to - [gen] appeler, nommer
TERM - [gen] terme *m*, limite *f*
[leg] clause *f*, session *f*
[gen] (a word) mot *m*, expression *f*, terme *m*
[math] terme *m*
~DISPLACEMENT - [nucl] (the displacement of the members of a compound quantity) déplacement *m* de termes
~OF MAINTENANCE - [leg] délai *m* d'entretien
TERMINAL - [el] (the parts of an electrical component to which the connecting wires are attached) borne *f*, poteau *m* d'arrêt
[el] (on an electrical machine) borne *f* de connexions
[mech] (of a belt conveyor etc) extrémité *f*
[railw] gare *f* terminus
[arch] (a pedestal, a terminal figure) piédestal *m*
[bot] distal *m*
[comput] (in a communication network) position *f* d'entrée et sortie
~BALLISTICS - [astronaut] balistique *f* de fin de trajectoire
~BAR - [el] (bar to which a group of plates of an accumulator is attached) borne *f* de connexions
~BLOCK - [el] bloc *m* de dérivation, bloc *m* de jonction, réglette *f* de raccordement
~BOARD - [el] plaquette *f* de connexions
[electron] planche *f* à bornes
~BOARD GUARD - [el] (of a motor) couvre-bornes *m*
~BOLT - [el] porte-bornes *m*
~BOX - [el] (box housing electric terminals) boîte *f* à bornes, plaque *f* à bornes
~BRACKET - [telecomm] console *f* d'arrêt
~BUSHING - [el] isolateur *m* de traversée
~CABLE - [el] câble *m* terminal
~CLAMP - [el] serre-fil *m*
~CONE - [astronaut] (of a rocket) cône *m* terminal
~CONNECTER - [el] borne *f*
~COVER - [el] cache-bornes *m*, cache-fils *m*
~CURRENT - [el] courant *m* aux bornes
~CURVATURE - [geol] (sudden local change in the dip of stratified rocks near a fault) courbure *f* frontale
~DENT - [mech] dent *f* de l'extrémité, dent *f*
~IMPEDANCE - [el] impédance *f* terminale de garde
~LIMIT-SWITCH - [el] commutateur *m* terminal
~LOADING COIL SECTION - [el] complément *m* d'une section de pupinisation
~-LUG - [el] (projection on a group of accumulator or plate-section, used for connecting it to an exter-

nal circuit) oreille *f* de plaque, queue *f* de plaque
TERMINAL NOSE-DIVE - [aero] (a dive at terminal velocity) piqué *m* limite
~PILLAR - [el] (or post-head; a post or pillar at which insulated cables may conveniently be terminated) boulon *m* polaire
~PLATE - s. terminal box
~POLE - [el] (or dead-end tower; a support at the end of a line, designed to withstand the longitudinal load of all the phase-conductors as well as the vertical and transverse loads) mât *m* d'arrêt, poteau *m* d'arrêt
~POSTS - [el] (USA only; terminals in GB) bornes *f*pl, pôles *m*pl
~PRINTER - [comput] imprimante *f* de sortie
~PULSE - [electron] impulsion *f* terminale
~PUNCHING - [el] broche *f* de contact
~RATE - [telecomm] (the portion of the charge due in the countries of origin and destination of an international telegram) taxe *f* terminale
~REPEATER - [telecomm] (repeater for use at the end of a trunk or a line) répéteur *m* terminal
~RESISTANCE - [telecomm] (resistance matched in such way to the wave resistance of a line or cable as not to reflect any energy) résistance *f* aux bornes
~-RETURN-LOSS - [telecomm] affaiblissement *m* aux bornes du circuit
~ROOM - [telecomm] salle *f* des appareils (de téléphonie automatique)
~SCREW - [el] (or clamping screw; a screw for holding a conductor to the terminal of a switch) borne *f* à vis
~SPEED - [aero] (the velocity at which the weight and drag of a body in free fall are exactly equal under given atmospheric conditions) vitesse *f* limite
~SPINDLE - [el] console *f* d'arrêt
~SPINE - [bot] épine *f* terminale
~STATION - [telecomm] station *f* terminale
~STRIP - [el] bloc *m* de connexion
~SUPPORT - [el] support *m* d'extrémité
~TOWER - s. terminal pole
~THRESHOLD TUBE - [electron] tube *m* à seuil terminal
~UNIT - [comput] unité *f* terminale
~VELOCITY - [electron] vitesse *f* finale
~VOLTAGE - [el] (voltage at the terminals of an electrical device) tension *f* aux bornes
~YOKE - s. terminal bar
TERMINATE, to - [gen] terminer, délimiter
TERMINATE - [math] exact
~DECIMAL FRACTION - [math] fraction *f* décimale exacte
TERMINATED LEVEL - [instr] (the reading of a level measuring set at a point in a system when terminated at that point by a resistance equal to the nominal impedance of the circuit) niveau *m* composé adapté
TERMINATING - [gen] terminaison *f*, fin *m*
~SET - [telecomm] (four-wire terminating set) termineur *m*
TERMINATION OF A LEASE - [leg] résiliation *f* du bail
~OF A LINE - [el] terminaison *f* d'une ligne
TERMINOLOGY - [gen] (the study or the use of terms) terminologie *f*
TERMINUS - [surv] borne *f*
[transp] point *m* terminus
[railw] gare *f* de tête
TERMITE - [zool] termite *m*
TERMS - [comm & leg] conditions *f*pl
~OF DELIVERY - [comm] conditions *f*pl de livraison
~OF PAYMENT - [comm] conditions *f*pl de payment
TERMOLECULAR - [phys] (a synonym for trimolecular) trimoleculaire
TERNARY - [math & chem] ternaire
[metall] (of an alloy) ternaire
~ALLOY - [metall] acier *m* ternaire
~CRITICAL POINT - [phys chem] (the point where, by adding a mutual solvent to two partially miscible liquids, the two solutions become consolute and one phase results) point *m* critique ternaire
~FISSION - [nucl] (the splitting of a nucleus into three nuclear fragments) fission *f* ternaire, tripartition *f*
~NOTATION - [comput] (scale of notation having the base 3) notation *f* ternaire
~STEEL - [metall] acier *m* ternaire
~SYSTEM - [metall] système *m* ternaire
TERNE, to - [metall] (to cover with a thin layer of lead and tin) plomber
TERNE - s. terne-plate
~-PLATE - [metall] tôle *f* matte, tôle *f* terne
TERPENE - [chem] (generic name for unsaturated hydrocarbons having the formula $(C_5H_8)_{\underline{n}}$) terpène *m*
TERPHENYL - [chem] (an intermediate for plastics) terphényle *m*
TERPIN HYDRATE - [chem] (a constituent of certain pharmaceuticals and a starting material for the production of terpineol) hydrate *m* de terpine
TERPINEOL - [chem] (a solvent for cellulose derivatives and a number of resins) terpinéol *m*
TERPINOLENE - [chem] (a solvent and intermediate in the manufacture of synthetic resins) terpinolène *m*
TERPINYL ACETATE - [chem] (a perfumery component) acétate *m* de terpinyle
TERRA ALBA - [chem] (powdered gypsum with uses as a plastics filler) terre *f* de pipe
~PONDEROSA - [min] spath *m* pesant
~VERTE - [paint] (alternative name for Green Earth, essentially silicate of iron, and used as a base for lakes) terre *f* verte
TERRACE, to - [constr] disposer en terrasse ,terrasser
TERRACE - [constr] terre-plein *m*, terrasse *f*
[constr] (the plane roof of a house) terrasse *f*, toiture *f* en plate-forme
[town planning] rangée *f* de maisons
~CROPPING - [agric] culture *f* en terrasses
~CULTIVATION - s. terrace cropping
~WALL - [constr] mur *m* de terrasse
TERRACED HOUSES - [constr] maisons *f*pl en bandes
TERRACOTTA - [gen] (hard and durable kiln-burnt clay) terre *f* cuite, argile *f* cuite
TERRAIN - [gen] terrain *m*
~CLEARANCE WARNING INDICATOR - [aero] (a radio warning device carried in an aircraft, which gives a signal when the distance between the aircraft and the ground directly below falls below a pre-set level) avertisseur *m* de marge d'altitude
TERRAZZO - [constr] (venetian mosaic) granito *m*
TERRESTRIAL - [gen] terrestre

TERRESTRIAL EQUATOR - [astr] équateur *m* de la terre
~LATITUDE - [geogr] latitude *f* géographique
~LONGITUDE - [geogr] longitude *f* géographique
~MAGNETIC FIELD - [el] (the field of terrestrial magnetism) champ *m* magnétique terrestre
~MAGNETIC POLES - [el] pôles *mpl* magnétiques terrestres
~MAGNETISM - [phys] (the magnetism exerted by the earth) magnétisme *m* terrestre
~POLES - [geogr] (the two diametrically opposed points in which the earth's axis cuts the earth's surface) pôles *mpl* terrestres
~RADIATION - [met] (the earth's loss of heat through radiation to the sky at night) rayonnement *m* terrestre
TERRIGENOUS SEDIMENTS - [geol] (sediments deposited on the shallower parts of the sea-floors) dépôts *mpl* terrigènes
TERRITORIAL - [gen] (pertaining to a territory) territorial
~SEA - [leg] (the stretch of sea under a sovereign jurisdiction) eaux *fpl* territoriales, territoire *m* maritime
~WATERS - s. territorial sea
TERRY - [text] velours *m* frisé
~CLOTH - s. terry
~LOOM - [text] métier *m* pour tissu éponge
~MOTION - [text] mouvement *m* du battant pour métier à tissu éponge
~PILE TOWELLING - [text] tissu *m* éponge
~VELVET - [text] velours *m* frisé, velours *m* épinglé
~WARP - [text] chaîne *f* à boucles
TERT - [gen] (an abbreviation for tertiary) tert-
~-BUTYL PERPHTHALIC ACID - [chem] acide *m* tert-butyle perphtalique
TERTIARY - [gen] (third in terms of time, number, degree etc) tertiaire
[chem] (as of an atom of carbon) tertiaire
[geol] étage *m* tertiaire
~AIR - [chem] (in coal combustion) air *m* tertiaire
~ALCOHOL - [chem] (an alcohol containing the $R_3C.OH$ group) alcool *m* tertiaire
~AMIDE - [chem] (an amide in which three hydrogen atoms have been replaced by acyl groups) amide *m* tertiaire
~CREEP - [nucl] (the rapid deformation sometimes occurring after the secondary creep) fluage *m* tertiaire
~EXCHANGE - [telecomm] bureau *m* automatique rural
~HOLES - [mech] (in gas turbine holes in a flame tube downstream of the secondary holes to provide air for dilution of the gases of combustion) orifices *mpl* tertiaires
~WINDING - [el] (additional winding in a transformer) enroulement *m* tertiaire
TERVALENT - [chem] (synonym for trivalent) trivalent
TESCHENITE - [geol] (coarse-grained igneous rock) teschénite *f*
TESSELATE, to - [constr] arranger en damier
TESSELATED - [constr] (of a floor etc) disposé en damier, en mosaique
TESSERA - [constr] tesselle *f*
TESSERAL SYSTEM - [phys] (in crystallography) système *m* terquaternaire, système *m* régulier

TEST, to - [gen] (to try by subjecting to an experiment) éprouver, essayer
[chem] (to analyze) analyser
[metall] (to refine by means of lead, as in the process of cupellation) coupeller, passer à la coupelle
TEST - [gen] (examination, analysis) épreuve *f*, experience *f*
[mech] (of a machine, a motor etc) essai *m*, test *m*
[metall] (for the refining of gold and silver) coupelle *f*, témoin *m*
[cin] film *m* d'essai
[chem] réactif *m*, coupelle *f*
[gen] (a series of questions, or other means to assess knowledge, aptitude etc) test *m*, examen *m*, test *m* de capacité
~AMPLIFIER - [electron] amplificateur *m* de mesure
~AT VARYING SPEEDS - [mech] épreuve *f* à régimes variables
~BAKING - [food] essai *m* de panification
~BALANCE - [meas] balance *f* d'essai
~BAR - [mech] éprouvette *f*, barreau *m* d'essai
[telecomm] barre *f* de test, fil *m* de fiche
~BED - [mech] (a solid mounting on which engines are supported during test) banc *m* d'essai
~-BENCH - [mech] (a special structure on which a unit is mounted for test; sometimes supporting or containing instruments and the like) banc *m* d'épreuve, band *m* d'essai
~BLOC - [metall] lingot-éprouvette *m*
~BOARD - [electron] (a panel provided with instruments, terminals and equipment for testing electronic apparatus) panneau *m* de contrôle
[gen] table *f* d'essais et de mesures
~BORES - [soil] forages *mpl* de reconnaissance
~BORING - [min] sondage *m* de recherche
~BOX - [telecomm] armoire *f* à protections, boîte *f* de coupure
~BUTTON - [metall] témoin *m*
~-BRUSH - [el] frotteur *m* privé, frotteur *m* d'essai
~BURNER - [th eng] brûleur *m* d'essai, brûleur-contrôleur *m*
~BY WEIGHING, to - [meas] trier par pesée
~CALL - [telecomm] communication *f* d'essai
~CAP - [el] isolateur *m* d'extrémité de câble
~CARD - s. test chart
~CELL - [instr] (instrument designed to determine the dielectric strength of a liquid) spintermètre *m*
~CERTIFICATE - [gen] (for machinery etc) certificat *m* d'épreuve
~CHAMBER - [aero] (the part of a wind tunnel in which the experimental materials is placed) chambre *f* d'expérience
~CHART - [med] (for optical examination) tabelle *f* optométrique
[telev] (chart showing geometrical patterns for checking purposes) image *f* test, mire *f*
~CIRCUIT - [telecomm] circuit *m* de contrôle
~COCK - [plumb] robinet *m* de jauge, robinet *m* de hauteur d'eau
~DECK - [comput] jeu *m* de cartes de test
~DESK - [el] (assembly of apparatus in the general form of a desk and comprising measuring instruments, devices for regulation of voltage, current and phase displacement and other accessories for calibration of meters and instruments) pupitre *m* d'étalonnage
~DRILLING - s. test boring

TEST DRIVER - [auto] contrôleur *m*
~-EQUIPMENT - [mech & instr] (any instruments or apparatus used in making tests) appareillage *m* de essai
~FAN - [aero] moulinet *m* d'essai
~FARM - [agric] ferme-pilote *f*
~FIELD - [agric] champ *m* d'essai
~FILM - [cin] film *m* d'essai
[telev] (film with patterns of varying gradation which makes possible the control of definition and contrast) film *m* d'essai
~FIRING - [astronaut] tir *m* d'essai
~FLIGHT - [aero] vol *m* d'essai
~FREQUENCY - [telecomm] fréquence *f* de mesure
~FURNACE - [th eng] fourneau *m* à essais
[metall] four *m* d'essai
~GAS - [gas ind] gaz *m* d'essai
~GLASS - [ind chem] éprouvette *f*
~HOLDER - [gas ind] gazomètre *m* de contrôle, gazomètre *m* d'expérience
~HOLE - [nucl] (opening in a reactor for the purpose of experiments) canal *m* expérimental
[gas ind] dispositif *m* fixe de détection de fuites
~HUT - [telecomm] guérite *f*, guérite *f* de coupure
~INSULATOR - [el] isolateur *m* de coupure
~JACK - [el] jack *m* d'essai
~LABEL - [text] étiquette *f* de titrage
~LEA - [text] échevette *f* d' épreuve
~LEVEL - [telecomm] (in a telephone circuit) niveau *m* composite
~LINE - [telecomm] ligne *f* d'essais
~LOAD - [mech] charge *f* d'épreuve, charge *f* de essai
~LOOP - [telecomm] boucle *f* de mesure
~MEAL - [med] repas *m* d'épreuve
~-METER - [instr] mesureur *m* de contrôle
~NEEDLE - [ind chem] touchaud *m*, aiguille *f* d'essai
~-OBJECT - [opt] test-objet *m*
~OUT, to - [gen] contrôler
~OVEN - s. test furnace
~PANEL - [telecomm] baie *f* de mesure
~PAPER - [chem] papier *m* à réactif
~PATTERN BEAM - [telev] (holding beam in the USA; electron beam of medium intensity holding the pattern of the screen) faisceau *m* d'accumulation
~PIECE - [gen] (a specially prepared piece of material for trial purposes) éprouvette *f*, coupon *m*
[mus] morceau *m* imposé
~PILE - [soil] pieu *m* d'essai
~PILOT - [aero] pilote *m* d'essais
~PIT - [mining] trou *m* de prospection, trou *m* d'exploration
~PLANT - [gen] (pilot plant) installation *f* pilote
~PLATE - [mech] timbre *m*
[el] disque *m* d'épreuve
~PLUG - [el] fiche *f* d'essai
~POINT - [el] point *m* de mesure du champ
~POLE - [telecomm] poteau *m* de coupure, poteau *m* d'essai
~PRINT - [cin] copie *f* étalon, positif *m* destiné aux producteurs
~PROGRAMME - [comput] (programme designed to show that a computer is functioning properly) programme *m* d'essai
~RECORD - [el acoust] disque *m* étalon
~REEL - [text] dévidoir *m* pour titrage
~REELING - [text] dévidage *m* pour l'essai

TEST RELAY - [el] relais *m* de contrôle, relais *m* de essai
~REPORT - [gen] certificat *m* d'épreuve
~ROD - [metall] barreau *m* d'épreuve
~ROOM - [gen] salle *f* d'expérimentation
~RUN - [railw] parcours *m* d'épreuve
[comput] passage *m* d'essai
~SAMPLE - [gen] échantillon *m*, épreuve *f* au hasard
~SCORING MACHINE - [comput] correctrice *f* électronique
~SECTION - [aero] (of a wind tunnel) chambre *f* d'essai (d'une souffleries)
~SELECTOR - [telecomm] sélecteur *m* d'essai
~SET - [el] boîte *f* de contrôle, boîte *f* de vérification
~SHEATH - s. test shield
~SHIELD - [el] gaine *f* métallique isolée
~SHOT - [cin & telev] prise *f* provisoire
~SHOT - [telev] essai *m* image, bout *m* d'essai
~SPECIMEN - [gen] (a piece of material specially cut or prepared for a test) éprouvette *f*
~STRIP - s. test shot
~TANK - [aero naut etc] bassin *m* d'expérimentation
~TAPE - [el acoust] (a means for measuring recording performance) ruban *m* d'essai
~TERMINAL BOX - [instr] (special terminal box placed beside a meter to facilitate its calibration without disturbing the operation of the installation in which it is included) boîte *f* à bornes d'essai
~THE YARN, to - [text] essayer les fils
~TRACK - [auto] piste *f* d'épreuve
~TRANSLATOR - [comput] programme *m* d'essai pour programmes d'assemblage
~-TUBE - [ind chem] (thin glass tube closed at one end, used for marking tests with small quantities of solutions and the like) tube *m* à essai, éprouvette *f*, burette *f*
~-TUBE BRUSH - [ind chem] (special cylindrical brush for cleaning test-tubes) brosse *f* à tubes
~-TUBE HOLDER - [ind chem] pince *f* pour tubes à essai
~-TUBE RACK - [ind chem] porte-tubes *m*
~-TUBE RACK - [photo] support *m* d'éprouvettes
~-TUBE STAND - s. test-tube rack
~VALUE - [el] valeur *f* d'essai
~WELL - [mining] sondage *m* d'exploration
~WELL - [hydr] puits *m* d'observation
TESTATOR - [leg] testateur *m*
TESTER - [gen] essayeur *m*, vérificateur *m*
[instr] appareil *m* contrôleur
[mech] machine *f* à essayer
TESTICLE - [anat] testicule *m*
TESTIFY, to - [leg] témoigner
TESTIMONIAL - [gen] lettre *f* testimoniale, attestation *f*
TESTING - [gen] essayage *m*, épreuve *f*, essai *m*
~BENCH - [gen] banc *m* d'essai
~CONDITIONS - [gen] conditions *f*pl de contrôle
[metall etc] (for the testing of materials) réglement *m* d'essai
~CREW - [telecomm] équipe *f* de mesure
~DATA - [gen] données *f*pl de contrôle
~DEPARTMENT - [gen] atelier *m* d'essai
~DRILL - [oil ind] sonde *f* de prospection
~FLAME - [mining] auréole *f*
~FOR WEIGHT - [fin] trébuchage *m*
~HOUSE - [text] atelier *m* de conditionnement

TESTING JOINTS - [el] jonctions *f*pl d'essai
~LOAD - [mech] charge *f* d'épreuve
~MACHINE - [mech] machine *f* à essayer, machine *f* d'essai
~OF FEEDING STUFFS - [agric] contrôle *m* des aliments
~OF MATERIALS - [ind] épreuve *f* des matériaux
~OVEN - [text] (for the moisture of silk) dessiccateur *m*, appareil *m* à conditionner
~PERIOD - [gen] période *f* d'épreuve
~PUMP - [mech] pompe *f* d'épreuve
~ROD - [el] (buzz stick in the USA) perche *f* d'essai
~ROOM - [ind] salle *f* d'expérimentation
~SET - [el] boîte *f* de vérification
~SHOP - [gen] atelier *m* d'essai
~STAND - [mech] banc *m* d'essai
~STANDARDS - [gen] normes *f*pl pour le contrôle
~TERMINALS - [el] (calibrating terminals) bornes *f*pl d'essai
~TICKET - [comm] fiche *f* de preuve
~TRACK - s. test track
~VAN - [telecomm] voiture *f* pour l'essai des câbles
~WIRE - [el] fil *m* pilote
TESTIS - s. testicle
TESTOSTERONE - [chem] (a natural steroid with applications in medicine) testostérone *f*
~CYCLOPENTYLPROPIONATE - [chem] (an ester of testerone having therapeutic uses) cyclopentylpropionate *m* de testostérone
~PROPIONATE - [chem] (odourless, white, crystalline powder with uses in medicine in replacement therapy and in the treatment of carcinoma) propionate *m* de testostérone
TETANIZATION - [med] tétanisation *f*
TETANUS - [med] tétanos *m*
TETARTOHEDRAL CRYSTALS - [crystall] cristaux *m*pl tétartoédriques
TETHER, to - [agric] pâturer au tière, pâturer au piquet
TETHERING - [agric] pâturage *m* au tière
TETRABROMOETHYLENE - [chem] (a synthesis intermediate) tétrabrométhylène *m*
TETRABUTYL TITANATE - [chem] (TBT; butyl titanate; titanium butylate) titanate *m* de tétrabutyle
~UREA - [chem] (plasticizer) tétrabutyl-urée *f*
~ZIRCONATE - [chem] (plastics crosslinking agent and condensation catalyst) zirconate *m* de tétrabutyle
TETRABUTYLTHIURAM DISULPHIDE - [chem] (a vulcanization agent and accelerator) bisulfure *m* de tétrabutylthiuram
TETRABUTYLTIN - [chem] (a polymerization catalyst and stabiliser for the silicones) tétrabutyl-étain *m*
TETRACAINE HYDROCHLORIDE - [chem] (amethocaine hydrochloride. A local anaesthetic) chlorhydrate *m* de tétracaîne
TETRACARBOXYBUTANE - [chem] (a curing agent for epoxy resins) tétracarboxybutane *m*
TETRACENE - [chem] (guanyl nitrosoaminoguanyl tetracene. A detonating explosive) tétracène *m*
TETRACHLORIDE - [chem] tétrachlorure *m*
TETRACHLOROBENZENE - [chem] (a synthesis intermediate) tétrachlorobenzène *m*
TETRACHLORODIFLUOROETHANE - [chem] (a nonflammable solvent) tétrachlorodifluoroéthane *m*
TETRACHLOROETHANE - [chem] (acetylene tetrachloride, applications as a solvent and in photographic film and protective coatings) tétrachloroéthane *m*
TETRACHLOROPHENOL - [chem] (a fungicide) tétrachlorophénol *m*
TETRACHLOROPHTHALIC ACID - [chem] (a synthesis intermediate) acide *m* tétrachlorophtalique
TETRACHLOROPHTHALIC ANHYDRIDE - [chem] (an intermediate for plasticizers, dyestuffs and drugs) anhydride *m* tétrachlorophtalique
TETRACOSANE - [chem] (a synthesis intermediate) tétracosane *m*
TETRACYANOETHYLENE - [chem] (a synthesis intermediate) tétracyanoéthylène *m*
TETRACYCLINE - [chem] (an antibiotic produced by varieties of Streptomyces, and having important therapeutic applications) tétracycline *f*
TETRAD - [chem] élément *m* tétravalent, tetrade *f* [comput] (a group of four; or a group of four pulses expressing a digit in the scale of I0 or I6) série *f* de quatre
TETRADECANE - [chem] (a synthesis intermediate) tétradécane *m*
TETRADECENE - [chem] (an intermediate for pharmaceuticals, perfumes, synthetic resins, and dyestuffs) tétradécène *m*
TETRADECYL MERCAPTAN - [chem] (a synthesis intermediate) tétradécyl-mercaptan *m*
TETRADECYLAMINE - [chem] (an intermediate for detergents and disinfectants) tétradécylamine *f*
TETRADYMITE - [min] (natural bismuth telluride, often found in gols quartz veins. A source of bismuth) tétradymite *f*
TETRAETHANOLAMMONIUM HYDROXIDE- [chem] (a rubber chemical catalyst; and dyestuffs solvent) hydrate *m* de tétraéthanolammonium
TETRAETHYLAMMONIUM CHLORIDE- [chem] (colourless crystals with uses in medicine as a ganglion-blocking agent) chlorure *m* de tétraéthanolammonium
TETRAETHYL DITHIOPYROPHOSPHATE - [chem] (an insecticide) dithiopyrophosphate *m* de tétraéthyle
TETRAETHYL LEAD - [chem] (colourless, poisonous, oily liquid. It is an important antiknock additive for motor fuels) plomb *m* tétraéthylique
TETRAETHYL PYROPHOSPHATE - [chem] (an insecticide and pesticide) pyrophosphate *m* de tétraéthyle
TETRAETHYLENE GLYCOL - (TEG) - [chem] (a plasticizer and solvent for cellulose compounds) tétraéthylène-glycol *m*
TETRAETHYLENE GLYCOL DICAPRYLATE - [chem] (plasticizer for vinyl chloride and vinyl chloride acetate) dicaprylate *m* de tétraéthylène-glycol
~GLYCOL DIMETHACRYLATE - [chem] (plasticizer) diméthacrylate *m* de tétraéthylène-glycol
~GLYCOL DISTEARATE - [chem] (plastics plasticizer) distéarate *m* de tétraéthylène-glycol
~GLYCOL MONOSTEARATE - [chem] (plastics plasticizer) monostéarate *m* de tétraéthylène-glycol
TETRAETHYLENEPENTAMINE - [chem] (intermediate in the production of synthetic rubbers) tétraéthylène-pentamine *f*
TETRAETHYLTHIURAM DISULPHIDE - [chem] (disulfiram. A pesticide, therapeutic agent in the treatment of alcoholism, and a component of vulcanization agents for rubber) bisulfure *m* de tétraéthylthiuram
~SULPHIDE - [chem] (monosulfiram, An insecticide and parasiticide in medicine) sulfure *m* de tétraé-

thylthiuram
TETRAFLUOROETHYLENE - [chem] (a starting material for a number of polytetrafluoroethylene polymers) tétrafluoroéthylène *m*
TETRAFLUOROHYDRAZINE - [chem] (a synthesis intermediate) tétrafluorohydrazine *f*
TETRAFLUOROMETHANE - [chem] (colourless, inert gas with uses as a refrigerant and rocket propellant) tétrafluorométhane *m*
TETRAGLYCOL DICHLORIDE - [chem] (a solvent and synthesis intermediate) bichlorure *m* de tétraglycol
TETRAGONAL - [math] tétragonal
~CRYSTAL - [phys] cristal *m* tétragonal
TETRAHEDRITE - [min] (natural copper antimony sulphide, often containing bismuth, mercury, silver, zinc and iron) tétraédrite *f*, tétrahédrite *f*
TETRAHEDRON - [geom] tétraèdre *m*
TETRAHYDROFURAN - (THF) - [chem] (flammable liquid with uses as a solvent for natural and synthetic resins, in polymerizations; and as an intermediate) tétrahydrofuranne *m*
TETRAHYDROFURFURYL ALCOHOL - [chem] (tetrahydrofuryl carbinol) alcool *m* tétrahydrofurfurylique
~LAURATE - [chem] (a plastics plasticizer) laurate *m* de tétrahydrofurfuryle
~LAEVULINATE - [chem] (plastics plasticizer) lévulinate *m* de tétrahydrofurfuryle
~OLEATE - [chem] (plastics plasticizer) oléate *m* de tétrahydrofurfuryle
~PALMITATE - [chem] (plasticizer for cellulose acetate butyrate) palmitate *m* de tétrahydrofurfuryle
~PHTHALATE - [chem] (plastics plasticizer) phtalate *m* de tétrahydrofurfuryle
TETRAHYDRONAPHTHALENE - [chem] (a solvent for cellulose derivatives, rubber and resins) tétrahydronaphtalène *m*
TETRAHYDROPHTHALIC ANHYDRIDE - [chem] (intermediate in the production of polyesters, alkyds and plasticizers) anhydride *m* tétrahydrophtalique
TETRAHYDROPYRAN - 2 - METHANOL - [chem] (a synthesis intermediate) tétrahydropyran-2-méthanol *m*
TETRAHYDROPYRIDINE - [chem] (a synthesis intermediate) tétrahydropyridine *f*
TETRAHYDROTHIOPHENE - [chem] (a solvent and synthesis intermediate) tétrahydrothiophène *m*
TETRAHYDROXYETHYLETHYLENEDIAMINE - [chem] (an intermediate in the production of synthetic resins and urethane foam crosslinking agents) tétrahydroéthyléthylènediamine *f*
TETRAIODOETHYLENE - [chem] (a fungicide) tétraiodoéthylène *m*
TETRAISOPROPYL TITANATE - [chem] (TPT; titanium isopropylate; isopropyl titanate, a condensation catalyst; also increases the adhesion of rubbers and plastics to metal surfaces) titanate *m* de tétraisopropyle
~ZIRCONATE - [chem] (a crosslinking agent and catalyst) zirconate *m* de tétraisopropyle
TETRAISOPROPYLTHIURAM DISULPHIDE - [chem] (a component of rubber vulcanizing agents) bisulfure *m* de tétraisopropylthiuram
TETRAL NOTATION - [comput] (quaternary notation) notation *f* quaternaire
TETRALIN - [chem] (a synonym for Tetrahydronaphthalene) tétraline *f*
TETRAMER - [chem] (a substance formed by the addition polymerization of four monomers) tétramère *m*
TETRAMETHYL - [chem] (having four methyl groups) tétramethyle *m*
~LEAD - [chem] (an anti-knock additive for motor fuels) plomb *m* tétraméthylique
TETRAMETHYLAMMONIUM CHLORODIBROMIDE - [chem] (a brominating agent) chlorodibromure *m* de tétraméthylammonium
TETRAMETHYLBUTANEDIAMINE - [chem] (a catalyst for synthetic resins) tétraméthylbutane-diamine *f*
TETRAMETHYLDIAMINOBENZHYDROL - [chem] (hydrol. An intermediate for dyestuffs and other organic compounds) tétraméthyldiaminobenzhydrol *m*
TETRAMETHYLDIAMINOBENZOPHENONE - [chem] (a dyestuffs intermediate) tétraméthyldiaminobenzophénone *m*
TETRAMETHYLDIAMINODIPHENYLEMETHANE - [chem] (an intermediate for dyestuffs) tétraméthyldiamino-diphényl-méthane *m*
TETRAMETHYLDIAMINODIPHENYLSULPHONE - [chem] (an analytical reagent and an intermediate for pharmaceuticals) tétraméthyl-diamino-diphénylsulfone *f*
TETRAMETHYLENEDIAMINE - [chem] (a synthesis intermediate) tétraméthylène-diamine *f*
TETRAMETHYLETHYLENEDIAMINE - [chem] (an intermediate in the production of polyurethanes and curing agents for epoxy resins) tétraméthyléthylène-diamine *f*
TETRAMETHYLTHIURAM DISULPHIDE - [chem] (a pesticide, disinfectant, and vulcanization accelerator) bisulfure *m* de tétraméthylthiuram
~MONOSULPHIDE - [chem] (a vulcanization accelerator in rubber compounding) monosulfure *m* de tétraméthylthiuram
TETRAMORPHOUS - [chem] (term used of a component which is capable of crystallization in four different forms) tétramorphe
TETRANITROANILINE - [chem] (an unstable substance used as an initiator for explosives) tétranitroaniline *f*
TETRANITROMETHANE - [chem] (an analytical reagent and a rocket propellant) tétranitrométhane *m*
TETRANOPSIA - [med] tétranopsie *f*, hémianopsie *f* en quadrant
TETRAPHENYLSILANE - [chem] (an intermediate for silicone polymers) tétraphénylsilane *m*
TETRAPHENYLTIN - [chem] (a synthesis intermediate) tétraphényl-étain *m*
TETRAPLOID - [bot] tétraploïde
TETRAPROPENYLSUCCINIC ANHYDRIDE - [chem] (a synthesis intermediate and a curing agent for synthetic resins) anhydride *m* tétrapropénylsuccinique
TETRAPROPYLENE - [chem] (dodecene. An intermediate for plasticizers and detergents) tétrapropylène *m*
TETRAPROPYLTHIURAM DISULPHIDE - [chem] (an ingredient of rubber accelerators and vulcanizing agents) bisulfure *m* de tétrapropylthiuram
TETRASTYLE - [arch] tétrastyle *m*
TETRATOMIC - [nucl] tétratomique
TETRAVALENT - [nucl] (capable of combining with four atoms of hydrogen or the equivalent) tétravalent
TETRODE - [electron] (a thermionic valve having four electrodes) tétrode *f*
~POINT-CONTACT TRANSISTOR - [electron] (a point-

contact transistor having three point contacts and one base connection) transistor *m* tétrode à contacts de pointe
TETRODE TRANSISTOR - [electron] (or transistor electrode; a four electrode transistor, usually a conventional junction triode having a second base electrode) transistor *m* tétrode
TETROXIDE - [chem] tétroxyde *m*
TETRYL - [chem] (an initiating explosive) tétryle *m*
TETTER - [med] herpès *m*, dartres *f*pl
TEXAS COTTON - [text] coton *m* du Texas
TEXASITE - [min] (emerald nickel) zaratite *f*
TEXT - [gen] texte *m*
TEXTILE - [gen] textile
[text] tissu *m*, étoffe *f*, matière *f* textile
~FABRICS - [text] tissus *m*pl, étoffes *f*pl
~FIBRE - [text] fibre *f* textile
~INDUSTRY - [text] industrie *f* textile
~INSERTION - [rubber ind] sous-couche *f* de textile, insertion *f* de toile
~MACHINE - [text] machine *f* textile
~MILL - [text] usine *f* textile
~OILS - [chem] (oils used to improve the physical characteristics of all types of fibres during processing) huiles *f*pl d'ensimage
~PLANT - [bot] plante *f* textile
~PROCESSING OILS - s.textile oils
~RAW MATERIAL - [text] matière *f* première textile
~SLEEVE - [mech] (of a mill) tuyau *m* de tissu filtrant
TEXTILITE PAPER YARN - [text] fil *m* de papier textilite
TEXTILOSE PAPER YARN - [text] fil *m* textilose
TEXTUAL - [gen] textuel
TEXTURE, to - [text] texturer
TEXTURE - [text] tissu *m*, tissure *f*
[text] (the arrangement of the threads) armure *f*
[biol] (the mode of union and disposition of elementary constituent parts) texture *f*, grain *m*
[gen] structure *f*
[geol] structure *f*, texture *f*
TEXTURED FABRIC - [text] tissu *m* en structures
~YARN - [text] fil *m* stretch
TEXTURELESS - [gen] (without texture) amorphe
TEXTURIZE, to - s. texture, to
TEXTURIZED YARN - [text] filé *m* high-bulk
THALASSEMIA - [med] anémie *f* méditerranéenne, syndrome *m* de Cooley
THALASSOTHERAPY - [med] thalassothérapie *f*
THALLINE - [chem] (a quinoline derivative with application in medicine as an antipyretic) thalline *f*
THALLIUM - [chem] (a metallic element, symbol Tl, A.N. 81, A.W. 204.39, with uses in metallurgy, pharmaceuticals, electronics, glass manufacture, analysis, and in the production of catalyst) thallium *m*
~ACETATE - [chem] (a component of ore flotation solutions) acétate *m* de thallium
~CARBONATE - [chem] (an analytical reagent) carbonate *m* de thallium
~CHLORIDE - [chem] (a catalyst) chlorure *m* de thallium
~HYDROXIDE - [chem] (an analytical reagent) hydrate *m* de thallium
~MONOXIDE - [chem] (a constituent of special glasses and an analytical reagent) monoxyde *m* de thallium
THALLIUM NITRATE - [chem] (an analitical reagent) nitrate *m* de thallium
~OXYSULFIDE - [chem] (compound of thallium oxygen and sulfur with photoconductive properties) oxysulfure *m* de thallium
~SULPHATE - [chem] (a pesticide and analytical reagent) sulfate *m* de thallium
~SULPHIDE - [chem] (a highly poisonous compound, used in photo-electric cells) sulfure *m* de thallium
THALLUS - [bot] thalle *m*
THALOFIDE CELL - [electron] (photoconductive cell in which the active light-sensitive material is thallium oxysulfide in a vacuum) cellule *f* thalofide
THANATOMANIA - [med] suicidomanie *f*
THANATOSIS - [med] gangrène *f*
THATCH, to - [constr] (the operation of covering a roof with a thatch) couvrir en chaume
THATCH - [constr] (covering of reeds, straw etc arranged on a roof so as to shed water) chaume *m* de toiture
~A STACK, to - [agric] couvrir en meulon
THATCHBOARD - [constr] panneau *m* de chaume
THATCHING - [constr] couverture *f* de chaume
THAW, to - [phys] (to change a frozen substance to a liquid by the application of heat) dégeler
THAW - [phys] (the act of thawing) dégel *m*, décongelation *f*
[met] fonte *f* des neiges
THAWING - [gen] dégèlement *m*, dégelage *m*
~POINT - [met] point *m* de rosée
~THE FIBRE - [text] décongélation *f* de la fibre
~TRANSFORMER - [el] transformateur *m* de dégèlement
T-HEAD CYLINDER - [mech] (type of i.c. engine cylinder having side-valves placed laterally) culasse *f* en T, culasse *f* à soupapes latérales
THEBAINE - [chem] (a poisonous alkaloid derived from opium) thébaïne *f*
THECA - [med] thèque *f*, gaine *f*, enveloppe *f*
THECOMA - [med] thécome *m*
THELAN PAN - [chem] (covered pan of semicircular cross section and mechanical scrapers, used in calcining sodium bicarbonate in the Solvay Process) pan *m* de Thelan
THELORRHAGIA - [med] thélorragie *f*
THENARDITE - [min] (natural sodium sulphate) thénardite *f*
THENARD'S BLUE - [paint] (alternative name for Cobalt Blue a blue pigment made by calcining cobalt oxide with alumina) bleu *m* de Thénard
THENYLDIAMINE HYDROCHLORIDE - [chem] (an anti-histamine drug used in medicine) chlorhydrate *m* de thényldiamine
THEOBROMINE - [chem] (an alkaloid obtained from Theobroma cacao and used in medicine in cardiac conditions and as a diuretic) théobromine *f*
~CALCIUM SALICYLATE - [chem] (a diuretic) salicylate *m* de théobromine et de calcium
~SODIUM SALICYLATE - [chem] (an approximately equimolar mixture of theobromine sodium and sodium salicylate used as a diuretic and in the treatment of angina pectoris) salicylate *m* de théobromine et de sodium
THEODOLITE - [instr] (instrument designed to measure horizontal and vertical angles) théodolite *m*
THEOMANIA - [med] folie *f* mystique, manie *f* religieuse

THEOPHYLLINE - [chem] (an alkaloid, obtained from tea or synthesized, and having application as a diuretic) théophylline *f*
THEOPHYLLINE-METHYLGLUCAMINE - [chem] (an approximately equimolecular mixture used in cardiovascular conditions) théophylline-méthylglucamine *f*
~SODIUM GLYCINATE - [chem] (a diuretic) théophylline-glycinate *m* sodique
THEOREM - [math] (a proposition which is demonstrably true) théorème *m*
~OF PARALLEL AXES - [phys] (the moment of inertia of a rigid body about any axis is equal to the moment of inertia about a parallel axis through the centre of mass plus the product of the mass of the body and the square of the perpendicular distance between the two axes) théorème *m* des axes parallèles
THEORETIC - s. theoretical
THEORETICAL - [gen] théorique
~AIR REQUIREMENT FOR COMBUSTION - [chem] (volume of air needed for complete combustion of unit mass of a fuel) air *m* théorique, facteur *m* de combustion
~CUT OFF FREQUENCY - [acoust] (a frequency at which the image attenuation constant changes from zero to a positive value or viceversa) fréquence *f* coupée théorique
~PLATE - [nucl] (hypothetical device for bringing two streams of material into so perfect a contact that they leave in equilibrium with each other) plateau *m* théorique
THEORIES OF PASSIVITY - [phys] (passivity is apparently due to a thin layer of oxide on the passive metal) théories *f*pl de passivité
THEORY - [gen] théorie *f*
~OF CHARGE TRANSFER - [phys] (or theory of charge transport; a theory used to understand the influence of the ion current in photoelectric processes) théorie *f* du transfert de la charge
~OF COLOUR VISION - [opt] théorie *f* de chromatopsie
~OF ELECTROLYTES - [el] (the fundamental idea underlying the modern view of electrolyte behaviour) théorie *f* des électrolytes
~OF EMULSIFICATION - [chem] (theory explaining the formation of oil-in-water and water-in-oil emulsions) théorie *f* d'émulsionnement
~OF INTERLACING - [text] théorie *f* des armures
~OF PROPORTION - [math] théorie *f* de proportionalité
~OF RELATIVITY - [phys] théorie *f* de la relativité
~OF VALENCY - [phys] (the theory that atoms combine with other atoms in determined proportions) théorie *f* de la valence
THERALITE - [geol] (coarse-grained igneous rock) théralite *f*
THERAPEUTIC - [med] (pertaining to medical treatment of diseases) thérapeutique
~RADIOLOGY - [med] radiothérapie *f*
THERAPEUTICS - [med] thérapeutique *f*
THERAPY - [med] (the curative method of treatment of disease) thérapie *f*
~TUBE - [radiat] (X-ray tube for use in X-ray therapy) tube *m* de thérapie
THERM - [meas] (a unit of quantity of heat equal 100,000 Btu. used as a unit for the sale of town gas) 100,000 B.T.U., 252 grandes calories *f*pl
THERM - [meas] (great calorie) grande calorie *f*
[meas] (small calorie) petite calorie *f*
[meas] (1000 great calories) thermie *f*, 1000 grandes calories *f*pl
THERMAL - [phys] thermal, thermique
[met] (an upward convection current in the atmosphere) vent *m* ascendant thermique
~ACTIVATION - [electron] (the role played by the thermal energy of an instrinsic semiconductor in maintaining the concentration of free electrons and holes in the crystals structure)activation *f* thermique
~AGITATION - [phys] (of the molecules of a body) agitation *f* thermique
~AGITATION NOISE - [el] (noise in a radio receiver due to random movement of electrons within the conducting elements) bruit *m* d'agitation thermique
~-AGITATION VOLTAGE - [electron] (the variations of voltage at the terminals of a conductor due to thermal agitation of the electrons in the conductor) tension *f* d'agitation thermique
~AMMETER - [el] (ammeter in which the deflection of the pointer depends on the sag of a fine wire carrying the current to be measured) ampèremètre *m* thermique
~ANALYSIS - [nucl] (a method of determining the temperatures of phase transformations by measuring the discontinuities in the slopes of the temperature-time curves obtained upon heating or cooling) analyse *f* thermique
[metall] (the use of cooling or heating curves in the study of changes in metals and alloys) analyse *f* thermique
~ANESTHESIA - s. thermanesthesia
~ANOMALY - [aero] anomalie *f* thermique
~BARRIER - [aero] (the limitation imposed by thermal effects on materials at high speeds and altitudes) barrière *f* thermique
~BLACK - [chem] (amorphous carbon used as a pigment) noir *m* de charbon
~BOND - [nucl] liaison *f* thermique
~BREAKDOWN - [electron] (or thermal runaway, of a semiconductor p-n junction) rupture *f* thermique
~BREEDER - [nucl] (a type of breeding reactor in which thermal neutrons are generated) surrégénérateur *m* à neutrons thermiques
~BUCKLING - [metall] (buckling arising in whole or part from thermal stress or deformation) tassement *m* thermique
~CAPACITY - [phys] (of water equivalent; the product of the specific heat of a body and its mass) capacité *f* thermique
~CAPTURE - [nucl] (capture of thermal neutrons) capture *f* thermique
~CHEMICAL REACTION - [nucl] (reaction which should be avoided by partial cooling of the graphite) réaction *f* chimique thermique
~CIRCUIT-BREAKER - [el] (device for automatic interruption of an electric circuit, operated by thermal action, e.g. expansion) interrupteur *m* thermique
~COEFFICIENT OF EXPANSION - [metall] coefficient *m* de dilatation
~COLUMN - [nucl] (a column of moderator material passing through a reactor shield to allow a controlled escape of thermal neutrons) colonne *f* thermique

THERMAL CONDUCTION - [phys] (conduction of heat) transmission *f* de la chaleur
~CONDUCTIVITY - [phys] (a measure of the facility with which a material transmits heat. Expressed in terms of the energy, (grammes-calories), transmitted per unit time (seconds) across unit area (sq. cm) through unit thickness (cm) for a specified temperature difference (I° C.) conductivité *f* thermique
~CONDUCTIVITY VACUUM GAUGE - [instr] jauge *f* thermique à vide
~CONTRACTION - [phys] (in rubber) rétrécissement *m* thermique
~CONVECTION - [phys] (convection of heat) convection *f* thermique
~CONVERTOR - [el] (electrically heated thermocouple) thermocouple *m*
~CROSS-SECTION - [nucl] (the cross-section as measured with thermal neutrons) section *f* efficace thermique
~CUT-OUT - s. thermal circuit-breaker
~CYCLE - [nucl] (a cycle of operation in which heat is transferred from one part of a system to another by changes in the temperature of a medium) cycle *m* thermique
~CYCLING - [metall] essai *m* à cycle thermique
~DEATH-POINT - [phys] (the temperature at which an organism is killed) limite *f* de résistance thermique
~DECOMPOSITION - [phys] (breaking down a compound by heat into simpler molecules) décomposition *f* thermique
~DEGRADATION - [plast ind] (deterioration of a plastic by the effect of heat) dégradation *f* thermique
~DEPRESSION - [met] (a depression developing over a land mass which has become substantially hotter than the sea areas round it) dépression *f* thermique
~DERATING FACTOR - [electron] (the factor by which the power dissipation rating must be reduced with increase of ambient or case temperature) facteur *m* thermique de réduction de puissance
~DETECTOR - [radio] détecteur *m* thermique
~DIFFUSION - [phys] (the phenomenon by which a temperature gradient in a mixture of two or more fluids tends to establish a concentration gradient) diffusion *f* thermique, effet *m* Soret
~DIFFUSION PLANT - [nucl] (for the separation of isotopes) installation *f* de diffusion thermique
~DIFFUSION RELAXATION - [phys] (a source of internal friction in solids. The heat generated at one point of the material by rapid compression diffuses into other areas before the strain is relaxed and is thus lost) perte *f* de chaleur par diffusion thermique
~DIFFUSION SEPARATION - [oil ind] séparation *f* par diffusion thermique
~DIFFUSITY - [phys] (constant determining the rate of rise of temperature at a point on a bar removed from the source of heat) diffusibilité *f*
~DISSOCIATION - [chem] dissociation *f* thermique
~EFFECT - s. thermal-agitation voltage
~EFFICIENCY - [phys] (the ratio of heat used to total energy consumed) rendement *m* thermique
~EMBRITTLEMENT - [phys] (increase in brittleness caused by rise of temperature) élévation *f* de la fragilité sous l'effet de la chaleur
~ENDURANCE - [glass man] (the relative ability of glass to withstand thermal shock) résistance *f* aux sauts thermiques
THERMAL ENERGY - [phys] (heat) chaleur *f*, énergie *f* thermique
~ENERGY REGION - [phys] (thermal range) domaine *m* d'énergie thermique
~ENERGY YIELD - [nucl] (the part of the total energy yield of the nuclear explosion which is radiated as thermal energy) rendement *m* thermique énergétique
~EQUATOR - [astr] (the belt of maximum temperature about the earth, which follows the sun's apparent seasonal motion) équateur *m* thermique
~EQUILIBRIUM - [phys] (the condition of a system in which the net rate of exchange of heat among its parts thas become zero) équilibre *m* thermique
~EXCITATION - [phys] (the acquisition of excess energy by atoms or molecules by collision processes with other particles) excitation *f* thermique
~EXPANSION - [phys] (the increase in volume due to heat) dilatation *f* thermique
~EXPANSION COEFFICIENT - [phys] (increase in volume per unit volume due a rise of I° C.) coefficient *m* de dilatation thermique
~EXPANSION INSTRUMENT - [instr] (hot-wire instrument) instrument *m* à fil chaud
~EXTENSION - s. thermal expansion
~FACTOR - [electron] (a measure of the frequency change in a magnetron caused by a given change in anode temperature) facteur *m* thermique
~FATIGUE - [mech] (fatigue caused wholly or in part by cyclic temperature variations) fatigue *f* due à la chaleur
~FATIGUE FAILURE - [electron] (in semiconductors) défaut *m* par fatigue thermique
~FISSION - [nucl] fission *f* provoquée par des neutrons thermiques
~GASES - [min] émanations *f*pl thermales
~GLASS - [glass man] (a glass having a low coefficient of expansion, making it thermally stable) verre *m* résistant à la chaleur
~GRADIENT - [geol] degré *m* géothermique
~IMPULSE WELDING - [mech] (welding in which thermal shock is applied to the material) soudage *m* par impulsions thermiques
~INERTIA - [phys] (the reciprocal of thermal response) inertie *f* thermique
~INSTABILITY - [phys] (a positive temperature coefficient, especially in a component of low heat capacity) instabilité *f* thermique
~-INSTABILITY REGION - [electron] (in magnetrons) zone *f* d'instabilité thermique
~INSTRUMENT - [instr] (an instrument the operation of which depends on the heating effect of a current) appareil *m* thermique
~INSULATING BOARD - [constr] (heat insulating material in board form) plaque *f* isolante
~INSULATION - [constr] isolation *f* thermique
~IONIZATION - [electron] (of a gas or a vapour; the ionization of atoms or molecules resulting from the thermal agitation of a gas or vapour) ionisation *f* thermique
~JET ENGINE - [aero] moteur *m* thermique à réaction
~LAG - [phys] inertie *f* thermique
~LEAKAGE FACTOR - [nucl] facteur *m* de fuite de neutrons thermiques

THERMAL LIMIT - [el] (the maximum permissible output of an electrical machine as governed by the considerations of temperature rise) limite *f* thermique de sécurité
~LOW - s. thermal depression
~METAMORPHISM - [geol] (the process by which the atoms forming the constituent parts of rocks are regrouped as a result of induced temperature changes) métamorphisme *m* thermique
~MICROPHONE - [el acoust] (a microphone which depends for its operation on the variation of a line wire on the passage of a sound-wave, the velocity of the particles in the sound-wave cooling the heated wire) microphone *m* à fil chaud, thermophone *m*
~MOTION - [phys] (effect caused by the motion of atoms and resulting in disturbing the regularity of a crystal lattice) mouvement *m* thermique
~MOTIONS - [phys] agitation *f* thermique
~NEUTRON - [nucl] (a neutron of which the kinetic energy is below about 10^{2} electron-volts) neutron *m* thermique
~NEUTRON BEAM - [nucl] faisceau *m* de neutrons thermiques
~NEUTRON CHAIN REACTION - [nucl] réaction *f* en chaîne à neutron thermiques
~NEUTRON CROSS-SECTION - s. thermal cross-section
~NEUTRON FISSION - [nucl] (slow neutron fission) fission *f* par neutrons thermiques
~NOISE - s. thermal agitation noise
~NOISE GENERATOR - [electron] (a generator using the inherent agitation of an electron tube to provide a calibrated noise source) générateur *m* de bruit thermique
~OHM - [el] (the thermal resistance of a body across the opposite faces of which there is a temperature difference of one centigrade when heat flows at the rate of I watt) résistance *f* thermique
~OVERLOAD CAPACITY - [el] (of a meter; the ratio between the current continuously applied for maximum permissible temperature rise and the rated current of a meter) capacité *f* de surcharge thermique
~OVERLOAD RELAY - [el] (a thermally operated device, connected in the operating motor-circuit, which served to limit the load on the motor by interrupting the supply to it) relais *m* thermique de surcharge
~POWER MEASURING ASSEMBLY - [meas] ensemble *m* de mesure de la puissance thermique
~PROCESS REGENERATION - s. thermal reclaiming process
~RADIATION - [phys] (radiation emitted by bodies which are not at absolute zero temperature) rayonnement *m* thermique
~RADIATION BURN - [radiat] lucite *f* par rayonnement thermique
~RANGE - s. thermal energy region
~REACTOR - [nucl] (slow reactor) réacteur *m* à neutrons thermiques
~RECLAIMING PROCESS - [rubber ind] régénération *f* à la vapeur surchauffée
~RELAY - [el] (temperature relay) relais *m* thermique
~RESISTANCE - [phys] (the property of a substance or a body which causes it to resist the transmission of heat) résistance *f* thermique

THERMAL RESISTANCE - [el] (of a cable; the resistance to the flow of heat, expressed in thermal ohms, of one cm length of cable, offered by the path from the conductor to the external surface of the cable) résistance *f* à chaud, résistance *f* contre la chaleur
~RESISTIVITY - [el] (specific thermal resistance; a measure of the property of a material which resists the flow of heat therein. It is usually expressed in thermal ohms per cubic centimetre) résistivité *f* thermique
~RESPONSE - [nucl] (rate of temperature rise in a reactor operating at the rated power if no heat is withdrawn by cooling) réponse *f* thermique
~SHIELD - [nucl] (a high-density heat-conducting portion of a shield placed near the reflector) bouclier *m* thermique
~SHOCK - [phys] choc *m* thermique
~SHOCK CONDITIONS - [mech] (conditions of wide temperature variations over very short periods of time, e.g. in air passing through a gas turbine during stopping and starting) changements *mpl* rapides de température
~SHOCK SHIELD - [nucl] écran *m* de choc thermique
~SHORT-TIME CURRENT RATING - [el] (USA only; rated short-circuit current) courant *m* de court-circuit nominal
~SHRINKAGE - [rubber ind] retrait *m* par refroidissement
~SHUNT - [el] (any device for rapid local abstraction of heat, as when a part is to be heated in one region and kept cool in another) dérivation *f* thermique
~SIPHON - [phys] (a close loop containing fluid, a vertical member of which is kept at different temperature from that of another vertical member) siphon *m* thermique
~SLEEVE - [nucl] (in a nuclear reactor) manchette *f* calorifuge
~SOARING FLIGHT - [aero] vol *m* plané thermique
~SPALLING - [metall] écaillement *m* thermique
~SPECTRUM - [phys] spectre *m* calorifique
~SPIKE - [phys] pointe *f* thermique
~SPRING - [min] source *f* thermale
~STABILITY - [phys] (stability of physical characteristics in respect to temperature changes) stabilité *f* thermique
~STABILITY UNDER LOAD - [mech] stabilité *f* thermique sous charge
~STATION - [el] (or thermal power-station; a power station in which electrical energy is produced by means of thermal prime movers) centrale *f* thermique
~STORAGE WATER HEATER - [el] chauffe-eau *m* à accumulation
~STREETS - [met] (narrow bands of vertically-moving air) émanations *fpl* thermiques d'air ascendant
~STRESS - [nucl] tension *f* thermique
~TIDE - [met] marée *f* atmosphérique d'origine thermique
~TRANSMITTANCE - [phys] (heat-loss coefficient) coefficient *m* de dispersion
~TRIP - [el] (tripping relay for a large circuit-breaker, which operates by thermal expansion) relais *m* à déclenchement à action thermique
~TUNER - [electron] (a microwave tuner using thermal tuning of a cavity resonator) syntonisateur *m*

thermique
THERMAL TUNING - [electron] syntonisation *f* thermique
~TUNING RATE - [electron] vitesse *f* de syntonisation thermique
~TUNING SENSITIVITY - [electron] sensibilité *f* pour syntonisation thermique
~TUNING TIME - [electron] durée *f* de syntonisation thermique
~UTILIZATION - [phys] (the probability that a thermal neutron, which is absorbed, is absorbed usefully) utilisation *f* thermique
~UNIT - [meas] (British Thermal Unit, or B.T.U. a unit of heat, that is the quantity of heat which is required to raise the temperature of one pound of water from 60° F. to 61° F.) unité *f* de chaleur
~VALUE - [phys] pouvoir *m* calorifique, puissance *f* calorifique
~VAPOURS - s. thermal gases
~WIND - [met] (the vector difference between winds at different levels which arises from a temperature gradient existing in the atmosphere between such levels) vent *m* thermique
~YIELD - s. thermal energy yield
THERMALIZATION - [phys] (reduction of the kinetic energy by neutrons to a point where they have approximately the same kinetic energy as the atoms or molecules of the medium in which the neutrons are undergoing elastic scattering) thermalisation *f*
~RANGE - [nucl] zone *f* de thermalisation
THERMALLY-ACTUATED ELEMENT - [gas ind] élément *m* thermo-sensible
~ATTACKED RUBBER - [rubber ind] caoutchouc *m* dégradé par la chaleur
~-STABLE OIL - [chem] (oil of which the physical characteristics do not vary seriously with temperature) huile *f* stabile à la chaleur
THERMALGESIA - [med] douleur *f* causée par la chaleur, thermo-esthésie *f*
THERMANALGESIA - [med] thermo-analgésie *f*
THERMANESTHESIA - s. thermoanalgesia
THERME - s. therm
THERMHYPESTHESIA - [med] hypoesthésie *f* à la chaleur
THERMIC - s. thermal
~BALANCE - [meas] bolomètre *m*
~CATALYTIC CRACKING - [ind chem] craquage *m* catalytique thermique
~INERTIA - [metall] (of incandescent metal etc) inertie *f* thermique
~WEIGHT - [phys] entropie *f*
THERMION - [electron] (an electron or ion liberated by thermionic emission) thermion *m*
THERMIONIC - [electron] thermoionique, électronique
~AMPLIFIER - [electron] amplificateur *m* thermoionique
~ARC - [electron] (electric arc characterized by the fact that the thermionic cathode is heated by the arc current itself) arc *m* thermoélectronique
~CATHODE - [electron] (or hot cathode; a cathode functioning primarily by the process of thermionic emission) cathode *f* à incandescence, cathode *f* thermionique
~CONVERTER - [electron] (converter converting heat energy directly into electric energy) convertisseur *m* thermionique
THERMIONIC CURRENT - [electron] (the electric current which is represented by the flow of electrons emitted from a thermionic cathode) courant *m* thermionique
~DETECTOR - [electron] (high-vacuum thermionic tube used for detecting radiofrequency signals) détecteur *m* thermionique
~DIODE - [electron] (a diode electron tube or valve having a heated cathode) diode *f* thermionique
~EMISSION - [electron] (electron emission resulting solely from the temperature of the emitter) émission *f* thermionique
~ENERGY CONVERSION - [electron] (the conversion of some form of energy into another form by means of thermionic equipment) transformation *f* thermionique d'énergie
~GENERATOR - s. thermionic converter
~GRID EMISSION - [electron] (or primary grid emission; currents produced by electrons thermionically emitted from a grid) émission *f* primaire de grille
~OSCILLATOR - [electron] oscillateur *m* thermionique
~RECTIFIER - [electron] (thermionic tube designed to produce a unidirectional current from an alternating input) redresseur *m* thermionique
~TUBE - s. thermionic valve
~VALVE - [electron] (electron valve in which the heating of one or more of the electrodes is designed to cause electron or ion emission) tube *m* thermionique
THERMISTOR - [electron] (electron device which makes use of the change of resistivity of a semiconductor with change in temperature) thermistor *m*
~MOUNT - [electron] (a waveguide termination in which a thermistor can be incorporated) monture *f* pour thermistor
~RELAY - [el] (relay device operated by a thermionic tube) relais *m* thermionique
THERMIT - [metall] (a mixture of a metallic oxide and aluminium powder with applications in metallurgy, as a method of welding in situ, and as a filling material for incendiary bombs) thermite *f*
~CRUCIBLE - [metall] (used in welding) creuset *m* pour aluminothermie
~MIXTURE - [metall] (used in welding) mélange *m* aluminothermique
~MOULD - [metall] (used in welding) moule *m* pour soudure à thermite
~PROCESS - [metall] aluminothermie *f*
~PROCESS BUTT-WELDING - [metall] soudure *f* avec apport de fer-thermite
~PROCESS INTERMEDIATE WELDING - [metall] soudure *f* avec apport de fer-thermite
~REACTION - [chem] (aluminothermic reaction) réaction *f* aluminothermique
[metall] réaction *f* à thermite
~REDUCTION - [metall] réduction *f* par réaction à thermite
~WELDING - [metall] soudure *f* à thermite
THERMOALGESIA - s. thermalgesia
THERMOANALGESIA - s. thermanalgesia
THERMOBAROMETER - [instr] thermobaromètre *m*
THERMOCAUTERY - [med] thermo-cautère *m*
THERMOCHEMICAL - [chem] thermo-chimique *m*
THERMOCHEMISTRY - [chem] (the science of the heat changes resulting from chemical action) thermo-chimie *f*

THERMOCOUPLE - [el] (device consisting of a junction of two dissimilar elements, the electrical resistance of which is related to the temperature to which they are subjected) thermo-couple *m*, couple *m* thermo-électrique
~AMMETER - [instr] (current-measuring device depending upon a junction between dissimilar metals) ampèremètre *m* à couple thermo-électrique
~INSTRUMENT - [instr] (an instrument designed to give temperature values by measuring the current obtained from a thermocouple) appareil *m* à thermocouple
~THERMOMETER - [instr] pyromètre *m* à couple thermoélectrique
~VACUUM GAUGE - [instr] manomètre *m* thermo-électrique, vacumètre *m* thermo-électrique
~VOLTMETER - [el] (combination of the thermocouple and sensitive millivoltmeter) pince *f* thermo-électrique
~WATTMETER - [instr] thermo-wattmètre *m*
THERMODYNAMIC - [phys] thermo-dynamique
~EQUATION OF STATE - [phys] (an equation derived by consideration of reversible energy changes, giving a relationship between pressure, volume and temperature for a state of matter) équations *f*pl thermo-dynamiques d'état
~EQUILIBRIUM - [phys] équilibre *m* thermo-dynamique
~POTENTIAL - [phys] potentiel *m* thermo-dynamique
~PROBABILITY - [phys] (the number equally probable states of a statistical assembly) probabilité *f* thermo-dynamique
~QUANTITIES - [phys] (macroscopic quantities affecting the internal state of a system which determines its internal energy) grandeurs *f*pl thermo-dynamiques
~SYSTEM - [phys] (system whose behaviour can be described by thermodynamic quantities) système *m* thermo-dynamique
THERMODYNAMICAL - s. thermodynamic
THERMODYNAMICS - [phys] (the mathematical aspect of the relation of heat to other forms of energy) thermo-dynamique *f*
THERMOELASTIC COEFFICIENT - [mech] (the modulus of elasticity which is defined for a three-dimensional system) coefficient *m* de thermo-élasticité
THERMOELECTRIC - [phys] thermo-électrique
~COOLER - [electron] (used for the local cooling of hot transistor below ambient conditions) dispositif *m* de refroidissement thermo-électrique
~COUPLE - [instr] (used to measure temperatures) couple *m* thermo-électrique
~EFFECT - [el] (the production of an e. m.f. due to a difference of temperature between two junctions of different metals or alloys forming part of the same circuit) effet *m* thermo-électrique, effet *m* Peltier, effet *m* Seebeck
~GENERATING SET - [el] groupe *m* thermique
~PAIR - s. thermoelectric couple
~POWER - [el] (the thermoelectric force per degree) puissance *f* thermo-électrique
~PYROMETER - s. thermoelectric thermometer
~THERMOMETER - [instr] (a pyrometer the operation of which depends on the variations of electrical resistance with temperature) pyromètre *m* à couple thermo-électrique
~TRACTION - [el] traction *f* thermo-électrique
THERMOELECTRICITY - [el] (the current flowing in a circuit due to an electromotive force generated by the difference in temperature between the junction of two dissimilar metal conductors and another part of the circuit) thermo-électricité *f*
THERMOELECTROMETER - [instr] thermo-électromètre *m*
THERMOELECTROMOTIVE FORCE - [el] (electromotive force developed by a thermo-couple) force *f* thermo-électromotrice
THERMOFORMING - [plast ind] (production of objects from plastic sheet with the aid of heat) thermoformage *m*
THERMOFORMS - [plast ind] (thermoformed products) produits *m*pl thermoformés
THERMOGALVANOMETER - [instr] (a galvanometer in which the current to be measured passes through a fine wire coil in which it produces a heating effect) thermogalvanomètre *m*
THERMOGRAM - [met] (the record made by a thermograph) thermogramme *m*
THERMOGRAPH - [instr] (a continuously recording thermometer) thermographe *m*
THERMOHARDENING - s. thermosetting
THERMOJUNCTION - [instr] (one of the surfaces of contact between the two conductors of a thermocouple) point *m* de contact
THERMOLABILE - [chem] (term used of a substance which has a tendency to decomposed under the action of heat) thermolabile
THERMOLOGY - [phys] thermologie *f*
THERMOLUMINESCENCE - [phys] (luminescence which occurs only at a certain temperature) thermoluminescence *f*
THERMOLYSIS - [chem] (decomposition or dissociation due to heat) thermolyse *f*, dissociation *f* par la chaleur
THERMOMECHANICAL RECLAIMING PROCESS - [rubber ind] (method for recovering rubber from scrap using a combination of heat and mechanical treatment) procédé *m* de régénération thermo-mécanique
THERMOMETAL - [metall] bimétal *m*
THERMOMETER - [instr] (instrument for measuring temperature) thermomètre *m*
~PROBE - [instr] (or pyrometer probe; a device comprising a thermocouple and so arranged that it can easily be inserted in the part of a machine of which the temperature is to be measured while in operation) sonde *f* pyrométrique
~RESISTOR - [instr] (resistor forming the temperature-sensitive part of a resistance thermometer) résistance *f* pyrométrique
~SCREEN - [mech] (shelter with louvred sides to protect meteorological instruments from the direct action of the weather) écran *m* du thermomètre
~TUBE - [instr] (thermocouple fitted into a protective tube) canne *f* pyrométrique
THERMOMETRIC - [instr] thermométrique
~COEFFICIENT - [phys] (the coefficient of expansion at constant pressure, or the coefficient of change of pressure at constant volume) coefficient *m* thermométrique
~COLUMN - [instr] colonne *f* du thermomètre
~CORRECTION - [instr] (temperature correction) correction *f* thermométrique
~CRAYON - [instr] (temperature-indicating crayon)

crayon *m* thermométrique
THERMOMETRIC SCALES - [meas] (the Centigrade scale, the Fahrenheit scale etc) échelles *f*pl thermométriques
THERMOMETRICAL - s. thermometric
THERMOMETRY - [meas] (the measurement of temperatures) thermométrie *f*
THERMOMOLECULAR PRESSURE - [phys] pression *f* thermo-moléculaire
THERMONUCLEAR - [nucl] (said of a nuclear reaction induced by heat) thermonucléaire
~ENERGY - [nucl] (ghe energy obtained by a nuclear reaction induced by heat) énergie *f* thermonucléaire
~POWER - s. thermonuclear energy
~REACTION - [nucl] (a nuclear reaction in which the energy necessary for the reaction is provided by colliding particles which have kinetic energy by virtue of their thermal agitation) réaction *f* de fusion nucléaire, réaction *f* thermonucléaire
THERMOPAIR - s. thermoelectric couple
THERMOPHILIC ORGANISM - [biol] (organism which thrives in temperature above room levels) organisme *m* thermophile
THERMOPHONE - [el acoust] (electro-acoustic transducer in which sound-waves of calculable magnitude result from the expansion and contraction of air adjacent to a conductor whose temperature varies in response to a current input) thermophone *m*
THERMOPILE - [instr] (an instrument in which the thermoelectric effect exhibited by two dissimilar metals is utilised to measure small temperature differences) thermopile *f*, pile *f* thermoélectrique
THERMOPLASTIC - [plast ind] (becoming plastic when heated) thermoplastique
~ADHESIVE - [ind chem] (a synthetic resin adhesive based on a thermoplastic resin) colle *f* thermoplastique
~RECORDING - [el acoust] (method of electron beam recording of information on a deformable plastic film) enregistrement *m* thermoplastique
~RESINS - [plast ind] (those which can be made plastic by heating and harden again on cooling, their properties remaining unchanged, e.g. vinyl, polystyrene, polyamide and acrylic resins) résines *f*pl thermoplastiques
THERMOPLASTICITY - [chem] thermoplasticité *f*
THERMOPLASTICS - [plast ind] (synthetic resins which may be formed by the application of heat and which regain their original properties on cooling) thermoplastique *f*
THERMOREGULATION - [th eng] thermo-régulation *f*
THERMOREGULATOR - [th eng] thermorégulateur *m*
THERMOS BOTTLE - [impl] bouteille *f* Thermos
~FLASK - s. thermos bottle
THERMOSCOPE - [instr] (instrument which is perceptive of change of temperature) thermoscope *m*
THERMOSETTING - [plast ind] (synthetic resins which can be formed by the application of heat which brings about a chemical change so that they cannot be remelted) thermodurcissable
~ADHESIVE - [ind chem] (adhesive which is set and hardened by the application of heat) colle *f* thermodurcissable
~COMPOSITIONS - [plast ind] (compositions in which a chemical reaction takes place while they are being moulded under heat and pressure) matières *f*pl plastiques thermodurcissables
TERMOSETTING PLASTIC - [plast ind] duroplaste *m* thermodurcissable *m*
THERMOSIPHON - [th eng] (an arrangement for the circulation of a liquid by the combination of siphon tubes and heat) thermosiphon *m*
~COOLING - [ind chem] refroidissement *m* par thermosiphon
~SYSTEM - [th eng] système *m* à thermosiphon
THERMOSTABLE - [phys] (term used of a substance which does not decomposed when subjected to moderate heat) thermostabile
THERMOSTAT - [contr] (controlling device actuated by temperature conditions) thermostat *m*
THERMOSTATIC - [instr] thermostatique
~CONTROL - [el] (the automatic regulation of temperature by means of a thermostat) commande *f* thermostatique
~OIL CONTROL VALVE - [contr] (an automatic valve actuated by a thermostat, to control the action of an oil cooler) valve *f* thermostatique de l'huile
~RELAY - [el] relais *m* thermostatique
~VALVE - [contr] valve *f* thermostatique, robinet *m* à commande thermostatique
THERMOSTATICS - [contr] thermostatique *f*
THERMOTANK - [th eng] (a heat exchanger) échangeur *m* thermique
THERMOTAXIS - [biol] (the response of an organism to the stimulus of heat) thermotropisme *m*
THERMOTHERAPY - [med] thermothérapie *f*
THERMOTOLERANT - [bot] (which can endure high temperatures) thermo-tolerant
THERMOTROPISM - s. thermotaxis
THETA - [gen] (in the Greek alphabet) thêta *m*
~FUNCTION - [math] fonction *f* thêta
~POLARIZATION - [el] (the state of an electromagnetic wave in which the E-vector is tangential to the meridian lines of a specified spherical frame of reference) polarisation *f* thêta
THI - s. thio
THIAMINE HYDROCHLORIDE - [chem] (vitamin B aneurine hydrochloride. The essential antineuritic vitamin, having important therapeutic and nutritional application) chlorhydrate *m* de thiamine
THIAMYLAL SODIUM - [chem] (a barbiturate of very short action with uses in medicine as an anaesthetic) thiamylal *m* sodique
THIAZINE - [chem] (a synthesis intermediate) thiazine *f*
THIAZOLE - [chem] (a synthesis intermediate for dyestuffs) thiazole *m*
THIAZOSULPHONE - [chem] (an antileprotic drug) thiazosulfone *f*
THIBET YARN - [text] fil *m* "Thibet"
THICK - [gen] épais, gros
[mech] de grosse épaisseur
[chem] consistant, visqueux, dense
[met] (of fog) épais, intense
[mining] (of a coal seam) puissant
~DIPOLE - [radio] (a dipole with a low ratio of length to diameter) dipôle *m* épais
~FABRIC - [text] tissu *m* épais
~FOG - [met] brouillard *m* épais, brume *f* intense
~HUSKED - [agric] à grosse cosse
~LEAF - [bot] plante *f* grasse
~LENS - [opt] lentille *f* épaisse
~MASH - [brew ind] trempe *f* épaisse

THICK-MASH METHOD - [brew ind] méthode *f* par décoction
~MUD - [soil] boue *f* épaisse
~NUT - [mech] (a nut of which the axial dimension is substantially greater than the standard) écrou *m* haut
~OIL - [chem] huile *f* dense
~PLATE - [metall] tôle *f* forte, tôle *f* grosse
~PULP - [paper man] pâte *f* épaisse
~SEAM - [mining] couche *f* puissante
~SOURCE - [nucl] source *f* épaisse
~SPACE - [print] espace *f* forte
~TARGET - [electron] (target of such thickness that there is considerable energy loss of the incident particles or photons traversing it) cible *f* épaisse
~THREAD - [text] fil *m* gros, fil *m* grossier
~-WALLED CELL - [bot] cellule *f* à paroi épaisse
~-WALLED COCOON - [text] cocon *m* à coque, cocon *m* à paroi épaisse
~YARN - [text] fil *m* gros
THICKEN, to - [gen] épaisser
[paint] épaissir
[tool] recharger
THICKENED BAST FIBRE - [bot] fibre *f* corticale épaisse
~CELL WALL - [bot] paroi *f* épaissie de la cellule
THICKENER - [ind chem] (device or machine for reducing the water content of a slurry) épaississant *m*
[min] (an ore-dressing apparatus) bassin *m* de décantation
[paper man] (an apparatus for reducing the water content of pulp stock) épaississant *m*
THICKENING DRUM FOR WOODPULP - [paper man] tambour *m* d'épaississement pour cellulose
~OF LATEX - [rubber ind] épaississement *m* du latex
THICKLY LIQUID IRON - [metall] fonte *f* froide
THICKNESS, to - [metall] (the operation of applying a temporary thickness to a mould) mettre une épaisseur provisoire
[carp] (the operation of making an even thickness) mettre d'épaisseur
THICKNESS - [gen] épaisseur *f*, couche *f*
[phys] densité *f*, consistance *f*
[mech] fausse pièce *f*, épaisseur *f*
[mech] (of a nut) hauteur *f*
[constr] (of a course) hauteur *f* d'assise
[mining] puissance *f*
~-CHORD RATIO - [aero] (the relation between the maximum thickness of an aerofoil section, taken perpendicularly to the cord and the length of its chord) épaisseur *f* relative
~GAUGE - [meas] calibre *m* d'épaisseur
~GAUGING - [metall] mesure *f* d'épaisseur
~LINES - [met] (lines on a map or chart passing through a series of points at which the altitude difference between given surfaces of constant pressure is the same) lignes *f*pl d'égale épaisseur
~OF A BOARD - [constr] rive *f* d'une planche
~OF A LIQUID - [ind chem] épaisseur *f* d'un liquide
~OF A PIPE - [plumb] épaisseur *f* de la paroi d'un tuyau
~OF A VAULT - [constr] force *f* d'une voûte, épaisseur *f* d'une voûte
~OF FALSE MOULD - [metall] largeur *f* de la chemise
THICKNESS OF FILLING - [text] épaisseur *f* de la trame
~OF FLANGE - [mech] épaisseur *f* de la bride
~OF SOIL - [geol] épaisseur *m* du sol
~OF THE CELL - [bot] épaisseur *f* de la cellule
~OF TOOTH - [mech] (of gearing) épaisseur *f* de la dent
~PIECE - [metall] mouche *f*
~RATIO - s. thickness-chord ratio
~STRICKLE - [metall] planche *f* secondaire
~TESTER - [instr] jauge *f* d'épaisseur
~TOLERANCE METER - [instr] mesurer *m* de tolérances dans l'épaisseur
THICKNESSING - [carp] mise *f* d'épaisseur
~MACHINE - [mach tool] (for wood working) machine *f* à raboter tirant des bois d'épaisseur
THICKSET VELVET - [text] velours *m* de coton rayé
THIEF - [gen] voleur *m*
[ind chem] (a device for extracting a sample of a liquid from a container) sonde *f*
[oil ind] tube *m* de repêchage
~GLASS - [ind chem] pipette *f* en verre à échantillons
~ROD - [oil ind] pipette *f* à échantillons
~SAMPLING - [oil ind] prise *f* d'échantillons par tarière
~SAND - [oil ind] sable *m* absorbant
~TUBE - s. thief [oil ind]
THIEMIA - [med] thiémie *f*
THIEVE, to - [gen] voler
[ind chem] échantillonner
THIEVING - [oil ind] échantillonnage *m*
THIGH - [anat] cuisse *f*
~BONE - [anat] fémur *m*
~BOOT - [shoe man] botte *f* cuissarde
THIGMOTAXIS - [biol] (the response of an organism to the stimulus of touch or contact) thigmotropisme *m*
THIGMOTROPISM - s. thigmotaxis
THILL - [min] (the floor in a coal mine) mur *m*
[impl] (of a horse-drawn vehicle) limon *m*, brancard *m*
THIMBLE - [impl] (for sewing) dé *m*
[mech] (a sleeve) bague *f*, bride *f*
[mech] (tubular cone used to expand tubes) mandrin *m* à arrondir les tubes de chaudière
[plumb] (a short blind tube, used in some thermal equipment, e.g. heat-recovery boilers) virole *f*, cosse *f*
[mech] (tubular distance-piece) pièce *f* d'écartement tubulaire
[mining] (a wall box) niche *f* murale
[metall] cuve *f* à laitier
[nucl] (tube closed at one end used to enclose control rods to be introduced into the reactor) chaussette *f*
[naut] (a metal antichafing ring forming a guard over a loop or eye in a sail) cosse *f* de câble
~IONIZATION CHAMBER - [electron] (small, usually cylindrical, ionization chamber, generally with walls of organic material) chambre *f* dé, chambre *f* d'ionisation à dé à coudre
~TUBE - [oil ind] tube *m* à gobelet
THIN, to - [gen] amincir, affiner
[paint] diluer, délayer
[agric] éplucher, effeuiller, éclaircir
THIN - [gen] mince, fin, menu

THIN - [gen] (not dense) peu épais
[photo] (said of a picture) faible
[chem] (of an oil, of low viscosity) fluide, peu consistant
[agric] (said of the soil) pauvre, appauvri
[gen] (pf air) rarefié, subtil
~CELL WALL - [bot] paroi *f* cellulaire mince
~FIBROUS GROWTH - [bot] poils *mpl* poussant espacés
~FILM STORAGE - [comput] mémoire *f* à pellicule mince magnétique
~-LAYER FILM - [photo] pellicule *f* à couche d'é-mulsion mince
~LENS - [opt] lentille *f* mince
~LENS RELATIONSHIP - [opt] relations *fpl* des len-tilles minces
~OUT, to - [agric] élaguer, émonder, ébrancher
~PILE - [text] poil *m* peu serré
~PULP - [paper man] pâte *f* fluide
~SINGLE FILAMENT - [text] fil *m* simple et fin
~SPACE - [print] espace *f* fine
~STRAW BLADE - [text] brin *m* de paille fétu
~TARGET - [electron] (target of such small thick-ness that there is very little loss or absorption of the incident particles or photons traversing it) cible *f* mince
~THREAD - [text] fil *m* fin, fil *m* mince
~-WALLED - [gen] à paroi mince
~WALLED CASTING - [metall] pièce *f* de fonte à pa-rois minces
~-WALLED COUNTER TUBE - [electron] tube *m* comp-teur à paroi mince
~-WINDOW COUNTER TUBE - s. thin-walled counter tube
THINDOWN - [astr] diffusion *f* de l'énergie des rayons cosmiques
THING - [gen] chose *f*, objet *m*
THINNED - [paint] dilué, aminci, effilé
THINNERS - [paint] (general term for any volatile liquid used to reduce the viscosity of a coating composition) diluants *mpl*
THINNESS - [gen] minceur *f*, peu *m* d'épaisseur, finesse *f*
[paint] dilution *f*
[chem] fluidité
[agric] état *m* clairsemé
THINNING - [paint] délayage *m*, dilution *f*
~DOWN - s. thinning out
~MEDIUM - [ind chem] diluant *m*
~OF THREADS BY DRAWING - [text] raffinement *m* des fils par laminage
~OUT - [chem] (reduction in viscosity of a lubricat-ing oil as temperature rises) dilution *f*
~RATIO - [paint] taux *m* de dilution
THIO - [chem] (prefix used in chemical terminology to indicate that a sulphur atom occupies the posi-tion normally held by an oxygen atom, in the com-pound designed) thio-
THIOACETIC ACID - [chem] (An analitycal reagent) acide *m* thioacétique
THIOACIDS - [chem] (acids in which the hydroxil of the carboxyl group has been replaced by SH) thioacides *mpl*
THIOALCOHOLS - [chem] (mercaptans) thioalcools *mpl*
THIOARSENIOUS - [chem] sulfarsénieux
THIOBENZOIC ACID - [chem] (a synthesis interme-diate) acide *m* thiobenzoïque
THIOCARBANILIDE - [chem] (a vulcanization accele-rator) thiocarbanilide *f*
THIOCARBONATE - [chem] thiocarbonate *m*
THIOCYANATE - [chem] (a salt or ester of thiocya-nic acid) thiocyanate *m*
THIOCYANIC ACID - [chem] (sulphocyanic acid. Volatile liquid which polymerizes readily) acide *m* thiocyanique
THIODIGLYCOL - [chem] (a synthesis intermediate) thiodiglycol *m*
THIODIGLYCOLIC ACID - [chem] (an analytical rea-gent) acide *m* thiodiglycolique
THIOETHERS - [chem] (alkyl sulphides, neutral, evil-smelling volatile liquids) sulfures *mpl* d'al-coyles
THIOGLYCEROL - [chem] (a plastics stabilizer) thioglycérol *m*
THIOGLYCOLIC ACID - [chem] (a component of hair waving solutions) acide *m* thioglycolique
THIOHYDANTOIN - [chem] (a synthesis intermedia-te) thiohydantoïne *f*
THIONYL CHLORIDE - [chem] (a catalyst and synthe-sis intermediate) chlorure *m* de thionyle
THIOPHEN - s. Thiophene
THIOPHENE - [chem] (colourless liquid hydrocarbon, found in coal tar) thiophène *m*
THIOPHENOL - [chem] (phenyl mercaptan. A synthe-sis intermediate) thiophénol *m*
THIOPHOSGENE - [chem] (a synthesis intermediate) thiophosgène *m*
THIOPROPOZATE HYDROCHLORIDE - [chem] (a tran-quilizer with application in the treatment of neuro-tic conditions) chlorhydrate *m* de thiopropazate
THIORIDAZINE HYDROCHLORIDE - [chem] (a pheno-phiazine derivative with application in medicine as a tranquillizer) chlorhydrate *m* de thioridazine
THIOSALICYLIC ACID - [chem] (an analytical rea-gent) acide *m* thiosalicylique
THIOSEMICARBAZIDE - [chem] (aminothiourea. A rodenticide an analytical reagent) thiosémicarba-zide *f*
THIOSEMICARBAZONE - [chem] (thiocetazone; ami-thiozone. A tuberculostatic and antileprotic drug) thiosémicarbazone *f*
THIOSULPHATE - [chem] (a salt of thiosulphuric acid) thiosulfate *m*
THIOSULPHURIC ACID - [chem] (unstable acid used in bleaching and photography) acide *m* thiosulfurique
THIOURACIL - [chem] (bitter-tasting, creamy, odourless powder used in the treatment of thyro-toxicosis) thio-uracile *m*
THIOUREA - [chem] (organic intermediate and rub-ber accelerator) thiourée *f*
~RESINS - [plast ind] (thermosetting resins of little commercial importance) résines *fpl* de thio-urée
THIRD - [gen & math] troisième
[math] soixantième *m* de seconde
[mus] tierce *f*
[auto] troisième vitesse *f*
~-ANGLE PROJECTION - [draw] (in mechanical drawing) projection *f* à l'américaine
~BRUSH - [el] (an additional brush fitted to a mo-tor or generator for regulation purposes) troi-sième balaim, balai *m* auxiliaire
~-BRUSH CONTROL - [el] (control of a motor or ge-nerator by the use of a third brush) réglage *m* par le troisième balai

THIRD HARMONIC BAND - [phys] (the spectral band produced when the vibrational energy of the molecule changes from an initial level in which the vibrational quantum number is 0 to level in which the vibrational quantum number is 3, or viceversa) bande *f* de la troisième harmonique
~MOTION SHAFT - [auto] (the main shaft) arbre *m* secondaire, arbre *m* de sortie de boîte
~-ORDER PRISM - [cryst] prisme *m* de troisième espèce, tritoprisme *m*
~ORDER PYRAMID - [cryst] pyramide *f* de troisième espèce, tritopyramide *f*
~ORDER THEORY - [opt] (for lens computation) théorie *f* du troisième ordre
~PICKINGS - [agric] récolte *f* postérieure
~PINIONE- [horol] roue *f* à pignon intermédiaire
~RAIL - [railw] (conductor rail) troisième rail *m*
~-RAIL CLEARANCE LINE - s. third-rail gauge
~-RAIL GAUGE - [railw] gabarit *m* de troisième rail
~-RAIL INSULATOR - [railw] isolateur *m* pour le troisième rail
~ROOT - [math] (cube root) racine *f* cubique
~TAP - [mech] taraud *m* finisseur, troisième taraud *m*, taraud *m* cylindrique
~WHEEL - [horol] roue *f* intermédiaire
THIRL - [mining] travers-bancs *mpl*
THIRTY-SIXMO - [print] in-trente-six *m*
~-TWOMO - [print] in-trente-deux *m*
THISTLE - [bot] chardon *m*
~FUNNEL - [ind chem] (a glass funnel with a long tube and bulbous head, used as a combined filling and safety device with chemical flasks) tube *m* de sûreté
~-SEED OIL - [ind chem] huile *f* de chardon
THIURAM - [chem] (a vulcanization accelerator) thiuram *m*
~SULPHIDE - [chem] (vulcanization accelerator) sulfure *m* de thiuram
THIXOTROPE - [chem] (colloid of which the properties are modified by mechanical treatment) thixotrope *m*
THIXOTROPIC - [chem] thixotrope
~FLUID - [phys] fluide *m* thixotrope
THIXOTROPY - [phys] (property of liquefying on agitation and of returning to the solid condition when at rest) thixotropie *f*
THOLE - [naut] (for an oar) tolet *m*, touret *m*
THOLEPIN - s. thole
THOLOBATE - [constr] tambour *m* de dôme
THOMAS-FERMI DIFFERENTIAL EQUATION - [nucl] (equation which occurs in studying the electron distribution in an atom) équation *f* différentielle de Thomas-Fermi
THOMAS-FERMI MODEL - [nucl] (a method for calculating atomic energy levels based on a statistical treatment of the assembly of electrons) modèle *m* de Thomas-Fermi
THOMAS CONVERTER - [metall] cornue *f* basique, convertisseur *m* Thomas
~METER - [instr] (instrument designed to measure the rate of flow of a gas in terms of the increase in temperature of the gas produced by a known quantity of heat) débitmètre *m* de Thomas
~PROCESS - [metall] (basic open-hearth process) procédé *m* basique, procédé *m* Thomas
~SLAG - [metall] scorie *f* Thomas, scorie *f* basique
~SLAG UTILIZATION WORKS - [metall] installation *f* pour l'utilisation du laitier Thomas
THOMAS STEEL - [metall] acier *m* Thomas
THOMPSON'S RULE - [el] (in a electric cell the heat of reaction is a direct measure of the electromotive force) loi *f* de Thompson
THOMSON EFFECT - [el] (a thermoelectric effect) effet *m* Thomson
~ISOTHERM - [phys] (an S-shaped isotherm showing the continuous transition from the gaseous to the liquid state) isotherme *m* de Thomson
~METER - [instr] (a meter in wich the driving torque is produced by the action of a series coil on a moving armature provided with a commutator and connected in shunt to the supply) compteur *m* électrodynamique à collecteur, compteur *m* Thomson
~PARABOLA METHOD - [electron] (in electron physics) méthode *f* de la parabole de Thomson
~SCATTERING - [electron] (the scattering of electromagnetic radiation by electrons) dispersion *f* de Thomson, diffusion *f* de Thomson
THOMSONITE - [min] (a natural zeolite) thomsonite *f*
THONZYLAMINE HYDROCHLORIDE - [chem] (histazylamine hydrochloride. An antihistamine drug) chlorhydrate *m* de thonzylamine
THORACALGIA - s. thoracodynia
THORACAORTA - s. thoracic aorta
THORACECTOMY - [med] thoracectomie *f*
THORACIC AORTA - [anat] aorte *f* thoracique
THORACODYNIA - [med] douleur *f* thoracique, pectoralgie *f*
THORACOLYSIS - [med] thoracolyse *f*
THORACOPLASTY - [med] thoracoplastie *f*
THORACOSCOPE - [med] (an instrument for viewing the pleura covering the lungs and the chest wall) thoracoscope *m*
THORACOSTENOSIS - [anat] étroitesse *f* des parois thoraciques
THORAX - [anat] thorax *m*
THORIA - [chem] (white, very heavy oxide of thorium) oxyde *m* de thorium
THORIANITE - [min] (mineral consisting largely of thorium oxides) thorianite *f*
THORIATED FILAMENT - [electron] (tungsten filament of an electron tube to which a small amount of thorium has been added) filament *m* thorié
THORIDES - [nucl] (a name proposed for the elements immediately following thorium) thorides *fpl*
THORITE - [min] (natural thorium silicate; an ore of thorium) thorite *f*
THORIUM - [chem] (a metallic, radioactive element, symbol Th, A.N.90, A.W. 232.12 with application in nuclear technology) thorium *m*
~CONTENT METER - [nucl] teneurmètre *m* en thorium
~DIOXIDE - [chem] (an X-ray contrast medium in medicine ; catalyst, and a component of special glasses and ceramics) bioxyde *m* de thorium
~EMANATION - [chem] émanation *f* de thorium, thoron *m*
~FISSION - [nucl] (nuclear process in which thorium is employed as fissile material) fission *f* de thorium
~FLUORIDE - [chem] (a constituent of special ceramics) fluorure *m* de thorium
~NITRATE - [chem] (an oxidizing agent) nitrate *m* de thorium
~ORE - [min] (thorium containing ore) minerai *m* de thorium

THORIUM OXALATE - [chem] (a component of special ceramics) oxalate *m* de thorium
~REACTOR - [nucl] (reactor containing thorium as fuel) réacteur *m* à thorium
~SERIES - [nucl] (the series of nuclides deriving from thorium 232) famille *f* du thorium
THOROGUMMITE - [min] (mineral containing lead and calcium) thorogummite *f*, nicolayite *f*
THORON - [chem] (a radio-active isotope of radon; it is produced by the disintegration of thorium. Half-life 54.5 secs) thoron *m*, émanation *f* du thorium
THOROTUNGSTITE - [miner](tungsten, thorium, cerium and zirconium containing mineral) thoritungstite *f*
THOROUGHBRED - [zool] (any animal bred from pure stock) pur sang *m*, de race
THOROUGHFARE - [gen] (a road or street through which the public has unobstructed passage) voie *f* de communication
[gen] (a passage) passage *m*
THOU - [meas] (coll. abbreviation for I/I000 in) un millième *m* de pouce
THOUSANDS FINE - [min] titre *m* en millièmes
THRASH, to - s, thresh
THRASHING - s. treshing
THRASHING FLAIL - [text] fléau *m*
~THE HEMP - [text] battage *m* en grange du chanvre
THREAD, to - [mech] (to cut a screw on a part) fileter, tarauder
[gen] (a needle) enfiler
[photo] (a film) mettre le film en place
THREAD - [gen] (very slender cord or line) fil *m*, filament *m*
[bot] (a filament of any substance) fil *m* (de plante) fibre *f*
[geol] (a vein) filet *m*, veinule *f* de minerai
[mech] (the spiral part of a screw which projects beyond the core) filet *m*, filetage *m*, pas *m*
~ANGLE - [mech] (of a screw thread) angle *m* du filetage
~BOARD - [text] planchette *f* de guide-fil
~BOLT - [text] boulon *m* fileté
~BRAKE - [text] frein *m* du fil
~BREAKAGE - [text] rupture *f* du fil
~CARRIER - [text] (of a knitting machine) guide-fil *m*, enrouleuse *f*, appareil *m* pour fil crépu
~CHAFING - [text] frottement *m* du fil
~CHASING TOOL - [mech] outil *m* de filetage extérieur
~CLEARER - [text] dispositif *m* de purgeage du fil
~CLEARER BRUSH - [text] brosse-nettoyeuse *f* du fil
~CLIP - [text] pince *f* pour fil
~COUNTER - [text] compte-fils *m*
~COVERING MACHINE - [text] machine *f* à guiper
~CUTTER - [mech] taraud *m*, filière *f*
~-CUTTING LATHE - [mach tool] tour *m* à fileter, taraudeuse *f*
~CUTTING MACHINE - s. thread cutting lathe
~-CUTTING SCREW - [mech] vis *f* coupante
~DIVIDER - [text] séparateur *m* de fils
~DOPE - [mech] lubrifiant *m* de filetage
~ENTANGLEMENT - [text] boucle *f* de fil, duvet *m* fixe
~EXTRACTOR - [text] machine *f* à retirer les fils durs des barbes
~EYE - [text] trou *m* pour le fil

THREAD FEED - [text] alimentation *f* en fils
~FEEDER - [text] donneur *m* de fil
~FORM - [mech] (of a screw thread) profile *m* du filetage
~FORMING INSERT - [mech] (a threaded plug inserted in a mould to provide a threaded socket in the finished article) broche *f* filetée
~FORMING TOOL - [tool] outil *m* à fileter
~GAUGE - [mech] (device for determining the pitch of a screw thread) calibre *m* pour la verification des vis, calibre *m* de mesure de pas de vis
~GRINDING MACHINE - [mach tool] rectifieuse *f* pour filets
~GRIPPER - [text] pince *f* du fil
~GUARD - [text] plaque *f* métallique de protection
~GUIDE - [text] guide-fil *m*
~-GUIDE ADJUSTING BAR - [text] barre *f* de réglage du guide-fil
~-GUIDE COUPLING - [text] embrayage *m* de guide-fil
~-GUIDE EYELET - [text] oeuillet *m* du guide-fil
~-GUIDE HOLDER - [text] support *m* du guide-fil
~-GUIDE LEVER - [text] levier *m* de guide-fil
~-GUIDE ROD - [text] arbre *m* de la tringle guide-fil
~GUIDE SCREW SPINDLE - [text] tringle *f* guide-fil en forme de vis
~INSERT - [mech] (insert for soft metals) filetage *m* rapporté
~LAPPET - [text] planchette *f* de guide-fil
~-LIKE STRIPS - [text] corps *mpl* filiformes
~MILLER - [mach tool] fraiseuse *f* pour filets
~MILLING MACHINE - s. thread miller
~MOUNT - [el] culot *m* à vis
~OF SCREW - [mech] spire *f* de vis
~OF THE WEFT - [text] duite *f*
~PICKER - s. thread extractor
~PICKING MACHINE - [text] machine *f* à séparer les fils durs des déchets
~PLATE - [text] (of a spinning machine) guide-mèche *m*
~PLUG GAUGE - [mech] calibre *m* pour filetage
~-PROTECTING CAP - s. thread protector
~PROTECTOR - [mech] embout *m* protecteur de filetage
~ROLLER - [tool] (for screw thread) rouleau *m* à fileter
~ROLLING DIE - [tool] poigne *m* à fileter
~SELECTOR - [text] sélecteur *m*
~SOCKET - [photo] raccord *m* fileté
~SPACING - [text] compte *m* de fils
~TENSION - [text] tension *f* du fil
~TENSION EQUALISER - [text] égaliseur *m* de tension du fil
~TENSION GEAR - [text] frein *m* pour le fil
~TENSIONER - [text] frein *m* du fil
~TESTER - [text] appareil *m* à contrôler les fils
~TURNING LEVER - [text] levier *m* pour tourner le fil
~UP, to - [plast ind] (to pass the beginning of an extruded filament through all the device designed to treat it before starting the process) enfiler
~WASTE - [text] déchets *mpl* de fil, fonds *mpl* de bobines
THREADBARE - [gen] râpe, élimé
[text] (so worn that the thread is visible) qui montre la corde
THREADED - [gen] enfilé

THREADED - [mech] fileté, à vis, taraudé
~AND COUPLED TUBE - [mech] tube *m* taraudé et manchonné
~BUSHING - [mech] (a bush provided with a thread for fixing purposes) manchonnage *m* fileté, coussinet *m* fileté
~CASING - [mech] tube *m* fileté
~DISK - [plumb] disque *m* fileté
~END - [plumb] extrémité *f* filetée
~JOINT - [mech] fermeture *f* à vis, filetage *m*, assemblage *m* par vis
~NIPPLE - [mech] mamelon *m* fileté
~PIPE - [mech] tube *m* fileté
~RING - [mech] anneau *m* fileté
~SLEEVE - [mech] manchon *m* fileté
~SOCKET - [mech] emboîtement *m* à vis
THREADER - [mach tool] (threading machine) tour *m* à fileter, taraudeuse *f*
THREADING - [mech] filetage *m*, taraudage *m* [gen] enfilement *m*
~CUTTER - [tool] outil *m* à fileter
~DIE - [tool] filière *f*
~HOOK - [text] crochet *m* d'enfilage
~MACHINE - s. threader [mach tool] [text] machine *f* d'enfilage
~MECHANISM - [text] appareil *m* d'enfilage
~MOTION - [text] dispositif *m* d'enfilage
~NEEDLE - [text] aiguille *f* d'enfilage
~OF THE FILM - [photo] insertion *f* de la pellicule
~OILS - [ind chem] huiles *fpl* pour filetage
~SLOT - [el acoust] fente *f* de mise en place
~TUBE - [text] tube *m* d'enfilage
~-UP - [cin] (the operation of inserting the start of the film into the mechanism of the camera or of the projector) mise *f* en place du film, placement *m* du film
THREATENING SKY - [met] (general appearance of the sky indicative of a marked change for the worse in weather conditions) ciel *m* orageux, ciel *m* menaçant
THREE - [math] trois
~-ADDRESS - [comput] à trois adresses
~-ADDRESS-INSTRUCTION - [comput] instruction *f* à trois adresses
~-AMMETER METHOD - [el] (a method for measuring the power carried by a single-phase circuit making use of three ammeters) méthode *f* des trois ampèremètres
~-AND-FOUR FLUTED DRILL - [mach tool] foret *m* à trois et quatre rainures
~AND ONE WARP - [text] chaîne *f* 3 et I
~AND ONE WEFT - [text] trame *f* 3 et I
~-AXLE TRUCK TRAILER - [auto] remorque *f* à trois essieux et six roues
~-BLADE PROPELLER - [mech] hélice *f* tripale
~BODY PROBLEM - [phys] problème *m* à trois cordes
~CARD SET - [text] assortiment *m* de trois cardes
~-CAVITY KLYSTRON - [electron] (klystron having three resonant cavities) klystron *m* à trois cavités
~-CENTRED ARCH - [arch] (an arch having the form of a false ellipse struck from three centres) voûte *f* à trois centres
~-COAT WORK - [constr] peinturage *m* en trois temps
~-COIL REGULATION - [el] (or four-busbar regulation) régulation *f* à trois inductances
~-COLOUR - [gen] (adjective) trichrome, trichromatique
THREE-COLOUR INKS - [print] encre *f* pour trichromie
~-COLOUR METHOD COLORIMETRY - [photo] colorimétrie *f* trichrome
~-COLOUR PHOTOGRAPHY - [photo] photographie *f* trichrome
~-COLOUR PROCESS - [photo] (any system of colour photography which analyzes the colours by three colour filters each giving a black-and-white record) procédé *m* trichrome
~-COMPONENT BALANCE - [aero] (device used in a wind tunnel to measure three components of moments and forces, usually lift, drag and pitching moment) balance *f* à trois composantes
~-CONDITION CABLE CODE - [telecomm] code *m* trivalent pour câble
~-CONE BIT - [mech] foret *m* à trois cônes
~-CORE CABLE - [el] (cable consisting of three insulated conductors) câble *m* tripolaire
~-COURSE - [agric] triennal
~-COURSE SYSTEM - [agric] assolement *m* triennal
~-CROP ROTATION SYSTEM - s. three-course system
~-CYLINDER LOCOMOTIVE - [railw] machine *f* à trois cylindres
~-DECKER - [naut] trois-ponts *m*
~-DECKER CAGE - [mining] cage *f* à trois étages
~-DIGIT GROUP - [comput] groupe *m* de trois chiffres
~-DIMENSIONAL - [phys] tridimensionnel
~-DIMENSIONAL CUTTING - [metall] coupe *f* tridimensionnelle
~ELECTRODE TUBE - [electron] triode *f*
~-ELEMENT AERIAL - [radio] (aerial consisting of a directing dipole and a receiving and a reflecting dipole) antenne *f* à trois éléments
~-ELEMENT ANTENNA - s. three-element aerial
~-END TWILL - [text] sergé *m* de trois
~FIELD SYSTEM - s. three course corporation system
~-FOLD CROSSING - [text] triple tour *m*
~-FOLD ROPE - [text] câble *m* en trois bouts
~-FORKED ROAD - [constr] chemin *m* fourchu
~-FURROW PLOUGH - [agric] charrue *f* trisoc
~-FURROW TRACTOR - [agric] tracteur *m* à trois sillons
~-GUN COLOUR PICTURE TUBE - [telev] (a colour TV tube in which three electron guns emit three electron beams, one for each primary colour) tube *m* image couleurs à trois canons électroniques
~-HEAD BATTERY - [constr] batterie *f* de trois pilons, bocard *m* à trois pilons
~-HIGH HOUSING - [metall] cage *f* à trio
~-HIGH MILL - [metall] (a stand of three rolls the upper-most running in the opposite sense to the lowermost, so that passes can be made in opposite directions in the same mill) cage *f* de laminoir trio
~-HIGH PLATE MILL - [metall] laminoir *m* à trio à tôles
~-HIGH PLATE MILL TRAIN - [metall] train *m* à trio à tôles
~-HIGH STAND - [metall] cage *f* trio
~-HINGED ARCH - [arch] (a continuous arch which is hinged at the crown and the abutments) arc *m* à trois rotules, arc *m* à trois articulations
~-HORN SET - [auto] jeu *m* de trois avertisseurs
~-IMPEDANCE STAR NETWEORK - [telecomm] réseau *m* en étoile à trois impédances
~-JAW AMERICAN SELF-CENTRING CHUCK - [mach

tool] mandrin *m* américain à trois mâchoires à serrage concentrique
THREE-JAW CONCENTRIC GRIPPING CHUCK - [mach tool] plateau *m* à trois griffes concentriques
~-JAW STEADY - s. three-jaw steady rest
~-JAW STEADY REST - [mach tool] lunette *f* à trois touches réglables
~-JUNCTION TRANSISTOR - [electron] transistor *m* à trois jonctions
~-LEAVED TWILL - [text] sergé *m* de trois
~-LEGGED - [gen] à trois branches
~-LEVEL MASER - [electron] (a solid state maser in which three energy levels are used for maser operations) maser *m* à trois niveaux
~-LIFT CONE PULLEY - [mech] cône-poulie *m* à trois gradins, cône-poulie *m* à trois étages
~-LIGHT WINDOW - [carp] (window with two mullions dividing the window-space into three compartments) fenêtre *f* triple
~-PART BOX - [metall] châssis *m* en trois parts
~-PART COMB - [text] peigne *m* en trois parties
~-PHASE - [el] (an a. c. system in which the displacement is one-third of a period) triphasé
~-PHASE ALTERNATING CURRENT - [el] courant *m* alternatif triphasé
~-PHASE CIRCUIT - [el] circuit *m* triphasé
~-PHASE CONVERTER - [el] convertisseur *m* triphasé
~-PHASE EQUILIBRIUM - [phys] équilibre *m* de trois phases
~-PHASE FOUR-WIRE SYSTEM - [el] (a system of distribution comprising four conductors of which three are connected to a three-phase supply and the fourth to a neutral point in the source of supply) distribution *f* triphasée quatre fils
~-PHASE MACHINE - [el] machine *f* triphasée
~-PHASE RECTIFIER - [electron] (a three-phase,half wave rectifier, also called zig-zag rectifier) redresseur *m* triphasé
~-PHASE STAR RECTIFIER - [el] redresseur *m* triphasé en étoile
~-PHASE SYSTEM - [el] (method of electric power supply in which there are three alternating currents displaced in phase by 120 degrees) système *m* triphasé
~-PHASE THREE-WIRE SYSTEM - [el] (a system of distribution comprising three conductors connected to a three-phase supply) distribution *f* triphasée trois fils
~-PHASE TRANSFORMER - [el] (a transformer, intended for three-phase working, in which three magnetic circuits have parts in common) transformateur *m* triphasé
~-PHASE WATER-GAS CYCLE - [gas ind] fabrication *f* à courant de vapeur alterné
~-PIN - [el] (with three contact pins) tripolaire
~-PIN CENTRAL BASE - [el] socle *m* à trois broches
~-PIN PLUG - [el] (plug fitted with two pins for the main circuit and one for an earth connexion) fiche *f* tripolaire
~-PINNED ARCH - s. three-hinged arch
~-PLY - [carp] bois *m* plaqué triplé
~-PLY BELTING - [mech] courroie *f*pl à trois plis
~-PLY CLOTH WEAVE - [text] armure *f* tissu triple
~-PLY ROPE - s. three-fold rope
~POINT LANDING - [aero] (a landing in which main and tail landing wheels touch the ground at the same instant) atterrissage *m* normal
THREE-POINT PROBLEM - [surv] problème *m* à trois points
~-POINT SUPPORT - [mech] triangle *m* de sustentation
~-POINTED ALTIMETER - [instr] (a type of altimeter in which three pointers indicate readings on a single dial graduated with three scales of different orders of magnitude) altimètre *m* à trois aiguilles
~-POLE - [el] tripolaire
~-PORT TWO-STROKE ENGINE - [mech] moteur *m* à deux temps et trois lumières
~-POSITION CONTROL - [mech]A réglage *m* pour tout-ou-rien
~-POSITION RELAY - [el] relais *m* à trois positions
~-POSITION VALVE - [mech] soupape *f* à trois positions
~-PRONGED CHUCK - [mach tool] mandrin *m* à trois pointes
~-QUARTER BAT - [constr] trois-quarts *m* de brique
~-QUARTER FLOATING AXLE - [auto] pont *m* trois-quarts flottant
~RIB SET - [text] boutage *m* à lignes à trois dents
~-RIBBED SLEEPER - [railw] traverse *f* à trois nervures
~-ROLLER TEMPLE - [text] templet *m* à trois cylindres
~-SACCHARIDES - [chem] trisaccharides *f*pl
~-SHEAVE BLOCK - [mech] moufle *f* à trois poulies
~-SPEED GEAR BOX - [auto] transmission *f* à trois vitesses
~-SQUARE FILE - [tool] lime *f* tiers-point, trois-carrés *m*, lime *f* triangulaire
~-STAGE COMPRESSOR - [mech] (air or gas compressor in which the fluid pressure is raised in three successive steps) compresseur *m* à trois étages
~-STAMP MILL - [constr] batterie *f* de trois pilons
~-START THREAD - [mech] triple pas *m*, vis *f* à triple pas
~-STEP CONTROL - [contr] action *f* a trois échelons
~-STICK SET - [mining] cadre *m* ordinaire
~-STRAND TWINE - [text] ficelle *f* à trois brins, merlin *m*
~-TERMINAL CONTACT - [el] contact *m* à trois bornes
~-THROW CRANK - [mech] manivelle *f* triple
~-THROW PUMP - [mech] pompe *f* à trois corps
~-TUBE TELECINE - [telev] télécinéma *m* à trois tubes
~-VOLTMETER METHOD - [meas] (a method of measuring single-phase power in which three voltmeters are used in conjunction with non-inductive resistances) méthode *f* des trois voltmètres
~-WATTMETER METHOD - [meas] méthode *f* de trois wattmètres
~-WAY - [mech] (of cock a valve which allows fluid to be directed into either of two pipe lines at will) à trois vois
~-WAY BIT - [oil ind] trépan *m* à trois ailettes
~-WAY COCK - [plumb] (plug cock with passages so arranged that fluid can be directed to either of two outlets) robinet *m* à trois voies
~-WAY DUMP-TRUCK - [auto] benne *f* basculante dans trois directions
~-WAY SWITCH - [el] (rotary-type single-pole switch with three independent contact positions) commutateur *m* à trois voies
~-WAY TIPPER - s. three-way dump-truck
~WAY VALVE - [mech] soupape *f* à trois voies [oil ind] soupape *f* à trois siéges
~WEFT BINDING - [text] liage *m* par triple duite
~-WHEEL DELIVERY VAN - [auto] fourgon *m* à trois

roues
THREE-WHEELER - [auto] véhicule m à trois roues
~-WING SHUTTER - [photo] obturateur m à trois ailettes
~-WIRE METER - [el] (electricity supply meter performing the simultaneous integration of the energy supplied by the two sides of a three-wire system) compteur m à trois fils
~-WIRE SYSTEM - [el] (a system of distribution with D.C. or single-phase A.C. comprising two conductors and a middle or neutral wire, the supply being taken from the middle wire and either outer conductor, the middle wire carries only the difference-current and is usually connected to earth) distribution f trois fils
~-YEAR ROTATION SYSTEM - [agric] assolement m triennal
THREEFOLD - [gen] triple
~TACKLE - [mech] palan m triple
~TRIPOD STAND - [impl] pied m à trois brisures
THREEWAY - [plumb] coude m double à branchement central
THREONINE - [chem] (an essential amino-acid with application in medicine and nutritional research) thréonine f
THRESH, to - [agric] battre le grain, défourrer
THRESHER - [agric] batteuse f
THRESHING - [agric] battage m
~CYLINDER - [agric] batteur m
~FLOOR - [agric] aire f
~MACHINE - s. thresher
THRESHOLD - [gen] seuil m
[el] (electrical treshold, the least current or voltage needed to obtain a minimal response) seuil m électrique, limite f
~AUDIOGRAM - [acoust] (a graph showing hearing loss, percent hearing loss, or percent hearing as a function of frequency) audiogramme m
~CIRCUIT - [electron] circuit m à valeur de seuil
~CONCENTRATION - [chem] concentration f limite
~CURRENT - [electron] (the current at which a gas discharge changes from- a non-sustained to a self-sustained discharge) courant m de suill
~DETECTOR - [nucl] détecteur m à seuil
~DIODE - [electron] (limiter diode) diode f de seuil
~DOSE - [nucl] (the minimum amount of irradiation which will produce a given result) dose f seuil
~ELEMENT - [electron] élément m de seuil
~ENERGY - [phys] (the energy limit for an incident particle or photon, below which a particular endothermic reaction will not occur) énergie f de seuil
~ENERGY OF FISSION - [nucl] seuil m d'énergie de fission
~EXPOSURE - [photo] (the minimum exposure to ensure discernibility) exposition f minimum
~FIELD - [comput] (in static magnetic storage) champ m de seuil
~FREQUENCY - [electron] (the frequency of incident radiant energy below which there is no photoemissive effect) fréquence f de seuil
~ILLUMINANCE - [opt] seuil m d'éclairement
~LIGHTS - [aero] (lights placed to show the limit of the usable length of a runway or the like) feux mpl de seuil
~OF AUDIBILITY - [acoust] (at a specified frequency under given conditions) seuil m d'audibilité
~OF DETECTABILITY - s. threshold of audibility
THRESHOLD OF DISCOMFORT - [astronaut] seuil m de malaise
~OF FEELING - [acoust] seuil m de douleur
~OF HEARING - s. threshold of audibility
~OF LUMINESCENCE - [electron] (the maximum wavelength of radiation capable of exciting a luminescent material) seuil m de luminescence
~OF RUNWAY - [aero] (the limit of the usable length of runway) seuil m de piste
~POTENTIAL - [nucl] (the minimum energy necessary in an electron beam in an ion source for the production of a given type of ions during the ionization of molecule) potentiel m d'apparition
~PRICE - [comm] prix m de seuil
~REACTION TEMPERATURE - [chem] seuil m thermique
~SENSITIVITY - [instr] (the lowest level of the measured variable which produces effective response of the instrument) sensibilité f de seuil
~TO RESPONSE TO PULSES - [instr] (of a counter) seuil m de réponse aux impulsions
~TREATMENT - [ind chem] (in water softening) traitement m limite
~TUBE - [radar & telev] tube m de seuil
~VALUE - [contr] (the minimum input producing a corrective action in the power element of an automatic controller) seuil m d'action, valeur f de seuil
~VOLTAGE - [electron] (the minimum voltage at which all pulses produced in the counter by any ionizing event are of the same size, independently of the size of the primary ionizing event) tension f de seuil
~WAVELENGTH - [electron] (the wavelength of the incident radiant energy above which there is no photoemissive effect) longueur f d'onde de seuil, seuil m de longueur d'onde photoélectronique
THRIBBLE - [oil ind] jeu m de trois
~-PLATFORM - [oil ind] plate-forme f intermédiaire pour trois tiges
THRIFT - [gen] économie f
[bot] statice m
THROAT - [anat] gorge f
[metall] (of a blast furnace; the opening at the highest part of the furnace) gueulard m
[mech] (of an extruder) goulotte f
[aero] (of a wind tunnel; the part of the duct at which the cross-section is least) entrée f
[naut] (of a stay sail) empointure f
[plumb] (of a pipe) étranglement m
[th eng] (of a burner) col m du mélangeur, col m du canon
[railw] (the inside angle of a wheel flange at the point at which it joins the tread) congé m
[mech] (in welding; the distance from the root of a fillet weld to its face) distance f utile
[glass man] (between melter and refiner of a tank; the submerged passage) barrage m
[naut] (of the anchor) collet m d'ancre
[constr] (a drip) cavet m droit
[geol] (of a volcano) cheminée f de volcan
[mach tool] portée f
[aero] (of jet engine) col m
~BRAIL - [naut] étrangloir m
~DEPTH - [mech] (the distance from the centre-line of the electrodes to the nearest interference point for sheet welding) distance f utile
~FLAME - [metall] (flame of burning gas at the top of a blast furnace) flamme f du gueulard

THROAT-LATCH - [agric] (the strap passing under the neck of a draft animal) sous-gorge *f*
~MICROPHONE - [el acoust] (a microphone designed to be worn over the larynx) laryngophone *m*
~OPENING - [metall] (blast furnace the mouth of the shaft at its highest point) ouverture *f* du gueulard
~PLATFORM - [metall] (the staging round the top from which the furnace is changed) plate-forme *f* du gueulard
~SPRAYER - [med] insufflateur *m*
~STOPPER - [metall] fermeture *f* du gueulard
~WASH - [med] gargarisme *m*
THROATING FILLET - [constr] regingot *m*
THROB, to - [gen] battre
[med] palpiter
[mech] vibrer
[mech] of an engine) vrombir
[geol] rejeter
THROB - [gen]battement *m*, pulsation *f*
[mech] (of an engine) vrombissement *m*
THROBBING - [med] palpitant, vibrant
THROES - [med] angoisse *f*, agonie *f*
THROMBASE - [biol] fibrin-ferment *f*, thrombine *f*
THROMBASTHENIA - [med] thrombasthénie *f*
THROMBIN - s. thrombase
THROMBOBLAST - [biol] thromboblaste *m*
THROMBOCYTE - [biol] (small colourless corpuscle found in large numbers in the blood of mammals) thrombocyte *m*, plaquette *f* sanguine
THROMBOEMBOLIA - [med] thrombo-embolie *f*
THROMBOLYSIS - [med] thrombolyse *f*
THROMBOPATHY - [med] thrombopathie *f*
THROMBOSIN - s. thrombin
THROMBOSIS - [med] (formation of a clot in a blood vessel) thrombose *f*
THROSTLE - [zool] grive *f* chanteuse
[text] métier *m* continu
~FRAME - [text] métier *m* continu
THROTTLE, to - [gen] étrangler
[mech] (the flow in a pipe etc) étrangler
THROTTLE - [mech] (device for controlling the flow of mixture to an i.c. engine) étrangleur *m*, obturateur *m* d'air
[anat] gosier *m*
[mech] robinet *m* modérateur, registre *m* de vapeur
~BUTTERFLY - [auto] papillon *m* des gaz
~CHAMBER - [mech] (in internal combustion engines) boisseau *m* d'étranglement
~CONTROL - [mech] commande *f* des gaz, poignée *f* de commande des gaz
~DOWN, to - [mech] mettre au ralenti
[mech] (in internal combustion engines) fermer le gaz
~FRICTION DEVICE - [mech] (device to ensure that a throttle lever remains in the position in which it is set) dispositif *m* d'assurance du clapet
~FRICTION LOCK - s. throttle friction device
~ICE - [aero & auto] (ice developing in the neighbourhood of an engine throttle, caused by the expansion of the air drawn in) givre *m* de carburateur
~LEVER - [mech] poignée *f* de commande des gaz
~ROD - [auto] tige *f* de commande des gaz
~SPINDLE - s. throttle rod
~VALVE - [mech] (of a jet engine) régulateur *m* du débit
[auto] s. throttle butterfly
~VALVE LEVER - [auto] levier *m* du papillon des gaz
THROTTLING - [metall] étouffement *m*
~GOVERNOR - [auto] régulateur *m* de carburateur
THROUGH-AND-THROUGH COAL - [min] charbon *m* tout venant
~-BAR - [mech] barre *f* d'assemblage
~-BOLT - [mech] boulon *m* libre, boulon *m* traversant
~BORE - [metall] trou *m* traversant
~BRIDGE - [constr work] pont *m* à tablier inférieur
~CARBURIZING - [metall] cémentation *f* totale
~CARRIAGE - [railw] voiture *f* directe
~CIRCUIT - [el] circuit *m* direct
~-COAL - [mining] charbon *m* tout-venant
~COMMUNICATION - [railw] intercommunication *f* entre wagons
~-COUPLING - [auto] (direct drive) prise *f* directe
~FAULT - [geol] fossé *m* d'effondrement
~HARDENING - [metall] endurcissement *m* total
~-HOLE - [mech] (a hole which passes completely through a part) trou *m* traversant
~JOINT - [tool] serre-fils *m*
~LEVEL - [meas] (the reading of a high-impedance level-measuring set at a point in a circuit) niveau *m* de tension absolu
~LINE REPEATER - [telecomm] répéteur *m* embroché
~MORTISE AND TENON - [carp] mortaise *f* passante
~-PIPE VEHICLE - [railw] (vehicle fitted with through brake pipe) véhicule *m* à conduite blanche
~POSITION CORD PAIR - [telecomm] dicorde *f* de transit
~-PUT - [gen] (a measure of a system efficiency) débit *m*
~RAIL - [railw] rail *m* contre-aiguille de la voie directe
~RETORT - [ind chem] (a retort with two mouthpieces) cornue *f* sans fond, cornue *f* à deux têtes
~ROAD - [constr] (a road serving as a connection between other roads) route *f* de transit
[railw] voie *f* directe
~ROLLER - [metall] rouleau *m* pour courber la bande
~-ROUTE LOCKING - [el] enclenchement *m* de transit rigide
~SHAFT - [mech] arbre *m* passant
~SIDING - [railw] garage *m* avec entrée directe des deux côtés
~STATION - [railw] gare *f* de passage
~STONE - [constr] (bondstone having a length equal to the thickness of the wall in which it is laid as a header) patpaing *m*
~SWITCHING CORD CIRCUIT - [telecomm] dicorde *f* de transit
~TENON - [carp] tenon *m* passant
~TRACK - [railw] (tha main track, or line) voie *f* directe
~TRAFFIC - [telecomm] trafic *m* de transit
~-TYPE INSERT - [plast ind] (a mould insert which is exposed on both sides of a moulded product) broche *f* traversante
THROUGHOUT - [gen] partout, d'un bout à l'autre
THROUGHPUT - [hydr] débit *m*
THROW, to - [gen] jeter, lancer
[cin] (a picture on a screen) projeter
[zool] (of animals) mettre bas
[pottery] tournasser
[text] (to twist) retordre
[zool] (of birds) muer
[zool] (of snackes) se dépouiller
THROW - [gen] lancement *m*, jet *m*

THROW - [gen] (of a missile) jet *m*
[geol] rejet *m*, cran *m*, coufflée *f*
[comput] (of a pinch roller in a magnetic tape transport) jeu *m*
[mech] (the radius of a crank) bras *m* de manivelle
[mech] (of a reciprocating piece, the maximum stroke) volée *f* du piston
[mech] (of a cam) excentricité *f*
[mech] (of a crankshaft) manivelle *f*, maneton *m* de vilebrequin
[el] portée *f*
[gas ind] pression *f* gazométrique
~A BELT OFF, to - [mech] débrayer une courroie
~A BELT ON, to - [mech] embrayer une courroie
~BACK, to - [biol] retourner à un type antérieur
~IN, to - [el] faire démarrer
~INTO GEAR, to - [mech] (of a clutch) engrener
[mech] (of the gears) embrayer
~OF HEAD - [text] course *f* du chariot
~OF PUMP - [mech] hauteur *f* de refoulement
~OF THE CRANK - [mech] course *f* de la manivelle
~OF THE NEEDLE FRAMES - [text] course *f* des peignes à broder
~OF THE REED - [text] coup *m* du peigne
~OF THE TAPPET - [mech] course *f* de l'excentrique
~-OFF CARRIAGE - [mech] (mobile device to provide for the discharge of material from a belt conveyor at any point in its length) wagonnet *m* à décharge automatique
~OFF CENTRE, to - [mech] excentrer, désaxer
~-OFF GEAR - s. through-off carriage
~-OFF POINT - [railw] aiguille *f* de déraillement
~ON, to - [el] (as a load on an electrical motor) appliquer
~OUT, to - [mech] (to disengage gears) débrayer
[el] désaccoupler
~-OUT BEARING - s. throw-out sliding muff
~-OUT COLLAR - [auot] anneau *m* de débrayage
~OUT OF BALANCE, to - [gen] déséquilibrer
~OUT OF GEAR, to - [mech] débrayer
[auto] (of a clutch) dégager une vitesse
~OUT OF LEVEL, to - [mech] déniveler
~-OUT SLIDING MUFF - [auto] manchon-guide *m* de la butée de débrayage
~OUT SPIRAL - [el acoust] (the blank spiral groove at the end of a pitch) sillon *m* de sortie
~OVER CONTACT - [el] contact *m* de commutation
~OVER RELAY - [el] relais *m* à deux directions
~-OVER SWITCH - [el] commutateur, inverseur *m* de courant
~OVER THE POINTS, to - [railw] renverser l'aiguille
THROWAWAYS - [cin] (of a film) déchet *m* de film
THROWER - [pottery] potier *m*, tournasseur *m*
~RING - [mech] (a ring fixed on a shaft to throw off oil which would otherwise creep along it) bague *f* de graissage, bague *f* de projection d'huile, segment *m* déflecteur
THROWING - [paint] (a defect in electrical insulating varnishes, which gives rise to the ejection of particles of the varnish by centrifugal action when used on high-speed rotary machines) éjection *f* de particules
~IN THE COUNTERSHAFT - [mech] ancrage *m* de l'arbre intermédiaire
~MACHINE - [text] métier *m* à retordre
~MILL - [text] retorderie *f*
~OFF CENTRE - [mech] excentration *f*, désaxage *m*
THROWING OILS - [ind chem] huiles *f* de retorderie
~OVER THE POINTS - [railw] manoeuvre *f* des aiguil
~POWER - [metall] (the capacity of a solution to effect an even deposition on a cathode of irregular form) pouvoir *m* couvrant, pouvoir *m* de pénétration
~UP OF ASHES - [geol] projection *f* de scories
~UP OF LAVA - s. throwing up of ashes
THROWN SIDE - [geol] lèvre *f* affaissée
~SILK - [text] soie *f* torse, soie moulinée
THROWOFF - [mach tool] mécanisme *m* d'arrêt de l'avance
THROWOUT - [mech] dispositif *m* de déclenchement
[mech] (the operation of disconnecting) débrayage *m*
THROWSTER - [text] tordeur *m* de soie, organsineur *m*
THRUM - [text] (the pringe of warp thread) penne *f*
[naut] landage *m*
THRUST, to - [gen] pousser
THRUST - [mech] (the propulsive force developed by a power unit) poussée *f*, butée
[mining] (of a gallery) charriage *m*
[aero] (of a jet engine) poussée *f*
[geol] chevauchement *m*, écroulement *m*
[mining] écrasement *m* des piliers
~BALL BEARING - [mech] butée *f* à billes
~BALL RACE - s. thrust bearing
~BEARING - [mech] (a special bearing to take the thrust of the feed screw) palier *m* de butée
~BEARING RING - [mech] bague *f* de palier de butée
~BEARING SHOE - s. thrust bearing ring
~BLOCK - s. thrust bearing
[constr] massif *m* d'ancrage
~-BORER - [impl] (pusher) pousse-tube *m*
~COLLAR - [mech] (a bearing designed to take axial loads) collier *m* de butée, collet *m* de palier de butée
~DEFLECTOR - [mech] (a device designed to change the direction of the gases escaping from the jet pipe) déflecteur *m* de jet
~FACE - [mech] surface *f* de poussée
~FAULT - [geol] pli-faille *m* inverse, faille *f* de compression
~HORSE POWER - [mech] (the product of thrust and speed reduced to terms of h.p.) puissance *f* de poussée
~IN A SETTING - [metall] poussée *f* d'un four
~INVERSER - [aero] inverseur *m* de poussée
~LINE - [aero] axe *f* de poussée
~LOAD - [mech] charge *f* axiale
~METER - [instr] (instrument designed to measure the thrust of a jet engine) mesureur *m* de la puissance de poussée
[meas] statimètre *m*
~OF AN ARCH - [constr] (the equal horizontal forces acting on the abutments of an arch, due to the load carried by it) poussée *f* d'un arc
~PLANE - [geol] plan *m* de charriage
~PLATE - [mech] (pressure plate) plaque *f* de butée
[th eng] plaque *f* d'ancrage
[metall] disque *m* de butée
~RACE - [mech] butée *f* à billes
~REVERSER - [aero] (a device designed to direct the flow from the jet pipe in a reverse sense, for deceleration on landing) inverseur *m* de poussée, déviateur *m* de jet
~RING - [mech] bague *f* de butée
~ROLLER - [mech] rouleau *m* de butée

THRUST RUNNER - [mech] rondelle *f* de butée
~SCREW - [mech] vis *f* de pression
~SHAFT - [mech] (portion of the propeller shaft fitted with collars) arbre *m* de butée
~SPOILER - [aero] (a device for reducing the thrust of a jet) déviateur *m* de jet
~TERMINATOR - [astronaut] dispositif *m* d'arrêt de poussée
~WASHER - [mech] (washer designed to receive a thrust or to act as a distance-piece) rondelle *f* de butée
~WEIGHT RATIO - [aero] (the thrust of the propulsion unit divided by the gross weight of the aircraft) rapport *m* de la poussée au poids total
THRUSTING - [geol] charriage *m*
THRUSTOR - [mech] servo-moteur *m*
THUJA OIL - [chem] (an oil obtained from Thuia occidentalis and used in perfumery) essence *f* de thuya
THUJONE - [chem] (a bicyclic monoterpene occurring in thuja and other essential oils) thuyone *f*
THULIUM - [chem] (a rare-earth element, symbol Tm, with application in nuclear technology) thulium *m*
THUMB - [anat] pouce *m*
[arch] quart *m* de rond d'arrête
~BOLT - [mech] boulon *m* à oreilles
~-LATCH - [mech] loquet *m* à poucier
~NUT - [mech] écrou *m* papillon, écrou *m* à oreilles
~-PIECE - [mech] (a handle) poucier *m*
~-SCREW - [mech] vie *f* ailée, papillon *m*
~STALL - [mech] doigtier *m* pour pouce
~-SWITCH - [mech] (a switch located on a control lever, and designed to be operated by the thumb) interrupteur *m* manuel
THUMBKIN - s. thumbscrew
THUMBSCREW - [mech] vis *f* à oreilles, vis *f* ailée
THUMBTACK - [impl] punaise *f*
THUMP, to - [gen] frapper à grands coups
THUMP - [gen] coup *m* sourd
[auto] (tyre thump) vibration *f* des pneus
[telecomm] (telegraph noise) buit *m* de télégraphe
THUNDER - [met] (the sound produced by a lightning discharge) tonnerre *f*
~-BALL - [met] éclair *m* en boule
~-CLOUD - [met] (cumuloninbus-type cloud associated with a thunderstorm) nuage *m* orageux
THUNDERBOLT - [met] (electric discharge accompanied by a clap of thunder) coup *m* de foudre
THUNDERHEAD - [met] (upstanding mass of cumulonimbus thundercloud) partie *f* supérieure d'un cumulus, coeur *m* de l'orage
THUNDERSTORM - [met] (local disturbance accompanied by heavy rain, wind and lightning) orage *m*
~CELL -[met] (localized low-pressure system not associated with a front) orage *m* local
~HIGH - [met] (small weak anticyclone developing immediately under a precipitating cumulonimbus cloud) anticyclone *m* dû à l'orage
THUNDERY DEPRESSIONS - [met] (non-frontal low-pressure cells commonly occurring in summer over Europe) orages *mpl* d'été
THWACK, to - [constr] battre à grands coups
THWART, to - [gen] contrarire, frustrer
THWART - [gen] transversal
[naut] banc *m* de nage
THYME OIL - [chem] (an essential oil obtained from species of Thymus and used in perfumery and as a flavourant) essence *f* de thym
THYMECTOMY - [med] thymectomie *f*
THYMIC ACID - [chem] (synonym of thymol) acide *m* thymique
THYMOL - [chem] (isopropyl metacresol. An antiseptic, disinfectant, perfumery component, and preservative) thymol *m*
~IODIDE - [chem] (an antiseptic) iodure *m* de thymol
THYMOMA - [med] thymome *m*
THYMOPHTHALEIN - [chem] (an analytical indicator) thymophtaléine *f*
THYMOPATHY - [med] trouble *m* de l'affectivité, thymopathie *f*
THYMOPRIVIC - [med] thymoprive
THYMOPRIVOUS - s. thymoprivic
THYMOQUINONE- [chem] (a fungicide) thymoquinone *f*
THYMUS GLAND - [anat] thymus *m*
THYRASTHENIA - [med] insuffisance *f* thyroïdienne
THYRATRON - [electron] (hot-cathode gas filled valve) thyratron *m*
THYRATRON CONVERTER - [electron] convertisseur *m* à thyratrons
~EXTINCTION - [electron] (the cutting off of the anode current which must be carried out by an external circuit if the thyratron is used in direct-current applications) extinction *f*
~TRANSISTOR - [electron] transistor *m* thyratron
THYRISTOR - [electron] (controlled semiconductor rectifier) redresseur *m* semiconducteur commandeé, thyristor *m*
THYROCARDITIS - [med] cardiothyréose *f*
THYROCHONDROTOMY - [med] thyrochondrotomie *f*
THYROID - [zool] (ductless gland in the pharyngeal region) glande *f* thyroïde, corps *m* thryroïde
~BODY - s. thyroid gland
~GLAND - [med] glande *f* thyroïde
THYROIDECTOMY - [med] thyroïdectomie *f*
THYROIDOTHERAPY - [med] thyroïdothérapie *f*
THYROTHERAPY - s. thyroidotherapy
THYROTROPIC HORMONE - [biol] hormone *f* thyréotrope
THYROXINE - [chem] (the hormone manufactured by the thyroid gland. It is used in medicine in the treatment of thyroid deficiency states) thyroxine *f*
TICK - [zool] tique *f*
~FEVER - [vet] fièvre *f* à tique
~-OVER SPEED - [mech](idling speed) marche *f* au grand ralenti
TICKER - [telecomm] télégraphe *m* imprimeur
~-TAPE - [telecomm] bande *f* de zone
TICKET - [gen] billet *m*
[comm] étiquette *f*
[comm] bulletin *m* de bagages
~COLLECTOR - [railw etc] contrôleur *m*
~OFFICE - [railw etc] guichet *m*
~-PRINTING MACHINE - [print] machine *f* à imprimes les tickets
~WINDOW - [railw etc] guichet *m*
TICKING - [horol] tic-tac *m*
~FREQUENCY - [telecomm] fréquence *f* de tic-tac
~-OVER - [mech] grand ralenti *m*
~WEAVE - [text] armure *f* coutil, armure *f* damier
TICKLER - [mech] titillateur *m*
[radio] bobine *f* de réaction
[auto] bouton *m* de noyage, titillateur *m*
[comm] (USA only) aide-mémoire *m*

TICKLER - [el] régulateur *m* d'intensité de courant
TIDAL - [gen] de marée, à marée
~AIR - [med] air *m* de respiration
~BASIN - s. tidal dock
~BORE - [hydr] barre *f* d'eau
~CURRENT - [hydr] courant *m* de marée
~DOCK - [constr] (a dock within which the level is the same as outside) bassin *m* à flot, darse *f*
~HARBOUR - [naut] port *m* accessible à haute mer
~MUD - [geol] boue *f* de marée descendante
~POWER STATION - s. tide power plant
~RIVER - [geogr] fleuve *m* à marée
~WAVE - [met] raz *m* de marée, vague *f* de fond, mascaret *m*, flot *m* de la marée
TIDE - [astr] (the effect of the gravitational attraction of the moon on the waters of the earth) marée *f*
~-DRIVEN - [naut] maremoteur *m*
~GAUGE - [instr] (apparatus designed to determine the variation of sea-level with time) marémètre *m*
~POWER PLANT - [el] centrale *f* marémotrice, usine *f* marémotrice
~REGISTER - s. tide gauge
~RIP - s. tidal wave
TIE, to - [gen] lier, attacher
TIE - [gen] lien *m*, attache *f*, cordon *m*, lacet *m*
[text] cravate *f*
[carp] tirant *m*
[railw] traverse *f*
[naut] itague *f*
[mus] liaison *f*
[constr] chaîne *f*, crampon *m*, harpon *m*
[aero] câble *m* d'haubannage
~ANGLE - [metall] (of an iron framework) cornière *f* de renfort
~BAND - [text] cordon *m* de rappondage
~BAR - [metall] tirant *m*, barre *f* d'entretoisement
[constr] moufle *f* noyée (dans la maçonnerie)
~BEAM - [constr] (structural member connecting the lower ends of a pair of principal refters) longrine *f*, moufle *f*, raineau *m* de pilotis, entrait *m*
~CUTTER - [impl] (for bales) pince *f* à ouvrir les balles
~FEEDER - [el] feeder *m* entre deux centrales
~-GUM - [ind chem] (adhesive for rubber-to-metal bonding) caoutchouc *m* d'ancrage
~-IN BUS - [el] barre *f* collectrice
~-IN WELD - [mech] soudure *f* de raccordement
~-IRONS - [constr] chaînage *m*
~LINE - [phys] (on an equilibrium diagram) ligne *f* de connexion
[telecomm] communication *f* directe
~OF A DRAWBRIDGE - [constr] tirant *m* d'un pont *m* à bascule
~PIECE - [carp] pièce *f* de renfort
[el] connexion *f*
~PLATE - [railw] (plate between rail and sleeper) semelle *f*, selle *f* d'arrêt
[shipbuild] (for strengthening the beams) virure *f* d'hiloire
[constr] plaque *f* d'assise, ancre *f*, tôle *f* de jonction
[metall] bande *f* de raccordement
~RAIL - [metall] rail *m* armé
~-ROD - [mech] (any rod in a structure which is designed to resist tensile stress in a particular direction, often fitted with clevises for length adjustment) tirant *m*
TIE-ROD - [auto] (of the steering system) barre *f* d'accouplement
[metall] armature *f*
[constr] (of the roof) entrait *m* de toit
[railw] tringle *f* de connexion des aiguilles
~-ROD END - [mech] tête *f* du tirant
~THREAD - [text] fil *m* à nouer
~UP, to - [gen] lier
~UP - [metall] armure *f*
~-WIRE - [el] (or binding-wire; wire used to attach a conductor to an insulator) cordelette *f* pour ligatures, fil *m* d'attache
TIED FLEECE - [text] toison *f* ficelée
~-UP HANK - [text] écheveau *m* lié en centaines
TIER - [gen] rangée *f*, étage *m*
[el] plan *m*
[naut] (of a rope) glène *f*
[naut] (a line of moored ships) plan *m* d'arrimage
[radio] (an aerial array) système *m* d'antennes
[text] (of carriages) rangée *f* (de chariots)
~OF CARRIAGES - [text] (in a lace machine) rangée *f* de chariots
TIERING RACK - [gen] étagère *f*
TIEROD - s. tie rod
TIES OF THE BALES - [text] cordes *fpl* d'emballage des balles
TIGER - [zool] tigre *m*
TIGER'S EYE - [min] (a form of silicified crocidolite) oeil-de-tigre *m*
TIGER SKIN PLUSH - [text] peluche *f* imitation tigre
TIGHT - [gen] imperméable, étanche
[gen] (closely held together) serré
[gen] (of thread, rope etc) raide, tendu
~ALIGNMENT - [radio] (alignment used to obtain the desired frequency response curve) alignement *m* serré
~BINDING APPROXIMATION - [electron] (in electron physics) approximation *f* d'une liaison serrée
~CASK - [gen] tonneau *m* étanche
~CHAIN - [text] chaîne *f* tendue
~COAT - [metall] revêtement *m* métallique sans défauts
~CORE CABLE - [el] câble *m* compact
~COUPLING - [radio] (a degree of coupling greater than the critical coupling) couplage *m* serré
~CURE - [rubber ind] vulcanisation *f* complète
~FIT - [mech] couplage *m* bien ajusté
~-FITTING SCREW - [mech] vis *f* de serrage
~GRAVEL - [constr] gravier *m* serré
~HOLE - [mining] trou *m* rétréci
~JOINT - [mech] joint *m* étanche
~MESHING - [mech] (of gears) engrenage *m* sans jeu
~PULLEY - [mech] poulie *f* fixe
~RING - [mech] segment *m* d'étanchéité
~RIVETING - [mech] rivure *f* étanche
~ROPE - [gen] corde *f* tendue
~SAND - [geol] sable *m* peu perméable
~SIDE - [mech] (of a transmission rope) brin *m* conducteur, brin *m* tendu
~STRAND - [text] toron *m* tendu
~YARN - [text] fil *m* tendu
TIGHTEN, to - [gen] serrer, resserrer, se tendre
[mech] (a screw, a nut) serrer
(the ropes) tendre
[mech] (of a spring) se bander
TIGHTEN A SEAM, to - [rubber ind] boucher une cou-

ture étanche
TIGHTENER - [mech] tendeur *m*, tenseur *m*
~PLATE - [text] plaque *f* de serrage
TIGHTENING - [gen] serrage *m*, raidissement *m*
[mech] (of a spring) bandage *m* (d'un ressort)
~CORD - [mech] corde *f* de tension
~FLAP - [auto] plaquette *f* basculante de fermeture
~NUT - [mech] écrou *m* de serrage
~PULLEY - [mech] poulie *f* de tension, galet *m* tendeur
~PULLEY ARM - [mech] bras *m* de poulie de tension
~PULLEY STUD - [mech] axe *m* pour la poulie de tension
~ROLLER - [text] rouleau *m* de tension
~SCREW - [mech] vis *f* de serrage, vis *f* de tension
~SLIDE - [text] coulisse *f* de tension
~STICK - [text] cheville *f* à tourniquet
~THE WARP - [text] tension *f* de la chaîne
~-UP - [mech] serrage *m*
~-UP DEVICE - [railw] (for coupling railway cars) dispositif *m* tendeur
TIGHTLY PACKED - [soil] très serré
~STRETCHED WARP - [text] chaîne *f* fortement tendue
~WOVEN STUFF - [text] tissu *m* serré, tissu *m* à corps plein
TIGHTNESS - [gen] étanchéité *f*, imperméabilité *f*
[mech] (tension) tension *f*
[fin] resserrement *m* (de l'argent)
~OF WINDING - [text] tension *f* de renvidage
~TEST - [plumb] essai *m* d'étanchéité
TIGLIC ACID - [chem] (a constituent of camomile oil) acide *m* tiglique
TIL - [bot] sésame *m* de l'Inde, till *m*
~OIL - [ind chem] huile *f* de sésame
TILE, to - [constr] couvrir de tuiles, carreler
TILE - [constr] (this slab of baked clay, terracotta, glass etc, used for roofing or covering walls or floors) brique *f* à paver, carreau *m*
[constr] (for roofing) tuile *f*
~CLADDING - [constr] carrelage *m*
~CLAY - [min] argile *f* téguline
~CONDUIT - [constr] tuyau *m* en poterie
~COVERING - [constr] pose *f* des tuiles
~-FIELD - [constr] tuilerie *f*
~FILTER BOTTOM - [hydr] plancher *m* de filtre en céramique, fond *m* de filtre en céramique
~FLOOR - [constr] carrelage *m*
~KILN - [constr] four *m* à carreaux, tuilerie *f*
~ORE - [min] (the earthy, brick-red variety or cuprite) oxyde *m* rouge de cuivre
~ROOF - [constr] toit *m* à tuiles
TILEMAKING - [constr] fabrication *f* de tuiles
~MACHINE - [mech] machine *f* à tuiles
TILERY - [constr] tuilerie *f*, fabrique *f* de carreaux
TILING - [constr] (of a roof) pose *f* des tuiles
[constr] (of a floor) carrelage *m*
TILL, to - [agric] labourer, cultiver
TILL - [gen] tiroir-caisse *m*
[geol] argile *f* à blocaux, terrain *m* erratique
~BY ROTARY HOE, to - [agric] travailler à la fraise
TILLAGE - [agric] labour *m*, labourage *m*, culture *f*
~-EQUIPMENT - [agric] matériel *m* de préparation du sol
~PAN - [soil] semelle *f* de labour, fonds *mpl* de raie du labour
TILLER - [naut] barre *f* franche
[bot] talle *f*, baliveau *m*
TILLER - [mining] tourne-à-gauche *m* de manoeuvre
~-BAR - [naut] barre *f* de direction
~ROPE - [naut] drosse *f* du gouvernail
~STEERING - [auto] (for trailers) direction *f* commandée par timon d'attelage
~WHEEL - [naut] roue *f* du gouvernail
TILLERING - [agric] tallage *m*
TILLITE - [geol] (boulder clay) tillite *f*
TILT, to - [gen] pencher, s'incliner, incliner, se renverser
[aero] glisser sur l'aile
[mining] (of coal etc) culbuter, déverser
[metall] ramper (un moule)
[metall] (iron) martiner, marteler
TILT - [gen] inclinaison *f*, pente *f*
[naut] gîte *f*
[photo] (of the camera) panoramique *f* verticale
[mech] carrossage *m* de l'essieu
[telev] (percentage difference between the initial and final value of the response of a rectangular pulse, due to the blocking of low frequencies) dénivellation *f*, déclivité *f*
[radar] (frame distortion) distorsion *f* d'image
~ADJUSTMENT - [photo] réglage *m* de l'inclinaison
~-AND-BEND SHADING - [telev] correction *f* de défauts de mosaïque
~BACK, to - [gen] basculer en arrière
~HAMMER - [mech] martinet *m*, marteau *m* à bascule marteau *m* à soulèvement
~INDICATOR - [photo] indicateur *m* d'inclinaison
~METER - [instr] clinomètre *m*
~MIXER - [telev] (a means to rectify the tilt) correcteur *m* de déformation de ligne
~-ROOF - [constr] toit *m* arrondi
~-TOP TABLE - [carp] table *f* à inclinaison
TILTED AERIAL - [radio] antenne *f* inclinée
~ANTENNA - s. tilted aerial
~CYLINDER MIXER - [mech] (mixer consisting of a plain cylinder revolving about a diagonal of its longitudinal section) mélangeur *m* à tonneau désaxé
~MIRROR - [photo] miroir *m* incliné
TILTER - [metall] culbuteur *m*
TILTING - [gen etc] incliné, inclinable, basculaire
~AMALGAMATOR - [oil ind] amalgamateur *m* incliné
~ATTACHMENT - [photo] dispositif *m* d'inclinaison
~BEARING - [mech] appui *m* à bascule, appui *m* à rotule
~CART - [impl] tombereau *m*
~DAM - [hydr] barrage *m* basculant
~DEVICE - [impl] bascule *f*, culbuteur *m*, dispositif *m* d'inclinaison
~-DRUM MIXER - [mech] malaxeur *m* à tambour inclinable
~FILLET - [constr] chanlatte *f*
~FURNACE - [metall] four *m* basculant
~HAMMER - s. tilt hammer
~HEAD - [photo] plate-forme *f* à bascule (du trépied)
~HEAD PRESS - [mech] (a hydraulic press so designed that the head of the press tilts during the opening movement) presse *f* à tête renversable
~INSULATOR SWITCH - [el] interrupteur *m* à isolateur basculant
~JOINT - [mech] rotule *f* d'inclinaison
~LADLE - [metall] poche *f* à bascule
~LEVER - [impl] queue *f* de lion
~OVER OF THE RAIL - [railw] (canting of the rail)

déversement *m* du rail
TILTING REFLECTOR - [auto] réflecteur *m* à bascul.
~SEAT - [aero, auto etc] (a seat designed to be folded back when not in use) strapontin *m*
~STAND - [impl] support *m* tournant
TIMBER, to - [gen] boiser, blinder
[constr] charpenter
[mining] cuveler (un puits)
TIMBER - [gen] bois *m* d'oeuvre
[constr] bois *m* de charpente
(a squared piece) poutre *f*, madrier *m*
[naut] membre *m*, allonge *f*
~ASSEMBLING - [constr] assemblage *m* des bois
~BEAM - [constr] poutre *f* en bois
~-CART - [transp] trique-balle *m*, fardier *m*
~CHUTE - [agric] (in forestry) lançoir *m*
~CLAIM - s. timber concession
~CONCESSION - [comm] concession *f* forestale
~CLIP - [carp] agrafe *f* (du chariot d'une scie à debiter en grumes)
~CRANE - [impl] chèvre *f*
~DAM - [hydr] barrage *m* en bois
~DOG - [impl] clampe *f*, clameau *m*
~DRAWING - [agric] déboisage *m*
~DRYING - [timber] dessiccation *f* du bois
~FORMS - s. timber shuttering
~FOUNDATION - [text] grillage *m* en charpente
~FRAMING - [constr] pan *m* de bois
~-HEAD - [naut] jambette *f*
~IN PLANKS - [carp] bois *m* méplat
~JACK - [carp] cric *m* de charpentier
~LINE - [geogr] limite *f* forestière
~MILL - [ind] scierie *f*
~PLANKING - [hydr] bordé *m* en bois
~RAFT - [forestry] train *m* de bois
~RIGHT - [forestry] droit *m* de coupe
~ROT - [bot] carie *f* du bois
~RUNNER - [constr] glissière *f* en bois
~SET - [mining] cadre *m* de boisage
~SHOOT - [forestry] lançoir *m*
~SHUTTERING - [hydr] coffrage *m* en bois
~SLIDE - [forestry] lançoir *m*
~-SPLITTING WEDGE - [impl] coin *m* à fendre le bois
~SUPPORT - [constr] support *m* en bois
[gen] charpente *f*
~-TREE - [bot] sarbre *m* de haute futaie
~TRESTLE - [constr] chevalet *m* en bois
~WAIN - s. timber shoot
~WORK - [constr] construction *f* en bois
~-YARD - [gen] chantier *m* de bois de charpente
TIMBERED SHAFT - [mining] puits *m* boisé
TIMBERING - [gen] boisage *m*, boisement *m*
[arch] forêt *f* de comble
[constr] armature *f* de bois
[mining] cuvelage *m* d'un puits
TIMBERWORK - s. timber work
TIMBERYARD - s. timber-yard
TIMBRE - [acoust] (distinctive quality of tone) timbre *m*
TIME, to - [gen] (to begin, to perform at the suitable time) fixer l'heure
[gen] (to ascertain or record speed etc) mesurer le temps, cronométrer
[gen] (to synchronize) synchroniser
[photo] calculer le temps de pose
[mech] (an engine) caler, ajuster l'allumage
[autom] (to determine the time for machining a piece etc) mesurer le temps de production
TIME - [gen] temps *m*
[an age] (a period in history) époque *f*, âge *m*
[gen] (a period of time) durée *f*
[gen] (a definite time) heure *f*
[mus] tempo *m*, mesure *f*
~-AND-PERCUSSION FUSE - [firearms] fusée *f* à percussion et à temps
~AVERAGE - [phys] (the time average of any time-dependent physical quantity over a given time interval is the time interval of the quantity taken over the internal, divided by the magnitude of the interval) moyenne *f* de temps
~AXIS - [radar] (the abscissa axis) axe *m* des temps
~-BASE - [electron] (network in a C.R.T. circuit to give a time-control to the trace) base *f* de temps
~-BASE OSCILLATOR - [electron] oscillateur *m* de base de temps
~-BASE VOLTAGE - [electron] tension *f* de base de temps
~BELT - [geogr] (a time zone) fuseau *m* horaire
~BILL - [railw] horaire *m*
~BOMB - [comm] bombe *f* à horlogerie
~-BOOK - [gen] livre *m* de pointage
~CARD - [gen] carte *f* de présence
~CHARTER - [naut] affrètement *m* à temps
~CHECK - [instr] horloge *f* de contrôle
~CLOCK - [instr] (a clock designed to record times) horloge *f* enregistreuse
~CONSTANT - [el] (the time in which the current in an inductance changes by the fraction of its ultimate change, when the applied voltage is changed) constante *f* de temps
~CONSTANT OF AN EXPONENTIAL QUANTITY - [el] (the time after which the quantity would reach its limit, if it maintained its initial rate of variation) constante *f* de temps d'une grandeur exponentielle
~CONSTANT OF THE APERIODIC COMPONENT - [el] (the time constant of that component when it is prapractically exponential or of the exponential which envelops it when it shows an appreciable periodicity) constante *f* de temps de la composante apériodique
~CONTROL - [contr] commande *f* à programme
~CONTROL PULSE - [radar] (the leading and trailing edge of the flat-topped pulse are used to start or stop an operation in radar control systems) impulsion *f* de contrôle des temps
~-CYCLE - [contr] en fonction du temps
~-CYCLE OPERATION - [contr] (the operation of one or more regulating units as a function of time only) opération *f* en fonction du temps
~DELAY - [el] retard *m* en temps, action *f* retardée, temporisation *f*
~-DELAY CIRCUIT - [el] circuit *m* retardateur
~-DELAY RELAY - [el] relais *m* à action retardée
~DEMODULATION - [radio] démodulation *f* par impulsions dans le temps
~-DEPENDENCE - [chem] (said of a reaction when it depends on time) en fonction du temps
~-DEPENDENT NUCLEAR REACTION - [nucl] réaction *f* nucléaire en fonction de temps
~DISCRIMINATOR - [el] (circuit indicating the time equality of two events, or the approximate magnitude of the inequality) discriminateur *m* de temps
~-DIVISION MULTIPLEX - [telecomm] (device for the transmission of two or more signals over a common

path by using successive time intervals for different signals) multiplex *m* dans le temps, transmission *f* par secteurs
TIME-DIVISION TELEGRAPH SYSTEM - [telecomm] système *m* télégraphique par partage dans le temps
~DOMAIN - [contr] domaine *m* des temps
~ELEMENT - [el] (the time-delay feature in the action of a circuit breaker) dispositif *m* de temporisation, relais *m* temporisé
~EXPOSURE - [photo] pose *f*
~FUEL GRAPH - [aero] (graph indicating the relation of time and fuel used during a flight) courbe *f* temps/carburant
~FUSE - [mining] fusée *f* fusante, fusée *f* à temps
~-FUSE FIRE - [firearms] tir *m* fusant
~GATE - [radio] (transducer which gives output only during chosen time intervals) porte *f* à intervalle de temps sélectionné
~INTERVAL BETWEEN STROKES - [el] intervalle *m* entre foudres
~-INTERVAL COUNTER - [electron] (electronic counter designed to measure a time interval) compteur *m* électronique d'intervalle de temps
~JITTER - [radar] instabilité *f* de la base de temps
~LAG - [el] retard *m*, temporisation *f*
[electron] retard *m* d'amorçage
~-LAG CUT-OUT - [el] coupe-circuit *m* à action différée
~-LAG DEVICE - [mech] dispositif *m* à action différée, appareil *m* à action temporisée
~-LAG ERROR - [el] erreur *f* sur le retard
~LAG IN COMPRESSION - [soil] lenteur *f* de la compression
~-LAG OF FLASHOVER - [el] retard *m* de contournement
~-LAG OF PUNCTURE - [el] retard *m* de percement
~-LAG RELAY - [el] automate *m* chronométrique, relais *m* temporisé
~-LAG OF RELEASE - [photo] durée *f* du déclenchement
~-LIMIT - [el] temps *m* limite, limite *f* de temps
[contr] à action différée
~-LIMIT ATTACHMENT - [el] (a mechanical device designed to open a circuit-breaker after a given time-delay) appareil *m* pour action différée
~-LIMIT PROTECTION - [el] (of electrical lines) protection *f* à action différée
~-LIMIT RELEASE - [el] action *f* différée
~-LIMIT RELEASE RELAY - [el] relais *m* à action différée
~METERING - [meas] comptage *m* à temps
~MODULATION - [radio] (modulation in which the time of a definite portion of a waveform, measured with respect to a reference time, is varied in accordance with a signal) modulation *f* par impulsions dans le temps
~OF ACTION - [hydr] temps *m* de contact, temps *m* de réaction
~OF CONTACT - s. time of action
~-OF-DAY TARIFF - [comm] tarif *m* double
~-OF-FLIGHT - [nucl] (of moving particles) temps *m* de vol
~OF FLIGHT METHOD - [nucl] (method for measuring cross-section as a function of energy in nuclear reactors) méthode *f* de temps de vol
~-OF-FLIGHT VELOCITY SELECTOR - [nucl] sélecteur *m* de vitesse à temps de vol

TIME OF FLOW - [hydr] temps *m* d'écoulement
~OF FLOWING THROUGH - s. time of passage
~OF LIBERATION - [phys] (the time necessary to liberate an electron from an emitting surface) temps *m* de libération
~OF PASSAGE - [hydr] durée *f* de passage
~OF REACTION - s. time of action
~OF RECOVERY - [metall] durée *f* de rétablissement, durée *f* de récupération
~OF RESPONSE - [contr] temps *m* de réponse
~PATTERN - [electron] (picture-tube presentation of horizontal and vertical lines or dot rows generated by two stable frequency sources operating at multiples of the line and field frequencies) diagramme *m* horaire
~QUENCHING - [metall] trempe *f* interrompue
~RATE - [el] (of a storage battery) ampérage *m* de décharge, régime *m* de décharge
~RATING - [mech] (of a machine) durée *f* de fonctionnement
~RECORDER - s. time clock
~RELEASE - [el] déclenchement *m* temporisé
~REVERSAL - [phys] (in equation of motion of a dynamic system) inversion *f* de temps
~RING - [firearms] (of a fuse) bague *f* des temps
~SCALE - [comput] échelle *f* des temps
[photo] échelle *f* de gris à éclairement constant et temps variable
~SCHEDULE - [autom] plan *m* d'échelonnement
~-SELECTION BAND - [comput] bande *f* de sélection adresses
~-SELECTOR TRANSDUCER - [radio] (a gating circuit) circuit *m* discriminateur de temps
~-SERIES - [gen] série *f* chronologique
~SETTING - [photo] réglage *m* du temps d'exposition
~-SETTING RANGE - [contr] domaine *m* d'ajustement au retard
~-SETTLEMENT CURVE - [soil] courbe *f* temps-tassement
~-SHARED AMPLIFIER - [radio] amplificateur *m* à subdivision dans le temps
~-SHARED OPERATION - [comput] opération *f* en parallèle
~SHARING - [comput] simultanéité *f* par partage du temps machine
~SHEET - [gen] feuille *f* de présence, semainier *m*
~SIGNAL - [radio] signal *m* horaire
~SIGNAL SET - [electron] générateur *m* électronique de signaux horaires
~SORTER - [phys] sélecteur *m* d'impulsions en dépendance du temps
~STUDY - [gen] (for factory organization) analyse *f* des temps de production
~SWITCH - [el] (switch designed to open or close a circuit at a predetermined time by means of an electrically wound clock) interrupteur *m* horaire
[meas] (a meter change-over clock) horloge *f* de commutation pour compteur
~-TEMPERATURE TRANSFORMATION CURVE - [metall] courbe *f* de transformation isothermique
~THE IGNITION, to - [auto] régler l'allumage
~TO IMPULSE FLASHOVER - [el] durée *f* de contournement
~TO TAKE-OFF - [aero] (the time consumed in taking off) temps *m* de décollage
~TICK - [meas] top *m*
~UNDERVOLTAGE PROTECTION - [el] protection *f*

contre sous-tension de longue durée
TIME-VARIABLE FIELD - [phys] champ *m* variable
~WAGES - [comm] salaire *m* à l'heure
~-WORK - [comm] travail *m* à l'heure
~-WORKER - [print] ouvrier *m* en conscience
~YIELD - [metall] durée *f* de charge
~ZONE - s. time belt
TIMEBASE - s. time-base
TIMED - [mech] (of an engine) au point
TIMEKEEPER - [horol] (general term denoting a watch or clock) chronométreur *m*
TIMEKEEPING - [gen] contrôle *m*, pointage *m* de présence
TIMER - [gen] chronométreur *m*
[instr] (a device, usually electronic, for automatic control of the time taken by an operation) compteur *m* de temps
[el] (a synchronizer) synchroniseur *m*
[gen] (a watch or clock) horloge *f*, chronomètre *m*
(int comb eng; of the spark) distributeur *m* d'allumage, rupteur *m*
[metall] (a device for regulating the times of a welding machine in a welding cycle) interrupteur *m* automatique de réglage
~SHAFT - [mech] (distributor shaft) axe *m* du dispositif de calage, axe *m* du distributeur
~SUBCHASSIS - [nucl] tiroir *m* programmateur
TIMETABLE - [gen] horaire *m*
TIMEWORK - [gen] s. time-work
TIMING - [auto etc] (in a reciprocating i.c. engine, the adjustment of ignition and valve action cycles in respect to crankshaft rotation) réglage *m* de l'allumage
[mech] calage *m*
[auto] distribution *f*
[telev] marquage *m*
[photo] calcul *m* du temps de pose
[electron] (of signals) timing *m*
~ADJUSTMENT - [auto] réglage *m* de l'allumage
~ANGLE - [mech] angle *m* de calage
~CAM - [text] doigt *m* du compteur
~CASE - [mech] (in motorbicycles) carter *m* de la distribution
~CHAIN - [mech] (a chain, driving the camshaft in a reciprocating i.c. engine) chaîne *f* de distribution
~CHEST - [mech] (of a motorbicylce) boîte *f* de la distribution
~CIRCUIT - [electron] circuit *m* de rythme
~COVER - [mech] (a cover fitted over timing gears or chain) couvercle *m* de la chaîne de distribution
~COVER GASKET - [mech] (an oil-tight seal placed between the timing cover and the body of the engine) joint *m* du couvercle de distribution
~DIAGRAM - [mech] (for the valve timing of an engine) diagramme *m* de la distribution
~ELEMENT - [el] relais *m* de temporisation
[meas] (a drum type counter) minuterie *f* de compteur
~GEAR - [mech] pignons *mpl* de distribution
~-GEAR CASE - [mech] (casing enclosing the camshaft drive) boîte *f* de distribution
~-GEAR HOUSING - [auto] couvercle *m* de distribution
~GEARS - [mech] (a train of gears driving the camshaft in a reciprocating i.c. engine) engrenages *mpl* di distribution
TIMING LEVER - [mech] levier *m* de calage
~MARK - [auto] repère *m* de distribution
~MARK - [comput] marque *f* de synchronisation
~MARK CHECK - [comput] contrôle *m* des marques de synchronisation
~PINION - [mech] pignon *m* de distribution
~POINTER HOLE - [auto] index *m* de calage, regard *m*
~PULSE - [electron] impulsion *f* de rythme
~PULSE GENERATOR - [comput] (a generator designed to preset the computing interval in electronic analog computers) générateur *m* d'impulsions d'horloge
~PULSES - [telev & radar] impulsions *fpl* de marquage
~RANGE - [auto] marge *f* de calage
~SHAFT - [auto] arbre *m* de distribution
~SIDE - [mech] (of a motorbicycle) côté *m* distribution
~SIGNAL - [electron] signal *m* de rythme
~SYSTEM - [mech] distribution *f*
~TAPE - [el acoust] (between two pieces of recorded tape) ruban *m* intermédiaire
~TRACK - [comput] piste *f* de rythme
~WASHER - [horol] rondelle *f* du balancier
TIMOTHY - [bot] fléole *f* des prés
TIN, to - [metall] (to coat metal with tin by immersing it in a bath of the liquid metal) étamer
[food] (to pack in tins) mettre en boîtes
TIN - [chem] (a metallic element, symbol Sn, A.N. 50, A.W. 118.7, with important metallurgical applications) étain *m*
[food] boîte *f* (à conserves), bidon *m*
[metall] étain *m*, fer *m* blanc
~ALLOYS - [metall] (used for soft solders, type metals etc) alliages *mpl* d'étain
~AMALGAM - [metall] amalgame *m* d'étain
~BAR - [metall] barre *f* d'étain
~-BASE ALLOY - [metall] alliage *m* à base d'étain
~BATH - [metall] tain *m*
~-BEARING - [min] stannifère
~BELT - [min] zone *f* stannifère
~BICHLORIDE - [chem] (stannous chloride) bichlorure *m* d'étain
~-BOX - [metall] boîte *f* en fer blanc
~BRONZE - [metall] bronze *m* d'étain
~-COATING GAUGE - [meas] jauge *f* d'épaisseur de la couche d'étain
~-COPPER PLATING - [metall] placage *m* à l'étain et le cuivre
~CRY - [min] cri *m* de l'étain
~DEPOSIT - [min] gîte *m* d'étain
~DISEASE - s. tin pest
~DISH FEED - [text] auge *f* alimentaire en fer blanc
~DREDGE - [mech] drague *f* à étain
~DREDGING - [min] dragage *m* stannifère
~-DRESSING - [min] préparation *f* mécanique du minerai d'étain
~FOIL - [metall] feuille *f* d'étain, clinquant *m* de étain
~-FOIL CAPACITOR - [el] (paper capacitor with tin foil as conducting material) condensateur *m* à feuilles d'étain
~GROUND - [min] terrain *m* stannifère
~GUARD - [mech] plaque *f* métallique de protection
~LEAF - [metall] feuille *f* d'étain
~LINING - [metall] étamage *m*
~LODE - [min] filon *m* d'étain, filon *m* stannifère

TIN-NICKEL BRASS - [metall] laiton *m* à étain et nickel
~-NICKEL PLATING - [metall] placage *m* à l'étain et le nickel
~OCTOATE - [chem] (a catalyst) octoate *m* d'étain
~ORE - [min] minerai *m* d'étain, cassitérite *f*
~PEST - [metall] peste *f* de l'étain
~PLACER - [min] placer *m* stannifère
~PLAGUE - s. tin pest
~PLATE - [metall] (thin steel-sheet covered with an adherent layer of tin) fer *m* blanc, ferblanterie *f*
~PLATE MILL - [metall] laminoir *m* pour fer blanc
~PLATE SHAFT - [mech] arbre *m* en fer blanc
~-PLATE WORKING - [metall] ferblanterie *f*
~PLATING - [metall] étamage *m*
~POWDER - [metall] poudre *f* d'étain
~PUTTY - [ind chem] potée *f* d'étain
~PYRITES - [min] (another name for Stannine, Stannite, Bell Metal Ore : a natural sulphide of copper, tin and iron) étain *m* pyriteux, stannine *f*
~RECLAMATION PROCESS - [metall] procédé *m* de régénération de l'étain
~ROLLER - [text] tambour *m* de friction
~ROLLER BRAKE - [text] frein *m* du tambour
~ROLLER PULLEY - [text] volant *m* sur l'axe des tambours
~ROLLER WHEEL - [text] pignon *m* du tambour en fer blanc
~SHEET - [metall] plaque *f* d'étain
~SHETT IRON - [metall] fer *m* blanc
~SHOP - [metall] ferblanterie *f*
~SMELTERY - [metall] fonderie *f* d'étain
~SMELTING - [metall] fusion *f* de l'étain
~SNIPS - [impl] cisaille *f* pour fer blanc
~SOLDER - [metall] (low temperature solder consisting of tin and lead) soudure *f* à l'étain
~SOLDERING - s. tin solder
~SPOT - [metall] tache *f* d'étain
~STRIP - [text] tôle *f* de guide, plaque *f* morte
~SWEAT - [metall] ressuage *m* d'étain
~TREE - [chem] arbre *m* de Jupiter
~WHITE COBALT - [min] smaltine *f*
~WORKS - s. tin smeltery
~-ZINC PLATING - [metall] placage *m* à l'étain et le zinc
TINCAL - [min] (crude natural borax, obtained from salt lakes, especially in Tibet) tincal *m*
TINCTORIAL POWER - [photo] (the measure of the depth of colour produced by a dye) pouvoir *m* colorant
~STRENGTH - s, tinctorial power
TINCTURE - [chem] (a solution, often of a drug, in alcohol or water) teinture *f*
TINDER - [gen] (any inflammable substance which will ignite on contact with a spark) mèche *f* de briquet
[th eng] (for boilers) filasse *f* inflammable
~BOX - [impl] boîte *f* d'amadou, briquet *m* à silex
TINE - [tool] (a spike or prong) dent *f*
[impl] (of a fork) fourchon *m*, pointe *f*
[zool] branche *f* de bois de cerf
~CULTIVATOR - [agric] cultivateur *m* à dents rigides
TINEA - [med] (ringworm) teigne *f*
TINFOIL, to - [metall] revêtir de feuille d'étain
TINFOIL - [metall] étain *m* battu, papier *m* d'argent, feuille *f* d'étain
TINGE, to - [gen] (to imbue with a faint trace of colour) teinter, nuancer
TINGE - [gen] (a slight colouration) teinte *f*, nuance *f*
[photo] (e. g. in underexposed picture) dominante *f*
TINGED COTTON - [text] coton *m* tacheté
TINGLE - [acoust] tintement *m*
[carp] (a tack, a small nail) broquette *f*
TINKER - [gen] rétameur *m*, chaudronnier *m*
TINKLING - [acoust] tintement *m*
TINMAN - [metall] ferblantier *m*
TINMAN'S SHEARS - [impl] cisailles *f*pl de ferblantier
~SNIPS - s, tinman's shears
TINMEN'S SOLDER - [metall] soudure *f* d'étain
TINNED - [metall] (coated with metallic tin) étamé
[food] conservé
~CONDUCTOR - [el] fil *m* étamé
~COPPER - [metall] cuivre *m* étamé
~FOODS - [food] conserves *f*pl alimentaires
~HEALD - [text] lisse *f* étamée
~IRON - [metall] (plate or sheet) tôle *f* étamée
~OIL - [oil ind] huile *f* en bidon
~STEEL - [metall] acier *m* étamé
~WIRE - [metall] fil *m* étamé
TINNER - s. tinman
TINNING - [metall] étamage *m*, avivage *m*
[metall] (in electro-typing) (the melting of lead-tin foil or tin plating on the back of shells) étamage *m*
~METAL - [metall] étamure *f*
~VAT - [metall] chaudière *f* à étain
TINNITUS - [med] (a persistent sensation of ringing noise in the ear) tintement *m* d'oreilles
TINPLATE - [metall] (thin sheet steel coated with tin) fer-blanc *m*
TINSEL - [text] (a silk fabric) fil *m* métallique aplati
[text] (fabric in which spangles or bits of metals are interwoven) lamé *m*
[text] (as used in embroidery) paillettes *f*pl
[el] filet *m*, clinquant *m*
~BRAID - [text] galon *m* lamé
~CORD - [el] cordon *m* souple à conducteur en hélice
~FOR TRIMMINGS - [text] lames *f*pl pour passementerie
TINSMAN - s. tinsmith
TINSMITH-[labour] étameur *m*, ferblantier *m*
TINSMITHING - [metall] ferblanterie *f*
TINSTONE - [min] (cassiterite,natural tin dioxide, the chief source of tin) cassitérite *f*, étain *m* oxydé
TINT, to - [gen] teinter
[print] (in engraving, any effect of light, shade etc) ombrer, hachurer
TINT - [gen] (tinge) teinte *f*, nuance *f*
[print] (the colour obtained by mixing white pigment with a small proportion of coloured pigment) grisé *m*
~OF PASSAGE - [opt] couleur *f* de transition
TINTED - [gen] teinté
~GLASS - [auto etc] (e. g. the windows of a car) verre *m* coloré
~GLASSES - [opt] lunettes *f*pl contre le soleil
TINTING - [gen] teinture *f*
[paint] (the operation for obtaining the desired paint colour) préparation *f* de la peinture
~POWER - s. tinting strength
~STRENGTH - [paint] (arbitrary scale indicating

the capacity of a pigment to produce a given tint) pouvoir *m* de coloration
TINTOMETER - [instr] (a colorimeter in which colours are compared with those of standard solutions or of specially prepared glass slides) tintomètre *m*
TINTYPE - [photo] amphitypie *f*
TINWARE - [gen] articles *mpl* en fer blanc
TIP, to - [gen] toucher, effleurer
[gen] (to tip over) renverser, verser
[aero] s'incliner
[mining] verser, culbuter, déverser
[mech] (e.g., a tool with carbide) rapporter la pointe
[shoe man] mettre un bout à
[gen] (to cap) ferrer, emboutir
TIP - [gen] bout *m*, extrémité *f*, pointe *f*
[gen] (the apex) sommet *m*
[aero] (of a wing) extrémité *f*
[mech] extrémité *f* de la came
[electron] (the small protuberance on the envelope of a tube) pointe *f*
[mining] terris *m*, dépôt *m* de déblais
[metall] crassier *m*
[impl] palette *f* à dorer
[transp](elevated runway) estacade *f*
[mech] (a tipping device) culbuteur *m*, basculeur *m*
~AND GUIDE POSITION - [el acoust] position *f* des pointes et du guide-bande
~BUCKET - [mech] benne *f*
~ELECTRODE - [el] (spot welding electrode) électrode *f* de soudage par points
~GROWTH - [bot] pousse *f* terminale
~OF PLUG - [el] tête *f* de la broche
~OF STAPLE - [text] pointe *f* de la mèche
~OF THE CARD TOOTH - [text] pointe *f* de l'aiguille
~-PATH PLANE - [el] (plane containing the path defined by the rotating blade tips) plan *m* du disque balayé
~PROJECTION - [el acoust] dépassement *m* des têtes
~RADIUS - [aero] rayon *m* de la pale
~SKID - [mech] (electrode skid in welding) glissement *m* de l'électrode
~TANK - [aero] réservoir *m* à l'extrémité de l'aile
~-TROUGH - [impl] goulotte *f* culbutante
~TRUCK - [transp] wagon *m* à bascule
~-UP SEAT - [aero auto etc] (tilting seat) strapontin *m*, siège *m* relevable
~WIRE - [el] fil *m* de pointe
TIPCART - [impl] tombereau *m*
TIPLESS BULB - [electron] (tipless envelope) ampoule *f* sans pointe
TIPPED - [mech] (of a tool) à plaquette rapportée
TIPPER - [mech] culbuteur *m*, basculeur *m*
[transp] (a tipping lorry) véhicule *m* à benne basculante
[railw] wagonnet *m* basculant
~TRUCK - [transp] benne *f* basculante
TIPPING - [gen] inclinaison *f*, renversement *m*
[naut] déjaugeage *m*
~BUCKET - [mech] (of earth-moving machinery) benne *f* à bascule
~CRIB - [agric] mangeoire *f* basculante
~DEVICE - [mech] (for unloading) appareil *m* de basculage, appareil *m* à renversement
~GRATE - [railw] (of a steam locomotive) grille *f* basculante, jette-feu *m* basculant
~HORSE CART - [agric] tombereau *m* attelé
TIPPING LEVER - [mech] levier *m* à bascule
~LORRY - s. tipper truck
~STAGE - [metall] plate-forme *f* de déversement
~TORCH - [electron] (or sealing-off burner; a component part of a machine for manufacturing electron tubes) chalumeau *m* de coupage
~TRAILER - [transp] remorque *f* à benne basculante
~TRAY - s. tip through
~TROUGH - s. tip trough
TIPPINGS - [mining] déblai *m* de mine
TIPPLE BOX - [text] boîte *f* à lin
TIPPLER - [mech] culbuteur *m*, verseur *m*
[mining] recette *f* supérieure
TIPULA - [zool] tipule *f*
TIRE - s. tyre
TIREFOND - [med] tire-fond *m*
TIRING - [med] cerclage *m*
T-IRON - [constr] (a T-shaped section) fer *m* T
TISANE - [gen] tisane *f*
TISSUE - [text] (woven fabric) tissu *m*, étoffe *f*
[paper man] (the lightest type of paper, e.g. for paper table napkins, toilet paper etc) papier *m* de soie
[biol] (aggregate of similar cells forming a definite continous fabric) tissu *m*
[photo] (carbon tissue) papier *m* au charbon
~DOSE - [radiat] (dose received by a tissue in the region of interest) dose *f* au tissu, dose *f* tissulaire
~-EQUIVALENT IONIZATION CHAMBER - [electron] (ionization chamber in which the material of the walls, electrodes and gas are such as to produce ionization essentially equivalent to the ionization which is characteristic of the tissue under consideration) chambre *f* d'ionisation équivalente au tissu
~-EQUIVALENT IONIZATION CHAMBER DOSE RATEMETER - [nucl] ictomètre *m* à chambre d'ionisation équivalente au tissu
~-EQUIVALENT MATERIAL - [radiat] (solid or liquid whose radiation-absorbing and scattering properties are similar to those of the human body) substance *f* équivalente au tissu
~PAPER - [paper man] papier *m* de soie
TIT - [physiol] tetin *m*
[mech] (small protrusion) teton *m*, téton *m*
[zool] cob *m*, cheval *m* de petite taille
TITAN CRANE - [mech] grue *f* titan
TITANATES - [chem] (compounds formed when titanium oxide, or titanium dioxide is fused with alkalies) titanates *mpl*
TITANIA - s. titanium oxide
TITANIC ACID - [chem] (a mordant in dyeing) acide *m* titanique
~OXIDE - s. titanium oxide
TITANIFEROUS - [min] titanifère, titané
~IRON ORE - [min] titane *m* oxydé ferrifère, ilménite *f*
TITANITE - [min] (natural titanium calcium silicate with uses as gemstone and as a source of titanium) titanite *f*
TITANIUM - [chem] (a metallic element, symbol Ti, A.N. 22. A.W.47.9, having excellent resistance to corrosion. Titanium has major application in the production of alloys for chemical plant, aircraft structural members, surgical equipment, and cermets) titane *m*
~BORIDE - [chem] (a refractory) borure *m* de titane

TITANIUM CARBIDE - [chem] (a component of cermets and carbide cutting tools) carbure *m* de titane
~CARBIDE CERMET - [ceramics] (material consisting of finely-divided ceramic bonded with titanium carbide, having high wear-resistance) cermet *m* de carbure de titane
~DIOXIDE - [chem] (white powder with high hiding power and used as a paint, plastics and rubber pigment and as a delustering agent for synthetic fibres) bioxyde *m* de titane
~OXIDE - [chem] (acidic and basic; forms titanates when fused with alkalies) oxyde *m* de titane
~PEROXIDE - [chem] (a constituent of ceramic glazes) peroxyde *m* de titane
~-POTASSIUM OXALATE - [chem] (a mordant in dyeing) oxalate *m* de titane et de potassium
~SULPHATE - [chem] (a textile processing agent) sulfate *m* de titane
~TETRACHLORIDE - [chem] (a colourless, corrosive liquid with uses as a polymerisation catalyst) tétrachlorure *m* de titane
~TRICHLORIDE - [chem] (anhydrous violet crystals with uses as an organic intermediate and as a polymerization catalyst for polyolefins) trichlorure *m* de titane
~WHITE - [dyes] (a non-toxic pigment with good hiding power, consisting of a mixture of titanium dioxide and barium sulphate) blanc *m* de titane
TITANOUS SULPHATE - [chem] (a dyeing assistant textiles) sulfate *m* de titane
TITANYL ACETYLACETONATE - [chem] (cross-linking agent for cellulosics) acétylacétonate *m* de titanyle
TITER - s. titre
TITILLATION - [gen] titillation *f*
TITILLOMANIA - [med] titillomanie *f*
TITLE - [gen] (of a book etc) titre *m*
[leg] (the means whereby the owner of lands has the just possession of his property) droit *m*, titre *m*
[chem] (of gold, silver etc) titre *m*
[cin] (subtitle) sous-titre *m*
~BLOCK - [print] cliché *m* d'un dessin
~DEED - [leg] titre *m* de propriété
~INSERT - [telev] carton *m* générique
~ON THE SPINE - [print] titre *m* au dos d'un livre
~PAGE - [print] grand titre *m*, frontispice *m*
~STRIP - [cin] bande *f* de tirage
~WRITER - [cin] auteur *m* du titrage
TITLING - [gen] titrage *m*
TITRANT - [chem] titrant *m*
TITRATE, to - [chem] (to determine the strength of a solution bv titration) titrer
TITRATE THE SILK , to - [text] titrer la soie
TITRATION - [chem] (the volumetric determination of the strength of a substance in a solution) titrage *m*
~OF SILK - [text] titrage *m* de la soie
TITRE - [chem] (the strength of a solution as determined by titration) titre *m* (d'une solution etc)
~BALANCE - [text] romaine *f* de titrage, balance *f* de titrage
TITREING PROCESS - [text] procédé *m* de titrage
TITRIMETER - [ind chem] appareil *m* pour titrages
TITRIMETRIC - [chem] titrimétrique
~ANALYSIS - [chem] (analysis carried out by electrometric titrations) analyse *f* titrométrique
TITRIMETRY - [chem] titrimétrie *f*
TITROMITE - [min] (a borosilicate containing some thorium) titromite *f*
TITUBATION - [med] titubation *f*
TITULAR - [gen] titulaire, nominalement
~HEAD - [comm] (of a firm) titulaire *m*
T-MATCHED AERIAL - [radio] antenne *f* avec adaptateur *m* à T
T-MATCHED ANTENNA - s, T-matched aerial
TM MODE - [electron] (transverse magnetic mode) mode *m* E, mode *m* transversal
T.M.S. - [telecomm] (transmission measuring set) hypsomètre *m*
TM WAVE - [electron] (transverse magnetic wave) onde *f* E, onde *f* transversale
T.N.B. - s. trinitrobenzene
TNT - s. trinitrotoluene
TO-AND-FRO - [gen] de long en large, aller et venir
[mech] va-et-vient
~MOTION - mouvement *m* de va-et-vient
~-AND-FRO MOVEMENT - s. to-and-fro motion
~-AND-FRO MOVING CARRIAGE - [text mach] chariot *m* à mouvement de va-et-vient
TOAD PIPE - [bot] prêle *f* des champs
TOADSTONE - [min] pierre *f* de crapaud, crapaudine *f*
TOAST, to - [gen] rôtir, griller
TOAST - [food] pain *m* grillé
TOASTER - [impl] grille-pain *m*
TOAT - [carp] (the handle of a bench plane) poignée *f*
T.O.B. - [aero] (initials of Take-Off Boost) pression *f* de suralimentation de décollage
TOBACCO - [bot] tabac *m*
~CUTTER - [impl] hacheuse *f* pour le tabac
TOBIAS ACID - [chem] (a synonym for 2-naphthylamine I - sulphonic acid) acide *m* de Tobias
TOCOGRAPHY - [med] enregistrement *m* des contractions utérines
TOCOLOGY - [med] (the science of midwifery) tocologie *f*
TOCOPHEROLS - [chem] (three hydroxy-compounds L, B, and L-tocopherol together constituting vitamin E. L-TOCOPHEROL, in the form of the acetate or the hydrogen succinate, has application in medicine) tocophérols *mpl*
TODDITE - [min] (niobium, iron, manganese and uranium containing mineral) toddite *f*
TOE, to - [gen] refaire la pointe
[mech] clouer transversalement
TOE - [anat] orteil *m*, doigt *m* de pied
[gen] (of a shoe or stocking) bout *m*, pointe *f*
[mech] patin *m*
[mech] ergot *m*, queue *f*, pivot *m*
[mech] (of a brake shoe) patin *m*
[mech] came *f*
[railw] (of a switch blade) patin *m*
[mech] (of a valve gear) touche *f*
[constr] empattement *m*, saillie *f*
[metall] (the junction between the face of a weld and the base metal) joint *m* de soudure
[instr] (of a sensitometric curve) pied *m*
[arch] empattement *m*
[zool] pince *f* de sabot
~BEARING TOOTH - [mech] (of gear) dent *m* à zone de contact à la base
~-BRAKE - [aero] (a landing gear brake controlled by a pedal moved by the pilot's toe) frein *m* à pied
~CHANGE - [auto] (the change of toe per unit verti-

cal displacement of both wheel centres relative to the sprung mass) modification *f* du parallélisme
TOE CIRCLE - [mech] cercle *m* de pied
~CLIP - [impl] (of a bicycle) cale-pied *m*
[agric] (in farriery) pinçon *m*
[metall] cale-pied *m*
~FILTER - [soil] tapis *m* filtrant
~-IN - [mech] (of the front wheel of an auto) pincement *m*
~LASTING - [shoe man] (in rubber shoes) mise *f* sur la forme du bout de semelle
~NAIL - [anat] ongle *f* d'orteil
~OF A DAM - [hydr] pied *m* d'une digue
~OF A HOLE - [mining] fond *m* du trou
~OF POINTS - [railw] pointe *f* de l'aiguille
~-OUT - [mech] (of the fron wheels of an auto) divergence *f* des roues avant
~PIECING OF SOLE - [shoe man] rénovation *f* du bout de semelle
~PUFF - [shoe man] (a box toe) bout *m* dur
~THE BALLAST, to - [railw] retrousser le ballast
~TRENCH - [soil] drain *m* de pied
TOED - [shoe man] à bout rapporté
TOENAIL ULCER - [med] onyxis *f* maligne
TO-FROM-INDICATOR - [radar] indicateur *m* entrée ou sortie
TOGGLE - [mech] (mechanical device consisting of two links pivoted together to form an elbow joint, the other ends being linked to, or butted against other parts) bras *m* articulé, joint *m* à genou
[naut] cabillot *m*, chevillot *m*
[radio] (by-stable trigger circuit) circuit *m* de déclenchement bistable
[horol] barrette *f*
[mech] clef *f*
~BOLT - [mech] boulon *m* à tête articulée
~CLAMP - [mech] (clamp in which the pressure is applied by a toggle mechanism) crampon *m* articulé
[metall] (in welding; a screw clamp fitted with a toggle action) dispositif *m* de blocage par bielle et coulisseau
~JOINT - [mech] joint *m* à genou, rotule *f*
~JOINT RIVETING MACHINE - [mech] riveuse *f* à rotule
~LEVER - [mech] levier *m* coudé
~LEVER PRESS - [mech] presse *f* à genouillère
~LINK - [mech] articulation *f* à genouillère
~LINKS - [aero] (the linking members in a toggle-type landing gear shock absorber, designed to prevent relative rotation of the telescopic elements) liaisons *f*pl articulées
~LOCK - [mech] fermeture *f* à garrot
~-OPERATED - [mech] à genouillère
~PIN - [mech] chevillot *m*
~PLATE - s.toggle link
~PRESS - [mech] (press actuated by a toggle action) presse *f* à genouillère, presse *f* à levier coudé
~SWITCH - [el] interrupteur *m* articulé
~SYSTEM - [mech] système *m* à leviers articulés
TOILET - [gen] toilette *f*
[constr] toilette *f*
[constr](USA only, bathroom) salle *f* de bain
[med] (of a wound) détersion *f*
TOILINET - [text] toilinette *f*
TOKEN - [gen] signe *f*, marque *f*, indication *f*
[print] dix mains *f*pl (de papier)
~SYSTEM - [el] système *m* du gage

TOLAN - [chem] (diphenylacetylene. A synthesis intermediate) tolane *m*
TOLAZOLINE HYDROCHLORIDE - [chem] (benzazoline hydrochloride; 2 - benzyliminoazoline hydrochloride. A sympatholytic and adrenolytic-agent with uses in the treatment of circulatory disturbances) chlorhydrate *m* de tolazoline
TOLERANCE, to - [draw] (to show the tolerances on a drawing) indiquer les tolerances
TOLERANCE - [mech] (permissible variation in dimensions to cover unavoidable inaccuracy in the manufacture of mating parts) tolérance *f*
[phys] (permissible variation from a specified value) tolérance *f*, limite *f*
~DOSE - [radiat] (maximum permissible dose) dose *f* maximale admissible
~LIMIT - [mech] limite *f* de tolérance
~ON FIT - [mech] tolérance *f* de la goupille
~ZONE - [mech] champ *m* de tolérance
TOLERANCING - [draw] indication *f* des tolérances
TOLIDINE - [chem] (an anatytical reagent and an intermediate for dyestuffs) tolidine *f*
TOLL, to - [gen] payer le droit de passage
[acoust] (of bells) tinter, sonner
TOLL - [gen] péage *m*, droit *m* de passage
[leg] octroi *m*
[acoust] (of a bell) tintement *m*
~AREA - [telecomm] zone *f* suburbaine
~BAR - s. toll gate
~BRIDGE - [gen] pont *m* à péage
~CABLE - [telecomm] câble *m* à long distance
~CALL - [telecomm] (a short-distance trunk call) appel *m* suburbain, appel *m* interurbain
~CENTRE - [telecomm] centre *m* de groupement
~EXCHANGE - [telecomm] bureau *m* central régional
~GATE - [gen] barrière *f* de péage
~HOUSE - [gen] bureau *m* de péage
~LINE - [telecomm] (short-haul trunk circuit; in the USA any long-distance circuit) ligne *f* interurbaine
~TELEPHONE CIRCUIT - [telecomm] circuit *m* téléphonique interurbain
~TEST PANEL - [telecomm] panneau *m* de coupure de la table d'essai des lignes interurbaines
TOLONIUM CHLORIDE - [chem] (toluidine Blue O. Green, crystalline powder with application in medicine for the control of bleeding) chlorure *m* de tolonium
TOLU BALSAM - [chem] (yellowish brown, resinous solid obtained from Toluifer balsamum and having application in perfumery and medicine) baume *m* de tolu
TOLUENE - [chem] (a colourless, flammable liquid with uses as an organic intermediate, solvent, and starting material for polyurethane resins) toluène *m*
~DIAMINE - [chem] (an intermediate for dyestuffs) toluène-diamine *f*
~DI-ISOCYANATE - [chem] (a starting material for a number of synthetic resins, elastomers and foamed plastics) toluène-di-isocyanate *m*
~SULPHANILIDE - [chem] (an intermediate for dyestuffs) toluène-sulfanilide *f*
~SULPHONAMIDE - [chem] (a synthesis intermediate) toluène-sulfonamide *f*
~-SULPHONIC ACID - [chem] (a catalyst and synthesis intermediate) acide *m* toluènesulfonique
TOLUHYDROQUINONE - [chem] (a polymerization

stabilizer and inhibitor) toluhydroquinone *f*
TOLUIC ACID - [chem] (a synthesis intermediate for plasticizers, resins, and insect repellents) acide *m* toluique
TOLUIDINE - [chem] (a dyeing mordant, dyestuffs intermediate, and rubber additive) toluidine *f*
~SULPHONIC ACID - [chem] (an intermediate for dyestuffs) acide *m* toluidine-sulfonique
TOLUOL - [chem] (a solvent for cellulose derivatives) toluol *m*
TOLUQUINONE - [chem] (quinaldine; 2-methyl-quinoline, A dyestuffs intermediate) toluquinone *f*
TOLUYLENE RED - [chem] (a synonym for neutral red and acid-base indicator) rouge *m* de toluylène
TOLYLALDEHYDE - [chem] (an intermediate for dyestuffs and pharmaceuticals and perfumery and flavourant component) tolylaldéhyde *m*
TOLYLDIETHANOLAMINE - [chem] (a dyestuffs intermediate) tolyldiéthanolamine *f*
TOLYLETHANOLAMINE - [chem] (an emulsifying agent) tolyléthanolamine *f*
TOTYL NAPHTHYLAMINE - [chem] (a dyestuffs intermediate) tolylnaphtylamine *f*
TOMATO - [bot] tomate *f*
~JUICE - [food] jus *m* de tomate
~LEAF MOULD - [bot] (plant disease) cladosporiose *f* de la tomate
~LEAF SPOT - [bot] septoriose *f* de la tomate
~MOTH - [zool] noctuelle *f* potagère
~PUREE - [agric] purée *f* de tomates
TOMBAC - [metall] (a copper-zinc alloy) tombac *m*
TOMBACK - s. tombac
TOMBAK - s. tombac
TOMBSTONE - [arch] pierre *f* tombale
TOMENTUM - [bot] (the covering of felted cottony hairs) duvet *m*
[anat] réseau *m* vasculaire de la pie-mère et du cortex
TOMMY - s. tommy bar
~BAR - [mech] (a metal rod passed through a hole in another part and used for turning it) broche *f* à visser
[impl] garrot *m*
~GUN - [firearms] mitraillette *f*
~-HEAD SCREW - [mech] vis *f* à tête cylindrique
~-HOLE - [mech] encoche *f*, entaille *f*
~NUT - [mech] écrou *m* à trous (qui se visse à la broche)
~SCREW - [mech] vis *f* à broche
TOMOGRAPHY - [radiat] (a radiography of a layer in the body, obtained by rotating the X-ray source and film a point in the plane of the layer) radiographie *f* en coupe, tomographie *f*, stratigraphie *f*
TON - [meas] (a measure of weight, 2,240 lb) tonne *f*
[naut] (100 cubic feet volume) tonneau *m* de jauge
~-KILOMETER - [meas] tonne/kilomètre
~-MILE - [railw] tonne *f* millénaire
~-MILE INDICATOR - [oil ind] enregistreur *m* du rendement des câbles
TONALITE - [geol] (coarse-grained igneous rock) tonalite *f*
TONALITY - [gen & mus] tonalité *f*
TONAPHASIA - [med] amusie *f*, tonaphasie *f*
TONE, to - [photo] (to impact colour to or to modify the shade of an existing colour) virer
[mus] (an instrument) accorder (un instrument)
TONE - [acoust] (sound in relation to quality, volume, duration and pitch) ton *m*
[gen] (characteristic strain) accent *m*, son *m*
[opt] (of a colour) (the prevailing effect) ton *m*, nuance *f*
[med] tonicité *f*
~-ARM - [el acoust] (pick-up arm) bras *m* de pick-up
~CHANNEL - [radio] (circuit utilizing an audio frequency as a means of transmission) canal *m* son
~CONTROL - [radio etc] correcteur *m* de tonalité
[telev] (in colour television) tréglage *m* de nuance de couleur
~FILTER - [acoust] filtre *m* de tonalité
~GENERATOR - [el acoust] (signal generator) générateur *m* de fréquences musicales
~KEYER - [comput] ondulateur *m*
~LOCALIZER - [radar] (radiobeacon with double modulation) balise *f* à deux modulations
~OPERATED NET LOSS ADJUSTER - [telecomm] tonlar *m*
~REDUCER SWITCH - [auto] commutateur *m* de réduction d'intensité du son de l'avertisseur
~REPRODUCTION - [opt] rendu *m* des valeurs, reproduction de la couleur
~-REPRODUCTION CURVE - [photo] courbe *f* du rendu tonal
~SENSE - [acoust] sens *m* musical
~SEPARATION - [photo] stylisation *f* par compression des valeurs moyennes
~SOURCE - [acoust] source *f* sonore
~SWITCH - [acoust] commutateur *m* de tonalité
~TEST - [telecomm] (in telephony) test *m* vibré
~WEDGE - [radio] (optical step wedge containing a number of steps of density between black and white) échelle *f* de nuances
TONER - [paint] (a concentrated, purely organic pigment) colorant *m* organique (précipité en forme d'un sel métallique)
[photo] bain *m* de virage
TONG CLAMP - [oil ind]clé *f* à étan
~DIE - [oil ind] peigne *m* à clés suspendues
~ROPE - [oil ind] câble *m* de tenaille
~TORQUE GAUGE - [oil ind] enregistreur *m* de couple de clé de serrage
~UP, to - [oil ind] serrer par la clé suspendue
TONGS - [tool] pincettes *f*pl, tenailles *f*pl
[tool] (a forge tool) goulue *f*, tenaille *f* creuse
[metall] (of a crucible) happe *f*
[oil ind] clés *f*pl suspendues
~-CURRENT-TRANSFORMER - [el] transformateur *m* de courant à pince
~FOR BLOCKS - [impl] tenailles *f*pl pour blocs
~HOLD - [mech] (used in forging) queue *f*
~HOLDER - [tool] poignée *f* de pince
TONGUE, to - [carp] langueter
TONGUE - [anat] langue *f*
[gen] (language) langue *f*
[mech] bouvetage *m*, ardillon *m*
[mech] (of scales) languette *f*, aiguille *f*
[mech] (of a knife etc) soie *f* (d un couteau etc)
[railw] (of a switch) pointe *f* de coeur, lame *f* d'aiguille
[carp] languette *f* de bois
[mus] (of reed pipes) anche *f*
[geogr] (a strip of land projecting into the sea) langue *f* (de terre)
[horol] battant *m*

TONGUE-AND-GROOVE BOARDING - [carp] planchéiage *m* à rainure et languette
~-AND-GROOVE JOINT - [carp] (joint formed between the butting edges of two boards) assemblage *m* à rainure et languette
[constr] tenon *m* à rainure
~-AND-GROOVE PLANE - [tool] rabot *m* à emboîter
~ATTACHMENT - [railw] (of points) patte *f* d'attache d'aiguille
~BONE - [anat] os *m* hyoïde
~DEPRESSOR - [med] abaisse-langue *m*
~GRAFTING - [agric] greffe *f* à cheval
~JOINT - [carp] assemblage *m* à rainure et languette
~-RAIL - [railw] pointe *f* de croisement
~TRACTION - [med] tractions *fpl* de la langue
TONGUED AND GROOVED - [constr] (of wood) à languette *f*
TONGUETIA - [med] ankyloglosse *f*, raccourcissement *m* du frein de la langue
TONGUING - [carp] languetage *m*
~AND GROOVING IRONS - [tool] fers *mpl* à bouveter
~AND GROOVING MACHINE - [mech] machine *f* à faire les rainures et languettes
~CUTTER - [mach tool] fraise *f* à bouveter
~IRON - [tool] fer *m* de bouvet double
~PLANE - [tool] bouvet *m* mâle, bouvet *m* à languette
[constr] bouvet *m* mâle
TONIC - [gen & med] tonique, fortifiant *m*
~ACCENT - [acoust] accent *m* tonique
~TRAIN SIGNALLING - [telecomm] signalisation *f* par ondes musicales
TONICITY - [gen & med] tonicité *f*
TONING - [photo] (the alteration of the colour of a silver print by chemical action) virage *m*
[chem] (chemical toning) virage *m* (chimique)
TONKA BEAN - [bot] (beans of various species of dipteryx. The are the source of coumarin having application as a flavourant and perfumery agent) tonka *m*
TONLAR - [radio] (tone operated net loss adjuster) tonlar *m*
TONNAGE - [meas] tonnage *m*
[naut] (the duty) jauge *f* de douane
[shipbuild] (the cubic content of a ship) capacité *f* de chargement
~ADMEASUREMENT - [naut] jauge *f* de registre
~DECK - [naut] pont *m* de tonnage
~STEEL - [comm] acier *m* courant
TONNE - [meas] (the metric ton, i.e. 1,000 kg) tonne *f* métrique
TONNEAU - [auto] tonneau *m*
TONOFIBRIL - [med] fibrille *f* épidermique
TONOMETER - [instr] (instrument designed to measure the pitch or the vibration rate of tones) tonomètre *m*
[instr] (instrument designed to measure tension or pressure, i. e. of the blood) sphygmomanomètre *m*
[instr] (instrument designed to measure the pressure of vapour) appareil *m* pour la mesure de la vapeur
TONOMETRY - [instr] (the study of the use of tonometers) tonométrie *f*
TONOPLAST - [anat] membrane *f* vacuolaire
TONSIL - [anat] amygdale *f*
TONSILLAR ABSCESS - [med] abcès *m* de l'amygdale
TONSILLITIS - [med] amygdalite *f*
TOOL, to - [mech] usiner, travailler
TOOL, to - [constr] (a stone) bretteler, layer
[bookbind] ciseler, dorer
[leather ind] ciseler
TOOL - [tool] outil *m*, ustensile *m*
[impl] instrument *m*
[bookbind] fer *m* de relieur
~ANGLE - [mach tool] angle *m* de taillant
~BAG - [mech] sac *m* à outils
~BASKET - [gen] cabas *m*
~BITS - [mech] forets *mpl*, barreaux *mpl* traités
~BOARD - [mech] tableau *m* porte-outils
~BOX - [mach tool] (of a planing machine) sabot *m* (d'une machine à raboter)
~CABINET - [mech & carp] casier *m* à outils
~CAR - [auto] dépanneuse *f*
~-CAR - [transp] camion *m* porte-outil
~CARRIAGE - [mach tool] chariot *m* porte-outil
~CARRIER - [tool] (of a machine tool) porte-outil *m*
~CASE - [tool] coffret *m* d'outils
~CHEST - s. tool bag
~CRIB - [mech] armoire *f* à outils
~CUTTING EDGE - [mech] taillant *m* d'un outil
~ENGINEERING - [gen] (the organization and equipment of industrial production) organisation *f* de la production industrielle
~FRAME FOR IMPLEMENTS - [mech] porte-outils *m*
~GAUGE - [mech] jauge *m* pour taillant d'outil
~GRINDER - [mach tool] machine *f* à affûter les outils
~GRINDING MACHINE - s. tool grinder
~HEEL - [tool] porte-outil *m* à charnière
~HOLDER - [mech] porte-outils *m*
~HOLDING SLIDE - [mech] coulisseau *m* porte-outil
~JACK - [mech] vérin *m* de serrage
~JOINT - [oil ind] (coupling with a conical thread, used for making a tight, leakproof connection between two joints of a drill pipe) accouplement *m* d'outils de sondage
~KIT - [gen] trousse *f* à outils
~MAKER - [mech] outilleur *m*, taillandier *m*
~-MAKER'S LATHE - [mach tool] tour *m* d'outillage
~MAKING - [mech] fabrication *f* d'outils
~MILLING MACHINE - [mach tool] fraiseuse *f* d'outillage
~OUTFIT - [gen] outillage *m*, jeu *m* d'outils
~PAD - [mech] nécessaire *m* d'outils
~POST - [mach tool] étrier *m* (d'outil)
~-PUSHER - [oil ind] (a drilling supervisor) contremaître *m* de forage
~REST - [mech] support *m* d'outil
~-ROLL - [impl] trousse *f* à outils
~SETTING - [mech] réglage *m* de l'outil
~SHARPENER - [mach tool] machine *f* à affûter les outils
~SHED - [agric] hangar *m* à outils, hangar *m* à matériel, resserre *f*
~SPINDLE - [mach tool] mandrin *m* porte-outil
~STEEL - [metall] acier *m* à outils
~TIP - [tool] pointe *f* de l'outil
~UP, to - [gen] organiser l'équipement
~WRAP - s. tool roll
~WRENCH - [oil ind] clés *fpl*
TOOLBAR - [agric] barre *f* d'attelage, barre *f* d'accrochage
TOOLBOX - s. tool box
TOOLED - [bookbind] ciselé
~ASHLAR - [constr] (a block of stone finished with

parallel vertical flutes) moellon *m* d'appareil layé
TOOLHEAD - [tool] (of a machine tool) tête *f* porte-outils
TOOLHOLDER - [tool] (any arrangement for gripping and supporting a cutter in a machine-tool) porte-outil *m*
TOOLING - [mech] usinage *m*
[bookbind] ciselure *f*
[leather ind] ciselage *m*
[constr] (of stones) bretture *f*
~CALF - [bookbind] veau *f* pour la reliure cilselée
~LAYOUT - [mech] disposition *f* des outils
~SHEEPSKIN - [bookbind] basane *f*
TOOLMARKER - [gen] outilleur *m*, taillandier, fabricant *m* d'outils
TOOLMARKER'S MICROSCOPE - [opt] microscope *m* d'usine
TOOLPLATE - s. toolholder
TOOLROOM - [gen] cabane *f* aux outils
[gen] (room used for storing and distributing tools) hangar *m* à outils
~INTERNAL GRINDING MACHINE - [mach tool] fraiseuse *f* à outils
~LATHE - [mach tool] tour *m* à outils
TOOLSLIDE - [mach tool] coulisseau *m* porte-outil
TOOLSTOCK - s. toolholder
TOOTH - [anat] dent *f*
[gen] (of a fork, or a comb) dent *f*
[mech] dent *f* (d'engrenage)
[tool] (of a file) taillant *m*
[el] (between two slots of an armature) dent *f*
[paper man] (grain of paper) grain *m* (du papier)
[carp] tenon *m* invisible
~BLOCK - [mech] modèle *m* de segment de denture
~BLOCK HOLDER - [metall] fouloir *m* porte-segment
~CHISEL - [tool] (for masons), marteau *m* plat, marteau *m* à parer
~CONTOUR - [mech] profil *m* de la dent
~CULTIVATOR - [agric] cultivateur *m* à dents rigides
~DEPTH - [mech] (of gears) hauteur *m* de la dent
~END ELEVATION - [mech] profil *m* d'extrémité de la dent
~EXTRACTION - [med] extraction *f* dentaire, avulsion *f* dentaire
~FACE - [mech] (of a gear) flanc *m* d'une dent, face *f* d'une dent
~FILLET - [mech] (of a gear tooth) congé *m* de la dent
~FLANK - [mech] (of a gear) flanc *m* de la dent
~FORM - [mech] forme *f* de la dent
~GEARING - [mech] transmission *f* à engrenages
~HOOD - [anat] capuchon *m* gingival
~OF COMB - [text] dent *f* du peigne, lame *f* du peigne
~OF RATCHET BAR - [mech] dent *f* de la barre
~OUTLINE - s. tooth contour
~PITCH - [el] (the interval between fixed points in corresponding positions in respect of two consecutive teeth) pas *m* dentaire
~PLANE - [tool] rabot *m* denté
~PROFILE - s. tooth outline
~RIPPLE - [el] ondulation *f* de la force électromotrice
~SPACE - [mech] (of a gear) écartement *m* des dents
~SPACING - [mech] (of a gear) pas *m* de denture
~SPASM - [med] convulsions *f*pl dentaires
TOOTH SURFACE - [mech] (of a gear) face *f* de la dent
~THICKNESS - [mech] (of a gear) épaisseur *f* de la dent
TOOTHED - [mech] denté, crénelé
[bot] dentelé
~BAR - [mech] barre *f* dentelée
~-BAR RACK - [mech] crémaillère *f*
~CIRCULAR KNIFE - [impl] couteau *m* circulaire dentelé
~CONVEYOR BELT - [mech] courroie *f* transporteuse dentée
~COUPLING - [mech] manchon *m* d'accouplement à dents
~DISK - [mech] disque *m* denté
~DISK MILL - [mech] (a grinding mill consisting of disks carrying teeth and running at a small angle to each other) broyeur *m* à couleurs
~FEED ROLLER - [text] cylindre *m* d'alimentation à dents
~FLUTING - [mech] cannelure *f* dentée
~FLYWHEEL CROWN - [mech] couronne *f* dentée du volant moteur
~GEARING - [mech] engrenage *m* à roues dentées
~GEARING HANDLE - [mech] manette *f* d'embrayage dentée
~KNIFE - [mech] couteau *m* dà dents
~MICA SPACER - [el] plaque *f* de mica entaillée
~PULLING ROLLER - [text] cylindre *m* briseur, cylindre *m* à pointes
~QUADRANT - [mech] secteur *m* à dents
~RACK STRUTTING DEVICE - [rubber ind] (for tyres) crémaillère *f* à étresillonner
~RIM - [mech] couronne *f* dentée
~SCALE - [text] écaille *f* dentée
~SECTOR - [mech] secteur *m* crénelé, secteur *m* denté
~SHUTTLE - [text] espolin *m* à crémaillère
~STRIP - [text] barreau *m* à dents
~SWIFT - [text] tambour *m* à dents
~V-BELT - [mech] courroie *f* trapézoïdale dentée
~WHEEL - [mech] roue *f* dentée, roue *f* d'engrenage, rouage *m*
~-WHEEL RIM - [mech] couronne *f* dentée d'une roue
TOOTHING - [impl] (in construction work) harpes *f*pl, arrachement *m*
[constr] appareil *m* en besace
[mech] (of a gear) denture *f*
[mech] (of a saw) taille *f* des dents
[constr] (of stone) bretture *f* (de la pierre)
~PLANE - [tool] rabot *m* denté
~STONE - [constr] attente *f*, pierre *f* d'attente, amorce *f*, pierre *f* d'arrachement
TOP, to - [gen] surmonter, couronner
[agric] étêter, écimer, élaguer
[agric] (in horticulture) pincer, décolleter
[gen] (to reach a summit) atteindre le sommet
[gen] (to exceed) excéder, dépasser
[metall] (to cut, e.g. the top of an ingot) ébouter
[dyes] (to enrich a tone by covering with another dye) recouvrir (avec une peinture)
[ind chem] (in chemical distillation, to take off the most volatile part) prédistiller
[naut] (of ships) s'élever (à la hauteur d'une lame)
[naut] apiquer

TOP - [gen] (summit) sommet *m*, cime *f*
[gen] (the upper limit) limite *f* supérieure, sommité *f*
[gen] (a lid, a covering) couvercle *m*
[auto] capote *f*
[text] (a combed sliver) peigné *m*, trait *m*
[bookbind] tête *f*
[impl] table *f* de travail, taque *f*
[mining] toit *m*
[naut] hune *f*
[constr] (of the roof) comble *m*
[metall] gueulard *m*
[constr] (of a slate-quarry) cosse *f* (d'une ardoisière)
~AERIAL - [radio] (aerial for a car radio fitted on top of the car) antenne *f* de toit (d'auto)
~AND BOTTOM CLEARANCE - [mech] jeu *m* au fond des dents
~AND BOTTOM RAM - [mech] piston *m* plongeur supérieur et inférieur
~AND BOTTOM REFLECTED REACTOR - [nucl] réacteur *m* à réflecteur supérieur et inférieur
~ANGLE - [mech] (of a gear tooth) cornière *f* de tête
~ANTENNA - s. top aerial
~BAND - [shoe man] bande *f* supérieure
~BAR JACQUARD - [text] Jacquard *m* supérieur, Jacquard *m* Menchester
~BARREL HOLE - [horol] trou *m* postérieur du barillet
~BEARING LEVER - [text] (of a loom) levier *m* du support supérieur de la broche
~BEVEL - [mech] (of a circular saw) (the transverse inclination of the cutting edge of the teeth, in relation to the centre-line of the saw spindle) coupe *f* au sommet
~BITE - [mech] (the opening between the upper pair of rollers in a calender or like machine) ouverture *f* des cylindres supérieurs, jeu *m* des cylindres supérieurs
~BLOWING - [metall] tuyère *f* supérieure
~BOBBIN - [text] bobine *f* de ruban peigné
~BOOM - [constr] membrure *f* supérieure
~BOOT - [shoe man] bottine *f* à haute tige, bottes *f*pl à revers
~BOWL - [text] poulie *f* supérieure
~BOX - [metall] contre-châssis *m*, contre-coussinet *m*
~BRACKET - [mech] support *m* supérieur
~-BRASS - [metall] contre-coussinet *m*
~BUCKLE - [mech] boucle *f* supérieure
~CALENDER ROLLER - [text] rouleau *m* d'appel supérieur
~CAM BOX RING - [text] boîte *f* à cames supérieure d'une machine double cylindre
~-CAP, to - [rubber ind] (of tyres) rechaper le sommet
~-CAP - [electron] (a small metal shell on the envelope of an electronic tube used to connect one electrode to an external circuit) téton *m* supérieur
~-CAPACITOR AERIAL - [radio] antenne *f* à charge terminale
~-CAPACITOR ANTENNA - s. top-capacitor aerial
~CASTING - [metall] (in foundry work) coulée *f* en chute directe
~CHAIN - [naut] suspente *f* de vergue
~CIRCLE - [mech] (of a gear) cercle *m* du diamètre externe
TOP CLAMP PLATE - [plast ind] plaque *f* d'entrée
~CLEARANCE - [mech] (of a gear) jeu *m* au fond des dents
~CLEARER - [text] chapeau *m* de propreté
~CLOTH - [text] toile *f* d'endroit, tissu *m* de dessus
~CLOTH WEAVE - [text] armure *f* du tissu de dessus
~COATING - [text] étoffe *f* pour paletot
~CONE - [mech] cône *m* supérieur
~CONE DRUM - [text] cône *m* supérieur
~CONE OF COP - [text] cône *m* supérieur de la canette
~CONTACT RAIL - [railw] (used in electric traction) troisième rail *m* supérieur
~COURSE - [rubber ind] couche *f* de couvrement, couche *f* supérieure
~COVER - [mech] couvercle *m* de cylindre
[auto] housse *f* de capote
~CROSS ROD - [mech] arbre *m* des secteurs
~CRUST - [metall] croûte *f*
~CUSHION - [rubber ind] (of tyres) souscouche *f* de la bande de roulement
~CUTTING - [agric] bouture *f* de tête
~DEAD CENTRE - [mech] (the dead centre at which the piston is nearest to the cylinder head) point *m* mort haut
~DIE - [metall] estampe *f* supérieure
~-DIP RATIO - [electron] (or peak-valley ratio; between the peak amplitude and the lowest amplitude in the response curve of a circuit) rapport *m* crête-creux
~DRAINAGE - [agric] drainage *m* de la surface
~DRAWSHEET - [print] feuille *f* de couverture, feuille *f* tendue
~-DRESSING - [agric] (the soil) engrais *m* en couverture, fumure *f* en surface, épandage *m* en surface
~DRIVING APPARATUS - [mach tool] transmission *f* supérieure
~EDGE RIB - [constr] solin *m*
~EJECTION - [plast ind] (system in which the moulding is ejected from the upper member of the mould) éjection *f* par le haut
~END - [text] tête *f*
~END PLUG - [mech] bouchon *m* de fermeture
~END SHOOT - [bot] poussée *f* terminale
~FILLER - [metall] chargeur *m* de haut fourneau
~FORCE - [plast ind] (upper part of a mould) matrice *f* supérieure
~FORK - [text] cheville *f* supérieure
~FRAME - [auto] châssis *m* de la capote
~FRICTION WASHER - [mech] disque *m* de friction supérieur
~FULLER - [tool] dégorgeoir *m* de dessus
~GAS - [metall] gaz *m* de gueulard
~GEAR - [auto] (of a gearbox) prise *f* directe
[naut] manoeuvres *f*pl autes
~GRAFTING - [bot] regreffage *m*
~HAMPER - [constr] superstructure *f*
~-HEAVINESS - [gen] excès *m* de charge en hauteur, manque *m* de stabilité
~-HEAVY - [gen] trop lourd du haut
[aero] jaloux
[naut] trop charge dans les hauts
~-HOLE - [mining] mine *f* de couronne
~IRON - [tool] contre-fer *m* de rabot
~JACK - [text] bricoteau *m*
~LAMINATION - [metall, plast ind etc] (in lamina-

ted material, the layer which will be exposed to view or to wear, in use) feuille *f* de couverture
TOP LAND - [mech] cordon *m* supérieur
~LANDING - [mining] recette *f* du jour, accrochage *m* du jour
~LAYER - [rubber ind] couche *f* de couvrement, revêtement *m* supérieur
[soil] couche *f* superficielle
~LEAF - [mech] (of a spring) lame-maîtresse *f*
~LIGHT - [naut] lanterne *f* de hune
~LINEARITY CONTROL - [telev] réglage *m* de la résistance de crête
~-LOADED AERIAL - s. top-capacitor aerial
~OF A MINE SHAFT - [mining] sommet *m* d'un puits de mine
~OF AN ARCH - [arch] clef *f* d'un arc
~OF CONNECTING ROD - [mech] (small end) pied *m* de bielle
~OF HOUSING - [mining] tête *f* de la cage
~OF STAPLE - [text] pointe *f* de la mèche
~OF STROKE OF PISTON - [mech] haut *m* de course de piston
~OF THE BASE - [mining] sommet *m* de la sole
~OFF, to - [oil ind] prélever les fractions supérieures
~OPENING FREEZER - [food] vitrine *f* frigorifique
~OVERHAUL - [auto] (the reconditioning of the cylinders) revision *f* partielle
~PALLET - [mech] (of a power hammer) frappe *f* supérieure, panne *f* du pilon
~PART OF FLASK - [metall] dessus *m* de châssis, chapeau *m* de châssis
~PIECE - [constr] poutre *f* de pignon
~PLATE - [metall & plast ind] (a steel plate bolted to the upper section of a mould) plateau *m* de dessus, plaque *f* supérieure
[metall] plaque *f* de recouvrement
[mech] (of a leaf spring) maîtresse-lame *f*
~PLUG - [oil ind] deuxième tampon *m* de cimentation
~POURING - [metall] coulée *f* en chute directe
~PRESSURE - [metall] pression *f* au gueulard
~PRINT - [metall] portée *f* de noyau
~RADDLE PLATE - [text] chapeau *m* du vautoir
~RAIL - [constr] (of a door frame etc) traverse *f* du haut
~RAKE - [mach tool] angle *m* de dégagement
~RAM PRESS - [mech] presse *f* à piston supérieur
~-ROAD BRIDGE - [constr] pont *m* à tablier supérieur
~ROD - [mining] tête *f* de sonde
~ROLL - [metall] (of a rolling mill) cylindre *m* mâle (d'un laminoir)
~ROLLER - [mech] (of a sugar mill; the highest roller of a mill, which nips the cane against the bagasse roller) cylindre *m* supérieur
~SHEDDING MACHINE - [text] mécanique *f* de levée
~SLEEVES - [th eng] (refractory material) viroles *fpl*
~SOIL - [soil] sol *m* superficiel
~SOIL PLOUGH - [agric] charrue *f* déchaumeuse
~SPEED - [mech] vitesse *f* maximum
~SPROCKET - [cin] (a film-feeding sprocket) tambour *m* denté supérieur
~STONE - [constr] couronnement *m*
~SURFACE - [aero] (of the wing) dessus *m* de l'aile
~SWAGE - [mech] (in die casting) dessus *m* d'étampe, étampe *f* de dessus
TOP TEXTURE - [text] tissu *m* de dessus
~TIMBERING - [mining] boisage *m* du toit
~-UP, to - [gen] (the operation of adding a relatively small quantity of liquid to the contents of a vessel so as to fill it to the normal limit, e.g. of a fuel tank) faire le plein, rétablir le niveau
[el] (to fill a battery with distilled water) reniveler, ramener à niveau
[aero] (a balloon) renflouer
~VIEW - [draw] vue *f* de haut en bas
~WALL - [mining] toit *m*
~WASTE - [text] bouts *mpl* de peigné
~WATER - [mining] eau *f* supérieure
~WEB - s. top texture
~WIRE - [mech] fil *m* collecteur
~YEAST - [brew ind] (yeast vegetating at the top of a fermenting vat) levure *f* haute
~YELLOWS - [bot] (a plant disease) mosaïque *f* du poix
TOPAZ - [min] (a gemstone, consisting of a luminium fluosilicate) topaze *f*
TOPESTHESIA - [med] sensibilité *f* tactile locale
TOPGALLANT - [naut] voile *f* de perroquet, mât *m* de perroquet
TOPHACEOUS - [med] (gritty, of the nature of tophus) tophacé
TOPHIC CONCRETION - s. tophus
TOPHUS - [med] tophus *m*, concrétions *fpl* tophacés
[geol] tuf *m* calcaire
TOPIARY TREE - [agric] arbre *m* formé
TOPINAMBUR - [bot] topinambour *m*
TOPMAKING - [text] étirage *m* des rubans de la peigneuse
TOPMAN - [naut] gabier *m*
TOPMAST - [naut] mât *m* de hune
TOPOCHEMICAL REACTION - [chem] (a reaction which takes place only at a specified point in a system) réaction *f* topochimique
TOPOGRAM - [comput] (schematic representation of the positions in an internal storage) topogramme *m*
TOPOGRAPHIC - [surv] (pertaining to topography) topographique
~MAP - [surv] carte *f* topographique
TOPOGRAPHICAL - s. topographic
TOPOGRAPHICAL ANATOMY - [med] anatomie *f* topographique
~SURVEY - [surv] levé *m* topographique
TOPOGRAPHY - [surv] (the art of representing the physical features of a place on a map) topographie *f*
TOPOLOGY - [math] (a branch of geometry dealing with the properties of solid bodies which do not vary under continuous deformation) topologie *f*
TOPOTAXIS - [biol] (reaction of an organism to a stimulus in which the organism orientates itself in relation to the stimulus and moves towards or away from it) topotaxie *f*
TOPPED CRUDE - [oil ind] (bottoms from atmospheric distillation for treatment in a vacuum still) residu *m* de première distillation
TOPPING - [ind chem] (in distillation) première distillation *f*
[food] (the milk) écrémage *m*
[text] (of hemp) épontage *m* du chanvre
[constr] (the mortar coat on walls and floors) finition *f*
[constr] (of road) couche *f* superficielle

TOPPING - [agric] décolletage *m*, écimage *m*, étêtement *m*
[agric] couverture *f* de terreau
~AND TAILING - [text] affinage *m* des deux bouts, épointage *m* du chanvre
~PLANT - [oil ind] unité *f* de première distillation
~THE FLAT - [text] montage *m* de la garniture du chapeau
~TOWER - [ind] tour *f* de première distillation
~TURBINE - [mech] turbine *f* à extraction de vapeur
~UNIT - [oil ind] unité *f* de première distillation
~-UP - [el] (the addition of distilled water to make up evaporation losses from the electrolyte) renivellement *m*
[aero] (of an aerostat) renflouage *m*
[gen] remplissage *m*
TOPPLE - [instr] composante *f* verticale de la précession d'un gyroscope
TOPS - [oil ind] (the lightest petrol fractions obtained when distilling crude oil. Also, in general, the top product of a fractionating column) distillats *mpl* de premier raffinage
TOPSAIL - [naut] hunier *m*
TOPSIDE - [naut] oeuvres *fpl* mortes
~STRINGER - [aero] contre-quille *f*
TOPSOIL - [agric] couche *f* arable
TOR - [geol] massif *m* de roche, pic *m*
TORBANITE - [min] (a dark-brown oil-shale, with a carbonaceous content of up to 80 p.c.) torbanite *f*
TORBERNITE - [min] (a natural hydrous phosphate of copper and uranium, usually occurring with autunite, and also with other uranium minerals) torbernite *f*, torbérite *f*
TORCH, to - [constr] rendre, araser
TORCH - [gen] torche *f*, flambeau *m*
[el] lampe *f* électrique de poche
[impl] (for welding or cutting) chalumeau *m*
[impl] (a vapourizing device used for heating or soldering) fer *m* à souder
~BRAZING - [metall] brasage *m* au chalumeau
~CUTTING - [metall] oxycoupage *m*, coupage *m* à l'autogène
~IGNITER - [aero] (in a jet engine a type of igniter for gas turbine burners, comprising a small injection nozzle and an electric ignition device, the whole giving a flame which initiates and maintains combustion in the main burners) allumeur
~HARDENING - [metall] (flame hardening) trempe *f* au chalumeau
~TIP - [impl] (for welding or cutting) pointe *f* du chalumeau
TORCHING - [mech] dégazage *m* à la flamme
TORE - [math] (geometrical surface generated by the revolution of a conic section about an axis in its plane) tore *m*
TORIC - [math] torique
~LENS - [opt] verre *m* ponctuel
TORMENTIL - [bot] tormentille *f*
TORMENTOR - [cin] (or flat; large portable wall) décor *m* insonorisé
TORN INTO SHREDS - [gen] arraché en morceaux
~TAPE RELAY - [telecomm] (manual tape relay) transit *m* manuel par bande perforée
TORNADO - [met] (violent rotary wind of small radius) tornado *m*, ouragan *m*
TOROID - [math] (a solid of annular form and circular cross-section) toroïde
TOROID - [electron] (or doughnut; a toroidal-shaped vacuum envelope in which electrons are accelerated) chambre *f* à vide
TOROIDAL - [gen] toroïdal
~COIL - [el] (a coil wound on an annular former or core) bobine *f* toroïdale
~COMBUSTION CHAMBER - [mech] chambre *f* de combustion toroïdale
~ELECTRON GUN - [electron] (cathode-ray tube so constructed as to generate a hollow beam) canon *m* électronique toroïdal
~INTAKE GUIDE-VALVE - [mech] (annular flaring guidevane to distribute air uniformly over the impeller intake) aubage *m* directeur toroïdal d'admission d'air
~PINCH DEVICE - [phys] (used in plasma physics) dispositif *m* de struction toroïdale
~PLASMOID - [phys] (a discrete piece of plasma like a toroid) plasmoïde *m* toroïdal
~TRANSFORMER - [electron] transformateur *m* toroïdal
~WINDING - [el] enroulement *m* en anneau
TORPEDO - [firearms] (a device containing an explosive to be fired by concussion etc) torpille *f*
[railw] (a detoning signal) pétard *m*
[plast ind] (the conical body in the centre of a tube extrusion die) torpille *f*
[oil ind] (for clearing a well in a drilling operation) torpille *f*, charge *f* explosive
~A WELL, to - [oil ind] torpiller un puits
~BOAT - [gen] torpilleur *m*
~FOG-SIGNAL - [railw] pétard *m* pour brouillard
~TUBE - [firearms] lance-torpille *m*
TORPIDITY - [med] torpidité *f*, engourdissement *m*
TORQUE - [mech] (the turning moment produced by a tangential force) couple *m*, couple *m* moteur
[phys] moment *m* de torsion, moment *m* de rotation
~AMPLIFIER - [radio] coupleur *m* synchronisé
~ARM - [mech] jambe *f* de réaction
~-ARM CENTRE - [auto] axe *m* de bras de réaction de couple
~AT RATED LOAD - [el] (of a motor) couple *m* normal
~COMPENSATION - [aero] compensation *f* du couple antagoniste
~CONVERTER - [mech] (a device, usually hydraulic, for modifying the value of torque in a mechanical transmission) convertisseur *m* de couple
~DYNAMOMETER - [instr] (an apparatus for measuring the torque of an engine) frein *m* dynamométrique
~GRADIENT - [el] gradient *m* de couple
~INCREASE - [mech] admission *f* de force
~LIMITER - [mech] (a device comprising friction discs or a shear-pin, designed to slip at a given preset value, to restrict torque to such value) limiteur *m* de torsion
~LINKS - [mech] (or toggle links; linkage used for preventing relative rotation between telescopic members) liaisons *fpl* articulées
~-METER - [instr] mesureur *m* du couple, compas *m*, torsiomètre *m*
~MOMENT - [mech] moment *m* de torsion
~MOTOR - [mech] (motor producing a torque when stationary over a relatively long period) moteur *m* à couple constant
~PLATE - [auto] (of the brake) plateau *m* porte-mâ-

choires du frein
ORQUE-ROD - [mech] jambe *f* de réaction
SPANNER - [tool] (spanner containing a device to limit the force which can be applied to a nut or the like) clef *f* de torsion
STAND - [mech] (a special type of test-bench for measuring engine torque) banc *m* d'essai de torsion motrice
STRESS - [phys] effort *m* de torsion
TENSOR - [el] tendeur *m* de couple
TESTER - [instr] vérificateur *m* de couple
TRANSMITTER - [el] transmetteur *m* de couple
TUBE - [mech] tube *m* de réaction
[auto] (around the propeller shaft) tube *m* d'arbre de transmission
-WEIGHT RATIO - [instr] (of a meter) couple *m* spécifique
WRENCH - s. torque spanner
WRENCH ADAPTER - [mech] (a device to enable a torque spanner to be used on nuts of various sizes) embour *m* dynamométrique
ƆRQUEMETER - [meas] (an instrument built into an engine to show the torque developed) mesureur *m* de couple, compas *m*
[instr] (electrical or optical device designed to measure torsion) torsiomètre *m*
ƆRQUER - [mech] moteur-couple *m*, dispositif *m* de production d'un couple
ƆRQUING - [instr] application *f* d'un couple à l'axe d'un gyroscope
ƆRREFACTION - [gen & min] torréfaction *f*
ƆRREFY, to - [min] (to dry or roast by exposure to heat) torréfier
ƆRRENT - [geogr] torrent *m*
CONTROL - [hydr] correction *f* d'un torrent
ƆRRID ZONE - [geogr] zone *f* torride
ƆRSIOGRAPH - [instr] (instrument designed to record torsional vibrations) torsiographe *m*
ƆRSION - [phys] (the state of strain set up in a part by twisting) torsion *f*
BALANCE - [instr] (apparatus for measuring deformation under torsion) balance *f* de torsion
BAR - [mech] (a bar of elastic properties, stressed in torsion and used to control the movement of the parts attached to it) barre *f* de torsion
BAR SUSPENSION - [auto] suspension *f* à barres de torsion
DYNAMOMETER - [instr] dynamomètre *m* de torsion
ELECTROMETER - [instr] balance *f* de Coulomb
HEAD - [instr] (a component of indicating measuring instruments) tête *f* de torsion
-HEAD WATTMETER - [instr] wattmètre *m* à tête de torsion
INDICATOR - [instr] (or transmission dynamometer a device for measuring the torque in a shaft and thus the power transmitted) indicateur *m* de torsion, torsiomètre *m*
LOAD - [phys] charge *f* de torsion
MACHINE - [constr] machine *f* à tordre
METER - s. torsion indicator
SPRING - [mech] ressort *m* de torsion
STRAIN GAUGE - [meas] résistance *f* extensométrique de torsion
TESTER - s. torsion indicator
TORQUE - [instr] (the torque produced by torsion) couple *m* de torsion
WIRE - [metall] fil *m* de torsion

TORSIONAL - [phys] (relating to torsion) de torsion
~ELASTICITY - [phys] élasticité *f* de torsion
~CRYSTAL - [phys] (a piezoelectric crystal designed to oscillate in the torsional mode of motion) cristal *m* à mode de torsion
~LOAD - [metall] force *f* de torsion
~MOMENT - [phys] moment *m* de torsion
~PENDULUM - [horol] (a pendulum in which the bob rotates by the twisting of the suspension ribbon) pendule *f* de torsion
~SHOCK ABSORBER - [mech] amortisseur *m* à torsion
~SPRING - [mech] ressort *m* de torsion
~STRAIN - [metall] déformation *f* due à la torsion
~STRENGTH - [metall] résistance *f* à la torsion
~STRESS - [phys] effort *m* de torsion
~VIBRATION - [phys] vibration *f* de torsion
~-VIBRATION BALANCER - [auto] antivibrateur *m*, amortisseur *m* de vibrations
TORTICOLLIS - [med] torticolis *m*
TORTOISE - [zool] tortue *f*
~SHELL - [gen] écaille *f* de tortue
TORTRIX MOTH - [bot] tordeuse *f*
TORTUOSITY - [gen] (e.g. of a road, abounding in irregular bends or turns) tortuosité *f*
TORTUOUS - [gen] tortueux
TORULOMA - [med] tumeur *f* cryptococcique
TORULOSE - [bot] (cylindrical, with swellings and contractions) toruleux
TORULOSIS - [med] (disease due to infection which affects especially the central nervous system) torulose *f*, exacose *f*
TORULOUS - s. torulose
TORUS - [arch] (large convex moulding) tore *m*
[math] s. tore
[med] tore *m*, protubérance *f*
TOSCANITE - [geol] (fine-grained acid igneous rock) toscanite *f*
TOSS, to - [gen] lancer, jeter
[gen] (to shake) secouer, agiter
[naut] (of a boat, ship etc) tanguer
[min] laver à la cuve
TOSSING - [metall] oxydation *f* par cuilérées
TOTAL - [gen] total, complet
~ABSORPTION - [radiat] (absorption arising from all causes) absorption *f* totale
~ABSORPTION COEFFICIENT - [phys] coefficient *m* d'absorption totale
~AMMONIA - [chem] (the sum of fixed and free ammonia in a product, especially ammonia liquor) ammoniac *m* total
~AMPLITUDE OF OSCILLATION OF A PERIODIC QUANTITY - [el] amplitude *f* totale d'une oscillation
~ANGULAR MOMENTUM - [phys] (the number which gives the resultant of the magnetic field engendered by the electron due to its orbital movement and to its revolving on its own axis) nombre *m* quantique interne
~ANODE POWER INPUT - [electron] puissance *f* de entrée anodique totale
~BODY RADIATION - [radiat] irradiation *f* globale, irradiation *f* totale
~BREAK TIME - [el] durée *f* totale de coupure
~BRILLIANCY - [opt] (in photometry) éclat *m* total
~CALORIFIC EFFICIENCY - [th eng] rendement *m* calorifique total
~CAPACITANCE - [el] capacité *f* globale
~CARBON - [metall] carbone *m* total

TOTAL CARBURIZING - [metall] cémentation *f* totale
~CARD - [comput] carte *f* résultat
~CHARGE NUMBER - [el] (the total number of coulombs of a charged body) nombre *m* de charge totale
~CROSS SECTION - [nucl] section *f* efficace totale
~-CURRENT REGULATION - [el] régulateur *m* du courant total
~CURVATURE OF LENS - [opt] courbure *f* totale de lentille
~CYANIDE - [chem] (total amount of the cyanide group present as a simple or complex cyanide) cyanure *m* total
~CYCLE - [comput] cycle *m* de prise de total
~DEVIATION - [comput] écart *m* actif de réglage, écart *m* de régulation
~ECLIPSE - [astr] éclipse *f* totale
~EFFECTIVE COLLISION CROSS-SECTION - [nucl] (the sum of the effective collision cross-sections of atoms and molecules contained in a unit volume of a gas) section *f* spécifique de choc
~ELECTRODE CAPACITANCE - [electron] (the capacitance of one electrode in relation to all other electrodes connected together) capacité *f* interélectrode totale
~ELECTRODE DISSIPATION - [electron] dissipation *f* totale des électrodes
~ELECTRON BINDING ENERGY - [electron] énergie *f* totale de liaison électronique
~EMISSION - [electron] (the maximum value of thermionic current which can be obtained from the cathode, all other electrodes being connected together and sufficient voltage being applied to them to raise the emission current to saturation value) émission *f* totale
~EMISSIVE POWER - [electron] (the rate at which the surface of a solid or a liquid emits electrons in the material by action of heat or radiant energy, or by the impact of other electrons on the surface) émissivité *f* totale
~EQUIVALENT BRAKE HORSE-POWER - [aero] (of a turbojet the sum of the propeller shaft B.H.P. and the equivalent h.p. developed by the jet thrust) puissance *f* totale équivalente au frein
~ERROR - [meas] erreur *f* totale
~FIELD DOSE - [radiat] (dose to a single given area during a course or treatment) dose *f* globale en surface
~FILTER - [radiat] (in X-rays) filtre *m* total
~HANGOVER TIME - [telecomm] temps *m* de retour au repos d'un suppresseur d'écho
~HARDNESS - [hydr] dureté *f* totale
~HARMONIC DISTORTION RATIO - [radio & telecomm] (the expression, in transmission units, of the reciprocal of the coefficient of harmonic distortion) affaiblissement *m* de distorsion totale harmonique
~HEAD - [mech] (the sum of static and dynamic pressures) chute *f* brute, hauteur *f* totale
~HEAD TUBE - [meas] (tube inserted in a flow to measure the stagnation pressure of the total head) tube *m* mesureur de la hauteur
~INPUT - [gas ind] (in a gasholder) volume *m* total de gaz injecté
~IONIZATION - [nucl] ionisation *f* totale
~LAST CARD - [comput] somme *f* dernière carte
~LOAD - [mech] charge *f* totale
~LOSS - [el] perte *f* totale

TOTAL MASS NUMBER - [nucl] (the total number of nucleons in the atomic configuration) nombre *m* massique total
~MEAN FREE PATH - [nucl] libre parcours *m* moyen total
~NEUTRON SOURCE DENSITY - [nucl] densité *f* totale d'une source de neutrons
~NUCLEAR BINDING ENERGY - [nucl] énergie *f* totale de liaison nucléaire
~OPERATING TIME - s. total break time
~OUTPUT - [gas ind] (of a gasholder) volume *m* total soutiré (pour une période donnée)
~OVERSHOOT - [contr] dépassement *m* total
~PITCH - [mech] (of a screw) pas *m* réel
~PLATE POWER INPUT - s. total anode power input
~POWER LOSS - [el] perte *f* totale de puissance
~PRESSURE - [mech] (the sum of static pressure and velocity pressure of a fluid) pression *f* totale
~PRESSURE HEAD - [hydr] charge *f* totale
~PUNCHING - [comput] perforation *f* récapitulative
~QUANTUM NUMBER - s. total angular momentum
~RADIATION PYROMETER - [instr] (a pyrometer using an optical device to concentrate on a thermocouple the radiation from a given area of the heat source, the temperature of which is to be measured) pyromètre *m* à rayonnement total
~RANGE - [meas] (of an instrument) échelle *f* totale
~-REFLECTING PRISM - [opt] prisme *m* à réflexion totale
~REFLECTION - [opt] réflexion *f* totale
~RESISTANCE - [el] effort *m* résistant total
~RESISTING EFFORT - [mech] résistance *f* totale à la traction
~ROLLING - s. total transfer
~SKIN DOSE - [radiat] dose *f* globale en surface
~SOLIDS - [gen] (dry matter) matières *f*pl solides
~SPECIFIC IONIZATION - [nucl] ionisation *f* spécifique totale
~SYSTEM DEVIATION - [contr] écart *m* actif de réglage
~THERMAL POWER - [nucl] (by all the fission produced in a nuclear reactor) puissance *f* thermique totale
~TIME CONSTANT - [el] constante *f* de temps globale
~TRANSFER - [comput] transfert *m* des totaux
~TRANSITION PROBABILITY - [nucl] probabilité *f* de transition totale
~VARIABLE RESISTANCE - [contr] résistance *f* variable totale
~WEIGHT - [auto] poids *m* total
~WIDTH - [nucl] largeur *f* du niveau d'énergie
~WHITE COUNT - [biol] (the number of white corpuscles per cubic mm of blood) formule *f* leucocytaire
~WITHDRAWAL - s. total output
~YIELD - [gen] rendement *m* total

TOTALIZATOR - [comput] (a device for indicating the number and amounts of bets staked on a race) totalisateur *m*

TOTALIZER - [comput] s. totalizator

TOTALLY ENCLOSED - [el] (said of an electrical machine) fermé
~ENCLOSED APPARATUS - [el] appareil *m* fermé
~-ENCLOSED MOTOR - [el] (a motor which has no provision for ventilation, but is not necessarily

gas-or water-tight) moteur *m* non ventilé
TOTE, to - [gen] (USA only) transporter
TOTE - [carp] (of a bench plane) poignée *f* (d'un riflard)
[comput] (short for totalizator) totaliseur *m*
~BOX - [impl] (container for hauling material in a factory) boîte *f* à transporter des matériaux
~PAN - [impl] collecteur *m* à boîte
TOUCH, to - [gen] toucher, se toucher
[gen] (to be in contact) toucher à
[naut] (a port) aborder à un port
TOUCH - [physiol] toucher *m*, tact *m*
[gen] (a slight quantity, a trace) pointe *f*, nuance *f*
[gen] (contact) contact *m*
[draw] touche *f* (de pinceau), coup *m* (de crayon)
~DOWN, to - [aero] (of an aircraft, to make contact with the ground and cease to be airborne) impacter
~-DOWN - [aero] (of an aircraft the action of making contact with the ground) impact *m*
~NEEDLE - [instr] touchaud *m*, aiguille *f* d'essai
~-UP - [paint] retouche *f*, rehaut *m*
TOUCHSTONE - [min] (fine-grained dark stone, formerly used to test the fineness of gold) pierre *f* de touche, basanite *f* jaspe *m* noir, lydite *f*
TOUGH - [gen] dur, tenace
~COPPER - [metall] cuivre *m* affiné
~PITCH - [metall] (said of copper in which the oxygen content has been correctly adjusted) cuivre *m* à oxyde cuivreux
~-RUBBER SHEATING - [el] gaine *f* isolante en caoutchouc
TOUGHENED GLASS - [glass man] (safety glass) verre *m* durci, verre *m* trempé
TOUGHENER - [metall] substance *f* pour renforcer un alliage
TOUGHNESS - [metall] (term denoting a condition which is intermediate between brittleness and softness) tenacité *f*, dureté *f*
[rubber ind] tenacité *f*, résistance *f*
~OF A METAL - [metall] ténacité *f* d'un métal
~TEST - [metall] essai *m* de fragilité
~VALUE - [metall] valuer *f* de ténacité à l'entaille
TOURER - [aero] avion *m* de tourisme
TOURING - [auto] de tourisme
~CAR - [auto] voiture *f* de tourisme
~PLANE - [aero] avion *m* de tourisme
~TOP LUGGAGE CARRIER - [auto] galerie *f* de toit pour les bagages
TOURISM - [gen] tourisme *m*
TOURMALINE - [min] (a complex native aluminium borosilicate with uses as a gemstone and in optical apparatus) tourmaline *f*
TONGS - [opt] pince *f* à tourmaline
TOURNADOZER - [mech] bulldozer *m* à roues
TOURNAROCKER - [mech] grande benne *f* basculante
TOURNATRAILER - [transp] semi-remorque *f* à fond glissant vers l'arrière
TOW, to - [transp] remorquer, touer
TOW - [naut etc] (the act of towing) remorque *f*
[naut] (a rope, a towline) câble *m* de remorque
[naut] (a barge) chaland *m*
[text] étoupe *f*, filasse *f*
[auto] (aero etc) (the tow rod) grelin *m* de remorque
~BAR - [mech] barre *f* de remorquage
~-BOAT - [naut] remorqueur *m*, toueur *m*
TOW BOX - [text] caisse *f* à étoupes
~CAR - [auto] remorque *f*
~CARPET - [text] tapis *m* en étoupe
~CATCHER - [text] peigne-détacheur *m* d'étoupes
~CHAIN - [transp] chaîne *f* de remorquage
~COMBING - [text] peignage *m* des étoupes
~COUPLING - [mech] accouplement *m* de traction
~DOFFER - [text] doffer *m* d'étoupes
~DOFFER COMB - s. tow doffer
~FIBRE - [text] fibre *f* d'étoupe
~HACKLING - s. tow combing
~HOOK - [auto etc] fourche *f* d'attelage
~LINEN - [text] toile *f* d'étoupe
~-OFF - [aero] (of a glider) lancement *m*
~-PATH - [naut] (narrow path used for towing boats along a canal) banquette *f* de halage, tirage *m*
~-PLANE - [aero] avion *m* remorqueur
~-ROPE - [gen] câble *m* de remorque, grelin *m* de remorque, corde *f* de halage
~ROVING - [text] mèche *f* de préparation d'étoupe
~SHAKING MACHINE - [text] secoueuse *f* à étoupes
~SORTER - [text] classeur *m* d'étoupes
~SPINNER - [text] fileuse *f* d'étoupes
~STRIPPER - [text] dépouilleur *m*
~WEAVER - [text] tisseur *m* d'étoupe
~YARN - [text] fil *m* d'étoupe
TOWAGE - [transp] remorquage *m*, touage *m*
[hydr] halage *m*
TOWED - [transp etc] remorqué, toué
~GLIDER - [aero] (a glider which is maintained in flight by towing) planeur *m* remorqué
TOWEL - [text] serviette *f*
~LOOM - [text] bâti *m* du métier pour tissu éponge
~WEAVE - [text] armure *f* pour serviette éponge
TOWELLING - [text] tissu-éponge *m*, toile *f* pour serviette
[paper man] papier *m* essuie-mains
TOWER - [ind chem] (a general term for a vertical hollow structure used to bring liquids and gases into contact, the former flowing downward against an upward current of the latter) tour *f*
[constr] tour *f*
[el] pylône *m*
[radio] (of an aerial) pylône *m*
[naut] (towing craft) remorqueur *m*, toueur *m*
~BODY - [el] fût *m* d'un pylône
~BOLT - [mech] (bolt working in a barrel-like case) verrou *m* à la capucine
[constr] verrou *m* à platine
~CRANE - [mech] grus *f* à pylône
~CROSS - [constr] croix *f* de tour
~DRYER - [ind chem] tour *f* de séchage
~GANTRY - [constr] (form of staging for the support of a derrick to be used in the construction of large buildings) portique *f* de grue à pylône
~LEGS - [el] montants *mpl* d'un pylône
~LINE - [el] ligne *f* sur pylônes
~LOADING - [el] charges *fpl* mécaniques totales d'un pylône
~PURIFICATION - [ind chem] épuration *f* en tour
~PURIFIER - [ind chem] tour *f* d'épuration, épurateur *m* en tour
~SCRUBBER - [ind chem] (a packed washing tower) laveur *m* statique
~SILO - [agric] silo-tour *m*
~WAGON - [railw] wagon *m* à plate-forme mobile
[transp] camion *m* à plate-forme élevée

TOWING - [naut] remorquage m, halage m, touage m
[transp] remorque f
~AMBULANCE - [auto] camion m dépanneur
~BAR - [mech] barre f de remorquage
~CAPACITY - [mech] puissance f de traction
~DOLLY - [aero] (a device used in towing aircraft on the ground by means of a tractor) remorqueur m d'avions
~EYE - [auto etc] anneau m de remorquage
~RING - s. towing eye
~STABILIZER - [transp] (a hydraulic damping device to keep the trailer in line with the towing vehicle) stabilisateur m de remorque
~TRACTOR - [mech] tracteur m de halage
TOWN - [gen] ville f, cité f
~DRAINAGE - [hydr] assainissement m urbain
~GAS - [gas ind] (gas made and supplied for industrial and domestic purposes, usually heating. It generally consists of a mixture of coal gas, carburetted water gas; gases obtained from oil deposits directly or indirectly are used increasingly as additions) gaz m de ville
~PLANNING - [surv] architecture f urbaine, urbanisme m
~SEWAGE - [hydr] eaux fpl d'égout urbaines
TOWNSEND AVALANCHE - [phys] (term denoting a cascade multiplication of ions) avalanche f de Townsend
~DISCHARGE - [electron] (non self-maintained discharge) décharge f de Townsend, décharge f non autonome
~IONIZATION - [nucl] (the process by which a single charged particle when accelerated by a strong electric field and colliding with neutral gas molecules, produced other charged particles) ionisation f de Townsend
TOXAPHENE - [chem] (a chlorinated camphene with uses as an agricultural pesticide) toxaphène m
TOXEMIA - [med] toxémie f
TOXIC - [chem] (poisonous, harmful life) toxique
TOXICITY - [chem] (the degree to which a substance possesses poisonous qualities) toxicité f
TOXICODERMA - [med] toxidermie f
TOXICODERMATITIS - [med] dermatite f toxique
TOXICOLOGY - [med] toxicologie f
TOXICOSIS - [med] toxicose f
TOXIN - [chem] (a specific physiologic poison) toxine f
TOXINOTHERAPY - [med] toxithérapie f
TOXIS - [med] intoxication f
TOY - [gen] jouet m
~TRANSFORMER - [el] petit transformateur m, transformateur m de jouets
TR BAND TUBE - [radio] (a type of gas filled switching tube, operating over a comparatively wide band of frequencies) tube m émission-réception à large bande
~CAVITY - [electron] (the resonant portion of a TR switch) cavité f d'un commutateur émission-réception
~SWITCH - [radio] (a switch, mostly of the gas discharged type, used when a common transmitting and receiving aerial is employed, which automatically uncouples the receiver from the aerial during the transmitting period) commutateur m émission-réception, duplexeur m, tube m commutateur
~TUBE - [radio] (transmission-reception-tube) tube m émission-réception
TRABEATE - [arch] à entablement, en poutres en saillie
TRABEATED - s. trabeate
~SYSTEM - [arch] système m à poutres en saillie
TRABEATION - [arch] (construction having beams or long stones as lintels instead of an arch) entablement m
TRACE, to - [draw] tracer
[draw] (to copy on a superimposed transparent sheet) tracer faire le tracer, calquer
[gen] (to follow the traces of) retrouver les vestiges
TRACE - [gen] (vestige, mark) trace f
[mech] (connecting rod) bielle f
[telev] (the forward motion of the scanning spot) ligne f
[radar] (the radial line made by the moving spot on the P.P.I.) ligne f
[impl] (of a harness) trait m
~-CHEMISTRY - [chem] (American term for microchemistry, i.e. the chemistry of substance present in exceedingly minute quantities) microchimie f
~CONCENTRATION - [chem] (a concentration of a substance below the usual limits of chemical detection) concentration f microscopique
~ELEMENT - [chem] oligoélément m
~ELEMENTS - [hydr] éléments mpl à l'état de traces
~INTERVAL - [electron] (in a cathode-ray tube) intervalle m de trace
~PROGRAMME - [comput] programme m de traçage
~ROTATION SYSTEM - [radar] système m de rotation de la trace
TRACER - [nucl] (a substance introduced into another substance to enable its subsequent movements to be followed by observation of radiation) traceur m isotopique, indicateur m
[impl] traçoir m
[tool] burin m
~BULLET - [firearms] balle f traceuse
~CHEMISTRY - [chem] (the use of isotopic tracers in chemical studies) chimie f des indicateurs isotopiques
~COMPOUND - [chem] (compound which, by its ease of detection, enables reaction or process to be studied conveniently) composé m traceur
~ELEMENT - [nucl] (radioactive element used as an indicator) traceur m radioactif
~GAUGING - [meas] juageage m par traceur
~METHOD - [nucl] méthode f d'analyse par traceurs isotopiques
~POINT - [mach tool] (of a copying machine) traçoir m de tour à copier
~STUDIES - [chem] (a technique for studying the role of an element, a group of elements or a compound in a biological, chemical or physical process) études fpl avec traceurs
~WHEEL - [impl] roue f traceuse
TRACERY - [arch] réseau m, entrelacs m, filigrane m
[constr] tracé m
~PANEL - [constr] ouvrage m en tracé
~VAULT - [constr] voûte f à réseau de nervures
TRACHEA - [anat] (air tube of the respiratory system) trachée f
TRACHEAL - [anat] trachéal
TRACHEITIS - [med] (inflammation of the mucous membrane of the trachea) trachéite f
TRACHELODYNIA - [med] douleur f cervicale

TRACHELOPEXY - [med] trachélopexie *f*
TRACHELOTOMY - [med] incision *f* du col de l'utérus
TRACHEOBRONCHITIS - [med] trachéobronchite *f*
TRACHEOSCOPY - [med] trachéoscopie *f*
TRACHEOSTENOSIS - [med] trachéosténose *f*
TRACHOMA - [med] trachome *m*, conjonctivite *f* granuleuse
~GLANDS - [anat] follicules *mpl* lymphatiques de la conjonctivite
TRACHYTE - [geol] (fine-grained igneous rock) trachyte *f*
TRACING - [draw] tracé *m*, dessin *m* calqué, calque *m*, décalque *m*
[gen] traçage *m*, calquage *m*
~BENCH - [impl] banc *m* à diviser
~CLOTH - [draw] toile *f* à calquer, papier *m* toile
~DISTORTION - [el acoust] bruit *m* indirect
~LINEN - s. tracing cloth
~PAPER - [draw] papier *m* à calquer, papier *m* translucide
~ROUTINE - [comput] (superroutine allowing a partial or complete monitoring by intervention in a programme) programme *m* d'analyse de programme, programme *m* de traçage
~TOOL - [carp] rénette *f*
TRACK, to - [gen] suivre à la trace
[aero] (by antiaircraft means etc) poursuivre
[railw] poser la voie
[mech] (to span between wheels) obtenir l'écartement de
[auto] (of a trailer, to be aligned) être en alignement
[radar] (the operation of continuously defining the coordinates of a moving target) poursuivre, guider
[cin] (the camera) prendre des vues en mouvement
[naut] haler à la cordelle
TRACK - [gen] (the trail left by the passage of anything) empreintes *fpl*, trace *f*
[gen] piste *f*
[mech] (of a ball bearing) piste *f* de roulement
[mech] (the metal belt for tractors or similar vehicles) chenille *f*, chemin *m* de roulement
[railw] (set of rails) voie *f* ferrée
[meas] (the width between the wheels of a vehicle) voie *f*
[mech] (of a crane, a pair parallel rails for an overhead travelling crane) voie *f* de roulement (de chariot de pont)
[aero] (the projection on the surface of the earth of the path followed by the C.G.of an aircraft) route *f* vraie
[aero] (of a landing gear, the distance between the outer points of contact with the ground of the main undercarriage wheels) distance *f* entre les roues
[naut] route *f* régulière
[comput] (a channel) piste *f* magnétique
[photo] chariot *m*
~ADHESION - [el acoust] adhésion *f* de la piste
~ADJUSTMENT - [mech] réglage *m* de la voie
~ALIGNMENT GAUGE - [auto] (a wheel tester and camber alignment tester) calibre *m* de réglage du pincement, calibre *m* de réglage de la voie
~AND CAMBER GAUGE - [auto] gabarit *m* de carrossage et de pincement

TRACK ANGLE - [aero] (the angle which the track of an aircraft makes with the meridian on which it lies at a given moment) angle *m* de route géographique
~ARM - [auto] levier *m* d'attaque de la fusée
~ASSEMBLY - [mech] chenilles *fpl*
~BED - [railw] assiette *f* de la voie
~BRAKING - [railw] (a method of braking in which a shoe is applied to the track-rails) freinage *m* de rail
~BUSHING - [mech] (of a tractor) bague *f* de chenille
~CAPACITY - [railw] capacité *f* d'une ligne
~CARRIER - [mech] (of a tractor) galet *m* de roulement de chenille
~CARRIER ROLLER - s. track carrier
~CENTRE LINE - [railw] axe *m* de la voie
~CHANNELER - [mining] trancheuse *f* montée sur rails
~CIRCUIT - [el] (the electric circuit formed by the two running rails of a traction system) circuit *m* d'occupation, circuit *m* de voie
~CIRCUIT SIGNALLING - [el] (electric signalling system making use of the change in resistance of a track circuit when an electric train passes over a section of the railway track) signalisation *f* par circuit de voie
~CLEARER - [railw] chasse-pierre *m*
[agric] sabot *m* séparateur
~CLOSED - [railw] voie *f* occupée
~CLUTCH CONTROL LEVER - [railw] levier *m* de commande
~CONTROLLER - [railw] distributeur *m* d'itinéraires
~CROSSING - [railw] croisement *m* de la voie
~DIAGRAM - [railw] panneau *m* indicateur des voies
~DISPLACEMENT - [telecomm] déplacement *m* du tracé
~DISTORTION - [railw] affaissement *m* de la voie, déformation *f* de la voie
~EDGE - [el acoust] délimitation *f* de la piste sonore
~ELEMENT - [railw] élément *m* de rail activateur
~ERROR - [el acoust] (fault in recording) erreur *f* de piste
~FASTENINGS - [railw] petit matériel *m* de voie
~FORMATION - [railw] plate-forme *f* de la voie, palier *m* de la voie
~GAUGE - [railw] gabarit *m* de voie, gabarit *m* d'écartement des rails
~GUIDE - [radio] (a radio beacon system to indicate to aircraft a specified track) indicateur *m* de route
~HUBS - [mech] (of a tractor etc) moyeux *mpl* des maillons de chenille
~IDLERS - [mech] (of a tractor etc) roues *fpl* porte-chenille
~IN, to - [cin] (to decrease the distance between the camera and the scene) travelling *m* en dedans
~INDICATOR - [radar] indicateur *m* de route
[railw] indicateur *m* de marche
~INSTRUMENT - [railw] contacteur *m* de rail
~LAYING - [railw] pose *f* de la voie
[telev] préparation *f* au montage des bandes à mixer
~LEVEL - [railw] règle *f* à dévers et niveau
~LEVEL AND GAUGE - s. track level
~LEVER - [railw] levier *m* d'itinéraire
~LINK - [mech] patin *m* de chenille
[auto] bielle *f* d'accouplement des roues avant

TRACK LOCKING - [railw] (locking by track circuit) enclenchement *m* par circuit de voie
~LOOSENER - [agric] effaceur *m* des traces de roue
~MAGNET - [el] aimant *m* de voie
~MAINTENANCE - [railw] entretien *m* de la voie
~MEASURING DEVICE - [railw] gabarit *m* universel de voie
~-MOTOR CAR - [railw] draisine *f*
~OCCUPANCY LIGHT - [railw] lampe *f* d'occupation de voie
~OCCUPATION DIAGRAM - [railw] tableau *m* d'occupation des voies
~OPEN - [railw] voie *f* praticable
~OUT, to - [cin] (or dolly out to increase the distance between the camera and the scene) travelling *m* en dehors
~PIN - [mech] (of a tractor) goujon *m* de chenille
~PINCH-BAR - [railw] pince *f* à rails
~PITCH - [el acoust] pas *m* de la piste
[comput] (in storage) entre-axe *m* des pistes, intervalle *m* entre pistes
~RAIL - [railw] rail *m*
~RAIL-BOND - [el] (rail bond designed to preserve the electrical continuity of the track rails) connexion *f* électrique des rails
~RAISING - [railw] relèvement *m* de la voie
~RECORDING COACH - [railw] wagon *m* d'auscultation de la voie
~RELAY - [el] (relay used in track-circuit signalling for controlling the electrically operated signals) relais *m* de voie
~RELEASE RELAY - [railw] relais *m* de contrôle de libération de la voie
~-RETURN CIRCUIT - [railw] circuit *m* de retour par la voie
~-RETURN SYSTEM - [railw] (a system in which the track-rails are used as an uninsulated portion of the system for the distribution of electric power to trains) système *m* à retour par la voie
~ROD - [auto] (of the steering assembly) barre *f* d'accouplement
~ROD LEVER - [auto] biellette *f* d'accouplement
~ROLLER - [mech] (of a tractor) chemin *m* de roulement
~ROLLER FRAME - [mech] (of a tràctor) longeron *m* du tendeur de chenille
~ROPE - [transp] (of a ropeway) câble *m* porteur
~SCALES - [railw] (USA only, a railway weighbridge) bascule *f* à wagons
~SECTIONING CABIN - [railw] (cabin for the switchgear by means of which the supply to different sections of an electrified railway may be disconnected) poste *m* de conduite
~SELECTION - [comput] sélection *f* de piste
~SHOE - [mech] (of a tractor) patin *m* de chenille
~SHOT - [cin] plan *m* pris en mouvement
~SLOTS - [mech] chemins *mpl* de roulement
~SPIKE - [railw] crampon *m*
~SUBSIDENCE - [railw] tassement *m* de la voie
~SWITCH - [el] (switch controlling the supply of current to a section of an electrified railway line) interrupteur *m* de feeder
~TANK - [railw] canal *m* d'alimentation
~TAPER - [phys] conicité *f* de trace, réduction *f* de trace
~TONGS - [railw] pince *f* à rails
~TREAD LANDING GEAR - [aero] train *m* d'atterrissage à chenilles
TRACK-TYPE TRACTOR - [agric] tracteur *m* à chenilles
~TYRE FOR TRACTOR - [rubber ind] pneu *m* de sillage pour tracteur
~VEHICLE - [auto] véhicule *m* à chenilles
~WHEEL - [railw] (the driving wheel) roue *f* motrice
~-WHILE-SCAN - [radar] (electronic device used to detect a radar target) dispositif *m* de traçage et balayage simultané
TRACKED LANDING GEAR - [aero] (landing gear in the form of a pair of tracks, similar in action to those of a crawler tractor) train *m* d'atterrissage à chenilles
~SHOVEL - [mech] (USA only, excavator, mechanical shovel) cuiller *f* droite
TRACKER - [mus] (in an organ) demoiselle *f*, abrégé *m*
TRACKING - [gen] (on a road) sillage *m*
[radar] (the continuous defining of the coordinates of a moving target) poursuite *f*
[el] (the formation of a carbonized conducting path across the surface of an insulating material between electrodes maintained at a potential difference) piste *f* carbonisée
[telev] commande *f* dynamique de la luminosité
[telev] (in video recording) centrage *m* de piste
[radio] (an arrangement by which one of a number of ganged circuits is maintained at a constant frequency difference from that of another circuit) réglage *m* exact
~ANTENNA - [radar] antenne *f* de poursuite
~APPARATUS - [el acoust] (apparatus designed to apply the magnetic sound tracks to the picture film) appareil *m* d'application de la piste sonore
~CIRCUIT - [radar] (circuit used in fire control) circuit *m* de contrôle de la direction du tir
~CONTROL - [telev] régleur *m* de la reproduction
~DEVICE - [radar] (a device for the automatic search of the target) dispositif *m* pour la recherche automatique de la cible
~ERROR - [el acoust] distorsion *f* de gravure
~FILTER - [radar] filtre *m* à affaiblissement de phase
~SHOT - s. track shot
~STATION - [radar] station *f* de poursuite
TRACKLESS TROLLEY SYSTEM - [el] (trolley system in which electrically equipped vehicles run on the ordinary roadway, the electric power is obtained from two overhead conductors, one positive and the other negative) système *m* trolleybus
TRACKMAN - [railw] garde-ligne *m*
TRACKWALKER - s. trackman
TRACT - [gen] espace *m*, étendue *f*
~OF LAND - [gen] terrain *m*, étendue *f* de pays
~OF THE RAIN - [met] zone *f* de pluie
TRACTION - [mech] (the act of drawing) traction *f*, tirage *m*
[phys] (the adhesive friction of a wheel on the ground) adhérence *f*
[med] (of the muscles) traction *f*, tirage *m*
~ACCUMULATOR - s. traction battery
~BATTERY - [el] (storage battery designed to supply power for vehicles) batterie *f* de traction
~BOND - [el] connexion *f* de traction
~COEFFICIENT - [railw] coefficient *m* d'adhérence
~DRIVE - [el] (a method of transmitting power to the suspension ropes by means of a sheave) trans-

mission *f* à poulie d'adhérence, transmission *f* à poulie motrice
TRACTION ENGINE - [mech] (a type of locomotive) machine *f* routière, tracteur *m*
~EQUIPMENT - [el] équipement *m* de traction
~MACHINE - [el] treuil *m*
~MOTOR - [el] (for electric locomotives) moteur *m* de traction
~OUTPUT OF A HEAT ENGINE - [el] puissance *f* de traction d'un moteur thermique
~PERMEAMETER - [instr] (apparatus for measuring the permeability of a sample of iron by weighing the mechanical force between the end of the sample and a surface forming part of the yoke of the apparatus) perméamètre *m* à attraction magnétique
~ROPE - [mech] (the endless rope used in an aerial ropeway system to effect the movement of the carries) câble *m* porteur, câble *m* de tirage
~SUBSTATION - [el] sous-station *f* pour traction
TRACTIVE - [mech] tractif, de traction
~EFFORT - [mech] (the total propelling-force measured at the rims of the driving wheels) effort *m* de traction
~FORCE - [mech] force *f* tractive, force *f* tirante
~FORCE METER - [instr] dynamomètre *m* de traction
~POWER - s, tractive effort
~REGISTRATION ARM - [railw] antibalançant *m* tirant
~RESISTANCE - [railw] (or train-resistance, the sum of frictional or atmospheric forces which resist vehicle or train movement) effort *m* résistant (en palier et en alignement)
TRACTOR - [mech] (machine for pulling or drawing) tracteur *m*, mototracteur *m*
[auto] tracteur *m* pour semi-remorque
[agric] agromotive *f*
~AEROPLANE - [aero] (an aeroplane driven by one or more tractor propellers) avion *m* à hélice tractive
~AND TRAILER - [auto] tracteur *m* pour remorque
~AND SEMITRAILER - [auto] tracteur *m* pour semi-remorque
~AIRSCREW - s. tractor propeller
~EXCAVATOR - [mech] excavateur *m* à chenilles
~LOADER - [mech] (a loading shovel) pelle *f* mécanique
~PLOUGH - [agric] charrue *f* à tracteur
~PROPELLER - [aero] (a propeller of which the shaft is normally in tension) hélice *f* tractive
~SHOVEL - s. tractor loader
~TRACK LINKS - [mech] patins *mpl* de chenille
~TRUCK - [auto] tracteur *m* pour semi-remorque
TRADE, to - [gen & comm] faire la commerce, trafiquer
TRADE - [comm] commerce *m*
[comm] (commercial exchange) trafic *m*, négoce *m*
[gen] emploi *m*, métier *m*
~AGREEMENT - [gen] accord *m* commercial
~EXPENSES - [comm] frais *mpl* de bureau
~-IN, to - [comm] (to give something in part payment for something else) reprendre en compte
~-IN - [comm] (something given or accepted in payment or part payment) reprise *f* en compte
[auto] (said of a second-hand car, given as part payment for a new car) vente *f* en reprise
~KIT - [gen] trousseau *m* professionnel
~MARGIN - [comm] marge *f* commerciale
~-MARK, to - [gen] appliquer la marque de fabrique
TRADE-MARK - [gen] marque *f* de fabrique, estampille *f*
~ROUTE - [comm] route *f* commerciale
~-SHOW - [cin] preview *m* pour distributeurs
~-UNION - [leg] syndacat *m* ouvrier, fédération *f* syndacale ouvrière
~WASTE WATER - [hydr] eaux *fpl* résiduaires industrielles
~WINDS - [met] (winds blowing consistently from N.E. in the Northern hemisphere and from the S.E. in the Southern, and prevailing between the horse latitudes and the equatorial belt of calms) alizés *mpl*
TRADER - [naut] navire *m* marchand
TRADING - [comm] commerce *m*, négoce *m*, trafic *m*
[gen] de commerce, marchand
~COMPANY - [comm] société *f* de commerce
~-STAMP - [comm] coupons *mpl* à prime
~YEAR - [fin] exercice *m*
TRAFFIC - [gen] trafic *m*, circulation *f*
[comm] trafic *m*, commerce *m*
[telecomm] trafic *m*
~ACCIDENT - [gen] accident *m* de rue, accident *m* de route
~BEAM - [auto] (of the headlights) éclairage *m* code
~CHANNEL - [telecomm] voie *f* d'écoulement du trafic
~CONGESTION - [gen] encombrement *m* de circulation
~CONTROL - [aero] (the supervision and direction of air traffic) contrôle *m* du trafic aérien
[gen] contrôle *m* de la circulation
~CONTROL POST - [gen] poste *m* de contrôle du trafic
~-CONTROL PROJECTOR - [aero] phare *m* d'aérodrome
~CONTROL TOWER - [aero] (of an airport) tour *f* de contrôle
~CURVE - [telecomm] courbe *f* du trafic, allure *f* de trafic
~DENSITY - [railw] débit *m* d'une ligne
~DIRECTOR - [aero] (a radar controller responsible for maintaining proper separation between aircraft and regulating traffic) directeur-contrôleur *m*
~DISTRIBUTOR - [telecomm] distributeur *m* d'appels
~EXCHANGE POINT - [railw] bureau *m* d'échange
~FLOW - [gen] écoulement *m* de la circulation
~INDICATOR - [auto] signalisateur *m*
~LANE - [gen] voie *f*
~LIGHTS - [auto] feux *mpl* de circulation, signalisation *f* routière lumineuse
~LINE - [road traffic] (the line dividing the lanes) ligne *f* de séparation des voies
~-LINE PAINT - [paint] vernis *m* pour signalisation routière
~LOAD - [telecomm] intensité *f* du trafic
~MANAGER - [railw] chef *m* du mouvement
~PILLAR - [road traffic] colonne *f* de signalisation routière
~PILOT - [comput] (multiplexer) multiplexeur *m*
~RECORDER - [instr] compteur *m* d'occupation totale
~RELAY - [telecomm] relais *m* de circulation
~ROAD - [gen] voie *f* de circulation
~ROUNDABOUT - [road traffic] sens *m* gyro
~SECTION - [telecomm] service *m* d'exploitation
~STREAM - [road traffic] courant *m* du trafic
~TILE - [road traffic] carreau *m* de signalisation
~-UNIT - [telecomm] communication-heure *f*
TRAFFICATOR - [auto] (traffic indicator) indicateur *m*

de direction, flèche *f* de direction
:RAFFICATOR SWITCH - [auto] interrupteur *m* du signalisateur
:RAGACANTH GUM - [chem] (a gum obtained from a species of Astragalus and having application as a suspending agent in medicine, in adhesives, and cosmetics, and in textile processing) gomme *f* adragante
'RAIL, to - [gen] traîner, suivre
[bot] (of plants) grimper, ramper
[railw] talonner
'RAIL - [gen] trace *f* de lumière, panache *f* de fumée
[firearms] (the part resting on the ground) fléche *f*, crosse *f*
[astronaut] (of a rocket) traînée *f*
[auto] (the distance between the point of contact of the front wheel with ground and the intersection of the kingpin axis and the ground) distance *f* de chasse
[aero] (of a landing gear, the distance by which the axis of a castoring wheel is offset to the rear of the swivelling axis) recul *m*
[arch] rinceau *m*
[astr] queue *f*
[gen] sentier *m*, piste *f*
BOARD - [shipbuild] frise *f* d'éperon
BOGIE - [railw] (of a locomotive) bogie *m* porteur
BRIDGE - [gen] bac *m* à traille
ROPE - [aero] (a rope trailed from a free balloon and in contact with the ground used to control speed and altitude) guide-rope *m*
RAILER - [transp] remorque *f*, baladeuse *f*
[cin] extrait *m* de film
[auto] (a caravan-like vehicle) roulotte *f*
[railw] (a trailer car) voiturette *f* remorque
[electron] (of a cathode-ray tube) strie *f* lumineuse
[railw] roue *f* porteuse d'arrière
[telev] amorce *f* de fin
BLOCK - [comput] (information block with data which do not appear in the normal blocks) bloc *m* complémentaire
BODY - [auto] carrosserie *f* de remorque
CARD - [comput] carte-suite *f*
CHASSIS - [auto] châssis *m* pour remorque
-COACH - [railw] (railway-vehicle not equipped with motors but intended for marshalling in trains) voiture *f* remorque
COUPLING - [auto] attelage *m* de remorque
COUPLING DEVICE - [mech] dispositif *m* d'attelage pour remorques
DUMP - [auto] (a vehicle combination of body and hoist mounted on a trailer chassis) remorque *f* à benne basculante
HITCH - [auto] (towing fork) fourche *f* d'attelage de remorque
LABEL - [comput] label-fin *m*
PLOUGH - [agric] charrue *f* traînée
PUMP - [hydraul] pompe *f* à remorque
STOCK - [railw] matériel *m* remorqué
WITH DUMPING MECHANISM - [auto] remorque *f* benne
RAILING - [gen] traîne *f*, poursuite *f*
[railw] talonnement *m*
ACTION - [auto] (caster action of a steering wheel) reversibilité *f*
AERIAL - [radio] (an aerial consisting of a wire trailed below the aircraft when in flight) antenne *f* pendante
TRAILING AXLE - [mech] essieu *m* porteur d'arrière
~CABLE - [el] (travelling cable, flexible cable providing electrical connection between the lift car and a fixed point) câble *m* flexible
[el] câble *m* de guidage
~CONTACTS - [el] contacts *mpl* à accompagnement, contacts *mpl* échelonnés
~EDGE - [aero] (the downstream edge of an aerofoil or a strut) bord *m* de fuite
[telev] (the rapid voltage alteration which together with the leading edge, forms a pulse) flanc *m* arrière
[electron] flanc *m* descendant
~EDGE CORD - [aero] (a length of wire or cord (sometimes of metal) attached to the trailing edge of a control surface and adjusted when the aircraft is on the ground, to modify, trim or balance) stabilisateur *m* bord de fuite
~EDGE RISE TIME - [radio] (the time at which the instantaneous amplitude last reaches a given fraction of the peak pulse-amplitude) temps *m* de descente du flanc arrière
~END - [cin] amorce *f* en fin de bobine
[comput]fin *m* de bande
~POLE EDGE - [el] arête *f* de sortie de la pièce polaire
~POLE HORN - [el] (the pole horn opposite to the leading pole horn) corne *m* polaire de sortie
~POLE TIP - s. trailing pole horn
~RIB - [aero] (of a wing] nervure *f* du bord de fuite
~ROPE - [rtransp] guide-rope *m*
~SWEEP - [aero] (deviation in a propeller blade in the direction of the trailing edge) déport *m* dans le plan de rotation
~THE POINTS - [railw] talonnage *m* d'aiguilles
~-TRUCK AXLE - s. trailing axle
~VORTEX - [aero] (a vortex extending downstream from its origin on the surface of a body) tourbillon *m* de bord de fuite
~-VORTEX DRAG - [aero] (drag arising from the presence of trailing vortices) traînée *f* de tourbillon de bord de fuite
~WHEEL - [railw] (of a locomotive) roue *f* porteuse d'arrière
TRAIN, to - [gen] (to bring to a specified standard) former, instruire, s'entraîner
[gen] (to render skilful, proficient) exercer, élever
[firearms] pointer
[instr] (to point]) pointer
[bot] (of flowers, climbing plants) faire grimper, conduire, diriger
[mining] tracer, suivre à la trace
TRAIN - [gen] traîne *f*, queue *f*, convoi *m*
[railw] train *m*
[mech] (a series of parts acting upon one another for transmitting motion) système *m* d'engrenages, rouage *m*
[metall] (of a rolling mill) train *m*, laminoir *m*
[firearms] (the variation of the axis of a gun in a horizontal plane) flèche *f*, crosse *f*
[mining] traînée *f* de poudre
[mining] rame *f* de wagons
~A CAMERA, to - [cin] diriger une chambre photographique

TRAIN ARRIVAL - [railw] arrivée *f* d'un train
~ARRIVAL INDICATOR - [railw] tableau *m* indicateur des arrivées de trains
~COMPOSITION - [railw] composition *f* d'un train
~CONTROL SECTION - [railw] poste *m* de régulation
~CONTROL TERRITORY - [railw] section *f* à commande automatique
~DEPARTURE - [railw] départ *m* d'un train
~DEPARTURE INDICATOR - [railw] tableau *m* indicateur des départs de trains
~DESCRIBER - [railw] appareil *m* enregistreur d'annonce des trains
~DIAGRAM - [railw] graphique *m* des trains, graphique *m* de marche
~FERRY - [railw] ferry-boat *m*, transbordeur *m* de trains, bac *m* transbordeur
~LINE - s. train pipe
~OF DEBRIS - [geol] traînée *f* de débris
~OF GEARING - [mech] équipage *m* d'engrenage, train *m* d'engrenages
~OF POWDER - [mining] traînée *f* de poudre
~OF ROLLS - [metall] équipage *m*, jeu *m* de cylindres
~OF WAVES - [phys] (a group of successive waves) train *m* d'ondes
~OF WHEELS - [mech] train *m* de roues
~-OIL - [zool] huile *f* de baleine
~PIPE - [railw] (for the braking system) conduite *f* du frein
~-PIPE HOSE - [railw] boyau *m* d'accouplement du frein
~RESISTANCE - [phys] (or tractive resistance, the sum of frictional and atmospheric forces resisting the movement of a train or any vehicle) résistance *f* des trains
~RESISTANCE ON THE LEVEL - [phys] (specific train resistance on level tangent track) résistance *f* en palier, résistance *f* spécifique en palier et en alignement
~RUNNING INDICATOR - [railw] enregistreur *m* de la marche des trains
~SHUNT - [railw] shunt *m* d'un train
~STAFF - [railw] bâton-pilote *m*
[gen] service *m* d'accompagnement
~TRAFFIC - [railw] mouvement *m* des trains
~UNIT - [railw] (motor train unit) élément *m* automoteur
~WHEEL BRIDGE - [horol] pont *m* du rouage
TRAINER AIRCRAFT - [aero] avion *m* d'entraînement
TRAINING-[gen]éducation *f*, instruction *f*
[gen] entraînement *m*
[bot] (of a plant) dressage *m*, palissage *m*
[firearms] pointage *m*, braquage *m*
~COLLEGE - s. training school
~COURSE - [gen] cours *m* de formation professionnelle
~PLANE - s. trainer aircraft
~REACTOR - [nucl] réacteur *m* d'entraînement
~SCHOOL - [gen] école *f* professionnelle
~SHIP - [naut] navire *m* école
~WALL - [hydr] chicane *f*, guideau *m*
~WHEEL - [mech] volant *m* de pointage
TRAIT - [gen] trait *m*
[biol] trait *m* de caractère
TRAJECTORY - [phys] (the path described by a body moving in space) trajectoire *f*
[nucl] (the path in space traversed by a particle) parcours *m*

TRAJECTORY BAND - [aero] (a reinforcing band over the upper surface, e.g. of a balloon envelope) bande *f* de renfort supérieure
~MEASURING SYSTEM - [meas] système *m* de trajectographie
TRAM, to - [mining] herscher, rouler
TRAM - [transp] (a tramway) tramway *m*
[text] (thick silk thread) soie *f* trame
[min] (for mine railway) berline *f*, herche *f*
[draw] compas *m* à verge
~BOY - [mining] suiveur *m* de rames
~LEVEL - [mining] galerie *f* de roulage
~RAIL - [mech] rail *m* à gorge, rail *m* de translation
TRAMCAR - [transp] voiture *f* de tramway
~MOTORCOACH - [transp] automotrice *f* de tramway
TRAMLINE - [transp] ligne *f* de tramway
TRAMMEL, to - [gen] entraver, empêtrer
TRAMMEL - [gen] entrave *f*
[fishing] (a three-layer net) trémail *m*, tramail *m*
[instr] (instrument having parts sliding on a rod for use as a compass) compas *m* à verge
[tool] (tool used for the alignment of components) outil *m* de centrage
[impl] (a device used to describe ellipse)ellipsographe *m*
~BAR - [impl] compas *m* à verge allongé
TRAMMING - [mining] herschage *m*, roulage *m*
TRAMONTANA - [met] (a northerly mountain wind blowing over Italy) tramontane *f*
TRAMP - [gen] marche *f*, bruit *m* de pas marqués
[naut] chemineau *m*
[shoe man] semelle *f* de fer
~IRON - [metall] (iron fragments from tools machines etc, mixed with other materials, causing damage and interference with processes) fragments *mpl* de fer dispersés
~METAL - [metall] (stray fragments of metal in a material and foreign to it) fragments *mpl* métalliques
~STEAMER - [naut] chemineau *m*, navire *m* sans ligne régulière
TRAMPING - [metall] serrage *m* à pied (du sable)
TRAMPLE MANURE, to - [agric] tasser le fumier
TRAMPLING - [agric] tassement *m* du fumier
TRAMROAD - [min] (USA only, a railroad in a mine) tramway *m* de mine
TRAMWAY - [transp] tramway *m*
~POLES - [el] (the steel poles fixed at the sides or at the centre of a road for supporting the overhead conductors) pylônes *mpl* pour les fils de contact, pylônes *mpl* pour le fils de trolley
TRANCE - [med] catalepsie *f*, extase *f*
TRANQUIL - [gen] calme, tranquille
TRANQUILLIZE,to - [gen & med] tranquilliser, calmer
TRANQUILLIZER - [chem] calmant *m*
TRANS - [chem] (a prefix denoting an isomer in which like atoms lie on opposite sides of the molecule) trans-
TRANSACTION - [comm] opération *f* commerciale
[fin] (a financial operation) opération *f* de bourse
~DATA - [comput] données *fpl* de mouvement
~RECORD - [comput] enregistrement *m* modificateur
TRANSACTIONAL - [leg] transactionnel
TRANSADMITTANCE - [electron] (from one electrode to another, the quotient of the sinusoidal component of the current of the second electrode by the sinusoidal component of the voltage of the first electro-

de, provided all other electrode voltages are kept constant) admittance *f* de transfert, transadmittance *f*
TRANSADMITTANCE COMPRESSION RATIO - [electron] taux *m* de compression de la transadmittance
TRANSAMINASE - [biol] transaminase *f*
TRANSANIMATION - [med] transanimation *f*
TRANSATLANTIC - [geogr] transatlantique
[naut] transatlantique *m*
~CABLE - [telecomm] câble *m* sousmarin transatlantique
TRANSCENDENTAL CURVE - [math] courbe *f* transcendante
~NUMBER - [math] numéro *m* transcendant
TRANSCEIVER - [radar] (transmitter-receiver) émetteur-récepteur *m*
TRANSCODER - [contr] transcodeur *m*
TRANSCONDUCTANCE - [electron] (or mutual conductance, transconductance between control grid and anode) pente *f*
[el] transconductance *f*
~METER - [instr] (mutual conductance meter) appareil *m* de mesure de pente
TRANSCRIBE, to - [gen] transcrire, copier
[radio] transmettre un programme enregistré
TRANSCRIBER - [comput] transcripteur *m*
TRANSCRIPTION - [gen] transcription *f*
[comput] transcription *f*
~MACHINE - [el acoust] dérouleur *m* son
TRANSCRYSTALLINE - [metall] (passing through the crystals) transcristalline
~FRACTURE - [metall] (the normal type of failure occurring in metals) cassure *f* transcristalline
TRANSCURIUM ELEMENTS - [chem] transcuriens *mpl*
TRANSDUCER - [radio] (device capable of being actuated by waves from a transmission system and of supplying related waves to one more other transmission system or media) transducteur *m*
~DISSIPATION LOSS - [radio] perte *f* de dissipation
~EQUIVALENT NOISE PRESSURE - [acoust] pression *f* de bruit équivalente
~GAIN - [acoust] gain *m* transductique
~LOSS - [el acoust] perte *f* de transducteur, affaiblissement *m* transductique
~PULSE DELAY - [radio] intervalle *m* entrée-sortie
TRANSDUCTOR - [el] transducteur *m* magnétique
~AMPLIFIER - [el acoust] amplificateur *m* magnétique
~ELEMENT - [el] élément *m* de transducteur
~REACTOR - [el acoust] transducteur -réactance *m*
~REGULATOR - [el acoust] régulateur *m* à transducteur
~RELAY - [el acoust] relais *m* à transducteur
TRANSEPT - [aech] (of a cruciform church) transept *m*
TRANSFER, to - [gen] transférer
[leg] (fin] céder, transmettre
[el acoust] (to transcribe) transcrire
[print] reporter
[comm] contre-passer
TRANSFER - [gen] transport *m*, translation *f*
[leg] (of shares etc) acte *m* de cession, transfert *m*, assignation *f*, virement *m*
[opt] (of images) transport *m*
[leg] (of property, stock etc) cession *f*
[draw] décalque *m*
[print] report *m* (sur la pierre)

TRANSFER ADDRESS - [comput] adresse *f* d'un ordre de transfert
~ADMITTANCE - [el] (the reciprocal of transfer impedance) transmittance *f* de transfert
~ATTENUATION CONSTANT - [el] atténuation *f* de transfert
~BAR - [el] barres *fpl* de transfert
~BOX - [el] boîte *f* dérivation
~BY HAND, to - [text] reproduire à la main
~CANAL - [nucl] (in a swimming-pool reactor) canal *m* de transfert
~CARD - [comput] carte *f* de transfert
~CASE - [mech] renvoi *m* à engrenages
~CHAMBER - [plast ind] (the chamber in transfer moulding in which the material is heated) chambre *f* de transfert
~CHAMBER RETAINER PLATE - [plast ind] plaque *f* contenant la chambre de transfert
~CHARACTERISTIC - [electron] caractéristique *f* de conversion (de tube à rayons cathodiques); caractéristique *f* interélectrode
~CHECK - [comput] (verification of transmitted information by temporary storing, retransmitting and comparing) contrôle *m* de double transfert
~CIRCUIT - [telecomm] (a circuit between two positions in an exchange, over which calls are extended) circuit *m* intermédiaire
~CONTACT - [electron] contact *m* inverseur
~CULL - [plast ind] (residual material in the transfer pot) culot *m*
~CURRENT - [electron] (of a cold cathode tube) courant *m* de déclenchement
~CURVE - [el acoust] (static characteristic of a transductor) caractéristique *f* de réglage d'un transducteur
~CYLINDER - [mech] cylindre *m* de transfert de pression
~DIE - [metall] moule *m* à dispositif de transfert
~ELECTRODES - [electron] (a series of supplementary cathodes placed between the true cathodes of a dekatron) électrodes *fpl* de transfert
~FUNCTION - [el] (transmittance) transmittance *f*, fonction *f* de transfert
~FUNCTION METER - [meas] transféromètre *m*
~HAMMER - [text] marteau-chargeur *m*
~IMPEDANCE - [el] (the ratio between the applied voltage at one part of a network and the current at another part of the same system) impédance *f* de transfert
~INK - [print] encre *f* lithographique
~INSTRUCTION - [comput] instruction *f* de transfert
~INTERPRETER - [comput] (interpreter making it possible for selected details from a set of punched cards to be reproduced in printed form on a separate set of unpunched cards) traductrice *f* de transfert
~JACK - [telecomm] jack *m* de renvoi, jack *m* de rupture, jack *m* de transfert
~KEY - [telecomm] clé *f* de renvoi
~LAG - [contr] retard *m* de transfert, retard *m* de transit
~LOCUS - [contr] lieu *m* géométrique des fonctions de transfert
~MACHINE - [mach tool] machine *f* à copier
~MECHANICAL IMPEDANCE - [el acoust] (complex quotient of the force applied at one point of a mechanical system and the resulting velocity at another point) impédance *f* mécanique de transfert

TRANSFER MOULD - [plast ind] (mould into which thermosets are forced in transfer moulding) moule *m* à transfert
~MOULDING - [plast ind] (process of moulding thermosets in which the softened material is forced into a closed hot mould) moulage *m* par transfert
~OF PROPERTY - [leg] translation *f* de propriété
~OPERATION - [comput] (operation designed to move from one store location to another) opération *f* de transfert
~PAPER - [draw] papier *m* à décalquer
[print] papier *m* à report
~PINION - [mech] pignon *m* satellite
~PLUNGER - [plast ind] (the ram which forces the material from the transfer chamber) piston *m* de transfert
~PLUNGER RETAINER PLATE - [plast ind] contre-plaque *f* de piston de transfert
~PORT - [mech] (in a two-stroke engine) canal *m* de transfert
~POT - [plast ind] pot *m* de transfert, chambre *f* de transfert
~PRINTING - [paint] (used in ceramics etc) décalque *m*
~PUMP - [mech] pompe *f* à transvaser
~SWITCH - [el] (a diverter switch) commutateur *m*, interrupteur *m* déviateur
~PLATFORM - [railw] quai *m* de transbordement
~THE DESIGN, to - [text] reproduire le dessin
~TIME - [electron] temps *m* de transfert
~UNIT - [ind chem] (in a chemical installation, one of the component parts in which the transfer of the treated substances occurs in a number of stages) unité *f* de transfert, élément *m* de transport
TRANSFERABLE - [comm & fin] transférable
TRANSFERASE - [biol] (an enzyme which brings about the transfer of a radical from one molecule to another) transférase *f*
TRANSFERENCE - [gen] transfèrement *m*
~ARRANGEMENT - [text] étaleur *m*
~NUMBER - s. transport number
TRANSFERMATICS - [mach tool] (USA only) machines *fpl* multiples en ligne
TRANSFERRED CHARGE CALL - [telecomm] conversation *f* payable à l'arrivée
TRANSFERRING BY MECHANICAL MEANS - [print] reproduction *f* mécanique du dessin
~BY PHOTOGRAPHIC METHODS - [print] mise *f* en carte photographique
~LEVER - [text] marteau-chargeur *m*
~MECHANISM - [mech] pignon *m* de commande
~ROLLERS - [text] cylindres *mpl* transporteurs
~ROPE - [text] câble *m* de transport
~THE SKETCH TO POINT PAPER - [text] reproduction *f* de l'esquisse sur papier quadrillé
TRANSFINITE - [math] transfini
TRANSFOCATOR - [opt] transfocateur *m*
TRANSFORM, to - [gen] transformer, convertir
~DOWN, to - [el] transformer en basse tension
~UP, to - transformer en haute tension
TRANSFORMATION - [gen] transformation *f*
[metall] (a constitutional change in a solid metal) transformation *f*
[chem] (a change from one phase to another) conversion *f*
[mech] conversion *f*
~CHAIN - [nucl] (a succession of nuclides, each of which transforms by radioactive disintegration into the next, until a stable nuclide results) famille *f* radioactive
TRANSFORMATION CIRCLE - [el] (in impedance matching) cercle *m* de transformation
~CONSTANT - [nucl] constante *f* de désintégration
~OF ELECTRICAL ENERGY - [el] transformation *f* de l'énergie électrique
~OF THE LAP INTO SLIVERS - [text] transformation *f* de la nappe en rubans
~POINTS - [metall] (critical points) points *mpl* d'arrêt
~RANGE - [metall] zone *f* critique
~RATIO - [el] (the ratio between the primary and the secondary voltages of a transformer on no load) rapport *m* de transformation
~SERIES - s. transformation chain
~TEMPERATURE - [metall] température *f* de transformation
TRANSFORMED PROGRAM - [comput] programme *m* transformé
TRANSFORMER - [el] (static apparatus for changing the voltage of an alternating current) transformateur *m*
~BANK - [el] (series of transformers grouped in line) groupe *m* de transformateurs en ligne
~BOX - [el] kiosque *m* de transformation
~CASE - [el] boîtier *m* du transformateur
~COIL - [el] bobine *f* du transformateur
~COMPOUND - [el] (material used for cooling, insulating and coating electrical equipment in contact with oil) huile *f* pour transformateurs
~CORE - [el] (the whole of the magnetic material forming the main magnetic circuit) noyau *m* du transformateur
~COUPLED CIRCUIT - [el] circuit *m* couplé par transformateur
~COUPLING - [el] (the transference in both directions of electrical energy, from one circuit to another by a transformer, of any degree of coupling, the primary being in one circuit, the secondary in the other) couplage *m* par transformateur
~ELECTROMOTIVE FORCE - [el] (in an alternating current commutator machine) force *f* électromotrice statique
~FOR REGULATING IN PHASE - [el] transformateur *m* variable en phase
~FOR REGULATING IN QUADRATURE - [el] transformateur *m* variable en quadrature
~KIOSK - [el] kiosque *m* de transformation
~LEG - [el] portion *f* du noyau
~LIMB - s. transformer leg
~LOSS - [el] (in communication practice) pertes *fpl* de transformateur
~OIL - [el] (a special type of oil of high dielectric strength, used as a cooling medium of electric power transformers) huile *f* pour transformateurs
~OVERCURRENT TRIPPING - [el] coupure *f* à courant de transformateur maximal
~PILLAR - s. transformator kiosk
~PLATE - [el] (sheet-iron of low magnetic loss used for transformer core laminations) tôle *f* pour transformateurs
~ROOM - [el] (in a factory, a building) cabine *f* de transformation
~SET - [el] groupe *m* transformateur
~SHEET - [metall] tôle *f* de transformateur, tôle *f*

dynamo, tôle *f* magnétique
TRANSFORMER SHELL - [el] cuirasse *f* du transformateur
~STAMPINGS - [el] (the laminations, stamped out of transformer plate, which are assembled to form the transformer core) paquet *m* de tôles
~STARTER - [el] démarreur *m* par auto-transformateur, démarreur *m* par transformateur
~STATION - [el] station *f* de transformation
~SUBSTATION - [el] sous-station *f* de transformation, poste *m* de transformation
~TANK - [el] (the steel tank encasing the core and windings of a transformer and holding the transformer oil) bac *m* du transformateur, cuve *f* du transformateur
~TAPE - [el] ruban *m* de transformateur
~TUBE - [el] (a steel tube on the outside of a transformer providing a vertical path of circulation for the transformer oil) tube *m* de circulation de l'huile du transformateur
~UNDERCURRENT TRIPPING - [el] coupure *f* à courant de transformateur minimal
~VAULT - [el] voûte *f* de transformateur
~WINDING - [el] (the electrically active part of a transformer, surrounding the magnetically active transformer core) enroulement *m* du transformateur
~WINDOW - [el] fenêtre *f* de transformateur
~WITH ADJUSTABLE LEAKAGE - [el] transformateur *m* à fuite ajustable
~WITH CONCENTRIC WINDINGS - [el] transformateur *m* à enroulements concentriques
~WITH EVAPORATIVE COOLING - [el] transformateur *m* refroidissement par liquide vaporisé
~WITH FORCED AIR COOLING - [el] transformateur *m* à refroidissement forcé par air
~WITH FORCED OIL CIRCULATION - [el] tranformateur *m* à circulation forcée de l'huile
~WITH SANDWICH WINDINGS - [el] transformateur *m* à bobines alternées
TRANSFORMING SECTION - [electron] (a length of waveguide of modified cross section) section *f* de adaptation d'impédance
TRANSFUSE, to - [gen & chem] (to transfer a liquid from one container to another) transfuser, transvaser
TRANSFUSION - [gen & ind chem] transfusion *f*, transvasement *m*
[med] (blood transfusion) transfusion *f* (de sang)
TRANSGRANULAR - [metall] transgranulaire
TRANSGRESSION - [leg] transgression *f*
[geol] discordance *f*, transgression *f*, envahissement *m* par la mer
TRANSHIP, to - [naut] transborder
TRANSHIPMENT - [comm & naut] transbordement *m*
TRANSHUMANCE - [agric] (of a herd) transhumance *f*, émigration *f* des animaux des plaines
TRANSIENT - [gen phys el] (term applied to short-time phenomena occurring in a system owing to sudden change of conditions) transitoire, phénomène *m* transitoire
~CONDITION - [el & radio] régime *m* transitoire
~CURRENT RATIO IN SATURATION - [electron] (of semi-conductors) rapport *m* de courant transitoire en saturation
~-DECAY CURRENT - [electron] (of a photo-electric device; the decreasing residual current occurring after the irradiation of the device has been abruptly cut off) traînage *m* d'un tube photoélectrique
TRANSIENT DISTORTION - [el] (delay distortion) distorsion *f* de retard
~EQUILIBRIUM - [nucl] (expression denoting that the rate of the amount of daughter to the amount of parent present is constant) équilibre *m* radioactif transitoire
~FAULT - [el] défaut *m* auto-extincteur
~MOTION - [phys] (the motion which has not reached or has ceased to be a steady state) mouvement *m* transitoire
~PHENOMENA - [phys] (phenomena appearing during the transition from one operating condition to another) phénomènes *mpl* transitoires
~POWER LIMIT - [el] limite *f* de stabilité momentanée
~RADIOACTIVE EQUILIBRIUM - [nucl] équilibre *m* radioactif transitoire
~RECOVERY VOLTAGE - [el] tension *f* transitoire de rupture
~RECOVERY VOLTAGE RATE - [el] courbe *f* de tension transitoire de rupture
~RESPONSE - [radio] (the evanescent part of the waveform response) réponse *f* du signal unité
~SINGLE-PHASE SHORT-CIRCUIT CURRENT - [el] courant *m* transitoire de court-circuit monophase
~STABILITY - [el] stabilité *f* transitoire
~STABILITY LIMIT - s. transient power limit
~STATE - [el] (a transition period, e.g. the variation of voltage when the load is suddenly taken off) état *m* transitoire
~THERMAL IMPEDANCE - [electron] (of semiconductors) impédance *f* thermique transitoire
~TIME-CONSTANT ON A GIVEN IMPEDANCE - [el] constante *f* de temps transitoire sur une impédance donnée
~TIME-CONSTANT ON OPEN CIRCUIT - [el] constante *f* de temps transitoire à circuit ouvert
TRANSIENTS - [electron] phénomènes *m* transitoires
TRANSILLUMINATION - [med] (the passing of a strong light through the walls of a cavity) transillumination *f*, diaphanoscopie *f*
TRANSILLUMINATOR - [instr] (apparatus for passing a strong light through the walls of a cavity) diaphanoscope *m*, transilluminateur *m*
TRANSISTOR - [electron] (a device in which a combination of semi-conductors and suitable contacts is made to perform some of the functions of thermionic valves) transistor *m*
~BATTERY - [electron] batterie *f* pour transistors
~CASE - [electron] (structure designed to contain and protect transistors) enveloppe *f* de transistor
~ELEMENT - [electron] (integral part of the semiconductor device contributing to its operation) élément *m* de transistor
~HEARING AID - [el acoust] appareil *m* de prothèse auditive à transistors, appareil *m* de correction auditive à transistors
~HOUSING - s. transistor case
~IGNITION SYSTEM - [electron] (fully transistorized ignition system) système *m* d'allumage transistorisé
~INPUT RESISTANCE - [electron] (the resistance presented by the input terminals of a transistor stage) résistance *f* d'entrée de transistor
~LEAD - [electron] (the supply conductor of a transistor) alimentation *f* de transistor

TRANSISTOR MUTUAL RESISTANCE - [electron] résistance *f* mutuelle de transistor
~OUTPUT RESISTANCE - [electron] (the resistance across the output terminals of a transistor stage) résistance *f* de sortie de transistor
~PENTODE - [electron] (a point-contact transistor with three emitters and one collector) transistor *m* pentode
~POWER GAIN - [electron] (in semiconductos) gain *m* de puissance de transistor
~RELAY - [el] relais *m* à transistor
~SOCKET - [electron] (socket designed for receiving transistors) douille *f* de transistor
~TETRODE - [electron] (four-electrode transistor, usually a conventional junction triode, having a second base electrode) transistor *m* tétrode
~TRANSIT TIME - [electron] (the time required for an injected carrier to diffuse across the barrier region) temps *m* de transit de transistor
~TRIODE - [electron] (a three-electrode transistor) transistor *m* triode
~VELOCIMETER - [instr] (transistorized instrument designed to measure the speed of sound in water) vélocimètre *m* transistorisé
TRANSISTORIZE, to - [electron] (to equip with transistors) transistoriser
TRANSISTORIZED - [electron] transistorisé
TRANSIT, to - [surv] (to rotate the telescope of a theodolite so that the positions of the ends of the telescope are reversed) faire une révolution complète, révolutionner complètement (un théodolite)
TRANSIT - [gen] (of a road etc) passage *m*
[transp] transport *m*
[astr] (the apparent passage of a heavenly body across the meridian of a place due to the earth's diurnal revolution) passage *m*, culmination *f*
[astr] (the passage of a smaller body across the disk of a larger body, as seen by an observer on the earth) passage *m* d'une planète (sur le disque d'un astre au méridien)
[instr] (a transit theodolite) théodolite *m* à lunette centrale
~BOOK - [surv] carnet *m* de levé au théodolite
~CIRCLE - [astr] cercle *m* méridien
~COMPASS - [instr] théodolite *m* à boussole
~DUTY - [comm] droit *m* de transit
~EXCHANGE - [telecomm] bureau *m* de transit, bureau *m* tandem
~GAUGE - [railw] (gauge for transit vehicles) gabarit *m* de transit, gabarit *m* pour véhicules de transit
~INSTRUMENT - [instr] (a telescope capable of being completely rotated about its horizontal axis) cercle *m* d'alignement, lunette *f* méridienne
~PHASE ANGLE - [electron] (the product of the transit time and the angular frequency when the current resulting from a flow of electrons is sinusoidal) angle *m* de transit
~RATE - [telecomm] (in telegraphy services) taux *m* de transit
~TIME - [phys] (the time taken by a charged particle in moving between two specified points) durée *f* de parcours, temps *m* de transit
[electron] (the time taken by an electron to move from the cathode to the anode) temps *m* de transit
[astr] (of a star across the meridian) temps *m* de passage
~-TIME CORRECTION - [telev] correction *f* des temps de transit
TRANSIT-TIME DAMPING - [telecomm] amortissement *m* dû au temps de transit
~-TIME EFFECT - [electron] (the acting of electrons as if they were out of phase with the grid signal, resulting in a much increased grid-cathode conductance) effet *m* de temps de transit
~-TIME MODE - [electron] (a condition of operation of an oscillator) mode *m* du temps de transit
~-TIME TUBE - [electron] tube *m* à modulation de vitesse
~-TIME VALVE - s. transit-time tube
TRANSITION - [gen] transition *f*, passage *m*
[cin] (between two sequences) passage *m*
[el] (the changing of the connections of electrical railway motors) transition *f*, transfert *m*
[mus] (modulation of transient value) transition *f*
[electron] (the process whereby a quantum mechanical system changes from one energy eigenstate to another) transition *f*
[hydr] conduite *f* de transition
~ANODE - [el] anode *f* de transfert
~APPARATUS - [el] transitionneur *m*
~COIL - [el] inductance *f* de passage
~CURVE - [surv] (curve of special form connecting a straight and a circular arc) courbe *f* de raccordement
~DISCHARGE PATH - [el] trajet *m* de décharge de transfert
~EFFECT - [electron] effet *m* de transition
~ELEMENT - [electron] (wave converter) organe *m* de raccord
~FACTOR - [electron] (reflection coefficient) coefficient *m* de réflexion
~FIT - [mech] (having a clearance or interference not exceeding tolerance limits) montage *m* transitoire
~FREQUENCY - [electron] (in semiconductors) fréquence *f* de transition
~LENGTH - [surv] (the length of a transition curve) longueur *m* de la courbe de raccordement
~LOSS - [electron] (of waveguides) perte *f* de transition
~METAL - [phys] (a member of one of several groups of elements in the periodic table having an incomplete inner shell) élément *m* de transition
~MULTIPOLE MOMENTS - [nucl] moments *mpl* multipôles de transition
~POINT - [phys] (the point of change from laminar to turbulent flow in a boundary layer) point *m* de transition, température *f* de transition
[el] (any point in a transmission system at which there is a change in the circuit constants) point *m* de transition
~PROBABILITY - [nucl] (the probability per unit that a system in state 'i' will undergo a transition to state 'f') probabilité *f* de transition
~RECTIFIER - [el] soupape *f* de transfert
~REGION - [electron] (the region between two homogeneous semiconductors, in which the impurity concentration changes) zone *f* de transition
~RESISTANCE - [el] résistance *f* de passage
~ROCK - [geol] roches *fpl* de transition
~SHOT - [cin] (shot uniting two different parts of a film) prise *f* de vue transitoire
~STRATA - [geol] terrains *mpl* de transition
~TEMPERATURE - [chem] (the temperature at which

both forms of a polymorphous substance can exist) température *f* de transition
TRANSITIONAL - [gen] transitionnel
TRANSITORY - [gen] transitoire, passager
TRANSITRON - [electron] (transitron oscillator; a negative-transconductance oscillator employing a screengrid valve with negative transconductance produced by a retarding field between the negative screen and the control grid serving as the anode) oscillateur *m* transitron
~CIRCUIT - [electron] circuit *m* transitron
TRANSLATABLE - [gen] traduisible
TRANSLATE, to - [gen] (from one language into another) traduire
[gen] (to convert) métamorphoser
[mech] imprimer un mouvement de translation
[telecomm] (in automatic telephony) répéter, retransmettre
TRANSLATING ROUTINE - [comput] programme *m* de traduction
TRANSLATION - [gen] traduction *f*
[mech] (motion in which all the parts follow the same direction) translation *f*, mouvement *m* de translation
[telecomm] (in automatic telephony; the alteration of the number and composition of the last two coded trains of impulses which are dialled by a subscriber and represent the wanted exchange) correction *f*
~FIELD - [telecomm] (in telegraphy) couronne *f* de contacts, arc *m* de broches
~GRATING - [phys] (a diffraction grating ruled on a transparent base) réseau *m* de diffraction transparente
~GROUP - [phys] (in crystallography) groupe *m* de translation
~LOSS - [el acoust] (the loss in the reproduction of a mechanical recording, in which the amplitude of motion of the reproducing stylus differs from the recorded amplitude in the medium) erreur *f* de piste
~OF AXIS - [math] transformation *f* de coordonnées
~OPERATION - [phys] (the geometrical process of displacing a body along a straight line, keeping lines fixed in the body always parallel to themselves) mouvement *m* de translation
~MOTION - [nucl] (a motion in which all particles move with the same velocity and acceleration in parallel paths) mouvement *m* de translation
~MOVEMENT - s. translation motion
~ROLLER - [photo] galet *m* de translation
TRANSLATIONAL ENERGY - [phys] énergie *f* de translation
TRANSLATOR - [comput] programme *m* de traduction
~KEY - [telecomm] clé *f* de translation
TRANSLATORY MOTION - [soil] mouvement *m* de translation
TRANSLUCENCE - [opt] translucidité *f*, diaphanéité *f*
TRANSLUCENT - [phys] (permitting the diffused transmission of light) translucide, transparent
~GLASS - [glass ind] verre *m* transparent
TRANSLUNAR - [astr] (situated beyond the moon) translunaire
~FLIGHT - [astronaut] vol *m* translunaire
TRANSMIGRATION - [med] transmigration *f*, diapédèse *f*, égarement *m*
TRANSMISSION - [gen] transmission *f*
TRANSMISSION - [auto] (the propeller shaft) système *m* d'arbres de transmission
[auto] (the gearbox) boîte *f* de vitesse
[auto] (gearbox and propeller shaft) transmission *f* et boîte de vitesse
[telecomm] transmission *f*, émission *f*
[mech] (of a belt etc) entraînement *m*
[el] transport *m*
~AND TRANSFER CASE - [auto] boîte *f* de transfert
~BAND - [electron] (of a uniconductor waveguide; the frequency range above the cut-off frequency) bande *f* de transmission
[electron] (of a filter circuit) bande *f* passante
~BELT - [mech] courroie *f* d'entraînement
~BRAKE - [auto] frein *m* sur la transmission
~BY LEVERS - [mech] transmission *f* par leviers
~CASE - [auto] carter *m* de boîte de vitesse
~CHANNEL - [telecomm] voie *f* de transmission
~CODE - [telecomm] code *m* de transfert
~COEFFICIENT - [electron] (of a waveguide) coefficient *m* de transmission
[nucl] (penetration probability) facteur *m* de pénétration, probabilité *f* de pénétration
~CURVE - [phys] courbe *f* d'absorption, courbe *f* d'atténuation
~DIAGRAM - [radio] (of an aerial) diagramme *m* de transmission
~DRIVE - [mech] commande *f* de transmission
~EXPERIMENT - [radiat] expérience *f* de transmission
~FAILURE - [telecomm] dérangement *m* de la transmission
~FEEDER - [el] feeder *m* de transmission
~FREQUENCY METER - [instr] filtre *m* de fréquence à transmission
~GAIN - [telecomm] coefficient *m* d'amplification, gain *m* de transmission
~GEAR - [mech] (mechanical arrangement for transferring power from one machine or assembly to another) organes *mpl* de transmission, engrenage *m* de transmission
~GREASE - [auto] graisse *f* pour engrenages
~HOUSING - s. transmission case
~IMPAIRMENT - [telecomm] réduction *f* de qualité de transmission
~LEVEL - [telecomm] (the power in a transmission circuit) niveau *m* de transmission
~LINE - [el] (the overhead conductor system by which electric power is transmitted at high voltage from one point to another) ligne *f* de transport d'énergie
[radio] (a material structure forming a continuous path from one point to another, so as to direct the transmission of electromagnetic energy along path) ligne *f* de transmission
~-LINE AMPLIFIER - [radio] amplificateur *m* distribué
~LINE CONTROL - [radio] (the control of the frequency of an oscillator by a resonant line) contrôle *m* à résonance
~LINE TOWER - [el] pylône *m* de ligne de transport d'énergie
~LOSS - [radio] (the power lost in transmission between one point and another) perte *f* de transmission
~MAIN - [gas ind] (a feeder main, a pipeline) canalisation *f* de transport, artère *f*, pipeline *m*

TRANSMISSION-MEASURING SET - [telecomm] (apparatus essentially consisting of a sending circuit and a level-measuring set) hypsographe *m*
~MODE - [electron] (a mode of propagation along a transmission line) mode *m* de transmission
~MONITOR - [electron] (master monitor in the USA final monitor controlling the image which is actually broadcast) moniteur *m* d'émission, moniteur *m* final
~NETWORK - [el] réseau *m* de transmission
~OF CRANK MOTION - [mech] transmission *f* du mouvement de la manivelle
~OF LOOM VIBRATION TO THE MACHINE - [text] transmission *f* de la vibration du métier à la mécanique
~OF POWER - [mech] transmission *f* de force (du moteur), transport *m* de force
~OF TELEGRAPHIC PICTURES - [telecomm]phototélégraphie *f*, radiotransmission *f* des images
~PERFORMANCE - [radio & telecomm] qualité *f* de transmission
~PERFORMANCE RATING - [radio & telecomm] indice *m* de qualité de transmission
~PLANE - [opt] (the plane of vibration of polarized light which will pass through a specified polarizer) plan *m* de transmission
~REGULATOR - [telecomm] (a device designed to maintain constant transmission over a transmission system) régulateur *m* de transmission
~ROD - [mech] barre *f* de renvoi
~ROUTE - [el] trajet *m* de transmission
~SHAFT - [auto] arbre *m* de transmission
~SPEED - [telecomm] vitesse *f* de transmission
~SYSTEM - [auto] système *m* de transmission
[el] canalisation *f*
~TARGET - [electron] (a target so arranged that the useful X-ray beam emerges from the surface remote from that on which the beam is incident) cible *f* à transmission
~TIME - [radio & telecomm] (the absolute time interval from transmission to reception of a signal) temps *m* de propagation
~TOWER - s. transmission line tower
~TROUBLE - [radio & telecomm] mauvaise audition *f*, difficulté *f* d'audition
~UNIT - [acoust] (obsolete for decibel) décibel *m*
TRANSMISSIVITY - [opt] (the ration between the emitted radiation and the normally-incident radiation, as radiation is passed through boundary between two media) nombre *m* de transmission
TRANSMISSOMETER - [instr] (a photoelectric visibility meter) transmissomètre *m*
TRANSMIT, to - [gen] transmettre
[radio] (to broadcast) émettre, transmettre
[mech] imprimer (un mouvement)
[el] transporter (la force)
~-RECEIVE TUBE - [electron] (gas-filled radiofrequency switching tube designed to protect the receiver in pulsed radiofrequency systems) tube *m* émission-réception
TRANSMITTANCE - [opt] (the ratio between the light flux transmitted by the medium and the light flux incident on it) transparence *f*
TRANSMITTANCY - [opt] (the ratio between the transmittance of a solution and that of the pure solvent in equivalent thickness) relation *f* de transmission interne

TRANSMITTED LIGHT - [opt] (light which has travelled through a medium without being absorbed) lumière *f* transmise
~WAVE - [electron] (that portion of a travelling wave which travels away from a transition point on the remote side) onde *f* transmise
TRANSMITTER - [telecomm] (any sending instrument) émetteur, installation *f* de translation, répétiteur *m*
[radio] (the complete assemblage of apparatus necessary for the production and modulation of radio frequency current) émetteur *m*
[telecomm] (in telephony) transmetteur *m*, microphone *m*
[telecomm] (in telegraphy) translateur *m*
[instr] (device for transmitting to the indicator) transmetteur *m*
[electron] (to convert sound waves into electrical oscillations) transducteur *m* acoustique
~AMPLIFIER - [radio] (in the transmitter circuit) amplificateur *m* d'émission
[telecomm] amplificateur *m* de l'émetteur
~INPUT POLARITY - [telev] polarité *f* du signal d'image
~NOISE - [radio] bruit *m* de microphone
~-RECEIVER - [radio] (apparatus designed for reception and transmission of radio signals, some parts of the circuits being common to both units) émetteur-récepteur *m*
~UNIT - [el acoust] capsule *f* de microphone
TRANSMITTING - [gen] transmetteur, émetteur, de transmission
~AERIAL - [radio] (aerial radiating electrical oscillations) antenne *f* d'émission
~ANTENNA - s. transmitting aerial
~CURRENT RESPONSE - [el acoust] (of an electroacoustic transducer used for sound emission) réponse *f* en transmission de courant
~DIRECTION - [telecomm] sens *m* du passage
~EFFICIENCY - [el acoust] rendement *m* de transducteur
~FILTER - [radio] filtre *m* passe-bande d'émission
~KEY - [telecomm] manipulateur *m*
~LOOP LOSS - [radio] pertes *fpl* à la transmission
~POWER RESPONSE - [el acoust] (of an electro-acoustic transducer used for sound emission) caractéristique *f* de fréquence à puissance constante, taux *m* de transmission
~SET - [radio] appareil *m* de radio-émission
~TUBE - s. transmitting valve
~VALVE - [electron] (electron valve used in broadcasting) tube *m* d'émission
~-VALVE COOLER - [electron] réfrigérateur *m* pour tubes
~VOLTAGE RESPONSE - [el acoust] (of an electro-acoustic transducer used for sound emission) réponse *f* en transmission de tension
TRANSMOUNTING - [auto] à support transversal
TRANSMUTATION - [phys] (the changing of one element into another. This may occur spontaneously or may be provoked by artificial means) transmutation *f*
~CONSTANT - [nucl] constante *f* radioactive
~EQUATION - [nucl] équation *f* de transmutation
TRANSMUTED ELEMENT - [chem] élément *m* transmuté
TRANSOM - [arch] traverse *f*, sommier *m*, linteau *m*
[naut] arcasse *f*

TRANSOM - [aero] entretoise *f*
[railw] (a crossbeam) traverse *f*
[mech] (of a vehicle) épart *m*, lisoir *m*
[constr] meneau *m* horizontal, tendière *f*
~BAR - [shipbuild] (a timber across the sternpost) barre *f* d'arcasse
~FRAME - [shipbuild] (framing supporting the projecting stern) arcasse *f*
~KNEE - [shipbuild] contrefort *m* d'arcasse
~PLATE - [shipbuild] tôle *f* d'arcasse
~WINDOW - [constr] fenêtre *f* à meneau horizontal, imposte *f* de porte
TRANSONIC - [aero] (said of a speed in which subsonic and supersonic speeds are both present) transonique
~FLOW - [aero] (flow in which regions of subsonic and supersonic speed are both present) écoulement *m* transonique
~RANGE - [aero] (the range of airspeed in which both super-and subsonic speeds of flow exist around the body in question) domaine *m* transonique
~TUNNEL - [aero] (transonic wind tunnel) soufflerie *f* transonique
TRANSPARENCE - [phys] transparence *f*
TRANSPARENCY - [phys] (offering no transmission of light or other electromagnetic vibrations) transparence *f*
[photo] (colour transparency) diapositif *m*, diapositive *f*
TRANSPARENT - [phys] transparent, diaphane
~ASBESTOS - [min] amiante *m* transparent
~COCOON - [text] cocon *m* transparent
~GAS STREAM - [oil ind] courant *m* de gaz transparent
~PACKING - [comm] emballage *m* transparent
~PARCHMENT - [paper man] papier *m* parchemin transparent
~SILK FIBRE - [text] fibre *f* de soie transparente
TRANSPIRATION - [aero] (the flow of gas along comparatively long passages, the flow being determined by the pressure difference and the viscosity of the gas) transpiration *f*
~COOLING - [astronaut] refroidissement *m* par sudation
TRANSPLANT, to - [agric] transplanter, déplanter
[med] (in surgery) transplanter, greffer (un tissu)
~THE SHEEP, to - [agric] transplanter le mouton
TRANSPLANTING - [agric] transplantation *f*
~MACHINE - [agric] repiqueuse *f* de plantes
TRANSPLANTS - [agric] plantes *fpl* repiquées
TRANSPONDER - [radar] (or racon, responder bacon; device emitting a signal when it is hit by the radiation of a radar transmitter) répondeur *m*
~BEACON - [radio] balise *f* répondeuse
TRANSPORT, to - [gen] transporter
TRANSPORT - [gen] transport *m*
[aero] transport *m* d'aviation
[naut] bâtiment *m* de transport
[nucl] transport *m* d'isotopes
~CHARGES - [comm] frais *mpl* de transport
~CROSS-SECTION - [nucl] (the reciprocal of transport mean path) section *f* efficace de transport
~EQUATION - [nucl] équation *f* de transport
~KERNEL - [nucl] noyau *m* intégral de transport
~LOCK - [comput] blocage *m* du transport
~MEAN FREE PATH - [nucl] (a modified mean free path used to correct for the persistence of velocities and anisotropy of scattering; also, when Fick's law is applicable, three times the diffusion coefficient of neutron flux) libre parcours *m* moyen de transport
TRANSPORT FACTOR - [electron] (of a point contact or junction transistor) taux *m* de transit
~MECHANISM - [el acoust] (of a tape recorder) platine *m* des moteurs
~NUMBER - [phys] (a number expressing the relatio between the quantity of electricity transported by the ions in question and the total quantity carried b all the ions of the electrolyte considered, during th same period) nombre *m* de transport des ions
~PLANE - [aero] avion *m* de transport
~SHAFT - [photo] rouleau *m* transporteur
~SHIP - [naut] bâtiment *m* de transport
~SPROCKET - [mech] tambour *m* entraînement
~THEORY - [nucl](a theory based on an approximation the Boltzmann equation for those conditions for which Fick's law is not applicable) théorie *f* du transport
~UNIT - [phys] unité *f* de transfert, élément *m* de transport
~WAGON - [railw] voiture *f* de transport
TRANSPORTABLE - [gen] transportable
~REACTOR - [nucl] réacteur *m* transportable
~SUBSTATION - [telecomm] sous-station *f* transportable
TRANSMITTER - [radio] émetteur *m* transportable
TRANSPORTATION - [gen] transport *m*, transportatio
[gen] (USA only) feuille *f* de route
~COSTS - [comm] frais *mpl* de transport
~FACILITIES - [transp] facilités *fpl* de transport
TRANSPORTER - [transp] entrepreneur *m* de transport
[mech] conveyeur *m*
~BRIDGE - [mech] pont *m* transporteur
TRANSPOSE, to - [gen] transposer
TRANSPOSE THE INTERLACING POINTS, to - [text] changer les points de croisure
TRANSPOSED SPOTS - [text] mouchetures *fpl* contresemplées
~TRANSMISSION LINE - [telecomm] ligne *f* croisée
TRANSPOSING POINT - [text] point *m* de transposition
TRANSPOSITION - [gen] transposition *f*
[el] (a spiralling of the line conductors of an overhead line in order to reduce the effects of asymmetry) rotation *f*, transposition *f*
[telecomm] (in telephony; interchange of circuit wires at intervals, in order to reduce crosstalk) croisement *m*, transposition *f*
[math] permutation *f*
[photo] transposition *f*
~BRACKET - [el] console *f* à bride, traverse *f* en porte-à-faux
~CYCLE - [el] pas *m* de transposition
~INSULATOR - [el] isolateur *m* de transposition
~OF PAIRS - [telecomm] croisement *m* de lacets
~OF WIRES - [telecomm] rotation *f*, croisement *m* des fils
~POLE - [el] appui *m* de croisement
~SECTION - [el] section *f* d'antiinduction complète de deux lignes l'une par rapport à l'autre
TRANSRECTIFICATION - [electron] (rectification occurring on one circuit when an alternating voltage is applied to another circuit) redressement *m*
~FACTOR - [electron] pourcent *m* de redressement

TRANSRECTIFIER - [electron] redresseur *m*
TRANSTAGE - [astronaut] étage *m* de changement de orbite
TRANSTRICTOR - [electron] (or field effect transistor; a transistor consisting in principle of a very small parallelepiped of semiconducting material) transistor *m* à effet de champ
TRANSUDE, to - [phys] (to pass through tissues or pores) transsuder
TRANSURANIC - [chem] transurien
~ELEMENTS - [nucl] (elements having atomic numbers higher than that of uranium) éléments *mpl* transuriens
TRANSVASATE - [chem] (the liquid decanted from a vessel) liquide *m* de transvasement
TRANSVERSAL - [gen] transversale, transversale *f*
~BLOCK PROFILE - [rubber ind] (for tyres) sculpture *f* à blocs transversaux
~EXTRUDER HEAD - [rubber & plast ind] tête *f* de équerre de boudineuse
~GROOVE PROFILE - [rubber ind] (for tyres) sculpture *f* à rainures transversales
~OR CROSS SECTION - [constr] section *f* transversale
~RIB - [rubber ind] (for tyres) nervure *f* transversale
TRANSVERSE - [gen] (lying across, being across) transversal, en travers
~AERIAL FUZE - [radio] (an electronic fuze in which the aerial is set at right angles to the longitudinal axis of the missile) fusée *f* transversale
~ANTENNA FUZE - s. transverse aerial fuze
~ARCHITRAVE - [constr] traverse *f*
~-BEAM TRAVELLING-WAVE TUBE - [electron] tube *m* à ondes progressives à faisceau transversal
~BRACING - [constr] entretoisement *m*
~CABLE - [el] câble *m* transversal
~CROSSTALK COUPLING - [telecomm] couplage *m* diaphonique
~DIFFERENTIAL PROTECTION - [el] protection *f* différentielle transversale
~DRAINING - [agric] drainage *m* transversal
~ELASTIC LIMIT - [phys] limite *f* d'élasticité transversale
~ELECTRIC AND MAGNETIC MODE - [electron] (or TEM mode, of waveguides) mode *m* transversal électrique et magnétique, mode *m* TEH
~ELECTRIC MODE - [electron] (or H mode, or TE mode; in a waveguide) mode *m* transversal électrique, mode *m* H, mode *m* TE
~ELECTRIC WAVE - [el] (in a homogeneous isotropic medium, an electromagnetic wave in which the electric field vector is everywhere perpendicular to the direction or propagation) onde *f* transversale électrique, onde *f* H, onde *f* TE
~ELECTROMAGNETIC WAVE - [el] onde *f* transversale électromagnétique, onde *f* TEM, onde *f* TEH
~FIBRE FEED - [text] alimentation *f* en travers
~FIELD - [el] (electric field not containing a magnetic-field component in the direction of propagation of electromagnetic waves) champ *m* transversal
~FIELD TRAVELLING-WAVE TUBE - [electron] tube *m* à ondes progressives à champ transversal
~FILM ATTENUATOR - [electron] (attenuator consisting of a conducting film placed transverse to the axis of a waveguide) affaiblisseur *m* transversal à couche conductrice

TRANSVERSE FINNING - [nucl] (a method of arrangi cooling fins) ailettes *fpl* transversales
~-FOCUSSING ELECTRIC FIELD - [electron] (a focussing electric field not containing a magnetic field component in the direction of propagation of particles) champ *m* électrique focalisateur transve sal
~FUZE - s. transverse aerial fuze
~MAGNETIC FIELD - [phys] champ *m* magnétique transversal
~MAGNETIC MODE - [electron] (or TM mode; a mode in which the longitudinal component of the magnetic field is everywhere zero and the longitudinal component of the electric field is not) mode *m* transversal magnétique, mode *m* TH, mode *m* TM
~MAGNETIC WAVE - [electron] (of E wave, or TM wave; of a waveguide) onde *f* transversale magnétique, onde *f* TH, onde *f* TM
~MAGNETIZATION - [el acoust] (magnetization of the recording medium in a direction perpendicular to the line of travel and parallel to the greater cross-sectional dimension) magnétisation *f* trans\ sale
~MEMBER - [auto] traverse *f*
~MOVEMENT OF CARRIAGE - [mach tool] course *f* transversale du chariot
~MOVING NEEDLE BOARD - [text] planche *f* à aiguilles double
~PHONON - [acoust & crystall] (longitudinal phonor phonon *m* longitudinal
~PITCH - [mech] pas *m* transversal
~PLATE - [electron] (of a wave-guide) cloison *f* transversale
~PRESSURE ANGLE - [mech] (of a bevel gear) angle *m* des pressions
~REINFORCING ROD - [constr] (used in concrete) ferrure *f* en fourchette
~RIB - [aero] nervure *f* de renfort
~SCAN - [telev] analyse *f* transversale
~SECTION - [mech] coupe *f* en travers, profil *m* en travers
~SEPTUM - s. transverse plate
~SPRING - [mech] ressort *m* transversal
~STRENGTH - [phys] résistance *f* à la flexion
~STRESS - [phys] charge *f* transversale
~STRIP - [text] loquettes *fpl* transversales
~STRIPE PATTERN - [text] dessin *m* à rayures transv sales
~TEST - [metall] (a test in which the test piece long tudinal axis is perpendicular to the direction of rol ling) essai *m* à charge transversale
~THRUST - [geol] décrochement *m* transversal, faille *f* transversale
~TRAVERSE - [mach tool] course *f* transversale
~WAVE - [electron] (or longitudinal wave; a wave characterized by a vector at right angle with the direction of propagation) onde *f* transversale
~WIRE - s. transverse cable
TRANSVERSECTOMY - [med] résection *f* de l'apophyse transverse
TRANSVERSELY-FINNED FUEL ELEMENT - [nucl] élément *m* combustible à ailettes transversales
TRANSVERSOSTOMY - [med] colostomie *f* du transverse
TRAP, to - [gen] prendre au piège
[mech] purger (la vapeur)
[plumb] (to set a hydraulic trap) mettre un siphor

disposer un siphon
TRAP, to - [ind chem] (to separate by traps, e.g. water from compressed air) séparer
TRAP - [gen] piège *m*, attrape *f*, trappe *f*
[ind chem] (of a diffusion pump) bouchon *m*
[plumb] (U-tube or chamber in a pipe-line, which always retains a quantity of the liquid passing and so prevents the passage of low-pressure gas, e.g. from a sewer) siphon *m*, pot *m* de purge
[transp] (a vehicle) charrette *f* anglaise
[geol](trappean rock) trapp *m*, roche *f* trappéenne
[electron] centre *m* de capture
[astronaut] grille *f*
[mining] porte *f* d'aérage
[geol] dislocation *f*
~-BAND - [bot] ceinture-piège *f*, ceinture *f* gluante
~-CELLAR - [constr] (of a theatre) dessous *mpl*
~-CROP - [bot] plante-piège *f*
~DOOR - [constr] abattant *m*, trappe *f*
[mining] porte *f* d'aérage
[agric] abat-foin *m*
~-NEST - [agric] nid-trappe *m*, pondoir-trappe *m*
~POINTS - [railw] aiguille *f* d'évitement
~ROCK - [geol] roche *f* verte
~SIDING - [railw] voie *f* de sûreté
TRAPEZE - [gen] trapèze *m*
~BAR - [aero] (of an airship) trapèze *m*
TRAPEZIFORM GROOVE - [text] rainure *f* trapéziforme
TRAPEZIUM - [gen] trapèze *m*
[anat] os *m* trapèze, os *m* trapézoïde
~DISTORTION - [electron] (in a cathode-ray tube) distorsion *f* en trapèze
~-SHAPED ROPE - [text] corde *f* de section trapézoïdale
TRAPEZOID - [math] quadrilatère *m* irrégulier
~BONE - [anat] os *m* trapézoïde
TRAPEZOIDAL - [gen] trapézoïdal
~DEFLECTION - [telev] déviation *f* trapézoïdale
~LOAD - [el] charge *f* à diagramme trapézoïdal
~RULE - [surv] planimètre *m* à réseau
~THREAD - [mech] filet *m* trapézoïdal
TRAPPED MODE - [electron] (of electromagnetic waves) mode *m* de propagation guidée
~ORBIT - [astronaut] orbite *f* d'un corps tournant toujours la même face vers le corps central
TRAPPING - [gen] piégeage *m*
[electron] (a process in which the electrons are held at an irregularity in the crystal lattice of a semiconductor until released by thermal agitation) piégeage *m*
~SPOT - [electron] (capture spot) zone *f* de capture
TRAPROCK - s. trap rock
TRASH - [agric] émondes *fpl*, bagasse *f*
~BOARD - [text] lit *m* de la navette, chemin *m* de la navette
~RACK - [impl] grille *f* grossière
TRASS - [geol] (material similar to pozzuolana) trass *m*, pouzzolane *f* en pierre
~CEMENT - [constr] ciment *m* de trass
~MORTAR - [constr] mortier *m* de trass
TRAUMA - [med] (any injury to the body or mind, caused by a shock) trauma *m*
TRAUMATIC - [med] traumatique
TRAUMATOPYRA - [med] fièvre *f* traumatique
TRAUZL TEST - [chem] (a technique for determining the power of an explosion. It consists of denoting a specific quantity of explosive in a cavity in a lead block of standard size and measuring the increase in volume of the cavity) essai *m* de Trauzl
TRAVAIL - [med] douleurs *fpl* de l'enfantement, enfantement *m*
TRAVEL, to - [gen] voyager
[mech] se déplacer, se mouvoir
TRAVEL - [gen] voyages *mpl*
[el] parcours *m*, trajectoire *f*
[mach tool] (term used for the distance through which a part of a machine moves in normal action, e.g. piston travel) course *f*
~GHOST - [cin] (a film defect) filage *m*
~OF KNIVES - [text] lève *f* des couteaux
~OF THE OIL - [oil ind] migration *f* du pétrole
~OF THE WARP - [text] marche *f* de la chaîne
~-REVERSING SWITCH - [el] commutateur *m* de renversement de marche
~SHOT - [cin] travelling *m*, plan *m* pris en mouvement
TRAVELLER - [gen] voyageur *m*
[comm] (a commercial traveller) commis *m* voyageur, représentant *m*
[text] (for cotton spinning) curseur *m* à filer
[naut] guide *m* de drisse
TRAVELLING - [gen] ambulant
[mech] roulant, volant
~APRON - [mech] tablier *m* transporteur, toile *f* transporteuse
~BAND CONVEYOR - [mech] bande *f* souple de transport, courroie *f* transporteuse, transporteur *m* mobile à bande
~BARREL - [mech] corps *m* de pompe mobile
~BELT SCREEN - [impl] tamis *m* roulant, tamis *m* à paniers
~BOARD - [cin] chariot *m*, plancher *m* roulant
~BOBBINS - [text] bobines *fpl* à mouvement tournant
~CABLE - [el] (flexible cable providing electrical connection between the lift-car and a fixed point) câble *m* souple de connexion
~CARRIAGE - [mech] (of a crane) chariot *m* roulant
~CARRIAGE SCAVENGER - [text] nettoyeur *m* automatique de chariot
~CLOTH CLEARER - [text] panne *f* sans fin nettoyant les cylindres
~CRAB - [mech] treuil *m* roulant
~CRADLE - [constr] échafaudage *m* volant
~CRANE - [mech] pont *m* roulant, grue *f* roulante
~CREEL - [text] râtelier *m* mobile
~DETECTOR - [electron] (a device for measuring electric field intensity in a waveguide as a function of distance along the guide) sonde *f* mobile
~DISTRIBUTOR - [mech] pulvérisateur *m* à va-et-vient, chariot *m* baladeur
~EXPENSES - [comm] frais *mpl* de route
~FLOOR LATTICE - [text] tablier *m* sans fin
~FRAME - [text] bâti *m* déplaçable
~GANTRY - [mech] (gantry of the platform gantry type but with a movable carriage on rails in place of the platform) portique *m* roulant
~-GANTRY CRANE - [mech] grue *f* roulante à portique
~GRATE - [th eng] grille *f* mobile
~-GRATE STOKER - [th eng] grille *f* sans fin
~HYDRAULIC RIVETTING PLANT - [mech] outillage *m* hydraulique à river monté sur roues
~JIB CRANE - [mech] grue *f* à volée

TRAVELLING LADDERWAY - [mech] échelle *f* mécanique
~LIFT - [mech] (used for loading and unloading shelves) élévateur-transporteur *m*
~MASKS - s. travelling matts
~MATTS - [cin] (film rolls used as light modulators in printing machines for obtaining transitional effects) bandes *fpl* d'essai de modulation
~OVEN - [metall] (a conveyor furnace) four *m* à passage
~OVERVOLTAGE - [el] surtension *f* mobile
~PADDLE MIXER - [mech] (paddle mixer assembly mounted to travel on rails above the mixing tank in which it operates) agitateur *m* mobile à pales
~-PLANE WAVE - [electron] onde *f* progressive plane
~PLATFORM - [railw] (traverse table) transbordeur *m* pour wagons, plateforme *f* roulante
~PLATFORM RUNNERS - [railw] chariot *m* de pont roulant
~RACK ON CARRIAGE - [text] crémaillère *f* voyageuse sur le chariot
~ROPE - [petr] câble *m* de transmission
~ROPE-HAULED DISTRIBUTOR - [mech] arroseur *m* à va-et-vient actionné par câble
~RUG - [text] couverture *f* de voyage
~STEEL TAPE - [text] lanière *f* mobile en acier
~VALVE - [oil ind] soupape *f* à piston
~-WAVE AERIAL - [radio] (progressive-wave aerial) antenne *f* à ondes progressives
~-WAVE ANTENNA - s. travelling-wave aerial
~-WAVE INTERACTION - [electron] interaction *f* d'onde progressive et faisceau d'électrons
~-WAVE MAGNETRON - [electron] magnétron *m* à ondes progressives
~-WAVE MASER - [electron] (a maser, i.e. microwave amplification by simulated emission of radiation, in which the interaction between paramagnetic material and radiation occurs in a non-resonant travelling wave-structure) maser *m* à ondes progressives
~-WAVE TUBE - [electron] tube *m* à ondes progressives
~-WAVE TUBE INTERACTION CIRCUIT - [electron] tube *m* à ondes progressives à retardement de champ
~YARN SKIP CARRIER - [text] panier *m* à fil roulant
TRAVERSE, to - [gen] traverser
[gen] (to move laterally) passer à travers
[carp] traverser avec un rabot
[leg] dénier, renier
[mech] charioter, surfacer
[surv] faire le levé, faire l'intersection
TRAVERSE - [gen] traverse oblique
[gen] (a road) passage *m* à travers
[mech] (of the moving part of a machine tool) translation *f* latérale, course *f* verticale, chariotage *m*
[constr] traverse *f*, entretoise *f*
[math] ligne *f* transversale
[naut] route *f* en zigzag
~ARC - [firearms] (of a gun) arc *m* de pointage
~CARRIAGE - s. travelling platform runners
~CHAIN - [text] chaîne *f* pour le mouvement des guide-fil
~CHAIN BOWL - [text] galet *m* de la chaîne de levée
~FEED - [mach tool] avance *f*
~GUIDE FORK - [text] fourchette *f* guide-mèche
TRAVERSE GUIDE RAIL - [text] barre *f* des guide-fil
~MOTION - [mech] mécanisme *m* de chariotage
~MOTOR - [mech] (of a machine tool etc) moteur *m* de l'avance
~NET - [text] réseau *m* divisible
~SHAPER - [tool] étau-limeur à outil mobile
~SURVEY - [surv] (a survey consisting of a set of connected lines, the length and directions of which are measured) levé *m* par intersection
~TABLE - [railw] pont *m* roulant, chariot *m* transbordeur
TRAVERSER CARRIAGE - [railw] chariot *m* de pont-roulant
~PIT - [railw] fossée *f* de roulement
TRAVERSING - [mach tool] translation *f* du chariot
[mach tool] chariotage *m*
[surv] s. traversing survey
~BINDER - [text] poil *m* protecteur traversant
~BRIDGE - [constr] (movable bridge which can be rolled forwards and backwards across an opening, e.g. a dock entrance, to allow a vessel through) pont *m* roulant
~-HEAD SHAPING MACHINE - [mach tool] étau-limeur *m* à outil mobile
~MANDREL - [mach tool] mandrin *m* à mouvement axial
~MOTION - [mach tool] mouvement *m* de chariotage
~SHAFT - [mech] arbre *m* transversal
~SIDE GRINDER - [mech] trotteuse *f* pour aiguisage latéral
~SPINDLE - s. traversing mandrel
~STRICKLE GRINDER - [text] appareil *m* d'aiguisage à toile d'émeri voyageante
~TABLE - [railw] (a transfer table) chariot *m* transbordeur, pont *m* roulant
[naut] table *f* de point
~THREAD GUIDE - [text] guide-fil *m* à va-et-vient
TRAVERTINE - [geol] (variety of calcareous tufa of light colour) travertin *m*
TRAWL, to - [naut] (the operation of fishing by dragging a net) pêcher à la traille, chaluter
TRAWL - [naut] (the dragnet for fishing) chalut *m*, chalon *m*
[fishing] (fishing line, a mile or so long, carrying a number of short lines fitted with hooks) palangre *f*, ligne *f* flottante, corde *f*
TRAWLER - [naut] chalutier *m*
TRAWCAVATOR - [mech] (a skimmer shovel) pelle *f* nivelleuse
TRAY - [gen] plateau *m*
[impl] (used in photography etc) cuvette *f*
[gen] casier *m*, châssis *m*
[firearms] planchette *f* de chargement
[el] (for storage batteries; a support to receive several batteries) caisse *f* de groupement
[mach tool] réservoir *m*, bac *m*
TREACLE - [food] (molasses. Syrup produced in sugar refining) mélasse *f*
~MUSTARD - [bot] giroflée *f* sauvage
TREAD, to - [gen] poser les pieds, marcher
[zool] (of a male bird) couvrir, côcher
TREAD - [gen] pas *m*
[auto] bande *f* de roulement
[rubber ind] (the part of a wheel which bears on the ground) chape *f*, surface *f* de roulement
[railw] (the part of the wheel which bears on the rail) table *f* de roulement

TREAD - [mech] (of a vehicle; the distance between the rails) largeur *f* de voie
[mech] (of a vehicle the distance between the points of contact with the ground of wheels pertaining to the same axle) écartement *m* des roues
[constr] semelle *f* de poutre
[constr] (of a step) giron *m* de marche d'escalier
[biol] cicatrice *f*
~BASE - [rubber ind] (for tyres) sous-couche *f* de la bande
~BUTT SPLICING MACHINE - [rubber ind] machine *f* à souder les bandes de roulement
~CHANGE - [auto] variation *f* de la voie
~COMPOUND - [rubber ind] mélange *m* du protecteur
~-CUT REPAIR GUN - [rubber ind] seringue *f* de solution pour la réparation des incisions
~CUTS - [rubber ind] (of tyres) incisions *fpl* de la chape
~DEPTH - [rubber ind] profondeur *f* de sculpture
~EXTRUDER - [rubber ind] boudineuse *f* pour bandes de roulement
~FILLER - s. tread compound
~HEAD - [rubber ind] (of an extruder) tête *f* de boudineuse pour bande de roulement
~LOOSENESS - [rubber ind] décollage *m* du protecteur
~OF THE WHEEL - [railw] surface *f* de roulement
~OF TYRE - [rubber ind] (the area of a tire which makes contact with the ground during revolution) chape *f* , surface *f* de roulement du pneu
~PATTERN - [rubber ind] dessin *m* de sculpture de la bande de roulement
~PROFILE - [rubber ind] sculpture *f* de bande de roulement
~RECLAIM - [rubber ind] régénéré *m* de chapes
~ROLLER - [rubber ind] roulette *f* à appuyer, rouleau *m* à convectionner la bande de roulement
~RUBBER STOCK - [rubber ind] caoutchouc *m* de rechapage
~SKIVE - [rubber ind] incision *f* en forme cratère de bande de roulement
~SKIVER - [impl] couteau *m* à débrider
~STRIP - [rubber ind] boudin *m* de mélange pour bande de roulement
TREADING - [rubber ind] (retreading, or recapping) rechapage *m*, regommage *m*
[rubber ind] (a stair carpet) tapis *m* d'escalier
~ROD - [text] pédale *f*, marche *f*
~TAPPET - [mech] pièce *f* excentrique (de pédale)
TREADLE, to - [mech] actionner la pédale, pédaler
TREADLE - [mech] (lever operated by the foot) pédale *f*
[text] marche *f*, pédale *f*
[el] (rail flexure) pédale *f* électromécanique à flexion de rail
~DRILLING MACHINE - [mach tool] machine *f* à percer à pédale
~BOWL - [text] galet *m* de marche
~BRACKET - [text] boîte *f* pour les marches
~CRANK - [mech] manivelle *f* à pédale
~DRIVE - [mech] commande *f* à pédale
~FOR FIGURING WARP - [text] marche *f* pour chaîné effet
~GUIDE - [text] grille *f* des marches
~HELL - s. treadle bracket
~LATHE - [mach tool] tour *m* à pédale, tour *m* marchant au pied
~LOOM - [text] métier *m* à marches
TREADLE MOTION - [text] mécanisme *m* des marches et lisses
~WORK - [text] façonnage *m* à la marche
TREADMILL - [mech] (obsolete type of mechanism rotated by the walking motion of one or more persons) moulin *m* de discipline
TREADPLATE - [gen] (used for vehicles etc. a non-slip plate) tôle *f* antidérapante
TREADWEAR - [rubber ind] usure *f* de roulement
TREASURE - [gen] trésor *m*
TREASURER - [fin etc] (in general, the person who has care of a treasury) trésorier *m*
TREASURY - [fin etc] (the body which receives and distributes public revenue, or, in the case of a corporation, the available funds) trésorerie *f*, trésor *m* publique
~BILL - [fin] billet *m* du Trésor
~RATING - [auto] puissance *f* fiscale
TREAT, to - [gen chem etc] traiter
[med] traiter, soigner
TREATED POLE - [el & telecomm] poteau *m* injecté
TREATER - [oil ind] installation *f* de traitement
TREATISE - [gen] traité *m*
TREATMENT - [gen] traitement *m*
[med] traitement *m* médical
~CONE - [radiat] (an attachement to an X-ray therapy-tube head, defining the cross-section of the radiation beam; a localizer) localisateur *m*, cône *m*
TREATY - [gen comm etc] traité *m*
TREBLE, to - [math] (to multiply by three) tripler
TREBLE - [gen] triple
[mus] soprano
~BLOCK - [mech] palan *m* triple
~KILN FLOOR - [brew ind] touraille *f* à trois plateaux
TREBLED ORGANZINE - [text] organsin *m* retordu
TREDDLE - [text] marche *f*, pédale *f*
TREE - [bot] arbre *m*
~AGATE - [min] agate *f* arborisée
~BARK BAST - [bot] filasse *f* d'écorce d'arbre, liber *m* d'arbre
~BLIGHT - [bot] blanc *m* sec, carie *f*
~BOX - [bot] buis *m* arborescent
~CABBAGE - [bot] chou *m* cavalier
~CALF - [bookbind] veau *m* raciné
~COTTON - [text] coton *m* de Maranham
~CULTURE - [agric] arboriculture
~-DOZER - [agric] abatteur-déracineur *m*
~NURSERY - [agric] pépinière *f* forestière
~PRUNER - [agric] échenilloir *m*
~PUTTY - [ind chem] mastic *m* antiseptique
~SCRAP - [rubber ind] sernamby *m*, scrap *m* d'écorce
~-STUMP - [bot] souche *f* d'arbre
~-SYSTEM - [el] système *m* de l'arbre
~-TRUNK - [bot] tronc *m* d'arbre
~WART - [bot] excroissance *f*
TREENAIL - [shipbuild] (wooden peg of dry wood swelling when wet) cheville *f* de bois
[naut] gournable *f*
TREES - [chem] (branching projections from the cathode formed during deposition) arborescences
~AND NODULES - [el chem] (macroscopic projections formed on a cathode during electrodeposition. Nodules are rounded and trees are branched) arborescences *fpl* et nodules
TREFOIL - [aero] (a group of three parachutes coupled together) trèfle *m*

TREFOIL - [bot] trèfle *m*
[arch] trèfle *m*
TRELLIS - [gen] treillis *m*, treillage *m*
[carp] treillis *m*
[agric] espalier *m*
~FRUIT - [agric] fruits *mpl* d'espalier
~-TRAINING - [agric] taille *f* en espalier
TRELLISWORK - [gen] treillage *m*
TREMATODE - [vet] trématode *m*
TREMBLE, to - [gen] trembler, vibrer
TREMBLER - [el] trembleur *m*
~BELL - [el] sonnerie *f* d'appel à courant continu, sonnerie *f* trembleuse
TREMBLING - [cin] (rapid movement of the film up down) saut *m* vertical
TREMIE - [constr] (a large metal funnel designed for the distribution of freshly mixed concrete over a site which is below water) trémie *f* pour béton
TREMOLITE - [min] (an amphibole consisting of magnesium and calcium silicate and with applications similar to those of asbestos) trémolite *f*, grammatite *f*
TREMOR - [med] tremblement *m*
[gen] (of glasses etc) trépidation *f*
TRENCH, to - [gen & constr] creuser une tranchée
[agric] planter dans une rigole
[soil] défoncer
[constr] arrimer (le lest)
TRENCH - [gen] tranchée *f*, fossé *m*
[agric] rigole *f*, saignée *f*
[constr] fossé *m*, fouille *f* de construction
~BACKFILL - [constr] comblement *m* de tranchée
~-COAT - [text] manteau *m* imperméable
~-CUTTING MACHINE - [agric] machine *f* à creuser les tranchées
~GUN - s. trench mortar
~HOE - s. trench plough
[impl] pelle *f* de tranchée
~LEG - s. trench shin
~MORTAR - [firearms] mortier *m*
~PLOUGH - [agric] charrue *f* fossoyeuse, charrue *f* rigoleuse
~PUMP - [hydr] pompe *f* portative
~RHEUMATISM - [med] rhumatisme *m* des tranchées
~SHIN - [med] fièvre *f* tibialgique, myofibrosite *f* des membre inférieurs
~SILO - [agric] silo-fosse *m*
TRENCHER - [impl] tanchoir *m*
[mech] machine *f* à creuser les fossés
TRENCHING - [el] caniveau *m* pour câbles
[mech] s.trencher
~PLOUGH - s. trench plough
TREND - [gen] (general course, direction) direction *f*, tendance *f*
[geogr] (of a river etc) direction *f* (d'un cours d'eau)
[naut] (of an anchor) bas *m* de la verge d'une ancre
~OF A FAULT - [gen] direction *f* d'une faille
TREPAN, to - [mech & med] (to use a trepan upon) trépaner
[mach tool] (e.g. on a lathe) tourner des rainures annulaires
[mech] (with a cylinder saw) percer
TREPAN - [impl] trépan *m*, tarière *f*
[tool] (used in mining, for boring) trépan *m* à double marteau
TREPANATION - [med] (operation by means of a trepan) trépanation *f*
TREPANNING - [metall] (in forging, a hollow punching) trépanation *f*
(the removal of a core from a piece of steel by means of a tubular cutter) poinçonnage *m* creux
~TOOL - [tool] outil *m* pour le poinçonnage creux
TREPHINATION - s. trapanation
TREPHINE - [impl] (a medical instrument used for trepanation, a trepan) tréphine *f*
TRESPASS, to - [gen & leg] transgresser, violer
[leg] (a boundary) passer sans autorisation
TRESPASS - [gen] transgression *f*
[leg] violation *f* des droits
TRESTLE - [impl] tréteau *m*, chevalet *m*
[constr] (a horse) chèvre *f*
[mech] tréteau *m*, treillage *m*
[carp] chevalet *m* de scieur
~BRIDGE - [constr] pont *m* sur chevalets, pont *m* sur tréteaux
~CRANE - [mech] grue *f* à chevalet
~SHORE - [constr] chevalement *m*
~TABLE - [impl] table *f* à tréteau
~WITH BOX - [text] (for silkworms) chevalet *m* avec claie
TRESTLETREE - [shipbuild] (one of a pair of pieces at right angles to a lower mast, to support the crosstree) élongis *m* de chouque
TRESTLEWORK - [constr] (a bridge made of braced framework) chevalets *mpl*
TREVET - [text] (trivet) rabot *m*
TREVETTE - s. trevet
TRIACETIN - [chem] (glyceral triacetate. A plasticizer, perfumery agent, and a camphor substitute) triacétine *f*
TRIACETYLOLEANDOMYCIN - [chem] (an ester of the antibiotic oleandromycin, having application in medicine) triacétyloléandomycine *f*
TRIACID - [chem] (containing three hydroxyl groups, which can be replaced by acid radicals) triacide *m*
TRIAD - [chem] (a trivalent element) triade *f*
[electron] (a phosphor dot trio; one of the tiny dots of phosphor materials used in groups of three one for each primary colour, on the screen of a cathode-ray tube) triade *f* à substance luminescente
TRIAL - [gen] essai *m*
[leg] jugement *m*
~-AND-ERROR - [gen] tâtonnement *m*
~-AND-ERROR CALCULATION - [math] calcul *m* par la règle de fausse position
~-AND-ERROR METHOD - [gen] procédé *m* par tâtonnement
[comput] méthode *f* d'expérimentation systématique
~BALANCE - [comm] bilan *m* aux fins de contrôle, balance de vérification
~-BALANCE BOOK - [comm] livre *m* des soldes
~BALLOON - [met] ballon *m* enregistreur
~BORE HOLE - [mining] forage *m* d'essai, sondage *m* de reconnaissance
~BORING - [mining] sondage *m* d'exploration
[soil] (of the ground) forage *m* d'essai
~BORING TOOL - [tool] sonde *f* d'exploration
~COCK - [hydr] robinet *m* de jauge
~CRUSHING - [min] broyage *m* d'essai
~DRIVING - [soil] battage *m* d'essai
~FIELD - [agric] champ *m* d'essai
~FLIGHT - [aero] vol *m* d'essai
~HEADING - [mining] galerie *f* de reconnaissance

TRIAL HOLE - [mining] sondage *m* de recherches
~JUDGE - [leg] (in the USA only) juge *m* de première instance
~LOADING - [constr] charge *f* d'essai
~PIT - [mining] trou *m* de prospection
~RUN - [auto] parcours *m* d'essai
[gen] (of a plant) marche *f* d'essai
~-SHAFT - [oil ind] puits *m* d'essai, puits *m* de recherches
~SPEED - [mech] vitesse *f* aux essais
~TEST - [mech] épreuve *f*, essai *m*
~WELL - [mining] puits *m* d'essai
TRIALLYL CYANURATE - [chem] (a synthesis intermediate) cyanurate *m* de triallyle
TRIAMINOTOLUENE TRIHYDROCHLORIDE - [chem] (a photographic chemical) chlorhydrate *m* de triaminotoluène
TRIAMINOTRIPHENYLMETHANE - [chem] (para-Leucaniline. An intermediate for dyestuffs) triaminotriphénylméthane *m*
TRIAMYL BORATE - [chem] (a component of protective coatings) borate *m* de triamyle
TRIAMYLAMINE - [chem] (a component of anti-corrosion systems) triamylamine *f*
TRIAMYLPHENYL PHOSPHATE - [chem] (a plastics plasticizer) phosphate *m* de triamylphényle
TRIANGLE - [geom] triangle *m*
~AERIAL - [radio] (an aerial shaped like a triangle) antenne *f* à triangle
~ANTENNA - s. triangle aerial
~CONNECTION - [el] montage *m* en triangle
~CURVE - [el] courbe *f* en triangles
~OF ERRORS - [surv] (the triangle formed in the trial-and error solution of the three-point problem) triangle *m* des erreurs
~OF ERRORS - [math] chapeau *m*
~OF FORCES - [phys] (three concurrent forces acting in such way that the common point is in translational equilibrium and can be represented by a triangle) triangle *m* de forces
TRIANGULAR - [geom] triangulaire
~BEACON - [aero] point *m* géosétique
[surv] point *m* topographique
~BREAKER BAR - [text] barre *f* briseuse triangulaire
~CAM - [mech] excentrique *m* triangulaire
~COMPASS - [impl] compas *m* à trois branches
~FILE - [tool] lime *f* triangulaire, lime *f* tiers-point, lime *f* trois-quarts, trois-carrès *m*
~FLUTES - [mech] cannelures *fpl* triangulaires
~LEVEL - [instr] niveau *m* triangulaire
~MESH - [text] maille *f* triangulaire
~PARACHUTE - [aero] (parachute which is approximately triangular when extended on the ground) parachute *m* triangulaire
~PASSAGEWAY - [mining] galerie *f* de forme triangulaire
~PLATE-SPRING - [mech] ressort *m* à lame en triangle
~PUNCHER - [cin] (a puncher used to eliminate sound splice noise) perforatrice *f* de défaut sonore
~RANDOM NOISE - [radio & telecomm] spectre *m* de puissance de bruit proportionnel à la fréquence de forme triangulaire
~RIGHT PRISM - [opt] prisme *m* droit triangulaire
~ROPE - [text] câble *m* triangulaire
~SET SQUARE - [impl] équerre *f* du dessinateur
~SILK FILAMENT - [text] fil *m* de soie triangulaire
TRIANGULAR THREAD - [mech] filet *m* triangulaire
~TRUSS - [constr] ferme *f* triangulaire
~WEIR - [hydr] déversoir *m* triangulaire
~WIPER - [text] broche *f* triangulaire
~WIRE - [text] fil *m* triangulaire
TRIANGULATED TRUSS OF A CUPOLA - [constr] charpente *f* triangulée de dôme
TRIANGULATION - [surv] (the process of dividing up a large area for survey purposes into a number of connected triangles) triangulation *f*
~BEACON - s. triangular beacon
~NET - [surv] réseau *m* de triangulation
~STATION - [surv] sommet *m* de triangulation
TRIASSIC - [geol] (pertaining to a geological system of rocks between the Permian and the Jurassic systems) triasique
TRIATOMIC - [phys] (containing three atoms) triatomique
[chem] (trivalent) trivalent
~GAS - [phys] (a gas in which the molecule contains three atoms) gas *m* triatomique
TRIAXIAL SHEARING TEST - [metall] essai *m* de cisaillement triaxial
~TEST - [mech] essai *m* de compression tridimensionnelle
TRI-BARREL ELECTRON GUN - [electron] canon *m* électronique à trois faisceaux
TRIBASIC - [phys] (a molecule containing three replaceable hydrogen satoms) tribasique
~CALCIUM PHOSPHATE - [chem] (a source of phosphorus and phoshoric acid; dyeing mordant, component in ceramic compounds; dietary supplement; and stabilizer for plastics) phosphate *m* tricalcique
~COPPER SULPHATE - [chem] (a fungicide) sulphate *m* tribasique de cuivre
~MAGNESIUM PHOSPHATE - [chem] (a plastics stabilizer; supplement, medicinal antacid, and mild abrasive) phosphate *m* tribasique de magnésium
~POTASSIUM PHOSPHATE - [chem] (used as an agricultural fertilizer) phosphate *m* tribasique de potassium
TRIBOELECTRIFICATION - [el] (the electrification occurring when two dissimilar substances are rubbed together) électricité *f* par frottement
TRIBOLUMINESCENCE - [phys] (luminescence exhibited by certain crystals when they are broken or subjected to friction) triboluminescence *f*
TRIBOMETER - [instr] (an instrument designed to determine the coefficients of friction) tribomètre *m*
TRIBROMOACETALDEHYDE - [chem] (bromal. A synthesis intermediate and, as the hydrate, a hypnotic used in medicine) tribromoacétaldéhyde *m*
TRIBROMACETIC ACID - [chem] (a synthesis intermediate) acide *m* tribromacétique
TRIBROMOETHANOL - [chem] (tribromoethyl alcohol. A basal anaesthetic) tribromoéthanol *m*
TRIBUNAL - [leg] tribunal *m*
TRIBUNE - [arch] (the raised floor for the chair of a Roman magistrate) tribune *f*
TRIBUTARY - [gen] (contributory, subsidiary) tributaire
[geogr] (of a stream) affluent *m*
~CHANNEL - [hydr] canal *m* secondaire
~WATER - [hydr] afflux *m*, affluence *f*
TRIBUTOXYETHYL PHOSPHATE - [chem] (a primary plasticizer, compatible with most resins, and giving improved flexibility) phosphate *m* de tribu-

toxyéthyle
TRIBUTYL ACONITATE - [chem] (a plasticizer and stabilizer for cellulose derivatives, synthetic rubbers and resins) aconitate *m* de tributyle
~BORATE - [chem] (a textile fireproofing agent and additive for protective coatings to give improved adhesion) borate *m* de tributyle
~CITRATE - [chem] (a solvent and plasticizer for cellulose derivatives)citrate *m* de tributyle
~PHOSPHATE - [chem] (a plasticizer and solvent for cellulose derivatives) phosphate *m* de trybutyle
~PHOSPHINE - [chem] (a polymerization and curing agent for synthetic resins) tributyl-phosphine *f*
~PHOSPHITE - [chem] (a fuel additive) phosphite *m* de tributyle
~TRICARBALLYATE - [chem] (a plastics plasticizer) tricarballylate *m* de tributyle
TRIBUTYLAMINE - [chem] (a solvent and synthesis intermediate) tributylamine *f*
TRIBUTYLPHENYL PHOSPHATE - [chem] (a plastics plasticizer) phosphate *m* de tributylphényle
TRIBUTYLTIN ACETATE - [chem] (a disinfectant) acétate *m* de tributyl-étail
~CHLORIDE - [chem] (a pesticide) chlorure *m* de tributyl-étain
~OXIDE - [chem] (a bacteriostatic agent) oxyde *m* de tributyl-étain
TRICAR - [transp] (a motor tricycle) véhicule *m* à trois roues
TRICE, to - [naut] hisser
TRICEPS - [anat] triceps *m*
TRICHIASIS - [med] (distortion of the eyelashes so that they rub across the eye) trichiasis *m*, trichosis *m*
TRICHINA - [zool] (a nematode worm whose larvae migrate to the muscles of the body) trichine *f*
TRICHINASIS - [med] (infestation of the human intestine with trichina as a result of eating raw or underdone pork) trichinose *f*
TRICHINELLOSIS - s. trichiniasis
TRICHINOSIS - s. trichinasis
TRICHITE - [geol] (hair-like crystallite occurring in volcanic rocks) trichite *f*
TRICHLOROACETIC ACID - [chem] (a synthesis intermediate; caustic and astringent in medicine and analytical reagent) acide *m* trichloroacétique
TRICHLOROBENZENE - [chem] (a solvent and synthesis intermediate) trichlorobenzène *m*
TRICHLOROBORAZOLE - [chem] (a catalyst and synthesis intermediate) trichloroborazol *m*
TRICHLOROCARBANILIDE - [chem] (a bacteriostat) trichlorocarbanilide *f*
TRICHLOROETHANE - [chem] (a solvent and synthesis intermediate) trichloroéthane *m*
TRICHLOROETHANOL - [chem] (a synthesis intermediate) trichloroéthanol *m*
TRICHLORO-ETHYL PHOSPHATE - [chem] (plasticizer for cellulose derivatives, polystyrene, polyvinyl acetate, polyvinyl butyral, vinyl chloride and vinyl chloride acetate)phosphate *m* de trichloroéthyle
TRICHLOROETHYLENE - [chem] (a solvent, synthesis intermediate, refrigerant, and fumigant) trichloroéthylène
TRICHLOROFLUOROMETHANE - [chem] (a solvent, aerosol propellant, refrigerant, and a filler for fire extinguishers) trichlorofluorométhane *m*
TRICHLOROISOCYANURIC ACID - [chem] (a bleaching agent and disinfectant) acide *m* trichloroisocyanurique
TRICHLOROMELAMINE - [chem] (a bleaching agent and disinfectant) trichloromélamine *f*
TRICHLOROMETHYL CHLOROFORMATE - [chem] (a synthesis intermediate and a military poison gas(diphosgene)) chloroformiate *m* de trichlorométhyle
TRICHLOROMETHYL PHENYL CARBINYL ACETATE - [chem] (a perfumery component) acétate *m* de trichlorométhyl-phényl-carbilyle
TRICHLOROMETHYLPHOSPHONIC ACID - [chem] (a condensation catalyst) acide *m* trichlorométhylphosphonique
TRICHLORONITROSOMETHANE - [chem] (a synthesis intermediate and a tear gas) trichloronitrosométhane *m*
TRICHLOROPHENOL - [chem] (a bacteriostat and fungicide) trichlorophénol *m*
TRICHLOROPHENOXYACETIC ACID - [chem] (a weedkiller) acide *m* trichlorophénoxyacétique
TRICHLOROPROPANE - [chem] (a solvent and degreaser) trichloropropane *m*
TRICHLOROSILANE - [chem] (an intermediate for other silanes) trichlorosilane *m*
TRICHLOROTRIFLUOROACETONE - [chem] (a solvent) trichlorotrifluoroacétone *m*
TRICHLOROTRIFLUOROETHANE - [chem] (an intermediate, fire extinguishing agent, refrigerant and solvent) trichlorotrifluoroéthane *m*
TRICHLOCEPHALIASIS - [chem] s. trichuriasis
TRICHOCEPHALOSIS -s. trichuriasis
TRICHOCLASIS - [med] trichoclasie *f*
TRICHOID - [zool] (hair-like) trichoïde
TRICHOPHAGY - [med] tricophagie *f*
TRICHOSIS - s. trichiasis
TRICHROISM - [opt] (the property of exhibiting three different colours when viewed in as many different directions) trichroïsme *m*
TRICHROMATIC - [opt] trichrome, trichromatique
~CAMERA - [telev] caméra *f* trichromatique
~COEFFICIENTS - [opt] coefficients *mpl* trichromatiques
~FILTER - [opt] (set of three filters arranged so as to suit a specified emulsion) filtre *m* trichromatique
~RESPONSE - [opt] courbes *fpl* d'égalisation chromatiques
TRICHROMATISM - [print] trichromie *f*
TRICHURIASIS - [med] (infestation of the human intestine with the nematode whip-worm trichuris trichiura) trichocéphalose *f*
TRICK - [gen] ruse *f*, tour *m*
~BUTTON - [cin] (in animation) bouton *m* de surimpression
~PHOTOGRAPHY - [photo] trucages *mpl* photographiques
~TABLE - [cin] (table for the accessories by means of which trick films are prepared) table *f* de trucages, table *f* de truquages
~VALVE - [mech] tiroir *m* trick, tiroir *m* à canal
TRICKLE, to - [gen] couler goutte à goutte, suinter, ruisseler
TRICKLE - [gen] filet *m*, filtrée *f*
~BATTERY - [el] batterie *f* tampon
~CHARGE - [el chem] (continuous charge at low rate for a considerable period, especially to make up small discharges or internal losses) charge *f* de compensation, charge *f* d'entretien

TRICKLE CHARGER - [el chem] appareil *m* à charger une petite batterie
~RECTIFIER - [el] redresseur *m* de charge continue
~SCALE - [metall] battitures *f*pl infiltrées
~COOLING PLANT - [th eng] réfrigérant *m* à ruissellement
~FILTER - [hydr] filtre *m* d'épuration des eaux d'égout, lit *m* percolateur
TRICKLING FILTER - [med] lit *m* bactérien
TRICLINIC - [phys] (crystals having neither axes nor planes of symmetry) triclinique
~SYSTEM - [crystall] système *m* triclinique
TRICOLOUR - [gen] tricolore
~TUBE - [electron] (electron tube used in the display of colour TV pictures) tube *m* à trois couleurs
TRICON RADAR SYSTEM - [radar] (or ratran; a radar system in which the receiver records the coincidence of received pulses from a group of three ground stations pulsed in variable time sequence) système *m* radar tricon
TRICOSANE - [chem] (a synthesis intermediate) tricosane *m*
TRICOT - [text] tricot *m*
~WEAVE - [text] armure *f* tricot
TRICRESYL PHOSPHATE - [chem] (a plasticizer; heat exchange medium, and fuel and lubricant additive) phosphate *m* de tricrésyle
~PHOSPHITE - [chem] (flame retardant, plasticizer and stabilizer for synthetic resins and plastics) phosphite *m* de tricrésyle
TRICYCLAMOL CHLORIDE - [chem] (an anticholinergic agent used in medicine) chlorure *m* de tricyclamol
TRICYCLE - [mech] tricycle *m*
~LANDING GEAR - [aero] (landing gear consisting of one main group of wheels and another single wheel under the nose of the aircraft) train *m* tricycle
~LAYOUT - s. tricycle landing area
~UNDERCARRIAGE - s. tricycle landing gear
TRIDECANE - [chem] (a synthesis intermediate) tridécane *m*
TRIDECYL ALCOHOL - [chem] (an intermediate for plasticizers, perfumery agents and detergents) alcool *m* tridécylique
~ALUMINIUM - [chem] (a polymerization catalyst) tridécylaluminium *m*
~PHOSPHITE - [chem] (a plastics stabilizer and synthesis intermediate) phosphite *m* de tridécycle
TRIDENT - [math] (in geometry) trident
~HADN - [med] main *f* en trident
TRIDIMENSIONAL - [gen] (having breadth, length and thickness) tridimensionnel
~PANTOGRAPH - [draw] pantographe *m* tridimensionnel
TRIDYMITE - [min] (high-temperature form of silíca) tridymite *f*
TRIETHANOLAMINE - [chem] (a dispersant and sotening agent) triéthanolamine *f*
TRIETHANOLAMINE LAURYL SULPHATE - [chem] (a detergent component) triéthanolamine-lauryl-sulfate *m*
TRIETHOXYHEXANE - [chem] (a synthesis intermediate) triéthoxyhexane *m*
TRIETHOXYMETHOXYPROPANE - [chem] (a synthesis intermediateand cross linking agent) triéthoxyméthoxypropane *m*
TRIETHYL ACONITANE - [chem] (a plastics plasticizer) aconitate *m* de triéthyle
~ALUMINIUM - [chem] (colourless, flammable liquid with uses as a plasticizer and solvent for cellulose, acetate and nitrate and natural resins, and as a plasticizer for vinyl and acetate resins) triéthyl-aluminium *m*
~CITRATE - [chem] (a perfumery agent and a solvent and plasticizer for cellulose derivates and other resins) citrate *m* de tryéthyle
~ORTHOFORMATE - [chem] (a synthesis intermediate) ortho-formiate *m* de triéthyle
~PHOSPHATE - [chem] (colourless, water-miscible liquid with uses as a catalyst, plastics and resins plasticizers, and solvent) phosphate *m* de triéthyle
~PHOSPHITE - [chem] (a synthesis intermediate) phosphite *m* de triéthyle
~PHOSPHOROTHIOATE - [chem] (a synthesis intermediate and plastics plasticizer) phosphorothioate *m* de triéthyle
~TRICARBALLYLATE - [chem] (a plasticizer) tricarballylate *m* de triéthyle
TRIETHYLAMINE - [chem] (colourless, flammable liquid with uses as a solvent in catalystic reactions and in the production of rubber accelerator activators) triéthylamine *f*
TRIETHYLBORANE - [chem] (a rocket fuel) triéthylborane *m*
TRIETHYLENE GLYCOL - [chem] (a synthesis intermediate, textile chemical, and a solvent for cellulose derivatives) triéthylène-glycol *m*
~GLYCOL DIACETATE - [chem] (plasticizer for cellulose derivatives and polymethyl methacrylate) diacétate *m* de triéthylène-glycol
~GLYCOL DICAPRYLATE - [chem] (a plasticizer for synthetic elastomers and resins) dicaprylate *m* de triéthylène-glycol
~GLYCOL DICAPRYLATE-CAPRATE - [chem] (a plasticizer for vinyl resins and cellulose derivatives) dicaprylate-caprate *m* de triéthylène-glycol
~GLYCOL DIDECANOATE - [chem] (a plastic plasticizer) didécanoate *m* de triéthylène-glycol
~GLYCOL DIETHYLHEZOATE - [chem] (a plasticizer di-éthylhexoate *m* de triéthylène-glycol
~GLYCOL DIHYDROABIETATE - [chem] (a plasticizer dihydroabiétate *m* de triéthylène-glycol
~GLYCOL DIMETHYL ETHER - [chem] (a coupling agent and an absorbent for gases) diméthyléther *m* de triéthylène-glycol
~GLYCOL DIPELARGONATE - [chem] (plasticizer for cellulose nitrate, ethyl cellulose, polyvinyl butyral, vinyl chloride and vinyl chloride acetate) dipélargonate *m* de triéthylène-glycol
~GLYCOL DIPROPIONATE - [chem] (a plastics plasticizer) dipropionate *m* de triéthylène-glycol
~GLYCOL PROPIONATE - [chem] (plasticizer for cellulose derivatives and polymethyl methacrylate) propionate *m* de tryéthylène-glycol
~THIOPHOSPHORAMIDE - [chem] (a cytotoxic agent used in medicine) triéthylène-thiophosporamide *f*
TRIETHYLENEMELAMINE - [chem] (tretamine; triethanomelamine. A cytotoxic agent used in medicine) triéthylènemélamine *f*
TRIETHYLENETETRAMINE - [chem] (a synthesis intermediate for rubber chemicals, pharmaceuticals, dyestuffs, and detergents) triéthylènetétramine *f*
TRIFLUOPROMAZINE HYDROCHLORIDE - [chem] (a

phenothiazine derivative with application in the treatment of psychotic conditions) chlorhydrate *m* de trifluoropromazine
'RIFLUOROACETIC ACID - [chem] (a catalyst and solvent) acide *m* trifluoroacétique
'RIFLUORONITROSOMETHANE - [chem] (a blue, irritant gas.It is a starting material in the production of synthetic elastomers) trifluoronitrosométhane *m*
'RIFURCATING BOX - [el] (a dividing-box for a three-core cable) boîte *f* de dérivation pour câble à trois conducteurs
JOINT - [el] boîte *f* de dérivation tri-mono
NERVE - [anat] trijumeau *m*
'RIGATRON - [electron] (electronic switch in which conduction is initiated by the breakdown of an auxiliary gap in a gas-filled envelope) trigatron *m*
'RIGGER, to - [gen] déclencher
'RIGGER - [firearms] (the fingerpiece used for releasing the hammer) détente *f*
[mech] (a catch, or a small lever actuating a mechanism) poussoir *m* à ressort, cliquet *m*, chien *m*, gâchette *f*
[el] (a tripping, or release device) débrayeur *m*, déclencheur *m*
ACTION - [mech] déclenchement *m*
AREA - [med] zone *f* reflexogène
BREAKDOWN VOLTAGE - [electron] (of a gasfilled tube the breakdown voltage of the main gap) tension *f* de rupture d'un relais électronique
CIRCUIT - [electron] (electron-tube circuit having two conditions of stability, with means for passing from one to the other when certain conditions are satisfied) circuit *m* à gâchette
[photo] (of a schutter) courant *m* de déclenchement
CRITERION - [comput] critère *m* de déclenchement
ELECTRODE - [electron] (auxiliary electrode forming part of the trigger gap) électrode *f* de déclenchement
ELEMENT - [el] organe *m* basculeur
FINGER - [med] doigt *m* à ressort
GAP - [electron] (of a gas-filled tube, a gap for carrying an auxiliary discharge which in turn initiates a discharge in a main gap) intervalle *m* de déclenchement
GUARD - [firearms] pontet *m*, sous-garde *f*
MAGNET - [el] (release magnet) aimant *m* de déclenchement
OFF, to - [gen & mech] (to initiate an action without subsequent control of it) déclencher
PULSE - [radar] (pulse used in a radar circuit) impulsion *f* de déclenchement
[electron] signal *m* de déclenchement
PULSE GENERATOR - [el] générateur *m* d'impulsions de déclenchement
RELAY - [electron] (a thermionic valve system so arranged that a given impulse will initiate discharge, but has no further control of it after such initiation) relais *m* électronique
[el] (a relay which, when operated, remains in its operated condition when the operating current is removed) relais *m* de déclenchement
RELEASE - [photo] déclenchement *m* au doigt
RETURN SPRING - [photo] ressort *m* du levier de déclenchement
SEMICONDUCTOR - [electron] (or trigistor, a silicon p-n-p-n- semiconductor device which can be controlled from the base in such way as to be either triggered on or off) semiconducteur *m* à déclenchement
TRIGGER SHARPENER - [radar] (circuit sharpening the shape of a wave or a pulse) circuit *m* à aiguiser une onde
~TIMING PULSE - [radar] impulsion *f* de contrôle de temps
~TUBE - [electron] (a three-electrode tube with a starting grid and filled with a rare gas, the electrons being released by ion bombardment) tube *m* à déclenchement, tube *m* à relais
~VALVE - [mech] (of an air compressor) soupape *f* pour la régulation de la pression
~ZONE - [med] zone *f* de déclenchement d'une douleur
TRIGGERED BLOCKING OSCILLATOR-[radio] oscillateur *m* de blocage déclenché
~SAWTOOTH GENERATOR [radio] générateur *m* en dents de scie de déclenchement
TRIGGERING LEVEL - [radar] niveau *m* de déclenchement
TRIGLYCOL DICHLORIDE - [chem] (an intermediate for dyestuffs and other organic compounds, and a solvent) bichlorure *m* de triglycol
TRIGLYPH - [arch] triglyphe *m*
TRIGONAL - [math] rhomboédrique, trigonal
~CRYSTAL - [phys] cristal *m* rhomboédrique
~REFLECTOR AERIAL - [radio] (a type of aerial array) antenne *f* à trièdre réflecteur
~REFLECTOR ANTENNA - s. trigonal reflector aerial
TRIGONE OF THE BLADDER - [med] trigone *m* de la vessie
TRIGONELLINE - [chem] (a widely distributed alkaloid, the methyl-betaine of nicotinic acid) trigonelline *f*
TRIGONOMETRIC - [math] trigonométrique
~POINT - [surv] point *m* trigonométrique
TRIGONOMETRICAL - s. trigonometric
~FUNCTIONS - [math] (functions of an angle or arc used in trigonometry) fonctions *f*pl trigonométriques
~RATIOS - [math] rapports *m*pl trigonométriques
~SETTING OUT - [surv] tracé *m* trigonométrique
~STATION - [surv] (a survey station used in a triangulation) point *m* trigonométrique
~SURVEY - [surv] levé *m* trigonométrique
~TABLES - [math] tables *f*pl trigonométriques
TRIGONOMETRY - [math] (that branch of mathematics which deals with the relations of the sides and angles of triangles and of the methods of applying these relations in the solution of problems involving triangles) trigonométrie *f*
TRIDHEDRAL BRACING ROD - [mech] (rod designed to resist stresses acting in three planes) fer *m* rond trièdre
TRIHEDRON - [geom] (having three intersecting surfaces as sides) trièdre *m*, angle *m* trièdre
TRIHEXYL PHOSPHITE - [chem] (a solvent and constituent of certain vinyl stabilizers) phosphite *m* de trihexyle
TRIHEXYLPHENIDYL HYDROCHLORIDE - [chem] (Benzhexol hydrochloride. An antispasmodic and parasympatholytic agent used in medicine) chlorhydrate *m* de trihexylphénidyle
TRIHYDRIC ALCOHOL - [chem] (an alcohol in which

hydraxy groups are attached to three different carbon atoms) trialcool *m*
TRIHYDROXYETHYLAMINE OLEATE- [chem] (an emulsificant) oléate *m* de trihydroxyéthylamine
TRIHYDROXYETHYLAMINE STEARATE - [chem] (triethanolamine stearate. A surfactant and emulsifying agent) stéarate *m* de trihydroxyéthylamine
TRIIODOTHYRONINE SODIUM - [chem] (liothyronine sodium. An active principle of the thyroid gland, with application in medicine and physiological research) tri-iodothyronine *f* sodique
TRIISOBUTYLALUMINIUM - [chem] (a polymerization catalyst) tri-isobutylaluminium *m*
TRIISOBUTYLENE - [chem] (a synthesis intermediate for elastomers and synthetic resins) triisobutylène *m*
TRIISOOCTYL PHOSPHITE - [chem] (a solvent, plastics stabilizer, and intermediate for pesticides) phosphite *m* de tri-isooctyle
TRIISOPROPANOLAMINE - [chem] (a synthesis intermediate) tri-isopropanolamine *f*
TRIISOPROPYL PHOSPHITE - [chem] (a solvent, synthesis intermediate and plastics plasticizer) phosphite *m* de tri-isopropyle
TRILATERAL - [geom] trilatéral
TRILAURYL AMINE - [chem] (a synthesis intermediate) trilauryl-amine *f*
TRILL - [mus] trille *m*
TRILOBAL - s. trilobed
TRILOBATE - [anat] trilobé
TRILOBATED - s. trilobate
TRILOBED - s. trilobate
TRIM, to - [gen] émonder, mettre en ordre
[gen] (to cut) couper, tondre
[naut] (to settle the position by arranging the cargo etc) balancer, redresser, équilibrer
[plast ind] ébavurer, détourer, couper les bords
[mech] (to remove flash or sharp edges and corners mechanically or by hand, as from a moulding) ébavurer, parer, corroyer, dresser
[aero] (to adjust the pitch angle) compenser
[bookbind] rogner les tranches (d'un livre)
[metall] (to clean a casting) ébavurer
[metall] (in sheet metal working) couper les bords
[constr] enchevêtrer, enclaver
TRIM - [gen] bon ordre *m*
[gen] (finishing work) affinage *m*
[auto] garniture *f* intérieure
[aero] (adjustment of the controls of an aircraft to maintain the required attitude in steady flight) trim *m*, équilibrage *m*, compensation *f*
[arch] garniture *f* en bois
[mech etc] (verticality) verticalité *f* de l'axe
[cin] coupe *f*
[naut] assiette *f*, arrimage *m*
~ACTUATOR - [aero] (servo-actuator operating a trim tab) servo-actionneur *m* de trim
~JACK - [aero] (a hydraulic cylinder used to operate a trim tab) actionneur *m* hydraulique de compensation
~SAILS, to - [naut] brasser
~TAB - [aero] ailette *f* de compensation
~THE CARD WIRES, to - [text] redresser les dents de carde
~THE END OF THE LAP, to - [text] égaliser la nappe
~THE FLASH, to - [metall] (in forging of a stamping) rogner, ébarder
TRIM THE GRINDING WHEEL, to - [mech] équilibrer la meule
TRIMER - [chem] (a molecule formed by the combination of three monomer molecules) trimère *m*
TRIMESIC ACID - [chem] (a derivative of mesitylene) acide *m* trimésique
TRIMETAPHAN CAMPHORSULPHONATE - [chem] (a hypotensive agent used in medicine) camphorsulfonate *m* de triméthaphane
TRIMETHADIONE - [chem] (troxidone. An anticonvulsant used in the treatment of epilepsy) triméthadione *f*
TRIMETHYL BORATE - [chem] (a solvent, fungicide, and intermediate) borate *m* de triméthyle
TRIMETHYL DIHYDROQUINOLINE POLYMER - [chem] (a polymerization inhibitor and antioxidane) triméthyl-dihydroquinone polymère *m*
~PHOSPHITE - [chem] (a synthetic intermediate) phosphite *m* de triméthyle
TRIMETHYLACETIC ACID - [chem] (a synthesis intermediate) acide *m* triméthylacétique
TRIMETHYLALUMINIUM - [chem] (a polymerization catalyst) triméthylaluminium *m*
TRIMETHYLAMINE - (TMA) - [chem] (colourless, highly flammable liquified gas with uses as an intermediate in the production of synthetic resins) triméthylamine *f*
TRIMETHYLBUTANE - [chem] (a synthesis intermediate) triméthylbutane *m*
TRIMETHYLCHLOROSILANE - [chem] (an intermediate for silicones) triméthylchlorosilane *m*
TRIMETHYLENE BROMIDE - [chem] (an intermediate for drugs and dyestuffs) bromure *m* de triméthylène
~GLYCOL - [chem] (a synthesis intermediate) triméthylène glycol *m*
TRIMETHYLHEXANE - [chem] (a synthesis intermediate) triméthylhexane *m*
TRIMETHYLNONANONE - [chem] (a synthesis intermediate and solvent) triméthylnonanone *f*
TRIMETHYLOLETHANE - [chem] (an intermediate for synthetic resins and drying oils) triméthyloléthane *m*
TRIMETHYLOLPROPANE - [chem] (a component of protective coatings and an intermediate for synthetic resins) triméthylolpropane *m*
~MONOOLEATE - [chem] (a low-temperature plasticizer) mono-oléate *m* de triméthylpropane
TRIMETHYLPENTANE - [chem] (a synthesis intermediate) triméthylpentane *m*
TRIMETHYLPENTENE - [chem] (a synthesis intermediate) triméthylpentène *m*
TRIMETHYLPENTYLALUMINIUM - [chem] (a polymerization catalyst) triméthylpentylaluminium *m*
TRIMMED - [gen] (e.g. of a car body) garni
[bookbind] rogné, ébardé
[metall plast ind] ébavuré, rogné
TRIMMER - [tool] (a forging tool) outil *m* à ébarder
[constr] (cross member framed between the full-length members to give intermediate support to the shorter joist in a trimming) chevêtrier *m*
[radio] (small adjustable capacitor connected in parallel with another capacitor so that the combined capacitances can be adjusted) trimmer *m*
[metall] (in a rolling mill) machine *f* à cisailler
[el] condensateur *m* shunt d'équilibrage
[photo] (a cutting implement) cisaille *f* à guillotine

TRIMMER SIGNAL - [railw] (USA only, shunting signal in GB) carré *m* violet, signal *m* de manoeuvre

TRIMMING - [gen] mise *f* en ordre, arrangement *m*
[constr] (the operation by which bridging joists are shortened and given intermediate support) enchevêtrement *m*, enclavement *m*
[mech] ébarbage *m*, rognage *m*
[plast ind] ébarbage *m*
[text] garnissage *m*
[metall] (in sheet metal working) finissage *m*
[agric] taille *f*, ébranchage *m*
[leather ind] échantillonnage *m* des peaux
[naut] arrimage *m*
[naut] (of the sails) balancement *m* (des voiles)
[carp] corroyage *m*, dressage *m*
[auto] garniture *f* intérieure

~-AXE - [impl] émondoir *m*

~CONDENSER - [radio] s. trimmer

~CUTTER - [mach tool] (a machine for removing irregular edges of sheet material) rogneuse *f*

~DIE - [tool] (tool used for cutting the contour of a drawn part on a press) outil *m* à ébarber
[metall] étampe *f* à ébarber

~FILTER - [telev] (in colour television) filtre *m* de ajustage

~JOIST - [constr] solive *f* d'enchevêtrure, chevêtrier *m*

~MACHINE - [metall] machine *f* à ébarber, ébarbeuse *f*

~MATERIALS - [text] passements *mpl*

~PRESS - [mech] (a press used in removing flash or defective edges) presse *f* à ébarber

~STRIP - [aero] (trailing edge cord) stabilisateur *m* bord de fuite

~TAB - s. trim tab

~TANK - [mech] caisse *f* d'assiette

~TESTBOARD - [radio] table *f* de réglage

~WITH WEFT LOOPS - [text] galon *m* avec picots par trame

~YARN - [text] fil *m* de passementerie

TRIMS - [aero] compensateurs *mpl* de gouverne

TRINITROBENZENE - [chem] (an explosive) trinitrobenzène *m*

TRINITROBENZOIC ACID - [chem] (a synthesis intermediate) acide *m* trinitrobenzoïque

TRINITROTOLUENE - [chem] (an explosive and a synthesis intermediate) trinitrotoluène *m*

TRINITROTOLUOL - [chem] s. trinitrotoluene

TRINOSCOPE - [telev] (an arrangement of three-picture tubes with colour filters and projection lenses, used in theatre television) trinoscope *m*

TRIOCTYL PHOSPHATE - [chem] (a solvent and plasticizer) phosphate *m* de trioctyle

TRIOCTYLTRIMELLITATE - [chem] (a plasticizer for vinyl plastics) tri-octyl-trimellitate *m*

TRIODE - [electron] (a three-electrode electronic valve containing an anode and a cathode) triode *f*

~-HEZODE - [electron] (an electron tube containing a triode and a hexode section and a common cathode) triode-hexode *f*

~-HEXODE CONVERTER - [electron] (a triode-hexode frequency changer) triode-hexode *f* convertisseuse de fréquence

~-PENTODE - [electron] (electron tube containing a triode and a pentode system and a common cathode) triode-pentode *f*

TRIODE-TRANSISTOR - [electron] (a three-electron transistor) transistor *m* triode

TRIOSES - [chem] (the simplest of the monosaccharoses, containing three atoms of oxygen in the molecule) trioses *mpl*

TRIOXANE - [chem] (metaformaldehyde. A synthesis intermediate and a non-luminous fuel) trioxane *m*

TRIOXIDE - [chem] trioxyde *m*

TRIP, to - [mech] déclencher, décliquer, culbuter, débrayer
[mech] (a compressed spring) déclencher
[el] déclencher

TRIP - [mech] (the action of a pawl, or similar device, which trips) déclic *m*, déclenchement *m*
[mech] (a catch) dent *f* d'arrêt, déclic *m*
[el] (of a circuit breaker) déclenchement *m*, interrupteur *m* de courant d'appel
[mining] (a convoy) train *m*, rame *f*, convoi *m*
[mining] (of the cage) trait *m*, cordée *f*
[oil ind] manoeuvre *f* de batterie

~-AND-RESET SWITCH - [el] interrupteur *m* à retour automatique

~BOB - [oil ind] barre *f* pour actionner un dispositif dans le puits

~CASING-SPEAR - [impl] arrache-tuyau *m* à déclic

~CATCH - [mech] couteau *m* de déclenche

~COIL - [electron] (a winding used to release another device, but not supplying energy to it) bobine *f* d'excitation

~DOG - [mech] déclic *m*

~-FREE - [el] (said of a circuit-breaker. Starter etc, to denote that the automatic release mechanism can function independently of the closing or operating mechanism) à déclenchement libre

~-FREE CIRCUIT-BREAKER - [el] disjoncteur *m* à déclenchement libre

~-FREE RELAY - [el] relais *m* à blocage d'ouverture

~GAS - [oil ind] gaz *m* échappé lors d'une manoeuvre

~GEAR - [mech] déclanche *f*, déclic *m*
[mech] (of a dredger bucket) culbuteur *m*

~HAMMER - [mech] marteau *m* à bascule, martinet *m*, marteau *m* à soulèvement

~IN, to - [mech] embrayer

~LEVER - [mech] levier *m* à déclic

~MARGIN - [nucl] (a power interval caused by the difference in power wich corresponds to the upper threshold of a safety circuit and the operating power of a reactor) intervalle *m* de puissance

~MILEAGE INDICATOR - [instr] compteur *m* de trajet, journalier *m*

~OF SIGNAL HAND LEVER - [railw] cran *m* actif du levier (d'un poste d'aiguillage)

~OVERWIND-GEAR - [mining] évite-molettes à déclic d'attelage

~PILE-DRIVER - [impl] sonnette *f* à déclic

~RELAY - [el] relais *m* interrupteur du courant d'appel

~RIDER- [mining] suiveur *m* de rames

~ROD - [el] choisisseur *m* de balais, déclenchement *m* de balais

~SHAFT - [mech] arbre *m* à déclic

~SPINDLE - s. trip rod

TRIPALMITIN -[chem] (a white crystalline solid, used in soap-making, medicine and leather manufacture) tripalmitine *f*

TRIPARANOL - [chem] (an agent inhibiting the synthesys of cholesterol in the body and having application

in medicine) triparanol *m*
TRIPHENYL PHOSPHITE - [chem] (a synthesis intermediate) phosphite *m* de triphényle
TRIPHENYLGUANIDINE - [chem] (a vulcanization accelerator) triphénylguanidine *f*
TRIPHENYLMETHANE - [chem] (a hydrocarbon occurring in colourless leaflets, used in the manufacture of dyes) triphénylméthane *m*
~DYES - [chem] colorants *mpl* au triphénylméthane
TRIPHENYLPHOSPHATE - [chem] (a plasticizer for cellulose derivatives) phosphate *m* de triphényle
TRIPHENYLSTIBINE - [chem] (a catalyst) triphénylstibine *f*
TRIPHENYLTIN CHLORIDE - [chem] (a synthesis intermediate) chlorure *m* de triphényl-étain
TRIPLANE - [aero] (an aeroplane having three main supporting surfaces arranged vertically one over the other) triplan *m*
TRIPLE, to - [math] tripler
TRIPLE - [gen & math] triple
~ACTING HORIZONTAL PUMP - [hydr] pompe *f* horizontale à triple effet
~ACTING UPRIGHT PUMP - [hydr] pompe *f* verticale à triple effet
~ACTION - [mech] à triple effet
[radio] instabilité *f*
~-ACTION PRESS - [mech] presse *f* à triple effet
~BOND - [chem] (indication of a state of unsaturation between two polyvalent atoms, showing that two hydrogen atoms can be attached to each atom connected by a triple bond before saturation is reached) liaison *f* triple
~CARPET - [text] tapis *m* triple, tapis *m* écossais
~CONCENTRIC CABLE - [el] (a concentring cable containing three conductors) câble *m* à trois conducteurs concentriques
~CONDUCTOR - [el] conducteur *m* triple
~-CYLINDER ENGINE - [mech] machine *f* à trois cylindres
~-CYLINDER MIXER - [mech] mélangeur *m* à trois cylindres
~-DETECTOR RECEPTION - [radio] réception *f* à double changement de fréquence
~-DIFFUSED TRANSISTOR - [electron] transistor *m* à triple diffusion
~DIODE - [electron] (electron tube including three cylindrical anodes arranged about a common cathode) triple diode *f*
~-EXPANSION ENGINE - [mech] machine *f* à triple expansion
~-EXPANSION UNDERGROUND PUMPING ENGINE - [mech] machine *f* d'épuisement souterrain à triple expansion
~EXTENSION - [photo] triple tirage *m*
~FEED RACK - [mech] rampe *f* de graissage à trois départs
~GEAR - [mech] triple harnais *m* d engrenage
~-GEAR SHAPING MACHINE - [mach tool] étau-limeur *m* à triple harnais d'engrenage
~HORN SET - [auto] jeu *m* de trois avertisseurs
~ISOMORPHISM - [phys] (or isotrimorphism, the condition in which two isomorphous substances are each trimorphous and each of the three pairs of form is isomorphous) isotrimorphisme *m*
~-LENGTH WORKING - [comput] opération *f* à triple longueur de mot
~-LENS TURRET - [opt] tourelle *f* à trois objectifs
TRIPLE MOLECULAR COLLISION - [phys] (a collision occurring when three molecules collide simultaneously) collision *f* trimoléculaire
~POINT - [phys] (the temperature and pressure at which the three physical states of a substance are in equilibrium) triple point *m*
~-POLE - [el] tripolaire
~-POLE REVERSER - [el] inverseur *m* tripolaire
~-POLE SWITCH - [el] commutateur *m* tripolaire
~ROCKET - [astronaut] fusée *f* à trois étages
~-ROLLER MILL - [min] (a special type of grinding mill for very fine grinding of pigmented compositions in which dispersion is effected by the crushing action of roller and shearing is promoted by the use of rollers running at different peripheral speeds) laminoir *m* triple
~-STUB TRANSFORMER - [electron] transformateur *m* de mode à trois bras de réactance
~WEFT MOQUETTE - [text] moquette *f* à triple trame
TRIPLET - [phys] (a bond which two atoms share three electrons) triplet *m*
[gen] (any combination of three) trio *m*
[opt] (a system of three lenses) triplet *m*
[mus] tercet *m*, triolet *m*
[physiol] trijumeau *m*, triplet *m*
~BIRTH - [med] accouchement *m* trigémellaire
TRIPLEX - [gen] à trois, en trois
~BOARDS - [paper man] cartons *mpl* triplex
~CABLE - [el] câble *m* à trois conducteurs câbles
~CONTROL - [contr] commande *f* triple
~CONTROLLER - [contr] régulateur *m* triple
~DRAWING BOARD - [draw] planche *f* à dessin à trois épaisseurs
~ENGINE - [mech] machine *f* à trois cylindres
~GLASS - [glass man] (a type of unbreakable glass) glace *f* triplex
~PAPERS - [paper man] cartons *mpl* triplex
~WINDING - [el] (a direct-current armature winding with three parallel paths per pole between positive and negative terminals) enroulement *m* triplex
TRIPLEXER - [telev] (a filter device) filtre *m* triplexeur
TRIPOD - [impl] trépied *m*
[mech] (of a rock-drill) affût-trépied
~BUSH - [photo] écrou *m* de pied
~CLAMP - [photo] entretoise *f* de trépied
~DERRICK - [oil ind] derrick *m* à trois pieds
~EXTENSION - [photo] extension *f* du pied
~HEAD - [photo] (the platform of a tripod on which the camera rests) tête *f* de pied, tête *m* de trépied
~JACK - [mech] vérin *m* à trépied
~LEG - [photo] branche *f* de trépied
~RIG - [oil ind] installation *f* à trépied
~STAND - [impl] (in a laboratory three-legged support for a flask or the like) trépied *m*, pied *m* à trois branches
~STRUT - [photo] branche *f* de trépied
TRIPOLI - [min] (a diatomaceous earth obtained from North Africa) terre *f* pourrie
~POWDER - s. tripoli
TRIPOLITE - [geol] tripolite *f*
TRIPPING - [el] déclenchement *m*, déclenche *m*
[mech] débrayage
~CURRENT - [el] courant *m* de déclenchement
~DEVICE - [el] déclencheur *m*
[mech] déclic *m*, dispositif *m* de déclenchement
~PULSE - [radar] (synchronizing pulse used as a

trigger in a saw-tooth generator valve of a time base) impulsion *f* de synchronisation et de déclenchement
TRIPPING TIME - [el] (e. g. of a circuit breaker) temps *m* de déclenchement
TRIPROLIDINE HYDROCHLORIDE - [chem] (a potent, short-acting-histamine drug) chlorhydrate *m* de triprolidine
TRIPROPYLENE - [chem] (a synthesis intermediate and lubricant additive) tripropylène *m*
~GLYCOL - [chem] (an intermediate for drugs, pesticides, synthetic resins and plasticizers, and dyestuffs) tripropylène-glycol *m*
TRIPSIS - [med] trituration *f*
TRIPSOMETER - [instr] (an instrument for assessing the resilience of a substance) tripsomètre *m*
TRIPTYCH - [art] triptyque *m*
TRISACCHARIDE - [chem] (any of a class of saccharides which yields three monosaccharide molecules when subjected to hydrolysis) trisaccharides *mpl*
TRIS(hydroxymethyl)AMINOMETHANE - [chem] (an emulsificant and synthesis intermediate) tris(hydroxyméthyl)aminométhane *m*
TRIS(diethylene glycol monoethyl ether)CITRATE - [chem] (a plasticizer) citrate *m* de tris(diéthyléne-glycol-mono-ethyl-ether)
TRIS(hydroxymethyl)NITROMETHANE - [chem] (an industrial bactericide) tris(hydroxyméthyl)nitrosométhane *m*
TRISETHYLHEXYL PHOSPHITE - [chem] (a synthesis intermediate) phosphite *m* de triséthylhexyle
TRISILICATE - [chem] trisilicate *m*
TRISILICIC - [chem] trisilicique
TRISMUS - [med] trismus *m*
TRISTIMULUS SIGNALS - [telev] (in colour television) signaux *mpl* RVB
~SPECIFICATIONS - s. tristimulus values
~VALUES - [opt] composantes *fpl* trichromatiques
TRITANOMALOUS VISION - [opt] tritanopie *f* partielle
TRITIATED WATER - [chem] (water which contains a small percentage of tritium) eau *f* tritiée
TRITIUM - [chem] (a hydrogen isotope of mass 3 : it has been produced by bombarding deuterons with deuterons) tritium *m*
~AIR MONITOR - [nucl] moniteur *m* atmosphérique de tritium
~LABILIZATION - [nucl] instabilisation *f* de tritium
~LUMINOUS COMPOUND - [nucl] composé *m* lumineux à tritium
~SURFACE CONTAMINATION METER - [nucl] contaminamètre *m* surfacique pour tritium
~TARGET - [nucl] cible *f* en tritium
TRITON - [phys] (the nucleus of tritium, i. e. of hydrogen of mass number 3) triton *m*
TRITONITE - [min] (borosilicate of cerium and other elements) tritonite *f*
TRITOPRISM - [phys] prisme *m* de troisième espèce, tritoprisme *m*
TRITOPYRAMID - [phys] pyramide *f* de troisième espèce, tritopyramide *f*
TRITURATE, to - [min] triturer
TRITURATION - [ind chem] (reduction of a substance to powder by grinding in a mortar) trituration *f*
~PLATES - [min] plaques *fpl* de trituration
TRITURATOR - [mech] triturateur *m*
TRIURANIUM OCTOXIDE - [chem] (a constituent of pitchblende with application in nuclear technology) octoxyde *m* de triuranium
TRIVALENCE - [chem] trivalence *f*
TRIVALENCY - s. trivalence
TRIVALENT - [chem] (capable of combining with three atoms of hydrogen or the equivalent) trivalent
TRIVET - [text] (a knife used for cutting the pile thread loops of velvet) taillerolle *f*
TROCAR - [med] (used in surgery, a sharp-pointed perforator) trocart *m*, trois-quarts *m*
TROCHE - [chem] tablette *f*, trochisque *f*
TROCHIN - [med] trochin *m*, petite tubérosité *f* de l'humérus
TROCHITER - [med] trochiter *m*, grosse tubérosité *f* de l'humérus
TROCHLEA - [anat] (any structure which is shaped like a pulley) trochlée *f*
TROCHLEAR - [anat] (pertaining to a trochlea) trochléen
TROCHOIDAL MASS ANALYZER - [instr] spectromètre *m* de masse à parcours trochoïde
TROCHOTRON - [electron] (a hot-cathode stepping tube) trochotron *m*
TROCTOLITE - [geol] (coarse-grained basic igneous rock) troctolite *f*
TROEGERITE - [min] (mineral containing uranylortho-arsenate) trögérite *f*
TROILITE - [min] troïlite *f*
TROLL, to - [fishing] pêcher à la cuiller
TROLLEY - [gen] (a vehicle) fardier *m*, chariot *m*, binard *m*
[mech] (of a travelling crane) chariot *m* de roulement
[transp] diable *m*
[el] (overhead device) trolley *m*
~ARM - [el] trolley *m*
~BASE - [el] base *f* trolley
~-BUS - [transp] trolleybus *m*
~-BUS SYSTEM - [el] système *m* trolleybus
~BUSH - [el] alvéole *m* en graphite
~CAP - s. trolley shield
~-CAR - [transp] tramway *m* à trolley
~COACH - [transp] voiture *f* trolley
~CORD - [el] corde *f* du trolley
~FROG - [el] (device used at a junction of a branch contact-wire with the main contact-wire to permit the passage of current-collectors along either wire) aiguillage *m* de trolley
~HARP - [el] archet *m* de trolley
~HEAD - [el] tête *f* de trolley
~JACK - [auto] cric *m* mobile
~LINE - [transp] ligne *f* de tramway
~LOCKER - [impl] coffret *m*
~MESSENGER - [el] câble *m* de suspension du trolley
~POLE - [el] perche *f* de trolley, tige *f* du trolley
~SHIELD - [el] chape *f* de trolley
~SHOE - [el] cuillère *f* de contact
~SYSTEM - [el] (in electric traction) système *m* par trolley, système *m* par fil aérien avec trolley
~WHEEL - [el] roulette *f* de trolley
~WIRE - [el] (a contact wire, an overhead conductor from which electric power is supplied) fil *m* de contact, fil *m* de trolley
TROLLEYBUS - s. trolley-bus
TROLLING - [fishing] pêche *f* à la cuiller
TROMBONE - [mus] trombone *m*
[radio] (a U-shaped length of waveguide of adjusta-

ble lenght) trombone *m*
TROMBONE TUNING BAR - [radio] (or hairpin tuning bar, a hairpin-shaped waveguide used for tuning a transmission line) barre *f* de syntonisation en U
TROMMEL, to - [min] trommeler, passer au trommel
TROMMEL - [min] (cylindrical inclined rotary screen) trommel *m*, cible *m* rotatif
~SCREEN - s. trommel
~TEST - [min] essai *m* au cible rotatif
TRONA - [min] (hydrated natural basic sodium carbonate, found in saline residues in desert areas. A source of sodium compounds) trona *m*
TROOP, to - [gen] s'assembler, s'attrouper
TROOP CARRIER - [aero] avion *m* pour le transport de troupes
TROOPSHIP - [naut] transport *m*
TROOPTRAIN - [railw] train *m* régimentaire
TROOSTITE - [metall] (term denoting structures in steel which consist of very fine aggregates of ferrite and cementite) troostite *f*
TROPACOCAINE HYDROCHLORIDE - [chem](benzoylpseudotropeine hydrochloride. A local anaesthetic) chlorhydrate *m* de tropacocaïne
TROPAEOLINES - [chem] (a group of dyes derived from p-hydroxy azobenzene) tropéolines *fpl*
TROPHEDEMA - [med] trophoedème *m*
TROPHIC - [zool] trophique
TROPHICITY - [med] trophicité *f*
TROPHOBLAST - [biol] trophoblaste *m*, couche *f* enveloppante
TROPHONEUROSIS - [med] trophonévrose *f*
TROPHOPLASM - [biol] (protoplasm which is mainly concerned with nutrition) trophoplasma *m*
TROPIC - [chem] tropique
[geogr] s. tropical
~ACID - [chem] (a monobasic unsaturated acid and a precursor of atropine) acide *m* tropique
~-PROOF BEER - [brew ind] (special beer to withstand tropical climates) bière *f* pour l'exportation aux régions chaudes
TROPICAL - [geogr] (pertaining to the tropics) tropique, tropical
~CLIMATE - [met] climat *m* tropical
~CONDITIONS - [gen] (relating to engines, electrical machines etc, which must be operated in the tropics) conditions *fpl* tropicales
~CYCLONE - [met] cyclone *m* tropical
~DISTURBANCE - [met] (term used by the U.S.Weather Bureau for a cyclonic wind system occurring in the tropical zone but not reaching the level of "storm" or "hurricane") perturbance *f* tropicale
~FRENZY - [med] délire *m* aigu tropical
~MARITIME AIR - [met] (air mass of high surface temperature and specific humidity, originating over ocean areas in the tropical zone) air *m* tropical maritime
~REVOLVING STORM - [met] (a small depression of cyclonic character originating over ocean areas in the tropical zone, giving rise to winds of great force blowing round a calm centre) cyclone *m* tropical
~SWITCH - [el] (a switch having feet or bosses so arranged as to provide an air space between the base and the mounting surface as a safeguard in excessively damp conditions) interrupteur *m* hydrofuge
TROPICAL YEAR - [astr] (the period on which the Civil Year is based (365.2425...mean solar days) année *f* tropicale
~ZONE - [geogr] (the belt of the earth's surface delimited by the N. and S. tropics) région *f* tropicale
TROPICALIZATION - [packag] (the treatment or packing of parts or units in a special manner to preserve them against the effects of a tropical climate) tropicalisation *f*
TROPICS - [astr] (the parallels of latitude marking the limits of the sun's declination (23 deg 27' N. or S. of the Equator) tropiques *mpl*
TROPINE DIPHENYLMETHYL ETHER - [chem] (benztropine methanesulphonate. An antihistaminic, anticholinergic and anaesthetic agent with application in medicine)éther *m* diphénylméthylique de tropine
TROPISM - [biol] (reflex response to an external stimulus, involving movements of the whole body) tropisme *m*
TROPOPAUSE - [met] (the region of transition between troposphere and stratosphere)tropopause *f*
TROPOSPHERE - [met] (the part of the atmosphere nearest to earth, in which the rate of change of temperature with respect to altitude is relatively great) troposphère *f*
TROPOSPHERIC ABSORPTION - [radio] (the loss of energy in the transmission of radio waves due to dissipation in the atmosphere) absorption *f* atmosphérique, absorption *f* troposphérique
~DUCT - [radio] (a stratum of the troposphere in which an abnormally large proportion of any radiation of sufficiently high frequency is confined and over part of which there exists a negative gradient of modified refractive index) conduit *m* troposphérique
~MODE - [radio] (a mode of propagation) mode *m* troposphérique
~SCATTERING - [radio] diffusion *f* troposphérique
~WAVE - [radio] onde *f* troposphérique
TROT - [zool] trot *m*
TROTTER - [zool] cheval *m* de trot, trotteur *m*
TROUBLE - [mech] trouble *m*, panne *f* , défaut *m* de fonctionnement
[gen] dérangement *m*, peine *f*
[gen] (labour troubles etc) conflits *mpl* sociaux
[med] dérangement *m*, trouble *m*
~-FREE - [mech] à l'épreuve de pannes
~HUNTER - [gen] dépanneur *m*
~HUNTING - s. trouble shooting
~LAMP - [auto] baladeuse *f*
~LOCATION - [comput etc] localisation *f* d'un défaut
~-LOCATION PROBLEM - [comput] problème *m* de la localisation d'un défaut
~-SHOOT, to - s. troubleshoot, to
~-SHOOTER - [gen] (an expert mechanic or electrician whose duty is to locate and rectify faults in equipment) ouvrier *m* spécialisé
[instr] (servicing instrument designed to measure voltages and current in electronic equipment) appareil *m* de dépannage
~-SHOOTING - [gen] (the process of locating and remedying faults in equipment of any kind) recherche *f* d'un défaut, dépistage *m* d'un défaut
~-SHOOTING DEVICE - [instr] appareil *m* pour la localisation de défauts
~TONE - [telecomm] (in telephony) signal *m* de dé-

rangement
TROUBLESHOOT, to - [comput] mettre au point
TROUGH - [met] (a belt of low atmospheric pressure) zone *f* dépressionnaire
[hydr] (a gutter) rigole *f*, pierrée *f*, cunette *f*, caniveau *m*
[gen] (an open receptacle) bac *m*, auge *f*, augette *f*
[naut] (of a wave) creux *m* d'une onde
[geol] fond *m* de bateau, auge *f*
[min] rigole *f* d'écoulement
[min] élément *m* de couloir oscillant
[geol] pli *m* synclinal
~-ACCUMULATOR - [el] accumulateur *m* à augets
~AND RIDGE FAULTS - [geol] failles *fpl* à rejet compensateur
~AXIS - [geol] axe *m* synclinal
~BATTERY - [el] pile *f* à auges
~BEND - [geol] charnière *f* synclinale
~BRIDGE - [constr] pont *m* à tablier inférieur, pont *m* en forme d'auge
~CASTING - [metall] coulée *f* avec cuve réfractaire intermédiaire
~CHARGING CRANE - [metall] grue *f* d'enfournement des récipients
~COMPASS - [instr] déclinatoire *m*
~CONVEYOR - [mech] transporteur *m* à palettes, convoyeur *m* à palettes
~CORE - [geol] noyau *m* synclinal
~FAULT - [geol] faille *f* en fond de bateau
~GIRDER - [constr] poutre *f* à ornière, poutre *f* en U
~GUTTER - [constr] (parallel gutter used along roof valleys) chéneau *m* encaissé
[constr] gouttière *f* en V
~GUTTER TILE - [constr] tuile *f* à onglet en auge
~LIMB - [geol] flanc *m* inverse, flanc *m* renversé
~LINE - [geol] arête *f* synclinale
~MIXER - [mech] malaxeur *m* à auge
~TIPPING WAGON - [constr] basculeur *m* à auge
~WASHER -[min] langelotte *f*
~WAVEGUIDE - [radio] (waveguide consisting essentially of a channel with a centre fin parallel to the outer walls) guide *m* d'ondes en gouttière
~WELDING - [mech] soudure *f* en U
TROUGHED BELT CONVEYOR - [mech] (type of conveyor in which the upper run of the belt is curved into the form of a channel or trough by suitably-arranged rollers) bande *f* transporteuse en V
~CORE - [metall] boîte *f* en auge
TROUGHING - [el] (a preformed channel in which cables are laid to protect them against external mechanical damage) caniveau *m*
TROUSERING - [text] étoffe *f* pour pantalons
TROUSSE - [gen] trousse *f*
TROUT - [zool] truite *f*
TROUTSTONE - s. troctolite
TROWAL - [met] (a Canadian term for a trough of warm air at some distance above the surface of the earth) trowal *m*, zone *f* de basse pression
TROWEL , to - [constr] (the operation of applying a pasty material with a trowel or spatula) étaler avec la truelle
TROWEL - [impl] truelle *f*, gâche *f*
[constr] (for plastering) plâtroir *m*, truelle *f* brettée
[metall] (for foundry work) spatule *f*
TROWELLING - [constr] (the application of a pasty material with a trowel or spatula) étalage *m* avec la truelle
TRUCE - [gen] trêve *f*
TRUCK - [gen] camion *m*, fardier *m*, binard *m*
[transp] diable-brouette *m*, diable *m* de camionneur
[railw] (of a locomotive or a railway coach) bogie *m*, truck *m*, bissel *m*
[el] (a self supporting structure for transformers etc) chariot *m*
[gen & comm] troc *m*, échange *m*
[comm] (payment in kind) payment *m* en nature
[mining] berline *f*, benne *f*
[naut] pomme *f* de mât
[mech] (in an overhead travelling crane) chariot *m* roulant
~AND TRAILER - [transp] train *m* routier
~-AXLE - [railw] essieu *m* porteur
~BOLSTER - [railw] traverse *f* dansante
~BRAKE - [mech] frein *m* du bogie
~CENTERING DEVICE - [railw] crapaudine *f* du bogie
~CENTRES - [railw] écartement *m* des pivots des bogies
~CRANE - [mech] grue *f* sur camion
~DRIVER - [transp] camionneur *m*
~EFFECT - [cin] (a stereoscopic effect) effet *m* de travelling
~FARM - [agric] (USA only) jardin *m* maraîcher, horticulture *f*
~FRAME - [mech] châssis *m* du bogie
~GARDEN - s. truck farm
~GENERATOR SUSPENSION - [el] suspension *f* de la génératrice
~HINGE - [auto] (a fitting attached to the truck frame which is pivoted to the body hinge) articulation *f* de carrosserie
~-LOAD - [transp] charge *f* de wagon
~SHOT - [cin] (shotting a moving object and keeping at the same distance) travelling *m*, plan *m* pris en mouvement
~SIDE - [railw] flanc *m* du bogie
~SWING - [railw] (angular displacement of bogey) angle *m* de braquage du bogie, braquage *m* du bogie
~SYSTEM - [comm] (a system whereby wages are paid in goods) paiement *m* des ouvriers en nature
~TRACTOR - [auto] tracteur *m* pour semiremorque
~TRAILER - [auto] remorque *f*
~-TYPE SWITCHBOARD - [el] (a switchboard embodying sections each of which is mounted on wheels or carried on a truck so that it may be completely disconnected from the remainder and taken away for adjustment or repair) tableau *m* roulant
~WHEEL - [railw] roue *f* de wagon, roue *f* porteuse
[mech] galet *m* de diable
TRUCKER - [transp] camionneur *m*
[agric] (USA only) horticulteur *m*
TRUCKING - [gen] roulage *m*
~EFFECT - s. truck effect
TRUE, to - [mech] ajuster, défausser, rectifier, dégauchir
[mech] (to dress a grinding wheel) dresser, redresser
TRUE - [gen] vrai, exact
[gen] (of manufactured goods) pur, authentique
[instr etc] exact, precis, réel
[mech] rectiligne, droit, ajusté
~A FLAT SURFACE, to - [mech] redresser une surface plane
~A ROD, to - [mech]défausser une tige, dégauchir

une tige
TRUE A WHEEL, to - [mech] assurer le centrage d'une roue
~ABSORPTION - [phys] (that part of the total absorption which derives from the complete or partial loss of energy of photons of the primary beam) absorption ℓ réelle
~ACTIVITY CONSTANT OF A SOURCE - [radiat] constante ℓ d'activité réelle d'une source
~ADHESION - [mech] adhérence ℓ réelle
~AIRSPEED - [aero] vitesse ℓ vraie
~ALTITUDE - [aero] (indicated altitude corrected for air temperature) altitude ℓ réelle
[astr] hauteur ℓ observée (d'un astre), altitude ℓ absolue
~ANGLE OF INCIDENCE - [aero] angle m d'incidence réel
~ARC VOLTAGE - [el] tension ℓ d'arc réelle
~AZIMUTH - [astr] (the azimuth measured from the true north) azimut m réel
~BEARING - [astr] (the angle between the plane of the meridian and a plane containing both the point of observation and the point observed) azimut m
~BEARING UNIT - [radar] appareil m de relèvement réel
~CENTRE SPINNING - [mech] centrifugation ℓ pure
~CENTRIFUGAL CASTING - s. true centre spinning
~COINCIDENCE - [phys] (in a counter tube, a coincidence due to the incidence of a single particle of several genetically related particles) coïncidence ℓ réelle
~COMPLEMENT - [math] complément m à B
~COPY - [comm & leg] copie ℓ certifiée, copie ℓ authentique
~COUNT - [astronaut] durée ℓ de phase ascensionnelle
~COURSE - [radar] route ℓ vraie
~DENSITY - [metall] densité ℓ théorique
~DIP - [geol] inclinaison ℓ maximum d'une couche
~HEADING - [aero] route ℓ géographique
~HEIGHT - [surv] (the vertical distance between a given point and a known datum, e.g. M.S.L.) altitude ℓ réelle
~HORIZON - [surv] horizon m réel
~JADE - [min] jadéite ℓ
~NORTH - [surv] (as referred to the earth's axis) nord m géographique
~OHM - [el] (the true value of the ohm) ohm m international
~PLOT - [radar] tracé m réel
~RADIO - [meas] (of a current or a potential instrument transformer) rapport m réel
~RESISTANCE - [el] résistance ℓ ohmique
~TENSILE STRESS - [phys] effort m maximal de traction
~TONE REPRODUCTION - [photo] rendu m correct des tons
~UP, to - [mech] dégauchir, ajuster, défausser, dégauchir
[mech] (in milling) ribler
~UP THE CUTTER, to - [mach tool] ajuster l'outil, rectifier l'outil
~WATTS - [el] puissance ℓ active
~ZENITH DISTANCE - [surf] (the zenith distance obtained by actual observation) distance ℓ zenithale réelle
TRUEING - [mech] dégauchissement m, redressement m
TRUEING ATTACHMENT - [mach tool] dispositif m à rectifier
~DEVICE - [mach tool] (for grinding wheels) meule ℓ à rectifier
TRUFFLE - [bot] truffe ℓ
TRUING - s. trueing
TRULY CONICAL SCREW - [mech] vis ℓ conique
TRUMPET, to - [mus] trompeter, sonner de la trompette
[zool] (of elephants) barrir
[mech] évaser
TRUMPET - [mus] trompette ℓ
[mus] (a reed organ stop) trompette ℓ
[mech] tuyau m évasé
[text] (a guide funnel) entonnoir m
[acoust] pavillon m
~ARCH - [constr] trompe ℓ
~GUIDE - [mech] entonnoir m
TRUMPETING - [mech] (the flaring of a a pipe etc) évasement m
[mech] (in motors and engines, carbon deposits) dépôts mpl de calamine
TRUNCATE, to - [gen] tronquer
[comput] (to drop digits of a number, or terms of a series) tronquer
[math] rompre
TRUNCATED CONE - [geom] tronc m de cône
~PICTURE - [telev] (term denoting that the upper part of the picture is cut off) image ℓ tronquée
TRUNCATION - [comput] tronquage m
~ERROR - [comput] erreur ℓ de tronquage
TRUNC, to - [min] débourber
TRUNK - [bot] (of a tree) tronc m
[arch] (column shaft) fût m (de colonne)
[gen] (a box-like container) coffre m, malle ℓ
[min] caisse ℓ à débourber le minerai
[el] bus m
[telecomm] (a circuit directly connecting two exchanges, also called, in GB, a junction circuit) ligne ℓ de jonction
[zool] (of an elephant) trompe ℓ (d'éléphant)
[naut] caisse ℓ
[railw] artère ℓ principale
[anat] tronc m
[hydr] (a large counduit) conduite ℓ maîtresse
[text] slip m
~CABLE - [telecomm] câble m à longue distance, câble m interurbain
~CIRCUIT - [telecomm] ligne ℓ auxiliaire, ligne ℓ de jonction, circuit m interurbain
~CONGESTION LAMP - [telecomm] lampe ℓ d'occupation interurbaine
~CONGESTION SIGNAL - [telecomm] signal m d'occupation interurbaine
~CONNECTION - [telecomm] communication ℓ interurbaine
~ENGINE - [mech] (a type of steam engine) machine ℓ à fourreau
~EXCHANGE - [telecomm] (in telephony, an exchange in a telephone area connected by long-distance lines to other trunk exchanges) bureau m central interurbain
~FEEDER - [el] feeder m d'interconnexion
~GROUP - [telecomm] faisceau m de lignes, faisceau m de jonctions
~GROUP AREA - [telecomm] réseau m téléphonique

urbain desservi par plusieurs bureaux
TRUNK LINE - [telecomm] (trunk circuit) ligne *f* jonction
~LINE CONDUIT - [telecomm] conduit *m* pour ligne urbaine
~LOSS - [telecomm] équivalent *m* de transmission effective du circuit interurbain
~MAIN - [gas ind etc] (a transmission main, a pipeline) canalisation *f* de transport, artère *f*, pipeline *m*
~OF CABLES - [el] groupe *m* de câbles
~OFFERING SELECTOR - [telecomm] sélecteur *m* pour l'offre des communications interurbaines
~PISTON - [mech] (the piston of a trunk engine) piston *m* à fourreau
~RECORD POSITION - [telecomm] position *f* d'enregistrement
~ROAD - [gen] (a main road) grande route *f*
~SERVICE OBSERVATION - [telecomm] observation *f* des circuits interurbains
~TRAFFIC - [telecomm] trafic *m* interdistrict
~WORKING - [telecomm] établissement *m* des communications interurbaines à la commande de l'opératrice locale
~ZONE - [telecomm] zone *f* interurbaine
TRUNKING - [telecomm] exploitation *f* urbaine par lignes auxiliaires
TRUNKING - [min] débourbage *m*
TRUNNEL - [shipbuild] (a treenail) gournable *f*
TRUNNION - [mech] tourillon *m*, goujon *m*
~BAND - [firearms] (of a cannon) frette *f* à tourillons
~JIG - [mech] calibre *m* à tourillons
~RING - [mech] ceinture *f*
~SHOULDER - [mech] embase *f* des tourillons
~TIP WAGON - [metall] culbuteur *m* à tourillons
~-TYPE FIXTURE - [mech] outil *m* oscillant
TRUNNIONS - [mech] (short shafts projecting from the sides of a part to provide a pivoting movement upon them, as in the elevation of an antique cannon) axes *mpl*, tourillons *mpl*
TRUSS, to - [constr] (to support by a truss) armer, renforcer, contre-ficher
[naut] latter
[agric] (the hay) botteler
TRUSS - [constr] (a built-up system forming a beam of the like stressed element) ferme *f*, armature *f*
[arch] (a large corbel, a modillion) encorbellement *m*
[med] (a surgical appliance) brayer *m*, bandage *m* herniaire
[naut] (iron band secured to a mast) drosse *f*
[agric] (of hay, straw etc) botte *f* (de foin), touffe *f*
[mech] (in autos) barre *f* de renfort
~-BEAM - [constr] poutre *f* armée
~BRIDGE - [constr] pont *m* métallique à poutres armées
~GIRDER - [constr] poutre *f* armée
~POINTER - [meas] (a measuring instrument pointer of reinforced construction) aiguille *f* renforcée
~ROD - [constr] étai *m*, tirant *m*
TRUSSED BLADE - [el] couteau *m* de contact renforcé
~POLE - [el & telecomm] appui *m* haubanné sur lui-même
TRUSSING - [constr] (the members of the truss) ferme *f* en arbalète, poutre *f* armée
TRUSSING - [constr] (reinforcement by struts etc) armature *f*
~OF FRAME - [auto] renforcement *m* du châssis
TRUST - [gen] confiance *f*
[comm] crédit *m*
[leg] fidéicommis *m*
[fin] trust *m*, syndicat *m*
TRUTH - [gen] vérité *f*, véracité *f*
[mech] (accuracy of position etc) centrage *m* de précision
~TABLE - [math] table *f* de définition
[comput] table *f* de fonction
TRY, to - [gen] éprouver
[gen] (to experiment) expérimenter
[gen] (to test) mettre à l'épreuve, essayer
TRY - [gen] (colloq. for trial) essai *m*, tentative *m*
~A TENON IN A MORTISE, to - [carp] présenter un tenon à une mortaise
~-COCK - [hydr] robinet *m* de hauteur d'eau, robinet *m* de jauge
~PLANE - [tool] varlope *f*, jointout *m*
~-SQUARE - [impl] équerre *f* à lame d'acier
TRYING PLANE - s. try-plane
~UP - [carp] varlopage *m*
~SQUARE - s. try-square
~-UP MACHINE - [mech] varlopeuse *f*
TRYOUT - [gen] essai *m* à fond
TRYPARSAMIDE - [chem] (a trypanocide used in medicine) tryparsamide *f*
TRYPSIN - [biochem] (enzyme occurring in plants and animals which converts protein to amino acids and peptones) trypsine *f*
TRYPTOPHANE - [chem] (an essential amino triptofano) tryptophane *m*
TRYSAIL - [naut] voile *f* goélette
T-SLOT - [mach tool] (a groove having a cross-section like an inverted "T", commonly formed in machine tables to aid in mounting workpieces) rainure *f* à T
T SQUARE - [impl] (instrument for laying out right angles or parallel lines) équerre *f* en T, té *m* de dessin, té *m*
TSUNAMI - [geol] (seismic sea-wave caused by earth-tremors on the sea floor, usually having a great translational velocity and sometimes great amplitude) onde *f* sismique maritime
TT PIECE - [plumb] croix *f* à quatre brides
TTT-CURVE - [metall] (Time-Temperature-Transformation curve) courbe *f* de transformation isothermique, courbe *f* TTT
TTV - [radiat] (Tenth-Value Thickness) couche *f* d'atténuation au dixième
TUB, to - [mining] (the lining of a circular shaft) cuveler, boiser
[agric] encaisser (une plante)
[ind chem] laver dans un bac
TUB - [impl] bac *m*, cuve *f*
[hydr] (in a bathroom) baigoire *f*
[mining] (a type of wagon) berline *f*, berlaine *f*
[min] (in coal mines, a hoisting bucket) cuffat *m*, benne *f*, tonne *f*
[naut] (colloq) barcasse *f*, vieux sabot *m*
~FILE - [comput] fichier *m* d'extraction
~SIZING - [paper man] (application of waterproofing agents after the formation of the paper) surfaçage *m*
~-SIZING-MACHINE - [paper man] machine *f* à surfacer

TUBA - [mus] (a reed organ stop) tuba *m*
[mus] (the brass wind instrument) tuba *m*
TUBATORSION - [med] torsion *f* de la trompe de Fallope
TUBBER - [impl] pic *m* à deux pointes
TUBBING - [mining] cuvelage *m*, boisage *m*
~PLATES - [mining] plaques *f*pl de cuvelage
TUBE, to - [gen] garnir de tubes
TUBE - [gen] (long, hollow cylindrical body) tube *m*, tuyau *m*
[auto] chambre *f* à air
[el] (a duct or pipe) conduite *f*
[firearms] tube *m*
[transport] (a tunnel, a subway) voie *f* souterraine tubulaire
[transp] (colloq) chemin *m* de fer métropolitain, métro *m*
[metall] (of a furnace) tube *m* de chaudière
[photo] barrillet *m* (d'un objectif)
[med] drain *m* d'une plaie
[anat] tube *m*, canal *m*
[mech] (of a lock) canon *m* (de serrure)
[electron] (electronic tube) tube *m*
[mus] (the lower portion of the reed pipe of an organ) partie *f* inférieure du tuyau
~BASE - [plumb] socle *m* du tube, base *f* du tube
~BEADER - [mech] mandrin *m*
~BENDER - [mech] cintreuse *f*
~BENDING MACHINE - [mech mach] cintreuse *f*
~-BOILER - [th eng] chaudière *f* tubulaire
~BRUSH - [tool] brosse *f* à tubes, brosse *f* à écouvillon
~BURR REMOVER - [tool] fraise *f* ébarbeuse
~BUSHING - [plumb] bouchon *m*
~CLEANER - [impl] passe-diable *m*
~CLIP - [impl] pince *f* pour tubes
~COEFFICIENT - [electron] (valve factor) constantes *f*pl de tubes électroniques
~COILING MACHINE - [mech] enrouleuse *f* pour tubes
~COLUMN - [hydr] colonne *f* de tubes
~COMPLEMENT - [electron] (or valve complement the number of electron tubes required in a piece of electronic equipment) garniture *f* de tubes électroniques
~CONVERTER - [radio] convertisseur *m* à tube
~CONVEYOR - [nucl] (passage in a nuclear reactor for a shuttle) tube *m* d'entrée
~COUNT - [electron] (a terminated discharge produced by an ionizing event) coup *m*, impulsion *f* de comptage
~CUTTER - [tool] coupe-tube *m*
~ELECTROMETER - [instr] (instrument for measuring small differences in potential by making use of capillary action) électromètre *m* à tube
~ENGINE - [metall] banc *m* à étirer les tubes
~EXPANDER - [tool] dudgeon *m*, extendeur *m*, expanseur *m*
~EXTRUDER - [rubber ind] boudineuse *f* à chambres à air
~EXTRUSION - [plast ind] (extrusion of material in pipe-like form) extrusion *f* de tubes
~FACTOR BRIDGE - [electron] pont *m* de mesure pour tubes électroniques
~FILTER - [radiat] (filter used as attachment to the X-ray tube shield) filtre *m* de gaine
~FOCUS - [radiat] (X-ray focal spot) foyer *m* de tube

TUBE FOR BORING WELLS - [mining] tube *m* pour l'exploitation par forage de gisements
~HEATING SURFACE - [th eng] surface *f* de chauffe des tubes
~HEATING TIME - [electron] temps *m* de préchauffage
~HOLDER - [ind chem] porte-tube *m*
~IN A VECTOR FIELD - s. tube of force
~INFLATION FORMER - [rubber ind] machine *f* à galber à air comprimé
~LIGHTER - [auto] (a tyre lever) démonte-pneu *m*, levier *m* de montage
~LOOM FOR CARPETS - [text] métier *m* à tubes pour tapis
~MILL - [min] (a grinding machine consisting of a cylindrical revolving barrel charged with pebbles or ceramic balls. Its action is similar to that of a ball mill, but as no part of the barrel is conical, there is no automatic classification of balls as wear reduces their size, nor are steel balls generally used) broyeur *m* à tube, tube *m* broyeur, tube *m* finisseur
~MOULD - [rubber ind] moule *m* pour chambres à air
~NOISE - [electron] bruit *m* de fond d'un tube électronique
~OF ELECTRIC FLUX - s. tube of force
~OF FORCE - [phys] (a tube through which unit flux passes) tube *m* de force
~OF MAGNETIC FLUX - [el] tube *m* de flux magnétique
~OF SPOOL - [text] tube *m* de la bobine
~-PLATE - [th eng] (the end-plate of a tubular boiler heat-exchanger or the like, which receives and supports a nest of tubes) plaque *f* tubulaire, plaque *f* de tête
~PRESSURE-GAUGE - [instr] manomètre *m* à tube
~PROTECTOR - [auto] protecteur *m* de chambre à air
~RAILWAY - [railw] chemin de fer *m* souterrain
~REACTOR - [radio] (a circuit incorporating a tube, or valve behaving as a reactance, the magnitude of which is a function of the tube and circuit parameter) circuit *m* à tube de réactance
~REACTOR MODULATOR - [radio] modulateur *m* à tube de réactance
~REAMER - [metall] fraise *f* ébarbeuse
~RECTIFIER - [electron] redresseur *m* à tube
~ROLLING - [metall] laminage *m* de tubes, procédé *m* dit "au pas de pélerin"
~ROLLING MILL - [metall] laminoir *m* à tubes
~SCRAPER - [impl] raclette *f* pour tubes de chaudière
[oil ind] racloir *m* de tubes
~SCREWING AND CUTTING OFF MACHINE - [mech] machine *f* à tarauder et à tronçonner les tubes
~-SCREWING MACHINE - [mech] machine *f* à tarauder les tubes
~SHEET - s. tube plate
~SHIELD - [electron] (or valve shield, shield placed around a tube) écran *m* de tube
~SOCKET - [radio] support *m* du tube
~SOLUTION - [rubber ind] dissolution *f* pour chambres à air
~SPANNER - [tool] (a box spanner) clé *f* à tube
~SPLICER - [rubber ind] machine *f* à joindre les bouts de chambres à air
~STAND - [ind chem] porte-éprouvette *m*
[radiat] (support for holding an X-ray tube in position) support *m* de tube

TUBE STEEL - [metall] acier *m* pour tubes
~TESTER - [electron] lampemètre *m*
~TESTING TUB - [rubber ind] cuveau *m* à essayer chambres à air
~-TONGS - [impl] pinces *fpl* pour tubes
~TRIMMING MACHINE - [mech] ébarbeuse *f* pour tubes
~VICE - [impl] étau *m* à tubes
~VOLTAGE DROP - [electron] (or valve voltage drop in an electronic tube, the anode voltage during conduction) chute *f* de tension
~VULCANIZER - [rubber ind] vulcanisateur *m* pour chambres à air
~WELD - [metall] soudure *f* de tube
~WELL - [mining] puits *m* instantané
~WINDING MACHINE - [mech] machine *f* à bandeler les tuyaux
[print] machine *f* à enrouler les étuis
~WITH DRAWN END - [glass man] tube *m* effilé
~WITH STOP-COCK - [ind chem] tube *m* à robinet
~-WORKS - [gen] tuyauterie *f*
~WRENCH - [impl] serre-tube *m*, clef *f* pour tubes
TUBEHEAD - s. tube plate
TUBELESS - [rubber ind] "tubeless", sans chambre à air
~TYRE - [rubber ind] (type of tyre having no inner airtube, a low leakage butyl rubber lining and a large diametre rubber seal against the wheel halves to retain the air pressure) pneu *m* tubeless, pneu *m* sans chambre à air
~TYRE MOUNTING BAND - [rubber ind] système *m* écarteur des talons
TUBER - [anat] tubérosité *f*
[bot] tubercule *m*, racine *f* tubéreuse
[med] tubercule *m*
TUBERCLE - [med] tubercule *m*
TUBERCULOCELE - [med] tuberculose *f* testiculaire
TUBERCULOSIS - [med] phtisie *f*, tuberculose *f*
~OF THE SKIN - [med] lupus *m*, tuberculose *f* cutanée
TUBEROSITY OF THE HUMERUS - [med] tubérosité *f* humérale
TUBES TYPE PLATE - [el chem] (plate in alkaline battery containing an assembly of metal tubes in which the active material is placed) plaque *f* à tube
TUBIFEROUS - s. tuberous
TUBING - [mech] tuyautage *m*
[metall] (a length of tube) tube *m*
[constr] tubage *m*
[mech] (good-quality pipes with threads and couplings) manchon *m*
[mining] tubage *m* d'un puits
[plast ind] (the making of tubes) boudinage *m* de tubes, extrusion *f* de tuyaux flexibles
[oil ind] tube *m* de pompage, tube *m* de production, tubing *m*
[oil ind] colonne *f* d'extraction
~A BOILER - [th eng] garnissage *m* de tubes de chaudière
~BLOCK - [oil ind] moufle *f* de tubing
~BORE-HOLES - [mining] tubage *m* des trous de sonde
~CATCHER - [oil ind] mâchoires *fpl* de suspension pour tubes
~CLAMP - [oil ind] collier *m* de serrage
~CUTTER - [tool] coupe-tubes *m*
[oil ind] coupe-tubing *m*

TUBING HEAD - [oil ind] (used in well drilling) tête *f* de tube
~HEAD SPOOL - [oil ind] bobine *f* de bridage du tubing
~MACHINE - [rubber ind] tubeuse *f*, boudineuse *f* à tuyaux
[print] machine *f* à emboutir les étuis
~SPIDER - [oil ind] (used in well drilling) potence *f* pour tubes
TUBINGLESS COMPLETION - [oil ind] complètement *m* sans colonne de production
TUBO-OVARIOTOMY - [med] résection *f* de l'ovaire
TUBOCURARINE CHLORIDE - [chem] (the chloride of tubocurarine, an alkaloid obtained from Chondodendron tomentosum, having application in medicine) chlorure *m* de tubocurarine
TUBOTORSION - s. tubatorsion
TUBULAR - [mech etc] (having the form of a tube containing, or provided with tubes, e.g."tubular boiler") tubulaire
~BOILER - [th eng] chaudière *f* tubulaire
~BRIDGE - [constr] pont *m* tubulaire
~CAPACITOR - [electron] condensateur *m* tubulaire
~CHASSIS - [auto] châssis *m* tubulaire
~CONSTRUCTION - [railw] (skinstressed structure for a vehicle body) construction *f* tubulaire
~COP - [text] canette *f* cocon, canette *f* tubulaire
~CROSS BEARER - [auto] traverse *f* tubulaire
~DIE - [plast ind] (an annular die used for extruding pipes) filière *f* annulaire
~FELT - [text] manchon *m* de feutre, feutre *m* tubulaire
~FILM - [plast ind] (film produced by extrusion as tube followed by slitting) feuille *f* extrudée en gaîne, feuille *f* soufflée
~FRAMEWORK - [metall] carcasse *f* tubulaire
~GIRDER - [constr] poutre *f* tubulaire, poutre-caisson *f*
~INSULATION - [plast ind] (provision of a heat-insulating sleeve within the gate of a mould to increase flowing time and reduce amount of material left in the gate) isolement *m* tubulaire
~LAMP - [el] lampe *f* tubulaire
~LANTERN - [impl] lanterne-tempête *f*
~MAST - [el & telecomm] mât *m* tubulaire
~RESISTANCE HEATER - [plast ind] (extruder barrel heater in the form of a tube formed to fit round the barrel and containing resistance wires) dispositif *m* de chauffage par résistance en tube
~RIVET - [mech] rivet *m* tubulaire
~SHAFT - [mech] arbre *m* tubulaire
~SPANNER - [impl] clé *f* à tube
~STREAMER - [impl] manche *f* à air
~THREAD GUIDE - [text] tube *m* courbe, guide-fil *m* tubulaire
~UNIT RADIATOR - [th eng] (column radiator) radiateur *m* tubulaire à éléments
~VECTOR FIELD - [el] (solenoidal field) champ *m* solénoïdal
~VIEWFINDER - [photo] viseur *m* tubulaire
TUBULATING MACHINE - [electron] (machine designed to seal the exhaust tube to the bulb) machine *f* à sceller les queusots
TUBULATION - [phys] (in the vacuum technique, a glass tube through which the air is pumped out of the envelope to be evacuated) queusot *m* de pompage

TUBULOUS BOILER - [th eng] chaudière *f* tubuleuse
TUBULURE - [ind chem etc] (of a retort etc) tubulure *f*
TUCK - [text] application *f* par couture
~AND WELT CLOTH - [text] rangée *f* de fil molleton
~CAM - [text] came *f* façom-métier
~IN , to - [print] mettre sous étuis
~NEEDLE - [text] aiguille *f* de retenue
~PATTERN - [text] effet *m* guilloché
~POSITION - [text] position *f* de cueillage
~PRESSER - [text] roue *f* de presse
~THE SAND, to - [metall] serrer le sable
~STITCH - [text] maille *f* double
TUCKED LOOP - [text] maille *f* chargée
TUCKING-IN MACHINE - [print] machine *f* à emboîter
TUE-IRON - [metall] tuyère *f*
TUFA - [min] (a variety of calcium carbonate with a cellular structure, deposited from streams and springs) tuf *m*, tuf *m* calcaire
TUFACEOUS - [min] tufacé
TUFF - s. tufa
TUFT, to - [gen] garnir d'une touffe
[pext] (to pad) capitonner, piquer
TUFT - [gen] touffe *f*
[text] houppe *f*, freluche *f*, flocon *m*, méche *f*
[anat] (of blood vessels) glomérule *m* (de vaisseaux sanguins)
~OF COTTON - [text] riste *f*, barbe *f* de fibre
~OF SILK FIBRES)- [text] touffe *f* de fibres de soie
~OF WOOL - [text] touffe *f* de laine
TUFTED FRACTURE - [text] surface *f* de rupture en forme de touffes
~SURFACE OF FRACTURE - s. tufted fracture
TUG, to - [gen] tirer avec un effort
[naut] remorquer
TUG - [naut] remorqueur *m*
[gen] traction *f*, saccade *f*
TUGBOAT - [naut] (generally a small compact vessel designed for towing) remorqueur *m*, toueur *m*
TUGGED WARP - [mech] chaîne *f* tiraillée
TUGGER HOIST - [mech] treuil *m* à air comprimé
TULE MINT - [bot] baume *m* des champs
TULIP - [bot] tulipe *f*
~-HEAD VALVE - [mech] (type of reciprocating i. c. engine valve, having a deeply-recessed and flared head of trumpet-shape) soupape *f* en forme de tulipe
~TREE - [bot] tulipier *m*, liriodendron *m*
~VALVE - s. tulip-head valve
TULLE - [text] tulle *m*
~CELL - [text] cellule *f* en tulle
TUMBLE, to - [gen] tomber, culbuter
[mech] (to treat objects by tumbling) broyer
[metall] dessabler au tonneau
~CARD - [comput] (a fractional card) carte *f* ventilée
~POLISHING - [mech] (process of polishing objects by placing them in a revolving barrel with other materials, e.g.pieces of cloth, wooden balls etc) polissage *m* au tonneau
~-TURN, to - [print] retourner
TUMBLER - [mech] tambour *m* de courroie, contrepoids *m*
[mining] culbuteur *m*
~SWITCH - [el] commutateur *m* à bascule, tumbler *m*
TUMBLING - [metall] (the operation of finishing, deburring or polishing objects by placing them in a revolving barrel) polissage *m* au tonneau
TUMBLING - [astronaut] basculement *m*
~BARREL - [metall] (a cylindrical vessel rotating on its horizontal axis, and charged with work-pieces and polishing agent) tonneau *m* de dessablage, tambour *m* dessableur
~SHAFT - [mech] (camshaft) arbre *m* à came, arbre *m* de la came
[mech] (a lay shaft) arbre *m* de relevage, arbre *m* de changement de marche
TUMBREL - [transp] topbereau *m*
[el] culbuteur *m* d'interrupteur
[paper man] cuve *f* de blanchiment, tambour *m* à décortication
TUMEFACTION - [med] (the swelling of a part, as in a tumour) tuméfaction *f*
TUMEFY, to - [med] (to swell, to puff up) tuméfier, de tuméfier
TUMID - [gen & med] gonflé, enflé
TUMOUR - [med] (a swelling on or in any part of the body) tumeur *f*
~DOSE - [radiat] (the depth dose at the tumour) dose *f* tumorale
TUN, to - [gen] mettre en tonneaux, entonner
TUN - [impl] (a large cask) tonneau *m*, fût *m*
[brew ind] cuve *f* de fermentation
~WAGON - [railw] wagon-foudre *m*
TUNA - [zool] thon *m*
~OIL - [ind chem] (an oil obtained from the livers of Thunnus vulgaris and a source of vitamins A and D. Tuna is also used in the manufacture of protective coatings) huile *f* de thon
TUNABLE ECHO BOX - [radio] (echo box consisting of an adjustable cavity operating in a single mode) résonateur *m* ajustable par échos artificiels
~ISOLATOR - [radio] (isolator for waveguides operating over the full waveguide frequency range) affaiblisseur *m* ajustable non-réciproque
~MAGNETRON - [electron] (a magnetron which can be tuned by electrical or mechanical devices over a range of frequencies) magnétron *m* réglable
TUNE, to - [gen and mus] accorder
[radio] syntoniser, accorder
[mech] (an engine) caler, mettre au point
[text] appareiller (un métier)
TUNE - [mus] (succession of musical tones forming a melody) air *m* (de musique)
[mus] (concord or unison) accord *m*
~INTO STATION, to - [radio] capter un poste, accrocher un poste
~UP, to - [mus] (to bring instruments to a common pitch) s'accorder
[mech] mettre au point
TUNED - [gen & mus] accordé
[radio] syntonisé
~AERIAL - [radio] (USA only, periodic or modulated antenna in the USA, standing aerial resonating to an applied sinusoidal E.M.F.) antenne *f* accordée
~AMPLIFIER - [el acoust] amplificateur *m* à résonance
~ANODE OSCILLATOR - [radio] oscillateur *m* à circuit anodique accordé
~CAVITY - [electron] (enclosure with a conductive inner wall whose resonant frequency is determined by its internal dimensions) cavité *f* accordée
~CIRCUIT - [el] circuit *m* accordé

TUNED DIPOLE - [radio] dipôle *m* accordé
~DOUBLET - s, tuned dipole
~-GRID OSCILLATOR - [radio] oscillateur *m* à circuit de grille accordé
~RADIO-FREQUENCY RECEIVER - [radio] récepteur *m* à amplification directe
~-REED RELAY - [el] relais *m* à lame vibrante, relais *m* à résonance
~TORSIONAL VIBRATION DAMPER - [contr] amortisseur *m* de vibrations à torsion accordé
~TRANSFORMER - [el] transformateur *m* accordé
~UP - [mech] (of an engine etc) mis au point
~WINDOW - [electron] (a window which is cut proportionally in dimensions so as reradiate waves of a given length or frequency) fenêtre *f* syntonisée
TUNER - [mus] accordeur *m*
[radio] (a device for tuning) dispositif *m* d'accord, syntonisateur *m*
[text] appareilleur *m* de métiers
~DRUM - [telev] tambour *m* sélecteur de canaux
TUNG OIL - [ind chem] (china-wood oil. A drying oil obtained from Aleurites cordata and used in the production of paints, varnishes, linoleum, and paint and varnish driers) huile *f* d'abrasin
TUNGAR TUBE - [electron] (a type of argon-filled tube or valve) redresseur *m* tungar
~VALVE - s, tungar tube
TUNGSTEN - [chem] (metallic element, symbol W, A.N.74, A.W.184 having a melting point of approximately 3400° C, the highest of all metals, has application in metallurgy in the production of corrosion-proof, magnetic and high-speed alloys, in electrical equipment, as filaments in electric lamps, and in X-ray technology) tungstène *m*, wolfram *m*
~ARC - [el] (an arc in which the main part of the radiation is emitted from incandescent tungsten electrodes) arc *m* à électrodes en tungstène
~BRASS - [metall] laiton *m* à tungstène
~BRONZE - [metall] bronze *m* à tungstène
~CARBIDE - [metall] (extremely hard compound with uses in the production of cutting tools and dies) carbure *m* de tungstène
~COBALT - [metall] alliage *m* de tungstène et cobalt
~COPPER - [metall] alliage *m* de tungstène et cuivre
~DISK - [radiat] (a part of a ray-tube targets) disque *m* en tungstène
~ELECTRODE - [el] électrode *f* en tungstène
~FILAMENT LAMP - [el] lampe *f* à filament de tungstène
~LAMP - s, tungsten filament lamp
~LIGHT - [el] (a tungsten-filament lamp light) lumière *f* au tungstène
~NICKEL - [metall] alliage *m* de tungstène et nickel
~STEEL - [metall] acier *m* au tungstène
TUNGSTIC ACID - [chem] (a mordant dyeing) acide *m* tungstique
~OCHRE - [min] tungstite *f*
~OXIDE - [chem] (a component of ceramic glazes and a textile flameproofing agent) oxyde *m* tungstique
TUNING - [mech] (i.c.engines) (the process of adjusting carburation, ignition, timing etc, to obtain maximum performance) mise *f* au point
[radio] (the adjustment of a radio receiver to obtain maximum response to the frequency it is desired to receive) syntonisation *f*
~ARCH - [mus] (in horns etc) corps *m* de rechange, ton *m* de rechange
TUNING BAND - [radio] gamme *f* de syntonie
~BOARD - [mus] (the wooden bar in a piano carrying the tuning pins) sommier *m* d'accordage
~COIL - [el acoust] bobine *f* syntonisatrice, self *m* d'accord
~CONDENSER - [radio] condensateur *m* d'accord
~CONTROL - [radio] (the optical control of the tuning of a receiver to sender by indicating instruments) indication *f* d'accord
~DIAL - [radio] (a graduated dial showing the frequency or wave-length to which a set is tuned) cadran *m* indicateur d'accord
~FORK - [acoust] (pitch-carrying instruments) diapason *m*
~-FORK CLOCK - [instr] horloge *f* à diapason
~-FORK OSCILLATOR - [radio] générateur *m* à diapason
~-FORK OSCILLATOR DRIVE - [radio] (electromechanical drive, the frequency of which is determined by the vibration of a tuning fork) pilotage *m* par diapason
~HAMMER - [tool] (used by a tuner) accordoir *m*, marteau *m* d'accordage
~IN - s, tuning (radio)
~INDICATOR - [radio] (a device usually a form of cathode ray tube which indicates visually whether the receiver is correctly tuned to an incoming signal) indicateur *m* d'accord
~KEY - s, tuning hammer
~KNOB - [radio] (a moulded head which can be turned with the fingers to adjust the tuning of a receiver) bouton *m* de syntonisation
~PIN - [mus] (USA only, peg in GB) cheville *f*
~PROBE - [radio] (for waveguides) sonde *f* ajustable
~RANGE - [electron] (in switching tubes) plage *f* de syntonisation
[radio] portée *f* d'un émetteur
~SCREW - [radio] (of a waveguides) vis *f* d'accord, vis *f* d'adaptation
~SHARPNESS - [radio] acuité *f* de syntonisation
~SLIDE - [mus] (of an organ pipe) coulisse *f* d'accord
~STRIP - [electron] (component part of the electrode structure of magnetrons) lamelle *f* d'accord
~STUB - [radio] (short length of transmission line connected to a transmission line for impedance matching purposes) bras *m* de réactance ajustable
~SUSCEPTANCE - [electron] (for special tubes) susceptance *f* de désaccord
~WAVE - [radio] onde *f* de syntonisation
~WIRE - [mus] (of reed pipes) rasette *f*
TUNNEL, to - [constr] percer un tunnel
TUNNEL - [constr] (artificial underground gallery) tunnel *m*, souterrain *m*
[mining] galerie *f*
[metall] (of a blast furnace) vide *m* d'un haut fourneau
~ACTION - [electron] (in a semi-conductor p-n junction, a process whereby conduction occurs through the potential barrier due to the tunnel affect and in which electrons pass in either directions between the conduction band in the n-region and the valence band in the p-region) action *f* tunnel
~BAR - [constr] affût *m* pour creusement de tunnels
~BURNER - [th eng] brûleur-tunnel *m*
~DIODE - [electron] (diode having a p-n junction in which tunnel action occurs) diode *f* tunnel
~DRIVING - [constr] percement *m* de tunnel

TUNNEL DRYER - [th eng] (tunnel heated from one end, through which bricks are passed on trucks to dry them) tunnel *m* de séchage
~EFFECT - [electron] (the piercing of a potential hill by a carrier, which would be impossible according to classical mechanics, but possible according to wave mechanics, if the width of the hill is small enough) effet *m* tunnel
~GUIDE - [ind chem] entonnoir *m*
~HEALDING - [constr] extrémité *f* de fendue
~KILN - [th eng] four *m* à tunnel
~MOUTH - [railw & constr] débouché *m* du tunnel
~OPENING - s. tunnel mouth
~OVEN - [metall] (a tubular furnace designed for annealing glass bulbs) four *m* à tunnel
~TRAY DRYER - [th eng] (apparatus consisting of a tunnel of sheet metal through which trucks carrying material on trays is passed) tunnel *m* de séchage à plateaux
~TYPE FURNACE - s. tunnel oven
~VAULT - [arch] voûte *f* en berceau
TUNNELING - s. tunnelling
TUNNELLING - [constr] percement *m* de tunnels
[mining] construction *f* d'une galerie (de mine)
~EFFECT - s. tunnel effect
TUP, to - [mech] (to strike with a power hammer) battre avec un mouton
[zool] (to copulate, of sheep) béliner
TUP - [mech] (the striking part of a power hammer) pilon *m*
[impl] (the striking part of a drop hammer) mouton *m*
[metall] (in foundry work, a cast iron mass used for breaking castings) poire *f* de casse-fonte
[zool] bélier *m*
~DIE - [mech] frappe *f*, panne *f* du pilon
~PALLET - s. tup die
TURBAN TUMOR - [med] épithélioma *m* bénin du cuir chevelu
TURBARY - [geol] tourbière *f*
TURBID - [gen] (cloudy, opaque (of a liquid) trouble, bourbeux
TURBIDIMETER - [instr] (an instrument for determining particle size) opacimètre *m*, turbidimètre *m*
TURBIDIMETRY - [phys chem] (the methods of analysis and measurement made by the turbidimeter) turbidimetrie *f*
TURBIDITY - [phys chem] (the cloudiness in liquid caused by the presence of finely-divided, suspended material) turbidité *f*, état *m* trouble
TURBINATE BODY - [med] os *m* turbiné, cornet *m*
~BONE - s. turbinate body
TURBINE - [mech] (an engine which consists of one or more rotary units mounted on a shaft and generally equipped with a series of curved vanes actuated by the reaction, impulse or suction of a fluid or gas under pressure) turbine *f*
~BLADE - [mech] aube *f* d'une turbine, palette *f* d'une turbine
~CHAMBER - [mech] réservoir *m* de turbine, chambre *f* d'eau, huche *f*, bâche *f* fermée
~DISK - [mech] (the rotating element carrying the blades of a turbine) disque *m* de turbine
~-DRIVEN LOCOMOTIVE - [railw] locomotive *f* à turbine
~-DRIVEN SET - [el] groupe *m* turbo-générateur
~-ELECTRIC DRIVE - [el] propulsion *f* turbo-électrique
TURBINE ENTRY DUCT - [mech] (the duct by which the gas stream passes from the combustion chamber to the turbine rotor) manche *f* à gaz
~ENTRY TEMPERATURE - [mech] (the temperature prevailing at the entrance to the turbine rotor casing) température *f* d'entrée de la turbine
~GENERATOR - s. turbogenerator
~IMPELLER - [mech] piston *m* tournant de turbine
~NOZZLE BLADES - [mech] palettes *fpl* directrices
~-PUMP - [mech] turbo-pompe *f*
~SET - [mech] groupe *m* de turbines
~SHELL - [mech] caisse *f* de la turbine
~SHROUD RING - [mech] (a ring designed to prevent gas escaping past the blade tips of a turbine) anneau *m* d'étanchéité de turbine
~STARTER - [mech] démarreur *m* à turbine
~-TYRE UNIT HEATER - [th eng] (a heater having a fan driven by a small turbine operated by the heating fluid) radiateur *m* soufflant à turbine
~WHEEL - [mech] (assembly consisting of the disk of a turbine with its blades) roue-turbine *f*
TURBINECTOMY - [med] turbinectomie *f*
TURBINED - [mech] à turbines
TURBINING - [oil ind] nettoyage *m* par cure-tubes rotatifs
TURBO-ALTERNATOR - s. turbo-generator
~-BLOWER - [mech] (used for compressed-air supply) turbo-souffleuse *f*
[metall] (for a cupola furnace) soufflerie *f* rotative
~-BOOSTER - s. turbo-blower
~-CHARGED - s. turbo-supercharged
~-CHARGER - s. turbo-supercharger
~-COMPRESSOR - [mech] turbo-compresseur *m*
~-DIESEL LOCOMOTIVE - [railw] locomotive *f* turbo-Diesel
~-DRILLING RING - [mech] (used in prospecting for oil and natural gas) turboforeuse *f*
~-DRYER - [mech] turbo-sécheur *m*
~-DYNAMO - [mech] turbo-dynamo *f*
~-ELECTRIC - [el] turbo-électrique
~-ELECTRIC PROPULSION - [mech] propulsion *f* turbo-électrique
~-EXHAUSTER - [gas ind] extracteur *m* centrifuge
~-FAN - [aero] (ducted-fan turbine engine) turboréacteur *m* à double flux
~GENERATOR - [el] (alternating-current generator with smooth rotor usually intended to be driven by a high-speed turbine) turbo-générateur *m*
~-GENERATOR SET - s. turbine-driven set
~-JET - [aero] (jet turbine engine) turboréacteur *m*
~-JET ENGINE - [aero] moteur *m* à turboréacteur
~-MOTOR - [mech] turbo-moteur *m*
~-PROPELLER ENGINE - [aero] (airscrew gas turbine) turbo-propulseur *m*
~-PUMP - [mech] pompe *f* turbine, turbo-pompe *f*
~-SUPERCHARGE, to - [mech] (to fit an engine with a supercharger operated by the exhaust gas) suralimenter par turbo-compresseur
~-SUPERCHARGER - [mech] turbo-compresseur *m*
TURBOPROP-JET ENGINE - s. turbopropeller engine
TURBOPUMP - s. turbo-pump
TURBORAM-JET ENGINE - [aero] (a jet engine with turbine and reheater) statoréacteur *m*
TURBOROCKET - [astronaut] fusée *f* à turboréacteur
TURBULENCE - [phys] (the existence of irregular eddy currents to a varying extent in a uniform flow)

turbulence *f*
TURBULENCE CLOUD - [met] (cloud resulting from turbulence in the flow of a wind over the surface of the earth) nuage *m* de turbulence
~COMBUSTION CHAMBER - [mech] (in internal combustion engines) chambre *f* de combustion à haute turbulence
TURBULENT - [phys chem] (of a liquid, disturbed, agitated) turbulent
[met] changeant, variable
~BOUNDARY LAYER - [phys] (boundary layer in which flow is turbulent) couche *f* limite turbulente
~BURNER - [th eng] brûleur *m* à turbulence
~FLOW - [phys] (non-laminar flow, in which eddies occur) écoulement *m* turbulent
TURF - [gen] gazon *m*
[soil] (peat) tourbe *f*
TURGESCENZE - [bot & med] (swelling, increasing in size) turgescence *f*
TURGESCENT - [gen] turgescent, enflé
TURGID - [gen & med] turgide, gonflé
TURGOR - [biol] (the state of being turgid) turgescence *f*
TURIN POUND - [text] livre *f* de Turin (368,8 g)
~TITRE - [text] titre *m* de Piémont
TURION - [bot] turion *m*, bourgeon *m*
TURKEY - [zool] dindon *m*
~CARPET - [text] tapis *m* de Smyrne, tapis *m* d'Orient
~-HEN - [zool] dinde *f*
~RED - [paint] (light red, pigment obtained by calcination of precipitated hydroxide or carbonate of iron) rouge *m* turc
~-RED OIL - [chem](sulphonated castor oil, A sulphonated castor oil with application in textile processing, soap manufacture, and medicine) huile *f* de ricin sulfonée
~STONE - [geol] novaculite *f*, pierre *f* du Levant
~UMBER - [paint] (a type of umber obtained from Cyprus) ombre *f* turque
TURKISH CARPET LOOM - [text] métier *m* pour tapis de Smyrne
~CRESCENT - [mus] chapeau *m* chinois
~JINGLE - s. turkish crescent
~KNOT - [text] noeud *m* de Smyrne
~TOWELLING - [text] tissu *m* éponge
TURMALINE - s. tourmaline
TURMERIC - [bot] (a spice and colouring matter) curcuma *m*, safran *m* des Indes
~PAPER - [chem] papier *m* curcuma
~YELLOW - [chem] jaune *m* curcuma
TURN, to - [gen] tourner
[gen] (to cause to change position) faire tourner
[gen] (to reverse direction or course) changer la direction, virer
[gen] (to transform) transformer, convertir
[gen] (to invert) invertir
[constr] faire pivoter
[mech] (the operation of machining in a lathe) tourner, cylindrer, façonner au tour
TURN - [gen] (of a road etc) tour *m*, révolution *f*
[gen] (change of direction) virage *m*
[mech] (a revolution about an axis) trour *m*, spire *f*
(of a rope) tour *m* d'une corde
[aero] virage *m*, virement *m*
[naut] giration *f*
[geogr] (of the tide) changement *m* de la marée, virement *m* d'eau
TURN - [print] caractère *m* retourné
[el] (of a winding) spire *f*
~ABOUT, to - [gen] se tourner, se retourner
~-AND-BANK INDICATOR - [instr] (a combined turn-indicator and cross-level) contrôleur *m* de virage
~-AND SIDESLIP INDICATOR - s. turn-and-bank indicator
~-AROUND TIME - [comput] temps *m* d'achèvement
~BENCH - [mech] tour *m* à archet, tour *m* d'horloger
~-BRIDGE - [constr] pont *m* tournant
~DOWN, to - [gen] baisser, atténuer, rabattre
[gen] (to reject, to refuse) refuser
~ERROR - [astronaut] erreur *f* de virage
~FOR A LANDING , to - [aero] virer pour l'atterrissage
~FULL ON, to - [mech] ouvrir en plein
~INDICATOR - [instr] (instrument to show the rate of turn about the vertical axis) indicateur *m* de virage
[auto] indicateur *m* de direction
~INTO, to - [gen] convertir
~LENGTH - [text] longueur *m* du tour
~OF A SPRING - [mech] spire *f* d'un ressort
~OF A WINDING - [el] spire *f* d'un enroulement
~OF CALL - [telecomm] (in telephony) tour *m* d'une demande de communication
~OF THE FLYER - [text] rotation *f* de l'ailette
~OF THE REEL - [text] tour *m* du dévidoir
~OF THE SLAY - [text] tour *m* du battant
~OFF, to - [railw] garer, aiguiller
~OFF - [gen] fermer, couper
[railw] garer, aiguiller, s'aiguiller
~-OFF THYRISTOR - [electron] (a thyristor which can be switched from the on-state to the off-state and vice-versa by applying control signals of suitable polarities to the gate terminal) thyristor *m* bloquable
~ON, to - [gen] ouvrir
[el etc] (the light) allumer
[mech] donner, lâcher
~ON A TAP, to - [hydr] lâcher un robinet
~OUT, to - [gen] mettre dehors, faire sortir, jeter
[naut] (a lifeboat etc) pousser en dehors (une embarcation)
[comm] produire, fabriquer
~-OUT TRACK - [railw] voie *f* de tiroir
~OVER, to - [mech] (to revolve the crankshaft of an engine slowly, e. g. for examination) tourner
[gen] retourner, verser, capoter
[metall] rabattre (les bords)
[print] (a letter etc) faire sauter
~OVER A MOULD, to - [metall] retourner un châssis de fonderie
~-OVER PLOUGH - [agric] charrue *f* brabant
~-OVER TABLE MOULDING-MACHINE - [metall] machine *f* à mouler à plaque rotative
~PICTURE CONTROL - [radar] (manual control for rotating the picture for correct orientation) commande *f* de rotation de l'image
~-PIN - [plumb] toupie *f*
~-PULLEY - [mech] poulie *f* de renvoi
~RATIO - [el] rapport *m* du nombre de spires
~ROUND, to - [gen] tourner, pivoter, tournoyer
[naut] virer de bord
~-SCREW - s. turnscrew
~SWITCH - [el] (a rotary switch) commutateur *m* rotatif, interrupteur *m* rotatif

TURN-TABLE - [metall] plaque *f* rotative
~THE EDGE, to - [mech] rabattre le bord
~THE FLAX, to - [text] remuer le lin, tourner le lin
~THE HEMP STRICK, to - [text] tourner la tresse de chanvre
~THE SCALES, to - [meas] trébucher
~THE SOIL, to - [agric] retourner
~THE TOW, to - [text] remuer l'étoupe
~-UNDER - [agric] enfouir le fumier
~-WREST PLOUGH - [agric] charrue *f* tourne-sous-age
TURNBUCKLE - [mech] (tensioning device for rods and ropes, consisting of two eyebolts screwed into a central body, one bolt having a L.H. thread and the other a R.H. one) tendeur *m*, lanterne *f* de serrage
[constr] lanterne *f*
[print] boîte *f* du preneur de matrices
~JOINT - [mech] assemblage *m* à lanterne
TURNED - [mech] (machined in a lathe) tourné, venu de tour
~BARS - [metall] barres *fpl* d'acier écroûtées
~EDGE - [metall] bord *m* tourné
~FINISH - [metall] finissage *m* brillant par tournage
~IN - [gen] rentré
~SORTS - [print] (characters turned face-downwards so as to obtain black marks and ensure that temporarily missing letters are later inserted) blocage *m*, lettre *f* bloquée
~UP FLANGE - [metall] collet *m* rabattu, collet *m* tombé
TURNER - [gen] tourneur *m*
TURNER'S YELLOW - [chem] oxychlorure *m* de plomb
TURNING - [gen] virage *m*
[gen] (adj) tournant
[mech] (machining in a lathe, i.e. by applying a cutting tool to a rotating workpiece) tournage *m*
[ceram] (finishing operation on pottery ware, using a machine somewhat similar to a wood-turning lathe) tournage *m*
~AND BORING MACHINE - [mach tool] tour *m* vertical
~BAR - [impl] tringle *m* de retournement
~BLADES - [mech] déflecteurs *mpl*
~BRIDGE - s. turn-bridge
~CARRIER - [impl] toc *m* (pour tourneur)
~CHIPS - [mech] copeaux *mpl* de tournage
~CHISEL - [tool] fermoir *m* de tour, ciseau *m* de tour
~CHUCK - [mech] chuck *m* d'un tour
~CIRCLE - [mech] cercle *m* de braquage
~CLIP - [el] virole *f*
~CRANE - [mech] grue *f* tournante
~ENGINE - [mech] tour *m* marchant au moteur
~FRICTION - [phys] friction *f* composée
~GEAR - [mech] dispositif *m* de virage
~GOUGE - [mech] gouge *f* de tour
~HANDLE - [mech] manivelle *f*
~KEY - [hydr] clé *f* à canon, clé *f* amovible
~MOMENT - [mech] moment *m* d'un couple
~OVER MOULDS - [metall] retournage *m* de moules
~POINT - [gen] moment *m* critique
[surv] (between two straight lines of a traverse) point *m* perdu
[instr] point *m* de virage
~RADIUS - [mech aero] (the normal radius needed for a given vehicle or aircraft to make a horizontal turn of 360 degrees) rayon *m* de braquage
~REST - [mach tool] support *m* à chariot
TURNING SAW - [tool] scie *f* à chantourner
~THE STRICK - [text] retournage *m* de la poignée
~TOOL - [tool] outil *m* de tour
~VANE - [mech] (e.g. of a ventilating duct) chicane *f*
TURNINGS - [mech] (fragmentary material detached in a turning operation) tournure *f*
TURNIP - [bot] navet *m*
TURNMETER - [instr] (gyroscopic instrument to measure the rate of turn of an aircraft about a given axis) indicateur *m* de virage
TURNOUT - [gen] production *f*
[railw] (a sidetrack) changement *m* de voie, voie *f* de garage
[railw] évitement *m*
[railw] (a switch) aiguillage *m*, branchement *m*
TURNOVER - [comm] chiffre *m* d'affaires
[gen] renversement *m*, culbute *f*
[print] (in a newspaper, a runover) fin *f* (d'article) sur une autre page
[med] cycle *m* métabolique
~BOARD - [metall] plaque *f* à fouler
~FREQUENCY - [acoust] (transition frequency of a disk recording system) fréquence *f* de recouvrement, fréquence *f* de transition
~MOULDING MACHINE - [mech] (used in foundry work) machine *f* à démouler avec plaque à modèle renversable
~PATTERN PLATE - [mech] (used in foundry work) plaque *f* renversable
~RATE - [med] (in radiobiology) vitesse *f* de renouvellement, vitesse *f* du cycle métabolique
~TIME - [biol] temps *m* de renouvellement
~UNIT - [metall] (in sheet-metal working) dispositif *m* de renversement
TURNPIKE - [gen] barrière *f* de péage
[gen] (a road) grande route *f*, route *f* à barrière de péage
TURNPLATE - s. turntable
TURNS-RATIO - [el] (of a transformer winding) rapport *m* du nombre de spires
TURNSCREW - [tool] (term used for a short strong screw-driver suitable for use where considerable force is needed) tournevis *m*
~BIT - [tool] tournevis *m* au fût, lame *f* de tournevis pour vilebrequins
TURNSICK - [vet] tournis *m*, coenurose *f*
TURNSOLE - [bot] tournesol *m*; héliotrope *m*
TURNSTILE AERIAL - [radio] (one or more systems of crossed horizontal half-wave dipoles giving a concentrated radially symmetrical field of radiation with horizontal polarization) antenne *f* croisée, antenne *f* en tourniquet, doublet *m* en croix
~ANTENNA - s. turnstile aerial
TURNTABLE - [metall] (a rotable plate supporting work on a machine) table *f* rotative
[railw] plaque *f* tournante
[el acoust] (of a record-player) plateau *m* de tourne-disques
[aero] (of an airship mooring) plate-forme *f* tournante
~PRESS - [mech] (type of press in which work is fixed on a revolving table and brought under the press in succession) presse *f* à plaque tournante
~RUMBLE - [el acoust] vibration *f* à basse fréquence, grognement *m*
TURPENTINE - [bot] (resinous oily mixture exuding from coniferous trees) térébenthine *f*, résine *f*

vierge
TURPENTINE - [chem] (the essential oil obtained by the steam distillation of turpentine, consisting of a mixture of terpenes) térébenthine *f*
~OIL - [chem] essence *f* de térébenthine
~SUBSTITUTE - [paint] (a cheaper substitute for turpentine, usually White Spirit, a petroleum fraction) essence *f* de térébenthine artificielle
TURPS - s. turpentine
TURQUOISE - [min] (natural hydrous aluminium and copper phosphate employed as a gemstone) turquoise *f*
~MATRIX - [min] turquoise *f* engagée dans sa gangue
TURRET - [gen & arch] tourelle *f*
[mach tool] (a toolholder) tourelle *f* revolver, porte-outil *m* revolver
[telev] (turret lens) tourelle *f* de lentilles
~COAL-CUTTING MACHINE - [min] haveuse *f* à colonne
~FEED LEVER - [mach tool] (of a lathe) levier *m* d'avance de la tourelle
~FEED SHAFT - [mach tool] (of a lathe) barre *f* de chariotage de la tourelle
~HEAD - [mach tool] tourelle-revolver *f*
~LATHE - [mach tool] tour *m* à revolver
~QUICK-TRAVERSE LEVER - [mach tool] levier *m* de traverse rapide de la tourelle
~SADDLE - [mach tool] traînard *m* du revolver
~SLIDE - [mach tool] chariot *m* porte-tourelle
~SLIDE CAM DRUM - [mach tool] tambour *m* de réglage du tambour porte-tourelle
~SPINDLE - [mach tool] mandrin *m* vertical
~TUNER - [telev] sélecteur *m* de canaux à tourelle
~-TYPE DRILLING MACHINE - [mach tool] perceuse *f* à tourelle
~WINDER - [mech] (a winding-up machine having several spindles mounted on rotating plate, so that one can be operating while others are being unloaded) enrouleur *m* à tourelle
TURTLE - [zool] tortue *f* de mer
[print] (strong frame shaped like a segment of a cylinder used to hold the type in a typerevolving web press) casse *f* courbe
~DECK - s. turtleback
~-DOVE - [zool] tourterelle *f*
TURTLEBACK - [shipbuild] (arched covering over the main deck of a ship as a protection against heavy seas) pont *m* en carapace de tortue
TUSK TENON - [carp] tenon *m* renforcé, tenon *m* de repos
~-TENON JOINT - [carp] assemblage *m* à tenon renforcé
TUSSOCK - [bot] touffe *f* d'herbe, tussack *m*
~MOTH - [zool] orgyie *f*
TUSSORE - [text] tussor *m*
~SCHAPPE YARN - [text] schappe *f* tussah
~SILK YARN - s. tussore schappe yarn
TUTOCAINE HYDROCHLORIDE - [chem] (butamin. A local anaesthetic) chlorhydrate *m* de tutocaïne
TUTTY - [chem] (impure zinc oxide obtained as a sublimate in the flues of zinc-smelting furnaces) tutie *f*, cadmie *f*
TUYERE - [mech] (nozzle to supply an air blast, as in a blast furnace or forge) tuyère *f*
~CAP - [metall] couverture *f* de tuyère
~COOLER - [metall] refroidisseur *m* de tuyère
~NOZZLE - [mech] (the orifice at the fire end of a tuyère) bec *m* de tuyère
TUYERE NOZZLE - [metall] nez *m* de la tuyère
~OPENING - [mech] (aperture in the wall through which the tuyere passes) chapelle *f*
~-PIPE - [metall] porte-vent *m*
~PLATE - [metall] taque *f* de varme
~STOCK - [metall] coude *m* du porte-vent
[mech] tubulure *f* extérieure de la tuyère
~WALL - [mech] (wall in the lower part of a blast furnace in which the tuyeres are set) paroi *f* de la tuyère
TV - s. television
TWANG, to - [acoust] résonner, vibrer
[mus] pincer, faire frémir
[gen] (in archery) lâcher
TWANG - [acoust] bruit *m* sec
[mus] (of a stringed instrument) son *m* aigu
TWEED - [text] (soft woollen fabric with a homespun surface) tweed *m*, cheviote *f* écossaise
TWEEL - [glass man] (counterweighted furnace door, opening vertically) porte *f* à guillotine
TWEET - [acoust] pépiement *m*
TWEETER - [el acoust] (a loudspeaker which only reproduces higher frequencies) haut-parleur *m* pour fréquences élevées
TWEEZER, to - [gen] extraire avec des pinces
TWEEZER - [text] pincettes *f*pl de tisserand
~PEG - [text] broche *f* fendue
TWEEZERS - [impl] (small pincers used for tiny objects) brucelles *f*pl, pinces *f*pl brucelles
T-WELD - [mech] soudure *f* en T
TWELVE-CHANNEL GROUP - [telecomm] (the assembly of twelve telephone channels) groupe *m* primaire
~END FLAT TWILL - [text] sergé *m* de douze à angle obtus
~-GAUGE - [firearms] (of a shotgun) (piéce) de douze
~PUNCH - [comput] perforation *f* I2, perforation *f* Y
TWELVEMO - [print] in-douze *m*
TWENTY-FOURMO - [print] in-vingt-quatre *m*
TWENTY-TWENTY - [opt] (the measure of visual acuity corresponding to a normal human eye) acuité *f* visuelle normale
TWIG - [bot] ramille *f*, brindille *f* de branche
~AND LEAF PATTERN - [text] dessin *m* de ramage
TWILIGHT - [met] (the interval between sunrise or sunset, and total darkness, when some light occurs because of scattering from particles in the atmosphere) crépuscule *m*
~ANAESTHESIA - [med] anesthésie *f* crépusculaire
~BLINDNESS - [med] amblyopie *f* crépusculaire
~STATE - [med] état *m* crépusculaire
~ZONE - [radio] (the narrow region on each side of a radio range course, in which A or N signals begin to predominate) zone *f* crépusculaire
TWILL, to - [text] croiser une étoffe
TWILL - [text] tissu *m* croisé, twill *m*
~BACKED CLOTH - [text] côte-cheval *f*
~COATING - [text] frise *f* croisée
~LEFT TO RIGHT - [text] croisé *m* de gauche à droite
~RIGHT TO LEFT - [text] croisé *m* de droite à gauche
~WEAVE - [text] (weave producing a pattern of diagonal lines) armure *f* sergé, armure *f* croisé
TWILLED CLOTH - [text] étoffe *f* croisée, étoffe *f* satinée
~FABRIC - [text] tissu *m* croisé

TWILLED MUSLIN - [text] mousseline *f* en croisé
~RIBBON - [text]ruban *m* en croisé
~SATIN - [text] sergé-satin *m*
~WARP BACK CLOTH - [text] côte-cheval *f* par chaîne
~WEFT BACK CLOTH - [text] côte-cheval *f* par trame
TWILLING BAR - [text] tringle *f*, réglette *f* de levé
TWIN - [gen & biol] jumeau
[phys] (in crystallography) macle *f*
~ACCOMMODATION - [cryst] (twinning) hémitropie *f*
~AERIALS - [radio] (a pair of two identical conductors) antenne *f* bifilaire
~ANTENNA - s, twin aerials
~-ARC LIGHT - [light] lampe *f* à arc double
~ARTICULATED VEHICLES - [railw] voitures *fpl* jumelées
~AXIS - [cryst] axe *m* d'hémitropie
~BIRTH - [biol] accouchement *m* gémellaire
~-BLADE WINDSCREEN WIPER - [auto] essuie-glace *m* à double balai
~BOUNDARY DIFFUSION - [cryst] (a source on internal friction in solids due to the displacement of the boundary between twinned crystallites) glissement *m* de joint de grains
~CABLE - [el] (a cable containing two cores not arranged concentrically) câble *m* bipolaire, câble *m* à paires torsadées
~CALORIMETER - [instr] (a differential steam calorimeter) calorimètre *m* differentiel
~-CARBON ARC LAMP - [el] lampe *f* à arc à électrodes de charbon jumelées
~-CAST CYLINDER - [auto] bloc *m* de deux cylindres
~-CATHODE RAY BEAM - [electron] (cathode-ray beam used in colour television) faisceau *m* jumelé de rayons cathodiques
~CHECK - [comput] contrôle *m* double
~COCK - [plumb] robinet *m* à deux faces
~COCOON - [text] cocon *m* double, doupion *m*
~COLUMN - [constr] colonnes *fpl* géminées
~CONDUCTOR - [el] conducteur *m* double
~CONTACT - [el] contact *m* jumelé
~-CONTACT TYRES - [aero] (a type of aircraft tyre designed to prevent shimmy, and having a double tread witha a deep depression between the ground-contact surfaces) pneus *mpl* jumelés
~-CONTACT WIRE - [el] fil *m* de contact double
~CRYSTAL - [cryst] (two crystals of the same substance having one common face) cristal *m* maclé, macle *f*
~CRYSTALLIZATION - [phys] hémitropie *f*
~-CYLINDER - [mech] à deux cylindres
~-CYLINDER MIXER - [mech] (mixer in which two cylinders are joined in V-form, and rotate about a line which is at 45 deg. to both their divergent axes) mélangeur *m* à tonneau en V
~-CYLINDER MOTOR - [mech] moteur *m* à deux cylindres
~DIODE - [electron] (rectifier comprising two nearly identical germanium diodes) diodes *fpl* jumelées
~DOUBLING SPINDLE - [text] broche *f* à retordre à double action
~ENGINES - [mech] machines *fpl* jumelles
~EXHAUST PIPES - [auto] double-tuyau *m* d'échappement
~EXTRUDER FOR TWO COLOURS - [rubber ind] boudineuse *f* jumelée pour tirer à deux couleurs
~FEEDER - [el] feeder *m* bifilaire
TWIN FLEXIBLE CORD - [el] (two flexible cords twisted together) cordon *m* flexible double
~FLUE - [th eng] carneaux *mpl* en épingle à cheveux
~GIRDER - [constr] poutres *fpl* jumelées
~HEATER - [rubber ind] presse *f* jumelle pour pneus
~HOLE - [mining] sondages *mpl* jumeaux
~IGNITION - [mech] (i.c.eng) double allumage *m*
~-JET - [aero] biréacteur *m*
~-JET CHARGE - [oil ind] charges *fpl* creuses accouplées
~-JET NOZZLE - [auto] injecteur *m* à deux jets
~LENS CAMERA - [photo] (camera with matched lenses, one for exposing, the other for focusing, usually with a reflex mirror) chambre *f* binoculaire
~LETTER-PRESS - [print] machine *f* double
~MAGAZINE - [cin] (a type of film magazine) chargeur-magazine *m*, chargeur-mécanisme *m*
~-MILL - [rubber ind] groupe *m* de deux malaxeurs
~MOTOR - [mech] bimoteur
~-ORIFICE BURNER - [th eng] (a type of burner having a pair of similar spray orifices) brûleur *m* à double orifice
~PLANAR TRANSISTOR - [electron] (twin transistor made up to planar silicon units) transistor *m* planaire jumelé
~PLANE - [cryst] plan *m* d'hémitropie
~POLES - [el & telecomm] poteau *m* jumelé
~-POINT FOCUSING - [photo] mise *f* au point conjuguée
~PROJECTORS - [photo] deux projecteurs *mpl* jumelés
~-RAIL CAMERA - [photo] chambre *f* à double rail
~ROLLING MILL - [metall] laminoir *m* due, train *m* duo
~-ROTOR - [aero] à deux rotors
~-ROTOR HELICOPTER - [aero] hélicoptère *m* à deux rotors
~-SCREW - [naut] à deux hélices (jumelles)
~-SCREW EXTRUDER - [rubber ind] (type of extruder in which two feed screws are used) boudineuse *f* à vis jumelée
~-SCREW LATHE - [mach tool] tour *m* à deux vis mères indépendantes et de pas différents
~-SCREW MOTORSHIP - [naut] vaisseau *m* à moteur à hélices jumelles
~SHAFTS - [mining] puits *mpl* jumeaux
~-SHELL DRY BLENDER - [rubber ind] mélangeur-sécheur *m* à double enveloppe
~-SIX - [mech] moteur *m* à douze cylindres en V
~SUCTION COUCH - [print] double-drague *f*
~-T NETWORK - [telecomm] réseau *m* en T en dérivation
~-TAIL - [aero] à double queue
~TAPE TRANSPORTER - [electron] (a type of magnetic tape transporting unit) transporteur *m* de bande magnétique à jumeaux
~TYRE - [auto] pneumatiques *mpl* jumelés
~TYRE PRESS - [rubber ind] pots *mpl* de cuisson jumeaux, presse *f* jumelle pour pneus
~-WIRE - [el] (a wire containing two cores not arranged concentrically) bifilaire
TWINE, to - [gen] tordre, tortiller
[text] retordre
TWINE - [text] ficelle *f*, fil *m* retors
[text] (harness cord) fil *m* d'arcade, arcade *f*
[gen] entrelacement *m*
[gen] ficelle
~BALL - [gtext] pelote *f* de ficelle

TWINE CUTTER - [impl] coupe-ficelle *m*
~EYE - [text] oeillet *m* en fil retors
~HEALD - [text] lisse *f* avec oeillet en fil retors
~HEDDLE - [text] lisse *f* tricotée
~HOLE - [text] trou *m* de laçage
~MANUFACTURING MACHINE - [text] machine *f* pour la fabrication de ficelles
~REEL - [text] rouleau *m* de ficelle
~SPINNER - s. twine manufacturing machine
TWINFLAT - [el] ligne *f* en ruban symétrique
TWINGE - [med] élancement *m* (de douleur)
TWINING MACHINE - [el] (for cables and wires) câbleuse *f*, assembleuse *f*
TWINKLE, to - [gen] scintiller
[gen] (of objects) papillonner
TWINKLE - [gen] scintillement *m*, clignotement *m*
TWINNING - [cryst] (a process in which a region in a crystal assumes an orientation which is symmetrically related to the basic orientation of the crystal) hémitropie *f*
[electron] (in a cathode-ray tube) jumelage *m*
[cryst] s. twin crystallization
~AXIS - [cryst] axe *m* d'hémitropie
TWINPLEX - [telecomm] (frequency-shift diplex operation) twinplex *m*
T-WIRE - [el] fil *m* de pointe
TWIST, to - [gen] tordre, tortiller
[text] (a thread on a reel etc) enrouler
[text ind] (of yarns) retordre
[aero] vriller
[mech] torsader
[text] (rope man) (to wind one strand round another) commettre
TWIST - [gen] cordon *m*
[text] (the twisting of yarns) tordage *m*
[text] (of the yarn) torsion *f*
[phys] (torsional stress) torsion *f*
[radio] (of a waveguide) section *f* en torsade
[aero] (of the wing) vrillage *m*
[comm] (in the cotton trade, the warp yarn) chaîne *f* à torsion ordinaire
~BIT - [tool] (used in carpentry) torse *f*, mèche *f* hélicoïdale
~CONSTANT - [phys] coefficient *m* de torsion
~COP - [text] canette-chaîne *f*
~DRILL - [tool] (a drill having helical grooves for the escape of the swarf) foret *m* hélicoïdal, foret *m* à hélice, mèche *f* americaine
~-DRILL CUTTER - [mach tool] fraise *f* à fraiser les forets hélicoïdaux
~-DRILL GRINDER - [mach tool] machine *f* à affûter les forets hélicoïdaux
~FACTOR - [text] coefficient *m* de torsion
~FACTOR AT BREAK - [text] torsion *f* de rupture
~GEAR - [mech] (helical gear) engrenage *m* à denture hélicoïdale
~GIMLET - [impl] vrille *f* à torsade
~-GRIP - [mech] (a control device consisting of a cylindrical grip which can be rotated to actuate a control mechanism) tenaille *f* giratrice
~HAND-REAMER - [tool] alésoir *m* à main en hélice
~IN OPPOSITE DIRECTION - [text] tors *m* croisé, commettage *m* croisé
~IN SAME DIRECTION - [text] tors *m* allongé
~IN THE YARN - [text] torsion *f* du fil
~MOTION LEVER - [text] levier *m* du compteur de torsion
TWIST MULTIPLIER - [text] coefficient *m* de torsion
~SETTING AGENTS - [chem] (melamine-formaldehyde resins used to retain twist during steaming of rayon and silk yarns) agents *mpl* empêchant la torsion
~TESTER - [instr] torsiomètre *m*, compteur *m* d'apprêt
~WELDING - [mech] (a method of plastics welding in which the welding rod is moved to and fro) soudure *f* à va-et-vient
TWISTED - [text] tors, tordu, retordu
~AUGER - [tool] tarière *f* torse, tarière *f* à double spire
~CHAIN - [mech] chaîne *f* torse, chaîne *f* en S
~CORD FABRIC - [text] tissu *m* corde retordue
~CURVE - [math] courbe *f* gauche, courbe *f* à double courbure
~FIBRE - [text] fibre *f* torse
~HEMP STRICK - [text] trasse *f* de chanvre
~JOINT - [el] torsade *f*
~STRAW YARN - [text] fil *m* de paille cordonné, cordonnet *m* en paille
~STRIP - [metall] bande *f* tordue
~THREAD - [text] fil *m* retors
~WINDING - [el] enroulement *m* torsadé
~WIRE HEALD - [text] lisse *f* en fil métallique torsionné
~YARN - [text] fil *m* retors
TWISTER - [met] (a tornado having a life of an hour or more and usually causing considerable damage) tornado *m* violent
[text] tordoir *m*, retordoir *m*
~FINGER - [text] doigt *m* fileur
~SPINDLE - [text] broche *f* câbleuse
TWISTING - [text] retordage *m*
[mech] (wrenching) vrillage *m*
[mech] (an operation by which forged crank-shafts are made by torsional displacement or individual webs) torsion *m* des vilebrequins
~APPARATUS - [text] (used for solidified silk filament) appareil *m* de retordage
~BOBBIN - [text] bobine *f* de retordage
~FRAME - [text] (doubling frame) métier *m* à retordre
~HOOK FRAME - [text] chariot *m* des crochets
~MACHINE - s. twisting frame
~MILL - [text] retorderie *f*
~MOMENT - [mech] moment *m* d'un couple
~-REELING MACHINE - [text] retordeuse *f* à écheveaux
~TEST - [gen] essai *m* à la torsion
~VICE - [impl] mâchoire *f* à tordre
TWISTOR - [electron] (storage device in which the information is stored on hairlike magnetic wire in the form of spirally-polarized magnetic zones) twistor *m*
TWITCHING - [med] secousse *f*, convulsion *f* clonique
~OF THE THREAD - [text] chocs *mpl* sur le fil
TWITTER - [acoust] (sound, as made by sparrows) gazouillement *m*
TWO-ADDRESS INSTRUCTION - [comput] instruction *f* à deux adresses
~-BEAM INSTABILITY - [phys] (a plasma instability) instabilité *f* par faisceaux entrecroisés
~-BODY PROBLEM - [phys] (the foundation of celestial mechanics)problème *m* à deux corps
~-BUSBAR REGULATION - [el] (single-coil regulation)

régulation *f* à deux barres
TWO-CAGE DISINTEGRATOR - [mining] (bar crusher) désintégrateur *m* à deux barres
~-CAVITY KLYSTRON - [electron] (klystron with two resonant cavities) klystron *m* à deux cavités
~-CHANNEL TRANSMITTER - [radio] émetteur *m* à deux canaux
~-CIRCUIT BAND-PASS FILTER - [telecomm] filtre *m* passe-bande à deux circuits
~-CIRCUIT PREPAYMENT METER - [el] compteur *m* à prépaiement à double circuit
~-COAT SIMULTANEOUS EXTRUSION - [plast ind] (extrusion of two concentric coats on to a wire at the same operation) extrusion *f* en enrobage double
~-COIL RELAY - [el] relais *m* à deux bobines
~-COLOUR DIRECT VIEW STORAGE TUBE - [electron] (storage tube capable of displaying information in either of two colours or in some intermediate hue, and selectively erasing such information) tube *m* mémoire bicolore à vision directe
~-COLOUR PRESS - [print] machine *f* pour impression en deux couleurs
~-COLOUR PRINTING - [print] bichromie *f*, impression *f* à deux couleurs
~-COLOUR PROCESS - [print photo] gravure *f* bichrome
~-COLOUR PYROMETER - [instr] pyromètre *m* à deux longueurs d'onde
~-CONDITION CABLE CODE - [telecomm] code *m* bivalent pour câble
~-CONDUCTOR EARTHED-RETURN WIRING SYSTEM- [el] distribution *f* deux fils à ligne de retour mise à la terre
~-CONDUCTOR EARTHED WIRING SYSTEM - [el] distribution *f* deux fils mise à la terre
~-CONDUCTOR INSULATED WIRING SYSTEM - [el] distribution *f* deux fils isolée
~-CORE-PER-BIT STORAGE - [comput] mémoire *f* à deux tores par bit
~-COURSE CROP ROTATION - [agric] assolement *m* à deux cultures, assolement *m* biennal
~-CUBE TWIN - [cryst] macle *f* de deux cubes
~-CYCLE - [mech] (of engines) à deux temps [phys] (in thermodynamics) cycle *m* à deux temps
~DAYLIGHT PRESS - [mech] (a press having three platens and thus two daylights) presse *f* à deux plateaux
~-DECKER - [naut] navire *m* à deux ponts [transp] autobus *m* à impériale
~-DECKER CAGE - [mining] cage *f* à deux étages
~-DIGIT GROUP - [comput] groupe *m* de deux digits
~-DIMENSIONAL GAS - [phys] (a layer of adsorbed gas of monomolecular thickness) gaz *m* à deux dimensions
~-ELEMENT AERIAL - [radio] antenne *f* à deux éléments
~-ELEMENT ANTENNA - s. two-element aerial
~-ELEMENT RELAY - [el] relais *m* à deux éléments
~-FIELD PICTURE - [telev] image *f* complète de télévision formée par deux trames
~-FIELD SYSTEM - s. two-course crop rotation
~-FLOOR KILN - [brew ind] touraille *f* à deux plateaux
~-FLUE BOILER - [metall] chaudière *f* à deux tubes-foyers
~-FLUID CELL + [el chem] (primary cell in which the electrodes are in contact with different electrolytes) pile *f* à deux liquides
TWO-FLUID MANOMETER - [instr] manomètre *m* à deux liquides
~-FREQUENCY SIGNALLING - [telecomm] téléphonie *f* sur deux fréquences locales
~-FURROW PLOUGH - [agric] charrue *f* à deux corps, charrue *f* bisoc
~-GAP SINGLE RESONATOR KLYSTRON - [electron] klystron *m* à un résonateur et deux espaces d'interaction
~-GROUP MODEL - [phys] (in nuclear physics) modèle *m* à deux groupes d'énergie
~-HANDED CROSSCUT SAW - [tool] scie *f* passe-partout
~HIGH MILL - [metall] laminoir *m* duo
~-HIGH REVERSING MILL - [metall] laminoir *m* trio
~-HIGH ROUGHENER STAND - [metall] cage *f* duo ébaucheuse
~-HIGH SLAY - [text] battant *m* à deux séries de navettes
~-HIGH STAND - [metall] cage *f* duo
~-HINGED ARCH - [constr] arc *m* à double articulation, arc *m* articulé aux naissances
~-HOLE DIRECTIONAL COUPLER - [electron] coupleur *m* directionnel à deux trous
~-INPUT ADDER - [comput] additionneur *m* à deux entrées
~-INPUT SUBTRACTER - [comput] unité *f* de soustraction à deux entrées
~-JAW CHUCK - [mach tool] mandrin *m* à deux mâchoires
~-LAMPS PROJECTOR - [cin] projecteur *m* à deux lampes
~-LAYER PLASTIC PIPE - [plast ind] (type of pipe in which one type of plastic is lined with another) tuyauterie *f* plastique de deux qualités
~-LAYER WINDING - [el] enroulement *m* à deux couches
~-LEAF DOOR - [constr] porte *f* à deux vantaux
~-LEVEL ACTION - [contr] action *f* à deux niveaux
~-LEVEL COACH - [railw] voiture *f* à deux étages
~-LEVEL CONTROLLER - [contr] régulateur *m* à action à deux échelons
~-LEVEL MASER - [electron] (a maser with only two energy levels) maser *m* à deux niveaux
~-LEVEL SUBROUTINE - [comput] sous-programme *m* à deux étages
~-LEVEL STORE - [comput] mémoire *f* à deux temps d'accès
~-LEVER PUNCHING MACHINE - [mech] poinçonneuse *f* à deux leviers
~-LINE BREVIER - [print] corps *m* 15
~-LINE ENGLISH - [print] corps *m* 26
~-LINE PICA - [print] corps 25
~-MASH METHOD - [brew ind] décoction *f* à deux trempes
~-PANEL DOOR - [constr] porte *f* à deux panneaux
~-PART CARRIER - [text] conducteur-entraîneur *m* double
~-PART CRANK ARM - [mech] bielle *f* en deux parties
~-PART DIE - [mech] coussinet *m* en deux pièces
~-PART FLASK - [metall] châssis *m* à deux parties
~-PART SCREW-PLATE - [metall] filière *f* double
~-PART TARIFF - [el etc] tarif *m* mixte
~-PEDAL CONTROL - [auto] commande *f* à deux pé-

dales
TWO-PHASE - [el] (of a system, in a two-phase system, the displacement is one quarter of a period) diphasé
~-PHASE CIRCUIT - [el] circuit *m* diphasé
~-PHASE CURRENT - [el] courant *m* diphasé
~-PHASE SYSTEM - [el] (a system comprising two sinusoidal quantities of the same effective value but of a relative phase displacement of a quarter of a period) distribution *f* diphasée
~-PHASE THREE-WIRE SYSTEM - [el] (a system of distribution with A. C. comprising three conductors, which form two circuits, supplied by two currents having a relative phase displacement of a quarter of a period) distribution *f* diphasée trois fils
~-PIN PLUG-SWITCH - [el] prise *f* de courant à deux broches
~-PLUS-ONE ADDRESS - [comput] à (2+I) adresses
~-PLY - [text] à deux brins
~-PLY BELT - [gen] courroie *f* à deux plis
~-PLY CARPETS - [text] tapis *mpl* doubles
~-PLY WOOD - [carp] contre-plaqué *m* à deux épaisseurs
~-POINT EMERGENCY CELL SWITCH - [telecomm] réducteur *m* double de batterie d'accumulateurs
~-POINT LANDING - [aero] (a landing in which the main undercarriage makes contact with the ground before the tail wheel does so) atterrissage *m* en deux points
~-POINT PROBLEM - [surv] problème *m* à deux points
~-PORT NETWORK - [electron] réseau *m* avec une entrée et une sortie
~-PORT REPRESENTATION - [electron] représentation *f* bipolaire
~-PRONG GRAB - [oil ind] harpon *m* de repêchage à deux jambres
~-REGION REACTOR - [nucl] réacteur *m* à deux régions
~-ROLL MILL FOR CRACKING - [rubber ind] broyeur *m* à deux cylindres
~-ROLLER MILL - [mech] (a type of machine similar to the Three-Roller Mill but having only two rollers) malaxeur *m* à deux cylindres
~-SEATER - [transp] à deux places
[auto] voiture *f* à deux places
~SET CARD - [text] carton *m* à deux séries de trous
~-SHOT - [cin] (USA only, close medium shot in GB) demi gros plan *m*, plan *m* italien
~SHUTTLE PLUSH - [text] peluche *f* par double trame
~-SIDED MOSAIC PICKUP TUBE - [electron] tube *m* image à mosaïque à double couche
~-SOURCE FREQUENCY KEYING - [radio] manipulation *f* par mutation des fréquences
~-SPEED - [mech] à deux vitesses
~-SPEED AXLE - [auto] pont *m* à deux vitesses
~-SPEED CONTROLLER - [contr] régulateur *m* à deux vitesses d'action
~-SPEED COUNTERSHAFT - [mech] arbre *m* à deux vitesses
~-SPEED SUPERCHARGER - [auto etc] compresseur *m* d'alimentation à deux vitesses
~-SPEED WINDSCREEN WIPER - [auto] essuie-glace *m* à deux vitesses
~-STAGE - [gen] à deux étages
~-STAGE COMPRESSOR - [mech] (air or gas compressor in which the fluid is compressed in two stages) compresseur *m* à deux étages
~-STAGE FILTER - [impl] filtre *m* double
TWO-STAGE GEAR-DRIVEN SUPERCHARGER - [auto] compresseur *m* d'alimentation à deux étages d'engrenage
~-STAGE TURBINE - [mech] (a turbine in which the gases of combustion are expanded in two consecutive stages) turbine *f* à expansion double
~-STAMP MILL - [min] bocard *m* à deux pilons
~-START THREAD - [mech] (a screw which has two threads the distance between which is half the true pitch) vis *f* à double pas
~-STATE PROCESS - [electron] (a process used in obtaining photo-ionization in pure gases, for light-sensitive devices) processus *m* à deux étages
~-STATION RANGE-FINDER - [meas] (range finder having a home base and two points of observation) télémètre *m* bistatique
~STEP CONE - [mech] bipoulie *f*
~-STEP RELAY - [el] relais *m* à deux seuils
~-STEP SATEEN - [text] satin *m* à double décochement
~-STICK SET - [mining] cadre *m* à un seul montant
~-STOREY CONTAINER - [hydr] réservoir *m* à double étage
~-STOREY TANK - [hydr] bassin *m* à deux étages
~STRAND TWINE - [text] ficelle *f* à deux brins, bitord *m*
~-STROKE CYCLE - [auto] (a working cycle in reciprocating i. c. engines, completed in two piston strokes only) cycle *m* à deux temps
~-STROKE ENGINE - [auto] moteur *m* à deux temps
~-TANGED FILE - [impl] lime *f* à deux soies
~-TERM CONTROLLER - [contr] (controller with proportional action and either integral or a derivative action) régulateur *m* PI
~-TERMINAL NETWORK - [el] (any group of impedances, possibly containing a source of electromotive force, connected with an external system by two terminals only) dipôle *m*
~-TONE DETECTOR - [radio] détecteur *m* à deux tons
~-TONE MOULDING - [plst ind] (a moulding having areas of different shades of its nominal colour, separated by a more or less definite demarcation line) moulage *m* en deux tons
~-TUBE CAMERA - [telev] caméra *f* à deux tubes
~-WAY - [mech] (of a valve etc) à deux voies
[el] (said of a circuit-breaker or similar apparatus to denote that it provides two alternative paths for current) à deux directions, à deux voies
[hydr] à deux eaux
~-WAY BREAK-BEFORE-MAKE CONTACT - [el] contact *m* à deux directions sans chevauchement
~-WAY COCK - [hydraul] robinet *m* à deux voies, robinet *m* à deux eaux
~WAY CONTACT - [el] contact *m* à deux directions
~-WAY CONTACT WITH NEUTRAL POSITION - [el] contact *m* à deux directions avec position neutre
~WAY MAKE-BEFORE-BREAK CONTACT - [el] contact *m* à deux directions avec chevauchement
~WAY SPRING - [mech] ressort *m* à deux voies
~-WAY SULKY PLOUGH - [agric] charrue *f* alternative
~-WAY SWITCH - [el] (single-pole change-over switch without off position, used when it is necessary to control a circuit from two or more positions) interrupteur *m* d'étage
~-WAY TURN-OVER PLOUGH - [agric] charrue *f* brabant
~-WHEEL GRINDING MACHINE - [mach tool] machine

f à meuler double
TWO-WHEEL TRAILER - [auto] remorque *m* à deux roues
~-WHEEL TRUCK - [railw] truck *m* à un seul essieu
~-WING SHUTTER - [cin] (a type of rotary shutter) obturateur *m* à deux lames
~-WIRE AERIAL - [radio] antenne *f* bifilaire
~-WIRE AMPLIFIER - [telecomm] (a telephone amplifier) amplificateur *m* simple pour circuit à deux fils, répéteur *m* pour circuit à deux fils
~-WIRE ANTENNA - s, two-wire aerial
~-WIRE CIRCUIT - [radio] circuit *m* à deux fils
~-WIRE TRUNK - [telecomm] ligne *f* auxiliaire à deux fils, ligne *f* de jonction à deux fils
~-WIRE WINDING TYPE - [el] type *m* d'enroulement à deux fils
TWOFOLD - [gen] double
~TACKLE - [mech] palan *m* double
TWYER - s; tuyere
TYING - [agric] accolage *m*, palissage *m*
[gen] nouage *m*
[constr] chaînage *m*, renforcement *m* avec des tirants
~CLIP - [mech] griffe *f* de serrage
~THE WARP ENDS - [text] nouage *m* des fils de chaîne
~UP - [text] encordage *m*, empoutage *m*
TYLOSIS - [med] tylose *f*, callosité *f*
TYMPAN - [print] (the sheet of paper between the impression cylinder and the paper) papier *m* de décharge, papier *m* intercalaire, marge *f*
[gen] (a tympanum) tympan *m*
~STONE - [metall] (of a blast furnace) tympe *f*
TYMPANECTOMY - [med] tympanectomie *f*
TYMPANIC CAVITY - [anat] cavité *f* du tympan
TYMPANISM - [med] colique *f* flatulente, météorisme *m*
TYMPANITIS - s. tympanism
TYMPANUM - [arch] (ornamental space over the doorway bounded by an arch) tympan *m*
[anat] (the middle ear) tympan *m*
[acoust] tympan *m*
TYNDALL EFFECT - [opt] (when a powerful beam of light is sent through a colloidal solution of high dispersity, the sol appears fluorescent and the light is polarized) effet *m* Tyndall
TYPE, to - [gen] représenter
[print] (short for typewrite,to) dactylographier
[med] (the blood group) déterminer le groupe
TYPE - [gen] type *m*, genre *m*
[print] caractère *m*
[gen] (of coins) empreinte *f*
[chem] composé type
~A 1 WAVES - [radio] (keyed continuous waves) ondes *fpl* entretenues manipulées, ondes *fpl* A 1
~A 2 WAVES - [radio] (keyed modulated waves) ondes *fpl* continues modulées manipulées, ondes *fpl* A 2
~A 3 WAVES - [radio] (sound-modulated waves) ondes *fpl* modulées à fréquence acoustique, ondes *fpl* A 3
~A 4 WAVES - [radio] (facsimile waves) ondes *fpl* fac-similé
~A 5 WAVES - [radio] (television waves) ondes *fpl* de télévision, ondes *fpl* A 5
~AO WAVES - [radio] (continuous waves) ondes *fpl* entretenues
~AREA - [print] justification *f*
TYPE B WAVES - [radio] (damped waves) ondes *fpl* amorties
~BAR - [print] (of a typewriter) tige *f* à caractères
[print] ligne-bloc *f*
~BED - [print] marbre *m*
~BODY - [print] corps *m* de la lettre
~BRUSH - [print] brosse *f* à battre les flans, brosse *f* à mouler
~CASE - [print] casse *f*
~CASTING - [print] fonte *f* de caractères
~-CASTING MACHINE - [print] fondeuse *f* de caractères
~-CASTING PUMP - [print] injecteur *m*
~CUTTER - [print] graveur *m* de caractères
~FACE - [print] oeil *m*
~FAMILY - [print] famille *f* de caractères
~FORM - [print] forme *f* à tirer
~FOUNDRY - [print] fonderie *f* de caractères
~FOUNT WEIGHT - [print] poids *m* de caractères
~GAUGE - s. typometer
~HEIGHT - [print] hauteur *f* de caractère
~HIGH - [print] (denoting the standard height of type from base to the level of the printing surface) à l'hauteur de caractère
~HIGH GAUGE - [print] calibre *m* de justification
~MATTER - [print] caractères *mpl* d'imprimerie
~METALL - [metall] alliage *m* pour caractères d'imprimerie, alliage *m* sans retrait
~-METAL MELTER - [print] creuset *m* pour alliages sans retrait
~MOULD - [print] matrice *f*
~OF DUTY - [el] (typical duty consisting of one or more constant operating conditions for given durations) service *m* type
~OF EMULSION - [chem] (emulsion are either lyophobic or lyophilic. The former are difficult to prepare and easily undergo phase reversal, while the latter are easy to prepare and difficult to precipitate) type *m* d'émulsion
~PLANER - [print] rabot *m* de caractères
~PUSHER - [print] expulseur *m* de types
~SCALE - s. typometer
~SETTER - [print] compositeur *m*
~SETTING - [print] composition *f*
~SETTING MACHINE - [print]machine *f* à composer
~SIZE - [print] corps *m* de la lettre
~STOCK - [print] réserve *f*, stock *m*
~TESTS - [mech] (of an engine, a motor etc)essais mpl d'homologation
[el] (prototype test which is only carried out on one unit or a few units of the same type) essais *mpl* de type
~TEST HORSE-POWER - [mech] (of an engine etc) essai *m* homologué
TYPESCRIPT - [print] manuscrit *m* dactylographié
TYPEWRITE, to - [gen] dactylographier, écrire à la machine
TYPEWRITER - [mech] machine *f* à écrire
TYPEWRITING - [gen] dactylographie *f*
TYPHA FIBRE - [text] fibre *f* de tipha
~WOOL - s. typha fibre
TYPHLECTOMY - [med] excision *f* du caecum
TYPHLITIS - [med] typhlite *f*
TYPHLOEMPYEMA - [med] abcès *m* du caecum
TYPHOID FEVER - [med] fièvre *f* typhoïde, tiphys *m* abdominal
TYPHOMALARIAL FEVER - [med] fièvre *f* typhomala-

rienne
TYPHOON - [met] (term used for a tropical revolving storm in China Sea) typhon *m*, toufan *m*
TYPHOPNEUMONIA - [med] pneumotyphus *m*
TYPHUS - [med] typhus *m*
TYPING OF BLOOD - [med] détermination *f* du groupe sanguin
TYPIST - [gen] dactylographe *m* & *f*
TYPOGRAPH - [print] (a type - setting machine) composeuse *f*
TYPOGRAPHER - [gen] typographe *m*
TYPOGRAPHIC - [print] typographique
TYPOGRAPHY - [print] typographie *f*
TYPOMETER - [instr] (instrument for the measurement of printing types) typomètre *m*
TYPOMETRY - [meas] (the measuring of types) typométrie *f*
TYRAMINE - [chem] (hydroxy-phenylethylamine, an organic base preset in ergot) tyramine *f*
TYRE - [rubber ind] pneu *m*, pneumatique *m*
[rubber ind] (the hoop of a wheel) enveloppe *f*
~BALANCING MACHINE - [rubber ind] dispositif *m* compensateur, machine *f* à vérifier l'équilibrage
~BEAD - [rubber ind] (the thickened edge of the tyre wall which engages the rim of the wheel) talon *m* du pneu
~-BENDING MACHINE - [rubber ind] machine *f* à cintrer les cercles des roues
~BORING AND TURNING MILL - [mach tool] tour *m* vertical pour jantes
~BRAKE - [auto] frein *m* sur pneu
~-BRUSH MACHINE - [rubber ind] machine *f* à brosser les pneus
~BUILDING MACHINE - [rubber ind] machine *f* à confectionner les pneus
~BUILDING ON DRUMS - [rubber ind] confection *f* des pneumatiques sur tambours
~BUILDING ON MANDRELS - [rubber ind] confection *f* sur noyau
~CANVAS - [rubber ind] tissu *m* enchevêtré pour pneus
~CARCASS - [rubber ind] carcasse *f*
~CARRIER - [auto] porte-pneu *m*
~CASING - [rubber ind] pneumatique *m*, enveloppe *f*
~CHAFING STRIP - [rubber ind] pare-clou *m*, flap *m*
~CHAINS - [auto] chaînes *fpl* anti-dérapantes
~-CLASP - s. tyre clip
~-CLEANING DRUM - [rubber ind] tambour *m* pour l'ébarbage des pneus
~-CLIP - [railw] agrafe *f* de bandage
~CORD - [text] toile *f* pour carcasse
~DEBAGGER - [rubber ind] éjecteur *m* d'airbag
~DESINTEGRATOR - [rubber ind] machine *f* pour le démembrement des pneus
~DESKIDDING MACHINE - [rubber ind] machine *f* à faire les incisions anti-dérapantes
~DRIVER - [rubber ind] batte *f* de pneu, fouloir *m* de pneu
~ENGRAVING - [rubber ind] sculpture *f* d'un pneu
~FABRIC - [rubber ind] tissu *m* cord pour pneu
~FLAP - s. tyre chafing strip
~GAITER - [auto] gaîne *f* à pneumatique, corset *m* de pneu, manchon-guêtre *m*
~GAUGE - [auto] manomètre *m* de contrôle de la pression des pneus, manomètre *m* de gonflage
~HOLDER - [auto] porte-pneu *m*
~INFLATING GUN - [auto] pistolet *m* à gonfler
TYRE IRON - [auto] démonte-pneu *m*, levier *m* de montage
~LEVER - s. tyre iron
~LIFE - [rubber ind] durée *f* du pneu
~MILL - [mech] machine *f* à refouler les cercles de roues
~MOULD - [rubber ind] moule *m* pour pneu
~MOUNTING AND DEMOUNTING DEVICE - s. tyre mounting rack
~MOUNTING PRESS - [rubber ind] presse-montage *f* pour bandage plein
~MOUNTING RACK - [auto] appareil *m* pour monter et démonter les pneus
~PATCH - [rubber ind] rustine *f*
~POCKET BUILDING MACHINE - [rubber ind] tambour *m* à confectionner les nappes de câbles, tambour *m* à croiser les toiles
~PORTATIVE FORCE - [rubber ind] portance *f* du pneu
~PRESS - [rubber ind] presse *f* à pneus
~PRESSURE - [auto] pression *f* de gonflage
~PRESSURE GAUGE - s. tyre gauge
~PUMP - [impl] gonfleur *m*
~RATE - [auto] (the static rate measured by the change of wheel load per unit vertical displacement of the wheel relative to the ground at a specified load and inflation pressure) composante *f* de charge statique
~RECAPPING - s. tyre reconditioning
~RECONDITIONING - [rubber ind] rechapage *m*, rajeunissement *m* de bande de roulement
~REMOVER - s. tyre mounting rack
~RIM - [auto] jante *f*
~SAFE LOAD - [rubber ind] charge *f* admissible du pneu
~SCULPTURE - s. tyre engraving
~SIDEWALL - [rubber ind] flanc *m* du pneu
~SIZE - [rubber ind] mesure *f* du pneu
~SPINNER - [rubber ind] support *m* tournant pour l'inspection des pneus
~SPREADER - [tool] écarte-pneu *m*, ouvre-pneu *m*
~SPRING-RING - s. tyre clip
~STOCK - [rubber ind] mélange *m* pneu
~-STOCK STRAINER - [rubber ind] filtreuse *f* pour gomme de chape
~TAPE - [rubber ind] ruban *m* adhésif pour pneumatique
~TOOL - s. tyre iron
~TREAD - [rubber ind] bande *f* de roulement, chape *f* de pneu
~VALVE - [auto] (inner tube valve) valve *f* de chambre à air
~VULCANIZER - [rubber ind] presse *f* pour la vulcanisation des pneus
~WIDTH - [rubber ind] largeur *f* de pneu
~WITH EXTENSIBLE HEELS - [rubber ind] pneu *m* à talons extensibles
~WITH REMOVABLE TREAD - [rubber ind] pneu *m* avec chape amovible
~WORKING LOAD - s. tyre safe load
~WRAPPING MACHINE - [rubber ind] machine *f* à enrouler les pneus
TYROCIDINE - [chem] (an antibiotic produced by Bacillus brevis. Tyrocidine and gramicidin are the constituents of tyrothricin and, as such are used in medicine) tyrocidine *f*
TYROMA - [med] masse *f* caséeuse, tumeur *f* tu-

berculeuse

TYROSINE - [chem] (a non-essential amino-acid with application in physiological research) tyrosine *f*

TYROTHRICIN - [chem] (an antibiotic) tyrothricine *f*

TYSONITE - [min] (mineral containing lanthanium) tysonite *f*

TYUYAMUNITE - [min] (natural hydrated vanadate of calcium and uranium) tyuyamunite *f*

U

U - [chem] (the symbol of Uranium) symbole *m* de l'uranium
U BAR - s. U iron
U-BOAT - [naut] sous-marin *m*
U-BOLT - [mech] (a bolt in the shape of the letter 'U', threaded at both extremities) étrier *m*, lien *m* en fer à U
U.C. - [print] (short for Upper Case) h.d.c., haut-de-casse *m*
U CHANNEL - s. U iron
U-CLAMP - [mech] bride *f* de serrage à U
U-HYBRID-RING-RATE-RACE - [electron] (hybrid junction of a waveguide) anneau *m* hybride
U IRON - [metall] fer *m* en U, fer *m* à U
U LEATHER - [leather ind] cuir *m* embouti
U-LINK - [mech] étrier *m* en U
U-STEEL - [metall] acier *m* en U
U.S. STANDARD THREAD - [mech] pas *m* américain (système Sellers)
U SECTION - [metall] profil *m* en U, section *f* en U
U-SHAPED BASE-PLATE - [mach tool] pied *m* à fourche
U.T. - s. Universal Time
U TUBE - [plumb] tube *m* en U
U-TUBE MANOMETER - [instr] (type of pressure-gauge in which the pressure to be measured is balanced against a column of liquid in a U-tube) manomètre *m* à tube en U
U-VALUE - [phys] (only USA, the heat transmission coefficient of a wall) coefficient *m* de transmission d'une paroi
UBEROUS - [zool] riche en lait
UBERTY - [biol] fertilité *f*
UDDER - [zool] mamelle *f*, pis *m* de vache
UDOGRAPH - [instr] (a type of self-registering rain gauge) udographe *m*, pluviographe *m*
UDOMETER - [instr] (instrument designed to measure the amount of rain which falls during a specified period) udomètre *m*, pluviomètre *m*
[ind chem] enregistreur *m* du niveau des liquides
UDOMETRIC - [met] pluviométrique, udométrique
UDOMETRY - [met] (the study of the measurement of rainfall) udométrie *f*
UHF - s. ultra-high frequency
ULALGIA - [med] douleur *f* gingivale
ULATROPHY - [med] rétraction *f* des gencives
ULCER - [med] (open sore generally accompanied by disintegration of tissue) ulcère *f*
ULCERATION - [med] (the forming of an ulcer) ulcération *f*
ULEDERMATITIS - s. ulerythema
ULERYTHEMA - [med] ulérythème *m*
ULETOMY - s. ulotomy
ULEXITE - [min] (natural hydrated sodium calcium borate) ulexite *f*
ULEXITE - [min] (boronatro-calcite) boronatrocalcite *f*, ulexite *f*
ULLAGE - [gen] (the volume in a cask not filled with its liquid contents) vidange *m*, perte *f*, freinte *f*
[gen] volume *m* libre
[print] (in engraving) copeaux *mpl* enlevés par le burin
~ ROCKET - [astronaut] fusée *f* d'équilibrage du propergol liquide dans le réservoir
ULLAGES - [brw ind] (waste beer) bière *f* de reste
ULLAGING - [meas] (a method of measuring the contents of a tank by measuring the height of the liquid surface from the top of the tank) estimation *f* du manquant, ouillage *m*
ULMIC ACID - [chem] acide *m* humique
ULMINS - [chem] ulmines *fpl*
ULMOUS SUBSTANCES - [chem] matières *fpl* humiques
ULNA - [anat] (the long bone of the forearm or fore-limb) cubitus *m*
ULNAR - [anat] cubital, ulnaire
ULOCARCINOMA - [med] carcinome *m* des gencives
ULONCUS - [med] tumeur *m* gingivale
ULOTOMY - [med] ulétomie *f*
ULTERIOR - [gen] (following) ultérieur
ULTIMATE - [gen] final, ultime
[mech] (denoting the maximum strength) limite
~ ANALYSIS - [chem] (in analytical chemistry) analyse *f* élémentaire
~ BENDING STRENGTH - [metall] résistance *f* à la rupture par flexion
~ CO_2 - [chem] carbone *m* total (dans un mélange gazeux)
~ CRUSHING STRENGTH - [metall] résistance *f* à la rupture par compression
~ ELONGATION - [rubber ind] allongement *m* à la rupture
~ FACTOR - [mech] (the factor of safety in respect of the ultimate load) facteur *m* de la résistance à la rupture
~ LIMIT SWITCH - [el] (in lifts) interrupteur *m* de fin de course auxiliaire
~ LINES - [phys] (the spectrum of a substance resulting from the most moderate excitation) lignes *fpl* ultimes
~ LOAD - [mech] (the greatest load which a structure must withstand) charge *f* limite
~ POSITION - [el] position *f* finale
~ PRESSURE - [phys] vide *m* limite, pression *f* limite
~ PRODUCTION - [gen] production *f* finale
~ RESERVOIR CAPACITY - [gas ind] (of a gasholder) capacité *f* totale théorique d'un réservoir

ULTIMATE RESERVOIR PRESSURE - [gas ind] (of a gasholder) pression *f* maximum de stockage
~SETTLEMENT - [soil] tassement *m* définitif, tassement *m* global
~STRENGTH - [mech] résistance *f* limite, résistance *f* à la rupture
~TENSILE STRENGTH - [phys] (the greatest tensile load which can be applied to a material before it fails) résistance *f* à la rupture par traction
ULTOR - [electron] (second anode, of a cathode-ray tube using electrostatic deflection, the electrode for accelerating the electron beam before deflection and after focusing the first anode) anode *f* accélératrice
ULTRA - [gen] (as a prefix, mainly of physical condition) ultra-
~-ACCELERATOR - [ind chem] (rubber chemical producing very rapid vulcanization) ultra-accélérateur *m*
~-ACID - [chem] ultra-acide
ULTRACENTRIFUGE - [mech] (centrifuge operating at very high speed) ultracentrifugeuse *f*
ULTRACONDENSED - [print] (of a type) ultra-étroit
ULTRAEXPANDED - [print] (of a type) ultra-large
ULTRAFST PINCH - [phys] striction *f* ultrarapide
ULTRAFILTER - [ind chem] (a filter used for colloidal solutions) ultra-filtre *m*
ULTRAFILTRATION - [chem] (filtration under pressure difference through a semipermeable membrane, to separate colloidal particles) ultrafiltration *f*
ULTRAFORMING - [ind chem] (reforming process in which the catalyst can be regenerated without interruption of the process) ultraforming *m*
ULTRAHIGH FREQUENCY - [radio] (Uhf radio frequencies between 300 and 3000 Mc/s, i.e. wavelengths between I0 cm and I meter) hyperfréquence *f*, UHF
~-SPEED EXTRUSION - [plast ind] (extrusion in machines in which the screw runs at I500 rpm or more) extrusion *f* à trés haute vitesse
~VACUUM - [phys] ultra-vide, vide *m* ultra-moléculaire
ULTRALIGHT ALLOY - [metall] alliage *m* ultra-léger
ULTRAMARINE - [dye] (a blue pigment, either obtained from lapis lazuli or made artificially from silica, chine clay, sulphur and soda, ash, in which case it is basically a complex aluminium-sodium sulpho-silicate) outremer *m*
~BLUE - [dye] (a pigment occurring naturally as lapis lazuli and also prepared artificially and used in paints, plastics, rubber, etc) bleu *m* d'outremer
ULTRAMICROBE - [biol] (microorganism which is invisible in the optical microscope) ultramicrobe *m*
ULTRAMICROMETER - [instr] ultramicromètre *m*
ULTRAMICROSCOPE - [instr] ultramicroscope *m*
ULTRAPHOTIC RAYS - [phys] (rays beyond the visible region of the spectrum) rayons *mpl* invisibles
ULTRARAPID - [gen] ultrarapide
[photo] à très grande ouverture
~LENS - [opt] objectif *m* utrarapide
~PICTURE - [photo] photographie *f* ultrarapide
ULTRARED - [phys] (infrared) infra-rouge
ULTRASENSITIVE - [gen instr etc] ultra-sensible
ULTRASHORT WAVES - [radio] (waves having a wavelength between I and I0 m) ondes *fpl* ultra-courtes
ULTRASONIC - [phys] (synonymous with Supersonic, but generally used of mechanical vibration and wave motion in solids and fluids) ultra-sonore
ULTRASONIC CLEANING - [phys] (method of cleaning metal parts by immersion in a solvent through which ultrasonic waves are made to pass) nettoyage *m* ultrasonore
~COAGULATION - [phys] (the bonding of small particles into larger aggregates by the action of ultrasonic waves) coagulation *f* par ondes ultrasonores
~CROSS GRATING - [phys] (space grating resulting from the crossing of beams of ultrasonic waves having different direction of propagation) réseau *m* de diffraction
~DELAY LINE - [electron] ligne *f* à retard pour ondes ultrasonores
~DETECTOR - [instr] (a device for the detection and measurement of ultrasonic waves) détecteur *m* supersonique
~DRILLING - [mech] (a method of drilling very hard and brittle materials by means of a tool vibrating at ultrasonic frequency) sondage *m* par ultrasons
~FLOW DETECTOR - [metall] détecteur *m* ultrasonore de fissures internes
~GENERATOR - [radio] (device for the production of sound waves of ultrasonic frequency) générateur *m* supersonique, générateur *m* ultrasonore
~GRATING CONSTANT - [phys] (the distance between diffracting centres of the sound wave which is producing particular light-diffraction spectra) constante *f* de diffraction
~INSPECTION - [metall] examen *m* ultrasonore
~LEVEL GAUGE - [instr] limnimètre *m* ultrasonore
~LIGHT DIFFRACTION - [phys] (the formation of optical diffraction spectra when a beam of light is passed through a longitudinal wave field) diffraction *f* de la lumière par les ondes ultrasonores
~MATERIAL DISPERSION - [phys] dispersion *f* des substances par les ondes ultrasonores
~SEALING - [plast ind] (sealing or welding of plastics by subjecting them to vibration above the frequency of sound, without the application of heat) soudure *f* par ultrasons
~SOUNDING - [meas] sondage *m* par ultrasons
~SPACE GRATING -[phys] (a periodical spatial variation of the index of refraction caused by the presence of acoustic waves within the medium) variation *f* de l'indice de réfraction
~STORAGE CELL - s. ultrasonic delay line
~STROBOSCOPE - [acoust] (light interruptor having an action based on the modulation of a light beam by an ultrasonic field) stroboscope *m* ultrasonore
~WAVE - [acoust] (elastic wave whose frequency is above the audible range) onde *f* supersonique, onde *f* ultrasonore
ULTRASONICS - [phys] (sound vibrations of very high frequency which the human ear is incapable of perceiving) ultra-sons *m.pl*
ULTRASOUND - [acoust] (acoustic oscillation whose frequency is too low to affect the sense of hearing) ultra-son *m*
ULTRAVIOLET - [phys] (radiations with a wavelength of I00-3900 A° and not visibly perceptible) ultraviolet
~ABSORBER - [chem] (a substance added to, e.g. rubber or plastics, to prevent degradation by ultraviolet light) adsorbant *m* d'ultraviolet
~DETECTOR TUBE - [electron] (of a light-sensitive

device) diode *f* détectrice sensible à l'ultraviolet
ULTRA VIOLET INHIBITOR - [chem] (a substance which enhances the resistance to fading from sunlight of e.g. plastics) inhibiteur *m* UV
~RADIATION - [phys] (electromagnetic radiation of frequency greater than that of the visible spectrum, in the wavelength band 3900 to 50 Angstrom units) rayonnement *m* ultraviolet
~RAYS - [phys] rayons *mpl* ultraviolets
~SPECTRUM - [phys] (portion of the electromagnetic spectrum) spectre *m* ultraviolet
~THERAPY - [med] thérapie *f* à rayons ultraviolets
ULTRAWHITE - [telev] dépassement *m* d'amplitude
~REGION - [telev] zone *f* de l'ultrablanc
UMBER - [min] (the primary material of Umber. It is a natural earth consisting essentially of oxides of iron and manganese) terre *f* d'ombre, ombre *f* [min] lignite *m* terreux
UMBILICAL - [anat] ombilical
~BELT - [med] ceinture *f* ombilicale
~CORD - [anat] cordon *m* ombilical
~RUPTURE BELT - [med] ceinture *f* d'hernie ombilicale
~TOWER - [astronaut] tour *f* ombilicale
UMBILICATION - [med] ombilication *f*
UMBILICUS - [anat] ombelic *m*, nombril *m*
UMBRA - [radiat] (region behind an object in a beam of radiation, so that a straight line drawn from any point in this region to any point in the source passes through the object) ombre *f*
UMBRELLA - [impl] parapluie *m*
~AERIAL - [radio] (top-loaded aerial, the shape of the top resembles an umbrella) antenne *f* en parapluie
~ANTENNA - s. umbrella aerial
~INSULATOR - [el] isolateur *m* en parapluie
~ROOF - [constr] comble *m* avec avant-toit
~-TYPE AERIAL - [radio] antenne *f* avec panneau réflecteur
~-TYPE ANTENNA - s. umbrella-type aerial
UMKLAPP PROCESS - [phys] (or flip-over process, a type of collision between phonons, or between phonons and electrons, where crystal momentum is not conserved) processus *m* de fustigation
UMOHOITE - [min] (rare mineral containing uranium) umohoite *f*
UNABLE - [gen] incapable
UNACCEPTABLE - [gen] inacceptable
[leg] irrecevable
UNACCOUNTABLE - [gen] inexplicable
UNADULTERATED - [gen & chem] pur, non falsifié, sans mélange
~OIL - chem] huile *f* non falsifié
~SOAP - [chem] savon *m* inaltéré
UNAFFECTED - [gen] veritable, sincère
[gen] (in industry) indemne, inaltérable
~ZONE - [metall] (in welding) zone *f* pas attaquée par la chaleur de soudure
UNALIENABILITY - [leg] inaliénabilité *f*
UNALLOTTED GROUND - [mining] terrain *m* non-concedé
UNALLOYED - [metall] pur, sans alliage
UNANNEALED WIRE - [metall] fil *m* cru
~WIRE ROPE - [metall] câble *m* métallique en fil non recuit
UNANSWERABLE - [gen] incontestable
UNARMOURED - [el] (said of a cable etc) sans armature
UNARMOURED CABLE - [el] câble sans armature extérieure
UNASSAILABLE - [gen] indiscutable, irréfutable
UNATTACKED BY ACIDS - [min] inattaquable par les acides
UNATTENDED OPERATION - [comput] opération *f* non asservie
~TIME - [comput] temps *m* de non-utilisation
UNAVALAIBLE ENERGY - [phys] (in an irreversible process) énergie *f* perdue
UNBALANCE, to - [gen & mech] (to deprive of balance) déséquilibrer
UNBALANCE - [mech] défaut *m* d'équilibrage, balourd *m*
~FACTOR - [el] (unsymmetry factor, in a three-phase system the ratio of the negative to the positive phase-sequence component) coefficient *m* de dissymétrie
~IN PHASE - [electron] discordance *f* de phase
UNBALANCED - [el] (of a load) non-équilibré
[mech] présentant un balourd, non-compensé
[comm] non soldé
~THREE-PHASED LOAD - [el] charge *f* triphasée non-équilibrée
~WIRE CIRCUIT - [el] (a circuit the two sides of which are electrically unlike) circuit *m* non-équilibré
UNBALLAST, to - [naut & aero] délester
UNBAK, to - [mech] (used in industry) découvrir le feu
UNBANKED - [constr] sans remblai, sans rampe
UNBAR, to - [gen] débarrer, débâcler
[mech] déboulonner
[naut] dessaisir
UNBATCHED JUTE - [text] jute *m* non ensimé
UNBENDING - [gen & mech] inflexible, rigide
UNBITT, to - [naut] débitter
UNBLANKING CIRCUIT - [telev] circuit *m* d'annulation de la suppression
~OF FORWARD SWEEP - [electron] annulation *f* de la suppression de l'aller
~PULSES - [electron] impulsions *fpl* d'annulation
~SIGNAL - [electron] impulsion *f* de commande de luminosité
UNBLEACHED - [text] écru, non blanchi
[paper man] non blanchi
~CELLULOSE - [text] cellulose *f* non blanchie
~CARDBOARD - [paper man] carton *m* nature, carton *m* non blanchi
~LINEN - [text] toile *f* écrue, toile *f* bise
~STUFF - [paper man] pâte *f* écrue
~TWINE - [text] ficelle *f* écrue
UNBLISTERED STEEL - [metall] acier *m* sans ampoules
UNBLOCK, to - [mech] décaler (une roue), enlever la cale
UNBOLT, to - [mech] déboulonner
[gen] déverrouiller
UNBOLTING - [mech] déboulonnage *m*
UNBOUND - [phys] (of particles, electrons) libre
[gen] délié
[print] (of a book) broché
UNBRACED - [constr] sans renforcement, sans lien, sans moise (d'une charpente), sans entretoise (d'un cadre)
UNBREAKABLE - [gen] incassable, imbrisable
~GLASS - [glass man] verre *f* imbrisable

UNBROKEN ORE - [min] minerai *m* non-abattu, minerai *m* en place
UNBUILD, to - [constr] démolir
[el] se désaimanter
UNBUNG, to - [brew ind] débondonner
UNBURNABLE - [gen] unbrûlable
UNBURNT - [th eng] (as fuel in combustion products of a furnace) non brulé
[constr] (of a brick) non cuit
~GYPSUM - [constr] plâtre *m* cru
~LIME - [constr] chaux *f* incuite
UNBUTTONING - [constr] démolition *f*, rasement *m*
UNCAGE, to - [contr] déverrouiller
UNCAGING - [mining] décagement *m*
UNCALCINED - [chem] incalciné
UNCANNED FUEL ELEMENT - [nucl] (a type of fuel element in a thermal breeder reactor) élément *m* combustible nu
UNCAP, to - [geol] découronner
UNCAPPING - [zool] désoperculation *f*
UNCASED HOLE - [mining] trou *m* non tubé
UNCEMENTED - [soil] sans cohésion
UNCERTAINTY -PRINCIPLE - [phys] (in quantum theory) principe *m* d'incertitude, principe *m* d'indétermination
UNCHARGED - [el] (without electric charge) sans charge
~PARTICLES - [phys] (a particle which has no electrical charge) particule *f* neutre
UNCIAL - [print] (form of letter found in ancient manuscripts) oncial
UNCIFORM - [anat] unciforme, onguiforme, crochu
~BONE - [anat] aos *m* unciforme, os *m* crochu
UNCINATUM - s. unciform bone
UNCINUS - [met] (longs wisps of cirrus cloud, hooked at the end) uncinus *m*
UNCLAMP, to - [mech] débrider
UNCLAMPING - [mech] déclavetage *m*
UNCLEAN - [gen] impure
UNCLUTCH, to - [mech] débrayer, désembrayer
UNCOATED - [el] sans couche magnétique
~LENS - [opt] objectif *m* non traité
UNCOCK, to - [firearms] désarmer
UNCOCKED - [agric] (of the hay) (foin) non emmeulé
UNCOIL, to - [gen] (a bobbin etc) dérouler, se dérouler
UNCOIL STAND - [text] (a frame to carry reels or bobbins from which material is to be unwound) appareil *m* à dérouler le tissu
UNCOILING - [el] (unwinding) déroulement *m*, débobinage *m*
~REEL - [text] dévidoir *m*
~SPOOL - [cin] (the spool delivering the film to be projected) bobine *f* de déroulement
UNCOMBED - [text] (of wool) non peigné
~SILK WASTE - [text] schappe *f* non peignée
UNCOMBINED - [chem] à l'état libre, non combiné
~FAT - [chem] graisse *f* non combinée
UNCOMPRESSED - [text] non abrégé
[gen] incomprimé
UNCOMPRESSIBLE - [phys] incompressible
UNCONDENSED - [chem] non-condensé
UNCONDITIONAL - [gen] inconditionnel
~CONTROL TRANSFER INSTRUCTION - [comput] instruction *f* de saut inconditionnelle
~JUMP - [comput] saut *m* inconditionnel
UNCONDITIONAL TRANSFER OF CONTROL - [comput] (transfer of control which is not subject to conditions external to the specific instruction) instruction *f* de saut inconditionnelle
UNCONFINED COMPRESSION TEST - [soil] essai *m* de compression sans étreinte latérale
UNCONFORMITY - [geol] discordance *f*
UNCONGEALABLE - [gen] incongelable
UNCONTROLLABLE - [gen] incontrôlable
UNCONTROLLED NUCLEAR TRANSFORMATION - [nucl] réaction *f* d'emballement, réaction *f* incontrôlée
~SPIN - [aero] (a spin in which recovery cannot be effected by the use of the controls) vrille *f* libre
UNCOOLED - [gen] non réfrigéré
~COMBUSTION CHAMBER - [mech] (in rockets) chambre *f* de combustion sans réfrigération
UNCOUPLE, to - [gen & railw] découpler, désaccoupler, dételer
[mining] désassembler
UNCOVER, to - [gen] découvrir, mettre à nu
UNCOVERED CARD - [text] carde *f* sans couvercle
~FLAT - [text] chapeau *m* dégarni
UNCOVERING - [metall] dépouille *f*, découverte *f*
UNCROSSED RIBBON FEED - [text] alimentation *f* en rubans à fibres en long
UNCRUSHED - [min] non-broyé
UNCTION - [chem] onction *f*, action *f* d'oindre
UNCTUOSUS CLAY - [metall] argile *f* grasse
UNCTUOUSNESS - [gen] (greasiness) onctuosité *f*
UNCUPPING - [mech] décrochage *m*
UNCURED - [rubber ind] (not vulcanized) non-vulcanisé
~COMPOUND - [rubber ind] mélange *m* non-vulcanisé
UNCURL, to - [text] lisser
UNCURLED WEFT YARN - [text] fil *m* de trame non frisé
UNCUT - [gen] non-taillé, non-coupé
~CARPET PLUSH - [text] peluche *f* de tapis non-coupée
~FLAX - [text] lin *m* non-coupé
~VELVET - [text] velours *m* frisé, velours *m* épinglé
UNDAMAGED - [gen] non-endommagé
~FLEECE - [text] toison *f* intacte
UNDAMPED - [phys&el] non amorti
[telecomm] entretenu
~WAVE - [radio] onde *f* entretenue
UNDATED - [comm] sans date
UNDECALACTONE - [chem] (a flavourant and perfumery component) undécalactone *f*
UNDECANE - [chem] (a synthesis intermediate) undécane *m*
UNDECANOIC ACID - [chem] (a synthesis intermediate) acide *m* undécandique
UNDECANOL - [chem] (a perfumery component, plasticizer, and synthesis intermediate) undécanol *m*
UNDECYLENIC ACID - [chem] (an antimycotic used in medicine and an intermediate for plastics additives) acide *m* undécylénique
~ALCOHOL - [chem] (a perfumery constituent) alcool *m* undécylénique
~ALDEHYDE - [chem] (a perfumery component) aldéhyde *m* undécylénique
UNDECYLENYL ACETATE - [chem] (a perfumery component) acétate *m* d'undécylényle
UNDEFINED - [photo] manquant de définition
UNDENTABLE - [gen] incabossable
UNDER - [gen] (preposition) sous, au-dessous de

UNDER - [gen] (in the process of) en voie de
[gen] (less than) insuffisant
~BODY - [aero] châssis *m*, chariot *m* d'atterrissage
[auto] plancher *m*
~-CAR AERIAL - [radio] antenne *f* sous le châssis
~-CAR ANTENNA - s. under-car aerial
~-CARRIAGE - s. under-body
~CONSTRUCTION - [gen] en construction
~-CONTACT RAIL - [railw] (a contact-rail so arranged that the collector shoe runs on the lower surface) rail *m* de contact avec la surface inférieure
~CONTROL - [gen] sous contrôle
~-FLOOR HOLD - [aero] (cargo or baggage space under the floor of an aircraft) soute *f* sous le plancher
~-NIPPER PLATE - [text] mâchoire *f* inférieure
~-PICK MOTION - [text] mécanisme *m* de chasse-navette inférieur
~PROOF - [chem] (term used to define spirit strength: Thus "20° under proof" means a spirit containing 20 p.c. of water and 80 p.c. of proof spirit by volume at 60° F. (15.5° C.) teneur *f* en alcool
~REPAIR - [gen] en réparation
~-WEB - [text] toile *f* d'envers
~-WEFT THREAD - [text] trame *f* d'envers, fil *m* de dessous
~-WING RADIATOR - [aero] (a radiator arranged on the underside of a wing) radiateur *m* d'intrados
UNDERBID, to - [comm] faire des soumissions plus avantageuses
UNDERBACK - [brew ind] reverdoir *m*
[brew ind] (filter set) batterie *f* de filtration
UNDERBODY - [auto] dessous *m* de la voiture, châssis *m*
~HOIST - [mech] vérin *m* sous le plancher, mécanisme *m* de benne sous plateau
UNDERBRIDGE - [railw] passage *m* en dessous
UNDERBUNCHING - [electron] (a condition representing less than optimum bunching) groupement *m* inférieur
UNDERCARPET - [text] thibaude *f*
UNDERCARRIAGE - [aero] (a main assembly of a landing gear) atterrisseur *m*
~FAIRING - [aero] carénage *m* de l'atterrisseur
~SPRINGING - [aero] suspensions *fpl* de l'atterrisseur
~STRUT - [aero] (a shock-absorber strut) mât *m* de l'atterrisseur
UNDERCASING BRACKET - [text] palier *m* à auge
UNDERCAST - [min] (for the ventilation of a shaft) puits *m* de ventilation
UNDERCLAY - [metall] couche *f* d'argile sous une couche de charbon
UNDERCLEARER - [text] plaque *f* nettoyeuse
UNDERCLOTH - [plast ind] (a belt of flexible material used to support sheets during an operation) feuille *f* ensouple
[text] doublier *m*
UNDERCLOTHES - [text] linge *m* de corps
UNDERCOAT - [paint] (a coating intermediate between primer and finishing coat, or sometimes a priming and undercoat combined) couche *f* de fond
[zool] (of a dog) sous-poil *m*
UNDERCOATING - [photo] sous-couche *f*
UNDERCOMPOUND EXCITATION - [el] excitation *f* hypercompound
UNDERCOOLED GRAPHITE - [metall] graphite *f* de surfusion
UNDERCOOLING - [phys chem] (the phenomenon which occurs when a substance is cooled without change of state below the temperature at which its state of aggregation normally changes) sous-refroidissement *m*
[metall] surfusion *f*
UNDERCOUPLING - [electron] (of a filter circuit etc) couplage *m* lâche
UNDERCROFT - [arch] crypte *f*
UNDERCURE - [rubber ind] sous-vulcanisation *f*, sous-cuisson *f*
[plast ind] (in moulding thermosetting plastics, a condition caused by insufficient time and/or temperature being allowed for adequate thermal hardening) sous-cuisson *f*
UNDERCURED - [rubber ind] (insufficiently vulcanized) sous-vulcanisé
UNDERCURRENT - [el] minimum *m* de courant
(in the sea) courant *m* sous-marin
~PROTECTION - [el] dispoditif *m* de protection à minimum de courant
~RELAY - [el] (minimum current relay) relais *m* à minimum
~RELEASE - [el] (a tripping device which operates when the current falls below the predetermined value at which the release has been adjusted to operate) déclenchement *m* à minimum de courant
UNDERCUT, to - [mining] haver, souchever
[gen] creuser, fouiller
[tool] dégager le tranchant
[comm] s. underbid,to
UNDERCUT - [gen] entaille *f* creusée
[metall] (the reserve taper in a mould) contre-dépouille *f*
[mining] havage *m*, sous-cave *f*, souchevage *m*
[mech] (a groove melted in the base metal adjacent to the toe of a weld and left unfilled by weld metal) gorge *f*, brèche *f*, ébréchure *f*
[metall] (in welding) caniveau *m*
~LAPPED SCARF - [constr] enture *f*
~STOPE OF A RIVER - [geogr] rive *f* concave d'une rivière
~TOOTH - [mech] (of a gear) dent *f* affaiblie à la base
UNDERCUTTER - [mining] (machine used in coal-mining) haveuse *f*, déhouilleuse *f*
UNDERCUTTING - [mech] (of a gear tooth) dégagement *m*
[mining] havage *m*
[soil] sapement *m* par érosion
~AND NICKING MACHINE - [mining] (machine used in coal-mining) haveuse-rouilleuse *f*
~MACHINE - s. undercutter
UNDERDAMPING - [el] (degree of damping sufficiently small, that, once the system has been subjected to a single disturbance, one or more cycles of oscillations are executed by the system) amortissement *m* sous-critique
UNDERDECK SPRAY - [astronaut] partie *f* de l'aspersion jaillissant sous la fusée
UNDERDEVELOPMENT - [photo] sous-développement *m*
UNDERDIGGER - [mining] burin *m* excentrique pour terrains tendres
UNDERDOUGH - [brew ind] "unterteig" *m*, pâte *f* inférieure

UNDERDRAIN - [agric] tuyau *m* de drainage
~SETTLING BASIN - [hydr] bassin *m* d'égouttage
UNDERDRAINAGE - [agric] drainage *m* profond
UNDERDRIVE, to - [mech] (to drive from beneath) commander par le bas
UNDERDRIVE - [mech] (reduction gear) réducteur *m*
UNDERDRIVEN - [mech] à commande par le bas
UNDERESTIMATE, to - [gen] sous-estimer
UNDEREXCITATION - [el] sous-excitation
UNDEREXPOSE, to - [photo] sous-exposer
UNDEREXPOSURE - [photo] (a too short time of exposure) sous-exposition *f*
UNDERFEED, to - [th eng] (a furnace) alimenter par en dessous
UNDERFEED FURNACE - [metall] foyer *m* à alimentation par en dessous
~STOKER - [metall] foyer *m* mécanique à alimentation par en dessous
UNDERFEEDING - [agric] malnutrition *f*, sous-alimentation *f*
[plast ind] (insufficient supply of material, especially to an injection moulding machine) dosage *m* insuffisant
UNDERFELT - [text] dessous *m* de tapis
UNDERFILL - [metall] (said of a material, parts of which have been incompletely filled during rolling) section *f* transversale incomplète
UNDERFLOOR - [mech etc] (e.g. engine) plancher *m*
~ENGINE - [auto] moteur *m* sous le plancher
~LOCATION - [mech] (of an engine) installation *f* sous le plancher
UNDERFLOW - [math] dépassement *m* de capacité inférieur
UNDERFOLD - [geol] pli *m* secondaire
UNDERFOOT - [cin] (or shuttle, part of a film-perforating machine) escomateur *m*
UNDERFRAME - [railw] (of a railway car) châssis *m* (de wagon)
[auto] soubassement *m* du châssis
UNDERFREQUENCY PROTECTION - [el] dispositif *m* de protection à minimum de fréquence
UNDERGAUGE HOLE - [mining] trou *m* sous-calibre
UNDERGLAZE DECORATION - [ceram] (ornamentation applied to biscuit before glazing) décoration *f* sous émail
UNDERGO, to - [gen] (to endure, to be subjected to) subit, passer par
~REAPIRS, to - [mech] être en réparation
UNDERGRADE BRIDGE - [constr] pont *m* à tablier supérieur
UNDERGRAZING - [agric] pâturage *m* insuffisamment poussé
UNDERGROUND - [gen] souterrain, sous terre
[gen] (a passage or space underneath the ground) souterrain *m*
[gen] (said of pipes) enterré
~BURST - [nucl] (the explosion of a nuclear weapon beneath the surface of the ground) explosion *f* nucléaire souterraine
~CANALIZATION - [el] canalisation *f* souterraine
~COLLECTOR - [el] (or plough, a device for maintaining a sliding contact between the conductors in a conduit-system and the electric circuit of a tramway car) patin *m* souterrain
~CONDUIT - [el] (for an electric tramway) caniveau *m* souterrain
~CONTACT RAIL - [el] rail *m* de contact souterrain
UNDERGROUND CONVEYANCE - [transp] transport *m* souterrain
~CORROSION - [metall] (of cast parts etc) corrosion *f* par enterrage
[soil] corrosion *f* du sol
~CROSSING - [railw] passage *m* en dessous
~DISTRIBUTION CHAMBER - [el & telecomm] sous-sol *m* de distribution des câbles
~FLOW - [hydr] (a subterranean stream) courant *m* d'eau souterraine
~FOREMAN - [mining] contremaître *m* du fond
~FURNACE - [mining] foyer *m* d'aérage
~HANDS - [mining] ouvriers *mpl* du fond
~HAULAGE - [mining] roulage *m* souterrain, transport *m* à l'intérieur des mines
~HYDRANT - [hydr] bouche *f* à clé sous trottoir
~LANDING STATION - [mining] recette *f* intérieure
~LIGHTING STATION - [mining] poste *m* de rallumage souterrain
~LINE - [el] (electric line situated in the ground) ligne *f* souterraine
~LINK BOX - [el] boîte *f* de distribution souterraine
~MILLING - [mining] exploitation *f* par tranches descendantes
~MINING - [mining] exploitation *f* en souterrain
~PASSAGE - [gen] (e.g. in a factory) passage *m* souterrain, chemin *m* souterrain
~RAILWAY - [railw] chemin *m* de fer souterrain
~RIVER - [geogr] rivière *f* souterraine
~ROAD - [mining] voie *f* souterraine
~SHEET OF WATER - [geol] nappe *f* d'eau souterraine
~STORAGE TANK - [auto] réservoir *m* souterrain
~STREAM - [geogr] cours *m* d'eau souterrain
~SURVEY - [surv] levé *m* souterrain
~TELEPHONE SYSTEM - [telecomm] réseau *m* téléphonique souterrain
~TROLLEY - [el] frotteur *m* souterrain
~WATER - [geol] nappe *f* phréatique, nappe *f* d'infiltration
~WORKINGS - [mining] exploitation *f* en souterrain
UNDERGROWTH - [bot] sous-bois *m*, broussailles *fpl*
UNDERHAND MOTION - [mech] renvoi *m* de mouvement se fixant sur le sol
~STOPE - [mining] gradin *m* droit
~STOPING - [mining] abattage *m* en gradins droits, abattage *m* descendant
UNDERHOLE, to - [mining] haver, sous-caver
UNDERHORN - [anat] corne *f* inférieure du ventricule latéral
UNDERINFLATED - [auto] (of tyres) insuffisamment gonflé
UNDERLAP - [radio] (a defect in reproduction occurring when the width of a scanning line is less than the scanning pitch) non-juxtaposition *f*
[telev] rétrécissement *m* de l'image
UNDERLAY, to - [print] (to insert a card or the like under a plate or type to bring it to type height) rehausser (la composition)
[min] s'incliner par rapport à la verticale
[gen] mettre en dessous
UNDERLAY - [print] hausse *f*
[min] inclinaison *f* par rapport à la verticale
[gen] (e.g. under a carpet) assise *f* de feutre
[geol] inclinaison *f*, plongement *m* de couches
UNDERLAYING - [print] mise *f* des hausses à
UNDERLIE OF PROPS - [mining] déversement *m* des montants

UNDERLOAD - [gen] charge *f* insuffisante
UNDERLYING - [gen] sous-jacent
UNDERMANTLE - [constr] faux jambage *m* de la cheminée
UNDERMINE, to - [gen] miner par le bas
UNDERMINING - [constr] affouillement *m*, sapement *m*
~BY SCOUR - s. undercutting
~BY WATER - [soil] affouillement *m*, éboulement *m*
~PITTING - [metall] formation *f* de piqûres sous la surface
UNDERMODIFIED MALT - [brew ind] malt *m* mal désagrégé
UNDERMODULATION - [electron] sousmodulation *f*
UNDERNEATH - [gen] au dessous de, dessous
~DRIVE - [mech] à commande par le bas
UNDERPAN - [mech] (of an engine) carter *m* inférieur
UNDERPASS - s. undercrossing
UNDERPICK LOOM - [text] métier *m* à chasse inférieure
~PICKER - [text] taquet *m* de chasse inférieure
UNDERPIN, to - [constr] (to support a structure from below) étayer, étançonner, enchevaler
[constr] (to insert masonry under a foundation, as a support) reprendre en sous-oeuvre
~WITH PILES, to - [constr] battre des pieux en dessous
UNDERPINNING - [constr] reprise *f* en sous-oeuvre
UNDERPOLING - [metall] perchage *m* incomplet
UNDERPOWER PROTECTION - [el] dispositif *m* de protection à minimum de puissance
UNDERPRODUCTION - [gen] production *f* déficitaire, sous-production *f*
UNDERPUNCH - [comput] sous-perforation *f*
UNDERQUOTE, to - [comm] faire une soumission plus avantageuse
UNDERREAM, to - [mining] (a well) élargir
UNDERREAMER - [mining] élargisseur *m*
~WITH CUTTERS - [tool] élargisseur *m* à couteaux
UNDERREAMING - [mining] élargissement *m*
UNDERRUN - [telev] termination *f* du programme en moins du temps prévu
UNDERRUNNING - [el] sous-voltage *m*
UNDERSCANNING - [telev] (reduced-amplitude scanning)balayage *m* à amplitude reduite
UNDERSEA - [gen] sous-marin
UNDERSEED - [agric] semis *m* sous couverture
UNDERSEEPAGE - [soil] écoulement *m* d'infiltration sous le barrage
UNDERSELL, to - [comm] vendre à bas prix
UNDERSET - [mining] déversé, déversement *m* des montants
UNDERSHOOT, to - [aero] (to alight short of the prescribed area, or to follow an approach path which would result in such alighting) atterrir trop court
UNDERSHOOT - [radio] (or precursor, the initial transient response which precedes the main transition and is opposite in sense) sous-vibration *f*
[aero] atterrissage *m* trop court
[telev] (distortion) sous-oscillation *f*, dépassement *m* balistique en sens négatif
UNDERSHOT - [hydr] (said of water wheel) en dessous
~WATER WHEEL - [hydr] roue *f* en dessous
UNDERSINTERED - [metall] fritté insuffisamment
UNDERSIZE - [gen] taille *f* au dessous de la moyenne
[phys] (in particle size classification, the term used for the material which passes through a screen of given mesh, as distinct from the oversize, which remains on it) trop petit
UNDERSIZE - [min] passé *m* de crible
~PARTICLE - [min] grain *m* surfin
UNDERSIZED - s. undersize
UNDERSLING, to - [mech] (a spring) surbaisser
UNDERSLUNG - [mech] (having the springs fixed to the axle from below) sous l'essieu, en cantilever, surbaisser
~FRAME - [auto] châssis *m* surbaissé
~SPRING - [auto] ressort *m* sous les essieux
UNDERSMOKES SHEETS - [rubber ind] feuilles *fpl* sous-fumées
UNDERSOLE - [shoe man] semelle *f* extérieure
UNDERSOWN - [agric] semé sous couverture
UNDERSPEED - [cin] vitesse *f* trop petite
UNDERSTAIRS - [constr] pièces *fpl* d'en bas
UNDERSTEEP, to - [brew ind] tremper insuffisamment
UNDERSTEERING - [auto] (said of car whose defective distribution of masses causes understeering) à rayon réduit de braquage
UNDERSTOCKING - s. undergrazing
UNDERSTUDY - [gen] (of an actor, also an assistant) suppléant *m*, doublure *f*
UNDERSWING - s. undershoot (telev)
UNDERSWUNG SLAY - [text] battant *m* inférieur
UNDERTABLE - [geol] hauteur *f* de la plaine du terrain
UNDERTAKE, to - [comm] entreprendre, entreprendre à forfait
UNDERTAKER - [comm] entrepreneur *m*
[gen] entrepreneur *m* de pompes funèbres
UNDERTAKING - [gen] engagement *m*, entreprise *f*
UNDERTHROW DISTORTION - s. undershoot
UNDERTONE - [acoust] demi-voix *f*, mi-voix *f*
[paint] demi-ton *m*
UNDERVALUATION - [gen] sous-évaluation *f*, mésestimation *f*
UNDERVOLTAGE - [el] tension *f* minimale
~PROTECTION - [el] protection *f* à tension minimale
~PROTECTIVE DEVICE - s. undervoltage protection
~RELAY - [el] (minimum voltage relay) relais *m* à minimum
~RELEASE - [el] (device designed to open an electric circuit when the voltage falls below a predetermined value) déclenchement *m* à tension minimale
~TRIPPING - [el] déclenchement *m* par manque de tension
UNDERWATER - [gen] sous l'eau, sous-marin
[geol] (underground water) eau *f* souterraine
~ANTENNA - [radio] (a submerged aerial) antenne *f* immergée
~BURST - [nucl] (the explosion of a nuclear weapon beneath the surface of the water) explosion *f* nucléaire sous-marine
~COMPLETION - [oil ind] achèvement *m* sous-marin
~SOUND PROJECTOR - [el acoust] (a transducer designed to produce sound in water) source *f* sonore immergée
UNDERWAY - [mech] (into motion from a position of rest) en mouvement
UNDERWEIGHT - [gen] manque *m* de poids
UNDERWINDING - [text] sous-renvidage *m*
UNDERWOOD - [agric] taillis *m*, sous-bois *m*, broussailles *fpl*
UNDERWRITING - [leg] garantie *f* d'émission, souscription *f* d'une police d'assurance

UNDETERMINABLE - [gen] indéterminable
UNDETERMINED - [gen] indéterminé
[chem] non-dosé
UNDINE - [med] gondole *f*, bassin *m* oculaire
UNDISSOLVED MATTER - [chem] matières *fpl* non dissoutes
UNDISTILLED - [chem] non distillé
UNDISTORTED - [el acoust] (without any distortion) sans distorsion
UNDISTURBED-ONE OUTPUT SIGNAL - [comput] signal *m* de sortie d'un un non-perturbé
~OUTPUT SIGNAL - [comput] signal *m* de sortie d'un élément de mémoire non-perturbé
~-ZERO OUTPUTS SIGNAL - [comput] (in static magnetic storage) signal *m* de sortie d'un zéro non-perturbé
UNDOCK, to - [naut] faire sortir du bassin
UNDRESSED STONE - [constr] pierre *f* non-taillée
UNDRIED COCOON - [text] cocon *m* vert
UNDRILLED - [mech] pas encore perforé, sans trou
[gen] inexercé
UNDRINKABLE - [gen] imbuvable
~WATER - [gen] eau *f* imbuvable, eau *f* non-potable
UNDULATE, to - [gen] (to move, or to cause to move like a wave) onduler, ondoyer
UNDULATE - [gen] ondulé
UNDULATED - [gen] ondulé, onduleux
~SHEET IRON - [metall] tôle *f* ondulée
UNDULATING LIGHT - [phys] (a light of which the intensity increases and decreases according to a fixed cycle without being extinguished at any time in such cycle) lumière *f* modulée
UNDULATION - [phys] (the continuous propagation of waves through a medium) ondulation *f*
~OF WEFT - [text] ondulation *f* de la trame
UNDULATOR - [telecomm] (in the Morse system, a high-speed tape recorder) ondulateur *m*
UNDULATORY - [gen & phys] ondulatoire
~CURRENT - [el] courant *m* ondulatoire
~ROUGHNESS - [plumb] (of pipes) rugosité *f* d'ondulation
~THEORY - [phys] théorie *f* des ondulations
UNEARNED - [gen & fin] non gagné par le travail
~INCOME - [fin] rente *f*
UNEARTH, to - [constr] déterrer
UNEATABLE - [gen] immangeable
UNEDGE, to - [mech & carp)] refouler
UNELASTIC - [gen] non élastique
[phys] (in construction theory) anélastique
UNEQUAL - [gen] inégal
~ANGLE - [metall] cornière *f* à ailes inégales
~IMPULSE - [comput] impulsion *f* d'inégalité
~-SIDED ANGLES - [metall] cornières *fpl* inégales, cornières *fpl* à branches inégales
~WEBS - [metall] ailes *fpl* inégales
UNEVEN - [gen] inégal, rugueux, raboteux
[gen] (of a road) dénivelé, accidenté
[math] impair
[geogr] (e.g. of a coastline) anfractueux
~FRACTURE - [min] cassure *f* raboteuse
~GROUND - [soil] terrain *m* accidenté
~LAP - [text] nappe *f* irrégulière
~ROAD - [gen] chemin *m* anfractueux, chemin *m* inégal
~SLIVER - [text] ruban *m* inégal
~YARN - [text] fil *m* inégal
UNEVENNES - [gen] inégalité *f*, irrégularité *f*
UNEVENNES - [soil] (of the ground) anfractuosité *f*, montuosité *f*
[constr] désaffleurement *m*
~OF THE WOOL FIBRE - [text] inégalité *f* dans les qualitée du brin de laine
UNEXCEPTIONABLE - [gen] inattaquable, irrécusable
UNEXCEPTIONAL - [gen] sans exception
UNEXCHANGEABILITY - [fin] impermutabilité *f*
UNEXCISED - [leg] exempt de droits de régie
UNEXHAUSTED - [mining] inépuisé
UNEXPOSED - [photo] (said of a film) virge
UNEXPLOITED - [mining] inexploité
UNFACTORED LOAD - [aero] (the greatest load which it is anticipated will be applied to an aircraft, or part of one, under given conditions) charge *f* critique
UNFADABLE - [paint] bon teint
UNFASTEN, to - [gen] détacher, délier
UNFELTED MATERIAL - [text] matières *fpl* non feutrées
UNFERMENTED - [gen] infermenté
~WORT - [brew ind] moût *m* non-fermenté
UNFINISHED - [gen] inachevé
[mech] non façonné, non usiné
~PRODUCTS - [gen] produits *mpl* demi-finis
UNFIRED TUBE - [electron] tube *m* non-amorcé
UNFIT - [gen] inapte, impropre
[med] en mauvaise santé
[constr] (of bricks) cru
~FOR HUMAN CONSUMPTION - [food] non comestible
UNFLASK, to - [metall] (in foundry work) décocher
UNFOLD, to - [gen] ouvrir, déplier
[agric] (the sheep) déparquer
UNFOLDED STRICK - [text] poignée *f* détordue
UNFREEZABLE - [chem] incongelable
UNFREEZE, to - [gen] dégeler, décongeler
UNFULLED CLOTH - [text] drap *m* brut, loden *m*
UNFURNISHED - [gen] (without furniture) démeublé, dégarni
UNFUSED - [metall] non fondu
~CHAPLET - [metall] support *m* mal soudé
UNGATHERED PARASHEET - [aero] (a parachute canopy not furnished with a hem cord) parachute *m* à bord non froncé
UNGOVERNABLE - [gen & mech] ingouvernable
UNGRADED - [min] tout-venant
~CARBIDE - [chem] carbure *m* tout venant
~COKE - [gas ind] coke *m* tout-venant
UNGROUND - [agric] non moulu
UNGROUNDED - [el] isolé
~SUPPLY SYSTEM - [el] système *m* isolé
UNGUENT - [chem] onguent *m*
UNGUIDED - [gen & mech] sans guide
UNGUIS - [anat] unguis *m*, os *m* lacrymal, ongle *f*
[bot] onglet *m* de pétale
UNGULA - [math] onglet *m*
UNHAIR, to - [leather ind] (the operation of removing the hair from a skin before tanning) dépiler, ébourrer
UNHAIRING - [leather ind] dépilage *m*, débourrage *m*, ébourrage *m*
~PROCESS - [leather ind] épilage *m*, ébourrage *m*
UNHANDLE, to - [mech] démancher
UNHANDLING - [mech] démanchement *m*
UNHARDENED - [metall] non trempé
UNHARNESS, to - [gen] dételer
UNHEAD A RIVET, to - [mech] dériver un clou

UNHEALTHINESS - [gen] insalubrité *f*
UNHEWN STONE - [constr] (undressed stone) pierre *f* brute, pierre *f* non-taillée
UNHINGE , to - [mech] décrocher
UNHOOK, to - [gen & mech] décrocher
UNHOOKING - [gen & mech] décrochage *m*, décrochement *m*
UNHUSKED BARLEY - [agric] orge *f* vêtue
UNIAXIAL - [gen] (having an unbranched main axis) uniaxe, à un axe
[cryst] (term denoting those crystalline minerals in which there is only one direction of single refraction) monoaxial
~CRYSTAL - [cryst] crystal *m* monoaxial
UNICELLULAR - [biol] unicellulaire
~ORGANISMS - [biol] protistes *mpl*
~PLANT - [bot] plante *f* unicellulaire
UNICONDUCTOR WAVEGUIDE - [electron] guide *m* d'ondes à uniconducteur
UNIDIAMETER JOINT - [el] (a joint between two similar-lead-covered cables, having approximately the same diameter as the cables, thus enabling the jointed lengths to be drawn into a duct like a normal length of cable) manchon *m* à diamètre du câble
UNIDIRECTIONAL - [gen] (moving in the same direction) unidirectionnel
[el] (of a current) unidirectionnel
~AERIAL - [radio] antenne *f* unidirectionnelle
~ANTENNA - s. unidirectional aerial
~CLOTH - [text] (cloth in which the strength is very much higher in one direction than in the other) tissu *m* unidirectionnel
~COUPLER - [electron] (of a waveguide, a coupler having connections for sampling one direction of transmission) coupleur *m* unidirectionnel
~CURRENT - [el] courant *m* unidirectionnel, courant *m* continu
~ELEMENT - [contr] (irreversible element) élément *m* unidirectionnel
~LOOP DIRECTION FINDER - [instr] (a D.F. loop designed to radiate or receive solely or mainly in one direction) radiogoniomètre *m* unidirectionnel à cadre
~MICROPHONE - [el acoust] (microphone which is responsive predominantly to sound incident within a solid angle not greater than one hemisphere) microphone *m* unidirectionnel
~PULSE - [radio] impulsion *f* unidirectionnelle
[electron] impulsion *f* unidirectionnelle
~TRANSDUCER - [radio] (transducer which cannot be actuated at its output by waves in such manner as to supply related waves as its input) transducteur *m* unidirectionnel
~TRANSMITTER - s, unidirectional microphone
~WEAVE - [text] tissu *m* unidirectionnel
UNIFICATION - [gen] unification *f*
[mech etc] (making a unit) assemblage *m*
UNIFIED - [gen] (made uniform) unifié
[mech etc] (made into a unit) à une seule pièce
~BODY - [auto] (a body having no separate frame) carrosserie *f* monocoque
~FIELD THEORIES - [phys] (term denoting any theory which unites two or more physical theories) théories *fpl* uniformes de champ
UNIFILAR SUSPENSION - [instr] (a type of suspension used in electrical instrument in which the moving part is suspended on a single wire or thread) suspension *f* unifilaire
UNIFORM - [gen] uniforme
[mech] (of a motion) uniforme
~ACCELERATION - [mech] accélération *f* uniforme
~COLOUR BLEND - [ind chem] mélange *m* de couleurs uniforme
~COLOUR SPACE - [opt] espace *m* chromatique uniforme
~CONVERGENCE - [math] convergence *f* uniforme
~FIELD - [el] (a field the intensity and direction of which are the same at all points of the space under consideration) champ *m* uniforme
~FLOW - [phys] (flow steady in time, also a flow which is the same at all points in space) effluent *m* égal
~GRAIN SIZE - [soil] granulométrie *f* uniforme
~LINE - [el] (a line having substantially identical electrical properties throughout its entire length) ligne *f* uniforme
~MOTION - [phys] (motion which is the same at all points in space) mouvement *m* uniforme
~PLANE WAVE - [radio] onde *f* plane uniforme
~RANDOM NOISE - [el acoust] (or white noise, noise distributed over the spectrum, so that the power per cycle per second is constant) bruit *m* à spectre continu et uniforme, bruit *m* blanc
~RAW SILK - [text] soie *f* grège uniforme
~SILK - [text] soie *f* grège uniforme
~SPUN YARN - [text] fil *m* égal
~VELOCITY - [mech] vitesse *f* uniforme
~WAVEGUIDE - [electron] guide *m* d'ondes uniforme
~WEB - [text] voile *m* uniforme
UNIFORMITY - [gen] uniformité *f*
[mech] régularité *f* de fonctionnement
~ANNEALING - [metall] recuit *m* d'homogénéité
~ERROR - [cin] (error which occurs in film perforating machines at the fourth space when the film is not moved correctly) défaut *m* de perforation
~FACTOR - [opt] coefficient *m* d'uniformité
UNIFORMLY - [gen] uniformément
~ACCELERATED MOTION - [mech] mouvement *m* uniformément accéléré
~LOADED BEAM - [constr] poutre *f* uniformément chargée
~SPUN COCOON - [text] cocon *m* confectionné régulièrement
UNIFY, to - [gen] (to cause to be unit) unifier
[mech] (to make into a unit) monter en une seule pièce
UNIJUNCTION TRANSISTOR - [electron] transistor *m* à une jonction
UNILATERAL - [gen] (one-sided) unilatéral
[mech] (with a nominal diameter equal to one of the limits of tolerance) unilatéral
~-AREA TRACK - [el acoust] (a sound-track in which only one edge of the opaque area is modulated in accordance with the recorded signal) piste *f* unilatérale
~CONDUCTIVITY - [el] conductivité *f* unilatérale
~GEAR - [mech] engrenage *m* unilatéral
~TRANSDUCER - s. unidirectional transducer
~TRANSMISSION - [el] transmission *f* unilatéral
UNILAYER - [phys] (or monolayer, a monomolecular layer, i.e. a film on a solid or liquid surface one molecule thick) couche *f* monomoléculaire
UNIMOLECULAR - [phys] (possessing one molecule) monomoléculaire

UNIMOLECULAR REACTION - [chem] (or monomolecular reaction, reaction whose speed is proportional to the concentration of only one reactant) réaction *f* monomoléculaire
UNIMPEDED MAGNETIZATION - [radio] (in a magnetic amplifier) magnétisation *f* libre
UNINFLAMMABLE - [phys] (incapable of supporting combustion) ininflammable, incombustible
~WOOD - [gen] bois *m* incombustible
UNINHABITABLE - [gen] inhabitable
UNINSULATED CONDUCTOR - [el] (a conductor at earth potential, such that it is unnecessary to insulate it from earth) conducteur *m* non-isolé
UNINTELLEGIBLE CROSSTALK - [telecomm] (inverted crosstalk in the USA) diaphonie *f* inintelligible
UNINTERRUPTED - [gen] ininterrompu
~DUTY - [el] service *m* ininterrompu
UNINVERTED CROSSTALK - [telecomm] (USA only, intelligible crosstalk in GB) diaphonie *f* intelligible
UNION - [gen] union *f*
[mech] (coupling or connection for pipes) raccord *m*, manchon *m*, raccordement *m*
[text] (fabric made of two or more materials) étoffe *f* mélangée
~ADAPTOR - [mech] (a form of screw fitting)joint *m* à manchon
~CLOTH - [text] drap *m* union, tissu *m* mi-laine
~-COCK - [hydr] robinet *m* à raccord
~ELBOW - [plumb] coude *m* de raccord
~LINEN - [text] tissu *m* mi-lin
~SCREW - [mech] tendeur *m* à vis, vis *f* de tension
~SILK - [text] mi-soie *f*
~SLEEVE - [mech] manchon *m* de raccordement
~SYSTEM - [brew ind] fermentation *f* en tonnes
~T - [plumb] raccord *m* à T
~YARN - [text] filé *m* mi-laine
UNIPIVOT BEARING - [instr] (in a moving-coil instrument to support the circular coil) support *m* à une seule pointe
UNIPOLAR - [el] (showing only one kind of polarity) monopolaire, unipolaire
~ELECTRODE SYSTEM - [el] (a pickup or stimulating system, consisting of one active and one dispersive electrode) système *m* unipolaire d'électrodes
~MACHINE - [el] machine *f* acyclique, machine *f* unipolaire
~TRANSISTOR - [electron] (or field effect transistor, a transistor consisting in principle of a very small parallelepiped of semiconducting material) transistor *m* à effet de champ
UNISELECTOR - [telev] carrousel *m*, rotaxteur *m*
UNISEXUAL - [biol] unisexuel, unisexué
UNISEXUALITY - [biol] unisexualité *f*
UNISON - [acoust] (simultaneous tones of the same pitch but not necessarily of the same timbre) unisson *m*
UNIT - [meas & math] (a standard quantity) unité *f*
[gen] (a single thing considered as an individual though belonging to a group) unité *f*, ensemble *m*, organe *m* complet
[chem] (the quantity of a drug, serum etc required to produce a given effect) unité *f*
[railw] (a motor train unit) élément *m* automoteur
[mech] élément *m*, bloc *m*
~AREA - [math] surface *f* unité
~AREA ACOUSTIC IMPEDANCE - [el acoust] impédance *f* acoustique par unité de surface, impédance *f* acoustique spécifique
UNIT AREA CAPACITANCE - [el] capacité *f* par surface unitaire
~-AREA IMPEDANCE - s. unit-area acoustic impedance
~-AREA REACTANCE - [el acoust] réactance *f* acoustique spécifique
~CELL - [cryst] (the basic unit of a crystal structure) cellule *f* élémentaire
~CHARGE - [el] (or unit quantity) charge *f*, quantité *f* d'électricité
[comm] (the running charge) tarif *m* à prix de base
~CONSTRUCTION - [mech] (of a motorcycle, motor and gearbox) moteur *m* et boîte de vitesse dans le même carter
~COST - [comm] prix *m* à la pièce
~COUNTER - [comput] compteur *m* d'unités
~ELECTRIC FLUX - [el] (the amount of electric flux associated with unit charge) flux *m* électrique unitaire
~ELEMENT - [telecomm] (alphabetic signal element having a duration equal to the unit interval) élément *m* unitaire
~FUNCTION RESPONSE - [radio] (the evanescent part of the waveform response) réponse *f* au signal unité
~HEATER - [th eng] (USA only, a unit warm-air) aérotherme *m*
~INTERVAL - [el] (in a winding) intervalle *m*
[radio] (signal interval) intervalle *m* de signal, élément *m* de signal
~INTERVAL AT THE COMMUTATOR - [el] (the distance on the commutator between two consecutive segments) intervalle *m* au collecteur
~LOAD - [mech] charge *f* unitaire
~KILOMETRE - [railw] (engine, or locomotive kilometre) kilomètre *m* élément, kilomètre *m* machine
~MOULD - [plast ind] élément *m* rapporté, pièce *f* complémentaire, élément *m* interchangeable
~OF ANGLE - [meas] unité *f* d'angle
~OF ANGULAR VELOCITY - [phys] unité *f* de vitesse angulaire
~OF AREA - [meas] unité *f* d'aire, unité *f* de surface
~OF CAPACITY - [el] unité *f* de capacité
~OF CURRENT - [el] unité *f* de courant
~OF DESIGN - [text] rapport *m* d'armure
~OF ENERGY - [phys] unité *f* d'énergie, unité *f* de travail
~OF FORCE - [phys] unité *f* de force
~OF HEAT - [heat] unité *f* de chaleur, unité *f* thermique
~OF ILLUMINATION - [opt] (used in photometry) unité *f* d'éclairement
~OF LIGHT - [phys] unité *f* d'intensité lumineuse
~OF LIGHT FLUX - [phys] unité *f* de flux lumineux
~OF LUMINOUS FLUX - s. unit of light flux
~OF MAGNETIC FLUX - [phys] unité *f* de flux magnétique, unité *f* de flux de force
~OF POWER - [mech] (horse power) unité *f* de puissance
~OF RESISTANCE - [el] unité *f* de résistance
~OF SURFACE - s. unit of area
~OF WEIGHT - [meas] unité *f* de poids
~OF WORK - [mech] unité *f* d'énergie
~OPERATION OF A CIRCUIT BREAKER - [el] action *f* sans retard

UNIT PLANE - [crystall) plan m cristallographique à indices de Miller
~PLANES - [opt] (principal planes) plans mpl principaux
~POLE - [el] (a magnetic pole which, when placed at a distance of I cm. from a like pole, experiences a force of repulsion equal to one dyne) unité f polaire étalon
~PRISM - [crystall] prisme m de première espèce, protoprisme m
~PROCESS - [chem] (a chemical process in which all reactions occur in a single apparatus) procédé m unitaire
~-PULSE SIGNAL - [contr] signal m d'impulsion unité
~PYRAMID - [crystall] protopyramide f, pyramide f de première espèce
~-RAMP SIGNAL - [contr] signal m unité de pente
~SENSITIVITY - [instr] (of an instrument) unité f de sensibilité
~-STEP FUNCTION - [contr] (a step function in which the constant finite value is unity on a specified scale) fonction f échelon unitaire
~-STEP RESPONSE - [contr] réponse f indicielle
~-STEP SIGNAL - [contr] signal m échelon unité
~STRESS - [phys] charge f
~STRING - [comput] chaîne f à un élément
~TIME - [meas] (the I/86400th part of a solar day) unité f de temps
~TUBE - [phys] (a tube through which unit flux passes) tube m unité
~TUBE OF FLUX - s. unit tube
~TUBE OF MAGNETIC FLUX - [el] (a tube of magnetic flux, the flux within which is one unit) tube m de flux magnétique unitaire
~VECTOR - [math] vecteur m unitaire, vecteur m unité
~VECTOR FIELD - [el] champ m directionnel
~VOLUME CHANGE - [soil] variation f du volume unitaire
UNITE, to - [gen] unir, amalgamer
[chem] s'unir, se combiner
UNITIZED CONSTRUCTION - [el & electron] construction f en sous-ensembles
UNITS OF ACCELERATION - [phys] unités fpl d'accélération
~OF FLUIDITY - [phys] (in the cgs system of units) unités fpl de fluidité
~OF TIME - [meas] (the fundamental unit of time in all standard system of physical units in the second) unités fpl de temps
UNITY COUPLING - [el] (perfect magnetic coupling between two coils) couplage m parfait
~GAIN - [comput] coefficient m d'amplification = I
UNIVALENT - [chem] (synonym for Monovalent) univalent, monovalent
~ALCOHOL - [chem] (synonym for monohydric alcohol) alcool m monovalent
UNIVARIANT - [phys & chem] (having one degree of freedom) monovariante f
~FUNCTION GENERATOR - [contr] générateur m de fonction à monovariante
UNIVERSAL - [gen] universel
~ANGULAR BIT-STOCK - [mech] vilebrequin m universel à genouillères
~BRIDGE - [el] (a bridge designed for all the usual d.c. and a.c. measurements) pont m universel
~CHUCK - [mach tool] plateau m universel, mandrin m universel
~COUPLING - [mech] raccord m universel
~FOUR-JAW CHUCK - [mach tool] plateau m universel à quatre griffes indépendantes
~GAUGE - [mech] gabarit m universel
~GRINDING MACHINE - [mach tool] machine f à rectifier universelle
~INDEX HEAD - [mach tool] tête f divisée universelle
~INSTABILITY - [phys] (instability of a non-uniform plasma in a magnetic field in the presence of a pressure gradient) instabilité f universelle
~JOINT - [mech] (a coupling connecting two shafts which allows of their relative angular movement in respect to the centre line while still transmitting power) joint m de Cardan, joint m universel, charnière f universelle
~JOINT HOUSING - [mech] carter m de joint de Cardan
~-JOINT SPIEDER - [mech] croisillon m du joint de Cardan
~MILL - [metall] laminoir m universel
~MILL PLATE - [metall] large-plat m
~MILL TRAIN - [metall] train m de laminoir universel
~MILLING MACHINE - [mach tool] fraiseuse f universelle
~MOTOR - [el] (a motor with a commutator, which can be operated with direct current or power-frequency alternating current) moteur m universel
~NETWORK - [el] (or all-pass network, or all-pass transducer, a two-terminal pair network having a loss which is independent of frequency) réseau m passe-tout, réseau m compensateur de phase
~PLIERS - [impl] pinces fpl universelles
~PROGRAMME TRANSMITTER - [comput] traceur m universel de programme
~RADIAL DRILL - [mach tool] machine f à percer radiale universelle
~RECEIVER - [radio] poste m recepteur tous courants, poste m récepteur universel
~ROLLING - [metall] laminage m universel
~SHUNT - [el] (a multiple shunt composed of a number or resistors connected in series in such way that all its ranges have a common connection) shunt m universel
~SLABBING MILL - [metall] laminoir m universel à brames
~STAND - [metall] cage f universelle
~THREAD COUNTER - [text] compte-fil m universel
~-TIME - [astr] (U.T. an internationally-agreed system of Greenwhich Mean Time, reckoned from Greenwich Mean Midnight as zero) temps m universel
~TIPPING WAGON - [transp] wagon m culbutant dans tous les sens
~TOOL GRINDER - [mach tool] machine f universelle à affûter les outils
~TRANSMISSION FUNCTION - [phys] fonction f représentant la propagation infrarouge dans l'atmosphère
UNIVERSE - [astr] (the totality of space surrounding, and including, the earth) univers m
UNIVOLTINE VARIETY - [text] (of silk-worms) variété f univoltine
UNKEY, to - [mech] décoincer, décaler
[mining] (a chute) déhourder (une cheminée)

UNKEYING - [mech] décoincement *m*, décalage *m*
UNKILLED STEEL - [metall] acier *m* effervescent
UNKNOWN - [gen] inconnu, ignoré
[math] inconnue *f*
~QUANTITY - [math] inconnue *f*
UNLATCHING - [mech] désengrenage *m*
UNLAWFUL - [leg] illégal
UNLAY, to - [text] (a rope) décommettre, détordre
UNLIKE POLES - [el] poles *mpl* de nom contraire
UNLIME, to - [ind chem] décalcifier
UNLIMITED - [gen] illimité
UNLINED - [gen] non revêtu
UNLOAD, to - [gen] décharger
UNLOADED Q - [electron] (in switching tubes, the Q of a tube unloaded either by the generator or the terminator) facteur *m* de qualité pour circuit ouvert
~RUBBER - [rubber ind] (pure stock) mélange *m* pure gomme
~VULCANIZATE - [rubber ind] (vulcanized rubber which does not contain loading material) vulcanisat *m* sans charge, vulcanisat *m* pure gomme
UNLOADER VALVE - [mech] (valve used to limit the maximum pressure in a hydraulic system by unloading the actuating pump) soupape *f* de réglage de la pression
UNLOADING - [soil] décompression *f*
~FACE - [nucl] (of a reactor) front *m* de décharge
~OF TROUGHS - [rubber ind] (mould stripping) démoulage *m*
~PIT - [gen] fosse *f* de déchargement
~STATION - [mining] station *f* de déchargement
UNLOCK, to - [mech] ouvrir, déverrouiller, dévisser
~A SWITCH, to - [railw] déverrouiller une aiguille
UNLOOSE, to - [mech] délier, détacher
UNLUTE, to - [ind chem] déluter
UNLUTING - [ind chem] délutage *m*
UNMACHINABLE - [mech] non-usinable
UNMAKE, to - [gen] défaire, démonter
UNMALLEABLE - [metall] non-malléable
UNMALTED GRAIN - [agric] (raw grain) grain *m* cru
UNMAN, to - [gen] dégarnir, désarmer
UNMANNED - [aero, astronaut etc] sans équipage
~ROCKET - [astronaut] fusée *f* non-habitée
UNMANURED - [agric] non fumé
UNMARKED POLE - [phys] pole *m* boréal
UNMARRIED PRINT - [cin] enregistrements *mpl* séparés
UNMASK, to - [comput] modifier la masque
UNMATCHED RECORDS - [comput] blocs *m* d'informations discordants
UNMATTED - [text] (of wool) non-feutré
~WOOL - [text] laine *f* non feutrée
UNMODIFIED INSTRUCTION - [comput] instruction *f* non-modifiée
UNMODIFIED SCATTER - [phys] (radiation scattered without change of photon energy) diffusion *f* non-modifiée
UNMODULATED GROOVE - [el acoust] (a blank groove) sillon *m* sans modulation
UNMONITORED CONTROL SYSTEM - [contr] système *m* de commande en boucle ouverte
UNMOOR, to - [naut] (to weight the anchors) démarrer, désamarrer
[naut] (to let the vessel ride on a single anchor) désaffourcher
UNMOOR A DREDGE, to - [mech] démarrer une drague
UNMOORING - [gen] démarrage *m*
UNMOUNTED - [gen] (e.g. of precious stones) hors d'oeuvre
UNNECESSARY OPERATION - [el] fonctionnement *m* intempestif
UNOBSTRUCTED DAYLIGHT - [photo] lumière *f* solaire dans un ciel dégagé
UNOBTAINABLE - [gen] non-procurable
UNOPENED - [gen] fermé
[min] (said of a seam etc) non exploité
~SEAM - [mining] filon *m* non exploité
UNPACK, to - [gen] déballer, décaisser
[comput] (to separate packed items of information into separate machine words) décomprimer, dégrouper
UNPADLOCK, to - [mech] décadenasser
UNPEELED - [bot] non épluché
[carp] (of timber) non écorcé
~FIBRE - [text] fibre *f* non pelée, fibre *f* non écorcée
UNPICKED - [gen] non choisi
[min] non trié
[agric] non épluché
UNPITCHED - SOUND - [acoust] (said of a complex sound devoid of any musical quality) son *m* inarticulé
UNPOLARIZED LIGHT - [opt] lumière *f* non polarisée
UNPREPARED SCRAP - [metall] mitraille f brute
UNPRIME A CARTRIDGE, to - [firearms] désamorcer une cartouche
UNPRINTED - s. unpublished
UNPROP, to - [constr] enlever les étais
UNPROPAGATED POTENTIAL - [el biol] (an evoked transient localized potential not necessarily associated with charged excitability) potentiel *m* non propagé
UNPUBLISHED - [print] inédit, non publié
UNRAM, to - [mining] débourrer
UNRAMMING - [mining] débourrage *m*
UNRAVEL, to - [gen] débrouiller
[text] (to disengage the threads of a fabric) effiler, effilocher
UNRECOVERABLE - [gen] irrécouvrable
UNREELING - [text] déroulement
UNREFINED - [gen] brut, non affiné, inépuré
UNRESTRICTED FLOW - [plast ind] (unhampered flow in a pressure mould) écoulement *m* libre
UNRIPE - [bot] vert
UNRIVET, to - [mech] dériver
UNROLL, to - [text] dérouler
UNROLLING CREEL - [text] ratelier *m* dérouleur
~DEVICE - [text] dispositif *m* d'ensouplage
UNSAFE - [gen] dangereux
UNSANITARY - [gen] malsain
UNSAPONIFIABLE - [chem] insaponifiable
UNSATURATED - [chem] (generic term for organic compounds capable of forming addition compounds) insaturé, non saturé
~ALCOHOL - [chem] (alcohols capable of forming addition compounds) alcool *m* non saturé
~COMPOUNDS - [chem] composés *mpl* non saturés
~STANDARD CELL - [chem] (standard cell in which the cadmium sulphate electrolyte is not saturated at ordinary temperatures) pile *f* étalon non saturée
UNSATURATION - [chem] (organic compounds with

the ability to react by addition) insaturation *f*
UNSCOURED SILK - [text] soie *f* écrue
UNSCREENED - [min] non criblé
UNSCREENED HORN BALANCE - [aero] (horn balance) corne *f* de compensation
UNSCREW, to - [mech] dévisser, se dévisser
UNSCREWING - [mech] dévissage *m*
~HEAD - [plast ind] (device for mechanically unscrewing moulded threaded parts) tête *f* pour dévisser
UNSEAL, to - [gen] desceller, décacheter
UNSEALED COWLING - [aero] (a cowling in which the air pressure is increased by a fan or by ramming) capotage *m* sous pression, capotage *m* non étanche
UNSEAM, to - [text] découdre
UNSEAWORTHY - [naut] incapable de tenir la mer
[leg] innavigable
UNSERVICEABLE - [gen] inutilisable
UNSET, to - [mining] dessertir
UNSET - [gen] (of precious stones) hors d'oeuvre
~CROWN - [tool] couronne *f* sans diamants
UNSEWERED - [hydr] sans égouts
UNSHARPENED - [gen] non-affûté
UNSHIELDED - [gen] non-protégé, exposé à
~ARC-WELDING - [metall] soudage *m* à arc non-protégé
UNSHIFT-ON-SPACE - [telecomm] blanc *m* des lettres automatique
UNSHIP, to - [naut] (to remove a part) démonter
[naut] (to unload) débarquer, décharger
UNSHIP A BELT, to - [mech] débrayer une courroie
UNSHOE, to - [gen] (a horse) déferrer (un cheval)
UNSHORE, to - [naut] enlever les accores
UNSHRINK, to - [mech] débloquer
UNSHRINKABLE - [text] irrétrécissable
UNSIFTED - [min] non-tamisé
UNSINKABLE - [naut] insubmersible
UNSKILLED - [gen] novice, inexpérimenté
~LABOUR - [gen] main-d'oeuvre *f* non spécialisée
UNSLAKED LIME - [chem] chaux *f* anhydre, oxyde *m* de calcium
UNSLING A STONE, to - [mining] débrider une pierre
UNSOIL, to - [mining] (in quarrying) enlever les terrains de recouvrement
UNSOILING - [mining] (in quarrying) enlèvement *m* des cosses
UNSOLDER, to - [mech] (to release a soldered connexion by heating it until the solder melts) dessouder
UNSOLDERING - [metall] dessoudure *f*
UNSORTABLE - [min] non-triable
UNSORTED FLAX - [text] lin *m* non classé
~ORE - [min] minerai *m* tout venant
UNSPILLABLE ACCUMULATOR - [el] accumulateur *m* à liquide immobilisé
UNSPINDLE, to - [text] débrocher
UNSPLICED INNER TUBE - [rubber ind] (air tube with two ends) chambre *f* à air à deux bouts
UNSPRUNG - [mech auto railw) (said of those mechanical parts which are not over springs) non suspendu, inélastique
~MASS - [mech] masse *f* inélastique
~WEIGHT - [auto] poids *m* non suspendu
UNSTABLE - [gen] instable
[chem] (said of easily decomposable compounds) instable
~EQUILIBRIUM - [phys] (an equilibrium state of a system of one or more particles such that the potential energy of the system is a maximum) équilibre *m* instable
UNSTABLE LAYER - [met] (an atmospheric layer in which convective movements tend to increase) couche *f* instable
~MULTIVIBRATOR - [radio] multivibrateur *m* instable
~OSCILLATION - [phys] (an oscillation which has a tendency to increase) oscillation *f* amplifiée
~POSITION - [el] position *f* instable
UNSTEADINESS OF WHEELS - [auto] flottement *m* des roues
UNSTEADY - [gen] peu stable, vacillant
[cin] (film defect) (image) qui danse
UNSTEEL, to - [metall] désaciérer
UNSTEM, to - [soil] débourrer
UNSTICK, to - [aero] (slag term meaning "to take off") décoller
UNSTRIPPED GAS - [gas ind] (benzolized gas) gaz *m* non débenzolé
~STEM - [bot] tige *f* non écorcée
UNSOPPORTED - [constr] en porte-à-faux
~SHEETING - [plast & rubber ind] (sheeting unprovided with backing material) feuilles *f*pl simples, feuilles *f*pl sans support
UNSYM (UNS) - (prefix denoting unsymmetrical structure in an organic compound) asymétrique, dissymétrique
UNSYMMETRICAL ALTERNATING CURRENT - [el] courant *m* alternatif asymétrique
~BENDING - [mech] flexion *f* asymétrique
~LOADING - [aero] (a loading condition for wings and connecting elements corresponding to a rolling attitude) charge *f* asymétrique
UNSIMMETRY FACTOR - [el] (or unbalance factor; in a three-phase system, the ratio of the negative to the positive phase-sequence component) coefficient *m* de dissymétrie
UNTAPPED - [mech] non-taraudé
UNTIE, to - [gen] (a knot) dénouer, délier
UNTIMBERING - [mining] déboisage *m*
UNTOP, to - s. unsoil, to
UNTREATED WATER - [hydr] eau *f* brute
UNTRUE - [gen] faux
[mech] (out of centre) faux, inexact
UNTRUSSED ROOF - [constr] comble *m* sans ferme
UNTUNED - [radio] syntonisé, non accordé
~AERIAL - [radio] (a periodic aerial) antenne *f* apériodique
UNTUNING - [radio] (deliberate disturbing of the tuned state of a receiver) désaccord
UNTWIST, to - [text] détordre
UNTWISTING - [gen] détorsion *f*
UNVENTED RADIATOR - [th eng] (flueless space-heating appliance) radiateur *m* sans dégagement
UNVITRIFIED - [min] invitré
UNVULCANIZED - [rubber ind] non vulcanisé
UNWASHED YARN - [text] fil *m* non lavé
UNWATER, to - [soil] dénoyer, assécher
UNWATERING - [soil] exhaure *f*, dénoyage *m*
UNWEATHERED - [geol] inaltéré
UNWEDGE, to - [mech] relâcher un coin
UNWELD, to - [mech] (of a welded seam etc) dessouder
UNWIELDY - [gen] immaniable, difficil à manier
UNWIND, to - [gen] dérouler, se dérouler
[text] t(the yarn from a bobbin etc) dévider

UNWIND, to - [el] (a winding) débobiner
[mech] (of a capstan) dévirer
[contr] expliciter, détailler
~THE COCOON FILAMENT, to - [text] dépelotonner le brin de cocon
UNWINDER - [mech](a dereeling device) débobineuse, dérouleuse *f*
UNWORKABLE SEAM - [mining] couche *f* inexploitable
UNWRAPPED CONSTRUCTION - [el chem] (non-lined construction a method of dry cell construction in which the polarizing mix is separated from the negative electrode only by a layer of paste) montage *m* sans habillage
UNWROUGHT - [metall] non-ouvré
~BAR - [metall] barre *f* non travaillée
UP AND DOWN - [railw] en amont et en aval
~-AND DOWN LOCK - [mech] (locking device to hold a retractable undercarriage in "up" or "down" position) verrou *m* de position
~-AND-DOWN MOTION - [mech] mouvement *m* ascendant et descendant
~-AND-DOWN WORKING - [oil ind] exploitation *f* à l'alternat
~DIP - [oil ind] amont pendage *m*
~DOWN AND STOP CONTROL - [el] (or semi-automatic control ; a method of control in which the momentary manual operation causes the car to start and to continue to travel in the direction corresponding to the button pressed, until stopped by pressing the "stop" button, or the terminal limit-switch) commande *f* semi-automatique
~-DRAFT - s. up-draft carburettor
~-DRAFT CARBURETTOR - [auto] carburateur *m* vertical
~ROAD - [railw] voie *f* montante
~-RIVER - [geogr] amont, d'amont
~-RUN - [gas ind] (in the production of water gas) fabrication *f* montante, fabrication *f* à courant de vapeur ascendant
~-RUN STEAM PURGE - [gas ind] purge *f* de fin de soufflage, purge *f* à la vapeur
~-STREAM - s. up-river
~-STREAM CUTWATER - [constr] avant-bec *m* d'une pile de pont
~THE WIND - s. up wind
~-TO-DATE - [gen] moderne
~-TRAVEL - [mech] (movement of a part, espec. of a press in an upward direction) course *f* ascendante
~-WIND - [aero naut] (towards the point from which the wind is blowing) contre le vent
UPCAST - [min] (a ventilating shaft) puits *m* de retour d'air
[geol] faille *f* inverse
~SHAFT - [mining] puits *m* de sortie d'air
UPDATE, to - [gen] (to bring up to date) mettre à jour
UPDRAW - [glass man] (the continuous drawing of glass canes or tubes of various cross-sections) tirage *m* vertical
UPENDING - [metall] (in forging, an operation by which the axial length of the work is shortened and the cross section increased) forgeage *m*
~TEST - [metall] (a forging test) essai *m* de forgeage
UPFLOW BAFFLE - [hydr] chicane *f* de retenue
UPFLOW CLARIFICATION - [hydr] clarification *f* par circulation ascendante
~OF THE WATER - [hydr] écoulement *m* ascendant de l'eau
UPGRADE, to - [gen] améliorer
UPGRADE - [gen] (an upward incline) rampe *f*
UPGRADED ORE - [min] minerai *m* affiné
UPGRANDING - [gen] montée *f* en grade
UPGRADING TREATMENT - [min] affinage *m*
~OF FARMS - [agric] agrandissement *m* des fermes trop petites
UPHAND SLEDGE - [tool] marteau *m* de forge
UPHEAVAL - [geol] soulèvement *m*, surrection *f*
~BEDS - [geol] couches *f*pl soulevées
~DOME - [geol] dôme *m* de soulèvement
UPHILL - [gen] montant, ardu
~CASTING - [metall] coulée *f* en source
~RUNNING - [metall] attaque *f* en remonte
~TEEMING - [metall] coulée *f* en source
UPHOLSTER, to -[gen] (to fit with coverings etc) tapisser, rembourrer, garnir
UPHOLSTERER - [gen] tapissier *m*
UPHOLSTERING MATERIAL - [text] tissu *m* d'ameublement, matière *f* de rembourrage
UPHOLSTERY - [gen] capitonnage *m*, rembourrage *m*
[auto] garniture *f* intérieure
~WEBBING - [text] sangle *f* de tapissier
UPKEEP - [gen] entretien *m*
UPLAND - [geogr] hautes terres *f*pl, région *f* montagneuse
~COTTON - [text] (a short-staple cotton) coton *m* à fibres courtes
~FARM - [agric] exploitation *f* agricole de montagne
~MEADOW - [agric] prairie *f* de montagne
UPLIFT, to - [gen] soulever, élever
UPLIFT - [constr] élévation *f*
[geol] surrection *f*, soulèvement *m*
[phys] force *f* ascensionnelle, poussée *f* de bas en haut
~PRESSURE - s. uplift
~PRESSURE OF A RETAINING WALL - [hydr] sous-pression *f* d'un barrage
UPLIFTED PENEPLAIN - [geol] pénéplaine *f* élevée
UPLOCK - [aero] (device to secure a retractable landing gear in the "up" position) verrou *m* de position de train rentré
UPLOCKING - [aero] (the operation of locking a retractable landing gear in the "up" position) verrouillage *m* en position "train rentré"
UPPER - [gen] supérieur, plus haut
~AIR TEMPERATURE REPORT - [met] bulletin *m* de températures aux couches supérieures de l'atmosphère
~BRACKET - [constr] console *f* supérieure
~BRIDGE - [horol] pont *m* supérieur
~CAMBER - [aero] courbure *f* supérieure
~CASE - [print] haut *m* de casse
~CHORD - [constr] arbalétrier *m*
~CULMINATION - [astr] passage *m* supérieur
~DEAD CENTRE - [mech] (top dead center in the USA) point *m* mort haut
~DECK - [naut] premier pont *m*
~EDGE OF LOADING AREA - [auto] bord *m* supérieur de la surface de chargement
~END-PIECE - [horol] plaque *f* de contre-écrou
~EGYPTIAN - [text] coton *m* de la Haute-Egypte

UPPER LEATHER - [shoe man] empeigne *f*
~LIMIT SWITCH - [el] (for lifts) contacteur *m* de fin de course vers le haut
~MOULD SECTION - [rubber ind] coquille *f* supérieure,demi-moule *m* supérieur
~PART OF A RIVER - s. upper waters of a river
~PITCH LIMIT - [acoust] limite *f* supérieure de tonalité
~PITCH OF A CURB ROOF - [constr] terrasson *m*
~PLATEN - [mech] (the upper of the two or more plates or tables between which work is placed in a press) plateau *m* supérieur
~-SURFACE - [aero] (aileron housed in the trailing edge of the wing and deflectable only upward, so as to modify the aerodynamic contour of the wing on the upper surface only) aileron *m* supérieur
~TEXTURE - [text] tissu *m* de dessu
~-TERMINAL - [constr] station-amont *f*
~TRESHOLD OF AUDIBILITY - [acoust] seuil *m* limite d'audibilité
~VALVE GEAR HOUSING - [auto] couvercle *m* de culbuterie, couvercle *m* de culasse
~WATERS OF A RIVER - [geogr] amont *m* d'une rivière
~WIND REPORT - [met] message *m* aérologique, message *m* de sondage du vent
UPPERBOARD - [mus] (of an organ) sommier *m*
UPPERS - [shoe man] empeignes *f*pl
UPRAISE - [mining] remontage *m*
UPRAISING - [mining] remontage *m*
UPRIGHT - [gen] vertical, perpendiculaire
[constr] (an upright, standing element) montant *m*, pied-droit *m*
[mech] (of a press etc) jumelle *f*
~BOILER - [th eng] chaudière *f* verticale
~BURNER - [gas ind] (for gas light) bec *m* droit
~COURSE OF BRICKS - [constr] rouleau-brique *m*
~CREEL - [text] cantre *m* vertical
~DRILLING MACHINE - [mach tool] machine *f* à percer sur colonne
~ENGINE - [mech] moteur *m* vertical
~EXTENSION - [constr] (of a scaffold) rallonge *f* d'une écoperche
~FLANGED BOBBINS - [text] bobines *f*pl verticales à plateaux
~LATTICE - [text] tablier *m* élévatoire, toile *f* sans fin
~MAST - [naut] mât *m* ciergé
~OF THE LOOM - [mech] montant *m* du métier
~POLE - [constr] écoperche *f*
~PROJECTION - [draw] élévation *f*
~PULLEY - [mech] renvoi *m* sur champ
~RUNNER - [metall] trou *m* de coulée, jet *m* de coulée
~SHAFT - [mech] arbre *m* de transmission vertical
~-SHAPE PICTURE - [photo] image *f* en hauteur
~SPINDLE - [text] broche *f* verticale
~SUPPORT - [carp] montant *m*
~TWILL - [text] croisé *m* à angle aigu
UPROAR - [acoust] (a badly defined mixture of loud sounds) vacarme *m*, tumulte *m*, tapage *m*
UPROOTING - [agric] arrachage *m* volontaires (de vignes)
UPSET, to - [gen] renverser, culbuter
[mech] (in forging) refouler
UPSET - [gen] renversé, versé
[overturning] renversement *m*
[comm] (at an auction, the opening price) mise *f* à prix
UPSET - [tool] (a smithing tool) refouleur *m*
[metall] (in forging) refoulement *m*
[mining] ouvrage *m* montant
~BUTT-WELDING - [mech] soudure *f* en bout refoulée
~CASING - [oil ind] tube *m* refoulé
~TUBING - [oil ind] tube *m* de pompage à refoulements extérieurs
~TYRE - [rubber ind] carcasse *f* plissée, pneu *m* refoulé
~WELDING - [mech] soudage *m* bout à bout
UPSETTER - [metall] machine *f* à refouler
UPSETTING - [metall] (the forging operation) refoulage *m*
~ALLOWANCE - [metall] (the allowance in stock length for the metal lost in upsetting, in upset welding) surépaisseur *f* pour l'usinage
~DIE - [metall] calibre *m* à refouler
~MACHINE - [metall] (a forging machine) machine *f* à refouler
~PRESS - [metall] presse *f* à refouler
~TEST - [metall] essai *m* d'aplatissement
UPSHAFT - [mining] (ventilation shaft) puits *m* d'aérage
UPSIDE-DOWN - [gen] sens dessus dessous
UPSLOPE FOG - [met] (fog resulting from the ascent of air up a hillside or mountain slope) brouillard *m* élévé
UPSTAIR - [gen] (pertaining to an upper floor) aux étages supérieurs, d'en haut
UPSTREAM - [phys] (in a fluid system, the region from which the flow is coming) amont
~WATER - [hydraulic] eaux *f*pl amont
UPSTROKE - [mech] course *f* montante
UPSURGE - [hydr] poussée *f*
UPSWEPT - [auto] ramassé, élevé
~FRAME - [auto] châssis *m* cintré
UPTAKE - [th eng] colonne *f* d'air montant
[mining] puits *m* de retour d'air
[metall] carneau *m* à gaz vertical
~FACTOR - [radiat] facteur *m* d'apport
~RATE - [nucl] (the velocity at which a radioactive isotope is absorbed by body) vitesse *f* d'apport
~SHAFT - [mining] puits *m* de sortie d'air
UPTHROW - [geol] rejet *m* en haut
~SIDE - [geol] lèvre *f* soulevée d'une faille
UPTHRUST - [soil] sous-pression *f* hydrostatique
UPWARD - [gen] montant, vers le haut
~EXHAUST PIPE - [mech] (of a motorcycle) tuyau *m* d'échappement rehaussé
~FORCE - [phys] force *f* ascendante
~PRESSURE - [soil] sous-pression *f* hydrostatique
UPWARDS CONVERSION - [telev] conversion *f* montante
UPWIND - s. up wind
URACIL - [chem] (a pyrimidine derivative with uses in physiological research) uracile *m*
URACONITE - [min] (mineral containing uranopilite) uraconite *f*
URANIDES - [chem] (a name proposed for the elements beyond actinium in the periodic system) uranides *m*pl
URANINE - [min] (a yellow dyestuff derived from fluorescein) uranine *f*
URANINITE - [min] (another name for Pitchblende, an important source of uranium and radium consisting essentially of uranyl uranate) uranite *f*, pechu-

rane *m*, uranine *f*

URANITE - [min] (or pitchblende, mineral largely consisting of uranium oxides) chalcolite *f*, torbérite *f*

URANIUM - [chem] (a radio-active metallic element, symbol U, A. N. 92, A. W. 238.07 half-life 4.5 x 10^9 years. Of great importance in nucleonics and nuclear engineering, e. g. in power generation) uranium *m*

URANIUM 233 - [nucl] (an isotope of uranium) uranium *m* 233

URANIUM 235 - [nucl] (an isotope of uranium, fissionable and a component of atomic bombs) uranium *m* 235

URANIUM 238 - [nucl] (an isotope or uranium with application as a catalyst, in ceramic glazes in electronics, and as an analytical reagent) uranium *m* 238

~ACETATE - s. uranyl acetate

~AGE - [min] âge *m* d'uranium

~BAR - [nucl] (a bar or uranium fuel for nuclear reactors) barre *f* d'uranium

~BARIUM OXIDE - [chem] (a constituent of ceramics glazes) oxyde *m* d'uranium et de barium

~CARBIDE - [chem] (a fuel for nuclear reactors) carbure *m* d'uranium

~CHALCOGENIDE - [chem] chalcogénure *m* d'uranium

~COMPOUND - [chem] (chemical compound of uranium) composé *m* d'uranium

~CONCENTRATE - [chem] concentré *m* uranifère

~CONTENT - [min] (the percentage of uranium in an ore) teneur *f* en uranium

~CONTENT METER BY BETA AND GAMMA RADIOACTIVITY - [instr] teneurmètre *m* en uranium par radioactivité bêta et gamma

~DIOXIDE - [chem] (a catalyst and a component of pigments and photographic chemicals. Uranium dioxide also has application in nuclear technology) bioxyde *m* d'uranium

~ENRICHED FUEL - [nucl] combustible *m* en uranium enrichi

~GRAPHITE LATTICE - [nucl] (a lattice in a reactor composed of uranium and graphite rods, slabs or slugs) réseau *m* uranium-graphite

~HALIDE - [chem] halogénure *m* d'uranium

~HEXAFLUORIDE - [chem] (the substance which, in its gaseous form, is used to separate the isotope U^{235} from the isotope) U^{238}) hexafluorure *m* de uranium

~HYDRIDE - [chem] (a reducing agent) hydrure *m* d'uranium

~MONOCARBIDE - [chem] (a fuel for nuclear reactors) monocarbure *m* d'uranium

~ORE - [min] (ore containing uranium) minerai *m* d'uranium

~OXIDE - [chem] (a constituent of ceramic glazes and pigments) oxyde *m* d'uranium

~OXOSALTS - [chem] sels *mpl* des oxo-acides de uranium

~-RADIUM SERIES - [nucl] famille *f* uranium-radium

~REACTOR - [nucl] (nuclear reactor using uranium as its principal fuel) réacteur *m* à uranium

~RECONNAISSANCE - [min] exploration *f* d' uranium

~ROD - s. uranium bar

~SERIES - [nucl] (the series of nuclides resulting from the decay of uranium ^{238}) famille *f* de l'uranium

URANIUM SLUG - [nucl] (small solid piece of uranium metal) lingot *m* d'uranium

~SULPHATE - s. uranyl sulphate

~TETRAFLUORIDE - [chem] (a source material for metallic uranium) tétrafluorure *m* d'uranium

~-THRIUM REACTOR - [nucl] réacteur *m* à uranium-thorium

~TRIOXIDE - [chem] (a constituent of pigments and ceramic glazes) trioxyde *m* d'uranium

URANOCIRCITE - [min] (a native uranium and barium phosphate) méta-uranocircite *f*

URANOGRAPHY - [astr] (a scientific description of the celestial bodies and the making of celestial globes and maps) uranographie *f*

URANOMETRY - [astr] (the making of maps of those celestial bodies which are visible to the naked eye also the measurement of the heavens) uranométrie *f*

URANOMOLYBDATE - [chem] (a uranium compound) uranomolybdate *m*

URANOPHANE - [min] (rare mineral containing uranium) lambertite *f*, uranophane *m*, uranitile *f*

URANOPILITE - [min] (mineral containing copper and uranium) uranopilite *f*

URANOSPHAERITE - [min] uranosphérite *f*

URANOTHORITE - [min] (mineral containing zinc and thorium) uranothorite *f*

URANYL - [chem] (the bivalent radical UO_2 found in many uranium compounds) uranyle *m*

~ACETATE - [chem] (an analytical reagent) acétate *m* d'uranyle

~AMMONIUM CARBONATE - [chem] (a component of ceramic glazes) carbonate *m* d'uranyle et ammonium

~NITRATE - [chem] (a photographic chemical and a constituent of ceramic glazes) nitrate *m* d'uranyle

~SULPHATE - [chem] (an analytical reagent) sulfate *m* d'uranyle

URATURIA - [med] uraturie *f*

URBAN - [gen] (pertaining to a city) urbain

~RAILWAY - [railw] (metropolitan railway) métro *m*

UREA - [chem] (carbamide. A diuretic, fertilizer, starting material for synthetic resins, and synthesis intermediate. Urea was the first organic compound to be prepared synthetically) urée *f*

~ADDUCTS - [chem] (inclusion complexes of urea and, usually, an unbranched aliphatic hydrocarbon) produits *mpl* d'insertion avec l'urée

~-FORMALDEHYDE RESINS - [chem] (an important class of synthetic thermosetting resins with good mechanical properties) résines *fpl* d'urée-formaldéhyde

~FROST - [med] givre *m* uréique

~PEROXIDE - [chem] (an oxidizing, bleaching and polymerizing agent, and a source of hydrogen peroxide) peroxyde *m* d'urée

UREASE - [chem] (an enzyme which hydrolyses urea, it has application in analysis) uréase *f*

URECCHYSIS - [med] infiltration *f* d'urine

UREDEMA - s. urecchysis

URENA FIBRE - [text] fibre *f* de l'urène

UREOMETER - [instr] (instrument designed to measure the amount of ureum in the urine) uréomètre *m*

URETER - [anat] uretère *m*

URETEROGRAPHY - [med] urétérographie *f*

URETEROLITH - [med] calcul *m* de l'uretère

URETEROSTOMY - [med] urétérostomie *f*
URETHANE - [chem] (ethyl carbamate. A cytotoxic agent in medicine and a synthesis intermediate) uréthane *m*
URETHRAL FEVER - [med] fièvre *f* urineuse
URETHRALGIA - s. urethrodynia
URETHREMPHRAXIS - [med] sténose *f* urétrale
URETHRITIS - [med] urétrite *f*
URETHROCELE - [med] urétrocèle *m*
URETHRODYNIA - [med] urétralgie *f*
URETHROPHRAXIS - s. urethremphraxis
URETHROSCOPE - [instr] (instrument for examining the urethra) urétroscope *m*
URGE, to - [gen] exhorter, presser
[phys] (e.g. a fire) activer
URGENCY SIGNAL - [telecomm] signal *m* d'urgence
URGENT - [gen] urgent, immédiat.
URIC ACID - [chem] (lithic acid. A synthesis intermediate) acide *m* urique
URIDINE - [chem] (a pyrimidine riboside with uses in biological research) uridine *f*
URIDYLIC ACID - [chem] (a nucleotide used in research) acide *m* uridylique
URINAL - [impl] urinal *m*
URINARY BLADDER - s. urocyst
URINE - [physiol] urine *f*
~BATH - [ind chem] bain *m* d'urine
~WASH - [text] lavage *m* à l'urine
URINOMA - [med] kyste *m* renfermant de l'urine
URINOMETER - [instr] (instrument designed to determine the specific gravity of urine) urinomètre *m*
URINOUS ABCESS - [med] abcès *m* urineux
URITIS - [med] dermatite *f* calorique
URN - [gen] (rounded or angular vase with a foot) urne *f*
[mining] (a water cistern) réservoir *m* d'eau
[ind chem] (a closed, heated container) récipient *m* chauffé
[hydr] (a water spring) source *f*
UROCYST - [med] vessie *f*
UROGRAPHY - [med] (the radiological examination of the urinary tract) urographie *f*
URONCUS - [med] tumeur *f* renfermant de l'urine
URSA - [astr] (Ursa Major and Ursa Minor) Ursa *f*
URTICA - [med] papule *f* ortiée, boule *f* d'oedème
USABLE HORSEPOWER - s. useful horsepower
USABLE RANGE - [instr] (the portion of the resistance element available for use after subtracting the end resistance) secteur *m* utile
USAGE - [gen] usage *m*, coutume *f*
USE, to - [gen] employer, user, se servir, emprunter
USE - [gen] emploi *m*
[leg] jouissance *f*
USED AIR OUTLET - [th eng] (ventilation outlet) ouverture *f* de sortie d'air vicié, ventouse *f*
~LYE - [chem] lessive *f* épuisée
~UP - [gen & mech] exténué, épuisé
USEFUL - [gen] utile
~BEAM - [electron] (the part of the primary radiation which passes through the aperture, cone or other collimator) faisceau *m* util
USEFUL CAPTURE - [nucl] capture *f* utile
~HEAD - [hydr] chute *f* exploitable
~HORSEPOWER - [auto] puissance *f* utile
~LIFE - [el etc] durée *f* utile
~LOAD - [aero etc] (the result of deducting the empty or tare weight from the gross weight) charge *f* utile
~OUTPUT - [th eng] puissance *f* thermique
~POWER - [mech] (the relation between applied energy to the resultant energy) puissance *f* utile
~SCREEN DIMENSIONS - [electron] (the dimensions of the luminescent part of the screen when viewing in a direction parallel with the tube axis) dimensions *fpl* utiles de l'écran
~WORK - [mech] (energy developed) travail *m* utile
USER-TO-USER CONNECTION - [comput] liaison *f* poste à poste
USHER - [gen] huissier *m*
USTION - [med] ustion *f*, brûlure *f*, cautérisation *f*
USUAL PRACTICAL UNITS - [el] (units which are not actually included in the practical units derived from the CGS system, but which are commonly used) unités *fpl* pratiques usuelles
USUFRUCT - [leg] usufruit *m*
USUFRUCTUARY - [leg] usufruitier
UTENSIL - [impl] ustensile *m*, outil *m*
UTERINE - [anat] (pertaining to the uterus) utérin
~SOUFFLE - [med] souffle *m* placentaire
UTEROPLASTY - [med] utéroplastie *f*
UTERUS - [anat] (the womb) utérus *m*, matrice *f*
UTILITIES - [town planning] entreprises *fpl* de service publique
UTILITY - [gen] utilité *f*
~CAR - [auto] voiture *f* utilitaire
~OPERATING METHOD - [telecomm] méthode *f* de attention continuelle
~PLANT - [ind] entreprise *f* de distribution publique
~VAN - [auto] camionnette *f*
~WAGON - [transp] chariot *m* à toutes fins
UTILIZATION - [gen] utilisation *f*, mise *f* en valeur
~COEFFICIENT - [radio] taux *m* d'utilisation
~FACTOR - [el] facteur *m* d'utilisation
~OF SEWER - [ind chem] utilisation *f* des eaux d'égout, mise *f* en valeur des eaux usées
~OF SLUDGE - [ind chem] utilisation *f* des boues
~TIME - [el biol] (the minimum duration which a stimulus of rheobasic strength must have to be just effective) temps *m* d'utilisation
UTILIZE, to - [gen] utiliser
UTRICLE OF THE EAR - [anat] utricule *m*
UVANITE - [min] (uranium and vanadium containing mineral) uvanite *f*
UVEA - [anat] (the inner, coloured layer of the iris) uvée *f*, membrane *f* irido-choroïdienne
UVEAL - [anat] uvéal
UVEITIS - [med] uvéite *f*
UVIOL - [phys] (short for ultraviolet) uviol
~GLASS - [opt] (a glass which is highly transparent to the ultraviolet) verre *m* uviol
~LAMP - [radiat] lampe *f* pour rayons ultraviolets
UVULA - [anat] (the pendent fleshy portion of the soft palate) uvula *f*

V

V - [chem] (the symbol of Vanadium) symbole *m* du Vanadium
[el] (volt, the practical unit of electromotive force and potential difference) volt *m*
[phys] symbole *m* du volume de gaz
[mech] en V
[met] (Beaufort letter for pure air) air *m* pur
V-AERIAL - [radio] antenne *f* à V
V ALIGNMENT - [mech] alignement *m* en V
V-ANTENNA - s. V-aerial
V-BLOCK - [draw] support *m* en V pour le traçage
V BOB - [impl] balancier *m* d'équerre, varlet *m*
V CROSSING - [railw] croisement *m* de changement
V-CUT - [gen & mech] incision *f* en V
V-DOOR SIDE - [oil ind] porte *f* centrale de la tour
V-ENGINE - [mech] (an engine in which the cylinders are arranged in two banks of cylinders in line, the banks being at an angle to each other) moteur *m* en V, moteur *m* à cylindres convergents
V F - [acoust] (Voice Frequency, frequency within the audible range) fréquence *f* vocale
V FORMAT - [comput] (a type of data set format) format *m* V
V.F.R. - [aero] (short for Visual Flight Rules) VFR
V GEAR - [mech] engrenage *m* hélicoïdal double
V-GROOVE CLUTCH - [mech] embrayage *m* à coins
V.H.F. - [radio] (short for Very High Frequency) hyperfréquence *f*, VHF
V.I. - [radio] (short for Volume Indicator) volumètre *m*
V.L.F. - [radio] (short for Very Low Frequency) très basse fréquence *f*
V-NOTCH - [carp] entaille *f* en V
V-NOTCH WEIR - [hydr] déversoir *m* triangulaire
V PARTICLE - [phys] particule *f* à trace en forme de V
V-POTENTIAL - [el] (or limb centre) potentiel *m* V, centre *m* des members, centre *m* de Wilson
V PULLEY - [mech] poulie *f* à corde
V REPONSE - [contr] (in remote processing) réponse *f*, réponse *f* V
V-RING - [el] (the metal V-ring of a commutator is the metal ring used in certain constructions of commutator for clamping the bars in position) anneau *m* métallique en V
V-SHAPED CHISEL - [tool] trépan *m* plat à tranchant en pointe de diamant
V-SHAPED TURNING TOOL - [tool] grain-d'orge *m* de tour
V-SHEAVE DRIVE - [el] (or traction drive, a method of transmitting power to the suspension ropes by means of a sheave) transmission *f* à poulie d'adhérence
V THREAD - [mech] filet *m* triangulaire, pas *m* triangulaire
V.T.O. - [aero] (short for Vertical Take Off) décollage *m* vertical
V.T.O.L. - [aero] (short for Vertical Take Off and Landing) (avion) capable d'atterrir et décoller verticalement
VA - [el] (volt-ampere, the unit expressing the product of the root-mean-square value of amperes and the root-mean-square value of volts) voltampère *m*
VACANCY - [gen] poste *m* vacant
[gen] (a gap etc) vide *m*, vacuité *f*
[crystall] (a site in the crystall lattice of an ionic crystal from which the ion which should be present is missing) lacune *f*, trou *m*
VACCINAL FEVER - [med] fièvre *f* vaccinale
VACCINATION - [med] vaccination *f*
VACCINE - [med] (an inoculabel immunizing agent) vaccin *m*
VACCINOTHERAPY - [med] vaccinothérapie *f*
VACUOLAR - [biol] (similar, or pertaining to a vacuole) vacuolaire
VACUOLE - [biol] (small space or cavity in a cytoplasm, generally containing fluid) vacuole *f*
VACUOLIZATION - [biol] (the formation of vacuoles) vacuolisation *f*
VACUOMETER - [instr] vacumètre *m*, manomètre *m*, jauge *f* à vide
VACUUM - [phys] (in theory, a space devoid of matter, in practice the region of space in which the atmospheric pressure has been reduced as much as possible) vide *m*
[mech] (operated by suction) à vide
~ADVANCE - [mech] (advance automatically operated by vacuum) avance *f* à l'allumage à vide
~ANNEALING - [metall] recuit *m* sous vide
~APPARATUS - [mech] (in vacuum technology) installation *f* de vide, appareil *m* pour le vide
~BACK FILM HOLDER - [print] porte-film à succion
~BELL JAR - [mech] cloche *f* à vide
~BOTTLE - [impl] bouteille *f* isolante, bouteille *f* thermos
~BOX - [instr] tambour *m* d'un baromètre
~BRAKE - [mech] (applied to railways) frein *m* à vide
~BREAK VALVE - [mech] robinet *m* d'entrée d'air
~BREAKER - [mech] (a valve used to break a vacuum by admitting air) vis *f* d'aérage, vis *f* d'admission d'air, casse-vide *m*
~BULB - [electron] ampoule *f* vide d'air
~CARBON TRAIN - [instr] appareil *m* de détermination de la teneur en carbone sous vide
~CASTING - [metall] coulée *f* sous vide

VACUUM CEMENT - [gen] ciment *m* à vide, mastic *m* pour le vide
~CHAMBER - [mech] chambre *f* à vide
~CHEMISTRY - [chem] chimie *f* du vide
~CLEANER - [el] (a domestic, or industrial appliance) aspirateur *m*
~COATING - [metall] métallisation *f* sous vide
~COMPONENT - [mech] (a type of fitting) raccord *m* à vide, pièce *f* de raccordement pour la vide
~COMPONENTS - s. vacuum equipment
~CONCRETE - [constr] vide *m* d'air
~CONTROL - [hydr] commande *f* de la dépression
~CONTROL UNIT - [auto] (accessory fitted to the distributor, controlled by the vacuum in the engine to advance or retard the ignition timing) commande *f* à dépression
~CONTROLLED ADVANCE - s. vacuum advance
~CONVEYOR TUBE - [mech] transporteur *m* pneumatique
~CORRECTOR - [instr] manomètre *m* régulateur
~DEGASSED STEEL - [metall] acier *m* dégazé sous vide, acier *m* moulé sous vide
~DEGASSING - [metall] dégazage *m* sous vide
~DEPOSITION - [metall] (coating with a film of metal under high vacuum in the vicinity of heated wires of the metal to be applied) évaporation *f* thermique sous vide poussé
[electron] pulvérisation *f* cathodique
~DISTILLATION - [chem] (method of distillation at reduced pressure, enabling lower temperature to be used) distillation *f* dans le vide, distillation *f* sous vide
~DRIER - [ind chem] (drying oven operating under vacuum) sécheur *m* à vide, étuve *f* à vide
~DRYING CABINET - [ind chem] armoire *f* de séchage sous vide
~DRYING OVEN - [ind chem] (laboratory drying oven which is designed to operate under vacuum) four *m* à sécher dans le vide
~DUST-REMOVING PLANT - [mech] installation *f* d'aspiration de la poussière
~ELECTRON - [electron] (in the Dirac electron theory, an electron in one of the negative energy states which are supposed to be all filled the case of a vacuum) électron *m* de vacuum
~ENCAPSULATION - [chem] (vacuum impregnation) imprégnation *f* sous vide
~EQUIPMENT - [mech] appareillage *m* pour le vide, équipement *m* de vide
~EVAPORATION - [phys] évaporation *f* dans le vide
~EVAPORATOR - [brew ind] évaporateur *m* à vide
~EXTRACTOR - [mech] (a type of vacuum pump) pompe *f* à vide
~EXTRUSION - [plast ind] extrusion *f* sous vide
~FAN - [mech] ventilateur *m* aspirant
~FEED - [mech] alimentation *f* à dépression
~FILAMENT LAMP - s. vacuum lamp
~FILTER - [ind chem] (a type of fibre in which the passage of liquid is accelerated by a vacuum) filtre *m* à vide
~FILTRATION - [ind chem] (filtration in which the pressure in the filtre vessel is maintained below that of the atmosphere with a view to accelerating the action) filtration *f* par le vide
~FITTING - s. vacuum component
~FLANGE - [mech] bride *f* à vide
~FLASK - s. vacuum bottle

VACUUM FOREPUMP - [mech] (in vacuum technology) pompe *f* à vide préliminaire
~FORMING - [plast ind] (a process for producing objects from sheet material, in which a blank cut from the sheet is placed in position over a female mould, heated by radiation and then, after the mould has closed, is drawn into the lower part by the effect of vacuum induced below the blank) formage *m* sous vide
~FREEZE DRYER - [ind chem] installation *f* de lyophilisation
~FUMIGATION - [chem] (used against infestation) fumigation *f* sous vide
~FURNACE - [metall] four *m* à vide
~FURNACING - [metall] traitement *m* à chaud sous vide
~FUSION - s. vacuum melting
~GAUGE - [instr] (instrument to show pressures below that of the atmosphere) vacuomètre *m*
~GAUGE CONTROL-BOX - [instr] coffret *m* d'alimentation d'un manomètre du vide
~GAUGE CONTROL CIRCUIT - [el] anneau *m* de réglage d'un indicateur de vide
~GREASE - [ind chem] graisse *f* à vide, graisse *f* pour le vide
~HEATING - [th eng] chauffage *m* sous vide
~HEATING SYSTEM - [th eng] système *m* de chauffage sous vide
~HOLDER - [photo] (used for film) porte-film *m* sous vide
~HOSE - [impl] (special type of hose for negative pressure) tuyau *m* à vide
~IMPREGNATION - [ch m] imprégnation *f* sous vide
~INGOT CASTING - [metall] coulée *f* de lingots sous vide
~INJECTION MOULDING - [metall] moulage *m* par injection sous vide
~JACKET - [mech] enveloppe *f* à vide, enveloppe *f* sous vide
~-JACKETED FEED TUBE - [ind chem] siphon *m* avec enveloppe sous vide, tube *m* d'ascension isothermique à double paroi évacuée
~JACKETED LIFT PIPE - s. vacuum jacketed feed tube
~JACKETED SIPHON - s. vacuum jacketed feed tube
~JET - [oil ind] éjecteur *m* par le vide
~KNEADER - s. vacuum gauge
~LAMP - [el] (short for vacuum filament-lamp, a filament lamp in which the filament operates in an inert gas) lampe *f* à vide
~LEAK DETECTOR - [instr] détecteur *m* de fuites à vide
~LIGHTNING ARRESTER - [el] parafoudre *m* à vide
~LIGHTNING PROTECTOR - [el] parafoudre *m* à gaz raréfié
~MELTING - [metall] fusion *f* sous vide
~METALLIZING - [metall] (process of metal coating by exposing the work to metal vapour under vacuum) métallisation *f* sous vide
~METALLURGY - [metall] métallurgie *f* sous vide
~METER - s. vacuum gauge
~METER WITH ALPHA EMITTER - [instr] manomètre *m* de vide à émetteur alpha
~MOTOR - [auto] (e.g.as used for windscreen wipers) moteur *m* à vide
~OPERATED - [mech] fonctionant par la dépression
~OPERATED CLUTCH - [auto] embrayage *m* commandé par dépression

VACUUM PACKAGE - s. vacuum packing
~PACKING - [packag] emballage *m* sous vide
~PAN - [ind chem] (vessel in which liquid is evaporated under reduced pressure) cuve *f* à vide
~PHOTOTUBE - [electron] tube *m* photoélectronique à vide
~PHYSICS - [phys] physique *f* du vide
~PIPE - [ind chem] tuyau *m* à vide, canalisation *f* pour le vide
~PIPING - [plumb] tuyauterie *f* d'une conduite à vide
~PLANT - [ind chem] installation *f* de vide, appareil *m* pour le vide
~PLATING - [metall] revêtement *m* par métallisation sous vide, revêtement *m* par évaporation thermique sous vide
~PLUMBING- [plumb] robinetterie *f* et jonctions pour le vide
~POLARIZATION - [electron] (a process by which an electromagnetic field generates virtual electron-positron pairs which modify the charge and current distribution producing the original electromagnetic field) polarisation *f* du vide
~PORT - [mech] (in vacuum technology) raccord *m* de vide, ajutage *m* d'évacuation
~POTTING - [ind chem] imprégnation *f* sous vide
~POWER BRAKE - s. vacuum servo brake
~PRINTING FRAME - [photo] châssis *m* pneumatique
~PROCESS - [ind chem] procédé *m* sous vide, procédé *m* à vide
~PROCESS TECHNOLOGY - [ind chem] technologie *f* des applications industrielles du vide, technique *f* du vide industriel
~PROCESSING - [ind chem] traitement *m* sous vide
~PUMP - [mech] (pumping device operated without pistons by steam) pumpe *f* à vide
~PUMP STATION - s. vacuum pump system [mech] groupe *m* de pompes à vide
~REGULATOR VALVE - [mech] (anti-suction valve) clapet *m* de sécurité à minimum de pression
~RESERVOIR - [mech] (large vessel inserted between the fore pump and high-vacuum pump in a vacuum system) réservoir *m* à vide
~SEAL - s. vacuum sealing
~SEALED - [mech] hermétique, étanche au vide
~SEALING - [mech] scellement *m* sous vide, bouchage *m* sous vide
~SERVO BRAKE - [auto] servo-frein *m* à dépression
~SHELF DRYER - [ind] (apparatus in which the product to be treated is placed on shelves in a vacuum chamber) armoire *f* de séchage sous vide, étuve *f* à vide
~SIDE - [mech] (of a pump) côté *m* vide, côté *m* d'aspiration
~SIDE PRESSURE - [mech] pression *f* côté vide, pression *f* d'entrée
~SIDE TRAP - [mech] (of a rotary pump) séparateur *m* côté vide (de pompe rotative)
~-SINTERING FURNACE - [metall] four *m* de frittage sous vide
~SPACE - [phys] espace *m* sous vide, volume *m* évacué
~SPARK CONTROL - s. vacuum control unit
~SPECTROGRAPH - [phys] (a spectrograph in which the entire light path is in vacuo) spectrographe *m* à vide
~SPRAYING - s. vacuum deposition
VACUUM STEAM - [phys] vapeur *f* sous dépression, vapeur *f* sous vide
~-STEAM HEATING - [th eng] chauffage *m* à la vapeur sous vide
~STEEL CASTING - [metall] coulée *f* de l'acier sous vide
~STEEL DEGASSING - [metall] dégazage *m* de l'acier sous vide
~STILL - [ind chem] (a distillation apparatus in which the pressure in the boiling vessel is maintained at a level below that of the atmosphere) alambic *m* à basse pression
~STOPCOCK - [mech] robinet *m* pour le vide
~SYSTEM - [ind] (used for cleaning etc) système *m* à vide, installation *f* de vide
~TANK - [hgen] récipient *m* à vide, enceinte *f* à vide, cuve *f* à vide
~TAP - s. vacuum stopcock
~TECHNOLOGY - [mech] technique *f* du vide, vacuotechnique *f*
~TESTER - [instr] appareil *m* inducateur de vide
~THERMOCOUPLE - [el] couple *m* thermoélectrique à vide
~TIGHT - s. vacuum sealed
~TIGHTNESS - [mech] étanchéité *f* au vide
~TREATMENT - s. vacuum processing
~TUBE - [electron] (electron tube evacuated to such an extent that its electrical characteristics are substantially unaffected by the ionization of residual gas or vapour) tube *m* à vide, kénotron *m*
~TUBE ELECTROMETER - [instr] (of valve electrometer) électromètre *m* à tubes
~TUBE HOLDER - [electron] (a valve holder) douille *f* de tube
~TUBE MODULATOR - [electron] (or valve modulator, a modulator employing a vacuum tube as modulating element) modulateur *m* à tube
~-TUBE NOISE - [telecomm] bruit *m* des répéteurs
~-TUBE RECTIFIER - [el] (a device which makes use of the unidirectional flow of current in a vacuum tube) redresseur *m* à lampe
~TUBE SWITCH - [electron] (electronic switch) interrupteur *m* électronique
~TUBE TRANSMITTER - [electron] (or valve transmitter, a transmitter in which vacuum tubes are used to convert the applied electric power into radiofrequency power) émetteur *m* à tubes
~TUBULATION - [ind chem] tubulure *f* d'évacuation
~TUNNEL - [aero] (a wind tunnel in which the pressure can be brought to vacuum point) soufflerie *f* à vide
~UNION - [plumb] élément *m* de raccordement pour le vide
~VALVE - s. vacuum tube
~VALVE - [mech] vanne *f* à vide
~VELOCITY - [astronaut] vitesse *f* dans le vide
~VESSEL - s. vacuum bottle
~VOLATILIZATION - [nucl] (a process used in treating irradiated fuel) volatilisation *f* sous vide
~WAX - [ind chem] cire *f* à vide, cire *f* d'étanchéité
~ZONE PURIFICATION - [ind chem] (a technique used in preparing single crystals of silicon) raffinage *m* en zones progressives sous vide
~ZONE REFINING - s. vacuum zone purification
VAGABOND CURRENT - [el] courant *m* vagabond
VAGINA - [anat] (the terminal part of the female

genital duct, leading from the uterus to the external genital opening) vagin *m*
VAGINA - [anat] (any sheath-like structure) gaine *f*
VAGINITIS - [med] (inflammation of the vagina) vaginite *f*
VAGINOMYCOSIS - [med] mycose *f* vaginale
VAGITUS - [med & gen] vagissement *m*
VAGOTONIA - [med] (a condition whereby the activity of the vagus nerve is heightened) vagotonie *f*, parasympaticotonie *f*
VAGOTONY - s. vagotonia
VAGOTROPISM - [med] vagotropisme *m*
VAGRANT - [gen & biol] errant, vagabond
VAGUS - [anat] (a cranial nerve supplying the viscera and the heart) (nerf) vague
VAH METER - [instr] (short for Volt-ampere hour meter) compteur *m* d'énergie apparente, voltampère-heuremètre *m*
VALANCE - [gen] (of an awning, or window) frange *f*, soubassement *m*, jupon *m*
~PANEL - [auto] bavolet *m*
VALE - [geogr] canal *m*
VALENCE - [chem] (the number of atoms of hydrogen an element will combine with or replace) valence *f*
~ANGLES - [phys] (the angles between the successive valence bonds of an atom) angles *mpl* de valence
~BAND - [phys] (the range of energy states in the spectrum of a solid crystal in which lie the energies of the valence electron which bind the crystals together) bande *f* de valence
~BOND - [phys] (the bond formed between electrons of two or more atoms) liaison *f* de valence
~CRYSTAL - [phys] (a crystal bound together by covalent bonds) cristal *m* de valence
~ELECTRON - [electron] (or outer-shell electron, or conduction electron, an electron belonging normally to the outer shell and concerned in light phenomena, also in the chemical properties of the atom) électron *m* de conduction, électron *m* de valence, électron *m* optique
~FORCE FIELD - [phys] (an assumed force field utilized in order to solve the equation in which the potential energy of vibration of a polyatomic molecule is expressed in terms of the energies of the restoring forces of each atom and the energies of their interaction terms) champ *m* de force de valence
~NUMBER - [phys] (number assigned to an atom or ion which is equal to its valence) nombre *m* de valence
~SHELL - [phys] (the group of electrons which constitute the outer electronic shell of an atom) couche *f* de valence
~STAGE - [phys] étage *m* de valence
VALENCY - s. valence
VALENTINITE - [min] (an ore of antimony) valentinite *f*, antimoine *m* oxydé
VALERIAN - [chem] (the roots and rhizome of Valeriana officinalis, an extract of which is used in medicine in the treatment of neurasthenia) valeriane *f*
~OIL - [chem] (an oil obtained from Valeriana officinalis and used in medicine, perfumery, and as a flavourant) essence *f* de valériane
VALERIC ACID - [chem] (intermediate in the production of plasticizers and vinyl stabilizers) acide *m* valérique, acide *m* valérianique
VALEROLACTONE - [chem] (a solvent and dyeing assistant) valérolactone *f*
VALID - [gen & leg] valide, régulier
VALIDATION - [statist] (the characteristic of having a high degree of correlation with its criterion) validation *f*
VALIDITY - [gen & leg] validité *f*
[instr] (the degree of correctness) validité *f*
~CHECK - [contr] contrôle *m* de validité
~CODE - [contr] (a list showing the code numbers and description of parts or subassemblies, with those of the major assemblies or units with which they can be used) code *m* de validité
VALINE - [chem] (an aliphatic amino-acid with application in medicine and nutritional research) valine *f*
VALLECULA - [anat] grande *f* scissure médiane du cervelet
VALLEY - [geogr] vallée *f*
[constr] (the gutter or the outer angle formed by two slopes of a roof) cornière *f*
[constr] (the material fixed in the valley of the roof) noue *f*, noulet *m*
~BEVEL - [constr] sauterelle *f* de noue
~BOARD - [constr] planche *f* de noue
~BREEZE - [met] (a light wind blowing up valleys and hillsides in daytime, when the ground is being warmed by the sun) brise *f* de vallée
~GLACIER - [geol] glacier *m* de vallée
~POINT - [electron] (of a semiconductor) point *m* de vallée
~POINT CURRENT - [electron] (the current value at the valley point) courant *m* de point de vallée
~PRINTING - [print] (application of ink to the upstanding area of an embossing roll, which deposits it in the depression formed in the sheet treated) impression *f* en vallée
RAFTER - [constr] arêtier *m* de noue, chevron *m* à noulet
VALORIZATION - [gen] valorisation *f*
VALUATION - [gen] évaluation *f*, estimation *f*
[comm] expertise *f*, prisée *f* et estimation
VALUE, to - [comm] évaluer, estimer
VALUE - [gen] valeur *f*, prix *m*
[min] qualité *f* industrielle
[chem] (constant value) indice *m*
[phys] pouvoir *m*
[math] valeur *f*
[mus] (the length of a note) valeur *f* (d'une note)
~PARCEL - [transp] colis-valeur *m*
VALVE, to - [mech] garnir de soupape
[hydr & mech] (to regulate the flow by a valve) commander par soupape
[rubber ind] appliquer la soupape
[phys] (to release) évacuer, laisser échapper
VALVE - [mech] (a device for controlling the movement of a fluid) soupape *f*, robinet *m*, clapet *m*, distributeur *m*
[electron] (an electron tube essentially designed to have unilateral conducting properties) tube *m*
[mech] (the device which controls the intake of explosive mixture or the escape of exhaust gas into or out of the cylinder usually of the mushroom type) valve *f*
[hydr] vanne *f*
[bot] valve *f*
~ACTION - [el chem] (said of an electrochemical

valve, the process involved in the operation of an electrochemical valve) action *f* électrochimique de soupape

VALVE ACTUATION - [mech] actionnement *m* de vanne

~ADJUSTMENT - [mech] réglage *m* de la soupape

~AMPLIFIER - [electron] (an electron-tube amplifier) amplificateur *m* à tubes électroniques

~AUGER - [mining] tarière *f* à soupape

~BALL - [mech] boulet *m* de soupape

~BODY - [mech] (the main structure or casing of a valve) corps *m* de vanne

~BONNET - [mech] (the upper part of a gate or globe valve, in which the stem is fitted) couvercle *m* de vanne

~BOX - [mech] chapelle *f* de soupape, chapelle *f* du tiroir

~CAGE - [mech] (the part of the valve assembly containing the operative parts of the valve) corps *m* de vanne

~CAP - [auto] bouchon *m* de valve, chapeau *m* de valve, capuchon *m*

~CAPACITY - [electron] capacité *f* de soupape

~CHAMBER - [meas] (in meters) boîte *f* de distribution
[mech] (in internal combustion engines) chapelle *f*

~CLAMPING WASHER - [mech] (for tyres) plaquette *f* de valve

~COCK - [mech] robinet-valve *m*

~COMPLEMENT - [electron] (the number of electron valves which is required for a piece of electronic equipment) garniture *f* de tubes électroniques

~CONTROL MECHANISM - [contr] (automatic valve control) commande *f* automatique de vanne, commande *f* par vannes

~-CORE - [mech] (of a tyre inner tube), the assembly of valve, valve-spring etc, inserted in the valve stem to permit inflation) obus *m* de valve

~COTTER - [mech] (a small retaining part fitted to a valve-stem in a reciprocating i.c. engine to transmit the effect of a spring) ressort *m* à spirale de soupape, ressort *m* à volute de soupape

~COVER - [mech] cache-soupape *m*

~CURING UNIT - [rubber ind] appareil *m* à vulcaniser pour valves, plateau *m* de chauffe pour valves caoutchouc

~CURRENT - [electron] courant *m* de tube

~CUTTER - [tool] fraise *f* pour la rectification des soupapes

~DETECTOR - [radio] détecteur *m* à lampe, détecteur *m* à tube électronique

~DIAGRAM - [mech] (of a steam engine) diagramme *m* de distribution, épure *f* de régulation

~DISC - [mech] disque *m* pour clapet

~DRIVE - [radio] (resonant-circuit drive) maître-oscillateur *m* à commande électronique

~ECCENTRIC - [mech] excentrique *m* du tiroir, excentrique *m* de commande du tiroir

~ELLIPSE - [mech] ellipse *f* du tiroir

~FACE - s. valve seat

~FACTOR BRIDGE - [electron] (a circuit adapted to the measurement of the dynamic characteristics of electron valves) pont *m* de mesure pour tubes électroniques

~FACTORS - [electron] (the constants describing the characteristics of an electron valve, such as the amplification factor and the transconductance) constantes *fpl* de tubes électroniques

VALVE FLANGE - [mech] (for tyres) épaulement *m* de la valve, plaquette *f* de pied de valve

~FLAP - [mech] clapet *m* d'une valve

~FOOTPLATE - s. valve flange

~GEAR - [mech] (of a steam engine) organes *mpl* de distribution de vapeur, distribution *f* par coulisse, coulisse *f*, détente *f*

~GLAND - [mech] presse-étoupe *f*

~GRATING - [mech] grille *f* de distribution

~GRID - s. valve grating

~GRINDING - [mech] (grinding a valve to improve its sealing qualities) rectification *f* des soupapes

~GRINDING PASTE - [mech] (abrasive paste used in grinding-in valves) pâte *f* abrasive pour rectifier les soupapes

~GUIDE - [mech] (device to ensure that a valve moves along the required path) guide-soupape *m*

~HEAD - [mech] (the enlarged obturating element of a mushroom similar valve) tête *f* de soupape

~HOOD - [mech] (a cover or housing to protect a valve, especially a tank vent valve) cache-soupape *m*
[aero] (of an airship envelope) capotage *m*

~HORN - [mus] cor *m* d'harmonie

~HOUSING - s. valve box

~-IN-HEAD - [auto] (overhead valve) à soupape en tête

~INSERT - [mech] (a valve seat shrunk into the head of i.c. engines) siège *m* de soupape rapporté

~INSIDE - s. valve core

~LAND - [mech] (of a piston valve) guide *m* de vanne

~LEVER - [mech] levier *m* de distribution
[auto] linguet *m* de soupape, culbuteur *m* de soupape

~LIFT - [mech] (the distance by which a valve is clear of its seat when in the open position) levée *f* de soupape, hauteur *mf* d'ouverture de la soupape

~LIFT DIAGRAM - [mech] (of an engine, a compressor etc) diagramme *m* de levée de soupape

~LIFT RESTRICTOR - [mech] (device for limiting the lift of a valve) limiteur *m* de levée de soupape

~LIFTER - [tool] poussoir *m* de soupape
[mech] (in internal combustion engines) lève-soupape *m*

~LIFTER GUIDE - [mech] guide *m* se poussoir de soupape

~LIFTER ROLLER - [auto] galet *m* de poussoir de soupape

~-MEASURING APPARATUS - s. valve tester

~NEEDLE - [mech] aiguille *f* de pointeau

~NOISE - [electron] (noise within a vacuum valve, e.g. microphonics and shot effects) bruit *m* de fond d'un tube électronique

~OSCILLATOR - [radio] oscillateur *m* à tubes à vide

~PLATE - [mech] (in meters) plafond *m*

~PLUG - [auto] obus *m*
[mech] clapet *m* de vanne

~PLUNGER - s. valve tappet

~PORT - [mech] lumière *f* de vanne

~POSITIONER - [contr] (positioner by which the valve takes up a position which corresponds to the output signal of the regulator) positionneur *m* de vanne

~PUSH ROD - s. valve tappet

~RATIO - [el] (the ratio, higher than unit, between the impedance of the valve to a current flowing in one direction and the impedance in the opposite direction) rapport *m* de soupape

VALVE REACTOR - [radio] (a reactance valve, tube reactor in the USA) circuit *m* à tube de réactance
~REACTOR MODULATOR - [radio] modulateur *m* à tube de réactance
~REFACER - [mach tool] rectifieuse *f* de soupape
~RECTIFIER - [electron] redresseur *m* à tube
~-RESEATER - [tool] (special tool for cutting new valve seats in reciprocating i. c. engines and the like) rectifieuse *f* des sièges de soupapes
~RESEATING - [mech] (the operation of forming or fitting a new seat to a valve) rectification *f* des sièges de soupape
~RIGGING - [mech] régulation *f* de soupape
~ROCKER - [mech] (the oscillating element which transmits the reciprocatng motion of a push-rod to an overhead valve in i. c. engine) culbuteur *m*
~ROCKER ARM - [mech] linguet *m* de soupape
~ROCKER ARM COVER - [mech] (a cover enclosing the valve rockers of an i. c. engine) couvercle *m* de culbuterie, couvercle *m* de culasse
~ROCKER COVER GASKET - [mech] (flat oil-seal placed under the joint-face of a valve-rocker cover) joint *m* de couvercle de culbuteríe
~ROCKER SHAFT - [mech] (the shaft which carries the valve rockers in an overhead valve engine) axe *m* de culbuterie
~ROD - [auto] (the links connecting the control handle to the operating elements of the valve) tige *f* de clapet, bielle *f* du tiroir, queue *f* de soupape
~ROTATOR - [mech] dispositif *m* pour la rotation des soupapes
~RUBBER - [rubber ind] (for tyres) caoutchouc *m* pour valve
~SEAT - [mech] (the fixed surface against which the obturating element of a valve closes) siège *m* de soupape, portée *f* de soupape
~SEAT GASKET - [mech] joint *m* du siège de vanne
~SEAT INSERT - [mech] siège *m* de soupape rapporté
~SEATING - [mech] (of a slide-valve) glace *f* du tiroir
~SETTING - [mech] réglage *m* des soupapes
~SHAFT - [mech] tige *f* de vanne, guide *m* de l'obturateur d'un robinet
~SHAPED - [gen] valviforme
~SHIELD - [electron] écran *m* de tube
~SPINDLE - [mech] tige *f* de soupape
[mech] (of a gate valve) arbre *m* de la soupape
~SPOOL - [mech] (the spindle element of the spool valve) boisseau *m* de clapet
~SPRING - [mech] (spring used to return a valve to the closed position after opening) ressort *m* de soupape
~SPRING CAP - [auto] cuvette *f* de ressort de soupape
~SPRING COVER - s. valve spring cap
~SPRING CUP WASHER - [mech] (recessed washer placed over a valve stem, and serving to transmit the thrust of the spring to the cotter) presse-étoupe *f* de cuvette de ressort de soupape
~SPRING KEY - [auto] clavette *f* de ressort de soupape
~SPRING TOOL - [tool] lève-soupape *m*
~STEM - [mech] (of an engine or pump the elongated part of a mushroom valve, below the head) queue *f* de soupape, tige *f* de soupape
[rubber ind] (a short metal pipe serving to house the valve-core, seated in the wall of the inner air tube and passing through the wheel to allow of infilation) armature *f* de valve, corps *m* de valve, contre-tige *f*
VALVE STEM RETREADER - [mech] (for tyres) taraudeur *m* de valve
~TABLE - s. valve plate
~TAPPET - [mech] (a reciprocating element in i.c. engine valve gear, which follows the cam contour and transmits linear movement to valve-stem or puch-rod) poussoir *m* de soupape
~TAPPET CLEARANCE - [auto] jeu *m* de poussoir
~TESTER - [electron] lampemètre *m*
~TIMING - [mech] (the process of setting i. c. engine valves to open and close at such point of crankshaft revolution as will ensure the required performance) réglage *m* de distribution
~TOGGLE - [mech] poignée *f* de robinet, garrot *m* de vanne
~TOMMY - s. valve toggle
~TOP - [mech] chapeau *m* de la vanne, couvercle *m* de la soupape
~TRAVEL - [mech] (the distance through which a valve can move when in action) course *f* de la soupape
~TUBING - [rubber ind] (for bicycles) tube *m* pour valves de pneus
~-TYRE ECHO SUPPRESSOR - [radio] suppresseur *m* d'écho à action continue
~VOLTAGE DROP - [electron] (the anode voltage during conduction) chute *f* de tension dans un tube électronique
~VOLTMETER - [instr] (type of voltmeter incorporating an electronic tube, used in electronic measurements, where it is necessary to avoid disturbance of circuit conditions etc) voltmètre *m* électronique
~WITH CONICAL SEAT - [mech] soupape *f* à siège conique
~WITH GUIDE WINGS - [mech] (of a pump etc) soupape *f* à guide à ailettes
~WITH INSERTED SEAT - [mech] siège *f* de soupape rapporté
~WITH VARIABLE SLOPE - [radio] lampe *f* à pente variable
VALVELESS - [mech] sans soupapes, sans valves
~MOTOR - [auto] moteur *m* sans valves
VALVES - [gas ind] robinetterie *f*
VALVOTOMY - [med] valvulotomie *f*
VALVULITIS - [med] (inflammation *f* of a valve of the heart) inflammation *f* des valves du coeur
VAMP, to - [shoe man] mettre une empeigne, remonter
VAMP - [shoe man] (the front upper part of a shoe or boot) empeigne *f*, claque *f*
[mus] (extemporized accompaniment to a song or an instrumental solo) accompagnement *m* improvisé
VAN - [transp] fourgon *m*, camion *m*
[mining] pelle *f* à vanner
[auto] boulangère *f*
[railw] fourfon *m* à bagages
~ALLEN BELTS - [phys] (Van Allen radiation belts) ceintures *f*pl de Van Allen
~DE GRAAFF ACCELERATOR - [electron] (a particle accelerator, an electrostatic generator employing a system of conveyor belt and spray point to charge an insulated electrode to a high potential) générateur *m* van de Graaff

VANADATE - [chem] (a salt or ester of vanadic acid) vanadate *m*
VANADIATE - s. vanadate
VANADIC ACID - [chem] (a synonym for vanadium pentoxide) acide *m* vanadique
VANADINITE - [min] (a native lead chlorovanadate and a source of vanadium) vanadinite *f*
VANADIUM - [chem] (metallic element, symbol V, with application in metallurgy, the production of catalyst, and in X-ray technology) vanadium *m*
~CARBIDE - [chem] (a component of alloys for machine tools) carbure *m* de vanadium
~CHLORIDE - [chem] (a textile dyeing mordant) chlorure *m* de vanadium
~DICHLORIDE - [chem] (a reducing agent) bichlorure *m* de vanadium
~DRIERS - [chem] (basically salts of vanadium, these are surface-drying catalysts) siccatifs *mpl* au vanadium
~ETHYLATE - [chem] (a polymerization catalyst) éthylate *m* de vanadium
~OXYTRICHLORIDE - [chem] (a polymerization catalyst) oxytrichlorure *m* de vanadium
~PENTOXIDE - [chem] (a catalyst photographic chemical, component of special glasses, and a textile chemical) pentoxyde *m* de vanadium
~STEEL - [metall] (generic term for steels containing varying proportions of vanadium, the inclusion of which improves tensile strength and other physical properties) acier *m* au vanadium
~SULPHIDE - [chem] (a starting material for other vanadium compounds) sulfure *m* de vanadium
~TETRACHLORIDE - [chem] (an intermediate for vanadium compounds) tétrachlorure *m* de vanadium
~TETRAOXIDE - [chem] (a catalyst) tétraoxyde *m* de vanadium
~TRICHLORIDE - [chem] (a catalyst in organic synthesis) trichlorure *m* de vanadium
~TRIOXIDE - [chem] (catalyst used in the production of ethyl alcohol) trioxyde *m* de vanadium
VANADYL SULPHATE - [chem] (a component of ceramic glazes, catalyst, mordant in dyeing, and a reducing agent) sulfate *m* de vanadyle
VANCOMYCIN HYDROCHLORIDE - [chem] (the hydrochloride of an antibiotic produced by Streptomyces orientalis and effective against streptococci, staphylococci, and pneumococci) chlorhydrate *m* de vancomycine
VANDYKE BROWN - [paint] (a brown earth consisting chiefly of organic material, chiefly used in artist colours) brun *m* foncé, brun *m* Van Dyke
VANE - [mech] (a flat or curved part, designed to produce motion in a fluid, to control such motion, or to be moved by a fluid stream) aube *f*, ailette *f*
[mech] (of a turbine) palette *f*, aubage *m*
[agric] (of a windmill) bras *m* (de moulin à vent)
[constr] (a weathercock) girouette *f*
[met] (of an anemometer) moulinet *m*, turbine *f*
[surv] (a disc attachment to a levelling staff) voyant *m*
[surv] (a sight) pinnule *f*
[zool] (of a feather) lame *f* (d'une plume)
~ANEMOMETER - [instr] (anemometer consisting of a single plate member against which the wind blows) anémomètre *m* à moulinet
~ANODE - [electron] (for magnetrons) anode *f* à ailettes
VANE ANODE MAGNETRON - [electron] (a magnetron incorporating an anode block of the vane design) magnétron *m* à ailettes
~ATTENUATOR - [electron] (in a rectangular waveguide) affaiblisseur *m* à cloison longitudinale
~MAGNETRON - s. vane anode magnetron
~PUMP - [mech] pompe *f* rotative à ailettes
~SUPERCHARGER - [aero] (a type of supercharger consisting of an excentric rotor fitted with radially-sliding blades, running in a cylindrical casing against which the outer edges of the blades or vanes bear) compresseur *m* à palettes
~-TYPE DRAUGHT GAUGE - [impl] (draught-gauge actuated by a moving vane exposed to the gas current) déprimomètre *m* à volet
~TYPE PUMP - [mech] pompe *f* à ailettes
~WATTMETER - [instr] (instrument designed to measure the power flow in a waveguide) wattmètre *m* à palettes
~WHEEL - [mech] roue *f* à palettes, roue *f* à ailettes
VANILIN - [chem] (vanilic aldehyde. A constituent of the vanilla bean, also prepared synthetically, and having application in perfumery and as a flavourant) vanilline *f*
VANILLA - [bot] vanille *f*
VANISHING POINT - [draw] (for perspectives) point *m* de fuite
VANNER - [min] (an ore dressing machine) vanneur *m*, vanoir *m*, crible *m*
VANNING - [min] (the assessment of the contents of an ore by washing on a flat shovel) vannage *m*
~SHOVEL - [impl] pelle *f* à vanner
VANOXITE - [min] (a mineral containing uranium and vanadium) vanoxite *f*
VAPORIMETER - [instr] (an apparatus for determining the volatility of an oil by heating it in a stream of air) vaporimètre *m*
VAPORIZATION - [phys] (the changing of a solid or liquid into the vapour state) vaporisation *f*
[chem] (of a liquid) pulvérisation *f*
[mech] (in internal combustion engines) carburation *f* du combustible
~HEAT - [phys] (latent heat of vaporization. The amount of heat, in calories, absorbed by one gram of liquid in passing from the liquid to the vapour phase without a change of temperature) chaleur *f* de vaporisation
~LATENT TEMPERATURE - [phys] chaleur *f* latente de vaporisation
~OF GETTER - [metall] vaporisation *f* du métal getter
~SPECIFIC TEMPERATURE - [phys] chaleur *f* spécifique de vaporisation
VAPORIZE, to - [phys] (to convert a liquid or a solid into vapour) vaporiser
VAPORIZED CARBON RESISTANCE - [radio] (or cracked-carbon resistance, component of radioreceivers) résistance *f* au carbone vaporisé
VAPORIZER - [impl] (an atomizer) vaporisateur *m*, atomiseur *m*
[mech] (in internal combustion engines) réchauffeur *m*
VAPORIZING - [phys] vaporisation *f*
~CHAMBER - [mech] (in internal combustion engines) chambre *f* de mélange
~ENGINE - [mech] (a kerosene engine) moteur *m* à kérosène
~OF GETTER - [electron] vaporisation *f* du getter

VAPORIZING OIL - [in chem] huile *f* pour moteurs
VAPOUR, to - [gen] s'évaporer, se vaporiser
[chem] déposer par vaporisation
VAPOUR - [phys] (a gas of which the temperature is below the critical temperature, hence it can be liquefied by compression only, without temperature reduction being essential) vapeur *f*
~BLASTING - [metall] (of metal surfaces) honage *m* au jet de vapeur
~BONNET - [ind chem] (chamber in a still, located above the pot, to trap spray rising with the vapour) capot *m* de vapeur
~BUBBLES - [phys] bulles *fpl* de vapeur
~CONDENSER - [ind chem] condensateur *m* de vapeur
~CONDUIT - [mech] (of a diffusion pump) conduite *f* d'amenée de vapeur (d'une pompe à fluide moteur)
~DEGREASING - [metall] (method of removing grease from metal parts by placing them while cold in a tank filled with the vapour of a solvent (e,g.trichloroethylene) which condenses on the part, dissolves the grease and drips off into the liquid solvent at the bottom of the tank) dégraissage *m* à vapeur de solvant
~DEGREASING CHAMBER - [metall] (closed vessel or tank containing liquid solvent which is heated in contact with its vapour phase, for cleaning metal parts) chambre *f* de dégraissage à la vapeur
~DENSITY - [phys] (a measure of the density of a vapour or gas in relation to the density of an equal volume of hydrogen) densité *f* de vapeur
~DENSITY BULB - [phys] ballon *m* de Dumas
~DEPOSITION - [metall] déposition *f* de la phase de vapeur
~DOME - [oil ind] dôme *m* à vapeur
~HEATING SYSTEM - [th eng] système *m* de chauffage à vapeur
~JET - s. vapour nozzle
~JET NOZZLE - [mech] diffuseur *m* de vapeur, tuyère *f* de vapeur
~JET OF AN EJECTOR PUMP - [mech] tuyère *f* à éjection d'une pompe à éjecteur
~JET PUMP - [impl] (a pump designed to obtain ultra-high vacua) pompe *f* à diffusion, pompe *f* à entraînement moléculaire
~LOCK - [mech] (interruption of normal flow in pipe system, caused by the presence of vapour, from the liquid conveyed by it) poche *f* de vapeur
~NOZZLE - [mech] injecteur *m* de vapeur, diffuseur *m* de vapeur
~PHASE - [phys] phase *f* de vapeur
~PRESSURE - [phys] (the pressure which a saturated vapour of a substance exerts in equilibrium with its solid or liquid form at any given temperature) pression *f* de vapeur
~PRESSURE EQUILIBRIUM - [phys] tension *f* de vapeur d'équilibre
~PRESSURE OF WATER - [phys] (steam tension) tension *f* de la vapeur d'eau
~PRESSURE THERMOMETER - [instr] (a temperature measuring instrument depending on the pressure exerted by a vapour in contact with its liquid phase) thermomètre *m* à pression de vapeur
~RECOVERY UNIT - [oil ind] installation *f* de récupération des vapeurs
~STREAM - [phys] flux *m* de vapeur
~STREAM DRYING - [ind chem] dessication *f* par courant de vapeur
VAPOUR STREAM PUMP - [mech] (a fluid entrainment pump) pompe *f* à jet de vapeur, pompe *f* à flux de vapeur
~TENSION - [phys] (the tendency of a liquid to enter the vapour state, balanced by the vapour pressure, and numerically equal to the latter) tension *f* de vapeur, force *f* élastique de la vapeur
~TESTING APPARATUS - [oil ind] installation *f* d'essai des gaz
~-TIGHT - [el] (of electrical machines or apparatus) étanche aux gaz, étanche à la vapeur
~TRAP - [mech] piège *m* à vapeur
~TRANSPORT - [mech] transfert *m* de vapeur
~VALVE - [constr] (in a hood) trappe *f* du conduit d'évacuation
VAPOURER MOTH - [zool] (a harmful insect) orgye *f* antique
VAPOURPROOF LAMP - [impl] (an antideflagrating device) lampe *f* étanche aux gaz
VAR - [radio] (short for visual-aural range) radiophare *m* d'alignement audio-visuel, VAR
[el] (short for reactive volt-ampere) unité *f* de puissance réactive
~-HOUR METER - or reactive-energy meter, integrating instrument designed to measure the reactive energy in var-hours) compteur *m* d'énergie réactive
VARACTOR - [electron] (a semiconductor device characterized by a variation of capacitance with voltage) varactor *m*
VARIABILITY - [phys] (the property of departing from an established value or standard) variabilité *f*, déviation *f*
VARIABLE - [gen] variable, changeant
[math] variable *f*
~ADDRESS - [comput] (address which may change during the execution of the program) adresse *f* modifiée
~APERTURE SHUTTER - [photo] obturateur *m* à largeur variable
~AREA EXHAUST NOZZLE - [aero] tuyère *f* d'échappement à section variable
~AREA PROPELLING NOZZLE - [mech] (a type of propelling nozzle in which the flow area can be varied) buse *f* propulsive à section variable
~AREA TRACK - [el acoust] (sound track divided longitudinally into transparent and opaque areas, by one or more sharp demarcation lines forming an oscillographic trace of the wave shape of the recorded signal) piste *f* à densité fixe
~BLOCK FORMAT - [comput] disposition *f* à bloc variable
~-BLOCK TARIFF -[el] (a tariff which is in form similar to the block tariff, but in which the size of the blocks are variable, according to maximum demand, capacity of apparatus, or the size of the premises) tarif *m* dégressif variable
~-CAMBER PROPELLER - [aero] (a type of propeller having two sets of blades in tandem, staggered at about 10 deg.and capable of individual pitch adjustment in flight) hélice *f* à courbure variable
~CAPACITOR - [el] (a capacitor whose capacitance can be varied, e.g.by moving one of the plates in relation to the other) condensateur *m* variable
~CAPACITY DIODE - [electron] (of a semiconductor) diode *f* à capacité variable
~CARRIER MODULATION - [radio] (controlled-carrier modulation) modulation *f* d'amplitude à taux de mo-

dulation constante
VARIABLE CONDENSER - s. variable capacitor
~COUPLING - [el] (inductive coupling of two or more coils which can be varied by moving one or more in relation to the others) couplage *m* variable
~CYCLE OPERATION - [comput] opération *f* à cycle variable
~-DATUM BOOST CONTROL - [mech] (a system of boost-control in which the manifold pressure is controllable, and is varied progressively with the movement of the hand throttle) limiteur *m* de pression d'admission progressif
~DELIVERY PUMP - [mech] pompe *f* à débit variable
~DENSITY TRACK - [el acoust] (sound track whose density varies in accordance with the recorded signal while remaining constant over the whole width of the track) piste *f* à densité variable
~DENSITY WIND TUNNEL - [aero] soufflerie *f* à densité variable
~ENERGY CYCLOTRON - [nucl] cyclotron *m* à puissance variable
~-FOCUS LENS - [opt] objectif *m* à focale variable
~-FOCUS PYROMETER - [instr] pyromètre *m* à focale variable
~INCIDENCE GUIDE VANE - [aero] (a guide-vane of which the angle of incidence can be adjusted) aube *f* distributrice à incidence variable
~- INDUCTANCE PICKUP - [el acoust] phonocapteur *m* à inductance variable, pick-up *m* à inductance variable
~INDUCTOR - s. variometer
~INJECTOR - [th eng] (adjustable injector) injecteur *m* réglable
~INTERMITTENT DUTY - [el] (of an electrical machine) service *m* intermittent variable
~LENGTH FIELD - [comput] alimentation *f* de cartes de longueur variable
~LOSSES - [el] (of an electrical machine or apparatus the losses which vary with the load of the machine or apparatus at a given speed and voltage) pertes *fpl* variables
~MOTION - [mech] mouvement *m* varié
~-MU CONDUCTANCE TUBE - s. variable mutual conductance tube
~-MU CONDUCTANCE VALVE - s. variable mutual conductance tube
~-MUTUAL CONDUCTANCE TUBE - [electron] (a vacuum tube, or valve, in which the amplification factor varies in a predetermined way with control-grid voltage) tube *m* à pente variable
~-MUTUAL CONDUCTANCE VALVE - s. variable mutual conductance tube
~PITCH - [electron] (of a grid when the turns of grid wires are not spaced at an equal distance) pas *m* variable
~-PITCH PROPELLER - [aero] (a propeller so designed that the pitch can be changed while it is rotating) hélice *f* à pas variable
~POINT - [math] point *m* variable
~POINT REPRESENTATION - [comput] notation *f* à point variable
~QUADRICORRELATOR - [telev] (in colour television) quadricorrélateur *m* à amplification variable
~-RATIO TRANSFORMER - [el] transformateur *m* à rapport variable
~RECORD LENGTH - [comput] longueur *f* variable d'un enregistrement
VARIABLE RELUCTANCE MICROPHONE - [el acoust] (microphone depending for its operation on variations of the reluctance of a magnetic circuit) microphone *m* électro-magnétique
~-RELUCTANCE PICKUP - [el acoust] (magnetic pick-up) pick-up *m* électro-magnétique
~RESISTANCE PICKUP - [el acoust] (pickup depending for its operation on the variation of resistance) pick-up *m* à résistance variable
~RESISTOR - [el] (wire-wound or composition resistor, the value of which may be changed) résistance *f* réglable, résistance *f* variable
~SPARK-GAP - [el] éclateur *m* réglable
~-SPEED - [mech] à vitesse variable, à vitesse réglable
~-SPEED AXLE DRIVEN GENERATOR - [el] (used for lighting railway coaches) génératrice *f* d'essieu à vitesse variable
~-SPEED AXLE GENERATOR - s. variable-speed axle driven generator
~-SPEED DRIVE - [mech] transmission *f* à vitesse réglable
~-SPEED MOTOR - [el] (a motor the speed of which can be varied gradually over a given range, but which, once adjusted, remains practically unaffected by the load) moteur *m* à vitesse réglable
~-SPEED SCANNING - [telev] analyse *f* à vitesse variable
~TEMPORARY DUTY - [el] (of an electrical machine) service *m* temporaire variable
~TRANSFORMER - [el] transformateur *m* variable
~VELOCITY - [mech] vitesse *f* variée
~VIDEO ATTENUATOR - [telev] potentiomètre *m* gain signal vidéo
~VOLTAGE CONTROL - [el] (in traction, a method of controlling the speed of motors, in which the applied voltage is varied by variable-voltage generator, Also a method of controlling a lift-motor by the use of a motor-generator in which a D.C. voltage applied to the lift-motor armature is varied by altering the strength and direction of the generator field) régulation *f* par variation de tension, régulation *f* par variation dans le champ de la génératrice
~VOLTAGE GENERATOR - [el] génératrice *f* de tension variable
~-VOLTAGE REGULATOR - [el] variateur *m* de tension
~-VOLTAGE WELDING SET - [el] (a welding set, open-circuit voltage of which can be varied) poste *m* de soudure à tension variable
~-VOLTAGE WELDING SOURCE - [el] génératrice *f* de soudage à tension variable
~WORD LENGTH COMPUTER - [comput] ordinateur *m* à longueur de mot variable
VARIABLES OF PERCEIVED COLOUR - [opt] variables *fpl* d'une couleur perçue
VARIANCE - [phys] (the number of degrees of freedom possessed by a system, also the degrees of freedom themselves) degrés *mpl* de liberté, nombre *m* de degrés de liberté
VARIATION - [geogr] (the azimuth angle between the true and the magnetic meridians) variation *f*, déclinaison *f* magnétique
[gen] variation *f*, changement *m*
~COMPASS - [instr] boussole *f* de variation
~OF THE COMPASS - [instr] déclinaison *f* de l'aiguille aimantée

VARIATOR - [mech] (e.g. of speed) variateur *m*
VARICOCELE - [med] varicocèle *m*, circocèle *m*
VARICOLE - s. varicocele
VARICOMPHALUS - [med] varices *f*pl ombilicales
VARICULA - [med] varice *f* de la conjonctive
VARIEGATED - [gen] (exhibiting different colours) varié, bariolé, versicolore
~COPPER ORE - [min] érubescite *f*, bornite *f*, cuivre *m* panaché
~SANDSTONE - [min] grès *m* bigarré
VARIETY - [gen] variété *f*, diversité *f*
[gen] (assortment) assortiment *m*
VARINDOR - [radio] (a type of inductor) varindor *m*
VARIOCOUPLER - [radio] (a type of transformer) variocoupleur *m*
VARIODE REGULATOR - [electron] (a semiconductor apparatus designed to improve and simplify current regulation of dynamos) régulateur *m* variode
VARIOMETER - [el] (electrical measuring instrument) variomètre *m*
[aero] (rate of climb indicator) variomètre *m*
VARISTOR - [radio] résistance *f* à variation automatique
VARMETER - [el] varmètre *m*
VARNISH, to - [paint] (to cover with varnish) vernir, vernisser
VARNISH - [paint] (transparent protective coating based on natural or synthetic resins) vernis *m*
~THINNER - [paint] diluants *m*pl pour vernis
VARNISHED - [paint] (covered with varnish) verni, vernissé
~FABRIC - [text] (fabric which has been coated or impregnated with a natural or synthetic resin) tissu *m* verni, toile *f* vernie, tissu *m* imprégné
~MATERIAL - [plast ind] feuille *f* imprégnée, bande *f* imprégnée
~PAPER - [paper man] (paper which has been impregnated or coated with natural or synthetic resins) papier *m* verni, papier *m* imprégné
~SHEET - [plast ind] (laminated material coated with varnish) feuille *f* vernie
~TUBE - [metall] (paper or fabric tube coated or impregnated with a natural or synthetic resin) tube *m* verni
~WEB - [paper man] (a continuous band of paper impregnated with varnish) bande *f* imprégnée
VARNISHER - [gen] vernisseur *m*
VARNISHING MACHINE - [mech] machine *f* à laquer, machine *f* à vernie
~TABLE - [impl] table *f* de vernissage
VARY, to - [gen] varier, changer
VARYING DUTY - [el] opération *f* à charge variable, service *m* interrompu à charge variable
~MODERATOR HEIGHT - [nucl] hauteur *f* de modérateur variable
VAS - [anat] vaisseau *m*
VASCULAR - [zool & bot] vasculaire, vasculeux
~AREA - [zool] aire *f* vasculaire
~BUNDLE - [bot] faisceau *m* fibro-vasculaire
~DISEASE - [med] (disease affecting vessels providing for the circulation of fluids) maladie *f* vasculaire
~TISSUE - [anat] tissu *m* vasculaire
VASCULARITIS - [med] vascularite *f*
VASE - [impl] vase *m*
[bot] vase *m*
VASELINE - [chem] (universally adopted tradename for various semi-solid hydrocarbons derived from petroleum) vaseline *f*
VASOCONSTRICTION - [med] vasoconstriction *f*
VASODILATATION - [med] vasodilatation *f*
VASOLIGATION - [med] s. vasorrhaphy
VASOLIGATURE - [med] s. vasorrhaphy
VASOMOTOR - [anat] vas-moteur *m*
VASOPRESSIN - [chem] (a hormone secreted by the posterior lobe of the pituitary gland and used in medicine in the treatment of diabetes insipidus) vasopressine *f*, pitressine *f*
VASOPUNCTURE - [med] ponction *f* du canal déférent, vasoponction *f*
VASORESECTION - [med] résection *f* du canal déférent
VASORRHAPHY - [med] ligature *f* du déférent, vasoligature *f*
VASOSPASM - [med] spasme *m* vasculaire, angiospasme *m*
VASOTRIPSY - [med] vasotripsie *f*, angiotripsie *f*
VAT, to - [agric] mettre en cuve
[brew ind] entonner, soutirer en fûts
[leather ind] mettre en fosse
VAT - [gen] cuve *f*, bac *m*
[leather ind] fosse *f* (de tannage)
[paper man] (a tank containing beater-pulp) cuve *f*, confit *m*
[metall] bain *m* (de trempage)
~DYEING - [text] teinture *f* à la cuve
~DYES - [chem] colorants *m*pl à la cuve
~-HOUSE - [paper man] salle *f* des cuves
~PADDING - [text] foulardage *m* en cuve
~PAPER - [paper man] (type of hand-made paper) papier *m* à la cuve
VATTED - [brew ind] (matured by storage in casks) mûr
VATTING PROPERTY - [text] réductibilité *f*
VAULT, to - [gen] sauter
VAULT - [arch] voûte *f*
[arch] (a burial chamber) caveau *m*
[constr] (underground room for storing valuables) chambre *f* forte, souterrain *m*
[plumb] (USA only, for service pipes) carter *m* de robinet, bouche *f* à clé
[gen] (for the wine etc) cave *f*, cellier *m*
[th eng] chapelle *f* de four
~-TYPE TRANSFORMER - [el] transformateur *m* de voûte
VAULTED - [constr] voûté
VAULTING - [constr] construction *f* de voûtes
[arch] voûte *f*
[constr] bombement *m*
VECTOR - [phys] (a quantity which has direction as well numerical value) vecteur *m*, grandeur *f* vectorielle
[astr] (radius vector) rayon *m* vecteur
[aero] (course direction) direction *f* de route
~ADDITION - s. vector sum
~ADMITTANCE - [el] admittance *f* vectorielle
~ALGEBRA - [math] algèbre *f* vectorielle
~ANALYSIS - [math] analyse *f* vectorielle
~CARDIOGRAPHY - [med] vectocardiographie *f*
~CURRENT - [el] (complex sinusoidal current) courant *m* sinusoïdal complexe
~ELECTROCARDIOGRAM - [med] (a loop pattern taken from leads placed orthogonally) cardiogramme *m* vectoriel, électrocardiogramme *m* vectoriel

VECTOR FIELD - [el] (a field of which the state at each point is represented by a vector) champ *m* vectoriel
~GROUP OF A TRANSFORMER - [el] (a designation of the whole of the connections of the windings, which determines the phase displacement between low and high voltage sides at the related terminals) couplage *m* d'un transformateur
~-GROUP SYMBOL OF A TRANSFORMER -[el] symbole *m* de couplage d'un transformateur
~INSECT - [vet] insecte *m* vecteur
~MODEL OF ATOM - [nucl] model *m* vectoriel d'un atome
~POTENTIAL - [el] potentiel *m* vectoriel
~POTENTIAL FIELD - [el] champ *m* du potentiel vectoriel
~POWER - [el] puissance *f* vectorielle
~PRODUCT - [el] (cross, or outer product) produit *m* vectoriel
~QUANTITY - [phys] (a physical quantity which may be represented by a vector) grandeur *f* vectorielle
~STEERING - [astronaut] guidage *m* par tuyères mobiles
~SUM - [math] somme *f* de vecteurs
VECTORCARDIOGRAM - s. vector electrocardiogram
VECTORIAL - [math & phys] vectoriel
~ANGLES - [math] angles *mpl* vectoriels
~COORDINATES - [math] coordonnées *fpl* vectorielles
VECTORSCOPE - [telev] (in colour television) vectorscope *m*
VEE - [gen] (V-shaped) syn.de V
~BELT - [mech] (usually of rubber and textile structure, having tapered sides so as to run in pulleys having grooves of V-section) courroie *f* trapézoïdale
~-BLOCK - [tool] (a block of cast-iron or steel machined to a true face on one surface and having an accurately-shaped V-shaped groove on the other (used in marking-off and inspection etc) cale *f* en V pour traçage, bloc *m* en V
~-FLIGHT-FORMATION - [aero] formation *f* de vol en V
~-GUTTER - [hydr] gouttière *f* en V
~-SHAPED RADIATOR FRONT - [auto] radiateur *m* en coupe-vent
~WELD - [metall] soudure *f* en V
VEER, to - [met] (of a wind, to change direction clockwise, formerly used to mean change of direction in the same sense as the apparent motion of the sun, but in scientific usage the meaning is now clockwise in both hemispheres) changer de direction vers la droite, tourner
[gen] virer, changer de direction
[naut] (to change the course) virer, changer de bord
[naut] (to let out) rôder
VEER - [gen] changement *m* de direction
[met] (of the wind) saute *f*
VEERING - [met] (change of direction of a wind in a clockwise sense) rotation *f* du vent vers la droite
VEGETABLE - [gen] (a plant of any kind) végétal
[bot] (the edible part of plants) légume *m*
~BIN - [th eng] (in a refrigerator) coffre *m* à légumes
~BLACK - [chem] (carbon black produced from vegetable matter) noir *m* végétal
VEGETABLE BUTTER - [agric] beurre *m* végétal
~CHARCOAL - [chem] (charcoal made from vegetable material) charbon *m* végétal
~CONTAINER - s. vegetable bin
~DOWNS - [text] duvets *mpl* de plantes
~DYE - [chem] (used for textiles) teinture *f* végétale
~FARMING - [agric] culture *f* légumière de plein champ
~FAT SOAP - [chem] savon *m* à l'huile végétale
~FATS - [food] graisses *fpl* végétales
~FIBRE - [text] fibre *f* végétale
~GLUE - [ind chem] colle *f* végétale
~GROWING - [agric] culture *f* maraîchère
~GUM - [chem] matière *f* gommeuse
~HORSEHAIR - [text] (used in upholstery) crin *m* végétal
~IMPURITIES - [text] débris *mpl* végétaux
~JUICE - [agric] jus *m* de légumes
~LAYER - [chem] substance *f* végétale en décomposition
~MANURE - [agric] engrais *m* végétal
~MOULD - [geol] roche-mère *f*
~MULCH - [agric] fumier *m*, engrais *m* végétal
~PARCHMENT - [paper man] papier *m* parchemin, papier *m* pergamine
~PITCH - [chem] poix *f* végétale
~POISON - [chem] poison *m* végétal
~RAW MATERIAL - [text] matière *f* première végétale
~ROUGE - [dyes)] rouge *m* de carthame
~SEEDLING - [agric] plant *m* de légume
~SILK - [text] soie *f* végétale
~SLICER - [agric] taille-racine *m*
~TALLOW - [bot] suif *m* végétal
~TANNAGE - [leather ind] tannage *m* végétal
~TAR - [ind chem] goudron *m* végétal
~WAX - [chem] cire *f* végétale
~WOOL - [text] laine *f* végétale
VEGETAL - [bot] (pertainig to plants, or vegetables) végétal
VEHICLE - [gen] véhicule *m*, voiture *f*
[chem] (the liquid by means of which the solids of a coating composition are conveyed to the surface to be covered) véhicule *m*
[paint] (the medium in which the pigment and filler etc of a coating composition is ground) milieu *m* de suspension, véhicule *m*
[chem] (in pharmaceutical products) excipient *m*
~-BORNE SCINTILLATOR - [instr] radiamètre *m* de prospection porté à scintillateur
~CONTROL SYSTEM - [astronaut] dispositif *m* de commande d'un véhicule
~GAUGE - [transp] gabarit *m* pour véhicules
~MASS RATIO - [phys] rapport *m* des masses d'un véhicule
~WEIGHT - [auto] poids *m* en ordre de marche
VEIL - [text] voile *m*
[photo] voile *m*
VEILING - [photo] voile *m* faible
VEIN - [anat] (the muscular tubular vessel conveying blood to the heart) veine *f*
[bot & zool] (the radiating support of the framework of a leaf or an insect wing) nervure *f*
[min] veine *f*, filon *m*, gîte *m* filonien
[min] (coloured streak in wood, marble etc) veine *f*
~GOLD - [min] or *m* filonien

VEIN MATERIAL - [min] matière *f* filonienne
~MATTER - s. vein material
~MINING - [mining] exploitation *f* des filons
~QUARTZ - [min] quartz *m* filonien
~ROCK - [geol] roche *f* filonienne
VEINED - [gen] (of marble) veiné, madré
VEINFILLING - [geol] remplissage *m* filonien
VEINING - [metall] gerces *fpl*, figures *fpl* de corrosion
[text] madrure *f*
VEINLET - [min] veinule *f*, filet *m*
VEINSTONE - [min] (the valueless part of a metalliferous vein) roche *f* de filon
VEINSTUFF - s. vein material
VELD - [geogr] (open ground in South Africa) veld *m*
VELDT - s. veld
VELLICATION - [med] tressaillement *m* musculaire
VELLUM - [gen] (fine parchment) vélin *m*
VELOCIMETER - [instr] vélocimètre *m*, mesureur *m* de vitesse radiale
VELOCITY - [gen] (the time rate of change of position) vitesse *f*
~ANTIRESONANCE - [mech] antirésonance *f* de vitesse
~CONTROL SERVO - [contr] (for the remote control of the change of speed) régulateur *m* de vitesse asservi
~DISTRIBUTION - [electron] (the selecting of electrons according to their velocity) distribution *f* de vitesse
~FACTOR - [radio] rapport *m* de démultiplication
~FEEDBACK - [contr] réaction *f* tachymétrique
~FOCUSING MASS SPECTROGRAPH - [phys] spectrographe *m* de masse à focalisation de vitesse
~HEAD - [hydr] charge *f* de la vitesse
[phys] énergie *f* cinétique
~-HEAD TACHOMETER - [instr] mètre *m* de vitesse d'eau
~LEVEL - [acoust] niveau *m* de la vitesse
~MICROPHONE - [el acoust] (pressure-gradient microphone) microphone *m* à gradient de pression
~MISALIGNMENT - [contr] alignement *m* faux de vitesse
~MISALIGNMENT COEFFICIENT - [contr] coefficient *m* d'écart angulaire par vitesse
~MODULATED AMPLIFIER - [radio] amplificateur *m* à modulation de vitesse
~MODULATED ELECTRON BEAM - [electron] faisceau *m* électronique modulé en vitesse
~MODULATED OSCILLATOR - [electron] oscillateur *m* à modulation de vitesse
~MODULATED TUBE - [electron] (an electron-beam tube in which the velocity of the electron stream is alternately increased and decreased with a period comparable with the total transit time) tube *m* à modulation de vitesse
~MODULATION - [electron] (of an electron beam, the process whereby a desired time variation in velocity is impressed on the electrons of a beam) modulation *f* de vitesse
~MODULATION TELEVISION SYSTEM - [telev] système *m* de télévision à modulation de vitesse du faisceau
~OF A WAVE - [el] vitesse *f* de propagation d'une onde
~OF AN ION - [phys] (the speed attained by an ion under the action of an electric field) vitesse *f* d'un ion
VELOCITY OF DEPOSITION - [hydr] vitesse *f* de décantation
~OF DISCHARGE - s. velocity of flow
~OF ENERGY TRANSMISSION - [el] (the energy flux per unit area divided by the energy density) vitesse *f* de transport d'énergie
~OF ESCAPE - [astronaut] vitesse *f* de libération
~OF EXHAUST - [mech] vitesse *f* d'échappement
~OF FLOW - [hydr] vitesse *f* d'écoulement, vitesse *f* de passage, vitesse *f* du courant
~OF PROPAGATION - [phys] vitesse *f* de propagation
~OF SOUND - [phys] vitesse *f* du son
~OF WAVE - s. velocity of a wave
~PRESSURE - [phys] (the component of the pressure of the moving fluid which is due to velocity) pression *f* cinétique
~PROFILE - [phys] (the graphical representation of the variation with displacement normal to the general direction of flow of the mean flow velocity in a shear flow) profil *m* de la vitesse
~RATE - s. velocity factor
~RATIO - [mech] rapport *m* de transmission
~RESONANCE - [phys] résonance *f* de phase
~SORTING - s. velocity distribution
~STAGE - [mech] (e.g. of a steam turbine) étage *m* de vitesse
VELODYNE - [contr] intégrateur *m* vélodyne
VELOUR - [ptext] (soft, closely woven cotton or wool fabric a velvet pile) velouté *m*, feutre *m* taupé
[text] (for hats) bichon *m*
~PAPER - [paper man] papier *m* velouté
~SABRE - [text] velours *m* sabre
~SPLITS - [mleather ind] croûtes *fpl* velours
~TRIMMING - [text] galon *m* velours
VELOURS - s. velour
VELVET - [text] velours *m*
[adj] velouté
~BENT - [bot] agrostide *f* canine
~CALFSKIN - [leather ind] veau *m* chamoisé
~CUTTING MACHINE - [text] coupeuse *f* à velours
~GRASS - [agric] houlque *f* laineuse
~KNIFE - [text] couteau *m* à velours
~PILE - [text] moquette *f*
~PILE CARPET - [text] tapis *m* velouté
~TRAP - [photo] (of a film magazine) revêtement *m* de velours formant chicane
VELVETEEN - [text] (a cotton fabric having a short close velvet-like pile) velours *m* lisse de coton, velours *m* de coton
~CORD - [text] velours *m* de Gênes
VELVETING - [paint] velouté *m*
VENA CONTRACTA - [phys] (the phenomenon of the contraction of the free jet of liquid issuing from a container through an orifice) contraction *f* d'une veine liquide
VENDER - [comm] vendeur *m*
VENDOR - s. vender
VENEER - [carp] (a thin sheet of wood used in building up plywood) feuille *f* de placage
~CHIPS FILLED MOULDING COMPOUND - [plast ind] matière *f* à mouler avec copeaux de bois
~CUTTING MACHINE - [mach] machine *f* à trancher le bois en feuilles de placage
~LAMINATES - [carp] feuilles *fpl* de bois
~PLANING MACHINE - [mach tool] raboteuse *f* à placage

VENEER SAW - [tool] (power saw designed to cut thin veneers) scie *f* à placage
~SMOOTHING MACHINE - [mach tool] machine *f* à lisser le placage
VENEERING - [carp] placage *m*
[carp] (the material) bois *m* de placage
~BAG - [impl] vessie *f* pour placage
VENEERS - [ind] feuilles *fpl* de bois
VENEPUNCTURE - [med] ponction *f* veineuse
VENESECTION - [med] phlébotomie *f*
VENETIAN BLIND - [constr] (flexible window screen which can be raised or lowered) jalousie *f* à lames mobiles
~CARPET - [text] (worsted carpet for stairs etc) tapis *m* d'escalier, courante *f*
~GLASS - [glass man] verre *m* de Venise
~GREEN - [dyes] vert *m* de Venise
~RED - [dye] (a natural red iron oxide pigment found in Italy) rouge *m* vénitien
~SHUTTERS - [constr] persiennes *fpl*
~WINDOW - [arch] fenêtre *f* à trois panneaux
VENICE TURPENTINE - [paint] (an oleo-resin obtained from Pinus larix, the Tyrolean larch) térébenthine *f* du mélèze
VENIPUNCTURE - s. venepuncture
VENOCLYSIS - [med] injection *f* intraveineuse
VENOFIBROSIS - [med] fibrose *f* veineuse
VENOGRAM - [med] phlébogramme *m*
VENOM - [chem] (poisonous fluid secreted by certain animals) venin *m*
VENOSTASIS - [med] stase *f* veineuse
VENOTOMY - s. venesection
VENT, to - [gen phys mech] (to permit to escape) décharger, faire exhaler
[metall] (to make a vent) ouvrir un trou d'air
[metall] (in die casting, to permit gas to escape) tirer l'air
VENT - [gen] reniflard *m*, évent *m*, trou *m*
[metall] (a hole made in a mould for the escape of gas or air) trou *m* d'air
[aero] (an opening in the centre of a parachute canopy) cheminée *f*
[geol] cheminée *f*
~-CAP - s. vent patch
~COCK - [plumb] robinet *m* de purge
~FLAP - [constr] clapet *m* d'air
~FORMER - [metall] tirette *f* d'air
~HEM - [aero] (the hem round the vent of a parachute) bord *m* de fuite
~HOLE - [metall] trou *m* d'air
~LINE - [plumb] tuyau *m* d'échappement
~PATCH - [aero] (a piece of material covering the vent of a parachute) chapeau *m* de cheminée
~PEG - [impl] fausset *m*
~PIPE - [hydr] (a pipe connecting a sanitary fixture with the vent stack) tuyau *m* d'évent, tube *m* aérateur
[th eng] tuyauterie *f* de mise à l'air libre
[ind chem] (of a retort) tuyauterie *f* d'aérage
[constr] ventilation *f*
[metall] lanterne *f*
~PLUG - s. vent peg
~PORT - [mech] lumière *f* de ventilation
~ROD - [metall] aiguille *f* à air
~STACK - [constr] aspirateur *m* de fumée
~VALVE - [mech] (of a fuel system) régulateur *m* d'aération

VENT WINDOW - [aero] déflecteur *m*
~WIRE - [metall] (a wire for punching small holes in the mould, so that gases can escape) aiguille *f* à air
VENTED MANIFOLD - [mech] distributeur *m* ouvert
VENTHOLE - [metall] trou *m* d'air
VENTIDUCT - [constr] conduit *m* d'air
VENTILATE, to - [gen] ventiler, aérer
[med] (the blood) oxygéner (le sang)
[mining] éventer
VENTILATED COMMUTATOR RISER MOTOR - [el] moteur *m* à radiales de collecteur ventilées
~FRAME - [el] (of a machine) à carcasse ventilée
~MOTOR - [el] moteur *m* ventilé
~RADIATOR - [el] (of an electrical machine or motor) à radiateurs ventilés
~RIBBED SURFACE - [el] (of an electrical machine) à nervures ventilées
VENTILATING - [gen] ventilation *f*, de ventilation
~BRICK - [constr] pierre *f* de ventilation
~COVER - [telecomm] tampon *m* avec grille d'aération
~DUCTS - [el] canaux *mpl* de ventilation
~FAN - [el] (of a motor) ventilateur *m* électrique
~FLAP - [constr] clapet *m* de registration
~GRILLE - [constr] grille *f* d'aération
~RIDGE TILE - [constr] faîtière *f* ventilatrice
~SHAFT - [mining] puits *m* d'aérage
VENTILATION - [gen] ventilation *f*, aérage *m*
~BAFFLE - [el] (of an electrical machine, or motor) cône *m* d'aérage
~FURNACE - [metall] foyer *m* d'aérage
~GARMENT - [astronaut] vêtement *m* ventilé
~HOLE - [metall] ouverture *f* aérante
~LOSS - [el] perte *f* par frottement de l'air
~PLUG - [mech] fausset *m*
~STACK - [constr] cheminée *f* d'aérage
VENTILATOR - [impl] ventilateur *m*
[mech] (e.g. in a Diesel locomotive) volet *m* d'aération
[constr] ventouse *f*
~COVER - [auto] couvercle *m* d'aération
~COWL - [auto] gaîne *f* de ventilateur
VENTING - [metall] tirage *m* d'air
~CHANNEL - [oil ind] canal *m* de pompage
VENTRAL - [bot] ventral
[aero] (relating to the under side of an aircraft fuselage) ventral
VENTRICLE - [anat] ventricule *m*
VENTRICULOGRAPHY - [med] (the radiological examination of the intercranial ventricles following direct introduction of air) ventriculographie *f*
VENTRILOQUY - [acoust] ventriloquie *f*
VENTURI - s. Venturi tube
~DISPERSION PLUG - [mech] (a plug formed with a conical orifice, used in the nozzle of an injection moulding machine to disperse colourant) poinçon *m* de dispersion Venturi
~GOVERNOR - [mech] régulateur *m* par Venturi
~METER - [instr] (a device designed to measure fluid or gases) compteur *m* Venturi
~TUBE - [mech] (a duct furnished with a constriction of such form and dimensions that the reduction of pressure at that point is a function of the speed of flow) Venturi *m*, diffuseur *m*
VENTURINE QUARTZ- [min] aventurine *f*
VERANDA - [arch] véranda *f*, galérie *f* à jour

VERANDAH - s. veranda
VERATRINE - [chem] (a mixture of poisonous alkaloids obtained from the dried ripe seeds of Schoenocaulon officinale and having application in medicine as a parasiticide) vératrine *f*
VERATROL - [chem] (an antiseptic) vératrol *m*
VERATRUM - [bot] (American hellebore. The rhizome and roots of Veratrum viride, an extract of which has been employed in medicine in the control of hypertension) vératre *m*
VERBENA OIL - [chem] (an oil obtained from the leaves of Verbena triphylla L. and used in perfumery) essence *f* de verveine
VERBIGERATION - [med] berbigération *f*
VERDIGRIS - [paint] (a pigment consisting of basic copper acetate, now little used) vert-de-gris *m*
[agric] verdet *m*
VERDUNIZATION - [med] verdunisation *f*
VERGE, to - [gen] être contigu à, toucher à
VERGE - [gen] bord *m*, extrémité *f*
[mech] (a stick or rod) tringle *f*, tige *f*
[mech] axe *m* du balancier
[constr] saillie *f* (de la couverture au dessus du pignon)
~BOARD - [arch] bordure *f* du pignon
~-PERFORATED CARD - [comput] carte *f* à bande perforée
VERIFICATION - [gen] (the process of checking the results of one or more operations) vérification *f*, contrôle *m*
VERIFIER - [comput] vérificatrice *f*
VERIFY, to - [gen & comput] vérifier
VERIFYING PAGEPRINTER - [comput] téléimprimeur *m* en page comparateur
VERMICIDE - [chem] vermicide *m*
VERMICULAR - [bot & zool] vermiculaire, vermoulu
VERMICULITE - [min] (a native hydrated aluminium magnesium-iron silicate, capable when heated, of considerable exfoliation. It is used in the production of lightweight concretes, as a thermal and acoustical insulant, as a refractory, adsorbent, catalyst carrier, as a paint, rubber, and plastics filler, and to lighten heavy solids) vermiculite *f*
VERMIFORM - [gen] vermiforme
VERMIFUGE - [med] vermifuge *m*
VERMILION - [paint] (a red pigment consisting essentially of mercuric sulphide) vermillon *m*
VERMIN - [zool] vermine *f*
VERMINATION - [med] vermination *f*
VERMINOUS BRONCHITIS - [med] broncho-pneumonie *f* vermineuse
VERNAL - [gen] printanier
[astr & bot] vernal
~EQUINOX - [astr] (the time when the sun is at the First Point of Aries, about the 21st of March) équinoxe *m* de printemps
VERNALIZATION - [bot] (the treatment of seeds before they are sown) printanisation *f*, vernalisation *f*, jarovisation *f*
VERNATION - [bot] (the manner in which the leaves are packed in a bud) vernation *f*, préfoliation *f*
VERNIER - [meas] (a small scale sliding in contact with another, which makes it possible to read the latter to fractions (usually tenths) of a division) vernier *m*
~ARM - [instr] (the part of an instrument carrying the verniers) bras *m* du vernier
VERNIER CALIPER - s. vernier gauge
~GAUGE - [tool] (a tool having a sliding jaw provided with a vernier for measuring diameters of parts and the like) jauge *f* micrométrique, calibre *m* à vernier
VERONA YELLOW - [paint] jaune *m* de Vérone
VERONAL - [chem] (diethylbarbituric acid, an important soporific) véronal *m*
VERONICA - [bot] véronique *f*
VERRUCA - [med] verrue *f*
VERRUCIFORM - [med] papillaire, d'aspect verruqueux
VERRUCOSE - [bot & med] verruqueux
VERSATILE - [gen] aux talents variés
[mech] (of a spindle) pivotant
[bot & zool] versatile, oscillant
VERSED SINE - [math] sinus *m* verse
VERSICOLOR - [med] versicolore
VERSINE - s. versed sine
VERSO - [print] verso *m*
VERTEBRA - [anat] vertèbre *f*
VERTEBRAL - [anat] vertébral
VERTEBRATE - [zool] vertébré *m*
~WAVEGUIDE - [radio] (a form of flexible waveguide) guide *m* d'ondes articulé
VERTEX - [gen math surv] sommet *m*
[astr] zénith *m*
[constr] (the crown of an arch etc) sommet *m*
~FIELD - [radio] (the feed with a conical aerial applied to the top of the cone) alimentation *f* au sommet
~OF A LENS - [opt] pôle *m* d'une lentille
~PLATE - [radio] (plate near the vertex of a reflector to prevent unwanted reflections back to the primary radiator) cache *f* sommet
VERTICAL - [gen & geom] vertical
~ADJUSTMENT - [mech] réglage *m* vertical
~ANCHOR - s. vertical bracing joist
~AND HORIZONTAL CORRECTION - [telev] correction *f* horizontale et verticale
~ANGLE -[astr] angle *m* vertical
~ANGLES - [math] angles *mpl* opposé par le sommet
~APERTURE CORRECTION - [telev] correction *f* verticale de l'ouverture
~AUTOMATIC TAPPING MACHINE - [mach] taraudeuse *f* verticale
~BAR GENERATOR - [telev] générateur *m* de barres verticales
~BEARING - [mech] palier *m* vertical
~BLANKING - [electron] (of a cathode-ray tube) suppression *f* verticale
~BOILER - [th eng] chaudière *f* verticale
~BORING MACHINE - [mach tool] machine *f* à aléser verticale
~BORING MILL - [mach tool] tour *m* vertical
~BRACE - [constr] ferrure *f* d'assemblage, ferrure *f* d'entretoisement
~BRACING JOIST - [th eng] fer *m* d'ancrage vertical
~BREAK SWITCH - [el] disjoncteur *m* à couteau vertical
~CASTING - [metall] (in foundry work) coulée *f* debout, coulée *f* verticale
~CENTRING - [telev] position *f* verticale
~CENTRING CONTROL - [telev] (control provided in a TV receiver or cathode-ray oscilloscope to shift the position of the centre image vertically on the screen) centrage *m* vertical

VERTICAL CHAMBER - [th eng] chambre *ℓ* verticale
~CIRCLE - [astr] (a great circle perpendicular to the plane of the horizon) vertical *m*
~CONDUCTOR - [el] conducteur *m* vertical
~CONVERGENCE CONTROL - [telev] réglage *m* de convergence verticale
~CONVERGENCE SHAPE CONTROL - [telev] réglage *m* de la configuration de la convergence de trame
~CONVERGENCE TILT CONTROL - [telev] réglage *m* linéaire de la convergence de trame
~CONVERGENCE YOKE - [telev] bobine *ℓ* de convergence de trame
~CROSS-HAIR - [instr] (in a telescope) fil *m* vertical, fil *m* collimateur
~CROSS TUBE BOILER - [metall] chaudière *ℓ* verticale à bouilleurs transversaux
~CURVE - [surv] (the parabolic curve introduced between two railway gradients to provide a gradual change from one to the other) raccordement *m* entre les rampes
~CYLINDER-GRINDING MACHINE - [mach tool] machine *ℓ* à rectifier en l'air les surfaces cylindriques
~DEFINITION - [telev] définition *ℓ* verticale
~DEFLECTION - [electron] (vertical tracing of the field on the screen of a cathode-ray tube) déviation *ℓ* verticale
~DEFLECTION ELECTRODES - [electron] (the pair of electrodes causing the electron beam to move up and down on the fluorescent screen of a cathode-ray tube employing electrostatic deflection) électrodes *ℓpl* de déviation verticale
~DEPTH - [mining] hauteur *ℓ* verticale
~DEVIATION - [astronaut] écart *m* en hauteur
~DRILLING MACHINE - [mach tool] foreuse *ℓ* verticale
~DRIVE - [mech] commande *ℓ* verticale
~DYNAMIC CONVERGENCE - [telev] (in colour television) convergence *ℓ* dynamique verticale
~DYNAMIC FOCUSING - [telev] (in colour television) focalisation *ℓ* dynamique verticale
~EASEMENT CURVE - [surv] arc *m* de raccordement
~EDGING ROLLS - [metall] cylindres *mpl* de refoulement verticaux
~ELEVATION - [astr] altitude *ℓ*
~ENGINE - [mech] (type of engine in which the cylindres are located above the crankshaft in a vertical position) moteur *m* vertical
~EXTRUDER - [mech] (type of extruder in which the barrel and screw are vertical and extrusion is downward) extrudeuse *ℓ* verticale
~FAULT - [geol] faille *ℓ* verticale
~FEED OF DRILLING SPINDLE - [mach tool] course *ℓ* verticale de l'arbre porte-foret
~FIN - [aero] (of the rudder) plan *m* fixe vertical
~FLOOR-BEARING - [constr] boitard *m*
~FLYBACK - [telev] (the return of the electron beam to the top of the screen at the end of each field in television) retour *m* vertical
~FREQUENCY - [telev] (the frame frequency) fréquence *ℓ* de balayage verticale, fréquence *ℓ* de trame
~GUIDES - s. vertical guiding
~GUIDING - [gas ind] guidage *m* vertical
~HAMMER - [mech] marteau *m* à chute verticale
~HOIST - [mech] (hoist which is mounted vertically in front of the body) treuil *m* vertical

VERTICAL HOLD-CONTROL - [telev] (a manual control for adjusting the vertical scanning synchronization) régleur *m* de la synchronisation de la trame
~HUNTING - [telev] (jumping of the picture due to faulty synchronization) instabilité *ℓ* verticale de l'image, saut *m* de l'image
~INJECTION PRESS - [plast ind] (injection moulding machine in which the material is injected in a vertical direction) presse *ℓ* d'injection verticale
~JOINT - [constr] joint *m* montant
~LIFTING TRUCK - [mech] monte-charge *m* sur chariot
~LIMB - [instr] (of a theodolite) limbe *m* vertical (d'un théodolite)
~LINEARITY CONTROL - [telev] réglage *m* de linéarité verticale
~LINING - [mining] boisage *m* vertical
~MAGNET - [el] aimant *m* de levage
~MILLING ATTACHMENT - [mach tool] appareil *m* pour fraiser verticalement
~MILLING MACHINE - [mach tool] fraiseuse *ℓ* verticale
~MOTION - [mech] ascension *ℓ*, soulèvement *m*
~MOVEMENT - [mach tool] course *ℓ* verticale
~PANNING - [cin] panoramique *ℓ* verticale
~PLANE DIRECTIONAL PATTERN - [radio] (the polar diagram obtained by radiation of the aerial in a vertical plane) diagramme *m* de rayonnement vertical
~PROJECTION - [draw] projection *ℓ* verticale
~RAM PUMP - [mech] pompe *ℓ* refoulante verticale
~RECORDING - [el acoust] (or hill and dale recording; a mechanical recording in which the modulation is in a direction perpendicular to the surface of the recording medium) enregistrement *m* en profondeur
~RESOLUTION - [telev] (the number of active lines in a television image) définition *ℓ* verticale
~RETRACE - s. vertical flyback
~RUDDER - [aero] gouvernail *m* de direction
~SCREW PRESS - [mech] presse *ℓ* à vis verticale
~SEISMOGRAPH - [instr] (instrument designed to record the vertical components of the earth's vibrations) sismographe *m* vertical
~SEPARATION - [aero] (the vertical distance which is ordered by a traffic control to be maintained between aircraft) espacement *m* vertical
~SHADING - [telev] canevas *m* à luminosité inégale verticale
~SHAFT - [mech] arbre *m* de transmission vertical
~-SHAFT TURBINE - [mech] turbine *ℓ* à axe vertical
~SHEAR - [soil] effort *m* tranchant
~SHORE - [constr] chandelle *ℓ*, étai *m* vertical
~SIZE CONTROL - [telev] régleur *m* de l'hauteur du cadre
~SLIP - [telev] décalage *m* vertical, glissement *m* vertical
~STAY - [mech] (of a metal structure) montant *m*
~STEP - [mech] pas *m* d'ascension
~STIFFENER - s. vertical stay
~SWEEP - [telev] balayage *m* vertical
~SWITCHBOARD - [el] tableau *m* vertical
~TAIL AREA - [aero] (the projection on the vertical plane of the full area of the rudder and fin) surface *ℓ* verticale de l'empennage
~TAKE-OFF AND LANDING AIRCRAFT - [aero] (an aircraft designed to leave and return to the ground with its longitudinal axis in a vertical position)

avion *m* capable de décoller et d'atterrir verticalement
VERTICAL TRAVERSE - [mach tool] chariotage *m* vertical
~WIND TUNNEL - [aero] (type of wind tunnel designed for vertical airflow) soufflerie *f* verticale
~WIRE - [comput] fil *m* vertical
VERTICALLY POLARIZED WAVE - [electron] (a linearly polarized wave whose direction of polarization is vertical) onde *f* à polarisation verticale
VERTIGO - [med] vertige *m*
VERVAIN - [bot] verveine *f*, herbe *f* sacrée
VERY CURLY WOOL - [text] laine *f* fortement ondulée
~GREASY WOOL - [text] laine *f* en suint riche en graisse
~HIGH FREQUENCY - [radio] (vhf, range of radio frequencies between 30 and 300 mc/s) hyperfréquence *f*, VHF
~HIGH FREQUENCY OMNIRANGE SYSTEM - s. VHF omnirange
~LIGHT - [signals] éyoile *f* éclairante
~LOW FREQUENCY - [radio] (V.L.F.; radio-frequencies below 30 kc/s (wavelengths exceeding 10.000 meters) très basse fréquence *f*
~PISTOL - [impl] (for signalling) pistolet *m* à fusée
VESICLE - [anat] vésicule *m*
VESICOPUSTULE - [med] vésicopustule *f*
VESICULA - s. vesicle
VESICULAR STRUCTURE - [metall] structure *f* vacuolaire
VESSEL - [impl] vase *m*, récipient *m*, vase *m*
[el] (of a storage battery) cuve *f*
[naut] vaisseau *m*
[anat] vaisseau *m*
[bot] trachée *f*
VESTIBULE - [constr] vestibule *m*, antichambre *f*
[metall] (of a gas-tight carburizing furnace) chambre *f* de précombustion
[railw] (luggage space) compartiment *m* à bagages
[railw] (US only, the gangway between coaches) à intercirculation
[anat] (of the ear) vestibule *m* (de l'oreille)
~CAR - s. vestibule coach
~COACH - [railw] voiture *f* à intercirculation
~OF THE EAR - [anat] vestibule *m* labyrinthique
~OF THE LARYNX - [anat] vestibule *m* du larynx
~TRAIN - [railw] train *m* à voitures à intercirculation
VESTIGE - [gen] vestige *m*, trace *f*
VESTIGIAL SIDEBAND - [telev] bande *f* latérale restante, bande *f* latérale résiduelle
~SIDEBAND AERIAL FILTER - [telev] (a filter for television transmitter) filtre *m* d'antenne pour émetteur de télévision à bande latérale restante
~SIDEBAND ANTENNA FILTER - s. vestigial sideband aerial filter
~SIDEBAND MODULATION - [radio] modulation *f* sur bande latérale résiduelle
~-SIDEBAND TRANSMISSION - [telev] transmission *f* à bande restante
VESTRY - [arch] sacristie *f*
VETCH - [bot] vesce *f*, orobe *m*
VETCHLING - [bot] gesse *f* des prés
VETERINARY - [vet] vétérinaire *m*
~MEDICINE - [vet] médecine *f* vétérinaire
~SURGEON - [vet] vétérinaire *m*

VETIVEROL - [chem] (a perfumery agent) vétivérol *m*
VETIVERT ACETATE - [chem] (a perfumery component) acétate *m* de vétiver
~OIL - [chem] (an aromatic oil obtained from the root of Vetiveria zizantoides and used in perfumery) essence *f* de vétiver
VETRIFIED BOND - [metall] agglomérant *m* céramique
V.F. - [acoust] (short for Voice Frequency, Telephone Frequency in the USA, any frequency within the part of the audio range required for a transmission of commercial quality, i.e. 300-3400 c/s) fréquence *f* vocale, fréquence *f* téléphonique
V.F. SIGNALLING RELAY SET - [telecomm] signaleur *m* à fréquence vocale
VFR - [aero] (Visual Flight Rules) VFR
~FLIGHT - [aero] (a flight carried out according to official visual flight rules) vol *m* VFR
V.G. RECORDER - [instr] (a recording accelerometer registering simultaneous values of airspeed and acceleration) enregistreur *m* V.g.
VH RECORDER - [instr] (an instrument which makes a graphic recorder of simultaneous values of indicated airspeed and altitude) enregistreur *m* V.H.
VHF - s. Very High Frequency
VHF OMNIRANGE - [radio, radar] (VOR, non directional radio range beacon) radiophare *m* VHF omnidirectionnel
VHF ROTATING TALKING BEACON - [radar] (an automatic VHF radio-telephone beacon with a continuously rotating radiation pattern) radiophare *m* parlant
VIA - [gen] via, par la voie de
[astr] voie *f*
VIABLE - [gen] (in a living state, e.g. live yeast) viable
VIAL - [impl] fiole *f*
VIAND - [gen] mets *m*
VIBRATE, to - [gen] (to give a rapid swinging motion) vibrer, osciller, faire vibrer
VIBRATED CONCRETE - [constr] béton *m* vibré
VIBRATILE MEMBRANE - [biol] membrane *f* vibratile
VIBRATING - [gen] vibrant, vibratoire, oscillant
~-BAR GRIZZLY - [impl] grille *f* à barres vibrantes
~CAPACITOR - [el] condensateur *m* vibrant
~CATCHER - [text] crochet *m* oscillant
~CIRCUIT - [telecomm] (auxiliary local timing circuit associated with the main line receiving relay of a telegraph circuit) circuit *m* vibratoire
~COMB - [text] vibrateur *m* denté
~CONDENSER - s. vibrating capacitor
~CONTACTOR - s. vibrator
~DRAIN - [transp] transporteur *m* par vibration
~FRAME - [paper man] appareil *m* de branlement
~LEVER - mech] levier *m* oscillant
~MACHINE - [soil] vibroleuse *f*
~PLATE - [auto] (of the horn) disque *m* de résonance (du cornet)
~PLATE COMPACTOR - [soil] plaque *f* vibrante
~RAMMER - [soil] dame *f* vibrante
~REED - [el] (device making and breaking contact in a cyclic fashion) lame *f* vibrante
~REED ELECTROMETER - [instr] (instrument taking negligible current for measuring unidirectional voltage which is converted into alternating voltage by means of a vibrating capacitor and subsequently amplified) électromètre *m* à vibration
~-REED INSTRUMENT - [instr] appareil *m* à lames

vibrantes
VIBRATING REED RELAY - [el] relais *m* à lame vibrante
~RELAY - [el] relais *m* vibreur
~SCREEN - [min] (a device for classifying finely-crushed material according to particle-size, and consisting of fabric woven frame) crible *m* oscillant
~SIEVE - s. vibrating screen
VIBRATION - [phys] (rapid swinging motion, broadly, any periodic physical process, e.g. a cyclic variation in electric or magnetic field intensity) vibration *f*, oscillation *f*
[gen] trépidation *f*, vibration *f*
~ABSORBER - [mech] amortisseur *m*
~DAMPER - [mech](for crankshafts etc) amortisseur *m* de vibrations
~-EXCITING EQUIPMENT - [meas] (equipment for the dynamic testing of constructions and material) appareillage *m* excitateur de vibrations
~-FREE SUSPENSION - [mech] suspension *f* antivibratile
~GALVANOMETER - [instr] galvanomètre *m* à vibration
~METER - [instr] (apparatus for the measurement of the displacement, velocity or acceleration of a vibrating body) vibromètre *m*
~-PROOF - s. vibrationless
~RAMMING - [metall] serrage *m* par vibration
~-ROTATION SPECTRUM - [phys] (spectrum in the infrared portion of the electromagnetic spectrum produced by vibrational and rotational transitions within a molecule) spectre *m* de vibration-rotation
~TEST - [mech] essai *m* de secousses
VIBRATIONAL PARTITION FUNCTION - [phys] (the contribution of the total partition function of molecules associated with their vibrational energy) fonction *f* de répartition vibrationnelle
~SUM RULE - [phys] règle *f* du total des vibrations
VIBRATIONLESS - [phys & mech] sans vibrations
VIBRATON - [electron] (a type of cavity resonator) vibraton *m*
VIBRATOR - [phys el] vibreur *m*, vibrateur *m*
[telecomm] (in wireless telegraphy) oscillateur *m*
[print] distributeur *m* d'encre
[mus] anche *f*
~COIL - [el] bobine *f* de vibreur
~FOR CONCRETE - [constr] vibrateur *m* pour béton
~INVERTER - [radio] déphaseur *m* à vibreur
~METHOD - [soil] méthode *f* du vibrateur
VIBRATORY - [gen] vibratoire
~CONVEYOR - [transp] transporteur *m* par secousses
~FEEDER - [mech] (a device for feeding a machine by rapid vibration) alimentation *f* par secousses
~ROLLER - [soil] rouleau *m* vibrant
VIBROFLOTATION - [constr] (a compacting method) vibro-flottation *f*
~SOIL COMPACTION - [constr] compactage *m* par vibro-flottation
VIBROGRAPH - [instr] (an instrument which shows diagrammatically the form, amplitude and frequency of mechanical vibrations) vibrographe *m*
VIBROMETER - s. vibration meter
VIBROSCOPE - [instr] vibroscope *m*
VIBROTAMPER - [constr] dame *f* pour vibro-flottation
VIBROTRON - [electron] (movable-anode triode) vibrotron *m*

VIBURNUM - [bot] (the dried bark of Viburnum opulus, having application in medicine) viorne *f*
VICE - [mech] étau *m*
~BENCH - [impl] étau *m* roulant, établi *m* roulant pour étaux
~CAP - [mech] (guards for soft materials) mordache *f*
~CLAMP - [mech] (brass or copper jaw fitted over the steel jaw of the vice) mordache *f*, estibois *m*
~PIN - [plumb] serre-joint *m*
~-PLATE - [mech] étau-plateau *m*
~PRESS - [mech] presse *f* à vis
~SLIDING BETWEEN PARALLEL BARS - [mech] étau *m* à barres parallèles
~-STAND - s. vice bench
~WITH INSERTED JAWS - [mech] étau *m* à machoîres rapportées
VICINAL - [gen] vicinal
~FACES - [crystall] plans *mpl* vicinaux
~ROAD - [gen] chemin *m* vicinal
VICKERS HARDNESS - [metall] dureté *f* Vickers
VICTUAL - [gen] approvisionnement *m*
VICUNA - [text] (a fine wool obtained from a small lama-like ruminant) vigogne *f*
~CLOTH - [text] tissu *m* de vigogne
~WOOL - [text] laine *f* de vigogne
~YARN - [text] fil *m* de vigogne
VIDEO - [telev] (synonym of television) télévision *f*
[telev] (relating to television) vidéo
~ADJUSTMENT - [telev] compensation *f* du signal d'image
~AMPLIFIER - [telev] amplificateur *m* vidéo
~CHANNEL - [telev] canal *m* vidéo
~CHIP - [comput] micro-image *f* magnétique
~CIRCUIT - [telev] circuit *m* vidéo
~COIL - [telev] bobine *f* vidéo
~CONVERTER - [telev] convertisseur *m* vidéo
~DEMODULATOR - [telev] démodulateur *m* vidéo
~DETECTION - [telev] détection *f* vidéo
~DISK APPARATUS - [telev] appareil *m* à disque vidéo
~DRUM APPARATUS - [telev] appareil *m* vidéo à tambour
~FREQUENCY - [telev] (the frequency in the video-signal spectrum) vidéofréquence *f*, fréquence *f* vidéo
~-GAIN CONTROL - [radar] contrôle *m* de gain vidéo
~HEAD - [telev] (in video recording) tête *f* vidéo
~IMPEDANCE - [telev] impédance *f* vidéo
~INSERTION - [telev] insertion *f* vidéo
~INTEGRATION - [telev] intégration *f* des signaux vidéo
~INTERCONNECTION POINT - [telev] point *m* de jonction de lignes de télévision locales
~MAPPING - [electron] (a procedure whereby a chart of an area is electronically superimposed on a radar display) représentation *f* topographique électronique
~MATRIX - [telev] grille *f* de distribution compensée de signaux vidéo
~MIXER - [telev] pupitre *m* de mélange image
~OUTPUT - [telev] signal *m* de sortie vidéo
~OUTPUT STAGE - [telev] étage *m* de sortie vidéo
~REPEATER - [telev] répéteur *m* vidéo
~SHEET APPARATUS - [telev] (in television recording) appareil *m* vidéo à feuille magnétique

VIDEO SIGNAL - [telev] (the signal consisting of the picture signal and the synchronizing signal) signal *m* d'image complet
~SIGNAL SEGMENTATION - [telev] segmentation *f* du signal vidéo
~SOCKET - [telev] (socked used to connect a TV set to a video recorder) connexion *f* vidéo
~SWITCHER - [telev] commutateur *m* vidéo
~-TAPE - [telev] (a television recording apparatus) bande *f* vidéo, bande *f* image
~TIME BASE - [telev] (image output) base *f* de temps vidéo
~TRANSMITTER - [telev] émetteur *m* d'images
VIDEOTAPE DUBBING - [telev] (in television recording) repiquage *m*
~REPRODUCER - [telev] magnétoscope *m* de lecture, vidéoscope *m*
~SPLICER - [telev] colleuse *f* pour bande image
VIDEOTRON - [electron] (electron-beam tube providing a picture signal derived from a fixed given image on an electrode inside the tube) monoscope *m*, tube *m* de mire
VIDICON - [electron] (camera tube in which a charge-density pattern is formed by photoconduction and stored on that surface of the photoconductor which is scanned by an electron beam) vidicon *m*
VIEW - [gen] vue *f*, regard *m*, champ *m* visuel
[gen] (exhibition) exposition *f*
[gen] (panorama) vue *f*
~-CAMERA - [photo] appareil *m* à pied, chambre *f* à pied
~FINDER - [photo] (a component for viewing the image-field of the camera) viseur *m*
~HOLE - [photo] trou *m* de visée
VIEWER - [telev] spectateur *m*, téléspectateur *m*
VIEWING - [gen] (term sometimes used for inspection of finished or partly-finished products) inspection *f*, contrôle *m*
[photo] visée *f*
~ANGLE - [opt] angle *m* visuel
~APPARATUS - [photo] appareil *m* d'examen
~DISTANCE - [telev] distance *f* de vision
~FILTER - [photo] filtre *m* de vision
~MIRROR - [telev] miroir *m* de vision
~RATIO - [telev] distance *f* optimale de vision
~SCREEN - [telev] écran *m*
~SYSTEM - [opt] système *m* de contrôle de visée
[nucl] système *m* de contrôle optique
VIGILANCE DEVICE - [el] dispositif *m* de vigilance
VIGNETTER - [photo] dégradateur *m*
VIGNETTING EFFECT - [opt] effet *m* dégradateur
VILLA - [arch] villa *f*
VILLAGE - [geogr] village *m*
VINAL - [chem] (man-made fibre based on vinyl alcohol and having good resistance to fungi, mildew and chemicals) vinal *m*
VINASSE - [ind chem] vinasses *f*pl de distillerie
VINBARBITAL - [chem] (a sedative and hypnotic) vinbarbital *m*
VINE - [agric] vigne *f*, pied *m* de vigne
~ARBOUR - [agric] treille *f*
~COTTON - [text] coton *m* grimpant
~CULTURE - [agric] viticulture *f*
~GROWING - [agric] (of a ground) vignoble
~-GROWING DISTRICT - [agric] région *f* viticole
~-GRUB - [zool] eumolpe *m* de la vigne
~KNIFE - [agric] couteau *m* à vendange
VINE-LOUSE - [bot] phylloxéra *m*
~MILDEW - [bot] (plant disease) mildiou *m*, oïdium *m*
~MOTH - [zool] cochylis *f*
~PESTS - [bot] parasites *m*pl de la vigne
~-PLANT - [bot] cep *m* de vigne
~SPRAYER - [agric] pulvérisateur *m* pour vignes
~WEEVIL - [zool] charançon *m* de la vigne
VINEGAR - [chem] (an aqueous solution of acetic acid with mineral salts and some esters obtained by alcoholic and acetic fermentation of fruit juice or malt extract) vinaigre *m*
~GENERATOR - [ind chem] (packed tower in which dilute alcohol solution is made to trickle over chips on which Acetobacter aceti have been cultivated) générateur *m* de vinaigre
~-MAKING - [agric] fabrication *f* du vinaigre
~-PLANT - [bot] sumac *m* de Virginie
VINELAND - [agric] pays *m* viticole
VINERY - [agric] grapperie *f*, serre *f* à raisin
VINEYARD - [agric] surface *f* complantée en vignes
~GRAFTING - [agric] greffe *f* sur place
VINICULTURE - [agric] viticulture *f*
VINOMETER - [instr] (an instrument for determining the alcohol content of wine) oenomètre *m*, vinomètre *m*
VINTAGE - [agric] récolte *f* du raisin, vendange *f*
VINYL - [chem] (the unsaturated group CH_2:CH-) vinyle *m*
~ACETATE - [chem] (colourless, flammable liquid, the main row material for vinyl plastics. It is also used in the manufacture of safety glass and as a textile finish) acétate *m* de vinyle
~BUTYL ETHER - [chem] (a synthesis intermediate for copolymers) vinyl-butyl-éther *m*
~BUTYRATE - [chem] (flammable liquid with application in polymers and water-based paint) butyrate *m* de vinyle
~CHLORIDE - [chem] (flammable, easily liquified gas and the starting material of major group of thermoplastics) chlorure *m* de vinyle
~COMPOUNDS - [chem] (highly reactive plastics intermediate containing the vinyl grouping) composés *m*pl vinyliques
~CYCLOHEXENE - [chem] (a synthesis intermediate for plastics and other organic compounds) vinylcyclohexène *m*
~CYCLOHEXENE MONOXIDE - [chem] (flammable liquid with application in organic synthesis and the production of polymers) monoxyde *m* de vinyl-cyclohexène
~ETHER - [chem] (divinyl ether. An inhalation anaesthetic) vinyl-éther *m*, éther *m* vinylique
~ETHOXYETHYL SULPHIDE - [chem] (a synthesis intermediate) sulfure *m* de vinyl-éthoxyéthyle
~ETHYL ETHER - [chem] (colourless, flammable liquid with uses as an intermediate and in copolymerization) vinyl-éthyléther *m*, éther *m* vinyl-éthylique
~ETHYL HEXOATE - [chem] (an intermediate for plastics) hexoate *m* de vinyléthyle
~ETHYLHEXYL ETHER - [chem] (an intermediate for pesticides, drugs, and adhesives) vinyl-éthyl-hexyl éther *m*
~ETHYLPYRIDINE - [chem] (an intermediate for plastics) vinyl-éthylpyridine *f*
~FLUORIDE - [chem] (a flammable gas to polyvinyl fluoride) fluorure *m* de vinyle
~ISOBUTYL ETHER - [chem] (isobutyl vinyl ether, IVE,

a colourless, flammable liquid with uses in adhesive protective coatings and as a plasticizer and modifier for polystyrene and alkid resins) éther *m* vinyl-isobutylique

VINYL METHYL ETHER - [chem] (methyl vinyl ether MVE, colourless, flammable gas which gives copolymers suitable for use in protective coatings, other applications include its use as a plastic plasticizer and as a modifier for a number of resins) éther *m* méthylvinylique

~PLASTICS - [chem] (synthetic organic thermoplastics with wide applications › Vinyl plastics include methyl methacrylate and acrylate, styrene, vinylidene chloride, vinyl acetate and chloride, acrylonitrile and a number of other polymers typified by a carbon double bond in the monomer molecule) plastiques *f*pl vinyliques

~PROPIONATE - [chem] (flammable liquid with applications in polymers and water-soluble paints) propionate *m* de vinyle

~PYRIDINE - [chem] (an intermediate for pharmaceuticals, polymers, and synthetic rubbers) pyridine *f* de vinyle

~RESIN - [chem] (synthetic resin formed by the polymerization of compounds containing the CH_2=CH- group) résine *f* vinylique

~STABILIZERS - [chem] (metallic soaps, amines or other substances added to vinyl chloride resins during processing to retard deterioration due to hydrogen chloride) stabilisants *m*pl de vinyle

~STEARATE - [chem] (a plasticizer and copolymerization agent) stéarate *m* de vinyle

~TOLUENE - [chem] (an intermediate and solvent) toluène *m* de vinyle

~TRICHLOROSILANE - [chem] (an intermediate for silicones) trichlorosilane *m* de vinyle

VINYLACETYLENE - [chem] (an intermediate for synthetic rubber and other organic compounds) vinylacétylène *m*

VINYLATION - [chem] (the reaction of alcohols amines or phenols with acetylene to give vinyl derivatives which in turn have uses as polymerization intermediates) préparation *f* vinylique

VINYLCARBAZOLE - [chem] (an intermediate for synthetic resins) vinylcarbazole *m*

VINYLIDENE CHLORIDE - [chem](VC) (colourless flammable liquid used with acrylonitrile or vinyl chloride in the production of saran) chlorure *m* de vinylidène

~FLUORIDE - [chem] (starting material in the production of a number of elastomers) fluorure *m* de vinylidène

~RESINS - [chem] (resins containing a chlorine, fluorine or cyanide radical in the structural unit) résines *f*pl à base de vinylidène

VINYON - [chem] (generic term for synthetic fibres containing a specific proportion of vinyl chloride units) vinyon *m*

VIOL - [mus] viole *f*

VIOLA - [mus] alto *m* à cordes

VIOLET - [dye] violet *m*

~RAY - [opt] rayon *m* violet

~-RED TOURMALINE - [min] tourmaline *f* violet-rougeâtre

VIOLIN - [mus] violon *m*

VIOLLE - [meas] (in photometry) violle *m*

~STANDARD - [meas] (in photometry) étalon *m* Violle

VIOLONCELL - [mus] (USA only, cello in GB) violoncelle *m*

VIOMYCIN - [chem] (an antibiotic produced by strains of Streptomyces puniceus and used in the treatment of tuberculosis) viomycine *f*

VIPER - [zool] vipère *f*

~-GRASS - [bot] scorsonère *f*, salsifis *m* noir

VIRDIAN GREEN - [paint] (a green pigment of complex composition derived primarily from hydrated chromium oxide) vert *m* de chrome

VIRGIN - [gen] pur, virginal, vierge
[metall] virginal, de première fusion
[ind chem] (of clay) (argile) crue

~FOREST - [geol] forêt *f* vierge

~IRON - [metall] fer *m* virginal

~LIQUOR - [ind chem] (in refrigeration process) produits *m*pl de première condensation des condenseurs à air

~MATERIAL - [plast ind] (new material as distinct from that which has been recovered and reground) produit *m* vierge

~NEUTRON FLUX - [phys] flux *m* de neutrons vierges

~NEUTRONS - [nucl] (neutrons from any source prior to collision) neutrons *m*pl vierges

~SOIL - [geol] terre *f* vierge

VERGINAL GROOVE - [el acoust] sillon *m* non-modulé

~TAPE - [el acoust] ruban *m* vierge

VIRGINIUM - [chem] (metallic element, symbol Fr atomic number 87) francium *m*

VIRILISM - [physiol] virilisme *m*

VIRILITY - [physiol] virilité *f*

VIROPEXIS - [med] fixation *f* des virus

VIROSE OF POTATO - [bot] (plant disease) virose *f* des pommes de terre

VIRTUAL - [.gen mech opt)] virtuel

~ADDRESS - [comput] adresse *f* virtuelle

~CATHODE - [electron] (a region in the space charge where there is a potential minimum which, by reason of the space-charge density, behaves as a source of electrons) cathode *f* virtuelle

~ENTROPY - [nucl] entropie *f* virtuelle

~FOCUS - [opt] foyer *m* virtuel

~GRAVITY - [phys] pesanteur *f* virtuelle

~IMAGE - [opt] image *f* virtuelle

~INERTIA - [aero] (in aeroelasticity, the portion of the effective inertia forces in an oscillation which results from the presence of the ambient air and varies with its density) inertie *f* virtuelle

~LEVEL - s. virtual state

~PARTICLE - [phys] (particle emitted and absorbed in a virtual process) particule *f* virtuelle

~PHOTON - [phys] (photon emitted and absorbed in a virtual process) photon *m* virutel

~PROCESS - [phys] (a process which may be represented as the emission of a particle or quantum followed so quickly by absorption, that the energy and momentum of the particle in this intermediate state cannot be clearly defined) processus *m* virtuel

~QUANTUM - [phys] (in second and higher order perturbation theory, a matrix element connecting an initial state with a final state, involves intermediate states in which energy is not conserved) quantum *m* virtuel

~REACTOR - [nucl] (reactor in which the method of images is used to solve the critically equations for a reactor) réacteur *m* virtuel

~SOURCE - [nucl] (theoretical source making use of

infinite kernels to solve the reactor equation) source *f* virtuelle
VIRTUAL STATE - [phys] (a quasi-stationary energy level, or state of a compound nuclear) niveau *m* virtuel
~TEMPERATURE - [phys] température *f* virtuelle
~WORK - [mech] travail *m* virtuel
VIRULENCE - [gen & med] virulence *f*
VIRUS - [med] virus *m*
~DISEASE - [med] maladie *f* à virus
VIRUSEMIA - [med] virémie *f*
VISBREAKING - s. viscosity breaking
VISCELLINE YARN - [text] fil *m* de viscelline
VISCOMETER - [instr] (an instrument for the determination of viscosity) viscomètre *m*
VISCOSE - [ind chem] (the solution from which viscose rayon is spun, this term is sometimes incorrectly used for the fibre itself, which is properly called "viscose rayon") viscose *f*
~FILAMENT - [text] fil *m* de viscose
~PROCESS - [ind chem] (process for the production of rayon involving cellulose xantate which, after conversion to fibres, is reconverted to cellulosa. Modifications of the process give high tenacity rayon) procédé *m* à la viscose
~PUMP - [text] pompe *f* à viscose
~RAYON - [text] (regenerated cellulose fibres produced by the viscose process) rayonne *f* de viscose
~SILK - [text] (artificial silk) soie *f* viscose
~SOLUTION - [text] solution *f* de viscose
VISCOSIMETER - [instr] (instrument to measure viscosity) viscosimètre *m*
VISCOSITY - [phys] (the ratio of the shear stress to the rate of shear of a fluid) viscosité *f*
~-BREAKING - [oil ind] (operation of reducing the viscosity of petroleum products by short-time decomposition) réduction *f* de viscosité
~DEPRESSANT - [chem] (a substance which when added to another, lowers its viscosity) réducteur *m* de viscosité
~INDEX - [chem] indice *m* de viscosité
~MANOMETER - [instr] (a type of vacuum meter) manomètre *m* à amortissement, jauge *f* à amortissement, manomètre *m* à viscosité
~METER - s. viscosimeter
VISCOUS - [phys] visqueux, gluant
~DAMPING - [auto] (a damping in which the force opposing the vibratory motion is proportional to the velocity) amortissement *m* visqueux
~DIFFUSITY - [phys] (the coefficient of viscosity divided by the density of a fluid) diffusivité *f* visqueuse
~FLOW - [phys] (low velocity flow of liquids characterized by absence of turbulence) flux *m* visqueux, écoulement *m* en régime visqueux, écoulement *m* en régime laminaire
~FLUID - [mech] fluide *m* visqueux
~OIL - [chem] huile *f* visqueuse
~SPINNING SOLUTION - [text] solution *f* à filer visqueuse
VISE - s. vice
VISIBILITY - [opt] (the maximum distance at which a standard object can be perceived and identified) visibilité *f*, champ *m* visuel
[met] visibilité *f*
~FACTOR - [opt] efficacité *f* lumineuse
~RANGE - [opt] marge *f* de visibilité
VISIBLE - [gen] visible
~HORIZON - [astr] (the line at which earth (or sea) and sky appear to meet) horizon *m* visible, horizon *m* apparent
~RADIATION - [opt] (the radiant energy perceived by the human eye) rayonnement *m* visible
~SIGNAL - [railw etc] signal *m* optique
[telecomm] (in telephony) signal *m* lumineux, voyant *m*
~SPEECH - [electron] (electronic method of translating spoken words into visible patterns) parole *f* visible
VISION - [gen] vision *f*, vue *f*
[opt] vue *f*, faculté *f* visuelle
~BANDWIDTH - [telev] largeur *f* de bande image
~CARRIER - [telev] porteuse *f* image
~CHANNEL - [telev] canal *m* image
~CONTROL - [telev] régie *f* image
~CONTROL SUPERVISOR - [telev] technicien *m* image
~CROSSTALK - [telev] diaphotie *f*
~FREQUENCY - [telev] (the frequency of the visual carrier of the television transmitter) fréquence *f* image
~MIXER - [telev] pupitre *m* régie image
~MIXTURE OPERATOR - [telev] opérateur *m* de mélange image
~MODULATION - [telev] modulation *f* image
~ON SOUND - [telev] interférence *f* d'image sur le son
~-PACKAGING - [pachag] (packing of articles in transparent containers, so that they can be seen without unpacking them) emballage *m* transparent
~SWITCHER - [telev] commutateur *m* image
VISIONPROOF GLASS - [glass man] (type of glass through which objects can be seen although transparent to light) verre *m* transparent
VISOR - [impl] (projecting piece on cap to shield the eyes) pare-soleil *m*
~SCREEN - [impl] (screen held or fixed before the eyes to pretect them during welding) masque *m*
VISTA SHOT - [telev] prise *f* de vues à distance
VISTAVISION - [cin] (a type of large screen) vistavision *f*
VISUAL - [opt] (pertaining to the sense of sight, ocular) visuel
~ACUITY - [opt] acuité *f* visuelle
~ANGLE - [opt] angle *m* visuel
~AURAL RANGE - [radio] (VAR, a VHF radio range, two tracks being shown visually in the aircraft and two aurally) radiophare *m* d'alignement audio-visuel *m*, VAR
~EXPOSURE METER - [photo] posemètre *m* optique
~FIELD - [opt] champ *m* visuel, champ *m* de vue, champ *m* de vision
~FIELD SLIDE RULE - [opt] régle *f* à calcul du champ visuel
~FLIGHT RULES - [aero] (code of regulations for visual flying) VFR
~INSPECTION - [gen] (inspection by observation with the eye alone, as distinct from physical tests or measurements) inspection *f* visuelle
~METEOROLOGICAL CONDITIONS - [met] (meteorological conditions observed visually , not by means of instruments, e.g. estimation of cloud amounts) conditions *fpl* météorologiques visuelles
~NERVE - [anat] nerf *m* optique
~RANGE - [nucl] (the value of the range of beta par-

ticles in an absorber, usually aluminium estimated by visual inspection of breaks in the aluminium absorption curve) portée *f* extrapolée
VISUAL RANGE - [opt] distance *f* de visibilité
~RECORD - [comput] enregistrement *m* visible
~SIGNAL - [electron] signal *m* optique
~SIGNALLING - [telecomm] signalisation *f* à voyant
~TRANSMITTER OUTPUT - s. visual transmitter power
~TRANSMITTER POWER - [telev] puissance *f* d'émission d'image
~TUNING INDICATOR - [radio] indicateur *m* optique de syntonie
VISUALIZATION - [med] évocation *f* à l'esprit, visualisation *f*
VITAL - [gen] vital, essentiel
VITAMIN A - [chem] (the vitamin essential for normal vision in poor light. It occurs, as the precursor, carotene, in plants and as the vitamin in cream, butter, fish liver, and eggs. Vitamin A is also prepared synthetically) vitamine *f* A, axérophtol *m*
~B - [chem] (synonym for thiamine, q.v.) vitamine *f* B, aneurine *f*, vitamine *f* antipolynévritique
~B_2 - [chem] (synonym for riboflavine, q.v.) vitamine B_2, riboflavine *f*, lactoflavine *f*
~B_6 - [chem] (synonym for pyridoxine, q.v.) vitamine *f* B_6, adermine *f*, pyridoxine *f*
~B_{12} - [chem] (cobalamin. Cyanocobalamin. An anti-anaemic substance obtained from liver or by the microbial metabolism of specific nutrients. Cyanocobalamin has therapeutic uses in medicine) vitamine *f* B_{12}, cyanocobalamine *f*
~C - [chem] (ascorbic acid) vitamine *f* C, vitamine *f* antiscorbutique
~CONTENT - [chem] teneur *f* en vitamines
~D - [chem] (calceferol. The antirachitic vitamin, occurring in eggs, milk, and fish, and also prepared by the UV irradiation of ergosterol) vitamine *f* D, ergostérol *m*
~DEFICIENCY - [med] avitaminose *f*
~H - [chem] (anti-egg-white factor) vitamine *f* H, biotine *f*
~K - [chem] (the antihaemorrhagic vitamin occurring in green vegetables and also synthesized) vitamine *f* K
~P - [chem] (permeability vitamin) vitamine *f* P, citrine *f*
~PP - [chem] (substance preventing pellagra) vitamine *f* PP, acide *m* nicotinique
VITAMINS - [chem] (accessory food factors of complex structure and present in raw foodstuffs. Vitamins are essential to health and have therapeutic uses in medicine) vitamines *fpl*
VITASCOPE - [cin] (a type of motion picture projector) vitascope *n*
VITIATE, to - [gen] vicier
VITICULTURE - [agric] viticulture *f*
VITREOUS - [gen] (glassy) vitreux
[chem] vitreux, hyalin
~BODY - [anat] humeur *f* vitrée
~CHINA - [gen] poterie *f* émaillée
~COATING - [metall] revêtement *m* vitreux
~ELECTRICITY - [el] (obsolete for positive electricity) électricité *f* vitrée
~ENAMELLING - [gen] émaillage *m* vitreux
~INCLUSION - [min] inclusion *f* vitreuse
~LUSTRE - [min] éclat *m* vitreux
~SILICA - [min] verre *m* quarteux
VITREOUS SILVER - [min] argentite *f*
~STATE - [phys] (crystals do not form at a definite temperature when liquids are cooled fairly rapidly, however, the viscosity of the liquid increases until a glassy substance is obtained) état *m* vitreux
VITRIFICATION - [ceramics] (phase of firing during which some of the constituents of the clay melt and assume a glassy condition) vitrification *f*
~POINT - [metall] point *m* de vitrification
VITRIFIED - [gen] (glazed, made vitreous, fused into glass) vitrifié
~BOND - [mech] agglutinant *m* vitrifié
[oil ind] mélange *m* vitrifié
~BOND GRINDING WHEEL - [tool] roue *f* à meuler vitrifiée
~BRICKS - [constr] (bricks fired to a stage of vitrification) briques *fpl* vitrifiées
~CLAY - s. vitrified tile
~DRAIN TILE - [constr] tuile *f* en grès verni
~TILE - [constr] grès *m*, porcelaine *f*, faïence *f*
VITRIFY, to - [gen] (to fuse into glass, to glaze) vitrifier
VITRIOL, to - [metall] (to pickle) décaper
VITRIOL - [chem] (sulphuric acid originally made from green vitriol, i.e. any sulphate from a heavy metal) vitriol *m*
VITRIOLIC ACID - [chem] acide *m* sulfurique, huile *f* de vitriol
VITROPRESSION - [med] vitropression *f*
VIVIANITE - [min] (a native hydrated ferrous phosphate) vivianite *f*
VIVIDIALYSIS - [med] dialyse *f* péritonéale
VIZOR - s. visor
VMC - s. Visual Meteorological Conditions
VOCAL CHORDS - [anat] (the organ which produces the human voice) cordes *fpl* vocales
VOCODER - [el acoust] (a system for the production of synthetic speech, using recorded voice signals to actuate the system instead of the mechanical keys used by the coder) vocoder *m*
VODAS - [telecomm] (a two-way radiotelephone transmission system employing the same frequency for transmission and reception) vodas *m*
VODER - [el acoust] (a system for the production of synthetic speech, employing a series of electronic tubes, controllable by mechanical keys, to vary the frequency, intensity, quality, duration, growth and decay of tones) voder *m*
VOGAD - [el acoust] (voice-operated gain adjusting device) vogad *m*, régulateur *m* vocal
VOICE, to - [gen] exprimer, voiser
[mus] (e.g. the organ pipes) harmoniser
VOICE - [gen] voix *f*
~COIL - [acoust] (the mobile loudspeaker-coil carrying the speed current) bobine *f* mobile
~CURRENT - [el acoust] (the current flowing through the winding on a cone of a loudspeaker) courant *m* de bobine mobile
~-EAR MEASUREMENT - [acoust] mesure *f* téléphonométrique
~FREQUENCY - [acoust] fréquence *f* vocale
~FREQUENCY DIALLING - [telecomm] sélection *f* à distance à fréquence vocale
~-FREQUENCY KEY PULSING - s. voice-frequency key
~-FREQUENCY KEY SENDING - [telecomm] envoi *m* par manipulation de signaux à fréquence vocale
~-FREQUENCY MULTICHANNEL TELEGRAPHY-[tele-

comm] télégraphie *f* à fréquence téléphonique, télégraphie *f* harmonique
VOICE-FREQUENCY TELEGRAPHY - s. voice-frequency multichannel telegraphy
~RECORDING EQUIPMENT - [el acoust] dispositif *m* d'enregistrement des conversations téléphoniques
VOICER - [mus] (organ voicing tuner) accordeur *m*
VOICING - [mus] accordage *m*
VOID, to - [gen] vider
[physiol] évacuer
[leg] annuler, résoudre
VOID - [gen] vide *m*
[adj] vide, inoccupé
[metall & plast ind] (a cavity in a moulding, generally due to imperfect filling) cavité *f*
~HOLE - [metall] retassure *f* interne
~RESULT - [math] résultat *m* indéterminé
VOIDANCE - [leg] annulation *f*
VOIDS - [constr] (spaces within a material, not filled with solid matter) vides *mpl*
VOILE - [text] voile *m*
~YARN - [text] fil *m* voile
VOLATILE - [chem phys] (having a low boiling or subliming point at normal atmospheric pressures) volatil, gazéifiable
~CARBON - [chem] carbone *m* volatil
~CONSTITUENTS - [chem] (those components of a mixture which will evaporate at normal temperatures) constituents *mpl* volatils
~HYDROCARBONS - [chem] hydrocarbures *mpl* volatils
~MATTER - [phys] (any component of a mixture which evaporates readily) matière *f* volatile
~-MATTER CONTENT - [min] (in coal testing) indice *m* de matière volatile
~MEMORY - [comput] mémoire *f* non-permanente
~OIL - [chem] (essential oil) huile *f* volatile,huile *f* essentielle
~SALTS - [chem] solution *f* de sels volatils
~SOLVENT - [chem] solvant *m* volatil, solvant *m* éthéré
~SOLVENT RECOVERY PLANT - [ind chem] installation *f* de récupération des solvants volatils
~STORAGE - s. volatile memory
~STORE - s. volatile memory
VOLATILITY - [chem & phys] (tendency of a substance to pass into the vapour phase also the product of the division, by its mole fractions, of the vapour pressure of a substance) volatilité *f*
~PRODUCT - [phys] (the product of the concentration of two or more ions or molecules which react to produce a volatile substance) produit *m* de volatilité
VOLATILIZE, to - [chem phys] volatiliser
VOLCANIC - [geol] volcanique
~ACTIVITY - [geol] activité *f* volcanique
~CHIMNEY - [geol] cheminée *f* volcanique
~GLASS - [geol] verre *m* volcanique, obsidiane *f*, silex *m* volcanique
~LAKE - [geol] lac *m* de barrage volcanique
~MUD - [geol] boue *f* volcanique
~NECK - s. volcanic chimney
~ROCK - [geol] (or igneous rock, a rock formed by the eruption of a volcano)roche *f* volcanique
~VAPOURS - [geol] émanations *fpl* volcaniques
VOLCANIZATION - [geol] volcanisation *f*
VOLCANO - [geol] volcan *m*
VOLCANOLOGY - [geol] (scientific study of volcanoes) volcanologie *f*
VOLITION - [gen & med] volition *f*, volonté *f*
VOLLEY - [gen] salve *f*, volée *f*
[mining] (simultaneous firing of holes) volée *f* (de coups de mine)
VOLT - [el] (the practical unit of electromotive force) volt *m*
~-AMPERE - [el] (of an a. c.) voltampère *m*, joule *m* par seconde
~-AMPERE-HOUR-METER - [instr] (apparent-energy meter) compteur *m* d'énergie apparente, voltampère heuremètre *m*
~-AMPERE METER - [instr] (instrument designed to measure apparent power in volt-amperes) voltampèremètre *m*, wattmètre *m*
~-BOX - [el] (voltage divider) boîte *f* de résistances
~EFFICIENCY - [el] (the relation between mean voltages of charge and discharge under fixed conditions) rendement *m* en tension
VOLTA EFFECT - [el] (the potential difference resulting when two dissimilar and insulated metals are brought into contact) effet *m* Volta, potentiel *m* de contact
VOLTAGE - [el] (the value of electromotive force or potential difference, expressed in volts) tension *f*, différence *f* de potentiel
~ADAPTER - [radio] (voltage adjustment switch) commutateur -adapteur *m* de tension
~AMPLIFICATION - [el] facteur *m* d'amplification en tension, rapport *m* des tensions
~-AMPLIFIER TUBE - [electron] (tube designed to produce voltage gains) tube *m* amplificateur de tension
~-AMPLIFIER VALVE - s. voltage-amplifier tube
~ATTENUATION - [radio] (of a transducer) atténuation *f* de la tension
~BETWEEN LINES - [el] (as of a three-phase system) tension *f* composée (d'un système polyphasé)
~BOOSTER - [el] survolteur *m*
~CHANGE - [el] variation *f* de tension
~CIRCUIT - [el] (or shunt circuit, the part of a measuring instrument, supplied by the voltage of the circuit it is to measure, or by a proportional voltage supplied by a transformer or a voltage divider) circuit *m* de tension, circuit *m* dérivé
~COMPARISON CONVERTER - [comput] traducteur *m* à sérvo-mécanisme
~CONNECTION - [el] montage *m* en tension
~DETECTOR - s. voltage indicator
~DIRECTIONAL RELAY - [el] (polarity directional relay) relais *m* directionnel de tension
~DIVIDER - [el] (a device, comprising, resistors capacitors or inductors, making it possible to obtain between two points a voltage proportional to the voltage to be measured) diviseur *m* de tension, réducteur *m* de tension
~-DOUBLER - [el] doubleur *m* de tension
~-DOUBLER RECTIFIER - [electron] redresseur *m* de doubleur de tension
~DROP - [el] (the diminution of potential along a conductor) chute *f* de tension, chute *f* de potentiel
~EFFICIENCY - [el] rendement *m* de tension
~FACTOR - [electron] facteur *m* d'amplification
~FEEDBACK - [el] réaction *f* de tension
~FLUCTUATIONS - [electron] (small variations in the electrode voltage in vacuum tube, resulting from thermal agitation, shot effect etc) fluctuations *fpl* de tension

VOLTAGE FUSE - [el] fusible *m* protecteur
~GAIN - [electron] gain *m* en tension
~GRADING ELECTRODE - [el] électrode *f* de répartition de potentiel
~INDICATOR - [el] (a device to show whether a circuit is alive) déceleur *m* de tension, détecteur *m* de tension
~JUMP - [electron] (in glow discharge tubes) saut *m* de tension
~LEVEL - [el] (the ratio between the voltage at any point in a transmission system existing at that point and an arbitrary voltage used as reference) niveau *m* de tension
~LIMITER - [el] limiteur *m* de tension
~MULTIPLIER RECTIFIER - [electron] redresseur *m* multiplicateur de tension
~OF FILAMENT BATTERY - [el] tension *f* de chauffage, tension *f* aux bornes du filament
~OF MAINS - [el] tension *f* au réseau
~PHASE-BALANCE PROTECTION - [el] disjoncteur *m* polyphasé de tension
~PROTECTION - [el] dispositif *m* de protection voltmétrique
~PULSE - [electron] (change in voltage in the central electrode system of a radiation counter) impulsion *f* de tension
~RATING - [electron] (the maximum sustained voltage which can be safely applied or taken from an electronic device) tension *f* nominale
~RATIO- [el] rapport *m* de transformation
~REFERENCE DIODE - [electron] diode *f* étalon de tension
~REFERENCE TUBE - [electron] tube *m* étalon de tension
~REGULATING RELAY - [el] relais *m* régulateur de tension
~REGULATOR - [el] (a device for the automatic control of the voltage produced by a generator) régulateur *m* de tension
~REGULATOR DIODE - [electron] (diode developing across its terminals an essentially constant voltage throughout a given current range) diode *f* régulatrice de tension
~REGULATOR TUBE - [electron] tube *m* régulateur de tension
~RELAY - [el] relais *m* de tension
~RESTRAINT - [el] relais *m* dépendant de la tension
~RISE - [el] surélévation *f* de tension
~SATURATION - [electron] saturation *f* anodique
~STABILIZER - [el] stabilisateur *m* de tension
~STABILIZING CIRCUIT - s. voltage stabilizing tube
~STABILIZING TUBE - [electron] tube *m* stabilisateur de tension
~STANDARDIZER - [instr] stabilisateur *m* de tension continue
~STANDING WAVE RATIO - [electron] (the ratio between the maximum voltage and the minimum voltage on the line) taux *m* d'ondes stationnaires
~TAPPING SWITCH - s. voltage adapter
~-TEMPERATURE COEFFICIENT - [electron] (for semiconductors) coefficient *m* tension-température
~TO EARTH - [el] tension *f* à la terre
~TO NEUTRAL - [el] (the voltage between a line conductor of a polyphase system and a neutral point) tension *f* étoilée
~TRANSFORMER - [el] transformateur *m* de tension
~TUNABLE TUBE - [electron] (oscillator tube whose operating frequency can be varied by changing one or more of the electrode voltages) tube *m* à syntonisation par variation d'une tension électrode
~TUNABLE VALVE - s. voltage-tunable tube
~WAVEFORM - [electron] allure *f* de la tension
VOLTAIC - [el] voltaïque, galvanique
~ARC - [el] arc *m* voltaïque
~BATTERY - [el] pile *f* électrique
~CELL - [el] (a system of two electrodes of different nature placed in a suitable electrolyte and designed to produce an electric current. Also under the names of various types, e.g. Bunsen Cell) pile *f* voltaïque
~COUPLE - [el] élément *m* voltaïque
~CURRENT - [el] (galvanic current) courant *m* galvanique, courant *m* voltaïque
~PILE - s. voltaic cell
VOLTAMETER - [instr] (instrument which measures a quantity of electricity by its electrochemical effects) voltamètre *m*, coulombmètre *m*
VOLTAMMETER - [instr] wattmètre *m*
VOLTMETER - [instr] (an instrument to read E.M.F. directly in volts) voltmètre *m*
VOLUME - [phys] (the space occupied by any body) volume *m*
[gen] (a book) livre *m*
[acoust] (the level of noise at a point in a sound channel) volume *m*, ampleur *f*, intensité *f*
~CHARGE DENSITY - [el] (the quantity of electric charge per unit volume of a three-dimensional charged insulator) densité *f* de charge de volume
~COMPRESSIBILITY - [rubber ind] compressibilité *f* tridimensionale
~CONTROL - [radio telev etc] régulateur *m* de volume, réglage *m* de volume
~CONTROL FADER - [el acoust] (a potentiometer designed to control the volume of sound in reproduction) potentiomètre *m* de réglage
~DILATOMETER - [instr] (a dilatometer designed to measure the variation of the volume of a liquid over a range of temperatures) dilatomètre *m* de volume
~DOSE - [radiat] dose *f* absorbée dans le volume, dose *f* volume
~EFFECT - [nucl] (isotopic effect for the heavy atoms which exceed the effect of the mutual motion) effet *m* de volume
~ENERGY - [chem] énergie *f* de volume
~EQUIVALENT - [acoust] équivalent *m* de référence
~IMPLANT - [radiat] (type of implant which is used when the volume to be treated approximates to a regular solid form) implant *m* en volume
~INTEGRAL - [math] intégrale *f* de volume
~IONIZATION - [phys] (average ionization density in a specified volume, independently of the specific ionization of the ionizing particles) densité *f* volumétrique d'ionisation, ionisation *f* volumique
~IONIZATION DENSITY - s. volume ionization
~LIFETIME - [electron] (of semiconductors) vie *f* moyenne du porteur minoritaire, vie *f* moyenne volumique
~LIMITER - [radio etc] limiteur *m* de volume
~PERCENTAGE - [chem] (the number of c.c. of absolute (q.v.) in 100 c.c. of the fluid in question at 60° F. (15.5° C.) pourcentage *m* en volume
~RANGE - [radio] (the range of intensities within which the volume of a programme fluctuates) gamme *f* dynamique
~RECOMBINATION - [phys] (occurring between po-

sitive and negative ions at low energies throughout the volume of an ionization chamber or counter) recombinaison *f* de volume

VOLUME RESISTIVITY - [el] (the electrical resistance between opposite faces of a unit cube of insulating material under prescribed conditions) résistivité *f* volumique

~SWELL - [rubber ind] augmentation *f* de volume par gonflement

~UNIT - [acoust] (a unit expressing the magnitude of a complex wave, such as that corresponding to speech or music) unité *f* de volume

~UNIT METER - [instr] volumemètre *m*

~VELOCITY - [elacoust] (across a surface element, the product of the area of the latter and the component of the sound particle velocity perpendicular to the surface) flux *m* de vitesse, vitesse *f* de volume

~VOLTAMETER - [instr] (a voltameter in which the quantity of electricity is determined by the volume of gas evolved) voltamètre *m* à volume

VOLUMENOMETER - [instr] (instrument for measuring volumes of solids, liquids or gases) volumé-nomètre *m*

VOLUMETER - [instr] (volumenometer) volumètre *m* [instr] (hydrometer) densimètre *m*

VOLUMETRIC - [gen] volumétrique

~ANALYSIS - [chem] (methods of quantitative analysis, using solutions of known concentration, the volumes of which are measured) analyse *f* volumétrique

~EFFICIENCY - [mech] (the ratio of the volume delivered by the feed unit to the maximum possible amount) rendement *m* volumétrique

~FLASK - [ind chem] ampoule *f* graduée

~GOVERNOR - [contr] régulateur *m* de débit

VOLUMINOUS - [gen] volumineux

VOLUMOMETER - s. volumenometer

VOLUNTEER PLANTS - [bot] plantes *fpl* accidentelles

VOLUTE - [arch] volute *f*, corne *f*

~CHAMBER - [mech] (of a centrifugal pump) canal *m* collecteur (de pompe centrifuge)

~PUMP - [mech] pompe *f* centrifuge à développantes

~SPRING - [mech] ressort *m* en volute

VOMIT - [med] vomissement *m*

VOR - s. VHF Omni Range

VORTA - s. vug

VORTEX - [phys] (a mass of fluid of which the motion is circulatory, with a core of intense vorticity) tourbillon *m*, tourbillonnement *m*

~BLADING - [mech] (e.g. of a gas turbine) système *m* d'aubes à tourbillon libre

~FILAMENT - [phys] filet *m* tourbillonnair

~LINE - [phys] (a line which is everywhere parallel to the local direction of the vorticity) ligne *f* de tourbillon
[aero] (a line of concentrated vorticity such as are found trailing from the free end of an aerofoil) filet *m* de tourbillon

~MOTION - [phys] (the motion of a fluid having non-zero vorticity) mouvement *m* tourbillonnaire

~RING - [phys] vortex *m*

~-RING STATE - [aero] (condition in a rotor in which the axial flow through the disk area is opposed in direction to that outside it, and also to the rotor thrust) fonctionnement *m* avec anneau tourbillonnaire

~SHEET - [phys] (a sheet composed of many vortex filaments streaming away from the trailing edge of an aerofoil) nappe *f* tourbillonnaire

VORTEX STREET - [phys] (a regular system of line vortices in two nearly parallel rows) rues *fpl* de tourbillons, alignement *m* de tourbillons

VORTICITY - [phys] (the measure of the rate of rotation of the fluid in a vortex) vecteur *m* de tourbillonnement

~EQUATION - [phys] équation *f* de variation du rotationnel

VORTICOSE - [gen] tourbillonnaire

VOUCHER - [comm] pièce *f* compatible, fiche *f* [leg] garant *m*

VOUSSOIR - [arch] (a tape red brick used to form an arch) voussoir *m*

VOWEL - [phonetics)] voyelle *f*

~ARTICULATION - [acoust] articulation *f* des voyelles, intelligibilité *f* des voyelles

VR TUBE - [electron] (glow-discharge voltage regulator) tube *m* régulateur

V.T.O. - s. vertical take-off

V.T.O.L. - s. vertical take-off and landing

VU METER - s. volume unit meter

VUG - [min] (or Vorta, a cavity in a rock lined with crystals) four *m* à cristaux, druse *f*, poche *f* à cristaux

VUGH - s. vug

VULCANITE - [rubber ind] (a hard rubber product made by the addition of sulphur i. e. by vulcanization) vulcanite *f*, ébonite *f*

VULCANIZATION - [rubber ind] (the combination under heat and pressure of rubber with sulphur to give improved properties) vulcanisation *f*

~ACCELERATOR - [rubber ind] accélérateur *m* de la vulcanisation

~BY HIGH FREQUENCY - [rubber ind] (electronic curing) radiovulcanisation *f*

~COEFFICIENT - [rubber ind] coefficient *m* de vulcanisation

~HEAT - [rubber ind] chaleur *f* de vulcanisation

~IN STAGES - [rubber ind] vulcanisation *f* échelonnée

~ON DRUM - [rubber ind] vulcanisation *f* sur tambour

~ON MANDREL - [rubber ind] vulcanisation *f* sur mandrin

~RATE - [rubber ind] (the speed at which vulcanization takes place) vitesse *f* de vulcanisation

VULCANIZE, to - [rubber man] vulcaniser, effectuer la cuisson

VULCANIZED - [rubber man] vulcanisé

~ASBESTOS - [ind] amiante *f* vulcanisée

~FIBRE - [ind chem] (a low-voltage insulant based on paper pulp treated with zinc chloride to give a product composed mainly of amyloid) fibre *f* vulcanisée

~ON-TYRE - [rubber ind] bandage *m* plein fixé à la jante par vulcanisation

~RUBBER - [rubber ind] (rubber which has been combined with sulphur and other additives under heat and pressure to give improved physical properties) caoutchouc *m* vulcanisé

~SCRAP - [rubber ind] (vulcanized rubber scrap) déchets *mpl* de caoutchouc vulcanisé

VULCANIZER - [rubber ind] vulcanisateur *m*, installation *f* de vulcanisation

VULCANIZING - [rubber ind] vulcanisation *f*

~AGENT - [rubber ind] agent *m* de vulcanisation

VULCANIZING APPARATUS - s. vulcanizer
~BOILER - [rubber ind] (a heating chamber used for the curing of rubber) chaudière *f* de vulcanisation
~HEATER - [rubber ind] appareil *m* de chauffage pour vulcanisation
~OVEN - [rubber ind] étuve *f* à vulcaniser

VULCANIZING PAN - s. vulcanizing heater
~PRESS - [rubber ind] presse *f* à cuisson, presse *f* à vulcaniser
~UNIT - s. vulcanizer
VULNERABILITY - [gen & med] vulnérabilité *f*
VULVA - [anat] vulve *f*
VULVITIS - [med] vulvite *f*

W - [chem] (the symbol for tungsten) symbole *m* du tungstène
[constr] (the symbol for total load) charge *f* totale (symbole)
[el] (symbol for electrical energy) énergie *f* électrique (symbole)
WAD, to - [gen] barrer, tamponner
[text] rembourrer
WAD - [text] (of cotton) pelote *f* d'ouate
[gen] (small compact mass of soft flexible substance) tampon *m*, bouchon *m*
[firearms] (piece of cloth or leather used to hold powder and shots in place in a shotgun) bourre *f*
[min] (hydrated oxide of manganese) wad *m*, hydroxyde *m* impur de manganèse
[metall] (the flash removed from a forging) barbe *f* intérieure
WADE, to - [gen] (to walk through water) marcher dans l'eau, passer à gué
[transp] (of a vehicle) faire passer à gué
WADDING - [text] ouatage *m*, ouate *f*
[firearms] (the material for wads) bourre *f*
~FILTER - [mech] filtre *m* d'ouate
~PICK - [text] trame *f* de remplissage
~THREAD - [text] fil *m* de fourrure
~WARP - [text] chaîne *f* de fourrure
~WASTE - [text] déchets *mpl* d'ouate
~WEFT - [text] trame *f* de fourrure
WADING - [gen] gué *m*
~TEST - [mech] (a test design to prove the watertightness of a craft) essai *m* de gué
WAFER - [food] gaufrette *f*
[chem] cachet *m*
[leg] (thin hardened disk of gelatine or other suitable substance, used for sealing letters etc) disque *m* de papier rouge
[electron] (a component part of semiconductor devices) plaquette *f* gaufrée
[comput] plaquette *f* microélément
~CORE - [metall] (in foundry work) galette *f*
~LOUDSPEAKER - [el acoust] (a flat loudspeaker) haut-parleur *m* plat
~SOCKET - [electron] (electron tube socket) culot *m* plat
WAFFLE - [food] (a type of batter cake) gaufre *f*
~INGOT - [metall] (a quarter of an inch thick aluminium ingot having a section of 3sq inches) lingot *m* mince d'aluminium
~IRON - [impl] (electrodomestic implement) gaufrier *m*
WAGE - [gen] salaire *m*, paie *f*
~CEILING - [gen] plafond *m* de rémunération, salaire *m* maximum
WAGE CLAIMS - [leg] revendications *fpl* de salaires
~-EARNER - [gen] salarié *m*
~-FREEZE - [leg] blocage *m* des salaires
~SCALE - [leg] indice *m* des salaires
~SHEET - [comm] feuille *f* de paie
~SLIDING SCALE - [leg] échelle *f* mobile
WAGER, to - [gen] parier, gager
WAGER - [gen] pari *m*
~-POLICY - [leg] (in insurance) police *f* gageuse
WAGGON - s. wagon
WAGNER EARTH - [el] (a bridge using an additional pair of ratio arms) prise *f* de terre de Wagner
WAGON - [transp] (USA only) voiture *f*, chariot *m*, camion *m*
[railw] wagon *m*
~BLOCK)- [railw] (a block wedged in for braking) cale *f* de roue
~BOILER - [th eng] chaudière *f* en tombeau
~BOX - [railw] caisson *m* du wagon
~-CARRYING TRAILER - [railw] remorque *f* porte-wagon
~CEILING - [constr] plafond *m* en berceau
~CORF - [mining] berline *f*
~DRILL - [mech] chariot *m* porte-manteau, jumbo *m*
~ELEVATOR - [mech] monte-wagon *m*
~HITCH - [mech] attelage *m* à chape
~-LOAD - [railw] charge *f* de wagon
[gen] charretée *f*
~MOVING DEVICE - s. wagon pinch bar
~PINCH BAR - [railw] pousse-wagon *m*
~REPLACER - [railw] rampe *f*
~SHOE - [railw] frein *m* à sabot du wagon
~VAULT - [constr] voûte *f* en berceau
~WAY - [mining] galerie *f* de roulage
~WITH BRAKE CABIN - [railw] wagon *m* à guérite
~WITH RADIAL BOLSTER - [railw] (used for the transport of very long loads, e. g. logs, machinery etc) wagon *m* à bascule
~ROOF - [constr] toit *m* cylindrique
~TIPPER - s. wagon-tipping mechanism
~TIPPING MECHANISM - [mech] culbuteur *m* pour wagons
~TIPPLER - s. wagon-tipping mechanism
WAILING - [min] triage *m* à la main
WAINSCOT, to - [constr] lambrisser, boiser
WAINSCOT - [constr] (facing for an inner wall) boiserie *f*
[constr] (the lower part of the interior wall showing a special finish) lambris *m*
~OAK - [constr] merrain *m*
~WOOD - [constr] bois *m* de lambrissage
WAINSCOTING - [constr] lambrissage *m*, boisage *m*
[constr](the material for wainscoting) bois *m* pour

lambrissage
WAINSCOTING - [build] (of a room) menuiserie *f* dormante d'une pièce habitable
~CAP - [constr] bord *m* supérieur du lambris
~PANELLING - [constr] lambrissage *m*
WAIST - [gen] (the part of the body between the chest and the hips) taille *f*, mi-corps *m*
[gen] (the middle part or section of an object) ceinture *f*, étranglement *m*
[auto] (of the body) ceinture *f*
[naut] embelle *f*
[text] (USA only) blouse *f*, chemisette *f*
~BELT - [text] ceinture *f*
~COAT STUFF - [text] étoffe *f* pour gilet
~COATING - s. waist-coat stuff
WAISTLINE - [auto] ligne *f* de ceinture
[text] (in tailoring) taille *f*
[shoe man] cambrure *f*
WAIT, to - [gen] attendre
[gen] (to serve) servir
[print] (copy) manquer (de copie)
~CONDITION - [comput] (in data processing) état *m* d'attente
WAITING - [gen] attente *f*
~-PASSENGER INDICATOR - [el] (in lifts) voyant *m* d'appel
~ROOM - [gen] salle *f* d'attente
~TIME - [comput] (or latency) temps *m* d'attente
WAIVE, to - [gen] renoncer, abandonner
[leg] (to surrender, to relinquish voluntarily) départir de, déroger
WAIVER - [leg] renonciation *f* (à un droit)
WAKE - [phys] (the region of disturbed fluid behind a body moving through it) houache *f*, sillage *m*
[mech] sillage *m* de viscosité
[naut] brassage *m*, remous *m*
~BLOCKAGE - [phys] (the proportion of blockage due to stream displacement by the wake of a model) blocage *m* dû au sillage d'un modèle
~FACTOR - [naut] facteur *m* de remous
WALE, to - [constr] moiser
[gen] marquer (de coups de fouet)
WALE - [text] (in a knitted fabric) rangée *f*
[text] (a rib, or a ridge on the surface of a cloth) côte *f* de drap
[constr] (horizontal timber used to bind together piles driven in a row) moise *f*
[shipbuild] (a strake of outer planking) plat-bord *m*
WALING - [constr] moisage *m*
~BOARD - [constr] traverse *f*
~PILES - [constr] moisage *m* des pieux
WALINGS - [constr] planches *f*pl de boisage
WALK, to - [gen] marcher,cheminer
WALK - [gen] marche *f*, démarche *f*
[gas ind] (a walkway, e.g. of a gasholder) passerelle *f*
[zool] pas *m*
~-AROUND BOTTLE - [astronaut] bouteille *f* d'oxygène portable
~-DOWN - [comput] (process in a magnetic cell) fuite *f*
WALKIE-LOOKIE - [telev] (USA only, a portable television transmitter) téléviseur *m* portatif
~-TALKIE - [radio] (radio communication unit providing a two-way service) émetteur-récepteur *m* portatif
WALKING - [gen] marche *f*
WALKING - [agric] à traction
[mech] (oscillating) oscillant
[mech] mobile
~BEAM - [mech] (in a vertical engine, a horizontal beam transmitting power to the crankshaft through the connecting rod) balancier *m*
[mining] levier *m* de battage
~BEAM DRILLING MACHINE - [mining] moteur *m* de battage
~BEAM FURNACE - [th eng] four *m* à longerons, four *m* à poutre mobile
[ovens] four *m* à sole oscillante
~BEAM SADDLE - [oil ind] support *m* du balancier
~CRANE - [mech] grue *f* vélocipède sur rail de translation
~HEARTH - [metall] sole *f* mobile
~LINE - [constr] (of stairs) ligne *f* de foulée, ligne *f* d'emmarchement
~STICK TRIPOD - [photo] pied-canne *m*
WALKOUT - [gen] (colloq.for strike) grêve *f*
WALKWAY - [mech] passerelle *f*
[oil ind] passerelle *f* de sonde
WALL, to - [constr] (to furnish with walls) murer, entourer de murs
[constr] (to enclose with walls) murailler
WALL - [constr] mur *m*, paroi *m*
[constr] (a kind of rampart) muraille *f*
[mech] (of a boiler) paroi *f*, mort mur *m*
[mining] (the side of a shaft etc) éponte *f*, ponte *f*
[mining] (of a mine-level) pied-droit *m*, paroi *f* latérale
[geogr] enceinte *f* de montagnes
[geol] lèvre *f*
[mining] fond *m* de taille
~ABSORPTION - [radiat] (the decrease in beta or gamma-ray output caused by the absorption in the walls of the container) absorption *f* aux parois
~ARCH - [constr] arcade *f* aveugle
~BAFFLE - [glass man] (in a furnace structure) chicane *f*
~BARLEY - [bot] (a weed) orge *f* des murs
~BEARING - [mech] (of a drive shaft) palier *m* mural
~BENCH - [constr] établi *m* mural
~BLOCK - [el] rosace *f* isolante
~BOARD - [constr] planche *f* pour cloison
~BOX - [constr] niche *f* murale, chaise-niche *f*
[mech] (for through shaft) caisson *m* mural
~BRACKET - [mech] chaise-console *f*
[gen] console *f* murale
~-BUILDING TEST - [oil ind] mesure *f* du panneau au filtre-presse
~CHASE - [constr] rainure *f* pour tuyaux
~CHASE - [oil ind] (for tubes) gorge *f* pour tubes
~CLAMPS - [constr] tirants *m* pour cloison double
~CLEANING GUIDES - [oil ind] grattoir *m* centralizateur
~CRAB - [mech] treuil *m* d'applique
~CRANE - [mech] grue *f* à potence, grue *f* murale
~CRIB - [mining] grand côté *f* du cadre de boisage
~CROSSING - [constr] croisement *m* de murs
~DISPLACEMENT - [phys] déplacement *m* de paroi
~DRILLING MACHINE - [mech] forerie *f* murale
~DRIVE - [mech] transmission *f* à paroi
~EFFECT - [phys] (the contribution to the ionization in ionization chambers by electrons from the walls) effet *m* de paroi
[soil] effet *m* de paroi

WALL ENCLOSURE - [constr] enceinte *f*
~ENERGY - [phys] (the energy per unit area of the boundary between oppositely oriented ferromagnetic domains) énergie *f* de paroi
~FACE - [mining] front *m* de taille
~FITTING - [el] applique *f*
~FRAME - s. wall box
~FRICTION - [constr] frottement *m* sur le mur
~HANGER - s. wall box
~HANGING - [text ind] tapisserie *f*
~HOLDFAST - [impl] clou *m* à patte
~HOOK - [impl] crampillon *m*
~IN, to - [constr] emmurer
~INSULATOR - [el] isolateur *m* de tranversée
~JIB CRANE - [mech] potence *f*
~LINING - [constr] revêtement *m* d'un mur
~MOUNTING - [el] montage *m* sur paroi
~PAPER - s. wallpaper
~PIECE - [mining] chapeau *m* sur murs de remblai
~PILLAR - [mining] pilier *m* du mur, pilier *m* de sûreté
~PLATE - [constr] sablière *f* de comble, plaque *f* d'assise, lambourde *f*
[mech] contre-plaque *f* de chaise
~-PLATE FOR BRACKETS - [mech] contre-plaque *f* pour chaises
~PEPPER - [bot] (a medical plant) poivre *m* de muraille, orpin *m* brûlant
~PIER - [constr] dosseret *m*, pilier *m* de mur
~POST - [constr] jambe-de-force *f* de console, poteau *m* accolé, poteau *m* contre un mur
~RECOMBINATION - [electron] (a basic mode of recombination) recombinaison *f* à la paroi
~ROCK - [geol] roche *f* encaissante
~SCATTERING - [nucl] (in a counter tube) diffusion *f* dans la paroi
~SCRAPER - [oil ind] grattoir *m* à expansion pour mur
~SLAB - [constr] plaque *f* murale
~SOCKET - [el] rosace *f*
~STRING - [constr] contre-limon *m*
~SWITCHBOARD - [el] tableau *m* mural
~TIE - [constr] ancre *f* de mur, ancrage *m*
~TIE CLOSER - [constr] parpaing *m*, pierre *f* d'ancrage
~TILE - [build] carreau *m*
~-TYPE - [el] (of a socket etc) sur paroi
~-TYPE RADIATION BEACON - [radiat] balise *f* de rayonnement murale
~WITH BUTTRESSES - [constr] mur *m* à pilastres, mur *m* à jambages
WALLED - [constr] muré, muraillé
~PLAIN - [atr] cirque *m* lunaire
WALLING - [constr] muraillement *m*, murage *m*
WALLING - [mining] revêtement *m* de puits
WALLMAN AMPLIFIER - [radio] (cascode amplifier) amplificateur *m* cascode
WALLOWING - [aero] instabilité *f*
WALLPAPER - [paper ind] papier *m* à tapisser
WALNUT - [bot] noix *f*
[bot] (the tree) noyer *m*
~BUTTER - [food] beurre *m* de noix
~DIE - [text] racinage *m*
~OIL - [chem] (oil expressed from the fruit of Juglans regia and used as a foodstuff and in the manufacture of paints) huile *f* de noix
WAND - [impl] baguette *f*
WANDER, to - [gen] errer
WANDER, to - [gen] (to deviate) dévier
[med] (to be delirious) délirer
WANDER - [gen] course *f* errante
[radar] (on a radar display, a rapid apparent displacement of the target from its mean position) déplacement *m* d'écho
[astronaut] dérive *f* apparente
~PLUG - [el] (a plug designed to connect a flexible conductor to one of several sockets forming terminals of different circuits) fiche *f* mobile
WANDERING CELLS - [biol] (leucocytes, migratory amoeboid cells) leucocytes *mpl*
~SEQUENCE - [metall] soudure *f* échelonnée
WANE - [carp] flache *f*
~-EDGED WOOD - [constr] bois *m* non-équarri
WANEY - [carp] flacheux
~EDGE - [constr] chanfrein *m*
WANTAGE - [comm] (USA only) manque *m*, déficit *m*
WANTED SIGNAL - [telecomm] signal *m* utile
WAP - [metall] (one turn in a wire coil) spire *f*
WAR DROPSY- [med] oedème *m* par carence, oedème *m* de famine
~GAS - [chem] gaz *m* de combat
~-HEAD - [firearms] (e.g. of a torpedo) cône *m* de charge
WARBLE FLY - [vet] oestre *m*, varron *m*
~-FLY INFESTATION - [vet] hypodermose *f* bovine
~TONE - [acoust] (a tone having a frequency continuously varying within fixed limits) ton *m* de huhulement, ton *m* de huhulé
WARBLER OSCILLATOR - [telecomm] (heterodyne) oscillateur *m* hétérodyne
WARBLING - [acoust] (sound made by singing birds) gazouillis *m*, ramage *m*
~CARRIER SYSTEM - [radio] (a method of increasing the degree of secrecy obtainable with a radiotelephone system using inversion) système *m* à onde porteuse et à variation de fréquence
WARD - [gen] guet *m*
[leg] pupille *m*
[leg] (custody) tutelle *f*
[med] (of a hospital) salle *f* d'hôpital
[mech] (of a key, a projection inside a lock, to avoid the use of any key other than the proper one) gardes *fpl*, bouterolles *fpl*, garniture *f*
~INFECTION - [med] infection *f* hospitalière
WARDED LOCK - [mech] serrure *f* à garnitures
WARDER - [gen] gardien *m*
[gen] (of a prison) gardien *m* de prison
WARDING FILE - [impl] (a small file used for delicate work) lime *f* à garnir, lime *f* à bouter
WARDROBE - [gen] armoire *f*
~TRUNK - [gen] malle-armoire *f*
WARDSHIP - [leg] tutelle *f*
WARE - [gen] articles *mpl*, marchandises *fpl*
[comm] (manufactured good) articles *mpl* fabriqués
~POTATO - [agric] pomme *f* de terre comestible
WAREHOUSE, to - [gen] magasiner, emmagasiner
[leg] (for excise purposes) entreposer
WAREHOUSE - [gen] magasin *m*, entrepôt *m*
WAREHOUSEMAN - [gen] magasinier *m*
[leg] (in a custom house) entreposeur *m*
WARFARIN - [chem] (a rodenticide) warfarine *f*
~SODIUM - [chem] (an anticoagulant used in medicine) warfarine *f* sodique
WARHEAD - [armament] (the ogive-shaped chamber containing the high explosive) tête *f* explosive

WARM, to - [gen] chauffer
WARM - [gen] chaud
~-AIR FURNACE - s. warm-air heater
~-AIR GRILLE - s. warm-air register
~-AIR HEATER - [th eng] générateur *m* d'air chaud
~-AIR MASS - [met] (an air mass originating in a latitude lower than that in which it is at the time of observation, hence warmer than the surface over which it is moving) masse *f* d'air chaud
~-AIR REGISTER - [th eng] bouche *f* d'air chaud
~BLEACH - [paper man] blanchiment *m* à chaud
~BLOOD - [zool] cheval *m* à sang chaud
~COLOURS - [opt] couleurs *fpl* chaudes
~FRONT - [met] (a line along which cold air being closely followed by warmer air) front *m* chaud
~SECTOR - [met] (a body of warm air lying between cold and warm fronts) secteur *m* chaud
~UP TIME - [electron] (the time which is required for all essential parts of a tube to attain an operating temperature) temps *m* de préchauffage
~-WATER RETTING - [text] rouissage *m* à l'eau chaude
WARMED CHILL - [metall] coquille *f* chauffée
~FLEECE - [text] toison *m* rechauffée
WARMING - [gen] chauffage *m*
~CHAMBER - [th eng] (of a domestic cooker) chauffe-assiette *m*
~OVEN - [th eng] étuve-table *f* chaude
~UP ALLOWANCE - [th eng] (of a heating system) marge *f* de puissance (d'une installation de chauffage)
WARN, to - [gen] avertir
[leg] (to give notice in advance) réprimander
WARNING - [gen] avertissement *m*
[gen] (alarm) signalisation *f*
[leg] congé *m*
[gen] (the act of notifying) action *f* d'avertir
[telecomm] annonce *f*
~BELL - [impl] cloche *f* d'alarme, sonnette *f* d'alarme
~BOARD - [auto] tableau *m* des lampes-témoin
~CIRCUIT - [telecomm] circuit *m* d'annonce
~DEVICE - [gen] (any device designed to inform an observer of a change in condition which may lead to danger.e.g.alteration in temperature, pressure etc beyond a set limit) dispositif *m* d'avertissement, avertisseur *m*
~HORN - [el acoust] (an acoustic warning device) sirène *f*
~LABEL - [nucl] (on an isotope container) étiquette *f* d'alarme
~LAMP - [mining] lampe *f* grisoumétrique
[gen] lampe *f* d'avertissement
~LIGHT - [el] (a luminous warning device) lampe-témoin *f*
~SHIELD - [nucl] (shield for the transport of radioactive materials) plaque *f* avertissante
WARP, to - [text] ourdir, empeigner (un métier)
[gen] (of timber) gauchir, déverser
[phys] (to change the form) fausser, se déformer
[naut] haler, touer
[aero] gauchir (les ailes)
[gen] s'enrouler
WARP - [gen] (the state of being twisted out of shape) courbure *f*, déformation *f*, gauchissement *m*
[text] (the threads running the long way of a fabric) chaîne *f*, lisse *f*
[rubber ind] (the heavy cords forming the carcass of a pneumatic tyre) chaîne *f* du tissu
WARP - [naut] (a light cable) amarre *f*, touée *f*, aussière *f* de halage
[rope man] (a length or rope yarn or rope) cordelle *f*
[geol] (alluvial soil formed by sediments) lais *m*, dépôt *m* alluvionnaire
[agric] colmate *f*
~BALL - [text] pelote *f* de chaîne, boule *f* de chaîne
~BALLING MACHINE - [text] pelotonneuse *f* pour chaînes
~BEAM - [text] ensouple *f* dérouleuse
~BEAM CARRIER - [text] porteur *m* d'ensouples
~BEAM CREEL - [text] support *m* d'ensouple
~BEAM FLANGE - [mech] plateau *m* d'ensouple, flasque *f* d'ensouple
~BEAMING - [text] montage *m* de la chaîne
~BEAMING MACHINE - [text] dresseuse *f*, machine *f* à monter sur ensouple
~BOBBIN - [text] bobine *f* de fil de chaîne
~BROCADED FABRIC - [text] broché *m* par chaîne
~CHAIN - [text] chaîne *f* d'ourdi
~CHEESE - [text] bobine *f* de chaîne croisée
~COP - [text] bobine *f* de chaîne
~COUNT - [mech] numéro *m* de la chaîne
~DIVIDER - [text] peigne *m* diviseur de chaîne, fausses lisses *fpl*
~DRYING FRAME - [text] cadre *m* de séchage pour chaîne
~EFFECT- [text] effet *m* de chaîne
~ENDS - [text] déchets *mpl* de la chaîne
~EYE - [text] (heddle eye) oeillet *m*, coulisse *f*
~FIGURING - [text] brochage *m* par chaîne
~FLOAT - [text] flotté *m* de chaîne
~FRAME - [text] ourdissoir *m*
~FRINGE - [text] frange *f* par chaîne
~GUIDE ROD - [text] tringle-guide *f* de la chaîne
~KNIT FABRIC - [text] tricot *m* de metièr chaîne
~KNITTING - [text] tricotage *m* sur metièr chaîne
~MACHINE - s. warp-frame
~PATTERN - [text] rapport *m* en chaîne
~PILE FABRICS - [text] velours *mpl* par chaîne
~PROTECTOR - [text] casse-chaîne *m*
~RIB - [text] cannelé *m* trame, reps *m* travers
~SATEEN - [text] satin *m* par chaîne
~SETTING - [text] compte *m* en chaîne
~SHEET - [text] nappe *f*
~SPOOLING MACHINE - [text] bobineuse *f* de fil de chaîne
~STOP - [text] (of a weaving machine) casse-chaîne *m*
~TENSION - [text] traction *f* sur la chaîne
~TENSION ROD - [text] tringle *f* de tension des fils
~THREAD - [text] fil *m* de chaîne
~THREAD ROW - [text] chaîne *f* supplémentaire sur l'envers
~TWILL - [text] sergé *m* par chaîne
~VELVET - [text] velours *m* par chaîne
~WINDER - [text] bobinoir *m*, bobineuse *f* de fil de chaîne
~WINDING - [text] bobinage *m*
~WINDING MACHINE - s. warp winder
~YARN - [text] fil *m* de chaîne
WARPAGE - [metall] déformation *f*
WARPED - [gen] gauchi, déjeté
[mech] fléchi, gauchi, voilé
[gen] (timber) déversé
[text] empeigné

WARPED CASTING - [metall] pièce *f* de fonte gauchie
~LAND - [agric] colmate *f*
~WHEEL - [mech] roue *f* voilée
~YARN - [text] fil *m* de chaîne
WARPER - [text] ourdisseur *m*
[text] (warping machine) ourdissoir *m*
WARPING - [gen & mech] gauchissement *m*, déjettement *m*, déversement *m*
[aero] (of a wing) gauchissement *m* des ailes
[text] ourdissage *m*
[metall] (of casting) déformation *f*, flexion *f*
[metall] bombement *m*
[agric] terrement *m* des champs, colmatage *m*
[naut] halage *m*
[electron] (the mechanical buckling of the grid wires of an electron tube owing to excessive heating) gondolage *m*
~BOBBIN - [mech] bobine *f* d'ourdissoir
~CAPSTAN - [naut] cabestan *m* de touage
~CHOCK - [naut] chaumard *m*
~COMB - [text] peigne *m* d'ourdissoir
~CREEL - [text] centre *m* d'ourdissage
~ENGINE - [naut] machine *f* à déhaler
~FRAME - [text] métier *m* à ourdir
~MACHINE - [text] ourdissoir *m*
~MILL - s. warping frame
WARPLANE - [aero] avion *m* de guerre
WARRANT, to - [gen & leg] garantir, certifier
[gen] (to authorize) autoriser
[gen] (to justify) justifier
WARRANT - [gen] garantie *f*, brevet *m*
[leg] mandat *m*
[comm] (a receipt for goods) certificat *m*
~OF ARREST - [leg] mandat *m* d'arrêt
~OF ATTORNEY - [leg] (power of attorney) procuration *f*
WARRANTED - [gen & comm] garanti
[leg] légitime, autorisé
WARRANTY - [comm] garantie *f*, attestation *f*
~PERIOD - [comm] durée *f* d'une garantie
WARREN - [gen] garenne *f*
[agric] (for rabbits) lapinière *f*
[impl] herse-bineuse *f*
WARSHIP - [gen] vaisseau *m* de guerre
WART - [med] verrue *f*
[vet] poireau *m*
[bot] excroissance *f*, loupe *f*
~DISEASE OF POTATOES - [bot] gale *f* verruqueuse
WARY FIBRED WOOD - [constr] bois *m* à fibre sineuse
WASH, to - [gen] laver, se laver
[metall] (of metals) métalliser
[gas ind] (of gas) laver
[constr] (walls, plasters) teinter
[geogr] (of the sea) baigner
[min] clairer, débourber
[text etc] blanchir, lessiver
WASH - [gen] lavage *m*, lessive *f*
[paint] badigeon *m*, badigeonnage *m*
[metall] (for a mould) noir *m* de fonderie
[phys] (the disturbance in air caused by the passage of an aerofoil through it) remous *m* d'air
(of the waves) remous *m* des vagues
[geol] (alluvion deposits) apports *mpl* d'alluvion
[naut] sillage *m*, souffle *m* (de l'hélice)
[naut] (of an oar) pale *f*
[med] lotion *f*
WASH - [ind chem] produits *mpl* du lavage
~BIT - [constr] trépan *m* à orifices d'évacuation de l'eau de curage
~BOTTLE - [ind chem] (flask fitted with double-bored stopper and tubes with open and jet ends, respectively, from which small jet of water can be sprayed by blowing into the open-ended tube) pissette *f*
~DIRT - [mining] alluvion *f* à laver
~DISSOLVE - [telev & cin] volet *m* ondulé
~DRAWING - [draw] dessin *m* lavé
~DRILL - [mech] perforatrice *f* à injection d'eau
~GRAVEL - s. wash dirt
~HEATING - [metall] chauffage *m* superficiel pendant le travail
~HOUSE - [constr] lavanderie *f*, buanderie *f*, blanchisserie *f*
~-IN - [aero] (increase in the angle of incidence in the direction of the wing tip) gauchissement *m* positif
~LEATHER - [leather ind] peau *f* de chamois
~MILL - [min] appareil *m* à analyser la fine granulométrie
~OIL - [gas ind] (oil used for the washing of gas) huile *f* de débenzolage
~OUT, to - [gen] laver
[print] dépouiller
~OUT - [el acoust] (the erasing of a recording from a magnetic tape or wire) démagnétisation *f*
[geol] (erosion, due to flood etc) ruissellement *m*
[aero] (decrease in the angle of incidence towards the wing tip) gauchissement *m* négatif
~PIPE - [oil ind] tube *m* de surforage
[constr] tige *f* de curage
~PLAN - [draw] plan *m* au lavis
~STUFF - [min] alluvion *f* à laver
~TANK - [paper man] cuve *f* de lavage
~TROUGH - [impl] (for the washing of minerals) battée *f*
~-UP, to - [gen] laver
[print] (the cylinders of a printing press) laver
~WATER - [gen] eau *f* de lavage
[ind chem] eau *f* de curage
~-WATER GUTTER - [hydr] (of a filter) goulotte *f* d'évacuation d'eau de lavage
~-WATER PUMP - [hydr] pompe *f* d'eau de lavage
WASHABILITY - [gen & paint] lavabilité *f*
WASHABLE - [gen] lavable
WASHAWAY - [geol] dégradation *f* du sol
WASHBASIN - [gen] cuvette *f* de lavabo
WASHBOARD - [carp] (a wood skirting) plinthe *f*
[glass man] (ripples on the surface of glass) ondulations *fpl*
[naut] (a plank) fargues *fpl* volantes
[naut] (a plank so adjusted that it turns the wash of the sea from a port) fargues *fpl* de sabord
WASHBOWL - s. washbasin
WASHBURN CORE - [metall] noyau *m* d'étranglement, noyau *m* de liaison
WASHDOWN - [gen] lavage *m*
~CLAY - [constr] (malm) mélange *m* de chaux et sable
~PUMP - [naut] (used on the boatdeck) pompe *f* de lavage
WASHED-OUT - [gen] (faded) fané, décoloré
WASHER - [mech] (annular, flat piece used under a nut to distribute pressure, or between surfaces to make a tight joint) rondelle *f*

WASHER - [ind chem] (for gases) laveur *m*
[mech] (a washing machine) machine *f* à laver
[min] (a washing machine for minerals) patouillet *m* pour laver les minerals
[paper man] (a washing engine) pile *f* laveuse
~CUTTER - [tool] coupe-rondelle *m*, coupe-cercle *m*, compas *m* à lame tranchante
~LOCK - [mech] blocage *m* par rondelle
~WASTE - [hydr] eau *f* résiduaire de lavage
~WITH PADDLES - [min] barboteur *m*
WASHERY - [min] laverie *f*, atelier *m* de lavage
WASHING - [gen] lavage *m*
[ind chem] (for purification) lavage *m*, épurage *m*
[geol] (an erosion which is due to running water) affouillement *m*, dégradation *f*
[min] (of minerals) débourbage *m*, lavée *f*, clairage *m*
[metall] raffinage *m* au bain de scories
~BOILER - [text] chaudière *f* pour lavage
~BOTTLE - [ind chem] flacon *m* laveur, barboteur *m* pour lavage
~CYLINDER - [text] cylindre *m* laveur
[min] patouillet *m* pour le minerai
~DRUM - [mech] tambour *m* laveur
~ENGINE - [paper man] pile *f* désagrégeante
~FLASK - s. wash bottle
~MACHINE - [constr] (used for washing gravel) laveuse *f* à sable et gravier
[text] (for pieces) appareillage *m* de lavage
[impl] (domestic implement) laveuse *f*
~MILL - [rubber ind] laminoir *m* à laver
~PAN - [min] cuve *f* de lavage
~PIT - [auto] fosse *f* de lavage
~ROOM - [ind chem] (room in a chemical laboratory fitted for the washing of apparatus) chambre *f* de lavage
~SODA - [chem] carbonate *m* de soude, soude *f* de commerce
~STUFF - [min] alluvion *f* à laver
~TANK - [mech] (used for mechanical parts) cuve *f* de lavage
~TOWER - [in d chem] tour *f* de lavage
~TROMMEL - [min] trommel *m* débourbeur
~TUBE - [ind chem] (laboratory implement) tube *m* laveur
~VAT - [ind chem etc] cuve *f* de lavage
~WATER - [gen] eau *f* de lavage
~WITH JETS OF WATER - [photo] lavage *m* sous jet
WASHMARKING - [metall] (a defect) surface *f* ondulée
WASHOUT - [geol] (through the action of water) poche *f* de dissolution
[oil ind] sortie *f* de liquide d'une conduite par cassure
~HOLE - [th eng] (for boilers) regard *m* de lavage, trou *m* de bras
~PLUG - [impl] tampon *m* de lavage, bouchon *m* de lavage
WASTAGE - [gen] gaspillage *m*
[chem mech etc] déperdition *f*, perte *f*
WASTE, to - [gen] gaspiller
[gen] (to wear away) dissiper
[mech] (to get lost) perdre
[gen] (to devastate) ravager, dévaster
[med] (e.g. for lack of food) amaigrir par une maladie
[leg] (a property etc) dégrader

WASTE - [gen] desert, désolé
[gen] (useless) de rebut
[gen] gaspillage *m*
[hydr] trop-plein *m*
[mech] gaspillage *m*, perte *f*, eau *f* de condensation
[text] rebuts *mpl*, bouts *mpl* veules
[text] (remnants of cotton or wool rags, used for cleaning) bourre *f* de coton, déchets *mpl* de coton
[min] (refuse ore) déblais *mpl*
[comm] (production in excess of consumption) production *f* non utilisée
[leg] dégradations *fpl*
[nucl] déchets *mpl* radioactifs
~ACID - [chem] acide *m* épuisé
~AREA - [gen] dépôt *m* d'ordures
~BILLY - [text] déchets *mpl* de coton
~BRINE - [ind chem] liqueur *f* résiduaire, lessive *f* résiduaire
~CHUTE - [min] cheminée *f* de remblai
~COCK - [mech] robinet *m* de purge
~COLLECTING COMB - [text] collecteur *m* de déchets
~DISPOSAL - [gen] destruction *f* des ordures ménagères
[nucl] (the operation of eliminating radioactive material) élimination *f* des déchets
~END - [text] déchets *mpl* de fil
[metall] chute *f*
~FILLING - [min] (the filling of cavities with waste) remblai *m* stérile
~FLOOR - [mining] niveau *m* de remblai
~GAS - [metall & th eng] flammes *fpl* perdues, gaz *mpl* perdus
~-GAS ENGINE - [mech] moteur *m* à flammes perdues
~GRINDER - [rubber ind] broyeur *m* à déchets
~HEAT - [mech] chaleur *f* perdue
~-HEAT BOILER - [th eng] (special type of boiler designed to utilise the heat contained in exhaust or other gases) chaudière *f* de récupération
~-HEAT FLUE - [th eng] conduit *m* d'évacuation des gaz chauds
~HOLDER - [text] caisse-collectrice *f* de déchets
~INSTRUCTION - [comput] (do nothing instruction) instruction *f* non-opération
~LAND - [agric] terrain *m* inculte
~LIQUOR - [hydr] eaux *fpl* d'égout, eaux *fpl* usées
~MATERIALS - [gen] résidus *mpl*
~PAPER - [paper man & gen] (paper which is thrown away as worthless) déchets *mpl* de papier, bardot *m*
~PICKLING LIQUOR - [ind chem] liqueur *f* de décapage épuisée
~PIPE - [plumb] tuyau *m* d'écoulement, tuyau *m* de décharge
~RECOVERY - [nucl] (the treatment of waste designed to obtain some useful material from it) récupération *f* des déchets
~SHAKER - [text] dispositif *m* à battre les déchets
~SILK - [text] déchets *mpl* de soie
~TREATMENT - [hydr] (sewage treatment) traitement *m* des eaux d'égout
~WADDING - [text] ouate *f* de déchets
~-WATER DRAIN - [hydr] égout *m* pour les eaux vannes
~-WATER GUTTER - [hydr] rigole *f* pour les eaux usées
~-WAX PROCESS - [metall] (in foundry work) système *m* à cire perdue
~WEIR - [hydr] barrage-déversoir *m*

WASTE WOOD - [agric] émondes *f*pl
WASTER PLATE - [metall] plaque *f* protectrice du paquet
WATCH, to - [gen] (to be vigilant, on the alert) veiller, garder
[gen] (to observe, to look on) observer
WATCH - [gen] (the act of watching) veille *f*, garde *f*, surveillance *f*
[naut] quart *m*
[horol] (small, portable time piece) montre *f*
~CASE RECEIVER - s. watch receiver
~-COAT - [naut] capote *f*
~DUTY - [gen] service *m* de garde
~-FIRE - [gen] feu *m* de bivouac
~GLASS - [ind chem] (shallow glass saucer to contain small quantities of liquids) verre *m* de montre
~-POST - [gen] poste *m* de garde
~RECEIVER - [telecomm] récepteur-montre *m*, récepteur *m* en forme de montre
~-TOWER - [gen] tour *f* d'observation
WATCHCASE - [horol] boîtier *m* de montre
WATCHDOG - [zool] chien *m* de garde, chien *m* d'attache
WATCHMAKER - [horol] horlogeur *m*
WATCHMAKER'S LATHE - [mech] tour *m* d'horloger
WATCHMAN - [gen] gardien *m*
WATCHWORK - [mech] mouvement *m* d'horlogerie
WATER, to - [gen] (to sprinkle with water) arroser
[chem] (to dilute in water) diluer, étendre d'eau
[naut] faire provision d'eau, faire de l'eau
[mech] alimenter d'eau
WATER - [chem] (hydrogen oxide, purified water obtained by de-ionisation or distillation) eau *f*
~ABSORPTION - [phys] (a measure of the water taken up by a material when immersed for a specified time) absorption *f* d'eau
~ADIT - [min] galerie *f* d'écoulement, galerie *f* de drainage
~ANALYSIS - [chem] analyse *f* d'eau
~ANCHOR - [naut] ancre *f* flottante
~AND MOISTURE PROOF SHEETING - [rubber & plast ind] feuilles *f*pl imperméables à l'eau et à l'humidité
~ANTILOPE - [zool] kob *m* singsing
~ASSOCIATION - [leg] association *f* des riverains
~ATOMISER - [mech] pulvérisateur *m* à eau
~BACK - [th eng] chaudière *f* postérieure
~BALANCE - [bot] (the ratio between the water which is taken in by a plant and the water which is lost by it) balance *f* hydraulique
[med] métabolisme *m* de l'eau
~BALLAST - [naut] lest *m* d'eau
~BAR - [constr] (a galvanized iron bar between the wood and the stone sills of a window) cassis *m*
~BASE - [gas ind] (of a counter) bâche *f*
~BASIN - [hydr] bassin *m* versant
~BATH - [ind chem] (device for heating chemical vessels without contact with flame, the heat being transmitted through water) bain-marie *m*
~BATTERY - [el] pile *f* hydroélectrique
~BEARING - [geol] aquifère, hydrofère
[mech] palier *m* glissant
~-BEARING GEOLOGICAL BED - [geol] couche *f* géologique aquifère
~-BEARING ROCK - [geol] roche *f* aquifère
~-BEARING STRATUM - [geol] niveau *m* aquifère, nappe *f* aquifère

WATER BED - [geol] couche *f* aquifère
~BLAST - [mining] trompe *f*, trompe *f* à eau
~BLOOM - [bot] (water flowers) fleur *m* d'eau, prolifération *f* d'algues
~BOILER REACTOR - [nucl] réacteur *m* à eau bouillante
~-BORNE - [gen] flottant, à flot, d'origine hydrique (maladie)
~-BORNE COAL - [transp] charbon *m* transporté par voie d'eau
~BOX - [gas ind] (of a counter) réservoir *m* de trop-plein
~BRASH - [med] pyrosis *m*, pituite *f*
~BRUSH - [impl] used for washing cars, machinery etc) brosse *f* à eau
~BUCKET - [mining] benne *f* à eau
~BUTT - [constr] tonneau *m* à eau
~CALORIMETER - [instr] (thermally-insulated metal cup containing water and furnished with a thermometer) calorimètre *m* à eau, calorimètre *m* à circulation d'eau
~CAN - [impl] broc *m* à eau, fontaine *f* d'arrosage
~-CARRIAGE SYSTEM - [constr] canalisation *f* d'évacuation d'une maison
~CARTRIDGE - [mech] cartouche *f* à chemise d'eau
~CATCHEMENT - [hydr] captage *m* d'eau, prise *f* d'eau
~-CEMENT - [constr] ciment *m* hydraulique
~-CEMENT RATIO - [constr] (in a concrete mix) rapport *m* ciment/eau
~CHAMBER - [auto] chambre *f* d'eau
~CHANNEL - [plast ind] (a continuous channel in a mould through which water can be circulated) passage *m* pour l'eau, canal *m* d'eau
~CHESTNUT - [bot] macre *f* commune, macre *f* flottante
~CIRCULATION - [gen & mech] circulation *f* d'eau
~-CIRCULATION PUMP - [mech] pompe *f* pour la circulation de l'eau
~CLARIFICATION - [hydr] clarification *f* de l'eau
~CLOSET - [hydr] cabinet *m* à chasse d'eau, cabinet *m* d'aisance
~-CLOSET BOWL - [hydr] cuve *f* de cabinet d'aisance
~-CLOSET FLUSH CISTERN - [hydr] bassin *m* de chasse
~-COCK - [hydr] robinet *m* pour conduite d'eau
~COLLAR - [plast ind] (a device surrounding a mould part and so designed that water can be circulated in a passage between the collar and the mould part) enveloppe *f* d'eau
~COLOURS - [draw] (artist's colours made by grinding appropriate pigments in aqueous solutions of gums) couleurs *f*pl à l'eau, couleurs *f*pl pour aquarelles
~COLUMN - [meas] (e.g. for measuring a pressure) colonne *f* d'eau
~-COLUMN COMPRESSOR - [mech] compresseur *m* à colonne d'eau
~-COLUMN LIGHTNING ARRESTER - [el] déchargeur *m* à écoulement
~CONCENTRATION - [min] (in ore dressing) concentration *f* humide, concentration *f* à l'eau
~CONDENSATION - [phys] (of steam) condensation *f*
~CONDITIONER - [hydr] installation *f* à conditionner l'eau
~CONDITIONING - [hydr] conditionnement *m* de l'eau
~CONING - [gas ind] (in a gas-holder) remontée *f* d'eau en pied de puits, formation *f* d'un cône d'eau

WATER CONSERVATION - [hydr] conservation *f* des eaux
~CONSUMPTION - [hydr] consommation *f* en eau
~CONTAINING LIME- [chem] eau *f* calcaire, eau *f* dure
~CONTENT - [chem] teneur *f* en eau, proportion *f* d'eau
~-COOL, to - [gen] refroidir par l'eau
~-COOLED - [mech etc] (term applied to any machine or unit from which excess heat is abstracted by circulating water round it in a jacket) refroidi par l'eau, à refroidissement d'eau
~COOLED HEAT-TRAP - [cin] (a device used for cooling a film in the projector) cuvette *f* à eau
~-COOLED MOTOR - [mech] moteur *m* à refroidissement d'eau
~COOLED SURFACE CONDENSER - [mech] condenseur *m* par surface au moyen de l'eau
~-COOLED TUBE - [electron] (electronic tube of great power, in which the heat generated is carried away by the circulation of cooling water) tube *m* à refroidissement par eau
~-COOLED VALVE - s. water-cooled tube
~COOLER - [ind] réfrigérateur *m* d'eau
~CORE - [metall] noyau *m* refroidi par l'eau
~COURSE - [geogr] course *f* d'eau, ruisseau *m*
~CRANE - [mech] grue hydraulique
[railw] grue *f* d'alimentation, bouche *f* d'eau
~CURE - [med] hydrothérapie *f*
[rubber ind] vulcanisation *f* à l'eau chaude
~CURTAIN - [gen] (used for fire fighting) arrêt-barrage *m* d'eau, rideau *m* d'eau
~-CUSHION - [rubber ind] coussin *m* à eau, matelas *m* à eau
~DEMAND - [hydr] besoins *mpl* en eau
~-DISCHARGE VENT - [hydr] robinet *m* d'écoulement d'eau
~-DIVINING - [hydr] sourcellerie *f*, radiesthésie *f*
~-DRAIN - [hydr] drain *m*
~-DRILL - [mech] perforatrice *f* à injection d'eau
~-DROPPING LIGHTNING ARRESTER - [el] déchargeur *m* à écoulement en gouttes
~ENGINEERING - [hydr] construction *f* hydraulique
~EQUIVALENT - [phys] (the product of the mass of a body and its specific heat. Synonymous with thermal capacity) équivalent *m* en eau
~EXTRACT - [chem] extrait *m* aqueux
~FEEDER - [mining] poche *f* d'eau
~-FEEDING RING - [hydr] anneau *m* d'arrosage
~FENCE - [leg] (only USA) fossé *m* limitant un domaine
~-FILLED TYRES - [transp] pneus *m* remplis d'eau
~FILTER - [hydr] filtre *m* à eau
~FINDING - [hydr] hydroscopie *f*, art *f* du sourcier
~-FINISHED - [paper man] (of paper which is supercalendered while still wet) calandré humide
~FLOW - [hydr] écoulement *m* libre
~-FLUSH DRILL - [mech] perforatrice *f* à injection d'eau
~-FLUSH DRILLING - [oil ind] sondage *m* avec curage hydraulique
~FLUSHING - [hydr] chasse *f* d'eau
~FOWL - [zool] oiseaux *mpl* aquatiques
~FRONT - [constr] faisant face à l'eau
~GAS - [chem] (a fuel gas and a starting material in synthesis. It is obtained by passing steam over incandescent coke and is composed chiefly of carbon monoxide and hydrogen) gaz *m* à l'eau, gaz *m* d'eau
WATER-GATE - [hydr] porte *f* d'écluse, robinet-vanne *m*, grille *f* d'accès
~GAUGE - [instr] tube *m* de niveau d'eau
[hydr] (to measure the level of water in a river etc) échelle *f* d'étiage
~-GAUGE PRESSURE - [meas] (pressure measured by the height of a column of water) pression *f* du robinet-jauge
~GILDING - [el metall] dorure *f* par immersion
~-GLASS - [chem] (common name for potassium silicate. Used as a catalyst and an ingredient in adhesives, as a binder and structural water-proofing agent, in textile chemistry and for preserving eggs) silicate *m* de potasse, verre *m* soluble
~-GRANULATED SLAG - [metall] laitier *m* granulé à l'eau
~GUIDE - [mech] (of a turbine) distributeur *m* (d'une turbine)
~HAMMER - [hydr] coup *m* de bélier
[phys] marteau *m* d'eau
~HAMMER ARRESTER - [constr] antibélier *m*
~HAMMERING - [hydr] coup *m* de bélier
~-HARDEN, to - [metall] tremper à l'eau
~HARDENING - [brew ind] burtonisation *f*
[metall] durcissement *m* à l'eau
~HEAD - [hydr] colonne *f* d'eau
[meas] ligne *f* de niveau de l'eau
~HEATER - [th eng] chauffe-eau *m*
~HEATER WITH FREE OUTLET - [th eng] chauffe-eau *m* à écoulement libre, chauffe-eau *m* sans pression
~HEMP AGRIMONY - [bot & med plant) eupatoire *f* chanvrine, chanvre *f* d'eau
~HOIST TANK - [mining] cage *f* à eau
~HOSE - [gen] tuyau *m* à eau
~HYDRANT - [hydr] hydrante *m*, bouche *f* d'eau
~INJECTION - [mech] (the process of injecting water (usually with methanol) into an engine to prevent detonation) injection *f* d'eau
~INFLOW - [hydr] venue *f* d'eau, coup *m* d'eau
~INLET - [hydr] orifice *m* d'introduction d'eau
~INTAKE - s. water inlet
~JACKET - [mech] (an enclosed space round any part which evolves heat, e.g. an i.c.engine cylinder, through which water is circulated to abstract such heat) chemise *f* d'eau, enveloppe *f* d'eau, chemise *f* à circulation d'eau
~JACKET PLUG - [auto] bouchon *m* de chemise d'eau
~JACKETED CORE - [metall] noyau *m* pour chambre à eau
~-JACKETED GENERATOR - s. water-jacketed producer
~-JACKETED PRODUCER - [gas ind] gazogène *m* à chemise d'eau
~JET - [gen] jet *m* d'eau
~JET AIR PUMP - s. water jet pump
~JET ASPIRATOR - s. water jet pump
~JET ELECTRIC HEATER - [th eng] (type of water heating in which water is sprayed from one electrode against another) chaudière *f* à jets d'eau
~JET HEATER - s. water jet electric heater
~-JET LIGHTNING ARRESTER - [el] déchargeur *m* à jet d'eau
~JET NOZZLE - [mech] buse *f* à jet d'eau
~JET PUMP - [hydr] pompe *f* à éjecteur d'eau, éjecteur *m* hydraulique

WATER JOINT - [gen] joint *m* d'eau
~JUMP - [gen] chute *f* d'eau
~KIBBLE - [impl] benne *f* à eau, cuffat *m* d'épuisement
~KNOCKOUT - [gas ind] (in the extraction of natural gas) séparateur *m* d'eau
~LEG - [th eng] (of a boiler) jambette *f*, culotte *f*, cuissard *m*
~LEVEL - [instr] niveau *m* d'eau
[naut] (water line) ligne *f* de flottement
[gen] niveau *m* de l'eau
[mining] galerie *f* de drainage
[geol] niveau *m* hydrostatique
~-LEVEL ALARM - [hydr] avertisseur *m* de niveau d'eau
~LEVEL GAUGE - [instr] indicateur *m* de niveau d'eau, limnimètre *m*
~-LEVEL RECORDER - [instr] limnigraphe *m*
~LEVEL REGULATOR - [mech] régulateur *m* du niveau d'eau
~LEVEL TUBE - [impl] tube *m* à niveller
~LEVEL VALVE - s. water level regulator
~LIME - [chem] chaux *f* hydraulique
~LINE - [naut] ligne *f* de flottaison
[meas] (a water level in a boiler, or tank) niveau *m*
[paper man] filigrane *f*
[shipbuild] (of a ship) ligne *f* d'eau
~LOAD - [electron] (matched waveguide termination in which the electromagnetic energy is absorbed in water) charge *f* à eau
~LOGGING - [nucl] (into a fuel element) pénétration *f* de l'eau
~LOSS - [hydr] déperdition *f* d'eau, perte *f* d'eau
~-LUBRICATED RUBBER BEARING - [mech] coussinet *m* en caoutchouc lubrifié avec de l'eau
~MAIN - [plumb] conduite *f* d'eau, conduite *f* principale d'eau, tuyau *m* principal
~MARK - [paper man] filigrane *f*
[naut] laisse *f*
~-MARKING ROLL - [paper man] cylindre *m* pour filigrane
~MELON - [bot] melon *m* d'eau, pastèque *f*
~METER - [instr] compteur *m* d'eau, compteur *m* à eau
~METER FOR RISING MAIN - [hydr] compteur *m* pour une colonne montante
~METER WITH OUTSIDE GEAR - [hydr] compteur *m* de type sec à cadran externe
~-METHANOL INJECTION - [mech] (the injection of water, with methanol as an anti-freeze, into an engine to suppress detonation) injection *f* d'eau et méthanol
~MILL- [agric] moulin *m* à eau
~MONITOR - [nucl] (a device for detecting and measuring waterborne radioactivity) moniteur *m* d'eau
~MONITORING - [nucl] contrôle *m* de l'eau
~OF CAPILLARITY - [constr] (the moisture drawn up by capillary action from the soil into the walls of a building) eau *f* capillaire
~OF COMPACTION - [soil] eau *f* de compactage
~OF CONSTITUTION - [chem] eau *f* de constitution
~OF CRYSTALLIZATION - [chem] (water chemically combined in a crystalline substance and which can be expelled at I00°C.) eau *f* de cristallisation
~OF DEHYDRATION - [hydr] eau *f* de déshydratation
~OF HYDRATION - [hydr] eau *f* d'hydratation
~OF SATURATION - [hydr] eau *f* de saturation

WATER ORGANISM - [biol] organisme *m* aquatique
~OUTLET - [mech] orifice *m* de sortie d'eau, sortie *f* de l'eau
~PACK - [mining] remblai *m* d'ensablage
~PACKER - [mining] garniture *f* étanche
~PACKING - [mining] remblayage *m* hydraulique, remblayage *m* par ensablage
~PARTING - [geogr] ligne *f* de faîte
~-PASSAGE - [mech] (a channel formed in an i. c. engine cylinder block or the like, through which cooling water passes) conduit *m* d'eau
~PILLAR - [railw] bâche *f*, grue *f* d'alimentation
~-PILLOW - s. water-cushion
~PIPE - [hydr] tuyau *m* d'eau, conduite *f* d'eau
[th eng] (of a boiler) tuyau *m* de prise d'eau
~PISTON - [mech] piston *m* à eau
~PISTON COMPRESSOR - [mech] compresseur *m* à piston liquide
~PLANE - [hydr] plan *m* d'eau
~PLANT - [bot] plante *f* aquatique
~PLUG - [hydr] hydrante *m*
~POLLUTION - [hydr] pollution *f* des eaux, contamination *f* des eaux
~POLLUTION CONTROL - [hydr] dépollution *f* de l'eau
~POLLUTION CONTROL WORKS - [hydr] ouvrages *mpl* d'épuration des eaux
~POWER - [hydr] force *f* hydraulique, houille *f* blanche
~POWER PLANT - [el] centrale *f* hydro-électrique, usine *f* hydro-électrique
~POWER STATION - s. water power plant
~POWER UTILIZATION - [hydr] utilisation *f* de l'énergie hydraulique
~-PRESSURE - [hydr] pression *f* d'eau, charge *f* d'eau
~PRESSURE INFLATION - [rubber ind] (for tyres) gonflage *m* à pression d'eau
~PRESSURE TEST - [plumb] essai *m* hydraulique à pression
~PUMP - [mech] (a device for delivering water against pressure, usually for cooling purposes) pompe *f* à eau
~PUMP PACKING - [mech] garniture *f* d'étanchéité de pompe à eau
~PUMP SHAFT - [auto] axe *m* de pompe à eau
~PUMPAGE - [hydr] élévation *f* de l'eau
~PURIFICATION - [hydr] épuration *f* d'eau
~PURITY METER - [instr] (electrical salinometer) salinomètre *m* électrique
~PUTTY - [carp] mastic *m* à l'eau
~QUENCH - [metall] durcissement *m* à l'eau
~RAISING - [agric] élévation *f* d'eau
~RAM - [impl] bélier *m* hydraulique
~RATE - [hydr] redevance *f* pour la fourniture d'eau
~RECOVERY - [mech] (the recovery of water from the exhaust gas for ballast) récupération *f* de l'eau
~REPELLENT - [gen] (capable of repelling water or moisture) impérméable
~RESISTANCE - [paint] (the property in a surface coating of opposing the passage of water in the liquid state) résistance *f* à l'eau
~RESISTING - [gen] hydrofuge
[constr] inaffouillable
~RESOURCES - [hydr] ressources *fpl* hydrauliques
~RESOURCES ENGINEERING - [hydr] génie *m* de l'hydro-économie
~RETENTION - [agric] rétention *f* de l'eau

WATER RETTED FLAX - [text] lin *m* roui à l'eau
~RETTING - [text] roui *m* à l'eau
~-RIB TILE - [build] (tile with a projecting rib to prevent entry of rain or snow) tuile *f* de bordure
~-ROLLED - [geol] roulé par les eaux
~-ROLLED PEBBLES - [geol] cailloux *mpl* roulés par les eaux
~SAPPHIRE - [min] saphir *m* d'eau
~SCOOP - [hydr] bec *m* de prise d'eau
~SCOOPING MACHINE - [hydr] installation *f* de pompage
~SCREW - [mech] vis *f* hydraulique
~-SEAL - [mech] fermeture *f* hydraulique, fermeture *f* à eau
[plumb] siphon *m* de tuyau de vidange
~-SEAL IMPELLER - [hydr] roue *f* à aubes à joint hydraulique
~-SEALED STUFFING BOX - [mech] presse-étoupe *m* à joint hydraulique
~SEPARATION - [min] triage *m* à l'eau
~SEPARATOR - [auto] séparateur *m* d'eau et de vapeur
~SERVICE - [hydr] distribution *f* d'eau, service *m* des eaux
~SHOCK - [hydr] coup *m* de bélier
~-SIDE - [gen] riverain
~SLIDE VALVE - [mech] vanne *f* à eau
~SOFTENER - [ind chem] adoucisseur *m* d'eau
~SOFTENING - [hydr] adoucissement *m* de l'eau
~-SOLUBLE - [chem] soluble dans l'eau
~SPACE - [th eng] (of a boiler) chambre *f* d'eau, réservoir *m* d'eau
~SPOUT - [constr] descente *f* d'eau, gouttière *f*
[met] s. waterspout
~SPRAY - [gen] jet *m* d'eau
[hydr] (for firefighting) rideau *m* d'eau
~SPRAYER - [hydr] pulvérisateur *m* d'eau, vaporisateur *m* d'eau
~SPRAYING - [agric] pulvérisation *f* de l'eau
~STAIN - [paper man] goutte *f*
[photo] tache *f* d'humidité
~STORAGE - [agric] emmagasinage *m* d'eau
~SUCTION PUMP - [mech] pompe *f* d'aspiration de l'eau
~SUIT - [astronaut] vêtement *m* à compression hydraulique
~SUPPLY - [hydr] approvisionnement *m* d'eau, réserve *f* en eau, alimentation *f* en eau
~-SUPPLY SYSTEM - s. water service
~SURFACE - [hydr] surface *f* de l'eau
~SURFACE TENSION - [phys] tension *f* superficielle de l'eau
~SWIVEL - [mining] touret *m* hydraulique
~SYSTEM - [hydr] (system of distribution) canalisation *f* d'eau, réseau *m* de canalisation
~TABLE - [geol] niveau *m* hydrostatique, niveau *m* de la nappe phréatique
~-TABLE FLUCTUATION - [geol] fluctuation *f* de la nappe aquifère
~TABLE PROFILE - [geol] profil *m* de la nappe d'eau
~TANK - [gen] réservoir *m* à eau
[railw] caisse *f* à eau, soute *f* à eau
~TAP - [hydr] robinet *m* hydraulique, robinet *m* pour eau
~TEMPER, to - [metall] tremper à l'eau
~TEMPERING - [metall] trempe *f* à l'eau
~-THISTLE - [bot/veed) (cirsium oleraceum) cirse *m* maraîcher

WATER THREAD - [hydr] filet *m* d'eau
~-TIGHT - [gen] étanche à l'eau, imperméable à l'eau
~-TIGHT BULKHEAD - [shipbuild] cloison *f* étanche
~-TIGHT CORE - [hydr] (of an impounding dam) noyau *m* d'étanchéité
~-TIGHT DIAPHRAGM - [hydr] rideau *m* d'injection, voile *m* d'étanchéité
~-TIGHT LINING - [mech etc] revêtement *m* étanche
~-TIGHT SCREEN - s. water-tight diaphragm
~TOOL-GRINDER - [mach tool] machine *f* à affûter les outil
~TOUGHENING - [metall] augmentation *f* de la ténacité par immersion dans l'eau
~TOWER - [hydr] château *m* d'eau, réservoir *m* sur tour
[railw] tour *f* réservoir
~TRAP - [plumb] siphon *m* de décantation
~TREATING PLANT - [ind chem] installation *f* pour le traitement de l'eau
~TREATMENT - [hydr] traitement *m* de l'eau
~TROUGH - [impl] (a blacksmith's implement) baquet *m*, auge *f* de forgeron
[mech] rigole *f* à eau
~TUBE - [gen] tube *m* d'eau, tuyau *m* d'eau
~-TUBE BOILER - [th eng] (type of boiler in which the water is circulated through tubes round which the products of combustion pass) chaudière *f* aquatubulaire, chaudière *f* à tubes d'eau
~TUNNEL - [aero] (a device similar to a wind tunnel, in which the working fluid is water instead of air) soufflerie *f* hydrodynamique
~TURBINE - [mech] turbine *f* à eau, turbine *f* hydraulique
~TWIST - [text] fil *m* de continu à anneau
~UTILIZATION - [hydr] hydrologie *f*
~VAPORIZATION - [phys] évaporation *f* de l'eau
~VAPOUR - [met] humidité *f* atmosphérique
[phys] vapeur *f* d'eau
~VAPOUR ABSORPTION - [phys] (the increase of weight of a substance exposed to a saturated atmosphere and due to the absorption of water vapour) absorption *f* par la vapeur d'eau
~-VASCULAR SYSTEM - [zool] système *m* hydrovasculaire
~VEIN - [hydr] veine *f* d'eau, veine *f* liquide
~WAVE - [acoust] (an optical effect on the surface of a record, caused by periodic alteration of the recording stylus with reference to the surface of the record, which varies the depth of cut) onde *f* dans l'eau
~WAVING - [text] moirage *m*, moirure *f*
~WELL - [hydr] puits *m* à eau
~WHEEL - [hydr] roue *f* hydraulique, roue *f* à aubes
[mech] (a turbine) turbine *f* hydraulique
[mech] (a noria) noria *f* à roue
~WITCHING - [geol] (the use of a divining rod to discover water) sourcellerie *f*
~WORKS - [ind] usine *f* de traitement des eaux, ouvrage *m* de purification de l'eau
WATERBAG - [rubber ind] (a curing bag) waterbag *m*, sac *m* à cuisson
WATERBRASH - [med] (heartburn) aigreurs *fpl* d'estomac
WATERCOURSE - [geogr] cours *m* d'eau
[hydr] canal *m*, conduite *f* d'eau
[hydr] (a drain) drain *m*
~LEVEL - [mining] galerie *f* de drainage

WATERCRAFT - [naut] embarcation *f*
WATERCRESS - [bot] cresson *m* de fontaine
WATERFALL - [geogr] chute *f* d'eau, saut *m*
~EFFECT - [phys] effet *m* de cascade, effet *m* Lénard
WATERFINDER - [gen] hydroscope *m*, tourneur *m* de baguette
~WAND - [impl] baguette *f* de sourcier
WATERFOG - [ind chem] (used in fire-fighting) nébulisation *f* de l'eau
WATERGLASS - s. water glass
WATERING - [agric] arrosage *m*, irrigation *f* des champs
[mech] alimentation *f* en eau
[ind chem] dilution *f*
[agric] (of animals) abreuvage *m*
~CAN - [agric] arrosoir *m*, chantepleure *f*
~CART - [impl] tonneau *m* d'arrosage
~POT - s. watering can
~ROLLER - [mech] cylindre *m* d'arrosage
~TROUGH - [gen] abreuvoir *m*
WATERISH - [gen] aqueux, humide
WATERLEAF - [paper man] (said of paper which is not sized) papier *m* métallisé
WATERLESS - [gen] sans eau
~GASHOLDER - [gas ind] (a piston-type gas-holder) gazomètre *m* sec, gazomètre *m* à piston
WATERLINE - s. water line
WATERLOGGED - [naut] (made heavy and unmanageable owing to the leakage of water into a hold) plein d'eau
[gen] (saturated with water) alourdi par absorption d'eau, envahi par les eaux
[soil] aqueux
WATERMARK - [paper man] (design formed in paper during passage over the wire screen of the machine, by means of a fine-wire patterns formed in the mesh of the screen) filigrane *f*
~POST - [hydr] échelle *f* fluviale, échelle *f* d'étiage
WATERPROOF, to - [gen] imperméabiliser
[text] cirer, caoutchouter
[hydr] hydrofuger
WATERPROOF - [gen] imperméable
[text] imperméable, imbrifuge, caoutchouté
[ind] étanche à l'humidité
[el] (said of an electrical machine or motor, when protected against infiltration of water) étanche à l'eau
~CEMENT - [constr] ciment *m* hydrofuge
~CLOTH - [text] tissu *m* imperméable
~COCOON - [text] cocon *m* étanche à l'eau
~FILAMENT - [text] fil *m* imperméable à l'eau
~GREASE - [ind chem] graisse *f* hydrofuge
~PAPER - [paper man] papier *m* hydrofuge
~PLASTERING - [constr] enduit *m* hydrofuge
~SERVING - [el] matelas *m* imperméable
WATERPROOFING - [text etc] (the process of rendering surfaces or materials impervious to water) imperméabilisation *f*, hydrofugeage
[gen] (the actual material) matériel *m* hydrofuge
WATERSHED - [geogr] (a drainage area) ligne *f* de partage des eaux
WATERSPOUT - [met] (a tornado occurring over a water surface, usually having a core of water precipitated by the fall in pressure at the centre) trombe *f* marine
[hydr] descente *f* d'eau
WATERSPROUT - [bot] gourmand *m*
WATERTIGHT - [gen mech etc] étanche à l'eau
~BULKHEAD - [shipbuild] cloison *m* étanche
~GLAND - [mech] presse-étoupe *m* étanche
~MOTOR - [el] moteur *m* étanche à l'eau
WATERTUBE - [th eng] à tube d'eau
WATERWALL - [hydr] digue *f*
[th eng] (of boilers) revêtement *m*, chemise *f*
~HEADER - [th eng] (in boilers) collecteur *m*
WATERWAY - [geogr] voie *f* navigable, cours *m* d'eau navigable
[hydr] cours *m* d'eau
WATERWORKS - s. water works
WATERWORN - [gen] usé par l'eau
WATT - [el] (the unit of electric power) watt *m*
~-HOUR - [el] (the unit of the electrical energy, i.e. the work done by one watt acting for an hour) watt-heure *m*
~-HOUR CAPACITY - [el] (the quantity of energy in watt-hours which can be delivered under given conditions of temperature, rate of the discharge and final voltage) capacité *f* en énergie
~-HOUR CONSTANT OF A METER - [el] constante *f* de watt-heure d'un compteur
~-HOUR EFFICIENCY - [el] (the ratio between average voltage during discharge and under given conditions of temperature, rate of discharge and final voltage) rendement *m* en énergie
~-HOUR METER - [instr] (energy meter) wattheuremètre *m*, compteur *m* d'énergie active
~-SECOND CONSTANT OF A METER - [el] constante *f* de watt-seconde d'un compteur
WATTAGE RATING - [el] (rated watt consumption) consommation *f* nominale, capacité *f* de charge
WATTLESS - [el] (in quadrature with the voltage or the current) déwatté
~COMPONENT - [el] (reactive voltage) composante *f* réactive du courant
[el] (reactive current) composante *f* active de la tension
WATTMETER - [el] wattmètre *m*
WAVE - [gen] (of the sea etc) vague *f*, lame *f*
[phys el] (modification of the physical state of a medium which is propagated as a result of a local disturbance) onde *f*
[glass man] (optical effect due by unevenness of glass distribution) ondulation *f*
~AERIAL - [radio] (the Beveridge aerial) antenne *f* de Beveridge
~AMPLITUDE - [phys] (the magnitude of the maximum change from equilibrium of the disturbance characterizing the wave) amplitude *f* d'onde
~ANALYZER - [radio] (an instrument designed to resolve a given wave-shape into its fundamental and harmonic components) analyseur *m* d'ondes
~ANTENNA - s. wave aerial
~-BUILT TERRACE - [geol] talus *m* littoral
~-CHANGE SWITCH - s. wave changer
~CHANGER - [radio] (wave-change switch, a switch used to produce a change in the range of frequencies over which a circuit can be tuned) commutateur *m* d'ondes
~CLUTTER - [radar] (picture on a screen caused by reflection on sea waves) réflexion *f* de la mer
~CONVERTER - [radio] (of a waveguide) organe *m* de raccord
~CREST - [gen] crête *f* d'une vague
[radio] (the position of maximum positive distur-

bance in a progressive wave) crête *f* d'onde
WAVE CYCLONE – [met] (extra-tropical cyclones developing along frontal surface, most commonly on polar fronts) cyclone *m* à ondulations
~DIFFRACTION – [phys] (the change in direction of a wave in its incidence at a given angle into a medium) diffraction *f* d'ondes
~DISTORTION – [phys] (a wave is distorted when it is not sine-shaped) distorsion *f* d'onde
~DRAG – [aero] (drag associated with shock waves) traînée *f* due aux ondes de choc
~DUCT – [radio] (waveguide with tubular boundaries) guide *m* d'ondes tubulaire
~EQUATION – [radio] (the partial differential equation of wave motion) équation *f* d'onde
~FILTER – [radio] (transducer for separating waves on the basis of the frequency) filtre *m* d'ondes
~-FORM – [phys] (the shape of a graph representing the instantaneous values of a periodically varying quantity plotted against time) forme *f* d'onde
~-FORM COMPONENTS – [telev] composantes *fpl* du signal de télévision
~-FORM CONVERTER – [radio] convertisseur *m* de la forme d'onde
~-FORM CORRECTOR – [radio] (for the elimination of a wave-form distortion) correcteur *m* de forme d'onde
~-FORM MONITOR – [electron] (a cathode-ray oscilloscope having a timebase suitable for viewing the waveforms of the video signal in a TV system) moniteur *m* de forme d'onde
~-FORM RESPONSE – [radio] réponse *f* de forme d'onde
~FRONT – [phys] (the part of the wave observed from the side towards which the wave is travelling) front *m* d'onde
~FUNCTION – [radio] (the solution of a partial or differential equation for wave propagation through a medium) fonction *f* d'onde
~GUIDE – s. waveguide
~IMPEDANCE – [el] (in a waveguide) impédance *f* d'onde
~INTENSITY – [radio] (the average rate of flow of energy in the direction of propagation per unit area of the wave front) intensité *f* d'onde
~INTERFERENCE – [radio] interférence *f* d'ondes
~LENGTH – s. wavelength
~LINE – [text] ligne *f* ondulée
~MECHANICS – [phys] (general theory which ascribes wave characteristics to the fundamental entities of atomic structures) mécanique *f* ondulatoire
~METER – s. wavemeter
~MOTION – [phys] ondulation *f*
~NORMAL – [radio] (a unit vector normal to an equiphase surface with its positive direction taken in the same side of the surface as the direction of propagation) normale *f* à la surface d'onde
~NUMBER – [radio] (the reciprocal of the wavelength in a harmonic wave) chiffre *m* d'onde
~PARAMETER – [radio] paramètre *m* d'onde
~PHASE – [radio] (the argument in the wave function) phase *f* d'onde
~QUENCHING – [radio] étouffement *m*
~RANGE SWITCH – [electron] commutateur *m* de gammes d'ondes
~RESISTANCE – [naut] (of a marine craft) résistance *f* de vague
WAVE SERIES WINDING – s. wave winding
~-SHAPE – [el] (the voltage time or current-time characteristic of a surge) forme *f* d'onde
~-SHAPING SET – [radio] (an electric network inserted in a telegraph circuit for improving the wave shape of the received signal) réseau *m* correcteur de forme
~SPACING – [telecomm] (signal spacing) espace *m* du signal
~SURFACE – s. wave front
~THEORY OF LIGHT)– [opt] (light is shown to have the properties of waves by the phenomena of interference and diffraction) théorie *f* ondulatoire de la lumière
~TILT – [radio] (the forward inclination of a radiowave caused by proximity to the ground) inclinaison *f* d'onde
~-TRAIN – [phys] (or train of waves, a group of successive waves) train *m* d'ondes
~TRAP – [radio] circuit *m* bouchon, piège *m* à ondes [instr] ondemètre *m* d'absorption
~TROUGH – [radio] (a point of minimum value of the disturbance in a progressive wave) creux *m* d'onde, vallée *f* d'onde
~VELOCITY – [radio] (the common value of the phase velocity and group velocity in a non-dispersive medium) vitesse *f* de propagation
~WINDING – [el] (winding of a drum armature in which the back and front pitches are in the same direction so that the winding pitch is the sum of the two) enroulement *m* ondulé
WAVED WHEEL – [mech] came *f* à montagne russe
~YARN – [text] fil *m* ondé
WAVEFORM – s. wave-form
WAVEGUIDE – [radio] (a system of material boundaries capable of guiding waves) guide *m* d'ondes
~ADAPTER – [radio] (a unit for adapting a coaxial cable to a waveguide) adapteur *m* de guide d'ondes
~ATTENUATOR – [radio] (attenuator designed for application to the microwave range of frequencies) affaiblisseur *m* de guide d'ondes
~AXIS – [radio] (the line of centres of a waveguide) ligne *f* des centres d'un guide d'ondes, ligne *f* médiane d'un guide d'ondes
~BEND – [radio] coude *m* progressif d'un guide des ondes
~COMPONENT – [radio] (a device designed to be connected at specified points in a waveguide system) pièce *f* détachée de guide d'ondes
~FILTER – [radio] filtre *m* de guide d'ondes
~FIN – [radio] ailette *f* de guide d'ondes
~HIGH-POWER LOAD – [radio] (attenuator used as a power-dissipating terminating section for a waveguide) affaiblisseur *m* au sable
~PAD – [radio] affaiblisseur *m* fixe
~PHASE-SHIFTER – [radio] déphaseur *m* de guide des ondes
~POST – [radio] saillie *f* de guide d'ondes
~RADIATOR – [radio] (open-ended waveguide used to radiate electromagnetic energy to a reflector or into space) guide *m* d'ondes rayonnant
~RESONATOR – [radio] (waveguide device for storing oscillating electromagnetic energy) section *f* de guide d'ondes résonnante
~SEAL – [radio] clôture *f* de guide d'ondes
~SHIM – [radio] joint *m* de contact de guide d'ondes
~SHUTTER – [radio] obturateur *m* de guide d'ondes

WAVEGUIDE SWITCH - [radio] commutateur m de guide d'ondes
~SYSTEM - [radio] (a system employing waveguides as the means of transmission, reception, or conveyance of electromagnetic energy) système m à guides d'ondes
~-TO-STRIPLINE COUPLER - [radio] (a small coupler capable of operating over a full waveguide bandwidth) coupleur m de guide d'ondes-ligne microbande
~TRANSFORMER - [radio] transformateur m de guide d'ondes
~TUNER - [radio] transformateur m ajustable de guide d'ondes
~WINDOW - [radio] (a thin conducting metal window placed transversely in a waveguide for impedance matching) fenêtre f de guide d'ondes
WAVELENGTH - [phys] (the distance between two successive points of a periodic wave in the direction of propagation, in which the oscillation has the same phase) longueur f d'onde
~METER - [instr] (instrument designed for measuring radio frequencies) ondemètre m
~OF LIGHT - [phys] longueur f d'onde de la lumière
~SHIFTER - [electron] (a photofluorescent compound used with a scintillator material to increase the wavelength of the optical photons, thus permitting a more efficient use of the photons by the photocell) agrandisseur m de longueur d'onde
WAVELLITE - [min] (native hydrated aluminium phosphate and a source of phosphorus) wavellite f
WAVEMETER - [instr] (instrument designed to measure the wavelength of electromagnetic waves) ondemètre m, cymomètre m
WAVER - [print] (a distributing roller) rouleau m distributeur
WAVETAIL - [phys] queue f d'onde
WAVETRAP - [el] (an acceptor circuit to short circuit signals or interference of a certain frequency) circuit m bouchon
WAVINESS - [metall] surface f ondulée
[text] ondulation f
WAVY EDGES - [metall] bords mpl ondulés
~FIBRE - [text] fibre f ondulée
WAX, to - [gen] cirer, enduire de cire
[el acoust] (to record) enregistrer
WAX - [chem] (generic term for a mixture of esters of monohydric alcohols and fatty acids, insoluble in water but soluble in most organic solvents) cire f
[el acoust] (in mechanical recording, a blend of waxes with metallic soaps) cire f
~BEAN - [bot] haricot m à cosse jaune
~-BERRY - [bot] myrte m bâtard
~BLOCK PHOTOMETER - [instr] (the Joly block screen) photomètre m de Joly
~BUDDING TAPE - [agric] ruban m ciré pour greffe
~CANDLE - [gen] bougie f de cire
~CLOTH - [text] (oil cloth) toile f cirée
~DISTILLATE - [oil ind] distillat m paraffineux
~END - [shoe man] chégros m, fil m poissé
~FINISH - [carp] cirage m
~MASTER - [acoust] (original recording on a wax surface for the purpose of preparing a master) enregistrement m en cire
~MOULD - [gen] empreinte f en cire, moule m en cire
WAX OPAL - [min] résinite f
~ORIGINAL - s. wax master
~PALM - [bot] arbre m à cire, céroxyle m
~PAPER - [paper ind] papier m ciré
~PENCIL - [photo] crayon m de cire
~POCKETS - [zool] glandes fpl cerifères
~POLISH - [photo] encaustique f
~RECORDING - [el acoust] (recording in wax, an obsolete technique for making a matrix) enregistrement m en cire
~SHEET - [print] papier m stencil
WAXED - [gen] ciré, enduit de cire
~HEALD - [text] lisse f cirée
~LEATHER - [leather ind] cuir m ciré
~THREAD - [text] fil m poissé
WAXING - [paper man] (operation of coating paper with wax for waterproofing, high-gloss or to give heat-sealing properties) cirage m
[gen] cirage m, encausticage m, empoissage m
WAY - [gen] voie f, chemin m, route f
[gen] (direction) direction f
[gen] (distance) distance f, parcours m
[gen] (a mode, a line or conduct) manière f, façon f
[gen] (condition) état m, condition f
[mech] (a track on which a part travels) guide m, glissière f
~AND STRUCTURES - [railw] (USA only, permanent way and fixed installations) voies fpl et installations fixes
~BILL - [comm] (a list of the goods, or passengers, carried by a common carrier, e. g. a train, a ship etc) lettre f de voiture, bordereau m d'expédition
~OUT - [gen] sortie f
~SHAFT - [mining] puits m intérieur
~STATION - [railw] (USA only, a railway station between main stations) gare f secondaire
~TRAIN - [railw] (USA only, a local train, a train which stops at way stations) train m omnibus
WAYBILL - s. way bill
WAYS - [shipbuild] (the structure on which the ship is built) couettes fpl
[shipbuild] (the structure which support the ship when it is launched) longrines fpl de lancement
WAYSIDE - [gen] bord m de la route
~ELEMENT - [railw] (the track element) élément m de rail activateur
WEAK - [gen] faible
[gen] (of a beverage) aqueux, dilué
[mech] (of a mixture) pauvre
~ACID - [chem] acide m faible
~BLOW - [text] chasse f douce
~COUPLING - [el] (of a circuit) couplage m faible
~CURRENT CABLE - [el] câble m à courant faible
~ELECTROLYTE - [el chem] électrolyte m faible
~FIBRE - [text] fibre f faible
~IGNITION - [mech] allumage m faible
~LENS - [opt] (a lens in which the focal lengths are long in comparison with the extent of the lens field) lentille f faible
~MIXTURE - [mech] (i. c. engines) (mixture containing a low proportion of fuel to air) mélange m pauvre
~-MIXTURE KNOCK RATING - [mech] (a number giving an anti-value for a given fuel under the best economic cruising conditions) indice m d'octane en mélange pauvre

WEAK ROASTING - [min] faible grillage *m*
~RUBBER - [rubber ind] (or "short rubber", a rubber which tears or breaks easily) weak-rubber *m*, caoutchouc *m* cassant
~SAND - [metall] (in foundry work) sable *m* maigre
~TENSION - [text] tension *f* souple, tension *f* élastique
~TIE - [aero] (of a parachute, a special cord or thread designed to yield under certain conditions and allow some special form of deployement) fil *m* à casser
WEAKEN , to [gen] affaiblir
[chem] (of a solution) appauvrir, diluer
WEAKENING RATIO - [el] (field weakening ratio) taux *m* de shuntage
WEALTH - [gen] richesse *f*, abondance *f*
WEAN, to - [zool] sevrer
WEANING - [zool] sevrage *m*
~BRASH - [med] diarrhée *f* de sevrage
WEANLING - [zool] nourisson *m* en sevrage
WEAPON - [gen] arme *f*
WEAR, to - [gen] porter
[mech etc] (to waste by use) user
WEAR - [gen] usage *m*, usure *f*, détérioration *f*
[text] (clothes) vêtements *mpl*
~AND TEAR - [gen mech phys etc] usure *f* normale
~AND TEAR TEST - [text] essai *m* d'usure
~-IN FAILURES - [comput] défaillances *f* prématurées
~MACHINE - [mech] (used in the textile industry to measure the resistance to wear) machine *f* à mesurer l'usure
~OF THE CYLINDER - [mech] (in internal combustion engines) ovalisation *f* du cylindre
~-OUT FAILURES - [comput] défaillances *f* par usure
~PLATE - [metall] plaque *f* d'usure, plaque *f* de friction
~RESISTANCE - [gen & mech] résistance *f* à l'usure
~-RESISTING - [mech] resistant à l'usure
~RING - [mech] (of the cylinder lining) gradin *m* d'usure
~TEST - [mech etc] essai *m* de résistance à l'usure
WEARING APPAREL - [text] (clothing) vêtements *mpl*
~ACTION - [geol] action *f* rongeante
~COAT - [soil] couche *f* de roulement, couche *f* d'usure
~COURSE - [constr] couche *f* superficielle, surface *f* frottante
~DEPTH - [mech] limite *f* d'usure admissible
~STRIPS - [el] (of a contact) bandes *fpl* de frottement, bandes *fpl* d'usure
~SURFACE - [constr] s. wearing course
[el] (sliding surface) surface *f* de frottement
[metall] couche *f* d'usure
WEATHER, to - [gen] (to expose to the action of the weather) altérer, se désagréger, exposer à l'intempérisme
[met] (to encounter successfully, e.g. a storm) étaler (une tempête)
[naut] doubler, tourner
[phys] (to undergo some alterations owing to weathering) s'altérer
[constr] tailler, en rejéteau
WEATHER - [gen & met] (the general atmospheric conditions) temps *m*, météorologique
[naut] du côté du vent
[mech] (of a windmill) airage *m*
~AREAS - [naut] surfaces *fpl* exposées aux vents

WEATHER BALLOON - [met] (balloons used for recording and transmitting atmospheric data) ballon-sonde *m*
~BAND - [hydr] gouttière *f*
~-BOUND - [gen & naut] (heald up by unfavourable weather) arrêté par le mauvais temps
~BOX - [instr] capucin *m* hydrométrique
~CHART - [met] (a chart in which are marked synchronous observations of atmospheric pressure, temperature, direction of winds etc) carte *f* météorologique
~CONDITIONS - [met] conditions *fpl* météorologiques
~DAMAGE - [gen] dommages *mpl* causés par le mauvais temps
~DECK - [naut] partie *f* du pont non recouverte par des roufs
~DOOR - [metall] porte *f* de ventilation
~FORECAST - [met] bulletin *m* météorologique
~GAUGE - [naut] avantage *m* du vent
~HELM - [naut] barre *f* au vent
~MAP - s. weather chart
~MOULDING - [constr] larmier *m*
~PARTING - [met] (meteorological limit) limite *f* météorologique
~-PROOF - [gen] à l'épreuve des intempéries
[el & telecomm] protégé contre les projections d'eau latérales
~REPORT - [met] (a statement of weather conditions at a given time in a specified erea) bulletin *m* météorologique
~SIDE - [naut] bord *m* du vent
[gen] côté *m* exposé au vent
~SLATING - s. weather tiling
~STATION - [met] station *f* météorologie
~STRIP - [gen] (for doors, windows etc) bourrelet *m* étanche, coupe-froid *m*
[auto] gouttière *f* d'étanchéité
~TILING - [constr] (tiles which eare hung vertically to the side of walls, to protect them against wet) tuilès *fpl* à recouvrement
~-VANE - [met] (a pivoted pointer so shaped as to indicate the direction of the wind) girouette *f*
WEATHERBOARD, to - [constr] garnir de planches à recouvrement
WEATHERBOARD - [constr] (board for the outside covering of wooden buildings) planche *f* à recouvrement
[naut] (the windward side) côté *m* du vent
WEATHERCOCK, to - [aero] (to turn into the wind like a weather-vane) virer face au vent
WEATHERCOCK - [met] (a vane, generally in the shape of a cock, which turns to show the direction of the wind) girouette *f*, flouette *f*
~STABILITY - [aero] (tendency to turn into the wind (either in pitch or yaw) stabilité *f* en girouette
WEATHERED - [geol] altéré
~CRUDE - [oil ind] pétrole *m* brut vieilli
~DISTILLATE - [oil ind] distillat *m* stabilisé
~OUTCROP - [geol] affleurement *m* altéré par l'action des agents atmosphériques
WEATHERGLASS - [instr] (a common barometer) baromètre *m* à cadran
WEATHERING - [phys] (the effect of exposure to the atmosphere and to meteorological conditions) intempérisme *m*, action *f* des agents atmosphériques
[mech] (weathering test) effritement *m*
[oil ind] (the loss of volatile fractions from a pe-

troleum fraction through slow evaporation during storage) vieillissement *m*
WEATHERING - [constr] (of a window) jet *m* d'eau
~TEST - [gen] (the exposure of a specimen to natural or artificial weathering to determine its resistance to this) essai *m* de vieillissement
~ZONE - [geol] zone *f* d'altération
WEATHEROMETER - [instr] (USA only, instrument designed to measure the life of film coatings by subjecting them to ultra violet rays, water etc) indicateur *m* de désagrégation
WEATHERPROOF CLOTHING - [text] habits *mpl* à l'épreuve des intempéries
WEATHERTIGHT - [gen] protégé contre la pluie
WEAVE, to - [text] tisser
WEAVE - [text] (any method or style of weaving) tissage *f*
[text] (a woven fabric) tissu *m*, étoffe *f*
[text] (a weaving pattern) armure *f*
[telev] manque *m* de fixité latérale
~BY HAND, to - [text] tisser à bras
~FOR BACK TEXTURE - [text] armure *f* du fond
WEAVER - [gen] tisseur *m*, tisserand *m*
WEAVER'S BEAM - [text] (for the warp) ensouple *f*
~COMB - [text] peigne *m* de tisserand
~NIPPERS - [text] pincettes *fpl* de tisserand
~SEAT - [text] banc *m* du tisseur
~SHUTTLE - [text] navette *f* de tissage
WEAVING - [text] tissage *m*
[gen] (of a road) serpentement *m*
~FIGURES BY EXTRA FLOATING WEFT - [text] brochage *m* au lancé
~MILL - [text] atelier *m* de tissage, usine *f* de tissage
~SHED - [text] salle *f* de tissage
~SHUTTLE - [text] navette *f* de tissage
~SPOOL - [text] bobine-trame *f*
~WIDTH - [text] empeignage *m* utile
~WASTE - [text] déchets *mpl* de tissage
~YARN - [text] filé *m* pour tissage
WEB - [text] (textile fabric being woven in the loom) tissu *m*
[text] (the textile sheet coming from the card) voile *m*
[paper man] (long roll of material formed like a web of cloth) rouleau *m* de papier
[print] cordon *m* de papier à imprimer
[paper man] (sheet of paper in full width) bande *f* (de papier)
[zool] (a cobweb) toile *f* d'araignée
[mech] (the disk which connects the rim and hub of certain wheels) disque *m*
[mech] (of a crankshaft) bras *m* de manivelle, flasque *f*, joue *f* de manivelle
[railw] (of a rail) âme *f* (d'un rail)
[mech] (the bit of a key) panneton *m* (d'un clef)
[naut] (of an oar, between blade and rowlock) bras *m*
[zool] (connective membrane) palmure *f*, membrane *f*
[mech & arch] nervure *f*
[plast ind] (endless sheeting) feuilles *fpl* en continu
[opt] réticule *m*
[mech] (of an anvil) estomac *m*, corps *m*
[constr] âme *f* d'une poutre
~BEAM - [text] ensouple *f* enrouleuse
~CALENDERED - [paper man] calandré en bande
WEB CONDUCTOR - [text] plaque-guide *f* du voile
~CORD - [rubber ind] (for tyres) nappe *f* corde sans trame
~-CORD PROCESS - [rubber ind] imprégnation *f* de nappes au trempé
~DIVIDER - [text] appareil *m* diviseur de voile
~DOFFING - [text] sortie *f* du voile
~-FOOT - [zool] (a foot with webbed toes) pied *m* palmé
~-FOOTED - [zool] (with the toes connected by a membrane) palmipède
~FRAME - [shipbuild] porque *f*
~GLAZING - s. web calendered
~GUIDE PLATE - s. web conductor
~MACHINE - [instr] (a web machine, in which the paper is fed from a continuous roll) presse *f* à bobines
~MEMBER - [constr] barre *f* de treillis
~PLATE - [constr] âme *f* d'une poutre
~PRINTING - [print] imprimerie *f* à bobines
~SAW - [tool] scie *f* à châssis
~STRIP - [text] ruban *m* de voile
~TRUMPET - [text] entonnoir *m* du voile
~WHEEL - [mech] roue *f* à disque
~WINDER - [text] enrouleuse *f* de voile
WEBBED BASE - [carp] (in upholstery) support *m* en sangles entrelacées
~EYEPIECE - [opt] oculaire *m* à réticule
WEBCORD PROCESS - [rubber ind] (for tyres) imprégnation *f* de nappes au trempé
WEBBING - [text] tissage *m*
[plast ind] (failure of sheet material to enter all the depressions in the mould) pont *m* de formage
[gen] sangles *fpl*, ruban *m* à sangles
[paint] (surface coating defect similar to gas checking, viz. the appearance of a network of fine cracks, often caused by the presence of town gas in the air) craquelure *f*
~SLING - [text] brassière *f*
WEBER - [phys] (the practical unit of magnetic flux) weber *m*
WEBSTERITE - [geol] (coarse-grained igneous rock consisting principally of ortho-and clino-pyroxenes) aluminite *f*, webstérite *f*
WEDGE, to - [gen] (to fixe in like a wedge) coincer, caler, enclaver
[gen] (to cleave apart with a wedge) fendre avec un coin
[mining] picoter, colleter
WEDGE - [gen] coin *m*
[mech] clavette *f*, coin *m*
[mech] (wedge-shaped duct) conduit *m* en coin
[met] (high-pressure region having its isobars shaped like a wedge) crête *f* de haute pression
[telev] (a beam of black diverging lines drawn in an angle to value the definition in TV) angle *m* de définition
[radio] (waveguide termination) coin *m* d'absorption
[mining] (a quarry wedge) quille *f*
[mining] (in timbering) picot *m*
[constr] ébuard *m*, clé *f*
~BAR - [mech] (for material testing) baguette *f* d'essai en coin
~BELT - [mech] sangle *f*
~BRICK - [build] brique *f* en coin
~CONTACT - [el] (contact consisting of two fingers between which a wedge-shaped contact on the moving

element is forced) contact *m* à coin
WEDGE-DRIVING - [mining] picotage *m*, colletage *m*
~-DRIVING MACHINE - [mining] machine *f* à picots
~FRICTION WHEEL - s. wedge wheel
~GATE - [metall] attaque *f* à languette
~GRIP GEAR - [el] (used for lifts) dispositif *m* d'arrêt à coin
~KEY - [mech] clef *f* de serrage
~LOCK - [mech] serrage *m* à coin
~PHOTOMETER - [instr] photomètre *m* à coin
PIECE - [metall] voussoir *m*
~PRESS - [mech] presse *f* à coin
~RANGEFINDER - [photo] télémètre *m* à prismes pivotants
~ROASTER - [metall] four *m* cylindrique multiple de grillage
~SECTION - [metall] profil *m* cunéiforme
~SECTION WIRE - [mech] fer *m* à biseau
~-SHAPED - [gen] en forme de coin, en coin
~SPANNER - [impl] clef *f* à clavette
~SPECTROGRAPH - [instr] spectrographe *m* à coin
~TRANSISTOR - [electron] (a transistor with a wedge shaped crystal) transistor *m* à coin
~-TYPE SAFETY GEAR - [el] parachute *f* à coin
~-TYPE VALVE - [mech] (a double-faced valve) vanne *f* à coin
~WHEEL - [mech] roue *f* de friction à coin
~-WIRE DECK - s. wedge-wire screen
~-WIRE SCREEN - [impl] (screen consisting of wires of bars drawn or roller to a wedge-shaped section) surface *f* criblante à fils métalliques
WEDGED ASSEMBLING - [constr] assemblage *m* à clavettes
WEDGING - [gen mech etc] (the operation of forcing in with or as with, a wedge) coinçage *m*, coincement *m*
[mining] picotage *m*, colletage *m*
[mech] clavage *m*, clavetage *m*
WEED, to - [agric] (to remove weeds) sarcler, désherber
WEED - [bot] mauvaise herbe *f*, plantes *f*pl adventices
~CUTTER - [impl] faucardeur *m*
~HARROW - [agric] scarificateur *m*
~INFESTATION - [bot] (plant disease) envahissement *m* par les mauvaises herbes
~KILLER - [agric] désherbant *m*, herbicide *m*
WEEDER - [agric] (an implement used to remove weeds) sarcleuse *f*, désherbeur *m*
WEEDING - [agric] sarclage *m*, désherbage *m*
~MACHINE - [agric] sarcleuse *f*
WEEDKILLERS - [chem] (materials, e.g. certain petroleum fractions and chemical compounds, which can be used to destroy the growth of plants) herbicides *m*pl
WEEDLESS - [naut] (term denoting a propeller whose blades are curved backwards in relation to the direction of rotation) anti-algue
~PROPELLER - [naut] hélice *f* antialgue
WEEDY - [bot] couvert de mauvaises herbes
WEEK - [gen] semaine *f*
WEEP HOLE - [oil ind] trou *m* de ressuage
[mining] chantepleure *f*
WEEPHOLES IN A RETAINING WALL - [constr] barbacanes *f*pl d'évacuation des eaux dans un mur de soutènement
WEEPING WILLOW - [bot] saule *m* pleureur
WEEPING WOUND - [med] plaie *f* baveuse
WEEVIL - [zool] charançon *m*
WEEVILLED - [agric] charançonné
WEFT - [text] (the cross-threads in a web of cloth) trame *f*
[gen] traînée *f* (de brume etc)
~-BACKED FABRIC - [text] tissu *m* double-face en trame
~BOBBIN - [text] bobine-trame *f*
~CARRIER - [text] chargeur *m*
~COP - [text] cannette *f* de trame
~COUNTER - [text] compteur *m* de duites
~CUTTING MOTION - [text] mécanisme *m* de coupage de la trame
~DAMPING - [text] mouillage *m* de la trame, vaporisage *m* de la trame
~DISTRIBUTOR - [text] distributrice *f* de trames
~EFFECT - [text] effet *m* de trame
~FEED - [text] alimentation *f* de trame
~FEEDING MOTION - [text] dispositif *m* d'alimentation de trame
~-FEELER - [text] (of a loom) tâteur *m*
~FLOAT - [text] flotté *m* de trame
~FORK - [text] fourchette *f* de casse-trame
~FRINGE - [text] frange *f* par trame, frange *f* latérale
~GRID - [text] grille *f* de la fourchette de trame
~HAMMER - [text] casse-trame *m* à marteau
~INSERTION - [text] insertion *f* de la trame
~LINE - [text] ligne *f* de trame
~LOOP - [text] galon *m* par trame
~MOISTENING - s. weft damping
~MULE - [text] renvideur *m* pour trame
~NEEDLE - [text] aiguille *f* à insérer la trame
~NIB - [text] nez *m* du casse-trame
~PILE CARPETS - [text] tapis *m*pl veloutés par trame
~PILE FABRICS - [text] velours *m*pl par trame
~PIRN - [text] fuseau *m* de trame
~RIB - [text] reps *m* en chaîne
~RIB EFFECT - [text] côte *f* par trame
~RING FRAME - [text] continu *m* pour chaîne
~SILK - [text] soie *f* trame
~SPACING - [text] duitage *m*
~SPOOLING - [text] cannetage *m* de la trame
~STOP - [text] (of a loom) casse-trame *m*
~STOP MOTION - s. weft stop
~TWIST - [text] torsion *f* trame
~WASTE - [text] déchets *m*pl de trame
~WINDER - [text] cannetière *f*
~WINDING - [text] bobinage *m* pour trame
~WINDING MACHINE - [text] bobinoir *m* pour trame
~YARN - [text] fil *m* de trame, trame *f*
~YARN SPINNER - [text] fileuse *f* de trame
WEHNELT CYLINDER - [electron] (cathode, or concentration cup, a cylindrical electrode for concentrating the beam of electrons emanating from the cathode) cupule *f* de concentration, cylindre *m* de Wehnelt
WEIGH, to - [gen] peser
[gen] (the action) faire la peser
[naut] (the anchor) lever (l'ancre)
~FEEDER - [plast ind] dispositif *m* de dosage
~FEEDING - [plast ind] dosage *m* pondéral
WEIGHBAR - [mech] (a rock shaft) arbre *m* basculant
WEIGHBRIDGE - [mech] pont *m* à bascule
WEIGHING - [gen] pesage *m*, pesée *f*
[comm] (the quantity weighed) pesée *f*

WEIGHING - [naut] levage *m* (de l'ancre)
~BOTTLE - [ind chem] (special bottle to contain substances for weighing) verre *m* de pesée
~MACHINE - [mech] bascule *f*
WEIGHSHAFT - s. weighbar
WEIGHT, to - [gen] alourdir, attacher un poids
[phys] (to express the precision probability of an observed phenomenon by a number) pondérer
[mining] donner charge
[text] (to add sizing material) charger, engaller
WEIGHT - [phys] (the force with which a body is attracted toward the earth) poids *m*
[meas] mesure *f* de poids, mesure *f* pondérale
[meas] (of a steelyard) poire *f*, poids *m* curseur
[constr & mining] charge *f*
[mech] (of a lever) résistance *f*
~CARRIED PER WHEEL - [auto] poids *m* porté par roue
~CASE ACCUMULATOR - [hydr] accumulateur *m* à caisse de contrepoids
~DISC - [text] disque-poids *m*
~EMPTY - [aero etc] (the measured weight of an aircraft etc less that of disposable load and of equipment which can be removed without infringing regulations) poids *m* à vide opérationnel
~FILL - [metall] remplissage *m* à poids
~HOOKS - [text] crochets *mpl*
~LEVER - [mech] levier *m* de pression
[text] levier *m* du contrepoids
~LIMIT - [transp] limite *f* de poids
~PER AXLE - [mech] (load per axle) chrage *f* par essieu
~PER HORSE-POWER - [mech aero] (the result obtained by dividing the dry weight of an engine by the maximum allowable power output) poids *m* au cheval, poids *m* par cheval
~PER POUND THRUST - [mech aero] (the dry weight of an engine divided by the highest allowable thrust under sea level conditions) poids *m* par kilo de poussée
~PERCENTAGE - [chem] (the number of grammes of absolute alcohol contained in I00 grammes of the spirit at I5 deg.C.) pourcentage *m* en poids
~RELIEVING MOTION - [mech] dispositif *m* pour relever le poids
~RING ACCUMULATOR - [hydr] accumulateur *m* à contrepoids par anneaux
~TRANSFER - [mech] déchargement *m* d'essieu
~VOLTAMETER - [instr] (a voltameter in which the quantity of electricity is determined by the weight of metal deposited) voltamètre *m* à poids
WEIGHTED - [gen] chargé d'un poids, alourdi
[text] chargé, engallé
~ACCUMULATOR - [contr] accumulateur *m* à poids
~AVERAGE - s. weighted mean
~BAND - [text] corde *f* à contrepoids
~BINARY SIGNAL - [comput] signal *m* pondéré binaire
~CODE - [comput] code *m* pondéré
~CORD - [mech] corde *f* de frein
~MEAN - [statist] moyenne *f* pondérée
~NOISE - [comput] bruit *m* pondéré
~RACK - [text] crémaillère *f* chargée
~ROPE - s. weighted band
~SIGNAL-TO-NOISE RATIO - [telev] rapport *m* signal-bruit subjectif
~SLEDGE - [text] traîneau *m* chargé
WEIGHTING - [text] (the operation of adding sizing materials) engallage *m*
WEIGHTING - [meas & statist] (the artificial adjustment of measurements in order to account for factors which would normally be different from the conditions during measurement) pondération *f*
[comput] pondération *f*
~FUNCTION - [nucl] (a measure of the relative effect on reactivity of localized changes in nuclear properties as a function of position and property change) fonction *f* de pondération
[contr] fonction *f* poids
~MATERIAL - [oil ind] matières *fpl* lourdes
~NETWORK - [el] (a network whose loss varies with frequency in a pre-determined manner) filtre *m* d'évaluation, réseau *m* à atténuation prédéterminée
~SUBSTANCE - [text] substance *f* pour la charge
~WASHER - [mech] plateau *m* de contrepoids
WEIGHTLESS - [gen & astronaut] sans poids
WEIGHTLESSNESS - [astronaut] apesanteur *f*
WEIGHTOMETER - [meas] (continuously weighing automatic device) pesée *f* automatique
WEIGHTS - [meas] (of a balance) (pieces of metal of known weight, used to counterpoise objects in weighing) poids *mpl*
[metall] plateau *m* de charge
WEIR - [hydr] (dam across a river to raise its level in dry weather) barrage *m*, déversoir *m*
[meas] (a device designed to measure the quantity of flowing water) déversoir *m* de jaugeage
~BOX - [hydr] (a device for measuring the quantity of irrigation water) caisson *m* de mesure
~CREST - [hydr] crête *f* d'un déversoir
~EDGE - [hydr] paroi *f* de déversoir
WELD, to - [mech] souder, se souder
[metall] souder, corroyer
WELD - [mech] (the consolidation of parts or pieces of metal by welding, that is by raising the temperature at the joints so that the metal becomes plastic and may be united by hammering or pressure) soudure *f*
~BEAD - [metall] cordon *m* de soudure
~DECAY - [metall] (the intercrystalline corrosion which develops in certain stainless steels) corrosion *f* intergranulaire de la soudure, attaque *f* de la soudure
~INGOT - [metall] (a nugget) perle *f* de soudure
~IRON SOCKET - [metall] emboîtement *m* à souder
~LINE - [metall] (a surface indication of the weld plane) ligne *f* de soudure
~MARKS - [plast ind] (marks caused by the reunion of divided streams of plastic in a mould) lignes *fpl* d'accollement
~METAL - [metall] métal *m* de fusion
~NUGGET - [metall] perle *f* de soudure
~SPATTER - s. weld nugget
WELDABILITY - [metall] soudabilité *f*
WELDED BOND - [el] connexion *f* soudée
~-CONTACT RECTIFIER - [electron] (a contact rectifier in which a point contact is permanently soldered to a semi-conductor) redresseur *m* à pointe soudée
~FLANGE - [mech] bride *f* soudée
~JOINT - [railw] (a joint between two rails made by electric welding) joint *m* soudé
~SEAL - [metall] scellement *m* soudé
~SEAM - [metall] soudure *f*, cordon *m* de soudure
~SOURCE - [nucl] (neutron source in a welded capsule) source *f* soudée
WELDER - [mech] machine *f* à souder

WELDER'S GOGGLES - [impl] device for protecting the eyes from U.V. radiation during welding) lunettes *f*pl protectrices pour soudeurs
WELDING - [mech] (the joining of two iron or steel pieces by raising the temperature at the joints, so that the metal becomes plastic and may be united by hammering or pressure) soudure *f*
~ARC VOLTAGE - [el] tension *f* de soudage à l'arc
~BLOWPIPE - [impl] chalumeau *m* soudeur
~BOOTH - [metall] (cabin or other enclosed structure in which welding is carried on, designed to prevent interference to other work in the neighbourhood by light radiation) cabine *f* de soudage
~BURNER - s. welding torch
~BY INDUCTION - [metall] (a welding process in which the necessary heat is generated by electromagnetic induction) soudure *f* à étincelles
~BY SPARKS - [metall] soudure *f* à énticelles
~CYCLE - [mech] (in resistance welding, the complete series of operations) cycle *m* de soudure
~ELECTRODE - [metall] (a rod of special metal used in electric welding to supply the material for the weld) électrode *f* de soudage
~ELECTRODE HOLDER - [metall] porte-électrode *m* de soudage
~FLUX - [metall] (a composition used to facilitate welding by chemical action) fondant *m* à souder, flux *m* de soudage
~GENERATOR - [el] (a direct-current generator providing electric energy to one or more arcs) génératrice *f* pour soudage à arc
~GROUND - [el] (USA only, backing electrode in GB) contre-électrode *m*
~GUN - [impl] pinces *f*pl à souder
~JIG - [metall] montage *m* de soudage
~JOINT - [metall] joint *m* soudé
~LEADS - [el] câbles *m*pl de soudage
~MACHINE - [mach tool] machine *f* à souder
~MACHINE TIMBER - [mech] (device to regulate the time during which a welding machine performs its cycle of operation) régulateur *m* de soudage
~MILL - [metall] laminoir-soudeur *m*
~NECK - [oil ind] col *m* à souder
~PASS - [metall] cordon *m* de soudure
~POSITIONER - [impl] dispositif *m* pour mettre en position pour soudage
~POSITIONS - [metall] positions *f*pl pour soudage
~POWDER - [metall] poudre *f* à souder
~PRESSURE - [mech] (the pressure which is exerted on the parts to be welded) pression *f* de soudure
~ROD - [metall] baguette *f* de soudure
~ROLLERS - [impl] (a pair of metal rollers used in seam welding) rouleaux *m*pl pour soudure continue
~SCREEN - [impl] écran *m* pour soudage
~SEAM - [metall] ligne *f* de soudure
~SEQUENCE - [metall] (the order of welding operations) ordre *m* de soudage
~SET - [el] (group of electrical equipment, e.g. motor-generator, transformer rectifier etc, used to supply or modify electric current for welding) appareil *m* d'alimentation pour soudage à l'arc, poste *m* de soudage
~SPATTER - [metall] (small drops of molten metal during welding) éclaboussures *f*pl
~STRESS - [metall] effort *m* interne dû a la soudure
~TEE - [oil ind] té *m* à souder
~TIMER - [metall] (a device for determining the time for a perfect welding) régulateur *m* de la durée de soudure
WELDING TORCH - [welding] (a device used in hot gas welding and designed to direct a jet of flame, heated air or inert gas on the area to be welded. The gas may be heated by electrical means or by allowing it to pass through a tube heated by means of gas flame) torche *f* à souder, chalumeau *m* soudeur
~WITH PRESSURE - [el heat] (welding which employs static or dynamic pressure to complete the union) soudure *f* à pression
~YOKE - s. welding gun
WELDLESS - [mech] sans soudure
~STEEL TUBE - [metall] tube *m* d'acier sans soudure
WELDMENT - [mech] (the operation of welding) soudage *m*
WELDON'S PROCESS - [chem] (a process for the industrial production of chlorine by the action of hydrochloric acid on manganese dioxide) procédé *m* Weldon
WELFARE - [gen] bien-être *m*
[leg] salut *m* public
~WORK - [gen] assistance *f* sociale
WELL - [gen] puits *m*, fontaine *f*, source *f*
[min] (the shaft sunk into the earth to obtain a fluid) puits *m*
[constr] (of the stair) jour *m* (d'escalier), cage *f* (d'ascenseur)
[naut] (the space in a vessel's hold containing the pumps) sentine *f*, puisard *m*, archipompe *f*
[auto] (of a spare-wheel) baignoire *f*
~BASE RIM - [auto] jante *f* à base creuse
~BEAM - [impl] poulie *f* de puits
~BORER - [mining] machine *f* pour forage de puits
~CASING - [mining] tubage *m* d'isolement, tubage *m* de puits, revêtement *m* de puits
~CASING STARTER - [oil ind] sabot *m* de tubage
~CHAMBER - [mining] chambre *f* du puits
~COUNTER - [instr] (a radiation counter with a heavy tubular shield closed at one end, in which the radiation detector and the radioactive sample are inserted to reduce the effect of background radiation) compteur *m* à canal pour échantillons
~CURB - [constr] margelle *f* (de puits)
~DRILL - [tool] sonde *f* de fontainier
~DRILLING - [mining] sondage *m* de puits
~GALLOW - [hydr] potence *f* de puits
~HOLE - [constr] (the shaft of a well) jour *m*, cage *f* d'escalier
[naut] archipompe *f*
~LOG - [mining] coupe *f* de sonde
~LOGGING - [oil ind] (investigation by nuclear methods) radiocarottage *f*
~LOWERING - [hydr] abaissement *m* d'un puits
~OF A RIM - [rubber ind] (of tyres) fond *m* de la jante, base *f* de la jante
~PIT - s. well hole
[hydr] abaissement *m* d'un puits
~PULLEY - s. well beam
~RIG - [mining] appareil *m* de sondage
~ROOM - [naut] sentine *f*
~SCREEN - [hydr] crépine *f*
~SHAFT - s. well hole
~SINKING - [mining] fonçage *m* de puits
~SPACING - [oil ind] espacement *m* entre puits
~STAIRCASE - [constr] escalier *m* tournant
~TOP - [hydr] avant-puits *m*, tête *f* de puits

WELL TRAILER - [transp] (for deep loading) remorque *f* à plate-forme abaissée
~TUBE - [mining] tube *m* de puits
~TUBING - [mining] tubage *m* d'extraction
~-TYPE IONIZATION CHAMBER - [nucl] (principally for the measure of activity of gamma-emitting sources) chambre *f* d'ionisation à puits
~-TYPE SCINTILLATOR DETECTOR - [radiat] détecteur *m* à scintillateur à puits
~-TYPE SODIUM IODIDE CRYSTAL - [crystall] cristal *m* d'iodure de sodium type puits
WELLHOLE - s. well hole
WELLINGTONS - [shoe man] bottes *f*pl en caoutchouc
WELT, to - [text] faire joindre
WELT - [shoe man] (strip of material applied to a seam to strengthen it) trépointe *f* de semelle
[text] (for stockings) bordure *f*
[carp] (for upholstery) cordon *m*
[mech] (a buttstrap)couvre-joint *m*, bande *f* de recouvrement
[auto] (USA only) pare-boue *m*, bavolet *m*
[text] (a rib) côte *f*
[text] bordure *f*
[med] (a wheal) papule *f*
[plumb] agrafe *f*
~BAR - [text] griffe *f* à revers
~FLOAT - [text] boucle *f*
~HOOK - [text] crochet *m* pour revers double
~SEAM - [text] couture *f* de passepoil
~YARN - [text] fil *m* pour revers double
WERNERITE - [geol] (a gem stone) wernérite *f*, scapolite *f*
WEST - [astr] (the point where the sun sets at the equinoxes) ouest *m*
[gen] occident *m*
WESTBOUND - [gen] (going westward) allant vers l'ouest
WESTER, to - [astr] (of stars etc, to move westward) se diriger vers l'ouest
[met] (of the wind] sauter à l'ouest
WESTERLY - [met] (of the wind] d'ouest
~WIND - [met] vent *m* d'ouest
WESTERN - [gen] occidental, ouest
[cin] western *m*
WESTON CADMIUM CELL - s. Weston normal cell
~NORMAL CELL - [el chem] (standard cell with an electrolyte consisting of a saturated solution of cadmium sulphate) pile *f* étalon Weston
WET, to - [gen] mouiller, humecter
[naut] empeser (une voile)
[ind chem] madéfier
WET - [gen] (humidity) humide, mouillé
[met] (rain) (temps) humide
[chem & metall] (of a process) voie *f* humide
~AGENT - [el chem] agent *m* mouillant, agent *m* tensio-actif
~AND DRY BULB PSYCHROMETER - [instr] (apparatus used for the measurement of humidity) psychromètre *m*
~ASSAY - [min] (determination of the metal content of an ore or other substance by solution methods) analyse *f* par voie humide
~BATTERY - s. wet cell
~-BEATEN PULP - [paper man] pâte *f* engraissée
~BEATING - [paper man] raffinage *m*
~BINDER - [plast ind] (aqueous solutions or emulsions sprayed on a preform to consolidate the chopped glass rovings) liant *m* humide
WET BLACKING - [metall] noir *m* liquide, noir *m* d'étuve
~BULB THERMOMETER - [instr] thermomètre *m* mouillé
~CELL - [el chem] (a primary cell having a liquid electrolyte which is contained in a vessel, as distinct from a dry cell, in which the electrolyte is absorbed in a paste) pile *f* liquide
~CLEANING - [ind chem] nettoyage *m* au moyen d'une solution aqueuse
~COLLODION PROCESS - s. wet-plate process
~COMPRESSOR - [mech] compresseur *m* humide
~CONCENTRATION - [min] concentration *f* à l'eau
~CRITICALITY - [nucl] (obtained with a liquid coolant) criticité *f* en présence de réfrigérant
~CRUSHING - [min] broyage *m* à l'eau
~CYLINDER LINE - [mech] (a type of cylinder liner the outer surface of which forms part of the water-jacket) chemise *f* humide de cylindre
~DOCK - [constr] (a dock in which water is impounded at a given level by means of dock gates) bassin *m* à flot
~DOUBLING FRAME - [text] métier *m* à retordre au mouillé
~-DRAWING - [metall] étirage *m* à lubrifiant liquide
~-DRAWING MACHINE - [metall] banc *m* à étirer à lubrifiant humide
~DRAWN WIRE - [metall] fil *m* écroui par voie humide
~END - [paper man] (part of a paper machine)partie *f* humide
~END OF PRESS - [print] presses *f*pl
~EXHAUST SYSTEM - [auto] système *m* d'échappement par voie humide
~EXPANSION - [paper man] allongement *m* à l'humidité
~FLASHOVER VOLTAGE - [el] (the voltage at which the air surrounding a clean wet insulator breaks down) tension *f* d'arc sous conditions humides
~FOUNDATION - [constr] fondation *f* enfoncée dans l'eau
~-FUEL ROCKET - [astronaut] fusée *f* à propergol liquide
~GAS - [min] (petroleum gas containing such quantities of the lower members of the paraffin hydrocarbon series that the recovery of liquid products from that gas may be economical) gaz *m* humide, gaz *m* riche en condensat
~GRINDING - [min] broyage *m* à l'eau
~LINER - [auto] (wet-cylinder liner) chemise *f* humide
~LUTE - [mech] (for chemical purification) joint *m* à garde hydraulique
~METHOD - [chem] voie *f* humide
~METHOD OF DRILLING - [oil ind] méthode *f* de forage avec courant d'eau
~MILL - [min] bocard à eau
~MILLING - [min] broyage *m* liquide
~MIX - [constr] (a concrete mix with an excess of water in it) mélange *m* aqueux
~MIXER FOR SOLUTIONS - [rubber ind] malaxeur *m* humide à dissolution
~NATURAL GAS - s. wet gas
~ON WET - [paint] (a painting system in which new coatings are applied before the previous ones have dried) mouillé-sur-mouillé *m*
~PERIMETER - [hydr] périmètre *m* mouillé
~PLATE - [photo] (a collodion plate) plaque *f* au collodion humide

WET-PLATE PROCESS - [photo] (wet collodion process, a process employing a glass plate coated with iodized collodion, sensitized in silver nitrate solution, physically developed with a solution of ferrous sulphate and acetic acid in water) procédé *m* au collodion
~PRESS MACHINE - [text] presse *f* humide
~PROCESS - [constr] (method of making P. C. in which the raw materials are ground together with water before burning) procédé *m* humide (pour Portland)
[chem] voie *f* humide
~PUDDLING - [metall] puddlage *m* gras
~QUENCING - [metall] extinction *f* humide
~REPROCESSING - [nucl] traitement *m* de combustible irradié par voie humide
~SANDING - [paint] (in industrial, especially automobile painting) polissage *m* par voie humide
~SEAL - [oil ind] étanchéité *f* hydraulique
~SEPARATION - [min] triage *m* par voie humide
~SORTING - [min] triage *m* à l'eau, triage *m* par la voie humide
~SPARKOVER VOLTAGE - [el] s. wet flashover
~SPINNER - [text] fileuse *f* au mouillé
~SPINNING - [text] (extrusion of synthetic fibres into a water bath) filage *m* au mouillé
~SPUN FLAX YARN - [text] fil *m* de lin de filage au mouillé
~STAMPING - [min] bocardage *m* à l'eau
~STAMPING MILL - s. wet mill
~STEAM - [phys] (steam in contact with the liquid phase from which it was generated, or formed by the cooling of dry saturated steam) vapeur *f* humide
~-STEAM COOLED REACTOR - [nucl] réacteur *m* refroidi par vapeur humide
~STRENGTH - [phys] (the strength of a material when saturated with water) résistance *f* en condition mouillé
~-SUMP LUBRICATION - [mech] (a lubrication system in which a certain quantity of oil is stored loose in the sump) lubrification *f* à carter humide
~SWELLING - [rubber ind] gonflement *m* par humidité
~TOOL-GRINDER - [mach tool] machine *f* à affûter les outils à l'eau
~TREATMENT - [ind chem] traitement *m* par voie humide, traitement *m* à l'eau
~TWISTING - [text] retordage *m* au mouillé
~-TYPE FILTER ELEMENT - [ind chem] (a type of element in which a liquid film acts as the filtering medium) élément *m* filtrant humide
~WAY - s. wet method
~WEAVING - [text] tissage *m* au mouillé
~WEIGHT - [gen] poids *m* humide
~WHEEL - [tool] meule *f* mouillé
WETBLAST MOULD CLEANING - [rubber ind] procédé *m* de sablage humide pour le nettoyage des moules
WETHER LAMB - [zool] bélier *m* châtré, mouton *m*
WETROT - [bot] (red root disease) carie *f* humide
WETTABILITY - [chem] (the extent to which a solid is wetted by a liquid, it is measured by the force of adhesion between the solid and the liquid phases) aptitude *f* d'être mouillé
WETTED - [gen] mouillé
~AREA - [gen] (that part of the surface of a body which is in contact with liquid) surface *f* mouillée
~ROVING - [text] mèche *f* humectée
WETTING AGENT - [chem] (substances capable of reducing surface tension and thus causing a liquid to "wet" bodies or particles, i. e. to spread over their surfaces in a thin film) agent *m* mouillant, agent *m* tensio-actif, agent *m* humectant
WETTING ANGLE - [soil] angle *m* de contact
~BOX - [print] appareil *m* de trempage
~OUT, to - [gen] (the operation of permeating a material completely with a liquid, usually water) imbiber à fond, impregner
~POWDER - [ind chem] poudre *f* mouillante
~POWER - [phys] (the ability of a specific liquid to make a contact with a substance) pouvoir *m* mouillant
~TABLE - [print] table *f* de mouillage
W. F. - [print] (wrong fount) coquille *f*, mastic *m*
W.G. - [meas] (abbreviation for water gauge, a measure of pressure equal to that exerted by a column of water of the specified height) indicateur *m* de niveau d'eau
WHALE - [zool] baleine *f*
~-BONE - [zool] fanon *m* de baleine
~OIL - [ind] (a non-drying oil obtained from the blubber of various whales and used in the production of margarine, edible oils, soap manufacture, metallurgy, lubricants, and leather processing) huile *f* de baleine
WHALEBOAT - [naut] baleinière *f*
WHALER - [naut] (vessel engaged in whaling) baleinière *f*
WHALING - [fishing] (the industry of capturing whales) pêche *f* à la baleine
~FACTORY - [shipbuild] baleinier *m*
~GUN - [fishing] canon *m* à harpon
WHARF - [naut] (a quay) appontement *m*, wharf *m*, quai *m*
WHARFAGE - [comm] droits *mpl* de quai
[gen] mise *f* en entrepôt
WHARFINGER - [comm] propriétaire *m* d'un quai
WHARVE - [text] (in a spinning machine, the flywheel of a spindle) noix *f*
WHEAL - [min] (a tin mine, GB only) mine *f* d'étain
WHEAT - [bot] blé *m*, froment *m*
~BRAN - [agric] son *m* de blé
~BULB FLY - [zool] mouche *f* grise du blé
~FLOUR - [agric] farine *f* de blé
~GRASS - [bot] chiendent *m*
~MALT - [brew ind] malt *m* de froment
~MEAL - [agric] farine *f* de blé fourragère
~MIDGE - [zool] cécidomyie *f* du blé
~STALK - [bot] chaume *m* de blé
~STRAW - [agric] paille *f* de blé, tige *f* de blé
WHEATSTONE BRIDGE - [el] (a bridge circuit in which the arms consist of resistors) pont *m* de Wheatstone
WHEEL - [gen] roue *f*
[mech] (of gear) roue *f*, engrenage *m*, galet *m* mécanique
[tool] (a grinding wheel) meule *f*
[aero] (the aileron control wheel) volant *m* (à poignée radiales)
[mech] (a pulley) poulie *f*
[mech] (of a turbine) roue *f* mobile, couronne *f* mobile
[mech] (of a fan) tourniquet *m*
~ALIGNER - [mech] (a track alignment gauge) dispositif *m* à vérifier l'alignement des roues
~ALIGNMENT - [aero] (the adjustment of landing gear wheels so that they revolve in a plane parallel to the direction of movement of the aircraft when on the

ground) parallélisme *m* des roues
WHEEL AND AXLE - [mech] (a simple machine the theoretical mechanical advantage of which is the ratio between the radius of the wheel and that of the axle) poulie *f*, galet *m* mécanique
~-AND-DISK INTEGRATOR - [contr] intégrateur *m* à disque et molette
~BALANCING - [auto] équilibrage *f* d'une roue
~BALANCING ARBOR - [mach tool] mandrin *m* pour l'équilibrage des ressorts
~BALANCING STAND - [impl] banc *m* pour l'équilibrage des ressorts
~BALANCING WEIGHT - [auto] poids *m* d'équilibrage des roues
~BAROMETER - [instr] baromètre *m* à cadran
~BASE - [aero & naut] (the distance between the centre of main-and tail-or nose-wheels, measured parallel to the longitudinal axis) empattement *m*
~BLADE - [mech] palette *f* de roue
~BODY - [mech] (of a gear) corps *m* de la roue
~BOND - [railw] connexion *f* de roue
~BORE - [railw] (axle seat) portée *f* de calage
~BOX - [auto] passage *m* de roue
~BRACE - [auto] vilebrequin *m* à roues
~CAMBER - [auto] (obliquity of the wheels) inclinaison *f* latérale
~CAMBER AND CASTER ALIGNMENT TESTER - s.wheel aligner
~CARRIAGE - [mech] chariot *m* porte-meule
~CASTER - [auto] chasse *f* de roue
~CENTRE - [mech] (flanged hub designed to support the grinding wheel) porte-meule *m*
[auto] (the point at which the axis of rotation of the wheel intersects the wheel plane) centre *m* de rotation de la roue
~CHOCK - [mech] cale *f* pour roues
~CLEARANCE - [auto] débattement *m* des roues
~COLLET - [horol] moyeu *m* de la roue
~CONTROL - [aero] commande *f* à volant
~COVER - [auto] couvre-roue *m*
~CRANK - [mech] manivelle *f* à plateau
~CUTTER - [mach tool] fraise *f* pour engrenages
~CUTTING ENGINE - [horol] (a machine designed for cutting the teeth of wheels) machine *f* à tailler les engrenages
~DISTANCE - s. wheel track
~DRESSER - s. wheel dressing roll
~DRESSING ROLL - [mech] décrasse-meule *m*, rhabilleur *m* pour meules
~-FIGHT - [auto] réaction *f* de direction
~FIT - [mech] (of an axle) portée *f* de calage
~FLANGE - [mech] (of railway carriages) boudin *m* de roue, bourrelet *m* de roue
~FLUTTER - [auto] flottement *m* des roues, dandinement *m* des roues
~FRICTION - [mech] frottement *m* des roues
~GAUGE - [auto] gabarit *m* à bandages
~GRIP - [phys] adhérence *f* des roues
~GUARD - [mach tool] couvre-meule *m*
[mech] couvre-boîte *m* à graisse
[mech] (a gear casing) carter *m* des engrenages, couvre-engrenages *m*
~HEAD - [mach tool] mandrin *m* porte-meule
~HOP - [auto] débattement *m* de la roue
~HOUSING - [auto] passage *m* de roue
~HUB - [auto] moyeu *m* de roue
~HUBS CUP - [auto] couvre-moyeu *m* de roue

WHEEL JACK - [impl] lève-roues *m*
~LANDING - [aero] atterrissement *m* sur les roues
~LATHE - [mach tool] tour *m* à roues
~LEVER - [mech] levier *m* de roue
~LEVER GRATE - [text] arrêt *m* du levier de roue
~MOULDING MACHINE - [mech] machine *f* à mouler les roues
~-MOUNTED - [gen & mech] monté sur roues
~NUT - [mech] écrou *m* de roue
~-ORE - [min] (bournonite, a natural sulphide of copper, antimony and lead. Orthorhombic : often appears in a form resembling a toothed wheel) bournonite *f*
~PLAN - [railw] schéma *m* de la disposition des roues
~PLANE - [auto] (the central plane of the tyre, normal to the axis of rotation) plan *m* de la roue
~PLOUGH - [agric] charrue *f* à avant-train
~PRESS - [railw] (a machine for pressing wheels on locomotive axles) presse *f* hydraulique pour la calage des roues
~PRINT TRIMMER - [photo] coupe-épreuves *m* à molette
~PRINTER - [comput] imprimante *f* à roues
~PULLER - [mech] extracteur *m* de roue
~QUARTERING MACHINE - [mach tool] (horizontal drilling machine with two opposed spindles at opposite ends of the bed, it is used to drill the crank-pin holes in both wheels of a locomotive coupled-axle simultaneously and in exact angular relationship) perceuse *f* horizontale à deux mandrins aux extrémités du banc
~RATE - [auto] (the change of wheel load, at the centre of tyre contact, per unit vertical displacement of the sprung mass relative to the wheel at a specified load) taux *m* de flexibilité rapporté à la roue
~RIM - [auto] jante *f* de roue
~ROLLING MILL - [metall] laminoir *m* à roues
~SEAT - [mech] (wheel spindle nose, e.g.in internal grinding) extrémité *f* de calage de la meule
[railw] (the axle section which is in contact with the railway wheel bore) portée *f* de calage (d'un essieu)
~SET - [railw] (an axle with fitted wheels) essieu *m* monté
~SHIELD - [auto] cache-roue *m*
~SLIDE - [mech] chariot *m* de la meule
~SLIP INDICATOR - [auto] indicateur *m* du glissement de la roue
~SPIDER - [auto] étoile *f* de roue
~SPINDLE - [mach tool] (of internal grinder) arbre *m* porte-meule
~SPINDLE NOSE - s. wheel seat
~SPOKE - [mech] rayon *m* de roue
~SPRAYER - [agric] pulvérisateur *m* à chariot, pulvérisateur *m* sur brouette
~STRETCHER - [horol] extenseur *m* du diamètre des roues
~SWARF - [mach tool] boue *f* de meule
~-TIE ARM STOP - [mech] butée *f* de bras de liaison de roue
~TIRE - [mech] cercle *m* de roue
~TOE - [auto] pincement *m* des roues
~TRACK - [auto] voie *f*
~TRAIN - [mech] train *m* de roues
~TREAD - [railw] surface *f* de roulement
~TROLLEY - [railw] (in electrical railway) trolley *m* à roue

WHEEL TRUING - [mech] rectification *f* de la meule
~TURNER - [text] (a ropemaker assistant) aide-fileur *m*
~TURNING LATHE - [mach tool] tour *m* à roues
~-TYPE EXCAVATOR - [mech] excavateur *m* à roues
~-TYPE TRACTOR - [agric] tracteur *m* à roues
~VALVE - [mech] robinet *m* à soupape à volant
~VANE - [mech] (an impeller) aube *f* de rotor
~WATER - [sugar ind] eau *f* de lavage des betteraves
~WEAR COMPENSATING DEVICE - [mach tool] (in a gear-grinding machine) dispositif *m* de compensation de l'usure de la meule
~WOBBLE - [auto] (to a lack of balance in the steering gear) flottement *m* des roues
~WRENCH - [auto] clé *f* pour écrous de roues
WHEELBARROW - [impl] brouette *f*
WHEELBASE - [auto] empattement *m*
WHEELCASE - [mech] carter *m* engrenages
WHEELED BUCKET - [mech] benne *f* roulante
~PUMP - [mech] pompe *f* roulante
WHEELHEAD - s. wheel head
WHEELHOUSE - [naut] kiosque *m* de la barre
[auto] passage *f* de roue
WHEELING STEP - [constr] (a winder) marche *f* tournante
WHEELSPIN - [auto etc] (a spin caused by insufficient adhesion) chasse *f* des roues
WHEELWORK - [mech] rouage *m*
WHEELWRIGHT - [mech] charron *m*
WHEELWRIGHT'S WORK - [mech] charronnage *m*
WHEEZE - [med] respiration *f* bruyante, respiration *f* asthmatique
WHET, to - [gen] (to sharpen) aiguiser, affûter, repasser
WHETSTONE - [impl] (a fine-grained stone used to sharpen the edge of knives, or tools) pierre *f* à aiguiser, affiloire *f*
WHETTING - [mech] repassage *m*, aiguisage *m*
WHEY - [agric] (the liquid which separates from the curd when milk is curdled by rennet or acids)petit lait *m*
~-CHEESE - [agric] fromage *m* de petit lait
~CONCENTRATE - [agric] petit lait *m* concentré
WHIFF - [gen] bouffée *f*, souffle *m*
WHIM - [mech] (an obsolete form of mine hoist) treuil *m* à manège
~SHAFT - [mining] puits *m* à cabestan
WHIMBLE - [tool] (a type of brace) vilebrequin *m*
WHIN - [bot] (furze, gorse) jenet *m* épineux
[mech] (a small winch) petit treuil *m*
[geol] grauwacke *f*
~DIKE - [geol] dyke *m* de basalte
WHINE, to - [acoust] geindre, glapisser
WHINING - [acoust] (the sound uttered by animals in pain) geignement *m*, glapissement *m*
WHINNYING - [acoust] (the sound made by horses) hennissement *m*
WHIP, to - [gen] (to strike with a lash, a whip) fouetter
[gen] (to agitate very rapidly, as in whipping cream) battre, faire mousser
[naut] garnir (un cordage)
[text] (a seam) surjeter (une couture)
[mech] (to vibrate, e,g.as a long spindle turning at high speed) vibrer
[naut] (to hoist) hisser
WHIP - [gen] fouet *m*
WHIP - [gen] (the stroke of a whip) fouettement *m*, coup *m* de fouet
[impl] (of a windmill) bras *m* (d'un moulin à vent), aile *f* (d'un moulin à vent)
[naut] (a very simple hoisting apparatus) cartahu *m*, palan *m*
[el] (vibrating spring closing different circuits) vibrateur *m* à ressort
~AERIAL - [radio] (aerial consisting of a flexible wire or rod conductor) antenne *f* flexible
~AND DERRY - [mining] chevalement *m* à palan
~ANTENNA - s. whip aerial
~CORD - [text] corde *f* à fouet
~GIN - [naut] (or whip, a hoisting apparatus) cartahu *m*
~GRAFTING - [agric] greffe *f* en fente anglaise
~PAN - [cin & telev] arraché *m*
~ROLL - [text] bobine *f* de trame brodeuse
[text] (the roll carrying the warp threads) cylindre *m* des fils de trame
~SAW - [tool] scie *f* à chantourner
~STALL - [aero] (a tail slide) glissade *f* sur la queue
~THREAD - [text] fil *m* brodeur, fil *m* à broder
WHIPCORD - [text] (a twill-weave fabric) whipcord *m*
[text] (strong hard-twisted hempen cord) corde *f* à fouet
WHIPPING - [mech] (the vibration of a long spindle when turning at a high r.p.m.) fouettement *m*, battement *m*
WHIPSAW - [tool] scie *f* de long, scie *f* à chantourner
WHIPSTOCK - [oil ind] (wedge-shaped device used in deviated drilling to deflect and guide the bit away from vertical) biseau *m* de déviation
WHIRL, to - [gen] tourbillonner, tournoyer
WHIRL - [mech] (a rapid revolving motion) mouvement *m* giratoire
[hydr & aerodyn] tourbillonnement *m*
[text] (in ropemaking) molette *f*
[el] tourniquet *m*
~BONE - [anat] rotule *f*
~GATE - [metall] piège *m* à crasses à canal tangentiel
~HEAD - [text] tête *f* du rouet
~-SINTERING - [ind chem] frittage *m* en lit fluidisé
WHIRLED THERMOMETER - [instr] thermomètre-fronde *m*
WHIRLER - [photo] (a whirling table used for coating photographic plates) tournette *f* d'essorage
WHIRLING - [gen] tourbillonnant
~ARM - [aero] (an arm revolving in a horizontal plane, used to carry a model for aerodynamic experiments) manège *m*
~MOTION - [hydr] mouvement *m* tourbulent
~RUNNER - [metall] chambre *f* d'épuration
~TABLE - [mech] dispositif *m* à accélération centrifuge
WHIRLPOOL - [hydr] (eddy or vortex where water moves with a rotating sweep) remous *m* d'eau,tourbillon *m* , gouffre *m*
WHIRLWIND - [met] (a local vortex in the atmosphere, in which air revolves round a core of low pressure) trombe *f* de vent, tourbillon *m* de vent
WHISKER - [zool] poustache *f*
[shipbuild] arc-boutant *m* de beaupré
[crystall] monocristal *m* sans dislocations
~RESISTANCE - [electron] (the resistance of the

whisker element of a semiconductor) résistance *f* de chercheur
WHISKING MACHINE - [rubber ind] (a beating or foaming machine) fouetteuse *f*, batteuse *f* à mousse
WHISPER - [acoust] chuchotement *m*
WHISTLE, to - [acoust] siffler
WHISTLE - [acoust] sifflement *m*
[impl] (the device for producing a whistle) sifflet *m*
~BOX - [cin] (a portable inductance for filtering the supply ripple in the electrical supply to arc lamps, which would otherwise emit noise) inductance *f* anti-sifflement
~VALVE - [el] clapet *m* de sifflet
WHISTLING BUOY - [acoust] (a buoy fitted with an acoustic warning device actuated by wave motion) bouée *f* à trompe
~KETTLE - [impl] bouilloire *f* à sirène
~METEOR - [astr] météore *m* sifflant
WHITE - [gen] blanc, de couleur claire
[anat] (of the eye) cornée *f*
[anat] (of tissues) (tissu) albuginé
~ACID - [chem] acide *m* hydrofluorique
~ADJUSTMENT - [telev] ajustage *m* du blanc
~AFTER BLACK - [telev] (black compression) blanc *m* après le noir
~ALLOY - [metall] antifriction *f*
~ANNEALING - [metall] recuit *m* blanc
~ANTIMONY - [paint] blanc *m* d'antimoine
~ARSENIC - [chem] (arsenic trioxide, obtained by condensing the fumes from the roasting of arsenical ores) arsenic *m* blanc
~ASH - [bot] frêne *m* américain
~BAR - [telev] (lines used as test signal) barre *f* blanche
~BASSWOOD - [bot] tilleul *m* américain
~BEARING METAL - [metall] métal *m* blanc (pour coussinets)
~BOLE - [min] terre *f* à porcelaine
~CAST IRON - [metall] fonte *f* blanche
~CEDAR - [bot] bois *m* blanchet, cèdre *m* blanc
~CELL - [biol] (leucocyte) leucocyte *m*
~CHALK - [geol] kaolin *m*
~CHEESEWOOD - [bot] pulai *m*
~CLIPPER - [telev] écrêteur *m* du blanc
~CLIPPING - [telev] limitation *f* du blanc
~COAL - [ind] (water power) houille *f* blanche
~COAT - [constr] (the finishing coat of plaster) couche *f* d'enduit
~COBALT - [min] smaltine *f*
~COLD BLAST PIG IRON - [metall] fonte *f* blanche à vent froid
~-COLLAR WORKER - [gen] (term denoting salaried workers, specifically clerical staff) employé *m* de bureau
~COMPRESSION - [telev] (compression in the signals corresponding to the white regions of the picture) écrasement *m* du blanc, tassement *m* du blanc
~CONTENT - [telev] contenu *m* en blanc
~COPPER - [metall] cuivre *m* blanche
~COPPERAS - [min] (goslarite) couperose *f* blanche
~CORPUSCLE - [biol] (a leucocyte, colourless blood corpuscle) leucocyte *m*
~CORUNDUM - [min] corindon *m* blanc
~COUNT - [med] numération *f* leucocytaire
~CRAG - [geol] crag *m* corallin
~CRUSHING - [telev] distorsion *f* de la crête du blanc

WHITE CRYSTALLINE IRON - [metall] fonte *f* blanche à facettes
~DAMP - [mining] (the carbon monoxide produced by the incomplete combustion of coal in a mine fire or by gas or dust explosion. Very poisonous) oxyde *m* de carbone
~EDGING - [telev] bord *m* blanc
~FIBRES - [zool] (inelastic fibres of connective tissue occurring in wavy bundles) fibres *fpl* de tissu conjonctif
~FIR - [bot] sapin *m* pectiné, sapin *m* argenté
~-FLAME HEAT - [phys] chaude *f* grasse
~FLOWERED FLAX - [text] lin *m* américain à fleur blanche
~FRENCH POLISH - [paint] (a clear type of French polish, made by dissolving bleached shellac in commercial alcohol) vernis *m* au tampon blanc
~FROST - [met] gelée *f* blanche
~GARNET - [min] leucite *f*, amphigène *m*
~GOLD - [metall] (an alloy of gold containing zinc and nickel) or *m* blanc
~-HEART MALLEABLE IRON - [metall] fonte *f* malléable à coeur blanc
~HEAT - [phys] chaleur *f* blanche, chaude *f* blanche
~-HOT - [phys] chauffé blanc
~IRON - [metall] (cast iron) fer *m* blanc, fonte *f* blanche
~IRON - [min] marcassite *f*
~IRON PYRITE - [min] (marcasite) pyrite *f* blanche, marcassite *f*
~LAC - [paint] (chemically-bleached shellac) laque *f* blanchie
~LEAD - [ind chem] (white pigment consisting of carbonate and hydroxide of lead) blanc *m* de céruse, plomb *m* blanc
~LEAD ORE - [min] (decomposition product of sphalerite) cérusite *f*
~LEVEL - [telev] (arbitrary level of the vision signal corresponding to the maximum brightness to be transmitted) niveau *m* du blanc
~LIGHT - [phys] (light containing all wavelengths in the visible range at the same intensity) lumière *f* blanche
~LIME - [constr] lait *m* de chaux
~LINE - [print] (a line of space) ligne *f* de blanc
[anat] ligne *f* blanche
~LINE PRINT - [photo] épreuve *f* en traits blancs
~LIQUOR - [ind chem] (solution obtained after precipitation of calcium carbonate from the green liquor in sulphate recovery) liqueur *f* blanche
~METAL - [metall] (a tin based alloy) métal *m* blanc, alliage *m* blanc, régule *m*
~MODIFYING RELAY - [telev] relais *m* modificateur de blanc
~MUNDIC - [min] arsénopyrite *f*
~NICKEL ORE - [min] cloanthite *f*
~NOISE - [acoust] (sound whose spectrum is continuous and uniform as a function of frequency, over a sufficiently large frequency range, as measured in a band of fixed width) bruit *m* à spectre continu et uniforme, bruit *m* blanc
~OBJECT - [opt] (an object which reflects all wavelengths of light) objet *m* blanc
~OILS - [oil ind] (petroleum products obtained by intense refining, e. g. with fuming sulphuric acid, used for pharmaceutics, cosmetics, special textile lubricants and other purposes) huiles *fpl* paraf-

finées
WHITE OUT , to - [print] espacer
~-PATTERN SIGNAL - [telev] signal *m* du canevas blanc
~PEAK - [telev] (the maximum excursion of the vision picture in the white direction) crête *f* du blanc
~PICKLING - [metall] décapage *m* supplémentaire
~PIG IRON - [metall] fonte *f* blanche ordinaire
~PRODUCT - [oil ind] produit *m* blanc
~PULP - [paper man] pâte *f* mécanique blanche
~RAMIE - [bot] ramie *f* blanche
~REINFORCING FILLER - [rubber ind] charge *f* renforçante blanche, pigment *m* renforçant non noir
~RUST - [metall] couche *f* d'oxyde de zinc
~SAPPHIRE - [min] (the colourless, pure variety of crystallized corundum) saphir *m* blanc
~SATURATION - [telev] (distortion of the picture signal consisting of extreme compression of the whitest range) saturation *f* du niveau de blanc
~SIDEWALL TYRE - [rubber ind] pneu *m* à flancs blancs
~SIZE - [paper man] colle *f* blanche
~SPIRIT - [paint] (a petroleum fraction with a boiling range of I40-I95 deg. C. and a flash point over 32.2 deg. C., very much used as a solvent for air-drying finishes and in general as a substitute for turpentine) white-spirit *m*, essence *f* de térébenthine artificielle
~SPRUCE - [bot] épinette *f* d'Engelmann
~STAR APPLE - [bot] mululu *m*
~TELLURIUM - [min] sylvanite *f*, tellure *f* graphique
~TO BLACK AMPLITUDE RANGE - [telev] amplitude *f* totale blanc-noir
~VITRIOL - [chem] (commercial name for goslarite) vitriol *m* blanc
~WATER - [paper] (liquid obtained from pulp thickening operations, from which fine fibre is recovered) eau *f* blanche
WHITECAP - [gen] (foam-crested waves) vague *f* à tête d'écume
WHITEHEART CAST IRON - s. white-heart cast iron
WHITEN, to - [paint] blanchir
WHITENING - [paint] clanchiment *m*
[ind chem] (the material used for whitening) blanc *m* de chaux
[metall] couche *f* de métal blanc, étamage *m*
[leather ind] écharnage *m*
WHITES - [med] pertes *f*pl blanches, leucorrhée *f*
WHITESMITH - [plumb] ferblantier *m*
WHITETHORN - [bot] aubépine *f*
WHITEWALL - [rubber ind] (a type of tyre) pneu *m* à flancs blancs
WHITEWASH, to - [paint] peindre à la chaux, badigeonner, chauler
[constr] (term used in the brick trade to denote white spots on the surface of a brick) montrer la carie
WHITEWASH - [paint] blanc *m* de chaux, badigeon *m*
WHITEWOOD - [timber] bois *m* blanc
WHITING - [min] (finely powdered native calcium carbonate used as a filler and pigment) craie *f*, carbonate *m* de calcium
~MIXTURE - [rubber ind] (a talcum mixture preventing the sticking of the tube) talc *m* pateux
WHITLOW - [bot] (a weed) lépidier *m*
[med] panaris *m*, tourniole *f*
WHITWORTH SPLINE - [mech] rainure *f* Whitworth
WHITWORTH SCREW-THREAD - [mech] (British standard Whitworth thread) pas *m* anglais Whitworth, filetage *m* normal anglais
WHIZ, to - [text] essorer
WHIZZER - [mech] (a machine consisting of a perforated cylinder containing loose material to be dried. This is flung outwards by revolving paddles, thus removing the water by centrifugal force) séchoir *m* centrifuge
WHIZZING - [text] centrifugation *f*
WHOLE - [gen] tout, entier
~-BODY RADIATION METER - [nucl] anthroporadiamètre *m*
~-BRICK WALL - [constr] (a wall having a thickness equal to the length of a whole brick, i.e. 9 inches) mur *m* de l'épaisseur d'une brique
~-CIRCLE BEARING - [surv] (the horizontal angle measured from 0 to 360° clockwise, from true north to a given survey line) orientation *f* par boussole
~-COILED WINDING - [el] (armature winding for an alternator having one armature coil per pole) enroulement *m* à pas correspondant au pas polaire
~DEPTH - [mech] (of a gear tooth) hauteur *f* totale (d'une dent)
~LATEX RUBBER - [rubber ind] caoutchouc *m* de latex total
~MALT - [brew ind] malte *m* entier
~MILK - [food ind] lait *m* intégral, lait *m* non écrémé
~NOTE - [mus] ronde *f*
~NUMBER - [math] nombre *m* entier
~PLATE - [photo] format *m* (I6,5 x 2I,6)
~SPIRAL TAPPING - [rubber ind] incision *f* en spirale entière
~STEP - [mus] s. whole note
[acoust] (the larger pitch interval between successive tones of a major scale) ton *m* entier
~TIMBER - [timber] bois *m* dégrossi
~TONE - [acoust] s. whole step
~-TONE SCALE - [mus] gamme *f* hexatone
~TYRE RECLAIM - [rubber ind] régénéré *m* de carcasse de pneu
~WORKING - [mining] traçage *m*
~WORKINGS - [mining] travaux *m*pl de traçage
WHOLESALE, to - [comm] vendre en gros
WHOLESALE - [comm] vente *f* en gros
WHOLESALER - [comm] commerçant *m* en gros
WHOOPING COUGH - [med] (pertussis) coqueluche *f*
WHORL - [gen] spire *f*, tour *m* d'une spirale
[text] (the flywheel of a spindle) volant *m* de fuseau
[bot] verticille *m*
[anat] circonvolution *f*, verticille *m*
WHORTLEBERRY - [bot] airelle *f*, myrtille *f*
WICK - [gen] mèche *f*
[med] mèche *f*
[railw] (a lubrication device) dispositif *m* de lubrification
~-FEED OILING - s. wick oiling
~OILED BEARINGS - [mech] palier *m* à rotins, palier *m* à graissage par mèche
~OILING - [mech] graissage *m* par mèche
~REEL - [text] dévidoir *m* pour fil de mèche
~TRIMMER - [impl] mouchette *f*
~TUBE - [impl] porte-mèche *m*
WICKER - [bot] osier *m*
~BASKET - [gen] panier *m* d'osier
WICKET - [constr] (small door, or opening, framed in a larger door) porte *f* à piétons

WICKET - [gen] (small opening) guichet *m*
[constr] (a small gate, framed in a larger gate) petite porte *f* à claire-voie
~GATE - [railw] (a level crossing side gate) portillon *m*, portillon *m* de passage à niveau
WICKING - [text] tissu *m* de mèche de lampe
WIDE - [gen] large
[text] (of a fabric) large
~ANGLE - [photo] grand angle *m*, à grand angle
~-ANGLE DEFLECTION - [telev] (a deflection occurring over so wide an angle that the whole of the useful screen area can be covered) déviation *f* à grand angle
~-ANGLE LENS - [photo] objectif *m* à grand angle, objectif *m* grandangulaire
~-ANGLE REFLECTOR - [photo] réflecteur *m* à large faisceau
~APERTURE - [photo] grande ouverture *f*
~-BAND AMPLIFIER - [radio] amplificateur *m* à large bande
~-BAND CHROMINANCE DECODING - [telev] décodage *m* de chrominance à large bande
~-BAND CHROMINANCE RECEIVER - [telev] récepteur *m* à décodage de chrominance à large bande
~-BAND DIPOLE - [radio] (dipole aerial radiating a wide frequency band) dipôle *m* à large bande
~-BAND IMPROVEMENT - [radio] amélioration *f* du rapport signal/bruit
~-BAND INTERPOLATION - [telev] interpolation *f* par filtre à large bande
~-BAND RATIO - [radio] facteur *m* d'utilisation de bande
~-BASE MORTAR - [constr] mortier *m* avec large pied
~-BASE RIM - [rubber ind] (of tyres, a tapered rim) jante *f* à repos de talon conique
~-CUT FILTER - [telev] (for colour television) filtre *m* chromatique à large bande
~-EARED - [agric] à large épis
~FILM - [cin] (a film which is more than 35 *mm* wide) film *m* grandeur, film *m* large
~GAUGE - [railw] voie *f* large
~-LONG SHOT - [photo] plan *m* de grandensemble
~LOOM - [text] métier *m* large
~-MESHED COCOON - [text] cocon *m* à grandes mailles
~-OPEN - [gen] grand ouvert
~-SPREAD - [gen] étendu
WIDELY SPACED LATTICE - [nucl] (fuel lattice with a large pitch) réseau *m* à pas large
WIDEN, to - [gen] élargir, agrandir, étendre
WIDENING - [mech] élargissement *m*, aggrandissement *m*
WIDTH - [gen] largeur *f*
[mech] (of a ball bearing etc) largeur *f*
[text] (of a fabric) laize *f*, largeur *f*, lé *m*
[aero] (of the wing) envergure *f*
[constr] (of a vault) ouverture *f* (d'une voûte)
~CHECK - s. width indicator
~CHOKE - [telev] bobine *f* d'ajustage de la largeur de l'image
~CONTROL - [electron] (a circuit arrangement designed to regulate the extent of the horizontal deflection of the beam in a cathode-ray tube) commande *f* de largeur
~INDICATOR - [auto] repère *m* de gabarit
~OF CLOTH - [text] largeur *f* d'une étoffe
~OF CREST - [hydr] largeur *f* au sommet
WIDTH OF CUT - [mech & carp] (obtained with a tool) largeur *f* de la coupe
~OF RESONANCE - [radio] largeur *f* de résonance
~OF SPACE - [mech] (in gearing) largeur *f* du vide, intervalle *m*
~OF TOOTH - [mech] (in gearing) largeur *f* de la dent
~OF TRANSITION STEEPNESS - [radar] (edge steepness) raideur *f* de flanc
~OF WARP BEAM - [text] largeur *f* de l'ensouple
WIELD, to - [gen] manier
WIEN EFFECT - [el] (the increase in conductance of an electrolyte at high potential gradients) effet *m* Wien
WIG - [gen] perruque *f*
~-WAG - [railw] signal *m* oscillant
~-WAG SIGNAL - [telecomm] signal *m* oscillant
WIGAN - [text] bougran *m*
WIGGING - [text] (the wool obtained from the sheep's head) laine *f* de tête
WIGNER EFFECT - [nucl] (decomposition effect) effet *m* Wigner
~GAP - [nucl] (between two graphite blocks) espace *m* de Wigner
~GROWTH - [nucl] croissance *f* par effet Wigner
WIIKITE - [min] (a mineral which contains niobates, titanates, silicate and rare earths) wiikite *f*
WILD - [gen] sauvage, désert
[agric] inculte
[met] (of the wind] violent, furieux
[zool] (of animals) sauvage
~BEER - [brew ind] bière *f* sauvage
~-CAT TRAIN - [railw] (only USA) train *m* hors horaire
~-ENGINE - [mech] machine *f* haut-le-pied
~FLOWING - [oil ind] éruption *f* d'un sondage
~GAS - [metall] (in foundry work, gas in a blast furnace caused by incomplete combustion) gaz *m* de haut-fourneau
~HEAT - [metall] (of steel) charge *f* effervescente
~JUTE PLANT - [bot] jute *m* sauvage
~MULBERRY SILK - [text] soie *f* du mûrier sauvage
~PING - [mech] (in internal combustion engines) préallumage *m* du carburant dans l'antichambre
~SILK - [text] soie *f* sauvage
~SILK-COCOON - [text] cocon *m* sauvage
~STEEL - [metall] (in a wild heat) acier *m* très effervescent
[metall] (made from a wild heat) acier *m* sauvage
~TRACK - [cin] (a sound track which is recorded independently of any photographic track) piste *f* sonore non-synchronisée
~WALL - [cin] (part of stage decorations) panneau *m* transportable
~WELL - [oil ind] sondage *m* éruptif
~YEAST - [brew ind] levure *f* sauvage
WILDCAT, to - [oil ind] (the operation of drilling to seek unproved oil possibilities) effectuer un forage de recherche
WILDCAT - [zool] chat *m* sauvage
[zool] (USA only) lynx *m*
[oil ind] (drilling operation seeking unproved oil possibilities) forage *m* de recherche
WILLEMITE - [min] (native zinc orthosilicate and an ore of zinc) willémite *f*
WILLEY - [text] loup *m* batteur
WILLOW, to - [text] (to clean cotton, wool etc with a willow, i.e. a machine designed to give preliminary

cleaning by means of long spikes projecting from a revolving cone or cylinder) louveter, effilocher
WILLOW, to - [paper man] battre, louveter
WILLOW - [bot] saule *m*
[text] loup *m*, effilocheuse *f*, batteuse *f*
~BAST - [text] liber *m* de saule
WILLOWED WOOL - [text] laine *f* louvetée
WILLOWING MACHINE - [text] loup *m*, diable *m*, effilocheuse *f*
WILLY-WILLY - [met] (tropical revolving storm) cyclone *m* tropical
WILSON CHAMBER - s. Wilson expansion chamber
~EXPANSION CHAMBER - [phys] (a cloud chamber) cambre *f* à brouillard, chambre *f* à nuage, chambre *f* de Wilson
WILT, to - [bot] se flétrir, se faner
WIMBLE, to - [mech & carp] forer
WIMBLE - [mech] vrille *f*
[mining] (a boring auger) tarière *f*
[mining] (a scoop) cloche *f* de curage, cuiller *f*
WIN, to - [gen] gagner
[mining] (to prepare for mining) creuser
[mining] (to mine, to extract) extraire
[metall] (metal from ore) séparer
WINCH, to - [gen & mech] (to hoist by means of a winch) hisser (avec un treuil)
WINCH - [mech] (a hoisting machine) treuil *m*
[aero] (special winding gear fitted to a helicopter for rescue or load handling) cabestan *m*, treuil *m*
[mech] (a crank) manivelle *f*
[mech] (of a derrick crane) singe *m* d'une chèvre
~BECK - [text] barque *f* à tourniquet
~FOR RAISING ORE - [mining] treuil *m* d'extraction
~ROPE - [text] câble *m* pour treuils
~SUSPENSION - [aero] (flying rigging of a kite balloon) haubannage *m* de vol
WIND, to - [gen] (to pass a thread, a cord, a twine, a wire etc around an object) enrouler
[el] (as a coil) bobiner
[text] (the yarn) dévider, envider
[mech] (to hoist, as with a windlass) remonter, enlever
[gen] (a movement) serpenter
WIND - [met] (air current in the atmosphere moving in a generally horizontal direction) vent *m*
[med] flatuosité *f*, tympanisme *m*
[metall] vent *m* de la soufflerie
~ACROSS - [naut & aero] (component of wind transverse to ship or catapult) vent *m* traversier
~ANGLE - [aero & naut] (the angle between the direction of the wind and the true course, or of the heading) angle *m* de vent
~AXES - [aero] (a system of co-ordinate axes, of which the direction is determined by that of the wind and having its origin in the aircraft) axes *mpl* liés au vent
[met] axe *m* de cordonnées lié au vent
~BEAM - [constr] poutre *f* de contre-ventement
~BELT - [metall] (of a cupola) boîte *f* à vent
[agric] (a row of trees forming a barrier against the wind) ceinture *f* de plantes
~BLOWN - [met] balayé par le vent, poussé par le vent
~BORE - [mech] (of a pump) crépine *f* d'aspiration
~BORNE - [gen] porté par le vent
~-BOUND - [naut] retenu par des vents contraires
~-BOX - [metall] (of a furnace) boîte *f* à vent
WIND-BRACE, to - [constr] contreventer
~-BRACE - [constr] contrevent *m*, entretoise *f* de contreventement
~-BRACE CONNECTION - [constr] assemblage *m* de contreventement
~-BRACING - [constr] (in telecommunication and electrical works) contreventement *m*
~BREAK - [agric] brise-vent *m*
[agric] (in forestry) volis *m*
~BURN - [med] dermatite *f* due au vent
~CHEST - [mus] (of an organ) sommier *m*
~CONE - [met] (a wind direction indicator in the form of an elongated truncated cone of textile material, flown from a mast) manche *f* à vent
~CORRECTION ANGLE - [aero] (the angle applied to the course to obtain heading) angle *m* de correction du vent
~-CUTTER - [railw] coupe-vent *m*
[mus] (of an organ) lèvre *f* supérieure
~DIRECTION - [met] (the true azimuth direction of the wind) direction *f* du vent
~-DIRECTION INDICATOR - [instr] (any device actuated by the wind to show the direction of the latter at the surface of the earth) indicateur *m* de direction du vent
~DISPERSAL - [bot] (the dispersal of spores and seeds by the wind) dispersion *f* anémophile
~DOWN - [aero & naut] (the horizontal wind component in the fore-and aft line of the ship) vent *m* longitudinal
~-DRIVEN GENERATOR - [el] (an electric generator driven by the wind) génératrice *f* actionnée par le vent
~ENGINE - [mech] moteur *m* à vent
~EROSION - [geol] érosion *f* éolienne
~FLOW - [gen] courant *m* d'air
~FORCE - [phys] force *f* du vent
~FURNACE - [metall] haut fourneau *m*
~GAG - [acoust] (bag of thin cloth placed over a microphone when this is used out of doors, to eliminate the noises due to wind) écran *m* brise-vent
~GAP - [geol] tronçon *m* de vallée
~GAUGE - [instr] anémomètre *m*, indicateur *m* de pression du vent
~INDICATOR - [instr] (a device to show the direction and force of the wind at the surface of the earth) indicateur *m* anémométrique
~INSTRUMENTS - [mus] instruments *mpl* à vent
~LOAD - [phys] (on a structure) charge *f* due au vent
~MACHINE - [mech] (a machine used in a theatre to reproduce the sound of wind) éolienne *f*
~OFF, to - [gen] (the operation of unwinding) dérouler, détourner
~ON BOBBINS, to - [text] enrouler sur les bobines
~ON THE LAP, to - [text] enrouler la nappe
~ON THE SILK, to - [text] bobiner la soie
~POLLINATION - [bot] (the conveyance of pollen from anthers to stigmas by the wind) anémophilie *f*
~POWER - [mech] énergie *f* empruntée au vent, houille *f* bleue
~POWER STATION - [el] centrale *f* éolienne, usine *f* éolienne
~PRESSURE - [phys met etc] pression *f* du vent
~PUMP - [mech] (a pump which is operated by the force of the wind rotating a multi-bladed propeller) pompe *f* à aéromoteur

WIND RIPPLES - [geol] rides *fpl* éoliennes
~ROAD - s. windway
~ROSE - [instr] (a diagram showing the frequency of winds of various directions and forces at a definite locality and usually over a considerable period) rose *f* de vents
~SAIL - [naut] manche *f* à vent
~SCALE - [met] échelle *f* des vents
~SCOOP - [naut] manche *f* à vent
[railw] (ventilating jack) prise *f* de ventilation
[auto etc] saut-de-vent *m*
~SHADOW - [gen] (an area sheltered from the wind by a large object such as a town or hill) aire *f* à l'abri du vent
~SHEAR - [met] (the change of wind velocity with distance along a line normal to its direction) cisaillement *m* du vent
~SLEEVE - s. wind cone
~SOCK - s. wind cone
~SPOUT - [met] trombe *f*
~STAR - [met] (a graphic solution for wind direction and speed obtained by plotting drift angles observed on two headings approximately at right angles) étoile *f* des vents
~STEEPLY, to - [text] croiser fortement le fil
~STOP - [auto] joint *m* d'étanchéité contre le vent
~SUCKING - [vet] cornage *m*
~TEE - [instr] (a large type of wind indicator shaped like the letter "T" and fixed on or near a landing ground) T *m* d'atterrissage
~-TIGHT - [gen] imperméable à l'air
~TRUNK - [mus] (the piping distributing the air under pressure to the wind chests in an organ) porte-vent *m*
~-TUNNEL - [aero] (an apparatus in which a controlled airstream can be produced, used for aerodynamic experiments) soufflerie *f* aérodynamique
~TUNNEL BALANCE - [aero] (device used in a wind tunnel for measuring aerodynamic forces and moments acting on the model) balance *f* de soufflerie aérodynamique
~TURBINE - [mech] éolienne *f*
~TURN - [constr] courbure *f*
~UP, to - [mech] monter, remonter
~-UP - [mech] (mechanism for winding up sheet into rolls) dispositif *m* d'enroulement
~VALVE - [mus] (of an organ) soupape *f* du sommier
[constr] éolipile *f* de cheminée
~VANE - [met] girouette *f*
[mech] (of a windmill) ailette *f*
~-WAY - s. windway
[mining] voie *f* d'aérage
~WHEEL - [mech] roue *f* à vent
WINDAGE - [el] (of an electrical machine) frottement *m* de l'air
[met] (air disturbance) déplacement *m* d'air
[mech] espace *m* libre, jeu *m*
[aero] dérive *f* due au vent
~LOSS - [el] (ventilation loss) perte *f* par ventilation, perte *f* par frottement de l'air
WINDBRACING - [constr] contreventement *m*
WINDCORD - s. wind stop
WINDER - [constr] (a step, usually triangular, for a winding stair) marche *f* dansante, marche *f* tournante
[text] bobinoir *m*, dévidoir *m*
[horol] remonteuse *f* (d'horloges)
WINDER - [paper man] bobineuse *f*
[auto] lève-glace *m*
[mining] (electrically-driven engine for hoisting a cage up a vertical shaft) machine *f* d'extraction
WINDING - [gen] sinueux
[constr] (of a stair) tournant, en vis
[gen] (a bend, a turning) détour *m*, sinueux, plein de détours
[carp] (warp from a plane surface) gauchissement *m*
[mining] extraction *f*, remonte *f*
[el] (assembly of conductors forming a circuit in a machine or a piece of electrical equipment) enroulement *m*, enroulage *m*
[el] (the act of making a winding) bobinage *m*
[el] (the single round of an electrical coil) spire *f*
[text] (of threads) embobinage *m*, bobinage *m*
[text] (of the yarn with a winding frame) renvidage *m*
[text] (twisting) retordage *m*
[mech] (e.g. the tightening of a spring) bandage *m* (de ressort)
[constr] lacets *mpl*, tournant *m*
[text] (rope man) écheveau *m* (de corde)
~-BAR - [impl] garrot *m* de scie
~BUTTON - [horol] couronne *f* de remontage
~COEFFICIENT - s. winding factor
~DEVICE - [horol] dispositif *m* de remontage
~DIAGRAM - [el] (diagram showing the arrangement and sequence of an armature winding and its circuit connections) diagramme *m* d'enroulement
~DRUM - [mech] (of a hoisting machine) tambour *m* d'enroulement, cylindre *m* enrouleur
~-DRUM MACHINE - [el] machine *f* à tambour d'enroulement
~ENDS - [el] extrémités *fpl* d'enroulement à phases
~ENGINE - [mech] (a hoisting apparatus for mines) machine *f* d'extraction
~FACTOR - [el] (a factor which takes into account the difference between the vector and arithmetic sums of the electromotive forces induced in a series of armature coils in successive positions round the periphery of the armature) facteur *m* de bobinage
~FRAME - [text] dévideuse *f*
~FROM THE COP - [mech] cannetage *m*
~GEAR - [el] (the mechanical gear of an electric winder) mécanisme *m* d'enroulement
~IN COP FORM - [text] renvidage *m* sous forme de canette
~LAYER - [text] couche *f* de renvidage
~MACHINE - [text] bobinoir *m*
[el] (a machine for electrical windings) machine *f* d'ascenseur
~OF BOBBINS - [text] renvidage *m*
~OFF FRAME - [text] bâti *m* de déroulement
~OFF REEL - [text] asple *f* de dévidage
~ON BEAM - [text] ensouple *f*
~ON BOBBIN - [text] bobinage *m*
~OPERATION - [el] (the complete operation which is necessary to make a winding) bobinage *m*
~PINION - [horol] pignon *m* de remontage
~PITCH - [el] pas *m* résultant d'un enroulement en tambour
~PLANT - [el] (the apparatus constituting an electrically driven winder) bobineuse *f*
~RATCHET - [text] rochet *m*
~RATIO - [el] (of a transformer) rapport *m* de transformation

WINDING SHAFT - [horol] arbre *m* de remontage
~SHIELD - [el] capot *m* protecteur
~SPACE - [el] (the cross-sectional space which is available in an armature slot for the insertion of the insulated conductors) espace *m* d'enroulement
~SPEED - [text] vitesse *f* d'enroulement
~SPINDLE - [text] brochette *f* de la bobine
~SPOOL - [text] bobine *f* tournante
~STAIRS - [constr] escalier *m* tournant, escalier *m* en vis
~STEM - s, winding shaft
~STOCK - [text] petite têtière *f*
~SURFACE - [text] surface *f* du renvidage, couche *f*
~TACKLE - [naut] moufle *m*
~TERMINAL - [el] borne *f* d'enroulement
~-TYPE CURRENT TRANSFORMER - [el] transformateur *m* de courant sans primaire
~-UP GEAR - [mech] dispositif *m* à enrouler
~-UP WHEEL - [horol] roue *f* de remontage
~UPWARDS - [text] renvidage *m* vers le haut, renvidage *m* du fil
~WHEEL - [text] roue *f* de commande du différentiel
WINDLACE - [auto] (packing round the door as a protection against draughts) profilé *m* étanche à l'air
WINDLASS - [mech] (hoisting machine) vindas *m*, treuil *m*
[mech] (a hand-operated hoist) vireveau *m*
[naut] (a hoisting apparatus turning on a horizontal axis) cabestan *m* horizontal
WINDMILL - [agric.] moulin *m* à vent
[aero] (an airscrew driven by the airstream produced by the motion of an aircraft and used to drive auxiliaries, e.g. an electric generator) moulinet *m* pour entraînement d'auxiliaires
~BRAKE STATE - [aero] (condition in which axial flow through, and outside the rotor disk area, is co-directional with the rotor thrust) fonctionnement *m* en moulinet-frein
WINDMILLING - [aero] (rotation of a propeller without engine power, by the effect of the forward movement of the aircraft) fonctionnement *m* en moulinet
WINDOW - [constr] (opening in a wall to provide access for light and ventilation) fenêtre *f*
(the assembly of shutter, framework etc) ferrures *f* pl de croisée
[railw] glace *f*
[comm] vitrine *f*
[phys] (aperture for the passage of particles or radiation) fenêtre *f*
[radar] (a flasher) réflecteur *m* radar, rubans *m* antiradar
[anat] fenêtre *f* du tympan
[comm] guichet *m*
[telev] (double limiter) circuit *m* à déclenchement périodique
[nucl] (energy range of high transparency in the total neutron cross section of a material) zone *f* d'énergie transparente
[electron] (channel width) largeur *f* de canal
~AMPLIFIER - [electron] amplificateur *m* à fenêtre
~BAR - [constr] (iron bars between the panes of a sash) petit bois *m* en fer, barreau *m* de fenêtre
[constr] (on a sill) barre *f* d'appui
[constr] (used to secure a shutter) barre *f* de fermeture

WINDOW BLIND - [carp] jalousie *f*
~BOARD - s. window sill
~BOX - [carp] bac *m* à fleurs
~CLOUD - [radar] (a large amount of metallized paper interference strips dropped by aircraft) nuage *m* de rubans antiradar
~COUNTER TUBE - [radiat] tube *m* compteur à fenêtre
~DRESSER - [gen] étalagiste *m*
~DRESSING - [comm] arrangement *m* de la vitrine, art *m* de l'étalage
~DROPPING - [radar] (flasher dropping) créneau *m* de lancement
~EFFICIENCY RATIO - [photo] (the day light factor) coefficient *m* de lumière naturelle
~EMBRASURE - [constr] embrasure *f* de fenêtre
~ENVELOPE - [comm] enveloppe *f* à fenêtre
~FASTENER - [mech] (a sash lock) fermeture *f* de cadre de fenêtre
~FRAME - [constr] dormant *m* de fenêtre, châssis *m* de fenêtre, châssis *m* dormant
~FRAME AERIAL - [radio & telev] antenne *f* de vitre
~GLASS - [glass man] (sheet glass) verre *m* à vitres
~GRATING - [gen] grille *f* de fenêtre
~GROOVE - [constr] feuillure *f* de fenêtre
~JAMB - [constr] montant *m* de fenêtre
~LATCH - s. window fastener
~LIFT - s. window raiser
~LINTEL - [constr] linteau *m* de fenêtre
~LOCK - [mech] crampon *m* de fermeture
~MOULDING - [auto] moulure *f* de glace
~OPENING- [constr] ouverture *f* de fenêtre
~-PANE - [glass man] vitre *f*, carreau *m*
~POST - [constr] poteau *m* d'huisserie
~RAISER - [auto] lève-glace *m*
~REGULATOR - [auto] lève-glace *m*
~SASH - [constr] châssis *m* de fenêtre, châssis *m* mobile
~SCATTERING - [radiat] (in a counter tube) diffusion *f* dans la fenêtre
~SHUTTER - [constr] persienne *f*
~SILL - [constr] rebord *m* de fenêtre, appui *m* de fenêtre
~TRIM - [comm] (used for publicity) étalage *m*
~WINDER - [auto] lève-glace *m*
WINDOWLESS PHOTOMULTIPLIER - [electron] photomultiplicateur *m* sans fenêtre
WINDSCREEN - s. windshield
WINDSHED - [met] (line of division between katabatic winds blowing down different slopes: analogous to a watershed) ligne *f* de séparation des vents
WINDSHIELD - [auto] (transparent screen fixed in front of the car) pare-brise *m*
~DEFROSTER - s. windshield de-icer
~DE-ICER - [auto & aero] (device for de-icing a windshield) dégivreur *m* de pare-brise
~FRAME PROFILE - [auto] cadre *m* du pare-brise
~HEATER - [auto & aero] dégivreur *m*
~REGULATOR HANDLE - [auto] poignée *f* de commande de l'ouverture du pare-brise
~VIZOR - [auto] pare-soleil *m*
~WASHER - [auto] lave-glace *m*
~WIPER - [auto] (device for wiping rain etc from a windshield) balai *m* d'essuie-glace, essuie-glace *m*
~WIPER ARM - [auto] (a mechanically-driven oscillating arm carrying a wiping blade and forming the active part of a windshield wiper) bras *m* d'essuie-glace

WINDSHIELD WIPER BLADE - [auto] balai *m* d'essuie-glace
WINDSOCK - s. wind cone
WINDSTORM - [met] tempête *f* de vent
WINDTIGHT - [mech etc] étanche à l'air
WINDWARD - [naut] au vent
WINDWAY - [mining] galerie *f* d'aérage, voie *f* d'air
[mus] (of a wind instrument) lumière *f*
WINE - [gen] (the fermented juice of grapes, or other fruit) vin *m*
~ACID - [chem] (tartaric acid) acide *m* tartarique
~BIN - [impl] porte-bouteilles *m*
~-BOTTLE POTENTIAL - [nucl] potentiel *m* à fond de bouteille
~CELLAR - [constr] cave *f* au vin
~GROWING - [agric] viniculture *f*, viticulture *f*
~MAKING - [agric] vinification *f*
~PRESS - [agric] pressoir *m*
~VINEGAR - [ind chem] (vinegar made by acetification of sour low grade wines) vinaigre *m* de vin
WINERY - [agric] caves *fpl*
WING - [zool] (organ of flight) aile *f*
[aero] (one lateral half of a main plane) aile *f*
[constr] (section of building projecting from the main part of it) aile *f*
[cin] (the sides of the stage which are unseen from the audience) secteur *m* d'obscuration
[auto] (the curved mudguard) garde-boue *m*
[constr] (the side section of a door) battant *m* (d'une porte)
[gen & zool] course *f* vol, essor *m*
~ALIGNMENT - [aero] alignement *m* alaire
~AREA - [aero] (the surface of a wing bounded by leading and trailing edges, the wing tips and the lines of intersection of wing and fuselage) surface *f* alaire
~AUGER - [oil ind] alésoir *m*
~AXIS - [aero] (a line passing through the aerodynamic centres of all the wing sections) axe *m* alaire
~BIT - s. wing auger
~BOLT - [mech] (a bolt of which the head has lugs for turning with the fingers, as in a wing-nut) boulon *m* à oreilles
~BOLT LOCK - [mech] serrure *f* à harpons
~CAM - [text] came *f* auxiliaire de descente
~CLEARANCE ANGLE - [aero] (the angle between the horizontal and a line drawn from the wing tip to the point of contact of the landing wheel on the same side and the ground, i.e. the maximum angle through which an aircraft on the ground can be tilted laterally) angle *m* d'inclinaison maximum de l'aile
~COMPASSES - [carp] (a form of quadrant dividers) compas *m* quart de cercle, compas *m* droit
~CONTOUR - [aero] profil *m* de l'aile
~CONTROL - [aero] contrôle *m* par déplacement des ailes
~COVERING - [aero] revêtement *m* d'aile
~FAN - [impl] ventilateur *m* à ailettes
~FENCE - [aero] (a chordwise projection from the surface of a wing, used to modify the distribution of pressure thereon) barrière *f* de décrochage
~FLAP - [aero] volet *m*
~FLUTTER - [aero] flutter *m* alaire, vibration *f* des ailes
~-FOOTED - [zool] chiroptères *mpl*
~GATE SLUICE - [hydr] écluse *f* à porte en éventail
~LAMP - [auto] feu *m* d'aile
WING LOAD - [aero] charge *f* alaire
~LOADING - [aero] (the weight carried by unit wing area) charge *f* alaire
~MIRROR - [auto] rétroviseur *m*
~MIXER - [metall] mélangeur *m* à ailettes
~NUT - [mech] (a nut provided with wings or lugs so as to be turned easily by hand) écrou *m* à oreilles, écrou *m* papillon
~OF CREEL - [text] bras *m* du cantre
~OF WARPING MILL - [text] montant *m* vertical du moulin
~-OVER - [aero] (a manoeuvre consisting of a climbing turn carried nearly to stalling point: the nose is then allowed to drop and the aircraft put into a dive, returning to normal flight on a course approximately reciprocal to the original one) renversement *m* sur l'aile
~PROFILE - [aero] (the outline of a wing section) profil *m* de l'aile
~RADIATOR - [aero] (a radiator fitted in a wing) radiateur *m* d'aile
~RESISTANCE - [aero] résistance *f* de volet
~RIB - [aero] (a cordwise member in a wing structure) nervure *f* de volet
~-ROOT FITTING - [aero] (a special part used to join the inner end of a wing to the fuselage structure) chape *f* de jonction demi-voilure
~SECTION - [aero] (a cross section of a wing at a given point, taken parallel to a plane of reference, usually that of symmetry) profil *m* de l'aile
~SETTING - [aero] (the angle between the plane of the chord and the plane in which the longitudinal axis of the aircraft lies) calage *m* de la voilure
~SHELL - [zool] avicule *f*
~SKID - [aero] (a skid placed near the wing tip to protect the latter during landing) patin *m* d'aile
~SLOT - [aero] fente *f* alaire
~SPAN - [aero] envergure *f* d'avion
~SPAR - [aero] (a principal spanwise member of a wing structure) longeron *m* d'aile
~SPAR BOX - [aero] (box structure part of wing spar) caisson *m* voilure
~STULL - [oil ind] plateforme *f* à aile
~TENSION - [text] frein *m* de tension à ailettes
~THICKNESS - [aero] épaisseur *f* de l'aile
~TIP - [aero] (the outer extremity of a wing) pointe *f* de l'aile
~-TIP FLARE - [aero] (a pyrotechnic device fixed to a wing tip to light the ground when landing at night) porte-fusée *m* de pointe d'aile
~-TIP FLOAT - [aero] (a small buoyant structure fixed near the wing tip of a flying boat or seaplane) flotteur *m* de pointe d'aile
~UNDERSIDE - [aero] ventre *m* de l'aile
~UNIT - [aero] cellule *f* d'avion
~VALVE - [mech] (a conical-seated valve guided by radial vanes or wings fitted inside the circular port) soupape *f* à oreilles
~WALKWAY - [aero] (a part of a wing strengthened to allow persons to walk upon it when the aircraft is on the ground) passerelle *f* de l'aile
~WALL - [constr] (lateral wall build on an abutment and designed to retain earth) mur *m* à ailes
WINGED DISC - [text] disque *m* à ailettes
~ROCKET - [astronaut] fusée *f* à ailes
WINGSPREAD - [aero] envergure *f*
WINK, to - [gen] cligner les yeux

WINK, to - [gen] (to flash intermittently) éclairer
WINKING - [auto etc] clignotement *m*
[telev] papillotement *m* partiel
~INDICATOR - [instr] clignotant *m*
WINKLE - [zool] bigorneau *m*
WINNING - [mining] (the operation of mining an ore) extraction *f*, abattage *m*
[mining] (the opening up of a new portion of a coal-seam) champ *m* d'exploitation
~HEADWAY - s. winning level
~LEVEL - [mining] galerie *f* de traçage
~PLACE - [mining] siège *m* d'extraction
WINNINGS - [mining] récolte *f*
WINNOW, to - [agric] (the grain) tarater (le grain), sasser, passer au van
WINNOWER - [agric] sasseur *m* mécanique, van *m* mécanique, vanneuse *f*
WINNOWING MACHINE - s. winnower
WINSEY - [text] (cotton flannelette of good quality) tissu *m* mi-lin mi-coton
WINTER, to - [gen] hiverner
WINTER - [astr] hiver *m*
~ANNUAL - [bot] (a short-lived plant growing and setting seeds during the winter) plante *f* annuelle hibernale
~-BERRY - [bot] apalachine *f*
~CLOTH - [text] étoffe *f* d'hiver
~CORN - [agric] céréales *fpl* d'automne
~COVER - [auto] écran *m* anti-gel
~FRONT - [auto] (a radiator cover) cache *m* de radiateur
~GRAIN - s. winter corn
~-PLOUGH, to - [agric] entre-hiverner
~PRUNING - [agric] taille *f* en sec, taille *f* d'hiver
~REST PERIOD - [biol] hibernation *f*
~WOOL - [text] laine *f* de printemps
WINTERGREEN OIL - [chem] (an oil obtained from Gaultheria procumbens and used in perfumery, as a flavourant, and in medicine) essence *f* de Gaulthéria, essence *f* de wintergreen
WINTERING - [agric] hivernage *m*, hibernation *f*, mise *f* en hivernage
~OF THE EGGS - [zool] hivernage *m* des graines
~OF THE PUPA - [zool] hivernage *m* de la chrysalide
WINTERIZATION - [auto aero etc] (to adjust an engine etc for running at low temperature) réglage *m* d'un moteur pour l'hiver
WINTERIZE, to - [auto aero etc] régler un moteur pour l'hiver
WINTERIZED EQUIPMENT - [auto] équipement *m* d'hiver
WINZE - [mining] (a shaft sunk from one level to another in a mine) descenderie *f*
WIPE, to - [gen] essuyer
[mech] (to apply solder with a piece of greased cotton or leather) ébarber (un joint)
[el acoust] (a record, or recorded tape) effacer
WIPE - [cin] (transition in which the new picture begins as a small area and subsequently grows until it spreads over the whole screen) fondu *m* masqué graduel
[telev] commutation *f* par volet, fondu *m* éffacé
~-JOINT - [metall] noeud *m* de soudure
~TEST - [radiat] (a smear test) essai *m* de frottement
WIPED GALVANIZED IRON - [metall] fil *m* galvanisé à chaud et passé entre rouleaux essuyeurs

WIPER - [gen] torchon *m*
[mech] came *f*, mentonnet *m*, virgule *f*
[telecomm] (in a selector, the conducting arm which rotated over a row of contacts and stops at an outlet) balai *m*, frotteur *m*, bras *m* porte-contacts
[el] (of a rheostat) contact *m* glissant
[text] (in weaving, a mechanism for converting a rotating to a reciprocating motion) excentrique *m*
[auto] (short for windshield wiper) essuie-glace *m*
~ARM - [auto] bras *m* d'essuie-glace
~BLADE - [auto] (the part of a screen wiper which presses against the glass and wipes off water or snow from it) balai *m* d'essuie-glace
~MOTION - [mech] mécanisme *m* des excentriques
~SHAFT - [mech] arbre *m* à cames, arbre *m* des cames
~WHEEL - [mech] roue *f* à cames
WIPING - [mech] glissant (d'un mouvement de came)
[el acoust] effacement *m*
~CONTACT - [electron] contact *m* de passage
~CURRENT - [telev] (in video recording) tension *f* d'effacement
~GLAND - [mech] (for cables, a projecting sleeve on a junction box) douille *f* soudée de jonction
~HEAD - [telev & el acoust] tête *f* d'effacement
WIRE, to - [gen] munir d'un fil métallique
[el] (to instal the electrical wiring) canaliser, poser des fils électriques dans
[telecomm] (to cable) télégraphier
WIRE - [gen & metall] fil *m* métallique, fil *m* de fer
[el] fil *m*, conducteur *m*
[aero] (for the structure) corde *f* à piano (de cellule)
[telecomm] (colloq. for telegram) télégramme *m*
[paper man] (term used for wire mesh thickening pulp) toile *f* (de machine à papier)
~-BAR - [el] (electrolytic wire-bar, a bar of electrolytically refined copper for rolling to form wire) barre *f* de cuivre électrolytique
[metall] fer *m* à tirer
~BASKET - [text] panier basculeur *m* en fils de fer
~BINDER - [bookbind] agrafe *f*
~BINDING - [text] liage *m* par fil métallique
~-BOUND RUBBER HOSE - [rubber ind] tuyau *m* en caoutchouc à gaine métallique
~-BREAKAGE LOCK - [railw] (a broken wire locking device) appareil *m* contrôleur de rupture de fil
~BRIDGE - [build] pont *m* suspendu
~BROADCASTING - [radio] (a radio programme sent over wires) télédiffusion *f*
~BRUSH - [impl] (a brush with bristles of steel wire, used for cleaning metal parts etc) brosse *f* en fil de fer
~BRUSHING - [metall] traitement *m* à brosse en fil de fer
~CARRIER - [mech] poulie *f* guide-fils
~CLAMP - [el] serre-fil *m*
~CLOTH - [paper man] (a continuous band of wire gauze) écran *m* en toile métallique
~CLOTH WEAVING - [text] tissage *m* de la toile métallique
~COATING - [el] revêtement *m* du conducteur
~COIL - [el] bobine *f* de fil
~COMPENSATOR - [railw] (for a suspended contact line) tendeur *m*
~CONTROL - [el] (remote control) télécommande *f*
~CORE - [el] noyau *m* en fil

WIRE COVERED ROLLER - [text] tambour *m* garni d'aiguilles
~COVERING COMPOUND - [plast ind] (elastomeric plastics compound suitable for use as a cable insulant) matière *f* d'isolement de câble
~CUTTER - [tool] coupe-fil *m*
~CUTTERS - [tool] cisailles *fpl*, pince *f* coupante
~DOG - [mech] chien *m* tendeur à pouce articulé, tendeur *m* articulé pour fils
~DRAWER'S PLATE - [metall] filière *f*, banc *m* à étirer, plaque-filière *f*
~DRAWING - [metall] (the process of reducing the diameter of a rod to a wire by pulling it through successively smaller holes in a herd-steel dieblock) étirage *m* de fil, tréfilerie *f*
~DRAWING MILL - [metall] tréfilerie *f*
~-EDGE - [tool] (used to sharpen tools) morfil *m*, marfil *m*, morflat *m*
~FAULT - [el] défaut *m* de fil
~FENCE - [gen] clôture *f* en fil de fer
~-GAUGE - [tool] (a tool formed with notches on its edge for determining the gauge diameter of wire) jauge *f* pour fils métalliques, jauge *f* de tréfilerie
~GAUZE - [ind chem] (square of wire mesh used to support vessels over a flame etc) tamis *m* métallique
[gen] toile *f* métallique
~GLASS - [glass man] verre *m* armé, verre *m* à fil de fer noyé
~GRATING - [electron] (of a waveguide) filtre *m* en grille
[mech] grille *f* en fil de fer
~GUARD - [gen] grille *f* de protection
~GUIDE - [el] fil-guide *m*
~GUY - [tool] hauban *m* en fil souple
~HEAD - [text] tête *f* du fer
~HEAD BRIDGE - [text] projection *f* de la tête du fer
~HEAD SPRING - [text] bande *f* en fer blanc élastique
~-INSULATING COMPOUND - [el] (rubber or plastic-based compound with good flexibility and dielectric properties used as an insulant in electric cables) matière *f* d'isolement de câbles
~JOINT - [mech] liage *m* par fil métallique
~LAMP - [el] lampe *f* à filament métallique
~LATTICE - [text] tablier *m* sans fin en fil métallique
~LOCKING - [mech] (for nuts) freinage *m* à fil
~MESH - [impl] (wire gauze used for straining liquids) filtre *m* en toile métallique
~-MESH OSCILLATING SCREEN - [mining] (for coke) grille *f* à secousses
~-MESH RECIPROCATING SCREEN - s. wire-mesh oscillating screen
~MILL - [metall] train *m* à serpenter, laminoir *m* à fil
~NETTING - [metall] treillis *m* métallique, grillage
~-NETTING CUTTER - [tool] coupe-net *m*
~PENETROMETER - [instr] (for radiations) pénétramètre *m* à fils
~PICK - [text] fer *m* placé sous la chaîne de poil
~PLIERS - [tool] pince *f* à tréfiler
~PRINTER - [comput] imprimante *f* à fils, imprimante *f* à bloc
~PROTECTION CAGE - [el] grille *f* de protection
~RECORDER - [el acoust] enregistreur *m* magnétique à fil
~REEL - [el] dérouleuse *f* pour dépose des fils
~RELEASE - [photo] déclencheur *m* métallique

WIRE ROD - [metall] fer *m* à étirer
~ROD PRODUCTS - [metall] produits *mpl* du fil machine
~ROD ROLLS - [metall] fil *m* machine en rouleaux
~RODS - [metall] fil *m* machine
~ROPE - [metall] (a steel rope made by twisting a number of strands over a central core) câble *m* métallique
~-ROPE CLAMP - [tool] serre-câble *m*, manchon-bride *m*
~ROPE CLIP - [mining] serre-câble *m*
~-ROPE HAULAGE - [mech] traction *f* par câble métallique
~-ROPE PULLEY - [mech] poulie *f* pour câble métallique
~-ROPE SOCKET - [mech] douille *f* pour câble métallique
~-ROPE SPEAR - [impl] harpon *m* pour câbles métalliques
~ROPEWAY - [transp] voie *f* à câble aérien
~SAW - [mining] (used in quarrying) scie *f* hélicoïdale, câble *m* helicoïdal
~SCRATCHER - [impl] (tool used to make scratch marks on a cement surface) grattoir *m*
~SCREEN - [paper man] (mesh of wire in the form of an endless belt or cylinder on which the paper web is formed from the wet pulp) tamis *m* de fil métallique
~-SHEATED CABLE - [metall] câble *m* revêtu d'une cuirasse de fils de fer
~SHED - [text] pas *m* du poil
~SIDE - [paper man] (of a sheet of paper) côté *m* toile
~SIEVE - [impl] crible *m* en toile métallique
~SPOOL - [text] bobine *f* de fil métallique
~SPRING RELAY - [el] relais *m* à fil élastique
~-STAPLE - [mech] clou *m* à deux pointes
~-STITCH, to - [bookbind] brocher
~STITCHING - [bookbind] (securing by means of wire staples) brocage *m* au fil métallique, piqûre *f* au fil métallique
~STRAIGHTENER - [mach tool] machine *f* à dresser les câbles
~STRAINER - [tool] cric-tenseur *m*, raidisseur *m*
~-STRAINING VICE - [telecomm] étau *m* tenseur, étau *m* raidisseur
~STRETCHER - [mech] cric-tenseur *m*, raidisseur *m*
~STRING - [mus] corde *f* métallique
~STRIPPING ROLLER - [text] rouleau *m* débourreur garni de fils métalliques
~TAPPING - [telecomm] captage *m* de messages télégraphiques
~TELEGRAPH NETWORK - [telecomm] réseau *m* télégraphique
~THREAD GUIDE - [text] queue *f* de cochon, guide-fil *m* métallique
~TRAIN - [plst ind] (the whole assembly of a wire-coating plant from extruder to take-up spool) circuit *m* du fil
~VIEWFINDER - [photo] viseur *m* à cadre
~WHEEL - [mech] (for motorcycles) roue *f* à rayons-fil
[tool] (a tool for cleaning metal pieces) brosse *f* circulaire métallique
~-WOUND ARMATURE - [el] (armature wound with insulated copper wire) induit *m* à enroulement de fils de cuivre isolés
~-WOUND HOSE - [mech] (hose protected with a

spiral wrapping of wire) tuyau *m* à armature de fil métallique fourré
WIRE-WOUND RESISTOR - [electron] résistance *f* bobinée
~WOVEN IN - [text] fer *m* passé dans le pas
~WRAPPED SCREEN - [min] crépine *f* à fil enroulé
WIRECORD FABRIC - [rubber ind] (for tyres) nappe *f* de fils d'acier
WIRED - [gen] (with a wire netting) monté sur fil de fer, armé de fil de fer
[auto] (of tyres) à tringles
~HOSE - s. wire-wound hose
~IN CHECK - [comput] contrôle *m* incorporé
~SYSTEM - [telev] filovision *f*, filodiffusion *f*
~TELEVISION RECEIVER - [telev] téléviseur *m* à filovision
WIREDRAW, to - [metall] tréfiler, fileter le métal, étirer en fil, laminer
WIREDRAWER - [mech mach] tréfile *f*
WIREDRAWING - s. wire drawing
[mech] (of steam) laminage *m* (de la vapeur)
~BLOCK - [metall] bobine *f* de tirerie
~MACHINE - [metall] étireuse *f*
WIREDRAWN STEAM - [mech] vapeur *f* laminée
WIRELESS, to - [telecomm] envoyer un message par radio
WIRELESS - [radio] (alternative term for radio) radio *m*
[telecomm] sans fil
~CONTROL - [mech] radiocommande *f*
~OPERATOR - [aero naut ect) sans-filiste *m*
~TELEGRAPHY - [telecomm] radiotélégraphie *f*
~TELEPHONY - [telecomm] radiotéléphonie *f*
WIREWORK - [metall] étirage *m*, tréfilerie *f*
[gen] grillage *m* métallique
WIREWORM - [zool] larve *f* de taupin
WIRING - [el & telecomm] (the system of wires) câblage *m*, filerie *f*, pose *f* de fils, montage *m*
[metall] repliage *m* du bord
~BOARD - [contr] tableau *m* de connexions
~DIAGRAM - [el] plan *m* de câblage
~HARNESS - [el] (the wires which are assembled and taped together for installation) faisceau *m* de fils
~JUNCTION BLOCK - [el] boîte *f* de connexion
~LUG - [mech] attache-tirant *m*
~PLATE - [aero] attache-fils *m*
~POINT - [el] (in an interior wire installation) accouplement *m* d'installation
WIREWAY - [el] tuyau *m* de protection pour câbles électriques
WISHBONE - [auto] (of a suspension) bras *m* oscillant
WITCHING - [hydr] (water divining) sourcellerie *f*
~ROD - [hydr] baguette *f* de sourcier
WITH CLOSED-CIRCUIT VENTILATION - [el] ventilé en circuit fermé
~NATURAL COOLING - [el] à refroidissement naturel
~OPEN-CIRCUIT VENTILATION - [el] ventilé en circuit ouvert
~THE GRAIN - [text] (in the same direction as that of the natural fibres of a material) parallèle à la direction des fibres
WITHDRAW, to - [gen] retirer, enlever
~A PLUG, to - [el] couper (le courant)
~THE CLUTCH - [mech] débrayer
~THE PATTERN - [metall] (in foundry work) retirer le modèle
~THE PLUG PIN - [metall] retírer les baguettes
WITHDRAWAL - [gen] retrait *m*
[comm] (of funds) retrait *m* (de monnaie)
~FLANGE - [mech] (of a clutch) bride *f*
~OF PATTERNS - [metall] démoulage *m* des modèles
~OF SLUDGE - [hydr] extraction *f* des boues
~OF THE ORE - [mining] enlèvement *m* du minerai
~SHAFT - [mech] arbre *m* de commande de débrayage
~SLEEVE - [mech] manchon *m* d'arbre de commande de débrayage
WITHDRAWING - [metall] retrait *m* de la matrice
[mining] déboisage *m*
~DEVICE - [metall] dispositif *m* d'enlèvement
~FORCE - [metall] force *f* de retrait
WITHER, to - [bot & gen] se flétrir, déperir, se dessécher
WITHERITE - [min] (native barium carbonate. It is a source of barium compounds) withérite *f*
WITHERS - [zool] (the region of the horse's back above the shoulders) garrot *m*
WITHOUT - [gen] (prep) sans, à l'extérieur
~CURRENT - [el] (de-energized) sans courant
WITHSTAND, to - [gen & phys] supporter, résister
~TEST - [phys] essai *m* de résistance
WITNESS, to - [gen & leg] témoigner
WITNESS - [leg] témoin *m*, témoignage *m*
~MARK - [mech] (of a mechanical operation) témoignage *m*
WOBBLE, to - [gen] vaciller, balloter
[auto] (to move irregularly and unsteadily) tanguer
[mech] (said of a wheel when it is inclined on its rotation angle) tourner à faux
[auto] flotter, se dandiner
WOBBLE - [gen] oscillation *f*, vacillation *f*
[mech] mouvement *m* à faux
[auto] (of the front wheels) shimmy *m*, dandinement *m*
~DRILL - [tool] mèche *f* à talon
~MODULATION - [radio] wobbulation *f*
~PLATE - [mech] (of a wobble-plate engine) plateau *m* oscillant d'un moteur
~-PLATE PUMP - [mech] pompe *f* à rotor à plateau oscillant
~PUMP - [mech] (used for refuelling an aircraft) pompe *f* auxiliaire à bras
~SEAL - [nucl] (flexible seal) scellement *m* flexible
WOBBLER - [metall] (of a rolling mill) branleur *m*, trèfle *m*
[radio] s. wobbulator
WOBBLING OF WHEELS - [auto] dandinement *m* des roues
WOBBULATION - [telev] (oscillator frequency variation in a frequency band, used in cathode-ray circuits) wobbulation *f*
WOBBULATOR - [radio] (a signal generator) wobbulateur *m*
[telev] générateur *m* wobbulé
WOLF - [zool] loup *m*
WOLFRAM - [chem] (tungsten) tungstène *m*, wolfram *m*
~CARBIDE - [chem] (USA for tungsten carbide) carbure *m* de tungstène
WOLFRAMATE - [chem] wolframiate *m*, tungstate *m*
WOLFRAMIC - [chem] tungstique
WOLFRAMITE - [min] (a major ore of tungsten) wolframite *f*
WOLLASTONITE - [min] (native calcium silicate, with uses in papermaking, paint manufacture, as a rein-

forcing agent for plastics and in rubber compounding) wollastonite *f*
WOMB - [anat] matrice *f*
WOMP - [cin] (USA colloq. for a flare spot) tache *f* hyperlumineuse
WOOD - [gen] (a forest, a large compact collection of trees) bois *m*, peuplement *m*
[gen] (the hard fibrous material of the tree) bois *m*
[gen] (a synonym of barrel) tonneau *m*, fût *m*
~ALCOHOL - [chem] (methanol, obtained by the destructive distillation of wood) alcool *m* de bois
~ASHES - [agric] (used as fertilizer) cendres *fpl* de bois
~BASE - [gen] embase *f*, patère *f*
~-BENDING MACHINE - [mech] (used in carpentry and joinery) cintreuse *f* pour bois
~BLOCK - [carp] (for floors etc) pavé *m* de bois
~BODY JACK - [mech] cric *m* fût bois
~-BORER - s, woodworm
~BRICK - [constr] brique *f* de bois
~BUD - [bot] bouton *m* à bois
~CASING - [el] baguette *f* électrique
[gen] revêtement *m* en bois
~CEMENT - [ind chem] béton *m* de bois
~CHIP - [gen] copeau *m* de bois
~CHIPPER - [paper man] coupeuse *f* à bois
~CHISEL - [tool] ciseau *m* de menuisier
~CHUCK - [mach tool] mandrin *m* pour tour à bois
~COAL - [ind] (charcoal) charbon *m* de bois
~COUNTERSINK - [tool] fraise *f* à bois
~-CUTTER - [gen] bûcheron *m*
~-CUTTER'S SAW - [impl] scie *f* à bûches
~ENGRAVING - [print] (the art of cutting a design on wood for printing, a woodcut) gravure *f* sur bois, xylographie *f*
~FELLING - [agric] coupe *f*, production *f* de bois
~FIBRE - [text] fibre *f* de bois
~FLOUR - [ind chem] (finely pulverized wood used as a filler for plastics, oil cloth etc, as an absorbent for nitroglycerine, and in the production of polishes) farine *f* de bois
~-FREE PAPER - [paper man] papier *m* sans bois, papier *m* sans pâte
~GAS - [ind] (a fuel gas obtained as a by-product of charcoal manufacture) gaz *m* de bois
~GRINDING STONE - [paper man] meule *f* de coupeuse à bois
~-HANDLED SPANNER - [impl] clef *f* à manche bois
~HURDLE SCRUBBER - [hydr] ruisseleur *m* à lattes de bois
~LAGGING - [gen] revêtement *m* en bois
~LATH CONE - [text] cône *m* à lattes de bois
~LINING - [text] tablette *f* de bois
~LOCK - [shipbuild] clef *f* de gouvernail
~MEADOW GRASS - [bot] pâturin *m* des bois
~MEAL - s. wood flour
~MOULDING - s. wood casing
~NAPHTA - [chem] méthanol *m*
~OPAL - [min] opale *f* xyloïde
~PACKING - [acoust] étoupage *m* en laine
~PANELLING - [constr] panneautage *m* en bois
~PAPER - [paper man] papier *m* de bois
~PASTE - [text] pâte *f* mécanique de bois
~PAVING - [constr] pavage *m* en bois, pavé *m* de bois
~PAVING BLOCK - [constr] pavé *m* de bois
[print] planche *f*
WOOD PIGEON - [zool] palombe *f*, ramier *m*
~PILING - [constr] palplanches *fpl* en bois
~PIRN - [text] fuseau *m* en bois, tube *m* en bois
~PLASTICS MATERIAL IRRADIATION - [radiat] irradiation *f* de bois plastique
~PULP - [paper ind] (disintegrated wood, used in paper manufacture) pâte *f* de bois
~PULP BLACK - [ind chem] (pigment made by carbonizing wood pulp) noir *m* de pâte de bois
~PULP BOARD - [paper man] carton *m* bois, carton *m* de pâte mécanique
~PULP YARN - [text] fils *mpl* de pâte de bois
~SAW - [impl] scie *f* à bois, scie *f* à bûches
~SCREW - [carp] vis *f* à bois
~SEASONING - [carp] séchage *m* du bois
~SHAVINGS - [gen] copeaux *mpl* de bois
~SHORING - s. wood piling
~SPIRIT - [chem] (a destructive-distillation product of wood, chiefly methanol) esprit *m* de bois
~SPLIT PULLEY - [mech] poulie *f* en bois en deux pièces
~-STICK BREAK - [el] section *f* isolatrice en bois
~STICK SECTION BREAK - s. wood-stick break
~STONE - [geol] bois *m* pétrifié
~STRICKLE - [text] plaque *f* d'aiguisage, planche *f* d'émeri
~TAR - [chem] goudron *m* végétal, goudron *m* de bois
~TIN - [min] (a variety of cassiterite) étain *m* de bois
~TRIMMER - [mach tool] machine *f* à trancher le bois en bout
~TURNER - [gen] tourneur *m* sur bois
~TURNING - [carp] tournage *m* du bois
~-TURNING LATHE - [mach tool] tour *m* à bois
~VINEGAR - [chem] (impure acetic acid from the distillation of wood) acide *m* pyroligneux
~-WIND - [mus] bois *mpl*
~WOOL - [constr] (shaving used for plaster) laine *f* de bois, paille *f* de bois
~-WOOL SLABS - [constr] (slabs made from long wood shavings with a cementing material) plaques *fpl* en laine de bois
WOODCUT - [print] (an engraving on wood, also a print from a block) gravure *f* sur bois, xylographie *f*, estampe *f*
WOODEN - [gen] (of wood) en bois, de bois
~BARREL - [brew ind] fût *m* en bois, tonneau *m* en bois
~BELLY - [med] ventre *m* de bois
~BLADE - [mech] couteau *m* de bois
~DOWEL - [carp] cheville *f* en bois
~DRILL-POLE - [mining] tige *f* de sonde en bois
~FUNNEL - [impl] entonnoir *m* en bois
~MALLET - [impl] maillet *m* en bois
~MAUL - [impl] mail *m* en bois
~NAIL - [carp] cheville *f*
~RACK - [photo] claie *f*
~-RAIL TRACK - [mining] chemin *m* de bois
~-SHOE HEART - [med] coeur *m* en sabot
~SLEEPER - [railw] traverse *f* de bois
~STAVE PIPE - [hydr] tuyau *m* en douves de bois
~TONGUE - [vet] (actino bacillosis of the tongue of cattle) actinomycose *f*
~VAT - [brew ind] cuve *f* en bois
WOODFORM FOR CONCRETE - [constr] coffre *m* pour béton coulé
WOODINESS - [brew ind] goût *m* de bois
WOODLAND AREA - [geogr] superficie *f* des forêts

WOODLAND MEADOW - [agric] prairie *f* forestière
~PASTURE - [agric] patûrage *m* forestier
WOODLOUSE - [zool] cloporte *m*
WOODMAN - [gen] bûcheron *m*
WOODROCK - [min] bois *m* pétrifié
WOODRUFF KEY - [mech] (a key in the form of a semi-circular flat piece of steel, the straight side of which engages a keyway in the bore of a wheel while the curved side lies a semi-circular keyseat cut in the shaft) clavette *f* Woodruff
WOODSMAN - s. wood-cutter
WOODWELDING - [carp] (very rapid glueing of wood by induction heating at very high frequency) collage *m* rapide du bois par induction à très haute fréquence
WOODWORK - [carp] charpente *f*, charpenterie *f*
[constr] (the wooden part of any structure) travail *m* du bois
WOODWORKER - [gen] (carpenter) charpentier *m*, menuisier *m*
~VICE - [impl] étau *m* de menuisier
WOODWORM - [zool] artison *m*
WOODY - [gen] boisé
[bot] (said of the parts of a plant consisting of wood) ligneux
[paper man] (said of paper containing mechanical wood pulp) à pâte de bois
~CORE - [bot] partie *f* ligneuse
WOOF - s. weft
WOOFER - [el acoust] (a bass loudspeaker) haut-parleur *m* de graves
WOOL - [text] (a staple fibre obtained from the fleece of the sheep and certain other species of Ungulata and used in the manufacture of clothing and fabrics) laine *f*
~BALE - [text] balle *f* de laine
~BASKET - [text] panier *m* à laine
~BEARING ANIMAL - [zool] bête *f* lanifère
~BLEND - [text] mélange *m* de laine
~BRAID - [text] galon *m* de laine
~CARDER - [text] cardeur *m* de laine
~CARDING - [text] cardage *m* de la laine
~CLIP - [text] rendement *m* de la tonte
~COMBING - [text] peignage *m* de la laine
~COMBING PLANT - [text] peignage *m* de laine
~CONSUMPTION - [text] consommation *f* de laine
~DRYING - [text] séchage *m* de la laine
~DUSTING MACHINE - [text] batteur *m* pour laine
~EATING SHEEP - [zool] mouton *m* mangeant la laine
~EXTRACT - [text] (the wool which is extracted from wool rags) laine *f* régénérée
~FAT - [chem] (lanolin) graisse *f* de laine
~FIBRE - [text] fibre *f* de laine
~GRASS - [bot] ériophoron *m*
~-GRASS LEAF - [bot] feuille *f* de l'ériophoron
~-GRASS PEAT - [bot] tourbe *f* de l'ériophoron
~GREASE - [chem] graisse *f* de laine
~KNOP - [text] bouton *m* de laine
~MILK - [chem] émulsion *f* de graisse de laine
~OIL - [text] huile *f* de laine
~OILING - [text] ensimage *m* de la laine, graissage *m* de la laine
~OPENER - [text] ouvreuse *f* à laine
~PARTICLES - [text] particules *fpl* de laine
~PROPELLER - [text] propulseur *m* de la laine, fourche *f* de transport
~REELER - [text] dévideur *m* de laine
WOOL RINSING MACHINE - [text] rinçeuse *f* pour laine
~SCOURER - [text] dessuinteur *m* de laine
~SHEARING - [text] tonte *f* du mouton
~SHEEP - [zool] mouton *m* à laine
~SKEP - [text] panier *m* à laine
~SORTER - [text] trieur *m* de laine
~SORTING - [text] triage *m* de la laine, détrichage *m* de la laine
~SPINNING - [text] filature *f* de la laine
~SPINNING MILL - [text] filature *f* de la laine
~STEEPING - [text ind] trempage *m* de laine, dessuintage *m* de la laine
~STEEPING APPARATUS - [text] appareil *m* à dessuinter la laine
~STOCK - [text] provisions *fpl* en laine
~TOP - [text] ruban *m* de laine peignée
~TESTING - [text] examen *m* de laine
~TUFT - [text] touffe *f* de laine
~WASHER - [text] vlaveur *m* de laine
~WASHING - [text] lavage *m* des laines
~WASHING MACHINE - [text] laveuse *f* à laine
~WASTE - [text] déchets *mpl* de laine
~WAX - [chem] cire *f* de laine
~WEAVING - [text] tissage *m* de la laine
WOOLEN - s. woollen
WOOLFF'S BOTTLE - [ind chem] (glass bottle with two necks, used for passing gases through liquids and the like) flacon *m* de Woolff
WOOLINESS - [cin] (only USA, reverberation) manque *m* de netteté, réverbération *f* trop forte
WOOLLEN - [text] de laine
~CARD - [text] carde *f* pour laine courte
~CARDING MACHINE ENGINE - s. woollen card
~CARPET YARN - [text] fil *m* de laine pour tapis
~CHEVIOT - [text] drap *m* cheviot
~CLOTH - [text] drap *m* de laine, flanelle *f*
~FELT - [text] feutre *m* de laine
~FLAKE YARN - [text] fil *m* flammé de laine cardée
~MUSLIN YARN - [text] fil *m* mousseline de laine
~TOP - [text] ruban *m* de laine
~YARN - [text] fils *mpl* de laine cardée
WOOLLENS - [text] tissus *mpl* de laine, laines *fpl*
WOOLPACK - [met] (a single large mass of cumulus cloud) cumulus *m*
WOOLWORK - [text] tapisserie *f*
WOOLY TAIL - [text] queue *f* laneuse
WORD, to - [gen] exprimer, formuler
WORD - [gen] mot *m*, parole *f*
~ADDRESS FORMAT - [comput] disposition *f* à adresse
~ARTICULATION - [acoust] (the intelligibility of words) intelligibilité *f* des mots
~FORMAT - [comput] structure *f* d'un mot
~MARK - [comput] marque *f* de mot
~-ORGANIZED MEMORY - [comput] mémoire *f* à courants coïncidents
~PERIOD - [comput] durée *f* de mot
~RATE - [telecomm] fréquence *f* de mots
~SEPARATION - [comput] séparateur *m* de mot
~STRUCTURE - [comput] structure *f* d'un mot
~-TIME - [comput] temps *m* de mot
WORK, to - [gen] travailler, faire travailler
[mech] (to function) fonctionner, faire fonctionner
[gen] (to manoeuvre) manoeuvrer, actionner, ouvrer
[gen] (to direct, to manage, e.g. a scheme, a farm)

diriger, administrer
WORK, to - [gen] (to produce) produire
[mining] exploiter
[brew ind] fermenter
WORK - [gen] travail *m*
[gen] (activity) emploi *m*
[mech] (any piece destined to be machined by a machine tool) pièce *f*
[phys] (the work done on a particle by a force during a given displacement) travail *m*
[gen] (a product) produit *m*
[naut] chemin *m* parcouri
~A MINE, to - [mining] exploiter une mine
~ARBOR - [mach tool] arbre *m* porte-pièce
~AREA - [comput] (intermediate storage) mémoire *f* de travail à accès très rapide
~-BENCH - [impl] banc *m*, établi *m*
~BY STEAM, to - [mech] marcher à vapeur
~-DAY - [gen] (working day) jour *m* ouvrable, jour *m* ouvrier
~DRIVING MOTOR - [mach tool] commande *f* de la pièce
~DUE TO FRICTION - [phys] travail *m* de friction
~ELECTRODES - [el] (a dielectric heating equipment) électrodes *fpl* de travail
~FUNCTION - [electron] (electron affinity) travail *m* de sortie
~-HARDEN, to - [metall] (to make metal hard by a purely mechanical operation, e. g. rolling or hammering) écrouir
~-HARDENING - [metall] (hardening which occurs in metal during mechanical treatment) écrouissage *m*
~HARDNESS - [metall] dureté *f* due au traitement mécanique
~HEAD TRANSFORMER - [el] (a transformer connected very close to the heating inductor in order to match its impedance to that of the generator) transformateur *m* d'adaptation, transformateur *m* de couplage
~HOLDING FIXTURE - [mach tool] outil *m* porte-pièce
~IN HAND - [gen] travail *m* en main, travail *m* en progrès
~-IN-PROCESS FILE - [comput] fichier *m* mouvementé
~IN PROGRESS - s. work in hand
~OF ADHESION - [phys] travail *m* d'adhesion
~OF COHESION - [phys & chem] (the work which is required to separate a column of liquid of one square centimetre in cross section into two) travail *m* de cohésion
~OUT, to - [mining] exploiter par méthode directe
~OUT A MINE, to - [mining] épuiser une mine
~-PIECE - [gen] (any piece which has to be worked) pièce *f* à usiner
~-PLATE - [mach tool] table *f* porte-pièce
~REST - [mach tool] (of a lathe) lunette *f*
~-ROOM - [gen] atelier *m*
~SHIFT - [gen] travail *m* par équipes
~SPINDLE - [mach tool] mandrin *m* porte-pièce
~SURFACE - [mech] surface *f* de la pièce
~-TABLE - [mach tool] plateau *m* porte-pièce
~THE BANKS - [mech] travailler en butte
~THE BUTTER - [agric] malaxer le beurre
~THE DREDGER - [mech] manoeuvrer la drague
~UNDERGROUND, to - [mining] travailler au fond
WORKABILITY - [gen mech etc] exploitabilité *f*
[agric] (of soil) propriétés *fpl* culturales
WORKABLE - [gen] exécutable, réalisable
WORKABLE - [min] exploitable
[carp & leather ind] ouvrable
WORKAWAY STOCK - [rubber ind] mélange *m* déchets
WORKBOOK - [gen] livret *m* de travail
WORKED - [gen & mech] usiné
~METAL - [metall] métal *m* usiné
~-OUT MINE - [mining] mine *f* épuisée
WORKER - [text] (the cylinder of a card) travailleur *m*
WORKING - [gen] travaillant, ouvrier
[mech] fonctionnant, ouvrier
[mech] (the actual functioning) fonctionnement *m*, travail *m*
[mech] jeu *m* d'un mécanisme
[metall] (of a furnace) marche *f*, allure *f*
[mining] chantier *m* de travail, galerie *f*
[el chem] (the process of stirring additional solid electrolyte into the fused electrolyte so as to produce a uniform solution) travail *m*
[agric] (of the wine) fermentation *f*
[math] calcul *m*
[agric] (of the butter) délaitement *m* (du beurre)
~ANGLE - [mach tool] angle *m* de travail
~APERTURE - [opt] (focal aperture) ouverture *f* utile (d'objectif)
~AREA - [mech] portée *f*, surface *f* d'appui
~BARREL - [mech] travaillante *f* (de pompe de mine)
~BEAM - [mech] balancier *m*
[mining] bascule *f* de battage, levier *m* de battage
~BLADE - [cin] (cutting, or master blade) lame *f* d'obturation
~CAPACITY - [gen] capacité *f* de travail
~CAPITAL - [fin] capital *m* de roulement
~CHARACTERISTIC - [el] caractéristique *f* dynamique, caractéristique *f* de régulation
~COMB - [text] peigne *m* de travail
~CONDITION TEST - [el] (proof test) essai *m* de fonctionnement
~CONDITIONS - [gen] conditions *fpl* de travail
~COST - [comm] coût *m* d'exploitation
~CURRENT - [el] courant *m* watté
~CYCLE - [mech] cycle *m* de fonctionnement
~DAY - [gen] jour *m* ouvrable, journée *f* de travail
~DEPTH - [mech] (of a gear) hauteur *f* de la dent
~DIAGRAM - [radio & telev] diagramme *m* de fonctionnement, démonstration *f* pratique
~DOOR - [metall] (in a furnace) porte *f* de travail
~DRAWING - [gen] épure *f*
~END OF AN ELECTRODE - [el] extrémité *f* de l'électrode
~FACE - [carp] plateau *m* de réglage
[mining] front *m* de taille, front *m* d'abattage
~FACE LOCOMOTIVE - [mining] (for use in mines) locomotive *f* pour front de taille
~FLUID - [phys] fluide *m* actif
~GAS - [gas ind] gaz *m* injecté utile
~GAUGE - [mech] (of threads) calibre *m* de travail
~IN THE BROKEN - [mining] dépilage *m* des piliers
~JOINT - [constr] (in concrete) joint *m* de reprise
~LEVEL - [mining] front *m* de taille en direction du chantier, chassage *m*
~LINE - [electron] (the load line) droite *f* de charge
[contr] conduite *f* de force motrice
~LOAD - [phys] (the maximum stress a machine or a structure is designed to withstand) charge *f* de travail
~MACHINE - [ind] machine *f* fonctionnante
~MOTION - [mach tool] mouvement *m* principal

WORKING PARTS - [mech] (of an engine) parties fpl ouvrières
~PIT - [mining] puits m d'extraction
[auto] fosse f
~PLACE - [mining] chantier m, atelier m
~PLAN - [gen] épure f
~PLATFORM - [metall] plancher m de manoeuvre
[mining] plate-forme f de manoeuvre
~POINT - [electron] (the point on a characteristic curve corresponding to the average voltage of an electrode) point m de fonctionnement
~PRESSURE - [gas ind] pression f d'exploitation
[mech] pression f de travail, pression f de marche
[el] tension f de service
~RANGE - [glass man] (the range of surface temperature in which glass is formed into articles in a given process) intervalle m de maniabilité
~REFERENCE SYSTEM - [telecomm] (or working standard, a secondary reference telephone system) système m étalon de travail
~SCHEDULE - [gen] programme m d'exploitation
~SHAFT - [mining] puits m d'extraction
~SPEED - [mach tool] vitesse f de régime, vitesse f de travail
~STANDARD - [el] (a standard for everyday use) fonctionnement m normal
~STORAGE - [comput] (internal storage locations reserved for intermediate and partial results) mémoire f de travail
~STRESS - [metall] effort m de travail
~-STROKE - [mech] (the stroke of a piston during which power is developed or work done) course f motrice, course f d'explosion
~SURFACE - [el] (the active surface of an electrode) surface f active d'une électrode, surface f utile d'une électrode
~TORQUE - [mech mach] couple m de travail
~TRIAL - [mech] essai m de fonctionnement
~VALUE - [el] tension f de service
~VALVE -[hydr] soupape f hydraulique de manoeuvre
~VOLTAGE - [el] (the voltage developed on closed circuit) tension f de fonctionnement
[el] (USA only) tension f en circuit fermé
~WINDOW - [metall] (of a furnace) ouverture f de travail
WORKINGS - [mining] exploitation f
WORKMANLIKE - [gen] bien travaillé
[comm] (of goods) fait de main d'ouvrier
WORKREST - [mach tool] (of a centreless grinding machine etc) support m d'outil, appui m
WORKROOM - [gen] atelier m
WORKS - [gen] usine f, atelier m
[mech] (the machinery and the structures of a plant) installation f
[horol] movement m
[naut] oeuvres fpl mortes
[mech] rouage m
~LABORATORY - [mech] laboratoire m de fabrication
~LOCOMOTIVE - [railw] (industrial locomotive) locomotive f d'usine
~SUPERINTENDENT - [gen] chef m du service des ateliers
WORKSHOP - [mech] (premises or building where manufacturing and/or repair work is carried out) atelier m, usine f
~MANUAL - [auto etc] manuel m de reparations, manuel m d'atelier
WORKSHOP RAILWAY - [railw] voie f ferrée d'atelier
WORKTABLE - [mach tool] table f porte-pièce
[mech] (a working bench) banc m de travail
WORM, to - [gen & zool] se glisser
[naut] (to insert yarn between the strands of a rope) congréer
[mech] fileter
WORM - [zool] ver m
[mech] (a specially-formed screw arranged to drive a gear wheel) vis f sans fin, filet m
[mech] (a conveyor) hélice f transporteuse
[ind chem] (worm pipe or tube) serpentin m d'alambic
~AND GEAR - [mech] (the assembly formed by a worm and the worm gear which it drives) engrenage m à vis sans fin
~-AND-GEAR STEERING - [auto] direction f à engrenage à vis sans fin
~-AND-PEG STEERING - [mech] direction f à vis et doigt
~AUGER - [mech] tarière f rubanée
~BIT - [tool] mèche f à vis
~DRIVE - [mech] (a gear drive consisting of a worm and gear) transmission f par vis sans fin
~GEAR - [mech] (a gear wheel designed to engage and be driven by a worm) engrenage m à vis sans fin, engrenage m à vis tangente
[mech] (a worm wheel) roue f à vis sans fin, engrenage m hélicoïdal
~GEAR HOB - [mech] fraise-mère f hélicoïdale pour roues hélicoïdales
~-GEARED LADLE - [metall] poche f à engrenage et vis sans fin
~GEARING - [mech] transmission f à vis sans fin
~HOB - [tool] fraise-mère f hélicoïdale
~KNOTTER - [paper ind] (screening machine fitted with slowly-rotating perforated plates through which pulp is forced by an helical element, under water spray) compresseur m rotatif
~LETOFF - [text] (of a loom) régulateur m de chaîne à vis sans fin
~PIPE - [plumb] serpentin m
~RIM - [mech] couronne f à vis sans fin
~SCREW - [mech] vis f sans fin
~SEGMENT - [auto] segment m de vis
~SHAFT - [mech] arbre m de vis sans fin
~STEERING GEAR - [auto] mécanisme m de direction à vis
~WHEEL - [mech] (the driven gear in a worm gear) roue f hélicoïdale
~-EATEN - [timber] (of wood) vermoulu, mouliné
WORM'S EYE VIEW - [photo] (a ground angle shot) contre-plongée f
WORMING - [naut] (the insertion of yarn between the strands of a rope) congréage m
WORMSEED - [bot] santonine f, ambroisie f
~MUSTARD - [bot] giroflée f sauvage
WORMWOOD - [bot] (a medical plant) armoise f amère, absinthe f
~OIL - [chem] (absinthe oil. An oil obtained from Artemisia absinthium and used as a flavourant) essence f d'absinthe
WORN - [gen] usé
~MASS - [rubber ind] masse f d'usure
~OUT - [gen & mech] hors d'usage, complètement usé

WORN OUT TREAD - [rubber ind] (of tyres) bande *f* de roulement lisse
~-OUT TYRE - [rubber ind] pneu *m* usé
~SHUTTLE - [text] navette *f* usée
WORSTED - [text] (fabric woven from long-staple wool) laine *f* peignée, tissu *m* de laine peignée
~BAND - [text] ruban *m* de laine
~BOBBIN - [text] bobine *f* de mèche peignée
~CLOTH - [text] étoffe *f* de laine peignée
~COP - [text] canette *f* en laine peignée
~COUNT - [text] (the number of 560-yard hanks of worsted which weigh one pound avoirdupois) numéro *m* de fils de laine peignée
~HEALD - [text] lisse *f* en câblé maine
~MULE - [text] renvideur *m* pour fil peigné
~PILE FABRICS - [text] tissus *mpl* veloutés en peigné
~PILE WARP - [text] chaîne *f* de poil en laine peignée
~POPLIN - [text] popeline *f* de laine
~SELFACTOR - [text] renvideur *m* pour fil peigné
~SPINNING - [text] filature *f* de la laine peignée
~SPINNING MILL - [text] usine *f* de filature de la laine peignée
~VELVET - [text] velours *m* de laine
~WARP - [text] peigné-chaîne *m*
~WEFT - [text] peigné-trame *m*
WORT - [brew ind] (the liquor run off from a malt tun) moût *m* (de bière)
~BOILING - [brew ind] cuisson *m* du moût
~COOLER - [brew ind] réfrigérant *m* du moût
~FROM THE LAUTER TUB - [brew ind] moût *m* clair
~PIPE - [brew ind] tuyau *m* de vidange du moût
WORTH - [gen] valeur *f*
[adj] valant
WORTLE - [metall] (small plate having a series of holes in a single row for drawing wire) plaque-filière *f*
WOUND, to - [gen] blesser
WOUND - [gen] blessure *f*
[gen] (the participle of wind, to) enroulé
~-ROTOR INDUCTION MOTOR - [el] (a slip-ring motor) moteur *m* à bagues
WOVE - [paper man] (wove paper) papier *m* tramé
WOVEN - [gen & text] tissé
~JACKET - [rubber ind] enveloppe *f* de tissage
~RESISTORS - [el] (heating elements made in the form of a net by intersecting thin conductors and fire-proof insulating threads) tissu *m* chauffant
~RUBBERIZED BELT - [rubber ind] (a belt of rubberized fabric) courroie *f* en tissu caoutchouté
~WIRE - [metall] (cloth woven from wire, used for filters, screens etc) toile *f* métallique
~WIRE BELT CONVEYOR - [mech] convoyeur *m* à courroie en tissu caoutchouté
~WIRE BRAIDING - [mech] tressage *m* de fils métalliques en tissu
WOW - [el acoust] (the effect of a change in pitch caused by variation in speed) pleurage *m*
[comput] scintillement *m*
~AND FLUTTER - [telev] scintillation *f* et pleurage
WRAITH - [text] peigne *m* d'avant, peigne *m* mobile
~HOLDER - [text] support *m* du peigne
WRAP, to - [gen] envelopper, se couvrir, enrouler
[packing & packaging] empaqueter
[gen] (of edges etc) s'enrouler
WRAP - [comput] (in static magnetic storage, one convolution of the tape about the axis) spire *f*
~REEL - [text] dévidoir *m* de numérotage
WRAP UP, to - [gen] empaqueter
~THE JOINT - [el] guiper l'épissure
~WIDTH - [comput] (tape width) largeur *f* du ruban
WRAPPED CONNECTION - [el] connexion *f* enroulée
~CURING - [rubber ind] vulcanisation *f* sous bandelage
~HOSE - [rubber ind] tuyau *m* bandelé, tuyau *m* enroulé de tissu
WRAPPER - [gen] chemise *f*
[gen] couverture *f*
[paper man] (packing paper) feuille *f* de papier d'emballage
WRAPPERS - [rubber ind] (of tyres) bandes *fpl* pour talon
WRAPPING - [gen] enveloppement *m*, papier *m* d'emballage
[hydr] (of pipes) bandage *m*
~CARD - [text] carton *m* à bobiner
~CLOTH - [gen] toile *f* d'enroulement, enrubannage *m*
~DRUM - [text] tambour *m* enrouleur
~MACHINE - [rubber ind] machine *f* à bandeler, machine à enrubanner, machine *f* à gainer les fils
~MACHINE FOR PIPES - [mech] machine *f* à bandeler les tuyaux
~OF A CABLE - [el] enroulement *m* d'un câble
~PAPER - [paper man] papier *m* d'emballage
~PROCESS - [chem] procédé *m* d'absorption
~REEL - s. wrap reel
~TEST - [mech] (of wire etc) essai *m* d'enroulement
WREATH, to - [carp] débillarder
WREATH - [gen] guirlande *f*
~FILAMENT - [el] filament *m* à couronne
WREATHED STRING - [constr] (of stairs) limon *m* débillardé
WREATHING - [carp & constr] débillardement *m*
WRECK, to - [gen] causer le naufrage, faire échouer
[gen] (to ruin) démolir, détruire
[railw] faire dérailler
WRECK - [gen] ruine *f*
[naut] naufrage *m*
[leg] épaves *fpl* de mer
~-BUOY - [naut] (a buoy of special form used to mark the position of a wreck) bouée *f* d'épave
WRECKAGE - [gen] débris *mpl*
[naut] naufrage *m*
WRECKER - [auto] dépanneuse *f*, véhicule *m* de secours
[naut] (a vessel employed to recover wrecked cargoes or disabled vessels) sauveteur *m* d'épaves
WRECKING BAR - [mech] pied-de-biche *m*
~CAR - [railw] (breakdown wagon) voiture *f* de secours
~FROG - [railw] rampe *f*
~LORRY - [auto] camion *m* de dépannage
~TRUCK - s. wrecker
WRENCH, to - [gen] tordre, forcer, tourner violemment
[med] (an ankle) se fouler (la cheville)
WRENCH - [gen] (a violent twist) mouvement *m* violent de torsion
[tool] (a tool for rotating screwed parts and the like. This term is often used for a spanner) clef *f*, tourne-à-gauche
~FLAT - [mech] (a flattened part for the purpose of allowing the use of a wrench) aplatissement *m* pour clé
~-HAMMER - [impl] clef *f* à marteau
WREST, to - [gen] arracher

WREST - [mus] (a tuning key) accordoir *m*, clef *f* d'accordeur
~-PIN - [mus] (the pins round which the wires of a piano are twisted and by which the wires are tuned) cheville *f* métallique, fiche *f* métallique
~-PLANK - [mus] (in a piano) sommier *m* de bois
WRETCH YEARS - [agric] années *f*pl de misère
WRING, to - [gen] tordre
[gen] (to squeeze out) exprimer, faire sortir
[metall] forcer, déformer (une plaque etc)
WRINGER - [mech] essoreuse *f* à rouleaux
~-ROLL - s. wringing roll
WRINGING PLATE - [text] plaque *f* à exprimer
~ROLL - [impl] cylindre *m* essoreur, rouleau *m* d'essoreuse
WRINKLE, to - [gen] rider, plisser
WRINKLE - [gen] ride *f*, pli *m*
[mech] (when bending a hot steel pipe) pli *m*
~FINISH - [paint] (special type of finish on painted surfaces, consisting of a pattern of small wrinkless) fini *m* ondulé
WRINKLING - [photo] (of a film gelatine) réticulation *f* (de la gélatine)
[paint] ridement *m*
WRINKLY CLOTH - [text] tissu *m* ondulé
WRIST - [anat] carpe *m*, os *m* du carpe
[mech] s. wrist-pin
[gen] poignet *m*
~BONE - s. wrist (anat
~DROP - [med] main *f* tombante
~GANGLION - [med] kyste *m* synovial du poignet
~JOINT - [anat] articulation *f* du carpe
~-PIN - [mech] (of a piston) tourillon *m* de crosse, bouton *m* de manivelle, maneton *m*, bouton *m* de bielle
[aero] (the pin which serves to couple an articulated to a master-connecting rod in a radial i.c.engine) axe *m* de tête de biellette
~-PIN END - [mech] (the end of an articulated connection rod which embraces the wrist-pin in the master-rod) tête *f* de biellette
~-PIN HEAD - [mech] pied *m* de bielle
~-PLATE - [mech] plateau *m* oscillant, plateau *m* conducteur
WRIT - [leg] mandat *m*, acte *m* judiciaire
WRITE, to - [gen] écrire
WRITE HEAD - [comput] tête *f* d'écriture
~LOCK OUT - [comput] barrage *m* d'écrire
~PULSE - [comput] impulsion *f* d'écriture, impulsion *f* écrire
~-READ HEAD - [comput] tête *f* d'inscription -lecture
~WINDING - [comput] (one of the drive windings of a magnetic core) enroulement *m* écrire
WRITER'S CRAMP - [med] chéirospasme *m*, crampe *m* des écrivains
WRITING BAR - [telecomm] (in continuous recorders) couteau *m* d'impression
WRITING BEAM - [electron] faisceau *m* écrivant
~BOOK - [gen] cahier *m*
~OFF - [comm] amortissement *m* (du capital)
~PAPER - [paper man] papier *m* à écrire
~SPEED - [gen] (the rate of writing on successive storage elements) vitesse *f* d'écriture
[telecomm] vitesse *f* d'exploration à la réception
WRONG - [gen] inexact, incorrect
[leg] tort *m*, injure *f*, lésion *f*
~FOUNT - s. wrong letter
~LETTER - [print] coquille *f*, mastic *m*
~SIDE - [text] côté *m* envers (du tissu)
WROUGHT - [gen] forgé
~BOARDING - [carp] planches *f*pl rabotées
~IRON - [metall] (iron containing only very small amounts of other elements) fer *m* forgé
~IRON JOINT - [metall] assemblage *m* du fer
~IRON OF SMITHING QUALITY - [metall] fer *m* de forge
~-IRON SCRAP - [metall] débris *m* de fer
~-IRON SPLIT PULLEY - [mech] poulie *f* en fer en deux pièces
~-IRON TAPPING SLEEVE - [hydr] (a malleable iron clip) collier *m* de prise en fer forgé, collier *m* de prise en fonte malléable
~STEEL - [metall] acier *m* soudé, acier *m* soudant
WULFENITE - [min] (natural lead molybdate, occurring with other ores of lead) wulfénite *f*
WULFF PROCESS - [ind chem] (a commercial process for the production of acetylene from a hydrocarbon gas and steam at elevated temperatures) méthode *f* Wulff
WURTZ SYNTHESIS - [ind chem] (the production of hydrocarbons from solutions of alkyl halides with metallic sodium) synthèse *f* de Wurtz
WURTZITE - [min] (natural zinc sulphide, having the same chemical composition as Sphalerite, but crystallizing in the hexagonal system) wurtzite *f*
WYE - [plumb] (a branch pipe with only one branch which is not at right angle to the main run) raccord *m* à 45°
~CONNECTION - [el] connexion *f* en étoile
~JUNCTION - [radio] (of a waveguide) jonction *f* en Y
~LEVEL - [surv] (a level whose characteristic is the support of the telescope which is similar to that of the wye theodolite) niveau *m* à lunette
~RECTIFIER - [electron] (a three-phase rectifier) redresseur *m* triphasé
~THEODOLITE - [surv] (a form of theodolite in which the telescope is not directly mounted on the trunnion axis but is supported on two Y-shaped forks, in which it may be reversed end to end so as to reverse the line of sight) théodolite *m* en Y
WYOMINGITE - [min] wyomingite *f*

X

X - [math] (the symbol for the principal unknown quantity) X *m*, symbole *m* d'une grandeur inconnue
[chem] (the general symbol for an electronegative atom or group) symbole *m* d'un atome électronégatif
[chem] (or Xe, symbol for Xenon) symbole *m* du Xénon
[met] (symbol for hoarfrost) symbole *m* de la gelée blanche
~ AXIS - [phys] (or horizontal axis, the horizontal axis on a cathode-ray oscilloscope screen or on a graph) axe *m* horizontal
~ AXIS DEFLECTION - [electron] (horizontal deflection) déviation *f* horizontale
~ -BIT - [mech] taillant *m* en couronne
~ -BRED - [zool] (term employed in the textile industry to denote a crossbreed) croisé
~ -CROSS MEMBER - [auto] traverse *f* cruciforme, traverse *f* en croix
~ -CUT CRYSTAL - [crystall] (a quartz plate obtained by cutting a slab from the mother-crystal normal to the X-axis, with major surfaces parallel to Y- and Z-axes) cristal *m* à coupe X
~ -ENGINE - [mech] (an engine in which the cylinders or banks of cylinders are so arranged as to resemble the letter "X" when viewed from the end of the crankshaft) moteur *m* en X
~ -FRAME - [auto] châssis *m* croisillonné, châssis *m* à traverse cruciforme
~ GUIDE - [radio] (of a waveguide, a surface wave transmission line with a dielectric structure whose section is X-shaped) ligne *f* de transmission de Goubau en X
~ -MEMBER FRAME - s, X frame
~ -OPERATION - [telecomm] (the advance operation of one of a group of contact units) opération *f* X
~ PLATES - [electron] (horizontal deflecting electrodes) électrodes *fpl* de déviation horizontale
~ -PUNCH - [comput] (or II punch, a punch in the first row above the zero-row on an IBM card) perforation *f* II, perforation *f* X
~ -RAY , to - [phys & med] (to expose to, or treat with, X-rays) radiographier
~ -RAY ANALYSIS - [crystall] (a technique based on the diffraction of X-rays by crystalline solids for the identification and solution of their crystal structures) analyse *f* par rayons X
~ -RAY APPARATUS FOR METAL INSPECTION - [metall] (X-ray apparatus providing an accurate means of determining the details of the internal structure) appareil *m* d'essai radiographique des métaux
~ -RAY CAMERA - [radiat] (an apparatus designed to obtain a photographic record of the diffraction beams produced when a crystalline specimen is irradiated in a beam of X-rays) chambre *f* à rayons X
X-RAY COVERAGE - [radiat] (the area covered by the X-rays emitted) champ *m* d'irradiation
~ -RAY CRYSTALLOGRAPHY - [phys] (the study of crystals by X-ray diffraction) cristallographie *f* à rayons X
~ -RAY DEFLECTION - [electron] (or horizontal deflection, horizontal tracing of picture lines on the screen of a cathode-ray tube) déviation *f* horizontale
~ -RAY DERMATITIS - [med] (inflammation of the skin caused by excessive exposure to X-rays or to rays emitted by radioactive substances) actinodermatose *f*, dermite *f* des rayons X, radiodermite *f*
~ -RAY DIFFRACTION - [phys] diffraction *f* des rayons X
[photo] radiodiffraction *f*
~ -RAY DIFFRACTION PATTERN - [phys] (in diffraction analysis) photographie *f* de la diffraction des rayons X
~ -RAY DIFFRACTOMETER - [instr] (instrument designed to determine diffraction and used in diffraction analysis) diffractomètre *m* à rayons X
~ -RAY EMISSION SPECTRA - [phys] (the relatively sharp K- and L-levels of atoms in a solid may be ionized by electron impact and filled by electrons dropping from the conduction band in a metal or from a filled band in an insulator) spectres *mpl* d'émission des rayons X
~ -RAY EXCITATION FLUORIMETER - [instr] fluorimètre *m* à excitation X
~ -RAY FILM - [radiat] radiofilm *m*
~ -RAY FLUORESCENCE THICKNESS METER - [instr] épaisseurmètre *m* à fluorescence X
~ -RAY FOCAL SPOT TUBE FOCUS - [electron] (that part of the target of an X-ray tube which is struck by the main electron stream) foyer *m* de tube
~ -RAY GAS TUBE - [radiat] (used in radiology) tube *m* à gaz à rayons X
~ -RAY HARDNESS - [radiat] (the penetrating power of X-rays which is in inverse proportion to the wavelength) dureté *f* de rayons X
~ -RAY INDUCED MUTATION - [radiat] mutation *f* induite par rayons X
~ -RAY METAL TUBE - [radiat] (used in radiology) tube *m* métallique à rayons X
~ -RAY MICROGRAPHY - [photo] microradiographie *f*
~ -RAY OUTPUT - [electron] (the quantity of X-rays emitted in relation to the energy supplied to the anode) rendement *m* des rayons X
~ -RAY OUTPUT AT FULL RATINGS - [electron] (the maximum output of an X-ray tube at full load) ren-

dement *m* optimal

X-RAY PHOTOGRAPH - [med & metall] radiographie *f*

~-RAY SPECTRA - [phys] (wavelength or frequency diagram in which a series of lines show by their position the particular X-rays emitted by a body as the result of cathode ray bombardment) spectres mpl de rayons X

~-RAY SPECTROGRAM - [phys] (a record of an X-ray diffraction pattern) spectrogramme *m* à rayons X

~-RAY SPECTROGRAPH - [phys] (an apparatus designed to record X-ray diffraction patterns) spectrographe *m* à rayons X

~-RAY SPECTROGRAPHY - [phys] (the study of X-ray spectra and their interpretation) spectrographie *f* par rayons X

~-RAY STRUCTURE - [phys] (the atomic or ionic structure of substances as determined by X-ray diffraction patterns obtained by the passage through it of X-rays) structure *f* aux rayons X

~-RAY THERAPY - [med] (radiotherapy by means of X-rays) röntgenthérapie *f*

~-RAY TUBE - [electron] (a vacuum tube designed for the production of X-rays) tube *m* à rayons X

~-RAYS - [phys] (electromagnetic rays of short wavelength produced when cathode rays impinge on matter) rayons *mpl* X

~-SHAPED CROSS MEMBER - s. X-cross member

~-UNIT - [meas] (a unit used in expressing the wavelength of X-or gamma-rays) unité *f* X

~-WAVE - [radio] (or extraordinary wave) onde *f* extraordinaire

~-Y-DIGITIZER - [comput] convertisseur *m* X-Y

~-Z AXES - [telev] (in colour television) axes *mpl* X-Z

~-Z MATRIX - [telev] (in colour television) matrice *f* X-Z

XANTHATES - [chem] (the salts of xanthic acid) xanthates *mpl*

XANTHATION - [text] (process in the manufacture of viscose rayon, alkali cellulose is converted into cellulose xanthate by mixing it with carbon bisulphide) xanthation *f*

XANTHENE - [chem] (diphenylene-methane oxide. Used in organic synthesis and for fungicides) xanthène *m*

~CARBOXYLIC ACID - [chem] (a synthesis intermediate) acide *m* carboxylique de xanthène

~DYESTUFFS - [chem] (a group of dyestuffs including the pyronines and phthaleins derived from xanthene) colorants *mpl* de xanthène

XANTHIC ACID - [chem] (ethyl hydrogen dithiocarbonate, the salts of which have application as flotation agents) acide *m* xanthique

XANTHINE - [chem] (dioxopurine. A synthesis intermediate) xanthine *f*

XANTHOCILIN - [chem] (an antibiotic complex produced by Penicilium notatum and used as an antiseptic) xanthocilline *f*

XANTHOPHYLL - [chem] (a yellow pigment occurring in plants) xanthophylle *f*

XANTHOPSIA - [med] xanthopsie *f*

XANTHOPTERIN - [chem] (a pterin occurring as a pigment in butterflies' wings) xanthoptérine *f*

XANTHOTIXIN - [chem] (methoxsalen. A compound which increases the formation of melanin pigments in the skin) xanthotoxine *f*

XANTHYDROL - [chem] (xanthenol. An analytical reagent) xanthydrol *m*

XENOCRYST - [geol] inclusion *f*

XENOGAMY - [bot] (pollination of a flower from a flower of the same species but on another plant) fécondation *f* croisée

XENOLITH - [min] enclave *f*, xénolithe *f*

XENOMORPHIC GRANULAR TEXTURE - [min] structure *f* granitoïde

XENON - [chem] (one of the noble gases, occurring in the atmosphere, symbol Xe, A. N. 54, A. W. 131.3) xénon *m*

~BUILD - UP - [nucl] (in a reactor) surempoisonnement *m* xénon

~EFFECT - [nucl] (reduction in reactivity caused by neutron capture in Xenon 135) empoisonnement *m* xénon, effet *m* xénon

~HIGH - PRESSURE LAMP - [photo] lampe *f* xénon à haute pression

~INSTABILITY - [nucl] (in a reactor) instabilité *f* xénon

~OVERRIDE - [nucl] (in a reactor) réserve *f* de réactivité par xénon

~-PLUS-TEMPERATURE - [nucl] (in a reactor) température *f* excessive par xénon

~POISONING COMPUTER - [nucl] calculatrice *f* analogique de l'effet d'empoisonnement

XENOTIME - [miner] (a native yttrium phosphate) xénotime *f*

XEROGEL - [chem] (a gel in a nearly dry state) xérogel *m*

XEROGRAPHIC PRINTER - [comput] imprimante *f* xérographique

XERORADIOGRAPHY - [metall] xéroradiographie *f*

XEROSIS - [med] xérose *f*

XEROSTOMIA - [med] xérostomie *f*, aptyalisme *m*

Xs - [radio] (atmospherics) bruits *mpl* parasites

XYLENE - [chem] (dimethylbenzene. Clear, inflammable, toxic liquid with uses as a synthesis intermediate, solvent, and fuel) xylène *m*

XYLENOL - [chem] (hydroxydimethol benzene, dimethyl phenol. A solvent and synthesis intermediate) xylénol *m*

~RESINS - [chem] (synthetic resins produced by reaction of a cresol with an aldehyde) résines *fpl* de xylénol

XYLIDINE - [chem] (aminodimethyl benzene. An intermediate for pharmaceuticals and dyestuffs) xylidine *f*

XYLITHOL - [chem] (pentanepentol. A synthesis intermediate) xylithol *m*

XYLOGEN - [min] lignin *m*

XYLOGRAPHY - [print] (a wood engraving, also printing with wood engravings) xylographie *f*, gravure *f* sur bois

XYLOID - [bot] (pertaining to, or resembling wood) xyloïde

XYLOL - [chem] (an aromatic hydrocarbon with uses as a solvent) xylol *m*

XYLOMETALOLINE HYDROCHLORIDE - [chem] (a vaso-constrictor and nasal decongestant used in medicine) chlorhydrate *m* de xylométaloline

XYLOMETER - [instr] (instrument designed to determine the specific gravity of wood) xylomètre *m*

XYLOPAL - [min] opale *f* xyloïde

XYLOPHONE - [mus] (graduated row of hard wooden bars played with two beaters) xylophone *m*, claque-

bois *m*
XYLOSE - [chem] (wood sugar, A dyeing assistant) xylose *f*, sucre *m* de bois

XYLYL BROMIDE - [chem] (a synthesis intermediate) bromure *m* de xylyle
XYSTER - [med] rugine *f* kystre *m*

Y

Y - [gen mech el] (adj Y-shaped) en étoile
[chem] (the symbol of Yttrium) symbole *m* de le yttrium
[el] (the symbol for admittance) symbole *m* de l'admittance
[telev] (luminance) symbole *m* de la luminance
[meas] (yard) yard *m*(0,9I4 m)
~-AERIAL - [radio] (a delta-matched impedance aerial) antenne *f* avec adapteur en delta
~-ANTENNA - s. Y-aerial
~-AXIS - [gen] (a vertical axis) axe *m* vertical
[electron] (the vertical axis on a cathode-ray oscilloscope screen) axe *m* vertical
~AXIS DEFLECTION - [electron] (vertical tracing of the field on the screen of a cathode-ray tube) déviation *f* verticale
~-BRANCH - [plumb] (a forked pipe fitting) culotte *f*, té *m* a tubulure oblique
~CIRCULATOR - [instr] (a microwave instrument for use with masers and parametric amplifiers, also in duplexing systems) circulateur *m* en Y
~CONNECTION - [el] (a method of connection in which three conductors meet at a common junction known as the star point) montage *m* en étoile, groupement *m* en étoile
~-CUT CRYSTAL - [crystall] (a quartz plate obtained by cutting a slab from the mother-crystal normal to the Y-axis) cristal *m* a coupe Y
~DEFLECTION - [electron] déviation *f* Y
~-EDGE LEADING - [comput] alimentation *f* côté Y
~GUIDE - [radio] (a surface waveguide with a cross section shaped as an inverted Y, used in railway safety installations) guide *f* d'ondes en Y renversé
~JOINT - [el] (a tree-joint, branch-joint in USA) té *m* de dérivation
~JUNCTION - [radio] (a junction of waveguides) jonction *f* en Y
~-LEVEL - [surv] (wye level) niveau *m* à lunette
[telev] niveau *m* du noir
~PLATES - [electron] (vertical deflecting electrodes) électrodes *f*pl de déviation
~-PUNCH - [comput] (or twelve punch, a punch in the second row above the zero-row on an IBM card) perforation *f* I2, perforation *f* Y
~RECTIFIER - [electron] (a three-phase rectifier, or zig-zag rectifier) redresseur *m* triphasé
~TUBE - [plumb] tube *m* en Y
~-VALVE - [mech] (fluid control device in which the bonnet and spindle are at an angle (commonly 45°) to the pipeline) vanne *f* inclinée
~VOLTAGE - [el] (voltage to neutral) tension *f* étoilée
YACHT - [naut] yacht *m*
YACHT VARNISH - [paint] (high quality varnish of excellent water resistance, suitable for application to wood) vernis *m* pour yachts
YAK - [zool] yak *m*, vache *f* de Tartarie
YAM - [bot] igname *f*
~BEAN - [agric] pois *m* patate
YAOURT - s. yoghurt
YAPPING - [acoust] (a shrill sound as made by puppies) jappement *m*
YARD - [meas] (British unit of length equal to 0.9I44 m) yard *m*
[gen] cour *f*
[railw] dépôt *m* de marchandises, gare *f* de triage
[constr] chantier *m*
[constr] (of materials) chantier *m* de matériaux
[naut] vergue *f*
~-ARM - [naut] fusée *f* de vergue
~DONKEY - [transp] (vehicle for the transport of logs) fardier *m*
~LOCOMOTIVE - [railw] locomotive *f* de manoeuvre, locomotive *f* de gare
~MANURE - [agric] fumier *m* d'étable, fumier *m* de ferme
YARDAGE - [meas] métrage *m*
[constr] (the volume of excavation in cubic yards) cubage *m*
YARDER - [transp] (engine for the haulage of felled logs) triqueballe *f*
YARDMASTER - [railw] (stationmaster) gareur *m*
YARDSTICK- [impl] (a stick of the length of I yard) yard-canne *m*
[gen] (sa standard for a comparison) aune *f*
YARN - [gen] (any spun material prepared for use in weaving, knitting etc) fil *m*
~BALANCE - [text] romaine *f* de numérotage, balance *f* pour échevettes
~BEAM - [text] ensouple *f* dérouleuse
~BRAKE - [text] frein *m* du fil
~BUNDLING - [text] empaquetage *m* des fils
~BUNDLING PRESS - [text] presse *f* à fils
~CHANGER - [text] appareil *m* rayeur
~CLEARER - [text] épurateur *m* de fil
~COUNTING - [text] numérotage *m* des fils, titrage *m* des fils
~COUNTS INDICATOR - [text] indicateur *m* du numéro du fil
~DAMPER - [text] assouplisseur *m* de trame, mouilleur *m* de trame
~DISTRIBUTOR - [text] livreur *m* de trames, distributrice *f* de trames
~DRESSER - [text] encolleuse *f* pour fils
~ELONGATION - [text] allongement *m* du fil
~FEED - [text] alimentation *f* du fil

YARN FILTER - [text] filtre *m* pour filés
~FINISHING - [text] apprêtage *m* du filé
~GUIDE - [text] guidage *m* du fil
~-GUIDE BRIDGE - [text] étrier *m* de porte-fil
~HURDLE - [text] claie *f* à bobines
~IN THE COP - [text] fil *m* en canette
~LAYER - [text] couche *f* de fils
~LEVER - [text] (lever used for feeding the yarn in a knitting machine) levier *m* d'alimentation du fil
~ON THE BOBBIN - [text] fil *m* en bobine
~ON TUBES - [text] fil *m* sur tube
~POLE - [text] perche *f* de séchage
~POURER - [text] porteur *m* de fuseaux
~REEL - [text] dévidoir *m* pour filé
~REST - [text] traverse *f* arrière, traverse *f* guide-fil
~SCALES - [text] romaine *f*, romaine *f* de numérotage
~SKIP - [text] panier *m* de fils
~SPEED GAUGE - [text] instrument *m* pour mesurer la vitesse du fil
~SPINNING - [text] filage *m* du fil simple
~SPOOLER - [text] bobineuse *f* de fil
~STEAMING CHAMBER - [text] chambre *f* de vaporisage pour fil
~TENSION BOWL - [text] galet *m* de tension
~TESTER - [text] dynamomètre *m* (pour filés)
~TESTING - [text] essai *m* des fils
~TWIST TESTER - [text] torsiomètre *m*
~UNTANGLER - [text] démêleuse *f* des fils
~WASTE - [text] déchets *mpl* de fils, bouts *mpl* durs
YARNING IRON - [impl] calfait *m*
~CHISEL - s. yarning iron
YAW, to-[aero] (to change direction about the normal axis) embarder, voler en lacet
[naut] (to steer abruptly) embarder
YAW - [aero] (angular motion of an aircraft about the normal axis) mouvement *m* de lacet
[naut] (abrupt and irregular steering) embardée *f*
~ANGLE - [aero] angle *m* de lacet
~AXIS - [aero] (the axis about which yawing takes place) axe *m* de lacet
~METER - [instr] (an instrument to show changes in the direction of an airstream) indicateur *m* de lacet
YAWING - [aero] (angular motion about the normal axis) lacet *m*
~AXIS - [aero] axe *m* vertical
~MOMENT - [aero] (the moment, taken about the normal axis, due to air flow) moment *m* de lacet
YAWNING - [physiol] baîllement *m*
YAWS - [med] pian *m*, framboesia *f*
YEAR - [astr & gen] an *m*, année *f*
~RING - [bot] anneau *m* d'accroissement
YEARBOOK - [gen comm statist etc] (a book published once a year, usually dealing statistical informations) annuaire *m*, almanach *m*
YEARLING - [zool] (animal) d'un an
~COLT - [zool] poulain *m* d'un an, laiteron *m*
YEARLING'S WOOL - [text] laine *f* d'agneau d'un an
YEARLY - [gen] annuel
YEAST - [chem] (generic term for unicellular organisms belonging to the family Saccharomycetaceae, and having application in the fermentation of sugars to produce alcohol, and as a source of Vitamin B) levure *f*, levain *m*
~BITE - [brew ind] (taste of yeast) goût *m* de levure
~CELL - [chem] cellule *f* de levure
YEAST CULTURE - [brew ind] culture *f* de levure
~FOOD - [chem] (nutritive substances added to increase the activity of a yeast) aliment *m* de la levure
~HEAD - [brew ind] (cover over a fermenting vat) couvercle *m*
~PRESS - [brew ind] presse *f* à levure
~PRESSINGS - [brew ind] (beer recovered by pressing yeast after fermentation) bière *f* de levure
~PROPAGATOR - [brew ind] (a plant for the production of pure yeast) appareil *m* à culture pure, appareil *m* à propagation
~RAKE - [brew ind] râteau *m*, fourquet *m*
~ROUSER - [brew ind] râteau *m* à levure
~STRAIN - [brew ind] (variety of yeast) race *f* de levure
~TUB - [brew ind] cuivier *m* à levure
YELLING - [acoust] (shrill sounding notes) cris *mpl* perçants
YELLOW - [dye] jaune
~AMBER - [min] succin *m*, ambre *f* jaune
~CHROME - [paint] (alternative name for Chrome Yellow, a lead chromate pigment) jaune *m* de chrome
~CLAY - [min] argile *f* jaune
~COCOON - [text] cocon *m* jaune
~COPPER - [metall] laiton *m*
~COPPER ORE - [min] (popular name for the sulphide of copper and iron, chalcopyrite) cuivre *m* pyriteux, chalcopyrite *f*
~DISCOLORATION - [photo] jaunissement *m*
~FEVER - [med] fièvre *f* jaune
~FLAME BURNER - [gas ind] (a diffusion-flame burner) brûleur *m* à flamme de diffusion, brûleur *m* à flamme blanche
~FOG - [photo] voile *m* jaune
~IRON ORE - [min] minerai *m* de fer jaune
~LEAD ORE - [min] (popular name for wulfenite) wulfénite *f*, plomb *m* jaune
~METAL - [metall] métal *m* jaune
~OCHRE - [paint] ocre *f* jaune
~ORE - s. yellow pyrites
~PHOSPHORUS - [chem] phosphore *m* ordinaire
~PRUSSIATE OF POTASH - [min] prussiate *m* jaune de potasse
~PYRITES - [min] cuivre *m* pyriteux, chalcopyrite *f*
~RIPENESS - [agric] maturité *f* jaune
~-SAND CLOVER - [bot] anthyllide *f* vulnéraire
~SCREEN - [photo] écran *m* jaune
~SPINEL - [min] rubicelle *f*
~SPOT - [telev] tache *f* jaune, macula *f* lutea
~TELLURIUM - [min] sylvanite *f*, tellure *m* graphique
YELLOWING - [paints] (the changing of a white pigment to a slightly yellow tint, especially by the action of light) jaunissement *m*
YELPING - [acoust] (the sound made by small animals) geignement *m*, glapissement *m*
YEW - (tree) [bot] if *m*
YIELD, to - [gen] rendre, produire, rapporter
[gen] (to give up, to relinquish) céder
[gen] (to give way) fléchir
YIELD - [gen] (the amount yielded) production *f*, rendement *m*
[fin] (the profit on invested capital) rendement *m*
[fin] (the proceeds of a tax) rapport *m*
[mining] débit *m*
[text] (of wool) rendement *m*
[rubber ind] commencement *m* d'écoulement
[min] rendement *m* des minerais

YIELD LOAD - [phys] (the load at which a part or test piece fails by a considerable degree of plastic deformation) poids *m* limite d'élasticité
~OF A WELL - [hydr] débit *m* d'un puits
~OF AN OIL WELL - [oil ind] débit *m* d'un puits à pétrole
~OF FIBRES - [text] rendement *m* en fibres
~OF HACKLING - [text] rendement *m* en fibres du peignage
~OF THE COCOON - [text] rendement *m* en soie du cocon
~POINT - [rubber ind] (the point at which considerable plastic deformation takes place under a constant load) limite *f* élastique
~STRENGTH - [phys] résistance *f* à l'écoulement
~STRESS - [phys] effort *m* d'écoulement
~VALUE - [rubber ind] seuil *m* de fluage, contrainte *f* critique d'écoulement
YIELDING - [of a beam etc) fléchissement *m*, affaissement *m*
~CAPACITY OF A WELL - s. yield of a well
~PROP - [min] étançon *m* élastique coulissant
YLANG YLANG OIL - [chem] (a volatile oil obtained from Cananga odorata and used in perfumery) essence *f* d'ylang-ylang
YLEM - [chem] (the primordial substance from which the chemical elements may have been derived) plasma *m* primordial, ylem *m*
YODEL - [acoust] (the abrupt change from chest voice to falsetto, as used by Tyrolean mountain people) ioulement *m*
YOGHOURT - [food] yahourt *m*
YOHIMBINE - [chem] (quebrachine. An alkaloid obtained from Pausinystalia yohimbe and used in medicine) yohimbine *f*
YOKE - [mech] (the lathe dog) toc *m*, doguin *m*
[mech] (a clamp) étrier *m*, chape *f*
[mech] (the slotted crosshead of a steam engine) joug *m*
[el] (those parts of the core, not surrounded by windings of the magnetic circuit and serving to complete the circuit) culasse *f*
[agric] (curved timber attachment used for coupling draft animals) joug *m*, attelage *m*
[impl] palanche *f*

YOKE - [constr] moufle *f*, longrine *f*
[el] (of a dynamo) bâti *m* de dynamo
~OF A RELAY - [el] culasse *f* d'un relais
~PRESSURE BOLT - [el] (of a transformer) boulon *m* de pression pour colliers
~PRESSURE PLATE - [el] (in a transformer, the core clamp) presse-culasse *m*
~SUSPENSION - [el] (bar suspension) suspension *f* à barre
YOLK - [zool] (the nutritive non living material in an ovum) jaune *m* d'oeuf
[chem] (wool fat) lanoline *f*
[text] (dried perspiration in raw wool) suint *m*
~-BAG - [zool] membrane *f* vitelline
~LOCKS - [text] abats *mpl* en suint
~WAX - [text] cire *f* du suint
YOUNG FUSTIC - [paint] (yellow pigment obtained from Rhus Cotinus, consisting basically of fustin) fustet *m*
YOUNG'S MODULUS - [phys] (the coefficient of tensile elasticity) module *m* d'Young
YPERITE - [chem] (mustard gas) ypérite *f*
YTTERBIUM - [chem] (a rare earth metal, symbol Yb, A. N. 70, A. W. 173.5 with application in metallurgy) utterbium *m*
YTTRIUM - [chem] (a metallic element, usually regarded as one of the rare earth metals. Symbol or Y or Yt, A. N. 39) yttrium *m*
~ACETATE - [chem] (an analytical reagent) acétate *m* d'yttrium
~CARBIDE - [metall] carbure *m* d'yttrium
~CHLORIDE - [chem] (an analytical reagent) chlorure *m* d'yttrium
~-90 ROD - [nucl] barre *f* en yttrium-90
~OXIDE - [chem] (a component of ceramic glazes and special glasses) oxyde *m* d'yttrium
~SULPHATE - [chem] (an analytical reagent) sulfate *m* d'yttrium
YTTROCRASITE - [min] (a mineral containing calcium, lead, thorium and uranium) yttrocrasite *f*
YTTROGUMMITE - [min] (decomposition product of cleveite) yttrogummite *f*
YTTROTANTALITE - [min] (mineral containing yttrium tantalum, calcium and iron) yttrotantalite *f*

Z

Z --[chem] (symbol for gram-equivalent weight) symbole *m* du poids gramme-équivalent
[phys] (the symbol of the number of molecular collisions per second) symbole *m* du nombre de collisions par seconde
[phys] (symbol for atomic weight) symbole *m* du poids atomique
[el] (the symbol for impedance) symbole *m* de l'impédance
~ARMATURE RELAX - [el] relais *m* à armature en Z
~AXIS MODULATION - [electron] modulation *f* dans l'axe des Z
~-BIT - [mech] taillant *m* en Z
~DISTORTION - [telev] distorsion *f* en Z
~MARKER BEACON - [radio] (a radio marker beacon giving a radiation pattern in which there is a vertical cone of silence) radiophare *m* Z
~MODULATION - [radio] modulation *f* de luminosité
~PISTON - [radio] (of a waveguide) piston *m* Z
~PLUNGER - s. Z piston
~POTENTIAL - s. zeta potential
~-TWIST - [text] (twisting of a yarn in a right-handed sense) torsion *f*
ZAFFER - [min] (a blue pigment made by roasting cobalt ores to yield an impure cobalt oxide used for enamel and for painting on glasses and porcelain) safre *m*, oxyde *m* bleu de cobalt
ZAPLA ORE - [min] hématite *f* rouge
ZARATITE - [min] (emerald nickel) zaratite *f*
ZAX - [tool] (or sax, an axe used for shaping slates) hache *f* d'ouvrage
ZEBRA - [zool] zèbre *f*
~COLOUR TUBE - [telev] tube *m* zèbre
~CROSSING - [gen] passage *m* clouté
~MARKING - [gen] zébrure *f*
~ROOF - [metall] toît *m* zébré
ZEBRAWOOD - [bot] (the wood of a large tree growing in Guiana, light brown in colour with dark stripes) gonçalo *m* alves
ZED - [gen] (the letter Z, commonly called zee in the USA) Z *m*
~IRON - [metall] fer *m* à Z
ZEIN - [chem] (a protein derived from corn used in the production of synthetic fibres, plastics, films and coatings) zéine *f*
ZELOTYPIA - [med] zèle *m* morbide
ZENER BREAKDOWN - [electron] (of a semiconductor diode, a breakdown caused by the field emission of holes and electrons in the depletion layer) rupture *f* de Zener
~CURRENT - [electron] (in a semiconductor) courant *m* de Zener
~DIODE - [electron] (in semiconductors) diode *f* de Zener
ZENER EFFECT - [electron] (the effect responsible for Zener breakdown in a semiconductor) effet *m* Zener
ZENITH - [astr] (geocentric) (the point at which a straight line drawn from the centre of the earth through the observer's position would, if produced, meet the celestial sphere) Zénith *m*
[astr] (true or geographical) (the point at which the normal to the tanget to the earth's surface at the observer's position would meet the celestial sphere) Zénith *m*
~ATTRACTION - [astr] attraction *f* zénithale
~DISTANCE - [aero] (the angular distance of a celestial body from the zenith measured along a vertical circle i. e. the complement of the altitude) distance *f* zénithale
ZENITHAL - [astr] zénithal
ZENOGRAPHIC - [astr] jovien
ZEOLITE SOFTENING - [hydr] adoucissement *m* par les zéolites
ZEOLITES - [min] (originally natural hydrated calcium and aluminium silicate but now also applied to artificial products used in ion-exchange processes) zéolithes *f*pl
ZEPHYR - [text] laine *f* zéphire
~YARN - [text] fil *m* zéphire
ZEPPELIN - [aero] (airship having a rigid, cigar-shaped body, originally designed and built by F. von Zeppelin) Zeppelin *m*
~AERIAL - [radio] (horizontal half-wave aerial with tuned feeder) antenne *f* de Zeppelin
~ANTENNA - s. Zeppelin aerial
ZERO, to - [instr] (to adjust an instrumant so that its indication is zero when the quantity to be measured is also zero) mettre à zéro
ZERO - [math] zéro *m*
~-ACCESS MEMORY - [comput] mémoire *f* à accès rapide
~-ACCESS STORE - s. zero-access memory
~-ACCESS STORAGE - s. zero-access memory
~-ADDRESS INSTRUCTION - [comput] instruction *f* sans part adresse
~ADJUSTER - [instr] (a device for bringing the pointer into the zero position) ajusteur *m* à zéro
~ADJUSTING DEVICE - s. zero adjuster
~ADJUSTMENT - [instr] réglage à zéro
~BALANCE - [comput] solde *m* zéro
~BALANCING - [comput] contrôle *m* à zéro
~-BASED LINEARITY - [constr] linéarité *f* à base zéro
~BEAT - [radio] (a system of reception or radiotelephony in which two frequencies in a mixing cir-

cuit are exactly the same) battement *m* nul
ZERO BIAS - [electron] (the condition in which the control grid and cathode of an electron tube are at the same direction current voltage) polarisation *f* nulle
~ CARRIER - [telev & radar] (zero amplitude of a waveform) porteuse *f* zéro
~ COMPRESSION - [comput] compression *f* de zéros
~ CONDITION - [comput] (of a magnetic cell) position *f* zéro
~ DELIVERY PRESSURE - [mech] (in a variable-delivery hydraulic pump, the pressure at which delivery falls to zero level) pression *f* de débit nul
~ DRIFT - [contr] dérive *f* zéro
~ ELIMINATION - [comput] suppression *f* des zéros
~-ENERGY LEVEL - [phys] (the energy level of a system of particles at absolute zero, also called zero-point energy) niveau *m* d'énergie zéro
~ ERROR - [instr] erreur *m* de zéro
~-FIELD EMISSION - [electron] (free-field emission current) courant *m* d'émission à champ nul
~ FLUE DRAUGHT - [th eng] absence *f* de tirage
~- g - [phys] (weightlessness) pesanteur *f* nulle
~ GRAVITY - s. zero-g
~ LAUNCH - [astronaut] lancement *m* à point fixe
~ LEAD - [el biol] (bioelectric null) zéro *m* bioélectrique, fil *m* neutre
~-LENGTH ROCKET - [astronaut] fusée *f* de lancement à point fixe
~ LEVEL - [surv] niveau *m* zéro
~ LEVEL ADDRESS - [comput] adresse *f* immédiate
~ LIFT - [aero] portance *f* nulle
~-LIFT ANGLE - [aero] (the angle of attack at which the lift of an aerofoil is zero) angle *m* de portance nulle
~-LIFT CHORD - [astronaut] corde *f* de portance nulle
~-LIFT LINE - [aero] axe *m* de portance nulle
~ LINE - [mech] axe *m* neutre
~ METHOD - [el] (a method of measuring an electric circuit quantity in which the correct value is given when the current flowing through the indicating instrument is zero) méthode *f* de zéro
~ OFFSET - [contr] déplacement *m* du zéro
~ OUTPUT - [comput] (in static magnetic storage) signal *m* de sortie zéro
~ PHASE-SEQUENCE - [el] (a three-phase vector system in which all three vectors are equal in magnitude and in phase with one another) séquence *f* nulle de phases
~ PHASE-SEQUENCE PROTECTION - [el] protection *f* homopolaire
~ PHASE-SEQUENCE REALY - [el] relais *m* fonctionnant au point nul de la phase
~ POINT - [instr] (of scale) point *m* zéro
~-POINT ENERGY - [phys] (the total energy of a system of particles at absolute zero temperature) énergie *f* au zéro absolu
~ POINT ENTROPY - [phys] entropie *f* au point zéro
~ POSITION - [photo] position *f* zéro
~ POTENTIAL - [el] (earth potential in electric circuits) potentiel *m* nul
~ POWER CHARACTERISTIC - [el] caractéristique *f* en courant réactif
~ POWER LENS COMBINATION - [opt] introduction *f* d'une lentille à pouvoir zéro
~-POWER REACTOR - [nucl] (a type of experimental reactor) réacteur *m* de puissance zéro
ZERO READER - [instr] (apparatus for finding the landing runway) chercheur *m* de la piste d'atterrissage
~-SEQUENCE COMPONENT - [el] (homopolar component, one of the quantities which constitute the homopolar coordinate) composante *f* homopolaire
~-SEQUENCE FIELD IMPEDANCE - [el] impédance *f* de champ homopolaire
~-SEQUENCE REACTANCE - [el] réactance *f* à séquence de phase zéro
~ SETTING - s. zero adjustment
~-SETTER - [instr] (a zero-setting control) appareil *m* à mettre à zéro
~ SHIFT - [contr] déplacement *m* du zéro
~ SIGNAL - [telev] signal *m* zéro
~ SPEED POSITION - [radar] position *f* de vitesse zéro
~ STATE - [comput] (of a static magnetic storage) état *m* "zéro"
~ SUBCARRIER CHROMATICITY - [opt] (the chromaticity which is intended to be displayed when the subcarrier amplitude is zero) couleur *f* correspondante à l'absence de sous-porteuse
~ SUPPRESSION - [comput] (the elimination of non-significant zeros to the left of the significant part of a quantity before printing begins) suppression *f* des zéros
~-THRUST PITCH - [aero] (the pitch of a propeller at which thrust is zero) pas *m* pour la traction nulle
~-VALENT - [chem] (incapable of combining with other atoms) zérovalent
~ VARIATION - [instr] (the part of the deflection of a measuring instrument having a restoring torque which remains after the cause producing it it has disappeared) déviation *f* résiduelle
ZEROFILL, to - [comput] remplir avec des zéros
ZETA POTENTIAL - [el boil] (electrokinetic potential, a set of four electric or velocity potentials accompanying relative motion between solids and liquids) potentiel *m* électrocinétique, potentiel *m* zéta
ZEUNERITE - [chem] (hydrous copper uranium arsenate) zéunérite *f*
ZIBELINE - [text] zibeline *f*
~ YARN - [text] fil *m* zibeline
ZIEGLER CATALYST - [chem] (stereospecific catalyst of specific composition) catalyseur *m* de Ziegler
~ PROCESS - [chem] (polymerization process for the production of linear polymers from propylene or ethylene. A feature of the process is the low pressure at which it is operated and the use of catalysts based on aluminium derivates) procédé *m* Ziegler
ZIGZAG, to - [gen] zigzaguer, disposer en zigzag
ZIGZAG - [gen] zigzag *m*
[mech etc] mouvement *m* à zigzag, disposition *f* en quinconce
[adj] en zigzag
~ AERIAL - [radio] (an aerial which is shaped like a row of sawteeth) antenne *f* en zigzag
~ ANTENNA - s. zigzag aerial
~ COMB - [text] peigne *m* en zigzag
~ CONNECTION - [el] (of transformer or reactor winding, a star connection) connexion *f* en zigzag
~ CLOVER - [bot] trèfle *m* intermédiaire
~ GAUZE - [text] gaze *f* en zigzag
~-HARROW - [agric] herse *f* zigzag
~-LINE COUPLER - [electron] (of a waveguide) cou-

plage *m* en zigzag
ZIGZAG RECTIFIER - [electron] (three-phase rectifier) redresseur *m* triphasé
~REED - s. zigzag comb
~RIBBON - [text] bande *f* en zigzag
~SEAM - [text] couture *f* en zigzag
~TWILL - [text] chevron *m*
ZIGZAGGING SUPPORTING WIRE - [el] câble *m* de support en zigzag, câble *m* entrelacé
ZINC, to - [metall] zinguer, galvaniser
ZINC - [chem] (a metallic element, symbol Zn, having application in alloys, as a protective coating for iron, in guttering, roofing, engraving, plates and in bearing metals) zinc *m*
~ACETATE - [chem] (a dyeing mordant, analytical reagent, and an astringent and emetic in medicine) acétate *m* de zinc
~AMMONIUM CHLORIDE - [chem] (an electrolyte in dry batteries and soldering flux) chlorure *m* d'ammonium et de zinc
~ANODES - [el] (used for protection against galvanic corrosion) zinc *m* anodique
~ARSENATE - [chem] (an insecticide) arséniate *m* de zinc
~ARSENITE - [chem] (a pesticide and wood preservative) arsénite *f* de zinc
~-BASE ALLOY - [metall] alliage *m* à base de zinc
~-BEARING ACCELERATOR - [rubber ind] accélérateur *m* contenant du zinc
~-BERYLLIUM ALLOY - [metall] alliage *m* de zinc et béryllium
~BLENDE - [min] (a generally used term for sphalerite) blende *f*, fausse galène *f*, sphalérite *f*
~BLOOM - [min] hydrozincite *f*, zincosine *f*
~BORATE - [chem] (a fungistat, textile fireproofing agent, and ceramic flux) borate *m* de zinc
~BOX - s. zinc precipitation box
~BROMIDE - [chem] (a photographic chemical and a sedative) bromure *m* de zinc
~BRONZE - [metall] bronze *m* au zinc
~-CADMIUM SULPHATE - [chem] (a fluorescent substance) sulfate *m* de cadmium et zinc
~CAPRYILATE - [chem] (a fungicide) caprylate *m* de zinc
~CARBONATE - [chem] (white, crystalline powder with uses in medicine as a mild astringent, as a pigment, textile flame-proofing agent, and in cosmetics) carbonate *m* de zinc
~CHLORIDE - [chem] (a catalyst, dyeing mordant, carbonising agent for wool, disinfectant, and constituent of galvanizing and electroplating baths. Zinc chloride is also used in medicine as a caustic and astringent) chlorure *m* de zinc
~CHLOROIODIDE - [chem] (a disinfectant) chloroiodure *m* de zinc
~CHROMATE - [chem] (a yellow pigment with uses in paint and linoleum manufacture) chromate *m* de zinc
~CHROME - [paint] (a yellow pigment, essentially chromate of zinc, used in anticorrosion paints and in primers) chromate *m* jaune de zinc, jaune *m* de zinc
~COATED STEEL - [metall] acier *m* zingué
~COATING - s. zinc plating
~CUTTINGS - [metall] rognures *fpl* de zinc
~CYANIDE - [chem] (an analytical reagent, constituent of plating baths, and an insecticide) cyanure *m* de zinc
ZINC DEPOSIT - [min] gîte *m* de zinc
~DIBUTYLDITHIOCARBAMATE - [chem] (a lubricant additive and vulcanization accelerator) dibutyl-dithiocarbamate *m* de zinc
~DICHROMATE - [chem] (an orange pigment) bichromate *m* de zinc
~DIETHYL - [chem] (diethylzinc. A catalyst and synthesis intermediate) zinc *m* diéthylique
~DIETHYLDITHIOCARBAMATE - [chem] (a vulcanization accelerator) diéthyl-dithiocarbamate *m* de zinc
~DIMETHYLDITHIOCARBAMATE - [chem] (rubber accelerator) diméthyl-dithiocarbamate *m* de zinc
~DRIERS - [paint] (nowadays usually zinc naphthenate, which also prevents rivelling) siccatifs *mpl* de zinc
~DUST - [chem] (a catalyst and polymerizing agent) limaille *f* de zinc, poudre *f* de zinc
~DUST PRIMER - [metall] peinture *f* riche en zinc
~ETHYLENEBISDITHIOCARBAMIDE - [chem] (a fungicide and pesticide) éthylène-bis-dithiocarbamide *f* de zinc
~ETHYLSULPHATE - [chem] (a synthesis intermediate) éthylsulfate *m* de zinc
~FERRITE - [min] ferrite *f* de zinc
~FLUORIDE - [chem] (a constituent of galvanizing baths and ceramics glazes) fluorure *m* de zinc
~FORMALDEHYDE SULPHOXYLATE - [chem] (a textile dye discharging agent) sulfoxylate *m* de formaldéhyde et de zinc
~FORMATE - [chem] (wood preservative, textile processing chemical, and a catalyst) formiate *m* de zinc
~FUMES - [chem] vapeurs *fpl* de zinc
~GREEN - [paint] (alternative name for cobalt green. It is a complex of zinc and cobalt oxides, made by calcination) vert *m* de zinc
[min] vert *m* de zinc
~GRIP - [metall] acier *m* à couche de zinc électrolytique
~HYDROSULPHITE - [chem] (a bleaching agent)hydrosulfite *m* de zinc
~INGOT METAL - [metall] lingot *m* de zinc
~IODIDE - [chem] (an analytical reagent) iodure *m* de zinc
~LAURATE - [chem] (a drier for paints and varnishes) laurate *m* de zinc
~LINOLEATE - [chem] (a paint and varnish drier) linoléate *m* de zinc
~NAIL - [mech] clou *m* en zinc
~NAPHTHENATE - [chem] (a fungicide and pesticide, driers for paints and varnishes, waterproofing agent for textiles, and a wood preservative) naphténate *m* de zinc
~NITRATE - [chem] (a mordant in dyeing, a catalyst, synthesis intermediate, analytical reagent, and a resin agent) nitrate *m* de zinc
~OCTOATE)- [chem] (a catalyst) octoate *m* de zinc
~OLEATE - [chem] (has application in medicine in the treatment of skin complaints) oléate *m* de zinc
~-ORE - [min] minerai *m* de zinc
~ORTHOSILICATE - [chem] (a compound used in making fluorescent screens) orthosilicate *m* de zinc
~OXALATE - [chem] (a synthesis intermediate) oxalate *m* de zinc
~OXIDE - [chem] (zinc white, Chinese white. A paint and linoleum pigment, reinforcing agent and vulcanisation accelerator for rubber, a mild astringent

in medicine, component of special cements, ceramic glazes and cosmetics, and a resist in textile dyeing) oxyde *m* de zinc
ZINC OXIDE SENSITIVITY - [rubber ind] sensibilité *f* à l'oxyde de zinc
~PALMITATE - [chem] (a plastics lubricant, rubber chemical, and a pigment suspending agent in paint systems) palmitate *m* de zinc
~PERBORATE - [chem] (an oxidizing agent) perborate *m* de zinc
~PERMANGANATE - [chem] (an oxidizing agent, an oxidant deodorant and astringent in medicine) permanganate *m* de zinc
~PEROXIDE - [chem] (an oxidizing agent, disinfectant and deodorant in medicine, and a rubber vulcanisation agent) peroxyde *m* de zinc
~PHENATE - [chem] (an insecticide) phénate *m* de zinc
~PHENOLSULPHONATE - [chem] (an astringent and antiseptic in medicine) phénolsulfonate *m* de zinc
~PHOSPHATE - [chem] (a component of dental cements, also a fluorescent substance) phosphate *m* de zinc
~PHOSPHATIZING - [metall] revêtement *m* en phosphate de zinc
~PHOSPHIDE - [chem] (a rodenticide) phosphure *m* de zinc
~PHOSPHITE - [chem] (white, crystalline powder with uses in medicine) phosphite *m* de zinc
~PLATING - [metall] zingage *m*
~PRECIPITATION BOX - [ind chem] caisse *f* de précipitation par le zinc, boîte *f* à zinc
~PROPIONATE - [chem] (a fungicide with application in medicine) propionate *m* de zinc
~PYROPHOSPHATE - [chem] (a white pigment) pyrophosphate *m* de zinc
~RESINATE - [chem] (a dispersing agent and a drier in paints and varnishes) résinate *m* de zinc
~RICINOLEATE - [chem] (a lubricant additive, fungicide, plastics plasticizer, and emulsificant) ricinoléate *m* de zinc
~SALICYLATE - [chem] (white, crystalline powder with uses in medicine) salicylate *m* de zinc
~SELENIDE - [chem] (a chemical substance used in the manufacture of semiconductors) séléniure *m* de zinc
~SHAVINGS - [metall] tournure *f* de zinc, copeaux *mpl* de zinc
~SHEET - [metall] tôle *f* de zinc
~SILICOFLUORIDE - [chem] (a mothproofing compound and a curing agent for concrete) silicofluorure *m* de zinc
~SMELTERY - [metall] fonderie *f* de zinc
~SMELTING - [metall] fusion *f* du zinc
~-SPAR - [min] calamine *f*, zinc *m* carbonaté
~-SPINEL - [min] gahnite *f*
~STEARATE - [chem] (a lubricant in rubber compounding and plastics extrusion, waterproofing agent, a mild antiseptic in medicine, also has application in cosmetics and in paint manufacture) stéarate *m* de zinc
~SULPHANILATE - [chem] (an astringent and antiseptic in medicine) sulfanilate *m* de zinc
~SULPHATE - [chem] (white vitriol. A fungicide, analytical reagent, astringent and emetic in medicine, dyeing mordant, hide preservative, and a bleaching agent for paper) sulfate *m* de zinc
ZINC SULPHIDE - [chem] (a white pigment for glass, rubber plastics and linoleum, and a textile dyeing assistant) sulfure *m* de zinc
~SULPHITE - [chem] (a preservative for biological specimens) sulfite *m* de zinc
~SULPHOXYLATE - [chem] (a dyeing assistant) sulfoxylate *m* de zinc
~TALLATE - [chem] (a drier for paints and varnishes) tallate *m* de zinc
~TELLURIDE - [chem] (a compound used in the manufacture of semiconductors, with a forbidden band gap of 2,2 electron volts) telluriure *m* de zinc
~THIOCYANATE - [chem] (an analytical reagent) thiocyanate *m* de zinc
~TURNINGS - [metall] tournure *f* de zinc
~UNDECYLENATE - [chem] (zinc undecanoate. An antimycotic agent in medicine and a synthesis intermediate) undécylénate *m* de zinc
~VALERATE - [chem] (white, unpleasant-smelling powder with application in the treatment of hysteria) valérianate *m* de zinc
~WHITE - [paint] (alternative name for zinc oxide q.v.) blanc *m* de zinc
~WORKS - [metall] zinguerie *f*
~YELLOW - [chem] (zinc chrome. Zin potassium chromate, used as a pigment and in corrosion-resistant paint systems) jaune *m* de zinc
~ZIRCONIUM SILICATE - [chem] (a component of ceramic glazes) silicate *m* de zinc et de zirconium
ZINCATE TREATMENT - [metall] traitement *m* au zincate
ZINCATES - [chem] (the salts of zinc) zincates *mpl*
ZINCIC - [min] zincique
ZINCIFEROUS - [min] zincifère
ZINCIFY, to - s. zinc, to
ZINCING - s. zinking
ZINCITE - [chem] (native zinc oxide. An ore of zinc) zincite *f*, spartalite *f*
ZINCOGRAPH - [print] zincogravure *f*, zincographie *f*
ZINCOGRAPHY - [print] (engraving process in which zinc is covered with wax and etched) zincographie *f*, photogravure *f* sur zinc
ZINKENITE - [min] (a native lead sulphantimonide) zinckénite *f*
ZINKING - [el metall] zincage *m*
ZIPFASTENER - s. zipper
ZIPPEITE - [min] (uranium and sulphur-containing mineral) zippéite *f*
ZIPPER - [mech] (a slide fastener) fermeture *f* éclair, fermeture *f* à curseur
~CLOSED CONVEYOR BELT - [mech] (type of material conveyor in which an endless belt provided with a zip fastener receives the material and is automatically closed over it into tubular form. A similar device opens the fastener when the belt has reached a discharging station at a higher level) courroie *f* transporteuse à fermeture-éclair
ZIRCON - [min] (native zirconium silicate, with uses as a gemstone, abrasive, and in the production of refractories) zircon *m*
~CEMENT - [metall] ciment *m* à base de zircon
ZIRCONATE - [chem] (a salt formed by the replacement of hydrogen in zirconium hydroxide by a metal) zirconate *m*
ZIRCONIA - [chem] (a zirconium dioxide, the opacifier used in radiography) zircone *f*
ZIRCONIUM - [chem] (a metallic element, symbol

Zr, with applications in nuclear technology, in metallurgy for special alloys, in the production of catalyst, and textile finishing compounds) zirconium *m*
ZIRCONIUM ACETATE - [chem] (a water repellant for textile and a resin curing agent) acétate *m* de zirconium
~ ACETYLACETONATE - [chem] (a crosslinking agent for a number of synthetic resins) acétylacétonate *m* de zirconium
~ ALLOY - [metall] alliage *m* de zirconium
~ BORIDE - [chem] (a high temperature refractory) borure *m* de zirconium
~ CARBIDE - [chem] (a refractory and abrasive material) carbure *m* de zirconium
~ CARBONATE - [chem] (a starting material for the production of zirconium oxide) carbonate *m* de zirconium
~ DIOXIDE - [chem] (heavy, white, odourless powder with uses in medicine and in cosmetics as a deodorant, as a paint pigment, catalyst, abrasive component of ceramic glazes and special glasses, dyestuffs stabilizer, and as a refractory) bioxyde *m* de zirconium
~ GLYCOLATE - [chem] (a sequestering agent and a deodorant in cosmetics) glycolate *m* de zirconium
~ HYDRIDE - [chem] (a catalyst and reducing agent) hydrure *m* de zirconium
~ HYDROXIDE - [chem] (white, amorphous powder with uses as a pigment and in the manufacture of special glasses) hydroxyde *m* de zirconium
~ LACTATE - [chem] (a deodorant in cosmetics) lactate *m* de zirconium
~ NAPHTHENATE - [chem] (a lubricant additive, paint additive to improve weathering characteristics, and a component of ceramic glazes) naphténate *m* de zirconium
~ NITRATE - [chem] (white, hygroscopic crystals with uses as a preservative) nitrate *m* de zirconium
~ NITRIDE - [chem] (a refractory) nitrure *m* de zirconium
~ OXIDE - s. zirconium dioxide
~ OXYCHLORIDE - [chem] (a lubricant additive, analytical reagent, and textile dyeing assistant) oxychlorure *m* de zirconium
~ PYROPHOSPHATE - [chem] (a polymerization catalyst) pyrophosphate *m* de zirconium
~ SULPHATE - [chem] (a lubricant additive and analytical reagent) sulfate *m* de zirconium
~ TETRAACETYLACETONATE - [chem] (a catalyst, crosslinking agent, analytical reagent, and lubricant additive) tétra-acétylacétonate *m* de zirconium
~ TETRACHLORIDE - [chem] (an analytical reagent, catalyst, and textile waterproofing agent) tétrachlorure *m* de zirconium
ZIRCONYL HYDROXYCHLORIDE - [chem] (a dyeing assistant and waterproofing agent for textiles, and a component of cosmetic deodorants) hydroxychlorure *m* de zirconyle
ZITHER - [mus] (plucked string instrument) cithare *f*
ZODIAC - [astr] (a belt on the celestial sphere extending 9 deg. on each side of the ecliptic) zodiaque *m*
~ LIGHT - [astr] (a faint glow extending along the zodiac in the neighbourhood of the sun) lumière *f* zodiacale
ZOISITE - [min] (thulite. An epidote mineral used as a decorative stone in building) zoïsite *f*
ZONAL ABERRATION - [opt] aberration *f* de zone
ZONDA WIND - [met] (a foehn wind occurring in Argentina) zonda *m*
ZONE, to - [town planning] répartir en zones
ZONE - [gen geogr geol] zone *f*
[gen] (a girdle) ceinture *f*
~ AXIS - [crystall] (the axis through the centre of a crystal parallel to the edge of a zone) axe *m* de zone
~ BIT - [comput] bit *m* de hors-texte
~ FOCUSING - [photo] mise *f* au point pour prises sur le vif
~ FOSSIL - [geol] fossile *m* de zone
~ -HARDENED - [metall] trempé localement
~ LEVELLING - [electron] (along a semiconductor body) nivellement *m* en zones
~ MARKER - [radar] radiophare *m* zonal à faisceau vertical
~ MELTING - [metall] (the process used in preparing pure semiconductors crystals) fusion *f* en zones
~ OF A CRYSTAL - [crystall] (a set of faces of a crystal meeting in a series of edges, all of which are parallel) zone *f* d'un cristal
~ OF ALARM - [med] zone *f* d'alarme
~ OF COMBUSTION - [metall] zone *f* de combustion
~ OF CONTACT - [mech] (of a gear) zone *f* de contact
~ OF DEPRESSION - [hydr] zone *f* de dépression, zone *f* d'influence
~ OF FOLDS - [geol] zone *f* de plissement
~ OF INTERSECTION - [aero] (that part of a civil airway which overlays any part of another such airway) zone *f* d'intersection
~ OF PLASTIC EQUILIBRIUM - [soil] zone *f* d'équilibre plastique
~ OF RADIAL SHEAR - [soil] zone *f* de cisaillement radial
~ OF REDUCTION - [metall] zone *f* de réduction, zone *f* réductrice
~ OF SEMICONTINUOUS SOIL MOISTURE - [soil] zone *f* funiculaire
~ OF WEATHERING - [geol] zone *f* de l'altération superficielle
~ POSITION INDICATOR - [radar] radar *m* auxiliaire à faisceau étroit
~ PUNCHING - [comput] perforation *f* hors texte
~ REFINING - [electron] (the passage of one or more molten zones along a semiconductor in order to reduce the impurity concentration of part of the ingot) raffinage *m* en zones
~ TELEVISION - [telev] (or rectilinear scanning, the process of scanning an area in a sequence of narrow straight parallel strips) analyse *f* par lignes
~ TIME - [astr] (U. S. S, only a system of standard time in which the surface of the earth is divided into zones of I5 deg, of longitude, the primary zone being centred on the meridian of Greenwich) temps *m* fuseau
~ WIND - [astr] vent *m* zonal
ZONED LENS - [radio] (of waveguides) lentille *f* à échelons
ZOOGENIC ROCK - [geol] roche *f* zoogène
ZOOGLEA - [hydr] zooglée *f*
ZOOGLEAL MATRIX - [hydr] matrice *f* zoogléale
ZOOM, to - [aero] (to gain height in a sudden climb, using the kinetic energy of the aircraft) monter en chandelle
[acoust] vrombir, bourdonner

ZOOM, to - [cin] (to change rapidly the physical position of the camera or camera lens reference to the fixed object being photographed) changer rapidement le plan
ZOOM - [acoust] vrombissement *m*
[photo cin telev] changement *m* rapide de plan
[aero] montée *f* en chandelle
~-AWAY SHOT - [cin] prise *f* de vue à augmentation de la distance focale
~AWAY, to - [photo & cin] ouvrir
~CAMERA - [photo & cin] caméra *f* avec objectif à distance focale variable
~IN, to - [photo & cin] serrer
~-IN SHOT - [cin] prise *f* de vue à diminution de la distance focale
~LENS - [cin] (lens with which the picture can be taken with warying focal length) objectif *m* à foyer réglable
ZOOMING - [aero] (the use of the kinetic energy of an aircraft to gain height in a sudden climb) chandelle *f*
ZOOPLANKTON - [hydr] (the animal life of the plankton) zooplancton *m*
ZOOTECHNY - [agric] zootechnie *f*
ZYGOMA - [anat] zygoma *m*, os *m* malaire
ZWITTERION - [nucl] (an ion which carries both positive and negative charges) ion *m* amphotérique
ZYGOMATIC - [anat] zygomatique
ZYGOTE - [biol] zygote *m*
ZYGOTOMERE - [biol] sporoblaste *m*
ZYMASE - [chem] (an enzyme present in yeast and responsible for the conversion of sugar to alcohol and carbon dioxide) zymase *f*
ZYMOGEN - [chem] (non-catalytic substance formed during the elaboration of an enzyme) zymogène *m*
ZYMOLOGY - [chem] (the science of ferments and associated substances) zymologie *f*
ZYMOMETER - [instr] (an instrument for measuring fermentation) zymosimètre *m*
ZYMOSCOPE - [instr] (an apparatus for determining the fermenting power of yeast) zymosiscope *m*
ZYMOSIMETER - s. zymometer
ZYMOSIS - [chem] (chemical changes produced by living micro-organisms or by enzymes) fermentation *f*
[med] (the morbid process, believed to be analogous to fermentation and constituting an infectious disease) zymose *f*
ZYMOTECHNICS - [chem] (the study of the fermentation processes) zymotechnie *f*
ZYMOTIC - [med] zymotique
ZYMURGY - [chem] (the chemistry of fermentation) enzymologie *f*